Hawley's
Condensed Chemical Dictionary

Hawley's

Condensed Chemical Dictionary

Sixteenth Edition

Michael D. Larrañaga
Richard J. Lewis, Sr.
Robert A. Lewis

WILEY

Published by John Wiley & Sons, Inc., Hoboken, New Jersey.
Published simultaneously in Canada.

For general information on our other products and services or for technical support, please contact our Customer Care Department within the United States at (800) 762-2974, outside the United States at (317) 572-3993 or fax (317) 572-4002.

Wiley also publishes its books in a variety of electronic formats. Some content that appears in print may not be available in electronic formats. For more information about Wiley products, visit our web site at www.wiley.com.

Library of Congress Cataloging-in-Publication Data is available.

ISBN: 978-1-118-13515-0

Printed in the United States of America.

SKY10057931_102323

Dedicated to Patricia and Ana Sofia for their support, sacrifice, patience, and most importantly, love.

Contents

Contents

Introduction

The First Edition of the *Condensed Chemical Dictionary* appeared in 1919 when the chemical industry in the United States was entering a huge expansion program as a result of World War I. The urgent need for such a reference book became apparent to Francis M. Turner, president of the Chemical Catalog Company, predecessor of the Reinhold Publishing Corporation. Under his supervision a succession of editors developed and expanded the *Condensed Chemical Dictionary* to meet the growing needs of the chemical industry. Since his death this development has continued, with the result that the work has achieved worldwide recognition in its field.

The *Condensed Chemical Dictionary* is a unique publication. It is not a dictionary in the usual sense of an assemblage of brief definitions, but rather a compendium of technical data and descriptive information covering many thousands of chemicals, chemical phenomena, and chem-biological materials organized in such a way as to meet the needs of those who have only minutes to devote to any given substance or topic.

Four distinct types of information are presented: (1) descriptions of chemicals, raw materials, processes, and equipment; (2) expanded definitions of chemical entities, phenomena, and terminology; (3) descriptions or identifications of a wide range of trademarked products used in the chemical industries; (4) definitions of bio-chemical materials, phenomena, and terminology. Supplementing these are listings of accepted chemical abbreviations used in the literature, short biographies of chemists of historic importance, and winners of the Nobel Prize in Chemistry. Also included are descriptions or notations of the nature and location of many U.S. technical societies and trade associations. In special cases, editorial notes have been supplied where it was felt necessary to clarify or amplify a definition or description. A few entries written by specialists are acknowledged by use of the author's name.

In a work of this nature, selection of topics for inclusion can hardly fail to be influenced by current interests and developing concerns within the topic area. The growing importance to scientists, public officials, and general public, of environmental and health hazards, which came to the forefront so quickly in the 1960s, was reflected in the Eighth Edition, which greatly increased its coverage of this aspect of chemistry.

Since then, the magnitude of the energy problem has been uppermost in the thinking of a broad spectrum of engineers, chemists, and physicists because it is certainly the most important technical problem confronting this country.

Both the Ninth and Tenth Editions, while retaining emphasis on environmental considerations, were expanded in the area of energy and its sources, as far as permitted by available information. The goal of the editors was to provide condensed, authoritative, factually oriented statements and descriptions, and to resist prognostications as to the future potential of any particular energy source. At the same time, continuing attention has been devoted to common hazards, such as flammable and explosive materials, poisons, pesticides, carcinogens, corrosive agents, and radioactive wastes, in line with the practice followed in earlier editions, and with the increasing public concern over these matters.

The Eleventh Edition added new chemicals, revised the format for chemical entries, and added new trademarked products and definitions. Chemical Abstract Services (CAS) numbers were included for many chemical entries to facilitate recourse to computerized databases.

In the Twelfth Edition, all trademarked entries were revised. The method of referencing was changed from superscripted numbers to the company names, facilitating access to the addresses of the manufacturers of trademarked products. Many additional definitions and cross-references were added to make the work current with the constantly growing field of chemistry.

In the Thirteenth Edition, all trademarked entries were revised and all CAS numbers verified. Many additional chemical entries, definitions, and cross-references were added to make the work current with the constantly growing field of chemistry.

In the Fourteenth Edition, links to the Internet were added. World Wide Web page addresses were added for manufacturers and associations in the appropriate entries and in Appendix III. Trademarked entries and their associated manufacturers were updated to reflect the constant flux in the modern chemical industry. Many additional chemical entries, definitions, and cross-references have been added to make the work current with the constantly growing field of chemistry.

In the Fifteenth Edition, over 4200 new or updated entries were included. Special effort was directed to add definitions and terms of art for biochemistry. Over 700 entries reflecting the great interest in biochemistry are now included. Over 90 terms relating to nanotechnology were added. Almost 3000 new chemicals were added, including trade named products. Links to World Wide Web page addresses for manufacturers and associations in the appropriate entries and in Appendix III were verified or added. Trademarked entries and their associated manufacturers were updated to reflect the constant flux in the modern chemical industry.

In the Sixteenth Edition, there is significant expansion of both chemical and biochemical terms. The reason for the additions of biochemical terms in the Sixteenth Edition is the emerging fields in biology and biological engineering such as synthetic biology, which highlight the merging of the sciences of chemistry and biology such that biology can be engineered to produce chemical substances and vice versa. In all, there are 1471 new definitions, 5236 revised or updated definitions, a new Chemical Abstract Number Index, and an update of all trademarks. Internet links were removed, except in certain instances; Internet search engines have become so powerful that it is no longer difficult to find relevant material about a particular chemical on the Internet.

In connection with certain classifications of substances, particularly pesticides and carcinogens, the statement "Use may be restricted" indicates that a state or local regulation may exist, even though a product has not been officially banned, or that a definitive ruling on its use is pending. When a product has been banned outright, the statement "Use has been prohibited" is used. A number of disputed cases have arisen in recent years; although some have been definitely settled, others are still being evaluated.

In such a work as this, in view of the many materials in various stages of evaluative testing, court procedures, appeals, hearings, and so forth, it is impossible to keep abreast of every development, every new chemical, and even every new biological organism. The user should check the current status of any questionable products/materials before making decisions that involve them (see Hazards below).

Chemicals, Raw Materials, and Biological Entries

The information in the categories listed next is given for each substance in the sequence indicated; where entries are incomplete, it may be presumed that no reliable data were provided by the reference systems utilized.

Name: The commonly accepted name is the key entry. Terminological variations are indicated where necessary. In virtually all cases, the name is given in the singular number. A name having initial caps and enclosed in quotes is a trademark (TM). The manufacturers of trademarked products are indicated directly after the name in italics and within brackets (e.g., *[Du Pont]*). The addresses of the manufacturers are given in Appendix III.

Synonym: Alternative names, as well as trivial names, are indicated. Obsolete and slang names have been eliminated as far as possible. Most synonyms are entered independently and cross-referenced, but space limitation has not permitted complete consistency in this regard.

Chemical Abstract Service Registry Number (CAS): This universally used number permits use and comparison of data on a given material even though published under different synonyms. It will, in fact, permit absolute identification of a compound with all of its synonyms. CAS numbers also facilitate

extraction of information from computerized data bases.

Formula: The molecular (empirical or atomic) formula is regularly given. Structural formulas are used in special cases of unusual importance or interest.

Properties: The properties typically given are physical state, atomic number, atomic weight, valence, isotopes, odor, taste, density, boiling point (at 760 mm Hg unless otherwise stated), melting point (freezing point), refractive index, and solubility or miscibility. Various other properties are given where pertinent: flash point, autoignition temperature, electrical properties, tensile strength, hardness, expansion coefficient, and so forth.

Source or Occurrence: Geographical origins of metals, ores, essential oils, vegetable oils, and other natural products may be given.

Derivation: The chemical reactions or other means of obtaining the product by current industrial methods are emphasized. Obsolete and "curiosity" methods have been largely eliminated.

Grade: Recognized grades are provided as reported in the industrial literature, including technical, CP, USP, refined, reactor, and semiconductor.

Hazard: This category includes flammability, toxicity characterization, explosion risks, and so on.

The ACGIH® classifies carcinogens in five categories. *confirmed human carcinogen*, indicating the evidence is compelling for causing human cancer; *suspected human carcinogen*, indicating the substance is carcinogenic in animals but convincing evidence for human carcinogenesis is lacking; *animal carcinogen*, indicating the substance is carcinogenic in animals; *not classifiable as a human carcinogen*, indicating inadequate data are available to make a decision; and *not suspected as a human carcinogen*, indicating the available data do not suggest the likelihood the substance causes cancer. The ACGIH does not classify all potential carcinogens and some carcinogens may be listed as possibly carcinogenic.

It was not considered practical to include occupational exposure recommendations made by the National Institute for Occupational Safety and Health or exposure standards established by the Occupational Safety and Health Administration, as these are readily available on the Internet.

The toxicity ratings are intended to be used only as indications of the industrial hazard presented by a given material, as most of them are based on tests made on laboratory animals. Qualified toxicologists, industrial hygienists, exposure scientists, or physicians should be consulted for specific evaluations, dosages, exposure times, and concentrations. For further information regarding these hazards, the reader is referred to the following entries: combustible material; flammable material; dust, industrial; corrosive material; oxidizing material; poison; toxicity; toxic materials; and carcinogens.

Use: These are primarily large-scale applications. Because of the rapidity of change in the chemical industries and the difficulty of obtaining reliable current data, no attempt has been made to list uses in the order of their tonnage consumption. The patent literature is not specifically represented.

General Entries

It is likely that no two editors would completely agree about what general subjects should be included in a dictionary of this kind. The major subdivisions of matter directly involved with chemical reactions, the various states of matter, biochemical materials, and the more important groups of compounds would almost certainly be regarded as essential, but beyond these, the area of selectivity widens rapidly. The topics either added or expanded by the present editor were chosen chiefly because of their interest and importance, both industrial and biochemical, and secondarily because of the terminological confusion evidenced in the literature and in industrial practice. Regarding the latter, the reader is referred to the entries on gum, resin, pigment, dye, filler, extender, reinforcing agent, and homogeneous and combustible materials. In some cases, a position has been taken that may not be accepted by all, but is defensible and certainly not arbitrary. Even editors must acknowledge that the meanings and uses of terms often change illogically, and that such changes are usually irreversible.

Among the general entries are important subdivisions of chemistry; short biographies of outstanding chemists of the past, including winners of the Nobel Prize in Chemistry; numerous group definitions (barbiturate, peroxide); major chemical and physical chemical phenomena (polymerization, catalysis); functional names (antifreeze, heat-exchange agent); terms describing special material forms (aerosol, foam, fiber); energy sources (solar cell, fuel cell, fusion); the more important chemical processes; and various types of machinery and equipment used in the process industries. No general entry is intended to be encyclopedic or definitive, but rather a condensation of essential information, to be supplemented by reference to specialized sources. To present all this in useful and acceptably complete form has been a challenging, though often frustrating task, which the editor leaves with the uneasy feeling that, like the breadcrumbs in the Hatter's butter, some mistakes are likely to have

got in as well. For any mistakes, we apologize in advance.

Trademarks

Continuing the policy of previous editions, an essential component of the Dictionary comprises descriptions of proprietary industrial products. The information was either provided by the manufacturers of these materials or taken from announcements or advertisements appearing in the technical press. Each proprietary name is enclosed in quotation marks, is stated to be a trademark (or brand name), and is followed by a portion of the name of the manufacturer of the trademarked product. Manufacturer names are displayed in italics and within brackets (e.g., *[Du Pont]*). The addresses of the manufacturers are given in Appendix III. We wish to thank the owners of these trademarks for making the information available. The space devoted to these is necessarily limited, as the constant proliferation and changes in trademarked products make it impossible to list more than a small fraction of them in a volume such as this.

The absence of a specific trademark designation does not mean that proprietary rights may not exist in a particular name. No listing, description, or designation in this book is to be construed as affecting the scope, validity, or ownership of any trademark rights that may exist therein. Neither the editor nor the publisher assumes any responsibility for the accuracy of any such description or for the validity or ownership of any trademark.

The editors would specifically like to acknowledge the contributions of the late Robert A. Lewis, who prepared much of the updated material in biochemistry and biology. For Robert's contributions, we are truly thankful.

A Request

Many corrections and suggestions have been made by readers from around the world during the long history of the earlier editions. The editors have always tried to acknowledge these to the best of their ability. They have welcomed this correspondence, for it has been an important source of information about the acceptance of the Dictionary by its readers. The present editor and publisher wish to encourage this reaction from the field, not only to permit corrections to be made in reprinted issues, but also to establish a basis for preparing future editions. All letters addressed to the publisher will be forwarded.

MICHAEL D. LARRAÑAGA

Abbreviations

ACS	American Chemical Society		i.e.	that is
ASTM	American Society for Testing and Materials		L, l	liter
			lb	pound
atm	atmosphere		m-	meta
autoign temp	autoignition temperature		mg	milligram
aw	atomic weight		mg/m^3	milligrams per cubic meter
bp	boiling point		μCi/mL	microcuries per milliliter
Btu	British thermal unit		μg/m^3	micrograms per cubic meter
C	degrees centigrade (Celsius)		min	minimum, minute
CAS	Chemical Abstract Service Registry Number		mm	millimeter
			mp	melting point
cc	cubic centimeter		mw, Mw	molecular weight
CC	closed cup		NF, N.F.	National Formulary grade of chemical
Ci	Curie		NIOSH	National Institute for Occupational Safety and Health
CI, C.I.	"Color Index" (a standard British publication giving official numerical designations to colorants)		nm	nanometers
			o-	ortho
CL	ceiling level		OC	open cup
COC	Cleveland open cup		OSHA	U.S. Occupational Safety and Health Administration
CP	chemically pure: a grade designation signifying a minimum of impurities, but not 100% purity		p-	para
			ppb	parts per billion
cP	centipoise		ppm	parts per million
cu	cubic		psi(a)	pounds per square inch (absolute)
d, D	density		%	percent
DOT	U.S. Department of Transportation		refr	refractive
e.g.	for example		sec	second
F	degrees Fahrenheit		sp vol	specific volume
FCC	"Food Chemical Codex"		sv	specific volume in volume/unit mass
FDA	U.S. Food and Drug Administration		TCC	Tagliabue closed cup
flash p	flash point		TLV	Threshold Limit Value
fp	freezing point		TM	trademark
ft	feet		TOC	Tagliabue open cup
FTC	U.S. Federal Trade Commission		USAN	United States Adopted Name
g	gram		USDA	U.S. Department of Agriculture
gal	gallon		USP	United States Pharmacopeia
g/L	grams per liter		UV	ultraviolet
g/mL	grams per milliliter		vap d	vapor density
Hg	mercury		vap press	vapor pressure
H$_2$O	water		wt/gal	weight per gallon
hr	hour			

A

α. See alpha.

A. Abbreviation for absolute temperature.

Å. Abbreviation for Ångstrom.

a. Abbreviation for atto-, a prefix meaning 10^{-18} unit.

AAAS. See American Association for the Advancement of Science.

AATCC. See American Associates of Textile Chemists and Colorists.

abaca. (Manila hemp). The strongest vegetable fiber, obtained from the leaves of a tree of the banana family. The fibers are 4–8 ft long, light in weight, soft, lustrous, and nearly white; they do not swell or lose strength when wet. Denier ranges from 300 to 500. Combustible, but self-extinguishing.
Source: Philippines, Central America, Sumatra.
Use: Heavy cordage and twine, especially for marine use; manila paper.
See hemp.

abate. (O,O-dimethyl phosphorothioate-O,O-diester with 4,4′-thiodiphenol; temephos).
CAS: 3383-96-8. $[(CH_3O)_2PSOC_6H_4]S$.
Properties: Colorless crystals. Mp 30C.
Hazard: Toxic by ingestion and inhalation. Cholinesterase inhibitor.
Use: Pesticide.

Abbé condenser. Substage two-lens condenser giving numerical aperture of 1:20–1:25. Three-lens versions give numerical aperture of 1:40.

Abbé number. (1)The reciprocal of dispersive power. (2) The measure of optical dispersion of a glass; the measure of two planes not displacing the axis.

Abegg's rule. The solubility of salts of alkali metals decreases in strong acids and increases in weak acids as atomic weight increases. This is an empirical rule; sodium chloride is an exception.

Abel-Pensky flash-point apparatus. Instrument used for the determination of the flash point of petroleum.

aberration. Deviation from the ideal in an optical system, the image points being imperfect or improperly located.

aberration, spherical. The aberration by which light passing through the edge of a lens with spherical surfaces comes to focus in a different image plane from that passing through the lens center.

abherent. Any substance that prevents adhesion of a material to itself or to another material. It may be in the form of a dry powder (a silicate such as talc, mica, or diatomaceous earth); a suspension (bentonite-water); a solution (soap-water); or a soft solid (stearic acid, tallow waxes). Abherents are used as dusting agents and mold washes in the adhesives, rubber, and plastics industries. Fats and oils are used as abherents in the baking industry. Fluorocarbon resin coatings on metals are widely used on cooking utensils.
See antiblock agent; dusting agent.

***Abies Siberica* oil.** See fir needle oil.

abietene. A volatile oil.
Derivation: Distilled from the resin or balsam of *Pinus sabiniana*.
Use: In varnishes and lacquers.

abietic acid. (abietinic acid; sylvic acid).
CAS: 514-10-3. $C_{19}H_{29}COOH$ (having a phenanthrene ring system). A major active ingredient of rosin, where it occurs with other resin acids. The term is often applied to these mixtures, separation of which is not achieved in technical grade material.
Properties: Yellowish, resinous powder. Mp 172–175C, optical rotation −106. Soluble in alcohol, ether, chloroform, and benzene; insoluble in water. Combustible.
Derivation: Rosin, pine resin, tall oil.
Method of purification: Crystallization.
Grade: Technical.
Use: Abietates (resinates) of heavy metals as varnish driers, esters in lacquers and varnishes, fermentation industries, soaps.

abietinic acid. $C_{20}H_{30}O_2$.
CAS: 514-10-3.
Properties: A crystalline acid with three isomeric forms.

Hawley's Condensed Chemical Dictionary, Sixteenth Edition. Michael D. Larrañaga, Richard J. Lewis, Sr., and Robert A. Lewis.
© 2016 John Wiley & Sons, Inc. Published 2016 by John Wiley & Sons, Inc.

Derivation: From *colophony rosin* and is a solid form or resin obtained from pines and other plants, mostly conifers.
Use: Stimulate growth of lactic and butyric bacteria.

"Abitol" *[Eastman].* TM for a colorless, tacky, very viscous liquid; mixture of tetra-, di-, and dehydroabietyl alcohols made from rosin.
Use: Plasticizers, tackifiers, adhesive modifiers.

ablation experiment. An experiment designed to produce an animal deficient in one or a few cell types in order to study cell lineage or cell function. The idea is to make a transgenic mouse with a toxin gene (often a diphtheria toxin) under control of a specialized promoter which activates only in the target cell type. When embryo development progresses to the point where it starts to form the target tissue, the toxin gene is activated, and the target tissue dies. Other tissues are unaffected.

ablative material. Any material that possesses a capability for rapidly dissipating heat from a substrate. Specialized ceramic tiles developed since 1980 for protection of the space shuttle have proved successful. The materials used are of two major types: (1) Fibers made from white silica, fused in an oven, cut into blocks, and coated with borosilicate glass; these are extremely efficient at temperatures up to 2300F. (2) An all-carbon composite (called reinforced carbon–carbon) made by laminating and curing layers of graphite fiber previously coated with a resin, which is pyrolyzed to carbon. The resulting tile is then treated with a mixture of alumina, silicon, and silicon carbide. Such composites are used for maximum-temperature (nose-cone) exposure up to 3000F. Both types are undamaged by the heat and are reusable. The tiles are adhered to the body of the spacecraft with a silicone adhesive. Ablative materials used on early spaceship trials were fluorocarbon polymers and glass-reinforced plastics, but these were wholly or partially decomposed during reentry.

Abram's law. The strength of concrete depends on the water/cement ratio.

abrasion. Gradual erosion of the surface of a material both by physical forces (simultaneous cutting, shearing, and tearing) and by chemical degradation, chiefly oxidation. Temperature is a significant factor: friction may raise the temperature of the surface layers to the point where they become subject to chemical attack. Abrasion causes deterioration of many materials, especially of rubber (tire treads), where it can be offset by a high percentage of carbon black; other materials subjected to abrasion in their service life are textiles (laundering), leather and plastics (shoe soles, belting), and house paints and automobile lacquers (airborne dust, grit, etc.).
See abrasive.

abrasive. A finely divided, hard, refractory material, ranging from 6 to 10 on the Mohs scale, used to reduce, smooth, clean, or polish the surfaces of other, less hard substances, such as glass, plastic, stone, wood, etc. Natural abrasive materials include diamond dust, garnet, sand (silica), corundum (aluminum oxide, emery), pumice, rouge (iron oxide), and feldspar; the more important synthetic types are silicon carbide, boron carbide, cerium oxide, and fused alumina. Abrasive in powder form may be (1) applied directly to the surface to be treated by mechanical pressure or compressed-air blast, as in cleaning building stone; (2) affixed to paper or textile backing after the particles have been coated with an adhesive; or (3) mixed with a bonding agent such as sodium silicate or clay, the particles being compressed into a wheel rotated by a power-driven shaft. Aluminum grinding wheels are fabricated by bonding industrial diamonds with fluorocarbon polymer ("Teflon"). The process involves reaction of fluorine with the surfaces of the diamonds, chemical bonding of the fluorinated diamonds to the fluorocarbon, and further chemical bonding of the resulting material to the aluminum, with application of heat and pressure.

abrasive, coated. See abrasive (2).

abric acid. A chemical that inhibits digestion and will show as undigested food during an autopsy.
Properties: Contains a tetanic glycoside.
Hazard: Poison.

abrin. (abrus agglutinin; jequiritin; toxalbumin; crab's eye; Indian licorice seed; jumble bead; prayer bead). Any of the five nearly identical proteinaceous phytotoxins whose a-chain inhibits ribosomal protein synthesis, killing the cell and the b-chain binds to the plasma membrane of the cells of the intestinal wall permitting the a-chain to enter the cytoplasm.
Properties: Chemically similar to botulinum toxin; inactivated by heating; composed of two disulfide-linked polypeptide chains. Molecular weight: 63,000–67,000Da; Yellowish-white powder; soluble in solutions of sodium chloride, usually with turbidity.
Derivation: Produced by the leguminous shrub, *Abrus precatorius.*
Hazard: Extreme irritant; lethal if ingested; toxin; poison.

Abrus. A genus of leguminous plants that produce poisonous lectins or phytotoxins.

ABS. Abbreviation for (1) alkyl benzene sulfonate (detergent); (2) acrylonitrile-butadiene-styrene copolymer.
See ABS resin.

abscisic acid.
CAS: 21293-29-8. $C_{15}H_{20}O_4$. A plant growth regulator that promotes detachment of leaves and fruit.

Properties: Colorless crystals. Mp 160C. Sublimes at 120C. Soluble in acetone, ether, chloroform; slightly soluble in water. Optically active.
Occurrence: In plants, fruits, and vegetables from which it can be extracted. Also made synthetically.
Use: In orchard sprays to facilitate fruit harvesting, defoliant, growth inhibitor.

absinthe. (absinth; *Artemisia absinthium*). A commercially important volatile oil or a highly poisonous alcoholic beverage.
Properties: Contains oil of wormwood, anise, and other herbs.
Derivation: Produced by plants of the genus *Artemisia*.

absinthium. (wormwood). $C_{30}H_{40}O_7$. An essential oil with intensely bitter taste due to presence of absinthin.
Hazard: Toxic by ingestion.
Use: A flavoring in liqueurs, vermouth.

absolute. (1) Free from admixture of other substances; pure. Example: absolute alcohol is dehydrated ethanol, 99% pure. (2) The pure essential oil obtained by double solvent extraction of flowers in the manufacture of perfumes.
See concrete (2). (3) absolute temperature.

absolute alcohol. (anhydrous alcohol; dehydrated alcohol; ethyl alcohol).
CAS: 64-17-5. C_2H_6O. A liquid rapidly absorbed from the gastrointestinal tract and distributed throughout the body. It has bactericidal activity.
Properties: Clear, colorless liquid; 99% ethanol, 1% water.
Use: Reagent in chemical reactions where water must be absent or nearly so; a topical disinfectant; as a solvent and preservative in pharmaceutical preparations; the primary ingredient in alcoholic beverages.

absolute configuration. The configuration of four different substituent groups around an asymmetric carbon atom. The absolute configurations of molecules in biochemistry are compared to the configuration of D- and L-glyceraldehyde.

absolute error. The actual difference between the approximate and the exact value in any calculation.

absolute temperature. The fundamental temperature scale used in theoretical physics and chemistry, and in certain engineering calculations, such as the change in volume of a gas with temperature. Absolute temperatures are expressed either in degrees Kelvin or in degrees Rankine, corresponding respectively to the centigrade and Fahrenheit scales. Temperatures in Kelvins are obtained by adding 273 to the centigrade temperature (if above 0C) or subtracting the centigrade temperature from 273 (if below 0C). Degrees Rankine are obtained by subtracting 460 from the Fahrenheit temperature.

absolute zero. Temperature at which the volume of a perfect gas theoretically becomes zero and all thermal motion ceases: −273.13C or −459.4F.

absorbent. (1) Any substance exhibiting the property of absorption, e.g., absorbent cotton, so made by removal of the natural waxes present. (2) A material that does not transmit certain wavelengths of incident radiation.
See absorption (1); absorption (2).

absorptiometer. An instrument for determining the solubility of a gas in a liquid.

absorption. (1) In chemical terminology, the penetration of one substance into the inner structure of another, as distinguished from adsorption, in which one substance is attracted to and held on the surface of another. Physicochemical absorption occurs between a liquid and a gas or vapor, as in the operation known as scrubbing, in which the liquid is called an absorption oil; sulfuric acid, glycerol, and some other liquids absorb water vapor from the air under certain conditions. Physiological absorption takes place via porous tissues, such as the skin and intestinal walls, which permit passage of liquids and gases into the bloodstream.
See adsorption; hygroscopic. (2) In physical terminology, retention by a substance of certain wavelengths of radiation incident upon it, followed either by an increase in temperature of the substance or by a compensatory change in the energy state of its molecules. The UV component of sunlight is absorbed as the light passes through glass and some organic compounds, the radiant energy being transformed into thermal energy. The radiation-absorptive capacity of matter is utilized in analytical chemistry in various types of absorption spectroscopy. (3) In physical chemistry, the ability of some elements to pick up or "capture" thermal neutrons produced in nuclear reactors as a result of fission. This is due to the capture cross-section of their atoms, which is measured in units called barns; elements of particularly high neutron absorption capability are cadmium and boron.

absorption band. The range of wavelengths absorbed by a molecule; for example, absorption in the infrared band from 2.3 to 3.2 μm indicates the presence of OH and NH groups, while in the band from 3.3 to 3.5 indicates aliphatic structure. Atoms absorb only a single wavelength, producing lines, such as the sodium D line.
See spectroscopy; resonance (2); ultraviolet absorber; excited state.

absorption (biology). Transport of the products of digestion from the intestinal tract into the blood.

absorption oil.   See absorption (1).

absorption spectroscopy.   An important technique of instrumental analysis involving measurement of the absorption of radiant energy by a substance as a function of the energy incident upon it. Absorption processes occur throughout the electromagnetic spectrum, ranging from the γ region (nuclear resonance absorption or the Mossbauer effect) to the radio region (nuclear magnetic resonance). In practice, they are limited to those processes that are followed by the emission of radiant energy of greater intensity than that which was absorbed. All absorption processes involve absorption of a photon by the substance being analyzed. If it loses the excess energy by emitting a photon of less energy than that absorbed, fluorescence or phosphorescence is said to occur, depending on the lifetime of the excited state. The emitted energy is normally studied. If the source of radiant energy and the absorbing species are in identical energy states (in resonance) the excess energy is often given up by the nondirectional emission of a photon whose energy is identical with that absorbed. Either absorption or emission may be studied, depending upon the chemical and instrumental circumstances. If the emitted energy is studied, the term *resonance fluorescence* is often used. However, if the absorbing species releases the excess energy in small steps by intermolecular collision or some other process, it is commonly understood that this phenomenon falls within the realm of absorption spectroscopy. The terms *absorption spectroscopy*, *spectrophotometry*, and *absorptiometry* are often used synonymously. Most absorption spectroscopy is done in the ultraviolet, visible, and infrared regions of the electromagnetic spectrum.
See emission spectroscopy; infrared spectroscopy.

absorption tower.   (scrubber; Paulson tower). A device used for gas purification by absorption of gas impurities in a liquid.

ABS resin.   Any of a group of tough, rigid thermoplastics that derive their name from the initial letters of the monomers which produce them.

abstraction reaction.   A reaction that removes an atom from a structure.

abundance.   The relative amount (% by weight) of a substance in the earth's crust, including the atmosphere and the oceans. (1) The abundance of the elements in the earth's crust is:

Rank	Element	% by wt
1	Oxygen	49.2
2	Silicon	25.7
3	Aluminum	7.5
4	Iron	4.7
5	Calcium	3.4
6	Sodium	2.6
7	Potassium	2.4
8	Magnesium	1.9
9	Hydrogen	0.9
10	Titanium	0.6
11	Chlorine	0.2
12	Phosphorus	0.1
13	Manganese	0.1
14	Carbon	0.09
15	Sulfur	0.05
16	Barium	0.05
	All others	0.51

(2) The percentages of inorganic compounds in the earth's crust, exclusive of water, are:

(1) SiO_2 55   (2) Al_2O_3 15   (3) $CaCO_3$ 8.8
(4) MgO 1.6   (5) Na_2O 1.6   (6) K_2O 1.9

(3) The most abundant organic materials are cellulose and its derivatives, and proteins.
Note: In the universe as a whole, the most abundant element is hydrogen.

"Abzol" *[Albemarle].*   TM for a solvent cleaner that can be used in place of chlorinated solvents. The main ingredient is n-propyl bromide and is an acceptable substitute for ozone-depleting substances.
Use: Electronic equipment and close tolerance metal parts.

AC.   Abbreviation for allyl chloride.

Ac.   Symbol for actinium; abbreviation for acetate.

acacia gum.   See arabic gum.

acaricide.   A type of pesticide effective on mites and ticks (acaricides).

ACC.   See the American Chemistry Council.

accelerator.   (1) A compound, usually organic, that greatly reduces the time required for vulcanization of natural and synthetic rubbers, at the same time improving the aging and other physical properties. Organic accelerators invariably contain nitrogen, and many also contain sulfur. The latter type are called ultra-accelerators because of their greater activity. The major types include amines, guanidines, thiazoles, thiuram sulfides, and dithiocarbamates. The amines and guanidines are basic, the others acidic. The normal effective concentration of organic accelerators in a rubber mixture is 1% or less depending on the rubber hydrocarbon present. Zinc oxide is required for activation, and in the case of acidic accelerators, stearic acid is required. The introduction of organic accelerators in the early twenties was largely responsible for the successful development of automobile tires and mechanical products for engineering uses. A few inorganic accelerators are still used in low-grade products, e.g., lime, magnesium oxide, and lead oxide.

See vulcanization; rubber. (2) A compound added to a photographic developer to increase its activity, such as certain quaternary ammonium compounds and alkaline substances. (3) A particle accelerator.

"Accepta" *[Accepta].* TM for a rig wash concentrated detergent miscible with fresh or sea waters.
Use: Cleaning in the shipping industry.

"Accepta 3538" *[Accepta].* TM for an emulsifying bilge cleaner and degreaser.

"Accepta 3547" *[Accepta].* TM for a water stain and scale remover.
Use: For steel, wood, glazed surfaces, toilet bowls, baths, and terrazzo.

acceptability (foods). See organoleptic.

acceptable risk. A concept that has developed in recent years, especially in connection with toxic substances (insecticides, mercurials, carcinogens), food additives, air and water pollution, and related environmental concerns. It may be defined as a level of risk at which a seriously adverse result is highly unlikely to occur, "but at which one *cannot* prove whether *or not* there is 100% safety. It means living with reasonable assurance of safety and acceptable uncertainty." (Schmutz, J. F., *Chemical and Engineering News*, Jan. 16, 1978). Examples of acceptable risk that might be cited are diagnostic X-rays, fluoridation of water, and ingestion of saccharin in normal amounts. The acceptability of the risks involved in nuclear power generation is controversial. The weight of the evidence has tended to shift toward the negative side since 1975 when an official safety study estimated the risk of a serious accident to be 1 in 20,000 years of reactor operation. An investigation made by the Oak Ridge National Laboratory based on data collected from 1969 to 1979 concluded that the risk of a major accident is 1 in 1000 years of reactor operation.

acceptor. See donor.

acceptor control. (electron transport chain). The regulation of the rate of respiration by the availability of ADP as a phosphate group acceptor.

accessory pigments. Visible light-absorbing pigments, such as carotenoids and xanthophyll in green plants and photosynthetic bacteria that trap energy from sunlight and pass it on to "special pairs."

"Accosoft" *[Stepan].* TM for a product that adds softening, lubricity, and heat resistance.

Use: In household and commercial textile industries.

"Accuchem" *[Accurate].* TM for a series of research biochemical compounds.

"Accudenz" *[Accurate].* TM for an autoclavable, universal centrifugation medium.

AccuGel. A native pea starch.
Use: Gives excellent gel strength, improved body and mouth feel without adding flavors.

"accutane"®. *[F. Hoffman].* (Isotretinoin, 13-*cis*-retinoic acid).
CAS: 4759-48-2. $C_{20}H_{28}O_2$. A chemical that is related to both retinoic acid and retinol (Vitamin A).
Properties: Yellow to orange crystalline powder; molecular weight: 300.44.
Hazard: Teratogen.
Use: Medicine to treat acne and other skin diseases.

aceclidine. (3-acetoxyquinuclidine glucostat; 3-quinuclidinyl acetate; 1-azabicyclo[2.2.2]octan-8-yl).
CAS:827-61-2. $C_9H_{15}NO_2$. The acetate ester of quinuclidinol.
Use: A cholinergic warfare agent.

acenaphthene. (1,8-dihydroacenaphthalene; ethylenenaphthalene).
CAS: 83-32-9. $C_{10}H_6(CH_2)_2$ (a tricyclic compound).
Properties: White needles. D 1.024 (99/4C), fp 93.6C, bp 277.5C, refr index (100C) 1.6048. Soluble in hot alcohol; insoluble in water. Combustible.
Derivation: From coal tar.
Grade: Technical, 98%.
Hazard: Questionable carcinogen.
Use: Dye intermediate, pharmaceuticals, insecticide, fungicide, plastics.

acenaphthenequinone. (1,2-acenapthene-dione). $C_{10}H_6(CO)_2$ (a tricyclic compound).
Properties: Yellow needles. Mp 261–263C. Insoluble in water; soluble in alcohol.
Derivation: By oxidizing acenaphthene, using glacial acetic acid and sodium or potassium dichromate.
Grade: Technical.
Use: Dye synthesis.

acenocoumarin. (3-(α-acetonyl-4-nitrobenzyl)-4-hydroxycoumarin).
CAS: 152-72-7. $C_{19}H_{15}NO_6$.
Properties: White, crystalline powder; tasteless and odorless. Mp 197C. Slightly soluble in water and organic solvents.
Use: Medicine (anticoagulant).

acephate. (acetylphosphoramidothioic acid ester).
CAS: 30560-19-1. $C_4H_{10}NO_3PS$.

Properties: White crystals. Mp 65C. Soluble in water; slightly soluble in acetone and alcohol.
Hazard: Moderately toxic by ingestion. Possible carcinogen.
Use: Insecticide.

Acephatemet. $C_2H_8NO_2PS$
CAS: 10265-92-6.
Properties: White crystalline structure; slightly water-soluble; melting point: 39–41C.
Use: Insecticide to control cutworms and borers.

acerola. The fruit of the West Indian cherry, *Malpighia punicifolia*, that is the most concentrated natural source of ascorbic acid known.

ACerS. See American Ceramic Society.

acetadol. See aldol.

acetal. (diethylacetal; 1,1-diethoxyethane; ethylidenediethyl ether).
CAS: 105-57-7. $CH_3CH(OC_2H_5)_2$.
Properties: Colorless, volatile liquid; agreeable odor; nutty aftertaste. D 0.831, bp 103–104C, vap press 20.0 mm (20C), flash p −5F (CC) (−20.5C), specific heat 0.520, refr index 1.38193 (20C), wt (lb/gal) 6.89, autoign temp 446F (230C). Stable to alkalies but readily decomposed by dilute acids. Forms a constant-boiling mixture with ethanol. Soluble in alcohol, ether, and water.
Derivation: Partial oxidation of ethanol, the acetaldehyde first formed condensing with the alcohol.
Grade: Technical.
Hazard: Highly flammable, dangerous fire risk. Explosive limits in air 1.65–10.4%. Moderately toxic and narcotic in high concentrations.
Use: Solvent, cosmetics, organic synthesis, perfumes, flavors.
See acetal resin.

acetaldehyde. (acetic aldehyde; aldehyde; ethanal; ethyl aldehyde).
CAS: 75-07-0. CH_3CHO.
Properties: Colorless liquid; pungent, fruity odor. D 0.783 (18/4C), bp 20.2C, mp −123.5C, vap press 740.0 mm (20C), flash p −40F (−40C) (OC), specific heat 0.650, refr index 1.3316 (20C), wt 6.50 lb/gal (20C). Miscible with water, alcohol, ether, benzene, gasoline, solvent naphtha, toluene, xylene, turpentine, and acetone.
Derivation: (1) Oxidation of ethylene; (2) vapor-phase oxidation of ethanol; (3) vapor-phase oxidation of propane and butane; (4) catalytic reaction of acetylene and water (chiefly in Germany).
Grade: Technical 99%.
Hazard: Highly flammable, toxic (narcotic). Dangerous fire, explosion risk, explosive limits in air 4–57%. Possible carcinogen, eye, and upper respiratory tract irritant.

Use: Manufacture of acetic acid and acetic anhydride, *n*-butanol, 2-ethylhexanol, peracetic acid, aldol, pentaerythritol, pyridines, chloral, 1,3-butylene glycol, and trimethylolpropane; synthetic flavors.

acetaldehyde ammonia. See aldehyde ammonia.

acetaldehyde cyanohydrin. See lactonitrile.

Acetaldehyde dehydrogenase.
CAS: 9028-86-8. An enzyme that rapidly converts acetaldehyde into the less harmful acetic acid.

acetal resin. (polyacetal). A polyoxymethylene thermoplastic polymer obtained by ionically initiated polymerization of formaldehyde + CH_2 to obtain a linear molecule of the type –O-CH_2-O-CH_2=CH_2–. Single molecules may have over 1500 –CH_2– units. As the molecule has no side chains, dense crystals are formed. Acetal resins are hard, rigid, strong, tough, and resilient; dielectric constant 3.7; dielectric strength 1200 volts/mil), 600 volts/mil (80-mil); dimensionally stable under exposure to moisture and heat; resistant to chemicals, solvents, flexing, and creep, and have a high gloss and low friction surface. Can be chromium plated, injection-molded, extruded, and blow-molded. Not recommended for use in strong acids or alkalies. They may be homopolymers or copolymers.
Properties: D 1.425, thermal conductivity 2.6 Btu-in/(hr $*$ ft^2 $*$ degF), coefficient of thermal expansion 4.5 × 10^{-5}/degree F, specific heat 0.35 Btu/(lb)(degree F), water absorption 0.41%/24 hour, tensile strength 10,000 psi, elongation 15%, hardness (Rockwell) R120, impact strength (notched) 1.4 ft-lb/inch, flexural strength 14,100 psi, shear strength 9500 psi. Combustible, but slow burning.
Use: An engineering plastic, often used as substitute for metals, as in oil and gas pipes; automotive and appliance parts; industrial parts; hardware; communication equipment; aerosol containers for cosmetics.
See "Delrin" *[Du Pont]*; "Celcon" *[CNA]*.

acetamide. (acetic acid amine; ethanamide).
CAS: 60-35-5. CH_3CONH_2.
Properties: Colorless, deliquescent crystals; mousy odor. D 1.159, mp 80C, bp 223C, refr index 1.4274 (78.3C). Soluble in water and alcohol; slightly soluble in ether. Combustible.
Derivation: Interaction of ethyl acetate and ammonium hydroxide.
Grade: Technical, CP (odorless), intermediate, reagent.
Hazard: A possible carcinogen.
Use: Organic synthesis (reactant, solvent, peroxide stabilizer), general solvent, lacquers, explosives,

soldering flux, hygroscopic agent, wetting agent, penetrating agent.

acetamidine hydrochloride. $C_2H_6N_2HCl$.
  Properties: Crystalline solid. Slightly deliquescent. Mp 166C. Soluble in water and alcohol; insoluble in acetone. Keep stoppered.
  Derivation: Alcohol solution of acetonitrile + HCl + ammonia.
  Hazard: Skin irritant, moderately toxic by ingestion.
  Use: Synthesis of pyrimidines and related groups of biochemically active compounds.

acetamido-.   Prefix indicating the group CH_3CONH-. Also called acetamino- or acetylamino-.

3-acetamido-5-aminobenzoic acid.
  Use: Intermediate in the manufacture of X-ray contrast media.

5-acetamido-8-amino-2-naphthalenesulfonic acid.   (acetyl-1,4-naphthalenediamine-7-sulfonic acid; acetylamino-1,6-Cleve's acid). $C_{10}H_5NHCOCH_3(NH_2)(SO_3H)$. A reddish-brown paste.
  Hazard: Toxic.
  Use: Chemical intermediate, dyes.

8-acetamido-5-amino-2-naphthalenesulfonic acid.   (acetyl-1,4 – napthalenediamine-6-sulfonic acid; acetylamino-1,7-Cleve's acid). $C_{10}H_5(NHCOCH)(NH_2)(SO_3H)$.
  Properties: Paste.
  Hazard: Toxic.
  Use: A chemical intermediate and in dyes.

p-acetamidobenzenesulfonyl chloride.   See N-acetylsulfanilyl chloride.

((p-acetamidobenzoyl)oxy)tributylstannane.
  See tributyltin-p-acetamidobenzoate.

acetamidocyanoacetic ester.   See ethyl acetamidocyanoacetate.

4-acetamido-2-etholxbenzoic acid methyl ester.   See ethopabate.

8-acetamido-2-naphthalenesulfonic acid magnesium salt.   (acetyl-1,7-Cleve's acid). $[C_{10}H_6(CH_3CONH)(SO_3)]_2Mg$.
  Properties: Brownish-gray paste containing approximately 80% solids.
  Use: Intermediate for dyes.

p-acetamidophenol.   See p-acetylaminophenol.

Acetamine.   A group of azo dyes and developers made for application to acetate yarn, and especially suited to nylon.

acetamino-.   See acetamido-.

acetaminophen.   See p-acetylaminophenol.

acetanilide.   (N-phenylacetamide).
  CAS: 103-84-4. $C_6H_5NH(COCH_3)$.
  Properties: White, shining crystalline leaflets or white, crystalline powder; odorless; slightly burning taste. Stable in air. D 1.2105, mp 114–116C, bp 303.8C. Soluble in hot water, alcohol, ether, chloroform, acetone, glycerol, and benzene. Flash p 345F (174C); autoign temp 1015F (545C). Combustible.
  Derivation: Acetylation of aniline with glacial acetic acid.
  Grade: Technical, CP.
  Hazard: Toxic by ingestion.
  Use: Rubber accelerator, inhibitor in hydrogen peroxide, stabilizer for cellulose ester coatings, manufacture of intermediates (p-nitroaniline, p-nitroacetanilide, p-phenylenediamine), synthetic camphor, pharmaceutical chemicals, dyestuffs, precursor in penicillin manufacture, medicine (antiseptic), acetanisole.
  See p-methoxyacetophenone.

acetate.   (1) A salt of acetic acid in which the terminal hydrogen atom is replaced by a metal, as in copper acetate, $Cu(CH_3COO)_2$. (2) An ester of acetic acid where the substitution is by a radical as in ethyl acetate, $CH_3COOC_2H_5$. In cellulose acetate the hydroxyl radicals of the cellulose are involved in the esterification.
  See cellulose acetate; vinyl acetate.

acetate dye.   One group consists of water insoluble azo or anthraquinone dyes that have been highly dispersed to make them capable of penetrating and dyeing acetate fibers. A second class group comprises water-insoluble amino azo dyes that are made water soluble by treatment with formaldehyde and bisulfite. After absorption by the fiber, the resulting sulfonic acids hydrolyze and regenerate the insoluble dyes.

acetate fiber.   A manufactured fiber in which the fiber-forming substance is cellulose acetate. Where not less than 92% of the hydroxyl groups are acetylated, the term *triacetate* may be used as a generic description of the fiber (Federal Trade Commission). This fiber was formerly called acetate rayon or acetate silk The term rayon is not permissible for this type.
  Properties: Thermoplastic; becomes tacky at 350F (176C). Moisture absorption 6%. Tenacity 1.4 g/denier (dry); about 1 g/denier (wet). Elongation 50% dry, 40% wet. Soluble in acetone and glacial acetic acid; decomposed by concentrated solutions of strong acids and alkalies. Combustible.
  Use: Wearing apparel, industrial fabrics.
  See cellulose acetate; cellulose triacetate.

acetate fiber, saponified. Regenerated cellulose fibers obtained by complete saponification of highly oriented cellulose acetate fibers.
Properties: Tensile strength (psi) 136,000–155,000; elongation 6%; d 1.5–1.6; moisture regain 9.6–10.7%; decomposes at about 149C. Similar to cotton in chemical resistance, dyeing, and resistance to insects and mildew. Combustible.
Available forms: Available in continuous filament form having a high degree of crystallinity and great strength.
Use: Cargo parachutes, typewriter ribbons, belts, webbing, tapes, carpet backing.

acetate film. A durable, highly transparent film with nondeforming characteristics produced from cellulose acetate resin. It is grease, oil, dust, and air proof and hygienic. Combustible.
Available forms: Rolls and cut-to-size sheets.
Use: Laminates, support for photographic film, document preservation, pressure-sensitive tape, magnetic sound-recording tape, window cartons, envelope packaging.

acetate kinase. (acetokinase). A phosphotransferase that catalyzes the formation of acetyl phosphate from ATP and acetate.

acetate of lime. Commercial term for calcium acetate made from pyroligneous acid and milk of lime. There are brown and gray acetates of lime.
See calcium acetate.

acetate process. See cellulose acetate.

acetate rayon. The yarn made from refined wood cellulose by the acetate process.
See acetate fiber.

acetate rayon process. In this process cellulose is combined with acetic anhydride to make cellulose acetate, which is dissolved in acetone and forced through spinnerette holes into a precipitating bath.
See acetate fiber.

acethydrazidepyridinium chloride. See Girard's reagent.

acetic acid. (ethanoic acid; vinegar acid; methanecarboxylic acid).
CAS: 64-19-7. (CH_3COOH). Glacial acetic acid is the pure compound (99.8% min), as distinguished from the usual water solutions known as acetic acid.
Properties: Clear, colorless liquid; pungent odor. Mp 16.63C, bp 118C (765 mm Hg), 80C (202 mm Hg), d 1.0492 (20/4C), wt/gal (20C) 8.64 lb, viscosity (20C) 1.22 cP, flash p 110F (43C) (OC), refr index 1.3715 (20C), autoign temp 800F (426C). Combustible. Miscible with water, alcohol, glycerol, and ether; insoluble in carbon disulfide.
Derivation: (a) Liquid- and vapor-phase oxidation of petroleum gases (with catalyst), (b) oxidation

of acetaldehyde, (c) reaction of methanol and carbon monoxide (with catalyst; this is the most cost-efficient method and has been in general use for some years), (d) fermentative oxidation of ethanol.
Grade: USP (glacial, 99.4 wt%, and dilute, 36–37 wt%), CP, technical (80, 99.5%), commercial (6, 28, 30, 36, 56, 60, 70, 80, and 99.5%), NF (diluted 6.0 g/100 mL).
Hazard: Moderate fire risk. Pure acetic acid is moderately toxic by ingestion and inhalation, but dilute material is approved by FDA for food use. Strong irritant to eyes, skin, and tissue; upper respiratory tract irritant and pulmonary function effects.
Use: Manufacturing of acetic anhydride, cellulose acetate, and vinyl acetate monomer; acetic esters; chloroacetic acid; production of plastics, pharmaceuticals, dyes, insecticides, photographic chemicals, etc.; food additive (acidulant); latex coagulant; oil-well acidizer; textile printing.
See vinegar.

acetic acid amine. See acetamide.

acetic acid benzyl ester. See benzyl acetate.

acetic acid, glacial. See acetic acid.

acetic acid, ((octylstannylidyne)trithio)tri-, tris(2-ethylhexyl) ester. See octyltris(2-ethylhexyloxycarbonylmethylthio)stannane.

acetic acid phenylmethyl ester. See benzyl acetate.

acetic aldehyde. See acetaldehyde.

acetic anhydride. (acetyl oxide; acetic oxide).
CAS: 108-24-7. $(CH_3CO)_2O$.
Properties: Colorless, mobile, strongly refractive liquid; strong odor. D 1.0830 (20/20C), bp 139.9C, fp −73.1C, flash p 121F (49.4C) (CC), autoign temp 732F (385C), wt/gal 9.01 lb (20C). Miscible with alcohol, ether, and acetic acid; soluble in cold water; decomposes in hot water to form acetic acid. Combustible.
Derivation: (1) Oxidation of acetaldehyde with air or oxygen with catalyst; (2) by catalyzed thermal decomposition of acetic acid to ketone; (3) reaction of ethyl acetate and carbon monoxide; (4) from carbon monoxide and methanol.
Grade: CP, technical (75, 85, 90–95%).
Hazard: Strong irritant and corrosive; may cause burns and eye damage. Moderate fire risk. Eye and upper respiratory tract irritant. Questionable carcinogen.
Use: Cellulose acetate fibers and plastics; vinyl acetate; dehydrating and acetylating agent in production of pharmaceuticals, dyes, perfumes, explosives; etc., aspirin. Esterifying agent for food starch (5% max).

acetic ester. See ethyl acetate.

acetic ether.   See ethyl acetate.

acetic oxide.   See acetic anhydride.

acetin.   (monoacetin; glyceryl monoacetate).
CAS: 102-76-1. $C_3H_5(OH)_2OOCCH_3$. Acetin may
also refer to glyceryl di- or triacetate, also known
as diacetin and triacetin.
Properties: Colorless, thick liquid. Hygroscopic. D
1.206 (20/4C), bp 158C (165 mm), 130C (3 mm).
Soluble in water, alcohol; slightly soluble in ether;
insoluble in benzene. Combustible.
Derivation: By heating glycerol and strong acetic
acid, distilling off the weak acetic acid formed, and
again heating with strong acetic acid and distilling.
Method of purification: Rectification.
Hazard: Moderately toxic, irritant.
Use: Tanning; solvent for dyes, food additive, gela-
tinizing agent in explosives.

acetoacetanilide.   (acetylacetanilide).
CAS: 102-01-2. $CH_3COCH_2CONHC_6H_5$.
Properties: White, crystalline solid. Mp 85C, d
1.26, flash p 325F (162.7C). Resembles ethyl acetoacetate in chemical reactivity. Slightly soluble in
water; soluble in dilute sodium hydroxide, alcohol,
ether, acids, chloroform, and hot benzene. Combustible.
Derivation: By reacting ethyl acetoacetate with aniline, and eliminating ethanol. Acetoacetanilide may
also be prepared from aniline and diketene.
Grade: Technical.
Use: Organic synthesis; dyestuffs (intermediate in
the manufacture of the dry colors generally referred
to as Hansa and benzidine yellows).

acetoacet-*o*-anisidide.
$CH_3COCH_2CONHC_6H_4OCH_3$.
Properties: White, crystalline powder. Mp 86.6C,
d 1.1320 (86.6/20C), flash p 325F (162.7C) (OC).
Combustible.
Use: Intermediate for azo pigments.

acetoacet-*o*-chloranilide.
CAS: 93-70-9. $CH_3COCH_2CONHC_6H_4Cl$.
Properties: White, crystalline powder. Mp 107C, bp
(decomposes), d 1.1920 (107/20C), flash p 350F
(176.6C) (OC). Almost insoluble in water. Com-
bustible.
Hazard: Toxic by ingestion.
Use: Intermediate for azo pigments.

acetoacet-*p*-chloranilide.
$CH_3COCH_2CONHC_6H_4Cl$.
Properties: White, crystalline powder. Mp 133C, bp
(decomposes), flash p 320F (160C) (OC). Com-
bustible. Very slightly soluble in water.
Hazard: Toxic by ingestion.
Use: Intermediate for azo pigments.

acetoacetic acid.   (acetylacetic acid; diacetic
acid; acetone carboxylic acid). CH_3COCH_2COOH.

Properties: Colorless, oily liquid. Soluble in water,
alcohol, and ether; decomposes below 100C into
acetone and carbon dioxide.
Hazard: Irritant to eyes and skin.
Use: Organic synthesis.

acetoacet-*p*-phenetidide.
$CH_3COCH_2CONHC_6H_4OCH_2CH_3$.
Properties: Crystalline powder. Mp 108.5C, bp
(decomposes), d 1.0378 (108.5/20C), flash p 325F
(162.7C) (OC). Combustible.
Hazard: Moderately toxic by ingestion.
Use: Intermediate for azo pigments.

acetoacet-*o*-toluidide.
$CH_3COCH_2CONHC_6H_4CH_3$.
Properties: Fine, white, granular powder. Mp 106C,
bp (decomposes), d 1.062 (106C). Slightly soluble
in water. Flash p 320F (160C). Combustible.
Grade: Technical.
Hazard: Moderately toxic.
Use: Intermediate in the manufacture of Hansa and
benzidine yellows.

acetoacet-*p*-toluidide.
$CH_3COCH_2CONHC_6H_4CH_3$.
Properties: White, crystalline powder. Mp 93.0–
96.0C, purity 99% min.
Hazard: Moderately toxic.
Use: Light-fast yellow pigment intermediate; diazo
coupler.

acetoacet-*m*-xylidide.   (AAMX).
$(CH_3)_2C_6H_3NHCOCH_2COCH_3$.
Properties: White to light-yellow crystals. Mp 89–
90C, d 1.238 (20C). Water solubility 0.5% (25C).
Flash p 340F (171C); combustible.
Use: Intermediate for yellow pigments.

acetoacetate.   (3-oxobutanoic acid).
CAS: 105-45-3. $C_4H_5O_3$.
Properties: Salt or ion of acetoacetic acid; can spon-
taneously release carbon dioxide, forming acetone.

acetoaminofluorene.   A pesticide. May not be
used in food products or beverages (FDA).
Hazard: Toxic by ingestion.

***p*-acetoanisole.**   See *p*-methoxyacetophenone.

acetoglyceride.   Usually an acetylated monoglyceride, although commercial acetoglycerides
will contain di- and triglycerides.
See acetostearin.

acetoin.   See acetylmethylcarbinol.

acetol.   See hydroxy-2-propanone.

acetolysis.   The procedure whereby acetyl compounds are treated with alcoholic or aqueous alkalies to remove the acetyl groups.

acetomeroctol.
$CH_3COOHgC_6H_3(OH)C(CH_3)_2CH_2C(CH_3)_3$.
Properties: White solid. Mp 155–157C. Freely soluble in alcohol; soluble in ether, chloroform; sparingly soluble in benzene; insoluble in water.
Hazard: Toxic by ingestion.
Use: Medicine (antiseptic, solution 1:1000).

acetone.   (dimethylketone; 2-propanone).
CAS: 67-64-1. CH_3COCH_3.

$$\underset{\underset{O}{\parallel}}{CH_3CCH_3}$$

Properties: Colorless, volatile liquid; sweetish odor. Mp −94.3C, bp 56.2C, refr index 1.3591 (20C), d 0.792 (20/20C), wt/gal 6.64 lb (15C), flash p 15F (−9.4C) (OC), autoign temp 1000F (537C). Miscible with water, alcohol, ether, chloroform, and most oils.
Derivation: (1) Oxidation of cumene; (2) dehydrogenation or oxidation of isopropyl alcohol with metallic catalyst; (3) vapor-phase oxidation of butane; (4) by-product of synthetic glycerol production.
Grade: Technical; CP; NF; electronic; spectrophotometric.
Hazard: Flammable, dangerous fire risk. Explosive limits in air 2.6–12.8%. Narcotic in high concentrations. Moderately toxic by ingestion and inhalation. Eye and upper respiratory tract irritant, central nervous system impairment, and hematologic effects. Questionable carcinogen.
Use: Chemicals (methyl isobutyl ketone; methyl isobutyl carbinol; methyl methacrylate; bisphenol-A); paint, varnish, and lacquer solvent; cellulose acetate, especially as spinning solvent; to clean and dry parts of precision equipment; solvent for potassium iodide and permanganate; delusterant for cellulose acetate fibers; specification testing of vulcanized rubber products.

acetone bromoform.   See tribromo-*tert*-butyl alcohol.

acetonecarboxylic acid.   See acetoacetic acid.

acetone chloroform.   See chlorobutanol.

acetone cyanohydrin.   (α-hydroxyisobutyronitrile; 2-methyllactonitrile).
CAS: 75-86-5. $(CH_3)_2COHCN$.
Properties: Colorless liquid. Bp 82C (23 mm), mp −20C, d 0.932 (19C), refr index 1.3996 (20C), flash p 165F (73.9C), autoign temp 1270F (685C). Soluble in alcohol and ether. Combustible.
Derivation: Condensing acetone with hydrocyanic acid.
Grade: Technical (97–98% pure).
Hazard: Toxic. Readily decomposes to hydrocyanic acid and acetone. Upper respiratory tract irritant, headache, hypoxia, and cyanosis.

Use: Insecticides; intermediate for organic synthesis, especially methyl methacrylate.

acetonedicarboxyllic acid.   See β-ketoglutaric acid.

acetone number.   The degree of polymerization of a bodied vegetable oil, measured by the amount of matter that is insoluble in acetone.

acetone oxime.   See acetoxime.

acetone semicarbazone. $(CH_3)_2CNNH$ $CONH_2$. A chemical intermediate.
Properties: White powder. Mp 188C.

acetone sodium bisulfate.   See sodium acetone bisulfate.

acetonitrile.   (methyl cyanide).
CAS: 75-05-8. CH_3CN.
Properties: Colorless, limpid liquid; aromatic odor. D 0.783; mp −41C; bp 82C; flash p 42F (5.56C). Soluble in water and alcohol; high dielectric constant; high polarity; strongly reactive.
Derivation: By-product of propylene-ammonia process for acrylonitrile.
Grade: Technical; nanograde; spectrophotometric.
Hazard: Flammable, dangerous fire risk. Toxic action by skin absorption and inhalation. Lower respiratory tract irritant. Questionable carcinogen.
Use: Solvent in hydrocarbon extraction processes, especially for butadiene; specialty solvent; intermediate; catalyst; separation of fatty acids from vegetable oils; manufacturing of synthetic pharmaceuticals.

acetonylacetone.   (1,2-diacetylethane; hexanedione-2,5; 2,5-diketohexane).
CAS: 110-13-4. $CH_3COCH_2CH_2COCH_3$.
Properties: Colorless liquid. D 0.9734 (20/20C); bp 192.2C; vap press 0.43 mm at 20C; fp −5.4C; flash p 185F (85C); bulk d 8.2 lb/gal (20C). Combustible; autoign temp 920F (493C). Soluble in water.
Derivation: By-product in the production of acetaldehyde from acetylene.
Grade: Technical.
Hazard: Irritant to eyes and skin.
Use: Solvent for cellulose acetate, roll-coating inks, lacquers, stains; intermediate for pharmaceuticals and photographic chemicals; electroplating.

acetonyl alcohol.   See hydroxy-2-propanone.

3-(α-acetonylbenzyl)-4-hydroxycoumarin.
See Warfarin.

3-(α-acetonylfurfuryl)-4-hydroxycoumarin.
(sodium salt also used). A rodenticide.
Hazard: Highly toxic by ingestion and inhalation.

3-(α-acetonyl-4-nitrobenzyl)-4-hydroxycoumarin.   See acenocoumarin.

(s)-acetophan.   See sinorphan.

acetophenetidin.   (*p*-acetylphenetidin; ace-tophenetidide; phenacetin; *p*-ethoxyacetanilide).
CAS: 62-44-2. $CH_3CONHC_6H_4OC_2H_5$.
Properties: White crystals or powder; odorless and stable in air. Mp 135C. Soluble in alcohol, chloroform, and ether; slightly soluble in water; has slightly bitter taste.
Derivation: By the interaction of *p*-phenetidin and glacial acetic acid or of ethyl bromide and *p*-acetaminophenol.
Method of purification: Crystallization.
Grade: Technical; USP, as phenacetin.
Hazard: Toxic by ingestion. Upper respiratory tract irritant, central nervous system impairment, and pregnancy loss. Confirmed carcinogen.
Use: Medicine (analgesic); veterinary medicine.

acetophenone.   (phenylmethylketone; hypnone; acetylbenzene).
CAS: 98-86-2. $C_6H_5COCH_3$.

Properties: Colorless liquid; sweet, pungent odor and taste. Bp 201.7, fp 19.7C, d 1.030 (20/20C), bulk d 8.56 lb/gal (20C), refr index 1.5363 (20C), flash p 180F (82.2C) (COC). Slightly soluble in water; soluble in organic solvents and sulfuric acid. Combustible.
Derivation: (1) Friedel–Crafts process with benzene and acetic anhydride or acetyl chloride; (2) by-product from the oxidation of cumene; (3) oxidation of ethylbenzene.
Method of purification: Distillation and crystallization.
Grade: Technical, refined, perfumery.
Hazard: Narcotic in high concentrations, hypnotic.
Use: Perfumery; solvent; intermediate for pharmaceuticals, resins, etc.; flavoring; polymerization catalyst; organic synthesis.

acetophenone oxime.
Properties: Crystals.
Use: Antiozonant properties, antioxidant, antiskinning agent, piezoelectric properties, emulsifier–water/oleoresinous systems, end blocker, polymerization short stopper.

acetostearin.
$CH_3(CH_3)_{16}COOCH_2CH_2OCH_2OOCCH_3$.
Acetylated glyceryl monostearate. Solid with peculiar combination of flexibility and nongreasiness. Derived from glyceryl monostearate or mixed glycerides by acetylation with acetic anhydride.
Use: Protective coating for food and as a plasticizer.

acetotoluidide.   See acetyl-*o*-toluidine; acetyl-*p*-toluidine.

acetoxime.   (acetone oxime; 2-propanone oxime). $(CH_3)_2CNOH$.

Properties: Colorless crystals; chloral-like odor. Both basic and acidic in properties. Volatilizes in air. D 0.97 (20/20C), bp 136.3C, mp 61C. Fairly easily hydrolyzed by dilute acids; soluble in alcohols, ethers, water. Combustible.
Derivation: Reaction of hydroxylamine in water solution with acetone, followed by ether extraction.
Grade: Technical.
Use: Organic synthesis (intermediate); solvent for cellulose ethers; primer for diesel fuels.

o-**acetoxybenzoic acid.**   See aspirin.

acetoxylation.   A method of synthesizing ethylene glycol in which ethylene is reacted with acetic acid in the presence of a catalyst, such as tellurium bromide, resulting in the formation of mixed mono- and diacetates; this is followed by hydrolysis to ethylene glycol and acetic acid, with up to 95% yield of the glycol. It is thus considerably more efficient than the ethylene oxide method.

acetoxydiethylphenylstannane.   See diethyl phenyltin acetate.

acetoxydiphenylethylstannane.   See ethyl-diphenyltin acetate.

4-(*p*-acetoxyphenyl)-2-butanone.   See cue-lure.

(*p*-acetoxyphenyl)methyl carbinol.   See 4-acetoxyphenyl methyl carbinol-4-acetoxyphenyl methyl carbinol.

4-acetoxyphenyl   methyl   carbinol-4-acetoxyphenyl methyl carbinol.
CAS: 53744-50-6. $C_{10}H_{12}O_3$.
Hazard: Toxic by ingestion. A severe skin irritant.

α-**acetoxytoluene.**   See benzyl acetate.

acetoxytriethylstannane.
CAS: 1907-13-7. $C_8H_{18}O_2Sn$.
Hazard: A poison by ingestion.

acetoxytriethyltin.   See acetoxytriethylstannane.

acetoxytrihexylstannane.
CAS: 2897-46-3. $C_{20}H_{42}O_2Sn$.
Hazard: A poison by skin contact. Moderately toxic by ingestion.

acetoxytrihexyltin.   See acetoxytrihexylstannane.

acetoxytriisopropylstannane.   See   triisopropyltin acetate.

acetoxytrimethylstannane.   See   trimethyltin acetate.

acetoxytripropylstannane.   See   tripropyltin acetate.

acetoxytris(β,β-dimethylphenethyl)stannane. See trineophyltin acetate.

acetozone.   See acetylbenzoyl peroxide.

"Acetulan" [Lubrizol].   TM for a special fraction of acetylated lanolin alcohols.
   Properties: Anhydrous; pale-straw color; nearly odorless, low-viscosity liquid. Miscible with ethanol, mineral oil, and other common formulating materials; insoluble in water. Combustible.
   Use: Hydrophobic penetrant, emollient, plasticizer, cosolvent, pigment dispersant, cosmetics.

acetylacetanilide.   See acetoacetanilide.

acetylacetic acid.   See acetoacetic acid.

acetylacetonate.   ((Z)-4-oxopent-2-en-2-olate). $C_5H_7O_2$. A chelate of acetylacetone with any number of metals such as aluminum or beryllium.
   Use: Extraction in organic solvents.

acetylacetonatobis(ethylene)rhodium(1).
   Properties: Orange crystal; air stable. Soluble in chloroform and ether.
   Use: Reactive ethylene complex.

acetylacetonatodicarbonylrhodium(1).
   Properties: Green crystal grades (dichroic red when crushed); air stable. Soluble in acetone, benzene, and chlorinated solvents.
   Use: Homogeneous catalyst for hydroformylation reactions.

acetylacetone.   (diacetylmethane; pentanedione-2,4).
   CAS: 123-54-6. $CH_3COCH_2OCCH_3$.
   Properties: Mobile, colorless or yellowish liquid. When cooled, solidifies to lustrous, pearly spangles. The liquid is affected by light, turning brown and forming resins. Bp 140.5C; d 0.9753 (20/20C); bulk d 8.1 lb/gal; fp −23.5C; flash p 105F (40.5C) (TOC). Soluble in water (acidified by hydrochloric acid); fairly soluble in neutral water, alcohol, chloroform, ether, benzene, acetone, and glacial acetic acid. Combustible.
   Derivation: By condensing ethyl acetate with acetone.
   Hazard: Moderately toxic; moderate fire risk.
   Use: Solvent for cellulose acetate, intermediate, chelating agent for metals, paint driers, lubricant additives, pesticides.

3-acetylacrolein.   See 4-ketopentenal.

N-acetyl-l-(+)-alanine.   See alanine.

acetylamino-.   See acetamido-.

p-acetylaminobenzenesulfonyl   chloride. See N-acetylsulfanilyl chloride.

o-acetylaminobenzoic   acid.   See acetylanthranilic acid.

p-acetylaminophenol.   (APAP;   N-acetyl-p-aminophenol; acetaminophen; p-acetamidophenol; p-hydroxyacetanilide).
   CAS: 103-90-2. $CH_3CONHC_6H_4OH$.
   Properties: Crystals; odorless; slightly bitter taste. D 1.293 (21/4C), mp 168C. Slightly soluble in water and ether; soluble in alcohol; pH saturated aqueous solution 5.5–6.5.
   Derivation: Interaction of p-aminophenol and an aqueous solution of acetic anhydride.
   Use: Intermediate for pharmaceuticals and azo dyes, stabilizer for hydrogen peroxide, photographic chemicals, medicine (analgesic).

p-acetylaminophenyl salicylate.   (phenetsal). $C_6H_4(NHCOCH_3)OOCC_6H_4OH$.
   Properties: Fine, white, crystal scales; odorless; tasteless. Mp 187–188C. Soluble in alcohol, ether, and hot water; insoluble in light hydrocarbon solvents; decomposed by strong alkalies.
   Derivation: By reducing p-nitrophenol salicylate to p-aminophenol salicylate and acetylating the latter.
   Hazard: See aspirin.
   Use: Medicine (analgesic).

p-acetylanisole.   See p-methoxyacetophenone.

N-acetylanthranilic   acid.   (o-acetylaminobenzoic acid). $CH_3CONHC_6H_4COOH$.
   Properties: Needles, plates, rhombic crystals (from glacial acetic acid). Mp 185C. Slightly soluble in water; soluble in hot alcohol, ether, and benzene. Combustible.
   Derivation: Oxidation of o-acetyltoluidine with potassium permanganate in the presence of magnesium sulfate or potassium chloride.
   Grade: Technical.
   Use: Chemical (organic synthesis, anthranilic acid).

acetylase.   (acetyltransacetylase;   acetyltransferase; transacetylase). Any enzyme that catalyzes acetylation or deacetylation and also reverses the reaction.

acetylated monoglycerides.
   Properties: Esters of glycerin with acetic acid and edible fat-forming fatty acids. (FCC III) May be white to pale yellow liquids or solids; bland taste. Sol in alc, acetone; insol in water.
   Use: Food additive.

acetylation. Introduction of an acetyl radical (CH_3CO-) into the molecule of an organic compound having OH or NH_2 groups. The usual reagents for this purpose are acetic anhydride or acetyl chloride. Thus, ethanol (C_2H_5OH) may be converted to ethyl acetate ($C_2H_5OCOCH_3$). Cellulose is similarly converted to cellulose acetate by treatment with a mix containing acetic anhydride. Acetylation is commonly used to determine the number of hydroxyl groups in fats and oils. See acetyl value.

acetylbenzene. See acetophenone.

acetylbenzoyl aconine. See aconitine.

acetylbenzoyl peroxide. (acetozone; benzozone).
CAS: 644-31-5. $C_6H_5CO•O_2•OCCH_3$.
Properties: White crystals. Decomposed by water, alkaloids, organic matter, and some organic solvents; decomposes slowly and evaporates when gently heated, and instantaneously (possibly explosively) if quickly heated, ground, or compressed. Mp 36.6C; bp 130C (19 mm). Moderately soluble in ether, chloroform, carbon tetrachloride, and water; slightly soluble in mineral oils and alcohol. The commercial product is mixed with a neutral drying powder and contains 50% acetyl benzoyl peroxide.
Hazard: Toxic by ingestion. Strong irritant to skin and mucus membranes. Strong oxidizing agent; dangerous in contact with organic materials. Moderate explosion risk when shocked or heated.
Use: Medicine (active germicide); disinfectant; bleaching flour.

acetyl bromide.
CAS: 506-96-7. CH_3COBr.
Properties: Colorless, fuming liquid; turns yellow in air. Bp 81C; mp −96C; d 1.663 (16/4C). Soluble in ether, chloroform and benzene.
Derivation: Interaction of acetic acid and phosphorus pentabromide.
Grade: Technical.
Hazard: Toxic by ingestion and inhalation; strong irritant to eyes and skin. Reacts violently with water or alcohol.
Use: Organic synthesis, dye manufacture.

α-acetylbutyrolactone. (α-acetobutyrolactone).

$CH_2CH_2CH(OCCH_3)C(O)O$.
Properties: Liquid; ester-like odor. D 1.18–1.19 (20C), bp 142–144 (30 mm). Partially soluble in water. Combustible.
Derivation: Sodium acetoacetate and ethylene oxide in absolute alcohol.
Hazard: Strong irritant to skin and mucus membranes.

Use: Organic synthesis.

acetyl carbinol. See hydroxy-2-propanone.

acetyl carbromal. (N-acetyl-N-bromodiethylacetylurea). $(C_2H_5)_2CBrCONHCONHCOCH_3$.
Properties: Crystals; slightly bitter taste. Mp 109C. Slightly soluble in water; freely soluble in alcohol and ethyl acetate.
Hazard: Overdoses may be fatal.
Use: Medicine (sedative).

acetyl chloride. (ethanoyl chloride).
CAS: 75-36-5. CH_3COCl.
Properties: Colorless, highly refractive, fuming liquid; strong odor. D 1.1051, mp −112C, bp 51–52C, flash p 40F (4.4C) (CC). Soluble in ether, acetone, acetic acid.
Derivation: By mixing glacial acetic acid and phosphorus trichloride in the cold and heating a short time to drive off hydrochloric acid. The acetyl chloride is then distilled.
Hazard: Corrosive to skin and mucous membranes; toxic; strong eye irritant. Flammable, dangerous fire risk. Reacts violently with water and alcohol.
Use: Organic preparations (acetylating agent); dyestuffs; pharmaceuticals.

acetylcholine. (acetylethanoltrimethylammonium hydroxide).
CAS: 51-84-3.

$$CH_3-\overset{\underset{|}{CH_3}}{\underset{|}{N}}-CH_2-CH_2-O-\overset{O}{\overset{\|}{C}}-CH_3$$

A derivative of choline; important because it acts as the chemical transmitter of nerve impulses in the autonomic system. It has been isolated and identified in brain tissue. The enzyme cholinesterase hydrolyzes acetylcholine into choline and acetic acid, and is necessary in the body to prevent acetylcholine poisoning.
Use: Medicine (as bromide and chloride); biochemical research.
See nerve gas; cholinesterase inhibitor.

acetylcholine chloride. (2-(acetyloxy)-N,N,N-trimethylethanaminium chloride; 2-acetyloxyethyl(trimethyl)azaniumchloride).
CAS: 60-31-1. $C_7H_{16}NO_2•Cl$. A neurotransmitter found at neuromuscular junctions, autonomic ganglia, parasympathetic effector junctions, a subset of sympathetic effector junctions and at many sites in the central nervous system.
Properties: Deliquescent crystalline powder; miotic agent; cholinergic drug.
Hazard: Toxic.

acetylcholinesterase. See cholinesterase.

acetylcholinesterase inhibitor. (anticholinesterase). A drug that blocks acetylcholinesterase, thereby increasing levels of acetylcholine in the synapse.

acetylcholinesterase reactivation. Acetylcholinesterase inhibition is spontaneously reversible by hydrolysis that removes the carbamylating or phosphorylating moiety from the enzyme.

acetyl-CoA acetyltransferase. (acetoacetyl-CoA thiolase; acetyl-CoA thiolase; thiolase). An acetyltransferase that produces acetoacetyl-CoA from two molecules of acetyl-CoA with the release of one CoA.

acetyl-CoA hydrolase. (acetyl-CoA acylase; acetylCoA deacylase). An enzyme that catalyzes the splitting of acetate from acetyl-CoA.

acetyl-CoA synthetase. A long-chain fatty acid-CoA ligase that catalyzes the formation of acyl-CoA.

acetylcoenzyme A. (acetyl CoA; active acetate).
CAS: 72-89-9. A high-energy ester of acetic acid that plays important roles in the tricarboxylic acid cycle in fatty acid biosynthesis, and participates in many metabolic acetylation reactions.
Derivation: It is a condensation product of coenzyme a and acetic acid that results mainly from the metabolism of glucose and fatty acids in the mitochondria resulting in the release of acetyl groups that combine with the thiol group of coenzyme a.

n-acetylcysteine conjugate. (mercapturic acid; (2R)-2-acetamido-3-sulfanylpropanoic acid; n-acetyl-l-cysteine; acetein; acetylcysteine, n-acetylcysteine; airbron; broncholysin; fluimucetin; fluimucil; flumicil; inspir; mucolyicum; mucolyticum lappe; mucomyst; mucosolvin; nac; nac-tb; parvolex; respaire).$C_5H_9NO_3S$. Any of a class of compounds that are terminal receptors in the detoxication of potentially harmful electrophiles. It is the *N*-acetyl derivative of cysteine. It has been shown to have antiviral effects in patients with HIV due to inhibition of viral stimulation by reactive oxygen intermediates.
Properties: Crystals from water.
Use: As a mucolytic agent to reduce the viscosity of mucous secretions.

N-acetyl-L-cysteine (USAN). $HSCH_2CH$ ($NHCOCH_3$)COOH. The *N*-acetyl derivative of the naturally occurring amino acid, L-cysteine.
Use: Medicine, biochemical research.

α-acetyldigitoxin (anhydrous). $C_{43}H_{66}O_{14}$.
Properties: Crystals. Sparingly soluble in water, ether, petroleum ether; soluble in most organic solvents.

Derivation: Obtained by enzymatic hydrolysis of a digitalis extract.
Hazard: Toxic.
Use: Medicine (heart disease).

acetyldigoxin. (Lantailin, Desglucolanatoside C, 3′′′-Acetyldigoxin, alpha-acetyldigoxin, sandolanid, Lanadigin, Digorid A, Acetyl-digoxin-alpha). $C_{43}H_{66}O_{15}$. Alpha- or beta-acetyl derivatives of digoxin or lanatoside C from *Digitalis lanata*.
Use: In congestive heart failure.

3-acetyl-2,5-dimethylfuran.
CAS: 10599-70-9. $C_8H_{10}O_2$.
Properties: Yellow liquid; strong roasted nutlike odor. D: 1.027–1.048, refr index: 1.475–1.496 (25 degree Celsius). Sol in alc, propylene glycol, fixed oils; sltly sol in water.
Use: Food additive.

acetyl dipterex. See chloracetophos.

acetylene. (ethene).
CAS: 74-86-2.   HC≡CH.
Properties: Colorless gas; ethereal odor. D 0.91 (air = 1), mp −81.8C (890 mm); bp −84C, flash p 0F (−17.7C) (CC), autoign temp 635F (335C). Soluble in alcohol and acetone; slightly soluble in water. An asphyxiant gas.
Derivation: (a) By the action of water on calcium carbide; (b) by cracking petroleum hydrocarbons with steam (Wulff process), or natural gas by partial oxidation (BASF process); (c) from fuel oil by modified arc process.
Grade: Technical, containing 98% acetylene and not more than 0.05% by volume of phosphine or hydrogen sulfide; 99.5%.
Hazard: Very flammable, dangerous fire risk; burns with intensely hot flame; explosive limits in air 2.5–80%. A asphyxiate. Forms explosive compounds with silver, mercury, and copper, which should be excluded from contact with acetylene in transmission systems. Copper alloys may be used with caution. Piping used should be electrically bonded and grounded.
Use: Manufacture of vinyl chloride; vinylidene chloride; vinyl acetate; acrylates; acrylonitrile; acetaldehyde; per- and trichloroethylene; cyclooctatetraene; 1,4-butanediol; carbon black; and welding and cutting metals.

10-acetyldithranol.
CAS: 117568-24-8. $C_{16}H_{12}O_5$.
Hazard: A poison. A skin irritant.

acetylene black. The carbon black resulting from incomplete combustion of or thermal decomposition of acetylene.
Properties: High liquid adsorption, retention of high bulk volume, high electrical conductivity.
Use: Dry cell batteries, component of explosives, reinforcing agent in rubber and in thermal and sound

insulation, gloss suppressor in paints, carburizing agent in hardening of steel, pigment in printing inks, filler in electroconductive polymers.
See carbon black.

(acetylenecarbonyloxy)triphenyltin.   See triphenyltin propiolate.

acetylene dichloride.   See *sym*-dichloroethylene.

acetylene hydrocarbon.   (alkyne). One of a class of unsaturated hydrocarbons of the homologous series having the generic formula C_nH_{2n-2} and a structural formula containing a triple bond.

acetylene polymer.   See polyacetylene; 1,3,5,7-cyclooctatetraene.

acetylene tetrabromide.   (*sym*-tetrabromethane).
CAS: 79-27-6. $CHBr_2CHBr_2$.
Properties: Yellowish liquid. D 2.98–3.00, bp 239–242C with decomposition, 151C (54 mm), mp 0.1C, refr index 1.638. Soluble in alcohol and ether, insoluble in water. Combustible.
Derivation: Interaction of acetylene and bromine and subsequent distillation.
Method of purification: Rectification.
Grade: Technical.
Hazard: Irritant by inhalation and ingestion. TLV: 1 ppm.
Use: Separating minerals by specific gravity; solvent for fats, oils, and waxes; fluid in liquid gauges; solvent in microscopy.

acetylene tetrachloride.   See *sym*-tetrachloroethane.

acetylenogen.   Calcium carbide.

N-acetylethanolamine.   (hydroxyethylacetamide). $CH_3CONHC_2H_4OH$.
Properties: Brown, viscous liquid. D 1.122 (20/20C); boiling range 150–152C (5 mm); decomposes (10 mm); bulk d 9.34 lb/gal; refr index 1.4730 (20C); flash p 350F (176.6C); fp 15.8C. Soluble in alcohol, ether, and water. Combustible.
Grade: Technical (75% solution in water).
Use: Plasticizer for polyvinyl alcohol and for cellulosic and proteinaceous materials; humectant for paper products, glues, cork, and inks; textile conditioner.

acetylethanoltrimethylammonium hydroxide.   See acetylcholine.

acetylethyltetramethyltetralin.   (1,1,4,4,-tetra-methyl-6-ethyl-7-acetyl-1,2,3,4-tetrahydronaphthalene; polycyclic musk; musk tetralin; aett; 1-(3-ethyl-5,5,8,8-tetramethyl-6,7-dihydronaphthalen-2-yl)ethanone).

CAS: 88-29-9. $C_{18}H_{26}O$. An organic compound.
Properties: White crystals.
Hazard: Very toxic; neurotoxic.
Use: Formerly in perfumes.

acetylferrocene.   (ferrocenyl methyl ketone). $C_5H_5FeC_5H_4COCH_3$.
Properties: Orange crystals. Mp 85–86C.
Use: Intermediate.
See ferrocene.

acetylformic acid.   See pyruvic acid.

n-acetylglutamate.   ((2S)-2-acetamidopentanedioic acid; (2S)-2-acetamidoglutaric acid). $C_7H_{11}NO_5$. An amino acid that activates carbamyl phosphate synthetase during urea synthesis. This amino acid causes a configurational change in the enzyme, therefore increasing its affinity for adenosine triphosphate.

6-acetyl-grayanotoxin i.   See grayanotoxin iii 6,14-diacetate.

acetylide.   A salt-like carbide formed by the reaction of acetylene and an alkali or alkaline-earth metal in liquid ammonia, or with silver, copper, and mercury salts in aqueous solution. The latter are explosive when shocked or heated.
See carbide; calcium carbide.

acetyl iodide.
CAS: 507-02-8. CH_3COI.
Properties: Colorless, transparent, fuming liquid; suffocating odor; turns brown on exposure to air or moisture. D 1.98; bp 105–108C; refr index 1.55. Soluble in ether and benzene; decomposed by water and alcohol.
Derivation: Reaction of acetic acid, iodine, and phosphorus; reaction of acetyl chloride and hydrogen iodide.
Method of purification: Distillation.
Grade: Technical.
Hazard: Strong irritant to eyes and skin.
Use: Organic synthesis.

acetylisoeugenol.   (isoeugenol acetate). $C_6H_3(CHCHCH_3)(OCH_3)(OCOCH_3)$.
Properties: White crystals; spicy, clove-like odor. Congealing point 77C, soluble in 27 parts of 95% alcohol. Combustible.
Method of purification: Crystallization.
Grade: Technical.
Use: Perfumery, particularly for carnation-type odors; flavoring.

acetyl ketene.   See diketene.

n-acetylloline.
CAS: 4914-36-7. $C_{10}H_{16}N_2O_2$.
Hazard: A poison.
Source: Natural product.

acetylmethionine. See *n*-acetyl-*l*-methionine.

n-acetyl-*l*-methionine.
CAS: 65-82-7. $C_7H_{13}NO_3S$.
Properties: Colorless or white crystals or powder; odorless. Sol in water, alc, alkali and mineral acids; insol in ether.
Hazard: A poison.
Use: Food additive; drug.

acetylmethylcarbinol.   (acetoin; 3-hydroxy-2-butanone; dimethylketol).
CAS: 513-86-0. $CH_3COCH_2OCH_3$.
Properties: Slightly yellow liquid or crystals (dimer), oxidizes gradually to diacetyl on exposure to air. D 1.016, bp 140–148C, mp 15C. Soluble in alcohol; miscible with water; slightly soluble in ether. Combustible.
Derivation: Reduction of diacetyl.
Grade: Technical, FCC (as acetoin).
Use: Aroma carrier; preparation of flavors and essences.

acetyl nitrate. CH_3COONO_2.
Properties: Colorless, fuming liquid. Hygroscopic. D 1.24.
Derivation: Reaction of acetic anhydride and nitrogen pentoxide.
Hazard: Corrosive to skin and mucous membranes. Explodes above 55C (130F) or in presence of mercuric oxide.
Use: Nitrating agent for organic compounds.

acetylnitro peroxide.   (Pan; peroxyacetyl nitrate, 2-benzofuran-1,3-dione). $C_8H_4O_3$. An organic peroxide.

acetyl oxide.   See acetic anhydride.

10-(acetyloxy)-1,8-dihydroxy-9(10h)-anthracenone.   See 10-acetyldithranol.

4(or 6)-(acetyloxy)-5(or 4)-hexenoic acid.
CAS: 83145-58-8. $C_8H_{12}O_4$.
Hazard: Moderately toxic by ingestion and skin contact. A mild skin and severe eye irritant.

o-(4-(1-((acetyloxy)imino)ethyl)-3-methylphenyl) o,o-diethyl-phosphorothioate.
CAS: 22936-44-3. $C_{15}H_{22}NO_5PS$.
Hazard: Moderately toxic by ingestion.
Use: Agricultural chemical.

acetyl peroxide.   (diacetyl peroxide).
CAS: 110-22-5. $(CH_3CO)_2O_2$.
Properties: Colorless crystals. Mp 30C. Slightly soluble in cold water; soluble in alcohol and ether, flash p 113 (45C) (OC), fp approx −8C, d 1.18 (20C). Combustible.
Available forms: A 25% solution in dimethyl phthalate.

Hazard: (25% solution) Moderate fire risk; strong irritant to skin and eyes. The pure material is a strong oxidizer and severe explosion hazard; should not be stored after preparation, nor heated above 30C.
Use: Initiator and catalyst for resins.

acetylphenetidin.   See acetophenetidin.

acetylphenol.   See phenyl acetate.

N-acetyl-*p*-phenylenediamine.   See *p*-aminoacetanilide.

acetyl phosphate.   (acetic phosphoric anhydride; phosphono acetate).
CAS: 590-54-5. $C_2H_3O_5P$. A high-energy phosphate that acts as an acetate donor in the metabolism of various bacteria.

acetylpropionic acid.   See levulinic acid.

acetyl propionyl.   (2,3-pentanedione; methyl ethyl glyoxal; methyl ethyl diketone).
CAS: 600-14-6. $CH_3COCOCH_2CH_3$.
Properties: Yellow liquid. Mp −52C, bp 106–110C, d 0.955–0.959 (15/4C). Partly soluble in water. Combustible.
Grade: 99%.
Use: Flavors of butterscotch and chocolate type.

2-acetyl pyrazine.
CAS: 22047-25-2. $C_6H_6N_2O$.
Properties: Colorless to pale-yellow crystals or liquid; sweet popcorn-like odor. Mp: 75–78C, d: 1.100–1.115 @ 20C, refr index: 1.530–1.540 @ 25C. Sol in acids, alc, ether, and water @ 230C.
Hazard: A skin and eye irritant.
Use: Food additive.

2-acetylpyrrole.
CAS: 1072-83-9. C_6H_7NO.
Properties: Light beige to yellow crystals from petroleum ether; bread-like odor. Mp: 90C, bp: 220C. Sol in acids, alc, ether, water @ 230C.
Use: Food additive.

4-acetylresorcinol.   (2,4-dihydroxyacetophenone). $C_6H_3(OH)_2COCH_3$.
Properties: Light tan crystals. Mp 146–148C. High absorptivity in UV region. Slightly soluble in water; soluble in most organic solvents except benzene and chloroform.
Use: UV absorber in plastics; dye intermediate; fungicide; plant growth promoter.

acetylsalicylate.   (2-acetyloxybenzoic acid). $C_9H_8O_4$. A prototypical analgesic that has anti-inflammatory and antipyretic properties and acts as an inhibitor of cyclooxygenase which results in the inhibition of the biosynthesis of prostaglandins.

Properties: Salt, colorless crystals; melting point: 143C; slightly soluble in water.
Use: In the treatment of mild to moderate pain; in the prevention of arterial and venous thrombosis.

acetylsalicylic acid. See aspirin.

N-acetylsulfanilyl chloride. (*p*-acetamidobenzenesulfonyl chloride; *p*-acetylaminobenzenesulfonyl chloride). $(CH_3CONH)C_6H_4(SO_2Cl)$.
Properties: Light tan to brownish powder or fine crystals. Mp 149C. Soluble in chloroform and ethylene dichloride.
Hazard: Irritant to skin and mucous membranes.
Use: Intermediate in the manufacture of sulfa drugs.

2-(acetylthioglycolic amide)benzothiazole.
CAS: 63123-39-7. $C_{11}H_{10}N_2O_2S_2$.
Hazard: A poison by ingestion.

2-acetylthiophene. (methyl-2-thienyl ketone). $CH_3COC_4H_3S$.
Properties: Yellowish, oily liquid. Mp 10–11C, bp 213.5C, 88–90C (10 mm). Very slightly soluble in ether.
Use: Organic intermediate.

acetyl-*o*-toluidine. (*o*-acetotoluidide). $CH_3CONHC_6H_4CH_3$.
Properties: Colorless crystals. Mp 110C, bp 296C, d 1.168 (15C). Soluble in alcohol, ether, benzene, chloroform, glacial acetic acid; slightly soluble in cold water; insoluble in hot water.
Derivation: By boiling glacial acetic acid with *o*-toluidine and distilling the product.
Grade: Technical.
Use: Organic synthesis.

acetyl-*p*-toluidine. (*p*-acetotoluidide). $CH_3CONHC_6H_4CH_3$.
Properties: Colorless needles. Mp 153C, bp 307C, d 1.212 (15/4C). Slightly soluble in water; soluble in alcohol, ether, ethyl acetate, glacial acetic acid.
Grade: Technical.
Use: Dyes.

n-acetyltransferase. A cytosolic enzyme isolated from rat liver, kidney, or certain other organs that catalyses the acetylation of xenobiotic amines.

acetyl triallyl citrate. $CH_3COOC_3H_4$ $(COOCH_2CH:CH_2)_3$.
Properties: Liquid. Boiling range 142–143C (0.2 mm), d 1.140 (20C), refr index 1.4665 (25C), flash p 365F (185C). Combustible.
Use: Cross-linking agent for polyesters; monomer for polymerization.

acetyl tributyl citrate. $CH_3COOC_3H_4$ $(COOC_4H_9)_3$.

Properties: Colorless, odorless liquid. Distillation range 172–174C (1 mm), pour p −60C, d 1.046 (25C), bulk d 8.74 lb/gal (25C), refr index 1.4408 (25C), viscosity 42.7 cP (25C), flash p 400F (204C). Insoluble in water. Combustible.
Derivation: Esterification and acetylation of citric acid.
Grade: Technical.
Use: Plasticizer for vinyl resins.
See "Citroflex A-4" *[Vertellus]*.

acetyl triethyl citrate. $CH_3COOC_3H_4$ $(COOC_2H_5)_3$.
Properties: Colorless liquid; odorless. Distillation range 131–132C (1 mm), pour p −47C, d 1.135 (25C), bulk d 9.47 lb/gal (25C), refr index 1.4386 (25C), viscosity 53.7 cP (25C), flash p 370F (187C). Slightly soluble in water. Combustible.
Derivation: Esterification and acetylation of citric acid.
Grade: Technical.
Use: Plasticizer for cellulosics, particularly ethyl cellulose.
See "Citroflex A-2" *[Vertellus]*; ATEC.

acetyl tri-2-ethylhexyl citrate. CH_3COOC_3 $H_4(COOC_8H_{17})_3$.
Properties: Liquid. Bp 225C (1 mm), flash p 430F (222C). Insoluble in water. Combustible.
Grade: Technical.
Use: Low-volatility plasticizer for vinyl resins.

acetyltri-*n*-hexyl citrate. See "Citroflex A-6" *[Vertellus]*.

N-acetyltryptophan. Available commercially as *N*-acetyl-*l*-tryptophan, mp 185–186C; *N*-acetyl-*dl*-tryptophan, mp 205C.
Use: Nutrition and biochemical research; medicine.

acetyl valeryl. (heptadione-2,3). CH_3CO COC_4H_9.
Properties: Yellow liquid. Combustible.
Grade: 92% pure.
Use: Cheese, butter, and miscellaneous flavors.

acetyl value. The number of milligrams of potassium hydroxide required for neutralization of acetic acid obtained by the saponification of 1 g of acetylated fat or oil sample. Acetylation is carried out by boiling the sample with an equal amount of acetic anhydride, washing, and drying. Saponification values on acetylated and on untreated fat are determined. From the results the acetyl value is calculated. It is a measure of the number of free hydroxyl groups in the fat or oil.

ACGIH. See American Conference of Governmental Industrial Hygienists.

achroglobin. A colorless respiratory protein compound found in certain invertebrate animals.

"Achromycin" *[Heritage].*　　TM for tetracycline hydrochloride.

achroodextrin.　　(achrodextrin). A low molecular weight dextrin produced during the digestion of starch by amylase.

acicular.　　Used to describe needle-shaped crystals or the particles in powders.

acid.　　One of a large class of chemical substances whose water solutions have one or more of the following characteristics: sour taste, ability to make litmus dye turn red and to cause other indicator dyes to change to characteristic colors, ability to react with and dissolve certain metals to form salts, and ability to react with bases or alkalies to form salts. All acids contain hydrogen. In water, ionization or splitting of the molecule occurs so that some or most of this hydrogen forms H_3O^+ ions (hydronium ions), usually written more simply as H^+ (hydrogen ion). Acids are referred to as strong or weak according to the concentration of H^+ ion that results from ionization. Hydrochloric, nitric, and sulfuric are strong or highly ionized acids; acetic acid (CH_3COOH) and carbonic acid (H_2CO_3) are weak acids. Tenth normal hydrochloric acid is 100 times as acid (pH = 1) as tenth normal acetic acid (pH = 3). The pH range of acids is from 6.9 to 1.
See pH. When dealing with chemical reactions in solvents other than water, it is sometimes convenient to define an acid as a substance that ionizes to give the positive ion of the solvent. The common definitions of acid have been extended as more detailed studies of chemical reactions have been made. The Lowry-Brønsted definition of an acid as a substance that can give up a proton is more useful in connection with an understanding of bases (see base). Perhaps the most significant contribution to the theory of acids was the electron-pair concept introduced by G. N. Lewis around 1915.
See Lewis electron theory. The terms *hard* and *soft* acids and bases refer to the ease with which the electron orbitals can be disturbed or distorted. Hard acids have a high positive oxidation state, and their valence electrons are not readily excited; soft acids and bases have little or no positive charge and easily excited valence electrons. Hard acids combine preferentially with hard bases, and soft acids with soft bases. Soft acids tend to accept electrons and form covalent bonds more readily than hard acids. For example, the halogen acids arranged in a series by increasing atomic weight (and decreasing chemical activity) show a progression from hard (HF) to soft (HI).
A brief outline of the major groups of acids is as follows:
Inorganic
Mineral acids: sulfuric, nitric, hydrochloric, phosphoric.
Hazard: All mineral acids are highly irritant and corrosive to human tissue.

Organic
Carboxylic (contain –COOH group)
aliphatic: acetic, formic
aromatic: benzoic, salicylic
Dicarboxylic (contain two –COOH groups)
oxalic, phthalic, sebacic, adipic
Fatty acids (contain –COOH group)
aliphatic: oleic, palmitic, stearic
aromatic: phenylstearic
Amino acids: N-containing protein components
See Lewis acid; carboxylic acid; fatty acid; amino acid; specific compounds.

1,2,4-acid.　　(1-amino-2-naphthol-4-sulfonic acid). $C_{10}H_5NH_2OHSO_3H$.
Properties: Pinkish-white to gray needles. Soluble in hot water; but almost insoluble in cold water.
Derivation: β-naphthol is nitrated to nitroso-β-naphthol by reaction with nitrous acid and the product treated with sodium bisulfite. Upon acidification the free sulfurous acid effects simultaneous reduction and sulfonation.
Use: Aniline dye intermediate.

1,8,2,4-acid.　　See Chicago acid.

acid aerosol.　　An aerosol that contains high concentrations of acidic substances that can irritate the lungs and other soft tissues.

acid alcohol.　　Ethanol that contains 1% hydrochloric acid.

acid amide.　　See amide.

acid ammonium sulfate.　　See ammonium bisulfate.

acid ammonium tartrate.　　See ammonium bitartrate.

acid anhydride.　　An oxide of a nonmetallic element or of an organic radical that is capable of forming an acid when united with water, or which can be formed by the abstraction of water from the acid molecule, or can unite with basic oxides to form salts.

acid anthraquinone brilliant blue.　　See C.I. acid blue 80.

AcidAway.　　A system to neutralize air conditioning and refrigeration systems that have experienced burnout.

Acid Black 2.
Use: Hair color, reagent, biological stain.

acid butyl phosphate.　　See *n*-butyl acid phosphate.

acid calcium phosphate.　　See calcium phosphate, monobasic.

acid chrome orange gr.   See C.I. mordant orange 6, disodium salt.

acid cleaner.   A cleaner whose action depends on the presence of caustic minerals.

acid deposition.   Acidic are pollutants (chiefly nitrogen oxides and sulfur oxides) deposited in the biogeosphere. These materials may be deposited as dry particulates or acidic gases or in rain, snow or fog.

acid dye.   An azo, triarylmethane, or anthraquinone dye with acid substituents such as nitro-, carboxy-, or sulfonic acid. These dyes are most frequently applied in acid soluble to wool and silk, and no doubt combine with the basic groups of the proteins of those animal fibers. Orange II (CI 15,510), black 10B, and acid alizarin blue B are examples.

acid ethylsulfate.   See ethylsulfuric acid.

acid ester.   A derivative of a polyvalent organic acid in which some of the acid H atoms are replaced by a radical.

acid ester sulfate.   (sulfuric acid ester). An ester sulfate in which one of the hydrogen atoms is replaced by a hydrocarbon. They are synthesized in phase II metabolic reactions releasing water-soluble products of xenobiotic compounds that are readily eliminated from the body.
Use: Industrial alkylating agent.

acid fungal protease.   (fungal protease enzyme).
Properties: Highly off-white powder.
Use: As a replacement for pepsin; chill-proofing agent for beer; in cereal treatment; feed supplement for baby pigs; rennet extender.

γ-acid.   (2-amino-8-naphthol-6-sulfonic acid; aminonaphtholsulfonic acid-2,8,6).

Properties: White crystals. Soluble in alcohol and ether; slightly soluble in water.
Derivation: β-naphthylamine sulfate is sulfonated with 66% oleum; the precipitate, 2-naphthylamine-6,8-disulfonic acid, is heated with 65% sodium hydroxide solution at 210C and 210 psi.
Use: Azo dye intermediate.

acid glaucine blue.   See peacock blue.

α-acid glycoprotein.   (acute-phase reaction protein; orosomucoid; AAG). An anionic polymorphic conjugated protein that is chiefly hepatic in origin. The carbohydrate moiety, made up of equal parts of hexosamine, hexose, and sialic acid, comprises 45% of the molecule. Plasma levels may rise during chronic inflammation or conditions of acute physiological stress; levels may fall as a consequence of severe hepatic damage, certain serious gastroenteropathies, and severe malnutrition. It combines electrostatically with many basic xenobiotics.

acid halide.   A compound that contains a carbonyl group bound to a halogen atom.

acid, hard.   See Lewis electron theory; acid.

acid hydrogen.   The hydrogen of the carboxyl ground in organic compounds that can be displaced by metals, alkyls, aryls, or basic radicals.

acid hydrolyzed proteins.
Properties: Liquid, paste or powder. Sol in water.
Use: Food additive.

acidic amino acid.   An amino acid that bears a net negative charge at neutral pH, that contain a second acid moiety and are hydrophilic.

acidic oxide.   An oxide of a nonmetal, e.g., SO_2, CO_2, P_2O_5, SO_3, that forms an acid when combined with water.
See acid anhydride.

acidic salt.   A salt that contains an ionizable hydrogen atom but does not necessarily produce acidic solutions.

acidimetry.   The determination of the concentration of acid solutions or of the quantity of acid in a sample or mixture. This is usually done by titration with a solution of base of known strength (standard solution); an indicator is used to establish the end point.
See pH.

acid lining.   Silica brick lining used in steelmaking furnaces.

acid magnesium citrate.   See magnesium citrate, dibasic.

acid magnesium phosphate.   See magnesium phosphate, monobasic.

acid methyl sulfate.   See methylsulfuric acid.

acid mine drainage.   (AMD). Water from both active and inactive coal mines that has become contaminated with sulfuric acid as a result of hydrolysis of ferric sulfate, the oxidation product of pyrite.

This is a factor in water and stream pollution, which can be corrected by use of appropriate ion-exchange resins.

acid mine water.   Acidic water in pyritic coal mines due to the presence of free sulfuric acid and acidic sulfates, especially ferrous sulfate, leached from iron pyrites.

acid number.   (acid value). Determined by the number of milligrams of potassium hydroxide required for the neutralization of free fatty acids present in 1 g of fat or oil. Also the measure of free acids present in a substance.

acidolysin.   See glycoprotease.

acidolysis.   A chemical reaction that is comparable to hydrolysis or alcoholysis, in which water or alcohol is used in place of the acid. It involves the decomposition of a molecule with the addition of the elements of an acid to the molecule.

acidophil.   (acidophile). An organism that grows and lives successfully in acidic environments.

acidophil granule.   (oxyphil granule). A granule that stains with an acid dye such as eosin.

acidosis.   A condition in which blood pH decreases, either for metabolic or respiratory reasons.

acid phosphatase.   An enzyme found in blood serum that catalyzes the liberation of inorganic phosphate from phosphate esters. Optimum pH 5; it is less active than alkaline phosphatase.
Use: Biochemical research.

acid phosphate.   An acid salt of phosphoric acid such as NaH_2PO_4, $CaHPO_4$. Also used to refer specifically to calcium phosphate monobasic, $Ca(H_2PO_4)_2$, or superphosphate of lime.

acid potassium oxalate.   See potassium binoxalate.

acid potassium sulfate.   See potassium bisulfate.

acid precipitation.   (acid rain). Any form of precipitation (wet deposition) having a pH of 5.6 or less, the most important deleterious components being the sulfur dioxide and oxides of nitrogen, either emitted as stack gases in highly industrialized areas or resulting from volcanic activity. The most sensitive sections of the U.S. are the East and the extreme Northwest; southeastern Canada is also affected. Acid precipitation is not only destructive to fish and other freshwater life, but also kills certain species of trees (especially spruce) and corrodes metal and cement structures. Industrial use of coal is largely responsible for the incidence of acid rain, especially in the Northeast.
See dry deposition.

acid rain.   See acid precipitation.

acid salt.   A type of compound derived from an acid or a base in which only part of the hydrogen of the acid is replaced by a basic radical.

acid, soft.   See Lewis electron theory; acid.

acid soil.   Soil that has a pH value of less than 6.6.

acidulant.   Any of a number of acids (chiefly organic) either occurring naturally in fruits and vegetables or used as additives in food processing. They function in the following ways: (1) as bacteriostats in processed foods; (2) as aids to the sterilization of canned foods by lowering the pH; (3) as chelating agents for metal ions such as iron and copper, which catalyze rancidity reactions in fats; (4) as flavor enhancers by offsetting excessive sweetness by their tart taste. Commonly used acidulants are citric, acetic, fumaric, ascorbic, propionic, lactic, adipic, malic, sorbic, and tartaric acids.

Acidulin.   Glutamic acid hydrochloride acid value. The number of milligrams of potassium hydroxide required to neutralize the free acids present in 1 g of oil, fat, or wax. The determination is made by titrating the sample in hot 95% ethanol using phenolphthalein as indicator.

acid value.   See acid number.

Acid Yellow 9.   See 4-aminoazobenzene-3,4'-disulfonic acid.

acifluorfen, sodium salt.   (Blazer).
Use: Postemergent herbicide for soybeans, peanuts, and rice.

"ACL 60" *[Occidental]*.   TM for a product line of chlorinated isocyanurates that are an effective economical disinfectant, germicide, and algicide.
Use: Sanitizing swimming pools and spas.

Aclar.   A series of fluorohalocarbon films.
Properties: Useful properties from -200 to $+198C$.
Use: In packaging applications where a transparent, vapor, and/or gas barrier is required, as in packaging of foods for astronauts. Used in electronic and electrical applications because of insulating and heat-resistant properties. Extreme chemical resistance and ability to seal make it useful as a tank lining, etc.

A-C-M.   A balanced mixture of ascorbic acid (vitamin C) and citric acid.
Use: As an antioxidant that protects flavor and prevents browning of fruits exposed to air. Used in home freezing and canning of fresh fruits.

Acofor.   A pale, distilled tall oil fatty acids.
Properties: D 0.907 (25/25C), refr index 1.471 (20C), flash p 380F (193C) (OC), acid number 192, saponification number 194, unsaponifiable matter 2.5%, rosin acids 4.5%. Combustible.
Use: Paint and varnish, inks, soaps, disinfectants, textile oils, core oils, etc.

Aconew.   A pale, distilled tall oil fatty acids with low rosin acid contents.

aconine. $C_{25}H_{41}NO_9$.
Properties: Amorphous shaped, monobasic, diterpendoid alkaloid solid; molecular weight: 499.6.
Heazard: Poison.

aconitase.   (citrate (isocitrate) hydrolyase; aconitate hydratase). The enzyme that catalyzes the reversible interconversion of citric acid, *cis*-aconitic acid, and isocitric acid in the tricarboxylic acid cycle. It is the target of fluoroacetate, a powerful rodenticide that condenses with oxaloacetate to produce an aconitase inhibitor.

aconite.   (monkshood; wolfsbane; friar's cowl).
Hazard: Antipyretic drug, alkaloid poison.

aconitic   acid.   (propene-1,2,3-tricarboxylic acid). H(COOH)C:C(COOH)CH$_2$(COOH).
Properties: White to yellowish, crystalline solid. Mp approx 195C (decomposes). Soluble in water and alcohol. Combustible.
Derivation: (a) By dehydration of citric acid with sulfuric acid; (b) extraction from sugar cane bagasse, *Aconitum napellus,* and other natural sources.
Use: Preparation of plasticizers and wetting agents; antioxidant; organic syntheses; itaconic acid manufacture; synthetic flavors.

aconitine.   (acetyl benzoyl aconine).
CAS: 302-27-2. $C_{34}H_{49}NO_{11}$.
Hazard: Highly toxic alkaloid, antipyretic drug, readily absorbed by skin.

Aconitum.   (aconite; friar's cowl; monkshood; wolfbane; wolf' s bane). A genus of more than 100 species of poisonous North temperature zone flowering plants of the family Ranunculaceae, some of which are violently poisonous.

acp-acetyltransferase.   (acetyl transacylase). An enzyme that initiates fatty acid biosynthesis by transferring an acetyl moiety from acetyl-CoA to acp.

acp-malonyltransferase.   (malonyl transacylase). An enzyme that transfers malonyl from malonyl-CoA to acp; a key step in the biosynthesis of fatty acids.

acridine dye.   Any of a class of polycyclic compounds derived from acridine, that are important as fluorochromes in cytochemistry, histology, and chemotherapy.

acrosin.   A serine proteinase that is active in spermatozoa.

acrylylcholine.   A choline ester toxin produced by the gastropod mollusk, *Buccinum undatum*. It has an action similar to that of muscarine and nicotine.
Hazards: cardiotoxic; hypotensive; stimulates gastric motility and secretion; increases respiration.

acquired genetic mutation.   See somatic cell genetic mutation.

acquired mutations.   See somatic cell genetic mutation.

"Acrafix" [Lanxess].   TM for a processing chemical for high-quality pigment prints.

acraldehyde.   See acrolein.

"Acrawax" [Lonza].   TM for N,N'-ethylenebisstearamide.
CAS: 110-30-5.
Grade: Atomized powdered, prilled, and beaded grades.
Use: In plastics and metal powder lubricants, varnishes, lacquers, defoamers, and mar-resistant antiblock paints.

Acree's reaction.   A test for protein in which a violet ring appears when concentrated sulfuric acid is introduced below a mixture of the unknown solution and a formaldehyde solution containing a trace of ferric oxide.

acridine   orange.   (N,N,N',N'-tetramethyl-3,6-acridinediamine monohydrochloride).
CAS: 494-38-2. $C_{17}H_{19}N_3$•HCl.
Hazard: An in vitro mutagen. Questionable carcinogen.
Use: Selective biological stain for tumor cells, *intravitam*; causes retardation of tumor growth.

acridine (tricyclic).
CAS: 260-94-6. $C_{13}H_9N$.
Properties: Small, colorless needles. Sublimes at 100C, mp 111C, bp above 360C. Soluble in alcohol, ether, or carbon disulfide; sparingly soluble in hot water.
Derivation: (a) By extraction with dilute sulfuric acid from the anthracene fraction from coal tar and adding potassium dichromate. The acridine chromate precipitated is recrystallized, treated with ammonia, and recrystallized. (b) Synthetically.
Hazard: Strong skin irritant.
Use: Manufacture of dyes; derivatives, especially acriflavine, proflavine; analytical reagent.

2-(1-(9-acridinyl)hydrazino)ethanol monohy-drochloride.
CAS: 28846-44-8. $C_{15}H_{15}N_3O \cdot ClH$.
Hazard: A poison by ingestion.
Use: Agricultural chemical.

4-(9-acridinyl)-2-methyl-3-thiosemicarbazide.
CAS: 28846-37-9. $C_{15}H_{14}N_4S$.
Hazard: A poison by ingestion.
Use: Agricultural chemical.

4-(9-acridinyl)-2-methyl-3-thiosemicarbazone acetone.
CAS: 29023-85-6. $C_{18}H_{18}N_4S$.
Hazard: A poison by ingestion.
Use: Agricultural chemical.

4-(9-acridinyl)-3-thiosemicarbazide.
CAS: 29023-84-5. $C_{14}H_{12}N_4S$.
Hazard: A poison by ingestion.
Use: Agricultural chemical.

acriflavine. $C_{14}H_{14}N_3Cl$. A mixture of 3,6-diamino-10-methylacridinium chloride and 3,6-diaminoacridine.
Properties: Brownish or orange, odorless, granular powder. Soluble in three parts of water; incompletely soluble in alcohol; nearly insoluble in ether and chloroform. The aqueous solutions fluoresce green on dilution. Also available as the hydrochloride.
Use: Antiseptic and bacteriostat.

"Acrilan" *[Solutia].* TM for an acrylic fiber.

acrinathrin. See 2,2-dimethyl-3-(3-oxo-3-(2,2,2-trifluoro-1-(trifluoromethyl)).

acroleic acid. See acrylic acid.

acrolein. (2-propenal; acrylaldehyde; allyl aldehyde; acraldehyde).
CAS: 107-02-8. CH_2CHCHO.

$$CH{=}O$$
$$|$$
$$CH$$
$$||$$
$$CH_2$$

Properties: Colorless or yellowish liquid; disagreeable, choking odor. Bp 52.7C, mp −87.0C, d 0.8427 (20/20C), bulk d 7.03 lb/gal (20C). Flash p below 0F (−17C) (COC), autoign temp 532F (277C). Soluble in water, alcohol, and ether. Polymerizes readily unless inhibitor (hydroquinone) is added. Very reactive.
Derivation: (1) Oxidation of allyl alcohol or propylene; (2) by heating glycerol with magnesium sulfate; (3) from propylene with bismuth-phosphorus-molybdenum catalyst.
Method of purification: Rectification.

Grade: Technical.
Hazard: Very irritating to eyes and skin. Toxic by inhalation and ingestion. Upper respiratory tract irritant, pulmonary edema, and pulmonary emphysema. Questionable carcinogen. Dangerous fire risk. Explosive limits in air 2.8–31%.
Use: Intermediate for synthetic glycerol, polyurethane and polyester resins, methionine, pharmaceuticals; herbicide; warning agent in gases.

acrolein dimer. (2-formyl-3,4-dihydro-2*H*-pyran).

$OCH{:}CHCH_2CH_2CHCHO.$

Properties: Liquid. D 1.0775 (20C), bp 151.3C, fp −100C, flash p 118F (47.7C) (OC), bulk d 8.96 lb/gal (20/20C). Soluble in water. Combustible.
Hazard: Moderate fire risk.
Use: Intermediate for resins, pharmaceuticals, dyestuffs.

acrolein test. A test for the presence of glycerine or fats. A sample is heated with potassium bisulfate. Acrolein is released for a positive test.

"Acronal" *[BASF].* TM for dispersions, solutions, and solids of acrylate homo- and copolymers.

ACR process. Abbreviation for advanced cracking reactor.
See ethylene.

acrylaldehyde. See acrolein.

acrylamide.
CAS: 79-06-1. $CH_2CHCONH_2$.
Properties: Colorless, odorless crystals. Mp 84.5C, bp 125C (25 mm), d 1.122 (30C). Soluble in water, alcohol, acetone; insoluble in benzene, heptane. The solid is stable at room temperature but may polymerize violently on melting.
Derivation: Reaction of acrylonitrile with sulfuric acid (84.5%) and neutralization.
Grade: Technical (approximately 97% pure).
Hazard: Toxic by skin absorption. Present in particular and vapor phases. Irritant to skin and mucous membranes. Central nervous system impairment. Probable carcinogen.
Use: Synthesis of dyes, etc., cross-linking agent, adhesives, paper and textile sizes, soil-conditioning agents, flocculants, sewage and waste treatment, ore processing, permanent-press fabrics.

acrylamide gels. A polymer gel used for electrophoresis of DNA or protein to measure their sizes (in daltons for proteins, or in base pairs for DNA). See "Gel Electrophoresis". Acrylamide gels are especially useful for high resolution separations of DNA in the range of tens to hundreds of nucleotides in length.

p-acrylamidobenzoic acid 2-(diethylamino)ethyl ester. See procaine acryloyl monomer.

2-acrylamido-2-methyl-1-propanesulfonic acid.
CAS: 15214-89-8. $C_7H_{13}NO_4S$.
Hazard: Low toxicity by ingestion.

acrylate.   (1) Any of the several monomers used for the manufacture of thermosetting acrylic surface coating resins, e.g., 2-hydroxyethyl acrylate (HEA) and hydroxypropyl acrylate (HPA). (2) Polymer of acrylic acid or its esters, used in surface coatings, emulsion paints, paper and leather finishes, etc.
See acrylic acid; acrylic resin.

acrylate-acrylamide resins.
Use: Food additive.

acrylic acid.   (acroleic acid; propenoic acid).
CAS: 79-10-7. $H_2C{:}CHCOOH$.

$$H_2C=\overset{\overset{\displaystyle H}{|}}{C}-\overset{\overset{\displaystyle O}{\|}}{C}-OH$$

Properties: Colorless liquid; acrid odor. Bp 140.9C, mp 12.1C, d 1.052 (20/20C), vap press 3.1 mm (20C), bulk d 8.6 lb/gal (20C), refr index 1.4224 (20C). Flash p 130F (54.5C) (OC). Polymerizes readily. Miscible with water, alcohol, and ether. Combustible.
Derivation: (1) Condensation of ethylene oxide with hydrocyanic acid followed by reaction with sulfuric acid at 320F; (2) acetylene, carbon monoxide, and water, with nickel catalyst; (3) propylene is vapor oxidized to acrolein, which is oxidized to acrylic acid at 300C with molybdenum-vanadium catalyst; (4) hydrolysis of acrylonitrile.
Grade: Technical (esterification and polymerization grades), glacial (97%).
Hazard: Irritant and corrosive to skin. Toxic by inhalation. May polymerize explosively. Upper respiratory tract irritant. Questionable carcinogen.
Use: Monomer for polyacrylic and polymethacrylic acids and other acrylic polymers.
See acrylic resin.

acrylic fiber.   A manufactured fiber in which the fiber-forming substance is any long-chain synthetic polymer composed of at least 85% (by weight) acrylonitrile units $-CH_2CH(CN)-$ (U.S. Federal Trade Commission).
Properties: Tensile strength 2–3 g/denier, water absorption 1.5–2.5%, d approx 1.17. Combustible.
Hazard: Fumes are toxic.
Use: Modacrylic fibers; blankets; carpets.
See modacrylic fiber; acrylic resin.

acrylic polymers.   See acrylic resin.

acrylic resin.   (acrylic fiber; nitrile rubber). Thermoplastic polymers or copolymers of acrylic acid, methacrylic acid, esters of these acids, or acrylonitrile. The monomers are colorless liquids that polymerize readily in the presence of light, heat, or catalysts such as benzoyl peroxide; they must be stored or shipped with inhibitors present to avoid spontaneous and explosive polymerization.
Properties: Acrylic resins vary from hard, brittle solids to fibrous, elastomeric structures to viscous liquids, depending on the monomer used and the method of polymerization. A distinctive property of cast sheet and extruded rods of acrylic resin is ability to transmit light.
Use: Bulk-polymerized: hard, shatterproof, transparent or colored material (glass substitute, decorative illuminated signs, contact lenses, dentures, medical instruments, specimen preservation, furniture components). Suspension-polymerized: beads and molding powders (headlight lenses, adsorbents in chromatography, ion-exchange resins). Solution polymers: coatings for paper, textiles, wood, etc. Aqueous emulsions: adhesives, laminated structures, fabric coatings, nonwoven fabrics. Compounded prepolymers: exterior auto paints, applied by spray and baked. Acrylonitrile-derived acrylics are extruded into synthetic fibers and are also the basis of the nitrile family of synthetic elastomers.
See acrylic acid; acrylonitrile; methyl methacrylate.

Acryloid coating resins.   Acrylic ester polymers in organic solvent solutions or 100% solid forms; water-white and transparent. Films range from very hard to very soft.
Use: Exceptionally resistant surface coatings, such as heat-resistant and fumeproof enamels; vinyl and plastic printing; fluorescent coatings; clear and pigmented coatings on metals.

Acryloid Modifiers.   Thermoplastic acrylic polymers in powder form. Various grades facilitate processing or improve physical properties of rigid or semi-rigid polyvinyl chloride formulations.

Acryloid oil additives.   Acrylic polymers supplied in special oil solution or in diester lubricant.
Use: Viscosity-index improvement, pour-point depression of lubricating oils and hydraulic fluids, sludge dispersancy in lubricating and fuel oils.

"Acrylon" [Stromdahl].   TM for a group of acrylic rubbers outstanding in resistance to oil, grease, ozone, and oxidation.
Hazard: Central nervous system impairment, and lower respiratory tract irritant.
Use: Gaskets and rubber parts for contact with oils and diester lubricants.

acrylonitrile.   (propenenitrile; vinyl cyanide).
CAS: 107-13-1. $H_2C{:}CHCN$.
Properties: Colorless, mobile liquid; mild odor. Fp −83C; bp 77.3–77.4C, d 0.8004 (25C), flash p 32F

(0C) (TOC). Soluble in all common organic solvents; partially miscible with water.

Derivation: (1) From propylene oxygen and ammonia with either bismuth phosphomolybdate or a uranium-based compound as catalysts; (2) addition of hydrogen cyanide to acetylene with cuprous chloride catalyst; (3) dehydration of ethylene cyanohydrin.

Hazard: Toxic by inhalation and skin absorption. A possible carcinogen. Flammable, dangerous fire risk. Explosive limits in air 3–17%.

Use: Monomer for acrylic and modacrylic fibers and high-strength whiskers; ABS and acrylonitrile styrene copolymers; nitrile rubber; cyanoethylation of cotton; synthetic soil blocks (acrylonitrile polymerized in wood pulp); organic synthesis; adiponitrile; grain fumigant; monomer for a semiconductive polymer that can be used like inorganic oxide catalysts in dehydrogenation of *tert*-butanol to isobutylene and water.

acrylonitrile-butadiene rubber.   See nitrile rubber.

acrylonitrile-butadiene-styrene resin.   Most contemporary ABS resins are true graft polymers consisting of an elastomeric polybutadiene or rubber phase, grafted with styrene and acrylonitrile monomers for compatibility, dispersed in a rigid styrene-acrylonitrile (SAN) matrix. Mechanical polyblends of elastomeric and rigid copolymers, e.g., butadiene-acrylonitrile rubber and SAN, historically the first ABS resins, are also marketed. Varying the composition of the polymer by changing the ratios of the three monomers and use of other comonomers and additives results in ABS resins with a wide range of properties.

Properties: Dimensional stability over temperature range from −40 to +71C. Attacked by nitric and sulfuric acids and by aldehydes, ketones, esters, and chlorinated hydrocarbons. Insoluble in alcohols, aliphatic hydrocarbons, and mineral and vegetable oils. Processed by conventional molding and extrusion methods. D 1.04; tensile strength about 6500 psi, flexural strength 10,000 psi, good electrical resistance, water absorption 0.3–0.4%. Combustible but slow-burning; flame retardants may be added. Can be vacuum-metallized or electroplated.

Grade: High-, medium-, and low-impact; molding and extrusion.

Use: Engineering plastics used for automobile body parts and fittings; telephones; bottles; heels; luggage; packaging; refrigerator door liners; plastic pipe and building panels (subject to local building codes); shower stalls; boats; radiator grills; machinery housings; business machines.

Note: Several trademarked types are "Cycolac" *[Sabic]*, Abson, "Lustran" *[INEOS ABS]*. See ABS resin.

acrylonitrile dimer.   See methylene glutaronitrile.

acrylonitrile-styrene copolymer.   A thermoplastic blend of acrylonitrile and styrene monomers having good dimensional stability and suitable for use in contact with foods. Among its numerous applications is that of bottles for soft drinks (Cyclesafe).

Hazard: FDA regulations limit to 0.3 ppm the amount of acrylonitrile monomer that will be allowed to migrate from the container to the contents.

acryloyl   chloride.   (acryloyl   chloride). $H_2CCHCOCl$.

Properties: Liquid. Bp 75C.

Use: Monomer, intermediate.

"Acrysol" *[Lawson].*   (1) TM for aqueous solutions of sodium polyacrylate or other polymeric acrylic salts. (2) TM for polyacrylic acid and copolymer products in aqueous solutions or dispersions. Some grades are solutions of sodium polyacrylate.

Use: (1) Thickeners in paints, fabric coatings and backings, adhesives. (2) Warp size for synthetic fibers, cotton and rayon, modifier of starch sizes.

ACS.   See American Chemical Society; American Carbon Society.

ACTH.   (adrenocorticotropic hormone; corticotropin).

CAS: 9002-60-2. One of the hormones secreted by the anterior lobe of the pituitary gland. It stimulates an increase in the secretion of the adrenal cortical steroid hormones. It is a polypeptide consisting of a 39-unit chain of amino acids, the sequence varying in certain positions with the biological species. ACTH was synthesized in 1960.

Properties: White powder with molecular weight approximately 3500. Freely soluble in water; soluble in 60–70% alcohol or acetone. Solutions are stable to heat.

Source: Extracted from whole pituitary glands of swine, sheep, and oxen. Normally isolated from swine.

Grade: Pure; USP as corticotropin injections.

Units: Based on comparison with USP Corticotropin Reference Standard.

Hazard: May have damaging side effects.

Use: Medicine; biochemical research.

"Acti-Brom" *[Nalco].*   TM for chemical program used to control microbiological deposits.

Use: In cooling water systems.

Acti-dione.   Antibiotic cycloheximide, an agricultural fungicide.

"Actimoist" *[Active].*   (hyaluronic acid).

CAS: 9067-32-7. TM for active natural mucopolysaccharide.

Properties: Clear, colorless, viscous liquid.

Use: To help maintain long-lasting moisture balance on the skin.

actin.   A protein making up the thin filaments of muscle and cytoskeleton of eukaryotic cells.

actinide series.   (actinoid series). The group of radioactive elements starting with actinium and ending with element 105. All are classed as metals. Those with atomic number higher than 92 are called transuranic elements. The series includes the following elements: actinium, 89; thorium, 90; protactinium, 91; uranium, 92; neptunium, 93; plutonium, 94; americium, 95; curium, 96; berkelium, 97; californium, 98; einsteinium, 99; fermium, 100; mendelevium, 101; nobelium, 102; and lawrencium 103. The isotopes of several of these elements are under study for possible use in such fields as radiography, neutron activation analysis, hydrology, and geophysics.

actinium.   Ac. A radioactive metallic element; first member of the actinide series. Similar to lanthanum.
Properties: Atomic number 89, aw 227 (most stable isotope), silvery white metal. Mp 1050C, bp (est) 3200C, oxidation state +3. Eleven radioactive isotopes; 227 has half-life of 21.8 years.
Derivation: Uranium ores; neutron bombardment of radium. Several compounds have been prepared. Available commercially at 98% minimum purity.
Hazard: Radioactive bone-seeking poison.
Use: Radioactive tracer (225 isotope).
See lanthanum.

actinohematin.   A red respiratory pigment found in certain sea anemones of the genus *Actinia*.

actinolite.   A greenish crystalline or fibrous variety of asbestos.

actinomycin.   A family of antibiotics produced by Streptomyces; reported to be active against *Escherichia coli*, other bacteria and fungi to have cytostatic and radiomimetic activity. There are many forms of actinomycin; two of commercial importance are cactinomycin and dactinomycin.

actinomycin D.   (2-amino-4,6-dimethyl-3-oxo-1-N,9-*N-bis*[7,11,14-trimethyl-2,5,9,12,15-pentaoxo-3,10-di(propan-2-yl)-8-oxa-1,4,11,14-tetrazabicyclo[14.3.0]nonadecan-6-yl]phenoxazine-1,9-dicarboxamide; act; actinomycindioic d acid; dilactone; actinomycin I; AD; cosmegen; dactinomycin; dilactone actinomycindioic d acid; HBF 386; lyovac cosmegen; meractinomycin;   NCI-C04682;   NSC-3053; oncostatin k).
CAS: 50-76-0. $C_{62}H_{86}N_{12}O_{16}$. It binds to DNA and inhibits RNA synthesis (transcription, with chain elongation more sensitive than initiation, termination, or release. As a result of impaired mRNA production, protein synthesis also declines after dactinomycin therapy.
Properties: Composed of two cyclic peptides attached to a phenoxazine.
Derivation: from Streptomyces parvullus.
Hazard: Poison; questionable carcinogen; neoplastigen; teratogen; tumorigen.

actinon.   Rn. An isotope of radon that has a half-life of about 4 seconds. It is a member of the noble gas family found in soil and is released during the decay of radium.

Actitex.   An activated carbon fiber.
Available forms: Rolls.
Use: Removal of pesticides from drinking water and VOCs from process gases.

activated alumina.   See alumina, activated.

activated carbon.   See carbon, activated.

activated sludge.   See sewage sludge (2).

activation.   The process of treating a substance or a molecule or atom by heat or radiation or the presence of another substance so that the first-mentioned substance, atom, or molecule will undergo chemical or physical change more rapidly or completely. Common types of activation are: (1) Processing of carbon black, alumina, and other materials to impart improved adsorbent qualities. Subjecting the material to steam or carbon dioxide at high temperatures is the method usually used.
See alumina, activated; carbon, activated. (2) Heating or otherwise supplying energy to a substance (e.g., ultraviolet or infrared radiation) to attain the necessary level of energy for the occurrence of a chemical reaction, or for emission of desired light waves, as in fluorescence or chemical lasers. The term *excitation* is also used. (3) An important variation of (2) is the process of making a material radioactive by bombardment with neutrons, protons, or other nuclear particles.
See activation analysis; energy, activation. (4) Catalytic processes in which the energy of activation for occurrence of a reaction is lowered by the presence of a nonreacting substance.

activation analysis.   An extremely sensitive technique for identifying and measuring very small amounts of various elements. A sample is exposed to neutron bombardment in a nuclear reactor, for the purpose of producing radioisotopes from the stable elements. The characteristics of the induced radiations are sufficiently distinct that different elements in the sample can be accurately identified. The technique is particularly useful when concentration of the elements is too small to be measured by ordinary means. Trace elements have thus been determined in drugs, fertilizers, foods, fuels, glass, minerals, dusts, water, toxicants, etc.

activation energy. (1) Distinct energy states corresponding to minima of a potential energy surface in a configuration space. If tunneling is considered, lower energy paths become possible, but an activation energy can be associated with the reaction (at a given temperature) via the relationship between temperature and reaction rate. (2) The amount of energy (SI unit of joules) required to convert 1 mole of a reactant from the ground state to the transition state.

activator. (1) A metallic oxide that promotes cross-linking of sulfur in rubber vulcanization. By far the most widely used is zinc oxide; in rubber mixes where no organic accelerator is used, oxides of magnesium, calcium, or lead are effective. (2) A fatty acid that increases the effectiveness of acidic organic accelerators; stearic acid is generally used, especially with thiazoles. (3) A substance necessary in trace quantities to induce luminescence in certain crystals. Silver and copper are activators for zinc sulfide and cadmium sulfide.
See initiator.

active amyl alcohol. See 2-methyl-1-butanol.

active carbon. See carbon, activated.

active metabolite. A biologically active substance produced by, or modified by, a metabolic process from a precursor.

active metal. A highly reactive metal, usually one that occupies a place in the first or second column of the periodic table. A metal that forms cations by readily giving up electrons.

active oxygen. A highly reactive species of oxygen that destroy microsomal cytochromes by initiating autoxidation and are formed in the animal body by successive additions of an electron and a hydrogen ion to a molecule of molecular oxygen, yielding OH_2^-.

active pharmaceutical ingredient. (API). The biologically active compound in a drug formutation that imparts the desired therapeutic effect.

active site. The region of an enzyme that binds a substrate molecule and catalytically transforms it, usually a small portion of the total enzyme molecule.

active transport. Energy-requiring transport of a solute across a membrane in the direction of increasing concentration. Contrast with passive transport.

activin.
CAS: 114949-22-3.
Hazard: Moderately toxic by ingestion and skin contact. A moderate skin and mild eye irritant.
Source: Natural product.

activity. (1) Chemical activity (thermodynamic activity): a quantity replacing actual molar concentration in mathematical expressions for the equilibrium constant so as to eliminate the effect of concentration on equilibrium constant. (2) Activity coefficient: a fractional number which when multiplied by the molar concentration of a substance in solution yields the chemical activity. This term gives an idea of how much interaction exists between molecules at higher concentration. (3) Activity of metals or elements: an active element will react with a compound of a less active element to produce the latter as the free element, and the active element ends up in a new compound. Thus, magnesium, an active metal, will displace copper from copper sulfate to form magnesium sulfate and free metallic copper; chlorine will liberate iodine from sodium iodide and sodium chloride is formed. See electromotive series. (4) Activity product: the number resulting from the multiplication of the activities of slightly soluble substances. This is frequently called the solubility product. (5) Catalytic activity: see catalysis. (6) Optical activity: the existence of optical rotation in a substance. (7) Radioactivity activity coefficient. (8) Biology: The true thermodynamic activity of a substance, as distinct from its molar concentration. Most of the time we call the activity equal to the molar concentration and ignore it.
See activity (2).

activity series. (displacement series; electromotive series). An arrangement of the metals in the order of their tendency to react with water and acids, so that each metal displaces from solution those below it in the series and is displaced by those above it. The arrangement of the more common metals is: K Na Mg Al Zn Fe Sn Pb H Cu Hg Ag Pt Au.

Actol. A series of polyoxypropylene diols, triols, and polyols. These vary in molecular weight from approximately 1000 to 3600; the diols and triols are almost insoluble in water, but the polyols are completely miscible with it.
Use: Urethane foams, elastomers, and coatings.

Acumer 1000. An acrylic homopolymer.
Use: For water treatment, scale inhibitor, and as a dispersant.

"Acumist" [Honeywell]. TM for homopolymer, and copolymer polyolefin waxes.
Use: For adhesives, coatings, color concentrates, cosmetics, inks, lubricants, paints, plastics, rubber, and textiles.

acyl. An organic acid group in which the OH of the carboxyl group is replaced by some other substituent (RCO–). Examples: acetyl, $CH_3CO–$; benzoyl, $C_6H_5CO–$.

n-acylamino acid.
CAS: 9012-37-7. An amino acid in which the nitrogen is linked to an acyl group in attached.

acyl carrier protein. (ACP). Any of a class of proteins that bind acyl intermediates during the synthesis of long-chain fatty acids. ACPs are importantly involved in the biosynthesis of fatty acids.

acyl chloride. A compound in which a chlorine atom is bound to the carbon atom of a carbonyl group.

acyl halide. Any compound composed of an acyl group bonded to a halogen.
Properties: Highly reactive with strong nucleophiles; fuming, irritant substance.
Derivation: They are derived from a carboxylic acid by replacement of the hydroxyl group by a halogen.

acyloin. (3-hydroxyoctadecan-2-one). $C_{18}H_{36}O_2$. A class of hydroxyl ketones.
Derivation: Produced by the condensation of aldehydes.

acyloxy radical. An oxygen-centered radical composed of an acyl radical bonded to an oxygen atom.

acyl phosphate. Any molecule with the general chemical form adenosine 3′,5′-cyclic monophosphate.

ADA. Abbreviation for acetonedicarboxylic acid. See β-ketoglutaric acid.

1-adamantanamine hydrochloride. See amantadine hydrochloride.

adamantane. (sym-tricyclodecane). $C_{10}H_{16}$. Has unique molecular structure consisting of four fused cyclohexane rings.
Properties: White crystals. Mp 270C (sublimes), approximately 99% pure. Derivatives (alkyl adamantanes) have potential uses in imparting heat, solvent, and chemical resistance to many basic types of plastics. Synthetic lubricants and pharmaceuticals are also based on adamantane derivatives. Adamantane diamine is used to cure epoxy resins.
See "Symmetrel" [Upsher-Smith].

adamsite. See diphenylamine chloroarsine.

Adams, Roger. (1889–1971). An American chemist, born in Boston; graduated from Harvard, where he taught chemistry for some years. After studying in Germany, he moved to the University of Illinois in 1916, where he later became chairman of the department of chemistry (1926–1954). During his prolific career, he made this department one of the best in the country, and strongly influenced the development of industrial chemical research in the U.S. His executive and creative ability made him an outstanding figure as a teacher, innovator, and administrator. Among his research contributions were development of platinum-hydrogenation catalysts, and structural determinations of chaulmoogric acid, gossypol, alkaloids, and marijuana.

He held many important offices, including president of the ACS and AAAS, and was a recipient of the Priestley medal.

addition polymer. A polymer formed by the direct addition or combination of the monomer molecules with one another. An example is the formation of polystyrene by stepwise combination of styrene monomer units (approximately 1000 per macromolecule).
See polymerization.

additive. A nonspecific term applied to any substance added to a base material in low concentrations for a definite purpose. Additives can be divided into two groups: (1) those which have an auxiliary or secondary function (antioxidants, inhibitors, thickeners, plasticizers, flavoring agents, colorants, etc.); and (2) those that are essential to the existence of the end product (leavening agents in bread, curatives in rubber, blowing agents in cellular plastics, emulsifiers in mayonnaise, polymerization initiators in plastics, and tanning agents in leather). It seems logical that the latter group should be regarded less as additives than as base materials, since the end products could not exist without them. In any case, a specific functional name is preferable to the neutral term additive.
See food additive.

additive genetic effects. When the combined effects of alleles at different loci are equal to the sum of their individual effects.
See anticipation.

adduct. See inclusion complex.

adenase. (adenine deaminase). An enzyme in animals that catalyzes the hydrolysis of adenine to hypoxanthine.

adenine. (6-aminopurine).
CAS: 73-24-5. $C_5H_5N_5$.

Properties: White, odorless, microcrystalline powder; sharp salty taste. Mp 360–365C (decomposes). Very slightly soluble in cold water; soluble in boiling water, acids, and alkalies; slightly soluble in alcohol; insoluble in ether and chloroform. Aqueous solutions are neutral.
Occurrence: Ribonucleic acids and deoxyribonucleic acids, nucleosides, nucleotides, and many important coenzymes.
Derivation: By extraction from tea; by synthesis from uric acid; prepared from yeast ribonucleic acid.
Use: Medicine and biochemical research.

adenosine. (adenine riboside; 9-β-D-ribofuranosyl-adenine).
CAS: 58-61-7. $C_5H_4N_5 \cdot C_5H_8O_4$. The nucleoside composed of adenine and ribose.
Properties: White, crystalline, odorless powder; mild, saline, or bitter taste. Mp 229C. Quite soluble in hot water; practically insoluble in alcohol.
Derivation: Isolation following hydrolysis of yeast nucleic acid.
Use: Biochemical research.

adenosine diphosphate. (5'-adenylphosphoric acid; ADP; adenosine-5'-pyrophosphate; adenosine diphosphoric acid).
CAS: 58-64-0. $C_{10}H_{15}N_5O_{10}P_2$. A nucleotide found in all living cells and important in the storage of energy for chemical reactions.
Derivation: (1) From adenosine triphosphate by hydrolysis with the enzyme adenosinetriphosphatase from lobster or rabbit muscle; (2) by yeast phosphorylation of adenosine.
Use: Biochemical research.
Available forms: Sodium or barium salt of adenosinediphosphoric acid.
See adenosine diphosphate.

adenosine kinase. An enzyme catalyzes the transfer of phosphate groups from ATP to adenosine, producing adenosine diphosphate (ADP) and adenosine monophosphate (AMP).

adenosine monophosphate. See adenylic acid.

adenosine monophosphate deaminase. (adenylic acid deaminase; AMP deaminase). An enzyme that catalyzes the conversion of adenylic acid to inosinic acid.

adenosine phosphate (USAN). (5'-adenyldiphosphoric acid; ATP; adenosine triphosphate; adenosine 5'-(tetra-hydrogen triphosphate)).
CAS: 56-65-5. $C_{10}H_{16}N_5O_{13}P_3$. A nucleotide that serves as a source of energy for biochemical transformations in plants (photosynthesis) and also for many chemical reactions in the body, especially those associated with muscular activity and replication of cell components. Composed of the purine base, adenosine, and D-ribose attached to three phosphate groups, two of which are high-energy groups. It is an essential energy-storage and energy-transfer component of all living cells. Metabolic hydrolysis of ATP releases pyrophosphate (O_7P_2)

and large amounts of free energy forming adenosine diphosphate (ADP). The large amount of free energy released in this reaction train drives numerous endergonic metabolic processes.
Properties: White, amorphous powder; odorless; faint sour taste. Soluble in water; insoluble in alcohol, ether, and organic solvents; stable in acidic solutions; decomposes in alkaline solution.
Derivation: Isolation from muscle tissue; yeast phosphorylation of adenosine. ATP is reconstituted mainly during photosynthesis and oxidative phosphorylation from ADP.
Use: Biochemical research.
Available forms: Disodium, dipotassium, and dibarium salts.

adenylate cyclase. (adenyl cyclase, 3',5'-cyclic adenosine monophosphate synthetase). The enzyme that catalyzes the conversion of ATP to adenosine 3',5'-cyclic monophosphate (cyclic AMP). It is sometimes a site of toxicant action that can have grave consequences. The active site of this enzyme is part of a complex polymer that includes hormone or neurotransmitter receptors, and guanine nucleotide-binding proteins.

adenylic acid. (adenosine monophosphate). Any of a group of isomeric nucleotides that contains the adenyl radical, $C_5H_4N_5$.

3'-adenylic acid. (yeast adenylic acid). $C_{10}H_{14}N_5O_7P$.
Properties: Crystals. Mp 197C (decomposes). Almost insoluble in cold water; slightly soluble in boiling water. Gives quantitative yield of furfural when distilled with 20% hydrochloric acid.
Derivation: Extracted from nucleic acids of yeast; also made synthetically.
Use: Biochemical research.

5'-adenylic acid. (adenosine monophosphate; AA; adenosine phosphate; AMP; adenosinephosphoric acid; cAMP).
CAS: 61-19-8. $C_{10}H_{14}N_5O_7P$. The monophosphoric ester of adenosine, i.e., the nucleotide containing adenine, d-ribose, and phosphoric acid. Adenylic acid is a constituent of many important coenzymes. Cyclic adenosine-3',5'-monophosphate is designated by biochemists as cAMP.

Properties: (Muscle adenylic acid) crystalline solid. Mp 196–200C. Readily soluble in boiling water. Gives only traces of furfural when boiled with 20% hydrochloric acid.
Derivation: Extracted from muscle tissue; phosphorylation of adenosine.
Use: Biochemical research.

adenylic acid deaminase.   (AMP deaminase; adenosine monophosphate deaminase). An enzyme that catalyzes the conversion of adenylic acid to inosinic acid and NH_3.

adenyl kinase.   (adenylic acid kinase; myokinase). A phosphotransferase that catalyzes the phosphorylation of one molecule of ADP by another, yielding adenosine triphosphate (ATP) and adenosine monophosphate (AMP).

adhesion.   The state in which two surfaces are held together by interfacial forces, which may consist of valence forces or interlocking action, or both (ASTM).

adhesion tension.   The decrease in free surface energy that occurs when a unit liquid–solid interface is substituted for a unit air–solid interface. It is equal to the product of the surface tension of the liquid and the cosine of the liquid–solid angle of contact.

adhesive.   Any substance, inorganic or organic, natural or synthetic, that is capable of bonding other substances together by surface attachment. A brief classification by type is as follows:

I. Inorganic
   (1) Soluble silicates (water glass)
   (2) Phosphate cements
   (3) Portland cement (calcium oxide-silica)
   (4) Other hydraulic cements (mortar, gypsum)
   (5) Ceramic (silica-boric acid)
   (6) Thermosetting powdered glasses ("Pyroceram")
II. Organic
   (1) Natural:
      (a) Animal: hide and bone glue, fish glue, blood and casein glues
      (b) Vegetable: soybean starch cellulosics, rubber latex, and rubber-solvent (pressure-sensitive); gums, terpene resins (rosin), mucilages
      (c) Mineral asphalt, pitches, hydrocarbon resins
   (2) Synthetic
      (a) Elastomer-solvent cements
      (b) Polysulfide sealants
      (c) Thermoplastic resins (for hot-melts): polyethylene, isobutylene, polyamides, polyvinyl acetate
      (d) Thermosetting resins: epoxy, phenoformaldehyde, polyvinyl butyral, cyanoacrylates
      (e) Silicone polymers and cements

For further information, refer to Case Western Reserve University in Cleveland, Ohio, which maintains a fundamental research center for adhesives and coatings.

adhesive, high-temperature.   (1) An organic polymer, e.g., polybenzimidazoles, that retains bonding strength up to 260C for a relatively long time (500–1000 hours); strength drops rapidly above 260C, 80% being lost after 10 minutes at 535C. (2) An inorganic (ceramic), e.g., silica-boric acid mixtures or cermets produce bonds having high strength above 2000F. Adhesive lap-bond strengths can be over 2000 psi at 1000F. These adhesives are used largely for aerospace service, and for metal–metal and glass–metal seals. A silicone cement is reported to have been used to adhere tiles to spacecraft.
See RTV.

adhesive, hot-melt.   A solid, thermoplastic material that melts quickly upon heating, and then sets to a firm bond on cooling. Most other types of adhesives set by evaporation of solvent. Hot-melt types offer the possibility of almost instantaneous bonding, making them well suited to automated operation. In general, they are low-cost, low-strength products, but are entirely adequate for bonding cellulosic materials. Ingredients of hot-melts are polyethylene, polyvinyl acetate, polyamides, hydrocarbon resins, as well as natural asphalts, bitumens, resinous materials, and waxes.
Use: Rapid and efficient bonding of low-strength materials, e.g., bookbinding, food cartons, sideseaming of cans, miscellaneous packaging applications.
See sealant.

adhesive, rubber-based.   (cement, rubber). (1) A solution of natural or synthetic rubber in a suitable organic solvent, without sulfur or other curing agent; (2) a mixture of rubber (often reclaimed), filler, and tackifier (pine tar, liquid asphalt) applied to fabric backing (pressure-sensitive friction tape); (3) a room-temperature curing rubber-solvent-curative mixture, often made up in two parts that are blended just before use; (4) rubber latex, especially for on-the-job repairing of such items as conveyor belts; (5) silicone rubber cement (see RTV and silicone).
Hazard: Those containing organic solvents, (1) and (3) above, are flammable.

adhesive working life.   (pot life). The length of time an adhesive is usable after being mixed.

adiabatic.   Descriptive of a system or process in which no gain or loss of heat is allowed to occur.
adigoside.
CAS: 14259-51-9. $C_{35}H_{54}O_9$.
Hazard: A poison.
Source: Natural product.

adipic acid. (hexanedioic acid; 1,4-butanedicarboxylic acid). CAS: 124-04-9.

$$CH_2CH_2CO|OH|$$
$$CH_2CH_2|COOH|$$

Properties: White, crystalline solid. Mp 152C, bp 265C (100 mm), d 1.360 (20/4C), flash p 385F (196C) (CC). Slightly soluble in water; soluble in alcohol and acetone. Relatively stable. Combustible.
Derivation: Oxidation of cyclohexane, cyclohexanol, or cyclohexanone with air or nitric acid.
Grade: Technical; FCC.
Hazard: Upper respiratory tract and autonomic nervous system impairment.
Use: Manufacture of nylon and of polyurethane foams; preparation of esters for use as plasticizers and lubricants; food additive (acidulant); baking powders; adhesives.

adipocere. The grease material formed when animal fats decompose.

adipocyte. An animal cell specialized for the storage of triacylglycerols.

adiponitrile. (1,4-dicyanobutane). CAS: 111-69-3. $NC(CH_2)_4CN$.
Properties: Water-white, odorless liquid. Mp 1–3C, refr index 1.4369 (20C), bp 295C, flash p 200F (93.3C) (OC). Slightly soluble in water, soluble in alcohol and chloroform. Combustible.
Derivation: Chlorination of butadiene to dichlorobutylene, which is reacted with 35% sodium cyanide soluble to yield 1,4-dicyanobutylene, which is hydrogenated to adiponitrile. Also by electroorganic synthesis from acrylonitrile.
Hazard: Toxic by ingestion and inhalation. Upper and lower respiratory tract irritant.
Use: Intermediate in the manufacture of nylon; organic synthesis.

Adipol. A series of adipate plasticizers.

adipose tissue. Connective tissue specialized for the storage of large amounts of triacylglycerols.

"Adiprene" *[Chemtura].* TM for a polyurethane rubber, the reaction product of diisocyanate and polyalkylene ether glycol. In its raw polymer form, it is a liquid of honey-like color and consistency, which is compounded chemically (to polymerize it further) and converted into products by casting and other techniques.
See polyurethane.

"Adiprene" *[Uniroyal].* See "Vibrathane."

adjuvant. A subsidiary ingredient or additive in a mixture (medicine, flavoring, perfume, etc.) which contributes to the effectiveness of the primary ingredient.

Adkins catalyst. A catalyst containing copper chromite and copper oxide. It is used for the reduction of organic compounds, usually at high temperatures and pressures. It is likewise used as a catalyst for dehydrogenation and for decarboxylation reactions.

Adkins–Peterson reaction. Formation of formaldehyde by air oxidation of methanol in the vapor phase over metal oxide catalysts. A 40% aqueous formaldehyde solution is obtained.

ADK STAB. A line of organic and inorganic industrial chemicals, synthetic resins, synthetic rubber, high-compound fertilizers, coating materials, latexes, pharmaceutical and food additives, explosives, photopolymers and platemaking systems, separation and ion-exchange membranes, systems, and equipment.

(-)-adlumidine. See *l*-capnoidine.

d-adlumidine.
CAS: 550-49-2. $C_{20}H_{17}NO_6$.
Hazard: A poison.
Source: Natural product.

"ADMA" *[Albemarle].* TM for a group of alkyldimethylamines composed of even-numbered carbon chains from C_8 to C_{18}.

ADME. The four steps a drug goes through when administered are absorption, dilution, metabolism, excretion.

admiralty metal. A nonferrous alloy containing 70–73% copper, 0.75–1.20% tin, remainder zinc.
Properties: Offers good resistance to dilute acids and alkalies, seawater, and moist sulfurous atmospheres. D 8.53 (20C), liquidus temp 935C, solidus temp 900C.
Use: Condenser, evaporator, and heat exchanger tubes, plates, and ferrules.

Admox. (lauryl dimethylamine oxide). A foam stabilizer and viscosity builder.
Use: In janitorial products.

adocain. A mixture of cocaine hydrochloride and adonidin (a glucoside from *Adonis vernalis*).
Use: Heart stimulant and diuretic.

"Adogen" *[Evonik].* TM for primary fatty amines.
Use: Corrosion inhibitors and sludge dispersants for lube oil additives, textile intermediates, antistatic agents, down well corrosion inhibitors, and bactericides.

"Adogen" *[Evonik].*   TM for dialkyl dimethyl quaternaries.
  Use: Paper debonders, dye retarders, organophilic clays, antistatic agents, car-wash rinse aids, sugar clarification, fabric softeners, creme rinses. A high performance cationic quaternary fabric softener for retail, industrial, laundry, textile, and paper softener applications.

"Adogen 442" *[Evonik].*   TM for ditallow dimethylammonium chloride, dihydrogenated.
  Grade: 75% paste, 96% powder.
  Use: High-performance cationic quaternary fabric softener for retail, industrial, laundry, textile, and paper softener applications.

adoMet.   See S-adenosylmethionine.

adonidin.   Any of the cardiotoxic phenanthrene glycosides that occur in various species of *Adonis*. Effects are similar to those of digitalin.
  Hazard: Poison.

adonitoxin.   (3-[96-deoxy-α-1-mannopyranosyl)-oxy}-14,16-dihydroxy-19-oxocard-20(22)-enolide).
  $C_{29}H_{42}O_{10}$. A phenanthrene glycoside.
  Derivation: Occurs in *Adonis vernalis*.
  Hazard: Toxic.

ADP.   Abbreviation for (1) adenosine diphosphate, (2) ammonium dihydrogen phosphate.
  See ammonium phosphate, monobasic.

adrenal cortex.   The outer portion of the adrenal gland that secretes various steroid hormones. The outer region of the cortex *zona glomerulosa* produces mineralocorticoids and glucocorticoids; the inner region, the *zona reticularis-fasciculata* secretes mainly cortisol and certain other glucocorticoids.

adrenal gland.   (adrenal body; suprarenal gland). Either of a pair of glandular, ductless glands in vertebrates. They are situated on or near the kidneys in mammals and are superior to the kidney in primates. Each is composed of an inner medulla that secretes adrenalin and noradrenalin, and an outer cortex that secretes steroid hormones (corticosteroids). In some vertebrates, the two types of adrenosecretory tissue form separate glands.

adrenaline.   (epinephrine). $C_9H_{13}O_3N$. A hormone having a benzenoid structure. It is obtained by extraction from the adrenal glands of cattle and is also made synthetically. Its effect on body metabolism is pronounced, causing an increase in blood pressure and rate of heart beat. Under normal conditions, its rate of release into the system is constant; but emotional stresses such as fear or anger rapidly increase the output and result in temporarily heightened metabolic activity.

Hazard: Toxic by ingestion and injection.

adrenal medulla.   The inner portion of the adrenal gland that secretes adrenalin and noradrenalin.

adrenergic antagonist B.   A drug that competitively inhibits the binding of catecholamines to β-adrenergic receptors. They are used to decrease cardiovascular activity, improve blood circulation, and inhibit hypertension, cardiac arrhythmias, and angina.
  Hazard: Bronchospasm; fatigue; hypoglycemia.

adrenocorticotropic hormone.   See ACTH; corticoid hormone.

adrenodoxin.   A ferredoxin involved in electron transfer from NADPH+ to cytochrome p-450. The process is catalyzed by a reductase in the adrenal gland.

Adripene.   A synthetic resin.
  Use: For soles of athletic footwear.

adsorbent.   A substance that has the ability to condense or hold molecules of other substances on its surface. Activated carbon, activated alumina, and silica gel are examples.

adsorption.   Adherence of the atoms, ions, or molecules of a gas or liquid to the surface of another substance, called the adsorbent. The best-known examples are gas–solid and liquid–solid systems. Finely divided or microporous materials presenting a large area of active surface are strong adsorbents and are used for removing colors, odors, and water vapor (activated carbon, activated alumina, silica gel). The attractive force of adsorption is relatively small, of the order of van der Waals forces. When molecules of two or more substances are present, those of one substance may be adsorbed more readily than those of the others. This is called preferential adsorption.
  See absorption; chemisorption.

adsorption indicator.   A substance used in analytical chemistry to detect the presence of a slight excess of another substance or ion in solution as the result of a color produced by adsorption of the indicator on a precipitate present in the solution. Thus, a precipitate of silver chloride will turn red in a solution containing even a minute excess of silver ion (silver nitrate solution) if fluorescein is present. In this example, fluorescein is the adsorption indicator.

Advacide.   TPLA; a free-flowing powder of triphenyl lead acetate containing 10% of a liquid aromatic hydrocarbon mixture.
  Use: Antifouling paints.

Advantage.   A variety of polymers.
  Use: Hair care resins.

advection. The transfer of air and its characteristics by horizontal motion.

aeolotropic. (eolotropic). Displaying change of physical properties with change of position or direction, as in the change of refractive index on changing position of double refracting crystals: not isotropic.

AEPD. Abbreviation for 2-amino-2-ethyl-1,3-propanediol.

aerate. To impregnate or saturate a material (usually a liquid) with air or some similar gas. This is usually achieved by bubbling the air through the liquid, or by spraying the liquid into air.

aerobe. An organism that uses oxygen as the terminal electron acceptor in respiration.

aerobic. Requiring or occurring in the presence of oxygen.
See bacteria.

aerobic dehydrogenase. An enzyme, usually a metalloflavoenyme such as xanthine oxidase, that catalyzes the transfer of hydrogen from a substrate molecule to oxygen with the formation of hydrogen peroxide.

"Aero" [Cytec]. TM; used as a combining form in naming a group of chemical products, e.g., "Aero-float." They include the following:

case-hardening mixtures	metal heat-treating salts
catalysts	metallurgical additives
fertilizer additives	anticaking agents
flotation agents	sizing emulsions
flocculants	wetting agents
frothing/foaming agents	reinforcing agents

aerogel. Dispersion of a gas in a solid or a liquid: the reverse of an aerosol. Flexible and rigid plastic foams are examples.
See foam; aerosol.

aerosol. A suspension of liquid or solid particles in a gas, the particles often being in the colloidal size range. Fog and smoke are common examples of natural aerosols; fine sprays (perfumes, insecticides, inhalants, antiperspirants, paints, etc.) are manufactured.

aerosols. Suspensions of various kinds may be formed by placing the components together with a compressed gas in a container (bomb). The pressure of the gas causes the mixture to be released as a fine spray (aerosol) or foam (aerogel) when a valve is opened. This technique is used on an industrial scale to spray paints and pesticides. It is also used in consumer items such as perfumes, deodorants, shaving cream, whipped cream, and the like. The propellant gas may be a hydrocarbon

(propane, isobutane) or dimethyl ether. Admixture of 15% of methyl chloride with the hydrocarbons reduces their fire risk; for this purpose water can be used with dimethyl ether. Carbon dioxide generated in situ is a propellant which does away with the flammability problem.

"Aerothene" [Dow]. TM for a group of chlorinated solvents used as vapor pressure depressants, or with compressed gases to replace fluorocarbon propellant systems.

Aerothenet TT.
CAS: 71-55-6. 1,1,1-trichloroethane solvent.
Use: Leather and suede cleaning.

aerotolerant anaerobe. Any organism such as a bacterium that grows and reproduces under both aerobic and anaerobic conditions. Energy is obtained by fermentation regardless of ambient conditions.

aerozine. A 1:1 mix of hydrazine and *uns*-dimethylhydrazine (UDMH).
Hazard: Flammable and explosive.
Use: Rocket fuels.
See hydrazine.

aerugidiol.
CAS: 116425-35-5. $C_{15}H_{22}O_3$.
Hazard: A poison by ingestion.

AES. Abbreviation for Auger electron spectroscopy.
See spectroscopy.

AET. See β-aminoethylisothiourea dihydrobromide.

AFCF. See ferric ammonium ferrocyanide.

affected relative pair. Individuals related by blood, each of whom is affected with the same trait. Examples are affected sibling, cousin, and avuncular pairs.
See avuncular relationship.

affinin. (*N*-isobutyl-2,6,8-decatrienamide). $C_{14}H_{23}NO$.
Properties: Yellowish, oily liquid. Bp 163C (0.5 mm), mp 23C, refr index 1.52. Soluble in alcohol; insoluble in alkalies and acids.
Derivation: From *Heliopsis longipes* or made synthetically.
Use: Insecticide activator.

affinity. The tendency of an atom or compound to react or combine with atoms or compounds of different chemical constitution. For example, paraffin hydrocarbons were so named because they are quite unreactive, the word *paraffin* meaning "very little affinity." The hemoglobin molecule has a much greater affinity for carbon monoxide than for oxygen. The free energy decrease is a quantitative measure of chemical affinity.

affinity constant. The reciprocal of the dissociation constant; a measure of the binding energy of a ligand in a receptor.

aflatoxins.
CAS: 1402-68-2. A group of polynuclear molds (mycotoxins) produced chiefly by the fungus *Aspergillus flavus*; they are natural contaminants of a wide range of fruits, vegetables, and cereal grains.
Hazard: Highly toxic to many species of animals, including fish and birds. The B_2 and G_1 strains are known carcinogens. Aflatoxins fluoresce strongly under UV, and are soluble in methanol, acetone, and chloroform, but only slightly soluble in water and hydrocarbon solvents. Prevention of mold growth is the most effective protection; removal or inactivation is possible by physical or chemical means (hand-sorting, solvent refining, etc.). Complete elimination of aflatoxins from foods is not feasible; FDA sets an upper limit of 20 ppb in foods and feeds, and 0.5 ppb in milk. Confirmed carcinogen.
See mycotoxin.

aflatoxin $C_{17}H_{12}O_6$. (2,3-6aα,9aα-tetrahydro-4-methoxycyclopenta-[c]-furo-[3n,2n:4,5]furo[2,3h][1]benzopyran-1,11-dione; afb1; aflatoxin b).
CAS: 1162-65-8. $C_{17}H_{12}O_6$.
Properties: Blue florescence; crystalline material.
Derivation: Secreted by Aspergillus flavus and Aspergillus parasiticus.
Hazard: Mycotoxin; liver poison; carcinogen; toxic; mutagen; tumorigen; neoplastigen; teratogen.

aflatoxin $C_{17}H_{14}O_6$. (2,3-6aα,8,9,9aα-tetrahydro-4-methoxycyclopenta-[c]-furo-[3n,2n:4,5]furo[2,3h][1]benzopyran-1,11-dione; dihydroaflatoxin B1).
CAS: 7220-81-7. $C_{17}H_{14}O_6$. The 8,9-dihydro derivative of aflatoxin b_1.
Properties: Blue florescence; yellow crystals.
Hazard: Toxic; mutagen; carcinogen; tumorigen.

aflatoxin G_1. (3,4,7aα,9,10,10aα-tetrahydro-5-methoxy-1h,12h-furo-[3n,2n:4,5,]-furo[2,3h]-pyrano[3,4-c][1]benzopyran-1,12-dione).
CAS: 1165-39-5. $C_{17}H_{12}O_7$.
Properties: Green florescence.
Derivation: Secreted by *Aspergillus parasiticus.*
Hazard: Mycotoxin; toxic; mutagen; carcinogen; neoplastigen.

aflatoxin $C_{17}H_{14}O_7$. (3,4,7aα,9,10,10aα-hexahydro-5-methoxy-1h,12h-furo-[3n,2n:4,5,]-furo[2,3h]-pyrano[3,4-c][1]benzopyran-1,12-dione; dihydroaflatoxin G1).
CAS: 7241-98-7. $C_{17}H_{14}O_7$. The 9,10-dihydro derivative of aflatoxin G_1.
Properties: Crystals with green florescence.

Derivation: Occur in foodstuffs contaminated with *Aspergillus flavus.*
Hazard: Toxic; mutagen; carcinogen.

aflatoxin $C_{17}H_{12}O_7$. (2,3,6a,9a-tetrahydro-9a-hydroxy-4-methoxycyclopenta[c]furo[2,3-h][1]-benzopyran-1,11-dione; 4-hydroxyaflatoxin B1).
CAS: 6795-23-9. $C_{17}H_{12}O_7$. The 4-hydroxylated derivative of aflatoxins B.
Properties: Crystals that exhibit blue-violet fluorescence.
Derivation: Occur in the milk of cattle that have been fed infested meal.
Hazard: Mutagen; carcinogen; extremely toxic (hepatotoxic).

aflatoxin M_2. (2,3,6a,8,9,9a-hexahydro-9a-hydroxy-4-methoxycyclorpenta[c]furo[3n,2n:4,5]-furo[2,3h][1]benzopyran-1,11-dione). The 4-hydroxylated derivative of aflatoxins B.
Derivation: Occurs in the milk of cattle that have been fed infested meal.
Hazard: Mutagen; carcinogen; toxic.

afterblow. In the Bessemer process, continuing the blast air flow in order to remove phosphorus after the removal of carbon.

after-chromed dye. A dye that is improved in color or fastness by treatment with sodium dichromate, copper sulfate, or similar materials, after the fabrics are dyed.

after-coppering dye. A dye that is improved in color or fastness by treatment with copper sulfate, after the fabrics are dyed.

after cure. Continuing the process of vulcanizing after the cure has been carried to the proper degree and the heat is cut off.

afterdamp. Carbon monoxide.
Derivation: Produced in coal mines by methane explosions.

after-flow. The action of the plastic flow in solids continuing after external forces have stopped.

after-glow. (1) The remaining luminescence in rarefied gas after electrodeless charge has passed through. (2) The glow that remains after the extinguishing of a flame.

Ag. Symbol for silver.

aga. (potassium-6-amino-4-sulfonaphthalene-2-sulfonatehydrate). $C_{10}H_{10}KNO_7S_2$. A codon of arginine that directs the placement of arginine into a polypeptide.

agar. (agar-agar). A phycocolloid derived from red algae such as *Gelidium* and *Gracilaria*; it is a polysaccharide mixture of agarose and agaropectin.

Properties: Thin, translucent membranous pieces or pale buff powder. Strongly hydrophilic, it absorbs 20 times its weight of cold water with swelling; forms strong gels at approximately 40C.

Grade: Technical, USP, FCC.

Use: Microbiology and bacteriology (culture medium); antistaling agent in bakery products, confectionery, meats, and poultry; gelation agent in desserts and beverages; protective colloid in ice cream, pet foods, health foods, laxatives, pharmaceuticals, dental impressions, laboratory reagents, and photographic emulsions.

See algae; alginic acid.

agarose.
CAS: 9012-36-6.

Properties: A neutral, linear polymer composed of alternating residues of D-galactose and 3,6-anhydro-1-galactose.

Derivation: Extracted from seaweed.

Use: The resolving medium in electrophoresis.

agarose gels. A polysaccharide gel used to measure the size of nucleic acids (in bases or base pairs). See "Gel Electrophoresis". The gel of choice for DNA or RNA in the range of thousands of bases in length, or even up to 1 megabase when employing pulsed field gel electrophoresis.

agate-ware. An enameled iron or steel frequently used for kitchenware.

agc. ((2S,3R,4S,5S,6R)-6-(hydroxymethyl) oxane-2,3,4,5-tetrol). $C_6H_{12}O_6$. A codon of serine that directs the placement of serine into a polypeptide.

age hardening. The spontaneous hardening of alloys at room temperature within a of couple days after quenching as a result of grain structures.

agent blue. (dimethylarsinic acid; herbicide blue). $C_2H_7AsO_2$.

Properties: A solution of cacodylic acid (0.37 kg/l) that is an arsenical.

Use: Has been used as a dermatologic agent and a herbicide.

agent orange. A toxic herbicide and defoliant containing 2,4,5-trichlorophenoxyacetic acid (2,4,5-T) and 2,4-dichlorophenoxyacetic acid (2,4-D), with trace amounts of dioxin. Its use has been restricted.

agent purple. (herbicide purple).

Properties: It is a mixture of the n-butyl esters of 2,4-d and 2,4,5-t and the isobutyl ester of 2,4,5-t, at concentrations of 0.50, 0.31, and 0.22 kg/l, respectively.

Use: A herbicide used by the U.S. military forces to control vegetation early in the Vietnam conflict, but it was replaced by agent orange in 1964.

agent white. (herbicide white).

Properties: It is a mixture of the triisopropanolamine salt of 2,4-d and picoram, at respective concentrations of 0.24 and 0.07 kg/l.

Use: A herbicide used by the U.S. military forces to control vegetation in the Vietnam conflict.

age-resister. See antioxidant.

agg. ((2S)-2-(butylsulfonylamino)-3-[4-(4-piperidin-4-ylbutoxy)phenyl]propanoic acid). $C_{22}H_{36}N_2O_5S$. A codon of arginine that directs the placement of arginine into a polypeptide.

agglomeration. (1) Combination or aggregation of colloidal particles suspended in a liquid into clusters or "flocs" of approximately spherical shape. It is usually achieved by neutralization of the electric charges which maintain the stability of the colloidal suspension. The terms *flocculation* and *coagulation* have a closely similar meaning. (2) The food industry uses "agglomeration" in the sense of increasing the particle size of powdered food products. Because such powders tend to be hydrophobic because of the high surface tension of water, agglomeration causes them to be more readily dispersible in water—a process known as "instantizing." The agglomerates have varying degrees of open spaces (voids) and are loosely bound, foam-like structures. They are formed by mechanical means in chamber spray dryers, tubes, or fluidized beds, usually in the presence of moisture.

See aggregation; agglutination; flocculation; coagulation.

agglutination. The combination or aggregation of particles of matter under the influence of a specific protein. The term is usually restricted to antigen–antibody reactions characterized by a clumping together of visible cells such as bacteria or erythrocytes. The antigen is called an agglutinogen, and the antibody an agglutinin because of an apparent gluing or sticking action.

See aggregation.

agglutinin. An antiantibody whose activity is directed against the idiotype of a specific immunoglobulin antibody that under certain circumstances will cause organic particles coated with a specific antigen or cells that contain the antigen to agglutinate.

aggregate. A collective term denoting any mixture of such particulates as sand, gravel, crushed stone, or cinders used in Portland cement formulations, road building, paving compositions, animal husbandry, trickle filters, horticulture, etc.

aggregation. A general term describing the tendency of large molecules or colloidal particles to combine in clusters or clumps, especially in solution. When this occurs, usually as a result of removal of the electric charges by addition of an appropriate electrolyte, by the action of heat, or by mechanical agitation, the aggregates precipitate or separate from the dissolved state. Included in this term are the more specific terms *agglutination*, *coagulation*, *flocculation*, *agglomeration*, and *coalescence*.

aggregation technique.   A technique used in model organism studies in which embryos at the 8-cell stage of development are pushed together to yield a single embryo (used as an alternative to microinjection).
See model organisms.

aging.   (1) *Deleterious*: gradual deterioration of a material due to long exposure to the environment. Aging characteristics of various materials:

(a) Vulcanized rubber and thermoplastic polymers lose strength and crack due to oxidation, sunlight, heat; this is retarded by antioxidants, e.g., phenyl-β-naphthylamine. For accelerated aging tests see bomb.
(b) Foods: spoilage and rancidity due to bacterial contamination, retarded by butylated hydroxyanisole and various propionates.
(c) Paints: cracking, fading, chalking due to exposure to weather and photochemical degradation. Retarded by proper selection of vehicles and pigments. Accelerated weathering tests are used.
(d) Metals: rusting, pitting, and scaling due to corrosion, especially in moist acid and alkaline environments. Avoided by use of alloys in which noncorrosive metals are incorporated (stainless steel) or by plating or cladding the base metal with chromium or nickel.

See corrosion; exposure testing.
(2) *Beneficial*: improvement of flavor by long storage. Cheeses develop a "sharp" flavor on ripening for 9–12 months, wines develop a "bouquet" after 2 or more years of storage. Whiskeys stored in oaken casks for several years modify their flavor by extracting components of the wood. Tobacco is aged from 3 to 5 years after curing to remove unpleasant odors and improve smoking qualities.

agitator.   Any rotating device that induces motion in fluid mixtures over a wide range of viscosities, thus effecting uniform dispersion of their components. An important class of agitators comprises impellers which produce turbulent flow in liquids of low viscosity; their diameter is much less than that of the container. They may be either top-entering (vertical) or side-entering (at an angle of approximately 45°); medium-viscosity liquids are agitated by paddles attached to a central rotating member. Pastes and high-viscosity mixtures in which no turbulent flow is possible require agitators that closely fit the mixing chamber so as to provide the necessary shearing and squeezing action throughout the mass. These are kneading devices utilizing curved or helical rotors or sigma blades, either single or double. Screw-type agitators permit continuous mixing by means of multiple shearing

and blending actions. The so-called ribbon agitators are effective for dry powders, slurries, etc. A number of ingenious modifications and combinations of these types are widely used in the process industries. Most are available in laboratory sizes.
See impeller; mixing; kneading; screw.

aglucon.   The nonglucose portion of a glucoside.

aglucone.   The nonsugarlike portion of a glucoside molecule.
See glycoside.

aglycon.   (aglycone; genin). The noncarbohydrate moiety resulting from hydrolysis of a glycoside molecule.

aglycone.   A nonsugar hydrolytic product of a glycoside.
See glycoside.

agmatine.   (2-(4-aminobutyl)guanidine;   4-(aminobutyl)guanidine;   1-amino-4-guanidobutane).
CAS: 306-60-5. $C_5H_{14}N_4$. Decarboxylated arginine.
Derivation: Occurs in ergot, sponges, the salivary glands of cephalopods, the pollen of *Ambrosia artemisiifolia*, herring sperm, and in octopus muscle.

Agre, Peter.   (1949–). An American born in Northfield, Minnesota who won the Nobel Prize for chemistry in 2003 for his pioneering work concerning discoveries about water channels in cell membranes. Agre received an undergraduate degree from Augsburg College and an M.D. from Johns Hopkins University, where he continues his research.

agricultural   chemical.   (agrichemical).   A chemical compound or mixture used to increase the productivity and quality of farm crops. Included are fertilizers, soil conditioners, fungicides, insecticides, herbicides, nematocides, and plant hormones.

agricultural waste.   See biomass; waste control; gasohol.

agrimek.   See avermectin b(sub 1).

"Agrisel Carbait 5 Insecticide" *[Agrisel]*.
TM for a turf insecticide for ornamental and agricultural grass.
Use: Controls 120 pests.

agromet.   See   *n*-(2,6-dimethylphenyl)-*n*-(methoxy)alanine methyl ester.

"Agsolex" *[ISP]*.   TM for chemical used as an agricultural solvent.

agstone.   (agricultural limestone; calcium carbonate).

CAS: 1317-65-3. $CCaO_3$. Carbonic acid calcium salt.
Properties: Odorless, tasteless powder or crystal.
Use: To lime soil; therapeutically as a phosphate buffer in hemodialysis patients and as a calcium supplement.

agu. (2-aminoguanidine).
CAS: 79-17-4. CH_6N_4. A codon of serine that directs the placement of serine into a polypeptide.

ah locus. A gene locus that regulates the induction of various enzymes by aromatic hydrocarbons.

AHMT (perfume). See: ethanone, 1-(5,6,7,8-tetrahydro-3,5,5,6,8,8-hexamethyl-2-naphthalenyl)-.

AI3-36161.
CAS: 108910-63-0.
Hazard: Moderately toxic by ingestion. Low toxicity by inhalation. A severe skin and moderate eye irritant.
Use: Agricultural chemical.

AIC. See American Institute of Chemists.

AIChE. See American Institute of Chemical Engineers.

air. A mixture (or solution) of gases, the composition of which varies with altitude and other conditions at the collection point. Following is the composition of dry air at sea level:

Substance	% by wt	% by vol
Nitrogen	75.53	78.00
Oxygen	23.16	20.95
Argon		0.93
Carbon dioxide		0.033*
Neon		0.0018
Helium		0.0005
Methane		0.0002
Krypton		0.0001
Nitrous oxide		0.000,05
Hydrogen		0.000,05
Xenon		0.000,008
Ozone		0.000,001

*The CO_2 content of air has increased 12–15% since 1900 due to combustion of fossil fuels.
See greenhouse effect. The density of dry air is 1.29 g/L at 0C and 760 mm Hg. It is noncombustible, but will support combustion. Liquid air is air that has been subjected to a series of compression, expansion, and cooling operations until it liquefies.
Use: Source of oxygen, nitrogen, and rare gases; coolant; power source (compressed); cryogenic agent (liquid); particle classification; blowing agent (asphalt, soap, ice-cream mixes, whipped cream, etc.); flotation.

air classification. The separation of solid particles according to weight and/or size, by suspension in and settling from an air stream of appropriate velocity, as in air-floated clays and other particulate products.

airedale brown md. See C.I. direct brown 2.

air floatation. See air classification.

air gas. See producer gas.

air knife. See doctor knife.

air pollution. (atmospheric pollution). Introduction of substances into the atmosphere that are not normally present therein and that have a harmful effect on humans, animals, or plant life. Photosynthesis is significantly inhibited by air pollutants, especially in urban areas. The worst offenders are sulfur dioxide (which forms sulfurous acid on contact with water vapor); automotive emission products; metal dusts from smelters, coal smoke, and other particulates; formaldehyde and acrolein; and radioactive emanations. Control of these is exercised by the Environmental Protection Agency. As conventionally used, the term does not apply to interior air spaces such as industrial workrooms. TLVs (Threshold Limit Values) for the latter are established by the American Conference of Governmental Industrial Hygienists (ACGIH) and by OSHA. See environmental chemistry.

aizen direct deep black eh. See apomine black gx.

"Ajicure" *[Ajinomoto].* TM for a line of epoxy curing agents.

"Ajisper" *[Ajinomoto].* TM for a pigment dispersing agent.

ajugarin. Any of a class of diterpenes that are effective antifeedants against the African Army Worm, a moth larva (family noctuidae) that feeds on alfalfa and various grains, sometimes in huge numbers.
Derivation: From the leaves of *Ajuga remota* (bugle weed) and related plants.
Use: Insecticide.

Akabori amino acid reactions. (1) Formation of aldehydes by oxidative decomposition of α-amino acids when heated with sugars. (2) Reduction of α-amino acids and esters by sodium amalgam and ethanolic hydrochloric acid to the corresponding α-amino aldehydes. (3) Formation of alkamines by heating mixtures of aromatic aldehydes and amino acids. No reaction has been observed with tertiary amino groups.

aklomide.
CAS: 3011-89-0. $C_7H_5ClN_2O_3$.

Properties: Crystals and gray scales from alc. Mp: 172C.
Use: Drug (veterinary); food additive.

"Akroflex" C [Du Pont]. TM for a rubber antioxidant containing 35% diphenyl-*p*-phenylenediamine $C_6H_4(NHC_6H_5)_2$ and 65% phenyl-α-naphthylamine ("Neozone" A).
Use: To improve the aging and service life of rubbers; anti-cross-linking agent for SBR (styrene-butadiene-rubber).

"Akucell" [Akzo Nobel].
CAS: 9004-32-4. TM for anionic water-soluble sodium carboxymethylcellulose polymer.
Use: As a binder, thickener, and suspension agent in food and pharmaceuticals.

Al.   Symbol for aluminum.

alabaster.   A fine-grained compact gypsum.

alachlor. (Lasso; 2-chloro-*N*-(2,6-diethylphenyl)-*N*-(methoxymethyl)acetamide; 2-chloro-2',6'-*di*-ethyl-*N*-(methoxymethyl)acetanilide; metachlor).
CAS: 15972-60-8. CH_6N_4. A preemergence selective anilide that is absorbed through the roots and inhibits root elongation. Alachlor persists in the soil for up to 3 months. It is a common groundwater pollutant in some agricultural areas.
Hazard: Toxic; questionable carcinogen; allergen; irritant; measured as inhalable fraction and vapor; may cause dermal sensitization; hemosiderosis of the liver, spleen and kidney.
Use: Herbicide to control weeds in corn and soybean fields.

"Alamine 304" [Henkel]. (trilaurylamine). TM for organic acid and metal extraction additive.
Use: For paints, plastics, and rubbers.

"Alanap" [Chemtura]. TM for *N*-1-naphthylphthalamic acid.
Use: A herbicide.

alanine. (α-alanine; α-aminopropionic acid; 2-aminopropanoic acid). $CH_3CH(NH_2)COOH$. A nonessential amino acid.
Properties: Colorless crystals. Soluble in water; slightly soluble in alcohol; insoluble in ether. Optically active.
DL-alanine, mp 295C (decomposes), sublimes at 200C.
L(+)-alanine, mp 297C (decomposes).
D(−)-alanine, mp 295C (decomposes).
L(+)-alanine hydrochloride; prisms, mp 204C (decomposes).
L(+)-alanine, *N*-acetyl-, crystals, mp 116C.
L(+)-alanine, *N*-benzoyl-, crystals, mp 152–154C.
Derivation: Hydrolysis of protein (silk, gelatin, zein), organic synthesis.
Grade: Reagent, technical.

Use: Microbiological research, biochemical research, dietary supplement.

β-alanine. (3-aminopropanoic acid; β-aminopropionic acid). $NH_2CH_2CH_2COOH$. A naturally occurring amino acid not found in protein.
Properties: White prisms. Mp 198C (decomposes). Soluble in water; slightly soluble in alcohol, insoluble in ether. The pH of a 50% solution is 6.0–7.3. Hydrochloride plates and leaflets: mp 122.5C. Platinichloride: yellow leaflets, mp 210C (decomposes).
Derivation: Addition of ammonia to β-propiolactone, other processes based on the reaction of ammonia with acrylonitrile, etc.
Use: Biochemical research, organic synthesis, calcium pantothenate production, buffer in electroplating.

dl-alanine.
CAS: 302-72-7. $C_3H_7NO_2$.
Properties: Needles or prisms, or white crystalline powder; odorless with a sweet taste. Mp: 295C (decomposes). Slightly sol in water, insol in Et_2O.
Hazard: Low toxicity by ingestion.
Use: Food additive.

alanine aminotransferase. (alt; alanine transaminase; glutamin-pyruvic transaminase).
CAS: 9000-86-6. An enzyme that catalyzes the transfer of amino groups from L-alanine to 2-ketoglutarate, or the reverse reaction from l-glutamate to pyruvate.

alanine-oxomalonate aminotransferase.
An enzyme that effects the transfer of the L-alanine group to oxomalonate.

alanosine. $C_3H_7N_3O_4$.
Properties: Finely divided crystals. Decomposes at 190C, optically active. Can be prepared in *l*-, *d*-, and *dl*-forms. Insoluble in most organic solvents; slightly soluble in water.
Derivation: Fermentation of *Streptomyces alanosinicus*; also synthetically.
Use: Inhibitor of insect reproduction, antineoplastic, antibiotic.

β-alanylhistidine.   See carnosine.

Alar. A plant growth regulator (succinic acid-2,2-dimethylhydrazide) that improves the color and texture of apples, grapes, and tomatoes, and prevents premature dropping; growth retardant; multiple-flower stimulator.

alarmona. One of a class of cell-growth regulating metabolites (nucleotides) that enable bacteria to respond to metabolic and environmental changes. They are thought to act by controlling or affecting several biochemical reactions simultaneously,

but the exact mechanism of their behavior has not been elucidated. One such nucleotide, discovered in 1982, is known as ZTP (5-amino-4-imidazole carboximide riboside-5'-triphosphate).
See nucleotide.

"Alathon" *[LyondellBasell].* TM for a polyethylene resin. "Alathon" G-0530, designated as a reinforced polyethylene, contains 30% by weight of glass fiber treated with a proprietary coupling agent that optimizes its reinforcing properties.

albendazole sulfone. See 5-(propylsulfonyl)-2-benzimidazolecarbamic acid methyl ester.

Alberti furnace. A reverberatory furnace for roasting mercury ores, the mercury being condensed in iron tubes and brick chambers.

Albert Precht effect. A technique used in photography to produce a reversed image. The exposed surface is immersed in chromic acid. It is then exposed to uniform light and developed over.

Albigen. A water-soluble polymer used in the textile industry for stripping vat and other dyestuffs. Has no affinity for the fiber and promotes the stripping effect of alkaline hydrosulfite solutions.

albino rat. A phenotype that sometimes occurs in various populations of *Rattus*. The fur is white and the eyes are pink due to a lack of pigment.

"Albone" *[Arkema].* TM for a series of hydrogen peroxide solutions which vary in hydrogen peroxide content from 35 to 90% by weight.
See hydrogen peroxide.

albumen. A 66.3 kDa protein comprising most of the protein in serum and largely responsible for buffering pH and volume of blood. Commercial term for dried egg white used in the food industry.
See albumin, egg.

albumin. Any of a group of water-soluble proteins of wide occurrence in such natural products as milk (lactalbumin), blood serum, and eggs (ovalbumin). They are readily coagulated by heat and hydrolyze to α-amino acids or their derivatives.

albumin, egg. (ovalbumin). Chief protein occurring in egg white as a viscous, colorless fluid; it becomes an amorphous solid when dried, which can be reconstituted with water. It is a heat-sensitive colloidal material that coagulates irreversibly at approximately 60C (140F). The dried product is available in commercial quantities.
Use: Protective colloid and emulsifying agent in bakery products (especially angel cake), clarification of wines, adhesives, paper coatings, pharmaceuticals,

enzyme activation, lithography, analytical reagent, antidote for mercury poisoning, mordant in dyes.
Note: A recombinant DNA technique has made possible the formation of ovalbumin by the bacterium *Escherichia coli.*

albumin, milk. (lactalbumin). A component of skim milk protein (2–5%). Can be crystallized. Exact function is not known, but probably aids in stabilization of the fat particles.
See milk.

ALCA. See American Leather Chemists' Association.

alchemy. The predecessor of chemistry, practiced from as early as 500 BC through the 16th century. Its two principal goals were transmutation of the baser metals into gold and discovery of a universal remedy. Modern chemistry grew out of alchemy by gradual stages.

alchlor process. To remove the unstable components of lubricating oil, this process used aluminum chloride instead of sulfuric acid.

alcian blue.
Properties: Greenish-black crystals with metallic sheen. Soluble in ethanol, cellosolve, ethylene glycol.
Use: Gelling agent for lubricating fluids, bacterial stain for histiocytes and fibroblasts.

Alcoa Mill Products. A wide variety of aluminum sheet and formed products.

Alcoa process. A more efficient method of producing aluminum from bauxite that requires one-third less electric power than the Hall process. Alumina is reacted with chlorine, the resulting aluminum chloride yielding the metal and chlorine on electrolysis. No fluorine is required in the process. Prototype plants are under development.

Alcoa Wheel Forged Products. A wide variety of aluminum forged products.

alcogel. A gel formed by the coagulation of a solution in which the liquid is an alcohol. The result is a mixture of liquid and solid parts within the same volume.

alcohol. A broad class of hydroxyl-containing organic compounds occurring naturally in plants and made synthetically from petroleum derivatives such as ethylene. Many are manufactured in tonnage quantities.
   The many types may be summarized as follows:

I. Monohydric (1 OH group)
   (1) Aliphatic
      (a) paraffinic (ethanol)
      (b) olefinic (allyl alcohol)

(2) Alicyclic (cyclohexanol)
(3) Aromatic (phenol, benzyl alcohol)
(4) Heterocyclic (furfuryl alcohol)
(5) Polycyclic (sterols)
II. Dihydric (2 OH groups): glycols and derivatives (diols)
III. Trihydric (3 OH groups): glycerol and derivatives
IV. Polyhydric (polyols) (3 or more OH groups)

Use: Organic synthesis for solvents, detergents, beverages, pharmaceuticals, plasticizers, and fuels.
For further information, see monohydric, dihydric, trihydric, polyol, and specific alcohol.

alcohol, absolute. See ethyl alcohol.

alcohol dehydrogenase. An enzyme found in animal and plant tissue that acts upon ethanol and other alcohols producing acetaldehyde and other aldehydes.
Use: Biochemical research.

alcohol, denatured.
CAS: 64-17-5. Ethanol to which another liquid has been added to make it unfit to use as a beverage (chiefly for tax reasons). In the U.S., it may be either Completely Denatured (CDA) or Specially Denatured (SDA). At least 50 formulations are officially authorized for making denatured alcohol. They include the following denaturants: SDA 40B must contain brucine, brucine sulfate, or quassin plus *tert*-butanol; SDA 40A must contain sucrose octaacetate plus *tert*-butyl alcohol; SDA 40B must contain "Bitrex" *[Johnson Matthey]* and *tert*-butyl alcohol; SDA40C must contain only *tert*-butyl alcohol. For exact formulas, consult 27CFR Part 21 and the Alcohol and Tobacco Tax Division of IRS, Washington D.C.
Properties: See ethanol.
Hazard: Flammable, dangerous fire risk.
Use: Manufacture of acetaldehyde and other chemicals, solvents, antifreeze and brake fluids, fuels.
See ethanol.

alcohol fermentation. The anaerobic conversion of glucose to ethanol via anaerobic glycolysis.
See also fermentation.

alcohol, grain. Ethanol made from grain.

alcohol, industrial. A mixture of 95% ethanol and 5% water, plus additives for denaturing or special solvent purposes.
See alcohol, denatured.

alcohols, C8-C10, ethoxylated propoxylated.
CAS: 68603-25-8.
Hazard: A severe eye irritant.

alcohols, C12-C6, ethoxylated.
CAS: 68551-12-2.
Hazard: A moderate eye irritant.

alcohols, C12-C13, ethoxylated.
CAS: 66455-14-9.
Hazard: Moderately toxic by ingestion and skin contact. A moderate skin irritant.

alcohols, C12-C15, ethoxylated.
CAS: 68131-39-5.
Hazard: Moderately toxic by ingestion.

alcohols, C14-C15, ethoxylated.
CAS: 68951-67-7.
Hazard: Moderately toxic by ingestion.

alcohols, C16-C18, ethoxylated.
CAS: 68439-49-6.
Hazard: Moderately toxic by ingestion and skin contact. A moderate skin and eye irritant.

alcohol, wood. See methyl alcohol.

alcoholate. (alkoxide).
Properties: Ionic organic compound.
Derivation: By replacement of the hydrogen of a hydroxyl group from an alcohol usually by an alkali metal.

alcoholic extract.
Properties: Solid.
Derivation: Extraction from the alcohol-soluble principles of a drug, followed by evaporation of the alcohol.

alcoholysis. A chemical reaction between an alcohol and another organic compound analogous to hydrolysis. The alcohol molecule decomposes to form a new compound with the reacting substance, the other reaction product is water. Both hydrolysis and alcoholysis may be considered as forms of solvolysis.
See solvolysis.

"Alconox" *[Alconox].* TM for a powdered precision cleaner.
Use: For manual and ultrasonic systems for glassware, metal, ceramic, and other materials in laboratories.

Aldactazide. A combination of spironolactone and hydrochlorothiazide.
Use: Drug.

"Aldactone" *[Pfizer].* TM for spironolactone.
Use: Drug.

aldazine. An azine of an aldehyde that contains two or more nitrogen atoms.

aldehol. An oxidation product of kerosene that is used as a denaturant of ethanol.

aldehydase. (aldehyde oxidase). An enzyme that catalyzes reactions that produce acids from aldehydes.

aldehyde. A broad class of organic compounds having the generic formula RCHO, and characterized by an unsaturated carbonyl group (C=O). They are formed from alcohols by either dehydrogenation or oxidation, and thus occupy an intermediate position between primary alcohols and the acids obtained from them by further oxidation. Their chemical derivation is indicated by the name *al*(cohol) + *dehyd*(rogenation). Aldehydes are reactive compounds participating in oxidation, reduction, addition, and polymerization reactions. For specific properties, see individual compounds.

aldehyde ammonia. (acetaldehyde ammonia; 1-amino-ethanol). $CH_3CH_2ONH_2$.

$$CH_3-CH\begin{array}{l} \diagup OH \\ \diagdown NH_2 \end{array}$$

Properties: White, crystalline solid. Mp 97C (partly decomposes). Stable in closed containers; resinifies on long exposure to air. Very soluble in water and alcohol.
Derivation: Action of acetaldehyde on ammonia.
Hazard: Irritant to eyes and skin; moderate fire risk.
Use: Accelerator for vulcanization of thread rubber, organic synthesis, source of acetaldehyde and ammonia.

aldehyde C-18. See δ-nonalactone.

aldehyde collidine. See 2-methyl-5-ethylpyridine.

aldehyde dehydrogenase.
CAS: 9028-86-8. An enzyme that catalyzes the formation of acids from aldehydes.

aldehyde hydrate. A compound produced by the oxidation of a hydrate of aldehyde.

aldehyde oxidase. A flavoprotein that occurs in the soluble fraction of hepatic cells that catalyzes chiefly the oxidation of endogenous aldehydes.

aldehydine. See 2-methyl-5-ethylpyridine.

Alder, Kurt. (1902–1958). A German chemist who won the Nobel Prize for chemistry along with Otto Diels in 1950 for a project involving a practical method for making ring compounds from chain compounds by forcing them to combine with maleic anhydride. This is known as the Diels–Alder reaction and provided a method for synthesis of complex organic compounds. Alder had degrees from the Universities of Berlin and Kiel.

Alder–Rickert rule. Adducts of 1,3-cyclohexadiene derivatives with acetylenedicarboxylic esters give phthalate ester and ethylene

on heating. Similar adducts of cyclopentadiene revert on heating to starting materials (retro-Diels–Alder).

Alder–Stein rules. Set of rules governing the stereochemistry of the Diels–Alder reaction. The most important are that (1) the stereochemical relationship of groups attached to the diene and the dienophile is maintained in the product (*cis*-addition); and (2) the product resulting from maximum accumulation of unsaturated centers in the transition state is favored (endo rule).

aldicarb.
CAS:116-06-3. $CH_3SC(CH_3)_2HC:NOCONHCH_3$.
Properties: Colorless crystals. Mp 100C (212F). Almost insoluble in water; slightly soluble in benzene and xylene; partly soluble in acetone and methylene chloride.
Hazard: Toxic by ingestion. Questionable carcinogen.
Use: Nematocide, insecticide.

aldimine. (methenaimine). CH_3N. Any of a class of amines derived from an aldehyde.

alditol. Any of a class of acyclic polyols that are derived from an aldose by reduction of the carbonyl group.

aldobiuronic acid. Any of a number of condensation products of an aldose and a uronic acid.

aldohexose. (pentahydroxyhexanal). $C_6H_{12}O_6$. A hexose that contains an aldehyde group. A primary source of energy for living organisms.
Derivation: Found in fruits and other parts of plants.
Use: Therapeutically in fluid and nutrient replacement.

aldoketose. Any of a subgroup of monosaccharides, saccharides, uronic acids, and aminosugars.

aldol. (acetaldol;   β-hydroxybutyraldehyde). $CH_3CH_2OCH_2CHO$.
Properties: Water-white to pale yellow, syrupy liquid. Decomposes into crotonaldehyde and water on distillation under atmospheric pressure. Miscible with water, alcohol, ether, organic solvents. D 1.1098 (15.6/4C), bp 83C (20 mm), vap press less than 0.1 mm (20C), specific heat 0.737, bulk d 9.17 lb/gal (20C), fp below 0C. Flash p 150F (65.5C) (OC), autoign temp 530F (276.6C). Combustible.
Derivation: By condensation of acetaldehyde in sodium hydroxide solution.
Grade: Technical (98%).
Hazard: Moderate fire risk.
Use: Synthesis of rubber accelerators and age resisters, perfumery, engraving, ore flotation, solvent, solvent mixtures for cellulose acetate, fungicides, organic synthesis, printer's rollers, cadmium plating, dyes, drugs, dyeing assistant, synthetic polymers.

aldolase. (zymohexase). An enzyme present in muscle involved in glycogenolysis and anaerobic glycolysis. It catalyzes production of dihydroxy-acetone phosphate and phosphoglyceric aldehyde from fructose-1,6-diphosphate.
Use: Biochemical research.

aldol condensation. A reaction between two aldehyde or two ketone molecules in which the position of one of the hydrogen atoms is changed in such a way as to form a single molecule having one hydroxyl and one carbonyl group. Since such a molecule is partly an alcohol (OH group) and partly an aldehyde (CHO group) and represents a union of two smaller molecules, the reaction is called an aldol condensation. It can be repeated to form molecules of increasing molecular weight. The condensation of formaldehyde to sugars in plants, which on repetition builds up the more complex carbohydrate structures such as starch and cellulose, is thought to be a reaction of this type. It occurs most effectively in an alkaline medium.

Aldo MO. (glycerol monoleate).
CAS: 25496-72-4. An antifog agent and dispersant.
Use: Bath oils, lotions, and creams; emulsifier for liquid and paste waxes, polish, and cleaners.

aldo-α-naphthylamine condensate.
Properties: Orange to dark red solid with characteristic odor. Softens at 64C min, d 1.16. Insoluble in water, gasoline; slightly soluble in alcohol and petroleum hydrocarbons; soluble in acetone, benzene, chloroform, and carbon disulfide.
Use: Antioxidant in tire carcasses, tubes, insulating tape, black soles.

aldonic acid. (glyconic acid; 2,3,4,5,6-pentahydroxyhexanoic acid). $C_6H_{12}O_7$. A monosaccharide derivative or polyhydroxy acid in which the aldehyde group has been oxidized to a carboxyl group.

aldose. Any of a group of sugars whose molecule contains an aldehyde group and one or more alcohol groups. An example is glyceraldehyde (HOCH$_2$•CH$_2$O•CHO), specifically called an aldotriose because it contains three carbon atoms.

aldoside. A glucoside in which the sugar moiety is an aldose.

aldosterone. (electrocortin). $C_{21}H_{28}O_5$.

An adrenal cortical steroid hormone that is the most powerful mineralocorticoid. Probably the chief regulator of sodium, potassium, and chlorine metabolism; approximately 30 times as active as deoxycorticosterone.
Properties: Crystals. Mp 108–112C.
Derivation: Isolated from adrenals; has been synthesized.
Use: Medicine.

aldoxime. The –CH:NOH radical resulting from reactions between aldehydes and hydroxylamine or by the oxidation of primary amines by persulfuric acid.
Properties: Colorless. Water soluble.

aldrin. (HHDN).
CAS: 309-00-2. $C_{12}H_8Cl_6$. The assigned common name for an insecticidal product containing 95% or more of 1,2,3,4,10,10-hexachloro-1,4,4a,5,8,8a-hexahydro-1,4,5,8-*endo,exo*-dimethanonaphthalene.
Properties: Brown to white, crystalline solid. Mp 104–105.5C. Insoluble in water; soluble in most organic solvents. Not affected by alkalies or dilute acids, compatible with most fertilizers, herbicides, fungicides, and insecticides.
Grade: Technical.
Hazard: Toxic by skin absorption. Central nervous system impairment, and liver and kidney damage. Questionable carcinogen.
Use: Insecticide.
See dieldrin; endrin.

Aldyl. A high-strength polyethylene pipe composed principally of "Alathon."

ale. See brewing.

alendronic acid. See 4-amino-1-hydroxybutane -1,1-diyldiphosphonic acid.

aleuritic acid. (9,10,16-trihydroxyhexadeca-noic acid).
CAS: 533-87-9. HOCH$_2$(CH$_2$)CH(OHCH(OH) (CH$_2$)COOH.
Use: Perfumes.

alfin. A catalyst obtained from alkali alcoho-lates derived from a secondary alcohol which is a methyl carbinol and olefins possessing the grouping –CH=CH-CH$_2$–, which may be part of a ring, as in toluene. The interaction of the alkali alcoholate (sodium isopropoxide) with the olefin halide (allyl chloride) gives a slurry of sodium chloride on which sodium isopropoxide and allyl sodium are adsorbed. This slurry is a typical alfin catalyst used to convert olefins into polyolefins. The elastomers produced are called alfin rubbers.

alfuzosin hydrochloride. See xatral.

algae. Chlorophyll-bearing organisms occurring in both salt- and freshwater; they have no flowers or seeds, but reproduce by unicellular spores. They range in size from single cells to giant kelp over 100 ft long, and include most kinds of seaweed. There are four kinds of algae: brown, red, green, and blue-green. Blue-green algae are said to be the earliest form of life to appear on earth. The photosynthetic activity of algae accounts for the fact that over two-thirds of the world total photosynthesis takes place in the oceans. Algae are harvested and used as food supplements (see carrageenan, agar), soil conditioners, animal feeds, and a source of iodine; they also contain numerous minerals, vitamins, proteins, lipids, and essential amino acids. Alginic acid is another important derivative. Blue-green algae are water contaminants and are toxic to fish and other aquatic life. Phosphorus compounds in detergent wastes stimulate the growth of algae to such an extent that overpopulation at the water surface prevents light from reaching many of the plants; these decompose, removing oxygen and releasing carbon dioxide, thus making the water unsuitable for fish. Algae are being used in treatment of sewage and plant effluent in a proprietary flocculation process.
See eutrophication; agar; biomass.

algae, brown.
Properties: Various seaweeds harvested in coastal waters of the northern Atlantic and the Pacific oceans.
Use: Food additive.

algae meal, dried.
Properties: Mixture of algae cells from *Spongiococcum*, molasses, corn steep liquor, and a maximum of 0.3% ethoxyquin.
Use: Food additive.

algae, red.
Properties: Various seaweeds harvested in coastal waters of Pacific ocean.
Use: Food additive.

alganet.
Use: Food additive.

Algar–Flynn–Oyamada reaction. Alkaline hydrogen peroxide oxidation of *o*-hydroxyphenyl styryl ketones (chalcones) to flavonols via the intermediate dihydroflavonols.

Algene. Quaternary ammonium compounds.
Use: As cationic surfactants.

algestone acetonide. See progesterone 16,17-acetonide.

algicide. Chemical agent added to water to destroy algae. Copper sulfate is commonly employed for large water systems.

algin. A hydrophilic polysaccharide (phycocolloid or hydrocolloid) found exclusively in the brown algae. It is analogous to agar. The seaweed (giant kelp) is sea harvested, water extracted, and refined. U.S. (California) and Great Britain are the chief producers.
See alginic acid; alginate.

alginate. Any of several derivatives of alginic acid (e.g., calcium, sodium, or potassium salts or propylene glycol alginate). They are hydrophilic colloids (hydrocolloids) obtained from seaweed. Sodium alginate is water soluble but reacts with calcium salts to form insoluble calcium alginate.
Use: Food additive (thickener, stabilizer), yarns and fibers, medicine (first-aid dressings), meat substitute, high-protein food analogs.

alginic acid. $(C_6H_8O_6)_n$. A polysaccharide composed of β,*d*-mannuronic acid residues linked so that the carboxyl group of each unit is free, while the aldehyde group is shielded by a glycosidic linkage. It is a linear polymer of the mannuronic acid in the pyranose ring form.
Properties: White to yellow powder possessing marked hydrophilic colloidal properties for suspending, thickening, emulsifying, and stabilizing. Insoluble in organic solvents, slowly soluble in alkaline solutions. Absorbs up to 300 times its weight of water.
Grade: Refined (food), technical (commercial), NF (sodium alginate), FCC.
Use: Food industry as thickener and emulsifier; protective colloid in ice cream, toothpaste, cosmetics, pharmaceuticals, textile sizing, paper coatings; waterproofing agent for concrete; boiler water treatment; oil-well drilling muds; storage of gasoline as a solid.

alicyclic. A group of organic compounds characterized by arrangement of the carbon atoms in closed ring structures sometimes resembling boats, chairs, or even birdcages. These compounds have properties resembling those of aliphatics and should not be confused with aromatic compounds having the hexagonal benzene ring. Alicyclics comprise three subgroups: (1) cycloparaffins (saturated); (2) cycloolefins (unsaturated, with two or more double bonds); and (3) cycloacetylenes (cyclynes) with a triple bond. The best-known cycloparaffins (sometimes called naphthenes) are cyclopropane, cyclohexane, and cyclopentane; typical of the cycloolefins are cyclopentadiene and cyclooctatetraene. Most alicyclics are derived from petroleum or coal tar. Many can be synthesized by various methods.

alicyclic acid. Any acid that contains a saturated ring.

aliphatic. One of the major groups of organic compounds, characterized by straight- or branched-chain arrangement of the constituent carbon atoms.

Aliphatic hydrocarbons comprise three subgroups: (1) paraffins (alkanes), all of which are saturated and comparatively unreactive, the branched-chain types being much more suitable for gasoline than the straight-chain; (2) olefins (alkenes or alkadienes), which are unsaturated and quite reactive; (3) acetylenes (alkynes), which contain a triple bond and are highly reactive. In complex structures, the chains may be branched or cross-linked.
See alicyclic; aromatic; chain.

aliphatic acid.   An organic acid.
  Derivation: From an aliphatic (nonaromatic) hydrocarbon.

aliphatic alcohol.   (alkyl alcohol). An alcohol in which the hydroxyl groups are attached to a carbon atom of a branched or straight-chain aliphatic hydrocarbon.
  Hazard: Toxic.
  Use: Industrial solvent.

aliphatic amine.   (methanamine). CH_5N. Any amine in which the substituted hydrocarbons are aliphatic.
  Hazard: Odorific; toxic.

aliphatic amino acid.   An amino acid that has a side chain with no aromatic components.

aliphatic esterase.   (alkyl ester hydrolase). An esterase that catalyzes the hydrolysis of alkyl esters.

aliphatic hydrocarbon.   A major class of hydrocarbons marked by arrangement of the constituent atoms into straight or branched-chain molecules. Aliphatic hydrocarbons do not contain aromatic rings. There are three subclasses: alkanes, alkenes, and alkynes.

aliquot.   A part that is a definite fraction of a whole, as in aliquot samples for testing or analysis.

alizarin.   (1,2-dihydroxyanthraquinone; CI 58000). $C_6H_4(CO)_2C_6H_2(OH)_2$. Parent form of many dyes and pigments including mordants.

Properties: Orange-red crystals; brownish-yellow powder. Mp 289C, bp 430C (sublimable). Soluble in aromatic solvents, hot methanol, and ether; sparingly soluble in water; moderately soluble in ethanol. Combustible.
  Derivation: Anthracene is oxidized to anthraquinone, the sulfonic acid of which is

then fused with caustic soda and potassium chlorate; the melt is run into hot water and the alizarin precipitated with hydrochloric acid. Occurs naturally in madder root.
  Grade: Technical.
  Use: Manufacture of dyes, production of lakes, indicators, biological stain.

alizarin blue.   (CI 67410). $C_{17}H_9NO_4$.
  Properties: Violet crystals. Mp 268C. Insoluble in water, soluble in glacial acetic acid and hot benzene.
  Use: Indicator.

alizarin yellow R.   (p-nitrophenylazosalicylate sodium; CI 14030). $O_2NC_6H_4NNC_6H_3OHCOONa$.
  Properties: Yellow-brown powder. Soluble in water.
  Use: Acid-base indicator, biological stain.

alkadiene.   See diolefin.

alkadiene.   Any unsaturated aliphatic hydrocarbon with two double bonds.

alkadiyne.   An alkyne that has two triple bonds.

alkali.   Any substance which in water solution is bitter and is irritating or caustic to the skin and mucous membranes, turns litmus blue, and has a pH value greater than 7.0. The alkali industry produces sodium hydroxide, sodium carbonate (soda ash), sodium chloride, salt cake, sodium bicarbonate, and corresponding potassium compounds.
See base; pH; alkali metal.

alkali blue.   Class name for a group of pigment dry powders prepared by the phenylation of p-rosaniline or fuchsine, followed by drowning in hydrochloride acid, washing, and sulfonating. Alkali blue on a weight basis has the highest tinting strength of all blue pigments. The presscake may be "flushed" with vehicle to replace the water in the pulp.
  Use: Printing inks, interior paints.

alkali cellulose.   The product formed by steeping wood pulp with sodium hydroxide, the first step in the manufacture of viscose rayon and other cellulose derivatives.
See carboxymethylcellulose.

alkalide.   A chemical compound that includes anions of alkali metals.
  Properties: Low ionization potentials; thermally labile; unusual stoichiometry.

alkali earth metal.   A term that refers to all six elements of Group II of the periodic table, that exhibit +2 oxidation state and are very reactive.

alkali metal.   A metal in group IA of the periodic table, i.e., lithium, sodium, potassium, rubidium, cesium, and francium. Except for francium, the alkali metals are all soft, silvery metals, which may be readily fused and volatilized; the melting and boiling points becoming lower with increasing atomic weight. The density increases with (but less rapidly than) the atomic weight, the atomic volume therefore becoming greater as the series is ascended. The alkali metals are the most strongly electropositive of the metals. They react vigorously, at times violently, with water; within the group itself, the basicity increases with atomic weight, that of cesium being the greatest.

alkalimetry.   The measurement of the concentration of bases or of the amount of free base present in a solution by titration or some other means of analysis.

alkaline battery.   (alkaline cell). A dry cell battery in which the electrolyte contains potassium hydroxide.

alkaline cleaner.   A cleaner whose action depends upon the presence of a caustic alkali, usually ammonium hydroxide or sodium hydroxide.

alkaline earth.   An oxide of an alkaline earth metal (lime).

alkaline-earth metals.   Calcium, barium, strontium, and radium (group IIA of the periodic table). In general they are white and differ by shades of color or casts; are malleable, extrudable, and machinable; may be made into rods, wire, or plate; are less reactive than sodium and potassium and have higher melting and boiling points.

alkaline phosphatase.   Any of a class of enzymes that catalyze the hydrolysis of phosphate monoesters under conditions of alkaline pH. Elevated activity in blood serum usually indicates bone cancer, obstructive jaundice or Paget's disease.

alkaline salt.   Any salt of a strong base.
  Hazard: toxic; paralyzer.

alkaline soil.   Soil that has pH value above 7.3.

alkaliphile.   (alkalophile). An alkaliphilic organism.

alkaliphobe.   An alkaliphobic organism.

alkali soil.   Any soil with a pH of 8.5 or more that contains salts injurious to plant life.

alkaloid.   A basic nitrogenous organic compound of vegetable origin. Usually derived from the nitrogen ring compounds pyridine, quinoline, isoquinoline, and pyrrole, designated by the ending -ine. Though some are liquids, they are usually colorless, crystalline solids, having a bitter taste, and which combine with acids without elimination of water. They are soluble in alcohol, insoluble or sparingly soluble in water. Examples are atropine, morphine, nicotine, quinine, codeine, caffeine, cocaine, and strychnine.

alkaloid nitrogen oxide.   Any alkaloid with a nitrogen oxide group that occurs in some plants.

alkaloidal reagent.   An alkaloid that precipitates protein and may affect salt bridges and hydrogen bonds.

alkalophile.   (alkaliphile). An organism that grows best under alkaline conditions (up to a pH of 10.5).

alkalosis.   A metabolic condition in which blood pH decreases, usually the result of a metabolic condition or vomiting.

alkamine.   ((2E,4E)-hexa-2,4-dienamide; amino alcohol). C_6H_9NO. An alcohol that contains an amine group.

alkane.   See paraffin (1).

alkane series.   A homologous series composed of methane, ethane, propane, butane, etc. The lower members of this series are gases while the high members are waxy solids.

alkanesulfonic acid, mixed. RSO_3H (R is methyl, ethyl, propyl, mixed). Trade designation for a mixture of methane-, ethane-, and propane sulfonic acids. A strong nonoxidizing, nonsulfonating liquid acid which is thermally stable at moderately high temperatures.
  Properties: Light amber liquid; sour odor. Mp below $-40C$, bp $120–140C$ (1 mm), d 1.38 (20C), pH (1% solution) 1.15. Very corrosive. Miscible with water and saturated fatty acids.
  Use: Catalyst; intermediate, reaction medium.

"Alkanol" [Du Pont].   TM for a series of fatty alcohol–ethylene oxide condensation products used as nonionic surface-active agents in detergents and dispersing and emulsifying agents in paper, leather, and textiles. These include grades OA, OE, OJ, OP, and HC. 189-S is a saturated hydrocarbon sodium sulfonate. B and BG are sodium alkylnaphthalene sulfonates. Sulfur is tetrahydronaphthalene sodium sulfonate.

alkanolamine.   (alkylolamine). A compound such as ethanolamine, $HOCH_2CH_2NH_2$, or triethanolamine, $(HOCH_2CH_2)_3N$, in which nitrogen is attached directly to the carbon of an alkyl alcohol. See specific compound.

alkatriene.   An acyclic hydrocarbon that contains three double bonds.

alkatriyne.   An alkyne that has three triple bonds.

alkene.   See olefin.

alkenyl halide.   (alkenyl organohalide). Any organic compound that contains at least one halogen atom and one carbon–carbon double bond. They are widespread environmental toxicants that exhibit a wide range of acute and chronic toxicities.

alkoxide.   Any of a class of compounds that are derivatives of alcohols.

alkoxy-alkylated mercury compound.   An organomercurial compound that contains an alkoxy radical.
Use: pesticide.

alkoxyaluminum hydrides.   ($H_n AlOR_{3n}$). A group of reducing agents especially useful in converting epoxides to alcohols. Derived by reaction of aluminum hydride with the corresponding alcohol in tetrahydrofuran.

alkoxylate.   An acyclic polyether.

alkyd resin.   A thermosetting coating polymer, chemically similar to polyester resins, conventionally made by condensation and polymerization of a dihydric or polyhydric alcohol (ethylene glycol or glycerol) and a polybasic acid (phthalic anhydride), usually with a drying oil modifier. The process requires heating at 230–250C for up to 12 hours. A new and quite different method utilizes epoxy addition polymerization, in which a mixture of glycidyl esters and organic acid anhydrides is heated with a metallic catalyst at 100C or less for only 2–4 hours. Cost and energy savings and improved application performance are realized by this process.
Use: Alkyd resins are used as vehicles in exterior house paints, marine paints, and baking enamels. Molded alkyd resins are used for electrical components, distributor caps, encapsulation, and a variety of similar applications.

alkyl.   A paraffinic hydrocarbon group which may be derived from an alkane by dropping one hydrogen from the formula. Examples are methyl CH_3-, ethyl C_2H_5-, propyl $CH_3CH_2CH_2-$, isopropyl $(CH_3)_2CH-$. Such groups are often represented in formulas by the letter R and have the generic formula $C_nH_2n_+1$.
See aryl.

alkyl(C12-C15) alcohol ethoxylated.
CAS: 58391-12-1.
Hazard: Low toxicity by ingestion and skin contact. A moderate skin and mild eye irritant.

alkylamine.   An alkane that contains an alkyl group attached to the nitrogen of an amine.

alkylaryl polyethyleneglycol ether.
Use: In surface-active agents.
See isooctylphenoxypolyoxyethylene ethanol for a typical example of this class of compound.

alkylaryl sulfonate.   An organic sulfonate of combined aliphatic and aromatic structure, e.g., alkylbenzene sulfonate.

alkylate.   (1) A product of alkylation. (2) A term used in the petroleum industry to designate a branched-chain paraffin derived from an isoparaffin and an olefin, e.g., isobutane reacts with ethylene (with catalyst) to form 2,2-dimethylbutane (neohexane). The product is used as a high-octane blending component of aviation and civilian gasolines. (3) In the detergent industry, the term is applied to the reaction product of benzene or its homologs with a long-chain olefin to form an intermediate, e.g., dodecylbenzene, used in the manufacture of detergents. It also designates a product made from a long-chain normal paraffin that is chlorinated to permit combination with benzene to yield a biodegradable alkylate. The adjectives *hard* and *soft* applied to detergents refer to their ease of decomposition by microorganisms.
See biodegradability; detergent.

alkylated mercury compound.   (alkylmercury compound; alkylmercury; alkyl mercury). Any mercury-containing alkane.

alkylated organotin compound.   (alkyltin compound; alkyltin). Any tin-containing alkane. A number of these compounds are immunotoxic and induce thymus atrophy without apparent effects on other organ systems.

alkylating   **agent.**   (1-methyl-1-nitroso-3-[(2S,3R,4R,5S,6R)-2,4,5-trihydroxy-6-(hydroxymethyl)oxan-3-yl]urea). $C_8H_{15}N_3O_7$. Any substance that is able to form covalent bonds and attach alkyl groups to nitrogen and oxygen atoms of the DNA bases and to other macromolecules.
Hazard: Highly reactive; carcinogen; mutagen; cytotoxic; immunotoxic; fetotoxic; teratogenic; abortifacient.
Use: An antineoplastic agent and to induce diabetes in experimental animals.

alkylation.   (1) The introduction of an alkyl radical into an organic molecule. This was one of the early chemical processes used in Germany to furnish intermediates for improved dyes, e.g., dimethylaniline. Other alkylation products are cumene, dodecylbenzene, ethylbenzene, and nonylphenol. (2) A process whereby a high-octane blending component for gasolines is derived from catalytic combination of an isoparaffin and an olefin.
See alkylate (2); neohexane.

alkylbenzene.   A compound that contains an alkyl group bonded to a benzene ring.

alkylbenzene sulfonate. (ABS). A branched-chain sulfonate type of synthetic detergent, usually a dodecylbenzene or tridecylbenzene sulfonate. Such compounds are known as "hard" detergents because of their resistance to breakdown by microorganisms. They are being replaced by linear sulfonates.
See alkyl sulfonate; linear molecule; detergent; sodium dodecylbenzene sulfonate.

alkyl carbonyl. A carbonyl in which a metal atom is bonded to an aromatic group and coordinated with several carbon monoxide molecules.

alkyl diaryl phosphate ester. See "Santicizer 141" *[Ferro]*.

alkyldimethylbenzylammonium chloride. General name for a quaternary detergent.
See benzalkonium chloride.

alkylene. A phosphated long-chain alcohol.

alkyl fluorophosphate. See diisopropyl fluorophosphate.

alkyl halide. (alkyl organohalide). Any organohalide in which one or more halogen atoms is substituted for hydrogen in an alkyl group.
Hazard: Central nervous system depressant; toxic.

alkylidene. Any compound of the alkylidene group.

alkylolamine. See alkanolamine.

alkyl polyamine. An amine in which two or more amino groups are bonded to alkane moieties.
Use: Chelating agents; industrial solvents; emulsifiers; epoxy resin hardeners; stabilizers; starting materials for dye synthesis.

alkyl sulfonate. (linear alkylate sulfonate; LAS). A straight-chain alkylbenzene sulfonate, a detergent specially tailored for biodegradability. The linear alkylates may be normal or iso (branched at the end only), but are C_{10} or longer.
See sodium dodecylbenzene sulfonate.

alkyne. See acetylene hydrocarbon.

all-*trans*-anhydroretinol.
CAS: 1224-78-8. $C_{20}H_{28}$.
Hazard: A reproductive hazard.

Allan–Robinson reaction. Preparation of flavones or isoflavones by condensing *o*-hydroxyaryl ketones with anhydrides of aromatic acids and their sodium salts.

allantoin. (glyoxyldiureide; 5-ureidohydantoin). $C_4H_6N_4O_3$. The end product of purine metabolism in mammals other than humans and other primates; it results from the oxidation of uric acid.

Properties: White to colorless powder or crystals; odorless; tasteless. Mp 230C (decomposes). 1 g is soluble in 190 cc water or 500 cc alcohol; readily soluble in alkalies. Optically active forms are known.
Derivation: Produced by oxidation of uric acid. Also present in tobacco seeds, sugar beets, wheat sprouts.
Use: Biochemical research, medicine.

allele. One of two or more types of genes that may occur at a given position on a strand of DNA.

allelochemical. A substance that inhibits the growth of animals and plants.
See allelopathic chemical.

allelopathic chemical. Any of a wide range of natural herbicides of varying toxicity produced by many species of plants, as well as by soil microorganisms (bacteria, fungi). These compounds adversely affect other plants in the vicinity, either inhibiting germination and growth or killing them outright. They are extracted from the growing plant by leaching of its leaves, root exudates, and decomposition of dead tissue. Examples of plants found to be sources of these toxic compounds are sunflowers, oats, and soybeans. Among the products that have been identified are amygdalin, caffeine, gallic acid, and arbutin. Many types of chemical structures are represented. Research is directed toward breeding and cultivation of allelopathic plants to utilize their weed-killing ability.

allelotoxin. An allelopathic substance that has toxic effects on species other than that which produces the substance.

allene. (propadiene; dimethylenemethane; propa-1,2-diene). $H_2C:C:CH_2$.
CAS: 463-49-0. C_3H_4. Any hydrocarbon or hydrocarbon derivative produced by substitution that has two double bonds from one carbon and to two others.
Properties: Colorless gas. Unstable. Fp -136.5C, bp 34.5C. Can be readily liquefied.
Derivation: (1) Action of zinc dust on 2,3-dichloropropene; (2) pyrolysis of isobutylene; (3) electrolysis of potassium itaconate.
Use: Organic intermediate.

Allen-O'Hara furnace. An horizontal double-hearth furnace that is used for calcining sulfide ores.

allergen. Any substance that acts in the manner of an antigen on coming into contact with body tissues by inhalation, ingestion, or skin adsorption. The allergen causes a specific reagin to be formed in the bloodstream. The ability to produce reagins in response to a given allergen is an inherited characteristic that differentiates an allergic from a nonallergic person. A reagin is actually an antibody. The specificity of the allergen–reagin reaction and its dependence on molecular configuration are similar to those of an antigen–antibody reaction.

The allergen molecule (often a protein such as pollen or wool) may be regarded as a key that precisely fits the corresponding structural shape of the reagin molecule. Allergies in the form of contact dermatitis can result from exposure to a wide range of plant products, some metals, and a few organic chemicals. Though they are alike in some ways, antigen–antibody reactions protect the individual, whereas allergen–reagin reactions are harmful.

See antigen–antibody.

allergenic serum.   A serum that induces hypersensitivity to antigen.

allethrin. $C_{19}H_{26}O_3$. Generic name for 2-allyl-4-hydroxtcyclopenten-1-one ester of chrysanthemummonocarboxylic acid. A synthetic insecticide structurally similar to pyrethrin and used in the same manner. For other synthetic analogs, see barthrin, cyclethrin, ethythrin, furethrin. Pyrethrin I differs in having a 2,4-pentadienyl group in place of the allyl of allethrin.
Properties: Clear, amber-colored, viscous liquid. D 1.005–1.015 (20/20C), refr index (20C) 1.5040. Insoluble in water; incompatible with alkalies; soluble in alcohol, carbon tetrachloride, kerosene, and nitromethane. Combustible.
Derivation: Synthetically (glycerol, acetylene, and ethyl acetoacetate are the major raw materials).
Grade: 90%, technical (approximately 90% pure with 10% of isomers or related compounds), 20% technical, 2.5% technical.
Use: Insecticides, synergist.

allethrin   I.   (2,2-dimethyl-3-(2-methyl-1-propenyl)cyclopropanecarboxylic acid 2-methyl-4-oxo-3-(2-propenyl)-2-cyclo-penten-1-yl   ester; allethrone ester of chrysanthemummonocarboxylic acid;   (2-methyl-4-oxo-3-prop-2-enylcyclopent-2-en-1-yl)2,2-dimethyl-3-(2-methylprop-1-enyl)cyclopropane-1-carboxylate). $C_{19}H_{26}O_3$. The commercial form of allethrin. Synthetic analog of the naturally occurring insecticides cinerin, jasmolin, and pyrethrin.
Hazard: Toxic.
Use: Insecticide.

allethrin   II.   (3-(3-methoxy-2-methyl-3-oxo-1-propenyl)-2,2-dimethylcyclopropane-carboxylic acid 2-methyl-4-oxo-3-(2-propenyl)-2-cyclopenten-1-yl ester; allenthrolone ester of chrysanthemumdicarboxylic acid monomethyl ester).
Properties: Oily, pale yellow mixture.
Hazard: Toxic.
Use: Insecticide.

allethrolone.   (2-methyl-4-oxo-3-(2-propenyl)-2-cyclopentenol;   4-hydroxy-3-methyl-2-prop-2-enylcyclopent-2-en-1-one). $C_9H_{12}O_2$. An analog of pyrethrolone in which 2-propanyl replaces the 2,4-pentadienyl group.
Use: Synthesis of allethrins.

Allianz OPT.   A polymer that makes sun screens water resistant.

allicin. $C_6H_{10}OS_2$. An antibacterial substance extracted from garlic (allium).
Properties: Yellow, oily liquid; sharp garlic odor. Unstable; decomposes rapidly when heated; slightly soluble in water; very soluble in alcohol, benzene, and ether.
Use: Medicine.

allidochlor.   (2-chloro-*N*,*n*-*di*-2-propenylacetamide; α-chloro-*n*.*n*-diallylacetaminde; *n*,*n*-diallyl-2-chloroacetamide; cdaa; 2-chloro-*N*,*N*-di(prop-2-enyl)acetamide).
CAS: 93-71-0. $C_8H_{12}ClNO$. An amide that modifies RNA and protein biosynthesis and inhibits cell division in primary meristems.
Hazard: Toxic.
Use: Pre-emergent herbicide.

alligatoring.   Formation of cracks on the surface of thick paint layers, the underlying material remaining soft.

allo-.   A prefix designating the more stable of two geometrical isomers.

allochromatic.   (1) Term used to refer to minerals whose color is a result of minute inclusions. (2) Term referring to crystals exhibiting photoconductivity.

allogeneic.   Variation in alleles among members of the same species.

allolactose.   (6-[(3,4,5-trihydroxy-6-(hydroxymethyl)oxan-2-yl)oxymethyl]oxane-2,3,4,5-tetrol).
CAS: 585-99-9. $C_{12}H_{22}O_{11}$. A disaccharide consisting of two glucose units in an alpha (1-6)glycosidic linkage. It is an effector molecule that permits activation of the lac operon by binding to the repressor molecule.
Derivation: Occurs in milk.

allomaleic acid.   See fumaric acid.

allomeric.   Having the same crystal form but different chemical composition.

allomerism.   A constancy of crystalline form or structure with a variation in chemical composition. See polyallomer.

allomorphism.   A physical change in a mineral without the gain or loss of components.

alloocimene.   (2,6-dimethyl-2,4,6-octatriene). $(CH_3)_2C(CH)_3CCH_3CHCH_3$.
Properties: Clear, almost colorless liquid. Boiling range (5–95%) 89–91C (20 mm), d 0.824 (15/15C), refr index 1.5278 (20C). Polymerizes and oxidizes readily. Combustible.

Derivation: Pyrolysis of α-pinene.
Use: Component of varnishes and a variety of polymers; fragrance.

allophanamide.   See biuret.

allophanate.   An unsaturated nitrogenous product made by reaction of an alcohol with two moles of isocyanic acid (a gas). Usually crystals, high-melting products that are easily isolated. Acid-sensitive and tertiary alcohols can be converted into allophanates.

Δ-allose.   (β-d-allopyranose). $C_6H_{12}O_6$.
Properties: Crystals. Mp 128C, mw 180.16. Soluble in water, insoluble in alcohol.
Derivation: Obtained from the leaves of *Protea rubropilosa*.

all-purpose cleaner.   Any of the numerous irritant formulations used in the home or the work place for general cleaning surface.
Properties: May contain ammonia water, artificial dyes, detergents, and/or fragrances.
Hazard: Toxic.

allosteric activator.   Any molecule which positively modulates the activity of an allosteric enzyme.

allosteric enzyme.   A regulatory enzyme, with catalytic activity modulated by the noncovalent binding of a specific molecule at a site other than the active site.

allosteric site.   The specific site on the surface of an allosteric enzyme molecule (distinct from the active site) to which a modulator molecule binds.

allothreonine.   See threonine.

allotrope.   One of the several possible forms of a substance.
See allotropy.

allotropy.   (polymorphism). The existence of a substance in two or more forms which are significantly different in physical or chemical properties. The difference between the forms involves either (1) crystalline structure; (2) the number of atoms in the molecule of a gas; or (3) the molecular structure of a liquid. Carbon is a common example of (1), occurring in several crystal forms (diamond, carbon black, graphite) as well as several amorphous forms. Diatomic oxygen and triatomic ozone are instances of (2); and liquid sulfur and helium of (3). Uranium has three crystalline forms, manganese four, and plutonium no less than six. A number of other metals also have several allotropic forms which are often designated by Greek letters, e.g., α-, γ-, and Δ-iron.

allouzarigenin.   See 17-α-uzarigenin.

alloxan.   (mesoxalylurea). $C_4H_2O_4N_2•H_2O$ and $4H_2O$.
Properties: White crystals, become pink on exposure to air; colorless aqueous solution which imparts pink color to skin. Mp 170C (decomposes) (various melting points in literature). Soluble in water and alcohol.
Derivation: Oxidation of uric acid in acid solution.
Use: Biochemical research, cosmetics, organic synthesis.

alloxan hydrate.   See 5,5-dihydroxy barbituric acid.

alloy.   A solid or liquid mixture of two or more metals, or of one or more metals with certain non-metallic elements, as in carbon steels. The properties of alloys are often greatly different from those of the components. The purpose of an alloy is to improve the specific usefulness of the primary component—not to adulterate or degrade it. Gold is too soft to use without a small percentage of copper. The corrosion and oxidation resistance of steel is markedly increased by incorporation of 15–18% of chromium and often a few percent of nickel (stainless steel). The presence of up to 1.5% carbon profoundly affects the properties of steels. Similarly, a low percentage of molybdenum improves the toughness and wear resistance of steel. The hundreds of special alloys available are the results of designs to meet specific operating conditions. Amorphous alloys for use in transformer coils are made by quick-quenching the melt.
See alloy, fusible; amalgam; superalloy.

alloy, fusible.   (low-melting alloy; fusible metal). An alloy melting in the range of approximately 51–260C, usually containing bismuth, lead, tin, cadmium, or indium. Eutectic alloys are the particular compositions that have definite and minimum melting points compared. The compositions of a few fusible alloys are given below:

System	Eutectic Composition	Temperature (C)
Cd-Bi	60 Bi-40 Cd	144
In-Bi	33.7 Bi-66.3 In	72
	67.0 Bi-33 In	109
Pb-Bi	56.5 Bi-43.5 Pb	125
Sn-Bi	58 Bi-42 Sn	139
Pb-Sn-Bi	52 Bi-16 Sn-32 Pb	96
Pb-Cd-Bi	52 Bi-8 Cd-40 Pb	92
Sn-Cd-Bi	54 Bi-20 Cd-26 Sn	102
In-Sn-Bi	58 Bi-17 Sn-25 In	79
Pb-Sn-Cd-Bi	*50 Bi-10 Cd-13.3 Sn-26.7 Pb	70
In-Pb-Sn-Bi	49.4 Bi-11.6 Sn-18 Pb-21 In	57
In-Cd-Pb-Sn-Bi	44.7 Bi-5.3 Cd-8.3 Sn-22.6 Pb-19.1 In	47

*Wood's metal.

alloy steel. A steel containing up to 10% of elements such as chromium, molybdenum, nickel, etc., usually with a low percentage of carbon. These added elements improve hardenability, wear resistance, toughness, and other properties. This term includes low-alloy steels in which the alloy content does not exceed 5%, but does not include stainless steel.
See steel.

allyl acetate. (acetic acid allyl ester; acetic acid-2-propenyl ester; 3-acetoxypropene; prop-2-enyl acetate).
CAS: 591-87-7. $C_5H_8O_2$.
Properties: Flammable; water insoluble liquid.
Hazard: Skin and eye irritant; poisonous.

allylacetone. (5-hexene-2-one).
$CH_2CHCH_2CH_2COCH_3$.
Properties: Colorless liquid. D 0.846 (20/4C), bulk d 6.99 lb/gal (20C), 5–95% distills between 127–129C. Soluble in water and organic solvents.
Use: Intermediate in pharmaceutical synthesis, perfumes, fungicides, insecticides, fine chemicals.

allyl acrylate. $CH_2CHCOOCH_2CHCH_2$. Liquid, bp 122–134C. Used as monomer for resins.

allyl alcohol. (2-propen-1-ol; propenyl alcohol).
CAS: 107-18-6. $CH_2=CH_2OH$.
Properties: Colorless liquid; pungent mustardlike odor. Bp 96.9C, mp −129C, d 0.8520 (20/4C), bulk d 7.11 lb/gal (20C), refr index 1.4131, flash p 70F (21C) (TOC), autoign temp 713F (375C). Miscible with water, alcohol, chloroform, ether.
Derivation: (1) Hydrolysis of allyl chloride (from propylene) with dilute caustic, (2) isomerization of propylene oxide over lithium phosphate catalyst at 230–270C, (3) dehydration of propylene glycol.
Hazard: Toxic by skin absorption. Eye and upper respiratory tract irritant. Questionable carcinogen.
Use: Esters for use in resins and plasticizers, intermediate for pharmaceuticals and other organic chemicals, manufacture of glycerol and acrolein, military poison, herbicide.

allylamine. (2-propenylamine). $C_3H_5NH_2$.
Properties: Colorless to light-yellow liquid; strong ammoniacal odor. Attacks rubber and cork. Bp 55–58C, d 0.759–0.761 (20/20C), refr index 1.4194 (22C). Soluble in water, alcohol, ether, and chloroform. Combustible.
Grade: CP, technical.
Use: Pharmaceutical intermediate, organic synthesis.

allyl bromide. (3-bromopropene; bromoallylene).
CAS: 106-95-6. $H_2C=CHCH_2Br$.
Properties: Colorless to light-yellow liquid; irritating, unpleasant odor. D 1.398 (20/4C), mp −199C; bp 71.3C, refr index 1.4654, flash p 30F (−1.1C), autoign temp 563F (295C). Soluble in alcohol, ether, chloroform, carbon tetrachloride, carbon disulfide; insoluble in water.

Grade: Technically pure (95% min purity via bromium titration).
Hazard: Strong irritant to skin and eyes, flammable, high fire risk. Upper respiratory tract irritant. Questionable carcinogen.
Use: Organic syntheses, preparation of resins, perfume intermediates.

allyl carbamate.
Use: Intermediate for fireproofing compound, emulsion, and solvent polymers.

allyl chloride.
CAS: 107-05-1. $H_2C=CHCH_2Cl$.
Properties: Colorless liquid with a disagreeable pungent odor. D 0.938 at 20C, mw 76.53, vap press 295 at 20C, flash p −25F (CC). Slightly soluble in water; miscible with alcohol, chloroform, ether, and petroleum ether.
Hazard: Skin and eye irritant. Upper respiratory tract irritant, liver and kidney damage. Questionable carcinogen.
Use: Synthesis of allyl compounds.

allyl diglycol carbonate. See diethylene glycol bis(allyl carbonate).

4-allyl-1,2-dimethoxybenzene. See methyl eugenol.

allylene. See methylacetylene.

2-allyl-2-ethyl-1,3-propanediol.
$C_3H_5C(CH_2OH)_2CH_2CH_3$.
Properties: Off-white solid. Fp 31C, d 0.976 (35C).
Use: Suggested as polymer additive and chemical intermediate.

allyl glycidyl ether. (AGE).
CAS: 106-92-3.
Properties: Colorless liquid of pleasant odor. D 0.9698 at 20C, mw 114, fp −100C, bp 153.9C (760 mm), vap press 4.7 mm at 25C, flash p 135F (OC). Slightly soluble in water.
Hazard: Skin, eye, and upper respiratory tract irritant, and dermatitis. Questionable carcinogen.
Use: Resin intermediate, stabilizer of chlorinated compounds, vinyl resins, and rubber.

1-allyl-4-hydroxybenzene. See chavicol.

allylic rearrangements. Migration of a C=C double bond in a three-carbon (allylic) system on treatment with nucleophiles under S_{N1} conditions (or under S_{N2} conditions when the nucleophilic attack takes place at the γ-carbon).

allyl-α-ionone. $C_{16}H_{24}O$. Yellow liquid with fruity odor. D 0.928–0.935 (25/25C). Stable, soluble in 70% alcohol. Made synthetically.
Properties: Combustible.
Use: A perfume and flavoring.

5-allyl-5-isobutylbarbituric   acid. $C_{11}H_{16}$ N_2O_3.
Properties: White, crystalline powder; odorless; slightly bitter taste. Mp 138–139C. Soluble in alcohol, ether, and chloroform; almost insoluble in water.
Use: Medicine (sedative).
See barbiturate.

allyl isocyanate. C_3H_5NCO.
Properties: Colorless liquid; turns yellow on standing. D 0.935–0.945 (20C), flash p 110F (43.3C), bp 85.5–86C, fp less than −80C. Combustible.
Grade: Purity, 98% min.
Hazard: Moderate fire risk. Toxic.
Use: Organic intermediate, cross-linking agent, polymer modifier.

allylisopropylacetaminde.   An amide that initiates autoxidation, destroying microsomal cytochromes, especially cytochrome P-450.

allyl isothiocyanate.   (allyl isosulfocyanate; mustard oil; 2-propenyl isothiocyanate).
CAS: 57-06-7. $H_2C=CHCH_2NCS$.
Properties: Colorless to pale yellow, oily liquid; pungent, irritating odor; sharp, biting taste. D 1.013–1.016 (25C), bp 152C, flash p 115F (46.1C), refr index 1.527; optically inactive. Soluble in alcohol, ether, carbon disulfide; slightly soluble in water. Combustible.
Derivation: Distillation of sodium thiocyanate and allyl chloride, or of seed of black mustard.
Grade: Technical, FCC.
Hazard: Toxic via ingestion, inhalation, skin contact; fire risk. Questionable carcinogen.
Use: Fumigant, ointments, and mustard plasters.

allyl isovalerate.   (isovaleric acid, allyl ester; allyl isovalerianate; allyl 3-methylbutyrate; 3-methyl-butanoic acid, 2-propanyl ester; 3-methylbutyric acid, allyl ester; 2-propenyl isovalerate; 2-propanyl 3-methylbutanoate).
CAS: 2835-39-4.
Properties: Organic compound.
Hazard: Toxic, irritant, questionable carcinogen.

allyl mercaptan.   See allyl thiol.

allyl methacrylate. $CH_2C(CH_3)COOC_3H_5$.
Use: As monomer and intermediate.

4-allyl-2-methoxyphenol.   See eugenol.

4-allyl-1,2-methylenedioxybenzene.   See safrole.

2-(3-allyloxyphenyl)imidazo(2,1-a)isoquinoline.
CAS: 61001-16-9. $C_{20}H_{16}N_2O$.
Hazard: A reproductive hazard.

allyl pelargonate. $C_3H_5OOC(CH_2)_7CH_3$.
Properties: Liquid, fruity odor, bp 87–91C (3 mm), refr index 1.4332 (20.5C). Combustible.
Use: Flavors and perfumes, polymers.

2-allylpentanoic acid.
CAS: 1575-72-0. $C_8H_{14}O_2$.
Hazard: Moderately toxic.

p-allylphenol.   See chavicol.

allyl phosphite.   See phosphorous acid, triallyl ester.

allyl propyl disulfide.
CAS: 2179-59-1. $CH_2=CHCH_2S_2C_3H_7$.
Properties: Liquid with a pungent, irritating odor. Mw 148.16, d 0.9289 at 15C. Soluble in ether, carbon disulfide, and chloroform.
Derivation: The chief volatile constituent of onion oil.
Hazard: Eye and upper respiratory tract irritant.

allyl resin.   A special class of polyester resins derived from esters of allyl alcohol and dibasic acids. Common monomers are allyl diglycol carbonate, also known as diethylene glycol bis(allyl carbonate), diallyl chlorendate, diallyl phthalate, diallyl isophthalate, and diallyl maleate. Polymerization occurs through the unsaturated allyl double bond to form thermosetting resins that are highly resistant to chemicals, moisture, abrasion, and heat, and have low shrinkage and good electrical resistivity.
Use: As laminating adhesives and coatings, especially by impregnation of layered materials with prepolymers (called prepregs); allylic glass cloth, applications requiring microwave transparency, encapsulation of electronic parts, vacuum impregnation of metal, casting of ceramics, molding compositions, heat-resistant furniture finishes.

allyl sulfide.   (diallyl sulfide; thioallyl ether).
$(CH_2CHCH_2)_2S$.
Properties: Colorless liquid with garlic odor. Bp 139C, d 0.888 (27/4C), refr index 1.4877 (27C). Insoluble in water; miscible with alcohol, ether, chloroform, and carbon tetrachloride. Combustible.
Use: Component of artificial oil of garlic.

allyl thiol.   (allyl mercaptan; 2-propene-1-thiol).
CH_2CHCH_2SH.
Properties: Water-white liquid (darkens on standing); strong garlic odor. D 0.925 (23/4C), bp 90C. Insoluble in water; soluble in ether and alcohol. Combustible.
Use: Pharmaceutical intermediate, rubber accelerator intermediate.

allylthiourea.   (allylsulfocarbamide; thiosinamine; allyl sulfourea). $C_3N_5NHCSNH_2$.

Properties: White, crystalline solid; slight garlic odor; bitter taste. D 1.22; mp 78C. Soluble in water, ether, and solutions of borax, benzoates, urethane; insoluble in benzene; slightly soluble in 70% alcohol.

Derivation: Warming a mixture of equal parts of allyl isothiocyanate and absolute alcohol with an equal amount of 30% ammonia.

Use: Medicine, corrosion inhibitor, organic synthesis.

allyltrichlorosilane. $CH_2CHCH_2SiCl_3$.

Properties: Colorless liquid; pungent, irritating odor. Bp 117.5C; d 1.217 (27C); refr index 1.487 (20C); flash p 95F (35C) (COC). Readily hydrolyzed by moisture with the liberation of hydrochloric acid, polymerizes easily. Fire hazard.

Derivation: Reaction of allyl chloride with silicon (copper catalyst).

Use: Intermediate for silicones, fiberous glass finishes.

allyltriphenyl stannane.   See allyltriphenyltin.

allyltriphenyltin.

CAS: 76-63-1. $C_{21}H_{20}Sn$.

Properties: Needle-like crystals from alc. Mp: 73–74.5C. Sol in most org solvs.

Hazard: A poison.

4-allyl veratrole.   See methyl eugenol.

almond emulsion.   (synaptase; amygdalase; β-glucosidase). An enzyme catalyzing the production of glucose from β-glucosides.

Properties: White powder; odorless; tasteless. Capable of hydrolyzing glucosides such as amygdalin to glucose and the other component substances. Soluble in water; insoluble in ether and alcohol.

Source: Sweet almonds.

Derivation: By extracting an emulsion of almonds with ether, filtering the clear solution, and precipitating the emulsion with alcohol.

Use: Food grade flavoring.

almond oil.   The volatile essential oil distilled from ground kernels of bitter almonds.

Use: Cosmetic creams, perfumes, liqueurs, food flavors (hydrocyanic acid-free).

See amygdalin.

Note: "Bitter almonds contain amygdalin together with an enzyme that catalyzes its hydrolysis. When the kernels are ground and moistened, a volatile oil produced by the hydrolysis can be distilled from them consisting mainly of benzaldehyde and hydrocyanic acid. This is the oil of bitter almond used in pharmacy as a food flavor after removal of the hydrocyanic acid)" (Eckey, *Vegetable Fats and Oils.*).

alnico.   Coined word for an alloy containing chiefly iron, aluminum, nickel, and cobalt which has outstanding properties as a permanent magnet.

aloe-emodin.   (1,8-dihydroxy-3-(hydroxymethyl)-9,10-anthracenedione; 1,8-dihydroxy-3-(hydro-xymethyl)anthraquinone; 3-hydroxy-methylchrysazin; bitter aloe; rhabarberone; 1,8-dihydroxy-3-(hydroxymethyl)anthracene-9,10-dione). $C_{15}H_{10}O_5$. A cathartic anthraquinoid that occurs free and as a glycoside.

Derivation: In the leaves of rhubarb, senna and certain species of *Aloe* (*lilaceae*).

aloin.   (barbaloin). A mixture of active principles obtained from aloe. Varies in properties according to variety used.

Properties: Yellow crystals; bitter taste; no or slight odor of aloe. Darkens on exposure to air. Approx 60% soluble in pyridine; slightly soluble in water and organic solvents.

Grade: Technical.

Use: Medicine, proprietary laxatives, electroplating baths, fermentation.

alpaca.   A natural fiber obtained from a South American animal similar to the llama. Properties resemble those of wool. Used for specialty clothing and also blended with polyester. Combustible.

Alpha.   Olefin (polypropylene) fiber.

Use: Automotive, home furnishings, and apparel.

alpha.   (α). (1) A prefix denoting the position of a substituting atom or group in an organic compound. The Greek letters α, β, γ, etc., are usually not identical with the IUPAC numbering system 1, 2, 3, etc., since they do not start from the same carbon atom. However, α and β are used with naphthalene ring compounds to show the 1 and 2 positions, respectively. α, β, etc., are also used to designate attachment to the side chain of a ring compound. (2) Both a symbol and a term used for relative volatility in distillation. (3) Symbol for optical rotation. (4) A form of radiation consisting of helium nuclei. See alpha particle. (5) The major allotropic form of a substance, especially of metals, e.g., α-iron.

alpha-cellulose.   See cellulose; α-cellulose.

Alpha helix.   A helical conformation of a polypeptide chain, predominantly right-handed, with maximal intrachain hydrogen bonding of the peptide bonds; one of the most common secondary structures in proteins.

alpha-hydroxy acids.   See: α-hydroxy acids.

alpha particle.   (α particle). A helium nucleus emitted spontaneously from radioactive elements both natural and manufactured. Its energy is in the range of 4–8 MeV and is dissipated in a very short path, i.e., in a few centimeters of air or less than 0.005 mm of aluminum. It has the same mass (4) and positive charge (2) as the helium nucleus.

Accelerated in a cyclotron alpha-particles can be used to bombard the nuclei of other elements.
See helium; decay; radioactive.

Alpha San.   A nylon carpet yarn, polyamide 6 for hospital, child care, and office locations.
Use: Antimicrobial.

Alrok process.   The immersion of aluminum-base metals in a hot solution of alkali carbonate and chromate in order to form a corrosion-resistant oxide film on the surface.

Alsilox.   A fusion product of 1% alumina, 65% litharge, and 34% silica, used in ceramics. Available in various particle sizes.

Alternaria.   A genus of fungi, at least some species of which are allergens and frequently infest laboratories. It is a common fungus commonly known as "black rot" on tomato fruit. Alternaria species is known as opportunistic pathogens in immunocompromised individuals.
Hazard: Pathogenic.

alternaric acid.   An acid that is a withering agent that wilts the leaves of solanaceous plants at concentrations as low as 5 μg/ml.
Derivation: Occurs in the fungus, *Alteraria solani.*

alternating polymer.   Polymer in which the molecules of different monomers alternate in the chain.

alternative splicing.   Different ways of combining a gene's exons to make variants of the complete protein.

altheine.   See asparagine.

Altman, Sidney.   (1939–). Awarded Nobel Prize in chemistry in 1989 jointly with Cech for the discovery that RNA acts as a biological catalyst as well as a carrier of genetic information. Doctorate in 1967 from the University of Colorado.

altretamine.   (2-*N*,2-*N*,4-*N*,4-*N*,6-*N*,6-*N*-hexamethyl-1,3,5-triazine-2,4,6-triamine).
CAS: 645-05-6. $C_9H_{18}N_6$. A hexamethyl-2,4,6-triamine derivative of 1,3,5-triazine.
Use: Antineoplastic alkylating agent and insect chemosterilant.
Hazards: Toxic.

altrose.   An aldohexose that is isomeric with glucose, tallose, allose, and others.

alugan.   See bromomethylhexachlorobicycloheptene.

alum.   See aluminum ammonium sulfate; aluminum potassium sulfate, aluminum sulfate.

alum, burnt.   (alum, dried). $AlK(SO_4)_2$. Aluminum ammonium sulfate or aluminum potassium sulfate heated just sufficiently to drive off the water of crystallization.
Properties: White, odorless powder; sweetish taste. Absorbs moisture on exposure to air. Soluble in hot water; slowly soluble in cold water; insoluble in alcohol.
Use: Medicine (astringent).

alum, chrome.   See chromium potassium sulfate.

alum, chrome ammonium.   See chromium ammonium sulfate.

alum-hematoxylin.   A purple nuclear histological stain that is a mixture of an aqueous solution of ammonium alum and an alcoholic solution of aged hematoxylin.

alumina.   See aluminum oxide and following entries.

alumina, activated.   A highly porous, granular form of aluminum oxide having preferential adsorptive capacity for moisture and odor contained in gases and some liquids. When saturated, it can be regenerated by heat (176–315C). The cycle of adsorption and reactivation can be repeated many times. Granules range in size from powder (7 microns for chromatographic work) to pieces approximately 1.5 inches in diameter. Average density approximately 50 lb/cu ft.
Use: An effective desiccant for gases and vapors in the petroleum industry. It is also used as a catalyst or catalyst carrier in chromatography, and in water purification.
See aluminum oxide.

alumina, fused.   See aluminum oxide.

alumina gel.   See aluminum hydroxide gel.

alumina-silica fiber.
Properties: Amorphous structure, excellent resistance to all chemicals except hydrochloric and phosphoric acids and concentrated alkalies. Available in both short and long staple. Low heat conductivity, high thermal shock resistance. Tensile strength 400,000 psi, elastic modulus 16 million psi, upper temperature limit in oxidizing atmosphere 800C. Noncombustible.
Derivation: The short fiber type is made by blasting a stream of molten alumina and silica with a steam jet. Long staple is spun from a molten mixture of alumina and silica modified with zirconium.
Available forms: Fibers, sheets, blankets.
Use: Nonwoven fabrics (short staple), woven fabric structures, cordage, thermal insulation, repair of furnace linings, piping molten metals, welding insulation (reusable), insulation for rocket and space applications.
Note: In finely divided form, alumina-silica is also used as a catalyst.

alumina trihydrate. (aluminum hydroxide; aluminum hydrate; hydrated alumina; hydrated aluminum oxide).
CAS: 21645-51-2. $Al_2O_3 \cdot 3H_2O$ or $Al(OH)_3$.
Properties: White crystalline powder, balls, or granules. D 2.42. Insoluble in water; soluble in mineral acids and caustic soda. Releases water on heating.
Derivation: From bauxite. The ore is dissolved in strong caustic and aluminum hydroxide precipitated from the sodium aluminate solution by neutralization (as with carbon dioxide) or by autoprecipitation (Bayer process).
Grade: Technical, CP.
Use: Glass, ceramics, iron-free aluminum and aluminum salts, manufacture of activated alumina, base for organic lakes, flame retardants, mattress batting. Finely divided form (0.1–0.6 microns) used for rubber reinforcing agent, paper coating, filler, cosmetics.

aluminium. British spelling of aluminum.

aluminon. (aurine tricarboxylic acid, ammonium salt).
Properties: Mw 473.44, mp 220.5C (decomposes).
Use: As a reagent for aluminum in solution.

aluminosilicate. A compound of aluminum silicate with metal oxides or other radicals.
Use: Catalyst in refining petroleum, to soften water, and in detergents. The sodium compound, $Na_{12}[(AlO_2)_{12}(SiO_2)_{12}]_xH_2O$, is typical.
See zeolite; molecular sieve.

aluminum. (Al; aluminium).
CAS: 7429-90-5. Metallic element of atomic number 13; group IIIA of the periodic table; aw 26.98154; valence 3; no stable isotopes. Monovalent in high-temperature compounds (AlCl and AlF). Most abundant metal in earth's crust; third most abundant of all elements. Does not occur free in nature.
Properties: Silvery white, crystalline solid. Tensile strength (annealed) 6800 psi, cold-rolled 16,000 psi. D 2.708, mp 660C, bp 2450C. Forms protective coating of aluminum oxide approximately 50 Å thick, which makes it highly resistant to ordinary corrosion. Attacked by concentrated and dilute solutions of hydrochloric acid, hot concentrated sulfuric acid, and perchloric acid. Also violently attacked by strong alkalies. Rapidly oxidized by water at 180C. Not attacked by dilute or cold concentrated sulfuric acid or concentrated nitric acid. Can ignite violently in powder form. Electrical conductivity approximately two-thirds that of copper. Aluminum qualifies as both a light metal and a heavy metal, according to their definitions.
Derivation: From bauxite by Bayer process and subsequent electrolytic reduction by Hall process. There are several processes for obtaining ultrapure aluminum: (1) electrolytic (three-layer), (2) zone refining, and (3) chemical refining. Impurities as low as 0.2 ppm are possible.
More efficient processes are the Alcoa and Toth processes which require much less electric power than the Hall process. Another method, using no electricity, involves heating a mixture of aluminum ores with a coal-derived fuel in a closed furnace. Still another process called calsintering, using fly ash as a source of alumina, has been described (1978). See calsintering.
Available forms: Structural shapes of all types, plates, rods, wire foil flakes, powder (technical and USP). Aluminum can be electrolytically coated and dyed by the anodizing process (see anodic coating); it can be foamed by incorporating zirconium hydride in molten aluminum, and it is often alloyed with other metals or mechanically combined (fused or bonded) with boron and sapphire fibers or whiskers. Strengths up to 55,000 psi at 500C have been obtained in such composites. A vapor-deposition technique is used to form a tightly adherent coating from 0.2 to 1 mil thick on titanium and steel.
Hazard: Fine powder forms flammable and explosive mixtures in air. Confirmed carcinogen.
Use: Building and construction, corrosion-resistant chemical equipment (desalination plants), die-cast auto parts, electrical industry (power transmission lines), photoengraving plates, permanent magnets, cryogenic technology, machinery and accessory equipment, miscible food-processing equipment, tubes for ointments, toothpaste, shaving cream, etc. Also as a powder in paints and protective coatings, as rocket fuel, as an ingredient of incendiary mixtures (thermite) and pyrotechnic devices, as a catalyst, and for foamed concrete vacuum metallizing and coating. Other uses are as foil in packaging, cooking, and decorative stamping, and as flakes for insulation of liquid fuels.
See aluminum alloy.

aluminum acetate. A salt obtained by reaction of aluminum hydroxide and acetic acid with subsequent recrystallization. Its neutral form $Al(C_2H_3O_2)_3$ is a white, water-soluble powder used in solution as an antiseptic, astringent, and antiperspirant. Its basic form is $Al(C_2H_3O_2)_2OH$, also known as aluminum diacetate and aluminum subacetate. It is a crystalline solid, insoluble in water, used as a mordant in textile dyeing, as a flame retardant and waterproofing agent, and in manufacture of lakes and pigments.
See mordant rouge.

aluminum acetylacetonate. $Al(C_5H_7O_2)_3$.
Properties: Solid. Mp 189C, bp 315C. Soluble in benzene and alcohol.
Use: Deposition of aluminum, catalyst.

aluminum alkyl. (Al trialkyl). Catalyst used in the Ziegler process.
Hazard: Pyrophoric liquid.
See triethylaluminum; triisobutylaluminum.

aluminum alloy. Aluminum containing variable amounts of manganese, silicon, copper, magnesium, lead, bismuth, nickel, chromium, zinc,

or tin. A wide range of uses and properties is possible. Alloys may be obtained for casting or working, heat-treatable or nonheat-treatable, with a wide range of strength, corrosion resistance, machinability, and weldability.
See duralumin.

aluminum ammonium chloride. (ammonium aluminum chloride). $AlCl_3 \cdot NH_4Cl$.
Properties: White crystals. Mp 304C. Soluble in water.
Use: Fur treatment.

aluminum ammonium sulfate. (ammonium alum; alum NF). $Al_2(SO_4)_3(NH_4)_2SO_4 \cdot 24H_2O$ or $AlNH_4(SO_4)_2 \cdot 12H_2O$.
Properties: Colorless crystals; odorless; strong astringent taste. D 1.645, mp 94.5C, bp loses 20 waters at 120C. Soluble in water and glycerol; insoluble in alcohol.
Derivation: By crystallization from a mixture of ammonium and aluminum sulfates.
Method of purification: Recrystallization.
Grade: Technical, lump, ground, powdered, CP, NF, FCC.
Use: Mordant in dyeing, water and sewage purification, sizing paper, retanning leather, clarifying agent, food additive, manufacture of lakes and pigments, and fur treatment.

aluminum, anodized. See anodic coating.

aluminum antimonide. AlSb.
Properties: Crystalline solid. Mp 1050C.
Derivation: Fusion of the elements followed by zone refining to purify.
Use: Semiconductor technology.

aluminum arsenide. AlAs. A semiconductor used in rectifiers, transistors, thermistors.
Hazard: Poisonous by ingestion.

aluminum borate. $Al_2O_3 \cdot B_2O_3$.
Properties: White, granular powder; decomposed by water.
Derivation: Interaction of aluminum hydroxide and boric oxide.
Grade: Technical, CP.
Use: Glass and ceramic industries, polymerization catalyst.

aluminum boride.
Properties: Powder. Apparent bulk density, (fully settled, light) 0.6–0.8 g/cc; dense, 1.2–1.4 g/cc; high neutron absorption.
Use: Nuclear shielding.

aluminum borohydride. $Al(BH_4)_3$.
Properties: Volatile, pyrophoric liquid. Bp 44.5C; fp −64.5C.
Derivation: (1) By reaction of sodium borohydride and aluminum chloride in the presence of a small

amount of tributyl phosphate; (2) by reaction of trimethylaluminum and diborane.
Hazard: Ignites spontaneously in air, reacts violently with water.
Use: Intermediate in organic synthesis, jet fuel additive, reducing agent.

aluminum brass. An alloy containing 76% copper, 21.5–22.25% zinc, and 1.75–2.50% aluminum.
Use: Condenser, evaporator, and heat exchanger tubes and ferrules.

aluminum bromide. (1) $AlBr_3$; (2) $AlBr_3 \cdot 6H_2O$.
Properties: White to yellowish, deliquescent crystals. Exists as double molecules Al_2Br_6 in the vapor. (1) D 3.01, mp 97.5C, bp 265C; (2) d 2.54, mp 93C (decomposes). Soluble in alcohol, carbon disulfide, or ether.
Derivation: (1) By passing bromine over heated aluminum; (2) reaction of hydrogen bromide with aluminum hydroxide.
Hazard: The anhydrous form reacts violently with water; corrosive to skin.
Use: Anhydrous: bromination, alkylation, and isomerization catalyst in organic synthesis.

aluminum bronze. An alloy containing 88–96.1% copper, 2.3–10.5% aluminum, and small amounts of iron and tin. Characterized by high strength, ductility, hardness, and resistance to shock, fatigue, most chemicals, and seawater.
Grade: Powder that is also called gold bronze powder. An alloy of 90% copper and 10% aluminum reduced from leaf form to powder, polished mechanically, and coated with stearic acid. Available in the following grades: litho, molding, printing-ink, and radiator.
Use: A pigment in paints and inks.

aluminum-n-butoxide. $Al(OC_3H_9)_3$.
Properties: Yellow to white crystalline solid. Mp 101.5C (pure) and 88–96C (commercial), d 1.0251 (20C), bp 290–310C (30 mm). Soluble in aromatic, aliphatic, and chlorinated hydrocarbons.
Use: Ester exchange catalyst, defoamer ingredient, hydrophobic agent, intermediate.

aluminum calcium hydride. (calcium aluminum hydride). Al_2CaH_8.
Properties: Grayish solid. Soluble in tetrahydrofuran, insoluble in benzene.
Derivation: Reaction of $AlCl_3$ and CaH_2 in tetrahydrofuran.
Hazard: Flammable in contact with water and alcohols. Spontaneous ignition in moist air: store and handle in nitrogen.
Use: Versatile reducing agent.

aluminum calcium silicate.
Hazard: A nuisance dust.
Use: Food additive.

aluminum carbide. Al_4C_3.
Properties: Yellow crystals or powder. D 2.36, stable to 1400C. Decomposes in water with liberation of methane.
Derivation: By heating aluminum oxide and coke in an electric furnace.
Grade: Technical.
Hazard: Dangerous fire risk in contact with moisture.
Use: Generating methane, catalyst, metallurgy, drying agent, reducing agent.

aluminum carbonate.   A basic carbonate of variable composition; formula sometimes given as $Al_2O_3{\cdot}CO_2$. White lumps or powder, insoluble in water; dissolves in hot hydrochloric acid or sulfuric acid. Formerly used as mild astringent, styptic. Normal aluminum carbonate $Al_2(CO_3)_3$ is not known as an individual compound.

aluminum chlorate. $Al(ClO_3)_3$.
Properties: Colorless crystals. Deliquescent; soluble in water and alcohol.
Hazard: Powerful oxidizing material: keep out of contact with combustibles.
Use: Disinfectant, color control of acrylic resins.

aluminum chloride, anhydrous. $AlCl_3$
Properties: White or yellowish crystals. D 2.44 (25C), mp 190C (2.5 atm), sublimes readily at 178C. The vapor consists of double molecules Al_2Cl_6. Soluble in water.
Derivation: (1) By reaction of purified gaseous chlorine with molten aluminum; (2) by reaction of bauxite with coke and chlorine at approximately 875C. (This product is used to make the hydrate).
Impurities: Ferric chloride-free aluminum, insolubles.
Grade: Technical, reagent.
Hazard: Powerful irritant to tissue; moderately toxic by ingestion. Reacts violently with water, evolving hydrogen chloride gas.
Use: Ethylbenzene catalyst, dyestuff intermediate, detergent alkylate, ethyl chloride, pharmaceuticals and organics (Friedel–Crafts catalyst), butyl rubber, petroleum refining, hydrocarbon resins, nucleating agent for titanium dioxide pigments.

aluminum chloride hydrate. $AlCl_3{\cdot}6H_2O$.
Properties: White or yellowish, deliquescent, crystalline powder; nearly odorless; sweet astringent taste. D 2.4, mp (decomposes). Soluble in water and alcohol. The water solution is acid.
Derivation: By crystallizing the anhydrous form from hydrochloric acid solution.
Grade: Technical, CP, NF. Aluminum chloride solution 32°Bé is a special grade consisting of a solution containing only 0.005% iron as impurity, and having an acid reaction but containing no free acid.
Use: Pharmaceuticals and cosmetics (antiperspirants), pigments, roofing granules, special papers, photography, textiles (wool).

aluminum chlorohydrate. $[Al_2(OH)_5Cl]_x$.
An ingredient of commercial antiperspirant and deodorant preparations. Also used for water purification and treatment of sewage and plant effluent.

aluminum cleaner.   A cleaner that contains aluminum salts.
Hazards: Moderately toxic.

aluminum dextran.   See aluminum monostearate.

aluminum diacetate.   See aluminum acetate.

aluminum diethyl monochloride.   See diethylaluminum chloride.

aluminum diformate.   (aluminum formate, basic). $Al(OH)(CHO_2)_2{\cdot}H_2O$.
Properties: White or gray powder. Soluble in water.
Derivation: Aluminum hydroxide is dissolved in formic acid and spray-dried. Solutions are also prepared by treating aluminum sulfate with formic acid, followed by lime.
Grade: Technical, solutions (12–20° Bé).
Use: Waterproofing, mordanting, antiperspirants, tanning leather, improving wet strength of paper.

aluminum distearate.
CAS: 300-92-5. $Al(OH)(C_{18}H_{35}O_2)_2$.
Properties: White powder. Mp 145C, d 1.009. Insoluble in water, alcohol, ether. Forms gel with aliphatic and aromatic hydrocarbons.
Use: Thickener in paints, inks, and greases; water repellent; lubricant in plastics and cordage; and in cement production.

aluminum ethylate.   (aluminum ethoxide). $Al(OC_2H_5)_3$.
Properties: Colorless liquid which gradually solidifies. Bp 200C (6 mm), mp 140C. Partly soluble in high-boiling organic solvents.
Derivation: Reaction of aluminum with ethanol, catalyzed by iodine and mercuric chloride.
Hazard: Strong irritant to eyes and skin.
Use: Reducing agent for aldehydes and ketones, polymerization catalyst.

aluminum ethylhexoate.   (aluminum octoate). A metallic salt of 2-ethylhexoic acid.
Use: A gelling agent for liquid hydrocarbons, and a paint additive.

aluminum fluoride, anhydrous. AlF_3.
Properties: White crystals. Sublimes at approximately 1250C, d 2.882. Slightly soluble in water; insoluble in most organic solvents.
Derivation: (1) Action of hydrogen fluoride gas on alumina trihydrate; (2) reaction of hydrogen fluoride on a suspension of aluminum trihydrate, followed by calcining the hydrate formed; (3) reaction of fluosilicic acid on aluminum hydrate.

Grade: Technical.
Hazard: Strong irritant to tissue.
Use: Production of aluminum to lower the melting point and increase the conductivity of the electrolyte, flux in ceramic glazes and enamels, manufacture of aluminum silicate, catalyst.

aluminum fluoride, hydrate. $AlF_3 \cdot 3.5H_2O$.
Properties: White, crystalline powder. Slightly soluble in water.
Derivation: Action of hydrofluoric acid on alumina trihydrate and subsequent recovery by crystallization.
Grade: Technical, CP.
Hazard: See fluorine.
Use: Ceramics (production of white enamel).

aluminum fluorosilicate.   See aluminum hexafluorosilicate.

aluminum fluosilicate.   (aluminum silicofluoride). $Al_2(SiF_6)_3$.
Properties: White powder. Slightly soluble in cold water, readily soluble in hot water.
Grade: Technical.
Hazard: See fluorine.
Use: Artificial gems, enamels, glass.

aluminum formate.   See aluminum triformate; aluminum diformate.

aluminum formate, basic.   See aluminum diformate.

aluminum formate, normal.   See aluminum triformate.

aluminum formoacetate. $Al(OH)(OOCH)$ $(OOCCH_3)$.
Properties: White powder. Soluble in water and alcohol.
Use: Textile water repellents.

aluminum hexafluorosilicate.
CAS: 17099-70-6. $F_{18}Si_3 \cdot 2Al$.
Hazard: Low toxicity by ingestion.
Source: Natural product.

aluminum hydrate.   See alumina trihydrate.

aluminum hydride. AlH_3.
Properties: White to gray powder. Decomposes at 160C (100C if catalyzed). Evolves hydrogen on contact with water.
Hazard: Dangerous fire and explosion risk.
Use: Electroless coatings on plastics, textiles, fibers, other metals; polymerization catalyst; reducing agent.

aluminum hydroxide.   See alumina trihydrate.

aluminum hydroxide gel.   (hydrous aluminum oxide; alumina gel). $Al_2O_3 \cdot xH_2O$.

Properties: White, gelatinous precipitate. Constants variable with the composition, d approx 2.4. Insoluble in water and alcohol, soluble in acid and alkali. Noncombustible.
Derivation: By treating a solution of aluminum sulfate or chloride with caustic soda, sodium carbonate, or ammonia, by precipitation from sodium aluminate solution, by seeding or by acidifying (carbon dioxide is commonly used).
Grade: Technical, CP, USP (containing 4% Al_2O_3), NF (dried containing 50% Al_2O_3).
Use: Dyeing mordant, water purification, waterproofing fabrics, manufacture of lakes, filtering medium, chemicals (aluminum salts), lubricating compositions, manufacture of glass, sizing paper, ceramic glaze, antacid.

aluminum hydroxystearate. $Al(OH)[OOC$ $(CH_2)_{10}CH_2O(CH_2)_5CH_3]_2$.
Properties: White powder. Mp 155C, d 1.045. Less soluble in nonpolar compounds than other aluminum stearates, and more soluble in polar compounds.
Use: Waterproofing of leather and cements; lubricant for plastics and ropes; paints and inks.

aluminum hypophosphite. $AlH_6O_6P_3$.
Properties: Crystalline solid. Decomposes at 218C. Insoluble in water; soluble in hydrochloric acid, weak sulfuric acid, and sodium hydroxide solution.
Derivation: By heating an aluminum salt solution with sodium hypophosphite.
Hazard: Decomposes to toxic phosphine.
Use: Finishing agent for polyacrylonitrile fiber.

aluminum iodide. AlI_3 (anhydrous).
Properties: Brown-black, crystalline pieces (white when pure). Mp 191C, bp 385C, d 3.9825. Soluble in alcohol, ether, carbon disulfide.
Derivation: Heating aluminum and iodine in a sealed tube.
Method of purification: Crystallization.
Grade: Technical.
Hazard: Reacts violently with water.
Use: Catalyst in organic synthesis.

aluminum isopropoxide.   See aluminum isopropylate.

aluminum isopropylate.   (aluminum isopropoxide). $Al(OC_3H_7)_3$.
Properties: White solid. D 1.035 (20C), mp 128–132C, bp 138–148C (10 mm). Soluble in alcohol and benzene; decomposes in water.
Derivation: From isopropanol and aluminum.
Grade: Distilled (purity approximately 100%).
Use: Dehydrating agent, catalyst, waterproofing textiles, organic synthesis, paints.

aluminum lactate. $C_9H_{15}AlO_9$.
Properties: Colorless, water-soluble powder.
Use: Fire foam.

aluminum metaphosphate. $Al(PO_3)_3$.
  Properties: White powder. Mp approximately 1527C. Insoluble in water.
  Use: As a constituent of glazes, enamels, and glasses, and as a high-temperature insulating cement.

aluminum monobasic stearate.   See aluminum monostearate.

aluminum monopalmitate.   See aluminum palmitate.

aluminum monostearate.   (aluminum dextran; aluminum monobasic stearate). $Al(OH)_2[OOC(CH_2)_{16}CH_3]$. A complex containing aluminum and dextran, a chain of molecular weight 2500, corresponding to a chain of 15 anhydroglucose units.
  Properties: Fine, white to yellowish-white powder; faint characteristic odor. Mw 344.48, mp 155C, d 1.020. Insoluble in water, alcohol, and ether. Forms a gel with aliphatic and aromatic hydrocarbons.
  Derivation: Mixing solutions of a soluble aluminum salt and sodium stearate.
  Grade: USP, which describes it as a mixture of the monostearate and monopalmitate containing 14.5–16.0% Al_2O_3.
  Use: Paints, inks, greases, waxes, thickening lubricating oils; waterproofing, gloss producer, stabilizer for plastics, food additive.

aluminum naphthenate.
  Properties: Yellow substance of rubbery consistency with high thickening power. Combustible.
  Derivation: Reaction of an aluminum salt with an alkali naphthenate in aqueous solution.
  Use: Paint and varnish drier and bodying agent, detergent in lube oils, the solution in organic solvents has been proposed for insecticides and siccatives.
  See soap (2).

aluminum nicotinate.
  CAS: 1976-28-9. $C_{18}H_{12}AlN_3O_6$.
  Properties: Solid.
  Use: Food additive.

aluminum nitrate. $Al(NO_3)_3 \cdot 9H_2O$.
  Properties: White crystals. Mp 73C, decomposes at 150C. Soluble in cold water, alcohol and acetone; decomposes in hot water.
  Derivation: Formed by the action of nitric acid on aluminum and crystallization.
  Grade: Technical, CP (99.75%).
  Hazard: Powerful oxidizing agent. Do not store near combustible materials.
  Use: Textiles (mordant), leather tanning, manufacture of incandescent filaments, catalyst in petroleum refining, nucleonics, anticorrosion agent, antiperspirant.

aluminum nitride.   AlN.

Properties: Crystalline solid. D 3.10, mp 2150C (4.3 atm), Mohs hardness 9+, decomposes when wet into aluminum hydroxide and ammonia.
  Derivation: From coal and bauxite when heated in a stream of nitrogen.
  Use: As semiconductor in electronics; nitriding of steel.

aluminum oleate. $Al(C_{18}H_{33}O_2)_3$.
  Properties: Yellowish-white, viscous mass. Insoluble in water; soluble in alcohol, benzene, ether, oil, and turpentine. Combustible.
  Derivation: By heating aluminum hydroxide, water, and oleic acid. The resultant mixture is filtered and dried.
  Use: Waterproofing, drier for paints, etc.; thickener for lubricating oils; medicine; as lacquer for metals; lubricant for plastics; food additive.

aluminum orthophosphate.   See aluminum phosphate.

aluminum oxalate. $Al_2C_6O_{12}$.
  Properties: Finely divided solid (as hydrate), soluble in nitric and sulfuric acids, almost insoluble in water and alcohol.
  Use: Mordant and dyeing assistant.

aluminum oxide.   (alumina).
  CAS: 1344-28-1. Al_2O_3. The mineral corundum is natural aluminum oxide, and emery, ruby, and sapphire are impure crystalline varieties. The mixed mineral bauxite is a hydrated aluminum oxide.
  Properties: Vary according to the method of preparation. White powder, balls, or lumps of various mesh. D 3.4–4.0, mp 2030C, insoluble in water, difficultly soluble in mineral acids and strong alkali. Noncombustible.
  Derivation: (1) Leaching of bauxite with caustic soda followed by precipitation of a hydrated aluminum oxide by hydrolysis and seeding of the solution. The alumina hydrate is then washed, filtered, and calcined to remove water and obtain the anhydrous oxide. (2) Aluminum sulfate from coal mine wastewaters is reduced to alumina.
  Grade: Technical, CP, fibers, high purity, fused, calcined.
  Hazard: Toxic by inhalation of dust. Confirmed carcinogen.
  Use: Production of aluminum, manufacture of abrasives, refractories, ceramics, electrical insulators, catalyst and catalyst supports, paper, spark plugs, crucibles and laboratory wares, adsorbent for gases and water vapors (see alumina, activated), chromatographic analysis, fluxes, lightbulbs, artificial gems, heat-resistant fibers, food additive (dispersing agent).
  See alumina trihydrate; aluminum hydroxide gel; alumina, activated.

aluminum oxide, hydrated.   See alumina trihydrate.

aluminum oxide, hydrous.   See aluminum hydroxide gel.

aluminum palmitate.   (aluminum monopalmitate). $Al(OH)_2(C_{16}H_{31}O_2)$.
Properties: White powder. Mp 200C, d 1.072. Insoluble in alcohol and water; forms gel with hydrocarbons. Combustible.
Derivation: By heating aluminum hydroxide and palmitic acid and water. The resultant mixture is filtered and dried.
Use: Waterproofing leather, paper, textiles; thickening for lubricating oils; thickening or suspending agent in paints and inks; production of high gloss on leather and paper; ingredient of varnishes; lubricant for plastics; food additive.

aluminum paste.   Aluminum powder ground in oil.
Use: Aluminum paints.

aluminum phosphate.   (aluminum orthophosphate). $AlPO_4$.
Properties: White crystals. D 2.566, mp 1500C. Insoluble in water and alcohol, slightly soluble in hydrochloric acid and nitric acid.
Derivation: Interaction of solutions of aluminum sulfate and sodium phosphate.
Hazard: Solutions are corrosive to tissue.
Use: Ceramics, dental cements, cosmetics, paints and varnishes, pharmaceuticals, pulp and paper.

aluminum phosphide.   AlP.
Properties: Dark gray or dark yellow crystals. D 2.85.
Hazard: Dangerous fire risk. It evolves phosphine.
Use: Insecticide, fumigant, semiconductor technology.

aluminum picrate. $Al[(NO_2)_3C_6H_2O]_3$.
Hazard: Toxic by ingestion and inhalation, dangerous fire risk in contact with combustibles, severe explosion risk when shocked or heated; strong oxidizer.
Use: Explosive compositions.

aluminum potassium sulfate.   (potash alum; alum NF; potassium alum).
CAS: 7784-24-9. $Al_2(SO_4)_3 \cdot K_2SO_4 \cdot 24H_2O$, sometimes written $AlK(SO_4)_2 \cdot 12H_2O$.
Properties: White; odorless crystals; astringent taste. D 1.75, mp 92C, bp loses 18 H_2O at 64.5C, anhydrous at 200C. Soluble in water; insoluble in alcohol. Solutions in water are acid. Noncombustible.
Derivation: (1) From alunite, leucite, or similar mineral. (2) Also derived by crystallization from a solution made by dissolving aluminum sulfate and potassium sulfate and mixing.
Grade: Technical, lump, ground, powdered, NF, FCC.

Use: Dyeing (mordant), paper, matches, paints, tanning agents, waterproofing agents, purification of water, aluminum salts, food additive, baking powder, astringent, cement hardener.

aluminum resinate.
Properties: Brown solid. Insoluble in water; soluble in oils.
Derivation: Heating together soluble aluminum salts and rosin.
Grade: Technical (fused, precipitated).
Use: Drier for varnishes.
Hazard: Flammable, dangerous fire risk.

aluminum rubidium sulfate.   (rubidium alum). $AlRb(SO_4)_2 \cdot 12H_2O$.
Properties: Colorless crystals. Soluble in hot water, insoluble in alcohol. D 1.867, mp 99C.

aluminum salicylate. $Al(C_6H_4OHCOO)_3$.
Properties: Reddish-white powder. Odorless. Soluble in dilute alkalies; insoluble in water, and alcohol. Decomposed by acid.
Use: Medicine.

aluminum silicate.   Any of the numerous types of clay which contain varying proportions of Al_2O_3 and SiO_2. Made synthetically by heating aluminum fluoride at 1000–2000C with silica and water vapor; the crystals or whiskers obtained are up to 1 cm long, have high strength, and are used in reinforced plastics.
Use: Same as for clay.
See mullite; kaolin; satintone.

aluminum silicofluoride.   See aluminum fluosilicate.

aluminum silicon.   A light-weight alloy available as ingots or in powder form and used for automotive parts, construction, etc. For new manufacturing method, see aluminum derivation.
Hazard: Powder is flammable, dangerous fire risk.

aluminum soaps.   See aluminum oleate; aluminum palmitate; aluminum resinate; aluminum stearate; soap (2).

aluminum sodium chloride.   See sodium tetrachloroaluminate.

aluminum sodium sulfate.   (SAS; sodium aluminum sulfate; soda alum; alum). $Al_2(SO_4)_3 \cdot Na_2SO_4 \cdot 24H_2O$ or $AlNa(SO_4)_2 \cdot 12H_2O$.
Properties: Colorless crystals; saline, astringent taste; effloresces in air. D 1.675, mp 61C. Soluble in water, insoluble in alcohol. Noncombustible.
Derivation: By heating a solution of aluminum sulfate and adding sodium chloride. The solution is allowed to cool with constant stirring. The alum meal deposited is washed with water and centrifuged.

Method of purification: Recrystallization.
Grade: Pure crystals, technical, CP, FCC.
Use: Textiles (mordant, waterproofing), dry colors, ceramics, tanning, paper size precipitant, matches, inks, engraving, sugar refining, water purification, medicine, confectionery, baking powders, food additive.

aluminum stearate.   (aluminum tristearate).
CAS: 637-12-7. $C_{54}H_{105}O_6 \cdot Al$.
Properties: White powder. D 1.070, mp 115C. Insoluble in water, alcohol, ether; soluble in alkali, petroleum, turpentine oil. Forms gel with aliphatic and aromatic hydrocarbons.
Derivation: Reaction of aluminum salts with stearic acid.
Grade: Technical.
Use: Paint and varnish drier, greases, waterproofing agent, cement additive, lubricants, cutting compounds, flatting agent, cosmetics and pharmaceuticals, defoaming agent in beet sugar and yeast processing.

aluminum subacetate.   See aluminum acetate.

aluminum sulfate.   (alum; pearl alum; pickle alum; cake alum; filter alum; papermakers' alum; patent alum).
$Al_2(SO_4)_3$; (2) $Al_2(SO_4)_3 \cdot 18H_2O$.
Properties: White crystals; sweet taste. Noncombustible. Stable in air, d (1) 2.71, (2) 1.62, mp decomposes at 770C. Soluble in water; insoluble in alcohol.
Derivation: (1) By treating pure kaolin or aluminum hydroxide or bauxite with sulfuric acid. The insoluble silicic acid is removed by filtration, and the sulfate is obtained by crystallization. (2) Similarly, from waste coal-mining shale and sulfuric acid.
Grade: Iron-free, technical, CP, USP, FCC. A liquid form (49.7 H_2O) is also available.
Use: Sizing paper, lakes, alums, dyeing mordant foaming agent in fire foams, cloth fireproofing, white leather tannage, catalyst in manufacturing ethane, pH control in paper industry, waterproofing agent for concrete, clarifier for fats and oils, lubricating compositions, deodorizer and decolorizer in petroleum refining, sewage precipitating agent and for water purification, food additive.

aluminum sulfide. Al_2S_3.
Properties: Yellowish-gray lumps. Odor of hydrogen sulfide. Decomposes in moist air to hydrogen sulfide and a gray powder. D 2.02, mp 1100C.
Hazard: Irritant to skin and mucous membranes.
Use: To prepare hydrogen sulfide.

aluminum tartrate. $C_{12}H_{12}Al_2O_{18}$.
Properties: Crystalline powder. Soluble in hot water and ammonia.
Use: Textile dyeing auxiliary.

aluminum thiocyanate.   (aluminum sulfocyanate). $Al(SCN)_3$.
Properties: Yellowish powder. Soluble in water; insoluble in alcohol and ether.
Use: Mordant for textile dyeing, manufacturing pottery.

aluminum trialkyl.   See triethylaluminum; triisobutylaluminum.

aluminum tributyl.   See *n*-tributyl aluminum.

aluminum triethyl.   See triethyl aluminum.

aluminum triformate.   (aluminum formate, normal). $Al(HCOO)_3 \cdot 3H_2O$.
Properties: White, crystalline powder. Soluble in hot water; slightly soluble in cold water.
Use: Textile (delustering rayon, mordanting, waterproofing, after-treatment of dyeings), paper (sizing), fur dyeing (mordant), and medicine.

aluminum trimethyl.   See trimethyl aluminum.

aluminum   triricinoleate. $Al(C_{17}H_{32}OHCOO)_3$.
Properties: Yellowish to brown plastic mass. Mp 95C. Limited solubility in most organic solvents. Combustible.
Derivation: Castor oil.
Use: Gelling agent, waterproofing, solvent-resistant lubricants.

aluminum tristearate.   See aluminum stearate.

alumite.   (alumstone). A hydrous sulfate of potassium and aluminum that occurs commonly as fine-grained masses.

alum, NF.   May be either aluminum ammonium sulfate or aluminum potassium sulfate.

alum, papermakers.   See aluminum sulfate.

alum, pearl.   Specially prepared aluminum sulfate for the papermaking industry.

alum, pickle.   Aluminum sulfate prepared to meet specifications of packers and preservers.

alum, porous.   See aluminum sodium sulfate.

alum, potash.   See aluminum potassium sulfate.

alum, soda.   See aluminum sodium sulfate.

alunite.   (alum stone).
CAS: 1344-28-1. $KAl_3(OH)_6(SO_4)_3$. A naturally occurring basic potassium aluminum sulfate usually found with volcanic and other igneous rocks.
Occurrence: Utah, Arizona, California, Colorado, Nevada, Washington, Italy, Australia.

Use: Production of aluminum potassium compounds, millstones, substitute for bauxite in aluminum manufacture, decolorizing and deodorizing agent, fertilizer.

aluorochrome. A natural or synthetic dye.
Properties: Fluoresce.

Am. Symbol for americium.

AMA-10.
CAS: 6317-18-6. Methylene bisthiocyanate.
Grade: 99% dry powder, 10% solution.
Available forms: Powder, liquid.
Use: Pesticide, raw materials for use in paper mills, cooling towers, petroleum drilling, water treatment, leather processing.

Amadori rearrangement. Conversion of N-glycosides of aldoses to N-glycosides of the corresponding ketoses by acid or base catalysis.

amagat unit. A compressibility factor, equal to pressure in atmospheres times volume under standard conditions.

amalgam. A mixture or alloy of mercury with any of a number of metals or alloys, including cesium, sodium, tin, zinc, lithium, potassium, gold, and silver as well as with some nonmetals. Dental amalgams are mixtures of mercury with a silver tin alloy. A sodium amalgam is formed in the preparation of pure sodium hydroxide by electrolysis of brine.
Use: Dental fillings, silvering mirrors, catalysis, analytical separation of metals; to facilitate application of active metals such as sodium, aluminum, and zinc in the preparation of titanium, etc., or in reduction of organic compounds.
See sodium amalgam.

amanitin. (α) $C_{39}H_{54}N_{10}O_{13}S$. ($\beta$) $C_{39}H_{53}N_9$ $O_{14}S$. Toxic principle from a species of mushroom (*Amanita phalloides*). The β form has been obtained as acicular crystals which are soluble in water, methanol, and ethanol.
Hazard: A poison. Ingestion may be fatal.

amantadine hydrochloride. (1-adamantana-mine hydrochloride). $C_{10}H_{17}\cdot HCl$. A derivative of adamantane. An anti-viral drug. Also used in treatment of Parkinson's disease.
See "Symmetrel" *[Endo]*.

amaranth. (FDC Red No. 2; Red Dye No. 2).
CAS: 915-67-3. $NaSO_3C_{10}H_5N=NC_{10}H_4(SO_3$ $Na)_2OH$. An azo dye derived from naphthionic and R acids.
Properties: Dark red to purple powder. D approximately 1.50. Soluble in water, glycerol, propylene glycol; insoluble in most organic solvents.

Hazard: A questionable carcinogen. May not be used in foods, drugs, or cosmetics.
Use: Formerly a certified food and drug colorant. (Replaced for some applications by FDC Red No. 40.) Textile dye, color photography.

amatol. An explosive mixture of ammonium nitrate and TNT. The 50:50 mixture can be melted and poured for filling small shells; the 80% ammonium nitrate mixture is granular.
Hazard: Highly explosive. A powerful irritant to mucous membranes and by skin absorption.

amber. A polymerized fossil resin derived from an extinct variety of pine. Readily accumulates static electrical charge by friction; good electrical insulator.

ambergris. A waxy, opaque mass containing 80% cholesterol formed in the intestinal tract of the sperm whale and found on beaches or afloat in the ocean.
Use: A fixative in perfumes, now largely replaced by synthetic products.

"Amberlac" *[Reichhold].* TM for modified alkyd-type resins for quick-drying lacquers.
Use: Metal primers, bottle cap coatings, food can coatings, appliance coatings.

"Amberlite" *[Dow].* TM for several types of ion-exchange resins. Insoluble cross-linked polymers of various types in minute bead form. Strong acid, weak acid, strong base, and weak base forms, each having various grades differing in exchange capacity and porosity, for removing simple and complex cations and anions from aqueous and nonaqueous solutions. Reversible in action; can be regenerated.
Grade: Laboratory, liquid, nuclear, mixed bed, pharmaceutical.
Use: Water conditioning (softening and complete deionization), recovery and concentration of metals, antibiotics, vitamins, organic bases, catalysis, decolorization of sugar, manufacture of chemicals, neutralization of acid mine-water drainage, analytical chemistry, water treatment in nuclear reactors, pharmaceuticals.

Amberol. Maleic-resin and resin-modified and unmodified phenolformaldehyde-type polymers in solid form. They react with various oils to produce fast-drying, high-gloss protective coatings and vehicles for printing inks.
Use: Varnishes, enamels, can liners, nitrocellulose sanding sealers, printing inks, tackifying and vulcanization of butyl rubber.

"Ambien" *[Sanofi].* TM for zolpidem tartrate.
Use: Drug.

ambient air.   The air that immediately surrounds an individual, a population, a community, or any real entity of interest and which potentially affects that entity.

ambient temperature.   The temperature of the environment in which an experiment is conducted or in which any physical or chemical event occurs. See room temperature.

ambithion.   See fenitrothion-malathion mixture.

"Ambitrol" *[Dow].*   TM for coolants used in stationary industrial engines.

amblygonite. $Li(AlF)PO_4$ or $AlPO_4 \cdot LiF$. A natural fluorophosphate of aluminum and lithium.
Properties: White to grayish-white. Contains up to 10.1% lithia, sometimes with partial replacement by sodium. D 3.01–3.09, Mohs hardness 6.
Occurrence: California, Maine, Connecticut, South Dakota, Germany, Norway, France.
Use: A source of lithium used in glazes and coatings.

ambomycin (USAN).   An antibiotic produced by *Streptomyces ambofaciens.*

ambrette seed oil.
Properties: Volatile oil from seeds of *Abelmoschus moschatus* Moench, syn. *Hibiscus Abelmoschus L.* (Fam. *Malvaceae*). Amber liquid; odor of ambrettolide. Sol in fixed oils; insol in glycerin, propylene glycol.
Use: Food additive.

ambrettolide.   (ω,6-hexadecenlactone; 6-hexadecenolide; 16-hydroxy-6-hexadecenoic acid; ω-lactone). $C_{16}H_{28}O_2$.
Properties: Colorless liquid having powerful musk-like odor. Found in ambrette seed oil.
Use: Flavoring, perfume fixative.

amebicide.   (amebacide; emetine; (2S,3R,11bS)-2{{[(1R)-6,7-dimethoxy-1,2,3,4-tetrahydroisoquinolin-1-yl]methyl}-3-ethyl-9,10-dimethoxy-2,3,4,6,7,11b-hexahydro-1H-pyrido[2,1-a]isoquinoline dihydrochloride). $C_{29}H_{41}ClN_2O_4$. The principle alkaloid of ipecac. It inhibits protein synthesis in eukaryotic cells but not prokaryotic cells.
Derivatives: From the ground roots of Uragoga (or Cephaelis) ipecacuanha or Uragoga acuminaata of the Rubiaceae.
Hazard: May cause cardiac, hepatic or renal damage, violent diarrhea and vomiting.
Use: Amebicide.

"Amerchol" *[Union].*   TM for a series of surface-active lanolin derivatives; most are soft solids.
Use: Emulsifiers and stabilizers for water and oil systems, emollients in pharmaceuticals and cosmetics.

American Association for the Advancement of Science.   (AAAS). Founded in 1848. The largest general scientific organization representing all fields of science. Membership includes 120,000 individuals and 262 societies. The interests of this association extend into all areas of natural and social science. Its leading publication is *Science*, established by Thomas Edison in 1880. It also publishes many symposium volumes based on papers presented at its annual meetings. One of its important activities in chemistry is sponsorship of the Gordon Research Conferences, which originated in 1931 under the leadership of Dr. Neil E. Gordon; these have since been expanded to more than 30 technical conferences attended by chemists from many foreign countries. Society headquarters are located at 1200 New York Ave NW, Washington, D.C. 20005. Website: http://www.aaas.org

American Association of Textile Chemists and Colorists.   (AATCC). Founded in 1921. It has over thousands of individual and corporate members in more than 60 countries world wide. A technical and scientific society of textile chemists and colorists in textile and related industries using colorants and chemical finishes. It is the authority for test methods. It is located at 1 Davis Drive, Research Triangle Park, NC 27709. Website: http://aatcc.org

American Carbon Society.   (ACS). The American Carbon Society (ACS), formally the American Carbon Committee, was established in 1957 with the express purpose of organizing U.S. conferences on carbon. The Society promotes interdisciplinary research and technology in the field of carbon science. Moreover, the Society sponsors a Biennial Conference on Carbon. The ACS promotes carbon science and technology through its Graffin Lecture series. It publishes *CARBON*, an international journal. Its mailing address is University of Kentucky, Center for Applied Energy Research, 2540 Research Park Drive, Lexington, KY 40511. Website: http://www.americancarbonsociety.org

American   Ceramic   Society.   (ACerS). Founded 1898. It has 9500 members. A professional society of scientists, engineers, researchers, manufacturers, plant personnel, educators, students, marketing and sales professionals interested in glass, ceramics-metal systems, cements, refractories, nuclear ceramics, white wares, electronics, and structural clay products. It is located at 600 N. Cleveland Ave., Suite 210, Westerville, OH 43082. Website: http://www.ceramics.org

American Chemical Society.   (ACS). Founded in 1876. It has over 161,000 members. The nationally chartered professional society for chemists in the U.S. One of the largest scientific organizations in the world. Its offices are at 1155 16th St., NW,

Washington D.C. 20036. For further information, see Appendix IIB.
Website: http://www.acs.org
See Chemical Abstracts.

American Chemistry Council. (ACC). Founded in 1863, and formerly called the Chemical Manufacturers Association, this nonprofit trade association of chemical manufacturers represents the voice of the US Chemical Industry. The American Chemistry Council represents the chemical industry on public policy issues, coordinates the industry's research and testing programs, and administers the industry's environmental, health, and safety performance improvement initiative, known as Responsible Care. The American Chemistry Council is located at 1300 Wilson Boulevard in Arlington, VA.
It has instituted an emergency telephone information service called ChemTrec to provide instant information for safety precautions in accidents involving chemicals.
Website: http://www.cmahq.com

American Conference of Governmental Industrial Hygienists. (ACGIH). Founded in 1938. Over 3000 members. A society of persons employed by official government units responsible for programs of industrial hygiene, education, and research. It was founded for the purpose of determining standards of exposure to toxic and otherwise harmful materials in workroom air. The standards are revised annually and are available from the secretary. It is located at 1330 Kemper Meadow Drive, Cincinnati, OH 45240. Website: http://www.acgih.org
See Threshold Limit Value.

American Institute of Chemical Engineers. (AIChE). Founded in 1908. There are approximately 45,000 members. The largest society in the world devoted exclusively to the advancement and development of chemical engineering. Its official publication is *Chemical Engineering Progress*. It has 110 local sections and many committees working in a wide range of activities. Its offices are located at 120 Wall Street, FL 23, New York, NY 10005. Website: http://www.aiche.org

American Institute of Chemists. (AIC). Founded in 1923, it is primarily concerned with chemists and chemical engineers as professional people rather than with chemistry as a science. Special emphasis is placed on the scientific integrity of the individual and on a code of ethics adhered to by all its members. It publishes a monthly journal, *The Chemist*. It is located at 315 Chestnut Street, Philadelphia, PA 19106. Website: http://www.theaic.org

American Leather Chemists' Association. (ALCA). Founded in 1903. This group works to devise and perfect methods for the analysis and testing of leathers and materials used in leather manufacturing. It is located at 1314 50th Street, Suite 103, Lubbock, TX 79412. Website: http://www.leatherchemists.org

American National Standards Institute. (ANSI). Founded in 1918. A federation of trade associations, technical societies, professional groups, and consumer organizations that constitutes the U.S. clearinghouse and coordinating body for voluntary standards activity on the national level. It eliminates duplication of standards activities and combines conflicting standards into single, nationally accepted standards. It is the U.S. member of the International Organization for Standardization and the International Electrochemical Commission. Over 125,000 companies are members of the ANSI. One of its primary concerns is safety in such fields as hazardous chemicals, protective clothing, welding, fire control, electricity and construction operations, blasting, etc. Its address is 1899 L Street, NW, 11th Floor, Washington, D.C. 20036. Website: http://www.ansi.org

American Oil Chemist's Society. (AOCS). Founded in 1909. It has over 4300 members. These members are chemists, biochemists, chemical engineers, research directors, plant personnel, and persons concerned with animal, marine, and regular oils and fats and their extraction, refining, safety, packaging, quality control and use. The address is 2710 S. Boulder, Urbana, IL 61802. Website: http://www.aocs.org

American Petroleum Institute. (API). Founded in 1919. It has 500 members. The members are the producers, refiners, marketers, and transporters of petroleum and allied products such as crude oil, lubricating oil, gasoline, and natural gas. The address is 1220 L Street, NW, Washington, D.C. 20005. Website: http://www.api.org

American Society for Testing and Materials. (ASTM). This society, organized in 1898 and chartered in 1902, is a scientific and technical organization formed for "the development of standards on characteristics and performance of materials, products, systems and services, and the promotion of related knowledge." There are over 30,000 members. It is the world's largest source of voluntary consensus standards. The society operates via more than 143 main technical committees that function in prescribed fields under regulations that ensure balanced representation among produces, users, and general-interest participants. Headquarters of the society is at 100 Barr Harbor Drive, P.O. Box C700, West Conshohocken, PA 19428. Website: http://www.astm.org

American Society of Pharmacology and Experimental Therapeutics. (ASPET).

Founded in 1908. It has over 4800 members. It is a scientific society of investigators in pharmacology and toxicology interested in research and promotion of pharmacological knowledge and its use among scientists and the public. The address is 9650 Rockville Pike, Bethesda, MA 20814. Website: http://www.aspet.org

americium.   Am. A synthetic radioactive element of atomic number 95, a member of the actinide series; aw 241; 14 isotopes of widely varying half-life; valence 3, but divalent, tetravalent, and higher valencies exist. α and γ emitter, forms compounds with oxygen, halides, lithium, etc. Metallic americium is silver-white crystalline, d 13.6, mp approximately 100C. Half-life of ^{241}Am is 458 years.
Derivation: Multiple neutron capture in plutonium in nuclear reactors, plutonium isotopes yield ^{241}Am and ^{243}Am on β decay. The metal is obtained by reduction of the trifluoride with barium in vacuum at 1200C.
Hazard: A radioactive poison.
Use: Gamma radiography, radiochemical research, diagnostic aid, electronic devices.

"Amerlate" phosphorus [Amerchol].   TM for the isopropyl ester of hydroxy, normal, and branched-chain acids of lanolin. A light-yellow, soft solid that liquefies on contact with the skin. A hydrophilic emollient, moisturizer, conditioning agent, lubricant, pigment dispersant, and nonionic auxiliary without emulsifier.

Ames dial.   A device used for measuring the thickness of paint film.

Ames test.   A simple bacterial test for carcinogens, based on the assumption that carcinogens are mutagens. Named after its inventor Bruce Ames. Somewhat less favored than it used to be, because of the idea that compounds causing cells to grow more rapidly will sometimes result in cancer.

amethopterin.   See methotrexate.

ametryn.   (2-thanmino-4-isopropylamino-6-methylmercapto-s-triazine; *N*/ethyl-*n'*-(1-methylethyl)-6-(methylthio)-1,3,5-triazine-2,4-diamine; ametryne; 4-*N*-ethyl-6-methylsulfanyl-2-*N*-propan-2-yl-1,3,5-triazine-2,4-diamine).
CAS: 834-12-8. $C_9H_{17}N_5S$. A methylthiotriazine similar in biological properties to atrazine and simazine but absorbed through foliage as well as the roots of green plants.
Use: Pre-emergence and post-emergence herbicides.

ametryn-atrazine mixture.   See herbifert.

amiben.   Generic name for 3-amino-2,5-dichlorobenzoic acid. $C_6H_2NH_2Cl_2COOH$.
Use: Herbicide or plant growth regulator.

amic acid.   (carbamoyl carboxylic acid). Any of a class of organic compounds that contain a carboxy and a carboxamide group.

amicyanin.   An electron transfer protein found in certain bacteria that contains a type 1 copper site.

amidase.   (acylase; acylamidase; amidohydrolase).
CAS: 9012-56-0. Any deamidizing hydrolase that catalyses the hydrolysis of mono carboxylic amides, yielding the free acid and ammonia.

amide.   A nitrogenous compound related to or derived from ammonia. Reaction of an alkali metal with ammonia yields inorganic amides, e.g., sodium amide ($NaNH_2$). Organic amides are characterized by an acyl group ($-CONH_2$) usually attached to an organic group (R=$CONH_2$); formamide ($HCONH_2$) and carbamide (urea) $CO(NH_2)_2$ are common examples.
See polyamide.

amide herbicide.   Any of a family of herbicides that inhibit seed germination and/or growth of seedlings, probably by blocking protein synthesis in the primary meristems. When applied to foliage, localized or general necrosis is seen at the loci of contact.
Use: Herbicide.

amide oxime.   Any of a class of organic compounds and derivatives formed by substitution of carboxamides.

amidinase.   A deamidinizing enzyme that enables the hydrolysis of a nonpeptide C-N bond of an amidine.

amidine.   Any strong monobasic organic compound resulting from the reaction of ammonia with nitriles or with imido esters.

amidinomycin. $C_9H_{18}N_4O$. An antibiotic that partially inhibits spore-forming bacteria.

4-amidino-1-(nitrosaminoamidino)-1-tetrazene.   See tetrazene.

amido black.   (disodium(6Z)-4-amino-3-(4-nitrophenyl)diaenyl-5-oxo-6-(phenylhydrazinylidene)naphtha-2,7-disulfonate).
$C_{22}H_{14}N_6Na_2O_9S_2$. A blue-black protein stain.
Use: In forensics to make bloody friction ridges more detailed.

amido black 10b.   (sodium-4-amino-3-(4-nitrophenyl)diazenyl-5-oxo-6-(phenylhydraziney-lidene)naphtha-2,7-disulfonic acid). $C_{22}H_{16}N_6O_9S_2$. An acid diazo dye.
Use: Protein stain in paper chromatography and electrophoresis; connective tissue stain.

amidohydrolase. (amidase; deamidase; deamidizing enzyme). Any of a class of enzymes that catalyze the hydrolysis of C-N bonds in amides.

amidol. See 2,4-diaminophenol hydrochloride.

amidopropylamine oxide.
Use: Foamer, foam booster, foam stabilizer; scour for household, cosmetic, and industrial applications.

amidoxime. (amide oxime; N-[amino-4[-5-[4-[amino(nitroso)methylidene]cyclo-2,5-dien-1-ylidene]furan-2-ylidene]cyclohexa-2,5-dien-1-ylidene]methyl]hydroxylamine). $C_{18}H_{16}N_4O_3$. Any oxime of an amide.

amidoxyl. The radical of an amidoxime in which the terminal H of the NOH group has been removed.

aminacrine hydrochloride. USAN name for 9-amino-acridine hydrochloride.

amination. The process of making an amine (RNH_2). The methods commonly used are (1) reduction of a nitro compound, and (2) action of ammonia on a chloro, hydroxy, or sulfonic acid compound.

amine. A class of organic compounds of nitrogen that may be considered as derived from ammonia (NH_3) by replacing one or more of the hydrogen atoms with alkyl groups. The amine is primary, secondary, or tertiary depending on whether one, two, or three of the hydrogen atoms are replaced. All amines are basic in nature and usually combine readily with hydrochloric or other strong acids to form salts.
See fatty amine.

amine 220. 2-(8-heptadecenyl)-2-imidazoline-1-ethanol).

$C_{17}H_{33}C:NC_2H_4NC_2H_4OH.$
Properties: D 0.9330 (20/20C), bulk d 7.76 (20C) lb/gal, bp 235C (1 mm), flash p 465F (240C). Combustible.
Use: Demulsifier used particularly in the recovery of tar from water-gas process emulsions. A powerful cationic wetting agent. Useful in flotation processes involving siliceous minerals and the formation of emulsions and dispersions under acidic conditions.

amine 248. Dark-colored liquid or paste consisting of a nonvolatile amine mixture with *bis*-(hexamethylene)triamine and its homologs as principal components. Disperses readily in water.
Use: Coagulant and flocculating agent, ingredient of antistripping agents for asphalt, corrosion inhibitor, anchoring agent, for making cationic amine salts.

amine absorption process. See Girbotol absorption.

aminimide. Any of a group of nitrogen compounds derived by reaction of 1,1-dimethylhydrazine with an epoxide in the presence of an ester of a carboxylic acid. A number of types of epoxides and esters may be used, providing a wide variety of products, including short- and long-chain aliphatics and aromatics. Their major uses are in tire-cord dips to increase adhesion of cord to rubber, in soil-removing detergents (nonionic), in coating formulations, and in cosmetic creams and shampoos. They are stated to be biodegradable and without toxic hazard. They also have elastomer and cross-linking applications from adhesives, caulks, and sealants to foams and mechanical goods. As isocyanate precursors, aminimides can be prepared in a large number of structural variations. Bisaminimides are especially valuable for producing stable, single-package prepolymer compositions for in situ generation of isocyanates in polyurethane applications.

***p*-aminoacetanilide.** (4'-aminoacetanilide; *N*-acetyl-*p*-phenylenediamine). $NH_2C_6H_4NHCOCH_3$.
Properties: Colorless or reddish crystals. Mp 162C, bp 267C. Soluble in alcohol and ether; slightly soluble in water. Combustible.
Derivation: Acetylation of *p*-phenylenediamine.
Hazard: Moderate toxicity by ingestion.
Use: Intermediate for azo dyes and pharmaceuticals.

aminoacetic acid. See glycine.

aminoacetophenetidide hydrochloride. See phenocoll hydrochloride.

amino acid. An organic acid containing both a basic amino group (NH_2) and an acidic carboxyl group (COOH). Amino acids are amphoteric and exist in aqueous solution as dipolar ions. The 25 amino acids that have been established as protein constituents are α-amino acids (i.e., the $-NH_2$ group is attached to the carbon atom next to the $-COOH$ group). Many other amino acids occur in the free state in plant or animal tissue. At least 22 amino acids with structures identical with those that exist today have been identified in pre-Cambrian sedimentary rock, indicating an age of at least 3 million years. Amino acids have been created in the laboratory by passing an electrical discharge through a mixture of ammonia, methane, and water vapor; it is believed that a similar reaction may have accounted for the original synthesis of amino acids on earth. Amino acids can be obtained by hydrolysis of a protein; or they can be synthesized in various ways, especially by fermentation of glucose. An essential amino acid is one that cannot be synthesized by the body and is necessary for survival, namely, isoleucine phenylalanine, leucine, lysine, methionine, threonine, tryptophan, and valine. Nonessential amino acids (alanine, glycine, and about a dozen others) can be synthesized by the body in adequate quantities. Arginine and histidine are essential during periods of intensive growth. All the essential and most of the nonessential amino acids have one or more asymmetric carbon atoms and are optically active. Amino acids are the building blocks of molecular biology. Various combinations of amino acids form the proteins, which are highly complex molecules present in all living things. *Note:* Use of amino acids as fortification additives to foods is restricted by the FDA to foods containing proteins. See genetic code; deoxyribonucleic acid; chromatin; protein; life, origin.

amino acid activation.   Enzymatic esterification of the carboxyl group of an amino acid to the 3'-hydroxyl group of its corresponding tRNA which requires 2 ATP-equivalents of energy.

amino acid alkaloid.   Usually an ergot alkaloid such as bromocriptine or ergotamine.

amino acid antagonist.   A substance that inhibits the absorption, metabolism, or biological activity of an amino acid.

amino acid antagonistic.   (antipathic). Having the action of an antagonist.

amino acid oxidase.   See amino oxidase.

amino-acid residue.   A structure (in addition to water) that results from the combination of two or more amino acids to form a peptide. Such a residue lacks a hydrogen atom of the amino group or the hydroxy group of the carboxy group or both. Thus each unit of the peptide chain are amino-acid residues.

aminoacyl-tRNA.   An aminoacyl ester of a tRNA.

aminoacyl-tRNA synthetases.   Enzymes that catalyze synthesis of an aminoacyl tRNA at the expense of ATP energy.

aminoamylene glycol.   See 2-amino-2-ethyl-1,3-propanediol.

5-amino-2-aniliobenzenesulfonic acid.   See 4-amino-diphenylamine-2-sulfonic acid.

o-**aminoanisole.**   See *o*-anisidine.

p-**aminoanisole.**   See *p*-anisidine.

aminoanthraquinone.
CAS: 117-79-3. $C_6H_4(CO)_2C_6H_3NH_2$.
Tricyclic: (1) 1-amino, (2) 2-amino.

Properties: (1) Red, iridescent needles. (2) Red or orange-brown needles. Soluble in alcohol, chloroform, benzene, and acetone; insoluble in water. Mp (1) 252C, (2) 302C, bp sublimes (both 1 and 2).
Derivation: By reduction of nitroanthraquinones, or by the substitution of the amino radical direct for the sulfonic acid.
Hazard: Questionable carcinogen.
Use: Dye and pharmaceutical intermediate.

4-aminoantipyrine.   (4-amino-1,5-dimethyl-2-phenyl-3-pyrazolone; 1,5-dimethyl-2-phenyl-4-aminopyrazolone). $C_{11}H_{13}N_3O$.
Properties: Mp 107–109C.
Use: Analytical reagent.

p-**aminoazobenzene.**   (aniline yellow; phenylazoaniline).
CAS: 60-09-3. $C_6H_5NNC_6H_4NH_2$.
Properties: Yellow to tan crystals. Mp 126–128C, bp above 360C. Soluble in alcohol and ether; slightly soluble in water.
Derivation: (1) Heating diazoaminobenzene with aniline hydrochloride as catalyst. (2) Diazotization of a solution of aniline and aniline hydrochloride with hydrochloric acid and sodium nitrite.
Hazard: Possible carcinogen.
Use: Dyes (chrysoidine, induline, solid yellow, and acid yellow), insecticide.

p-**aminoazobenzene.**   ((*p*-phenylazo)aniline; C.I. 11000; C.I. solvent blue 7; 4-phenyldiazenylaniline).
CAS: 60-09-3. $C_{12}H_{11}N_3$.
Properties: Aromatic amine; high lipophilicity.
Hazard: Bioaccumulate; toxic; possible carcinogen.
Use: Dye in lacquers, varnishes, stains and resins; an intermediate in the manufacture of acid yellow and indulines.

4-aminoazobenzene-3,4'-disulfonic acid.   (Acid Yellow 9). $C_6H_4(SO_3H)NNC_6H_3NH_2(SO_3H)$.
Properties: Bright, violet needles.
Use: For synthesizing dyes for dyeing wool.

p-**aminoazobenzene hydrochloride.**   (aminoazobenzene salt). $C_6H_5NNC_6H_4NH_2 \cdot HCl$.
Properties: Steel-blue crystals. Soluble in alcohol; slightly soluble in water.
Derivation: By passing dry hydrogen chloride gas into a solution of aminoazobenzene.
Hazard: Suspected carcinogen.
Use: Dyes, coloring lacquers, intermediate.

aminoazobenzenemonosulfonic acid. NH_2 $C_6H_4NNC_6H_4SO_3H$.
Properties: Yellowish-white, microscopic needles. Barely soluble in water; almost insoluble in alcohol, ether, and chloroform.

Derivation: By sulfonating aminoazobenzene.
Hazard: Suspected carcinogen.
Use: Dyestuff manufacture.

_o_-aminoazotoluene. (Solvent Yellow; 3,2-amino-5-azotoluene; toluazotoluidine).
CAS: 97-56-3. $CH_3C_6H_4N_2C_6H_3NH_2CH_3$.
Properties: Reddish-brown to yellow crystals. Mp 100–117C. Soluble in alcohol, ether, oils, and fats; slightly soluble in water.
Derivation: From *o*-toluidine by treatment with nitrite and hydrochloric acid.
Hazard: Possible carcinogen.
Use: Dyes, medicine.

6,*p*-(*p*-aminobenzamido)benzamido-1-naphthol-3-sulfonic   acid.
$H_2NC_6H_4CONHC_6H_4CONHC_{10}H_5(OH)(SO_3H)$.
Gray paste containing approximately 35% solids.
Use: Intermediate.

aminobenzene.   See aniline.

_p_-aminobenzenearsonic acid.   See arsanilic acid.

2-amino-*p*-benzenedisulfonic acid. (aniline-2,5-disulfonic acid). $C_6H_3NH_2(SO_3H)_2 \cdot 4H_2O$.
Properties: Crystals. Very soluble in water and alcohol.
Derivation: Boiling sodium salt of 4-chloro-3-nitrobenzene sulfonate with sodium sulfite, resulting in formation of sodium-2-nitrobenzene disulfonate, which is reduced with iron and acetic acid to aniline-2,5-disulfonic acid.
Use: Intermediate.

4-amino-*m*-benzenedisulfonic   acid. (aniline-2,4-disulfonic acid). $C_6H_3NH_2(SO_3H)_2 \cdot 2H_2O$.
Properties: Needles decompose when heated above 120C. Very soluble in water and alcohol.
Derivation: By heating sulfanilic acid with fuming sulfuric acid at 170–180C.
Use: Dye intermediate.

_m_-aminobenzenesulfonic acid.   See metanilic acid.

_p_-aminobenzenesulfonic acid.   See sulfanilic acid.

_m_-aminobenzoic acid. $C_6H_4NH_2CO_2H$.
CAS: 99-05-8.
Properties: Yellowish or reddish crystals, sublimes easily, sweet taste. Mp 173–174C. Slightly soluble in water, alcohol, and ether.
Use: Dye intermediate.

_o_-aminobenzoic acid.   See anthranilic acid.

_p_-aminobenzoic acid.   (PABA).

CAS: 150-13-0. $NH_2C_6H_4CO_2H$. Required by many organisms as a vitamin for growth, active in neutralizing the antibacteriostatic effect of some sulfonamide drugs.

Properties: Light buff, odorless crystals; white when pure; discolor on exposure to light and air. Mp 186–187C. Sparingly soluble in cold water; soluble in hot water, glacial acetic acid, ethyl acetate. Unstable to ferric salts and oxidizing agents.
Derivation: Reduction of *p*-nitrobenzoic acid. Commercially available as the calcium, potassium, and sodium salts.
Food source: Widely distributed, especially in yeast.
Grade: Technical, NF.
Hazard: Questionable carcinogen.
Use: Dye intermediate, pharmaceuticals, nutrition, UV absorber in suntan lotions.

2-aminobenzothiazole. $C_6H_4NC(NH_2)S$ (bicyclic).
Use: As an azo dye intermediate and in photographic chemicals.

_m_-aminobenzotrifluoride.
CAS: 98-16-8. $CF_3C_6H_4NH_2$.
Properties: Colorless to oily yellow liquid.
Grade: Technical (88% min), purified (98% min).
Use: Pharmaceutical intermediate.

_o_-aminobenzoylformic   anhydride.   See isatin.

***N*-(*p*-aminobenzoyl)glycine.**   See *p*-aminohippuric acid.

_o_-aminobiphenyl. (*o*-phenylaniline; *o*-biphenylamine).
CAS: 90-41-5. $C_6H_5C_6H_4NH_2$.
Properties: Colorless or purplish crystals. Mp 49.3C, bp 299C. Slightly soluble in water.
Derivation: Reduction of *o*-nitrobiphenyl.
Hazard: Toxic by ingestion, inhalation, and skin absorption. A carcinogen.
Use: Research, analytical chemistry.

1-aminobutane.   See *n*-butylamine.

2-aminobutane.   See *sec*-butylamine.

aminobutanoic acid.   See aminobutyric acid.

2-amino-1-butanol.
CAS: 96-20-8. $CH_3CH_2CHNH_2CH_2OH$.

Properties: Colorless liquid. D 0.944 at 20/20C, mp −2C, bp 178C (at 10 mm Hg = 79–80C), flash p 164F (73.3C), bulk wt 7.85 lb/gal (20C); pH (0.1 M aqueous solution) 11.11, refr index 1.453 (20C). Completely miscible with water at 20C; soluble in alcohols. Corrosive to copper, brass, aluminum. Combustible.

Use: Emulsifying agent (in soap form) for oils, fats, and waxes; absorbent for acidic gases; organic synthesis.

aminobutyric acid. (aminobutanoic acid; GABA).
CAS: 56-12-2. An unusual amino acid having the following isomers: α: $CH_3CH_2CH(NH_2)COOH$. Isolated from a bacterium *(Corynebacterium diphtheriae)*. The *dl* form is a crystalline solid, mp 305C, soluble in water, slightly soluble in alcohol, insoluble in ether. The *l*(+) form is solid, mp 270C, sweetish taste, soluble in water. β: $CH_2CH(NH_2)CH_2COOH$. The *dl* form is a tasteless solid, mp 190C, water soluble, insoluble in ether and alcohol. The *d*(−) form decomposes at 220C. γ: (GABA) $H_2N(CH_2)_3COOH$. Obtained from bacteria, yeast, and plant life. Crystalline solid, mp 202C, soluble in water, insoluble in organic solvents. Decomposes to pyrrolidone and water on quick heating. This substance is reported to be a neurotransmitter that activates or retards nervous reactions in the cells of the brain, including the sense of pain. All three isomers have been synthesized by various reaction sequences, the first reported in 1880.

ã-aminobutyric acid. (4-aminobutanoic acid; ã-amino-*n*-butyric acid; piperidic acid; GABA).
CAS: 56-12-2. $C_4H_9NO_2$. An amino acid that is not a component of protein. It is an inhibitory neurotransmitter of the cns and may mediate the inhibitory actions of local interneurons. Inhibition of the synthesis or transmission of ã-aminobutyric acid can cause convulsions.
Derivation: It occurs chiefly in nervous tissue.
Use: Hypertensive agent.

ß-aminobutyric acid. (3-aminobutanoic acid; ß-amino-*n*-butyric acid).
Properties: Tasteless; water-soluble.
Hazard: Highly toxic.

α-aminocaproic acid. See norleucine.

aminocaproic lactam. See caprolactam.

aminocarb. (4-(dimethylamino)-3-methylphenol-methylcarbamate ester; methyl-carbamic acid 4-(dimethylamino)-*m*-tolyl ester; 4-dimethyl-amino-*m*-tolyl methylcarbamate; [4-(dimethylamino)-3-methylphenyl]*N*-methylcarbamate). $C_{11}H_{16}N_2O_2$.
Properties: Tan; crystalline; slightly water-soluble; melting point 93–94C.

Hazard: Highly poisonous.
Use: Insecticide.

**1-(aminocarbonyl)propyl   *n*-
(((methylamino)carbonyl)oxy)
ethanimidothioate.**
CAS: 92065-18-4. $C_8H_{15}N_3O_3S$.
Hazard: A poison by ingestion.
Use: Agricultural chemical.

aminochlorobenzene.   See chloroaniline.

3-amino-5-chlorobenzoic acid.
CAS: 21961-30-8. $C_7H_6ClNO_2$.
Hazard: Low toxicity by ingestion.

4-amino-4′-chlorodiphenyl. $C_{12}H_{10}ClN$.
Properties: Crystalline solid. Mp 128C. Insoluble in water; soluble in acetone, benzene, glacial acetic acid, and alcohol.
Derivation: Chlorination of 4-nitrodiphenyl and reduction of the product with iron and hydrochloric acid in alcohol.
Use: Sulfur determination in various materials, e.g., rubber, coal, etc.

2-amino-4-chlorophenol. (*p*-chloro-*o*-aminophenol). $C_6H_3OHNH_2Cl$.
Properties: Light brown crystals. Mp 138C (decomposes).
Derivation: Reduction of *p*-chloro-*o*-nitrophenol.
Use: Intermediate.

2-amino-4-chlorotoluene. 5-chloro-2-methylaniline (NH_2 = 1); 4-chloro-*o*-toluidine (CH_3 = 1).
CAS: 95-69-2. $ClNH_2C_6H_3CH_3$.
Properties: Off-white solid or light-brown oil that tends to darken on storage. Mp 20–22C.
Hazard: Toxic by ingestion and inhalation. Probable carcinogen.
Use: Intermediate.

2-amino-5-chlorotoluene. $ClNH_2C_6H_3CH_3$.
Properties: Crystalline solid. Mp 26–27C. Sparingly soluble in water; soluble in dilute acids.
Use: Intermediate.

2-amino-6-chlorotoluene. 6-chloro-*o*-toluidine (CH_3 = 1); 3-chloro-2-methylaniline (NH_2 = 1). $ClNH_2C_6H_3CH_3$.
Properties: Liquid. Mp 0–2C.
Hazard: Toxic by ingestion or inhalation.
Use: Intermediate.

4-amino-2-chlorotoluene. 2-chloro-*p*-toluidine (CH_3 = 1); 3-chloro-4-methylaniline (NH_2 = 1).
CAS: 95-69-2. $ClNH_2C_6H_3$.
Properties: Liquid. Mp 21–24C.
Hazard: Probable carcinogen.
Use: Intermediate.

4-amino-2-chlorotoluene-5-sulfonic acid. (brilliant toning red amine; permanent red 2B amine). $ClNH_2CH_3C_6H_2SO_3H$.
Properties: White to buff powder. Essentially insoluble as free acid; soluble as sodium or ammonium salt.
Grade: 98.5% min purity.
Use: Intermediate for azo pigments.

5-amino-2-chlorotoluene-4-sulfonic acid. (lake red carbon amine). $ClNH_2CH_3C_6H_2SO_3H$.
Properties: White to pink powder. Essentially insoluble as free acid; soluble as sodium or ammonium salt.
Grade: 98.5% min purity.
Use: Intermediate for azo pigments.

m-amino-*p*-cresol methyl ether. See 5-methyl-*o*-anisidine.

aminocyclohexane. See cyclohexylamine.

1-aminocyclopropane-1-carboxylase. (ACC synthase). An enzyme that catalyzes the rate limiting step in the ethylene biosynthetic pathway. It contributes significantly in the ripening of fruit. Plants generally bear a number of distinct ACC synthase genes that are differentially regulated in response to various developmental, environmental, and chemical factors.

l-amino-dehydrogenase. See amino oxidase.

3-amino-2,5-dichlorobenzoic acid. See amiben.

2-amino-4,6-dichlorophenol. See 2,4-dichloro-6-aminophenol.

p-aminodiethylaniline. (*N,N*-diethyl-*p*-phenylenediamine; diethylaminoaniline). $(C_2H_5)_2NC_6H_4NH_2$.
Properties: Liquid. Bp 260–262C. Insoluble in water; soluble in alcohol and ether.
Derivation: Treatment of diethylaniline with nitrous acid and subsequent reduction.
Hazard: See aniline.
Use: Dye intermediate, source of diazonium compounds in diazo copying process.

p-aminodiethylaniline hydrochloride. $C_{10}H_{16}N_2 \cdot HCl$.
Properties: Colorless needles. Soluble in water and alcohol; insoluble in ether.
Hazard: See aniline.
Use: Color photography.

2-amino-4,6-dihydroxypteridine. See xanthopterin.

p-aminodimethylaniline. (dimethylaminoaniline; dimethyl-*p*-phenylenediamine). $(CH_3)_2NC_6H_4NH_2$.

Properties: Colorless; asbestos-like needles; stable in air when pure. If impure, the crystals liquefy. Mp 41C, bp 257C. Soluble in water, alcohol, and benzene.
Derivation: By reduction of *p*-nitrosodimethylaniline with zinc dust and hydrochloric acid.
Method of purification: Recrystallization from mixture of benzene and ligroin.
Hazard: Toxic by ingestion or inhalation of vapor.
Use: Base for production of methylene blue, photodeveloper, reagent for detection of hydrogen sulfide, reagent for cellulose, organic synthesis, reagent for certain bacteria.

3-amino-*n*-(2-(dimethylamino)ethyl) naphthalimide. See nafidimide.

aminodimethylbenzene. See xylidine.

2-amino-4,6-dimethylpyridine. $(CH_3)_2C_5H_2NNH_2$.
Properties: Solid. Mp 65.2–68.5C, bp 235C. Water soluble.
Derivation: Prepared from 2-aminopyridine.
Use: Organic intermediate.

2-amino-4,6-dinitrophenol. See picramic acid.

p-aminodiphenyl. (4-aminodiphenyl; *p*-xenylamine).
CAS: 92-67-1. $C_6H_5C_6H_4NH_2$.
Properties: Colorless crystals. Mw 169.23, mp 53C, bp 302C.
Hazard: Toxic by ingestion, inhalation, skin absorption. Confirmed carcinogen. Bladder and liver cancer.
Use: Organic research.

p-aminodiphenylamine. (*N*-phenyl-*p*-phenylenediamine). $NH_2C_6H_4NHC_6H_5$.
Properties: Purple powder. Mp 75C. Insoluble in water; soluble in alcohol and acetone.
Derivation: Reduction of the coupling product of diazotized sulfanilic acid and diphenylamine.
Use: Dye intermediate, pharmaceuticals, photographic chemicals.

4-aminodiphenylamine-2-sulfonic acid. (5-amino-2-anilinobenzenesulfonic acid). $C_6H_5NHC_6H_3NH_2SO_3H$.
Properties: Needle like crystals. Barely soluble in water.
Derivation: From *p*-nitrodiphenylamine-*o*-sulfonic acid by reduction with iron and hydrochloric acid.
Use: Synthesis of dyestuffs.

aminodithioformic acid. See dithiocarbamic acid.

aminoethane. See ethylamine.

2-aminoethanesulfonic acid.   See taurine.

2-aminoethanethiol.   (cysteamine;   mercaptamine; thioethanolamine).
CAS: 60-23-1. $HSCH_2CH_2NH_2$.
Properties: Crystals with unpleasant odor. Oxidizes on contact with air. Mp 97C. Soluble in water. Combustible.
Use: Medicine (believed to offer protection against radiation).

1-aminoethanol.   See aldehyde ammonia.

2-aminoethanol.   See ethanolamine.

2-(2-aminoethoxy)ethanol.   (Diglycolamine; DGA).
CAS: 929-02-6. $NH_2CH_2CH_2OCH_2CH_2OH$.
Properties: Colorless, slightly viscous liquid; mild amine odor. Bp 221C, d 1.0572 (20/20C), flash p 260F (126.6C), fp −12.5C. Miscible with water and alcohols. Combustible.
Hazard: Strong irritant to tissue.
Use: Removal of acid components from gases, especially carbon dioxide and hydrogen sulfide from natural gas; intermediate.

4-amino-*n*-(6-ethoxy-3-pyridazinyl) benzenesulfonamide.   See sulfaethoxypyridazine.

aminoethoxyvinylglycine.
Use: Inhibitor of ethylene biosynthesis in plants.

1-[(aminoethyl)amino]-2-propanol.
$C_5H_{14}N_2O$.
Properties: Thick, colorless liquid. Bp 112C (11 mm Hg), d 0.984, refr index 1.743. Slight odor of ammonia.
Use: Curing epoxy resins.

3-amino-9-ethylcarbazole.   (3-amino-*N*-ethylcarbazole hydrochloride; 9-ethylcarbazol-3-amine).
CAS: 132-32-1. $C_{14}H_{14}N_2$. A hydrochloride salt and free amine solid compound.
Hazard: Questionable carcinogen; poison.

4-(2-aminoethyl)diethylenetriamine.   See 2,2′,2″-nitrilotris(ethylamine).

aminoethylethanolamine.   See   hydroxyethylethylenediamine.

4-aminoethylglyoxaline.   See histamine.

4-(2-aminoethyl)imidazole.   See histamine.

2-aminoethylisothiouronium diacetate.
CAS: 63680-10-4. $C_3H_9N_3S•2C_2H_4O_2$.
Hazard: Moderately toxic.

β-aminoethylisothiourea   dihydrobromide.
(2-(2-aminoethyl)-2-thiopseudourea dihydrobromide; AET). $C_3H_9N_3S•2HBr$.
Properties: Crystals, hygroscopic. Mp 194–195C.
Derivation: Thiourea is refluxed with 2-bromoethylamine hydrobromide in isopropanol.
Use: Enzyme activator, free radical detoxifier (believed to offer protection against radiation).

4-(1-aminoethyl)phenol.
CAS: 134855-87-1. $C_8H_{11}NO$.
Hazard: A severe eye irritant.

N-aminoethylpiperazine.
CAS: 140-31-8.

$H_2NC_2H_4NCH_2CH_2NHCH_2CH_2$.
An amine combining a primary, secondary, and tertiary amine in one molecule.
Properties: Liquid. D 0.9837, bp 222.0C, flash p 200F (93.3C), fp 17.6C. Soluble in water. Combustible.
Hazard: Strong irritant to tissue.
Use: Epoxy curing agent, intermediate for pharmaceuticals, anthelmintics, surface-active agents, synthetic fibers.

2-amino-2-ethyl-1,3-propanediol.   (AEPD; aminoamylene   glycol). $CH_2OHC(C_2H_5)NH_2$ CH_2OH.
Properties: Solid or viscous liquid. Mp 38C. Soluble in water and alcohol.
Use: Emulsifying agent (in soap form) for oils, fats, and waxes, absorbent for acidic gases CO_2 and H_2S, organic synthesis.

s-(2-aminoethyl) phosphorothioate.
CAS: 5746-40-7. $C_2H_8NO_3PS$.
Hazard: A reproductive hazard.

2-aminoethylsulfuric   acid. $NH_2CH_2CH_2$ OSO_3H.
Properties: White, crystalline powder; noncorrosive. Mp 274–280C; sinters at 274C and darkens without complete melting at 280C; d 1.782; bulk d 1.007. Soluble in water; insoluble in most organic solvents; pH (1% aqueous solution) 4.0 (20C), (5% aqueous solution) 3.3 (39C).
Use: Organic synthesis of ethyleneimine and various other compounds, amination of cotton.

3-amino-α-ethyl-2,4,6-triiodohydrocinnamic acid.   See iopanoic acid.

4-aminofolic acid.   See aminopterin.

aminoform.   See hexamethylenetetramine.

N-[p-[[(2-amino-5-formyl-5,6,7,8-tetrahydro-4-hydroxy-6-pteridinyl)methyl]amino] benzoyl]glutamic acid.   (folinic acid; (2S0-2-{[4-{[2-amino-5-formyl-4-oxo-1,6,7,8-tetrahydropteridin-6-yl]methylamino}benzoyl]amino} peacid; leucovorin).

CAS: 58-05-9.　$C_{20}H_{23}N_7O_7$. A calcium salt.
Use: Antidote to folic acid antagonists.

amino-G acid.　　(2-naphthylamine-6,8-disulfonic acid; 7-amino-1,3-naphthalene-disulfonic acid). $C_{10}H_5(NH)_2(SO_3H)_2$.
Properties: White, crystalline solid. Soluble in water.
Derivation: (1) Form G acid by heating sodium salt with ammonia and sodium bisulfite solution in an autoclave under pressure. (2) Sulfonation by β-naphthylamine.
Use: Azo dye intermediate.

α-aminoglutaric acid.　　See glutamic acid.

amino-4-guanidovaleric acid.　　See arginine.

7-aminoheptanenitrile.
CAS: 23181-80-8.　$C_7H_{14}N_2$.
Hazard: Moderately toxic by ingestion.

aminohexamethyleneimine.

$\overline{}$
$NH_2NCH_2CH_2CH_2CH_2CH_2CH_2$.

Properties: Liquid. Bp 170C. Soluble in water and most organic solvents. Combustible.
Use: Intermediate for dyes, pharmaceuticals, and photographic chemicals.

2-aminohexanoic acid.　　See norleucine.

6-aminohexanoic acid.　　See norleucine.

p-aminohippuric acid.　　[N-(p-aminobenzoyl) glycine; PAHA]. $NH_2C_6H_4CONHCH_2COOH$.
Properties: White, crystalline powder. Discolors on exposure to light. Soluble in alcohol and most organic solvents. Very soluble in dilute hydrochloric acid and alkalies. Forms a water-soluble sodium salt. Mp 197–199C.
Grade: USP.
Use: Medicine (diagnostic agent), intermediate.

aminohydroxybenzoic acids.　　See aminosalicylic acids.

4-amino-1-hydroxybutane-1,1-diyldiphosphonic acid.
CAS: 66376-36-1.　$C_4H_{13}NO_7P_2$.
Hazard: A reproductive hazard.

α-amino-β-hydroxybutyric acid.　　See threonine.

2-amino-2-hydroxymethyl-1,3-propanediol.
See tris(hydroxymethyl)aminomethane.

α-amino-β-hydroxypropionic acid.　　See serine.

6-amino-n-hydroxy-3-pyridinecarboxamide.
See 6-aminonicotinohydroxamic acid.

α-amino-β-imidazolepropionic acid.　　See histidine.

α-aminoisocaproic acid.　　See leucine.

α-aminoisovaleric acid.　　See valine.

amino-J acid.　　(2-naphthylamine-5,7-disulfonic acid; 6-amino-1,3-naphthalene-disulfonic acid). $C_{10}H_5(NH_2)_2(SO_3H)_2$.
Properties: Crystallizes in white lustrous leaflets from water and in long needles from hydrochloric acid solution.
Derivation: By sulfonation of either 2-naphthylamine-5-sulfonic acid or 2-naphthylamine-7-sulfonic acid.
Use: Azo dye intermediate.

2-amino-6-mercaptopurine.
Use: Pharmaceutical end product for treatment of leukemia.

aminomercuric chloride.　　See mercury, ammoniated.

aminomethane.　　See methylamine.

3-amino-4-methoxybenzanilide.
$CH_3OC_6H_3(NH_2)CONHC_6H_5$.
Properties: Gray powder.
Use: Dyes, pharmaceuticals, and other organic chemicals.

1-amino-2-methoxy-5-methylbenzene.　　See 5-methyl-o-anisidine.

m-(4-amino-3-methoxyphenylazo) benzenesulfonic acid.　　$H_2NC_6H_3(OCH_3)NNC_6H_4SO_3H$.
Properties: Maroon paste containing approximately 38% solids.
Use: Intermediate.

4-aminomethyl-1-benzylpyrrolidin-2-one fumarate (2:1).
CAS: 97205-35-1.　$C_{24}H_{32}N_4O_2 \cdot C_4H_4O_4$.
Hazard: Moderately toxic by ingestion.

3-amino-3-methyl-1-butyne.　　See 2-methyl-3-butyn-2-amine.

4-amino-4'-methyldiphenylamine-2-sulfonic acid.　　(aminotoluidinobenzenesulfonic acid). $CH_3C_6H_4NHC_6H_3NH_2SO_3H$.
Properties: Light to dark gray paste with characteristic odor.
Use: Intermediate.

4-amino-10-methylfolic acid.   See methotrexate.

3-amino-5-methylisoxazole.
Use: Analog of the nucleic acid constituent guanine.

2-amino-1-methyl-2-oxoethyl-*n*-(((methylamino)carbonyl)oxy) ethanimidothioate.
CAS: 92065-77-5. $C_7H_{13}N_3O_3S$.
Hazard: A poison by ingestion.
Use: Agricultural chemical.

2-amino-3-methylpentanoic   acid.   See isoleucine.

2-amino-4-methyl-5-phosphono-3-pentenoic acid.
CAS: 137424-81-8. $C_6H_{12}NO_5P$.
Hazard: A poison.

2-amino-2-methyl-1,3-propanediol.
(AMPD; aminobutylene glycol; butanediolamine). $CH_2OCH(CH_3)NH_2CH_2OH$.
Properties: Colorless crystals. Mp 110C. Soluble in water and alcohol. Corrosive to copper, brass, and aluminum.
Use: Emulsifying agent (in soap form) for oils, fats, and waxes; absorbent for acidic gases; organic synthesis; cosmetics.

2-amino-2-methyl-1-propanol.   (isobutanolamine; AMP). $CH_3(CH_3)NH_2CH_2OH$.
Properties: Solid or viscous liquid. Mp 30C, bp 165C, d 0.93, refr index 1.45.
Hazard: Toxic by ingestion.
Use: Emulsifying agent (in soap form) for oils, fats, and waxes, absorbent for acidic gases, organic synthesis, cosmetics.

2-amino-3-methylpyridine.   (2-amino-3-picoline).

N:C(NH$_2$)C(CH$_3$):CHCH:CH.
Properties: Liquid. Bp 221C, mp 29.5–33.3C. Soluble in water.
Derivation: From 2-aminopyridine.
Hazard: Toxic by ingestion.
Use: Intermediate.

2-amino-4-methylpyridine.   (2-amino-4-picoline).

N:C(NH$_2$)CH:C(CH$_3$)CH:CH.
Properties: Crystals. Bp 230.9C (115–117C at 11 mm), mp 96–99.0C. Sublimes on slow heating. Soluble in water and lower alcohols.
Derivation: Prepared from 2-aminopyridine.
Use: Intermediate, medicine.

2-amino-5-methylpyridine.   (2-amino-5-picoline).

N:C(NH$_2$)CH:CHC(CH$_3$):CH.
Properties: Crystals. Bp 227.1C, mp 76.6C. Soluble in water.
Derivation: Prepared from 2-aminopyridine.
Use: Intermediate.

2-amino-6-methylpyridine.   (2-amino-6-picoline).

N:C(NH$_2$)CH:CHCH:C(CH$_3$).
Properties: Crystals. Bp 214.4C, mp 43.7C. Soluble in water.
Derivation: Prepared from 2-aminopyridine.
Use: Intermediate.

2-amino-4-(methylthio)butyric   acid.   See methionine.

α-amino-β-methylvaleric acid.   See isoleucine.

α-amino-γ-methylvaleric acid.   See leucine.

***dl*-2-amino-3-methylvaleric acid.**   See *dl*-isoleucine.

aminonaphthalenesulfonic acid.   See naphthylaminesulfonic acid.

aminonaphthol.
Derivation: From nitronaphthols by reduction.

4-amino-1-naphthol.   (4-hydroxy-α-naphthylamine). $C_{10}H_9NO$.
Properties: Acicular crystals. Soluble in water.
Derivation: Rearrangement of α-naphthylhydroxylamine in acetone.
Use: Inhibitor of polymerization, chemical intermediate. Must be kept dry during storage to avoid oxidation and discoloration.

aminonaphtholsulfonic acid.   Any of several sulfonated aromatic acids derived from naphthol or naphthylamine and used as azo dye intermediates.

6-aminonicotinic acid.   (6-aminopyridine-3-carboxylic acid). $C_6H_6N_2O_2$.
Properties: Crystals. Mp decomposes above 300C.

6-aminonicotinohydroxamic acid.
CAS: 76706-59-7. $C_6H_7N_3O_2$.
Hazard: A poison by ingestion.
Use: Agricultural chemical.

amino oxidase.   (L-amino acid oxidase; D-amino acid oxidase; L-amino dehydrogenase). An enzyme that catalyzes the deamination of α-amino acids by dehydrogenation to keto acids and ammonia. Two types are recognized, acting on the D- and L-amino acids. Recent emphasis has been on characterization of the D-amino oxidase, which is known to contain the flavin isoalloxazine as coenzyme. Both types are found in animal tissue, especially in liver and kidney, as well as in snake venom and certain bacteria.

**2-amino-2-oxoethyl-2,2-dimethyl-
n-(((methylamino)carbonyl)oxy)
propanimidothioate.**
CAS: 92065-91-3. $C_9H_{17}N_3O_3S$.
Hazard: A poison by ingestion.
Use: Agricultural chemical.

**2-amino-2-oxoethyl
n-(((methylamino)carbonyl)oxy)
ethanimidothioate.**
CAS: 92065-82-2. $C_6H_{11}N_3O_3S$.
Hazard: A poison by ingestion.
Use: Agricultural chemical.

2-amino-6-oxypurine.   See guanine.

1-aminopentane.   See *n*-amylamine.

2-aminophenetole.   See *o*-phenetidine.

4-aminophenetole.   See *p*-phenetidine.

***m*-aminophenol.**   (*m*-hydroxyaniline).
CAS: 591-27-5. $C_6H_4NH_2OH$.

Properties: White crystals. Mp 122C. Soluble in water, alcohol, and ether.
Derivation: Fusion of *m*-sulfanilic acid with caustic soda and subsequent extraction of the melt with ether.
Hazard: Toxic by ingestion.
Use: Dye intermediate, intermediate for *p*-aminosalicylic acid.

***o*-aminophenol.**   (*o*-hydroxyaniline).
CAS: 95-55-6. $C_6H_4NH_2OH$.
Properties: White crystals; turn brown with age. Mp 172–173C, sublimes on further heating. Soluble in cold water, alcohol, benzene; freely soluble in ether.
Derivation: By reduction of *o*-nitrophenol mixed with aqueous ammonia by means of a stream of hydrogen sulfide. Also available as the hydrochloride.
Grade: Technical, 99% min.
Use: Dyeing furs and hair; dye intermediate for azo and sulfur dyes; pharmaceuticals.

***p*-aminophenol.**   (*p*-hydroxyaniline).
CAS: 123-30-8. $C_6H_4NH_2OH$.
Properties: White or reddish-yellow crystals turn violet on exposure to light. Mp 184C (decomposes). Soluble in water and alcohol.
Derivation: (1) By reduction of *p*-nitrophenol with iron filings and hydrochloric acid; (2) by electrolytic reduction of nitrobenzene in concentrated sulfuric acid and treatment with an alkali to free the base. Also available as the hydrochloride.
Grade: Technical, photographic.
Use: Dyeing textiles, hair, furs, feathers, photographic developer, pharmaceuticals, antioxidants, oil additives.

4-amino-1-phenol-2,6-disulfonic   acid. $C_6H_2OHNH_2(SO_3H)_2$.
Properties: Fine needles. Soluble in water; slightly soluble in alcohol; insoluble in ether.
Derivation: Action of sulfur dioxide on *p*-nitrophenol.
Use: Dyes.

2-amino-1-phenol-4-sulfonic   acid.   (*o*-amino-phenol-*p*-sulfonic acid). $C_6H_3OHNH_2SO_3H$.
Properties: Brown crystals, No mp. Decomposes on heating. Fairly soluble in hot water, very soluble in alkaline solution.
Derivation: (1) Sulfonation and nitration of chlorobenzene followed by hydrolysis to phenol with caustic soda with subsequent reduction by sodium sulfide. (2) Sulfonation of *o*-aminophenol. (3) Sulfonation of phenol followed by nitration and reduction.
Use: Intermediate for dyes.

***p*-aminophenylarsonic acid.**   See arsanilic acid.

***m*-aminophenylboronic acid hemisulfate.**
(3-aminobenzeneboronic acid). $H_2NC_6H_4B(OH)_2$•$1/2H_2SO_4$.
Properties: Mw 186, mp above 300C.
Use: Adsorbent additive for the chromatographic separation of 3′-terminal polynucleotides from RNA.

1-amino-2-phenylethane.   See   2-phenylethylamine.

***o*-aminophenylglyoxalic lactim.**   See isatin.

***p*-aminophenylmercaptoacetic   acid.** $NH_2C_6H_4SCH_2CO_2H$.
Properties: Mp 186–187C. Insoluble in water, alcohol, benzene, chloroform; soluble in aqueous acid or alkali solutions.
Use: Synthetic intermediate for dyes and pharmaceuticals.

2-(p-aminophenyl)-6-methylbenzothiazole.
See dehydrothio-p-toluidine.

m-aminophenyl   methyl   carbinol.
$NH_2C_6H_4CH(OH)CH_3$.
Properties: Solid. D 1.12, bp 217.3C (100 min), mp 66.4C, flash p 315F (157C). Soluble in water. Combustible.
Use: Carrier for dyeing synthetic fibers; intermediate for perfume, chemicals, and pharmaceuticals.

1-(m-aminophenyl)-3-methyl-5-pyrazolone.

$NH_2C_6H_4NNC(CH_3)CH_2CO$.
Properties: Light tan paste containing approximately 45% solids.
Use: Intermediate.

aminophenylnorharman.
CAS: 219959-86-1. $C_{17}H_{13}N_3$.
Hazard: A poison by ingestion.

1-amino-2-phenylpropane.   See 2-phenylpropylamine.

α-amino-β-phenylpropionic   acid.   See phenylalanine.

aminopherase.   See transaminase.

aminophylline.   (3,7-dihydro-1,3-dimethyl-1H-purine-2,6-dione compounded with 1,2-ethanediamine (2:1)). $C_{16}H_{24}N_{10}O_4$.
Properties: White or slightly yellowish granular powder; slight ammonia odor; bitter taste. Mw 420.44.
Derivation: Prepared from theophylline and aqueous ethylenediamine.
Hazard: Cardiovascular and respiratory collapse.
Use: Small animal muscle relaxant for heaves in horses, and diuretic in dogs with congestive heart failure.

aminopicoline.   See aminomethylpyridine.

aminoplast resin.   (amino resin). A class of thermosetting resins made by the reaction of an amine with an aldehyde. The only such aldehyde in commercial use is formaldehyde, and the most important amines are urea and melamine.
Use: Molding, adhesives, laminating, textile finishes, permanent-press fabrics, wash-and-wear apparel fabrics, protective coatings, paper manufacture, leather treatment, binders for fabrics, foundry sands, graphite resistors, plaster-of-paris fortification, foam structures, and ion-exchange resins.
See dimethylol urea; methylol urea; melamine resin; urea-formaldehyde resin.

2-aminopropane.   See isopropylamine.

2-aminopropanoic acid.   See alanine.

3-aminopropanoic acid.   See β-alanine.

1-amino-2-propanol.   See isopropanolamine.

2-amino-1-propanol.   (2-aminopropyl alcohol; β-propanolamine). $CH_3CH(NH_2)CH_2OH$.
Properties: Colorless to pale yellow liquid. Both l and dl forms are available. dl-Form: fish odor. Bp 173–176C. Freely soluble in water, alcohol, ether. l-Form: refr index 1.4480–1.4495 (26C), distillation range approximately 114C at 100 mm Hg. Combustible.
Use: Organic synthesis and chemical intermediate.

3-amino-1-propanol.   (propanolamine). $H_2NCH_2CH_2CH_2OH$.
Properties: Colorless liquid. Mp 12.4C, bp 184–186C (168C), flash p 175F (79.4C), d 0.9786 (30C). Miscible with alcohol, water, acetone, and chloroform. Combustible.
Grade: 99% pure.
Hazard: Irritant to tissue.
Use: Organic intermediate.

3-aminopropionitrile. $H_2NCH_2CH_2CN$.
Properties: Colorless liquid. Bp 185C, refr index 1.44. May polymerize if stored in presence of air.
Derivation: Reaction of acrylonitrile with ammonia.
Use: Production of β-alanine and pantothenic acid.

2-aminopropyl   alcohol.   See 2-amino-1-propanol.

N-aminopropylmorpholine.   (4-(3-aminopropyl)morpholine).

$CH_2CH_2OCH_2CH_2NC_3H_6NH_2$.
Properties: Colorless liquid. D 0.9872 (20/20C), bp 224.5C, flash p 220F (104.4C) (OC), fp −15C. Soluble in water and alcohol. Combustible.
Hazard: Strongly irritant to tissue.
Use: Fiber synthesis; chemical intermediate.

n-(3-aminopropyl)-1,3-propanediamine polymer with (chloromethyl)oxirane.
CAS: 51961-45-6. $(C_6H_{17}N_3 \cdot C_3H_5ClO)_x$.
Hazard: A severe eye irritant.

(3-aminopropyl)triethylsilane.
CAS: 17887-09-1. $C_9H_{23}NSi$.
Hazard: A poison.

γ-aminopropyltrimethoxysilane.
$NH_2(CH_2)_3Si(OCH_3)_3$.
Properties: 100% active, water-white liquid. D 1.01; refr index 1.42.
Use: Glass fabric sizing, binder, adhesion promoter.

aminopterin. (4-aminofolic acid; aminopteroyl-glutamic acid). $C_{19}H_{20}N_8O_5 \cdot 2H_2$. Differs slightly in structure from folic acid and antagonizes the utilization of folic acid by the body, an antimetabolite.
Properties: Occurs as clusters of yellow needles that are soluble in aqueous sodium hydroxide solutions.
Use: Medicine, rodenticide.

aminopteroylglutamic acid. See aminopterin.

aminopurine. (2-aminopurine; 1-H-purin-2-amine; SQ 22451). $C_5H_5N_5$. An analog of adenine that pairs with cytosine.
Properties: Crystals from water.
Hazard: Mutagen.

6-aminopurine. See adenine.

2-aminopyridine. (α-pyridylamine).
CAS: 504-29-0. $C_5N_4NNH_2$.
Properties: White leaflets or large colorless crystals. Fp 58.1C; bp 210.6C. Soluble in water, alcohol, benzene, ether.
Hazard: Toxic.
Use: Intermediate for antihistamines and other pharmaceuticals.

3-aminopyridine. (β-pyridylamine).
$C_6H_4NNH_2$.
Properties: White crystals. Mp 64C; bp 250–252C. Soluble in water, alcohol, benzene, ether.
Use: Intermediate in preparation of drugs and dyestuffs.

4-aminopyridine.
CAS: 504-24-5. $C_5H_4NNH_2$.
Properties: Crystals. Mp 158.9C, bp 273.5C. Soluble in water.
Derivation: From 2-aminopyridine.
Grade: 95% (minimum).
Use: Intermediate.

3-amino-pyridine-2-carboxaldehyde.
CAS: 143621-35-6. $C_7H_9N_5S$.
Hazard: A poison.

3-aminopyridine-2-carboxaldehyde thiosemi-carbazone. See 3-amino-pyridine-2-carbo-xaldehyde.

4-aminosalicylic acid. (PASA; PAS; p-amino-salicylic acid; 4-amino-2-hydroxybenzoic acid).
CAS: 65-49-6. $NH_2C_6H_3(OH)COOH$.
Properties: White or nearly white bulky powder. Odorless or has slight acetous odor. Mp 150–151C (decomposes). Affected by light and air. Darkened solutions should not be used. Soluble in dilute sodium hydroxide and dilute nitric acid, slightly soluble in ether; practically insoluble in benzene. Water solutions decompose with evolution of carbon dioxide.

Derivation: From m-aminophenol and potassium bicarbonate, soluble under pressure.
Grade: USP.
Use: Medicine, industrial preservative.

5-aminosalicylic acid. (m-aminosalicylic acid; 5-amino-2-hydroxybenzoic acid).
CAS: 89-57-6. $NH_2C_6H_3(OH)COOH$.
Properties: White crystals, sometimes pinkish. Mp 260–280C (decomposes). Soluble in hot water or alcohol.
Derivation: From the corresponding nitrosalicylic acid by reduction.
Use: Dyes, intermediate.

α-aminosuccinamic acid. See asparagine.

l-α-aminosuccinamic acid. See l-asparagine.

aminosuccinic acid. See aspartic acid.

dl-aminosuccinic acid. See dl-aspartic acid.

amino sugar. Any of a class of monosaccharides in which an alcoholic hydroxyl group has been replaced by an amino group.

aminotetrazole. (5-aminotetrazole; 5-amino-1h-tetrazole; 1h-tetrazol-5-amine).
CAS: 4418-61-5. CH_3N_5.
Properties: Crystals. Mp: 206C (decomp), mw: 85.09.
Hazard: Moderately toxic. An unstable material; explodes with KOH.

5-aminotetrazole. See aminotetrazole.

amino-terminal residue. The amino acid residue in a polypeptide chain which has a free alpha-amino group.

5-amino-1h-tetrazole. See aminotetrazole.

2-aminothiazole. (2-thiazylamine).
$SCHCHNCNH_2$.
Properties: Light-yellow crystals. Mp 90C, distills at 3 mm without decomposition. Slightly soluble in cold water, alcohol, and ether; soluble in hot water and dilute mineral acids.
Derivation: Chlorination of vinyl acetate and condensation with thiourea.
Use: Intermediate in synthesis of sulfathiazole, medicine (thyroid inhibitor).

α-amino-β-thiolpropionic acid. See cysteine.

aminothiourea. See thiosemicarbazide.

aminotoluene. See o-toluidine; m-toluidine; p-toluidine; benzylamine.

aminotransferases. Enzymes that catalyze the transfer of amino groups from a-amino to a-keto acids; commonly called transaminases.

6-amino-*s*-triazine-2,4-diol. See ammelide.

aminotriazole. (amitrole; 3-amino-1,2,4-triazole; 3-amino-1H-1,2,4-triaole; 1,2,4-riazol-3-ylamine; ata).
CAS: 61-82-5. $C_2H_4N_4$. A powerful plant growth suppressant and cotton defoliant that is readily absorbed by leaves and roots. It inhibits many aspects of growth and differentiation including cell division in the primary root meristems in some plants.
Properties: Colorless, crystalline solid; soluble in water and ethanol.
Hazard: Toxic; carcinogen.
Use: Herbicide banned in the U.S. in 1971.

3-amino-1,2,4-triazole. (amitrole).
CAS: 61-82-5.

NHNC(NH$_2$)NCH.
Properties: White, crystalline solid. Mp 156–159C. Soluble in water and alcohol.
Hazard: Use on food crops not permitted. Questionable carcinogen.
Use: Herbicide, defoliant.

4-amino-3,5,6-trichloropicolinic acid. See picloram.

3-(3-amino-2,4,6-triiodophenyl)-2-ethylpropanoic acid. See iopanoic acid.

aminourea hydrochloride. See semicarbazide hydrochloride.

Amino Vital Fast Charge. Pharmaceutical grade amino acids drink powders.

Amino Vital Mix and Shake. Pharmaceutical grade amino acids drink powders.

Amino Vital RTD Drink. Pharmaceutical grade amino acids drink powders.

"Aminox" *[Chemtura].* TM for a low-temperature reaction product of diphenylamine and acetone.
Properties: Light-tan powder. D 1.13, mp 85–95C. Soluble in acetone, benzene, and ethylene dichloride; insoluble in water and gasoline.
Use: Antioxidant for nylon and light-colored rubber products.

aminoxylene. See xylidine.

amiton. (*s*-[2-(diethylamino)ethyl]phosphorothioic acid *o,o*-diethyl ester; *o,o*-diethyl *s*-(β-diethylamino)ethylphosphorothiolate;

2-diethoxyphosphorylsulfanyl-*N,N*-diethylethanamine). $C_{10}H_{24}NO_3PS$
Hazard: Extremely toxic contact insecticide and miticide; a cholinesterase inhibitor.

amitrole. Generic name for 3-amino-1,2,4-triazole.

amlodipine.
CAS: 88150-42-9. $C_{20}H_{25}ClN_2O_5$.
Hazard: Human systemic effects.

amlure. See propyl 1,4-benzodioxan-2-carboxylate.

Ammate. Ammonium sulfamate in various grades.
Hazard: See ammonium sulfamate.
Use: Nonselective herbicide.

ammelide. (6-amino-*s*-triazine-2,4-diol; cyanuramide).

NC(OH)NC(OH)NC(NH$_2$).
Properties: Crystalline solid. Mp (decomposes). Insoluble in alcohol; slightly soluble in hot water. Similar to melamine and suggested for melamine-type (amino) resins.

ammeline. (4,6-diamino-*s*-triazine-2-ol; cyanurdiamide).

NC(OH)NC(NH$_2$)NC(NH$_2$.
Properties: Crystalline solid. Mp decomposes. Insoluble in water and alcohol. A compound similar to melamine.
Use: In melamine-type resins and in special high-temperature lubricants.

ammine. A coordination compound formed by the union of ammonia with a metallic substance in such a way that the nitrogen atoms are linked directly to the metal. Note the distinction from amines in which the nitrogen is attached directly to the carbon atom.
See cobaltammine; coordination compound.

ammonia, anhydrous.
CAS: 7664-41-7. NH_3.
Properties: Colorless gas (or liquid); sharp, intensely irritating odor; lighter than air; easily liquefied by pressure. Bp −33.5C, fp −77C, vap press of liquid 8.5 atm (20C), sp vol 22.7 cu ft/lb (70C), d (liquid) 0.77 at 0C and 0.6819 at bp. Very soluble in water, alcohol, and ether. Autoign temp 1204F (650C). Combustible. *Note:* Ammonia is the first complex molecule to be identified in interstellar space. It has been observed in galactic dust clouds in the Milky Way and is believed to constitute the rings of the planet Saturn.

Derivation: From synthesis gas, a mixture of carbon monoxide, hydrogen, carbon dioxide, and nitrogen (from air) obtained by steam reforming or by partial combustion of natural gas (U.S.), or from the action of steam on hot coke (Haber–Bosch process). The latter method is used in South Africa. After removal of the carbon oxides, the gas composition is adjusted to a ratio of three parts H_2 to one part N_2 and passed to the synthesis unit over a catalyst at pressures of about 300 atm and temperature of approximately 500C. The catalyst most widely used is produced by fusion of iron oxide (Fe_3O_4) containing aluminum oxide and potassium oxide as promoters, followed by reduction of the oxide. Chemisorption of the nitrogen on the catalyst surface is the rate-controlling step.

Other methods include use of refinery off-gases, coke-oven gas, electrolytic hydrogen, and calcium cyanimide. Ammonia has been made experimentally using solar energy to activate the reaction

$$N_2 + 3H_2O \overset{h\nu}{\underset{TiO_2}{\rightarrow}} 2NH_3 + 1.5O_2.$$

Ammonia is formed as an end product of animal metabolism by decomposition of uric acid.

Grade: Commercial 99.5%, refrigerant 99.97%.

Hazard: Inhalation of concentrated fumes may be fatal. Moderate fire risk, explosive limits in air 16–25%. Forms explosive compounds in contact with silver or mercury. Eye damage and upper respiratory tract irritant.

Use: Fertilizers, either as such or in the form of compounds, e.g., ammonium nitrate; manufacture of nitric acid, hydrazine hydrate, hydrogen cyanide, urethane, acrylonitrile, and sodium carbonate (by Solvay process); refrigerant; nitriding of steel; condensation catalyst; synthetic fibers; dyeing; neutralizing agent in petroleum industry; latex preservative; explosives; nitrocellulose; ureaformaldehyde; nitroparaffins; melamine; ethylenediamine; sulfite cooking liquors; fuel cells; rocket fuel; yeast nutrient; developing diazo films.

See ammonium hydroxide; Haber, Fritz; synthesis gas.

ammonia, aromatic spirits. A mixture of 10% of ammonia in alcohol. Strong, suffocating odor.

Hazard: Irritant to mucous membranes. Flammable, keep tightly sealed.

Use: Medicine (respiratory stimulant).

ammonia-soda process. See Solvay process.

ammoniated glycyrrhizin.
CAS: 1407-03-0.
Properties: From roots of *Glycyrrhiza glabra*.
Use: Food additive.

ammoniated mercury chloride. See mercury, ammoniated.

ammoniated ruthenium oxychloride. See ruthenium red.

ammoniated superphosphate. Fertilizer produced by mixing ammonia with superphosphate in the ratio of 5 parts to 100.

ammonia water. (ammonia solution; ammonium hydroxide; aqueous ammonia; household ammonia; aqua ammonia; aqua ammoniae; azanium hydroxide).
CAS: 1336-21-6. H_5NO. The hydroxy salt of ammonium ion.
Properties: Highly volatile, extremely irritant liquid.
Derivation: Formed when ammonia reacts with water molecules in solution.
Hazard: Irritant; poison; burns.
Use: All-purpose cleaner; cleaner used in dry cleaning.

ammonio-cupric sulfate. See copper sulfate, ammoniated.

ammonio-ferric oxalate. See ferric ammonium oxalate.

ammonio-ferric sulfate. See ferric ammonium sulfate.

ammonium acetate.
CAS: 631-61-8. $NH_4(C_2H_3O_2)$.
Properties: White, deliquescent, crystalline mass. Mp 114C, d 1.073. Soluble in water and alcohol. Mp 114C, d 1.073. Combustible.
Derivation: By the interaction of glacial acetic acid and ammonia.
Use: Reagent in analytical chemistry, drugs, textile dyeing, preserving meats, foam rubbers, vinyl plastics, explosives.

ammonium acid carbonate. See ammonium bicarbonate.

ammonium acid fluoride. See ammonium bifluoride.

ammonium acid methanearsonate. $CH_3AsO(OH)ONH_4$. A postemergent herbicide, available as "Ansar" 157 *[KMG-Bernuth]*, a clear solution containing 9.54% As.

ammonium acid phosphate. See ammonium phosphate, monobasic.

ammonium alginate. (ammonium polymannuronate). $C_6H_7O_6 \cdot NH_4$. A hydrophilic, colloid.
Properties: Filamentous, grainy, granular or powder, colorless or slightly yellow, may have a slight characteristic smell and taste. Slowly soluble in water forming a viscous solution. Insoluble in alcohol.
Grade: Technical, FCC.

Use: Thickening agent and stabilizer in food products.
See algin.

ammonium alum.   See aluminum ammonium sulfate.

ammonium aluminum chloride.   See aluminum ammonium chloride.

ammonium arsenate.
CAS: 7784-44-3. $(NH_4)_2HAsO_4$.
Properties: White crystals or powder efflorescing in air with loss of ammonia. D 1.99. Soluble in water; decomposes in hot water.

ammonium   aurin   tricarboxylate.
$C(C_6H_3OHCOONH_4)_2(C_6H_3(O)COONH_4)$.
Forms colored lakes with aluminum, chromium, iron, and beryllium.

ammonium benzenesulfonate.   (ammonium sulfonate). $C_6H_5SO_3NH_4$.
Properties: Mp 271C (decomposes), d 1.34.
Grade: 35% solution in kerosene.
Hazard: Flammable.

ammonium benzoate. $C_6H_5COONH_4$.
Properties: White crystals or powder. Soluble in water, alcohol, and glycerol. Decomposes at 198C, d 1.260, sublimes at 160C.
Use: Medicine, latex preservative.

ammonium biborate.   See ammonium borate.

ammonium bicarbonate.   (ammonium acid carbonate; ammonium hydrogen carbonate).
CAS: 1066-33-7. NH_4HCO_3.
Properties: White crystals. D 1.586, mp (decomposes at 36–60C). Soluble in water; insoluble in alcohol. Noncombustible.
Derivation: By heating ammonium hydroxide with an excess of carbon dioxide and evaporation.
Impurities: Ammonium carbonate.
Grade: Technical, CP, FCC.
Hazard: Evolves irritating fumes on heating to 35C.
Use: Production of ammonium salts, dyes; leavening agent for cookies, crackers, cream-puff doughs; fire-extinguishing compounds; pharmaceuticals; degreasing textiles; blowing agent for foam rubber; boiler scale removal; compost treatment.

ammonium bichromate.   See ammonium dichromate.

ammonium bifluoride.   (ammonium acid fluoride; ammonium hydrogen fluoride).
CAS: 1341-49-7. NH_4HF_2.
Properties: White crystals. Deliquescent, d 1.211. Soluble in water and alcohol.
Derivation: Action of ammonium hydroxide on hydrofluoric acid with subsequent crystallization.

Hazard: Corrosive to skin.
Use: Ceramics, chemical reagent, etching glass (white acid), sterilizer for brewery, dairy, and other equipment; electroplating processing beryllium, laundry sour.

ammonium binoxalate. $(NH_4)HC_2O_4 \cdot H_2O$.
Properties: Colorless crystals. D 1.556, decomposes on heating. Soluble in water.
Derivation: Action of ammonium hydroxide on oxalic acid with subsequent crystallization.
Use: Analytical reagent, ink removal from fabrics.

ammonium biphosphate.   See ammonium phosphate, monobasic.

ammonium bisulfate.   (acid ammonium sulfate; ammonium hydrogen sulfate). NH_4HSO_4.
Properties: Colorless, deliquescent powder. Mp 145C; d 1.79. Soluble in water; insoluble in acetone and alcohol.
Use: Catalyst in organic synthesis, hair wave formulation.

ammonium bisulfide.   See ammonium sulfide.

ammonium bitartrate.   (acid ammonium tartrate). $(NH_4)HC_4H_4O_6$.
Properties: White crystals. D 1.636. Soluble in water, acids, and alkalies; insoluble in alcohol.
Derivation: By the action of ammonium hydroxide on tartaric acid.
Use: Baking powder.

ammonium boranecarboxylate.
CAS: 74861-59-9. $CH_2BO_2 \cdot H_4N$.
Hazard: Moderately toxic.

ammonium borate.   (ammonium biborate). $NH_4HB_4O_7 \cdot 3H_2O$.
Properties: Colorless crystals, efflorescent with loss of ammonia. Soluble in water. D 2.38–2.95. Noncombustible.
Derivation: Action of ammonium hydroxide on boric acid with subsequent crystallization.
Hazard: Evolves irritating fumes, especially when heated.
Use: Fireproofing compounds, electrical condensers, herbicide.

ammonium bromide. NH_4Br.
Properties: Colorless crystals or yellowish white powder, soluble in water and alcohol, somewhat hygroscopic. D 2.43, mp sublimes. Noncombustible.
Derivation: Action of hydrobromic acid on ammonium hydroxide with subsequent crystallization.
Grade: Technical, pure, CP, NF.
Use: Precipitating silver salts for photographic plates, medicine (for its bromide ion), analytical chemistry, process engraving, textile finishing, fire retardant, anticorrosive agents.

ammonium cadmium bromide. See cadmium ammonium bromide.

ammonium cadmium chloride. $4NH_4Cl \cdot CdCl_2$.
Properties: Colorless, rhombic crystals. Mw 397.3, d 2.01. Soluble in water.
Hazard: A poison and confirmed carcinogen.

ammonium caprylate. (octanoic acid ammonium salt). $C_8H_{19}NO_2$.
Properties: Hygroscopic crystals. Decomposed at room temperature, mp approximately 75C. Hydrolyzes readily. Soluble in alcohol and glacial acetic acid; partly soluble in acetone; insoluble in benzene.
Use: Pesticide, photographic emulsions, chemical intermediate.

ammonium carbamate. $NH_4CO_2NH_2$.
Properties: White, rhombic, crystalline powder; very volatile; forms urea upon heating. Soluble in water and alcohol. Sublimes at 60C, decomposes in air to evolve ammonia.
Derivation: Interaction of dry ammonia gas and carbon dioxide from ammonia liquor with ammonia and ammonium carbonate.
Grade: Technical.
Hazard: Evolves irritating fumes when heated.
Use: Fertilizer.

ammonium carbazotate. See ammonium picrate.

ammonium carbonate. (crystal ammonia; ammonium sesquicarbonate; hartshorn).
CAS: 506-87-6. $(NH_4)HCO_3 \cdot (NH_4)CO_2NH_2$. A mixture of ammonium acid carbonate and ammonium carbamate.
Properties: Colorless, crystalline plates or white powder; unstable in air, being converted into the bicarbonate. Strong odor of ammonia, sharp ammoniacal taste. Soluble in water, decomposes in hot water, yielding ammonia and carbon dioxide. Noncombustible.
Derivation: Ammonium salts are heated with calcium carbonate.
Method of purification: Sublimation.
Grade: Technical, lumps, cubes, powder, CP, NF, FCC.
Hazard: Evolves irritating fumes when heated.
Use: Ammonium salts, medicine (expectorant), baking powders, smelling-salts, fire-extinguishing compounds, pharmaceuticals, textiles (mordant), fermentation accelerator in wine manufacture, organic chemicals, ceramics, washing wool.

ammonium ceric nitrate. See ceric ammonium nitrate.

ammonium chlorate.
CAS: 10192-29-7. NH_4ClO_3.

Properties: Colorless or white crystals, water soluble.
Derivation: Reaction of ammonium chloride with sodium chlorate in solution.
Hazard: Spontaneous chemical reaction with reducing agents. Powerful oxidizer. When contaminated with combustible materials, it can ignite. Shock sensitive; can detonate when exposed to heat or vibration, especially when contaminated.
Use: Explosives.

ammonium chloride. (sal ammoniac).
CAS: 12125-02-9. NH_4Cl.
Properties: White crystals; cool, saline taste; somewhat hygroscopic. Sublimes at 350C, d 1.54. Soluble in water, glycerol; slightly soluble in alcohol.
Derivation: (1) As a by-product of the ammonia-soda process, (2) reaction of ammonium sulfate and sodium chloride solutions.
Grade: Technical (lumps or granulated), CP, USP, FCC.
Hazard: Eye and upper respiratory tract irritant.
Use: Dry batteries, mordant (dyeing and printing), soldering flux, manufacturing of various ammonia compounds, fertilizer, pickeling agent in zinc coating and tinning, electroplating, washing powders, melt-retarding snow treatment, production of urea-formaldehyde resins and adhesives, bakery products.

ammonium chloroosmate. See ammonium hexachloroosmate.

ammonium chloroplatinate. See ammonium hexachloroplatinate.

ammonium chloroplatinite. (platinous ammonium chloride; platinum ammonium chloride).
CAS: 16919-58-7. $(NH_4)_2PtCl_4$.
Properties: Dark ruby-red crystals. Decomposes 140–150C. Soluble in water; insoluble in alcohol.
Use: In photography.

ammonium chromate. (chromic acid, diammonium salt).
CAS: 7788-98-9. $(NH_4)_2CrO_4$.
Properties: Yellow crystals. D 1.866, mp (decomposes). Soluble in cold water; insoluble in alcohol.
Derivation: Addition of ammonium hydroxide to a solution of ammonium bichromate, recovery by crystallization.
Impurities: Dichromates.
Grade: Technical, CP.
Hazard: Toxic by inhalation, strong irritant. Confirmed carcinogen.
Use: Mordant in dyeing; photography (sensitizer for gelatin coatings); analytical reagent; catalyst; corrosion inhibitor.

ammonium chrome alum. See chromium ammonium sulfate.

ammonium chromium sulfate. See chromium ammonium sulfate.

ammonium citrate, dibasic. $(NH_4)_2HC_6H_5O_7$.
Properties: White granules, soluble in water, very slightly soluble in alcohol.
Use: Pharmaceuticals, rustproofing, cotton printing, plasticizer, analytical reagent in determination of phosphate in fertilizer.

ammonium cobaltous phosphate. See cobaltous ammonium phosphate.

ammonium decaborate. See ammonium pentaborate.

ammonium dichromate. (ammonium bichromate).
CAS: 7789-09-5. $(NH_4)_2Cr_2O_7$.
Properties: Orange needles. D 2.152 (25C), decomposes with slight heating. Soluble in water and alcohol.
Derivation: Action of chromic acid on ammonium hydroxide with subsequent crystallization.
Hazard: Dusts and solutions are toxic, irritating to eyes and skin; dangerous fire risk. Strong oxidizing agent may explode in contact with organic materials. TLV: 0.05 mg(Cr)/m^3; Confirmed human carcinogen.
Use: Mordant for dyeing, pigments, manufacturing of alizarin, chrome alum, oil purification, pickling, manufacture of catalysts, leather tanning, synthetic perfumes, photography, process engraving and lithography (sensitizer for photo-chemical insolubilization of albumin, etc.), chromic oxide, pyrotechnics.

ammonium dihydrogen phosphate. See ammonium phosphate monobasic.

ammonium dimethyldithiocarbamate. $(CH_3)_2NCS_2NH_4$.
Properties: Yellow crystals. Soluble in water, decomposes in air.
Grade: 42% solution in water.
Use: Fungicide.

ammonium dinitro-*o*-cresolate.
Hazard: Dangerous in contact with combustible materials. Strong oxidizing agent. Flammable.
Use: Herbicide.

ammonium dithiocarbamate. $CH_6N_2S_2$.
Properties: Yellow crystals. Mp 99C (decomposes), d 1.45. Soluble in water.
Derivation: Reaction of ammonia with carbon disulfide.
Use: Organic synthesis, especially heterocyclic compounds; analytical reagent.

ammonium-ferric-cyano-ferrate(II). See ferric ammonium ferrocyanide.

ammonium fluoride.
CAS: 12125-01-8. NH_4F.
Properties: White crystals. D 1.31, decomposed by heat. Soluble in cold water.
Derivation: Interaction of ammonium hydroxide and hydrofluoric acid with subsequent crystallization.
Method of purification: Recrystallization.
Grade: Technical, CP.
Hazard: Corrosive to tissue.
Use: Fluorides, analytical chemistry, antiseptic in brewing, etching glass, textile mordant, wood preservation, mothproofing.

ammonium fluoroberyllate.
CAS: 14874-86-3. $BeF_4 \cdot 2H_4N$.
Hazard: A poison by ingestion and inhalation.

ammonium fluosilicate. (ammonium silicofluoride).
CAS: 16919-19-0. $(NH_4)_2SiF_6$.
Properties: White, crystalline powder. D 2.01. Soluble in alcohol and water.
Hazard: Strong irritant to eyes and skin.
Use: Laundry sours, mothproofing, disinfectant in brewing industry, glass etching, light metal casting, electroplating.

ammonium formate. $HCOONH_4$.
Properties: Deliquescent, crystalline powder. D 1.26; mp 115C. Soluble in water and alcohol.
Derivation: Reaction of formic acid with ammonia.
Use: Analytical chemistry (metal precipitant).

ammonium gluconate. $NH_4C_6H_{11}O_7$.
Properties: White powder. Soluble in water, insoluble in alcohol. Optical rotation +11.6 (in water).
Derivation: From gluconic acid by neutralization with ammonia.
Use: Emulsifying agent for cheese and salad dressings.

ammonium glutamate. (monoammonium glutamate).
See sodium glutamate.

ammonium hexachloroosmate. (ammonium chlorosmate; osmium ammonium chloride). $(NH_4)_2OsCl_6$.
Properties: Red powder. Contains 43.5% osmium. Soluble in alcohol and water.
Hazard: See osmium.

ammonium hexachloroplatinate. (ammonium chloroplatinate; platinic-ammonium chloride; platinic sal ammoniac; platinum ammonium chloride). $(NH_4)_2PtCl_6$.
Properties: Orange-red crystals or yellow powder. D 3.06, mp (decomposes). Slightly soluble in water; insoluble in alcohol.
Grade: Technical, CP.
Use: Plating, platinum sponge.

ammonium hexafluorogermanate. $(NH_4)_2$ GeF_6.
Properties: White crystalline solid. D 2.564, mp 380C (sublimes). Soluble in cold water; insoluble in alcohol.

ammonium hexafluorosilicate. (cryptohalite; diammonium hexafluorosilicate; ammonium fluosilicate; ammonium silicofluoride; diazanium hexafluorosilicon(2-)).
CAS: 16919-19-0. $F_6H_8N_2Si$.
Properties: Odorless, crystalline powder.
Hazard: Very toxic; fatal.
Use: In pesticides.

ammonium hexanitratocerate. See ceric ammonium nitrate.

ammonium hydrate. See ammonium hydroxide.

ammonium hydrogen carbonate. See ammonium bicarbonate.

ammonium hydrogen fluoride. See ammonium bifluoride.

ammonium hydrogen sulfate. See ammonium bisulfate.

ammonium hydrosulfide. See ammonium sulfide.

ammonium hydroxide. (ammonia solution; aqua ammonia; ammonium hydrate). NH_4OH.
Properties: Colorless liquid, strong odor. Concentration of solutions range up to approximately 30% ammonia.
Grade: Technical, CP, 16%, 20%, 26%, NF (strong), FCC.
Hazard: Liquid and vapor extremely irritating, especially to eyes.
Use: Textiles, manufacture rayon, rubber, fertilizers, refrigeration, condensation polymerization, photography (development of latent images), pharmaceuticals, ammonia soaps, lubricants, fireproofing wood, ink manufacture, explosives, ceramics, ammonium compounds, saponifying fats and oils, organic synthesis, detergents, food additives, household cleanser.

ammonium hypophosphite. $NH_4N_2PO_2$.
Properties: Deliquescent crystals or white powder, decomposes when heated. Soluble in water and alcohol.
Hazard: Evolves flammable and toxic fumes on heating.
Use: Catalyst in nylon manufacture.
See phosphine.

ammonium ichthosulfonate. See ichthammol.

ammonium iodate. NH_4IO_3.
Properties: White, odorless, granular powder.
Hazard: Fire risk in contact with organic materials.
Use: Oxidizing agent.

ammonium iodide. NH_4I.
Properties: White, hygroscopic crystals or powder. D 2.56, mp sublimes with decomposition. Soluble in water or alcohol. Affected by light.
Use: Iodides, medicine (expectorant), photography.

ammonium laurate, anhydrous. $C_{11}H_{23}COONH_4$.
Properties: Tan, waxlike material; free from ammonia odor. D 0.88 (25C), pH (5% dispersion) 7.6–7.8, mp 42–56C, neutralization value 120–125. Soluble in ethanol, methanol, cottonseed oil, mineral oil, (hot) in naphtha, toluene, and vegetable oil. Combustible.
Use: Production of oil-in-water emulsions with high oil content; cosmetics.

ammonium lignin sulfonate. See lignin sulfonate.

ammonium linoleate. $C_{17}H_{31}COONH_4$.
Properties: Yellow paste with ammoniacal odor. D 1.1, pH (5% dispersion) 9.5–9.8, total solids 82%. Soluble in water, ethanol and methanol; emulsifies in naphtha, toluene, mineral oil, and vegetable oil. Combustible.
Grade: Technical, 80%.
Use: Emulsifying agent for oils, waxes, and hydrocarbon solvents; surface tension reducer; detergent; water-repellent finishes.

ammonium magnesium arsenate. See magnesium ammonium arsenate dihydrate.

ammonium metavanadate. (ammonium vanadate). NH_4VO_3.
Properties: White crystals. D 2.326, mp decomposes at 210C. Insoluble in saturated ammonium chloride solution; slightly soluble in cold water. Nonflammable.
Derivation: Alkali solutions of vanadium pentoxide and precipitation with ammonium chloride.
Use: For catalyst as vanadium pentoxide, dyes, varnishes, indelible inks, drier for paints and inks, photography, analytical reagent.

ammonium molybdate. (molybdic acid 85%). $(NH_4)_6Mo_7O_{24} \cdot 4H_2O$.
Properties: White, crystalline powder. D 2.27, mp decomposes. Soluble in water; insoluble in alcohol. Nonflammable.
Derivation: Dissolving molybdenum trioxide in aqueous ammonia.
Grade: Technical, CP, reagent (contains 85% MoO_3).

Hazard: Irritant.

Use: Analytical reagent, pigments, catalyst for dehydrogenation and desulfurization in petroleum and coal technology, production of molybdenum metal, source of molybdate ions.

ammonium-12-molybdophosphate.
(ammonium phosphomolybdate). $(NH_4)_3PO_4 \cdot 12MoO_3 \cdot 3H_2O$, or $(NH_4)_3PMo_{12}O_{40} \cdot xH_2O$.
Properties: Yellow, crystalline powder. Soluble in alkali; insoluble in alcohol and acids; very slightly soluble in water. Nonflammable.
Derivation: Interaction of ammonium molybdate with phosphoric and nitric acids.
Grade: 91% MoO_3.
Use: Reagent, ion exchange columns, photographic additives, imparting water resistance.

ammonium-12-molybdosilicate. (ammonium silicomolybdate). $(NH_4)_4SiMo_{12}O_{40} \cdot xH_2O$.
Properties: Crystalline, yellow granules; thermally stable. Only slightly soluble in water, ethanol, and ethyl acetate. Nonflammable.
Grade: Technical, reagent.
Use: Catalysts, reagents, in atomic energy as precipitants and inorganic ion-exchangers, in photographic processes as fixing agents and oxidizing agents, in plating processes as additive, and in plastics, adhesives, and cement industries for imparting water resistance.

ammonium nickel chloride. See nickel ammonium chloride.

ammonium nickel sulfate. See nickel ammonium sulfate.

ammonium nitrate. (Norway saltpeter). NH_4NO_3.
Properties: Colorless crystals. D 1.725, mp 169.6C, bp (decomposes at 210C with evolution of nitrous oxide). Soluble in water, alcohol, and alkalies.
Derivation: Action of ammonia vapor on nitric acid.
Grade: Usually expressed in percent of nitrogen, as 20.5% N, 33.5% N. FGAN is a fertilizer grade, prilled and usually coated with kieselguhr. Also available as an 83% solution. A temperature-stabilized grade is also available that inhibits breakdown of prills due to crystalline changes.
Hazard: May explode under confinement and high temperatures, but not readily detonated. Ventilate well. To fight fire, use large amounts of water. The material must be kept as cool as possible and removed from confinement and flooded with water in event of fire. Explodes more readily if contaminated with combustibles. Strong oxidizing agent. May be made resistant to flame and detonation by proprietary process involving addition of 5–10% ammonium phosphate.
Use: Fertilizer, explosives especially as prills/oil mixture, pyrotechnics, herbicides and insecticides,

manufacture of nitrous oxide, absorbent for nitrogen oxides, ingredient of freezing mixtures, oxidizer in solid rocket propellants, nutrient for antibiotics and yeast, catalyst.
See explosives.

ammonium nitrate-carbonate mixtures.
See calcium ammonium nitrate.

ammonium nitroso-β-phenylhydroxylamine.
See cupferron.

ammonium oleate. $C_{17}H_{33}COONH_4$.
An ammonium soap.
Properties: Yellow to brownish ointmentlike mass; ammonia odor. Decomposes on heating. Soluble in water and hot alcohol. Combustible.
Use: Emulsifying agent, cosmetics.

ammonium oxalate. $(NH_4)_2C_2O_4 \cdot H_2O$.
Properties: Colorless crystals. D 1.502, decomposed by heat. Soluble in water.
Use: Analytical chemistry, safety explosives, manufacture of oxalates, rust and scale removal.

ammonium palmitate. $C_{15}H_{31}COONH_4$. An ammonium soap.
Properties: Yellowish granules. Mp 20C. Soluble in water, hot alcohols, and benzene. Combustible.
Derivation: Reaction of palmitic acid and ammonium carbonate.
Use: Thickening agent for petroleum-derived solvents, lubricants, etc.; waterproofing.

ammonium paratungstate. See ammonium tungstate.

ammonium pentaborate. (ammonium decaborate). $(NH_4)_2B_{10}O_{16} \cdot 8H_2O$.
Properties: Crystals or powder. Soluble in water.
Use: Intermediate for boron chemicals; as a power-level control in atomic submarines.

ammonium pentadecafluorooctanoate. See ammonium perfluorooctanoate.

ammonium perchlorate. (AP; APC). NH_4ClO_4.
Properties: White crystals. D 1.95, mp (decomposes on heating). Soluble in water.
Derivation: Interaction of ammonium hydroxide, hydrochloric acid, and sodium chlorate. Recovery by crystallization.
Hazard: Strong oxidizing agent; ignites violently with combustibles. Shock sensitive; may explode when exposed to heat or by spontaneous chemical reaction. Sensitive, high explosive when contaminated with reducing materials. Skin irritant.
Use: Explosives, pyrotechnics, analytical chemistry, etching and engraving agent, smokeless rocket and jet propellant.

ammonium   perfluorooctanoate. (ammonium pentadecafluorooctanoate).
CAS: 3825-26-1. $C_8F_{15}COONH_4$.
Properties: White powder. Mw 431, d 0.6–0.7 g/cc, bp 125C (sublimes). Highly soluble in water.
Hazard: Toxic by inhalation and skin contact. Liver damage. Possible carcinogen.
Use: For polymerization of fluorinated monomers.

ammonium permanganate. NH_4MnO_4.
Properties: Crystal or powder form, having a metallic sheen in rich violet-brown or dark-purple shades, soluble in water.
Hazard: May explode on shock or on exposure to heat. Toxic by ingestion or by inhalation of dust or fume. Powerful oxidizer.

ammonium peroxychromate. $(NH_4)_3CrO_2$.
Properties: Red-brown crystals. Mp @ 40C (decomp), bp: explodes @ 50C, mw 234.1.
Hazard: A poison. Confirmed carcinogen. Moderately flammable by chemical reaction with reducing agents. A powerful oxidizer. Moderately explosive when heated.

ammonium perrhenate. NH_4ReO_4
Properties: Colorless, weakly oxidizing solid. Stable to heat; decomposes at 365C. Moderately soluble in hot water; slightly soluble in cold water.
Derivation: Liquid ion-exchange.
Hazard: Moderate fire risk in contact with reducers.

ammonium persulfate.
CAS: 7727-54-0. $(NH_4)_2S_2O_8$.
Properties: White crystals. D 1.98; mp decomposes. Strong oxidizer. Water soluble.
Derivation: Electrolysis of concentrated solution of ammonium sulfate. Recovered by crystallization.
Hazard: Fire risk in contact with reducers.
Use: Oxidizer, bleaching agent; photography; etchant for printed circuit boards, copper; electroplating; manufacture of other persulfates; deodorizing and bleaching oils; aniline dyes; preserving foods; depolarizer in batteries; washing infected yeast.

ammonium phosphate, dibasic. (ammonium phosphate, secondary; diammonium hydrogen phosphate; diammonium phosphate; DAP). $(NH_4)_2HPO_4$.
Properties: White crystals or powder. D 1.619, mildly alkaline in reaction. Soluble in water; insoluble in alcohol. Noncombustible.
Derivation: Interaction of ammonium hydroxide and phosphoric acid in proper proportions.
Grade: Technical, CP, fertilizer, feed, dentifrice, highly purified for phosphors, FCC.
Use: Flameproofing of wood, paper, and textiles; coating vegetation to retard forest fires; to prevent afterglow in matches and smoking of candle wicks; fertilizer (high analysis phosphate type); plant nutrient solutions; manufacture of yeast, vinegar, and

bread improvers; feed additive; flux for soldering tin, copper, brass, zinc; purifying sugar; in ammoniated dentifrices; halophosphate phosphors.

ammonium phosphate, hemibasic.
$NH_4H_2PO_4 \cdot H_3PO_4$.
Properties: White, crystalline material; somewhat hygroscopic. Strongly acid in reaction. Soluble in water. Noncombustible.
Use: Nutrient for truck gardens, yeast food, buffer for adjustment of pH values, metal cleaning.

ammonium phosphate, monobasic. (ammonium acid phosphate; ammonium biphosphate; ammonium dihydrogen phosphate; ammonium phosphate, primary). $NH_4H_2PO_4$.
Properties: Brilliant white crystals or powder. D 1.803. Mildly acid in reaction. Moderately soluble in water. Noncombustible.
Derivation: Interaction of phosphoric acid and ammonia in proper proportions.
Grade: Technical, CP, FCC, single crystals.
Use: Fertilizers; flameproofing agent; to prevent afterglow in matches; plant nutrient solutions; manufacturing of yeast, vinegar, yeast foods, and bread improvers; food additive; analytical chemistry.

ammonium phosphite. (neutral ammonium phosphite). $(NH_4)_2HPO_3 \cdot H_2O$.
Properties: Colorless, crystalline mass. Hygroscopic. Soluble in water.
Grade: Technical.
Use: Chemical (reducing agent), lubricating grease (corrosion inhibitor).

ammonium phosphomolybdate. See ammonium-12-molybdophosphate.

ammonium phosphotungstate. (ammonium phosphowolframate). $2(NH_4)_3PO_4 \cdot 24WO_3 \cdot xH_2O$.
Properties: White powder. Soluble in alkali, insoluble in acid, sparingly soluble in water.
Derivation: Interaction of ammonium tungstate, ammonium phosphate, and nitric acid.
Use: Chemical reagent, ion-exchange.

ammonium phosphowolframate. See ammonium phosphotungstate.

ammonium picrate. (ammonium carbazoate; ammonium picronitrate).
CAS: 131-74-8. $C_6H_2(NO_2)_3ONH_4$.
Properties: Yellow crystals. D 1.72, mp (decomposes). Slightly soluble in water and alcohol.
Hazard: A high explosive when dry, flammable when wet.
Use: Pyrotechnics, explosive compositions.

ammonium polymannuronate. See ammonium alginate.

ammonium polyphosphate. See ureaammonium polyphosphate; Poly-N.

ammonium polysulfide. $(NH_4)_2S_x$.
  Properties: Known only in solution. Yellow, unstable; H_2S odor. Decomposed by acids with evolution of hydrogen sulfide.
  Derivation: Passing hydrogen sulfide into 28% ammonium hydroxide and dissolving an excess of sulfur in the resulting solution.
  Hazard: Evolves toxic and flammable gas on contact with acids.
  Use: Analytical reagent, insecticide spray.

ammonium potassium hydrogen phosphate.
  Use: Food additive.

ammonium reineckate.   See Reinecke salt.

ammonium ricinoleate. $C_{17}H_{32}OHCOONH_4$.
  Properties: White paste. Combustible.
  Grade: Technical.
  Use: Detergent, emulsifying agent.

ammonium saccharin.
  CAS: 6381-61-9. $C_7H_8N_2O_3S$.
  Properties: White crystals or crystalline powder; intense sweet taste. Sol in water.
  Hazard: A severe eye irritant.
  Use: Food additive.

ammonium salts.   Salts formed by neutralization of ammonium hydroxide with acids. Usually white and water soluble; usually decomposed by heat into ammonia and the corresponding acid, which may also decompose. All ammonium salts liberate ammonia (NH_3) when heated with a strong base, e.g., sodium hydroxide or calcium hydroxide.

ammonium selenate. $(NH_4)_2SeO_4$.
  Properties: Colorless crystals. D 2.194. Soluble in water, insoluble in alcohol.
  Use: Mothproofing agent.

ammonium selenite. $(NH_4)_2SeO_3 \cdot H_2O$.
  Properties: Colorless or slightly reddish crystals. Keep away from dust or light. Soluble in water.
  Grade: Technical.
  Use: Analysis (test for alkaloids), glass colorant.

ammonium sesquicarbonate.   See ammonium carbonate.

ammonium silicofluoride.   See ammonium fluosilicate.

ammonium silicomolybdate.   See ammonium-12-molybdosilicate.

ammonium soap.   A soap resulting from the reaction of a fatty acid with ammonium hydroxide. Has an appreciable vapor pressure of ammonia and decomposes on continued exposure, leaving the fatty acid residue. Usually not sold as detergents, but used in toilet preparations and emulsions.

ammonium stearate.   (octadecanoic acid; ammonium salt).
  CAS: 1002-89-7. $C_{17}H_{35}COONH_4$.
  Properties: Tan, waxlike solid; free from ammonia odor. Mw 301.58, d 0.89 (22C), pH (3% dispersion) 7.6, mp 73–75C, neutralization value 70–80. Soluble in boiling water and hot toluene; partly soluble in hot butyl acetate and ethanol. Combustible.
  Grade: Available as anhydrous solid or as paste.
  Hazard: A nuisance dust.
  Use: Vanishing creams, brushless shaving creams, other cosmetic products; waterproofing of cements, concrete, stucco, paper and textiles, etc.

ammonium sulfamate.
  CAS: 7773-06-0. $NH_4OSO_2NH_2$.
  Properties: White, hygroscopic solid. Mp 130C, decomposes at 160C. Soluble in water and ammonia solution. Nonflammable.
  Derivation: Hydrolysis of the product obtained when urea is treated with fuming sulfuric acid.
  Hazard: Hot acid solutions when enclosed may explode.
  Use: Flameproofing agent for textiles and certain grades of paper, weed and brush killer, electroplating, generation of nitrous oxide.

ammonium sulfate.
  CAS: 7783-20-2. $(NH_4)_2SO_4$.
  Properties: Brownish-gray to white crystals according to degree of purity. D 1.77, mp 513C with decomposition. Soluble in water, insoluble in alcohol and acetone. Nonflammable.
  Derivation: (1) Ammoniacal vapors from destructive distillation of coal react with sulfuric acid, followed by crystallization and drying. (2) Synthetic ammonia is neutralized with sulfuric acid. (3) By-product of manufacture of caprolactam. (4) From gypsum by reaction with ammonia and carbon dioxide.
  Method of purification: Recrystallization or sublimation.
  Grade: Commercial, technical, CP, enzyme (no heavy metals), FCC.
  Use: Fertilizers, water treatment, fermentation, fireproofing compositions, viscose rayon, tanning, food additive.

ammonium sulfate nitrate.
  Properties: A double salt of approximately 60% ammonium sulfate and 40% ammonium nitrate, 26% nitrogen content. White to light gray granules, soluble in water.
  Hazard: Oxidizer, dangerous in contact with organic materials.

ammonium sulfide.
  CAS: 12124-99-1. $(NH_4)_2S$. The true sulfide is stable only in the absence of moisture and below

0C. The ammonium sulfide of commerce is largely ammonium bisulfide or hydrosulfide, NH_4HS.
Properties: Yellow crystals. Mp (decomposes). Soluble in water, alcohol, and alkalies. Evolves hydrogen sulfide on contact with acids.
Grade: Technical, CP, liquid, 40–44%.
Hazard: Strong irritant to skin and mucous membranes.
Use: Textile industry, photography (developers), coloring brasses, bronzes, iron control in soda ash production, synthetic flavors.

ammonium sulfite.
CAS: 10192-30-0. $(NH_4)_2SO_3 \cdot H_2O$.
Properties: Colorless crystals; acrid, sulfurous taste. D 1.41. Hygroscopic, sublimes at 150C with decomposition. Soluble in water.
Use: Chemical (intermediates, reducing agent), medicine, permanent wave solutions, photography, metal lubricants.

ammonium sulfocyanate.   See ammonium thiocyanate.

ammonium sulfocyanide.   See ammonium thiocyanate.

ammonium sulfonate.   See ammonium benzenesulfonate.

ammonium sulforicinoleate.
Properties: Yellow liquid. Soluble in alcohol. Combustible.
Grade: Technical.
Use: Medicine, furniture polish.

ammonium tartrate.
CAS: 3164-29-2. $(NH_4)_2C_4H_4O_6$.
Properties: White crystals. D 1.601, decomposes on heating. Soluble in water and alcohol.
Use: Textile industry, medicine.

ammonium tetrathiocyanodiammonochromate.   See Reinecke salt.

ammonium tetrathiotungstate. $(NH_4)_2WS_4$.
Properties: Orange-colored, crystalline powder; sensitive to heat; hydrogen sulfide odor. Mp (decomposes). Soluble in water, ammoniacal, and amine solutions.
Use: Source of high purity tungsten disulfide for catalysts, lubricants, semiconductors.

ammonium thiocyanate.   (ammonium sulfocyanide; ammonium sulfocyanate).
CAS: 1762-95-4. NH_4SCN.
Properties: Colorless, deliquescent crystals. D 1.3057; mp 149.6C; decomposes at 170C. Soluble in water, alcohol, acetone, and ammonia.
Derivation: By boiling an aqueous solution of ammonium cyanide with sulfur or polysulfides, or by the reaction of ammonia and carbon disulfide.

Grade: Technical, CP, 50–60% solution.
Use: Analytical chemistry; chemicals (thiourea); fertilizers; photography; ingredients of freezing solutions, especially liquid rocket propellants; fabric dyeing; zinc coating; weed killer and defoliant; adhesives; curing resins; pickling iron and steel; electroplating; temporary soil sterilizer; polymerization catalyst; separator of zirconium and hafnium, and of gold and iron.

ammonium thioglycolate. $HSCH_2COONH_4$.
Properties: Colorless liquid, repulsive odor. Evolves hydrogen sulfide. Combustible.
Use: Solutions of various strengths are used for hair waving and for hair removal.

ammonium thiosulfate.
CAS: 7783-18-8. $(NH_4)_2S_2O_3$.
Properties: White crystals decomposed by heat. PH of 60% solution 6.5–7.0. Very soluble in water.
Grade: Pure crystals (97%), 60% photographic solution.
Use: Photographic fixing agent, especially for rapid development; analytical reagent; fungicide; reducing agent; brightener in silver plating baths; cleaning compounds for zinc-base die-cast metals; hair waving preparations; fog screens.

ammonium titanium oxalate.   (titanium ammonium oxalate). $(NH_3)_2TiO(C_2O_4)_2$.
Properties: A water-soluble powder.
USE: Mordant in dyeing cellulosic fibers, leather, etc.

ammonium tungstate.   (ammonium wolframate;   ammonium   paratungstate). $(NH_4)_6W_7O_{24} \cdot 6H_2O$.
Properties: White crystals. Soluble in water; insoluble in alcohol.
Derivation: Interaction of ammonium hydroxide and tungstic acid with subsequent crystallization.
Use: Preparation of ammonium phosphotungstate and tungsten alloys.
See ammonium metatungstate.

ammonium valerate.   (pentanoic acid, ammonium salt; valeric acid, ammonium salt). $C_5H_{13}NO_2$.
Properties: Very hygroscopic crystals. Mp 108C, mw 119.16. Very soluble in water, alcohol, and ether.
Grade: Food and flavor codex.
Use: Flavoring material.

ammonium vanadate.   See ammonium metavanadate.

ammonium wolframate.   See ammonium tungstate.

ammonium zirconifluoride.   See zirconium ammonium fluoride.

ammonium zirconyl carbonate.
$(NH_4)_3ZrOH(CO_3)_3 \cdot 2H_2O$. D 1.238 (24C). Stable up to approximately 68C; decomposes in dilute acids, alkalies.
Grade: Aqueous solution.
Use: Ingredient in water repellents for paper and textiles, catalyst, stabilizer in latex emulsion paints, ingredient in floor wax to aid in resistance to detergents, lubricant in fabrication of glass fibers.

ammonobasic mercuric chloride. See mercury, ammoniated.

ammonolysis. The procedure that is analogous to hydrolysis, with ammonia substituted for water.

ammonotelic. Organisms that excrete excess nitrogen in the form of ammonia.

Ammo-Phos. A high-analysis ammonium phosphate-containing fertilizers.

amniote egg. The type of egg laid by reptiles and birds, having a nutritious yolk and a hard outer shell to protect the embryo from the dry environment. The amniote egg is named for the amnion, a sac that contains the embryo.

amobarbital. (5-ethyl-5-isoamylbarbituric acid). $C_{11}N_{18}N_2O_3$.
Properties: White, crystalline powder; odorless with bitter taste. Mp 156–161C. Solutions are acid to litmus. Very slightly soluble in water; soluble in alcohol.
Grade: USP.
Hazard: May be a habit forming drug of abuse.
Use: Medicine (also as sodium salt), hypnotic.

Amoco Performance Products. Specialty engineered polymers to help reduce pollution.

amodiaquine hydrochloride.
$C_{20}H_{22}ON_3Cl \cdot 2HCl \cdot 2H_2O$.
Properties: Yellow crystalline solid; odorless; bitter. Mp 150–160C (decomposes), pH (1% solution) 4.0–4.8. Soluble in water; sparingly soluble in alcohol; very slightly soluble in benzene, chloroform, and ether.
Grade: NF.
Use: Medicine (antimalarial).

amorphous. Noncrystalline, having no molecular lattice structure which is characteristic of the solid state. All liquids are amorphous. Some materials that are apparently solid, such as glasses, or semisolid, such as some high polymers, rubber, and sulfur allotropes, also lack a definite crystal structure and a well-defined melting point. They are considered high-viscosity liquids. The cellulose molecule contains amorphous as well as crystalline areas. Carbon derived by thermal decomposition or partial combustion of coal, petroleum, and wood is amorphous (coke, carbon black, charcoal), though other forms (diamond, graphite) are crystalline. Amorphous metallic alloys for transformer coils are made by extremely rapid cooling of the molten mixture. They are composed of iron, nickel, phosphorus, and boron.
See liquid; liquid crystal; glass, metallic.

amorphous phosphorus. (red phosphorus).
CAS: 7723-14-0. P.
Properties: Nontoxic, less flammable, dark red mass or powder.
Derivation: Produced by heating ordinary phosphorus to 260C in the absence of oxygen.

amorphous wax. See microcrystalline wax.

amosite. A type of asbestos.
See asbestos.

AMP. (1) Abbreviation for 2-amino-2-methyl-1-propanol. (2) Abbreviation for adenosine monophosphate.
See adenylic acid.

AMP resistance. See antibiotic resistance.

A5MP. Abbreviation for adenosine-5-monophosphoric acid.
See 5′-adenylic acid.

"AMP-95" [Dow]. TM for 2-amino-methyl-1-propanol.
Grade: 95% and 99+%.
Available forms: Liquid.
Use: Multifunctional amine, used as codispersant, solubilizer, stabilizer, neutralizer, buffer, and catalyst.

Ampco. A series of aluminum-iron-copper alloys containing 6–15% aluminum, 1.5–5.25% iron, balance copper. Resistant to fatigue, corrosion, erosion, wear, and cavitation pitting.
Use: For bushings, bearings, gears, slides, etc.

Ampcoloy. A series of industrial copper alloys including low-iron-aluminum bronzes, nickel-aluminum bronzes, tin bronzes, manganese bronzes, lead bronzes, beryllium-copper, and high-conductivity alloys.

Ampco-Trode. A series of aluminum-bronze arc-welding electrodes and filler rod containing 9.0–15.0% aluminum, 1.0–5.0% iron, balance copper, for joining like or dissimilar metals and overlaying surfaces resistant to wear, corrosion, erosion, and cavitation-pitting.

AMPD. Abbreviation for 2-amino-2-methyl-1,3-propanediol.

amphetamine. (1-phenyl-2-aminopropane; methylphenethylamine; Benzedrine). $C_6H_5CH_2CH(NH_2)CH_3$.
Properties: Colorless, volatile liquid; characteristic strong odor; slight burning taste. Bp 200–203C (decomposes), flash p 80F (26.6C). Soluble in alcohol and ether; slightly soluble in water.
Grade: Dextro-, dextrolevo-. Also available as phosphate and sulfate.
Hazard: Flammable, moderate fire risk. Basis of a group of hallucinogenic (habit-forming) drugs that affect the central nervous system. Sale and use restricted to physicians. Production limited by law.
Use: Medicine.

amphibole. A type of asbestos.
See asbestos.

amphibolic pathway. A metabolic pathway used in both catabolism and anabolism.

amphipathic. Containing both polar and nonpolar domains.

amphipathy. The simultaneous attraction and repulsion in a single molecule or ion consisting of one or more groups having an affinity for the phase in which they are dissolved together with groups that tend to be expelled by the medium.

amphiphilic. Molecule having a water-soluble polar head (hydrophilic) and a water-insoluble organic tail (hydrophobic), e.g., octyl alcohol, sodium stearate. Such molecules are necessary for emulsion formation and for controlling the structure of liquid crystals.
See emulsion; liquid crystal.

amphiprotic. Capable of donating and accepting protons, thus able to serve as an acid or a base.

ampholyte. A substance that can ionize to form either anions or cations and thus may act as either an acid or a base. An ampholytic detergent is cationic in acid media and anionic in base media. Water is an ampholyte.
See amphoteric.

amphora catalyst. See catalyst, amphora.

"Amphosol" *[Stepan].* (cocamidopropyl betaine). TM for chemical used in shampoos, liquid hand soap, bubble bath, thickeners, and foam enhancers.

amphoteric. Having the capacity of behaving either as an acid or a base. Thus, aluminum hydroxide neutralizes acids with the formation of aluminum salts, $Al(OH)_3 + 3HCl \rightarrow AlCl_3 + 3H_2O$, and also dissolves in strongly basic solutions to form aluminates, $Al(OH)_3 + 3NaOH \rightarrow Na_3AlO_3 + 3H_2O$. Amino acids and proteins are amphoteric, i.e., their molecules contain both an acid group (COOH) and a basic group (NH_2). Thus, wool can absorb both acidic and basic dyes.

amphoteric dye. A dye that contains both acidic and basic chromophores.

amphotericin B. A polyene antifungal antibiotic. $C_{47}H_{73}NO_{17}$.
Properties: Pale yellow, semicrystalline powder. Mp >170C (gradual decomposition). Insoluble in water; slightly soluble in methanol; somewhat more soluble in dimethylsulfoxide.
Derivation: Fermentation with *Streptomyces nodosus.* Commercially available as a deoxycholate complex.
Grade: USP.
Hazard: May have undesirable side effects.
Use: Medicine (meningitis treatment).

amphoteric ion-exchange resin. Any ion-exchange resin that contains both positive and negative groups. Such resins are most useful in ion retardation where all ionic materials can be removed from solution.

amphoteric oxide. An oxide that shows some acidic and some basic properties insofar as it can accept or donate protons.

ampicillin (USAN). (6,D,α-aminophenyl-acetamido penicillanic acid). $C_{16}H_{19}N_3O_4S$. A semisynthetic antibiotic, active against some Gram-negative infections.

amplification. An increase in the number of copies of a specific DNA fragment; can be in vivo or in vitro.
See cloning; polymerase chain reaction.

amprolium. (1-[(4-amino-2-propyl-5-pyrimidinyl)-methyl]-2-picolinium chloride). $C_{14}H_{19}ClN_4$. A coccidiostat used in veterinary medicine.

amprotropine phosphate. (phosphate of the *dl*-tropic acirdiethylamin*o*-2,2-dimethyl-*l*-propanol). $C_{18}H_{29}NO_3 \cdot H_3PO_4$.
Properties: Bitter crystals. Mp 142–144C. Soluble in water; slightly soluble in alcohol.
Use: Medicine (antispasmodic).

AMPS (sulfonic acid). See 2-acrylamido-2-methyl-1-propanesulfonic acid.

ampule. (ampul; ampulla). A small-sealed glass container used for sterile liquids. It protects contents from air or contamination.

amsonic acid. (4,4′-diamino-2,2′-stilbene-disulfonic acid). $C_{14}H_{14}N_2O_6S_2$.

Properties: Acicular crystals. Slightly soluble in water.
Use: Manufacture of bleaching agents and organic dyes.

amsulosin hydrochloride. See r-(-)-5-(2-(2-ethoxyphenoxy)ethyl)amino)propyl)-2-methyl-benzene-sulfanomide hydrochloride.

amygdalic acid. See mandelic acid.

amygdalin. (mandelonitrile-β-gentiobioside; amygdaloside). $C_6H_5CHCNOC_{12}H_{21}O_{10}$. A glycoside found in bitter almonds, peaches, and apricots.
Properties: White crystals; bitter taste. Mp 214–216C (anhydrous). Soluble in water and alcohol; insoluble in ether.
See almond oil (note).

amyl. The 5-carbon aliphatic group C_5H_{11}, also known as pentyl. Eight isomeric arrangements (exclusive of optical isomers) are possible. The amyl compounds occur (as in fusel oil) or are formed (as from the petroleum pentanes) as mixtures of several isomers, and, since their boiling points are close and their other properties similar, it is neither easy nor usually necessary to purify them. See amyl alcohol.

amyl acetate. (amylacetic ester; banana oil; pear oil).
CAS: 628-63-7. $CH_3COOC_5H_{11}$. Commercial amyl acetate is a mixture of isomers, the composition and properties depending upon the grade and derivation. The main isomers are isoamyl, normal, and secondary amyl acetates.
Properties: Colorless liquid; persistent banana-like odor. Flash p 65–95F (18.3–35C) (CC) depending on grade, autoign temp approximately 714F (380C).
Derivation: Esterification of amyl alcohol (often fusel oil) with acetic acid and a small amount of sulfuric acid as catalyst.
Method of purification: Rectification.
Grade: Commercial (85–88%), high test (85–88%), technical (90–95%), pure (95–99%), special antibiotic grade. Amyl acetate is also sold by original source as from fusel oil, pentane, or Oxo process.
Hazard: Flammable, high fire risk. Explosive limits in air 1.1–7.5%.
Use: Solvent for lacquers and paints, extraction of penicillin, photographic film, leather polishes, nail polish, warning odor, flavoring agent, printing and finishing fabrics, solvent for phosphors in fluorescent lamps.

***n*-amyl acetate.**
CAS: 628-63-7. $CH_3COOCH_2CH_2CH_2CH_2CH_3$.
Properties: Colorless liquid. Bp 148.4C, mp −70.8C, d 0.879 (20/20C), wt/gal (20C) 7.22 lb, flash p 77F (25C) (CC). Very slightly soluble in water; miscible with alcohol and ether. Vapor heavier than air. Autoign temperature 714F (380C).

Derivation: Esterification of *n*-amyl alcohol with acetic acid.
Hazard: Flammable, dangerous fire risk.
Use: See amyl acetate.

***sec*-amyl acetate.**
CAS: 626-38-0. $CH_3CO_2C_5H_{11}$.
Properties: Colorless liquid. May be mixture of secondary isomers. Distillation range 123–145C, mild odor, nonresidual, purity of ester content as amyl acetate 85–88%, d 0.862–0.866 (20/20C), flash p 89F (31.6C) (CC), wt/gal (20C) approximately 7.19 lbs.
Derivation: Esterification of *sec*-amyl alcohol and acetic acid.
Grade: Technical.
Hazard: Flammable, moderate fire risk. Toxic. Upper respiratory tract irritant.
Use: Solvent for nitrocellulose and ethyl cellulose, cements, coated paper, lacquers, leather finishes, nail enamels, plastic wood, textile sizing, and printing compounds.

amylacetic ester. See amyl acetate.

amyl acid phosphate. $(C_5H_{11})_2HPO_4$ and $C_5H_{11}H_2PO_4$. A mixture of primary and amyl isomers. Water-white liquid, d 1.070–1.090, flash p 245F (118.3C) (COC). Insoluble in water, soluble in alcohol. Combustible.
Hazard: Strong irritant to tissue.
Use: Curing catalyst and accelerator in resins and coatings, stabilizer, dispersion agent, lubricating and antistatic agent in synthetic fibers.

amyl alcohol. (amyl hydrate). Eight isomers of amyl alcohol, $C_5H_{11}OH$, are possible, exclusive of several optical isomers, and six are offered commercially. In addition, definite mixtures of the isomers are sold under a variety of names (unfortunately, some of them identical with the names of the pure isomers, see (1) *n*-amyl alcohol, primary; (2) 2-methyl-1-butanol (active amyl alcohol from fusel oil); (3) isoamyl alcohol, primary; (4) 2-pentanol; (5) 3-pentanol; (6) *tert*-amyl alcohol. The other two isomers not described in detail are (7) *sec*-isoamyl alcohol; (8) 2,2-dimethyl-1-propanol. (1), (2), (3), and (8) are primary alcohols; (4), (5), and (7) are secondary alcohols; and (6) is a tertiary alcohol. (1), (4), and (5) are normal, and (2), (3), (6), (7), and (8) are branched-chain compounds. (2), (4), and (7) are asymmetric and have optically active forms.

***n-sec*-amyl alcohol.** See 2- and 3-pentanol.

***tert*-amyl alcohol.** (dimethylethylcarbinol; 2-methyl-2-butanol; amylene hydrate; *tert*-pentanol).
CAS: 75-85-4. $(CH_3)_2C(OH)CH_2CH_3$.
Properties: Colorless liquid; camphor odor; burning taste. D 0.81 (20/20C), fp −11.9C, bp 101.8C, refr

index 1.4052 (20C), wt/gal 6.76 lb, flash p 70F (21.2C) (OC), Autoign temp 819F (437C). Slightly soluble in water; miscible with alcohol and ether. Solutions neutral to litmus.
Derivation: Fractional distillation of the mixed alcohols resulting from the chlorination and alkaline hydrolysis of pentanes.
Grade: Technical, CP, NF.
Hazard: Flammable, dangerous fire risk.
Use: Solvent, flotation agent, organic synthesis, medicine (sedative).

sec-amyl alcohol, active.   See 2-pentanol.

amyl alcohol, fermentation.   See fusel oil.

amyl alcohol, primary.   A mixture of primary amyl alcohols made from normal butenes by the Oxo process is sold under this name. It consists of 60% primary n-amyl alcohol, 35% 2-methyl-1-butanol, and 5% 3-methyl-1-butanol.
Hazard: Flammable, moderate fire risk.
Use: A solvent.

n-amyl alcohol, primary.   (1-pentanol; n-butyl carbinol).
CAS: 71-41-0. $CH_3(CH_2)_4OH$.
Properties: Colorless liquid; mild odor. Bp 137.8C, fp −78.9C, d 0.812–0.819 (20/20C), wt/gal (20C) 6.9 lb, flash p 123F (50.5C) (OC). Autoign temp 572F (300C). Slightly soluble in water; miscible with alcohol, benzene, and ether. Combustible.
Derivation: Fractional distillation of the mixed alcohols resulting from the chlorination and alkaline hydrolysis of pentane.
Grade: Technical, CP, 98%.
Hazard: Lower explosive level in air 1.2% by volume. Moderate fire risk.
Use: Raw material for pharmaceutical preparations; organic synthesis solvent.

amyl alcohol, primary active.   See 2-methyl-1-butanol.

amyl aldehyde.   See n-valeraldehyde.

n-amylamine.   (pentylamine; 1-aminopentane). $C_5H_{11}NH_2$.
Properties: Colorless liquid, d 0.75 (20/20C), fp −55.0C, bp 104.4C, flash p 45F (7.2C) (OC). Soluble in water, alcohol, and ether.
Derivation: Reaction of ammonia and amyl chloride which gives a mixture of mono-, di-, and triamyl amines.
Grade: Technical.
Hazard: Flammable, dangerous fire risk. Strong irritant.
Use: Chemical intermediate, dyestuffs, rubber chemicals, insecticides, synthetic detergents, flotation agents, corrosion inhibitors, solvent, gasoline additive, pharmaceuticals.

sec-amylamine.   (2-aminopentane). $CH_3(CH_2)_2CH(CH_3)NH_2$.
Properties: Colorless liquid. Bp 198F, d 0.7; flash p 20F (65.5C).
Hazard: Flammable, dangerous fire risk.
Use: See n-amylamine.

amylase.   A class of enzymes which convert starch into sugars. Fungal and bacterial amylases from specific fungi and bacteria have been suggested for commercial fermentation processes.
Use: Textile desizing, conversion of starch to glucose sugar in syrups (especially corn syrups), baking (to improve crumb softness and shelf life), dry cleaning (to attack food spots and similar stains).
See amylopsin; diastase; ptyalin.

n-amylbenzene.   (1-phenylpentane). $C_6H_5CH_2(CH_2)_3CH_3$. Water-white liquid, mild odor, fp −75C, bp 205C, d 0.8585 (20/4C), flash p 150F (65.5C) (OC). Insoluble in water, soluble in hydrocarbons and coal tar solvents. Combustible.
Hazard: Irritant to skin and eyes, narcotic in high concentrations. Moderate fire risk.

sec-amylbenzene.   (2-phenylpentane). $CH_3CH(C_6H_5)CH_2CH_2CH_3$.
Properties: Clear liquid. Fp −75C, bp 190.3C, d 0.861 (20/4C).
Hazard: Moderate fire hazard. Combustible. Irritant to skin and eyes, narcotic in high concentrations.
Grade: Pure, 99.0 mole %, technical, 95.0 mole %.
Use: Weed control, chemical intermediate.

amyl benzoate.   See isoamyl benzoate.

amyl butyrate.   See isoamyl butyrate.

amyl carbinol.   See hexyl alcohol.

n-amyl chloride.   (1-chloropentane). CAS: 543-59-9. $CH_3(CH_2)_3CH_2Cl$.
Properties: Colorless liquid. Bp 107.8C, fp −99C, d 0.883 (20/4C), refr index 1.4128 (20C), flash p 54F (12.2C) (OC). Miscible with alcohol and ether; insoluble in water. Autoign temp 500F (260C).
Derivation: (1) Distillation of amyl alcohol with salt and sulfuric acid, (2) addition of hydrochloric acid to α-amylene.
Grade: Technical.
Hazard: Flammable, dangerous fire risk; lower explosive level 1.4%, upper explosive level 8.6%. May be narcotic in high concentrations.
Use: Chemical intermediate.

amyl chlorides, mixed.
Properties: Straw- to purple-colored liquid. D 0.88 (20C), 95% distills between 85 and 109C, wt/gal 7.33 lb, refr index (20C) 1.406. Insoluble in water; water azeotrope at 77–82C approximately 90% $C_5H_{11}Cl$, miscible with alcohol and ether. Flash p 38F (3.3C) (OC). Components:

1-chloropentane, bp 107.8C; 2-chloropentane, bp 96.7C; 3-chloropentane, bp 97.3C; 1-chloro-2-methylbutane, bp 99.9C; 1-chloro-3-methylbutane, bp 98.8C; 3-chloro-2-methylbutane, bp 93.0C; 2-chloro-2-methylbutane, bp 86.0C.
Derivation: Vapor phase chlorination of mixed normal pentane and isopentane.
Hazard: Flammable, dangerous fire risk. Explosive limits in air 1.4–8.6%. May be narcotic in high concentrations.
Use: Synthesis of other amyl compounds, solvent, rotogravure ink vehicles, soil fumigation.

α-amylcinnamic alcohol. (2-benzylidene-1-heptanol). $C_6H_5CH{:}C(CH_2OH)C_5H_{11}$.
Properties: Yellow liquid; floral odor. D 0.954–0.962 (25/25C). Soluble in 3 parts 70% alcohol. Combustible.
Derivation: Synthetic.
Use: Perfumery, flavoring.

α-amylcinnamic aldehyde. (jasmine aldehyde; α-pentylcinnamaldehyde). $C_6H_5CH{:}C(CHO)C_5H_{11}$.
Properties: Clear, yellow, oily liquid; jasmine-like odor. D 0.962–0.968, refr index 1.554–1.559. Aldehyde content 97%. Soluble in 6 volumes of 80% alcohol. Combustible.
Derivation: Synthetic.
Grade: Technical, FCC.
Use: Perfumery, flavoring.

6,*n*-amyl-*m*-cresol. $C_{12}H_{18}O$.
Properties: Liquid at room temperature. Mp 24C. Insoluble in water; soluble in acetone and ethanol.
Use: Prevention of molds, bactericide.

α,*n*-amylene. Legal label name for 1-pentene.

β,*n*-amylene. Legal label name for 2-pentene.

amylene dichloride. See dichloropentane.

amylene hydrate. See *tert*-amyl alcohol.

n-amyl ether. (diamyl ether). $C_{10}H_{22}O$.
Properties: Colorless liquid. Bp 186C, fp −70C, d 0.783, flash p 135F (57C), autoign temp 340F (171C), refr index 1.41. Insoluble in water; soluble in alcohol and ether.
Hazard: Narcotic in high concentration.
Use: General solvent for fats, oils, waxes, resins, etc.

amyl formate. $HCOOC_5H_{11}$.
Properties: Colorless liquid composed of a mixture of isomeric amyl formates with isoamyl formate in predominance. Plum-like odor. Less odoriferous and more active solvent than amyl acetate. It also has both a lower boiling point and a higher rate of evaporation. D 0.880–0.885, bp 123.5C, flash p 80F (26.6C). Miscible with oils, hydrocarbons, alcohols, ketones; slightly soluble in water.

Grade: Technical, FCC.
Hazard: Flammable, dangerous fire risk. Toxic by ingestion and inhalation.
Use: Solvent for cellulose esters, resins; solvent mixtures; films and coatings; perfume for leather; flavoring.

n-amylfuroate. (amyl pyromucate). $C_4H_3OCO_2C_5H_{11}$.
Properties: Colorless oil. Decomposes on standing, d 1.0335 (20/4C), bp 233C. Insoluble in water; soluble in alcohol. Combustible.
Derivation: Esterification of furoic acid.
Use: Perfumes, lacquers.

amyl heptanoate. $C_{12}H_{24}O_2$.
Properties: Colorless to pale-yellow liquid; fruity taste. D: 0.859, refr index: 1.422.
Use: Food additive.

amyl hydrate. See amyl alcohol.

amyl hydride. See *n*-pentane.

amyl mercaptan. Legal label name for pentanethiol.

tert-amyl mercaptan. See 2-methyl-2-butanethiol.

6-amylmercaptopurine. See 6-(pentylthio) purine.

amyl methyl alcohol.
CAS: 105-30-6. $C_6H_{14}O$.
Properties: Liquid. Bp: 130C, flash p: 114F (CC), d: 0.804, vap d: 3.52.
Hazard: Moderately toxic by ingestion and skin contact. A skin and severe eye irritant. Human systemic irritant by inhalation. A flammable liquid; can react with oxidizing materials. To fight fire, use CO_2, dry chemical.

amyl nitrate (mixed isomers). $C_5H_{11}NO_3$.
Properties: Colorless liquid; ethereal odor. Bp 145C, flash p 118F (47.8C), d 0.99 (20C). Flammable.
Hazard: Oxidizing agent, moderate fire risk.
Use: Additive to increase cetane number of diesel fuels.

amyl nitrite. (isoamyl nitrite).
(CAS: 463-04-7. $CH_3)_2CHCH_2CH_2NO_2$.
Properties: Yellowish liquid; peculiar ethereal, fruity odor; pungent aromatic taste. D 0.865–0.875 (25C); bp 96–99C; autoign temp 405F (207C). Soluble in alcohol; almost insoluble in water. Decomposes on exposure to air, light, or water.
Derivation: Interaction of amyl alcohol with nitrous acid.
Grade: NF, (75% min), technical.
Hazard: Flammable, dangerous fire risk, a strong oxidizer. Vapor may explode if ignited.
Use: Perfumes, diazonium compounds.

amyloglucosidase.   An enzyme used commercially to convert starches to dextrose.

amylograph.   A device designed to record the action of heat on a mixture of flour and water. It is used in measuring diastatic activity through viscosity change.

amyloid.   A gelatinous hydrate that results from diluting a solution of concentrated sulfuric acid and cellulose with water.

amylopectin.   The outer, almost insoluble portion of starch granules. It is a hexosan, a polymer of glucose, and is a branched molecule of many glucose units. It stains violet with iodine and forms a paste with water.

amylopsin.   (animal diastase). The starch-digesting enzyme of pancreatic juice, the most powerful enzyme of the digestive tract. It is an amylase that converts starches through the soluble-starch stage to various dextrins and maltose. It acts in neutral, slightly acid, and slightly alkaline environments, with an optimum pH of 6.3–7.2. It requires the presence of certain negative ions for activation.
Use: Biochemical research.

amylose.   The inner, relatively soluble portion of starch granules. Amylose is a hexosan, a polymer of glucose, and consists of long, straight chains of glucose units joined by a 1,4-glycosidic linkage. It stains blue with iodine. Microcrystalline amylose is available chiefly as a food ingredient and dietary energy source.

***o-sec*-amylphenol.** $C_5H_{11}C_6H_4OH$.
Properties: Clear, straw-colored liquid. D 0.955–0.971 (30/30C), initial bp over 235.0C, final bp below 250.0C, wt/gal 8.0 lbs. Very slightly soluble in water; soluble in oil and organic solvents, flash p 219F (OC). Combustible.
Use: Dispersing and mixing agent for paint pastes, antiskinning agent for paint, varnish and oleoresinous enamels, organic synthesis.

***p-tert*-amylphenol.** $(CH_3)_2(C_2H_5)CC_6H_4OH$.
Properties: White crystals. Mp 93C, bp 265–267C (138C at 15 mm). Slightly soluble in water; soluble in alcohol and ether, flash p 232F (111C) (OC). Combustible.
Use: Manufacture of oil-soluble resins, chemical intermediate.

***p-tert*-amylphenyl   acetate.** $C_5H_{11}C_6H_4OOCCH_3$.
Properties: Colorless liquid. D 0.996 (20C), boiling range 253–272C, fruity odor, flash p 240F (115.5C). Combustible.
Use: Perfumes, flavorings.

amyl propionate. $CH_3CH_2COOC_5H_{11}$. Probably the isoamyl isomer.
Properties: Colorless, high-boiling liquid; apple-like odor. D 0.869–0.873 (20/20C); wt/gal (20C) approximately 7.25 lb; distillation range 135–175C; flash p 106F (41.1C) (OC); autoign temp 712F (377C). Miscible with most organic solvents. Flammable.
Derivation: By reacting amyl alcohol with propionic acid in the presence of sulfuric acid as a catalyst, followed by neutralization, drying, and distillation.
Hazard: Fire hazard.
Use: Perfumes, lacquers, flavors.

amyl pyromucate.   See *n*-amyl furoate.

amyl salicylate.   See isoamyl salicylate.

amyl sulfide.   See diamyl sulfide.

amyltrichlorosilane. $C_5H_{11}SiCl_3$. A mixture of isomers.
Properties: Colorless to yellow liquid. Bp 168C, d 1.137 (25/25C), refr index 1.4152 (20C), flash p 145F (62.8C) (COC). Readily hydrolyzed by moisture with the liberation of hydrogen chloride. Combustible.
Derivation: By Grignard reaction of silicon tetrachloride and amyl magnesium chloride.
Hazard: Toxic and corrosive.
Use: Intermediate for silicones.

amyl valerate.   See isoamyl valerate.

amyl valerianate.   See isoamyl valerate.

amyris oil, West Indian type.
Properties: Extracted from *Amyris balsamifera L.* (Fam. *Rutaceae*). Clear, pale yellow viscous liquid; odor of sandalwood. Sol in mineral oil, propylene glycol; insol in glycerin.
Use: Food additive.

"Amytal" *[Marathon].*   TM for amobarbital (USP).

anabaena, aphanizomenon, and microcystis. The three genera of blue-green bacteria that are toxic to animals and are known as "annie, fannie and mike" by water treatment professionals.

anabasine.   (neonicotine; 2-(3-pyridyl)piperidine). $C_{10}H_{14}N_2$. A naturally occurring alkaloid.

Properties: Colorless liquid, darkens on exposure to air. Bp 270C, fp 9C, d 1.046 (20/20C), refr index 1.5430 (20C). Miscible with water; soluble in alcohol and ether.
Derivation: (1) Extraction from *Anabasis aphylla* and *Nicotiana glauca*, (2) synthetic.
Use: Insecticide.

anabolism. The phase of intermediary metabolism concerned with the energy-requiring biosynthesis of cell components from smaller precursors.

anacardic acid. $C_{22}H_{32}O_3$. The chief component of cashew nutshell oil.

anacardium gum. See cashew gum.

Anacyclus pyrethrum. A prostrate perennial shrub (family Asteraceae) that is endemic to morocco, the flowers are a source of pyrethrin.

anaerobe. An organism that lives without oxygen and uses another chemical species as a terminal electron acceptor. Obligate anaerobes die when exposed to oxygen; facultative anaerobes can live in both aerobic and anaerobic environments.

anaerobic. Descriptive of a chemical reaction or a microorganism that does not require the presence of air or oxygen. Examples are the fermentation of sugars by yeast and the decomposition of sewage sludge by anaerobic bacteria. It is also applied to certain polymers that solidify when kept out of contact with air.
See "Loctite" *[Henkel]*; bacteria; botulism.

anaerobic bacteria. Bacteria that grow only in the absence of molecular oxygen.

aneuploid. An organism or somatic cell that has a chromosome number that is not an exact multiple of the normal haploid set of chromosomes.

anagenesis. Evolutionary change along an unbranching lineage; change without speciation.

analcite. (analcime). $Na_2O \cdot Al_2O_3 \cdot 4SiO_2 \cdot 2H_2O$. A mineral, one of the zeolites.
Properties: Colorless, white, sometimes greenish-grayish, yellowish, or reddish white. Hardness 5–5.5; d 2.22–2.29.
Occurrence: Europe, U.S., Nova Scotia.

analgin-tempidone mixture. See tempalgin.

analytical chemistry. The subdivision of chemistry concerned with identification of materials (qualitative analysis) and with determination of the percentage composition of mixtures or the constituents of a pure compound (quantitative analysis).

The gravimetric and volumetric (or "wet") methods (precipitation, titration, and solvent extraction) are still used for routine work; indeed, new titration methods have been introduced, e.g., cryoscopic methods, pressure-metric methods (for reactions that produce a gaseous product), redox methods, and use of an F-sensitive electrode. However, faster and more accurate techniques (collectively called instrumental) have been developed in the last few decades. Among these are infrared, ultraviolet, and X-ray spectroscopy, where the presence and amount of a metallic element are indicated by lines in its emission or absorption spectrum; colorimetry, by which the percentage of a substance in solution is determined by the intensity of its color, chromatography of various types, by which the components of a liquid or gaseous mixture are determined by passing it through a column of porous material or on thin layers of finely divided solids, separation of mixtures in ion-exchange columns; and radioactive tracer analysis. Optical and electron microscopy, mass spectrometry, microanalysis, nuclear magnetic resonance (NMR), and nuclear quadrupole resonance (NQR) spectroscopy all fall within the area of analytical chemistry. New and highly sophisticated techniques have been introduced in recent years, in many cases replacing traditional methods.
See spectroscopy; nuclear magnetic resonance; nuclear quadrupole resonance spectroscopy; chromatography; fiber, optical; supercritical fluid.

anandamide. A fat produced by the brain that stimulates the same receptors as THC (1-*trans*-Δ9-tetrahydrocannabinol), the psychoactive substance in marijuana and hashish.
See n-(2-hydroxyethyl)-5,8,11,14-eicosatetraenamide (all-z)-.

anaphoresis. The migration of dispersed particles under the influence of an electrical field toward an anode.

anaphylaxis. Abnormal reaction to a second injection of a foreign protein, e.g., penicillin. It is an extreme form of allergy which often has serious consequences (swelling of tissues) and has been known to be fatal.

anaplerotic reaction. An enzyme-catalyzed reaction that can replenish the supply of intermediates in a metabolic pathway, most commonly the citric acid cycle.

anatase. (octahedrite). A natural crystallized form of titanium dioxide, d 3.8, refr index 2.5, mp 1560C.

anavenin. (anavenom). A toxoid of snake venom. A venom that retains its antigenic activity following detoxification by formaldehyde.

andalusite. Al_2OSiO_4. A natural silicate of aluminum.
Properties: Gray, greenish, reddish, or bluish in color. D 3.1–3.2, hardness 7–7.5.
Occurrence: Massachusetts, Connecticut, California, Nevada, Europe, South Africa, Australia.
Use: Constituent of sillimanite refractories, spark-plug insulators, laboratory ware, superrefractories.

androgen.   A male sex hormone. The androgenic hormones are steroids and are synthesized in the body by the testis, the cortex of the adrenal gland, and, to a slight extent, by the ovary.

androstenolone.   See DHEA.

androsterone. $C_{19}H_{30}O_2$. An androgenic steroid, metabolic product of testosterone. The international unit (IU) of androgenic activity is defined as 0.1 mg androsterone.

Properties: Crystalline solid. Mp 185–185.5C, sublimes in high vacuum. Dextrorotatory in solution. Not precipitated by digitonin. Practically insoluble in water; soluble in most organic solvents.
Derivation: Isolation from male urine, synthesis from cholesterol.
Use: Medicine, biochemical research.

Andrussow oxidation.   Ammonia and methane are oxidized with air in the presence of platinum catalyst to form hydrogen cyanide. Side reactions are hydration of methane to carbon dioxide and hydrogen, and oxidation of methane and ammonia to carbon monoxide and nitrogen. The reaction is strongly exothermic. The process has been elaborated wherever natural gas is abundant.

anesthetic.   A chemical compound that induces loss of sensation in a specific part or all of the body. A brief classification of the more important agents is as follows:
(A) General
  (1) Hydrocarbons
    (a) Cyclopropane (USP). Effective in presence of substantial proportions of oxygen; flammable.
    (b) Ethylene (USP). Rapid anesthesia and rapid recovery; flammable.
  (2) Halogenated hydrocarbons
    (a) Chloroform. Nonflammable. Its use is being abandoned because of its high toxicity.
    (b) Ethyl chloride. A gas at room temperature, liquefies at relatively low pressure. Applied as a stream from container directly on tissue. Sometimes used in gaseous form as inhalation-type general anesthetic. Flammable.
    (c) Trichloroethylene. Toxic and flammable. Used as general anesthetic since 1934.
  (3) Ethers
    (a) Ethyl ether (USP). First anesthetic used in surgery (1846), now largely replaced with less dangerous types. Highly flammable, explodes in presence of spark or open flame.
    (b) Vinyl ether. A liquid having many of the physiological properties of ethylene and ethyl ether. Highly flammable.
  (4) Miscellaneous
    (a) Tribromoethanol. Basal anesthetic, supplemented by an inhalation type when general anesthesia is needed. Ingredient of Avertin.
    (b) Nitrous oxide. Originally prepared by Priestley in 1772 (laughing gas); first used as anesthetic by Humphry Davy in 1800. Used (with oxygen) largely for dental surgery. Nonflammable.
    (c) Barbiturates.
(B) Local
  (1) Alkaloids (cocaine).
  (2) Synthetic products (procaine group, e.g., "Novocain" *[Hospira]*); alkyl esters of aromatic acids (topical).
  (3) Quinine hydrochloride.

anesthetic gas.   (nitrous oxide; dinitrogen monoxide; hyponitrous acid anhydride; laughing gas).
CAS: 10024-97-2. N_2O. Any respirable gas that effects general anesthesia upon controlled inhalation.
Properties: Nonflammable, colorless gas with a slightly sweet odor; Molecular weight of 44.0; boiling point of −127F.
Hazard: Dyspnea; drowsiness, headache; asphyxia; reproductive effects; liquid frostbite.
Use: Anesthetic, analgesic, a food aerosol in the preparation of whipped cream.

anethole.   (anise camphor; *p*-methoxypropenylbenzene; *p*-propenylanisole). $CH_3CH:CHC_6H_4OCH_3$.
Properties: White crystals; sweet taste; odor of oil of anise. Affected by light. D 0.983–0.987, refr index 1.557–1.561, optical rotation 0.08, mp 22–23C, distillation range 234–237C. Soluble in 8 volumes of 80% alcohol, 1 volume of 90% alcohol; almost immiscible with water.
Derivation: By crystallization from anise or fennel oils; synthetically from *p*-cresol.
Grade: USP, technical, FCC.

Use: Perfumes, particularly for dentifrices, flavors, synthesis of anisic aldehyde, licorice candies, color photography (sensitizer in color-bleaching process), microscopy.

Anfinsen, Christian B. (1916–1995). An American biochemistry who won the Nobel Prize for chemistry in 1972. His work involved the molecular basis of evolution and the chemistry of enzymes. He works with Moore and Stein. His doctorate was granted from Harvard.

ANFO. A high explosive based on ammonium nitrate.
See explosive, high.

angelic acid. (*cis*-2-methyl-2-butenoic acid; α-methyl-crotonic acid). $CH_3CH:C(CH_3)COOH$. The *cis* isomer of tiglic acid.
Properties: Colorless needles or prismatic crystals; spicy odor. D 0.9539 (76/4C), mp 45C, bp 185C, refr index 1.4434 (47C). Soluble in alcohol, ether, and hot water.
Derivation: From the root of *Angelica archangelica* or from the oil of *Anthemis nobilis* by distillation.
Use: Flavoring extracts.

angelica oil.
Properties: Essential oil; strong aromatic odor; spicy taste. D 0.853–0.918, optical rotation +16 to +41. Soluble in alcohol. Chief known constituents: phellandrene, valeric acid. Combustible.
Derivation: Distilled from the roots and seeds of *Angelica archangelica* found principally in Europe.
Grade: Technical, FCC.
Use: Preparation of liqueurs, perfumery.

angelica seed oil.
Properties: Extracted from seeds of *Angelica archangelica L.* A light-yellow liquid; sweet taste. Sol in fixed oils; sltly sol in mineral oil; insol in glycerin, propylene glycol.
Use: Food additive.

Angio-Conray. An 80% solution of sodium iothalamate used in diagnostic medicine.

angiotensin. (angiotonin; hypertensin). A peptide found in the blood, important in its effect on blood pressure. Both a decapeptide and an octapeptide are known. Their amino acid sequences and hence the complete structures have been established.

angiotensin II. ((3S)-3-amino-4-[[(2S)-5-(diaminomethylideneamino)-1-[[(2S)-1-[[(2S)-1-[[(2S,3S)-1-[[(2S)-1-[(2S)-2-[(2S)-1-hydroxy-1-oxo-3-phenylpropan-2-yl]carbamoyl]pyrrolidin-1-yl]-3-(1H-imidazol-5-yl)-1-oxopropan-2-yl]amino]-3-methyl-1-oxopentan-2-yl]amino]-3-(4-hydroxyphenyl)-1-oxopropan-2-yl]amino]-3-methyl-1-oxobutan-2-yl]amino-3-methyl-1-oxobutan-2-yl]amino]-1-oxopentan-2-yl]amino]-1-oxopentan-2-yl]amino]-4-oxobutanoic acid.

CAS: 53-73-6. $C_{49}H_{70}N_{14}O_{11}$. A octopeptide with eight aminoacid residues that is a powerful vasoconstrictor and stimulator of aldosterone secretion from the adrenal cortex.
Derivation: Formed from anginotensin I, which is produced by cleavage of the protein angiotensinogen in a reaction catalyzed by renin secreted by the kidneys.

angiotonin. See angiotensin.

Ångstrom. (Å). A unit of length almost one one-hundred millionth (10^{-8}) centimeter. The Ångstrom is defined in terms of the wavelength of the red line of cadmium (6438.4696 Å). Used in stating distances between atoms, dimensions of molecules, wavelengths of short-wave radiation, etc.
See nanometer.

angular aldehyde. The aldehyde group linked to carbon 13 between rings C and D of the steroid nucleus in aldosterone.

anhydrase. An enzyme that catalyzes the removal of water from a compound; most such enzymes are now known as hydrases, hydrolyases, or dehydrateases.

anhydride. A chemical compound derived from an acid by elimination of a molecule of water. Thus, sulfur trioxide (SO_3) is the anhydride of sulfuric acid (H_2SO_4), carbon dioxide (CO_2) is the anhydride of carbonic acid (H_2CO_3), and phthalic acid $[C_6H_4(CO_2H)_2]$ minus water gives phthalic anhydride $[C_6H_4(CO_2)O]$. Not to be confused with anhydrous.

anhydrite. $CaSO_4$. A natural calcium sulfate usually occurring as compact granular masses and resembling marble in appearance. Differs from gypsum in hardness and lack of hydration.

8,9-anhydro-4″-deoxy-3′-*n*-desmethyl-3′-*n*-ethylerythromycin b-6,9-hemiacetal.
CAS: 150785-53-8. $C_{38}H_{67}NO_{10}$.
Hazard: A poison.

anhydroenneaheptitol. (AEH; 4-hydroxy-2H-pyran-3,3,5,5(4H,6H)tetramethanol).

$OCH_2(CH_2OH)_2CH_2OC(CH_2OH)_2CH_2$.

Use: Alkyd resins, rosin esters, urethane coatings and foams, surfactants, lubricating oil additives.

anhydrous. Descriptive of an inorganic compound that does not contain water either adsorbed on its surface or combined as water of crystallization. Do not confuse with anhydride.

anhydrous ammonia. (aqua ammonia; synthetic ammonia; azine).
CAS: 7664-41-7. H_3N.

Properties: Colorless gas with a pungent, suffocating odor, usually liquefied under high pressure; molecular weight of 17.0; boiling point of −28F.
Use: Nitrogen fertilizer; chemical feed stock for industrial production of all other nitrogen fertilizers.

anhydrous borax. (dehydrated borax; anhydrous sodium borate; anhydrous sodium borax; 4-octylbenzoic acid).
CAS: 1330-43-4. $B_4Na_2O_7$.
Properties: White, hygroscopic, nonflammable, free-flowing crystalline substance.
Hazard: Toxic.
Use: Herbicide, manufacture of glass, enamels and ceramics.

anhydrous cupric sulfate. CuO_4S
Properties: Hygroscopic; grayish-white substance consisting of rhombic crystals or powder; water soluble.
Hazard: Toxic; highly irritant.
Use: Fungicide, to detect and remove water from organic compounds.

anhydrous hydrogen fluoride. (anhydrous hydrofluoric acid; aqueous hydrogen fluoride).
CAS: 7664-39-3. HF.
Properties: Extremely acidic substance that forms a weak acid in aqueous solution; colorless gas or fuming liquid below 67F with a strong, irritating odor; molecular weight of 20.0; boiling point of 67F.
Hazard: Toxic; corrosive; can cause painful burns.

anidex. A synthetic fiber designated by the FTC as a cross-linked polyacrylate elastomer.

anileridine. (ethyl-1-(p-aminophenethyl)-4-phenylisonipecotate). $C_{22}H_{28}N_2O_2$.
Properties: White crystalline powder; odorless. Oxidizes and darkens in air and on exposure to light. Exhibits polymorphism of two crystalline forms observed, one melts at approximately 80C and the other at approximately 89C. Soluble in alcohol and chloroform; very slightly soluble in water.
Grade: NF.
Hazard: Addictive.
Use: Medicine (narcotic).

anilide. (2-chloro-N-[2-(4-chlorophyenyl) phenyl]pyridine-3-carboxamide). $C_{18}H_{12}Cl_2N_2O$. Any of a class of amines in which the hydrogen of the amido group is replaced by a phenyl group. These are N-acyl derivatives of aniline.

anilide herbicide.
Use: Pre- and postemergence herbicides to control annual grasses and broadleaved weeds in crops.

aniline. (aniline oil; phenylamine; aminobenzene).

CAS: 62-53-3. $C_6H_5NH_2$. One of the most important of the organic bases, the parent substance for many dyes and drugs.

Properties: Colorless, oily liquid; characteristic odor and taste; rapidly becomes brown on exposure to air and light. Vapors will contaminate foodstuffs and damage textiles. D 1.0235, solidifies at −6.2C, bp 184.4C, wt/gal 8.52 lb (20C), refr index 1.5863 (20C), flash p 158F (70C) (CC), autoign temp 1140F (615C). Soluble in alcohol, ether, and benzene; soluble in water.
Derivation: By (1) catalytic vapor-phase reduction of nitrobenzene with hydrogen; (2) reduction of nitrobenzene with iron filings using hydrochloric acid as catalyst; (3) catalytic reaction of chlorobenzene and aqueous ammonia; (4) ammonolysis of phenol (Japan).
Grade: Commercial, CP.
Hazard: An allergen. Toxic if absorbed through the skin. Combustible. Skin irritant. Questionable carcinogen.
Use: Rubber accelerators and antioxidants, dyes and intermediates, photographic chemicals (hydroquinone), isocyanates for urethane foams, pharmaceuticals, explosives, petroleum refining, diphenylamine, phenolics, herbicides, fungicides.

aniline acetate. $C_6H_5NH_2 \cdot CH_3COOH$.
Properties: Colorless liquid, becomes dark with age; on standing or heating is converted gradually to acetanilide. D 1.070–1.072. Miscible with water and alcohol. Combustible.
Use: Organic synthesis.

aniline, arsenate. See arsenic acid, aniline salt.

aniline black. A black dye developed on cotton and other textiles from a bath containing aniline hydrochloride, an oxidizing agent (usually chromic acid), and a catalyzer (usually a vanadium or copper salt).
Hazard: See aniline.

aniline blue. (disodium 3-(aminomethyl)-2-[[4-(2-sulfoanilino)phenyl]-4-(2-sulfonatophenyl)iminocyclohexa-2,5-dien-1-ylidene]methyl]benzenesulfonate). $C_{32}H_{25}N_3Na_2O_9S_3$
Properties: A mixture of sulfonated triphenylmethane dyes.
Use: A connective tissue stain and counterstain.

aniline chloride. See aniline hydrochloride.

aniline-2,4-disulfonic acid. See 4-amino-m-benzenedisulfonic acid.

aniline-2,5-disulfonic acid.   See 2-amino-*p*-benzene-3,4'-disulfonic acid.

aniline dye.   Any of a large class of synthetic dyes made from intermediates based on or made from aniline. Most are somewhat toxic and irritating to eyes, skin, and mucous membranes. They are generally much less toxic than the intermediates from which they are derived.

aniline hydrochloride.   (aniline salt; aniline chloride). $C_6H_5NH_2 \cdot HCl$.
  Properties: White crystals; commercial product frequently greenish in appearance; darkens in light and air. D 1.2215, mp 198C, bp 245C. Soluble in water, alcohol, and ether.
  Derivation: By (1) passing a current of dry hydrogen chloride gas into an ethereal solution of aniline, (2) neutralizing aniline at 100C with concentrated hydrochloric acid and subsequent crystallization.
  Hazard: See aniline.
  Use: Intermediates, dyeing and printing, aniline black.

aniline ink.   A fast-drying printing ink used on kraft paper, cotton fabric, cellophane, polyethylene, etc. The name is due to the fact that original inks for this purpose were solutions of coal tar dyes in organic solvents. Modern inks usually employ pigments rather than dyes and are of two types: spirit inks containing organic solvent as the vehicle, and emulsion inks in which water is the main vehicle.

aniline point.   The lowest temperature at which equal volumes of aniline and the test liquid are miscible. Cloudiness occurs on phase separation. Used as a test for components of hydrocarbon fuel mixtures.

aniline salt.   See aniline hydrochloride.

***p*-anilinesulfonic acid.**   See sulfanilic acid.

aniline yellow.   See *p*-aminoazobenzene.

1-anilino-4-hydroxyanthraquinone.
  $C_6H_4(CO)_2C_6H_2(OH)NHC_6H_5$ (tricyclic).
  A chemical intermediate.

anilinophenol.   See *p*-hydroxydiphenylamine.

anilophos.
  CAS: 64249-01-0. $C_{13}H_{19}ClNO_3PS_2$.
  Hazard: Moderately toxic by ingestion and skin contact.
  Use: Agricultural chemical.

animal black.   See bone black.

animal diastase.   See amylopsin.

animal oil.   See bone oil.

animal starch.   See glycogen.

anion.   An ion having a negative charge, anions in a liquid subjected to electric potential collect at the positive pole or anode. Examples are hydroxyl, OH^-; carbonate, CO_3^{2-}; phosphate, PO_4^{3-}.

anionic detergent.   Any detergent in which the surface activity moiety bears a negative charge at neutral pH.

***o*-anisaldehyde.**   (*o*-methoxybenzaldehyde; *o*-anisic aldehyde).
  CAS: 135-02-4. $C_6H_4(OCH_3)CHO$.
  Properties: White to light-tan solid; burned, slightly phenolic odor. Bp 238C, mp (two crystalline forms) 38–39C and 3C, d (liquid) 1.1274 (25/25C), (solid) 1.258 (25/25C), refr index 1.5608 (20C), flash p 244F (117C). Insoluble in water; soluble in alcohol. Combustible.
  Grade: 95% (min).
  Use: Intermediate.

***p*-anisaldehyde.**   (aubepine; *p*-anisic aldehyde; *p*-methoxybenzaldehyde).
  CAS: 123-11-5. $C_6H_4(OCH_3)CHO$.

$$CH_3O - \langle \ \rangle - CHO$$

  Properties: Colorless to pale yellow liquid; odor of hawthorn. D 1.119–1.122, refr index 1.570–1.572, mp OC, bp 248C. Insoluble in water; soluble in 5 volumes of 50% alcohol. Combustible.
  Derivation: Obtained from anethole or anisole by oxidation.
  Grade: Liquid and crystals, the latter being the disulfite compound.
  Use: Perfumery, intermediate for antihistamines, electroplating, flavoring.

(-)-anisatin.
  CAS: 5230-87-5. $C_{15}H_{20}O_8$.
  Hazard: Moderately toxic by ingestion.
  Source: Natural product.

anise alcohol.   See anisic alcohol.

anise alcohol.   (4-methoxybenzenmethanol; *p*-methoxybenzyl alcohol; anisyl alcohol; 4-methoxyphenyl)methanol).
  CAS: 105-13-5. $C_8H_{10}O_2$.
  Properties: Organic liquid; nearly insoluble in water; freely soluble in ethanol and ethyl ether.
  Hazards: Extremely toxic.

anise camphor.   See anethole.

anise oil.   (anise seed oil; aniseed oil). See anethole.

anisic acid. (p-methoxybenzoic acid). $CH_3OC_6H_4COOH$.
Properties: White crystals or powder. D 1.385 (4C), mp 184C, bp 275–280C. Soluble in alcohol and ether; almost insoluble in water.
Derivation: Oxidation of anethole.
Use: Repellent and ovicide.

anisic alcohol. (anisyl alcohol; anise alcohol; p-methoxybenzyl alcohol). $CH_3OC_6H_4CH_2OH$.
Properties: Solidifies at room temperature; floral odor. D 1.111–1.114, refr index 1.541–1.545, boiling range 255–265C. Soluble in 1 volume of 50% alcohol; insoluble in water. Combustible.
Derivation: Obtained from anisic aldehyde by reduction.
Grade: Technical, FCC.
Use: In perfumery for light floral odors, pharmaceutical intermediate, flavoring.

anisic aldehyde. See anisaldehyde.

o-anisidine. (o-methoxyaniline; o-amino-anisole).
CAS: 90-04-0. $CH_3OC_6H_4NH_2$.
Properties: Reddish or yellowish oil; becomes brownish on exposure to air, volatile with steam. D 1.097 (20C), bp 225C, fp 5C. Soluble in dilute mineral acid, alcohol, and ether; insoluble in water.
Derivation: (1) Reduction of o-nitroanisole with tin or iron and hydrochloric acid; (2) heating o-aminophenol with potassium methyl sulfate.
Method of purification: Steam distillation.
Grade: 99% (1% max moisture).
Hazard: Strong irritant. Toxic when absorbed through the skin. Possible carcinogen.
Use: Intermediate for azo dyes and for guaiacol.

p-anisidine. (p-methoxyaniline; p-aminoanisole).
CAS: 104-94-9. $CH_3OC_6H_4NH_2$.
Properties: Fused, crystalline mass. Crystallizing point 57.2C (min), d 1.089 (55/55C), bp 242C. Soluble in alcohol and ether; slightly soluble in water.
Derivation: (1) Reduction of p-nitroanisole with iron filings and hydrochloric acid; (2) methylation of p-aminophenol.
Grade: Technical.
Hazard: Strong irritant. Toxic when absorbed through the skin. Questionable carcinogen.
Use: Azo dyestuffs, intermediate.

anisindione. (2,p-anisyl-1,3-indandione). $C_{16}H_{12}O_3$.
Properties: Pale yellow crystals. Mp 156–157C.
Grade: ND.
Use: Anticoagulant (blood).

anisole. (methylphenyl ether; methoxybenzene).
CAS: 100-66-3. $C_6H_5OCH_3$.
Properties: Colorless liquid; aromatic odor. D 0.999 (15/15C), fp −37.8C, bp 155C, refr index 1.5170 (20C), flash p 125F (51.6C) (OC). Soluble in alcohol and ether; insoluble in water. Combustible.
Derivation: From sodium phenate and methyl chloride, heating phenol with methanol.
Use: Solvent, perfumery, vermicide, intermediate, flavoring.

anisomycin. $C_{14}H_{19}NO_4$.
Properties: Needles. Mp 140C. Slightly soluble in water; soluble in low-molecular weight alcohols, ketones, and chloroform; slightly soluble in organic solvents. Water solutions are stable at room temperature over a broad range of pH.
Use: Fungus inhibitor, mildew preventive in vegetables.

anisotropic. Descriptive of crystals whose index of refraction varies with the direction of the incident light. This is true of most crystals, e.g., calcite (Iceland spar); it is not true of isometric (cubic) crystals, which are isotropic.

anisotropy factor. The coefficient by which the contribution of one absorption band to the ordinary refraction has to be multiplied in order to obtain the rotatory contribution of the same band.

anisoyl chloride.
CAS: 100-07-2. $CH_3OC_6H_4COCl$.
Properties: Clear crystals or amber liquid. Mp 22C, bp 262–263C. Soluble in acetone and benzene; decomposed by water or alcohol. Fumes in moist air.
Use: Intermediate for dyes and medicines.
Hazard: Solutions corrosive to tissue. Explosion risk when in closed containers due to pressure caused by decomposition at room temperature.

n-anisoyl-gaba.
CAS: 72432-14-5. $C_{12}H_{15}NO_4$.
Hazard: A poison by ingestion.

anisyl acetate. (p-methoxybenzyl acetate). $CH_3OC_6H_4CH_2COCH_3$.
Properties: Colorless liquid; lilac odor. D 1.104—1.107, refr index 1.514–1.516. Soluble in 4 volumes of 60% alcohol. Combustible.
Derivation: Reaction of anisic alcohol with acetic anhydride.
Grade: Technical, FCC.
Use: Perfumery, flavoring.

anisylacetone. (generic name for 4-(p-methoxyphenyl)-2-butanone). $CH_3OC_6H_4C_2H_4COCH_3$.
Properties: A colorless to pale-yellow liquid. Mp 8C. Combustible.
Use: In insect attractants, organic synthesis, flavoring.

anisyl alcohol. See anisic alcohol.

p-anisylchlorodiphenylmethane. (*p*-methoxytriphenylmethyl chloride). $CH_3OC_6H_4C(C_{62}H_5)_2Cl$.
Properties: Mp 122–124C, mw 308.81.
Grade: Research.
Hazard: Skin and eye irritant.
Use: NH_2- protecting reagent for amino acids in oligonucleotide synthesis.

anisyl formate. (*p*-methoxybenzyl formate). $CH_3OC_6H_4CH_2OCOH$.
Properties: Colorless liquid; lilac odor. D 1.139–1.141. Soluble in 5.5 volumes of 70% alcohol. Combustible.
Use: Perfumery, flavoring.

2,p-anisyl-1,3-indandione. See anisindione.

annatto. $C_{27}H_{34}O_4$. Vegetable dye containing ethyl bixin.
Derivation: From the seeds of *Bixa orellana*.
Occurrence: South America, West Indies, India.
Use: (As extract) coloring margarine, sausage casings, etc.; food-product marking inks. For details, see regulations of Meat Inspection Division, USDA and FDA regulations.
See bixin.

anneal. Generally synonymous with "hybridize."

annealing. Maintenance of glass or metal at a specified temperature for a specific length of time (at least 3 days for plate glass) and then gradual cooling at a predetermined rate. This treatment removes the internal strains resulting from previous operations and eliminates distortions and imperfections. A clearer, stronger, and more uniform material results.
See temper.

annealing point, glass. The temperature at which the internal stress in glass is substantially relieved within 15 minutes.

annellation. A chemical reaction in which one cyclic or ring structure is added to another to form a polycyclic compound.

annotation. Adding pertinent information such as the gene coded for, the amino acid sequence, or other commentary to the database entry of raw sequence of DNA bases.
See bioinformatics.

Ano. A series of dyestuffs used for coloring anodized aluminum.
See anodic coating.

anode. The positive electrode of an electrolytic cell, to which negatively charged ions travel when an electric current is passed through the cell. Such anodes are usually made of graphite or other forms of carbon, although titanium has been successfully introduced in the chloralkali industry. In a primary cell (battery or fuel cell), the anode is the negative electrode.
See cathode; electrode.

anode mud. Residue obtained from the bottom of a copper or other plating bath. In the electrolytic refining of copper, the anode mud contains the relatively inert metals platinum, silver, and gold and is usually collected and treated for the recovery of these metals and other rare elements.

anode process. See electrophoresis.

anodic coating. (anodizing). The electrolytic treatment of aluminum, magnesium, and a few other metals as a result of which heavy, stable films of oxides are formed on their surfaces. A thin oxide film will form on an aluminum surface without special treatment on exposure of the metal to air. This provides excellent resistance to corrosion. This fact led to the development of electrochemical processes to produce much thicker and more effective protective and decorative coatings. The chief electrolytes used are sulfuric, oxalic, and chromic acids. The metal acts as the anode. Such anodic coatings are hard and have good electrical insulating properties. Their ability to absorb dyes and pigments makes it possible to obtain finishes in a complete range of colors, including black. The luster of the underlying metal gives them a metallic sheen; colorants can be used to reproduce the color of any metal with which the aluminum might be used. Anodized coatings can be used as preparatory treatments to electroplating; copper, nickel, cadmium, silver, and iron have been successfully deposited over oxide coatings.
See aluminum.

anodizing. See anodic coating.

anomalous dispersion. The inversion of the usual change of refractive index with wavelength near an absorption band.

anomer. A special kind of diastereoisomer (or epimer) occurring in some sugars and other substances having asymmetric carbon atoms.

anomeric. Denoting sugar modifications that differ only in that the H and OH, or H and OMe on the reducing carbon are reversed.

anomers. Two stereoisomers (α and β) of a given sugar that differ only in the configuration about the carbonyl (anomeric) carbon atom (carbon 1 for pyranoses and carbon 2 for furanoses).

anoxomer.
CAS: 60837-57-2.

Properties: A polymer consisting of 1,4-benzenediol, 2-(1,1-dimethylethyl)-polymer with diethylbenzene, 4-(1,1-dimethylethyl)phenol, 4-methoxyphenol, 4,4'(1-methylethylidene)bis(phenol) and 4-methylphenol.
Use: Food additive.

ANPO. Abbreviation for α-naphthylphenyloxazole.

Anrade's theory. This states that viscosity is the result of transportation of momentum in collisions of molecules.

ANSI. See American National Standards Institute.

antabuse.
Use: Drug for treatment of alcoholism.
See tetraethylthiuram disulfide.

antacid. Any mildly alkaline substance, such as sodium bicarbonate, taken internally or in water solution to neutralize excess stomach acidity.

antagonist, structural. (antimetabolite). An organic compound that is structurally related to a biologically active substance (enzyme, nucleic acid, amino acid, etc.) and that acts as an inhibitor of its growth and development. Such biological antagonism exists between sulfa drugs and *p*-aminobenzoic acids, and also between histamine and a group of compounds collectively called antihistamines. One of the most important of these from an agricultural standpoint is imidazole, which, together with similar compounds, is used to "antagonize" the metabolism of insects, especially those attacking fabrics; it is also being used in irrigation waters to protect plants from pests. Structural antagonists have important medical applications in the complex field of allergic disease.
See antihistamine; anticoagulant; antigen–antibody.

Antarcticine. A high molecular weight glycoprotein obtained from an Antarctic strain of bacteria.
Use: Provides stability to proteins and lipid membranes exposed to extreme cold and dryness.

Antaron FC-34. A high-foaming, water-soluble, amphoteric surfactant with soaplike qualities, a complex fatty amido compound 40% active.
Use: Fulling agent and detergent for woolen and worsted fabrics; effective under neutral, acid, and alkaline conditions; recommended for use in bubble baths, detergents, and in soaps for dedusting purposes.

antazoline. $C_{17}H_{19}N_3$.
Properties: White, odorless crystalline powder with bitter taste. Mp 120C. Sparingly soluble in

alcohol and water; practically insoluble in benzene and ether.
Use: Medicine (antihistamine). Available as hydrochloride and phosphate.

anteiso-. Prefix denoting an isomer (usually a fatty acid or derivative) that has a single, simple branching attached at the third carbon from the end of a straight chain, in distinction to an iso-compound, where the attachment is to the second carbon from the end. For example, isododecanoic acid would be $CH_3CH(CH_3)(CH_2)_8COOH$, while anteisododecanoic acid would be $CH_3CH_2CH(CH_3)(CH_2)_7COOH$.

anthelmintic. An agent used in veterinary medicine as a vermifuge.

anthelmycin. USAN name for an antibiotic substance produced by *Streptomyces longissimus*.

antheridiol. $C_{29}H_{42}O_5$.
Properties: Colorless, fine crystals. Mp 250C. Slightly soluble in water; soluble in warm methanol.
Use: A plant hormone having a specific sex function, it is secreted by certain water molds. It has been used to modify plant fertility. Said to be the first plant sex hormone to be discovered (1942).

anthocyanin. A flavonoid plant pigment that accounts for most of the red, pink, and blue colors in plants, fruits, and flowers. Water soluble.
See flavonoid.

anthonaphthol as. See C.I. azoic coupling component 2.

anthopyllite. $(Mg,Fe)_7Si_8O_{22}(OH)_2$. A natural magnesium-iron silicate.
See asbestos.

anthopyllite. $(Mg,Fe)_7Si_8O_{22}(OH)_2$. A natural magnesium-iron silicate.
See asbestos.

anthracene.
CAS: 120-12-7. $C_6H_4(CH)_2C_6H_4$.
Properties: Yellow crystals with blue fluorescence. D 1.25 (27/4C), mp 217C, bp 340C, fp 250F (121C) (CC). Soluble in alcohol and ether; insoluble in water. Combustible. It has semiconducting properties.

Derivation: (1) By salting out from crude anthracene oil and draining. The crude salts are purified by pressing and finally phenanthrene and carbazole are removed by the use of various solvents. (2) By distilling crude anthracene oil with alkali carbonate in iron retorts, the distillate contains only anthracene and phenanthrene. The latter is removed by carbon disulfide.

Impurities: Phenanthrene, carbazole, and chrysene.

Method of purification: By sublimation with superheated steam or by crystallization from benzene followed by sublimation; for very pure crystals, zone melting of solid anthracene.

Grade: Commercial (90–95%), pure crystals.

Hazard: A questionable carcinogen.

Use: Dyes, alizarin, phenanthrene, carbazole, anthraquinone, calico printing, a component of smoke screens, scintillation counting crystals, organic semiconductor research.

anthracene oil.   A coal tar fraction boiling in the range 270–360C, a source of anthracene and similar aromatics. Also used as a wood preservative and pesticide, except on food crops.

Hazard: A carcinogen.

1,8,9-anthracenetriol.   See anthralin.

anthracite.   See coal.

anthragallic acid.   See anthragallol.

anthragallol.   (1,2,3-trihydroxyanthraquinone; anthragallic acid). $C_6H_4(CO)_2C_6H(OH)_3$. Tricyclic.

Properties: Brown powder. Soluble in alcohol, ether, and glacial acetic acid; slightly soluble in water and chloroform. Sublimes at 290C.

Derivation: Product of the reaction of benzoic, gallic, and sulfuric acids.

Use: Dyeing.

Anthragen.   A series of lake colors. Used for printing inks, wallpaper, coated paper, paint, rubber, and organic plastics.

Anthralan.   A series of acid dyestuffs. Used on wool.

anthralin.   (1,8,9-anthracenetriol; 1,8-dihydroxy anthranol). $C_{14}H_{10}O_3$.

Properties: Odorless, tasteless, yellow powder. Mp 176–181C. Filtrate from water suspension is neutral to litmus. Soluble in chloroform, acetone, benzene, and solutions of alkali hydroxide; slightly soluble in alcohol, ether, and glacial acetic acid; insoluble in water. Combustible.

Derivation: By catalytic reduction of 1,8-dihydroxyanthraquinone with hydrogen at high pressure.

Grade: NF, 95%.

Hazard: Very irritating. Do not use on scalp or near eyes.

Use: Medicine (treatment of psoriasis).

anthranilic acid.   (*o*-aminobenzoic acid).
CAS: 118-92-3. $C_6H_4(NH_2)(CO_2H)$.

Properties: Yellowish crystals; sweetish taste. Mp 144–146C (sublimes). Soluble in hot water, alcohol, and ether. Combustible.

Derivation: Phthalimide plus an alkaline hypobromite solution.

Grade: Technical (95–98%), 99% or better.

Hazard: Questionable carcinogen.

Use: Dyes, drugs, perfumes, and pharmaceuticals.

anthranol.   (9-hydroxyanthracene). $C_{14}H_9OH$.

Properties: Crystals. Mp 120C. Soluble in organic solvents with a blue fluorescence. Changes in solution to anthrone. Combustible.

Use: Dyes.

anthranone.   See anthrone.

anthrapurpurin.   (1,2,7-trihydroxy anthraquinone; isopurpurin; purpurin red). $C_6H_3OH(CO)_2C_6H_2(OH)_2$. Tricyclic.

Properties: Orange-yellow needles. Mp 369C, bp 462C. Soluble in alcohol and alkalies; slightly soluble in ether and hot water.

Derivation: By fusion of anthraquinonedisulfonic acid with caustic soda and potassium chlorate; the melt is run into hot water and the anthrapurpurin precipitated by hydrochloric acid.

Use: Dyeing, organic synthesis.

anthraquinone.
CAS: 84-65-1. $C_6H_4(CO)_2C_6H_4$.

Properties: Yellow needles. D 1.419–1.438, mp 286C, bp 379–381C, flash p 365F (185C) (CC). Soluble in alcohol, ether, and acetone; insoluble in water. Combustible.

Derivation: (1) By heating phthalic anhydride and benzene in the presence of aluminum chloride and dehydrating the product; (2) by condensation of 1,4-naphthoquinone with butadiene.

Method of purification: Sublimation.

Grade: Sublimed, 30% paste (sold on 100% basis), electrical 99.5%.

Hazard: Possible carcinogen.

Use: Intermediate for dyes and organics, organic inhibitor, bird repellent for seeds.

See anthraquinone dye.

anthraquinone-1,5- and 1,8-disulfonic acids.
(rho acid, chi acid, respectively). $C_{14}H_8O_8S_2$.

Properties: (in pure state) Slightly yellow to white crystals. The technical grade is grayish-white.

Soluble in water and strong sulfuric acid. The 1,8-isomer is the more soluble. The 1,5-disulfonic acid melts with decomposition at 310–311C. The 1,8-isomer melts and decomposes at 293–294C.

Derivation: Anthraquinone is sulfonated with fuming sulfuric acid in the presence of mercury or mercuric oxide to a mixture of the 1,5- and 1,8-disulfonic acids, which are separated by fractional crystallization.

Method of purification: Fractional crystallization from strong sulfuric acid, or in the form of their alkali salts from either acid or alkaline solutions.

Grade: Technical.

Use: Dyes.

anthraquinone   dye. $C_6H_4(CO)_2C_6H_4$.   A dye whose molecular structure is based on anthraquinone. The chromophore groups are $=C=O^-$ and $=C=C=$. The benzene ring structure is important in the development of color. CI numbers from 58000 to 72999. These are acid or mordant dyes when OH or HSO_3 groups, respectively, are present. The anthraquinone dyes that can be reduced to an alkaline solution leuco (vat) derivative having affinity for fibers and that can be reoxidized to the dye are known as anthraquinone vat dyes. They are largely used on cotton, rayon, and silk, and have excellent properties of color and fastness.

anthrarufin.   See 1,5-dihydroxyanthraquinone.

anthrax.   A notifiable disease caused by *Bacillus anthracis* toxin. It occurs most commonly among domestic cattle and sheep but pigs and horses can also be affected. Humans are sometimes infected by contact with infect meat products or by direct contact with diseased animals.

Hazards: Hemorrhage and serous effusions in various organs and body cavities; extreme prostration; death.

anthrax toxin.   (*Bacillus anthracis* toxin). A culture filtrate of *Bacillus anthracis* that contains at least three different substances: an edema factor, a lethal factor, and a protective antigen.

anthrone.   (anthranone;   9,10-dihydro-9-oxoanthracene). $C_{14}H_{10}O$. The keto is the more stable form of anthranol.

Properties: Colorless needles. Mp 156C. Insoluble in water; soluble in alcohol, benzene, and hot sodium hydroxide solutions.

Derivation: Reduction of anthraquinone with tin and hydrochloric acid.

Use: Rapid determination of sugar in body fluids, and of animal starch in liver tissue; general reagent for carbohydrates; organic synthesis.

anti-.   (1) A prefix used in designating geometrical isomers in which there is a double bond between the carbon and nitrogen atoms. This prevents free rotation, so that two different spatial arrangements of substituent atoms or groups are possible. When a given pair of these are on opposite sides of the double bond, the arrangement is called *anti-*; when they are on the same side, it is called *syn-*, as indicated below:

$$C_6H_5-\underset{HO-N}{\overset{\|}{C}}-H \qquad C_6H_5-\underset{N-OH}{\overset{\|}{C}}-H$$
$$\text{anti} \qquad\qquad \text{syn}$$

These prefixes are disregarded in alphabetizing chemical names. (2) A prefix meaning "against" or "opposed to," as in antibody, antimalarial, etc.

antianxiety agent.   See psychotropic drug.

antiarsenin.   A nonarsenical substance thought to be secreted by the body in response to the administration of doses of arsenic trioxide (as arsenic acid) that are protective against toxic responses to a higher dosage.

antiauxin.   A substance that inhibits the growth-regulating effects of an auxin, sometimes by preventing auxin transport.

antiauxin herbicide.   Any of a number of growth-regulating compounds that are chemically related to auxins. These compounds have effects on cell elongation that are generally opposite to those of auxins. They stimulate root elongation and inhibit that of coleoptile segments.

antibiotic.   A chemical substance produced by microorganisms that has the capacity in dilute solutions to inhibit the growth of other microorganisms or destroy them. Only approximately 20 out of several hundred known have proved generally useful in therapy. Those that are used must conform to FDA requirements. The most important groups of antibiotic-producing organisms are the bacteria, lower fungi, or molds, and actinomycetes. These antibiotics belong to very diverse classes of chemical compounds. Most of the antibiotics produced by bacteria are polypeptides (such as tyrothricin, bacitracin, polymyxin). The penicillins are the only important antibiotics produced by fungi. Actinomycetes produce a wide variety of compounds (actinomycin, streptomycin, chloramphenicol, tetracycline). The antimicrobial activity (antibiotic spectrum) of antibiotics varies greatly; some are active only on bacteria, others on fungi, still others on bacteria and fungi; some are active on viruses, some on protozoa, and some are also active on neoplasms. An organism sensitive to an antibiotic may, on continued contact with it, develop resistance and yet remain sensitive to other antibiotics. Certain antibiotics are used as direct food additives to inhibit growth of bacteria and fungi; among these are nisin, pimaricin, nystatin, and tylosin. *Note:* A

number of antibiotics have been restricted by FDA, both for direct use by humans and as additives to animal feeds. Among those are streptomycin, chloramphenicol, tetracycline, and penicillin.
See penicillin; cephalosporin; plasmid; Waksman; Fleming.

antibiotic resistance.   Plasmids generally contain genes that confer on the host bacterium the ability to survive a given antibiotic. If the plasmid pBR322 is present in a host, that host will not be killed by (moderate levels of) ampicillin or tetracycline. By using plasmids containing antibiotic resistance genes, the researcher can kill off all the bacteria that have not taken up this plasmid, thus ensuring that the plasmid will be propagated as the surviving cells divide.

"Antiblaze" *[Hickory].*   TM for a trichloropropyl phosphate additive flame retardant.
Use: On either side of two-component rigid urethane systems and in rebonded flexible foam, coatings, and elastomers.

antiblock agent.   A substance (e.g., a finely divided solid of mineral nature) that is added to a plastic mix to prevent adhesion of the surfaces of films made from the plastic to each other or to other surfaces. They are of particular value in polyolefin and vinyl films. The hard, infusible particles tend to roughen the surface and so maintain a small air space between adjacent layers of the film, thus preventing adhesion. Silicate minerals are widely used for this purpose. Another type of antiblock function is performed by high-melting waxes, which bloom to the surface and form a layer that is harder than plastic.

antibody.   See antigen–antibody.

anticaking agent.   An additive used primarily in certain finely divided food products that tend to be hygroscopic to prevent or inhibit agglomeration and thus maintain a free-flowing condition. Such substances as starch, calcium metasilicate, magnesium carbonate, silica, and magnesium stearate are used for this purpose in table salt, flours, sugar, coffee, whiteners, and similar products.

anticancer drug.   See antineoplastic.

antichlor.   Any product that serves to neutralize and remove hypochlorite or free chlorine after a bleaching operation. For many years the term was considered synonymous with sodium thiosulfate, but it may equally be applied to sodium disulfite or any other product used for the purpose.

anticholinergic.   A drug or pharmaceutical that inhibits the action of acetylcholine by reactivating the cholinesterase. Examples are atropine sulfate and pralidoxime iodide.

anticipation.   Each generation of offspring has increased severity of a genetic disorder; e.g., a grandchild may have earlier onset and more severe symptoms than the parent, who had earlier onset than the grandparent.
See additive genetic effects; complex trait.

anticholinesterase.   (acetylcholinesterase inhibitor; cholinesterase inhibitor). Any substance that inhibits or inactivates acetylcholinesterase, thereby preventing the hydrolysis of acetylcholine. Their action leads to an accumulation of endogenous acetylcholine with a resultant hyperactivity of cholinergic neurons, which can prove lethal.
Hazard: Tremors; convulsions; respiratory failure; death.
Use: Pesticides.

anticoagulant.   A complex organic compound, often a carbohydrate, that has the property of retarding the clotting or coagulation of blood. The most effective of these is heparin, which acts by interfering with the conversion of prothrombin to thrombin and by inhibiting the formation of thromboplastin. In addition to their specific clinical uses, anticoagulants have been applied with limited success to rodenticides (see warfarin). They are regarded as cumulative poisons, requiring multiple ingestions to be lethal. One type (diaphenadione) has been found to reduce blood cholesterol in experimental animals.

anticoagulant poison.   An anticoagulant.
Use: To control small mammal pests.

anticoagulant rodenticide.   A rodenticide that causes massive internal bleeding by virtue of its action as an anticoagulant. These rodenticides are potentially hazardous to all mammals and birds.

anticodon.   A specific sequence of three nucleotides in a tRNA, complementary to a codon for an amino acid in an mRNA.

anticonvulsant.   An agent used to prevent, control, or to relieve convulsions states.

anticrotalus serum.   An antivenomous serum which is protective against envenomation by crotalids.

anticurare.   A chemical agent that neutralizes the myotoxic action of curare.

anticytotoxin.   A substance, usually a specific antibody, that inhibits the action of a cytotoxin.

antidepressant.   See psychotropic drug.

antidote.   Any substance that inhibits or counteracts the effects of a poison which has entered the body by any route. Mild acids or alkalies (except sodium bicarbonate) exert neutralizing action if corrosive materials have been swallowed; for noncorrosive poisons, warm salt water or milk may be given to cause vomiting. Activated charcoal in water is effective in protecting the throat and stomach linings, except for corrosive poisons. Nothing should be administered by mouth if the subject is unconscious. Artificial respiration may be necessary. In no case should alcohol be used as an antidote. Atropine sulfate and pralidoxime iodide have been successfully used as antidotes for poisoning by cholinesterase inhibitors.

antiemetic agent. A compound usually classed as a pharmaceutical that inhibits or prevents nausea. A well-known type is dimenhydrinate ("Dramamine" *[Prestige]*) which is useful to counteract motion sickness, nausea due to pregnancy, etc. A more recent antiemetic is a benzoquinolizine derivative used for nausea resulting from anesthesia (Emete-con).

antiendotoxin. An antibody that counteracts the effects of an endotoxin.

antienzyme. A substance present in the substrate that restricts or negates the catalytic activity of the enzyme on that substrate.

antiestrogen. Any substance that chemically reduces or inhibits the biological action of an estrogen.

antifertility agent. A synthetic steroid sex hormone of the type normally produced by the body during pregnancy. Said to act by simulating the conditions of pregnancy and thus suppressing ovulation, which automatically prevents conception. The hormones used are basically of two kinds: (1) progestin (synthetic progesterone) and (2) synthetic estrogen. There are also several other derivatives, all of which are much stronger than natural progesterone and estrogen when taken orally. The chemical modifications are necessary for oral potency. Injectable contraceptives that remain effective over much longer periods than the oral type are also available. The active ingredient is medoxyprogesterone acetate. Experimental work is also being done with N,N'-octaethylenediamine bis(dichloroacetamide) (Fertilysin). An estrogen-free type is available in England (chloromadinone acetate); it is said to be less effective, but with less tendency to cause blood clotting than estrogenic types.
Hazard: FDA requires that oral contraceptives carry a label warning of the tendency of these agents to form blood clots. There is also a possibility that they have other adverse side effects. *Note:* Proteins and peptides that act as enzyme inhibitors have been identified in the semen of some mammals.

antifoam agent. See defoaming agent.

antifouling paint. An organic coating formulated especially for use on the hulls and bottoms of ships, boats, buoys, pilings, and the like to protect them from attack by barnacles, teredos, and other marine organisms. The chief specific ingredient is a metallic naphthenate, e.g., copper naphthenate; mercury compounds are also used.

antifreeze. (1) Water additive. Any compound that lowers the freezing point of water. Both sodium chloride and magnesium chloride were once used, but their extreme corrosive properties made them a liability in automotive cooling systems. Methanol requires only 27% by volume for protection to $-17.7C$. Due to its tendency to evaporate rapidly at operating temperatures, its flammability, and its low boiling point (63.9C), it has been replaced by glycol derivatives which are relatively noncorrosive, nonflammable, have very low evaporation rates, and are effective heat-exchange agents. A concentration of 35% protects against freezing to $-17.7C$. Ethylene and propylene glycol antifreezes can be carried in an automotive cooling system for several years without damage, and are satisfactory coolants at summer operating temperatures. Methoxypropanol has been introduced as an antifreeze-coolant for diesel engines. (2) Gasoline additive. A proprietary preparation, TM "Drygas," *[Cristy]* consists of methanol, isopropanol, or mixtures of these, that lowers the freezing point of water enough to inhibit ice formation in feed lines and carburetors. It is added directly to the gasoline.
Hazard: Poisonous. Flammable.
See coolant.

Antigen–antibody. An antibody is a blood serum protein of the globulin fraction that is formed in response to introduction of an antigen. It has a molecular weight of approximately 160,000. An antigen is an infective organism (protein) with molecular weight of at least 10,000; it is able to induce formation of an antibody in an organism into which it is introduced (by injection). Thus an animal is able to resist infections to which it has previously been exposed. The entire science of immunology is based on antigen–antibody reactions, the most outstanding feature of which is their specificity. The antibodies produced in the bloodstream can react only with the homologous antigen or with those of a similar molecular structure. As a result, the animal can destroy a particular virus or bacterium and become immune. The specificity of antigens is due not so much to the composition of the molecule as to its configuration. Certain radicals (polar and quaternary ammonium groups) seem to "mate" with corresponding complementary structures in the antibody molecule. A precipitate or agglutinate is formed by the reaction which is analogous to colloidal (catalytic) reactions in some respects, e.g., surface configuration.
See immunochemistry; Pasteur; allergen; antagonist, structural.

antiglobulin. An agent used to coagulate globulin.

antihistamine. A synthetic substance structurally analogous to histamine whose presence in minute amounts prevents or counteracts the action of excess histamine formed in body tissues. These compounds are usually complex amines of various types and also have other physiological effects and medical uses. Examples are chlorpheniramine maleate, dimenhydrinate, diphenhydramine hydrochloride, imidazole, pheniramine maleate, pyrilamine maleate, thonzylamine hydrochloride, tripelennamine hydrochloride.
See antagonist, structural.

antihypertensive agent. An organic compound having the property of lowering blood pressure in animals and humans. Among the better-known types are the alkaloid reserpine and its derivative syrosingopine, guanethidine sulfate, α-methyl dopa (α-methyl-3,4-dihydroxyphenylalanine), and hydralazine. They function by various nerve-blocking mechanisms involving structural antagonism. They should be taken only by prescription.

antiinflammatory agent. Any of a number of drugs that prevent or inhibit inflammation of tissue. Most common of these is aspirin.

antiinvasin. Either of the two enzymes present in normal blood plasma that are antagonistic to hyaluronidase.

antiknock agent. Any of a number of organic compounds that increase the octane number of a gasoline when added in low percentages by reducing knock, especially in high-compression engines. Knock is caused by spontaneous oxidation reactions in the cylinder head, resulting in loss of power and characteristic ignition noise. Branched-chain hydrocarbon gasolines ameliorate this problem, and antiknock additives virtually eliminate it. Tetraethyl lead, the most effective of these, has been used for many years, but its contribution to air pollution has almost eliminated its use in automotive fuels. Lead-free gasolines are now used in conjunction with catalytic converters. The antiknock agents used in them are nonmetallic compounds such as methyl-*tert*-butyl ether (MTBE) or a mixture of methanol and *tert*-butyl alcohol.
See octane number; gasoline.

antilymphocytic serum. (ALS). An immunological suppressant for use in organ transplants. It acts to control the buildup of rejection factors in the blood that result from introduction of foreign organs into the body.
See immunochemistry.

antimatter. See antiparticle.

antimetabolite. See antagonist, structural; antihistamine; metabolite.

antimonate. Salt in which antimony has a valence of 5.

antimonial lead alloy. (hard lead). Lead alloy containing approximately 6–28% antimony. Common grades are as follows: (1) 15% antimony, resistant to sulfuric acid used in type metal; (2) national stock pile specification, 10.7–11.3% antimony; (3) battery grids, 5–11% antimony; (4) battery terminals, 4% antimony; (5) cable sheaths, 1% antimony.

antimonic. Adjective used for compounds in which antimony has a valence of 5, e.g., antimony pentachloride, pentasulfide, etc.

antimonic acid. See antimony pentoxide.

antimonic anhydride. See antimony pentoxide.

antimonite. See stibnite.

antimonous. (antimonious). Adjective used for compounds in which antimony has a valence of 3, as in antimony tribromide, antimony trichloride, antimony trioxide, antimony trisulfide.

antimonous sulfide. See antimony trisulfide.

antimony.
CAS: 7440-36-0. Sb (from Latin stibium). Metallic element of atomic number 51, group VA of the periodic table, aw 121.75, valences 3, 4, and 5. Two stable isotopes.
Properties: Silver-white solid. Mp 630.5C, bp 1635C. Low thermal conductivity, Mohs hardness 3 to 3.5. Oxidized by nitric acid; not attacked by hydrochloric acid in absence of air; reacts with sulfuric acid and aqua regia. Combustible. A semiconductor.
Available forms: Besides the stable metal, there are two allotropes: yellow crystals and amorphous black modifications.
Source: Ores: Stibnite, kermasite, tetrahedrite, livingstonite, jamisonite.
Occurrence: Algeria, Bolivia, China, Mexico, South Africa, Peru, Yugoslavia.
Derivation: Reduction of stibnite with iron scrap; direct reduction of natural oxide ores. About half the antimony used in the U.S. is recovered from lead base battery scrap metal.
Grade: Up to 99.999% pure, technical, powder, commercial grade in 55-lb cakes $10 \times 10 \times 2.5$ inches.
Hazard: Use with adequate ventilation. Soluble salts are toxic.
Use: Hardening alloy for lead, especially storage batteries and cable sheaths; bearing metal; type metal; solder; collapsible tubes and foil; sheet and

pipe; semiconductor technology (99.999% grade); pyrotechnics.

antimony-124. Radioactive antimony isotope.
Properties: Half-life 60 days, radiation β and γ. The chemical form used is often antimony trichloride and oxychloride in hydrochloric acid solution.
Use: As a tracer, especially in solid-state studies and marker of interfaces between products in pipelines. The γ ray has sufficient energy to eject neutrons from beryllium. Convenient portable neutron sources that may be reactivated in a nuclear reactor are made by irradiation of an antimony pellet encased in a beryllium shell.
Hazard: Radioactive poison.

antimony-125. Radioactive antimony isotope. Half-life 2.4 years, emits β and γ rays.
Hazard: Radioactive poison.

antimony black. Metallic antimony in the form of a fine powder produced by electrolysis or chemical action on an antimony salt solution.
See antimony trisulfide.

antimony bromide. See antimony tribromide.

antimony, caustic. See antimony trichloride.

antimony chloride. See antimony trichloride.

antimony chloride, basic. See antimony oxychloride.

antimony dichlorotrifluoride. $SbCl_2F_3$. A thick liquid stored in iron drums. Used as catalyst for fluorocarbon manufacture.

antimony fluoride. See antimony trifluoride.

antimony hydride. (stibine).
CAS: 7803-52-3. SbH_3.
Properties: Colorless gas. Mp −88C, bp −17C.
Derivation: Action of hydrogen chloride on antimony-metal compounds such as Zn_3Sb_2; also released by reduction of antimony compounds in hydrochloric acid solutions with zinc or other reducing metal.
Hazard: Toxic.

antimony iodide. See antimony triiodide.

antimonyl. The radical or group SbO which occurs commonly in formulas of antimony compounds. Thus SbOCl is often named antimonyl chloride, and numerous other antimony compounds are sometimes named in a similar manner.

antimony lactate.
CAS: 58164-88-8. $Sb(C_3H_5O_3)_3$.
Properties: Tan-colored mass. Soluble in water.
Grade: Technical.

Hazard: Toxic.
Use: Mordant in fabric dyeing.

antimony oxide. See antimony trioxide.

antimony oxychloride. (antimony chloride, basic; antimonyl chloride). SbOCl.
Properties: White powder. Mp 170C (decomposes). Soluble in hydrochloric acid and alkali tartrate solutions; insoluble in alcohol, ether, and water.
Derivation: Interaction of water and antimony chloride.
Hazard: Toxic.
Use: Antimony salts, flameproofing textiles.

antimony pentachloride. (antimony perchloride).
CAS: 7647-18-9. $SbCl_5$.
Properties: Reddish-yellow, oily liquid; offensive odor; hygroscopic. Mp 2.8C, d 2.34, bp 92C (30 mm). Solidifies by absorption of moisture. Decomposed by excess water into hydrochloric acid and antimony pentoxide. Soluble in an aqueous solution of tartaric acid in hydrochloric acid and chloroform.
Derivation: Action of chlorine on antimony powder.
Hazard: Corrosive, fumes in moist air, reacts strongly with organics.
Use: Analysis (testing for alkaloids and cesium), dyeing intermediates, as chlorine carrier in organic chlorinations.

antimony pentafluoride.
CAS: 7783-70-2. SbF_5.
Properties: Viscous, hygroscopic liquid. D 2.99 (23C), mp 7C, bp 149.5C, hydrolyzed by water. Soluble in potassium fluoride (KF) and liquid sulfur dioxide.
Derivation: Antimony pentachloride and anhydrous hydrogen fluoride.
Hazard: Corrosive to skin and tissue.
Use: Catalyst and/or source of fluorine in fluorination reactions.

antimony pentasulfide. (antimony red; antimony persulfide; antimony sulfide golden).
CAS: 1315-04-4. Sb_2S_5.
Properties: Orange-yellow powder; odorless. Insoluble in water; soluble in concentrated hydrochloric acid with evolution of hydrogen sulfide; soluble in alkali. Decomposes on heating.
Hazard: Flammable, dangerous fire risk near oxidizing materials.
Use: Red pigment, rubber accelerator.

antimony pentoxide. (antimonic anhydride; antimonic acid; stibic anhydride).
CAS: 1314-60-9. Sb_2O_5.
Properties: White or yellowish powder. D 3.78, mp 450C, loses oxygen above 300C. Slightly soluble in water; soluble in strong bases forming antimonates; insoluble in acids except concentrated hydrochloric acid.

Derivation: Action of concentrated nitric acid on the metal or the trioxide.
Use: Preparation of antimonates and other antimony compounds, flame retardant for textiles.

antimony persulfide.   See antimony pentasulfide.

antimony potassium tartrate.   (tartar emetic; potassium antimonyl tartrate; tartrated antimony). $K(SbO)C_4O_6 \cdot 1/2H_2O$.
Properties: Transparent, odorless crystals efflorescing on exposure to air or white powder; sweetish metallic taste. D 2.6, at 100C loses all its water. Soluble in water, glycerol; insoluble in alcohol. Aqueous solution is slightly acid.
Derivation: By heating antimony trioxide with a solution of potassium bitartrate and subsequent crystallization.
Grade: Technical, crystals, powdered, CP, USP.
Hazard: Toxic.
Use: Textile and leather mordant, medicine, insecticide.

antimony salt.   (deHaens salt). Mixture of antimony trifluoride and either sodium fluoride or ammonium sulfate.
Properties: White crystals. Soluble in water.
Hazard: Toxic.
Use: Dyeing and printing textiles.

antimony sodiate.   See sodium antimonate.

antimony sulfate.   (antimony trisulfate).
CAS: 7446-32-4. $Sb_2(SO_4)_3$.
Properties: White powder or lumps. Deliquescent, decomposes in water. D 3.62 (4C), flammable.
Hazard: A poison.
Use: Matches, pyrotechnics.

antimony tribromide.   (antimony bromide). $SbBr_3$.
Properties: Yellow, deliquescent, crystalline mass. Decomposed by water, d 4.148, mp 96.6C, bp 280C. Soluble in carbon disulfide, hydrobromic acid, hydrochloric acid, ammonia.
Hazard: Toxic.
Use: Analytical chemistry, mordant, manufacturing antimony salts.

antimony trichloride.   (antimonous chloride; antimony chloride; caustic antimony).
CAS: 10025-91-9. $SbCl_3$.
Properties: Colorless, transparent, very hygroscopic, crystalline mass. Fumes slightly in air. D 3.14, bp 223.6C, mp 73.2C. Soluble in alcohol, acetone, acids. With water forms antimony oxychloride.
Derivation: Interaction of chlorine and antimony or by dissolving antimony sulfide in hydrochloric acid.
Grade: Technical, CP.

Hazard: Corrosive liquid or solid. Very irritating to eyes, skin.
Use: Antimony salts, bronzing iron, mordant, manufacturing lakes, chlorinating agent in organic synthesis, pharmaceuticals, fireproofing textiles, analytical reagent.

antimony trifluoride.   (antimony fluoride).
CAS: 7783-56-4. SbF_3.
Properties: White to gray hygroscopic crystals. Mp 292C, d 4.58. Soluble in water.
Grade: 99–100%.
Hazard: Strong irritant to eyes and skin.
Use: Porcelain, pottery, dyeing, fluorinating agent.

antimony triiodide.   (antimony iodide). SbI_3.
Properties: Red crystals. Volatile at high temperatures. D 4.768, mp 167C, bp 420C. Soluble in carbon disulfide, hydrochloric acid, and solution of potassium iodide; insoluble in alcohol and chloroform; decomposes in water with precipitation of oxyiodide.
Derivation: Action of iodine on antimony.

antimony trioxide.   (antimony white; antimony oxide).
CAS: 1309-64-4. Sb_2O_3. Occurs in nature as valentinite.
Properties: White, odorless, crystalline powder. D 5.67; mp 655C. Insoluble in water; soluble in concentrated hydrochloric and sulfuric acids, strong alkalies. Amphoteric.
Derivation: Burning antimony in air, adding ammonium hydroxide to antimony chloride, directly from low-grade ores.
Grade: Technical, pigment.
Hazard: Possible carcinogen during production.
Use: Flameproofing of textiles, paper, and plastics (chiefly polyvinyl chloride); paint pigments; ceramic opacifier; catalyst; intermediate; staining iron and copper; phosphors; mordant; glass decolorizer.

antimony trisulfate.   See antimony sulfate.

antimony trisulfide.   (antimony orange; black antimony; antimony needles; antimonous sulfide; antimony sulfide).
CAS: 1345-04-6. Sb_2S_3.
Properties: (1) Black crystals; (2) orange-red crystals. D 4.562, mp 546C. Insoluble in water; soluble in concentrated hydrochloric acid and sulfide solutions.
Derivation: (1) Occurs in nature as black crystalline stibnite; (2) As precipitated from solutions of salt of antimony, the trisulfide is an orange-red precipitate, which is filtered, dried, and ground.
Grade: Technical.
Hazard: Explosion risk in contact with oxidizing materials. Questionable carcinogen.

Use: Vermilion or yellow pigment, antimony salts, pyrotechnics, matches, percussion caps, camouflage paints (reflects infrared radiation in same way as green vegetation), ruby glass.

antimycin A$_1$.   (C$_{28}$H$_{40}$O$_9$N$_2$). An antibiotic substance said to have strong fungicidal properties.
Properties: Crystals. Mp 139–140C. Soluble in alcohol, ether, acetone, and chloroform; slightly soluble in benzene, carbon tetrachloride, and petroleum ether; insoluble in water.
Derivation: From *Streptomyces*.
Use: Active against a large group of fungi but, in general, not against bacteria; possible insecticide and miticide.

antineoplastic.   A drug that inhibits the formation of tumors (neoplasms). Many of these are antibiotics used in treatment of cancer.
See bleomycin.

antineutron.   See antiparticle.

antioxidant.   An organic compound added to rubber, natural fats and oils, food products, gasoline, and lubricating oils to retard oxidation, deterioration, rancidity, and gum formation, respectively. Rubber antioxidants are commonly of an aromatic amine type, such as di-β-naphthyl-*p*-phenylenediamine and phenyl-β-naphthylamine 1% or less based on the rubber content of a mixture affords adequate protection. Many antioxidants are substituted phenolic compounds (butylated hydroxyanisole, di-*tert*-butyl-*p*-cresol, and propyl gallate). Food antioxidants are effective in very low concentration (not more than 0.01% in animal fats) and not only retard rancidity but protect the nutritional value by minimizing the breakdown of vitamins and essential fatty acids. Sequestering agents, such as citric and phosphoric acids, are frequently employed in antioxidant mixtures to nullify the harmful effect of traces of metallic impurities.
Note: Max concentration of food antioxidants approved by FDA is 0.02%.

antiozonant.   (antiozidant). A substance used to reverse or prevent the severe oxidizing action of ozone on elastomers both natural and synthetic. Among antiozonant materials used are petroleum waxes, both amorphous and microcrystalline, secondary aromatic amines (such as *N,N*-diphenyl-*p*-phenylenediamine), quinoline, and furan derivatives.
See ozone.

antiparallel.   Describing two linear polymers that are opposite in polarity or orientation.

antiparticle.   (antimatter). Any of several species of subatomic particles that are identical in mass to ordinary particles, but opposite in electrical charge or (in the case of the neutron) in magnetic moment.

Thus, a positron is an electron with a positive charge, an antiproton is a proton with a negative charge, and an antineutron has no charge but has a magnetic moment opposite to that of a neutron. A photon is its own antiparticle. When an antiparticle collides with its opposite particle (e.g., a collision of an electron and a positron) both particles are annihilated and their masses are converted to photons of equivalent energy. The same is true of other subatomic particles (neutrinos, mesons, etc.) some of which are fantastically short-lived (of the order of billionths of a second).

antiperspirant.   Any substance having a mild astringent action that tends to reduce the size of skin pores and thus restrain the passage of moisture on local body areas. The most commonly used antiperspirant agent is aluminum chlorohydrate. Use of zirconium compounds in antiperspirant sprays has been virtually discontinued because of their suspected carcinogenicity, though they are permissible in creams. Antiperspirants exert a neutralizing action that gives them deodorant properties. The FDA classified them as drugs rather than as cosmetics.

antiport.   Co-transport of two solutes across a membrane in opposite directions.

antiprecipitant.   Substances that prevent precipitation in chemical processes.

antiproton.   See antiparticle.

antipsychotic agent.   See psychotropic drug.
antipyretic.
Use: Any of a group of drugs that reduce fever or inflammation, e.g., aconite.

antipyrine.   (phenazone; 2,3-dimethyl-1-phenyl-3-pyrazolin-5-one).
CAS: 60-80-0.   (CH$_3$)$_2$(C$_6$H$_5$)C$_3$HN$_2$O.
Properties: Colorless crystals or powder; odorless; slightly bitter taste. D 1.19, mp 110–113C, bp 319C. Soluble in water, alcohol, and chloroform; slightly soluble in ether.
Derivation: Condensation of methylphenylhydrazine and ethyl acetoacetate.
Method of purification: Crystallization.
Grade: Technical, NF.
Use: Medicine (analgesic); analytical reagent for nitrous acid, nitric acid, and iodine number.

antipyrine chloral hydrate.   See chloral hydrate antipyrine.

antipyrine iodide.   See iodophenazone.

antirheumatic.   Any of various drugs used in the treatment of rheumatoid arthritis. Among these are certain gold salts, e.g., disodium aurothiomalate and gold sodium thiosulfate. Penicillamine is also reported to be effective. Cortisone is no longer widely used because of deleterious side effects.

antiscorbutic.   Tending to prevent scurvy. See ascorbic acid.

antisense.   Nucleic acid that has a sequence exactly opposite to an mRNA molecule made by the body; binds to the mRNA molecule to prevent a protein from being made. See transcription.

antisense DNA.   One of the two strands of double-stranded DNA, usually that which is complementary to the mRNA, that is the nontranscribed or non-coding strand.

antisense molecule.   (antisense nucleotide). An oligonucleotide fragment or an analog that is complementary to, and binds with, an RNA or a DNA segment and inhibits its normal function.

antisense RNA.   An RNA sequence that is complementary to all or part of a functional mRNA molecule, with which it hybridizes, thereby blocking translation.

anti-sense strand.   See discussion under "Sense strand."

antiseptic.   A substance applied to humans or animals that retards or stops the growth of microorganisms without necessarily destroying them, e.g., alcohol; boric acid and borates; certain dyes, as acriflavine; menthol; hydrogen peroxide; hypochlorites; iodine; mercuric chloride; and phenol. Many of these are corrosive and poisonous, and should be used with great caution. Among the newer antiseptics are hexachlorophene, which is also toxic, and some quaternary ammonium compounds. See disinfectant; sanitizer; fumigant.

antiskinning agent.   A liquid antioxidant used in paints and varnishes to inhibit formation of an oxidized film on the exposed surface in cans, pails, or other open containers.

antistatic agent.   The marked tendency of thermoplastic polymers to accumulate static charges which result in adherent particles of dust and other foreign matter has required study of possible means of eliminating or reducing this property. The following have been tried: (1) Development of more electrically conductive polymers, e.g., tetracyanoquinodimethane. (2) Incorporation of additives that migrate to the surface of the plastic or fiber and modify its electrical properties. (Examples of these are fatty quaternary ammonium compounds, fatty amines, and phosphate esters.) Other types of antistatic additives are hygroscopic compounds, such as polyethylene glycols and hydrophobic slip additives, that markedly reduce the coefficient of friction

of the plastic. (3) Copolymerization of an antistatic resin with the base polymer.

Antistine phosphate.   Antazoine phosphate.

antitussive.   A medicinal preparation for suppressing coughs, often containing codeine. Chloroform is no longer permitted as an ingredient.

antizymotic.   Property of preventing fermentation. A material that prevents fermentation.

Antonow's rule.   The rule states that the interfacial tension of two liquids in equilibrium is equal to the difference between the surface tensions.

Antox.   A rubber antioxidant, a condensation product of butyraldehyde-aniline. Amber liquid.

Antron.   Nylon textile fibers in the form of continuous filament yarns and staple.

ANTU.   (α-naphthylthiourea). CAS: 86-88-4. **Hazard:** Questionable carcinogen.

A-number.   A number that indicates the amount of fatty acids precipitated from edible fats.

AOAC International.   (AOAC). Formerly called the Association of Official Agricultural Chemists, and the Association of Official Analytical Chemists. It was founded in 1884. There are 4000 members. These scientists develop, test, and study methods for analyzing fertilizer, foods, feeds, drugs, pesticides, cosmetics and other products related to agricultural and public health. It is located at 481 North Frederick Avenue, Suite 500, Gaithersburg, MD 20877-2417. Website: http://www.aoac.org

AOCS.   See American Oil Chemist's Society.

AOD process.   Injection of a mixture of argon and oxygen into molten steel to reduce carbon impurities.

AP.   Abbreviation for ammonium perchlorate.

ap-.   A prefix denoting formation from or relationship to another compound, e.g., apomorphine.

apamin.   (apamine). $C_{79}H_{131}N_{31}O_{24}S_4$. Occurs in the venom of the honey bee, *Apis mellifera*; molecular weight of 2027.38. **Properties:** Small, highly potent, highly basic polypeptide. **Hazard:** Central nervous system poison able to cross the blood–brain barrier; neurotoxic; poison.

AP-1 site.   The binding site on DNA at which the transcription "factor" AP-1 binds, thereby altering the rate of transcription for the adjacent gene. AP-1 is actually a complex between c-fos protein and c-jun protein, or sometimes is just c-jun dimers. The AP-1 site consensus sequence is (C/G)TGACT(C/A)A. Also known as the TPA-response element (TRE). [TPA is a phorbol ester, tetradecanoyl phorbol acetate, which is a chemical tumor promoter.

APAP. Abbreviation for acetyl-*p*-amino-phenol. See *p*-acetylaminophenol.

apatite. A natural calcium phosphate (usually containing fluorine) occurring in the earth's crust as phosphate rock. It is also the chief component of the bony structure of teeth.
Properties: Color variable. D 3.1–3.2, hardness 5, frequently in hexagonal crystals.
Occurrence: Eastern U.S., California, the former U.S.S.R., Canada, Europe.
Use: Source of phosphorus and phosphoric acid, manufacture of fertilizers, laser crystals.
See fluoridation.

apazone. (5-(dimethylamino-9-methyl-2-propyl-1H-pyrazolo[1,2-a][1,2,4]benzotriazine-1,3(2h)-dione; 3-dimethylamino-7-methyl-1,2-(*n*-pro-pylmalonyl)-1,2-dihydro-1,2,4-benzotriazine; azapropazone; 5-(dimethylamino)-9-methyl-2-propylpyrazolo[1,2a][1,2,4]benzotriazine-1,3-dione; azapropazon; azapropazone; cin-namon; cinnopropazone; 1,2-dihydro-3-dimethylamino-7-methyl-1,2-(propylmalonyl)-1,2,4-benzotriazine; 3-dimethylamino-7-methyl-1,2-(*n*-propylmalonyl)-1,2-dihydro-1,2,4-benzotriazine; MI 85; MSC-102824;prolixan; rheumox; sinnamin). $C_{16}H_{20}N_4O_2$
Properties: Nearly colorless, crystalline solid.
Hazard: Nausea, vomiting, abdominal pain, gastric ulcers; moderately toxic.
Use: Anti-inflammatory agent; analgesic; antipyretic; to treat rheumatoid arthritis and gout.

APC. Abbreviation for ammonium perchlorate.

apertomer. A device for measuring the angular aperture of an objective. It was invented by Abbé.

aphicide. An insecticidal agent that kills aphids.

aphid. (plant louse). Any homopterous insect, aphids are nearly cosmopolitan in distribution and many are among the most serious pests of crops and orchards. They feed on plant juices, causing developmental anomalies, distorted growth, wilt-ing, and/or gall formation; they transmit certain important viral diseases of plants. Aphids secrete a sweet liquid that is attractive to ants.
apholate.
CAS: 52-46-0. $C_{12}H_{24}N_9P_3$. Generic name for 2,2,4,4,6,6-hexakis(1-aziridinyl)-2,2,4,4,6,6-hexahydro-1,3,5,2,4,6-triazatriphosphorine.

Prevents reproduction in certain insects by inhibiting formation of DNA in eggs.
Hazard: Questionable carcinogen.
Use: Insect sterilant.

aphrodine. See yohimbine.

API. See active pharmaceutical ingredient.

API. See American Petroleum Institute.

apigenin 8-c-glucoside. See 8-β-d-glucopyranosyl-apigenin.

API gravity. A scale of density measurement adopted by the API. It runs from 0.0 (equivalent to d 1.076) to 100.0 (equivalent to d 0.6112). The API values as used in the petroleum industry decrease as density increases.

aplysiatoxin.
CAS: 52659-57-1. $C_{32}H_{47}BrO_{10}$.
Hazard: A poison.

APO. See triethylenephosphoramide.

apoatropine. (apoatropin; atropamin; atropamine; atropyltropene; apatropine; endo-α-methylenebenzeneacetic acid 8-methyl-8-azabicyclo[3,2,1]oct-3-yl ester; 1-α-*h*, 5-α-*h*-tropan-3-α-ol atropate; atropamine; atropyl-tropeine; (8-methyl-8-azabicyclo[3,2,1]octan-3-yl)2-phenylprop-2-enoate). $C_{17}H_{21}NO_2$.
Properties: Alkaloid composed of colorless crys-talline prisms.
Derivation: Occurs in the roots of *Atropa belladonna* and is derived from atropine by the action of nitric acid.
Hazard: Highly toxic; poison.
Use: Antispasmodic agent.

apocarotenal. Food color supplied in dark purplish-black beadlets. Vitamin A activity 120,000 units/g. Dispersible in warm water. Approved for food use by FDA.
See β-apo-8′-carotenal.

β-apo-8′-carotenal.
CAS: 1107-26-2. $C_{30}H_{40}O$.
Properties: Fine crystalline powder with dark metal-lic sheen or violet crystals. Mp: 139°. Sol in chlo-roform; sltly sol in acetone; insol in water.
Use: Food additive.

apoenzyme. The protein portion of an enzyme, absent any organic or inorganic cofactors or pros-thetic groups that might be required for catalytic activity.

apolar adsorption. (homopolar adsorption). Adsorption in which molecules of the adsorbed sub-stance are concentrated at the adsorption layer with-out any fission.

apomine black gx.
CAS: 1937-37-7. $C_{34}H_{25}N_9O_7S_2 \cdot 2Na$.
Hazard: Confirmed carcinogen. Moderately toxic by ingestion and inhalation. An eye irritant.

apomorphine.   (apomorfin; apomorphine; 6-α-β-aprophine-10,11-diol).
CAS: 58-00-4. $C_{17}H_{17}NO_2$. A derivative of morphine that is a dopamine D2 agonist.
Properties: Alkaloid; white crystalline mass; turns green on exposure; weakly soluble in water.
Hazard: Poison; central nervous system effects.
Use: To treat acute poisoning; in the diagnosis and treatment of parkinsonism; a weak sensitizer and a powerful emetic.

apopino oil.   See shiu oil.

apoplectic anthrax.   A type of anthrax that affects livestock and other animals. Death may ensue within a few minutes to a few hours following infection.

apoprotein.   A polypeptide chain or protein that has not yet complexed with the prosthetic group or metal to form the active holoprotein.

apoptosis.   Programmed cell death, the body's normal method of disposing of damaged, unwanted, or unneeded cells.
See cell.

apparent density.   See density.

Appert, Nicolas.   (1752–1841). A French pioneer in the science of food preservation. Though not a chemist, his work on application of heat to food products led to a form of home preserving that eventually developed into the canning industry. The idea of destroying bacteria by heat treatment was later applied more exhaustively by Pasteur.

applaud.   See buprofezine.

apple acid.   See malic acid.

apple oil.   See isoamyl valerate.

applied research.   The experimental investigation of a specific practical problem for the immediate purpose of creating a new product, improving an older one, or evaluating a proposed ingredient. The experimental program is set up to answer the question "What happens?" rather than "Why does it happen?" Examples are the determination of the value of a new rubber antioxidant, the substitution of one drying oil for another in a paint formulation, and the development of a new synthetic product. While applied research has produced the multitude of new materials that have revolutionized industry, agriculture, and medicine in the last 50 years or so,

its achievements have usually been practical outgrowths of prior fundamental research.

approved name.   The official name for a chemical adopted by a government bureau, trade, or medical association.

apramycin.
CAS: 37321-09-8. $C_{21}H_{41}N_5O_{11}$.
Properties: Crystals. Mp: 245–247C. Sol in water; sltly sol in lower alcs.
Use: Drug (veterinary); food additive.

aprotic solvent.   A type of solvent that neither donates nor accepts protons. Examples: dimethylformamide, benzene, dimethyl sulfoxide.

APS.   Abbreviation for appearance potential spectroscopy.
See spectroscopy.

aptation.   Change in an organism resulting from natural selection; a structure which is the result of such selection.

apurinic acid.   DNA from which the purine bases have been removed by mild acid treatment.

araban.   A polysaccharide and a constituent of some pectins. It yields arabinose on hydrolysis.

arachnida.   A large class of invertebrates that are predaceous terrestrial forms such as scorpions, spiders, harvestmen, mites, ticks, and related forms. They are characterized by a cephalothorax that bears four pairs of walking appendages, a pair of usually sensory pedipalps and a pair of prehensile buccal chelicerae that are often served by poison glands; the abdomen lacks appendages. The sexes are separate, development is direct, compound eyes are lacking, most species breathe via tracheae, book lungs, or gill lungs, they eat only fluids and excrete by coxal glands and malpighian tubules.

ardrox.
Properties: Fluorescent yellow dye.
Use: With UV light to visualize cyanoacrylate ester fumed friction ridge detail in human digits.

arene carbonyl.   A carbonyl in which a metal atom is bonded to an aromatic group and coordinated with several carbon monoxide molecules.

arene epoxide.   Any of a class of epoxides derived from arenes by the 1,2-addition of an oxygen atom to a double bond.

arginine ester hydrolase.   Any of a class of enzymes that catalyze the hydrolysis of the ester or peptide linkage of an arginine residue. They are constituents of many crotalid and viperid venoms and some hydrophid venoms.

arginine glutamate. (2-amino-5-(diamino methylideneamino)pentaacid; 2-aminopentanedioic acid). $C_{11}H_{23}N_5O_6$. The L(+)-arginine salt of glutamic acid.
Use: Intravenous ammonia detoxicant in cases of liver failure.

a ribonucleoside. 5'-triphosphate functioning as a phosphate group donor in the cell energy cycle as a carrier of chemical energy between metabolic pathways by serving as a shared intermediate coupling endergonic and exergonic reactions.

aroclor. Any of a variety of commercial mixtures of polychlorinated biphenyls and/or triphenyls. Each mixture is identified by a four-digit number, the first of which indicates the presence of biphenyls (12), triphenyls (54), or both (25, 44). The last two digits represent the average percentage by weight of chlorine in the mixture.
Hazard: Immunotoxic; carcinogen.
Use: Originally used in heat exchangers, transformers, capacitors, and other types of electrical equipment.

aroclor 1242. (chlorodiphenyl 42% chlorine; 2,4-dichloro-1-(2,4-dichlorophenyl)benzene). CAS: 53469-21-9. $C_{12}H_6Cl_4$.
Properties: Clear; mobile liquid; approximately 3.1 chlorine atoms per molecule.

aroclor 1254. (chlorodiphenyl 54% chlorine; 1,2,3-trichloro-4-(2,3-dichlorophenyl)benzene). CAS: 11097-69-1. $C_{12}H_5Cl_5$. A mixture of polychlorinated biphenyls that induces hepatic microsomal UDP-glucuronyl transferase activity toward thyroxine.
Properties: Light-yellow viscous liquid; approximately 4.96 chlorine atoms per molecule.
Hazards: Toxic, carcinogen.

aroclor 1260. (1,2,3-trichloro-4-(2,3,4-trichlorophenyl)benzene). CAS: 11096-82-5. $C_{12}H_4Cl_6$.
Properties: Soft, sticky, light-yellow resin; approximately 6.3 chlorine atoms per molecule.

aromatic alcohol. (aryl alcohol). An alcohol in which the hydroxyl group is a substituent of an alkyl side chain of an aromatic compound.

aromatic amine. (arylamine; aniline). C_6H_7N. Any amine derived from an aromatic hydrocarbon by the replacement of at least one hydrogen on the benzene ring by an amino group.
Properties: Somewhat unpleasant odor of rotten fish, ignites readily, colorless, slowly oxidizes and resinifies in air giving a red-brown tint to aged samples.
Hazard: Carcinogen; bioaccumulate.
Use: In the manufacture of precursors to polyurethane.

aromatic amino acid. An amino acid that contains a benzene ring.
Hazard: Hydrophogic.

aromatic ring. (benzene ring). An exceptionally stable six-membered planar ring of carbon atoms in which the Π electrons are delocalized and the ring is sometimes said to have three conjugated double bonds.

arsane. H_3As. Any of a class of saturated hydrides of trivalent arsenic.
Hazard: Toxic.

arsenate. (arsoric acid). AsO_4. Any salt or ester of arsenic acid. They are distributed widely in nature and contaminate coal and metal ores. Arsenates uncouple oxidative phosphorylation by replacing inorganic phosphorous in the ATP.
Hazard: Carcinogen; toxic.
Use: Pesticides; herbicides; fungicides; algicides.

arsenilic acid. (4-aminophenylarsonic acid; *p*-aminobenezenearsonic acid; atoxylic acid). An organic arsenical compound.
Derivation: Prepared by the reaction of aniline with arsenic acid.
Hazard: Toxic.
Use: Manufacture of other arsenicals; additive to the rations of poultry and swine to promote growth and control dysentery in swine.

arsenide. (arseniuret). Any compound of arsenic.
Derivation: From arsines in which one or more hydrogen atoms are replaced by a metal.

arsenite. (arsorous acid). AsO_3. Any salt or ester of arsenous acid or any salt that contains trivalent arsenic.
Derivation: Occur widely in nature.
Hazard: Corrosive to tissues; carcinogen; poisonous.
Use: Pesticides.

arsenopyrite. An arsenic-containing mineral that can be smelted to produce elemental arsenic.

Artemisia californica. A sagebrush that secretes volatile terpenes that absorb to the soil and inhibit the growth of grasses found in the soft chaparral communities of southern California.

artemisia. (sage; sagebrush; mugwort). A genus of more than 200 species of plants (Family asteraceae). They produce commercially important volatile oils.
Properties: Chief constituent is Thujone.
Derivation: Found in the plant *Artemisia absinthium L.*
Hazard: Moderately toxic; an allergen.

aqua ammonia. See ammonium hydroxide.

"Aquacar" *[Gibson].*　TM for water treatment microbiocides that are aqueous solutions of glutaraldehyde.
Use: For controlling slime-forming bacteria, sulfate-reducing bacteria, and algae in water cooling towers, air washers, pasteurizers, and other recirculation water systems.
See glutaraldehyde.

"Aquaclean Alkaline" *[Geberit].*　TM for a synthetic detergent hold cleaner.
Use: To remove stains and discoloring after transporting coke and coal cargoes.

"Aquaclean Clean Up" *[Geberit].*　TM for a cleaning agent.
Use: After transport of cargoes where the hold did not have barrier or release agent installed prior to loading.

"Aquaclean HD" *[Geberit].*　TM for a cargo hold cleaning material.
Use: To clean cargo containers after transport of oily cargos.

"Aquaflow" *[Flow].*　TM for rheology cellulose.

"Aquapel" *[Nano-Tex].*　TM for an alkylketene dimer that is reactive with alkaline cellulose.
Available forms: Liquid.
Use: For sizing agents.

"Aquaprint" *[Aquasol].*　TM for a resin-bonded pigment color for printing on textiles. The vehicle, an oil-in-water emulsion, contains a water-insoluble binder that adheres to the fibers and anchors the color permanently to the cloth.

aqua regia.　(nitrohydrochloric acid; chloronitrous acid; chlorazotic acid).
Properties: Fuming yellow, volatile, suffocating liquid.
Derivation: A mixture of nitric and hydrochloric acids, usually one part of nitric acid to three or four parts of hydrochloric acid.
Grade: Technical.
Hazard: A powerful oxidizer, toxic, corrosive liquid.
Use: Metallurgy, testing metals, dissolving metals (platinum, gold, etc.).

"Aquarex" *[Windmoller].*　TM for a series of wetting agents for elastomers. They act as stabilizers and mold lubricants.

"Aquasorb" *[Consolidated].*　TM for a cellulose gum.

"Aquasorb AR" *[Consolidated].*　TM for a phosphorus pentoxide (P_2O_5)-based desiccant.
Hazard: Powerful oxidizer and caustic.

"Aquatreat" *[Akzo Nobel].*　TM for water-soluble dithiocarbamate salt bactericides.

Use: In cooling towers, paper and pulp mills, air washers, petroleum production, and sugar processing.

aqueous tension.　Water vapor pressure at a given temperature over pure water or an aqueous solution.

Ar.　Symbol for argon.

ara-A.　(α-β-D-arabinofuranosyladenine; vidarabine).
CAS: 24356-66-9. A biologically active pharmaceutical product having both antitumor and antiviral properties. It was originally prepared (1959) by chemical synthesis at Stanford Research Institute, and later isolated from a fermentation beer of *Streptomyces antibioticus.*

arabic gum.　(acacia gum).
CAS: 9000-01-5. The dried, water-soluble exudate from the stems of *Acacia senegal* or related species.
Properties: Thin flakes, powder, granules, or angular fragments; color white to yellowish white; almost odorless; mucilaginous consistency. Completely soluble in hot and cold water, yielding a viscous solution of mucilage; insoluble in alcohol. The aqueous solution is acid to litmus. Combustible.
Occurrence: Sudan, West Africa, Nigeria.
Derivation: A carbohydrate polymer, complex and highly branched. The central core or nucleus is D-galactose and D-glucuronic acid (actually the calcium, magnesium, and potassium salts), to which are attached sugars such as L-arabinose and L-rhamnose.
Grade: USP, FCC (both grades of acacia).
Use: Pharmaceuticals, adhesives, inks, textile printing, cosmetics, thickening agent, and colloidal stabilizer in confectionery and food products, binding agent in tablets, emulsifier.

arabinogalactan.　A water-soluble polysaccharide extracted from timber of the western larch trees. It is a complex, highly branched polymer of arabinose and galactose.
Properties: Dry, light-tan powder. Readily soluble in hot and cold water; both powder and solutions fairly stable. Combustible.
Use: Dispersing and emulsifying agent, lithography.

arabinose.　(pectin sugar; gum sugar). $C_5H_{10}O_5$. Both the D- and L-enantiomers occur naturally. L-Arabinose is common in vegetable gums, especially arabic.

$$
\begin{array}{c}
\text{CHO} \\
|\\
\text{HCOH} \\
|\\
\text{HOCH} \\
|\\
\text{HOCH} \\
|\\
\text{CH}_2\text{OH}
\end{array}
$$

Properties: White crystals. Mp 158.5C, d 1.585 (20/4C). Soluble in water and glycerol; insoluble in alcohol and ether. Combustible.
Use: Culture medium.

Ara-C. Abbreviation for cytosine arabinoside.

arachidic acid. (eicosanoic acid).
$CH_3(CH_2)_{18}COOH$. A widely distributed but minor component of the fats of peanut oils and related plant species.
Properties: Shining, white, crystalline leaflets. Mp 75.4C, d 0.2840 (100/4C), bp 328C (decomposes), refr index 1.4250. Soluble in ether; slightly soluble in alcohol; insoluble in water. Combustible.
Derivation: From peanut oil.
Grade: Technical, 99%.
Use: Organic synthesis, lubricating greases, waxes and plastics, source of arachidyl alcohol, biochemical research.

arachidonic acid. (5,8,11,14-eicosatetraenoic acid).
CAS: 506-32-1. $CH_3(CH_2)_4(CH:CHCH_2)_4(CH_2)_2$ COOH. A C_{20} unsaturated fatty acid. Combustible. An essential fatty acid.
Source: Liver, brain, glandular organs; also made synthetically.
Use: Biochemical research, source of prostaglandins and other pharmacologically active compounds. See eicosanoid.

arachidyl alcohol. (1-eicosanol).
$CH_3(CH_2)_{18}CH_2OH$. A long-chain saturated fatty alcohol much like stearyl alcohol.
Properties: White, waxlike solid. Mp 66.5C; bp 369C (220C at 3 mm Hg); refr index 1.455. Soluble in hot benzene. Combustible.
Derivation: Ziegler synthesis (trialkylaluminum process).
Grade: Technical, 99%.
Use: Lubricants, rubber, plastics, textiles, research.

arachin. (arachine). A protein from peanuts, a globulin containing arginine, histidine, lysine, cystine; yellow-green syrup, soluble in water and alcohol, insoluble in ether. Combustible.

aragonite. A form of calcium carbonate appearing in pearls.
See nacre.

aragonite needles. Slender crystals of the mineral aragonite that constitute most carbonate muds in the modern ocean. Some of the needles form by direct precipitation from seawater, and some by the collapse of the skeletons of organisms.

aralkonium chloride. $C_{21}H_{36}Cl_3N$.
Properties: Water-soluble solid; bitter taste.
Use: As sanitizer and deodorant.

aralkyl. See arylalkyl.

aramid. Generic name for a distinctive class of highly aromatic polyamide fibers that are characterized by their flame-retardant properties. Some types are also suitable for protective clothing, dust-filter bags, tire cord, and bullet-resistant structures. They are derived from *p*-phenylenediamine and terephthaloyl chloride.
See "Nomex" *[Du Pont]*; "Kevlar" *[Du Pont]*; polyamide.

"Aramite" *[ExcelAg]*.
CAS: 140-57-8. $(CH_3)_3CC_6H_4OCH_2CH(CH_3)$ $OSOOC_2H_4Cl$. TM for 2-(*p-tert*-butylphenoxy)isopropyl-2-chloroethyl sulfite.
Properties: Clear, light-colored oil. D 1.148–1.152 (20C); bp 175C (0.1 mm Hg). Very soluble in common organic solvents; insoluble in water. Noncorrosive.
Grade: Technical (90% min), wettable powder, emulsifiable concentrate restricted to postharvest application on fruit trees.
Hazard: A possible carcinogen. Irritant to eyes and skin, toxic by ingestion.
Use: Antimicrobial agent, miticide.

"Aranox" *[Chemtura]*. TM for *p*-(*p*-tolysulfonylamido)-diphenylamine.
$CH_3C_6H_4SO_2NHC_6H_4NHC_6H_5$.
Properties: Gray powder. D 1.32; mp 135C (min). Soluble in acetone, benzene, and ethylene dichloride; insoluble in gasoline and cold water; slightly soluble in hot water or hot alkaline solutions.
Use: Antioxidant for light-colored rubber products.

"Arasan" *[Du Pont]*. TM for seed disinfectants based on thiram.

"Arazate" *[Chemtura]*. TM for zinc dibenzyl dithiocarbamate.

arbutin. (ursin). $C_{12}H_{16}O_7$. Available commercially in both natural and synthetic forms. Pure synthetic arbutin is hydroquinone-β,D-glucopyranoside.

Properties: White powder (pure synthetic). Mp 199–200C. Soluble in water and alcohol. Stable in storage.
Derivation: A glucoside found in the leaves of the cranberry, blueberry, and manzanita shrubs, and in the roots, trunks, and leaves of most pear species.

Pure arbutin can be prepared synthetically from ace-tobromoglucose and hydroquinone in the presence of alkali.
Use: Oxidation inhibitor, polymerization inhibitor, color stabilizer in photography, intermediate.

arc furnace. A furnace that is heated by an electric arc with carbon or graphite as one electrode, and a similar unit, or the furnace charge, as the other electrode.

archaometry. Application of chemical and physical analytical methods to archaeology. Among those used are microanalytical methods, spectroscopic analysis, X-rays, and other types of nondestructive tests. For age determination, ^{14}C measurement (chemical dating) is one of the most valuable techniques.

arene. See aromatic.

Arens-van Dorp synthesis. The preparation of alkoxyethynyl alcohols from ketones and ethoxy-acetylene. In the Isler modification, β-chlorovinyl ether is reacted with lithium amide to give lithium ethoxyacetylene which is then condensed with the ketone. This avoids the tedious preparation of ethoxyacetylene.

argentite. (silver glance). Ag_2S. Lead-gray to black or blackish-gray mineral. A natural silver sulfide. Contains 87.1% silver. Differs from other soft black minerals in cutting like wax. Soluble in nitric acid, d 7.2–7.36, Mohs hardness 2–2.5.
Occurrence: Nevada, Colorado, Montana, Mexico, Chile, Canada.
Use: An important ore of silver.

argentum. The Latin name for silver, hence the symbol Ag.

arginase. An enzyme that produces ornithine and urea by splitting arginine. It is found in liver.
Use: Biochemical research.

arginine. (guanidine aminovaleric acid; amino-4-guanidovaleric acid).

$$H_2N \diagdown \atop HN \diagup C-N-CH_2CH_2CH_2CH-C-OH \atop \underset{H}{|} \ \ \ \ \ \ \underset{O}{\|}, \ NH_2$$

An essential amino acid for rats, occurring naturally in the L(+) form. Available as glutamate and hydrochloride.
Properties: Prisms from water containing two molecules of H_2O, anhydrous plates from alcohol solution. Dehydrates at 105C, decomposes at 244C. Sparingly soluble in alcohol; insoluble in ether.

Derivation: Widely found in animal and plant proteins. It is precipitated as flavianate from gelatin hydrolyzate in industry.
Use: Biochemical research, medicine, pharmaceuticals, dietary supplement.

5-*l*-argininecyanoginosin la.
CAS: 101043-37-2. $C_{49}H_{74}N_{10}O_{12}$.
Hazard: A poison by ingestion and inhalation.
Source: Natural product.

arginine vasopressin.
CAS: 113-79-1. $C_{46}H_{65}N_{15}O_{12}S_2$.
Hazard: A poison.

argon.
CAS: 7440-37-1. Ar. A nonmetallic element of atomic number 18, in the noble gas group of the Periodic System. Aw 39.948. Present in atmosphere to 0.94% by volume.
Properties: Colorless monatomic gas; odorless; tasteless. It is not known to combine chemically with any element, but forms a stable clathrate with β-hydroquinone. Fp −189.3C, bp −185.8C, d 1.38 (air = 1), sp vol 9.7 cu ft/lb (21.1C @ 1 atm). Slightly soluble in water. Noncombustible; an asphyxiant gas.
Derivation: (1) By fractional distillation of liquid air. (2) By the treatment of atmospheric nitrogen with metals such as magnesium and calcium to form nitrides. (3) Recovery from natural gas oxidation bottoms-steam in ammonia plant. (4) Originally formed by radioactive decay of ^{40}K.
Method of purification: (1) Highly purified argon is obtained by passing the gas through a bed of titanium at 850C. (2) Synthetic zeolite molecular sieves separate oxygen from argon to give high purity gas.
Grade: Technical, highest purity (99.995%).
Use: Inert-gas shield in arc welding, furnace brazing, plasma jet torches (with hydrogen), electric and specialized light bulbs (neon, fluorescent, sodium vapor, etc.; titanium and zirconium refining; flushing molten metals (steel) to remove dissolved gases; in Geiger-counter tubes; lasers; inert gas or atmosphere in miscible applications; decarburization of stainless steel (AOD process).

Argyrol. TM for an organic compound of silver and a protein, used in medicine for its specific antiseptic and bacteriostatic action.

Aridye. A product and process for printing colors on textiles using permanent and insoluble pigments suspended in an organic vehicle into which water is emulsified to give printing consistency. The vehicle contains a water-insoluble binder that adheres to the clot and anchors the color permanently to the fibers.

aripiprazole.
CAS: 129722-12-9. $C_{23}H_{27}Cl_2N_3O_2$.
Hazard: A poison.

aristolochic acid, sodium salt.   See sodium aristolochate I.

arkose.   A rock consisting primarily of sand-sized particles of feldspar. Most arkose accumulates close to the source area of the feldspar, because feldspar weathers quickly to clay and seldom travels far.

Armalon.   TFE-fluorocarbon fiber felt and also for TFE-fluorocarbon resin-coated glass fabrics, tapes, and laminates.

Armstrong's acid.   (naphthalene-1,5-disulfonic acid). $C_{10}H_6(SO_3H)_2$.
Properties: White, crystalline solid. Soluble in water.
Derivation: Sulfonation of naphthalene with fuming sulfuric acid at low temperature followed by separation from the 1,6-isomer.
Use: Dye intermediate.

Arndt-Eistert synthesis.   Procedure for converting an acid to its next higher homolog.

Arnel.   An acetate fiber made from cellulose triacetate. It has a higher melting point, and is less soluble than cellulose acetate.
See acetate fiber; cellulose triacetate.

aromatic.   (arene). A major group of unsaturated cyclic hydrocarbons containing one or more rings, typified by benzene, which has a 6-carbon ring containing three double bonds. The vast number of compounds of this important group, derived chiefly from petroleum and coal tar, are rather highly reactive and chemically versatile. The name is due to the strong and not unpleasant odor characteristic of most substances of this nature. Certain 5-membered cyclic compounds such as the furan group (heterocyclic) are analogous to aromatic compounds.
Note: The term "aromatic" is often used in the perfume and fragrance industries to describe essential oils that are not aromatic in the chemical sense.

aromaticity.   A stable electron shell configuration in organic molecules, especially those related to benzene.
See resonance; orbital theory.

aromatic oils (hydrocarbons).   See aromex.

aromatization.   See hydroforming.

aromex.
CAS: 78308-32-4.
Hazard: A poison by ingestion.
Use: Agricultural chemical.

Arosurf TA100.   (distearyldimethylammonium chloride).
CAS: 107-64-2. A powder cationic-quaternary fabric softener.

Use: For retail powdered detergent-softener, industrial laundry and pap-softening formulas, and cosmetic formulations.

arrack.   An oriental distilled liquor which is obtained from palm or rice juice.

arrayed library.   Individual primary recombinant clones (hosted in phage, cosmid, YAC, or other vector) that are placed in two-dimensional arrays in microtiter dishes. Each primary clone can be identified by the identity of the plate and the clone location (row and column) on that plate. Arrayed libraries of clones can be used for many applications, including screening for a specific gene or genomic region of interest.
See library; genomic library; gene chip technology.

arrest point.   The temperature at which a system of more than one component, being heated or cooled absorbs or yields heat without changing temperature.

Arrhenius, Svante.   (1859–1927). A native of Sweden, he won the Nobel Prize in chemistry in 1903. He is best known for his fundamental investigations on electrolytic dissociation of compounds in water and other solvents, and for his basic equation stating the increase in the rate of a chemical reaction with rise in temperature:

$$\frac{d \ln k}{dT} = \frac{A}{RT^2}$$

in which k is the specific reaction velocity, T is the absolute temperature, A is a constant usually referred to as the energy of activation of the reaction, and R is the gas-law constant.

arrowroot.   (maranta). The starch that is obtained from the roots of the maranta plant, which has many uses, including food ingredients, cosmetics, glues, and starches.

arsacetin.   (sodium   acetylarsanilate; sodium *p*-acetyl   aminophenylarsonate). $CH_3CONHC_6H_4AsO(OH)ONa$.
Properties: White, crystalline powder; odorless; tasteless. Free of arsenous or arsenic acid. Solutions will admit of thorough sterilization. Soluble in cold water, but more so in warm water.
Use: Medicine (antisyphilitic).

9-arsafluoreninic acid.
CAS: 5687-22-9. $C_{12}H_9AsO_2$.
Hazard: A poison by ingestion.
Use: Agricultural chemical.

arsanilic acid.   (atoxylic acid; *p*-aminobenzenearsonic acid; *p*-aminophenylarsonic acid). $C_6H_4 \cdot C_6H_8AsNO_3$.

Properties: White, crystalline powder; practically odorless. Mp 232C. Soluble in hot water; slightly soluble in cold water, alcohol, and acetic acid; insoluble in acetone, benzene, chloroform, and ether.
Derivation: By condensing aniline with arsenic acid, removing the excess of aniline by steam distillation in alkaline solution, and setting the acid free using hydrochloric acid.
Hazard: A poison. Yields flammable vapors on heating above melting point.
Use: Arsanilates, manufacture of arsenical medicinal compounds such as arsphenamine, veterinary medicine, grasshopper bait.

arsenic.
CAS: 7440-38-2. As. A nonmetallic element of atomic number 33, group Va of periodic table, aw 74.9216, valences of 2, 3, 5; no stable isotopes.
Properties: Silver-gray, brittle, crystalline solid that darkens in moist air. Allotropic forms: black, amorphous solid (β-arsenic); yellow, crystalline solid, d 5.72 (commercial product ranges from 5.6 to 5.9), mp 814C (36 atm), sublimes at 613C (1 atm), Mohs hardness 3.5, insoluble in water and in caustic and nonoxidizing acids. Attacked by hydrochloric acid in presence of oxidant. Reacts with nitric acid. Low thermal conductivity; a semiconductor.
Derivation: Flue dust of copper and lead smelters from which it is obtained as white arsenic (arsenic trioxide) in varying degrees of purity. This is reduced with charcoal. The commercial grade is not made in the U.S.
Grade: Technical, crude (90–95%), refined (99%), semiconductor grade (99.999%), single crystals.
Hazard: Confirmed carcinogen and mutagen. OSHA employee exposure limit:
Use: (Metallic form) Alloying additive for metals, especially lead and copper as shot, battery grids, cable sheaths, boiler tubes. High-purity (semiconductor) Grade: Used to make gallium arsenide for dipoles and other electronic devices; doping agent in germanium and silicon solid state products; special solders; medicine.
See arsenic trioxide.

arsenic acid.   (orthoarsenic acid).
CAS: 7778-39-4. $H_3AsO_4 \cdot 1/2H_2O$. Arsenic pentoxide is also sometimes called arsenic acid.
Properties: White, translucent crystals. D 2–2.5, mp 35.5C, bp (loses water at 160C). Soluble in water, alcohol, alkali, glycerol.
Derivation: By digestion of arsenic with nitric acid.
Grade: Pure, technical, CP.
Hazard: Confirmed carcinogen.
Use: Manufacture of arsenates, glass making, wood treating process, defoliant (regulated), desiccant for cotton, soil sterilant.

arsenic acid, aniline salt.
CAS: 63957-41-5. $C_{18}H_{21}N_3 \cdot AsH_3O_4$.
Hazard: A poison by ingestion.
Use: Agricultural chemical.

arsenical Babbitt.   See Babbitt metal.

arsenical nickel.   See niccolite.

arsenic anhydride.   See arsenic pentoxide.

arsenic, black.   (β-arsenic).
See arsenic.

arsenic bromide.   Legal label name for arsenic tribromide.

arsenic chloride.   See arsenic trichloride.

arsenic disulfide.   (arsenic monosulfide; ruby arsenic; red arsenic glass; red arsenic sulfide; red arsenic).
CAS: 56320-22-0. As_2S_2 or AsS. Occurs as the mineral realgar.
Properties: Orange-red powder. D 3.4–2.6, mp 307C. Soluble in acids and alkalies; insoluble in water.
Derivation: By roasting arsenopyrite and iron pyrites and sublimation.
Grade: Technical.
Use: Leather industry, depilatory agent, paint pigment, shot manufacture, pyrotechnics, rodenticide, taxidermy.

arsenic hydride.   See arsine.

arsenic pentafluoride. AsF_5. A gas, bp −52.8C, fp −79C. Readily hydrolyzed, soluble in alcohol and benzene.
Use: Doping agent in electroconductive polymers.

arsenic pentasulfide. As_2S_5.
Properties: Yellow or orange powder. Soluble in nitric acid and alkalies; insoluble in water. Decomposes to sulfur and the trisulfide when heated.
Derivation: By precipitation from arsenic acid in a hydrochloric acid solution with hydrogen sulfide. It is filtered, then dried.
Grade: Technical.
Use: Paint pigments, light filters, other arsenic compounds.

arsenic pentoxide.   (arsenic oxide; arsenic anhydride; arsenic acid).
CAS: 1303-28-2. As_2O_5.
Properties: White, amorphous solid; deliquescent; forms arsenic acid in water. D 4.086, mp 315C. Soluble in water, alcohol.
Derivation: By action of oxidizing agent, such as nitric acid, on arsenious oxide.
Hazard: Confirmed carcinogen.
Use: Arsenates, insecticides, dyeing and printing, weed killer, colored glass, metal adhesives.

arsenic sesquioxide.   See arsenic trioxide.

arsenic sulfide.   Legal label name for arsenic disulfide.

arsenic thioarsenate. $As(AsS_4)$.
  Properties: Dry, free-flowing yellow powder, stable, high melting. Insoluble in water and organic solvents, but soluble in aqueous caustics.
  Use: Scavenger for certain oxidation catalysts and thermal protectant for metal-bonded adhesives and coating resins.

arsenic tribromide. (arsenic bromide; arsenious bromide; arsenous bromide).
  CAS: 7784-33-0. $AsBr_3$.
  Properties: Yellowish-white, hygroscopic crystals. D 3.54 (25C), mp 33C, bp 221C. Decomposed by water.
  Derivation: Direct union of arsenic and bromine.
  Use: Analytical chemistry, medicine.

arsenic trichloride. (arsenic chloride; arsenious chloride; arsenous chloride; caustic arsenic chloride; fuming liquid arsenic).
  CAS: 7784-34-1. $AsCl_3$.
  Properties: Colorless or pale yellow oil. Bp 130.5C, fp −18C, d 2.163 (14/4C). Soluble in concentrated hydrochloric acid and most organic solvents, decomposed by water. Fumes in moist air. Noncombustible.
  Derivation: (1) By action of chlorine on arsenic; (2) by distillation of arsenic trioxide with concentrated hydrochloric acid.
  Grade: Technical.
  Hazard: Strong irritant to eyes and skin.
  Use: Intermediate for organic arsenicals (pharmaceuticals, insecticides), ceramics.
  See arsenic.

arsenic trifluoride. (arsenious fluoride).
  CAS: 7784-35-2. AsF_3.
  Properties: Mobile liquid that fumes in air.
  Hazard: Extremely toxic.
  Use: Fluorinating reagent, catalyst, ion implantation source, and dopant.

arsenic trioxide. (crude arsenic; white arsenic; arsenious acid; arsenious oxide; arsenous anhydride).
  CAS: 1327-53-3. As_2O_3.
  Properties: White powder; odorless; tasteless. D 3.865. Slightly soluble in water; soluble in acids, alkalies, and glycerol. Sublimes on heating.
  Derivation: Smelting of copper and lead concentrates. Flue dust to which pyrite or galena concentrations are added yields As_2O_3 vapor. Condensation gives product of varying purity called crude arsenic (90–95% pure). A higher-purity oxide called white arsenic (99+% pure) is obtained by resubliming the crude As_2O_3.
  Hazard: A confirmed carcinogen.
  Use: Pigments, ceramic enamels, aniline colors, decolorizing agent in glass, insecticide, rodenticide, herbicide, sheep and cattle dip, hide preservative, wood preservative, preparation of other arsenic compounds.

arsenic trisulfide. (arsenious sulfide; arsenic sulfide [yellow]; arsenous sulfide; arsenic tersulfide).
  CAS: 1303-33-9. As_2S_3.
  Properties: Yellow crystals or powder, changes to a red form at 170C. D 3.43, mp 300C. Insoluble in water and hydrochloric acid; dissolves in alkaline sulfide solutions and nitric acid.
  Derivation: Occurs in nature as the mineral orpiment. May be precipitated from arsenious acid solution by the action of hydrogen sulfide.
  Grade: Technical, pigment, single crystals.
  Use: Pigment, reducing agent, pyrotechnics, glass used for infrared lenses, semiconductors, hair removal from hides.

arsine. (arsenic hydride).
  CAS: 7784-42-1. AsH_3.
  Properties: Colorless gas. Fp −113.5C, bp −62C, decomposes 230C. Soluble in water; slightly soluble in alcohol and alkalies.
  Derivation: Reaction of aluminum arsenide with water or hydrochloric acid, electrochemical reduction of arsenic compounds in acid solutions.
  Grade: Technical, 99% pure or in mixture with other gases.
  Hazard: Highly poisonous by inhalation. Peripheral nervous system and vacular system impairment, kidney and liver impairment.
  Use: Organic synthesis, military poison, doping agent for solid-state electronic components.

arsphenamine. A specific for syphilis originally developed by Ehrlich, but no longer in use. It was a derivative of arsenic and benzene.
  See Ehrlich.

Artic. A refrigeration grade of methyl chloride.

artificial cinnabar. See mercuric sulfide, red.

artificial snow. A copolymer of butyl and isobutyl methacrylate, often dispersed from an aerosol bomb or other atomizing device, used in decorative window displays, etc. Man-made snow is crystallized water vapor made by mechanical means.

aryl. A compound whose molecules have the ring structure characteristic of benzene, naphthalene, phenanthrene, anthracene, etc., (i.e., either the 6-carbon ring of benzene or the condensed 6-carbon rings of the other aromatic derivatives). For example, an aryl group may be phenyl C_6H_5 or naphthyl $C_{10}H_9$. Such groups are often represented in formulas by "R."
  See alkyl.

arylalkyl. A compound containing both aliphatic and aromatic structures, e.g., alkyl benzenesulfonate. Also called aralkyl.

aryl carbamate herbicide. (aryl carbamic ester). Any of a class of herbicides that are absorbed via the roots and have colchicine-like actions.
Hazard: Toxic to monocotyledons; inhibition of oxidative phosphorylation and of RNA and protein synthesis; inhibition of photosynthesis.
Use: Herbicide used to control grasses in crops such as peas and beets.

arylesterase.
CAS: 9032-73-9. Any enzyme that enables the hydrolysis or arylesters.

aryl halide. An aromatic compound with two benzene rings that contain substituted halogens. The two major classes of aryl halides are those based upon maphthalene and the other upon biphenyl.

aryl hydrocarbon hydroxylase. Any of certain isozymes that catalyzes the hydroxylation of aromatic hydrocarbons. Aryl hydrocarbon hydroxylase activity resides in isozymes of cytochrome P-450.

aryne. A hydrocarbon derived from an arene by abstraction of two hydrogen atoms from adjacent carbon atoms. Arynes are commonly represented with a formal triple bond. The analogous heterocyclic compounds are called heteroarynes or hetarynes.

As. Symbol for arsenic.

as-. Abbreviation for asymmetrical, same as *uns-*.

ASA. (1) American Standards Association. (2) Abbreviation for acrylic ester-modified styrene acrylonitrile terpolymer.
See Luran Sulfur.

asarone. See 2,4,5-trimethoxy-1-propenylbenzene.

asbestine. A soft, fibrous magnesium silicate.
Use: As a filler in paper, rubber, and plastics.

asbestos.
CAS: 1332-21-4. A group of impure magnesium silicate minerals that occur in fibrous form.
Properties: White, gray, green, brown. D 2.5. Noncombustible. (1) Serpentine asbestos is the mineral chrysotile, a magnesium silicate. The fibers are strong and flexible, so that spinning is possible with the longer fibers. A microcrystalline form, TM "Avibest," has been developed. (2) Amphibole asbestos includes various silicates of magnesium, iron, calcium, and sodium. The fibers are generally brittle and cannot be spun, but are more resistant to chemicals and to heat than serpentine asbestos. (3) Amosite. (4) Crocidolite.
Occurrence: Vermont, Arizona, California, North Carolina, Africa, Italy, Yukon, Quebec, Mexico,

Hazard: A confirmed carcinogen. Highly toxic by inhalation of dust particles. Pneumoconiosis, lung cancer, mesothelioma.
Use: Fireproof fabrics, brake lining, gaskets, roofing compositions, electrical and heat insulations, paint filler, chemical filters, reinforcing agent in rubber and plastics, component of paper dryer felts, diaphragm cells, cement reinforcement.
Note: A promising substitute for asbestos for cement reinforcement is glass fiber made from slate and limestone.

ascaridida. (ascaridorida; ascarididea; ascaridata; ascarididae). An order of large, intestinal, parasitic nematode worms that includes species such as *Ascaris lumbricoides*, that are important to humans. Important genera include *Ascaridia, Ascaris, Subuluris, Heterakis,* and *Anisakis.*

ascaridole. (1,4-peroxido-*p*-menthene-2). $C_{10}H_{16}O_2$.
Properties: A liquid, naturally occurring peroxide. Bp 84C (5 mm), d 1.011 (13/15C), refr index 1.4743 (20C).
Derivation: By vacuum distillation of chenopodium oil.
Hazard: Strong oxidizing agent, explodes on heating to 130C or in contact with organic acids.
Use: Initiator in polymerization, medicine.

"Ascarite II" [Thomas]. TM for a sodium hydroxide-nonfibrous silicate formulation.
Use: Quantitative absorption of carbon dioxide in the determination of carbon in steel and organic compounds by direct combustion and other analysis.

ascaron. A peptone of certain helminthes, especially ascarids.
Hazard: Toxic.

ascorbic acid. (*l*-ascorbic acid; vitamin C).
CAS: 50-81-7. $C_6H_8O_6$.

$$\begin{array}{c} \overline{-O}\qquad H \\ || \\ C-C=C-C-C-CH_2OH \\ \parallel||| \\ OOH\,OH\,HOH \end{array}$$

A dietary factor that must be present in the diet of humans to prevent scurvy. It cures scurvy and increases resistance to infection. Ascorbic acid presumably acts as an oxidation-reduction catalyst in the cell. It is readily oxidized; citrus juices should not be exposed to air for more than a few minutes before use.
Properties: White crystals (plates or needles). Mp 192C. Soluble in water; slightly soluble in alcohol; insoluble in ether, chloroform, benzene, petroleum ether, oils and fats. Stable to air when dry. One international unit is equivalent to 0.05 milligram of *l*-ascorbic acid.

Source: Food source: acerola (West Indian cherry); citrus fruits; tomatoes; potatoes; green, leafy vegetables. Commercial sources: Synthetic product made by fermentation of sorbitol.
Grade: USP, FCC.
Use: Nutrition, color fixing, flavoring, and preservative in meats and other foods, oxidant in bread doughs, abscission of citrus fruit in harvesting, reducing agent in analytical chemistry. The iron, calcium, and sodium salts are available for biochemical research.

ascorbic acid oxidase.   An enzyme found in plant tissue that acts upon ascorbic acid in the presence of oxygen to produce dehydroascorbic acid.
Use: Biochemical research.

ascorbyl palmitate. $C_{22}H_{38}O_7$. A white or yellowish-white powder having a citrus-like odor. Mp 116–117C; soluble in alcohol and in animal and vegetable oils; slightly soluble in water.
Derivation: Palmitic and *l*-ascorbic acids.
Grade: FCC.
Use: Antioxidant for fats and oils, source of vitamin C, stabilizer, emulsifier.

ascorbyl stearate.
Use: Food additive.

-ase.   A suffix characterizing the names of many enzymes, e.g., diastase, cellulase, cholinesterase, etc. However, the names of some enzymes end in *-in* (i.e., pepsin, rennin, papain).

asexual reproduction.   A type of reproduction involving only one parent that usually produces genetically identical offspring. Asexual reproduction occurs without meiosis or syngamy, and may happen through budding, by the division of a single cell, or the breakup of an entire organism into two or more parts.

ash.   (1) In analytical chemistry, the residue remaining after complete combustion of a material. It consists of mineral matter (silica, alumina, iron oxide, etc.) the amount often being a specification requirement. (2) The end product of large-scale coal combustion as in power plants; now said to be the sixth most plentiful mineral in the U.S. It consists principally of fly ash, bottom ash, and boiler ash. Some of its values are recoverable, and there are a number of industrial uses of fly ash, e.g., in cement products and road fill.
See fly ash.

askarel.   A generic descriptive name for synthetic electrical insulating (dielectric) material which, when decomposed by the electric arc, evolves only nonexplosive gases or gaseous mixtures, i.e., chlorinated aromatic derivatives, particularly pentachlorodiphenyl and trichlorobenzene, but also including pentachlorodiphenyl oxide, pentachlorophenylbenzoate, hexachlorodiphenylmethane, pentachlorodiphenyl ketone, and pentachloroethylbenzene. Nonflammable.
Use: Insulating medium in transformers, dielectric fluid.
See dielectric; transformer oil.

ASM International.   (ASM). Formally organized in 1935, this society actually had been active under other names since 1913, when the name for standards of metal quality and performance in the automobile became generally recognized. Most recent previous name was American Society for Metals. ASM International has over 30,000 members and publishes *Metals Review* and the famous *Metals Handbook*, as well as research monographs on metals. It is active in all phases of metallurgical activity, metal research, education, and information retrieval. Its headquarters is at 9639 Kinsman Road, Materials Park, OH 44073. Website: http://www.asminternational.org

asparagic acid.   See aspartic acid.

L-asparaginase.   (L-asnase; colaspase; elspar). CAS: 9015-68-3. An enzyme used in the treatment of certain types of leukemia. Produced by biochemical activity of certain bacteria, yeasts, and fungi. Yields are in excess of 3500 units/g of source.

asparagine.   (α-aminosuccinamic acid; β-asparagine; althein; aspartamic acid; aspartamide). $NH_2COCH_2CH(NH_2)COOH$. The β amide of aspartic acid, a nonessential amino acid, existing in the D(+)- and L(−)-isomeric forms as well as the DL-racemic mixture. L(−)-asparagine is the most common form.
Properties: L(−)-Asparagine monohydrate: White crystals. Mp 234–235C. Acid to litmus. Nearly insoluble in ethanol, methanol, ether and benzene; soluble in acids and alkalies.
Derivation: Widely distributed in plants and animals both free and combined with proteins.
Use: Biochemical research, preparation of culture media, medicine.

l-asparagine.
CAS: 70-47-3. $C_4H_8N_2O_3 \cdot H_2O$.
Properties: White crystalline powder or rhombic hemihedral crystals; sltly sweet taste. Mp: 234C. Sol in water; insol in alc, ether.
Use: Food additive.

asparaginic acid.   See aspartic acid.

Aspartame.   (3-amino-n-(α-carboxyphenethyl)succinamic acid n-methyl ester, stereoisomer; aspartylphenylalanine methyl ester; n-l-α-aspartyl-l-phenylalanine 1-methyl ester; canderel; dipeptide sweetener; Equal; methyl aspartylphenylalanate; 1-methyl n-l-α-aspartyl-l-phenylalanine; Nutrasweet; sweet dipeptide). $C_{14}H_{18}N_2O_5$.

Properties: White crystalline powder from water or alcohol; odorless; sweet taste. Mw: 294.34; mp 190. Slightly soluble in water, alcohol. A synthetic non-nutritive sweetener approved by FDA for tabletop use and as a packaged food additive. The U.S., Canada, and South Africa permit its use in carbonated beverages. A combination of aspartic acid and L-phenylalanine, it is said to be 200 times sweeter than sugar.
See sweetener, nonnutritive.

aspartamic acid.   See asparagine.

aspartamide.   See asparagine.

aspartate aminotransferase. (aminotransferase; aminotransaminase; aspartate transaminase; AST; glutamic-aspartic transaminase; (serum) glutamic-oxaloacetic transaminase). An enzyme that catalyzes reversibly the transfer of an amino group from glutamic acid to oxaloacetic acid, yielding α-ketoglutaric acid and aspartic acid. The codons for aspartate are GAC and GAU.

aspartic acid. (asparaginic acid; asparagic acid; aminosuccinic acid). $COOHCH_2CH(NH_2)COOH$. A naturally occurring nonessential amino acid. The common form is L(+)-aspartic acid.

$$HO-\overset{\overset{O}{\|}}{C}-CH_2\underset{\underset{NH_2}{|}}{C}HC-OH$$

Properties: Colorless crystals. Soluble in water; insoluble in alcohol and ether. Optically active. DL-aspartic acid. Mp 278–280C (decomposes), d 1.663 (12/12C). L(+)-aspartic acid. Mp 251C. D(−)-aspartic acid. Mp 269–271C (decomposes), d 1.6613.
Source: Young sugar cane, sugar beet molasses.
Derivation: Hydrolysis of asparagine, reaction of ammonia with diethyl fumarate.
Use: Biological and clinical studies, preparation of culture media, organic intermediate, ingredient of aspartame, detergents, fungicides, germicides, metal complexation. Available commercially as DL(−)-, L(+)-, and DL-aspartic acid.

dl-aspartic acid.
CAS: 617-45-8. $C_4H_7NO_4$.
Properties: Colorless to white monoclinic crystals; acid taste. Mp: 280C (decomp). Sltly sol in water; insol in alc, ether.
Hazard: Very low toxicity.
Use: Food additive.

l-aspartic acid.
CAS: 56-84-8. $C_4H_7NO_4$.
Properties: Colorless to white crystals or leaflets; acid taste. Mp: 270C. Sltly sol in water; insol in alc, ether.

Hazard: Low toxicity.
Use: Food additive; drug.

l-aspartic acid, n-acetyl-, dilithium salt.   See dilithium n-acetyl-l-aspartate.

aspartocin.   USAN for antibiotic produced by *Streptomyces griseus*.

aspergillic acid. (2-hydroxy-3-isobutyl-6-(1-methylpropyl)pyrazine-1-oxide). $C_{12}H_{20}N_2O_2$. An antibiotic from strains of *Aspergillus flavus*.
Properties: Yellow crystals. Mp 97C. Insoluble in cold water; soluble in common organic solvents and dilute acids. Hydrochloride melts at 178C and is soluble in water.
Use: Antibiotic.

aspergillin. (Gliotoxin; Gladiocladium fimbriatum). $C_{13}H_{14}N_2O_4S_2$. A broad-spectrum antibacterial antibiotic.
Properties: Monoclinic crystals from MeOH.
Derivation: Obtained from cultures of the molds *Aspergillus flavus* and *Aspergillus fumigatus*.
Hazard: Poison.

Aspergillus. A genus of small molds and fungi used in industry to ferment carbohydrates for producing citric and other organic acids.

Aspergillus. A genus of imperfect fungi, at least 20 species are known to cause opportunistic infections in humans. *Aspergillus* species are ubiquitous in the environment and are the second most commonly recovered fungus from opportunistic mycoses following those of *Candida* (commonly known as yeast) species. Many *Aspergillus* produce mycotoxins, which are potent toxins.
Hazard: Allergenic, pathogenic, toxic.

ASPET.   See American Society of Pharmacology and Experimental Therapeutics.

asphalt. (petroleum asphalt; Trinidad pitch; mineral pitch).
CAS: 8052-42-4. A dark-brown to black cementitious material, solid or semisolid in consistency, in which the predominating constituents are bitumens that occur in nature as such or are obtained as residua in petroleum refining (ASTM). It is a mixture of paraffinic and aromatic hydrocarbons and heterocyclic compounds containing sulfur, nitrogen, and oxygen.
Properties: Black solid or viscous liquid. D approximately 1.0. Soluble in carbon disulfide, flash p 450F (132C), autoign temp 900F (482C), solid softens to viscous liquid at approximately 93C, penetration value (paving) 40–300, (roofing) 10–40. Good electrical resistivity. Combustible.
Occurrence: California, Trinidad, Venezuela, Cuba, Canada (Athabasca tar sands).

Hazard: Toxic by inhalation of fume. Upper respiratory tract and eye irritant. Probable carcinogen.
Use: Paving and road-coating, roofing, sealing and joint filling, special paints, adhesive in electrical laminates and hot-melt compositions, diluent in low-grade rubber products, fluid loss control in hydraulic fracturing of oil wells, medium for radioactive waste disposal, pipeline and underground cable coating, rust-preventive hot-dip coatings, base for synthetic turf, water-retaining barrier for sandy soils, supporter of rapid bacterial growth in converting petroleum components to protein.
See bacteria; protein; oil sands.

asphalt (blown).   (mineral rubber; oxidized asphalt; hard hydrocarbon).
Properties: Black, friable solid obtained by blowing air at high temperature through petroleum-derived asphalt with subsequent cooling. Penetration value 10–40, softening point 85 to 121C. Combustible.
Use: Primarily roofing, as diluent in low-grade rubber products and as thickener in oil-based drilling fluids.

asphalt (cut-back).   A liquid petroleum product produced by fluxing an asphaltic base with suitable distillates. (ASTM).
Properties: Flash p 50F(10C) (OC).
Grade: Solution of residue from distillation in carbon tetrachloride, 99.5%.
Hazard: Flammable, dangerous fire hazard.
Use: Road surfaces.

asphaltene.   A component of the bitumen in petroleums, petroleum products, malthas, asphalt cements and solid native bitumens. Soluble in carbon disulfide but insoluble in paraffin naphthas. (ASTM). It is composed of polynuclear hydrocarbons of molecular weight up to 20,000, joined by alkyl chains.

asphalt, liquid.   See residual oil; asphalt (cut-back).

asphalt (oxidized).   See asphalt (blown).

asphalt paint.   Asphaltic base in a volatile solvent with or without drying oils, resins, fillers, and pigments. Ground asbestos was frequently used as a component of heavy asphaltic paints for roofing and waterproofing purposes.
Hazard: Flammable, dangerous fire risk.

asphyxiant gas.   A gas that has little or no positive toxic effect but that can bring about unconsciousness and death by replacing air and thus depriving an organism of oxygen. Among the so-called asphyxiant gases are carbon dioxide, nitrogen, helium, methane, and other hydrocarbon gases.

aspidospermine. $C_{22}H_{30}O_2N_2$.

Properties: White to brownish-yellow crystalline alkaloid. Mp 208C, bp 220C (1–2 mm Hg), sublimes at 180C under reduced pressure. Soluble in fats, fixed oils, absolute alcohol, and ether. Its sulfate and hydrochloride are soluble in water.
Use: Medicine (respiratory stimulant).

aspirin.   (acetylsalicylic acid; o-acetoxybenzoic acid).
CAS: 50-78-2. $CH_3COOC_6H_4COOH$.

Properties: White crystals or white, crystalline powder; odorless; slightly bitter taste. Stable in dry air, slowly hydrolyzes in moist air to salicylic and acetic acids. Mp 132–136C, bp 140C (decomposes). Soluble in water, alcohol, chloroform, and ether; less soluble in absolute ether. Dissolves with decomposition in solutions of alkali hydroxides and carbonates.
Derivation: Action of acetic anhydride on salicylic acid.
Method of purification: Crystallization.
Grade: Technical, USP.
Hazard: An allergen; may cause local bleeding especially of the gums; 10-g dose may be fatal. May cause excessive biosynthesis of prostaglandins. Dust dispersed in air is serious explosion risk. Skin and eye irritant.
Use: Medicine (analgesic, antiinflammatory, antipyretic).

Assam Milk Tea.   A less astringent tea brewed by boiling leaves in milk.

assay.   (1) The quantitative determination by chemical analysis of metallic values in an ore. Ores are assayed by heat fractionation. (2) Dry assay refers to the determination of metal values in ores by chemical methods not involving the use of solutions. (3) Assay value refers to the proportion of precious metal indicated in ounces of metal per ton of ore. (4) Wet assay is the chemical determination of metal per ton of ore. (5) Pharmaceutical products are assayed to validate the amount of drug present in a given unit. (6) Assays on organisms (bacteria) to determine their reactions to an antibiotic or insecticide. This is called bioassay. Organic materials are assayed by solvent extraction and chemical separation.

assembly.   Putting sequenced fragments of DNA into their correct chromosomal positions.

assimilation.   The activity whereby different parts of the substrate in a fermentation are changed into cell substance.

assistant. A term loosely used in the textile industry for any chemical compound that aids in a processing step, e.g., scouring, dyeing, bleaching, finishing, etc.
See auxiliary; dyeing assistant.

association. A reversible chemical combination due to any of the weaker classes of chemical bonding forces. Thus, the combination of two or more molecules due to hydrogen bonding, as in the union of water molecules with one another or of acetic acid molecules with water molecules, is called association; also, combination of water or solvent molecules with molecules of solute or with ions, i.e., hydrate formation or solvation. Formation of complex ions or chelates, as copper ion with ammonia or copper ion with 8-hydroxyquinoline is another example. Aqueous solutions of soaps or synthetic detergents are often called association colloids.

association polymer. A polymer formed by simple addition of molecules or by rearrangement of bonds.

A-stage resin. (resole; one-step resin). An alkaline catalyzed thermosetting phenol-formaldehyde-type resin consisting primarily of partially condensed phenol alcohols. At this stage, the product is fully soluble in one or more common solvents (alcohols, ketones) and is fusible at less than 150C. On further heating and without use of a catalyst or additive, the resin is eventually converted to the insoluble, infusible, cross-linked form (C-stage). The A-stage resin is a constituent of most commercial laminating varnishes and is also used in special molding powders.
See B-stage resin; C-stage resin; novolak; phenol-formaldehyde resin.

astatine. At. Nonmetallic element of atomic number 85. Group VIIA of periodic table. aw 211. Heaviest member of the halogen family, has 20 isotopes, all radioactive; derived by α-bombardment of bismuth. The two most stable isotopes have half-lives of approximately 8 hours. Astatine occurs in nature to the extent of approximately one ounce in the entire earth's crust. Like iodine, it concentrates in the thyroid gland. Its use in medicine is still experimental.

aster lake. Red lake pigment having high oil absorption properties.

ASTM. See American Society for Testing and Materials.

Aston, Francis William. (1877–1945). This noted English chemist and physicist carried out much of his work with J. J. Thomson at Cambridge. He was the pioneer investigator of isotopes, and his method of separating the lighter from the heavier atomic nuclei provided the technique that later developed into the mass spectroscope, which utilizes a magnetic field for this purpose. Aston received the Nobel Prize for this discovery in 1922, just three years after Rutherford performed the first transmutation of elements. Aston also correctly estimated the energy content of a hydrogen atom and predicted the controlled release of this energy.

"Astracel" [IP]. TM for a group of fast dyes for union shades on polyester-carbon blended fabrics.

Astragal. Dyeing auxiliaries.
Use: In dye processes to ensure level uptake of dyes and give color fastness.

astrazon golden yellow gl. See C.I. basic yellow 28.

astrochemistry. Application of radioastronomy (microwave spectroscopy) to determination of the existence of chemical entities in the gas clouds of interstellar space and of elements and compounds in celestial bodies, including their atmospheres. Such data are obtained from spectrographic study of the light from the sun and stars, from analysis of meteorites, and from actual samples from the moon. Hydrogen is by far the most abundant element in interstellar space, with helium a distant second. Over 25% of the elements, including carbon, have been identified, as well as molecules of water, carbon monoxide, carbon dioxide, ammonia, ethane, methane, acetylene, formaldehyde, formic acid, methyl alcohol, hydrogen cyanide, and acetonitrile. When applied to the planets only, the science is called chemical planetology.
See nucleogenesis.

Astrol. A group of fast alizarin direct blues.

Astro Starch. Cationic potato starches.
Use: As an alkaline-size emulsion stabilizer.

asulam potassium salt. See potassium asulam.

asymmetric carbon atom. One having its four valences held by four different groups or atoms. Compounds possessing asymmetric carbon atoms have optically active isomers.

asymmetric induction. The action of a force, arising in an optically active molecule, that influences adjacent symmetrical molecules in such a way that they become asymmetric.

asymmetric syntheses. Processes that produce optically active compounds from symmetrically constituted molecules by the intermediate use of optically active reagents, but without the use of any of the methods of resolution.

asymmetry. A molecular structure in which an atom having four tetrahedral valences is attached to four different atoms or groups. The commonest cases involve the carbon atom, though they may exist also with other elements such as nitrogen and sulfur, i.e., lactic acid which contains one asymmetric carbon (indicated by *). In such cases, two optical isomers (L and D enantiomers) result that are nonsuperposable mirror images of each other.

$$
\begin{array}{ccc}
& COOH & & COOH \\
OH\!-\!C^{*}\!-\!H & & H\!-\!C\!-\!HO \\
& CH_3 & & CH_3 \\
& L & & D
\end{array}
$$

Amino acids are also characterized by asymmetric carbons. Many compounds have more than one asymmetric carbon, e.g., tartaric acid, sugars, terpenes, etc. This results in the possibility of many optical isomers, the number being determined by the formula 2^n, where n is the number of asymmetric carbons.
See optical isomer; enantiomer; glyceraldehyde.

At.   Symbol for astatine.

atactic.   A type of polymer molecule in which substituent groups or atoms are arranged randomly above and below the backbone chain of atoms when the latter are all in the same plane, as shown below. See polymer, stereospecific.

$$
\begin{array}{c}
\ \ H\ \ R\ \ H\ \ R\ \ H\ \ H\ \ H\ \ R\ \ H\ \ H\ \ H\ \ H \\
\ \ |\ \ \ |\ \ \ |\ \ \ |\ \ \ |\ \ \ |\ \ \ |\ \ \ |\ \ \ |\ \ \ |\ \ \ |\ \ \ | \\
-C\!-\!C\!-\!C\!-\!C\!-\!C\!-\!C\!-\!C\!-\!C\!-\!C\!-\!C\!-\!C\!-\!C- \\
\ \ |\ \ \ |\ \ \ |\ \ \ |\ \ \ |\ \ \ |\ \ \ |\ \ \ |\ \ \ |\ \ \ |\ \ \ |\ \ \ | \\
\ \ H\ \ H\ \ H\ \ H\ \ R\ \ H\ \ H\ \ H\ \ R\ \ H\ \ R\ \ H
\end{array}
$$

ATBC.   (acetyltri-*n*-butyl citrate).
CAS: 77-90-7. Aqueous-based pharmaceutical coatings.

ATE.   Abbreviation for aluminum triethyl.
See triethyl aluminum.

-ate.   A suffix having two different meanings. (1) In inorganic compounds, it indicates a salt whose metal or radical is in the highest oxidation state, as in calcium sulfate, ammonium nitrate, etc. (2) In engineering terminology, it means "result of," as in precipitate, condensate, alkylate, distillate, etc.

ATEC.   (acetyltriethyl citrate).
CAS: 77-89-4. Aqueous-based pharmaceutical coatings.

atenolol.   (1,*p*-carbamoylmethylphenoxy-3-isopropylamino-2-propanol). $C_{14}H_{22}N_2O_3$.
Properties: Colorless crystals. Mp 147C.
Use: An adrenergic blocker used in treatment of hypertension. FDA approved.

$$CH_3 \searrow$$
$$\qquad CHNHCH_2CH(OH)CH_2O\!-\!\!\bigcirc\!\!-\!CH_2CNH_2$$
$$CH_3 \nearrow \qquad\qquad\qquad\qquad\qquad\qquad\ \ \overset{\|}{O}$$

ATG or AUG.   The codon for methionine; the translation initiation codon. Usually, protein translation can only start at a methionine codon (although this codon may be found elsewhere within the protein sequence as well). In eukaryotic DNA, the sequence is ATG; in RNA it is AUG. Usually, the first AUG in the mRNA is the point at which translation starts, and an open reading frame follows, i.e., the nucleotides taken three at a time will code for the amino acids of the protein, and a stop codon will be found only when the protein coding region is complete.

atlantic black gac.   See apomine black gx.

ATM.   Abbreviation for aluminum trimethyl.
See trimethyl aluminum.

atmolysis.   The separation of admixed gases by means of a porous partition. Use is made of the different speeds at which the gases diffuse.

atmosphere.   (1) The gaseous envelope that surrounds the earth. It comprises of four major divisions: the troposphere (from sea level to approximately 10 km), the stratosphere (ozone region) which extends from approximately 10 to 50 km, the mesosphere which extends from approximately 50 to 100 km, and the thermosphere which ranges from approximately 100 to 1000 km or more. There is no sharp boundary between the layers. The pressure drops rapidly as altitude increases (from 1 atm at sea level to 10^{-13} atm at 1000 km). The chemical entities and reactions that occur in these spheres are the subject of extensive research. (2) The pressure exerted by the air at sea level (14.696 psi), which will support a column of mercury 760 mm high (approximately 30 inches). This is standard barometric pressure, though it varies slightly with local meteorological conditions. It is often used to indicate working pressures of steam. The accepted abbreviation is *atm*. (3) Any environmental gas or mixture of gases, e.g., an atmosphere of nitrogen or an inert atmosphere.

atmosphere (controlled).   As used in the technology of food preservation and storage, a gaseous environment in which the concentrations of oxygen, carbon dioxide, and nitrogen are held constant at a specific level, the temperature also being controlled. Controlled atmosphere storage techniques are used on a commercial scale in the U.S.

atmospheric pollution.   See air pollution.

Atnul.   (glycerol monostearate).
CAS: 31566-31-1. An intermediate chemical.
Use: For food processing and baking, snack goods, pharmaceuticals, plastics, paints, coatings, and cosmetics.

atom. The smallest possible unit of an element, consisting of a nucleus containing one or more protons and (except hydrogen) one or more neutrons, and one or more electrons which revolve around it. The protons are positively charged, the neutrons have no charge, the electrons are negatively charged. As each atom contains the same number of protons as electrons, the atom is electrically neutral. Atoms in general are characterized by stability. One might wonder why the negatively charged electrons are not attracted into the positively charged nucleus in response to the law of opposite charges, causing the atom to collapse on itself. That this does not occur is due to the nature of the electron, which is not only a particle but also a standing quantum wave. As explained by Dr. W. V. Houston, "The normal state of an atom is balance between the attraction of the nucleus for the electron wave and what might be called the elastic resistance to compression of the wave itself."

Atoms of the various elements differ in mass (weight), that is, in the number of neutrons and protons and also in the number of electrons. Atoms of a given element are identical, except that an element may have atoms of different masses, called isotopes. Individual atoms of uranium and thorium have been resolved at 5 Å in the scanning electron microscope. Motion pictures of uranium atoms at magnification of 7.5 million times have been made at the Enrico Fermi Institute at the University of Chicago.

Atoms of the same or different elements combine to form molecules. When the atoms are of two or more different elements, these molecules are called compounds. Atoms remain essentially unchanged in chemical reactions except that some of the outermost electrons may be removed, shared, or transferred as occurs in oxidation, ionization, and chemical bonding. A few atomic species disintegrate as a result of nuclear changes and thus are radioactive. Heavy unstable atoms such as uranium-235 and plutonium can be split by bombardment with high-energy particles yielding tremendous energy. See electron; proton; bonds, chemical; orbital theory; ionization; radioactivity; fission.

atom-cavity microscope. A technique that traps single atoms in a magneto-optical trap and are dropped through a high-finesse optical cavity. A laser probe measures atomic motions.
Use: Study of atomic-scale motions of single atoms. See scanning tunneling microscope; chemical force microscopy.

atom form factor. (atomic scattering factor; atomic structure factor; f-value). Quantity in the expression for the intensity of an X-ray beam reflected by a crystal, whose value is dependent on the varying configuration of the electrons in the crystal atoms relative to the nuclei, as well as on the incident angle and the wave length of the X-rays.

atomic absorption coefficient. Ratio of absorption coefficient of an element to the number of atoms per unit volume.

atomic absorption spectroscopy. An analytical technique in which the substance to be analyzed is converted into an atomic vapor by spraying a solution into an acetylene-air flame. Some types of compounds require a reducing flame, such as acetylene-nitrous oxide. The absorbance at a selected wavelength is measured and compared with that of a reference substance. The absorbance measured is proportional to the concentration.

atomic energy. See nuclear energy.

atomic force microscopy. /D A technique similar to scanning tunneling microscopy employing an instrument that uses a sharply tipped electrode in close proximity to a surface. As the electrode is moved, the change in electrode signal relates to the shape of the surface being scanned.
Use: Study of atomic-scale structure of surfaces, transport of atoms and molecules. See scanning tunneling microscope; chemical force microscopy.

atomic hydrogen welding. A method of welding in which hydrogen gas is passed through an arc between two tungsten electrodes. The arc breaks down the molecules to form atomic hydrogen. The recombination of the atoms to form molecules and the combustion of the molecular hydrogen in atmospheric oxygen produce a flame temperature of 4000–5000C.

atomic number. The number of protons (positively charged mass units) in the nucleus of an atom, upon which its structure and properties depend. This number represents the location of an element in the periodic table. It is normally the same as the number of negatively charged electrons in the shells. Thus, an atom is electrically neutral except in an ionized state, when one or more electrons have been gained or lost. Atomic numbers range from 1, for hydrogen, to 109 for meitnerium. See periodic table; atomic weight; mass number.

atomic radius. One-half the distance between two adjacent atoms in crystals of elements. It varies according to interatomic forces.

atomic scattering factor. See atom form factor.

atomic structure factor. See atom form factor.

atomic susceptibility. Product of the specific or mass susceptibility of an element and its atomic weight.

atomic theory. See Dalton, John.

atomic volume. The atomic weight of an element divided by its density.

atomic weight. (aw). The average weight or mass of all the isotopes of an element as determined from the proportions in which they are present in a given element, compared with the mass of the 12 isotope of carbon (taken as precisely 12.000), which is the official international standard. The true atomic weight of carbon when the masses of its isotopes are averaged is 12.01115; that of oxygen is 15.9994. The total mass of any atom is the sum of the masses of all its constituents (protons, neutrons, and electrons). Official atomic weight determinations are released periodically by the IUPAC.
See atomic number; mass number.

ATP. Abbreviation for adenosine triphosphate.

ATPase. An enzyme that hydrolyzes ATP to yield ADP and phosphate; usually coupled to some process requiring energy such as the sodium potassium ATPase.

ATP synthase. An enzyme complex that forms ATP from ADP and phosphate during oxidative phosphorylation in the inner mitochondrial membrane or the bacterial plasma membrane, and during photophosphorylation in chloroplasts. Uses a proton gradient to chemiosmotically drive the synthesis.

atracurium. (tracium; benzenesulfonate; benzenesulfonate; 5-[3-[1-[(3,4-dimethoxyphenyl)methyl]-6,7-dimethoxy-2-methyl-3,4-dihydro-1H-isoquinolin-2-ium-2-yl]propanoyloxy]pentyl-3-[1-[(3,4-dimethoxyphenyl)methyl]-6,7-dimethoxy-2-methyl-3,4-dihydro-1H-isoquinolin-2-ium-2-yl]propanoate). $C_{65}H_{82}N_2O_{18}S_2$. A nondepolarizing neuromuscular blocking agent with short duration of action.
Hazard: Extremely toxic.

atranorin.
CAS: 479-20-9. $C_{19}H_{18}O_8$.
Hazard: A poison by skin contact. A skin irritant.

atrazine. (2-chloro-4-ethylamino-6-isopropylamino-s-triazine).
CAS: 1912-24-9.
Hazard: Hematologic, preproductive and developmental effects. Questionable carcinogen.
Use: Herbicide, plant growth regulator, and weed-control agent for corn, etc., and for noncrop and industrial sites. Reported to inhibit photosynthesis of algae in streams.

atropine. (daturine).
CAS: 51-55-8. $C_{17}H_{23}NO_3$. An alkaloid obtained from species of *Atropa, Datura,* or *Hyoscyamus.*

Properties: White crystals or powder. Optically inactive (but usually contains levorotatory hyoscyamine). Mp 114–116C. Soluble in alcohol, ether, chloroform, and glycerol; slightly soluble in water.
Derivation: By extraction from *Datura stramonium* or by synthesis.
Grade: Technical, NF.
Use: Medicine (antidote for cholinesterase-inhibiting compounds, organophosphorus insecticides, nerve gases); artificial respiration may also be necessary.

Attack All. A laundry detergent containing bleach, disinfectant, deodorizer, and softener in one.

attapulgite.
CAS: 12174-11-7. $(MgAl)_5Si_8O_{22}(OH)_4 \cdot 4H_2O$. A hydrated aluminum-magnesium silicate, the chief ingredient of fuller's earth.
Hazard: Possible carcinogen.
Use: Drilling fluids, decolorizing oils, filter medium.
See clay.

attar. (otto). An essential oil (fragrance) made by steam distillation of flowers, especially roses.
See essential oil; perfume.

attenuator. An RNA sequence involved in regulating the expression of some prokaryotic genes.

atto-. (a).
Prefix meaning 10^{-18} unit (abbreviated a), e.g., 1 ag = 1 attogram (10^{-18} g).

attrition mill. (burr mill). A grinding machine consisting of two metal plates or discs with small projections (burrs). One plate may be stationary while the other rotates, or both may rotate in opposite directions. Feed enters through a hopper above the plates, and ground product emerges at the bottom. There are numerous variations in design.

Au. Symbol for gold, from Latin *aurum.*

aua. ((1R,2R)-2-[(1S,2S)-2-[[(2S)-2-[[(2S)-2-acetamido-4-methylpentanoyl]amino]-4-methylsulfanylbutanoyl]amino]-1-hydroxy-4-methylpentyl]-N-[(2S)-1-(butylamino)-3-methyl-1-oxobutan-2-yl]-4-oxocyclopentane-1-carboxamide). A codon of isoleucine that directs the placement of isoleucine into a polypeptide.

AUG. See initiation codon.

auger. See screw.

Auger electron. Low-energy conversion electron produced by absorption of X-ray quanta by an electron of an outer shell.
See AES; spectroscopy.

auramine.   [4,4′-(imidocarbonyl)bis(*N,N*-dimethylaniline)].
CAS: 492-80-8. $(CH_3)_2NC_6H_4(C:NH)C_6H_4N$ $(CH_3)_2{\cdot}HCl$.
Properties: Yellow flakes or powder. Soluble in water, alcohol, and ether.
Hazard: Confirmed carcinogen.
Use: Yellow dye for paper, textiles, leather; antiseptic; fungicide.

auramine hydrochloride.   (yellow pyoktanin; 4-[4-(dimethylamino)benzenecarboxi]-*N,N*-dimethylaniline hydrochloride).
CAS: 2465-27-2. $C_{17}H_{22}ClN_3$. An aniline dye. It is weakly fluorescing and binds specifically to certain proteins.
Properties: A yellow, water-soluble, crystalline compound, soluble in ethanol.
Hazard: Probable carcinogen.
Use: A commercial dye, a disinfectant and antiseptic agent.

aureolin.   See Indian yellow.

"Aureomycin" *[Zoetis].*   TM for chlortetracycline hydrochloride. An antibiotic. Must conform to FDA requirements.

p. aureus.   (mung bean; black gram; golden gram; green gram; gram; mungo bean). An annual herb (Family Fabaceae) with green or yellow seeds, that is indigenous to Asia. Both the seeds and sprouts are edible.
Use: Forage or green manure.

Auric.   A ferric oxide brown pigment.

auric compounds.   See gold compounds.

aurin (*p*-rosolic acid). $(C_6H_4OH)_2CC_6H_4O$. A triphenylmethane derivative.
Properties: Reddish-brown pieces with greenish metallic luster; easily powdered. Insoluble in water, benzene, and ether; soluble in alcohol.
Use: Indicator, dye intermediate.

aurous compounds.   See gold compounds.

austenite.   A component of steel, a nonmagnetic solid solution of carbon or ferric carbide in γ-iron. Very unstable below its critical temperature, but may be obtained in high-carbon steels by rapid quenching from high temperatures. Addition of manganese and nickel lowers critical transition temperature, and stable austenite may be obtained at room temperature. Characterized by a face-centered cubic lattice.

austenitic alloys.   (austenitic steels). Alloys of iron, chromium, and nickel noted for their resistance to corrosion.

Australian bark.   See wattle bark.

authentic protein.   A recombinant protein that has all the properties of its naturally occurring counterpart.

autocatalysis.   A catalytic reaction induced by a product of the same reaction. This occurs in some types of thermal decomposition, in autoxidation, and in many biochemical systems, as when an enzyme activates its own precursor.
See autoxidation.

autoclave.   A chamber, usually of cylindrical shape, provided with a door or gate at one end which can be securely closed during operation. It is built heavily enough to accommodate steam pressures of considerable magnitude. It is used to effect chemical reactions requiring high temperature and pressure, such as open-steam vulcanization of rubber. Sizes vary from laboratory units to production size, which may be over 50 ft long and three or more feet in diameter. The latter are provided with baffles to ensure equal distribution of the entering steam. Autoclaves are also used in certain sterilization processes.

"Autofloc" *[Nalco].*   TM for computerized apparatus in mining.
Use: For monitoring and control of clarification/thickening operations.

autohemplysin.   An autoantibody that lyses erythrocytes in the presence of complement in the same individual in whose body the lysine is produced.

autohesion.   The formation of a bond between two contiguous surfaces of the same material, when they are pressed together.

autoignition point.   (autoignition temperature; autoign temp). The minimum temperature required to initiate or cause self-sustained combustion in any substance in the absence of a spark or flame. This varies with the test method. Some approximate autoignition temperatures follow:

phenol	715C (1319F)
acetone	537C (1000F)
aniline	537C (1000F)
toluene	537C (1000F)
magnesium powder	472C (883F)
butane	430C (806F)
amyl acetate	398C (750F)
pine shavings	265C (507F)
cotton batting	230C (446F)
ethyl ether	180C (356F)
nitrocellulose film	137C (279F)
carbon disulfide	100C (212F)

See flash point for additional ignition properties.

autointoxicant. Any endogenous substance, metabolic waste or toxicant in the excreta of an organism to which it or its progeny may be exposed.

autoionization. Spontaneous dissociation of a molecule.

autoisolysin. An antibody that lyses cells in the presence of complement in the individual in whose body the lysine is formed and in other individuals of the same species.

autolysin. (autocytolysin). An antibody that is responsible for autolysis.

autolysis. Hydrolysis by tissue proteases; self digestion.

autolyte. A substance that conducts an electric current by itself without requiring a dissolved salt to carry the current.

autolytic enzyme. An enzyme that causes lytic disintegration of the cell that produced it.

autolyzed yeast extract. See baker's yeast extract.

automatic control. Maintenance of desired process conditions (temperature, pressure, etc.) by means of sensing devices that function either electromechanically (thermostat) or electronically (feedback). Applicable to many operations and processes in the chemical industries, such as petroleum refining, evaporation, distillation, heat transfer, electroplating, calendering, extrusion, and many others. Automatic control is not identical with automation.
See instrumentation.

automation. Substitution of specially designed machines for manual labor in such mechanical operations as wrapping, packaging of small units, filling and capping bottles, sealing cans and containers, and materials feeding and proportioning. Automated procedures are much more efficient than manual and effect notable cost savings provided that the machinery is reliable. Do not confuse automation with automatic control.

automotive exhaust emission. See air pollution.

autonomic agent. A substance that inhibits or intensifies the rate of nerve impulse transmission across synaptic junctions, especially those of the autonomic nervous system.

autonomic poison. A poison that acts on the autonomic nervous system.
Hazard: Causes miosis, bronchoconstriction, salivation, lacrimation, respiratory failure, cardiovascular collapse, and death.

autoradiography. A technique that uses X-ray film to visualize radioactively labeled molecules or fragments of molecules.
Use: In analyzing length and number of DNA fragments after they are separated by gel electrophoresis.

autosomal dominant. A gene on one of the non-sex chromosomes that is always expressed, even if only one copy is present. The chance of passing the gene to offspring is 50% for each pregnancy.
See autosome; dominant; gene.

autosome. A chromosome not involved in sex determination. The diploid human genome consists of a total of 46 chromosomes: 22 pairs of autosomes, and 1 pair of sex chromosomes (the X and Y chromosomes).
See sex chromosome.

autotroph. An organism that can synthesize its own complex molecules from simple carbon and nitrogen sources, such as CO_2 and NH_3.

autoxidation. (autooxidation). A spontaneous, self-catalyzed oxidation occurring in the presence of air. It usually involves a free-radical mechanism. It is initiated by heat, light, metallic catalysts, or free-radical generators. Industrial processes, such as manufacture of phenol and acetone from cumene, are based on autoxidation. Other instances are the drying of vegetable oils, the spoilage of fats, gum formation in lubricating oils, and the degradation of high polymers exposed to sunlight for long periods.
See autocatalysis.

auu. A codon of isoleucine that directs the placement of isoleucine into a polypeptide.

Auwers–Skita rule. In its original form, the rule stated that in *cis*-trans isomeric hydroaromatic compounds, the *cis* had the higher density and refractive index and the lower molecular refractivity. A more modern statement of the von Auwers–Skita rule (which has undergone several modifications since it was first enunciated) is that, among alicyclic epimers not differing in dipole moment, the isomer of highest heat content has the higher density, index of refraction, and boiling point.

Auwers synthesis. Expansion of coumarones to flavonols by treatment of 2-bromo-2-(α-bromobenzyl)coumarones with alcoholic alkali.

auxiliary. Any of a number of chemical compounds used in some phase of textile processing. They may be classified as follows: (1) fats, oils, and waxes; (2) starches, gums, and glues (sizing); (3) soaps and detergents; (4) inorganic chemicals (bleaching, mercerizing); (5) organic solvents; (6)

special-purpose products (flameproofing, mildew-proofing, repellent and decorative coatings, and permanent-press resins). They are sometimes also called assistants.

auxin.   A natural or synthetic plant growth hormone that regulates longitudinal cell structure so as to permit bending of the stalk or stem in phototropic response. The natural materials are formed in small amounts in the green tips of growing plants, in root tips, and on the shaded side of growing shoots. 3-Indoleacetic acid is the most important natural auxin.
See plant growth regulator.

auxin herbicide.   Any of a number of synthetic auxins used as herbicides. While these herbicides often cause dedifferentiation and the initiation of cell division in mature cells, they usually inhibit cell division in the primary meristems of intact plants.

auxochrome.   A radical or group of atoms whose presence is essential in enabling a colored organic substance to be retained on fibers. The best examples are the groups $-COOH$, $-SO_3H$, $-OH$, and $-NH_2$.

auxoflore.   An atom or group of atoms present in a molecule, that shifts the fluorescent radiation of the atom toward a shorter wavelength, or increases the fluorescence.

Auxogluc.   An atomic group within a molecule that heightens sweetness.

auxotox.   An atomic group which, when present in the molecules of a compound, confers or intensifies the toxicity of the compound.

auxotroph.   See auxotrophic mutant.

auxotrophic mutant.   (auxotroph). A mutant organism defective in the synthesis of a given biomolecule, which must therefore be supplied for the organism's growth.

"Avadex *[Gowan].*   TM for a series of liquid or granular herbicides containing 2,3-dichloroallyl diisopropylthiocarbamate. Widely used to control growth of wild oats in agricultural crops.
Hazard: By ingestion and inhalation.

Availaphos.   A mineral supplement supplying phosphorus and calcium in readily available form for animal and poultry feeds.

avalite.
Properties: Claylike silicate containing chromium.

Avantia.   A pharmaceutical tablet-coating systems.

"Avelox Tabs" *[Bayer].*   TM for film coated tablets.
Use: Treatment of bacterial bronchitis, pneumonia and sinusitis.

avermectin. $C_{48}H_{72}O_{14}$. Any of a group of broad spectrum antiparasitic antibiotics produced by the actinomycete, *Streptomyces avermitilis*.
Hazard: Toxic.
Use: Sprayed on wool as a protection against moths and beetles.

avermectin b_1.
CAS: 71751-41-2.
Hazard: A poison by ingestion. Moderately toxic by inhalation and skin contact.
Use: Agricultural chemical.

avermectin b_{1a}.
CAS: 65195-55-3. $C_{48}H_{72}O_{14}$.
Hazard: A poison by ingestion. Moderately toxic by skin contact.
Use: Food additive; insecticide; miticide; agricultural chemical.

aviation gasoline.   See gasoline.

Avibest.   A microcrystalline form of asbestos.

"Avicel" *[FMC].*   TM for microcrystalline cellulose, a highly purified particulate form of cellulose.
Properties: Particle size ranges from less than 1 to 150 microns (average varies with grade), density 1.55 (bulk density 0.3–0.5). Insoluble in dilute acids, organic solvents, oils; swells in dilute alkali. Dispersible in water to form stable gels or pourable suspensions. Adsorbs oily and syrupy materials.
Use: Aid to stabilization and emulsification, ingredient in foods, suspending agent, binder and hardening agent in tableting, separatory medium in column and thin-layer chromatography, pure cellulose raw material.

avidin.   A protein occurring in egg white, where it constitutes approximately 0.2% of the total protein. It has the property of combining firmly with biotin and rendering it unavailable to organisms, since proteolytic enzymes do not destroy the avidin–biotin complex. Avidin loses its ability to combine with biotin when subjected to heat; hence cooked egg white does not lead to biotin deficiency.

"Avitene" *[C. R. Bard].*   TM for a microcrystalline form of collagen.

"Avitex" *[Du Pont].*   TM for a group of textile softeners, lubricants, and antistatic agents. Both anionic and cationic types are available.

Avitex.   A group of textile softeners, lubricants, and antistatic agents. Both anionic and cationic types are available.

"Avitone" *[Du Pont].*   TM for a group of chemical compounds based on hydrocarbon sodium sulfonates that are used principally as softening lubricating and finishing agents for textiles, leather, and paper.

avocado oil.   An edible oil high in unsaturated fatty acids.
  Properties: Greenish oil; faint odor, bland taste. D 0.91, acid value 1–7, saponification value 177–198, iodine value 71–95, fp 7–9C, refr index 1.461–1.465 (40C).
  Use: Cosmetic creams, hair conditioners, suntan preparations, salad oils.

Avogadro's law.   A principle stated in 1811 by the Italian chemist Amadeo Avogadro (1776–1856) that equal volumes of gases at the same temperature and pressure contain the same number of molecules regardless of their chemical nature and physical properties. This number (Avogadro's number) is 6.023×10^{23}. It is the number of molecules of any gas present in a volume of 22.41 L and is the same for the lightest gas (hydrogen) as for a heavy gas such as carbon dioxide or bromine. Avogadro's number is one of the fundamental constants of chemistry. It permits calculation of the amount of pure substance in a mole, the basis of stoichiometric relationships. It also makes possible determination of how much heavier a simple molecule of one gas is than that of another; as a result the relative molecular weights of gases can be ascertained by comparing the weights of equal volumes. Avogadro's number (conventionally represented by "N" in chemical calculations) is now defined to be the number of atoms present in 12 g of the carbon-12 isotope (one mole of carbon-12) and can be applied to any type of chemical entity.
See mole.

avuncular relationship.   The genetic relationship between nieces and nephews and their aunts and uncles.

"Axell" *[Accurate].*   TM for polyclonal antibodies.

"AxellPrep" *[Accurate].*   TM for products for the isolation of human mononuclear cells.

axis of symmetry.   An imaginary line in a crystal. The crystal, when rotated around this line through 360°, shows the same crystal face, at least twice.

"Axyll" *[Accurate].*   TM for monoclonal antibodies.

aza-.   Prefix indicating the presence of nitrogen in a heterocyclic ring.

3-azabicyclo(3,2,2)nonane. $C_8H_{15}N$.

Properties: White-tan solid. Mp 180C (sublimes). Partly soluble in water; solubility decreases with an increase in temperature. Readily soluble in alcohol, bulk d 4.67 lb/gal (20C).
Use: Intermediate for the preparation of pharmaceuticals and rubber chemicals.

1-azabicyclo(2.2.2)octane.   See quinuclidine.
azabicyclooctanol methyl bromide diphenylacetate.
CAS: 69766-47-8. $C_{22}H_{26}NO_2 \cdot Br$.
Hazard: A poison.

azaguanine.   (5-amino-2,3-dhydrotriazolo[4,5-*d*]pyrimidin-7-one).
CAS: 134-58-7. $C_4H_4N_6O$. One of the early purine analogs showing antineoplastic activity. It functions as an antimetabolite and is easily incorporated into ribonucleic acids.
Hazard: Mitotoic poison.

8-azaguanine.   (5-amino-1,4-dihydro-7H-1,2,3-triazolo[4,5-d]pyrimidin-7-1).
CAS: 134-58-7. $C_4H_4N_6O$.
Properties: Crystals from dilute aqueous sodium hydroxide. Insoluble in water, alcohol, and ether.
Grade: Refined.
Use: Inhibitor of purine synthesis.

azane. H_3N. Any of a class of saturated acyclic nitrogen hydrides.
Properties: Colorless, alkaline gas.
Derivation: Formed in the body during decomposition of organic materials during a large number of metabolically important reactions.

azaserine.   (azaserin; *L*-azaserine; AZS; CI-337; CN-15,757; diazoacetate (ester)-*L*-sterine; *L*-diazoacetate (ester) serine; diazo-acetic acid ester with Serine; NSC-742; P-165; RCRA water Number U015; *L*-serine diazoacetate(ester); serine diazoacetate; 0-0-diazoacetyl-*L*serine; 0-diazoacetyl-*L*-serine; (E)-1-(2-amino-3-hydroxy-3-oxopropoxy)-2-diazonioethenolate).
CAS: 115-02-6. $C_5H_7N_3O_4$. An inhibitor of purine synthesis.
Properties: Light-yellow needles from EtOH.
Derivation: Produced by the strain *Streptomyces fragilis*.
Hazard: Toxic; possible carcinogen; neoplastigenic; tumorigenic; poison; teratogen; mutagen.
Use: Antibiotic; antifungal; antineoplastic agent.

azaspiracid.
CAS: 214899-21-5. $C_{47}H_{71}NO_{12}$.
Hazard: A poison by ingestion.

2-azaspiro(5.5)undec-7-ene.
CAS: 6671-96-1. $C_{10}H_{17}N$.
Hazard: A poison by ingestion and skin contact. A severe skin and eye irritant.

azathioprine. (Imuran).
CAS: 446-86-6. An immunosuppressive drug administered for the purpose of inhibiting the natural tendency of the body to reject foreign tissues by one or more types of immunizing reactions, i.e., formation of leucocytes or antibodies. It has been used with some success in cases of kidney and liver transplants.
Hazard: Confirmed carcinogen.

6-azauridine. (6-azauracil riboside; *as*-triazine-3,5(2*H*,4*H*)dione riboside).
CAS: 54-25-1. $C_8H_{11}N_3O_6$.
Derivation: Microbiological fermentation.
Use: Research on cell formation and cancer.

azelaic acid. (nonanedioic acid; 1,7-heptanedicarboxylic acid).
$HOOC(CH_2)_7COOH$.
Properties: Yellowish to white, crystalline powder. Mp 106C, bp 365C (decomposes). Soluble in hot water, alcohol, and organic solvents.
Derivation: Oxidation of oleic acid by ozone.
Grade: Technical.
Use: Organic synthesis, lacquers, product of hydrotropic salts, alkyd resins, polyamides, polyester adhesives, low-temperature plasticizers, urethane elastomers.

azelaoyl chloride.
$ClOC(CH_2)_7COCl$.
Properties: Bp 125–130C (3 mm). Slowly decomposes in cold water; soluble in hydrocarbons and ethers.
Use: Organic synthesis.

azeotrope. See azeotropic mixture.

azeotropic distillation. A type of distillation in which a substance is added to the mixture to be separated in order to form an azeotropic mixture with one or more of the components of the original mixture. The azeotrope or azeotropes thus formed will have boiling points different from the boiling points of the original mixture and will permit greater ease of separation.

azeotropic mixture. (azeotrope). A liquid mixture of two or more substances which behaves like a single substance in that the vapor produced by partial evaporation of liquid has the same composition as the liquid. The constant boiling mixture exhibits either a maximum or minimum boiling point as compared with other mixtures of the same substances.

azide. Any of a group of compounds having the characteristic formula $R(N_3)_x$. R may be almost any metal atom, a hydrogen atom, a halogen atom, the ammonium radical, a complex ([CO(NH3)6], [Hg(CN)$_2$M] with M = Cu, Zn, Co, Ni) an organic radical like methyl, phenyl, nitrophenol, dinitrophenol, *p*-nitrobenzyl, ethyl nitrate, etc., and a variety of other groups or radicals. The azide group has a chain structure rather than a ring structure. All the heavy metal azides, such as hydrogen azide, and most if not all of the light metal azides (under appropriate conditions) are explosive. They should be handled with utmost care and protected from light, shock, and heat. Many of the organic azides are also explosive.
See lead azide; hydrazoic azide.

aziminobenzene. See 1,2,3-benzotriazole.

azimino compounds. Stable heterocyclic compounds having three adjacent nitrogen atoms in one ring.

azimsulfuron.
CAS: 120162-55-2. $C_{13}H_{16}N_{10}O_5S$.
Hazard: Low toxicity by ingestion, inhalation, and contact. A mild eye irritant.

azine dye. A class of dyes derived from phenazine. $(C_6H_4)N_2(C_6H_4)$ (tricyclic). The chromophore group may be =C=N–, but the color is more probably due to the characteristic unsaturation of the benzene rings. The members of the group are quite varied in application. The nigrosines (CI 50415–50440) and safranines (CI 50200–50375) are examples of this group.
See dye, synthetic.

azinphos methyl. (*O,O*-dimethyl-*S*-4-oxo-1,2,3-benzotriazin-3(4*H*)-yl methyl phosphorodithioate; Guthion).
CAS: 86-50-0. $C_{10}H_{12}N_3O_3PS_2$.
Properties: Brown, waxy solid. Mp 73C. Slightly soluble in water; soluble in most organic solvents.
Hazard: A poison, cholinesterase inhibitor. Absorbed by skin. Questionable carcinogen.
Use: Insecticide for fruit. Use may be restricted.

aziridine. A compound based on the ring structure.
See ethyleneimine; ethylethyleneimine; propyleneimine; polypropyleneimine; 1-aziridineethanol.

1-aziridineethanol. (*N*-(2-hydroxyethylethyleneimine)).
CAS: 1072-52-2. $C_2H_4NCH_2CH_2OH$.
Properties: Colorless liquid. Bp 167.9C, flash p 185F (OC). Combustible.
Handling: Inhibited with 1–3% dissolved sodium hydroxide.
Hazard: Irritant to skin and eyes.
Use: Chemical intermediate.

1-aziridinepropionic acid, cinnamyl ester.
CAS: 99900-94-4. $C_{14}H_{17}NO_2$.
Hazard: A poison.

A-Z Lite. A corrosion inhibitor.
Use: In cooling tower systems.

azlon. Generic name for a manufactured fiber in which the fiber-forming substance is composed of any regenerated naturally occurring protein (FTC). Proteins from corn, peanuts, and milk have been used. Azlon fiber has a soft hand, blends well with other fibers, and is used like wood. Combustible.

azobenzene. (diphenyldiimide; benzeneazobenzene).
CAS: 103-33-3. $C_6H_5N_2C_6H_5$.
Properties: Yellow or orange crystals. Mp 68C, bp 297C, d 1.203 (20/4C). Soluble in alcohol and ether; insoluble in water. Combustible.
Derivation: Reduction of nitrobenzene with sodium stannite.
Hazard: Toxic; may cause liver injury. Questionable carcinogen.
Use: Manufacture of dyes and rubber accelerators, fumigant, acaricide.

azobenzene-*p*-sulfonic acid. $C_{12}H_{10}O_3N_2S$.
Properties: Orange crystals. Mp 129C.
Use: Intermediate and reagent chemicals.

4,4′-azobis(4-cyanovaleric acid).
CAS: 2638-94-0. $[=NC(CH_3)(CN)CH_2CH_2CO_2H]_2$.
Properties: Light sensitive. Mp 90C (decomposes), mw 280.28.
Grade: 80% solution in water.
Hazard: Toxic.

azobisdimethylvaleronitrile.
Properties: White, crystalline solid.
Use: Initiator for suspension polymerization of vinyl chloride; solution polymerization of various monomers such as acrylonitrile, MMA, vinyl acetate.

1,1′-azobisformamide. (azodicarbonamide).
$H_2NCONNCONH_2$.
Properties: Yellow powder. D 1.65 (20/20C), mp above 180C (decomposes). Insoluble in common solvents; soluble in dimethyl sulfoxide. Hydrolyzes at high temperatures to nitrogen, carbon dioxide, and ammonia.
Derivation: From hydrazine.
Grade: Technical, FCC.
Use: Blowing agent for plastics and rubbers, maturing agent for flours.

2,2′-azobisisovaleronitrile.
CAS: 13472-08-7. $C_{10}H_{16}N_4$.
Hazard: Moderately toxic by ingestion.

azobisisobutyronitrile.
$(CH_3)_2C(CN)NNC(CN)(CH_3)_2$.
Properties: White powder. Mp 105C (decomposes). Insoluble in water; soluble in many organic solvents and in vinyl monomers.

Hazard: Toxic by ingestion.
Use: Catalyst for vinyl polymerizations and for curing unsaturated polyester resins, blowing agent for plastics.

azocarmine dye. A dye that stains tissues a dark purplish red.

azocholoramid. (n.n-dichloroazodi carbonamidine).
Properties: Water-soluble bacteria.

azocyclotin. See (1h-1,2,4-triazolyl-1-yl)tricyclohexylstannane.

azodicarbonamide. See 1,1′-azobis-formamide.

azodine. (benzeneazonaphthylethylenediamine).
CAS: 136-40-3. $C_{18}H_{18}N_4$.
Properties: Red crystals. Mp 107–108C.
Use: Reagent for rapid determination of penicillin in blood, urine, and other media.

Azodrin. [3-(dimethoxyphosphinyloxy)n-methyl-cis-crotonamide); monocrotophos].
CAS: 6923-22-4. Dimethylphosphate of 3-hydroxy-*N*-methyl-*cis*-crotonamide.
$(CH_3O)_2P(O)OC(CH_3):CHC(O)NHCH_3$.
Properties: Reddish-brown solid with a mild ester odor. Bp 125C. Soluble in water and alcohol; almost insoluble in kerosene and diesel fuel. Commercially available as a water-miscible solution.
Hazard: Flammable, dangerous fire risk. Use may be restricted. Toxic via ingestion, inhalation, and skin absorption. Questionable carcinogen. Cholinesterase inhibitor.
Use: Controls certain insects which attack cotton plants.

azo dye. Any of a broad series of synthetic dyes that have –N=N– as a chromophore group. They are produced from amino compounds by diazotization and coupling. Over half of the commercial dyestuffs are in this general category. By varying the chemical composition it is possible to produce acid, basic, direct, or mordant dyes. This general group is subdivided as monoazo, disazo, trisazo, and tetrazo according to the number of –N=N– groups in the molecule. Examples are Chrysoidine Y, Bismarck Brown 2R, and Direct Green B.
See dye, synthetic; azoic dye.

azo dye intermediate. Any of various sulfonated aromatic acids derived from α- and β-naphthol, naphthalene, and α- and β-napththylamine. Besides their systematic names, some are named after their discoverers, while others have letter designations. All have a fused ring structure with amino, hydroxyl, or sulfonic groups at various locations. For details, see the following entries:

1,2,4-acid
amino-G acid
amino-J acid
Armstrong's acid
B acid
Broenner' acid
Casella's acid
Chicago acid (SS acid)
chromotropic acid
Cleve's acid
crocein (Bayer's acid)
epsilon acid
G acid

γ acid
J acid
Kochs acid
L acid
Laurent's acid
M acid
Neville-Winter acid
peri acid
R acid
R R acid (2R acid)
Schaeffer's acid
Schoelkopf's acid
sultam acid
Tobias acid

azoic dye. There is no fundamental chromophoric difference between azo and azoic dyes. The differentiation is made to characterize a group of azo pigments which are precipitated within the cellulosic fiber by carrying out the dye coupling on the fiber. With the advent of the equally brilliant but more easily applied fiber-reactive dyes, the azoics have lost some of their importance.

azolitmin. A purplish-red colorant.
Derivation: Natural litmus synthesized by oxidation of orcinol in the presence of ammonia, lime, and potash.
Use: Broad indicator of pH.

azophenylene. See phenazine.

Azosol. A series of dyestuffs; soluble in organic solvents.
Use: Coloring spirit lacquers and spirit inks.

azosulfamide. (disodium 2-(4'-sulfamylphenylazo)-7-acetamido-1-hydroxynaphthalene-3,6-disulfonate). $C_{18}H_{14}N_4Na_2O_{10}S_3$.
Properties: Dark red powder; odorless; tasteless. Soluble in water with an intense red color; practically insoluble in organic solvents.

Use: Medicine (antibacterial).

azote. The French word for nitrogen (a = not, plus zoo = alive as in $zoon$ = animal). Nitrogen-bearing compounds can be recognized by this root word in such terms as azo, azide, azobenzene, carbazole, thiazole, etc. The derivation is due to the chemical inertness of nitrogen.

azotic acid. See nitric acid.

azoxybenzene. (diphenyldiazene oxide).
CAS: 495-48-7. $C_6H_5NO=NC_6H_5$.
Properties: Yellow crystals. D 1.16, mp 36C. Soluble in alcohol; insoluble in water.
Use: Intermediate in organic synthesis.

azoxystrobin.
CAS: 131860-33-8. $C_{22}H_{17}N_3O_5$.
Hazard: Moderately toxic by inhalation.

azoxytoluidine. See diaminoazoxytoluene.

azulene.
CAS: 275-51-4. $C_{10}H_7$. Aromatic hydrocarbon with a 7-carbon ring fused to a 5-carbon ring.
Properties: Blue to greenish-black leaflets. Mw 128.19, mp 99–100C, bp 242C, decomposes at 270C, bp 115–135C at 10 mm Hg. Soluble in alcohol, ether, acetone.

azure blue. See cobalt blue.

azuresin.
Properties: Moist, irregular, dark blue or purple granules; slightly pungent odor.
Derivation: Carbacrylic cation-exchange resin, in reversible combination with 3-amino-7-dimethylaminophenazathionium chloride (azure A dye).
Grade: NF.
Use: Medicine (diagnostic test).

B

β. See beta.

B. Symbol for boron.

Ba. Symbol for barium.

^{137}Ba. See barium-137.

babassu oil. A nondrying, edible oil expressed from the kernels of the babassu palm, which grows in profusion in Brazil. Composition: 44% lauric acid, 15% myristic acid, 16% oleic acid, balance mixed acids. Usable in foods and soapmaking, but supply is limited by cost of exploitation of the large quantities potentially available. Combustible.

Babbitt metal. One of a group of soft alloys used widely for bearings. They have good bonding characteristics with the substrate metal, maintain oil films on their surfaces, and are nonseizing and antifriction. Used as cast, machined, or preformed bimetallic bearings in the form of a thin coating on a steel base; the main types are lead base, lead-silver base, tin base, cadmium base, and arsenical. The latter contains up to 3% arsenic.
Hazard: Dust is toxic by inhalation.

Babcock test. A rapid test for butterfat in milk introduced by Stephen M. Babcock in 1890 and now in worldwide use in the dairy industry.

Babinet principle. Identical diffraction patterns of a substance are produced by two diffraction screens one of which is the exact negative of the other.

Babo's law. The lowering of vapor pressure is proportional to the mole fraction of nonvolatile solute in a solution.

BAC. See (1) Biologically activated C; (2) blood alcohol concentration; (3) bromoacetyl cellulose; (4) bacterial artificial chromosome.

B acid. (1-amino-8-naphthol-3,5-disulfonic acid). $C_{10}H_4NH_2OH(SO_3H)_2$.
Derivation: Sulfonation of 1-amino-8-naphthol-3-sulfonic acid.
Use: Azo dye intermediate.

bacillomycin. See bacillus subtilis BPN.

bacillus. A type of bacteria characterized by a rodlike shape.

Bacillus anthracis. A bacterium, the causative agent of the disease anthrax.
Use: A potential biological warfare and terrorism agent.
Hazard: Pathogenic.

Bacillus stearothermophilus. A biological indicator.
Use: To ensure that an autoclave has been thoroughly sterilized.

bacillus subtilis BPN. (bacillomycin; fungocin; subtilisins).
CAS: 1395-21-7.
Hazard: A severe eye irritant.
Use: Laundry detergents.

bacitracin. (4-[[2-[[2-(1-amino-2-methylbutyl)-4,5-dihydro-1,3-thiazole-4-carbonyl]amino]-4-methylpentanoyl]amino]-5-[[1-[[19-(2-amino-2-oxoethyl)-4-(3-aminopropyl)-10-benzyl-7-butan-2-yl-16-(carboxymethyl)-13-(4H-imidazol-4-ylmethyl)-2,5,8,11,14,17,20-heptaoxo-3,6,9,12,15,18,21-heptazacyclopentacos-1-yl]amino]-3-methyl-1-oxopentan-2-l]amino]-5-oxopentanoic acid; zinc; ayfivin; baciguent; baci-jel; baciliquin; bacitek ointment; fortracin; parentracin; pentracin; topitracin; USAF CB-7; zutracin).
CAS: 1405-87-4. $C_{66}H_{100}N_{17}O_{16}5Z_{12}$. A complex of cyclic peptide antibiotic of known chemical structure isolated from the Tracy-I strain of *Bacillus subtilis.*
Properties: White to pale-buff, hygroscopic powder; odorless or slight odor; freely soluble in water, alcohol, methanol, and glacial acetic acid; insoluble in acetone, chloroform, and ether.
Hazard: Poison; moderately toxic; mutagen.
Use: Antibiotic against *Staphylococci,* hemolytic *Streptococci,* and aerobic, Gram-positive, rod-shaped bacteria, and to treat eczema and infected dermal ulcers.

bacitracin methylene disalicylate.
Properties: White to gray-brown powder. Slight unpleasant odor, less bitter than bacitracin. Soluble in water, pyridine, ethanol; less soluble in acetone, ether, chloroform, benzene; pH of saturated

aqueous soluble 3.5–5.0. Available also as the sodium salt.
Use: Antibiotic, feed additive.

bacitracin zinc.
Use: Drug (veterinary); feed additive.

backcross. A cross between an animal that is heterozygous for alleles obtained from two parental strains and a second animal from one of those parental strains. Also used to describe the breeding protocol of an outcross followed by a backcross.
See model organisms.

back filling. Process of applying starch, with or without other filling or weighting materials, to the back of a cloth.

Back–Goudsmit effect. Effect of a weak magnetic field on the spectral lines of an element having a nuclear magnetic moment.

back-mutation. A mutation that causes a mutant gene to revert bacto its wild-type genotype.

bacteria. Microorganisms often composed of a single cell in the form of straight or curved rods (bacilli), spheres (cocci), or spiral structures. Their chemical composition is primarily protein and nucleic acid. Chlorophyll molecules are also present, enabling bacteria to carry out photosynthesis. Some types, called anaerobic, are able to live and reproduce in the absence of oxygen; aerobic types require oxygen. Bacteria that can live either with or without oxygen are called facultative. Filamentous bacteria are related to blue-green algae. Molds that yield antibiotics (*Actinomycetes*) are of this type. Pathogenic bacteria are infectious organisms that cause such diseases as pneumonia, tuberculosis, syphilis, and typhus.
The staining of bacteria for microscopic identification was originated by Koch, a German physician and bacteriologist (1843–1910). Bacteria are often classified as Gram-positive or Gram-negative.
Food spoilage is often induced by bacterial contamination. There are many beneficial types of bacteria in the body, e.g., intestinal flora that aid in metabolism. Bacteria rich in proteins can be produced by fermentation of animal wastes for feed supplements.
The outstanding development in this field is the laboratory modification of bacteria by gene-splicing techniques. This noteworthy achievement has an enormous future potential in the chemical, agricultural, food, and pharmaceutical industries. The Supreme Court has ruled it to be a patentable invention.
See recombinant DNA; biotechnology.
Use: (1) Fermentation processes used in baking and the manufacture of alcohol, wine, vinegar, beer (yeast), and antibiotics (molds). (2) Fixation of atmospheric nitrogen in the soil. (3) Reaction with hydrocarbons (methane and other paraffins) to yield proteins (yeasts). (4) Purification of sewage sludge activated by bacteria (see sewage sludge). (5) Reaction with cellulose to form biopolymers and high-protein foodstuffs. (6) Reaction with waste materials (coal and cement dusts, gasworks effluent) to release plant nutrients for inexpensive fertilizers (the former U.S.S.R.). (7) Precipitation and concentration of uranium and some other metals by compounds obtained from bacteria grown on carbonaceous materials such as lignin and cellulose. (8) Formation of azo compounds in soil treated with the herbicide propanil. (9) Synthesis of hormones by recombinant DNA methods (*E. coli*).
See insulin.
(10) Miscellaneous reactions, e.g., oxidation of pentaerythritol to tris(hydroxymethyl)acetic acid; conversion of the sulfur in gypsum to elemental sulfur via hydrogen sulfide; clean-up of oil spills.
See fermentation; virus; enzyme; biotechnology.

bacterial artificial chromosome. (BAC). A vector used to clone DNA fragments (100- to 300-kb insert size; average, 150 kb) in Escherichia coli cells. Based on naturally occurring F-factor plasmid found in the bacterium *E. coli*.
See cloning vector.

bactericide. (germicide). Any agent that will kill bacteria, especially those causing disease. Bactericides vary greatly in their potency and specificity. They may be other organisms (bacteriophages), chemical compounds, or shortwave radiation.
See virus; antibiotic; biocide.

bacterioagglutinin. An antibody that agglutinates bacteria.

bacteriochlorin. (bacteriochlorophyll; 7,8,17,18-tetrahydroporphyrin; 2,3,12,13,22,24-hexahydroporphyrin). $C_{20}H_{16}N_4$. A light-absorbing bacterial pigment that is closely related to chlorophyll.
Properties: Has two pairs of non-fused saturated carbon atoms in two of the pyrrole rings, c-7, c-8 and c-17, c-18. The tetrapyrrole ring bears a central manganese as opposed to magnesium in chlorophyll.
Derivation: A reduced porphyrin.

bacteriochlorophyll. A light-absorbing pigment, closely related to chlorophyll. The magnesium at the center of the tetrapyrrole ring is replaced by manganese.
Derivation: Occurs in photosynthesizing green sulfur and purple sulfur bacteria.

bacteriocin. Any of a class of antibacterial proteins of bacterial origin that selectively kill or inhibit but do not lyse closely-related bacteria other than the type that produced it. They are generally more potent, but have a narrower range of activity than antibiotics. They act by blocking a specific metabolic pathway.

bacteriohodopsin. A naturally-occurring carotenoid pigment. This substance can absorb light and generate a proton gradient across the membrane of halobacterium and initiating synthesis of ATP in the absence of chlorophyll.

bacteriolysin. An antibacterial antibody that lyses bacteria.

bacteriophage. A type of virus that attacks and destroys bacteria by surrounding and absorbing them.

bacteriophage lambda. A virus which infects *E. coli*, and which is often used in molecular genetics experiments as a vector, or cloning vehicle. Recombinant phages can be made in which certain non-essential DNA is removed and replaced with the DNA of interest. The phage can accommodate a DNA "insert" of about 15–20 kb. Replication of that virus will thus replicate the investigator's DNA.

bacteriophage M13. A single-stranded DNA bacteriophage used as a vector for DNA sequencing.

bacteriopsonin. An opsonin that acts upon bacteria.

bacteriostat. A substance that prevents or retards the growth of bacteria. Examples are quaternary ammonium salts and hexachlorophene. See antiseptic.

bacteriotoxin. A toxin produced by bacteria. They are classes as bacterial endotoxins and exotoxins and play important roles in a large number of disease processes and in food poisoning.

baddeleyite. (zirconia). ZrO_2. A natural zirconium oxide.
Properties: Black, brown, yellow to colorless; streak white, luster submetallic to vitreous to greasy. D5.5–6.0, mp 2500–2950C. Highly resistant to chemicals.
Grade: Crude (53%, 73–75%), purified (98%).
Occurrence: Brazil, Ceylon.
Use: Corrosion- and heat-resistant applications, source zirconium.

Badische acid. (2-naphthylamine-8-sulfonic acid). $C_{10}H_9NO_3S$.
Properties: Colorless needles, partially soluble in water and alkalies, slightly soluble in alcohol.
Derivation: Sulfonation of 2-naphthylamine.
Use: Azo dye intermediate.

Baekeland (Bakelite) process. Condensation of phenol and formaldehyde to *o*-hydroxymet-hylphenol (Lederer–Manasse), which undergoes further arylation yielding a polymeric structure.

Baekeland, Leo Henricus Arthur. (1863–1944). Born in Ghent, Belgium. He did early research in photographic chemistry and invented Velox paper (1893). After working for several years in electrolytic research, he undertook fundamental study of the reaction products of phenol and formaldehyde, which culminated in his discovery in 1907 of phenol-formaldehyde polymers originally called "Bakelite." The reaction itself had been investigated by Bayer in 1872 but Baekeland was the first to learn how to control it to yield dependable results on a commercial scale. The Bakelite Co. was founded in 1910 and now is Bakelite AG. See phenol-formaldehyde resins.

Baeyer–Drewson indigo synthesis. Formation of indigos by an aldol addition of *o*-nitrobenzaldehydes to acetone, pyruvic acid, or acetaldehyde. Of interest mainly as a method of protecting *o*-nitrobenzaldehydes.

Baeyer–Villiger reactions. The oxidation of aromatic, open-chain, and cyclic ketones to esters and lactones by peracids.

baffle. A flow-regulating device consisting of a perforated metal plate placed horizontally in liquid-mixing tanks, distillation columns, and the like to restrict or divert the passage of liquid, thus providing a uniformly dispersed flow. Baffles are also used in open-steam autoclaves to ensure even distribution of the entering steam.

bagasse. A form of cellulose (biomass) derived as a by-product of the crushing of sugar cane or guayule plants. Contains a high proportion of hemicellulose. After pulping with either soda or kraft cooking liquor, it can be made into a low grade of paper. It is also used in compressed form as an insulating board in construction, as a medium for growth of nutritive bacteria, in animal feeds, in manufacturing of furfural; and as on-site fuel for cane-sugar mills. In Hawaii, it is being used as a fuel for electric power generation.
Hazard: Dust is flammable.
See biomass.

baghouse. (bag filter). A large-scale dust-collecting device composed of a series of large cotton or nylon bags assembled in a heavy metal frame or housing. The bags may be as much as 10 ft high, each "bag" being made up of three units sewn together as one element approximately 18 inches in diameter. Discharge hoppers are located beneath the bags. A suction or blower system forces dust-laden air through an inlet port on one side of the frame just above the hopper space. It enters the bags, where it deposits its suspended solids, while the cleaned air is drawn through and leaves by an outlet port. A

motor-driven shaker mechanism agitates the bags periodically, dislodging the accumulated layer of dust, which falls into the hoppers. Installations of this type are often of impressive size, some containing over 300 bags.
See "Nomex" *[Du Pont]*.

bait. An insecticide or rodenticide placed in such a way as to attract the pest. Arsenic compounds and Bordeaux mixture are typical insect baits. All types are highly toxic.
See pesticide.

Bakelite. Polyethylene, polypropylene, epoxy, phenolic, polystyrene, phenoxy, perylene, polysulfone, ethylene copolymers, ABS, acrylics, and vinyl resins and compounds.

Baker–Nathan effect. Effect originally observed in the reaction of *p*-substituted benzyl bromides with pyridine and other processes in which the observed rates are opposite to those predicted by the electron-releasing inductive effect of alkyl groups, i.e., $CH_3 > CH_3CH_2 > (CH_3)_2CH > C(CH_3)_3$. To explain it, a type of electron delocalization involving σ electrons was proposed, termed hyperconjugation, which manifests itself in systems in which a saturated carbon atom attached to an unsaturated carbon or one with an empty orbital bears at least one hydrogen atom.

baker's yeast extract.
Properties: From ruptured cells of *Saccharomyces cerevisiae.* Liquid, paste or powder. Water sol.
Use: Food additive.

Baker–Venkataraman rearrangement.
Base-catalyzed rearrangement of *o*-acyloxyketones to β-diketones, important intermediates in the synthesis of chromones and flavones.

Bake Smart. A carb-based powder comprised of glucose syrup, dextrose and fiber.
Use: Delivers moisture control, binds moisture in a broad range of foods.

baking finish. A paint or varnish that requires baking at temperatures greater than 66C for the development of desired properties (ASTM). Such finishes are based on oil-modified alkyd, melamine, epoxy, nitrocellulose, or urea resins, or combinations of these. Baking is often done by infrared radiation, producing high-molecular weight coatings that are dense and tough.

baking powder. A synthetic leavening agent widely used in the baking industry. There are several types, all of which are composed of a carbonate, a weak acid or acidic compound, and a filler. A typical composition is sodium bicarbonate, tartaric acid or monobasic calcium phosphate, and cornstarch. Ingredients sometimes used are ammonium carbonate and potassium bitartrate. Upon contact with moisture and heat, the active ingredients react to evolve carbon dioxide, which "raises" the dough in the early minutes of heat exposure, thus producing a stable solid foam. Wheat-flour gluten is sufficiently elastic to retain the bubbles of carbon dioxide.

baking soda. See sodium bicarbonate.

BAL. Abbreviation for British Anti-Lewisite. See 2,3-dimercaptopropanol.

balagrin.
CAS: 71330-43-3.
Hazard: Moderately toxic by ingestion.
Use: Agricultural chemical.

balance. (1) Exact equality of the number of atoms of various elements entering into a chemical reaction and the number of atoms of those elements in the reaction products. For example, in the reaction $NaOH + HCl \rightarrow NaCl + H_2O$, the atoms in the input side are H[2], Na[1], O[1], and Cl[1]. Each of these is also present in the products, though in different combination. The atoms of catalysts (when present) do not enter into reactions and therefore are not involved. The balance of chemical reactions follows the law of conservation of mass. The term *material balance* is used by chemical engineers in designing processing equipment. It denotes a precise list of all the substances to be introduced into a reaction and all those that will leave it in a given time, the two sums being equal. (2) A precision instrument designed for weighing extremely small amounts of material with high accuracy. An analytical balance or microbalance for weights from about 1 g to 0.1 mg is standard equipment in chemical laboratories. Its essential feature is a one-piece metal beam (lever) pivoted on a knife-edge or flexure at its exact center (fulcrum) so that it is free to oscillate. From it are suspended two scale pans approximately 2 inches in diameter, each of which is also positioned on a knife-edge on the lower arms of the beam. Exact balance is indicated by a pointer attached to the beam. Either an aluminum rider or a chain and vernier is provided for maximum accuracy. Highly sophisticated balances operating electronically with built-in microprocessors have become available in recent years.

Balanites aegyptiaca.
A genus of trees indigenous to the Near East, the berries of which contain a biologically active principle that is lethal to fish, tadpoles, mollusks, and various larvae.
Use: Prophylacitic against schistosomiasis in drinking water.

balata. See rubber, natural (note).

balb/3t3. A type of cultured mouse embryo fibroblast used to test the capacity of a chemical in vitro to induce the transformation of mammalian cells.

Baldwin rules for ring closure.   Set of empirical rules, stereochemical in nature, predicting the relative facility of ring closure reactions.

Baldwin's phosphorus.   Fused nitrate of lime, which emits light for several hours after exposure to sunlight.

ball-and-ring method.   The melting point of a material is determined by filling the aperture of a metal ring with the material, placing a metal ball on it, and heating to the temperature at which the material softens and is then pushed out of the ring by the weight of the ball. It is used for substances that have no definite melting point, but soften before melting.

ball clay.   A clay that has good plasticity, strong bonding power, high refractoriness, and fires to a white or cream-colored product. Used as bonding and plasticizing agent or chief ingredient of whiteware, porcelains, stoneware, terra-cotta, glass refractories, and floor and wall tile.
See clay.

Balling hydrometer.   A device for determining the approximate percentage of sugar in solution at 60F.

ball mill.   A jacketed steel cylinder rotating on a horizontal axis and containing steel balls of varying diameter; the interior walls are usually equipped with baffle bars to impart a rolling and cascading action to the balls. The total weight of the balls may be 2000 lb or more. The grinding efficiency depends on the number of contacts between any two balls; thus the greater the number of balls, the more effective the grinding action. The material is introduced through an opening in the axis of the cylinder, which is then hermetically closed. Discharge is by the same opening after replacement of the cover plate with a grill to retain the balls. Ball mills can be adapted to continuous operation in which the feed enters at one end and is discharged at the other. Products ground are dry chemicals, paint pigments, etc.
See pebble mill; jar mill.

Bally–Scholl synthesis.   Formation of mesobenzanthrones by the action of glycerol or a derivative, and sulfuric acid on anthraquinones or anthranols.

balsam.   A resinous mixture of varying composition obtained from several species of evergreen trees or shrubs. Contains oleoresins, terpenes, and usually cinnamic and benzoic acids. All types are soluble in organic liquids and insoluble in water. Some have a penetrating, pleasant odor. They are combustible and in general nontoxic. The best-known types are as follows:
(1) Peru balsam, from Central America, is a thick, viscous liquid (d 1.15) containing vanillin. Used in flavoring, chocolate manufacturing, as an ingredient in expectorants and cough syrups, and as a fragrance in shampoos and hair conditioners. A mild allergen. Shipped in drums.
(2) Tolu balsam, a plastic solid, is derived from a related tree in Colombia. Its uses are similar to Peru balsam. Source of tolu oil. Odorless.
(3) Copaiba balsam from Brazil and Venezuela (d 0.94–0.99) is a viscous liquid used in varnishes and lacquers as an odor fixative and in manufacture of photographic paper. It is the source of copaiba oil.
(4) Balm of Gilead is from a Middle Eastern shrub and is used in perfumery and medicine.
(5) Canada balsam, from the North American balsam fir, is a liquid, d 0.98, used in microscopy, in fine lacquers, as a flavoring, and as a fragrance.
(6) Benzoin resin (Benjamin gum).
See benzoin resin.

balsam fir oil.   See fir needle oil, Canadian type.

Bamberger's formula.   A structural formula for naphthalene that shows the valences of the benzene rings pointing to the centers.

bamboo.   A grass native to southeast Asia having a rather high cellulose content, which makes possible its use for specialty papers. Its fibers are longer than those of most other plants of this type and are comparable to those of coniferous woods. Has a composition of total cellulose 58%, α-cellulose 35%, pentosans 28%, lignin 23%. Also used for making light furniture, fishing rods, etc. Combustible.

bambuterol.
CAS: 81732-65-2. $C_{18}H_{29}N_3O_5$.
Hazard: A poison.

Bamford–Stevens reaction.   Formation of olefins by base-catalyzed decomposition of *p*-toluenesulfonylhydrazones of aldehydes and ketones.

banana oil.   (1) (Banana liquid.) A solution of nitrocellulose in amyl acetate or similar solvent; so termed because of its penetrating banana-like odor.
(2) Synonym for amyl acetate.

Banbury mixer.   A batch-type internal-mixing machine, named after its inventor, that has been widely used in the rubber industry since 1920 for high-volume production. It will also accept plastic molding powders. Its chief feature is an enclosed, barrel-shaped chamber in which two rotors with oppositely curved contours rotate rapidly on a horizontal axis, first masticating the rubber and then efficiently incorporating the dry ingredients. Both steam and water jacketing are provided. Batches may be up to 1000 lb. A plunger at the entrance port rides on top of the batch to furnish enough pressure for proper mixing. A hydraulically operated discharge gate is located below the mixing chamber.

band, absorption.   See absorption band.

Bandane.   Polychlorodicyclopentadiene isomers.
Use: As an herbicide.

banded iron formation.   An iron formation that consists of alternating iron-rich and iron-poor layers. Most rocks of this type are older than about two billion years.

bandwidth.   Difference between limiting frequencies of a frequency band.

banod shift assay.   See gel shift assay.

Banox.   A series of dry, powdered, phosphate-type corrosion inhibitors. No. 1 is artificially colored. Nos. 1-P and WT are colorless.
Use: Refrigerator cars, refrigeration brine, cooling towers, and small water systems.

Banting, Sir Frederick.   (1891–1941). A native of Ontario, Canada, Banting did his most important work in endocrinology. His brilliant research culminated in the preparation of the antidiabetic hormone that he called insulin, derived from the isles of Langerhans in the pancreas. He received the Nobel Prize in medicine for this work together with MacLeod of the University of Toronto. In 1930, the Banting Institute was founded in Toronto. He was killed in an airplane crash.

Banvel D.   An herbicide containing 2-methoxy-3,6-dichlorobenzoic acid (dimethylamine salt).

BAP.   Abbreviation for benzyl-p-aminophenol.

Barafene.   Coatings used to prevent permeation of volatile ingredients, oils, and oxygen through polyolefin containers.

"Barak" [Technopharma].   TM for dibutylammonium oleate $(C_4H_9)_2NH_2COOC_{17}H_{33}$.
Properties: Translucent, light brown liquid. Combustible.
Use: To activate accelerators and improve processing of rubber and synthetic rubbers.

barban.   Generic name for 4-chloro-2-butynlyl-m-chlorocarbanilate. $C_6H_4(Cl)NHCOOCH_2C$:CCH_2Cl.
Herbicide and plant growth regulator.

***o*-barene.**   See 1,2-dicarbadodecaborane(12).

barberite.   A nonferrous alloy containing 88.5% copper, 5% nickel, 5% tin, 1.5% silicon.
Properties: D 8.80, mp 1070C. It offers good resistance to sulfuric acid in all dilutions up to 60%, seawater, moist sulfurous atmospheres, and mine waters.

Barbier–Wieland   degradation.   Stepwise carboxylic acid degradation of aliphatic acids (particularly in sterol side chains) to the next lower homolog. The ester is converted to a tertiary alcohol that is dehydrated with acetic anhydride, and the olefin oxidized with chromic acid to a lower homologous carboxylic acid.

barbital.   (diethylmalonylurea; diethylbarbituric acid; Veronal).
CAS: 57-44-3. $C_8H_{12}N_2O_3$.

Properties: White crystals or powder; bitter taste; odorless. Stable in air. Mp 187–192C. Soluble in hot water, alcohol, ether, acetone, and ethyl acetate.
Derivation: By the interaction of diethyl ester or diethylmalonic acid and urea.
Grade: Technical, CP.
Hazard: See barbiturates.
Use: Medicine (sedative), stabilizer for hydrogen peroxide.
See barbiturates.

barbital sodium.   (barbital Na; barbital soluble; barbitone sodium; diethylbarbiturate monosodium; diethylmalonylurea sodium; embinal; natriumbarbitals; nervoseton; 2,4,6(1H,3H,5H)-Pyrimidinetrione, 5,5-diethyl-, monosodium salt (9CI); sodium barbital; bodium barbitone; sodium diethylbarbiturate; sodium ethylbarbital; sodium malonylurea; sodium veronal; soluble barbital; soprinal; thyalone; veronal sodium; sodium 5,5-diethyl-4,6-dioxo-1H-pyrimidin-2-olate).
CAS: 144-02-5. $C_8H_{11}N_2NaO_3$.   The soluble monosodium salt of barbital. It is a long-acting barbiturate that depresses most metabolic processes at high doses.
Properties: Bitter crystals or powder.
Hazard: Poison; causes marked depression, prolonged coma, and death; teratogen; questionable carcinogen; tumorigen; neoplastigen; mutagen.
Use: A hypnotic and sedative, and in veterinary practice for central nervous system depression.

barbiturate.   A derivative of barbituric acid that produces depression of the central nervous system and consequent sedation. Used by prescription only for sedative and anesthetic purposes.

Hazard: Habit-forming. Several types, including amo-, seco-, and pentabarbital, are under government restriction.

barbituric acid.   (malonylurea; pyrimidinetrione; 2,4,6-tri-oxohexahydro pyrimidine).

OCNHCOCH$_2$CONH•2H$_2$O.

Properties: White crystals; efflorescent; odorless. Mp 245C with some decomposition. Slightly soluble in water and alcohol; soluble in ether. Forms salts with metals.
Derivation: By condensing malonic acid ester with urea.
Grade: Technical.
Use: Preparation of barbiturates, polymerization catalyst, dyes.

Bardhan–Sengupta phenanthrene synthesis.
Formation of octahydrophenanthrene derivatives by cyclodehydration of derivatives of 2-(β-phenethyl)-1-cyclohexanol and consequent dehydration to phenanthrenes with selenium.

bar disintegrator.   See cage mill.

Barff process.   The deposition of a rust-resistant coating on iron-base metals by oxidation with superheated steam.

Barfoed's reagent.   Aqueous solution of copper acetate.
Use: To distinguish monosaccharides from disaccharides (red cuprous oxide forms in presence of glucose).

barite.   (BaSO$_4$). Natural barium sulfate, barytes, heavy spar.

barium.
CAS: 7440-39-3. Ba. Alkaline-earth element of atomic number 56, group IIA of periodic table; aw 137.34; valence 2; 7 stable isotopes.
Properties: Silver-white, somewhat malleable metal. D 3.6. Values for melting and boiling points are reported as 704–850C and 1140–1637C, respectively (The most acceptable values, based on reliable original work, appear to be mp 710C and bp 1500C). Extremely reactive, reacts readily with water, ammonia, halogens, oxygen, and most acids. Gives green color in flame. Extrudable and machinable.
Occurrence: Ores of barite and witherite are found in Georgia, Missouri, Arkansas, Kentucky, California, Nevada, Canada, Mexico.
Derivation: Reduction of barium oxide with aluminum or silicon in a vacuum at high temperature.
Available forms: Rods, wire, plate, powder.
Grade: Technical, pure.

Hazard: Flammable (pyrophoric) at room temperature in powder form; store under inert gas, petroleum, or other oxygen-free liquid. When heated to approximately 200C in hydrogen, barium reacts violently, forming BaH$_2$. Eye, skin, and gastrointestinal irritant, and muscular stimulant. Questionable carcinogen.
Use: Getter alloys in vacuum tubes, deoxidizer for copper, Frary's metal, lubricant for anode rotors in X-ray tubes, spark-plug alloys.

barium-137.   Radioactive isotope of barium. See cesium-137.

barium acetate.
CAS: 543-80-6.   Ba(C$_2$H$_3$O$_2$)$_2$•H$_2$O.
Properties: White crystals. D 2.02, mp (decomposes). Soluble in water, insoluble in alcohol.
Derivation: Acetic acid is added to a solution of barium sulfide. The product is recovered by evaporation and subsequent crystallization.
Grade: Technical, CP.
Use: Chemical reagent, acetates, textile mordant, catalyst manufacturing, paint, and varnish driers.
See barium.

barium aluminate.   3BaO•Al$_2$O$_3$.
Properties: Gray, pulverized mass. Soluble in water, acids.
See barium.

barium azide.
CAS: 18810-58-7.   Ba(N$_3$)$_2$.
Crystalline solid, d 2.936, loses nitrogen at 120C, soluble in water, slightly soluble in alcohol.
Hazard: Explodes when shocked or heated.
Use: High explosives.
See barium.

barium binoxide.   See barium peroxide.

barium borotungstate.   (barium borowolframate).
2BaO•B$_2$O$_3$•9WO$_3$•18H$_2$O.
Properties: Large, white crystals. Effloresces in air. Keep well stoppered!! Soluble in water.
Hazard: A poison.
Use: Making borotungstates.

barium borowolframate.   See barium borotungstate.

barium bromate.
CAS: 13967-90-3.   Ba(BrO$_3$)$_2$•H$_2$O.
Properties: White crystals or crystalline powder. D 3.820, decomposes at 260C. Slightly soluble in water, insoluble in alcohol.
Derivation: By passing bromine into a solution of barium hydroxide; barium bromide and barium bromate are formed, which are separated by crystallization.
Grade: Pure, reagent.

Hazard: A poison. Moderate fire risk in contact with organic materials.
Use: Analytical reagent, oxidizing agent, corrosion inhibitor.

barium bromide. $BaBr_2 \cdot 2H_2O$.
Properties: Colorless crystals. D 3.852, mp (anhydrous) 847C. Soluble in water and in alcohol.
Derivation: Interaction of barium sulfide and hydrobromic acid with subsequent crystallization.
Grade: Technical, CP.
Hazard: A poison.
Use: Manufacturing bromides, photographic compounds, phosphors.

barium carbonate.
CAS: 513-77-9. $BaCO_3$.
Properties: White powder; found in nature as the mineral witherite. D 4.275, mp 174C at 90 atmospheres, 811C at one atmosphere. Insoluble in water; soluble in acids (except sulfuric).
Derivation: Precipitated barium carbonate is made by reaction of sodium carbonate or carbon dioxide with barium sulfide.
Grade: Technical, CP, reagent 99.5%.
Hazard: A poison.
Use: Treatment of brines in chlorine-alkali cells to remove sulfates, rodenticide, production of barium salts, ceramic flux, optical glass, case-hardening baths, ferrites, in radiation-resistant glass for color television tubes.

barium chlorate.
CAS: 13477-00-4. $Ba(ClO_3)_2 \cdot H_2O$.
Properties: Colorless prisms or white powder. D 3.179, mp 414C. Soluble in water. Combustible.
Derivation: Electrolysis of barium chloride.
Grade: Technical, CP, reagent.
Hazard: A poison. Strong oxidizer, fire risk in contact with organic materials.
Use: Pyrotechnics, explosives, textile mordant, manufacture of other chlorates.

barium chloride.
CAS: 10361-37-2. $BaCl_2 \cdot 2H2O$.
Properties: Colorless, flat crystals. D 3.097, mp 960C (anhydrous). Soluble in water; insoluble in alcohol, Combustible.
Derivation: (1) By the action of hydrochloric acid on barium carbonate or barium sulfide; (2) by heating a mixture of barium sulfate, carbon, and calcium chloride.
Grade: Technical (crystals or powdered), 99%, crystals, powdered, CP.
Hazard: Ingestion of 0.8 g may be fatal.
Use: Chemicals (artificial barium sulfate, other barium salts), reagents, lubrication oil additives, boiler compounds, textile dyeing, pigments, manufacture of white leather.

barium chromate.   (lemon chrome; ultramarine yellow; baryta yellow; Steinbuhl yellow).

CAS: 10294-40-3. $BaCrO_4$.
Properties: Heavy, yellow, crystalline powder. D 4.498. Soluble in acids; insoluble in water. Combustible.
Derivation: Interaction of barium chloride and sodium chromate. The precipitate is washed, filtered, and dried.
Grade: Technical, CP.
Hazard: Confirmed carcinogen.
Use: Safety matches, corrosion inhibitor in metal-joining compounds, pigment in paints, ceramics, fuses, pyrotechnics, metal primers, ignition control devices.
See chrome pigment.

barium citrate. $Ba_3(C_6H_5O_7)_2 \cdot H_2O$.
Properties: Grayish-white crystalline powder. Soluble in water, hydrochloric, and nitric acids.
Hazard: See barium.
Use: Manufacture of barium compounds, stabilizer for latex paints.

barium cyanide.
CAS: 542-62-1. $Ba(CN)_2$.
Properties: White, crystalline powder. Soluble in water and alcohol.
Derivation: By the action of hydrocyanic acid on barium hydroxide with subsequent crystallization.
Hazard: Questionable carcinogen.
Use: Metallurgy, electroplating.

barium cyanoplatinite.   (platinum barium cyanide; barium platinum cyanide). $BaPt(CN)_4 \cdot 4H_2O$.
Properties: Yellow or green crystals. Mp 100C (loses $2H_2O$), d 2.08. Soluble in water; insoluble in alcohol.
Grade: CP.
Hazard: See barium and cyanides.
Use: X-ray screens.

barium cyclohexanesulfamate.
CAS: 64011-64-9.
Hazard: A poison.
Use: Agricultural chemical.

barium dichromate.   (barium bichromate). $BaCr_2O_7 \cdot 2H_2O$.
Properties: Brownish-red needles or crystalline masses. Soluble in acids; decomposed by water.
Hazard: See barium.

barium dioxide.   See barium peroxide.

barium diphenylamine sulfonate.
$(C_6H_5NHC_6H_4SO_3)_2Ba$.
Properties: White crystals. Soluble in water.
Hazard: See barium.
Use: Indicator in oxidation-reduction titrations.

barium di-*o*-phosphate.   See barium phosphate, secondary.

barium   dithionate.   (barium   hyposulfate). $BaS_2O_6 \cdot 2H_2O$.
Properties: Colorless crystals. D 4.536. Soluble in hot water; slightly soluble in alcohol.
Derivation: Action of manganese dithionate on barium hydroxide.
Hazard: See barium.

barium ethylsulfate. $Ba(C_2H_5SO_4)_2 \cdot 2H_2O$.
Properties: Colorless crystals. Soluble in water and alcohol. Combustible.
Derivation: Interaction of barium hydroxide and ethylsulfuric acid.
Hazard: See barium.
Use: Organic preparations.

barium ferrite.
Grade: Powder.
Use: Permanent-magnet material.

barium fluoride.
CAS: 7787-32-8. BaF_2.
Properties: White powder. D 4.828, mp 1354C. Sparingly soluble in water.
Derivation: Interaction of barium sulfide and hydrofluoric acid, followed by crystallization.
Grade: Technical, CP, single pure crystals, and 99.98%.
Hazard: See barium.
Use: Ceramic flux, carbon brushes for electrical equipment, glass making, manufacture of other fluorides, crystals for spectroscopy, electronics, dry-film lubricants.

barium fluosilicate.   (barium silicofluoride). $BaSiF_6H$.
Properties: White, crystalline powder. Insoluble in water.
Grade: Technical.
Hazard: See barium.
Use: Ceramics, insecticidal compositions.

barium fructose diphosphate.   See fructose-1,6-diphosphate, calcium, barium salts.

barium   hexafluorogermanate. $BaGeF_6$. White crystalline solid, mp approximately 665C, dissociates to barium fluoride and germanium fluoride, d 4.56.

barium hydrate.   See barium hydroxide.

barium hydrosulfide. $Ba(SH)_2$.
Properties: Yellow crystals. Hygroscopic. Soluble in water.
Hazard: See barium.

barium hydroxide (anhydrous). $Ba(OH)_2$.
Available commercially.
See barium hydroxide hydrates.

barium hydroxide, monohydrate.   (barium monohydrate). $Ba(OH)_2 \cdot H_2O$.

Properties: White powder. Soluble in dilute acids; slightly soluble in water.
Hazard: See barium.
Use: Manufacturing of oil and grease additives, barium soaps, and chemicals. Refining of beet sugar, alkalizing agent in water softening, sulfate removal agent in treatment of water and brine, boiler scale removal, dehairing agent, catalyst in manufacture of phenol-formaldehyde resins, insecticide and fungicide, sulfate-controlling agent in ceramics, purifying agent for caustic soda, steel carbonizing agent, glass, refining edible oils, elastomer vulcanization.

barium hydroxide, octahydrate.   (barium hydrate; barium octahydrate; caustic baryta). $Ba(OH)_2 \cdot 8H_2O$.
Properties: White powder or crystals. D 2.18, mp 78C (losing its water of crystallization), mp (anhydrous $Ba(OH)_2$) 408C. Absorbs carbon dioxide from air. Keep well stoppered!! Soluble in water, alcohol, and ether.
Derivation: (1) By dissolving barium oxide in water with subsequent crystallization, (2) By precipitation from an aqueous solution of the sulfide by caustic soda, (3) By heating barium sulfide in earthenware retorts into which a current of moist carbonic acid is passed, after which superheated steam is passed over the resulting heated carbonate.
Impurities: Iron and calcium in commercial grades.
Grade: Technical (crystals or anhydrous powder), CP, ACS reagent.
Hazard: See barium.
Use: Organic preparations, barium salts, analytical chemistry.
See barium hydroxide monohydrate.

barium hydroxide pentahydrate.   (barium pentahydrate). $Ba(OH)_2 \cdot 5H_2O$.
Properties: Translucent, free-flowing, white flakes. Approximately 65 lb/cu ft.
Hazard: See barium.
Use: Same as the octahydrate.

barium hypophosphite. $BaH_4(PO_2)_2$.
Properties: White, crystalline powder; odorless. Soluble in water; insoluble in alcohol.
Hazard: See barium.
Use: Nickel plating.

barium hyposulfate.   See barium dithionate.

barium hyposulfite.   See barium thiosulfate.

barium iodate. $Ba(IO_3)_2$.
Properties: White, crystalline powder. D 5.23, mp decomposes at 476C. Slightly soluble in water, hydrochloric, and nitric acids; insoluble in alcohol.
Hazard: See barium.

barium iodide. $BaI_2 \cdot 2H_2O$.
Properties: Colorless crystals, decomposes and reddens on exposure to air. D 5.150, mp loses $2H_2O$

and melts at 740C. Soluble in water; slightly soluble in alcohol.

Derivation: Action of hydriodic acid on barium hydroxide or of barium carbonate on ferrous iodide solution.

Hazard: See barium.

Use: Preparation of other iodides.

barium manganate. (manganese green; Cassel green). $BaMnO_4$.

Properties: Emerald-green powder. D 4.85. Insoluble in water; decomposed by acids.

Hazard: See barium.

Use: Paint pigment.

barium mercury bromide. See mercuric barium bromide.

barium mercury iodide. See mercuric barium iodide.

barium metaphosphate. $Ba(PO_3)_2$.

Properties: White powder. Slowly soluble in acids; insoluble in water.

Hazard: See barium.

Use: Glasses, porcelains, and enamels.

barium metasilicate. See barium silicate.

barium molybdate. $BaMoO_4$.

Properties: White powder. Absolute d 4.7, approximate mp 1600C. Slightly soluble in acids and water.

Grade: Crystal, 99.84% pure.

Hazard: See barium.

Use: Electronic and optical equipment, pigment in paints and other protective coatings.

barium monohydrate. See barium hydroxide monohydrate.

barium monosulfide. See barium sulfide.

barium monoxide. See barium oxide.

barium nitrate.

CAS: 10022-31-8. $Ba(NO_3)_2$.

Properties: Lustrous, white crystals. D 3.244, mp 575C. Soluble in water; insoluble in alcohol.

Derivation: By the action of nitric acid on barium carbonate or sulfide.

Grade: Technical, crystals, fused mass or powder, CP.

Hazard: Strong oxidizing agent. See barium.

Use: Pyrotechnics (gives green light), incendiaries, chemicals (barium peroxide), ceramic glazes, rodenticide, electronics.

barium nitrite. $Ba(NO_2)_2 \cdot H_2O$.

Properties: White to yellowish, crystalline powder. D 3.173, decomposes at 217C. Soluble in alcohol, water.

Hazard: See barium.

Use: Diazotization, corrosive inhibitor, explosives.

barium octahydrate. See barium hydroxide octahydrate.

barium oxalate. $BaC_2O_4 \cdot H_2O$.

Properties: White, crystalline powder. D 2.66. Slightly soluble in water; soluble in dilute nitric or hydrochloric acid.

Hazard: See barium.

Use: Analytical reagent, pyrotechnics.

barium oxide. (barium monoxide; barium protoxide; calcined baryta).

CAS: 1304-28-5. BaO.

Properties: White to yellowish-white powder. D 5.72, mp 1923C. Absorbs carbon dioxide readily from air. Soluble in acids and water. Reacts violently with water to form the hydroxide.

Derivation: Decomposition of carbonate at high temperature in presence of carbon, oxidation of barium nitrate.

Grade: Technical (regular grind) 208 lb/cu ft, technical fine grind (175 lb/cu ft), porous, carbide-free, and 97%.

Hazard: Toxic by ingestion. See barium.

Use: Dehydrating agent for solvents, detergent for lubricating oils.

barium pentahydrate. See barium hydroxide pentahydrate.

barium perchlorate. $Ba(ClO_4)_2 \cdot 4H_2O$.

Properties: Colorless crystals. D 2.74, mp 505C. Soluble in methanol and water.

Hazard: Oxidizer, fire and explosion risk in contact with organic materials. Toxic by ingestion. See barium.

Use: Manufacture of explosives, experimentally in rocket fuels.

barium permanganate. $Ba(MnO_4)_2$.

Properties: Brownish-violet crystals. Soluble in water.

Hazard: Oxidizing material. Fire and explosion risk in contact with organic materials. Toxic by ingestion.

See barium.

Use: Strong disinfectant, manufacture of permanganates, depolarizing dry cells.

barium peroxide. (barium binoxide; barium dioxide; barium superoxide).

CAS: 1304-29-6. BaO_2 and $BaO_2 \cdot 8H_2O$.

Properties: Grayish-white powder. D 4.96, mp 450C, decomposes 800C. Slightly soluble in water.

Derivation: By heating barium oxide in oxygen or air at approximately 1000F.

Grade: Technical, reagent.

Hazard: Oxidizing material. Fire and explosion risk in contact with organic materials. Keep cool and dry. Toxic by ingestion, skin irritant.

Use: Bleaching, decolorizing glass, thermal welding of aluminum, manufacture of hydrogen peroxide, oxidizing agent, dyeing textiles.

barium phosphate, secondary.   (barium di-*o*-phosphate). $BaHPO_4$.
Properties: White powder. D 4.16. Soluble in dilute nitric acid or dilute hydrochloric acid; slightly soluble in water.
Hazard: See barium.
Use: Flame retardant, phosphors.

barium phosphosilicate.
Use: Anticorrosive pigment for solvent-based epoxies and as auxiliary pigment for 1-package zinc-rich coatings. Also used in water-based coatings.

barium platinum cyanide.   See barium cyanoplatinite.

barium potassium chromate.   (Pigment E). $BaK(CrO_4)_2$.
Properties: Pale-yellow pigment. D 3.65. Compared with other chromate pigments, it has a low chloride and sulfate content and forms stronger, more elastic paint films.
Derivation: By a kiln reaction at 500C between potassium dichromate and barium carbonate.
Hazard: See barium.
Use: Component of anticorrosive paints for use on iron, steel, and light metal alloys.

barium protoxide.   See barium oxide.

barium pyrophosphate. $Ba_2P_2O_7$.
Properties: White powder. Soluble in acids and ammonium salts; very slightly soluble in water.
Hazard: See barium.

barium reineckate.
CAS: 22708-05-0. $C_4H_6CrN_6S_4 \cdot 1/_2Ba$
Hazard: Moderately toxic by ingestion.
Use: Agricultural chemical.

barium selenide.   BaSe.
Properties: Crystalline powder. D 5.0. Decomposes in water.
Hazard: See barium.
Use: Semiconductors, photocells.

barium silicate.   (barium metasilicate). $BaSiO_3$.
Properties: Colorless powder. D 4.4, bp 1604C. Insoluble in water; soluble in acids.
Use: In ceramics.
Hazard: See barium.

barium silicide. $BaSi_2$.
Properties: Light-gray solid. Evolves hydrogen on exposure to moisture.
Hazard: See barium.
Use: Metallurgy to deoxidize steel, etc.

barium silicofluoride.   See barium fluosilicate.

barium-sodium niobate.   A synthetic electro-optical crystal used to produce coherent green light in lasers, also to make such devices as electro-optical modulators and optical parametric oscillators. The crystal undergoes no optical damage from laser irradiation at high power levels.

barium stannate. $BaSnO_3 \cdot 3H_2O$.
Properties: White, crystalline powder. Sparingly soluble in water; readily soluble in hydrochloric acid.
Hazard: See barium.
Use: Production of special ceramic insulations requiring dielectric properties.

barium stearate. $Ba(C_{18}H_{35}O_2)_2$.
Properties: White crystalline solid. Mp 160C, d 1.145. Insoluble in water or alcohol. Combustible.
Use: Waterproofing agent; lubricant in metalworking, plastics, and rubber; wax compounding; preparation of greases; heat and light stabilizer in plastics.

barium sulfate.   (barytes [natural]; blanc fixe [artificial, precipitated]; basofor).
CAS: 7727-43-7. $BaSO_4$.
Properties: White or yellowish powder; odorless; tasteless. D 4.25–4.5, particle size 2–25 microns, mp 1580C. Soluble in concentrated sulfuric acid. Noncombustible.
Derivation: (1) By treating a solution of a barium salt with sodium sulfate (salt cake), (2) by-product in manufacture of hydrogen peroxide, (3) occurs in nature as the mineral barite (Arkansas, Missouri, Georgia, Nevada, Canada, Mexico).
Grade: Technical, dry, pulp, bleached, ground, floated, natural, CP, USP, X-ray.
Hazard: Pneumoconiosis.
Use: Weighting mud in oil-drilling, paper coatings, paints, filler, and delustrant for textiles, rubber, plastics, and lithograph inks, base for lake colors, X-ray photography, opaque medium for gastrointestinal radiography, in battery plate expanders.

barium sulfide.   (barium monosulfide; black ash). BaS.
Properties: Yellowish-green or gray powder or lumps. D 4.25. Soluble in water, decomposes to the hydrosulfide.
Derivation: Barium sulfate (crude barite) and coal are roasted in a furnace. The melt is lixiviated with hot water, filtered, and evaporated.
Impurities: Iron, arsenic.
Hazard: See barium.
Use: Dehairing hides, flame retardant, luminous paints, barium salts, generating pure hydrogen sulfide.

barium sulfite. $BaSO_3$.
Properties: White powder, decomposed by heat. Soluble in dilute hydrochloric acid; insoluble in water.
Grade: Technical, CP.

Hazard: See barium.
Use: Analysis, paper manufacturing.

barium sulfocyanide.   See barium thiocyanate.

barium superoxide.   See barium peroxide.

barium tartrate. $BaC_4H_4O_6$.
Properties: White crystals. D 2.98. Soluble in water; insoluble in alcohol.
Hazard: See barium.
Use: Pyrotechnics.

barium thiocyanate.   (barium sulfocyanide). $Ba(SCN)_2 \cdot 2H_2O$.
Properties: White crystals. Soluble in water and in alcohol. Deliquescent.
Derivation: By heating barium hydroxide with ammonium thiocyanate and subsequent crystallization.
Hazard: See barium.
Use: Making aluminum or potassium thiocyanates, dyeing, photography.

barium thiosulfate.   (barium hyposulfite). $BaS_2O_3 \cdot H_2O$.
Properties: White, crystalline powder. D 3.5, decomposed by heat. Slightly soluble in water; insoluble in alcohol.
Hazard: See barium.
Use: Explosives, luminous paints, matches, varnishes, photography.

barium titanate. $BaTiO_3$.
Properties: Light gray-buff powder. Mp 3010F, d 5.95. Insoluble in water and alkalies; slightly soluble in dilute acids; soluble in concentrated sulfuric and hydrofluoric acids.
Use: Ferroelectric ceramics (single crystals either pure or doped with iron) are used in storage devices, dielectric amplifiers, and digital calculators.

barium tungstate.   (barium wolframate; barium white; tungstate white; wolfram white). $BaWO_4$.
Properties: White powder. D 5.04. Insoluble in water.
Use: Pigment and in X-ray photography for manufacturing of intensifying and phosphorescent screens.

Barium XA.   A product used by manufacturers of high quality tool steels. Eliminates chain-type occlusions and degasifies the steel.

barium zirconate. $BaZrO_3$.
Properties: Light gray-buff powder. D 5.52, bulk d 140 lb/cu ft, mp 4550F. Insoluble in water and alkalies; slightly soluble in acid.
Use: Manufacture of a white, easily colored silicone rubber compound having good heat stability at temperatures up to 500F; electronics.

barium zirconium silicate.   A complex of BaO, ZrO_2, SiO_2.

Properties: White powder. Bulk d 118 lb/cu ft, mp 2800F. Insoluble in water and alkalies; slightly soluble in acids; soluble in hydrofluoric acid.
Use: Production of electrical resistor ceramics, glaze opacifiers, and stabilizer for colored ground coat enamels.

bark.   The cellulosic outer layer or cortex of trees and other woody plants. The bark of certain species such as oak, hemlock, etc., is a source of tannic acid; medicinal products, e.g., quercitrin and quillaja, are also derived from barks, especially cinchona, from which quinine is obtained. Phenolic-rich bark extracts mixed with epichlorohydrin are reported useful as adhesive compounds. An unusual form of bark is cork from the oak species *Quercus suber*. In the pulp industry, bark is removed from logs with high-pressure jets of water.
See hydraulic barking; cork; quinine.

barking, hydraulic.   See hydraulic barking.

bark tannage.   To tan leather by using vegetable tannins found in bark, wood, or plant matter instead of tanning via minerals.

barn.   A unit of measurement equal to 10^{-24} cm^2, for the cross-section (target area) of the nucleus of an atom.

Barnett acetylation method.   Acetylation of hydroxy compounds such as cellulose with acid anhydrides in the presence of chlorine and sulfur dioxide. With cellulose, the process yields the diacetate below 65C and the triacetate above this temperature.

barometric pressure.   The pressure of the air at a particular point on or above the surface of the earth. At sea level, this pressure is sufficient to support a column of mercury approximately 29.9 inches in height (760 mm), equivalent to 14.7 lb/inch^2 absolute (psia) or 1 atm.

"Barosperse" *[Layfayette].*   TM for a special barium sulfate formulation used in radiographic examinations of the gastrointestinal tract.

barostat.   An instrument for regulation or maintenance of pressure at a constant value.

barrel finishing.   Cleaning, smoothing, and polishing of metal or plastic items by mechanical friction obtained by placing them in drums or barrels that rotate on their horizontal axis. An abrasive medium and water are usually added. The barrels are six- or eight-sided, and often contain vertical dividers to make two or more compartments that can be individually loaded and unloaded. Such treatment is widely used for large-scale cleaning and burnishing of metal parts, which it finishes to exact dimensional tolerances much more economically than is possible by manual methods.

barrier layer.   The electrical double layer formed at interface between a metal and a semiconductor or between two metals.

barrier, moisture.   Any substance that is impervious to water or water vapor. Most effective are high-polymer materials such as vulcanized rubber, phenolformaldehyde resins, polyvinyl chloride, and polyethylene, which are widely used as packaging films. The chief factors involved are polarity, crystallinity, and degree of cross-linking. Water-soluble surfactants and protective colloids increase the susceptibility of a film to water penetration. Any pigments and fillers must be completely wetted by the polymer. Properly formulated paints are effective moisture barriers.

barrier substance.   A substance applied to skin for protection against exposure to irritants.

Bartell cell.   Displacement cell for determining adhesion tension between liquids and solids.

barthrin.   Generic name for a synthetic analog of allethrin described as the 6-chloropiperonyl ester of chrysanthemummonocarboxylic acid.
Use: As insecticide with applications similar to allethrin and other analogs as furethrin, ethythrin, and cyclethrin. Relatively nontoxic to humans.

Barton, Derek H. R.   (1918–1998). An English organic chemist who won the Nobel Prize for chemistry in 1969 with Hassel. The field of conformational analysis in organic chemistry was initiated through his research in the terpene and steroid fields. He did extensive research in the area of carbanion autoxidations. He was instrumental in research concerning the relationship of molecular rotation to structure in complex organic molecules. His education took place in London, France, and Ireland.

Barton reaction.   Conversion of a nitrite ester to a γ-oximino alcohol by photolysis involving the homolytic cleavage of an N–O bond followed by hydrogen abstraction.

Bart reaction.   (Scheller modification; Starkey modification). Formation of aromatic arsonic acids by treating aromatic diazonium compounds with alkali arsenites in the presence of cupric salts or powdered silver or copper; in the Scheller modification, primary aromatic amines are diazotized in the presence of arsenious chloride and a trace of cuprous chloride.

baryon.   A type of strongly interacting elementary particle that is actually a massive composite hadron that consists of three quark triplets.

baryta.   (oxobarium).
CAS: 1304-28-5.   BaO. A barium-containing ore.

baryta, calcined.   See barium oxide.

baryta, caustic.   See barium hydroxide.

baryta water.   A solution of barium hydroxide.

baryta yellow.   See barium chromate.

barytes.   See barium sulfate.

Basacryl.   A series of cationic dyestuffs for the dyeing and printing of polyacrylonitrile fiber.

basal group.   The earliest diverging group within a clade; for instance, to hypothesize that sponges are basal animals is to suggest that the lineage(s) leading to sponges diverged from the lineage that gave rise to all other animals.

basal metabolic rate.   The rate of oxygen consumption by an animal's body at complete rest under fasting conditions.

basal metabolism.   See metabolism.

"Basazol" [BASF].   TM for dyes used in printing and dyeing paper composed of cellulosic fibers.

base.   Any of a large class of compounds with one or more of the following
Properties: bitter taste, slippery feeling in solution, ability to turn litmus blue and to cause other indicators to take on characteristic colors, ability to react with (neutralize) acids to form salts. Included are both hydroxides and oxides of metals.
Water-soluble hydroxides such as sodium, potassium, and ammonium hydroxide undergo ionization to produce hydroxyl ion (OH^-) in considerable concentration, and it is this ion that causes the previously mentioned properties common to bases. Such a base is strong or weak according to the fraction of the molecules that breaks down (ionizes) into positive ion and hydroxyl ion in the solution. Base strength in solution is expressed by pH. Common strong bases (alkalies) are sodium and potassium hydroxides, ammonium hydroxide, etc. These are caustic and corrosive to skin, eyes, and mucous membranes. The pH range of basic solutions is from 7.1 to 14.
Modern chemical terminology defines bases in a broader manner. A Lowry–Brønsted base is any molecular or ionic substance that can combine with a proton (hydrogen ion) to form a new compound. A Lewis base is any substance that provides a pair of electrons for a covalent bond with a Lewis acid. Examples of such bases are hydroxyl ion and most anions, metal oxides, and compounds of oxygen, nitrogen, and sulfur with nonbonded electron pairs (such as water, ammonia, hydrogen sulfide).
See Lewis electron theory for hard and soft bases.

base pair.   (bp).
  Two nitrogenous bases (adenine and thymine or guanine and cytosine) held together by weak bonds. Two strands of DNA are held together in the shape of a double helix by the bonds between base pairs.

base saponification number.   The number of milligrams of KOH equivalent to the amount of acid required to neutralize the alkaline constituents present after saponifying 1 g of sample.

base sequence.   The order of nucleotide bases in a DNA molecule; determines structure of proteins encoded by that DNA.

base sequence analysis.   A method, sometimes automated, for determining the base sequence.

BASF process.   A process for producing acetylene by burning a mixture of low-molecular weight hydrocarbons (as natural gas) with oxygen to produce a temperature of 1485C. The combustion products and cracked gases are quickly chilled by scrubbing with water, and the acetylene is separated by distillation and solvent extraction from ethylene, carbon monoxide, hydrogen, and other reaction products. The Sachsse process is similar.

basic.   Descriptive of a compound that is more alkaline than other compounds of the same name, e.g., lead carbonate, basic; basic salt.

basic amino acid.   (dibasic amino acid).
  Properties: Bears a net positive charge at neutral pH; hydrophilic; weakly basic polar compound; contains a second basic moiety.

basic anhydride.   An oxide of a metal that forms a base when reunited with water.

basic bismuth salicylate.   See salicylic acid, bismuth basic salt.

basic brown 4.   See C.I. basic brown 4.

basic chemicals.   See heavy chemicals.

basic dichromate.   See bismuth chromate.

basic dye.   (cationic dye). A basic dye that ionizes in solution, usually uniting with negatively charged moieties of the material to be colored.

basic fuchsin.   (CI 42500).
  CAS: 569-61-9. A mixture of three parts pararosaniline acetate and one part pararosaniline hydrochloride.
  Grade: Certifiable.
  Use: For staining *Tubercle bacillus* and in distinguishing between the *coli* and *aerogenes* types of bacteria in the Endo medium. Also used in the periodic acid-Schiff (PAS) method, in the Feulgen stain, in the Gomoris aldehyde-function method for staining elastic tissue.

basic lead carbonate.   (lead carbonate hydroxide; lead subcarbonate; white lead; flake lead; ceruse; cerussa; lead(2+)dicarbonate dihydroxide). CAS: 1319-46-6. $(PbCO_3)_2 \cdot Pb(OH)_2$.
  Properties: Acid-soluble, heavy, white powder or crystalline substance, insoluble in water and alcohol.
  Derivation: Occurs naturally as the mineral cerussite.
  Use: Pigment in paints.

basic lime phosphate.   Superphosphate neutralized with 6% excess calcium carbonate.

basic lining.   A furnace lining containing basic compounds that decompose under furnace conditions to give basic oxides. The usual basic linings contain calcium and magnesium oxides or carbonates.

basic nitrogen.   The nitrogen of a protein that are precipitated by phosphotungstic acid.

basic oxide.   An oxide that is a base or that forms a hydroxide when combined with water and/or that will neutralize acidic substances. Basic oxides are all metallic oxides, but there is a great variation in the degree of basicity. Some basic oxides, such as those of sodium, calcium, and magnesium, combine with water vigorously or with relative ease and also neutralize all acidic substances rapidly and completely. The oxides of the heavy metals are only weakly basic, do not dissolve or react with water to any extent, and neutralize only the more strongly acidic substances. There is a gradual transition from basic to acidic oxides, and certain oxides, such as aluminum oxide, show both acidic and basic properties.
  See base.

basic protein.   (alkaline protein). A protein in solution in which the pH is 7.0 to 7.5.

basic research.   See fundamental research.

basic salt.   A compound belonging in the category of both salt and base because it contains OH (hydroxide) or O (oxide) as well as the usual positive and negative radicals of normal salts. Among the best examples are bismuth subnitrate, often written $BiONO_3$, and basic copper carbonate, $Cu_2(OH)_2CO_3$. Most basic salts are insoluble in water, and many are of variable composition.

basic slag.   A slag produced in the manufacturing of steel. It contains a variable amount of tricalcium phosphate, calcium silicate, lime, and oxides of iron, magnesium, and manganese. Used as a fertilizer for its phosphorus and lime.
  See slag.

basic yellow 11.   See C.I. basic yellow 11.

basic yellow 40.
Properties: Fluorescent yellow dye
Use: In forensics with light to visualize cyanoacrylate ester fumed friction ridge detail in human digits.

basil oil, comoros type.
Properties: From steam distillation of *Ocimum basilicum* L. Light yellow liquid; spicy odor. Sol in fixed oils, mineral oil; slightly soluble in propylene glycol; insoluble in glycerin.
Use: Food additive.

basin, tub, and tile cleaner.   Any cleaner used to clean ceramic surfaces in the bathroom.
Hazard: Toxic.

basis metal.   In electroplating, the metal that is being coated constitutes the cathode. It may be any of a large number of metals.

Basogal phosphorus.   A leveling agent for vat dyeing.

basophil.   (basophil cell; basophile). A cell, the cytoplasm, or inclusions of which stain with basic dyes.

bastnasite.   An ore from which all nine of the lanthanide minerals (rare earths) are obtained. The only large deposit in the U.S. is in southwest California.
See monazite.

batch distillation.   Distillation in which the entire sample of the material to be distilled (the charge) is placed in the still before the process is begun, and product is withdrawn only from the condenser of the apparatus.

bating.   In leather processing, the treatment of delimed skins with pancreatin or other tryptic enzyme to give a softer and smoother-grained product. The extent of bating varies from none for sole leather to 10 hours or more for soft kid skins. The chemical mechanism is not clearly defined.

batrachotoxin.   (batrachotoxinin A 20-(2,4-dimethyl-1h-pyrrole-3-carboxylate);   3α,9α-epoxy-14β,18β-(epoxyethano-n-methylamino)-5 β-pregna-7,16-diene-3 β,11 α,20 α-triol; 20 α-ester with 2,4-dimethylpyrrole-3-carboxylic acid; btx). $C_{31}H_{42}N_2O_6$.
Properties: Noncrystal.
Derivation: The active principle steroidal alkaloid isolated from poison-dart (poison-arrow) frogs of the genus *Phyllobates*.
Hazard: Extremely toxic; deadly poison; neurotoxic; cardiotoxic activity; bara-chotoxin that blocks neuromuscular transmission; respiratory paralysis; death.

batrachotoxinin A. $C_{24}H_{35}NO_5$. An isomeric component of batrachotoxin; the strongest neurotoxin among venoms. It is a steroidal alkaloid. The A form is only 1/500 as strong as the complete venom, but is still as toxic as strychnine. It is found in the so-called poison dart frog of Colombia. Its structure has been elucidated; when synthesized it may prove useful in medicine.
See snake venom.

battery.   An electrochemical device that generates electric current by converting chemical energy to electrical energy. Its essential components are positive and negative electrodes made of more or less electrically conductive materials, a separating medium, and an electrolyte. There are four major types: (1) primary batteries (dry cells), which are not reversible and in which the anode (zinc) is the negative plate and the cathode (graphite) is the positive plate with ammonium chloride as electrolyte; (2) secondary or storage batteries, which are reversible and can be recharged and in which lead sponge is the negative plate (anode) and lead oxide the positive plate (cathode), with sulfuric acid as electrolyte; (3) nuclear and solar cells, or energy converters; and (4) fuel cells. So-called superbatteries of high charge density have been developed using solid electrolytes of lithium-titanium dioxide and trilithium nitride in which lithium atoms are intercalated in the crystal structure.
See dry cell; storage battery; voltaic cell; fuel cell; solar cell; intercalation compound.

battery acid.   (electrolyte acid). Sulfuric acid of strength suitable for use in storage batteries.
Properties: Water-white; odorless. Practically free from iron.
Derivation: By diluting high-grade commercial sulfuric acid with distilled water to standard strengths.
Hazard: Corrosive to skin and tissue.
See sulfuric acid.
Use: Storage batteries.

battery limit.   That portion of a chemical plant in which the actual processes are carried out, as distinguished from storage buildings, offices, and other subordinate structures called offsites.

batu.   A variety of East India copal resin.
See East India.

Baudisch reaction.   Synthesis of *o*-nitrosophenols from benzene or substituted benzenes, hydroxylamine, and hydrogen peroxide in the presence of copper salts.

Baumé.   (Bé). An arbitrary scale of specific gravities devised by the French chemist Antoine Baumé and used in the graduation of hydrometers. The relations of specific gravity (at 60/60F) are as follows: Bé = 145–145/*d* for liquids heavier than water, Bé = 140/*d* − 130 for liquids lighter than water.

bauxite. A natural aggregate of aluminum-bearing minerals, more or less impure, in which the aluminum occurs largely as hydrated oxides. It is usually formed by prolonged weathering of aluminous rocks. Contains 30–75% Al_2O_3, 9–31% H_2O, 3–25% Fe_2O_3, 2–9%, SiO_2, 1–3% TiO_2.

Properties: White cream, yellow, brown, gray, or red. D 2–2.55, Mohs hardness 1–3. Insoluble in water; decomposed by hydrochloric acid. Noncombustible.

Occurrence: Australia, Jamaica, France, Guiana, Guinea, U.S. (Arkansas), Brazil.

Use: Most important ore of aluminum, aluminum chemicals, abrasives, aluminous cement, refractories, decolorizing and deodorizing agent, catalysts, filler in rubber, plastics, paints, cosmetics, hydraulic fracturing.

See Bayer process; Hall process.

Note: Due to increasing cost of bauxite the use of other aluminum-containing minerals is under active investigation.

"Baybond" [Bayer]. TM for aqueous polyurethane dispersions.

Bayer process. Process for making alumina from bauxite. The main use of alumina is in the production of metallic aluminum. Bauxite is mixed with hot concentrated sodium hydroxide, which dissolves the alumina and silica. The silica is precipitated, and the dissolved alumina is separated from the solids, diluted, cooled, and then crystallized as aluminum hydroxide. The aluminum hydroxide is calcined to anhydrous alumina, which is then shipped to reduction plants.

See Hall process.

Bayer's acid. See crocein acid.

"Baygon" [S. C. Johnson]. TM for *o*-isopropoxyphenyl methylcarbamate

bay oil. See myrcia oil.

baypival. See 1-(*p*-chlorophenoxy)-3,3-dimethyl-1-(1-imidazolyl)-2-butanone.

"Baytex" [Bayer]. TM for *o,o*-dlN(methylthio)-*m*-tolyl)phosphorothioate.
See fenthion.

BBO. See 2,5-dibiphenylyloxazole.

BBP. Abbreviation for butyl benzyl phthalate.

b cell. A major type of cell in the immune system that can differentiate to form memory cells or antibody-producing cells.

BCG. See bromocresol green.

BCWL. Abbreviation for basic carbonate white lead.
See lead carbonate, basic.

BDA. An inhibited hydrochloric acid solution containing surfactants.

Use: In limestone and dolomite formations, in oil-well fracturing and acidizing.

4-BDAF. See 4,4'-((2,2,2-trifluoro-1-(trifluoromethyl)ethylidene)bis(4,1-phenyleneoxy))bisbenzenamine.

Be. Symbol for beryllium.

Bé. Abbreviation for Baumé.

bead. (1) In a rubber-fabric composite (tires, transmission belts), the point at which the cut edges of the fabric meet after being folded over. A length of pure gum rubber, called a bead strip, is used to seal the joint. The bead must be removed from tires before reclaiming-an operation called debeading. (2) See microsphere.

beamhouse. The part of a tannery where raw animal skins are washed, soaked, dehaired, and prepared for tanning.

Bearite. A series of lead-base bearing metals containing about 17% antimony and fractional percentages of copper and bismuth; babbitt.

"Bearium" [Metal Tek]. TM for high-lead bronze with 18–26% lead that is used for bearings and similar items.

beat. Maximum or minimum of intensity caused by interference of two wave series of slightly different frequencies. The number of beats per second is equal to the difference in frequency between the two tones.

beater. An open, oval tank into which digested paper pulp is fed together with water and other processing ingredients such as clay, rosin, pigments, etc. The resulting slurry (approximately 5% solids content) is strongly agitated by a rotating drum equipped with closely spaced, horizontal, finlike projections which effectively disintegrate and macerate the wood fibers while the slurry is continuously circulated. The time of beating varies with the type and quality of the paper being made. Conical refiners (called Jordans) are often used to complete the beating operation for high-grade papers. The beating operation is critical in converting wood pulp to paper.

beat frequency. The difference in frequencies of two interfering waves.

Bechamp reaction. Formation of *p*-amino or *p*-hydroxyphenylarsonic acids by heating aromatic amines or phenols with arsenic acid. The arsonylation requires an active hydrogen atom, and is practically limited to the benzene series. Apart from para substitution, a small amount of ortho arsonylation is observed, particularly when the para position is blocked.

Beckacite.   Fumaric, maleic, and modified phenolic resins.

Beckamine.   Urea-formaldehyde resins.

Beckmann rearrangement.   The conversion of a ketone oxime to a substituted amide by an intermolecular rearrangement brought about by a catalyst. For example, the oxime of cyclohexanone is converted into caprolactam with sulfuric acid as catalyst.

$$R'C-R \longrightarrow RC-OH \longrightarrow R-C=O$$
$$\overset{\|}{NOH} \qquad \overset{\|}{R'N} \qquad \overset{|}{R\ N-H}$$

Beckmann thermometer.   A specific form of mercury thermometer that because of its large bulb has greater sensitivity but smaller range. It is used to measure small changes very accurately.

"Beckosol" [Reichhold].   TM for alkyd resins used at coating vehicles.

Becquerel, Antoine Henri.   (1852–1908). A French physicist who shared the Nobel Prize in physics with the Curiers for the discovery of the radioactivity of uranium salts. He also discovered the defelction of electrons by a magnetic field, as well as the existence and properties of γ-radiation.

Becquerel effect.   The electromotive force resulting from unequal illumination of two similar electrodes immersed in an electrolyte.

beer.   See brewing.

Beer's law.   States that the degree of absorption of light is dependent on the thickness of the layer crossed and on the molecular concentration of colored substances in that layer.

beerstone.   A deposit occurring on containers during brewing operations and consisting of calcium oxalate and organic material.

beeswax.   Wax from the honeycomb of the bee. Beeswax consists largely of myricyl palmitate, cerotic acid and esters, and some high-carbon paraffins.
Properties: Brown or white (bleached) solid; faint odor. D 0.95, melting range 62–65C. Insoluble in water, slightly; soluble in alcohol; soluble in chloroform, ether, and oils. Combustible.
Grade: Technical, crude, refined, NF, FCC, USP (white).

Use: Furniture and floor waxes, shoe polishes, leather dressings, anatomical specimens, artificial fruit, textile sizes and finishes, church candles, cosmetic creams, lipsticks, adhesive compositions.

beeswing.   A light, gummy sediment deposited in bottled wines.

beet sugar.   See sucrose.

behavioral genetics.   The study of genes that may influence behavior.

behenic acid.   (docosanoic acid). $CH_3(CH_2)_{20}COOH$. A saturated fatty acid, a minor component of the oils of the type of peanut and rapeseed.
Properties: Solid. Mp 80.0C, bp 306C (60 mm Hg), 265C (15 mm Hg), d 0.8221 (100/4C), refr index 1.4270 (100C). Combustible.
Derivation: Occurs in bean oil, hydrogenated mustard oil, and rapeseed oil.
Grade: Technical, 99%.
Use: Cosmetics, waxes, plasticizers, chemicals, stabilizers.
See "Hystrene" [PMC].

behenone. $C_{22}H_{44}O$. An aliphatic ketone. Insoluble in water, inert, compatible with high-melting waxes, fatty acids. Incompatible with resins, polymers, and organic solvents at room temperature but compatible with them at high temperature.
Use: As an antiblocking agent.

behenyl alcohol.   (1-docosanol). $CH_3(CH_2)_{20}CH_2OH$. A long-chain, saturated fatty alcohol.
Properties: Colorless waxy solid. Mp 71C, bp 180C (0.22 mm Hg). Insoluble in water; soluble in ethanol and chloroform. Combustible.
Derivation: Reduction of behenic acid with lithium aluminum hydride as catalyst.
Grade: Technical; 99%.
Use: Synthetic fibers, lubricants, evaporation retardant on water surfaces.

Beilstein, F. P.   (1838–1906). A German chemist noted for his compilation "Handbuch der Organischen Chemie," the first edition of which appeared in 1880. A multivolume compendium of the properties and reactions of organic compounds, it has been revised several times and remains a unique and fundamental contribution to chemical literature.

Beilstein's test.   A test to detect halogens in organic compounds. Copper gauze is heated in a flame until the flame shows no green color; if the addition of an organic compound produces a green flame, a halogen is present.

BEK.   See butyl ethyl ketone.

bel. (1) An interval corresponding to a factor of 10 on a logarithmic scale showing the relationship of two quantities of power. (2) A unit of difference in sound sensation degrees.

belladonna. (deadly nightshade; banewort). A herbaceous perennial bush (*Atropa belladonna*) of which the leaves and roots are used for their content of hyoscyamine and atropine.
Occurrence: Southern and central Europe, Asia Minor, Algeria; cultivated in North America, England, France.
Grade: Belladonna leaf, USP; belladonna root.
Hazard: Very toxic when high in atropine.
Use: Medicine (gastrointestinal relaxant).

belladonna alkaloid. Any organic ester compounded from tropic acid and an organic base such as tropine or scopine. These alkaloids are produced by solanaceous plants such as *Atropa belladonna*, *Datura stramonium*, *Hyoscyamus niger*, and *Scopolia carniolica*.

belladonna extract. (deadly nightshade).
Properties: A pill or powder which contains 1.15–1.35 grams of belladonna alkaloids per 100 grams.
Hazard: Deadly poison.
Use: Anticholinergic.

Bellier index. (Bellier number). The temperature at which solid fatty acids begin to crystallize from 70% alcohol solution, acidulated with acetic acid, when the solution is slowly and progressively cooled with constant agitation.

Bellier number. See Bellier index.

bell jar. A bell-shaped glass jar that is used for the protection of instruments, etc.

bell metal. A copper-based hard alloy that contains 15–40% tin and optional additions of zinc, iron, and lead. Frequently used in bells and musical chimes.

belonesite. A molybdenum ore comprised of monoclinic crystals.

bemberg. A cuprammonium rayon fiber. Flammable, not self-extinguishing.

Bemul. A practically odorless emulsifying agent; a pure white, edible glycerol monostearate in bead form; mp 58–59C; completely dispersible in hot water; completely soluble in alcohols and hot hydrocarbons.
Use: Pharmaceuticals; cosmetics; foodstuffs; protective coating for edible hygroscopic powders, tablets, and crystals; pour-point depressant for lubricating oils; textile sizes.

"Benadryl." [Johnson & Johnson]. Proprietary name of diphenhydramine hydrochloride.

Benary reaction. Action of Grignard reagents on enamino ketones or aldehydes yields β-substituted α,β-unsaturated ketones or aldehydes.

benazolin ethyl ester. See ethyl 4-chloro-2-oxo-3(2h)-benzothiazoleacetate.

bench gas. See coal gas.

bench test. A simulated test in which the conditions are approximated, but the equipment is not necessarily identical, with that in which the item will be used.

bendiocarb. $C_{11}H_{13}NO_4$.
Properties: White powder. Mp 130C. Slightly soluble in water.
Hazard: Poison by ingestion, skin absorption.
Use: Contact insecticide.

"Benecel" [Hercules].
CAS: 9004-67-5.
TM for methylcellulose.
Use: In food, pharmaceutical, and personalcare items.

Benedict solution. A water solution of sodium carbonate, copper sulfate, and sodium citrate. The blue color changes to a red, orange, or yellow precipitate or suspension in the presence of a reducing sugar such as glucose, and is therefore used in testing for such materials, especially for urinalysis in the treatment of diabetes.
See Fehling's solution.

beneficiation. A process used in extractive metallurgy whereby an ore, either metallic or nonmetallic, is concentrated in preparation for further processing (smelting). Calcination is often an important step in beneficiation; others steps include physical separation of high-grade ore from impurities (gangue) by screening, washing, milling, or magnetic means. A process for removing sulfur from coal by chemical comminution has been developed.

benefin. (*N*-butyl-*N*-ethyl-α,α,α-tri-fluoro-2,6-dinitro-*p*-toluidine). $C_6H_2(NO_2)_2CF_3NC_6H_{14}$.
Properties: Yellow-orange solid. Mp 65–66.5C, bp 121–122C (0.5 mm Hg). Slightly soluble in water; readily soluble in acetone and xylene.
Hazard: Highly toxic.
Use: Herbicide.

"Ben-ex" [Kelco]. TM for specialty drilling-fluid additives.

Bengal fire. Material used in pyrotechnics composed of realgar, potassium nitrate, and sulfur.

Benjamin gum. See benzoin resin.

benomyl. (Benlate; methyl-1-(butylcarbamoyl)-2-benzimidazole-carbamate).
CAS: 17804-35-2. Generic name for a postharvest fungicide for peaches, apples, etc. Also used as oxidizer in sewage treatment.
Hazard: High toxicity by ingestion. Upper respiratory tract irritant, male reproductive, testicular, and embryo/fetal damage. Possible carcinogen.

bensulide. (*N*-(2-mercaptoethyl)benzenesulfonamide).
CAS: 741-58-2.
Use: Herbicide.

benthiocarb. (*s*-[(4-chlorophenyl)methyl] *N,N*-diethylcarbamothioate).
CAS: 28249-77-6. $C_{12}H_{16}ClNOS$.
Properties: Amber-colored, slightly water-soluble, liquid.
Hazard: Toxic.
Use: Thiocarbamate herbicide used to control sedges and grasses in rice paddies.

benthos. The bottom-dwelling life of an ocean or freshwater environment.

bentonite. A colloidal clay (aluminum silicate) composed chiefly of montmorillonite. There are two varieties: (1) sodium bentonite (Wyoming or western), which has high swelling capacity in water; and (2) calcium bentonite (southern), with negligible swelling capacity.
Properties: (Wyoming) Light to cream-colored impalpable powder; forms colloidal suspension in water, with strongly thixotropic properties.
Occurrence: Wyoming, Mississippi, Texas, Canada, Italy, the former U.S.S.R.
Use: Oil-well drilling fluids; cement slurries for oil-well casings; bonding agent in foundry sands and pelletizing of iron ores; sealant for canal walls; thickener in lubricating greases and fireproofing compositions; cosmetics; decolorizing agent; filler in ceramics, refractories, paper coatings; asphalt modifier; polishes and abrasives; food additive; catalyst support.
See clay, "Flo-Fre" *[Oil-Dri]*.

benzaconine. (picraconitine; benzoylaconine; napelline; isaconitine; pikraconitin). $C_{32}H_{45}NO_{10}$.
An alkaloid produced by the partial hydrolysis of aconitine.

benzalacetone. See benzylidene acetone.

benzalazine. (benzylidene azine). $C_6H_5CH:NN:CHC_6H_5$.

Properties: Yellow crystals. Mp 91–93C. Insoluble in cold water; soluble in benzene, and hot alcohol.
Use: Stabilizer; polymerization catalyst; UV absorbent; reagent and intermediate.

benzal chloride. See benzyl dichloride.

benzaldehyde. (benzoic aldehyde; synthetic oil of bitter almond).
CAS: 100-52-7. C_6H_5CHO.

Properties: Colorless or yellowish, strongly refractive, volatile oil; odor resembling oil of bitter almond; burning aromatic taste. Oxidizes readily. D 1.0415 (25/4C), refr index 1.5440–1.5464 at 20C, fp −56C; bp 178C, flash p 145F (62.7C) (CC). Miscible with alcohol, ether, fixed and volatile oils; slightly soluble in water. Oxidizes in air to benzoic acid. Combustible. Autoign temp 377F (191.6C).
Derivation: (1) Air oxidation of toluene with uranium or molybdenum oxides as catalysts; (2) reaction of benzyl dichloride with lime; (3) extraction from oil of bitter almond.
Impurities: Usually chlorides.
Method of purification: Rectification.
Grade: Technical, NF. *Note:* The specifications, especially regarding impurities, vary considerably for the grades used for dye manufacture from those used in perfumery.
Hazard: Highly toxic.
Use: Chemical intermediate for dyes, flavoring materials, perfumes, and aromatic alcohols; solvent for oils, resins, some cellulose ethers, cellulose acetate and nitrate; flavoring compounds; synthetic perfumes; manufacturing of cinnamic acid, benzoic acid; pharmaceuticals; photographic chemicals.

benzaldehyde cyanohydrin. See mandelonitrile.

benzaldehyde green. See Malachite green.

benzalkonium chloride. (alkyldimethylethylbenzyl ammonium chloride; alkyl(ethylphenyl) methyl)dimethyl quarternary ammonium chlorides; bentrol; BTC 471; cequartyl; drapolex; enuclen; germinal; germitol; octyl-octadecyl dimethyl ethylbenzyl ammonium chlorides; paralkan; roccal; rodalon; zephiran chloride; zephirol; benzyldodecyl-dimethylazaniumchloride).
CAS: 8001-54-5. Chemical formula varies.
Properties: Quaternary ammonium salt, very bitter taste, white or yellowish-white powder or gelatinous pieces; aromatic odor; highly alkaline

in aqueous solution; mixture of alkyl dimethylbenzylammonium chlorides.

Grade: USP.

Hazard: Highly toxic; poison.

Use: Cationic detergent; surface antiseptic; fungicide; bacericide.

benz(e)acephenanthrylene. (3,4-benz(e)acephenanthrylene; 2,3-benzfluoranthene; benzo(b)fluoranthene; b(b)f).

CAS: 205-99-2. $C_{20}H_{12}$.

Properties: Needles from C_6H_6 or EtOH. Mw 252.32, mp 168C.

Hazard: Confirmed carcinogen.

benzamide. (benzoylamide). $C_6H_5CONH_2$.

$$C_6H_5C \overset{NH_2}{\underset{O}{<}}$$

Properties: Colorless crystals. Mp 130C, bp 288C, d 1.341. Soluble in hot water, hot benzene, alcohol, and ether. Combustible.

Derivation: From benzoyl chloride and ammonia or ammonium carbonate.

Grade: Technical.

Use: Organic synthesis.

benzaminoacetic acid. See hippuric acid.

benzanilide. (benzoylaniline; phenylbenzamide). $C_6H_5NH(COC_6H_5)$.

Properties: White to reddish crystals and powder. Related to acetanilide, containing benzoyl in place of acetyl radical. D 1.306, mp 160–162C. Soluble in alcohol; insoluble in water; slightly soluble in ether.

Derivation: From benzoic anhydride and aniline with sodium hydroxide.

Use: Intermediate in the synthesis of dyes, drugs, and perfumes.

benz(a)anthracene. (benzanthracene; 1,2-benzanthracene; benzo(b)phenanthrene; tetraphene).

CAS: 56-55-3. $C_{18}H_{12}$.

Properties: Colorless leaflets or plates from EtOH/AcOH. Mw 228.30, mp 160C, bp 400C.

Hazard: Confirmed carcinogen. Found in oils, waxes, smoke, food, drugs.

benzanthrone. $C_{17}H_{10}O$. A four-ring system.

Properties: Pale-yellow needles. Mp 170C. Soluble in alcohol and other organic solvents.

Derivation: (1) From anthranol and glycerol via condensation via sulfuric acid (anthranol is made from anthraquinone), (2) from anthracene in sulfuric acid solution by addition of glycerol and heating to 100–110C until the anthracene disappears. The reaction mass is then diluted with water, salted out, and purified.

Method of purification: Crystallization from toluene.

Use: Dyes.

benzathine penicillin G. (*N,N'*-dibenzylethylenediamine dipenicillin G). $2C_{16}H_{18}N_2O_4S \cdot C_{16}H_{20}N_2 \cdot 4H_2O$.

Properties: White crystalline powder; odorless. The pH of a saturated solution is 4.5–7.5. Slightly soluble in alcohol; almost insoluble in water.

Grade: USP.

Use: Medicine (antibiotic).

benzazimide. See 4-ketobenzotriazine.

Benzedrine. An amphetamine sulfate.

benzene.

CAS: 71-43-2. C_6H_6.

Structure: I. Complete ring showing all elements.

II. Standard ring showing double bonds only.

III. Simple ring without double bonds, with numerals indicating position of carbon atoms to which substituent atoms or groups may be attached (2 = ortho, 3 = meta, 4 = para).

IV. Generalized structure with enclosed circle suggesting the resonance of this compound. This structure is now in general use.

These structures are also referred to as the benzene nucleus.

Properties: Colorless to light-yellow; mobile; nonpolar liquid of highly refractive nature; aromatic odor. Bp 80.1C, fp 5.5C, d 0.8790 (20/4C), wt/gal 7.32 lb, refr index 1.50110 at 20C, flash p 12F (−11C) (CC), surface tension 29 dynes/cm; autoign temp 1044F (562C). Miscible with alcohol, ether, acetone, carbon tetrachloride, carbon disulfide, acetic acid; slightly soluble in water. Vapors burn with smoky flame.

Derivation: (1) Hydrodealkylation of toluene or pyrolysis of gasoline; (2) transalkylation of toluene by disproportionation reaction; (3) catalytic reforming of petroleum; (4) fractional distillation of coal tar.

Grade: Crude, straw color, motor, industrial pure (2C), nitration (1C), thiophene-free, 99 mol%, 99.94 mol%, nanograde.

Hazard: A confirmed carcinogen. Highly toxic. Flammable, dangerous fire risk. Explosive limits in air 1.5 to 8% by volume.
Use: Manufacturing of ethylbenzene (for styrene monomer), dodecylbenzene (for detergents), cyclohexane (for nylon), phenol, nitrobenzene (for aniline), maleic anhydride, chlorobenzene, diphenyl, benzene hexachloride, benzene-sulfonic acid, and as a solvent.
See aromatic.

benzene azimide. See 1,2,3-benzotriazole.

benzeneazoanilide. See diazoaminobenzene.

benzeneazobenzene. See azobenzene.

benzeneazo-p-benzeneazo-β-naphthol.
(Sudan III; tetraazobenzene-β-naphthol). $C_{22}H_{16}ON_4$. A red dye; CI 26100.444.
Properties: Brown powder. Mp 195C. Insoluble in water; soluble in alcohol, oils, chloroform, glacial arctic acid.
Use: Coloring oils red; biological stain.

benzeneazonaphthylethylenediamine. See azodine.

benzenecarboxylic acid. See benzoic acid.

benzenediazonium chloride. $C_6H_5N(N)Cl$.
Properties: Ionic salt. Very soluble in water; insoluble in most organic solvents.
Hazard: Highly toxic. Can explode on heating.
Use: Dye intermediate.

benzene dibromide. See dibromobenzene.

1,3-benzenedicarbonitrile. See m-phthalodinitrile.

1,2-benzenedicarboxylic acid, 4,4′-carbonylbis-, ar,ar′-dietyl estercompd. with 1,3-benzenediamine.
CAS: 65701-07-7. $C_{21}H_{18}O_9 \cdot C_6H_8N_2$.
Hazard: Moderately toxic by ingestion. A severe skin and eye irritant.

1,2-benzenedicarboxylic acid, di-c7-c9-branched alkyl ester.
CAS: 68515-41-3.
Hazard: A reproductive hazard.

1,2-benzenedicarboxylic acid dipentyl ester, branched and linear.
CAS: 84777-06-0.
Hazard: A reproductive hazard.

benzene-o-dicarboxylic acid. See phthalic acid.

benzene-p-dicarboxylic acid. See terephthalic acid.

benzene hexachloride. (BHC). A commercial mixture of isomers of 1,2,3,4,5,6-hexachlorocyclohexane.
Hazard: The γ-isomer is highly toxic. Use may be restricted.
Use: An insecticide.
See lindane.

benzene, 1-isocyanato-2-methyl-. See 1-isocyanato-2-methylbenzene.

benzenemonosulfonic acid. See benzenesulfonic acid.

benzene-1,2-oxide. An epoxide, it is an intermediate in the metabolic oxidation of benzene.
Hazard: Toxic.

benzenephosphinic acid. (phenylphosphinic acid). $C_6H_5H_2PO_2$.
Properties: Colorless crystals. Mp 82–84C, d 1.376 (29C). Decomposes at 200C. Stable in air. Soluble in water, alcohol, acetone; slightly soluble in ether; insoluble in benzene, hexane, CCl_4. Combustible.
Use: Antioxidant, intermediate for metallic-salt formation, accelerator for organic peroxide catalysts.

benzenephosphonic acid. (phenylphosphonic acid). $C_6H_5H_2PO_3$.
Properties: Colorless crystals. Mp 158C, d 1.475 (4C), decomposes at 275C. Soluble in water, alcohol, CCl_4. Combustible.
Hazard: Highly toxic.
Use: Intermediate in antifouling paint agents; catalyst in organic reactions.

benzenephosphorus dichloride. $C_6H_5PCl_2$.
Properties: Highly reactive, colorless liquid. Mp −51C, bp 224.6C, d 1.315 (25C), refr index 1.5958 (25C). Soluble in common inert organic solvents. Fumes in air; hydrolyzes in water.
Hazard: Highly corrosive to skin, tissue. Flammable.
Use: Organic synthesis, for derivation of plasticizers, polymers, antioxidants; oil additives.

benzenephosphorus oxydichloride. $C_6H_5POCl_2$.
Properties: Reactive colorless liquid. Mp 3.0C, bp 258C, d 1.197 (25C), refr index 1.5585 at 25C. Soluble in common inert organic solvents; hydrolyzes in water. Combustible.
Hazard: Strong irritant to skin, mucous membranes.
Use: Organic synthesis, for derivation of plasticizers, polymers, antioxidants, oil additives.

benzenesulfinic acid, zinc salt. See zinc benzenesulfinate.

benzenesulfonic acid. (benzenemonosulfonic acid; phenylsulfonic acid).
CAS: 42615-29-2. $C_6H_5SO_3H$.

Properties: Fine, deliquescent needles or large plates. Mp 65–66C when anhydrous; with 1.5 molecules water, mp 43–44C. Soluble in water, alcohol; slightly soluble in benzene; insoluble in ether and carbon disulfide.
Derivation: By reacting benzene with fuming sulfuric acid.
Hazard: Irritant to skin, eyes, mucous membranes.
Use: Manufacturing of phenol, resorcinol, and other organic syntheses, and as a catalyst.

benzenesulfonic acid, dodecyl-, compd. with isopropylamine.
CAS: 26264-05-1. $C_{18}H_{30}O_3S \cdot C_3H_9N$.
Hazard: A severe eye irritant.

benzene, toluene, ethyl benzene, and xylene.
(BETX).
Properties: Volatile organic compounds.
Derivation: Occur in petroleum derivatives such as gasoline.
Hazard: Effects the central nervous system.

benzene-1,3,5-tricarboxylic acid chloride.
See trimesoyl trichloride.

benzenoid.   Any organic compound containing or derived from the benzene ring structure, e.g., phenol, nitrobenzene, anthracene, styrene. This large array of unsaturated compounds, derived chiefly from petroleum and coal tar, provides a broad base for the synthesis of polymers, dyes, and intermediates.
See aromatic.

benzenyl trichloride.   See benzotrichloride.

benzethonium chloride. $C_{27}H_{42}ClNO_2$.   A synthetic quaternary ammonium compound.
Properties: Colorless plates; odorless; very bitter taste. Mp 164–166C. Soluble in water, alcohol, acetone. Aqueous solution yields flocculent white precipitate with soap solutions.
Grade: NF.
Hazard: An oral poison.
Use: Antiseptic; cationic detergent.

2,3-benzfluoranthene.   See benz(e)acephenanthrylene.

benzhydrol.   (benzohydrol; diphenylmethanol; diphenylcarbinol).
CAS: 91-01-0. $(C_6H_5)_2CH_2O$.

Properties: Needlelike, colorless crystals. Mp 69C, bp 298C, 176C (13 mm Hg). Slightly soluble in water; easily soluble in alcohol, ether, chloroform, and carbon disulfide. Combustible.
Derivation: Reduction of benzophenone with magnesium or zinc dust.
Use: Preparation of certain antihistamines; insecticide.

benzhydryl bromide.   See diphenylmethyl bromide.

benzhydryl chloride. $(C_6H_5)_2CHCl$.
Properties: Water-white to light straw-colored liquid. Refr index 1.596, mp 16C, bp 140C (3 mm Hg). Combustible.
Use: Organic synthesis.

benzidine.   (benzidine base; *p*-diaminodiphenyl).
CAS: 92-87-5. $NH_2(C_6H_4)_2NH_2$.

Properties: Grayish-yellow, white, or reddish-gray crystalline powder. Mp 127C, bp 400C. Soluble in hot water, alcohol, ether; slightly soluble in cold water. Combustible. Also available as the hydrochloride.
Derivation: (1) By reducing nitrobenzene with zinc dust in alkaline solution followed by distillation; (2) by electrolysis of nitrobenzene, followed by distillation; (3) by nitration of diphenyl followed by reduction of the product with zinc dust in alkaline solution, with subsequent distillation.
Grade: Technical (paste; powder 80–85%).
Hazard: Highly toxic by ingestion, inhalation, and skin absorption. Confirmed carcinogen.
Use: Organic synthesis; manufacture of dyes, especially of Congo red; detection of blood stains; stain in microscopy; reagent; stiffening agent in rubber compounding.

benzidinedicarboxylic acid.   See diaminodiphenic acid.

benzidine dye.   Any of a group of azo dyes derived from 3,3'-dichlorobenzidine; they include yellow and orange colors claimed to be lightfast and alkali-resistant. Congo red is derived from benzidine and naphthionic acid.
Hazard: These compounds are highly toxic and carcinogenic. Physical contact with them should be avoided.

benzidine rearrangment; semidine rearrangement. The acid-catalyzed rearrangement of hydrazobenzenes to 4,4'-diaminobiphenyls. If the hydrazobenzene contains a *para* substituent, the product is a *p*-aminodiphenylamine (semidine rearrangement).

benzidine sulfate.
CAS: 531-86-2. $C_{12}H_{12}N_2 \cdot H_2SO_4$.
Properties: White, crystalline powder. Soluble in ether; sparingly soluble in water, alcohol, dilute acids.
Derivation: Action of sulfuric acid or sodium sulfate on benzidine with subsequent recovery by precipitation.
Hazard: Poison by ingestion, skin absorption. A carcinogen.
Use: Organic synthesis.

benzil.   (dibenzoyl). $C_6H_5CO \cdot COC_6H_5$.

Properties: Yellow needles. Mp 95C, bp 346–348C, d 1.521. Soluble in alcohol, ether; insoluble in water.
Derivation: From benzoin by oxidation with HNO_3.
Use: Organic synthesis; insecticide.

benzilic acid.   (diphenylglycolic acid). $(C_6H_5)_2C(OH)COOH$.
Properties: White to tan powder with a characteristic odor. Mp 148–151C. Soluble in hot water and alcohol. Combustible.
Use: Chemical intermediate.

benzilic acid rearrangement.   Rearrangement of benzyl to benzilic acid on treatment with base.

α-benzil monoxime.
CAS: 14090-77-8. $C_{14}H_{11}NO_2$.
Hazard: Moderately toxic by ingestion.
Use: Agricultural chemical.

benzimidazole.   (1,3-benzodiazole; azindole; benzoglyoxaline).
CAS: 51-17-2. $C_7H_6N_2$.
Properties: Tabular crystals. Mp 172–174C, mw 118.13, bp >360C. Weak base sparingly soluble in cold water and ether. Freely soluble in alcohol; practically insoluble in benzene and petroleum ether.

benzine.   The name *benzine* is archaic and misleading and should not be used. (ASTM Petroleum Definitions D-288.) Do not confuse with benzene. See ligroin.

1,2-benzisothiazolin-3-one 1,1-dioxide calcium salt.   See calcium saccharin.

benzocaine.   See   ethyl-*p*-aminobenzoate hydrochloride.

benzoctamine.
CAS: 17243-39-9. $C_{18}H_{19}N$.
Hazard: Moderately toxic by ingestion.

benzodihydropyrone.   (dihydrocoumarin). $C_6H_8O_2$ (bicyclic).
Properties: White to light yellow, oily liquid with a sweet odor. Congeals at 23C. Insoluble in water; soluble in alcohol, chloroform, ether. Combustible.
Use: Perfumery.

1,3-benzodioxole-4-carboxaldehyde.
CAS: 7797-83-3. $C_8H_6O_3$.
Hazard: A poison by ingestion.

1,3-benzodioxol-4-ol, 2,2-dimethyl-, acetyl-methylcarbamate.
CAS: 22791-33-9. $C_{13}H_{15}NO_5$.
Hazard: A poison by ingestion.
Use: Agricultural chemical.

1,3-benzodioxol-4-yl methylcarbamate.
CAS: 22791-23-7. $C_9H_9NO_4$.
Hazard: A poison by ingestion.
Use: Agricultural chemical.

BenzoFlex.   A battery cleaner.

"Benzoflex" [Eastman].   TM for a series of plasticizers that are dibenzoate esters of dipropylene glycol or any of several polyethylene glycols.
Use: Primary plasticizer for vinyl resins; adhesive formulations; some grades in food-packaging adhesives.

8,9-benzofluoranthene.   See   benzo(k)fluoranthene.

benzo(k)fluoranthene.
CAS: 207-08-9. $C_{20}H_{12}$.
Properties: Yellow prisms from C_6H_6 or AcOH. Mp 217C, bp 480C.
Hazard: Possible carcinogen.

benzofuran.   See coumarone.

benzoglycolic acid.   See mandelic acid.

benzoguanamine.   (2,4-diamino-6-phenyl-*s*-triazine). $C_6H_5C_3N_3(NH_2)_2$.
Properties: Crystals. D 1.40, mp 227–228C. Soluble in methyl "Cellosolve," alcohol, dilute hydrochloric acid; partially soluble in dimethylformamide, acetone; practically insoluble in chloroform, ethyl acetate; insoluble in water, benzene, ether. Combustible.
Derivation: Benzonitrile and dicyandiamide in the presence of sodium and liquid ammonia.

Use: Thermosetting resins, resin modifiers; chemical intermediate for pesticides, pharmaceuticals, and dyestuffs.

benzohydrol. See benzhydrol.

benzoic acid. (carboxybenzene; benzenecarboxylic acid; phenylformic acid).
CAS: 65-85-0. C_6H_5COOH. It occurs naturally in benzoin resin.
Properties: White scales or needle crystals; odor of benzoin or benzaldehyde. D 1.2659, mp 121.25C, partially sublimes at 100C, p 249.2C, flash p 250F (121.1C) (CC). Freely volatile in steam. Soluble in alcohol, ether, chloroform, benzene, carbon disulfide, carbon tetrachloride, turpentine; slightly soluble in water. Combustible.
Derivation: (1) Decarboxylation of phthalic anhydride in the presence of catalysts; (2) chlorination of toluene to yield benzotrichloride, which is hydrolyzed to benzoic acid; (3) oxidation of toluene; (4) from benzoin resin.
Method of purification: Sublimation.
Grade: Technical, CP, USP, FCC.
Hazard: Moderately toxic by ingestion. Use restricted to 0.1% in foods.
Use: Sodium and butyl benzoates, plasticizers, benzoyl chloride, alkyd resins, food preservative, seasoning tobacco, flavors, perfumes, dentifrices, standard in analytical chemistry, antifungal agent.

benzoic aldehyde. See benzaldehyde.

benzoic anhydride. $(C_6H_5CO)_2O$.
Properties: Colorless prisms. D 1.198, mp 42C, bp 360C, refr index 1.576. Soluble in most organic solvents; insoluble in water.
Use: Dyes, intermediates, pharmaceuticals (benzoylating agent), organic synthesis.

benzoic trichloride. See benzotrichloride.

benzoin. (bitter almond-oil camphor; benzoylphenylcarbinol; 2-hydroxy-2-phenylacetophenone; phenylbenzoylcarbinol).
CAS: 119-53-9. $C_6H_5CH_2OCOC_6H_5$.
Properties: White or yellowish crystals; slight camphor odor. Mp 137C. Slightly soluble in water and ether; soluble in acetone and hot alcohol. Optically active. Combustible.
Derivation: Condensation of benzaldehyde in an alkaline cyanide solution.
Hazard: Highly toxic.
Use: Organic synthesis, intermediate, photopolymerization catalyst.
Note: Do not confuse with benzoin resin.

benzoin condensation. Cyanide ion-catalyzed condensation of aromatic aldehydes to give benzoins (acyloins).

α-benzoin oxime. (benzoin antioxime).
$C_6H_5CH_2OC:NOHC_6H_5$.
Properties: Solid. Mp 150–152C.
Use: Organic intermediates and photographic chemicals, analytical reagent for determination of metals.

benzoin resin. (gum benzoin; Benjamin gum).
Properties: Reddish-brown globules; balsamic, vanilla-like odor. Brittle at room temperature but softened by heat. Soluble in warm alcohol and carbon disulfide; insoluble in water.
Source: Obtained from the styrax tree in Southeast Asia and Sumatra. The Sumatran grade is higher melting and only 75% soluble in alcohol.
Grade: Technical, tincture USP.
Chief constituents: Benzoic acid, cinnamic acid, vanillin.
Use: Source of benzoic acid; perfumery; cosmetics; medicine (antiseptic and expectorant).
Note: Do not confuse with benzoin.

benzol. Obsolete name for benzene, no longer in approved use.

benzonitrile. (phenyl cyanide).
CAS: 100-47-0. C_6H_5CN.
Properties: Colorless oil; almond-like odor; sharp taste. D 1.0051; bp 190.7C; fp −13.1C, viscosity 1.054 centistokes (100F), refr index 1.5289. Soluble in boiling water, alcohol, ether; slightly soluble in cold water.
Derivation: From benzoic acid by heating with lead thiocyanate.
Hazard: High toxicity; absorbed by skin.
Use: Manufacture of benzoguanamine; intermediate for rubber chemicals; solvent for nitrile rubber, specialty lacquers, and many resins and polymers, and for many anhydrous metallic salts.

benzo(b)phenanthrene. See benz(a)anthracene.

benzophenol. See phenol.

benzophenone. (diphenylketone).
CAS: 119-61-9. $(C_6H_5)_2CO$.
Properties: White prisms with sweet, roselike odor. Mp 47.5C; bp 305C. Partially soluble in alcohol, ether; soluble in chloroform; insoluble in water. Combustible.
Method of purification: Crystallization from alcohol.
Grade: Free from chlorine (FFC), also FCC.
Use: Organic synthesis; odor fixative; derivatives are used as ultraviolet absorbers; flavoring; soap fragrance; pharmaceuticals; polymerization inhibitor for styrene.

benzophenone oxide. See xanthone.

3,3′,4,4′-benzophenonetetracarboxylic dianhydride. (BTDA). $C_{17}H_6O_7$.
Properties: Free-flowing powder. Mp 228C.
Use: Epoxy curing agent, heat-resistant polymers, specialty alkyd resins, polyesters, and plasticizers.

12h-benzo(b)phenoselenazine.
CAS: 64050-25-5. $C_{16}H_{11}NSe$.
Hazard: Moderately toxic.
Use: Agricultural chemical.

benzopyrene. (3,4-benzypyrene).
CAS: [a] form 50-32-8. $C_{20}H_{12}$.
A polynuclear (five-ring) aromatic hydrocarbon. Found in coal tar and cigarette smoke, and in the atmosphere as a product of incomplete combustion.
Derivation: Occurs as benzo[a]pyrene and benzo[e]pyrene.
Properties: (Benzo[a]pyrene) Yellowish crystals. Mp 179C, bp 310–312C (10 mm Hg). Insoluble in water; slightly soluble in alcohol; soluble in benzene, toluene, xylene.
Hazard: Highly toxic, confirmed carcinogen by inhalation.

benzo[α]pyrene hydroxylase. Any of certain isozymes of cytochrome P-450 that exhibit aryl hydrocarbon hydroxylase activity.

benzo[α]pyrene-7,8-diol-9,10-epoxide.
Any of four mutagenic steriosomers resulting from the metabolism of benzo[α]pyrene.
Hazard: Toxic.

benzopyrone. See coumarin.

benzoquinone. See quinone.

benzosulfimide. See saccharin.

s-(2-(2-benzothiazolylamino)-2-oxoethyl) ethanethioate. See 2-(acetylthioglycolic amide)benzothiazole.

benzothiazole. C_6H_4SCHN (bicyclic).
Properties: Yellow liquid; unpleasant odor. D 1.246, refr index 1.637, bp 227C. Slightly soluble in water; soluble in alcohol. Combustible.
Hazard: Highly toxic by ingestion.
Use: Derivatives used as rubber accelerators.

benzothiazolyl disulfide. See 2,2′-dithiobis-(benzothiazole).

benzothiazyl-2-cyclohexylsulfenamide.
See N-cyclohexyl-2-benzothiazole-sulfenamide.

2-benzothiazyl-N,N-diethylthiocarbamyl sulfide. (diethyldithiocarbamic acid-2-benzothiazoyl ester). $(C_6H_4SCN)SSCN(C_2H_5)_2$.

Properties: Free-flowing, light-yellow to tan powder. D 1.27, mp 69C (min).
Use: Rubber accelerator.

benzothiazyl disulfide. See 2,2′-dithiobis-(benzothiazole).

1-benzothiophene-4-ol. See 4-hydroxybenzothiophene.

1,2,3-benzotriazole. (aziminobenzene; benzene azimide). $C_6H_4NHN_2$.
Properties: White to light tan crystalline powder; odorless. Boiling range 201–204C (15 mm Hg). Very stable toward acids and alkalies and toward oxidation and reduction. Its basic characteristics are very weak, but it forms stable metallic salts. Can exist in two tautomeric forms. Soluble in alcohol and benzene; slightly soluble in water. Derivatives are ultraviolet absorbers.
Hazard: Highly toxic by ingestion. May explode under vacuum distillation.
Use: Photographic restrainer, chemical intermediate.

benzotrichloride. (toluene trichloride; benzenyl trichloride; benzoic trichloride; phenylchloroform).
CAS: 98-07-7. $C_6H_5CCl_3$.
Properties: Colorless to yellowish liquid; fumes in air; hydrolyzes in water; penetrating odor. D 1.38; bp 220C, fp −5C, refr index 1.5584. Soluble in alcohol and ether; insoluble in water.
Derivation: Chlorination of boiling toluene.
Method of purification: Rectification.
Hazard: Highly toxic by inhalation, fumes highly irritant. Eye, skin, and upper respiratory tract irritant. Probable carcinogen.
Use: Synthetic dyes, organic synthesis.

benzotrifluoride. (toluene trifluoride; trifluoromethylbenzene). $C_6H_5F_3$.
Properties: Water-white liquid; aromatic odor. Bp 102.1C; fp −29.1C, d 1.1812 (25/4C), refr index 1.4146, flash p 54F (12.2C) (CC). Miscible with alcohol, acetone, benzene, carbon tetrachloride, ether, n-heptane; insoluble in water.
Hazard: Highly toxic by inhalation. Flammable, dangerous fire risk.
Use: Intermediate for dyes and pharmaceuticals, solvent and dielectric fluid, vulcanizing agent, insecticides.

trans-β-benzoylacrylic acid. $C_6H_5COCH:CHCOOH$.
Properties: Straw-yellow needles or plates. Mp 99C. Soluble in most solvents; only slightly soluble in cold water and ligroin. Combustible.
Use: Reagent for characterizing phenols; intermediate in the manufacturing of bactericides, insecticides, surface-active agents, and the upgrading of drying oils.

N-benzoyl-L(+)-alanine.   See alanine.

benzoylamide.   See benzamide.

benzoylamine.
CAS: 55-21-0.   C₇H₇NO.
A alkaloid present in various species of Aconitum.
Hazard: Depresses the central nervous system; toxic.

benzoylaminoacetic acid.   See hippuric acid.

benzoylaniline.   See benzanilide.

p-benzoylbenzoic acid.
CAS: 611-95-0.   C₁₄H₁₀O₃.
Hazard: A poison.

benzoyl chloride.
CAS: 98-88-4.   C₆H₅COCl.

Properties: Transparent, colorless liquid; pungent odor; vapor causes tears. D 1.2188, fp −0.5C, bp 197.2C, refr index 1.5536 (20C), flash p 162F (72.2C). Soluble in ether and carbon disulfide; decomposes in water. Combustible.
Derivation: (1) Interaction of benzoic acid and sulfuryl chloride, (2) benzotrichloride and water in the presence of zinc chloride, (3) phosphorus tri- or pentachloride and benzoic acid.
Grade: Technical, CP.
Hazard: Highly toxic. Strong irritant to skin, eyes, and mucous membranes, and via ingestion, inhalation. Upper respiratory tract irritant. Probable carcinogen.
Use: Introduction of benzoyl group, dye intermediates, benzoyl peroxide manufacturing, analytical reagent.

benzoyl-2,5-diethoxyaniline.
C₆H₅CONHC₆H₃(OC₂H₅)₂.
Properties: Gray pellets. Mp 83–84C.
Hazard: Possibly toxic.
Use: Intermediate for pharmaceuticals, dyestuffs, and other organic chemicals.

benzoylferrocene.   (phenyl ferrocenyl ketone).
C₅H₅FeC₅H₄COC₆H₅.
Properties: Dark red, crystalline solid. Mp 107–108C.
Hazard: Possibly toxic.
Use: Intermediate.

benzoyl fluoride.   C₆H₅COF.
Hazard: High toxicity.
Use: Manufacturing of acyl and other fluorides.

benzoylglucuronic acid.   The conjugation product of benzoic and glucuronic acids. It is the detoxified form of benzoic acid.

benzoylglycin.   See hippuric acid.

benzoylglycocoll.   See hippuric acid.

6-benzoylheteratisine.   See heteratisine 6-benzoate.

2-benzoyl-2-hydroxypropane.   See α-hydroxy-α-methylpropiophenone.

1-benzoylnapelline.
CAS: 198126-85-1.   C₂₉H₃₇NO₄.
Hazard: A poison.
Source: Natural product.

3-(3-(6-benzoyloxy-3-cyano-2-pyridyloxycarbonyl)benzoyl)-1-ethoxymethyl-5-fluorouracil.
CAS: 110690-43-2.   C₂₈H₁₉FN₄O₈.
Hazard: A poison by ingestion. A reproductive hazard.

benzoyloxytriphenylstannane.   See triphenylstannyl benzoate.

benzoyl peroxide.   (dibenzoyl peroxide).
CAS: 94-36-0.   (C₆H₅CO)₂O₂.

Properties: White, granular, crystalline solid; tasteless; faint odor of benzaldehyde. Active oxygen, approximately 6.5%. Mp 103–105C, decomposes explosively above 105C, autoign temp 176F (80C), d 1.3340 (25C). Soluble in nearly all organic solvents; slightly soluble in alcohols, vegetable oils; slightly soluble in water.
Grade: Technical, wet or dry; FCC.
Hazard: Highly toxic via inhalation. May explode spontaneously when dry (<1% of water). Never mix unless at least 33% water is present. Skin and upper respiratory tract irritant. Questionable carcinogen.
Use: Bleaching agent for flour, fats, oils, and waxes; polymerization catalyst; drying agent for unsaturated oils; pharmaceutical and cosmetic purposes; rubber vulcanization without sulfur; burnout agent for acetate yarns; production of cheese; embossing vinyl flooring (proprietary).
See peroxides.

benzoylphenyl carbinol.   See benzoin.

2-benzoylpyridine.   C₆H₅COC₅H₄N.
Properties: Colorless liquid. Fp 42.7C. Insoluble in water.

Grade: 98% (minimum).
Use: Organic synthesis.

4-benzoylpyridine. $C_6H_5COC_5H_4N$.
Properties: Colorless liquid. Fp 71.4C. Insoluble in water.
Grade: 98% (minimum).
Use: Organic synthesis.

benzoylsulfonic imide. See saccharin.

benzoyloxytributylstannane.
CAS: 4342-36-3. $C_{19}H_{32}O_2Sn$.
Hazard: A poison by ingestion.
Use: Agricultural chemical.

benzozone. See acetyl benzoyl peroxide.

1,2-benzphenanthrene. See chrysene.

benzphetamine. (*n*,α-dimethyl-*n*-
(phenylmethyl)-benzeneethanamine; *n*-benzyl-
n,α-dimethylphenethylamine; *d*-*n*-methyl-*n*-
benzyl- β-phenylisopropylamine; didrex;
N-methyl-1-phenyl-*N*-(phenylmethyl)propan-
2-amine).
CAS: 156-08-1. $C_{17}H_{21}N$. A sympathomimetic
agent with properties similar to dextroamphetamine.
Hazard: Elevate blood pressure, disturb heart
rhythm, cause restlessness, insomnia, hyperactivity,
headache, euphoria, depression, psychosis, tremor,
dryness of mouth, unpleasant tastes, diarrhea, stomach upset, altered sex drive, impotence, aggressive-
ness, hallucinations, panic.
Use: Appetite suppressant; used to assay certain
isozymes of cytochrome P-450.

benzpyrene. See benzopyrene.

benzyl abietate. $C_{19}H_{29}COOCH_2C_6H_5$.
Properties: Nonvolatile, viscous liquid that resembles Canada balsam. Soluble in most anhydrous
solvents.
See balsam.

benzyl acetate. (acetic acid benzyl ester; acetic
acid phenylmethyl ester; α-acetoxytoluene; benzyl
ethanoate; phenylmethyl acetate).
CAS: 140-11-4. $C_6H_5CH_2OOCCH_3$.
Properties: Water-white liquid; floral odor. Mw
150.19, d 1.059–1.062 (15C); mp −51.5C, bp 212C;
refr index 1.5015–1.5035, flash p 216F (102.2C)
(CC), autoign temp 862F (460C), vap press 1 mm
(45C), vap d 5.1, refr index 1.501. Soluble in alcohol, most fixed oils, propylene glycol; insoluble in
glycerin and water at 214C.
Derivation: (a) By treating benzyl chloride with
sodium acetate in various solvents; (b) by esterification of benzyl alcohol with acetic anhydride or
acetic acid.
Method of purification: Distillation.

Grade: Free-from-chlorine grade, which should
have an ester content of 97% but for which lower-
grade material is sometimes substituted; technical
grade, which is not free from chlorine and for which
ester content varies considerably; FCC.
Hazard: A poison by inhalation. Moderately toxic
by ingestion. Combustible. Upper respiratory tract
irritant. Questionable carcinogen.
Use: Artificial jasmine and other perfumes, soap perfume, flavoring, solvent and high boiler for cellulose
acetate and nitrate, natural and synthetic resins, oils,
lacquers, polishes, printing inks, varnish removers.

benzyl alcohol. (α-hydroxytoluene; phenylmethanol; phenylcarbinol).
CAS: 100-51-6. $C_6H_5CH_2OH$.

Properties: Water-white liquid; slight odor; sharp,
burning taste. Bp 206C, flash p 220F (105C) (OC),
d 1.040–1.050 (25/25C), refr index 1.5385–1.5405
(20C). Somewhat soluble in water; miscible with
alcohol, ether, chloroform. Autoign temperature
817F (436C). Combustible.
Derivation: (a) By hydrolysis of benzyl chloride;
(b) from benzaldehyde by catalytic reduction or
Cannizzaro reaction.
Method of purification: Distillation and chemical
treatment.
Grade: Free from chlorine (FFC), technical, NF, textile, photographic, reagent, FCC.
Hazard: Highly toxic.
Use: Perfumes and flavors; photographic developer for color movie films; dyeing nylon filament, textiles, and sheet plastics; solvent for
dyestuffs, cellulose esters, casein, waxes, etc.; heat-
sealing polyethylene films; intermediate for benzyl esters and ethers; bacteriostat; cosmetics, ointments, emulsions; ball point pen inks; stencil inks.

benzylamine. (aminotoluene).
$C_6H_5CH_2NH_2$.

Properties: Light amber liquid; strongly alkaline
reaction. D 0.9813; bp 184.5C; refr index 1.540 at
20C. Soluble in alcohol, ether, water. Combustible.
Derivation: From benzyl chloride and ammonia.
Hazard: Highly toxic, strong irritant to skin and
mucous membranes.
Use: Chemical intermediate for dyes, pharmaceuticals, and polymers.

N-benzyl-*p*-aminophenol. (BAP).
$C_6H_5CH_2NHC_6H_4OH$.

Properties: Light brown powder. Mp 84–90C, 96–99% pure. Solubility of 50% in anhydrous methanol, 50% in 95% ethanol, 0.06% in water; 0.1–0.5% in gasoline, varying with chemical nature of gasoline.
Use: In cracked gasoline, in concentration of 0.001–0.004% by weight to prevent gum formation.

2-benzylamino-1-propanol.
$C_6H_5NHCH(CH_2OH)CH_3$.
Properties: White to yellow solid. Both *l* and *dl*-forms are available. Mp (*dl*- form) 70–73C. Specific rotation (*l*- form) +38° to +44° (1.0% solution in alcohol) at 25C. Combustible.

6-benzylaminopurine.   (6-benzyladeine; BAP; *N*-(phenylmethyl)-7H-purin-6-amine).
CAS: 1214-39-7. $C_{12}H_{11}N_5$.
A synthetic cytokinin; one of the most active of those substituted adenines that show cytokinin activity.

benzylaniline. $C_6H_5NHCH_2C_6H_5$.
Properties: Colorless prisms. Mp 33C, bp 310C. Soluble in alcohol and ether; insoluble in water.
Use: Organic synthesis.

benzylbenzene.   See diphenylmethane.

benzyl benzoate.
CAS: 120-51-4. $C_6H_5CH_2OOCC_6H_5$.
Properties: Water-white liquid; sharp, burning taste; faint aromatic odor. D 1.116–1.120 (25/25C), bp 325C, mp 18.8C, refr index 1.568–1.569 (20C), flash p 298F (147.7C). Supercools easily. Insoluble in water, glycerol; soluble in alcohol, chloroform, ether. Combustible.
Derivation: (a) By a Cannizzaro reaction from benzaldehyde, (b) by esterifying benzyl alcohol with benzoic acid, (c) by treating sodium benzoate with benzyl chloride.
Method of purification: Distillation and crystallization.
Grade: Technical, USP, FCC.
Hazard: Irritant to eyes, skin.
Use: Fixative and solvent for musk in perfumes and flavors; medicine (external); plasticizer for nitrocellulose and cellulose acetate; miticide.

benzyl bromide.   (α-bromotoluene).
$C_6H_5CH_2Br$.
Properties: Clear, refractive liquid; pleasant odor. Not easily hydrolyzed. D 1.438 at 16C, bp 198–199C, fp −3.9C, vap d 5.8. Soluble in alcohol, benzene, ether; insoluble water.
Derivation: (1) Bromination of toluene, (2) interaction of benzyl alcohol and hydrobromic acid.
Hazard: Highly toxic. Corrosive to skin and tissue. A lachrymator.
Use: Making foaming and frothing agents, organic synthesis.

benzyl butyrate. $C_3H_7COOCH_2C_6H_5$.
Properties: Liquid; fruity odor. Bp 240C, d 1.016 (17.5C). Soluble in alcohol. Combustible.
Grade: Technical, FCC.
Use: Plasticizer, odorants, flavoring.

benzyl carbinol.   See phenethyl alcohol.

benzyl "Cellosolve" [Union].   See ethylene glycol monobenzyl ether.

benzyl chloride.   (α-chlorotoluene).
CAS: 100-44-7. $C_6H_5CH_2Cl$.
Properties: Colorless liquid; pungent odor. D 1.090–1.111 (25/25C), fp −43C, bp 179C, refr index 1.5365 (25C), flash p 153F (67.2C) (OC), autoign temp 1085F (525C). Soluble in alcohol, ether; insoluble in water. Combustible.
Derivation: By passing chlorine over boiling toluene until it has increased 38% in weight. The product is washed with water and separated by fractional distillation.
Grade: Technical; CP; 95%; redistilled.
Available forms: Anhydrous; stabilized (with aqueous sodium carbonate solution).
Hazard: Highly toxic, intense eye and skin irritant. A lachrymator. Upper respiratory tract irritant. Probable carcinogen.
Use: Dyes; intermediates; benzyl compounds; synthetic tannins; perfumery; pharmaceuticals; manufacture of photographic developer; gasoline gum inhibitors; penicillin precursors; quaternary ammonium compounds.

benzyl chlorocarbonate.   (benzyl chloroformate). C_6H_5OCOCl.
Properties: Oily liquid with lachrymatory properties; odor of phosgene. Decomposes above 100C, reacts with water to form hydrochloric acid.
Hazard: Highly toxic, emits very toxic phosgene fumes at 100C. Irritant to eyes.
Use: Peptide synthesis.

benzyl chloroformate.   See benzyl chlorocarbonate.

o-**benzyl-*p*-chlorophenol.**   (chlorophene, USAN; Santophen; septiphene; 4-chloro-α-phenyl-*o*-cresol).
$C_6H_5CH_2C_6H_3OHCl$.
Properties: White to light tan or pink flakes; slight phenolic odor. Crystallizing point 45C min, D 1.202–1.206 (55/55C). Insoluble in water; highly soluble in alcohol, other organic solvents; dispersible in aqueous media with the aid of soaps or synthetic dispersing agents; noncorrosive to most metals. Combustible.
Hazard: Highly toxic, an irritant.
Use: Active principle or enhancing agent for disinfectants.

benzyl cinnamate. (cinnamein). $C_9H_7O_2\ C_7H_7$.
Properties: White crystals; aromatic odor. Mp 39C; congeal point (min) 34C, bp 244C (25 mm Hg). Insoluble in water; soluble in alcohol.
Hazard: Moderately toxic.
Grade: Technical, FCC.
Use: Perfumery and flavors.

benzyl cyanide. (phenylacetonitrile; α-tolunitrile). $C_6H_5CH_2CN$.
Properties: Colorless, oily liquid; aromatic odor. D 1.0157, fp −24C, bp 230C, refr index 1.5211 (25C). Soluble in alcohol and ether; insoluble in water.
Derivation: Interaction of benzyl chloride and potassium cyanide.
Grade: Technical.
Hazard: Highly toxic, absorbed by skin.
Use: Organic synthesis, especially penicillin precursors.

benzyl dichloride. (benzylidene chloride; benzal chloride; chlorobenzal; α,α-dichlorotoluene).
CAS: 98-87-3. $C_6H_5CHCl_2$.
Properties: Colorless, oily liquid; faint aromatic odor. D 1.295 (16C), fp −16.1C, bp 207C, refr index 1.5502 (20C). Soluble in alcohol, ether, dilute alkali; insoluble in water. Combustible.
Derivation: Chlorination of toluene, until two formula weights of chlorine are absorbed, in absence of catalysts but presence of light.
Hazard: Strong irritant and lachrymator. Probable carcinogen.
Use: Dyes; manufacture of benzaldehyde and cinnamic acid.

benzyldimethyloctadecyl ammonium 3-nitrobenzenesulfonate.
CAS: 124088-59-1. $C_{27}H_{50}N \cdot C_6H_4NO_5S$.
Hazard: Moderately toxic by ingestion and skin contact. A severe eye and mild skin irritant.

4-(benzyl-(2-((2,5-diphenyloxazole-4-carbonyl)amino)ethyl)carbamoyl)-2-decanoylaminobutyric acid.
CAS: 219905-91-6. $C_{40}H_{48}N_4O_6$.
Hazard: A poison.

benzyl disulfide. See dibenzyl disulfide.

benzyl ethanoate). See benzyl acetate.

***N*-benzylethanolamine.**
$C_6H_5CH_2NH(C_2H_4OH)$.
Properties: Colorless to light yellow liquid. D 1.044 (27C); refr index 1.5400–1.5430; distillation range 240–255C. Combustible.
Use: Corrosion inhibitor; intermediate.

benzyl ether. (dibenzyl ether).
CAS: 103-50-4. $(C_6H_5CH_2)_2O$.
Properties: Colorless, unstable liquid. Bp 295C, d 1.001, flash p 275F (135C), refr index 1.54. Insoluble in water; soluble in alcohol, ether, acetone.

Derivation: Benzaldehyde is reduced with cobalt complex, $[Co(CO)_4]_2$.
Hazard: Moderately toxic by ingestion. A skin irritant.
Use: Solvent, plasticizer for nitrocellulose.

6-benzyl-1-(ethoxymethyl)-5-isopropyluracil. See emivirine.

benzyl ethyl ether. $C_6H_5CH_2OC_2H_5$.
Properties: Colorless, oily liquid; aromatic odor. Bp 185C, d 0.949, refr index 1.4955 (20C). Volatile in steam; insoluble in water; miscible with alcohol, ether. Combustible.
Derivation: By boiling benzyl chloride with either sodium or potassium ethylate.
Hazard: Narcotic in high concentration. May be skin irritant.
Use: Organic synthesis, flavoring.

benzyl ethylsalicylate.
$C_6H_5CH_2OOCC_6H_4OC_2H_5$.
Use: As fixative and solvent in perfumes.

benzyl fluoride. $C_6H_5CH_2F$.
Properties: Colorless liquid. Forms acicular crystals on prolonged cooling. D 1.022 at 25C, bp 139.8C (753 mm), fp −35C.
Derivation: By decomposing benzyltrimethylammonium fluoride.
Hazard: Very irritant.
Use: Organic synthesis.

benzyl formate. $C_6H_5CH_2OOCH$.
Properties: Colorless liquid; fruity-spicy odor. Resembles benzyl acetate in many respects but differs in its greater volatility. D 1.083–1.087, refr index 1.511–1.513, bp 203C. Miscible with alcohols, ketones, oils, aromatic, aliphatic and halogenated hydrocarbons; insoluble in water.
Hazard: May be narcotic in high concentration.
Use: Perfumes, flavoring, solvent for cellulose esters.

benzyl fumarate.
$C_6H_5CH_2OOCCH=CHCOOCH_2C_6H_5$.
Properties: White powder. Mp 59C, bp 210C (5 mm Hg). Insoluble in water; soluble in alcohol and ether.
Derivation: Reaction of fumaric acid and benzyl alcohol.
Use: Spray deodorant.

5′-benzyl-3′-furylmethyl α-ethyl-phenylacetate.
CAS: 51628-36-5. $C_{22}H_{22}O_3$.
Hazard: Moderately toxic by ingestion.
Use: Agricultural chemical.

5′-benzyl-3′-furylmethyl α-isopropyl-4-methoxyphenyl acetate.
CAS: 51628-56-9. $C_{24}H_{26}O_4$.
Hazard: Moderately toxic by ingestion.
Use: Agricultural chemical.

benzylhexadecyldimethylammonium chloride.   See cetalkonium chloride (USAN).

benzylhydroquinone.   See *p*-benzyloxyphenol.

benzylidene acetone.   (benzalacetone; methyl styryl ketone; 4-phenyl-3-buten-2-one). $C_6H_5CH=CHCOCH_3$.
Properties: Colorless crystals; odor of coumarin. Mp 72C, congeal point 39C (min), bp 260–262C, refr index 1.5836 (46C), d 1.0377 (15/15C). Soluble in alcohol, ether, benzene, chloroform; insoluble in water. Combustible.
Derivation: Condensation of benzaldehyde and acetone.
Use: Organic synthesis, perfumery, fixative, flavoring.

benzylidene azine.   See benzalazine.

benzylidene chloride.   See benzyl dichloride.

2-benzylidene-1-heptanol.   See α-amylcinnamic alcohol.

***N*-benzylidiethanolamine.**
$(C_6H_5CH_2N(C_2H_4OH)_2$.
Properties: Colorless to light yellow liquid. D 1.073, refr index 1.5345–1.5375, distilling range 240–255C. Miscible with water. Combustible.
Use: Corrosive inhibitor; intermediate.

***N*-benzylidimethylamine.**
$C_6H_5CH_2N(CH_3)_2$.
Properties: Colorless to light-yellow liquid. D 0.894 (27C); refr index 1.4985–1.5005 (25C); bp 180–182C; distilling range 65–68C (18 mm Hg). Combustible.
Use: Intermediate, especially for quaternary ammonium compounds; dehydrohalogenating catalyst; corrosion inhibitor; acid neutralizer; potting compounds; adhesives; cellulose modifier.

benzyl iodide. $C_6H_5CH_2I$.
Properties: Colorless crystals or liquid. D 1.7335, mp 34.1C, bp (decomposes). Soluble in alcohol, carbon disulfide, ether; insoluble in water.
Derivation: Interaction of benzyl chloride and hydriodic acid.
Hazard: Powerful irritant.

benzyl isoamyl ether.   See isoamyl benzyl ether.

benzyl isobutyl ketone.   See 4-methyl-1-phenyl-2-pentanone.

benzyl isoeugenol.   (1,α-phenyl-4-propenylveratrole).
$CH_3CHCHC_6H_3(OCH_3)OCH_2C_6H_5$.
Properties: White, crystalline solid; floral odor of the carnation type. Soluble in alcohol, ether. Combustible.
Use: Perfumery, fixative.

***N*-benzylisopropylamine.**
$C_6H_5CH_2NH(CH_3CHCH_3)$.
Properties: Colorless to yellow liquid. D 0.895 (25C), refr index 1.4995–1.5015 (25C). Combustible.
Use: Rust inhibitor; intermediate.

benzyl isothiocyanate. $C_6H_5CH_2NCS$.
Properties: Colorless to slightly yellow liquid; a lachrymator.
Hazard: Very irritant to tissues.
Use: Chemical intermediate.

benzyl mercaptan.   See benzyl thiol.

benzylmethylamine. $C_6H_5CH_2NHCH_3$.
Properties: Colorless to light yellow liquid. D 0.936 (25C), refr index 1.5185–1.5220 (25C), distillation range 183–188C. Combustible.
Use: Organic synthesis.

***N*-benzyl-*N*,*N*-methylethanolamine.**
$C_6H_5CH_2NCH_3(C_2H_4OH)$.
Properties: Colorless to light-yellow liquid. D 1.006 (27C), refr index 1.5250–1.5270 (25C), distillation range 95–105C (2 mm Hg). Combustible.
Use: Corrosive inhibitor, intermediate.

3-benzyl-4-methyl umbelliferone.
$C_6H_5CH_2CH_3C_9H_4O_3$.
Properties: Tan, crystalline powder. Mp 255C min. Slightly soluble in ethanol; insoluble in water.
Use: Optical whitening agent; intermediate.

5-benzyloxy-8-chloro-*n*,*n*-dimethyl-1,2,3,4-tetrahydro-1-naphthylamine hydrochloride.
CAS: 63978-98-3. $C_{19}H_{22}ClNO•ClH$.
Hazard: A poison.

***p*-benzyloxyphenol.**   (benzylhydroquinone; "Agerite Alba" *[Vanderbilt]*). $C_6H_5CH_2OC_6H_4OH$.
Properties: Light-tan powder; faint odor. D 1.26, mp 121–122C. Slightly soluble in water; practically insoluble in petroleum hydrocarbons; very soluble in benzene and alkalies. Combustible.
Use: Rubber antioxidant; stabilizer; polymerization inhibitor; chemical intermediate.

***p*-(benzyloxy)phenyl bis(1-aziridinyl)-phosphinate.**
CAS: 41920-59-6. $C_{17}H_{19}N_2O_3P$.
Hazard: Moderately toxic.

***n,n'*-(3-benzyloxy-1,2-propanedioxysulfinyl)bis(3-methylphenylmethylcarbamate).**
CAS: 81862-10-4. $C_{28}H_{32}N_2O_9S_2$.
Hazard: Moderately toxic by ingestion.
Use: Agricultural chemical.

n,n'-(3-benzyloxy-1,2-propanedioxysulfinyl)bis(1-naphthylmethylcarbamate).
CAS: 81877-67-0. $C_{34}H_{32}N_2O_9S_2$.
Hazard: Moderately toxic by ingestion.
Use: Agricultural chemical.

benzyl pelargonate. $C_6H_5CH_2OOCC_8H_{17}$.
Properties: Liquid; mild odor. D 0.962 (15.5/15.5C), bp 315C.
Use: In flavors and perfumes, bactericides and fungicides, organic synthesis.

p-benzylphenol.   (4-hydroxydiphenylmethane).
$C_6H_5CH_2C_6H_4OH$.
Properties: White crystals from ethanol. Mp 84C, bp 320–322C. Soluble in ethanol, ether, chloroform, benzene, acetic acid, caustic alkalies; moderate solubility in hot water. Combustible.
Hazard: Toxic by ingestion.
Use: Antiseptic and germicide; organic synthesis.

benzyl phenylacetate.
$C_6H_5CH_2COOCH_2C_6H_5$.
Properties: Colorless liquid; honeylike odor. D 1.097–1.099, refr index 1.554–1.556. Soluble in alcohol. Combustible.
Use: Perfumery and flavors.

benzyl phenyl ketone.   See deoxybenzoin.

benzylphosphonic acid dibutyl ester.
CAS: 3762-27-4. $C_{15}H_{25}O_3P$.
Hazard: A poison by ingestion.

benzyl propionate. $C_2H_5COOCH_2C_6H_5$.
Properties: Similar to benzyl acetate but has sweeter odor. Liquid. Bp 220C, d 1.036 (17.5C). Insoluble in water. Combustible.
Grade: Technical; FCC.
Use: Perfumes; flavoring.

benzylpyridine. $C_6H_5CH_2C_5H_4N$.
Properties: Liquid. Bp 276.8C, mp 13.6C, d 1.061 (20C), refr index 1.5797. Insoluble in water. Combustible.
Hazard: Toxic by ingestion.

benzyl salicylate. $C_6H_4(H_4OH)COOCH_2C_6H_5$.
Properties: Colorless liquid; faint sweet odor. D 1.176–1.179, refr index 1.580–1.581, mp 24C min, bp 208C(26 mm). Soluble in 9 vols of 90% alcohol. Combustible.
Derivation: Reaction of sodium salicylate and benzyl chloride.
Grade: Technical, FCC.
Use: Perfume fixative, solvent for synthetic musk, sunscreening lotions, soap odorant.

benzyl   succinate.   (dibenzyl   succinate).
$C_6H_5CH_2OOCCH_2CH_2COOCH_2C_6H_5$.

Properties: White crystalline powder; almost tasteless. Mp 45C. Soluble in alcohol, ether, chloroform, also in fixed and volatile oils; insoluble in water. Combustible.
Use: Medicine (antispasmodic).

benzyl sulfide. $(CH_2C_6H_5)_2S$.
Properties: Colorless plates. D 1.0712, mp 49C. Soluble in alcohol and ether; insoluble in water.
Derivation: Action of potassium sulfide on benzyl chloride and subsequent distillation.
Use: Organic synthesis.

7-benzyl-3-thia-7-azabicyclo(3.3.1)nonane perchlorate.
CAS: 89398-07-2. $C_{14}H_{19}NS•ClHO_4$.
Hazard: A poison by ingestion.

benzyl thiocyanate. $C_6H_5CH_2CNS$.
Properties: Colorless crystals. Mp 41C, bp 230C. Insoluble in water; soluble in alcohol and ether.
Hazard: Strong irritant to skin, tissue. Moderate fire hazard.
Use: Insecticide.

benzylthiol.   (benzyl mercaptan; α-toluenethiol).
$C_6H_5CH_2SH$.
Properties: Colorless liquid; strong odor. Bp 195C, flash p 158F (70C) (CC), d 1.05. Insoluble in water; soluble in alcohol, carbon disulfide. Combustible.
Hazard: Toxic by inhalation and ingestion; irritant to tissue.
Use: Odorant, flavors.

2-benzyl-6-thiouracil. $C_6H_5CH_2C_3N_2OS$.   A drug intermediate.

benzyltrimethylammonium chloride.
$C_6H_5CH_2N(CH_3)_3Cl$.
Properties: A quaternary ammonium salt. Colorless crystals; stable up to 135C, above which benzyl chloride and trimethylamine are formed. Properties of 60% solution: d 1.07 (20/20C), wt/gal 8.90 lb, fp below −50C. Readily soluble in water, ethanol, and butanol; slightly soluble in butyl phthalate and tributyl phosphate.
Grade: 60–62% aqueous solution.
Use: Solvent for cellulose, gelling inhibitor in polyester resins, intermediate.

benzyltrimethylammonium hexafluorophosphate. $C_6H_5CH_2N(CH_3)_3PF_6$.
Properties: Crystals. Mp 160C.
Hazard: Toxic by ingestion, irritant to skin.

4-benzyltrimethylammonium methoxide.
$C_6H_5CH_2(CH_3)_3NOCH_3$.
Properties: A quaternary ammonium salt. Yellow liquid; decomposes on distillation.
Hazard: Toxic by ingestion, irritant to skin.
Use: Catalyst, organic-soluble strong base.

Benzyl Tuex.   Tetrabenzoylthiuram disulfide.

benzyl violet.   (benzyl violet 4B; sodium-3-[[4-[(4-dimethylazaniumylidenecyclohex-2,5-dien-1-ylidene)-4[4-[ethyl-[(3-sulfonatophenyl)methyl]amino]*p*]-N-ethylanilino]methyl]benzenesulfo).
CAS: 1694-09-3. $C_{39}H_{40}N_3NaO_6S_2$.
Hazard: Possible carcinogen.

benzyne. C_6H_4.
Properties: An unsaturated, cyclic hydrocarbon with a structure similar to benzene, in which one of the double bonds is replaced by a triple bond. It may be prepared from benzenediazonium-2-carboxylate or from isatoic anhydride.

bepridil hydrochloride monohydrate.
See   3-isobutoxy-2-pyrrolidino-*n*-phenyl-*n*-benzylpropylamine.

berberine. $C_{20}H_{18}NO_4^+$.
Properties: White to yellow crystals. Mp 145C (anhydrous). Iinsoluble in water; soluble in ether, alcohol. Salts of berberine are berberine bisulfate, berberine sulfate, and berberine hydrochloride. All three are yellow crystals, slightly soluble in water.
Derivation: From the root of Berberis vulgaris or Hydrastis canadensis.
Hazard: Toxic via ingestion, inhalation, skin absorption.
Use: Medicine (antipyretic).

"Berbond 8200" [Cranston].   TM for a strippable wallpaper saturant.
Use: In many paper manufacturing areas.

"Bercohem 4842" [Cranston].   TM for a low molecular weight polyacrylate-based dispersant.
Use: For pigmented paper coating formulations.

bergamot oil.
Properties: An essential oil. Brownish-yellow to green liquid; agreeable odor; bitter taste.
Use: Perfumery.

Bergenia crassifolia.   A hardy, perennial garden herb (family Saxifragaceae) with large, fleshy evergreen leaves that contain arbutin.

Bergius, Friedrich Karl Rudolf.   (1884–1949). A German chemist who won a Nobel Prize in 1931 with Bosch for chemical high-pressure methods. He invented a method of converting coal dust into oil via pressurized hydrogen. He also invented a method for production of cattle feed and sugar from wood by hydrolysis. He was educated in Poland and Germany.

Bergius process.   Formation of petroleum-like hydrocarbons by hydrogenation of coal at high temperatures and pressures (e.g., 450C and 300 atm) with or without catalysts; production of toluene by subjecting aromatic naphthas to cracking temperatures at 100 atm with a low partial pressure of hydrogen in the presence of a catalyst.

Bergius–Willstatter saccharification process.
Process for industrial production of fermentable sugar from wood by hydrolysis of tannin and xylan-free cellulose with 40–45% hydrochloric acid. The use of concentrated acid requires acid-resistant equipment and recovery of acid. The sugar produced must be rehydrolyzed prior to fermentation.

Bergmann azlactone peptide synthesis.
Conversion of an acetylated amino acid and an aldehyde into an azlactone with an alkylene side chain; reaction with a second amino acid with ring opening and formation of an acylated unsaturated dipeptide, followed by catalytic hydrogenation and hydrolysis to the dipeptide.

Bergmann degradation.   Stepwise degradation of polypeptides involving benzoylation, conversion to azides, and treatment of the azides with benzyl alcohol; this treatment yields, via rearrangement to isocyanates, carbobenzoxy compounds which undergo catalytic hydrogenation and hydrolysis to the amide of the degraded peptide.

Bergmann–Zervas carbobenzoxy method.
Formation of the *N*-carbobenzoxy derivative of an amino acid for use in peptide synthesis and liberation of the amino group at an appropriate stage of synthesis by hydrogenolysis of the activated CO bond.

Berg, Paul.   (1926–). An American molecular biologist who won the Nobel Prize for chemistry in 1980 with Sanger and Gilbert. Berg's research concerned the biochemistry of nucleic acid, particularly regarding recombinant DNA, that is combining DNA from another species into a molecule. His Doctorate was attained at Western Reserve, and later he performed research at Stanford University.

berkelium.   Bk. Synthetic radioactive element with atomic number 97, first produced (1949) as the 243 isotope by bombarding americium with helium ions in a cyclotron. The chemical properties of berkelium have been studied by tracer techniques and are similar to those of the other transuranium elements. Its oxidation behavior is similar to that of the rare earth cerium. It has a mp of 986C. There are 8 isotopes ranging from 243 to 250; the 249 isotope has been made by neutron bombardment of curium 244. Atomic weight is generally accepted as 249. The following compounds have been identified by X-ray diffraction: berkelium dioxide (BkO_2), berkelium sesquioxide (Bk_2O_3), berkelium trifluoride (BkF_3), berkelium trichloride ($BkCl_3$), and berkelium oxychloride (BkOCl).

Berlin blue. Any of a number of the varieties of iron blue pigments.
See iron blue.

Berlin red. A red pigment consisting, essentially, of red-iron oxide.

"Bersize 6125" *[Cranston].* TM for a surface size for paper and paperboard.
Use: To give a good printing surface to fine papers.

Berthelot, Pierre Eugene Marcellin. (1827–1907).
A French chemist who did fundamental work on the organic synthesis of hydrocarbons, fats and carbohydrates. Opposed the then current idea that a "vital force" is responsible for synthesis. Did important work on explosives for French government. He was one of the first to prove that all chemical phenomena depend on physical forces that can be measured.

bertholite. A name given to chlorine when used as a poison gas.

Berthollet, Claude Louis. (1748–1822). A French chemist. Followed Lavoisier, but did not accept the latter's contention that oxygen is the characteristic constituent of acids. He was the first to propose chlorine as a bleaching agent. His essay on chemical physics (1803) was the first attempt to explain this subject. His speculations on stoichiometry, especially as regards relative masses of reacting atoms, profoundly affected later theories of chemical affinity.

bertrandite. (Berylium silicate hydrate). $Be_4Si_2O_7(OH)_2$.
Properties: Transparent, colorless to pale-yellow, beryllium silicate with orthorhombic crystals.
Hazard: Carcinogen.

beryl. $Be_3Al_2(SiO_3)_6$.
Sometimes with replacement of beryllium by sodium, lithium, cesium. A natural silicate of beryllium and aluminum.
Hazard: A confirmed carcinogen.
See beryllium.

beryllia. See beryllium oxide.

beryllides. Intermetallic compounds made by chemically combining beryllium with such metals as zirconium and tantalum.

beryllium.
CAS: 7440-41-7. Be.
Metallic element of atomic number 4, group IIA of the periodic table; aw 9.0121; valence 2; no stable isotopes.
Properties: A hard, brittle, gray-white metal. D 1.85, mp 1280C. Soluble in acids (except nitric) and alkalies. Resistant to oxidation at ordinary temperatures. High heat capacity and thermal conductivity. It is the lightest structural metal known; can be fabricated by rolling, forging, and machining. Joining is chiefly by shrink-fitting; brazing and welding are difficult. Highly permeable to X-rays.
Occurrence: Beryl, the ore of beryllium, is found chiefly in South Africa, Zimbabwe, Brazil, Argentina, and India. Principal sources in the U.S. are Colorado, Maine, New Hampshire, and South Dakota. There are undeveloped deposits in Canada.
Derivation: The ore is converted to the oxide or hydroxide, then to the chloride or fluoride. The halide may be (1) reduced in a furnace by magnesium metal, or (2) reduced by electrolysis. Liquid–liquid extraction with an organophosphate chelating agent can be used as a method of purification, or as an alternative process on the ore itself.
Grade: Technical, over 99.5% pure.
Available forms: Hot-pressed or cold-pressed and sintered blocks; sheet (0.04 inch); tube; rods; wire; powder.
Hazard: A confirmed carcinogen. Very high toxicity, especially by inhalation of dust.
Use: Structural material in space technology; moderator and reflector of neutrons in nuclear reactors; source of neutrons when bombarded with α-particles; special windows for X-ray tubes; in gyroscopes, computer parts, inertial guidance systems; additive in solid-propellant rocket fuels; beryllium-copper alloys.

beryllium acetate.
CAS: 543-81-7. $Be_4O(C_2H_3O_2)_6$.
Properties: White crystals. Mp 285–286C, bp 330–331C. Insoluble in water; hydrolyzed by hot water, dilute acids. Soluble in chloroform and other organic solvents. Can be crystallized from hot acetic acid in very pure form.
Hazard: Toxic by inhalation and ingestion.
Use: Source of pure beryllium salts.
See beryllium.

beryllium acetylacetonate. $Be(C_5H_7O_2)_2$.
Properties: Crystalline powder. Mp 108C, bp 270C. Freely soluble in alcohol and ether; slightly soluble in water. Resistant to hydrolysis. A chelating nonionizing compound.
Hazard: Toxic.
See beryllium.

beryllium carbide. Be_2C.
Properties: Fine, hexagonal, hard, refractory crystals; attacked vigorously by strong, hot alkali solutions forming methane gas and alkali beryllate. D 1.91; decomposes above 2100C.
Derivation: By direct interaction of elemental beryllium and carbon; by reduction of beryllium oxide with carbon above 1500C.
Use: Nuclear-reactor cores.

beryllium chloride.
CAS: 7787-47-5. $BeCl_2$.
Properties: White or slightly yellow deliquescent crystals; sweetish taste. Mp 440C, bp 520C, d 1.90.

Very soluble in water; soluble in alcohol, benzene, ether, carbon disulfide. Readily hydrolyzed.
Derivation: By passing chlorine over a mixture of beryllium oxide and carbon.
Hazard: Very toxic.
Use: In dry form, as catalyst for organic reactions.
See beryllium.

beryllium-copper.   A precipitation-hardenable alloy; often also contains nickel or cobalt and has relatively high electrical conductivity, high strength, and high hardness.
Properties: D 8.22. Tensile strength of heat-treated sheet 175,000 psi, elongation 5% in 2 inches; Brinell hardness 350; good electrical conductivity. Typical analysis: copper 97.4; beryllium 2.25; nickel 0.35.
Hazard: Avoid inhalation.
See beryllium.
Use: Electrical switch parts, watch springs, optical alloys, electronic equipment, valves and parts, spot-welding electrodes, nonsparking tools, springs and diaphragms, shims, cams, and bushings.
Note: A comparatively recent development is an 85 copper, 9 nickel, 6 tin alloy reported to be 15% stronger than Be–Cu.

beryllium fluoride. BeF_2.
Properties: Hygroscopic solid. Mp 800C, d 1.986. Readily soluble in water; sparingly soluble in alcohol.
Derivation: By the thermal decomposition (900–950C) of ammonium beryllium fluoride.
Hazard: A known carcinogen. Toxic by inhalation and ingestion.
Use: Production of beryllium metal by reduction with magnesium metal; nuclear reactors; glass manufacturing.

beryllium hydrate.   See beryllium hydroxide.

beryllium hydride. BeH_2.
Properties: White solid. Reacts with water, dilute acids, and methanol, liberating hydrogen. When heated to 220C it liberates hydrogen rapidly.
Hazard: For toxicity see beryllium. Fire risk when exposed to water, organic materials, and heat.
Use: Experimentally in rocket fuels.

beryllium hydroxide.   (beryllium hydrate).
CAS: 13327-32-7. $Be(OH)_2$.
Properties: White powder. Decomposes to the oxide at 138C. Soluble in acids and alkalies; insoluble in water.
Derivation: By precipitation with alkali from pure beryllium acetate.
Grade: Technical.
Hazard: Very toxic.
See beryllium.

beryllium metaphosphate. $Be(PO_3)_2$.
Properties: White porous powder or granular material. High melting point. Insoluble in water.

Hazard: Very toxic.
Use: Raw material for special ceramic compositions; catalyst carrier.
See beryllium.

beryllium nitrate.
CAS: 13597-99-4. $Be(NO_3)_2 \cdot 3H_2O$.
Properties: White to faintly yellowish, deliquescent mass. Mp 60C, decomposes 100–200C. Soluble in water, alcohol.
Derivation: Action of nitric acid on beryllium oxide, with subsequent evaporation and crystallization; reaction of beryllium sulfate with barium nitrate.
Grade: Technical, CP.
Use: Chemical reagent, gas-mantle hardener.
Hazard: Very toxic. Oxidizing material, dangerous fire risk.
See beryllium.

beryllium nitride. Be_3N_2.
Properties: Hard, refractory, white crystals. Mp 2200C, reacts with mineral acids to form the corresponding salts of beryllium and ammonia. Readily attached by strong alkali solutions, liberating ammonia.
Derivation: By heating beryllium metal powder in a dry, oxygen-free nitrogen atmosphere at 700–1400C.
Use: Atomic energy, production of the radioactive carbon isotope ^{14}C for tracer uses.
See beryllium.

beryllium oxide.   (beryllia).
CAS: 1304-56-9.   BeO.
Properties: White powder. A unique ceramic material. D 3.016, mp 2570C. Hardness (Mohs) 9. Soluble in acids and alkalies; insoluble in water. High electrical resistivity and thermal conductivity; transparent to microwave radiation; undamaged by nuclear radiation. High heat-stress resistance. Can be fabricated into finished shapes.
Derivation: By heating beryllium nitrate or hydroxide.
Grade: Technical, CP, pure, single crystals.
Hazard: Highly toxic by inhalation. Keep container tightly closed and flush out after use.
Use: Electron tubes; resistor cores; windows in klystron tubes; transistor mountings; high-temperature reactor systems; additive to glass, ceramics, and plastics; preparation of beryllium compounds; catalyst for organic reactions.
See beryllium.

beryllium potassium fluoride.   (potassium beryllium fluoride). $BeF_2 \cdot 2KF$.
Properties: White, crystalline masses. Soluble in water; insoluble in alcohol.
Hazard: Toxic by inhalation and ingestion.

beryllium potassium sulfate. $BeSO_4K_2SO_4$.
Properties: Shiny crystals, insoluble in alcohol, soluble in water and concentrated potassium sulfate solution.

Use: Metal plating especially chromium and silver.
Hazard: See beryllium.

beryllium sodium fluoride. (sodium beryllium fluoride).
$BeF_2•2NaF$.
Properties: White, crystalline mass. Mp approximately 350C. Soluble in water.
Hazard: Toxic by inhalation and ingestion.
Use: Making pure beryllium metal.

beryllium sulfate.
CAS: 13510-49-1. $BeSO_4•4H_2O$.
Properties: Colorless crystals. D 1.713, decomposes at 540C. Soluble in water; insoluble in alcohol.
Hazard: A confirmed carcinogen.
See beryllium.

Berzelius, Jöns Jacob. (1779–1848). A native of Sweden, Berzelius was one of the foremost chemists of the 19th century. He made many contributions to both fundamental and applied chemistry; coined the words *isomer* and *catalyst*; classified minerals by chemical compound; recognized organic radicals which maintain their identity in a series of reactions; discovered selenium and thorium, and isolated silicon, titanium, and zirconium; did pioneer work with solutions of proteinaceous materials which he recognized as being different from "true" solutions.

Be Square. A hard or plastic grade of petroleum microcrystalline wax.

Bessel function. Solution of the Bessel's equation.

Bessel's equation. Linear differential equation $x2y'' + xy' + (x^2 - n^2)y = 0$, whose solutions are expressible as power series in x.

Bessonoff's reagent. (phospho-tungstic-phosphomolybdic acid). It gives a deep blue reaction with ascorbic acid.
Use: For the detection of ascorbic acid.

Best, Charles H. (1899–1978). Born in Maine, Best was educated at the University of Toronto, where he distinguished himself as a student of biochemistry. He collaborated with the late Dr. Frederick Banting in the isolation of the hormone insulin. He later became head of the insulin division of the Connaught Laboratories of the University as well as of the Banting and Best Research Institute. He also developed histaminase (an antiallergic enzyme) and the anticoagulant heparin.
See Banting, Sir Frederick.

beta. (β). A prefix having meanings analogous to those of α.
(1) It indicates (a) the position of a substituent atom or radical in a compound; (b) the second position in a naphthalene ring; or (c) the attachment of a chemical unit to the side chain of an aromatic compound.
(2) It refers to a secondary allotropic modification of a metal or compound.
(3) It designates a type of radioactive decay.
See beta particle.

beta battery. Alternative name for the sodium-sulfur battery under development.
See storage battery; electrolyte.

beta-emitter. Radioactive substance that decays with the release of beta particles.

betaine. (1-carboxy-*N,n,n*-trimethylmethanaminium hydroxide inner salt; (carbodymethyl)trimethylammonium hydroxide inner salt; glycine betaine; glycocoll betaine; lycine; oxyneurine; trimethylglycine hydroxide inner salt; trimethylglycocoll anhydride; 2-(trimethylazaniumyl)acetate).
CAS: 107-43-7. $C_5H_{11}NO_2$.
Properties: Alkaloid; soluble in water, ethanol, and methanol.
Use: Organic synthesis; soldering and resin-curing flux; therapeutically as a hepatoprotectant.; as a source of hydrochloric acid in the treatment of hypochlorhydria; in the treatment of liver disorders, for hyperkalemia, for homocystinuria, and gastrointestinal distrubances.

betaine-aldehyde dehydrogenase. Enzyme that catalyzes the oxidation of betaine aldehyde to betaine. It is a component of the choline oxidase system.

betaine hydrochloride. (lycine hydrochloride).
CAS: 107-43-7. $C_5H_{11}O_2N•HCl$.
Properties: Colorless crystals. Mp 227–228C (decomposes). Soluble in water and alcohol; insoluble in chloroform and ether. Aqueous solutions are strongly acid. Liberates hydrogen chloride at the mp.
Grade: Technical.
Use: Source of hydrogen chloride in solders and fluxes, organic synthesis.

betaine phosphate. $C_5H_{11}O_2N•H_3PO_4$.
Properties: White, odorless granules; acid taste. Mp 198–200C. Very soluble in water.
Grade: Technical.
Use: Source of phosphoric acid.

beta iron. A nonmagnetic allotrope of pure iron that is stable between 770C and 910C.

Betanox Special. A low-temperature reaction product of phenyl-β-naphthylamine and acetone.
Properties: Tan powder. D 1.16, mp above 120C. Soluble in acetone, benzene, and ethylene dichloride; insoluble in water and gasoline.

Use: Antioxidant for wire insulation, tire treads, tire carcass, inner tubes, dark-colored footwear, proofing, and mechanical goods.

beta particle.   A high-speed electron or positron (accompanied by a neutrino) emitted by a nucleus during radioactive decay or nuclear fission.

beta particle.   (β particle). A charged particle emitted from a radioactive atomic nucleus either natural or manufactured. The energies of β particles range from 0 to 4 MeV. They carry a single charge; if this is negative, the particle is identical with an electron; if positive, it is a positron. β-rays (streams of these particles) may cause skin burns and are harmful within the body. Protection to the skin can be afforded by a thin sheet of metal.
See electron; decay; radioactive.

Betaprene.   Olefinic hydrocarbon resins used as coating vehicles.

betatron.   An electromagnetic device for accelerating electrons (β particles). Its action is similar in principle to that of an electric transformer in which the secondary windings are replaced with focusing magnets. The electrons travel around the core in a vacuum tube placed between the magnets. At each revolution around the core the electrons pick up the same energy as the voltage that would have been induced in one turn of wire at that point. The betatron can generate electron beams up to 320 MeV. Invented by D. W. Kerst in 1940, it is used chiefly for basic physical research.

"bethanechol   chloride".   [TEVA].   (2-[(aminocarbonyl)oxy]-*N,N,N*-trimethyl-1-propanaminium chloride; duboid; urecholine® [TEVA]). C$_7$H$_{17}$ClN$_2$O$_2$.
Properties: Colorless or white crystalline substance or powder.
Hazard: Headache, flushing, gastrointestinal distress, diarrhea, hypotension, excessive salivation, sweating, hypersensitivity.
Use: Parasymtaphomimetic agent to stimulate the parasympathic nervous system.

Bethanizing.   The process of electrodepositing very pure zinc on iron or steel.

Bettendorf's reagent.   A reagent used for the detection of arsenic in presence of bismuth and antimony compounds. It consists of a concentrated solution of stannous chloride in fuming hydrochloric acid.
Hazard: Powerful tissue irritant.

Better Locks Lite Almond Conditioner.   A hair treatment.
Use: Adds sheen, softens, and absorbs oils.

Betterton–Kroll process.   A process for obtaining bismuth and purifying desilverized lead

that contains bismuth. Metallic calcium or magnesium is added to the molten lead to cause formation of high-melting intermetallic compounds with bismuth. These separate as a surface scum and are skimmed off. The excess calcium and magnesium are removed from the lead by use of chlorine gas as mixed molten chlorides of lead or zinc. Bismuth of 99.995% purity is produced in this way.

Betti reaction.   The reaction of aromatic aldehydes, primary aromatic or heterocylic amines, and phenols leading to α-aminobenzylphenols.

Betts   process.   An electrolytic process for removing impurities from lead in which pure lead is deposited on a thin cathode of pure lead from an anode containing as much as 10% of silver, gold, bismuth, copper, antimony, arsenic, selenium, and other impurities. The electrolyte is lead fluosilicate and fluosilicic acid. The scrap anodes and the residues of impurities associated with them are either recast into anodes or treated to recover antimony lead, silver, gold, bismuth, etc.

betula oil.   See methyl salicylate.

betweenanenes.   Class of bicyclic alkenes joined by a double bond between the bridgehead atoms and a trans attachment of each branch to this bond.

Betzig, Eric.   (1960–). An American Physicist who won the 2014 Nobel Prize jointly with Stefan W. Hell and William E. Moerner. The scientists developed a groundbreaking method to bypass the natural resolution limit in optical microscopy by using fluorescent molecules. In Betzig and Moerner's method, light is used to excite or deactivate fluorescent molecules of proteins and an image is created by stitching images of different activated molecules of the protein. This method enables to track process of active viruses and molecules in living cells. Betzig received an M.S. and Ph.D. from Cornell University, followed by an illustrious career that included the William L. McMillan Award (1992) and William O. Baker Award for Initiatives in Research (1993).

"Beutene"   [Chemtura].   TM   for   a butyraldehyde-aniline reaction product.
Properties: Reddish-brown, free-flowing liquid. D 0.95. Soluble in acetone, benzene, and ethylene dichloride; slightly soluble in gasoline; insoluble in water.
Use: Rubber accelerator.

bevantolol hydrochloride.
CAS: 42864-78-8.   C$_{20}$H$_{27}$NO$_4$•ClH.
Hazard: Moderately toxic by ingestion.

BFE.   Abbreviation for bromotrifluoroethylene.

BF$_3$-ether complex.   See boron trifluoride etherate.

BFI powder.   Proprietary preparation of bismuth formic iodide.
Use: Skin antiseptic.

BF₃-MEA.   See boron trifluoride monoethylamine.

BF₃MeOH.   See boron trifluoride-methanol.

BFPO.   Abbreviation for bis(dimethylamino)-fluorophosphine oxide.
See dimefox.

Bh.   Symbol for bohrium.

BHA.   Abbreviation for butylated hydroxyanisole.

BHC.   Abbreviation for benzene hexachloride.

BHMT amine. $NH_2(CH_2)_6NH(CH_2)_6NH_2$. A liquid polyalkylene polyamine.
Use: In asphalt additives and corrosive inhibitors.

BHN.   Brinell hardness number.

BHT.   Abbreviation for butylated hydroxytoluene.
See 2,6-di-*tert*-butyl-*p*-cresol.

Bi.   Symbol for bismuth.

bi-.   Prefix meaning two; di- is preferred in chemical nomenclature. Exceptions are bicarbonate, bisulfate, bitartrate, in which it indicates the presence of hydrogen in the molecule, e.g., $NaHCO_3$ (sodium bicarbonate).
See bis-.

bialamicol.   (3,3′-bis[(diethylamino)methyl]-5,5′-diallyl-α,α′-bis(diethylamino)-*m,m*′-bitolyl-4,4′-diol; 6,6′-diallyl-α,α′-bis(diethylamino)-4,4′-bi-*o*-cresyl; biallylamicol; 2-(diethylaminomethyl)-4-[3-(diethylaminomethyl)-4-hydroxy-5-prop-2-enylphenyl]-6-prop-2-enylphenol). $C_{28}H_{43}N_2O_2$. Antiamebic agent.

bibenzyl.   See *sym*-diphenylethane.

bicalcium phosphate.   See calcium phosphate, dibasic.

bicarburetted hydrogen.   See ethylene.

2,2′-bichavicol.
CAS: 528-43-8. $C_{18}H_{18}O_2$.
Hazard: Moderately toxic by ingestion.
Source: Natural product.

bicyclic.   An organic compound in which only two ring structures occur. They may or may not be the same type of ring.
See naphthalene.

bicyclic alkene.   Any of a class of hydrocarbons in which the bridgehead atoms are joined by a double bond. There is also a trans attachment of each branch to the double bond.

bicyclobutylidine.
CAS: 6708-14-1. C_8H_{12}.
Hazard: Moderately toxic by ingestion and inhalation.

bicyclohexyl.   (dicyclohexyl). $C_{12}H_{22}$.
Properties: Colorless, mobile liquid with pleasant odor. Bp 238.5C, fp 1–3C, d 0.883 (25/16C), wt/gal 7.37 lb, refr index 1.480 (20C), flash p 165F (73.9C), autoign temp 471F (244C). Combustible.
Derivation: Hydrogenation of diphenyl.
Use: High-boiling solvent and penetrant.

"Bidrin" *[AMVAC].*   (TM for dimethyl phosphate of 3-hydroxy-*N,N*-dimethyl-*cis*-crotnamide; dicrotophos).
CAS: 141-66-2. $(CH_3O)_2P(O)OC(CH_3):CHC(O)N(CH_3)_2$.
Properties: Brown liquid with a mild ester odor. Bp 400C. Miscible with water and xylene; slightly soluble in kerosene and diesel fuel. Commercially available water-miscible solution.
Hazard: Cholinesterase inhibitor. Toxic by skin absorption. Questionable carcinogen.
Use: Insecticide.

biformin. $C_9H_6O_2$. An antibiotic produced by the fungus *Polyporus biformis*, reported to be active against various bacteria and fungi.

biformychlorazin.   See triforine.

Biginelli reaction.   Synthesis of tetrahydropyrimidinones by the acid-catalyzed condensation of an aldehyde, a β-keto ester, and urea.

B-I-K.   A surface-coated urea.
Properties: White powder. D 1.32, melting range 129–134C. Soluble in water. Surface coating not soluble in water but soluble in rubber. Slightly soluble in acetone; insoluble in benzene, gasoline, and ethylene dichloride.
Use: Promoter for azodicarbonamide, a nitrogen blowing agent; activator for thiazoles, sulfenamides, and thiurams; odor reducer when used with nitrosoamine-type blowing agents.

bilayer.   A double layer of amphipathic lipid molecules that orient themselves so the hydrocarbon tails face inward to form a continuous nonpolar phase, and the polar head groups face outward. In this way, they form the basic structure of membranes.

bile.   A secretion produced by the vertebrate liver. It is stored temporarily in the gall bladder and is conducted from the gall bladder to the duodenum of the small intestine via the bile duct under pressure

by contractions of the gall bladder. The components may be excreted or recycled facilitating digestion.
Properties: Dark yellow or brown-green alkaline digestive fluid; contains bile salts, bile pigments, cholesterol, lecithin, phospholipids, electrolytes, urea, various xenobiotics, toxicants, and metabolites.

bile acid. An acid found in bile (secretion of the liver). Bile acids are steroids having a hydroxyl group and a five-carbon-atom side chain terminating in a carboxyl group. Cholic acid is the most abundant bile acid in human bile. Others are deoxycholic and lithocholic acids. The bile acids do not occur free in bile but are linked to the amino acids glycine and taurine. These conjugated acids are water soluble. Their salts are powerful detergents and as such aid in the absorption of fats from the intestine.

bile salts. Sodium salts of glycocholic and taurocholic acids important for physiological fat absorption.

bilirubin. (bilifulvin). $C_{33}H_{36}O_6N_4$.
Red coloring matter of bile. Also occurs in blood serum as decomposition product of hemoglobin.
Properties: Orange-red powder. Mp 192C. Soluble in acids, alkalies, chloroform, and benzene; insoluble in water; very slightly soluble in alcohol and ether.
Derivation: From bile pigment.
Use: Analytical chemistry, biochemical research.

biliverdin. (3-[2[[(Z)-[(5Z)-3-(2-carboxyethyl)-5-[(4-methylpyrrol-2-ylidene]methyl]-5[(Z)-(3-ethenyl-4-methyl-5-oxopyrrol-2-ylidene)methyl]-4-methyl-1H-pyrrol-3-yl]propanoic acid).
CAS: 114-25-0. $C_{33}H_{34}N_4O_6$. A precursor of bilirubin.
Derivation: Produced from hemoglobin during the normal metabolic destruction of erythrocytes.

bimetal. A type of thermometer in which the sensing element consists of two thin strips of metals having different expansion coefficients bonded together in a helical or spiral structure. The extent of deflection or bending induced by temperature change is indicated by a pointer on a dial. Reasonably accurate readings are obtained in this way, the range being from −185 to 425C. Bimetals are used in both laboratory and industry.
See thermometer.

binapacryl. Generic name for 2-sec-butyl-4,6-dinitrophenyl-3-methyl-2-butenoate. $C_{15}H_8O_6N_2$.
Hazard: Toxic by ingestion and inhalation.
Use: Acaricide and fungicide.

binary. Descriptive of a system containing two and only two components. Such a system may be a chemical compound composed of two elements,

an element and a group (hydroxyl, methyl, etc.), or two groups, (e.g., oxalic acid). It may also be a two-component solution or alloy.

binary acid. An acid containing no oxygen, e.g., hydrofluoric acid.

binary alloy. Alloy containing two major elements, exclusive of impurities.

binary diagram. Constitution diagram for a binary metal alloy system.

bind. To exert a strong physiochemical attraction, as often occurs between various proteins and water in hydrophilic gels, between organic dyes and fabrics, or between acids or bases and various chemical complexes.

binder. (1) The film-forming ingredient in paint, usually either a drying oil or a polymeric substance. (2) In the food industry, a material used in sausage manufacture that absorbs moisture at high temperatures, e.g., various flours, dried milk, and soy protein. (3) Any cementitious material that is soft at high temperatures and hard at room temperature, used to hold dry powders or aggregate together, e.g., asphalt and sulfur in paving compositions, and resins used in sand casting.
See paint.

binding energy. The energy that holds the protons together in an atomic nucleus. Since protons are positively charged, they exert strong mutually repulsive forces and tremendous energy is required to keep them from flying apart. This energy is so great that it results in a slightly lower value for the mass of a nucleus taken as a whole than for the sum of its constituents taken individually. This phenomenon is of vast significance, for it means that a small fraction of mass has been converted into energy within the nucleus, as shown by Einstein's equivalence equation $E = mc^2$. Thus, when a ^{235}U nucleus (92 protons) is split, as in the fission process, a portion of its binding energy (equivalent to the mass difference) is released. It amounts to approximately 200 million electron volts per nucleus.
Binding energy may also be defined as the minimum energy required to dissociate a nucleus into its component neutrons and protons. The neutron or proton binding energy is that required to remove a neutron or a proton from a nucleus; the electron binding energy is that required to remove an electron from an atom or molecule.
See mass defect; fission.

binding site. A place on cellular DNA to which a protein (such as a transcription factor) can bind. Binding sites may be found in the vicinity of genes and are involved in activating transcription of that gene (promoter elements), in enhancing the transcription of that gene (enhancer elements), or in reducing the transcription of that gene (silencers). *Note*: Whether the protein in fact performs these functions may depend on some condition, such as the presence of a hormone, or the tissue in which the gene is being examined. Binding sites could also be involved in the regulation of chromosome structure or of DNA replication.

Bingham body. Fluid that does not exhibit Newtonian flow, but moves in plugs.

Bingham plastic. Substance that will not flow until its yield value is reached, then will flow normally.

Bioaoctive. A wine additive.
Use: Enhances yeast survival.

"Bioban" [Dow]. TM for monocyclic oxazolidines.
Grade: Industrial.
Available forms: Liquid.
Use: Antimicrobial, preservative for coating formulations, latexes, emulsions, oil recovery, and metalworking fluids.

Bio-Care. Hyaluronic acid.
Use: Conditioner for hair and skin.

"Biocheck" [GeneSystem]. TM for a family of biocides, fungicides, and slimicides.
Use: Controlling and eliminating microbiological growth in pulp- and paper-mill water systems as well as for antibacterial papers.

biochemical oxygen demand. (BOD). A standardized means of estimating the degree of contamination of water supplies, especially those that receive contamination from sewage and industrial wastes. It is expressed as the quantity of dissolved oxygen (in mg/L) required during stabilization of the decomposable organic matter by aerobic biochemical action. Determination of this quantity is accomplished by diluting suitable portions of the sample with water saturated with oxygen and measuring the dissolved oxygen in the mixture both immediately and after a period of incubation, usually 5 days.
See sewage sludge; biodegradability; dissolved oxygen (DO); oxygen consumed.

biochemistry. Originally a subdivision of chemistry but now an independent science, biochemistry includes all aspects of chemistry that apply to living organisms. Thus, photochemistry is directly involved with photosynthesis, and physical chemistry with osmosis—two phenomena that underlie all plant and animal life. Other important chemical mechanisms that apply directly to living organisms are catalysis, which takes place in biochemical systems by the agency of enzymes; nucleic acid and protein constitution and behavior, which are known to control the mechanism of genetics; colloid chemistry, which deals in part with the nature of cell walls, muscles, collagen, etc.; acid-base relations, involved in the pH of body fluids; and such nutritional components as amino acids, fats, carbohydrates, minerals, lipids, and vitamins, all of which are essential to life. The chemical organization and reproductive behavior of microorganisms (bacteria and viruses) and a large part of agricultural chemistry are also included in biochemistry. Particularly active areas of biochemistry are nucleic acids, cell surfaces (membranes), enzymology, peptide hormones, molecular biology, and recombinant DNA.
See biotechnology.

biochrome. The colored matter that can be extracted from plants or animals.

biocide. General name for any substance that kills or inhibits the growth of microorganisms such as bacteria, molds, slimes, fungi, etc. Many of them are also toxic to humans. Biocidal chemicals include chlorinated hydrocarbons, organometallics, halogen-releasing compounds, metallic salts, organic sulfur compounds, quaternary ammonium compounds, and phenolics.
See antiseptic; disinfectant; fungicide; bactericide.

biocolloid. An aqueous colloidal suspension or dispersion produced by or within a living organism. Blood, milk, and egg yolk are examples.

biocomputer. A computer in which the silicon in the microchips is replaced by a synthetic protein or polypeptide coated with a silver compound, the combination behaving as a metallic semiconductor. Such chips have been made experimentally; they have the potential of improving the storage capacity and operating efficiency of silicon chips substantially. The materials used in the experimental chips were polylysine on a glass substrate coated with an acrylate polymer and treated with silver nitrate.

bioconversion. Utilization of animal manures, garbage, and similar organic wastes for production of fuel gases by digestion, gasification, or liquefaction.
See biogas; biomass.

biocytin. (ε-*N*-biotinyl-L-lysine). $C_{16}H_{28}N_4O_4S$.
Properties: A naturally occurring complex of biotin isolated from yeast. Mp 228.5C. Water-soluble crystals.
The molecule arising from covalent attachment of biotin to a Lys residue via an amide linkage.

biocytinase. Enzyme in blood that catalyzes the hydrolysis of biocytin to biotin and lysine.

biodegradability. The susceptibility of a substance to decomposition by microorganisms, specifically the rate at which detergents and pesticides

2

and other compounds may be chemically broken down by bacteria and/or natural environmental factors. Branched-chain alkylbenzene sulfonates (ABS) are much more resistant to such decomposition than are linear alkylbenzene sulfonates (LAS), in which the long, straight alkyl chain is readily attacked by bacteria. If the branching is at the end of a long alkyl chain (isoalkyls), the molecules are about as biodegradable as the normal alkyls. The alcohol sulfate anionic detergents and most of the nonionic detergents are biodegradable. Among pesticides, the organophosphorus types, while highly toxic, are more biodegradable than DDT and its derivatives. Tests on a number of compounds gave results as follows. Easily biodegraded: *n*-propanol, ethanol, benzoic acid, benzaldehyde, ethyl acetate. Less easily biodegraded: ethylene glycol, isopropanol, *o*-cresol, diethylene glycol, pyridine, triethanolamine. Resistant to biodegration: aniline, methanol, monoethanolamine, methyl ethyl ketone, acetone. Additives that accelerate biodegradation of polyethylene, polystyrene, and other plastics are available.

bioelectrochemistry.   Application of the principles and techniques of electrochemistry to biological and medical problems. It includes such surface and interfacial phenomena as the electrical properties of membrane systems and processes, ion adsorption, enzymatic clotting, transmembrane pH and electrical gradients, protein phosphorylation, cells, and tissues.

bioengineering.   Application of the principles and methods of chemical engineering to biotechnology.

bioethics.   An interdisciplinary science for which research facilities were established in 1971, encompassing the ethical and social issues resulting from advances in medicine and the biosciences. Its scope includes a number of areas of importance to chemistry, e.g., reproductive and genetic phenomena, organ transplants, gerontology and antiaging techniques, biological warfare, contraception, etc. The Kennedy Institute at Georgetown University, Washington, DC, is the chief center for information about this developing aspect of biomedical science.

biofilm.
Properties: Coating comprised of populations or communities of microorganisms; include extracellular polysaccharides synthesized and secreted by the microbiota that form the biofilms.
Hazard: May cause certain human bacterial infections.

bioflavonoid.   A group of naturally occurring substances thought to maintain normal conditions in the walls of the small blood vessels. The bioflavonoids are widely distributed among plants,

especially citrus fruits, black currants, and rose hips (hesperidin, rutin, quercitin). They have little or no medicinal value.

biofuel.   Any fuel derived from a biological source.

biogas.   Methane generated from animal manure by bacterial anaerobic digestion. Small-scale units have been in use for some years, and the possibilities of utilizing the tremendous quantities of manure available in the U.S. as an energy source have stimulated investigation of large-scale production. One installation utilizing a thermophilic fermentation technique at 55–60C has been operating in Florida since 1979, and another in Colorado since 1981. This energy source is also being exploited in China and India.
See biomass.

biogenesis.   See life, origin.

biogenic amine.   Any of a large group of naturally occurring, biologically active compounds and neurotransmitters that contain a primary amine group.

biogenic emission.   A substance discharged by living organisms into the atmosphere as products of respiration or fermentation.

biogenic sediment.   Sediment consisting of mineral grains that were once parts of organisms.

biogeochemistry.   A branch of geochemistry dealing with the interactions between living organisms and their mineral environment. It includes, among other studies, the effect of plants on weathering of rocks, of the chemical transformations that produced petroleum and coal, of the concentration of specific elements in vegetation at some time in the geochemical cycle (iodine in sea plants, uranium in some forms of decaying organic matter), and of the organic constituents of fossils.

bioinformatics.   The science of managing and analyzing biological data using advanced computing techniques. Especially important in analyzing genomic research data.
See informatics.

bioinorganic chemistry.   Study of the mechanisms involved in the behavior of metal-containing molecules in living organisms, e.g., biological transport of iron, the effect of copper on nucleic acid and nucleoproteins, molybdenum, and manganese complexes, etc.

bioisotere.   (nonclassical isostere). A compound resulting from the exchange of an atom or a group of atoms with another, broadly similar, atom or group of atoms.

biolarvicide. An endotoxin of *Bacillus thuringiensis israelensis* that is toxic to mosquito larvae.

Biologically activated Carbon. supports active microbial growth, in order to aid in the degradation of organics that have been absorbed on its surface and in its pores.

biological availability. Blood level or similar tests that establish a significant concentration of a drug in the bloodstream or other body systems where its presence is understood to be effective.

biological molecule. Any molecule produced by living organisms even though alternate pathways of production are known.

biological pesticide. Any microorganism that is effective in controlling target pests.

biological pigment. Any pigment that occurs in living organisms that play diverse roles in life processes.

biological rhythm. Any of numerous periodic phenomena associated with living organisms that are partly endogenous, requiring only an external signal to bring the cycle into phase with environmental cycles.

biologicals. Medical products produced from living organisms or their products. These include antigens, antitoxins, serums, and vaccines.

biological stain. A dye for determining microscopic structure of cells and tissues. The usefulness depends on selective adsorption of stain by bacteria or tissue.

biological warfare agent. Any of a large number of biological entities such as microbial toxins and viral, bacterial, and fungal pathogens, that harm or kill humans or plants and other animals on which humans use in the conduct of warfare.
Hazard: Death; toxins; food poisoning; plant or animal diseases, foot-and-mouth disease; fowl plague; black stemrust of cereal grasses; potato blight; rice blast disease.

biological weapon. Any potentially injurious or lethal biological weapon developed for or suitable for application in biological warfare.

bioluminescence. (cold light).
See chemiluminescence.

biomarker. Specific biochemical compounds that are detected within the body, and which have a particular molecular feature that makes it useful for measuring a specific process such as flux through a pathway, the progress of a disease, or the effects of treatment of a disease.

biomass. Any organic source of energy or chemicals that is renewable. Its major components are (1) trees (wood) and all other vegetation; (2) agricultural products and wastes (corn, fruit, garbage ensilage, etc.); (3) algae and other marine plants; (4) metabolic wastes (manure, sewage); and (5) cellulosic urban waste. Conversion of these is performed in several ways: (1) by combustion (heat); (2) by fermentation (alcohol); (3) by gasification (synthesis gas); and (4) by anaerobic digestion (methane).
In terms of energy, wood is by far the most important component of biomass. It has become a significant source of industrial heat, e.g., in paper mills and power plants, and intensive cultivation of trees for this purpose is under way. Wood is also a potential source of alcohols; ethyl alcohol is produced from wood on large scale in Brazil as a gasoline substitute. Agricultural wastes are fermented or gasified to synthesis gas; manures and municipal waste yield methane (biogas) on digestion. In 1981, biomass supplied 3.5% of U.S. energy requirements, and this is expected to increase substantially.

biomaterial. Any material suitable for use as a surgical implant within the body to replace or support joints or tissues. Included are such metals as aluminum, stainless steels, titanium, various forms of carbon, and especially plastics (polycarbonate, polyurethane, nylon, silicones). They have been used successfully in many areas of the body, from hip and knee replacements to mastectomies. They must be compatible with the interior environment, noncorrosive, and nondegradable, and must duplicate as closely as possible the properties of the tissues they replace. A notable breakthrough was made in this field in 1982 when a complete artificial heart was implanted successfully in a living human. Its chief component was polyurethane, the base was aluminum, and the valves were pyrolytic graphite together with polycarbonate and titanium. Other materials under consideration are styrene-butadiene copolymers for the diaphragms. The polyurethane used in the first implanted heart is Biomer.

Biomer. A unique type of polyurethane used in heart implants.

bioMeT 12. A rodent-repellent coating for cables in which the active ingredient is a mixture of tributyl tin salts. It has proved 95% effective in preventing destruction of telephone cables by rats and other rodents. Flexible, transparent, and effective for 6 months or more, it is applied mechanically over the plastic cable sheathing. Can also be used to protect other types of wiring, shipping containers, and similar products.

biomethylated arsenic. Methanearsonic acid and cacodylic acid compounds.
Derivation: Arsenic in plaster and wallpaper that was converted under humid conditions.
Hazard: Arsenic poison.

biomimetic chemistry. An interdisciplinary approach to biochemistry including both organic and inorganic aspects of this field. The term means imitation or mimicry of natural organic processes in living systems, and encompasses such subjects as enzyme systems, vitamin B_{12} and flavins, oxygen binding and activation, bioorganic mechanisms, and nitrogen and small-molecule fixation. The technique was utilized in the synthesis of the bleomycin molecule. A notable example of biomimetic chemistry is the development of model synthetic catalysts that imitate the action of natural enzymes. The behavior of chymotrypsin has been duplicated by a manufactured catalyst that can accelerate certain reaction rates by the incredible factor of 100 billion.

biomineral. A mineral made by biological organisms, characterized by intricate composite microarchitectures.

biophyl. A highly refined form of verxite, consisting of more than 99% pure hydrobiotite (magnesium, iron, aluminum silicate).
Use: In foods and pharmaceuticals.

biopolymer. A water-soluble polymer resulting from the action of bacteria (genus *Xanthomonas*) on carbohydrates. The viscosity is almost as low as that of water. Chromium ion can be added to increase viscosity if desired. Such polymers are being used as viscosity builders in oil-well drilling muds and as thickeners and gel-strength additives in aqueous media.
See fermentation.

biopterin. An enzymatic cofactor derived from pterin and involved in certain oxidation-reduction reactions.

Biore. A marshmallow whip facial foam.

bioremediation. The use of biological organisms such as plants or microbes to aid in removing hazardous substances from an area.

bioresmethrin. $C_{22}H_{26}O_3$. A synthetic insecticide of the pyrethrin type. Biodegradable and has low toxicity, it is nonpersistent and can act as a synergist.

biorex.
CAS: 205943-18-6. $C_{22}H_{19}Cl_2NO_3 \cdot C_{19}H_{30}O_5 \cdot C_8H_{10} \cdot C_5H_9NO \cdot C_4H_6O_3$.
Hazard: A poison by ingestion.

biostat. Any product that regulates growth of microorganisms.

biosynthesis. (1) Natural synthesis of organic compounds by plants and animals. For plants, this includes not only photosynthetic formation of carbohydrates and protein synthesis by means

of nitrogen-fixing bacteria, but also a wide range of specialized organic compounds that are specific to individual species; many of these are poisonous. Animals and humans synthesize certain amino acids, hormones, cholesterol, etc. Some types (snakes, fish, toads, etc.) synthesize unique and powerful toxic principles.
(2) This term is also applied to such research achievements as synthesis of edible single-cell proteins by fermentation and gene-splicing techniques.
See protein, single-cell; genetic engineering; biotechnology.

biota. A collective term for all the animals and plants of an ecosystem.

biotechnology. A definition prepared by a committee of British scientists in a report issued by the Organization for Economic Cooperation and Development (Paris) may be considered official and definitive. It states that biotechnology is "application of scientific and engineering principles to the processing of any organic or inorganic substance by biological agents to provide goods and services; the biological agents include a wide range of biological catalysts, particularly microorganisms, enzymes, and animal and plant cells." This involves commercial production of chemical compounds from either (1) renewable resources (biomass) or (2) nucleic acids (DNA). Examples of (1) are production of antibiotics, alcohols, and single-cell proteins by fermentation, and of (2) production of insulin, interferon, and synthetic bacteria by gene splicing. Biotechnology constitutes a major worldwide technological revolution, and may prove to be the most important development in the chemical industries since the plastics explosion of the 1930s.

biotin. (vitamin H; 2'-keto-3,4-imidazolido-2-tetrahydrothiophene-*n*-valeric acid). $C_{10}H_{16}N_2O_3S$. Biotin, frequently referred to as a member of the vitamin B complex, is necessary for the maintenance of health in animals and for growth of many microorganisms.

It influences fat metabolism, decarboxylation and carbon dioxide fixation, and deamination of some amino acids. It is closely related metabolically to pantothenic acid and folic acid. A biotin deficiency may be induced by ingestion of avidin, a raw-egg protein, because of the formation of

a nonabsorbable biotin–avidin complex. Biotin is synthesized in the intestinal tract of humans; therefore, normally it is not essential in the diet.
Properties: White crystals. Mp 230–232C. Soluble in water and alcohol; insoluble in naphtha and chloroform. Stable to heat, stable in neutral or acid solution, destroyed by strong alkali or oxidizing agents. Amounts are expressed in milligrams or micrograms of biotin.
Source: Egg yolk, kidney, liver, yeast, milk, molasses.
Grade: Practical, FCC.
Use: Medicine, nutrition.

biotinidase. An enzyme that catalyzes the hydrolysis of biotin amide, biocytin, and other biotinides to biotin.

***N*-biotinyl-*l*-lysine.** See biocytin.

biotite. A component of igneous rocks and of soil, similar to mica. It is a silicate of magnesium, iron, potassium, and aluminum.

Biozan. Welan gum.

biphenol AF. See hexafluoroacetone bisphenol a.

biphenyl. See diphenyl.

***o*-biphenylamine.** See *o*-aminobiphenyl.

***o*-biphenyl biguanide.**
$NH_2(CNHNH)_2C_6H_4C_6H_5 \cdot H_2O$.
Properties: White to faintly pink powder. Mp greater than 150C on dried material, ash less than 0.5%. Soluble in alcohol, "Carbitol" *[Union]*, and "Cellosolve" *[Union]*; very slightly soluble in water.
Use: Soap, antioxidant.

2,4′-biphenyldiamine. See 2,4′-diphenyldiamine.

2,4-biphenyldiol.
CAS: 134-52-1. $C_{12}H_{10}O_2$.
Hazard: A poison by skin contact.

2-(4-biphenylyl)-5,6-dihydro-*s*-traizolo(5,1-a)isoquinoline.
CAS: 75318-64-8. $C_{22}H_{17}N_3$.
Hazard: Moderately toxic. A reproductive hazard.

(4,4′-biphenylylenebis(2-oxoethylene))bis(3-iodopyridinium) dibromide.
CAS: 63906-07-0. $C_{26}H_{20}I_2N_2O_2 \cdot 2Br$.
Hazard: A poison.

4-(2-(1,1′-biphenyl)-4-ylethoxy)quinazoline.
CAS: 124428-11-1. $C_{22}H_{18}N_2O$.

Hazard: Moderately toxic by ingestion. A reproductive hazard.

(2-biphenyloxy)tributyltin.
CAS: 3644-37-9. $C_{24}H_{37}OSn$.
Hazard: A poison.

α-(4-biphenylyloxy)propionic acid.
CAS: 5555-13-5. $C_{15}H_{14}O_3$.
Hazard: A poison.

((2-biphenylyloxy)tributyl)stannane. See (2-biphenyloxy)tributyltin.

2-(4-biphenylyl)-5h-*s*-triazolo(5,1-a)-isoindole.
CAS: 75318-65-9. $C_{21}H_{15}N_3$.
Hazard: Moderately toxic. A reproductive hazard.

biphosphoglycerate. (2,3-biphosphoglycerate; 2,3-dpg). An important regulator of the affinity of hemoglobin for oxygen.

Birch reduction. Reduction of aromatic rings by means of alkali metals in liquid ammonia to give mainly unconjugated dihydro derivatives.

birefringent. Descriptive of a type of crystal that separates an impinging light ray into two components that are polarized at right angles to each other; as a result, two images appear, each of which is caused by a light ray vibrating in only one direction (plane-polarized light). Such anisotropic crystals (e.g., Iceland spar) are used in nicol prisms. See nicol; anisotropic.

birth defect. Any harmful trait, physical or biochemical, present at birth, whether a result of a genetic mutation or of some other nongenetic factor.
See congenital; gene; mutation; syndrome.

bis-. Prefix meaning "twice" or "again." Used in chemical nomenclature to indicate that a chemical grouping or radical occurs twice in a molecule, e.g., bisphenol A, where two phenolic groups appear: $(CH_3)_2C(C_6H_5OH)_2$.

***N,N-bisacetoxethylaniline*.**
Grade: Brown liquid.
Hazard: Combustible.
Use: Coupling agent for disperse dyes for synthetic fibers.

bis(acetoxydibutylstannane) oxide.
CAS: 5967-09-9. $C_{20}H_{42}O_5Sn_2$.
Hazard: A poison.

2,2-bis(acetoxymethyl)propyl acetate.
$CH_3C(CH_2OOCCH_3)_2CH_2OOCCH_3$.
Properties: Colorless liquid. Refr index 1.4359 (20C).
Use: As a plasticizer.

2,3-bis(acetoxymethyl)quinoxaline di-*n*-oxide.   See quinoxidine.

1,5-bis(4-aldoximinopyridinium)diethylether bibromide.   See 1,1'-(oxydiethylene)bis(4-formylpyridinium bromide), dioxime.

bisamides.   General   formula   RCONHR'NHCOR.
Properties: When R and R' have high molecular weight, they are hard, light-colored waxes.

bis(2-aminoethyl)sulfide. $(H_2NCH_2CH_2)_2S$. An ethyleneimine derivative.
Properties: Colorless liquid. Fp 2.6C, bp 238C, d 8.7 lb/gal, refr index 1.5277, flash p 246F (118.9C). Very soluble in water, benzene, and ethanol. Combustible.
Use: See ethyleneimine.

bis(4-amino-3-methylcyclohexyl)methane.
CAS: 6864-37-5. $C_{15}H_{30}N_2$.
Hazard: Moderately toxic by inhalation.

1,4-bis(3-aminopropoxy)butane.
CAS: 7300-34-7. $C_{12}H_{24}N_2O_2$.
Hazard: A poison by skin contact. Moderately toxic by ingestion and inhalation.

N,N-bis(3-aminopropyl)methylamine.
$CH_3N(C_3H_6NH_2)_2$.
Properties: Liquid. D 0.9307 (20/20C), bp 240C, fp −29.6C, flash p 220F (104.4C). Miscible with water. Combustible.
Hazard: Irritant.
Use: Chemical intermediate.

bis(3-aminopropyl)methylamine-epichlorohydrin copolymer.   See epichlorohydrin-bis(3-aminopropyl)methylamine copolymer.

bisazo compound.   Any compound that contains two azo groups.
Use: Dyes.

1,3-bis(2-benzothiazolylmercaptomethyl)-urea. $(C_6H_4NCS•SCH_2NH)_2CO$.
Properties: Buff to light tan powder. Mp 220C, d 1.38 (25C).
Use: Rubber accelerator.

***p*-bis[2-(5-*p*-biphenylyloxazoyl)]benzene.**
(BOPOB). $C_{36}H_{25}O_2N_2$.
Properties: Shiny, yellow flakes. Mp 327–328C, fluorescence peak 4400 Å. Sparingly soluble in toluene.
Grade: Purified.
Use: Scintillation counter, wavelength shifter in liquid scintillators.

2,2-bis(*p*-bromophenyl)-1,1,1-trichloroethane. $C_{14}H_9Br_2Cl_3$. The bromine analog of DDT.
Hazard: Toxic by ingestion.
Use: Insecticide.

bis(butoxymaleoyloxy)dibutylstannane.
CAS: 15546-16-4. $C_{24}H_{40}O_8Sn$.
Hazard: A poison by ingestion.

bis(butoxymaleoyloxy)dioctylstannane.
CAS: 29575-02-8. $C_{32}H_{56}O_8Sn$.
Hazard: Moderately toxic by ingestion.

bisaniline-p.
CAS: 2616-10-1. $C_{24}H_{28}N_2$.
Hazard: Low toxicity by ingestion and skin contact.

***cis*-bisascorbato(racemic-1,2-diaminocyclohexane)platinum(II) hydrate.**
CAS: 92784-30-0. $C_{18}H_{28}N_2O_{12}Pt$.
Hazard: A poison.

bis(*tert*-butylperoxy)-2,5-dimethylhexane.
$C_{16}H_{34}O_4$. A cross-linking agent for polymers.

bis(butylthio)dimethyl stannane.   See bis(butylthio)dimethyltin.

bis(butylthio)dimethyltin.
CAS: 1000-40-4. $C_{10}H_{24}S_2Sn$.
Hazard: A poison.

2,3-bis(carbomethoxymercapto)quinoxaline.
CAS: 58705-49-0. $C_{12}H_{10}N_2O_4S_2$.
Hazard: A poison.

Bischler–Mohlau indole synthesis.   Formation of 2-substituted indoles by heating *o*-halogeno- or *o*-hydroxy-ketones with excess aniline via cyclization of the intermediate 2-arylaminoketone.

Bischler–Napieralski   reaction.   Cyclodehydration of β-phenethylamides to 3,4-dihydroisoquinoline derivatives by means of condensing agents such as phosphorus pentoxide or zinc chloride.

bis(2-chloroethoxy)methane.   See dichloroethyl formal.

4-(((*p*-(bis(2-chloroethyl)amino)phenyl)-imino)methyl)-5-hydroxy-6-methyl-3-pyridinemethanol.
CAS: 79967-32-1. $C_{18}H_{21}Cl_2N_3O_2$.
Hazard: A poison.

5-[bis(2-chloroethyl)amino]uracil.   (uracil mustard).

┌─────────────────────────┐
CONHCONHCHCN(C$_2$H$_4$Cl)$_2$.
└─────────────────────────┘

Properties: A cream-white, odorless, crystalline compound. Moderately soluble in methanol and acetone.
Use: In medicine.

bis(2-chloroethyl) (2-chloroethyl)-phosphonate.
CAS: 6294-34-4. C$_6$H$_{12}$Cl$_3$O$_3$P.
Hazard: Moderately toxic by ingestion. Low toxicity by skin contact. Experimental reproductive effects. A mild skin and eye irritant.
Hazard: A reproductive hazard.

9-(2,2-bis(2-chloroethyl)hydrazino)acridine monohydrochloride.
CAS: 29023-83-4. C$_{17}$H$_{17}$Cl$_2$N$_3$•ClH.
Hazard: A poison by ingestion.
Use: Agricultural chemical.

bis(2-chloroethyl) phthalate.
CAS: 6279-87-4. C$_{12}$H$_{12}$Cl$_2$O$_4$.
Hazard: Moderately toxic by ingestion and skin contact. A mild skin irritant.

bis(2-chloroethylsulfonylmethyl)ether.
See 1,1'-(oxybis(methylenesulfonyl))bis(2-chloroethane).

1,1'-bischloromercuriferrocene. [1,1-di(chloromercuri)ferrocene]. (ClHgC$_5$H$_4$)$_2$Fe.
Use: Inorganic polymers.

bis-1,2-(chloromethoxy)ethane.
CAS: 13483-18-6. C$_4$H$_8$Cl$_2$O$_2$.
Properties: Viscous liquid. Bp 99–100C @ 22 mm, d 1.2879 @ 14–15C.
Hazard: Questionable carcinogen.

bis(chloromethyl)ether. (dichloromethyl ether).
CAS: 542-88-1. (CH$_2$Cl)O(CH$_2$Cl).
Properties: Reported to form spontaneously from formaldehyde and chloride ions in moist air.
Hazard: A carcinogen. Toxic by ingestion.
Use: Intermediate for ion-exchange resins; laboratory reagent.

3,3-bis(chloromethyl)oxetane. See Penton.

bis(p-chlorophenoxy)methane.
(ClC$_6$H$_4$O)$_2$CH$_2$.
Properties: Solid. Mp 65C. Insoluble in water and oils; soluble in ether and acetone.
Use: Acaricide.

2,7-bis(4-chlorophenyl)benzo(lmn)(3,8)-phenanthroline-1,3,6,8(2h,7h)-tetrone.
CAS: 64005-91-0. C$_{26}$H$_{12}$Cl$_2$N$_2$O$_4$.
Hazard: A poison.

2,2-bis(p-chlorophenyl)-1,1-dichloroethane.
See TDE.

1,1-bis(p-chlorophenyl)ethanol. See di(p-chlorophenyl)ethanol.

bis(p-chlorophenylthio)dimethyl stannane.
See bis(p-chlorophenylthio)dimethyltin.

bis(p-chlorophenylthio)dimethyltin.
CAS: 55216-04-1. C$_{14}$H$_{14}$Cl$_2$S$_2$Sn.
Hazard: A poison.

1,1'-bis(p-chlorophenyl)-2,2,2-trichlor-ethanol. (4,4'-dichloro-α-trichloromethyl-benzhydrol; Kelthane). CCl$_3$C(C$_6$H$_4$Cl)$_2$OH. An alcohol analog of DDT.
Hazard: Toxic by inhalation and ingestion.
Use: Miticide.

bis(cumene)chromium. See dicumene chromium.

n,n-bis(2-cyanoethyl)-n-4-hydroxy-1-anthraquinonylsulfonilamide.
CAS: 66903-22-8. C$_{26}$H$_{20}$N$_4$O$_5$S.
Hazard: A poison.

n,n'-bis-(1-(2-cyanoethyl)thioacetaldehyde o-(n-methylcarbamoyl)oxime)disulfide.
CAS: 68789-93-5. C$_{14}$H$_{20}$N$_6$O$_4$S$_4$.
Hazard: A poison by ingestion.
Use: Agricultural chemical.

n,n'-bis(2-cyano-2-methylpropionaldehydeo-(n-methylcarbamoyl)oxime)sulfide.
CAS: 63942-43-8. C$_{14}$H$_{20}$N$_6$O$_4$S.
Hazard: A poison by ingestion.
Use: Agricultural chemical.

bis(cyclohexyl)carbodiimid. See n,n'-methanetetrayl biscyclohexanamine.

bis-cyclopentadienyliron. See dicyclopentadienyliron.

bis(l-cysteinato)mercury.
CAS: 12550-82-2. C$_6$H$_{12}$HgN$_2$O$_4$S$_2$.
Hazard: Moderately toxic by ingestion.

bis(decanoyloxy)di-n-butylstannane.
CAS: 3465-75-6. C$_{28}$H$_{56}$O$_4$Sn.
Hazard: A poison by ingestion. A severe skin and eye irritant.

bis(decanoyloxy)di-n-butyltin. See bis(decanoyloxy)di-n-butylstannane.

bis(dibutylacetoxytin)oxide. See bis(acetoxydibutylstannane) oxide.

bis(dibutyldithiocarbamato)copper.   See copper bis(dibutyldithiocarbamate).

bis(dibutyldithiocarbamato)dibenzylstannane.
CAS: 64653-03-8. $C_{32}H_{50}N_2S_4Sn$.
Hazard: A poison.

bis((dibutyldithiocarbamoyl)oxy)-dibenzylstannane.   See bis(dibutyldithiocarbamato)dibenzylstannane.

bis(dibutyldithiocarbamato)-dimethylstannane.
CAS: 66009-08-3. $C_{20}H_{42}N_2S_4Sn$.
Hazard: A poison.

bis((dibutyldithiocarbamoyl)oxy)-dimethylstannane.   See bis(dibutyldithiocarbamato)dimethylstannane.

2,7-bis(3,4-dichlorophenyl)benzo(lmn)(3,8)-phenanthroline-1,3,6,8(2h,7h)-tetrone.
CAS: 222420-34-0. $C_{26}H_{10}Cl_4N_2O_4$.
Hazard: A poison.

bis(2,2-dichlorovinyl) sulfoxide.   See 1,1'-sulfinylbis(1,2-dichloroethane).

bis(diethyldithiocarbamato)copper.   See copper(II) diethyldithiocarbamate.

bis(1,3-dithiocyanato-1,1,3,3-tetrabutyldistannoxane).
CAS: 38998-91-3. $C_{36}H_{72}N_4O_2S_4Sn_4$.
Hazard: A poison.

bis(dodecanoloxy)dioctylstannane.   See dioctyldi(lauroyloxy)stannane.

bis((2-(ethyl)hexyloxy)maleoyloxy)   di(*n*-butyl)stannane.
CAS: 15546-12-0. $C_{32}H_{56}O_8Sn$.
Hazard: A poison by ingestion. A skin and eye irritant.

bis(2,6-diethylphenyl)carbodiimide.
$(CH_2H_5)_2C_6H_3NCNC_6H_3(C_2H_5)_2$.
Properties: Light-yellow to red-brown liquid; faintly acrid odor. D 1.007 (20/20C), bp 192–194C (0.4 mm Hg), refr index 1.591 (23C). Soluble in organic solvents. Combustible.
Hazard: Toxic by inhalation. Damaging to eyes.
Use: Stabilizers in polyester and urethane systems, intermediate for textile chemicals and pharmaceuticals.

bis(2,6-diisopropylphenyl)carbodiimide.
CAS: 2162-74-5. $C_{25}H_{34}N_2$.
Hazard: A poison by ingestion.

2,6-bis(dimethylaminomethyl)cyclohexanone.
$[(CH_3)_2NCH_2]_2C_6H_8O$.

Properties: D 0.95 (20C).
Use: Preservative for aqueous paint systems, casein, pigment dispersions, and adhesives.

2,6-bis(dimethylaminomethyl)cyclohexanone dihydrochloride. $C_{12}H_{24}N_2O•2HCl$.
Properties: Free-flowing, white to off-white crystalline salt.
Use: Preservative for aqueous systems, latex paints, adhesives coatings, wax emulsions, casein, and starch solution.

bis(4-dimethylamino)triphenylcarbinol.
$C_6H_5C(OH)[C_6H_4N(CH_3)_2]_2$.
Properties: Solid. Mp 121–123C. Very soluble in ether and hot benzene; soluble in acids.
Use: Dyestuffs.
See malachite green.

bis(1,3-dimethylbutyl)amine.
$[(CH_3)_2CH_2CH_2CH(CH_3)]_2NH$.
Properties: Liquid. D 0.772–0.778 (20/20C), distillation range 179–205C, bulk d 6.5 lb/gal, flash p 160F (71.1C). Combustible.

((3,5-bis(1,1-dimethylethyl)-4-hydroxyphenyl)methyl)phosphonic   acid, monoethyl ester, nickel(2+) salt (2:1).
CAS: 30947-30-9. $C_{34}H_{56}O_8P_2•Ni$.
Hazard: Confirmed human carcinogen. Moderately toxic by ingestion.

2,4-bis(1,1-dimethylethyl)-6-(1-(4-methoxyphenyl)ethyl)phenol.
CAS: 71712-04-4. $C_{23}H_{32}O_2$.
Hazard: Moderately toxic by ingestion.
Use: Agricultural chemical.

bis((4-(1,1-dimethylethyl)phenyl)methyl)   3-pyridinylcarbonimidodithioate.
CAS: 51308-76-0. $C_{28}H_{34}N_2S_2$.
Hazard: Moderately toxic by ingestion.
Use: Agricultural chemical.

bis((5,5-dimethyl-2-isopropylimino-4-(*o*-(*n*-methylcarbamoyl)oximino)-1,3-dithiolane))sulfide.
CAS: 71108-02-6. $C_{20}H_{32}N_6O_4S_5$.
Hazard: A poison by ingestion.
Use: Agricultural chemical.

1,1'-bis(dimethyloctoxysilyl)ferrocene.
CAS: 32613-12-0. $C_{30}H_{54}FeO_2Si_2$.
Hazard: A poison by inhalation. Low toxicity by ingestion.

***N,N'*-bis(1,4-dimethylpentyl)-*p*-phenylenediamine.**   (diheptyl-*p*-phenylenediamine). $C_6H_4(CNH_7H_{15})_2$.
Properties: Amber to red liquid. D 0.90, fp 7.2C. Combustible.
Use: Gasoline antioxidant and sweetener.

bis(dimethyl(vinyl)silyl)amine.   See   1,1,3,3-tetramethyl-1,3-divinyldisilazane.

n,n'-bis-(1,4-dithiane-2-o-(n-methylcarbamoyl)oximino)sulfide.
CAS: 63956-71-8. $C_{12}H_{18}N_4O_4S_5$.
Hazard: Moderately toxic by ingestion.
Use: Agricultural chemical.

bis(ethane-1,2-diamine)copper(2+)   **diperchlorate.**
CAS: 36407-48-4. $C_4H_{16}CuN_4$•$2ClO_4$.
Hazard: Moderately toxic by ingestion and inhalation.

bis(ethenyldimethylsilyl)   **ether.**   See   *sym*-tetramethyldivinyldisiloxane.

bis(2-ethylhexyloxycarbonylmethylthio)-dibutylstannane.   See   bis(isooctyloxycarbonylmethylthio)dibutyl   stannane.

bis(2-ethylhexyloxycarbonylmethylthio)-dimethylstannane.   See   bis(isooctyloxycarbonylmethylthio)dimethylstannane.

bis(2-ethylhexyl)phthalate. $C_{24}H_{38}O_4$.
Properties: A liquid.
Use: In vacuum pumps.

bis(2-ethylhexyl)thiodipropionate.
CAS: 10526-15-5. $C_{22}H_{42}O_4S$.
Hazard: Moderately toxic by ingestion. A mild skin and eye irritant.

bis(2-ethylhexylthioglycolate)dibutyltic.
See   dibutyldi(2-ethylhexyloxycarbonylmethylthio)-stannane.

bis(2-ethylhexylthioglycolate)dioctyltin.
See di-n-octyltin bis(2-ethylhexyl) mercaptoacetate.

N,N-bis(1-ethyl-3-methylpentyl)-p-phenylenediamine. $C_6H_4(NHC_8H_{17})_2$.
Properties: Dark, reddish-brown liquid. D approximately 0.90; bulk d 7.5 lb/gal. Combustible.
Use: Antioxidant for polyunsaturated elastomers.

2,2-bis(p-ethylphenyl)-1,1-dichloroethane.
See 1,1-dichloro-2,2-bis(p-ethylphenyl)ethane.

bis(ethylthio)methylene malononitrile.
CAS: 18771-38-5. $C_8H_{10}N_2S_2$.
Hazard: A poison by ingestion and skin contact. A severe eye irritant.

bisethylxanthogen. $(C_2H_5OCSS)_2$.
Properties: Yellow needles; onionlike odor. Mp 28–32C. Insoluble in water; freely soluble in benzene, ether, petroleum fractions.
Grade: 58% soluble in oil.
Use: Weed control, rubber accelerator, fungicide.

bis(2-fluoro-2,2-dinitroethoxy)methane.
See formaldehyde bis(2-fluoro-2,2-dinitroethyl) acetal.

1-(2-(bis(4-fluorophenyl)methoxy)ethyl)-4-(3-phenylpropyl)piperazine   **dihydrochloride.**
CAS: 67469-78-7. $C_{28}H_{32}F_2N_2O$•$2ClH$.
Hazard: A poison.

bisfuranoid mycotoxin.   Any mycotoxin that contains the dihydrofurobenzofuran system.
Properties: Include either 7,8-di-hydrofurano-(2,3-b)furan or 2,3,7,8-tetra-hydro(2,3-b)furan.
Hazard: Hepatotoxic; tumorigen.

1,3-bis(3-glycidoxypropyl)tetramethyldisiloxane. $[OCH_2CHCH_2O(CH_2)_3Si(CH_3)_2]_2O$.
Properties: Liquid. D 0.99 (25C), refr index 1.4500 (25C), bp approximately 185C (2 mm Hg). Soluble in acetone and benzene; insoluble in water. Flash p 300F (149C). Combustible.
Use: Chemical intermediate.

bis(guanidinium) chromate.
CAS: 5188-42-1. $C_2H_{10}N_6$•CrH_2O_4.
Hazard: Poison. Confirmed carcinogen

bis(hexanoyloxy)di-n-butylstannane.
CAS: 19704-60-0. $C_{20}H_{40}O_4Sn$.
Hazard: A poison by ingestion. A skin and eye irritant.

bis(hexanoyloxy)di-n-butyl-tin.   See bis(hexanoyloxy)di-n-butylstannane.

bis(hydrogen   **maleato)dibutyl-tin**   **bis(2-ethylhexyl) ester.**
See   bis((2-(ethyl)hexyloxy)maleoyloxy)   di(n-butyl)stannane.

bis(hydrogen   **maleato)dioctyltin**   **bis(2-ethylhexyl) ester.**   See di-n-octyltin bis(2-ethylhexyl maleate).

bishydroxycoumarin.   [3,3'-methylenebis(4-hydroxycoumarin)dicoumarol]. $C_{19}H_{12}O_6$.
Properties: White or creamy-white crystalline powder; faint pleasant odor; slightly bitter taste. Mp 287–293C. Readily soluble in solutions of fixed alkali hydroxides; slightly soluble in chloroform; almost insoluble in water, alcohol, and ether.
Derivation: (1) Originally extracted from spoiled sweet clover; (2) synthetically from methyl acetylsalicylate sodium and formaldehyde.

Grade: USP.
Hazard: May cause hemorrhage.
Use: Anticoagulant for blood.

bis(1-hydroxycyclohexyl)peroxide.
  $C_6H_{10}(OH)_2O_2$.
  Properties: Fine, white powder. Mp 66–68C. Active oxygen 6.6% min.
  Hazard: Dangerous fire risk in contact with organic materials. Strong oxidizing agent.
  Use: Catalyst for polymerization of polyester resins.

N,N-bis(2-hydroxyethyl)alkylamine.   Clear liquid used as an antistatic for blow molding applications for polyolefins. Approved for use in food packaging films.

bis(hydroxyethyl)butynediol   ether.
  $HO(CH_2)_2OCH_2C•CCH_2C•CCH_2O(CH_2)_2OH$.
  Properties: Dark brown liquid. D 1.136 (25/15C), solidifies below −15C, distillation range 116–235C (10 mm Hg).
  Hazard: May explode under alkaline conditions at high temperature.
  Use: Intermediate for polyesters, plasticizers, and plastics, nickel brightener in electroplating, corrosive inhibitor, pickling inhibitor prior to copper plating.

n,n-bis(2-hydroxyethyl)cocoamide.   See coconut oil acid diethanolamine.

bis(hydroxyethyl) cocoamine oxide.   A derivative coconut oil claimed to be useful as a gasoline additive to inhibit rust formation and icing of carburetors. Also used as foaming agent in shampoos, detergents, tooth pastes, and the like.

N,N-bis(hydroxyethyl)oleamide.
  $CH_3(CH_2)_7HC:CH(CH_2)_7CON(CH_2CH_2OH)_2$.
  A technical grade containing 25% excess amine. Light amber liquid with faint odor.
  Use: As a surface-active agent.

bis[hydroxyethylpoly(ethyleneoxy)ethylpropyleneglycol.   See poloxalene; poloxalene free-choice liquid type c feed.

β-bishydroxyethyl sulfide.   See thiodiglycol.

1,7-bis(hydroxymethyl)carborane.   See *m*-carboranedimethanol.

bis(hydroxymethyl)-*m*-carborane.   See *m*-carboranedimethanol.

bis(hydroxymethyl)-*o*-carborane.   See *o*-carboranedimethanol.

1,3-bishydroxymethylurea.   See dimethylolurea.

bis(4-hydroxyphenyl)methanone (2,4-dinitrophenyl)hydrazone.
  CAS: 2675-35-6. $C_{19}H_{14}N_4O$.
  Hazard: Moderately toxic by ingestion.

4,4-bis(4-hydroxyphenyl)pentanoic acid.
  (diphenolic acid; DPA).
  $CH_3C(C_6H_4OH)_2CH_2CH_2COOH$.
  Properties: Light-tan granules. Mp 170–173C, d 1.30–1.32. Soluble in acetic acid, acetone, and ethanol; insoluble in benzene, carbon tetrachloride, and xylene. Slightly soluble in water.
  Use: Paint formulations, coatings, and finishes.

bishydroxyphenyl sulfone.   See dihydroxydiphenyl sulfone.

1,4-bis(2-hydroxypropyl)-2-methylpiperazine. $C_{11}H_{24}O_2N_2$.
  Properties: Liquid; odorfree. Bp 145C (3 mm Hg), d 1.0013 (25/25C), refr index 1.4803 (20C), flash p 300F (149C). Miscible with water. Combustible.
  Use: Catalyst, chemical intermediate.

bis-intercalator.   A unique type of natural antibiotics that have antitumor and antimicrobial properties. They function by interposing two symmetrical groups between the nucleotide bases of the DNA molecule. Some are quite toxic and may be mutagenic.
  See carzinophillin A.

bis(isooctyloxycarbonylmethylthio)dibutyl stannane.
  CAS: 25168-24-5. $C_{28}H_{56}O_4S_2Sn$.
  Hazard: Moderately toxic by ingestion.

bis(isooctyloxycarbonylmethylthio)-dimethylstannane.
  CAS: 26636-01-1. $C_{22}H_{44}O_4S_2Sn$.
  Hazard: Moderately toxic by ingestion.

bis(isooctyloxymaleoyloxy)dioctylstannane.
  CAS: 33568-99-9. $C_{40}H_{72}O_8Sn$.
  Hazard: Moderately toxic by ingestion.

bis(isopropylbenzene)chromium.   See dicumene chromium.

2,4-bis(isopropylamino)-6-methoxy-*s*-triazine.   (Gesafram;   ontrach;   promitol; prometon).
  CAS: 1610-18-0. $C_{10}H_{19}N_5O$.
  Properties: White, extruded pellets.
  Hazard: Toxic by ingestion.
  Use: Industrial herbicide.

bis-keto-triazine.   (Permafresh 110). Water-white liquid, 40% active. Can be cured with magnesium chloride to provide chlorine resistance; compatible with optical brighteners. Used as a wash-and-wear finish for cellulosics and blends of cellulosics.

bis(lauroyloxycarbonylmethylthio)-dioctylstannane.   See di-*n*-octyltin bis(laurylthioglycolate).

bismanol.   (MnBi). An alloy or compound of bismuth and manganese which has an exceptionally high coercive force (3400 oersteds) and a high energy product. Produced by U.S. Naval Ordnance Laboratory by powder metallurgy methods and used as a permanent magnet.

Bismarck brown.   See C.I. basic brown 4.

Bismarck Brown Y.   (toluene-2,4-diazo-bis-*m*-toluylenediamine hydrochloride; CI 21010). $CH_3C_6H_3[NNC_6H_2(CH_3)(NH_2)_2]_2 \cdot 2HCl$.
Properties: Dark brown powder. Soluble in water and alcohol.
Derivation: Action of nitrous acid on toluylene diamine.
Use: Dye for wool and leather, biological stain.

Bismarck Brown Y.   (4,4'-[*m*-phenylenebis(azo)]bis(*m*-phenylenediamine)dihydrochloride; CI 21000). C_6H_4-1,3-$[NNC_6H_3$-2,4-$(NH_2)_2 \cdot HCl]_2$.
Properties: Black to brown powder. Soluble in water; insoluble in benzene and carbon tetrachloride.
Use: Dyeing textiles, biological stain.

bis(2-methoxyethoxy)ethyl   ether.   See dimethoxytetraglycol.

2,12-bis(1-(methoxyimino)ethyl)-5,9-dioxo-n,n,n',n',6,8-hexamethyl-4,10-dioxa-7-thia-3,6,8,11-tetraazatrideca-2,11-dienediamide.
CAS: 90293-56-4. $C_{18}H_{30}N_8O_8S$.
Hazard: A poison by ingestion.
Use: Agricultural chemical.

bis(methoxymaleoyloxy)dibutylstannane.
CAS: 15546-11-9. $C_{18}H_{28}O_8Sn$.
Hazard: A poison by ingestion.

bis(methoxymaleoyloxy)dioctylstannane.
CAS: 60494-19-1. $C_{26}H_{44}O_8Sn$.
Hazard: Moderately toxic by ingestion.

2,7-bis(4-methoxyphenyl)benzo(lmn)(3,8)-phenanthroline-1,3,6,8(2h,7h)-tetrone.
CAS: 64005-84-1. $C_{28}H_{18}N_2O_6$.
Hazard: A poison.

bis(1-methylamyl)sodium   sulfosuccinate.
(dihexyl sodium sulfosuccinate). $C_{16}H_{29}NaO_7S$.

Properties: White, waxy particles. Readily soluble in hot water; slowly in cold water; soluble in benzene, carbon tetrachloride, acetone, and glycerol. Hydrolyzes in alkaline media.
Use: Surfactant, wetting agent.

***N,N'*-bis(1-methylheptyl)-*p*-phenylenediamine.**
$C_6H_4[NHCH(CH_3)C_6H_{13}]_2$.
Properties: Dark reddish-brown liquid. D approximately 0.90 at (26.6C), wt/gal 7.5 lb. Combustible.
Use: Antiozonant in polyunsaturated elastomers.

bis(2-methylallyl) diglycolate.
CAS: 63917-25-9.
Hazard: Moderately toxic by ingestion.
Use: Agricultural chemical.

***n,n'*-bis-(2-(*o*-(*n*-methylcarbamoyl)oximino)-1,4-dithiane)disulfide.**
CAS: 68789-89-3. $C_{12}H_{18}N_4O_4S_6$.
Hazard: A poison by ingestion.
Use: Agricultural chemical.

2,3:4,5-bis-*o*-(1-methylethylidene)-β-d-fructopyranose,   methyl((2-(1-methyl-ethoxy)phenoxy)carbonyl)amidosulfite.
CAS: 81897-50-9. $C_{23}H_{33}NO_{10}S$.
Hazard: Moderately toxic by ingestion.
Use: Agricultural chemical.

1,2:4,5-bis-*o*-(1-methylethylidene)-β-d-fructopyranose, methyl((3-methylphenoxy) carbonyl) amidosulfite.
CAS: 81862-13-7. $C_{21}H_{29}NO_9S$.
Hazard: Moderately toxic by ingestion.
Use: Agricultural chemical.

2,3:4,5-bis-*o*-(1-methylethylidene)-β-d-fructopyranose, methyl((3-methylphenoxy) carbonyl) amidosulfite.
CAS: 81862-12-6. $C_{21}H_{29}NO_9S$.
Hazard: Moderately toxic by ingestion.
Use: Agricultural chemical.

1,2:4,5-bis-*o*-(1-methylethylidene)-β-d-fructopyranose,   methyl((1-naphthaleny-loxy) carbonyl)amidosulfite.
CAS: 81862-22-8. $C_{24}H_{29}NO_9S$.
Hazard: Moderately toxic by ingestion.
Use: Agricultural chemical.

1,2:5,6-bis-*o*-(1-methylethylidene)-α-*d*-glucofuranose,   ((((2-(dimethylamino)-2-oxo-1-(methylthio)ethylidene)amino)oxy)-carbonyl)methylamidosulfite.
CAS: 81877-66-9. $C_{19}H_{31}N_3O_{10}S_2$.
Hazard: A poison by ingestion.
Use: Agricultural chemical.

bis((4-(1-methylethyl)phenyl)methyl) 3-pyridinylcarbonimidodithioate.
CAS: 51308-75-9. $C_{26}H_{30}N_2S_2$.
Hazard: Moderately toxic by ingestion.
Use: Agricultural chemical.

2,7-bis(4-methylphenyl)benzo(lmn)(3,8)-phenanthroline-1,3,6,8(2h,7h)-tetrone.
CAS: 989-74-2. $C_{28}H_{18}N_2O_4$.
Hazard: A poison.

1,4-bis[2-(4-methyl-5-phenyloxazolyl)]-benzene. (dimethyl-POPOP).
CAS: 3073-87-8. $C_6H_4[C_3NO(CH_3)C_6H_5]_2$. Yellow needles with a bluish fluorescence, mp 231–234C.
Use: As a scintillation phosphor.

***n,n'*-bis(2-methylsulfonyl-2-methylpropionaldehyde-*o*-(*n*-methylcarbamoyl)oxime)sulfide.**
CAS: 63942-42-7. $C_{14}H_{26}N_4O_8S_3$.
Hazard: A poison by ingestion.
Use: Agricultural chemical.

bis((methylsulfonyl)oxy)dibutylstannane.
See di-*n*-butyltin bismethanesulfonate.

bis((methylsulfonyl)oxy)dipropylstannane.
See di-*n*-propyltin bismethanesulfonate.

***n,n'*-bis-(1-methylthioacetaldehyd-*o*-(*n*-methylcarbamoyl)oxim)-disulfide.**
CAS: 68789-90-2. $C_{10}H_{18}N_4O_4S_4$.
Hazard: A poison by ingestion.
Use: Agricultural chemical.

***n,n'*-bis(1-methylthio-1-(n,n-dimethylcarbonyl)formaldehyde-*o*-(*n*-methylcarbamoyl)oxime)sulfide.**
CAS: 63942-44-9. $C_{14}H_{24}N_6O_6S_3$.
Hazard: A poison by ingestion.
Use: Agricultural chemical.

bismite. See bismuth trioxide.

bismuth. Bi. Metallic element of atomic number 83, group VA of the periodic table. Aw 208.9804. Valences 3, 5; no stable isotopes; four naturally radioactive isotopes.
Properties: Brittle metal with reddish tinge. Soluble in nitric and hydrochloric acids. Highly diamagnetic (mass susceptibility -1.35×10^6). Expands 3.3% on solidification. Electrical resistivity higher in solid than in liquid state. Extrudable at 437F, not fabricable at room temperature, d 9.8 (20C), mp 271C, bp 1560C, Brinell hardness 7. Thermal conductivity (0.018 cal/sec cm (250C)) is lowest of all metals except mercury. On heating it burns to form the oxide.

Source: (1) Metallurgical by-products (often lead bullion) obtained chiefly from smelting ores of lead, silver, copper, and gold; (2) ores used chiefly for their bismuth and one or two other metals as tin and tungsten.
Derivation: Debismuthizing of lead bullion by (1) fractional crystallization; (2) electrolytic (Betts) refining; or (3) addition of calcium or magnesium (Betterton–Kroll process), which removes bismuth.
Method of purification: By addition of molten caustic, zinc, and finally chlorine (to make removable chlorides of the impurities).
Impurities: Lead, iron, copper, arsenic, antimony, selenium.
Available forms: Rods, wire, lump, powder.
Grade: 99.5+% pure, high purity (less than 10 ppm impurities), single crystals.
Hazard: Flammable in powder form.
Use: Pharmaceuticals and medicine, cosmetics (eye shadow, lipstick), component of low-melting (fusible) alloys, catalyst in making acrylonitrile, additive to improve machinability of steels and other metals, coating selenium, thermoelectric materials, permanent magnets, semiconductors.
See bismuthinite.

bismuth ammonium citrate.
Properties: Pearly, shining, transparent scales or white powder; slightly acid, metallic taste. Composition varies. Soluble in water; slightly soluble in alcohol.
Derivation: Interaction of bismuth subnitrate, citric acid, and ammonium hydroxide.
Use: Medicine.

bismuthane. BiH_3. Any of a class of saturated hydrides of tervalent bismuth.

bismuth antimonide. BiSb. Single crystals used as semiconductors.

bismuth bromide. (bismuth tribromide). $BiBr_3$.
Properties: Yellow, crystalline powder. D 5.7, bp 453C, mp 218C. Hygroscopic. Decomposed by water with formation of bismuth oxybromide. Soluble in either hydrochloric acid (dilute) or solutions of potassium iodide, bromide, and chloride. Insoluble in alcohol.

bismuth bromide oxide. See bismuth oxybromide.

bismuth carbonate, basic. See bismuth subcarbonate.

bismuth chloride. (bismuth trichloride). $BiCl_3$.
Properties: White, very deliquescent crystals; volatilized by heat. D 4.56, mp 227C, bp decomposes at 300C. Soluble in acids; insoluble in alcohol, decomposes in water to the oxychloride.
Derivation: Action of hydrochloric acid on bismuth.
Use: Bismuth salts, catalyst.

bismuth chloride, basic.   See bismuth oxychloride.

bismuth chromate.   (basic   dichromate). $Bi_2O_3 \cdot 2CrO_3$.
Properties: Orange-red, amorphous powder. Soluble in alkalies and acids; insoluble in water.
Derivation: Interaction of bismuth nitrate and potassium chromate.

bismuth citrate. $BiC_6H_5O_7$.
Properties: White powder. D 3.458, mp (decomposes). Soluble in ammonia or alkali citrates; insoluble in water; slightly soluble in alcohol.
Derivation: Boiling bismuth subnitrate with citric acid.
Use: Medicine.

bismuth dimethyldithiocarbamate.
CAS: 21260-46-8. $Bi[(CH_3)_2NC(S)S]_3$.
Properties: Lemon-yellow powder. D 2.04, mp greater than 230C (decomposes). Soluble in chloroform; slightly soluble in benzene and carbon disulfide; insoluble in water.
Use: Accelerator for natural rubber and SBR, especially in cable covers and mechanical items.

bismuth ditannate.   See bismuth tannate.

bismuth ethyl chloride. $BiHC_2H_5Cl$. White powder.
Hazard: Ignites spontaneously in air, dangerous fire risk.
See bismuth.

bismuth gallate, basic.   See bismuth subgallate.

bismuth glance.   See bismuthinite.

bismuth hydrate.   See bismuth hydroxide.

bismuth hydroxide.   (bismuth hydrate; bismuth oxyhydrate; bismuth trihydroxide; bismuth trihydrate; hydrated bismuth oxide). $Bi(OH)_3$.
Properties: White, amorphous powder. D 4.36. Soluble in acids, insoluble in water.
Derivation: Action of sodium hydroxide on a solution of bismuth nitrate.
Use: Plutonium separation, hydrolysis of ribonucleic acid, absorbent.

bismuthinite.   (bismuth glance). Bi_2S_3. May contain copper or iron.
Properties: Lead-gray mineral, often with yellow tarnish; metallic luster. Contains 81.2% bismuth, 18.8% sulfur, soluble in nitric acid, d 6.4–6.5, Mohs hardness 2.
Occurrence: Utah, Bolivia, Mexico.
Use: Ore of bismuth.

bismuth iodide.   (bismuth triiodide). BiI_3.
Properties: Grayish-black, metallic, glistening crystals. D 5.778 (15C); sublimes at 438C. Soluble in alcohol, hydriodic acid, and potassium iodide solutions; insoluble in water. Decomposes in hot water.
Derivation: By the interaction of bismuth and iodine.
Hazard: Toxic by ingestion.
See bismuth.
Use: Analytical chemistry, manufacturing bismuth oxyiodide.

bismuth-β-naphthol. $Bi_2O_2(OH) \cdot C_{10}H_7O$.
Properties: Brown to gray powder. Almost insoluble in water or other solvents.
Derivation: By treating a solution of sodium β-naphtholate with acetic acid solution of bismuth nitrate and adding caustic soda solution to neutralize excess acid.
Use: Medicine.

bismuth nitrate.   (bismuth ternitrate; bismuth trinitrate). $Bi(NO_3)_3 \cdot 5H_2O$.
Properties: Lustrous, clear, colorless, hygroscopic crystals; acid taste. D 2.83; bp 75–80C (decomposes). Soluble in dilute nitric acid, alcohol, and acetone; slowly decomposed by water to the subnitrate.
Derivation: Action of nitric acid on bismuth with subsequent recovery by evaporation and crystallization.
Hazard: Oxidizing material; fire risk near organic materials.
Use: Preparation of other bismuth salts, bismuth luster on tin, luminous paints and enamels, precipitation of alkaloids.

bismuth nitrate, basic.   See bismuth subnitrate.

bis(octanoyloxy)di-*n*-butyl stannane.
CAS: 4731-77-5. $C_{24}H_{48}O_4Sn$.
Hazard: Low toxicity. A severe skin and eye irritant.

bis(octanoyloxy)di-*n*-butyltin.   See bis(octanoyloxy)di-*n*-butyl stannane.

bismuth oleate.
Properties: Yellowish-brown, soft, granular mass. Soluble in ether; insoluble in water.
Derivation: A mixture of bismuth trioxide and oleic acid.
Use: Catalyst in oxo process.

bismuth oxide.   See bismuth trioxide.

bismuth oxide, hydrated.   See bismuth hydroxide.

bismuth oxybromide.   (bismuth bromide, basic; bismuth bromide oxide). BiBrO.
Properties: Dry powder. Insoluble in water and alcohol; soluble in hydrochloric and hydrobromic acids.
Use: Cathodes for dry cells.

bismuth oxycarbonate.   See bismuth subcarbonate.

bismuth oxychloride.   (bismuth chloride, basic; bismuth subchloride).
BiOCl.
Properties: White lustrous crystalline powder. D 7.717. Soluble in acid; insoluble in water.
Derivation: By action of water on bismuth chloride, interaction of dilute nitric acid solution of bismuth nitrate with sodium chloride.
Use: Cosmetics, pigment, dry-cell cathodes.

bismuth oxyhydrate.   See bismuth hydroxide.

bismuth oxynitrate.   See bismuth subnitrate.

bismuth pentafluoride. BiF_5.
Properties: Crystals. Sublimes at 550C.
Hazard: Reacts violently with water and petrolatum above 50C. Strong irritant to eyes and skin.
Use: Fluorinating agent.

bismuth pentoxide. Bi_2O_5. An acid anhydride, its salts have not been prepared in pure state. Made by oxidation of bismuth trioxide, giving a scarlet precipitate.

bis(*p*-phenoxyphenyl)diphenyltin.
CAS: 17601-12-6. $C_{36}H_{28}O_2Sn$.
Hazard: A poison.

bis(*p*-phenoxyphenyl)diphenylstannane.
See bis(*p*-phenoxyphenyl)diphenyltin.

bismuth phosphate. $BiPO_4$.
Properties: Odorless crystals that do not melt when heated. D 6.32 (15C). Soluble in nitric and hydrochloric acids; insoluble in acetic acid and alcohol; slightly soluble in water and weak acids.
Use: Plutonium recovery, optical glass.

bismuth potassium iodide. $BiI_3 \cdot 4KI$.
Properties: Red crystals, decomposed by water. Soluble in potassium iodide solution.
Use: Precipitation of vitamins and antibiotics from solution.

bismuth salicylate, basic.   See bismuth subsalicylate.

bismuth   selenide.   (bismuth   triselenide).
Bi_2Se_4.
Properties: Black crystals. Mp 710C. Insoluble in water; decomposes on heating.
Use: Semiconductor technology.

bismuth stannate. $Bi_2(SnO_3)_3 \cdot 5H_2O$.
Properties: Light-colored crystals. Insoluble in water. Temperature of dehydration is approximately 140C.
Hazard: Toxic material.
See bismuth.

Use: Component of ceramic capacitors, useful with barium titanate.

bismuth subcarbonate.   (bismuth oxycarbonate; bismuth carbonate, basic). $(BiO)_2CO_3$ or $Bi_2O_3 \cdot CO_2 \cdot 1/2H_2O$.
Properties: White, odorless powder; tasteless. D 6.86. Stable in air but slowly affected by light. Insoluble in water and alcohol; soluble in nitric or hydrochloric acid with effervescence.
Derivation: By adding ammonium carbonate to a solution of a bismuth salt.
Grade: Technical, CP, USP (90% Bi_2O_3 min).
Use: Bismuth compounds, cosmetics, opacifier in X-ray diagnosis, enamel fluxes, and ceramic glazes.

bismuth subchloride.   See bismuth oxychloride.

bismuth subcitrate.
CAS: 57644-54-9. $C_{12}H_8O_7 \cdot Bi \cdot 3K$.
Hazard: A poison by ingestion.

bismuth subgallate.   (basic bismuth gallate).
$C_6H_2(OH)_3COOBi(OH)_2$.
Properties: Saffron-yellow powder; odorless; tasteless. Soluble in dilute alkalies; insoluble in water, alcohol, and ether. Stable in air but affected by light.
Derivation: Interaction of bismuth nitrate, glacial acetic acid, and gallic acid in aqueous solution.
Use: Medicine (treatment of alimentary canal).

bismuth subnitrate.   (basic bismuth nitrate; bismuth oxynitrate). $4BiNO_3(OH)_2 \cdot BiO(OH)$.
Properties: White, heavy, slightly hygroscopic powder that shows acid to moistened litmus paper. D 4.928, mp 260C (decomposes). Soluble in acids; insoluble in water and alcohol.
Derivation: Hydrolysis of bismuth nitrate, filtering and drying.
Grade: Technical, CP, NF.
Use: Cosmetics, ceramic glazes, enamel fluxes.

bismuth subsalicylate.   (basic bismuth salicylate). $Bi(C_7H_5O_3)_3Bi_2O_3$.
Properties: White, bulky, crystalline powder; tasteless; odorless. Soluble in acids and alkalies; insoluble in water, alcohol, and ether. Stable in air but affected by light.
Derivation: By treating freshly prepared bismuth hydroxide with salicylic acid.
Use: Surface-coating plastics and copying paper.
See salicylic acid, bismuth basic salt.

bismuth sulfate. $Bi_2(SO_4)_3$.
Properties: White needles or powder. Contains approximately 68.5% bismuth. D 5.08, decomposes at 405C. Soluble in dilute hydrochloric or nitric acid; insoluble in alcohol, water.
Use: Analysis of metallic sulfates.
See bismuth trisulfate.

bismuth sulfide.   (bismuth trisulfide). Bi_2S_3.
  Properties: Blackish-brown powder. D 7.6–7.8, mp (decomposes). Soluble in nitric acid; insoluble in water.
  Derivation: (1) By melting bismuth and sulfur together. (2) By passing hydrogen sulfide into a soluble of a bismuth salt. (3) Occurs as the mineral bismuthinite.
  Use: Manufacturing bismuth compounds.

bismuth tannate.   (bismuth ditannate).
  Properties: Light brownish-yellow powder containing approximately 36% bismuth. Insoluble in water and alcohol, soluble in mineral acids.
  Derivation: From freshly prepared bismuth hydroxide and tannin.
  Use: Medicine (astringent).

bismuth telluride.   (bismuth tritelluride).
  CAS: 1304-82-1. Bi_2Te_3.
  Properties: Gray, hexagonal platelets. Mp 573C, d 7.642.
  Derivation: Stoichiometric combination of the elements.
  Grade: Ingot, single crystals.
  Hazard: Toxic.
  Use: Semiconductors for thermoelectric cooling and power generation.

bismuth tetraoxide. Bi_2O_4.
  Properties: Heavy, yellowish-brown powder. D 5.6, mp 305C. Soluble in acids; insoluble in water.
  Derivation: By further oxidation of bismuth trioxide.
  Use: Lubricant for metal-extrusion dies.

bismuth tribromide.   See bismuth bromide.

bismuth trichloride.   See bismuth chloride.

bismuth trihydrate.   See bismuth hydroxide.

bismuth trihydroxide.   See bismuth hydroxide.

bismuth triiodide.   See bismuth iodide.

bismuth trinitrate.   See bismuth nitrate.

bismuth trioxide.   (bismuth oxide; bismuth yellow; bismite). Bi_2O_2.
  Properties: Heavy, yellow powder. D 8.8, mp 820C. Soluble in acids; insoluble in water.
  Derivation: Heating bismuth nitrate in air, ignition of bismuth hydroxide.
  Use: Enameling cast iron, ceramic and porcelain colors.

bismuth trisulfate.
  CAS: 7787-68-0. $Bi \cdot 3/_2H_2O_4S$.
  Hazard: Moderately toxic by ingestion.

bismuth trisulfide.   See bismuth sulfide.

bismuth tritelluride.   See bismuth telluride.

bismuth yellow.   See bismuth trioxide.

p-bis[2,5(5-α-naphthyloxazolyl)]benzene.
  (NOPON). $C_{32}H_{20}O_2N_2$.
  Properties: Crystals. Mp 215–217C.
  Grade: Purified.
  Use: Scintillation counting.

bis(3-nitrophenyl)   disulfide.   See nitrophenide.

bis(nonyloxymaleoyloxy)dioctylstannane.
  See di-n-butyltin di(monononyl)maleate.

bisphenol A.   (4,4′-isopropylidenediphenol; 2,2-bis(4-hydroxyphenol)propane).
  CAS: 80-05-7. $(CH_3)_2C(C_6H_4OH)_2$.

  Properties: White flakes with a mild phenolic odor. Bp 220C (4 mm), mp 153C, d 1.195 (25/25C), flash p 175F (79.4C), insoluble in water; soluble in alcohol and dilute alkalies; slightly soluble in carbon tetrachloride. Combustible.
  Derivation: Condensation reaction of phenol and acetone catalyzed by hydrochloric acid at 65C.
  Use: Intermediate in manufacture of epoxy, polycarbonate, phenoxy, polysulfone, and certain polyester resins; flame retardants, rubber chemicals, fungicide.

bisphenol a.
  (4-[2-(4-hydroxyphenyl)propan-2-yl]phenol; 2,2-bis-4′-hydroxyfenylpropan; bis (4-hydroxyphenyl) dimethylmethane; bis(4-hydroxyphenyl)propane; 2,2-bis(p-hydroxyphenyl)prpane; 2,2-bis(4-hydroxyphenyl)propane; dian; p,p'-dihydoxydiphenyldimethylmethane; 4,4′-dihydroxydiphenyldimethylmethane; p,p'-dihydroxydiphenylpropane; 2,2-(4,4′-dihydroxydiphenyl)propane; 4,4′-dihydroxydiphenylpropane; 4,4′-dihydroxydiphenyl-2,22-propane; 4,4′-dihydroxy-2,2′-diphenylpropane; β-di-p-hydroxyphenylpropane; 2,2-di(4-hydroxyphenyl)propane; dimethyl dis(p-hydroxyphenyl)methane; dimethylmethylene-p,p'-diphenol; 2,2-di(4-phenylol)propane; p,p'-isopropylidenebisphenol; 4,4′-isopropylidenebisphenol; p,p'-isopropylidenediphenol; NCI-C50635).
  CAS: 80-05-7. $C_{15}H_{16}O_2$.
  Any of a class of organic compounds, the methylenediphenols and their substitution products.

Properties: White flakes; mild phenolic odor; insoluble in water; soluble in alcohol and dilute alkalies.
Derivation: Condensation of two equivalent amounts of phenol with an aldehyde or ketone.
Hazard: Poison; moderately toxic; teratogen; irritant.

bisphenol A disodium salt.
CAS: 2444-90-8. $C_{15}H_{16}O_2 \cdot 2Na$.
Hazard: Moderately toxic by ingestion and skin contact. A moderate skin and severe eye irritant.

2,7-bis(2-phenylethyl)benzo(lmn)(3,8)-phenanthroline-1,3,6,8(2h,7h)-tetrone.
CAS: 222420-31-7. $C_{30}H_{22}N_2O_4$.
Hazard: A poison.

2,7-bis(phenylmethyl)benzo(lmn)(3,8)-phenanthroline-1,3,6,8(2h,7h)-tetrone.
CAS: 106897-63-6. $C_{28}H_{18}N_2O_4$.
Hazard: A poison.

1,4-bis-2(5-phenyloxazoyl)benzene.
(POPOP). $(C_6H_5HNO)_2C_6H_4$.
Properties: Light yellow, cottony needles. Mp 245–246C, fluorescence max 4200 Å. Solubilities (g/100 g at 25C): water 0.00, 95% ethanol 0.00, toluene 0.12, hexane 0.02. Combustible.
Grade: Purified.
Use: Band shifter in scintillation counting.

2,5-bis(phenylthio)benzoquinone.
CAS: 17058-53-6. $C_{18}H_{12}O_2S_2$.
Hazard: A poison.

bis(phenylthio)dimethyltin.
CAS: 4848-63-9. $C_{14}H_{16}S_2Sn$.
Hazard: A poison.

bis(2-pyridinethiol 1-oxide)copper.
CAS: 14915-37-8. $C_{10}H_8CuN_2O_2S_2$.
Hazard: Moderately toxic by ingestion, inhalation, and skin contact.

bisteroid. A molecule comprised of two molecules of a given steroid liked by a carbon-to-carbon bond.

bis(tetrachloroethyl) disulfide. $C_4H_2Cl_8S_2$.
Properties: Liquid. D 1.785 (23.3C), bp 185C (3 mm Hg). Soluble in benzene, hexane, ethanol. Combustible.
Hazard: Toxic.
Use: Agricultural chemicals, additives.

bis(tetradecanoyloxy)dibutylstannane.
CAS: 28660-67-5. $C_{36}H_{72}O_4Sn$.
Hazard: A poison by ingestion. A severe skin and eye irritant.

7-(3,5-bis((tetrahydro-2h-pyran-2-yl)oxy)-2-(4-phenoxy-3-((tetrahydro-2h-pyran-2-yl)oxy)-1-butenyl)cyclopentyl)-2-(phenylseleno)-5-heptenoic acid, methyl ester.
CAS: 62524-93-0. $C_{44}H_{60}O_9Se$.
Hazard: A reproductive hazard.

1,2-bis(thiocyanato)ethane. See ethylenedithiocyanate.

bis(tribenzylstannyl)sulfide.
CAS: 10347-38-3. $C_{42}H_{42}SSn_2$.
Hazard: A poison by ingestion. A skin and eye irritant.

bis(tribenzyltin) sulfide. See bis(tribenzylstannyl)sulfide.

bis(tribromophenoxy)ethane.
Properties: White, crystalline powder.
Use: Flame retardant for many thermoplastic and thermoset systems.

bis(tributyl(sebacoyldioxy))tin.
CAS: 30099-72-0. $C_{34}H_{70}O_4Sn_2$.
Hazard: A poison.

bis(tributyltin) itaconate.
CAS: 25711-26-6. $C_{29}H_{58}O_4Sn_2$.
Hazard: A poison.

bis(tri-*n*-butyltin) oxide. (hexabutyldistannoxane; TBTO; tributyltin oxide).
CAS: 56-35-9. $(C_4H_9)_3SnOSn(C_4H_9)_3$.
Properties: Slightly yellow liquid. Bp 180C (2 mm Hg), solidifies below −45C, d 1.17 (25C), flash p >212F (100C) (TCC), viscosity 4.8 centistokes at 25C. Almost insoluble in water; miscible with organic solvents. Forms compounds with cellulosic and lignin-containing materials not easily decomposed or dissolved in water. Combustible.
Derivation: Hydrolysis of tributyl tin chloride.
Hazard: Toxic via ingestion and inhalation.
Use: Fungicide and bactericide in underwater and antifouling paints, pesticide.

bis(tributyltin)sulfide. See 1,1,1,3,3,3-hexabutyldistannthiane.

1,3-bis(2,4,5-trichlorophenoxy)-1,1,3,3-tetrabutyldistannoxane. See oxybis(dibutyl(2,4,5-trichlorophenoxy)tin).

bis(trichlorosilyl)ethane. (1,1,1,4,4,4-hexachloro-1,4-disilabutane). $Cl_3SiCH_2CH_2SiCl_3$.
Properties: Colorless liquid. Bp 202.9C, flash p 190F (COC). Readily hydrolyzed with liberation of hydrogen chloride. Combustible.
Derivation: Reaction of acetylene and trichlorosilane in presence of a peroxide catalyst.
Grade: Technical.

Hazard: Corrosive when exposed to moisture.
Use: Intermediate for silicones.

bistridecyl phthalate. $C_6H_4(COOC_{13}H_{27})_2$.
Properties: Liquid. D 0.9497 (25C), boiling range 280–290C (4 mm Hg), fp −35C, flash p 485F (251C), refr index 1.483 (25C). Combustible.
Use: Primary plasticizer for most PVC resins.

bis(triethyltin) sulfate.
CAS: 57-52-3. $C_{12}H_{30}O_4SSn_2$.
Hazard: A poison by ingestion.

bis(trifluoroacetoxy)dibutyltin.
CAS: 52112-09-1. $C_{12}H_{18}F_6O_4Sn$.
Hazard: A poison by ingestion. Moderately toxic by skin contact.

1,3-bis(trifluoromethyl)-5-isocyanobenzene.
CAS: 141206-73-7. $C_9H_3F_6N$.
Hazard: Low toxicity by ingestion.
Use: Agricultural chemical.

bistrifluoromethylmethane.   See 1,1,1,3,3,3-hexafluoropropane.

bis(trifluoromethylthio)mercury.
CAS: 21259-75-6. $C_2F_6HgS_2$.
Hazard: A poison.

bis(triisobutylstannane).   See   hexaisobutylditin.

bis(trimethylhexyl)tin dichloride.
CAS: 64011-39-8. $C_{18}H_{38}Cl_2Sn$.
Hazard: A poison.

bis(trimethylsilyl)trifluoroacetamide.
(BSTFA). $CF_3C[=NSi(CH_3)_3]OSi(CH_3)_3$.
Properties: Bp 45–50C (14 mm Hg), refr index 1.3839 (20C), mw 257.4.
Use: Preparation of volatile derivatives of a wide range of biologically active compounds for gas-liquid chromatographical analysis.

bis(triphenylsilyl)chromate.
CAS: 1624-02-8. $C_{36}H_{30}CrO_4Si_2$.
Hazard: Moderately toxic by ingestion and skin contact.

bis(triphenyltin)acetylenedicarboxylate.
CAS: 73940-87-1. $C_{40}H_{30}O_4Sn_2$.
Hazard: A poison.

bis(triphenyltin)sulfate.
CAS: 3021-41-8. $C_{36}H_{30}Sn_2 \cdot O_4S$.
Hazard: A poison.

bis(triphenyltin)sulfide.
CAS: 77-80-5. $C_{36}H_{30}SSn_2$.

Properties: Colorless crystals. Mp 144C. Sol in org solvs.
Hazard: Moderately toxic by ingestion.

bis(tripropyltin)oxide.
CAS: 1067-29-4. $C_{18}H_{42}OSn_2$.
Properties: Air-sensitive liquid. Bp 154.5C @ 3.5 mm.
Hazard: A poison.

bisulfate.   (hydrogen sulfate; acid sulfate).
CAS: 14996-02-2. HO_4S
A salt of sulfuric acid that contains the HSO_4 radical from sulfuric acid.

bisulfite.   (hydrogen sulfite; acid sulfite).
CAS: 15181-46-1. HO_3S.
A salt that contains the HO_3S radical from sulfurous acid.
Hazard: Questionable carcinogen.

bisulfite compound.   An addition compound of sodium bisulfite and an aldehyde or ketone.

bithionol.   (bis[2-hydroxy-3,5-dichlorophenyl]-sulfide; 2,2′-thiobis[4,6-dichlorophenol]).
CAS: 97-18-7. $HOCl_2C_6H_2SC_6H_2Cl_2OH$.
Properties: White or grayish-white, crystalline powder; odorless or with slight aromatic or phenolic odor. Mp 187C. Insoluble in water; freely soluble in acetone, alcohol, and ether; soluble in chloroform and dilute solution of fixed alkali hydroxides.
Grade: NF.
Hazard: Skin irritant, may not be used in cosmetics (FDA).
Use: Deodorant, germicide, fungistat, pharmaceuticals.

"Bitrex" *[Johnson Matthey]*.   (benzyldiethyl-[(2,6-xylylcarbamoyl)methyl]ammonium   benzoate; denatonium benzoate [USAN]).
CAS: 3734-33-6. TM for a substance for use as a denaturant for alcohol.
Properties: Bitter taste. Mp 165C; soluble in water and alcohol; insoluble in ether.
See denatonium benzoate.

bitter almond oil.   See almond oil.

bittern.   The solution of bromides and magnesium and calcium salts that remains after sodium chloride has been crystallized by concentration of seawater or brines.

bitter orange oil.   See orange oil, bitter, cold-pressed.

bitumen.   A mixture of hydrocarbons occurring both in the native state and as residue from California petroleum distillation. Soluble in carbon disulfide. Solid to viscous, semisolid liquid. Used in hot-melt adhesives, coatings, paints, sealants, roofing, and road coating. Bitumens are found in asphalt,

mineral waxes, and lower grades of coal. Combustible.
See asphalt; gilsonite; glance; shale oil; oil sands.

bituminous coal.   See coal.

biuret.   (allophanamide;   carbamylurea). $NH_2CONHCONH_2 \cdot H_2O$.
Properties: White needles; odorless. Mp 190C (decomposes). Soluble in water and alcohol; very slightly soluble in ether. Loses water of crystallization at approximately 110C.
Derivation: From urea by heat.
Method of purification: Crystallization.
Use: Analytical reagent, especially for proteins.

biuret reaction.   A reaction that produces characteristic rose color when certain substances in alkaline solution are treated with copper sulfate solution.

bixin. $C_{25}H_{30}O_4$. A carotenoid obtained from seeds of *Bixa orellana.*
Properties: Orange crystals. Decomposes at 217C.
Derivation: Methyl ester (methyl bixin) $C_{26}H_{32}O_4$: blue crystals, mp 203C. Ethyl ester (ethyl bixin), $C_{27}H_{34}O_4$: Red crystals, mp 138C.
Use: The ethyl ester is used as a food coloring.
See annatto.

Bk.   Symbol for berkelium.

black.   Any of several forms of finely divided carbon, either pure or admixed with oils, fats, or waxes. See acetylene black; bone black; carbon black; ivory black.

black, aniline.   See aniline black.

black antimony.   See antimony trisulfide.

black ash.   (1) (Papermaking) The product obtained by heating black liquor in furnaces. The organic material is reduced to carbon. The alkaline components are leached out and used again in papermaking. The carbon may be treated to obtain activated carbon.
(2) See barium sulfide.

blackbody.   In radiation physics, an ideal blackbody is a theoretical object that absorbs all the radiant energy falling upon it and emits it in the form of thermal radiation. The power radiated by a unit area of a blackbody is given by Planck's radiation law, and the total power radiated is expressed by the Stefan–Boltzmann law.

black, bone.   See bone black.

black cyanide.   A mixture containing 45% calcium cyanide made from calcium cyanamid by heating it with sodium chloride and carbon.

black-eyed susan.   A flowering plant. *Abrus precatorius.*

black lead.   See graphite.

Black Leaf 40.   A pesticide consisting of a 40% solution of nicotine sulfate.

black liquor.   (1) The liquor resulting from cooking pulpwood in an alkaline solution in the soda or sulfate (kraft) papermaking process. It is a source of lignin and tall oil and is said to be effective in removal of mercury from industrial effluents (USDA). (2) Iron acetate liquor (black mordant).

black oil.   See residual oil.

black phosphorus.   See phosphorus.

black plate.   Thin sheet steel obtained by rolling and usually used for containers. It is not coated with any metal, but a special lacquer or baked enamel finish is usually applied by the manufacturer.

black, platinum.   See platinum black.

black powder.   (blasting powder). A low explosive composed of potassium nitrate, charcoal, and sulfur. In some cases sodium nitrate is substituted for potassium nitrate. Typical proportions are 75%, 15%, and 10%. Gunpowder is probably the oldest variety.
Hazard: Sensitive to heat, deflagrates rapidly. Does not detonate but is a dangerous fire and explosion hazard.
Use: Time fuses for blasting and shell, in igniter and primer assemblies for propellants, pyrotechnics, mining, and blasting.

black rouge.   See iron oxide, black.

black sand.   A deposit of dark minerals with a high density found in stream beds and on beaches. Magnetite and ilmenite are usually present and also monazite and other minerals.

blackstrap.   See molasses.

Bladen.   (hexaethyl tetraphosphate). An insecticide.
Use: Insecticide.

Blaise reaction.   Formation of β-oxoesters by treatment of α-bromocarboxylic esters with zinc in the presence of nitriles. The intermediate organozinc compound reacts with the nitrile, and the complex is hydrolyzed with 30% potassium hydroxide.

Blanc (chloromethylation) reaction.   Introduction of the CH_2Cl group into aromatic rings on treatment with formaldehyde and hydrochloric acid in the presence of zinc chloride.

blanc fixe. Precipitated barium sulfate.

blanch. To immerse vegetables or fruits in either hot water (80–100C) or steam in preparation for cooking, freezing, or canning. The times and temperatures vary among processors; the higher the temperature, the shorter the time required. The operation loosens the skin or peel when present, tends to remove the color, and decreases the volume. It also causes some loss of nutrient value, especially vitamins.

Blancophor. Optical whitening agents. FFG: A comarin derivative used as whitening agent for wool, nylon, acetate rayon, and mixed fibers. "HS Brands": Stilbene derivatives used on cellulosic fibers, cotton and rayon fabrics, paper, and in household and industrial detergents.

Blanc reaction-Blanc rule. Cyclization of dicarboxylic acids on heating with acetic anhydride to give either cyclic anhydrides or ketones, depending on the positions of the carboxyl groups: 1,4- and 1,5-diacids give anhydrides, while diacids in which the carboxy groups are in 1,6- or further-removed positions give ketones.

"Blandol" [Sonneborn]. TM for white mineral oil (NF).
Use: Pharmaceutical and cosmetic formulations, plasticizers, paper penetrants, foam depressants.

blank. (1) A piece of material of any desired shape cut by a stamping die-prepared for further processing. (2) See control (1).

BLAST. A computer program that identifies homologous (similar) genes in different organisms, such as human, fruit fly, or nematode.

blast furnace. A vertical coke-fired furnace used for smelting metallic ores, e.g., iron ore.

blast-furnace gas. By-product gas from smelting iron ore obtained by the passage of hot air over the coke in blast furnaces. A typical gas will analyze 12.9% carbon dioxide, 26.3% carbon monoxide, 3.7% hydrogen, 57.1% nitrogen.
Hazard: Toxic by inhalation. See carbon monoxide.
Use: Heating blast-furnace stoves, boilers, or as a gas-engine fuel.

blasting agent. See black powder; ammonium nitrate; explosive, high; explosive, permissible; explosive, low.

blasting gelatin. (SNG). A type of gelatinized dynamite containing approximately 7% of nitrocellulose.
Hazard: High explosive.

blasting powder. See black powder.

blastoderm. (chloroethane).
CAS: 75-00-3. C_2H_5Cl.
A highly reactive manmade volatile organic compound that is highly reactive in the atmosphere. It readily reacts with oxidizing agents to release the chlorine atoms which, circulate and cause tropospheric ozone to decompose.
Properties: Gas at room temperature low boiling point.
Use: Manufacture of drugs, dyes, and insecticides; and a local pain releaver.

B-L-E. A high-temperature reaction product of diphenylamine and acetone.
Properties: Darkbrown, viscous liquid. D 1.087. Soluble in acetone, benzene, and ethylene dichloride; insoluble in gasoline and water. Combustible.
Use: General-purpose rubber antioxidant.

bleach. To whiten a textile or paper by chemical action. Also the agent itself. Bleaching agents include hydrogen peroxide (the most common), sodium hypochlorite, sodium peroxide, sodium chlorite, calcium hypochlorite, hypochlorous acid, and many organic chlorine derivatives. Chlorinated lime is a bleaching powder used on an industrial scale. Household bleaching powders are sodium perborate and dichlorodimethylhydantoin.
Hazard: See calcium hypochlorite; lime, chlorinated. Some bleaching agents are toxic and strong oxidizing agents.

bleached lard. See lard (unhydrogenated).

bleach, household. (sodium hypochlorite). ClNaO.
Properties: Liquid; contains hypochlorite salts, chlorine, lye, artificial dyes, detergents, fluorescent brighteners, synthetic fragrances.
Hazard: Strong irritant; harmful if swallowed; poison.
Use: An oxidizing and bleaching agent, and a disinfectant.

bleaching assistant. A material added to bleaching baths to secure more rapid and complete penetration of the bleach or improved regulation of the bleaching action, e.g., compounds of sulfonated oils and solvents, soluble pine oils, fatty alcohol salts, sodium silicate, sodium phosphate, magnesium sulfate, and borax.

bleaching clay. A clay it its natural state or after chemical activation that is able to adsorb dyes and other colorants.

bleaching, fluorescent. The use of colorless fluorescent organic compounds to produce a whitening effect on textiles.

bleaching powder. See bleach; calcium hypochlorite.

bleach liquor.   A solution of either sodium or calcium hypochlorite and water.

bleed.   (1) When a dye runs. (2) To release pressure gradually, as via a valve.

blend.   A uniform combination of two or more materials either of which could be used alone for the same purpose as the blend. For example, a fabric may be a blend of wool and nylon, either of which is itself usable as fabric. Instances of materials that are often blended are

plastics (polyblends)	grains
whiskeys	coffees
fabrics	paints
colors	tobaccos
metal powders	solvents
fertilizers	

See mixture; mixing; kneading.

bleomycin.
CAS: 11056-06-7.
A glycopeptide antibiotic produced by *Streptomyces verticillus*; it functions as an antineoplastic and diagnostic agent. The molecule is exceedingly complex, but synthesis was achieved in 1982. It is a colorless to yellowish powder, soluble in water and methanol but insoluble in acetone and ether. It induces rupture of DNA strands.
Hazard: Possible carcinogen.

bleomycin A2.
(3-[[2-[2-[[(2S,3R)-2-[[(2S,3S,4R)-4-[[(2S,3R)-2-[[6-amino-2-[(1S)-3-amino-1-[[(2S)-2,3-diamino-3-oxopropyl]amino]-3-oxopropyl]-5-methylpyrimidine-4-carbonyl]amino]-3-[3-[4-carbamoyloxy-3,5-dihydroxy-6-(hydroxymethyl)oxan-2-yl]-oxy-4,5-dihydroxy-6-(hydroxymethyl)oxan-2-yl]oxy-4,5-dihydroxy-6-(hydroxymethyl)oxan-2-yl]oxy-3-(1H-imidazol-5-yl)propanoyl]amino]-3-hydroxy-2-methylpentanoyl]amino]-3-hydroxybutanoyl]amino]ethyl]-1,3-thiazol-4-yl]-1,3-thiazole-4-carbonyl]amino]propyl-dimethylsulfanium).
CAS: 11056-06-7. $C_{55}H_{84}N_{17}O_{21}S_3$. A species of bleomycin noted for its adverse pulmonary effects in humans. It is a complex of related glycopeptide antibiotics from *Streptomyces verticillus* consisting of bleomycin A2 and B2.
Hazard: Poison; mutagen; possible carcinogen; causes adverse pulmonary effects.
Use: Antineoplastic for solid tumors.

blinding.   (blister copper).
Properties: Copper (96–99% purity) produced by the reduction and smelting of copper ores. It has a blistered appearance, probably caused by gas pockets. It is usually further refined electrolytically.

blister agent.   (vesicant agent). A chemical warfare agent.
Hazard: Inflammation and blistering of exposed tissue on contact; blisters on the eyes, lungs, mucous membranes, skin, and blood-forming organs such as the spleen; persistent damage; death.
Use: In warfare to weaken and demoralize an opposing force by producing casualties and forcing enemy troops to wear bulky protective equipment which degrades their fighting ability.

blister gas.   See dibromodiethyl sulfide.

blister packaging.   A type of packaging used widely in the food and pharmaceutical industries, consisting of a hollow cavity of various shapes and capacities in which the material is enclosed. Polyester and polyethylene resins are often used.

block.   (1) Undesirable cohesion of films or layers of plastic.
See antiblock agent. (2) A type of polymer. See block polymer.

block copolymer.   Polymer containing long stretches of two or more monomeric units linked together by chemical valences in one single chain. See block polymer.

α-blocker.   Alpha-adrenergic blocking agent.

β-blocker.   Beta-adrenergic blocking agent.

blocking agent.   (antiadrenergic agent blocking agent; sympatolytic agent; sympathicolytic agent; sympathoparalytic agent).
An agent that blocks or inhibits the action of, or transmission by, adrenergic nerves.

blocking antibody.   An antibody molecule that does not cause precipitation of specific antigen at certain concentrations following combination but also blocks the effects of additional antibodies on the antigen.

block polymer.   A high polymer whose molecule is made up of alternating sections of one chemical composition separated by sections of a different chemical nature or by a coupling group of low molecular weight. An example is blocks of polyvinyl chloride interspersed with blocks of polyvinyl acetate. Such polymer combinations are made synthetically. They depend on the presence of an active site on the polymer chain that initiates the necessary reactions.
See graft polymer; stereoblock polymer.

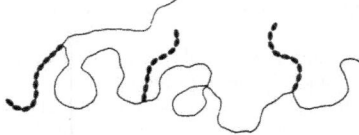

blood.   A complex, liquid tissue of d 1.056 and pH 7.35–7.45. It is comprised of erythrocytes (red cells), leucocytes (white cells), platelets, plasma, proteins, and serum. The plasma fraction (55–70%) is whole blood from which the red and white cells and the platelets have been removed by centrifuging. Hemoglobin is a protein found in the erythrocytes. It contains the essential iron atom and functions as the transport agent for oxygen from the lungs (arterial blood) and of carbon dioxide to the lungs from the heart (venous blood). Experimental work has been reported on the effectiveness of fluorocarbon compounds in carrying out the essential transport functions of blood, especially of the red cells.
Use: Plasma is used to restore liquid volume and thus osmotic pressure in the body where blood loss has been extensive. Animal blood is used as a component of adhesive mixtures. In dried or powder form it is a component of fertilizers, poultry feeds, and deer repellents.
See hemoglobin; plasma; platelet; rhesus factor.

blood agar.   Agar plus sterile blood.

blood agent.   (cyanogen agent). A chemical compound, or the cyanide group, that interferes with bodily function by preventing the normal transfer of oxygen from blood to the respiring tissues.

blood alcohol concentration.   amount of alcohol in a person's blood.

blood coagulant.   An agent that causes blood to clot following damage to a blood vessel.

blood coagulation.   A complicated process that involves the sequential activation of a number of soluble proteins in blood plasma with consequent production of a fibrin network.

blood group antigen.   (blood group substance). Any of a number of inherited antigens that determine a blood grouping reaction with specific antiserum.
Derivation: Occur on the surface of erythrocytes.

blood plasma.   The liquid portion of blood which remains when blood cells have been removed.

blood serum.   (bogomolet's serum; antireticular cytotoxic serum).
Properties: Thin, watery, proteinaceous fraction of blood remaining following removal of fibrin, coagulation factors, and formed elements of whole blood.
Use: Tests; preparation and use of antisera where clotting might interfere with the process or results.

blood–brain barrier.   (bbb; blood–cerebral barrier; blood–cerebrospinal fluid barrier; menatoencephalic barrier).
A barrier that separates systemic blood from the parenchyma of the central nervous system. It consists of endothelials of the capillaries and surround glial membranes to which the capillaries are closely joined. It is impermeable to most ions and protein-bound chemicals, but allows passage of lipid-soluble molecules. It effectively protects tissues of the brain and spinal cord from the intrusion of most toxicants but can be damaged or reduced.

blood–testis barrier.   A permeability barrier to the passage of blood-borne materials into the seminiferous tubules of the testis.

blood–vascular system.   A network or a continuous series of vessels and/or spaces within a multicellular animal that conducts blood through the body in response to a pressure gradient generated by one or more hearts. Such a system transports respiratory gases, nutrients, and wastes to and/or from the tissues of the body.

bloom.   (1) A thin coating of an ingredient of a rubber or plastic mixture that migrates to the surface, usually within a few hours after curing or setting. Sulfur bloom in vulcanized rubber products is most common; it is harmless but impairs the eye appeal of the product. Paraffin wax is often included purposely; when it migrates to the surface it provides an efficient barrier to sunchecking and oxidation. (2) A piece of steel made from an ingot. (3) An arbitrary scale for rating the strength of gelatin gels. When so used the word is capitalized. (4) Efflorescence of phytoplankton in seawater causing discoloration of the surface water.
See red tide.

blotting.   A technique for detecting one RNA within a mixture of RNAs (a Northern blot) or one type of DNA within a mixture of DNAs (a Southern blot). A blot can prove whether that one species of RNA or DNA is present, how much is there, and its approximate size. Basically, blotting involves gel electrophoresis, transfer to a blotting membrane (typically nitrocellulose or activated nylon), and incubating with a radioactive probe. Exposing the membrane to X-ray film produces darkening at a spot correlating with the position of the DNA or RNA of interest. The darker the spot, the more nucleic acid was present there.
The DNA is first transferred from the gel to a membrane by capillary action. Fluid wicks from the gel through the blotting membrane to several layers of absorbent paper, but the nucleic acids stick to the membrane. Baking the filter fixes the DNA or RNA to the filter.
Specific bands are detected by hybridization. The filter membrane is incubated with radioactive probe, which hybridizes to some bands. After the filter is

washed (to remove unused probe), an X-ray film exposed to the filter will show which bands have hybridized.

blow-down. The cyclic or constant removal of water from a boiler to deter the collection of solids.

blowing agent. A substance incorporated into a mixture for the purpose of producing a foam. One type decomposes when heated to processing temperature to evolve a gas, usually carbon dioxide, which is suspended in small globules in the mixture. Typical blowing agents of this kind are baking powder (bread and cake), sodium bicarbonate or ammonium carbonate (cellular or sponge rubber), halocarbons and methylene chloride in urethane, pentane in expanded polystyrene, and hydrazine and related compounds in various types of foamed plastics. Another type is air used at room temperature as a blowing agent for rubber latex; it is introduced mechanically by whipping, after which the latex is coagulated with acid. Air is also used for this purpose in ice cream, whipped cream, and other food products, as well as in blown asphalt and blown vegetable oils.
See foam.

blow molding. A technique for production of hollow thermoplastic products. It involves placing an extruded tube (parison) of the thermoplastic in a mold and applying sufficient air pressure to the inside of the tube to cause it to take on the conformation of the mold. Polyethylene is usually used, but a number of other materials are adaptable to this method, e.g., cellulosics, nylons, polypropylene, and polycarbonates. It is an economically efficient process and especially suitable for production of toys, bottles, and other containers, as well as air-conditioning ducts and various industrial items. The method is not limited to hollow products; housings can be made by blowing a unit and sawing it along the parting line to make two housings.

blown asphalt. See asphalt (blown).

blown oil. (oxidized oil; base oil; thickened oil; polymerized oil). Vegetable and animal oils that have been heated and agitated by a current of air or oxygen. They are partially oxidized, deodorized, and polymerized by the treatment and are increased in density, viscosity, and drying power. Common blown oils are castor, linseed, rape, whale, and fish oils.
Use: Paints, varnishes, lubricants, and plasticizers.

blue copperas. See copper sulfate.

blue copper protein. A protein in the electron transport chain that contains a type one copper site. Coordination of the copper by a cysteine sulfur moiety accounts for both the strong absorption in the visible portion of the electromagnetic spectrum, and

an electron paramagnetic resonance signal marked by an uncommonly small hyperfine coupling to the copper nucleus.

blue cross gas. See diphenylchloroarsine.

blue gas. See water gas.

blueing. (1) Producing a blue oxide film on steel by low-temperature heating; (2) heat treatment of steel springs to relieve cold-working strains.

blueing, laundry. The material used in washing to give the fabric a bluish tint, thereby neutralizing the yellow coloring to produce a whiter shade.

blueing, steel. The process for shading the surface of steel by heating to blueing temperature in the presence of air and steam.

blue lead. See lead sulfate, blue basic.

blueprint. See Turnbull's Blue.

blueprint paper. A paper that has been dipped in ammonium ferric citrate and potassium ferricyanide solution. When it is exposed to light and a wash, all black marks show as white, and the white background is colored blue.

bluestone. (copper sulfate pentahydrate; blue vitriol).
CAS: 7758-98-7. $CuH_{10}O_9S$.
The pentahydrate salt of cupric sulfate. It is a potent emetic.
Use: As an antidote for poisoning by phosphorus, and to prevent the growth of algae.

blue verdigris. See copper acetate, basic.

blue vitriol. See copper sulfate.

blush. Precipitation of water vapor in the form of colloidal droplets on the surface of a varnish or lacquer film, caused by lowering of the temperature immediately above the coated surface due to solvent evaporation. This results in unsightly graying of the dried film and can be avoided by use of a less-volatile solvent.

4-BOC-styrene. See *p-tert*-butoxycarbonyloxystyrene monomer.

BOD. See biochemical oxygen demand.

Bodroux–Chichibabin aldehyde synthesis. Formation of aldehydes by treatment of orthoformates with Grignard reagents.

Bodroux reaction. Formation of substituted amides by reaction of a simple aliphatic or aromatic ester with an amino magnesium halide obtained by treatment of a primary or secondary amine with a Grignard reagent at room temperature.

body. (1) A nonspecific term approximately synonymous with consistency or viscosity, usually descriptive of liquids, e.g., a heat-bodied oil (linseed oil that has been polymerized by heating).
See consistency.
(2) In biochemistry, an agglutinous substance present in the blood or tissues, e.g., antibody.
See agglutination.
(3) An object having a unique physical property, e.g., blackbody.

body-centered structure. (cubic centered). The internal crystal structure of substances in which the equivalent lattice points, as determined by X-ray, are at the corners of the cell and at the center of the cube.

bodying. Gaining body, or thickening after standing, or by mixing with another ingredient.

Boeseken's method. Determination of the relative configurations of the hydroxyl groups on the reducing and vicinal carbon atoms in a cyclic sugar. Boric acid forms complexes with *cis* hydroxyl groups on vicinal carbon atoms, and the electrical conductivity of the solution is enhanced. Since there is no complex formation with *trans* hydroxyl groups, no increase in conductivity is observed.

Bogert–Cook synthesis. Condensation of β-phenylethylmagnesium bromide with cyclohexanones, followed by cyclodehydration of the tertiary alcohol with concentrated sulfuric acid, with formation of octahydrophenanthrene derivatives and a small amount of spiran.

boghead coal. A variety of bituminous or subbituminous coal resembling cannel coal in appearance and behavior during combustion, characterized by a high percentage of algal remains and volatile matter. Upon distillation it gives exceptionally high yields of tar and oil (ASTM).

Bohn–Schmidt reaction. Hydroxylation of anthraquinones containing at least one hydroxyl group by fuming sulfuric acid or sulfuric acid and boric acid in the presence of a catalyst such as mercury.

bohrium. Bh. A transfermium element. Atomic number 107. Very short half-life.

boiled oil. See linseed oil, boiled.

boiler compound. Material added to boiler feed water to reduce corrosion, foaming, or deposit formation.

boiler scale. A rocklike deposit occurring on boiler walls and tubes in which hard water has been heated or evaporated. Consists largely of calcium carbonate, calcium sulfate, or similar materials, depending on the mineral content of the water. Boiler scale decreases the rate of heat transfer through the boiler and tube walls, resulting in increased heating costs and shortening of boiler life. Most boiler feed water is softened (treated to remove calcium and magnesium ions) before use. Scale may be removed by treatment with ammonium bicarbonate solution.
See water, hard; zeolite.

boiling point. The temperature of a liquid at which its vapor pressure is equal to or very slightly greater than the atmospheric pressure of the environment. For water at sea level it is 212F (100C).

boiling-point elevation, molecular. The rise in boiling point produced by 1 gram-molecule of dissolved material in 1 kg solvent. Equal to 0.52C with water as the solvent.

boiling point, initial. The temperature at which the first drop of distillate falls from a condenser.

bois de rose oil.
Properties: From steam distillation of chipped wood of *Aniba rosaeodora* var. *amazonica* Ducke, (Fam. Lauraceae). Colorless to pale yellow liquid; slt pleasant floral odor. Sol in fixed oils, propylene glycol, mineral oil; sltly sol in glycerin.
Use: Food additive.

bolaffinine. (bolaffinin). A protein isolated from the mushroom, *Boletus affinis* (family Bolet-aceae).
Hazard: Toxic; poison.

bolesatine. A protein isolated from *Boletus satanas* that inhibits in vitro protein synthesis.

Bolstar. Sulprofos.

Bolton's reagent. A composite formed by 78% picric acid, 2% nitric acid, 20% water.

Boltzmann constant. The ratio of the molar ideal-gas constant to Avogadro's number.

Boltzmann engine. Ideal thermodynamic device working in cycles.

Boltzmann entropy hypothesis. The entropy of a system of material particles is proportional to the logarithm of the statistical probability of the distribution.

Boltzmann factor. A correction for thermal excitation in calculation of spectral-line intensities.

Boltzmann, Ludwig. (1844–1906). Born in Vienna, Boltzmann was interested primarily in physical chemistry and thermodynamics. His work has importance for chemistry because of his development of the kinetic theory of gases and the rules governing their viscosity and diffusion. The mathematical expression of his most important generalizations is known as Boltzmann's law, still regarded as one of the cornerstones of physical science.

bomb. A small metal container that can contain gases or liquids under varying degrees of pressure. An aerosol bomb contains liquids that are emitted as an atomized spray on release of pressure, the gases used being carbon dioxide, nitrous oxide, butane, etc. at relatively low pressure. An oxygen bomb is used for accelerated aging tests for rubber and plastic products; oxygen under high pressure is used. This device must be handled by a trained technician.

bombardment. Impingement upon an atomic nucleus of accelerated particles such as neutrons or deuterons for the purpose of inducing fission or of creating unstable nuclei. This operation was first accomplished with positively charged particles in the cyclotron in the early 1940s and subsequently in nuclear reactors. Neutrons are commonly used in reactors because their lack of electrical charge permits easier penetration of the target nucleus. See radioisotope; fission; fast-atom bombardment.

bomb calorimeter. An instrument to obtain the caloric or thermal value of fuel or foods.

bombesin. (2-*l*-glutamine-6-*l*-asparaginealytesin; alytesin). $C_{71}H_{110}N_{24}O_{18}S$. A tetradecapeptide.
Derivation: From the skin secretions of a genus of venomous disc-tongued frogs (family Discoglossidae).
Hazard: Toxic.

bomb tube. See Carius tube.

BON. Abbreviation for β-oxynaphthoic acid. See 3-hydroxy-2-naphthoic acid; BON red.

bonaid. See buquinolate.

Bonaril. A hydrolyzed polyacrylamide for use in foundry sands.

bond, chemical. An attractive force between atoms strong enough to permit the combined aggregate to function as a unit. A more exact definition is not possible because attractive forces ranging upward from 0 to those involving more than 250 kcal/mole of bonds are known. A practical lower limit may be taken as 2–3 kcal/mole of bonds, the work necessary to break approximately 1.5 ×

10^{24} bonds by separating their component atoms to infinite distance.
All bonds appear to originate with the electrostatic charges on electrons and atomic nuclei. Bonds result when the net coulombic interactions are sufficiently attractive. Different principal types of bonds recognized include metallic, covalent, ionic, and bridge.
Metallic bonding is the attraction of all the atomic nuclei in a crystal for the outer shell electrons which are shared in a delocalized manner among all available orbitals. Metal atoms characteristically provide more orbital vacancies than electrons for sharing with other atoms.
Covalent bonding results most commonly when electrons are shared by two atomic nuclei. Here the bonding electrons are relatively localized in the region of the two nuclei, although frequently a degree of delocalization occurs when the shared electrons have a choice of orbitals. The conventional *single* covalent bond involves the sharing of two electrons. There may also be *double* bonds with four shared electrons, *triple* bonds with six shared electrons, and bonds of intermediate multiplicity.
Covalent bonds may range from *nonpolar*, involving electrons evenly shared by the two atoms, to extremely *polar*, where the bonding electrons are very unevenly shared. The limit of uneven sharing occurs when the bonding electron spends full time with one of the atoms, making the atom into a negative ion and leaving the other atom in the form of a positive ion. Ionic bonding is the electrostatic attraction between oppositely charged ions.
Bridge bonds involve compounds of hydrogen in which the hydrogen bears either a + or − charge. When hydrogen is attached by a polar covalent bond to one molecule, it may attract another molecule, bridging the two molecules together. If the hydrogen is +, it may attract an electron pair of the other molecule. This is called a *protonic bridge*. If the hydrogen is −, it may attract through a vacant orbital the nucleus of an atom of a second molecule. This is called a *hydridic bridge*. Such bridges are at the lower range of bond strength but may have a significant effect on the physical properties of condensed states of those substances in which they are possible.
See hydrogen bond.

bond energy. The energy required to disrupt the bonds in one gram-mole of a chemical compound.

"Bonderite" [Henkel]. TM for chemical compositions for producing a corrosion-inhibiting finish on metals, preparing metal surfaces for the subsequent application of finish coats, and conditioning metal surfaces to facilitate metal-deformation operations.

"Bonderlube" [Henkel]. TM for soaplike chemical composition for treating metal surfaces that have been pretreated with phosphatizing coating chemicals in order to form a lubricant layer adapted to cold forming and in order to retard the formation of rust.

Bonding Agent M-3P.   1-aza-5-methylol-3,7-doxabicyclo-(3,3,0)-octane,   a   4%   partitioning agent.
Use: A methylene donor for improving rubber-to-fabric adhesion.

Bonding Agent R-6.   A resorcinol-acetaldehyde condensate.
Use: A resorcinol donor for improving rubber-to-fabric adhesion.

Bonding   Agent   TZ.   2-chloro-4,6-bis(N-phenyl-p-phenylene diamino)-1,3,5-triazine.
Use: A rubber-to-metal bonding agent that replaces the traditional cobalt salt.

bone ash.   An ash composed principally of tribasic calcium phosphate but containing minor amounts of magnesium phosphate, calcium carbonate, and calcium fluoride. Noncombustible.
Derivation: By calcining bones. A synthetic product is also available.
Use: Cleaning and polishing, ceramics, animal feeds. The better grades are used in coating molds for copper wire, bars, slabs, and other metals.

bone black.   (bone char; bone charcoal).
Black pigment made by carbonizing bones. Carbon content is usually approximately 10%. Nonflammable in bulk.
Hazard: Flammable as suspended dust.
Use: Manufacturing activated carbon; decolorizing agent and filtering medium, cementation reagent, adsorptive medium in gas masks, paint and varnish pigment; clarifying shellac, water purification.

bone china.   Ceramic tableware of high quality in which a small percentage of bone ash is incorporated. Made chiefly in England.

bone meal.   A product made by grinding animal bones. Raw meal is made from bones that have not been previously steamed. If pressure steaming has been used, the meal is called *steamed*. The fertilizer grade contains 43–55% tricalcium phosphate, 20–25% phosphoric acid, and 4–5% ammonia. The feed grade, according to Bureau of Animal Industry specifications, must contain 65–75% tricalcium phosphate and only approximately 2% ammonia. Much of the latter grade is imported.
Use: Fertilizer (raw); animal feeds (steamed).

bone oil.   (animal oil; Dippel's oil; hartshorn oil).
CAS: 8001-85-2.
Properties: Dark-brown, fixed oil; repulsive odor. D 0.900–0.980. Soluble in water. Combustible.

Chief constituents: Hydrocarbons, pyridine bases, and amines.
Derivation: Destructive distillation of bones or other animal substances. Note distinction from bone tallow.
Grade: Technical.
Hazard: Evolves very toxic ammonium cyanide when heated to 180C.
Use: Organic preparations, source of pyrrole, denaturant for alcohol.

bone phosphate.   (BPL). Phosphoric acid occurring in bones in the form of tribasic calcium phosphate.

bone seeker.   An element or radioisotope that tends to lodge in the bones when absorbed into the body. Examples are fluorine, calcium, and strontium.

bone tallow.   (bone fat). Fat obtained from animal bones by boiling in water, treating with steam under pressure, or solvent extraction. It is a glyceryl ester.

BON maroon.   A calcium or manganese precipitated compound of 3-hydroxy-2-naphthoic acid and 2-naphthylamine-1-sulfonic acid.
See BON red.

BON red.   Class name for a group of organic azo pigments made by coupling 3-hydroxy-2-naphthoic acid to various amines and forming the barium, calcium, strontium, or manganese salts. They have bright shades ranging from yellow-red to deep maroon, good light and heat resistance, nonbleeding in vehicles and solvents, and good opacity. They are widely used in printing inks, paints, enamels, lacquers, rubber, plastics, wallpaper, textiles, floor coverings, and crayons.

Boord olefin synthesis.   Regiospecific synthesis of olefins from aldehydes and Grignard reagents.

boort.   (bort).
See diamond, industrial.

BOPOB.   See p-bis[2-(5-p-biphenylyloxazoyl)]-benzene.

boracic acid.   See boric acid.

boral.   A composite material consisting of boron carbide crystals in aluminum with a cladding of commercially pure aluminum. Concentration of up to 50% boron carbide can be obtained.
Use: Reactor shields, neutron curtains, shutters for thermal curtains, safety rods, containers for fissionable material.
See composite.

"Boran" [LaMotte].   TM for diaminochrysazin used in the colorimetric determination of boron.

borane. One of a series of boron hydrides (compounds of boron and hydrogen). The simplest of these, BH_3, is unstable at atmospheric pressure and becomes diborane (B_2H_6) as gas at normal pressures. This is converted to higher boranes, i.e., penta-, deca-, etc., by condensation. This series progresses through a number of well-characterized crystalline compounds. Hydrides up to $B_{20}H_{26}$ exist. Most are not very stable and readily react with water to yield hydrogen. Many react violently with air. As a rule, they are highly toxic. Their properties have suggested investigation for rocket propulsion, but they have not proved satisfactory for this purpose. There are also a number of organoboranes used as reducing agents in electroless nickel-plating of metals and plastics. Some of the compounds used are di- and triethylamine borane and pyridine borane. See carborane; diborane; pentaborane; organoborane.

boranecarboxylic acid, ammonium salt. See ammonium boranecarboxylate.

borates, tetra, sodium salt, pentahydrate. See sodium tetraborate pentahydrate.

borax. (sodium borate; tincal; borax decahydrate). $Na_2B_4O_7 \cdot 10H_2O$. A natural hydrated sodium borate found in salt lakes and alkali soils. Also the commercial name for sodium borate. **Hazard:** Toxic.

borax, anhydrous. (borax, dehydrated; sodium borate, anhydrous). CAS: 1303-43-4. $Na_2B_4O_7$. **Properties:** White, free-flowing crystals; hygroscopic; forms partial hydrate in damp air. Mp 741C, d 2.367. Slightly soluble in cold water. Noncombustible. **Grade:** Technical, 99% $Na_2B_4O_7$, standard, fine granular form, glass or fused. **Hazard:** Toxic. **Use:** Manufacturing of glass, enamels, and other ceramic products; herbicide.

borax glass. Fused anhydrous borax used as a metal flux.

borax pentahydrate. $Na_2B_4O_7 \cdot 5H_2O$. **Properties:** Free-flowing powder. Begins to lose water of hydration at 122C, d 1.815. Noncombustible. **Grade:** Crude, technical (99.5% $Na_2B_4O_7 \cdot 5H_2O$). **Hazard:** Toxic. **Use:** Weed killer and soil sterilant, fungus control on citrus fruits (FDA tolerance 8 ppm of boron residue).

borazole. (borazine). $B_3N_3H_6$. Inorganic analog of benzene.

Properties: Colorless liquid. Fp −58C, bp 53C, d 0.824 (0C), hydrolyzes to evolve boron hydrides. **Hazard:** Dangerous fire risk. Toxic via ingestion and inhalation, strong irritant to tissue.

Borcher's metal. A group of alloys of chromium with nickel and cobalt or of chromium and iron with a small proportion of molybdenum and/or silver or gold. Heat and corrosion resistant. **Use:** Chemical apparatus, crucibles, pyrometer tubes, heat treating, and annealing pots.

Borcher's process. An electrolytic method for refining silver by using silver sheet cathodes and dilute nitric acid electrolyte.

bordeaux direct. See C.I. direct red 13, disodium salt.

bordeaux mixture. **Properties:** Mixture of cupric sulfate, calcium hydroxide and water. **Use:** Fungicide on vines and fruit trees.

Bordeaux mixture. A fungicide and insecticide mixture made by adding slaked lime to a copper sulfate solution. It is either made by the user or bought as a powder ready for dissolving. Stabilizing agents are sometimes added to delay settling. Used especially for potato bugs and similar garden pests. **Hazard:** Irritant and corrosive by ingestion.

boric acid. (boracic acid; orthoboric acid). CAS: 10043-35-3. H_3BO_3. **Properties:** Colorless, odorless scales or white powder; stable in air. D 1.4347 (15C), mp indeterminate since it loses water in stages through metaboric acid, HBO_2, to pyroboric acid, $H_2B_4O_7$, and to the oxide, B_2O_3. Soluble in boiling water, alcohol, and glycerol. Noncombustible. **Derivation:** (1) By adding hydrochloric or sulfuric acid to a solution of borax and crystallizing. (2) From weak borax brines by extraction with a kerosene solution of a chelating agent such as 2-ethyl-1,3-hexanediol or other polyols. Borates are stripped from the chelate by sulfuric acid. **Method of purification:** Recrystallization. **Grade:** Technical, 99.9%, CP, USP. **Hazard:** Toxic via ingestion. Use only weak solutions. Irritant to skin in dry form. **Use:** Heat-resistant (borosilicate) glass, glass fibers, porcelain enamels, boron chemicals, metallurgy (welding flux, brazing copper), flame-retardant in cellulosic insulation, mattress batting and cotton textile products, fungus control on citrus fruits (FDA tolerance 8 ppm boron residue), ointment and eye wash (water solution only), nickel electroplating baths.

boric acid esters. (borate ester; trimethyl borate; tri-*n*-butyl borate; tricyclohexyl borate; tridodecylborate; tri-*p*-cresyl borates).

Trihexylene glycol biborate compounds that are readily hydrolyzed to boric acid and the respective alcohols.
Properties: Colorless to yellow liquids. Bp 230–350C. Combustible.
Use: Dehydrating agents, catalysts, sources of boric oxide, special solvents, stabilizers, plasticizers or adhesion additives to latex paints, ingredients of soldering and brazing fluxes.

orthoboric acid. See boric acid.

boric anhydride. See boric oxide.

boric oxide. (boric anhydride; boron oxide).
CAS: 1303-86-2. B_2O_3.
Properties: Colorless powder or vitreous crystals; slightly bitter taste. D 2.46, mp approximately 450C, bp 1500C. Soluble in alcohol and hot water; slightly soluble in cold water. Noncombustible.
Derivation: By heating boric acid.
Grade: Glass or fused form, powder, technical, or highpurity 99.99+%.
Hazard: Eye and upper respiratory tract irritant.
Use: Production of boron, heat-resistant glassware, fire-resistant additive for paints, electronics, liquid encapsulation techniques, herbicide.

boride. An interstitial compound of boron and another metal (transition, alkaline-earth, or rare-earth). Such compounds are not stoichiometric, the boron atoms being linked together in zigzag chains, two-dimensional nets, or three-dimensional structures throughout the crystal.
Properties: Color varies from gray (transition-metal) to black (alkaline-earth) to blue (rare-earth). Highly refractory, with mp from 2000 to 3000C; Mohs hardness from 8 to 10; thermally and electrically conductive. High chemical stability. Does not react with hydrochloric or hydrofluoric acids but is attacked by hot alkali hydroxides.
Derivation: (1) Sintering mixtures of metal powder and boron at 2000C; (2) reduction of mixture of the metal oxide and boric oxide with aluminum, silicon, or carbon; (3) fused-salt electrolysis; (4) vapor-phase deposition.
Use: High-temperature service such as rocket nozzles, turbines, etc.

boron compound.
CAS: 7440-42-8.
Any of a class of inorganic or organic compounds that contain boron as an integral part of the molecule.
Hazard: Very toxic; industrial poison; causes depression of the circulation; persistent vomiting; diarrhea; shock and coma.

borneol. (bornyl alcohol; 2-camphanol; 2-hydroxycamphane).
CAS: 507-70-0. $C_{10}H_{17}OH$.
Properties: White, translucent lumps; sharp, camphorlike odor; burning taste. Optically active in natural form, racemic form made synthetically. D 1.011, mp 208C, bp 212C. Soluble in alcohol and ether; insoluble in water. Flammable.
Derivation: Natural form from a species of tree in Borneo and Sumatra. Synthesized from camphor by hydrogen reduction or from α-pinene.
Grade: Technical.
Hazard: Fire risk in presence of open flame.
Use: Perfumery, esters.

Born equation. An equation representing the free energy of solvation of an ion.

bornyl acetate.
$C_{10}H_{17}OOCCH_3$.
Properties: Colorless liquid; piney-camphoraceous odor. D 0.980–0.984, refr index 1.463–1.465, mp 29C. Solidifies to crystals at approximately 50F. Soluble in 3 volumes of 70% alcohol; miscible with 95% alcohol and ether. Combustible.
Derivation: Interaction of borneol and acetic anhydride in the presence of formic acid.
Grade: Technical, FCC.
Use: Perfumery, flavoring, nitrocellulose solvent.

bornyl alcohol. See borneol.

2-bornyl chloride.
CAS: 464-41-5. $C_{10}H_{17}Cl$.
Hazard: A poison by inhalation. Moderately toxic by ingestion.

bornyl formate. $C_{10}H_{17}OOCH$.
Properties: Colorless liquid having a piney odor. D 1.007–1.009. Combustible.
Grade: Technical.
Use: Perfuming of soaps, disinfectants, and sanitary products; flavoring.

bornyl isovalerate.
$C_{10}H_{17}OOC_5H_9$.
A constituent of valerian oil.
Properties: Limpid fluid; aromatic, valerian-like odor. D 0.951 (20C), bp 255–260C. Soluble in alcohol and ether; insoluble in water. Combustible.
Use: Medicine, essential-oil intermediate, flavoring.

boroethane. See diborane.

boron.
CAS: 7440-42-8. B.
Nonmetallic element of atomic number 5; group IIIA of the periodic table. Aw 10.81. Valence 3. Two stable isotopes: 11 (approximately 81%) and 10 (approximately 19%).
Properties: Black, hard solid; brown, amorphous powder; crystals. D 2.45, mp 2300C, Mohs hardness 9.3. Highly reactive. Soluble in concentrated nitric and sulfuric acids; insoluble in water, alcohol, and ether. High neutron absorption capacity. Amphoteric. A plant micronutrient.
Source: Borax, kernite, colemanite, ulexite.

Derivation: (1) By heating boric oxide with powdered magnesium or aluminum; (2) by vapor-phase reduction of boron trichloride with hydrogen over hot filaments (80–2000C); (3) by electrolysis of fused salts.

Available forms: Filament, powder, whiskers, single crystals.

Grade: Technical (90–92%), 99% pure, high-purity crystals.

Hazard: Dust ignites spontaneously in air; severe fire and explosion hazard. Reacts exothermally with metals above 900C; explodes with hydrogen iodide.

Use: Special-purpose alloys, cementation of iron, neutron absorber in reactor controls, oxygen scavenger for copper and other metals, fibers and filaments in composites with metals or ceramics, semiconductors, boron-coated tungsten wires, rocket propellant mixtures, high-temperature brazing alloys.

See boron alloy; boron fiber; boron-10.

boron-10. Nonradioactive isotope of boron of mass number 10.

Properties: Has marked capacity for absorbing slow neutrons, emitting a high-energy α-particle in the process.

Derivation: Constitutes approximately 19% of natural boron.

Available forms: Crystalline powder, dry amorphous powder, colloidal suspension of dry amorphous powder in oil; in boron trifluoride-calcium fluoride, in potassium borofluoride, in boron trifluoride ethyl etherate, in boric acid.

Use: Neutron counter, radiation shielding (in the form of boral), medicine.

boron alloy. A uniformly dispersed mixture of boron with another metal or metals. Ferroboron usually contains 15–25% boron, manganese boron usually contains 60–65% manganese.

Use: Degasifying and deoxidizing agents, to harden steel (in trace quantities), to increase conductivity of copper, turbojet engines.

boron bromide. See boron tribromide.

boron carbide. B_4C.

Properties: Hard, black crystals. D 2.6, Mohs hardness 9.3, mp 2350C, bp 3500C. Soluble in fused alkali; insoluble in water and acids. High capture cross-section for thermal neutrons.

Derivation: Heating boron oxide with carbon in an electric furnace.

Available forms: Powder, crystals, rods, fibers, whiskers.

Hazard: Avoid inhalation of dust or particles.

Use: Abrasive powder, abrasion resister and refractory, control rods in nuclear reactors, reinforcing agent in composites for military aircraft, and other special applications.

See boral.

boron chloride. See boron trichloride.

boron fiber. A vapor-deposited filament made by deposition of boron on a heated tungsten wire. The filament is 0.004 inch in diameter, while the wire is only 0.0005 inch. Tensile strength 350,000–450,000 psi, elastic modulus 55 million psi, upper temperature limit in oxidizing atmosphere 250C. Used in composites with epoxy resins for aircraft and space applications. The fibers can be woven into fabrics.

boron fluoride. See boron trifluoride.

boron fuel. See rocket fuel.

boron hydride. See borane; diborane; decaborane; pentaborane.

boron nitride. BN.

Properties: White powder; particle diameter approximately 1 micron. Mp 3000C (sublimes). Graphite-like, hexagonal plate structure. High electrical resistance. Compressed at 10^6 psi, it becomes hard as diamond. Excellent heat-shock resistance, low mechanical strength, hygroscopic. Noncombustible.

Derivation: Heating a mixture of boric acid and tricalcium phosphate in ammonia atmosphere in an electric furnace.

Available forms: Powder, compressed solid, fibers, whiskers.

Use: Refractory, furnace insulation, crucibles, rectifying tubes, dielectric, chemical equipment, self-lubricating bushings, molten-metal pump parts, transistor and rectifier mounting wafers, heat-shield for plasma, nose-cone windows, heat-resistant fibers stable to 870C in oxidizing atmosphere for military composites, metalworking abrasive, high-temperature insulator, high-strength fibers.

boron oxide. See boric oxide.

boron phosphate. (Sometimes called borophosphoric acid).

BPO_4.

Properties: White, nonhygroscopic crystals. D 1.873. Soluble in water; pH (1% solution) 2.0.

Use: Special glasses, ceramics, acid cleaner, dehydration catalyst.

boron phosphide. BP.

A refractory, maroon powder; noncorrosive; Mohs hardness 9.5.

Derivation: Direct union of boron and phosphorus at approximately 1000C in a reducing atmosphere.

Hazard: Evolves toxic fumes in contact with water and acids. Ignites spontaneously at 390F (199C).

boron steel. See ferroboron.

boron tribromide. (boron bromide).

CAS: 10294-33-4. BBr_3.

Properties: Colorless, fuming liquid. Decomposed by alcohol and water. D 1.69 (15C), bp 90C, fp −46C.
Derivation: (High purity.) Direct bromination of boron followed by rectification in quartz columns.
Grade: Technical, high purity.
Hazard: Corrosive to tissue. May explode when heated. Upper respiratory tract irritant.
Use: Catalyst in organic synthesis manufacture of diborane.

boron trichloride. (boron chloride).
CAS: 10294-34-5. BCl_3.
Properties: Colorless, fuming liquid. D 1.35 (25C), bp 12.5C, fp −107C. Decomposed by alcohol and water. Reacts with hydrogen at 1200C.
Derivation: (1) Heating boric oxide and carbon with chlorine; (2) combining boric oxide with phosphorus pentachloride.
Grade: Technical (99%), CP (99.5%).
Hazard: Strong irritant to tissue. Fumes are corrosive and toxic.
Use: Catalyst in organic syntheses; source of many boron compounds; refining of alloys; soldering flux; making electrical resistors; extinguishing magnesium fires in heat-treating furnaces; manufacturing of diborane.

boron trifluoride. (boron fluoride).
CAS: 7637-07-2. BF_3.
Properties: Colorless gas. D 3.076 g/L, fp −126.8C, bp −101C. Does not support combustion. Soluble in cold water; hydrolyzes in hot water; soluble in concentrated sulfuric acid and most organic solvents. Easily forms double compounds such as that with ether, known as boron trifluoride etherate or BF_3-ether complex.
Derivation: From borax and hydrofluoric acid or from boric acid and ammonium bifluoride. The complex formed is then treated with cold-fuming sulfuric acid.
Grade: Pure (99% min).
Hazard: Toxic by inhalation, corrosive to skin and tissue. Lower respiratory tract irritant, and pneumonitis.
Use: Catalyst in organic synthesis, production of diborane, instruments for measuring neutron intensity, soldering fluxes, gas brazing.

boron trifluoride etherate. (BF_3-ether complex).
$CH_3CH_2O(BF_3)CH_2CH_3$.
A relatively stable coordination complex formed by the combination of diethyl ether with boron trifluoride, in which the boron atom is bonded to the oxygen of the ether.
Properties: Fuming liquid. Bp 259C (126C), flash p 147F (63.8C) (OC). Combustible.
Hazard: Toxic by inhalation, corrosive to skin and tissue.
Use: Catalyst in organic synthesis.

boron trifluoride-methanol.
(BF_3-MeOH).
A solution of boron trifluoride in methanol.
Properties: Concentrated 14 g/100 cc, d 0.90 (20C).
Hazard: Moderate fire risk. Combustible.
Use: Esterification reagent for fats and oils.

boron trifluoride monoethylamine. (boron fluoride monoethylamine; BF_3-MEA).
BF_3-$C_2H_5NH_2$.
Properties: White to pale-tan flakes. D 1.38, mp 88–90C. Soluble in furfuryl alcohol, polyglycol, acetone. Releases boron trifluoride above 110C.
Hazard: Moderate fire risk. Combustible.
Use: Elevated-temperature cure of epoxy resins.

borophosphoric acid. See boron phosphate.

borosilicate glass. See glass, heat-resistant.

Boro-Silicone. Fire- and heat-resistant field-castable elastomer with high hydrogen content. A solid material with resiliency to minimize impact due to secondary missile formation.

borotungstic acid. (borowolframic acid).
Various formulas and properties given.
Grade: Technical.
Use: Mineralogic assay.

Borsche–Drechsel cyclization. Acid-catalyzed rearrangement of cyclohexanone phenylhydrazone to tetrahydrocarbazole. Subsequent oxidation yields carbazole.

Bosch, Karl. (1874–1940) A German chemist who was the 1931 recipient of the Nobel Prize with Friederick Bergius. In World War I, his catalyst study led to the production of synthetic gasoline. He also worked in the area of chemical high-pressure methods. His research in ammonia synthesis aided in the manufacture of fertilizers and explosives. His doctorate was awarded in Liepzig, Germany.

Bosch–Meiser urea process. Industrial process for formation of urea by reaction of carbon dioxide with ammonia at elevated temperatures and pressures.

Bosch process. A method of recovering hydrogen from water gas wherein carbon monoxide is reacted with steam at 500C in the presence of catalysts to form carbon dioxide and hydrogen.

Bose–Einstein statistics. The statistical analysis of interspin particles such as radiation quanta.

bosentan.
CAS: 147536-97-8. $C_{27}H_{29}N_5O_6S$.
Hazard: A reproductive hazard.

boson.
A particle, such as a photon, with integral spin. Any number of bosons can occupy the same state and propagate an intense monochromatic light beam.

"Botran" [Pfizer].   TM for an agricultural fungicide, 2,6-dichloro-4-nitroaniline.

botulism.   An intense and often fatal poisoning caused by ingestion of the anaerobic bacterium *Clostridium botulinum*, a Gram-positive bacillus that proliferates in many kinds of inadequately sterilized canned food products those preserved at home. Death from respiratory paralysis occurs in from 30 to 65% of the cases. Heating to 80C or higher for 25 to 30 minutes before serving is effective protection against this powerful toxin. Extensive tests have indicated that meats such as bacon can be made resistant to botulism by treatment with a combination of potassium sorbate and sodium nitrite.

Bouchardat's solution.   Iodine dissolved in aqueous potassium iodide.

Bouin's fluid.   A tissue preservative composed of picric acid, formalin, and acetic acid.

bound water.   Water molecules that are tightly held by various chemical groups in a larger molecule. Carboxyl, hydroxyl, and amino groups are usually involved; hydrogenbonding is often a factor. Proteins tend to bind water in this way, and in meats it will remain unfrozen as low as −40C.

Bourdon gauge.   A pressure gauge using a flattened tube (Bourdon tube), that straightens out under pressure.

Bouveault aldehyde synthesis.   Action of Grignard reagents on *N,N*-disubstituted formamides yields aldehydes.

Bouveault–Blanc reduction.   Formation of alcohols by reduction of esters with sodium and an alcohol.

Boyer, Paul D.   (1918-). A chemist who won the Nobel Prize in 1997 for his work on elucidating the enzymatic mechanism by which ATP synthase (ATPase) catalyzes the synthesis of adenosine triphosphate (ATP), the energy source of living cells. He is a professor of biochemistry at the University of California, Los Angeles. He received a Ph.D. in Biochemistry in 1943 from the University of Wisconsin, Madison, U.S.A.
See Walker, John E., Skou, Jens C.

Boyle, Robert.   (1627–1691). A native of Ireland, Boyle devoted his life to experiments in what was then called "natural philosophy," i.e., physical science. He was influenced early by Galileo.

His interest aroused by a pump that had just been invented, Boyle studied the properties of air, on which he wrote a treatise (1660). Soon thereafter, he stated the famous law that bears his name (see following entry). Boyle's group of scientific enthusiasts was known as the "invisible college", and in 1663 it became the Royal Society of London. Boyle was one of the first to apply the principle that Francis Bacon had described as "the new method"—namely, inductive experimentation as opposed to the deductive method of Aristotle—and this became and has remained the cornerstone of scientific research. Boyle also investigated hydrostatics, desalination of seawater, crystals, electricity, etc. He approached but never quite stated the atomic theory of matter; however, he did distinguish between compounds and mixtures and conceived the idea of "particles" becoming associated to form molecules.

Boyle's law.   The volume of a sample of gas varies inversely with the pressure if the temperature remains constant. The relation is strictly true only for a perfect or ideal gas, but the law is satisfactory for practical calculations except when pressures are high or temperatures are approaching the liquefaction point. The van der Waals equation is a refinement that is adequate over a wider range.
See ideal gas.

B.   (bp).
Abbreviation for base pair(s). Double stranded DNA is usually measured in bp rather than nucleotides (nt).

BPIC.   Technical grade of terbutylperoxyisopropyl carbonate, a polymerization initiator for acrylic, ethylene, styrene, and other monomers, and a cross-linking agent for silicone and ethylene propylene elastomers.

BPL.   (1) Abbreviation for bone phosphate of lime.
See bone phosphate.
(2) Abbreviation for β-propiolactone.

BPR.   An insecticidal mixture containing varying proportions of pyrethrin, piperonyl butoxide, and rotenone in liquid or dust base.

Br.   Symbol for bromine.

BRA.   Abbreviation for β-resorcylic acid.

brackish water.   Water that is lower in salinity than normal seawater and higher in salinity than freshwater, ranging from 30 to 0.5 parts salt per 1000 parts water.

Bradsher reaction.   Acid-catalyzed cyclodehydration of *o*-acyldiarylalkanes to polycyclic hydrocarbons and their heterocyclic analogs.

bradykinin. ((2S)-2-[[(2S)-2-[[(2S)-1-[(2S)-2-[[(2S)-[[(2S)-2-[[2-[[(2S)-1-[(2S)-1-[(2S)-2-amino-5-(diaminomethylideneamino)penta-2-carbonyl]pyrrolidine-2-carbonyl]amino]acetyl]amino]-3-phenylpropanoyl]amino]-3-hydroxypropanoyl]pyrrolidiene-2-carbonyl]amino]-3-phenylpropanoyl]amino]-5-(diaminomethylideneamino)penta acid; BRS 640; Kallidin; PRS 640; syntetic bradykinin). CAS: 58-82-2. $C_{50}H_{73}N_{15}O_{11}$.
A nonapeptide messenger that is a potent but short-lived agent of arteriolar dialation.
Properties: Amorphous solid.
Derivation: Enzymatically produced from kallidin in the blood, released from mast cells during asthma attacks, from gut walls and from damaged tissues as a pain signal.
Hazard: Powerful vasodilator; increased capillary permeability; stimulates pain receptors; contraction of smooth muscle; teratogen; mutagenic.

Bragg angle. Characteristic angle of the diffraction of X-rays from planes of a crystal.

Bragg law. The equation describing how a system of parallel atomic layers in a crystal reflects a beam of X-rays with great intensity.

Bragg X-ray method. The X-ray examination of crystals using a single large crystal rotated through a small angle around an axis in a crystal face.

brake fluid. See hydraulic fluid.

branched chain. See chain.

branch migration. See Holliday intermediate.

bran oil. See furfural.

brasilin. (brazilin; brazilwood extract). $C_{16}H_{14}O_5$.
The crystalline, colorizing principle of brazilwood.
Properties: White or pale-yellow, rhombic needles from alcohol; turns orange in air or light. Soluble in water, alcohol, and ether; in alkalis giving a carmine-red color. Decomposes above 130C.
Use: Dyeing wood, ink, textiles, etc. red and purple shades. Acid-base indicator, turning yellow in acid and carmine red in alkali. Biological stain.

brass. Copper-zinc alloys of varying composition. Low-zinc brasses (below 20%) are resistant to stress-corrosion cracking and are easily formed. Red brass (15% zinc) is highly corrosionresistant. Yellow brasses contain 34–37% zinc, have good ductility and high strength, and can withstand severe cold-working. Cartridge brass contains 30–33% zinc. Muntz metal (40% zinc) is primarily a hot-working alloy used where cold-forming operations are unnecessary. Some brasses also contain low percentages of other elements, e.g., manganese, aluminum, silicon, lead, and tin (admiralty metal, naval brass).
Hazard: Flammable in powder or finely divided form.
Use: Condenser tube plates, piping, hose nozzles and couplings, oil gauges, flow indicators, air cocks, drain cocks, marine equipment, jewelry, fine arts, stamping dies.
See admiralty metal; aluminum brass; red brass; yellow brass; Muntz metal.

brassidic acid. (*trans*-13-docosenoic acid). $CH_3(CH_2)_7HC=CH(CH_2)_{11}COOH$.
An isomer of erucic acid.
Properties: White crystals. Mp 61–62C, bp 282C (30 mm), d 0.859, refr index 1.448 (57C). Insoluble in water; slightly soluble in alcohol; soluble in ether. Combustible.
Derivation: By treating erucic acid with nitrous acid (catalyst).

von Braun reaction. Reaction of tertiary amines with cyanogen bromide to form disubstituted cyanamides and an alkyl halide.

Brazil wax. See carnauba wax.

brazing. A welding method in which a nonferrous filler alloy is inserted between the ends or edges of the metals to be joined.
See welding.

Bredt's rule. A restriction applying to bridged systems, that states that in some bridged systems the branching points (the bridgeheads) cannot be involved in a double bond. As a corollary, reactions that should lead to such compounds will be hindered or will take a different course.

breeder. A particularly efficient type of nuclear reactor that is able to utilize the tremendous energy latent in ^{238}U. This cannot be exploited in conventional (thermal) nuclear reactors, which are fueled with enriched uranium or plutonium, for these eventually become depleted and must be replaced. The fuel used for the breeder reactor is a mixture of nonfissionable ^{238}U and ^{239}Pu sealed in long, thin hexagonal metal tubes, which are in turn contained in cans called subassemblies. These constitute the reactor core. Around it are placed several layers of U^{238}, also in subassemblies. When criticality is reached, the unmoderated neutron flux from the core permeates the entire system and thus "breeds" fissionable ^{239}Pu in the surrounding ^{238}U. The amount of fissionable material thus made available is about 100 times as great as that obtainable with a conventional reactor, since all the energy potential of the ^{238}U can be released. Twenty pounds of uranium has the potential of delivering approximately 52×10^6 kWh of electricity; only a small fraction of this would be extractable without breeding.

The breeder utilizes fast neutrons that are much more efficient than the slow (thermal) neutrons used in conventional reactors. Liquid sodium is the coolant in breeder reactors, as it has no retarding effect on the neutrons: 2.9 neutrons per fission are produced in the breeder compared with only 2.4 in water-moderated reactors. This excess of neutrons makes it possible for the fast breeder to produce more fuel than it consumes. Breeders have been operating on a commercial scale in several European countries for some years. The NRC authorized construction of the Clinch River breeder, but funding was canceled as a result of opposition by environmentalist and antinuclear groups. The only operating breeder reactor in the U.S. is Argonne's EBR-11 in Idaho.

breeze.   Coke particles less than one-half inch in diameter. This occurs to the extent of approximately 100 lb/ton of coal processed.

bretonite.   (iodoacetone).
CH_3COCH_2I.
Properties: Bp 102C.

brevetoxin.
CAS: 98225-48-0.
Hazard: A poison by ingestion.
Source: Natural product.

brewing.   The production of beer, ale, and malt liquors by a process involving a complex series of enzymatic reactions. The most important of these is the conversion of starch to a malt extract (wort), which in turn is fermented with yeast. Mashing is the preparation of wort from malt and cereals by enzymatic hydrolysis, after which the product is boiled with hops, which impart the characteristic taste and aroma of beer. The malt extract must contain the nutrients required for yeast growth. Mashing involves a complex interplay of chemical and enzymatic reactions that are not fully understood. Few changes have been made in the basic brewing processes for more than a century, but increased automatic operation and quality-control techniques ensure a consistently good product.
See fermentation; yeast; wort.

Brewster angle.   The angle of incidence at which a wave, polarized in the plane of the angle of incidence, undergoes a phase shift of one quadrant on reflection at the surface.

Brewster process.   A method for the extraction of acetic acid from the acid distillate of the destructive distillation of wood. Isopropyl ether is used as the solvent for the acetic acid.

brick, refractory.   A highly heat-resistant and nonconductive material used for furnace linings, as in the glass and steel industries and other applications where temperatures above 1600C are involved. Some types are made of quartzite or high-silica clay, others of metallic ores such as chromite, magnesite, and zirconia.
See refractory.

bridge.   See bond, chemical; hydrogen bond.

brightener.   A compound that when added to a nickel-plating formulation of the Watts type (nickel sulfate and nickel chloride in a six-to-one ratio plus boric acid) will yield a bright, reflective finish. There are two types: (1) naphthalenedisulfonic acids, diphenyl sulfonates, aryl sulfonamides, etc., which give bright deposits on polished surfaces; and (2) metal ions having high hydrogen overvoltage in acid solution (zinc, cadmium, selenium, etc.) and unsaturated organic compounds such as thiourea, acetylene derivatives, azo dyes, etc., which give mirror brightness as a result of their "leveling" action. Usually both types are used for maximum effectiveness.
See leveling (2); optical brightener.

brightening agent.   (optical bleach).
A colorless substance that absorbs UV radiation and produces a bluish hue complementary to the yellow tint of an off-white substrate.

bright stock.   Lubricating oil of high viscosity obtained from residues of petroleum distillation by dewaxing and treatment with fuller's earth or similar material. Sometimes also applied to viscous petroleum distillates.
Use: For blending with neutral oils in preparing automotive engine lubricating oils.

Brij.   A series of emulsifiers and wetting agents developed for use in emulsions of high alkalinity or acidity. They are polyoxyethylene ethers of higher aliphatic alcohols. Soluble in water and lower alcohols. Insoluble in coal tar hydrocarbons.

brilliant crocein.   (CI 17190; Crocein Scarlet MOO).
$C_6H_5N_2C_6H_4N_2C_{10}H_4OH(S_3Na)_2$.
Properties: Light-brown powder, cherry-red solution in water.
Use: To dye wool and silk red from acid solution and cotton and paper with aid of a mordant. Also used for red lakes, biological stain.

brilliant green.   (CI 42040; Malachite Green G).
$C_{27}H_{34}N_2O_4S$.
Properties: Yellow crystals. Soluble in water and alcohol.

Derivation: Condensation of benzaldehyde with diethylaniline in presence of sulfuric acid, followed by oxidation of the triphenylmethane product formed and conversion to sulfate.

Use: Dyeing textiles, inks, etc.; indicator; staining bacteria; antiseptic.

See Malachite Green.

brilliant toning red amine. See 4-amino-2-chlorotoluene-5-sulfonic acid.

brimonidine.
CAS: 59803-98-4. $C_{11}H_{10}BrN_5$.
Hazard: A poison by ingestion.

brimstone. Lumps or blocks of sulfur obtained in refining of sulfur. It collects on the floor of the condensing chamber. where it is cast into sticks.
See sulfur.

brine. Any solution of sodium chloride and water, usually containing other salts also. The most industrially important brines are (1) in subterranean wells as in Michigan; (2) in desert lakes such as the Great Salt Lake, Searles Lake, Salton Sea, and Dead Sea; and (3) in the ocean. These are the sources of many inorganic chemicals such as soda ash, sodium sulfate, potassium chloride, bromine, chlorine, borax, etc. Brines are also used for the preservation and pickling of certain vegetables, meat curing, and freezing mixtures. Concentrations range from 3% (ocean) to 20% or more.

Large areas of sand and shale containing brines under high pressure exist along the Gulf Coast. These are reported to be an important undeveloped source of natural gas and other hydrocarbons suitable for fuel or petrochemical feedstocks.
See desalination; demineralization.

Brinell hardness test. The standard method of measuring the hardness of metals. The smooth surface of the metal is dented by a steel ball under force. The standard load and time are 500 kg for 60 seconds for soft metals and 3000 kg for 30 seconds for steel and other hard metals. The size (diameter) of the resulting dent is measured, and the hardness determined from a chart or formula.

brisance. The shattering power of an explosive measured by the ratio of the weight of graded sand shattered when a charge of the test explosive is detonated in a standard manner to the weight of sand shattered by TNT detonated in the same manner.

britannia metal. See pewter; white metal.

"Britesorb" *[PQ].* (silica gel).
CAS: 7631-86-9.
TM for adsorbent for proteins and other materials.
Use: Beer, wine, and cooking oil.

British anti-Lewisite. See 2,3-dimercapto-propanol.

British thermal unit. See Btu.

brittle point. The temperature at which a sample shatters on application of pressure. This is slightly above the transition point.

Brix degree. A measure of the density or concentration of a sugar solution. The number of degrees Brix equals the percentage by weight of sucrose in the solution and is related empirically to the density.

brodifacoum. (3-[3-(4'-bromo[1,1'-biphenyl]-4-yl)-1,2,3,4-tetrahydro-1-maphthalenyl]-4-hydroxy-2h-1-benzopyran-2-one; 3-[3-[4-(4-bromophenyl)phenyl]-1,2,3,4-tetrahydronaphthalen-1-yl]-2-hydroxychromen-4-one).
CAS: 56073-10-0. $C_{31}H_{23}BrO_3$.
Properties: Whitish, water-insoluble, anticoagulant powder.
Hazard: Poison, anticoagulant.
Use: Rodenticide.

Broenner acid. (2-naphthylamine-6-sulfonic acid).
$C_{10}H_6(NH)_2SO_3H$.
Properties: Colorless needles. Soluble in boiling water.
Derivation: Heating sodium-2-naphthol-6-sulfonate with concentrated ammonia at 180C in an autoclave.
Grade: Technical. Available as the sodium salt, an odorless gray-to-pink powder.
Use: Azo dye intermediate.

Brom 55.
CAS: 77-48-5.
1,3-dibromo-5,5-dimethylhydantoin.
Use: Organic synthesis and as a disinfectant or sanitizer.

bromacil. (5-bromo-3-*sec*-butyl-6-methyluracil).
CAS: 314-40-9.
Substitute approved by EPA for some uses of 2,4,5-T.
Hazard: Possible carcinogen. Thyroid effects.
Use: Herbicide.

bromadialone. (3-[3-(4N-bromo[1,1n-biphenyl]-4-yl)-3-hydroxy-1-phenylpropyl]-4-hydroxy-2h-1-benzopyran-2-one; 3-[α-{p-(p-bormophenyl)-β-hydroxyphenethyl]-benzyl]-4-hydroxycoumarin; 3-[3-[4-(4-bromophenyl)phenyl]-3-hydrox-1-phenylpropyl]-2-hydroxychromen-4-one; bromadiolone; canadien 2000; contrac; (hydroxyl-4-columbarinyl 3)-3-phenyl-3-(bromo-4-biphenylyl-4)-1-pro-panol-1; LM-637; MAKI; Ratimus; Super-caid; super-rozol; sup'operats; temus).
$C_{30}H_{23}BrO_4$.
Properties: Whitish, slightly water-soluble powder.

Hazard: Deadly poison, vitamin K antagonist.
Use: Rodenticide.

bromal.   See tribromoacetaldehyde.

bromate.
CAS: 15541-45-4. BrO_3.
The anion, BrO_3, of bromic acid.
Hazard: Toxic; flammable; neurotoxic; likely to produce cancer.

bromatotoxin.   (bormatotoxismus).
A toxicant in food.
Derivation: Results from fermentation.

bromcresol green.   (tetrabromo-*m*-cresolsulfonphthalein).
An acid-base indicator showing color change from yellow to blue over the pH range 3.8–5.4
Properties: Yellow crystals. Mp 218C. Slightly soluble in water; soluble in alcohol.
See indicator.

bromcresol purple.   (dibromo-*o*-cresolsulfonphthalein).
An acid-base indicator that changes from yellow to purple between pH 5.2 and 6.8.
Properties: Yellow crystals. Mp 241C. Insoluble in water; soluble in alcohol.
See indicator.

bromelin.   (bromelain).
CAS: 9001-00-7.
A milk-clotting proteolytic enzyme. It is precipitated from pineapple juice with acetone or ammonium sulfide.
Use: Biochemical research, meat-tenderizing formulations, texturizer in baking, medicine.

bromeosin.   See eosin.

bromethalin.   (benzenamine;   2,4-dinitro-*n*-methyl-*n*-(2,4,6-tribromophenyl)-6-(trifluoromethyl;   bromoethaline;   4,6-dinitro-*n*-methyl-*N*-(2,4,6tribromophenyl)-α,α,α-trifluor *o-o*-toluidene; EL 614; gold crest vengeance; lilly 126714;   *o*-toluidene;   4-6-dinitro-*N*-methyl-*N*-(2,4,6-tribromophenyl)-α,α,α-trifluor   *o*; vengeance;   *n*-methyl-2,4-dinitron-*n*-(2,4,6-tribromophenyl)-6-(trifluoromethyl)benzenamine; *N*-methyl-2,4-dinitro-*N*-(2,4,6-tribromophenyl)-6-(trifluoromethyl)aniline).
CAS: 63333-35-7. $C_{14}H_7Br_3F_3N_3O_4$.
Properties: Pale yellow, water-insoluble crystalline substance.
Hazard: Poisonous; moderately toxic; causes headache, confusion, personality changes, seizures, coma, death.
Use: Rodenticide.

bromic acid. $HBrO_3$.
Properties: Colorless or slightly yellow liquid; turns yellow on exposure; unstable except in very dilute solution. D 3.28, bp decomposes at 100C. Exists only in water solution.
Derivation: Sulfuric acid is added to a solution of barium bromate, and the product is recovered by subsequent distillation and absorption in water.
Hazard: By ingestion and inhalation. Strong irritant to tissue.
Use: Dyes, intermediates, pharmaceuticals, oxidizing agent.

bromide compound.
Any salt or ester of hydrobromic acid, or any binary compound of bromine in which bromine has a valence of −1. It is a cholinesterase inhibitor.
Hazard: Central nervous system depressant, elevated spinal fluid pressure, nausea, vomiting, drowsiness, irritability, ataxia, vertigo, confusion, mania, hallucinations, coma, death.
Use: In the treatment of myasthenia gravis and to reverse the effects of muscle relaxants.

bromide paper.
Photographic paper coated with an emulsion of silver bromide that usually contains a small amount of silver iodide.

brominate.
To introduce bromine into an organic molecule.

brominated camphor.   See camphor bromate.

bromine.
CAS: 7726-95-6.   Br.
Nonmetallic halogen element of atomic number 35, group VIIA of the periodic table. Aw 79.904. Valences 1, 3, 5 (valence of 7 also reported). There are two stable isotopes.
Properties: Dark, reddish-brown liquid; irritating fumes.Bp 58.8C, fp −7.3C, d 3.11 (20/4C), vap d vs. air (at 15C) 5.51, wt/gal 25.7 lb, specific heat 0.107 cal/g, refr index 1.647, dielectric constant 3.2. Soluble in common organic solvents; very slightly soluble in water. Attacks most metals, including platinum and palladium; aluminum reacts vigorously and potassium explosively. Dry bromine does not attack lead, nickel, magnesium, tantalum, iron, zinc, or (below 300C) sodium.
Derivation: From seawater and natural brines by oxidation of bromine salts with chlorine; solar evaporation (Great Salt Lake); from salt beds at Stassfurt and the Dead Sea.
Method of purification: Distillation.
Grade: Technical, CP, 99.8%, 99.95%.
Hazard: Toxic by ingestion and inhalation, severe skin irritant. Strong oxidizing agent, may ignite combustible materials on contact. Upper and lower respiratory tract irritant, and lung damage.
Use: Manufacture of ethylene dibromide (antiknock gasoline), organic synthesis, bleaching,

water purification, solvent, intermediate for fumigants (methyl bromide), analytical reagent, fireretardant for plastics, dyes, pharmaceuticals, photography, shrink-proofing wool.

bromine azide. (bromoazide).
CAS: 13973-87-0. BrN_3.
Properties: Crystals or red liquid. Mp approximately 45C, bp explodes. A strong oxidizing agent.
Hazard: Explosive when heated or shocked. Will ignite combustible materials on contact.
Use: Detonators and other explosive devices.

bromine chloride. $BrCl$.
Properties: Reddish-yellow, mobile liquid. Fp −66C, decomposes with evolution of chlorine at 10C. Soluble in water, carbon disulfide, ether. Readily hydrolyzes. Reacts with ammonia to form bromamines.
Hazard: Irritant. Oxidizing agent.
Use: Industrial disinfectant, especially for wastewaters.

bromine cyanide. See cyanogen bromide.

bromine iodide. See iodine monobromide.

bromine pentafluoride.
CAS: 7789-30-2. BrF_5.
Properties: Colorless, fuming liquid. D 2.466 (25C), fp −61C, bp 40.5C, vap press (21.1C) 7 psi. Reacts with every known element except inert gases, nitrogen, and oxygen.
Derivation: By reacting bromine, diluted with nitrogen and fluorine, in a copper vessel at 200C.
Grade: 98% min.
Hazard: Corrosive to skin and tissue. Explodes on contact with water. Eye and upper respiratory tract irritant.
Use: Synthesis, oxidizer in liquid rocket propellants.

bromine trifluoride.
CAS: 7787-71-5. BrF_3.
Properties: Colorless liquid. D 2.80, mp 9C, bp 125C, vap press (21.1C) 0.15 psi. Decomposed violently by water.
Derivation: See bromine pentafluoride.
Grade: 98% min.
Hazard: Corrosive to skin. Very reactive and dangerous.
Use: Fluorinating agent, electrolytic solvent.

bromine water. A mixture of 3.2 g bromine in 100 g water.
Use: A laboratory reagent.

bromkal 80.
CAS: 61288-13-9.
Hazard: Confirmed carcinogen.

bromlost. (blister gas).
See dibromodiethyl sulfide.

N-bromoacetamide. (NBA). $CH_3CONHBr$.
Properties: White powder with bromine odor. Mp 105–108C. Contains approximately 57% active bromine, decomposes appreciably above 26.6C.
Hazard: Emits very toxic fumes of bromine on heating.
Use: Brominating and oxidizing agent in organic synthesis.

2-bromoacetamide.
CAS: 683-57-8. C_2H_4BrNO.
Hazard: A poison by ingestion. Moderately toxic by skin contact. A severe skin and eye irritant.

bromoacetic acid.
CAS: 79-08-3. $CH_2BrCOOH$.
Properties: Colorless, deliquescent crystals. Keep from air and moisture. Mp 51C, bp 208C, d 1.93. Soluble in water, alcohol, and ether.
Derivation: By heating acetic acid and bromine.
Hazard: Strong irritant to skin and tissue.
Use: Organic synthesis, abscission of citrus fruit in harvesting.

bromoacetate. (2-bromoacetate).
$C_2H_2BrO_2$.
A salt of bromoacetic acid.

bromoacetone.
CAS: 598-31-2. $CH_2BrCOCH_3$.
Properties: Colorless liquid when pure, rapidly becomes violet even in absence of air. D 1.631, bp 136C (partial decomposition), fp −54C, vap d 4.75, vap press 9 mm Hg (20C). Soluble in acetone, alcohol, benzene, and ether; slightly soluble in water.
Derivation: By treating aqueous acetone with bromine and sodium chlorate at 30–40C.
Grade: Technical.
Hazard: Toxic by inhalation and skin contact. A lachrymator gas, strong irritant.
Use: Organic synthesis, tear gas.

bromoacetone cyanohydrin.
$CH_2BrC(OH)(CN)CH_3$.
Properties: Colorless liquid. D 1.584 (13C); bp 94.5C (5 mm Hg). Soluble in alcohol, ether, and water.
Derivation: Interaction of bromoacetone and hydrogen cyanide at approximately 0C.
Use: Organic synthesis.

bromoacetyl cellulose. used as a quantitative micro-method for the measurement of antibody in small amounts of serum (0.1–0.2 ml).

bromoallylene. See allyl bromide.

4-bromoaniline. (p-bromoanaline; 4-bromobenzeneamine).
CAS: 106-40-1. C_6H_6BrN.
Properties: Colorless, rhombic crystals. Mp 66C. Soluble in alcohol and ether; insoluble in cold water.

Derivation: Steam distillation of *p*-bromoacetanilide and sodium hydroxide or bromination of aniline.
Use: Azo dye manufacturing, preparation of dihydroquinazolines (with formaldehyde).

5-bromoanthranilic acid.   (2-amino-5-bromo-benzoic acid). $C_7H_6BrNO_2$.
Properties: Colorless crystals. Mp 217C. Soluble in acetone; partially soluble in alcohol, benzene, and acetic acid.
Use: Analytical reagent for metal determination (cobalt, copper, nickel, zinc).

bromoauric acid.   (gold tribromide acid). $HAuBr_4 \cdot 5H_2O$.
Properties: Dark, red-brown, needle crystals or granular masses; odorless; metallic and acidic taste. Mp 27C. Stable in air if pure, but deliquescent if chloride is present. Soluble in water and alcohol.
Derivation: By dissolving auric bromide in hydrobromic acid, concentration, and crystallization.

p-bromobenzaldehyde. BrC_6H_4CHO.
Properties: Solid. Mp 58C.
USE: Chemical intermediate.

bromobenzene.   (phenyl bromide).
CAS: 108-86-1. C_6H_5Br.
Properties: Heavy, mobile, colorless liquid. Pungent odor. D 1.499, wt/gal 12.51 lb, bp 156.6C, fp −30.5C, flash p 124F (51.1C), refr index 1.5625. Miscible with most organic solvents; insoluble in water. Autoign temp 1051F (566C). Combustible.
Derivation: Bromination of benzene in presence of iron.
Grade: Technical, pure.
Hazard: Skin irritant. Moderate fire risk.
Use: Solvent, top-cylinder compounds, crystallizing solvent, organic synthesis, lubricating- oil additive.

p-bromobenzenesulfonic acid.
$BrC_6H_4SO_3H$.
Properties: Crystallizes in needles. Mp 102–103C, bp 155C (25 mm Hg). Soluble in hot water and hot alcohol.

p-bromobenzoic acid. $C_6H_4BrCOOH$.
Properties: Colorless or reddish crystals. Mp 254C. Soluble in alcohol and ether; very slightly soluble in water.
Derivation: From *p*-bromotoluene by oxidation.
Use: Organic synthesis, detection of strontium.

α-bromobenzyl cyanide.   (2-bromo-2-phynl-acetonitrile). C_8H_6BrN. A tear gas.
Hazard: Eye irritant.
Use: Military gas.

o-bromobenzyl cyanide.   (*o*-bromophenyl-acetonitrile; 2-bromo-α-cyanotoluene). $BrC_6H_4CH_2CN$.

Properties: Colorless solid or liquid. Mp 29C, d 1.519, bp 242C (decomposes). Soluble in organic solvents. Nonflammable.
Hazard: Strong lachrymator, irritant to tissue.

1-bromobutane.   See *n*-butyl bromide.

2-bromobutane.   See *sec*-butyl bromide.

5-bromo-3-*sec*-butyl-6-methyluracil.   See bromacil.

α-bromobutyric acid. $CH_3CH_2CHBrCOOH$.
Properties: Colorless, oily liquid. D 1.54, bp 181C (760 mm Hg), 214–217C (760 mm Hg) with decomposition, fp −4C. Soluble in alcohol and ether; sparingly soluble in water. Combustible.
Derivation: By heating bromine and butyric acid.
Hazard: Toxic by ingestion.
Use: Organic synthesis.

bromocarnallite.   An artificial carnallite in which chlorine is replaced by bromine.

bromochlorodifluoromethane.   See Halon 1211.

1-bromo-3-chloro-5,5-dimethylhydantoin.
CAS: 16079-88-2. $C_5H_6BrClN_2O_2$.
Hazard: Moderately toxic by ingestion and skin contact.
Use: Agricultural chemical.

3-bromo-1-chloro-5,5-dimethylhydantoin.
$BrCl(CH_3)_2C_3N_2O_2$.
Properties: Free-flowing, white powder; faint halogen odor. Mp 163–164C. Soluble in benzene, methylene dichloride, chloroform. Active bromine 33% min, active chlorine 14% min.
Hazard: See bromine, chlorine.
Use: Germicide and fungicide in treatment of water, disinfectant, halogenating agent, catalyst of ionic type, selective oxidant.

sym-bromochloroethane.   (ethylene chlorobromide).
CH_2BrCH_2Cl.
Properties: Colorless, volatile liquid; chloroform-like odor. D 1.70, bp 107–108C, wt/gal 14.9 lb (0C), fp −16.6C. Soluble in alcohol and ether; insoluble in water. Nonflammable.
Derivation: By action of bromine and chlorine on ethylene gas.
Hazard: By ingestion and inhalation, skin irritant.
Use: Solvent, especially for cellulose esters and ethers; organic synthesis; fumigant for fruits and vegetables.

7-bromo-6-chlorofebrifugine hydrobromide.
See halofuginone hydrobromide.

bromochloromethane.   (methylene chlorobromide; chlorobromomethane; Halon 1011).

CAS: 74-97-5. $BrCH_2Cl$.
Properties: Clear, colorless, volatile liquid; chloroform-like odor. D 1.93 (25C), bp 67.8C, fp −88C, refr index 1.48 (25C), vap d 4.46. Soluble in organic solvents; insoluble in water. Nonflammable.
Hazard: By inhalation.
Use: Fire extinguishers, organic synthesis.

2-(3-bromo-4-chlorophenyl)-4-chloro-5-((6-chloro-3-pyridinyl)methoxy)-3(2h)-pyridazinone.
CAS: 122322-22-9. $C_{16}H_9BrCl_3N_3O_2$.
Hazard: A poison by ingestion.
Use: Agricultural chemical.

4-bromo-2-(4-chlorophenyl)-1-ethoxymethyl-5-trifluoromethylpyrrole-3-carbonitrile.
CAS: 122453-73-0. $C_{15}H_{11}BrClF_3N_2O$.
Hazard: A poison by ingestion. Moderately toxic by inhalation.
Use: Agricultural chemical.

1-bromo-3-chloropropane. (trimethylene chlorobromide). $BrCH_2CH_2CH_2Cl$.
Properties: Colorless liquid. Fp <−50C, bp 143–145C, d 1.594 (25/25C), wt/gal 13.27 lb (25C), refr index 1.484 (25C). Insoluble in water; soluble in methanol and ether. Nonflammable.
Hazard: Toxic. Avoid inhalation of fumes.
Use: Organic synthesis, pharmaceuticals.

3-bromo-1-chloropropene.
CAS: 3737-00-6. C_3H_4BrCl.
Hazard: A poison by ingestion. Low toxicity by inhalation. Human systemic effects.

2-bromo-2-chloro-1,1,1,-trifluoroethane.
See halothane.

bromocresol green.
CAS: 76-60-8. $C_{21}H_{14}Br_4O_5S$.
Hazard: Moderately toxic.
See tetrabromo-m-cresolphthalein sulfone.

bromocriptine. (bromocriptin; α-bromoergocriptine; bromoergocryptine; 2-bromoergocryptine; 2-bromo-α-ergokryptin; 2-bromo-12′-hydroxy-2′-(1-methylethyl)-5′-α-(2-methylpropyl)ergoramin-3′,6′,18-trione; CB-154; (5nα)2-bromo-12n-hydroxy-2n-(1-methylethyl)-5n-(2-methylpropyl)ergotaman-3n,6n,18-trione; 2-bromoergocryptine; 2-bromo-α-ergokryptine).
CAS: 25614-03-3. $C_{32}H_{40}BrN_5O_5$.
An semisynthetic ergotamine alkaloid derivative and powerful dopamine D2 agonist. It inhibits prolactin secretion and release from the pituitary and retards tumor growth.
Properties: Crystals.
Use: in the treatment of endocrine disorders; as an antiparkinsonian.
Hazard: Poison; teratogen; developmental abnormalities of the respiratory system, musculoskeletal system, rogenital system, craniofacial area, and body wall; teratogen; mutagen; questionable carcinogen; tumorigen; causes nausea, vomiting, orthostatic hypotension; constipation, dyskinesias, psychoses, digital spasm, erythromelalgia.

2-bromo-α-cyanotoluene. See o-bromobenzyl cyanide.

bromocyclopentane. See cyclopentyl bromide.

bromodan. See bromomethylhexachlorobicycloheptene.

bromoderma. An acneform or granulomatous skin eruption due to hypersensitivity to, or prolonged use of, bromides.

4-bromo-2-(3,4-dichlorophenyl)-5-((6-iodo-3-pyridinyl)methoxy)-3(2h)-pyridazinone.
CAS: 122322-26-3. $C_{16}H_9BrCl_2IN_3O_2$.
Hazard: A poison by ingestion.
Use: Agricultural chemical.

(z)-1-(2-bromo-1,2-diphenylethenyl)-4-ethylbenzene. See cis-broparestrol.

1-bromododecane. See lauryl bromide.

bromoethane. See ethyl bromide.

2-bromoethyl alcohol. See ethylene bromohydrin.

2-bromoethylamine hydrobromide.
$BrCH_2CH_2NH_2 \bullet HBr$.
Use: Intermediate, suggested as a soldering flux.

bromoethyl chlorosulfonate.
$BrCH_2CH_2OSO_2Cl$.
Properties: Liquid. Bp 100–105C (18 mm Hg).
Derivation: Interaction of sulfuryl chloride and glycol bromohydrin.
Hazard: Irritant to skin and tissue.

bromofenoxim. (benzaldehyde; 3,5-dibromo-4-hydroxy-, (2,4-dinitrophenylbenzaldoxim-o-(2′,4′-dinitrophenyl)-aether; 3,5-dibromo-4-hydroxybenaldehyde 2,4-dinitrophenyl oxime; 3,5-dibromo-4-hydroxybenzaldehyde (2′,4′-dinitrophenyl) oxine; 3,5-dibromo-4-hydroxybenzaldehyde-o-(2′,4′-dinitrophenyl)oxime; faneron; 2,6-dibromo-4-[[(2,4-dinitrophenoxy)amino]mehylide-2,5-dien-1-one). $C_{13}H_7Br_2N_3O_6$. A nitrophenol.
Hazard: Moderately toxic.
Use: Herbicide for selective weed control in cereal crops.

p-bromofluorobenzene. C_6H_4BrF.
Properties: Colorless liquid. Bp 151–152C, fp −17.4C, d 1.593 (15C), refr index 1.5245 (25C). Insoluble in water.
Use: Intermediate, production of p-fluorophenol.

bromoform. (tribromomethane; methyl tribromide).
CAS: 75-25-2. $CHBr_3$.
Properties: Colorless, heavy liquid; odor and taste similar to those of chloroform. D 2.887, bp 151.2C, wt/gal 24 lb, boiling range 150.3–151.2C, fp 9C, surface tension 41.53 dynes/cm (20C), dielectric constant 4.5 (20C), refr index 1.6005. Soluble in alcohol, ether, chloroform, benzene, solvent naphtha, fixed and volatile oils; slightly soluble in water. Nonflammable.
Derivation: By heating acetone or ethanol with bromine and alkali hydroxide and recovery by distillation (similar to acetone process of chloroform).
Grade: Technical, pharmaceutical, spectrophotometric.
Hazard: A questionable carcinogen. By ingestion, inhalation, and skin absorption. Liver damage, eye and upper respiratory tract irritant.
Use: Intermediate in organic synthesis, geological assaying, solvent for waxes, greases, and oils, medicine (sedative).

bromobenzyl cyanide. (α-bromobenzeneacetonitrile; α-bromobenzyl cyanide; α=bromophenyacetonitrile)
CAS: 5798-79-8 C_8H_6BrN
Properties: Light yellow, gas, slightly water soluble, soluble in organic solvents.
Hazard: Toxic, lacrimator.
Use: Tear gas in riot control and a war gas, and agricultural fungicide to control certain fungal diseases of stone fruit.

1-bromohexane. See n-hexyl bromide.

(4-bromo-3-hydroxy-2-naphthoato)(8-quinolinolato)copper.
CAS: 14039-99-7. $C_{20}H_{12}BrCuNO_4$.
Hazard: A poison.

5-bromo-6-(2-imidazolin-2-ylamino)-quinoxaline. See brimonidine.

bromol. (2,4,6-tribromophenol).
CAS: 118-79-6. $C_6H_2Br_3OH$.
Properties: Soft, white needles; sweet taste; penetrating bromine odor. Sublimation point 96C, d 2.55 (20/20C), bp 244C. Soluble in alcohol, chloroform, ether, and caustic alkaline solution; almost insoluble in water.
Derivation: Action of bromine on phenol.
Hazard: By ingestion, inhalation, skin absorption. Strong skin irritant.

bromomethane. See methyl bromide.

bromomethylethyl ketone. $BrCH_2COC_2H_5$.
Properties: Colorless to pale-yellowish liquid. Affected by light. D 1.43, bp 145–146C (decomposes). Soluble in alcohol, benzene, ether, insoluble in water.

Derivation: Reaction of sodium bromide and methyl ethyl ketone in the presence of sodium chlorate.
Hazard: Strong irritant to skin and eyes.
Use: Organic synthesis.

bromomethylhexachlorobicycloheptene.
CAS: 1715-40-8. $C_8H_5BrCl_6$.
Hazard: Low toxicity by ingestion.
Use: Agricultural chemical.

α-bromonaphthalene. $C_{10}H_7Br$.
Properties: Colorless, thick liquid; pungent odor. D 1.4870, solidifies at 6.2C, bp 279C, refr index 1.6601. Soluble in alcohol, ether, and benzene; slightly soluble in water.
Derivation: Bromination of naphthalene.
Use: Organic synthesis, microscopy, refractometry of fats.

1-bromo-2-naphthol. $BrC_{10}H_6OH$.
Solid, mp 121–125C. Used as a dye intermediate.

bromonitropropanediol. (2-bromo-2-nitrol-1,3-propanediol; bndp, 2-bromo-2-nitropropan-1,3-diol; β-bromo-β-nitrotri-methyleneglycol; bronocot; bronopol; bronosol).
CAS: 52-51-7. $C_3H_6BrNO_4$.
Hazard: Toxic by all routes of exposure; skin irritant.
Use: An antiseptic.

2-bromopentane. $CH_3CH_2CHBrCH_3$.
Properties: Colorless to yellow liquid; strong odor. D 1.1850 (25/25C).

bromopheniramine maleate. (2-[p-bromo-α(2-dimethylaminoethyl)benzyl]pyridine bimaleate). $C_{16}H_{19}BrN_2 \cdot C_4H_4O_4$.
Properties: Crystals. Mp 130–135C. Soluble in water, less soluble in alcohol.
Grade: NF.
Use: Medicine (antihistamine).

p-bromophenol.
CAS: 106-41-2. $HO(C_6H_4)Br$.
Properties: Crystals. D 1.840 (15C), 1.5875 (80C), mp 64C, bp 238C. Slightly soluble in water; soluble in alcohol, chloroform, ether, and glacial acetic acid.
Use: Disinfectant.

bromophenol blue. (tetrabromophenolsulfonaphthalein).
Properties: An acid-base indicator, showing color change from yellow to purple over the range pH 3.0–4.6.

o-bromophenylacetonitrile. See o-bromobenzyl cyanide.

4-((3-((4-bromophenyl)amino)-4,5-dihydro-2h-benz(gndazol-2-yl)acetyl)morpholine.
CAS: 301644-27-9. $C_{23}H_{23}BrN_4O_2$.
Hazard: A poison by ingestion.

3-((4-bromophenyl)amino)-n-(2-ethoxy-ethyl)-4,5-dihydro-2h-benz(g)indazole-2-acetamide.
CAS: 301644-26-8. $C_{23}H_{25}BrN_4O_2$.
Hazard: A poison by ingestion.

2-(2-bromophenyl)-1h-benzimidazole.
CAS: 13275-42-8. $C_{13}H_9BrN_2$.
Hazard: Moderately toxic by ingestion.
Use: Agricultural chemical.

2-(4-bromophenyl)-4-chloro-5-((4-chloro-phenyl)methoxy)-3(2h)-pyridazinone.
CAS: 107359-69-3. $C_{17}H_{11}BrCl_2N_2O_2$.
Hazard: A poison by ingestion.
Use: Agricultural chemical.

3-(3-(4-(2-(4-bromophenyl)ethyl)phenyl)-1,2,3,4-tetrahydro-1-naphthalenyl)-4-hydroxy2h-1-benzopyran-2-one.
CAS: 90035-11-3. $C_{33}H_{27}BrO_3$.
Hazard: A poison by ingestion.
Use: Agricultural chemical.

2-(p-bromophenyl)imidazo(2,1-a)isoqui-noline.
CAS: 61001-06-7. $C_{17}H_{11}BrN_2$.
Hazard: A reproductive hazard.

5-((4-bromophenyl)methoxy)-4-chloro-2-(4-chloro-2-fluorophenyl)-3(2h)-pyridazinone.
CAS: 107359-76-2. $C_{17}H_{10}BrCl_2FN_2O_2$.
Hazard: A poison by ingestion.
Use: Agricultural chemical.

5-((4-bromophenyl)methoxy)-4-chloro-2-(4-chlorophenyl)-3(2h)-pyridazinone.
CAS: 107359-42-2. $C_{17}H_{11}BrCl_2N_2O_2$.
Hazard: A poison by ingestion.
Use: Agricultural chemical.

3-(3-(4-((4-bromophenyl)methoxy)phenyl)-1,2,3,4-tetrahydro-1-naphthalenyl)-4-hydroxy2h-1-benzopyran-2-one.
CAS: 90035-06-6. $C_{32}H_{25}BrO_4$.
Hazard: Moderately toxic by ingestion.
Use: Agricultural chemical.

(4-bromophenyl)methyl butyl 3-pyridin-ylcarbonimidodithioate.
CAS: 51308-80-6. $C_{17}H_{19}BrN_2OS$.
Hazard: Moderately toxic by ingestion.
Use: Agricultural chemical.

s-((4-bromophenyl)methyl) o-butyl 3-pyri-dinylcarbonimidothioate.
CAS: 51308-79-3. $C_{17}H_{19}BrN_2OS$.
Hazard: Moderately toxic by ingestion.
Use: Agricultural chemical.

(4-bromophenyl)methyl 1-methylethyl 3-pyridinylcarbonimidodithioate.
CAS: 34763-45-6. $C_{16}H_{17}BrN_2S_2$.
Hazard: Moderately toxic by ingestion.
Use: Agricultural chemical.

2-bromo-4-phenylphenol. $C_6H_5C_6H_3BrOH$.
Properties: Light-colored solid; faint characteristic odor. D 1.536 (25/4C), mp 93.6–95.6C, flash p 405F (207C), bp (decomposes at 18 mm Hg), 195–200C. Soluble in alkalies, most organic solvents; insoluble in water. Combustible.
Hazard: Toxic by ingestion and inhalation.
Use: Germicide.

3-(4-bromophenyl)-n-(4-propylcyclohexyl)-2-propenamide.
CAS: 315706-76-4. $C_{18}H_{24}BrNO$.
Hazard: A poison by ingestion.

4-bromophenyl trifluoroacetate.
CAS: 5672-84-4. $C_8H_4BrF_3O_2$.
Hazard: Moderately toxic by ingestion.

bromophosgene.   (carbonyl bromide; carbon oxybromide). $COBr_2$.
Properties: Heavy, colorless liquid; strong odor. D 2.5 (approximately 15C), bp 64–65C. Hydrolyzed by water. Decomposed by light and heat.
Derivation: Action of sulfuric acid on carbon tetra-bromide.
Hazard: Toxic by ingestion and inhalation.
Use: Military poison (toxic suffocant); making crystal-violet-type coloring agents.

bromopicrin.   (nitrobromoform; tribromoni-tromethane). CBr_3NO_2.
Properties: Prismatic crystals. Decomposes with explosive violence if heated rapidly. D 2.79 (18C), bp 127C (118 mm Hg), mp 103C. Soluble in alco-hol, benzene, and ether; slightly soluble in water.
Derivation: Action of picric acid on an aqueous solu-tion of bromine and calcium oxide followed by dis-tillation under reduced pressure.
Hazard: Powerful irritant. Severe explosion hazard when heated.
Use: Organic synthesis, military poison.

3-bromopropene.   See allyl bromide.

α-bromopropionic acid.   (2-bromopropionic acid). $CH_3CHBrCOOH$.
Properties: Colorless liquid. D 1.69, mp 24.5C, bp 203C (decomposes). Soluble in water, alcohol, and ether.
Derivation: By heating propionic acid with bromine.
Method of purification: Distillation.

4-bromo-n-(4-propylcyclohexyl)benzamide.
CAS: 315706-68-4. $C_{16}H_{22}BrNO$.
Hazard: A poison by ingestion.

3-bromo-1-propyne. See propargyl bromide.

2-bromopyridine. C_5H_4NBe.
Properties: Liquid. Bp 195C, d 1.627 (20C), refr index 1.5714 (20C). Solubility in 100 g water 2.08 (20C).
Use: Synthesis of pyridine compounds.

3-bromopyridine. C_5H_4NBr.
Properties: Needles. Bp 174.4C, d 1.628 (20/20C), refr index 1.5710 (20C). Slightly soluble in water; readily soluble in common organic solvents.

5-((6-bromo-3-pyridinyl)methoxy)-4-chloro-2-(4-chlorophenyl)-3(2h)-pyridazinone.
CAS: 122322-20-7. $C_{16}H_{10}BrCl_2N_3O_2$.
Hazard: A poison by ingestion.
Use: Agricultural chemical.

5-((6-bromo-3-pyridinyl)methoxy)-4-chloro-2-(3,4-dichlorophenyl)-3(2h)-pyridazinone.
CAS: 122322-21-8. $C_{16}H_9BrCl_3N_3O_2$.
Hazard: A poison by ingestion.
Use: Agricultural chemical.

5-bromosalicylhydroxamic   acid.
$BrC_6H_3(OH)CONH)OH$.
Properties: Crystals. Decomposes at 232C. Very slightly soluble in water, forming a water-soluble sodium salt.
Use: Medicine.

β-bromostyrene. (bromostyrol).
CAS: 103-64-0. $C_6H_5CHCHBr$.
Properties: Yellowish liquid; strong floral odor. D 1.395–1.424, refr index 1.602–1.608, mp min −2C. Soluble in 4 volumes of 90% alcohol.
Use: Perfumery.

bromosuccinic acid. $HOOCCH_2CHBrCOOH$.
Properties: Colorless crystals. D 2.073, mp 159–161C. Soluble in water and alcohol; insoluble in ether.
Derivation: By heating bromine and succinic acid.
Use: Organic synthesis.
Note: The above are properties of the *dl* form. Optically active forms are also known.

N-bromosuccinimide. (NBS) $(CH_2CO)_2NBr$.
Properties: Fine crystals; white to cream in color. Melting range 172–178C (decomposes). Soluble in carbon tetrachloride. 44.5% min active bromine.
Hazard: Use a respirator when handling dry material, which evolves toxic fumes of bromine. Strong irritant to eyes and skin.
Use: Controlled, low-energy bromination.

5-bromosulfamethazine. See sulfabromomethazine sodium.

6-bromo-2-thio-2h-1,3-benzoxazine-2,4(3h)-dione.
CAS: 23611-66-7. $C_8H_4BrNO_2S$.
Hazard: A poison by ingestion.
Use: Agricultural chemical.

4-bromothiophenol. BrC_6H_4SH.
Properties: White solid. Fp 73C, bp 239C. Almost insoluble in water; soluble in methanol and in alkaline solution (with which it reacts).
Hazard: Toxic by ingestion.
Use: Intermediate.

α-bromotoluene. See benzyl bromide.

p-bromotoluene. (*p*-tolyl bromide).
$BrC_6H_4CH_3$.
Properties: Crystals. Mp 28.5C, bp 184–185C, d 1.3898 (20C), refr index 1.5490, flash p 185F (85C). Combustible. Insoluble in water, soluble in alcohol, ether, and benzene.
Hazard: Toxic by ingestion.
Use: Intermediate.

bromotributylstannane. See tri-*n*-butyltin bromide.

bromotrichloromethane. (trichlorobromomethane).
CAS: 75-62-7. CCl_3Br.
Properties: Colorless, heavy liquid; chloroform-like odor. D 2.0; bp 104C; refr index 1.5051 (20C). Miscible with many organic liquids.
Hazard: Toxic by ingestion and inhalation of fumes.
Use: Organic synthesis.

bromotriethylstannane.
CAS: 2767-54-6. $C_6H_{15}BrSn$.
Properties: Colorless liquid. D 1.630, mp −13.5C, bp 221C Sol in org solvs.
Hazard: Moderately toxic by inhalation. A reproductive hazard.

bromotrifluoroethylene. (BFE).
$BrFC:CF_2$.
Gas (monomer) or liquid (polymer). The latter are usually clear oils at room temperature and solids at −55C. Viscosities and densities vary widely. Monomer: High-purity gas (97%). Shipped in cylinders.
Hazard: Flammable (gas or liquid). Dangerous fire risk.
Use: (BFE polymers). Flotation fluids for gyros or accelerometers used in inertial guidance systems. BFE polymers can also be used like CFE polymers.

bromotrifluoromethane. $CBrF_3$.
Properties: Colorless gas. Noncorrosive. Fp −168C, bp −58C, d at bp 8.71 g/L. Nonflammable.
Derivation: Bromination of fluoroform or perfluoropropane in nonmetallic reactor.
Hazard: Toxic by inhalation.

Use: Chemical intermediate, refrigerant, metal hardening, fire extinguishing.

bromotripentylstannane.
CAS: 3091-18-7. $C_{15}H_{33}BrSn$.
Hazard: A poison.

bromotripropylstannane.
CAS: 2767-61-5. $C_9H_{21}BrSn$.
Properties: Liquid. D 1.426 @ 25C/4C, mp −49C.
Hazard: Moderately toxic by inhalation.

5-bromouracil.   (5-bromo-2,4(1H,3H)-
pyrimidine-dione; bru; 5-bromo-1H-pyrimidine-
2,4-dione).
CAS: 51-20-7. $C_4H_3BrN_2O_2$.
A synthetic analog of thymine, in which a bromine
atom has replaced the methyl group. It is a potent
mutagenic pyrimidine analog that modifies the
base-pair sequencing of DNA by replacing thymine.
bp 133C.
Properties: Prisms from H_2O.
Hazard: Moderately toxic, alters DNA by replacing
thymine.
Use: An experimental mutagen.

bromoxynil.   (3,5-dibromo-4-hydroxybenzonit-
rile;   3,5-dibromo-4-hydroxyphenyl   cyanide;
2,6-dibromo-4-cyanophenol; broxynil).
CAS: 1689-84-5. $C_7H_3Br_2NO$.
Use: Post-emergence herbicide.
Hazard: Toxic.

bromoxynil octanoate.   See 2,6-dibromo-4-
cyanophenyl octanoate.

bromphenol blue. $C_{19}H_{10}Br_4O_5S$.
Properties: Prisms or crystals. Decomposes at 280C.
Soluble in sodium hydroxide solution; partially soluble in alcohol and benzene; almost insoluble in
water.
Use: Indicator (yellow = pH 3.0, purple = pH 4.6).

bromthymol blue. $C_{27}H_{28}Br_2O_5S$.
Properties: Yellow-white crystals. Soluble in alcohol and alkaline solution; slightly soluble in water.
Use: Indicator (yellow = pH 6.0, blue = pH 7.6).

brontin.   See elargin.

bronze.   An alloy of copper and tin usually containing from 1 to 10% tin. Special types contain from 5 to 10% aluminum (Al bronze), fractional percentages of phosphorus (phosphor bronze) as deoxidizer, or low percentages of silicon (Si bronze).
Hazard: Powder is flammable.
Use: Spark-resistant tools, springs, fourdrinier wire, paint, cosmetics (as powder), electrical hardware, vacuum dryers, blenders, water gauges, flow indicators, valves, drain cocks, fine arts.
See brass; phosphor bronze.

bronze blue.   Any of a number of varieties of
iron-blue pigments.

bronze orange.   See red lake C.

bronzing liquid.   (1) A solution of pyroxylin in
amyl acetate together with a bronze powder, usually aluminum bronze. (2) Gloss oils and aluminum
bronze. (3) Spirit varnishes and aluminum bronze.

cis-**broparestrol.**
CAS: 22393-63-1. $C_{22}H_{19}Br$.
Hazard: A reproductive hazard.

brosylate ester.   An ester of *p*-bromoben-
zenesulfonic acid.

broth.   A liquid nutrient medium, usually containing agar, used to promote the growth of bacteria in
the fermentation industry and to prepare cultures
for microbiological research.

brown algae.   See algae, brown.

Brown–Boverti test.   A way of determining the
oxidation resistance of oils.

Brown, Herbert C.   (1912–2004). An English-
born chemist who was the recipient of the Nobel
Prize for chemistry with Wittig in 1979. Via his
work in organic synthesis, Brown discovered new
routes to add substituents to olefins selectively. His
early education was irregular and disjointed as a
result of family circumstances and the economic
depression of the 1930s. He eventually received his
Ph.D. from the University of Chicago. The reduction of carbonyl compounds with diborane was the
topic of his thesis. The bulk of his career has been
spent at Purdue University.

Brownian movement.   The continuous zigzag
motion of the particles in a colloidal suspension,
e.g., rubber latex particles. The motion is caused
by impact of the molecules of the liquid upon the
colloidal particles. Named after the British botanist
Robert Brown, who first noted this phenomenon.

browning reaction.   (Maillard reaction).
A complicated and not completely evaluated
sequence of chemical changes occurring without
the involvement of enzymes during heat exposure
of foods containing carbohydrates (usually sugars)
and proteins, as well as during storage. It is responsible for the surface color change of bakery products and meats. It begins with an aldol condensation
reaction involving the carbonyl groups of the proteins, and ends with formation of furfural, which
produces the dark-brown coloration. Besides color
change, the reaction is accompanied by alterations
in flavor and texture, as well as in nutritive value. It
was first noted by the French chemist Maillard.

brucine. (dimethoxystrychnine).
CAS: 357-57-3. $C_{23}H_{26}O_4N_2 \cdot 2H_2O$ or $4H_2O$.
Properties: White, crystalline alkaloid; very bitter taste. Loses water at 100C, mp 178C. Soluble in alcohol, chloroform, and benzene; slightly soluble in water, ether, glycerol, and ethyl acetate. Forms brucine sulfate, hydrochloride, and nitrate (mp 230C). Also available as the sulfate.
Derivation: By extraction and subsequent crystallization from nux vomica or ignatia seeds.
Hazard: Poison by ingestion and inhalation.
Use: Denaturing alcohol, lubricant additive, separation of racemic mixtures.

brucite. (nemalite).
CAS: 1317-43-7. $Mg(OH)_2$. Natural magnesium hydroxide.
Properties: Colorless, white, gray, greenish; luster pearly or waxy. D 2.39; Mohs hardness 2.5.
Occurrence: Nevada, Washington, Canada.
Use: Refractories.

Brugma process. A distillation procedure, which is used in petroleum refining, wherein the number of columns equals the power to which 2 must be raised to give the desired number of fractions.

bryonia. The dried roots of *Bryonia spp* that contains several toxic glycosides including bryonin and bryonidin.

B-stage resin. (resitol). A thermosetting phenolformaldehyde-type resin that has been thermally reactive beyond the A stage so that the product has only partial solubility in common solvents (alcohols, ketones) and is not fully fusible even at 150–180C. The B stage resin has limited commercial use.

BT. (*Bacillus thuringiensis*). A species of bacteria used as a pesticide for agricultural crops. It is of the stomach-poison type and has been approved for commercial use.

BT-31 ASHAI DENKA. A line of organic and inorganic industrial chemicals, synthetic resins, synthetic rubber, high-compound fertilizers, coating materials, latexes, pharmaceutical and food additives, explosives, photopolymers and platemaking systems, separation and ion-exchange membranes, systems, and equipment.

BTDA. See 3,3′,4,4′-benzophenonetetracarboxylic dianhydride.

Btu. (British thermal unit).
The quantity of heat required to raise the temperature of 1 pound of water 1 degree F (usually from 39 to 40F). This is the accepted unit for the comparison of heating values of fuels. For example, fuel gases range from 100 (low producer gas) to 3200 (pure butane) Btu/cu ft. The usual standard for a city gas is approximately 500 Btu/cu ft.

BTX. Commercial abbreviation for benzene, toluene, xylene, the three major aromatic compounds.

Bu. Informal abbreviation for butyl.

bubble cap column. See tower, distillation.

bubbler cap plate. A part of distillation equipment for obtaining efficient contact between gases and liquids. The liquid flows over the surface of a perforated plate whereas the gas flows through the perforations.

"Bubreak" [Buckman]. TM for a defoamer.
Use: For leather processing.

Bucherer–Bergs reaction. Preparation of hydantoin from carbonyl compound by reaction with potassium cyanide and ammonium carbonate, or from the corresponding cyanohydrin and ammonium carbonate.

Bucherer carbazole synthesis. Formation of carbazoles from naphthols or naphthylamines, arylhydrazines, and sodium bisulfite.

Bucherer reaction. A procedure for preparation of β-naphthylamine by heating β-naphthol with a water solution of ammonium sulfite. "A sulfite solution is prepared by saturating concentrated ammonia solution with sulfur dioxide and adding an equal volume of concentrated ammonia solution, β-naphthol is added and the charge is heated in an autoclave provided with a stirrer or a shaking mechanism." (L. F. Fieser). This reaction is also involved in the preparation of several azo dye intermediates, e.g., Tobias acid.

Buchner–Curtius–Schlotterbeck reaction. Formation of keto compounds from aldehydes and aliphatic diazo compounds; ethylene oxides may also be formed.

Buchner, Eduard. (1860–1917). A German chemist who was awarded the Nobel Prize for chemistry in 1907. His works included the synthesis of diiodoacetamid through alcoholic fermentation caused by enzymes, as well as the discovery of zymase, the first enzyme to be isolated. He received his Ph.D. at the University of Munich, where he became a lecturer. Later, he taught and performed research at Tubingen, Berlin, and Wurzburg.

Buchner method of ring enlargement. Diazoacetic acid ester reacts with benzene and homologs to give the corresponding esters of non-caradienic acid, transformed at high temperatures to derivatives of cycloheptatriene, phenylacetic acid, and β-phenylpropionic acid (when one or more methyl groups are present in the initial hydrocarbon).

bucket elevator. See conveyor (5).

buckminsterfullerene. (buckyballs). C_{60}.

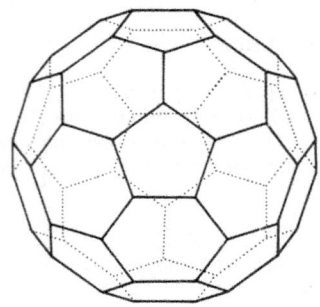

Spherical aromatic molecule with a hollow truncated-icosahedron structure, similar to a soccer ball. First reported in the mid-1980s. Capable of enclosing ions or atoms in a host-guest relationship.

buckthorn. (coyotillo; karwinskia humbold-tiana; tullidora).
Properties: A shrub which grows to 6 feet with small elliptical leaves 1 to 2 inches long. It produces a berr which turns black when mature and has a pit. It grows wild in western Texas and New Mexico.
Hazard: Berries contain poisonous anthracenones, which can cause loss of function in the peripheral nervous system including respiratory paralysis and death.

buckthorn toxins. A mixture of anthracenones in the fruit of Karwinskia humboldtiana that produce an often fatal demyelinating neuropathy of peripheral nerves in humans and all forms of livestock.

buckyballs. See buckminsterfullerene.

buckytube.
CAS: 7440-44-0.
Properties: Sequence of carbon atoms augmented with other compounds during manufacture.
Use: Superconductor of electrical current.

buclizine hydrochloride. $C_{28}H_{33}ClN_2 \cdot 2HCl$.
1-p-chlorobenzhydryl-4-(p-(*tert*)-butylbenzyl)-piperazinedihydrochloride.
Use: Medicine (antihistamine).

Budde effect. The increase in volume of halogen vapors on exposure to light.

Budium. A polybutadiene finish for application to tin plate.

bufadienolide. (5-[(8R,9S,10S,13S,14S,17S)-10,1-dimethyl-2,3,4,5,6,7,8,9,11,12,14,15,16-tetradecahydro-1H-cyclopenta[a]phenanthren-17-yl]pyran-2-one). $C_{24}H_{34}O_2$.
Any of a subclass of triterpenoids composed of 24 carbon atoms that are derivatives of bufanollide.

bufagin. (bufagenin). $C_{24}H_{34}O_5$.
Any of a group of steroid toxins found in toads of the family Bufonidae. Their action is similar to that of digitalis.
Hazard: Cardiotoxic.

bufaline. (3,14-dihydroxybufa-20,22-dienolide; 5-[(3S,5R,10S,13R,14S,17R)-3,14-dihydroxy-10,13-dimethyl-1,2,3,4,5,6,7,8,9,11,12,15,16,17-tetradecahydrocyclopenta[a]phe-17-yl]pyran-2-one).
CAS: 465-21-4. $C_{24}H_{34}O_4$.
Cardiotonic steroid component of ch'an su, the dried venom of the toad, *Bufo b. gargarizans*.
Hazard: Toxic.

bufanolide. (5R)-5-[(5S,8R,9S,10S,13S,14R,17R)-1-dimethyl-2,3,4,5,6,7,8,9,11,12,14,15,16-tetradecahydro-1H-cyclopenta[a]phenanthren-17-yl]oxan-2-one).
$C_{24}H_{38}O_2$.
Any of a group of basic steroid lactones of certain squill-toad (family Bufonidae) toxins and plant glycosides.
Hazard: Toxic.

bufencarb. ((3-(1-ethylpropyl)phenol methylcarbamate mixture with 3-(1-methylbutyl)-phenyl methylcarbamate (1:3); methylcarbamic acid m-(1-ethylpropyl)phenyl ester mixture with m-(1-methylbutyl)phenyl ester; metalkamate; (3-pentan-2-ylphenyl)N-methylcarbamate; (3-pentan-3-ylphenyl)N-methylcarbamate). $C_{26}H_{38}N_2O_4$. A dangerous cholinesterase inhibitor.
Hazard: Toxic.
Use: Insecticide.

buffer. A solution containing both a weak acid and its conjugate weak base, whose pH changes only slightly on addition of acid or alkali. The weak acid becomes a buffer when alkali is added, and the weak base becomes a buffer when acid is added. This action is explained by the reaction

$$A + H_2O \rightarrow B + H_3O_n$$

in which the base B is formed by the loss of a proton from the corresponding acid A. The acid may be a cation such as NH_4^+, a neutral molecule such as CH_3COOH, or an anion such as $H_2PO_4^-$. When alkali is added, hydrogen ions are removed to form water, but as long as the added alkali is not in excess of the buffer acid, many of the hydrogen ions are replaced by further ionization of A to maintain the equilibrium. When acid is added, this reaction is reversed as hydrogen ions combine with B to form A. The pH of a buffer solution may be calculated by the mass-law equation, pH = pK$'$ + log C_b/C_a in which pK$'$ is the negative logarithm of the apparent ionization constant of the buffer acid and the concentrations are those of the buffer base and its conjugate acid.

Bufin.   A leather finishing product.

Bufo.   A genus of some 100 species of toads (family Bufonidae), about 18 of which occur in North America. Nearly all are terrestrial and active at night, or at twilight, remaining in burrows or otherwise hidden and protected from high temperature and dehydration during the day. A few are diurnal. They typically enter water only to spawn. Some common or biologically important species that occur in North America are *B. americanus* (American toad), *B. marinus* (giant toad),*B. terrestris* (southern toad), *B. cognatus* (great plains toad), *B. boreas* (eastern toad), *B. punctatus* (red-spotted toad), and *B. woodhouse* (Woodhouse's toad).

bufogenin.   Any of a class of toxic principles in toad secretions. They have marked effects on involuntary muscle, including the heart.
Hazard: Toxic; deadly poison.

bufogenin.   (desacetylbufotalin;   3-β-14,16,-β-trihydroxy-5-β-bufa-20,22-dienolide;   3,14,16-tri-hydroxybufa-20,22-dienolide; desacetylbufotalin; 5-[(3S,5R,10S,13R,14S,16S,17R)-trihydroxy-10,13-dimethyl-1,2,3,4,5,6,7,8,9,11,12,15,16,17-tetradecahydrocyclopenta[a]phe-17-yl]pyran-2-one).
$C_{24}H_{34}O_5$. A toxicant isolated from the Chinese drug, ch'an su.
Properties: Elongated crysta prisms from methanol; slightly soluble in chloroform, methanol and acetone.
Hazard: Toxic; deadly poison.

Bufonidae.   A family of anuran amphibians that includes many of the true toads. They are nearly cosmopolitan in distribution, although they are absent from the Madagascar and Australian regions. Maxillary teeth are usually absent and the parotid gland is conspicuous. The dermal glands secrete a number of pharmacologically active agents, some of which are toxic. Bufonid toxins are chemically diverse and include biogenic amines, bufogenins, and indolalkyyamines. Several of these toxins have an action similar to that of digitalis.

bufonin.   A toxin of toads of the family Bufonidae.
Hazard: Toxic.

bufotaline.   (bufotalin;   3β,14,16-β-trihydroxy-5-β-bufa-20,22-dienolide-16-acetate;16-(acetyloxy)-3,14-dihydroxybufa-20,22-dienolide;

3β,14,16β-trihydroxy-5β-bufa-20,22-dienolide-16-acetate;   [(3S,5R,10S,13R,14S,16S,17R)-dihydroxy-10,13-dimethyl-17-(6-oxopyran-3-yl_-1,2,3,4,5,6,7,8,9,11,12,15,16-tetradecahydrocyclopenta[a]phe-16-yl]acetate).
CAS: 471-95-4 $C_{26}H_{36}O_6$.
Properties: Crystalline genin.
Derivation: In the skin and saliva of the common European toad, *bufo vulgaris*.
Hazard: Deadly poison.

bufotenine.   [3-(2-dimethylaminoethyl)-5-indolol]. $C_{12}H_{16}N_2O$.
Properties: Colorless prisms. Insoluble in water; soluble in alcohol, dilute acids and alkalies; slightly soluble in ether.
Derivation: From toads and toadstools; also made synthetically.
Hazard: A hallucinogenic agent.
Use: Medicine (experimental).
See hallucinogen.

builder detergent.   A substance that increases the effectiveness of a soap or synthetic detergent by acting as a softener and a sequestering and buffering agent. Phosphate-silicate formulations, once widely used, have been restricted for environmental reasons. They have largely been replaced by EDTA or zeolites, sometimes combined with nitrolotriacetic acid. Certain starch derivatives can be used as builders.
See zeolite.

bulan.   (2-nitro-1,1-bis(*p*-chlorophenyl)butane).
CAS: 76-20-0. $C_{16}H_{15}Cl_2NO_2$.
Hazard: A toxic chlorinated nitrogenous compound used as an insecticide. When mixed with Prolan, the product is called Dilan.
See Prolan; Dilan.

bulk density.   See density.

bulking agent.   Chemically inert material used for increasing volume.

bullion.   Bulk precious metals as produced at refineries, or gold-silver alloy produced in refining.

bunamiodyl.   [3(3-butyrylamino-2,4,6-tri-iodophenyl)-2-ethyl   sodium   acrylate].
$C_3H_7CONHC_6HI_3CH:C(C_2H_5)COONa$.
Use: Medicine (radiopaque contrast medium, diagnostic aid).

buna rubbers.   German vulcanizable synthetic rubbers from butadiene with sodium as a catalyst.
See rubber.

bungarotoxin.   Any of a class of proteins contained in the venom of the southeast Asian banded krait, *Bungarus multicinctus* (family Elapidae), and other snakes of this genus.
Hazard: Highly toxic.

α-bungarotoxin.
CAS: 11032-79-4. $C_{338}H_{529}N_{97}O_{105}S_{11}$.
A protein that binds specifically to acetylcholine receptors.
Properties: Single polypeptide weight; molecular weight of 8000.
Hazard: Nuerotoxic.

β-bungarotoxin. A presynaptic neurotoxin that inhibits acetylcholine release at myoneural junctions of voluntary muscles without affecting the postsynaptic membrane.
Properties: Material comprised of two subunits with molecular weights of 13,000 and 7,000, connected by sulfide bonds.
Hazard: Neurotoxic.

bunker fuel oil. A heavy residual oil used as fuel by ships, by industry, and for large-scale heating installations.

Bunsen burner. Common laboratory burner that allows regulation of the air to be mixed with the gas before burning.

Bunsen, Robert Wilhelm. (1811–1899). Born in Germany, Bunsen is remembered chiefly for his invention of the laboratory burner named after him. He engaged in a wide range of industrial and chemical research, including blast-furnace firing, electrolytic cells, separation of metals by electric current, spectroscopic techniques (with Kirchhoff), and production of light metals by electrical decomposition of their molten chlorides. He also discovered two elements, rubidium and cesium.

buoyancy balance. Balance, made of silica, capable of extreme accuracy.
Use: To determine the density of gases.

buprofezine.
CAS: 69327-76-0. $C_{16}H_{23}N_3OS$.
Hazard: Moderately toxic by ingestion. Low toxicity by skin contact.
Use: Agricultural chemical.

buquinolate.
CAS: 5486-03-3. $C_{20}H_{27}NO_5$.
Properties: Crystals. Mp 288–291C.
Use: Drug (veterinary); food additive.

burette. A liquid-measuring device used extensively in chemical laboratories. It is a vertical glass tube, open at the top, supported on a bracket, and equipped with scale graduation marks and a hand-operated stopcock at or near the bottom. The liquid to be dispensed is flowed in at the open end and can then be withdrawn in measured amounts by operating the stopcock.

Burgundy pitch. A resin obtained from Norway spruce or European silver fir. Other types, for example, that from various species of pines, are also offered under this name. Characterized by extreme tackiness, soluble in acetone and alcohol. Used to some extent in surgeons' tape and various special adhesive compositions.

burlap. A coarse, loose-woven fabric made from jute or similar fiber, used in low-cost laminated composites; as liner or backing in upholstery, carpets, etc.; and as a bagging material. It is often impregnated with hot-melt adhesive.

burnable poison. A neutron absorber (poison) such as boron that when incorporated in the fuel or fuel cladding of a nuclear reactor, gradually "burns up" (is changed into nonabsorbing material) under neutron irradiation. This process compensates for the loss of reactivity that occurs as fuel is consumed and fission-product poisons accumulate and keeps the overall characteristics of the reactor nearly constant during use.

burnt lime. See calcium oxide.

burnt sienna. See iron oxide red.

burnt umber. See umber.

burr mill. See attrition mill.

Burton Water Salts. An additive containing papain.
Use: In brewing to prevent chill haze.

Busan. A liquid bactericide, fungicide, and preservative.
Use: For leather processing.

bushy stunt virus. A viral protein present in tomato-plant infections.
Properties: Mw 7,600,000, pH 4.1.
See virus.

busulfan. (1,4-butanediol dimethanesulfonate esters; 1,4-bis(methanesulfonoxy)butane; 1,4-di(methanesulfonyloxy)butane; 1,4-di(methylsulfonoxy)butane; methanesulfonic acid tetramethylene ester; tetramethylene bis(methanesulfonate); busulfan; bsf; 4-methylsulfonyloxybutylmethanesulfonate).
CAS: 55-98-1. $C_6H_{14}O_6S_2$.
An alkylating agent having a selective immunosuppressive effect on bone marrow.
Properties: Crystalline, acetone-soluble alkylsulfonate.
Hazard: Extremely toxic, carcinogen, clastogenic, teratogenic, immunosuppressive, delayed bone marrow aplasia, cataracts, pigmentation, pulmonary thrombosis, cardiotoxic effects, thrombocytopenia.

Use: Antineoplastic alkylating agent, the palliative treatment of chronic myeloid leukemia, and insect sterilant.

butachlor. ((N-butoxymethyl)-2-chloro-n-(2,6-diethylphenyl)acetamide; n-(butoxymethyl)-2-chloro-2′,6′-diethylacetanilide; 2-chloro-2,6-diethyl-n-(butoxymethyl)acetanilide).
CAS: 23184-66-9. $C_{17}H_{26}ClNO_2$.
Hazard: Moderately toxic.
Use: Pre-emergence selective anilide herbicide used to control grass and broad-leaved weeds in seeded and transplanted rice paddies.

butadiene. (pliolite).
CAS:106-99-0. C_4H_6.
Properties: Flammable, gaseous hydrocarbon.
Hazard: Carcinogen.
Use: Manufacture of synthetic rubbers.

1,2,3,4-butadiene epoxide. The oxidation product of 1,3-butadiene.
Hazard: Carcinogen.

Butacite. A polyvinyl butyral resin, available as soft, pliable sheeting in 750-ft rolls, 10–84 inches wide.
See polyvinyl acetal; resins.

1,3-butadiene. (vinylethylene; erythrene; bivinyl; divinyl).
CAS: 106-99-0. $H_2C:CHHC:CH_2$.

$$HC \stackrel{\textstyle CH_2}{\underset{\textstyle CH_2}{\lessgtr}}$$

Properties: Colorless gas; mild aromatic odor. Easily liquefied. Bp −4.41C, d 0.6211 (liquid at 20C), fp −108.9C, flash p −105F (−76C), specific volume 6.9 cu ft/lb (700F), autoign temp 780F (414C), vap press 17.65 psi (0C). Soluble in alcohol and ether; insoluble in water. The material polymerizes readily, particularly if oxygen is present, and the commercial material contains an inhibitor to prevent spontaneous polymerization during shipment or storage.
Derivation: (1) Catalytic dehydrogenation of butenes or butane; (2) oxidative dehydrogenation of butenes.
Method of purification: Extractive distillation in the presence of furfural, absorption in aqueous cuprous ammonium acetate, or use of acetonitrile.
Grade: Technical (98.0%), CP (99.0%), instrument (99.4%), research (99.8%).
Hazard: A confirmed carcinogen. Irritant in high concentration. Highly flammable gas or liquid, explosive limits in air 2–11%. May form explosive peroxides on exposure to air. Must be kept inhibited during storage and shipment. Inhibitors often used are di-n-butylamine or phenyl-β-naphthylamine.

Storage is usually under pressure or in insulated tanks <35F (<1.67C).
Use: Synthetic elastomers (styrene-butadiene, polybutadiene, neoprene, nitriles), ABS resins, chemical intermediate.

butadiene-acrylonitrile copolymer. See nitrile rubber.

butadiyne. See diacetylene.

butaldehyde. See butyraldehyde.

butanal. See butyraldehyde.

butane. (n-butane).
CAS: 106-97-8. $CH_3CH_2CH_2CH_3$.
Properties: Colorless gas; natural-gas odor. Bp −0.5C, fp −138.3C, condensing pressure approximately 30 lb at 32.5C, d (liquid at 0C) 0.599, d (vapor at 0C (air = 1)) 2.07, critical temp 153.2C, critical press (absolute) 525 psi, heating value (25C) 3266 Btu/cu ft, sp vol (21.1C), 6.4 cu ft/lb, flash p −76F (−60C), autoign temp 761F (405C). Very soluble in water; soluble in alcohol and chloroform. Extremely stable, has no corrosive action on metals, does not react with moisture. An asphyxiant gas.
Derivation: A by-product in petroleum refining or gasoline manufacture.
Grade: Research 99.99 mol%, pure 99 mol%, technical 95 mol%, also available in various mixtures with isobutane, propane, pentanes, etc.
Hazard: Highly flammable, dangerous fire and explosion risk. Explosive limits in air 1.9–8.5%. Narcotic in high concentration. Central nervous system impairment.
Use: Organic synthesis, raw material for synthetic rubber and high-octane liquid fuels, fuel for household and many industrial purposes, manufacture of ethylene, solvent, refrigerant, standby and enricher gas, propellant in aerosols, pure grades used in calibrating instruments, food additive.
Note: Butane in liquid form may be stored both above and below ground. Besides storage in liquefied form under its vapor pressure at normal atmospheric temperatures, refrigerated liquid storage at atmospheric pressure may be used. Such systems are closed and insulated, and the liquid petroleum gas vapor is circulated through pumps and compressors to serve as the refrigerant for the system. Butane may be stored in pits in the earth capped by metal domes and in underground chambers (Compressed Gas Association). The foregoing also applies to propane.

butanedial. See succinaldehyde.

1,4-butanedicarboxylic acid. See adipic acid.

butanedioic anhydride. See succinic anhydride.

1,3-butanediol. See 1,3-butylene glycol.

1,4-butanediol.   See 1,4-butylene glycol.

2,3-butanediol.   See 2,3-butylene glycol.

butanediolamine.   See 2-amino-2-methyl-1,3-propanediol.

butanedione.   See diacetyl.

2,3-butanedione oxime thiosemicarbazone. $CH_3C(NOH)C(CH_3)N_2HCSNH_2$. A test reagent for manganese in very dilute solution made from dimethylglyoxime and thiosemicarbazide.

butane dioxime.   See dimethylglyoxime.

butanediperoxoic acid di-*tert*-butyl ester.
CAS: 16580-04-4. $C_{12}H_{22}O_6$.
Hazard: Low toxicity by ingestion.

butanenitrile.   See butyronitrile.

1-butanethiol.   (*n*-butyl mercaptan).
CAS: 109-79-5. C_4H_9SH.
Properties: Colorless liquid; strong, obnoxious odor. D 0.8412 (20/4C), refr index 1.4427 (20C), flash p 35F (1.67C), bp 97.2–101.7C. Slightly soluble in water; very soluble in alcohol and ether.
Grade: 95%.
Hazard: Toxic by inhalation. Flammable, dangerous fire risk. Upper respiratory tract irritant.
Use: Intermediate, solvent.

2-butanethiol.   (*sec*-butyl   mercaptan).
$C_2H_5CH(SH)CH_3$.
Properties: Colorless liquid; obnoxious odor. Boiling range 73–89C, d 0.8288 (20/4C), refr index 1.4363 (20C), flash p −10F (−23.3C).
Grade: 95%.
Hazard: Toxic by inhalation. Flammable, dangerous fire risk.

1,2,4-butanetriol. $HOCH_2CH_2OCH_2CH_2OH$.
Properties: Almost colorless; odorless liquid. Hygroscopic, bp 312C (extrapolated), d 1.184, refr index 1.473, flash p 332F (166C). Miscible with water and ethanol. Combustible.
Derivation: Reaction of 2-butyne-1,4-diol with water, followed by reduction.
Grade: Technical, nitration.
Use: Intermediate for alkyd resins and explosives, cellulose plasticizer, emulsifier for cosmetics, inks, finishes, paper, cork, textiles.

butanoic acid.   See butyric acid.

butan-3-one-2-yl butyrate. $C_8H_{14}O_3$.
Properties: White to slightly yellow liquid; red berry odor. D 0.972–0.992, refr index 1.408–1.429. Sol in alc, propylene glycol, most oils; insol in water.
Use: Food additive.

1-butanol.   See *n*-butyl alcohol.

2-butanol.   (*sec*-butanol; *sec*-butyl alcohol; 2-butyl alcohol; butylene hydrate; secondary butyl alcohol; ethyl methyl carbinol; methyl ethyl carbinol; 2-hydroxybutane; butan-2-ol; 2-buten-1-ol).
CAS: 78-92-2. C_4H_{100}.
Properties: Colorless, flammable liquid.
Hazard: Toxic, mutagenic, upper respiratory tract irritant, central nervous system impairment.
Use: In the extraction of fish meal; a flavoring agent.

2-butanol acetate.   See *sec*-butyl acetate.

***tert*-butanol.**   (*tert*-butyl alcohol; 2-methyl-2-propanol; tertiary butanol; *tert*-butyl hydroxide; trimethyl carbonol; trimethyl methanol; 1,1-dimethylethanol; NCI-C55367).
CAS: 75-65-0. $C_4H_{10}O$.
Properties: Colorless liquid or white crystalline solid rhombi prisms or plates with camphoraceous odor, soluble in water and ethanol.
Hazard: Alcoholic intoxicant, skin irritant, physical dependence, slightly to moderately toxic; teratogen. TLV: 100 ppm.
Use: Denaturant of ethanol, solvent for pharmaceuticals, octane booster in unleaded gasoline.

2-butanone.   See methyl ethyl ketone.

butanoyl chloride.   See butyroyl chloride.

"Butazate" [Chemtura].   TM for zinc dibutyl-dithiocarbamate.

Butazolidin.   Phenylbutazone.

butea frondosa, flower petals, alcoholic extract.   See butea mono-spra (lam.) kuntze, flower extract.

butea mono-spra (lam.) kuntze, flower extract.
CAS: 93333-82-5.
Hazard: Moderately toxic. A reproductive hazard.
Use: Reproductive effector.

2-butenal.   See crotonaldehyde.

Butenandt, Adolf.   (1903–1995). A German biochemist who won a Nobel Prize with Ruzicka in 1939. His work involved insecticides for plants and hormones. He received his doctorate at the University of Marburg, Germany. He received a multitude of honorary degrees and awards.

butene-1.   (ethylethylene; α-butylene).
CAS: 25167-67-3. $CH_2{:}CHCH_2CH_3$.
Properties: A colorless liquefied petroleum gas. Bp −6.3C, d 0.5951 (20/4C), fp approximately −185C, specific volume 6.7 cu ft/lb (70F), flash p −110F

(−79C), autoign temp approx 700F (371C). Soluble in most organic solvents; insoluble in water.

Derivation: (1) Gases containing appreciable content of butene-1, along with other butene and butane hydrocarbons, are obtained by fractional distillation of refinery gas. (2) Can be produced directly from ethylene.

Grade: Technical 95%, CP 99.0%, research 99.4%.

Hazard: Asphyxiant gas. Highly flammable, flammable limits in air 1.6–9.3% by volume. Dangerous fire and explosion risk.

Use: Polymer and alkylate gasoline; polybutenes; butadiene; intermediate for C_4 and C_5 aldehydes, alcohols, and other derivatives; production of maleic anhydride by catalytic oxidation.

***cis*-butene-2.** (dimethylethylene; β-butylene; also called the "high-boiling" butene-2). $CH_3HC:CHCH_3$.

$$CH_3 \diagdown_{C=C} \diagup^{CH_3}$$
$$H \diagup \qquad \diagdown H$$

Properties: A clorless liquefied petroleum gas. Bp 3.7C, d 0.6213 (10/4C), fp −139C, specific volume 6.7 cu ft/lb (21.1C), flash p −100F (−72C), autoign temp 615F (323C). Soluble in most organic solvents; insoluble in water.

Derivation: Gases containing appreciable content of *cis*-butene-2, along with other butene and butane hydrocarbons, are obtained by fractional distillation of refinery gas.

Grade: Technical 95%, CP 99%, research 99.8%.

Hazard: Asphyxiant gas. Highly flammable. Flammable limits in air 1.8–9.7% by volume. Dangerous fire and explosion risk.

Use: Solvent, cross-linking agent, polymerization of gasoline, butadiene synthesis, synthesis of C_4 and C_5 derivatives.

***trans*-butene-2.** (dimethylethylene; β-butylene; also called the "low-boiling" butene-2). $CH_3HC:CHCH_3$.

$$CH_3 \diagdown_{C=C} \diagup^{H}$$
$$H \diagup \qquad \diagdown CH_3$$

Properties: A colorless liquified petroleum gas. Bp 0.88C, fp −105.8C, d 0.6042 (20/4C), specific volume 6.7 cu ft/lb (21.1C), flash p −100F (−72C), autoign temp 615F (324C). Soluble in organic solvents; insoluble in water.

Derivation: Gases containing appreciable content of *trans*-butene-2, along with other butene and butane hydrocarbons, are obtained by fractional distillation of refinery gas.

Grade: Technical 95%, CP 99.0%, research 99.9%.

Hazard: Asphyxiant gas. Highly flammable. Flammable limits in air 1.8–9% by volume. Dangerous fire and explosion risk.

Use: See *cis*-butene-2.

***trans*-butenedioic acid.** See fumaric acid.

2-butene-1,3-diol. $HOCH_2CH:CHCH_2OH$.

Properties: Almost colorless; odorless liquid. Fp range 4.0–7.0C, bp range 232–235C, d 1.067–1.074, refr index 1.476–1.478 (25C), flash p 263F (128C). Miscible with water, ethanol, and acetone; sparingly soluble in benzene. Technical butenediol is predominantly the *cis* isomer. Combustible.

Derivation: Reduction of 2-butyne-1,4-diol, by high-pressure synthesis from acetylene and formaldehyde.

Hazard: Primary skin irritant.

Use: Intermediate for alkyd resins, plasticizers, nylon, pharmaceuticals; cross-linking agent for synthetic resins; fungicides.

butenoic acid. See crotonic acid.

butenolide. (2H-furan-5-one).
CAS: 497-23-4. $C_4H_4O_2$.
A mycotoxin that contains an α,β-unsaturated lactone moiety.

Derivation: Produced by *Fusarium sporotrichioides*.

Hazard: Extremely toxic.

3-buten-2-one. See vinyl methyl ketone.

butesin. See *n*-butyl-*p*-aminobenzoate.

butethal. (5-butyl-5-ethylbarbituric acid). $C_{10}H_{16}N_2O_3$.

Properties: White crystals or powder; odorless; bitter taste. Mp 124–127C. Fairly soluble in alcohol or ether; practically insoluble in water.

Hazard: May cause addiction.

Use: Medicine (hypnotic).
See barbiturate.

butethamine hydrochloride. [2-(isobutyl-amino)ethyl-*p*-aminobenzoate hydrochloride]. $NH_2C_6H_4COOCH_2CH_2NHCH_2CH(CH_3)_2 \cdot HCl$.

Properties: Whitecrystals or crystalline powder; odorless; bitter taste and local anesthetizing effects on tongue. Mp 192–196C, pH (1% solution) approximately 4.7, stable in air. Soluble in water; slightly soluble in alcohol and chloroform; very slightly soluble in benzene; practically insoluble in ether.

Grade: NF.

Use: Medicine (local anesthetic).

buthalitone sodium. See sodium buthalital.

    ***l*-buthionine sulfoximine.**
CAS: 83730-53-4. $C_8H_{18}N_2O_3S$.

Hazard: A reproductive hazard.

butinol-*N*-(3-chlorophenyl)carbamate.
Highly selective aryl carbamate.

Use: Herbicide to control *Avena fatua* (wild oats) in cereal crops.

butonate. $(CH_3O)_2P(O)CH(CCl_3)OOCC_3H_7$. (generic name for *o,o*-dimethyl-2,2,2,-trichloro-1-*n*-butyryloxyethyl phosphonate).
Properties: Colorless, somewhat oily liquid; slight ester odor. D 1.3742, refr index 1.4707, bp 129C, wt/gal 11.4 Miscible with most organic solvents; stable in neutral or acid aqueous solution; unstable in aqueous alkali. lb.
Hazard: Restrict use, toxic, cholinesterase inhibitor.
Use: Insecticide.

butopyronoxyl. (butylmesityl oxide oxalate; *n*-butyl-3,4-dihydro-2,2-dimethyl-4-oxo-1,2H-pyran-6-carboxylate).
CAS: 532-34-3. $C_{12}H_{18}O_4$
Properties: Yellow to pale reddish-brown liquid; aromatic odor. Reasonably stable in air. Slowly affected by light. D 1.052–1.060 (25/25C); refr index 1.4745–1.4755 (25C); distilling range 256–270C. Insoluble in water; miscible with alcohol, chloroform, ether, glacial acetic acid.
Derivation: Condensation of mesityl oxide and dibutyl oxalate in the presence of sodium ethoxide.
Grade: Technical.
Hazard: Toxic by ingestion. May cause liver damage.
Use: Insect repellent.

***p-tert*-butoxycarbonyloxystyrene monomer.**
CAS: 87188-51-0. $C_{13}H_{16}O_3$.
Hazard: Moderately toxic by skin contact. A mild skin irritant.

2-butoxyethanol. See ethylene glycol monobutyl ether.

2-(2-butoxyethoxy)ethyl thiocyanate. See β-butoxy-β'-thiocyanodiethyl ether.

1-butoxyethoxy-2-propanol.
$CH_3CH_2OCH_2OC_2H_4OC_4H_9$.
Properties: Colorless liquid, d 0.9310 (20/20C), bp 230.3C, fp −90C, soluble in water, wt/gal 7.8 lb, flash p 250F (121C). Combustible.
Use: Solvent, hydraulic-fluid components, antistall additive for automotive fuels, plasticizer, intermediate.

butoxyethyl laurate. See ethylene glycol monobutyl ether laurate.

butoxyethyl oleate. See ethylene glycol monobutyl ether oleate.

butoxyethyl stearate. See ethylene glycol monobutyl ether stearate.

Butoxyne 497. Hydroxyethyl esters of butyne-diol mixtures.

Grade: 100% active.
Available forms: Liquid.
Use: Nickel brightener in electroplating, copper pickling inhibitor, corrosion inhibitor in aerosol cans.

butoxypentachlorobenzene.
CAS: 90842-58-3. $C_{10}H_9Cl_5O$.
Hazard: Moderately toxic by ingestion and skin contact. A severe skin irritant.

***p*-butoxyphenol.** $HOC_6H_4OC_4H_9$.
Properties: White to faint-yellow, crystalline powder. Mp 61–65C. Soluble in alcohol, acetone, ether, benzene, aqueous alkali; insoluble in water. Combustible.
Grade: 93% pure.
Use: Synthesis.

butoxy polypropylene glycol. (generic name for polypropylene glycol monobutyl ether).
$CH_3CH_2O(CH_2OCHCH_3)_nCH_2OC_4H_9$.
Colorless liquid.
Use: An insect repellent.

***n*-butoxypropanol.**
CAS: 63716-40-5.
Properties: Colorless liquid, d 0.8801 (20/20C), bp 170.2C, fp −80C (sets to glass), flash p 154F (67.7C). Soluble in water. Combustible.
Use: Solvent for water-based enamels.

β-butoxy-β'-thiocyanodiethyl ether. [2-(2-butoxyethoxy)ethyl thiocyanate].
$H_3(CH_2)_3OCH_2OCH_2CH_2SCN$.
Hazard: Toxic by ingestion and skin absorption. Skin irritant.
Use: Insecticide.

butoxytriglycol. (triethylene glycol monobutyl ether).
CAS: 143-22-6. $C_4H_9O(C_2H_4O)_3H$.
Properties: Liquid. D 1.0021 (20/20C), bp (decomposes), fp −47.6C, flash p 290F (143C). Miscible with water. Combustible.
Use: Plasticizer, intermediate.

butter. (1) A colloidal system (emulsion) in which the continuous phase is composed of liquid fat from fat globules disintegrated by mechanical agitation and the dispersed phase is composed of finely divided water droplets and undamaged fat globules. (2) Outmoded term for hygroscopic metallic chlorides of viscous consistency, e.g., butter of zinc, etc.

buttercup yellow. See zinc yellow.

butter-yellow. (*p*-dimethylaminoazobenzene; *N,N*-dimethyl-4-phenyldiazenylaniline).
CAS: 60-11-7. $C_{14}H_{15}N_3$.
A banned food coloring.

Hazard: Carcinogen.
Use: A reagent to induce experimental liver cancer.

butterfat.   The oily portion of the milk of mammals.
Properties: Composition is largely glycerides of oleic, stearic, and palmitic acids, with smaller amounts of the glycerides of butyric, caproic, caprylic, and capric acids. D range 0.910–0.914. Cow's milk contains approximately 4% butterfat. See milk.

butter starter distillate.   See starter distillate.

butter yellow.   See dimethylaminoazobenzene.

Butvar.   Polyvinyl butyral resins, with various hydroxy content, whose solutions provide a range of viscosities.
Use: For film clarity, flexibility, abrasion, and water resistance; for use in primers, structural hot-metal adhesives, inks, and waterproof coatings.

butyl.   (1) The group C_4H_9; (2) butyl rubber.

_n_-butyl acetate.
CAS: 123-86-4. $CH_3COOCH_2CH_2CH_2CH_3$.
Properties: Colorless liquid; fruity odor. Vapor is heavier than air, d 0.8826 at 20/20C, bp 126.3C, vap press 8.7 mm Hg (20C), fp −75C, refr index 1.2951 (20C), wt/gal 7.35 lb (20C), flash p 98F (36.6C) (TOC), autoign temp 790F (421C). Soluble in alcohol, ether, and hydrocarbons; slightly soluble in water.
Derivation: Esterification and then distillation after contact of butyl alcohol with acetic acid in the presence of a catalyst such as sulfuric acid.
Hazard: Skin irritant, toxic. Flammable, moderate fire risk. Eye and upper respiratory tract irritant.

sec-butyl acetate.
CAS: 105-46-4. $CH_3COOCH(CH_3)(C_2H_5)$.
Properties: Colorless liquid. Bp 112.2C, d 0.8905 at 0/4C, 0.870 at 20/4C, refr index 1.389 (20C), wt/gal 7.21 lb, flash p 88F (31C) (OC). Miscible with alcohol and ether; insoluble in water.
Derivation: Esterification of *sec*-butyl alcohol.
Hazard: Flammable, dangerous fire risk. Eye and upper respiratory tract irritant.
Use: Solvent for nitrocellulose lacquers, thinners, nail enamels, leather finishes.

tert-butyl acetate.
CAS: 540-88-5. $CH_3COOC(CH_3)_2$.
Properties: Colorless liquid. Bp 96C, d 0.896 (20C). Insoluble in water; soluble in alcohol and ether.
Hazard: Flammable, moderate fire risk. Eye and upper respiratory tract irritant.
Use: Solvent, gasoline additive.

butyl acetate dilution ratio.   Measure of approximate tolerance of nitrocellulose solutions

for petroleum diluents. The higher the ratio, the better the solvent.

butyl acetoacetate.
CAS: 591-60-6. $CH_3COCH_2COOCH_2CH_2CH_2$
CH_3.
Properties: Colorless liquid. D 0.9694 (20/20C), bp 213.9C, vap press 0.19 mm Hg (20C), flash p 185F (85C), wt/gal 8.1lb (20C). Insoluble in water; soluble in alcohol and ether. Combustible.
Grade: Technical.
Use: Intermediate in synthesis of metal derivatives, dyestuffs, pharmaceuticals, flavoring.

butyl acetoxystearate.
$CH_3(CH_2)_5CH(CH_3COO)(CH_2)_{10}COOC_4H_9$.
Properties: See butyl acetyl ricinoleate.
Derivation: From castor oil, butylalcohol, and acetic anhydride with hydrogenation.
Use: Plasticizer, textile oils, adhesives.

butyl acetylene.   See 1-hexyne.

butyl acetyl ricinoleate. $C_{24}H_{44}O_4$.
Properties: Yellow, oily liquid; mild odor. D 0.940 (20/20C), sapon number 125, fp indefinite, becomes cloudy at −32C, solidifies at −65C, flash p 230F (110C) (OC), refr index 1.4614 (20C), Saybolt viscosity 123 sec at 100F, wt/gal 7.8 lb (20C), autoign temp 725F (385C). Miscible with most organic solvents. Almost insoluble in water. Combustible.
Derivation: From castor oil, butanol, and acetic anhydride.
Grade: Technical.
Use: Plasticizer, emulsifier, lubricant, detergent, protective coatings, special cleansing compounds, quick-breaking emulsions.

_n_-butyl acid phosphate.   (*n*-butylphosphoric acid; acid butyl phosphate).
CAS: 12788-93-1.
Properties: Water-white liquid. D 1.120–1.125 (25/4C), refr index 1.429 (25C), flash p 230F (110C) (COC). Soluble in alcohol, acetone, and toluene; insoluble in water and petroleum naphtha. Combustible.
Hazard: Strong irritant to skin and tissue.
Use: Esterification catalyst and polymerizing agent, curing catalyst and accelerator in resins and coatings, special detergents.

N-tert-butylacrylamide.
$H_2C{:}CHCONHC(CH_3)_3$.
Properties: White, crystalline solid. Mp 128–130C, d 1.015 (30C). Soluble in methanol, ethanol, chloroform, and acetone. Combustible.
Hazard: Toxic by ingestion and inhalation. Irritant to skin.
Use: Monomer, organic intermediate.

_n_-butyl acrylate.
CAS: 141-32-2. $CH_2{:}CHCOOC_4H_9$.

Properties: Colorless liquid. Fp −64C, boiling range 145.7–148.0C, polymerizes readily on heating, vap press (20C) 3.2 mm Hg, d 0.9015 (20/20C), wt/gal 7.5 lb (20C), flash p 120F (49C) (OC). Nearly insoluble in water. Flammable.

Derivation: Reaction of acrylic acid or methyl acrylate with butanol.

Grade: Technical (inhibited).

Hazard: Moderate fire risk. Questionable carcinogen.

Use: Intermediate in organic synthesis, polymers and copolymers for solvent coatings, adhesives, paints, binders, emulsifiers.

See acrylic resin.

tert-butyl acrylate.

CAS: 1663-39-4. CH_2:$CHCOOC(CH_3)_3$.

Properties: Liquid. Bp 120C, d 0.879 (25C), refr index 1.4080 (25C), flash p 66F (18.8C) (TOC). Commercial grade contains 100 ppm hydroquinone monomethyl ether as stabilizer.

Hazard: Toxic by ingestion and inhalation. Flammable, dangerous fire risk.

Use: Monomer for acrylic resins.

butyl acrylate-vinylidine chloride copolymer.

See vinylidene chloride-butyl acrylate copolymer.

n-butyl alcohol.   (1-butanol; butyric alcohol).

CAS: 71-36-3. $CH_3(CH_2)_2CH_2OH$.

Properties: Colorless liquid; vinous odor. Bp 117.7C, fp −89.0C, d (20/20C) 0.8109, wt/gal (20C) 6.76 lb, refr index 1.3993 (20C), flash p 95F (35C), autoign temp 689F (365C). Soluble in water 7.7 wt% (20C). Solution of water in *n*-butanol 20.1%. Miscible with alcohol and ether.

Derivation: (1) Hydrogenation of butyraldehyde, obtained in the Oxo process; (2) condensation of acetaldehyde to form crotonaldehyde, which is then hydrogenated (aldol condensation).

Hazard: Toxic on prolonged inhalation, irritant to eyes. Toxic when absorbed by skin. Flammable, moderate fire risk. Eye and upper respiratory tract irritant.

Use: Preparation of esters, especially butyl acetate; solvent for resins and coatings; plasticizers; dyeing assistant; hydraulic fluids; detergent formulations; dehydrating agent (by azeotropic distillation); intermediate; "butylated" melamine resins; glycol ethers; butyl acrylate.

sec-butyl alcohol.   (SBA; 2-butanol; methylethylcarbinol).

CAS: 78-92-2. $CH_3CH_2CH_2OCH_3$.

Properties: Colorless liquid; strong odor. Bp 99.5C, fp −114C, d 0.808 (20/4C), wt/gal 66.74 lb (20C), refr index 1.3949 (25C), flash p 75F (23.8C) (CC), autoign temp 763F (406C). Moderately soluble in water; miscible with alcohol and ether.

Derivation: Absorption of butene from cracking petroleum or natural gas in sulfuric acid with subsequent hydrolysis by steam.

Grade: Technical.

Hazard: Toxic on prolonged inhalation, irritating to eyes and skin. Flammable, dangerous fire risk. Central nervous system impairment and upper respiratory tract irritant.

Use: Preparation of methyl ethyl ketone, solvent, organic synthesis, paint removers, industrial cleaners.

tert-butyl alcohol.   (2-methyl-2-propanol; trimethyl carbinol).

CAS: 75-65-0. $(CH_3)_3COH$.

Properties: Colorless liquid or crystals; camphor odor. Fp 25.5C, bp 82.9C, d 0.779 (liquid 26C), refr index 1.3878 (20C), flash p 52F (11.1C) (CC), autoign temp 892F (477C). Miscible with water, alcohol, and ether.

Derivation: Absorption of isobutene from cracking petroleum or natural gas in sulfuric acid with subsequent hydrolysis by steam.

Grade: Technical.

Hazard: Irritant to eyes and skin. Flammable, dangerous fire risk. Central nervous system impairment. Questionable carcinogen.

Use: Alcohol denaturant, solvent for pharmaceuticals, dehydration agent, perfumery, chemical intermediate, paint removers, manufacture of methyl methacrylate, octane booster in unleaded gasoline (EPA approved).

n-butyl aldehyde.   See butyraldehyde.

n-butylamine.   (1-aminobutane).

CAS: 109-73-9. $C_4H_9NH_2$.

Properties: Colorless, volatile liquid; amine odor. Bp 77.1C, fp −49.1C, d 0.7385 (20/20C), wt/gal 6.2 lb (20C), refr index 1.401 (20C), flash p 30F (1.1C) (OC). Miscible with water, alcohol, ether.

Derivation: Reaction of butanol or butyl chloride with ammonia.

Grade: Technical.

Hazard: Skin irritant. Flammable, dangerous fire risk. Eye and upper respiratory tract irritant.

Use: Intermediate for emulsifying agents, pharmaceuticals, insecticides, rubber chemicals, dyes, tanning agents.

sec-butylamine.   (2-aminobutane).

CAS: 13952-84-6. $CH_3CHNH_2C_2H_5$.

Properties: Colorless liquid; amine odor. D 0.725 (20C), boiling range 63–68C, refr index 1.395 (20C), solidification point −104C, flash p 15F (−9.4C), wt/gal 6.0 lb (20C).

Hazard: Flammable, dangerous fire risk.

Use: Fungicide.

tert-butylamine.

CAS: 75-64-9. $(CH_3)_3CNH_2$.

Properties: Colorless liquid. Bp 44–46C, fp −72C, d 0.700 (15C), refr index 1.3794 (18C), flash p approx 50F (10C). Miscible with water; soluble in common organic solvents.

Grade: Technical.

Hazard: Skin irritant. Flammable, dangerous fire risk.

Use: Intermediate for rubber accelerators, insecticides, fungicides, dyestuffs, pharmaceuticals.

butyl-*o*-aminobenzoate.   See butyl anthranilate.

n-butyl-*p*-aminobenzoate.

$H_2NC_6H_4COOC_4H_9$.

Properties: White powder; odorless; tasteless. Mp 57–59C, bp 174C (8 mm Hg). Soluble in dilute acids, alcohol, chloroform, ether, and fatty oils; almost insoluble in water.

Grade: NF.

Hazard: Toxic by ingestion.

Use: Medicine (local anesthetic), treatment of burns, ointments, UV absorber in suntan preparations.

N-*n*-butylaminoethanol. $C_4H_9NHC_2H_4OH$.

Properties: Liquid. D 0.88–0.99 (20/20C), distillation range 192–210C, wt/gal 7.4 lb, flash p 170F (76.6C). Combustible.

tert-butylaminoethyl methacrylate.

$CH_2{:}C(CH_3)COOCH_2CH_2NHC(CH_3)_3$.

Properties: Liquid. Bp 100–105C (12 mm Hg), d 0.914 (25C), wt/gal 7.61 lb, refr index 1.4440 (25C), flash p 205F (96.1C) (COC). Combustible.

Use: Coatings, textile chemicals, dispersing agent for nonaqueous systems, antistatic agent, stabilizer for chlorinated polymers, ion-exchange resins, emulsifying agent, cationic precipitating agent.

N′-*n*-butyl-3-amino-4-methoxybenzenesulfonamide. $CH_3OC_6H_3(NH_2)SO_2NHC_4H_9$.

Properties: Pink powder. Mp 96–100C. Insoluble in water; partially soluble in alcohol and acetone.

USE: An intermediate.

N-*n*-butylaniline. $C_6H_5NHC_4H_9$.

Properties: Amber liquid; aniline odor. D 0.932 (20C), boiling range 236–242C, refr index 1.534 (20C), flash p 225F (107C). Very soluble in alcohol and ether; insoluble in water. Combustible.

Use: Organic synthesis, dyes.

butyl anthranilate.   (butyl-*o*-aminobenzoate). $C_4H_9OOCC_6H_4NH_2$.

Use: Flavoring.

butylated   hydroxymethylphenol.   mf: $C_{15}H_{24}O_2$.

Properties: White crystalline powder. Mp 140–141C. Sol in alc; insol in water, propylene glycol.

Use: Food additive.

2-*tert*-butylanthraquinone. $C_{18}H_{16}O_2$.

Properties: Yellow powder. Mp 102–104C. Soluble in alcohol and acetone. Combustible.

Grade: Technical (98%).

Use: Organic synthesis.

butylated hydroxyanisole.   (BHA).

CAS: 25013-16-5. $(CH_3)_3CC_6H_3OH(OCH_3)$.

A mixture of 2- and 3-*tert*-4-methoxyphenol.

Properties: White or slightly yellow, waxy solid having a faint characteristic odor. Melting range 48–63C. Not naturally water soluble but can be made so by special treatment. Freely soluble in alcohol and propylene glycol. Combustible.

Grade: FCC, water soluble.

Hazard: Toxic by ingestion. Use in foods restricted; consult FDA regulations.

Use: Antioxidant for fats and oils; food packaging.

butylated   hydroxytoluene.   See 2,6-di-*tert*-butyl-*p*-cresol.

(e)-1-t-butylazo-1-hydroxycyclopentane.

CAS: 50265-78-6. $C_9H_{18}N_2O$.

Hazard: Low toxicity by inhalation.

n-butylbenzene.   (1-phenylbutane).

CAS: 104-51-8. $C_6H_5C_4H_9$.

Properties: Colorless liquid. Bp 183.2C, fp −87.9C, d 0.860 (20C), refr index 1.489 (20C), flash p 160F (71.1C) (OC), autoign temp 774F (412C). Combustible

Grade: Technical, pure, research.

Hazard: Toxic by ingestion.

Use: Organic synthesis, especially of insecticides.

sec-butylbenzene.   (2-phenylbutane).

CAS: 135-98-8. $C_6H_5C(CH_3)C_2H_5$.

Properties: Colorless liquid. Bp 170.65C, vap press 15 mm Hg (60C), fp −75.68C, d 0.8618 (204C), wt/gal 7.2 lb (20C), refr index 1.4901 (20C), flash p 145F (62.7C) (OC), autoign temp 784F (417C). Combustible.

Grade: Technical 95%, pure, research.

Hazard: Toxic by ingestion.

Use: Solvent for coating compositions, organic synthesis, plasticizer, surface-active agents.

tert-butylbenzene.   (2-methyl-2-phenylpropane).

CAS: 98-06-6. $C_6H_5C(CH_3)_3$.

Properties: Colorless liquid. Bp 169.1C, fp −57.8C, d 0.866 (20C), refr index 1.492 (20C), flash p 140F (60C) (OC). Combustible. Autoign temp 842F (450C). Insoluble in water; soluble in alcohol.

Grade: Technical, pure, research.

Hazard: Toxic by ingestion.

Use: Organic synthesis, polymerization solvent, polymer linking agent.

butylbenzenesulfonamide.   (n,n-butylbenzenesulfonamide). $C_6H_5SO_2NHC_3H_9$.

Properties: Liquid; pleasant odor; amber to straw color. D 1.148 (25/25C), refr index 1.5235 (25C), bp 189–190C (4.5 mm Hg).

Hazard: Toxic by ingestion.
Use: Synthesis of dyes, pharmaceuticals, other organic chemicals, in resin manufacturing, as plasticizer for some synthetic polymers.

butyl benzoate. (*n*-butyl benzoate). $C_6H_5COOC_4H_9$.
Properties: Colorless, oily liquid. D 1.00 (20C), bp 247.3C, fp −22C, flash p 225F (107C) (OC). Insoluble in water; miscible with alcohol or ether. Combustible.
Grade: Technical.
Use: Solvent for cellulose ether, plasticizer, perfume ingredient, dyeing of textiles.

N-*tert*-butyl-2-benzothiazolesulfenamide.
$C_6H_4NCS(SNH)C_4H_9$.
Properties: Light-buff powder or flakes (sometimes colored blue). Mp 104C min, d 1.29 (25C). Soluble in most organic solvents. Combustible.
Use: Rubber accelerator.

di-*n*-butyl benzylphosphonate. See benzylphosphonic acid dibutyl ester.

butylbenzyl phthalate. (BBP).
$C_4H_9OOCC_6H_4COOC_7H_7$.
Properties: Clear, oily liquid; slight odor. D 1.113–1.121 (25/25C), flash p 390F (198C). Combustible.
Grade: Technical.
Hazard: Questionable carcinogen.
Use: Plasticizer for polyvinyl and cellulosic resins, organic intermediate.

butylbenzyl sebacate.
$C_4H_9OOC(CH_2)_8COOC_7H_7$.
Properties: Light-straw-colored liquid. Bp 245–285C (10 mm Hg), d 1.023 (25C), wt/gal 8.6 lb, flash p 395F (201C). Combustible.
Use: Plasticizer for resins.

butyl borate. See tributyl borate.

n-butyl bromide. (1-bromobutane).
CAS: 109-65-9. C_4H_9Br.
Properties: Colorless liquid. D 1.279 (20/20C), bp 101.6C, fp −112.4C, flash p 75F (23.9C) (OC), autoign temp 509F (265C). Insoluble in water; soluble in alcohol and ether.
Grade: 99%.
Hazard: Flammable, dangerous fire risk.
Use: Alkylating agent.

sec-butyl bromide. (2-bromobutane).
CAS: 78-76-2. $CH_3CHBrCH_2CH_3$.
Properties: Clear, colorless liquid; pleasant odor. Bp 91.2C, fp −112C, d 1.2425 (25/25C), refr index 1.4320–1.4344 (25C), flash p 70F (21.1C) (OC). Soluble in alcohol and ether; insoluble in water.
Hazard: Narcotic in high concentration. Flammable, dangerous fire risk.
Use: Synthesis, alkylating agent.

butyl butanoate. See *n*-butyl butyrate.

n-butyl butyrate. (butyl butanoate).
CAS: 109-21-7. $CH_3(CH_2)_2COOC_4H_9$.
Properties: Colorless liquid. D 0.8721 (20/20C), refr index 1.4059 (20C), fp −91.5C, bp 165.7C (736 mm Hg), flash p 128F (53.5C) (OC). Insoluble in water; soluble in alcohol and ether. Flammable.
Grade: FCC.
Hazard: Irritant and narcotic. Moderate fire risk.
Use: Flavoring.

n-butyl carbinol. See *n*-amyl alcohol, primary.

sec-butyl carbinol. See 2-methyl-1-butanol.

butyl "Carbitol" [Union]. TM for diethylene glycol monobutyl ether.

butyl "Carbitol" acetate [Union].
CAS: 124-17-4.
TM for diethylene glycol monobutyl ether acetate.

p-*tert*-butylcatechol. (4-*tert*-butyl-1,2-dihydroxybenzene).
CAS: 98-29-3. $(CH_3)_3CC_6H_3(OH)_2$.
Properties: Colorless crystals. Mp 56–57C, d 1.049 (60/25C), bp 285C, flash p 265F (129C). Combustible. Soluble in ether, alcohol, acetone; slightly soluble in water at 80C.
Hazard: Toxic by ingestion and skin absorption.
Use: Polymerization inhibitor for styrene-butadiene and other olefins.

butyl "Cellosolve" [Union]. TM for ethylene glycol monobutyl ether.

butyl "Cellosolve" acetate [Union]. TM for ethylene glycol monobutyl ether acetate.

butyl chloral hydrate. (trichlorobutyraldehyde hydrate). $CH_3CHClCCl_2CH(OH)_2$.
Properties: Colorless leaflets. D 1.693 (20/4C), mp 78C. Slightly soluble in water, soluble in alcohol and ether.
Derivation: Action of chlorine on paraldehyde.
Use: Medicine (hypnotic, anticonvulsant).

n-butyl chloride. (1-chlorobutane).
CAS: 109-69-3. $CH_3CH_2CH_2CH_2Cl$ or C_4H_9Cl.
Properties: Colorless liquid. D 0.8875 (20/20C), bp 78.6C, wt/gal 7.35 lb (20C), refr index 2.4015 (20C), vap press 80.1 mm Hg (20C), fp −122.8C, viscosity 0.0045 cP (20C), flash p 15F (−9.4C) (OC), autoign temp 860F (460C). Insoluble in water; miscible with alcohol and ether.
Grade: NF, technical.
Hazard: Toxic on prolonged inhalation. Flammable, dangerous fire risk.
Use: Organic synthesis (alkylating agent), solvent, anthelmintic.

sec-butyl chloride.   (2-chlorobutane).
CAS: 78-86-4. $CH_3CHClCH_2CH$.
Properties: Colorless liquid. Bp 68C, d 0.875 (20/4C), flash p 32F (0C), refr index 1.39. Miscible with alcohol and ether; sparingly soluble in water.
Hazard: Flammable, dangerous fire risk.
Use: Organic synthesis.

tert-butyl chloroacetate.
CAS: 107-59-5. $C_6H_{11}ClO_2$.
Hazard: A poison by ingestion. Moderately toxic by inhalation skin contact. Severe skin and moderate eye irritant.

tert-butylchlorodimethylsilane.
CAS: 18162-48-6. $C_6H_{15}ClSi$.
Hazard: Moderately toxic.

butyl(4-chlorophenyl)methyl   3-pyridinyl-carbonimidodithioate.
CAS: 34763-20-7. $C_{17}H_{19}ClN_2S_2$.
Hazard: Moderately toxic by ingestion.
Use: Agricultural chemical.

tert-butyl chromate.   (chromic acid di-*t*-butyl ester).
CAS: 1189-85-1. $[(CH_3)_3CO]_2CrO_3$.
Properties: Liquid. Mp −5 to 0C.
Hazard: Toxic by skin absorption. A very powerful oxidizer and dangerous fire hazard. Skin and lower respiratory tract irritant.

butyl citrate.   See tributyl citrate.

6-tert-butyl-*m*-cresol.   (MBMC; 6-*tert*-butyl-3-methylphenol). $(CH_3)_3CC_6H_3(OH)CH_3$.
Properties: Clear liquid. Solidifies slightly below room temperature, fp 23.1C, bp 244C, d 0.922 (80C), flash p 116F (47C). Soluble in organic solvents and aqueous potassium hydroxide. Flammable.
Hazard: Irritant to skin. Moderate fire risk.
Use: Germicide, disinfectant, synthesis of antioxidants and rubber-processing chemicals, additives to lubricating oils, synthetic resins, perfumes (fixative).

butyl crotonate. $CH_3CH:CHCOOC_4H_9$.
Properties: Water-white liquid; persistent odor. D 0.9037 (20/20C), bp 180.5C, wt/gal 7.52 lb (20C). Soluble in alcohol and ether; insoluble in water. Combustible.

butylcyclohexyl phthalate.
$C_4H_9OOCC_6H_4COOC_6H_{11}$.
Properties: Clear liquid; very mild odor. D 1.078, saponification number 369, acidity (as phthalic acid) 0.01 max. Miscible with most organic solvents. Combustible.
Use: Plasticizer for polymers and elastomers, nitrocellulose lacquers.

butyldecyl phthalate.
$C_4H_9OOCC_6H_4COOC_{10}H_{21}$.
Properties: Clear, oily liquid. D 0.977–0.987 (25/25C). Combustible.
Use: Primary plasticizer for polyvinyl chloride and copolymer resins.

n-butyldiamylamine. $C_4H_9N(C_5H_{11})_2$.
Properties: Straw-colored liquid; amine odor. D 0.788 (20C), boiling range 229–241C, flash p 200F (93.3C). Combustible.

n-butyldichlorarsine. $C_4H_9AsCl_2$.
Properties: Oily liquid; somewhat agreeable odor. Bp 192–194C. Decomposed by water.
Hazard: Toxic by inhalation and ingestion.
Use: Military poison.

tert-butyldichloroamine.
CAS: 2156-72-1. $C_4H_9Cl_2N$.
Hazard: Moderately toxic by ingestion and skin contact. Low toxicity by inhalation. A severe eye irritant.

butyldichlorophenoxyacetate.   See 2,4-D.

butyl(3,4-dichlorophenyl)methyl   3-pyridinylcarbonimidodithioate.
CAS: 34763-43-4. $C_{17}H_{18}Cl_2N_2S_2$.
Hazard: Moderately toxic by ingestion.
Use: Agricultural chemical.

1-*n*-butyl-3-(3,4-dichlorophenyl)-1-methylurea.   (neburon).
$Cl_2C_6H_3NHCONCH_3(C_4H_9)$.
Properties: White, crystalline solid. Mp 102C. Very low solubility in water and hydrocarbon solvents. Stable toward oxidation and moisture.
Use: Weed killer.

5-tert-butyl-3-(2,4-dichloro-5-propargyloxyphenyl)-1,3,4-oxadiazol-2(3h)-one.
CAS: 39807-15-3. $C_{15}H_{14}Cl_2N_2O_3$.
Hazard: A reproductive hazard.

n-butyl diethanolamine.
$C_4H_9N(CH_2CH_2OH)_2$.
Properties: Liquid; very light-straw color; faint amine odor. D 0.97 (20C), bp 272C, wt/gal 18.08 lb (20C), flash p 245F (118C). Combustible.
Use: Organic synthesis.

tert-butyl diethanolamine.
$C_4H_9N(CH_2CH_2OH)_2$.
Properties: Liquid, similar to normal compound.
Use: Organic synthesis, epoxy curing agent, catalyst for polyester resins, inhibitor for printing inks.

n-butyl diethyl malonate.
$C_4H_9CH(COOC_2H_5)_2$.

Properties: Colorless liquid with a fruity odor. D 0.972–0.974 (25/25C), refr index 1.420–1.422 (25C). Soluble in alcohols, ketones, esters. Combustible.
Use: Intermediate.

n-butyldiethyltin iodide.
CAS: 17563-48-3. $C_8H_{19}ISn$.
Hazard: A poison.

butyl diglycol carbonate.   (diethylene glycol bis(*n*-butylcarbonate)). $(C_4H_9OCO_2CH_2)_2O$.
Properties: Colorless liquid of low volatility. D 1.07 (20/4C), boiling range 164–166C (2 mm Hg), flash p 372F (188C), Saybolt viscosity 21 cP (20C), refr index 1.435 (20C), evaporation rate 0.59 mg/sq cm/hour (100C). Insoluble in water (very stable to hydrolysis); widely soluble in organic solvents; compatible with many resins and plastics. Combustible.
Use: Plasticizer; high-boiling-point solvent and softening agent; manufacture of pharmaceuticals and lubricant compositions.

butyl diglyme.   See diethylene glycol dibutyl ether.

4-*tert*-butyl-1,2-dihydroxybenzene.   See *p-tert*-butylcatechol.

butyldimethylamine.
CAS: 927-62-8. $C_6H_{15}N$.
Hazard: Moderately toxic by ingestion. Low toxicity by inhalation and skin contact. A moderate eye irritant.

o-butyl   s-((4-(1,1-dimethylethyl)phenyl)-methyl)-3-pyridinylcarbonimidothioate.
CAS: 51308-64-6. $C_{21}H_{28}N_2OS$.
Hazard: Moderately toxic by ingestion.
Use: Agricultural chemical.

n-butyl-2,6-dimethyl-1-piperidinecarboxamide.
CAS: 67626-66-8. $C_{12}H_{24}N_2O$.
Hazard: Moderately toxic by ingestion. A moderate skin and eye irritant.

butyl "Dioxitol" *[Shell]*.   TM for diethylene glycol monobutyl ether.

4-butyl-1,2-diphenyl-3,5-pyrazolidinedione.
See phenylbutazone.

butyl dodecanoate.   See butyl laurate.

butylene.   (butene). One of the liquefied petroleum gases butene-1, *cis*-butene-2,*trans*-butene-2, and isobutene.

butylene dimethacrylate. $C_{12}H_{18}O_4$
Properties: Liquid. Bp 110C (3 mm Hg), d 1.011, (25/15.6C), refr index 1.4502 (25C), flash p >150F (65C). Combustible.
Use: Monomer for resins.

1,3-butylene glycol.   (1,3-butanediol).
$HOCH_2CH_2CH(OH)CH_3$. Can exist in optical isomeric forms.
Properties: Practically colorless; viscous liquid. Hygroscopic. D 1.0059 (20/20C), wt/agl 8.4 lb (20C), bp 207.5C, vap press 0.06 mm Hg (20C), refr index 1.4401 (20C), flash p 250F (121C) (COC), autoign temp 741F (393C). Completely soluble in water and alcohol; slightly soluble in ether. Combustible.
Derivation: Reduction of aldol.
Use: Polyesters, polyurethanes, surface-active agents, plasticizers, humectant, coupling agent, solvent, food additive, flavoring.

1,4-butylene glycol.   (1,4-butanediol; tetramethylene glycol).
CAS: 110-63-4. $HOCH_2CH_2CH_2CH_2OH$.
Properties: Colorless, oily liquid. Bp 230C, mp 16C, d 1.020 (20/4C). Flash p >250F (>121C). Miscible with water; soluble in alcohol; slightly soluble in ether. Combustible.
Derivation: From acetylene and formaldehyde by high-pressure synthesis.
Grade: Technical.
Hazard: Toxic by ingestion.
Use: Solvents, humectant, intermediate for plasticizers, pharmaceuticals, cross-linking agent in polyurethane elastomers, manufacture of tetrahydrofuran, terephthalate plastics.

2,3-butylene glycol.   (2,3-dihydroxybutane; 2,3-butanediol; pseudobutylene glycol; *sym*-dimethylethylene glycol). $CH_3CH_2OCH_2OCH_3$.
Can exist in optical isomeric forms.
Properties: Nearly colorless, crystalline solid or liquid. Hygroscopic. D 1.045 (20/20C), mp 23–27C, bp 179–182C, refr index 1.438 (20C), flash p 185F (85C) (OC). Soluble in alcohol and ether; miscible with water. Combustible.
Derivation: From corn sugar by acid hydrolysis, also from fermentation of sugar by acid hydrolysis, also from fermentation of sugar-beet molasses.
Grade: 99%.
Use: Resins, solvent for dyes, intermediate, blending agent.

1,2-butylene oxide.   (1,2-epoxybutane).

CAS: 106-88-7. $H_2COCHCH_2CH_3$.

Properties: Colorless liquid. D 0.8312 (20/20C), bp 63C, sets to a glass below −150C, flash p approx 0F (−17C) (CC). Soluble in water; miscible with most organic solvents.

Grade: Approximately 97.5% purity.
Hazard: Toxic concentration of vapors occurs at room temperature. Highly flammable, dangerous fire risk. Possible carcinogen.
Use: Intermediate for various polymers, stabilizer for chlorinated solvents.

2,3-butylene oxide.   (2,3-epoxybutane).

$$CH_3HCOCHCH_3.$$

Two forms, *cis* and *trans*, are known.
Properties: *cis*: fp −80C, bp 59.7C (742 mm Hg), d 0.8266 (25/4C). *trans*: fp −85C, bp 53.5C (742 mm Hg), d 0.8010 (25/4C). Flash p approx 0F (−17C). Very soluble in ether, benzene, organic solvents; decomposes in water.
Hazard: Toxic concentration of vapors occurs at room temperature. Highly flammable, dangerous fire risk.
Use: Intermediate.

butyl epoxystearate.   (butyl-9,10-epoxyoctadecanoate).

$$CH_3(CH_2)_7CHOCH(CH_2)_7COOC_4H_9.$$

Properties: Colorless liquid; mild, slightly fatty, slightly fruity odor. D 0.910 (20C), wt/gal 7.59 lb. Combustible.
Use: Plasticizer for low-temperature flexibility improvement of vinyl resins.

n-butylethanolamine. $C_4H_9NHCH_2CH_2OH$.
Properties: Colorless liquid; very faint amine odor. D 0.892 (20C), boiling range 194–204C, flash p 170F (76.6C). Combustible.

t-butylethanolamine.
CAS: 4620-70-6. $C_6H_{15}NO$.
Hazard: Moderately toxic by ingestion and skin contact. A severe eye irritant.

butyl ether.   (*n*-dibutyl ether).
CAS: 142-96-1. $C_4H_9OC_4H_9$.
Properties: Colorless liquid; stable; mild ethereal odor. D 0.7694 (20/20C), bp 142.2C, vap press 4.8 mm Hg (20C), flash p 77F (25C), fp −95.2C, latent heat of vaporization 67.8 cal/g at 140.9C, refr index 1.3992 (20C), wt/gal 6.4 lb (20C), viscosity 0.0069 cP (20C), autoign temp 382F (194C). Miscible with most common organic solvents; immiscible with water.
Grade: Technical, spectrophotometric.
Hazard: Toxic on prolonged inhalation. Flammable, moderate fire risk. May form explosive peroxides, especially in anhydrous form.

Use: Solvent for hydrocarbons, fatty materials; extracting agent used especially for separating metals, solvent purification, organic synthesis (reaction medium).

butylethylacetaldehyde.   See 2-ethylhexaldehyde.

5-butyl-5-ethylbarbituric acid.   See butethal.

n-butyl ethyl ether.   See ethyl-*n*-butyl ether.

butyl ethyl ketone.   (3-heptanone).
CAS: 106-35-4. $(C_4H_9)(C_2H_5)CCO$.
Properties: Clear liquid, d 0.8198 (20/20C), bp 148C, fp −36.7C, refr index 1.4224 (20C), flash p 115F, vap d 3.93.
Grade: Available as 20% solution in hexane.
Hazard: Fumes are irritating. Flammable, dangerous fire risk. Store solution under nitrogen.
Use: Reactive chemical intermediate.

2-butyl-2-ethylpropanediol-1,3.   See 2-ethyl-2-butylpropanediol-1,3.

tert-butylformamide. $(CH_3)_3CNHCOH$.
Properties: Colorless, high-boiling liquid. Soluble in water and common hydrocarbon solvents.
Use: Solvent and in petroleum additives.

butyl formate.
CAS: 592-84-7. $HCOOC_4H_9$.
Properties: Colorless liquid. D 0.885–0.9108, bp 107C, fp −90C, flash p 64F (17.7C) (CC), autoign temp 612F (322.5C). Miscible with alcohols, ethers, oils, hydrocarbons; slightly soluble in water.
Grade: Technical.
Hazard: Narcotic and irritating in high concentration. Flammable, dangerous fire risk.
Use: Solvent for nitrocellulose, some types of cellulose acetate, many cellulose ethers, many natural and synthetic resins, lacquers, perfumes, organic synthesis (intermediate), flavoring.

n-butyl furfuryl ether. $C_4H_9OCH_2C_4H_3O$.
Properties: Colorless liquid, turning dark on exposure to air. Extremely hygroscopic, unstable with moisture. D 0.955 (10/0C), bp 189–190C (765 mm Hg), refr index 1.4522 (20C).

n-butyl furoate. $C_4H_3OCO_2C_4H_9$.
Properties: Colorless oil. Decomposes on standing. D 1.055 (20/5C), bp 83–84C (1 mm Hg), 118–120C (25 mm Hg). Insoluble in water, soluble in alcohol and ether.

n-butyl glycidyl ether.   (glycidylbutylether; BGE).
CAS: 2426-08-6. $C_4H_9OH_2CHOCH_2$.
Properties: Clear, colorless liquid; irritating odor. Bp 164C, vapor press 3.2 mm Hg (25C), vap d 3.78, d 0.908 (25/4C). Soluble in water.

Hazard: A mild skin and eye irritant. Sensitization and reproduction effects.

n-butyl glycol phthalate. See dibutoxyethyl phthalate.

tert-butyl hydroperoxide.
CAS: 75-91-2. $(CH_3)_3COOH$.
A highly reactive peroxy compound.
Properties: Water-white liquid. Fp −8C, decomposes at 75C, d 0.896 (20/4C), refr index 1.396 (25C) (90% pure), flash p (90%) 130F (54.4C). Moderately soluble in water; very soluble in organic solvents and alkali-metal hydroxide solutions. Combustible.
Grade: 70%, 90% pure.
Hazard: Moderate fire risk. Oxidizer.
Use: Polymerization, oxidation, sulfonation catalyst, bleaching, deodorizing.

tert-butylhydroquinone.
$C_6H_3(OH)_2C(CH_3)_3$.
Properties: Intermediate. Mp 125C. Insoluble in water; soluble in alcohol, acetone, and ethyl acetate.

butylhydroxyoxostannane. See butyl stannoic acid.

tert-butyl hypochlorite. $(CH_3)_3COCl$.
Properties: Yellowish liquid.
Hazard: Toxic by ingestion and inhalation. May explode at room temperature.
Use: Organic chlorinations, oxidation of alcohols to ketones and sulfides to sulfoxides.

4,4′-butylidenebis(6-tert-butyl-m-cresol).
$[(CH_3)_3CC_6H_2(OH)(CH_3)]_2CHC_3H_7$.
Properties: White powder. Mp 209C (min), d 1.03 (25C).
Use: Antioxidant for rubber, dry or latex.

n-butylisocyanate.
CAS: 111-36-4. C_4H_9NCO.
Properties: Colorless liquid. Bp 115C, d 0.88 (20/4C).
Use: Intermediate for pesticides, herbicides, pharmaceuticals.
Hazard: Strong irritant to eyes and skin.

butyl isodecyl phthalate.
Properties: Clear liquid; mild odor; color (Hazen) 50 max. D 0.9 (20/20C), saponification number 310, acidity (as phthalic acid) 0.01 max.
Use: Plasticizer for polyvinyls.

tert-butylisopropylbenzene hydroperoxide.
Properties: White crystals.
Hazard: Dangerous fire risk. Reacts strongly with reducing materials. Oxidizing agent.

n-butyl lactate.
CAS: 138-22-7. $CH_3CH_2OCOOC_4H_9$.

Properties: Water-white, stable liquid; mild odor. D 0.974–0.984 (20/20C), flash p 168F (75.5C) (TOC), mp −43C, wt/gal 8.15 lb, bp 188C, refr index 1.4216 (20C), vap press 0.4 mm Hg (20C), latent heat of vaporization 77.4 cal/g (20C), autoign temp 720F (382C). Miscible with many lacquer solvents, diluents, oils, slightly soluble in water, hydrolyzed in acids and alkalies. Combustible.
Grade: Technical, 95% min.
Hazard: Toxic. Upper respiratory tract irritant.
Use: Solvent for nitrocellulose, ethyl cellulose, oils, dyes, natural gums, many synthetic polymers, lacquers, varnishes, inks, stencil pastes, antiskinning agent, chemical (intermediate), perfumes, dry-cleaning fluids, adhesives.

N-n-butyl lauramide. $C_{11}H_{23}CONHC_4H_9$.
Properties: White solid; lauric acid odor. Boiling range 200–225C (2 mm Hg), flash p 375F (190C). Combustible.

butyl laurate. (butyl dodecanoate).
$C_{11}H_{23}COOC_4H_9$.
Properties: Liquid. D 0.855 (25C), bp 130–180C (5 mm Hg), fp −10C. Insoluble in water. Combustible.
Derivation: Alcoholysis of coconut oil with butyl alcohol followed by fractional distillation.
Use: Plasticizer, flavoring.

butyllithium. (*n*-butyllithium; *sec*-butyllithium; *tert*-butyllithium). $CH_3CH_2CH_2CH_2Li$; $CH_3CHLiCH_2CH_3$; $(CH_3)_3CLi$.
Properties: Available usually in solution in one of the C_5 to C_7 hydrocarbons, in which it is quite stable. *sec*-Butyllithium solution must be kept at or below 15.5C. Sold by percent butyllithium in the solution.
Hazard: Irritant. Solid and solution highly flammable, ignites in moist air.
Use: Polymerization of isoprene and butadiene; intermediate in preparation of lithium hydride; rocket fuel component; metalating agent.

n-butylmagnesium chloride. C_4H_9MgCl.
Properties: Liquid. D 0.88.
Derivation: From magnesium and butyl chloride.
Grade: Available in solution in ethyl ether or in tetrahydrofuran.
Hazard: Flammable, dangerous fire risk.
Use: Grignard reagent, as an alkylating agent.

n-butylmelamine.
CAS: 5606-24-6. $C_7H_{14}N_6$.
Hazard: A poison.

n- and sec-butyl mercaptan. See 1-butanethiol; 2-butanethiol.

tert-butylmercaptan. Legal label name for 2-methyl-2-butanethiol.

butyl mesityl oxide. See butopyronoxyl.

n-butyl methacrylate.   (methacrylic acid, butyl ester).
CAS: 97-88-1. $H_2C:C(CH_3)COOC_4H_9$.
Properties: Colorless liquid. Bp 163.5–170.5C, fp <−75C, d 0.895 (25/25C), flash p 130F (54.4C) (OC), refr index 1.4220. Readily polymerized. Insoluble in water. Combustible.
Derivation: Reaction of methacrylic acid or methyl methacrylate with butanol.
Grade: Technical (inhibited).
Hazard: Toxic by ingestion. Moderate fire risk.
Use: Monomer for resins, solvent coatings, adhesives, oil additives; emulsions for textiles, leather, and paper finishing.
See acrylic resin.

tert-butyl methacrylate.
$H_2C:C(CH_3)COOC(CH_3)_3$.
Properties: Colorless liquid. Bp 66C (57 mm Hg), d 0.877 (25C), refr index 1.4124 (24C), flash p 92F (33.3C) (TOC).
Grade: Technical containing 100 ppm hydroquinone monomethyl ether as inhibitor.
Hazard: Toxic by ingestion. Flammable, dangerous fire risk.
Use: Monomer for resins.
See acrylic resin.

butyl methacrylate resins.   See "Elvacite" [Lucite].

tert-butyl-4-methoxyphenol.   See butylated hydroxyanisole.

sec-butyl-6-methyl-3-cyclohexene-1-carboxylate.   See siglure.

p-tert-butyl-α-methylhydrocinnamaldehyde.
(α-methyl-β-(p-tert-butylphenyl)-propionaldehyde).
$(CH_3)_3CC_6H_4CH_2CH(CH_3)CHO$.
Properties: Light-yellow liquid; strong floral odor. D 0.942–0.949 (25/25C), refr index 1.503–1.510 (20C), flash p 204F (95.5C) (TCC). Soluble in 1-part 90% alcohol. Stable, nondiscoloring. Combustible.
Grade: 93%, 85% purity.
Use: Perfume.

6-tert-butyl-3-methylphenol.   See 6-tert-butyl-m-cresol.

n-butyl myristate. $CH_3(CH_2)_{12}COOC_4H_9$. The butyl ester of myristic acid.
Properties: Water-white oily liquid. Saponification number 193–203, fp 1–7C, boiling range 167–197C (5 mm Hg), d 0.850–0.858 (25C). Insoluble in water; soluble in acetone, castor oil, chloroform, methanol, mineral oil, toluene. Combustible.
Derivation: Alcoholysis of coconut oil with butyl alcohol followed by fractional distillation.

Use: Plasticizer, lubricant for textiles, paper stencils, cosmetic preparations.

n-butyl nitrate. $C_4H_9NO_3$.
Properties: Water-white liquid; ethereal odor. D 1.103 (20C), bp 123C, flash p 97F (36C). Insoluble in water; soluble in alcohol and ether.
Hazard: Flammable, moderate fire risk in contact with reducing materials. Oxidizing agent, may explode from shock or heating.

tert-butyl nitrite.   (nitrous acid tert-butyl ester). $(CH_3)_3CONO$.
Properties: Yellowish liquid; pleasant odor. D 0.867, bp 63C, refr index 1.36. Soluble in alcohol, carbon disulfide, chloroform; slightly soluble in water.
Derivation: Reaction of tert-butyl alcohol, sodium nitrite, and sulfuric acid.
Use: Jet fuel.

butyl nonanoate.   See butyl pelargonate.

butyl octadecanoate.   See n-butyl stearate.

butyloctyl phthalate.
$C_4H_9OOCC_6H_4COOC_8H_{17}$.
Properties: Water-white liquid; mild characteristic odor. D 0.991–0.997 (20/20C), saponification number 298–308. Miscible with most organic solvents. Combustible.
Use: Plasticizer for vinyl resins.

butyl oleate.
CAS: 142-77-8. $CH_3(CH_2)_7CH:CH(CH_2)_7$ $COOC_4H_9$.
Properties: Light-colored, oleaginous liquid; mild odor. D 0.873 (20/20C), iodine value 76.8, fp opaque at 12C, solid at −26.4C, wt/gal 7.26 lb (20C), boiling range 173–227C (2 mm Hg), flash p 356F (180C). Insoluble in water; miscible with alcohol, ether, vegetable and mineral oils. Combustible.
Derivation: Alcoholysis of olein or esterification of oleic acid with butanol.
Use: Plasticizer (particularly for PVC), solvent, lubricant, water-resisting agent, coating compositions, polishes, waterproofing compounds.

butyl Oxitol.   An ethylene glycol monobutyl ether.
Hazard: Slight irritant to skin and eyes.
Use: Solvent in various types of surface-coating formulations to improve gloss and leveling, component of hydraulic fluids, coupling agent in various types of cleaning and cutting oils.

butylparaben.   (n-butyl hydroxybenzoate). $C_{11}H_{14}O_3$.
Properties: Finely divided solid. Mp 68C. Soluble in acetone, alcohol, and propylene glycol.
Available forms: Calcium and magnesium salts are available.

Use: Pharmaceutical preservative, fungistat. See Parabens.

n-butyl pelargonate. (*n*-butyl nonanoate). $C_4H_9OOCC_8H_{17}$.
Properties: Liquid with fruity odor. Combustible. D 0.865 (15.5/15.5C), bp 270C.
Use: Flavors and perfumes, chemical intermediate.

sec-butyl pelargonate.
$C_2H_5CH(CH_3)OOCC_8H_{17}$.
Properties: Liquid. D 0.8608 (20/4C), bp 123C (15 mm Hg), refr index 1.4220. Combustible. Shows optical activity.
Use: Chemical intermediate.

tert-butyl peracetate. See *tert*-butyl peroxyacetate.

tert-butyl perbenzoate. See *tert*-butyl peroxybenzoate.

tert-butyl perisobutyrate. See *tert*-butyl peroxyisobutyrate.

tert-butyl permaleic acid. See *tert*-butyl peroxymaleic acid.

tert-butyl peroxide. See di-*tert*-butyl peroxide.

tert-butyl peroxyacetate. (*tert*-butyl peracetate).
CAS: 107-71-1. $(CH_3)_3COOOCCH_3$.
Properties: Flash p <80F (<26.6C) (COC), d 0.923.
Grade: Available as a 72–76% solution in benzene.
Hazard: Flammable, dangerous fire risk. Oxidizer.
Use: Polymerization initiator for vinyl monomers, manufacture of polyethylene and polystyrene.

tert-butyl peroxybenzoate. (*tert*-butyl perbenzoate).
CAS: 614-45-9. $(CH_3)_3COOOCC_6H_5$.
Properties: Colorless liquid; mild aromatic odor. D 1.04 (25/25C), fp 8.5C, vap press 0.33 mm Hg (50C), flash p 200F (93.3C). Soluble in alcohols, esters, ethers, ketones; insoluble in water.
Grade: 98% min.
Hazard: Oxidizing material; do not store near combustible materials.
Use: Polymerization initiator for polyethylene, polystyrene, polyacrylates, and polyesters; chemical intermediate.

tert-butyl peroxy-2-ethylhexanoate. (*tert*-butyl peroctoate). $(CH_3)_3COOOCCH(C_2H_5)C_4H_9$.
Properties: Colorless liquid; faint odor. D 0.895 (25/25C), fp <−30C, refr index 1.426 (25C), decomposes at 89C, flash p 190F (87.7C). Insoluble in water; miscible with most organic solvents.
Hazard: Oxidizing material. Do not store near combustibles.

tert-butyl peroxyisobutyrate. (*tert*-butyl perisobutyrate).
CAS: 109-13-7. $(CH_3)_3COOOCCH(CH_3)CH_3$.
Properties: Flash p <80F (<26.6C).
Grade: Available as a 72–75% solution in benzene.
Hazard: Flammable, dangerous fire risk. Oxidizing agent.
Use: Polymerization catalyst.

tert-butylperoxyisopropyl carbonate.
(BPIC). $(CH_3)_3COOOCOCH(CH_3)_2$.
Properties: Liquid. Fp −3C, d 0.945, refr index 1.4050 (20C), flash p 112–118F (44–47C) (TOC). Almost insoluble in water; miscible with hydrocarbons, esters, and ethers. Relatively stable under ordinary conditions. Flammable.
Grade: Technical (8.6% active oxygen).
Hazard: Moderate fire risk. Oxidizing agent.
Use: Polymerization initiator, cross-linking agent.

tert-butylperoxymaleic acid. (*tert*-butylpermaleic acid). $(CH_3)_3COOOCCH:CHCOOH$. An unsaturated peroxide.
Properties: White crystals. Mp 114–116C (decomposes). Slightly soluble in water, cool 5% alkaline solution, and alcohols; moderately soluble in oxygenated organic solvents and polyester monomers; slightly soluble in naphtha, carbon tetrachloride, and chloroform; insoluble in benzene.
Grade: 95% pure.
Hazard: Oxidizing agent. Do not store near combustible materials.
Use: Polymerization catalyst, bleaching, pharmaceuticals.

tert-butylperoxyphthalic acid. (*tert*-butylperphthalic acid). $(CH_3)_3COOOCC_6H_4COOH$.
Properties: White crystals. Mp 96–99C. Insoluble in water; soluble in cool 5% alkaline solutions and in alcohols; moderately soluble in oxygenated organic solvents, chlorinated hydrocarbons and polyester monomers; slightly soluble in petroleum hydrocarbons.
Grade: 95% pure.
Hazard: Oxidizing agent. Do not store near combustible materials.
Use: Polymerization catalyst and oxidizing agent.

tert-butylperoxypivalate.
CAS: 927-07-1. $(CH_3)_3COOOCC(CH_3)_3$.
Properties: Colorless liquid. D 0.854 (25/25C). Solidifies below −19C, refr index 1.410 (25C), decomposes at 70C, flash p 155–160F (68–71C). Insoluble in water and ethylene glycol; soluble in most organic solvents.
Grade: Available as 75% soluble in mineral spirits.
Hazard: (Solution) Flammable, dangerous fire risk. Oxidizing agent. May explode on heating.
Use: Polymerization initiator.

tert-butyl perphthalic acid. See *tert*-butyl peroxyphthalic acid.

o-sec-**butylphenol.**
CAS: 89-72-5. $C_2H_5(CH_3)CHC_6H_4OH$.
Properties: A slightly volatile liquid. Mw 150.22, bp 226–228C, flash p 225F, d 0.891. Insoluble in water; slightly soluble in alcohol, ether, and alkalies. Combustible.
Hazard: Skin, eye, and upper respiratory tract irritant.
Use: Chemical intermediate in preparation of resins, plasticizers, surface-active agents.

o-tert-**butylphenol.** $(CH_3)_3CC_6H_4OH$.
Properties: Light-yellow liquid. Fp −7C, d 0.982 (20C), bp 224C, flash p 230F (110C) (OC). Soluble in isopentane, toluene, and ethanol; insoluble in water. Combustible.
Hazard: Toxic by ingestion, moderate irritant to eyes and skin.
Use: Chemical intermediate for synthetic resins, plasticizers, surface-active agents, perfumes, and other products. A permissible antioxidant for aviation gasoline (ASTM D910-64T).

p-tert-**butylphenol.**
CAS: 98-54-4. $(CH_3)_3CC_6H_4OH$.
Properties: White crystals; distinctive odor. D (crystals) 1.03, d (molten) 0.908 (114/4C), bp 239C, mp 100C. Insoluble in water, soluble in alcohol and ether. Combustible.
Derivation: Catalytic alkylation of phenol with olefins.
Hazard: Irritant to eyes and skin.
Use: Plasticizer for cellulose acetate; intermediate for antioxidants, special starches, oil-soluble phenolic resins; pour-point depressors and emulsion breakers for petroleum oils and some plastics; synthetic lubricants; insecticides; industrial odorants; motor-oil additives.

n-**butylphenyl acetate.** $C_4H_9OOCCH_2C_6H_5$.
Properties: Colorless liquid; rose-honey odor. D 0.991–0.994 (25/25C), bp 135–141C (18 mm Hg), refr index 1.488–1.490 (20C). Soluble in 2 volumes 80% alcohol. Combustible. Made synthetically.
Grade: 98% min.
Use: Perfumes, flavoring.

n-**butylphenyl ether.** $C_4H_9OC_6H_5$.
Properties: Water-white liquid; aromatic odor. D 0.929 (20C), boiling range 202–212C, flash p 180F (82C). Combustible.
Hazard: Toxic by ingestion.

1-(3-(4-*tert*-butylphenyl)-2-methylpropyl)-piperidine.   See fenpropidine.

4-*tert*-butylphenyl salicylate.
$(CH_3)_3CC_6H_4OOCC_6H_4OH$.
Properties: Off-white, odorless crystals. Mp 62–64C. Soluble in alcohol, ethyl acetate, toluene; insoluble in water.
Use: Light absorber, best at 2900–3300 Å.

p-tert-**butylphenyl salicylate.**
CAS: 87-18-3. $C_{17}H_{18}O_3$.
Hazard: Moderately toxic by ingestion.
Use: Food additive.

n-**butylphosphoric acid.**   See *n*-butyl acid phosphate.

n-**butyl phthalate.**   See dibutyl phthalate.

butylphthalylbutyl glycolate.
$C_4H_9OOCC_6H_4COOCH_2COOC_4H_9$.
Properties: Colorless liquid; odorless liquid. D 1.093–1.103 (25/25C), bp 219C (5 mm Hg), solidifies below −35C, darkens on heating above 290C, flash p 390F (199C) (OC). Insoluble in water. Extremely light stable. Combustible.
Use: Plasticizer for polyvinyl chloride. FDA approved for use in vinyl food wrappings.

n-**butyl phthalyl-*n*-butyl glycolate.**   See "Morflex 190" *[Reilly]*.

n-**butyl propionate.**
CAS: 590-01-2. $C_2H_5CO_2C_4H_9$.
Properties: Water-white liquid; applelike odor. D 0.875 (20C), 0.874 (15.5C), wt/gal 7.3 lb, bp 146C (commercial grades boil over a range of 130–150C due to presence of butyl alcohol and esters), fp −89C, flash p 90F (32.2C), autoign temp 800F (426C). Soluble in alcohol and ether; miscible with all coal tar and petroleum distillates; very slightly soluble in water.
Derivation: Esterification of propionic acid with butyl alcohol and sulfuric acid catalyst.
Grade: Technical (85–90% to 95% ester content).
Hazard: Skin and eye irritant. Flammable, moderate fire risk.
Use: Solvent for nitrocellulose, retarder in lacquer thinner, ingredient of perfumes, flavors.

2-*t*-butylpyrimidine.   See 2-(1,1-dimethylethyl)pyrimidine.

butyl ricinoleate. $C_{17}H_{32}(OH)COOC_4H_9$.
Properties: Yellow to colorless oleaginous liquid. D 0.916 (20/20C), bp approx 275C (13 mm Hg), flash p 220F (104.4C), Saybolt viscosity 112 (100F), fp indefinite, slightly opaque at −30C, very viscous at −50C, wt/gal 7.62 lb (20C). Soluble in alcohol and ether; insoluble in water. Combustible.
Derivation: Castor oil and butyl alcohol.
Use: Plasticizer, lubricant.

butyl rubber.   A copolymer of isobutylene (97%) and isoprene (3%). Polymerized below −95C with aluminum chloride catalyst.
Properties: D 0.92. Vulcanizates have tensile strength up to 2000 psi (unreinforced) and 3000 psi (reinforced). Service temperatures range from −55 to +204C. Good abrasion resistance, excellent impermeability to gases, high dielectric constant,

excellent resistance to aging and sunlight, superior shock-absorbing and vibration-damping qualities. Resistance to oils and greases only fair. Will support combustion.

Grade: Stabilized, latex, chlorine-containing elastomer, low molecular weight (liquid).

Use: Tire carcasses and linings, especially for tractors and other outsize vehicles; electric wire insulation; encapsulating compounds; steam hose and other mechanical rubber goods; pond and reservoir sealant. Latex is used for paper coating, textile and leather finishing, adhesive formulations, air bags, tire vulcanization, self-curing cements, pressure-sensitive adhesives, tire-cord dips, sealants.

See isobutylene-isoprene copolymer.

butyl sebacate.   See dibutyl sebacate.

butyl sorbate. $C_4H_9OOCC_5H_7$.

An insect attractant for the European chafer.

n-butylstannic acid. $[(C_4H_9)Sn(OH)_3]$.

Properties: White, infusible, and insoluble free-flowing powder. Exists as a polymer of undetermined chemical structure. Further dehydration results in polymers with molecular weights of 1000–5000 that are soluble in organic solvents.

Hazard: Toxic by ingestion and skin absorption.

Use: Polymerization catalyst, antioxidant, and heat stabilizer for PVC (not approved by FDA for food containers), electrically conducting tin oxide coatings, alkaline-earth-metal phosphates in fluorescent light bulbs, intermediate for silicones.

butyl stannoic acid.

CAS: 2273-43-0. $C_4H_{10}O_2Sn$.

Properties: White infusible solid. Sol in Me_2CO.

Hazard: A poison.

butyl stearamide. $C_{17}H_{35}CONHC_4H_9$.

Properties: Light-straw-colored liquid. D 0.869 (20/20C), boiling range 195–200C (2 mm Hg), flash p 430F (221C), amide odor. Combustible.

Use: Plasticizer and intermediate for the synthesis of insecticides, surface-active agents, pharmaceuticals, and textile assistants.

butyl stearate.   (butyl octadecanoate).

$C_{17}H_{35}COOC_4H_9$.

Properties: Colorless, stable, oleaginous liquid; practically odorless, sometimes with faint fatty odor. D 0.855–0.860 (25/25C), mp 19.5–20C, flash p 320F (160C) (CC). Wt/gal 7.14 lb (20C), refr index 1.4430 (20C), bp 350C. Miscible with mineral and vegetable oils; soluble in alcohol and ether; insoluble in water. Combustible.

Derivation: Alcoholysis of stearin or esterification of stearic acid with butanol.

Grade: Technical, cosmetic, chemically pure.

Use: Ingredient of polishes, special lubricants, and coatings; lubricants for metals and in textile and

molding industries; in wax polishes as dye solvent; plasticizer for laminated fiber products' rubber hydrochloride; chlorinated rubber and cable lacquers; carbon paper and inks; emollient in cosmetic and pharmaceutical products; lipsticks; damp proofer for concrete; flavoring.

butyl sulfide.   See di-*tert*-butyl sulfide.

4-*tert*-butyl-*o*-thiocresol.   (2-methyl-4-*tert*-butylthiophenol). $(CH_3)_3CC_6H_3(CH_3)SH$.

Properties: Colorless liquid; mild (nonmercaptan) odor. D 0.983 (25C), fp −4C, bp 250C, refr index 1.546 (25C). Insoluble in water; soluble in hydrocarbons. Combustible.

Grade: Available as 98% pure, supplied under nitrogen atmosphere, and as 55% solution in hydrocarbon.

Use: Peptizer for rubbers, polymer modifier, lubricating-oil additive.

4-*tert*-butylthiophenol. $(CH_3)_3CC_6H_4SH$.

Properties: Colorless liquid; mild (nonmercaptan) odor. D 0.986 (25C), fp −11C, bp 238C, refr index 1.546 (25C). Insoluble in water; soluble in hydrocarbons.

Grade: 98%, supplied under nitrogen atmosphere.

Use: Lubricating-oil additives, polymer modifiers, antioxidants, dyes.

(butylthio)trioctylstannane.

CAS: 70303-47-8. $C_{28}H_{60}SSn$.

Hazard: A poison.

Use: Drug.

(butylthio)tripropylstannane.   See tripropyl(butylthio)stannane.

n-butyltin trichloride. $C_4H_9SnCl_3$.

Properties: Colorless liquid. Fumes in contact with air. D 1.71 (25/4C), bp 102C (12 mm Hg), refr index 1.5190 (25C). Soluble in organic solvents; sparingly soluble in water with partial hydrolysis.

Hazard: Toxic by ingestion, strong irritant to skin. Avoid exposure to liquids or vapors.

Use: Intermediate, catalyst, stabilizer.

n-butyltin tris(dibutyldithiocarbamate).

CAS: 73927-88-5. $C_{31}H_{63}N_3S_6Sn$.

Hazard: A poison.

butyl titanate.   See tetrabutyl titanate.

p-*tert*-butyltoluene.   (1-methyl-4-*tert*-butylbenzene).

CAS: 98-51-1. $(CH_3)_3CC_6H_4CH_3$.

Properties: Colorless liquid. D 0.857–0.863 (20/20C), bp 192.8C. Insoluble in water. Combustible.

Grade: Technical.

Hazard: Toxic by inhalation, ingestion, and skin absorption. Eye and upper respiratory tract irritant.

Use: Solvent, intermediate.

n-butyltrichlorosilane. $C_4H_9SiCl_3$.
Properties: Colorless liquid. Bp 142C, d 1.1608 (25/25C), refr index 1.4363 (25C), flash p 126F (52C) (COC). Readily hydrolyzed with liberation of hydrogen chloride. Soluble in benzene, ether, heptane.
Derivation: Grignard reaction of silicon tetrachloride.
Grade: Technical, 95%.
Hazard: Corrosive to skin and tissue. Moderate fire risk.
Use: Intermediate for silicones.

butyl trichloro stannane.
CAS: 1118-46-3. $C_4H_9Cl_3Sn$.
Properties: Liquid. D 0.85 @ 20C/4C, bp 93C @ 10 mm.
Hazard: Moderately toxic by ingestion. A severe skin and eye irritant.

N-n-butylurea.
CAS: 592-31-4. $C_4H_9HNCONH_2$.
Properties: White solid; odorless. Decomposes on heating, mp 96C. Soluble in water, alcohol, and ether.

n-butyl vinyl ether. See vinyl-n-butyl ether.

1-butyne. See ethylacetylene.

2-butyne. See crotonylene.

butynediol.
CAS: 110-65-6. $HOCH_2C:CCH_2OH$.
Properties: White, orthorhombic crystals. Mp 58C, bp 238C, refr index 1.450 (25C). Soluble in water, aqueous acids, alcohol, and acetone; insoluble in ether and benzene. Combustible.
Derivation: High-pressure synthesis from acetylene and formaldehyde.
Grade: Crystalline solid, 97%, aqueous soluble 35%.
Hazard: Toxic by ingestion. May explode on contamination with mercury salts, strong acids, and alkaline earth hydroxides and halides at high temperatures.
Use: Intermediate, corrosion inhibitor, electroplating brightener, defoliant, polymerization accelerator, stabilizer for chlorinated hydrocarbons, cosolvent for paint and varnish removal.

3-butyn-1-ol. (β-ethynyl ethanol).
$HC:CCH_2CH_2OH$.
Properties: Water-white liquid; characteristic odor. D 0.9257 (20/4C), refr index 1.4409 (20C), bp 128.9C, fp −63.6C. Combustible.
Use: Preparation of perfume bases, acetylenic esters, plastics, plasticizers, pharmaceuticals, wetting agents, medicinals, and organic synthesis.

1,1'-(2-butynylene)dipyrrolidine dihydrochloride. See tremorine dichlorohydrate.

butyraldehyde. (butaldehyde; n-butanal; n-butylaldehyde; butyric aldehyde).
CAS: 123-72-8. $CH_3(CH_2)_2CHO$.
Properties: Water-white liquid; characteristic pungent aldehyde odor. D 0.8048 (20/20C), bp 75.7C, vap press 91.5 mm Hg (20C), flash p 20F (−6.6C), wt/gal 6.7 lb (20C), coefficient of expansion 0.00114 (20C), fp −99C, viscosity 0.043 cP (20C), autoign temp 446F (230C). Slightly soluble in water; soluble in alcohol and ether.
Derivation: (1) Oxo process; (2) dehydrogenating butanol vapors over a catalyst, the butyraldehyde being separated by distillation; (3) partial reduction of crotonaldehyde.
Grade: Technical (93% min).
Hazard: Flammable, dangerous fire risk.
Use: Plasticizers, rubber accelerators, solvents, high polymers.

butyric acid. (n-butyric acid; butanoic acid; ethylacetic acid; propylformic acid).
CAS: 107-92-6. $CH_3CH_2CH_2COOH$.
Properties: Colorless liquid; penetrating and obnoxious odor. Refr index 1.3981 (20C), d 0.9583 (20/4C), fp −5.0 to −8C, bp 163.5C (757 mm Hg) and 75C (25 mm Hg), vap press 0.84 mm Hg (20C), autoign temp 846F (452C). Miscible with water, alcohol, and ether. Combustible.
Derivation: Occurs as glyceride in animal milk fats. Produced as a by-product in hydrocarbon synthesis, by oxidation of butyraldehyde, and by butyric fermentation of molasses or starch.
Grade: 90%, 95%, 99%, edible, synthetic, reagent, technical, FCC.
Hazard: Strong irritant to skin and tissue.
Use: Synthesis of butyrate ester perfume and flavor ingredients, pharmaceuticals, deliming agent, disinfectants, emulsifying agents, sweetening gasolines.

butyric alcohol. See n-butyl alcohol.

butyric aldehyde. See butyraldehyde.

butyric anhydride.
CAS: 106-31-0. $(CH_3CH_2CH_2CO)_2O$.
Properties: Water-white liquid. Hydrolyzes to butyric acid. D 0.9681 (20/20C), fp −75C, bp 199.5C, vap press 0.3 mm Hg (20C), flash p 190F (87.7C), wt/gal 8.1 lb (20C). Combustible.
Grade: Technical, 98%.
Use: Manufacture of butyrates, drugs, and tanning agents.

butyrin. See glyceryl tributyrate.

β-butyrolacetone. (3-hydroxybutanoid acid β-lactone; hydroxybutyric acid lactone;

3-hydroxybutyric acid lactone; 4-methyl-2-oxetanone; 4-methyloxetan-2-one).
CAS: 3068-88-0. $C_4H_6O_2$.
Hazard: Possible carcinogen, neoplastigen; tumorigenmutagen, irritant, slightly toxic.

butyrolactam.   See 2-pyrrolidone.

butyrolactone.   (γ-butyrolactone).
CAS: 96-48-0. $OCH_2CH_2CH_2CO$.

$$CH_2 \cdot CH_2 \cdot CH_2 \cdot$$
$$O \text{———} CO$$

Properties: Colorless liquid; pleasant odor. Bp 204C, fp −44C, d 1.144, flash p 209F (98.3C) (OC). Miscible with water, alcohol, and ether. Combustible.
Derivation: High-pressure synthesis from acetylene and formaldehyde.
Grade: Technical.
Hazard: Toxic by ingestion. Questionable carcinogen.
Use: Intermediate for synthesis of butyric acid compounds, polyvinylpyrrolidone, methionine, solvent for acrylate and styrene polymers, ingredient of paint removers and textile assistants.

butyrone.   See dipropyl ketone.

butyronitrile.   (propyl cyanide; butanenitrile).
CAS: 109-74-0. $CH_3(CH_2)_2CN$.
Properties: Colorless liquid. D 0.796 (15C), fp −112.6C, bp 116–117C, flash p 79F (26.1C) (OC). Slightly soluble in water; soluble in alcohol and ether.
Hazard: Flammable, dangerous fire risk.
Use: Basic material in industrial, chemical, and pharmaceutical intermediates and products; poultry medicines.

butyroyl chloride.   (butyryl chloride; butanoyl chloride).
C_3H_7COCl.

Properties: Colorless liquid; pungent acid chloride odor. Reacts with alcohol and water; miscible with ether. Fp −89C, distillation range 100–110C, d 1.028 (15C), refr index 1.4121 (20C).
Hazard: Toxic by ingestion and inhalation, strong irritant to tissue.
Use: Organic synthesis.

butyryl chloride.   See butyroyl chloride.

10-butyryldithranol.
CAS: 75464-11-8. $C_{18}H_{16}O_4$.
Hazard: A poison by ingestion. A mild skin irritant.

n-**butyryltri-*n*-hexyl citrate.**   See "Citroflex B-6" *[Vertellus]*.

buxine.   A neurotoxin isolated from the leaves of *Buxus sempervirens* (common boxwood).

BVE.   Abbreviation for butyl vinyl ether. See vinyl-*n*-butyl ether.

B-X-A.   A diarylamine-ketone-aldehyde reaction product.
Properties: Brown powder. D 1.10, melting range 85–95C. Store in a cool place. Soluble in acetone, benzene, and ethylene dichloride; insoluble in water and gasoline.
Use: Antioxidant for rubber and nylon.

"Bynel" *[DuPont].*   TM for resins developed as coextrudable adhesives for multilayer packaging structures.
Use: Food packaging, sporting goods, wire and cable, building construction and automotive. More recent uses are extrusion laminations, in extrusion coatings, and as adhesive films and webs for thermal laminating and other new uses not related to packaging.

BZ.   A nonlethal gas that causes temporary disability. It is a derivative of lysergic acid.

C

C. Symbol for carbon.

^{14}C. (carbon-14). The naturally occurring radioactive isotope of carbon used in chemical dating, tracer studies, etc.

C12-C14-*tert*-alkyl amines.
CAS: 68955-53-3.
Hazard: A poison by ingestion. Moderately toxic by skin contact. Low toxicity by inhalation. A severe skin irritant.

C-22. A high-strength, nickel-based alloy with corrosion resistance to oxidizing or reducing medias, excellent resistance to localized corrosion while maintaining ease of welding and fabrication.
Available forms: Sheet, plate, bars, rods, welding electrodes, and wire.
Use: Fabrication into all types of process equipment.

CA. Abbreviation for cellulose acetate and cortisone acetate; also for controlled atmosphere.

C$_3$A. Abbreviation for tricalcium aluminate as used in cement.
See cement, Portland.

Ca. Symbol for calcium.

CAA. (2-cyanoaccetamide).
CAS: 107-91-5. $C_3H_4N_2O$.
A codon of glutamine that directs the placement of glutamine into a polypeptide.

CAB. Abbreviation for cellulose acetate butyrate.

"Cab-O-Sil" [Cabot]. TM for colloidal silica particles sintered together in chainlike formations. Surface area ranges from 50 to 400 m^2/g, depending on grade.
Grade: Standard M-5, L-5, SD-20.
Use: Thickening and emulsifying agent for oil–water systems, drilling muds, cattle-feed supplements, tile cleaners, dispersion of oil slicks on seawater, plastics, solar-heated ceiling tiles.

"Cab-O-Sperse" [Cabot]. TM for aqueous dispersions of pyrogenic silica for use in the paper and textile industries.

cacao butter. (cocoa butter).
See theobroma oil.

C acid. (2-naphthylamine-4,8-disulfonic acid).
$C_{10}H_5(NH)_2(SO_3H)_2$.
Properties: White, crystalline solid. Slightly soluble in water.
Derivation: Reduction of 2-nitronaphthalene-4,8-disulfonic acid. The sodium salt is recrystallized from water.
Use: Azo dye intermediate.

cacodylic acid. (dimethylarsinic acid).
CAS: 75-60-5. $(CH_3)_2AsOOH$.
Properties: Colorless, deliquescent crystals; odorless. Mp 200C. Soluble in water, alcohol, and acetic acid; insoluble in ether.
Derivation: By distilling a mixture of arsenic trioxide and potassium acetate, and oxidizing the resulting product with mercuric oxide.
Hazard: Toxic by ingestion.
Use: Herbicide, especially for control of Johnson grass on cotton; soil sterilant; chemical warfare; timber thinning.

cactinomycin. USAN for an antibiotic produced from *Streptomyces*, which is 10% dactinomycin and 90% of two forms of actinomycin C.

Cadalyte. A series of compounds for cadmium electroplating.

cadaverine. (1,5-diaminopentane; pentamethylenediamine).
$NH_2(CH_2)_5NH_2$.
A ptomaine formed in the decay of animal proteins after death, also made synthetically.
Properties: Syrupy, colorless liquid; fuming; odorous. Mp 9C, bp 178–179C. Soluble in water and alcohol; slightly soluble in ether.
Hazard: Toxic by ingestion, absorbed by skin, a skin and eye irritant.
Use: Preparation of high polymers, intermediate, biological research.

Cade oil. See juniper tar oil.

cadiene. See sesquiterpene.

Cadminate. A turf fungicide containing 60% cadmium succinate and 40% inert matter.
Hazard: See cadmium.
See cadmium.

cadmium.
CAS: 7440-43-9.   Cd.

Hawley's Condensed Chemical Dictionary, Sixteenth Edition. Michael D. Larrañaga, Richard J. Lewis, Sr., and Robert A. Lewis.
© 2016 John Wiley & Sons, Inc. Published 2016 by John Wiley & Sons, Inc.

Metallic element of atomic number 48, group IIB of the periodic table. Aw 112.40. Valence 2. There are eight stable isotopes.

Properties: Soft, blue–white, malleable metal or grayish-white powder. D 8.642, mp 320.9C, bp 767C, refr index 1.13, Mohs hardness 2.0. Soluble in acids, especially nitric, and in ammonium nitrate solution. Lowers melting point of certain alloys when used in low percentages. Tarnishes in moist air. Corrosion resistance poor in industrial atmospheres. Becomes brittle at 80C. Resistant to alkalies. High neutron absorber. Combustible.

Occurrence: A greenockite (cadmium sulfide) ore containing zinc sulfide; also in lead and copper ores containing zinc. Canada, Central and Western U.S., Peru, Australia, Mexico, Zaire.

Derivation: (1) Dust or fume from roasting zinc ores is collected, mixed with coal or coke and sodium or zinc chloride, and sintered. The cadmium fume is collected in an electrostatic precipitator, leached, fractionally precipitated, and distilled. (2) By direct distillation from cadmium-bearing zinc. (3) By recovery from electrolytic zinc process (approximately 40%).

Grade: Technical, powder, pure sticks, ingots, slabs, high-purity crystals (<10 ppm impurities).

Hazard: Flammable in powder form. Toxic by inhalation of dust or fume. A carcinogen. Cadmium plating of food and beverage containers has resulted in a number of outbreaks of gastroenteritis (food poisoning). Soluble compounds of cadmium are highly toxic; however, ingestion usually induces a strong emetic action that minimizes the risk of fatal poisoning. Use as fungicide may be restricted. Kidney damage. Confirmed carcinogen.

Use: Electrodeposited and dipped coatings on metals, bearing and low-melting alloys, brazing alloys, fire-protection systems, nickel–cadmium storage batteries, power transmission wire, TV phosphors, basis of pigments used in ceramic glazes, machinery enamels, baking enamels, Weston-standard-cell control of atomic fission in nuclear reactors, fungicide, photography and lithography, selenium rectifiers, electrodes for cadmium-vapor lamps and photoelectric cells.

cadmium acetate.
CAS: (2) 543-90-8.   (1) $Cd(OOCCH_3)_2 \cdot 3H_2O$; (2) $Cd(OOCCH_3)_2$.
Properties: Colorless crystals. (1) d 2.01, loses water at 130C; (2) d 2.341, mp 256C. Soluble in water and alcohols.
Derivation: Interaction of acetic acid and cadmium oxide.
Hazard: See cadmium. Confirmed carcinogen.
Use: Ceramics (iridescent glazes), manufacture of acetates, assistant in dyeing and printing textiles, electroplating baths, laboratory reagent.
See cadmium.

cadmium ammonium bromide. (ammonium–cadmium bromide).

$CdBr_2 \cdot 4NH_4Br$.
Properties: Colorless crystals. Soluble in alcohol and water.
Hazard: See cadmium.
See cadmium.

cadmium antimonide.   A semiconductor used in thermoelectric devices.
Hazard: See cadmium and antimony.

cadmium-base Babbitt.   See Babbitt metal.

cadmium borotungstate.
$Cd_5(BW_{12}O_{40}) \cdot 18H_2O$.
Properties: Yellow, heavy crystals. Mp 75C. Soluble in water. The solution is yellow to light brown.
Grade: Technical.
Hazard: See cadmium.
Use: Separating minerals.
See cadmium.

cadmium bromate.
$Cd(BrO_3)_2 \cdot H_2O$.
Properties: White crystals or crystalline powder. D 3.758, mp (decomposes). Soluble in water; insoluble in alcohol.
Derivation: By adding cadmium sulfate to a solution of barium bromate.
Hazard: Strong oxidizer, dangerous in contact with organics. Highly toxic, irritant, oxidizer.
Use: Analytical reagent.

cadmium bromide.
CAS: 7789-42-6. $CdBr_2$ or $CdBr_2 \cdot 4H_2O$.
Properties: White to yellowish efflorescent crystalline. Mp (anhydrous) 568C, bp 863C. Soluble in water, acetone, alcohol, and acids.
Derivation: By heating cadmium in bromine vapor.
Grade: Technical, reagent.
Hazard: See cadmium.
Use: Photography, process engraving, lithography.
See cadmium.

cadmium carbonate.
CAS: 513-78-0. $CdCO_3$.
Properties: White, amorphous powder. D 4.258, decomposes at <500C. Soluble in dilute acids and in concentrated solution of ammonium salts; insoluble in water.
Grade: Reagent.
Hazard: See cadmium.
See cadmium.

cadmium chlorate.
$Cd(ClO_3)_2 \cdot 2H_2O$.
Properties: Colorless, prismatic crystals. Hygroscopic, d 2.28 (18C), mp 80C. Soluble in alcohol, water, and acetone.
Grade: Technical.
Hazard: Dangerous in contact with organic materials.

cadmium chloride.
CAS: 10108-64-2.   (1) $CdCl_2$; (2) $CdCl_2 \cdot 2.5H_2O$.
Properties: Small, white, odorless crystals. D (1) 4.05, (2) 3.327, mp (1) 568C, bp (1) 960C. Soluble in water and acetone.
Derivation: Action of hydrochloric acid on cadmium with subsequent crystallization.
Grade: Technical, reagent.
Use: Preparation of cadmium sulfide, analytical chemistry, photography, dyeing and calico printing, ingredient of electroplating baths, addition to tinning solution, manufacture of special mirrors, manufacture of cadmium yellow.

cadmium cyanide.
$Cd(CN)_2$.
Obtained as a white precipitate when potassium or sodium cyanide is added to a concentrated solution of a cadmium salt. A complex ion is formed when it is dissolved in an excess of the precipitating agent, and a solution of this complex is used as an electrolyte for electrodeposition of cadmium.
Hazard: See cadmium and cyanide.
Use: Electroplating copper.

cadmium diethyldithiocarbamate.
$Cd[SC(S)N(C_2H_5)_2]_2$.
Properties: White to cream-colored rods. D 1.39, melting range 68–76C. Mostly soluble in benzene, carbon disulfide, chloroform; insoluble in water and gasoline.
Hazard: See cadmium.
Use: Accelerator for butyl rubber.
See cadmium.

cadmium fluoride.
CAS: 7790-79-6. CdF_2.
Available as pure crystals, 99.89%, d 6.6, mp approximately 1110C, soluble in water and acids, insoluble in alkalies.
Hazard: See cadmium.
Use: Electronic and optical applications, high-temperature dry-film lubricants, starting material for crystals for lasers, phosphors.
See cadmium.

cadmium halide.   An inorganic halide that forms white, lustrous, hexagonal, flake-like scales comprised of two water-soluble allotropes.
Hazard: Suspected human carcinogen.
Use: Menatocide, lubricant, the electrodeposition of cadmium, photography, lithography, process engraving, analytic chemistry.

cadmium hydroxide.   (cadmium hydrate).
$Cd(OH)_2$.
Properties: White, amorphous powder. D 4.79, mp loses H_2O at 300C. Soluble in ammonium hydroxide and in dilute acids; insoluble in water and alkalies; absorbs carbon dioxide from air.
Derivation: By the action of sodium hydroxide on a cadmium–salt solution.

Grade: Technical, CP.
Hazard: See cadmium.
Use: Cadmium salts, cadmium plating, storage-battery electrodes.
See cadmium.

cadmium iodate.
$Cd(IO_3)_2$.
Properties: Fine, white powder. D 6.48; mp (decomposes). Slightly soluble in water; soluble in nitric acid or ammonium hydroxide.
Grade: Technical.
Hazard: Fire risk in contact with organic materials.
Use: Oxidizing agent.

cadmium iodide.
CdI_2.
Properties: White flakes or crystals; odorless. Becomes yellow on exposure to air and light. Occurs in two allotropic forms. D (α) 5.67, (β) 5.30; mp (α) 388C, (β) 404C; bp (α) 796C. Soluble in water, alcohol, ether, acetone, ammonia, and acids.
Derivation: By the action of hydriodic acid on cadmium oxide.
Hazard: See cadmium.
Use: Photography, process engraving and lithography, analytical chemistry, electroplating, lubricants, phosphors, nematicide.
See cadmium.

cadmium molybdate.
$CdMoO_4$.
Properties: Yellow crystals. D 5.347, mp approximately 1250C. Slightly soluble in water; soluble in acids.
Grade: Technical, crystals, 99.98% pure.
Hazard: See cadmium.
Use: Electronic and optical applications.
See cadmium.

cadmium nitrate.
CAS: 10325-94-7.   (1) $Cd(NO_3)_2 \cdot 4H_2O$; (2) $Cd(NO_3)_2$.
Properties: White, amorphous pieces or hygroscopic needles. (1) d 2.455, mp 59.5C, bp 132C; (2) mp 350C. Soluble in water, ammonia, and alcohol.
Derivation: Action of nitric acid on cadmium or cadmium oxide and crystallization.
Grade: Technical, reagent.
Hazard: Dangerous fire and explosion hazard.
Use: Cadmium salts, photographic emulsions, coloring glass and porcelain, laboratory reagent, cadmium salts.

cadmium oxalate.
$Cd(COO)_2 \cdot 3H_2O$.
Properties: White, amorphous powder. D 3.32 (dehydrated), mp (decomposes at 340C). Soluble in dilute acids, ammonium hydroxide; insoluble in alcohol and water.

cadmium oxide.
CAS: 1306-19-0. CdO (forms 1 and 2).
Properties: (1) Colorless, amorphous powder; d 6.95. (2) Brown or red crystals; d 8.15. Both decompose on heating at 900C. The crystals are soluble in acids and alkalies; insoluble in water.
Derivation: Cadmium metal is distilled in a retort, the vapor reacted with air, and the oxide collected in a baghouse.
Hazard: Inhalation of vapor or fume may be fatal. A confirmed carcinogen.
Use: Cadmium plating baths, electrodes for storage batteries, cadmium salts, catalyst, ceramic glazes, phosphors, nematicide.

cadmium pigment. A family of pigments based on cadmium sulfide or cadmium selenide, used chiefly where high color retention is required. They are lightfast and have good alkali resistance. Red shades are obtained with cadmium selenide, yellow with cadmium sulfide. Used in paints and high-gloss baking enamels, they are often extended with barium sulfate and are then called cadmium lithopone.

cadmium potassium iodide.
$CdI_2 \cdot 2KI \cdot 2H_2O$.
Properties: White powder, becomes yellowish with age. Deliquescent. D 3.359; mp 76C (decomposes). Soluble in water, alcohol, ether, and acid.
Derivation: By combining cadmium iodide and potassium iodide in solution in proportion of their combining weights, and subsequent crystallization.
Hazard: See cadmium.
Use: Analytical chemistry, medicine.
See cadmium.

cadmium propionate.
$Cd(OOCC_2H_5)_2$.
A solid used in scintillation counters.

cadmium ricinoleate.
$Cd[CH_3(CH_2)_5CH_2OCH_2CH:CH(CH_2)_7CO_2]_2$.
Properties: Odorless, fine, white powder derived from castor oil. Mp 104C, d 1.11.
Hazard: See cadmium.
Use: Solution used to stabilize polyvinyl chloride and copolymers against light and heat.
See cadmium.

cadmium selenide.
CdSe.
Properties: Red powder. D 5.81 (15/4C), mp >1350C. Insoluble in water; stable at high temperature.
Use: Red pigment, semiconductors, phosphors, photoelectric cells.

cadmium selenide lithopone. See cadmium pigment.

cadmium stearate.
Hazard: See cadmium.

Use: As a lubricant and stabilizer in plastics.
See cadmium.

cadmium succinate.
$Cd(OOCCH_2)_2$.
Properties: A white powder. Slightly soluble in water; insoluble in alcohol.
Hazard: See cadmium.
Use: Fungicide.
See cadmium.

cadmium sulfate.
CAS: 10124-36-4. (1) $CdSO_4$; (2) $3CdSO_4 \cdot 8H_2O$; and (3) $CdSO_4 \cdot 4H_2O$.
Properties: Colorless, odorless crystals. D (1) 4.69, (2) 3.09, and (3) 3.05; mp (1) 1000C. Soluble in water; insoluble in alcohol.
Derivation: By the action of dilute sulfuric acid on cadmium or cadmium oxide.
Grade: Technical, CP.
Hazard: A confirmed carcinogen.
Use: Pigments, fluorescent screens, electrolyte in Weston standard cell, electroplating.
See cadmium.

cadmium sulfate (1:1) hydrate (3:8). (cadmium sulfate octahydrate; cadmium (2+) trisulfate octahydrate; sulfuric acid, cadmium salt, hydrate).
$Cd_3H_{16}O_{20}S_3$.
Properties: Crystals from aqueous solution.
Hazard: Suspected carcinogen; tumorigen; neoplastigen; teratogen; mutagen.

cadmium sulfide. (orange cadmium).
CAS: 1306-23-6. CdS.
Properties: Yellow or brown powder. D 4.82, mp 1750C (100 atm), sublimes (in nitrogen) 980C. Insoluble in cold water; forms a colloid in hot water; soluble in acids and ammonia. Can be polished like a metal. It is an n-type semiconductor.
Derivation: (1) By passing hydrogen sulfide gas into a solution of a cadmium salt acidified with hydrochloric acid. The precipitate is filtered and dried. (2) Occurs naturally as greenockite.
Grade: Technical, NF, high purity (single crystals).
Hazard: A confirmed carcinogen, highly toxic. See cadmium.
Use: Pigments and inks, ceramic glazes, pyrotechnics, phosphors, fluorescent screens, scintillation counters, rectifiers, photoconductor in xerography, transistors, photovoltaic cells, solar cells, catalyst in photodecomposition of hydrogen sulfide.
See cadmium pigment.

cadmium telluride.
CAS: 1306-25-8. CdTe.
Properties: Brownish-black, cubic crystals. Mp 1090C, d 6.2 (25/4C). Oxidizes on prolonged exposure to moist air. Insoluble in water and mineral acids except nitric, in which it is soluble with decomposition.

Derivation: Fusion of the elements, reaction of hydrogen telluride and cadmium chloride.
Grade: High purity crystals, 99.99+%.
Hazard: Toxic by inhalation.
Use: Semiconductors, phosphors.

cadmium tungstate.
$CdWO_4$.
Properties: White or yellow crystals or powder. Soluble in ammonium hydroxide, alkali cyanides; almost insoluble in water.
Derivation: By the interaction of cadmium nitrate and ammonium tungstate.
Available forms: Single crystal rods, broken crystals (crackle).
Hazard: Toxic by inhalation.
Use: Fluorescent paint, X-ray screens, scintillation counters, catalyst, phosphors.

Cadox PS.
CAS: 94-17-7.
A mixture of 50% bis(*p*-chlorobenzoyl)peroxide in silicone fluid.
Available forms: Paste.
Use: Curing agent for silicone rubbers.
See *p*-chlorobenzoyl peroxide.

Cardura E Ester.
Glycidyl ester of Versatic 911 Acid.
Hazard: Irritant to skin.
Use: Modifier for alkyd resins and thermosetting acrylic systems, reactive diluent for epoxy resins.

Caenorhabditis elegans.
A species of roundworm. It was the first animal whose genome was completely sequenced and its genes were characterized.

caerulein.
CAS: 17650-98-5. $C_{58}H_{73}N_{13}O_{21}S_2$.
Hazard: A poison.

caesium.
See cesium.

C₄AF.
Abbreviation for tetracalcium aluminoferrite as used in cement.
See cement, Portland.

caffeine.
(theine; methyltheobromine; 1,3,7-trimethylxanthine).
CAS: 58-08-2. $C_8H_{10}N_4O_2 \cdot H_2O$.

Properties: White, fleecy masses or long, flexible, silky crystals; odorless; bitter taste. An alkaloid. Loses water at 80C. Efflorescent in air. Mp 236.8C. Soluble in chloroform; slightly soluble in water and alcohol; very slightly soluble in ether. Solution neutral to litmus.
Derivation: By extraction of coffee beans, tea leaves, or kola nuts; also synthetically. Much of the caffeine of commerce is a by-product of decaffeinated-coffee manufacture.
Method of purification: Recrystallization.
Grade: Technical, USP, FCC.
Hazard: One grain or more is toxic, 200 µg/mL has been found to inhibit activity of the enzyme DNA polymerase. Use in soft drinks not to exceed 0.02%. Questionable carcinogen.
Use: Beverages, medicine.

CAG.
([(2R, 3S, 4R,5R)-5-(2-amino-6-oxo-7,8-dihydro-3H-purin-9-yl)-3,4-dihydroxyoxolan-2-yl]methyl[hydoxy]phosphoryl]oxyhydrogenphosphate).
$C_{18}H_{25}N_6O_{16}P_3$.
A codon of glutamine that directs the placement of glutamine into a polypeptide.

cage compound.
See inclusion complex; clathrate compound.

cage mill.
(bar disintegrator, squirrel-cage disintegrator).
A comminuting device that may consist of two rotating structures similar to water wheels, one fitting inside the other. They are provided with horizontal crossbars or breaker plates. The assembly is covered with a close-fitting housing. The two wheels or cages rotate at high speed in opposite directions on a horizontal axis. The material to be reduced is fed into the smaller cage from a hopper. It is ejected at speeds up to 12,000 ft/min and is fragmented by contact with the bars. As the pieces are thrown back and forth within the cages, they are disintegrated further by mutual impact. Some types of cage mills have only a single cage, but others have more than two. They are used for size reduction of niter cake, fertilizers, coal, and other friable materials. Particle size can be varied by adjusting the space between the crossbars.

cage zeolite.
(sodalite).
A structure of sodium aluminosilicates often arranged in combined tetrahedra at the intersections of which are sodium atoms, with oxygen atoms at the midpoints. Such sodalite units often combine to form supercages. They are common and extremely effective catalysts. Many zeolites occur naturally, but synthetic types are tailor-made for special purposes and have particular catalytic functions.
See zeolite.

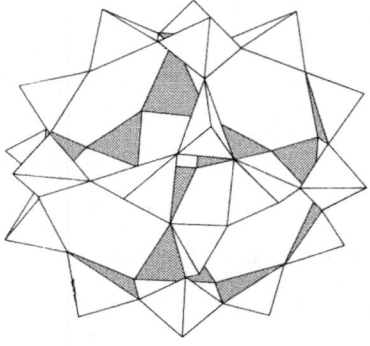

"Cairox" *[Carus].* TM for potassium permanganate.

cajeputene. (dipentene; *dl-p*-mentha-1,8-diene, inactive; limonene).
CAS: 138-86-3. $C_{10}H_{16}$.
Properties: Bp 178C, flash p 109F (42C), refr index 1.4739 (20C), d 0.840.

Cake alum. See aluminum sulfate.

calabarine. See physostigmine.

Caladryl. Proprietary name for diphenhydramine, camphor, and calamine.
Use: Dermatological.

calamine. (1) A hydrated zinc silicate containing 67.5% zinc oxide; d 3.5, Mohs hardness 4.5–5. Pyroelectric. Occurs in U.S. and Europe. Source of metallic zinc. (2) Zinc oxide with low percentage of ferric oxide; zinc oxide must be 98%. Soluble in mineral acids, insoluble in water. Pharmaceutical preparation (USP).

calamus oil.
Properties: Yellow to brownish-yellow essential oil. D 0.959–0.970 (15C), refr index 1.503–1.510, saponification value 6–20. Slightly soluble in water.
Derivation: By steam distillation of calamus, the stem or root of the sweet flag. Chief known constituents are asarone (see 2,4,5-trimethoxy-1-propenylbenzene) and eugenol.
Use: Perfumery, flavoring agent.

calandria. The heating unit of a vacuum-evaporating system.

Calan SR. Verapamil hydrochloride in sustained-release oral caplet form.
Use: Drug.

Calan Tablets. Verapamil hydrochloride in a tablet form.
Use: Drug.

calaverite.
$AuTe_2$.
One of the gold telluride group of minerals. Corresponds to the same general formula as sylvanite and krennerite. Pale bronze-yellow color or tin-white, tarnishing to bronze-yellow on exposure. Metallic luster. Contains 40–43% gold, 1–3% silver; d 9.0; Mohs hardness 2.5.
Occurrence: U.S. (California, Colorado), Australia, Canada.
Use: Important source of gold.

Calcene. A specially prepared precipitated calcium carbonate for use in compounding rubber, paints, and plastics. The particles of grade TM are coated with stearic acid to aid in dispersion. Grade NC is not coated.

calciferol. See ergocalciferol.

calcimine. (kalsomine).
Essentially chalk and glue in a powdered form ready to mix with water. Used as temporary decoration for interior plaster walls. Will not withstand washing.

calcination. Heating of a solid to a temperature below its melting point to bring about a state of thermal decomposition or a phase transition other than melting. Included are the following types of reactions: (1) thermal dissociation, including destructive distillation of organic compounds, e.g., concentration of aluminum by heating of bauxite; (2) polymorphic phase transitions, e.g., conversion of anatase to rutile form of TiO_2; and (3) thermal recrystallization, e.g., devitrification of glass. Calcination is often used in the beneficiation of ores.
See destructive distillation; pyrolysis.

calcine. Radioactive waste prepared for disposal in the form of granular powder made by drying liquid wastes at high temperature (approximately 800C).

calcined. Processed by burning or incinerating. Examples are chalk converted to lime and zinc sulfide converted to zinc oxide.

calciotropic hormone. Any of a group of hormones that contribute to the remodeling of bone by stimulating growth and inhibiting the resorption of bone.

calcite.
$CaCO_3$.
The most common form of natural calcium carbonate. Dogtooth spar, Iceland spar, nailhead spar, and satin spar are varieties of calcite. Essential ingredient of limestone, marble, and chalk.
Properties: Colorless, white, and various-colored crystals; vitreous to earthy luster; good cleavage in three directions. D 2.72, Mohs hardness 2. May

contain small amounts of magnesium, iron, manganese, and zinc. Reacts with acids to evolve carbon dioxide.

Use: Phosphor. Iceland spar is used in optical instruments.

calcitic dolomite. (calcium salt of carbonic acid; calcium carbonate).

$CaCO_3$.

A carbonate rock that contains from 10–50% calcite and 50–90% dolomite.

Properties: Odorless, tasteless powder or crystal.

Use: Therapeutically as a phosphate buffer in hemodialysis patients and as a calcium supplement.

calcitonin. A thyroid hormone controlling the proportion of calcium in circulating blood. May be used in calcium balance control and possibly also in the treatment of bone fractures, hypervitaminosis, and other calcium-related diseases. It is obtained in a purified form from pig thyroid; has been made synthetically.

calcium.

CAS: 7440-70-2. Ca.

Alkaline-earth element of atomic number 20, group IIA of the periodic table. Aw 40.08. Valence 2. Six stable isotopes.

Properties: Moderately soft, silver-white, crystalline metal. D 1.57, mp 845C, sublimes above mp in vacuum, bp 1480C, Brinell hardness 17. Oxidizes in air to form adherent protective film. Can be machined, extruded, or drawn. Soluble in acid. Decomposes water to liberate hydrogen.

Derivation: Electrolysis of fused calcium chloride, by thermal process under high vacuum from lime reduced with aluminum. Does not occur free in nature.

Available forms: Crowns, nodules, ingots, crystals up to 99.9% pure.

Hazard: Evolves hydrogen with moisture. Flammable in finely divided state. Fire and explosion hazard when heated or on contact with strong oxidizing agents.

Use: Alloying agent for aluminum, copper, and lead; reducing agent for beryllium; deoxidizer for alloys. Dehydrating oils. Decarburization and desulfurization of iron and its alloys, getter in vacuum tubes. Separation of nitrogen from argon. Reducing agent in preparation of chromium metal powder, thorium, zirconium, and uranium. Fertilizer ingredient.

Note: Calcium is an essential component of bones, teeth, shells, and plant structures. It occurs in milk in trace amounts and is necessary in animal and human nutrition. Vitamin D aids in the deposition of calcium in bones.

calcium 45. Radioactive calcium of mass number 45.

Properties: Half-life 164 days; emits β-radiation.

Derivation: By reactor irradiation of calcium carbonate, by neutron bombardment of scandium, or as a by-product of the irradiation of calcium nitrate for the preparation of ^{14}C.

Available forms: Calcium chloride in hydrochloric acid solution and solid calcium carbonate.

Hazard: Dangerous radiation hazard, MPC in air 3×10^{-8} μCi/mL. This isotope is a bone seeker and may cause damage to the blood-forming organs.

Use: Research aid for studying water purification, calcium exchange in clays, detergency, surface wetting and other surface phenomena, calcium uptake and deposition in bone, soil characteristics as related to soil utilization of fertilizer and crop yield, diffusion of calcium in glass, etc.

calcium abietate.

$(C_{20}H_{29}O_2)_2Ca$.

Product of the action of lime on rosin or resin acids. See calcium resinate.

calcium acetate. (vinegar salts; gray acetate; lime acetate; calcium diacetate).

$Ca(CH_3COO)_2 \cdot H_2O$.

Properties: Brown, gray, or white (when pure) powder, amorphous or crystalline; slightly bitter taste, slight odor of acetic acid; decomposes on heating. Soluble in water; slightly soluble in alcohol. Combustible.

Derivation: Action of pyroligneous acid on calcium hydroxide; the solution, on being filtered and evaporated to dryness, yields gray acetate of lime.

Grade: Technical (80% basis), reagent, CP, pure, brown, gray, FCC.

Use: Manufacture of acetone, acetic acid, acetates, mordant in dyeing and printing of textiles, stabilizer in resins, additive to calcium soap lubricants, food additive (as antimold agent in bakery goods, sausage casings), corrosion inhibitor.

calcium acetylsalicylate. (aspirin, soluble).

$(CH_3COOC_6H_4COO)_2 \cdot 2H_2O$.

Properties: White powder. Aqueous solutions are unstable. Soluble in water.

Derivation: (1) Action of acetylsalicylic acid upon calcium carbonate in the presence of a small amount of water; (2) by passing carbon dioxide into an aqueous solution of calcium carbonate and acetylsalicylic acid.

Use: Medicine (antipyretic).

calcium acid sulfite. See calcium hydrogen sulfite.

calcium acrylate.

$(H_2C:CHCOO)_2Ca$.

Properties: Free-flowing, white powder. Soluble in water; deliquescent. Solution polymerizes to form hydrophilic resin.

Use: Ion-exchange clay-soil stabilizer, binder, and sealer (laboratory scale only).

calcium alginate.
Properties: White or cream-colored powder or filaments, grains, or granules; slight odor and taste. Insoluble in water and acids; soluble in alkaline solution.
Grade: FCC.
Hazard: Flammable but self-extinguishing.
Use: Pharmaceutical products; food additive; thickening agent and stabilizer in ice cream, cheese products, canned fruits, and sausage casings; synthetic fibers.
See algin; alginic acid.

calcium aluminate. (tricalcium aluminate).
$3CaO \cdot Al_2O_3$.
Properties: Crystals or powder. D 3.038 (25C), mp (decomposes at 1535C). Soluble in acids. A refractory and an important ingredient of cements, especially of aluminous cement. Fused calcium aluminate (a glass) can be used for infrared transmission and detection. Noncombustible.

calcium aluminum hydride. See aluminum calcium hydride.

calcium aluminum silicate. See aluminum calcium silicate.

calcium ammonium nitrate. A uniform mixture of approximately 60% ammonium nitrate and 40% limestone and/or dolomite. A fertilizer containing approximately 20% N.
Hazard: Oxidizer, fire risk in contact with organic materials.

calcium arsenate. (tricalcium orthoarsenate).
CAS: 7778-44-1. $Ca_3(AsO_4)_2$.
Properties: White powder. Slightly soluble in water; soluble in dilute acids. Decomposes on heating.
Derivation: Interaction of calcium chloride and sodium arsenate.
Grade: Technical, CP.
Hazard: Toxic by ingestion and inhalation.
Use: Insecticide, germicide.

calcium arsenite.
$CaAsO_3H$.
Properties: White, granular powder. Insoluble in water; soluble in acids.
Grade: Technical.
Hazard: Toxic by inhalation and ingestion.
Use: Germicide, insecticide.

calcium arsonate (1:1).
CAS: 52740-16-6. $AsH_2O_3 \cdot Ca$.
Hazard: A confirmed human carcinogen

calcium ascorbate.
$Ca(C_6H_7O_6)_2 \cdot 2H_2O$.
Properties: A white to slightly yellow crystalline powder; odorless. Soluble in water; slightly soluble in alcohol; insoluble in ether. The pH of a 10% solution is between 6.8 and 7.4.
Grade: FCC.
Use: Food preservative.

calcium biphosphate. See calcium phosphate, monobasic.

calcium bisulfide. See calcium hydrosulfide.

calcium bisulfite. See calcium hydrogen sulfite.

calcium borate.
CaB_4O_7.
Properties: White powder. Soluble in dilute acids; partially soluble in water.
Use: Fire-retardant compositions, antifreeze compounds, metallurgical flux, porcelain.

calcium bromate.
$Ca(BrO_3)_2 \cdot H_2O$.
Properties: White, crystalline powder. D 3.329, loses its water at 180C. Very soluble in water.
Grade: FCC.
Hazard: Oxidizing agent, fire risk in contact with organic materials.
Use: Maturing agent, dough conditioner.

calcium bromide.
CAS: 7789-41-5. (1) $CaBr_2 \cdot 6H_2O$; (2) $CaBr_2$.
Properties: White powder or crystals; odorless; sharp saline taste. (1) D 2.295 (25C), mp 38C, bp 149C (decomposes); (2) d 3.353 (25C), mp 730C (slight decomposition), bp 806–812C, very deliquescent. Very soluble in water; soluble in alcohol and acetone.
Derivation: By the action of hydrobromic acid on calcium oxide, carbonate, or hydroxide and subsequent crystallization.
Grade: Technical, CP.
Use: Photography, medicine, dehydrating agent, food preservative, road treatment, freezing mixtures, sizing compounds, wood preservative, fire retardant.

calcium carbide.
CAS: 75-20-7. CaC_2.
Properties: Grayish-black, irregular, hard solid; garlic-like odor. Must be kept dry. D 2.22, mp approximately 2300C. Decomposes in water with formation of acetylene and calcium hydroxide and evolution of heat.
Derivation: Interaction of pulverized limestone or quicklime with crushed coke or anthracite in an electric furnace.
Grade: Technical, lumps, powder.
Hazard: Forms flammable and explosive gas and corrosive solid with moisture.
Use: Generation of acetylene gas for welding, chloroethylenes, vinyl acetate monomer, acetylene chemicals, reducing agent.

calcium carbonate.
CAS: 1317-65-3. $CaCO_3$.

Properties: White powder or colorless crystals; odorless; tasteless. D 2.7–2.95, decomposes at 825C. Very slightly soluble in water (a few ppm); soluble in acids with evolution of carbon dioxide. Noncombustible.

Occurrence: Calcium carbonate is one of the most stable, common, and widely dispersed materials. It occurs in nature as aragonite, oyster shells, calcite, chalk, limestone, marble, marl, and travertine, especially in Indiana (structural limestone), Vermont (marble), Italy (travertine), and England (chalk).

Derivation: (1) Mined from natural surface deposits. (2) Precipitated (synthetic) by reaction of calcium chloride and sodium carbonate in water solution or by passing carbon dioxide through a suspension of hydrated lime ($Ca(OH)_2$) in water.

Hazard: A nuisance particulate dust.

Use: Source of lime; neutralizing agent, filler, and extender in rubber, plastics, paints; opacifying agent in paper; fortification of bread; putty; tooth powders; antacid; whitewash; Portland cement; sulfur dioxide removal from stack gases; metallurgical flux; analytical chemistry; carbon dioxide generation (laboratory).

See chalk, calcite, marble, limestone, whiting.

Note: Calcium carbonate is a major cause of boiler scale when hard water is used in heating systems.

calcium caseinate.

Properties: White or slightly yellow powder; nearly odorless. Insoluble in cold water; forms a milky solution when suspended in water, stirred, and heated.

Use: Medicine, special foods.

See casein.

calcium chlorate.

CAS: 10137-74-3. $Ca(ClO_3)_2 \cdot 2H_2O$.

Properties: White to yellowish crystals. Keep well stoppered. Melts when rapidly heated at 100C, mp (anhydrous) 340C, hygroscopic, d 2.711. Soluble in water and alcohol.

Derivation: By the action of chlorine on hot calcium hydroxide slurry.

Hazard: Oxidizer, dangerous fire risk, forms explosive mixtures with combustible materials.

Use: Photography, pyrotechnics, dusting powder to kill poison ivy, herbicide.

calcium chloride.

CAS: 10043-52-4.   (1) $CaCl_2$; (2) $CaCl_2 \cdot H_2O$; (3) $CaCl_2 \cdot 2H_2O$; and (4) $CaCl_2 \cdot 6H_2O$.

Properties: White, deliquescent crystals, granules, lumps, or flakes. (1) D 2.15 (25C), mp 772C, bp alone 1600C; (2) mp 260C; (3) USP grade, d 0.835 (25C); and (4) d 1.71 (25C), bp loses $4H_2O$ at 30C and $6H_2O$ at 200C. All forms soluble in water and alcohol. Water solution is neutral or slightly alkaline.

Derivation: (1) Action of hydrochloric acid on calcium carbonate and subsequent crystallization. (2) Commercially obtained as a by-product in the Solvay soda and other processes. (3) Recovery from brines.

Grade: Technical, CP, FCC, USP (dihydrate), various forms and purities, solution.

Use: De-icing and dust control of roads; drilling muds; dust proofing; freeze proofing and thawing coal, coke, stone, sand, ore; concrete conditioning; paper and pulp industry; fungicides; refrigeration brines; drying and desiccating agent; sequestrant in foods; firming agent in tomato canning; tire weighting; pharmaceuticals; electrolytic cells. The hexahydrate (4) has been evaluated for solar-heat storage.

calcium chlorite.

CAS: 14674-72-7. $Ca(ClO_2)_2$.

Properties: White crystals. D 2.71. Decomposes in water.

Hazard: Strong oxidizer, fire risk in contact with organic materials.

calcium chromate.

CAS: 13765-19-0. $CaCrO_4$; $CaCrO_4 \cdot 2H_2O$.

Properties: Bright-yellow powder. Hydrate loses water at 200C. Anhydrous: d 2.89. Soluble in dilute acids; slightly soluble in water.

Hazard: Probable carcinogen.

Use: Pigment, corrosion inhibitor, oxidizing agent, depolarizer for batteries, coating for light metal alloys.

calcium citrate.   (lime citrate; tricalcium citrate).

$Ca_3(C_6H_5O_7)_2 \cdot 4H_2O$.

Properties: White powder; odorless. Loses most of its water at 100C and all of it at 120C. Almost insoluble in water; insoluble in alcohol.

Grade: Reagent, technical, FCC.

Use: Dietary supplement, sequestrant, buffer, and firming agent in foods.

calcium cyanamide.   (lime nitrogen; calcium carbimide).

CAS: 156-62-7. $CaCN_2$.

Properties: Colorless crystals or powder. D 1.083, mp 1300C, sublimes above 1150C. Decomposes in water, liberating ammonia and acetylene.

Derivation: Calcium carbide powder is heated in an electric oven into which nitrogen is passed (24–26 hours). Any uncombined calcium carbide is leached out after removal.

Grade: Fertilizer, 21% N, industrial.

Hazard: Fire risk with moisture or combined with calcium carbide. Skin, eye, and upper respiratory tract irritant. Questionable carcinogen.

Use: Fertilizer, nitrogen products, pesticide, hardening iron and steel.

calcium cyanide.

CAS: 592-01-8. $Ca(CN)_2$.

Properties: Colorless crystals or white powder, gray-black (technical). Decomposes in moist air liberating hydrogen cyanide. Decomposes above 350C. Soluble in water and very weak acid with evolution of hydrogen cyanide.

Hazard: Toxic by ingestion and skin absorption.

Use: Rodenticide; fumigant for greenhouses, flour mills, grain, seed, citrus trees under tents for control of scale insects; leaching of gold and silver ores; manufacture of other cyanides.

calcium cyclamate. (calcium cyclohexylsulfamate; calcium cyclohexanesulfamate).

CAS: 139-06-0. $(C_6H_{11}NHSO_3)_2Ca \cdot 2H_2O$.

Properties: White crystals or powder; nearly odorless; very sweet taste. Freely soluble in water (solutions are neutral to litmus); almost insoluble in alcohol, benzene, chloroform, and ether; pH (10% solution) 5.5–7.5. Sweetening power approximately 30 times that of sucrose.

Grade: NF, FCC.

Hazard: Not permitted for use in foods and soft drinks, because of suspected carcinogenicity.

Use: Nonnutritive sweetener.

See cyclamate.

calcium cyclamate dihydrate.

CAS: 5897-16-5. $C_{12}H_{24}N_2O_6S_2 \cdot Ca \cdot 2H_2O$.

Hazard: Low toxicity by ingestion.

calcium dehydroacetate.

$(C_8H_7O_4)_2$.

Properties: White to cream powder. Almost insoluble in water and organic solvents.

Grade: 96% min.

Use: Fungicide.

See dehydroacetic acid.

calcium diacetate. See calcium acetate.

calcium dichromate.

$CaCr_2O_7 \cdot 3H_2O$ (or $\cdot 4H_2O$).

Properties: Brownish-red crystals. Deliquescent, d (4.5H_2O variety) 2.136. Soluble in water.

Grade: Technical, CP.

Use: Corrosion inhibitor, catalyst, manufacture of chromium compounds.

calcium dihydrogen sulfite. See calcium hydrogen sulfite.

calcium dioxide. See calcium peroxide.

calcium disodium edetate. (USAN) (calcium disodium EDTA; edathamil calcium disodium; calcium disodium ethylenediaminetetraacetate).

$CaNa_2C_{10}H_{12}N_2O_8 \cdot xH_2O$.

For other variations of the name, see ethylenediaminetetraacetic acid. The calcium disodium salt is a mixture of the dihydrate and trihydrate.

Properties: White, odorless powder or flakes; slightly hygroscopic; faint saline taste. Stable in air. Soluble in water; insoluble in organic solvents. It acts as a chelating agent for heavy metals.

Grade: USP, FCC.

Use: Medicine (antidote in heavy-metal poisoning), in foods to "complex" trace heavy metals, as a preservative, and to retain color and flavor; antigushing agent in fermented malt beverages. (For restrictions on food uses, see FDA regulations.)

calcium distearate. See calcium stearate.

calcium ethylhexoate. See soap (2).

calcium ferrocyanide.

$Ca_2Fe(CN)_6 \cdot 12H_2O$.

Properties: Yellow crystals. D 1.68, decomposes on heating. Soluble in water; insoluble in alcohol.

Use: Removal of metallic impurities in the manufacture of citric, tartaric, and other acids.

calcium fluoride.

CAS: 7789-75-5. CaF_2.

Properties: White powder occurring in nature as fluorite (pure form) or fluorspar (mineral). Mp 1402C, bp approximately 2500C, d 3.18, Mohs hardness 4. Reacts with hot, concentrated sulfuric acid to liberate hydrogen fluoride. Insoluble in water; soluble in ammonium salts.

Derivation: (1) By powdering pure fluorite or fluorspar; (2) by the interaction of soluble calcium salt and sodium fluoride.

Grade: See fluorspar.

Hazard: An irritant.

Use: See fluorspar. Single pure (99.93%) crystals of calcium fluoride are also produced for use in spectroscopy, electronics, lasers, and high-temperature dry-film lubricants.

calcium fluorophosphate. (fluorapatite; FAP). $CaPO_3F$.

A laser crystal, said to have the lowest energy threshold of any room-temperature crystal.

calcium fluosilicate. See calcium silicofluoride.

calcium formate.

$Ca(OOCH)_2$.

Properties: White powder. Mp >300C, d 2.015. Soluble in water; insoluble in alcohol.

Use: Briquet binder, drilling fluids, lubricants, chrome tanning.

calcium 4-(β-D-galactosido)-D-gluconate. See calcium lactobionate.

calcium gluconate.

CAS: 299-28-5. $Ca(C_6H_{11}O_7)_2 \cdot H_2O$.

Properties: White fluffy powder or granules; odorless; practically tasteless. Stable in air. Loses water at 120C. Soluble in hot water; less soluble in cold water; insoluble in alcohol, acetic acid, and other organic solvents; specific rotation (20/D) approximately +6 degrees. Solution neutral to litmus.

Derivation: Neutralization of gluconic acid with lime or calcium carbonate.
Grade: Technical, USP, FCC, special for ampules.
Use: Food additive, buffer and sequestering agent, vitamin tablets.

calcium glutamate.   Similar to sodium glutamate.

calcium glycerophosphate.   (calcium glycerinophosphate).
$CaC_3H_7O_2PO_4$.
Properties: White, crystalline powder; odorless; almost tasteless. Slightly hygroscopic; decomposes above 170C. Slightly soluble in water; insoluble in alcohol.
Derivation: By esterification of phosphoric acid with glycerol and conversion of glycerophosphoric acid to the calcium salt.
Grade: Technical, pure, FCC.
Use: Stabilizer for plastics, nutrient and dietary supplement.

calcium glycolate.
$(CH_3OHCOO)_2Ca$.
Properties: White solid.
Grade: Technical.
Use: Source of glycolic acid and of the glycolic acid radical in chemical synthesis.

calcium hexametaphosphate.
Hazard: A nuisance dust.
Use: Food additive.

calcium hexasilicofluorate.   See calcium silicofluoride.

calcium hydrate.   See calcium hydroxide.

calcium hydride.
CAS: 7789-78-8. CaH_2.
Properties: Grayish-white lumps or crystals. Forms calcium hydroxide in moist air and evolves hydrogen. D 1.7, decomposes at 675C. Decomposed by water, organic acids, and lower alcohols.
Grade: Technical, 94% pure.
Hazard: Evolves highly flammable hydrogen when wet; solid product is slaked lime. Irritating to skin.
Use: Reducing agent, drying agent, analytical reagent in organic chemistry, easily portable source of hydrogen, cleaner for blocked-up oil wells.

calcium hydrogen phosphate.
CAS: 7789-77-7. $CaHPO_4 2H_2O$.
Properties: White crystals. Mp 36C. Slowly sol in water.
Use: Food additive.

calcium hydrogen sulfite.   (calcium bisulfite; calcium dihydrogen sulfite; calcium acid sulfite).
$Ca(HSO_3)_2$.

A solution of calcium sulfite in aqueous sulfur dioxide.
Properties: Yellowish liquid; strong sulfur dioxide odor. D 1.06. Corrosive to metals.
Derivation: Action of sulfur dioxide on calcium hydroxide (solution).
Hazard: Irritating and corrosive to skin and tissue.
Use: Antichlor in bleaching textiles, paper pulp (dissolving lignin) preservative, bleaching sponges, hydroxylamine salts, germicide, disinfectant.

calcium hydrosulfide.   (calcium bisulfide; calcium sulfhydrate).
$Ca(HS)_2·6H_2O$.
Properties: Colorless, transparent crystals. Soluble in alcohol and water. Decomposes in air (15–18C).
Use: Leather industry.

calcium hydroxide.   (calcium hydrate; hydrated lime; caustic lime; slaked lime).
CAS: 1305-62-0. $Ca(OH)_2$.
Properties: Soft, white, crystalline powder; alkaline, slightly bitter taste. D 2.34, loses its water at 580C, pH of water solution (25C) 12.4. Slightly soluble in water; soluble in glycerol, syrup, and acids; insoluble in alcohol. Absorbs carbon dioxide from air.
Derivation: Action of water on calcium oxide.
Impurities: Calcium carbonate, magnesium salts, iron.
Grade: Technical, chemical lime (insoluble matter less than 2%, magnesium less than 3%), building lime, USP, CP, FCC.
Hazard: Skin, eye, and upper respiratory tract irritant, avoid inhalation.
Use: Mortar, plasters, cements, calcium salts, causticizing soda, depilatory, unhairing of hides, whitewash, soil conditioner, ammonia recovery in gas manufacture, disinfectant, water softening, purification of sugar juices, accelerator for low-grade rubber compounds, petrochemicals, food additive as buffer and neutralizing agent, shell-forming agent (poultry).

calcium hypochlorite.   (calcium oxychloride).
CAS: 7778-54-3. $Ca(OCl)_2$.
Properties: White, crystalline solid. D 2.35, decomposes at 100C. Decomposes in water and alcohol; not hygroscopic. Practically clear in water solution. Stable chlorine carrier. An oxidizer.
Derivation: Chlorination of a slurry of lime and caustic soda with subsequent precipitation of calcium hypochlorite dihydrate, dried under vacuum.
Grade: Commercial (70%), high purity (99.2% available chlorine as calcium hypochlorite).
Hazard: Dangerous fire risk in contact with organic materials.
Use: Algicide, bactericide, deodorant, potable-water purification, disinfectant for swimming pools, fungicide, bleaching agent (paper, textiles).
See lime, chlorinated.

calcium hypophosphite.
$Ca(H_2PO_2)_2$.

Properties: White powder. Evolves phosphine at 300C. Soluble in water; insoluble in alcohol.
Use: Medicine, corrosion inhibitor.

calcium iodate.
$Ca(IO_3)_2$.
Properties: White crystals or powder; odorless. D 4.5 (15C), decomposes at 540C. Soluble in water and nitric acid; insoluble in alcohol. Oxidizer.
Grade: Technical, CP, FCC.
Hazard: Fire risk in contact with organic materials.
Use: Deodorant, mouth washes, food additive, dough conditioner (up to 0.0075 parts per 100 of flour used).

calcium iodide.
$CaI_2 \cdot 6H_2O$.
Properties: Yellowish-white crystals, deliquescent, decomposes in air by absorption of carbon dioxide. D 2.55 (anhydrous 4.0 at 25C), loses $6H_2O$ at 42C, mp 783C, bp approximately 1100C. Soluble in water, ethanol, and pentanol.
Derivation: Action of hydriodic acid on calcium carbonate.
Use: Photography, medicine.

calcium iodobehenate.
$Ca(OOCC_{21}H_{42}I)_2$.
Properties: White, odorless powder. Insoluble in alcohol; slightly soluble in cold water.
Use: Medicine and pharmaceuticals.

calcium lactate.
CAS: 814-80-2. $C_6H_{10}CaO_6 \cdot xH_2O$.
Properties: White crystalline powder with up to 5 H_2O. Sol in water; insol in alc.
Hazard: A poison.
Use: Food additive.

calcium lactobionate.
CAS: 5001-51-4. $C_{24}H_{42}CaO_{24}$.
Properties: White powder. Mp 120C (decomp). Sol in water; insol in alc, ether.
Use: Food additive.

calcium lignosulfonate.
CAS: 8061-52-7.
Properties: Brown, amorphous polymer obtained from the spent sulfite pulping liquor of wood. Sol in water.
Use: Dispersing agent for pesticides; food additive.

calcium metasilicate. (calcium silicate).
CAS: 1344-95-2. $CaSiO_3$.
Properties: White powder. D 2.9. Insoluble in water.
Hazard: Irritating dust. Use in foods restricted to 5% in baking powder, 2% in table salt. Upper respiratory tract irritant. Questionable carcinogen.
Use: Absorbent, antacid, filler for paper and paper coatings, cosmetics, food additive (anticaking agent), manufacture of glass and Portland cement.
See dicalcium silicate.

calcium molybdate.
CAS: 7789-82-4. $CaMoO_4$.
Properties: White, crystalline powder. Mp approximately 1250C, d 4.35. Soluble in mineral acids; insoluble in alcohol, ether, or water. Noncombustible.
Derivation: Fusion of calcium oxide and a molybdenum ore.
Grade: Technical, single crystals, 99.97%.
Use: Molybdic acid, alloying agent in production of iron and steel, crystals in optical and electronic applications, phosphors.

calcium naphthenate.
Properties: Sticky, tenacious mass. Insoluble in water; soluble in ethyl acetate, carbon tetrachloride, gasoline, benzene, and ether. Combustible.
Derivation: Precipitation from aqueous solutions of calcium salts and sodium naphthenate.
Use: Waterproofing compositions, adhesives, driers, wood fillers, grafting waxes, cements, varnishes, color lakes.
See soap (2).

calcium nitrate. (lime nitrate; nitrocalcite; lime saltpeter; Norwegian saltpeter).
CAS: 10124-37-5. (1) $Ca(NO_3)_2 \cdot 4H_2O$; (2) $Ca(NO_3)_2$.
Properties: White, deliquescent mass. D (1) 1.82, (2) 2.36; mp (1) 42C, (2) 561C. Soluble in water, alcohol, and acetone.
Grade: Technical, pure, CP, reagent.
Hazard: Strong oxidizer, dangerous fire risk in contact with organic materials, may explode if shocked or heated.
Use: Pyrotechnics, explosives, matches, fertilizers, other nitrates, source of ^{14}C by nuclear irradiation.

calcium nitride.
Ca_3N_2.
Properties: Brown crystals. D 2.63 (17C), mp 1195C. Soluble in water with evolution of ammonia (irritating gas), soluble in dilute acids, insoluble in absolute alcohol.

calcium nitrite.
$Ca(NO_2)_2 \cdot H_2O$.
Properties: Colorless or yellowish crystals, hygroscopic. D 2.23 (34C) (anhydrous), mp loses its water at 100C. Soluble in water; slightly soluble in alcohol.
Grade: Technical.
Use: Corrosion inhibitor in lubricants and steel-reinforced concrete.

calcium novobiocin. See novobiocin.

calcium octoate. See soap (2).

calcium oleate. (oleic acid calcium salt).
$Ca(C_{18}H_{33}O_2)_2$.

Properties: Yellowish crystals. Soluble in benzene and chloroform; almost insoluble in water, alcohol, and acetone; decomposes at approximately 140C.
Use: Grease-thickening agent, emulsifying agent, waterproofing concrete.

calcium orthophosphate.   See calcium phosphate, tribasic.

calcium   orthotungstate.   See   calcium tungstate.

calcium oxalate.
CaC_2O_4.
Properties: White, crystalline powder. D 2.2. Soluble in dilute hydrochloric and nitric acids; insoluble in acetic acid and water.
Grade: Technical, CP.
Hazard: An irritant.
Use: Making oxalic acid and organic oxalates, glazes, rare-earth-metal separations.

calcium oxide.   (lime; quicklime; burnt lime; calx; unslaked lime; fluxing lime).
CAS: 1305-78-8.   CaO.
Properties: White or grayish-white hard lumps; sometimes with a yellowish or brownish tint due to iron; odorless. Crumbles on exposure to moist air, d 3.40, mp 2570C, bp 2850C. Soluble in acid; reacts with water to form calcium hydroxide with evolution of heat.
Derivation: Calcium carbonate (limestone) is roasted in kilns until all the carbon dioxide is driven off.
Impurities: Calcium carbonate; magnesium, iron, and aluminum oxides.
Grade: Technical, refractory, agricultural, FCC.
Hazard: Evolves heat on exposure to water. Dangerous near organic materials. Upper respiratory tract irritant.
Use: Refractory, flux in steel manufacture, pulp and paper, manufacture of calcium carbide, sulfur dioxide removal from stack gases, sewage treatment (phosphate removal, pH control), poultry feeds, neutralization of acid waste effluents, insecticides and fungicides, dehairing of hides, sugar refining, food additive, glass manufacture, sodium carbonate by Solvay process, carbon dioxide absorbent.
Note: Like other high-melting solids (tungsten, zirconia, carbon), lime (CaO) becomes incandescent when heated to near its mp (2500C). Of both historical and etymological interest is the use of lime as an illuminant in stage lighting for some years before the advent of electricity (1850–1880). Invented in 1816, this technique involved an oxyhydrogen flame impinging on a cylinder of lime, causing it to emit a brilliant white light that was concentrated to a beam by a lens. The light was powerful enough to spotlight actors or simulate sunshine. It became known in theatrical circles as limelight and was the origin of the familiar phrase "in the limelight."
See lime.

calcium oxychloride.   See calcium hypochlorite.

calcium palmitate.
$Ca(C_{15}H_{31}CO_2)_2$.
Properties: White or pale yellow powder produced by reacting a soluble palmitate with a soluble calcium salt. Insoluble in water; slightly soluble in alcohol or ether. Combustible.
Use: Waterproofing agent, thickener for lubricating oils, manufacture of solidified oils. Available only as technical grade.
See soap (2).

calcium pantothenate.
$(C_5H_{16}NO_5)_2Ca$.
The calcium salt of pantothenic acid, available in both the dextro and racemic forms. Only the dextro form has vitamin activity.
Properties: (Both forms identical) white, slightly hygroscopic powder; odorless; sweetish taste. Mp 170–172C; decomposes at 195–196C; specific rotation (5% aqueous solution) +28.2 (25C). Stable in air. Solutions have a pH of 7–9. Soluble in water and glycerol; insoluble in alcohol, chloroform, and ether.
Source: Same as pantothenic acid.
Grade: USP (both forms), FCC.
Use: Medicine, animal feeds, dietary supplement. See pantothenic acid.

calcium pectate.   A material developed in plants, such as beans, peas, potatoes, which enables them to seal off fungus-infected areas.
See pectic acid; pectin.

calcium perborate.
$Ca(BO_3)_2 \cdot 7H_2O$.
Properties: Gray-white lumps or powder. Soluble in acids, and in water with evolution of oxygen.
Use: Medicine, bleach, tooth powders.

calcium perchlorate.
$Ca(ClO_4)_2$.
Properties: White crystals. D 2.651, decomposes at 270C. Soluble in water and alcohol.
Hazard: Strong oxidizer, dangerous fire risk in contact with organic materials.

calcium permanganate.
CAS: 10118-76-0. $Ca(MnO_4)_2 \cdot 4H_2O$.
Properties: Violet, deliquescent crystals. D 2.4. Soluble in water and ammonia; decomposed by alcohol.
Grade: Technical, pure.
Hazard: Strong oxidizer, dangerous fire risk in contact with organic materials.
Use: Textile industry, sterilizing water, dentistry, disinfectant, deodorizer, an additive (with hydrogen peroxide) in liquid rocket propellants, in binders for welding electrode coatings.

calcium peroxide. (calcium superoxide; calcium dioxide).
CAS: 1305-79-9. CaO_2.
Properties: White or yellowish powder; odorless; almost tasteless. Decomposes at approximately 200C. Almost insoluble in water; soluble in acids with formation of hydrogen peroxide. Available oxygen 22.2% (min 13.3% in technical grade).
Derivation: Interaction of solution of a calcium salt and sodium peroxide with subsequent crystallization.
Grade: 60–75%, FCC.
Hazard: Strong oxidizing agent. Dangerous fire risk in contact with organic materials. Irritating in concentrated form.
Use: Seed disinfectant, dentifrices, dough conditioners, antiseptic, bleaching of oils, modification of starches, high-temperature oxidation.

calcium phenylate. (calcium phenoxide; calcium phenate).
$Ca(OC_6H_5)_2$.
Properties: Finely divided red crystals. Slightly soluble in water and alcohol.
Use: Lubricating-oil detergent, emulsifier.

calcium phosphate. See calcium phosphate, dibasic; calcium phosphate monobasic; calcium phosphate tribasic.

calcium phosphate, dibasic. (dicalcium orthophosphate; bicalcium phosphate; secondary calcium phosphate).
$CaHPO_4 \cdot 2H_2O$ and $CaHPO_4$.
Properties: White, crystalline powder; tasteless; odorless. (Hydrate) d 2.306, loses its water at 109C. Soluble in dilute hydrochloric, nitric, and acetic acids; insoluble in alcohol; slightly soluble in water. Nonflammable.
Derivation: Interaction of fluorine-free phosphoric acid with milk of lime.
Grade: USP, FCC, dentifrice grade, feed grade, 18.5 or 21% P.
Use: Animal-feed supplement, food supplement, dentifrice, medicine, glass, fertilizer, stabilizer for plastics, dough conditioner, yeast food.

calcium phosphate, monobasic. (calcium biphosphate; acid calcium phosphate; calcium phosphate primary; monocalcium phosphate).
$CaH_4(PO_4)_2 \cdot H_2O$.
Properties: Colorless, pearly scales or powder; deliquescent in air. Loses water at 100C, decomposes at 200C, d 2.20. Soluble in water and acids. Aqueous solutions are acid. Nonflammable.
Derivation: By dissolving either dicalcium or tricalcium phosphate in phosphoric acid and allowing the solution to evaporate spontaneously.
Grade: FCC, ceramic, anhydrous, hydrated.
Use: Baking powders, fertilizers, mineral supplement, stabilizer for plastics, to control pH in malt, glass manufacture, buffer in foods, firming agent.
See superphosphate.

calcium phosphate, precipitated. See calcium phosphate, tribasic.

calcium phosphate, primary. See calcium phosphate, monobasic.

calcium phosphate, secondary. See calcium phosphate, dibasic.

calcium phosphate, tertiary. See calcium phosphate, tribasic.

calcium phosphate, tribasic. (calcium orthophosphate; tricalcium phosphate; precipitated calcium phosphate; tricalcium orthophosphate; tertiary calcium phosphate).
$Ca_3(PO_4)_2$.
True $Ca_3(PO_4)_2$ can be prepared thermally but is rare. Precipitated "tricalcium phosphate" is a hydroxyapatite with the approximate formula $Ca_5OH(PO_4)_3$.
Properties: White, crystalline powder; odorless; tasteless. D 3.18, mp 1670C, refr index 1.63. Soluble in acids; insoluble in water, alcohol, and acetic acid. Nonflammable.
Derivation: (1) Phosphate rock, apatite, and phosphorite; (2) by the interaction of soluble of calcium chloride and sodium triphosphate with excess of ammonia; and (3) by interaction of hydrated lime and phosphoric acid.
Grade: Granular, technical, CP, NF, pure precipitated, FCC.
Use: Ceramics, calcium acid phosphate, phosphorus and phosphoric acid, polishing powder, cattle foods, clarifying sugar syrups, medicine, mordant (dyeing textiles with Turkey red), fertilizers, dentifrices, stabilizer for plastics, in meat tenderizers, in foods as anticaking agent, buffer, nutrient supplement, removal of ^{90}Sr from milk.
See bone ash.

calcium phosphide. (photophore).
CAS: 1305-99-3. Ca_3P_2 (or Ca_2P_2).
Properties: Red-brown crystals or gray granular masses. D 2.51 (15C), mp approximately 160C, insoluble in alcohol and ether.
Derivation: By heating calcium phosphate with aluminum or carbon, by passing phosphorus vapors over metallic calcium.
Grade: Technical.
Hazard: Dangerous fire risk; decomposed by water to phosphine, which is highly toxic and flammable. See phosphine.
Use: Signal fires, torpedoes, pyrotechnics, rodenticide.

calcium phosphite. (dicalcium orthophosphite).
$CaHPO_4 \cdot 2H_2O$.

Properties: White powder, loses its water at 200–300C (decomposes), forming phosphine. Slightly soluble in water; insoluble in alcohol.
Use: Catalyst for polymerization reactions.

calcium phytate.   (hexacalcium phytate).
$C_6H_6(CaPO_4)_6$.
Properties: Free-flowing, white powder. Slightly soluble in water. The pH of saturated solution is neutral.
Derivation: Corn steep liquor.
Use: Sequestering agent to remove excess metals from wine and vinegar, source of calcium in pharmaceuticals and nutrition, source of phytic acid and its salts.

calcium plumbate.
Ca_2PbO_4.
Properties: Orange to brown crystalline powder. Decomposed by hot water or carbon dioxide, d 5.71. Soluble in acids (decomposes); insoluble in cold water.
Hazard: Fire risk in contact with organic materials. Toxic by ingestion.
Use: Oxidizing agent, pyrotechnics and safety matches, glass, storage batteries.

calcium polysilicate.
$CaO \cdot 12SiO_2$.
A powder used as an anticaking agent.

calcium propionate.
$Ca(OOCCH_2CH_3)_2$.
Occurs also with one water. White powder, soluble in water, slightly soluble in alcohol.
Grade: FCC.
Use: Mold-inhibiting additive in bread, other foods, tobacco, pharmaceuticals, medicine (antifungal agent).

calcium propyl arsenate.
$C_3H_7AsO_3Ca$.
Properties: Crystals. Soluble in water.
Hazard: Toxic by ingestion.
Use: Preemergence control of crabgrass.

calcium pyrophosphate.
$Ca_2P_2O_7$.
Properties: White powder. D 3.09, mp 1230C. Soluble in dilute hydrochloric and nitric acids; insoluble in water.
Grade: FCC.
Use: Polishing agent in dentifrices, mild abrasive for metal polishing, nutrient and dietary supplement.

calcium resinate.
CAS: 9007-13-0.
Properties: Yellowish-white, amorphous powder or lumps; rosin odor. Insoluble in water; soluble in acid, amyl acetate, butyl acetate, ether, amyl alcohol.

Derivation: By boiling calcium hydroxide with rosin and filtering; fusion of hydrated lime and melted rosin.
Grade: Technical, fused.
Hazard: Flammable, dangerous fire risk, spontaneous heating.
Use: Waterproofing; manufacturing paint driers, porcelains, perfumes, cosmetics, enamels; coating for fabrics, wood, paper; tanning leather.
See soap (2).

calcium ricinoleate.
$Ca[CH_3(CH_2)_5CH_2OCH_2CHCH(CH_2)_7CO_2]_2$.
Properties: White powder; slight odor of fatty acids. Mp 84C, d 1.04. Combustible.
Derivation: Derived from castor oil.

calcium *d*-saccharate.
$CaC_6H_8O_8 \cdot 4H_2O$.
Properties: White, crystalline powder; odorless; tasteless. Insoluble in water and alcohol; soluble in calcium gluconate solution.
Derivation: Oxidation of *d*-gluconic acid and neutralization with lime.
Use: Plasticizer for cement, mortar, etc.

calcium saccharin.
mf: $C_{14}H_8CaN_2O_6S_2 \cdot 3.5H_2O$.
Properties: White crystalline powder; faint aromatic odor. Sol in water.
Use: Food additive.

calcium selenate.
$CaSeO_4$.
Properties: A powder. D 2.7 (dihydrate). Water soluble.
Use: General pesticide.

calcium silicate.   See calcium metasilicate; wollastonite; Portland cement.

calcium silicide.
$CaSi_2$.
Properties: Solid. D 2.5. Insoluble in cold water; decomposes in hot water; soluble in acids and alkalies.
Hazard: Flammable, may ignite spontaneously in air.

calcium silicofluoride.   (calcium hexafluorosilicate).
$CaSiF_6$.
Properties: (Dihydrate) Finely divided solid. D 2.25. Insoluble in cold water and acetone; decomposed by hot water.
Hazard: Toxic to experimental animals.
Use: Flotation agent, insecticide, rubber compounding, ceramic glazes.

calcium–silicon alloy.   Contains 30% calcium.
Hazard: Flammable, may ignite spontaneously in air.

calcium sorbate.
$Ca(OOCC_5H_7)_2$.
Use: Chemical preservative in foods.

calcium stannate.
$CaSnO_3 \cdot 3H_2O$.
Properties: White, crystalline powder. Insoluble in water; dehydrates at approximately 350C.
Hazard: Toxic by ingestion and inhalation.
Use: Additive to ceramic capacitors, production of ceramic colors.

calcium stearate. (calcium distearate; octadecanoic acid, calcium salt).
CAS: 1592-23-0. $Ca(C_{18}H_{35}O_2)_2$.
Variable proportions of calcium stearate and calcium palmitate.
Properties: White powder; slight characteristic odor. Mp 179C. Insoluble in water; slightly soluble in hot alcohol. Decomposed by many acids and alkalies.
Derivation: Interaction of sodium stearate and calcium chloride, then filtration.
Grade: Technical, FCC.
Hazard: A nuisance dust.
Use: Water repellent, flatting agent in paints, lubricant in making tablets, emulsions, cements, wax crayons, stabilizer for vinyl resins, food additive, mold-release agent, cosmetics.

calcium stearoyl lactate.
Properties: Cream-colored powder; caramel odor. Slightly sol in hot water.
Use: Food additive.

calcium stearoyl lactylate.
$C_{48}H_{86}CaO_{12}$.
Properties: Finely divided, nonhygroscopic solid. Almost insoluble in water.
Use: Food additive (flour mixing, egg whites, etc.).

calcium strontium sulfide.
$CaSrS_2$ or $CaS \cdot SrS$.
Use: Phosphorescent pigment; a phosphor.

calcium sulfamate.
$Ca(SO_3NH_2)_2 \cdot 4H_2O$.
Properties: White crystals. Soluble in water. Aqueous solution is stable on boiling. Nonflammable.
Grade: Technical.
Use: Flame-proofing agent for textiles and certain grades of paper.

calcium sulfate.
CAS: 10101-41-4. $CaSO_4$ or $CaSO_4 \cdot 2H_2O$ or $CaSO_4 \cdot 0.5H_2O$.
Occurs in nature as anhydrite and in hydrated form as gypsum (plaster of Paris, the hemihydrate $CaSO_4 \cdot 0.5H_2O$).
Properties: (Pure anhydrous) White, odorless powder or crystals. D 2.964, mp 1450C. Slightly soluble in water. (Dihydrate, pure precipitated) D 2.32, loses 1.5 waters at 128C, becomes anhydrous at 163C. Noncombustible. Neither the anhydrous nor the dihydrate form can set with water.
Derivation: From natural sources and as a by-product in many chemical operations.
Grade: Technical, pure precipitated (as the dihydrate), FCC.
Hazard: Nasal symptoms.
Use: Portland-cement retarder; tile and plaster; source of sulfur and sulfuric acid; polishing powders; paints (white pigment, filler, drier); paper (size filler, surface coating); dyeing and calico printing; metallurgy (reduction of zinc minerals); drying industrial gases, solids, and many organic liquids; in a granulated form as a soil conditioner; quick-setting cements, molds, and surgical casts; wallboard; food additive; desiccant.

calcium sulfhydrate. See calcium hydrosulfide.

calcium sulfide.
CaS.
Properties: Yellow to light-gray powder; odor of hydrogen sulfide in moist air; unpleasant alkaline taste. D 2.6. Gradually decomposes in moist air or in weak acids; decomposed by acids. Slightly soluble in water with partial decomposition; insoluble in alcohol.
Derivation: Strong heating of pulverized calcium sulfate and charcoal.
Hazard: Irritating to skin and mucous membranes.
Use: Luminous paint, depilatory, preparation of arsenic-free hydrogen sulfide, lubricant additive, ore dressing and flotation agent, phosphors.

calcium sulfite.
$CaSO_3 \cdot 2H_2O$.
Properties: White powder. Loses its water at 100C. Soluble in sulfurous acid; slightly soluble in water.
Derivation: Action of sulfurous acid on calcium carbonate.
Use: Textiles (antichlor), disinfectant in sugar industry, brewing, biological cleansing, food preservative and discoloration retarder, paper manufacture.

calcium sulfocyanate. See calcium thiocyanate.

calcium superoxide. See calcium peroxide.

calcium tannate.
Properties: Yellowish-gray powder. Soluble in dilute acids; slightly soluble in water.
Use: Pharmaceuticals, adhesives.

calcium tartrate.
$CaC_4H_4O_6 \cdot 4H_2O$.
Properties: White crystals. Soluble in dilute acids; slightly soluble in water or alcohol.
Derivation: Interaction of calcium salt and crude cream of tartar.
Grade: Technical, CP.
Use: Tartaric acid, food preservative, antacid.

calcium thiocyanate.   (calcium sulfocyanate). $Ca(SCN)_2 \cdot 3H_2O$.
Properties: White, hygroscopic crystals. Soluble in water and alcohol.
Use: Solvent for cellulose and polyacrylate, acrylonitrile polymers; stiffening and swelling of textiles.

calcium thioglycollate.
$CaCH_2COOC \cdot 3H_2O$.
Properties: White powder. Loses water at 100C. Decomposes at 250C. Slightly soluble in water; insoluble in alcohol.
Use: Depilatories and hair-waving preparations.

calcium titanate.
$CaTiO_3$.
Properties: Powder. D 3.98, mp 1800C.
Use: Electronics.

calcium trisodium pentetate (USAN).   [tririsodium[(carboxymethyl)imino]bis(ethylenenitrolo) tetraacetate]. $CaNa_3C_{14}H_{18}N_3O_{10}$.
A chelating agent, antidote for lead poisoning.

calcium tungstate.   (calcium orthotungstate; calcium wolframate normal).
$CaWO_4$.
Properties: White crystals. D 6.062. Soluble in ammonium chloride; insoluble in water; decomposed by hot acids.
Derivation: (1) Interaction of calcium chloride and sodium tungstate. (2) Occurs in nature as scheelite, in Nevada, California, Arizona, Utah, Colorado, New Zealand, Europe.
Method of purification: A slurry of powdered scheelite is treated with soda ash to form the soluble sodium tungstate. Insoluble impurities are filtered off, and calcium tungstate is precipitated with lime.
Use: Luminous paints, fluorescent lamps, photography, medicine (radiopaque agent).
Note: Synthetic crystals are available for use as scintillation counters and possible application in lasers.

calcium undecylenate.
$[CH_2CH(CH_2)_8COO]_2Ca$.
Properties: A fine, white powder. Mp 155C. Of limited solubility.
Use: Bacteriostat and fungistat in cosmetics and pharmaceuticals.

calcium zirconate.
$CaZrO_3$.
Properties: Solid. Mp 2550C, d 4.78. Soluble in nitric and other acids. Noncombustible.

calcium zirconium silicate.
$CaZrSiO_5$ or $CaOZrO_2SiO_2$.
Properties: White solid. Mp 1593C. Insoluble in water, alkalies; slightly soluble in acids; soluble in hydrofluoric acid. Noncombustible.
Use: Electrical resistor ceramics, glaze opacifier.

Calcocid.   A series of acid dyestuffs used in the dyeing of wool and worsted goods, natural silk, and jute; and in coloring diversified materials.

Calcofast.   A series of metallized dyes containing chemically combined chromium used for dyeing wool. They can also be applied to leather, nylon, etc.

Calcofluor.   A series of direct-dyeing dyes that possess fluorescent properties. Used for dyeing cotton, linen, viscose, acetate, nylon, wool, and certain synthetics. Used also in soaps as a brightener for textiles.

caldopentamine.   See tetrapropylenepentamine.

caldron process.   Procedure for the recovery of silver in which a slurry of ore in a copper vessel is agitated with salt.

calender.   A machine in which material is passed between heated steel rolls for any of several purposes: (1) to convert it into a sheet of uniform thickness; (2) to cause it to impregnate a textile fabric; or (3) to increase its surface gloss and hardness. Calenders (i.e., cylinders) are composed of from 3 to as many as 10 or 12 hollow cast-iron rolls up to 84 inches in width, set vertically in a frame. The standard rubber or plastics calender has three steam-heated rolls that turn in opposing directions, at either the same speed or different speeds. When moving at the same speed, they deliver the mixture fed between the top and center rolls in a smooth sheet that can be as thin as 0.005 inches, from between the center and bottom rolls. If fabric impregnation is desired, the center roll runs faster than the other two, thus pressing the soft and tacky mixture, called friction, into the textile material (tire carcasses, electrical tape, etc.). In paper manufacturing a high-speed calender "stack" imparts a smooth finish to the sheet as it leaves the drying unit. Smaller calenders are used to apply coating compositions. Laboratory sizes of all types are available. See supercalender.

"Calgon" *[Calgon].*   TM for a sodium phosphate glass cleaner (commonly called sodium hexametaphosphate). It has a molecular ratio of 1:1 $Na_2O:P_2O_5$, with a guaranteed min of 67% P_2O_5. Several specialized compositions are available.
Derivation: From food-grade phosphoric acid and commercial soda ash by a thermal process.
Available forms: Powder, agglomerated particles, and broken glassy plates, either pure or adjusted with mild alkalies.
Properties: Miscible with water but insoluble in organic solvents. It possesses sequestering, dispersing, and deflocculating properties and precipitates proteins. In very low concentration, it inhibits corrosion of steel and prevents the precipitation of

slightly soluble, scale-forming compounds such as calcium carbonate and calcium sulfate.

Use: Softening water without precipitate formation, as in dyeing, laundering, textile processing, and washing operations; corrosion inhibitor in deicing salt preparations; frozen desserts; pretanning hides in the manufacture of leather; dispersing clays and pigments; threshold treatment for scale; and corrosion prevention.

Calgosil. A metaphosphate silicate compound, used to provide corrosion reduction in low-hardness waters.

caliche. See sodium nitrate.

Califlux. A series of oils composed principally of nitrogen bases and acidaffins. Used in plasticizers, in extenders, and as reclaiming agents for dark-colored compounds.

californium.
Cf.
A synthetic radioactive element of the actinide group with atomic number 98 and aw 252. It has several isotopes, two of which (^{252}Cf and ^{249}Cf) are available in milligram amounts. The pure metal has not yet been obtained. Several compounds are known: trioxide, trifluoride, trichloride, and sesquioxide. The 252 isotope has potential uses in neutron-activation analysis for continuous materials testing, mineral prospecting, oil-well logging, etc. Biologically, it is a bone-seeking element and has specialized applications in medicine.

calixarenes.
$C_{56}H_{72}O_{12}$.
Phenolic metacyclophanes and macrocyclic compounds capable of assuming a basket-shaped conformation.
Derivation: Formed from *p*-hydrocarbyl phenols and formaldehyde.

Calmette's serum. Snake antivenin.

calmodulin. A ubiquitous calcium binding eukaryotic protein involved in the regulation of many metabolic functions. By binding calcium ions, it is responsible for many of the known cellular effects of calcium ions.

Calo-Clor. A mercurial turf fungicide containing 73% mercury in chemical combination (principally mercuric and mercurous chlorides).
Hazard: Toxic by ingestion or inhalation.

Calocure. A mercurial turf fungicide having a mercury content of 36%, mainly mercuric and mercurous chlorides.
Hazard: A poison. See mercury.
See mercury.

Calogran. A turf fungicide composed of an extremely finely divided form of mercurous chloride. Contains 85% mercury (insoluble in water).
Hazard: A poison. See mercury.
See mercury.

calomel. See mercurous chloride.

calorie. (1) The amount of heat needed to raise 1 g of water 1C at 1 atm. (2) A kilogram calorie is the amount of heat required to raise 1 kg of water 1C (equal to 4184×10^7 ergs). In the latter case, the word *Calorie* should be capitalized when used alone, or the abbreviation kcal may be used. Kilogram calories are used in connection with foods and beverages.

calorific value. The number of heat units obtained by the complete combustion of unit mass of a fuel or other material.

calorimeter. Device for determining heat evolution or absorption in a chemical reaction.

calorizing. The process by which steel is coated with aluminum by heating it in aluminum powder. The aluminum forms an alloy with the steel surface and produces a thin, tightly adherent coating. See cementation.

calotropagenin.
CAS: 24211-64-1. $C_{23}H_{32}O_6$.
Hazard: A poison.
Source: Natural product.

calsintering. A method for recovering alumina from fly ash developed in 1978 by Oak Ridge National Laboratories. The essential steps are pelletizing and sintering a mixture of fly ash and lime (from either limestone or gypsum) at 1000–1200C for 20 min. The product is ground to 40 mesh and leached with dilute sulfuric acid. The leached solution is then solvent extracted, crystallized, and calcined to alumina, which then can be converted to aluminum by standard methods. The high recovery of alumina suggests that fly ash is a viable alternative raw material for aluminum.

calspar. See calcite.

calutron. A type of electromagnetic separator used for radioisotope purification.
See electromagnetic separation.

Calvin cycle. The cyclic pathway used by plants to fix carbon dioxide and produce triose phosphates. Named after Melvin Calvin, an early worker in the field.

Calvin, Melvin. (1911–1997). An American chemist who won the Nobel Prize for chemistry in

1961. Much of his work involved the study of photosynthesis, biophysics, and application of physics and chemistry of molecules to some of the basic problems of biology. His doctorate was from the University of Minnesota. He did postgraduate work in England and at Northwestern University and the University of Notre Dame.

"Cambrelle" *[Camtex]*.   TM for a nonwoven, melded fabric used for carpet backing, road reinforcement, upholstery, interlinings, tablecloths, and other household applications. It is said to be composed of two different polymers, one lying within the other. Upon heating to the melting point of the external polymer, the fibers soften and unite to form a fabric. The term *melded* refers to this type of fusion.
See nonwoven fabric.

cAMP.   Biochemical designation for cyclic adenosine-3′,5′-monophosphate, an activator of hormones and initiator of prostaglandin synthesis.

Campbell synthesis.   Ketoxime plus aryl Grignard reagent yields arylated ethylenimine.

2-camphanol.   See borneol.

2-camphanone.   See camphor.

camphene.
CAS: 79-92-5. $C_{10}H_{16}$.
A terpene.
Properties: Colorless crystals. Mp 48–52C, bp 159–162C. Soluble in ether; slightly soluble in alcohol; insoluble in water.
Derivation: (1) By heating pinene hydrochloride with alkalies, aniline, or alkali salts such as sodium acetate. (2) A constituent of certain essential oils.
Grade: Technical (mp 46C).
Hazard: Toxic by ingestion.
Use: Manufacture of synthetic camphor, camphor substitute.

camphor.   (gum camphor; 2-camphanone).
CAS: 76-22-2. $C_{10}H_{16}O$.
A ketone occurring naturally in the wood of the camphor tree (*Cinnamomum camphora*).

Properties: Colorless or white crystals, granules, or easily broken masses; penetrating aromatic odor. D 0.99, mp 174–179C, sublimes slowly at room temperature, flash p 150F (65.5C) (CC), autoign temp 871F (466C). Slightly soluble in water; soluble in alcohol, ether, chloroform, carbon disulfide, solvent naphtha, and fixed and volatile oils. Combustible.
Derivation: Steam distillation of camphor-tree wood and crystallization. This product is called natural camphor and is dextrorotatory. Synthetic camphor, most of which is optically inactive, may be made from pinene, which is converted into camphene; by treatment with acetic acid and nitrobenzene it becomes camphor, turpentine oil is also used.
Grade: Technical (synthetic, mp 163–168C), USP (mp 174–179C).
Hazard: Evolves flammable and explosive vapors when heated. Eye and upper respiratory tract irritant, and anosmia. Questionable carcinogen.
Use: Medicine (internal and external), plasticizer for cellulose nitrate, other explosives and lacquers, insecticides, moth and mildew proofings, tooth powders, flavoring, embalming, pyrotechnics, intermediate.
Note: A liquid form (camphor oil) is produced almost exclusively in Taiwan and was formerly used in the manufacture of sassafras oil; the available supply is used chiefly as a fragrance or flavoring material and to some extent as a pharmaceutical product.

camphor   bromate.   (α-bromo-1-camphor; brominated camphor).
$C_{10}H_{15}BrO$.
Properties: Colorless crystals; slight camphor odor and taste. Also available as powder. Discolors in light and should be stored in cool, dark place. Mp 76C, bp 274C, d 1.449. Soluble in alcohol, ether, chloroform, and oils; insoluble in water.
Derivation: By heating camphor with bromine.
Use: Medicine, manufacture of camphor derivatives.

camphoric acid.
$C_8H_{14}(COOH)_2$.
Properties: Colorless, odorless needles or scales. D 1.0–1.86 (20/4C), mp 186–188C. Soluble in alcohol, ether, fatty oils, chloroform. Partially soluble in water.
Derivation: By oxidizing camphor with nitric acid.
Use: Pharmaceuticals, medicine.

camphor, Malayan.   See borneol.

dl-**camphoroquinone.**   (2,3-bornanedione).
Properties: Mw 166.22, mp 198–200C.
Use: Synthetic intermediate.

camphor peppermint.   See menthol.

Campillit.   See cyanogen bromide.

Camp's   quinoline   synthesis.   Formation of hydroxyquinolines from *o*-acylaminoacetophenones in alcoholic sodium hydroxide. The relative proportions of the isomeric products are mainly determined by the acyl residue on the amino nitrogen.

Canada balsam.   See balsam.

Canadian Society for Mass Spectrometry.
(CSMS).
The Canadian Society for Mass Spectrometry is
a society dedicated to the promotion of mass
spectrometry. The members are involved in mass
spectrometric research and development in university, industrial and governmental laboratories. The
CSMS publishes a newsletter every 6 months and
hosts the Tandem Mass Spectrometry Workshop in
Lake Louise, Alberta. It can be reached at 10 Marie
Curie, Ottawa, ON K1N 6N5, Canada. Website:
http://www.csms-scsm.ca

canal ray.   A stream of cations formed when
electrons are removed from the gas particles in a
cathode-ray tube. Due to their positive charge, canal
rays flow toward the cathode in the cathode ray tube
whereas the negatively charged cathode rays move
toward the anode. Canal rays are so-named because
they have been observed to pass through canals in
the cathode.

cananga oil.   An essential oil similar in odor to
ylangylang oil, strongly levorotatory. Used for floral
odors in perfumery and as a flavoring agent.

canavanine.
$NH_2CNHNHOCH_2CH_2CHNH_2COOH.$
A nonprotein amino acid obtained from jack bean
meal. It is found naturally in the $l(+)$ form.
Properties: Crystals from dilute alcohol. Sulfate:
crystals from dilute alcohol, mp 172C (decomposes). Soluble in water; nearly insoluble in alcohol.
Use: Biochemical research.

cancer.   (malignant tumor; malignant neoplasm).
Any of various types of malignant neoplasms, characterized by malignant, uncontrolled proliferation
of cells resulting in a tumor or other abnormal
condition, regardless of the tissues or body parts
involved. Most are able to invade adjacent tissues,
and have a tendency toward metastasis. Many cancers are initiated by xenobiotic substances, both
natural and anthropogenic, or by viruses. 5–20%
of deaths from cancer may be caused by exposure
to toxic substances.

candelilla wax.
Properties: Yellowish-brown, opaque to translucent
solid. D 0.983, mp 67–68C, saponification value
65, iodine number 37, refr index 1.4555. Soluble in chloroform, turpentine, carbon tetrachloride,
trichloroethylene, toluene, hot petroleum ether, and
alkalies; insoluble in water. Combustible.
Occurrence: Mexico, Texas.
Grade: Crude, refined, powdered.
Use: Leather dressing, polishes, cements, varnishes,
candles, electric insulating composition, sealing
wax, waterproofing and insect-proofing containers,

paint removers, dentistry, paper sizes, stiffener for
soft waxes.

candicidin (USAN).   An antifungal antibiotic
produced by *Streptomyces griseus.*

Candida.   A genus of fungi commonly known
as yeast. *Candida* species are the most common
cause of opportunistic mycoses worldwide. It is a
widespread environmental contaminant. The spectrum of candidiasis is extremely diverse, and almost
any human organ or human system can be affected.
Hazard: Pathogenic.

candida lipolytica.
Properties: Derived from *Candida lipolytica* Fam.
Cryptococcaceae.
Use: Food additive.

candidate gene.   A gene located in a chromosome region suspected of being involved in a disease.
See positional cloning; protein.

candida guilliermondii.
Properties: Derived from *Candida guilliermondii*
Fam. Cryptococcaceae.
Use: Food additive.

candidin.   An antifungal antibiotic produced by
Streptomyces viridoflavus.

canescegenin.   See nigrescigenin.

cane sugar.   See sucrose.

cannabis.   (marijuana).
CAS: 8063-14-7.
Its principle, tetrahydrocannabinol, can be made syn-
thetically.
Derivation: Dried flowering tops of pistillate plants
of *Cannabis sativa.*
Source: Iran, India; cultivated in Mexico and Europe.
Hazard: A mild hallucinogen. Sale is illegal in U.S.
Use: Medicine, ophthalmology (treatment of glau-
coma).

cannel coal.   A variety of bituminous or subbituminous coal of uniform and compact fine-grained
texture. Dark gray to black; has a greasy luster. It
is noncaking, yields a high percentage of volatile
matter, and burns with a luminous, smoky flame.
(ASTM definition, ASTM D493-39.) Combustible.
Hazard: Explosion risk in form of dust.

Cannizzaro reaction.   Base catalyzed dismuta-
tion of aromatic aldehydes or aliphatic aldehydes
with no α-hydrogen into the corresponding acids
and alcohols. When the aldehydes are not identical, the reaction is called the "crossed Cannizzaro
reaction."

Cannizzaro, Stanislao. (1826–1910). Born in Italy, he extended the research of Avogadro on the molecular concentration of gases and thus was able to prove the distinction between atoms and molecules. His investigations of atomic weights helped to make possible the discovery of the periodic law by Mendeleyev. His research in organic chemistry led to the establishment of the Cannizzaro reaction involving the oxidation reduction of an aldehyde in the presence of concentrated alkali.

canola. (rapeseed; colza oil).
Properties: Yellow-brown liquid. D 0.913–0.917, mp 17–22C.
The oil obtained from rapeseed.

canonical form. The basic or general form.

cantharidine. (hexahydro-3a,7a-dimethyl-4,7-epoxyisobenzofuran-1,3-dione; 2,3-dimethyl-7-oxabicyclo-[2.2.1]heptane-2,3-dicarboxylic anhydride; exo-1,2-cis-dimethyl-3,6-epoxy-hexahydrophthalic anhydride; cantharides camphor; cantharone, cantharidin, kantaridin).
CAS: 56-25-7. $C_{10}H_{12}O_4$.
A potent and specific inhibitor of protein phosphatases 1 (PP1) and 2A (PP2A).
Properties: White powder or colorless, crystalline compound, slightly soluble in water, acetone, chloroform, and ethanol.
Derivation: Secreted by certain beetles of the family *meloidae*.
Hazard: Extremely toxic; questionable carcinogen; powerful irritant to all cells and tissues; deadly poison; respiratory failure; death.
Use: In veterinary medicine as a vesicant, rubefacient, and counterirritant.

canthaxanthin.
$C_{40}H_{52}O_2$.
A carotenoid colorant occurring in many natural products. It has been isolated from a variety of edible mushrooms. It has also been synthesized. The *trans* form is a violet, crystalline solid that is soluble in chloroform and various oils. It is a permissible colorant for foods, drugs, etc. It is used to produce artificial tanning of the skin but is not approved for this use.
Hazard: Oral intake may cause loss of night vision.

CAO-1 and CAO-3. Technical and food-grade BHT (butylated hydroxytoluene), also known as 2,6-di-tertiary butyl para cresol.

CAO-5. The hindered phenolic antioxidant 2,2'-methylene-bis(4-methyl-6-tertiary-6-butyl phenol).

Caoutchouc. See rubber.

CAP. Abbreviation for catabolite gene activator protein.

CAP. Abbreviation for chloramphenicol.

cap. All eukaryotes have at the 5' end of their messages a structure called a "cap," consisting of a 7-methylguanosine in 5'-5' triphosphate linkage with the first nucleotide of the mRNA. It is added post-transcriptionally and is not encoded in the DNA.

capillarity. The attraction between molecules, similar to surface tension, which results in the rise of a liquid in small tubes or fibers or in the wetting of a solid by a liquid. It also accounts for the rise of sap in plant fibers and of blood in capillary (hair-like) vessels.

capillary. A small thin-walled, one cell layer thick, blood vessel from which oxygen can diffuse from the blood into nearby cells and carbon dioxide can diffuse into the capillary from the cells. They are endothelial tubes comprised of flat pavement cells, which are separated from cells of the surrounding tissue cells by a layer of connective tissue.

capillary array. Gel-filled silica capillaries used to separate fragments for DNA sequencing. The small diameter of the capillaries permit the application of higher electric fields, providing high-speed, high-throughput separations that are significantly faster than traditional slab gels.

capillator. An instrument for colorimetrically comparing pH values in capillary tubes.

***l*-capnoidine.**
CAS: 485-50-7. $C_{20}H_{17}NO_6$.
Hazard: A poison.
Source: Natural product.

"Capoten" [Par]. TM for captopril.

"Capow" [Kenrich]. TM for a series of powdered forms of coupling agents based on titanium, zirconium, or aluminum.

Capracyl. A group of neutral-dyeing, premetalized acid colors that produce the highest possible degree of lightfastness on nylon. Also suitable for dyeing wool, particularly in blends with cellulosic fibers.

capraldehyde. See *n*-decanal.

"Capran" [Honeywell]. TM for transparent nylon 6 thermoplastic film used for food packaging.

capreomycin (USAN). Antibiotic produced by *Streptomyces capreolus*. Used in medicine.

capric acid. (decanoic acid; decoic acid; decylic acid).
CAS: 334-48-5. $CH_3(CH_2)_8COOH$.
Occurs as a glyceride in natural oils.

Properties: White crystals, unpleasant odor. D 0.8858 (40C), bp 270C, 172.6C (30 mm Hg), mp 31.5C, refr index 1.4288 (40C), acid number 308-315. Soluble in most organic solvents and dilute nitric acid; insoluble in water. Combustible.
Derivation: Fractional distillation of coconut-oil fatty acids.
Grade: Technical, 90%, FCC.
Use: Esters for perfumes and fruit flavors, base for wetting agents, intermediates, plasticizer, resins, intermediate for food-grade additives.

caproic acid.   (hexanoic acid; hexylic acid; hexoic acid).
CAS: 142-62-1. $CH_3(CH_2)_4COOH$.
Present in milk fats to extent of approximately 2%.
Properties: Oily, colorless or slightly yellow liquid; odor of Limburger cheese. D 0.9276 (20/4C), fp −4.0C, bp 205C, refr index 1.4168 (20C), wt/gal 7.7 lb, viscosity 0.031 cP (20C), flash p 215F (101C OC). Soluble in alcohol and ether; slightly soluble in water. Combustible.
Derivation: From crude fermentation of butyric acid, fractional distillation of natural fatty acids.
Grade: Technical, reagent to 99.8%, FCC.
Hazard: Strong irritant to tissue.
Use: Analytical chemistry, flavors, manufacture of rubber chemicals, varnish driers, resins, and pharmaceuticals.

caproic aldehyde.   See *n*-hexaldehyde.

caprolactam.   (aminocaproic lactam; 2-oxohexamethyleneimine).
CAS: 105-60-2.

$$CH_2 \begin{cases} CH_2-CH_2-C=O \\ CH_2-CH_2-NH \end{cases}$$

Properties: White flakes or fused. Mp 68–69C; bp 180C (50 mm Hg); d (70% solution) 1.05; refr index 1.4935 (40C); 1.4965 (31C), heat of fusion 29 cal/g, heat of vaporization 116 cal/g; viscosity 9 cP 78C; vap press 3 mm Hg 100C, 50 mm Hg 180C. Soluble in water, chlorinated solvents, petroleum distillate, and cyclohexene.
Derivation: (1) Catalytic oxidation of cyclohexane to cyclohexanol, reacting with peracetic acid to form caprolactone, and further reaction with ammonia; (2) catalytic hydrogenation of phenol to cyclohexanone, reaction with ammonia to cyclohexanone oxime with Beckmann rearrangement with sulfuric acid catalyst; (3) catalytic oxidation of cyclohexane to cyclohexanone, reaction with hydroxylamine sulfate and ammonia to cyclohexanone oxime followed by sulfuric acid–catalyzed Beckmann rearrangement; and (4) UV-catalyzed reaction of cyclohexane with nitrosyl chloride to cyclohexanone oxime hydrochloride, followed by Beckmann rearrangement. Method (1) was never used commercially.

Method (3) has been modified to minimize formation of by-product ammonium sulfate.
Available forms: Flake, molten.
Hazard: Toxic by inhalation. Upper respiratory tract irritant.
Use: Manufacture of synthetic fibers (especially nylon 6), plastics, bristles, film, coatings; synthetic leather, plasticizers and paint vehicles, crosslinking agent for polyurethanes, synthesis of amino acid lysine.

caprolactone.
Derivation: Reaction product of peracetic acid and cyclohexanone, an intermediate product in the manufacture of caprolactam.
Use: Intermediate in adhesives, urethane coatings, and elastomers; solvent; diluent for epoxy resins; synthetic fibers; organic synthesis.

"Caprolan" [General Electric].   TM for a polyamide fiber made from polymerized caprolactam. Has excellent dyeability and a wide variety of end uses, maintains a superior level of dimensional stability after heat setting, and has outstanding mechanical qualities.
See caprolactam; nylon.

"Capron" [General Electric].   TM for type of nylon 6 resin.
Use: Injection or blow molding, compounding extrusion, and film production.

capryl compounds.   The term *approximately capryl* is generally but erroneously used in the trade to refer to octyl compounds. Thus, the definition of capryl and caprylic compounds will be found under the corresponding octyl entry, e.g., for capryl alcohol. See *sec-n*-octyl alcohol; for caprylic halides, see corresponding octyl halide; for caprylic acid see octanoic acid.

caprylic acid.   (octanoic acid).
CAS: 124-07-2. $C_8H_{16}O_2$.
An eight-carbon straight chain fatty acid produced by the hydrolysis that occurs in butter, coconuts, and breast milk.
Properties: Oily liquid, slightly unpleasant rancidlike smell, minimally soluble in water.

caprylyl peroxide.   Legal label name for octyl peroxide.

capsanthin.
CAS: 465-42-9. $C_{40}H_{56}O_3$.
Hazard: A poison by skin contact.

capsid.   The protein coat of a virus particle.

cap site.   (a) In eukaryotes, the cap site is the position in the gene at which transcription starts, and really should be called the "transcription initiation site." The first nucleotide is transcribed from this

site to start the nascent RNA chain. That nucleotide becomes the 5′ end of the chain, and thus the nucleotide to which the cap structure is attached.
(b) In bacteria, the CAP site (note the capital letters) is a site on the DNA to which a protein factor (the Catabolite Activated Protein) binds.

captafol. See *cis-n*-[1,1,2,2-(tetrachloroethyl) thio]-4-cyclohexene-1,2-dicarboximide.

captan. (*N*-trichloromethylmercapto-tetrahydrophthalimide).
CAS: 133-06-2. $C_9H_8Cl_3NO_2S$.
Properties: White to cream powder. Mp 158–164C, d 1.5. Practically insoluble in water; partially soluble in acetone, benzene, and toluene, slightly soluble in ethylene dichloride and chloroform.
Derivation: Reaction product of tetrahydrophthalimide and trichloromethylmercaptan.
Hazard: A questionable carcinogen. A skin irritant. Avoid inhalation of dust or spray mist. Avoid contamination of feed and foodstuffs.
Use: Seed treatment; fungicide in paints, plastics, leather, fabrics; fruit preservation; bacteriostat.

captopril. (2,*d*-methyl-3-mercaptopropanoyl-*l*-proline).
An orally active hypertensive drug. Its use is limited by FDA. Said to be free from usual side effects of commonly used hypertensive drugs.

capture. The process in which an atomic or nuclear system acquires an additional particle, e.g., the capture of electrons by positive ions or capture of electrons or neutrons by nuclei.
See cross-section (1); neutron; fission.

caramel. (1) A sugar-based food colorant made from liquid corn syrup by heating in the presence of catalysts to approximately 250F (121C) for several hours, cooling to 200F (93C), and filtering. The brown color results from either Maillard reactions, true caramelization, or oxidative reactions. Caramels are colloidal in nature, the particles being held in solution by either positive or negative electric charges.
See caramelization; browning reaction.
(2) A low-enriched uranium reactor fuel containing 6.8% ^{235}U (approximately 10 times as much as natural uranium). It has sufficient neutron density to yield plutonium.

caramelization. A type of nonenzymic browning reaction occurring during exposure of food products to heat when the products contain no nitrogen compounds, e.g., sugars.
See browning reaction.

caraway oil.
Properties: Yellowish liquid. D 0.900–0.910. Insoluble in water; soluble in alcohol.

The oil obtained from the caraway seed. The chief components are carvone and *d*-limonene.
Use: Medicinels and food flavoring.

Carbachol chloride. (2-carbamoyloxyethyl (trimethyl)azachloride; 2-((aminocarbonyl)oxy)-*N,N,N* – trimethylethanaminium chloride; carbachol; carbacholin; carbacholine chloride; carbacolina; carbamic acid, ester, with choline chloride; carbaminocholine chloride; carbaminoylcholine chloride; carbamiotin; carbamoylcholine chloride; γ-carbamoyl choline chloride; carbamylcholine chloride; carbochol; carbocholin; carbyl; carcholin; choline carbamate chloride; choline chlorine carbamate; choline, chloride carbamate(ester); coletyl; doryl (pharmaceutical); (2-hydroxyethyl) trimethylammonium chloride carbamate; isopto carbachol; jestryl; lentin; lentine; moistat; mistura c; moryl; p.V. carbachol; TL 457; vasoperif).
CAS: 51-83-2. $C_6H_{15}CIN_2O_2$.
An acetylcholine counterfit molecule that is not inactivated by acetylcholinesterase and is a slowly hydrolyzed cholinergic agonist that acts at both muscarinic and nicotinic receptors.
Properties: Hard prisms; insoluble in $CHCl_3$ and Et_2O.
Hazard: Deadly poison; may cause lowered blood pressure, venous dilation, nausea or vomiting, sweating and lachrymation; a parasympathetic nerve stimulant.

carbamate. A compound based on carbamic acid (NH_2COOH), which is used only in the form of its numerous derivatives and salts.

carbamate herbicide. Any of a number of carbamates. They act selectively on monocotyledonous seedlings in which they strongly inhibit cell division by interrupting mitotic cycles in root and shoot meristems.
Use: Herbicide.

carbamate insecticide. Any salt or ester derivatives of carbamic acid used to kill insect larvae, nymphs, and adults on contact. These inhibit cholinesterase at nerve junctions by prompting the accumulation of acetylcholine at nerve terminals, thus interrupting nerve conduction. They break down more quickly and are less hazardous to humans and most nontarget organisms.
Use: Insecticide.

carbamic acid, ester with salicylaldehyde dimethyl acetal. See salicylaldehyde dimethyl acetal carbamate.

carbamide. See urea.

carbamide peroxide. See urea peroxide.

carbamide phosphoric acid. (urea phosphoric acid).
$CO(NH_2)_2 \cdot H_3PO_4$.

Properties: White, rhombic crystals. Very soluble in water and alcohol.
Hazard: Evolves toxic fumes when heated.
Use: Catalyst for acid-setting resins, flame-proofing compositions, cleaning compounds, acidulant.

carbamidine. See guanidine.

carbamite. See *sym*-diethyldiphenylurea.

carbamoyl carboxylic acid. Any of a group of organic compounds, such as 5-carbamoylnicotinic acid that contains a carboxy and carboxamide group.

carbamoylglutamic acid.
$C_6H_{10}O_5N_2$.
An intermediate in the carbamoylation of ornithine to citrulline in the urea cycle.

carbamoyl-phosphate synthase. A phosphotransferase that catalyzes the condensation of 2ATP, NH_3, COY, and H_2O to yield 2 ADP + Pi + carbamoyl phosphate. The reverse reaction is catalyzed by carbamate kinase.

carbamylguanidine sulfate. See guanylurea sulfate.

carbamylhydrazine hydrochloride. See semicarbazide hydrochloride.

carbamylurea. See biuret.

carbanil. See phenyl isocyanate.

carbanilide. See diphenylurea.

carbanion. A negatively charged organic ion, such as H_3C^- or RC^-, having one more electron than the corresponding free radical. Carbanions are short-lived but important intermediates in base-catalyzed polymerization and alkylation reactions. See carbonium ion; carbene; free radical.

carbaryl. ("Sevin" *[TechPac]*; generic name for 1-naphthyl-*N*-methylcarbamate).
CAS: 63-25-2. $C_{10}H_7OOCNHCH_3$.
Properties: Solid. Mp 142C, d 1.23. Insoluble in water.
Derivation: Synthesized directly from 1-naphthol and methyl isocyanate or from naphthyl chloroformate (1-naphthol and phosgene) plus methylamine.
Hazard: Toxic by ingestion, inhalation, and skin absorption; irritant. A reversible cholinesterase inhibitor. Use may be restricted. Questionable carcinogen. Male reproductive and embryo damage.
Use: Insecticide.

carbazide. See carbodihydrazide.

carbazole. (dibenzopyrrole; diphenylenimine).
CAS: 86-74-8. $(C_6H_4)_2NH$ (tricyclic).

Properties: White crystals with characteristic odor. Mp 244–246C, bp 352–354C. Partially soluble in alcohol and ether; insoluble in water.
Derivation: (1) From crude anthracene cake by selective solution of the phenanthrene with crude solvent naphtha, removal of the anthracene by conversion into a sulfonic derivative, and extraction by means of water. (2) Synthetically from *o*-aminobiphenyl.
Grade: Technical, 97%.
Hazard: Possible carcinogen.
Use: Manufacture of dyes, reagents, explosives, insecticides, lubricants, rubber antioxidants; odor inhibitor in detergents, UV sensitizer for photographic plates.

carbazotic acid. See picric acid.

carbene. (methylene).
An organic radical containing divalent carbon. Some divalent-carbon derivatives where the carbon is multiple bonded to oxygen or nitrogen are stable compounds (carbon monoxide); most are highly reactive units known only as reaction intermediates. Carbenes, carbonium ions, carbanions, and free radicals are the four most important classes of organic reaction intermediates containing carbon in an unstable valence state. A typical synthesis involving a carbene is that of cyclopropanes by the addition of carbenes to olefins.

carbenoid. Class of materials precipitated when certain crude oils or bitumens are dissolved in naphtha.

2-carbethoxycyclohexanone.
$OC_6H_9COOC_2H_5$.
Properties: Colorless liquid with a characteristic ester odor. Bp 106–107C (11 mm Hg), d 1.074 (25C), refr index 1.4750 (17.5C). Soluble in dilute alkali; insoluble in water. Combustible.
Use: Intermediate.

2-carbethoxycyclopentanone. (ethyl cyclopentanone-2-carboxylate; ethyl-2-oxocyclopentanecarboxylate).
$OC_5H_7COOC_2H_5$.
Properties: Colorless liquid with characteristic ester odor. Bp 122–124C (25 mm Hg), flash p 191F (88.3C), refr index 1.451 (25C), d 1.0976 (0C). Soluble in equimolar amounts of dilute alcohol; insoluble in water. Combustible.
Hazard: Vapors are toxic, as is skin contact.
Use: Pharmaceutical intermediate.

β-carbethoxyethyltriethoxysilane.
$C_2H_5OOC(CH_2)_2Si(OC_2H_5)_3$.
Properties: Colorless liquid. Bp 246C. Combustible.
Hazard: An irritant.
Use: Intermediate.

N-carbethoxypiperazine.
$C_7H_{14}N_2O_2$.

Properties: Colorless, somewhat viscous liquid; slight odor. Bp 116–117C (12 mm Hg), 237C, refr index 1.4756 (25C). Miscible with water and common organic solvents.
Use: Intermediate.

β-carbethoxypropylmethyldiethoxysilane.

$C_2H_5OOC(C_3H_6)CH_3Si(OC_2H_5)_2$.
Properties: Colorless liquid. Bp 228C. Combustible.
Hazard: An irritant.
Use: Intermediate.

carbide. A binary solid compound of carbon and another element. The most familiar carbides are those of calcium, tungsten, silicon, boron, and iron (cementite). Two factors have an important bearing on the properties of carbides: (1) the difference in electronegativity between carbon and the second element and (2) whether the second element is a transition metal. Saltlike carbides of alkali metals are obtained by reaction with acetylene. Those obtained from silver, copper, and mercury salts are explosive.
See acetylide; carbide, refractory; carbide, cemented.

carbide, cemented. A powdered form of refractory carbide united by compression with a bonding material (usually iron, nickel, or cobalt), followed by sintering. Tungsten carbide is bonded with cobalt at 1400C; from 3 to 25% of cobalt is used, depending on the properties desired. Used chiefly in metal-cutting tools that are hard enough to permit cutting speeds in rock or metal up to 100 times that is obtained with alloy steel tools.

carbide, refractory. A carbide characterized by great hardness, thermal stability, high melting point, and chemical resistance. Decomposed by fusion with alkali and attacked by mixtures of nitric and hydrofluoric acids. The best-known refractory carbides are those of silicon, boron, tungsten, and tantalum. Used as abrasives, furnace linings, and in other high-temperature applications. Some types are bonded.

carbinol. (1) Synonym for methanol, CH_3OH; (2) any compound of similar structure retaining the COH radical and in which hydrocarbon radicals may be substituted for the hydrogen originally attached to the carbon. Thus, isopropanol, $(CH_3)_2CH_2O$, and benzyl alcohol, $C_6H_5CH_2OH$, may be named dimethylcarbinol and phenylcarbinol, respectively.

carbinoxamine maleate. (2-[p-chloro-α-(2-dimethylaminoethoxy)-benzyl]pyridinemaleate).
$ClC_6H_4CH(C_5H_4N)OCH_2CH_2N(CH_3)_2·C_4H_4O_4$.
Properties: White crystalline powder; odorless; bitter. Mp 116–121C, pH (1% solution) 4.6–5.1. Very soluble in water; freely soluble in alcohol and chloroform; very slightly soluble in ether.

Grade: NF.
Use: Medicine.

"Carbitol" [Union]. TM for a group of mono- and dialkyl ethers of diethylene glycol and their derivatives, specialized solvents with a wide variety of properties and uses. Specific types are as follows:
butyl "Carbitol."
See diethylene glycol monobutyl ether.
butyl "Carbitol" acetate.
See diethylene glycol monobutyl ether acetate.
"Carbitol" acetate.
See diethylene glycol monoethyl ether acetate.
N-hexyl "Carbitol."
See diethylene glycol monohexyl ether.
methyl "Carbitol."
See diethylene glycol monomethyl ether.

carbitol acrylate. See ethylcarbitol acrylate.

carbobenzyloxy-L-alanine.
$CH_3CH(NHCO_2CH_2C_6H_5)CO_2H$.
Properties: Mw 223.23, mp 82–84C, optical rotation −14.2 degrees (23C).
Grade: 98+% pure.

Carbocaine. Proprietary name for mepivacaine hydrochloride.
Use: Local anesthetic in dentistry.

carbocalcitonin.
CAS: 60731-46-6. $C_{148}H_{244}N_{42}O_{47}$.
Hazard: A poison by ingestion.

carbocation. A positively charged ion whose charge resides at least partially on a carbon atom or group of carbon atoms.
See carbonium ion.

carbocoal. A fuel, obtained by distilling a mix of coal pitch and residue, which produces dark-gray briquettes.

carbocyclic. Any organic compound whose skeleton is in the form of a closed ring of carbon atoms. This includes both alicyclic and aromatic structures.

carbodihydrazide. (carbazide).
$CO(NHNH_2)_2$.
Properties: Colorless crystals. Mp 154C, d 1.1616 (−5C). Very soluble in water and alcohol.
Use: Organic intermediate and photographic chemical.

carbodiimide. See cyanamide (1).

-carbodithioic. Suffix for an organic acid in which sulfur replaces both oxygen atoms of the carboxyl.

carbofuran. See "Furadan" [FMC].

carbohydrase. An enzyme whose catalytic activity is directed toward the breaking down of complex carbohydrates to simpler units. Illustrations are amylase, invertase, and maltase.

carbohydrase, aspergillus.
Properties: From fermentation of *Aspergillus oryzae* var. Tan amorphous powder or liquid. Sol in water.
Use: Food additive.

carbohydrase and cellulase.
Properties: Derived from *Aspergillus niger.*
Use: Food additive.

carbohydrase and protease, mixed.
Properties: From controlled fermentation of *Bacillus licheniformis* var. Brown amorphous powders or liquid. Sol in water; insol in alc, chloroform, ether.
Use: Food additive.

carbohydrase, rhizopus.
Properties: Derived from *Rhizopus oryzae.*
Use: Food additive.

carbohydrate.
A compound of carbon, hydrogen, and oxygen that contains the saccharose unit or its first reaction product and in which the ratio

$$H-\underset{\underset{OH}{|}}{C}-\underset{\underset{O}{||}}{C}-$$

of hydrogen to oxygen is the same as in water. Carbohydrates are the most abundant class of organic compounds, constituting three-fourths of the dry weight of all vegetation. They are also widely distributed in animals and lower forms of life. They comprise: (1) monosaccharides, simple sugars such as fructose (levulose) and its isomer glucose (dextrose), both having the formula $C_6H_{12}O_6$; (2) disaccharides, sucrose ($C_{12}H_{22}O_{11}$), maltose, cellobiose, and lactose; and (3) polysaccharides (high polymeric substances). The last group includes all starch and cellulose families, as well as pectin, the seaweed products agar and carrageenan, and natural gums. The simple sugars are crystalline and water soluble, with a sweet taste; starches are water soluble, tasteless, and amorphous; cellulose is insoluble in water and organic solvents and is only partially crystalline. Galactose, sorbose, xylose, arabinose, and mannose are constituents of more complex sugars. The natural gums are water soluble plant products composed of monosaccharide units joined by glycosidic bonds (arabic, tragacanth).
Carbohydrates are an important natural source of ethanol now in extensive use in gasohol and other energy applications.
See energy sources; gasohol; fermentation.

carbohydrate binding protein. (CBP).
Proteins that mediate cellular processes that are central to immune regulation and disease.

carbohydrate metabolism. The catabolic and anabolic processes that involve carbohydrates. Long, complicated, reaction sequences and cycles are components of this process. Carbohydrate biosynthesis also takes place, resulting in the production of storage glycogen. Toxicants can interfere with carbohydrate catabolism and biosynthesis.

Carbohydrogen. See oil gas.

carboid. An asphaltic material insoluble in trichlorethylene.

Carbolac. High-color channel blacks, used for paint, varnish, and lacquer.

carbolated camphor. See phenol, camphorated.

carbolfuchsin. (Ziehl's stain).
A staining solution of fuchsin in alcohol and aqueous phenol used in the study of microorganisms.

carbolic acid. Legal label name for phenol.

carbolic oil. (middle oil).
The fraction, having a boiling range of about 190–250C, obtained from distillation of coal tar and containing naphthalene, phenol, and cresols.

β-carboline. (9H-pyrido[3,4-b]indole).
CAS: 244-63-3. $C_{11}H_8N_2$.
Any of a number of derivatives of tryptoline or 1,2,3,4-tetrahydro-β-carboline produced by the Pictet–Spengler isoquinoline synthesis.
Derivation: Occur in foods.
Hazard: Toxic; nervous system effects.

carbomer. A polymer composed of acrylic acid cross-linked with allyl sucrose.

carbometer. Device for measuring the quality of carbon in steel by determining the magnetic characteristics of a standard-size sample as cast at the furnace.

carbomethoxy malathion.
CAS: 3700-89-8. $C_8H_{15}O_6PS_2$.
Hazard: Moderately toxic by ingestion. Human systemic effects. A severe skin and eye irritant.

3-carbomethoxy-1-methyl-4-piperidone hydrochloride.
$C_8H_{13}NO_3 \cdot HCl.$
Properties: White, crystalline solid. Mp 165C. Soluble in water, alcohols; insoluble in ether, hydrocarbons.
Use: Pharmaceutical intermediate.

2-carbomethoxy-1-methylvinyl dimethyl phosphate. The β-isomer is also α pesticide. See mevinphos for the α-isomer.

carbomycin.
CAS: 4564-87-8. $C_{42}H_{67}NO_{16}$.
An antibiotic isolated from products of *Streptomyces halstedii* when grown in suitable media by the deep-culture method. It inhibits growth of certain gram-positive bacteria such as staphylococci, pneumococci, and hemolytic streptococci. Mp 214C.

carbon.
CAS: 7440-44-0. C.
A nonmetallic element, atomic number 6, aw 12.011, group IVA of the periodic table, normal valence 4, but divalent forms are known (carbenes). Carbon has two stable and four radioactive isotopes. The ^{12}C isotope, which comprises 99% of the element, is the standard to which atomic weights of all other elements are referred (i.e., $^{12}C = 12.00$ exactly). One mole of carbon atoms (6.02×10^{23}) is contained in 12 g of ^{12}C. Carbon has two crystalline allotropes (diamond and graphite) and several amorphous allotropes (coal, coke, carbon black, charcoal). Carbon is present in all organic and in a few inorganic compounds (carbon oxides, carbon disulfide, and metallic carbonates such as calcium carbonate). It is the active element in photosynthesis and thus occurs in all plant and animal life. The radioisotope ^{14}C is used in tracer research and chemical dating. Carbon is a strong reducing agent and is used as such in purifying metals. It is one of the few elements capable of forming four covalent bonds. Its strong electrical conductivity is used to advantage in electrodes and other electrical devices. Its presence in small proportions in steel has a pronounced effect on the properties of the metal.
Carbon forms binary compounds called carbides with many metals and some nonmetals. A few compounds are known that contain divalent carbon (carbenes or methylenes).
Since its major properties and uses vary widely with its form, the following entries should be consulted: diamond, graphite, activated carbon, carbon black, industrial carbon, charcoal, wood, coke, steel, carbon cycle.

carbon-11. (^{11}C).
A short-lived radioactive isotope of carbon that emits positrons, which in turn become a source of γ-rays when they collide with an electron within the body. It is used experimentally in nuclear medicine for labeling pharmaceuticals.

carbon-12. (methane)
CAS: 7440-44-0. CH_4.
The standard of atomic mass comprising 98.89% of natural carbon.

carbon-13. (^{13}C).
A stable, nonradioactive carbon isotope used for special analytical research. Commercially available in gram quantities.

carbon-14. (^{14}C; radiocarbon).
Naturally occurring, radioactive carbon isotope of mass number 14, a special case of radioactivity induced by cosmic rays in the upper atmosphere. Neutrons produced by cosmic radiation impact nitrogen atoms to yield ^{14}C and a proton. Half-life 5580 years; β-radiation. Can be made by reactor irradiation of calcium nitrate.
Use: Radiation source in thickness gauges and other instruments, elucidation of mechanisms in organic chemistry, metallurgy, and biochemical reactions, radiocarbon dating in geology and archaeometry.
See chemical dating; archaeometry.

Carbona. An obsolete proprietary mixture of petroleum ether (bp 70–72C) and carbon tetrachloride.
Use: Cleaning fluid.

carbon, activated. (active carbon; activated charcoal). An amorphous form of carbon characterized by high adsorptivity for many gases, vapors, and colloidal solids. The carbon is obtained by the destructive distillation of wood, nutshells, animal bones, or other carbonaceous material. It is "activated" by heating to 800–900C with steam or carbon dioxide, which results in a porous internal structure (honeycomb-like). The internal surface area of activated carbon averages approximately 10,000 square feet per gram. The density is from 0.08 to 0.5. It is not effective in removing ethylene.
Grade: Technical, USP, as activated charcoal.
Hazard: Flammable. Toxic by inhalation of dust.
Use: Decolorizing of sugar, water and air purification, solvent recovery, waste treatment, removal of sulfur dioxide from stack gases and "clean" rooms, deodorant, removal of jet fumes from airports, catalyst for natural-gas purification, brewing, chromium electroplating, air-conditioning.

carbonado. See diamond, industrial.

-carbonal. A suffix that indicates the presence of the –CHO group.

carbon, amorphous. See carbon, activated; carbon black.

carbonate. A compound resulting from the reaction of either a metal or an organic compound with carbonic acid. The reaction with a metal yields a salt (calcium carbonate) and that with an aliphatic or aromatic compound forms an ester, e.g., diethyl carbonate, diphenyl carbonate. The latter are liquids used as solvents and in synthesizing polycarbonate resins.
See carbonic acid.

carbonate hardness of water.   Hardness as a result of the introduction of carbonate, bicarbonate of calcium, and magnesium.

carbonate mineral.   A mineral in which the basic building block is a carbon atom linked to three oxygen atoms. Calcite, aragonite, and dolomite are the most abundant examples found in sediments and sedimentary rocks.

carbonate rock.   A sedimentary rock that consists primarily of carbonate minerals. The dominant mineral is nearly always either calcite, in which case the rock is limestone, or dolomite, in which case the rock is dolomite.

carbonate sediment.   Unconsolidated sediment that consists primarily of carbonate minerals, usually aragonite or calcite.

carbonation.   (1) Saturation of a liquid with carbon dioxide. (2) Production of a carbonate by means of carbon dioxide. (3) Removal of excess lime in sugar refining by carbon dioxide.

carbon atom.   (1) Primary: a carbon atom having not more than one other carbon directly attached to it. (2) Secondary: a carbon atom having two other carbon atoms directly attached to it. (3) Tertiary: a carbon atom that is joined to three other carbon atoms. (4) Quaternary: a carbon atom having four other carbon atoms directly attached to it.

carbon bisulfide.   See carbon disulfide.

carbon black.
CAS: 1333-86-4.
A finely divided form of carbon, practically all of which is made by burning vaporized heavy-oil fractions in a furnace with 50% of the air required for complete combustion (partial oxidation). This type is also called furnace black. Carbon black can also be made from methane or natural gas by cracking (thermal black) or direct combustion (channel black), but these methods are virtually obsolete. All types are characterized by extremely fine particle size, which accounts for their reinforcing and pigmenting effectiveness.
Grade: (Furnace black) conducting (CF), fine (FF), high modulus (HMF), high elongation (HEF), reinforcing (RF), semireinforcing (SRF), high abrasion (HAF), super abrasion (SAF), fast extruding (FEF), general purpose (GPF), intermediate super abrasion (ISAF), channel replacement (CRF), easy-processing furnace black (EPF).
Hazard: Possible carcinogen. Bronchitis.
Use: Tire treads, belt covers, and other abrasion-resistant rubber products; plastics as a reinforcing agent, opacifier, electrical conductor, UV-light absorber; colorant for printing inks; carbon

paper; typewriter ribbons; paint pigment; nucleating agent in weather modification; expanders in battery plates; solar-energy absorber (see note).
Note: A suspension of finely divided carbon particles in compressed air has been researched as a solar-energy absorber. The heat is absorbed until the particles vaporize, yielding energy that can be used directly or for power production. The suspension is placed in a transparent container located in a solar concentrator.

carbon-black oil.   A heavy refinery fraction similar to fuel oil, used as a feedstock for furnace black.

carbon, combined.   A metallurgical term for carbon that has combined chemically with iron to form cementite, as distinct from graphitic carbon in iron or steel.
See ferrite.

carbon cycle.   (1) The progress of carbon from air (carbon dioxide) to plants by photosynthesis (sugar and starches), then through the metabolism of animals to decomposition products that ultimately return it to the atmosphere in the form of carbon dioxide. (2) One of the processes by which the sun and other self-luminous astronomical bodies are thought to derive their energy. The net process is the combination (fusion) of four hydrogen atoms to form helium. The carbon cycle involves successive additions of hydrogen atoms, followed by β decay, to an initial carbon-12 atom until a final step is reached in which the new nucleus breaks down to a helium atom and regenerated carbon-12. The carbon thus functions as a catalyst for the process. At the temperatures prevailing in the sun, all atoms are stripped of their electrons and the reaction is between the nuclei of the atoms (thermonuclear reaction). Symbolically the set of reactions is written
$$^{12}C + {}^1H \rightarrow {}^{13}N, {}^{13}N \rightarrow {}^{13}C + e;$$
$$^{13}C + {}^1H \rightarrow {}^{14}N, {}^{14}N + {}^1H \rightarrow {}^{15}O.$$
$$^{15}O \rightarrow {}^{15}N + e, {}^{15}N \rightarrow {}^{12}C + {}^4He.$$
See fusion.

carbon dating.   Radiocarbon dating is a method of determining quite accurately the age of a carbon-bearing material derived from living plants or animals within the last 70,000 years. It is based on determining the ratio of carbon-14 in the material to that in a modern reference sample by measuring the radioactivity of the carbon-14 in the material. Since the half-life of carbon-14 is 5730 ±30 years and the living precursor utilized carbon dioxide from the atmosphere or some other part of the earth's dynamic carbon reservoir, a process that ceased when the original plant or animal died, the amount of carbon-14 now present gives directly the age of the material. The carbon-14 in the reservoir is constantly being replaced by the sequence $^{14}N \rightarrow {}^{14}C + O \rightarrow {}^{14}CO_2$. This has maintained the constant ratio of carbon isomers during the ages; however, burning of fossil fuels since the Industrial Revolution has lowered somewhat the fraction of carbon-14 in the atmosphere during the last few centuries, which does not affect measurements on older objects. The sample to be tested must be carefully prepared to prevent contamination by younger carbon.

The radiocarbon technique was discovered by Willard F. Libby (1908–1980), who won a Nobel Prize (1960), and has been applied with great success in the fields of archaeology, geology, geochemistry, and geophysics. Its accuracy has been checked and verified by use of tree-ring counts (dendrochronology) and with the known ages of objects from ancient cultures, such as Egyptian and Chinese. The former shows that for the 2400–6000-year age of bristlecone-pine tree rings, 5200 ^{14}C years equal 6000 calendar years.

carbon dichloride.   See perchloroethylene.

carbon dioxide.
CAS: 124-38-9. CO_2.
Properties: (1) Gas: colorless; odorless. D 1.97 g/L (0C, 1 atm), d 1.53 (air = 1.00). (2) Liquid: volatile, colorless; odorless. D 1.101 (−37C), sp volume 8.76 cu ft/lb (70F). (3) Solid (dry ice): white, snowlike flakes or cubes. D 1.56 (−79C), mp −78.5C (sublimes). All forms are noncombustible. Miscible with water (1.7 volumes per volume at 0C and 0.76 volume per volume at 25C and 760 mm Hg partial pressure of CO_2). Also miscible with hydrocarbons and most organic liquids. An asphyxiant gas in concentrations of 10% or more; low concentrations (1–3%) increase lung ventilation and are used admixed with oxygen in resuscitation equipment.
Derivation: (1) Gas: for industrial use, carbon dioxide is recovered from synthesis gas in ammonia production, substitute-natural gas production, cracking of hydrocarbons, and natural springs or wells. For laboratory purposes, it is obtained by the action of an acid on a carbonate. It is also a by-product of the fermentation of carbohydrates and an end product of combustion and respiration. Air contains 0.033% of carbon dioxide (see greenhouse effect). (2) Liquid: by compressing and cooling the gas to approximately −37C. (3) Solid (dry ice): by expanding the liquid to vapor and snow in presses that compact the product into blocks. The vapor is recycled.
Grade: Technical, USP, commercial and welding, 99.5%, bone dry (99.95%).
Hazard: Solid damaging to skin and tissue; keep away from mouth and eyes. Asphyxia.
Use: Refrigeration, carbonated beverages, aerosol propellant, chemical intermediate (carbonates, synthetic fibers, p-xylene, etc.), low-temperature testing, fire extinguishing, inert atmospheres, municipal water treatment, medicine, enrichment of air in greenhouses, fracturing and acidizing of oil wells, mining (Cardox method), miscible pressure source, hardening of foundry molds and cores, shielding gas for welding, cloud seeding, moderator in some types of nuclear reactors, immobilization for humane animal killing, special lasers, blowing agent, as demulsifier in tertiary oil recovery, possible source of methane, (liquid) carrier for powdered-coal slurry.
Note: Carbon dioxide is the source of the carbon utilized by plants to form organic compounds in the photosynthetic reaction catalyzed by chlorophyll.
See photosynthesis; carbon cycle (1).

carbon disulfide.   (carbon bisulfide).
CAS: 75-15-0. CS_2.
Properties: Clear, colorless, or faintly yellow liquid; almost odorless when pure; usually strong, disagreeable odor. D 1.260 at 25/25C, bp 46.3C, fp −111C, wt/gal 10.48 lb (25C), refr index 1.6232 (25C), flash p −22F (−30C), autoign temp 212F (100C). Soluble in alcohol, benzene, and ether; slightly soluble in water. Classed as an inorganic compound.
Derivation: (1) Reaction of natural gas or petroleum fractions with sulfur. (2) From natural gas and hydrogen sulfide at very high temperature (plasma process). (3) By heating sulfur and charcoal and condensing the carbon disulfide vapors.
Method of purification: Distillation.
Impurities: Sulfur compounds.
Grade: 99.9%, spectrophotometric.
Hazard: A poison. Toxic by skin absorption. Highly flammable, dangerous fire and explosion risk, can be ignited by friction. Explosive limits in air 1–50%. Questionable carcinogen. Peripheral nervous system impairment.
Use: Viscose rayon, cellophane, manufacture of carbon tetrachloride and flotation agents, solvent.

carbon, divalent.   See carbene.

Carbone primary cell.   This device uses a porous, cup-shaped carbon element as the positive electrode and a zinc ring as the negative. The electrolyte is a caustic-soda solution with a special mineral oil floating on its surface. The oil prevents evaporation of the electrolyte.

carbon fiber.   See graphite fiber.

carbon fixation reactions.   (dark reactions).
The light-independent enzymatic reactions involved in the synthesis of glucose from CO_2, ATP, and NADPH.

carbon fluoride.
$(CF)_x$, C_4F.
A solid, nonconductive material formed on carbon anodes during electrolysis of molten potassium fluoride-hydrogen fluoride mixtures to yield elemental fluoride. C_4F is unstable above 60C, $(CF)_x$ forms only at high temperatures.

carbon, graphitic.   A metallurgical term referring to practically pure carbon that forms in pig iron during cooling because the absorbing power of iron for carbon decreases as its temperature falls. It exists in the form of tiny flakes distributed throughout the mass. The tendency of graphitic carbon is to weaken the metal, whereas combined carbon, up to the limit of approximately 0.90%, strengthens it. See pearlite; cementite; ferrite.

carbon hexachloride. See hexachloroethane.

carbonic acid. H_2CO_3.

$$\begin{array}{c} HO \\ \diagdown \\ C=O \\ \diagup \\ HO \end{array}$$

A weak acid formed by reaction of carbon dioxide with water. Both organic and inorganic carbonates are formed from it by reaction with organic compounds or metals, respectively. Thus inorganic carbonates ($CaCO_3$, K_2CO_3, Na_2CO_3, etc.) are salts of carbonic acid, and organic carbonates are esters of carbonic acid.

carbonic anhydrase. An enzyme in red blood cells that catalyzes the production of carbon dioxide and water from carbonic acid.
Use: Biochemical research.

carbon, industrial. Any form of pure carbon used for industrial purposes, exclusive of fuel. Coke is one of the most important. Besides its use (combined with coal tar pitch) for refractories, furnace linings, electrodes, fibers, etc., it has tremendous volume consumption for reduction of iron in blast furnaces (see coke). Graphite in its many applications is another form; activated carbon for decolorizing and solvent recovery, carbon black for rubber and printing inks, industrial diamonds as abrasives and drilling bits, compressed carbon for electrodes and other electrical uses, and carbon fibers and whiskers are all included in this term.

carbonium ion. A positively charged organic ion, such as H_3C^+, H_2RC^+, R_3C^+, etc., having one less electron than the corresponding free radical and acting in subsequent chemical reactions as though the positive charge were localized on the carbon atom. Such ions can exist only when corresponding negative ions are also present. An electron-deficient carbon atom is extremely reactive and has only a transitory existence in most cases, but many organic rearrangement and replacement reactions are effectively explained in terms of a carbonium-ion intermediate, including acid-catalyzed polymerization of propylene and other olefins. In this case, propylene and hydrogen ion form a carbonium ion as follows:

$$H_3C-HC=CH_2 + H^+ \rightarrow H_3C-HC^+-CH_3$$

The latter then combines with another molecule of $HC_3-HC=CH_2$ to start chain growth.

The difference between a carbonium ion, a free radical, and a carbanion may be illustrated as follows:

$$\begin{bmatrix} R \\ R:\ddot{C} \\ \ddot{R} \end{bmatrix}^+ \quad \begin{bmatrix} R \\ R:\ddot{C}\cdot \\ \ddot{R} \end{bmatrix}^0 \quad \begin{bmatrix} R \\ R:\ddot{C}: \\ \ddot{R} \end{bmatrix}^-$$

Carbonium   Free   Carbanion
ion   radical

carbonium salt. Alkyl halide in which three hydrogen atoms in the methyl group have been replaced by aryl radicals. It acts as an electrolyte when dissolved.

carbonization. See destructive distillation.

carbon monoxide.
CAS: 630-08-0. CO.
Discovered by Priestly in America in 1799.
Properties: Colorless gas or liquid; practically odorless. Burns with a violet flame. D 0.96716 (air = 1.0), bp −190C, fp −207C, sp vol 13.8 cu ft/lb (21.1C), autoign temp (liquid) 1128F (609C). Slightly soluble in water; soluble in alcohol and benzene. Classed as an inorganic compound.
Derivation: (1) Made almost pure by placing a mixture of oxygen and carbon dioxide in contact with incandescent graphite, coke, or anthracite. (2) Action of steam on hot coke or coal (water gas) or on natural gas (synthesis gas). In the latter case, carbon dioxide is removed by absorption in amine solution, and the hydrogen and carbon monoxide separated in a low-temperature unit. (3) By-product in chemical reactions. (4) Combustion of organic compound with limited amount of oxygen, as in automobile cylinders. (5) Dehydration of formic acid.
Grade: Commercial (98%), CP (99.5%).
Hazard: Highly flammable, dangerous fire and explosion risk. Flammable limits in air 12–75% by volume. Toxic by inhalation. *Note:* Carbon monoxide has an affinity for blood hemoglobin over 200 times that of oxygen. A major air pollutant.
Use: Organic synthesis (methanol, ethylene, isocyanates, aldehydes, acrylates, phosgene), fuels (gaseous), metallurgy (special steels, reducing oxides, nickel refining), zinc white pigments.

carbon nanotubes. Carbon atoms formed into extended hollow tubes instead of closed, hollow spheres as in fullerenes. They also form as a series of nested, concentric tubes. Can be used as nanometer-scale syringe needles for injecting molecules into cells and as nanoscale probes for making fine-scale measurements.
Use: Drug delivery, chemical reactors, electronic devices.
See nanophase carbon materials.

carbon nitride.
(C_2N_2).
Properties: Mw 22.02, d 0.87, mp −34.4, bp −20.5.

carbon oxide. (formaldehyde).
CAS: 50-00-0. CH_2O.
A highly reactive aldehyde gas formed by the oxidation or incomplete combustion of hydrocarbons.
Hazard: Toxic.
Use: In the manufacture of resins and textiles, as a disinfectant, as a laboratory fixative or preservative.

carbon oxybromide.   See bromophosgene.

carbon oxychloride.   See phosgene.

carbon oxycyanide.   See carbonyl cyanide.

carbon oxyfluoride.   See carbonyl fluoride.

carbon oxysulfide.   See carbonyl sulfide.

carbon ratio.   (1) Fixed carbon percentage coal. (2) Ratio of isotopes ^{12}C to ^{13}C.

carbon steel.   See steel.

carbon suboxide. C_3O_2. Molecular structure: $O=C=C=C=O$.
Properties: Colorless gas or liquid; strong, pungent odor. Bp 7C, fp −110C, refr index 1.45, d 1.12. Forms malonic acid with water. Polymerizes on storage, even under pressure of 7 atm.
Derivation: From malonic acid by destructive distillation.
Hazard: Explosive limits in air 6–30%. Strong irritant to eyes and mucous membranes, cause lachrymation and impaired breathing.
Use: Dyeing auxiliary, chemical intermediate.

carbon tetrabromide.   (tetrabromomethane).
CAS: 558-13-4. CBr_4.
A brominated hydrocarbon.
Properties: Colorless crystals. D 3.42, mp 90.1C, bp 189.5C. Insoluble in water; soluble in alcohol, ether, and chloroform. Noncombustible.
Hazard: A poison; narcotic in high concentration. Liver damage, eye, skin, and upper respiratory tract irritant.
Use: Organic synthesis.

carbon   tetrachloride.   (tetrachloromethane; perchloromethane).
CAS: 56-23-5. CCl_4.
A chlorinated hydrocarbon.
Properties: Colorless liquid; sweetish, distinctive odor. Vapor 5.3 times heavier than air, d 1.585 (25/4C), bp 76.74C, fp −23.0C, refr index 1.4607 (20C), vap press 91.3 mm Hg (20C), wt/gal 13.22 lb (25C), flash p none. Miscible with alcohol, ether, chloroform, benzene, solvent naphtha, and most of the fixed and volatile oils; insoluble in water. Noncombustible.
Derivation: (1) Interaction of carbon disulfide and chlorine in the presence of iron; (2) chlorination of methane or higher hydrocarbons at 250–400C.

Method of purification: Treatment with caustic alkali solution to remove sulfur chloride, followed by rectification.
Grade: Technical, CP, electronic.
Hazard: Toxic by ingestion, inhalation, and skin absorption. Do not use to extinguish fire. Narcotic. A possible carcinogen. Liver damage. Decomposes to phosgene at high temperatures.
Use: Refrigerants. Metal degreasing, agricultural fumigant, chlorinating organic compounds, production of semiconductors, solvent (fats, oils, rubber, etc.).
Note: Not permitted in products intended for home use.

carbon tetrafluoride.   See tetrafluoromethane.

carbon trichloride.   See hexachloroethane.

carbonyl.   A compound of carbon monoxide with a metal, as in $Co(CO)_3$.

N,N'-carbonylbis(4-methoxymetanilic acid) disodium salt.   (sodium methoxymetanilate urea).
$[C_6H_3(OCH_3)(SO_3Na)NH]_2CO$.
Properties: Gray paste, solids approximately 70%.
Grade: Technical.
Use: Intermediate.

carbonyl bromide.   See bromophosgene.

carbonyl chloride.   See phosgene.

carbonyl cyanide.   (carbon oxycyanide).
$CO(CN)_2$.
Properties: Colorless liquid. Unstable in the presence of water. Bp −83C, d 1.139 (−114C), mp 114C.
Hazard: Toxic.
Use: Organic synthesis.

1,1′-carbonyldiimidazole.   (N,N-carbonyldiimidazole; 1,1′-carbonylbis-1H-imidazole).
$C_7H_6N_4O$.
Should be handled in absence of atmospheric moisture to avoid release of carbon dioxide.
Properties: Off-white powder or crystals. Mw 162.15, mp 118–120C.
Use: Enzyme cross-linking agent, condensing agent for nucleoside triphosphate synthesis.

carbonyl fluoride.   (fluoroformyl fluoride; carbon oxyfluoride).
CAS: 353-50-4. COF_2.
Properties: Colorless, hygroscopic gas. Unstable in the presence of water. Bp −83C, d 1.139 (−114C), fp −114C. Min purity 97 mole %. Nonflammable.
Derivation: Action of silver fluoride on carbon monoxide.
Grade: Technical.

Hazard: Toxic by inhalation, strong irritant to skin. Lower respiratory tract irritant. Bone damage.
Use: Organic synthesis.

carbonyl group. The divalent group =C=O, which occurs in a wide range of chemical compounds. It is present in aldehydes, ketones, organic acids, and sugars and in the carboxyl group, i.e.,

$$-C\overset{\displaystyle O}{\underset{\displaystyle OH}{}}$$

In combination with transition metals, it forms coordination compounds that are highly toxic, because they decompose to release carbon monoxide when absorbed by the body, e.g., nickel carbonyl. Several metal carbonyls have antiknock properties. The carbonyl group is also found in combination with nonmetals, as in phosgene (carbonyl chloride); these compounds are also poisonous.

carbonyl sulfide. (carbon oxysulfide).
CAS: 463-58-1. COS.
Properties: Colorless gas with typical sulfide odor except when pure. D gas 2.1 (air = 1), fp −138.8C, bp −50.2C (1 atm). Soluble in water and alcohol.
Derivation: Hydrolysis of ammonium or potassium thiocyanate.
Hazard: Narcotic in high concentrations. Flammable, explosive limits in air 12–28.5%. Central nervous system impairment.

carbophenothion. (generic name for S-[{(p-chlorophenyl)thio}methyl]-O,O-diethyl phosphorodithioate; O,O-diethyl-S-(p-chlorophenylthiomethyl) phosphorodithioate).
$(C_2H_5O)_2P(S)SCH_2S(C_6H_4)Cl$.
Properties: Amber liquid. Bp 82C (0.1 mm Hg), d 1.29 (20C). Essentially insoluble in water; miscible with common solvents.
Hazard: Use may be restricted. A cholinesterase inhibitor.
Use: Insecticide, acaricide.

carboquone. (2[2-[(aminocarbonyl)oxy]-1-methoxyethyl]-3,6-bis(1-aziridinyl)-5-methyl-2,5-cyclohexadiene-1,4-dione; 2,5-bis(1-aziidinyl)-3-(2-hydroxy-1-methxyethyl)-6-methyl-p-benzoquinone carbamate (ester); 2,5-bis(1-aziridinyl)-3-(2-carbamoyloxy-1-methoxy-ethyl)-6-methyl-1,4-benzoquinone; carbazilquinone; [2-[2,5-bis(azinidin-1-yl)-4-methyl-3,6-dioxocyclohexa-1,4-dien-1-yl]-2-methoxyethyl]carbamate).
$C_{15}H_{19}N_3O_5$.
An alkylating agent structurally similar to mitomycin and found to be effective in the treatment of leukemia and various other neoplasms in mice.
Properties: Red to reddish-brown crystalline compound, nearly insoluble in water, slightly soluble in most organic solvents.

Hazard: Extremely toxic; causes leukemia and thrombocytopenia in almost all human patients.
Use: Antineoplastic alkylating agent.

carborane. A crystalline compound composed of boron, carbon, and hydrogen. It can be synthesized in various ways, chiefly by the reaction of a borane (penta- or deca-) with acetylene, either at high temperature in the gas phase or in the presence of a Lewis base. Alkylated derivatives have been prepared. Carboranes have different structural and chemical characteristics and should not be confused with hydrocarbon derivatives of boron hydrides. The predominant structures are the cage type, the nest type, and the web type, these terms being descriptive of the arrangement of atoms in the crystals. Active research on carborane chemistry has been conducted under sponsorship of the U.S. Office of Naval Research.

m-carboranedimethanol.
CAS: 23924-78-9. $C_4H_{16}B_{10}O_2$.
Hazard: Low toxicity by ingestion. A moderate eye irritant.

o-carboranedimethanol.
CAS: 19610-37-8. $C_4H_{16}B_{10}O_2$.
Hazard: Moderately toxic by ingestion and inhalation. A moderate eye irritant.

Carborundum. Abrasives and refractories of silicon carbide, fused alumina, and other materials.
Properties: For silicon carbide, crystalline form ranges from small to massive crystals in the hexagonal system, the crystals varying from transparent to opaque, with colors from pale green to deep blue or black; hardness Mohs 9.17; d 3.06–3.20. Noncombustible, not affected by acids, slowly oxidizes at temperatures above 1000C, good heat dissipator, highly refractory.
Use: Abrasive grains and powders for cutting, grinding, and polishing; valve-grinding compounds; grinding wheels; coated abrasive products; antislip tiles and treads; refractory grains.

carbosand. Fine sand that has been treated with an organic solution and roasted to produce a material that can be sprayed onto oil slicks to aid in sinking or dispersing them.

"Carboset" [Lubrizol]. TM for water-soluble acrylic thermoset and thermoplastic products.
Use: Protective metal coatings, paints, ceramics, adhesives, textiles, paper, leather, and cosmetics.

-carbothioic. A suffix denoting an organic acid in which an atom of sulfur replaces an atom of oxygen.

-carbothionic. A suffix of organic acids in which the oxygen of the CO group has been replaced by sulfur.

"Carbowax" *[Union].* TM for polyethylene glycols and methoxypolyethylene glycols.

Grade: Available in various numbered grades, i.e., 200, 400, 1000, 4000, 6000. Usually designated by approximate molecular weight of polymer.

Use: Water-soluble lubricants; solvents for dyes, resins, proteins; plasticizers for casein and gelatin compositions, glues, zein, cork, and special printing inks; solvent and ointment bases for cosmetics and pharmaceuticals; intermediates for nonionic surfactants and alkyd resins.

carboxamide hydrochloride.
CAS: 158681-13-1. $C_{22}H_{21}Cl_3N_4O \cdot ClH$.
Hazard: A poison.

carboxamidoacetamide. See malonamide.

carboxybenzene. See benzoic acid.

1-carboxyethane-2-phosphonic acid. See 3-phosphonopropanoic acid.

2-carboxyethyl acrylate. See hydracrylic acid, acrylate.

6-(2-carboxyethylthio)purine.
CAS: 608-10-6. $C_8H_8N_4O_2S$.
Hazard: A poison.

carboxyhemoglobin. (HbCO; carboxyhemoglobin; carbon monoxide hemoglobin).
Carbon monoxide-contain hemoglobin. It is a much more stable complex than oxyhemoglobin. It is formed in the presence of carbon monoxide, which binds to hemoglobin much more readily than oxygen. The affinity of hemoglobin for carbon monoxide is 200–300 times greater than that of oxygen. In carbon monoxide poisoning, this gas displaces much of the oxygen that is complexed to hemoglobin in a process that is nearly irreversibly forming carboxyhemoglobin, thereby preventing the exchange of carbon dioxide released from the cells and oxygen during the circulation of blood, with increasingly severe asphyxiation and eventually death.
Hazard: Asphyxiation, death.

2-carboxy-2′-hydroxy-5′-sulfoformazylbenzene. (*o*-[α-(2-hydroxy-5-sulfophenyl)-azobenzylidene]-hydrazino benzoic acid).
$HO_3SC_6H_3(OH)N:NC(C_6H_5):NNC_6H_4COOH$.
Use: Reagent used for the colorimetric determination of zinc and copper.

carboxylase. A decarboxylase enzyme, found in plant tissues that act on pyruvic acid, producing acetaldehyde and carbon dioxide.
Use: Biochemical research.

carboxyl group. The chemical group characteristic of carboxylic acids that include fatty acids and amino acids. It usually occupies the terminal position in the molecule and is capable of assuming a negative charge, which makes the end of the molecule water soluble. Though it is customarily shown as either COOH or CO_2H, the structure of the group is

$$-C{\overset{\displaystyle O}{\underset{\displaystyle OH}{}}}$$

Thus it is composed of a carbonyl group and a hydroxyl group bonded to a carbon atom. The carbon–oxygen unsaturation within the carboxyl group is of a different order from the carbon-to-carbon unsaturation in the alkyl chain in unsaturated fatty acids. For this reason, fatty acids in which no double bond is present except that in the carboxyl group are called saturated.
See fatty acid.

carboxylic acid. Any of a broad array of organic acids comprised chiefly of alkyl (hydrocarbon) groups (CH_2, CH_3), usually in a straight chain (aliphatic), terminating in a carboxyl group (COOH). Exceptions to this structure are formic acid (HCOOH) and oxalic acid (HOOCCOOH). The number of carbon atoms ranges from 1 (formic) to 26 (cerotic), the carbon of the terminal group being counted as part of the chain. Carboxylic acids include the large and important class of fatty acids and may be either saturated or unsaturated. A few contain halogen atoms (chloroacetic). There are also some natural aromatic carboxylic acids (benzoic, salicylic), as well as alicyclic types (abietic, chaulmoogric).
See amino acid.

carboxylic acid hydrochloride.
CAS: 164150-85-0. $C_{20}H_{18}F_2N_4O_3 \cdot ClH$.
Hazard: A reproductive hazard.

carboxylic esterase. (carboxylesterase; carboxylic ester hydrolase).
CAS: 9016-18-6.
An esterase that catalyzes the hydrolysis of esters of carboxylic acids.

carboxyltransferase. (transcarboxylases).
Any enzyme that catalyzes the transfer of carboxyl groups from one compound to another.

carboxymethoxylamine hemihydrochloride.
$(H_2NOCH_2CO_2H)_2 \cdot HCl$.
Demonstrates anticonvulsant activity by inhibiting glutamic acid decarboxylase and γ-aminobutyric-α-ketoglutaric aminotransaminase, increasing the brain γ-aminobutyric acid concentrations. Increases the sugar content of sugarcane, sugar beets, and sorghum. Forms digitoxin derivatives.

Properties: Off-white crystals. Mw 218.59, mp 156C (decomposes), hygroscopic.
Grade: 98% research.

carboxymethylcellulose. (acetic acid; 2,3,4,5,6-pentahydroxyhexanal; sodium).
CAS: 9004-32-4. $C_8H_{16}NaO_8$.
A cellulose derivative which is a beta-(1,4)-D-glucopyranose polymer.
Use: Bulk laxative; a stabilizer for reagents; and an emulsifier and thickener in cosmetics and pharmaceuticals.

carboxymethylcellulose. (CMC; sodium carboxymethylcellulose; CM cellulose).
CAS: 9004-32-4.
A semisynthetic, water-soluble polymer in which CH_2COOH groups are substituted on the glucose units of the cellulose chain through an ether linkage. Mw ranges from 21,000 to 500,000. Since the reaction occurs in an alkaline medium, the product is the sodium salt of the carboxylic acid R-O-CH_2COONa.
Properties: Colorless, odorless, nontoxic, water-soluble powder or granules. D 1.59, refr index 1.51, tensile strength 8000–15,000 psi, pH (1% solution) 6.5–8.0, stable in pH range 2–10. Viscosity of 1% solution varies from 5 to 2000 cP, depending on the extent of etherification. Insoluble in organic liquids. Reacts with heavy-metal salts to form films that are insoluble in water, transparent, relatively tough, and unaffected by organic materials. Many of its colloidal properties are superior to those of natural hydrophilic colloids. It also has thixotropic properties and functions as a polyelectrolyte.
Derivation: By reaction of alkali cellulose and sodium chloroacetate.
Grade: Crude, technical (approximately 75% pure), high viscosity, low viscosity, semirefined, refined (99.5+% pure), USP, FCC.
Use: Detergents; soaps; food products (dietetic foods and ice cream), where it acts as water binder, thickener, suspending agent, and emulsion stabilizer; textile manufacturing (sizing); coating paper and paperboard to lower porosity; drilling muds; emulsion paints; protective colloid; pharmaceuticals; cosmetics.
See cellulose, modified.

carboxymethyl cellulose sodium. (carboxymethyl; methyl ether cellulose sodium salt; CMC; sodium carboxymethylcellulose; sodium cellulose glycolate).
CAS: 9004-32-4.
Properties: White granular colloid, variable solubility in water.
Use: A vehicle for insoluble toxicants or drugs.

carboxymethylmercaptosuccinic acid.
$HOOCCH_2SCH(COOH)CH_2(COOH)$.

Properties: White powder. Melting range 135–138C. Water solubility of 137 g/100 g (25C); ethanol solubility of 76 g/100 g (25C).
Use: Heavy-metal chelator and deactivator.

carboxyl-terminal residue. The amino acid residue in a polypeptide chain with a free a-carboxyl group.

carboxymethylpyridinium chloride hydrazide. See Girard's reagent.

1-carboxymethyl-1-methylpyrrolidinium iodide methyl ester.
CAS: 22041-28-7. $C_8H_{16}NO_2 \cdot I$.
Hazard: A poison.

carboxymethyltrimethylammonium chloride hydrazide. See Girard's reagent.

α-(3-carboxy-1-oxosulfopropyl)-ω-hydroxy-poly(oxy-1,2-ethanediyl), C10-C16 alkyl ethers, disodium salts.
CAS: 68815-56-5.
Hazard: A severe skin irritant.

carboxypeptidase. A proteolytic enzyme found in the pancreas that catalyzes the hydrolysis of native food proteins. It acts on polypeptides, producing simpler peptides and amino acids.
Use: Biochemical research.

carboxypeptidase.
CAS: 11075-17-5.
An enzyme that catalyzes the cleavage of amino acids from a protein at the C-terminus of a polypeptide chain.

(o-carboxyphenyl)hydroxymercury. See o-hydroxymercuriobenzoic acid.

(p-carboxyphenyl)hydroxymercury. See p-hydroxymercuribenzoic acid.

carboxyplast. A high-molecular-weight synthetic polymer in which the principal chain contains only carbon and oxygen.

4-carboxyresorcinol. See β-resorcylic acid.

6-carboxyuracil. See orotic acid.

carburet of sulfur. See carbon disulfide.

carburetted hydrogen. See ethylene.

carbylamine. (isocyanide; isonitrile).
$N \equiv C-C_6H_5$
An unpleasantly odored compound containing the radical $N \equiv C$.

carbylic acid. A general designation for carboxylic, carbaazylic, or carboxazylic.

carcinogen. Any substance that causes the development of cancerous growths in living tissue. Such substances are usually grouped in two classifications: (1) those that are known to induce cancer in humans or animals, either by operational exposure in industry or by ingestion in feedstuffs, e.g., asbestos particulates, nickel carbonyl, trichloroethylene, benzidine and compounds, vinyl monomer, benzopyrene, aflatoxin, chloromethyl ether, β-naphthylamine, as well as anthracene, phenanthrene, chrysene, and other polynuclear hydrocarbons of coal tar origin; (2) experimental carcinogens that have been found to cause cancer in animals under experimental conditions, namely, by external applications, feeding, or injection of the substance. Among these are dimethyl sulfate, cyclamate compounds, ethyleneimine, and 4-dimethylaminoazobenzene. The Delany amendment to the Food, Drug, and Cosmetic Act forbids the use in human foods of any substance falling in this group.

The substances mentioned above by no means comprise a complete list of carcinogens. There are approximately 3500 known and suspected carcinogenic compounds, and new ones are constantly being discovered.

carcinolytic. (cancericidal; cancerocidal). Destructive to malignant tumors.

carcinoma. Any malignant, invasive tumor that arises in epithelial tissue or any malignant neoplasm of any tissue arising from embryonic ectoderm and endoderm.

cardamom oil.
Properties: Colorless or pale-yellow essential oil; strongly aromatic, camphoraceous odor and taste. Strongly dextrorotatory.
Use: Flavoring for foods, confectionery, liqueurs, pharmaceuticals.

cardanolide. A fully saturated system of digitaloid lactones in which the configuration at C_{20} is the same as in cholesterol.

Cardio-Green. Indocyanine green, a diagnostic dye used in medicine.

cardiotoxin. A cytolytic and cardiotoxic polypeptide component of cobra venom comprised of a single chain of 60 amino acid residues cross-linked by disulfide bridges. The N- and C-terminal residues of the chain are leucine and asparagine, respectively. It irreversibly depolarizes the nerve cell membrane, with contraction of both striated and smooth muscle. The toxicity of cardiotoxin is potentiated by phospholipase a and is inhibited by the action of heparin and RNA gangliosides.

CARDIPOL. An oxidized hydrocarbon wax.

CARDIS. An oxidized hydrocarbon wax.

Cardosol Brand Resin. A water-soluble ketone formaldehyde condensate that can be gelled and cured by alkali or heat.
Use: Water-resistant adhesives for box board, coatings, and glass fibers and as a ceramic binder.

carene. See (+)car-3-ene.

(+)car-3-ene.
CAS: 74806-04-5. $C_{10}H_{16}$.
Hazard: Low toxicity by ingestion and skin contact. A mild skin and eye irritant.

Δ-3-carene. 3,7,7-trimethylbicyclo[4.1.0]-hept-3-ene.
CAS: 13466-78-9. $C_{10}H_{16}$.
A terpene hydrocarbon with both a 6-member and 3-member ring.
Properties: Clear, colorless liquid. D 0.8668 (15C), bp 170C, refr index 1.4723 (20C). Stable to approximately 250C, resinifies with oxygen. Insoluble in water; miscible with organic solutions. Combustible.
Derivation: From wood turpentine.
Use: Solvent, intermediate.

carfentrazone-ethyl.
CAS: 128639-02-1. $C_{15}H_{14}Cl_2F_3N_3O_3$.
Hazard: Moderately toxic.

Carinthiam process. A procedure for reduction of lead ores; a small charge is roasted slowly at a low temperature, the lead being collected outside the furnace by means of an inclined hearth.

Carius (wet combustion) method. Decomposition of organic compounds containing halogen or sulfur in a sealed tube in the presence of red, fuming nitric acid at 250–300C in such a manner that halogen is converted to ionic halide and sulfur to sulfate.

Carius tube. (bomb tube).
A hard, glass thick-walled tube, sealed at one end, that is approximately 40 cm long.

Carmethose. Sodium carboxymethylcellulose.

carmine. An aluminum lake of the pigment from cochineal. Bright-red pieces; easily powdered; soluble in alkali solution, borax; insoluble in dilute acids; slightly soluble in hot water.
Grade: Technical.
Use: Dyes, inks, indicator in chemical analysis, coloring food materials, medicines, etc.

carminic acid.
$C_{22}H_{20}O_{13}$.

The essential constituent of carmine. A tricyclic compound.
Properties: Dark, purplish-brown mass or bright-red powder. Mp (decomposes at 136C), pH 4.8 (yellow), pH 6.2 (violet). Soluble in water, alcohol, concentrated sulfuric acid; insoluble in ether, benzene, chloroform. Combustible.
Derivation: By extraction from the insects *Coccus cacti* (cochineal).
Use: Stain in microscopy, indicator in analytical chemistry, coloring proprietary medicines, pigment for fine oil colors, color photography, dyeing.

carmustine. (*N,N'*-bis(2-chloroethyl)-*N*-nitrosourea; BCNU; BiCNU; 1,3-bis(2-chloroethyl)-1-nitrosourea).
CAS: 154-93-8. $C_5H_9Cl_2N_3O_2$.
A chloroethylnitrosourea derivative. It is a cell-cycle phase nonspecific alkylating antineoplastic agent.
Hazards: Extremely toxic, central nervous system depression, pulmonary fibrosis, renal and hepatic damage, cytotoxic, immunosuppressive, carcinogen.
Use: An antineoplastic agent in the treatment of Hodgkin's disease, other lymphomas, meningeal leukemia, metastatic brain tumors; and other malignant neoplasms.

carnallite.
$KCl\cdot MgCl_2\cdot 6H_2O$ or $KMgCl_3\cdot 6H_2O$.
Properties: A natural hydrated double chloride of potassium and magnesium, white, brownish, and reddish; streak white; shining, greasy luster; strongly phosphorescent; bitter taste. Deliquescent, d 1.62, Mohs hardness 1.
Occurrence: West Germany, Alsace, New Mexico.
Use: A commercial source of manufactured potash salts.

carnauba wax. (Brazil wax).
The hardest and most expensive commercial wax.
Properties: Hard solid in form of yellow to greenish-brown lumps; slight odor. D 0.995 (15/15C), mp 84–86C, acid number 2-9, iodine number 13.5. Soluble in ether, boiling alcohol, and alkalies; insoluble in water. Combustible.
Derivation: Exudation from leaves of the wax palm *Copernica cerifera* (Brazil).
Grade: By numbers and sources; crude and refined; powdered; FCC.
Use: Shoe polishes, leather finishes, varnishes, electric insulating compositions, furniture and floor polishes, carbon paper, waterproofing, to prevent sunchecking of rubber and plastic products, confectionery, cosmetics.

carnosine. (β-alanylhistidine; ignotine).
$C_9H_{14}N_4O_3$.
An amino acid occurring in the muscle of many animals and humans. Occurs naturally in the L(+)form.
Properties: Mp 245–250C (decomposes). Soluble in water. Nitrate: crystals. Mp 222C (decomposes).

Soluble in water. Hydrochloride: crystals. Mp 245C (decomposes). Soluble in water. D(−)carnosine: crystals, mp 260C.
Use: Biochemical research.

Carnot cycle. An ideal closed cycle of reversible changes with which the performance of heat engines may be compared. It consists of four reversible operations: isothermal and adiabatic expansions followed by isothermal and adiabatic compressions.

carnotite.
$K_2(UO_2)_2(VO_4)_2\cdot 3H_2O$.
A natural hydrated vanadate of uranium and potassium usually found in sandstones and other sedimentary rocks.
Properties: Bright lemon-yellow, dull, or earthy luster (pearly or silky when coarsely crystalline). Soluble in acids. Radioactive. Usually occurs as a powder or in fine-grained aggregates.
Occurrence: Colorado, Utah, Arizona, New Mexico, South Dakota, Australia, Zaire, the former U.S.S.R.
Hazard: A radioactive poison.
Use: Ore of uranium, source of radium.

Carnot's reagent. A reagent for the determination of potassium, an alcoholic solution of sodium bismuth thiosulfate made from sodium thiosulfate and bismuth subnitrate.

Carnot theorems. (1) No heat engine working between two temperatures can have a greater efficiency than a reversible engine working between those temperatures. (2) The efficiency of any reversible heat engine working between two temperatures is independent of the nature of the engine and of the working substances and depends only on the temperatures.

carob-seed gum. (locust-bean gum).
A polysaccharide plant mucilage that is essentially galactomannan (carbohydrate). Mw approximately 310,000. Swells in cold water, but viscosity increases when heated. Insoluble in organic solvents. Combustible.
Derivation: Extracted from carob seeds, from the tree *Ceratonia siliqua*.
Grade: Technical, FCC (as locust-bean gum).
Use: In foods as stabilizer, thickener, emulsifier, and packaging material; cosmetics; sizing and finishes for textiles; pharmaceuticals; paints, bonding agent in paper manufacture; drilling fluids.

Caro's acid. (peroxysulfuric acid; persulfuric acid).
H_2SO_5 or $HOSO_2OOH$.
Properties: White crystals. Mp 45C (decomposes).
Derivation: Action of hydrogen peroxide on concentrated sulfuric acid; action of 40% sulfuric acid on potassium persulfate.

Hazard: Strong irritant to eyes, skin, and mucous membranes. Strong oxidizer, may explode in contact with organic materials.

Use: Caro's reagent, a pasty mass of great oxidizing power for testing aniline, pyridine, and alkaloids; dye manufacture; oxidizing agent; bleaching.

Caro's reagent. A thick mixture used in testing for aniline, composed of ammonium or potassium persulfate, which is dissolved in sulfuric acid, has great oxidizing power.

carotene. (provitamin A).
$C_{40}H_{56}$.
A precursor of vitamin A occurring naturally in plants. It consists of three isomers, approximately 15% α, 85% β, and 0.1% γ. Carotene is a member of a large class of pigments called carotenoids. It has the same basic molecular structure as vitamin A and is transformed to the vitamin in the liver.

Properties: Ruby-red crystals, easily oxidized on contact with air. Mp (α) 188C, (β) 184C, and (γ) 178C. Insoluble in water; slightly soluble in alcohol; soluble in chloroform, carbon disulfide, ether, and benzene.

Source: Orange-yellow pigment in plants, algae, and some marine animals, especially in leaves, vegetation, and root crops, in trace concentrations. Notably present in butter and carrots.

Derivation: By extraction from carrots and palm oil; concentration by a chromatographic process from alfalfa. β-Carotene is also made by a microbial fermentation process from corn and soybean oil.

Grade: According to USP, units of vitamin A, sold as pure crystals, as solutions in various oils, as colloidal dispersions. Also FCC.

Use: Pharmaceuticals, coloring margarine and butter, feed and food additive.

β-carotene.
$(C_{40}H_{56})$.
Properties: Red, shining crystals. Mw 536.44, mp 181C. Insoluble in water; slightly soluble in alcohol.
Occurrence: In vegetables.
Use: Food additive.

carotene cochineal.
Use: Food additive.

carotenoid. A class of pigments occurring in the tissues of higher plants, algae, and bacteria, as well as in fungi. Also present in some animals, as squalene in shark liver oil. They include the carotenes and xanthophylls.

Properties: Yellow to deep-red, crystalline solids. Soluble in fats and oils; insoluble in water; high melting; stable to alkali but unstable to acids and to oxidizing agents; color easily destroyed by hydrogenation or by oxidation; some are optically active.

carotenoids. Lipid-soluble photosynthetic pigments made up of isoprene units.

Carothers, Wallace H. (1896–1937). Born in Iowa, Carothers obtained his doctorate in chemistry at the University of Illinois. He joined the research staff of Du Pont in 1928, where he undertook the development of polychloroprene (later called neoprene) that had been initiated by Nieuland's research on acetylene polymers. Carothers's crowning achievement was the synthesis of nylon, the reaction product of hexamethylenetetramine and adipic acid. Carothers's work in the polymerization mechanisms of fiber like synthetics of cyclic organic structures was brilliant and productive, and he is regarded as one of the most original and creative American chemists of the early 20th century.

carrageenan. (3,6-anhydro-d-galactan; carrageen).
CAS: 9000-07-1.
A sulfur phycocolloid: the aqueous, usually gel-forming, cell-wall polysaccharide mucilage found in red algae (*Chondrus crispus* and several other species). It is water-extracted from a seaweed called carrageen or Irish moss (east coast of southern Canada, New England, and south to New Jersey). It is a mixture of polysaccharide fractions: (1) The λ fraction is cold-water soluble, contains d-galactose and 35% esterified sulfate, and does not gel. (2) The κ fraction contains d-galactose and 3,6-anhydro-d-galactose (1.4:1 ratio) and 25% esterified sulfate. The κ form does not gel without addition of a solute; the properties of the gel depend on the amount and nature of the added solute. Another species of seaweed produces 100% κ from North Carolina to the tropics. Carrageenan is a hydrophilic colloid that absorbs water readily and complexes with milk proteins.

Available forms: Dehydrated, purified powder.
Grade: Technical, FCC.
Hazard: Questionable carcinogen.
Use: Emulsifier in food products, chocolate milk, toothpastes, cosmetics, pharmaceuticals, protection colloid, stabilizing aid in ice cream (0.02%).

Carrel–Dakin solution. A medicinal solution composed of sodium chloride and sodium hypochlorite solution.

carrier. (1) A neutral material such as diatomaceous earth used to support a catalyst in a large-scale reaction system. (2) A gas used in chromatography to convey the volatilized mixture to be analyzed over the bed of packing that separates the components. (3) An atomic tracer carrier: a stable isotope or a natural element to which radioactive atoms of the same element have been added for purposes of chemical or biological research.
See tracer.

carrier ampholyte. A mixture of small natural or synthetic, soluble electrolytes that can act ether as an acid or as a base in standard ISO-DALT 2d gel electrophoresis.

carrier gas. The mobile phase in gas elution chromatography.

carrier protein. (membrane transport protein). A protein in the plasmalemma (cell membrane) that binds to a specific type of solute molecule and carries it across the membrane.

Carroll reaction. Preparation of γ,Δ-unsaturated ketones by base-catalyzed reaction of allylic alcohols with β-ketoesters or thermal rearrangement of allyl acetoacetates.

cartap. (carbamothioc acid; S,s'-[2-(dimethyl-amino)-1,3-propanediyl] ester; 1,3-bis(carbamoyl-thio)-2-n,n-(dimethylamino)propane; S-[3-carba-moylsulfanyl-2-(dimethylamino)propyl]carbamo-thioate).
$C_7H_{15}N_3O_2S_2$.
Properties: Colorless, crystalline, synthetic.
Hazard: Very toxic.
Use: Insecticide.

carthamin. (carthamic acid; safflor carmine; safflor red).
$C_{21}H_{22}O_{11}$.
Properties: Dark-red powder with green luster. Slightly soluble in water; soluble in alcohol; insoluble in ether; solutions rapidly decompose.
Derivation: A glucoside from *Carthamus tinctorius*.
Use: Colorant in food products, cosmetics.

"Carulite" [Carus]. TM for a family of oxidation catalysts.
Use: Applications requiring control of carbon monoxide, ozone, and hydrocarbons.

Carusorb. An air-filter medium.
Use: Purification of air in homes, commercial buildings, and industry.

carvacrol. (isopropyl-o-cresol; 2-methyl-5-isopropylphenol; 2-hydroxy-p-cymene).
CAS: 499-75-2.

Properties: Thick, colorless oil; thymol odor. D 0.976 (20/4C); bp 237C; fp 0C; refr index 1.523 (20C). Insoluble in water; soluble in alcohol, ether, and alkalies. Combustible.
Derivation: From p-cymene by sulfonation followed by alkali fusion.
Use: Perfumes, fungicides, disinfectant, flavoring, organic synthesis.

carvone.

$CH_3C{:}CHCH_2CH[C(CH_3){:}CH_2]CH_2CO.$
CAS: 99-49-0.
A ketone derived from the terpene dipentene. It is optically active, occurring naturally in both d- and l-forms.
Properties: Pale-yellowish or colorless liquid with a strong characteristic odor. D 0.960 (20C), bp 227–230C, refr index 1.4999 (18C). Soluble in alcohol, ether, chloroform, propylene glycol, and mineral oils; insoluble in glycerol and water. Combustible.
Derivation: The d-form is the main constituent of caraway and dill oils; the l-form occurs principally in spearmint oil and may be synthesized from d-limonene.
Method of purification: Rectification.
Grade: FCC (both d- and l-forms), technical.
Use: Flavoring, liqueurs, perfumery, soaps.

Carwinate. Isocyanates used to make urethane elastomers, coatings, foam adhesives, rigid plastics, sealants, and one-shot flexible, semiflexible, and semirigid foams.

caryophyllene.
$C_{15}H_{24}$.
A mix of sesquiterpenes occurring in many essential oils. It forms the chief hydrocarbon component of clove oil.

caryophyllic acid. See eugenol.

carzinophillin A.
CAS: 1403-29-8.
A natural antibiotic produced by a *Streptomyces* strain. Discovered in Japan in the 1950s, it has strong antitumor potential, though its toxicity may preclude medical use. It is a member of the so-called bis-intercalator group of antibiotics that act by interlocking between the nucleotide bases of DNA; thus, it may have mutagenic activity. The molecule contains carbon, oxygen, hydrogen, and nitrogen. The structure was elucidated in 1982.

CAS. Chemical Abstracts Registry or Chicago Academy of Science.

Casabond TX. A textile bonding agent. Bonds nylon, rayon, cotton, and terylene to natural and synthetic rubber. Also for rubber and metal bonding, PVC bonding, and plasticized variety of synthetic textiles.
Use: Technical goods, footwear, and automotive.

cascade. (1) A series of operational units or stages so arranged that the heat produced in the first unit serves as the heat source for the second unit and so on. An example of this is a triple-effect evaporator in which the latent heat of condensation is passed from one unit to another. This principle is also used in distillation, each column plate representing one stage in the cascade. In flash distillation, used in the desalination of seawater, the heat of condensation is used to warm the incoming water. (2) Coined term used to describe a large number of compounds derived from a common source, e.g., the arachidonic acid cascade.

"Cascamite" [Momentive].
CAS: 9011-05-6.
TM for urea-formaldehyde resins.
Use: Adhesives for particle board, plywood, various wood-gluing applications, wet-strength paper.

cascarilla oil.   The volatile oil obtained by steam distillation of the dried bark of *Croton eluteria* Bennet. Light-yellow to brown-amber liquid having a pleasant, spicy odor. Soluble in most fixed oils and mineral oil; almost insoluble in glycerol and propylene glycol.
Use: Flavoring agent in foods, medicine, tobacco.

"Cascowax" [Momentive].   TM for wax emulsion.
Use: Sizing agent for composite wood panels.

case hardening.   A process that imparts a hard surface to steel while the interior remains soft and tough. This is accomplished by heating the steel out of contact with air while packed in carbonaceous material, cooling it to black heat, reheating to a high temperature, and quenching. The materials are usually wood charcoal with sodium, potassium, or barium carbonates, cyanides, etc.

casein.
CAS: 9005-46-3.
Though commonly regarded as the principal protein in milk (approximately 3%), casein is actually a colloidal aggregate composed of several identifiable proteins together with phosphorus and calcium. It occurs in milk as a heterogeneous complex called calcium caseinate, which can be fractionated by a number of methods. It can be precipitated with acid at pH 4.7 or with the enzyme rennet (rennin). The product of the latter method is called paracasein, the term being applied to any of the casein fractions involved, i.e., α, β, κ, etc.
Properties: White, amorphous solid; tasteless; odorless. D 1.25–1.31, hygroscopic. Stable when kept dry but deteriorates rapidly when damp. Soluble in dilute alkalies and concentrated acids; almost insoluble in water; precipitates from weak acid solutions.
Derivation: Acid casein: warm skim milk is acidified with dilute sulfuric, hydrochloric, or lactic acid, the whey drawn off, and the curd washed, pressed, ground, and dried. Rennet casein: warm skim milk is treated with an extract of the enzyme rennin (rennet). The curd contains combined calcium and calcium phosphate.
Source: Midwestern U.S., Australia, Argentina, New Zealand, Poland.

Grade: Acid-precipitated (domestic edible, imported inedible), paracasein.
Use: Cheese making, plastic items, paper coatings, water-dispersed paints for interior use, adhesives (especially for wood laminates), textile sizing, foods and feeds, textile fibers, dietetic preparations, binder in foundry sands.

casein-sodium.   See sodium caseinate.

cashew gum.   (anacardium gum).
The exudation from the bark of the cashew-nut tree, *Anacardium occidentale*. Hard, yellowish-brown gum; partly soluble in water. Used for inks, insecticides, pharmaceuticals, mucilage, tanning agent, natural varnishes, bookbinders' gum.

cashew nutshell oil.   (cashew nutshell liquid).
The oil obtained from the spongy layer between the inner and outer shells of cashew nuts. The raw liquid contains approximately 90% anacardic acid, $C_{22}H_{32}O_3$, and a blistering compound containing sulfur. Most of the liquid used in commerce has been heated or treated with chemicals to make it safe to handle. The liquid is nondrying but can be made drying by proper treatment. It polymerizes on heating and forms condensation products with aldehydes.
Hazard: (untreated) Strong irritant.
Use: Varnishes and impregnating materials, modifier for phenol-based resins, plasticizers, germicides and insecticides, coloring materials and indelible inks, lubricants, and preservatives.

casing-head gasoline.   See gasoline.

caspase.   An enzyme that catalyzes the breakdown of cellular proteins during apoptosis.

cassava.   See tapioca.

casse.   The decolorization of wine due to action of enzymes.

Cassel brown.   See Van Dyke brown.

Cassel green.   See barium manganate.

Cassella's acid.   (2-naphthol-7-sulfonic acid; β-naphtholsulfonic acid F; F acid).
$C_{10}H_6(OH)SO_3H$.
Properties: White crystals. Mp 89C. Soluble in water and alcohol.
Derivation: Fusion of naphthalene-2,7-disulfonic acid with caustic soda.
Grade: Technical. Also available as the sodium salt (F salt).
Use: Intermediate for azo dyes.

Cassella's acid F.   (2-naphthylamine-7-sulfonic acid; delta acid).
$C_{10}H_6(NH_2)SO_3H$.

Properties: Colorless crystals. Soluble in water, alcohol, and ether.
Derivation: Heating sodium 2-naphthol-7-sulfonate with aqueous ammonia and ammonium acid sulfate in an autoclave.
Use: Azo-dye intermediate.

cassia oil. (Chinese cinnamon oil; cinnamon; cassia oil; cinnamon oil; USP).
CAS: 8007-80-5.
Properties: See cinnamic aldehyde, its chief constituent.
Derivation: Distilled from leaves and twigs of *Cinnamomum cassia*.
Grade: USP (as cinnamon oil), redistilled, technical, lead free.
Use: Flavoring, perfumery, medicine, soaps.

cassiterite. (tinstone; wood tin; stream tin).
SnO_2.
Natural tin dioxide, usually in igneous rocks.
Properties: Brown, black, yellow, white color; luster adamantine or dull submetallic; streak white. Mohs hardness 6–7, d 6.8–7.1.
Occurrence: Malaya, Bolivia, Indonesia, Africa.
Use: Principal ore of tin.

Castertrak. A program used to control addition of water-treatment chemicals.
Use: Caster cooling systems.

castile soap. A soap made from olive oil.
Use: Detergent or antiseptic. Fabric finishing and washing.

casting. (1) In the metal industries, conversion of a molten metal (iron, steel) into bars (pigs) or products by pouring it into open troughs or channels made of foundry sand (sand casting) or into products of precise shape by forcing it into steel dies or molds under pressure (die casting).
See foundry sand; cast iron; die casting; investment casting.
(2) In the plastics industry, formation of a product either by filling an open mold with liquid monomer and allowing it to polymerize in situ or (for film and sheet) by pouring the liquid mixture onto a moving flat surface.
See molding.

cast iron. Generic term for a group of metals that basically are alloys of carbon and silicon with iron. Relative to steel, cast irons are high in carbon and silicon, carbon ranging from 0.5 to 4.2% and silicon from 0.2 to 3.5%. All these metals may contain other alloys added to modify their properties.
Iron castings are produced in an exceptionally wide range of sizes and weights, from piston rings a fraction of an inch in diameter and weighing less than 1 oz to steam-turbine bases 20 ft long and weighing 180,000 lb.

Most cast iron is manufactured by melting a mixture of steel scrap, cast-iron scrap, pig iron, and alloys in a cupola using coke as a fuel. A small percentage is melted in electric furnaces. It is poured into molds of silica sand bonded with bentonite, fireclay, and water. A small percentage is cast into metal molds or into molds of baked or fired ceramics. Internal cavities are formed by hard but collapsible cores of sand bonded with drying oils or synthetic resins. Small molds and cores usually are made by machine, using patterns of wood or metal.

Castner process. A method of producing sodium metal from fused sodium hydroxide by means of an electrolytic cell with heavy iron anodes surrounding the cathode in the bath. Sodium is collected on an iron gauze diaphragm midway between the electrodes. Hydrogen and oxygen are also collected during the process.

castor. Perfume fixative obtained from secretions of the beaver. Synthetic types are available.

castor oil. (ricinus oil).
CAS: 8001-79-4.
Properties: Pale-yellowish or almost colorless, transparent, viscous liquid; faint, mild odor; nauseating taste. A nondrying oil. D 0.945–0.965 (25/25C), saponification value 178, iodine value 85, fp −10C. Flash p 445F (229C), autoign temp 840F (448C). Combustible. Soluble in alcohol, benzene, chloroform, and carbon disulfide; dextrorotatory.
Derivation: From the seeds of the castor bean, *Ricinus communis* (Brazil, India, the former U.S.S.R., U.S.). They are cold-pressed for the first grade of medicinal oil and hot pressed for the common qualities, approximately 40% of the oil content of the bean being obtained. Residual oil in the cake is obtained by solvent extraction.
Chief constituent: Ricinolein (glyceride of ricinoleic acid).
Grade: USP No. 1; No. 3; refined; FCC.
Hazard: Undergoes spontaneous heating.
Use: Plasticizer in lacquers and nitrocellulose, production of dibasic acids, lipsticks, polyurethane coatings, elastomers and adhesives, fatty acids, surface-active agents, hydraulic fluids, pharmaceuticals, industrial lubricants, electrical insulating compounds, manufacture of Turkey Red oil, source of sebacic acid and of ricinoleates, medicine (laxative).
See castor oil, dehydrated; blown oil.

castor oil, acetylated. See glyceryl 1,2-tri(acetylricinoleate).

castor oil acid. See ricinoleic acid.

castor oil, blown. See blown oil.

castor oil, dehydrated. (DCO).
A castor oil from which approximately 5% of the chemically combined water has been removed and

that as a result has drying properties similar to those of tung oil. Dehydration is carried out commercially by heating the oil in the presence of catalysts such as sulfuric and phosphoric acids, clays, and metallic oxides. The commercial product is offered in a wide range of viscosities and analytical constants. Used in protective coatings and alkyd resins.

castor oil, hydrogenated.   Principally glyceryl-tri(12)-hydroxystearate. A hard, waxy product. It is insoluble in water and organic solvents. Mp 85C.
Use: Hydroxystearic acid, waterproofing fabrics, cosmetics, lubricant, mold-release agent, candles, carbon paper.

castor oil, polymerized.   A rubber-like polymer results from combination of castor oil with sulfur or diisocyanates; this can be blended with polystyrene to give a tough, impact-resistant product.

castor oil, sulfonated.   See Turkey red oil.

castor seed oil meal.   (castor cake; castor meal). The residue from extraction of oil from the castor seed (ricinus). The normal product contains 29.5% crude protein, 35.8% crude fiber, 13.2% N-free extract, and 1.0% crude fat. The total digestible nutrients are approximately 25%. The ash content of 7.5% is high in potash and phosphate.
Hazard: Contains ricin, which must be removed before internal use.
Use: Animal feeds (after removal of toxic ingredients), fertilizer.

catabolism.   The   phase   of   intermediary metabolism concerned with the energy-yielding degradation of nutrient molecules.

catabolite gene activator protein.   (CAP). A specific regulatory protein that controls initiation of transcription of the genes producing the enzymes required for a bacterial cell to use some other nutrient when glucose is lacking.

CAT box.   CCAAT box.

"Cata-Chek" *[Ferro]*.   TM for catalysts.

Catacobal.   Organic and inorganic-cobalt compounds.

Catadry.   A paint drying catalyst.

Catal.   Accelerators of unsaturated polyester.

catalase.   An oxidizing enzyme occurring in both plant and animal cells. It decomposes hydrogen peroxide. It can be isolated and is used in food preservation (removing oxygen in packaged foods) and in decomposing residual hydrogen peroxide in bleaching and oxidizing processes.

catalysis.   One of the most important phenomena in nature, catalysis is the loosening of the chemical bonds of two (or more) reactants by another substance in such a way that a fractionally small percentage of the latter can greatly accelerate the rate of the reaction while remaining unconsumed. See catalyst. Thus, one part by volume of catalyst can activate thousands of parts of reactants. Though the mechanism of their action is not completely known, the electronic configuration of the surface molecules of the catalyst is often the critical factor. The surface irregularities give rise to so-called active points at which intermediate compounds can form. Most industrial catalysis is performed by finely divided transition metals or their oxides.
Solid catalysts may combine chemically (bond) at the surface with one or more of the reactants. This is known as chemisorption and occurs on only a small portion of the catalyst surface (i.e., at the active points); it results in changing the chemical nature of the chemisorbed molecules. Catalysis of chemical reactions by surfaces must proceed by chemisorption of at least one of the reactants.
Catalysis and catalytic mechanisms permeate almost every aspect of chemistry and are of such wide-ranging importance that they have long been the subject of continuing research. Many interrelated disciplines are involved, among them organometallic reactions (stereospecific catalysts), electrochemistry, colloid and surface chemistry, coordination chemistry, and biochemistry. One example of comparatively recent investigation is so-called cluster catalysis, in which a metal is bonded to carbon monoxide to form a metallic cluster anion. Such aggregations or crystallites may be up to 12 Å in diameter. Metal cluster research is motivated by the need to develop selective catalysts for C_1 compounds, in view of the likelihood that the basic raw materials for chemistry in the future will be derived from coal or biomass rather than oil. Another research development is the creation of a uniquely shaped catalyst particle that permits greater efficiency because of the increase in surface-to-volume ratio. Such particles are known as amphora catalysts.

catalysis, heterogeneous.   A catalytic reaction in which the reactants and the catalyst comprise two separate phases, e.g., gases over solids, or liquids containing finely divided solids as a disperse phase.

catalysis, homogeneous.   A catalytic reaction in which the reactants and the catalyst comprise only one phase, e.g., an acid solution catalyzing other liquid components.

catalyst.   Any substance of which a small proportion notably affects the rate of a chemical reaction without itself being consumed or undergoing a chemical change. Most catalysts accelerate reactions, but a few retard them (negative catalysts, or inhibitors). Catalysts may be inorganic, organic, or a complex of organic groups and metal halides (see catalyst, stereospecific). They may be gases, liquids, or solids. In some cases, their action is destructive and undesirable, as in the oxidation of iron to its oxide, which is catalyzed by water vapor, and similar types of corrosion. The life of an industrial catalyst varies from 1000 to 10,000 hours, after which it must be replaced or regenerated.

Though it is not a substance, light in both the visible and ultrashort wavelengths can act as a catalyst, as in photosynthesis and other photochemical reactions, e.g., as polymerization initiator and cross-linking agent.

Catalysts are highly specific in their application. They are essential in virtually all industrial chemical reactions, especially in petroleum refining and synthetic organic chemical manufacturing. For details of application, see the following list. Since the activity of a solid catalyst is often centered on a small fraction of its surface, the number of active points can be increased by adding promoters that increase the surface area in one way or another, e.g., by increasing porosity. Catalytic activity is decreased by substances that act as poisons that clog and weaken the catalyst surface, e.g., lead in the catalytic converters used to control exhaust emissions.

Besides inorganic substances, there are many organic catalysts that are vital in the life processes of plants and animals. These are called enzymes and are essential in metabolic mechanisms, e.g., pepsin in digestion. Synthetic organic catalysts have been developed that imitate the action of enzymes such as chymotrypsin. Such model catalysts are examples of biomimetic chemistry. They approach the catalytic activity of natural enzymes.

Following is a partial list of catalysts; an asterisk indicates a destructive effect.

Substance	Reaction Type
aluminum alkyl + titanium chloride	Ziegler catalyst for stereo-specific polymers
aluminum chloride	condensation (Friedel-Crafts)
aluminum oxide	hydration, dehydration
ammonia	condensation (polymers)
chromic oxide	methanol synthesis, aro-matization, polymerization
cobalt (Oxo process)	hydrocarbon synthesis
copper salts	oxidation (of rubber)*
ferric chloride	Friedel-Crafts
hydrogen fluoride	alkylation, condensation, dehydration, isomerization
iodine	condensation, alkylation
iron	ammonia synthesis, hydrocarbon synthesis
iron oxide	dehydrogenation (oxidation)
manganese dioxide	oxidation

Substance	Reaction Type
molybdenum oxide	dehydrogenation, polymerization aromatization, partial oxidation
nickel	hydrogenation (oils to fats), methanation
phosphoric acid	polymerization, isomerization,
platinum metals	hydrogenation, aromatization, oxidation
silica-alumina	cracking hydrocarbons
silver	hydration, oxidation
sulfuric acid	isomerization, corrosion*
triethylaluminum	polymerization (stereospecific)
vanadium pentoxide	oxidation (sulfuric acid)
water (esp. + NaCl)	oxidation (corrosion)*
zeolites	cracking hydrocarbons

See catalysis; enzyme.

catalyst, amphora. Catalyst particles made from a slurry of critical viscosity by compressing it into spheres or droplets followed by unidirectional heating and air-drying on a moving belt, the spheres being supported by a powder bed material. This process results in particles from 1.5 to 6 mm in diameter. The unique feature is the formation of an internal cavity having an orifice at one point, so that the particle roughly resembles a doughnut from which a bite has been taken. Since the shape of the cavity suggests an amphora (Greek vase), the catalyst was so named. This shape affords a higher surface-to-volume ratio than is possible with solid spheres and other conventional forms, with consequent greater efficiency. A variety of materials can be used, e.g., alumina, zeolites, metallic oxides, etc. Amphora catalysts are effective in a wide range of chemical processing applications (oxidation and reforming of hydrocarbons, hydrotreating).

catalyst, negative. See inhibitor.

catalyst, organic. See enzyme.

catalyst poison. A substance that interferes with the action of a catalyst.

catalyst promoter. A compound that reduces or activates fresh catalysts at polymerization temperatures, reactivates spent catalysts, and scavenges and reacts with catalyst poisons.

catalyst, shape-selective. A catalytic solid (transition metal) introduced into a crystalline aluminosilicate (zeolite) having pores or openings of approximately 5 Å. This permits the catalyst to

exert a selective effect between molecules that differ in shape rather than in the reactivity of their chemical groups. The cage structure of the zeolite effectively prevents contact of the catalyst with all molecules whose shapes and sizes exclude their entry. The ability of a catalyst to discriminate among molecules on the basis of their shapes is of great value in the cracking of straight-chain hydrocarbons and has attractive possibilities in other types of catalytic reactions.
See zeolite; cage zeolite.

catalyst, stereospecific.   An organometallic catalyst that permits control of the molecular geometry of polymeric molecules. Examples are Ziegler and Natta catalysts derived from a transition metal halide and a metal alkyl or similar substances. There are many patented catalysts of this general type, and most of them developed in connection with the production of polypropylene, polyethylene, or other polyolefins.
See polymer, stereospecific; Natta catalyst; Ziegler catalyst.

catalyst, thermonuclear.   See carbon cycle (2).

catalytic RNA.   RNA which contains an intron sequence that has an enzyme-like catalytic activity. This intron sequence can fold up to form a complex surface that acts like an enzyme in reactions with other RNA molecules.

catalytic site.   See active site.

cataphoresis.   The migration of colloidal particles toward an electrode under the influence of an electric current.

CAT assay.   An enzyme assay. CAT stands for chloramphenicol acetyl transferase, a bacterial enzyme that inactivates chloramphenicol by acetylating it. CAT assays are often performed to test the function of a promoter. The gene coding for CAT is linked onto a promoter (transcription control region) from another gene, and the construct is "transfected" into cultured cells. Largely supplanted by the reporter gene luciferase.

catechol.   See pyrocatechol.

catecholamines.   Hormones, such as epinephrine, that are amino derivatives of catechol.

catecholborane.   (1,3,2-benzodioxaborole). A monofunctional hydroborating agent.
Properties: A liquid. Mw 119.92, mp 12C, bp 50C (50 mm Hg), optical rotation 1.5070 degrees (20C).
Use: Preparation of alkaneboronic acid and esters from olefins.

catechol *o*-methyltransferase.   A type of enzyme, typically found in the soluble fraction of a number of tissues, that catalyzes the methylation of catecholamines and other derivatives of pyrocatechol.

catenane.   A compound with interlocking rings that are not chemically bonded but that cannot be separated without breaking at least one valence bond. The model would resemble the links of a chain.

catenyl.   An ester that has been reacted with an alkylene oxide or its polymer.

"Cat-Floc" *[Calgon].*   (diallyldimethylammonium chloride).
TM for a quaternary ammonium polymer.
Derivation: Monomer in water solution is mixed with a catalytic amount of butylhydroperoxide and kept at 50–75C for 48 h. The solid formed is taken up in water, precipitated, and washed with acetone.
Use: Flocculating agent, textile spinning aid, antistatic agent, wet-strength improvers in paper, rubber accelerators, curing epoxy resins, surfactants, bacteriostatic and fungistatic agents.

catharometer.   Device for determining rate of flow or change in composition of gases.

cathetometer.   A device for exact measurement or observation of short vertical distances, which consists of a horizontal-reading telescope or microscope movable along a vertical scale.

cathode.   The negative electrode of an electrolytic cell, to which positively charged ions migrate when a current is passed as in electroplating baths. The cathode is the source of free electrons (cathode rays) in a vacuum tube. In a primary cell (battery), the cathode is the positive electrode.
See anode; electrode.

cathode ray.   A stream of electrons that emanates from the cathode of a partially evacuated discharge tube.

cathode sputtering.   See sputtered coating.

cathodic protection.   The reduction or prevention of corrosion of a metallic surface by making it cathodic, e.g., by the use of sacrificial anodes for impressed currents bringing a metal, by an external current, to a potential where it is thermodynamically stable.

catholyte.   The solution surrounding the cathode in an electrolytic cell.

cation.   An ion having a positive charge. Cations in a liquid subjected to electric potential collect at the negative pole, or cathode.

cation exchange.   See ion exchange.

cation-exchange resin. A polymeric substance with fixed negative charges used in the chromatographic separation of cationic substances, e.g., CM (carboxymethyl)-cellulose.

cationic detergent. Any detergent in which the surface active moiety bears a positive charge at neutral ph.
Use: Dishwashing detergents and fabric softeners.

cationic reagent. One of several surface-active substances in which the active constituent is the positive ion. Used to flocculate and collect minerals that are not flocculated by oleic acid or soaps (in which the surface-active ingredient is the negative ion). Reagents used are chiefly quaternary ammonium compounds, e.g., cetyltrimethylammonium bromide.

catlinite. (pipestone).
A fine-grained silicate mineral related to pyrophyllite, which is easily compressible, has high surface friction, and is used for gaskets in very high-pressure equipment.

CAU. ((2S)-(9H-carbazol-4-yloxy)-3-(propan-2-ylamino)propan-2-ol).
$C_{18}H_{22}N_2O_2$.
A codon of histidine that directs the placement of histidine into a polypeptide.

caulking compound. See sealant.

caulophylline. (1,2,3,4,5,6-hexahydro-3-methyl-1,5-methano-8H-pyrido[1,2-a][1,5]diazocin-8-one; 12-methylcytisine; n-methylcytisine).
$C_{12}H_{16}N_2O$.
An alkaloid.
Derivation: Isolated from blue cohosh, *Caulophyllum thalictroides*, a perennial herb that is native to the Eastern U.S.
Hazard: Extremely toxic; cardiotoxic; poisonous.

caustic. (1) Unqualified, this term usually refers to caustic soda (NaOH). (2) As an adjective, it refers to any compound chemically similar to NaOH, e.g., caustic alcohol (C_2H_5ONa). (3) Any strongly alkaline material that has a corrosive or irritating effect on living tissue.

caustic alkali. (corrosive alkali).
Any strongly corrosive base, especially a metallic hydroxide or a carbonate that can neutralize acids.
Properties: High degree of ionization; water soluble
Hazard: Cause rapid, deep, very painful destruction tissues giving it a grayish color and a slippery or soapy texture.

caustic baryta. See barium hydroxide.

caustic embrittlement. The corrosion resulting in cracking of steel stressed beyond its yield point, due to localized concentration of hydroxide ions breaking down the cohesion between the ferrite grains.

causticized ash. Combinations of soda ash and caustic soda in definite proportions and marketed for purposes where an alkali is needed ranging in causticity between the two materials. Causticized ash is usually designated by its caustic-soda content, and the range of standard marketed products embraces 7, 10, 15, 25, 36, 45, and 67% of caustic soda.

caustic lime. See calcium hydroxide.

caustic potash. See potassium hydroxide.

caustic soda. See sodium hydroxide.

cavitands. Three-dimensional stacked-layer polycyclic compounds that maintain a rigid structure and bind a variety of molecules in the cavities produced by the structure of the molecules.

cavitation. Formation of vapor bubbles in a liquid such as saltwater when subjected to tension, causing severe mechanical damage to the surfaces of metals exposed to it, e.g., ship propellers, steam condensers, pumps, and piping systems. The erosive effect is due to the shock waves created by collapse of the bubbles. The pressures exerted by cavitation have been calculated to be in the range of 30,000 psi. This phenomenon plays a part in corrosion of metals and in emulsion formation.
See corrosion; homogenization.

Cavitron. Cyclodextrins, alpha, beta, gamma, and their derivatives.
CAS: α (1006-20-3), α (7585-39-9), γ (17465-86-0).
Available forms: Crystalline powders.
Use: Host inclusion complexation agents for stabilization and property modification of guest molecules.

Cb. Symbol for columbium, an obsolete name for niobium.

CBM. Abbreviation for chlorobromomethane (see bromochloromethane); also for constant-boiling mixture.
See azeotropic mixture.

cc. (1) Abbreviation for cubic centimeter.
See milliliter.

ccc. (2-chloroethyl(trimethyl)azanium chloride; Chlormequat chloride).
CAS: 999-81-5. $C_5H_{13}C_{12}N$.
A codon for proline that directs the placement of proline into a polypeptide.

CC. Abbreviation for closed cup.
See flash point, TCC.

CCAAT box.   (CAAT box; CAT box).
A sequence found in the 5' flanking region of certain genes that is necessary for efficient expression.

CCD.   See charge-coupled device.

ccg.   (S-((2R,3S,4S,6S)-6-[[(2R,3S,4S,5R,6R)-5-[[(2S,4S,5S)-5-(ethylamino)-4-methoxyoxan-2-yl]oxy-4-hydroxy-6-[[(2S,5Z,9S,13E)-9-hydroxy-12-(methoxycarbonylamino)-13-(2-methylsulfanyldisulfanylethylideni-oxo-2-bicyclo[7.3.1]trideca-1(12),5-dien-3,7-diynl]oxy-2-methyloxan-3-yl]amino]oxy-4-hydroxy-2-methyloxan-3-yl]-4-[(2S,3R,4R,5S,6S)-3,5-dihydroxy-4-methoxy-6-methyloxan-2-yl]oxy-5-iodo-2,3-dimethoxy-6-methylbenzenecarbothioate).
$C_{55}H_{74}IN_3O_{21}S_4$.
A codon for proline that directs the placement of proline into a polypeptide.

ccu.   ((2Z,4Z)-hexa-2,4-dienedioic acid).
$C_6H_6O_4$.
A codon for proline that directs the placement of proline into a polypeptide.

Cd.   Symbol for cadmium.

CDA.   Abbreviation for completely denatured alcohol.
See alcohol, denatured.

CDAA.   See α-chloro-*N,N*-diallylacetamide.

CDEC.   See 2-chloroallyldiethyldithiocarbamate.

CDMA.   Abbreviation for the Commercial Development and Marketing Association.

cDNA.   (complementary DNA).
A DNA, usually made by reverse transcriptase, which is complementary to given mRNA and used in cloning.

cDNA clone.   (complementary DNA).
A piece of DNA copied from an mRNA. The term "clone" indicates that this cDNA has been spliced into a plasmid or other vector in order to propagate it. A cDNA clone may contain DNA copies of such typical mRNA regions as coding sequence, 5'-untranslated region, 3' untranslated region or poly(A) tail. No introns will be present, nor any promoter sequences (or other 5' or 3' flanking regions). A "full-length" cDNA clone is one that contains all of the mRNA sequence from nucleotide #1 through to the poly(A) tail.

cDNA library.   A collection of DNA sequences that code for genes. The sequences are generated in the laboratory from mRNA sequences.
See messenger RNA.

CDP.   (1) Abbreviation for cytidine diphosphate. (2) Abbreviation for cresyl diphenyl phosphate. See cytidine phosphates.

CDTA.   See *trans*-1,2-diaminocyclohexanetetraacetic acid monohydrate.

Ce.   Symbol for cerium.

Cech, R. Thomas.   (1947–). Awarded Nobel Prize in chemistry in 1989 jointly with Altman for the discovery that RNA acts as a biological catalyst, as well as a carrier of genetic information. Doctorate awarded in 1975 by the University of California.

cedar leaf oil.   An essential oil distilled from the leaves of *Juniperus virginiana*. Strongly dextrorotatory. Used in microscopy, perfumery, flavoring.

cedrol.
$C_{15}H_{26}O$.
A tertiary terpene alcohol.
Properties: Colorless crystals; cedarwood odor. Mp 86C, soluble in 11 parts of 95% alcohol. Combustible.
Use: Perfumery, for woody and spicy notes; odorant for disinfectants.

cedryl acetate.
$CH_3COOC_{15}H_{25}$.
Properties: Colorless liquid; light cedar odor. D 0.975–0.995, refr index 1.496–1.510. Soluble in one volume of 90% alcohol. Combustible.
Use: Perfumery.

CEELS.   Abbreviation for characteristic electron energy-loss spectroscopy.

cefodizime disodium.
CAS: 86329-79-5. $C_{20}H_{18}N_6O_7S_4$·2Na.
Hazard: Low toxicity by ingestion.

cefprozil.
CAS: 92665-29-7. $C_{18}H_{19}N_3O_5$.
Hazard: Moderately toxic by ingestion. Human systemic effects.

ceftiofur.
Properties: Powder.
Use: Drug (veterinary); food additive.

ceftriaxone sodium hydrate.
CAS: 74578-69-1. $C_{18}H_{16}N_8O_7S_3$·2Na·$7/_2$H$_2$O.
Hazard: Moderately toxic. Low toxicity by ingestion. Human systemic effects.

Celanar.   A polyester film made from polyethylene terephthalate.
Properties: Transparent, biaxially oriented crystals. D 1.395, tensile strength 30,000 psi, dielectric strength 7000 volts/mil, mp 260C, service temp

−60 to 150C. Outstanding dimensional stability and chemical resistance.

Use: Magnetic recording tape, drafting and engineering reproduction materials, metallic yarn, roll leaf, pressure-sensitive tapes, packaging, dielectric material in capacitors, wire and cable, motors, generators, transformers, and oils.

Celanese CL. A series of polyvinyl acetate emulsions.
Available forms: 102: Fine particle size, water-resistant homopolymer emulsion. 202: Fine particle size, water-resistant copolymer emulsion. 203: Vinyl-acrylic copolymer emulsion. 204: Vinyl copolymer emulsion.
Use: Paints, adhesives, and paper-coating specialties.

Celanese Solvent. A series of special solvents. Replacement for butanol and methylisobutyl carbinol in lacquers and brake fluids, distillation range 125–155C, flash p 120F (48.9) (OC).
Available forms: 203: Replacement for normal butyl alcohol in nitrocellulose lacquers, alkyd resin formulations, and thinners, distillation range 115–120C, flash p 100F (37.7C) (OC). 601: Replacement for methyl ethyl ketone in vinyl and nitrocellulose applications, distillation range 74–84C, flash p 10F (−12.2C) (OC).
Hazard: Flammable, dangerous fire risk.

Celanthrene. A group of anthraquinone disperse dyes designed especially for acetate, also suitable for application to nylon.

"Celatom" [EP]. TM for a group of diatomaceous silicas (diatomite) of high quality and uniformity.
Use: Filter aid; foods and beverages; absorbents, as in insecticides and fertilizers; catalyst supports; fillers for paper; paints; explosives; concrete and asphalt; chromatography.

Celcon. A highly crystalline acetal copolymer based on trioxane.
See acetal resin.

celery seed oil.
Properties: From steam distillation of fruit and seed of *Apium graveolens* L. Yellow to green-brown liquid; aromatic odor. D: 0.870–0.910. Sol in fixed oils, mineral oil; slightly sol in propylene glycol; insol in glycerin.
Use: Food additive.

celestine blue. (CI 51050).
$C_{17}H_{18}ClN_3O_4$.
Properties: Dark-green powder. Slightly soluble in water, ethylene glycol, and alcohol. Water solution gives purple color, alcoholic solution blue.
Use: Biological stain, mordant in dyeing.

celestite.
$SrSO_4$.
Natural strontium sulfate, usually found in sedimentary rocks.
Properties: Colorless, white, pale-blue or red, luster vitreous to pearly. Resembles barite. D: 3.95, Mohs hardness 3–3.5.
Occurrence: U.S., Canada, Europe, Mexico.
Use: Strontium chemicals, oil-well drilling mud, sugar refining, ceramics, production of very pure sodium hydroxide.

celiac disease. (CD; celiac sprue disease).
Also referred to as gluten sensitive enteropathy (GSE). Considered to be the most undiagnosed common disease, affecting 1 in 133 people in the U.S. It is a chronic inherited disease. Symptoms can include diarrhea, bloating, fatigue, constipation, and osteoporosis and can ultimately result in intestinal lymphoma. Frequently misdiagnosed as irritable bowel syndrome or Crohn's disease. Strict adherence to a gluten free diet is the only treatment currently available. This involves elimination from the diet of wheat, rye, barley, and sometimes oats and the derivatives of these grains.

Celiac Sprue Association. (CSA).
A member-based nonprofit organization to help individuals with Celiac Sprue and Dermatitis Herpetformes through education and research. Its mailing address is P.O. Box 31700, Omaha, NE 68131.
Website: http://www.csaceliacs.org

Celite. Diatomaceous earth and related products.

"Celkate" [Imerys]. TM for silicates with good liquid-absorption capacities.
Use: Gloss control for coatings; decolorizing specific acid absorbers.

cell. (1) The fundamental unit of biological structure, comprised of (a) an outer membrane, or wall, approximately 100 Å thick, which, being semipermeable, maintains by osmosis the biochemical equilibrium of the intracellular fluids; (b) the cytoplasm, containing mitochondria, ribosomes, and other structures; and (c) the nucleus, in which lie the chromosomes and genes. An extremely complex biochemical organization, the cell is the dynamic unit of all life. Its ability to reproduce itself and to control its functions systematically is of basic importance to maintenance of life and growth. All organic matter is either found in cells or produced by cellular activity and was ultimately derived from photosynthesis. Cells are comparatively large units that can be resolved in an optical microscope; the largest single cells are represented by the eggs of oviparous animals.
See mitosis.

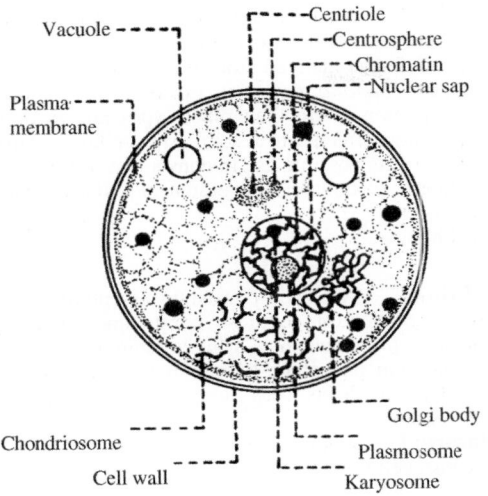

Vacuole
Centriole
Centrosphere
Chromatin
Nuclear sap
Plasma membrane
Chondriosome
Cell wall
Golgi body
Plasmosome
Karyosome

(2) Any self-contained unit having a specific functional purpose, as (a) voltaic cell (battery) to generate electric current, (b) electrolytic cell to effect electrolysis, (c) fuel cell to convert chemical energy into electricity, and (d) solar cell to capture heat from sunlight. All except the last involve use of electrodes and electrolytes.
(3) Any completely enclosed hollow unit, as in a honeycomb or cellular plastic.

cell-bound antibody.   An antibody that forms on the surface of a sensitized lymphocyte and effects cell-mediated (delayed) sensitivity to antigens and cells.

cell, concentration.   An electrolytic cell whose emf is due to a difference in concentration of the electrolyte or active metal at the anode and cathode.

cell constant.   The factor by which the observed resistance of an electrolyte is multiplied to obtain its specific resistance.

Cell Guard NRT.   A magnesium hydroxide slurry.
Use: For refinery bleaching replacing caustic soda and sodium silicate.

cell, dry.   See dry cell.

cell, electrolytic.   See electrolytic cell.

cell, fuel.   See fuel cell.

cellulase and carbohydrase.   See carbohydrase and cellulase.

Cellitazol.   A series of developed acetate dyestuffs.

Celliton.   A series of disperse dyestuffs characterized by good fastness to light, washing, etc. Used for dyeing and printing acetate fibers.

cell, local.   A cell whose emf is due to differences of potential between areas on a metallic surface in an electrolyte.

cell membrane.   The outer membrane of a cell, which separates it from the environment. Also called a plasma membrane or plasmalemma.

cellobiose.
$C_{12}H_{22}O_{11}$.
The product of the partial hydrolysis of cellulose, composed of two d-glucose molecules.
Properties: Colorless crystals. Mp 225C (decomposes). Soluble in water; slightly soluble in alcohol; nearly insoluble in ether; insoluble in acetone.
Use: Bacteriology.

celloidin.   A pure form of pyroxylin; used for embedding sections in microscopy.

Cellolube.   A finisher.
Use: To give dyed and printed textiles final characteristics for water and soil repellency.

"Cellolyn" [Pinova].   TM for a series of pale, low-melting hydroabietyl ester resins. Used for lacquers, inks, and adhesives.

cellophane.   (regenerated cellulose).
Film produced from wood pulp by the viscose process.
Properties: Transparent, strong, flexible, and highly resistant to grease, oil, and air. The base cellulose film is modified by softeners, flame-resisting materials, and dyes, also by coating with other materials. On exposure to heat the untreated film loses strength at 149C, decomposes at 176–204C, does not melt, burns readily, and is not self-extinguishing.
Hazard: Flammable, moderate fire risk.
Use: Wrapper or protective package for fabricated articles and industrial applications.
See rayon.

"Cellosize" [Union].   TM for hydroxyethyl cellulose and carboxymethylcellulose.
See cellulose, modified.

"Cellosolve" [Union].   TM for mono- and dialkyl ethers of ethylene glycol and their derivatives widely used as industrial solvents.
butyl "Cellosolve"
See ethylene glycol monobutyl ether.
butyl "Cellosolve" acetate
See ethylene glycol monobutyl ether acetate.
"Cellosolve" acetate
See ethylene glycol monoethyl ether acetate.
"Cellosolve" solvent
See ethylene glycol monohexyl ether.
methyl "Cellosolve"
See ethylene glycol monomethyl ether.

cell, photovoltaic.   See solar cell.

cell, primary.   A device producing an electrical current from a nonreversible electrochemical reaction.

cell, reduction.   An electrolytic unit in which free metal is produced by electrolysis of one of its compounds.

cell, reversible.   A galvanic cell in which chemical and electrical energy can be interconverted reversibly.

cell, solar.   See solar cell.

cell, storage.   An electrolytic cell for the generation of electric energy that may be recharged after use by an electric current flowing in a direction opposite to the current flow by which the cell discharges.

cellular plastic.   A thermosetting or thermoplastic foam composed of cellular cores with integral skins having high strength and stiffness. The cells result from the action of a blowing agent, either at room temperature or during heat treatment of the plastic mixture. The resulting product may be either flexible or rigid, the latter being machinable.

The foaming action in some cases may occur in situ (foamed-in-place plastics). Cellular plastics are combustible. For details, see foam, plastic.
Use: (Flexible) Furniture, automobile interiors, mattresses, etc., where softness and resiliency are desired. (Rigid) Insulating material, boat building and similar light construction, salvage of water-logged ships.
See foam; plastic; rubber sponge.
For further information, refer to Cellular Plastics Division, Society of the Plastics Industry.

cellulase.   An enzyme complex, produced by the fungi *Aspergillus niger* and *Trichoderma viride* that is capable of decomposing cellulosic polysaccharides into smaller fragments, primarily glucose. It has been used as a digestive aid in medicine and in the brewing industry. Research has been devoted to experimental application of cellulase to disposal of cellulosic solid wastes. The resulting glucose can be

fermented to ethanol, used to grow yeast for animal-feed proteins, or used as a chemical feedstock.
Note: Cellulase derived from the thermophilic soil fungus *Thielatia terrestris* functions at a much higher temperature than other types and is thus much more effective in decomposing cellulose. This indicates its possible use in conversion of biomass to energy. Commercial development of this product is expected.
See biomass.

Celluloid.   A plastic consisting essentially of a solid solution of cellulose nitrate and camphor or other plasticizer plus a flame-retardant such as ammonium phosphate to minimize flammability; available in sheets, rods, tubes, films. Also called pyroxylin.
Hazard: Flammable, dangerous fire risk.
See nitrocellulose.

cellulose.
CAS: 9004-34-6. $(C_6H_{10}O_5)_n$.
A natural carbohydrate high polymer (polysaccharide) consisting of anhydroglucose units joined by an oxygen linkage to form long molecular chains that are essentially linear. It can be hydrolyzed to glucose. The degree of polymerization is from 1000 for wood pulp to 3500 for cotton fiber, giving a molecular weight from 160,000 to 560,000.

Cellulose is a colorless solid, d approximately 1.50, insoluble in water and organic solvents. It will swell in sodium hydroxide solution and is soluble in Schweitzer's reagent. It is the fundamental constituent of all vegetable tissues (wood, grass, cotton, etc.) and the most abundant organic material in the world. Cotton fibers are almost pure cellulose; wood contains approximately 50%.
The physical structure of cellulose is unusual in that it is not a single crystal but consists of crystalline areas embedded in amorphous areas. Chemical reagents penetrate the latter more easily than the former. Cellulose is virtually odorless and tasteless and is combustible, with an ignition point of approximately 450F. In some forms, it is flammable. For example, railroad shipping regulations require a flammable label on such items as burnt fiber, burnt cotton, wet waste paper, and wet textiles. Fires have been known to occur in warehouses in which telephone books were stored. These were undoubtedly due

to heat buildup in the paper caused by microbial activity and self-sustaining oxidation.
See flammable material.
The most important uses of cellulose are bulk woods of many kinds; paper, most of which is made from wood pulp; cotton products (clothing, sheeting, industrial fabrics); packaging, ranging from wooden barrels to candy pats; and as a source of ethanol (enzymatic hydrolysis) and methanol (destructive distillation of wood). Specialized uses include nonwoven fabrics, medical equipment (artificial kidney), insulation and soundproofing, sausage casings, etc. Cellulose has approximately 60% of the energy content of bituminous coal; its use as a fuel has increased, especially in rural locations.
See biomass.
There are many chemical modifications of cellulose, including its esters (cellulose acetate) and its ethers (methylcellulose), the nitrated product (nitrocellulose), and rayon and cellophane (from cellulose xanthate). Thus, it is the basis of many plastics, fibers, coatings, lacquers, explosives, and emulsion stabilizers. Alkali cellulose is an intermediate made by the action of sodium hydroxide solution on cellulose and is used for making cellulose ethers and viscose.
See cellulose, modified.
Cellulose exists in three forms—α, β, and γ. α-Cellulose has the highest degree of polymerization (DP) and is the chief constituent of paper pulp. It is insoluble in strong sodium hydroxide solution. The β and γ forms have much lower DP and are known as hemicellulose. Methods of determining the α content of pulps are detailed in TAPPI Method T203 and ASTM D-588-42.
See pulp, paper.
Cellulose has been prepared in microcrystalline form.
See Avicel.
Cellulose can be decomposed to glucose by the enzyme cellulase or by hydrolysis.
Hazard: Cellulosic materials (paper, cotton, and textile wastes), when wet with water, are a dangerous fire hazard and subject to shipping regulations. Upper respiratory tract irritant.
See rayon; cellophane; nitrocellulose; carboxymethylcellulose.

α-cellulose.   The major component of wood and paper pulp. It is that portion of holocellulose that is insoluble in strong sodium hydroxide solution.
See cellulose.

p-cellulose.   See phosphorylation.

cellulose acetate.   (CA).
A cellulose ester in which the cellulose is not completely esterified by acetic acid.
Properties: White flakes or powder. A thermoplastic resin, softening at approximately 60–97C and melting at approximately 260C. D: 1.27–1.34. Insoluble in acetone, ethyl acetate, cyclohexanol, nitropropane, ethylene dichloride. Notable for toughness, high-impact strength, and ease of fabrication. Subject to dimensional change due to cold flow, heat, or moisture absorption (1–7%). Fibers weaken above 80C and are difficult to dye uniformly; not attacked by microorganisms.
Derivation: Reacting cellulose (wood pulp or cotton linters) with acetic acid or acetic anhydride, with H_2SO_4 catalyst. The cellulose is fully acetylated (three acetate groups per glucose unit), and at the same time the sulfuric acid causes appreciable degradation of the cellulose polymer so that the product contains only 200–300 glucose units per polymer chain. At this point in the process the cellulose acetate ordinarily is partially hydrolyzed by the addition of water until an average of 2–2.5 acetate groups per glucose unit remain. This product is thermoplastic and soluble in acetone. Fibers are produced by forcing an acetone solution through orifices of a spinnerette into a stream of warm air, which evaporates the solvent. Fibers are also produced in a similar manner from cellulose triacetate, which is insoluble in acetone but soluble in methylene chloride.
See acetate fiber; acetate film.
Available forms: Sheet, film or fiber, molded items.
Grade: Filtered and unfiltered. Also graded by percent combined acetic acid content: plastic 52–54%, lacquer 54–56%, film 55.5–56.6%, water resisting 56.5–59%, triacetate 60.0–62.5%.
Hazard: Flammable, not self-extinguishing, moderate fire risk.
Use: Acetate fiber, lacquers, protective-coating solution, photographic film, transparent sheeting, thermoplastic molding composition, cigarette filters, magnetic tapes, osmotic-cell membrane.

cellulose acetate butyrate.   (CAB; cellulose acetobutyrate).
Properties: White pellets or granules. A thermoplastic resin. D 1.2. Soluble in ketones; organic acetates; lactates; methylene, ethylene, and propylene chlorides; and high-boiling solvents. Excellent weathering properties, low water absorption, low heat conductivity, high dielectric strength, high resistance to oil and grease. Combustible.
Derivation: Reaction of purified cellulose with acetic and butyric anhydrides, with sulfuric acid as catalyst and glacial acetic acid as solvent. The ratio of acetic to butyric components may be varied over a wide range.
Grade: According to percent butyryl content (17, 27, 38, 50%).
Use: Molding compositions, film and sheet, photographic film, lacquers, protective-coating solutions, taillight lenses, water-desalination membrane, piping and tubing, covering for aluminum fibers, toys, packaging, brush handles, hydrometers, miscellaneous consumer products.

cellulose acetate phthalate. A reaction product of phthalic anhydride and cellulose acetate, used for coating of tablets and capsules.

cellulose acetate propionate. (cellulose propionate).
Similar to cellulose acetate butyrate but made with propionic anhydride instead of butyric anhydride. Unusually stable; requires less plasticizer and is compatible with more plasticizers than the butyrate.

cellulose acetobutyrate. See cellulose acetate butyrate.

cellulose, colloidal. Cellulose hydrosol, solidifying to a dry, hardened material that is soluble in cupriethylenediamene.

cellulose ether. See ethylcellulose; methylcellulose.

cellulose gel. See cellulose, microcrystalline.

cellulose gum. A purified grade of carboxymethylcellulose.

cellulose, hydrated. (hydrocellulose).
Cellulose that has been caused to react with water (approximately 8–12%), forming a gelatinous mass. Combustible.
Derivation: By mechanical pulverization and agitation with water; by the action of strong salt solutions, alkalies, or acids.
Use: In the manufacture of paper, vulcanized fiber, mercerized cotton, viscose rayon.
See hydration.

cellulose methyl ether. See methylcellulose.

cellulose, microcrystalline.
Properties: Fine white crystalline powder from treatment of α-cellulose with mineral acids. Insol in water, most org solvs.
Hazard: A nuisance dust.
Use: Food additive.

cellulose, modified. One of many derivatives of cellulose, formed by substitution of appropriate radicals (carboxyl, alkyl, acetate, nitrate, etc.) for hydroxyl groups along the carbon chain. Such reactions are usually not stoichiometric. Some of these products (carboxymethyl and hydroxyalkyl cellulose) are water-soluble ethers; others are organosol esters (cellulose acetate), nitrates (nitrocellulose), or xanthates (viscose).
Biodegraded cellulose has been used as a microbial growth medium for protein formation. A tobacco substitute based on a cellulose modification is Cytrel.

cellulose nitrate. See nitrocellulose.

cellulose, oxidized. (cellulosic acid).
Derivative of cotton. Cellulose produced by treatment with nitrogen dioxide. Soluble in alkali but may be made to retain original form of the cellulose and much of its tensile strength.
Properties: Slightly off-white gauze, lint, or powder; slight charred odor; acid taste. Soluble in aqueous organic bases, in dilute alkali, and in ammonium hydroxide, forming salts and esters; insoluble in water, acids, and common organic solvents. It slowly degrades at room temperatures and should be kept cool. Combustible.
Grade: USP, technical.
Use: Surgery and medicine, ion-exchange medium, thickening agent.

cellulose propionate. See cellulose acetate propionate; Forticel.

cellulose, regenerated. See cellophane; rayon.

cellulose solubility number. The percentage by weight of a sample of cellulose that is soluble in cold caustic-soda solution under standard conditions. This is a measure of degradation.

cellulose sponge. A sponge of regenerated cellulose; highly absorbent, soft, and resilient when wet; long lasting. It will not scratch, can be sterilized by boiling, and is not affected by ordinary cleaning compounds. The pores vary in size from coarse (the size of a pea) to fine (the size of a pinhead). Yarn made of cellulose sponge consists of cotton fiber covered with the sponge product. Combustible.
Use: Washing automobiles and trucks, walls and painted surfaces, windows, etc.; general cleaning; sponge used in photographic laboratories; wet mops, cleaning pads, etc.

cellulose triacetate. A cellulose resin in which the cellulose is almost completely esterified by acetic acid.
Properties: White flakes. D 1.2. Soluble in chloroform, methylene chloride, tetrachloroethane. Combustible.
Derivation: Reaction of purified cellulose with acetic anhydride in the presence of sulfuric acid as catalyst and glacial acetic acid as solvent, followed by very slight hydrolysis.
Grade: Flake.
Use: Protective coatings resistant to most solvents, textile fibers; base for magnetic tape.

cellulose xanthate. See viscose process.

cellulosic plastic. One of a number of semisynthetic polymers based on cellulose.
See cellophane; cellulose acetate; cellulose, modified; nitrocellulose; rayon; carboxymethylcellulose.

cellulosic thiocarbonate. A reactive intermediate in the graft polymerization of certain synthetic

polymers to cellulosic fibers. The latter are treated with sodium hydroxide and a sulfur-containing compound, and the cellulose thus activated is placed in an emulsion or solution of monomers. Polymerization at 50C occurs with a peroxide catalyst to form the graft.
See graft polymer.

cell, voltaic.   See voltaic cell.

cell wall.   Rigid structure deposited outside the cell membrane. Plants are known for their cell walls of cellulose, as are the green algae and certain protists, while fungi have cell walls of chitin.

"Celogen" *[CelChem].*   TM for a series of blowing agents: AZ azodicarbonamide; OT phosphorus-*p*-oxybis-benzenesulfonylhydrazide; RA *p*-toluenesulfonyl semicarbazide.

Celotex.   Structural building and insulation board produced in large sheets. Made from bagasse or wood fiber and treated to be resistant to fungi, termites, and water penetration. The name also includes roofing products, gypsum, wallboard, lath, plasters, mineral wool, and hardboard.

Celsius, A.   (1701–1744). Swedish physicist who proposed the use of the centigrade temperature scale. His name is now generally applied to this scale (degrees centigrade = degrees Celsius).
See centigrade.

cement, aluminous.   (high alumina cement).
A hydraulic cement that contains at least 30–35% alumina (in contrast to Portland cement, which contains less than 5%). The alumina is usually supplied by inclusion of bauxite. Aluminous cement attains its maximum strength more rapidly than Portland cement. It is also more resistant to solutions of sulfates. It exists in two modifications, sintered and fused.

cementation.   A process in which steel or iron objects are coated with another metal by immersing them in a powder of the second metal and heating to a temperature below the melting point of any of the metals involved. Zinc, chromium, aluminum, copper, and other metals are applied to iron or steel in this fashion. The process is basically diffusion of one metal into the other so that intermetallic alloy layers are formed at the interface of the basis and coating metals.
See sherardizing.

cemented carbide.   See carbide, cemented.

cement, hydraulic.   Any mixture of fine-ground lime, alumina, and silica that will set to a hard product by admixture of water that combines chemically with other ingredients to form a hydrate.
See cement; aluminous; cement, Portland.

cementite.
Fe$_3$C.
A carbide of iron formed in the manufacture of pig iron and steel. Composed of 93.33% iron and 6.67% carbon, it is very hard and brittle and will scratch glass and feldspar but not quartz. It is about two-thirds as magnetic as pure iron under an exciting current. It occurs in ordinary steels of more than 0.85% carbon and takes its name from cement steel, made by the cementation process, which contains a great deal of this carbide.
See carbide.

cement, organic.   Any of various types of rubber cement, silicone adhesives, deKhotinsky cement.
See adhesive, rubber-based; silicone; deKhotinsky cement.

cement, Portland.
CAS: 65997-15-1.
A type of hydraulic cement in the form of finely divided, gray powder composed of lime, alumina, silica, and iron oxide as tetracalcium aluminoferrite (4CaO·Al$_2$O$_3$·Fe$_2$O$_3$), tricalcium aluminate (3CaO·Al$_2$O$_3$), tricalcium silicate (3CaO·SiO$_2$), and dicalcium silicate (2CaO·SiO$_2$). These are abbreviated, respectively, as C$_4$AF, C$_3$A, C$_3$S, and C$_2$S. Small amounts of magnesia, sodium, potassium, and sulfur are also present. Hardening does not require air and will occur under water. Cement may be modified with various plastic latexes in proportions up to 0.2 part latex solids to one part cement to improve adhesion, strength, flexibility, and curing properties. Water evaporation can be retarded by adding such resins as methylcellulose and hydroxyethylcellulose. For further information, refer to the Portland Cement Association, 5420 Old Orchard Road, Skokie, IL 60077-1083.
Hazard: Pulmonary function effects, asthma, respiratory symptoms.
See concrete (1).

cement, rubber.   See adhesive, rubber-based.

CE Methyl Esters.   A series of methyl esters of straight chain (normal), even-numbered fatty acids ranging from C$_8$ (octanoate) to C$_{18}$ (octadecanoate) and including mixtures of these.
Properties: Colorless to light-yellow liquids or white solids. D approximately 0.87.
Use: Chemical intermediate, lubricants, cosmetic ingredients, formulating aids (rubber, wax, etc.).

C&EN.   *Chemical & Engineering News.* Weekly publication of the American Chemical Society.

Chemical Heritage Foundation.   (CHF).
Formerly the Beckman Center for History of Chemistry was established in 1982 to discover and disseminate information about historical resources and

to encourage research scholarship and popular writing in the history of chemistry, chemical engineering, and the chemical process industries. The activities of the Foundation include organizing interviews and oral histories on major developments in modern chemistry; locating historical manuscript and archival records of individuals, societies, trade associations, and companies important in the history of the chemical sciences; and encouraging academic research in the field. Typically, graduate students in History and Sociology of Science participate most actively in its programs. It is located at 315 Chestnut Street, Philadelphia, PA 19106-2702.
Website: http://www.chemheritage.org
See Appendix II.

"Chemigum" *[Goodyear].*   TM for NBR elastomers, copolymers of acronylite and butadiene.
Use: Adds good mechanical properties and resistance to fuels, fats, and oils.

chemiosmotic coupling.   Coupling of ATP synthesis to electron transfer via an electrochemical pH gradient across a membrane (inner mitochondrial membrane, bacterial plasma membrane, or thylakoid membrane).

Chemmer Epoxy.   An encapsulating resin for mini electronics.

CenPrem.   A tableting sugar processed into layer particle size.
Use: For mixes and tablets.

Centara III.   A dietary pea fiber.
Use: Adds excellent moisture retention capacity and fiber enrichment while being devoid of color.

"Centellin" *[Sabinsa].*   TM for an extract of Centella asiatica skin conditioning agent.
Use: For wound healing; supports circulation.

centigrade.   (Celsius).
The internationally used scale for measuring temperature, in which 100 degrees is the boiling point of water at sea level (1 atm) and 0 degree is the freezing point. A temperature given in centigrade degrees may be converted to the corresponding Fahrenheit temperature by multiplying it by 9/5 (or 1.8) and adding 32 to the product. A temperature given in Fahrenheit degrees is converted to the corresponding centigrade temperature by subtracting 32 and multiplying the remainder by 5/9. The centigrade scale was devised by the Swedish scientist Celsius; his name is officially used to designate it, even though centigrade is more meaningful.

centigrade heat unit.   See chu.

centimorgan.   (cM).
A unit of measure of recombination frequency. One centimorgan is equal to a 1% chance that a marker at one genetic locus will be separated from a marker at a second genetic locus due to crossing over in a single generation. In human beings, one centimorgan is equivalent, on average, to one million base pairs. See megabase.

centipoise.   (cP).
1/100 poise. The poise is the cgs-metric system unit of viscosity and has the dimensions of dyne-seconds per square centimeter or grams per centimeter-second.

centistoke.   (cS).
1/100 stoke, the unit of kinematic viscosity. The kinematic viscosity in stokes is equal to the viscosity in poises divided by the density of the fluid in grams per cubic centimeter, both measured at the same temperature.

central dogma.   The organizing principle of molecular biology: genetic information flows from DNA to RNA to protein.

centrifugation.   A separation technique based on the application of centrifugal force to a mixture or suspension of materials of closely similar densities. The smaller the difference in density, the greater the force required. The equipment used (centrifuge) is a chamber revolving at high speed (10,000 rpm or more) to impart a force up to 17,000 times gravity. The materials of higher density are thrown toward the outer portion of the chamber while those of lower density are concentrated in the inner portion. This technique is used effectively in a number of biological and industrial operations, such as separation of the components of blood, concentration of rubber latex, and separation of fat particles from other milk components. Separation of isotopes, e.g., those of uranium, by this method is now practicable for producing enriched uranium. This method is economically superior to the gaseous diffusion process.
See ultracentrifuge.

centrifuge.   See centrifugation; ultracentrifuge.

centroid diagram.   A chart showing relationship of atomic electron energy levels in successive elements.

centromere.   A specialized region of a chromosome that serves as the attachment point for the mitotic or meiotic spindle.

cephalin.   (kephalin; phosphatidyl ethanolamine; phosphatidylserine).
$CH_2OR_1CHOR_2CH_2OP(O)(OH)OR_3$.
Properties: Yellowish, amorphous substance; characteristic odor and taste. Insoluble in water and acetone; soluble in chloroform and ether; slightly soluble in alcohol.

A group of phospholipids in which two fatty acids (R_1 and R_2) form ester linkages with the two hydroxyl groups of glycerophosphoric acid, and either ethanolamine or serine (R_3) forms an ester linkage with the phosphate group. Cephalins are therefore either phosphatidylethanolamine or phosphatidylserine. They are associated with lecithins found in brain tissue, nerve tissue, and egg yolk.
Use: Medicine, biochemical research.

cephaloridine. (1-[[2carboxy-8-oxo-7-[(thienyl-acetyl)amino]-5-thia-1-azabicyclo[4.2.0] oct-2-en-3-yl]methyl]pyridinium hydroxide inner salt; 1-[7'-β-[2-(2-thienyl)acetamido-8'-oxo-1'-aza-5'-thiabicyclo[4.2.0]oct-2'-en-3'-yl)-methyl] pyridinium-2'-carboxylate; N-[7-[(2-thienyl) acetamido)ceph-3-em-3-ylmethyl]pyridinium-4-carboxylate; n-[7-(2'-thienylacetamidoceph-3-yl-methyl)]pyridinium-2-carboxylate; cefaloridin; (6R,7R)-8-oxo-3-(pyridine-1-ium-1-ylmethyl)-7-[(2-thiophen-2-ylacetyl)amino]-5-thia-1-azabicyclo[4.2.0]oct-2-ene-2-carboxylate).
CAS: 50-59-9. $C_{19}H_{17}N_3O_4S_2$.
Hazard: Moderately to very toxic.
Use: Antibacterial agent.

cephalosporin. Any of a family of antibiotics related to penicillin, discovered in 1953; an important member of this group was synthesized by Woodward in 1966. Several cephalosporins are used clinically (cephalothin, cephaloridine, and cephalexin). The molecule contains a fused β-lactam-dihydrothiazine ring system with an N-acyl side chain and an acetoxy group attached to the dihydrothiazine ring. The formula for cephalosporin (C) is $C_{16}H_{21}N_3O_8S$. Cephalosporins are reported to be free from the allergic reactions common with penicillin. Development of new cephalosporin derivatives is being actively pursued.
See penicillin; antibiotic.

cephamycin. Any of a group of antibiotics related to cephalosporins and produced by several species of *Streptomyces*.

"Ceramer" [Celanese]. TM for a modified hydrocarbon wax.

ceramic. A product, manufactured by the action of heat on earthy raw materials, in which silicon and its oxide and complex compounds known as silicates occupy a predominant position (American Ceramic Society). The chief groups of the ceramics industry are as follows: (1) structural clay products (brick, tile, terra-cotta, glazed architectural brick); (2) whitewares (dinnerware, chemical and electrical porcelain, e.g., spark plugs, sanitary ware, floor tile); (3) glass products of all types; (4) porcelain enamels; (5) refractories (materials that withstand high temperatures); (6) Portland cement, lime, plaster, and gypsum products; (7) abrasive materials such as fused alumina, silicon carbide, and related

products; and (8) aluminum silicate fibers. A wide range of ceramics are available as ultrafine particles (10–150 microns), and ceramic foams are offered commercially.

ceramic, ferroelectric. A unique type of polycrystalline ceramic having properties that make possible the production of reliable, high-density optical memories for computers that are more efficient than conventional types. Lead zirconate titanate, heated and pressed into thin plates, is one of the compounds used. As a result of its ferroelectric properties, an applied voltage aligns the electric charges in the molecules of ceramic in the direction of the field and the polarization so induced remains indefinitely. Thus, the material accommodates itself to the requirements of the digital system, namely, binary 0 and binary 1.
See ferroelectric.

ceramic, glass. See glass ceramic.

ceramide. (E,2S,3R)-2-aminooctadec-4-ene-1,3-diol).
$C_{20}H_{39}NO_3$.
An amino alcohol with a long unsaturated hydrocarbon chain. It is the common lipid component of glycosphingolipids, composed of a long-chain basic alcohol (sphingosine) and an amide-linked fatty acid, bonded to 4-sphingenine).

"Ceramix" [Unimin]. TM for a technical grade of barium carbonate used in the ceramic industry.
Hazard: See barium.

"Ceraphyl" [ISP]. TM for octyldodecyl stearoyl stearate.
Available forms: Liquid.
Use: An emollient, pigment dispersant and binder. Imparts long-lasting lubricity and rich cushioned feel to the skin.

Ceratak. A grade of petroleum microcrystalline wax, min mp 73.8C.

Cerathane 63-L. An emulsifiable microcrystalline wax, min mp 93.3C.

Ceraweld. A grade of petroleum microcrystalline wax, min mp 73.8C.

Cercor. Thin-walled, cellular ceramic structures that can be used for a wide range of high-temperature applications.
Use: Gaseous heat exchangers, burner plates, acoustics, flame arresting, filtering, and insulation.

cerebral anthrax. A form of anthrax in which the bacilli have invaded the capillaries of the brain. It most often occurs in individuals with pulmonary or intestinal anthrax. In such cases, hemorrhagic meningitis complicates the effects of the pulmonary or intestinal anthrax.

cerebrosides. Derivatives of sphingosine in which the amino group is connected in an amide linkage to a fatty acid and the terminal hydroxyl group is connected to a molecule of sugar, usually galactose, in glycosidic linkage. They are found in brain and nervous tissue, usually in association with sphingomyelin.

Ceresan. A series of mercury compounds used as seed disinfectants.

ceresin wax. (purified ozocerite; earth wax; mineral wax; cerosin; cerin).
Properties: White or yellow, waxy cake; white is odorless, yellow has a slight odor. D 0.92–0.94, mp 68–72C. Soluble in alcohol, benzene, chloroform, naphtha; insoluble in water. Combustible.
Derivation: Purification of ozocerite by treatment with concentrated sulfuric acid and filtration through animal charcoal.
Grade: White, yellow.
Use: Candles, sizing, bottles for hydrofluoric acid, electrical insulation, shoe and leather polishes, impregnating and preserving agent, lubricating compounds, wood filler, floor polishes, antifouling paints, waxed papers, cosmetics, ointments, matrix compositions, waterproofing textile fabrics.

ceria. See ceric oxide; rare earth.

ceric ammonium nitrate. (cerium ammonium nitrate; ammonium hexanitratocerate).
$Ce(NO_3)_4 \cdot 2NH_4NO_3$.
Properties: Small, prismatic, yellow crystals. Soluble in water and alcohol; almost insoluble in concentrated nitric acid; soluble in other concentrated acids.
Derivation: By electrolytic oxidation of cerous nitrate in nitric acid solution and subsequently mixing solutions of cerium nitrate and ammonium nitrate followed by crystallization.
Hazard: Strong oxidizer, dangerous fire risk in contact with organic materials.
Use: Analytical chemistry, oxidant for organic compounds, polymerization catalyst for olefins, scavenger in the manufacture of azides.

ceric hydroxide. (ceric oxide, hydrated; cerium hydrate).
$CeO_2 \cdot xH_2O$.
Properties: Whitish powder when pure. Commercially a hydrated oxide containing 85–90% ceric oxide. Soluble in concentrated mineral acids; insoluble in water.
Derivation: By treating a solution of a ceric salt with strong alkali. Reagent grade is prepared by adding a saturated solution of ceric ammonium nitrate to an excess of ammonium hydroxide.
Grade: Commercial, high purity, reagent.

Use: Production of cerium salts and ceric oxide, opacifier in glasses and enamels (imparts yellow color), shielding glass.

ceric oxide. (cerium dioxide; cerium oxide; ceria).
CeO_2.
Properties: Pale-yellow, heavy powder (white when pure). Commercial product is brown. D 7.65, mp 2600C. Soluble in sulfuric acid; insoluble in water and dilute acid. Requires reducing agent with acid to dissolve the anhydrous oxide. Noncombustible.
Derivation: By decomposition of cerium oxalate by heat. Hardness depends on firing temperature.
Grade: Technical, high purity (99.8%).
Use: Ceramics; abrasive for glass polishing; opacifier in photochromic glasses; retarder of discoloration in glass, especially radiation shielding and color TV tubes; catalyst; enamels and ceramic coatings; phosphors; cathodes; capacitors; semiconductors; refractory oxides; diluent in nuclear fuels; heat stabilizer (alumina catalyst).

ceric sulfate. (cerium sulfate).
$Ce(SO_4)_2 \cdot 4H_2O$.
Properties: White or reddish-yellow crystals. D 3.91. Soluble in water (decomposes); soluble in dilute sulfuric acid. Strong oxidizer.
Derivation: Action of sulfuric acid on cerium carbonate.
Hazard: Fire risk in contact with organic materials.
Use: Dyeing and printing textiles, analytical reagent, waterproofing, mildew-proofing.

ceric sulfide. (cerium sulfide).
A high-temperature thermoelectric material that is stable and efficient up to 1100C.

cerin. See ceresin wax.

ceritamic acid. See ascorbic acid.

cerite. A rare-earth ore found chiefly in Sweden. A minor source of cerium.

cerium.
CAS: 7440-45-1. Ce.
A rare-earth element of the lanthanide group of the periodic table. Atomic number 58, aw 140.12, valences 3, 4. Four stable isotopes.
Properties: Gray, ductile, highly reactive metal. D 6.78, mp 795C, bp 3257C. Attacked by dilute and concentrated mineral acids and by alkalies. Readily oxidizes in moist air at room temperature. It has four allotropic forms. It is the second most reactive rare-earth metal. Cerium forms alloys with other lanthanides (see misch metal); it also forms a non-metal with hydrogen, as well as carbides and intermetallic compounds. Decomposes water.
Source: Cerite (Sweden), bastnasite (California, New Mexico), monazite (beach sands in Florida, Brazil, India, and South Africa).

Derivation: Chemical processing and separation of ores.

Grade: Granules, ingots, rods (99.9% pure).

Hazard: May ignite on heating to 300F (148.9C). Strong reducing agent.

Use: Cerium salts, cerium–iron pyrophoric alloys, ignition devices, military signaling, illuminant in photography, reducing agent (scavenger), catalyst, alloys for jet engines, solid-state devices, rocket propellants, getter in vacuum tubes, diluent in plutonium nuclear fuels.
See misch metal.

Note: Cerium compounds have been found to have antiknock properties, e.g., cerium (2,2,6,6-tetramethyl-3,5-heptanedionate).

cerium-141.   Radioactive cerium of mass number 141.

Properties: Half-life 21.5 days; radiation β and γ.

Derivation: From cerium-140 by capture of a neutron and emission of a γ photon.

Available forms: $CeCl_3$ in hydrochloric acid solution.

Hazard: Radioactive poison.

Use: Biological and medical research.

cerium-ammonium nitrate.   See ceric ammonium nitrate.

cerium carbonate.   See cerous carbonate.

cerium chloride.   See cerous chloride.

cerium dioxide.   See ceric oxide.

cerium hydrate.   See ceric hydroxide; cerous hydroxide.

cerium naphthenate.

Properties: A rubbery material, very difficult to dry. Almost insoluble without small quantities of organic stabilizers.

Derivation: By saponifying naphthenic acids and treating the sodium naphthenate formed with a suitable cerium salt. The commercial product is a mixture of rare-earth soaps.

Use: See soap (2).

cerium nitrate.   See cerous nitrate.

cerium oxalate.   See cerous oxalate.

cerium oxide.   See ceric oxide.

cerium sulfate.   See ceric sulfate; cerous sulfate.

cermet.   (ceramic + metal).
A semisynthetic product consisting of a mixture of ceramic and metallic components having physical properties not found solely in either one alone, e.g., metal carbides, borides, oxides, and silicides. They combine the strength and toughness of the metal with the heat and oxidation resistance of the ceramic material. The composition may range from predominantly metallic to predominantly ceramic; e.g., SAP-sintered aluminum contains 85% aluminum and 15% Al_2O_3. The most important industrial cermets are titanium carbide-based, aluminum oxide-based, and special uranium dioxide types. Cermets are made by powder metallurgy techniques involving use of bonding agents such as tantalum, titanium, and zirconium. High stress-to-rupture rate; operate continuously at 982C, for short periods at 2200C.

Use: Gas turbines; rocket motor parts; turbojet engine components; nuclear fuel elements; coatings for high-temperature resistance; sensing elements in instruments; seals; bearings, etc.; in special pumps and other equipment.

Cer-O-Cillin.   An antibacterial substance that differs from penicillin G in that the benzyl group is replaced with an allylmercaptomethyl group. In concentrations up to 500,000 units per cc in sterile water for injection, sterile sodium chloride for injection, or sterile 5% dextrose.

Use: Medicine.

"Cerophyl" *[Pines].*   TM for a lipid analogue that delivers long-lasting moisture to rinsed off hair products.

cerosin.   See ceresin wax.

cerotic acid.   (hexacosanoic acid; cerinic acid). $CH_3(CH_2)_{24}COOH$.
A fatty acid obtained from beeswax, carnauba wax, or Chinese wax.

Properties: White, odorless crystals or powder. D 0.8198 (100/4C), mp 87.7C, refr index 1.4301 (100C). Insoluble in water; soluble in alcohol, benzene, ether, acetone. Combustible.

cerous carbonate.   (cerium carbonate). $Ce_2(CO_3)_3 \cdot 5H_2O$.

Properties: White powder. Soluble in mineral acids (dilute); insoluble in water.

Derivation: By adding an alkali carbonate to a solution of a cerous salt.

cerous chloride.   (cerium chloride).
CAS: 7790-86-5. $CeCl_3 \cdot xH_2O$.

Properties: White crystals. Deliquescent, d 3.88 (anhydrous), mp 848C (anhydrous), bp 1727C. Soluble in water, alcohol, and acids.

Derivation: Action of hydrochloric acid on cerium carbonate or hydroxide.

Use: Incandescent gas mantles, spectrography, preparation of cerium metal, polymerization catalyst.

cerous fluoride.   (cerium fluoride).
CAS: 7758-88-5. $CeF_3 \cdot xH_2O$.

Properties: Off-white powder. D 6.16 (anhydrous), mp 1460C, bp 2300C. Insoluble in water and acids.
Derivation: By treating cerous oxalate with hydrofluoric acid.
Hazard: An irritant.
Use: In arc carbons to increase their brilliance; preparation of cerium metal.

cerous hydroxide. (cerium hydrate).
Approximate formula $Ce(OH)_3$.
Properties: White, gelatinous precipitate; yellow, brown, or pink when impurities are present. Soluble in acids; insoluble in water and alkali.
Derivation: Chief source is monazite sand.
Grade: Pure, crude.
Use: Pure form: to produce cerium salts, impart yellow color to glass, opacifying agent in glazes and enamels. Crude form: flaming arc lamp.

cerous nitrate. (cerium nitrate).
$Ce(NO_3)_3 \cdot 6H_2O$.
Properties: Colorless crystals, deliquescent. Mp loses $3H_2O$ at 150C; bp (decomposes at 200C). Soluble in water, alcohol, and acetone.
Derivation: Action of nitric acid on cerous carbonate.
Hazard: Strong oxidizer, fire risk in contact with organic materials.
Use: Separation of cerium from other rare earths, catalyst for hydrolysis of phosphoric acid esters.

cerous oxalate. (cerium oxalate).
$Ce_2(C_2O_4)_3 \cdot 9H_2O$.
Properties: Yellowish-white crystalline powder; odorless; tasteless. Decomposes on heating. Soluble in dilute sulfuric and hydrochloric acids; insoluble in water, oxalic acid solution, alkalies, alcohol, and ether.
Derivation: By extraction from monazite sand with oxalic or hydrochloric acid and conversion into the oxalate, followed by crystallization.
Grade: Pure; the commercial product is a complex mixture of oxalates of cerium, lanthanum, and didymium.
Hazard: Toxic by ingestion.
Use: Medicine, isolation of cerium metals.

cerous sulfate. (cerium sulfate).
$Ce_2(SO_4)_3 \cdot 8H_2O$.
Properties: White crystals or powder. Mp 630C (dehydrated), d 2.886. Soluble in water and acids.
Derivation: Reagent grade is prepared by reducing a solution of ceric sulfate in sulfuric acid with hydrogen peroxide.
Grade: Technical and purified (reagent).
Use: Developing agent for aniline black.

certified color. See food color; FD&C colors.

cerulean blue. A light-blue pigment, essentially cobaltous stannate, $CoO \cdot x(SnO_2)$.

cerulein. See caerulein.

ceruloplasmin.
CAS: 9031-37-2.
A copper-containing alpha globulin in blood plasma that contains type 1, type 2, and type 3 copper centers in which the type 2 and type 3 centers are near each other, constituting a trinuclear copper cluster. It may play a part in red blood cell production and oxygen reduction.

cerussite.
$PbCO_3$.
Natural lead carbonate, found in the upper zone of lead deposits.
Properties: Colorless, white, gray, adamantine luster. Mohs hardness 3–3.5, d 6.55. Effervesces in nitric acid.
Occurrence: Colorado, Arizona, New Mexico, Idaho, Australia, Europe.
Use: An ore of lead.

Cer-Vit. A glass ceramic having linear expansion coefficient near zero. Used for specialty products, such as telescope mirrors, where minimum distortion is essential.

ceryl alcohol.
$C_{26}H_{53}OH$.
An alcohol obtained from Chinese insect wax.
Properties: Colorless crystals. Mp 79C. Insoluble in water; soluble in alcohol and ether. Combustible.

ceryl cerotate.
$C_{26}H_{53}OOCC_{25}H_{51}$.
The chief constituent of Chinese insect wax and typical of natural waxes. Colorless crystals with mp 84C.

CES. Abbreviation for cyanoethyl sucrose.

cesium. (caesium).
CAS: 7440-46-2. Cs.
An alkali-metal element of group IA of the periodic table, atomic number 55, aw 132.9054, valence 1. No stable isotopes.
Properties: Liquid at slightly greater than room temperature, soft solid below the melting point. Highly reactive. Decomposes water with evolution of hydrogen, which ignites instantly. Also reacts violently with oxygen, the halogens, sulfur, and phosphorus with spontaneous ignition and/or explosion. D 1.90, mp 28C, bp 705C, Mohs hardness 0.2. Cesium has the highest position in the electromotive series; it also has the lowest melting point of any alkali metal and the lowest ionization potential of any element. Soluble in acids and alcohol.
Derivation: By thermochemical reduction of cesium chloride with calcium or by electrolysis of the fused cyanide. Its chief ore is pollucite, found in Maine, South Dakota, Manitoba, Elba, South Africa.
Grade: Technical, 99.9%.

Hazard: Dangerous fire and explosion risk, ignites spontaneously in moist air, may explode in contact with sulfur or phosphorus, reacts violently with oxidizing materials, causes burns in contact with skin.
Use: Photoelectric cells, getter in vacuum tubes, hydrogenation catalyst, ion propulsion systems, plasma for thermoelectric conversion, atomic clocks, rocket propellant, heat-transfer fluid in power generators, thermochemical reactions, seeding combustion gases for magnetohydrodynamic generators.

cesium-137.   Radioactive cesium of mass number 137.
Properties: Half-life 33 years; radiation β.
Hazard: Radioactive poison. The β decay of ^{137}Cs produces ^{137}Ba, which in turn is radioactive, emitting a 0.662 Mev γ ray with a 2.6-min half-life.
Use: Most applications of ^{137}Cs depend on the fact that any ^{137}Cs preparation has an equivalent amount of the γ-emitting ^{137}Ba daughter. Approved for sterilization of wheat, flour, potatoes, and sewage sludge.

cesium alum.   See cesium aluminum sulfate.

cesium aluminum sulfate.   (cesium alum). $CsAl(SO_4)_2 \cdot 12H_2O$.
Properties: Colorless crystals. D 2.0215, mp 117C. Soluble in water; insoluble in alcohol.
Derivation: By adding a solution of cesium sulfate to a solution of potassium alum, concentrating, and crystallizing.
Grade: Pure.
Use: Mineral waters, purification of cesium by fractional crystallizing, preparation of cesium salts.

cesium antimonide.
Hazard: Toxic by ingestion.
Use: As a high-purity binary semiconductor.

cesium arsenide.
Hazard: See arsenic and cesium.
Use: As a high-purity binary semiconductor.

cesium bromide.
CsBr.
Properties: Colorless, crystalline powder. D 4.44, mp 636C, bp 1300C. Soluble in water; slightly soluble in alcohol.
Grade: Technical, single crystals.
Use: Crystals for infrared spectroscopy, scintillation counters, fluorescent screens.

cesium carbonate.
Cs_2CO_3.
Properties: White, hygroscopic, crystalline powder. Very stable. Can be heated to high temperature without loss of CO_2. Soluble in water, alcohol, and ether.
Derivation: By passing carbon dioxide into a solution of cesium hydroxide and subsequent crystallization.

Use: Brewing, mineral waters, ingredient of specialty glasses, polymerization catalyst for ethylene oxide.

cesium chloride.
CsCl.
Properties: Colorless crystals. D 3.972, mp 646C, sublimes at 1290C. Soluble in water and alcohol; insoluble in acetone.
Derivation: Action of hydrochloric acid on cesium oxide and crystallization.
Grade: Pure; low optical density; single crystals.
Use: Brewing; preparation of other cesium compounds; mineral waters; evacuation of radio tubes (positive ions supplied at surface of filament); for a density gradient in ultracentrifuge separations; fluorescent screens; contrast medium.

cesium dioxide.
Cs_2O_2.
Properties: Yellow needles. D 4.25, mp 400C, decomposes at 650C. Soluble in water and acids.
Grade: Technical, pure.
Use: Cesium salts.

cesium fluoride.
CAS: 13400-13-0.   CsF.
Properties: Deliquescent crystals. D 4.115, mp 682C, bp 1251C. Very soluble in water; soluble in methanol, insoluble in dioxane and pyridine.
Grade: 99% min, single crystals.
Hazard: A poison.
Use: Optics, catalysis, specialty glasses.

cesium hexafluorogermanate.
Cs_2GeF_6.
Properties: White, crystalline solid. Mp approximately 675C, d 4.10. Slightly soluble in cold water and acids; very soluble in hot water.
Hazard: An irritant; see fluorides.

cesium hydrate.   See cesium hydroxide.

cesium hydroxide.   (cesium hydrate).
CAS: 21351-79-1.   CsOH.
Properties: Colorless or yellowish, fused, crystalline mass. Strong alkaline reaction. Hygroscopic. Keep well stoppered. D 3.675, mp 272.3C. Very soluble in water. The strongest base known.
Derivation: By adding barium hydroxide to an aqueous solution of cesium sulfate.
Grade: Technical, 50% aqueous solution.
Hazard: A poison. Skin, eye, and upper respiratory tract irritant.
Use: Recommended as electrolyte in alkaline storage batteries at subzero temperatures, polymerization catalyst for siloxanes.

cesium iodide.
CsI.
Properties: Colorless, crystalline powder. Deliquescent. D 4.510, mp 621C, bp 1280C. Soluble in alcohol and water.

Grade: Technical, single crystals.
Use: Crystals for infrared spectroscopy, scintillation counters, fluorescent screens.

cesium nitrate.
CAS: 7789-18-6. $CsNO_3$.
Properties: Crystalline powder, saltpeter taste. D 3.687, mp 414C, bp (decomposes). Soluble in water and acetone; slightly soluble in alcohol.
Derivation: Action of nitric acid on cesium oxide and crystallization.
Grade: Pure, 99.0% min.
Hazard: Dangerous, may ignite organic materials on contact.
Use: Cesium salts.

cesium oxide.
Cs_2O.
Properties: Orange-red crystals. D 4.36, mp 360–400C (decomposes). Very soluble in water; soluble in acids.
Grade: Technical, pure.
Use: Cesium salts.

cesium pentachlorocarbonylosmium(III).
Properties: Orange-red crystal. Air stable. Slightly soluble in water, hydrochloric acid; and insoluble in organic solvents.

cesium pentachloronitrosylosmium(II).
Properties: Dark-red crystals. Air-stable, sparingly soluble in water.
Use: Starting material for other cesium nitrosyl complexes.

cesium perchlorate.
$CsClO_4$.
Properties: Crystals. D 3.327 (4C), mp 250C (decomposes). Soluble in water (much more in hot than cold); slightly soluble in alcohol and acetone.
Grade: 99% min.
Hazard: Dangerous fire risk, strong oxidizing agent, may ignite organic materials on contact.
Use: Optics, catalysis, specialty glasses, power generation.

cesium peroxide.   See cesium tetroxide.

cesium phosphide.
Hazard: Fire risk by decomposition to phosphine, which is very toxic.
Use: As a high-purity binary semiconductor.

cesium sulfate.
Cs_2SO_4.
Properties: Colorless crystals. D 4.2434, mp 1010C. Soluble in water; insoluble in alcohol.
Derivation: Action of sulfuric acid on cesium carbonate.
Grade: Pure; low optical density.
Use: Brewing, mineral waters, for density gradient in ultracentrifuge separation.

cesium tetroxide.   (cesium peroxide).
Cs_2O_4.
Properties: Yellow crystals. D 3.77, mp 600C. Decomposes violently in water; soluble in acids. Strong oxidizing agent.
Hazard: Fire risk in contact with organic materials.

cesium trioxide.
Cs_2O_3.
Properties: Chocolate-brown crystals. D 4.25, mp 400C, decomposes in water. Soluble in acids.

cetalkonium chloride (USAN).
(cetyldimethylbenzylammonium chloride; benzyl-hexadecyldimethylammonium chloride).
$C_6H_5CH_2N(CH_3)_2(C_{16}H_{33})Cl$.
A quaternary ammonium germicide.
Properties: Colorless crystalline powder; odorless. Mp 58–60C. Soluble in water to form colorless; odorless solution with pH 7.2. Compatible with alkalies and antihistamines. Soluble in alcohol, acetone, esters, carbon tetrachloride.
Use: Germicide, fungicide, surface-active agent, mildew preventive.

cetane.   See *n*-hexadecane.

cetane number.   A rating for diesel fuel comparable to the octane-number rating for gasoline. It is the percentage of cetane ($C_{16}H_{34}$) that must be mixed with heptamethylnonane to give the same ignition performance under standard conditions as the fuel in question.

cetearyl alcohol.   See "Lanette O" *[BASF Care]*.

cetene.   See 1-hexadecene.

cetin.   (cetyl palmitate; palmitic acid cetyl ester).
$C_{15}H_{31}COOC_{16}H_{33}$.
Properties: White, crystalline, waxlike substance. Chief constituent of commercial purified spermaceti. Mp 50C, bp 360C, d 0.832, refr index 1.4398 (70C). Soluble in alcohol and ether; insoluble in water. Combustible.
Derivation: By solution from spermaceti.
Use: Base for ointments, cerates and emulsions, manufacture of candles, soaps, etc.

cetrimonium pentachlorophenoxide.
$C_{25}H_{42}Cl_5NO$.
A crystalline solid melting at 68C, used as a fungicide and as preservative for agricultural products.

cetyl alcohol.   (alcohol C-16; cetylic alcohol; 1-hexadecanol, normal; primary hexadecyl alcohol; palmityl alcohol).
CAS: 36653-82-4. $C_{16}H_{33}OH$.
A fatty alcohol. Combustible.
Properties: White, waxy solid; faint odor. D 0.8176 (49.5C), mp 49.3C, bp 344C, refr index 1.4283

(79C). Partially soluble in alcohol and ether; insoluble in water.
Derivation: By saponifying spermaceti with caustic alkali, reduction of palmitic acid.
Method of purification: Crystals, distillation.
Grade: Technical, cosmetic, NF.
Use: Perfumery, emulsifier, emollient, foam stabilizer in detergents, face creams, lotions, lipsticks, toilet preparations, chemical intermediate, detergents, pharmaceuticals, cosmetics, base for making sulfonated fatty alcohols, to retard evaporation of water when spread as a film on reservoirs or sprayed on growing plants.

c-cetylbetamine. An amphoteric product showing cationic properties in acid solutions, anionic properties in alkaline solutions.
Use: Alkalone peroxide bleaching systems.

cetyl bromide.
$C_{16}H_{33}Br$.
Properties: Dark-yellow liquid. Fp 15C, bp 186–197C (10 mm Hg), d 0.991 (25/25C), w/gal 8.25 lb (25C), refr index 1.460 (25C), flash p 350F (176C). Soluble in ether; very slightly soluble in water and methanol. Combustible.
Use: Synthesis.

cetyldimethylbenzylammonium chloride.
See cetalkonium chloride.

cetyldimethylethylammonium bromide.
$C_{16}H_{33}(CH_3)_2H_5NBr$.
A quaternary ammonium salt.
Properties: Paste.
Use: Disinfectant, deodorant, germicide, fungicide, detergents.

cetyldimethylethylammonium chloride.
$C_{16}H_{33}(CH_3)_2C_2H_5NCl$.
A quaternary ammonium salt.

cetylic acid. See palmitic acid.

cetylic alcohol. See cetyl alcohol.

cetyl mercaptan. (hexadecyl mercaptan; 1-hexadecanethiol).
CAS: 2917-26-2. $C_{16}H_{33}SH$.
Properties: Liquid; strong odor. Fp 18C, bp 185–190C (7 mm Hg), d 0.8474 (20/4C), refr index 1.474 (20C), flash p 275F (135C). Combustible.
Grade: 95% (min) purity.
Use: Intermediate, synthetic-rubber processing, surface-active agent, corrosion inhibitor.

cetyl palmitate. See cetin.

cetyl pyridinium bromide.
$C_{16}H_{33}C_5H_5NBr$.
A quaternary ammonium compound.

Properties: Cream-colored, waxy solid. Soluble in acetone, ethanol, and chloroform.
Use: Germicide, deodorant, laboratory reagent, surfactant.

cetylpyridinium chloride. (Ceepryn; Cepacol; Cetamium; Dobendan; 1-hexadecylpyridinium chloride; Medilave; Pristacin; Pyrisept).
CAS: 123-03-5. $C_{21}H_8ClN \cdot H_2O$.
A quaternary ammonium salt.
Properties: (Monohydrate) White powder. Mp 77–83C, surface tension (25C) 43 dynes/cm (0.1% aqueous solution). Soluble in water, alcohol, chloroform; very slightly soluble in benzene and ether.
Use: Antibacterial in cough lozenges and syrups; emulsifier.

cetyltrimethylammonium bromide. (hexadecyltrimethylammonium bromide).
CAS: 57-09-0. $C_{16}H_{33}(CH_3)_3NBr$.
A quaternary ammonium salt.
Properties: White powder. Soluble in water, alcohol, and chloroform.
Grade: Technical.
Use: Surface-active agent, germicide.

cetyltrimethylammonium chloride.
$C_{16}H_{33}(CH_3)_3NCl$.
A quaternary ammonium salt.
See hexadecyltrimethylammonium chloride.

cetyltrimethylammonium tosylate.
$[C_{16}H_{33}(CH_3)_3N]OSO_2C_6H_4CH_3$.
A high-temperature stable quaternary ammonium compound.
Use: Germicide, surfactant.

cetyl vinyl ether. (vinyl cetyl ether).
$C_{16}H_{33}OCH:CH_2$.
Properties: Colorless liquid. D 0.822 (27C), mp 16C, bp 142C (1 mm Hg), 173C (5 mm Hg), flash p 325F (162C) (OC), refr index 1.444 (25C). Combustible.
Grade: 97%.
Hazard: Toxic by inhalation, skin irritant. Reacts strongly with organic materials.
Use: Reactive monomer that may be copolymerized with a variety of unsaturated monomeric materials, including acrylonitrile, vinyl chloride, vinylidene chloride, and vinyl acetate to yield internally plasticized resins.

ceylon cinnamon bark oil. See cinnamon leaf oil.

Cf. Symbol for californium.

CF. Abbreviation for citrovorum factor. See folinic acid.

C-Fatty Acids. Fatty acids derived from coconut oil. The major component acids are lauric and myristic. They differ primarily in amount of unsaturated acid components and color.

Properties: Light-yellow solids that liquefy at approximately 25C. Obtained from naturally occurring triglycerides. Combustible.

Use: Intermediate, rubber compounding, cosmetic ingredients, buffing compounds, alkyd resins, emulsifiers, grease manufacture, and candles.

cfc-11. (cfc13; trichlorofluorocarbon; trichloro(fluoro)methane).
CAS: 75-69-4. CCl_3F.
It is one of two primary chlorofluorocarbons that undergo photolysis at high altitudes with the release of chlorine. It is thus an important contributor to ozone depletion.

Use: In aerosols.

CFE. Abbreviation for chlorotrifluoroethylene. Also used for polychlorotrifluoroethylene resins.

CFNP. See fluoronitrofen.

cga. (2-chloroethyl-tris(2-methoxyethoxy)silane).
$C_{11}H_{25}ClO_6Si$.
A codon of arginine that directs the placement of arginine into a polypeptide.

cgc. A codon of arginine that directs the placement of arginine into a polypeptide.

CGMP. (Current Good Manufacturing Practices).
Refers to the body of regulations that describe the methods, equipment, facilities, and controls required for producing human and veterinary products, medical devices, and processed food.

cgs. Abbreviation of centimeter-gram-second, a system of measurement used internationally by scientists.

cgu. A codon of arginine that directs the placement of arginine into a polypeptide.

chabazite.
$CaAl_2Si_4O_{12} \cdot 6H_2O$.
Essentially a natural hydrated calcium aluminum silicate, usually containing some sodium and potassium. A zeolite.

Properties: White, reddish, yellow, or brown color; vitreous luster. D 2.1. Mohs hardness 4–5.

Occurrence: New Jersey, Colorado, Oregon, Europe.

Use: Water treatment.

Chadwick-Goldhaber effect. Dissociation of an atomic nucleus due to absorption of γ-rays.

Chadwick, Sir James. (1891–1974). A British physicist who was awarded the Nobel Prize in 1935 for his discovery of the neutron (1932), the existence of which had been predicted by Rutherford. See neutron.

Chaetomium. A genus of fungi found typically in soil and plant debris and is a source of contamination in water-affected indoor environments. Some species of *Chaetomium* are causative agents of infections in humans and produce potent mycotoxins.

Hazard: Pathogenic, mycotoxins are highly toxic.

chain. A series of atoms of a particular element directly connected by chemical bonds, which constitutes the structural configuration of a compound. Such chains are usually composed of carbon atoms and are often shown without their accompanying hydrogen. Carbon chains may be of the following types:

(1) Open or straight chain: a sequence of carbon atoms extending in a direct line, characteristic of paraffins and olefins, the former being saturated

```
    H  H  H  H
    |  |  |  |
  H-C--C--C--C-H
    |  |  |  |
    H  H  H  H
```

and the latter unsaturated:.

```
    H  H  H  H
    |  |  |  |
    C=C--C=C
    |        |
    H        H
```

(2) Branched chain: a linear series of carbon atoms occurring in paraffinic hydrocarbons and some alcohols that is isomeric with its straight chain counterpart and has a subordinate chain of one or more carbon atoms:

```
    H       H       H
    |       |       |
  H-C-------C-------C-H
    |       |       |
    H    H-C-H      H
            |
            H
```

Such compounds are designated by the prefix *iso-*; the *iso*-paraffins are much more efficient in combustion than their straight-chain isomers, especially in gasoline. For example, octane has a low antiknock rating, whereas that of isooctane is high.

(3) Closed chain or ring: a cyclic arrangement of carbon atoms giving a closed geometric structure, i.e., a pentagon or other form, characteristics of alicyclic, aromatic, and heterocyclic compounds. See cyclic compound.

(4) Side chain: a group of atoms attached to one or more of the locations in a cyclic or heterocyclic compound, e.g., tryptophan:

chain initiation.   The activation of one molecule in a mass of inert molecules subject to a chain reaction.

chain isomerization.   Isomerization resulting from differences in arrangement of carbon atoms.

chain mechanism.   See free radical.

chain reaction.   See fission; nuclear energy.

chain stopper.   A substance or event that stops the growth of a chain polymerization.

chain termination.   The end of a chain reaction, when an activated molecule takes part in some reaction other than the one required to continue the chain.

chain transfer.   The interruption of growth of a molecule chain by formation of a new radical that may act as a nucleus for forming a new chain.

chalcocite.   (copper glance).
Cu_2S.
Natural cuprous sulfide, occurring with other copper minerals.
Properties: Lead-gray color, tarnishing dull black, metallic luster. D 5.5–5.8, Mohs hardness 2.5–3.
Occurrence: Montana, Arizona, Utah, Nevada, Alaska, Chile, Mexico, Europe.
Use: Important ore of copper.

chalcopyrite.   (copper pyrites; yellow copper).
$CuFeS_2$.
Natural copper–iron sulfide found in metallic veins and igneous rocks. Also made synthetically.
Properties: Yellow or bronze iridescent crystals with metallic luster and greenish streak. D 4.1–4.3, Mohs hardness 3.5–4. May carry gold or silver or mechanically intermixed pyrite.
Occurrence: Montana, Utah, Arizona, Tennessee, Wisconsin, Europe, Chile, Canada.
Use: Important ore of copper, semiconductor research.

Chalfie, Martin.   (1947–). An American chemist who was awarded the 2008 Nobel Prize jointly with Osamu Shimomura and Roger Y. Tsien. The scientists discovered and researched the green fluorescent protein (GPF), naturally produced in jellyfish (*Aequorea victoria*), which is used to illuminate certain cells to study their biological processes. Chalfie's research showed how the gene responsible GPF production could be used in other organisms so they could make their own GPF. He was awarded his Ph.D. from Harvard University and became a professor of biological sciences at Columbia University in 1982.

chalk.   A natural calcium carbonate composed of the calcareous remains of minute marine organisms. Decomposed by acids and heat. Odorless, tasteless.

See calcite; calcium carbonate; whiting; chalk, prepared.

chalk, drop.   See chalk, prepared.

chalk, French.   A variety of soapstone or steatite. See talc.

chalking.   A natural process by which paints develop a loose, powdery surface formed from the film. Chalking results from decomposition of the binder, due principally to the action of UV rays.

chalk, prepared.   (drop chalk; calcium carbonate, prepared).
Properties: Fine, white to grayish-white impalpable powder; often formed in conical drops. Odorless; tasteless; stable in air. Mp (decomposes at 825C with evolution of carbon dioxide). Decomposed by acids; practically insoluble in water; insoluble in alcohol. Noncombustible.
Derivation: By grinding native calcium carbonate to a fine powder, agitating with water, allowing the coarser particles to settle, decanting the suspension, and allowing the fine particles to settle slowly.
Use: Medicine (antacid), tooth powders, calcimine, polishing powders, silicate cements.
See whiting; calcium carbonate.

chamber acid.   Sulfuric acid made by the chamber process; 50°Bé sulfuric acid (62.18%).

chamber process.   An obsolete method for manufacturing sulfuric acid in lead-lined chambers from sulfur dioxide, air, and steam in the presence of NO_x as catalysts. It is no longer used in the U.S.

chamois.   A very soft, flexible leather made from the flesh layer of a split sheepskin by treating with fish oils, piling in contact with similarly treated skins, and allowing the fish oils to oxidize.

Chance–Claus process.   The process recovers sulfur from sulfide waste by treatment with carbon dioxide and oxidation of the resulting hydrogen sulfide with air in the presence of a catalyst.

channel black.   See carbon black.

channeling.   (1) Establishment of flow paths in a bed of solid particles through which a disproportionate quantity of the introduced liquid passes. (2) The direct transfer of a reaction product from the active site of one enzyme directly to the active site of a different enzyme, where it serves as a substrate.

Channing's solution.   (mercury potassium iodide).
Properties: Yellow prisms. Mw 620.47. Soluble in alcohol.
Use: Medicine.

Chapman rearrangement.   Thermal rearrangement of aryl imidates to *N,N*-diaryl amides.

charcoal, activated.   See carbon, activated.

charcoal, animal.   See bone black.

charcoal, bone.   See bone black.

charcoal, vegetable.   See vegetable black; carbon, activated.

charcoal, wood.   A highly porous form of amorphous carbon.
Derivation: Destructive distillation of wood.
Grade: Technical, in lumps, powder briquettes.
Hazard: Dangerous fire risk in briquette form or when wet. May ignite spontaneously in air.
Use: Chemical (precipitant in the cyanide process, precipitant of iodine and lead salts from their solutions, catalyst, calcium carbide); decolorizing and filtering medium; gas adsorbent; component of black powder and other explosives; fuel; arc-light electrodes; decolorizing and purifying oils; solvent recovery; deodorant.

Chardonnet, Hilaire Bernigaud, count de.
(1839–1924).
A native of France, he has been called the father of rayon because of his successful research in producing what was then called artificial silk from nitrocellulose. He was able to extrude fine threads of this semisynthetic material through a spinnerette-like nozzle, and the textile product was made on a commercial scale in several European countries. He was awarded the Perkin medal for this work.

charge-coupled device.   (CCD).
A solid-state detector that is capable of generating a measurable electric charge after being struck by a single photon. It is used in small-molecule X-ray diffractometers.

charge-mass ratio.   The ratio of change of electrified particle or ion to its mass, expressed usually as e/m.

Charles' law.   At constant pressure, the pressure of a confined gas is proportional to its absolute temperature.
See Gay-Lussac's law.

Charpy.   A standard testing device for impact strength.

Chattock gauge.   A differential manometer that uses the pressure difference of two columns of liquid of nearly equal density.

chaulmoogra oil.   (gynocardia oil; hydnocarpus oil).
Properties: Brownish-yellow oil or soft fat; characteristic odor; somewhat acrid taste. D 0.940; iodine value 85–105 (based on type), optically active. Soluble in ether, chloroform, benzene, solvent naphtha; sparingly soluble in cold alcohol; almost entirely soluble in hot alcohol, carbon disulfide.

Chief constituents: Glycerides of chaulmoogric and hydnocarpic acids.
Use: Largely superseded for treatment of leprosy and other infective skin diseases.

chaulmoogric acid.   (hydnocarpyl acetic acid).

$CH_2CH_2CHCHCH(CH_2)_{12}COOH.$ A cyclic fatty acid.
Properties: Colorless, shiny leaflets. Mp 68.5C. Soluble in ether, chloroform, and ethyl acetate.
Source: Chaulmoogra oil.
Use: Medicine, biochemical research.

Chauvin, Yves.   (1930–2015). Born in France, Chauvin won the Nobel Prize for chemistry in 2005 for his pioneering work concerning the development of the metathesis method in organic synthesis. Chauvin received his degree from the Lyons School of Chemistry, Physics and Electronics in 1954. He is honorary research director at the Institut Francis du Petrole and a member of the French Academy of Science.

chavicol.   (p-allylphenol; 1-allyl-4-hydroxybenzene).
$C_3H_5C_6H_4OH.$
Properties: Liquid. Mp 16C, bp 230C, d 1.033 (18/4C). Soluble in water and alcohol. Occurs in many essential oils.

CHDM.   See 1,4-cyclohexanedimethanol.

checking.   Development of small cracks in the surface of a material such as rubber, paint, or ceramic glaze.

Chel.   Chelating agents based on polyaminocarboxyllic acids.
Use: To reduce the harmful effects of trace metals in the blood.

chelidonine.
CAS: 476-32-4. $C_{20}H_{19}NO_5.$
Properties: White, crystalline powder alkaloid.
Derivation: Isolated from poppies such as *Chelidonium majus.*
Hazard: Poison; central nervous system depressant causing sleepiness, depression, slowing of the pulse, coma and circulatory failure.

chelate.   The type of coordination compound in which a central metal ion such as Co^{2+}, Ni^{2+}, Cu^{2+}, or Zn^{2+} is attached by coordinate links to two or more nonmetal atoms in the same molecule, called ligands. Heterocyclic rings are formed with the central (metal) atom as part of each ring. Ligands offering two groups for attachment to the metal are termed *bidentate* (two-toothed); three groups, *tridentate*; etc.
A common chelating agent is ethylenediaminetetraacetic acid (EDTA). Nitrilotriacetic acid $N(CH_2COOH)_3$ and ethyleneglycol-bis(β-aminoethyl

ether)-*N*,*N*-tetraacetic acid ($HOOCCH_2$)$_2$NCH$_2$ CH$_2$OCH$_2$CH$_2$OCH$_2$CH$_2$N(CH$_2$COOH)$_2$ are used in analytical chemical titrations and to remove ions from solutions and soils. Metal chelates are found in biological systems, e.g., the iron-binding porphyrin group of hemoglobin and the magnesium-binding chlorophyll of plants. Medicinally, metal chelates are used against gram-positive bacteria, fungi, viruses, etc.
See ammine; sequestration.

Chem-Hoe.   Isopropyl-*N*-phenylcarbamate (IPC). A selective herbicide.

***Chemical Abstracts*.**   A weekly publication of the American Chemical Society that consists of research articles and patents in all major fields of chemistry throughout the world. It is completely computerized, including the ability to draw chemical structures as the basis for a search of the database, and available in various forms from several computerized services. It is the most indispensable information source in chemical literature and is the largest scientific abstract journal in the world. For further information, see Appendix II.

Chemical Abstracts Service.   Division of the American Chemical Society responsible for the production of Chemical Abstracts.
See *Chemical Abstracts*.

chemical bond.   See bond, chemical.

chemical change.   Rearrangement of the atoms, ions, or radicals of one or more substances resulting in the formation of new substances often having entirely different properties. Such a change is called a chemical reaction. In some cases, energy in the form of heat, light, or electricity is required to initiate the change; then the reaction is called endothermic. When energy is given off as a result of rupture of chemical bonds, the change is said to be exothermic. Chemical changes should be distinguished from physical changes, in which only the state or condition of a substance is modified, its chemical nature remaining the same. A physicochemical change has some of the characteristics of both. Examples of the three types are

Chemical Changes
fuel + oxygen → CO_2 + water + heat (exothermic)
water + CO_2 + energy → sugar + oxygen (endothermic)

Physical Changes
water to ice or steam; crystallization; coagulation of latex distillation processes

Physiochemical Changes
cooking of food, vulcanization of rubber, tanning of leather, drying of oil- or plastic-based paints.

chemical data storage.   The recording, coding, storage, and retrieval of chemical information. Early methods involved the use of punched-card systems, of which there were several types. The advent of digital computers made it possible to store and retrieve such data as molecular structure, physical and chemical properties, and other relevant facts with much greater efficiency. The state of the art in chemical documentation is the sophisticated and extensive information bank assembled by Chemical Abstracts Service, a division of ACS. It features on-line call-up and display of the most complex organic formulas and property data for more than 12 million compounds. Personal computers are used for laboratory instrument control, data collection, and analysis.

chemical dating.   Estimation of the age of geologic structures and events by measuring the amount of radioactive decay products in existing samples. The age of a uranium-containing material can be determined by measuring the percentage of lead (or helium) formed as a result of disintegration of the uranium. Uranium decays to both helium and the 206 lead isotope, but measurement of helium content is inaccurate because of its strong tendency to escape. By determining the ratio of the percentage of lead in a sample to the percentage of uranium, the age can be calculated. A more recent method, applicable to events within about 10,000 years, involves the use of the natural radioactive carbon isotope (^{14}C). The percentage of this isotope determined in a carefully prepared sample is an index of its age, based on the half-life of ^{14}C (5700 years), which was present in the atmospheric carbon dioxide absorbed by plants centuries ago. This method has yielded valuable results in the study of archaeological specimens, deep-sea sedimentation, and dates of volcanic and glacier activity.

chemical deposition.   The precipitation of one metal from a solution of one of its salts by adding to it another metal higher in the electromotive series or a reducing agent.

chemical economics.   The principles and practical application of industrial economics, in particular reference to the manufacture of chemicals and chemically derived products. Economic factors that apply mainly to the chemical industries are (1) heavy capital investment; (2) vast range of raw materials and products; (3) complex and varied production methods; (4) high-level R&D for new products, plus extensive testing for safety and acceptability; (5) market research to develop new product uses, etc.; (6) substitution of synthetics for natural products or (in some cases) the reverse; (7) utilization of by-product materials; and (8) waste reclamation and pollution control.
See chemical process industry.

chemical education.   The instruction and training of students at secondary, college, and graduate levels in both the theoretical and practical aspects of chemistry. Well-balanced courses include a substantial amount of laboratory experimentation

in addition to lectures and textbook study. Comprehensive 1-year courses are offered by most colleges for nonscience majors. The more advanced curricula emphasize quantum-mechanical and thermodynamic considerations, stoichiometry, and spectroscopy. Mechanical models of molecular structure are useful in teaching organic chemistry. A few schools provide courses in industrial chemistry. Student employment in a chemical process plant for part of the semester has proved successful. Tape recordings of symposia, analytical methods and lectures in general chemistry, as well as a broad group of high-level short courses in special subjects are offered by the ACS, which also sponsors radio broadcasts on topics of popular interest in chemistry. The field is well served by two publications the—*Journal of Chemical Education* and *Chemistry*. Among the leaders in the development of chemical education may be mentioned Ira Remsen, James B. Conant, Gilbert N. Lewis, Louis Fieser, Roger Adams, and Joel Hildebrand. In recent years, the pervasive influence of computer technology has penetrated almost all of the physical sciences. Thus basic training in its applications to chemistry has become an essential adjunct to chemical education. The publication of data and scientific papers on the World Wide Web and collaboration via the Internet is of increasing importance.
See chemical literature; model; computational chemistry.

chemical efficiency. The output of a process divided by its net input.

chemical energy. The energy given off in a chemical reaction.

chemical engineering. That branch of engineering concerned with the production of bulk materials from basic raw materials by large-scale application of chemical reactions worked out in a laboratory. The unit operations of chemical engineering are those common to all chemical processing: fluid flow, heat transfer, filtration, evaporation, distillation, drying, mixing, adsorption, solvent extraction, and gas absorption, based on the fundamentals involved in every technical problem in chemical engineering: conservation, equilibrium, kinetics, and control. Conservation involves the laws of the conservation of matter and energy. The limits of any chemical process are established by its equilibrium conditions. The principles underlying the equilibrium concept (the law of mass action) are of fundamental importance (see chemical laws). Kinetics means time-dependent processes or rate processes (momentum, thermal, mass transfer, and chemical kinetics). Control involves either the systems approach or feedback or closed-loop methods, as well as considerations of the stability of the system.
See equilibrium constant; stoichiometry; kinetics, chemical.

chemical equation. A representation of a chemical reaction using symbols to show the weight relationship between the reacting substances and the products.

chemical equivalent. (1) The atomic weight of an element divided by its valence. (2) The number of parts by weight of an element or radical that will combine with or displace eight parts by weight of oxygen or one part by weight of hydrogen. (3) The molecular weight of a salt divided by the valence of the particular element considered.

chemical fallout. Pollution that results from chemicals in the air being deposited on land or water surfaces.

chemical flooding. A method of enhanced oil recovery in which a mixture of detergents, alcohols, and other solvents is pumped into the porous oil-bearing strata, followed by much brine, which furnishes the pressure necessary to drive the mixture through the rocky structure. Field tests show this process to be no more than 50% efficient, and it will probably not be important in secondary oil recovery for some years. Another technique, called micellar flooding, involves the use of surfactants and a mixture of water and a high polymer. It is a two-step process: the surfactants (petroleum sulfonate and alcohol) are injected into the oil-bearing strata, followed by the water-polymer mixture. Volumes of these materials are required.
See hydraulic fracturing.

chemical force microscopy. A technique in which specific chemical groups are attached to the probe tip of an atomic force microscope.
Use: Study of adhesion and lubrication; mapping functional group microstructure in polymers and binding-recognition interactions in biological systems.
See: atomic force microscopy; scanning tunneling microscope.

chemical laws. A group of basic principles governing the combining power and reaction characteristics of elements. Among the more important are:
(1) *Law of mass action*: The rate of a homogeneous (uniform) chemical reaction at constant temperature is proportional to the concentration of the reacting substances.
(2) *Law of definite or constant composition*: any chemical compound contains the same elements in the same fixed proportions by weight. Exceptions to this law occur in solid compounds, such as silicates, that are known as nonstoichiometric compounds.
(3) *Law of multiple proportions*: When two elements unite to form two or more compounds (e.g., nitrogen and oxygen can form five different oxides), the weight of one element that combines with a given weight of the other is in the ratio of small whole

numbers. Hydrogen and oxygen unite in the ratio of 1 to 8 in water and of 1 to 16 in hydrogen peroxide. Thus, the weights of oxygen that unite with 1 g of hydrogen are in the ratio of 1 to 8.

(4) *Law of conservation of mass*: Any chemical reaction between two or more elements or compounds leaves the total mass unchanged, the reaction products having exactly the same mass as present in the reactants, regardless of the extent to which other properties are changed.

(5) *Law of Avogadro*: Equal volumes of gases at constant temperature and pressure contain the same number of molecules whether the gases are the same or different. 22.4 L of any gas contains 6.02×10^{23} molecules.

See Avogadro's law; mole.

chemical literature. Worldwide chemical information available in the form of journals, patents, and books (handbooks, dictionaries, encyclopedias, reports of conferences and symposia, review volumes, research treatises, and textbooks). The output of journal articles has increased almost exponentially in the last 70 years. Of major importance in making this information explosion available to scientists are the abstract journals, particularly *Chemical Abstracts*, described in Appendix II.

Outstanding among periodicals is the *Journal of the American Chemical Society*, in which more than 2000 articles appear yearly. Hundreds of other journals throughout the world add to the vast stream of chemical literature.

In book form, there are four monumental multi-volume compendia (the dates are for the original edition, but in most cases there have been several revisions): Abegg and Gmelin's *Handbuch der Anorganischen Chemie* (1870), Beilstein's *Handbuch der Organischen Chemie* (1881), Mellor's *Inorganic and Theoretical Chemistry* (1922), and Heilbron's *Dictionary of Organic Compounds* (approximately 1920). Other noteworthy reference sources are *Merck Index* (1889), *Handbook of Chemistry and Physics* (1918), Lange's *Handbook of Chemistry* (1934), *Chemical Engineers Handbook* (1934), *Ring Index* (1940), Hackh's *Chemical Dictionary* (1929), and Kirk-Othmer's *Encyclopedia of Chemical Technology* (1947). The ACS Monograph Series, established in 1919, is devoted to high-level treatises in developing areas of chemistry. Many useful series of "Advances," reviewing progress in specific fields, have appeared from time to time. Many of these publications are available for computerized information retrieval.

See chemical education.

chemically pure. (CP). Refers to an element or compound that has been purified for fine chemical work.

Chemical Mace. See Mace.

chemical microscopy. See microscopy, chemical.

chemical milling. The process of producing metal parts of predetermined dimensions by removing metal from the surface with chemicals. Acid or alkaline pickling or etching baths are used for this purpose. Immersion of a metal part will result in uniform removal of metal from all surfaces exposed to the solution. This is used by the aircraft industry for weight reduction of large parts. It is also used in the manufacturing of instruments and other components where exact tolerances are required.

chemical nomenclature. The origin and use of the names of elements, compounds, and other chemical entities, as individuals and as groups, as well as various proposals for systematizing them. It may be considered to include three major aspects: (1) the gradual sporadic development of these names, which may go back to the alchemists of the Middle Ages; (2) the proliferation of terminology due to rapid extension of organic chemistry in the mid-19th century, which led to the recommendations of the Geneva System in 1892; and (3) the additional reforms adopted by the International Union of Pure and Applied Chemistry in 1930. There is yet no clear-cut elimination of the older names, in spite of changes introduced in these reformed systems. Thus, present nomenclature is in some respects a hybrid; for example, the earlier terms *paraffin* and *olefin* are still widely used instead of the more modern *alkane* and *alkene*, and *methyl alcohol* is still in common use for *methanol*.

A comparatively recent development in the nomenclature of inorganic and complex compounds is use of the Stock system, in which Roman numerals indicate the oxidation state or coordination value. For example, iron(II) chloride stands for ferrous chloride ($FeCl_2$), and iron(III) chloride for ferric chloride ($FeCl_3$).

See Geneva System; benzene.

chemical oxygen demand. (COD). Refers to the amount of oxygen, expressed in parts per million, consumed under specified conditions in the oxidation of the organic and oxidizable inorganic matter contained in an industrial wastewater, corrected for the influence of chlorides.

See oxygen consumed.

chemical planetology. Application of various branches of chemistry (analytical, physical, and geochemistry) to study the composition of the surface and atmosphere of the planets, mainly Venus, Mars, and Jupiter. Much information has been obtained by spectrographic methods, and valuable additional data have resulted from space probes.

See astrochemistry.

chemical potential. The rate at which the total free energy of a phase in a system changes as the amount of a particular component changes, keeping pressure, temperature, and amounts of all other components of the phase the same.

chemical process industry. An industry whose products result from (1) one or more chemical or physicochemical changes; (2) extraction, separation, or purification of a natural product with or without chemical reactions; and (3) preparation of specifically formulated mixtures of materials, either natural or synthetic. Examples are as follows (with allowance for some overlapping): (1) the plastics, rubber, leather, food, dye, and synthetic organic industries; (2) the petroleum, paper, textile, and perfume industries; and (3) many of these involve one or more unit operations of chemical engineering, as well as such basic processes as polymerization, oxidation, reduction, hydrogenation, etc., usually with the aid of a catalyst. This definition may be interpreted to include ore processing, separation, and refinement, as well as the manufacture of metal products; however, these are usually considered to comprise the metal and metallurgical industries.

chemical reaction. A chemical change that may occur in several ways, e.g., combination, replacement, by decomposition, or by some modification of these. Reactions are endothermic when heat is needed to maintain them and exothermic when they evolve heat. All chemical reactions are in balance; i.e., the numbers of atoms of the various elements in the reacting substances are always equal to the numbers of atoms in the reaction products. Common types of reactions are oxidation, reduction, ionization, combustion, polymerization, hydrolysis, condensation, enolization, saponification, rearrangement, etc. Chemical reactions involve rupture of only the bonds that hold the molecules together and should not be confused with nuclear reactions, where the atomic nucleus is involved. A reversible reaction is one in which the reaction product is unstable and thus changes back into the original substance spontaneously. In a complete reaction, the activity goes to completion and is indicated by an arrow →; if heat or a catalyst is used, it is indicated by a symbol or word, usually placed in small type above the arrow as:

$$\Delta \qquad \text{catalyst}$$
$$\rightarrow \qquad \rightarrow$$

A reversible reaction is shown by ↔.

chemical research. See applied research; fundamental research.

chemical sediment. A sediment created by precipitation of one or more minerals from natural waters.

chemical smoke. Chemically generated aerosols, used primarily for military purposes. They are of four types: (1) FS, a mixture of sulfuric anhydride and chlorosulfonic acid, used in shells and bombs and sprayed from airplanes; (2) FM, titanium tetrachloride, the same as FS but brilliant white and will drop like a curtain when sprayed; (3) HC, a mixture of hexachloroethane, aluminum, and zinc oxide, burns to yield a white cloud; and (4) WP, a white phosphorus, burns to form a white cloud of phosphoric acid, an excellent smoke producer.
See fog; smoke; chemical warfare.

chemical specialty. A chemically formulated product manufactured from chemical components and used without further processing by household and industrial customers for specific and specialized purposes.

Chemical Specialties Manufacturers Association. (CSMA).
Founded in 1914, the Chemical Specialties Manufacturers Association represents an array of companies engaged in the manufacture, formulation, distribution and sale of chemical specialty products for household, institutional and industrial use. The Association's mission is to: foster high standards for the industry; and concern for the health, safety and environmental impacts of its products; address legislative and regulatory challenges at the federal, state and local levels; meet the needs of industry for technical and legal guidance; and provide a forum to share ideas for scientific and marketing excellence. Its headquarters are at 1667 K Street, NW, Suite 300, Washington, DC 20006. Website: http://www.cspa.org

chemical stoneware. (brick, chemical).
A clay pottery product widely employed to resist acids and alkalies. It is used for utensils, pipes, stopcocks, ball mills, laboratory sinks, etc.

chemical technology. A general term covering a broad spectrum of physicochemical knowledge of the materials, processes, and operations used in the chemical process industries. It includes (1) basic phenomena such as activation, adsorption, oxidation, catalysis, corrosion, surface activity, polymerization, etc.; (2) the properties, behavior, and handling of industrial materials and products (plastics, textiles, coatings, soap, foods, metals, pharmaceuticals, etc.); and (3) their formulation, fabrication, and testing (compounding, extruding, molding, assembly, and the like).
See chemical process industry.

chemical thermodynamics. That aspect of thermodynamics concerned with the relationship of heat, work, and other forms of energy to equilibrium in chemical reactions and changes of state.
See thermodynamics; thermochemistry; kinetics, chemical; equilibrium constant.

Chemical Transportation Emergency Center.
See ChemTrec.

chemical warfare.   The employment of a chemical agent directly for military purposes, i.e., to cause casualties by irritating, burning, asphyxiation, or poisoning; to contaminate the ground; to screen action by smoke; or to cause incendiary damage. Includes use of all forms of toxic or irritating gases, including nerve gases, smoke-inducing agents, flammable gels such as napalm, and incendiary materials such as magnesium and thermite. Biological warfare involves the use of bacteria and other infective agents against enemy forces or populations. Chemical warfare agents derived from biological organisms can be classified as biological or chemical warfare agents.

chemical warfare agent.   (warfare agent; war poison; military poison; military gas; warfare poison).
Any warfare substance intended for use in military operations developed to incapacitate, injure, kill, or demoralize enemy forces or used to deprive the enemy of resources such as food, water, or shelter. Such agents have been used on civilian populations and may cause substantial collateral damage.
Hazard: Toxic; poison; nerve damage; incapacitation; radiation; blisters; choking; psychological effects.
Use: In military operations as a warfare agent.

chemical waste.   Unusable by-products from many chemical and metal-processing operations, which often contain toxic or polluting materials, such as digester streams from paper manufacture, fluorides from aluminum manufacturing, mercury from chlor-alkali cells, tailings from asbestos production, and insecticidal wastes that become environmental threats if improperly disposed of. Disposal techniques approved by EPA are (1) landfill in which an impermeable barrier is placed between the waste and ground water, such as a layer of hard-packed clay; (2) incineration of organic materials plus use of specially equipped ships in which liquid wastes are incinerated at sea; and (3) such chemical and biological methods as neutralization of acidic and basic wastes, oxidation, and activated sludge treatment of organic wastes. Dumping of wastes into lakes and watercourses is strictly forbidden. Storage in metal drums is inadequate because of corrosion and subsequent leakage. Federal authority to provide for control and safe disposal of hazardous wastes is provided by the Resource Conservation and Recovery Act (RCRA), passed by the U.S. Congress in 1976.
See radioactive waste.

chemical weapon.   A device that disperses or disseminates a chemical agent intended to cause harm to humans or living systems.

Chemico process.   A technique used for extracting sulfur from low-grade ores (25–50% sulfur) by means of hot water.

chemiluminescence.   The emission of absorbed energy (as light) due to a chemical reaction of the components of the system. It includes the subclasses bioluminescence and oxyluminescence, in which light is produced by chemical reactions involving organisms and oxygen, respectively. Chemiluminescence occurs in thousands of chemical reactions covering a wide variety of compounds, both organic and inorganic. Emission of light by fireflies is a common example of bioluminescence.
See luminescence.

chemisorption.   The formation of bonds between the surface molecules of a metal (or other material of high surface energy) and another substance (gas or liquid) in contact with it. These bonds are comparable in strength to ordinary chemical bonds and much stronger than the Van der Waals type characterizing physical adsorption. Chemisorbed molecules are often altered. Hydrogen is chemisorbed on metal surfaces as hydrogen atoms. Chemisorption of hydrocarbons may result in formation of chemisorbed hydrogen atoms and hydrocarbon fragments. Even when dissociation does not occur, the properties of the molecules are changed by the surface in important ways. This mechanism is the activating force of catalysis.
An example of chemisorption is the boundary lubrication of moving metal parts in machinery. An oil film forms a chemisorbed layer at the interface and averts the high frictional forces that would otherwise exist. Solids with high surface energies are necessary for chemisorption to occur, e.g., nickel, silver, platinum, iron.
See catalysis.

chemistry.   A basic science whose central concerns are (1) the structure and behavior of atoms (elements); (2) the composition and properties of compounds; (3) the reactions between substances, with their accompanying energy exchange; and (4) the laws that unite these phenomena into a comprehensive system. Chemistry is not an isolated discipline; it merges into physics and biology. The origin of the term is obscure. Chemistry evolved from the medieval practice of alchemy. Its bases were laid by such men as Boyle, Lavoisier, Priestly, Berzelius, Avogadro, Dalton, and Pasteur.
See inorganic chemistry; organic chemistry; physical chemistry; and Appendix II A–E.
Note: Chemistry has been variously defined, to the point where definition has become a semantic exercise of questionable, if not negative, value. "Chemistry is the science of matter" and "Chemistry is a branch of physics" are two instances of such definitions. The first relegates physics to the background while the second accords it supremacy.

chemistry history.   See Appendix II A–E.

chemistry in space.   See space, chemistry in.

chemodynamics. A comprehensive, interdisciplinary study of chemicals in the environment with special reference to pesticides.

chemokine. Any of a class of pro-inflammatory cytokines that are able to attract and activate leukocytes.

chemometrics. Application of computer data-analysis techniques to the classification, assimilation, and interpretation of chemical information. Its major purpose is to correlate data in such a way that trends or patterns are indicated. Molecular spectra, thermodynamic functions, and distribution of chemicals in the atmosphere in relation to rainfall are some fields in which this science has been utilized. Formal programs in chemometrics are under way at the University of Washington and Umea University in Sweden.

"Chemonite" *[Arch].* (copper arsenite, ammonical).
TM for a wood-preservative solution prescribed by Federal Specification TT-W-549 to contain copper hydroxide $(Cu(OH)_2)$ 1.84%, arsenate, arsenic pentoxide (As_2O_5) 1.3%, ammonia (NH_3) 2.8%, acetic acid 0.05%, and water as necessary to 100.0%.
Hazard: Toxic by ingestion.

chemonuclear production. Manufacture of chemicals using the energy of a nuclear reactor. Feasibility studies on making hydrogen cyanide from nitrogen and methane using fission fragments as the energy source for the heat of reaction indicated that this process cannot yet compete economically with standard methods. In the case of hydrazine, the economic aspect is more favorable because of the high cost of conventional methods.

chemosetting. Hardening through a chemical reaction.

chemosmosis. A chemical reaction with the use of a semipermeable membrane.

chemosorption. Adsorption in which chemical energy causes an accumulation of the disperse constituent.

chemosterilant. A term coined by the USDA for materials or processes that sterilize insects, usually the males, thus preventing their reproduction. γ-irradiation is one method used. The males are brought to the radiation by sex attractants.

chemotaxis. The tendency of certain bacteria, especially *E. coli* and *Salmonella*, to move toward nutrients and other attractants and away from repellents and toxic chemicals. Such bacteria live in a liquid medium and propel themselves through it by means of threadlike appendages called flagellae. They apparently can determine differences in the concentration of chemicals in the medium as they move through it, thus exhibiting a memory in respect to their environment.

chemotherapy. The treatment or prevention of a disease by administration of a chemical. The term was first used by Paul Ehrlich, discoverer of the arsphenamine treatment for syphilis (1910). He said that chemotherapy results from the interaction of chemically reactive groups on drugs with chemically active receptor groups on parasitic cells and that an effective drug must be of quite low molecular weight. His achievement was one of the great triumphs of biomedical science. More-recent important achievements are the development of antimalarials, the synthesis of sulfa drugs, the discovery and proliferative development of antibiotics (penicillin, streptomycin, etc.), and the synthesis of cortisone. Much research effort has been devoted to the chemotherapeutic investigation of cancer. An antibiotic, adriamycin, is said to be effective against certain types of cancer. Treatment with ^{60}Co, either by irradiation or as a tissue implant, has been successfully used. A broad spectrum of antidepressant drugs to treat acute mental depression is another outstanding development of chemotherapy.

chemotroph. An organism that obtains energy by metabolizing organic compounds derived from other organisms.

chemotropism. The reaction of living things to chemicals.

CHEMRAWN. Abbreviation of Chemical Research Applied to World Needs, a program established by the International Union of Pure and Applied Chemistry and cosponsored by the American Chemical Society and the Chemical Institute of Canada to discuss and study the problems of long-range conversion to a postpetroleum economy. The subject range includes all forms of natural raw materials, particularly those used for fuel, food, and the manufacturing of organic chemicals. The conference's first meeting was held in Toronto in 1978.
Website: http://www.iupac.org/standing/chemrawn.html

ChemTrec. Abbreviation of Chemical Transportation Emergency Center was founded in 1971 by the American Chemistry Council, formerly known as the Chemical Manufacturers Association, as a public service to provide 24-hour information to emergency responders at the site of emergencies involving hazardous materials. The offices are at 1300 Wilson Blvd., Arlington, VA 22209.
Website: http://www.chemtrec.org/

chemurgy. Development of nonfood uses for farm products, especially such waste materials as

301

CHIRASELECT

corncobs, nutshells, etc. This term was introduced in 1935 but has since become obsolete.
See biomass; biotechnology.

chenopodium oil. (wormseed oil; American goose-foot oil).
CAS: 8006-99-3.
Properties: Colorless or yellowish oil; characteristic penetrating odor; bitterish burning taste. D 0.965–0.990 (15C), optical rotation −4 to −8 degrees, refr index 1.4740–1.4790 (20C). Soluble in 3–10 volumes of 70% alcohol (inferior and adulterated oils do not yield a clear solution).
Chief known constituents: Ascaridole ($C_{10}H_{16}O_2$); o-cymene; l-limonene.
Derivation: Distilled from the seeds and leaves of *Chenopodium ambrosioides anthelminticum*.
Use: Medicine (anthelmintic).

chert. An impure rock, often gray in color, that consists primarily of extremely small quartz crystals precipitated from water solutions.

(CHF). See Chemical Heritage Foundation.

chi acid. See anthraquinone-1,8-disulfonic acid.

Chicago acid. (1-amino-8-naphthol-2,4-disulfonic acid; 8-amino-1-naphthol-5,7-disulfonic acid; SS acid; 1,8,2,4-acid; 2-sulfur acid).

Properties: Gray paste, white when pure. Soluble in water and sodium hydroxide solution.
Derivation: 1-Naphthylamine-8-sulfonic acid is reacted with sulfuric acid to yield 1-naphthylamine-4,8-disulfonic acid; further reaction with 25% oleum gives 1,8-naphthosulfam-2,4-disulfonic acid, which melts at 155C with 40% NaOH.
Use: Azo dye intermediate.

Chichibabin pyridine synthesis. Condensation of carbonyl compounds with ammonia or amines under pressure to form pyridine derivatives; the reaction is reversible and produces different pyridine derivatives and by-products.

Chichibabin reaction. Amination of pyridines and other heterocyclic nitrogen compounds with alkali-metal amides.

chicle. A thermoplastic, gumlike substance obtained from the latex of the sapodilla tree native to Mexico and Central America. Softens at 32.3C. Insoluble in water, soluble in most organic solvents.

Chief use is as a chewing gum, after incorporation of sugar and specific flavoring.
Hazard: Ingestion should be avoided.

Chilean nitrate. See sodium nitrate.

Chilean saltpeter. See sodium nitrate.

Chile saltpeter. (sodium nitrate; nitric acid sodium salt).
CAS: 7631-99-4. $HNaO_3$.
Properties: Natural, impure, sodium nitrate.
Derivation: Occurs naturally in large quantities in South America.

chimera. An organism that contains cells or tissues with a different genotype. These can be mutated cells of the host organism or cells from a different organism or species.

chimeraplasty. An experimental targeted repair process in which a desirable sequence of DNA is combined with RNA to form a chimeraplast. These molecules bind selectively to the target DNA. Once bound, the chimeraplast activates a naturally occurring gene-correcting mechanism. Does not use viral or other conventional gene-delivery vectors.
See gene therapy; cloning vector.

China clay. See kaolin.

China-wood oil. See tung oil.

Chinese insect wax. A wax secreted on the leaves of plants in China by a louselike insect. Its chief ingredient is ceryl cerotate.

chinoxidin. See quinoxidine.

chips. (DNA chips; microarrays).
Identified expressed gene sequences of an organism can, as ESTs or synthesized oligonucleotides, be placed on a matrix. This matrix can be a solid support like glass. If a sample containing DNA or RNA is added, those molecules that are complementary in sequence will hybridize. By making the added molecules fluorescent, it is possible to detect whether the sample contains DNA or RNA of the respective genetic sequence initially mounted on the matrix.

chiral. In chemistry, this term describes asymmetric molecules that are mirror images of each other; i.e., they are related like right and left hands. Such molecules are also called enantiomers and are characterized by optical activity.
See optical isomer; enantiomer.

ChiraSelect. Products of very high and guaranteed enantiomeric purity. The TM is restricted to compounds used as chiral derivatizing agents or as chiral standards.

chitin.
$(C_8H_{13}NO_5)_n$.
A glucosamine polysaccharide. Contains approximately 7% nitrogen and is structurally similar to cellulose. Principal constituent of the shells of crabs, lobsters, and beetles. It is also found in some fungi, algae, and yeasts.
Properties: White, amorphous, semitransparent mass. Insoluble in the common solvents; soluble in concentrated hydrochloric, sulfuric, and nitric acids.
Use: Biological research, source of chitosan.

chitosan. Deacylated derivative of chitin. Absorbs heavy metals from water and industrial waste streams; also used as a dyeing assistant and in photographic emulsions.

clodronate sodium. See sodium clodronate.

chlophedianol hydrochloride (USAN). (α-(2-dimethylaminoethyl)-o-chlorobenzyhydrol hydrochloride).
$C_{17}H_{20}ClNO \cdot HCl$.
Use: In medicine.

chloracetofon. See *O,O*-Dimethyl (2,2,2-trichloro-1-(chloroacetoxy)ethyl)phosphonate.

chloracetophos.
CAS: 5952-41-0. $C_6H_{10}Cl_3O_5P$.
Hazard: Moderately toxic.
Use: Agricultural chemical.

chloracetyl chloride. See chloroacetyl chloride.

chloral. (trichloroacetaldehyde).
CAS: 75-87-6. CCl_3CHO.
Properties: Colorless, mobile, oily liquid; penetrating odor. D 1.505 (25/4C), fp −57.5C, bp 97.7C, vap press 35 mm Hg (20C), refr index 1.4557 (20C), latent heat of vaporization 97.1 Btu/lb. Soluble in water, alcohol, ether, and chloroform; combines with water, forming chloral hydrate.
Derivation: (1) By chlorination of ethyl alcohol, addition of sulfuric acid, and subsequent distillation; (2) by chlorination of acetaldehyde.
Grade: Technical, 94% min.
Hazard: Toxic by ingestion. Probable carcinogen.
Use: Manufacture of chloral hydrate and DDT.

chloral hydrate. (knockout drops; trichloroacetaldehyde, hydrated; trichloroethylidene glycol).
CAS: 302-17-0. $CCl_3CH(OH)_2$.
Properties: Transparent, colorless crystals; aromatic, penetrating, slightly acrid odor; slightly bitter, sharp taste. D 1.901, mp 52C, bp 97.5C. Slowly volatilizes when exposed to air. Soluble in water, alcohol, chloroform, ether, olive oil, and turpentine oil.

Derivation: Action of one-fifth of its volume of water on chloral.
Grade: Technical, USP.
Hazard: Overdose toxic, hypnotic drug, dangerous to eyes. Probable carcinogen.
Use: Medicine (sedative), manufacture of DDT, liniments.

chloral hydrate antipyrine. (antipyrine chloral hydrate).
$C_{11}H_{12}N_2OCl_3CH(OH)_2$.
Properties: Colorless crystals. Mp 67C. Moderately soluble in water; soluble in alcohol.
Use: Medicine (sedative).

chlor-alkali cell. See electrolytic cell.

α-chloralose.
CAS: 15879-93-3. $C_8H_{11}Cl_3O_6$.
Properties: Acicular solid. Mp 186C. Soluble in water, ether, and glacial acetic acid.
Derivation: By heating a mixture of chloral and glucose.
Hazard: May cause addiction, highly toxic.
Use: Coating seeds to protect them from birds.

Chloramben. (3-amino-2,5-dichlorobenzoic acid).
CAS: 133-90-4.
Use: Herbicide.

chloramine.
CAS: 10599-90-3. NH_2Cl.
A colorless, unstable, pungent liquid; soluble in water; decomposes (slowly in dilute solution) to form nitrogen plus hydrochloric acid and ammonium chloride. Mp −66C, soluble in alcohol and ether. (Do not confuse with chloramine-T.) Chloramine is an intermediate in the manufacturing of hydrazine.
Hazard: Questionable carcinogen.

chloramine-B. (sodium benzenesulfochloramide).
$C_6H_5SO_2NClNa$.
Properties: White powder; faint chlorine odor. Soluble in water.
Use: Medicine (antiseptic).

chloramine brown 2me. See C.I. direct brown 2.

chloramine-T. (sodium *p*-toluenesulfochloramine).
$CH_3C_6H_4SO_2NNaCl \cdot 3H_2O$.
Properties: White or slightly yellow crystals or crystalline powder, containing more than 11.5% and less than 13% active chlorine. Slight odor of chlorine. Decomposes slowly in air, liberating chlorine. (Not to be confused with NH_2Cl, which is also termed chloramine.) Soluble in water; insoluble in benzene, chloroform, ether; decomposed by alcohol.

Derivation: Reaction of ammonia and *p*-toluenesulfochloride under pressure. The product is reacted with sodium hypochlorite in the presence of an alkali, and the chloramine produced by crystallization.
Use: Medicine (antiseptic), reagent.
See dichloramine-T.

chloramphenicol.   (*d*(−)*threo*-1-(*p*-nitrophenyl)-2-dichloroacetamido-1,3-propandiol).
CAS: 56-75-7. $C_{11}H_{12}Cl_2N_2O_5$.

An antibiotic derived from *Streptomyces venezuelae* or by organic synthesis, it was the first substance of natural origin shown to contain an aromatic nitro group.
Properties: Fine, white to grayish-white or yellowish-white, needlelike crystals or elongated plates; bitter taste. Neutral to litmus and reasonably stable in neutral or slightly acid solutions. Mp 149–153C. The alcohol solution is dextrorotatory, while the ethyl acetate solution is levorotatory. Very slightly soluble in water; freely soluble in alcohol, propylene glycol, acetone, and ethyl acetate.
Grade: USP.
Hazard: Has deleterious and dangerous side effects. Must conform to FDA labeling requirements. Use is closely restricted. Probable carcinogen.
Use: Medicine (antibiotic), antifungal agent.

chloramphenicol alcohol.
CAS: 23885-72-5. $C_{11}H_{14}N_2O_6$.
Hazard: Moderately toxic.

chloranil.   (tetrachloroquinone; tetrachloro-*p*-benzo-quinone).
CAS: 118-75-2. $C_6Cl_4O_2$.
Properties: Yellow crystals. Mp 290C, d 1.97. Soluble in ether; insoluble in water. Good storage stability.
Derivation: From phenol, *p*-chlorophenol, or *p*-phenylenediamine by treatment with potassium chlorate and hydrochloric acid.
Hazard: Skin irritant.
Use: Agricultural fungicide, dye intermediate, electrodes for pH measurements, reagent.

chloranthrene yellow.   See flavanthrene.

chlorapatite.   See apatite.

chlorauric acid.   See gold trichloride.

chlorazine.   (generic name for 2-chloro-4,6-bis(diethylamino)-*s*-triazine).
CAS: 580-48-3.

$[(C_2H_5)_2N]_2CNC(Cl)NCN.$

Properties: Solid. Mp 15–18C, d 1.096 (20C). Soluble in hydrocarbons, alcohols, ketones; insoluble in water.
Use: Herbicide.

chlorazol black e (biological stain).   See apomine black gx.

chlorbenside.   (*p*-chlorobenzyl-*p*-chlorophenyl sulfide).
CAS: 103-17-3. $ClC_6H_4CH_2SC_6H_4Cl$.
Generic name for an agricultural toxicant.
Properties: Crystals; almond-like odor (technical grade). Mp 75–76C. Insoluble in water; soluble in aromatic hydrocarbons, acetone. Resistant to acid and alkaline hydrolysis.
Grade: Technical.
Hazard: Skin irritant.
Use: Acaricide.

chlor- compounds.   Most organic compounds of chlorine retain the letter *o* in accepted chemical terminology, e.g., chlorobenzene, chloroacetic, etc., although the form without the *o* is sometimes used. Therefore, for chlor- compounds, see also chloro-.

chlorcyclizine hydrochloride.
$ClC_6H_4CH(C_6H_5)C_4H_8N_2CH_3·HCl$ and ether. Solutions acid to litmus, pH (1–100 solution) 4.8–5.5.
Grade: USP.
Use: Medicine (antihistamine).

chlordane.   (1,2,4,5,6,7,8,8-octachloro-4,7-methano-3a,4,7,7a-tetrahydroindane).
CAS: 57-74-9. $C_{10}H_6Cl_8$.
Properties: Colorless, viscous liquid. D 1.57–1.67 (60/60F), viscosity SSU 100 sec (38C), organic chlorine 64–67% by weight, purity 98%, bp 175C (2 mm Hg), refr index 1.56–1.57 (25C). Soluble in many organic solvents; insoluble in water; miscible with deodorized kerosene; decomposes in weak alkalies.
Grade: Technical and pure.
Hazard: A possible carcinogen. Toxic by ingestion, inhalation, and skin absorption. Liver damage.
Use: Insecticide, fumigant.

chlordiazepoxide hydrochloride (USAN).   (7-chloro-2-methylamino-5-phenyl-3,H,1,4-benzodiazepine-4-oxide hydrochloride).
$C_{16}H_{14}ClN_3O·HCl$.
Properties: Crystals. Mp 212–218C. Soluble in water; sparingly soluble in alcohol; insoluble in ether and chloroform.

Hazard: CNS depressant. Manufacture and dosage controlled by law.
Use: Medicine (tranquilizer).

chlordimeform. (Galecron; [*N'*-(4-chloro-*o*-tolyl)-*N,N*-dimethylformamine]).
CAS: 6164-98-3.
Ovicide, insecticide, and miticide designed for use on cotton and vegetable crops. Available in a concentrated emulsion form, it is stated to be less toxic than organophosphates and is biodegradable.
Hazard: Questionable carcinogen.

chlorendic anhydride. (hexachloroendomethylenetetrahydrophthalic anhydride; 1,4,5,6,7,7-hexachlorobicyclo-(2,2,1)-5-heptene-2,3-dicarboxylic anhydride).
$C_9H_2Cl_6O_3$.
Properties: Fine, white, free-flowing crystals. Mp 239–240C, d 1.73. Readily soluble in acetone, benzene, toluene; slightly soluble in water, *n*-hexane, and carbon tetrachloride. Nonflammable.
Derivation: By Diels-Alder reaction of maleic anhydride and hexachlorocyclopentadiene.
Grade: Technical, pure.
Use: Flame-resistant polyester resins, hardening epoxy resins, chemical intermediate, source of chlorendic acid.

"Chlorez" *[Dover].* TM for a chlorinated paraffin additive.
Use: Flame-retardant, plasticizers, and lubricants in metal-working industries.

chlorfenac. (2,3,6-trichlorobenzeneacetic acid; 2,3,6-trichlorophenylacetic acid; fenac; 2-(2,3,6-trichlorophenyl)acetic acid).
$C_8H_5Cl_3O_2$.
Hazard: moderately toxic.
Use: Herbicide to control annual weeds, *Agropyron repens*, and seedling perennials in maize and *beta vulgaris*.

chlorfenapyr. See 4-bromo-2-(4-chlorophenyl)-1-ethoxymethyl-5.

chlorfenethol. (4-chloro-α-(4-chlorophenyl)-α-methylbenzenemethanol; 4,4'-dichloro-α-methylbenzhydrol; *p,p'*-dichlorodiphenylmethyl carbinol; 1,1-bis(*p*-chlorophenyl)ethanol; 1,1-bis-(*p*-chlorophenyl)methyl carbinol; dmc; 1,1-bis(4-chlorophenyl)ethanol).
$C_{14}H_{12}Cl_2O$.
Properties: Colorless, crystalline, water-insoluble compound.
Use: Miticide for ornamental plants.

chlorfenvinphos. (*O,O*-diethyl-*O*-[2-chloro-1-(2,4-dichlorophenyl)vinyl]phosphate).
CAS: 470-90-6. $C_{12}H_{14}Cl_3O_4P$.
Properties: Yellowish liquid. Soluble in most organic solvents; slightly soluble in water.

Hazard: A cholinesterase inhibitor.
Use: Insecticide, nematicide, parasiticide.

"Chlorhydrol" *[Reheis].* TM for aluminum chlorohydrate.
Use: Excellent antiperspirant properties.

chloric acid.
CAS: 7790-93-4. $HClO_3 \cdot 7H_2O$.
Occurrence: Only in aqueous solution.
Derivation: Reaction of barium chlorate and sulfuric acid.
Hazard: Toxic by ingestion and inhalation. Strong oxidizer, ignites organic materials on contact.
Use: Catalyst in polymerization of acrylonitrile.

chloride. Any salt of hydrochloric acid that contains the chloride ion. It is the most common extracellular ion and functions significantly in the maintenance of water balance and distribution, the maintenance of osmotic pressure, and normal anion–cation balance of extracellular fluid.

chlorine. (bertholite; chloor; chlor; chlore; chlorine mol., cloro; molecular chlorine).
CAS: 7782-50-5. Cl_2.
The 15th most abundant element in the earth's crust, occurring only in the combined state, mainly in common salt. A strong corrosive acid.
Properties: Pungent, greenish-yellow gas, liquid or rhombic crystals; atomic number 17; atomic weight 35.4527; density $3.124 \, kg/m^3$; melting point −101.03°C; boiling point −33.9°C.
Derivation: Formed by dissolving hydrogen chloride in water.
Hazard: Moderately toxic; eye and upper respiratory tract irritant.
Use: A laboratory reagent.

chloridizing. Heating in the presence of chlorine as a step in the recovery of certain metals from their oxides or other compounds.

chlorinated acetone. See chloroacetone.

chlorinated camphene. See toxaphene.

chlorinated diphenyl. See chlorodiphenyl.

chlorinated hydrocarbon. See hydrocarbon, halogenated.

chlorinated isocyanuric acid.
Use: Dry bleach.
See dichloroisocyanuric acid; trichloroisocyanuric acid; potassium dichloroisocyanurate; sodium dichloroisocyanurate.

chlorinated lime. See lime, chlorinated.

chlorinated naphthalene. (chloronaphthalene).
$C_{10}H_7Cl$.

From the chlorination of naphthalene. Physical state varies from oily liquids to crystalline solids, depending on the extent of chlorination.
Hazard: (Tri- and higher) Toxic by ingestion, inhalation, and skin absorption. Strong irritants.
Use: Solvent, immersion liquid in microscopy.
See oil; wax; tetrachloronaphthalene.

chlorinated paraffin.   See paraffin, chlorinated.

chlorinated para red.   A modification of para red that contains some chlorine. Much lighter than para or toluidine red; has excellent brilliance but poorer heat resistance.

chlorinated polyether.   See Penton.

chlorinated polyolefin.   See rubber, chlorinated; polypropylene, chlorinated.

chlorinated rubber.   See rubber, chlorinated.

chlorinated trisodium phosphate.   See trisodium phosphate, chlorinated.

chlorination.   To introduce chlorine into an organic compound or into a liquid.

chlorine.
CAS: 7782-50-5.   Cl.
Nonmetallic halogen element of atomic number 17, group VIIA of the periodic table. Aw 35.453, valences of 1, 3, 4, 5, 7. Two stable isotopes ^{35}Cl (75.4%) and ^{37}CL (24.6%).
Properties: (1) Dense, greenish-yellow, diatomic gas; pungent, very irritating odor. Noncombustible, but supports combustion (oxidizing agent). Liquefaction press 7.86 atm (25C), 1 atm (−35C). Water solubility 0.64 g Cl_2 per 100-g water. D 3.21 g/L (0C, 1 atm) (air = 1.29). Thermodynamic properties: (a) critical temp 144.0C; (b) critical press 78.525 atm absolute; (c) critical vol 1.763 L/kg. Strongly electronegative.
(2) Liquid; clear amber; very irritating odor. D 1.56 (−35C), fp −101C, 1 L liquid = 456.8 L gas at 0C and 1 atm. Very low electrical conductivity. Soluble in chlorides and alcohols. Extremely strong oxidizing agent. Slightly soluble in cold water.
Occurrence: Not free in nature; component of minerals halite (rock salt), sylvite, and carnallite; chloride ion in seawater.
Derivation: (1) Electrolysis of sodium chloride brine in either diaphragm or mercury-cathode cells; chlorine is released at the anode. (2) Fused-salt electrolysis of sodium or magnesium chloride. (3) Electrolysis of hydrochloric acid. (4) Oxidation of hydrogen chloride with nitrogen oxide as catalyst and absorption of steam with sulfuric acid (KeloChlor process). No by-product caustic is produced.
Grade: Technical (gas and liquid), pure (99.9%).
Hazard: A military poison. Dangerous in contact with turpentine, ether, ammonia, hydrocarbons,

hydrogen, powdered metals, and other reducing materials. Eye and upper respiratory tract irritant. Questionable carcinogen.
Use: Manufacture of carbon tetrachloride, trichloroethylene, chlorinated hydrocarbons, polychloroprene (neoprene), polyvinyl chloride, hydrogen chloride, ethylene dichloride, hypochlorous acid, metallic chlorides, chloroacetic acid, chlorobenzene, chlorinated lime; water purification, shrinkproofing wool, flame-retardant compounds, special batteries (with lithium or zinc); processing of meat, fish, vegetables, and fruit. For information, refer to the Chlorine Institute, 2001 L St., NW, Suite 506, Washington, DC 20036.

chlorine-36.   Radioactive chlorine of mass number 36. Half-life approximately 440,000 years; radiation: β.
Derivation: Separated from various isotopes produced during irradiation of potassium chloride.
Available forms: As hydrochloric acid solution and as solid potassium chloride.
Hazard: MPC 4×10^{-7} mCi/mL of air.
Use: Tracer in studying the saltwater corrosion of metals, especially mechanism of steel reaction with chlorinated hydrocarbons, location and flow of saltwater in porous media, etc.

chlorine, available.   The weight of free chlorine that would exert the same oxidizing action as the chlorine compound in question.

chlorine bromide.   See bromine chloride.

chlorine dioxide.
CAS: 10049-04-4. ClO_2.
Properties: Red-yellow gas. Fp −59.5C, bp 10C. Very reactive, unstable, strong oxidizer. Decomposes in water. Dissolves in alkalies, forming a mixture of chlorite and chlorate.
Derivation: Usually made at point of consumption from sodium chlorate, sulfuric acid, and methanol or from sodium chlorate and sulfur dioxide. Concentration of gas is limited to 10% to reduce explosion hazard.
Grade: Sold as hydrate in frozen form.
Hazard: Explodes when heated or by reaction with organic materials. Very irritating to skin and mucous membranes. Lower respiratory tract irritant. Broncitis.
Use: Bleaching wood pulp, fats, and oils; controversial maturing agent for flour; water treatment (purification and taste removal); swimming pools; odor control; biocide.

chlorine heptoxide.
Cl_2O_7.
Properties: Colorless, viscous liquid. Fp −91C; bp 82C; d 1.86. Hydrolyzes to form perchloric acid.
Hazard: Explodes on contact with iodine or flame or by shock. Strong irritant to tissue and very toxic.
Use: Cellulose esterification catalyst.

chlorine monofluoride. (chlorine fluoride).
ClF.
Properties: Colorless gas, slightly yellow when liquid. Mp −155.6C, bp −100.1C, critical temp −14C.
Hazard: Extremely reactive. Destroys glass instantly, attacks quartz readily in presence of moisture. Organic matter bursts into flame on contact. Violent reaction with water. Extremely corrosive to skin, eyes, mucous membranes, and respiratory tissues.
Use: Fluorinating reagent.

chlorine monoxide.
Cl_2O.
Properties: Yellow gas; strong, unpleasant odor. Fp −120C; bp 2.2C. Soluble in water and carbon tetrachloride.
Derivation: Reaction of mercuric oxide and chlorine.
Hazard: Explodes on contact with organic materials. Strong irritant to eyes, skin, and mucous membranes.
Use: Chlorination.

chlorine nitrate. (nitryl hypochlorite).
CAS: 14545-72-3 $ClNO_3$.
A compound that slows the rate of destruction of ozone.
Hazard: Explosive reaction with metals, metal chlorides, alcohols, ethers, and most organic materials.

chlorine number. The number of grams of chlorine or bleaching powder absorbed by 100-g oven-dry cellulose pulp in a definite time under certain conditions. Also an indication of bleach consumption.

chlorine, residual. The amount of chlorine present in water at any time before the addition of chlorine.

chlorine trifluoride.
CAS: 7790-91-2. ClF_3.
Properties: Nearly colorless gas, pale-green liquid or white solid. Bp 11.4C, fp −76.3C, d of gas 3.14 g/L (air = 1.29). Very reactive, comparable to fluorine.
Derivation: By reaction of chlorine and fluorine at 280C and condensation of the product at −80C. Obtained 99.0% pure.
Hazard: Explodes in contact with organic materials or with water. Dangerous fire risk. A poison, very toxic, corrosive to skin. Lung damage, eye, and upper respiratory tract irritant. Questionable carcinogen.
Use: Fluorination, cutting oil-well tubes, reprocessing reactor fuels, oxidizer in propellants.

chlorine war gas. (chlorine warfare gas).
Chlorine gas in containers for release against enemy troops in combat.

chlorine water. Clear, yellowish liquid; deteriorates on exposure to air and light. Made by saturating water with approximately 0.4% chlorine.
Use: Deodorizer, disinfectant, antiseptic.

chlorinolysis. The chlorination of organic compounds under conditions that rupture the carbon–carbon bonds to yield chloro compounds with fewer carbon atoms than in the original.

chloriodized oil. Chlorinated and iodinated vegetable oil. Contains 26.0–28.0% iodine in organic combination.
Properties: Pale yellow, viscous, oily liquid; faint, bland taste. Practically insoluble in water; slightly soluble in alcohol; freely soluble in benzene, chloroform, and ether.
Derivation: Formed by chemical addition of iodine monochloride to a vegetable oil.
Hazard: A poison. Very toxic by ingestion.
Use: Medicine (radiopaque medium).

chlorisondamine chloride. (4,5,6,7-tetrachloro-2-(2-dimethylaminoethyl)isoindoline dimethylchloride).
$C_{14}H_{20}Cl_6N_2$, a quaternary ammonium compound.
Properties: Crystals. Decomposes 258–265C. Soluble in water and alcohol.
Use: Medicine (blood-pressure control).

chlormequat. See 2-chloroethyltrimethyl ammonium chloride.

chlormequat chloride. (2-chloro-N,N,N-trimethylethanaminium chloride; (2-chloroethyl) trimethylammonium chloride; chlorocholine chloride; choline dichloride; 2-chloroethyl (trimethyl)azanium chloride).
CAS: 999-81-5. $C_5H_{13}Cl_2N$.
A phytocidal bipyridilium quaternary ammonium salt.
Properties: White, water soluble, hygroscopic, crystalline solid, fish like odor.
Hazard: Toxic; corrosive.
Use: Plant growth regulator on ornamental plants.

chlormethazanone. (2-(4-chlorophenyl-3-methyl)-4-meta-thiazanone-1-dioxide).
$C_{11}H_{12}ClNO_3S$.
Properties: Crystals. Mp 117C. Insoluble in water; slightly soluble in alcohol.
Use: Medicine (tranquilizer and muscle relaxant).

chloroacetaldehyde.
CAS: 107-20-0. $ClCH_2CHO$.
Properties: (of 40% aqueous solution) Clear, colorless liquid; pungent odor. Bp 85C, fp −16.3C, d 1.19 (25/25C), refr index 1.397 (25C), wt/gal 0.9 lb (25C). Soluble in water, acetone, methanol; at greater than 50% concentration in water, it forms an insoluble hemihydrate. Pure substance flash p 190F (87.7C).

Hazard: Corrosive to skin and mucous membranes. TLV: ceiling 1 ppm.
Use: Intermediate, fungicide.

chloroacetaldehyde dimethyl acetal. See dimethyl chloroacetal.

chloroacetamide. (α-chloroacetamide; 2-chloroethanamide).
$ClCH_2CONH_2$.
Properties: Colorless to pale-yellow crystals; characteristic odor. Mp 117–119C, bp 220C (decomposes). Soluble in water and alcohol; insoluble in ether.
Hazard: Strong irritant to skin and tissue.
Use: Intermediate.

chloroacetic acid. (chloroacetic acid; MCA; monochloroacetic acid).
CAS: 79-11-8. $CH_2ClCOOH$.
Properties: Colorless to light-brownish crystals. Deliquescent, d 1.58. Crystallizing point: α-form 61.0–61.7C, β-form 55.5–56.5C, γ-form 50C. The commercial material melts at 61–63C, boiling range 186–191C. Soluble in water, alcohol, ether, chloroform, carbon disulfide.
Derivation: Action of chlorine on acetic acid in the presence of acetic anhydride, phosphorus, or sulfur.
Grade: Technical, medicinal, 99.5% pure.
Hazard: Use in foods prohibited by FDA. Irritating and corrosive to skin. Upper respiratory tract irritant. Questionable carcinogen.
Use: Herbicide, preservative, bacteriostat, intermediate in production of carboxymethylcellulose; ethyl chloroacetate, glycine, synthetic caffeine, sarcosine, thioglycolic acid, EDTA, 2,4-D, 2,4,5-T.

chloroacetic acid *tert*-butyl ester. See *tert*-butyl chloroacetate.

chloroacetic anhydride. (chloroethanoic anhydride; symmetrical dichloroacetic anhydride).
$(ClCH_2CO)_2O$.
Properties: Colorless to slightly yellow crystals; pungent odor. Mp 51–55C, bp 203C, d 1.55. Soluble in chloroform and ether. Hydrolyzes to chloroacetic acid.
Hazard: Irritating to skin and eyes, moderately toxic by inhalation.
Use: Intermediate for acetylation of amino acids, cellulose chloroacetates.

***o*-chloroacetoacetanilide.**
$CH_3COCH_2CONHC_6H_4Cl$.
Properties: White crystals resemble ethyl acetoacetate in chemical reactivity. Mp 107C, vap press 0.1 mm Hg (20C). Nonflammable. Insoluble in water.
Use: Organic synthesis, dyestuffs.

chloroacetone. (monochloroacetone; 1-chloro-2-propanone; chloracetone; chlorinated acetone).
CAS: 78-95-5. CH_3COCH_2Cl.
A lachrymator.

Properties: Colorless liquid; pungent, irritating odor. D 1.162 (16C), bp 119C, fp −44.5C. Soluble in alcohol, ether, chloroform, and water.
Derivation: Chlorination of acetone.
Hazard: Strong irritant to tissue, eyes, and mucous membranes; toxic by ingestion and skin contact. Upper respiratory tract irritant.
Use: Couplers for color photography, enzyme inactivator, insecticides, perfumes, intermediate, organic synthesis, tear gas, polymerization of vinyl monomers.

chloroacetonitrile. (chloroethane nitrile; chloromethyl cyanide).
CAS: 107-14-2. $ClCH_2CN$.
Properties: Colorless liquid; pungent odor. D 1.202–1.2035 (25/25C), refr index 1.4210–1.4240 (25C), 5–95% distills between 124 and 129C. Soluble in hydrocarbons, alcohols; insoluble in water.
Hazard: Irritant. Questionable carcinogen.
Use: Fumigant, intermediate.

α-chloroacetophenone. (chloroacetophenone; phenacylchloride; phenyl chloromethyl ketone).
CAS: 532-27-4. $C_6H_5COCH_2Cl$.
A strong lachrymator.
Properties: (ω isomer) White crystals; floral odor. Mp 56C, bp 247C. The para isomer has mp 20C, bp 237C. Insoluble in water; soluble in acetone, benzene, carbon disulfide.
Derivation: From chloroacetylchloride, benzene, and aluminum chloride.
Hazard: Strong irritant to eyes and tissue as gas or liquid. Skin and upper respiratory tract irritant. Questionable carcinogen.
Use: Pharmaceutical intermediate, riot-control gas. See Mace.

(chloroacetoxy)tributylstannane. See tributyltin chloroacetate.

chloroacetyl chloride. (chloracetyl chloride).
CAS: 79-04-9. $ClCH_2COCl$.
A lachrymator.
Properties: Water-white liquid; pungent odor. D 1.495 (0C), bp 105–110C, decomposes in water.
Derivation: (1) Action of chlorine on acetyl chloride in sunlight. (2) Dropping phosphorous trichloride on chloroacetic acid.
Hazard: Irritant to eyes, corrosive to skin. Upper respiratory tract irritant.
Use: Preparation of chloroacetophenone, intermediate, tear gas.

chloroacetylurethane.
$ClCH_2CONHCOOC_2H_5$.
Properties: Crystals. Mp 129C. Soluble in alcohol; sparingly soluble in water.
Derivation: By interaction of a urethane derivative and ethyl chloracetate.

chloroacrolein.
$H_2C:CClCHO$.

Properties: Colorless liquid. D 1.205 (15C), bp 29–31C (17 mm Hg).
Derivation: Chlorination of acrolein.
Hazard: Irritant to eyes and skin.
Use: Tear gas.

α-chloroacrylonitrile.

Properties: Readily polymerizes and copolymerizes with other unsaturated monomers. High cross-linking ability.
Derivation: Chlorination of acrylonitrile and dehydrohalogenation by cracking.
Use: Synthetic fibers, coatings and films, acrylic polymers, treatment of cotton fiber, intermediate.

2-chloroallyl diethyldithiocarbamate.

(CDEC).
$(C_2H_5)_2NCSSCH_2CCl:CH_2$.
Properties: Amber liquid. Bp 128–130C (1 mm Hg). Very slightly soluble in water; soluble in benzene, alcohol, acetone, chloroform, and ether.
Available forms: Liquid and granular.
Hazard: Dry preparations are irritating to eyes and skin.
Use: Herbicide, pesticide.

1-(3-chloroallyl)-3,5,7-triaza-1-azoniaadamantane chloride.

$C_6H_{12}N_4(CH_2CHCHCl)Cl$.
Properties: White- to cream-colored powder. Soluble in water and methanol; almost insoluble in acetone.
Use: Bactericide used as preservative in latexes, paints, floor polishes, joint cements, adhesives, inks, starches, etc.

chloroaluminum diisopropoxide.

$[(CH_3)_2CHO]_2AlCl$.
Properties: White crystals. Mp 160C (decomposes). Soluble in most organic solvents; hydrolyzes.
Use: Catalyst and intermediate.
Hazard: An irritant.

chloroamino-.   See aminochloro-.

2-chloro-5-aminobenzoic acid.   Grade: Technical.

Use: Intermediate in manufacture of azo dyes for textiles and plastics and as coupling agent of color pigments in color photography.

p-chloro-aminophenol.   See 2-amino-4-chlorophenol.

2-chloro-4-*tert*-amylphenol.

$C_5H_{11}C_6H_3ClOH$.
Properties: Water–white liquid; aromatic odor. D 1.11 (20C), boiling range 253–265C, flash p 225F (107C). Combustible.

m-chloroaniline.   (m-aminochlorobenzene).

CAS: 108-42-9. $ClC_6H_4NH_2$.

Properties: Colorless to light-amber liquid, tends to darken during storage. Boiling range 228–231C, fp −10.6C. Insoluble in water; soluble in organic solvents.
Grade: Technical.
Use: Intermediate for azo dyes and pigments, pharmaceuticals, insecticides, agricultural chemicals.

o-chloroaniline.   (o-aminochlorobenzene).

CAS: 95-51-2. $ClC_6H_4NH_2$.
Properties: Amber liquid; amine odor; darkens on exposure to air. Distillation range 208–210C, fp −2.3C, d 1.213 (20/4C), refr index 1.5896 (20C). Miscible with alcohol and ether, insoluble in water.
Grade: Technical.
Hazard: Toxic by ingestion.
Use: Dye intermediate, standards for colorimetric apparatus, manufacture of petroleum solvents and fungicides.

p-chloroaniline.   (p-aminochlorobenzene).

CAS: 106-47-8. $ClC_6H_4NH_2$.
Properties: White or pale-yellow solid. Mp 69.5C, distilling range 229–233C, d 1.17. Soluble in hot water and organic solvents.
Grade: Technical.
Hazard: Toxic by inhalation and ingestion. Possible carcinogen.
Use: Dye intermediate, pharmaceuticals, agricultural chemicals.

4-chloroaniline-3-sulfonic acid.

$HSO_3C_6H_3ClNH_2$.
Properties: White to light-gray powder.
Use: Intermediate for dyes.

2-chloroanthraquinone.

$C_{14}H_7ClO_2$.
Properties: Mp 208–211C. Insoluble in water; soluble in hot benzene.
Derivation: Condensing phthalic anhydride and chlorobenzene in the presence of anhydrous aluminum chloride to form p-chlorobenzoylbenzoic acid. Ring closure of the intermediate acid is brought about by heating in sulfuric acid solution.
Use: Starting material for certain vat dyes.
See anthraquinone; 2-methylanthraquinone.

chloroauric acid.   See gold trichloride.

chloroazotic acid.   See aqua regia.

chlorobenzal.   See benzyl dichloride.

chlorobenzaldehyde.

C_6H_4CHOCl.
Properties: Colorless to yellowish liquid (o-) or powder (p-). Boiling range 209–214C, fp 8.0C (min), d 1.240–1.245 (25/25C). Soluble in alcohol, ether, and acetone; insoluble in water. Combustible.
Use: Intermediate in the preparation of triphenyl methane and related dyes, organic intermediate.

3-chloro-4-benzamido-6-methylaniline.
ClC$_6$H$_2$NH$_2$CH$_3$(NHCOC$_6$H$_5$).
Properties: White solid. Mp 198–199C.
Use: Azoic dyes, pigments.

chlorobenzanthrone.
C$_{17}$H$_9$ClO.
Properties: All isomers: yellow needles. Soluble in alcohol, benzene, toluene, acetic acid.
Derivation: From benzanthrone by treatment with chlorine.

chlorobenzene. (monochlorobenzene; phenyl chloride).
CAS: 108-90-7.   C$_6$H$_5$Cl.

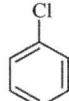

Properties: Clear, volatile liquid; almond-like odor. D 1.105 (25/25C), bp 131.6C, fp −45C, wt/gal 9.19 lb (25C), refr index 1.5216 (25C), flash p 85F (29.4C) (CC), autoign temp 1180F (637C). Miscible with most organic solvents; insoluble in water.
Derivation: By passing dry chlorine into benzene with a catalyst.
Grade: Technical.
Hazard: A possible carcinogen. Avoid inhalation and skin contact. Moderate fire risk. Explosive limits 1.8–9.6%.
Use: Phenol, chloronitrobenzene, aniline, solvent carrier for methylene diisocyanate, solvent, pesticide intermediate, heat transfer.

p-chlorobenzenesulfonamide.
ClC$_6$H$_4$SO$_2$NH$_2$.
Properties: White; odorless powder. Mp 145–148C. Soluble in alcohol.
Grade: 98–99% purity.
Use: Intermediate for pharmaceuticals and resins.

o-chlorobenzenethiol. See *p*-chlorothiophenol.

p-chlorobenzhydrol. (*p*-chlorobenzohydrol).
ClC$_6$H$_4$C(C$_6$H$_5$)H$_2$O.
Properties: White to off-white, crystalline powder. Mp 57–61C. Insoluble in water; soluble in ether, alcohol, and benzene.
Use: Organic synthesis.

"Chlorobenzilate" *[Novartis].* TM for ethyl-4,4'-dichlorobenzilate.

CAS: 510-15-6. (C$_6$H$_4$Cl)$_2$C(OH)COOC$_2$H$_5$.

Properties: Viscous, yellow liquid. Bp 141–142C (0.06 mm Hg); d (technical 90%) 1.2816 (20/4C). Slightly soluble in water; soluble in acetone, benzene, methanol.

Hazard: Questionable carcinogen. Caused testicular damage in farm workers. Use has been restricted.
Use: Pesticide, acaricide.

p-chlorobenzohydrol. See *p*-chlorobenzhydrol.

2-(2-chlorobenzyl)-4,4-dimethyl-1,2-oxazolidin-3-one.
CAS: 81777-89-1.   C$_{12}$H$_{14}$ClNO$_2$.
Hazard: Moderately toxic by ingestion, inhalation, and skin contact. A reproductive hazard.
Use: Agricultural chemical.

o-chlorobenzylidene malononitrile. (CS; OCBM).
CAS: 2698-41-1.   ClC$_6$H$_4$CH=C(CN)$_2$.
Properties: White crystals; odor of pepper. Mw 189, mp 93–95C, bp 310–315C. Insoluble in water.
Available forms: Available both unground and ground with 5% silica aerogel or treated "Cab-O-Sil" *[Cabot]*.
Hazard: Toxic by inhalation and skin contact. Strong irritant to eyes and mucous membranes.
Use: An incapacitating agent used by the military and law-enforcement officers.

4-(4-chlorobenzyl)pyridine.
CAS: 4409-11-4.   C$_{12}$H$_{10}$ClN.
Hazard: A poison.

chlorobenzoic acid.
ClC$_6$H$_4$COOH.
Properties: Nearly white, coarse powder. Mp (*o*-) 142C, (*m*-) 158C, (*p*-) 243C. Soluble in methanol, alcohol, ether, and hot water.
Use: Intermediate for the preparation of dyes, fungicides, pharmaceuticals, and other organic chemicals; preservative for adhesives and paints.

p-chlorobenzophenone.
ClC$_6$H$_4$COC$_6$H$_5$.
Properties: White to off-white, crystalline powder. Mp 73–78C; bp 332C. Soluble in acetone, benzene, carbon tetrachloride, ether, and hot alcohol; insoluble in water.
Use: Intermediate.

chlorobenzotriazole.
ClC$_6$H$_4$NHN:N.
Properties: White solid. Mp 157–159C.
Use: Intermediate and photographic chemical.

o-chlorobenzotrichloride.
ClC$_6$H$_4$CCl$_3$.
Properties: Colorless liquid or solid. Mp 29.37C, bp 264.3C, d 1.5131 (25/4C), refr index 1.5836 (20C). Soluble in alcohol, ether, and acetone; decomposed by water.
Use: Intermediate for pharmaceuticals, dyes, and other organic chemicals.

p-chlorobenzotrichloride.
CAS: 5216-25-1. $ClC_6H_4CCl_3$.
Properties: Water–white liquid. Boiling range 248–257C, fp approximately 3.8C, d 1.480–1.490 (25/25C). Soluble in alcohol, ether, and acetone; insoluble in water.
Use: Same as *o*-chlorobenzotrichloride.

m-chlorobenzotrifluoride.   (*m*-chlorotrifluoromethylbenzene; *m*-chloro-α,α,α-trifluorotoluene).
CAS: 98-15-7. $ClC_6H_4CF_3$.
Properties: Water–white, aromatic liquid. Bp 138C, fp −56C, refr index 1.446 (20C), flash p 122F (50C) (CC), d 1.351 (15.5/15.5C). Flammable.
Hazard: Moderate fire risk.
Use: Intermediate in manufacturing of dyes and pharmaceuticals, dielectrics, insecticides.

o-chlorobenzotrifluoride.   (*o*-chlorotrifluoromethylbenzene; *o*-chloro-α,α,α-trifluorotoluene).
CAS: 88-16-4. $ClC_6H_4CF_3$.
Properties: Colorless liquid with aromatic odor. D 1.379 (15.5/15.5C), refr index 1.456 (20C), bp 152C, fp −7.4C, flash p 138F (58.8C) (CC), wt/gal 11.50 lb (15.5C).
Hazard: See *m* form.
Use: Dye intermediate, chemical intermediate, solvent, and dielectric fluid.

p-chlorobenzotrifluoride.   (*p*-chlorotrifluoromethylbenzene; *p*-chloro-α,α,α-trifluorotoluene).
CAS: 98-56-6. $ClC_6H_4CF_3$.
Properties: Water-white liquid, aromatic odor. Bp 139.3C, fp −36C, refr index 1.446 (20C), flash p 116F (46.6C) (CC), d 1.3533 (15.5/15.5C), wt/gal 11.28 lb (15.5C).
Hazard: See *m*-form.
Use: See *o*-chlorobenzotrifluoride.

chlorobenzoyl chloride.
ClC_6H_4COCl.
Properties: Colorless liquid. Boiling range 227–239C; fp −4 to −6C, fp (*o*-) −4 to −6C, (*p*-) 10–12C; d 1.374–1.376 (25/15C). Soluble in alcohol, ether, and acetone; insoluble in water.
Use: Intermediate for pharmaceuticals, dyes, and other organic chemicals.

p-chlorobenzoyl peroxide.
CAS: 94-17-7. $(ClC_6H_4CO)_2OO$.
Properties: White, odorless powder. Decomposes violently on heating or contamination. Insoluble in water; soluble in organic solvents.
Hazard: Dangerous fire and explosion risk, explodes when heated to 38C, strong oxidizer, will ignite on contact with organic materials. Store in dark, cool locality. Toxic.
Use: Bleaching agent, polymerization catalyst.

1-(o-chlorobenzoyl)-2-thiobiuret.
CAS: 127019-52-7. $C_9H_8ClN_3O_2S$.
Hazard: A poison.

3-(2-(4-chlorobenzylamino)ethyl)indole monohydrochloride.
CAS: 63938-62-5. $C_{17}H_{17}ClN_2 \cdot ClH$.
Hazard: A poison.

chlorobenzylate.   (4,4′-dichlorobenzilic acid ethyl ester; 4,4′-dichlorobenzilate; ethyl-4-chloro-α-(4-chlorophenyl)-α-hydroxybenzeneacetate; ethyl-*p*,*p*′-dichlorobenzilate; ethyl-4,4′-dichlorodiphenyl glycollate; ethyl-4,4′-di-chlorophenyl glcollate; ethyl ester of 4,4′-dichlorobenzilic acid; ethyl-2-hydroxy-2,2-bis(4-chlorophenyl)acetate; benz-*o*-chlor; chlorbenzilate; ethyl-2,2-bis(4-chlorophenyl)-2-hydroxyacetate).
$C_{16}H_{14}Cl_2O_3$.
Properties: Viscous, slightly water soluble, liquid.
Hazards: Skin and eye irritant, moderately toxic, nephrotoxic, carcinogen.
Use: Pesticide.

chlorobenzyl chloride.
CAS: 104-83-6. $ClC_6H_4CH_2Cl$.
Properties: Colorless liquid. Boiling range 216–222C, mp (*o*-) −17C, (*p*-) 31C, d 1.270–1.280 (25/15C). Soluble in alcohol, ether, and acetone; insoluble in water.
Hazard: Irritating to skin, eyes.
Use: Intermediate for organic chemicals, pharmaceuticals, and dyes.

p-chlorobenzyl-p-chlorophenyl sulfide.   See chlorbenside.

p-chlorobenzyl cyanide.
$ClC_6H_4CH_2CN$.
Properties: Colorless to pale-yellow solid. Mp 27C. Soluble in acetone and alcohol.
Hazard: A poison.
Use: Organic synthesis.

p-chlorobenzyl-p-fluorophenyl sulfide.   See fluorbenside.

o-chlorobenzylidene malononitrile.   (([(2-chlorophenyl)methylene]propanedinitrile; *o*-chlorobenzalmalononitrile, β,β-dicyano-*o*-chlorostyrene; cs; 2-[(2-chlorophenyl)methylidene]propane; 2-chlorobenzal malononitrile; *o*-chlorobenzylidene malononitrile; 2-chlorobenzylidene malononitrile; 2-chlorobmn; NCI-C55118; propanedinitrile((2-chlorophenyl)methylene); USAF KF-11).
$C_{10}H_5ClN_2$.
Properties: White, slightly water-soluble, crystalline solid.
Hazard: Extremely toxic; poison; causes temporary irritation of the eyes and the mucosal surface of the respiratory tract.
Use: Riot-control agent.

2-(p-chlorobenzyl)pyridine.   (2-(4-chlorobenzyl)pyridine).
$ClC_6H_4CH_2C_5H_4N$.

Properties: Liquid. Bp 310.5C, fp 8.4C, d 1.168 (25C), refr index 1.5865 (20C). Insoluble in water.
Use: Organic synthesis.

chlorobromo-.   See bromochloro-.

2-chlorobutadiene-1,3.   See β-chloroprene.

1-chlorobutane.   See *n*-butyl chloride.

chlorobutanol.   (trichloro-*tert*-butyl alcohol; 1,1-trichloro-2-methyl-2-propanol; acetone chloroform).
CAS: 57-15-8. $Cl_3CC(CH_3)_2OH$.
Properties: Colorless to white crystals; characteristic odor and taste. Mp (anhydrous form) 97C, mp (hemihydrate) 78C, bp 167C, sublimes easily. Soluble in alcohol and glycerol, hot water, ether, chloroform, and volatile oils.
Derivation: By action of potassium hydroxide on a solution of chloroform and acetone.
Grade: USP.
Hazard: Action similar to chloral hydrate. Combustible.
Use: Plasticizer for cellulose esters and ethers, preservative for biological fluids and solutions, antimicrobial agent, anesthetic in dentistry.

4-chloro-1-butene.
CAS: 927-73-1. C_4H_7Cl.
Hazard: Moderately toxic by ingestion.

4-chloro-2-butynyl-*m*-chlorocarbanilate.
See barban.

s-(4-chloro-2-butynyl) *O,O*-diethyl phosphorothioate.
CAS: 76706-97-3. $C_8H_{14}ClO_3PS$.
Hazard: A poison.
Use: Agricultural chemical.

s-(4-chloro-2-butynyl) diphenylphosphinothioate.
CAS: 76706-99-5. $C_{16}H_{14}ClOPS$.
Hazard: A poison.
Use: Agricultural chemical.

s-(4-chloro-2-butynyl) *o*-ethyl phenylphosphonothioate.
CAS: 76706-98-4. $C_{12}H_{14}ClO_2PS$.
Hazard: A poison.
Use: Agricultural chemical.

4-chlorobutyric acid tributylstannyl ester.
See tributyltin-γ-chlorobutyrate.

2-chlorocamphane.   See 2-bornyl chloride.

chlorocarbon.   A compound of carbon and chlorine or carbon, hydrogen, and chlorine, such as carbon tetrachloride, chloroform, tetrachloroethylene, etc.

chlorocarbonyl ferrocene.   (ferrocenoyl chloride).
$C_5H_5FeC_5H_4COCl$.
Properties: Orange-red solid. Mp 48–49C.
Use: Intermediate.

3-chloro-*n*-(5-chloro-2,6-dinitro-4-trifluoromethylphenyl)-5-trifluoromethyl-2-pyridinamine.
CAS: 79622-59-6. $C_{13}H_4Cl_2F_6N_4O_4$.
Hazard: Moderately toxic by ingestion and inhalation. A reproductive hazard.
Use: Agricultural chemical.

4-chloro-2-(4-chloro-2-fluorophenyl)-5-((4-chlorophenyl)methoxy)-3(2h)-pyridazinone.
CAS: 107359-74-0. $C_{17}H_{10}Cl_3FN_2O_2$.
Hazard: A poison by ingestion.
Use: Agricultural chemical.

4-chloro-2-(4-chlorophenyl)-5-((4-chlorophenyl)methoxy)-3(2h)-pyridazinone.
CAS: 107359-39-7. $C_{17}H_{11}Cl_3N_2O_2$.
Hazard: A poison by ingestion.
Use: Agricultural chemical.

4-chloro-2-(4-chlorophenyl)-5-((6-iodo-3-pyridinyl)methoxy)-3(2h)-pyridazinone.
CAS: 122322-19-4. $C_{16}H_{10}Cl_2IN_3O_2$.
Hazard: A poison by ingestion.
Use: Agricultural chemical.

chlorochromic anhydride.   See chromyl chloride.

chlorocosane.   See paraffin, chlorinated.

3-chlorocoumarin.
$C_9H_5O_2Cl$.
Properties: Slightly yellow, crystalline solid. Mp 118C.
Grade: Technical.
Use: Tin-plating solutions.

p-chloro-*m*-cresol.   See 4-chloro-3-methylphenol.

chlorodecone.   [1,2,3,4,5,5,6,7,8,9,10,10-dodecachlorooctahydro-1,3,4-metheno-2-cyclobuta-(c,d)-pentalone].
CAS: 143-50-0.
Properties: Mp 349C (decomposes). Solubility in water 0.4% at 100C.
Use: Insecticide.

α-chloro-*N,N*-diallylacetamide.   (CDAA).
$ClCH_2CON(CH_2CH{:}CH_2)_2$.

Properties: Amber liquid or granules. Bp 74C (0.3 mm Hg). Slightly soluble in water; soluble in alcohol, hexane, and xylene.
Hazard: Toxic by ingestion. Dry formulations are irritating to eyes and skin.
Use: Herbicide.

chlorodiazepoxide. (7-chloro-*N*-methyl-5-phenyl-3h-1,4-benzodiazepin-2-amine-4-oxide; 7-chloro-2-methylamino-5-phenyl-3h-1,4-benzodiazephine-4-oxide; metaminodiazpoxide; methaminodiazpoxide; clopoxide; 7-chloro-4-hydroxy-*N*-methyl-5-phenyl-3H-1,4-benzodiazepin-2-imine).
CAS: 58-25-3. $C_{16}H_{14}ClN_3O$.
A benzodiazepine derivative.
Hazard: Anticonvulsant, sedative, and amnesic properties.
Use: Anxiolytic and veterinary tranquilizer; in the symptomatic treatment of alcohol withdrawal.

1-chloro-2-dichloroarsinoethene. See β-chlorovinyldichloroarsine.

4-chloro-2-(3,4-dichlorophenyl)-5-((6-iodo-3-pyridinyl)methoxy)-3(2h)-pyridazinone.
CAS: 122322-18-3. $C_{16}H_9Cl_3IN_3O_2$.
Hazard: A poison by ingestion.
Use: Agricultural chemical.

2-chloro-1-(2,4-dichlorophenyl)vinyldiethyl phosphate. See diethyl-1-(2,4-dichlorophenyl)-2-chlorovinyl phosphate.

2-chloro-4-(diethylamino)-6-(isopropylamino)-*s*-triazine. See isodiazine.

chlorodifluoroacetic acid.
$CClF_2 \cdot COOH$.
Properties: Colorless, pungent liquid. Bp 122C, fp 23C. Miscible with water and most organic solvents. Strong acid, dissolves cellulose and proteins.
Use: Catalyst, particularly for esterification and condensation reactions; herbicides; intermediate.

1-chloro-1,1-difluoroethane. (chlorodifluoroethane; difluoromonochloroethane; Freon 142).
CAS: 75-68-3. CH_3CClF_2.
Properties: Colorless gas; nearly odorless. Bp −9.5C, fp −130.8C, d 1.194 (−9C), lel 9.0%, uel 14.8%. Insoluble in water.
Derivation: Chlorinating 1,1-difluoroethane in UV light.
Grade: Technical.
Hazard: Flammable gas. Explosive limits in air 9.0–14.8%.
Use: Refrigerant, solvent, intermediate.

chlorodifluoromethane. (monochlorodifluoromethane; difluorochloromethane; difluoromonochloromethane; Freon 22).
CAS: 75-45-6. $CHClF_2$.
Properties: Colorless gas; nearly odorless. D (gas at its bp) 4.82 g/L, bp −40.8C, fp −160C. Partly soluble in water.
Derivation: Reaction of chloroform with anhydrous hydrogen fluoride with antimony chloride catalyst.
Grade: Technical, 99.9% pure.
Hazard: Asphyxiant. Central nervous system impairment, cardiac sensitization. Questionable carcinogen.
Use: Refrigerant, low-temperature solvent, fluorocarbon resins, especially tetrafluoroethylene polymers.
See chlorofluorocarbon.

2-chloro-4,5-dihydro-1,3,2-dithiarsenole. See ethylene chlorothioarsenate(III).

8-chloro-3,4-dihydrospiro(naphthalene-2(1h),4'(5'h)-oxazol)-2'-amine.
CAS: 162280-52-6. $C_{12}H_{13}ClN_2O$.
Hazard: A poison by ingestion.

3-chloro-5-(((((4,6-dimethoxy-2-pyrimidinyl)amino)carbonyl)amino)sulfonyl)-1-methyl-1h-pyrazole-4-carboxylic acid, methyl ester.
CAS: 100784-20-1. $C_{13}H_{15}ClN_6O_7S$.
Hazard: Moderately toxic by ingestion.

5-chloro-2-dimethylaminobenzoxazole.
CAS: 64037-20-3. $C_9H_9ClN_2O$.
Hazard: Moderately toxic by ingestion.

2-chloro-9-(2,2-dimethylhydrazino)acridine.

CAS: 29023-82-3. $C_{15}H_{14}ClN_3$.
Hazard: A poison by ingestion.
Use: Agricultural chemical.

2-chloro-*n,n*-dimethyl-3-oxobutanamide.
CAS: 5810-11-7. $C_6H_{10}ClNO_2$.
Hazard: A poison by skin contact. Moderately toxic by ingestion. A severe skin and eye irritant.

1-chloro-*n,n*-dimethyl-2-propanamine hydrochloride.
CAS: 17256-39-2. $C_5H_{12}ClN \cdot ClH$.
Hazard: A poison by ingestion. Moderately toxic by skin contact. A severe eye irritant.

1-chloro-2,4-dinitrobenzene. (dinitrochlorobenzene).
CAS: 97-00-7. $C_6H_3(NO_2)_2Cl$.
Properties: Pale-yellow needles; almond odor. D 1.69, mp 53C, bp 315C, flash p 382F (194C). Soluble in hot alcohol, ether, benzene, carbon disulfide; insoluble in water.

Derivation: Chlorination of dinitrobenzene.
Grade: Technical, fused.
Hazard: Toxic by ingestion, inhalation, and skin absorption. Combustible. Upper explosive limit 22%. A skin irritant.
Use: Dyes, organic synthesis.

2-chloro-1,5-dinitro-3-(trifluoromethyl) benzene.
CAS: 392-95-0. $C_7H_2ClF_3N_2O_4$.
Hazard: Low toxicity by inhalation. A severe skin and eye irritant.

chlorodiphenyl.
CAS: (42% chlorine) 53469-21-9; (54% chlorine) 11097-21-9.
Properties: Colorless, mobile liquid. Bp 340–375C, flash p 383F (195C) (OC). Resistant to acids and alkalies.
Grade: 54% chlorine, 42% chlorine.
Hazard: A possible carcinogen. Toxic by ingestion, inhalation, and skin absorption; strong irritant. Liver damage, chloroacne, eye and upper respiratory tract irritant.
Use: Plasticizer for cellulosics, vinyl resins, and chlorinated rubbers.
See diphenyl.

4-chlorodiphenyl sulfone.   See *p*-chlorophenyl phenyl sulfone.

1-chloro-2,3-epoxypropane.   See epichlorohydrin.

2-chloroethanamide.   See chloroacetamide.

chloroethane.   See ethyl chloride.

chloroethane nitrile.   See chloroacetonitrile.

2-chloroethanesulfonyl fluoride.
CAS: 762-70-9. $C_2H_4ClFO_2S$.
Hazard: A poison by ingestion and skin contact. Low toxicity by inhalation. A severe eye irritant.

chloroethanoic anhydride.   See chloroacetic anhydride.

2-chloroethanol.   See ethylene chlorohydrin.

chloroethene.   See vinyl chloride.

(1-(((2-chloroethoxy)(2-chloroethyl)phosphinyl)oxy)ethyl)-phosphonic acid, 1-(bis (2-chloroethoxy)phosphinyl)ethyl 2-chloroethyl ester.
CAS: 4351-70-6. $C_{14}H_{28}Cl_5O_9P_3$.
Hazard: Moderately toxic by ingestion. Low toxicity by inhalation and skin contact. A mild skin and eye irritant.

chloroethyl alcohol.   See ethylene chlorohydrin.

2-chloro-4-ethyl-amino-*s*-triazine.   See atrazine.

β-chloroethylchloroformate.
CH_2ClCH_2OOCCl.
Properties: Colorless liquid. D 1.3825 (20C), bp 152.5C (752 mm Hg). Decomposed by alkaline solutions and hot water. Insoluble in cold water.
Derivation: By bubbling gaseous phosgene into ethylene chlorohydrin at 0C.
Hazard: Very irritating to eyes and skin. Toxic by ingestion and inhalation.

β-chloroethyl chlorosulfonate.
$ClCH_2CH_2OSO_2Cl$.
Properties: Colorless liquid; chloropicrin-like odor. Bp 101C (23 mm Hg). Darkens on long storage and decomposes with evolution of hydrogen chloride.
Derivation: Interaction of sulfuryl chloride and ethylene chlorohydrin. Also from action of sulfur trioxide on ethylene chloride below 45C.
Hazard: Toxic by ingestion and inhalation. Strong irritant to tissue.

2-chloroethyldiisopropylamine hydrochloride.
CAS: 4261-68-1. $C_8H_{18}ClN·ClH$.
Hazard: A poison by inhalation, and skin contact. A severe skin and eye irritant.

chloroethylene.   See vinyl chloride.

n-(2-chloroethyl)-n-ethyl-2-bromobenzylamine.
CAS: 62078-98-2. $C_{11}H_{15}BrClN$.
Hazard: A poison.

2-chloroethylphosphonic acid.   (ethephon; "Ethrel").
$ClCH_2CH_2PO_3H_2$.
A plant-growth regulator that increases rate of ripening of plants by promoting release of ethylene, stimulates rubber latex formation; ripening of sugarcane; flowering agent for pineapple; color enhancer.

2-chloroethyltrimethyl ammonium chloride.   (chlormequat).
$ClCH_2CH_2N(CH_3)_3Cl$.
A plant-growth regulator stated to be effective in shortening the height of wheat and as a ripening agent in sugarcane.

2-chloroethyl vinyl ether.   (2-chloroethoxy ethane).
CAS: 110-75-8. $CH_2ClCH_2OCHCH_2$.
Properties: Colorless liquid. Bp 110C, d 1.052. Stable in caustic solution; hydrolyzes in acid solutions.

CHLOROFENETHOL

314

Derivation: Reaction of a mixture of sodium hydroxide and triethanolamine with 2,2'-dichlorodiethyl ether.
Hazard: Moderate fire risk. Combustible.
Use: Manufacture of cellulose ethers.

chlorofenethol. (di(p-chlorophenyl)methylcarbinol; 1,2-bis(p-chlorophenyl)ethanol).
Properties: White solid. Mp 69.5–70C. Insoluble in water; soluble in petroleum ether, ethanol, toluene.
Use: Miticide.

chlorofenvinphos. (Birlane; 2-chloro-1-(2,4-dichlorophenyl)-vinyldiethyl phosphate).
CAS: 470-90-6.
Properties: Bp 168–170C (67C) (0.5 mm Hg).
Use: General-purpose stored-product insecticide.

chlorofluorocarbon. Any of several compounds composed of carbon, fluorine, chlorine, and hydrogen, the best known of which are trichlorofluoromethane and dichlorodifluoromethane. Their use was prohibited in 1979 except for a few specialized items because of their depleting effect on stratospheric ozone.

5-(2-chloro-6-fluoro-4-(trifluoromethyl) phenoxy)-n-(ethylsulfonyl)-2-nitrobenzamide.
CAS: 77227-69-1. $C_{16}H_{11}ClF_4N_2O_6S$.
Hazard: A reproductive hazard.
Use: Agricultural chemical.

chloroform. (trichloromethane).
CAS: 67-66-3. $CHCl_3$.
Properties: Colorless, highly refractive, heavy, volatile liquid; characteristic odor; sweet taste. D 1.485 (20/20C), bp 61.2C, fp −63.5C, wt/gal 12.29 lb (25C), refr index 1.4422 (25C). Keep from light. Miscible with alcohol, ether, benzene, carbon disulfide, carbon tetrachloride, fixed and volatile oils; slightly soluble in water.
Derivation: (1) Reaction of chlorinated lime with acetone, acetaldehyde, or ethanol; (2) by-product from the chlorination of methane.
Method of purification: Extraction with concentrated sulfuric acid and rectification.
Grade: Technical, CP, ACS, NF, reagent.
Hazard: A possible carcinogen. Toxic by inhalation; anesthetic; prolonged inhalation or ingestion may be fatal. It has been prohibited by FDA from use in drugs, cosmetics, and food packaging, including cough medicines, toothpastes, etc. Nonflammable. Will burn on prolonged exposure to flame or high temperature. Liver and embryo/fetal damage, and central nervous system impairment.
Use: Fluorocarbon plastics, solvent, analytical chemistry, fumigant, insecticides.

chloroformoxime.
ClHCNOH.

Properties: Needles; odor resembles that of hydrocyanic acid. Stable at 0C, unstable at normal temperature. Small quantities volatile, large quantities decompose. Aqueous solutions slowly decompose. Soluble in water, alcohol, ether, benzene; slightly soluble in carbon disulfide.
Derivation: Interaction of hydrochloric acid and sodium cyanate.
Hazard: Toxic by inhalation; strong irritant to tissue.
Use: Organic synthesis, tear gas, and vesicant.

chloroformyl chloride. See phosgene.

4-chloro-N-furfuryl-5-sulfamoylanthranilic acid. (aisemide; aluzine; 5-(aminosulfuranyl)-4-chloro-2-((2-furnaylmethyl)amino)benzoic acid; beronald; chlor-N-(2-furylmethyl)-5-sulfamylanthranilsaeure; 4-chloro-N-(2-furylmethyl)-5-sulfamoylanthranilic acid; frusemide; 4-chloro-2-(furan-2-ylmethylamino)-5-sulfamoylbenzoic acid; desdemin; diural; dryptal; errolon; eutensin; frusemide; frusemin; frusid; fulsix; fuluvamide; furanthril; furanthryl; furantril; furesis; furosedon; furosemide; furosemide; furosemide "mita"; fursemid; fursemide; fusid; hydro-rapid; katlex; lasex; Lasix; LB 502; lowpstron; macasirool; NCI-C55936; nicorol; prefemin; profemin; radonna; posemide; salix; seguril; transit; trofurit; urex; urosemide).
CAS: 54-31-9. $C_{12}H_{11}ClN_2O_5S$.
A benzoic-sulfonamide-furan that is a diuretic with fast onset and short duration.
Properties: Crystals from EtOH.
Hazard: Poison; moderately toxic; teratogen; questionable carcinogen; mutagen.
Use: For edema and chronic renal insufficiency.

chlorogenic acid. (3-(3,4-dihydroxycinnamoyl) quinic acid).
$(HO)_2C_6H_3CHCHCOOC_6H_7(OH)_3COOH$.
Important metabolic factor in many plant tissues.
Properties: Crystals. Mp 208C. Slightly soluble in cold water; soluble in hot water, alcohol, acetone.

chlorohydrin. (α-chlorohydrin; 1-chloropane-2,3-diol; glyceryl α-chlorohydrin).
CAS: 96-24-2. $CH_2OHCHCH_2OCH_2Cl$.
Properties: Colorless, heavy liquid; unstable; hygroscopic. The commercial grade is a mixture of the two isomers, α and β, of which α is in a greater proportion. D 1.326 (18C), bp 213C (decomposes), wt/gal 11.102 lb, fp −40C, viscosity 2.388 cP (20C). Soluble in water, alcohol, and ether; immiscible with oils. Nonflammable.
Derivation: By passing hydrogen chloride gas into glycerol containing 2% acetic acid.
Grade: Technical.
Hazard: Toxic by ingestion, inhalation.
Use: Solvent for cellulose acetate, glyceryl phthalate resins; partial solvent for gums; intermediate in organic synthesis; antifreeze agent for dynamite.

chlorohydrin rubber. An elastomer made from epichlorohydrin. Both a homopolymer and a copolymer with ethylene oxide are available.

chlorohydroquinone. (2-chloro-1,4-dihydro-xybenzene; 2,5-dihydroxychlorobenzene [Cl = 1]). $ClC_6H_3(OH)_2$.
Properties: White to light-tan fine crystals. Mp 100C, bp 263C. Very soluble in water and alcohol; slightly soluble in ether.
Grade: Photographic, commercial.
Use: Photographic developer, organic intermediate, dyestuffs, bactericide.

chlorohydroxybenzene. See chlorophenol.

5-chloro-2-hydroxybenzophenone.
$C_6H_5COC_6H_3OHCl$.
Properties: Yellow crystals; nearly odorless. Mp 93–95C. Soluble in alcohol, ethyl acetate, methyl ethyl ketone; insoluble in water.
Use: Light absorber, best at 320–380 nm.

2-chloro-4-(hydroxymercuri)phenol. See hydroxymercurichlorophenol.

4-chloro-1-hydroxy-3-methylbenzene. See 4-chloro-3-methylphenol.

6-chloro-3-hydroxytoluene. See 4-chloro-3-methylphenol.

chloro-IPC. (isopropyl-*n*-(3-chlorophenyl) carbamate; isopropyl-3-chlorocarbanilate; CIPC; chlorpropham).
CAS: 101-21-3. $C_6H_4ClNHCOOC_3H_7$.
Properties: Light-tan powder. Mp 41.4C, vap press 2 mm Hg (149C), d 1.18 (30C). Very slightly soluble in water.
Hazard: Toxic by ingestion.
Use: Preemergence herbicide, prevents sprouting of potatoes.

2-chloro-*N*-isopropylacetanilide. (*N*-isopropyl-α-chloroacetanilide).
CAS: 1918-16-7. $C_6H_5N[CH(CH_3)_2]COCH_2Cl$.
Properties: Light-tan powder or granules. Mp 67–76C, bp 110C (0.03 mm Hg). Very slightly soluble in water; soluble in acetone, alcohol, benzene, xylene, and carbon tetrachloride.
Hazard: Toxic by ingestion and skin absorption.
Use: Herbicide.

chloroisopropyl alcohol. See propylene chlorohydrin.

s-4-chloro-*n*-isopropylcarbanilolylmethyl *O,O*-dimethyl phosphorodithioate. See anilophos.

chloro isopropyl ketone. See isobutyric acid chloride.

6-chloro-4-isopropyl-1-methyl-3-phenol. See chlorothymol.

7-chlorolincomycin hydrochloride.
CAS: 17431-55-9. $C_{18}H_{33}ClN_2O_5S\cdot ClH$.
Hazard: Moderately toxic by subcutaneous and ingestion routes. A poison by intraperitoneal and intravenous routes. When heated to decomposition, it emits vapors of NO_x, SO_x, HCl, and Cl^-.

chloromadinone acetate. A nonestrogenic sex hormone used in oral contraception. See antifertility agent.

chloromaleic anhydride.

CH:CClC(O)OC(O).
Properties: Yellow liquid. D 1.5, mp 10–15C, bp 192C.
Use: Catalyst for epoxy resins, intermediate.

chloromercuriferrocene.
$C_5H_5FeC_5H_4HgCl$.
Orange, crystalline solid; mp 193–194C. Used as an intermediate and for inorganic polymers.
Hazard: Toxic by ingestion and inhalation.

chloromercuriphenol. See 2-hydroxy-phenylmercuric chloride.

chloromethane. See methyl chloride.

chloromethapyrilene citrate. See chlorothen citrate.

2-chloro-*n*-(2-methoxy-3,6-dimethylphenyl)-*n*-((1-methylethoxy)methyl)acetamide.
CAS: 78194-09-9. $C_{15}H_{22}ClNO_3$.
Hazard: Moderately toxic by ingestion.
Use: Agricultural chemical.

chloromethoxynil. See 5-(2,4-dichlorophe-noxy)-2-nitroanisole.

3-chloro-2-methylaniline. See 2-amino-6-chlorotoluene.

3-chloro-4-methylaniline. See 4-amino-2-chlorotoluene.

5-chloro-2-methylaniline. See 2-amino-4-chlorotoluene.

chloromethylated aminated stryene-divinylbenzene resin.
CAS: 60177-39-1.
Use: Food additive.

chloromethylated diphenyl oxide.
$C_6H_5OC_6H_5$.

Up to three -CH$_2$Cl radicals can be substituted for hydrogens.
Properties: Straw-colored liquids or white solids. Mp up to 55C, d 1.19–1.30 (25/25C), flash p 307F (152C). Insoluble in water; very soluble in ether. Combustible.
Use: Intermediate, resins, plasticizers.

chloromethylbenzene.   See chlorotoluene.

1-chloro-3-methylbutane.   See isoamyl chloride.

chloromethylchloroformate.
ClCOOCH$_2$Cl.
Properties: Mobile, colorless liquid; penetrating, irritating odor. Hydrolyzed by water. Decomposed by alkalies. D 1.465 at 15C, bp 106.5–107C, vap d 4.5, vap press 5.6 mm Hg (20C). Soluble in most organic solvents.
Hazard: Toxic by ingestion and inhalation.

chloromethylchlorosulfonate.
ClCH$_2$OClSO$_2$.
Properties: Colorless liquid. D 1.63, bp 49–50C (14 mm Hg).
Derivation: By protracted boiling of chlorosulfonic acid with chloromethylchloroformate; also from paraformaldehyde and chlorosulfonic acid.
Hazard: Toxic by ingestion and inhalation; strong irritant to tissue.

chloromethyl cyanide.   See chloroacetonitrile.

1-chloromethylethylbenzene.   See ethylbenzyl chloride.

4-chloro-α-(1-methylethyl)benzeneacetic acid, (2,6-dimethyl-4-(2-propynyl)phenyl) methyl ester.
CAS: 51629-58-4.   C$_{23}$H$_{25}$ClO$_2$.
Hazard: Moderately toxic by ingestion.
Use: Agricultural chemical.

1-chloromethylnaphthalene.   (α-naphthylmethyl chloride).
C$_{10}$H$_7$CH$_2$Cl.
Properties: Colorless to greenish-yellow liquid; sharp pungent odor. D 1.182 (25/25C), coagulation p 23C, refr index 1.6354–1.6360 (25C). Insoluble in water; soluble in usual organic solvents. Very reactive.
Hazard: Vapor irritating to eyes.
Use: A lachrymator, intermediate.

4-chloro-17-α-methyl-19-nortestosterone.
CAS: 3415-90-5.   C$_{19}$H$_{27}$ClO$_2$.
Hazard: A reproductive hazard.
Use: Hormone.

4-chloro-3-methylphenol.   (4-chloro-1-hydroxy-3-methylbenzene; 6-chloro-3-hydroxytoluene; 4-chloro-*m*-cresol; so-called *p*-chloro-*m*-cresol).
CAS: 59-50-7.   C$_6$H$_3$CH$_3$OHCl.
Properties: White or slightly pink crystals; phenolic odor. Mp 64–66C, bp 235C, volatile with steam. Soluble 1:250 in water at 25C; soluble in alkalies, organic solvents, fats, and oils.
Hazard: Irritant to skin.
Use: External germicide; preservative for glues, gums, paints, inks, textile, and leather goods.

4-chloro-2-methylphenoxyacetic acid.   See MCPA.

8-chloro-4-(2-methylphenoxy)quinoline.
CAS: 124496-00-0.   C$_{16}$H$_{12}$ClNO.
Hazard: Moderately toxic by ingestion.

chloromethylphosphonic acid.
ClCH$_2$PO(OH)$_2$.
Properties: White, hygroscopic solid. Mp 85–95C.
Use: Intermediate for flame-proofing agents, resins, lubricants, additives, plasticizers.

chloromethylphosphonic dichloride.
ClCH$_2$POCl$_2$.
Properties: Water-white to light-straw liquid. Highly reactive. D 1.638 (25C), refr index 1.4960–1.4970 (25C).
Hazard: Toxic by inhalation; irritant to eyes, lungs, and mucous membranes.
Use: Intermediate for flame-proofing agents, resins, lubricants, additives, and plasticizers.

chloromethylpropane.   (1-chloro-2-methylpropene; α-chloroisobutylene; α,α-dimethylvinyl chloride; isocrotyl chloride; 1-chloro-2-methylprop-1-ene).
CAS: 563-47-3.   C$_4$H$_9$Cl.
Properties: Colorless, volatile liquid.
Hazard: Irritant; poison; anesthetic; questionable carcinogen; toxic.
Use: In organic synthesis.

3-chloro-2-methyl-1-propene.   See β-methylallyl chloride.

chloronaphthalene.   See chlorinated naphthalene.

chloronaphthalene oil.
Properties: Almost colorless, thin, mobile liquid. D 1.20–1.25 (68F), liquid down to −25F (−31.6C), congealing p −30F, flash p 350F (176C), volatile at 212F (100C), bp 480–550F (248–287C). Soluble in practically all organic solvent liquids and oils. Combustible.
Derivation: By chlorinating naphthalene.
Hazard: By inhalation, strong skin irritant.
Use: Plasticizer, carbon softener and remover, heat-transfer medium, solvent for rubber, aniline and

other dyes, mineral and vegetable oils, varnish gums and resins, waxes.

α-chloro-*m*-nitroacetophenone.

$NO_2C_6H_4COCH_2Cl$.

Properties: Off-white, free-flowing granules. Mp approximately 95–100C. Soluble in chlorinated solvents; insoluble in water.

Use: Bacteriostat and fungistat in cutting oils, water systems, paint, plastics, textiles; chemical intermediate.

2-chloro-4-nitroaniline. (*o*-chloro-*p*-nitroaniline).

CAS: 121-87-9. $C_6H_3ClNO_2NH_2$.

Properties: Yellow needles. Mp 107C. Soluble in alcohol, benzene, ether; slightly soluble in water and strong acids.

Derivation: (1) From 1,2-dichloro-4-nitrobenzene by heating with alcoholic ammonia. (2) From the chlorination of *p*-nitroaniline in acid solution.

Hazard: Toxic by ingestion and inhalation.

Use: Intermediate in manufacture of dyes.

4-chloro-2-nitroaniline. (*p*-chloro-*o*-nitroaniline).

CAS: 89-63-4. $C_6H_3ClNO_2NH_2$.

Properties: Orange crystals. Mp 163C. Insoluble in water, methanol, and ether.

Hazard: Toxic by ingestion and inhalation.

Use: Dye and pigment intermediate.

4-chloro-3-nitroaniline.

CAS: 635-22-3. $C_6H_3ClNO_2NH_2$.

Properties: Yellow to tan powder. Mp 95–97C. Soluble in alcohol and acetone; partially soluble in hot water.

Hazard: Toxic by ingestion and inhalation.

Use: Intermediate in the manufacture of azo dyes, pharmaceuticals, and other organic compounds.

2-chloro-4-nitrobenzamide. See aklomide.

m-chloronitrobenzene.

CAS: 121-73-3. $C_6H_4ClNO_2$.

Properties: Yellowish crystals. D 1.534, mp 44C, bp 236C. Soluble in most organic solvents; insoluble in water.

Derivation: By chlorinating nitrobenzene in the presence of iodine and recrystallizing.

Hazard: Toxic by inhalation and ingestion. Combustible. Questionable carcinogen.

Use: Intermediate for dyes.

o-chloronitrobenzene.

CAS: 88-73-3. $C_6H_4ClNO_2$.

Properties: Yellow crystals. D 1.368, bp 245.5C, mp 32C, flash p 261F (127C). Soluble in alcohol and benzene; insoluble in water

Derivation: By nitrating chlorobenzene and purifying by rectification.

Hazard: Toxic by inhalation and ingestion. Combustible. Questionable carcinogen.

Use: Intermediate, especially for dyes.

p-chloronitrobenzene. (*p*-nitrochlorobenzene).

CAS: 100-00-5. $C_6H_4ClNO_2$.

Properties: Yellowish crystals. D 1.520, mp 83C, bp 242C. Soluble in organic solvents; insoluble in water.

Derivation: Nitration of chlorobenzene and recrystallization.

Hazard: A questionable carcinogen. Very toxic by inhalation and ingestion. Absorbed via skin. Combustible. Methemoglobinemia.

Use: Intermediate, especially for dyes; manufacture of *p*-nitrophenol from which parathion is made; agricultural chemicals; rubber chemicals.

2-chloro-5-nitrobenzenesulfonamide.

$ClNO_2C_6H_3SO_2NH_2$.

Properties: Grayish-white solid. Insoluble in water; soluble in benzene.

Hazard: Toxic by ingestion.

Use: Dye and pharmaceutical intermediates.

6-chloro-3-nitrobenzenesulfonic acid, sodium salt.

$NaSO_3C_6H_3NO_2Cl$.

Properties: Off-white, moist crystals.

Hazard: Toxic by ingestion.

Use: Intermediate for dyes and pharmaceuticals.

4-chloro-3-nitrobenzoic acid.

$ClC_6H_3NO_2COOH$.

Properties: Light-gray or white powder. Mp 170–174C.

Hazard: Toxic by ingestion and inhalation.

Use: Intermediate for dyes, perfumes, flavors, pharmaceuticals.

4-chloro-3-nitrobenzotrifluoride. (*p*-chloro-*m*-nitrotrifluorotoluene).

$C_6H_3CF_3NO_2Cl$.

Properties: Thin, oily liquid. D 1.542 (15.5/15.5C), fp −7.5C, flash p 275F (135C), refr index 1.491 (20C), bp 222C. Combustible. Soluble in organic solvents; insoluble in water.

Grade: 97.5%.

Hazard: Very toxic by inhalation and ingestion.

Use: Intermediate for dyestuffs, agricultural chemicals, pharmaceuticals.

4-chloro-2-nitrophenol, sodium salt.

$ClC_6H_3NO_2ONa$.

Properties: Red needles with one water of crystallization. Soluble in hot water.

Derivation: Nitration of *p*-dichlorobenzene followed by hydrolysis.

Grade: 90% anhydrous sodium salt, containing 80% base.

Use: Dye intermediate, manufacture of 2-amino-4-chlorophenol.

1-chloro-1-nitropropane.   See korax.

2-chloro-6-nitrotoluene.
ClNO$_2$C$_6$H$_3$CH$_3$.
Properties: Solid. Mp 36.5–40C. Insoluble in water. Intermediate.
Hazard: Dangerous fire risk.

4-chloro-2-nitrotoluene.
ClNO$_2$C$_6$H$_3$CH$_3$.
Properties: Solid. Mp 35–37C. Insoluble in water; soluble in alcohol and ether.
Hazard: Dangerous fire risk.

p-chloro-m-nitrotrifluorotoluene.   See 4-chloro-3-nitrobenzotrifluoride.

chloronitrous acid.   See aqua regia.

chloropentafluoroacetone.
ClF$_2$CCOCF$_3$.
Properties: Colorless, hygroscopic, highly reactive gas. Bp 7.8C, fp −133C, d 1.43 (25C). Non-flammable.
Hazard: Evolves heat on contact with water. Causes cardiac sensitivity.
Use: Intermediate.

chloropentafluoroethane.   (monochloropenta-fluorethane; fluorocarbon 115).
CAS: 76-15-3.   CClF$_2$CF$_3$.
Properties: Colorless gas. Bp −37.7F, fp −106C. Insoluble in water; soluble in alcohol and ether. Has good thermal stability. Nonflammable.
Use: Dielectric gas.

1-chloropentane.   See n-amyl chloride.

chlorophacinone.
C$_{23}$H$_{15}$ClO$_3$.
Properties: Crystals. Mp 139C. Almost insoluble in water; soluble in common organic solvents.
Hazard: Inhibits blood coagulation.
Use: Rodenticide.

chlorophene.   USAN for o-benzyl-p-chloro-phenol.

m-chlorophenol.   (3-chloro-1-hydroxyben-zene).
CAS: 108-43-0.   C$_6$H$_4$OHCl.
Properties: White crystals; odor similar to phenol; discolors on exposure to air. D 1.245, mp 33C, bp 214C. Soluble in alcohol, ether, and aqueous alkali; slightly soluble in water.
Derivation: From m-chloraniline through the diazonium salt.
Hazard: Toxic by skin absorption, inhalation, or ingestion.
Use: Intermediate in organic synthesis.

o-chlorophenol.   (2-chloro-1-hydroxybenzene).
CAS: 95-57-8.   C$_6$H$_4$OHCl.
Properties: Colorless to yellow-brown liquid; unpleasant, penetrating odor. Bp 175C, fp 9.3C, d 1.265 (15.5C), flash p 225F (107C). Very soluble in water; soluble in alcohol, ether, and aqueous sodium hydroxide; volatile with steam. Combustible.
Derivation: Chlorination of phenol.
Hazard: Toxic by skin absorption, inhalation, or ingestion. Strong irritant to tissue.
Use: Organic synthesis (dyes).

p-chlorophenol.   (4-chloro-1-hydroxybenzene).
CAS: 106-48-9.   C$_6$H$_4$OHCl.
Properties: White crystals (yellow or pink when impure); unpleasant penetrating odor. Bp 217C, mp 42–43C, d 1.306, refr index 1.5579 (40C), flash p 250C (121C). Slightly soluble in water; soluble in benzene, alcohol, and ether. Volatile with steam. A 1% solution is acid to litmus. Combustible.
Derivation: Chlorination of phenol, from chloraniline through the diazonium salt.
Grade: NF, technical.
Hazard: Toxic by skin absorption, inhalation, or ingestion; strong irritant to tissue.
Use: Intermediate in synthesis of dyes and drugs, denaturant for alcohol, selective solvent in refining mineral oils, antiseptic.

chlorophenoxyacetic   acid.   (p-chlorophen-oxyacetic acid; 4-chlorophenoxyacetic acid; 4-CP; 4-CPA; PCPA; 2-(4-chlorophenoxy)acetic acid; sure-set; tomato fix concentrate tomato hold; tomatotone).
CAS: 122-88-3.   C$_8$H$_7$ClO$_3$.
Properties: Needles or prisms from H$_2$O.
Hazard: Moderately toxic; mutagen.
Use: Herbicide and growth regulator for tomatoes and peaches.

1-(p-chlorophenoxy)-3,3-dimethyl-1-(1-imidazolyl)-2-butanone.
CAS: 38083-17-9.   C$_{15}$H$_{17}$ClN$_2$O$_2$.
Hazard: Moderately toxic by ingestion. Low toxicity by skin contact.
Use: Agricultural chemical.

chlorophenoxy herbicide.
Use: Herbicide to control broadleaved weeds.

chlorophenoxypropionic acid.
C$_9$H$_9$O$_3$Cl.
Use: Growth regulator and fruit thinner for plums and prunes.

4-((3-((4-chlorophenyl)amino)-4,5-dihydro-2h-benz(g)indazol-2-yl)acetyl)morpholine.
CAS: 301644-25-7.   C$_{23}$H$_{23}$ClN$_4$O$_2$.
Hazard: A poison by ingestion.

3-((4-chlorophenyl)amino)-4,5-dihydro-*n*-(1-methylethyl)-2h-benz(g)indazole-2-acetamide.
CAS: 301644-22-4. $C_{22}H_{23}ClN_4O$.
Hazard: A poison by ingestion.

3-((4-chlorophenyl)amino)-4,5-dihydro-*n*-(phenylmethyl)-2h-benz(g)indazole-2-acetamide.
CAS: 301644-24-6. $C_{26}H_{23}ClN_4O$.
Hazard: A poison by ingestion.

3-((4-chlorophenyl)amino)-*n*-(2-ethoxyethyl)-4,5-dihydro-2h-benz(g)indazole-2-acetamide.
CAS: 301644-23-5. $C_{23}H_{25}ClN_4O_2$.
Hazard: A poison by ingestion.

_p_-chlorophenyl benzenesulfonate.   See fenson.

2-(*o*-chlorophenyl)benzimidazole.
CAS: 3574-96-7. $C_{13}H_9ClN_2$.
Hazard: A poison by ingestion.
Use: Agricultural chemical.

2-(4-chlorophenyl)-1h-benz(de)isoquinoline-1,3(2h)-dione.
CAS: 6915-00-0. $C_{18}H_{10}ClNO_2$.
Hazard: A poison.

_p_-chlorophenyl-_p_-chlorobenzenesulfonate.
See ovex.

2-(4-chlorophenyl)-5-((4-chlorophenyl)methoxy)-4-iodo-3(2h)-pyridazinone.
CAS: 128758-36-1. $C_{17}H_{11}Cl_2IN_2O_2$.
Hazard: A poison by ingestion.
Use: Agricultural chemical.

2-(4-chlorophenyl)-5-((6-chloro-3-pyridinyl)methoxy)-4-iodo-3(2h)-pyridazinone.
CAS: 122322-23-0. $C_{16}H_{10}Cl_2IN_3O_2$.
Hazard: A poison by ingestion.
Use: Agricultural chemical.

4-chloro-α-phenyl-*o*-cresol.   See *o*-benzyl-*p*-chlorophenol.

3,*p*-chlorophenyl-1,1-dimethylurea.   See monuron.

3-(3-(4-(2-(4-chlorophenyl)ethyl)phenyl)-1,2,3,4-tetrahydro-1-naphthalenyl)-4-hydroxy2h-1-benzopyran-2-one.
CAS: 90035-12-4. $C_{33}H_{27}ClO_3$.
Hazard: A poison by ingestion.
Use: Agricultural chemical.

_n_-(2-(4-chlorophenyl)ethyl)-1,3,4,5-tetrahydro-7,8-dihydroxy-2h-2-benzazepine-2-carbothioamide.
CAS: 138977-28-3. $C_{19}H_{21}ClN_2O_2S$.
Hazard: A poison.

2-(3-chlorophenyl)imidazo(2,1-a)isoquinoline.
CAS: 61001-11-4. $C_{17}H_{11}ClN_2$.
Hazard: A reproductive hazard.

4-(4-chlorophenyl)-1-(1h-indol-3-ylmethyl)piperidin-4-ol.
CAS: 81226-60-0. $C_{20}H_{21}ClN_2O$.
Hazard: A poison.

_m_-chlorophenyl isocyanate.
ClC_6H_4NCO.
Properties: Clear water-white to light-yellow liquid. Fp −4.4C, flash p 215F (101.6C) (COC). Combustible.
Hazard: Toxic via strong irritant to skin, eyes, and mucous membranes.
Use: Intermediate for pharmaceuticals, herbicides, and pesticides.

_p_-chlorophenyl isocyanate.
ClC_6H_4NCO.
Properties: Colorless to slightly yellow liquid or white crystals. Fp 29.9C, flash p 230F (110C) (COC). Combustible.
Hazard: Strong irritant to skin, eyes, and mucous membranes.
Use: Intermediate for pharmaceuticals, herbicides, and pesticides.

5-((4-chlorophenyl)methoxy)-2-(3,4-dichlorophenyl)-4-iodo-3(2h)-pyridazinone.
CAS: 128758-37-2. $C_{17}H_{10}Cl_3IN_2O_2$.
Hazard: A poison by ingestion.
Use: Agricultural chemical.

(2-chlorophenyl)methyl   (4-chlorophenyl)methyl-3-pyridinylcarbonimidodithioate.
CAS: 34763-34-3. $C_{20}H_{16}Cl_2N_2S_2$.
Hazard: Moderately toxic by ingestion.
Use: Agricultural chemical.

(4-chlorophenyl)methyl   1,1-dimethylpropyl   3-pyridinylcarbonimidodithioate.
CAS: 34763-25-2. $C_{18}H_{21}ClN_2S_2$.
Hazard: Moderately toxic by ingestion.
Use: Agricultural chemical.

(4-chlorophenyl)methyl dodecyl 3-pyridinylcarbonimidodithioate.
CAS: 40199-26-6. $C_{25}H_{35}ClN_2S_2$.
Hazard: Moderately toxic by ingestion.
Use: Agricultural chemical.

(4-chlorophenyl)methyl   1-ethyl-1-methyl-propyl-3-pyridinylcarbonimidodithioate.
CAS: 34763-27-4. $C_{19}H_{23}ClN_2S_2$.
Hazard: Moderately toxic by ingestion.
Use: Agricultural chemical.

s-((4-chlorophenyl)methyl) o-ethyl 3-pyridinylcarbonimidothioate.
CAS: 34763-52-5. $C_{15}H_{15}ClN_2OS$.
Hazard: Moderately toxic by ingestion.
Use: Agricultural chemical.

(4-chlorophenyl)methyl heptyl 3-pyridinylcarbonimidodithioate.
CAS: 34763-28-5. $C_{20}H_{25}ClN_2S_2$.
Hazard: Moderately toxic by ingestion.
Use: Agricultural chemical.

3-(o-chlorophenyl)-2-methyl-4-quinazolone.
See 2-methyl-3-(2-chlorophenyl)chinazolon-4.

3-(p-chlorophenyl)-5-methylrhodanine.
$C_{10}H_8ClNOS_2$.
Properties: Yellow crystals. Mp 106–110C. Insoluble in water; soluble in acetone.
Hazard: Toxic by ingestion.
Use: Fungicide and for nematode control.

(4-chlorophenyl)methyl hexadecyl 3-pyridinylcarbonimidodithioate.
CAS: 41643-23-6. $C_{29}H_{43}ClN_2S_2$.
Hazard: Moderately toxic by ingestion.
Use: Agricultural chemical.

(4-chlorophenyl)methyl hexyl 3-pyridinylcarbonimidodithioate.
CAS: 34763-26-3. $C_{19}H_{23}ClN_2S_2$.
Hazard: Moderately toxic by ingestion.
Use: Agricultural chemical.

(4-chlorophenyl)methyl 1-methylethyl 3-pyridinylcarbonimidodithioate.
CAS: 34763-19-4. $C_{16}H_{17}ClN_2S_2$.
Hazard: Moderately toxic by ingestion.
Use: Agricultural chemical.

(4-chlorophenyl)methyl 1-methylpropyl 3-pyridinylcarbonimidodithioate.
CAS: 34763-22-9. $C_{17}H_{19}ClN_2S_2$.
Hazard: Moderately toxic by ingestion.
Use: Agricultural chemical.

o-((4-chlorophenyl)methyl) s-(2-methylpropyl)-3-pyridinylcarbonimidothioate.
CAS: 51308-77-1. $C_{17}H_{19}ClN_2OS$.
Hazard: Moderately toxic by ingestion.
Use: Agricultural chemical.

s-((4-chlorophenyl)methyl) o-(2-methylpropyl)-3-pyridinylcarbonimidothioate.
CAS: 51308-78-2. $C_{17}H_{19}ClN_2OS$.
Hazard: Moderately toxic by ingestion.
Use: Agricultural chemical.

(2-chlorophenyl)methyl methyl 3-pyridinylcarbonimidodithioate.
CAS: 34763-36-5. $C_{14}H_{13}ClN_2S_2$.
Hazard: Moderately toxic by ingestion.
Use: Agricultural chemical.

1-((4-chlorophenyl)methyl)-2-(nitromethylene)imidazolidine.
CAS: 69840-61-5. $C_{11}H_{12}ClN_3O_2$.
Hazard: A poison.
Use: Agricultural chemical.

(4-chlorophenyl)methyl octadecyl 3-pyridinylcarbonimidodithioate.
CAS: 41643-24-7. $C_{31}H_{47}ClN_2S_2$.
Hazard: Moderately toxic by ingestion.
Use: Agricultural chemical.

(4-chlorophenyl)methyl octyl 3-pyridinylcarbonimidodithioate.
CAS: 34763-29-6. $C_{21}H_{27}ClN_2S_2$.
Hazard: Moderately toxic by ingestion.
Use: Agricultural chemical.

(4-chlorophenyl)methyl tetradecyl 3-pyridinylcarbonimidodithioate.
CAS: 41643-22-5. $C_{27}H_{39}ClN_2S_2$.
Hazard: Moderately toxic by ingestion.
Use: Agricultural chemical.

4-(4-chlorophenyl)-6h-1,3,5-oxathiazine.
CAS: 58955-81-0. C_9H_8ClNOS.
Hazard: Moderately toxic by ingestion.
Use: Agricultural chemical.

chloro-o-phenyl-phenol.   (chloro-2-phenylphenol).
$C_6H_3(OH)ClC_6H_5$.
Properties: Clear, colorless to straw-colored, viscous liquid with faint characteristic odor. D 1.228 (20/4C), fp <−20C, boiling range (5–95%) 146–158.7C (5 mm Hg), flash p 273F (134C). Readily soluble in most organic solvents. Combustible.
Composition: (80%)   4-chloro-2-phenylphenol, (20%) 6-chloro-2-phenylphenol.
Hazard: Toxic by ingestion and inhalation.
Use: Fungicide.

m-chlorophenyl phenyl sulfide.
CAS: 38700-88-8. $C_{12}H_9ClS$.
Hazard: Moderately toxic by ingestion and skin contact. A mild skin and eye irritant.

p-chlorophenyl   phenylsulfone.   (4-chloro-diphenyl sulfone; sulphenone).
$ClC_6H_4SO_2C_6H_5$.

Properties: Dimorphic crystals; slight aromatic odor; tasteless. Insoluble in water; soluble in most organic solvents. Relatively stable in acids and alkalies.
Hazard: Toxic by ingestion.
Use: Insecticide and acaricide for pests (harmful to most grapes and pears).

m-chlorophenylpiperazine.
CAS: 6640-24-0. $C_{10}H_{13}ClN_2$.
Hazard: A poison.

3-(4-chlorophenyl)-n-(4-propylcyclohexyl)-2-propenamide.
CAS: 315706-74-2. $C_{18}H_{24}ClNO$.
Hazard: A poison by ingestion.

3-((4-chlorophenyl)sulfonyl)-2-propenenitrile.
CAS: 1012-71-1. $C_9H_6ClNO_2S$.
Hazard: Moderately toxic by ingestion. A moderate eye irritant.

5-(p-chlorophenyl)-2,3,5,6-tetrahydroimidazo(1,2-c)quinazolin.
CAS: 23597-98-0. $C_{16}H_{14}ClN_3$.
Hazard: A poison by ingestion.

o-(4-((4-chlorophenyl)thio)phenyl) o-ethyl s-propyl phosphorothioate.
CAS: 59010-86-5. $C_{17}H_{20}ClO_3PS_2$.
Hazard: Moderately toxic by ingestion and skin contact.

2-(p-chlorophenyl)-5h-s-triazolo(5,1-a)isoindole.
CAS: 57312-03-5. $C_{15}H_{10}ClN_3$.
Hazard: A reproductive hazard.

2-p-chlorophenyl-2-(1h-1,2,4-triazol-1-ylmethyl)hexanenitrile.   See myclobutanil.

chlorophenyltrichlorosilane.
CAS: 26571-79-9. $ClC_6H_4SiCl_3$.
A mixture of isomers.
Properties: Colorless to pale-yellow liquid. Bp 230C, d 1.439 (25/25C), refr index 1.5414 (20C), flash p 255F (124C) (COC). Readily hydrolyzed with liberation of hydrogen chloride. Combustible.
Derivation: By Grignard reaction of silicon tetrachloride and chlorophenylmagnesium chloride.
Grade: Technical.
Hazard: Corrosive to wet skin.
Use: Intermediate for silicones.

3-β-(4-chlorophenyl)tropane-2-β-carboxylic acid phenyl ester hydrochloride.
CAS: 141807-57-0. $C_{21}H_{22}ClNO_2 \cdot ClH$.
Hazard: A poison.

4-chlorophthalic acid.
$C_6H_3Cl(COOH)_2$.

Properties: Colorless crystals. Mp 150C. Decomposes on further heating; soluble in alcohol and ether; insoluble in water.
Derivation: Chlorination of phthalic acid.

chlorophyll.   The green pigment essential to photosynthesis. It is present in all plants except fungi and bacteria. It occurs in three forms (a, b, and c), all of which are magnesium-centered porphyrins containing a hydrophilic carbocyclic ring with a lipophilic phytyl tail. Chlorophyll is a photoreceptor for wavelengths up to 700 nm. It can readily transfer radiant energy to its chemical environment and thus acts as a transducer in photosynthesis. It is structurally analogous to the red blood pigment hemin. Chlorophyll has been synthesized by two different routes. Its derivatives are relatively unstable to light, oxidizing agents, and chemical reagents.

Properties: Chlorophyll a: $C_{55}H_{72}MgN_4O_5$. Blue-green, microcrystalline wax; approximately three times as plentiful as chlorophyll b: mp 117–120C. Freely soluble in ether, ethanol, acetone, chloroform, carbon disulfide, benzene; sparingly soluble in cold methanol; insoluble in petroleum ether. The alcoholic solution is blue-green with a deep-red fluorescence.
Chlorophyll b: $C_{55}H_{70}MgN_4O_6$. Yellow-green, microcrystalline wax. Sparingly soluble in absolute alcohol, ether. Ether solution has a brilliant green color. Solutions with other organic solvents are usually green to yellow-green with red fluorescence.
Chlorophyll c: Occurs in marine organisms and may be as important as chlorophyll b.
Derivation: Alcoholic extraction of green plants; isolation by chromatography.
Grade: Aqueous, alcoholic, or oil solutions; water solutions are prepared by saponification of oil-soluble chlorophyll.
Use: Colorant for soaps, oils, fats, waxes, liquors, confectionery, preserves, cosmetics, perfumes; dentistry, source of phytol, sensitizer for color film, toothpaste additive, deodorant.
See photosynthesis; porphyrin; chelate.
Note: An experimental use of chlorophyll to act as energy converter in a synthetic photovoltaic cell has aroused interest in connection with solar-energy research. The cell is termed a *synthetic leaf*, because

it is an attempt to imitate the energy-trapping function of a natural leaf.

chlorophyll A.   (phenophytinate a).
CAS: 479-61-8. $C_{55}H_{72}MgN_4O_5$.
Occurs in all photosynthetic organisms and is the major species in all of those that evolve oxygen.

chlorophyllase.   (chlorophyll esterase).
A hydrolyzing enzyme that catalyzes the removal of the phytyl group from chlorophyll, leaving a chlorophyllide.

chlorophyll B.   (magnesium pheophytinate b).
CAS: 519-62-0. $C_{55}H_{70}MgN_4O_6$.
A species that is generally characteristic of higher plants and is an accessory chlorophyll that occurs in green algae. Its salts are essential in nutrition, being required for the activity of many enzymes, especially those concerned with oxidative phosphorylation. It is a component of both intra- and extracellular fluids and is excreted in the urine and feces.
Properties: Light, silvery, metallic element.

chloropyll C.   An accessory chlorophyll found in some protistans, especially brown algae, diatoms, and flagellates.

chlorophyll D.
$C_{54}H_{70}MgN_4O_6$.
A species of chlorophyll.

chlorophyllin.   Reaction product of alcoholic potassium or sodium hydroxide and alcoholic leaf extracts. The methyl and phytyl groups are replaced by alkali, but the magnesium is not replaced. Used in food coloring, dyes, deodorants, and medicine.

chloropicrin.   (chlorpicrin; nitrotrichloromethane; trichloronitromethane; nitrochloroform).
CAS: 76-06-2. CCl_3NO_2.
Properties: Pure product slightly oily, colorless, refractive liquid. D 1.692 (0C), bp 112C, fp −69.2C. Relatively stable, no decomposition by water or mineral acids. Soluble in alcohol, benzene, ether, carbon disulfide; slightly soluble in water. Nonflammable.
Derivation: (1) Action of picric acid on calcium hypochlorite; (2) nitrification of chlorinated hydrocarbons.
Hazard: Very toxic by ingestion and inhalation; strong eye irritant; pulmonary edema. Questionable carcinogen.
Use: Organic synthesis, dyestuffs (crystal violet), fumigants, fungicides, insecticides, rat exterminator, tear gas.

1-chloropinacolone.   See 1-monochloropinacoline.

chloroplast.   A chlorophyll-containing plastid found in algal and green plant cells (eukaryotic).

chloroplast chromosome.   Circular DNA found in the photosynthesizing organelle (chloroplast) of plants instead of the cell nucleus where most genetic material is located.

chloroplatinic acid.
CAS: 16941-12-1. $H_2PtCl_6 \cdot 6H_2O$.
Properties: Red-brown crystals. D 1.431, mp 60C. Soluble in water, alcohol, and ether.
Derivation: By solution of platinum in aqua regia, evaporation, and crystallization.
Hazard: See platinum chloride.
Use: Electroplating, platinizing pumice and the like for catalysts, etching zinc for printing, platinum mirrors, indelible ink, ceramics (producing fine color effects on high-grade porcelain), microscopy.

β-chloroprene.   (2-chlorobutadiene-1,3).
CAS: 126-99-8. $H_2C:CHCCl:CH_2$.

$$H_2C=CH-\overset{\overset{\displaystyle CH_2}{\|}}{C}Cl$$

Properties: Colorless liquid. Bp 59.4C, d 0.9583 (20/20C), flash p −4F (−20C). Soluble in alcohol; slightly soluble in water.
Derivation: Addition of cold hydrochloric acid to vinylacetylene, chlorination of butadiene.
Grade: Pure, 95% min.
Hazard: Toxic by ingestion, inhalation, and skin absorption. Flammable, dangerous fire risk, explosive limits in air 4.0–20%. Eye and upper respiratory tract irritant. Possible carcinogen.
Use: Manufacture of neoprene.

1-chloropropane.   See propyl chloride.

3-chloropropane-1,2-diol.   See chlorohydrin.

1-chloro-2-propanol.   See propylene chlorohydrin.

1-chloro-2-propanone.   See chloroacetone.

1- or 3-chloropropene.   See allyl chloride.

2-chloropropene.   (isopropenyl chloride).
CAS: 75-29-6. $CH_3CCl:CH_2$.
Properties: Colorless gas or liquid. D 0.918 (9C), fp −137.4C, bp 22.65C (1 atm).
Derivation: By treating propylene dichloride with alcoholic potassium hydroxide and fractionating from the simultaneously formed 1-chloropropene.
Grade: 95% purity.
Hazard: Toxic by ingestion and inhalation. Flammable, dangerous fire risk.
Use: Intermediate in organic synthesis, formulation of copolymers.

2-chloropropionic acid. (α-chloropropionic acid).
CAS: 598-78-7. $CH_3CHClCOOH$.
Properties: Crystals. D 1.260–1.268 (20C), bp 183–187C. Soluble in water.
Hazard: Combustible. Toxic by skin contact. Male reproductive damage.
Use: Intermediate for weed killers.

3-chloropropionic acid. (β-chloropropionic acid).
CAS: 107-94-8. CH_2ClCH_2COOH.
Properties: Crystals. Mp 41C, bp 200C. Soluble in water, alcohol, chloroform. Combustible.
Use: Intermediate.

3-chloropropionitrile.
CAS: 542-76-7. $ClCH_2CH_2CN$.
Properties: Colorless liquid. Fp −51C, flash p 168F (75.5C) (CC), refr index 1.4341 (25C), d 1.1363 (25C), bp 176C (decomposes). Miscible with acetone, benzene, carbon tetrachloride, alcohol, and ether.
Hazard: Toxic by ingestion, inhalation, and skin contact. Combustible.
Use: Intermediate in polymer synthesis.

4-chloro-*n*-(4-propylcyclohexyl)benzamide.
CAS: 315706-66-2. $C_{16}H_{22}ClNO$.
Hazard: A poison by ingestion.

α-chloropropylene. See allyl chloride.

chloropropylene oxide. See epichlorohydrin.

3-chloropropyl mercaptan.
$ClCH_2CH_2CH_2SH$.
Properties: Liquid. Distillation range 141.1–145.6C, d 1.131 (15.5C), refr index 1.492 (20C), flash p approximately 110F (43C).
Hazard: Moderate fire risk. Combustible.
Use: Intermediate.

3-chloro-1-propyne. See propargyl chloride.

2-chloropyridine.
CAS: 109-09-1. ClC_5H_4N.
Properties: Oily liquid. D 1.205 (15C), bp 170C. Slightly soluble in water; soluble in alcohol and ether.
Hazard: Toxic by ingestion.
Use: Production of antihistamines, germicides, pesticides, and agricultural chemicals.

5-((6-chloro-3-pyridinyl)methoxy)-2-(3,4-dichlorophenyl)-4-iodo-3(2h)-pyridazinone.
CAS: 122322-24-1. $C_{16}H_9Cl_3IN_3O_2$.
Hazard: A poison by ingestion.
Use: Agricultural chemical.

(1-((6-chloro-3-pyridinyl)methyl)-4,5-dihydro-1h-imidazol-2-yl)cyanamide.
CAS: 111988-43-3. $C_{10}H_{10}ClN_5$.
Hazard: A poison.
Use: Agricultural chemical.

n-((6-chloro-3-pyridinyl)methyl)-1,2-ethanediamine.
CAS: 101990-44-7. $C_8H_{12}ClN_3$.
Hazard: A poison.
Use: Agricultural chemical.

((6-chloro-3-pyridinyl)methyl)guanidine.
CAS: 200258-65-7. $C_7H_9ClN_4$.
Hazard: A poison.
Use: Agricultural chemical.

1-((6-chloro-3-pyridinyl)methyl)hexahydro-2-(nitromethylene)pyrimidine.
CAS: 101336-64-5. $C_{11}H_{13}ClN_4O_2$.
Hazard: A poison.
Use: Agricultural chemical.

1-((6-chloro-3-pyridinyl)methyl)-2-imidazolidinone.
CAS: 120868-66-8. $C_9H_{10}ClN_3O$.
Hazard: A poison by ingestion.
Use: Agricultural chemical.

1-((6-chloro-3-pyridinyl)methyl)-2-imidazolidinone hydrazone.
CAS: 131206-84-3. $C_9H_{12}ClN_5$.
Hazard: A poison.
Use: Agricultural chemical.

(1-((6-chloro-3-pyridinyl)methyl)-2-imidazolidinylidene) acetonitrile.
CAS: 141631-47-2. $C_{11}H_{11}ClN_4$.
Hazard: A poison.
Use: Agricultural chemical.

1-((6-chloro-3-pyridinyl)methyl)-3-methyl-*n*-nitro-2-imidazolidinimine.
CAS: 117906-15-7. $C_{10}H_{12}ClN_5O_2$.
Hazard: A poison.
Use: Agricultural chemical.

n-((6-chloro-3-pyridinyl)methyl)-*n'*-nitroguanidine.
CAS: 131748-56-6. $C_7H_8ClN_5O_2$.
Hazard: A poison.
Use: Agricultural chemical.

1-((6-chloro-3-pyridinyl)methyl)-*n*-nitro-1h-imidazol-2-amine.
CAS: 115086-54-9. $C_9H_8ClN_5O_2$.
Hazard: A poison.
Use: Agricultural chemical.

(3-((6-chloro-3-pyridinyl)methyl)-2-thiazolidinylidene)cyanamide.
CAS: 111988-49-9. $C_{10}H_9ClN_4S$.

Hazard: Moderately toxic by ingestion and inhalation. A reproductive hazard.
Use: Agricultural chemical.

((6-chloro-3-pyridinyl)methyl)urea.
CAS: 200258-66-8. $C_7H_8ClN_3O$.
Hazard: A poison.
Use: Agricultural chemical.

6-chloroquinaldine.
$C_9H_5N(CH_3)Cl$.
Properties: Brownish-black, oily, crystalline mass.
Grade: Technical.
Use: Intermediate.

chloroquine.
CAS: 54-05-7. $C_9H_5NClNHCH(CH_3)(CH_2)_3$ $N(C_2H_5)_2$.
(7-chloro-4-(4-diethylamino-1-methylbutylamino)-quinoline).
Properties: Colorless crystals; bitter taste. Insoluble in alcohol, benzene, chloroform, ether.
Derivation: Condensation of 4,7-dichloroquinoline with 1-diethylamino-4-aminopentane.
Hazard: Toxic by ingestion. Questionable carcinogen.
Use: Medicine (antimalarial). Usually dispensed as the phosphate.

5-chlorosalicylanilide.
$ClC_6H_3OHCONHC_6H_5$.
Properties: White crystals. Mp 209–211C. Slightly soluble in water; soluble in alcohol, ether, chloroform, and benzene.
Hazard: Toxic via ingestion.
Use: Fungicide, antimildew agent, intermediate for pharmaceuticals, dyes, pesticides.

5-chlorosalicylic acid.
CAS: 321-14-2. $ClC_6H_3OHCOOH$.
Properties: White crystals. Mp 174–176C. Slightly soluble in water; soluble in alcohol, ether, chloroform, and benzene.
Hazard: Toxic by ingestion.
Use: Moth-proofing, insecticide, intermediate for pharmaceuticals, dyes, pesticides.

o-chlorostyrene.
CAS: 2039-87-4. C_8H_7Cl.
Properties: Mp 138.6C, d 1.1 (20C), bp 188.7C. Soluble in alcohol, ether, and acetone.
Hazard: Central nervous system impairment and peripheral neuropathy.

n-chlorosuccinimide.   (NCS).
$COCH_2CH_2CONCl$.
Properties: White crystals. Mp 148–149C. Soluble in water; sparingly soluble in chloroform and carbon tetrachloride.
Use: Chlorinating agent, disinfectant for swimming pools, bactericide.

3-chloro-6-sulfanilamidopyridazine.   See sulfachlorpyridazine.

chlorosulfonic acid.   (sulfuric chlorohydrin).
CAS: 7790-94-5. $ClSO_2OH$.
Properties: Colorless to light yellow, fuming, slightly cloudy liquid; pungent odor. D 1.76–1.77 (20/20C, fp −80C, bp 158C. Decomposes in water to sulfuric and hydrochloric acids. Decomposed by alcohol and acids.
Derivation: By treating sulfur trioxide or fuming sulfuric acid with hydrochloric acid.
Grade: Technical.
Hazard: Toxic by inhalation; strong irritant to eyes and skin; causes severe burns. Can ignite combustible materials. Evolves hydrogen on contact with most metals.
Use: Synthetic detergents, pharmaceuticals, sulfonating agent for dyes, pesticides, intermediates, ion-exchange resins, anhydrous hydrogen chloride, and smoke-producing chemicals.

4-chlorosulfonylbenzoic acid.
$ClSO_2C_6H_4COOH$.
Properties: Light-tan powder. Soluble in benzene; slightly soluble in ether.
Use: Intermediate.

3-(chlorosulfonyl)benzoyl chloride.
CAS: 4052-92-0. $C_7H_4Cl_2O_3S$.
Hazard: Moderately toxic by ingestion and skin contact. A moderate eye irritant.

chlorosulfuric acid.   See sulfuryl chloride.

4-chlorotestosterone 17-acetate.
CAS: 855-19-6. $C_{21}H_{29}ClO_3$.
Hazard: A poison by ingestion. A reproductive hazard.
Use: Hormone.

chlorotetracycline.   See chlortetracycline.

chlorotetrafluoroethane.   (monochlorotetrafluoroethane).
CAS: 2837-89-0. C_2CF_4Cl.
Properties: Gas; odorless; colorless. Much heavier than air. Nonflammable.

6-chloro-2,3,4,5-tetrahydro-3-methyl-1-(3-methylphenyl)-1h-3-benzazepine-7,8-diol, hydrobromide.
CAS: 67287-95-0. $C_{18}H_{20}ClNO_2 \cdot BrH$.
Hazard: A poison.

6-chloro-2,3,4,5-tetrahydro-1-phenyl-1h-3-benzazepine-7,8-diol, hydrobromide.
CAS: 67287-39-2. $C_{16}H_{16}ClNO_2 \cdot BrH$.
Hazard: A poison.

6-chloro-2,3,4,5-tetrahydro-1-phenyl-3-(2-propenyl)-1h-3-benzazepine-7,8-diol, hydrobromide.
CAS: 74115-01-8. $C_{19}H_{20}ClNO_2 \cdot BrH$.
Hazard: A poison.

1-chloro-2,2,5,5-tetramethyl-4-imidazolidinone.
CAS: 38951-85-8. $C_7H_{13}ClN_2O$.
Hazard: A poison by ingestion. Low toxicity by skin contact.

chlorothen citrate.   (chloromethapyrilene citrate).
$C_{14}H_{18}ClN_3S \cdot C_6H_8O_7$.
Properties: White, crystalline powder; practically odorless. Mp 112–116C. On further heating solidifies and remelts; 125–140C decomposes. Slightly soluble in alcohol and water; practically insoluble in chloroform, ether, and benzene. One percent solution is clear and colorless; pH (1% solution) 3.9–4.1.
Grade: NF.
Use: Medicine (antihistamine).

Chlorothene.   A series of chlorinated organic solvents.
Grade: NU and industrial are both inhibited 1,1,1-trichloroethane.
Use: Cold cleansing of metal parts; industrial solvents.

6-chloro-2-thio-2h-1,3-benzoxazine-2,4(3h)-dione.
CAS: 7672-94-8. $C_8H_4ClNO_2S$.
Hazard: A poison by ingestion.
Use: Agricultural chemical.

6-chloro-4-thiochromanyl-*O,O*-diethyl dithiophosphate.
CAS: 41219-31-2. $C_{13}H_{18}ClO_2PS_3$.
Hazard: A poison by ingestion.
Use: Agricultural chemical.

6-chloro-4-thiochromanyl   *O,O*-dimethyl dithiophosphate.
CAS: 41219-30-1. $C_{11}H_{14}ClO_2PS_3$.
Hazard: A poison by ingestion.
Use: Agricultural chemical.

_p_-chlorothiophenol.   (*p*-chlorobenzenethiol).
CAS: 106-54-7. ClC_6H_4SH.
Properties: Moist white to cream crystals. Mp 52–55C, bp 205–207C. Soluble in most organic solvents.
Hazard: Toxic by ingestion.
Use: Oil additives, agricultural chemicals, plasticizers, rubber chemical, dyes, wetting agents, and stabilizers.

chlorothymol.   (6-chloro-4-isopropyl-1-methyl-3-phenol).
$CH_3C_6H_2(OH)(C_3H_7)Cl$.

Properties: White crystals or granular powder; characteristic odor; aromatic, pungent taste. Becomes discolored with age, affected by light. Mp 59–61C. Soluble in benzene, chloroform, dilute caustic soda, alcohol; slightly soluble in water.
Derivation: Action of sulfuryl chloride on thymol in a solution of carbon tetrachloride.
Grade: NF.
Hazard: Irritant to skin and mucous membranes in concentrated solution.
Use: Bactericide, component of antiseptic solutions.

α-chlorotoluene.   See benzyl chloride.

_m_-chlorotoluene.   (3-chloro-1-methylbenzene).
CAS: 108-41-8. $CH_3C_6H_4Cl$.
Properties: Colorless liquid. D 1.07218 (20/4C), bp 161.6C, fp −48.0C, refr index 1.52 (20C).
Derivation: Diazotization of *m*-toluidine followed by treating with cuprous chloride.
Hazard: Narcotic in high concentration. Avoid inhalation.
Use: Solvent, intermediate.

_o_-chlorotoluene.   (2-chloro-1-methylbenzene).
CAS: 95-49-8. $CH_3C_6H_4Cl$.
Properties: Colorless liquid. Bp 159.2C, fp −35.1C, d 1.0776 (25/4C), refr index 1.5268 (20C). Miscible with alcohol, acetone, ether, benzene, carbon tetrachloride, and *n*-heptane; slightly soluble in water.
Derivation: By catalytic chlorination of toluene.
Hazard: Toxic by inhalation. Eye, skin, and upper respiratory tract irritant.
Use: Solvent and intermediate for organic chemicals and dyes.

_p_-chlorotoluene.   (4-chloro-1-methylbenzene).
CAS: 106-43-4. $CH_3C_6H_4Cl$.
Properties: Colorless liquid. Boiling range 162–166C, fp approximately 6.5C, d 1.065–1.067 (25/15C), refr index 1.5184 (22C). Soluble in alcohol, ether, acetone, benzene, and chloroform; slightly soluble in water.
Hazard: Avoid inhalation; strong irritant.
Use: Solvent and intermediate for organic chemicals and dyes.

2-chlorotoluene-4-sulfonic acid.   (*o*-chlorotoluene-*p*-sulfonic acid).
$CH_3C_6H_3(SO_3H)Cl$.
Properties: White, glistening plates. Soluble in hot water.
Derivation: Chlorination of toluene-*p*-sulfonic acid.
Method of purification: Recrystallization from water.
Use: Dye intermediate.

2-chloro-*p*-toluidine.   See 4-amino-2-chlorotoluene.

3-chloro-*p*-toluidine. (CPT; 3-chloro-4-methylaniline; 1-amino-3-chloro-4-methylbenzene; 4-amino-2-chlorooluene; 2-chloro-4-aminotoluene; DKC 1347; DRC 1339; NCI-C02040; 3-Chloro-4-methylaniline).
CAS: 33240-95-8. C_7H_8ClN.
Properties: A solid or liquid.
Hazard: Nephrotoxic; hepatotoxic; poison; mutagen.
Use: Avicide.

4-chloro-*o*-toluidine. See 2-amino-4-chlorotoluene.

6-chloro-*o*-toluidine. See 2-amino-6-chlorotoluene.

4-chloro-*o*-toluidine hydrochloride.
CAS: 3165-93-3. $CH_3C_6H_3(Cl)NH_2 \cdot HCl$.
Hazard: Toxic by ingestion, inhalation, and skin absorption.

2-chloro-5-toluidine-4-sulfonic acid. (6-chloro-*m*-toluidine-4-sulfonic acid).
$CH_3C_6H_2(NH_2)(SO_3H)Cl$.
Properties: Fine, white crystals. Soluble in dilute caustic solution.
Derivation: From *o*-chlorotoluene-*p*-sulfonic acid by nitration and subsequent reduction.
Use: Intermediate.

chlorotriazinyl dye. A fiber-reactive dye, both mono- and di-derivatives have been developed. They react readily and permanently with cellulose under alkaline conditions with 70–80% efficiency.

chlorotribenzylstannane.
CAS: 3151-41-5. $C_{21}H_{21}ClSn$.
Properties: Colorless needles from EtOAc. Mp: 142–144C.
Hazard: A poison by ingestion. A skin and severe eye irritant.

chlorotributylstannane.
CAS: 1461-22-9. $C_{12}H_{27}ClSn$.
Properties: Liquid. D: 1.2105 @ 20°, bp 171–173C @ 25 mm.
Hazard: A poison by ingestion and skin contact. A severe eye irritant. Tributyl tin compounds are extremely toxic to marine life.
Use: Agricultural chemical.

2-chloro-6-(trichloromethyl)pyridine. See nitrapyrin.

2-chloro-1,1,3-triethoxy propane.
CAS: 10140-99-5. $C_9H_{19}ClO_3$.
Hazard: Moderately toxic by ingestion and skin contact. Low toxicity by inhalation. A mild skin irritant.

chlorotriethylstannane. See triethyltin chloride.

2-chloro-1,1,2-trifluoroethyl difluoromethyl ether. See enflurane.

chlorotrifluoroethylene. (CFE; CTFE; trifluorochloroethylene).
CAS: 79-38-9. $ClFC:CH_2$.
Properties: Colorless gas; faint ethereal odor. Bp −27.9C, fp −157.5C, d (liquid) 1.305 (20C). Decomposes in water.
Derivation: From trichlorotrifluoroethane and zinc.
Grade: Technical 99.0%.
Hazard: Dangerous fire risk. Flammable limits in air 8.4–38.7%.
Use: Intermediate, monomer for chlorotrifluoroethylene resins.

chlorotrifluoroethylene polymer. (polytrifluorochloroethylene resin; fluorothene).
A polymer of chlorotrifluoroethylene, usually including vinylidine fluoride, characterized by the repeating structure $(CF_2\text{-}CCl)$.
Properties: Colorless. D 2.10–2.15, refr index 1.43. Impervious to corrosive chemicals, resistant to most organic solvents, heat resistant. Nonflammable. Thermoplastic. Zero moisture absorption; high-impact strength; transparent films and thin sheets.
Use: Chemical piping, gaskets, tank linings, connectors, valve diaphragms, wire and cable insulation, electronic components.
See "Kel-F"; fluorocarbon polymer.

chlorotrifluoromethane. (monochlorotrifluoromethane; trifluorochloromethane).
CAS: 75-72-9. $CClF_3$.
Properties: Colorless gas; ethereal odor. Bp −81.4C, fp −181C, heavier than air. Nonflammable
Derivation: From dichlorodifluoromethane in vapor phase with aluminum chloride catalyst.
Grade: 99.0% min purity.
Hazard: Toxic by inhalation; slightly irritant.
Use: Dielectric and aerospace chemical, hardening of metals, pharmaceutical processing.
See chlorofluorocarbon.

2-chloro-5-trifluoromethylaniline.
$C_6H_3ClCF_3NH_2$.
Properties: Amber-colored oil. Fp 6–8C. Insoluble in water; soluble in alcohol and acetone. Forms water-soluble salts with mineral acids.
Hazard: Irritant to eyes and skin.
Use: Intermediate for dyes and other organic chemicals.

chlorotrifluoromethylbenzene. See chlorobenzotrifluoride.

chloro-α,α,α-trifluorotoluene. See chlorobenzotrifluoride.

chloro(triisobutyl)stannane.
CAS: 7342-38-3. $C_{12}H_{27}ClSn$.

Properties: Solid. D: 1.1290 @ 34°, mp 30.2C, bp: 174C @ 13 mm.
Hazard: A poison. Tributyl tin compounds are very toxic to marine life.

chlorotriisopropylstannane. See triisopropyltin chloride.

chlorotrimethylstannanechlorotrimethyl stannane.
CAS: 1066-45-1. C_3H_9ClSn.
Properties: Colorless needles. Mp: 42C, bp: 154–156C.
Hazard: A deadly poison. A reproductive hazard.

chlorotrioctylstannane. See tri-*n*-octyltin chloride.

chlorotripropylstannane.
CAS: 2279-76-7. $C_9H_{21}ClSn$.
Properties: Colorless liquid. D: 1.2678 @ 28°, mp −23.5C. Sol in org solvs.
Hazard: A poison.

chloro(trivinyl)stannane.
CAS: 10008-90-9. C_6H_9ClSn.
Properties: Colorless liquid. Bp: 59–60C @ 6 mm.
Hazard: A poison.

chlorous acid. (chlorite).
CAS: 14998-27-7. ClO_2^-.
An acid known only in solution and as its salts.

β-chlorovinyldichloroarsine. (1-chloro-2-dichloroarsinoethane; dichloro[2-chlorovinyl]arsine; chlorovinylarsinedichloride; Lewisite). Two isomers, probably *cis* and *trans*, are known. ClCH:CHAsCl₂.
Properties: Colorless liquid when pure; geranium-like odor. Impurities lead to colors ranging from violet to brown, decomposed by water and alkalies. Inactivated by bleaching powder. Antidote is dimercaptopropanol.
Derivation: Condensation of arsenic trichloride with acetylene in the presence of aluminum, copper, or mercury chloride. The mixed arsines are separated by fractionating.
Hazard: Vesicant gas, a poison. See arsenic.
Use: Poison gas, skin-blistering agent.

β-chlorovinylethylethynylcarbinol. See ethchlorvynol.

β-chlorovinylmethylchloroarsine.
ClCH:CHAsClCH₃.
Properties: Liquid. Bp 112–115C (10 mm Hg). Decomposed by water.
Derivation: Interaction of acetylene and methyldichloroarsine in the presence of aluminum chloride.
Hazard: Strong irritant poison, absorbed by skin.

chloroxine. (5,7-dichloro-8-quinolinol).
Properties: Colorless crystals. Mp 180C. Soluble in benzene, acetone, acids, and sodium hydroxide solution.
Use: Analytical reagent, antibacterial.

Chloroxuron. (3-(*p*-(*p*-chlorophenoxy)phenyl)-1,1-dimethylurea).
CAS: 1982-47-4.
Use: Herbicide.

***p*-chloro-*m*-xylenol.** (4-chloro-3,5-dimethylphenol).
$C_6H_2(CH_3)_2OHCl$.
Crystals with phenolic odor.
Hazard: Toxic by ingestion; strong irritant, absorbed by skin.
Use: Active ingredient in germicides, antiseptics, etc., fungistat; mildew preventive; preservative; chemical intermediate.

6-chloro-3,4-xylyl methylcarbamate. (6-chloro-3,4-dimethylphenyl-*N*-methylcarbamate).
$ClC_6H_2(CH_3)_2OCONHCH_3$.
Properties: Solid. Mp 120–133C.
Hazard: Toxic by ingestion and inhalation.
Use: Pesticide.

chloroxyphenamine.
CAS: 16726-46-8. $C_{23}H_{31}ClN_2O_3 \cdot 2ClH$.
Hazard: Moderately toxic by ingestion.

chlorpheniramine maleate. (chlorprophenpyridamine maleate).
$C_{16}H_{19}ClN_2 \cdot C_4H_4O_4$.
(1-(*p*-chlorophenyl)-1-(2-pyridyl)3-dimethylaminopropane maleate).
Properties: White, odorless crystals. Mp 130–135C. Slightly soluble in ether; soluble in alcohol, chloroform, and water; pH (1% solution) approximately 4.8.
Grade: USP.
Use: Medicine (antihistamine).

chlorphenol red. (dichlorosulfonphthalein).
$C_6H_4SO_2OC(C_6H_3ClOH)_2$.
An acid–base indicator showing color change from yellow to red over the pH range 5.2–6.8.

chlorpromazine. (2-chloro-10-(3-dimethylaminopropyl)-phenothiazine).
CAS: 50-53-3. $C_{17}H_{19}ClN_2S$.

Properties: Oily liquid; amine odor. Alkaline reaction, bp 200–205C (0.8 mm Hg).
Hazard: Toxic by ingestion.
Use: Medicine (antipsychotic drug).

chlorpromazine sulfone.   See oxochlorpromazine.

chlorpropham.   See chloro-IPC.

chlorpyrifos.   See Dursban.

chlorquinaldol.   (5,7-dichloro-8-hydroxy quinaldine; 5,7-dichloro-2-methyl-8-quinolinol). $CH_3C_9H_3N(OH)Cl_2$.
Properties: Yellow, crystalline powder; tasteless; pleasant medicinal odor. Mp 114C. Soluble in benzene, alcohol, chloroform; insoluble in water.
Use: Medicine (bactericide and fungicide).

chlortetracycline.   (CTC; chlorotetracycline). $C_{22}H_{23}ClN_2O_8$.
An antibiotic produced by growth of *Streptomyces aureofaciens* in submerged cultures. It has a wide antimicrobial spectrum, including many gram-positive and gram-negative bacteria, rickettsiae, and several viruses. Its chemical structure is that of a modified naphthacene molecule. Also available as the hydrochloride.
Properties: Golden-yellow crystals. Mp 168–169C. Very soluble in aqueous solutions above pH 8.5; freely soluble in the "Cellosolves" dioxane and "Carbitol"; slightly soluble in water, methanol, ethanol, butanol, acetone, ethyl acetate, and benzene; insoluble in ether and naphtha.
Derivation: By submerged aerobic fermentation, filtration, solvent extraction, and crystallization.
Use: Medicine (antibiotic), feed supplement, preservative for raw fish.

chlorthion.   (generic name for *O,O*-dimethyl-*O*-(3-chloro-4-nitrophenyl)thiophosphate).
A phosphoric acid ester containing chlorine.
Hazard: Toxic by inhalation and ingestion, cholinesterase inhibitor, absorbed via intact skin. Use may be restricted.
Use: Insecticide.

CHOC.   Abbreviation for Center of History of Chemistry.
See Appendix II D.

chocolate agar.   Agar and heated blood that is brown in color.

choking agent.   (lung-damaging agent).
A chemical warfare agent
Hazard: Damage to lung tissue causing severe edema.

cholaic acid.   See taurocholic acid.

cholane.   ((5R,8R,10S,13R,17R)-10,13-dimethyl-17-[(2R)-pentan-2-yl]-2,3,4,5,6,7,8,9,11,12,14,15,16,17-tetradechahydro-1H-cyclopenta[a]phenanthrene).
$C_{24}H_{42}$.
A parent compound of steroids, certain other hormones, bile acids, and toad poisons.
Properties: Pentacyclic hydrocarbon.
Hazard: Carcinogen.

cholanthrene.   (1,2-dihydrobenz[j] aceanthrylene).
$C_{20}H_{14}$.
The structural precursor of 3-methylcholanthrene.
Properties: Pale yellow leaflets of pentacyclic hydrocarbons.
Hazard: Questionable carcinogen; tumorigen.

cholate.   ((4R)-4-[(3R,5S,7R,8R,9S,10S,12S,13R)-trihydroxy-10,13-dimethyl-2,3,4,5,6,7,8,9,11,12,14,15,16-tetradecahydro-1H-cyclopenta[a]phenanthren-17-yl]pentanoic acid.
CAS: 81-25-4. $C_{24}H_{40}O_5$.
A salt or ester of cholic acid. A major primary bile acid produced in the liver and usually conjugated with glycine or taurine. It facilitates fat absorption and cholesterol excretion.

cholecalciferol.   (5,7-cholestadien-3-β-ol; 7-dehydrocholesterol; activated vitamin D_3).
CAS: 67-97-0. $C_{27}H_{44}O$.
A free vitamin D_3, isolated in crystalline state from the 3,5-dinitrobenzoate, produced by irradiation, and equivalent in activity to vitamin D_3 of tuna-liver oil.
Properties: Colorless crystals. Melting range 84–88C, specific rotation +105–112 degrees. Unstable in light and air. Insoluble in water; soluble in alcohol, chloroform, and fatty oils.
Grade: USP, FCC.
Available forms: Hermetically sealed under nitrogen.
Use: Medicine (antirachitic vitamin).

cholecystokinin.
$C_{166}H_{261}N_{51}O_{52}S_4$.
A vertebrate hormone peptide of about 33 amino acids, secreted in the duodenum of the intestinal tract (upper intestinal mucosa) and in the central nervous system that stimulates the gallbladder to contract releasing pancreatic exocrine enzymes and affects other gastrointestinal functions.

cholecystokinin tetrapeptide.   See gastrin tetrapeptide amide.

choleic acid.   A general term applied to the coordination complexes formed by deoxycholic acid (a bile acid) with fatty acids or other lipids and with a variety of other compounds including such aromatics as phenol and naphthalene. These complexes are similar to those used in separation processes such as the urea adducts for large-scale purification. See cholic acid.

cholestane. (10,13-dimethyl-17-(6-methyl-heptan-2-yl)-2,3,4,5,6,7,8,9,11,12,14,15,16,17-tetradecahydro-1H-cyclopenta[a]phenanthrene; 5α-cholestane).
$C_{27}H_{48}$.
The parent hydrocarbon of cholesterol.

cholesteric. A molecular structure found in some liquid crystals, so called because it was first noted in cholesteryl alcohol (in 1888). It occurs in some optically active compounds and in mixtures of chiral compounds and nematic liquid crystals.

cholesterol. (cholesterin; 5-cholesten-3-β-ol).
CAS: 57-88-5. $C_{27}H_{45}OH$.
The most common animal sterol, a monohydric secondary alcohol of the cyclopentenophenanthrene (4-ring fused) system, containing one double bond. It occurs in part as the free sterol and in part esterified with higher fatty acids as a lipid in human blood serum. The primary precursor in biosynthesis appears to be acetic acid or sodium acetate. Cholesterol itself in the animal system is the precursor of bile acids, steroid hormones, and provitamin D_3.

Properties: White or faintly yellow pearly granules or crystals; almost odorless. Affected by light. Mp 148.5C, bp 360C (decomposes), d 1.067 (20/4C), levorotatory, specific rotation (25C) −34 to 38 degrees. Sparingly soluble in water; moderately soluble in hot alcohol; soluble in benzene, oils, fats, and aqueous solutions of bile salts.
Occurrence: Egg yolk, liver, kidneys, saturated fats, and oils. All body cells contain cholesterol produced by the liver (approximately 1000 mg a day).
Source: Prepared from beef spinal cord by petroleum ether extraction of the nonsaponifiable matter; purification by repeated bromination.
Grade: Technical, USP, SCW (standard for clinical work).
Hazard: Questionable carcinogen.
Use: Emulsifying agent in cosmetic and pharmaceutical products, source of estradiol.

cholestyramine. A synthetic anion-exchange polymer in which quaternary ammonium groups are attached to a copolymer of styrene and divinylbenzene. A white- to buff-colored powder having a particle size of 50–100 mesh, it is stable to 150C. Insoluble in water and organic solvents. It is effective in binding bile salts such as cholesterol. Research on test animals and in clinical trials on humans indicates that it is effective in eliminating from the body such toxic organochlorine compounds as "Kepone."

cholestyramine resin. (cholestyramine chloride; cholestyramine resin; colestyramin; cuemid; quantalan; questran).
CAS: 11041-12-6.
A synthetic, strongly basic anion exchange resin that contains functional quaternary ammonium groups linked to a styrene-divinylbenzene copolymer.
Hazard: Low toxicity; tumorigen; questionable carcinogen; teratogen.

cholic acid.
CAS: 81-25-4. $C_{23}H_{49}O_3COOH$.

The most abundant bile acid. In bile, it is conjugated with the amino acids glycine and taurine as glycocholic acid and taurocholic acid, respectively, and does not occur free.
Properties: The monohydrate crystallizes in plates from dilute acetic acid. Bitter taste with sweetish aftertaste. Anhydrous form. Mp 198C. Not precipitated by digitonin. Soluble in glacial acetic acid, acetone, and alcohol; slightly soluble in chloroform; practically insoluble in water and benzene.
Derivation: From glycocholic and taurocholic acids in bile; organic synthesis.
Grade: FCC.
Use: Biochemical research, pharmaceutical intermediate, emulsifying agent in foods (up to 0.1%).

choline. (choline base; β-hydroxyethyltrimethylammonium hydroxide).
CAS: 62-49-7. $(CH_3)_3N(OH)CH_2CH_2OH$.
Member of the vitamin B complex. Essential in the diet of rats, rabbits, chickens, and dogs. In humans, it is required for lecithin formation and can replace methionine in the diet. There is no evidence of disease in humans caused by choline deficiency. It is a dietary factor important in furnishing free methyl groups for transmethylation; has a lipotropic function.
Source: Egg yolk, kidney, liver, heart, seeds, vegetables, and legumes; synthetic preparation from trimethylamine and ethylene chlorohydrin or ethylene oxide.

Properties: Viscous, alkaline liquid. Soluble in water and alcohol. Amounts are expressed in milligrams of choline.
Use: Medicine, nutrition, feed supplement, catalyst, curing agent, control of pH, neutralizing agent, solubilizer.

choline bicarbonate.   See hydroxyethyltrimethylammonium bicarbonate.

choline bitartrate.
CAS: 87-67-2. $(C_5H_{14}NO)C_4H_5O_6$.
Properties: White, crystalline powder; odorless or faint trimethylamine-like odor; acidic taste. Hygroscopic; soluble in water and alcohol; insoluble in ether, chloroform, and benzene.
Grade: FCC.
Use: Medicine, dietary supplement, nutrient.

choline chloride.
CAS: 67-48-1. $(CH_3)_3N(Cl)CH_2CH_2OH$.
Animal-feed additive derived from agricultural waste or made synthetically. Available as 50% dry feed grade and 70% solution.

cholinesterase.   (1)   (Acetylcholinesterase) Enzyme specific for the hydrolysis of acetylcholine to acetic acid and choline in the body. It is found in the brain, nerve cells, and red blood cells and important in the mechanism of nerve action.
See nerve gas; parathion; insecticide.
Derivation: From bovine erythrocytes.
Use: Biochemical research, determination of phosphorus in insecticides and poisons.
(2) "Pseudo" or nonspecific cholinesterase; prepared from horse serum. This esterase hydrolyzes other esters, as well as choline esters. It occurs in blood serum, the pancreas, and the liver.

cholinesterase inhibitor.   A chemical compound that deactivates the enzyme cholinesterase, thus preventing or retarding hydrolytic breakdown of the highly toxic acetylcholine formed in the body by the nervous system. Nerve gases act in this way, as do a number of insecticides, usually organic esters of phosphoric acid derivatives. Serious poisoning and death may occur on ingestion or prolonged inhalation of such compounds. Cholinesterase can be reactivated by administration of atropine sulfate or pralidoxime iodide.
See parathion; nerve gas.

cholytaurine.   See taurocholic acid.

chondroitin sulfate.   A major constituent of the cartilaginous tissue in the body.

chondrus.   See carrageenan.

chorionic gonadotropin.   (HCG).
A hormone isolated from blood and urine of pregnant women; secreted by the placenta. It is a glycoprotein containing approximately 11% galactose and having a molecular weight of approximately 100,000.
Properties: Rods or needlelike crystals. Soluble in water and glycols. Unstable in aqueous solution, stable in dry form. It enhances estrone and progesterone production.
Units: One international unit equals the activity of 0.1 mg of a standard preparation.
Use: Medicine, veterinary medicine.

Chromacyl.   A group of dyes that contain chromium in the molecule. Suitable for wool and nylon.

chromate.   Compound in which chromium has a valence of six.

chromated zinc chloride.   See zinc chloride, chromated.

chromate pigment.
Hazard: Toxic.
Use: In pigments.

chromaticity.   A qualitative description of color based on hue and saturation independent of brilliance.

chromaticity coordinates.   The ratios of each tristimulus value of a color to the sum of tristimulus values.

chromatic scale.   The arrangement of the colors of the visible spectrum.

chromatin.   A deoxyribonucleoprotein complex consisting of (1) double-stranded DNA molecules; (2) a basic protein called histone; and (3) other proteins. The latter protect the DNA from attack by enzymes. Chromatin occurs in the cell nucleus, where it forms chromosomes, the carriers of genes. Its name is derived from its sensitivity to biological stains.
See deoxyribonucleic acid; chromosome; gene.

chromatogram.   A column of adsorbent gas- or liquid-carrying bands or zones of adsorbed molecules.
See chromatography.

chromatography.   A group of laboratory separation techniques based on selective adsorption by which components of complex mixtures (vapors, liquids, solutions) can be identified. Its discoverer, Tswett (1906), named the procedure chromatography because the plant pigments used in his early experiments produced bands of characteristic color. Since the 1930s, the method has been widely applied in many variations to the analysis of colorless mixtures such as hydrocarbons, metallic salts, etc. Separation is due to redistribution of the molecules of the mixture between the thin phase (adsorption layer) and the bulk phase (adsorbent) with which it is in contact. Because the thin phase sometimes approaches molecular dimensions, the size and shape of the molecules of the mixture are of great significance.

Chromatography involves the flow of a mobile (gas or liquid) phase over a stationary phase (which may be a solid or a liquid). Liquid chromatography is used for soluble substances and gas (vapor-phase) chromatography for volatile substances. As the mobile phase moves past the stationary phase, repeated adsorption and desorption of the solute occurs at a rate determined chiefly by its ratio of distribution between the two phases. If the ratio is large enough, the components of the mixture will move at different rates, producing a series of bands (chromatographs) by which their identity can be determined.

See liquid chromatography; gas chromatography; paper chromatography; thin-layer chromatography; ion-exchange chromatography; gel filtration.

chromatoplate.   A thin plastic or glass plate that has a thin coat of a substance that is suitable for use in thin-layer chromatographic analysis.

chrome.   (chromium).
CAS: 7440-47-3.   Cr.
A dye or pigment that contains chromium, a trace element that plays a role in glucose metabolism.
Hazard: Carcinogen; tumorigen.

chrome alum.   See chromium potassium sulfate.

chrome ammonium alum.   See chromium ammonium sulfate.

chrome cake.   A green form of salt cake (sodium sulfate) containing a low percentage of chromium. A by-product of sodium dichromate manufacture used in the paper industry.

chrome dye.   A mordant dye, most frequently one in which sodium dichromate is used as the mordant.
Use: A mordant for wool dyeing.

chrome fast orange g.   See C.I. mordant orange 6, disodium salt.

chrome green.   See chrome pigment.

chrome leather scarlet se.   See C.I. direct red 23.

chrome liquor.   A solution of basic chromic salt used in chrome tanning. It is usually a strong solution of sodium dichromate treated with sulfur dioxide or glucose and sulfuric acid.

chrome-molybdenum steel.   Steel, made by any accepted method of quality steelmaking, containing both chromium and molybdenum, usually

in the ranges of chromium 0.35–1.10% and molybdenum 0.08–0.35%.

chrome-nickel steel.   See steel, stainless.

chrome orange.   Basic lead chromate.
Use: Pigments.

chrome pigment.   An inorganic pigment containing chromium. The most important types are (1) chrome oxide green, one of the most permanent and stable pigments known, the pure grade consisting of 99% Cr_2O_3, used in paints applied to cement and lime-containing surfaces; (2) chrome green, chrome yellow, and chrome red, consisting chiefly of lead chromate and used in paints, rubber, and plastic products; and (3) miscellaneous pigments such as molybdate orange and zinc yellow, based on lead and zinc compounds of chromium, respectively. All these are more stable to sunlight, weathering, and chemical action than the brighter organic dyes.
Hazard: Toxic by ingestion.

chrome potash alum.   See chromium potassium sulfate.

chrome red.   See chrome pigment.

chrome steel.   A steel, made by any accepted method of quality steelmaking, containing chromium as alloying element, usually in the range 0.20–1.60%, although the content may be as high as 25% in specialized heat-resistant and wear-resistant steels and in stainless steels.
See steel, stainless.

chrome tanning.   See tanning.

chrome tape.
Magnetic recording tape that has a coating of chrome dioxide.

chrome-vanadium steel.   A steel, made by any accepted method of quality steelmaking, containing both chromium and vanadium, usually in the ranges of chromium 0.50–1.10% and vanadium 0.10–0.20%.

chrome yellow.   See lead chromate; chrome pigment.

chromia.   See chromic oxide.

chromic.   Designating compounds in which the chromium is positive trivalent.

chromic acetate.   (chromium acetate).
CAS: 1066-30-4. $Cr(C_2H_3O_2)_3 \cdot H_2O$.
Properties: Grayish-green powder or bluish-green, pasty mass. Soluble in water; insoluble in alcohol.

Derivation: Action of acetic acid on chromium hydroxide. The solution is evaporated and crystallized.

Hazard: Toxic by ingestion.

Use: Textile mordant, tanning, polymerization and oxidation catalyst, emulsion hardener.

chromic acid. (chromium trioxide; chromic anhydride).
CAS: 7738-94-5. CrO_3.
The name is in common use, although the true chromic acid, H_2CrO_4, exists only in solution.

Properties: Dark purplish-red crystals. Deliquescent, d 1.67–2.82, mp 196C. Soluble in water, alcohol, and mineral acids.

Derivation: (1) Sulfuric acid is added to a solution of sodium dichromate and the product is crystallized out; (2) chromite is fused with soda ash and limestone and then treated with sulfuric acid; and (3) electrolysis.

Grade: Technical, CP.

Hazard: A human carcinogen. A poison. Corrosive to skin. Powerful oxidizing agent, may explode on contact with reducing agents, may ignite on contact with organic materials. Upper respiratory tract irritant.

Use: Chemicals (chromates, oxidizing agents, catalysts), chromium-plating intermediate, medicine (caustic), process engraving, anodizing, ceramic glazes, colored glass, metal cleaning, inks, tanning, paints, textile mordant, etchant for plastics.

chromic bromide.
$CrBr_3$.
Properties: Black crystals. Soluble in boiling water; insoluble in cold water unless chromous salts are added.

Derivation: Passage of bromine vapor over pulverized chromium at 1000C.

Use: Olefin polymerization catalyst.

chromic chloride. (chromium chloride; chromium trichloride; chromium sesquichloride).
CAS: 10025-73-7. (1) $CrCl_3$. (2) $CrCl_3 \cdot 6H_2O$.
Properties: (1) Violet crystals. D 1.76, mp 1150C, sublimes at approximately 1300C. Insoluble in water and alcohol. (2) Greenish-black or violet deliquescent crystals, depending on whether chlorine is coordinated with the chromium. D 1.76, mp 83C. Soluble in water and alcohol; insoluble in ether.

Derivation: (1) By passing chlorine over a mixture of chromic oxide and carbon. (2) By the action of hydrochloric acid on chromium hydroxide.

Hazard: A poison.

Use: Chromium salts, intermediates, textile mordant, chromium plating including vapor plating, preparation of sponge chromium, catalyst for polymerizing olefins, waterproofing.

chromic fluoride. (chromium fluoride; chromium trifluoride).
CAS: 7788-97-8. $CrF_3 \cdot 4H_2O$ or $CrF_3 \cdot 9H_2O$.

Properties: Fine, green crystals. D (anhydrous): 3.8, mp >1000C, bp (sublimes) 1100–1200C. Insoluble in water and alcohol; soluble in hydrochloric acid.

Derivation: Interaction of chromium hydroxide and hydrofluoric acid.

Grade: Technical, high purity (CrF_3).

Hazard: Irritant to skin and eyes, especially in solution.

Use: Printing and dyeing woolens, moth-proofing, halogenation catalyst.

chromic hydroxide. (chromic hydrate; chromium hydroxide; chromium hydrate).
CAS: 1308-14-1. $Cr(OH)_3$.
Properties: Green, gelatinous precipitate. Decomposes to chromic oxide by heat. Insoluble in water; soluble in acids and strong alkalies.

Derivation: By adding a solution of ammonium hydroxide to the solution of a chromium salt.

Use: Guignet's green, catalyst, tanning agent, mordant.

chromic nitrate. (chromium nitrate).
CAS: 13548-38-4. $Cr(NO_3)_3 \cdot 9H_2O$.
Properties: Purple crystals. Mp 60C, decomposes at 100C. Soluble in alcohol and water.

Derivation: By the action of nitric acid on chromium hydroxide.

Hazard: May ignite organic materials on contact; may be explosive when shocked or heated; powerful oxidizer. Very toxic.

Use: Catalyst, corrosion inhibitor.

chromic oxide. (chromium(III) oxide; chromia; chromium sequioxide; green cinnabar).
CAS: 1308-38-9. Cr_2O_3.
Properties: Bright-green, extremely hard crystals. D 5.2, mp 2435C, bp 4000C. Insoluble in water, acids, and alkalies.

Derivation: (1) By heating chromium hydroxide, (2) by heating dry ammonium dichromate, and (3) by heating sodium dichromate with sulfur and washing out the sodium sulfate.

Hazard: Toxic by ingestion and inhalation.

Use: Metallurgy, green paint pigment, ceramics, catalyst in organic synthesis, green granules in asphalt roofing, component of refractory brick, abrasive.

chromic phosphate. (chromium phosphate).
CAS: 7789-04-0. (1) $CrPO_4 \cdot 6H_2O$. (2) $CrPO_4 \cdot 4H_2O$.
Properties: (1) Violet crystals. D 2.12 (14C); (2) green crystals. Soluble in acids; insoluble in water.

Derivation: (1) Interaction of solutions of chromium chloride and sodium phosphate; (2) by mixing chrome alum and disodium hydrogen phosphate. Violet, amorphous powder (not the hexahydrate) is formed that becomes crystalline on contact with water. On boiling, it is converted into green crystalline hydrate.

Use: Paint pigment, catalyst.

chromic sulfate. (chromium sulfate).
CAS: 10101-53-8. (1) $Cr_2(SO_4)_3$; (2) $Cr_2(SO_4)_3 \cdot 15H_2O$; and (3) $Cr_2(SO_4)_3 \cdot 18H_2O$.
Properties: (1) Violet or red powder; (2) dark-green amorphous scales; and (3) violet cubes. D (1) 3.012, (2) 1.867, and (3) 1.70. (1) Insoluble in water and acids; (2) soluble in water, insoluble in alcohol; and (3) soluble in water and alcohol.
Derivation: Action of sulfuric acid on chromium hydroxide with subsequent crystallization.
Use: Chrome plating, chromium alloys, mordant, catalyst, green paints and varnishes, green ink, ceramics (glazes). The basic form (reduction of sodium dichromate) is used in tanning.

chrominance. The colorimetric difference between any color and a reference color of equal luminance, the reference color having a specific chromaticity.

chromite. (chrome iron ore).
CAS: 1308-31-2. $FeCr_2O_4$.
A natural oxide of ferrous iron and chromium, sometimes with magnesium and aluminum present. Usually occurs in magnesium and iron-rich igneous rocks.
Properties: Iron-black to brownish-black color, dark-brown streak, metallic to submetallic luster. D 3.6, Mohs hardness 5.5.
Occurrence: The former U.S.S.R., South Africa, Zimbabwe, Philippines, Cuba, Turkey.
Grade: Metallurgical, refractory, chemical.
Hazard: A confirmed carcinogen.
Use: Only commercial source of chromium and its compounds.

chromium.
CAS: 7440-47-3. Cr.
Metallic element of atomic number 24, group VIB of the periodic table. Aw 51.996, valences of 2, 3, 6; four stable isotopes. Name derived from Greek for *color*.
Properties: Hard, brittle, semigray metal. D 7.1, mp 1900C, bp 2200C. Compounds have strong and varied colors. Cr ion forms many coordination compounds. Exists in active and passive forms, the latter giving rise to its corrosion resistance due to a thin surface oxide layer that passivates the metal when treated with oxidizing agents. Active form reacts readily with dilute acids to form chromous salts. Soluble in acids (except nitric) and strong alkalies, insoluble in water.
Occurrence: The former U.S.S.R., South Africa, Turkey, Philippines, Zimbabwe, Cuba.
Derivation: From chromite by direct reduction (ferrochrome), by reducing the oxide with finely divided aluminum or carbon, and by electrolysis of chromium solutions.
Available forms: (1) Chromium metal as lumps, granules, or powder; (2) high- or low-carbon ferrochromium; and (3) single crystals, high-purity crystals, or powder run 99.97% pure.

Grade: (Ore) Chromium ores are classified as (1) metallurgical, (2) refractory, and (3) chemical, and their consumption in the U.S. is in that order. (1) Must contain a minimum of 48% Cr_2O_3 and have Cr–Fe ratio of 3:1; (2) must be high in Cr_2O_3 and Al_2O_3 and low in iron; and (3) must be low in SiO_2 and Al_2O_3 and high in Cr_2O_3.
Hazard: Hexavalent chromium compounds are questionable carcinogens and corrosive on tissue, resulting in ulcers and dermatitis on prolonged contact.
Use: Alloying and plating element on metal and plastic substrates for corrosion resistance, chromium-containing and stainless steels, protective coating for automotive and equipment accessories, nuclear and high-temperature research, constituent of inorganic pigments.

chromium-51. Radioactive chromium of mass number 51.
Properties: Half-life 26.5 days, radiation g (0.32 MeV).
Grade: USP (as sodium chromate ^{51}Cr injection).
Hazard: Radioactive poison.
Use: Diagnosis of blood volume (as tracer).

chromium acetate. See chromic acetate.

chromium acetylacetonate.
CAS: 13681-82-8. $[CH_3COCHC(CH_3)O]_3Cr$.
Properties: Purple powder or red-violet crystals. Mp 216C, bp 340C. Insoluble in water; soluble in acetone and alcohol.
Derivation: Reaction of chromium chloride, acetylacetone, and sodium carbonate.
Use: Reduction of detonation of nitromethane.

chromium ammonium sulfate. (ammonium chromium sulfate; chrome ammonium alum).
CAS: 10141-00-1. $CrNH_4(SO_4)_2 \cdot 12H_2O$.
Properties: Green powder or deep-violet crystals. D 1.72, mp 94C. Soluble in water; slightly soluble in alcohol. The aqueous solution is violet when cold, green when hot.
Grade: Technical.
Use: Mordant, tanning.

chromium boride. One of several compounds of chromium and boron, e.g., CrB, CrB_2, and Cr_3B_2. They have high melting points, are very hard and corrosion resistant, and may be suitable for use in jet and rocket engines.
Properties: CrB may be crystals. D 6.2, mp 1550C, Mohs hardness 8.5, resistivity 67 $\mu\Omega$ cm (20C). CrB_2: d 5.15, mp 1850C, hardness 2010 (Knoop), resists oxidation up to 1100C. Cr_3B_2: may be crystals, d 6.1, Mohs hardness 9+.
Use: Metallurgical additives, high-temperature electrical conductors, cermets, refractories, coatings resistant to attack by molten metals.

chromium bromide. See chromous bromide.

chromium carbide.
CAS: 12012-35-0. Cr_3C_2.
Properties: Orthorhombic crystals. D 6.65, micro-hardness 2700 kg/sq mm Hg (load 50 g), mp 1890C, bp 3800C, resistivity 95 $\mu\Omega$ cm (room temperature). Highest oxidation resistance at high temperature of all metal carbides also resistant to acids and alkalies.
Use: Gage blocks and hot extrusion dies, in powder form as spray-coating material, components for pumps and valves.

chromium carbonate. See chromous carbonate.

chromium carbonyl. See chromium hexacarbonyl.

chromium chloride. See chromic chloride; chromous chloride.

chromium-copper. A copper-chromium alloy containing 8–11% chromium. Used in the manufacture of hard steels for increased elasticity.

chromium dioxide.
CrO_2.
Properties: Black, semiconducting material. D 4.9. Acicular crystals having strong magnetic properties. Also in powder form.
Derivation: By heating chromic acid.
Use: Magnetic component in recording tapes, catalyst.

chromium fluoride. See chromic fluoride.

chromium hexacarbonyl. (chromium carbonyl).
CAS: 13007-92-6. $Cr(CO)_6$.
Properties: White, crystalline solid. D 1.77, mp 150–151C (inert atmosphere), 210C (explodes). Slightly soluble in iodoform and carbon tetrachloride; insoluble in water, alcohol, ether, or acetic acid. More stable than most metal carbonyls, decomposed by chlorine and fuming nitric acid, but resists bromine, iodine, water, and cold concentrated nitric acid. Decomposes photochemically when solutions are exposed to light.
Grade: 98% pure.
Hazard: Explodes at 400F (204C). Toxic by inhalation and ingestion.
Use: Isomerization and polymerization catalyst, gasoline additive, intermediate.

chromium hydrate. See chromic hydroxide.

chromium hydroxide. See chromic hydroxide.

chromium manganese antimonide. Brittle gray solid having magnetic properties when above a definite temperature, depending on the composition.

chromium naphthenate.
Properties: Dark-green liquid or violet powder.
Derivation: By addition of chromium salts to solution of sodium naphthenate and recovery of the precipitate.
Grade: 6% chromium.
Hazard: Toxic by ingestion.
Use: Paints (antichalking agent).

chromium nitrate. See chromic nitrate.

chromium oxide. See chromic oxide.

chromium oxychloride. See chromyl chloride.

chromium oxyfluoride. See chromyl fluoride.

chromium phosphate. See chromic phosphate.

chromium potassium sulfate. (chrome alum; potassium chromium sulfate; chrome potash alum).
CAS: 10141-00-1. $CrK(SO_4)_2 \cdot 12H_2O$.
Properties: Dark, violet-red crystals; efflorescent. D 1.813, mp 89C, loses $10H_2O$ at 100C. Soluble in water.
Derivation: By reducing potassium dichromate in dilute sulfuric acid with sulfurous acid.
Hazard: Toxic by ingestion.
Use: Tanning (chrome-tan liquors), textile dye (mordant), photography (fixing bath), ceramics.

chromium sesquichloride. See chromic chloride.

chromium sesquioxide. See chromic oxide.

chromium steel. See steel, stainless; iron, stainless.

chromium sulfate. See chromic sulfate.

chromium trichloride. See chromic chloride.

chromium trifluoride. See chromic fluoride.

chromium trioxide. See chromic acid.

chromocyte. A pigmented cell such as an erythrocyte.

chromogen. See chromophore.

Chromogene. Mordant dyestuffs used on wool and leather. Characterized by very good fastness to light, fulling, etc.

chromomere. One of the serially aligned beads or granules of a eukaryotic chromosome, resulting from local coiling of a continuous DNA thread.

chromophore. A chemical grouping that, when present in an aromatic compound (the chromogen), gives color to the compound by causing a displacement of, or appearance of, absorbent bands in the visible spectrum. Dyes are sometimes classified on the basis of their chief chromophores, e.g., –NO, nitroso dyes; –NO$_2$, nitro dyes; –N=N–, azo dyes; etc.

chromosomal deletion.   The loss of part of a chromosome's DNA.

chromosomal inversion.   Chromosome segments that have been turned 180 degrees. The gene sequence for the segment is reversed with respect to the rest of the chromosome.

chromosomal protein.   Any protein that binds selectively to chromosomal DNA and facilitates transcription. Such proteins may also respond to various hormones or phytomitogens by undergoing structural or functional changes.

chromosome.   The heredity-bearing gene carrier of the living cell derived from chromatin and consisting largely of nucleoproteins (DNA) together with other protein components (histones).
See deoxyribonucleic acid; gene.

chromosome painting.   Attachment of certain fluorescent dyes to targeted parts of the chromosome. Used as a diagnostic for particular diseases, e.g., types of leukemia.

chromosome region p.   A designation for the short arm of a chromosome.

chromosome region q.   A designation for the long arm of a chromosome.

chromosome walking.   A technique for cloning everything in the genome around a known piece of DNA (the starting probe). A screen is executed against a genomic library for all clones hybridizing with the probe looking for one that extends furthest into the surrounding DNA. The most distal piece of this most distal clone is then used as a probe, so that ever more distal regions can be cloned.

"Chromosorb" *[Imerys].*   TM for a series of screened calcined and flux-calcined diatomite aggregates. Available in non–acid-washed, acid-washed, and acid-washed dimethyldichlorosilane treatments.
 Grade: Chromosorb phosphorus, W, G for analytical use; Chromosorb A for preparative chromatography.
 Use: Supports in gas or liquid chromatography.

chromotropic acid.
 (1,8-dihydroxynaphthalene-3,6-disulfonic acid).

Properties: White needles. Soluble in water; insoluble in alcohol and ether.
Derivation: Reaction of H-acid with 10% NaOH at 280C.
Use: Azo-dye intermediate, analytical reagent.

chromous bromide.   (chromium bromide).
 CrBr$_2$.
 Properties: White crystals that change to yellow on heating. Oxidizes in moist air but stable in dry air. D 4.356, mp 842C. Soluble in water (blue color).
 Hazard: Toxic by ingestion, irritant to skin and tissue.

chromous carbonate.   (chromium carbonate).
 CrCO$_3$.
 Properties: Grayish-blue, amorphous mass. D 2.75. Soluble in mineral acids; slightly soluble in water containing carbon dioxide; insoluble in alcohol.
 Hazard: Toxic by ingestion; irritant to skin and tissue.

chromous chloride.   (chromium chloride).
 CrCl$_2$.
 Properties: White, deliquescent needles. D 2.878 (25C), mp 824C, active reducing agent. Very soluble in water; insoluble in alcohol and ether.
 Derivation: Reaction of the metal with anhydrous hydrogen chloride.
 Hazard: A poison.
 Use: Reducing agent, catalyst, reagent, chromizing.

chromous fluoride.
 CAS: 10049-10-2.   CrF$_2$.
 Properties: Greenish, shiny crystals. Mp 890C, d 3.80. Slightly soluble in water; insoluble in alcohol; soluble in boiling hydrochloric acid.
 Derivation: Reaction of chromous chloride with hydrofluoric acid.
 Hazard: A poison. Strong irritant to eyes and skin.
 Use: Catalyst for alkylation and hydrocarbon cracking.

chromous formate.
 CAS: 4493-37-2.   Cr(HCOO)$_2$.
 Properties: (Monohydrate) Reddish, acicular crystals. Soluble in water.
 Derivation: Reaction of chromous chloride with sodium formate.
 Use: Chromium electroplating solutions, catalyst for organic reactions.

chromous oxalate.
 CrC$_2$O$_4$·H$_2$O.
 Properties: Yellow, crystalline powder. D 2.468. Soluble in water; active reducing agent.

Hazard: Toxic by ingestion; irritant to skin and tissue; a poison.

chromous sulfate.
CAS: 13825-86-0. $CrSO_4·5H_2O$.
Properties: (Pentahydrate) Blue crystals. Soluble in water; insoluble in acetone; slightly soluble in alcohol. Solutions are subject to atmospheric oxidation.
Use: Oxygen scavenger, reducing agent, analytical reagent.

Chromoxane. Mordant dyestuffs. Used on wool, characterized by fairly good fastness to light, very good fastness to fulling, etc., and by a relatively bright shade.

"Chromspun" [Kodak]. TM for solution-dyed acetate fiber.

chromyl chloride. (chromium oxychloride; chlorochromic anhydride).
CAS: 14977-61-8. CrO_2Cl_2.
Properties: Mobile, dark-red liquid. Bp 116C, fp −96.5C, d 1.911. Fumes in air, reacts vigorously with water to form chromic acid, chromic chloride, hydrochloric acid, and chlorine. Miscible with carbon tetrachloride, tetrachloroethane, carbon disulfide. Nonconductor of electricity. Protect from light and moisture.
Derivation: By heating sodium dichromate and sodium chloride with sulfuric acid.
Hazard: Corrosive to tissue. Strong oxidizing agent. Skin and upper respiratory tract irritant. Probable carcinogen.
Use: Organic oxidations and chlorinations, solvent for chromic anhydride, chromium complexes and dyes, catalyst.

chromyl fluoride. (chromium oxyfluoride).
CAS: 7788-96-7. CrF_2O_2.
Properties: Black crystals. Sublimes at 29.6C, polymerizes on exposure to light.
Hazard: Highly reactive, ignites hydrocarbon gases at high temperatures.
Use: Fluorination catalyst, dyeing polyolefins.

chrysamine G.
$C_{26}H_{16}N_4O_6Na_2$.
Properties: Yellowish-brown powder. Very sparingly soluble in water.
Derivation: By coupling diazotized benzidine with salicylic acid.
Use: Yellow direct-cotton dyestuff.

chrysanthemummonocarboxylic acid ethyl ester.
$(CH_3)_2CCH(COOC_2H_5)CHCH:C(CH_3)_2$.
Properties: Colorless to pale yellow; pleasant ester odor. D 0.924–0.927 (25/25C). Insoluble in water; soluble in alcohols and ketones.
Hazard: Toxic by ingestion.

Use: Synthesis of perfumes, pharmaceuticals, insecticides, other organic chemicals.
See cinerin I; pyrethrin I.

chrysarobin. (chrysophanic acid anthranol; 3-methylanthralin; 1,8,9-trihydroxy-3-methylanthracene; purified Goa powder; purified araroba; 1,8-dihydroxy-3-methyl-9-anthrone; 3-methyl-1,8,9-anthracenetriol; 1,8-dihydoxy-3-methyl-10H-anthracen-9-one).
$C_{15}H_{12}O_3$.
A reduction product of chrysophanic acid and a commercial mixture comprised of 15–30% pure chrysarobin and 70–85% of various neutral anthraquinone derivatives of Goa powder.
Properties: Flammable, odorless, tasteless, slightly water-soluble, brownish to yellow-orange microcrystalline powder.
Hazard: Irritant; allergen; mutagen; can cause kidney damage.
Use: Treat chronic, noninfectious skin diseases.

chrysazin. See 1,8-dihydroxyanthraquinone.

chrysene. (1,2-benzphenanthrene).
CAS: 218-01-9. $C_{18}H_{12}$ (a tetracyclic hydrocarbon).
Properties: Crystals. D 1.274 (20/4C), mp 254C, bp 448C, sublimes easily in a vacuum. Slightly soluble in alcohol, ether, glacial acetic acid; insoluble in water.
Derivation: Distillation of coal tar.
Hazard: Possible carcinogen.
Use: Organic synthesis.

chrysoidine. (2,4-diaminoazobenzene hydrochloride; 4-phenyldiazenylbenzene-1,3-diamine hydrochloride).
CAS: 532-82-1. $C_{12}H_{13}ClN_4$.
A dye made from aniline.
Hazard: Questionable carcinogen.
Use: Histological stain, indicator changing from orange to yellow at pH 4.0–7.0.

chrysoidine hydrochloride. (m-diaminoazobenzene hydrochloride).
CAS: 532-82-1. $(C_6H_5NNC_6H_3)(NH_2)_2·HCl$.
Properties: Red-brown powder or large black shiny crystals; green luster. Mp 117C. Soluble in alcohol and water, giving orange-brown solutions; insoluble in ether.
Use: Orange dye for cotton and silk.

chrysolite. See olivine.

chrysomycin.
Properties: Yellow flakes.
Use: Antibiotic.

chrysotile. See asbestos.

chu.   Abbreviation for centigrade heat unit. It is the amount of heat required to raise the temperature of one pound of water, one centigrade degree, from 15 to 16C. It is sometimes called a PCU (pound centigrade unit).

"Chuck-Eez" *[Specialty Lubricants].*   TM for a heavy duty boundary lubricant.
Use: To increase the clamping pressure of Power Chucks.

Chugaev reaction.   Formation of olefins from alcohols without rearrangement through pyrolysis of the corresponding xanthates via *cis* elimination.

Churchill's caustic iodine solution.   A solution of 25-g iodine in an aqueous solution of 50-g potassium iodide in 100 cc of water.

chylomicron.   A plasma lipoprotein composed of a triacylglycerol core and a shell of protein and phospholipid. The chylomicron carries lipid from the intestine to the tissues.

chymosin.   See rennin.

chymotrypsin.   A proteolytic enzyme found in the intestine that catalyzes the hydrolysis of various proteins (especially casein) and protein-digestion products to form polypeptides and amino acids.
Properties: White to yellowish-white crystals or amorphous powder; odorless. Soluble in water or saline solution.
Derivation: Crystallized extract of the pancreas gland of the ox.
Use: Biochemical research, medicine.

chymotrypsinogen.   A crystalline enzyme occurring in the pancreas that gives rise to chymotrypsin.
Use: Biochemical research.

Ci.   Abbreviation for curie.

C.I. acid blue 80.
CAS: 4474-24-2. $C_{32}H_{28}N_2O_8S_2 \cdot 2Na$.
Hazard: Moderately toxic by ingestion.

Ciamician–Dennstedt rearrangement.
Expansion of the pyrrole ring by heating with chloroform or other halogeno compounds in the alkaline solution. The intermediate dichlorocarbene, by addition to the pyrrole, forms an unstable dihalogenocyclopropane that rearranges to a 3-halogenopyridine.

ciaphos.   See cyanophos.

C.I. azoic coupling component 2.
CAS: 92-77-3. $C_{17}H_{13}NO_2$.
Hazard: Low toxicity by ingestion.

"Cibacet" *[Ciba].*   TM for disperse dyes for polyamide and cellulose acetate fibers.

Cibacron.   Reactive dyes for cellulosic fibers.

Cibanone.   Anthraquinone vat dyes.

Cibaphasol.   Coacervate-forming colloidal product used in textile dyeing.

C.I. basic blue 16.
CAS: 4569-88-4. $C_{30}H_{24}N_5O \cdot Cl$.
Hazard: A poison.

C.I. basic brown 4.
CAS: 8005-78-5.
Hazard: A severe eye irritant.

C.I. basic yellow 11.
CAS: 4208-80-4. $C_{21}H_{25}N_2O_2 \cdot Cl$.
Hazard: Moderately toxic.

C.I. basic yellow 28.
CAS: 54060-92-3. $C_{20}H_{24}N_3O \cdot CH_3O_4S$.
Hazard: A moderate eye irritant.

C.I. direct black 32, trisodium salt.
CAS: 6428-38-2. $C_{48}H_{40}N_{13}O_{13}S_3 \cdot 3Na$.
Hazard: Moderately toxic. Low toxicity by ingestion.

C.I. direct brown 2.
CAS: 2429-82-5. $C_{29}H_{19}N_5O_7S \cdot 2Na$.
Hazard: Low toxicity by ingestion.

C.I. direct red 13, disodium salt.
CAS: 1937-35-5. $C_{32}H_{22}N_6O_7S_2 \cdot 2Na$.
Hazard: Low toxicity by ingestion.

C.I. direct red 23.
CAS: 3441-14-3. $C_{35}H_{25}N_7O_{10}S_2 \cdot 2Na$.
Hazard: Low toxicity by ingestion.

cidofovir.   See (s)-1-(3-hydroxy-2-phosphonyl-methoxypropyl)cytosine.

Ciechanover, Aaron.   (1947–). Born in Haifa, Israel, Ciechanover won the Nobel Prize for chemistry in 2004 for his pioneering work concerning the discovery of ubiquitin-mediated protein degradation. He was awarded an M.D. from Hadassah Medical School and a doctorate in medicine in 1982 from the Technion. In 2000, he received the Albert Lasker Award for Basic Medical Research.

C.I. fluorescent brightener 85.
CAS: 17958-73-5. $C_{36}H_{34}N_{12}O_8S_2 \cdot 2Na$.
Hazard: Low toxicity by ingestion.

C.I. fluorescent brightener 208.   See disodium bisethylphenyltriaminotriazine stilbenedisulfonate.

C.I. fluorescent brightener 220.
CAS: 16470-24-9. $C_{40}H_{40}N_{12}O_{16}S_4 \cdot 4Na$.
Hazard: Low toxicity by ingestion and skin contact.

C.I. mordant orange 6, disodium salt.
CAS: 3564-27-0. $C_{19}H_{12}N_4O_6S \cdot 2Na$.
Hazard: Low toxicity by ingestion.

cigar.   A tubular roll of tobacco leaf designed for smoking that burns at a lower temperature than cigarettes producing far fewer carcinogenic hydrocarbons.

cigarette.   A roll of finely cut tobacco wrapped in paper. Combustion produces a smoke that is toxic, dangerously carcinogenic, and addictive when inhaled.

cigarette smoke.   Tobacco smoke, especially from cigarette smoke, that is a major indoor pollutant and affects anyone who shares a smoker's environment. In closed rooms, the gaseous toxicants and the lighter particles remain in the air for substantial periods of time that may be inhaled by smokers and nonsmokers alike.
Hazard: Toxic, dangerously carcinogen, addictive when inhaled.

cigarette tar.   The comparatively nonvolatile residue from the burning of cigarette tobacco that appears in a finely divided form in the smoke. Cigarette tar is known to contain various aromatic ring compounds (especially benzo[a]pyrene), also found in coal tar, that are carcinogens. Various forms of activated carbon are used in filters to try to adsorb the toxic components of the tars that find their way into the smoke.
See smoke (5).

ciguatoxin.
$C_{28}H_{52}NO_5Cl$.
A complex toxic principle in bony fishes; has both fat- and water-soluble fractions. Ninhydrin test positive. It is a type of quaternary ammonium compound, and one fraction is said to be an irreversible anticholinesterase. The pharmacology is unknown.

CIIT Centers for Health Research.   (CIIT).
Founded in 1974 and formerly called the Chemical Industry Institute of Toxicology, it transitioned to The Hamner Institutes for Health Sciences in 2007 as an independent, nonprofit organization that offers and open, collaborative, and cross-disciplinary approach to translational biomedical research, biopharmaceutical safety, oncology, and nanosafety. It is located at 6 Davis Drive, P.O. Box 12137, Research Triangle Park, NC 27709.
Website: http://www.thehamner.org

"Cimcool Cimperial 1070" *[Milacron].*   TM for a soluble oil metalworking fluid with extreme-pressure lubrication additive for multimetal, multi-operation applications.

"Cimcool Cimstar 60" *[Milacron].*   TM for a general purpose semisynthetic metalworking fluid with extreme pressure additive for a variety of multimetal applications.

"Cimcool Cimtech 310" *[Milacron].*   TM for a heavy duty synthetic metalworking fluid for machining and grinding of aerospace aluminum alloys.

"Cimcool Cimtech 410C" *[Milacron].*   TM for a heavy duty synthetic metalworking fluid for ferrous metal removal operations.

"Cimcool Milform 8050" *[Milacron].*   TM for a synthetic patented metalforming fluid for heavy duty forming, deep-drawing, punching and piercing operations of ferrous metals.

cincholepidine.   See lepidine.

cinchona bark.   The bark of one of several species of cinchona trees native to South America and cultivated in Indonesia, Peru, Ecuador, and western Africa. The best-known types are calisaya, loxa, and succirubra. It is used primarily as the source of natural quinine, quinidine, chinchonidine, cinchonine, and related alkaloids formerly used as antimalarials.

cinder.   See slag.

cinene.   See dipentene.

cineol.   See eucalyptol.

cinerin I.
CAS: 8003-34-7. $C_{20}H_{28}O_3$.
One of the four primary active insecticidal principles of pyrethrum flowers. It is the 3-(2-butenyl)-4-methyl-2-oxo-3-cyclopenten-1-yl ester of chrysanthemummonocarboxylic acid.
Properties: A viscous liquid. Quickly oxidized in air. Bp 200C (0.1 mm Hg). Insoluble in water; soluble in organic solvents. Incompatible with alkalies.
Hazard: Toxic by ingestion.
Use: Household insecticide.
See cinerin II; pyrethrin I; pyrethrin II.

cinerin II.
CAS: 8003-34-7. $C_{21}H_{28}O_5$.
One of the four primary active insecticidal principles of pyrethrum flowers. It is the 3-(2-butenyl)-4-methyl-2-oxo-3-cyclopenten-1-yl ester of chrysanthemumdicarboxylic acid monomethyl ester.
Properties: A viscous liquid. Quickly oxidized in air. Bp 200C (0.1 mm Hg). Insoluble in water; soluble in organic solvents.
Hazard: Toxic by ingestion.
Use: Household insecticide.
See cinerin I; pyrethrin I; pyrethrin II.

cinnabar.   (natural vermilion; liver ore). HgS.

Natural mercuric sulfide occurring in veins near recent volcanic rocks and hot springs.

Properties: Red, scarlet, reddish-brown to blackish solid, streak scarlet; luster adamantine to dull earthy when impure. D 8.10, Mohs hardness 2.5. Soluble in aqua regia. Has greater optical rotation than any other substance (+325 degrees).

Occurrence: California, Nevada, Spain, Italy, Mexico, Yugoslavia.

Hazard: See mercuric sulfide.

Use: The only important ore of mercury.

cinnamaldehyde.   See cinnamic aldehyde.

cinnamein.   See benzyl cinnamate.

cinnamene.   See styrene, monomer.

cinnamic acid.   (β-phenylacrylic acid; 3-phenylpropenoic acid; cinnamylic acid).
CAS: 621-82-9. $C_6H_5CH{:}CHCOOH$.

Properties: White, crystalline scales. Congealing p 133C (min), bp 300C. Soluble in benzene, ether, acetone, glacial acetic acid, carbon disulfide, oils; insoluble in water. Combustible.

Derivation: By heating benzaldehyde with sodium acetate in the presence of a dehydrating agent (acetic anhydride) or by heating benzyl chloride with sodium acetate in an autoclave. Occurs naturally in tobacco and some balsams.

Grade: Technical, refined.

Use: Medicine (anthelmintic), perfumes, intermediate.

cinnamic alcohol.   (cinnamyl alcohol; phenyllallylic alcohol; 3-phenyl-1-propen-1-ol; styryl carbinol).
CAS: 104-54-1. $C_6H_5CH{:}CHCH_2OH$.

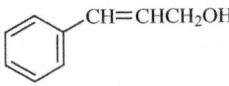

Properties: White needles or crystals; hyacinth-like odor. Congealing p 33C (min pure) as low as 24C (technical); bp 257C. Soluble in water, alcohol, glycerol, and organic solvents. Combustible.

Derivation: (1) From oil of cassia or oil of cinnamon; occurs as an ester. (2) Reduction of cinnamic aldehyde.

Method of purification: Recrystallization.

Grade: Technical, FCC.

Use: Perfumery, particularly for lilac and other floral scents; flavoring agent; soaps; cosmetics.

cinnamic aldehyde.   (cinnamaldehyde; 3-phenylpropenal; cinnamyl aldehyde).
CAS: 104-55-2. $C_6H_5CH{:}CHCHO$.

Properties: Yellowish oil; cinnamon odor; sweet taste. Thickens on exposure to air. D 1.048–1.052,

refr index 1.618–1.623, fp −8C, bp 248C. Soluble in five volumes of 60% alcohol, very slightly soluble in water. Combustible.

Derivation: (1) From Ceylon and Chinese cinnamon oils; (2) by condensation of benzaldehyde and acetaldehyde.

Method of purification: Rectification.

Use: Flavors, perfumery.

See cassia oil.

cinnamon leaf oil.
CAS: 8015-91-6.

Properties: Extracted by steam distillation of leaves from *Cinnamomum zeylanicum* Nees. Light to dark brown liquid; spicy cinnamon, clove odor, and taste. Sol in fixed oils, propylene glycol, mineral oil; insol in glycerin.

Hazard: Moderately toxic by ingestion. Low toxicity by skin contact.

Use: Food additive.

cinnamon oil.   See cassia oil.

cinnamoyl chloride.   (phenylacrylyl chloride).
CAS: 102-92-1. $C_6H_5CH{:}CHCOCl$.

Properties: Yellow crystals. Mp 35C, bp 170C (58 mm Hg), d 1.1617 (45/4C). Decomposes in water. Soluble in hydrocarbons and esters.

Hazard: Skin irritant.

Use: Reagent for determination of water, chemical intermediate.

cinnamyl acetate.
CAS: 103-54-8. $C_6H_5CH{:}CHCH_2OOCCH_3$.

Properties: Colorless liquid; floral spicy odor. D 1.048–1.052, refr index 1.539–1.542. Soluble in four volumes of 70% alcohol. Combustible.

Use: Perfumery (fixative), flavoring.

cinnamyl alcohol.   See cinnamic alcohol.

cinnamyl aldehyde.   See cinnamic aldehyde.

cinnamyl cinnamate.   (styracin).
CAS: 122-69-0. $C_9H_7O_2C_9H_9$.

Properties: Rectangular, prismic crystals. Mp 40C (min). Soluble in alcohol, ether, benzene.

Derivation: Esterification of cinnamic acid with cinnamic alcohol.

Use: Perfumery, flavoring.

α-cinnamyl-*n*-cyclopropylmethyl-α-ethyl-*n*-methyl-2-furfurylamine.
CAS: 137246-17-4. $C_{21}H_{27}NO$.

Hazard: Moderately toxic by ingestion.

cinnamylic acid.   See cinnamic acid.

cinnamyl anthranilate.   (anthranilic acid; cinnamyl ester; 2-aminobenzoic acid-3-phenyl-2-propenyl ester; cinnamyl alcohol anthranilate;

cinnamyl-2-aminobenzoate; cinnamyl-*o*-amino-benzoate; 3-phenyl-2-propenylanthranilate; 3-phenyl-2-propen-1-ylanthranylate; 3-phenylprop-2-enyl 2-aminobenzoate).
CAS: 87-29-6. $C_{16}H_{15}NO_2$.
Properties: Water insoluble, reddish-yellow powder, soluble in ethanol, ethyl ether and chloroform.
Hazard: Questionable carcinogen.

cinobufagin acetate.
CAS: 4026-97-5. $C_{28}H_{36}O_7$.
Hazard: A poison.
Source: Natural product.

CIPC.   See chloro-IPC.

cis-.   (Latin: "on this side.") A prefix used in designating geometrical isomers in which there is a double bond between two carbon atoms This prevents free rotation, and thus two different spatial arrangements of substituent groups or atoms are possible. When two atoms, or radicals are positioned on one side of the carbon axis or backbone, the isomer is called *cis-*; when they are on the opposite sides, the isomer is called *trans-* (Latin: "on the other side"). This is exemplified in the following formulas:

This is often called *cis–trans* isomerism. These prefixes are disregarded in alphabetizing chemical names.

cistron.   A unit of DNA or RNA corresponding to one gene.

citalopram hydrobromide.
CAS: 59729-32-7. $C_{20}H_{21}FN_2O\cdot BrH$.
Hazard: A poison.

citiolone.   (*N*-acetylhomocysteinethiolactone).
CAS: 1195-16-0. $C_6H_9NO_2S$.
Properties: Acicular crystals. Mp 112C, maximum UV absorption 238 nm.
Use: Fogging preventive in photography.

citraconic anhydride.   (methylmaleic anhydride).
CAS: 616-02-4. $C_5H_4O_3$.
Properties: Colorless liquid. Mp 7–8C, bp 213–214C, d 1.25 (15/4C). Soluble in ether.
Grade: Reagent.
Use: Reagent for alkalies, alcohols, and amines.

citral.   (geranial; geranialdehyde; 3,7-dimethyl-2,6-octadienal).
CAS: 5392-40-5. $(CH_3)_2CCHC_2H_4C(CH_3)CHCHO$.
Commercial material is a mixture of α and β isomers.

Properties: Mobile, pale-yellow liquid; strong lemon odor. D 0.891–0.897 (15C); refr index 1.4860–1.4900 (20C); not optically active. Soluble in five volumes of 60% alcohol; soluble in all proportions of benzyl benzoate, diethyl phthalate, glycerol, propylene glycol, mineral oil, fixed oils, and 95% alcohol; insoluble in water. Combustible.
Derivation: Principal component of lemongrass oil, which can be isolated by fractional distillation. Obtained synthetically by oxidation of geraniol, nerol, or linalool by chromic acid.
Grade: Technical, pure, FCC.
Hazard: Questionable carcinogen.
Use: Perfumes, flavoring agent, intermediate for other fragrances, vitamin A synthesis.

citrate.   (2-hydroxypropane-1,2,3-tricarboxylic acid).
CAS: 13754-17-1. $C_6H_8O_7$.
A chelating agent that is a salt or ester of citric acid and a base. It is a key intermediate in metabolism and is an acid compound found in citrus fruits.
Derivation: Naturally-occurring.
Use: Food additives, in pharmaceuticals, household detergents, anticoagulants.

citrate lyase.   (citratase; citrase; citridesmolase; citrate aldolase).
An enzyme that catalyzes the cleavage of citric acid to acetic acid and oxaloacetic acid in the absence of coenzyme A.

citrate synthase.   (cirogenase; condensing enzyme; oxaloacetate transacetase).
CAS: 9027-96-7.
An enzyme that catalyzes the condensation of oxaloacetic acid and acetyl-CoA to yield citric acid and coenzyme A.

citric acid.   (2-hydroxy-1,2,3-propanetricarboxylic acid).
$HOOCCH_2C(OH)(COOH)CH_2COOH\cdot H_2O$.
Properties: Colorless, translucent crystals or powder; odorless; strongly acidic taste; hydrated form is efflorescent in dry air. D 1.542, mp 153C (anhydrous form), decomposes before boiling. Very soluble in water and alcohol, soluble in ether. Combustible.
Occurrence: In living cells, both animal and plant.
Derivation: By mold fermentation of carbohydrates, including deep fermentation, from lemon, lime, and pineapple juice and molasses.
Grade: Both hydrous and anhydrous, technical, CP, USP, FCC.
Use: Preparation of citrates, flavoring extracts, confections, soft drinks, effervescent salts; acidifier, dispersing agent; medicines; acidulant and antioxidant in foods (for details see regulations of Meat Inspection Division of USDA), sequestering agent, water-conditioning agent and detergent builder; cleaning and polishing stainless steel and other metals; alkyd resins; mordant; removal of sulfur dioxide

from smelter waste gases, abscission of citrus fruit in harvesting; cultured dairy products. See TCA cycle.

citric acid cycle.   See TCA cycle.

"Citri-Fi 100" [Fiberstar].   TM for pure citrus fiber.
Use: For fat replacement, increase of moisture content, and addition of structure to food products.

"Citri-Fi 200" [Fiberstar].   TM for an all-natural citrus fiber product.
Use: Oil replacement and thickening in foods.

"Citri-Fi 200 FG" [Fiberstar].   TM for finely granulated citrus product.
Use: In bakery and meat products.

"Citrofil" [Fiberstar].   TM for a plasticizer that is a carrier solvent and a food carrier solvent. It reduces hardness, increases plasticity, and induces elasticity.
Properties: High boiling point.
Use: Dried egg whites, film coated tablets, aerosol hair spray, deodorants, perfumes, flavors, cellulose acetate films, and cigarette filters.

"Citroflex" [Vertellus].   TM for a series of organic citrates.

"Citroflex 2" [Vertellus]. (triethyl citrate).
CAS: 77-93-0.
TM for a plasticizer.
Use: Cellulosics.

"Citroflex 4" [Vertellus].   (tri-n-butyl citrate).
CAS: 77-94-1.
TM for a plasticizer.
Use: Vinyls and cellulosics.

"Citroflex A-2" [Vertellus].   (acetyltriethyl citrate).
CAS: 77-89-4.
TM for a plasticizer.
Use: Cellulosics.

"Citroflex A-4" [Vertellus].   (acetyltri-n-butyl citrate).
CAS: 77-90-7.
TM for a plasticizer.
Use: Vinyls, adhesives, and coatings.

"Citroflex A-6" [Vertellus].   (acetyltri-n-hexyl citrate).
TM for a plasticizer.
Use: Polymeric medical articles.

"Citroflex B-6" [Vetellus].   (n-butyryltri-n-hexyl citrate).
TM for a plasticizer.
Use: Polymeric medical articles.

citronellal.   (3,7-dimethyl-6(or 7)-octenal).
CAS: 8000-29-1.
Has both d- and l-isomers. $C_9H_{17}CHO$.
Derivation: From citronella oil, of which it is the main component; also from lemongrass oil.
Use: Soap perfumery, manufacture of hydroxycitronellal, insect repellent.

citronellal hydrate.   See hydroxycitronellal.

citronella oil.
CAS: 8000-29-1.
Properties: Light-yellowish, essential oil; rather pungent, citruslike odor. D 0.887–0.906, refr index 1.468–1.483; solutions are levorotatory. Soluble in 80% alcohol. Combustible.
Hazard: May not be used on food crops.
Use: Insect repellent; perfumes for soaps and disinfectants; manufacture of cironellal and geraniol; denaturant for alcohol.

citronellol.   (3,7-dimethyl-6(or 7)-octen-1-ol).
CAS: 106-22-9. $CH_2:C(CH_3)(CH_2)_3CH(CH_3)CH_2CH_2OH$.
The 7-octene form, which predominates.
Occurrence: Citronella, geranium, rose, savin, and other essential oils.
Use: Perfumery (floral odors, mainly rose types).
See rhodinol.

citronellyl acetate.
CAS: 150-84-5. $C_{10}H_{10}OOCCH_3$.
Properties: Colorless liquid; odor somewhat like that of bergamot oil. D 0.884–0.891, bp 119–121 (15 mm Hg), optical rotation usually slightly dextrorotatory, up to +1 degree, refr index 1.450–1.452. Soluble in nine volumes of 70% alcohol. Combustible.
Derivation: Action of acetic anhydride on citronellol.
Use: Perfumery, flavoring.

citronellyl formate.
CAS: 105-85-1. $C_{10}H_{19}OOCH$.
Properties: Colorless liquid; floral odor. D 0.890–0.903 (25/25C), refr index 1.4430–1.4490 (20C). One volume dissolves in three volumes 80% alcohol; soluble in most oils. Combustible.
Grade: FCC.
Use: Flavoring.

citron yellow.   See zinc yellow.

citrovorum factor.   See folinic acid.

citrulline.
$NH_2CONH(CH_2)_3CHNH_2COOH$.
An arginine derivative. An amino acid found in watermelon juice in the l(+) form.
Properties: Crystals from methanol–water mixture. Mp 222C. Soluble in water; insoluble in methanol and ethanol.
Use: Biochemical research.

Citrus Burst 7.   A citrus and vegetable derived solvent that is environmentally friendly.
Use: For parts cleaning, industrial and janitorial uses.

citrus red 2.   (1-[(2,5-dimethylxyphenyl)azo]-2-naphthol;   (1Z)-1-[(2,5-dimethoxyphenyl)hydrazineylidene]-2-one).
CAS: 6358-53-8. $C_{18}H_{16}N_2O_3$.
Hazard: Possible carcinogen.
Use: Food color to color the skin of mature oranges.

citrus-peel oils.   Edible oils expressed from the peel or rind of grapefruit, lemon, lime, orange, and tangerine.
Properties: Color, odor, and taste characteristic of source. D 0.84–0.89, refr index 1.473–1.478, optically active, unsaponifiable. Soluble in alcohol, vegetable oils, mineral oil; orange and lemon oils are soluble in glacial acetic acid. Combustible.
Chief constituents: Limonene, citral, and terpenes in varying percentages.
Use: Flavoring agents in desserts, soft drinks, ice cream; odorants in perfumery and cosmetics, furniture polish (lemon oil).
Note: Terpene-free grades are available at much higher concentration (15–30 times the original oil). These grades have much lower optical rotation and less tendency to spoil on storage, since terpenes tend to oxidize to undesirable components such as carvone and *p*-cymene.
See terpeneless oil.

citrus-seed oils.   Edible oils expressed from the seeds of grapefruit, orange, lemon, lime, and tangerine reclaimed from cannery processing.
Properties: Nondrying; odor, color, and taste characteristic of source; bitter taste removed by alkali refining. D 0.91–0.92, saponification number 190–195, iodine number 100–110, optically inactive. Combustible. May be bleached, deodorized, or hydrogenated.
Chief constituents: Chiefly palmitic, oleic, and linoleic acids.
Use: Flavoring, food products, cosmetics, odorants in special soaps.

civet.   (zibeth).
Unctuous secretion from the civet cat; used as a perfume fixative. Synthetic types are available.

civettal.   See   1,2,3,4-tetrahydro-6-methylquinoline.

Civex process.   See reprocessing.

Cl.   Symbol for chlorine; the molecular formula is Cl_2.

C. laburnum.   (laburnum, golden chain).
A plant whose leaves and seeds are toxic and produce effects similar to nicotine.

Hazard: Nausea, intense vomiting, dysphagia, diarrhea, prostration, irregular pulse and respiration, delirium, twitching, coma, renal damage.

cladding.   The process in which two metals are bonded by being rolled together at suitable pressure and temperature. Controlled explosion is also used. At the interface, each metal diffuses sufficiently into the other to form an alloy. Cladding is generally from 5 to 20% of total thickness, but may be heavier depending on the properties desired. The base metal is usually carbon or low-alloy steel clad with stainless steel, nickel, or other protective metal. Nonferrous metals are also clad; copper with cupronickel is used for coinage.

clade.   A group of organisms that includes the most recent common ancestor of all of its members and all of the descendants of that most recent common ancestor.

cladogenesis.   The development of a new clade; the splitting of a single lineage into two distinct lineages; speciation.

cladogram.   A diagram, resulting from a cladistic analysis, that depicts a hypothetical branching sequence of lineages leading to the taxa under consideration. The points of branching within a cladogram are called nodes. All taxa occur at the endpoints of the cladogram.

Claisen condensation.   A condensation reaction discovered in 1890 in which aldehydes react with esters of the type RCH_2COOR' in the presence of metallic sodium or sodium ethylate to form unsaturated esters $R''CH=C(R)COOR'$. The mechanism of the reaction parallels that of acetoacetic ester condensation. An addition compound of the ester and sodium ethylate is first formed and then transformed into an unsaturated sodio ether; an additional compound between this ether and the aldehyde forms next; this changes to the unsaturated ester through a molecular rearrangement and cleavage of sodium hydroxide. Amides of alkali metals are excellent catalysts for the reaction, but they tend to form some amide by-products. Union with the carbonyl carbon takes place at the α-carbon in the acid.

Claisen rearrangement.   Thermal rearrangement of allyl ethers of enols or phenols to γ,Δ-unsaturated ketones or omicron-allylphenols, respectively, by a 3,3-sigmatropic shift.

Claisen-Schmidt condensation.   Condensation of an aromatic aldehyde with an aliphatic aldehyde or ketone in the presence of a relatively strong base (hydroxide or alkoxide ion) to form an α,β-unsaturated aldehyde or ketone.

Clapeyron equation.   The equation $dp/dT = \Delta H/T\Delta V$. It states that the rate of change of vapor pressure of a liquid (expressed in ergs per cubic centimeter) with absolute temperature (in Kelvins) equals the heat of vaporization (in ergs per gram) divided by the product of the absolute temperature and the increase in volume (V, in cubic centimeters) when a gram of the liquid changes to vapor. Other consistent units may be used. The approximate Clapeyron equation, $d\ln p/dT = \Delta H/RT^2$, expresses the same relation in a less-exact form because in its derivation, it is assumed that ΔV is equal to the volume of vapor. This assumption (that the volume of liquid is negligibly small) is usually true within a few percent under ordinary conditions of temperature and pressure.

claphnin.   A natural glucoside classified as a derivative of styrene.

clarification.   Removal of bulk water from a dilute suspension of solids by gravity sedimentation, aided by chemical flocculating agents. Large circular tanks equipped with revolving plows or rakes are used for this purpose. Important applications of clarification are for separation of wood-pulp waste from paper-mill effluents and excess water from activated sludge.
See dewatering.

clarithromycin.   See 6-o-methylerythromycin.

Clark cell.   Standard cell for measuring electrical potential with mercury and zinc amalgam electrodes in zinc sulfate solution.

Clark-Lubs indicators.   Phenol red, cresol red, bromophenol blue, bromocresol purple, thymol blue, bromthymol blue, and methyl red.

clary oil.
   Properties: From steam distillation of flowering tops and leaves of *Salvia sclarea L.* (Fam. *Labiatae*). Pale yellow liquid; herbaceous odor. Sol in fixed oils, mineral oil; insol in glycerin, propylene glycol.
   Use: Food additive.

clary sage oil.   See sage oil.

class A fire.   Paper, wood, and rubbish fires, which can be extinguished by water.
See fire extinguishment.

class B fire.   Flammable liquids and grease fires, which require smothering.
See fire extinguishment.

class C fire.   Fires in electrical equipment, which require smothering and cooling with a nonconducting extinguishing agent such as liquefied CO_2.
See fire extinguishment.

classification.   In chemical-engineering parlance, the mechanical process of separating subdivided solids (crushed stone, cement, mineral aggregate, and the like) into two (or more) *classes*, each containing a specific size range. They may vary from an inch or more in diameter to powders of considerable fineness. Customary classifying equipment includes so-called grizzlies, perforated metal and vibrating screens, sifters, sieves, and similar devices. A magnetic separator is often used in conjunction with the screen to remove tramp metal. A similar method is used in the food industry for size grading of certain fruits and vegetables, e.g., peas, berries, etc.

clathrate compound.   An inclusion complex in which molecules of one substance are completely enclosed within the other, as argon within hydroquinone crystals. Urea adducts are inclusion complexes of the channel or canal type. In these, the complexing urea crystals wrap around the molecule of the other substance, usually a straight-chain unbranched aliphatic hydrocarbon. Similar complexes are formed with thiourea.
See inclusion complex.

Claude system.   A process for the production of liquid air in which the compressed gas is made to perform work in an expansion engine and thus cool itself.

Clausius–Mosotti law.   A relation between density and dielectric constant of dielectrics.

clay.   A hydrated aluminum silicate. Generalized formula $Al_2O_3SiO_2 \cdot xH_2O$. Component of soils in varying percentages.
   Properties: Fine, irregularly shaped crystals ranging from 150 microns to less than 1 micron (colloidal); reddish-brown to pale buff color, depending on iron oxide content; odorless. D approximately 2.50. Insoluble in water and organic solvents. Absorbs water to form a plastic, moldable mass and in some cases a thixotropic gel (bentonite). Refractory material; strong ion-exchange capability; important in soil chemistry and construction engineering.
   Derivation: Weathering of rocks.
   Occurrence: Southeastern U.S., Wyoming, Texas, Canada, England, France, the former U.S.S.R.
   Available forms: Kaolinite, montmorillonite, atapulgite, illite, bentonite, halloysite.
   Grade: Natural, refined, air floated.
   Hazard: Dusts may be irritating to nose and throat. Suspensions of dust are a fire hazard.
   Use: Ceramic products, refractories, colloidal suspensions, oil-well drilling fluids, filler for rubber and plastic products, films, paper coating, decolorizing oils, temporary molds, filtration, carrier in insecticidal sprays, catalyst support.
   See Fuller's Earth; bentonite; ceramic; refractory; kaolin; slip clay; polyorganosilicate graft polymer.

cleaner. (cleaning agent).
Any formulation designed for use in cleaning materials, objects, or surfaces.

Cleanwiz. Mold cleaners and strippers.

clearer. A chemical agent.
Properties: Miscible with dehydrating or fixing solution and the embedding material.
Use: Histological preparations.

cleave. (1) Of a crystal, to break or separate along definite planes defined by the crystalline structure. It may cleave in one direction, as in mica, or in several. (2) Of an alkene molecule, to divide into two compounds (aldehydes or ketones) at the double bond. This is usually done by ozone, followed by hydrolysis in the presence of powdered zinc.

Cleland's reagent. See dithiothreitol.

Clemmensen reaction. The Clemmensen method of reduction (1913) consists in refluxing a ketone with amalgamated zinc and hydrochloric acid. Acetophenone, for example, is reduced to ethylbenzene. The method is applicable to the reduction of most aromatic–aliphatic ketones to at least some aliphatic and alicyclic ketones, to the γ-keto acids obtainable by Friedel-Crafts condensations with succinic anhydride (succinolylation), and to the cyclic ketones formed by intramolecular condensation.

clenoliximab.
Properties: Monkey/human chimeric CD4 monoclonal antibodies
CAS: 182912-58-9.
Hazard: A poison.

cleroden. (vermifuge).
A naturally occurring epoxide.
Use: Vermifuge.

Cleveland Open Cup. See COC.

Cleve's acid. (1-naphthylamine-6-sulfonic acid).
Properties: Colorless needles. Mp >330C. Slightly soluble in water.
Derivation: Nitration of naphthalene-β-sulfonic acid. On reduction with iron, this yields a mixture of 1-naphthylamine-6-sulfonic acid (Cleve's acid) and 1-naphthylamine-7-sulfonic acid (or Cleve's acid-1,7). The latter is separated by crystallization as the sodium salt, the 1,6-acid precipitates on acidification.
Use: Azo dye intermediate.

Cleve's acid-1,7. (1-naphthylamine-7-sulfonic acid).
See Cleve's acid.

cliffstone Paris white. A special grade of whiting made from a hard grade of English chalk.

climbazol. See 1-(p-chlorophenoxy)-3,3-dimethyl-1-(1-imidazolyl)-2-butanone.

clindamycin hydrochloride. See 7-chloro-lincomycin hydrochloride.

clinical chemistry. A subdivision of chemistry that deals with the behavior and composition of all types of body fluids, including the blood, urine, perspiration, glandular secretions, etc. It involves analysis and testing of these for content of numerous metabolic constituents, as well as foreign materials; thus, it also includes toxicological factors.

clinoptilolite. A natural, inorganic zeolite used as a selective ion-exchange medium for removal of ammonia from plant wastewater.

clofibrate. (2-(4-chlorophenoxy)-2-methylpropanoic acid ethyl ester; ethyl 2-(4-chlorophenoxy)-2-methylpropanoate).
CAS: 637-07-0. $C_{12}H_{15}ClO_3$.
An oil that accelerates the excretion of neutral steroids and inhibits cholesterol synthesis and a fibric acid derivative.
Properties: Water-insoluble oil, miscible with acetone, chloroform, ethanol, and ether.
Hazard: Questionable carcinogen; toxic; causes nausea, vomiting, diarrhea, weakness, stiffness, cramps, and muscle tenderness.
Use: Therapeutically as an antihyperlipoproteinemic drug; in the treatment of hyperlipoproteinemia type III and severe hypertriglyceridemia.

clomazone. See 2-(2-chlorobenzyl)-4,4-dimethyl-1,2-oxazolidin-3-one.

clonazepam. (5-(2-chloropheny)-7-nitro-1,3-dihydro-1,4-benzodiaepin-2-one).
CAS: 1622-61-3. $C_{15}H_{10}ClN_3O_3$.
A benzodiazepine.
Properties: Anticonvulsant; antiepileptic; anxiolytic; antimanic.
Use: To treat myotonic seizures, atonic seizures, photosensitive epilepsy, absence seizures
Hazard: Moderately toxic.

clone. An identical copy of an organism. Most plants, fungi, algae, and many other organisms naturally reproduce by making clones of themselves as a form of asexual reproduction.

cloning. Using specialized DNA technology to produce multiple, exact copies of a single gene or other segment of DNA to obtain enough material for further study. This process, used by researchers in the Human Genome Project, is referred to as cloning DNA. The resulting cloned (copied) collections of DNA molecules are called clone libraries. A second type of cloning exploits the natural process of cell division to make many copies of an entire cell. The genetic makeup of these cloned cells, called a cell line, is identical to the original cell. A third type of cloning produces complete, genetically identical animals such as the famous Scottish sheep, Dolly. See cloning vector.

cloning vector.   DNA molecule originating from a virus, a plasmid, or the cell of a higher organism into which another DNA fragment of appropriate size can be integrated without loss of the vector's capacity for self-replication; vectors introduce foreign DNA into host cells, where the DNA can be reproduced in large quantities. Examples are plasmids, cosmids, and yeast artificial chromosomes; vectors are often recombinant molecules containing DNA sequences from several sources.

clopidol.   (3,5-dichloro-2,6-dimethyl-4-pyridinol; Coyden).
CAS: 2971-90-6. $C_7H_7Cl_2NO$.
A penta-substituted pyridine derivative.
Properties: A solid. Mw 192.06, mp >320C. Insoluble in water.
Hazard: Toxic. Mutagenic effects. Questionable carcinogen. Clopyralid.
Use: Food additive; herbicide.

clorox.   (sodium hypochlorite).
CAS: 7681-52-9.   ClNaOH.
Properties: A solution of sodium carbonate and sodium perborate.
Use: An oxidizing and bleaching agent, and as a disinfectant.

clorprenaline   hydrochloride.   See isoprophenamine hydrochloride.

clorsulon.
Use: Drug (veterinary); food additive.

closed system.   A system that exchanges neither matter nor energy with the surroundings.

clostebol acetate.   See 4-chlorotestosterone 17-acetate.

cloud point.   In petroleum technology, the temperature at which a waxy solid material appears as a diesel fuel is cooled. This material is harmful to engine performance.

cloud seeding.   See nucleation.

clove oil.   (caryophyllus oil).
An essential oil distilled from cloves. Optically active.
Use: Medicine (local), flavoring, dentistry, perfumery, confectionery, soaps.

clupanodonic acid.
$C_{21}H_{35}COOH$.
Derived from herring oil.

clupeine.   A protamine (simple protein) from herring. Contains no sulfur.
Properties: Water soluble.

cluster catalysis.   See catalysis.

CM.   A flame-retardant composition based on ammonium sulfamate and modified to prevent afterglow and improve penetration.
Properties: Fine, white granules. Soluble in water; insoluble in dry-cleaning solvents.
Use: Treatment of fabrics, paper, paper products, and other cellulosic materials.

Cm.   Symbol for curium.

cM.   See centimorgan.

CMA.   Abbreviation for Chemical Manufacturers Association.

CMC.   See carboxymethylcellulose.

CM-cellulose.   Abbreviation for carboxymethylcellulose, used especially by biochemists.

CMHEC.   Abbreviation for carboxymethyl hydroxyethyl cellulose.

CMP.   Abbreviation for cytidine monophosphate. See cytidylic acid.

CMPP.   Abbreviation for 2-(4-chloro-2-methylphenoxyl)propionic acid. See mecoprop.

CMRA.   Abbreviation for Chemical Marketing Research Association.

CMU.   Abbreviation for chlorophenyl dimethylurea.
See monuron.

CNS.   Abbreviation for central nervous system (applied to the actions of certain drugs).

Co.   Symbol for cobalt.

coacervation.   An important equilibrium state of colloidal or macromolecular systems. It may be defined as the partial miscibility of two or more optically isotropic liquids, at least one of which is in the colloidal state. For example, gum arabic shows the phenomenon of coacervation when mixed with gelatin.

coagulant.   A substance that induces coagulation. Coagulants are used in precipitating solids or semisolids from solution, as casein from milk, rubber particles from latex, or impurities from water. Compounds that dissociate into strongly charged ions are normally used for this purpose. Blood contains the natural coagulant thrombin.

coagulase. (hemocoagulate; heptocoagulase; plasma coagulase; plasmocoagulase; reptilases; RP-093; staphylocoagulase; thrombin coagulase; coagulating enzyme; clotting enzyme; curdling enzyme).
An enzyme that catalyzes coagulation processes.

coagulation. Irreversible combination or aggregation of semisolid particles, such as fats or proteins, to form a clot or mass. This can often be brought about by addition of appropriate electrolytes, e.g., by the addition of an acid to milk or of aluminum sulfate to turbid water. Mechanical agitation and removal of stabilizing ions, as in dialysis, also cause coagulation. The clotting of blood by thrombin, of rubber particles in latex by acetic acid, and of egg white by heat are additional instances.
See agglomeration (1); flocculation.

coagulation value. (flocculation value).
Concentration of a coagulent that effects a standard change.

Coahran process. Recovery of acetic acid from pyroligneous acid by extracting with ether. It is an improved version of the Brewster process but is basically the same.

coal. A natural, solid, combustible material formed from prehistoric plant life, it occurs in layers or veins in sedimentary rocks. It is far more plentiful than petroleum in the U.S., an important source of heat and energy. It occurs chiefly in West Virginia and Kentucky, as well as in Wyoming and other western states. Much of it is too high in sulfur content to meet desirable pollution standards unless sulfur is removed by scrubbing.
Chemically, coal is a macromolecular network composed of groups of polynuclear aromatic rings, to which are attached subordinate rings connected by oxygen, sulfur, and aliphatic bridges. This extended open structure is conducive to catalytic reactions, which in effect subdivide it into smaller molecules that can be defined readily.
Coal is an important source of chemical raw materials. Pyrolysis (destructive distillation) yields coal tar and hydrocarbon gases that can be upgraded by hydrogenation or methanation to synthetic crude oil and fuel gas, respectively. Catalytic hydrogenation yields hydrocarbon oils and gasoline. Gasification produces carbon monoxide and hydrogen (synthesis gas), from which ammonia and other products can be made. Numerous processes for adapting these reactions to large-scale production of fuel oil and gasoline have been developed in the U.S., but none has yet proved economically successful. The process is being used abroad (Lurgi and Sasol methods).

See gasification; hydrogenolysis; hydrosolvation; peat; lignite.

coal, conversion. See gasification.

coal dust. (anthracite particles; coal facings; coal; ground bituminous; coal-milled; coal slag-milled; sea coal).
Properties: Black powder or dust.
Derivation: Produced during the use of coal.
Hazard: Questionable carcinogen; lung damage; pulmonary fibrosis; chronic pneumoconiosis; turmorigen; variable toxicity.

coalescence. (1) The combination of globules in an emulsion as the result of molecular attraction of the surfaces. (2) The combination of one or multiple crystals of aggregates into a simple larger unit.

coalescer. Unit containing glass fibers or other suspended material to facilitate separation of immiscible liquids of similar density.

coalescing agent. A material that assists in filmmaking by helping polymer particles flow together to form a continuous film.

coal gas. (bench gas; coke-oven gas).
A mixture of gases produced by the destructive distillation of bituminous coal in highly heated fire-clay or silica retorts or in by-product coke ovens.
Hazard: Flammable.
Use: Directly on open-hearth furnaces.

coal gasification. See gasification.

coal hydrogenation. See hydrogenolysis, gasification.

coal oil. The crude oil obtained by the destructive distillation of bituminous coal, or the distillate obtained from this oil.
Hazard: Flammable, moderate fire risk.

coal tar. (coal tar pitch volatiles).
CAS: 65996-93-2.
Properties: Black, viscous liquid (or semisolid); naphthalene-like odor; sharp, burning taste. D 1.18–1.23. Soluble in ether, benzene, carbon disulfide, chloroform; partially soluble in alcohol, acetone, methanol; only slightly soluble in water.
Coal tar may be hydrogenated under pressure to form a petroleum-like fuel suitable for residual use. Coal tar fractions obtained by distillation and the chemicals found in each are as follows: (1) light oil (up to 200C): benzene, toluene, xylenes, cumenes, coumarone, indene; (2) middle oil (200–250C); and (3) heavy oil (250–300C): naphthalene, acenaphthene, methylnaphthalenes, fluorene, phenol, cresols, pyridine, picolines; (4) anthracene oil (300–350C): phenanthrene, anthracene, carbazole, quinolines; and (5) pitch. Typical yields: 5% light oil, 17%

middle oil, 7% heavy oil, 9% anthracene oil, 62% pitch. Treatment with alkalies, acids, and solvents is necessary to separate the individual chemicals.

Derivation: Obtained by destructive distillation of bituminous coal as in coke ovens. One ton of coal yields 8.8 gal of coal tar.

Grade: Crude, refined, USP.

Hazard: A human carcinogen. Toxic by inhalation. Combustible.

Use: Raw material for plastics, solvents, dyes, drugs, and other organic chemicals. The crude or refined product or fractions thereof are also used for waterproofing, paints, pipe coating, roads, roofing, and insulation and as pesticides and sealants.

coal tar creosote. Low-cost, water-insoluble, low-volatility wood preservative.

coal tar distillate. The lighter fractions of coal tar. The terms *coal tar light oil*, *coal tar naphtha*, and *coal tar oil* are loosely defined and sometimes regarded as synonymous.

Hazard: Flammable, dangerous fire risk. Toxic by inhalation and skin absorption.
See naphtha.

coal tar dye. A dye produced from the coal tar hydrocarbons or their derivatives, such as benzene, toluene, xylene, naphthalene, anthracene, aniline, etc.
See dye, synthetic.

coal tar light oil. See coal tar distillate; light oil.

coal tar naphtha. See coal tar distillate; naphtha, solvent.

coal tar oil. See coal tar distillate.

coal tar pitch. Dark-brown to black amorphous residue left after coal tar is redistilled. It is composed almost entirely of polynuclear aromatic compounds and constitutes 48–65% of the usual grades of coal tar. Different grades have different softening points: roofing pitch softens at 65C, electrode pitch at 110–115C. Combustible.

Hazard: Volatile components (anthracene, phenanthrene, acridine) are confirmed carcinogens.

Use: Binder for carbon electrodes, base for paints and coatings. Impregnation of fiber pipe for electrical conduits and drainage, foundry core compounds, briquetting coal, tar-bonded refractory brick, paving and roofing, plasticizers for elastomers and polymers, extenders, saturants, impregnants, sealants.

coating. A film or thin layer applied to a base material, called a substrate. The coatings most commonly used in industry are metals, alloys, resin solutions, and solid–liquid suspensions on various substrates (metals, plastics, wood, paper, leather, etc.). They may be applied by electrolysis, vapor deposition, vacuum evaporation, or mechanical means

such as brushing, spraying, calendering, and roller coating. Products such as cables and power cord are coated by extrusion. Such thermosetting resins as acrylics, epoxies, and polyesters appropriately compounded are available in powder form and are applied to auto bodies, machinery, and other industrial products by electrostatic spraying techniques. This process is claimed to minimize the pollution problems and waste encountered with solvent-based sprayed coatings.
See protective coating; film; paint; vacuum deposition.

cobalamin. Imprecise name for vitamin B_{12}; it refers to the molecule cyanocobalamin without the cyano- group (CN).
See cyanocobalamin.

cobalt.
CAS: 7440-48-4. Co.
Metallic element of atomic number 27, group VIII of the periodic table. Aw 58.9332. Valences 2, 3; no stable isotopes. There are several artificial radioactive isotopes, the most important being ^{60}Co. Coordination number: 4 (divalent), 6 (trivalent).

Properties: Steel-gray, shining, hard, ductile, somewhat malleable metal; ferromagnetic, with permeability two-thirds that of iron; has exceptional magnetic properties in alloys. D 8.9, mp 1493C, bp 3100C. Attacked by dilute hydrochloric and sulfuric acids, soluble in nitric acid. Corrodes readily in air. Hardness: cast 124 Brinell, electrodeposited 300 Brinell. An important trace element in soils and necessary for animal nutrition. Cobalt has unusual coordinating properties, especially the trivalent ion. Noncombustible except as powder.

Occurrence: Principal ores are smaltite, cobaltite, chloanthite, linnaeite (Canada, Zaire, Zambia).

Derivation: From ore concentration by roasting followed (1) by thermal reduction by aluminum; (2) by electrolytic reduction of solutions of metal; or (3) by leaching with either ammonia or acid in an autoclave under elevated temperatures and pressures and subsequent reduction by hydrogen.

Available forms: Rondels (1 × 3/4 in), shot, anodes, 150 and finer mesh powder to 99.6% purity, ductile strips (95% cobalt, 5% iron) and high-purity strips, 99.9% pure, single crystals.

Hazard: A possible carcinogen. Toxic by inhalation. Dust is flammable. Asthma, myocardial effects, and pulmonary function impairment.

Use: Chemical (cobalt salts, oxidizing agent); electroplating ceramics; lamp filaments; catalyst (sulfur removal from petroleum, Oxo process, organic synthesis); trace element in fertilizers; glass; drier in printing inks, paints, and varnishes; colors; cermets. Principal use in alloys, especially cobalt steels for permanent and soft magnets and cobalt-chromium high-speed tool steels. Cemented carbides, jet engines, coordination, and complexing agent.
See cobaltammine.

cobalt-57.   Radioactive cobalt of mass number 57.
Properties: Half-life 267 days; radiation γ, κ, X ray.
Derivation: Bombardment of a nickel target with protons.
Available forms: Cobaltous chloride in hydrochloric acid solution, cyanocobalamin cobalt-57 solution (USP).
Hazard: Radioactive poison.
Use: Biological research.

cobalt-58.   Radioactive cobalt of mass number 58.
Properties: Half-life 72 days; radiation positron, γ, κ, X ray.
Available forms: Cobaltous chloride in hydrochloric acid solution.
Hazard: Radioactive poison.
Use: Biological and medical research.

cobalt-60.   Radioactive cobalt of mass number 60. One of the most common radioisotopes.
Properties: Half-life 5.3 years; radiation β and γ. Radiocobalt is available in larger quantities and is cheaper than radium.
Derivation: Irradiation of cobalt oxide, Co_2O_3, or of cobalt.
Available forms: Encapsulated pellets or wire needles, cobaltous chloride in hydrochloric acid solution, solid cobaltic oxides, labeled compounds such as cyanocobalamin (USP).
Hazard: Radioactive poison.
Use: Radiation therapy (cancer), radiographic testing of welds and castings, as a source of ions in gas-discharge devices, as the radiation source in liquid-level gauges, for locating buried telephone and electrical conduits, portable radiation units, γ-irradiation for wheat and potatoes, as a research aid in studying the permeability of porous media to flow of oil, wearing quality of floor wax, oil consumption in internal-combustion engines, wool dyeing, etc.

cobalt acetate.   See cobaltous acetate.

cobaltammine.   A coordination or complexing compound containing the group $[Co(NH_3)_6]_3{}^+$ or its derivatives in which some of the ammonia has been replaced by other groups or ions. The names *hexammine*, *pentammine*, etc., are used to indicate the number of ammonia groups present. Prepared by adding excess ammonia to a cobaltous salt, exposing to air so that oxygen is absorbed, and boiling to oxidize the cobalt.
These compounds show none of the ordinary properties of cobalt. Different types of salts with various acid radicals are known. The ammonia in the amines may be replaced, molecule for molecule, by other nitrogen compounds such as hydroxylamine or ethylene diamine, by water, by ions such as hydroxyl, chloride, nitrate, etc., or by groups such as nitro (NO_2).
See coordination compound.

cobalt ammonium sulfate.   See cobaltous ammonium sulfate.

cobalt arsenate.   See cobaltous arsenate.

cobalt black.   See cobaltic oxide.

cobalt bloom.   See erythrite (2).

cobalt blue.   (Thenard's blue; cobalt ultramarine; azure blue).
Properties: Blue to green pigment of variable composition, consisting essentially of mixtures of cobalt oxide and alumina, approximating cobaltous aluminate, $Co(AlO_2)_2$. Cobalt blue is said to be the most durable of all blue pigments, being resistant to both weathering and chemicals.
Derivation: By heating alumina with (1) cobaltous oxide, or a material yielding this oxide on calcination; (2) cobalt phosphate; and (3) cobalt arsenate. Greenish shades may be made by incorporating zinc oxide.
Grade: Technical (called genuine to distinguish it from the imitation, which is ultramarine blue).
Use: Pigments in oil or water, cosmetics (eye shadows, grease paints).

cobalt bromide.   See cobaltous bromide.

cobalt caprylate.
Use: Food additive.

cobalt carbonate.   See cobaltous carbonate, basic.

cobalt carbonyl.
(cobalt tetracarbonyl; dicobalt octacarbonyl).
CAS: 10210-68-1. $C_8O_8Co_2$.

Properties: Orange-red crystals. Mw 341.94, mp 51C, vap press 1.5 mm Hg (25C). Insoluble in water, soluble in ether, alcohol, carbon disulfide, and naphtha.
Hazard: Possible carcinogen.
Use: Catalyst for hydroformulation, hydrogenation, hydrosilation, isomerization, carboxylation, carbonylation, and polymerization.

cobalt carbonyl hydride.   See cobalt hydrocarbonyl.

cobalt chloride.   See cobaltous chloride.

cobalt chromate.   See cobaltous chromate.

cobalt chromate, basic.   See cobaltous chromate.

cobalt difluoride.   See cobaltous fluoride.

cobalt-2-ethylhexoate.   (cobalt octoate).
$C_4H_9CH(C_2H_5)COOH$.
Probably the cobaltous salt of 2-ethylhexoic acid.
Properties: Blue liquid. D 1.013 (25C).
Use: Paint drier, whitener, catalyst.

cobalt-gold alloy.   Made by a vapor deposition technique into magnetic films. Compositions range from 25 to 60% gold.

cobalt hydrate.   See cobaltic hydroxide; cobaltous hydroxide.

cobalt hydrocarbonyl.   (cobalt carbonyl hydride; tetracarbonylhydrocobalt).
CAS: 16842-03-8. C_4HCoO_4.
Properties: Gas; offensive odor. Mw 171.98, mp −26C. Slightly soluble in water.
Hazard: Flammable gas. Possible carcinogen.
Use: Catalyst.

cobalt hydroxide.   See cobaltic hydroxide; cobaltous hydroxide.

cobaltic acetylacetonate.
CAS: 21679-46-9. $Co[CH_3COCHC(CH_3)O]_3$.
Properties: Dark-green or black crystals. D 1.43, mp 241C.
Derivation: Reaction of cobaltous carbonate with acetylacetone and peroxide.
Use: Catalyst for olefins, diens, polyesters, combustion, solid propellants; polymerization; vulcanizing; coloring for synthetic resins; deposition of metal and/or oxide as memory storage for computers; intermediate for synthesis; glass tinting.

cobaltic boride.
CoB.
Properties: Crystal prisms. D 7.25 (18C), mp >1400C. Decomposes in H_2O, soluble in nitric acid.
Use: Ceramics.

cobaltic fluoride.   See cobalt trifluoride.

cobaltic hydroxide.   (cobalt hydroxide; cobalt hydrate).
$Co(OH)_3$, actually considered to be $Co_2O_3 \cdot 3H_2O$.
Properties: Dark-brown powder. D 4.46, loses water at 100C. Soluble in cold concentrated acids; insoluble in water and alcohol.
Derivation: Addition of sodium hydroxide to a solution of cobaltic salt, action of chlorine on a suspension of cobaltous hydroxide, action of sodium hypochlorite on a cobaltous salt.
Use: Cobalt salts, catalyst.

cobaltic oxide.   (cobalt oxide; cobalt black).
CAS: 1308-04-9. Co_2O_3.
Sometimes incorrectly called cobalt peroxide.

Properties: Steel-gray or black powder. D 4.81–5.60, mp (decomposes at 895C). Soluble in concentrated acids; insoluble in water.
Derivation: By heating cobalt compounds at low temperature with excess of air.
Use: Pigment, coloring enamels, glazing pottery.

cobaltic potassium nitrite.   (cobalt yellow).
$CoK_3N_6O_{12}$ or $CoK_3(NO_2)_6$.
Properties: Yellow, crystalline solid. Insoluble in alcohol; slightly soluble in water and acetic acid; decomposed by sulfuric acid.
Derivation: By adding potassium nitrite to a cobalt salt solution.
Use: Colorant for rubber, glass, and ceramic products; separation of cobalt from nickel; analytical chemistry.

cobalt iodide.   See cobaltous iodide.

cobaltite.   (cobalt glance).
CoAsS.
Properties: Silver-white to gray mineral, metallic luster. Contains 35.5% cobalt, d 6–6.3, Mohs hardness 5.5.
Occurrence: Canada, Zaire, Sweden.
Hazard: Dust or powder toxic when inhaled.
Use: An important cobalt ore; ceramics.

cobalt linoleate.   See cobaltous linoleate.

cobalt molybdate.
CAS: 13762-14-6. $CoMoO_4$.
A molybdenum catalyst (a gray–green powder) used in petroleum technology, in reforming and desulfurization.

cobalt monoxide.   See cobaltous oxide.

cobalt naphthenate.   See cobaltous naphthenate.

cobalt neodecanoate.   Deep-blue paste containing 12% cobalt. Used as a drying additive for printing inks.

cobalt nitrate.   See cobaltous nitrate.

cobaltocene.   (dicyclopentadienylcobalt).
CAS: 1277-43-6. $(C_5H_5)_2Co$.
A metallocene.
Properties: Purple crystals. Mp 172–173C. Soluble in hydrocarbons; highly reactive compound that is readily oxidized by air, water, and dilute acids.
Hazard: Toxic by ingestion.
Use: Polymerization inhibitor of olefins up to 200C, Diels–Alder reaction, catalyst, paint drier, oxygen stripping agent.

cobaltocobaltic oxide.   (tricobalt tetraoxide).
CAS: 1308-06-1. Co_3O_4.

Properties: Steel-gray to black in anhydrous form. D 6.07; changes to cobaltous oxide at 900–950C. Insoluble in water, hydrochloric and nitric acids; soluble in sulfuric acid and fused sodium hydroxide; hygroscopic.
Derivation: By heating strongly other cobalt oxides in air. Thus, the commercial oxides contain a substantial quantity of Co_3O_4.
Use: Ceramics, pigments, catalyst, preparation of cobalt metal, semiconductors.

cobalt octoate.   See cobalt-2-ethylhexoate; soap (2).

cobalt oleate.   See cobaltous oleate.

cobaltous acetate.   (cobalt acetate).
CAS: 71-48-7. $Co(C_2H_3O_2)_2 \cdot 4H_2O$.
Properties: Reddish-violet, deliquescent crystals. D 1.7043, mp loses H_2O at 140C. Soluble in water, acids, and alcohol.
Derivation: Action of acetic acid on cobalt carbonate with subsequent crystallization.
Grade: Technical, pure crystals, CP.
Hazard: May not be used in food products (FDA).
Use: Sympathetic inks, paint and varnish driers, catalyst, anodizing, mineral supplement in feed additives, foam stabilizer.

cobaltous aluminate.   See cobalt blue.

cobaltous ammonium phosphate.   (ammonium cobaltous phosphate; cobalt violet).
CoH_4NO_4P.
Properties: Violet to dark-red crystals. Insoluble in water; soluble in common acids.
Use: Plant nutrient; colorant in glass, glazes, enamels; analytical chemistry.

cobaltous ammonium sulfate.   (cobalt ammonium sulfate).
$CoSO_4 \cdot (NH_4)_2SO_4 \cdot 6H_2O$.
Properties: Ruby-red crystals. D 1.902. Soluble in water; insoluble in alcohol.
Derivation: Crystals of cobaltous sulfate with ammonium sulfate.
Use: Ceramics, cobalt plating, catalyst.

cobaltous arsenate.   (cobalt arsenate).
CAS: 7785-24-2. $Co_3(AsO_4)_2 \cdot 8H_2O$.
Properties: Violet-red powder. D 3.178 (15C). Soluble in acids, insoluble in water.
Derivation: Interaction of solutions of sodium arsenate and a cobalt salt.
Grade: Technical.
Use: Painting on glass and porcelain in light-blue colors, coloring glass.
See erythrite.

cobaltous bromide.   (cobalt bromide).
CAS: 7789-43-7. $CoBr_2 \cdot 6H_2O$.
Properties: Red-violet crystals. D 2.46, mp 47–48C, loses $6H_2O$ at 130C. Soluble in water, alcohol, and ether. Anhydrous crystals: bright green.
Derivation: By the action of bromine on cobalt or of hydrobromic acid on cobaltous hydroxide or carbonate followed by crystallization.
Grade: Technical, CP.
Use: In hygrometers, catalyst.

cobaltous carbonate.
CAS: 7542-09-8. $CoCO_3$.
Properties: Red crystals. D 4.13, mp (decomposes). Insoluble in water and ammonia; soluble in acids. The cobalt carbonate of commerce is usually the basic salt (see following entry).
Derivation: By heating cobaltous sulfate with a solution of sodium bicarbonate.
Use: Ceramics, trace element added to soils and animal feed, temperature indicator, catalyst, pigments.

cobaltous carbonate, basic.
$2CoCO_3 \cdot 3Co(OH)_2 \cdot H_2O$.
The cobalt carbonate of commerce.
Properties: Red-violet crystals. Mp (decomposes). Soluble in acids; insoluble in cold water; decomposes in hot water.
Derivation: By adding sodium carbonate to a solution of cobaltous acetate followed by filtration and drying.
Use: Manufacturing cobaltous oxide, cobalt pigments, cobalt salts; intermediate.

cobaltous chloride.   (cobalt chloride).
CAS: 7646-79-9.   (1) $CoCl_2$, (2) $CoCl_2 \cdot 6H_2O$.
Properties: (1) Blue, (2) ruby-red crystals. D (1) 3.348, (2) 1.924; mp (1) sublimes, (2) 86.75C. Soluble in water, alcohol, and acetone.
Derivation: By the action of hydrochloric acid on cobalt; its oxide, hydroxide, or carbonate. Concentration gives (2) and dehydration (1).
Hazard: May not be used in food products (FDA). Can cause blood damage.
Use: Absorbent for ammonia, gas masks, electroplating, sympathetic inks, hygrometers, manufacture of vitamin B_{12}, flux for magnesium refining, solid lubricant, dye mordant, catalyst, barometers, laboratory reagent, fertilizer additive.

cobaltous chromate.   (basic cobalt chromate; cobalt chromate).
CAS: 13455-25-9. $CoCrO_4$.
Properties: Brown or yellowish-brown powder. Variable composition. (Pure cobaltous chromate is $CoCrO_4$, gray-black crystals.) Soluble in mineral acids and solutions of chromium trioxide; insoluble in water.
Hazard: A carcinogen.
Use: Ceramics (tinting).

cobaltous citrate.
CAS: 866-81-9. $Co_3(C_6H_5O_7)_2 \cdot 2H_2O$.

Properties: Rose-red, amorphous powder. Mp 150C (loses $2H_2O$). Slightly soluble in water; soluble in dilute acids.
Use: Vitamin preparations, therapeutic agents.

cobaltous cyanide.
(1) $Co(CN)_2 \cdot 2H_2O$, (2) $Co(CN)_2$.
Properties: (1) Buff crystals, (2) blue-violet powder; d (2) 1.872, mp (2) 280C. Insoluble in water; soluble in potassium cyanide, hydrochloric acid, ammonium hydroxide.
Hazard: Highly toxic.

cobaltous ferrite.
$CoOFe_2O_3$.
A constituent of magnetically soft ferrites that have high permeability, low coercive force, low magnetic saturation, and high resistivity.
See ferrite.

cobaltous fluoride.   (cobalt difluoride).
CAS: 10026-17-2. CoF_2.
Properties: Rose-red crystals or powder. D 4.46, mp approximately 1200C, bp 1400C. Soluble in cold water and hydrofluoric acid. Decomposes in hot water. Ammine complexes can be prepared from the hydrate.
Hazard: Highly toxic.
Use: Catalyst.

cobaltous formate.
$Co(CHO_2)_2 \cdot 2H_2O$.
Properties: Red crystals. D 2.129 (22C), mp 140C (loses $2H_2O$), decomposes at 175C. Soluble in cold water; insoluble in alcohol.
Use: Catalyst.

cobaltous hydroxide.   (cobalt hydroxide; cobalt hydrate).
CAS: 21041-93-0. $Co(OH)_2$.
Properties: Rose-red powder. D 3.597. Soluble in acids and ammonium salt solutions; insoluble in water and alkalies.
Derivation: Addition of sodium hydroxide to a solution of cobaltous salt.
Use: Cobalt salts, paint and varnish driers, catalyst, storage-battery electrodes.

cobaltous iodide.   (cobalt iodide).
CAS: 15238-00-3. $CoI_2 \cdot 6H_2O$.
Properties: Brownish-red crystals. Loses iodine on exposure to air, d 2.90, loses $6H_2O$ at 27C. Soluble in water and alcohol. Anhydrous cobaltous iodide, CoI_2, in the form of black crystals, d 5.68, or in a yellow β-modification that gives a colorless aqueous solution.
Derivation: Heating cobalt powder with hydriodic acid; anhydrous cobaltous iodide is prepared by heating cobalt in iodine vapor.
Use: Hygrometers; determination of water in organic solvents; catalyst.

cobaltous linoleate.   (cobalt linoleate).
CAS: 14666-96-7. $Co(C_{18}H_{31}O_2)_2$.
Properties: Brown, amorphous powder. Soluble in alcohol, ether, and acids; insoluble in water. Combustible.
Derivation: By boiling a cobalt salt and sodium linoleate.
Use: Paint and varnish drier, especially enamels and white paints.
See soap (2).

cobaltous naphthenate.
Properties: Brown, amorphous powder or bluish-red solid. Insoluble in water; soluble in alcohol, ether, oils. Composition indefinite. Combustible.
Derivation: By treating cobaltous hydroxide or cobaltous acetate with naphthenic acid.
Use: Paint and varnish drier, bonding rubber to steel and other metals.

cobaltous nitrate.   (cobalt nitrate).
CAS: 10141-05-6. $Co(NO_3)_2 \cdot 6H_2O$.
Properties: Red crystals. Deliquescent in moist air. D 1.88 (25C), mp 56C, decomposes at 75C. Soluble in most organic solvents.
Derivation: Action of nitric acid on metallic cobalt or cobalt oxide, hydroxide, or carbonate, with subsequent crystallization.
Use: Sympathetic inks, cobalt pigments, catalysts, additive to soils and animal feeds, vitamin preparations, hair dyes, porcelain decoration.
Hazard: Oxidizing agent, dangerous fire risk in contact with organic materials.

cobaltous oleate.   (cobalt oleate).
$Co(C_{18}H_{33}O_2)_2$.
Properties: Brown, amorphous powder. Mp 235C. Soluble in alcohol and ether; insoluble in water. Combustible.
Derivation: By heating cobaltous chloride and sodium oleate followed by filtration and drying.
Use: Paint and varnish driers.
See soap (2).

cobaltous oxide.   (cobalt oxide; cobalt monoxide).
CAS: 1307-96-6.   CoO.
Properties: Grayish powder under most conditions, can form green–brown crystals. D 6.45, mp 1935C. Soluble in acids and alkali hydroxides; insoluble in water and ammonium hydroxide.
Derivation: Calcination of cobalt carbonate or its oxides at high temperature in a neutral or slightly reducing atmosphere.
Grade: Technical, ceramic.
Use: Pigment in paints and ceramics, preparation of cobalt salts, catalyst, porcelain enamels, coloring glass, feed additive, cobalt metal powder.

cobaltous perchlorate.
CAS: 13478-33-6. $Co(ClO_4)_2$.

Properties: Red needles. D 3.327. Soluble in water, alcohol, acetone.
Hazard: Fire and explosion risk in contact with organic materials. Strong oxidizing agent.
Use: Chemical reagent.

cobaltous phosphate.   (cobalt phosphate).
CAS: 13455-36-2. $Co_3(PO_4)_2 \cdot 8H_2O$.
Properties: Reddish powder. D 2.769 (25C), loses $8H_2O$ at 200C. Slightly soluble in cold water; soluble in mineral acids.
Derivation: Interaction of solutions of cobalt salts and sodium phosphate.
Use: Cobalt pigments, coloring glass, painting porcelain, animal-feed supplement.

cobaltous resinate.   (cobalt resinate).
Principally cobalt abietate.
Properties: Brown-red powder. Insoluble in water; soluble in oils.
Derivation: See soap (2).
Grade: Fused, precipitated. The precipitated grade is higher in cobalt, more expensive, and more effective.
Hazard: Spontaneously flammable in air, reacts strongly with oxidizing materials.
Use: Varnish drier.
See abietic acid.

cobaltous silicofluoride.
$CoSiF_6 \cdot 6H_2O$.
Properties: Pale-red crystals. D 2.087. Soluble in water.
Hazard: Irritating to tissue.
Use: Ceramics.

cobaltous succinate.
CAS: 3267-76-3. $Co(C_4H_4O_4) \cdot 4H_2O$.
Properties: Violet crystals. Slightly soluble in cold water; soluble in alkalies; insoluble in alcohol.
Use: Vitamin preparations, therapeutic agents.

cobaltous sulfate.   (cobalt sulfate).
CAS: 10124-43-3.   (1) $CoSO_4$, (2) $CoSO_4 \cdot 7H_2O$.
Properties: Red powder. D (1) 3.472, (2) 1.948; mp (1) 735C, (2) loses $7H_2O$ at 420C. Soluble in water.
Derivation: Action of sulfuric acid on cobaltous oxide.
Hazard: May not be used in food products (FDA). Possible carcinogen.
Use: Ceramics, pigments, glazes, in plating baths for cobalt, additive to soils, catalyst, paint and ink drier, storage batteries.

cobaltous tungstate.   (cobalt tungstate; cobalt wolframate).
CAS: 10101-58-3. $CoWO_4$.
Properties: Reddish-orange powder. D 8.42. Insoluble in water; soluble in hot concentrated acids.
Derivation: By adding a sodium tungstate solution to a solution of a cobalt salt.

Use: Pigment, drier for enamels, inks, paints, electronic devices, antiknock agents.

cobalt oxide.   The commercial cobalt oxides are not usually definite chemical compounds but are mixtures of two or more cobalt oxides.
See cobaltic oxide; cobaltous oxide; cobaltocobaltic oxide.

cobalt phosphate.   See cobaltous phosphate.

cobalt phthalocyaninesulfonate.
CAS: 30638-08-5. $C_{32}H_{15}CoN_8O_3S \cdot H$.
Hazard: Moderately toxic by ingestion and skin contact. A mild skin and severe eye irritant.

cobalt potassium cyanide.
CAS: 13963-58-1. $K_3Co(CN)_6$.
Properties: Yellow crystals. D 1.878 (25C), mp (decomposes). Soluble in water; slightly soluble in alcohol.
Grade: Pure, electronic.
Hazard: A poison.
Use: Electronic research.

cobalt potassium nitrite.   (cobalt yellow; potassium cobaltinitrate; Fischer's salt; potassium hexanitrocobaltate III).
CAS: 13782-01-9. $K_3Co(NO_2)_6$.
Properties: Yellow, microcrystalline powder. Mp (decomposes at 200C). Slightly soluble in water; insoluble in alcohol.
Derivation: By adding potassium nitrate and acetic acid to a solution of a cobalt salt.
Use: Medicine, yellow pigment, painting on glass or porcelain.

cobalt resinate.   See cobaltous resinate.

cobalt selenite.
$CoSe_2O_3 \cdot 2H_2O$.
A blue–red powder, insoluble in water.

cobalt silicide.   A semiconductor reported to have as much as 15% efficiency in converting heat to electricity in the temperature range 20–800C.

cobalt soap.   See cobaltous linoleate; cobaltous naphthenate; cobaltous oleate; cobaltous resinate; cobalt tallate.

cobalt sulfate.   See cobaltous sulfate.

cobalt tallate.   Cobalt derivative of refined tall oil of varying composition. Used as a drier in paints and varnishes. Combustible.
See soap (2).

cobalt tetracarbonyl.   (dicobalt octacarbonyl).
CAS: 10210-68-1. $Co_2(CO)_8$.
Properties: Orange or dark-brown crystals; white when pure. D 1.78, mp 51C (decomposes above this

temperature). Insoluble in water; soluble in alcohol, ether, and carbon disulfide.

Derivation: Combination of finely divided cobalt with carbon monoxide under pressure.

Hazard: Toxic by ingestion and inhalation.

Use: Catalyst in Oxo process, high-purity cobalt salts.

cobalt titanate.
Co_2TiO_4.
Properties: Greenish-black crystals. D 5.07–5.12. Soluble in concentrated hydrochloric acid.

cobalt trifluoride. (cobaltic fluoride).
CAS: 10026-18-3. CoF_3.
Properties: Light-brown, free-flowing powder. D 3.88 (25C), no odor except hydrogen fluoride odor developed in moist air; stable in sealed containers; reacts readily with moisture in the atmosphere to form a dark, almost black powder; reacts with water to form a black, finely divided precipitate (cobaltic hydroxide). Insoluble in alcohol and benzene. As a fluorinating agent, yields one atom of fluorine and reverts to the difluoride. The spent cobalt difluoride may be regenerated with elemental fluorine.
Hazard: Strong irritant to tissue.
Use: Fluorination of hydrocarbons.

cobalt tungstate. See cobaltous tungstate.

cobalt ultramarine. See cobalt blue.

cobalt violet. See cobaltous ammonium phosphate.

cobalt wolframate. See cobaltous tungstate.

cobalt yellow. See cobalt potassium nitrite.

cobamide. (cobalt(2+);
[(2R,3S,4R,5S)-4,5-dihydroxy-2-(hydroxymethyl)oxolan-3-yl][(2R)-1-[3-[(1R,2R,3R,4Z,7S,9Z,12S,13S,14Z,17S,18S,19R)-2,13,18-tris(2-amino-2-oxoethyl)-7,12,17-tris(3-amino-3-oxopropyl)-3,5,8,8,13,15,18,19-octamethyl-2,7,12,17-tetrahydro-1H-corrin-21-id-3-yl]propanoylamino]propan-2-yl]phosphate).
$C_{53}H_{80}CoN_{11}O_{15}P$.
The hexaamide of cobamic acid. It is a component of vitamin B_{12}.

Cobon. 2-Nitroso-1-naphthol used for the colorimetric determination of cobalt. Sensitivity: 0.005 ppm cobalt.

Cobrate. Corrosion inhibitors used primarily for copper and copper alloys.

"Cobratec" [PMC]. TM for corrosion, staining, and tarnish inhibitors for copper and brass.
Use: Cutting oils, lubricating oils, and antifreezes.

cobric acid.
Properties: White, microcrystals, contain calcium sulfate.
Derivation: From cobra venom.

COC. Abbreviation for Cleveland Open Cup, a standard technique of flash-point determination.

cocaine. (methylbenzoylecgonine).
CAS: 50-36-2. $C_{17}H_{21}NO_4$.
An alkaloid.

Properties: Colorless to white crystals or white powder. Mp 98C. Soluble in alcohol, chloroform, and ether; slightly soluble in water (solution is alkaline to litmus). The hydrochloric acid solution is levorotatory.
Derivation: By extraction of the leaves of *Erythroxylon coca* with sodium carbonate solution, treatment of the latter with dilute acid and extraction with ether, evaporation of the solvent, resolution of the alkaloid, and subsequent crystallization. Also synthetically from the alkaloid ecgonine.
Method of purification: Recrystallization.
Grade: Technical, NF.
Hazard: CNS stimulant, a poison. Possession is illegal in U.S.
Use: Local anesthetic (medicine, dentistry), usually as the hydrochloride.

cocaine. (methyl (2R,3S)-3-(benzoyloxy)-8-methyl-8-azabicyclo[3.2.1]octane-2-carboxylate; benyolmethylecconine; bernice; bernies; burese; 2-β-carbomethoxy-3-β-benzoxytropane; "c" carrie; cecil; cholly; (-)-cocaine; β-cocaine; -cocaine; coke; corine; ecgonine, methyl ester, benzoate (ester); eritroxilina; erytroxylin; girl; gold dust; happy dust; 3-β-hydroxy-1-α-H,5-α-H-tropane-2-β-carboxylic acid methyl ester, benzoate; kokain; kokan; kokayeen; methyl-3-β-hydroxy-1-α-H,5-α-H-tropane-2-β-carboxylate benzoate (ester); neurocaine; star dust; 2-β-tropanecarboxylic acid, 3-β-hydroxy-, methyl ester, benzoate (ester); 3-tropanylbenzoate-2-carboxylic acid methyl ester).
CAS: 50-36-2. $C_{17}H_{21}NO_4$.
An alkaloid ester.
Properties: Colorless to white crystals or prisms from alcohol.
Derivation: Extracted from the leaves of several South American shrubs of the genus *erythroxylon*, especially *e. coca*.
Hazard: Poison; powerful central nervous system effects.

Use: A local narcotic anesthetic applied topically to mucous membranes and a vasoconstrictor.

cocamidopropyl betaine. See "Amphosol" *[Stepan].*

cocarboxylase. (TPP; thiamine pyrophosphate chloride).
$C_{12}H_{19}ClN_4O_7P_2S\cdot H_2O$.
The coenzyme of the yeast enzyme carboxylase. The key substance in decarboxylation, an energy-producing reaction in the body.
Properties: Crystals from alcohol containing some hydrochloric acid. Mp 240–244C (decomposes). Soluble in water. Dry substance very stable.
Use: Biochemical research.

P. coccineus. (scarlet runner bean).
An ornamental vine with pods that are considered edible.
Hazard: Contains traces of poisonous lectin and phytohaemagglutinin.

cocculin. See picrotoxin.

cocculus, solid. Legal label name (air) for picro-toxin.

coccus. (1) A type of bacteria characterized by a spherical shape, e.g., pneumococcus. (2) Synonym for cochineal.

cochineal. (coccus).
A red coloring matter consisting of the dried bodies of the female insects *Coccus cacti*. The coloring principle is carminic acid.
CAS: 1343-78-8. $C_{22}H_{20}O_{13}$.
Grade: Technical, NF, silver grain, black grain.
Use: Coloring food, medicinal products, toilet preparations, manufacture of red and pink lakes.

cochliodinol. (3.6-bis(5-(3-methyl-2-butenylindol-3-yl)-2,5-dihydroxy-*p*-benzoquinone; 2,5-dihydroxy-3,6-bis(5-(3-methyl-2-butenyl)-1H-indol-3-yl)2,5- cyclohexadiene-1,4-dione)
$C_{32}H_{30}N_2O_4$.
From the ascomycetous fungus, *Chaetomium* sp.
Properties: Purple crystals.
Hazard: Extremely toxic; mycotoxin; poison.

cocoa. (cacao).
Powder prepared from roasted, cured kernels of ripe seed of *Theobroma cacao*.
Grade: Commercial, USP.
Use: Flavoring agent for pharmaceuticals, food products, beverages.

cocoa butter. See theobroma oil.

coco amido betaine.
CAS: 61789-40-0.
Hazard: A severe eye irritant.

cocoa oil. See theobroma oil.

"Cocoloid" *[Kelco].* TM for an algin-carrageenan composition, a hydrophilic colloid.
Use: Stabilizer for chocolate-milk products, sterilized cream, and other milk products.

coconut acid. Mixture of fatty acids derived from hydrolysis of coconut oil. Acid chain lengths vary from 6 to 18 carbons but are mostly 10, 12, and 14. Combustible.
Grade: Distilled, double distilled.
Use: Soaps, detergents, source of long-chain alkyl groups.

coconut amine oil condensate.
CAS: 68154-34-7.
Hazard: A severe skin irritant.

coconut cake. (coconut palm cake; copra cake).
The residual product from expression of oil from the seed of the coconut.
See coconut oil meal.

coconut diethanolamide. See coconut oil acid diethanolamine.

coconut oil.
CAS: 8001-31-8.
Properties: White, semisolid fat containing C_{12} to C_{15}; slight odor. D 0.92, saponification number 250-264, iodine number 7-10, sharp mp at 25C. Soluble in alcohol, ether, chloroform, carbon disulfide; immiscible with water. Highly digestible, resists oxidative rancidity but is susceptible to that induced by molds and other microorganisms. Nondrying. Combustible.
Chief constituents: Glycerides of lauric acid and of capric, myristic, palmitic, and oleic acids.
Derivation: Hydraulic press or expeller extraction from coconut meat followed by alkali-refining, bleaching, and deodorizing.
Grade: Crude, refined, Ceylon, Manila.
Use: Food products (margarine, hydrogenated shortenings); synthetic cocoa butter; soaps; cosmetics; emulsions; cotton dyeing; synthetic detergents; source of fatty acids, fatty alcohols, and methyl esters; base for laundering and cleaning preparations for soft leathers.

coconut oil acid diethanolamine.
CAS: 68603-42-9.
Hazard: A poison by ingestion. A moderate skin irritant.

coconut oil, esters with polyethylene glycol nonylphenyl ether.
CAS: 68333-98-2.
Hazard: A severe eye irritant.

coconut oil meal. The dried and crushed form of coconut cake recovered from the hydraulic or expeller process of extraction of oil from the meat. The usual product of commerce contains 24.2% crude protein, 13.3% crude fiber, 35.7% nitrogen-free extract, 7.4% ether-soluble fat, and 6.0% ash. The total digestible nutrients approximate 72%.
Use: Animal feeds, fertilizer ingredient.

cocoonase. A proteolytic enzyme derived from silk moths, very similar to trypsin, isoelectric point approximately 9.5 (slightly less than for trypsin), pH approximately 8.0, mw approximately 25,000. As secreted by the insect, it is 80% pure. Identical with trypsin in catalytic specificity and mechanism of action.

cocoyl sarcosine.
$C_{11}H_{23}CON(CH_3)CH_2COOH$.
Yellow liquid, d 0.970, mp 22–28C. Combustible. Used as a detergent emulsifier.

"C-O-C-S" *[Loveland].* TM for insecticides containing copper oxychloride sulfate.
Use: Insecticide for fruits and vegetables.

COD. Abbreviation for chemical oxygen demand.

codehydrogenase 1. See nicotinamide adenine dinucleotide phosphate.

codeine. (methylmorphine).
CAS: 76-57-3. $C_{18}H_{21}NO_3·H_2O$.
A narcotic alkaloid.
Properties: Colorless or white crystals or powder. Effloresces slowly in dry air, affected by light. Mp 154.9C. Slightly soluble in water; soluble in alcohol and chloroform. Levorotatory in acid and alcohol solutions.
Derivation: From opium by extraction and subsequent crystals; also by the methylation of morphine.
Grade: Technical, NF.
Hazard: Habit-forming narcotic, sale legally restricted.
Use: Medicine (analgesic).

coding sequence. The portion of a gene or an mRNA which actually codes for a protein. Introns are not coding sequences; nor are the 5′ or 3′ untranslated regions (or the flanking regions, for that matter—they are not even transcribed into mRNA). The coding sequence in a cDNA or mature mRNA includes everything from the AUG (or ATG) initiation codon and inclusion of the stop codon.

coding strand. An ambiguous term intended to refer to one specific strand in a double-stranded gene.
See: Sense strand.

cod-liver oil. (morrhua oil).

Properties: Pale-yellow, viscous liquid; fixed, nondrying oil; slightly fishy odor and taste. D 0.918–0.927, saponification number 180-192, iodine number 145-180, maumene test 102-113, acid value 204-207. Soluble in ether, chloroform, ethyl acetate, and carbon disulfide; slightly soluble in alcohol. Combustible.
Chief constituents: Glycerides of palmitic and stearic acids, cholesterol, butyl alcohol esters, etc.
Derivation: From the livers of codfish (*Gadus morrhua*) and other species of Galidae. These are rendered by steam heat, and the oil separated and chilled until the stearin solidifies; then it is pressed and the clear oil collected.
Method of purification: Filtration.
Grade: Pale, light brown, dark brown, NF.
Use: Medicine (for its vitamin A and D content, now largely is replaced by synthetic products), chamois-leather tanning.

codominance. Situation in that two different alleles for a genetic trait are both expressed.
See autosomal dominant; recessive gene.

codominant gene. A set of two or more alleles such that each is expressed regardless of the presence of one of the others.

codon. In an mRNA, a codon is a sequence of three nucleotides which codes for the incorporation of a specific amino acid into the growing protein. The sequence of codons in the mRNA unambiguously defines the primary structure of the final protein. Of course, the codons in the mRNA were also present in the genomic DNA, but the sequence may be interrupted by introns.

Cohen law. Materials of greater dielectric constant assume a higher potential than those of lower dielectric constant to an extent that is proportional to their dielectric constants.

coenocytic. Condition in which an organism consists of filamentous cells with large central vacuoles, and whose nuclei are not partitioned into separate compartments. The result is a long tube containing many nuclei, with all the cytoplasm at the periphery.

coenzyme. A comparatively low-molecular-weight organic substance that can attach itself to, and supplement, a specific protein to form an active enzyme system. Generally synonymous with the term *prosthetic group*.

coenzyme A. (CoA).
CAS: 85-61-0. $C_{21}H_{36}O_{16}N_7P_3S$.
Essential for the formation of acetylcholine and for acetylation reactions in the body. It has been synthesized and is built up from pantothenic acid, cysteamine, adenosine, and phosphoric acid.
Properties: White powder. Water soluble; insoluble in alcohol and ether.

coenzyme b12. ((2R,3R,4S,5R)-2-(6-aminopurin-9-yl)-5-methanidyloxolane-3,4-diol; cobalt(3+); [(2R,3S,4R,5S)-5-(5,6-dimethylbenzimidazol-1-yl)-4-hydroxy-2-(hydroxymthyl)oxolan-3-yl]; [(2R)-1-[3-[(1R,2R,3R,4Z,7S,9Z,12S,13S,14Z,17S,18S,19R)-2,13,18-tris(2-amino-2-oxoethyl)-7,12,17-tris(3-amino-3-oxopropyl)-3,5,8,8,13,15,18,19-octamethyl-2,7,12,17-tetrahydro-1H-corrin-21-id-3-yl]propanoylamino]propan-2-yl]phosphate. CAS: 13870-90-1. $C_{72}H_{100}CoN_{18}O_{17}P$.
A cofactor derived from the cobalamin that is involved in alkyl group transfers.

coenzyme I. See nicotinamide adenine dinucleotide phosphate.

coenzyme Q.
CAS: 303-98-0. $CH_3C_6(O)_2(OCH_3)_2[CH_2CH:C(CH_3)CH_2]_nH$.
Found in animal organs and yeast. Active in the citric acid cycle in carbohydrate metabolism. The n in the formula varies according to the source.

coercivity. The magnetic intensity required to reduce the magnetic induction in a substance from saturation to zero.

cofactor. An inorganic ion or a coenzyme required for enzyme activity.

coffearine. See trigonelline.

coffee.
$C_{20}H_{25}N_3O$.
The seed of the coffee tree, genus *Coffea*. These beans, when roasted, are pulverized and brewed with hot water to form a widely used beverage. Coffee is grown chiefly in countries where highly toxic pesticides are still used. A semisynthetic derivative of ergot (*Claviceps purpurea*). It has complex effects on serotonergic systems including antagonism at some peripheral serotonin receptors, both agonist and antagonist actions at central nervous system serotonin receptors, and possible effects on serotonin turnover.
Hazard: Teratogen; mutagen; long-term usage may cause insomnia, depression, stomach upset, ulcers, headache, heart attacks; large quantities can promote miscarriage, premature birth and birth defects; linked to cancer of the pancreas.

coffinite.
$U(SiO_4)_{1-x}(OH)_{4}x$, (or $USiO_4$, with appreciable $(OH)_4$ in place of some SiO_4).
A naturally occurring uranium mineral.
Properties: Black color; adamantine luster; commonly fine grained, mixed with organic matter and other minerals. D 5.1.
Occurrence: Colorado, Utah, Wyoming, Arizona.
Use: Ore of uranium (Colorado).

cogeneration. Simultaneous production of electricity and steam from the same energy source. Research has indicated a potential fuel savings of 30% by use of this process.

cognate. Describing two biomolecules that normally interact; for example, an enzyme and its normal substrate, or a receptor and its normal ligand.

coherence. The attraction between molecules of a substance that prevents separation of the substance into parts if acted upon by superficial forces.

coherent light. Light having a single wavelength, frequency, and phase. See laser.

cohesion. The force that holds adjacent molecules of a single material together.

cohesive ends. Two DNA ends in the same DNA molecule, or in different molecules, with short overhanging, single-stranded segments that are complementary to one another, facilitating ligation of the ends; also known as sticky ends. Usually generated by the action of restriction enzymes.

cohune oil. An edible, nondrying oil with properties similar to coconut and babassu oils. Its composition is 46% lauric acid, 16% myristic acid, 10% oleic acid, and the balance mixed acids. Obtained from a palm native to Mexico and Central America. Combustible.

Coilife. Special epoxy resin encapsulation of random-wound stators using solventless epoxy-resin formulations and rotational seasoning.

cointegrate. An intermediate in the migration of certain DNA transposons in which the donor DNA and target DNA are covalently attached.

coion. Small ions, entering a solid ion exchanger, that have the same charge as that of the fixed ions.

coisogenic. (congenic).
Nearly identical strains of an organism varying only at a single locus.

colicin. A bacteriocin.
Derivation: Produced by coliform bacteria.

coliform bacteria. Bacteria that obtain energy by fermentation or aerobic respiration and may contaminate water via fecal pollution.
Properties: Gram-negative, peritrichous, rod-shaped eubacteria
Hazard: Toxic.

collagen.
CAS: 9007-34-5.

Protein of connective tissues that comprises approximately 30% of total body protein from which body tissues are formed.
Properties: Fibrous insoluble.

collagenase. (2-hydroxy-2-methylbuanedioic acid).
CAS: 9001-12-1. $C_5H_8O_5$.
A rare proteinase that catalyzes the hydrolysis of collagen. Collagenase activity has been demonstrated in several venoms of crotalid and viperid snakes and it occurs in certain species of *clostridium*.

collidin. Any of the alkyl substituted pyridines isolated from sources such as the sticky remains of decaying animal bodies.

collidine. (2,4,6-trimethylpyridine).
CAS: 108-75-8. $C_8H_{11}N$.
Methyl, ethyl, propyl, and trimethyl homologs of pyridine.
Properties: Oily, acrid, organic bases.
Derivation: By-products of the coking process or are synthesized.
Hazard: Toxic.

colloid. Any mixture of particles with diameters less than 10 μm.

colocynthin. (colocynthis; bitter apple; bitter cucumber; 25-(acetyloxy-2-(β-D-gluco-pyranosyloxy)-16,20-dihydroxy-9-methyl-19-norlanosta-1,5,23-triene-3,11,22-trione; 2-o-β-D-glucopyranosylcucurbitacine; [(E,6R)-6-hydroxy-6-[(8S,9R,10R,13R,14S,16R,17R)-hydroxy-4,4,9,13,14-pentamethyl-3,11-dioxo-2-[(3R,4S,5S,6R)-3,4,5-trihydroxy-6-(hydroxymethyl)oxan-2-yl]oxy-8,10,12,15,16,17-hexahydro-7H-cyclopenta[a]phenanthren-17-yl]-2-methyl-5-oxohept-3-en-2-yl]acetate).
$C_{38}H_{54}O_{13}$.
A glycoside.
Derivation: Isolated from the dried unripe fruit of *Citrullus colocynthis*.
Use: Insecticide; purgative; abortifacient.

coke. The carbonaceous residue of the destructive distillation (carbonization) of bituminous coal, petroleum, and coal tar pitch. The principal type is that produced by heating bituminous coal in chemical recovery or beehive coke ovens (metallurgical coke), one ton of coal yielding approximately 0.7 tons of coke. It is used chiefly for reduction of iron ore in blast furnaces and as a source of synthesis gas. Petroleum yields coke during the cracking process. Coke from petroleum residues and coal tar pitch is used for refractory furnace linings in the electrorefining of aluminum and other high-temperature service and for electrodes in electrolytic reduction of Al_2O_3 to aluminum, as well as in electrothermal production of phosphorus, silicon carbide, and calcium carbide.

coke oven gas. 53% hydrogen, 26% methane, 11% nitrogen, 7% carbon monoxide, 3% heavier hydrocarbons.
Hazard: Confirmed carcinogen.
Use: To produce hydrogen.

coking. The residue of a procedure of distillation to dryness of a product containing complex hydrocarbons that break down in structure during the process.

cola. (kola; kola nuts; kola seeds; Soudan coffee; guru).
Contains caffeine, theobromine.
Derivation: Seeds of *Cola nitida* or other species of *Cola*.
Source: West Africa, West Indies, India.
Use: Soft drinks.

"Colar" [Du Pont]. TM for chemical-resistant finishes having a base of polyamide-catalyzed epoxy resins.

colchicine.
CAS: 64-86-8. $C_{22}H_{25}NO_6$.
An alkaloid plant hormone.

Properties: Yellow crystals or powder; odorless, or nearly so. Affected by light, mp 135–150C. Soluble in water, alcohol, and chloroform. Solutions are levorotatory.
Derivation: From *Colchicum autumnale* by extraction and subsequent crystallization. Has been synthesized.
Grade: Technical, USP.
Hazard: As little as 20 mg may be fatal if ingested.
Use: To induce chromosome doubling in plants; phytopathology.

colcothar. Red ferric oxide produced by heating ferrous sulfate in air.
Use: Pigment; abrasive in polishing glass.

cold flow. The permanent deformation of a material that occurs as a result of prolonged compression or extension at or near room temperature. Some plastics and vulcanized rubber exhibit this behavior. In metals, it is known as creep.

cold light. See bioluminescence.

cold rubber. Synthetic rubber produced by polymerization at relatively low temperature, specifically SBR or butadiene-styrene elastomers produced by polymerization at approximately 4.4C rather than the usual temperature of approximately 49C. A special catalyst system is required.

cold-short.   See short.

colemanite.
($Ca_2B_6O_{11}\cdot5H_2O$).
The ore of calcium borate. Used to replace boric acid in the manufacture of glass fibers. Mined in Turkey, it began to be imported into the U.S. in large volume in 1965 and is competitive with domestically produced B_2O_3.
Properties: D 2.26–2.48.
Derivation: From kernite.

colestipol.   An anion exchange resin, highly cross-linked and insoluble. A specific formula has not been ascertained. Said to be a copolymer of diethylene triamine and chloroepoxypropane. It is used as a cholesterol-sequestering agent in medicine.

"Coleus Forskohlli oil" *[Sabinsa].*   TM for a refined oil.
Use: As an antimicrobial, anti-acne preparation, and in perfumery.

colistin.
$C_{45}H_{85}N_{13}O_{10}$.
Antibiotic produced by a soil microorganism. Probably identical to polymyxin E and closely related chemically to polymyxin B since it is a polypeptide composed of amino acids and a fatty acid.
See polymyxin.

collagen.   A fibrous protein constituting most of the white fiber in the connective tissues of animals and humans, especially in the skin, muscles, and tendons. The most abundant protein in the animal kingdom, it is rich in proline and hydroxyproline. The molecule is analogous to a three-strand rope, in which each strand is a polypeptide chain. It has a molecular weight of approximately 100,000. Glue made from the collagen of animal hides and skins is still widely used as an adhesive. So-called "soluble" collagen is that first formed in the skin; upon aging it becomes increasingly cross-linked and less hygroscopic. Soluble collagen is used in the cosmetic industry as the basis for face creams, lotions, and hair preparations. Special forms of collagen have been developed for dialysis membranes. Microcrystalline collagen is used in prosthetic devices and other medical and surgical applications. Regenerated collagen, used in sausage casings, is made by neutralizing with acid collagen that has been purified by alkaline treatment. Collagen is converted to gelatin by boiling water, which causes hydrolytic cleavage of the protein to a mixture of degradation products.
See gelatin.

collection trap.   Cooled container that collects gas-chromatographic eluant, preserving the eluant for the compound-identification step.

2,4,6-collidine.   (2,4,6-trimethylpyridine).
CAS: 29611-84-5. $(CH_3)_3C_5H_2N$.
Properties: Colorless liquid. Bp 170.4C, fp −44.5C, d 0.913 (20/20C), refr index 1.4981 (20C). Soluble in alcohol; slightly soluble in water. Combustible.
Grade: Technical (97.5% purity).
Use: Chemical intermediate, dehydrohalogenating agent.

colligative property.   A property independent of the chemical nature of the molecules of a substance, resulting only from the number of molecules present.

collodion.   A solution of pyroxylin (nitrocellulose) in ether and alcohol. USP specifications are pyroxylin 40 g, ether 750 ml, and alcohol 250 ml.
Properties: Pale-yellow, syrupy liquid; odor of ether. Immiscible with water; flash p approximately 0F (−17.7C).
Grade: Technical, USP.
Hazard: Flammable, dangerous fire risk.
Use: Cements, coating wounds and abrasions, solvent for drugs, corn removers, process engraving, lithography, photography.
See nitrocellulose.

colloidal solution.   A system intermediate between a true solution and a suspension. A dispersion where the particle size is between 1 and 100 nm. Colloids have little or no tendency to dialyze and small or no freezing-point depression.

colloid, association.   See association.

colloid chemistry.   A subdivision of physical chemistry comprising the study of phenomena characteristic of matter when one or more of its dimensions lie in the range between 1 millimicron (nanometer) and 1 micron (micrometer). It thus includes not only finely divided particles but also films, fibers, foams, pores, and surface irregularities. Dimension, rather than the nature of the material, is characteristic. Colloidal particles may be gaseous, liquid, or solid, and occur in various types of suspensions (imprecisely called solutions), e.g., solid–gas (aerosol), solid–solid, liquid–liquid (emulsion), gas–liquid (foam). In this size range, the surface area of the particle is large with respect to its volume so that unusual phenomena occur, e.g., the particles do not settle out of the suspension by gravity and are small enough to pass through filter membranes. Macromolecules (proteins and other high polymers) are at the lower limit of this range; the upper limit is usually taken to be the point at which the particles can be resolved in an optical microscope. The first specific observations were made by Thomas Graham in approximately

1860 and were extended by Ostwald, Hatchek, and Freundlich. Though the term is often used synonymously with *surface chemistry*, in a strict sense it is limited to the size range noted in at least one dimension, whereas surface chemistry is not. Natural colloid systems include rubber latex, milk, blood, egg white, etc.
See surface chemistry; colloid, protective; emulsion.

colloid mill.   See homogenization.

colloid, protective.   A hydrophilic high polymer whose particles (molecules) are of colloidal size, such as protein or gum. It may be either naturally present in such systems as milk and rubber latex or intentionally added to mixes to stop coagulation or coalescence of the particles of fat or other dispersed material. Protective colloids are also called stabilizing, suspending, or thickening agents; they also act as emulsifiers. Examples are (1) hydrocarbon particles of latex, which are covered with a layer of protein that keeps them from cohering as a result of the impact due to their Brownian motion; (2) gelatin, sodium alginate, or gum arabic, which are added to ice cream to inhibit formation of ice particles and to confectionery and other food products to obtain a smooth, creamy texture. They are readily adsorbed by the suspended particles and reinforce the protective effect of proteins that may be naturally present.
See thickening agent; gum, natural; gelatin.

Colloisol.   A series of vat dyes for dyeing and printing textiles of cellulosic fibers.

cologne.   (toilet water).
A scented, alcohol-based liquid used as a perfume, after-shave lotion, or deodorant. Combustible.

Cologne brown.   See Van Dyke brown.

colonial.   Condition in which many unicellular organisms live together in a somewhat coordinated group. Unlike true multicellular organisms, the individual cells retain their separate identities and, usually, their own membranes and cell walls.

colophony.   A rosin residue that remains after the volatiles have been removed by distillation of crude turpentine from any of the *Pinus* species.

colorant.   Any substance that imparts color to another material or mixture. Colorants are either dyes or pigments and may be (1) naturally present in a material (chlorophyll in vegetation); (2) admixed with it mechanically (dry pigments in paints); or (3) applied to it in a solution (organic dyes to fibers).
Note: There is no generally accepted distinction between dyes and pigments. Some have proposed one on the basis of solubility or physical form and method of application. Most pigments, so called, are insoluble inorganic powders, the coloring effect

being a result of their dispersion in a solid or liquid medium. Most dyes, on the other hand, are soluble synthetic organic products that are chemically bound to and actually become part of the applied material. Organic dyes are usually brighter and more varied than pigments but tend to be less stable to heat, sunlight, and chemical effects. The term *colorant* applies to black and white, as well as to actual colors. Instruments for measuring, comparing, and matching the hue, tone, and depth of colors are called colorimeters.
See dye; pigment; colorimetry; food color; FD&C color.

colorimeter.   An analytical device used to measure the comparative intensity of color in solutions by comparison with standard solutions.
See photoelectric colorimeter.

colorimetric analysis.   Analysis based on the law that intensity of color of certain solutions is proportional to the amount of substance in the solution.

colorimetric purity.   The ratio of luminance of spectrally pure light that must be mixed with reference achromatic (white) light to produce a color match for the specimen.

colorimetry.   An analytical method based on measuring the color intensity of a substance or a colored derivative of it. For example, the yellow-carotene content of butter is determined by saponifying a sample of butter in an alkaline solution, extracting the carotene with ether, and measuring the intensity of yellow color in the ether extract. Colorimetric methods are used to determine very minute amounts. They are used in hospital laboratories for blood and urine analysis; in food laboratories for determination of vitamins, preservatives, coloring matter, etc.; and in metallurgical laboratories for traces of metals in raw materials and finished products.

Color Index.   (CI; C.I.).
A listing of all commercial dyes, each of that has an individual number that identifies it on specifications.

color lake.   See lake.

colorless dye.   Synonym for optical brightener.

Colorundum.   A balanced mixture of abrasive aggregates, mineral oxide color, and stearate for surfacing concrete floors, in a choice of colors.

Columbite.   An ore of tantalum.
See tantalite.

columbium.
Cb.

Alternate name for the element niobium. The latter name became official in 1949; columbium is still used by metallurgists.

column, distillation. See tower, distillation.

"Coly-Mycin" *[Par Sterile].* TM for colistin in the form of sulfate or methane sulfonate (colistimethate sodium).

colza oil. See canola.

Combes quinoline synthesis. Formation of quinolines by condensation of β-diketones with primary arylamines followed by acid-catalyzed ring closure of the intermediate Schiff base.

combination. A chemical reaction in which two substances unite to form a third; the reacting substances may be elements or compounds, but the product is always a compound, e.g., $Cu + Cl_2 \rightarrow CuCl_2$ and $2NH_3 + H_2SO_4 \rightarrow (NH_4)_2SO_4$.
Polymerization is a special case of combination, where complex organic molecules of the same kind unite to form chains or clusters of high molecular weight.
See polymerization.

combining number. See valence.

combining volumes, law of. When gases combine, they do so in volumes that bear a simple relationship to each other and to that of the product if gaseous.

combining weight. See equivalent weight.

combustible material. Any substance that will burn, regardless of its autoignition temperature or whether it is a solid, liquid, or gas. Although this definition necessarily includes all flammable materials as well, this fact is disregarded in official classifications. As usually defined, the term *combustible* refers to solids that are relatively difficult to ignite and that burn relatively slowly and to liquids having a flash p greater than 100F (37.7C). It is difficult to generalize about the combustibility of solids. The rate and ease of combustion may depend as much on their state of subdivision as on their chemical nature. Many metals in powder or flake form will ignite and burn rapidly, whereas most are noncombustible as bulk solids. Cellulose is combustible as a textile fabric or as paper and is flammable as fine fibers (cotton linters). A plastic that burns at flame temperature will be a greater fire hazard as a foam than as a bulk solid because of the large surface area exposed to air and the thinness of the cell walls. Some polymers, such as nylon and polyvinylidene chloride, will melt and burn but not propagate flame; others, e.g., polyvinyl chloride and polyurethane, ignite at high temperature and evolve toxic fumes. Acrylics

and cellulose-derived plastics, such as rayon and cellulose acetate, are readily combustible. This may be partially offset by use of fire-retardant chemicals. Glass is noncombustible in all forms.
See flammable material; hazardous material; combustion.

combustion. An exothermic oxidation reaction that may occur with any organic compound, as well as with certain elements, e.g., hydrogen, sulfur, phosphorus, magnesium. The end products of elemental combustion are oxides; end products of organic compounds are carbon dioxide and water. Examples are (1) for an element: $2H_2 + O_2 \rightarrow 2H_2O$; (b) for an organic compound (carbohydrate): $C_6H_{12}O_6 + 6O_2 \rightarrow 6CO_2 + 6H_2O$. Here, combustion is the reverse of photosynthesis. The heat of combustion is due to rupture of chemical bonds and formation of new compounds. Substances differ greatly in their combustibility, i.e., in their ignition points (solids and gases) or their flash points (liquids). Carbon disulfide burns almost explosively at 100C, whereas rubber hydrocarbon and nylon are difficult to ignite at any temperature. Oxygen is not combustible but actively supports combustion; no oxygen is needed if another oxidizing agent is present, as in the combustion of a mixture of hydrogen and chlorine to form hydrogen chloride.
Spontaneous combustion may occur at or even below room temperature (1) by exposure to air of substances that are highly sensitive to oxidation (e.g., phosphorus); (2) by heat buildup from bacterial activity (compost, sewage sludge) or oxidation catalyzed by moisture, as in wet waste materials (paper, cotton, wool); and (3) by internal heat accumulation caused by autoxidation (fish oils, linseed oil).
See oxidation; pyrophoric material; autoignition point; flash point.

combustion tube. A tube used in the combustion method of chemical analysis.

comicellization. The phenomenon wherein a surfactant in a concentration considerably below its own critical micelle concentration solubilizes an insoluble compound in a comparatively large quantity.

comirin. Exact formula undetermined. An antifungal agent produced by *Bacterium antimyceticum* composed mostly of amino acids. A gray to white powder that resists heat and acids but decomposes in the presence of alkalies; soluble in organic acids and bases; slightly soluble in water. A fungicide in paints, textile products, plant growth.

Commercial Development and Marketing Association. (CDMA).
The Commercial Development and Marketing Association, a Special Interest Group of PDMA, is the world's leading professional association dedicated

to fostering, promoting, and sharing business processes for long-term growth and value creation in the chemical and allied industries. It provides information on world-class practices of business development, corporate growth, business strategy, marketing, and relating functional areas. The focus is to conduct workshops, business conferences, networking forums, and local section activities. The offices of the organization are at 330 N. Wabash Avenue, Suite 2000, Chicago, IL 60611. Website: http://www.pdma.org

comminution. Size reduction of materials by grinding, cutting, shredding, chopping, etc. Solids can be reduced to a particle size approaching 1 micron in special fine-grinding equipment. Comminution of coal by chemical means is possible via a low-molecular-weight compound such as sodium hydroxide or anhydrous ethanol, which penetrates the natural fault system of the coal, causing fragmentation without mechanical crushing. This permits removal of sulfur from coal without burning or grinding. Approximately 100 lb of chemical are needed per ton of coal.

commodity chemicals. Low-value, high-volume compounds produced in dedicated plants and used for a wide variety of applications.

common acetic acid.
  Properties: Clear, colorless, aqueous solution of 35% acetic acid, completely miscible with water or ethanol.

common intermediate. Referring to a metabolite: a chemical compound common to two enzymatic reactions, as a product of one and a substrate in the other.

common-ion effect. The reversal of ionization when a compound is added to a solution of another compound with which it has a common ion, the volume being kept constant.

comparative genomics. The study of human genetics by comparisons with model organisms such as mice, the fruit fly, and the bacterium *E. coli.*

compatibility. The ability of two or more materials to exist in close and permanent association indefinitely. Liquids (solvents) are compatible if they are miscible and do not undergo phase separation on standing. Thus, water is compatible with alcohol but not with gasoline. Liquids and solids are compatible only if the solid is soluble in the liquid. Solids are compatible if they can exist in intimate contact for long periods with no adverse effect of one on the other.

"Compazine" *[PBM].* TM for a brand of prochlorperazine, as the maleate or the edisylate.

competent bacteria. A bacterial culture so managed that the bacteria are able to take up DNA molecules without transduction or conjugation is heightened.

competitive inhibition. A type of enzyme inhibition characterized by the inhibitor binding to the active site. The inhibition can be reversed by increasing the concentration of substrate. A competitive inhibitor is usually a structural analog of the substrate.

complement. In immunochemistry, any of a number of blood proteins that act in conjunction with antibodies to cause disintegration of invading cells. They are an essential component of immune serum.

complementary. Having a molecular surface with chemical groups arranged to interact specifically with chemical groups on another molecule.

complementary DNA. See cDNA.

complementary sequence. Nucleic acid base sequence that can form a double-stranded structure with another DNA fragment by following base-pairing rules (e.g., A pairs with T and C with G). The complementary sequence to GTAC, for example, is CATG.

complex compound. See coordination compound.

"Complex Gel" *[Centerchem].* TM for a formulated gel.
  Use: To stabilize either free vitamin A or vitamin C, when packed in dual packaging with end user activating.

complexing agent. See ligand; chelate; ethylenediaminetetraacetic acid.

complex ion. An ion that has a molecular structure consisting of a central atom bonded to other atoms by coordinate covalent bonds.
See coordination compound.

complex mixture. A mixture in which the component parts cannot be separated out by convenient or cost-efficient means. Such mixtures, if toxic, are extremely difficult to assess. During the course of testing or experimentation, the makeup of such a mixture may change. Furthermore, such mixtures may produce complex and confusing combinations of signs and symptoms. The set of symptoms may vary complexly with concentration and age of the mixture due to differential stabilities, solubilities, toxicities, and mode of action and other properties of the component materials.

complex protein. A complex protein usually has more than one folding domain that comprises a sequence of 100–300 amino acids. The entire folding architecture is precisely constructed such that the protein is functional. While such proteins may consist of a single polypeptide chain or of multiple chains linked by disulfide bonds, some complex proteins contain nonpolypeptide organic structures.

complex trait. Trait that has a genetic component that does not follow strict Mendelian inheritance. May involve the interaction of two or more genes or gene–environment interactions.
See Mendelian inheritance; additive genetic effects.

component. One of the minimum sets of substances required to generate the composition of all phases of a system in the absence of chemical reaction of any substances in a mixture.
See constituent.

composite. A mixture or mechanical combination on a macroscale of two or more materials that are solid in the finished state, are mutually insoluble, and differ in chemical nature. The major types are (1) laminates of paper, fabric, or wood (veneer) and a thermosetting material (resin, rubber, or adhesive); examples are tire carcasses, plywood, and electrical insulating structures; (2) reinforced plastics, principally of glass fiber and a thermosetting resin; other types of fibers such as boron, aluminum silicate, and silicon carbide may be used. See whiskers. (3) Cermets, which are mixtures of ceramic and metal powders, heat treated and compressed. (4) Fabrics, e.g., woven combinations of wool or cotton and a synthetic fiber. (5) Filled composites in which a bonding material, i.e., linseed oil, resin, or asphalt, is loaded with a filler in the form of flakes or small particles; examples are linoleum, glass flake-plastic mixtures for battery cases, and asphalt-gravel road-surfacing mixtures.

composting. Aerobic bacterial decomposition of solid organic wastes, both agricultural and urban, including sewage sludge. As many as 500 tons a day can be handled in the larger installations, the waste degrading quickly without external heating. Decomposition is accelerated by adding ammonium bicarbonate. The product can be used as a soil conditioner and for landfill. The waste is piled and turned frequently to provide aeration and to maintain a high temperature in the pile to destroy pathogenic organisms. The volume of composted waste is from 20 to 60% of original volume.

compound. (1) A substance composed of atoms or ions of two or more elements in chemical combination. The constituents are united by bonds or valence forces. A compound is a homogeneous entity where the elements have definite proportions by weight and are represented by a chemical formula. A compound has characteristic properties quite different from those of its constituent elements. It is decomposed by energy in the form of a chemical reaction, heat, or electric current. Example: water is a *liquid* formed by chemical combination of two *gases*; it can be separated into hydrogen and oxygen by an electric current (electrolysis); in certain reactions, it is split into its constituent ions (H, OH) (hydrolysis); it is not chemically changed by heat or cold. See mixture; homogeneous; chemical reaction. (2) Loosely, a product formula (often proprietary) of various types, e.g., pharmaceuticals (a vegetable compound), rubber (a fast-curing compound), etc. (3) Having two sets of lenses (compound microscope).

compound 1080. Use may be restricted.
See sodium fluoroacetate.

compreg. A hardwood impregnated with a phenolformaldehyde resin under heat and pressure.

compressed gas. Any material or mixture that, when enclosed in a container, has an absolute pressure exceeding 40 psi at 21.1C or, regardless of the pressure at 21.1C, has an absolute pressure greater than 104 psi(a) at 54.4C, or any flammable material having a vapor pressure greater than 40 psi abs at 37.7C (vapor pressure determined by Reid method (ASTM)). Compressed gases include liquefied petroleum gases and oxygen, nitrogen, anhydrous ammonia, acetylene, nitrous oxide, and fluorocarbon gases. Some of these are shipped in tonnage volume. For details on properties, containers, and shipping regulations, see the entries for specific gases.

compression molding. Formation of a rubber or plastic article to a desired shape, by either placing the raw mixture in a specially designed cavity or bringing it into contact with a contoured metal surface. After the material is in place, heat and pressure are supplied by a hydraulic press, the time and temperature varying with the nature of the material. For rubber products, vulcanization occurs simultaneously. Most plastic molding is now done by the injection method, which is more economically efficient.
See injection molding.

Compton effect. One of the principal processes by which high-energy electromagnetic radiation (γ-rays) interacts with or is absorbed by matter. In the Compton process, the γ-ray frees an electron in matter as if the electron were unbound, dividing the momentum of the γ-ray between the ejected electron and a new γ-ray of lower energy going off in a new direction.

computational chemistry. Use of computers in organic synthesis and in chemical engineering as a more efficient means of research than conventional laboratory experimentation. The capacity of sophisticated computers for fast mathematical calculations has made them an invaluable aid

in exploring and evaluating the more likely pathways for a given organic synthesis, for which there may be innumerable possible sequences. The term *heuristic* is applied to such procedures. Computers can also handle the vast complexity of quantum-mechanical calculations and aid in the elucidation of the complicated molecular structures that occur in pharmaceutical compounds and recombinant-DNA research. The Quantum Chemistry Program Exchange at Indiana University offers many programs in this field, from subroutines to major computational systems. Chemical engineers utilize computers to develop more thermodynamically efficient procedures and to consolidate overall plant operations, especially in the areas of energy consumption, reaction rates, and hazardous waste problems.
See retrosynthesis.
Note: Notwithstanding the immense capability of computers to point the way to solutions of chemical and engineering problems, experimentation will remain the ultimate proof of theory. It is interesting to speculate how much time and effort such empirical scientists as Goodyear and Edison could have saved had computers been available to them.

concanavalin A. (Con A; NSC-143504; ricin–toxin con a; 1-[(2R,4S,5S)-4-azido-5-(hydroxymethyl)oxolan-2-yl]-5-methylpyrimidine-2,4-dione; 4-(dipropylsulfamoyl)benzoic acid).
CAS: 11028-71-0. $C_{23}H_{32}N_6O_8S$.
A lectin mitogen that binds specifically to and agglutinates transforms into tumors. It binds to glucose and mannose residues in the surface.
Properties: Unlike most lectins, it lacks a covalently bound carbohydrate and therefore is not a glycoprotein.
Derivation: Isolated from *canavalia ensiformis* (jack bean).
Hazard: Poison; teratogen; mutagen.

Conant, James Bryant. (1893–1978). An American chemist and educator, born in Boston, who received his doctorate in chemistry from Harvard in 1916 and was president of Harvard for 20 years (1933–1953). His major scientific activities included pioneering research on chlorophyll and important contributions to the Manhattan Project. Perhaps his greatest achievements lay in the educational field, in which he exerted a strong liberalizing influence at both the collegiate and secondary-school levels. He also was ambassador to postwar Germany and educational adviser to Berlin. He wrote many books on science and education, including basic chemical texts, and received a number of scientific and educational awards.

concave. See gyratory crusher.

concentration. The amount of a given substance in a stated unit of a mixture, solution, or ore. Common methods of stating concentration are percentage by weight or by volume, normality, or weight per unit volume (as grams per cubic centimeter or pounds per gallon). The concentration of an atom, ion, or molecule in a solution may be symbolized by the use of square brackets, as $[Cl^-]$. For radioactivity, the concentration is usually expressed as millicuries per milliliter (mCi/mL) or millicuries per millimole (mCi/mM).

conchiolin.
$C_{32}H_{98}N_2O_{11}$.
A natural bonding agent in the calcium carbonate structure of pearls.

conchoidal. A term adopted from mineralogy by chemists to describe a type of surface formed by fracturing a hard solid by impact. Certain materials present involutely curved fracture surfaces suggestive of the shape of the shells of bivalves (conch), from which the term is derived. Examples are glass, blown asphalt, and numerous minerals.

conclyte calcium. See calcium lactate.

concrete. (1) A conglomerate of gravel, pebbles, sand, broken stone, and blast-furnace slag or cinders, termed the aggregate, embedded in a matrix of either mortar or cement, usually standard Portland cement in the U.S. Reinforced concrete and ferro-concrete contain steel in various forms.
See cement, Portland.
Use: Building and road construction, radiation shielding.
(2) A waxy solid obtained from roses by extraction with nonpolar solvents (benzene) after trace quantities of solvent have been removed. When the concrete is dewaxed by a properly chosen second solvent (alcohol), the desired essential oil remains. This is called an absolute.
See absolute; perfume.

concrete, cellular. A lightweight concrete foam that may be made in several ways: (1) by adding aluminum powder to the concrete mix and applying heat, which releases hydrogen; (2) by whipping air into the mix containing an entraining agent; and (3) by adding preformed foam to the mix. Such foams are made from a foaming agent such as dried blood, a stabilizer, organic solvents, and a germicide.
See foam.

concrete, reinforced. See concrete.

condensation. (1) A type of chemical reaction in which two or more molecules combine with the separation of water, alcohol, or other simple substance. The distinction from an addition reaction is not sharp, because reactions can be either stopped at the addition stage or carried a step beyond it. If a polymer is formed by condensation, the process is called polycondensation, e.g., phenolformaldehyde resin.

See Claisen condensation; aldol.
(2) The change of state of a substance from the vapor to the liquid (or solid) form.

conductance. An electrical property of a solution, defined as the reciprocal of the resistance. It is usually used in connection with electrolytic solutions.
See conduction (2).

conductometric method. A method of analysis in which the end point of a reaction is determined by conductance measurements.

conduction. (1) Transfer of heat through a substance or from one substance to another when the two substances are in physical contact (thermal conduction). Crystalline solids (especially metals and alloys) are good thermal conductors; liquids such as water and glass and high polymers such as rubber and cellulose usually are not.
(2) Transfer of an electric current through a solid or liquid (electrical conduction). In metallic or electronic conductors, the current is carried by a flow of electrons, the atomic nuclei remaining stationary. This type of conduction is common to all metals and alloys, carbon and graphite, and certain solid compounds (manganese dioxide, lead sulfide). In electrolytic or ionic conductors, the current is carried by ions, as in solutions of acids, bases, and salts and in many fused compounds. In electrolytic conduction, as in metallic conduction, heat is generated and a magnetic field is formed around the conductor; a transfer of matter also occurs. In a few materials, such as solutions of alkali and alkaline earth metals in anhydrous liquid ammonia, both types of conduction take place simultaneously; such conductors are called mixed conductors.
See semiconductor; transference number.

conduction band. A partly filled energy band whereby the electrons can move freely, allowing the material to carry an electric current.

conductivity. The property of a substance or mixture that describes its ability to transfer heat or electricity. It is the reciprocal of resistivity.
See conduction.

Condy's fluid. A solution of potassium permanganate.
Use: Disinfectants.

confidentiality. In genetics, the expectation that genetic material and the information gained from testing that material will not be available without the donor's consent.

configuration. The spatial arrangement of an organic molecule that is conferred by the presence of either a double bond, about which there is no freedom of rotation, or chiral centers, around which

four different substituent groups are arranged in a stereospecific fashion. The arrangement in either of these cases cannot be changed without breaking at least one covalent bond.

conformation. (1) A shape or arrangement in three-dimensional space that an organic molecule can assume by rotating carbon atoms or their substituents around single covalent bonds. The conformation of a molecule is not fixed, though one or another shape may be more likely to occur. The number of conformational isomers is infinite. Conformational analysis involves the study of the preferred (or most likely) conformations of a molecule in the ground, transition, and excited states. Research on the conformations of cyclohexane and various sugars has contributed much to this aspect of stereochemistry.
(2) An extended, zigzag arrangement of a polypeptide chain that can be parallel or antiparallel. One of the common secondary structure in proteins.

congenic. See coisogenic.

congenital. Any trait present at birth, whether the result of a genetic or nongenetic factor.
See birth defect.

Congo red. (sodium diphenyl-bis-α-naphthylamine sulfonate; CI 2120).
CAS: 573-58-0. $C_{32}H_{22}O_6N_6S_2Na_2$.
Properties: Brownish-red powder; odorless. Decomposes on exposure to acid fumes. Soluble in water and alcohol; insoluble in ether.
Derivation: Combination of tetraazotized benzidine and naphthionic acid.
Use: Dye, medicine (diagnostic aid), indicator, biological stain.

Congo resin. A variety of copal fossil resin. The natural product is insoluble in organic solvents but forms transparent gels with some alcohols and hot solvents. When thermally processed (cracked), it is soluble in all organic solvents, fatty acids, and vegetable oils. Its high acid number prevents its use in paints containing reactive pigments.
Use: High-gloss varnishes, paints for metal surfaces, wrinkle finishes.

coniferin.
$C_{16}H_{22}O_8$.
A glucoside contained in pine bark and other conifers. When decomposed, it yields coniferyl alcohol, which can be oxidized to vanillin. Used as a raw material for manufacture of synthetic vanillin.

coning oil. Usually an emulsified mineral oil used as lubricant for textile fibers in processing to the finished yarn. Fatty-acid esters are often used for the oil-in-water emulsions.

conjugated double bonds. Two or more double bonds that alternate with single bonds in an unsaturated compound, as in the formula for butadiene-1,3 ($H_2C=CH-CH=CH_2$) or maleic acid (the $O=C-C=C-C=O$ skeleton).

conjugated fatty acid. A fatty acid containing conjugated double bonds.

conjugated protein. A protein containing one or more prosthetic groups.

conjugated system. A chemical system characterized by a transmission of chemical reactivity from one atom to another.

conjugate acid–base pair. A proton donor and its corresponding deprotonated species, e.g., acetic acid (donor) and acetate (acceptor).

conjugate layers. Two layers of a liquid system, each composed of a different ternary mixture and in equilibrium with one another.

conjugate redox pair. An electron donor and its corresponding electron acceptor form, e.g., NADH (donor) and NAD+ (acceptor).

connate water. The watery substance entrapped in the interstices of a rock at the time of deposition. It is expressed as a percentage of the pore volume occupied by such water.

connine. (2-propylpiperidine; cicutine; conicine).
A liquid alkaloid and derivative of pyridine, it is the chief toxic agent of poison hemlock, *conium maculatum* (family apiaceae).
Hazard: Toxic; weakness, nervousness, drowsiness, nausea, vomiting, trembling, dyspnea, cardiac arrhythmia, bradycardia, decrease of body temperature, fatal paralysis, death due to asphyxia, affects to skeletal muscle tone, narcosis.

Conrad–Limpach reaction. Thermal condensation of arylamines with β-ketoesters followed by cyclization of the intermediate Schiff bases to 4-hydroxyquinolines.

consensus sequence. A nominal sequence inferred from multiple, imperfect examples. Multiple lanes of shotgun sequence can be merged to show a consensus sequence. The optimal sequence of nucleotides recognized by some factor. A DNA binding site for a protein may vary substantially, but one can infer the consensus sequence for the binding site by comparing numerous examples.

conservation of energy law. (First Law of Thermodynamics).
See energy.

conserved sequence. A base sequence in a DNA molecule (or an amino acid sequence in a protein) that has remained essentially unchanged throughout evolution.

conservative substitution. Replacement of an amino acid residue in a polypeptide by another residue with similar properties. For example, substitution of Lys by Arg.

consistency. Resistance to flow of a material, usually a liquid. For Newtonian liquids, consistency and viscosity are synonymous. For non-Newtonian liquids, it qualitatively represents plastic flow. The term is used by food technologists.
See body (1); viscosity.

consolute. A liquid, added to a mixture of two slightly miscible liquids, which is soluble in each of these liquids.

consolute temperature. The critical solution temperature.

constantan. Generic name for an alloy containing from 40 to 45% nickel and from 55 to 60% copper. Used in thermocouples and specialized heat-measuring devices.

constant-boiling mixture. See azeotropic mixture.

constant composition law. See chemical laws (2).

constituent. Any of the elements or subgroups in a molecule of a compound. For example, nitrogen is a constituent of proteins; the carboxyl group is a constituent of fatty acids.
See component.

constitutional diagram. See equilibrium diagram.

constitutive ablation. Gene expression that results in cell death.

constitutive enzymes. Enzymes required at all times by a cell and present at some constant level, e.g., many enzymes of the central metabolic pathways. Sometimes called "housekeeping" enzymes.

contact acid. Sulfuric acid made by the contact process.

contact herbicide. One that damages or kills a plant on contact.

contact insecticide. A poison that kills insects or other arthropods on contact with the outer surface of the body. The chemical enters the body either by penetrating the surface or by entering the tracheal system.

Use: Insecticide to control phytophagous insects that have suck mouth parts and are not killed by ingestion of insecticides applied to plant surfaces.

contact pesticide.   A chemical that kills pests through contact with the body surface.

contact potential.   The difference between the work functions of two metals.

contact process.   A process for the manufacture of sulfuric acid and oleum in which the sulfur dioxide from combustion of sulfur, pyrites, or other sulfur sources is oxidized with air to sulfur trioxide by contact with a vanadium pentoxide catalyst. Most of the sulfuric acid produced in the U.S. is made by this process.
See sulfuric acid.

contact resin.   (impression resin; low-pressure resin).
A synthetic thermosetting resin characterized by cure at relatively low pressure. The usual components are an unsaturated, high-molecular-weight monomer, such as an allyl ester, or a mixture of styrene or other vinyl monomer with an unsaturated polyester or alkyd. Cure requires heat and a catalyst, as well as some pressure. The curing does not result in water formation as with phenol-formaldehyde resins.

container.   This term refers not only to units used to pack chemical products for shipment but also to those used to transport them. Factors that dictate their selection include size of shipment, compatibility of the product with container material, ease of storage and handling, and cost. Common containers are:
tank cars, tank trucks, hopper cars (bulk chemicals)
barges (petroleum products, petrochemicals, other chemicals)
tankers (crude oil, refined products)
pipelines (gases, liquid chemicals, etc.)
55-gal drums of steel, plastic, or fiberboard, with or without polyethylene lining.
5-, 15-, 30-gal metal cans and pails
5-gal plastic carboys
1-gal metal cans
steel cylinders (compressed gases)
multiwall paper bags (dry powders)
glass and plastic bottles, vials, etc.
wooden kegs, barrels, boxes, bales; burlap bags (bulk imports, such as gums, etc.)
76-lb flasks (mercury)

contaminant.   Any substance, unintentionally introduced into air, water, or food products, that has the effect of rendering them toxic or otherwise harmful. Examples are sulfur dioxide resulting from combustion of high-sulfur fuels; pesticide residues in vegetables, fish, or other food products; industrial dusts; and radioactive materials resulting from nuclear explosions.
See fallout; decontamination.

Note: Pesticide residues in foods are often referred to as unintentional additives though they are actually contaminants.

contig.   A group of copied pieces of DNA representing overlapping regions of a particular chromosome.

contig map.   A map depicting the relative order of a linked library of overlapping clones representing a complete chromosomal segment.

Continental.   A series of channel blacks used in natural and synthetic rubber, paints, inks, and plastics.

Continex.   Furnace blacks. Used in rubber, plastics, paints, paper.

continuous distillation.   Distillation in which a feed, usually of nearly constant composition, is supplied continuously to a fractionating column, and the product is continuously withdrawn at the top, the bottom, and sometimes at the intermediate points.

continuous phase.   See phase (2).

contractile vacuole.   In many protists, a specialized vacuole with associated channels designed to collect excess water in the cell. Microtubules periodically contract to force this excess water out of the cell, regulating the cell's osmotic balance.

control.   (1) In any chemical or other scientific experiment, the reference base with which the results are compared. This base is invariably a sample of identical constitution and prepared under the same conditions from which all experimental variations are omitted. Thus, the control represents known values as far as any specific experiment is concerned. Such a sample is often called a "blank." The use of a control is vital to significant interpretation of experimental data.
(2) Automatic.
See automatic control.

Control GP.   Corrosion inhibitor and stabilizer for glycol natural-gas dehydrators.

controlled atmosphere.   See atmosphere, controlled.

controlled release.   Descriptive of a compound manufactured in such a way that its effect will be kept uniform over an extended time period; this not only provides more effective control, but also reduces the waste involved in using unnecessarily high concentrations. This principle has been applied successfully to fertilizers, pesticides, pharmaceuticals, flavors, and fragrances. It may involve (1) encapsulation of the agent; (2) incorporating it into a neutral matrix such as a rubber or plastic; (3)

coating the particles with sulfur (fertilizers); and (4) absorbing the agent into substrates of various types.

controlled substance. A material whose purchase or use is legally controlled. These are usually drugs of abuse.

convallatoxin. (convallaotoxin; covallaton; convallatoxoside; convallotoxin; corglycon; corglycone; corglykon; korglykon; manopyrantoside, strophanthidin-3-6-deoxy-,α-l-; rhamnoside, strophanthidin-3,α-l-; strophanthidin-3-(6-deoxy-α-l-mannopyranoside); strophanthidin-α-l-rhamnoside; 3β-[(6-deoxy-α-l-mannopyranosyl)oxy]-5,14-dihydroxy-19-oxo-5β-card-20(22)-enolide; strophanthidin α-l-rhamnoside; (3S,5S,8R,9S,10S, 13R, 14S)-5,14-dihydroxy-13-methyl-17-(5-oxo-2H-furan-3-yl)-3-[(3R,4R,5R,6S)-3,4,5-trihydroxy-6-methyloxan-2-yl]oxy-2,3,4,6,7,8,9,11,12,15,16,17-dodecahydro-1H-cyclopenta[a]phenanthrene-10-carbaldehyde). $C_{29}H_{42}O_{10}$.
A phytotoxin from the flowers of *Adonis vernalis*, convallaria majalis, ornthogalum umbellatum and antiaris toxicaria. The aglycone is convallatoxigenin and the sugar is a rhamnose.
Properties: Crystals or prisms from methanol or ether; soluble in acetone; slightly soluble in chloroform, ethyl acetate and water; insoluble in ether.
Hazard: Extremely potent cardiotoxin.
Use: Therapeutically as a cardiotoxic.

convection. The transfer of heat from one place to another by a moving gas or liquid. Natural convection results from differences in density caused by temperature differences. Warm air is less dense than cool air; the warm air rises relative to the cool air, and the cool air sinks. Forced convection involves motion caused by pumps, blowers, or other mechanical devices.

convergence. Similarities that have arisen independently in two or more organisms that are not closely related.
See homologies.

conversion ratio. In nucleonics, the ratio of the number of atoms of fissionable fuel generated to the number of atoms of fissionable fuel consumed in a reactor.

converting. A term used specifically in the paper industry to refer to (1) modification of raw paper by coating, impregnating, laminating, and corrugating (wet converting); and (2) fabrication of a multitude of finished products such as bags, cartons, packaging materials, napkins, facial tissues, cups, plates, and the like (dry converting). Specialized equipment is required for these operations, e.g., roll coaters, extrusion coaters, embossers, crepers, carton and bag-making machines, etc.

conveyor. Any device for continuous transport of raw materials, assembly units, or finished products. The more important types are as follows, all except (4), (6), and (7) being mechanically driven.
(1) Chain
(a) Plain chain: four or more parallel chains whose tops are a little higher than the tracks on which they move (used for flat objects, cartons, metal bars, etc.).
(b) In floor: two parallel chains operating in metal channels about 16 inches apart and set into the floor (used for conveying cartons of finished goods to storage or loading dock).
(c) Enclosed chain: steel sprockets or drags, wholly enclosed; horizontal, vertical, or inclined movements are possible (used for powders, grain, vegetable products).
(2) Belt
(a) Light: flat, rubber/fabric structure for carrying units through a packaging or assembly line; also for feeding rubber or plastic mixes to extruder.
(b) Heavy: rubber/fabric structure with thick cover, for long-distance conveying of bulk solids (coal, coke, gravel, ores).
(3) Screw: an auger rotating in a metal trough or channel; used for feeding particulates to mixing equipment, handling light solids (wood chips, sawdust, etc.).
(4) Gravity roller: cylindrical metal rollers attached to metal track or frame; used for conveying cartons or boxes where plant layout permits uniform gravity movement; the angle of incline is critical.
(5) Bucket: metal scoops mounted on belts or steel frames (for vertical carriage of gravel, crushed stone, ores, etc.).
(6) Pneumatic: metal tubes of varying diameters operated by positive or negative air pressure; used for light particulates (unloading bins and hopper cars and intraplant transport).
(7) Pipeline: plastic or welded steel tubes for long-distance movement of crude oil, refined products, ammonia, natural gas, water, steam. Positive pressure required.

cooking. (1) Conversion of a foodstuff from the raw state to a palatable and more readily digestible condition by application of heat, as in boiling, frying, roasting, or baking. One or more chemical and physical changes occur: (a) hydrolysis of collagen in the connective tissue of meats; (b) softening and partial hydrolysis of cellulose and starches to sugars in fruits and vegetables; (c) denaturation and coagulation of proteins in meat, resulting in a less tightly ordered structure and improved texture; (d) coagulation of egg albumin and wheat gluten (bakery products); (e) modification or "shortening" of wheat flour by added fats that coat the particles to form a laminar structure, thus preventing formation of chains of gluten; (f) evolution of carbon dioxide by the action of leavening agents (yeast, baking powder), which cause *rising* of gas bubbles that are retained by the gluten to form a stable, solid foam;

(g) browning of meats and bakery products due to reaction of sugars and amino acids (nonenzymatic); (h) deactivation of enzymes and destruction of bacteria; and (i) loss or deactivation of water-soluble vitamins in meat and vegetables.

(2) In the paper industry, digestion of wood pulp with such compounds as sodium sulfate, sodium hydroxide, and sodium sulfide to separate the lignin content from the cellulose.

See pulp, paper.

coolant. (heat-transfer medium; thermofor).
Any liquid or gas having the property of absorbing heat from its environment and transferring it effectively away from its source. Coolants are used in all types of automobiles, as well as in chemical processing and nuclear engineering equipment. One of the most effective and cheapest coolants is water, which is almost universally used in automotive and ordinary reaction equipment. Air is also used. Where intense heating requires a more efficient medium, special coolants are used: liquid sodium in nuclear reactors, liquid hydrogen in high-thrust nuclear rocket engines; carbon dioxide, propylene glycol, and "Dowtherm" in chemical-processing reactors. Methoxy propanol has been introduced for diesel engines. Some coolants provide antifreeze protection.
See antifreeze.

Coolidge, William D. (1873–1975). An American physical chemist born in Massachusetts. He received a degree in electrical engineering at M.I.T. (1896) and a doctorate in physics at Leipzig (1899). In 1905, he joined the General Electric Research Laboratory, which had been established 5 years earlier. Here he invented the ductile tungsten filament and developed the use of tungsten in electrical switches and medical X-ray tubes. He did pioneer research in experimental metallurgy and powder metallurgy. He also had a prominent part in evaluating uranium research (1941) and in setting up the Manhattan Project. He was the recipient of many honors and awards, including induction into the National Inventors Hall of Fame in 1975.

coomassie blue. (4-[4-anilino)-5-sulfonaphthalen-1-yl]diazenyl-5-hydroxynaphthalene-2,7-disulfonic acid).
CAS: 3861-73-2. $C_{26}H_{19}N_3NaO_{10}S_3$
Blue protein stain.
Use: To enhance bloody friction ridge detail.

cooperage. The manufacture of barrels, formerly of wood but now including drums made of various materials such as resin-bonded fiber, metal, and plastic.

coordinate bond. (dative bond).
A covalent bond consisting of a pair of electrons donated by only one of the two atoms it joins.

coordination compound. (complex compound).
A compound formed by the union of a metal ion (usually a transition metal) with a nonmetallic ion or molecule called a ligand or complexing agent. The ligand may be either positively or negatively charged (such ions as Cl^- or $NH_2NH_3^+$), or it may be a molecule of water or ammonia. The most common metal ions are those of cobalt, platinum, iron, copper, and nickel, which form highly stable compounds. When ammonia is the ligand, the compounds are called ammines. The total number of bonds linking the metal to the ligand is called its coordination number. It is usually 2, 4, or 6 and often depends on the type of ligand involved. All ligands have electron pairs on the coordinating atom (e.g., nitrogen) that can be either donated to or shared with the metal ions. The metal ion acts as a Lewis acid (electron acceptor) and the ligand as a Lewis base (electron donor). The bonding is neither covalent nor electrostatic but may be considered intermediate between the two types. The charge on the complex ion is the sum of the charges on the metal ion and the ligands; for example, $4NH_3 + 2Cl^- + Co^{3+}$ forms the complex $[Co(NH_3)_4Cl_2]^+$. The brackets enclose the metal ion and the coordinated ligands.
See chelate; sequestration; metallocene.

coordination isomerism. Isomerism due to two central atoms, one in the positive complex and one in the negative complex.

coordination number. (CN).
The number of points at which ligands are attached to the metal ion in a complex. Common coordination numbers are 2, 4, and 6, exemplified by the ions $[Ag(NH_3)_2]^+$, $[Ni(CN)_4]^{2-}$, and $[PdCl_6]^{2-}$.

copaiba oil. A levorotatory, essential oil obtained from a tree native to Brazil; a component of copaiba resin (a type of balsam). The sap of this tree is a turpentine-like liquid that can be used as an automotive fuel; intensive cultivation of copaiba trees for this purpose is under experimentation in Brazil.

copaiba resin. See balsam (3); copaiba oil.

copaibic acid.
$C_{20}H_{30}O_2$.
A monobasic acid derived from copaiba balsam.

copal. A group of fossil resins still used to some extent in varnishes and lacquers. Insoluble in oils and water. The most important types are Congo, kauri, and manila.

Cope elimination reaction. Formation of an olefin and a hydroxylamine by pyrolysis of an amine oxide.

Cope rearrangement. Thermal isomerization of 1,5-dienes by 1 3,3 shift.

"Copherol" *[Cognis].*
CAS: 7695-91-2.
TM for natural tocopheryl acetate source of vitamin E.
Use: Antioxidant and moisturizer for topical skincare and sunscreen products.

copigment. A material that forms an unstable addition compound with the anthocyanin pigment of flowers, such as tannin.

copolymer. An elastomer produced by the simultaneous polymerization of two or more dissimilar monomers, as SBR synthetic rubber from styrene and butadiene.

copper.
CAS: 7440-50-8.   Cu.
Metallic element of atomic number 29, group IB of the periodic table, aw 63.546, valences 1, 2; two stable isotopes.
Properties: Distinctive reddish color. D 8.96, mp 1083C, bp 2595C. Ductile, excellent conductor of electricity. Complexing agent, coordination numbers 2 and 4. Dissolves readily in nitric and hot concentrated sulfuric acids; in hydrochloric and dilute sulfuric acids slowly, but only when exposed to the atmosphere. More resistant to atmospheric corrosion than iron, forming a green layer of hydrated basic carbonate. Readily attacked by alkalies. A necessary trace element in human diet, and a factor in plant metabolism. Essentially nontoxic in elemental form. Noncombustible, except as powder.
Source: Azurite, azurmalachite, chalcocite, chalcopyrite (copper pyrites), covellite, cuprite, malachite from Michigan, Arizona, Utah, Montana, New Mexico, Nevada, Tennessee, Chile, Canada, the former U.S.S.R., Zambia, Zaire.
Derivation: Varies with the type of ore. With sulfide ores, the steps may be (1) concentration (of low-grade ores) by flotation and leaching, (2) roasting, (3) formation of copper *matte* (40–50% Cu), (4) reduction of matte to *blister* copper (96–98%), and (5) electrolytic refining to 99.9+% copper.
Available forms: Ingot, sheet, rod, wire, tubing, shot, powder, high purity (impurities <10 ppm) as single crystals or whiskers.
Hazard: Flammable in finely divided form. Gastrointestinal irritant and metal fume fever.
Use: Electric wiring; switches; plumbing; heating; roofing and building construction; chemical and pharmaceutical machinery; alloys (brass, bronze, Monel metal, beryllium-copper); electroplated protective coatings and undercoats for nickel, chromium, zinc, etc., cooking utensils; corrosion-resistant piping; insecticides; catalyst; antifouling paints. Flakes used as insulation for liquid fuels. Whiskers used in thermal and electrical composites.

copper abietate. (cupric abietate).
CAS: 10248-55-2. $Cu(C_{20}H_{29}O_2)_2$.

Properties: Green scales. Soluble in alcohol and in oils (producing a fine-green color); insoluble in water.
Derivation: Heating copper hydroxide with abietic acid.
Use: Preservative metal paint, fungicide.

copper acetate. (cupric acetate; crystals of Venus; verdigris, crystallized).
CAS: 142-71-2. $Cu(C_2H_3O_2)_2 \cdot H_2O$.
Properties: Greenish-blue, fine powder. D 1.9, mp 115C, decomposes at 240C. Soluble in water, alcohol, and ether.
Derivation: Action of acetic acid on copper oxide and subsequent crystallization.
Use: Pesticide, catalyst, fungicide, pigments, manufacture of Paris green.

copper acetate, basic. (copper subacetate; verdigris; verdigris blue; verdigris green).
Properties: Masses of minute, silky crystals, either pale green or bright blue in color. Blue variety approximate formula $(C_2H_3O_2)_2Cu_2O$. Green variety approximate formula $CuO \cdot 2Cu(C_2H_3O_2)_2$. Coppery taste. The green rust with which uncleaned copper vessels become coated and which is commonly termed verdigris is a copper carbonate and must not be confused with true *verdigris*. Apart from its impurities, verdigris is a variable mixture of the basic copper acetates. Soluble in acids; very slightly soluble in water and alcohol.
Derivation: Action of acetic acid on copper in the presence of air.
Use: Paint pigment, insecticide, fungicide, mildew preventive, mordant in dyeing and printing.

copper acetoarsenite. (cupric acetoarsenite; Paris green; king's green; Schweinfurt green; imperial green).
CAS: 12002-03-8. $(CuO)_3As_2O_3 \cdot Cu(Cu_2H_3O_2)_2$.
Properties: Emerald-green powder. Soluble in acids; insoluble in alcohol and water. Phytotoxic.
Derivation: By reacting sodium arsenite with copper sulfate and acetic acid.
Hazard: Toxic by ingestion.
Use: Wood preservative, larvicide, marine antifouling paints.

copper acetylacetonate. (copper-2,4-pentanedione).
CAS: 13395-16-9. $Cu(C_5H_7O_2)_2$.
Crystalline powder, slightly soluble in water and alcohol, soluble in chloroform, mp >230C. Resistant to hydrolysis. A chelating nonionizing compound.

copper amalgam.
Properties: Hard, brown leaflets. Contains approximately 74% mercury and approximately 24% copper. Soluble in nitric acid.
Use: Dental cement.

copper aminoacetate. See copper glycinate.

copper aminosulfate.   See copper sulfate, ammoniated.

copper ammonium acetate.   (cuprous acetate, ammoniacal).
Use: In an absorption process for separating butadiene from C_4 streams at refineries or from by-products of ethylene production.

copper arsenate.   Probably the basic cupric arsenate.
Properties: Light-blue, blue, or bluish-green powder. Variable composition. Contains approximately 33% copper and approximately 29% arsenic. Soluble in dilute acids, ammonium hydroxide; insoluble in alcohol, water.
Use: Insecticide, fungicide.

copper arsenite.   (cupric arsenite; copper orthoarsenite; Scheele's green).
CAS: 10290-12-7. $CuHAsO_3$ or $Cu_3(AsO_3)_2 \cdot 3H_2O$; variable.
Properties: Fine, light-green powder. Mp (decomposes). Soluble in acids; insoluble in water and alcohol.
Use: Insecticide, fungicide, pigment, wood preservative, rodenticide.

copper arsenite, ammoniacal.   See "Chemonite" [Arch].

copperas.   See ferrous sulfate.

copperas, blue.   See copper sulfate.

copperas, green.   See ferrous sulfate.

copperas, white.   See zinc sulfate.

copper benzoate.
CAS: 533-01-7. $(C_6H_5COO)_2Cu \cdot 2H_2O$.
Properties: Blue, crystalline powder; odorless. Mp (loses water) 110C. Slightly soluble in cold water, acids, and alcohol.
Derivation: Interaction of solutions of a benzoate and copper salt.

copper-beryllium.   See beryllium-copper.

copper bis(dibutyldithiocarbamate).
CAS: 13927-71-4. $C_{18}H_{36}N_2S_4 \cdot Cu$.
Hazard: Moderately toxic by ingestion.
Use: Agricultural chemical.

copper, bis(1-hydroxy-2(1h)-pyridinethionato)-.   See bis(2-pyridinethiol 1-oxide)copper.

copper, blister.   See blinding copper.

copper blue.   See mountain blue.

copper borate.   See copper metaborate.

copper bromide.   (cupric bromide).
$CuBr_2$.
Properties: Black powder or crystals, deliquescent. Mp 498C; d 4.77 (25C). Soluble in acetone, alcohol, water.
Use: Photography (intensifier), organic synthesis (brominating agent), battery electrolyte, wood preservative.

copper carbonate.   (cupric carbonate; copper carbonate, basic; artificial malachite; mineral green. For the native mineral see malachite).
CAS: 12069-69-1. $Cu_2(OH)_2CO_3$.
Properties: Green powder. D 3.7–4.0, decomposes at 200C. Soluble in acids; insoluble in water.
Derivation: By adding sodium carbonate to a solution of copper sulfate, filtering, and drying.
Grade: Technical, CP.
Hazard: Toxic by ingestion.
Use: Pigments, pyrotechnics, insecticides, copper salts, coloring brass black, astringent in pomade preparations, antidote for phosphorus poisoning, smut preventive, fungicide for seed treatment, feed additive (in small amounts).

copper chloride.   (cupric chloride).
CAS: 1344-67-8.   (1) $CuCl_2$, (2) $CuCl_2 \cdot 2H_2O$.
Properties: (1) Brownish-yellow powder, hygroscopic, (2) green, deliquescent crystals, soluble in water and alcohol. D (1) 3.386 (25C), (2) 2.54 (25C); (1) mp 620C decomposes at 993C to cuprous chloride (2) loses $2H_2O$ at 100C.
Derivation: (1) By the union of copper and chlorine. (2) Copper carbonate is treated with hydrochloric acid and the product crystallized.
Grade: Technical, CP, reagent.
Hazard: Toxic by ingestion and inhalation.
Use: Isomerization and cracking catalyst, mordant in dyeing and printing fabrics, sympathetic ink, disinfectant, pyrotechnics, wood preservation, fungicides, metallurgy, preservation of pulpwood, deodorizing and desulfurizing petroleum distillates, photography, water purification, feed additive, electroplating baths, pigment for glass and ceramics, acrylonitrile manufacturing.
See cuprous chloride.

copper chromate.   (cupric chromate, basic).
CAS: 27787-63-9. $CuCrO_4 \cdot 2CuO \cdot 2H_2O$.
Properties: Light chocolate-brown powder. Loses water at 260C. Soluble in nitric acid; insoluble in water.
Derivation: Action of chromic acid on copper hydroxide.
Hazard: A carcinogen.
Use: Mordant in dyeing, wood preservative, seed treatment fungicides.

copper cyanide.   (cupric cyanide).
CAS: 544-92-3. $Cu(CN)_2$.
Properties: Green powder. Keep well stoppered. Soluble in acids and alkalies; insoluble in water.

Derivation: Addition of potassium cyanide to a solution of copper sulfate; cupric cyanide is precipitated. This can be dried but is not stable.
Grade: Technical.
Hazard: Poison.
Use: Electroplating copper on iron, intermediate (introduction of the cyanide group in place of the amino radical in aromatic organic compounds).
See cuprous cyanide.

copper, deoxidized.   (copper, oxygen-free).
Copper metal specially treated (as by addition of phosphorus) to remove all or a part of the 0.05% oxygen normally present. It is more ductile than ordinary copper.

copper(II) diethyldithiocarbamate.
CAS: 13681-87-3. $C_{10}H_{20}N_2S_4 \cdot Cu$.
Hazard: Moderately toxic by ingestion.
Use: Agricultural chemical.

copper dihydrazine sulfate.
CAS: 33271-65-7. $H_{10}N_4O_8S_2 \cdot Cu$.
Hazard: Moderately toxic by ingestion.
Use: Agricultural chemical.

copper dihydrazinium sulfate.
CAS: 33271-65-7. $(CuSO_4(N_2H_4)_2 \cdot H_2SO_4)$.
Properties: Bluish powder. Mp >300C, starts to decompose at 140C, very slightly soluble in water (250 ppm at 80C).
Hazard: Toxic by ingestion or inhalation. Skin and eye irritant.
Use: Foliage fungicide.

copper, electrolytic.   Copper refined by electrolysis. The purest form of copper available commercially.

copper ethylacetoacetate.
CAS: 14284-06-01. $Cu(C_6H_9O_3)_2$.
Properties: Blue-green powder. Mp 192–193C. Insoluble in water; soluble in most organic solvents.
Use: Fungicide, intermediate.

copper ferrocyanide.   (cupric ferrocyanide).
CAS: 13601-13-3. $Cu_2Fe(CN)_6 \cdot 7H_2O$.
Properties: Reddish-brown powder. Insoluble in water and acids; soluble in ammonium hydroxide and potassium cyanide solutions.
Use: Pigment in paints and enamels, analytical test for traces of copper, inorganic osmotic membranes.

copper fluoride.   (cupric fluoride).
CAS: 7789-19-7. $CuF_2 \cdot 2H_2O$.
The anhydrous form, CuF_2, is also available.
Properties: Blue crystals. D 2.93 (25C). Slightly soluble in water; soluble in acids and alcohol.
Derivation: (1) By decomposing copper carbonate with hydrofluoric acid and subsequent crystallization; (2) fluorination of copper hydroxyfluoride at 525C.

Grade: Technical.
Hazard: A poison.
Use: Ceramics, enamels, flux in metallurgy, high-energy batteries, fluorinating agent.

copper(II) fluoroacetate.
CAS: 20424-95-7. $C_4H_4F_2O_4 \cdot Cu$.
Hazard: A poison.
Use: Agricultural chemical.

copper fluosilicate.   (copper silicofluoride; cupric fluosilicate; cupric silicofluoride).
CAS: 12062-24-7. $CuSiF_6 \cdot 4H_2O$.
Properties: Blue, hygroscopic crystals. D 2.158, decomposed by heat. Soluble in water; slightly soluble in alcohol.
Derivation: Interaction of copper hydroxide and hydrofluosilicic acid.
Method of purification: Crystallization.
Grade: Technical.
Hazard: A poison.
Use: Dyeing and hardening white marble, treating grapevines for "white disease."

copper fungicide.   Any inorganic fungicide that contains copper. Copper fungicides are often based on copper oxychlorides or copper oxide; they may also be blended with other types of fungicide.

copper glance.   See chalcocite.

copper gluconate.   (cupric gluconate).
CAS: 527-09-3. $[Cu_2OH(CH_2O)_4COO]_2Cu$.
Properties: Odorless, light-blue, fine, crystalline powder. Soluble in water; insoluble in acetone, alcohol, and ether.
Grade: Pharmaceutical, FCC.
Use: Feed additive, dietary supplement, mouth deodorant.

copper glycinate.   (copper aminoacetate).
CAS: 13479-54-4. $(NH_2CH_2COO)_2Cu$.
Properties: Blue, triboluminescent crystals. Mp 130C. Slightly soluble in water and alcohol; insoluble in hydrocarbons, ethers, and ketones.
Grade: Anhydrous, hydrate.
Use: Catalyst for rapid biochemical assimilation of iron, electroplating baths, photometric analysis, feed additive.

copper hemioxide.   See copper oxide red.

copper hydrate.   See copper hydroxide.

copper hydroxide.   (cupric hydroxide; copper oxide hydrated; copper hydrate).
CAS: 20427-59-2. $Cu(OH)_2$.
Properties: Blue powder. D 3.368, mp (decomposes). Soluble in acids and ammonium hydroxide; insoluble in water.
Derivation: Interaction of a solution of a copper salt with an alkali.

Grade: Technical.
Hazard: Toxic by ingestion and inhalation.
Use: Copper salts, mordant, cuprammonium rayon, pigment, staining paper, feed additive, pesticide and fungicide, catalyst.

copper-8-hydroxyquinoline.   See   copper-8-quinolinolate.

copper indium selenide.
CAS: 12018-95-0. $CuInSe_2$.
Hazard: A reproductive hazard.

Copper Inhibitor 50.   50% disalicylalpropylenediamine and 50% aromatic solvent. Used to prevent catalytic action of copper on oxidation of natural and synthetic rubbers.
Hazard: Flammable, moderate fire risk.

copper iodide.   See cuprous iodide.

copper lactate.   (cupric lactate).
CAS: 16039-52-4. $Cu(C_3H_5O_3)_2 \cdot 2H_2O$.
Properties: Greenish-blue crystals or granular powder. Soluble in water and ammonium hydroxide.
Use: Source of copper in copper-plating fungicides.

copper mercury iodide.   See mercuric cuprous iodide.

copper metaborate.   (copper borate; cupric borate).
$Cu(BO_2)_2$.
Properties: Bluish-green, crystalline powder. D 3.859. Insoluble in water; soluble in acids.
Derivation: Interaction of copper sulfate and sodium borate.
Grade: Technical, CP.
Hazard: Toxic by ingestion.
Use: Dehydrogenation catalyst, wood preservative, fire retardant, paint and ceramic pigment, insecticides (especially wheat-rust compounds).

copper methane arsenate.
CH_3AsO_3Cu.
Properties: Greenish solid.
Derivation: Reaction of disodium methyl arsenate with copper salts.
Hazard: Toxic by ingestion.
Use: Algicide.

copper methane arsonate.
CAS: 63869-12-5. $CH_3AsO_2 \cdot Cu$.
Hazard: A poison.

copper molybdate.
CAS: 108844-49-1. $CuMoO_4$.
Properties: Crystals. D 3.4, mp approximately 500C. Insoluble in water.
Grade: 99.98% pure. Electronic and optical equipment.

Use: Paint pigment and protective coatings, corrosion inhibitor.

copper monoxide.   See copper oxide black.

copper naphthenate.
CAS: 1338-02-9.
Properties: Green–blue solid. High germicidal power. Soluble in gasoline, benzene, and mineral-oil distillates.
Derivation: Addition of solution of cupric sulfate to aqueous solution of sodium naphthenate.
Grade: 6, 8, 11.5% copper.
Hazard: Flammable, moderate fire risk (solution). Toxic by ingestion and inhalation.
Use: Wood, canvas, and rope preservative, insecticide, fungicide, antifouling paints.
See soap (2).

copper nitrate.   (cupric nitrate).
CAS: 3251-23-8.   (1) $Cu(NO_3)_2 \cdot 3H_2O$, (2) $Cu(NO_3)_2 \cdot 6H_2O$.
Properties: Blue, deliquescent crystals. Soluble in water and alcohol. D (1) 2.32 (25C), (2) 2.074; mp (1) 114.5C, (2) loses $3H_2O$ at 26.4C; (1) decomposes 170C.
Derivation: By treating copper or copper oxide with nitric acid. The solution is evaporated and product recovered by crystallization.
Grade: Technical, CP.
Hazard: Oxidizer, causes violent combustion or explosion with organic materials.
Use: Light-sensitive papers; analytical reagent; mordant in textile dyeing; nitrating agent; insecticide for vines; coloring copper black; electroplating; production of burnished effect on iron; paints; varnishes, enamels; pharmaceutical preparations; catalyst.

copper nitrite.   (copper nitrite basic; cupric nitrite).
$Cu(NO_2)_2 \cdot 3Cu(OH)_2$, variable.
Properties: Green powder. Decomposes at 120C. Soluble (with decomposition) in dilute acids, ammonium hydroxide; slightly soluble in water.

copper nitrite basic.   See copper nitrite.

copper octoate.   See soap (2).

copper oleate.   (cupric oleate).
CAS: 10402-16-1. $Cu(C_{18}H_{33}O_2)_2$.
Properties: Brown powder or greenish-blue mass. Soluble in ether; insoluble in water. Combustible.
Derivation: Interaction of copper sulfate and sodium oleate.
Use: Preserving fish nets and marine lines, fungicide, insecticide, ore flotation, lubricating oil antioxidant, emulsifying agent, fuel-oil ignition improver, catalyst.
See soap (2).

copper oxalate.   (cupric oxalate).
CAS: 814-91-5. CuC_2O_4.
  Properties: Bluish-green powder, decomposes at approximately 300C to copper oxide. Insoluble in water, alcohol, and acetic acid; soluble in ammonium hydroxide.
  Derivation: Reaction of oxalic acid with copper sulfate.
  Hazard: Toxic by ingestion; tissue irritant.
  Use: Catalyst in organic synthesis, rodent repellent in seed coatings.

copper oxide black.   (cupric oxide; copper monoxide).
CAS: 1344-70-3.   CuO.
  Properties: Brownish-black powder. D 6.32, decomposes at 1026C. Soluble in acids; difficultly soluble in water.
  Derivation: Ignition of copper carbonate or copper nitrate.
  Hazard: Toxic by ingestion.
  Use: Ceramic colorant, reagent in analytical chemistry, insecticide for potato plants, catalyst, purification of hydrogen, batteries and electrodes, aromatic acids from cresols, electroplating, solvent for chromic iron ores, desulfurizing oils, rayon, metallurgical and welding fluxes, antifouling paints, phosphors.

copper oxide hydrated.   See copper hydroxide.

copper oxide red.   (cuprous oxide; copper protoxide; copper hemioxide; copper suboxide).
CAS: 1317-39-1. Cu_2O.
For the native ore see cuprite.
  Properties: Reddish-brown, octahedral crystals. D 5.75–6.09, mp 1235C, bp 1800C. Soluble in acids and ammonium hydroxide; insoluble in water.
  Derivation: (1) Oxidation of finely divided copper. (2) Addition of bases to cuprous chloride. (3) Action of glucose on cupric hydroxide.
  Grade: Technical, CP, 97% min (for pigments), also USN Type I (97%), USN Type II (90%).
  Hazard: Toxic by ingestion.
  Use: Copper salts, ceramics, porcelain red glaze, red glass, electroplating, antifouling paints, fungicide, catalyst, brazing preparations, photocells.

copper oxinate.   See copper-8-quinolinolate.

copper oxychloride.   (cupric oxychloride).
Composition variable, possibly $3CuO·CuCl_2·3.5H_2O$.
  Properties: Bluish-green powder. Soluble in acids, ammonia; insoluble in water.
  Hazard: Toxic by ingestion and inhalation.
  Use: Pigment, pesticide, fungus control in grapevines.

copper-2,4-pentanedione.   See copper acetylacetonate.

copper phenolsulfonate.   (copper sulfocarbolate).
$[C_6H_4(OH)SO_3]_2Cu·6H_2O$.
  Properties: Green, prismatic crystals. Soluble in water and alcohol. Combustible.
  Derivation: Interaction of barium phenolsulfonate and copper sulfate.
  Use: Esterification catalyst, electroplating.

copper phosphate.   (copper orthophosphate).
CAS: 7798-23-4. $Cu_3(PO_4)_2·3H_2O$.
  Properties: Light-blue powder. Soluble in acids, ammonium hydroxide; insoluble in water.
  Use: Analysis, fungicide, catalyst, oxidation inhibitor for metals.

copper phosphide.
CAS: 12019-57-7. Cu_3P_2.
  Properties: Grayish-black, metallic powder. D 6.67. Insoluble in water and hydrochloric acid; soluble in nitric acid; insoluble in hydrochloric acid.
  Derivation: By heating copper and phosphorus.
  Hazard: Dangerous, spontaneously flammable, and toxic phosphine evolved on reaction with water. May explode when mixed with potassium nitrate.
  Use: Manufacturing phosphor-bronze.

copper phthalate.
$C_8H_4O_4Cu$.
  Properties: Fine, blue powder; assay min 95%; very slightly soluble in common organic solvents or water. Combustible.
  Hazard: Toxic by ingestion.
  Use: Fungicide.

copper phthalocyanine blue.   See Pigment Blue 15.

copper phthalocyanine green.   See Pigment Green 7.

copper potassium ferrocyanide.   (potassium copper ferrocyanide).
$K_2CuFe(CN)_6·H_2O$.
  Properties: Brownish-red powder. Insoluble in water.
  Use: Pigment.

copper protoxide.   See copper oxide red.

copper pyrites.   See chalcopyrite.

copper pyrophosphate.   See "Unichrome" [ATOTECH].

copper-8-quinolinolate.   (copper-8; copper oxinate; copper-8-hydroxyquinoline).
CAS: 10380-28-4. $Cu(C_9H_6ON)_2$.

Properties: Yellow-green, nonhygroscopic, odorless powder. Insoluble in water; somewhat soluble in weak acids; soluble in strong acids. Insoluble in most organic solvents but somewhat soluble in pyridine and quinoline. Soluble copper-8 refers to the product formed by heating copper-8-quinolinolate with certain organic acids (naphthenic, lactic, stearic, etc.) or their salts. In such products, the copper-8-quinolinolate does not settle out on standing, even after dilution with various solvents.
Derivation: From 8-quinolinol and copper salt such as copper acetate.
Grade: 10% active salt (1.8% Cu) solution.
Hazard: Toxic by ingestion. Questionable carcinogen.
Use: Fungicide and mildew-proofing of fabrics, analysis for copper.

copper resinate.   (cupric resinate).
Properties: Green powder. Soluble in ether and oils; insoluble in water. Combustible.
Derivation: By heating copper sulfate and rosin oil and filtering and drying the precipitate.
Use: Antifouling paints, insecticide.
See soap (2).

copper ricinoleate.
$Cu(C_{17}H_{32}OHCOO)_2$.
Properties: Green plastic solid. Soluble in water, aliphatic hydrocarbons, ketones, and aromatic hydrocarbons; partially soluble in alcohols and glycols; soluble in Softening p 64C.
Hazard: Toxic by ingestion. Combustible.
Use: Fungicides, insecticides.

copper scale.   A coating formed on copper after heating, composed of cupric and cuprous oxides.

copper selenate.   (cupric selenate).
CAS: 15123-69-0. $CuSeO_4 \cdot 5H_2O$.
Properties: Light-blue crystals. D 2.559; loses $4H_2O$ at 50–100C. Soluble in acids, ammonium hydroxide, water; insoluble in alcohol.
Use: Black colorant for copper.

copper silicate.   A complex mixture precipitated by solutions of copper salts from sodium silicate solutions. Used in pigments, catalysts, and insecticides.

copper silicide.   See silicon-copper.

copper silicofluoride.   See copper fluosilicate.

copper sodium chloride.   (sodium copper chloride).
$CuCl_2 \cdot 2NaCl \cdot 2H_2O$.
Properties: Light-green crystals. Soluble in water.

copper sodium cyanide.   See sodium copper cyanide.

copper stearate.   (cupric stearate).
CAS: 660-60-6. $Cu(C_{18}H_{35}O_2)_2$.
Properties: Light-blue powder. Soluble in ether, chloroform, benzene, turpentine, hot benzene, toluene, carbon tetrachloride; insoluble in water and alcohol. Mp 125C. Combustible.
Derivation: By the interaction of copper sulfate and sodium stearate.
Use: Preservative for cellulosic materials, antifouling paints, catalyst.

copper subacetate.   See copper acetate, basic.

copper suboxide.   See copper oxide, red.

copper sulfate.   (cupric sulfate; blue vitriol; blue stone; blue copperas).
$CuSO_4 \cdot 5H_2O$.
Properties: Blue crystals or blue, crystalline granules or powder; slowly efflorescing in air; white when dehydrated; nauseous metallic taste. Soluble in water, methanol; slightly soluble in alcohol and glycerol; d 2.284.
Derivation: Action of dilute sulfuric acid on copper or copper oxide (often as oxide ores) in large quantities with evaporation and crystallization.
Method of purification: Recrystallization.
Grade: Technical, CP, NF, also sold as monohydrate. Available as crystals or powder.
Hazard: Toxic by ingestion, strong irritant.
Use: Agriculture (soil additive, pesticides, Bordeaux mixture), feed additive, germicides, textile mordant, leather industry, pigments, electric batteries, electroplated coatings, copper salts, reagent in analytical chemistry, medicine, wood preservative, preservation of pulp wood and ground pulp, process engraving and lithography, ore flotation, petroleum industry, synthetic rubber, steel manufacture, treatment of natural asphalts. The anhydrous salt is used as a dehydrating agent.

copper sulfate, ammoniated.   (cupric ammonia sulfate; ammonio-cupric sulfate; copper aminosulfate).
$Cu(NH_3)_4SO_4 \cdot H_2O$.
Properties: Dark-blue, crystalline powder; decomposes in air; soluble in water; insoluble in alcohol.
Derivation: By dissolving copper sulfate in ammonium hydroxide and precipitating with alcohol.
Hazard: Toxic by ingestion.
Use: Calico printing, manufacturing copper arsenate, insecticide, treating fiber products.

copper sulfate, tribasic.
CAS: 7758-98-7. $CuSO_4 \cdot 3Cu(OH)_2 \cdot H_2O$.
Properties: Aqua-colored powder of extremely fine particle size. Water soluble. Stable in storage, forms essentially neutral water dispersion.
Hazard: Toxic by ingestion.
Use: A fixed copper fungicide. Also micronutrient for plants. Compatible with arsenicals, organic

insecticides, sulfur, and cryolite. Used as spray or dust. Does not inhibit photosynthesis.

copper sulfide.   (cupric sulfide).
CAS: 1317-40-4. CuS.
Properties: Black powder or lumps. D 3.9–4.6, decomposes at 220C. Soluble in nitric acid; insoluble in water. Occurs as the mineral covellite.
Derivation: By passing hydrogen sulfide gas into a solution of a copper salt.
Hazard: Toxic by ingestion.
Use: Antifouling paints, dyeing with aniline black, catalyst preparation.
See cuprous sulfide.

copper sulfocarbolate.   See copper phenolsulfonate.

copper sulfocyanide.   See cuprous thiocyanate.

copper tallate.   See soap (2).

copper trifluoroacetylacetonate.
Cu[OC(CH$_3$):CHCO(CF$_3$)]$_2$.
Properties: Solid. Mp 188–190C.
Use: Metal analysis standards, vapor-phase deposition of metals, laser studies.

copper tungstate.   (cupric tungstate).
CAS: 13587-35-4. CuWO$_4$·2H$_2$O.
Properties: Light-green powder. Soluble in ammonium hydroxide; slightly soluble in acetic acid; insoluble in alcohol and water.
Use: Polyester catalyst, semiconductors, nuclear reactors.

copper yellow.   See chalcopyrite.

copper zinc chromate.   Variable in composition. Used as a fungicide.
Hazard: Toxic by ingestion and inhalation. A carcinogen.

Coppralyte.   A group of products for electroplating copper.

Coprantine.   Dyes requiring after-treatment with copper compounds.

copra oil.   The name applied to lower grades of coconut oil.

coprine.   (1-cyclopropanol-1-n^5-glutamine; n^5-(1-hydroxycyclopropyl)-l-glutamine, (2S)-2-amino-5-[(1-hydroxycyclopropyl)amino]-5-oxopentanoic acid).
C$_8$H$_{14}$N$_2$O$_4$.
A mycotoxin of the mushrooms *Coprinus atramentarius*, *Clitocybe clavipes*, and a few other congeners. It is a water-soluble Y-glutamyl conjugate of 1-aminocyclopropanol.
Hazard: Toxic.

coprinus   atramentarius.   (alcohol   inky; smooth ink cap mushroom).
An edible mushroom (Family *coprinaceae*) that occurs in grass, woody rubble, and on soil near buried wood throughout much of North America. It is not toxic in the usual sense but can be if taken with alcohol.

coptisine.
CAS: 3486-66-6. C$_{19}$H$_{14}$NO$_4$.
Hazard: A poison.

coral.   Skeletons of the coral polyps found in the warmer oceans and consisting mainly of calcium carbonate colored with ferric oxide. Coral rock is porous. It occurs in the form of reefs east of Australia and at other locations in the southwest Pacific. It has been found to be the habitat of organisms that are rich in prostaglandins.
Hazard: Will infect open wounds.
Use: Building stone, cement, road and airfield construction, jewelry.

Coral.   A stereospecific, polyisoprene rubber consisting essentially of *cis*-1,4-polyisoprene.
Properties: Similar to those of natural rubber, both unvulcanized and vulcanized.
Use: Replacement for natural rubber.

cordial.   See liqueur.

cordite.   A smokeless powder that is a mixture of nitrocellulose and nitroglycerin, with approximately 5% petrolatum added to thicken and stabilize the mixture. Materials are dissolved in acetone and mixed. Evaporation of the excess acetone leaves a gelatinous mass, which is extruded into cords.

"Cordobond" [Ferro].   TM for epoxy repair systems, liquid coating, and dispersions.
Use: For industrial, marine applications giving strength and abrasion resistance.

core of the earth.   The central part of the earth below a depth of 2900 km. It is thought to be composed largely of iron and to be molten on the outside with a solid central region.

Corey, Elias James.   (1928–). An American who won the Nobel Prize for chemistry in 1990 for development of novel methods for the synthesis of complex natural compounds (retrosynthetic analysis). He specialized in synthesis of terpines. Awarded a doctorate by M.I.T. in 1951.

Corey-Winter olefin synthesis.   Synthesis of olefins from 1,2-diols and thiocarbonyldiimidazole. Treatment of the intermediate cyclic thionocarbonate with trimethylphosphite yields the olefin by *cis* elimination.

"CoRezyn" *[Interplastic].*　TM for a variety of polymer resins tailored for specific uses and for materials used with the polymers in making finished products.

"Corgard" *[King].*　TM for a borosilicate glass armored with an opaque, filament-wound laminate of glass fiber and a modified polyester resin.

coriander oil.　An essential oil with strong, spicy taste; may be irritating when ingested. Dextrorotatory. Used chiefly to flavor gin.

coriandrol.　See linalool.

Coriolis effect.　The tendency of a current of air or water flowing over the surface of the earth to bend to the right in the northern hemisphere and to the left in the southern hemisphere.

cork.　A form of cellulose constituting the light outer bark of the oak tree known as *Quercus suber.* It grows naturally in southern Europe and northern Africa and has been cultivated in the southwestern U.S. Its special properties are extreme lightness, relative imperviousness to water, resilient structure, and low rate of heat transfer. These account for its usefulness as bottle stoppers, insulation, wallboard, life preservers, gaskets, and sound-deadening insertions. D 0.1–0.25. Combustible.

corkboard.　A mixture of ground cork and paper pulp formed into thick sheets for insulating purposes.

(+)-corlumine.
CAS: 485-51-8.　$C_{21}H_{21}NO_6$.
Hazard: A poison.
Use: Agricultural chemical.

corn endosperm oil.
Properties: Reddish-brown liquid.
Use: Food additive.

Cornforth, Sir John Warcup "Kappa" Jr. (1917–2013).
An Australian-born chemist who won the Nobel Prize for chemistry in 1975 with V. Prelog for work on the chemical synthesis of organic compounds. Although deaf since childhood, he attained his doctorate from Oxford and held prestigious posts all over the world, as well as authoring many papers on organic and biochemical subjects.

corn oil.　(maize oil).
Properties: Pale yellow liquid; characteristic taste and odor. D 0.914–0.921, saponification number 188-193, iodine number 102-128, flash p 490F (254C). Insoluble in water; soluble in ether, chloroform, amyl acetate, benzene, and carbon disulfide; slightly soluble in alcohol. Combustible. Nontoxic and nondrying. Moderate tendency to spontaneous heating.
Chief constituents: Linoleic and oleic acids (unsaturated), palmitic and stearic (saturated).
Derivation: The germ of common corn (Indian corn, *Zea mays*) is removed from the grain and pressed.
Grade: Crude, refined, USP, technical.
Use: Foodstuffs, soap, lubricants, leather dressing, factice, margarine, salad oil, hair dressing, solvent.

corn silk and corn silk extract.
Properties: Extracted from the fresh styles and stigmas of *Zea mays L.*
Use: Food additive.

cornstarch.　A carbohydrate polymer derived from corn of various types, composed of 25% amylose and 75% amylopectin. A white powder that swells in water, it is the most widely used starch in the U.S. The so-called waxy variety (made from waxy corn) contains only branched amylopectin molecules. Its chief uses are as a source of glucose, in the food industry as a filler in baking powder and a thickening agent in various food products, and in adhesives and coatings. It has been proposed as an additive to plastics to promote rapid degradation in such products as bottles and waste containers. See starch.

corn steep liquor.　The dilute aqueous solution obtained by soaking corn kernels in warm 0.2% sulfur dioxide solution for 48 hours as the first step in the recovery of cornstarch, corn oil, and gluten from corn. The solution contains mineral matter and soluble organic material extracted from the corn. It is used as a growth medium for penicillin and other antibiotics, and it is also concentrated and used as an ingredient of cattle feeds.

corn sugar.　See glucose.

corn syrup.　See glucose syrup.

corn syrup, high-fructose.　See high-fructose corn syrup.

corona.　An electrical discharge effect that causes ionization of oxygen and the formation of ozone. It is particularly evident near high-tension wires and in spark-ignited automotive engines. The ozone formed can have a drastic oxidizing effect on wire insulation, cable covers, and hose connections. For this reason, such accessories are made of oxidation-resistant materials such as nylon, neoprene, and other synthetics.

corotoxigenin-rhamnose.
CAS: 58917-39-8.　$C_{29}H_{42}O_9$.
Hazard: A poison.
Source: Natural product.

coroxon. (*O,O*-diethyl-*O*-(3-chloro-4-methylcoumarin-7-yl)phosphate).
Hazard: A cholinesterase inhibitor.
Use: Insecticide, fungicide.

correlation table. See scatter diagram.

corresponding states. (reduced states).
Two substances are in corresponding states when their pressures, volumes (or densities), and temperatures are proportional, respectively, to their critical pressures, volumes (or densities), and temperatures. If any two of these ratios are equal, the third must also be equal. This principle has been useful in the development of physical and thermal properties of substances.

corrosion. (1) The electrochemical degradation of metals or alloys caused by reaction with their environment, which is accelerated by the presence of acids or bases. In general, the corrodability of a metal or alloy depends on its position in the activity series. Corrosion products often take the form of metallic oxides. This is actually beneficial in the case of aluminum and stainless steel, because the oxide forms a strongly adherent coating that effectively prevents further degradation. Hence, these metals are widely used for structural purposes. The rusting of iron is a familiar example of corrosion that is catalyzed by moisture. Acidic soils are highly corrosive. Sulfur is a corrosive agent in automotive fuels and in the atmosphere (as SO_2). Sodium chloride in the air at locations near the sea is also strongly corrosive, especially at temperatures above 21C. Copper, nickel, chromium, and zinc are among the more corrosion-resistant metals and are widely used as protective coatings for other metals. Excellent corrosion-resistant alloys are stainless steel (18 Ni-8 Cr), Monel metal (66 Ni-34 Cu), and duralumin.
(2) The destruction of body tissues by strong acids and bases.
See corrosive material; protective coating; paint; tarnish.

Corrosion Inhibitor CS. A synergistic combination of sodium nitrate, borax, and organic inhibitors, used to prevent corrosion of ferrous and nonferrous metal and alloy surfaces in low-makeup closed cooling and heating systems.

corrosive acid. A strong acid, especially a mineral acid that has a high degree of ionization and solubility in water.
Hazard: Rapid, deep, and painful destruction of tissues.

corrosive alkali. Any strongly corrosive base, especially a metallic hydroxide or a carbonate that has a high degree of ionization and solubility in water.

Hazard: Rapid, deep, and very painful destruction of tissues; gelatinize tissues, causing a grayish color and a slippery or soapy texture.

corrosive material. Any solid, liquid, or gaseous substance that attacks building materials or metals or that burns, irritates, or destructively attacks organic tissues, most notably the skin and, when taken internally, the lungs and stomach. Among the more widely used chemicals that have corrosive properties are the following:

acetic acid, glacial	hydrofluoric acid
acetic anhydride	nitric acid
bromine	potassium hydroxide
chlorine	sodium hydroxide
fluorine	sulfuric acid hydrochloric acid

See toxic substances.

corrosive sublimate. Obsolete term for mercuric chloride.

cortexone. ((8S,9S,10R,13S,14S,17S)-17-(2-hydroxyacetyl)-10,13-dimethyl-1,2,6,7,8,9,11,12,14,15,16,17-dodecahydrocyclopenta[a]phenanthren-3-one).
CAS: 64-85-7. $C_{21}H_{30}O_3$.
A pregnene-derived steroid metabolite that is the 11-deoxy derivative of corticosterone and the 21-hydroxy derivative of progesterone.
Hazard: Toxic.

corticoid hormone. A hormone produced or isolated from the cortex (external layer) of the adrenal gland. Corticoid hormones now used in medicine include cortisone, hydrocortisone, deoxycorticosterone, fluorcortisone, prednisone, prednisolone, methyl prednisolone, triamcinolone, dexamethasone, corticotropin (ACTH), and aldosterone. Some occur naturally in adrenal extract; others are modifications of the natural hormones. All are now made synthetically. They are derivatives of cyclopentanophenanthrene.
See cortisone; ACTH.

corticosteroids. Steroid hormones formed by the adrenal cortex.
See: corticoid hormone.

corticosterone.
CAS: 50-22-6. $C_{21}H_{30}O_4$.
One of the less active adrenal cortical steroid hormones.
Properties: Crystalline plates. Mp 180–182C. Soluble in organic solvents; insoluble in water.
Derivation: Isolation from adrenal cortex extract, synthesis from deoxycholic acid.
Use: Biochemical research, medicine.

corticotropin. See ACTH.

cortisol. See hydrocortisone.

cortisone. (11-dehydro-17-corticosterone).
CAS: 53-06-5. $C_{21}H_{28}O_5$.
An adrenal, cortical, steroid hormone. It affects carbohydrate and protein metabolism.
Properties: White, crystalline solid. Mp 220–224 (decomposes). Dextrorotatory in solutions. Slightly soluble in water; sparingly soluble in ether, benzene, and chloroform; fairly soluble in methanol, ethanol, and acetone.
Derivation: From adrenal gland extract (usually from cattle) (historical method), synthetically from bile acids, from other steroids or sapogenins.
Hazard: Damaging side effects, e.g., sodium retention from ingestion.
Use: An anti-inflammation drug used in treatment of acute arthritis, Addison's disease, inflammable diseases of eyes and skin.

corundum. (emery).
CAS: 1302-74-5. Al_2O_3.
Properties: A varicolored mineral with transparent crystals that are very hard and resistant to attack by acids. D 3.95–4.10, mp 2050C, bp 2977C.
Natural aluminum oxide, sometimes with small amounts of iron, magnesium, silica, etc.
Occurrence: New York, Greece, Asia Minor.
Use: Various polishing and abrasive operations, grinding wheels.
See aluminum oxide; diaspore; sapphire.

corynine. See yohimbine.

cosmetic. Any preparation in the form of a liquid, semiliquid, paste, or powder applied to the skin to improve its appearance, and for cleaning, softening, or protecting the skin or its adjuncts but without specific medicinal or curative effects. Cosmetics include hairsprays, shampoos, nail polish, deodorants, shaving creams, facial creams, dusting powders, rouge, etc. Detergents, common soap, and bactericidal agents are not themselves classed as cosmetics although they may be components of cosmetic mixtures. A partial list of cosmetic ingredients follows:
animal fats (lanolin)
vegetable oils, waxes
alcohols (glycerol, glycols)
surfactants (alkyl sulfonates)
UV blocking agents (PABA)
phenylene diamine
aluminum chlorohydrate
FDC organic dyes
talc (magnesium silicate)
essential oils
inorganic pigments
chlorophyllins
nitrocellulose lacquers
steroid hormones
Knowledge of the structure and function of the skin is essential for proper cosmetic formulation (cosmetology). All ingredients must be tested for possible toxic effects since the skin is an important means of access for poisons. Addition of proteins to cosmetic preparations is of questionable value.

cosmic rays. The nuclei and electrons of atoms, mostly hydrogen, that impinge upon the earth from all directions of space with nearly the speed of light. Also referred to as cosmic radiation or primary cosmic rays.

cosmid. Artificially constructed cloning vector containing the cos gene of phage lambda. Cosmids can be packaged in lambda phage particles for infection into *E. coli*; this permits cloning of larger DNA fragments (up to 45 kb) than can be introduced into bacterial hosts in plasmid vectors.

cosmochemistry. See astrochemistry; chemical planetology.

"Cosmoperine" *[Sabinsa].* TM for tetrahydropiperine preparation.
Use: To enhance permeation through skin surface.

Cosol. High-boiling coal tar solvents for use in alkyd resin enamels and synthetic lacquers.

costus root oil.
Properties: From the dried roots of *Saussurea lappa* Clarke (Fam. *Compositae*). Light yellow to brown liquid; persistent odor. D: 0.995–1.039, refr index: 1.515 @ 20°. Sol in fixed oils, mineral oil; insol in glycerin, propylene glycol.
Use: Food additive.

cotoneral abs. See C.I. direct black 32, trisodium salt.

cotransport. The simultaneous transport, by a single transporter, of two solutes across a membrane. It is either antiport or symport, depending on whether the solutes travel in opposite or same direction.

cotton. Staple fibers surrounding the seeds of various species of *Gossypium*. Both *Egyptian* and Sea Island cotton have unusually long staple (approximately 2 inches). Cotton is the major textile fiber and an important source of cellulose, which constitutes 88–96% of the fiber. So-called *absorbent cotton* is almost pure cellulose.
Properties: Tenacity, 3–6 g/denier (dry), 4–8 g/denier (wet); elongation 3–7%; d 1.54; moisture regain 7% (21C, 65% relative humidity); yellows slowly at 121C; decomposes at approximately 148C; low permanent set; decomposed by acids; swells in caustic but is undamaged. Soluble in cuprammonium hydroxide. Subject to mildew. May be dyed by direct, vat, azoic, sulfur, and basic dyes. Combustible.
Source: U.S., Brazil, Egypt, India.
Hazard: Toxic by inhalation. Moderately flammable in the form of dust or linters; fiber ignites readily. In

the form of dust or linters, exposure of workers in textile mills may cause "brown lung." Byssinosis, bronchitis, and pulmonary function. Questionable carcinogen.
Use: Apparel, industrial and household fabrics, upholstery, medicine, thread.
See cellulose.

cotton, acetylated. Cotton fibers, threads, or fabrics treated with acetic anhydride, acetic acid, perchloric acid, and catalyst to improve the heat, rot, and mildew resistance by forming a surface coating of cellulose acetate.
Hazard: Ignites readily, not self-extinguishing.

cotton, aminized. A cotton fabric produced by reacting 2-aminoethylsulfuric acid with the cellulose of the fabric in a strongly alkaline solution. The treated cotton can take acid wool dyes and can be made rot-resistant and water-repellent.

cotton, cyanoethylated. Cotton treated with acrylonitrile. It is passed through a caustic bath that induces mild swelling of the fiber and catalyzes the subsequent reaction with acrylonitrile. The fabric is then neutralized with acetic acid, washed, and dried. The treatment leaves 3–5% nitrogen attached to the cellulose polymer. The cyanoethylated fiber is claimed to have permanent rot and mildew resistance, greater retention of strength after exposure to heat, improved receptiveness to dyes, and higher abrasion and stretch resistance.

cotton linters. Short, fleecy fibers that adhere to cottonseed after it has been passed once through a cotton gin. They are removed from the seed by a second ginning.
Hazard: Flammable, dangerous fire risk.
Use: Rayon manufacture, cellulosic plastics, nitrocellulose lacquers, soil–cement binder in road construction, explosives.

cotton, mercerized. Cotton that has been strengthened by passing through 25–30% solution of sodium hydroxide under tension and then washed with water, while under tension. This causes the fibers to shrink and increases their strength and attraction for colors, as well as imparting luster. A process using liquid ammonia for this purpose has been introduced in the U.K.

Cotton–Mouton effect. (magnetic double refraction).
Double refraction produced in some pure liquids by a magnetic field transverse to the light beam.

cotton oil. (cotton spraying oil).
A compounded oil sprayed (in the form of a fine mist) onto cotton to condition the fibers for yarn-making operations. Used to lubricate the fibers, to reduce static, "fly," and dust, and generally to improve the suppleness and strength of the fibers.

cottonseed. The seed of the cotton plant, *Gossypium hirsutum*. It contains about 22% crude fiber, 20% protein, 20% oil, 10% moisture, and 24% N-free extract; also contains from 1 to 2% of the toxic pigment gossypol; specially processed kernels are free of this.
Hazard: Generates dangerous amounts of heat if piled or stored wet or hot. Powerful allergen; may cause asthma and other respiratory difficulties on inhalation.
Use: Source of cottonseed oil and meal, cotton linters; source of nutritional protein after removal of gossypol by centrifugation.

cottonseed meal. (cottonseed cake).
The pulverized cottonseed press cake. Depending on the extractive process, varying percentages of protein will remain in the meal, and it is normally sold with 36–45% protein content. The 42% product contains approximately 42% crude protein, 6% crude fiber, 25% N-free extract, 10% other extract (fat), and 7% ash. The total digestible nutrient averages 79%. The ash is high in potash and phosphate; some types contain gossypol.
Use: Animal feeds, fertilizer ingredient, filler for plastics.

cottonseed, modified products.
Properties: Extracted from decorticated, partially defatted, cooked, ground cottonseed kernels.
Use: Food additive.

cottonseed oil.
Properties: Pale-yellow or yellowish-brown to dark-ruby-red or black-red, fixed, semi-drying oil depending on the nature and condition of the seed. The pure oil is odorless and has a bland taste. D 0.915–0.921, iodine value 109–116, flash p 486F (252C). Soluble in ether, benzene, chloroform, and carbon disulfide; slightly soluble in alcohol. Combustible.
Chief constituents: Glycerides of palmitic, oleic, and linoleic acids.
Derivation: From cottonseeds by hot pressing or solvent extraction.
Grade: Crude, refined, prime summer yellow, bleachable, USP.
Use: Leather dressing, soap stock, lubricant, glycerol, base for cosmetic creams; hydrogenated to semisolid for use in food products; waterproofing compositions, dietary supplement.

Cottrell, Frederick G. (1877–1948). A native of California, Cottrell obtained his doctorate from Liebig in 1902. His major contribution to industrial chemistry was his discovery of a practical method of dust elimination by electrical precipitation. Used in factory stacks and other large units, this process has contributed greatly to purifying the atmosphere of industrial areas. The principle involves charging a suspended wire with electricity. This creates a field that ionizes the surrounding air, the particles assuming the charge on contact and then moving to the wall of the stack, where they are electrically discharged and precipitated.

couch roll.   (Pronounced *cooch*.) A hollow suction roll on a fourdrinier paper machine over which the formed sheet or web passes as it leaves the wire. The suction is provided by a vacuum or suction box inside the roll, whose face is perforated to offer as large a vacuum area as possible. The chief feature of the couch roll is its great water-removing capacity; this gives the sheet enough strength to enable it to hold together as it passes to the pickup felts.

coumaphos.   Generic name for *O,O*-diethyl-*O*-(3-chloro-4-methyl-2-oxo-2H-1-benzopyran-7-yl)-phosphorothioate.
CAS: 56-72-4. $C_9H_3O_2(CH_3)ClOPS(OC_2H_5)_2$.
Properties: Crystals. Mp 91C. Insoluble in water; soluble in aromatic solvents.
Hazard: Use may be restricted; cholinesterase inhibitor. Questionable carcinogen.
Use: Insecticide, anthelmintic.

***p*-coumaric acid.**   (4-coumaric acid; *p*-cumaric acid; 4-hydroxycinnamic acid; 4'-hydroxycinnamic acid; *p*-hydroxyphenylacrylic acid; (3-(4-hydroxyphenyl)-2-propenoic acid; *p*-hydroxycinnamic acid; β-[4-hydroxyphenyl]-acrylic acid; (E)-3-(4-hydroxyphenyl)prop-2-enoic acid).
CAS: 7400-08-0. $C_9H_8O_3$.
Properties: Crystalline needle-like substance, soluble in hot water, ethanol, and ethyl ether; slightly soluble in cold water; nearly insoluble in benzene and ligroin.
Hazard: Moderately toxic.
Use: Plant growth inhibitor.

coumarin.   (cumarin; benzopyrone; tonka bean camphor).
CAS: 91-64-5. $C_9H_6O_2$. A lactone.
Properties: Colorless crystals, flakes, or powder; fragrant odor similar to vanilla; bitter, aromatic burning taste. Mp 69C, bp 290C. Soluble in 10 vols of 95% alcohol and in ether, chloroform, and fixed volatile oils; slightly soluble in water. Combustible.
Derivation: (1) By heating salicylic aldehyde, sodium lactate, and acetic anhydride; (2) fine grades are isolated from tonka beans.
Hazard: Toxic by ingestion; carcinogenic. Use in food products prohibited (FDA). Questionable carcinogen.
Use: Deodorizing and odor-enhancing agent, pharmaceutical preparations.

courmarin anticoagulant.   Any anticoagulant that contains coumarin as the active component.
Hazard: Anemia, weakness, pale mucous membranes, dyspnea, moist rales, bloody feces, scleral,

conjunctival and intraocular hemorrhage, staggering, ataxia, blood-tinged froth around mouth and nose, death.

coumarone.   (benzofuran).
CAS: 271-89-6. C_8H_6O.
A bicyclic ring compound derived from coal tar naphtha, the parent substance of coumarone-indene resins.
Properties: Colorless, oily liquid. D 1.09, bp 165–175C. Insoluble in water.
Hazard: Possible carcinogen.

coumarone-indene resin.   A thermosetting resin derived by heating a mixture of coumarone and indene with sulfuric acid, which induces polymerization. It is soft and sticky at room temperature; it hardens on heating to a resinous solid. Soluble in hydrocarbon solvents, pyridine, acetone, carbon disulfide, and carbon tetrachloride; insoluble in water, alcohol. Combustible. Said to have been the first synthetic polymer.
Use: Adhesives, printing inks, floor-tile binder, friction tape.
See "Cumar" *[Neville]*; Nevindene; Paradene.

count.   (1) The external indication given by a radiation detector, such as a Geiger counter, of the amount of radioactivity to which the detector is exposed. The background counts are those that come from a source external to that being measured. (2) The number of warp and filler threads in a linear inch of a textile fabric, e.g., the count of a sheeting may be 80 × 60.

countercurrent.   Descriptive of a process in which a liquid and a vapor stream, or two streams of immiscible liquids, or a liquid and a solid are caused to flow in opposite directions and past or through one another with more or less intimate contact so that the individual substances present are more or less completely transferred to that stream in which they are more soluble or stable under the conditions existing. The streams leaving such a process are usually of higher purity that can be attained otherwise at equal cost. Distillation with a fractionating column is also a typical countercurrent process, in which rising vapor is purified by contact with descending liquid (reflux). Leaching, washing, and chemical reaction are frequently carried out in a countercurrent manner.
See liquid–liquid extraction.

coupled reactions.   Two chemical reactions that have a common intermediate and thus a means of energy transfer from one to the other.

coupling.   (1) The combination of an amine or phenol with a diazonium compound to give an azo

compound, the reaction by which azo dyes are prepared. Thus m-phenylenediamine $C_6H_4(NH_2)_2$ couples with benzene diazonium chloride $C_6H_5N_2Cl$ to produce the dye chrysoidine $C_6H_5N_2C_6H_3(NH_2)_2$.
See azo-dye intermediate.
(2) Oxidative coupling.
(3) An agent, e.g., a vinyl silane, used to protect fiberous glass laminates from effects of water absorption.
(4) A condensation polymerization of amino acids to form proteins; it can be done synthetically only by suppressing certain active sites on the amino acid molecules.

covalent bond.   (homopolar).
Sharing of electrons by a pair of atoms.
See bond, chemical.

covalent radius.   The radius for each element such that the actual bond length between any two elements that form a covalent single bond is roughly equal to the sums of their covalent radii.

covering power.   See opacity.

Cowles process.   The direct manufacture of aluminum alloys, such as copper aluminum, from aluminum ores by reacting with carbon in an electric furnace in the presence of the alloying metal.

Cox chart.   A special semilogarithmic plot of vapor pressure versus temperature especially useful for the petroleum hydrocarbons. The graph corresponding to each separate hydrocarbon is a straight line. All the lines appear to intersect at a point outside the chart.

coxistac.   See salinomycin.

CP.   Abbreviation for chemically pure, an accepted grade of drugs and fine chemicals that contain a minimum of impurities.
See chemically pure.

CP-40.   Chlorination derivatives of paraffin wax.
Use: Rendering fabrics waterproof and fire-retardant; fire-retardant paints; plasticizer and extender for certain plastic materials, etc.

C-P-B.   Dibutyl xanthogen disulfide.
Properties: Amber-colored, free-flowing liquid. D 1.15. Soluble in acetone, benzene, gasoline, and ethylene dichloride; insoluble in water.
Use: Accelerator for pure-gum handmade drugs, sundries and medical supplies, bathing shoes, bathing caps, novelties, and cold-cure cements.

CPR.   Abbreviation for cyclonene–pyrethrin–rotenone. Applied to various insecticide formulations containing as active ingredients approximately 10 parts piperonyl cyclonene, 5 parts rotenone, and 1 part pyrethrin.

CPVC.   See critical pigment volume concentration.

Cr.   Symbol for chromium.

"CR-39" *[PPG].*   TM for allyl diglycol carbonate, an optical plastic.
Properties: Clear optical plastic highly resistant to impact and abrasion. Furnished as a clear liquid. Thermosetting.
Use: Ophthalmic lenses, shields, instrument panels, marine, and aircraft glazing.

cracking.   A refining process involving decomposition and molecular recombination of organic compounds, especially hydrocarbons obtained by means of heat, to form molecules suitable for motor fuels, monomers, petrochemicals, etc. A series of condensation reactions takes place, accompanied by transfer of hydrogen atoms between molecules which brings about fundamental changes in their structure. Thermal cracking, the older method, exposes the distillate to temperatures of approximately 540–650C (1000–1200F) for varying periods of time; it is no longer used for gasoline, but is still of value in producing hydrocarbon gases for plastics monomers. The development of premium fuels for airplanes and automobiles resulted from the use of catalysts in cracking. In this process, hydrocarbon vapors are passed at approximately 400C (750F) over a metallic catalyst (e.g., silica-alumina or platinum); the complex recombinations (alkylation, polymerization, isomerization, etc.) occur within seconds to yield high-octane gasoline. Among the chemical changes induced are conversion of alicyclic compounds (cyclohexane) to aromatic compounds, and of straight-chain to branched-chain structures (isomerization). Cracking reactions are exothermic. Free radicals, carbonium ions, and other chain-initiating agents are involved in these rearrangements.
Catalytic cracking is carried out by either the moving-bed or the fluid-bed technique. In the former the catalyst is pelleted, while in the latter it is finely divided. Instances in which cracking does not involve production of gasoline are the steam cracking of methane or naphtha to form synthesis gas, thermal cracking of naphtha to ethylene, and thermal decomposition of methane to carbon black and hydrogen.
See catalysis; synthesis gas; fluidization; pyrolysis.

Crag.   (Sesone).
CAS: 136-78-7. $C_8H_7Cl_2O_5S\cdot Na$.
TM for agricultural chemicals, including: (1) Fly repellent (active ingredient, butoxypolypropylene glycol). Colorless liquid, 100% active material. (2) Fungicide 974 (active ingredient, 3,5-dimethyltetrahydro-1,3,5-2H-thiadiazine-2-thione). Wettable powder, 85% active material.

Irritant. (3) Glyodin solution (active ingredient, 2-heptadecyl glyoxaldine acetate), 34% active solution. (4) Herbicide-1 (SES) (active ingredient, sodium-2,4-dichlorophenoxyethyl sulfate). Water-soluble powder, 90% active material.
Hazard: Toxic by inhalation.

Craig method.   Introduction of a halogen into the α-position of pyridine by treatment of a solution of α-aminopyridine with sodium nitrite in hydrogen halide, followed by warming.

crambe-seed oil.   A vegetable oil obtained from the seeds of the crambe, a plant related to mustard and rape. Growth and processing techniques have been studied by the Agricultural Research Service of USDA. The plant can be grown on marginal and strip-mined land. Its high content of erucic acid and the high protein content of its meal make it economically attractive. The oil itself is useful as an industrial lubricant, especially for molds for continuous steel casting.

Cram, Donald James.   (1919–2001). Awarded the Nobel Prize for chemistry, together with Lehn, in 1987 for work in elucidating mechanisms of molecular recognition, which are fundamental to enzymic catalysis, regulation, and transport. Cram also studied three-dimensional cyclic compounds that maintained a rigid structure, accepting substrates in a structurally preorganized cavity. He called these compounds cavitands, while Lehn named them cryptands. Cram was awarded a doctorate by Harvard University in 1947.

Cram's rule of asymmetric induction.   Rule governing the stereochemistry of addition to carbonyl compounds containing an asymmetric center. The diastereomer that is formed by approach of the reagent from the less hindered side of the carbonyl group predominates in the product. The α-substituents are schematically represented as S (small), M (medium), and L (large).

crazing.   Development of minute cracks in the surface of a material, such as ceramic glaze, varnish, paint, etc., often as a result of exposure to sunlight or weathering.

c-reactive protein.
A serum protein, the concentration of which closely follows an individual's response to inflammation, thereby offering an assessment of the state of many disease processes. It is present in acutely ill patients and disappears during recovery. C-reactive protein has been detected in bacterial, viral, and other infections, as well as in certain noninfectious diseases. C-reactive protein appears to activate neutrophils and thereby facilitating cell-mediated cytotoxic reactions against cells infected with microbes and stimulates monocyte-macrophage tumoricidal activity.

cream of tartar.   See potassium bitartrate.

creatine.   (N-methyl-N-guanylglycine; (α-methyllguanido)acetic acid).
CAS: 60-27-5. $HN:C(NH_2)N(CH_3)CH_2COOH$.
A nitrogenous acid widely distributed in the muscular tissue of the body.
Properties: (Monohydrate) Prisms from water, anhydrous at 100C. Decomposes 303C. Slightly soluble in water; insoluble in ether.
Source: Commercially isolated from meat extracts.
Grade: Technical, CP.
Use: Biochemical research.

creatinine.
CAS: $C_4H_7N_3O$.
The anhydride of creatine, a metabolic waste product.
Properties: Colorless to yellow liquid. D 1.092 (25C), mp 5.5C, bp 220C. Slightly soluble in water, alcohol, benzene, chloroform, ether, acetic acid.

Creighton process.   Electrochemical reduction of sugars, e.g., reduction of glucose, xylose, and galactose to sorbitol, xylitol, and dulcitol.

creosote, coal tar.   (creosote oil; liquid pitch oil; tar oil).
CAS: 8001-58-9.
Properties: Yellowish to dark-green-brown, oily liquid; clear at 38C or higher; naphthenic odor; frequently contains substantial amounts of naphthalene and anthracene. Autoign temp 637F (335C), d 1.06–1.10, distilling range 200–400C, flash p 165F (74C) (CC). Soluble in alcohol, benzene, and toluene; immiscible with water.
Derivation: Fractional distillation of coal tar.
Method of purification: Rectification.
Grade: Technical, crude, refined.
Hazard: Toxic by inhalation of fumes, skin, and eye irritant. Use may be restricted. Probable carcinogen.
Use: Wood preservative (ties, telephone poles, marine pilings, etc.), disinfectants, fungicide, biocide.

p-cresidine.   See 5-methyl-o-anisidine.

Creslan.   An acrylic fiber.

cresol.   (methylphenol; hydroxymethylbenzene; cresylic acid).
CAS: 1319-77-3. $CH_3C_6H_4OH$.
A mixture of isomers obtained from coal tar or petroleum.
Properties: Colorless, yellowish, or pinkish liquid; phenolic odor. D 1.030–1.047, wt/gal 8.66–8.68 lb, flash p approximately 180F (82C), mp 11–35C, bp 191–203C. Soluble in alcohol, glycol, dilute alkalies, and water.
Derivation: Coal tar (from coke and gas works), also from toluene by sulfonation or oxidation.

Grade: Various, depending on phenol content or other properties. NF grade contains not more than 5% phenol.

Hazard: Irritant, corrosive to skin and mucous membranes, absorbed via skin. Questionable carcinogen.

Use: Disinfectant, phenolic resins, tricresyl phosphate, ore flotation, textile scouring agent, organic intermediate, manufacture of salicylaldehyde, coumarin, and herbicides, surfactant, synthetic food flavors (*para* isomer only).

See cresylic acids.

m-cresol.   (*m*-cresylic acid; 3-methylphenol).
CAS: 108-39-4. $CH_3C_6H_4OH$.

Properties: Colorless to yellowish liquid; phenol-like odor. D 1.034, mp 12C, bp 203C, wt/gal 8.66 lb, flash p 187F (86C), autoign temp 1038F (558C). Soluble in alcohol, ether, and chloroform; soluble in water.

Derivation: By fractional distillation of crude cresol (from coal tar), also synthetically.

Method of purification: Rectification.

Grade: Technical (95–98%).

Hazard: Questionable carcinogen.

See cresol.

o-cresol.   (*o*-cresylic acid; 2-methylphenol).
CAS: 95-48-7. $CH_3C_6H_4OH$.

Properties: White crystals; phenol-like odor. D 1.047; mp 30.9C; flash p 178F (81C); autoign temp 1110F (598C); bp 191C; wt/gal 8.68 lb. Soluble in alcohol, ether, chloroform, and hot water.

Derivation: (1) By fractional distillation of crude cresol from coal tar. (2) Interaction of methanol and phenol.

Method of purification: Crystals.

Grade: According to fp: 25, 29, 30, 30.5C, etc.

Hazard: Questionable carcinogen.

See cresol.

p-cresol.   (*p*-cresylic acid; 4-methylphenol).
CAS: 106-44-5. $CH_3C_6H_4OH$.

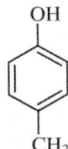

Properties: Crystalline mass; phenol-like odor. Wt/gal 8.67 lb, d 1.039, bp 202C, mp 35.25C, flash p 187F (86C), autoign temp 1038F (558C). Soluble in alcohol, ether, chloroform, and hot water.

Derivation: (1) By fractional distillation of crude cresol; (2) from benzene by the cumene process (see phenol).

Method of purification: Crystallization.

Grade: Technical, 98%, 99.0% min purity or 34C min fp.

Hazard: Questionable carcinogen.

See cresol.

cresolphthalein.

$$C_6H_4COOC(C_6H_3(OH)CH_3)_2.$$

An acid–base indicator, changes from colorless to red between pH 8.2 and 9.8, reagent.

See indicator.

cresol purple.

$$C_6H_4SO_2OC(C_6H_3(OH)CH_3)_2.$$

m-Cresolsulfonphthalein, an acid–base indicator, showing color change from red to yellow over the range pH 1.2–2.8, and from yellow to purple over the range pH 7.4–9.0.

See indicator.

cresol red.

$$C_6H_4SO_2OC(C_6H_3(OH)CH_3)_2.$$

o-Cresol-sulfonphthalein, an acid–base indicator, changes from yellow to red between pH 7.0 and 8.8.

See indicator.

cresotic acid.   (cresotinic acid; hydroxytoluic acid).
CAS: 83-40-9. $CH_3C_6H_3(OH)COOH$.

Ten possible isomers; most common is 2-hydroxy-3-methylbenzoic acid, also known as *o*-cresotic acid or *o*-homosalicylic acid. The description that follows is of this isomer.

Properties: White crystals or powder. Mp 166C, bp approximately 250C. Partially soluble in hot water; soluble in alcohol and ether. Combustible.

Derivation: Treatment of *o*-cresol with caustic and carbon dioxide under pressure.

Use: Dye intermediate, research on plant growth inhibition.

m-cresyl acetate.   (*m*-tolyl acetate).
CAS: 140-39-6. $CH_3C_6H_4OCOCH_3$.

Properties: Colorless, oily liquid; odor similar to phenol. Bp approximately 112C, distillation with steam. Insoluble in water; soluble in common organic solvents. Combustible.

Use: Medicine (antiseptic, fungicide).

o-cresyl acetate.   (*o*-tolyl acetate).
CAS: 533-18-6. $CH_3COOC_6H_4CH_3$.

Properties: Liquid. Bp 208C. Nearly insoluble in cold water; soluble in hot water and organic solvents. Combustible.

Use: Flavoring.

p-cresyl acetate.   (*p*-tolyl acetate).
CAS: 140-39-6. $CH_3C_6H_4COCH_3$.

Properties: Colorless liquid; floral odor. D 1.0532 (15C), optical rotation 0 degrees (100 mm Hg), refr index 1.500–1.504, acid value 0.7, ester value 341.6. Soluble in 2.5 volumes of 70% alcohol and in most fixed oils; insoluble in glycerol. Combustible.
Grade: Technical, FCC.
Use: Perfumery, flavoring.

cresyldiglycol carbonate.   (diethylene glycol bis(cresylcarbonate)).
$C_{20}H_{22}O_7$.
Properties: Colorless liquid of low volatility. D 1.19 (20/4C), bp approximately 250C (2 mm Hg), flash p 475F (246C), refr index 1.523 (20C). Insoluble in water (very stable to hydrolysis). Widely soluble in organic solvents. Compatible with many resins and plastics. Combustible.
Use: Plasticizer.

cresyldiphenyl phosphate.   (cresyl phenyl phosphate).
CAS: 26444-49-5. $(CH_3C_6H_4)(C_6H_5)_2PO_4$.
Probably seldom a pure compound, but a mixture of o-, m-, and p-cresyl and phenyl phosphates.
Properties: Colorless, transparent liquid; very slight odor. D 1.20 (20/20C), fp −38C, boiling range 235–255C (4 mm Hg), flash p 450F (232C). Insoluble in water; soluble in most organic solvents except glycerol. Combustible.
Use: Plasticizer, extreme-pressure lubricant, hydraulic fluids, gasoline additive, food packaging.

cresylic acids.   Commercial mixtures of phenolic materials boiling above the cresol range. An arbitrary standard in use for cresylic acids is that 50% must boil above 204C. If the boiling point is less than 204C, the material is called cresol. Cresylic acid varies widely according to its source and boiling range.
A typical commercial cut, bp 220–250C, has the composition m- and p-cresols 0–1%; 2,4- and 2,5-xylenols 0–3%; 2,3- and 3,5-xylenols 10–20%; 3,4-xylenol 20–30%; and C_9 phenols 50–60%. Excellent electrical insulators.
Derivation: Petroleum, coal tar. Imported cresylic acid is derived from coal tar (gasworks), also made synthetically.
Hazard: Corrosive to skin, absorbed via skin.
Use: Phosphate esters, phenolic resins, wire enamel solvent, plasticizers, gasoline additives, laminates, coating for magnet wire for small electric motors. Disinfectants, metal-cleaning compounds, phenolic resins, flotation agents, surfactants, chemical intermediates, oil additives, solvent refining of lubricating oils, scouring compounds, pesticides.

cresylate.
Any salt of a cresol or cresylic acid.

p-cresyl isobutyrate.   (p-tolyl isobutyrate).
CAS: 103-93-5. $CH_3H_6H_4OCOCH(CH_3)_2$.
Use: Flavoring.

cresylphenyl phosphate.   See cresyldiphenyl phosphate.

cresyl silicate.
$(CH_3C_6H_4O)_4Si$.
Properties: Colorless liquid. Bp 450C.
Derivation: Reaction of cresol and silicon tetrachloride.
Use: Heat-transfer fluid.

cresyl-p-toluene sulfonate.   (tolyl-p-toluene sulfonate).
$CH_3C_6H_4SO_3C_6H_4CH_3$.
Properties: Brown, oily liquid; faint odor. D 1.207, flash p 365F (185C), mp 68.70C. Combustible.
Derivation: From reaction of p-toluenesulfonyl chloride with p-cresol.
Use: Plasticizer.

Crick-Watson structure.   See deoxyribonucleic acid.

cricondentherm.   The maximum temperature at which two phases can coexist.
See phase.

Criegee reaction.   Oxidative cleavage of vicinal glycols by lead tetraacetate.

crinotoxic fish.   More than 60 species of fish that release a poisonous, toxic substance from epidermal glandular activity that is lethal to other fish and possibly other marine animals.

cristae.   Infoldings of the inner mitochondrial membrane.

criteria pollutant.   Any of seven pollutants that are sulfur dioxide, carbon monoxide, nitrogen dioxide, hydrocarbons, oxidants, particulate matter, and lead, that are regulated by the U.S. Environmental Protection Agency based on national ambient quality standards. The term "criteria pollutant" stems from the requirement that the Environmental Protection Agency must describe the characteristics and potential health and welfare effects of these pollutants. This information provides the criteria by which standards are set or revised.

critical assembly.   A system of fissionable material (enriched uranium) and moderator sufficient to sustain a chain reaction at a low and controllable power level, as in a nuclear reactor.
See fission; nuclear reactor.

critical constant.   A maximum or minimum value for a physical constant that is characteristic of a substance; for example, the critical temperature of a gas is the temperature above which it cannot be liquefied by an increase in pressure.

criticality. The state of a nuclear reactor when it is sustaining a chain reaction.

critical mass. The minimum mass of a fissionable material (^{235}U or ^{239}Pu) that will initiate an uncontrolled chain reaction as in an atomic bomb. The critical mass of pure ^{239}Pu is about 10 lb, and of ^{235}U about 33 lb. This phenomenon was unknown before 1940.

Critical pigment volume concentration. (CPVC).
The transition point above and below which there are substantial differences in the appearance and behavior of a paint film.

critical point. (1) The transition point between the liquid and gaseous states of a substance. (2) The temperature above which a gas cannot be liquefied, however high the pressure. (3) The temperature at which internal changes take place within a metal.

critical potential. The amount of energy needed to raise an electron from a lower to a higher level.

critical solution temperature. The temperature above or below which two liquids are miscible with all proportions. Some pairs of liquids have both an upper and a lower critical solution temperature; that is, they can exist in two phases only in a medium temperature range.

critical volume. The volume of a unit mass of a substance at critical temperature and pressure.

crocein acid. (croceic acid; Bayer's acid; 2-naphthol-8-sulfonic acid).

Derivation: Sulfonation of β-naphthol with 94% sulfuric acid at 95C and recrystallization from a salt.
Use: Azo-dye intermediate.

Crocein Scarlet MOO. See Brilliant Crocein.

crocetin.
CAS: 27876-94-4. $C_{20}H_{24}O_4$.
A dicarboxylic carotenoid derived from saffron.
Properties: Red, rhomboid crystals. Mp 285C. Soluble in pyridine and dilute sodium hydroxide; slightly soluble in water and organic solvents. Combustible.
Use: Experimental treatment of arteriosclerosis by increasing oxygen diffusion through arterial walls, thus decreasing buildup of cholesterol.

crocidolite. A type of asbestos.
See asbestos.

crocking. Removal of a dye or pigment from the surface of a paint or textile by rubbing or attrition.

Crodamal. Fatty acid esters.
Grade: In liquid, solid, and flake forms.
Use: As emollient esters for skin care, sun care, and stick products.

"Crodamide" *[Croda]*. (Erucamide)
TM for slip and mold release agents.
Use: Plastics casting.

"Crodamide OR" *[Croda]*. TM for oleamide slip and mold release additive.
Available forms: Beads, powder, and pastilles.
Use: In manufacture of plastics.

"Croda Oleochemicals" *[Croda]*. TM for chemicals used in hair care, sun screen, polymer medical additives, and cleaning products.

Crome of Nature. A no-lye hair relaxer.
Use: A color changing formula that signals when Crome is ready.

"Cronox 861 OS" *[Baker Hughes]*. TM for a batch treatment drilling corrosion inhibitor.
Use: As a corrosion inhibitor

crospovidone. See polyvinylpolypyrrolidone.

Cross–Bevan (viscose) process. Production of rayon by treatment of cellulose with alkali and carbon disulfide to yield cellulose xanthate, solution in dilute caustic, and extrusion of the viscous "Viscose" into a coagulating bath, a 7–10% sulfuric acid solution containing 1–5% zinc sulfate and an active surface agent.

crosshead. A device attached to the head of an extrusion machine that permits the material to be extruded in opposite directions simultaneously at right angles to the barrel. It is applicable chiefly to coating of wire, cable, and small-diameter hose.

crossing over. The breaking during meiosis of one maternal and one paternal chromosome, the exchange of corresponding sections of DNA, and the rejoining of the chromosomes. This process can result in an exchange of alleles between chromosomes.
See recombination.

cross-linking. Attachment of two chains of polymer molecules by bridges, composed of either an element, a group, or a compound, that join certain carbon atoms of the chains by primary chemical bonds, as indicated in the schematic diagram.

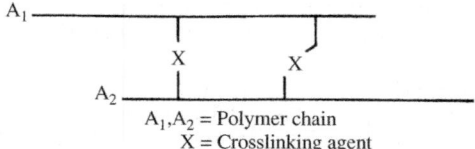

A₁, A₂ = Polymer chain
X = Crosslinking agent

Cross-linking occurs in nature in substances made up of polypeptide chains that are joined by the disulfide bonds of the cystine residue, as in keratins, insulin, and other proteins. Polysaccharide molecules can also cross-link to form stable gel structures (dextran). Cross-linking can be effected artificially, either by adding a chemical substance (cross-linking agent) and exposing the mixture to heat, or by subjecting the polymer to high-energy radiation. Examples are (1) vulcanization of rubber with sulfur or organic peroxides; (2) cross-linking of polystyrene with divinylbenzene; (3) cross-linking of polyethylene by means of high-energy radiation or with an organic peroxide; and (4) cross-linking of cellulose with dimethylol carbamate (10% solution) in durable-press cotton textiles. Cross-linking has the effect of changing a plastic from thermoplastic to thermosetting. Thus, it also increases strength, heat and electrical resistance, and especially resistance to solvents and other chemicals.
See vulcanization; polyethylene; keratin.

cross section.   (1) A measure of the probability that a nuclear reaction will occur. Usually measured in barns, it is the apparent (or effective) area presented by a target nucleus (or particle) to an oncoming particle or other nuclear radiation, such as a photon or γ-radiation. Also called capture cross section.
(2) A section made by a plane cutting through a solid. Tissue cross sections are widely used for microscopic observation.

Crosultaines.   (sulfobetaines).
Tallow and coconut versions.
Use: Mild surfactants for baby shampoos, excellent for foam boosting, stabilization, and thickening.

croton.   A mixture of toxic proteins contained in the seeds of *Croton tiglium*, a small Asiatic tree (family *euphorbaceae*). It is the commercial source of the highly toxic croton oil.
Hazard: Fatally toxic when ingested.

crotonaldehyde.   (2-butenal; crotonic aldehyde; β-methyl acrolein).
CAS: 4170-30-3. $CH_3CH:CHCHO$.
Commercial crotonaldehyde is the *trans* isomer.
Properties: Water-white, mobile liquid; pungent, suffocating odor. Turns to a pale-yellow color in contact with light and air. A lachrymator. D 0.8531 (20/20C), bp 102C, flash p 55F (12.7C), fp −69C, vap press 30 mm Hg (20C). Very soluble in water;

miscible with all proportions with alcohol, ether, benzene, toluene, kerosene, gasoline, solvent naphtha.
Derivation: Aldol condensation of two molecules of acetaldehyde.
Grade: Technical, 87% water-wet form.
Hazard: An animal carcinogen. Irritating to eyes, skin, and upper respiratory tract irritant. Flammable, dangerous fire risk. Explosive limits in air 2.9–15.5% by volume. Questionable carcinogen.
Use: Intermediate for *n*-butyl alcohol and 2-ethylhexyl alcohol, solvent, preparation of rubber accelerators, purification of lubricating oils, insecticides, tear gas, fuel-gas warning agent, organic synthesis, leather tanning, alcohol denaturant.

crotonic acid.   (2-butenoic acid; β-methacrylic acid).
CAS: 3724-65-0. $CH_3CH:CHCOOH$.
Exists in *cis* and *trans* isomeric forms, the latter being the stable isomer used commercially. The *cis* form melts at 15C and is sometimes called isocrotonic acid.
Properties: White, crystalline solid. D 0.9730, mp 72C, bp 185C, flash p 190F (87.7C) (COC). Soluble in water, ethanol, toluene, acetone. Combustible.
Derivation: Oxidation of crotonaldehyde.
Grade: 97%.
Hazard: Strong irritant to tissue.
Use: Synthesis of resins, polymers, plasticizers, drugs.

crotonic aldehyde.   See crotonaldehyde.

croton oil.   (tiglium oil).
CAS: 8001-28-3.
Properties: Brownish-yellow liquid. D 0.935–0.950 (25C), refr index 1.470–1.473 (40C). Soluble in ether, chloroform, and fixed or volatile oils; slightly soluble in alcohol.
Chief constituents: Glycerides of stearic, palmitic, myristic, lauric, and oleic acids and croton resin, a vesicant.
Derivation: By expression from the seeds of *Croton tiglium*.
Hazard: Strong skin irritant, ingestion of small amounts may be fatal.
Use: Medicine (counterirritant, cathartic).

crotonylene.   (2-butyne; dimethylacetylene).
CAS: 503-17-3. $CH_3C:CCH_3$.
Properties: Liquid. Bp 27C.
Hazard: Flammable, dangerous fire risk. Moderate explosion hazard.

crotoxin.   Exact formula undetermined. A toxic principle of rattlesnake venom having a polypeptide structure, one component being basic and the other acidic. Classed as a neurotoxin. Injection may be fatal.

crotoxyphos. (dimethyl-2-(α-methylbenzyloxy-carbonyl)-1-methylvinyl phosphate; *O,O*-dimethyl-*O*-(1-methyl-2-(1-phenylcarbethoxy)vinyl) phosphate; α,α-methyl-benzyl-3-(dimethoxy-phosphinyloxy)-isocrotonate).
CAS: 7700-17-6. $C_{14}H_{19}O_6P$.
Properties: Yellowish liquid. D 1.19, bp 135C at 4.0 (0.03 mm Hg), refr index 1.50. Soluble in acetone, alcohol, chlorinated hydrocarbons; slightly soluble in saturated hydrocarbons.
Hazard: A cholinesterase inhibitor.
Use: Insecticide used externally on farm animals.

crotyl alcohol. (2-buten-1-ol; 3-methylallyl alcohol).
CAS: 6117-91-5. $CH_3CH:CHCH_2OH$.
Properties: Clear, stable liquid. D 0.8550 (20/20C), boiling range 121–126C, flash p 113F (45C) (TOC). Partially soluble in water (17%); wholly soluble in alcohol and ether. Combustible.
Hazard: Toxic by ingestion, strong eye and skin irritant. Moderate fire risk.
Use: Chemical intermediate, source of monomers, herbicide, and soil fumigant.

crown ether. A cyclic molecule in which ether groups (polyethers) connected by dimethylene linkages are coordinated to a centrally located metal atom via the oxygen atoms of the ethers, which function as ligands (electron donors). Such compounds have strong complexing or chelating capabilities. In some types, silicon replaces the dimethylene linkages. They were so named because their molecular models resemble a crown.
See porphyrin; chelate.

crown filler. A mineral filler, usually calcium sulfate or carbonate or a mixture thereof, used in paper manufacture.

crown glass. See glass, optical.

crown group. All the taxa descended from a major cladogenesis event, recognized by possessing the clade's synapomorphy. See stem group.

crucible. (1) A cone-shaped container having a curved base and made of a refractory material, used for laboratory calcination and combustibles. Some types are equipped with a cover. A gooch crucible has openings in its base to permit filtration with suction; named after its inventor, an American chemist. (2) In the steel industry, a special type of furnace provided with a cavity for collecting the molten metal.

crude oil. Petroleum in its natural state prior to refining. A complex combination of aliphatic, alicyclic, and aromatic hydrocarbons that may also contain small amounts of nitrogen, oxygen, and sulfur compounds.
Hazard: Toxic; volatile; flammable.

crufomate. (*O*-methyl-*O*-(4-*tert*-butyl-2-chlorophenyl)methylphosphoramidate; 4-*t*-butyl-2-chlorophenylmethyl methyl phosphoramidate; ruelene).
CAS: 299-86-5.
Properties: Crystals. Mw 291.71, mp 61C. Insoluble in water; soluble in alcohol, benzene, and carbon tetrachloride. Commercial product is a yellow oil, bp 117C.
Hazard: Cholinesterase inhibitor. Questionable carcinogen.
Use: Systemic insecticide and anthelmintic.

crusher, gyratory. See gyratory crusher.

crust. The outermost layer of the lithosphere, consisting of felsic and mafic rocks less dense than the rocks of the mantle below.

Crutzen, Paul. (1933–). From the Netherlands who won the Nobel Prize for chemistry along with Mario Molina and Frank Sherwood Rowland in 1995 for their work in atmospheric chemistry, particularly concerning the formation and decomposition of ozone.
See Molina, Mario; Rowland, Frank Sherwood.

cryochemistry. That branch of chemistry devoted to the study of reactions occurring at extremely low temperature (−200C and lower). It permits synthesis of compounds that are too unstable or too reactive to exist at normal temperature.

cryogen. A substance, often a liquid, that is used to reach very low temperatures, usually below −150°C.

cryogenic gas. A gas that has been liquefied by lowering its temperature, usually to less than −110°C.

cryogenic liquid. A liquid cryogen that is usually defined as having a boiling point at atmospheric pressure below −150°C.

cryogenics. Study of the behavior of matter at temperatures below −200C. The use of the liquefied gases oxygen, nitrogen, and hydrogen at approximately −260C is standard industrial practice. Examples: Use of liquid nitrogen for quick-freezing of foods and of liquid oxygen in steel production. Some electronic devices and specialized instruments, such as the cryogenic gyro, operate at liquid-helium temperature (approximately 4 K). Many lasers and computer circuits require low temperature. Original research in this field was carried out by W. F. Giauque in the U.S. and by Kamerlingh–Onnes in Holland.
See superconductivity.

cryolite. (Greenland spar; icestone).
CAS: 15096-52-3. Na_3AlF_6.

A natural fluoride of sodium and aluminum or made synthetically from fluorspar, sulfuric acid, hydrated alumina, and sodium carbonate.

Properties: Colorless to white, sometimes red, brown, or black; vitreous to greasy luster. Hardness 2.5, d 2.95–3.0. Refr index 1.338, mp 1000C. Soluble in concentrated sulfuric acid and in fused aluminum and ferric salts.

Occurrence: Colorado; the former U.S.S.R., Greenland (only commercial source).

Derivation: Synthetic product is made by fusing NaF and aluminum fluoride.

Use: Electrolyte in the reduction of alumina to aluminum; ceramics; insecticide; binder for abrasives; electric insulation; explosives; polishes.

Cryovac. A light, shrink-film, transparent packaging material based on polyvinylidene chloride. Used especially for meats and other perishables.

cryptands. See cavitands.

Cryptococcus. A fungi (encapsulated yeast) common in soil worldwide and often associated with pigeon droppings, eucalyptus trees, and decaying wood forming wood hollows in trees. There are over 30 species of *Cryptococcus. Cryptococus neoformans* and *Cryptococcus gattii* are responsible for nearly all cryptococcal infections in humans in animals and have been known to infect those with healthy immune systems.

Hazard: Pathogenic

cryptocyanine. (1,1′-diethyl-4,4′-carbocyanine iodide).
CAS: 4727-50-8. $C_{25}H_{25}N_2I$.
Properties: Solid. Mp 250–255C.
Use: Organic dye, soluble, used as a chemical shutter in laser operation.
See cyanine dye.

cryptostegia rubber. Rubber from leaves of *Cryptostegia grandiflora* and *C. madagascariensis*.

cryptoxanthin. (provitamin A; hydroxy-β-carotene).
CAS: 472-70-8. $C_{40}H_{56}O$.
A carotenoid pigment with vitamin A activity.
Properties: Garnet-red prisms with metallic luster. Mp 170C. Soluble in chloroform, benzene, and pyridine; slightly soluble in alcohol and methanol.
Occurrence: In many plants, egg yolk, butter, blood serum. Can be made synthetically.
Use: Nutrition, medicine.

crystal. The normal form of the solid state of matter. Crystals have characteristic shapes and cleavage planes due to the arrangement of their atoms, ions, or molecules, which form a definite pattern called a lattice. Crystals may be face-centered, body-centered, cubic, orthorhombic, monoclinic, prismatic, etc. They have flat surfaces, sharp edges, and a definite angle between a given pair of surfaces. The form of a crystal is called its "habit." Among the most important features of a crystal are its optical properties, chief of which is the index of refraction, i.e., the extent to which a beam of light is slowed on passing through the crystal. With respect to light transmission, a crystal may be isotropic or anisotropic. Anisotropic crystals can polarize light. See optical isomer; optical rotation.

Crystals also have electrical and magnetic properties now being used in computers and other electronic devices. Crystals are almost always imperfect and contain impurities (atoms of other elements). These are utilized in semiconductors. For methods of growing crystals, see nucleation.

Single crystals are used in masers, lasers, semiconductors, miniaturized components, and computer memory systems, and as "whiskers." Many metals are now available in large, single-crystal, form and such natural crystals as ruby, garnet, sapphire, etc., are used in these applications.
See crystallization; nucleation; liquid crystal; hole; vacancy.

crystal face. The recurring characteristic surface of a crystal or a plane parallel to it.

crystal face, indices. Reciprocals of intercepts of a crystal plane on reference axes based on a chosen system of coordinates.

crystal-growth step. A ledge on the surface of a crystal, one or more lattice spacings high, where crystal growth occurs.

crystalline rocks. Igneous or metamorphic rock.

crystal, liquid. See liquid crystal.

crystallite. That portion of a crystal whose constituent atoms, ions, or molecules form a perfect lattice, without strains or other imperfections. Single crystals may be quite large, but crystallites are usually microscopic.
See crystal.

crystallization. The phenomenon of crystal formation by nucleation and accretion. The freezing of water into ice is one of the commonest examples of crystallization in nature. Industrially, it is used as a means of purifying materials by evaporation and solidification. The sugar of commerce is made in this way. Similarly, salt cake is derived from crystallization of natural brines (e.g., Searles Lake). Nucleated crystallization is also used to form polycrystalline ceramic structures.
See crystal.

crystallogram. A photograph of an X-ray diffraction pattern of a crystal.

crystallographic systems. (crystal systems). A categorization of crystals according to their degree of symmetry. They are cubic, hexagonal, orthorhombic, tetragonal, monoclinic, and triclinic. The cubic form has the highest symmetry.

crystallography. The study of the crystal formation of solids, including X-ray determination of lattice structures, crystal habit, and the shape, form, and defects of crystals. When applied to metals, this science is called metallography.

crystals of Venus. See copper acetate.

crystal systems. See crystallographic systems.

crystal violet. See methyl violet.

"Crystamet" [Crosfield]. TM for sodium metasilicate pentahydrate.
Grade: Granular.
Use: Detergent, industrial cleaner, paper, and textiles.

C_2S. Abbreviation for dicalcium silicate as used in cement.
See cement, Portland.

C_3S. Abbreviation for tricalcium silicate as used in cement.
See cement, Portland.

cS. Abbreviation for centistoke.

Cs. Symbol for cesium.

C. scoparius. (Sarothamnus scoparius; broom; common broom; scotch broom).
An attractive rather hardy European shrub, naturalized in parts of North America, that is often planted as cover along roadsides or other dry gravelly embankments.
Properties: Contains the alkaloids quinolizidine and sparteine.
Hazard: Toxic; nausea, diarrhea, headache, vertigo, paralysis of the ileus, tachycardia, and circulatory collapse.

CS gas. See o-chlorobenzylidne malononitrile.

CSMA. Abbreviation for the Chemical Specialties Manufacturers Association.

C-stage resin. (resite).
The fully cross-linked phenol-formaldehyde resin, which is infusible and insoluble in all solvents.
See A-stage resin; phenol-formaldehyde resin.

CTFE. See chlorotrifluoroethylene.

CTP. Abbreviation for cytidine triphosphate.

c-type lectin. Any of a class of Ca^{++}-dependent lectins that have a distinctive sequence composed of their carbohydrate recognition domain.

Cu. Symbol for copper.

cua. A codon of leucine that directs the placement of leucine into a polypeptide.

cuam. Abbreviation for cuprammonium, the copper ammonium radical.

cubeb oil.
CAS: 8007-87-2.
Properties: From steam distillation of mature, unripe fruit of piper cubeba L. (Fam. Piperaceae). Colorless to light green liquid; spicy odor, slt acrid taste. D: 0.898–0.928, refr index: 1.492–1.502 @ 20°. Sol in fixed oils, mineral oil; insol in glycerin, propylene glycol.
Hazard: A skin irritant.
Use: Food additive.

cube root. A powdered insecticidal preparation containing 5% rotenone.

Cubex. An oriented silicon–iron alloy in rolled sheet form for use as cores for transformers and other inductive devices. The alloy sheet comprises cubic grains with faces parallel to the sheet surface. In one form, the sheet is doubly oriented with two directions of easy magnetization parallel to the surface to the sheet. One direction of easy magnetization is parallel to the rolling direction, and the second is perpendicular to the rolling direction.

cubic-centered. See body-centered structure.

cubic centimeter. (cc; mL).
Unit of volume for liquids and finely divided solids.
See milliliter.

Cubidow. Compacted salt comprising either or both calcium and sodium chloride.

cucumber alcohol. See trans,cis-2,6-nonadienol.

cucurbitacin a.
CAS: 6040-19-3. $C_{32}H_{46}O_9$.
Hazard: A poison.
Source: Natural product.

cue-lure. Generic name for 4-(p-hydroxphenyl)-2-butanone acetate. $CH_3COCH_2CH_2C_6H_4OOCH_3$.
Properties: Liquid. Boiling range 117–124C. Insoluble in water; soluble in most organic solvents.
Use: Insect attractant.

cuen. Abbreviation for cupriethylenediamine.

cullet. In ancient glass manufacturing, chunks of glass of varying sizes and colors furnished to artisans for shaping and finishing. In modern practice, the term refers to fragments of scrap glass from production operations that are collected and recycled to the furnace for remelting.

culm. Anthracite tailings, especially prevalent in eastern Pennsylvania. They represent a considerable source of energy that could be used, for example, in fluidized-bed boilers.

cumaldehyde. See cuminic aldehyde.

"Cumar" *[Neville].* TM for a series of neutral, stable, synthetic resins of the coumarone-indene type, manufactured from selected distillates of tar.
Use: Softener and tackifier in varnishes, floor tile, rubber products, printing ink, adhesives, and waterproofing materials, and in the leather, electrical, radio, paper, and other industries.
See coumarone-indene resin.

cumene. (isopropylbenzene).
CAS: 98-82-8. $C_6H_5CH(CH_3)_2$.

Properties: Colorless liquid. D 0.8620, bp 152.7C, wt/gal 7.19 lb (25C), fp −96C, refr index 1.489 (25C), flash p 115F (46C), autoign temp 795F (424C). Soluble in alcohol, carbon tetrachloride, ether, and benzene; insoluble in water. Combustible.
Derivation: (1) Alkylation of benzene with propylene (phosphoric acid catalyst) and (2) distillation from coal tar naphtha fractions or from petroleum.
Grade: Technical, research, pure.
Hazard: Toxic by ingestion, inhalation, and skin absorption; a narcotic. Moderate fire risk. Eye, skin, and upper respiratory tract irritant, and central nervous system impairment. Possible carcinogen.
Use: Production of phenol, acetone, and α-methylstyrene; solvent.

cumene hydroperoxide. (α,α-dimethylbenzyl hydroperoxide).
CAS: 80-15-9. $C_6H_5C(CH_3)_2OOH$.

Properties: Colorless to pale-yellow liquid. Slightly soluble in water; readily soluble in alcohol, acetone, esters, hydrocarbons, chlorinated hydrocarbons. Flash p 175F (79.4C). Combustible.
Derivation: A solution or emulsion of cumene is oxidized with air at approximately 130C.

Hazard: Toxic by inhalation and skin absorption. Strong oxidizing agent; may ignite organic materials.
Use: Production of acetone and phenol; polymerization catalyst, particularly in redox systems, used for rapid polymerization.

cumerone.
C_8H_{60}.
Properties: Colorless oily liquid. D 1.09.6, bp 165–175C. Insoluble in water.

cumic alcohol. See cuminic alcohol.

cumic aldehyde. See cuminic aldehyde.

cumidine. (*o*-isopropylaniline).
CAS: 643-28-7. $(CH_3)_2CHC_6H_4NH_2$.
Properties: Colorless liquid. D 0.957 (20/4C), fp −63C, bp 225C. Insoluble in water.
Use: Reagent in determination of tungsten.
See pseudocumidine.

cuminic alcohol. (*p*-isopropylbenzyl alcohol; cuminyl alcohol; cumic alcohol).
CAS: 536-60-7. $CH_2OH(C_6H_4)CH(CH_3)_2$.
Found in caraway seed.
Properties: Colorless liquid; caraway-like odor; aromatic taste. D 0.981 (15C), bp 248C, refr index 1.522 (24C). Insoluble in water; miscible with alcohol and ether. Combustible.
Use: Flavoring.

cuminic aldehyde. (cumic aldehyde; cumaldehyde; *p*-isopropylbenzaldehyde).
CAS: 122-03-2. $(CH_3)_2CHC_6H_4CHO$.
Properties: Colorless to yellow liquid; cumin odor. D 0.986 (22C), bp 235C. Insoluble in water; soluble in alcohol and ether. Combustible.
Use: Perfumery, flavoring, synthesis.

cumin oil. A dextrorotatory essential oil used in perfumery and flavoring.

cumminum cyminum linn., extract.
CAS: 84775-51-9.
Hazard: Moderately toxic. A reproductive hazard.
Source: Natural product.

cumulene. A chain of up to six double-bonded carbon atoms derived from acetylene.

cumylphenol.
CAS: 599-64-4. $C_6H_5C(CH_3)_2C_6H_4OH$.
Properties: White to tan crystals; characteristic phenol odor. Fp 72.0C, d 1.115 g/mL (25C), distillation range 188.9–190.9C (10 mm Hg), flash p 320F (160C). Combustible.
Use: Intermediate for resins, insecticides, lubricants.

cupellation process. A process for freeing silver, gold, or other nonoxidizing metals from base metals that can be oxidized. The metallic mixture is placed in a cupel, which is a shallow, porous cup, and roasted in a blast of air. The base-metal oxides are absorbed in the cupel, leaving the pure metal to be decanted.

cupferron. (ammonium nitroso-β-phenylhydroxylamine).
CAS: 135-20-6. $C_6H_5N(NO)ONH_4$.
Properties: Creamy-white crystals. Mp 163–164C. Soluble in water and alcohol.
Derivation: By treating an ether solution of β-phenylhydroxylamine with dry ammonia gas and amyl nitrite.
Use: Analytical reagent, especially for separation and precipitation of metals, e.g., copper, iron, vanadium.

cupiennius. The banana spider, which is sometimes found in produce or other materials imported into the U.S.

cupola. A vertical furnace, similar to a blast furnace, used for melting iron or other metals for casting.

cuprammonium process. A minor process for making rayon by dissolving cellulose in an ammoniacal copper solution and reconverting it to cellulose by treatment with acid.

cupreine. (hydroxycinchonine).
CAS: 524-63-0. $C_{19}H_{22}O_2N_2 \cdot 2H_2O$.
One of the cinchona alkaloids.
Properties: Colorless crystals. Mp (anhydrous) 198C. Soluble in alcohol; slightly soluble in water, chloroform, ether, benzene.
Derivation: From cuprea bark *Remijia pedunculata*.
Use: Medicine (antimalarial).

cupric. Form of the word *copper* used in naming copper compounds in which the copper has a valence of 2.
Look for the corresponding compound under copper.

cupric acetoarsenite. ((acetate)trimetaarsenitodicopper; acetoarsenite de cuivre; copper ethanoacetoarsenate; basle green; CI 77410; CI pigment green 21; copper acetoarsenite, solid; Paris green; copper acetate arsenite; emerald green; ENT 884; French green; genuine paris green; imperial green; king's green; meadow green; mineral green; mitis green; moss green; mountain green; neuwild green; new green; ortho p-g bait; parrot green; patent green; powder green; schweinfurterguren; Schweinfurt green; sowbug & cutworm bait; Swedish green; Vienna green).
$C_4H_6As_6Cu_4O_{16}$.
Properties: Bright blue-green, crystalline powder.
Hazard: Toxic; poison; gastrointestinal disturbances, muscular cramps, tremors, peripheral neuritis.

Use: Insecticide; in antifouling compositions and wood preservatives.

cupric arsenite. (arsenic acid copper(2+) salt (1:1); arsenious acid coper (2+) salt (1:1); Scheele's green; copper hydrogen arsorite).
CAS: 10290-12-7. $AsCuHO_3$
Properties: Yellowish-green, highly water-insoluble, crystalline powder, varying composition.
Use: Insecticide, fungicide, wood preservative.

cupric chloride. (copper chloride; copper (II) chloride; dichlorocopper).
CAS: 7447-39-4. Cl_2Cu.
Properties: Yellowish-brown, deliquescent powder, soluble in water, ethanol and ammonium chloride.

cupric chromate basic. See copper chromate.

cupric diethyldithiocarbamate. See copper(ii) diethyldithiocarbamate.

cupric cyanide. (cupric cyanamide; copper dicyanide).
CAS: 14763-77-0. C_2CuN_2
Properties: Yellowish-green powder, reacts violently with magnesium.
Hazard: Toxic.

cupric sulfate. (copper sulfate; copper monosulfate; hydrous copper sulfate; sulfuric acid, copper (2+) salt (1:1)).
CAS: 7758-98-7. CuH_2O_4S.
A sulfate salt of copper.
Properties: Water-soluble, hygroscopic salt, grayish-white to greenish-white crystals or amorphous powder.
Hazard: Strong irritant, systemic poison, toxic.
Use: Herbicide, insecticide, wood preservative, food and fertilizer additive, paints and dyes, electroplating solutions, copper-plating baths, therapeutically as an antidote to phosphorus poisoning, topical antifungal, emetic, anthelmintic, additive in copper deficiency of ruminants, algicide.

cupriethylene diamine. Purple liquid, ammoniacal odor. Dissolves cellulose products.
Hazard: Strong irritant to tissue.

cuprinol. A copper naphthenate or sodium pentachlorophenate preparation used as a wood and fabric preservative.

cuprite. (copper ore ruby; red oxide of copper).
Cu_2O.
Crimson, scarlet, vermilion, deep- or brownish-red; secondary mineral; adamantine or dull luster; brownish-red streak. Soluble in nitric and concentrated hydrochloric acids, d 5.85–6.15, Mohs hardness 3.5–4.

Occurrence: U.S., England, Germany, France, Siberia, Australia, China, Peru, Bolivia.
Use: Source of copper.

cupronickel. An alloy of copper and nickel used in coinage, and in condenser and heat-exchanger tubes. Most types contain from 10 to 30% nickel. Strongly corrosion resistant, especially to seawater.

Cuprophenyl. After-coppering direct dyes for cellulosic fibers.

cuprophyte. A plant adapted to, or having a high tolerance to copper in the soil.
Use: Indicator of copper-containing soils.

cuprotungsten. An alloy of copper and tungsten.

cuprous. Form of the word *copper* used in naming copper compounds in which the copper has a valence of 1.

cuprous acetate ammoniacal. See copper ammonium acetate.

cuprous acetylide.
Cu_2C_2.
Properties: Amorphous, red powder. A salt of acetylene.
Derivation: Reaction of acetylene with aqueous soluble of cuprous salts.
Hazard: Severe explosion risk when shocked or heated.
Use: Detonators and other explosive devices.

cuprous bromide.
CAS: 7787-70-4. CuBr or Cu_2Br_2.
Properties: White, crystalline solid becoming green in light. Mp 500C; d 4.71. Soluble in hydrochloric acid and ammonium hydroxide; decomposes in hot water; insoluble in acetone and sulfuric acid. Protect from light when stored.
Use: Catalyst in organic reactions.

cuprous chloride.
CAS: 7758-89-6. CuCl or Cu_2Cl_2.
Properties: White, cubic crystals. D 4.14. Mp 430C; bp 1490C. Becomes greenish on exposure to air, and brown on exposure to light. Slightly soluble in water; soluble in acids, ammonia, ether; insoluble in alcohol and acetone.
Derivation: Copper and cupric chloride solution or copper and hydrochloric acid in air.
Grade: Technical, reagent, single crystals.
Use: Catalyst, preservative and fungicide, desulfurizing and decolorizing agent in petroleum industry, absorbent for carbon monoxide.

cuprous cyanide.
CAS: 544-92-3. $Cu_2(CN)_2$ or CuCN.

Properties: Cream-colored powder. D 1.9, mp 475C. Insoluble in water; soluble in sodium and potassium cyanides, hydrochloric acid, and ammonium hydroxide.
Use: Electroplating, antifouling paints, insecticide, catalyst.

cuprous iodide. (copper iodide).
CAS: 7681-65-4. CuI.
Properties: White to brownish-yellow powder. D 5.653 (15C), bp 1290C, mp 606C. Soluble in ammonia and potassium iodide solutions; insoluble in water.
Derivation: Interaction of solutions of potassium iodide and copper sulfate.
Hazard: Toxic.
Use: Feed additive, in table salt as source of dietary iodine (up to 0.01%), catalyst, cloud seeding.

cuprous mercuric iodide.
CAS: 13876-85-2. Cu_2HgI_4.
Properties: Dark-red crystals that become dark brown to black when heated to about 65C. Soluble in water and alcohol.
Hazard: Toxic by ingestion.
Use: Temperature indicator for bearings and other moving machinery parts.

cuprous oxide. See copper oxide red.

cuprous potassium cyanide.
CAS: 13682-73-0. $KCu(CN)_2$.
Properties: Crystalline solid. D 2.38. Insoluble in water; soluble in dimethylsulfoxide. Decomposed by heating in water.
Derivation: By evaporating a water solution of cuprous cyanide and potassium cyanide.
Hazard: A poison by ingestion.
Use: Copper-electroplating vats.

cuprous selenide.
CAS: 20405-64-5. Cu_2Se.
Properties: Dark-blue to black crystalline solid. Mp 1100C, d 6.84. Soluble in sulfuric acid (sulfur dioxide evolved); also in potassium cyanide solution.
Use: Semiconductor research.

cuprous sulfide.
CAS: 22205-45-4. Cu_2S.
Properties: Black powder or lumps. D 5.52–5.82, mp approximately 1100C. Soluble in nitric acid and ammonium hydroxide; insoluble in water. Occurs as the mineral chalcocite.
Derivation: By heating cupric sulfide in a stream of hydrogen.
Grade: Technical, single crystals.
Use: Antifouling paints, solar cells, electrodes, solid lubricants, luminous paints, catalyst.

cuprous sulfite.
CAS: 35788-00-2. $Cu_2SO_3 \cdot H_2O$.

Properties: White, crystalline powder. D 3.83. Soluble in ammonium hydroxide, hydrochloric acid (decomposes); insoluble in water.
Use: Catalyst, fungicide, textile dyeing.

cuprous thiocyanate. (copper sulfocyanide). CAS: 1111-67-7. CuSCN.
Properties: Yellow-white powder. D 2.843, mp 1084C. Insoluble in water; soluble in ammonia.
Use: Manufacture of organic chemicals, antifouling paints, printing textiles, primers of explosives.

Curalon. A series of urethane curatives. Curalon L, a mixture of hindered aromatic primary diamines. Curalon M, p,p'-methylene-bis(o-chloroaniline).

curare.
CAS: 8063-06-7.
A highly toxic mixture of approximately 40 alkaloids occurring in several species of South American trees. It acts on the central nervous system, and derivatives are used to some extent in medicine as a muscle relaxant.
See snake venom.

curative. (1) Any substance or agent that effects a fundamental and desirable change in a material to make it suitable for practical use, e.g., meats, rubber, tobacco. (2) Any substance that combats disease by killing bacteria or that restores health by chemical means.
See curing.

curcumenone.
CAS: 100347-96-4. $C_{15}H_{22}O_2$.
Hazard: A poison.

curcumin. (tumeric yellow; 1,7-bis(4-hydroxy-methoxyphenyl)-1,6-heptadiene-3,5-dione; CI 75300).
CAS: 458-37-7. $[CH_3OC_6H_3(OH)CH:CHCO]_2CH_2$.
Properties: Orange-yellow needles. Mp 183C. Soluble in water and ether; soluble in alcohol.
Derivation: The coloring principle from curcuma.
Use: Analytical reagent, food dye, biological stain. As an acid–base indicator it is brownish-red with alkalies, yellow with acids (pH range 7.4–8.6); also an indicator for boron.

cure. (1) The length of time required for a plastic compound to stay in the mold for complete reaction so it becomes infusible and chemically inert. (2) To change the physical properties of a material by chemical reaction or vulcanization. This is usually accomplished by the action of heat and catalysts.

curie. (Ci).
The official unit of radioactivity, defined as exactly 3.70×10^{10} disintegrations per second. This decay rate is nearly equivalent to that exhibited by 1 of radium in equilibrium with its disintegration products. A millicurie (mCi) is 0.001 curie. A microcurie (μCi) is 1-millionth curie.

Curie law. The magnetic susceptibility of a paramagnetic substance varies inversely as the absolute temperature.

Curie, Marie S. (1867–1934). Born in Warsaw, Poland, she and her husband Pierre made an intensive study of the radioactive properties of uranium. They isolated polonium in 1898 from pitchblende ore. By devising a tedious and painstaking separation method, they obtained a salt of radium in 1912, receiving the Nobel Prize in physics for this achievement in 1903 jointly with Becquerel. In 1911, Mme. Curie alone received the Nobel Prize in chemistry. Her work laid the foundation of the study of radioactive elements, which culminated in control of nuclear fission.
See Rutherford, Sir Ernest.

curie point. Transition temperature above which ferromagnetism ceases to exist.

Curie-Weiss law. The magnetic susceptibility of a paramagnetic substance is inversely proportional to the increase of its temperature above a certain fixed temperature characteristic of the substance.

curing. Conversion of a raw product to a finished and useful condition, usually by application of heat and/or chemicals that induce physicochemical changes. Many food products require aging under specified temperature conditions. The more common types of curing are as follows:
(1) *Meats*: Use of sodium chloride, sugars, sodium nitrite, sodium nitrate, ascorbic acid. These not only act as preservatives, but also aid in color retention. Some types are subsequently smoked. Conversion of collagen to gelatin occurs as a result of "hanging" meat for several days.
(2) *Leather*: Treatment of hides and skins with tanning agents of vegetable or mineral origin. This converts the protein structure into a firm and durable product as a result of complexing reactions.
See tanning.
(3) *Tobacco*: Exposure for 3–5 days to temperatures from 37 to 65C to reduce moisture content, convert starches to reducing sugars, and discharge the chlorophyll, followed by aging from 1 to 5 years to remove odors and improve smoking quality.
(4) *Cheese*: Aging for 9–12 months at 4.5–10C to develop sharp flavor; the process is also called ripening.
(5) *Rubber*: Addition of sulfur and accelerator, followed by exposure to heat, which effects crosslinking. This converts the material from a thermoplastic to a thermosetting product. High-energy radiation can also be used.
See vulcanization.

curium.
Cm.
Synthetic radioactive element of atomic number 96, aw 244, valences 3, 4. Isotopes available: 244 and 242 (gram quantities).
Properties: Silvery-white metal. D 13.5, mp 1340C. Chemically reactive. More electropositive than aluminum. An α emitter. Biologically it is a bone-seeking element. Forms compounds such as CmO_2, Cm_2O_3, CmF_3, CmF_4, $Cm(OH)_3$, $CmCl_3$, $CmBr_3$, $CmI_3 \cdot Cm_2(C_2O_4)_3$.
Use: Thermoelectric power generation for instrument operation in remote locations on earth or in space vehicles.
See actinide series.

Curl, Robert F., Jr. (1933–). An American who won the Nobel Prize for chemistry along with Sir Harold W. Kroto and Richard E. Smalley in 1996, the 100th anniversary of Alfred Nobel's death. The trio won for the discovery of the C_{60} compound called buckminsterfullerene. He graduated from Rice University and received a Ph.D. from the University of California, Berkeley, in 1957.
See buckminsterfullerene; Kroto, Sir Harold W.; Smalley, Richard E.

current density. In an electroplating bath or solution, the electric current per unit area of the object or surface being plated. Expressed in amperes per square centimeter or, more usually, amperes per square decimeter.

Current Good Manufacturing Practices. See CGMP.

cuscohygrine. (1,3-bis(1-methyl-2-pyrrolidinyl-2-propanone; cuskhygrine; cuskohygrine; bellaradine; 1-[(2S)-1-methylpyrrolidin-2-yl]-3-[(2R)-1-methylpyrrolidin-2-yl]propan-2-one). $C_{13}H_{24}N_2O$.
Isolated with crude hygrine from the leaves of the common cinchona or cusco-bark tree, *cinchona pubescens*, and from those of *erythroxylum coca*.
Properties: Oily alkaloid.

Curtius rearrangement. Formation of isocyanates by thermal decomposition of acyl azides.

cutaneous anthrax. The form of the disease most common in humans causing more than 95% of reported cases. It is acquired by handling infected material.
Hazard: Lesions that open turning into a bluish-black necrotic mass, vomiting, profuse sweating, high fever, prostration, death.

cutback. A coating substance or varnish that has been diluted or thinned.

cuticle. An outer protective layer of pellicle or skin that is a secreted material that covers the free surface of epithelial cells and is impervious to water. In many plants and animals, it serves as a barrier to the penetration of water and many chemicals.

"Cutless" [*International Paper*]. TM for a turf growth regulator.
Use: To reduces the leaf blade elongation for a more compact plant form.

cutting fluid. A liquid applied to a cutting tool to assist in the machining operation by washing away the chips or serving as a lubricant or coolant. Commonly used cutting fluids are water, water solutions or emulsions of detergents and oils, mineral oils, fatty oils, chlorinated mineral oils, sulfurized mineral oils, and mixtures of the foregoing oils. Transparent grades are available.

cuu. A codon of leucine that directs the placement of leucine into a polypeptide.

Cyana. Used in connection with the textile finishes obtained by applying Aerotex resins and similar products.

Cyanamer. Water-soluble polymers.
Use: Antiscalants, antiprecipitants, viscosity modifiers, corrosion inhibitors, and dispersants in a wide range of water treatment applications.

cyanamide. (1) (cyanogenamide; carbodiimide).
CAS: 420-04-2. $HN:C:NH$ or $N:CNH_2$.
Properties: Deliquescent crystals. Mp 43C, d 1.08. Very soluble in water, alcohol, ether, phenols, ketones.
Hazard: Strong irritant to skin and mucous membranes; avoid inhalation or ingestion.
See calcium cyanamide.

cyanamide process. See nitrogen fixation; ammonia, anhydrous.

cyanatotributylstannane.
CAS: 4027-17-2. $C_{13}H_{27}NOSn$.
Hazard: A poison. Tributyl tin compounds are very toxic to marine life.

cyanatryn.
CAS: 21689-84-9. $C_{10}H_{16}N_6S$.
Hazard: A reproductive hazard.
Use: Agricultural chemical.

cyanazine. (2-[[4-chloro-6-(ethylamino)-1,3,5-triazin-2-yl]amino]-2-methylpropanenitrile; 2-[[4-chloro-6-(ethylamino)-s-triazin-2-yl]amino]-2-methylpropionitrile).
CAS: 21725-46-2. $C_9H_{13}ClN_6$.
Properties: White solid.
Hazard: Very toxic.
Use: Selective pre- and postemergence herbicide to control weeds in crops such as alfalfa, corn, cotton, sorghum, soybeans, and wetland.

cyanic acid.   See isocyanic acid.

cyanide.   (carbon nitride ion; cyanide, dry; cyanide anion; cyanide ion; cyanide solutions; cyanure; formonitril; hydrocyanic acid, ion; isocyanide; rcra wate number P030).
CAS: 57-12-5. CN^+.
Properties: Colorless gas, highly water-soluble, various forms, numerous compounds that contain the $-CN$ group; slight odor of bitter almonds.
Derivation: Occurs naturally in many plants usually in the form of glycosides; found in the smoke of various tobacco products and released by combustion of nitrogen-containing organic materials.
Hazard: Cellular asphyxiation, respiration inhibition, highly toxic; very poisonous.
Use: Fumigants, soil sterilizers, fertilizers, rodenticides.

cyanide pulp.   The mixture obtained by grinding crude gold and silver ore and dissolving the precious metal content in sodium cyanide solution.

cyanine dye.   One of a series of dyes consisting of two heterocyclic groups (usually quinoline nuclei) connected by a chain of conjugated double bonds containing an odd number of carbon atoms. Example: cyanine blue $C_2H_5NC_9H_6:CHC_9H_6NC_2H_5$. They include the isocyanines, merocyanines, cryptocyanines, and dicyanines.
Use: Sensitizers for photographic emulsions.

cyanmethemoglobin.   (cyanide methemoglobin).
A compound of cyanide and methemoglobin that results from the administration of methylene blue in cyanide poisoning.
Hazard: Somewhat toxic.

cyanoacetamide.   (malonamide nitrile; propionamide nitrile).
CAS: 107-91-5. $CNCH_2CONH_2$.
Properties: White crystals. Mp 119C, bp (decomposes). Soluble in water and alcohol. Combustible.
Derivation: Ammonolysis of cyanoacetic ester or dehydration of ammonium cyanoacetate.
Hazard: Toxic by ingestion.
Use: Organic pharmaceutical synthesis, plastics.

cyanoacetic acid.   (malonic nitrile).
CAS: 372-09-8. $CNCH_2COOH$.
Properties: White crystals, hygroscopic. Mp 66.1–66.4C; decomposes at 160C. Soluble in water, alcohol, and ether.
Derivation: Interaction of sodium chloroacetate and potassium cyanide solution.
Hazard: Toxic by ingestion.
Use: Organic synthesis.

cyanoacrylate adhesive.   An adhesive based on the alkyl 2-cyanoacrylates (see, e.g., methyl 2-cyanoacrylate). The latter are prepared by pyrolyzing the poly(alkyl)-2-cyanoacrylates produced when formaldehyde is condensed with the corresponding alkyl cyanoacetates. These adhesives have excellent polymerizing and bonding properties. To prevent premature polymerization, inhibitors are added. Supplied commercially as Eastman 910.

cyanoacrylate ester.   (superglue; CA or CAE).
An adhesive used in a fuming method to develop friction ridge detail.

cyanobacteria.   (singular, cyanobacterium; blue-green bacteria; formerly blue-green algae; cyanophyceae; cyanophyta; myxophyta).
A homogenous group of closely related unicellular, prokaryotic, photosynthetic, colonial or filamentous nonmotile microorganisms. Formerly considered algae, they are now usually considered a phylum of the kingdom monera. Most occur in surface water bodies or in moist terrestrial situations as free-living or symbiotic associates of lichens or fungi.
Hazard: Poison; toxic.

"Cyanobrik" [Du Pont].
CAS: 143-33-9.
TM for sodium cyanide granular briquettes.
Use: For metal extraction, cleaning, intermediate for pharmaceuticals, dyes, inks, pigments, insecticides.

cyanocarbon.   Any of a class of compounds in which the cyanide radical ($-CN$) replaces hydrogen in organic compounds, as in tetracyanoethylene, $(CN)_2C:C(CN)_2$. The compounds are quite reactive and form colored complexes with aromatic hydrocarbons.
See nitrile.

Cyanocel.   Chemically modified (cyanoethylated) cellulose.

cyanocobalamin.   (vitamin B_{12}).
CAS: 68-19-9. $C_{63}H_{88}CoN_{14}O_{14}P$.

The anti-pernicious-anemia vitamin. All vitamin B_{12} compounds contain the cobalt atom in its trivalent state. There are at least three active forms: cyanocobalamin, hydroxocobalmin, nitrocobalamin. Vitamin B_{12} is a component of a coenzyme that takes part in the shift of carboxyl groups within molecules. As such it has an influence on nucleic acid synthesis, fat metabolism, conversion of carbohydrate to fat, and metabolism of glycine, serine, methionine, and choline.

Properties: Dark-red crystals or red powder; odorless; tasteless. Very hygroscopic. Slightly soluble in water; soluble in alcohol; insoluble in acetone and ether.

Source: (Food) Liver, eggs, milk, meats, and fish. Commercial.

Source: produced by microbial action on various nutrients (spent antibiotic liquors, sugar-beet molasses, whey), also from sewage sludge.

Grade: USP, radioactive.

Use: Medicine (blood and nerve treatment), nutrition, animal-feed supplements.

n'-cyano-*n,n*-dimethylguanidine.
CAS: 1609-06-9. $C_4H_8N_4$.
Hazard: Moderately toxic by ingestion and skin contact. A mild eye irritant.

2-cyanoethyl acrylate.
CAS: 106-71-8. CH_2:$CHCOOCH_2CH_2CN$.
Properties: Liquid. D 2.0690, bp (polymerizes when heated), fp −16.9C, wt/gal 8.9 lb, flash p 255F (124C) (COC). Soluble in water. Combustible.
Hazard: Toxic by ingestion and inhalation.
Use: Forms polymers and copolymers for viscosity-index improvers, adhesives, textile finishes, and sizes.

cyanoethylation.
Process for introducing the group −OCH_2CH_2CH into an organic molecule by reaction of acrylonitrile with a reactive hydrogen such as that on a hydroxyl or amino group.

cyanoethyl sucrose.   (CES).
CAS: 18304-13-7. $C_{12}H_{14}O_3(OH)_{0.7}$ $(OC_2H_4CN)_{7.3}$.
Properties: Clear, very viscous, pale-yellow liquid. D 1.20 (20/20C), fp (sets to a glass at −10C), bp >300C, flash p >375F (190C), refr index 1.615 (25C). Combustible. Viscosity decreases very rapidly when heated; has high volume resistivity, low power factor, unusually high dielectric constant.
Use: Capacitor impregnation, phosphor binding in electroluminescent panels, modification of electrical properties in coatings.

cyano(4-fluoro-3-phenoxyphenyl)methyl-3-(2,2-dichloroethenyl)-2,2-dimethylcyclopropanecarboxylate.
Use: Food additive; insecticide.

cyanoformic chloride.
$CNCOCl$.

Properties: Oily liquid. Bp 126–18C (750 mm Hg).
Derivation: Reaction between phthaloyl chloride and the amide of ethyl oxalate.

Cyanogas.
A pesticide containing not less than 42% calcium cyanide; evolves hydrogen cyanide gas on exposure to atmospheric moisture.
Hazard: Poisonous.

cyanogen.   (dicyan; oxalonitrile).
CAS: 460-19-5. $N\equiv C-C\equiv N$
Properties: Colorless gas; pungent penetrating odor. Specific gravity 1.8064 (air = 1), fp −28C, bp −20.7C. Burns with a purple-tinged flame. Soluble in water, alcohol, and ether.
Derivation: (1) Potassium cyanide solution is slowly dropped into copper sulfate solution; (2) mercury cyanide is heated.
Grade: Technical, pure.
Hazard: Flammable limits in air 6–32%. Store away from light and heat. A very toxic material. Eye and lower respiratory tract irritant.
Use: Organic synthesis, welding and cutting metals, fumigant, rocket propellant.

cyanogenamide.   See cyanamide (1).

cyanogen azide.   (carbonpernitride).
$N=N^+=N-C\equiv N$.
Properties: Colorless, oily liquid. Unstable at room temperature.
Hazard: Explodes when shocked or heated; store and handle in solvents, e.g., acetonitrile.
Use: Organic synthesis; has wide range of reactivity.

cyanogen bromide.   (bromine cyanide).
CAS: 506-68-3. $BrC\equiv N$.
Properties: Crystals; penetrating odor. D 2.02, bp 61.2C, mp 52C, vap d 3.6. Slowly decomposed by cold water. Corrodes most metals. Soluble in water, alcohol, benzene, ether.
Derivation: (1) Action of bromine on potassium cyanide. (2) Interaction of sodium bromide, sodium cyanide, sodium chlorate, and sulfuric acid.
Hazard: A poison. Strong irritant to skin and eyes. Pulmonary edema and upper respiratory tract irritant.
Use: Organic synthesis, parasiticide, fumigating compositions, rat exterminants, cyaniding reagent in gold extraction processes.

cyanogen chloride.
CAS: 506-77-4. $CNCl$.
Properties: Colorless gas or liquid. D 1.2, bp 12.5C, fp −6C, vap d 2.1. Min purity 97 mole %. Soluble in water, alcohol, and ether.
Derivation: Action of chlorine on moist sodium cyanide suspended in carbon tetrachloride and kept cooled to −3C, followed by distillation.
Hazard: Toxic by ingestion and inhalation, strong irritant to eyes and skin. TLV: ceiling 0.3 ppm.

Use: Organic synthesis, tear gas, warning agent in fumigant gases.

cyanogen fluoride.   (fluorine cyanide).
CNF.
Properties: Colorless gas. Forms a white, pulverulent mass if cooled strongly, and sublimes at −72C. Insoluble in water.
Derivation: Interaction of silver fluoride and cyanogen iodide.
Hazard: Inhalation or ingestion poison, strong irritant to eyes and skin.
Use: Organic synthesis, tear gas.

cyanogen iodide.   (iodine cyanide).
CAS: 506-78-5.   CNI.
Properties: Colorless needles; very pungent odor; acrid taste. Mp 146.5C, d 1.84. Soluble in water, alcohol, and ether.
Derivation: By heating a metal cyanide with iodine.
Hazard: Strong irritant to eyes and skin. A poison.
Use: Taxidermists' preservatives.

"Cyanogran" [Du Pont].   TM for a 98% sodium cyanide in granular form. White, crystalline solid crushed to pass 100% through 10 mesh, retained on 50 mesh.
Hazard: A deadly poison by inhalation and ingestion, strong irritant to eyes and skin.

cyanoguanidine.   See dicyandiamide.

cyanomethyl acetate.   (methyl cyanoethanoate).
$CNCH_2COOCH_3$.
Properties: Colorless liquid. Fp −22C, bp 200C, d 1.12.
Hazard: Highly toxic.
Use: Organic synthesis, pesticide.

cyano(methylmercuri)guanidine.
(methylmercury dicyandiamide).
$CH_3Hg(NHC(:NH)NHCN$.
Properties: Crystals. Mp 156C. Soluble in water.
Hazard: A poison by inhalation or ingestion, strong skin irritant.
Use: Seed fungicide and disinfectant.

5-cyano-n-methyl-1-phenyl-1h-pyrazole-4-carboxamide.
CAS: 98477-03-3. $C_{12}H_{10}N_4O$.
Hazard: Moderately toxic by ingestion.

Cyanophenfos.
$C_{15}H_{14}NO_2PS$.
Properties: Colorless crystals. Mp 80C, refr index 1.58. Partially soluble in aromatic solvents and ketones; almost insoluble in water.
Hazard: A poison.
Use: Insecticide.

cyano(4-phenoxyphenyl)methyl 4-chloro α-(1-methylethylbenzeneacetate).
CAS: 66827-38-1. $C_{25}H_{22}ClNO_3$.
Hazard: Moderately toxic by ingestion.
Use: Agricultural chemical.

(+)cyano(3-phenoxyphenyl)methyl(ñ)-1-(difluoromethoxy)-α-(1-methylethyl)benzeneacetate.
Use: Food additive; insecticide.

Cyanophos.   (ciafos; O,O-dimethyl-O-(4-cyanophenyl)phosphorothioate).
CAS: 2636-26-2. $C_9H_{10}NO_3PS$.
Properties: Yellowish liquid. Bp 120C (0.1 mm Hg), fp 14C, refr index 1.54. Soluble in alcohol, methanol, acetone; slightly soluble in water. Decomposed by light and in alkaline environment.
Hazard: Poison. A cholinesterase inhibitor.
Use: Pesticide.

3-cyanopyridine.   (3-azabenzonitrile; nicotinonitrile).
CAS: 100-54-9. C_5H_4NCN.
Properties: Colorless liquid. Bp 206.2C, mp 49.6C. Soluble in water.
Use: Organic synthesis.

4-cyanopyridine.
CAS: 100-48-1. C_5H_4NCN.
Properties: Colorless liquid. Bp 195.4C, mp 78.5C. Partially soluble in water; soluble in most organic solvents.
Use: Organic synthesis.

(3ar,9bs)-n-(4-(8-cyano-1,3a,4,9b-tetrahydro-3h-benzopyrano(3,4-c)pyrrol-e-2-yl)butyl)-4-.
CAS: 273203-30-8. $C_{29}H_{29}N_3O_2$.
Hazard: A poison.

cyano (type II) pyrethroid.
A pyrethroid that contains an alpha-cyano group in the alcohol moiety.
Hazard: Profuse salivation and choreoathetosis.

cyanuramide.   See ammelide.

cyanurdiamide.   See ammeline.

cyanuric acid.   (tricarbimide; tricyanide).
CAS: 108-80-5. $HOCHC(OH)NC(OH)N·2H_2O$.
Properties: White crystals; odorless; slight bitter taste. D 1.768, decomposes to cyanic acid at 320C. Soluble in hot water and concentrated mineral acids, insoluble in alcohol and acetone.
Use: Intermediate for chlorinated bleaches, selective herbicide, whitening agents.
See isocyanuric acid.

cyanuric   chloride.   (2,4,6-trichloro-1,3,5-triazine).
CAS: 108-77-0. $C_3N_3Cl_3$ (cyclic).

Properties: Crystals; pungent odor. D 1.32, mp 146C, bp 194C (764 mm Hg). Soluble in chloroform, carbon tetrachloride, hot ether, dioxane, ketones; very slightly soluble in water (hydrolyzes in cold water).
Hazard: Toxic by ingestion and inhalation.
Use: Chemical synthesis, dyestuffs, herbicides, optical brighteners.

cyanurtriamide. See melamine.

cycasin. (methyl-Onn-azoxy)methyl β-D-glucopyranoside; methylazoxymethanol β-D-glucoside; β-D-glucosyloxyazoxymethane; (Z)-methyl-oxido-[[(2S,3R,4S,5S,6R)-3,4,5-trihydroxy-6-(hydroxymethyl)oxan-2-yl]oxymethylimino]azanium).
CAS: 14901-08-7. $C_8H_{16}N_2O_7$.
A naturally occurring alkylating agent that is produced by sago palms (*Cycas*), and is found in flour made from the sago nut.
Hazard: Very toxic, possible carcinogen, hepatotoxic, teratogen, neurotoxic.

cyclain. See osmocaine.

cyclamate. Group name for synthetic nonnutritive sweetening agents derived from cyclohexylamine or cyclamic acid. The series includes sodium, potassium, and calcium cyclamates. As a result of a study made on laboratory animals in 1970, which indicated that these compounds cause genetic damage in chick embryos and cancer in rats at high dosage, their use in beverages and food products was banned in the U.S. More recent research has failed to confirm the carcinogenicity of these compounds in laboratory animals even at levels up to 240 times human intake. Notwithstanding these results, FDA has not yet withdrawn its ban on use of cyclamates as food additives or as table-top sweeteners, in view of the continuing uncertainty about their safety.
See sweetener, nonnutritive.

cyclamate calcium dihydrate. See calcium cyclamate dihydrate.

cyclamen alcohol. The alcohol corresponding to cyclamen aldehyde, used as a stabilizer of cyclamen aldehyde.

cyclamen aldehyde. (methyl-*p*-isopropylphenylpropyl aldehyde).
CAS: 103-95-7. $(CH_3)_2CHC_6H_4CH(CH_3)CH_2CHO$.
Properties: Colorless liquid; floral odor. D 0.949–0.959, refr index 1.507–1.520. Soluble in 1 vol of 80% alcohol, and in most oils.
Grade: FCC.
Use: Perfumery, soap perfumes, flavoring.

cyclamic acid. (USAN name for cyclohexanesulfamic acid; cyclohexylsufamic acid).
CAS: 100-88-9. $C_6H_{11}NHSO_3H$.

Properties: Odorless, white, crystalline solid; sweet-sour taste. Mp 170C. Strong, stable acid; soluble in water and alcohol; insoluble in oils.
Hazard: Suspected carcinogen.
Use: Nonnutritive sweetener, acidulant.
See cyclamate.

cyclanilide.
CAS: 113136-77-9. $C_{11}H_9Cl_2NO_3$.
Hazard: A poison. Moderately toxic by skin contact and inhalation.

cycle compound. See carbocyclic.

cyclethrin. (3-(2-cyclopentenyl)-2-methyl-4-oxo-2-cyclopentenyl ester of chrysanthemummonocarboxylic acid).
Properties: Viscous, brown liquid. Soluble in petroleum solvents and other common organic solvents. Formulated principally as liquid for spray applications corresponding to natural pyrethrins.
Hazard: Toxic by inhalation and ingestion.
Use: Insecticide with applications similar to allethrin and other analogs.
See furethrin; barthrin; ethythrin.

cyclic amp. (adenosine cyclic 3′,5′-(hydrogen phosphate); adenosine 3′,5′-cyclic monophosphate; adenosine 3′,5′-monophosphate; adenosine 3′,5′-phosphate; cyclic adenosine 3′,5′-monophosphate; acrasin; 3′,5′-AMP; cAMP; (1S,6R,8R,9R0-8-(6-aminopurin-9-yl)-3-hydroxy-3-oxo-2,4,7-trioxa-3,5-phosphabicyclo[4.3.0]nonan-9-ol).
CAS: 60-92-4. $C_{10}H_{12}N_5O_6P$.
A substance that is widely distributed among living organisms at concentrations from 10^{-7} to 10^{-6} moles/kg body weight. An adenine nucleotide containing one phosphate group that is esterified to both the 3′- and 5′-positions of the sugar moiety. It is a second messenger and a key intracellular regulator, functioning as a mediator of activity for a number of hormones. It is the conversion product of ATP catalyzed by adenylate cyclase.

cyclic compound. An organic compound whose structure is characterized by one or more closed rings, it may be mono-, bi-, tri-, or polycyclic, depending on the number of rings present. There are three major groups of cyclic compounds: (1) alicyclic, (2) aromatic (also called arene), and (3) heterocyclic. For more detailed information, consult specific entries.

cyclic electron flow. In chloroplasts, the light-induced flow of electrons originating from and returning to photosystem I to produce ATP without production of NADPH.

cyclic gmp. (guanosine cyclic 3′,5′-(hydrogen phosphate); cyclic guanosine-3′,5′-monophosphate; guanosine 3,5′-monophosphate; guanosine 3′,5′-cyclic monophosphate; guanosine

3′,5′-cyclic phosphate; guanosine 3′,5′-gmp; cgmp; 2-amino-9-[(1S,6R,8R,9R)-3,9-dihydroxy-3-oxo-2,4,7-trioxa-3,5-phosphabicyclo[4.3.0]nonan-8-yl]-3H-purine-6-one).
CAS: 7665-99-8. $C_{10}H_{12}N_5O_7P$.
A guanine nucleotide containing one phosphate group, which is esterified to the sugar moiety in both the 3′- and 5′-positions. It is the conversion product of GTP catalyzed by the enzyme guanylate cyclase. It is a regulatory agent of many intracellular processes and is sometimes referred to as "second messenger." Its levels increase in response to a variety of hormones and it has been found to activate specific protein kinases.

cyclic photophosphorylation.   ATP synthesis driven by cyclic electron flow through photosystem I (no NADPH produced).

cyclizine hydrochloride.   (1-diphenylmethyl-4-methylpiperazine hydrochloride).
CAS: 303-25-3. $(C_6H_5)_2CHC_4H_8N_2CH_3·HCl$.
Properties: White, crystalline powder or small colorless crystals; odorless or nearly so; bitter taste. Mp 285C (decomposes). Slightly soluble in water, alcohol, chloroform; insoluble in ether; pH (2% solution) 4.5–5.5.
Grade: USP.
Use: Medicine (antiemetic).

cycloaliphatic epoxy resin.   (cycloalkenyl epoxides).
A polymer prepared by epoxidation of multicycloalkenyls (polycyclic aliphatic compounds containing carbon–carbon double bonds) with organic peracids such as peracetic acid. Resistant to high temperatures.
Use: Space vehicles, outdoor electrical installations in polluted and humid atmospheres, high-temperature adhesives.

cycloate.   (S-ethyl-N-cyclohexyl-N-ethlcarbamothioate).
CAS: 1134-23-2. $C_{11}H_{21}NOS$.
A thiocarbamate.
Use: Herbicide.

cyclobarbital.   [5-(1-cyclohexenyl)-5-ethylbarbituric acid; tetrahydrophenobarbital].
CAS: 52-31-3. $C_{12}H_{16}N_2O_3$.
Properties: White crystals or crystalline powder; odorless; bitter taste. Mp 170–174C. Soluble in alcohol or ether; very slightly soluble in cold water or benzene.
Derivation: Hydrogenation of phenobarbital with colloidal palladium in alcohol as a catalyst.
Hazard: See barbiturates.
Use: Medicine (hypnotic, sedative).

cyclobutane.   (tetramethylene).
CAS: 287-23-0. C_4H_8.

$$CH_2{-}CH_2$$
$$CH_2{-}CH_2$$

Properties: Colorless gas. D 0.7083 (11C), bp 13C, fp −80C, flash p < 50F. Insoluble in water; soluble in alcohol and acetone.
Derivation: Catalytic hydrogenation of cyclobutene.
Hazard: Flammable, dangerous fire risk.

cyclobutene.   (cyclobutylene).
CAS: C_4H_6,
Properties: Gas. D 0.733, bp 2.0C.
Derivation: From petroleum.
Hazard: Flammable, dangerous fire risk.

cyclobutylidinecyclobutane.   See bicyclobutylidine.

cyclochlorotine.   (chlorine-containing peptide; 17,18-dichloro-13-eithyl-3,10-bis(hydroxymethyl)-7-phenyl-1,4,8,11,14-pentazabicyclo[14.3.0]nonadeca-2,5,9,12,15-pentone; islanditoxin).
CAS: 12663-46-6 $C_{24}H_{31}Cl_2N_5O_7$.
Properties: Chlorinated cyclic peptapeptide; white needles.
Derivation: Isolated from *Penicillium islandicum*.
Hazard: Extremely toxic; deadly poison; hepatotoxic; mycotoxic; tumorigen.

cyclocitrylideneacetone.   See ionone.

cyclocumarol.
CAS: 518-20-7. $C_{20}H_{18}O_4$.
A synthetic blood anticoagulant.
Properties: White, crystalline powder; slight odor. Mp 164–168C. Soluble in water; slightly soluble in alcohol.

cyclodiene insecticide.   Any of a class of chlorinated insecticides based on a cyclodiene ring structure.
Hazard: Highly toxic.

cyclodisone.   See 1,5,2,4-dioxadithiepane-2,2,4,4-tetraoxide.

cyclofenil diphenol.   See 4-(cyclohexylidene(4-hydroxyphenyl)methyl)phenol.

cycloheptane.   (heptamethylene; suberane).
CAS: 291-64-5. C_7H_{14}.
Properties: Colorless liquid. D 0.809, bp 117, fp −12C, aniline equivalent −6, flash p >70F (20C). Soluble in alcohol; insoluble in water.
Grade: Technical.
Hazard: Flammable, dangerous fire risk. Narcotic by inhalation.
Use: Organic synthesis.

cycloheptanone.   (suberone).
  CAS: 502-42-1. $C_7H_{12}O$.
  Properties: Colorless liquid; peppermint odor. Bp 179C, d 0.95. Insoluble in water; soluble in ether and alcohol. Combustible.
  Use: Research, intermediate.

cyclohexane.   (hexamethylene; hexanaphthene; hexalhydrobenzene).
  CAS: 110-82-7. C_6H_{12}.

Structure: A typical alicyclic hydrocarbon. It exists in two modifications called the "boat" and the "chair," as shown. This is due to slight distortion of the bond angles in accordance with the modified version of Baeyer's strain theory. Cyclohexane has been studied extensively on a theoretical basis in a branch of advanced chemistry called conformational analysis. See conformation.

chair   boat

Properties: Colorless, mobile liquid; pungent odor. D 0.779 (20/4C), bp 80.7C, mp 6.5C, refr index 1.4263, vap press 100 mm Hg at 60.8C, vap d 2.90, aniline equivalent 7. Insoluble in water; soluble in alcohol, acetone, benzene; flash p (98% grade) −1F (−18.3C) (CC); autoign temp 473F (245C). Flammable limits in air 1.3–8.4%.
  Derivation: (1) Catalytic hydrogenation of benzene. (2) Constituent of crude petroleum.
  Grade: 85, 98, 99.86%, spectrophotometric.
  Hazard: Flammable, dangerous fire risk. Moderately toxic by inhalation and skin contact. Central nervous system impairment.
  Use: Manufacture of nylon; solvent for cellulose ethers, fats, oils, waxes, bitumens, resins, crude rubber; extracting essential oils; chemicals (organic synthesis, recrystallizing medium); paint and varnish remover; glass substitutes; solid fuels; fungicides; analytical chemistry.

1,4-cyclohexanebis(methylamine).
  CAS: 2549-93-1. $C_6H_{10}(CH_2NH_2)_2$.
  The commercial product is about 40% *cis* and 60% *trans*.
  Properties: Clear liquid. D 0.9419 (20/4C), bp 239–244C. Miscible with water, alcohol, and most other organic solvents. Combustible.
  Use: Intermediate, resins.

cyclohexanecarboxylic acid.   See hexahydrobenzoic acid.

1,2-cyclohexanedicarboxylic anhydride. See hexahydrophthalic anhydride.

1,4-cyclohexanedimethanol.   (CHDM).
  CAS: 105-08-8. $C_6H_{10}(CH_2OH)_2$.
  cis and *trans* isomers are known and are present in the commercial product in about 30–70%.
  Properties: Liquid. Bp 286.0C (735 mm Hg *cis* isomer), mp 41–61C, d (super cooled) 1.0381 (25/4C), flash p 330F (165C) (COC), refr index 1.4893 (20C). Soluble in water and ethyl alcohol. Combustible.
  Use: Polyester films and protective coatings, reduction of reaction time in esterification.

cyclohexanesulfamic acid.   See cyclamic acid.

cyclohexanol.   (hexahydrophenol).
  CAS: 108-93-0. $C_6H_{11}OH$.
  Properties: Colorless, oily liquid; camphorlike odor; hygroscopic. D 0.937 (37.4C), mp 23C, bp 160.9C, wt/gal approximately 8 lb, flash p 154F (67.7C), refr index 1.465 (22C). Sparingly soluble in water; miscible with most organic solvents and oils. Combustible, autoign temp 572F (300C).
  Derivation: Phenol is reduced with hydrogen over active nickel at 71–76.6C. The cyclohexanone is removed by condensing with benzaldehyde in the presence of alkali.
  Grade: Technical (contains freezing inhibitor).
  Hazard: Toxic by skin absorption and inhalation; narcotic. Eye irritant, and central nervous system impairment.
  Use: In soap making to incorporate solvents and phenolic insecticides; source of adipic acid for nylon; textile finishing; solvent for alkyd and phenolic resins, cellulosics; blending agent for lacquers, paints, and varnishes; finish removers; emulsified products; leather degreasing; polishes, plasticizers, plastics; germicides.

cyclohexanol acetate.   (cyclohexanyl acetate).
  CAS: 622-45-7. $CH_3COOC_6H_{11}$.
  Properties: Colorless liquid; odor resembling that of amyl acetate. D 0.966, bp 177C, flash p 136F (57C), autoign temp 633F (333C). Miscible with most lacquer solvents and dilutions and with halogenated and hydrogenated hydrocarbons; soluble in alcohol; insoluble in water. Combustible.
  Hazard: Narcotic.
  Use: Solvent for nitrocellulose, cellulose ether, bitumens, metallic soaps, basic dyes, blown oils, crude rubber, many natural and synthetic resins and gums, lacquers.

cyclohexanone.   (pimelic ketone; ketohexamethylene).

CAS: 108-94-1.   C₆H₁₀O.

Properties: Water-white to pale-yellow liquid; acetone- and peppermint-like odor. Bp 156.7C, fp −32C, d 0.948, flash p 111F (44C), refr index 1.4507 (20C), vap press 135 mm Hg (100C), autoign temp 788F (420C). Slightly soluble in water; miscible with most solvents. Combustible.
Derivation: By passing cyclohexanol over copper with air at 280F, also by oxidation of cyclohexanol with chromic acid or oxide.
Hazard: Moderate fire risk. Toxic via inhalation and skin contact. TLV: 25 ppm.
Use: Organic synthesis, particularly of adipic acid and caprolactam (about 95%), polyvinyl chloride and its copolymers, and methacrylate ester polymers; wood stains; paint and varnish removers, spot removers; degreasing of metals; polishes; leveling agent dyeing and delustering silk; lubricating oil additive; solvent for cellulosics; natural and synthetic resins, waxes, fats, etc.

cyclohexanone peroxide.   (1-hydroperoxycyclohexyl-1-hydroxycyclohexyl peroxide).
CAS: 78-18-2.   C₆H₁₀(OOH)OOC₆H₁₀OH.
Properties: Grayish paste. Insoluble in water; soluble in most organic solvents.
Hazard: Dangerous fire risk, powerful oxidizer.

cyclohexanyl   acetate.   See   cyclohexanol acetate.

cyclohexene.   (1,2,3,4-tetrahydrobenzene).
CAS: 110-83-8.   C₆H₁₀.

H₂C—CH=CH—CH₂—CH₂—CH₂ (ring structure)

Properties: Colorless liquid. D 0.811 (20/4C), bp 83C, fp −1037C, refr index 1.445 (25C), flash p 11F (−11.6C), aniline equivalent 10, wt/gal 6.7 lb (25C), autoign temp 590F (310C). Soluble in alcohol; insoluble in water.
Grade: Technical 95%, 99%, research 99.9 mole %.
Hazard: Flammable, dangerous fire risk. Toxic by inhalation. Eye and upper respiratory tract irritant.
Use: Organic synthesis, catalyst solvent, oil extraction, manufacturing of adipic and maleic acids.

3-cyclohexene-1-carboxaldehyde.   (1,2,3,6-tetrahydrobenzaldehyde).

CH₂CH:CHCH₂CH₂CHCHO.

Properties: Liquid. D 0.9721, bp 164.2C, fp −100C, wt/gal 8.1 lb (20C), flash p 135F (57.2C). Slightly soluble in water. Combustible.
Hazard: Strong irritant to tissue.
Use: Intermediates, improves water resistance of textiles.

cyclohexene-1-dicarboximidomethylchrysanthemate.   See neopinamine.

cyclohexene oxide.
CAS: 286-20-4.   C₆H₁₀O.
Properties: Colorless liquid; a strong odor. Bp 129–130C, d 0.967 (25/4C), refr index 1.4503 (25C), flash p 81F (27.2C). Soluble in alcohol, ether, acetone; insoluble in water.
Grade: 98% pure.
Hazard: Flammable, dangerous fire risk.
Use: Chemical intermediate.

cyclohexenylethylbarbituric acid.   See cyclobarbital.

cyclohexenylethylene.   See vinylcyclohexene.

cyclohexenyltrichlorosilane.
CAS: 10137-69-3.   C₆H₉SiCl₃.
Properties: Colorless liquid. Bp 202C, d 1.263 (25/25C), refr index 1.488 (25C), flash p 200F (93.3C) (COC). Readily hydrolyzed by moisture, with liberation of hydrogen chloride. Combustible.
Grade: Technical.
Hazard: Strong irritant, corrosive to skin.
Use: Intermediate for silicones.

cycloheximide.   (generic name for 3-[2-(3,5-dimethyl-2-oxocyclohexyl)-2-hydroxyethyl] glutarimide).
CAS: 66-81-9.

CH₂CH(CH₃)CH₂CH(CH₃)COCHCH(OH)

CH₂CHCH₂CONHCOCH₂.

A plant growth regulator. By-product in the manufacture of streptomycin.
Properties: Crystals. Mp 115.5–117C. Slightly soluble in acetone, alcohol, and chlorinated solvents.
Hazard: Toxic by ingestion.
Use: Fungicide, antibiotic, abscission of citrus fruit in harvesting, turf disease control.

cyclohexylamine.   (hexahydroaniline; aminocyclohexane).
CAS: 108-91-8.   C₆H₁₁NH₂.
Properties: Colorless liquid; unpleasant odor. Bp 134.5C, d 0.8647 (25/25C), fp −18C. Strong organic base, pH of 0.01% aqueous solution 10.5, forms an azeotrope with water, bp 96.4C. Miscible with most solvents, flash p 90F (32.2C) (OC), autoign temp 560F (293C).
Grade: Technical (98%).

Hazard: Flammable, moderate fire risk. Toxic by ingestion, inhalation, skin absorption. Eye and upper respiratory tract irritant. Questionable carcinogen.

Use: Boiler-water treatment, rubber accelerator, intermediate in organic synthesis.

cyclohexylbenzene.   See phenylcyclohexane.

n-cyclohexyl-2-benzothiazolesulfenamide.
(benzothiazyl-2-cyclohexylsulfenamide).
CAS: 95-33-0. $C_6H_4SNCSNHC_6H_{11}$.
Properties: Cream-colored powder. D 1.27, melting range 93–100C. Insoluble in water; soluble in benzene.
Use: Rubber accelerator.

cyclohexyl bromide.
CAS: 108-85-0. $C_6H_{11}Br$.
Properties: Liquid, not more than faintly yellow, penetrating odor, d 1.32–1.34 (25/25C), refr index 1.4926–1.4936 (25C).

((cyclohexylcarbonyl)oxy)tributylstannane.
See tributyltin cyclohexanecarboxylate.

1-(cyclohexylcarbonyl)-1,2,3,6-tetrahydropyridine.
CAS: 63697-52-9. $C_{12}H_{19}NO$.
Hazard: Moderately toxic by ingestion. A moderate eye irritant.
Use: Agricultural chemical.

cyclohexyl chloride.
CAS: 542-18-7. $C_6H_{11}Cl$.
Properties: Colorless liquid. Fp −43C, bp 142C, flash p 89F (31.6C), d 0.992.
Hazard: Flammable, moderate fire risk.

2-cyclohexylcyclohexanol.
CAS: 6531-86-8. $C_6H_{11}H_{10}OH$.
Properties: Colorless liquid. Fp 29C, bp 271–277C, d 0.977 (25/25C), wt/gal 8.13 lbs, refr index 1.495 (25C), flash p 255F (124C). Soluble in methanol and ether; slightly soluble in water. Combustible.

cyclohexyl 4-(1,1-(dimethylethyl)phenyl) methyl-3-pyridinylcarbonimidodithioate.
CAS: 42754-23-4. $C_{23}H_{30}N_2OS_2$.
Hazard: Moderately toxic by ingestion.
Use: Agricultural chemical.

2-cyclohexyl-4,6-dinitrophenol.   See dinitrocyclohexylphenol.

1-cyclohexylethanol.
CAS: 1193-81-3. $C_8H_{16}O$.
Hazard: Moderately toxic by ingestion and skin contact. A severe skin and mild eye irritant.

4-(cyclohexylidene(4-hydroxyphenyl) methyl)phenol.
CAS: 5189-40-2. $C_{19}H_{20}O_2$.
Hazard: A reproductive hazard.

cyclohexylidenemalononitrile.
CAS: 4354-73-8. $C_9N_{10}N_2$.
Hazard: A poison.

cyclohexyl isocyanate.
CAS: 3173-53-3. $C_6H_{11}NCO$.
Use: To form cyclohexyl carbamates or ureas for agricultural chemicals or pharmaceutical use.

cyclohexyl methacrylate.
CAS: 101-43-9. $H_2C:C(CH_3)COOC_6H_{11}$.
Properties: Colorless monomeric liquid; pleasant odor. Bp 210C, refr index 1.4578 (20C), d 0.9626 (20/20C), viscosity 5.0 cP (25C). Insoluble in water. Combustible.
Use: Optical lens systems, dental resins, encapsulation of electronic assemblies.

2-cyclohexyl-4-methylphenol.
CAS: 1596-09-4. $C_{13}H_{18}O$.
Hazard: Moderately toxic by ingestion. A severe eye irritant.

p-cyclohexylphenol.
CAS: 1131-60-8. $C_6H_{11}C_6H_4OH$.
Properties: Crystals. Mp 120C (min). Combustible.
Grade: Technical.
Use: Intermediate for resins and organic synthesis.

cyclohexyl phenyl phosphate.
CAS: 4281-67-8. $C_{18}H_{21}O_4P$.
Hazard: Moderately toxic by ingestion. A mild eye irritant.

n-cyclohexylpiperdine.
$C_6H_{11}NC_5H_{10}$.
Properties: Yellow liquid. Refr index 1.4856 (20C). Combustible.
Use: Intermediate.

cyclohexyl stearate.
CAS: 104-7-4. $C_6H_{11}OOCC_{17}H_{35}$.
Properties: Pale-yellow powder. D 0.882 at 30/15.5C, mp 26–28C. Soluble in benzene, toluene, and acetone; insoluble in water.
Use: Plasticizer for natural and synthetic resins.

cyclohexylsulfamic acid.   See cyclamic acid.

n-cyclohexyl-p-toluenesulfonamide.
CAS: 80-30-8. $C_6H_{11}NHSO_2C_6H_4CH_3$.
Properties: Yellow-brown fused mass, relatively light stable. Mp 86C, bp 350C. Soluble in alcohol, esters, ketones, aromatic hydrocarbons, and vegetable oils; insoluble in water. Compatible with a wide variety of resins including most of the cellulosic and vinyl resins. Combustible.
Use: Resin plasticizer.

cyclohexyl trichlorosilane.
CAS: 98-12-4. $C_6H_{11}SiCl_3$.
Properties: Colorless to pale-yellow liquid. Bp 206C, d 1.226 (25/25C), refr index 1.4759 (25C), flash p 185F (85C) (COC). Readily hydrolyzed by moisture with liberation of hydrogen chloride. Combustible.
Derivation: By Grignard reaction of silicon tetrachloride and cyclohexylmagnesium chloride.
Grade: Technical.
Hazard: Toxic by ingestion and inhalation, strong irritant to tissue.
Use: Intermediate for silicones.

cycloleucine.   (1-aminocyclopentanecarboxylic acid; Acpc; 1-aminocyclopentane-1-carboxylic acid).
CAS: 52-52-8. $C_6H_{11}NO_2$.
A synthetic amino acid formed by cyclization of leucine, which acts as an immunosuppressive agent and as a valine antagonist.
Hazard: Very toxic.
Use: Therapeutically as an antineoplastic agent.

"Cyclolube" [Tricor].   TM for a series of oils composed principally of cycloparaffins. Used as plasticizers for nonpolar polymers, lubricants for polar polymers, and extenders for relatively saturated polymers.

cyclomethylcaine.   (3-(2-methylpiperidine) propyl-*p*-cyclohexyloxybenzoate).
CAS: $C_{22}H_{33}NO_3$.
Properties: White, crystalline powder; odorless. Sparingly soluble in water, alcohol, and chloroform; very slightly soluble in acetone, ether, and dilute acids.
Use: Medicine (topical anesthetic).

cyclooxygenase.   An enzyme that catalyzes the conversion of arachidonic acid to one of a number of prostaglandins.

cyclone.   A dust-collecting device consisting of a cylindrical chamber the lower portion of which is tapered to fit into a cone-shaped receptacle placed below it. The dust-laden air enters through a vertical slotlike duct on the upper wall of the chamber at the rate of at least 100 f/sec. Since the particles enter at a tanget, they whirl in a circular or cyclonic path within the chamber. The centrifugal force exerted on the particles is proportional to their weight and to the square of their velocity. The particles slide along the walls of the chamber and gradually circulate down into the conical receptor while the clean air escapes through a central pipe at the bottom. The dust accumulates in the cone and is discharged at intervals or continuously. The larger the particles, the more efficient the removal; in simple cyclones particles those less than 50 microns in diameter are not retained, but improved models retain particles

as small as 20 microns. Cyclones are also used in cleaning and firing pulverized coal.

cyclonite.   (*sym*-trimethylenetrinitramine; hexahydro-1,2,5-trinitro-*sym*-triazine; trinitrotrimethylenetriamine; cyclotrimethylenetrinitramine; RDX).
CAS: 121-82-4. $C_3H_6N_6O_6$.
Properties: White, crystalline solid. D 1.82, mp 203.5C. Soluble in acetone; insoluble in water, alcohol, carbon tetrachloride, and carbon disulfide; slightly soluble in methanol and ether.
Derivation: Reaction of hexamethylenetetramine with concentrated nitric acid.
Hazard: High explosive, easily initiated by mercury fulminate. Toxic by inhalation and skin contact. Liver damage. Questionable carcinogen.
Use: Explosive 1.5 times as powerful as TNT.

3-cyclooctadiene.

$\overline{HC{:}CH(CH_2)_2CH{:}CHCH_2CH_2}$. Intermediate for such compounds as suberic acid, 1,5-cyclooctadiene. A butadiene dimer.
Properties: Liquid. Fp −56.39C, distillation range 301–303F (technical), bp 149.34C (pure), d 0.88328 (20/4C), wt/gal 7.38 lbs, vap press 0.50 psia (37.7C), refr index 1.4933 (20C), flash p 100F (37.7C). Combustible.
Derivation: Catalytic dimerization of butadiene.
Grade: Technical 95%, 99%, 99.8 mole %.
Hazard: Moderate fire risk.
Use: Resin intermediate, third monomer in EPT rubber.

cyclooctane.
CAS: 292-64-8. C_8H_{16}.
Properties: Colorless liquid. D 0.835, bp 148C, mp 14C. Combustible.

1,3,5,7-cyclooctatetraene.
CAS: 629-20-9. C_8H_8.
Properties: Colorless liquid. Fp −7C, bp 140C, d 0.943 (0/4C), refr index 1.5394 (20C). It behaves like an aliphatic hydrocarbon, is relatively reactive, and resinifies on standing in air. Combustible.

$$\begin{array}{ccc} & CH{=}CH & \\ HC & & CH \\ \| & & \| \\ HC & & CH \\ & CH{=}CH & \end{array}$$

Derivation: Nickel-catalyzed polymerization of acetylene, a reaction discovered by Reppe in Germany about 1940. The mechanism has several possible pathways.
Use: Organic research
See polyacetylene.

cycloolefin.   An alicyclic hydrocarbon having two or more double bonds, e.g., the very reactive and widely used cyclopentadiene derived from coal tar, as well as cyclohexadiene and cyclooctatetraene, containing six and eight carbon atoms, respectively. The latter has four double bonds and is a polymer of acetylene.

cycloparaffin. An alicyclic hydrocarbon in which three or more of the carbon atoms in each molecule are united in a ring structure and each of these ring carbon atoms is joined to two hydrogen atoms or alkyl groups. The simplest members are cyclopropane (C_3H_6), cyclobutane (C_4H_8), cyclopentane (C_5H_{10}), cyclohexane (C_6H_{12}), and derivatives of these such as methylcyclohexane ($C_6H_{11}CH_3$).
Hazard: All members of the cycloparaffin series are narcotic and may cause death through respiratory paralysis. For most of the members, there appears to be a narrow range between the concentration causing deep narcosis and those causing death.

cyclopentadiene. (1,3-cyclopentadiene; cyclopenta-1,3-diene; cyclopentadiene; pentole; pyropentylene; r-pentine).
CAS: 542-92-7. C_5H_6.
Properties: Colorless, cyclic, liquid dialkene.
Derivation: Produced by distillation during the carbonization of coal or the cracking of petroleum hydrocarbons.
Hazard: Low toxicity.
Use: Manufacture of resins, synthetic alkalpoids, camphors, sesquiterpenes.

cyclopenta[c,d]pyrene. (acepyrene; acepyrylene; cyclopenteno(c,d)pyrene; cpp).
CAS: 27208-37-3. $C_{18}H_{10}$.
Properties: Polycyclic aromatic hydrocarbon solid.
Derivation: Component of fossil fuel combustion and carbon black.
Hazard: Mutagen; probable carcinogen; neoplastigen; tumorigen.

1,3-cyclopentadiene.
CAS: 542-92-7. C_5H_6.

$$H_2C\diagdown \begin{array}{c} CH=CH \\ | \\ CH=CH \end{array}$$

Properties: Colorless liquid. D 0.805, bp 42.5C. Insoluble in water; soluble in alcohol, ether, and benzene.
Derivation: From coal tar and cracked petroleum oils.
Hazard: Decomposes violently at high temperature. Toxic. Eye and upper respiratory tract irritant.
Use: Chemical intermediate, organic synthesis (Diels–Alder reaction), starting material for synthetic prostaglandin, chlorinated insecticides, formation of sandwich compounds by chelation, e.g., cyclopentadienyl iron dicarbonyl dimer $[C_5H_5Fe(CO)_2]_2$.

cyclopentamine hydrochloride. (1-cyclopentyl-2-methyl-aminopropane hydrochloride).
CAS: 538-02-3. $C_5H_9CH_2(CH_3)NHCH_3$:HCl.
Properties: White, crystalline powder; mild characteristic odor; bitter taste. Mp 113.0–116.0C. Freely soluble in water, alcohol, and chloroform; soluble in benzene; slightly soluble in ether; pH (1% solution) approximately 6.2.
Grade: NF.
Use: Medicine (vasoconstrictor).

cyclopentane. (pentamethylene).
CAS: 287-92-3. C_5H_{10}.

$$\begin{array}{c} CH_2 \\ CH_2 \quad CH_2 \\ | \qquad | \\ CH_2 - CH_2 \end{array}$$

Properties: Colorless liquid. D 0.7445 (20/4C); bp 49.27C; fp −94C; refr index 1.406 (20/D), flash p −35F (−37.2C). Soluble in alcohol; insoluble in water.
Derivation: Catalytic cracking of cyclohexane.
Grade: Technical, 95%, 99%, research.
Hazard: Flammable, dangerous fire risk. Moderately toxic by ingestion and inhalation. Skin, eye and upper respiratory tract irritant, and central nervous system impairment.
Use: Solvent for cellulose ethers, motor fuel, azeotropic distillation agent.

1,2,3,4-cyclopentanetetracarboxylic acid.
CAS: 3786-91-2. $C_5H_6(COOH)_4$.
Properties: Crystalline powder. Mp 195–196C. Soluble in water but insoluble in most organic solvents.
Use: Curing agent for resins; imparts thermal stability and high-temperature properties.

1,2,3,4-cyclopentanetetracarboxylic acid dianhydride.
CAS: 6053-68-5. $C_5H_6(C_2O_3)_2$.
Properties: White solid. Mp 220–221C. Moderately soluble in dimethyl formamide, butyrolactone, and dimethylsulfoxide; only slightly soluble in acetone; insoluble in hydrocarbons.
Use: Curing agent for epoxy resins.

cyclopentanol. (cyclopentyl alcohol).

$$\overline{CH_2CH_2CH_2CH_2CH_2O}.$$

Properties: Colorless, viscous liquid; pleasant odor. D 0.946 (20/4C), refr index 1.4575 (20C), fp −19C, bp 139–140C, flash p 124F (51C). Slightly soluble in water; soluble in alcohol. Combustible.
Hazard: Moderate fire risk.
Use: Perfume and pharmaceutical solvent, intermediate for dyes, pharmaceuticals, and other organics.

cyclopentanone.
CAS: 120-92-3. C_5H_8O.

CH₂−CH₂ structure:

$$\begin{array}{c} CH_2-CH_2 \\ | \quad\quad\; >CO \\ CH_2-CH_2 \end{array}$$

Properties: Water-white, mobile liquid; distinctive ethereal odor somewhat like peppermint. Bp 131C; d 0.943; refr index 1.437; flash p 87F (30.5C) (CC); insoluble in water; soluble in alcohol and ether.
Hazard: Flammable, moderate fire risk. Narcotic in high concentration.
Use: Intermediate for pharmaceuticals, biologicals, insecticides, and rubber chemicals.

cyclopentanone oxime.
CAS: 1192-28-5. C₅H₈NOH.
Nearly colorless and odorless, crystalline solid; mp 56C; bp 196C; soluble in water, alcohol. Used as intermediate in synthesis of the amino acids, proline, and ornithine.

cyclopentene.
CAS: 142-29-0.

CH:CHCH₂CH₂CH₂.

Properties: Colorless liquid. D 0.772, bp 44C, fp −135.21C, refr index 1.4225 (20/D), flash p −20F (−29C).
Grade: Technical, research (99.89 mole %).
Hazard: Flammable, dangerous fire risk. Moderate narcotic action.
Use: Organic synthesis, polyolefins, epoxies, cross-linking agent.

cyclopentenylacetone. [1-(1-cyclopentenyl)-2-propanone].
C₅H₇CH₂COCH₃.
Properties: Clear, colorless liquid; ketone odor. Bp 170C; refr index 1.4545–1.4550 (25C). Combustible.
Use: Organic synthesis.

cyclopentylacetone. (1-cyclopentyl-2-propanone).
CAS: 1122-98-1. C₅H₉CH₂COCH₃.
Properties: Liquid. D 0.893 (25/25C), bp 180–184C, refr index 1.4420 (25C). Combustible.
Use: Organic synthesis.

cyclopentyl alcohol. See cyclopentanol.

cyclopentyl bromide. (bromocyclopentane).
CAS: 137-43-9. C₅H₉Br.
Properties: Clear, mobile liquid; sweet aromatic odor. Bp 137–138C, d 1.3866 (20/4C), wt/gal 11.6 lb (20C), refr index 1.4885 (n 20/D), flash p 108F (42.2C) (CC). Insoluble in water. Combustible.
Hazard: Moderate fire risk; moderately toxic.
Use: Organic synthesis (pharmaceuticals).

1-cyclopentyl-2-methylaminopropane hydrochloride. See cyclopentamine hydrochloride.

cyclopentyl phenyl ketone.
CAS: 5422-88-8. C₅H₉COC₆H₅.
Properties: Colorless to light-yellow liquid. Bp 145–146C (15 mm Hg). Soluble in most common organic solvents; insoluble in water. Combustible.
Use: Pharmaceutical intermediate.

1-cyclopentyl-2-propanone. See cyclopentylacetone.

cyclopentylpropionic acid.
CAS: 140-77-2. C₅H₉CH₂CH₂COOH.
Properties: Liquid. Bp 130–132C (12 mm Hg), flash p 116F (46.6C). Insoluble in water. Combustible.
Hazard: Moderate fire risk.
Use: Intermediate, wood preservatives.

cyclopentylpropionyl chloride. (cyclopentylpropionic acid chloride).
CAS: 104-97-2. C₅H₉CH₂CH₂COCl.
Properties: Liquid. Bp 81–82C (10 mm Hg), flash p 104F (40C). Soluble in water. Combustible.
Hazard: Moderate fire risk.
Use: Intermediate.

cyclophane. A term applied in organic research to two molecules, e.g., of benzene, that are connected by covalent bonds involving all their carbon atoms (superphane). Such molecules exhibit a high degree of strain. In other types, some of the carbon atoms are not involved. Cyclophanes can be made by a reaction mechanism in which dimerization of benzocyclobutenes plays a part.

cyclophorase. Any of a set of enzymes and coenzymes that catalyze the complete oxidation of pyruvic acid to carbon dioxide and water. They occur in mitochondria and are basically those enzymes and coenzymes of the tricarboxylic acid cycle.

cyclophosphamide hydrate. (1-bis(2-chloroethyl)amino-1-oxa-2-aza-5-oxaphosphoridine monohydrate; 2-(bis(2-chloroethyl)amino)-1-oxa-3-aza-2-phosphocyclohexane 2-oxide monohydrate; (bis(chloro-2-ehtyl)amino)-2-tetrahydro-3,4,5,6-oxazaphosphorine; 1,3,2-oxide-2-monohydrate; bis(2-chloroethyl)phosphoramide cyclic porpanolamide ester monohydrate; N,N-bis(β-chloroethyl)-N',O-propylenephosphoic acid ester amine monohydrate; N,N-bis(2-chloroethyl)-1-oxo-6-oxa-2-aza-1,5-phosphacyclohexan-1-amine hydrate; N,N-bis(β-chloroethyl)-N',O-trimethylenephosphoric acid ester diamide monohydrate; CB 4564; clafen; cyclic N',o-propylene ester of N,N-bis(2-chloroethyl)phosphordiamidic acid monohydrate; cyclophosphamide monohydrate; cyclophosphamidum; cyclophosphan; cyclophosphane; cyclophosphanum, cytophosphan; Cytoxan; 2-(di(2-chloroethyl)amino)-1-oxa-3-aza-2-phosphacyclohexane-2-oxide monohydrate; n,n-di(2-chloroethyl)amino-N,O-propylene phosphoric acid ester diamide monohydrate;

endoxana; endoxan-asta; endoxan monohydrate; endoxan R; enduxan; genoxal; mitoxan; NSC-26271; procytox; semdoxan; sendoxan; senduxan). CAS: 6055-19-2. $C_7H_{17}Cl_2N_2O_3P$.
The monohydrate of cyclophosphamide. Precursor of an alkylating nitrogen mustard antineoplastic and immunosuppressive agent that must be activated in the liver to form the active aldophosphamide.
Hazard: Sterility; antineoplast, carcinogen, mutagen.
Use: In the treatment of lymphoma and leukemia.

cyclophosphamide.
CAS: 50-18-0. $C_7H_{15}Cl_2N_2O_2P$.
A nitrogen mustard.
Properties: Crystalline solid. Mp 41C. Slightly soluble in benzene, alcohol, carbon tetrachloride; very slightly soluble in ether and acetone.
Hazard: A confirmed carcinogen.
Use: Antineoplastic for treatment of leukemia, etc., tested for use in chemical shearing of sheep, and as an insect chemosterilant.

cyclopiazonic acid.
$C_{20}H_{20}N_2O_3$.
An indole produced by penicillium cyclopium and related species and by several species of aspergillus. It is a common contaminant of cheese, corn, and peanuts and occurs widely in mammalian tissues.
Properties: A solid.
Hazard: Extremely toxic; poison; teratogen; mutagen; myotoxic; hepatotoxic; nephrotoxic; cytotoxic; catalepsy, hypothermia, hypokinesia, incoordination, ptosis, sedation without loss of right reflex, tremor, convulsions.
Use: Antifungal agent.

cyclopropane.   (trimethylene).
CAS: 75-19-4. C_3H_6.

$$CH_2$$
$$/ \quad \backslash$$
$$CH_2-CH_2$$

Properties: Colorless gas; characteristic odor resembling that of solvent naphtha; pungent taste. D 0.72–079, bp −32.9C, fp −126.6C. Soluble in alcohol and ether; partially soluble in water. Autoign temp 928F (497C).
Derivation: Reduction of dibromocyclopropane with zinc dust.
Grade: Technical, USP, 99.5% min.
Hazard: Highly flammable. Forms flammable and explosive mixtures with air or oxygen. Explosive limits in air 2.4–10.3% by volume. Moderately toxic by inhalation. Narcotic in high concentration.
Use: Organic synthesis, anesthetic.

cyclopropanespirocyclopropane.   See spiropentane.

cyclopropanone.   A metabolite of coprine that inhibits the metabolic detoxification of alcohol and accounts for the toxic effects of alcoholic beverages when consumed with the mushrooms *coprinus atramentarius* and *clitocybe clavipes*.

n-cyclopropyl-*n'*-(2,5-difluorophenyl) urea.
CAS: 81356-60-7. $C_{10}H_{10}F_2N_2O$.
Hazard: A poison. Moderately toxic by skin contact. A mild skin and eye irritant.

cyclosilane.   See silane.

Cyclo Sol.   A series of hydrocarbon solvents composed of 50–99% aromatic hydrocarbons.
Hazard: Flammable, dangerous fire risk.

cyclosporine A.   (antibiotic S 7481F1; ciclosporin; cyclosporin; cyclosporine; cyclosporine a; OL 27-400; S 7481F1; sandimmun(e); 32-ethyl-2-[(E)-1-hydroxy-2-methylhex-4-enyl]-3,6,9,12,14,17,21,27,30-nonamethyl-8,11,20,26-tetrakis(2-methylpropyl)-5,23-di(propan-2-yl)-3,6,9,12,15,18,21,24,27,30,33-undecazacyclotritriacontane-1,4,7,10,13,16,19,22,25,26,31-undecone).
CAS: 59865-13-3. $C_{62}H_{111}N_{11}O_{12}$.
A cyclosporine undecapeptide that has antifungal action and inhibits T-cell activity.
Derivation: An extract of soil fungi.
Hazard: Highly toxic; nephrotoxic; carcinogen; poison; mutagen.
Use: Immunosuppressant to prevent the rejection of organ transplants.

cyclotrimethylenetrinitramine.   See cyclonite.

cyclotron.   A circular electromagnetic device for accelerating positively charged particles (protons, deuterons, alpha particles). It was invented in 1929 by E. O. Lawrence (1901–1958) at the University of California at Berkeley. It is now used chiefly for basic nuclear research. The acceleration is achieved by successive applications of small accelerations at low voltage synchronized with the rotational period of the particles in a magnetic field. Energies of over 700 MeV for protons and over 900 MeV for alpha particles have been attained. The energized particles emerging from the cyclotron impinge upon a target nucleus, resulting in formation of radioactive isotopes, neutrons, and ionizing radiation. The first plutonium was made in a cyclotron in 1939. Powerful cyclotrons with huge electromagnets are in use in physical research laboratories throughout the world.
See bombardment.

cycloversion.   A process using bauxite as a catalyst for (1) desulfurization, (2) reforming, and (3) cracking of petroleum to form high-octane gasoline.

cyclyd.   A coined term referring to the cyclic alkyd coatings prepared from "Polycyclol 1222," used in baking metal primers and air-drying maintenance paints.

"Cycocel" *[BASF]*.   TM for a plant growth regulator (2-chloroethyltrimethyl ammonium chloride) said to be effective for cereal grains, tomatoes, and peppers.

"Cydril" *[Cytec]*.   TM for a drilling mud additive, shale stabilizer, and viscofier.
Use: Reduces drilling costs by stabilizing the well bore.

cyhexatin.   (Plictran; TCHH; TCHTH; trichlohexyltin hydroxide).
CAS: 13121-70-5. $C_{18}H_{34}OSn$.
Properties: White crystals. Mw 385.16, mp 195–198C, vap press nil. Insoluble in water; soluble in acetone, chloroform, and methanol.
Hazard: Upper respiratory tract irritant, body weight effects, kidney damage. Questionable carcinogen.
Use: An acaricide.

cymarose.   (2,6-deoxy-3-methoxyaldohexose; 2,6-dideoxy-3-*o*-methylribohexose;   3-methyldigitoxose;   (2R,4S,5R,6R)-4-methoxy-6-methyloxane-2,5-diol).
$C_7H_{14}O_4$.
A hexose
Derivation: From the hydrolysis of cardiac glycosides that occur in members of the family *Apocynaceae*.

cymene.   (cymol;   isopropyltoluene;   methylpropylbenzene).
CAS: *m*- 535-77-3; *o*- 527-84-4; *p*- 99-87-6; mixed 25155-15. $CH_3C_6H_4CH(CH_3)_2$.
The *o*-, *m*-, and *p*-isomers are known.

Properties: Colorless, transparent liquids with aromatic odor. Combustible. D: *o*- 0.8748, *m*- 0.862, *p*- 0.8551. Fp: *o*- −71C, *m*- −64C, *p*- −68C. Bp: *o*- 177C, *m*- 175.6C, *p*- 176.5C. Refr index: *p*- 1.489 (20C). Soluble in alcohol, ether, and chloroform; insoluble in water. Flash p (*p*-) 127F (51C) (CC).
Derivation: Mixed cymenes are produced from toluene by alkylation. *p*-Cymene occurs in several essential oils and is made from monocyclic terpenes by dehydrogenation. These terpenes can be made from turpentine or obtained as a by-product from the sulfite digestion of spruce pulp in paper manufacture.
Method of purification: Washing with sulfuric acid, water, and alkali.
Grade: Technical.
Hazard: Moderate fire risk. Moderately toxic by ingestion.

Use: Solvents, synthetic-resin manufacture, metal polishes, organic synthesis (oxidation to hydroperoxides used as catalysts for synthetic-rubber manufacture; cymene alcohols are made by hydrogenating the hydroperoxides). Pure *p*-cresol and carvacrol are made from *p*-cymene.

cynapine.   The   active   principle   of   *Aethusa cynapium*.
Properties: Volatile alkaloid.
Hazard: Nausea, vomiting, toxic.

cynaustine hydrochloride.
CAS: 17958-39-3. $C_{15}H_{25}NO_4 \cdot ClH$.
Hazard: A poison.
Source: Natural product.

cypermethrin.   (cyano(3-phenoxyphenyl) methyl; 3-(2,2-dichlorovinyl)-2,2-dimethylcyclopropanecarboxylate).
CAS: 52315-07-8.
Use: Insecticide.

cyperquat.
CAS: 48134-75-4. $C_{12}H_{12}N$.
Hazard: Moderately toxic.

5-α-cyprinol.
CAS: 2952-70-7. $C_{27}H_{48}O_5$.
Hazard: A poison.

cyproconazole.
CAS: 94361-06-5. $C_{15}H_{18}ClN_3O$.
Hazard: Moderately toxic by ingestion, inhalation, and skin contact. A mild eye irritant.
Use: Agricultural chemical.

cypromid.   (cipromid; clobber; 3,4'-dichlorocyclopropanecarboxanilide; *N*-(3,4-dichlorophenyl)cyclopropanecarboxamide).
$C_{10}H_9Cl_2NO$.
Properties: Amide.
Hazard: Poison; moderately toxic.
Use: Herbicide.

"Cyracure" *[Union]*.   TM for cycloaliphatic epoxides.
Use: For UV-light-cured coatings such as varnishes, inks, industrial coatings, and sealants.

Cys.   Abbreviation for cysteine.

cysteamine.   See 2-aminoethanethiol.

cysteamine   hydrochloride.   (2-aminoethanethiol).
CAS: 156-57-0. C_2H_7NS.
A radiation-protective agent that oxidized in air to form cystamine.
Properties: Water- and ethanol-soluble crystalline compound.
Use: To treat radiation sickness.
Hazard: Very toxic.

cysteine. (α-amino-β-thiolpropionic acid; β-mercaptoalanine).
CAS: 52-90-4. $HSCH_2CH(NH_2)COOH$.
A nonessential amino acid derived from cystine, occurring naturally in the L(+) form.
Properties: Colorless crystals. Soluble in water, ammonium hydroxide, and acetic acid; insoluble in ether, acetone, benzene, carbon disulfide, and carbon tetrachloride.
Derivation: Hydrolysis of protein; degradation of cystine. Found in urinary calculi.
Available forms: Available commercially as L(+)-cysteine hydrochloride.
Use: Biochemical and nutrition research, reducing agent in bread doughs (up to 90 ppm).

cysteic acid. (3-sulfoalanine; 2-amino-3-sulfopropanoic acid).
CAS: 498-40-8. $C_3H_7NO_5S$.
An amino acid with a C-terminal sulfonic acid group that has been isolated from human hair oxidized with permanganate.
Derivation: Oxidation product of cysteine and a precursor of taurine and isethionic acid; outer part of a sheep's fleece where the wool is exposed to light and weather.

cystine. (β,β'-dithiobisalanine; di[α-amino-β-thiolpropionic acid]).
CAS: 56-89-3. $HOOCCH(NH_2)CH_2SSCH_2$ $CH(NH_2)COOH$.
A nonessential amino acid.
Properties: White, crystalline plates. Soluble in water; insoluble in alcohol. Optically active. DL-cystine, mp 260C. D(+)-cystine, mp 247–249C. L(−)-cystine, mp 258–261C with decomposition.
Derivation: Hydrolysis of protein (keratin), organic synthesis. Occurs as small hexagonal crystals in urine.
Grade: FCC.
Use: Biochemical and nutrition research, nutrient and dietary supplement.

cysteine conjugate β-lyase.
A type of enzyme that metabolizes cysteine conjugates, producing the thiol derivative of pyruvic acid and ammonia.

Cystokon. A 30% solution of sodium acetrizoate.

cytase. An enzyme that dissolves cellulose.
Derivation: Occurs in grass seeds.

cythion. Proprietary malathion.
Use: Insecticide.

cytidine.
CAS: 65-46-3. $C_9H_{13}N_3O_5$.
The nucleoside consisting of D-ribose and cytosine.
Properties: White, crystalline powder. Soluble in water, acid, alkali; insoluble in alcohol.
Derivation: From yeast ribonucleic acid. Also available as the hemisulfate, $(C_9H_{13}N_3O_5)_2 \cdot H_2SO_4$.
Use: Biochemical research.

cytidine phosphates. Nucleotides used by the body in growth processes. Important in biochemical and physiological research. Those isolated and commercially available (as sodium salts) are the monophosphate (CMP; see cytidylic acid), diphosphate (CDP), and triphosphate (CTP).

cytidylic acid. (cytosylic acid; cytidinephosphoric acid; cytidine monophosphate; CMP).
$C_9H_{14}N_3O_8P$.
The monophosphoric ester of cytosine, i.e., the nucleotide containing cytosine D-ribose and phosphoric acid. The phosphate may be esterified to the 2, 3, or 5 carbon of ribose, yielding cytidine-2'-phosphate, cytidine-3'-phosphate, or cytidine-5'-phosphate, respectively.
Properties: (Cytidine-3'-phosphate) White, crystalline powder; odorless; mild sour taste. Mp: crystals from 50% alcohol 230–233C (with decomposition); crystals from water 227C (with decomposition). Slightly soluble in water and dilute alkalies; insoluble in alcohol and other organic solvents.
Derivation: (commercial product) From yeast nucleic acid by hydrolysis. The 5'-monophosphate is made synthetically by phosphorylation and hydrolysis of isopropylidene cytidine.
Use: Biochemical research.

cytisine. (1,2,3,4,5,6-hexahydro-1,5-methano-8H-pyrido-[1,2-a][1,5]diazocin-9-one; baptitoxine; cytisine; cytitone; sophorine; ulexine; 2-amino-3,7-dihydropurin-6-one).
CAS: 485-35-8. $C_{11}H_{14}N_2O$.
An alkaloid.
Properties: Bitter, crystalline alkaloid solid.
Derivation: Found in *Laburnum anagyroides* and other plants of the family *fabaceae*.
Use: Formerly as a cathartic and diuretic.
Hazard: Highly toxic; poison.

cytisus. (broom; cytisus).
A genus of stiff or spiny deciduous and evergreen shrubs (family *fabaceae*) that are native to Europe, Western Asia, and Northern Africa but widely cultivated elsewhere. They have showy racemose flowers, a two-lipped calyx, and grow to a height of 3 meters.

cytochalasin. (cyclical lactone).
$C_{30}H_{39}NO_4$.
Any of a large class of macrolide mycotoxins with a highly substituted hydrogenated isoindole ring fused to a macrocyclic ring comprised of 11–14 atoms and is a carbocycle or a lactone.
Derivation: Originally isolated from *helminthosporium dematioideum, metarrhizium anisopliae,* and *rosellina necatrix,* but are known from a wide variety of *phoma*-type fungi.
Hazard: Cytotoxic; interfere with cytokinesis; induce extrusion of nuclei; reversibly inhibit movement; inhibit such processes as glucose transport, phagocytosis, aggregation of platelets, clot contraction, release of growth hormone, thyroid secretion.
Use: Medical research in cell biology.

cytochalasin B. (7,20-dihydroxy-16-methyl-10-phenyl-24-oxo[14]cytochalasa-6,12,13,21-triene-1,23-dione; (e,e)-16-benzyl-6,7,8,9,10,12a,13,14,15,15a,16,16-dodecahydro-5,13-dihydroxy-9,15-dimethyl-14-methylene-2H-oxycyclotetradec-[2,3-d]isoindole-2,18(5h)-dione; phomin).
CAS: 14930-96-2. $C_{29}H_{37}NO_5$.
Properties: Needles.
Derivation: Initially obtained from *Phoma* spp. as phomin.
Hazard: Poison; teratogen; mutagen; in vitro effects include cytokinesis, single-cell movement, morphogenesis of epithelia, and contractile processes of smooth muscle and cardiacs.
Use: A tool in cytological research.

cytochalasin D.
$C_{30}H_{37}NO_6$.
A mycotoxin.
Derivation: From *Chaetomium globusum*.
Hazard: Extremely toxic.

cytochalasin E.
$C_{28}H_{32}NO_7$.
Properties: A solid; food storage mold metabolite.
Derivation: Metabolite of *Aspergillus clavatus*.
Hazard: Toxic; poison; ataxia, drowsiness, cyanosis, coma, death, circulatory collapse, brain edema, pulmonary hemorrhage, injury to the vascular wall, congestive degenerative changes and necrosis of liver, kidney, spleen, and small intestines.

cytochemistry. The branch of biochemistry devoted to study of the chemical composition of cells and cell membranes, including chromosomes, genes, and the complex reactions involved in cell growth and replication, as well as the mechanism of enzyme activity.
See molecular biology.

cytochrome. A class of iron-porphyrin proteins of great importance in cell metabolism. They are pigments occurring in the cells of nearly all animals and plants. Several types have been identified. Cytochrome carbon is the most abundant and has been obtained in pure forms. The cytochromes and cytochrome oxidase have important functions in cell respiration. The latter is an iron-porphyrin–containing protein that is an important enzyme in cell respiration. It catalyzes the oxidation of cytochrome carbon and is reduced itself in the reaction. It is then reoxidized by oxygen.
See porphyrin.

cytochrome b5. A microsomal cytochrome that oarticipates in the desaturation of fatty acids and usually in the reduction of cytochrome P-450.

cytochrome c. (cromoci; cytorest; ferricytochrome c; ferrocytochrome c; hematin-protein; horse-cytochrome c; horse heart cytochrome c; landrax; myohematin; nitrosylferricytocrome

c; hematin protein; 3-[7,12-bis[1-[2-amino-3-(methylamino)-3-oxopropyl]sulfanylethyl]-18-(2-carboxyethyl)-3,8,13,17-tetramethylprophyrin-21,23-diid-2-yl]propanoic acid).
CAS: 9007-43-6. $C_{42}H_{52}FeN_8O_6S_2$.
Any hemeprotein enzyme derived from protoporphyrin IX and present in cells of aerobic organisms. They play an important role in the electron transport system of plants and animals. They serve as redox intermediates that accept electrons from mitochondrial electron transport complex III and transfer them to mitochondrial electron transport complex IV.
Properties: Reduced from crystallizes as separate needles; oxidized form as rosettes.
Hazard: Mutant.
Derivation: Found in eukaryotic mitochondria.

cytochrome c oxidase. (cytochrome a).
A cytochrome that catalyzes the transfer of electrons from cytochrome oxidase to oxygen in respiration.

cytochrome oxidase. (cytochrome a_3; Warburg's respiratory enzyme).
CAS: 9001-16-5.
Any of a family of heme- and copper-containing cytochromes in the electron transfer chain that react directly with reduced oxygen. They accept electrons from cytochrome c oxidase transferring them to oxygen with the production of water and the release of sufficient free energy to support the formation of ATP. Cytochrome oxidase is inhibited by cyanide or carbon monoxide which effectively halts electron transport and prevents the formation of ATP.

cytochrome P-420. A degradation product of cytochrome P-450, often seen in microsomal preparations, that binds carbon monoxide and a number of type II ligands, but lacks oxidative activity.

cytochrome P-448. A cytochrome complex that often activates carcinogens and other toxic chemicals.

cytochrome P-450. Any of a number of cytochromes in the electron transport chain that are active in enzymatic hydroxylation, demethylation, and N-oxidation reactions in liver microsomes and adrenal mitochondria. P-450s act as a monooxygenase in catalyzing metabolic functions; they affect the rate of metabolism of drugs and other xenobiotics.
Derivation: Most abundant in the liver and adrenal microsomes of vertebrates but also in bacteria, plants, yeasts, and insects.

cytogenetics. The study of the physical appearance of chromosomes.
See karyotype.

cytokinesis. The final separation of daughter cells following mitosis.

cytokinin. (phytokinin; kinin; *N*-(furan-2-ylmethyl)-7H-purin-6-amine).
CAS: 525-79-1. $C_{10}H_9N_5O$.
Any of a naturally occurring class of plant growth promoters produced in root tissue and derived from adenine. They stimulate cell division in the presence of auxin and induce flowering in some plants, delay senescence, and induce the breaking of dormancy in axillary buds and some seeds. Unusually high endogenous concentrations of cytokinins occur in plant diseases such as grown gall and witches broom.

cytokinins. See kinin.

cytological band. An area of the chromosome that stains differently from areas around it.
See cytological map.

cytological map. A type of chromosome map whereby genes are located on the basis of cytological findings obtained with the aid of chromosome mutations.

cytolysin. An antibody that acts in association with a complement to destroy an animal cell. They typically exhibit specificity for particular types of cells and are thus usually named accordingly.

cytoplasm. The extra nuclear components of the living cell, containing mitochondria, plastids, spherosomes, etc. This, together with the nucleus, constitutes the protoplasm. The chemical constituents are chiefly proteins, plus a high percentage of water.

cytoplasmic trait. A genetic characteristic in which the genes are found outside the nucleus, in chloroplasts or mitochondria. Results in offspring inheriting genetic material from only one parent.

cytophilic antibody. (agglutinating antibody; cytotropic antibody).
Antibody that has an affinity for certain kinds of cells in addition to its affinity for a specific antigen due to the properties of the Fe portion of the heavy chain.

Cytosar. Cytosine arabinoside.

cytosine. (2-oxo-4-aminopyrimidine).
CAS: 71-30-7. $C_4H_5N_3O$.
A pyrimidine found in both ribonucleic and deoxyribonucleic acids and certain coenzymes. One member of the base pair GC (guanine and cytosine) in DNA.

Properties: (Monohydrate) Lustrous platelets. Decomposes at 320–325C. Slightly soluble in water and alcohol; insoluble in ether.
Derivation: Isolation following hydrolysis of nucleic acids; organic synthesis.
Use: Biochemical research.
See base pair; nucleotide.

cytosine arabinoside. (Ara-C; 1-β-D-arabino-furanosyl cytosine).
CAS: 147-94-4. $C_9H_{13}N_3O_5$.
A drug synthesized in 1969 at Salk Institute. It is useful in combating myelocytic leukemia in adults and has been approved by the FDA as a prescription drug.

cytosine monophosphate. See cytidylic acid.

cytoskeleton. Integrated system of molecules within eukaryotic cells that provides them with shape, internal spatial organization, and motility, and it may assist in communication with other cells and the environment. Red blood cells, for instance, would be spherical instead of flat if it were not for their cytoskeleton.

cytosol. The aqueous phase of the cytoplasm (including dissolved solutes and enzymes, but excluding all organelles). It is most commonly the product of centrifugation at $100,000 \times g$ for 1 h.

cytosylic acid. See cytidylic acid.

"Cytotec" *[Searle].* TM for misoprostol.
Use: Drug.

cytotoxin. Any substance produced by a living organism that inhibits cellular processes or is otherwise toxic to or destroys living cells.

"Cytox" *[Cytec].* TM for microbiocide. Controls the formation of slime in effluents, boiler waters, and processing streams.
Use: This microbiocide is active against a broad spectrum of microorganisms, including bacteria, fungi, yeasts, and algae.

Cytrel. A tobacco substitute derived from cellulose. It is free from nicotine and has 15–30% of the tar content of tobacco.

Cytrol. 3-Amino-1,2,4-triazole liquid-formulation herbicide.

Cyzine. 10% 2-acetylamino-5-nitrothiazole feed supplement.

D

D.   Symbol for deuterium.

D.   (D-)
Prefix indicating the right-handed enantiomer of an optical isomer. Usually printed as a small capital letter.
See L; glyceraldehyde; asymmetry; enantiomer.

d-.   (d-)
Prefix indicating that a substance is dextrorotatory. A plus sign (+) is now preferred.

2,4-D.   (2,4-dichlorophenoxyacetic acid).
CAS: 94-75-7. $Cl_2C_6H_3OCH_2COOH$.
Properties: White to yellow crystalline powder. Stable, mp 138C, bp 160C (0.4 mm Hg). Difficultly soluble in water or oils; soluble in alcohols.
Derivation: Reaction of 2,4-dichlorophenol and chloroacetic acid in aqueous sodium hydroxide.
Available forms: Sodium salt (60–85% acid), amine salts (10–60% acid), esters (10–45% acid). These forms are dispersible in water or oils (esters) and can be applied as sprays.
Grade: Technical.
Hazard: Irritant. Use may be restricted. Thyroid effects and kidney tubular damage. Questionable carcinogen.
Use: Selective weed killer and defoliant, fruit drop control.

DAA.   Abbreviation for diacetone acrylamide.

DACE.   See S,N-diacetylcysteine monoethyl ester.

"Dacron" *[Du Pont].*   TM for a polyester fiber made from polyethylene terephthalate. Available as filament yarn, staple, tow, and fiberfill.
Properties: D 1.38, mp 250C, tensile strength (psi) 4–5 g/denier (about 50–60 thousand psi), break elongation 10–36%, moisture regain 0.4%, high elastic recovery, good insect resistance. Difficult to ignite; self-extinguishing. Soluble in *m*-cresol (hot), trifluoroacetic acid, and *o*-chlorophenol.
Derivation: Reaction of dimethyl terephthalate and ethylene glycol. The resulting polymer is melt-extruded through a spinnerette and stretched.
Use: Textile fabrics and suitings, often combined with wool and other fibers; cordage; fire hose.
See dimethyl terephthalate; terephthalic acid.

dactimicin sulfate.
CAS: 73245-91-7. $C_{18}H_{36}N_6O_6 \cdot 2H_2O_4S \cdot H_2O$.
Hazard: A poison.

dactinomycin.   (USAN for actinomycin D).
$C_{62}H_{86}N_{12}O_{16}$.
An antibiotic produced from *Streptomyces*. Used in medicine.

Dakin reaction.   Replacement of the aldehyde or acetyl groups in phenolic aldehydes or ketones by a hydroxyl group by means of hydrogen peroxide.

Dakin's solution.   An aqueous solution containing 0.5% sodium hypochlorite, used as an antiseptic, especially for wound treatment.

Dakin-West reaction.   Reaction of α-amino acids with acetic anhydride in the presence of base to give α-acetamido ketones. The reaction occurs via the intermediate azlactone.

"Dalamar" *[Clariant].*   TM for an azo yellow pigment.

dalapon.
CAS: 75-99-0. CH_3CCl_2COOH.
Generic name for 2,2-dichloropropionic acid. The sodium salt (sodium 2,2-dichloropropionate) is commonly used.
Properties: (free acid) Liquid. Bp 185–190C (760 mm Hg), 90–92C (14 mm Hg), d 1.389 (22.8 4C). Very soluble in water and alcohol; soluble in ether. (sodium salt): Crystals; salty taste. Decomposes 174–176C. Corrosive to iron. Soluble in water; aqueous solutions hydrolyze above 70C.
Hazard: Strong irritant to eyes and skin. Upper respiratory tract irritant. Questionable carcinogen.
Use: Herbicide.

"Dalpad" *[Dow].*   TM for a coalescing agent, a stable, low-odor, low-temperature film-forming aid for polyvinyl acetate and acrylic latex paints.

dalton.   A unit of mass equal to 1/12 the mass of the ^{12}O.
See atomic weight.

Dalton, John.   (1766–1844). The first theorist since the Greek philosopher Democritus to conceive

Hawley's Condensed Chemical Dictionary, Sixteenth Edition. Michael D. Larrañaga, Richard J. Lewis, Sr., and Robert A. Lewis.
© 2016 John Wiley & Sons, Inc. Published 2016 by John Wiley & Sons, Inc.

of matter in terms of small particles. The founder of the atomic theory on which all succeeding chemical investigation has been based (1807). His essential concept of the indivisibility of the atom was not called into question until 1910 when radioactive decay was established by Rutherford. Dalton's theories relating to pressures of gases and atomic combinations led to the basic generalizations stated in the law of multiple proportions, the law of constant composition, and the law of conservation of matter.
See Priestley, Joseph; chemical laws.

Dalton's law of partial pressures.   In any mixture of gases, each constituent exerts its pressure independently as if the other constituents were absent, and the solubility of mixed gases in liquid is proportional to the partial pressure of each.

DAMA.   Fatty dialkyl methylamines.

dammar.   A group of tree-derived resins soluble in hydrocarbon and chlorinated hydrocarbon solvents, partially soluble in alcohols, insoluble in water. Used in colorless and overprint varnishes, cellulosic lacquers, alkyd baking enamels, and paper and textile coatings.

dandy roll.   A light roller covered with wire cloth that rides on the formed sheet near the dry end of the fourdrinier wire. Its purpose is to provide a closer finish to the sheet as well as to impress a screen pattern on the upper surface of the sheet similar to that made by the wire on the under surface. Paper so marked is called "wove" in the trade. The dandy roll may also impress a pattern of parallel lines, in which case the paper is called "laid." It is also used to apply watermarks.

dansyl.   1-dimethylaminonaphthalene-5-sulfonyl, derivatives of *N*-terminal amino acid residues.

dansyl   chloride.   (5-dimethylamino-1-napthalenesulfonyl   chloride;   dimethlaminonaphthalenesulfonyl chloride)
Properties: Mw 269.8, mp 69–71C.

danthron.   See 1,8-dihydroxyanthraquinone.

DAP.   (1) Abbreviation for diallyl phthalate. (2) Abbreviation for diammonium phosphate.
See ammonium phosphate, dibasic.

Dapon.   A series of diallyl phthalate resins. Used for molding compounds, prepregs, and coatings.

dapsone (USAN and USP name).   (4,4'-sulfonyldianiline).
CAS: 80-08-0. $C_{12}H_{12}N_2O_2S$.
Properties: Crystals. Mp 176C, vap d 8.3. Nearly insoluble in water; soluble in acetone and alcohol.

Hazard: A poison. Questionable carcinogen.
Use: Used in leprosy treatment and veterinary medicine.

"Daran" [Owensboro].   TM for polyvinylidene chloride latexes used in packaging materials.
Use: Barrier coatings for packaging papers, paperboards, plastic films, and specialty saturants.

Daratax.   Specialty vinyl acetate acrylic copolymers.
Available forms: Latex emulsions.
Use: For paint, coating, adhesive, textile, and industrial applications.

Darcy's law.   The volumetric rate of flow of water through a sand filter bed is directly proportional to the cross-sectional area of the bed and the pressure difference across the bed and inversely proportional to the thickness of the bed.

Darex.   Styrene-butadiene latexes and related vehicles.
Use: Textile coatings, rug backings, saturants, shoe products, coatings, paints, and adhesives.

Dariloid.   Sodium alginate blends.

Dariloid KB.   Propylene glycol alginate blends.

Darling Ladies.   Five supplements including collagen, herbs, vitamins, minerals, and calcium.

dark reactions.   See carbon fixation reactions.

Darzens condensation.   Formation of α- and β-epoxy esters (glycidic esters) by the condensation of aldehydes or ketones with esters of α-haloacids; the corresponding thermally unstable glycidic acids yield aldehydes or ketones on decarboxylation.

Darzens-Nenitzescu synthesis of ketones.   Acylation of olefins with acid chlorides or anhydrides catalyzed by Lewis acids. When performed in the presence of a saturated hydrocarbon, the product is the saturated ketone.

Darzens synthesis of tetralin derivatives.   Cyclization of compounds such as α-benzyl-α-allylacetic acid by moderate heating in concentrated sulfuric acid to yield tetralin derivatives.

DAS.   Abbreviation   for   4,4'-diamino-2,2'-stilbenedisulfonic acid.

data warehouse.   A collection of databases, data tables, and mechanisms to access the data on a single subject.

dating chemical.   See chemical dating.

dative bond.   See coordinate bond.

daturic acid.   See *n*-heptadecanoic acid.

daughter element.   The element formed when another element undergoes radioactive decay. The latter is called the parent. The daughter may or may not be radioactive.

daunomycin.   (daunorubicin).
CAS: 20830-81-3. $C_{27}H_{29}NO_{10}$.
An antibiotic.
Properties: Reddish needles. Bp (decomposes) 190C. Soluble in water and methyl alcohol; insoluble in chloroform and benzene.
Hazard: A possible carcinogen.

Davy, Sir Humphry.   (1778–1829). Born in Cornwall, Davy was the first to isolate the alkali metals and recognize the identity of chemical and electrical energy. A pioneer in the science of electrochemistry, he carried out basic studies of electrolysis of salts and water, and his application of electricity to the decomposition of molten caustic potash led to the isolation of metallic potassium.

"Daxad" *[GEO].*   TM for anionic, polymer-type dispersing agents. Supplied as light-colored powders or aqueous solutions. Effective dispersant for aqueous suspensions of insoluble dyestuffs, polymers, clays, tanning agents, and pigments.
Use: Manufacture of dyestuff pastes, textile backings, latex paints and paper coatings, retanning and bleaching of leather, dye resist in leather dyeing, dispersion of pitch in paper manufacture, prefloc prevention in the manufacture of synthetic rubber.

Daypro.   Oxaprozin.
Use: Drug.

Dazzle.   A dishwasher stable rinse agent not compatible with strong oxidizing agents.

Db.   Symbol for dubnium.

2,4-DB.   Abbreviation for 2,4-dichlorophenoxybutyric acid.

DBC.   Abbreviation for 1,4-dichlorobutane.

DBCP.   Abbreviation for 1,2-dibromo-3-chloropropane.

DBM.   Abbreviation for dibutyl maleate.

DBMC.   Abbreviation for 4,6-di-*tert*-butyl-*m*-cresol.

DBP.   Abbreviation for dibutyl phthalate.

DBPP.   See dibutylphenyl phosphate.

DBS.   Abbreviation for dibutyl sebacate.

D & C.   Product certified for use in drugs and cosmetics.

DCA.   Abbreviation for deoxycorticosterone acetate.

DCCD.   See *N,N'*-methanetetrayl biscyclohexanamine.

DCHP.   Abbreviation for dicyclohexyl phthalate.

DCO.   Abbreviation for dehydrated castor oil. See castor oil, dehydrated.

DCP.   Abbreviation for dicapryl phthalate.

DCPA.   Abbreviation for dimethyl-2,3,5,6-tetrachloroterephthalate.

DCPC.   Abbreviation for dichlorophenyl methyl carbinol. See di(*p*-chlorophenyl)ethanol.

DDB.   Abbreviation for dodecylbenzene.

DDBSA.   Abbreviation for dodecylbenzenesulfonic acid. See sodium dodecylbenzenesulfonate.

DDD.   Abbreviation for dichlorodiphenyldichloroethane. See TDE.

DDDA.   Abbreviation for dodecanedioic acid.

DDDM.   Abbreviation for 2,2'-dihydroxy-5,5'-dichlorodiphenylmethane. See dichlorophene.

DDE.
CAS: 72-55-9.
Abbreviation for dichlorodiphenyldichloroethylene, $(ClC_6H_4)_2C{:}CCl_2$. It is a degradation product of DDT found as an impurity in DDT residues.
Hazard: Questionable carcinogen.

2,4'-DDE.   See 1,1-dichloro-2-(*o*-chlorophenyl)-2-(*p*-chlorophenyl)ethylene.

DDFC.   See gemcitabine.

DDH.   Abbreviation for dichlorodimethylhydantoin.

DDM.   (1) Abbreviation for diaminodiphenylmethane. (2) Abbreviation for *n*-dodecylmercaptan.

DDNP.   Abbreviation for diazodinitrophenol.

DDP.   Abbreviation for dodecyl phthalate.

DDQ. See 2,3-dichloro-5,6-dicyanobenzoquinone.

DDS. Abbreviation for diaminodiphenylsulfone.

DDT. (dichlorodiphenyltrichloroethane; dicophane; chlorophenothane; 1,1,1-trichloro-2,2-bis(chlorophenyl)ethane).
CAS: 50-29-3. $(ClC_6H_4)_2CHCCl_3$.

Properties: Colorless crystals or white to slightly off-white powder; odorless or with slight aromatic odor. Insoluble in water; soluble in acetone, ether, benzene, carbon tetrachloride, kerosene, dioxane, and pyridine. Not compatible with alkaline materials.
Derivation: Condensing chloral or chloral hydrate with chlorobenzene in the presence of sulfuric acid.
Grade: Technical, purified, aerosol, USP, as chlorophenothane.
Hazard: Toxic by ingestion, inhalation, and skin absorption, especially in solution. Lethal dosage for humans estimated to be 500 mg/kg of body weight (solid material). Since DDT is not biodegradable and is ecologically damaging, its agricultural use in the U.S. was prohibited in 1973 (though its manufacturing for export is permitted). A confirmed human carcinogen. DDT can be used for a few specialized purposes, e.g., to combat the tussock moth. FDA tolerance: 5 ppm in foods. Causes liver damage. Possible carcinogen.
Use: Insecticide for tobacco and cotton, pesticide (tussock moth).

DDVP. Abbreviation for dimethyl dichlorovinyl phosphate.
See dichlorvos.

DE. Abbreviation for dextrose equivalent.

DEA. Abbreviation for diethanolamine, also abbreviation for Drug Enforcement Administration, a government agency that replaced the Bureau of Narcotics.

DEAC. Abbreviation for diethylaluminum chloride.

deacetylandromedotoxin. See grayanotoxin iii.

deacetylanhydroandromedotoxin. See grayanotoxin ii.

deacetyllyoniatoxin.
CAS: 28894-74-8. $C_{20}H_{32}O_6$.
Hazard: A poison.

Deacon process. A method of converting hydrogen chloride to chlorine by oxidation of hydrogen chloride with oxygen at 400–500C over a copper–salt catalyst, $2HCl + O \rightarrow Cl_2 + H_2O$. It is a means of producing chlorine without caustic and of utilizing the large amounts of by-product hydrogen chloride from the chlorination of organic compounds. When conducted in the presence of an organic compound that reacts with the chlorine formed, it is known as oxychlorination, e.g., $CH_2=CH_2 + 2HCl + O \rightarrow CH_2ClCH_2Cl + H_2O$.

DEAE. Abbreviation for diethylaminoethyl.

DEAE-cellulose. (diethylaminoethyl cellulose). A cellulose ether containing the group $(C_2H_5)_2NCH_2CH_2$ bound to the cellulose in an ether linkage. An anionic ion exchange material.
Use: Chromatography.

DEAE-dextran. A diethylaminoethyl ether of dextran, an electropositively charged polymer.

deamination. The enzymatic removal of amino groups from biomolecules, such as amino acids or nucleotides.

deanol. See 2-dimethylaminoethanol.

deblooming agent. A substance added to mineral oils to mask fluorescence. Nitronaphthalene and yellow coal tar dyes are among such products.

Debye characteristic temperature. A parameter relating to the lattice specific heat of a solid. The temperature at which the specific heat of a simple specific cubic crystal equals 5.67 calories per degree per mole.

Debye-Huckel theory. A theory advanced in 1923 for quantitatively predicting the deviations from ideality of dilute electrolytic solutions. It involves the assumption that every ion in a solution is surrounded by an ion atmosphere of opposite charge. Results deduced from this theory have been verified for dilute solutions of strong electrolytes, and it provides a means of extrapolating the thermodynamic properties of electrolytic solutions to infinite dilution.

Debye, Peter J. M. (1884–1966). A Dutch chemist and physicist who received the Nobel Prize in 1936 for his pioneer studies of molecular structure by X-ray diffraction methods. The interference patterns are still called Debye-Sherrer rings. He also made outstanding contributions to knowledge of polar molecules and to fundamental electrochemical theory.
See Debye-Huckel theory.

DEC. Abbreviation for β-diethylaminoethyl chloride hydrochloride.

decaborane.
CAS: 17702-41-9. $B_{10}H_{14}$.
Properties: Colorless crystals. Stable indefinitely at room temperature; decomposes slowly into boron and hydrogen at 300C. D 0.94 (25/4C), d 0.78 (100C), mp 99.7C, bp 213C. Slightly soluble in cold water; hydrolyzes in hot water; soluble in benzene, hexane, alcohol, carbon tetrachloride, and toluene.
Derivation: By-product of the pyrolysis of diborane.
Grade: Technical 95%, high purity 99%.
Hazard: May explode in contact with heat or flame or with oxygenated and halogenated solvents. Ignites in contact with oxygen. Absorbed by skin. Central nervous system convulsant and cognitive decrement.
Use: Catalyst, corrosion inhibitor, fuel additive, stabilizer, rayon delustrant, mothproofing agent, dye-stripping agent, reducing agent, fluxing agent, oxygen scavenger, propellant.

decachlorooctahydro-1,2,4-metheno-2H-cyclobuta[cd]-pentalen-2-one.   See kepone.

decafentin.
CAS: 15652-38-7. $C_{28}H_{36}P \cdot C_{18}H_{15}BrClSn$.
Hazard: A poison by skin contact. Moderately toxic by ingestion.
Use: A pesticide.

decaglycerol.   See polyglycerol.

decahydronaphthalene.
CAS: 91-17-8. $C_{10}H_{18}$.
cis and *trans* forms are known.
Properties: Colorless liquid; aromatic odor. *cis*: d 0.8927 (20/4C), fp −43.2C, bp 194.6C, refr index 1.48113 (20C). *trans*: d 0.8700 (20/4C), fp −31.5C, bp 185.5C, refr index 1.46968 (20C). Flash p 136F (57.7C) (CC), autoign temp 482F (250C). Insoluble in water; soluble in alcohol and ether. Combustible.
Derivation: By treatment of naphthalene in a fused state (above 100C) with hydrogen in the presence of a copper or nickel catalyst.
Grade: Technical.
Hazard: Moderate fire risk. Irritant to eyes and skin.
Use: Solvent for oils, fats, waxes, resins, rubber, etc. Substitute for turpentine; cleaning machinery; stain remover; shoe creams; floor waxes; cleaning fluids; lubricants; motor-fuel additive.

Δ-decalactone.   Artificial flavoring for margarine. Approved by FDA.

"Decalin" [Sigma-Aldrich Biotechnology].
TM for decahydronaphthalene.

decamethrin.   (1R-(1-α(S*),3-α))-cyano(3-phenoxyphenyl)methyl-3-(2,2-dibromovinyl)-2,2-dimethylcyclopropanecarboxylate).
CAS: 52918-63-5.
Use: Insecticide.

decamethyltetrasiloxane.
CAS: 141-62-8. $C_{10}H_{30}O_3Si_4$.
Properties: Colorless liquid. Bp 195C, fp −70C, d 0.853, refr index 1.34. Soluble in light hydrocarbons and benzene; slightly soluble in alcohol. Stable over wide temperature range.
Use: Silicone oils, antifoam agent in lubricating oils.

n-decanal.   (capraldehyde; capric aldehyde; n-decyl aldehyde; aldehyde C-10).
CAS: 112-31-2. $CH_3(CH_2)_8CHO$.
Properties: Colorless to light-yellow liquid; floral-fatty odor. D 0.831-0.838 (15C), refr index 1.427–1.431 (20C). Soluble in 80% alcohol, fixed oils, volatile oils, mineral oil; insoluble in water and glycerol. Combustible.
Derivation: Occurs in lemon grass, citronella, orange, and many other oils. Synthetically by oxidation of the corresponding alcohol or reduction of the acid.
Grade: Technical, FCC.
Use: Perfumery, flavoring.

n-decane.   (decyl hydride).
CAS: 124-18-5. $CH_3(CH_2)_8CH_3$.
Properties: Colorless liquid. D 0.7298, bp 174C, fp −30C, refr index 1.4114 (20C), flash p 111F (44C) (CC), autoign temp 482F (250C). Soluble in alcohol; insoluble in water. Combustible.
Grade: Technical, 95%, 99%, research.
Hazard: Moderate fire risk. Narcotic.
Use: Organic synthesis, solvent, standardized hydrocarbon, jet-fuel research.

decanedioic acid.   See sebacic acid.

decanoic acid.   See capric acid.

1-decanol.   (n-decyl alcohol; alcohol C-10).
CAS: 112-30-1. $CH_3(CH_2)_8CH_2OH$.
Properties: Colorless, water-white liquid; sweet odor. D 0.829, bp 232.9C, mp 6C, flash p 180F (OC) (82.2C); refr index 1.4372 (20C). Insoluble in water (25C); soluble in alcohol and ether. Combustible.
Derivation: Reduction of coconut oil fatty acids, from C_9 olefin and synthesis gas, by the Oxo process.
Grade: Technical, high purity.
Use: Plasticizers, detergents, synthetic lubricants, solvents, perfumes, flavorings, antifoam agent.

decanoyl chloride.   (sometimes called caproyl chloride).
CAS: 112-13-0. $CH_3(CH_2)_8COCl$.
Available in bottles, carboys, and drums. Intermediate, polymerization initiator.

decanoyl peroxide.
CAS: 762-12-9. $CH_3(CH_2)_8C(O)OOC(O)(CH_2)_8CH_3$.

Properties: Soft, white granules. Mp 38–42C (decomposes). Insoluble in water and alcohol; soluble in ether and benzene.
Hazard: Strong oxidizer, fire risk in contact with organic materials.
Use: Polymerization catalyst.

decarboxylase. One of a group of enzymes in the living cell that removes carbon dioxide from various carboxylic acids without oxidation. An enzyme that catalyzes a decarboxylation reaction.

decay. Spontaneous disintegration of an unstable atomic nucleus (e.g., uranium, radium), with emission of α, β, and γ radiation, and eventual formation of another element of lower atomic weight.
See radioactivity; half-life.

***cis*-4-decenal.**
CAS: 21662-09-9. $C_{10}H_{18}O$.
Properties: Colorless to slightly yellow liquid; fatty, orangelike odor. D: 0.847, refr index: 1.442–1.444. Sol in alc, fixed oils; insol in water.
Hazard: Low toxicity by ingestion and skin contact. A skin irritant.
Use: Food additive.

1-decene. See decylene.

decision theory. A branch of applied mathematics based on probability theory and concerned with choice in a nondeterministic environment.

deck. The platen of a compression molding press.

deckle. A strip or bar placed along the sides of a fourdrinier wire to equalize the flow of the pulp slurry and give the sheet a straight edge. When a ragged or "deckle" edge is desired on certain specialty papers, the strip is removed.

"Declomycin" *[Abbvie].* TM for dimethyl-chlortetracycline hydrochloride.

decoction. Pharmaceutical term for a liquid produced by boiling one or more drugs in water and filtering.

decoic acid. See capric acid.

decolorizing agent. Any material that removes color by a physical or chemical reaction. Charcoals, blacks, clays, earths, or other materials of highly adsorbent character used to remove undesirable color, as from sugar, vegetable and animal fats and oils, etc. Also refers to bleaches involving a chemical reaction for removing color.

decomposition. A fundamental type of chemical change. In simple decomposition, one substance breaks down into two simpler substances, e.g., water yields hydrogen and oxygen. In double decomposition, two compounds break down and recombine to form two different compounds, e.g., $2HCl + CaCO_3 \rightarrow CaCl_2 + H_2CO_3$. In some cases, heat is absorbed, and in others it is evolved.
Decomposition may occur as a result of (1) reaction at room temperature ($NaOH + HCl \rightarrow NaCl + H_2O$), (2) heating in air ($C + H_2O \rightarrow CO + H_2$), (3) electrolysis (inorganic compounds), ($NaCl_{aq} + e \rightarrow Na^+ + Cl^-$), (4) bacterial or enzymic action (fermentation, $C_6H_{12}O_6 \rightarrow 2C_2H_5OH + 2CO_2$), (5) radiation (photodecomposition), as in the breakdown of chlorofluorocarbons in the upper atmosphere and of biodegradable polymers exposed to sunlight, and (6) heating in absence of air (thermal decomposition), in which carbonaceous raw materials such as coal and natural gas are converted into carbon and volatile organic compounds without undergoing combustion (coal $\rightarrow$ coke, coal, tar, and coal gas). The term thermal decomposition is virtually synonymous with pyrolysis and destructive distillation.
See degradation; pyrolysis; destructive distillation.

decontamination. Removal of radioactive poisons from skin, clothing, equipment, etc. Skin can often be decontaminated by washing with soap and water; application of titanium dioxide paste or a saturated solution of potassium permanganate, followed by a rinse of 5% sodium bisulfite is approved procedure. Contaminated clothing should not be sent to commercial laundries or burned in open incinerators. Water, steam, and detergents are effective on painted or metal surfaces.

decoquinate.
CAS: 18507-89-6. $C_{24}H_{35}NO_5$.
Properties: Crystals. Mp: 86–87C.
Use: Drug (veterinary); food additive.

decortication. Removal of the hard coating (cortex) from certain vegetables, nuts, fruits, etc., by either mechanical or manual means.
See hydraulic barking.

decumbin. ((1R,2R,3E,7S,11E,15S)-2,15-dihydroxy-7-methyl-6-oxabicyclo[11.3.0]hexadeca-3,11-dien-5-one).
CAS: 20350-15-6. $C_{16}H_{24}O_4$.
A mycotoxin.
Derivation: Produced by *Penicillium decumbens.*
Hazard: Very toxic; respiratory distress and hemorrhage.

decyl acetate. (acetate C-10).
CAS: 112-17-4. $CH_3(CH_2)_9OOCCH_3$.
Properties: Liquid; floral orange–rose odor. Bp 187–190C, d 0.862–0.864, refr index 1.426. Soluble in

80% alcohol, ether, benzene, glacial acetic acid; insoluble in water. Combustible.
Grade: Technical.
Use: Perfumery.

_n_-decyl alcohol. See 1-decanol.

_n_-decyl aldehyde. See _n_-decanal.

_n_-decylamine.
CAS: 2016-57-1. $CH_3(CH_2)_9NH_2$.
Properties: Water-white liquid; amine odor. Boiling range 215–221C, d 0.797 (20/20C), refr index 1.437 (20C), flash p 210F (99C). Combustible.
Hazard: An irritant.

decyl carbinol. See 1-undecanol.

decylene. (1-decene).
CAS: 872-05-9. $C_{10}H_{20}$ or $H_2C{:}CH(CH_2)_7CH_3$.
Properties: Colorless liquid. D 0.7396 (20/4C), bp 172C, fp −66.3C, refr index 1.4220 (20C), flash p 130F (55C), autoign temp 455F (235C). Soluble in alcohol; insoluble in water. Combustible.
Grade: Technical, high purity.
Use: Organic synthesis of flavors, perfumes, pharmaceuticals, dyes, oils, resins.

decyl-hydride. See _n_-decane.

decylic acid. See capric acid.

decyl mercaptan.
CAS: 143-10-2. $C_{10}H_{21}SH$.
Properties: Liquid; strong odor. Fp −26C, bp 114C (13 mm Hg), d 0.8410 (20/4C), refr index 1.4536 (20C). Combustible.
Grade: 95% (min) purity.
Use: Intermediate, synthetic rubber processing.

decyl-octyl methacrylate.
$H_2C{:}C(CH_3)COO(CH_2CH_3)$.
Use: Monomer for plastics, molding powders, solvent coatings, adhesives, oil additives, emulsions for textile, leather, and paper finishing.

decylthiuronium chloride.
CAS: 5392-26-7. $C_{11}H_{24}N_2S{\cdot}ClH$.
Hazard: A poison.

decyltriphenylphosphonium bromochlorotriphenylstannate. See decafentin.

DEET. (_N,N'_-diethyl-_n_-toluamide).
CAS: 134-62-3.
A broad-spectrum insect repellent.

DEF. _S,S,S_-tributyl phosphorotrithioate.

defecation. Purification, used specifically in the industrial clarification of sugar solutions.

"Defenz" [Danisco]. TM for a specially formulated homeopathic solution of iodized sodium chloride containing biologically important trace minerals.
Use: Food supplement.

deferoxamine. (_N'_-[5-[[4-[[5-(acetylhydroxyamino)pentyl]amino]-1,4-dixoxbutyl]hydroxylamino]pentyl]-_n_-(5-aminopentyl)-_n_-hydroxybutanediamide; n-[5-[3-[(5-aminopentyl)hydroxycarbamoyl]propionamido]pentyl-3-[[5-(_n_-hydroxyacetamido)pentyl]carbamoyl]propionohydroxamic acid; 1-amino-6,17-dihydroxy-7,10,18,21-tetraoxo-27-(n-acetylhydroxylamino)-6,11,17,22-tetraazaheptaeicosane; desferrioxamine b; _N'_-[5-[acetyl(hydroxyl)amino]amino]pentyl]-_N_-[5[[[4-[5-aminopentyl(hydroxyl)amino]-4-oxobutanoyl]amino]pentyl]-_N_-hydroxybutanediamide; 30-amino-3,14,25-trihydroxy-3,9,14,20,25-pentaazatriacontane-2,10,13,21,24-pentaone; _N_-benzoylferrioxamine b; deferoxaminum; deferrioxamine; deferrioxamine b; desferal; desferral; desferrin; desferrioxamine; desferrioxamine b; Df B; DFO; DFOA; DFOM; NSC-52760).
CAS: 70-51-9. $C_{25}H_{48}N_6O_8$.
Properties: Crystals from EtOH.
Derivation: Isolated from _Streptomyces pilosus_.
Hazard: Poison; moderately toxic; mutagen; causes changes in hearing acuity, eye hemorrhage, optic nerve neuropathy, thrombocytopenia, visual field changes.
Use: Therapeutically as a parenteral chelating agent for iron and aluminum; to treat dialysis encephalopathy, and to promote excretion of iron in patients with a secondary iron overload from multiple transfusions.

deflagration. Very rapid autocombustion of particles of explosive as a surface phenomenon. Usually initiated by contact with a flame or spark but may be caused by impact or friction. Deflagration is characteristics of low explosives.
See detonation.

defoaming agent. (antifoaming agent).
A substance used to reduce foaming due to proteins, gases, or nitrogenous materials that may interfere with processing. Examples include 2-octanol, sulfonated oils, organic phosphates, silicone fluids, dimethylpolysiloxane, etc. For restrictions on their use in foods, see FDA regulations.

defoliant. An herbicide that removes leaves from trees and growing plants. They may be either organic or inorganic. Some examples: (organic) phenoxyacetic acids, trichloropicolinic acid, carbamates, and nitro compounds; (inorganic) arsenic compounds, cyanides, thiocyanates, and chlorates. Several of the more persistent types have been

used in military operations. Many defoliants are toxic.
See 2,4-D; 2,4,5-T; Agent Orange.

DEG. (1) Abbreviation for diethylene glycol. (2) Abbreviation for diethanolglycine.

degenerate code. A code in which a single element in one language is specified by more than one element in a second language.

degenerate codon. A codon that specifies the same amino acid as another.

degenerate electron gas. A gas whose properties are chiefly dependent on the behavior of free electrons. Such a gas supplies the pressure that supports white dwarfs against collapse.

degenerate gas.
Properties: Gas with free electrons or free neutrons as densely spaced as permitted by the laws of quantum mechanics; unusually high, temperature-independent density; resistive to further compression.

degenerate neutron gas. A gas whose properties are chiefly dependent on the behavior of free neutrons. Such a gas supplies the pressure that supports neutron stars against collapse.

degenerate system. A system having several distinct wave functions corresponding to the same energy level.

DEGN. Abbreviation for diethylene glycol dinitrate.

degradation. A type of decomposition characteristic of high-molecular-weight substances such as proteins, polymers, branched-chain sulfonates, etc. It may result from oxidation, heat, sunlight, solvents, bacterial action, or, in the case of body proteins, infectious microorganisms.
See biodegradability; decomposition.

degras. Crude wool grease obtained by solvent washing of wool. It is a dark-brown semisolid with strong unpleasant odor and high water-absorbing capacity. A type known as moellen degras is a by-product of tanning chamois leather with various fish oils. The chief use of degras is as the source of lanolin; minor uses are in leather dressing and printing inks.
Available forms: Neutral, common, technical.

degree of polymerization. (DP).
The number of monomer units in an average polymer molecule in a given sample. For natural cellulose, it is about 3000, but in most polymers, it is still higher. It can be controlled by appropriate processing techniques. DP is an important factor in plastics technology, as it directly affects the viscosity of solutions and properties of the end product.
See polymerization; shortstopping agent.

degree of substitution. The average number of (alcohol) groups per polymer that have been replaced.

degrees of freedom. The number of variables (e.g., pressure, temperature, or concentration) that can be changed without producing a change in the number of phases in a system. The number of these variables must be fixed arbitrarily to completely define the system.

deguelin.
CAS: 522-17-8. $C_{23}H_{22}O_8$.
Properties: Greenish powder. Mp 170C. Soluble in alcohol; insoluble in water.
Hazard: Toxic by inhalation; skin irritant.
Use: Insecticide.

DEH. A variety of polyamines and polyamides suitable for curing epoxy resins.

dehairing. See unhairing.

dehalogenase. Any enzyme that catalyzes the removal of halogen atoms from organic halides.

dehumidification. The removal of moisture (water vapor) from air. Also sometimes extended to analogous processes of removing a vapor from a gas mixture.

"Dehydran 1208" [Cognis]. TM for coatings based on polyester polyacrylates, PUR, and epoxy resins.
Use: Solvent-free waterborne and solvent-borne coatings, contains silicones.

"Dehydran 1293" [Cognis]. TM for a defoamer for coatings based on polymer emulsions.
Use: Application to curtain coating.

"Dehydratine" [Euclid]. TM for bituminous water barrier coatings.

dehydration. + (1) Removal of 95% or more of the water from a material, usually a foodstuff, by exposure to high temperature by various means. Its primary purpose is to reduce the volume of the product, increase its shelf life, and lower transportation costs. Special equipment for dehydration includes tunnel dryers, vacuum (shelf) dryers, drum dryers, etc., in which the bulk product is exposed to a hot air environment. Another method is spray drying, in which a liquid product is ejected from a nozzle into hot air; dried milk and egg white are prepared in this way. The term *dehydration* is not applied to loss of water by evaporation or sun drying.
See drying.

(2) Removal of one or more molecules of H_2O from a chemical compound, e.g., of ethanol to ethylene.

Dehydrite.   Anhydrous granular magnesium perchlorate.
Use: A desiccant.

dehydroabietic acid.
CAS: 1740-19-8. $C_{20}H_{28}O_2$.
Solid, used as a basis for thermoplastic resins.

dehydroacetic acid.   (DHA, methylacetopyranone).
CAS: 520-45-6

$$CH_3C{:}CHC(O)CH(COCH_3)C(O).$$

Properties: Colorless crystals; odorless; tasteless. Mp 108.5C, bp 270C. Partially soluble in acetone and benzene; insoluble in water. Highly reactive. Combustible.
Derivation: (1) By action of N-bromosuccinimide on ketene dimer, (2) by strong heating of acetoacetic ester.
Grade: Technical, FCC.
Hazard: Toxic by ingestion.
Use: Fungicide and bactericide, plasticizer, chemical intermediate, medicated toothpastes.

5,6-dehydro-n-acetylloline.
CAS: 194205-01-1. $C_{10}H_{14}N_2O_2$.
Hazard: A poison.
Source: Natural product.

dehydroascorbic acid.

$$OCOCOCOCHCH_2OCH_2OH.$$ The oxidized form of ascorbic acid with the same vitamin activity.
Properties: Needles. Mp 225C (decomposes). Acid soluble in water at 60C.
Derivation: Synthesized from ascorbic acid.
Use: Nutrition, medicine.

7-dehydrocholesterol.   (provitamin D_3).
CAS: 434-16-2. $C_{27}H_{44}O \cdot H_2O$.
A sterol found in the skin of humans and animals that forms vitamin D_3 upon UV irradiation

Properties: Slender platelets from ether–methanol. Mp 150C. Insoluble in water; soluble in organic solvents.
Use: Nutrition, medicine, biochemical research.
See cholecalciferol.

dehydrocholic acid.
$C_{24}H_{34}O_5$.
A polycyclic compound.
Properties: White, fluffy powder, odorless; bitter taste. Mp 231–240C. Almost insoluble in water; slightly soluble in ether and alcohol; soluble in chloroform, glacial acetic acid, and solutions of alkali hydroxides and carbonates.
Derivation: Oxidation of cholic acid.
Grade: NF.
Use: Medicine, pharmaceutical intermediate.

dehydrocorticosterone.   (21-hydroxypregn-4-ene-3,11,20-trione; $\bar{A}^4$-pregen-21-ol-3,11,20-trione;   17-(1-keto-2-hydroxyethyl-$\bar{a}^4$-androsten-3,11-dione;   kendallns   compound   a;   17-desoxycortisone;   17-(2-hydroxyacetyl)-10,13-dimethyl-2,6,7,8,9,12,14,15,16,17-decahydro-1H-cyclopenta[a]phenanthrene-3,11-dione).
$C_{21}H_{28}O_4$.
A steroid hormone secreted by the adrenal cortex.
Derivation: Produced synthetically.
Use: Antiallergic agent.

dehydrocurdione.
CAS: 38230-32-9. $C_{15}H_{22}O_2$.
Hazard: A poison.

dehydrocyclodimerization.   A method of converting paraffin (straight-chain) hydrocarbons containing from three to five carbon atoms into aromatic (ring-type) hydrocarbons. Its main steps are (1) removal of hydrogen from the paraffins; (2) dimerization of the resulting olefins; (3) aromatization of the dimerized olefins and diolefins; and (4) isomerization or transalkylation to C_8 to C_{10} alkylbenzene isomers. Metallic catalysts are essential in some or all of these steps. The process is not in large-scale use.

dehydroepiandrosterone acetate.   See 3-β-hydroxyandrosten-17-one acetate.

dehydrogenases.   Enzymes catalyzing the removal of pairs of hydrogen atoms from their substrates.
See oxidase.

dehydrogenation.   The process whereby hydrogen is removed from compounds by chemical means. Dehydrogenation of primary alcohols yields the group of compounds called aldehydes. It is considered to be a form of oxidation, as two hydrogen atoms, each of which contains an electron, have been removed. An example is the reaction
$CH_3CH_2OH \rightarrow CH_3CHO + H_2$

11-dehydro-17-hydroxycorticosterone.   See cortisone.

dehydroisoandrosterone.   See DHEA.

dehydrothio-*p*-toluidine.

$$CH_3C_6H_3SC(C_6H_4NH_2)N.$$

Properties: Long, yellowish, iridescent needles. Mp 191C, bp 434C. Solutions have a violet-blue fluorescence. Soluble in alcohol; very slightly soluble in water.
Derivation: By heating *p*-toluidine and primuline base with sulfur and separation from the primuline base by distillation in vacuo.
Use: Dyestuffs, intermediate.

de-icing compound.   See calcium chloride; sodium chloride; alcohol.

de-inking.   The removal of printing inks from paper by use of strong alkaline solutions such as soda–ash liquor, caustic soda, or lime which dissolve, varnish, and free the ink carbon. Removal of the carbon is accomplished by use of colloidal agents such as talc or bentonite and by mechanical agitation with water.

deionized water.   (oxidane).
CAS: 7732-18-5. H_2O.
Water from which most salts have been removed by ion exchange.

deionizing.   (demineralizing).
A method for purifying water that involves two steps. First, soluble salts are converted into acids by passing through a hydrogen exchanger. Second, they are removed by an acid adsorbent or synthetic resin.

Deisenhofer, Johann.   (1943–). Awarded Nobel Prize for chemistry in 1988, along with Huber and Michel, for work that revealed the three-dimensional structure of closely linked proteins that are essential to photosynthesis. Doctorate awarded in 1974 by Max Planck Institute for Biochemistry, Germany.

deKhotinsky cement.   A thermoplastic adhesive mixture of shellac and pine tar. It is not attacked by water, sulfuric acid, nitric acid, hydrochloric acid, carbon disulfide, benzene, gasoline, or turpentine; very little affected by ether, chloroform, alkalies, but readily dissolved by ethanol.

"Delac" *[Chemtura]*.   TM for a series of delayed action rubber accelerators.

Delepine reaction.   Preparation of primary amines by reaction of alkyl halides with hexamethylenetetramine, followed by acid hydrolysis of the formed quaternary salts.

deletion.   A loss of part of the DNA from a chromosome; can lead to a disease or abnormality. See chromosome; mutation.

deletion map.   A description of a specific chromosome that uses defined mutations—specific deleted areas in the genome—as biochemical signposts, or markers for specific areas.

deletion mutation.   A mutation resulting from the deletion of one or more nucleotides from a gene.

delhi hard.   A ferrous alloy (d 7.75, mp 500C) containing, in addition to iron, 16.5–18% chromium, 1.1% carbon, 0.75–1% silicon, and 0.35–0.5% manganese. It is resistant to cold ammonium hydroxide in all concentrations and to mine and seawaters and moist sulfurous atmospheres.

deliquescent.   Tending to absorb atmospheric water vapor and become liquid. The term refers specifically to water-soluble chemical salts in the form of powders that dissolve in the water absorbed from the air. Such salts should be kept closely stoppered or otherwise enclosed.
See hygroscopic.

delphinine.
CAS: 561-07-9. $C_{33}H_{45}NO_9$.
Properties: Polycyclic diterpene alkaloid.
Derivation: Produced by plants of the genera *Aconitum* and *delphinium*.

delphinoidine.
Properties: Diterpenoid alkaloid.
Derivation: Isolated from seeds of *Delphinium staphisagria*.

delphisine.
$C_{28}H_{42}NO_8$.
Properties: Alkaloid isomer of delphinine.
Derivation: Isolated from seeds of *Delphinium staphisagria*.

Delrin.   A type of acetal resin. White and colors available. Also supplied as pipe and fittings. Thermoplastic.
Use: Injection-molded and extruded parts, door handles, bushings, other mechanical items; underground pipe; automotive parts.

Delsan.   Fungicide–insecticide seed treatment containing 60% thiram and 15% dieldrin.
Hazard: Toxic by ingestion and inhalation.

delta acid.   See Casella's acid F.

delta iron.
  Properties: Allotrope of iron, stable between 1400C and melting point of iron; structurally identical to alpha iron.

delta ray.   A particle, especially an electron, ejected from matter by ionizing radiation.

delustrant.   A substance used to produce dull surfaces on a textile fabric. Chiefly used are barium sulfate, clays, chalk, etc. They are applied in the finishing coat.

De Mayo reaction.   Synthesis of 1,5-diketones by photoaddition of enol derivatives of 1,3-diketones to olefins, followed by a retro-aldol reaction.

demecolcine.   (6,7-dihydro-1,2,3,10-tetramethoxy-7-(methylamino)benzo[a]heptalen-9(5H)-one; *n*-deacetyl-*n*-methylcolchicine; *n*-desacetyl-*n*-methylcolchicine;   *n*-methyl-*n*-desacetylcolchicine; colchamine; (7S0-1,2,3,10-tetramethoxy-7-(methylamino)-6,7-dihydro-5H-benzo[a]heptalen-9-one).
  CAS: 477-30-5. $C_{21}H_{25}NO_5$.
  Properties: Alkaloid.
  Derivation: derived from *Colchicum autumnale* (family *liliaceae*).
  Use: Antineoplastic agent.

4-demethoxydaunorubicin hydrochloride.
  CAS: 57852-57-0. $C_{26}H_{27}NO_9 \cdot ClH$.
  Hazard: A poison.

5-*o*-demethylavermectin ala.   See avermectin b_{1a}.

1-demethyltoxoflavine.
  CAS: 5016-18-2. $C_6H_5N_5O_2$.
  Hazard: A poison.

demeton.   (Systox).
  CAS: 8065-48-3. $C_8H_{19}O_3PS_2$.
  A mixture of *O,O*-diethyl-*O*-2-(ethylthio)ethyl phosphorothioate (demeton-*O*) and *O,O*-diethyl-*S*-2-(ethylthio)ethyl phosphorothioate (demeton-*S*).
  Properties: (Mixture) Pale-yellow liquid. Bp 134C (2 mm Hg), d 1.118. Slightly soluble in water; soluble in most organic solvents.
  Hazard: Toxic by skin absorption; cholinesterase inhibitor. Use may be restricted. Cholinesterase inhibitor.
  Use: Systemic insecticide (absorbed by plant, which then becomes toxic to sucking and chewing insects).

demeton methyl.   (*O,O*-dimethyl-*S*,2-(ethylthio)ethyl phosphorothiolate).
  CAS: 8022-00-2.
  Use: Systemic insecticide.

demeton-*o*-methyl   sulfoxide.   (bay 21097; demeton-*s*-methyl sulfoxide; *o,o*-dimethyl-2-(2–aethylsulfinyl-aethyl)-thiolphosphat;   demeton-methyl   sulphoxide;   *o,o*-dimethyl-*s*-(2-ethionylethyl)phosphorothioate;   dimethyl-*s*-(2-ethionylethyl) thiophosphate; *o,o*-dimethyl-*s*-(2-ethylsulfinyl)-ehtyl   thiophosphate;   *o,o*-dimethyl-s-ethylsulfinylethyl   phosphorothiolate; *s*-[2-(ethylsulfinyl)ethyl]-*o,o*-dimethyl   phosphorothioate;   2-ethylsulfinylethoxy-dimethoxy-sulfanylidene-1,5-phosphane;   dimethyl-*s*-(2-ethionylethyl)thiophosphate;   ENT   24,964; *s*-(2-(ethylsulfinyl)ethyl)-o,o-dimethyl   phosphorothioate; isomethylsystox sulfoxide; metaisosystoxsulfoxide; metasystemox; metasystox-r; methyl demeton-*o*-sulfoxide;   metilmercaptofosoksid; oxydemetonmethyl;   oxydementon-metile;   R 2170;   thiophosphate   de   *o,o*-dimethyle et de *s*-2-ethylsulfinylethyle).
  $C_6H_{15}O_3PS_2$
  Properties: Yellow liquid; soluble in water and most organic solvents; insoluble in ether.
  Hazard: Toxic; poison; mutagen.
  Use: Insecticide.

demeton-*S*.   (diethyl-*s*-(2-ethioethyl) thiophosphate;   *o,o*-diaethyl-*s*-(2-aethylthio-aethyl)-monothiophosphat;   diaethylthiophosphorsaeureester des aethylthioglykol; *o,o*-diethyl-*s*-(2-ethioethyl)phosphorothioate;   *o,o*-diethyl-*s*-ethyl-2-ethylmercaptophosphorothiolate; *o,o*-diethylphosphorothioate;   1-diethoxyphosphorylsulfanyl-2-ethylsulfanylethane;   *o,o*-diethyl-*s*-(2-ethylthio-ethyl)-monothiofosfaat; *o,o*-diethyl-*s*-2-(ethylthio) ethyl   phosphorothioate; *o,o*-diethyl-*s*-(2-(ethylthio) ethyl)   phosphorothiolate; *o,o*-dietil-*s*-(2-etiltioetil)-monotiofosfato;   *o,o*-diethyl-s-2-etyl-merkaptoetyltiofosfat;   2-(ethylthio)-ethanethiol *s*-ester   with   *o,o*-diethyl   phosphorothioate; isodemeton; izosystox; po-systox; thioldemeton; thiol systox; thiophosphate de *o,o*-diethyle et de *s*-(2-ethylthio-e-thyle).
  CAS: 126-75-0. $C_8H_{19}O_3PS_2$ $C_8H_{19}O_3PS_2$.
  An organophosphate that produces PO_x and SO_x fumes on decomposition by heating.
  Hazard: Toxic; poison.
  Use: Insecticide.

demeton-*S*-methyl-sulfone.   (demeton-*s*-methyl-sulphone;   *o,o*-dimethyl-*s*-(2-ethsulfonylethyl)-phosphorothioate;   dimethyl-*s*-(2-ethsulfonyl-ethyl)thiophosphate; *o,o*-dimethyl-*s*-ethyl-2-sulfonylethyl   phosphorothiolate; *o,o*-dimethyl-*s*-ethylsulfonylethyl   phosphorothiolate; dixoydemeton-*s*-methyl;   1-dimethyoxyphosphorylsulfanyl-2-ethylsulfonylethane).
  CAS: 17040-19-6. $C_6H_{15}O_5PS_2$.
  Very toxic, PO_x and SO_x fumes are released on decomposition by heating.

Hazard: Poison; moderately toxic; mutagen.
Use: Insecticide.

demineralization. Removal from water of mineral contaminants, usually present in ionized form. The methods used include ion exchange techniques, flash distillation, or electrodialysis. Acid mine wastes may be purified in this way, thus alleviating the pollution problem.
See desalination; deionizing.

Demjanov rearrangement. Deamination of primary amines by diazotization to give rearranged alcohols.

Democritus. A Greek philosopher (approximately 465 BC). The first thinker of record to conceive of matter as existing in the form of small indivisible particles, which he called atoms. However, this concept was overshadowed by Aristotle's theories, and it was not until some 2000 years later that it was developed by John Dalton in England—an astonishing length of dormancy for one of the most creative ideas in the history of science.
See Dalton, John.

demulsification. The process of destroying or "breaking" an unwanted emulsion, especially water-in-oil types occurring in crude petroleum. Both chemical and physical means are used. Chemical means include addition of polyvalent ions to neutralize electrical charges or of a strong acid; physical means include heating, centrifuging, or use of high-potential alternating current.
See emulsion; nonylphenol.

demurrage. A fee imposed on shippers of chemicals and other products by the railroads for retaining freight cars at loading docks for more than a given period of time (usually 24 hours).

DEN. A series of epoxy novolacs for multifunctional resins for all uses where maximum chemical or heat resistance is required.

denatonium benzoate. See "Bitrex" [Johnson Matthey].

denaturant. See alcohol, denatured.

denaturation. A change in the molecular structure of globular proteins that may be induced by bringing a protein solution to its boiling point or by exposing it to acids or alkalies or to various detergents. Denaturation reduces the solubility of proteins and prevents crystallization. It involves rupture of hydrogen bonds so that the highly ordered structure of the native protein is replaced by a looser and more random structure. It is usually irreversible but in some cases is reversible, depending on the protein and the treatment involved.

Alteration of the specific native conformation of a polypeptide chain, protein, or nucleic acid.
See degradation.

denatured alcohol. See alcohol, denatured.

denatured DNA. Double-stranded DNA that has been converted to single strands by breaking the hydrogen bonds that couple complementary nucleotide pairs (usually by heating), a process that is often reversible.

dendrimer. A tree-like highly branched polymer molecule (Greek dendra = tree). Dendrimers are synthesized from monomers with new branches added in discrete steps ("generation") to form a tree-like architecture. A high level of synthetic control is achieved through stepwise reactions and purifications at each step to control the size, architecture, functionality, and monodispersity.

dendrolasin. (3-[(3E)-4,8-dimethylnona-3,7-dienyl]furan).
CAS: 23262-34-2. $C_{15}H_{22}O$.
A component of anti-venom.
Derivation: Occurs in certain plants.

denier. A unit used in the textile industry to indicate the fineness of a filament. If 9000 m of a filament weighs 1 g, the filament is 1 denier; if 10,000 m weighs 1 g, the filament is 1 grex. Sheer women's hosiery usually runs from 15 to 10 denier.

Denomega. Nutritional oils from marine raw materials.
Use: In foods and supplements.

de novo pathway. Pathway for synthesis of a biomolecule from simple precursors, such as synthesis of purine nucleotides from Asp, CO2, Gly, formate, and Glu: Distinguish from a salvage pathway.

"Denox" [Grace]. TM for "300 Highlighter" series, a group of products.
Use: Treating denim with stone washing and other finishes.

density. Mass per unit volume expressed in grams per cubic centimeter for solids and liquids and usually as grams per liter for gases. Densities of some common substances follow:

g/cc	g/L
Sulfur	2.06
Aluminum	3.7
Sodium	0.967
Glycerol	1.27
Water*	1.0

g/cc	g/L
Chlorine	3.214
Carbon dioxide	1.977
Air**	1.293
Oxygen	1.429
Hydrogen	0.0899

*Basis of comparison for solids and liquids.
**Basis of comparison for gases.

For discussion of density vs. specific gravity, see specific gravity. Apparent density is the mass of a unit volume of powder, usually expressed in grams per cubic centimeter, determined by a specified method (MPA definition, MPA Standard 9-50T). *Bulk density* is an alternative term for *apparent density.*
See current density.

"Deo-Base" *[Chemtura].* TM for light petroleum distillate, superfine grade of kerosene without its objectionable odor.

deodorant. A substance used to remove or mask an unpleasant odor. It may or may not have a distinctive odor of its own. Deodorants act (1) by adsorption (activated carbon, charcoal, chlorophyllin), (2) by replacement (pine oil or other perfume), (3) by neutralization (aluminum chlorohydrate), and (4) by oxidation or hydrogenation, e.g., of fish oils. The cosmetic industry supplies a wide variety of deodorants and antiperspirants, chiefly based on neutralization. Mouthwashes and breath "sweeteners" often contain calcium iodate, thymol, peppermint, or a similar substance to mask or replace odors.
See odor; cosmetic.

deoxidizer. An agent that removes oxygen from a compound or from a molten metal.

deoxy-. Preferred prefix indicating replacement of hydroxyl by hydrogen in the parent compound. The meaning is the same as that of desoxy, and the two prefixes are used interchangeably.

deoxyandenosine-5′-triphosphate. (dATP; [(2R,3S,5R)-5-(6-aminopurin-9-yl)-3-hyroxy-oxolan-2-yl]methyl[hydroxyl(phos-phonooxy)phospho]hydrogen phosphate).
CAS: 1927-31-7. $C_{10}H_{16}N_5O_{12}P_3$.
A direct precursor molecule of DNA.

deoxyanisoin. (4′-methoxy-2-(p-methoxyphenyl)acetophenone).
CAS: 120-44-5. $CH_3OC_6H_4COCH_2C_6H_4OCH_3$.
Properties: Off-white to buff, crystalline powder; sweet, faint, cinnamon-like odor. Mp 110–112C.
Use: Intermediate.

deoxybenzoin. (α-phenylacetophenone, benzyl phenyl ketone).
CAS: 451-40-1. $C_6H_5CH_2COC_6H_5$.
Properties: Colorless crystals. Mp 53–60C. Slightly soluble in hot water; soluble in alcohols and ketones.
Use: Intermediate.

deoxycholic acid. (desocycholic acid).
CAS: 83-44-3. $C_{24}H_{40}O_4$.
A bile acid, contains one less hydroxyl group than cholic acid.
Properties: Crystals. Mp 172–173C. Not precipitated by digitonin. Practically insoluble in water and benzene, slightly soluble in chloroform and ether, and soluble in acetone and solutions of alkali hydroxides and carbonates, freely soluble in alcohol. Also available as sodium salt. Forms coordination compounds with fatty acids.
Derivation: Isolation from bile, organic synthesis.
Grade: Technical, FCC (as desoxycholic acid).
Use: Medicine, precursor for organic synthesis of cortisone, emulsifying agent in foods (up to 0.1%).

deoxycorticosterone. (4-pregnen-21-ol-3,20-dione; 11-deoxycorticosteroid).
CAS: 64-85-7. $C_{21}H_{30}O_3$.
An adrenal cortical steroid hormone. Active in causing the retention of salt and water by the kidney.
Properties: Crystalline plates. Mp 141–142C. Freely soluble in alcohol and acetone.
Derivation: From adrenal cortex extract, synthesis from other steroids.
Use: Medicine (usually as acetate).

deoxyguanosine-5′-triphosphate. (dGTP; [(2R,3S,5R0-5-(2-amino-6-oxo-3H-purin-9-yl)-3-hydroxyoxolan-2-yl]methyl[hydroxyl(phosphonooxy)phospho]hydrogen phosphate).
CAS: 2564-35-4. $C_{10}H_{14}N_5Na_2O_{13}P_3$.
An immediate precursor of DNA, that is required for DNA synthesis.

deoxyguanylic acid. (dGmp; guanine deoxyn-bonucleotide; deoxyguanosine phosphate; [(2R,3S,5R)-5-(2-amino-6-oxo-3H-purin-9-yl)-3-hydroxyoxolan-2-yl]methyl dihydrogen phosphate).
CAS: 902-04-5. $C_{10}H_{14}N_5O_7P$.
A hydrolysis product of DNA.

deoxyribonuclease. One of a group of enzymes that cause the splitting of deoxyribonucleic acids. Pancreatic deoxyribonuclease, the most widely studied, cleaves the acid at the 3′-phosphate bond. Other deoxyribonucleases cleave the 5′-phosphate bond.

deoxyribonucleic acid. (DNA).

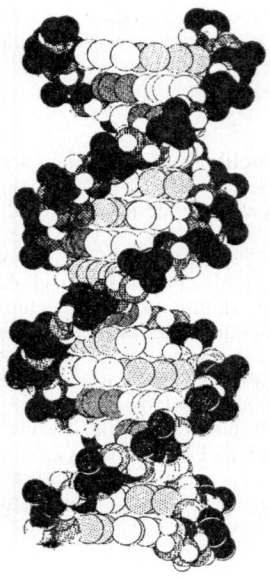

A complex sugar–protein polymer of nucleoprotein that contains the complete genetic code for every enzyme in the cell. It occurs as a major component of the genes, which are located on the chromosomes in the cell nucleus. The DNA molecule is a unique and intricate structure first elucidated in England by the chemists Crick and Watson in 1953. It is composed of 3000 to several million nucleotide units arranged in a double helix containing phosphoric acid, 2-deoxyribose, and the nitrogenous bases adenine, guanine, cytosine, and thymine. The spiral consists of two chains of alternating phosphate and deoxyribose units in continuous linkages. The nitrogenous bases project toward the axis of the helix and are joined to each other by hydrogen bonds. Adenine always unites with thymine, and cytosine with guanine. The complementarity of the bases on the joined chains allows each chain to act as a template for replication of the other when the chains are separated, thus producing two new strands of DNA. The sequence of the bases on the chains varies with the individual, and it is this sequence that expresses the genetic code. DNA works in conjunction with ribonucleic acid (RNA). Synthesis of self-replicating DNA was reported in 1967. Elucidation of the structure of the DNA molecule is under continuing research. Studies on synthetic DNA indicate that the helix may have a left-handed rather than a right-handed form. See ribonucleic acid; gene; nucleic acid; genetic code; replication; and recombinant DNA.

deoxyribonucleotides. Nucleotides containing 2-deoxy-D-ribose as the pentose component.

D-deoxyribose.
$CH_2OHCH_2OCH_2OCH_2CHO$.
A five carbon–atom sugar that is unusual in that there is no oxygen atom attached to the second carbon atom. It is a constituent of deoxyribonucleic acid.

DEP. Abbreviation for diethyl phthalate.

2,4-DEP.
(tris-(2,4-dichlorophenoxy)ethyl phosphite).
An herbicide.

"Depakene" *[Sanofi].* TM for sodium valproate.

Department of Transportation (DOT). The Federal agency that has been responsible since 1967 for the regulation and control of transportation of hazardous materials.
Website: http://www.dot.gov/

DEPC. (1) Abbreviation for diethyl pyrocarbonate. (2) Abbreviation for γ-diethylaminopropyl chloride hydrochloride.

dephlegmation. Partial condensation of vapor from a distillation operation to produce a liquid richer in higher-boiling constituents than the original vapor. The residual vapor is richer in the lower-boiling constituents.

depilatory. A substance used to remove hair from skin. Sulfides are largely used for this purpose. The leather industry uses large amounts of sodium sulfide for unhairing hides. The cosmetic industry also offers various sulfide preparations for removing unwanted body hair.

deptropine citrate. See elargin.

depurator. A system of filters, scrubbers, or electrostatic devices.
Use: For removing impurities from the air or from solvents.

dequalinium chloride.
$C_{30}H_{40}Cl_2N_4$.
Properties: Crystals. Mp 326C. Soluble in water.
Use: Bacteriostat, antiseptic.

DER. A series of epoxies including liquid resins, solid resins and solutions, flexible resins, and flame-retardant resins.

Deraspan. A group of epoxy resins and curing agents.

derived unit. A unit derived from the fundamental units of time, length, and mass, such as units of speed or density.

Dermasoft. Skin care products.

derris root.   The root of the shrubs *Derris ellip-tica* and *Derris malaccensis*. Chief active con-stituent is rotenone. Used as an insecticide.

DES.   Abbreviation for diethylstilbestrol.

16-desacetyl-16-anhydroacoschimperoside p.
CAS: 20819-47-0. $C_{30}H_{44}O_8$.
Hazard: A poison.
Source: Natural product.

desalination.   (Desalting).
Any of several processes for removing dissolved mineral salts from ocean water and other brines. The most important are as follows: (1) Distillation with reuse of vapors by compressive distillation or multiple-effect evaporation. Solar distillation has been in use on the Greek islands for some years. (2) Electrodialysis, an ion exchange method more efficient for purification of brackish water than sea-water (see demineralization). (3) Reverse osmosis, which uses pressure applied to the surface of a saline solution that is separated from pure water by a semipermeable membrane that ions cannot easily penetrate.
See osmosis.
The pressure forces the water component of the solu-tion through the membrane, thus effectively sepa-rating the components of the solution. Membranes used are cellulose acetate or graphitic oxide. This method is planned for use in a desalination plant proposed for the brackish waters of the lower Col-orado River that is said to be the world's largest. It is also used in a Potomac River installation. (4) Flash distillation appears to be the most effective method so far developed for seawater desalina-tion, accounting for about 90% of world production capacity.
There are approximately 350 desalination plants in the U.S., producing over 65 million gallons of fresh water a day. Development is under control of the Office of Saline Water, Dept. of Interior.

desaturases.   Enzymes (dehydrogenases) that catalyze the introduction of double bonds into the hydrocarbon portion of fatty acids.

"Descote" *[Particle Dynamics].*
CAS: 98-92-0.
TM for niacinamide tablet formulations.
Use: As a taste masker.

desert.   A terrestrial environment that receives less than about 25 cm (10 inches) of rain per year, and consequently supports only a few kinds of plants.

desiccant.   A hygroscopic substance such as acti-vated alumina, calcium chloride, silica gel, or zinc chloride. Such substances adsorb water vapor from the air and are used to maintain a dry atmosphere in containers for food packaging, chemical reagents, etc.
See molecular sieve.

desiccator.   A tightly closed vessel containing a desiccant. Used in the laboratory for drying test materials. Some types have partial vacuum.

"Desicote" *[Glaxo-SmithKline].*   TM for a mixture of hydrophobic monomers stabilized in chlorinated hydrocarbon and aromatic solvents. Rapidly decomposes on contact with sorbed water on glass surfaces, leaving surface water repellent.

desiodothyroxine.   See thyronine.

desmethylcyproheptadine.
CAS: 14051-46-8. $C_{20}H_{19}N$.
Hazard: A reproductive hazard.

***n*-desmethyltamoxifen.**
CAS: 31750-48-8. $C_{25}H_{27}NO$.
Hazard: A reproductive hazard.
s-desmethylzopiclone.
CAS: 151776-26-0. $C_{16}H_{15}ClN_6O_3$.
Hazard: A poison.

desmetryne.   (4-*N*-methyl-6-methylsulfanyl-2-*N*-propan-2-yl-1,3,5-triazine-2,4-diamine).
CAS: 1014-69-3. $C_8H_{15}N_5S$.
An atrazine that is similar in biological activity to atrazine and simazine but is absorbed through foliage as well as the roots.
Use: A pre-emergence and postemergence herbicide.

"Desmodur" *[Bayer AG].*   TM for a group of isocyanates and isocyanate prepolymers for ure-thane coatings, foams, adhesives, etc.

"Desmophen" *[Bayer AG].*   TM for a group of polyesters and polyethers for cross-linking with isocyanates.

"Desmophen A" *[Bayer AG].*   (urethane acrylics).
TM for polyacrylate resins.
Use: In chemical- and weather-resistant polyurethane coatings.

"DeSolite" *[DSM IP].*   TM for a product for application to artificial grass.

desolvation.   In aqueous solution, the release of bound water surrounding a solute, such as an enzyme.

desorption.   The process of removing an adsorbed material from the solid on which it is adsorbed. See adsorption. Desorption may be accomplished by heating, by reduction of pressure, by the presence of another more strongly adsorbed substance, or by a combination of these means.

desoxy-. See deoxy-

desoxycholate amphotericin b. See fungizone intravenous.

desoxycholic acid. FCC name for deoxycholic acid.

desoxycorticosterone. (4-pregnen-21-ol-3,20-dione; 11-deoxycorticosteroid; 11-desoxycorticosteroid; (8S,9S,10R,13S,14S,17S)-17-(2-hydroxyacetyl)-10,13-dimethyl-1,2,6,7,8,9,11,12,14,15,16,17-dodecahydrocyclopenta[a]phena-3-one).
CAS: 64-85-7. $C_{21}H_{30}O_3$.
A steroid metabolite that is the 11-deoxy derivative of corticosterone and the 21-hydroxy derivative of progesterone.
Properties: Phlogistic adrenal corticoid.
Hazard: Retention of salt and water by the kidney.
Use: In medicine as the acetate.

d-**desoxyephedrine.** See *d*-1-phenyl-2-methylaminopropane.

Despretz law. States that the temperature of maximum density of water is lowered from 4C on the addition of a solute by an amount proportional to the concentration of the solution.

destructive distillation. An operation in which a highly carbonaceous material, such as coal, oil shale, or tar sands, is subjected to high temperature in the absence of air or oxygen, resulting in decomposition to solids, liquids, and gases. As the solid end product is carbon, the term *carbonization* is often used. Other terms with the same general meaning as destructive distillation are *pyrolysis* and *thermal decomposition*. Destructive distillation of coal is carried out in the temperature range of 350–1000C, yielding coal tar, coal gas, and char (coke, carbon).

destruxin.
Properties: Cyclodepsipeptides with a number of N-CH$_3$ groups.
Use: Insecticide.

detergent. Any substance that reduces the surface tension of water, specifically a surface-active agent that concentrates at oil–water interfaces, exerts emulsifying action, and thus aids in removing soils. The older and still widely used types are the common sodium soaps of fatty acids, which are relatively weak. The much stronger synthetic detergents are classed as anionic, cationic, or nonionic, depending on their mode of chemical action. The latter functions by a hydrogen-bonding mechanism. The most widely used group comprises linear alkyl sulfonates (LAS), often aided by "builders." LAS are preferable to alkyl benzene sulfonates (ABS) because they are readily decomposed by microorganisms (biodegradable). LAS are straight-chain compounds having 10 or more carbon atoms in the chain. The branched-chains characteristic of ABS resists decomposition; these have been largely replaced by LAS because of water pollution.
See surface tension; emulsion; wetting agent; soap (1); alkylate (3); biodegradability; eutrophication; and builder detergent.

Detergent 8. Aqueous detergents.
Properties: Liquid.
Use: A detergent that removes rosins, fluxes, and resins from printed circuit boards prior to coating. It is phosphate-free and biodegradable.

determinate error. An error that can be identified and thus corrected or reduced.

"Detojet" *[Alconox].* TM for a low foaming, heavy-duty alkaline detergent.
Use: For use in labware washers, parts washers, sprays, and ultrasonic systems.

detonation. The extremely rapid, self-propagating decomposition of an explosive accompanied by a high-pressure temperature wave that moves at 1000–9000 msec. May be initiated by mechanical impact, friction, or heat. Detonation is a characteristic of high explosives, which vary considerably in their sensitivity to shock, nitroglycerin being one of the most dangerous in this respect.
See explosive, high; deflagration.

deuterium. (heavy hydrogen).
Symbol D. An isotope of hydrogen whose nucleus contains one neutron and one proton and is, therefore, twice as heavy (aw 2.014) as the nucleus of normal hydrogen. The ratio in nature is 1 part deuterium to 6500 parts normal hydrogen.
See deuteron.
Properties: Almost identical with hydrogen. D 2.0 (H = 1), fp −254.4C (121 mm Hg), bp −249.5C, autoign temp 1085F. Noncorrosive.
Derivation: Electrolysis of high-purity heavy water, fractional distillation of liquid hydrogen.
Grade: 98, 99.5 atom %.
Hazard: Highly flammable and explosive. Explosive range 4–74%.
Use: Bombardment of atomic nuclei, tracer element, thermonuclear reactions.
See deutero-; heavy water.

deuterium oxide. See heavy water.

deutero-. (deuterated).
Prefix indicating that one or more of the hydrogens in a compound is the deuterium isotope. Example: deuteroborane solution, used for labeling olefinic unsaturation. The adjective form, deuterated, has

the same meaning. Deuterated ethylene, sometimes written ethylene-1,1-D_2, has the formula $CH_2:CD_2$.

deuteron. (deuton).
A nuclear particle having mass and a positive charge of 1, identical with the nucleus of the deuterium atom.

Devarda's metal. (Devarda's alloy).
Properties: Gray powder. Contains copper, aluminum, and zinc in the proportion of 50:45:5. Slightly soluble in hydrochloric acid.
Grade: Reagent (20-mesh and finer).
Use: Analysis (testing for nitrogen).

developer. (1) A term applied in the dyeing industry to certain organic compounds that in combination with some other organic compound already deposited upon the fiber will develop a colored compound, or if united with a dye already upon the fiber will form a new coloring matter possessing a more desirable or faster color.
(2) A substance used in photography to convert a latent image to a visible one by chemical reduction of a silver compound to metallic silver more rapidly in the portions exposed to light than in those not exposed. Such reducing agents as hydroquinone, pyrogallol, and p-phenylenediamine are used.
See photographic chemistry.

devitrification. Formation of unwanted crystals of silica on heating or cooling. The term is used largely in the glass industry. The tendency to devitrify results from the unstable nature of glasses. It usually occurs if the melt is cooled too slowly.

De Vry's reagent. Contains molybdate ion.
Use: Test reagent for alkaloids.

devulcanization. Technically a misnomer, since vulcanization is irreversible. The term is used to describe the softening of a vulcanizate caused by heat and chemical additives during reclaiming.

dewatering. Removal of gross water from a suspension or sludge by filtration, expression, centrifugation, or clarification. Paper pulp is dewatered by the fourdrinier wire; rubber latex may be concentrated in a centrifuge, in which half or more of the water is removed. Sludges and organic wastes are also dewatered centrifugally.
See drying; dehydration; centrifugation; clarification.

dew of death. See β-chlorovinyldichloroarsine.

dew point. The temperature at which air is saturated with moisture, or in general the temperature at which a gas is saturated with respect to a condensable component.

"Dexedrine" *[Ameda].* TM for dextroamphetamine sulfate.

dexniguldipine hydrochloride. See (-)-nigaldipine hydrochloride.

"Dexon" *[Covidien].* TM for p-dimethylaminobenzenediazo sodium sulfonate.

dextran. (macrose).
Certain polymers of glucose that have chain-like structures and molecular weights up to 200,000. Produced from sucrose by *Leuconostoc* bacteria. Occurs as slimes in sugar refineries, on fermenting vegetables, or in dairy products. Clinical dextran is standardized to a low molecular weight (75,000); made by partial hydrolysis and fractional precipitation of the high-molecular-weight particles.
Properties: Stable to heat and storage. Soluble in water making very viscous solutions. Solutions can be sterilized. Combustible.
Use: Blood plasma substitute or expander, confections, lacquers, oil-well drilling muds, filtration gel, food additive.

dextranase. An enzyme reported to be effective in reducing dental caries.

dextran sulfate. See sodium dextran sulfate.

dextrin. (starch gum).
A group of colloidal products formed by the hydrolysis of starches. Industrially, it is made by treatment of various starches with dilute acids or by heating dry starch. The yellow or white powder or granules are soluble in boiling water and insoluble in alcohol and ether.
Use: Adhesives, thickening agent, sizing paper, and textiles, substitute for natural gums, food industry, glass-silvering compositions, printing inks, felt manufacture, substitute for lactose in penicillin manufacture, fuel in pyrotechnic devices.

dextrorotatory. Having the property when in solution of rotating the plane of polarized light to the right or clockwise. Dextrorotatory compounds are given the prefix d or (+) to distinguish them from their levorotatory, l or (−) isomers. The plus (+) and minus (−) signs are preferred.
See optical rotation.

dextrose. Glucose is the preferred term.
See glucose.

dextrose equivalent. (DE).
The total amount of reducing sugars expressed as dextrose that is present in a corn syrup, calculated as a percentage of the total dry substance. The usual technique for determining DE in the corn products industry is the volumetric alkaline copper method.
See glucose syrup.

df. (methylphosphonyl difluoride; difluorophos-phorylmethane).
CAS: 99517-98-3. $C_{12}H_{13}N_3O_5$ CH_3F_2OP.
An organophosphate.
Use: Chemical warfare agent.

DFDD. (difluorodiphenyldichloroethane).
CAS: 603-55-4. $C_{14}H_{10}Cl_2F_2$.
Properties: Colorless crystals. Mp 75C.
Hazard: Toxic by ingestion and skin contact.
Use: Contact insecticide.

DFDT. (difluorodipheynltrichloroethane).
CAS: 475-26-3. $(FC_6H_4)_2CHCCl_3$.
Fluorine analog of DDT.
Properties: A low-melting white solid; odor resem-bling ripe apples. Mp 45.5C. Insoluble in water; soluble in organic solvents.
Derivation: By condensing chloral and fluoroben-zene in the presence of sulfuric acid or chlorosul-fonic acid.
Hazard: Toxic by ingestion, inhalation, and skin absorption. Use may be restricted.
Use: Contact insecticide.

"DFL No. 3" [Sysmex]. A solution of buffered phosphate esters, used as a lubricant release agent and corrosive inhibitor for synthetic rubber driers.

DFP. Abbreviation for diisopropyl fluorophos-phate.

d gene. A small segment of immunoglobulin heavy-chain and T-cell receptor DNA that codes for the third hypervariable region of most receptors.

4-α-D-glucantotransferase. (dextrin transgly-cosylase; dextrin glycosyltransferase; D enzyme; disproportionating enzyme; amylomaltase).
A 4-glycosyltransferase that converts maltodextrins into amylose by transferring portions of 1,4-glucan chains to new 4-positions on glucose or other 1,4-glucans.

d(+)-glucuronic acid. (glucuronic acid; 3,4,5,6-tetrahydroxyoxane-2-carboxylic acid).
CAS: 6556-12-3. $C_6H_{10}O_7$
An oxidation product of glucose and glucuronic acid that is widely distributed in both plants and animals, usually as part of a larger molecule.
Properties: Crystalline compound, soluble in water and ethanol.

DHA. Abbreviation for (1) dihydroxyacetone, (2) DHEA.

DHEA. (androstenolone; dehydroisoandros-terone; DHA; diandron; 17-hormoforin; pras-terone).
CAS: 53-43-0. $C_{19}H_{28}O_2$.

An androgenic steroid, a metabolic product of the adrenal steroid hormones with about one-third of the androgenic activity of androsterone.
Properties: Dimorphous needles: mp 140–14CC. Leaflets: mp 152–153C. Mw 288.47. Soluble in benzene, alcohol, and ether; sparingly soluble in chloroform and petroleum ether.
Derivation: Isolated from male urine, synthesis from cholesterol or sitosterol, extracted from wild yams.
Use: A hormone used in health food products, medicine, biochemical research.

D-homo rearrangement of steroids. Origi-nally discovered in 17β-hydroxy-20-ketosteroids, but thoroughly studied in the 17α-hydroxy-20-keto series, this reaction involves an acid- or base-catalyzed acyloin rearrangement that yields a 6-membered D-ring.

DHP-MP. 1,4-bis(2-hydroxpropyl)-2-methylpiperazine.

DHS. Abbreviation for dihydrostreptomycin.

Di. Symbol for didymium.

di-. Prefix meaning two.
See bi-.

diabetes mellitus. A complex endocrine-metabolic disease resulting from insulin deficiency; characterized by a failure in glucose transport from the blood into cells at normal glucose concentra-tions, resulting in hyperglycemia.

diacetic acid. See acetoacetic acid.

diacetin. (glyceryl diacetate).
CAS: 25395-31-7. $CH_2O(OCCH_3)CH_2OCH_2O(OCCH_3)$.
Properties: Hygroscopic liquid. A mixture of iso-mers. D 1.18, bp approximately 259C, refr index 1.44. Miscible with water, benzene, and alcohol. The commercial mixture gels approximately −30C. Combustible.
Derivation: Heating one mole of glycerol with two moles of glacial acetic acid.
Grade: Technical.
Use: Plasticizer and softening agent, solvent for cel-lulose derivatives, "Glyptal" resins [Glyptal, Inc.], shellac.

diacetone acrylamide. (DAA).
CAS: 2873-97-4. $C_9H_{15}NO_2$.
A vinyl monomer.
Properties: White, crystalline solid; purity 99+%. Mp 57C, bp (8 mm Hg) 120C. Highly soluble in water and most organic solvents. (The DAA homopolymer is insoluble in water); polymerizes readily.
Use: Imparts water tolerance and vapor permeability to copolymer films; latex and water-based coating

compositions; adhesion improver for cellulosics, concrete, glass; cross-linking agent in polyester resins; color photography.

diacetone alcohol. (diacetone; 4-hydroxy-4-methylpentanone-2; 4-hydroxy-2-keto-4-methylpentane).
CAS: 123-42-2. $CH_3COCH_2C(CH_3)_2OH$.
Properties: Colorless liquid; pleasant odor. D 0.9406 at 20/20C, bp 169.1C, flash p varies from <73F (23C) to 100F (38C) or higher depending on grade, wt/gal 7.8 lb (20C), viscosity 0.032 cP (20C), fp −42.8C, refr index 1.42416 (20C), autoign temp 1118F (603C). Miscible with alcohols, aromatic and halogenated hydrocarbons, esters, and water. A constant boiling mixture with water has bp 99.6C and contains approximately 13% diacetone alcohol.
Derivation: Condensation of acetone.
Grade: Technical, acetone-free, reagent.
Hazard: Flammable, dangerous fire risk, explosive limits in air 1.8–6.9. An eye and upper respiratory tract irritant.
Use: Solvent for nitrocellulose, cellulose acetate, various oils, resins, waxes, fats, dyes, tars, lacquers, dopes, coating compositions, wood preservatives, stains, rayon and artificial leather, imitation gold leaf, dyeing mixtures, and antifreeze mixtures; extraction of resins and waxes; preservative for animal tissue; metal-cleaning compounds; hydraulic compression fluids; stripping agent (textiles); and laboratory reagent. The technical grade containing acetone has greater solvent power.

diacetonyl sulfide.
$(CH_3COCH_2)_2S$.
Properties: Crystals. Bp 136–137C (15 mm Hg), mp 47C.
Derivation: Interaction of chloroacetone and hydrogen sulfide gas.

diacetoxytetrabutyldistannoxane. See bis(acetoxydibutylstannane) oxide.

diacetyl. (biacetyl; butanedione; diketobutane; dimethyl diketone; dimethylglyoxal).
CAS: 431-03-8. $CH_3COCOCH_3$.
Properties: Yellow liquid; strong odor. D 0.990 (15/15C), mp >3C to >4C, bp 88–91C, refr index 1.3933 (18C), flash p >80F (26C). Soluble in water, alcohol, and ether.
Derivation: Special fermentation of glucose, synthesis from methyl ethyl ketone.
Grade: Technical, flavor grade, FCC.
Hazard: Flammable, dangerous fire risk. Liver damage. Questionable carcinogen.
Use: Aroma carrier in food products.

diacetylaminoazotoluene. (4-o-tolylazo-o-diacetotoluide).
CAS: 83-63-6. $[CH_3C_6H_4NNC_6H_3(CH)N(CH_3CO)_2]$.

Properties: Crystalline powder. Color varies from yellowish-red through rose to red. Mp 74–76C. Acted upon by atmospheric water vapor. Soluble in alcohol, chloroform, ether; fats, oils, and greases; insoluble in water.
Hazard: Questionable carcinogen.
Use: Medicine (external).

S,N-diacetylcysteine monoethyl ester.
CAS: 19547-89-8. $C_9H_{15}NO_4S$.
Hazard: Moderately toxic.

diacetylene.
CAS: 460-12-8. $HC{\equiv}C\text{-}C{\equiv}CH$.
An unsaturated hydrocarbon containing two triple bonds with the type formula C_nH_{2n-6}. The simplest is butadiyne or biacetylene, a gas that boils at 10C. Combustible.
Hazard: Ignites spontaneously in contact with moist silver salts, may explode at −25C.

1,2-diacetylethane. See acetonylacetone.

1,1′-diacetylferrocene.
CAS: 1273-94-5. $(C_5H_4COCH_3)_2Fe$.
Red, crystalline solid, mp 122–124C. Used as an intermediate.
See ferrocene.

diacetylmethane. See acetylacetone.

diacetylmorphine. (diamorphine; heroin).
CAS: 561-27-3. $C_{17}H_{17}NO(C_2H_3O_2)_2$.
Properties: White, crystals, or crystalline powder; odorless; bitter taste. Mp 173C. Soluble in alcohol.
Derivation: By acetylization of morphine.
Hazard: Addictive narcotic; ingestion of less than one grain may be fatal. Cannot be legally sold in the U.S.

diacetyl peroxide. See acetyl peroxide.

diacetyl tartaric acid esters of mono- and diglycerides.
Properties: Vary from sticky, viscous liquid to waxy solid; faint acid odor. Sol in oil, methanol, acetone, acetic acid, and water.
Use: Food additive.

diacolation. To percolate and extract drugs using a solvent under pressure.

diacylamine. (secondary amide).
Any of a class of compounds that have two acyl groups substituted on ammonia or a primary amine.

diagenesis. The set of processes, including solution, that alter sediments at low temperatures after burial.

"Diak" *[Vanderbilt].*   TM for a series of rubber accelerators used to vulcanize Viton fluoroelastomer and polyacrylate elastomers.

dialdehyde starch.   See starch dialdehyde.

dialifor.   (*S*-(2-chloro-1-phthalmidoethyl)-*O*,*O*-diethylphosphorothionate).
CAS: 10311-84-9.
Properties: White crystals. Mp 167–169C. Insoluble in water; soluble in common organic solvents.
Hazard: Toxic by ingestion, cholinesterase inhibitor.
Use: Acaricide; pesticide against codling moth, red spider mite, etc., of deciduous fruit.

dialkylchloroalkylamine hydrochloride.   A group of amine salts having the formula RCl·HCl, when R represents such groups as $(CH_3)_2NCH_2CH_2$ (β-dimethylaminoethyl chloride hydrochloride), $(CH_3)_2NCH_2CH(CH_3)$ (β-dimethylaminoisopropyl chloride hydrochloride), etc. Used in organic synthesis.

di-allate.   See 2,3-dichloroallyl diisopropylthiocarbamate.

diallyl adipate.
CAS: 2998-04-1. $C_3H_5OOC(CH_2)_4COOC_3H_5$.
Properties: Liquid. Color-maximum #100 Pt-Co, characteristic odor, d 1.025 (20C). Combustible.
Use: Monomer.

diallylamine.   (di-2-propenylamine).
CAS: 124-02-7. $(CH_2:CHCH_2)_2NH$.
Properties: Liquid. D 0.7889 (20C), bp 112C, fp −100C, refr index 1.4404 (20C). Soluble in water. Combustible.
Derivation: From allylamine and allylbromide.
Hazard: Toxic by inhalation and skin absorption.
Use: Intermediate.

diallylbarbituric acid.   (5,5-diallylbarbituric acid).
CAS: 52-43-7. $C_{10}H_{12}N_2O_3$.
Properties: White, crystals or crystalline powder; odorless; slightly bitter taste. Mp 171–173C. Soluble in alcohol or ether; slightly soluble in water.
Hazard: See barbiturates.
Use: Medicine (sedative).

5,5′-diallyl-2,2′-biphenyldiol.   See   2,2′-bichavicol.

diallyl chlorendate.
CAS: 3232-62-0. $(C_3H_5OOC)_2C_7H_2Cl_6$.
Solid, fp 29.5C, viscosity 4.0 cP (20C), d 1.47 (20C). Used as a monomer for allyl resins especially in flame-retardant compositions.

diallyl cyanamide.
CAS: 538-08-9. $(H_2C:CHCH_2)_2NCN$.

Properties: Liquid. Fp <−70C, bp 222C, d 0.90. Insoluble in water; soluble in organic solvents.
Derivation: Reaction of allyl bromide and disodium cyanamide.
Hazard: Yields very toxic cyanide fumes on heating.
Use: Organic intermediate, polymers.

diallyldibromo stannane.
CAS: 17381-88-3. $C_6H_{10}Br_2Sn$.
Hazard: A poison.

diallyl diglycolate.   See diglycolic acid, diallyl ester.

diallyl diglycollate.
$(C_3H_5OOCCH_2)_2O$.
Properties: Liquid. Color-maximum #100 Pt-Co, characteristic odor, d (20C) 1.1113.
Use: Monomer.

diallyldimethylammonium   chloride.   See "Cat-floc" *[Calgon].*

diallyl isophthalate.
CAS: 1087-21-4. $C_6H_4(COOH_2C:CHCH_2)_2$.
Properties: Monomer is liquid. Color-maximum #175 Pt-Co, mild characteristic odor, d 1.124 (20C). Prepolymer is solid. D 1.256 (25C).
Use: Molding and laminating, cross-linker for polyesters.

diallyl maleate.
$C_3H_5OOCCH:CHCOOC_3H_5$.
CAS: 999-21-3.
Properties: Colorless or straw-colored liquid. Bp 109–110C (3 mm Hg), d 1.077 (20C), refr index 1.4699 (20C). Polymerizes readily when exposed to light or temperature above approximately 50C. Combustible.
Hazard: Toxic by ingestion, irritating to skin.
Use: Polymers and copolymers, insecticide formulations.

diallylmelamine.

$$(C_3H_5)_2NCNC(NH_2)NC(NH_2)N.$$

Properties: White, crystalline solid. Mp 142C, d 1.24 (30C). Combustible.
Hazard: Toxic by ingestion, irritating to skin, evolves cyanide on heating.
Use: Monomer for resins.

diallyl phosphite.
CAS: 23679-20-1. $(CH_2:CHCH_2O)_2PHO$.
Properties: Water-white liquid. Fp 0C, bp 62C (1 mm Hg), refr index 1.444 (25C), d 1.080 (25/15C). Combustible.
Use: Synthesis of organophosphorus compounds.

diallyl phthalate. (DAP).
CAS: 131-17-9. $C_6H_4(COOCH_2CH:CH_2)_2$.
The name is also used for the polymer.
Properties: Nearly colorless, oily liquid. D 1.120 (20/20C), fp −70C (viscous liquid), boiling range 158–165C (4 mm Hg), odor mild lachrymatory, flash p 330F (165.5C), viscosity 13 cP (20C). Limited solubility in gasoline, mineral oil, glycerol, glycols, and certain amines. Soluble in most other organic liquids. Insoluble in water. Combustible.
Hazard: Toxic by ingestion.
Use: Primary plasticizer that will polymerize if not inhibited, a monomer that will polymerize with heat and catalyst. Forms low-pressure laminates with various fillers, such as glass cloth, paper, etc., for electrical insulation.

diallyl sulfide. See allyl sulfide.

diallyltin dibromide. See diallyldibromo stannane.

dialysis. The separation of small molecules from macromolecules in a solution by means of a semipermeable membrane such as parchment or collodion. The rates of diffusion of the small and the large molecules are so widely different that the former will readily pass through the membrane, whereas the latter will penetrate with extreme difficulty. For example, the diffusion rates are about 2.3 for sodium chloride, 7 for cane sugar, and from 50 to 100 for proteins and other macromolecules. This differential led Thomas Graham to define substances that would pass through the membrane easily as crystalloids and those having a tendency to be retained by the membrane as colloids.
See colloid chemistry; electrodialysis; Graham, Thomas.

diamide hydrate. See hydrazine hydrate.

diamidino phenylindole. (4,6-diamidino-2-phenylindole-2Hci; dapi; 2-(4-carbamimidoyl-phenyl)-1H-indole-6-carboximidamide).
$C_{16}H_{15}N_5$
A fluorescent probe for DNA.

diamine. See hydrazine.

3,6-diaminoacridine. See acriflavine.

***m*-diaminoazobenzene hydrochloride.** See chrysoidine hyrochloride.

diaminoazoxytoluene. (azoxytoluidine).
$C_6H_3(CH_3)(NH_2)N_2OC_6H_3(NH_2)(CH_3)$.
Properties: Yellow or orange crystals. Mp 168C. Soluble in alcohol; insoluble in water, Combustible.
Derivation: By alkaline reduction of *p*-nitro-*o*-toluidine.
Use: Dye intermediate.

diaminobenzidine. (DAB; *N,N*-dimethyl-4-phenyldiazenylaniline).
CAS: 91-95-2. $C_{12}H_{14}N_4$ $C_{14}H_{15}N_3$.
A reagent.
Hazard: Carcinogen.
Use: To detect or enhance the detail of bloody friction ridges of human digits; to induce experimental liver cancer.

3,3′-diaminobenzidine. (3,3′4,4′-biphenyl-tetramine).
CAS: 91-95-2. $(H_2N)_2C_6H_3C_6H_3(NH_2)_2$.
Properties: Solid. Mp 178–180C.
Use: Copolymerized with diphenyl isophthalate to make high-temperature resistant polybenzimidazoles.

1,3-diaminobutane.
CAS: 590-88-5. $NH_2CH_2CH_2CHNH_2CH_3$.
Properties: Water-white liquid; amine odor. Boiling range 143–150C, d 0.858 (20/20C), refr index 1.450 (20C), flash p 125F (51.6C). Combustible.
Hazard: Toxic by ingestion and skin absorption.

2,6-diamino-4-butylamino-*s*-triazine. See *n*-butylmelamine.

α,ε-diaminocaproic acid. See lysine.

diaminochrysazin.
$(NH_2)_2(OH)_2C_{14}H_4O_2$.
(1,8-diamino-4,5-dihydroxyanthraquinone).
Use: Colorimetric determination of boron.

***trans*-1,2-diaminocyclohexanetetraacetic acid monohydrate.** (CDTA).
$C_6H_{10}[N(CH_2COOH)_2]_2·H_2O$.
Properties: White, crystalline solid. Mp 200–220C. Very slightly soluble in water and insoluble in most common organic solvents. Partially soluble in dimethyl formamide and dimethyl sulfoxide upon heating. Forms stable complexes.
Use: Chelating agent similar to ethylenediaminetetraacetic acid.

diaminodiethyl sulfide.
$S(CH_2CH_2NH_2)_2$.
Properties: Mobile, colorless liquid; amine-like odor. Bp 230–240C, d 1.054 (25C). Miscible with water and benzene, insoluble in aliphatic hydrocarbons. Combustible.

1,8-diamino-4,5-dihydroxyanthraquinone. See diaminochrysazin.

diaminodihydroxyarsenobenzene dihydrochloride. See arsphenamine.

di-*p*-aminodimethoxydiphenyl. See dianisidine.

diaminodimethyl acridine. (3,6-diamino-2,7-dimethylacridine; acridine yellow base; 2,8-diamino-3,7-dimethylacridine; diaminoacridine sulfate proflavine).
$C_{15}H_{15}N_3$.
Properties: A yellow crystalline dye.
Hazard: Poison; mutagen.

4,4′-diamino-3,3′-dimethyldicyclo-hexylmethane. See bis(4-amino-3-methylcyclohexyl)methane.

diaminodiphenic acid. (benzidine dicarboxylic acid).
$C_6H_3(CO_2H)NH_2C_6H_3(CO_2H)NH_2$.
Properties: White crystals. Soluble in alcohol and ether; insoluble in water.
Derivation: By boiling *m*-nitrobenzaldehyde with caustic soda, reducing with zinc dust, and acidifying.
Hazard: See benzidine.
Use: Dyestuff.

p-diaminodiphenyl. See benzidine.

diaminodiphenylamine.
$HN(C_6H_4NH_2)_2$.
Properties: Yellowish crystals. Mp 158C. Soluble in alcohol and ether; insoluble in water.
Use: Dye intermediate, detection of hydrogen cyanide.

diaminodiphenylethylene. See *p*-diaminostilbene.

p,p′-diaminodiphenylmethane. (4,4′-methylenedianiline; MDA).
CAS: 101-77-9. $H_2NC_6H_4CH_2C_6H_4NH_2$.
Properties: Light-brown crystals. Mp 92–93C, bp 398–399C, flash p 440F. Slightly soluble in cold water; very soluble in alcohol, benzene, ether. Combustible.
Hazard: A possible carcinogen. Toxic by inhalation and skin contact. Liver damage.
Use: Determination of tungsten and sulfates, polymer and dye intermediate, corrosion inhibitor, epoxy-resin hardening agent, isocyanate resins, polyamides.

diaminodiphenylthiourea. (diaminothiocarbanilide).
$(NH_2C_6H_4NH)_2CS$.
Properties: Colorless plates or crystalline solid. Mp 195C. Soluble in alcohol and ether; sparingly soluble in water.
Derivation: By boiling *p*-phenylenediamine with carbon disulfide.

diaminodiphenylureadisulfonic acid.
$CO(NHC_6H_3NH_2SO_3H)_2$.

Properties: Colorless, needlelike crystals. Slightly soluble in water.
Derivation: Action of phosgene upon either *p*-phenylenediaminesulfonic acid or 4-nitroaniline-3-sulfonic acid.
Use: Dye manufacture.

3,3′-diaminodipropylamine. See 3,3′-iminobispropylamine.

diaminoditolyl. See *o*-tolidine.

p,p′-diaminoditolylmethane.
$NH_2C_7H_6CH_2C_7H_6NH_2$.
Properties: Glistening, crystalline plates. Mp 149C. Soluble in alcohol and ether.
Derivation: By heating formaldehyde and *o*-toluidine.

1,2-diaminoethane. See ethylenediamine.

6,9-diamino-2-ethoxyacridine lactate monohydrate. See ethodin.

diaminoethyl ether tetraacetic acid.
$(HOOCCH_2)_2NCH_2CH_2OCH_2CH_2N(CH_2COOH)_2$.
Properties: Slightly soluble in water. Purity 98% min.
Use: A chelating agent.

1,6-diaminohexane. See hexamethylenediamine.

3,6-diamino-10-methylacridinium chloride. See acriflavine.

diaminonaphthalene. See naphthalenediamine.

1,5-diaminopentane. See cadaverine.

2,3-diaminophenazine.
CAS: 655-86-7. $C_{12}H_{10}N_4$.
Properties: Brownish, needlelike crystals. Mp 265C. Soluble in alcohol and benzene; sublimes on heating.
Use: Analytical reagent for detection of metals.

2,5-diaminophenol.
CAS: 636-25-9. $C_6H_3OH(NH_2)_2$.
Properties: Colorless crystals. Mp 68C. Soluble in water.
Derivation: By reduction of 2,5-dinitrophenol.
Hazard: May be skin irritant.
Use: Organic synthesis.

2,4-diaminophenol hydrochloride. (amidol).
CAS: 137-09-7. $C_6H_3(NH_2)_2OH\cdot2HCl$.
Properties: Grayish-white crystals. Soluble in water; slightly soluble in alcohol.

Derivation: By interaction of dinitrophenol with iron and hydrochloric acid.
Use: Photographic developer, dyeing furs and hair, analytical reagent.

2,4-diamino-6-phenyl-*s*-triazine.   See benzoguanamine.

1,2-diaminopropane.   (propylenediamine; 1,2-propanediamine).
CAS: 78-90-0. $NH_2CH_2CH(NH_2)CH_3$.
Properties: Colorless, very hygroscopic, strongly alkaline liquid. D 0.8732 (20/20C), refr index 1.4460 (20C), flash p 92F (33C), bp 117C. Very soluble in water. Ammoniacal odor.
Grade: Technical, 75%, 90%, 98% solution.
Hazard: Dangerous fire risk. Toxic by ingestion and skin absorption.
Use: Synthesis of medicinals, dyes, rubber accelerators, electroplating, analytical reagent.

1,3-diaminopropane.   (1,3-propanediamine).
CAS: 109-76-2. $NH_2CH_2CH_2CH_2NH_2$.
Properties: Water-white mobile liquid; amine odor. D 0.8881 (20/20)C, bp 139.7C, fp −12C, refr index 1.459 (20C), flash p 120F (49C) (OC). Completely soluble in water, methanol, and ether. Combustible.
Hazard: Moderate fire risk. Strong irritant to eyes and skin.
Use: Intermediate.

***N,N'*-diaminopropylethylenediamine.**   See 1,5,8,12-tetraazadodecane.

2,6-diaminopyridine.
CAS: 141-86-6. $NC_5H_3(NH_2)_2$.
Properties: Crystals. Mp 120.8C, bp 285C. Soluble in water. Combustible.
Derivation: From 2-aminopyridine.

2,6-diamino-4-(2-pyridyl)-*s*-triazine.
CAS: 25007-79-8. $C_8H_8N_6$.
Hazard: A poison.

***p*-diaminostilbene.**   (diaminodiphenylethylene).
$C_6H_4(NH_2)CHCHC_6H_4(NH_2)$.
Properties: Colorless needles or plates. Mp 227C. Soluble in alcohol and ether; insoluble in water. Combustible.
Derivation: Reduction of dinitrostilbene.

4,4'-diamino-2,2'-stilbenedisulfonic acid.
(DAS).
CAS: 81-11-8. $C_6H_3(NH_2)(SO_3H)CHCHC_6H_3$ $(SO_3H)(NH_2)$.
Properties: Yellowish, microscopic needles. Soluble in alcohol and ether; insoluble in water.
Derivation: Boiling sodium salt of *p*-nitrotoluene-*o*-sulfonate in water and caustic soda and reduction with zinc dust.
Hazard: Toxic by ingestion.
Use: Dyestuffs.

diaminothiocarbanilide.   See diaminodiphenylthiourea.

di-α-amino-β-thiolpropionic acid.   See cystine.

diaminotoluene.   See toluene-2,4-diamine.

4,6-diamino-*m*-toluenesulfonic acid.   See *m*-tolylenediaminesulfonic acid.

4,6-diamino-*s*-triazine-2-ol.   See ammeline.

2,5-diaminovaleric acid.   See ornithine.

diammine(benzylmalonato)platinum (II).
CAS: 63919-17-5. $C_{10}H_{14}N_2O_4Pt$.
Hazard: A poison.

***cis*-diamminedibromoplatinum(II).**
CAS: 15978-91-3. $Br_2H_6N_2Pt$.
Hazard: A poison.

diammonium beryllium tetrafluoride.   See ammonium fluoroberyllate.

diammonium ethylenebisdithiocarbamate.
CAS: 3566-10-7. $NH_4S_2CNH(CH_2)_2NHCS_2NH_4$.
Properties: Mp 72.5C. Very soluble in water.
Grade: 42% solution in water.
Use: Fungicide, intermediate, corrosion inhibitor. See nabam.

diammonium hydrogen phosphate.   See ammonium phosphate, dibasic.

diammonium phosphate.   See ammonium phosphate, dibasic.

diamond.   An allotropic form of carbon that crystallizes isometrically and consists of carbon atoms covalently bound by single bonds only in a predominantly octahedral structure.

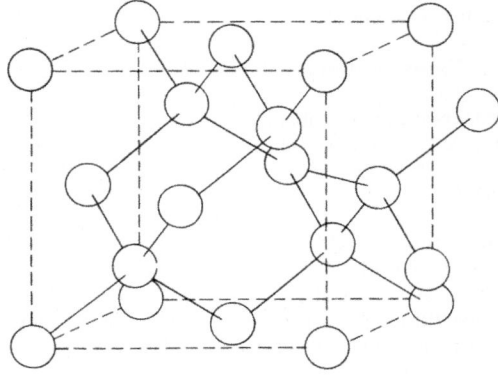

This accounts for its extreme hardness (Mohs 10) and great stability.

Properties: High refractive index (2.42), d 3.50, coefficient of friction 0.05, highest thermal conductivity of any substance, transparent to infrared, mp 3700C, bp 4200C. The purest diamonds used for gems are mined in South Africa, lower grades in Brazil, Venezuela, India, Borneo, Arkansas. Also made synthetically by heating carbon and a metal catalyst in an electric furnace at about 3000F under high pressure.

Use: Special surgical knives, windows in space probes, high-capacity transmitters.

See diamond, industrial.

diamond, industrial. Low-grade diamonds (bort and carbonado) as well as those made synthetically in an electric furnace (3000F, 1.3 million psi).

Use: Oil-well drill bits, primary grinding of steel, wire-drawing dies, glass and metal cutting, grinding wheels.

See abrasive.

diamond pyramid hardness.   See hardness.

diamthazole dihydrochloride.
$C_{15}H_{23}N_3OS \cdot 2HCl$.
(6-(2-diethylaminoethoxy)-2-dimethylaminobenzothiazole dihydrochloride).

Properties: Crystals. Decomposes 269C. Soluble in water, ethanol, and methanol.

Use: Topical therapy (medicine), antifungal agent.

di-*n*-amylamine.   (di-*n*-pentylamine).
CAS: 2050-92-2. $(C_5H_{11})_2NH$.

Properties: Colorless liquid. Bp 202–3C (745 mm Hg), d 0.77–0.78 (20C), refr index 1.430 (20C), flash p 124F (51.1C). Very slightly soluble in water; soluble in alcohol and ether. Combustible.

Derivation: From reaction of amyl chloride and ammonia.

Hazard: Moderate fire risk. Toxic by ingestion and inhalation.

Use: Rubber accelerators, flotation reagents, dyestuffs, and corrosion inhibitors; solvent for oils, resins, and some cellulose esters.

***N,N*-diamylaniline.**   (mixed isomers).
$C_5H_5N(C_5H_{11})_2$.

Properties: Dark-amber liquid. D 0.898 (20C), boiling range 276–292C, faint aniline odor, flash p 260F (126C). Combustible.

Use: Organic dyes.

di-*tert*-amyl disulfide.
CAS: 34965-30-5. $CH_3CH_2C(CH_3)_2$
$SSC(CH_3)_2CH_2CH_3$.

Properties: Liquid. D 0.931 (15.5/15.5C), vacuum distillation range 86–102C, refr index 1.495 (20C), flash p 220F (104.4C). Combustible.

α-(diamylaminomethyl)-1,2,3,4-tetrahydro-9-phenathrene methanol.
CAS: 27074-70-0. $C_{26}H_{39}NO$.

Hazard: A poison.

diamylene.
$C_{10}H_{20}$.

Properties: Colorless liquid, fp <−50C, bp 150C, flash p 118F (47.7C) (OC), d 0.77. Combustible.

Hazard: Moderate fire risk.

Use: Solvent, organic synthesis.

2,5-di(*tert*-amyl)hydroquinone.   (DAHQ, 2,5-di[*tert*-pentyl]hydroquinone).
CAS: 79-74-3. $(C_5H_{11})_2C_6H_2OH)_2$.

Properties: Buff powder. Mp 176C, d 1.05 (25C). Slightly soluble in water; soluble in alcohol and benzene.

Use: Antioxidant for uncured rubber and for unsaturated resins and oils; food packaging, polymerization inhibitor.

diamyl maleate.
$(CHCOOC_5H_{11})_2$.

Properties: Water-white liquid. D 0.981 (20C), boiling range 263–300C, odor faintly alcoholic, flash p 270F (132C). Combustible.

diamyl phenol.   (1-hydroxy-2,4-diamyl-benzene).
$(C_5H_{11})_2C_6H_3OH$.
Commercial form is a mixture of isomers including both secondary amyl and tertiary amyl groups, mainly in 2,4 positions.

Properties: Light-straw–colored liquid with mild phenolic odor; miscible with both aliphatic and aromatic hydrocarbons, insoluble in water and 10% aqueous alkalies. Boiling range (ASTM 5–95%) 280–295C, d 0.930 (20C), wt/gal 7.8 lb (20C), flash p 260F (126)(TOC). Combustible.

Hazard: Irritant to skin.

Use: Synthetic resins, lubricating-oil additives, rust preventives, plasticizers, synthetic detergents, antioxidants and antiskinning agents, rubber chemicals, rodenticide, fungicide.

diamyl phthalate.
CAS: 131-18-0. $C_6H_4(COOC_5H_{11})_2$.

Properties: Colorless, oily liquid; nearly odorless. D 1.022 (20C), wt/gal 8.52 lb (20C), refr index 1.488 (25C), bp 342C, fp <−55C, flash p 357F (180C) (CC). Combustible.

Derivation: By esterification of phthalic anhydride with amyl alcohol in the presence of approximately 1% concentrated sulfuric acid as catalyst.

Use: Plasticizer.

diamyl sodium sulfosuccinate.
CAS: 922-80-5. $C_{14}H_{25}NaO_7S$.

Properties: White powder. Soluble in water, acetone, carbon tetrachloride, glycerol.

Derivation: Action of alcohol on maleic anhydride, followed by addition of sodium bisulfite.
Use: Wetting agent, emulsion polymerization.

diamyl sulfide. (amyl sulfide).
CAS: 872-10-6. $(C_5H_{11})_2S$.
A mixture of isomers.
Properties: Yellow liquid; obnoxious odor. D 0.85–0.91 (20/20C), distillation range 170–180C, flash p 185F (85C) (OC), refr index 1.477 (19C). Combustible.
Hazard: Irritating by inhalation, ingestion, and skin absorption.
Use: Organic sulfur compounds by addition reactions, flotation agent in metallurgical processes, odorant.

diandron. See DHEA.

dianhydrosorbitol. See sorbide.

1,4,3,6-dianhydrosorbitol. See isosorbide.

1,2-dianilinoethane.
CAS: 150-61-8. $C_{14}H_{16}N_2$.
Properties: Colorless crystals. Mp 67C, bp 228C (12 mm Hg). Soluble in alcohol and ether.
Derivation: Heating aniline with dichloro- or dibromoethane.
Use: Manufacture of antihistamines, resin, and rubber stabilizer, reagent for aldehydes.

dianisidine. (di-*p*-aminodi-*m*-methoxydiphenyl; 3,3'-dimethoxybenzidine).
CAS: 119-90-4. $[C_6H_3(OCH_3)NH_2]_2$.
Properties: Colorless crystals. Mp 137C, flash p 403F (206C). Soluble in alcohol and ether; insoluble in water. Combustible.
Derivation: The methyl ether of *o*-nitrophenol is reduced by zinc dust and caustic soda to the hydrazo compound, which is then rearranged with hydrochloric acid.
Hazard: See anisidine, benzidine.
Use: Azo-dye intermediate.

dianisidine diisocyanate. (3,3'-dimethoxybenzidine 4,4'-diisocyanate).
CAS: 91-93-0. $[OCN(CH_3O)C_6H_3]_2$.
Properties: Gray to brown powder. Mp 112C min. Soluble in ketones and esters.
Hazard: Toxic by inhalation and ingestion, skin irritant. Questionable carcinogen.
Use: Polymers and adhesive systems, high-strength backbone or cross-linking intermediate.

diaphragm cell. A type of electrolytic cell for the production of sodium hydroxide and chlorine from sodium chloride brine. The cell contains anode and cathode compartments separated by a porous diaphragm or membrane to prevent mixing of the solutions. Asbestos fibers are usually used for this diaphragm, though a recent development is a plastic material made from perfluorosulfonic acid (see "Naflon"). The brine is fed continuously to the anode compartment, where chlorine is released at the graphite anode and flows through the diaphragm to the steel cathode, where hydrogen is liberated. Sodium hydroxide accumulates in the liquid and is continuously drained from the cathode compartment. The Hooker cell and the Vorce cell are two widely used types of diaphragm cell.

"Diaron" *[Reichhold].* TM for powdered melamine adhesives.

diasolysis. The process of diffusion of organophilic compounds in organic solvent through gum, plastic, or rubber membranes for separating hydrophilic and colloidal substances that do not dialyze.

diaspore.
$Al_2O_3 \cdot H_2O$.
A natural hydrous aluminum oxide occurring in bauxite and with corundum and dolomite.
Properties: White, gray, yellowish, and greenish; luster vitreous to pearly. D 3.35-3.45, Mohs hardness 6.5-7.
Occurrence: Arkansas, Missouri, Pennsylvania, Switzerland, the former U.S.S.R., Czechoslovakia.
Use: Refractory, abrasive.

diastase malt. A commercial mixture containing amylolytic enzymes.
Properties: Yellowish-white, amorphous powder or syrupy liquid. Soluble in water; almost insoluble in alcohol.
Derivation: The filtrate from the mash of malted grain is concentrated at low temperature in vacuum. The sugar acts as preservative. The diastase hydrolyzes starch to malt sugar.
Use: Desizing of textiles, bread making, malted foods, converting starch to sugar.
See amylase.

diastereoisomer. (diastereomer).
In any group of four optical isomers occurring in compounds containing two asymmetric carbon atoms, such as 4-carbon sugars, there are two pairs of enantiomers (structures that are mirror images of each other) indicated by the letters D and L. The two D and the two L isomers are not mirror images, and these are called diastereoisomers. For example, in the structures below, (1) and (2) are enantiomers and so are (3) and (4); (1) and (3) are diastereoisomers, as are (2) and (4)

CHO
H—C—OH
HO—C—H
CH₂OH
L
(1)

CHO
HO—C—H
H—COH
CH₂OH
D
(2)

CHO
HO—C—H
HO—C—H
CH₂OH
(3)

CHO
H—C—OH
H—C—OH
CH₂OH
D
(4)

See optical isomer; enantiomer; epimer; anomer.

diatomaceous earth. (diatomite; kieselguhr; infusorial earth).
CAS: 61790-53-2.
Properties: Soft, bulky, solid material (88% silica) composed of skeletons of small prehistoric aquatic plants related to algae (diatoms). They have intricate geometric forms. Available as light-colored blocks, bricks, powder, etc. True d 1.9–2.35, bulk density from 5 to 15 lb/cu ft. Insoluble in acids except hydrogen fluoride, soluble in strong alkalies. Absorbs 1.5–4 times its weight of water, has high oil absorption capacity. Poor conductor of sound, heat, and electricity. Noncombustible.
Occurrence: Western U.S., Europe, Algeria, the former U.S.S.R.
Grade: Natural, chemical, airfloated.
Use: Filtration; clarifying and decolorizing; insulation; absorbent; mild abrasive; drilling-mud thickener; extender in paints, rubber, and plastic products; ceramics; paper coating; anticaking agent in fertilizers; asphalt compositions; chromatography; refractories; acid-proof liners; and catalyst carrier.

diatomic. Descriptive of a gas whose molecules are composed of two atoms, e.g., O_2, N_2, Cl_2, H_2. Gases in which the element is present as single atoms are called monatomic, e.g., Ar, Ne.

diatomite. See diatomaceous earth.

diatrizoate sodium. See sodium diatrizoate.

1,4-diazabicyclo[2.22]octane.

CH₂CH₂NCH₂CH₂NCH₂CH₂.

Properties: Crystals. Hygroscopic. Mp 158C, bp 174C, forms crystalline hydrate, sublimes easily. Soluble in water and organic solvents.

Use: Possible catalyst for urethane foams and coatings, chemical intermediate.

diazepam. (USAN for 7-chloro-1,3-dihydro-1-methyl-5-phenyl-2*H*-1,4-benzodiazepin-2-one).
CAS: 439-14-5. $C_{16}H_{13}ClN_2O$.
Properties: Slightly yellowish, crystalline powder; practically no odor. Mp 131.5–134.5C. One g of diazepam dissolves in about 350 mL of water, in approximately 15 mL of 95% ethanol, or in approximately 2 mL of chloroform.
Hazard: Central nervous system depressant; addictive; manufacture and usage restricted. Questionable carcinogen.
Use: Medicine (tranquilizer).

1,3-diazine. (pyrimidine; miazine).

CHN(CH)₃N.

Properties: Liquid or crystalline mass; penetrating odor. Mp 20–22C, bp 123–124C. Soluble in water, alcohol, and ether.
See pyrimidine.

Diazinon.
CAS: 333-41-5. [C(CH₃)₂CHC₄N₂H(CH₃)O]PS (OC₂H₅)₂.
O,O-diethyl-*O*-(2-isopropyl-4-methyl-6-pyrimidinyl) phosphorothioate.
Properties: Colorless liquid. Bp 83–84C (0.002 mm Hg). Slightly soluble in water; freely soluble in petroleum solvents, alcohol, and ketones. More stable in alkaline than neutral or acid solutions.
Hazard: Toxic by ingestion, inhalation, and skin absorption; cholinesterase inhibitor; use may be restricted. Questionable carcinogen.
Use: Insecticide (use against fire ants permitted by EPA).

diazoaminobenzene. (1,3-diphenyltriazene; benzeneazoanilide).
CAS: 136-35-6. $C_6H_5NNNHC_6H_5$.
Properties: Golden-yellow scales. Mp 96C. Soluble in alcohol, ether, and benzene; insoluble in water.
Derivation: Interactions of nitrous acid and an alcoholic solution of aniline.
Hazard: Explodes on heating to 150C, dangerous.
Use: Organic synthesis, dyes, insecticide.

p-diazobenzenesulfonic acid.
CAS: 305-80-6. $C_6H_4SO_3N_2$.
Properties: White or slightly red crystals or white paste. Soluble in water and ether; insoluble in alcohol.
Derivation: From sulfanilic acid, sodium nitrite, and sulfuric acid.
Hazard: Explodes when shocked or heated.
Use: Dyestuffs, reagent.

diazo compound.　A compound of the general formula Ar-N=N-X, in which Ar is an aromatic radical and X an inorganic group, such as halogen or nitrate.

_p_-diazodimethylaniline zinc chloride double salt.　(_p_-dimethylaminobenzene diazonium chloride, zinc chloride double salt; _p_-diazotized aminodimethylaniline zinc chloride double salt). $(CH_3)_2NC_6H_4N_2Cl \cdot ZnCl_2$.
Properties: Yellow-to-orange (light-sensitive) crystals. Specification: Moisture content 5-20%, zinc 17–23%, chloride 31–35%.
Hazard: Irritant.
Use: Rapid diazotype coupler used in coatings for light-sensitive paper.

2-diazo-4,6-dinitrobenzene-1-oxide.　See diazodinitrophenol.

diazodinitrophenol.　(2-diazo-4,6-dinitrobenzene-1-oxide; 5,7-dinitro-1,2,3-benzoxadiazole; DDNP).
CAS: 87-31-0.　$(NO_2)_2C_6H_2ON_2$ (bicyclic).
Properties: Yellow, crystalline compound; darkens rapidly on exposure to sunlight. D 1.6. Soluble in nitrobenzene, acetone, acetic acid, and nitroglycerin; desensitized by water.
Derivation: Diazotization of picramic acid in aqueous solution with sodium nitrite and hydrochloric acid.
Hazard: Explodes when shocked or heated to 180C; dangerous, an initiating explosive.
Use: Primary charge in blasting caps.

_p_-diazodiphenylamine sulfate.
CAS: 27990-92-7.　$(C_6H_5NHC_6H_4N_2)_2SO_4$.
Properties: Yellow–green solid; unpleasant odor. Sensitive to light. Soluble in water.
Use: Used as a light-sensitive diazo compound for coating on reproduction paper, giving direct positive prints of various colors with different developers or coupling agents.

1,2-diazole.　See pyrazole.

diazomethane.　(azimethylene).
CAS: 334-88-3.　$H_2C=N^+=N^-$.
Properties: Yellow gas at room temperature. Fp −145C, bp −23C, d 1.45.
Hazard: Severe explosion risk when shocked; may explode on contact with alkali metals, rough surfaces, or heat (100C); toxic by inhalation, a questionable carcinogen. Eye and upper respiratory tract irritant.
Use: Organic synthesis (methylating agent).

1-diazo-2-naphthol-4-sulfonic acid.
CAS: 887-76-3.　$C_{10}H_5N_2OSO_3H$.
Properties: Yellow needles in paste or dry form. Slightly soluble in water. Also available as sodium salt. Mp 168C. Combustible.

Derivation: Diazotization of 1-amino-2-naphthol-4-sulfonic acid and filtering of the diazo compound.
Hazard: May explode if heated above 100C.
Use: Azo dyes, valuable chrome-dyestuff component.

diazo paper.　Paper that is coated with a diazo light-sensitive dye.

diazotization.　The reaction of a primary aromatic amine with nitrous acid in the presence of excess mineral acid to produce a diazo (-N=N-) compound. Widely used in organic synthesis, especially production of dyes.

diazotizing salt.　See sodium nitrite.

DIBA.　See diisobutyl adipate.

DIBAC.　Abbreviation for diisobutylaluminum chloride.

DIBAL-H.　Abbreviation for diisobutylaluminum hydride.

dibasic.　See monobasic.

dibenalacetone.　(dibenzylidene acetone). $C_6H_5HC=CHCOHC=CHC_6H_5$.
Properties: (_cis_-trans) Yellow crystals. Mp 60C.
Derivation: Reaction of benzaldehyde and acetone. There are three geometric isomers (_trans_-trans, _cis_-trans, _cis_-cis).
Use: Suntan lotions, cosmetics.

dibenz(a,h)acridine 3,4-diol-1,2-epoxide.
CAS: 125276-72-4.　$C_{21}H_{15}NO_3$.
Hazard: Experimental carcinogenic data reported.

1,2:5,6-dibenzanthracene.　See dibenz(a,h)anthracene.

dibenz(a,h)anthracene.
CAS: 53-70-3.　$C_{22}H_{14}$.
Properties: Silvery leaflets from AcOH. Mp: 266-267C.
Hazard: A probable carcinogen. A poison.

dibenzanthrone.　(volanthrone).
CAS: 116-71-2.　$C_{34}H_{16}O_2$.
Violet-blue vat dye.
Properties: Bluish black powder. Soluble in nitrobenzene and concentrated sulfuric acid.
Derivation: From benzanthrone.
Use: Intermediate.

3,3′-dibenzanthronyl.　(also called the 13,13′-compound).
$C_{34}H_{18}O_2$.
Properties: Dark-yellow needles. Mp 412C. Soluble in concentrated sulfuric acid.
Use: Intermediate.

4,4′-dibenzanthronyl.
(also called the 2,2′-compound).
CAS: 116-90-5. $C_{34}H_{18}O_2$.
Properties: Yellow needles. Mp 320C. Soluble in nitrobenzene; slightly soluble in benzene, alcohol, ether.
Use: Intermediate.

dibenzarsenolic acid. See 9-arsafluoreninic acid.

2,3,6,7-dibenzoanthracene. See pentacene.

dibenzocycloheptadienone.
CAS: 1210-35-1. $C_{15}H_{12}O$.
A tricyclic compound.
Properties: Light-yellow to amber solid. Mp 28.5C, bp 203–204C (7 mm Hg), d 1.1635 (20C), refr index (20C) 1.6324. Soluble in alcohol and most organic solvents; insoluble in water. Combustible.
Use: Intermediate.

dibenzo(b,jk)fluorene. See benzo(k)fluoranthene.

dibenzofuran. See diphenylene oxide.

Dibenzo G-M-F.
$(C_6H_5COON)_2C_6H_4$.
Dibenzoyl-p-quinonedioxime.
Properties: Brownish-gray powder. D 1.37, starts to decompose above 200C, good storage stability. Insoluble in acetone, benzene, gasoline, ethylene dichloride, and water.
Grade: Available in superdispersing grades.
Use: Nonsulfur vulcanizing agent, in tire-curing bags, gaskets, and wire insulation to impart heat resistance.
See benzoyl peroxide.

dibenzopyran. See xanthene.

(+-)-dibenzo(a,l)pyrene-11,12-dihydrodiol.
CAS: 153857-27-3. $C_{24}H_{16}O_2$.
Hazard: Experimental carcinogenic data reported.

(-)-(11r,12r)-dibenzo(a,l)pyrenedihydrodiol.
CAS: 189880-63-5. $C_{24}H_{16}O_2$.
Hazard: Experimental carcinogenic data reported.

dibenzopyrone. See xanthone.

dibenzopyrrole. See carbazole.

dibenzothiophene.
CAS: 132-65-0

$C_6H_4C_6H_4S$.

Properties: Colorless crystals. Mp 97–98C. Combustible.
Hazard: Questionable carcinogen.
Use: Cosmetics and pharmaceuticals, intermediate.

dibenzoxazepine.
CAS: 12770-99-9. $C_{13}H_9NO$.
Hazard: A severe human skin irritant.

dibenzoyl. See benzil.

trans-1,2-dibenzoylethylene.
CAS: 4070-75-1. $C_6H_5COCHCHCOC_6H_5$.
Properties: Yellow–orange crystals. Mp 111C. Soluble in glacial acetic acid, ethyl acetate, benzene, and chloroform; sparingly soluble in alcohol; insoluble in water and petroleum ether. Combustible.
Use: Enzyme inhibitor, bactericide, and intermediate.

dibenzoylmethane. (1,3-diphenyl-1,3-propanedione).
CAS: 120-46-7. $C_6H_5COCH_2COC_6H_5$.
Properties: Crystals. Mp 80C, bp 219–221C (18 mm Hg). Partially soluble in alcohol; soluble in ether, chloroform, and aqueous sodium hydroxide. Combustible.
Use: Colorimetric determination of uranium.

dibenzoyl peroxide. See benzoyl peroxide.

dibenzoyl-p-quinonedioxime. See "Dibenzo G-M-F."

2,4-dibenzoylresorcinol.
$C_6H_5COC_6H_2(OH)_2COC_6H_5$.
Properties: Light-yellow crystals; nearly odorless. Mp 125–128C. Soluble in alcohol, ethyl acetate, methyl ethyl ketone; insoluble in water.
Use: Light absorber, best at 280–370 microns.

dibenzyl. See sym-diphenylethane.

N,N-dibenzylamine.
CAS: 103-49-1. $HN(CH_2C_6H_5)_2$.
Properties: Colorless to light-yellow, oily liquid; ammonia-like odor. Bp 300C (partial decomposition), mp −26C, d 1.017 (20C), refr index 1.5730–1.5740 (25C), distilling range 168–172C (10 mm Hg). Insoluble in water, soluble in alcohol and ether. Combustible.
Hazard: Toxic by ingestion and inhalation.
Use: Intermediate, rubber activator, reagent for metals.

N,N-dibenzyl-p-aminophenol.
$(C_6H_5CH_2)_2NC_6H_4OH$.
Properties: Brown powder. Mp >110C. Soluble in acetone, benzene, anhydrous methanol.

N,N-dibenzylaniline.
CAS: 91-73-6. $C_6H_5N(CH_2C_6H_5)_2$.

Properties: Yellowish-white crystals. Mp 70C, bp >300C. Soluble in alcohol and ether; insoluble in water.
Hazard: See aniline.
Use: Manufacture of dyes.

dibenzyl disulfide.   (benzyl disulfide).
CAS: 150-60-7. $C_6H_5CH_2SSCH_2C_6H_5$.
Properties: Pink solid; odor similar to benzaldehyde. Mp 70–72C. Dissolves in most organic solvents; very slightly soluble in water. Combustible.
Use: Antioxidant and antisludging agent for petroleum oils, extreme pressure lubricating oils and greases, and silicone oils.

dibenzyl ether.   (benzyl ether).
CAS: 103-50-4. $C_6H_5CH_2OCH_2C_6H_5$.
Properties: Colorless, unstable liquid; faint, almond odor. D 1.035, bp 298–300C, flash p 275F (135C). Insoluble in water; soluble in most organic solvents. Combustible.
Grade: Technical.
Hazard: Irritant and narcotic.
Use: Plasticizer for various resins, perfumery (solvent for nitro-musks), flavoring.

dibenzylidene acetone.   See dibenalacetone.

_N_,_N_-dibenzylmethylamine.
CAS: $CH_3N(CH_2C_6H_5)_2$.
Properties: Colorless to light-yellow liquid. D 0.99 (25C), refr index 1.5560–1.5590 (25C), distilling range 152–158C (11 mm Hg). Partially soluble in water; soluble in organic solvents. Combustible.
Use: Intermediate, oil-soluble rust inhibitor, cutting oils, hydraulic fluids, lubricants.

dibenzyl sebacate.
CAS: 140-24-9 140-24-9. $C_6H_5CH_2OOC(CH_2)_8$ $COOCH_2C_6H_5$.
Properties: Light-straw–colored liquid. Mp 28C, bp 265C (4 mm Hg), d 1.055 (30/20C), complete non-volatility, gives excellent low-temperature flexibility, flash p 450F (232C). Combustible.
Use: Plasticizer, especially for plastic linings for containers.

dibenzyl succinate.   See benzyl succinate.

dibenzyltin bis(dibutyldithiocarbamate).
See bis(dibutyldithiocarbamato)dibenzylstannane.

2,5-dibiphenylyloxazole.   (BBO).
$C_{27}H_{21}NO$.
Properties: Crystalline solid. Mp 237–239C.
Grade: Purified.
Use: Scintillation counter or a wavelength shifter in soluble scintillators.

dibismuth trisulfate.   See bismuth trisulfate.

diborane.   (diboron hexahydride; boroethane).
CAS: 19287-45-7. B_2H_6.
Properties: Colorless gas; repulsive odor. Bp −92.5C, fp −165C, density 0.18 g/mL (17C), flash p −130F (−90C), autoign temp 100–125F (38–51C). Soluble in carbon disulfide; decomposes in water. Highly reactive.
Derivation: Reaction of lithium hydride and boron trifluoride catalyzed by ether at 25C.
Grade: Technical 95%, high purity 99+%.
Hazard: Highly flammable, dangerous fire risk, reacts violently with oxidizing materials including chlorine. Toxic by inhalation, strong upper respiratory tract irritant. Can cause headaches.
Use: Synthesis of organic boron compounds and metal borohydrides, polymerization catalyst for ethylene, fuel for air-breathing engines and rockets, reducing agent, doping agent for p-type semiconductors.

dibromoacetic acid.   (2,2-dibromoacetic acid; acetic acid, dibomo-).
$C_2H_2Br_2O_2$.
Properties: Colorless, crystalline, water-soluble compound.
Hazard: Experimental reproductive effects; possible carcinogen.

dibromoacetylene.   (dibromomethyne).
CAS: 624-61-3.   BrC≡CBr.
Properties: Heavy, colorless liquid; disagreeable odor. D approximately 2, mp 76–76.5C. Soluble in most organic solvents.
Derivation: (1) Interaction of magnesium dibromoacetylene and an ethereal solution of cyanogen bromide; (2) Interaction of tribromoethylene and alcoholic potash.
Hazard: Explodes on contact with oxygen or on heating. Dangerous!! Toxic by inhalation and injection, strong irritant.
Use: Organic synthesis (halogenated ethylene).

9,10-dibromoanthracene.
CAS: 523-27-3. $C_6H_4C_2Br_2C_6H_4$ (tricyclic).
Properties: Yellow crystals. Mp 226C, sublimes. Soluble in chloroform, slightly soluble in alcohol and ether, insoluble in water.
Derivation: Bromination of anthracene.
Use: Organic synthesis.

_o_-dibromobenzene.   (benzene dibromide).
CAS: 26249-12-7. $C_6H_4Br_2$.
Properties: Heavy liquid; pleasant, aromatic odor. Bp 225.5C, fp 7.13C, d 1.9767 (25/4C), refr index 1.6155 (20C). Combustible. Miscible with alcohol, acetone, ether, benzene, carbon tetrachloride, and *n*-heptane; insoluble in water.
Derivation: Interaction of benzene with an excess of bromine in presence of iron.
Use: Solvent for oils, motor fuels, top-cylinder compounds, organic synthesis, ore flotations.

p-dibromobenzene.   (benzene dibromide).
CAS: 26249-12-7. $C_6H_4Br_2$.
Properties: Colorless crystals. Mp 89C, bp 219C, d 2.261, refr index 1.5743 (99C). Soluble in alcohol and ether.
Use: Organic synthesis of dyestuffs and drugs, manufacture of intermediates, fumigant.

N,N-dibromobenzenesulfonamide.
$C_6H_5SO_2NBr_2$.
Properties: Solid. Mp 109–111C. Active bromine 50.4%.
Use: Halogenating agent.

dibromochloromethane.
CAS: 124-48-1. $CHBr_2Cl$.
Properties: Clear, colorless, heavy liquid. D 2.38, bp 116C.
Hazard: Irritant and narcotic.
Use: Organic synthesis.

1,2-dibromo-3-chloropropane.   (DBCP).
CAS: 96-12-8. $CH_2BrCHBrCH_2Cl$.
Properties: Colorless (when pure) liquid. D 2.05 (20C), bp 195.5C, fp 6.7C, refr index 1.5530 (20C), flash p 170F (76.6C) (TOC). Slightly soluble in water, miscible with oils. Combustible.
Hazard: A possible carcinogen, reported to cause sterility. Use regulated by EPA. OSHA standard of 1 ppb. No longer made in the U.S.
Use: Pesticide, nematocide, soil fumigant.

2,6-dibromo-4-cyanophenyl octanoate.
CAS: 1689-99-2. $C_{15}H_{17}Br_2NO_2$.
Hazard: A poison by ingestion.
Use: Agricultural chemical.

dibromodibutylstannane.
CAS: 996-08-7. $C_8H_{18}Br_2Sn$.
Properties: Mp: 20°.
Hazard: A poison by ingestion. Moderately toxic by skin contact.

dibromodiethyl sulfide.
$(CH_2CH_2Br)_2S$.
The bromine analog of mustard gas.
Properties: White crystals. D 2.05 (15C), bp 240C (decomposes), mp 31–34C. Soluble in alcohol, benzene, ether; insoluble in water.
Derivation: Action of hydrobromic acid on an aqueous solution of thiodiglycol.
Hazard: Toxic by inhalation, strongly irritant poison gas.
Use: Organic synthesis.

1,1-dibromoethane.   (ethylidene   bromide; ethylidene dibromide).
$C_2H_4Br_2$
Properties: Heavy liquid; insoluble in water; soluble in organic solvents.
Hazard: Moderately toxic; mutagen.

Use: Grain and soil fumigant for insect control; a component of leaded gasoline for scavenging lead from engine cylinders.

dibromodiethyl sulfone.
$(BrCH_2CH_2)_2SO_2$.
Properties: Plates. Mp 111–112C. Soluble in alcohol, benzene, and ether.
Derivation: Interaction of dibromodiethyl sulfide, chromic anhydride, and dilute sulfuric acid.

dibromodiethyl sulfoxide.
$(BrCH_2CH_2)_2SO$.
Properties: Glittering crystals. Mp 100–101C. Soluble in alcohol, benzene, and ether.
Derivation: Interaction of benzoyl peroxide and a hot solution of dibromodiethyl sulfide in chloroform.

dibromodifluoromethane.   (difluorodibromomethane).
CAS: 75-61-6. CF_2Br_2.
Properties: Colorless, heavy liquid. Fp −141C, bp 24.5C, d 2.288 (15/4C), refr index 1.399 (12C). Insoluble in water; soluble in methanol and ether. Nonflammable.
Derivation: Vapor-phase bromination of difluoromethane.
Grade: Pure (95.0% min).
Hazard: Irritant. Upper respiratory tract irritant, central nervous system impairment, and liver damage.
Use: Synthesis of dyes, pharmaceuticals, quaternary ammonium compounds, fire-extinguishing agent.

6,8-dibromo-dihydro-1,3-benzoxazine-2-thione-4-one.
CAS: 23611-68-9. $C_8H_3Br_2NO_2S$.
Hazard: A poison by ingestion.
Use: Agricultural chemical.

1,3-dibromo-5,5-dimethylhydantoin.

$$\overline{BrNCONBrCOC(CH_3)_2}.$$

Properties: Free-flowing, cream-colored powder; slight bromine odor. Mp 187–191C (decomposes). Quite stable at 75C; soluble in benzene, chloroform, glacial acetic acid; slightly soluble in water and carbon tetrachloride; insoluble in hexane. Contains 55% active bromine, which is slowly released in aqueous solution.
Derivation: Bromination of dimethylhydantoin.
Use: Controlled bromination and oxidation of organic compounds, water treatment, polymerization catalyst, potential germicide and sanitizer.

dibromodimethyl stannane.   See dimethyltin dibromide.

dibromodiphenylstannane.
CAS: 4713-59-1. $C_{12}H_{10}Br_2Sn$.
Properties: White or colorless crystals. Mp: 38C; bp: 230C at 42 mm. Sol in alc and ether.
Hazard: A poison.

1,2-dibromoethane.   See ethylene dibromide.

4,5′-dibromofluorescein.
$C_{20}H_{10}Br_2O_5$.
Properties: Yellow powder (clean shade). Mp 265–267C.
Grade: 99+%.
Use: Lipstick dye.

2,4-dibromofluorobenzene.
$C_6H_3Br_2F$.
Properties: Colorless liquid. D 2.047 (20C), bp 214C, refr index 1.5790 (25C). Insoluble in water; soluble in alcohol, acetone, ether, benzene, chloroform, ethyl acetate, and glacial acetic acid.
Hazard: Irritant to skin and eyes.
Use: Intermediate for agricultural and pharmaceutical chemicals.

dibromoformoxime.
CBr_2NOH.
Properties: Crystals. Mp 70–71C, distills between 75 and 85C (3 mm Hg).
Hazard: Evolves highly toxic fumes on heating. A military poison.

dibromoiodoethylene.
Br_2CCHI.
Properties: Liquid. D 2.952 (24C), bp 91C (15 mm Hg).
Derivation: Reaction of iodine and dibromoacetylene.

dibromomalonic acid.
$HOOC \cdot CBr_2COOH$.
Properties: Light-yellow needles or prisms. Mp 147C (decomposes).
Use: Intermediate for drugs and fine chemicals.

dibromomalonyl chloride.
$ClOCCBr_2COCl$.
Properties: Yellowish, oily liquid. Bp 75–77C (15 mm Hg).
Use: Chemical intermediate.

dibromomethane.   See methylene bromide.

dibromomethyl ether.
$(CH_2Br)_2O$.
Properties: Colorless liquid. D 2.2, bp 154–155C, fp −34C. Decomposed by water; soluble in acetone, benzene, ether.
Derivation: (1) The reaction product of paraformaldehyde and sulfuric acid is treated with ammonium bromide; (2) interaction of hydrobromic acid and paraformaldehyde.
Hazard: Evolves highly toxic fumes on heating, strong irritant to eyes. Fire risk in contact with oxidizers.
Use: Military poison (lachrymator).

9,10-dibromooctadecanoic acid. (9,10-dibromostearic acid).
$CH_3(CH_2)_7CHBrCHBr(CH_2)CO_2H$.
Properties: Yellow solid or liquid. D 1.2458 (30/4C), mp 29–30C, refr index 1.4893 (42C). Insoluble in water; soluble in alcohols, ketones, and aromatic and chlorinated hydrocarbons. Also available as methyl ester.
Grade: Technical (amber liquid).
Use: Chemical intermediate.

1,5-dibromopentane.   See pentamethylene dibromide.

1,3-dibromopropane.   See trimethylene bromide.

dibromopropanol. (2,3-dibromo-1-propanol).
CAS: 96-13-9. $CH_2BrCHBrCH_2OH$.
Properties: Colorless liquid. D 2.120 (20/4C), bp 219C. Soluble in acetone, alcohol, ether, and benzene.
Hazard: A possible carcinogen.
Use: Intermediate in preparation of flame retardants, insecticides, and pharmaceuticals.

dibromoquinonechlorimide.
$C_6H_2Br_2ClNO$.
A reagent used for spot visualizations in chromotographic systems.
Properties: Yellow powder. Slightly soluble in water; moderately soluble in hot alcohol.
Hazard: Highly sensitive to heat; explodes at 120C, decomposes with rapid heat evolution at 60C.

3,5-dibromosalicylaldehyde. (3,5-dibromo-2-hydroxybenzaldehyde).
$Br_2(OH)C_6H_2CHO$.
Properties: Pale-yellow crystals. Mp 86C. Slightly soluble in water; soluble in ether, benzene, chloroform, alcohol, and acetic acid.
Use: Medicine (external), fungicide.

4′,5-dibromosalicylanilide.   See dibromsalan.

9,10-dibromostearic acid.   See 9,10-dibromooctadecanoic acid.

2,5-dibromoterephthalic acid.
$C_6H_2Br_2(COOH)_2$.
A flame-retardant monomer for production of polyester fibers that are made by reacting this acid with dimethyl terephthalate and ethylene glycol. Permanent lowering of flammability is said to be

gained by this method of incorporating bromine in molecular combination.

sym-dibromotetrafluoroethane.
$CBrF_2CBrF_2$.
Properties: Liquid. Bp 47.3C, d 2.18 (21.1C). Non-flammable.
Use: Refrigerant, fire-extinguishing agent, control fluid.

1,2-dibromo-1,1,5-trichloropentane.
CAS: 19792-94-0. $C_5H_7Br_2Cl_3$.
Hazard: Moderately toxic by ingestion.

dibromsalan.   (USAN for 4′,5-dibromosalicylanilide).
$BrC_6H_3(OH)CONHC_6H_4Br$.
Hazard: A suspected carcinogen; use in cosmetics prohibited (FDA).
Use: Disinfectant.

dibucaine.   (2-*n*-butoxy-*N*-(2-diethylaminoethyl) cinchoninamide).
CAS: 85-79-0. $C_{20}H_{29}N_3O_2$.
Properties: Colorless or almost colorless powder; odorless. Mp 62–65C. Somewhat hygroscopic; affected by light. Soluble in alcohol and acetone; slightly soluble in water. Also available as the hydrochloride.
Grade: NF.
Hazard: Possibly allergenic.
Use: Medicine (anesthetic).

dibutoline sulfate.
$(C_{15}H_{33}N_2O_2)_2SO_4$.
Bisdibutylcarbamate of ethyl(2-hydroxyethyl)dimethylammonium sulfate.
Properties: Hygroscopic powder. Decomposes 166C. Soluble in water and benzene.
Use: Surface-active agent in medicine.

2,5-dibutoxyaniline.
$C_6H_3(OC_4H_9)_2NH_2$.
Properties: Mp 18C. Insoluble in water; soluble in organic solvents.
Hazard: See aniline.
Use: Dyes, synthesis.

1,4-dibutoxybenzene.   See hydroquinone di-*n*-butyl ether.

dibutoxyethyl adipate.
$(C_2H_4COOC_2H_4OC_4H_9)_2$.
Properties: Colorless, oily liquid; mild, butyl-like odor. D 0.997 (20/20C); fp −34C; boiling range 205–215 (4 mm Hg); flash p 370F (187C); refr index 1.442 (25C); wt/gal 8 lb. Soluble or only slightly soluble in mineral oil, glycerol, glycols, and some amines; soluble in most other organic liquids. Combustible.

Use: Primary plasticizer for most resins, imparting flexibility at very low temperature as well as stability to UV light.

dibutoxyethyl phthalate.   (*n*-butylglycol phthalate).
CAS: 117-83-9. $C_6H_4(COOC_2H_4OC_4H_9)_2$.
Properties: Colorless liquid. Fp −55C, d 1.06 (20C), bp 270C, wt/gal 8.86 lb, fast to light, water-resistant, flash p 407F (208C). Soluble in organic solvents. Combustible.
Use: Plasticizer for polyvinyl chloride, polyvinyl acetate, and other resins.

dibutoxymethane.
$CH_2(OC_4H_9)_2$.
Properties: Colorless liquid. Wt/gal 6.97 lb (20C), refr index 1.40615 (20C), d 0.838 (20/20C), flash p 140F (60C) (CC), boiling range 164–186C. Insoluble in water. Combustible.
Hazard: Moderate fire risk.

dibutoxytetraglycol.   (tetraethylene glycol dibutyl ether).
$(C_4H_9OC_2H_4OC_2H_4)_2O$.
Properties: Practically colorless liquid; characteristic odor. D 0.9436 (20/20C), wt/gal 7.85 lb (20C), bp 237C (50 mm Hg), 330C, vap press <0.01 mm Hg (20C), fp −20C, flash p 355F (180C), solubility of water in product 4.8% by wt (20C), refr index 1.4357. Slightly soluble in water (1.3% by weight). Combustible.
Use: Solvent, especially for DDT.

N,*N*-di-*n*-butyl acetamide.
$CH_3CON(C_4H_9)_2$.
Properties: Colorless liquid. D 0.890 (20C), boiling range 245–250C, faint odor, flash p 225F (107C). Combustible.

di-*n*-butylamine.
CAS: 111-92-2. $(C_4H_9)_2NH$.
Properties: Colorless liquid; amine odor. Bp 159.6C, fp −62C, d 0.7613 (20/20C), wt/gal 6.33 lb (20C), refr index 1.4175 (20C), flash p 125F (51.6C) (OC). Partially soluble in water; soluble in alcohol and ether; miscible with hydrocarbons. Combustible.
Derivation: By reaction of butanol or butyl chloride with ammonia.
Grade: Technical.
Hazard: Moderate fire risk. Toxic by ingestion and inhalation.
Use: Corrosion inhibitor; intermediate for emulsifiers, rubber accelerators, dyes, insecticides, and flotation agents; inhibitor for butadiene.

di-*sec*-butylamine.
$(CH_3CHCH_2CH_3)_2NH$.
Properties: Water-white liquid; amine odor. Boiling range 132–135C, d 0.754 (20/20C), refr index 1.412 (20C), flash p 75F (23.9C).

Hazard: Flammable, dangerous fire risk. Toxic by ingestion and inhalation.

dibutylamine pyrophosphate.
Grade: Available as 40% solution in ethanol–acetone.
Hazard: Flammable, dangerous fire risk.
Use: Anticorrosion agent in lacquers and cotton solutions, for light and heat stabilization of vinyl chloride and vinyl copolymer resins.

***N*,*N*-di-*n*-butylaminoethanol.**   (2-*N*-dibutylaminoethanol).
CAS: 102-81-8. $(C_4H_9)_2NCH_2CH_2OH$.
Properties: Colorless liquid; faint amine-like odor. D 0.859 (20C), boiling range 224–232C, flash p 200F (93.3C). Combustible.
Hazard: Toxic by ingestion and skin absorption. Eye and upper respiratory tract irritant.
Use: Synthesis.

dibutylammonium oleate.   See "Barak" [*Technopharma*].

di-*n*-butylammonium tetrafluoroborate.
$(C_4H_9)_2NH_2BF_4$.
Properties: Solid. Mp 266C.
Use: Lubricating, surface treating, fluxing of aluminum.

***N*,*N*-di-*n*-butylaniline.**
$C_6H_5N(C_4H_9)_2$.
Properties: Amber liquid; faint aniline odor. D 0.904 (20C), boiling range 267–275C, refr index 1.519 (20C), flash p 230F (110C). Soluble in alcohol and ether; insoluble in water. Combustible.
Hazard: See aniline.

2,5-di-*tert*-butylbenzoquinone.
$[C(CH_3)_3]_2C_6H_2O_2$.
Properties: Yellow crystals. Mp 149–151C. Insoluble in water; soluble in ethyl acetate, acetone, and benzene; slightly soluble in ethyl alcohol.
Hazard: See quinone.
Use: Oxidant, polymerization catalyst.

dibutylbutyl phosphonate.
$C_4H_9P(O)(OC_4H_9)_2$.
Properties: Colorless liquid; mild odor. D 0.948 (20/4C), bp 127–128C (2.5 mm Hg), flash p 310F (154.4C) (COC). Stable, insoluble in water, miscible with most common organic solvents. Combustible.
Use: Heavy metal extraction and solvent separation, gasoline additives, antifoam agent, plasticizer, textile conditioner, and antistatic agent.

dibutyl chlorophosphate.
$(C_4H_9O)_2P(O)Cl$.
Properties: Water-white liquid. Bp 103–106C (1.5 mm Hg), d 1.0742 (25C), refr index 1.4289 (25C). Soluble in alcohols; hydrolyzes slowly.
Use: Intermediate in organic synthesis.

4,6-di-*tert*-butyl-*m*-cresol.   (DBMC; 4,6-*tert*-butyl-3-methylphenol).
$[C(CH_3)_3]_2CH_3C_6H_2OH$.
Properties: Crystalline solid. Mp 62.1C, bp 282C, d 0.912 (80/4C), viscosity 9.9 centistokes (80C), 1.42 centistokes (160C), flash p 262F (127.7C) (OC), Combustible. Very soluble in ethanol, benzene, carbon tetrachloride, ethyl ether, and acetone; essentially insoluble in water, ethylene glycol, and 10% aqueous sodium hydroxide.
Use: Rubber reclaiming, surface-active agents, resins and plasticizers, antioxidants, and perfumes.

2,6-di-*tert*-butyl-*p*-cresol.   (2,6-*tert*-butyl-4-methylphenol; butylated hydroxytoluene; BHT).
CAS: 128-37-0. $[C(CH_3)_3]_2CH_3C_6H_2OH$.
Properties: White, crystalline solid. Fp 70C, bp 265C, d 1.048 (20/4C), viscosity 3.47 centistokes (0C), 1.54 centistokes (120C), refr index 1.4859 (75C), flash p 275F (135C) (COC). Soluble in methanol, ethanol, isopropanol, "Cellosolve" (12C), naphtha, benzene, methyl ethyl ketone, and linseed oil; insoluble in water and 10% sodium hydroxide. Combustible.
Grade: Technical, feed, FCC.
Hazard: Use in foods restricted. Upper respiratory tract irritant. Questionable carcinogen.
Use: Antioxidant for petroleum products, jet fuels, rubber, plastics, and food products, food packaging, and animal feeds. Satisfies ASTM D910-64T for use in aviation gasoline.

di-*n*-butyl(dibutyryloxy)stannane.
CAS: 28660-63-1. $C_{16}H_{32}O_4Sn$.
Hazard: A poison by ingestion. A severe eye and skin irritant.

***N*,*N*-di-*tert*-butylethylenediamine.**
CAS: 4062-60-6. $C_{10}H_{24}N_2$.
Hazard: A poison by ingestion. A mild skin irritant.

dibutyldi(2-ethylhexyloxycarbonylmethylthio)
stannane.
CAS: 10584-98-2. $C_{28}H_{56}O_4S_2Sn$.
Hazard: Moderately toxic by ingestion.

dibutyldifluorostannane.
CAS: 563-25-7. $C_8H_{18}F_2Sn$.
Hazard: A poison by ingestion.

dibutyl(diformyloxy)stannane.
CAS: 7392-96-3. $C_{10}H_{20}O_4Sn$.
Hazard: A poison by ingestion. A severe skin and eye irritant.

2,2-dibutyldihydro-6h-1,3,2-oxathiastannin-6-one.   See dibutyltin mercaptopropionate.

dibutyldiiodostannane.
CAS: 2865-19-2. $C_8H_{18}I_2Sn$.

Hazard: A poison by ingestion. Moderately toxic by skin contact.

2,6-di-*tert*-butyl-α-dimethylamino-*p*-cresol.
(Ethyl Antioxidant 703).
$(C_4H_9)_2C_6H_2OH[CH_2N(CH_3)_2]$.
Properties: Light-yellow, crystalline solid. Mp 93.9C, flash p 280F (137.7C) (OC). Insoluble in water and 10% sodium hydroxide; soluble in organic solvents. Combustible.
Use: Antioxidant in gasoline and oils, including jet engine oils.

dibutyldipentanoyloxystannane.
CAS: 3465-74-5. $C_{18}H_{36}O_4Sn$.
Hazard: A poison by ingestion. A severe skin and eye irritant.

di-*tert*-butyl diperphthalate.
$(CH_3)_3COOCOC_6H_4COOOC(CH_3)_3$.
Available as a 50% solution in dibutyl phthalate.
Properties: Clear liquid. Flash p 145F (62.7C) (OC). D 1.056. Insoluble in water; soluble in most organic solvents. Combustible.
Hazard: May ignite organic materials, strong oxidizer, may explode when shocked or in contact with reducing materials.
Use: High-temperature polymerization catalyst for vinyl and polyester resins.

dibutyldiphenyl tin.
$(CH_3CH_2CH_2CH_2)_2Sn(C_6H_5)_2$.
Properties: Clear, slightly greenish liquid. Contains 30.7% Sn. Bp 175C (2 mm Hg), refr index 1.563 (17.5C), d 1.19. Combustible.
Hazard: Toxic.

dibutyldipropionyloxystannane.
CAS: 3465-73-4. $C_{14}H_{28}O_4Sn$.
Hazard: A poison by ingestion. A severe skin and eye irritant.

di-*tert*-butyl disulfide.
$C(CH_3)_3SSC(CH_3)_3$.
Properties: Liquid. D 0.9291, boiling range 190–207C, refr index 1.491 (20C), flash p approximately 170F (76.6C). Combustible.
Use: Intermediate, diesel- and jet-fuel additive, lubricant additive.

dibutyldithiocarbamic acid-s-tributylstannyl ester.
CAS: 67057-34-5. $C_{21}H_{45}NS_2Sn$.
Hazard: A poison.

((dibutyldithiocarbamoyl)oxy)
tributylstannane. See dibutyldithiocarbamic acid-s-tributylstannyl ester.

***n*-dibutyl ether.** See butyl ether.

dibutyl fumarate.
$(C_4H_9)OOCCH:CHCOO(C_4H_9)$.
Properties: Colorless liquid. D 0.9873 (20C), bp 285.2C, fp −15.6C, refr index 1.4466 (20C), flash p 300F (149C). Insoluble in water. Combustible.
Use: Monomeric plasticizers, copolymers, intermediate.

dibutyl hexahydrophthalate.
$C_6H_{10}(COOC_4H_9)_2$.
Properties: Liquid. D 1.005, bp 185–190C, flash p 305F (151.6C). Combustible.
Use: Plasticizer for nitrocellulose.

9-(2,2-dibutylhydrazino)acridine monohydrochloride.
CAS: 28846-39-1. $C_{21}H_{27}N_3 \cdot ClH$.
Hazard: A poison by ingestion.
Use: Agricultural chemical.

2,5-di-*tert*-butylhydroquinone.
$[C(CH_3)_3]_2C_6H_2(OH)_2$.
Properties: White powder. Mp 210–212C. Soluble in acetone, alcohol, and benzene; insoluble in water, aqueous alkali.
Use: Polymerization inhibitor, antioxidant, stabilizer against UV deterioration of rubber.

dibutyl 2-hydroxyethyl phosphate. See 2-hydroxyethyl dibutyl phosphate.

di-*n*-butyl itaconate.
$CH_2:C(COOC_4H_9)CH_2(COOC_4H_9)$.
Properties: Clear, colorless liquid; slight odor. Bp 145C (10 mm Hg), d 0.9833 (22C), refr index 1.442 (25C). Insoluble in water. Combustible.
Use: Resins, lubricating oil additives, plasticizers.

***N,N*-di-*n*-butyl lauramide.**
$C_{11}H_{23}CON(C_4H_9)_2$.
Properties: Straw-colored liquid; odor of lauric acid. D 0.861 (20C), boiling range 200–230C (3 mm Hg), flash p 375F (190C). Combustible.

dibutyl maleate. (DBM).
CAS: 105-76-0. $C_4H_9OOCCH:CHCOOC_4H_9$.
Properties: Colorless, oily liquid. Bp 280.6C, sets to a glass below −85C, d 0.9964 (20/20C), wt/gal 8.3 lb (20C), flash p 285F (140.5C) (OC). Insoluble in water. Combustible.
Use: Copolymers, plasticizers, intermediate.

4,6-di-t-butyl-2-(α-methyl-4-methoxybenzyl)phenol. See 2,4-bis(1,1-dimethylethyl)-6-(1-(4.

2,6-di-*tert*-butyl-4-methylphenol. See 2,6-di-*tert*-butyl-*p*-cresol.

4,6-di-*tert*-butyl-3-methylphenol. See 4,6-di-*tert*-butyl-*m*-cresol.

dibutyl oxalate.
$(COOC_4H_9)_2$.
Properties: Water-white, high-boiling liquid; mild odor. Bp 240–250C, refr index 1.425, fp −30C, wt/gal approximately 8.24 lb (20C), flash p 220F (104.4C) (CC). Miscible with most alcohols, ketones, esters, oils, hydrocarbons. Combustible.
Derivation: By the standard esterification process using normal butyl alcohol and oxalic acid.
Grade: According to ester content 90%, 95%, 99–100%.
Hazard: Toxic by ingestion and inhalation, strong skin irritant.
Use: Organic synthesis, solvent.

2,2-dibutyl-1,3,2-oxathiastannolane.
CAS: 27371-95-5. $C_{10}H_{22}OSSn$.
Hazard: A poison by ingestion. Moderately toxic by skin contact.

di-*tert*-butyl peroxide. (*tert*-butyl peroxide; DTBP).
CAS: 110-05-4. $(CH_3)_3COOC(CH_3)_3$.
Properties: Clear, water-white liquid. D 0.791 (25/25C), fp −40C, bp 111C, refr index 1.389 (20C), flash p 65F (18.3C) (CC). Soluble in styrene, ketones, and most aliphatic and aromatic hydrocarbons; insoluble in water.
Hazard: Flammable, dangerous fire hazard. Strong oxidizer, may ignite organic materials or explode when shocked or in contact with reducing materials.
Use: Polymerization catalyst for resins, including olefins, styrene, styrenated alkyds, and silicones; ignition accelerator for diesel fuel; organic synthesis; intermediate.

2,4-di-*tert*-butylphenol.
CAS: 96-76-4. $[(CH_3)_3C]_2C_6H_3OH$.
Properties: Tan, crystalline solid. Mp 52C, bp 152–157C (25 mm Hg), d 0.907 (60/4C), wt/gal 7.57 lb (60C), flash p 265F (129C). Soluble in methanol and ether; very slightly soluble in water. Combustible.
Hazard: See phenol.
Use: Intermediate, antioxidant, stabilizer, germicide.

2,6-di-*tert*-butylphenol.
CAS: 128-39-2. $[(CH_3)_3C]_2C_6H_3OH$.
Properties: Light-straw colored, crystalline solid. Mp 37C, d 0.914 (20C), bp 253C, flash p 245F (118C). Soluble in alcohol and benzene; insoluble in water. Combustible.
Hazard: See phenol.
Use: Intermediate, antioxidant, satisfies ASTM D910-64T for use as antioxidant in aviation gasoline.

***N,N′*-di-*sec*-butyl-*p*-phenylenediamine.**
$C_{14}H_{24}N_2$.
Properties: Amber to red liquid (normally supercooled below 18C). D 0.94 (15.5/15.5C), fp 20C, flash p 290F (COC). Soluble in gasoline, absolute alcohol, and benzene; insoluble in water or caustic solutions. Combustible.
Hazard: Toxic by ingestion, inhalation, and skin absorption; strong irritant to tissue, causes skin burns.
Use: Oxidation inhibitor and stabilizer in gasoline (satisfies ASTM D910-64T as antioxidant in aviation gasoline), prevents decomposition of tetraethyl lead in gasoline.

dibutyl phenyl phosphate. (DBPP; phosphoric acid, dibutyl phenyl ester).
CAS: 2528-36-1. $C_{14}H_{23}PO_4$.
Properties: Clear liquid; butanolic odor. Mw 286.26, sp g 0.0691 (23/25C), bp 131–132C, vap press 0.007 mm Hg 25C, flash p 129C. Insoluble in water.
Hazard: Combustible. Cholinesterase inhibitory and upper respiratory tract irritant.
Use: Hydraulic fluids.

dibutyl phosphate. (di-*n*-butyl phosphate).
CAS: 107-66-4.
Properties: Pale-amber liquid. Mw 210.21, mp (decomposes above 100C), vap press <1 mm Hg (20C).
Hazard: Bladder, eye, and upper respiratory tract irritant.
Use: Organic catalyst and antifoaming agent.

dibutyl phosphite.
CAS: 1809-19-4. $(C_4H_9O)_2PHO$.
Properties: Water-white liquid. Bp 95C (1 mm Hg), d 0.9860 (25C), refr index 1.4228 (25C), flash p 120F (49C). Soluble in common organic solvents. Combustible.
Hazard: Moderate fire risk.
Use: Solvent, antioxidant, intermediate.

dibutyl phthalate.
CAS: 84-74-2. $C_6H_4(COOC_4H_9)_2$.
Properties: Colorless, stable, oily liquid; odorless. D 1.0484 (20/20C), fp −35C, viscosity 0.203 cP (20C), distillation range 227–235 (37 mm Hg), flash p 340F (171C) (COC), wt/gal 8.72 lb (20C), refr index 1.4915 (25C), bp 340.0C, vap press 1.11 mm Hg (150C). Miscible with common organic solvents; insoluble in water. Autoign temp approximately 750F (398.8C). Combustible.
Derivation: By treating *n*-butyl alcohol with phthalic anhydride, followed by purification, which results in a product unusually free from odor and color.
Grade: Technical, 99–100% dibutyl phthalate.
Hazard: Toxic. Testicular damage, eye and upper respiratory tract irritant.
Use: Plasticizer in nitrocellulose lacquers, elastomers, explosives, nail polish, and solid rocket propellants; solvent for perfume oils; perfume fixative; textile lubricating agent; safety glass; insecticides; printing inks; resin solvent; paper coatings; adhesives; insect repellent for textiles.

2,5-di-*tert*-butylquinone.
$[C(CH_3)_3]_2C_6H_2O_2$.
Properties: Yellow powder. Mp 149–151C. Insoluble in water; soluble in alcohol, acetone, ethyl acetate, and benzene.
Hazard: Fire risk in contact with organic materials.
Use: Oxidizing agent.

dibutylsebacate. (DBS).
$C_4H_9OCO(CH_2)_8OCOC_4H_9$.
Properties: Clear, colorless, liquid; odorless. Bp 349C (760 mm Hg), 180C (3 mm Hg), fp −11C, d 0.936 (20/20C), wt/gal 7.81 lb (20C); refr index 1.4395 (25C); flash p 350F (176C). Insoluble in water. Combustible.
Grade: Technical.
Use: Plasticizer, rubber softener, dielectric liquid, cosmetics and perfumes, sealing food containers, flavoring.

dibutylstannane. See dibutyltin dihydride.

N,N-dibutylstearamide.
$C_{17}H_{35}CON(C_4H_9)_2$.
Properties: Yellow liquid. D 0.860 (20/20C), boiling range 173–175C (0.4 mm Hg), flash p 420F (215C), fatty acid odor. Combustible.

di-*n*-butyl succinate.
$C_{12}H_{22}O_4$.
Properties: Colorless liquid. Bp 120C, fp −29C, d 0.977, refr index 1.43.
Derivation: Reaction of butyl alcohol with succinic acid.
Use: Insect repellent.

di-*tert*-butyl sulfide. (butyl sulfide).
$[(CH_3)_3C]_2S$.
Properties: Liquid. Fp −11C, boiling range 297–303F, d 0.8316, wt/gal 6.93 lb, refr index 1.451 (20C), flash p 125F (51.6C). Combustible.
Hazard: Moderate fire risk.
Use: Intermediate, flavoring.

dibutyl tartrate.
$C_4H_9OOCCH_2OCH_2OCOOC_4H_9$.
Properties: Light-tan liquid. Mp 21C, bp approximately 204C (26 mm Hg), refr index 1.4463 (20C), flash p 195F (90.5C) (CC), combustible, autoign temp 544F (284.4C), wt/gal 9.07 lb (20C). Miscible with the common organic solvents, oils, and hydrocarbons.
Use: Solvent and plasticizer for cellulose esters and ethers, elastomers, and lubricant; rubberized fabrics; lacquers; dopes; transfer inks.

dibutyl(tetrachlorophthalato)stannane.
CAS: 23535-89-9. $C_{16}H_{18}Cl_4O_4Sn$.
Hazard: A poison.

dibutylthiourea.
$C_4H_9NHCSNHC_4H_9$.

Properties: White to light-tan powder. Mp 59–69C. Slightly soluble in water; soluble in methanol, ether, acetone, benzene, and ethyl acetate; insoluble in gasoline.
Use: Corrosion inhibitor for pickling cast iron or carbon steel, reducing corrosion of ferrous metals and aluminum alloys in brine, intermediate.

dibutyl-tin bis(isooctylthioglycollate). See bis(isooctyloxycarbonylmethylthio)dibutyl stannane.

dibutyltinbis(lauryl) mercaptide.
$(C_4H_9)_2Sn(SC_{12}H_{25})_2$.
Properties: Yellow liquid, tin content 18.5%, soluble in toluene and heptane.
Hazard: Toxic by skin absorption.
Use: Antioxidant and metal cleaning (or protective) agent.

di-*n*-butyltin bismethanesulfonate.
CAS: 73927-86-3. $C_{10}H_{24}O_6S_2Sn$.
Hazard: A poison.

dibutyltin bis(methyl maleate). See bis(methoxymaleoyloxy)dibutylstannane.

dibutyltin bis(trifluoroacetate). See bis(trifluoroacetoxy)dibutyltin.

dibutyltin diacetate.
CAS: 1067-33-0. $(C_4H_9)_2Sn(C_2H_3O_2)_2$.
Properties: Clear, yellow liquid. Bp 130C (2 mm Hg), fp <12C, flash p 290F (143C). Soluble in water and most organic solvents. Combustible.
Derivation: Reaction of acetic acid with dibutyltin oxide.
Hazard: Toxic by skin absorption.
Use: Stabilizer for chlorinated organics, catalyst for condensation reactions.

dibutyltin dibromide. See dibromodibutyl-stannane.

dibutyltin dicaprylate. See bis(octanoyloxy)di-*n*-butyl stannane.

dibutyltin dichloride.
CAS: 683-18-1. $(C_4H_9)_2SnCl_2$.
Properties: White, crystalline solid. Mp 43C, bp 135C (10 mm Hg), d 1.36 (24/4C), refr index 1.4991 (51C), flash p 335F (168C). Insoluble in cold water; hydrolyzed by hot water; soluble in many organic solvents. Combustible.
Derivation: Reaction of butylmagnesium chloride with tin tetrachloride.
Hazard: Toxic by skin absorption.
Use: Organotin intermediate.

dibutyltin di-2-ethylhexoate.
CAS: 2781-10-4. $(C_4H_9)_2Sn(OCOC_{10}H_{20}CH_3)_2$.

Properties: Waxy, white solid. Mp 52.54C. Insoluble in water; soluble in most organic solvents.
Derivation: Reaction of dibutyltin oxide with 2-ethyl-hexoic acid.
Hazard: Toxic by skin absorption.
Use: Catalyst for silicone curing, polyether foams.

dibutyltin difluoride.   See dibutyldifluorostannane.

di-*n*-butyltin diformate.   See dibutyl (diformyloxy)stannane.

di-*n*-butyl tin di(hexadecylmaleate).
CAS: 19706-58-2. $C_{48}H_{88}O_8Sn$.
Hazard: A poison by ingestion. An eye irritant.

dibutyltin dihydride.
CAS: 1002-53-5. $C_8H_{20}Sn$.
Hazard: Moderately toxic by ingestion. Questionable carcinogen.

dibutyltin diiodide.   See dibutyldiiodostannane.

dibutyltin dilaurate.
CAS: 77-58-7. $(C_4H_9)_2Sn(OCOC_{10}H_{20}CH_3)_2$.
Properties: Clear, pale-yellow liquid. D 1.066, mp 23C, fp 8.0C, wt/gal 8.84 lb (20C), flash p 440F (226C). Soluble in acetone and benzene; insoluble in water. Combustible.
Hazard: Toxic by skin absorption.
Use: Stabilizer for vinyl resins, lacquers, and elastomers; catalyst for urethane and silicones.

di-*n*-butyltin di(monon-onyl)maleate.
CAS: 69239-37-8. $C_{42}H_{76}O_8Sn$.
Hazard: A poison by ingestion.

di-*n*-butyltin dipentanoate.   See dibutyldipentanoyloxystannane.

di-*n*-butyltin dipropionate.   See dibutyldipropionyloxystannane.

di-*n*-butyltin di(tetradecanoate).   See bis (tetradecanoyloxy)dibutylstannane.

dibutyltin maleate.
CAS: 15535-69-0. $[(C_4H_9)_2Sn(OOCCH)_2]_x$.
Properties: White, amorphous powder. Mp 110C, flash p 400F (204C). Insoluble in water; soluble in benzene and organic esters. Combustible.
Derivation: Reaction of maleic acid with di-butyltin oxide.
Hazard: Toxic by skin absorption.
Use: Stabilizer for polyvinyl chloride resins, condensation catalyst.

dibutyltin mercaptopropionate.
CAS: 78-06-8. $C_{11}H_{22}O_2SSn$.
Hazard: A poison.

dibutyltin-*o,s*-mercaptopropionate.   See dibutyltin mercaptopropionate.

dibutyltin methyl maleate.   See bis (methoxymaleoyloxy)dibutylstannane.

dibutyltin octanoate.   See bis(octanoyloxy)di-*n*-butyl stannane.

dibutyltin oxide.
CAS: 818-08-6. $[(C_4H_9)_2SnO^-]_x$.
Properties: White powder. Mp (decomposes). Insoluble in water. Combustible.
Derivation: Hydrolysis of dibutyltin dichloride with caustic.
Hazard: Toxic by skin absorption.
Use: Condensation catalyst, intermediate for other organotins.

dibutyltin sulfide.
CAS: 4253-22-9. $[(C_4H_9)_2SnS]_3$.
Properties: Colorless, oily liquid. Combustible.
Derivation: Reaction of dibutyltin oxide with hydrogen sulfide.
Hazard: Toxic by skin absorption.
Use: Vinyl stabilizer, antioxidant, lubricating additive.

dibutyltin tetrachlorophthalate.   See dibutyl (tetrachlorophthalato)stannane.

2,6-di-*t*-butyl-*p*-tolyl methylcarbamate.
See terbutol.

2,6-di-*tert*-butyl-*p*-tolyl-*N*-methylcarbamate.
$[(CH_3)_3C]_2CH_3C_6H_2OOCNHCH_3$.
White solid, mp 200C, insoluble in water, soluble in alcohol.
Use: Herbicide.

1,1-dibutylurea.   (*N,N*-dibutylurea).
$NH_2CON(C_4H_9)_2$.
Properties: Liquid. Mp 22–25C, boiling range 118–119C (2–3 mm Hg), flash p 279F (137C). Soluble in alcohol and ether. Combustible.
Use: Thermoplastic resins by copolymerization with urea with formaldehyde catalyst.

dibutylxanthogen disulfide.   See C-P-B.

DIC.   Abbreviation for β-diisopropylaminoethyl chloride hydrochloride.

dicalcium magnesium aconitate.   (calcium magnesium aconitate).
$[C_3H_3(COO)_3]_2Ca_2Mg$.
Properties: White, crystalline powder or lumps.
Derivation: Precipitation from molasses with lime.
Use: Conversion to aconitic acid, tributyl aconitate, and similar ester plasticizers.

dicalcium orthophosphate. See calcium phosphate, dibasic.

dicalcium orthophosphite. See calcium phosphite.

dicalcium phosphate dihydrate. See calcium phosphate, dibasic.

dicalcium silicate.
$2CaO \cdot SiO_2$.
One of the components of cement. Noncombustible.
Occurrence: Obtained as a by-product in electric furnace operation.
Use: Used to neutralize acid soils.
See cement, Portland.

"Dicalite" [Raymond G]. TM for a group of products made from either diatomite or perlite, used in filters and filter aids.
See diatomaceous earth.

dicamba. Generic name for 3-6-dichloro-o-anisic acid (2-methoxy-3,6-dichlorobenzoic acid).
$HOOC(Cl)C_6H_2Cl(OCH_3)$.
Properties: Crystals. Mp 114–116C. Slightly soluble in water; moderately soluble in xylene; very soluble in alcohol.
Use: Herbicide, pest control.

dicapryl adipate.
$C_8H_{17}OOC(CH_2)_4COOC_8H_{17}$.
Properties: Almost water-white liquid. Bp 213–216.5C (4 mm Hg), flash p 352F (177C). Combustible.
Use: Plasticizer for vinyl resins and cellulose esters.

dicapryl phthalate. (DCP, di-(2-octyl) phthalate).
$(C_8H_{17}OOC)_2C_6H_4$.
Properties: Colorless, viscous liquid. Bp 227–234C (4.5 mm Hg), fp −60C, refr index 1.480 (20C), d 0.965 (25C), flash p 395F (201C). Insoluble in water; compatible with vinyl chloride resins and some cellulosic resins. Combustible.
Use: Monomeric plasticizer for vinyl and cellulosic resins.

dicapryl sebacate.
$C_8H_{17}OOC(CH_2)_8COOC_8H_{17}$.
Properties: Light straw-colored liquid. Bp 231.5–239C (4 mm Hg). Nonvolatile, gives excellent low-temperature flexibility, flash p 445F (229C). Combustible.
Use: Plasticizer for vinyl resins and acrylonitrile rubber.

dicapthon. (generic name for O-(2-chloro-4-nitrophenyl) O,O-dimethyl phosphorothioate).
$(CH_3O)_2P(:S)OC_6H_3(Cl)NO_2$.

Properties: White solid. Mp 51–52C. Insoluble in water; soluble in acetone, cyclohexanone, ethylacetate, toluene, and xylene.
Hazard: Cholinesterase inhibitor. Use may be restricted.
Use: Insecticide.

1,2-dicarbadodecaborane(12).
CAS: 16872-09-6. $C_2H_{12}B_{10}$.
Hazard: Moderately toxic by ingestion and inhalation.

dicarboxylate. See sodium (r,r)-5-(2-((2-(3-chlorophenyl)-2.

dicarboxylic acid. A carboxylic acid containing two -COOH groups, e.g., adipic, oxalic, phthalic, sebacic, and maleic acids.

dicatechol borate 1,3-di(o-tolyl)guanidine salt.
CAS: 16971-82-7. $C_{15}H_{17}N_3 \cdot Cl_2H_8BO_4 \cdot H$.
Hazard: Moderately toxic by ingestion.

di-(C9-C11 alkyl) phthalate.
CAS: 68515-49-1.
Hazard: Experimental reproductive effects.

dicetyl. See dotriacontane.

dicetyl ether. (dihexadecyl ether).
Properties: Crystals. Mp 54C, bp decomposes at 300C, d 0.8117 (54/4C).
Use: Electrical insulators, water repellents, lubricants in plastic molding and processing, antistatic substances, chemical intermediate.

dicetyl sulfide. (dihexadecyl thioether; dihexadecyl sulfide).
$(C_{16}H_{33})_2S$.
Properties: Solid. Mp 57–58C, bp (decomposes), d 0.8253 (60/4C).
Use: Organic synthesis (formation of sulfonium compounds).

dichlobenil. (2,6-dichlorobenzonitrile).
CAS: 1194-65-6. $C_7H_3Cl_2N$.
An amide solution applied to the soil that disrupts many aspects of growth and differentiation including blocking of cell division, which results in a blackening and death of the growing points in many plant species.
Hazard: Slightly toxic.
Use: A selective pre- and postemergence, amide herbicide.

dichlobenil. (generic name for 2,6-dichlorobenzonitrile).
$Cl(Cl)C_6H_3CN$.
Properties: White solid. Mp 144C. Almost insoluble in water; soluble in organic solvents.

Hazard: Toxic by ingestion and inhalation.
Use: Herbicide.

dichlofenthion. (*O,O*-diethyl-*O*-2,4-dichlorophenyl phosphorothioate).
CAS: 97-17-6. $C_{10}H_{13}Cl_2O_3PS$.
Properties: Colorless liquid. Bp 165C (0.3 mm Hg), d 1.30, refr index 1.52. Slightly soluble in water; soluble in organic solvents.
Hazard: Cholinesterase inhibitor, use may be restricted.
Use: Pesticide.

dichlone. (generic name for 2,3-dichloro-1,4-naphthoquinone).
CAS: 117-80-6. $C_{10}H_4Cl_2O_2$.
Properties: Yellow needles. Mp 193C. Soluble in xylene and *o*-dichlorobenzene; slightly soluble in ethyl alcohol, glacial acetic acid, and carbon tetrachloride; almost insoluble in water.
Hazard: Toxic by ingestion and inhalation; irritant to skin and eyes.
Use: Seed disinfectant, fungicide for foilage and textiles, insecticide, organic catalyst.

dichloramine-T. (*p*-toluenesulfondichloroamide).
$CH_3C_6H_4SO_2NCl_2$.
Properties: Pale-yellow crystals containing neither <28% nor >30% active chlorine. Chlorine odor; stable when pure, decomposed slowly by air, rapidly by impurities, petrolatums, kerosene, olive oil, and alcohol. Soluble in glacial acetic acid, chlorinated paraffin hydrocarbons, eucalyptol, benzene, chloroform, and carbon tetrachloride; almost insoluble in water; mp 80C.
Derivation: Product of the reaction between toluene-*p*-sulfonamide and calcium hypochlorite solution is acidified with acetic acid and extracted with chloroform. The chloroform solution is dried chemically, filtered, and evaporated.
Use: Germicide, antibacterial.
See chloramine-T.

dichlorimipramine.
CAS: 3589-22-8. $C_{19}H_{22}Cl_2N_2$.
Hazard: A poison.

dichloroacetaldehyde.
$CHCl_2CHO$.
Properties: Colorless liquid; penetrating pungent odor. D 1.436 (25C), wt/gal 12.1 lb, flash p 140F (60C) (CC). Combustible.
Hazard: Moderate fire hazard. Toxic by ingestion and inhalation; strong skin irritant.
Use: Manufacture of insecticides.

dichloroacetic acid.
CAS: 79-43-6. $CHCl_2COOH$.
Properties: Colorless liquid. D 1.5724 (13C), fp −4C, bp 193–194C. Soluble in water, alcohol, and ether.

Derivation: Chlorination of acetic acid in presence of iodine.
Hazard: Toxic by ingestion and inhalation; strong eye, skin, and upper respiratory tract irritant. Testicular damage. Possible carcinogen.
Use: Intermediate, pharmaceuticals, medicine.

α,α-dichloroacetophenone.
$C_6H_5COCHCl_2$.
Properties: Crystals. D 1.34 (15C), bp 247C (decomposes) mp 20–21.5C.
Hazard: Toxic by ingestion and inhalation; strong irritant to eyes and skin.

dichloroacetyl chloride.
$Cl_2CHCOCl$.
Properties: Fuming liquid; acrid odor. D 1.5315 (16/4C), bp 107–108C, refr index 1.4638 (16C); soluble in ether; decomposes in water and alcohol.
Hazard: Strong irritant to skin and eyes.
Use: Intermediate.

dichloroacetylene.
CAS: 7572-29-4. $ClC{\equiv}CCl$.
Properties: A liquid. Mw 94.93, mp −66C, bp (explodes). Soluble in alcohol and acetone.
Occurrence: Thermal decomposition of trichloroethylene; closed-circuit anesthesia using trichloroethylene and soda lime absorption of carbon dioxide. May be synthesized by passing trichloroethylene over alkaline materials at 70F.
Hazard: Questionable carcinogen.

2,3-dichloroallyl diisopropylthiolcarbamate. (diallate).
CAS: 2303-16-4. $[(CH_3)_2CH]_2NCOSCH_2C(Cl){:}CHCl$.
Properties: Brown liquid. Bp 159C (9 mm Hg). Almost insoluble in water; soluble in acetone, benzene, chloroform, kerosene, and xylene.
Hazard: Toxic. Absorbed by skin. Use may be restricted. Nausea, peripheral nervous system impairment, and questionable carcinogen.
Use: Herbicide.

2,4-dichloro-6-aminophenol. (2-amino-4,6-dichlorophenol).
$C_6H_5Cl_2NO$.
Properties: Acicular crystals. Mp 95C. Soluble in benzene; less so in carbon disulfide. The hydrochloride is soluble in water and alcohol.
Use: Azo-dye intermediate.

2,5-dichloroaniline.
$C_6H_3NH_2Cl_2$.
Properties: Light-brown or amber-colored crystalline mass. Mp 47–50C, bp 251–252C. Slightly soluble in water; soluble in alcohol, benzene, and dilute hydrochloric acid. Combustible.
Derivation: Nitration of *p*-dichlorobenzene with subsequent reduction.

Hazard: See aniline.
Use: Dye intermediate.

3,4-dichloroaniline.

CAS: 95-76-1. $C_6H_3NH_2Cl_2$.
Properties: Crystals. Mp 68–72C, bp 272C, flash p 331F (166C). Slightly soluble in water; soluble in most organic solvents. Combustible.
Derivation: Nitration of o-dichlorobenzene with subsequent reduction.
Hazard: See aniline.
Use: Intermediate for manufacture of dyes and pesticides.

dichlorobenzaldehyde.

$C_6H_3CHOCl_2$.
Includes 2,4-, 2,5-, and 3,4-isomers.
Properties: White, crystalline solid. Bp 233C, mp 65–67C (2,4-), 35C (3,4-). 2,4-Isomer soluble in methanol, ethanol, ether, and acetone; insoluble in water. 2,5-Isomer soluble in ethanol and ether. 3,4-Isomer soluble in ethanol, ether, and acetone; slightly soluble in methanol and amyl ether; insoluble in water.
Use: Intermediate in the manufacture of pharmaceuticals, dyes, and other organic chemicals.

dichlorobenzalkonium chloride.

$C_{21}H_{36}Cl_3N$.
Properties: Colorless crystals; very bitter taste. Soluble in alcohol and water.
Derivation: Reaction of N-dimethyldodecylamine with 3,4-dichlorobenzyl chloride.
Hazard: Skin irritant.
Use: Algicide, antiseptic, sterilant.

m-dichlorobenzene.   (1,3-dichlorobenzene).

CAS: 541-73-1. $C_6H_4Cl_2$.
Properties: Colorless liquid. D 1.288 (20/4C), bp 172C, fp −24C, refr index 1.547 (20.9C). Soluble in alcohol and ether; insoluble in water. Combustible.
Derivation: Chlorination of monochlorobenzene.
Hazard: Questionable carcinogen.
Use: Fumigant and insecticide.

o-dichlorobenzene.   (1,2-dichlorobenzene).

CAS: 95-50-1. $C_6H_4Cl_2$.
Properties: Colorless liquid; pleasant odor. A mixture of isomers containing at least 85% o- and varying percentages of p- and m-. Autoign temp 1198F (647C), d 1.284, wt/gal 10.7 lb, bp 172–179C, fp −17C, flash p 150F (65.5C). Miscible with most organic solvents; insoluble in water. Combustible.
Derivation: Chlorination of monochlorobenzene.
Method of purification: Rectification.
Grade: Purified, technical.
Hazard: Toxic by inhalation and ingestion. Liver damage; eye and upper respiratory tract irritant. Questionable carcinogen.
Use: Manufacture of 3,4-dichloroaniline; solvent for a wide range of organic materials and for oxides of nonferrous metals; solvent carrier in production of toluene diisocyanate; dye manufacture; fumigant and insecticide; degreasing hides and wool; metal polishes; industrial odor control; heat transfer.

p-dichlorobenzene.   (1,4-dichlorobenzene; PDB).

CAS: 106-46-7. $C_6H_4Cl_2$.
Properties: White crystals; volatile (sublimes readily); penetrating odor. D 1.458, bp 173.7C, mp 53C, flash p 150F (65.5C) (CC). Soluble in alcohol, benzene, and ether; insoluble in water. Combustible.
Derivation: Chlorination of monochlorobenzene.
Grade: Technical.
Hazard: Toxic by ingestion, irritant to eyes, kidney damage. Possible carcinogen.
Use: Moth repellent, general insecticide, germicide, space odorant, manufacture of 2,5-dichloroaniline, dyes, intermediates, pharmacy, agriculture (fumigating soil).

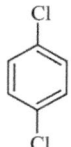

See Santochlor.

N,N-dichlorobenzenesulfonamide.

$C_6H_5SO_2NCl_2$.
Properties: Solid, fine, white crystals. Mp 68–71C.
Use: A source of positive chlorine.

3,3′-dichlorobenzidine.

CAS: 91-94-1. $C_6H_3ClNH_2C_6H_3ClNH_2$.
Properties: Gray to purple, crystalline solid. Mp 165C. Insoluble in water; soluble in alcohol and ether.
Hazard: A tumorigen and carcinogen; absorbed by skin. Possible carcinogen.
Use: Intermediate for dyes and pigments, curing agent for isocyanate-terminated resins for urethane plastics.

2,4-dichlorobenzoic acid.

$Cl_2C_6H_3COOH$.
Properties: White to slightly yellowish powder. Mp 158–162C. Soluble in alcohol, ether, acetone, 5% caustic; insoluble in water and heptane.
Use: Intermediate for antimalarials, dyes, fungicides, pharmaceuticals, and other organic chemicals.

3,4-dichlorobenzoic acid.

$Cl_2C_6H_3COOH$.
Properties: White to slightly yellowish powder. Mp 202–204C. Soluble in alkali, alcohol, ether, and acetone; slightly soluble in diacetone; insoluble in water, ethylene dichloride, and toluene.
Use: Intermediate for pharmaceuticals, dyes, etc.

2,6-dichlorobenzonitrile.   See dichlobenil.

3,4-dichlorobenzotrichloride.
$Cl_2C_6H_3CCl_3$.
Properties: Water-white liquid. Boiling range 276–285C, fp approximately 24.0C, d 1.585–1.590 (25/15C). Soluble in alcohol, ether, and acetone; insoluble in water. Combustible.
Use: Intermediate for pharmaceuticals, dyes, etc.

dichlorobenzoyl chloride.
$Cl_2C_6H_3COCl$.
Properties: Colorless liquid. Boiling range 250–260C, fp 15–16C, d 1.500–1.510 (25/15C). Soluble in alcohol, ether, and acetone; slightly soluble in heptane; insoluble in water. Combustible.
Use: Intermediate in the manufacture of pharmaceuticals, dyes, and other organic chemicals.

3,5-dichlorobenzoyl chloride.
CAS: 2905-62-6. $C_7H_3Cl_3O$.
Hazard: Moderately toxic by ingestion. A moderate skin and mild eye irritant.

dichlorobenzyl chloride.
$Cl_2C_6H_3CH_2Cl$.
Properties: Colorless liquid. Boiling range 245–252C; d 1.415–1.420 (25/15C). Soluble in alcohol, ether, and acetone; insoluble in water. Combustible.
Use: Intermediate for organic chemicals, pharmaceuticals and dyes, insecticide.

1,1-dichloro-2,2-bis(p-chlorophenyl)ethane.
See TDE.

3,3'-dichlorobiphenyl.
CAS: 2050-67-1. $C_{12}H_8Cl_2$.
Hazard: A poison by ingestion.
Use: Agricultural chemical.

1,1-dichloro-2,2-bis(p-ethylphenyl)ethane.
(diethyldiphenyldichloroethane; Perthane).
$C_{18}H_{20}Cl_2$.
Properties: Crystals. Mp 56C. Insoluble in water; soluble in acetone, kerosene, and diesel fuel.
Hazard: Toxic. Absorbed by skin.
Use: Insecticide, formulated as emulsifiable concentrate or wettable powder. Used against insects, especially moths.

1,4-dichlorobutane.   (tetramethylene dichloride; DCB).
$ClCH_2(CH_2)_2CH_2Cl$.
Properties: Colorless, mobile liquid; pleasant odor. D 1.141 (20/4C), bp 155C, flash p 104F (40C) (OC), refr index 1.4542 (20C). Insoluble in water; soluble in most common organic solvents. Combustible.
Hazard: Moderate fire risk.
Use: Organic synthesis, including adiponitrile.

1,3-dichlorobutene-2.
$ClH_2CCH:CClCH_3$.

Properties: Clear to straw-colored liquid. Bp 125–130C, flash p 80F (26.6C) (COC). Insoluble in water; soluble in organic solvents.
Hazard: Flammable, moderate fire risk. Toxic by inhalation.

1,4-dichlorobutene-2.
CAS: 764-41-0. $ClH_2CCH:CHCH_2Cl$.
Properties: Colorless liquid; distinct odor. Bp 158C (60C) (20 mm Hg), mp 3.5C, d 1.1858 (25/4C), refr index 1.4863 (25C). Miscible with benzene, alcohol, and carbon tetrachloride; immiscible with ethylene glycol, glycerol, and water. Combustible.
Grade: Available as 95–98% *trans* isomer, 2–5% *cis* isomer. Above constants are for the pure *trans*-isomer.
Hazard: Irritant to skin and eyes, causes blisters. Moderate fire hazard. Upper respiratory tract irritant. Possible carcinogen.
Use: Intermediate.

dichlorocarbene.
CCl_2.
Properties: Exists only at low temperature and pressure, fp −114C, bp −20C, decomposes on distillation at normal pressure to hexachloroethane and hexachlorobenzene.
Derivation: Reaction of carbon tetrachloride vapor with carbon at 1300C and 10^{-3} mm Hg.
Hazard: Explosive reaction with carbon, forms phosgene on reaction with oxygen.
Use: Research.

1,1'-dichlorocarbonylferrocene.   (ferrocenoyl dichloride).
$[C_5H_4(COCl)]_2Fe$.
Properties: Red, crystalline solid. Mp 93–95C.
Use: Intermediate.

1,1-dichloro-2-(o-chlorophenyl)-2-(p-chlorophenyl)ethylene.
CAS: 3424-82-6. $C_{14}H_8Cl_4$.
Hazard: Moderately toxic by ingestion.
Use: Agricultural chemical.

4,6-dichloro-N-(2-chlorophenyl)-1,3,5-triazin-2-amine.
$ClC_6H_4NHC_3N_3Cl_2$.
Properties: Tan, crystalline solid. Mp 159–160C. Insoluble in water.
Hazard: See aniline.
Use: Foliage fungicide.

dichloro(2-chlorovinyl)arsine.   See β-chlorovinyldichloroarsine.

3,3'-dichloro-4,4'-diaminodiphenylmethane.
See 4,4'-methylenebis(2-chloroaniline).

dichlorodibenzofuran.
CAS: 43047-99-0. $C_{12}H_6Cl_2O$.

Hazard: A poison.
Use: Agricultural chemical.

1,3-dichloro-2-(3,4-dichlorophenoxy)benzene.
CAS: 130892-66-9. $C_{12}H_6Cl_4O$.
Hazard: A reproductive hazard.

2,3-dichloro-5,6-dicyanobenzoquinone.
(DDQ).

OCC(Cl):C(Cl)COC(CN):C(CN).

Properties: Bright, yellow-orange solid. Mp 213–215C.
Use: Highly selective oxidizing agent for organic compounds.

2,2′-dichlorodiethyl ether. See *sym-*dichloroethyl ether.

dichlorodiethylformal. See dichloroethylformal.

dichlorodiethylstannane.
CAS: 866-55-7. $C_4H_{10}Cl_2Sn$.
Properties: Water-white crystals. Mp: 85C, bp: 277C.
Hazard: A poison by ingestion.

dichlorodiethyl sulfide. (mustard gas; dichloroethyl sulfide).
CAS: 505-60-2. $S(CH_2CH_2Cl)_2$.
Properties: Yellow liquid. Bp 228C, fp 14C, d 1.27, flash p 220F (104C).
Derivation: Bubbling ethylene through sulfur chloride, also from thiodiglycol and hydrogen chloride.
Grade: Pure, technical (containing excess sulfur as a polysulfide).
Hazard: Vesicant war gas, causes conjunctivitis and blindness. Can be decontaminated by chloroamines or bleaching powder. Vapor is extremely poisonous and can be absorbed via skin. Confirmed carcinogen.
Use: Organic synthesis, poison gas, medicine.

dichlorodiethyl sulfone.
$(ClCH_2CH_2)_2SO_2$.
Properties: Colorless crystals. Bp 179–181C (14–15 mm Hg), mp 52C. Soluble in alcohol, chloroform, and ether; slightly soluble in water.
Hazard: Strong irritant to eyes and skin.

dichlorodiethyltin. See dichlorodiethylstannane.

2,2-dichloro-1,1-difluoroethyl methyl ether.
(methoxyflurane).
CAS: 76-38-0. $HCCl_2CF_2OCH_3$.
Properties: Clear, colorless liquid; fruity odor. Bp 104.65C, fp −35, d 1.4223 (25C). Completely stable

in the presence of alkali, air, light, or moisture. Slightly soluble in water. Combustible.
Grade: ND.
Use: Anesthetic.

dichlorodifluoromethane. (difluorodichloromethane; fluorocarbon-12).
CAS: 75-71-8. CCl_2F_2.
Properties: Colorless, odorless, noncorrosive gas. Bp −29.8C, fp −158C, critical pressure 43.2 atm. Insoluble in water; soluble in most organic solvents. Nonflammable.
Derivation: (1) Reaction of carbon tetrachloride and anhydrous hydrogen fluoride, in the presence of an antimony halide catalyst; (2) high-temperature chlorination of vinylidene fluoride (vinylidene fluorides made by addition of hydrogen fluoride to acetylene).
Grade: 99.9% min purity.
Hazard: Narcotic in high concentration. Cardiac sensitivity. Questionable carcinogen.
Use: Refrigerant in air conditioner; plastics, blowing agent, low-temperature solvent, leak-detecting agent, freezing of foods by direct contact, chilling cocktail glasses.
See chlorofluorocarbon.

dichlorodihexylstannane.
CAS: 2767-41-1. $C_{12}H_{26}Cl_2Sn$.
Hazard: A poison by ingestion.

6,8-dichloro-dihydro-1,3-benzoxazine-2-thione-4-one.
CAS: 23611-67-8. $C_8H_3Cl_2NO_2S$.
Hazard: A poison.
Use: Agricultural chemical.

dichlorodiisononyl stannane. See bis(trimethylhexyl)tin dichloride.

dichlorodiisopropylstannane. See diisopropyltin dichloride.

1,3-dichloro-5,5-dimethylhydantoin.
(DDH).
CAS: 118-52-5.

ClNCONClOC(CH_3)$_2$.

Properties: White powder with mild chlorine odor. Mp approximately 130C, sublimes approximately 100C without decomposition. Contains approximately 36% active chlorine. Slightly soluble in water with gradual liberation of hypochlorous acid; soluble in benzene, chloroform, ethylene dichloride, and alcohol. Combustible with evolution of chlorine at 210C.
Derivation: Chlorination of dimethylhydantoin.
Grade: Technical.
Hazard: Toxic by inhalation, skin and upper respiratory tract irritant.

Use: Household laundry bleach, water treatment, mild chlorinating agent, pharmaceutical intermediate, catalyst.

2,3-dichloro-*N,N*-dimethyl-6-quinoxaline sulfonamide.
CAS: 2347-47-9. $C_{10}H_9Cl_2N_3O_2S$.
Hazard: Moderately toxic by ingestion.
Use: Agricultural chemical.

dichlorodimethylsilane.   See dimethyldichlorosilane.

2,5-dichloro-4,6-dinitrophenyl crotonate.
CAS: 24291-70-1. $C_{10}H_6Cl_2N_2O_6$.
Hazard: Low toxicity by ingestion.
Use: Agricultural chemical.

3,6-dichloro-2,4-dinitrophenyl methacrylate.
CAS: 24291-69-8. $C_{10}H_6Cl_2N_2O_6$.
Hazard: Moderately toxic by ingestion.
Use: Agricultural chemical.

dichlorodioctylstannane.   See di-*n*-octyltindichloride.

dichlorodipentylstannane.   See   dipentyltin dichloride.

dichlorodiphenyldichloroethane.   See TDE.

dichlorodiphenyldichloroethylene.   See DDE.

dichlorodiphenylstannane.   See   diphenyltin dichloride.

dichlorodiphenyltrichloroethane.   See DDT.

dichlorodipropylstannane.
CAS: 867-36-7. $C_6H_{14}Cl_2Sn$.
Properties: Colorless crystals. Mp: 82.5–83C, bp: 118–121C at 10 mm.
Hazard: A poison by ingestion.

dichlorodipropyltin.   See   dichlorodipropylstannane.

dichlorodivinylstannane.   See divinyltin dichloride.

1,1-dichloroethane.   See ethylidene chloride.

1,2-dichloroethane.   See ethylene dichloride.

3-(2,2-dichloroethenyl)-2,2-dimethylcyclopropanecarboxylic acid, cyano(3-phenoxyphenyl) methyl ester, (1-α(S*),3-β)-(+−)-.
CAS: 71697-59-1. $C_{22}H_{19}Cl_2NO_3$.
Hazard: A poison.
Use: Agricultural chemical.

dichloroether.   See *sym*-dichloroethyl ether.

1,2-dichloro-1-ethoxyethane.
CAS: 623-46-1. $C_4H_8Cl_2O$.
Hazard: Human systemic effects.

dichloroethoxymethane.   See   dichloroethylformal.

1,2-dichloroethyl acetate.
$CH_3COOCHClCH_2Cl$.
Properties: Water-white liquid. D 1.296 (20C), boiling range, 58–65C (13 mm Hg), fp <−32C, refr index 1.444 (20C), bp (decomposes), flash p 307F (152C). Combustible. Miscible with alcohol and ethyl ether; immiscible with water.
Hazard: Toxic by inhalation.
Use: Organic synthesis.

***p*-di(2-chloroethyl)aminophenylalamine.**
See melphalan.

dichloroethylarsine.   See ethyldichloroarsine.

dichloroethyl carbonate.
$(ClH_2CCH_2O)_2CO$.
Properties: Colorless liquid. Slowly hydrolyzed by alkalies; volatile in steam. D 1.3506 (20C), bp 240C (partial decomposition). Insoluble in water.
Derivation: By heating ethylene chlorohydrin and trichloromethylchloroformate together (under reflux).

***sym*-dichloroethylene.**   (1,2-dichloroethylene; acetylene dichloride; dichloracetylene).
CAS: 540-59-0.   ClHC:CHCl.
Exists as *cis* and *trans* isomers.
Properties: Colorless, low-boiling liquid; pleasant odor. Decomposes slowly on exposure to air, light, and moisture. Soluble in most organic solvents; slightly soluble in water. *trans* isomer: d 1.257, bp 47–49C. *cis* isomer: d 1.282, bp 58–60C, flash p 39F (3.9C), fp −80C.
Derivation: Two stereoisomeric compounds made by the partial chlorination of acetylene.
Grade: Technical, as *cis*, *trans*, and mixture of both.
Hazard: Flammable, dangerous fire hazard. Toxic by ingestion, inhalation, and skin contact; eye irritant and narcotic in high concentration. Central nervous system impairment.
Use: General solvent for organic materials, dye extraction, perfumes, lacquers, thermoplastics, and organic synthesis.

***sym*-dichloroethyl   ether.**   (dichloroether; dichloroethyl oxide; bis(2-chloroethyl) ether; 2,2′-dichlorodiethyl ether).
CAS: 111-44-4. $ClCH_2CH_2OCH_2CH_2Cl$.
Properties: Colorless liquid; odor like that of ethylene dichloride. Bp 178.5C, d 1.2220 (20/20C), wt/gal 10.2 lb (20C), refr index 1.457 (20C), flash p (CC) 131F (55C), fp −51.8C, autoign temp 696F

(368C). Miscible with most organic solvents; insoluble in water. Combustible.

Derivation: Chlorination of ethyl ether.

Grade: Technical.

Hazard: Moderate fire hazard. Toxic by inhalation, ingestion, and skin absorption; strong eye and upper respiratory tract irritant. Nausea. Questionable carcinogen.

Use: General solvent, selective solvent for production of high-grade lubricating oils, textile scouring and cleansing, fulling compounds, wetting and penetrating compounds, paints, varnishes, lacquers, finish removers, spotting and dry cleaning, soil fumigant, and intermediate and cross-linker in organic synthesis.

dichloroethylformal. (dichlorodiethylformal).
$CH_2(OCH_2CH_2Cl)_2$.

Properties: Colorless liquid. Bp 218.1C, fp −32.8C, d 1.2339 (20/20C), wt/gal 10.3 lb (20C), flash p 230F (110C) (OC). Slightly soluble in water; decomposed by mineral acids. Combustible.

Hazard: Toxic by inhalation and ingestion, strong irritant.

Use: Solvent, intermediate for polysulfide rubber.

1,1-N-(2,2-dichloroethylidene)bis[4-chlorobenzene]. (1,1-dichloro-2,2-bis(p-chlorophenyl)et2,3-dimerocapto-1-propanesulfonic acid; 2,3-dithiolpropanesulfonic acid).

A chelating agent.

Hazard: Toxic.

Use: Antidote for heavy metal poisoning.

dichloroethyl oxide. See *sym*-dichloroethyl ether.

di-2-chloroethyl phthalate. See bis(2-chloroethyl) phthalate.

dichloroethyl sulfide. See dichlorodiethyl sulfide.

2,4-dichloro-6-fluorophenyl *p*-nitrophenyl ether. See fluoronitrofen.

dichloroformoxime.
CCl_2NOH.

Properties: Colorless, prismatic crystals; disagreeable, penetrating odor. High vap press; slowly decomposes at normal temperature, the rate depending on temperature and humidity. Bp 53–54C (18 mm Hg), mp 39–40C. Soluble in water, alcohol, ether, and benzene.

Derivation: (1) Action of chlorine on fulminic acid, HONC, (2) reduction of trichloronitrosomethane with either aluminum amalgam or hydrogen sulfide.

Hazard: Strong irritant to skin and eyes.

Use: A military poison.

1,3-dichloro-1,1,2,2,3,3-hexafluoropropane.
CAS: 662-01-1. $C_3Cl_2F_6$.

Hazard: Low toxicity by inhalation.

α-dichlorohydrin. (α-propenyldichlorohydrin; 1,3-dichloro-2-propanol; *sym*-dichloroisopropyl alcohol).
CAS: 96-23-1. $CH_2ClCH_2OCH_2Cl$.

Properties: Colorless, slightly viscous, unstable liquid; faint chloroform-like odor. D 1.36-1.39, fp −4C, bp 174C, refr index 1.47–1.48, flash p 165F (73.9C), vap press 7 mm Hg. The commercial product is a mixture of two isomers. Miscible with most organic solvents, vegetable oils; slightly soluble in water. Combustible.

Derivation: Interaction of glycerol and dry hydrogen chloride gas and subsequent distillation.

Hazard: Moderate fire risk. Toxic by inhalation and ingestion. Possible carcinogen.

Use: General solvent, intermediate in organic synthesis, paints, varnishes, lacquers, watercolor binder, and photographic lacquers.

5,7-dichloro-8-hydroxyquinaldine. See chlorquinaldol.

1,3-dichloroisobutylene.
CAS: 3375-22-2. $C_4H_6Cl_2$.

Hazard: Low toxicity by inhalation.

1,3-dichloro-5-isocyanobenzene.
CAS: 60357-67-7. $C_7H_3Cl_2N$.

Hazard: Moderately toxic by ingestion.

Use: Agricultural chemical.

dichloroisocyanuric acid. (dichloro-*s*-triazine-2,4,6-trione).

OCNClCONClCONH.

Properties: White, slightly hygroscopic, crystalline powder, granules. D (loose bulk, approximate) powder 34 lb/cu ft, granular 53 lb/cu ft. Active ingredient approximately 70% available chlorine, decomposes 225C.

Hazard: Oxidizer, may ignite organic materials on contact. Irritant to eyes.

Use: Household dry bleaches, dishwashing compounds, scouring powders and detergent sanitizers, replacement for calcium hypochlorite.

dichloroisocyanuric acid potassium salt. See potassium dichloroisocyanurate.

sym-dichloroisopropyl alcohol. See α-dichlorohydrin.

dichloroisopropyl ether.
CAS: 108-60-1. $[ClCH_2C(CH_3)H]_2O$.

Properties: Colorless liquid. D 1.1135 (20/20C), bp 187.4C (760 mm Hg), vap press 0.10 mm Hg (20C), flash p 185F (85C), wt/gal 9.3 lb (20C), coefficient of expansion 0.00096 K^{-1} (20C), viscosity 0.0230 cP (20C). Miscible with most oils and organic solvents; immiscible with water. Combustible.

Hazard: Toxic by ingestion and inhalation, strong irritant.

Use: Solvent for fats, waxes, greases, extractant, paint and varnish removers, spotting agents, and cleaning solutions.

dichloromethane.   Legal label name for methylene chloride.

dichloromethane diphosphonate.   See sodium clodronate.

3′,4′-dichloro-2-methylacrylanilide.   See dicryl.

dichloromethylchloroformate.
ClCOOCHCl$_2$.
Properties: Colorless liquid. D 1.56 (15C), bp 110–111C, vap d 5.7 (air = 1.29). Decomposed by water and alkalies; soluble in alcohol, benzene, and ether.
Derivation: (1) By chlorinating methyl formate, (2) by chlorinating methyl chloroformate. In both methods, the mixture of chloro derivatives is then separated by fractionation.
Hazard: Strong irritant to eyes and skin.

***sym*-dichloromethyl ether.**
O(CH$_2$Cl)$_2$.
Properties: Colorless, volatile liquid; suffocating odor. D 1.315 (20C); bp 105C. Decomposed by heat and water; soluble in acetone, benzene, ethyl alcohol, and methyl alcohol; insoluble in water.
Derivation: (1) Action of chlorine on methyl ether, (2) interaction of hydrochloric acid and formaldehyde with subsequent dehydration of the chloromethyl alcohol formed.
Hazard: A carcinogen by ingestion and inhalation. Strong irritant to eyes and mucous membranes.

5,7-dichloro-2-methyl-8-quinolinol.   See chlorquinaldol.

dichloromethylsilane.
CH$_3$SiHCl$_2$.
Properties: Colorless liquid. Bp 41C, d 1.113 (25/25C), refr index 1.3983 (20C). Combustible.
Hazard: Corrosive to tissue.
Use: Intermediate.

dichloromethyl sulfate.
(ClCH$_2$O)$_2$SO$_2$.
Properties: Colorless, liquid; odorless. D 1.60 (20C), bp 96–97C (14 mm Hg). Soluble in alcohol, benzene, and ether. Combustible.
Derivation: (1) By bubbling sulfur trioxide through (cooled) dichloromethyl ether, (2) by heating chlorosulfonic acid with formaldehyde.

dichloromonofluoromethane.   (dichlorofluoromethane; fluorodichloromethane; fluorocarbon 21).
CAS: 75-43-4.   CHCl$_2$F.
Properties: Colorless heavy gas; nearly odorless. Bp 8.9C, fp −135C, d 1.426 (0C), critical pressure 51.0 atm. Soluble in alcohol and ether; insoluble in water. Nonflammable.
Derivation: Reaction of chloroform and hydrogen fluoride.
Grade: Technical.
Hazard: Toxic by inhalation.
Use: Solvent, refrigerant.
See chlorofluorocarbon.

dichloronaphthalene.   See chlorinated naphthalenes.

2,3-dichloro-1,4-naphthoquinone.   See dichlone.

1,2-dichloro-4-nitrobenzene.
Cl$_2$C$_6$H$_3$NO$_2$.
Properties: Solid. Mp 43C, bp 255–256C, d 1.4266 (100/4C);. Insoluble in water; soluble in hot alcohol and ether.
Hazard: Fire risk by spontaneous reaction.
Use: Intermediate.

2,5-dichloronitrobenzene.
Cl$_2$C$_6$H$_3$NO$_2$.
Properties: Pale-yellow crystals. D 1.669 (22C), mp 55C, bp 266C. Insoluble in water; soluble in chloroform and hot alcohol.
Hazard: Fire risk by spontaneous reaction.
Use: Intermediate.

1,1-dichloro-1-nitroethane.
CAS: 594-72-9.   H$_3$CC(Cl)$_2$NO$_2$.
Properties: Bp 124C, d 1.4153 (20/20C), flash p 168F (75.5C) (OC). Combustible.
Hazard: Strong irritant. Upper respiratory tract irritant.
Use: Grain fumigant, solvent.

4,6-dichloro-3-((1e)-3-oxo-3-(phenylamino)-1-propenyl)-1h-indole-2-carboxylic acid.
CAS: 153436-22-7.   C$_{18}$H$_{12}$Cl$_2$N$_2$O$_3$.
Hazard: A poison.

1,3-dichloro-1,1,2,2,3-pentafluoropropane.
CAS: 507-55-1.   C$_3$HCl$_2$F$_5$.
Hazard: Low toxicity by ingestion, inhalation, and skin contact.

3,3-dichloro-1,1,1,2,2-pentafluoropropane.
CAS: 422-56-0.   C$_3$HCl$_2$F$_5$.
Hazard: Low-toxicity ingestion, inhalation, and skin contact.

dichloropentane.
CAS: 30586-10-8.   C$_6$H$_{10}$Cl$_2$.

A mixture of the dichloro derivatives of both normal pentane and isopentane. About 40% are amylene dichlorides having two chlorine atoms attached to adjacent carbon atoms.

Properties: Light-yellow liquid. D 1.06–1.08 (20C), acidity as hydrochloric acid not over 0.025% (distillation = 95%, 130–200C), flash p < 80F (26C), wt/gal 8.94 lb. Water solubility negligible; water azeotrope at 80–97C = approximately 66% $C_6H_{10}Cl_2$.

Hazard: Flammable, dangerous fire risk.

Use: Solvent for oils, greases, rubber resins, and bituminous materials; removal of tar; reclaiming rubber; paint and varnish removers; degreasing of metals; insecticide; soil fumigant; removal of wax deposits on oil-well equipment.

dichlorophenarsine hydrochloride. (3-amino-4-hydroxyphenyldichloroarsine hydrochloride).

$C_6H_3(AsCl_2)(OH)NH_2 \cdot HCl$.

Properties: White, hygroscopic powder. Mp 200C. Soluble in water and solutions of alkali hydroxides and carbonates, and in dilute mineral acids.

Hazard: Toxic by ingestion.

Use: Medicine (syphilis treatment).

dichlorophene. (2,2'-dihydroxy-5,5'-dichlorodiphenylmethane; DDDM).

CAS: 97-23-4. $(C_6H_3ClOH)_2CH_2$.

Properties: Light-tan, free-flowing powder; weakly phenolic odor. Mp 177C, vap press 10^{-4} mm Hg (100C) and about 10^{-10} mm Hg (25C) (extrapolated value). Soluble in acetone and alcohols; slightly soluble in benzene, toluene, and carbon tetrachloride; insoluble in water.

Derivation: Condensation of p-chlorophenol with formaldehyde in the presence of sulfuric acid.

Grade: Pure and technical.

Hazard: Toxic by ingestion.

Use: Fungicide and bactericide, textile preservative, some dermatological and cosmetic applications, veterinary medicine.

2,4-dichlorophenol.

CAS: 120-83-2. $Cl_2C_6H_3OH$.

Properties: White, low-melting solid. Bp 210C, mp 45C, flash p 237F (113C). Soluble in alcohol and carbon tetrachloride; slightly soluble in water. Combustible.

Derivation: By chlorination of phenol.

Hazard: Toxic by ingestion, strong irritant to tissue. Possible carcinogen.

Use: Organic synthesis.

2,4-dichlorophenoxyacetic acid. See 2,4-D.

2,4-dichlorophenoxybutyric acid. (2,4-DB). $Cl_2C_6H_3O(CH_2)_3COOH$.

Mp 118–120C, used as a herbicide.

2-(2,4-dichlorophenoxy)ethyl benzoate.

See sesin.

di(4-chlorophenoxy)methane.

$(ClC_6H_4O)_2CH_2$.

Properties: Solid. Mp 65C. Insoluble in water and oils; soluble in ether and acetone.

Use: Acaricide.

5-(2,4-dichlorophenoxy)-2-nitroanisole.

CAS: 32861-85-1. $C_{13}H_9Cl_2NO_4$.

Hazard: Low toxicity by ingestion and skin contact.

Use: Agricultural chemical.

(r)-2-(2,4-dichlorophenoxy)propanoic acid.

CAS: 15165-67-0. $C_9H_8Cl_2O_3$.

Hazard: A poison by ingestion. Low toxicity by inhalation and skin contact.

Use: Agricultural chemical.

2-(2,4-dichlorophenoxy)propionic acid. (dichloroprop).

$Cl_2C_6H_3OCH(CH_3)COOH$.

Properties: Solid. Mp 117–118C. Soluble in acetone, alcohol, and ether; insoluble in water.

Use: Herbicide.

(2,4-dichlorophenoxy)tributylstannane.

CAS: 39637-16-6. $C_{18}H_{30}Cl_2OSn$.

Hazard: A poison. Tributyltin compounds are extremely toxic to marine life.

2,4-dichlorophenyl benzenesulfonate.

$Cl_2C_6H_3OSO_2C_6H_5$.

Properties: Waxy solid. Vap press 2.7×10^{-4} mm Hg (30C). Insoluble in water; soluble in most organic solvents.

Use: Acaricide, insecticide.

2-(3,4-dichlorophenyl)-1h-benz(de)isoquinoline-1,3(2h)-dione.

CAS: 222420-28-2. $C_{18}H_9Cl_2NO_2$.

Hazard: A poison.

O-(2,4-dichlorophenyl)-O,O-diethyl phosphorothioate.

$Cl_2C_6H_3OP(S)(OC_2H_5)_2$.

Properties: (Pure compound) Liquid. Bp 120–123C (0.2 mm Hg). Slightly soluble in water; miscible with most organic solvents.

Hazard: Cholinesterase inhibitor. Absorbed by skin, use may be restricted.

n-(2,4-dichlorophenyl)-n-(4,5-dihydro-2-thiazolyl)-3-pyridinemethanamine.

CAS: 89985-01-3. $C_{15}H_{13}Cl_2N_3S$.

Hazard: Moderately toxic by ingestion.

3-(3,4-dichlorophenyl)-1,1-dimethylurea.

See diuron.

di(p-chlorophenyl)ethanol. (1,1-bis(p-chlorophenyl)ethanol; di(p-chlorophenyl) methylcarbinol; DMC; DCPC).

$CH_3C(C_6H_4Cl)_2OH$.

Properties: Colorless crystals. Mp 70C. Insoluble in water, soluble in common organic solvents.

Derivation: Reaction of 4,4′-dichlorobenzophenone with methylmagnesium bromide, followed by treatment with water.
Use: Insecticide.

2-(3,4-dichlorophenyl)imidazo(2,1-a)isoquinoline.
CAS: 61001-12-5. $C_{17}H_{10}Cl_2N_2$.
Hazard: A reproductive hazard.

3,4-dichlorophenyl isocyanate.
$Cl_2C_6H_3NCO$.
Properties: White to yellow crystals. Strong irritant to tissue, especially eyes and mucous membranes.
Use: Chemical intermediate, organic synthesis.

3-(3,4-dichlorophenyl)-1-methoxy-1-methylurea.   See linuron.

di(p-chlorophenyl)methylcarbinol.   See di(p-chlorophenyl)ethanol.

o-(2,4-dichlorophenyl)-o-methyl isopropylphosphoramidothioate.   (DMPA).
$Cl_2C_6H_3OP(S)(OCH_3)NHCH(CH_3)_2$.
Properties: Solid. Mp 51.4C, vap press 2 mm Hg (150C). Slightly soluble in water (5 ppm); freely soluble in acetone, benzene, and carbon tetrachloride.
Hazard: Cholinesterase inhibitor, use may be restricted.
Use: Herbicide, insecticide.

(2,6-dichlorophenyl)methyl methyl 3-pyridinylcarbonimidodithioate.
CAS: 34763-39-8. $C_{14}H_{12}Cl_2N_2S_2$.
Hazard: Moderately toxic by ingestion.
Use: Agricultural chemical.

2,4-dichlorophenyl-4-nitrophenyl ether.
$Cl_2C_6H_3OC_6H_4NO_2$.
Properties: Dark-brown solid. Setting point 62.5C. Soluble in acetone, methanol, and xylene.
Hazard: Toxic by ingestion.
Use: Herbicide.

3-(2,4-dichlorophenyl)-n-(4-propylcyclohexyl)-2-propenamide.
CAS: 315706-77-5. $C_{18}H_{23}Cl_2NO$.
Hazard: A poison.

2-(4-(2,4-dichlorophenoxy)phenoxy)-methyl-propionate.   See illoxan.

((2,5-dichlorophenyl)thio)methylcarbamic acid, 2,3-dihydro-2,2-dimethyl-7-benzofuranyl ester.
CAS: 50673-11-5. $C_{18}H_{17}Cl_2NO_3S$.
Hazard: A poison by ingestion.
Use: Agricultural chemical.

dichlorophenyltrichlorosilane.
$Cl_2C_6H_3SiCl_3$.
A mixture of isomers.
Properties: Straw-colored liquid. D 1.562, bp 260C, refr index 1.5638 (20C), flash p 286F (141C) (COC). Readily hydrolyzed with liberation of hydrogen chloride; soluble in benzene and perchloroethylene. Combustible.
Derivation: Chlorination of phenyltrichlorosilane.
Grade: Technical.
Hazard: Strong irritant to skin and eyes.
Use: Intermediate for silicones.

3,6-dichlorophthalic acid.
$C_6H_2Cl_2(COOH)_2$.
Properties: Colorless, thick crystals. Soluble in hot water.
Derivation: By oxidizing dichloronaphthalene tetrachloride (see chloronaphthalene) with nitric acid.

dichloroprop.   See 2-(2,4-dichlorophenoxy) propionic acid.

1,2-dichloropropane.   See propylene dichloride.

1,3-dichloro-2-propanol.   See α-dichlorohydrin.

1,3-dichloropropene.   (1,3-dichloropropylene).
CAS: 542-75-6. $CHCl:CHCH_2Cl$.
Properties: Exists in *cis* and *trans* isomeric forms, both colorless liquids. D 1.225 (10/4C), flash p 95F (35C) (OC). Insoluble in water; soluble in acetone, toluene, and octane. Bp *cis* 104C, *trans* 112C, refr index (20C) *cis* 1.469, *trans* 1.475.
Derivation: Chlorination of propylene.
Hazard: Confirmed carcinogen. Toxic by skin absorption. Flammable, moderate fire risk. Strong irritant. Kidney damage.
Use: Organic synthesis, soil fumigants.

dichloropropene-dichloropropane   mixture.
The technical mixture of dichloropropenes and dichloropropane is a clear amber liquid with a pungent odor. It is soluble in halogenated solvents, esters, and ketones. It was widely used as a soil nematocide before planting.
Hazard: A poison via ingestion, inhalation, or absorbed through the skin; corrosive to tissue (skin and eyes).
Use: Pesticide, insecticide.

3,4-dichloropropionanilide.   See propanil.

2,2-dichloropropionic acid.   See dalapon.

2,4-dichloro-n-(4-propylcyclohexyl) benzamide.
CAS: 315706-69-5. $C_{16}H_{21}Cl_2NO$.
Hazard: A poison.

3-(2,4-dichloro-5-(2-propynyloxy)phenyl)-5-(1,1-dimethylethyl)-1,3,4-oxadiazol-2(3h)-one.   See 5-*tert*-butyl-3-(2,4-dichloro-5-propargyloxyphenyl)-1,3,4.

3,6-dichloro-2-pyridinecarboxylic   acid. See clopyralid.

2,6-dichloroquinonechlorimide.   Reagent used for spot visualizations in chromatographic systems.
Hazard: Explodes readily on slight heating.

2,6-dichlorostyrene.
CAS: 6607-45-0. $C_6H_3(CH:CH_2)Cl_2$.
Properties: Colorless liquid. Bp 92–94C (5 mm Hg), flash p 225F (107C) (OC). Insoluble in water; soluble in most organic solvents. Polymerizes slowly on standing unless inhibited. Combustible.
Use: Monomer and comonomer in plastic research.

***p-N,N*-dichlorosulfamylbenzoic   acid.**   See halazone.

dichlorosulfonphthalein.   See chlorphenol red.

dichlorotetrafluoroacetone.
$CClF_2COCClF_2$.
Properties: Colorless liquid. Bp 45.2C, fp < −100C. Soluble in water and most organic solvents; stable to acids but not alkalies.
Hazard: Toxic by ingestion and inhalation.
Use: Solvent in acidic media, complexing agent for active hydrogen-compound separation.

***sym*-dichlorotetrafluoroethane.**
(fluorocarbon-114; tetrafluorodichloroethane).
CAS: 76-14-2. $CClF_2CClF_2$.
Properties: Colorless, nonflammable gas; nearly odorless. Bp 3.55C, fp −94C, critical press 32.3 atm. Insoluble in water.
Derivation: By treating perchloroethylene with hydrogen fluoride.
Grade: Technical 95%.
Hazard: An asphyxiant. Inhibits pulmonary function. Questionable carcinogen.
Use: Solvent, fire extinguishers, refrigerant and air conditioner fluid, blowing agent, dielectric fluid. See chlorofluorocarbon.

2,5-dichlorothiophene.
$C_4H_2Cl_2S$ (cyclic).
Properties: Colorless to light-yellow liquid. Bp 161C. Combustible.
Use: Intermediate.

dichlorotoluene.   (chlorobenzyl chloride).
$C_6H_3CH_3Cl_2$.
Properties: Colorless liquid. Boiling range 200–300C, fp approximately −13C, d 1.245–1.247

(25/15C), refr index 1.5480 (22C). Soluble in alcohol, ether, and acetone; insoluble in water. Exists as *o*- and *p*- isomers. Combustible.
Hazard: Irritant, see benzyl chloride.
Use: High-boiling solvent, intermediate for organic synthesis.

α,α-dichlorotoluene.   See benzyl dichloride.

dichloro-*sym*-triazine-2,4,6-trione.   See dichloroisocyanuric acid.

β,β'-dichlorovinylchloroarsine.
$(ClCH:CH)_2AsCl$.
Properties: Yellowish liquid. Bp 230C (decomposes), d 1.70.
Hazard: Toxic by inhalation and ingestion, strong irritant to skin and mucous membranes.
Use: Military poison.

2,2-dichlorovinyldimethyl   phosphate.   See dichlorovos.

β,β'-dichlorovinylmethylarsine.
$(ClCH:CH)_2AsCH_3$.
Properties: Liquid. Bp 140–145C (10 mm Hg).
Derivation: Interaction   of   acetylene   and methyldichloroarsine in the presence of aluminum chloride.
Hazard: Toxic by inhalation and ingestion, strong irritant to skin and mucous membranes.
Use: Military poison.

dichlorovos.   (2,2-dichlorovinyldimethyl phosphate; DDVP).
CAS: 62-73-7. $(CH_3O)_2P(O)CH:CCl_2$.
Properties: Liquid. Bp 120C (14 mm Hg). Slightly soluble in water and glycerol; miscible with aromatic and chlorinated hydrocarbon solvents and alcohols.
Hazard: Cholinesterase inhibitor, a poison. Toxic by skin absorption. Possible carcinogen.
Use: Insecticide, fumigant.

dichloroxylenol.   (2,4-dichloro-3,5-dimethylphenol).
$C_8H_8Cl_2O$.
Properties: Crystals. Mp 95C. Partially soluble in acetone, diethyl ketone, and chloroform; almost insoluble in water.
Use: Mold inhibitor, antibacterial agent, preservative.

2,4-dichlorphenoxyacetic   acid   octyl   ester. See octyl 2,4-dichlorophenoxyacetate.

dichroic.   A term used in crystallography to denote crystals that refract incident light in two directions, thus displaying two colors when observed from different angles, e.g., calcite. See anisotropic; birefringent.

dichroism. The property of some crystals of exhibiting different colors according to the direction of the light ray due to difference in extent of absorption of the various rays.

dichromate. A compound in which two chromium atoms with a valency of 6 are present, as in potassium dichromate.

dichromatic. Characterizing certain dyes and indicators for which different colors may be seen, depending on the thickness of the solution viewed.

dicobalt octacarbonyl. See cobalt tetracarbonyl.

dicofol. (generic name for 4,4′-dichloro-α-trichloromethylbenzhydrol).
CAS: 115-32-2.
Hazard: Questionable carcinogen.

dicumarol. See bishydroxycoumarin.

dicopper(I) acetylide.
CAS: 1117-94-8. CuC≡CCu.
Properties: Mw 151.10.
Hazard: Explodes on impact or heating to 100C.

dicresyl glyceryl ether. (glyceryl ditolyl ether). This may be a mix of *o*-, *m*-, and *p*-isomers.
$CH_3C_6H_4OCH_2CH_2OCH_2OC_6H_4CH_3$.
Properties: Similar to cresyl glyceryl ether. D 1.136, refr index 1.549, boiling range 328–340C.

dicresyl glyceryl ether acetate.
$CH_3C_6H_4OCH_2CHOOCCH_3OC_6H_4CH_3$.
Properties: Fairly stable liquid. D 1.115, bp 360C, refr index 1.53. Combustible.

dicrotophos. See "Bidrin" *[AMVAC]*.

dicryl. (generic name for 3′,4′-dichloro-2-methylacrylanilide).
CAS: 2164-09-2. $Cl_2C_6H_3NHCOC(CH_3):CH_2$.
Properties: Solid. Mp 128C. Insoluble in water; soluble in acetone, alcohol, isophorone, and dimethyl sulfoxide.
Use: Herbicide, pest control.

"Dicrylan" *[Novartis].* TM for a group of products that includes silicone elastomers and other polymeric chemicals used for coating textile fabrics.

dicobalt edetate. (Ba 2724; cobalt(2)-edathamil; dicobalt edta; ((ethylenedinitrilo) tetraacetato(2-))-cobaltate(2-) cobalt(2+) salt; kelocyanor; kobalt-edta; 2-[2-[bis(2-oxido-2-oxoethyl)amino]ethyl-(2-oxido-2-oxoethyl)amino]acetate; cobalt (2+)).
$C_{10}H_{12}Co_2N_2O_8$

A chelating agent that sequesters a variety of polyvalent cations.
Hazard: Deadly poison.
Use: Antidote for cyanide poisoning.

dicumarol. (3,3′-methylenebis[4-hydroxy-2H-1-benzopyran-2-one]; bishydroxycoumarin; dicoumarol; dicoumarin; dicumol; dufalone; melitoxin; 2-hydroxy-3-[(2-hydroxy-4-oxochromen-3-yl)methyl]chromen-4-one).
CAS: 66-76-2. $C_{19}H_{12}O_6$.
A vitamin K antagonist that decreases the activity of vitamin K-dependent clotting factors in blood plasma. It inhibits the production of prothrombin in the liver and increases prothrombin time.
Hazard: Cumulative poison; therapy may cause nausea, abdominal pain, and diarrhea.
Use: Therapeutically as a long-acting, oral anticoagulant.

dicumene chromium.
CAS: 12001-89-7. $C_{18}H_{24}$·Cr.
Hazard: A confirmed carcinogen. Poison by skin contact. Moderately toxic by ingestion. A skin and eye irritant. Confirmed carcinogen.

dicumyl peroxide.
$[C_6H_5C(CH_3)_2O]_2$.
Hazard: Strong oxidizer, may ignite organic materials on contact.
Use: Polymerization catalyst and vulcanizing agent.

"Di-cup" *[GEO].* TM for a series of vulcanizing and polymerization agents containing dicumyl peroxide.

dicyan. See cyanogen.

dicyandiamide. (cyanoguanidine).
$NH_2C(NH)(NHCN)$.
Properties: Pure-white crystals. D 1.400 (25C), stable when dry, melting range 207–209C. Soluble in liquid ammonia; partly soluble in hot water. Nonflammable.
Derivation: Polymerization of cyanamide in the presence of bases.
Grade: 99% pure, technical.
Use: Fertilizers; nitrocellulose stabilizer; organic synthesis, especially of melamine, barbituric acid, and guanidine salts; pharmaceutical products; dyestuffs; explosives; retarding rancidity in fats and oils; fireproofing compounds; case-hardening preparations; cleaning compounds; soldering compounds; accelerator; thinner for oil-well drilling muds; stabilizer in detergent compositions; modifier for starch products; catalyst for epoxy resins.

dicyanine. See cyanine dye.

***m*-dicyanobenzene.** See *m*-phthalodinitrile.

***o*-dicyanobenzene.** See phthalonitrile.

1,3-dicyanobenzene. See *m*-phthalodinitrile.

1,4-dicyanobutane. See adiponitrile.

2,4-dicyanobutane-1. See methylene glutaronitrile.

dicyclohexyl. See bicyclohexyl.

dicyclohexyl adipate.
$(CH_2CH_2COOC_6H_{11})_2$.
Properties: White crystals; odorless. D 0.913-0.919 (15/15C), bp 256C, fp −0.1C, d 45 lb/cu ft, acidity (as adipic acid) <0.05%. Compatible with most natural and synthetic resins. Soluble in most organic solvents; insoluble in water.
Use: Plasticizer.

dicyclohexylamine.
CAS: 101-83-7. $(C_6H_{11})_2NH$.
Properties: Colorless liquid; faint amine odor. D 0.913-0.919 (15/15C), bp 256C, fp −0.1C, refr index 1.4823 (25C), flash p 210F (98.9C). Slightly soluble in water; miscible with organic solvents. Strongly basic.
Hazard: Toxic by ingestion, strong irritant to skin and mucous membranes. Combustible.
Use: Intermediate; insecticides; plasticizer; corrosion inhibitors; antioxidants in rubber, lubricating oils, and fuels; catalyst for paint, varnishes, and inks; detergents, extractant.

dicyclohexyl carbodiimide.
CAS: 538-75-0. $C_6H_{11}NCNC_6H_{11}$.
Properties: White crystals; heavy sweet odor. Set point 29–30C, bp 138–140C (2 mm Hg). Soluble in organic solvents.
Use: Chemical intermediate, coupling agent in peptide synthesis.

dicyclohexylphenyltin hydroxide.
CAS: 53413-47-1. $C_{18}H_{28}OSn$.
Hazard: A poison by ingestion. A moderate skin irritant.

dicyclohexyl phthalate. (DCHP).
$C_6H_4(COOC_6H_{11})_2$.
Properties: White, granular solid; nonvolatile; mildly aromatic odor. D 1.20 (25/25C), mp 62–65C, flash p 405F (207C). Soluble in most organic solvents; insoluble in water; compatible with a large number of polymers. Combustible.
Use: Plasticizer for nitrocellulose, ethyl cellulose, chlorinated rubber, polyvinyl acetate, polyvinyl chloride, and other polymers.
See "Morflex 150" *[Vertellus]*.

dicyclopentadiene.
CAS: 77-73-6. $C_{10}H_{12}$.
Properties: Liquid. D 0.979 (20/20C), bp 172C, mp 33.6C, bulk d 8.2 lb/gal (60F), refr index 1.5073 (31C), flash p 90F (32.2C). Soluble in alcohol; insoluble in water.
Derivation: Olefin manufacture.
Hazard: Flammable, moderate fire risk. Toxic by ingestion. Eye, upper and lower respiratory tract irritant.
Use: Chemical intermediate for insecticides, EPDM elastomers, metallocenes, paints and varnishes, flame retardants for plastics.

dicyclopentadiene dioxide.
$C_{10}H_{12}O_2$.
Properties: White, crystalline powder. Mp 180–184C, d 1.331 (25C). Slightly soluble in water; soluble in acetone and benzene.
Use: Intermediate for epoxy resins; plasticizers; protective coatings.

dicyclopentadienyl cobalt. See cobaltocene.

dicyclopentadienyl iron. (biscyclopentadienyl iron).
CAS: 102-54-5.
The first organometallic "sandwich" compound (synthesized in 1951), which served as a prototype for metallocenes. Such compounds are based on cyclic unsaturates combined with a transition metal or its halide.
Hazard: Toxic; liver damage.
See ferrocene.

dicyclopentadienyl nickel. See nickelocene.

dicyclopentadienyl osmium. See osmocene.

dicyclopentadienyl titanium dichloride.
See titanocene dichloride.

dicyclopentadienyl zirconium dichloride.
See zirconocene dichloride.

DIDA. See diisodecyl adipate.

didecyl adipate.
$(CH_2CH_2COOC_{10}H_{21})_2$.
Properties: Light-colored liquid. D 0.9181 (20/20C), bulk d 7.7 lb/gal (20C), bp 245C (5 mm Hg), vap press (0.58 mm Hg (100C), viscosity 26.3 cP (20C), flash p 425F (218C). Insoluble in water. Combustible.
Use: Plasticizer.

didecylamine.
$[CH_3(CH_2)_9]_2NH$.
Properties: Light-straw-colored liquid; faint amine odor. Boiling range 195–215C (12 mm Hg), d 0.840 (20/20C) (solid). Combustible.

didecyl ether.
$(C_{10}H_{21})_2O$.

Properties: Liquid. Mp 16C, bp 170–180C (6 mm Hg), d 0.819 (20/4C), refr index 1.4418 (20C). Combustible.
Grade: 95% (min) purity.
Use: Electrical insulators, water repellent, lubricant in plastic molding and processing, antistatic agent, intermediate.

didecyl phthalate.   (DDP).
CAS: 84-77-5. $C_6H_4(COOC_{10}H_{21})_2$.
Properties: Light-colored liquid. D 0.9675 (20/20C) bulk d 8.05 lb/gal (20C), bp 261C (5 mm Hg), vap press 0.3 mm Hg (200C), viscosity 113.2 cP (20C), flash p 445F (229C). Insoluble in water. Combustible.
Use: Plasticizer, especially for vinyl resins.

didecyl sulfide.   (didecyl thioether).
$(C_{10}H_{21})_2S$.
Properties: Liquid. Solidifies at 22.2C, bp 205–206C (4 mm Hg), d 0.831 (20/4C), refr index 1.4569 (33.5C). Combustible.
Grade: 95% (min) purity.
Use: Organic synthesis (formation of sulfonium compounds).

didecyl thioether.   See didecyl sulfide.

2′,3′-didehydro-3′-deoxythymidine.   See stavudine.

didodecylamine.   See dilaurylamine.

didodecyl ether.   See dilauryl ether.

didodecyltin dichloride.
CAS: 5827-58-7. $C_{24}H_{50}Cl_2Sn$.
Hazard: Low toxicity by ingestion.

didodecyl-3,3′-thiodipropionate.   See dilauryl thiodipropionate.

didodecyl thioether.   See dilauryl sulfide.

DIDP.   See diisodecyl phthalate.

didymium.
Di.
Commercial mixture of rare earth elements obtained from monazite sand by extraction, followed by the elimination of cerium and thorium. The name is used like that of an element in naming mixed oxides and salts. The approximate composition of didymium from monazite, expressed as rare earth oxides, is 46% La_2O_3, 10% Pr_6O_{11}, 32% Nd_2O_3, 5% Sm_2O_3, 0.4% yttrium earth oxides, 1% CeO_2, 3% Gd_2O_3, and 2% others. Commercially used didymium salts are acetate, carbonate, chloride, fluoride, nitrate, etc.

didymium oxide.
Di_2O_3.

Properties: Brown powder. Insoluble in water; soluble in acids.
Use: (As salts) Coloring and decolorizing glass, in temperature-compensating capacitors for radio, television, and radar. In carbon-arc cores (fluoride), in stainless steel (oxide), metallurgical research, textile treatment.

die.   A device, usually of steel, having a specific shape or design that it imparts to such materials as metals and plastics by impact (stamping), by the contour of a negative cavity (casting), or by passing the material through it (extrusion). Diamond dies may be used for wiredrawing. The terms "die" and "mold" are virtually synonymous in the sense of a negative cavity into which a molten metal or plastic is introduced under pressure, the former being used in reference to metals and the latter for plastics, rubber, etc.
See die casting; investment casting; injection molding; and extrusion.

die casting.   Shaping of metal products by forcing a molten metal or alloy under high pressure into a negative-cavity die by means of a hydraulic ram. The die is usually made of an alloy steel. Metals commonly used for die casting are zinc, aluminum, copper, lead, and their alloys, some of which also include silicon. Die castings can be held to tolerances as low as 0.001–0.0015 inches, and sizes 75–100 lb (Al) are possible. The largest end-use area for die castings is automobile and airplane parts. They are also used in washing and drying machines, in electrical equipment and appliances, and for various military applications.

Dieckmann reaction.   Base-catalyzed intramolecular cyclization of esters of dicarboxylic acids to give β-ketoesters.

dieldrin.   (HEOD).
CAS: 60-57-1. $C_{12}H_{10}OCl_6$.
(Not <85% of 1,2,3,4,10,10-hexachloro-6,7-epoxy-1,4,4a,5,6,7,8,8a-octahydro-1,4-*endo*,*exo*-51,2,3,4,10,10-hexa-8-dimethanonaphthalene and not <15% active related compounds.) Endrin is a stereoisomer of dieldrin.
Properties: Light-tan, flaked solid. Mp 175C. Insoluble in water, methanol, and aliphatic hydrocarbons; soluble in acetone and benzene. Compatible with most fertilizers, herbicides, fungicides, and insecticides.
Derivation: By oxidation of aldrin with peracids.
Hazard: Toxic by ingestion, inhalation, and skin absorption. Use restricted to nonagricultural applications. Liver damage, reproductive effects, and central nervous system impairment. Questionable carcinogen.
Use: Insecticide.
See endrin.

dielectric.   A substance with very low electrical conductivity, i.e., an insulator. Such substances have electrical conductivity of <1 millionth mho/cm. Those with a somewhat higher conductivity (10^{-6} to 10^{-3} mho/cm) are called semiconductors. Among the more common solid dielectrics are glass, rubber, and similar elastomers, and wood and other cellulosics. Liquid dielectrics include hydrocarbon oils, askarel, and silicone oils.
See transformer oil.

dielectric cohesion.   The molecular force of a dielectric that opposes an electric field tending to ionize the molecules.

dielectric constant.   An index of the ability of a substance to attenuate the transmission of an electrostatic force from one charged body to another, as in a condenser. The lower the value, the greater the attenuation. The standard measurement apparatus utilizes a vacuum whose dielectric constant is 1. In reference to this, various materials interposed between the charged terminal have the following values at 20C: air 1.00058, glass 3, benzene 2.3, acetic acid 6.2, ammonia 15.5, ethanol 25, glycerol 56, and water 81. The exceptionally high value for water accounts for its unique behavior as a solvent and in electrolytic solutions. Most hydrocarbons have high resistance (low conductivity). Dielectric-constant values decrease as the temperature rises.

dielectric strength.   The maximum electric field that an insulator or dielectric can withstand without breakdown, usually measured in kV/cm. At breakdown, a considerable current passes as an arc, usually with more or less decomposition of the material along the path of the current.

Diels-Alder reaction.   An important organic reaction for the synthesis of six-membered rings discovered in 1928. It involves the addition of an ethylenic double bond to a conjugated diene, i.e., a compound containing two double bonds separated by one single bond, as in 1,3-butadiene (CH_2=CH-CH=CH_2) or cyclopentadiene. The ease of addition of the ethylenic compound is greatly enhanced by adjacent carbonyl groups; hence maleic anhydride reacts quantitatively with hexachlorocyclopentadiene to form chlorendic anhydride.

Diels, Otto P. H.   (1876–1954). A German chemist who won a Nobel Prize in chemistry with Alder in 1950. He was awarded the prize for diene synthesis work that led to improved methods of analyzing and synthesizing organic compounds. His research resulted in the discovery of carbon suboxide, methods of dehydrating cyclical hydrocarbons using selenium, and determination of the structure of steroids. A student of Fischer's, Diels graduated from the University of Berlin.

dien.   Abbreviation of diethylenetriamine as used in formulas for coordination compounds. See *en*; *pn*; *py*.

diene.   See diolefin.

Diene.   A solution-polymerized polybutadiene rubber.
Use: In tires, high-impact polystyrene, and ABS.

dienestrol.
$HOC_6H_4C(CHCH_3)C(CHCH_3)C_6H_4OH$.
(3,4-bis(*p*-hydroxyphenyl)-2,4-hexadiene).
A synthetic with estrogenic activity.
Properties: Colorless, needles or powder; odorless. Mp 227C. Soluble in alcohol; practically insoluble in water; sensitive to light.
Grade: NF.
Use: Medicine (estrogenic hormone).

dienone-phenol rearrangement.   Transformation of a 4,4-disubstituted cyclohexadienone into a 3,4-disubstituted phenol upon acid treatment.

diepoxybutane.   (*di*-butadiene dioxide; 1,2,3,4-dianhydro-*di*-threitol; (±)-1,2,3,4-diepoxybutane; *dl*-1,2,3,4-deipoxybutane; 2-(oxiran-2-yl)oxirane). $C_4H_6O_2$.
Properties: Colorless liquid.
Hazard: Probable carcinogen; neoplastigen; tumorigenic; poison; mutagen.

diesel ignition improver.   A substance such as amyl nitrate that is added to diesel fuels to improve fuel ignition and to raise the cetane number.

diesel oil.   (fuel oil no. 2).
CAS: 68476-34-6.
Fuel for diesel engines obtained from distillation of petroleum. Its efficiency is measured by the cetane number. It is composed chiefly of unbranched paraffins; its volatility is similar to that of gas oil. Flash p 110–190F, d <1. Combustible.
Hazard: Moderate fire risk. Environmental hazard. Dermatitis. Possible carcinogen.
Use: Fuel for trucks, ships, and other automotive equipment, drilling muds, mosquito control (coating on breeding waters).
See fuel oil.

Diesel-Treat.   A dry, granular, orange sodium dichromate used as a corrosion inhibitor. Sold in 50-lb drums.
Use: Closed cooling systems; particularly diesel engines, cooling tower systems.

dietary food supplement.   Any food product to which enough vitamins and minerals have been added to furnish >50% of the recommended daily allowance in a single serving (FDA). Such foods must have added ingredients identified on labels.

diethanolamine. (DEA; di(2-hydroxyethyl) amine).
CAS: 111-42-2. $(HOCH_2CH_2)_2NH$.
Properties: Colorless crystals or liquid. Active base. Mp 28.0C, bp 269C, d 1.092 (30/20C), flash p 306F (152C) (OC). Very soluble in water and alcohol; insoluble in ether, benzene. Combustible.
Derivation: Ethylene oxide and ammonia.
Hazard: Toxic. Liver and kidney damage. Possible carcinogen.
Use: Liquid detergents for emulsion paints, cutting oils, shampoos, cleaners, and polishes; textile specialties; absorbent for acid gases; chemical intermediate for resins, plasticizers, etc.; solubilizing 2,4-D; humectant; dispersing agent.

diethanololeamide. See oleic acid diethanolamide.

N,N-diethanolglycine. (DEG).
$(HOCH_2CH_2)_2NCH_2COOH$.
Use: Chelating agent. Also available as the sodium salt.

diethofencarb. See isopropyl 3,4-diethoxycarbanilate.

2,5-diethoxyaniline.
$NH_2C_6H_3(OC_2H_5)_2$.
Properties: White to gray powder. Mp 83–85C. Insoluble in water; soluble in organic solvents.
Hazard: See aniline.
Use: Intermediate.

1,4-diethoxybenzene. See hydroquinone diethyl ether.

1,1-diethoxyethane. See acetal.

1,2-diethoxy ethylene.
CAS: 16484-86-9. $C_6H_{12}O_2$.
Hazard: Moderately toxic.

diethoxyethyl phthalate. See diethyl glycol phthalate.

diethylacetal. See acetal.

diethyl acetaldehyde. See 2-ethylbutyraldehyde.

N,N-diethylacetamide.
$CH_3CON(C_2H_5)_2$.
Properties: Colorless liquid. D 0.920 (20C), boiling range 182–186C, faint odor, flash p 170F (76.6C). Combustible.
Hazard: Toxic by ingestion.

diethylacetic acid. See 2-ethylbutyric acid.

diethylberyllium. (beryllium diethyl; beryllium ethane).
$C_4H_{10}Be$.
Properties: Volatile, colorless oily liquid; reacts violently with water, HCl, and alcohols; boils at 63C, combusts spontaneously in air releasing beo as dense white smoke
Hazard: Extremely hazardous; very poisonous; carcinogen.

N,N-diethylacetoacetamide.
$CH_3COCH_2CON(C_2H_5)_2$.
Properties: Liquid. D 0.9950 (20/20C), bp (decomposes), fp −70C (sets to glass below this temperature), flash p 250F. Miscible with water. Combustible.
Use: Intermediate for pigments.

diethyl adipate.
$C_2H_5OCO(CH_2)_4OCOC_2H_5$.
Properties: Colorless liquid. D 1.002 (25C), refr index 1.426 (25C), bp 245C, fp −21C. Insoluble in water. Combustible.
Use: Plasticizer.

4′-(N,N-diethylalanyl)methanesulfonanilide hydrochloride.
CAS: 60735-64-0. $C_{14}H_{22}N_2O_3S·ClH$.
Hazard: Moderately toxic by ingestion.

diethylaluminum chloride.
CAS: 96-10-6.
(aluminum diethyl monochloride; DEAC).
$(C_2H_5)_2AlCl$.
Properties: Colorless pyrophoric liquid. Bp 208C, fp −50C.
Derivation: Reaction of triethylaluminum with ethylaluminum sesquichloride.
Hazard: Pyrophoric and highly flammable in air, reacts violently with water. Dangerous fire and explosion hazard.
Use: Polyolefin catalyst, intermediate in production of organometallics.

diethylaluminum hydride.
$(C_2H_5)_2AlH$.
A pyrophoric mix with triethylaluminum.
Derivation: Action of ethylene and hydrogen on aluminum.
Hazard: Highly flammable in air, reacts violently with water. Dangerous fire and explosion hazard.
Use: Catalyst, reducing agent.

diethylamine.
CAS: 109-89-7. $(C_2H_5)_2NH$.
Properties: Colorless liquid; ammoniacal odor; alkaline reaction. Bp 55.5C, fp −49.8C, d 0.7062 (20/20C), wt/gal 5.91 lb (20C), autoign temp 594F (312C), flash p <−15F (<−26C). Miscible with water, alcohol, and most organic solvents.
Derivation: From ethyl chloride and ammonia under heat and pressure.

Grade: Technical.
Hazard: Highly flammable, dangerous fire risk. Flammable limits in air 1.8–10.1%. Toxic by ingestion; strong eye, skin, and upper respiratory tract irritant. Questionable carcinogen.
Use: Rubber chemicals, textile specialties, selective solvent, dyes, flotation agents, resins, pesticides, polymerization inhibitors, pharmaceuticals, petroleum chemicals, electroplating, corrosion inhibitors.

α-diethylaminoaceto-2,6-xylidide.   See lidocaine.

4-(diethylamino)-2-butanone.
CAS: 3299-38-5. $C_8H_{17}NO$.
Hazard: Moderately toxic by ingestion.
Use: Agricultural chemical.

5-diethylamino-2-aminopentane.   (1-diethylamino-4-aminopentane).
$CH_3CH(NH_2)(CH_2)_2CH_2N(C_2H_5)_2$.
Properties: Liquid; amine odor. D 0.82, bp 142–144C. Soluble in water, alcohol, and ether. Combustible.
Use: Pharmaceuticals.

diethylaminoaniline.   See   *p*-aminodiethylaniline.

diethylaminocellulose.   (DEAE cotton).
A cellulose derivative containing a tertiary amine group that acts as a catalyst for epoxide reactions. It is also used in ion exchange fractionations. It is made by adding β-chloroethyldiethylamine hydrochloride to cellulose in sodium hydroxide. Repeated treatments increase the nitrogen content of the cotton (cellulose) to over 1%, with beneficial effect on crease resistance. Combustible.
See epoxide.

2-diethylaminoethanethiol hydrochloride.
$(C_2H_5)_2NCH_2CH_2SH \cdot HCl$.
Properties: Solid. Mp 170C. Soluble in water and alcohol; insoluble in benzene.
Use: Pharmaceutical intermediate, pesticides, polymerization promoter.

diethylaminoethanol.   (*N,N*-diethylethanolamine).
CAS: 100-37-8. $(C_2H_5)_2NCH_2CH_2OH$.
Properties: Colorless, hygroscopic liquid base combining the properties of amines and alcohols. Bp 161C, d 0.88–0.89 (20/20C), vap press 21 mm Hg (20C), flash p 140F (60C) (OC), wt/gal 7.14 lb (20C), fp −70C. Soluble in water, alcohol, and benzene. Combustible.
Grade: Technical.
Hazard: Moderate fire risk. Toxic by ingestion and skin absorption. Central nervous system impairment and upper respiratory tract irritant.

Use: Water-soluble salts, fatty acid derivatives, textile softeners, pharmaceuticals, antirust compositions, emulsifying agents in acid media, derivatives containing tertiary amine groups, curing agent for resins.

diethylaminoethoxyethanol.
$(C_2H_5)_2NC_2H_4OC_2H_4OH$.
Properties: Colorless liquid. D 0.930–0.950 (20/20C), boiling range (95%) 215.0–228.0C. Combustible.
Use: Intermediate.

2-(2-(diethylamino)ethoxy)ethanol.
CAS: 140-82-9. $C_8H_{19}NO_2$.
Hazard: A poison by ingestion and skin contact. A severe eye and mild skin irritant.

β-diethylaminoethyl chloride hydrochloride.
(DEC).
CAS: 869-24-9. $(C_2H_5)_2NCH_2CH_2Cl \cdot HCl$.
Intermediate in the manufacture of pharmaceuticals and as an organic intermediate for attaching the diethylaminoethyl radical.

N,N′-diethyl-3-amino-4-methoxybenzene-sulfonamide.
$NH_2(CH_3O)C_6H_3SO_2N(C_2H_5)_2$.
Properties: White to pink crystals. Mp 100–103C. Insoluble in water and ether; partially soluble in alcohol.
Use: Intermediate.

1-diethylamino-2-methylbenzene.   See *N,N*-diethyl-*o*-toluidine.

7-diethylamino-4-methylcoumarin.
(MDAC; 4-methyl-7-[diethylamino]coumarin).
$CH_3C_9H_4O(O)N(C_2H_5)_2$.
Properties: Granular, light-tan color. Mp 68–72C. Gives a bright blue–white fluorescence in very dilute solutions. Soluble in aqueous acid solutions, resins, varnishes, vinyls, and nearly all common organic solvents; slightly soluble in aliphatic hydrocarbons.
Use: Optical bleach in textile industry, in coatings for paper, labels, book covers, etc.; to lighten plastics, resins, varnishes, and lacquers; invisible marking agent.

5-diethylamino-2-pentanone.   (1-diethylamino-4-pentanone).
$CH_3CO(CH_2)_3N(C_2H_5)_2$.
Properties: Liquid; an amine odor. Combustible.
Use: Pharmaceuticals.

9-diethylaminoethyl-2-phenylimidazo(1,2-a)benzimidazole dihydrochloride.
CAS: 23572-32-9. $C_{21}H_{24}N_4 \cdot 2ClH$.
Hazard: Moderately toxic by ingestion.

m-**diethylaminophenol.**
$C_6H_4OHN(C_2H_5)_2$.
Properties: White, crystalline solid. Mp 78C, bp 276–280C. Soluble in alcohol, caustic soda, ether.
Derivation: Diethylaniline is sulfonated with oleum, and the resulting diethylaniline-*m*-sulfonic acid fused with caustic soda.
Hazard: See phenol.
Use: Dyes.

3-diethylaminopropylamine.
$(C_2H_5)_2NCH_2CH_2CH_2NH_2$.
Properties: Water-white liquid; amine odor. Bp 169C, fp −100C (sets to a glass), d 0.82 (20/20C), refr index 1.442 (10C), flash p 138F (58.9C) (OC), Combustible.
Hazard: Moderate fire risk. Irritant to skin.
Use: Curing agent for epoxy resins, intermediate.

γ-**diethylaminopropyl chloride hydrochloride.**   (DEPC).
$(C_2H_5)_2NCH_2CH_2CH_2Cl·HCl$.
Use: Manufacture of pharmaceuticals, intermediate for attaching the diethylaminopropyl radical.

3-(diethylamino)propyl   isopropyl(phenyl)glycolate hydrochloride.
CAS: 15422-00-1. $C_{18}H_{29}NO_3·ClH$.
Hazard: A poison.

10-(2-(diethylamino)propyl)-10h-pyrido(3,2-b)(1,4)benzothiazine.
CAS: 67465-66-1. $C_{18}H_{23}N_3S$.
Hazard: A poison by ingestion.

N,N-**diethylaniline.**
CAS: 91-66-7. $(C_2H_5)_2NC_6H_5$.
Properties: Colorless to yellow liquid. D 0.9351, fp −38 to −39C, bp 215–216C, flash p 185F (85C). Slightly soluble in alcohol and ether; soluble in water. Combustible.
Derivation: By heating aniline hydrochloride with alcohol at 180C under pressure.
Grade: Technical.
Hazard: See aniline.
Use: Organic synthesis, dyestuff intermediate.

diethylbarbituric acid.   See barbital.

diethylbenzene.
CAS: 25340-17-4. $(C_6H_4(C_2H_5)_2$.
Grade: The commercial product is either a mixture of isomers or the *p*-isomer.
Available forms: *p*-Isomer available in both pure and technical grades.
Properties: Colorless liquid. Boiling range 179.8–184.8C, d 0.865 (25/25C), flash p 132F (55.5C). Soluble in alcohol, benzene, carbon tetrachloride, and ether; insoluble in water; wt/gal 7.22 lb; refr index 1.49; autoign temp 806F (430C). Combustible.

Hazard: Moderate fire risk. Moderately toxic.
Use: Intermediate, solvent.

diethylbis(octanoyloxy)stannane.
CAS: 2641-56-7. $C_{20}H_{40}O_4Sn$.
Hazard: A poison by ingestion.

o,o-**diethyl-*o*-(4-(1-((((butylamino) carbonyl)oxy)imino)ethyl)phenyl) phosphorothioic acid ester.**
CAS: 22936-20-5. $C_{17}H_{27}N_2O_5PS$.
Hazard: Moderately toxic by ingestion.
Use: Agricultural chemical.

di(2-ethylbutyl) azelate.
$C_6H_{13}OOC(CH_2)_7COOC_6H_{13}$.
Properties: Pale-yellow to water-white liquid. D 0.9340 (20/20C), viscosity 56 seconds Saybolt (100F), flash p 385F (196C), fp <−40C, acid number <1.0, faint odor. Stable to heat, light, and hydrolysis. Combustible.
Use: As plasticizer for polyvinyl chloride and its copolymers and for cellulose esters.

di(2-ethylbutyl) phthalate.   (dihexyl phthalate).
$C_6H_4(COOC_6H_{13})_2$.
Properties: Oily, slightly aromatic liquid. D 1.010–1.016 (20/20C), bp 350C (735 mm Hg), acidity (as phthalic acid) 0.01% max, fp −50C, ester content 98% max, flash p 381F (193C). Combustible.
Grade: Technical.
Use: Plasticizer for cellulose ester and vinyl plastics.

diethylcadmium.
$(C_2H_5)_2Cd$.
Properties: Colorless, oily liquid. Bp 64C (19 mm Hg), fp −21C. Pyrophoric.
Derivation: Reaction of cadmium acetate with triethylaluminum.
Hazard: Ignites spontaneously in air, dangerous fire hazard.
See cadmium.
Use: TEL production, synthesis of ketones from acid chlorides.

diethylcarbamazine citrate.
CAS: 1642-54-2. $C_{10}H_{21}N_3O·C_6H_8O_7$.
(1-diethylcarbamyl-4-methylpiperazine dihydrogen citrate).
Properties: White, crystalline powder; odorless or slight odor. Mp 135–138C. Slightly hygroscopic; very soluble in water; sparingly soluble in alcohol; practically insoluble in acetone, chloroform, and ether.
Grade: USP.
Use: Medicine (anthelmintic).

N,N'-**diethylcarbanilide.**   See *sym*-diethyldiphenylurea.

diethyl carbinol.   See 3-pentanol.

1,1′-diethyl-4,4′-carbocyanine iodide.   See cryptocyanine.

diethyl carbonate.   (ethyl carbonate).
CAS: 105-58-8. $(C_2H_5)_2CO_3$.
Properties: Colorless liquid; mild odor. Stable, d 0.975 (20/4C), bp 125C, fp −43C, flash p 77F (25C) (OP). Miscible with alcohols, ketones, esters, aromatic hydrocarbons, and some aliphatic solvents; insoluble in water. Combustible.
Derivation: The steps in its manufacture are (1) reacting chlorine and carbon monoxide to produce phosgene $(COCl_2)$; (2) reacting phosgene with ethanol to make ethyl chlorocarbonate $(ClCO_2C_2H_5)$; and (3) reacting ethyl chlorocarbonate with anhydrous ethanol to produce diethyl carbonate.
Grade: Technical.
Hazard: Flammable, dangerous fire risk.
Use: Solvent for nitrocellulose, cellulose ethers, many synthetic and natural resins; organic synthesis; adhering rare earths to cathodes.

diethyl chlorophosphate.
$(C_2H_5O)_2P(O)Cl$.
Properties: Water-white liquid. Bp 60C (2 mm Hg), d 1.1915 (25C), refr index 1.4153 (25C). Soluble in alcohols. Combustible.
Grade: Technical.
Hazard: Toxic by ingestion, inhalation, and skin absorption; cholinesterase inhibitor.
Use: Intermediate in organic synthesis.

diethylcyclohexane.
$(C_2H_5)_2C_6H_{10}$.
Properties: Liquid. D 0.8037 (20/20C), bp 174C, fp −100C, insoluble in water, flash p 125F (51.6C) (OC), autoign temp 465F (240.5C). Combustible.
Hazard: Moderate fire risk.

***N,N*-diethylcyclohexylamine.**
$C_6H_{11}N(C_2H_5)_2$.
Properties: Colorless liquid. Bp 194.5C. Soluble in ether and benzene; slightly soluble in water. Combustible.
Grade: Technical.
Use: Solvent, intermediate.

diethyl-1-(2,4-dichlorophenyl)2-chlorovinyl phosphate.
$(C_2H_5O)_2PO·OC(C_6H_3Cl_2):CHCl$.
Properties: Liquid. Bp 167–170C (5 mm Hg). Insoluble in water; miscible with acetone, ethanol, kerosene, and xylene.
Hazard: Cholinesterase inhibitor.
Use: Insecticide.

diethyldichlorosilane.
CAS: 1719-53-5. $(C_2H_5)_2SiCl_2$.
Properties: Colorless liquid. Bp 130.4C, d 1.053 (25/25C), refr index 1.4309 (25C), flash p (COC)

77F (25C). Readily hydrolyzed with liberation of hydrogen chloride.
Derivation: Reaction of powdered silicon and ethyl chloride at 300C, in presence of copper powder.
Grade: Technical.
Hazard: Flammable, dangerous fire risk. Corrosive to tissue.
Use: Intermediate for silicones.

diethyl (2-(diethoxymethylsilyl)ethyl) phosphonate.
CAS: 18048-06-1. $C_{11}H_{27}O_5PSi$.
Hazard: A poison by ingestion and skin contact. A moderate skin and severe eye irritant.

***O,O*-diethyl-*S*,2-diethylaminoethyl phosphorothioate hydrogen oxalate.**   See tetram.

diethyl diethylmalonate.
$(C_2H_5)_2C(COOC_2H_5)_2$.
Properties: Colorless liquid; sweet odor. D 0.984 (25/25C).
Use: Intermediate.

diethyldiiodostannane.
CAS: 2767-55-7. $C_4H_{10}I_2Sn$.
Properties: Very slightly sol white crystals or white needles. Mp: 44C, bp: 240–245C (decomp).
Hazard: A poison by ingestion.

diethyl 2,3-dimercaptosuccinate.
CAS: 17660-58-1. $C_8H_{14}O_4S_2$.
Hazard: Moderately toxic.

***o,o*-diethyl *o*-(4-(1-((((dimethylamino) carbonyl)oxy)imino)ethyl)phenyl) phosphorothioic acid ester.**
CAS: 22942-43-4. $C_{15}H_{23}N_2O_5PS$.
Hazard: A poison by ingestion.
Use: Agricultural chemical.

diethyl 2,6-dimethyl-4(2-pyridyl)-1,4-dihydro-3,5-pyridine-dicarboxylate.
CAS: 23125-28-2. $C_{18}H_{22}N_2O_4$.
Hazard: A poison.

diethyldiphenyldichloroethane.   See 1,1-dichloro-2,2-bis(*p*-ethylphenyl)ethane.

diethyl diselenide.   See ethyl diselenide.

***sym*-diethyldiphenylurea.**   (*N,N′*-diethylcarbanilide; ethyl centralite; carbamite).
$C_2H_5(C_6H_5)NCON(C_6H_5)C_2H_5$.
Properties: White crystalline, solid; peppery odor. Mp 79C, bp 325–330C, d 1.12 (20C), flash p 302F (150C). Insoluble in water; soluble in organic solvents.
Use: Stabilizer for nitrocellulose-based smokeless powder and in solid rocket propellants.

diethylenediamine.   See piperazine.

1,4-diethylene dioxide.   See 1,4-dioxane.

diethylene disulfide.   See 1,4-dithiane.

diethylene ether.   See 1,4-dioxane.

diethylene glycol.   (dihydroxydiethyl ether; diglycol; DEG).
CAS: 111-46-6. $CH_2OHCH_2OCH_2CH_2OH$.
Properties: Colorless, syrupy liquid; practically odorless; sweetish taste. Bp 245.0C, fp −80C, d 1.1184 (20/20C), wt/gal 9.35 (15C), refr index 1.446 (25C), surface tension 48.5 dynes/cm (25C), viscosity 0.30 cP (25C), vap press 0.01 mm Hg (30C), flash p 255F (123.9C), autoign temp 444F (228.9C). Extremely hygroscopic. Noncorrosive. Lowers freezing point of water; miscible with water, ethanol, acetone, ether, and ethylene glycol; immiscible with benzene, toluene, and carbon tetrachloride. Combustible.
Derivation: By-product of manufacture of ethylene glycol.
Grade: Technical.
Hazard: Hazardous for household use in concentration of 10% or more (FDA).
Use: Production of polyurethane and unsaturated polyester resins, triethylene glycol; textile softener; petroleum-solvent extraction; dehydration of natural gas, plasticizers, and surfactants; solvent for nitrocellulose and many dyes and oils; humectant for tobacco, casein, synthetic sponges, and paper products; cork compositions, bookbinding adhesives, dyeing assistant, cosmetics, antifreeze solutions.

diethylene glycol acetate.   See diethylene glycol monoacetate.

diethylene glycol bis(allyl carbonate).   (allyl diglycol carbonate).
CAS: 142-22-3. $O[CH_2CH_2OCOO(C_3H_5)]_2$.
Properties: Liquid. Fp −4C, bp 160C (4 mm Hg), d 1.143 (20C), viscosity 9 cP (20C).
Use: Monomer for allyl resins, particularly in optically clear castings.

diethylene glycol bis(n-butyl carbonate). See butyl diglycol carbonate.

diethylene glycol bis(chloroformate).   See diglycol chloroformate.

diethylene glycol bis(cresyl carbonate).   See cresyl diglycol carbonate.

diethylene glycol bis(2,2-dichloropropionate).
A herbicide.
See "Garlon" *[Dow Agrosciences]*.

diethylene glycol bis(phenyl carbonate).
See phenyl diglycol carbonate.

diethylene glycol diacetate.   (diglycol acetate).
$(CH_3COOCH_2CH_2)_2O$.
Properties: Colorless liquid. D 1.1159, bp 250C, mp 19.1C, flash p 275F (135C), vap press 0.02 mm Hg. Miscible with water. Combustible.
Grade: Technical.
Use: Solvent for cellulose esters, printing inks, lacquers.

diethylene glycol dibenzoate.
$C_6H_5COO(CH_2)_2O(CH_2)_2OOCC_6H_5$.
Properties: Liquid. Bp 225–227C (3 mm Hg), flash p 450F (232C). Combustible.
Use: Plasticizer.

diethylene glycol dibutyl ether.   (dibutyl "Carbitol" *[Union]*).
CAS: 112-73-2. $C_4H_9O(C_2H_4O)_2C_4H_9$.
Properties: Almost colorless liquid. D 0.8853 (20/20C), bulk d 7.36 lb/gal (20C), bp 256C, vap press 0.02 mm Hg (20C), fp −60.2C, viscosity 2.39 cP (20C), flash p 245F (118C). Slightly soluble in water. Combustible.
Use: High-boiling, inert solvent with application in extraction processes and in coatings and inks; diluent in vinyl chloride dispersions; extractant for uranium ores.

diethylene glycol dicarbamate.   See diglycol carbamate.

diethylene glycol diethyl ether.   (diethyl "Carbitol" *[Union]*; ethyl diglyme).
CAS: 112-36-7. $C_2H_5O(C_2H_4O)_2C_2H_5$.
Properties: Colorless liquid. Very stable. D 0.9082 (20/20C), bp 189C, flash p 180F (82.2C), wt/gal 7.56 lb (20C), fp −44.3C. Soluble in hydrocarbons and water. Combustible.
Use: Solvent for nitrocellulose, resins, and lacquers; high-boiling medium and solvent for organic synthesis.

diethylene glycol dimethyl ether.   (diglyme; diglycol methyl ether).
CAS: 111-96-6. $CH_3(OCH_2CH_2)_2OCH_3$.
Properties: Colorless liquid; mild odor. Bp 162.0C, fp −68.0C, d 0.9451 (20/20C), flash p 153F (67.2C), viscosity 1.089 cP (20C). Miscible with water and hydrocarbons. Combustible.
Grade: Technical.
Use: Solvent, anhydrous reaction medium for organometallic synthesis.

diethylene glycol dinitrate.   (DEGN; diglycol nitrate; dinitroglycol).
CAS: 693-21-0. $(O_2NOCH_2CH_2)_2O$.
Properties: Liquid. D 1.377 (25/4C), fp −11.3C, bp 161C. Slightly soluble in water and alcohol; soluble in ether.
Hazard: Severe explosion hazard when shocked or heated. Toxic by ingestion.
Use: Plasticizer in solid rocket propellants.

diethylene glycol dipelargonate.
$(C_8H_{17}COOCH_2CH_2)_2O$.
A simple ester of pelargonic acid. Acid number 2.0, bp 229C (5 mm Hg), pour point 10F, viscosity (SUV at 110C) 36 seconds, flash p 410F (210C). Combustible.
Use: Secondary plasticizer for vinyl resins.

diethylene glycol distearate. See diglycol stearate.

diethylene glycol ethyl ether acrylate. See ethylcarbitol acrylate.

diethylene glycol isopropyl ether. See isopropyl carbitol.

diethylene glycol monoacetate. (diethylene glycol acetate).
$HO(CH_2)_2(CH_2)_2OOCCH_3$.
Miscible with water and aromatic hydrocarbons. Solvent for nitrocellulose, cellulose acetate, camphor, and rosin.

diethylene glycol monobutyl ether. (butyl "Carbitol" [Union]).
CAS: 112-34-5. $C_4H_9OCH_2CH_2OCH_2CH_2OH$.
Properties: Colorless liquid; faint butyl odor. Bp 230.6C, d 0.9536 (20/20C), wt/gal 7.94 lb (20C), refr index 1.4316 (20C), viscosity 0.0649 cP (20C), vap press 0.01 mm Hg (20C), specific heat 0.546 cal/g K (20–25C), flash p 172F (77.7C), autoign temp 442F (227.7C), coefficient of expansion 0.00088 K^{-1} to 20C, fp −68.1C. Soluble in oils and water. Combustible.
Grade: Technical.
Hazard: Hematologic, liver and kidney effects.
Use: Solvent for nitrocellulose, oils, dyes, gums, soaps, and polymers; plasticizer intermediate.

diethylene glycol monobutyl ether acetate.
(butyl "Carbitol" acetate [Union]).
CAS: 124-17-4. $CH_3CO(OC_2H_4)_2OC_4H_9$.
Properties: Colorless liquid. D 0.9810 (20/20C), bp 246.7C, vap press <0.01 mm Hg (20C), flash p 240F (115.5C), wt/gal 8.16 lb (20C), nitrocellulose-xylene dilution ratio 1:8, coefficient of expansion 0.0010 K^{-1} (20C), fp −32.3C, viscosity 0.0356 cP (20C), autoign temp 570F (298.8C). Miscible with most organic liquids. Combustible.
Grade: Technical.
Use: Solvent for oils, resins, and gums, also for cellulose nitrate and polymeric coatings; plasticizer in lacquers and coatings.

diethylene glycol monoethyl ether. ("Carbitol" solvent [Union]).
CAS: 111-90-0. $CH_2OHCH_2OCH_2CH_2OC_2H_5$.
Properties: Colorless, hygroscopic liquid; mild, pleasant odor; slightly viscous; stable. Bp 195–202C, d 1.0272 (20/20C), refr index 1.425 (25C),

flash p 205F (96.1C), wt/gal 8.55 lb (20C). Miscible with water and the common organic solvents. Combustible.
Grade: Technical.
Use: Solvent for dyes, nitrocellulose, and resins; mutual solvent for mineral-oil–soap and mineral-oil–sulfonated-oil mixtures; nonaqueous stains for wood, for setting the twist and conditioning yarns and cloth; textile printing, textile soaps, lacquers, organic synthesis; brake–fluid diluent.

diethylene glycol monoethyl ether acetate.
("Carbitol" acetate [Union]).
CAS: 112-15-2. $CH_3COOCH_2OCH_2CH_2OC_2H_5$.
Properties: Colorless liquid. D 1.0114 (20/20C), bp 217.4C, vap press 0.05 mm Hg (20C), flash p 230F (110C), wt/gal 8.4 lb (20C), coefficient of expansion 0.00105, fp −25C, refr index 1.418 (30C), viscosity 0.0279 cP (20C). Soluble in water; miscible with most organic solvents. Combustible.
Grade: Technical.
Use: Solvent for cellulose esters, gums, resins; coatings and lacquers; printing inks.

diethylene glycol monohexyl ether. (n-hexyl "Carbitol" [Union]).
CAS: 112-59-4. $C_6H_{13}OC_2H_4OC_2H_4OH$.
Properties: Water-white liquid. D 0.9346 (20/20C), 7.8 lb/gal (20C), bp 259.1C, vap press <0.01 mm Hg (20C), fp −33C, viscosity 8.6 cP (20C), flash p 285F (140.5C). Combustible.
Use: High-boiling solvent.

diethylene glycol monolaurate. See diglycol laurate.

diethylene glycol monomethyl ether. [(2-β-methyl "Carbitol" [Union]), methoxyethoxy ethanol].
CAS: 111-77-3. $CH_3OCH_2CH_2OCH_2CH_2OH$.
Properties: Colorless liquid. Refr index 1.4264 (27C), d 1.0211 (20/4C), bp 194C, soluble in water, flash p 200F (93.3C), wt/gal 8.51 lb (20C). Combustible.
Use: Solvent, brake–fluid component, intermediate.

diethylene glycol monomethyl ether acetate.
(methyl "Carbitol" acetate [Union]).
CAS: 629-38-9. $CH_3COOC_2H_4OC_2H_4OCH_3$.
Properties: Colorless liquid. Bp 209.1C, flash p 180F (82.2C) (OC), d 1.04 (20/20C), vap press 0.12 mm Hg (20C). Combustible.
Use: Solvent.

diethylene glycol monooleate. See diglycol oleate.

diethylene glycol monoricinoleate. See diglycol ricinoleate.

diethylene glycol monostearate. See diglycol monostearate.

diethylene glycol phthalate. See diglycol phthalate.

diethylene glycol stearate. See diglycol stearate.

diethylene oxide. See 1,4-dioxane.

diethylenetriamine.
CAS: 111-40-0. $NH_2C_2H_4NHC_2H_4NH_2$.
Properties: Yellow liquid; ammoniacal odor. Bp 206.7C, fp −39C, d 0.9542 (20/20C), vap press 0.37 mm Hg (20C), flash p 215F (101.6C), wt/gal 7.9 lb (20C), viscosity 0.0714 cP (20C), coefficient of expansion 0.00088 K^{-1}. Strongly alkaline, hygroscopic, somewhat viscous; soluble in water and hydrocarbons. Corrosive to copper and its alloys.
Grade: Technical.
Hazard: Toxic by ingestion, inhalation, and skin absorption. Strong irritant to eyes, skin, and upper respiratory tract.
Use: Solvent for sulfur, acid gases, various resins, and dyes; saponification agent for acidic materials; fuel component.
See hydyne.

diethylenetriamine pentaacetic acid.
CAS: 67-43-6. $HOOCCH_2N[CH_2CH_2N(CH_2COOH)_2]_2$.
Properties: White, crystalline solid. Mp 230C (decomposes). Slightly soluble in cold water; soluble in hot water.
Grade: Technical.
Use: Chelating agent.

N,N-diethylethanolamine. See diethylaminoethanol.

diethyl ether. See ethyl ether.

diethyl ethoxymethylenemalonate.
$C_2H_5OCH:C(COOC_2H_5)_2$.
Properties: Liquid. D 1.0855 (15/15C), refr index 1.4625 (20C), bp 279–281C (decomposes), flash p 190F (87.7C). Insoluble in water.
Grade: 98% (min purity). Combustible.
Use: Synthesis.

uns-diethylethylene. See 2-ethyl-1-butene.

N,N-diethylethylenediamine.
CAS: 100-36-7. $(C_2H_5)_2NC_2H_4NH_2$.
Properties: Colorless liquid. Bp 145.2C, sets to a glass below −100C, d 0.8211 (20/20C), wt/gal 6.8 lb (20C), flash p 115F (46.1C) (OC). Miscible with water. Combustible.
Hazard: Moderate fire risk. Moderately toxic.
Use: Intermediate.

p,p′-(1,2-diethylethylene)diphenol. See hexestrol.

diethyl ethylmalonate.
$C_2H_5CH(COOC_2H_5)_2$.
Properties: Colorless liquid; ester odor. D 0.9994 (25/25C). Combustible.
Use: Intermediate.

diethyl ethylphosphonate.
$C_2H_5P(O)(OC_2H_5)_2$.
Properties: Colorless liquid; mild odor. D 1.025 (20/4C), bp 82–83C (11 mm Hg), flash p 220F (104.4C) (COC). Stable; miscible with most common organic solvents; slightly soluble in water; soluble in alcohol. Combustible.
Use: Heavy metal extraction and solvent separation, gasoline additives, antifoam agent, plasticizer, textile conditioner, and antistatic agent.

o,o-diethyl s-(2-(ethylthio)-6-methyl-4-pyrimidinyl) phosphorodithioate.
CAS: 32522-68-2. $C_{11}H_{20}N_2O_2PS_3$.
Hazard: A poison by ingestion.
Use: Agricultural chemical.

1,1′-diethyl ferrocenoate. (1,1′-ferrocene dicarboxylic acid diethyl ester).
$(C_2H_5OOCC_5H_4)_2Fe$.
Properties: Orange crystals. Mp 38–40C.
Use: Intermediate, high-temperature plasticizer.

diethylgermanium dichloride.
$(C_2H_5)_2GeCl_2$.
Properties: Colorless liquid. Fp −38C, bp 175C. Decomposed by water.
Use: Biocide, intermediate.

diethylglycol phthalate. (diethoxyethyl phthalate).
$(C_2H_5OCH_2CH_2OOC)_2C_6H_4$.
Properties: Water-white to pale-straw liquid. D 1.115–1.120 (20/20C), wt/gal 9.31 lb, flash p 343F (172C). Combustible.
Use: Plasticizer.

di(2-ethylhexyl) adipate. (DOA; dioctyl adipate).
CAS: 103-23-1. $[CH_2CH_2COOCH_2CH(C_2H_5)C_4H_9]_2$.
Properties: Light-colored, oily liquid. D 0.9268 (20/20C), refr index 1.4472, flash p 385F (196C), pour point −75C, bp 417C (214C at 5 mm Hg), vap press 2.60 mm Hg (200C), viscosity 13.7 cP (20C), wt/gal 7.7 lb (20C). Insoluble in water. Combustible.
Grade: 99% min.
Hazard: Questionable carcinogen.
Use: Plasticizer, commonly blended with general purpose plasticizers such as DOP and DIOP; in processing polyvinyl and other polymers; solvent; aircraft lubricants.

di(2-ethylhexyl)amine. (dioctylamine).
CAS: 20830-75-5. $[C_4H_9CH(C_2H_5)CH_2]_2NH$.

Properties: Water-white liquid; slightly ammoniacal odor. D 0.8062 (20/20C), 6.7 lb/gal (20C), bp 281.1C, vap press 0.01 mm Hg (20C), viscosity 3.70 cP (20C), refr index 1.4420 (20C), flash p 270F (132C). High solubility in hydrocarbons; low solubility in water. Combustible.
Hazard: Moderately toxic.
Use: Synthesis of dyestuffs, insecticides, emulsifying agents, etc.

di(2-ethylhexyl)aminoethanol. See di(2-ethylhexyl)ethanolamine.

di(2-ethylhexyl) azelate. (DOZ; dioctyl azelate).
$(CH_2)_7[COOCH_2CH(C_2H_5)C_4H_9]_2$.
Properties: Colorless liquid; odorless. D 0.919 (20/20C), refr index 1.4472, bp 376C, flash p 430F (221C). Combustible.
Grade: 99% pure.
Use: Plasticizer for vinyls, especially used as low-temperature plasticizer; base for synthetic lubricants.

di(2-ethylhexyl)ethanolamine. (di(2-ethylhexyl)aminoethanol; dioctylaminoethanol).
$[C_4H_9CH(C_2H_5)CH_2]_2N(CH_2)_2OH$.
Properties: Colorless liquid. Wt/gal 7.2 lb, flash p 280F (137C). Insoluble in water. Combustible.
Use: Emulsifier, acid-stable wetting agent.

di(2-ethylhexyl) ether.
CAS: 10143-60-9. $[C_4H_9CH(C_2H_5)CH_2]_2O$.
Properties: Colorless, stable liquid; mild odor. D 0.8121 (20/20C), 6.6 lb/gal (20C), bp 269.4C, vap press <0.01 mm Hg (20C), sets to glass below −95C, viscosity 2.89 cP (20C), refr index 1.4525 (20C). Almost insoluble in water. Combustible.
Use: High-boiling, inert reaction medium; component of certain foam breakers.

di(2-ethylhexyl)-2-ethylhexyl phosphonate.
(bis(2-ethylhexyl)-2-ethylhexyl phosphonate).
$C_8H_{17}PO(OC_8H_{17})_2$.
Properties: Colorless liquid; mold odor. D 0.908 (20/4C), bp 160–161C (0.26 mm Hg), flash p 420F (215C). Insoluble in water; miscible with most common organic solvents. Combustible.
Use: Heavy metal extraction, solvent separation, gasoline additive, antifoam agent, plasticizer, stabilizer, textile conditioner, and antistatic agent.

di(2-ethylhexyl) fumarate. (dioctyl fumarate; DOF).
CAS: 141-02-6. $C_8H_{17}OOCCH:CHCOOC_8H_{17}$.
Properties: Clear, mobile liquid. D 0.937–0.940 (25/25C), bp 211–220C, flash p 365F (185C). Combustible.
Use: Monomer for polymerization and copolymerization.

di(2-ethylhexyl) hexahydrophthalate.
(dioctyl hexahydrophthalate).
$C_6H_{10}[COOCH_2CH(C_2H_5)C_4H_9]_2$.
Properties: Light-colored liquid. D 0.9586 (20/20C), 8.0 lb/gal (20C), bp 216C (5 mm Hg), vap press 2.2 mm Hg (200C), viscosity 42.1 cP (20C), flash p 425F (218C). Insoluble in water. Combustible.
Use: Plasticizer.

di(2-ethylhexyl) hydrogen phosphate.
(bis(2-ethylhexyl) hydrogen phosphate).
$(C_8H_{17})_2HPO_4$.
Properties: Solid. D 0.972 (20/4C), flash p 340F (171C) (COC). Insoluble in water.
Use: Heavy metal extraction.

di(2-ethylhexyl) isophthalate. (dioctyl isophthalate).
$C_6H_4[COOCH_2CH(C_2H_5)C_4H_9]_2$.
Properties: Colorless liquid. Bp 258C (10 mm Hg), d 0.984 (20/20C), 8.2 lb/gal, pour point −46C, viscosity 86.5 cP (20C). Insoluble in water. Combustible.
Use: Plasticizer.

di(2-ethylhexyl) maleate. (dioctyl maleate; DOM).
CAS: 142-16-5. $C_8H_{17}OOCCH:CHCOOC_8H_{17}$.
Properties: Liquid. Bp 209C (10 mm Hg), fp sets to glass below −60C, d 0.9436 (20/20C), wt/gal 7.9 lb (20C), flash p 365F (185C) (OC). Insoluble in water. Combustible.
Use: Copolymers, intermediate.

di(2-ethylhexyl) phosphite. (bis(2-ethylhexyl) phosphite).
$(C_8H_{17}O)_2PHO$.
Properties: Mobile, colorless liquid; mild odor. High degree of thermal stability. D 0.937 (20/4C), bp 163–164C (3 mm Hg), refr index 1.444 (25C), flash p 330F (165C). Insoluble in water (hydrolyzes very slowly); miscible with most common organic solvents. Combustible.
Use: Lubricant additive, intermediate, adhesive.

di(2-ethylhexyl) phosphoric acid. (di-*n*-octyl phosphoric acid; di[2-ethylhexyl] hydrogen phosphate).
CAS: 298-07-7. $[C_4H_9CH(C_2H_5)CH_2]_2HPO_4$.
Properties: Strongly acidic liquid. D 0.973 (25/25C), fp −60C, refr index 1.4420 (25C), flash p 385F (196C), wt/gal 8.2 lb. Insoluble in water; soluble in organic solvents. Combustible.
Use: Metal extraction and separation, intermediate for wetting agents and detergents.

di(2-ethylhexyl) phthalate. (di-*sec*-octyl phthalate; dioctyl phthalate; DOP).
CAS: 117-81-7. $C_6H_4[COOCH_2CH(C_2H_5)C_4H_9]_2$.
Properties: Light-colored, odorless liquid. D 0.9861 (20/20C), pour p −46C, refr index 1.4836, flash p

425F (218C), 8.20 lb/gal (20C), bp 231C (5 mm Hg) vap press 1.32 mm Hg (200C), viscosity 81.4 cP (20C). Insoluble in water; miscible with mineral oil. Combustible.
Derivation: Reaction of 2-ethylhexanol and phthalic anhydride.
Hazard: Lower respiratory tract irritant. Possible carcinogen.
Use: Plasticizer for many resins and elastomers.

di(2-ethylhexyl) sebacate.   (dioctyl sebacate).
CAS: 122-62-3. $C_4H_8(COOC_8H_{17})_2$.
Properties: Pale straw-colored liquid. D 0.91 (25C), refr index 1.447 (28C), bp 248C (4 mm Hg), fp −55C, flash p 410F (210C) (COC). Insoluble in water; partially compatible with cellulose acetate and cellulose acetate butyrate; compatible with ethyl cellulose, polystyrene, polyethylene, vinyl chloride, and vinyl chloride acetate. Combustible.
Use: Plasticizer.

di(2-ethylhexyl) sodium sulfosuccinate.   See dioctyl sodium sulfosuccinate.

di(2-ethylhexyl) succinate.   (dioctyl succinate).
$C_8H_{17}OCOCH_2CH_2COOC_8H_{17}$.
Properties: Liquid. Bp 257C (50 mm Hg), fp sets to glass below −60C, d 0.9346 (20/20C), wt/gal 7.8 lb (20C), flash p 315F (157C) (OC), vap press <0.01 mm Hg (20C). Solubility in water <0.01% by wt (20C). Combustible.
Use: Plasticizer, intermediate.

di-2-ethylhexyltin dichloride.
CAS: 25430-97-1. $C_{16}H_{34}Cl_2Sn$.
Properties: Crystals.
Hazard: A poison.

diethylhydroxylamine.
CAS: 3710-84-7. $(C_2H_5)_2NOH$.
Properties: Liquid. Refr index 1.4238 (20C). Combustible.
Grade: 85%.
Hazard: Upper respiratory tract irritant.
Use: Photographic developer, antioxidant, corrosion inhibitor.

diethyl isoamylethylmalonate.
$(C_2H_5)(C_5H_{11})C(COOC_2H_5)_2$.
Properties: Colorless liquid; sweet odor. D 0.950 (25/25C). Combustible.
Use: Intermediate.

diethyl ketone.   (metacetone; propione; 3-pentanone; ethyl propionyl).
CAS: 96-22-0. $C_2H_5COC_2H_5$.
Properties: Colorless, mobile liquid; acetone-like odor. Autoign temp 846F (452C), bp 101C, d 0.816, fp −42C, flash p 55F (12.78C) (OC). Soluble in alcohol and ether; slightly soluble in water.

Derivation: By distilling sugar with an excess of lime.
Method of purification: Rectification.
Grade: Technical.
Hazard: Flammable, dangerous fire hazard. Upper respiratory tract irritant and central nervous system impairment.
Use: Medicine, organic synthesis.

diethylmagnesium.   (magnesium ethane).
$C_4H_{10}Mg$.
Properties: Pyrophoric solid compound; reacts violently with water and steam; combusts spontaneously in air or in carbon dioxide.
Hazard: Toxic.

diethyl maleate.
CAS: 141-05-9. $C_2H_5OOCCH:CHCOOC_2H_5$.
Properties: Water-white liquid. D 1.0687, bulk d 8.92 lb/gal (20C), refr index 1.4400 (20C), bp 225C, fp approximately −115C, viscosity 3.567 cP (20C), flash p 200F (93.3C) (COC), dielectric constant 2.18 (calc) (25C), surface tension 3.70 dynes/cm (20C). Readily soluble in alcohol, diethyl ether, paraffinic hydrocarbons, and common organic solvents; soluble in water; readily hydrolyzed by alkaline solutions. Combustible.
Derivation: Reaction of maleic anhydride with ethanol in the presence of a catalyst.
Hazard: Irritant to eyes and skin.
Use: Organic synthesis, flavoring.

diethyl malonate.   See ethyl malonate.

diethylmalonylurea.   See barbital.

***o,o*-diethyl s-((5-methoxy-1,3,4-thiadiazol-2-yl)methyl) phosphorothioate.**
CAS: 38090-84-5. $C_8H_{15}N_2O_4PS_2$.
Hazard: A poison by ingestion.

***o,o*-diethyl o-(4-(1-((((methylamino)carbonyl)oxy)imino)ethyl)phenyl) phosphorothioic acid ester.**
CAS: 22941-83-9. $C_{14}H_{21}N_2O_5PS$.
Hazard: Moderately toxic by ingestion.
Use: Agricultural chemical.

diethyl(1-methylbutyl) malonate.
$[C_3H_7CH(CH_3)]CH(COOC_2H_5)_2$.
Properties: Colorless liquid; ester odor. D 0.969 (25/25C).
Use: Intermediate, organic synthesis.

***o,o*-diethyl o-(4-(1-(((((1-methylethyl)amino)carbonyl)oxy)imino)ethyl) phenyl) phosphorothioic acid ester.**
CAS: 22936-17-0. mf: $C_{16}H_{25}N_2O_5PS$.
Hazard: A poison by ingestion.
Use: Agricultural chemical.

diethylmethylmethane.   See 3-methylpentane.

***O,O*-diethyl-*O-p*-nitrophenyl   phosphoroth-
ioate.**   See parathion.

**N,N-diethyl-*n*-(1,2,3,4,4a,5,10,10a-
octahydro-6-hydroxy-1-propylbenzo(g)qui
nolin-3-yl)-sulfamide,   (3-α,4a-α,10a-β)-
(+−)-.**
CAS: 87056-78-8. $C_{20}H_{33}N_3O_3S$.
Hazard: Human systemic effects reported. A reproductive hazard.

diethyl oxalate.   See ethyl oxalate.

diethyl oxide.   See ethyl ether.

di(*p*-ethylphenyl)dichloroethane.   See
1,1-dichloro-2,2-bis(*p*-ethylphenyl)ethane.

N,N-diethyl-*p*-phenylenediamine.   See
p-aminodiethylaniline.

diethyl phenyltin acetate.
CAS: 64036-46-0. $C_{12}H_{18}O_2Sn$.
Hazard: A poison by ingestion.

**diethylphosphinic acid anhydride with diethyl
phosphorothionate.**
CAS: 7506-77-6. $C_8H_{20}O_4P_2S$.
Hazard: Moderately toxic.

diethyl phosphite.
CAS: 762-04-9. $(C_2H_5O)_2HPO$.
Properties: Water-white liquid. Bp 138C, d 1.069
(25C), refr index 1.4061 (25C), flash p 195F (90.5C)
(COC). Soluble in water and common organic solvents. Combustible.
Use: Paint solvent, lubricant additive, antioxidant,
reducing agent, intermediate for flame retardants
and insecticides, phosphorylating agent.

***O,O*-diethyl phosphorochloridothioate.**
(ethyl PCT).
CAS: 2524-04-1. $(C_2H_5O)P(S)Cl$.
Properties: Colorless to light-amber liquid. D 1.196
(25/25C), fp less than −75C, bp 49C (below 1 mm
Hg), refr index 1.4705 (25C). Insoluble in water;
soluble in most organic solvents. Stable at room
temperature, slowly isomerizes at 100C.
Hazard: Cholinesterase inhibitor, irritant to eyes and
lungs.
Use: Intermediate for pesticides, oil, and gasoline
additives, flame retardants, flotation agents.

**4-(*o*-(*o,o*-diethylphosphorothioyl))-
benzaldoximino-*n*-butylcarbamate.**
CAS: 22942-02-5. $C_{16}H_{25}N_2O_5PS$.
Hazard: A poison by ingestion.
Use: Agricultural chemical.

**4-(*o*-(*o,o*-diethylphosphorothioyl))-
benzaldoximino-*n*-morpholinylcarbamate.**
CAS: 22935-72-4. $C_{16}H_{23}N_2O_6PS$.

Hazard: A poison by ingestion.
Use: Agricultural chemical.

diethyl phthalate.   (ethyl phthalate; DEP).
CAS: 84-66-2. $C_6H_4(CO_2C_2H_5)_2$.
Properties: Water-white, stable liquid; odorless; bitter taste. Fp −40.5C, refr index 1.5002 (25C), surface tension 37.5 dynes/cm (20C), viscosity 31.3
centistokes (OC), vap press 14 mm Hg (163C), 30
mm Hg (182C), 734 mm Hg (295C), bp 298C, flash
p 325F (162.7C) (OC), wt/gal approximately 9.31
lb (20C), d 1.120 (25/25C). Miscible with alcohols,
ketones, esters, and aromatic hydrocarbons; partly
miscible with aliphatic solvents; insoluble in water.
Combustible.
Derivation: By reacting phthalic anhydride with
ethanol, followed by careful purification.
Grade: Technical.
Hazard: Toxic by ingestion and inhalation, strong
irritant to eyes, mucous membranes, and upper respiratory tract. Narcotic. Questionable carcinogen.
Use: Solvent for nitrocellulose and cellulose acetate,
plasticizer, wetting agent, insecticidal sprays, camphor substitute, plastics, perfumery as fixative and
solvent, alcohol denaturant, mosquito repellents,
plasticizer in solid rocket propellants.

2,2-diethyl-1,3-propanediol.
$HOCH_2C(C_2H_5)_2CH_2OH$.
Properties: Colorless liquid. Mp 61.3C, bp 160C
(50 mm Hg), d 0.949 (at melting point), wt/gal 8.2
lb (60C), flash p 215C (101.6C) (OC). Soluble in
water. Combustible.
Grade: Technical, pharmaceutical.
Hazard: Toxic by ingestion.
Use: Emulsifying agent, intermediate, medicine.

diethyl s-propyl phosphorothiolate.   See
ethyl propyl phosphorothioate.

***O,O*-diethyl-*O*-2-pyrazinyl phosphorothioate.**
See thionazin.

diethylpyrocarbonate.   (DEPC).

$$C_2H_5O-\overset{\overset{O}{\|}}{C}-O-\overset{\overset{O}{\|}}{C}-OC_2H_5$$

Properties: Colorless liquid; sweet esterlike odor.
Refr index 1.395–1.398 (25C). Miscible with
ethanol and methanol.
Grade: FCC.
Hazard: Toxic. Use in food products prohibited
(FDA); irritant to eyes and skin.
Use: Fermentation inhibitor.

diethylstannium diiodide.   See diethyldiiodostannane.

diethylstilbestrol.   (stilbestrol; DES; 3,4-bis(*p*-
hydroxyphenyl)-3-hexene).

CAS: 56-53-1.

A nonsteroid, synthetic estrogen, always in the *trans* form. It is the most active of the commonly used stilbene compounds.

Properties: White, crystalline powder; odorless. Mp 169–172C. Almost insoluble in water; soluble in alcohol, chloroform, ether, fatty oils, and dilute alkali hydroxide.

Derivation: From anethole hydrobromide, from anisole, from anisoin.

Grade: USP.

Hazard: A confirmed carcinogen. Under USDA regulations, no residues are permitted in tissues of slaughtered animals; not permitted in cattle feeds (FDA).

Use: Biochemical research, medicine (prevents spontaneous abortion), veterinary medicine

diethyl succinate.

CAS: 123-25-1.   (-CH$_2$COOC$_2$H$_5$)$_2$.

Properties: Colorless liquid; faint pleasant odor. Bp 216.2C, fp −21C, d 1.0418 (20/20C), wt/gal 8.7 lb (20C), refr index 1.4201 (20C), flash p 230F (OC) (110C). Miscible with alcohol and ether; slightly soluble in water, Combustible.

Use: Plasticizer, intermediate, flavoring.

diethyl sulfate.   (ethyl sulfate).

CAS: 64-67-5.   (C$_2$H$_5$)$_2$SO$_4$.

Properties: Colorless liquid; faint ethereal odor; irritating aftereffects. D 1.1803, bp 208C (decomposes), vap press 0.19 mm Hg (20C), flash p 220F (104.4C), autoign temp 817F (436C), wt/gal 9.8 lb (20C), fp −24.4C, viscosity 1.79 cP (20C). Noncorrosive. Soluble in alcohol and ether; insoluble in water. Combustible.

Derivation: Action of fuming sulfuric acid on ethanol.

Method of purification: Rectification in vacuo.

Grade: Technical.

Hazard: Toxic by ingestion and inhalation, strong irritant. Probable carcinogen.

Use: Ethylating agent in organic synthesis.

diethyl sulfide.   See ethyl sulfide.

diethyl tartrate.

C$_4$H$_4$O$_6$(C$_2$H$_5$)$_2$.

Properties: Colorless, thick, oily liquid. Bp 280C, mp 17C, D 1.204 (20/4C). Soluble in water and alcohol. Combustible.

Use: Plasticizer for automobile lacquers; solvent for nitrocellulose, gums, and resins.

o,o-diethyl o-tetrahydrofurfuryl ester phosphorothioic acid.

CAS: 3513-92-6.   C$_9$H$_{19}$O$_4$PS.

Hazard: A poison by ingestion.

Use: Agricultural chemical.

1,3-diethylthiourea.

CAS: 105-55-5.   C$_2$H$_5$NHCSNHC$_2$H$_5$.

Properties: Buff solid. Mp 68–71C. Slightly soluble in water; soluble in methanol, ether, acetone, benzene, and ethyl acetate; insoluble in gasoline.

Hazard: Questionable carcinogen.

Use: Inhibitor of corrosion in metal pickling solutions; accelerator; activator in elastomers.

diethyltin chloride.   See dichlorodihexylstannane.

diethyltin di(10-camphorsulfonate).

CAS: 73940-85-9.   mf: C$_{24}$H$_{40}$O$_8$S$_2$Sn.

Hazard: A poison. TWA 0.1 mg(Sn)/m^3; STEL 0.2 mg(Sn)/m^3 (skin).

diethyltin dicaprylate.   See diethylbis(octanoyloxy)stannane.

diethyltin diiodide.   See di-2-ethylhexyltin dichloride.

diethyltin dioctanoate.   See diethylbis(octanoyloxy)stannane.

N,N-diethyl-m-toluamide.   (deet).

CAS: 134-62-3.   CH$_3$C$_6$H$_4$CON(C$_2$H$_5$)$_2$.

Properties: Colorless liquid; mild bland odor. Bp 160C (19 mm Hg), d 0.996–1.002 (25/25C), refr index 1.5200–1.5235 (25C). Soluble in water, alcohol, ether, and benzene. Combustible.

Grade: USP.

Hazard: Toxic by ingestion, irritant to eyes and mucous membranes.

Use: Insect repellents, resin solvent, film formers.

N,N-diethyl-m-toluidine.

CH$_3$C$_6$H$_4$N(C$_2$H$_5$)$_2$.

Properties: Light-amber oil. Bp 231C, refr index (20C) 1.5361.

Hazard: See toluidine.

Use: Dye intermediate.

N,N-diethyl-o-toluidine.   (1-diethylamino-2-methylbenzene).

CH$_3$C$_6$H$_4$N(C$_2$H$_5$)$_2$.

Properties: Prisms from water. Mp 72.3C, bp 209C. Soluble in water, alcohol, and ether.

Derivation: From *o*-toluidine.
Hazard: See toluidine.
Use: Dye intermediate.
See DEET.

O,O-diethyl-*O*-3,5,6-trichloro-2-pyridyl phosphorothioate.

$Cl_3C_5NHOP(S)(OC_2H_5)_2$.
Properties: Solid. Mp 41.5–43C. Soluble in acetone, benzene, and ether; almost insoluble in water.
Hazard: Cholinesterase inhibitor.
Use: Insecticide.

3,9-diethyl-6-tridecanol. (heptadecanol).

$C_4H_9CH(C_2H_5)C_2H_4CH(OH)C_2H_4CH(C_2H_5)_2$.
Properties: White solid. D 0.8475, bp 309C, flash p 310F (154C) (OC), refr index 1.4531 (20C). Insoluble in water. Combustible.
Use: Intermediate for synthetic lubricants, defoamers, and surfactants.

1,1-diethylurea.

$NH_2CON(C_2H_5)_2$.
White solid; mp 75C; soluble in water, alcohol, and benzene. When copolymerized with simple urea by the use of formaldehyde, it yields modified resins that differ in nature from those made from mono-substituted ureas. These resins tend to be permanently thermoplastic.

diethylzinc. (ethylzinc; zinc ethyl; zinc diethyl).

CAS: 557-20-0. $Zn(C_2H_5)_2$.
The first known organometallic compound.
Properties: Colorless, pyrophoric liquid. D 1.207 (20C), fp −28C, bp 118C. Soluble in most hydrocarbons.
Derivation: Action of ethyl iodide on zinc and sodium-zinc, or by interaction of zinc chloride with triethyl aluminum.
Grade: Technical.
Hazard: Ignites spontaneously on contact with air, dangerous fire hazard, decomposes violently in water.
Use: Organic synthesis, catalyst for polymerization of olefins, high-energy aircraft and missile fuel, production of ethyl mercuric chloride.

differential centrifugation. Separation of cell

organelles or other particles of different density by their different rates of sedimentation in a centrifugal field.

differential gravimetric analysis. (DGA).

A variation of differential thermal analysis in which additional information is obtained by determining the rate of change in weight during the heating process.

differential scanning calorimetry. (DSC).

Determines the electrical energy input rate necessary to establish zero temperature difference between a substance and a reference material against either time or temperature as they both are subjected to a controlled temperature change.

differential thermal analysis. (DTA).

The method of precisely measuring the temperature and the rate of temperature change as heat is added to or abstracted from a sample of material that is in a controlled constant environment. The method determines whether the sample is a pure substance or a mixture and yields information about its composition and thermal properties.

differentiation. Specialization of cell structure

and function during embryonic growth and development.

diffraction, neutron. An analytical technique

analogous to X-ray diffraction, in which an incident beam of neutrons is scattered by the atoms of a crystal. Because elements that are close to each other in the periodic table differ considerably in their neutron-scattering ability, neutron diffraction is capable of distinguishing between them. For example, carbon, nitrogen, and oxygen atoms can be readily identified by neutron diffraction whereas they appear almost identical by the X-ray method. More accurate determination of the bond lengths of light atoms and distribution of molecular bonding electrons is also possible. Details of molecular structure that can be inferred only by other techniques can often be observed directly by neutron diffraction. Investigations using this method include hydrogen bonding, so-called metal cluster compounds (C-H-M relationships), and electronic charge distributions.

diffraction, X-ray. A method of spectroscopic

analysis involving the reflection or scattering of x-radiation by the atoms of a substance (lattice) as the rays pass through it. The rays are reflected by the atoms at an angle that is characteristic of the substance, yielding a spectrum that indicates its atomic or molecular structure. The spectra thus obtained are well defined and specific; from them the properties of elements and the structure of both crystalline and amorphous materials can be obtained. For example, unvulcanized rubber gives an amorphous pattern, while vulcanized rubber is crystalline; the cellulose macromolecule has alternating crystalline and amorphous areas. X-ray diffraction was one of the earliest and most successful methods of instrumental analysis; developed by Bragg and van Laue early in this century, it was used with dramatic effect by Moseley (1912) in establishing the location of several elements in the periodic system.
See lattice; crystal; x-radiation.

diffusion. The spontaneous mixing of one sub-

stance with another when in contact or separated by a permeable membrane or microporous barrier. The rate of diffusion is proportional to the concentration of the substances and increases with temperature.

Diffusion occurs most readily in gases, less so in liquids, and least in solids. The theoretical principles are stated in Fick's laws. In gases, diffusion takes place counter to gravity, and the rate at which different gases diffuse into a particular gas (e.g., air) is inversely proportional to the square root of the density. For example, carbon dioxide and chlorine vapor will diffuse in air until a uniform mixture results. Diffusion occurs in the cell walls of plants and animals (osmosis). Many substances diffuse through a parchment membrane.
See dialysis; diffusion; gaseous; osmosis.

diffusion current. Limiting current reached by electrolytic migration of ions, in a solution, under the application of a potential difference to the electrodes.

diffusion, gaseous. A technique used for separating the light isotope of uranium (^{235}U) from the heavy isotope (^{238}U). The uranium is allowed to diffuse through a series of microporous barriers whose apertures are of molecular dimensions in the form of the gas uranium hexafluoride; this is a mix of ^{238}UF$_6$ and ^{235}UF$_6$ in a ratio code of 140:1. Because of the vastly greater number of the heavier molecules and the extremely small difference in their masses, the mix must pass through a barrier a great many times to obtain a high concentration of the 235 isotope. Assuming that the diffusion rate of two gases through a porous barrier is inversely proportional to the square root of their molecular weights, the ideal separation factor is the square root of the product M_1 times M_2, where M_2 is the molecular weight of ^{238}UF$_6$, and M_1 that of ^{235}UF$_6$. This method is still in use for uranium enrichment for nuclear fuel.

diffusion layer. The liquid layer around an electrode, within which the electrolyte concentration changes.

diffusion length. A property of materials used in reactors for moderators or reflectors. It is a measure of the distance a thermal neutron diffuses after it is thermalized until it is captured. It is related to the density of the material and to the scattering and absorption cross sections.

diflufenzopyr. See 2-(1-((((3,5-difluorophenyl) amino)carbonyl)hydrazono)ethyl).

diflubenzuron. (N-((4-chlorophenyl)aminocarbonyl)-2,6-difluorobenzamide; 1-(4-chlorophenyl)-3-(2,6-difluorobenzoyl)urea).
CAS: 35367-38-5.
Use: Insecticide.

difluophosphoric acid. See difluorophosphoric acid.

2,4-difluoroaniline.
C$_6$H$_5$F$_2$N.
Properties: Liquid. Bp 170C (753 mm Hg), fp −7.5C, density 10.7 lb/gal, flash p 158F (70C).
Use: Organic synthesis.

1,1,1-difluorochlorethane. See 1,1,1-chlorodifluoroethane.

difluorochloromethane. See chlorodifluoromethane.

difluorodiazine.
FN:NF.
Properties: Gas. Can exist as *cis* and *trans* isomers.
Grade: All *trans* isomer, 95–99.8%.
Hazard: See fluorine.
Use: *trans* form: preparation of ionic fluorine compounds; *cis* form: polymerization initiator.

difluorodibromomethane. See dibromodifluoromethane.

difluorodichloromethane. See dichlorodifluoromethane.

difluorodimethylstannane.
CAS: 3582-17-0. C$_2$H$_6$F$_2$Sn.
Properties: White crystals. Bp: decomp at <360C. Sol in water.
Hazard: A poison.

difluorodiphenyltrichloroethane. See DFDT.

1,1-difluoroethane. (ethylidene fluoride).
CAS: 75-37-6. CH$_3$CHF$_2$.
Properties: Colorless, gas; odorless. Bp −24.7C, fp −117C, d 1.004 (−25C), refr index 1.255 (20C). Insoluble in water.
Derivation: By adding hydrogen fluoride to acetylene.
Grade: Technical, 98%.
Hazard: Flammable, dangerous fire risk. Flammable limits in air 3.7–18%. Narcotic in high concentration.
Use: Intermediate.

1,1-difluoroethylene. (Air) Legal label name for vinylidene fluoride.

difluoromethane.
CAS: 75-10-5. CH$_2$F$_2$.
Properties: Gas. Bp −51.6C. Soluble in alcohol; insoluble in water. High thermal stability.
Use: Refrigeration, organic synthesis.

difluoromonochloroethane. Legal label name for 1,1,1-chlorodifluoroethane.

difluoromonochloromethane. See chlorodifluoromethane.

2-(1-(((((3,5-difluorophenyl)amino)carbonyl) hydrazono)ethyl)-3-pyridinecarboxylic acid, 98.1%.
CAS: 109293-97-2. $C_{15}H_{12}F_2N_4O_3$.
Hazard: A reproductive hazard.
Use: Agricultural chemical.

difluorophosphoric acid. (difluophosphoric acid).
HPO_2F_2.
Properties: Mobile, strongly fuming, colorless liquid. D 1.583 (25/4C), fp −75C, bp 116C. Corrosive to glass and fabric. Noncombustible.
Hazard: When heated, corrosive to tissue.
Use: Chemical polishing agent, protective coatings for metal surfaces, catalyst.

difluron.
CAS: 35367-38-5. $C_{14}H_9ClF_2N_2O_2$.
Properties: Solid. Mp 239C. Sparingly soluble in water.
Hazard: Toxic by ingestion.
Use: Larvicide.

Difolatan. See *cis-n*-[1,1,2,2-tetrachloroethyl) thio]-4-cyclohexene-1,2-dicarboximide.

difurylmethane. See 2-(2-furfuryl)furan.

digester. A cylindrical metal vessel, used chiefly in the preparation of wood pulp for papermaking, in which lignin is separated from cellulose by chemical means. It operates at approximately 150 psi and 170C. The wood is fed to the digester in the form of chips, and cooking liquor is added. Standard digesters are 12 ft in diameter and 45 ft high, with a wall thickness of 2 inches. These hold about 20 cords of wood, and some are even larger. The cooking cycle varies from 2.5 hours for board stock to 5 hours for bleached paper. Heat supply is by circulating steam and heat exchanger, though some types have direct steam injection. Digesters are designed for both batch and continuous operation. They are also used in reclaiming fabricated rubber products (tires, boot, shoe, etc.).
See pulp, paper; digestion (2).

digestion. (1) The physiological processes involved in the assimilation of nutrients from ingested foods by the animal organism. Hydrochloric acid in the gastric juice plays a prominent part, aided by the saliva that initiates carbohydrate breakdown, and bile and pancreatic secretions in the intestine. Numerous types of enzymes catalyze these processes. (2) In chemical engineering, the term refers to several processes: (a) the preferential dissolution of certain mineral constituents in some ore concentrates, (b) the liquefaction of organic waste materials by microbiological action as in activated sludge, (c) the removal of lignin from wood by hot chemical solutions in the manufacture of chemical cellulose and paper pulp, and (d) the separation

of fabric from scrap tires by hot sodium hydroxide solution in the reclaiming of rubber. The equipment for (c) and (d) is called a digester.
See metabolism; nutrition.

digitalin. ((3β,5β,16β)-3-[(6-deoxy-4-o-β-d-glucopyranosyl-3-o-methyl-β-d-galactopyranosyl) oxy]-14,16-dihydroxy-card-20,22-enolide; digitalinum vernum; digitalinum true; schmiedeberg's digitalin; 3-[3-[5-[5-(4,5-dihydroxy-6-methyloxan-2-yl)oxy-4-hydroxy-6-methyloxan-2-yl]oxy-4-hydroxy-6-methyloxan-2-yl]oxy-14-hydroxy-10,13-dimethyl-1,2,3,4,5,6,7,8,9,11,12,15,16,17-tetradecahydrocyclopenta[a]phenanthren-17-yl]-2H-furan-5-one).
CAS: 71-63-6. $C_{41}H_{64}O_{13}$.
A cardiac glycoside.
Properties: White, crystalline, steroid alkaloid.
Derivation: Isolated from the seeds and leaves of *Digitalis purpurea*.
Hazard: Toxic.

digitalis. A drug obtained from dried leaves of the purple foxglove, native to southern Europe but grown in the U.S. Used in the treatment of cardiac diseases, both human and animal. Allergic reactions are infrequent. Contains both digitonin and digitoxin.
Hazard: An overdose can be fatal.

digitonin.
CAS: 11024-24-1. $C_{56}H_{92}O_{29}$.
A saponin derived from digitalis seeds.
Use: Determination of cholesterol (an insoluble addition compound is formed), analytical reagent.

digitoxigenin. (3,14-dihydrocard-20,22-enolide; $\bar{A}^{20:22}$-3,14,21-trihydroxynorcholenic acid lactone; cerberigenin; evonogenin; 3-[(3S,5R,8R, 9S,10S,13R,14S,17R)-3,14-dihydroxy-10,13-dimethyl-1,2,3,4,5,6,7,8,9,11,12,15,16,17-tetradecahydrocyclopenta[a]phenanthren-17-yl]-2H-furan-5-one).
CAS: 143-62-4. $C_{23}H_{34}O_4$.
A cardenolide which is the aglycone of digitoxin.
Properties: Crystalline steroid lactone.

digitoxin.
CAS: 71-63-6. $C_{41}H_{64}O_{13}$.
Most active glycoside of *Digitalis purpurea*.
Properties: White, leaflets or powder; odorless; bitter taste. Mp 255–256C. Slightly soluble in water or ether; soluble in alcohol.
Derivation: From digitalis leaves, usually *Digitalis purpurea* (foxglove).
Grade: USP.
Hazard: Toxic by ingestion; overdose can be fatal.
Use: Medicine (cardiac treatment).

diglycerol. See polyglycerol.

diglycidyl ether. (DGE; di(2,3-epoxypropyl) ether).
CAS: 2238-07-5. $C_6H_{10}O_3$.
Properties: A colorless liquid; strong, irritating odor. Mw 130.14, d 1.262 (25C), bp 260C, vap d 3.78 (25C), vap press 0.09 mm Hg.
Hazard: A questionable carcinogen. Severe eye, skin, and respiratory tract irritant. Male reproductive damage.
Use: Intermediate.

1,3-diglycidyloxybenzene. See resorcinol diglycidyl ether.

diglycoaldehyde. See inosine dialdehyde.

diglycol. See diethylene glycol.

diglycol acetate. See diethylene glycol diacetate.

diglycol carbamate. (diethylene glycol dicarbamate).
$O(CH_2CH_2OCONH_2)_2$.
Properties: White, crystalline solid. Relatively stable to acid hydrolysis, but less stable to basic conditions.
Use: Manufacture of resins.

diglycol chloroformate. [diethylene glycol bis(chloroformate)]. $O(CH_2CH_2OCOCl)_2$.
Properties: Liquid. Bp 125–127C (5 mm Hg), flash p 295F (146C). Soluble in acetone, alcohol, ether, chloroform, and benzene. Combustible.
Use: Preparation of nonvolatile plasticizers or modifying agent.

diglycol chlorohydrin.
CAS: 628-89-7. $ClCH_2CH_2OCH_2CH_2OH$.
Properties: Colorless liquid. D 1.1698, bp 196.8C, flash p 225F (107C), vap press 0.17 mm Hg. Miscible with water. Combustible.
Hazard: Toxic by ingestion and inhalation; an irritant.

diglycolic acid.
$O(CH_2COOH)_2$.
Properties: White, crystalline solid. Mp 148C. Soluble in water and alcohol; pH of 10% aqueous solution 1.4. Forms a nonhygroscopic monohydrate at relative humidities above 72% at 25C.
Use: Manufacture of resins and plasticizers, organic synthesis, sequestering agent, emulsion breaker in petroleum.

diglycolic acid, diallyl ester.
CAS: 5441-63-4. $C_{10}H_{14}O_5$.
Hazard: A poison by ingestion. Moderately toxic by skin contact.
Use: Agricultural chemical.

diglycol laurate. (diethylene glycol monolaurate).
$C_{11}H_{23}COOC_2H_4OC_2H_4OH$.
Properties: Light-straw-colored, oily liquid; practically odorless. Edible. D 0.96, flash p 290F (143C). Insoluble in water; soluble in methanol, ethanol, toluene, naphtha, and mineral oil; miscible in certain proportions with cottonseed oil, acetone, and ethyl acetate. Combustible.
Derivation: Lauric acid ester of diethylene glycol.
Use: Emulsifying agent for oils and hydrocarbon solvents; emulsions for lubrication, sizing and finishing of textiles, paper, and leather; fluid emulsions of oils for hand lotions, hair dressings, etc.; cutting and spraying oils; dry cleaning soap base; antifoaming agent.

diglycol methyl ether. See diethylene glycol dimethyl ether.

diglycol monostearate. (diethylene glycol monostearate).
CAS: 106-11-6. $C_{17}H_{35}COOC_2H_4OC_2H_4OH$.
Properties: Small, white flakes.
Available forms: Regular or water-dispersible types.
Use: Emulsifier and thickener in cosmetics, mold release lubricant for die casting, temporary binder for ceramics and grinding wheels.

diglycol nitrate. See diethylene glycol dinitrate.

diglycol oleate. (diethylene glycol monooleate).
CAS: 106-12-7. $C_{17}H_{33}COOC_2H_4OC_2H_4OH$.
Properties: Light-red, oily liquid; fatty odor. D 0.93, iodine value 65–75, titer below 0C, pH 7.7–8.2 (25C) (5% aqueous dispersion). Soluble in ethanol, naphtha, ethyl acetate, and methanol; partly soluble in cottonseed oil; insoluble in water. Combustible.
Derivation: Oleic acid ester of diethylene glycol.
Use: Emulsifying agent for fluid water-in-oil emulsions for the manufacture of furniture and automobile polish, water-emulsion paints, agricultural sprays.

diglycol phthalate. (diethylene glycol phthalate).
$C_6H_4(COOC_2H_4OH)_2$.
Properties: Pale-yellow, liquid resin. D 1.29, saponification value 430–450, acid value 170–175. Soluble in methanol, ethanol, acetone, and ethyl acetate; partly soluble in toluene, naphtha, mineral oil, and cottonseed oil; insoluble in water. Combustible.
Use: Plasticizer, emulsifier.

diglycol ricinoleate. (diethylene glycol monoricinoleate).
CAS: 5401-17-2. $C_{17}H_{32}(OH)COOC_2H_4OC_2H_4OH$.
Properties: Light-yellow liquid. Fp less than −60C, d 0.980 (25C). Soluble in alcohol, acetone, and ethyl acetate; insoluble in water. Combustible.
Use: Plasticizer for high polymers and elastomers.

diglycol stearate.   (diethylene glycol distearate). CAS: 109-30-8. $C_{17}H_{35}COOC_2H_4)_2O$.
Properties: White, waxlike solid; faint fatty odor. Mp 54–55C, d 0.9333 (20/4C). Disperses in hot water; soluble in hot alcohol, oils, and hydrocarbons. Combustible.
Derivation: Stearic acid ester of diethylene glycol.
Grade: Technical, cosmetic.
Use: Emulsifying agent for oils, solvents, and waxes; lubricating agent for paper and cardboard; suspending medium for powders in the manufacture of polishes, cleaners, and textile delusterants; temporary binder for abrasive powders and clays for ceramic insulation; protective coating for hygroscopic powders; thickening agent; pharmaceuticals.

diglyme.   See diethylene glycol dimethyl ether.

digoxin.
CAS: 20830-75-5. $C_{41}H_{64}O_{14}$.
A cardiotonic digitalis glycoside.
Properties: Colorless to white crystals, or white, crystalline powder; odorless. Melts at approximately 235C (decomposes). Insoluble in water, chloroform, and ether; soluble in pyridine and dilute alcohol.
Derivation: From the leaves of *Digitalis lanata*.
Grade: USP.
Hazard: See digitalis.
Use: Medicine (cardiac diseases).

diheptyl-*p*-phenylenediamine.   See   *N,N'*-bis(1,4-dimethylpentyl)-*p*-phenylenediamine.

dihexadecylamine.   See dipalmitylamine.

dihexadecyl ether.   See dicetyl ether.

dihexadecyl sulfide.   See dicetyl sulfide.

dihexadecyl thioether.   See dicetyl sulfide.

dihexyl.   See *n*-dodecane.

di-*n*-hexyl adipate.
(-$CH_2CH_2COOC_6H_{13})_2$.
Properties: Liquid, colorless, water-white to maximum 100 Pt-Co. D 0.939 (20C), refr index 1.438 (25C), surface tension 32.7 dynes/cm (20C), viscosity 8.8 cP (20C), bp 183–192C (4 mm Hg) (midpoint 191C), flash p 325F (162.7C). Water solubility 0.1% (25C); gasoline and oil solubility complete. Combustible.
Use: Low-temperature plasticizer for SBR elastomers.

di-*n*-hexylamine.
CAS: 143-16-8. $[CH_3(CH_2)_5]_2NH$.
Properties: Water-white liquid. Bp 233–243C, d 0.788 (20/20C), refr index 1.434 (20C), flash p 220F (104.4C). Combustible.
Hazard: Toxic by ingestion and skin absorption.

di-*n*-hexyl maleate.
CAS: 105-52-2. $C_6H_{13}OOCCH:CHCOOC_6H_{13}$.
Properties: Liquid. D 0.9602 (20/20C), bp 179C (10 mm Hg), refr index 1.449 (20C), vap press <0.01 mm Hg (20C), fp −70C, viscosity 10.2 cP (20C). Soluble in water <0.01% by wt (20C). Combustible.
Use: Preparation of resins.

dihexyl phthalate.   See di(2-ethylbutyl) phthalate.

dihexyl sebacate.
CAS: 2449-10-7. (-$CH_2CH_2CH_2CH_2COOC_6H_{13})_2$.
Properties: Light-straw–colored liquid. Bp 203C (4 mm Hg), flash p 415F (212C). Combustible.
Derivation: By reacting dodecyl alcohol with sebacic acid.
Use: Plasticizer for vinyl resins.

dihexyltin dichloride.   See dichlorodihexylstannane.

dihydrazine sulfate.
$(N_2H_4)_2 \cdot H_2SO_4$.
Properties: White, crystalline flakes. Mp approximately 104C, decomposes at approximately 180C. Soluble in water; insoluble in most organic solvents.
Grade: 95% is available commercially.
Hazard: An irritant.
Use: Reducing agent.

dihydric.   Containing two hydroxyl groups connected to different carbon atoms, e.g., a dihydric alcohol. Dihydric alcohols are collectively called glycols (diols).
See ethylene glycol.

dihydroabietyl alcohol.   (hydroabietyl alcohol).
CAS: 26266-77-3. $C_{19}H_{31}CH_2OH$.
Properties: Solid. D 1.007–1.008, refr index 1.5280, vap press 1.5×10^{-5} mm Hg (25C), mp 32–33C, flash p 375F (190C). Insoluble in water. Combustible.
Use: Plasticizer.

1,8-dihydroacenaphthylene.   See acenaphthene.

4,5-dihydro-2-amino-1-((6-chloro-3-pyridinyl)methyl)-1h-imidazole-4,5-diol.
CAS: 200258-64-6. $C_9H_{11}ClN_4O_2$.
Hazard: A poison.
Use: Agricultural chemical.
2,3-dihydro-1,3-benzoxazine-4h-2-thione-4-one
See 2-thio-2h-1,3-benzoxazine-2,4(3h)-dione.

dihydrochalcone.   Any of a group of disaccharide sugars having a sweet taste. They are derived from such flavanones as naringin and hesperidin. Research on the conversion of these compounds

into acceptable synthetic sweeteners has been in progress for some years.
See flavanone.

dihydrocholesterol.   (β-cholestanol; 3-β-hydroxycholestane).
CAS: 360-68-9. $C_{27}H_{47}OH$.
A sterol found in the feces.
Properties: White crystals. Mp (monohydrate) 142C, optical rotation α +23 (25C) degrees. Soluble in fat solvents; insoluble in water.
Derivation: By a series of oxidation and reduction reactions from cholesterol.
Use: Biochemical research, pharmaceutical preparations.

4,5-dihydro-3-(4-chlorophenyl)-4-methyl-1-((((4-(trifluoromethyl)phenyl)amino) carbonyl)-1h-pyrazole-4-carboxylic acid methyl ester.
CAS: 99832-61-8. $C_{20}H_{17}ClF_3N_3O_3$.
Hazard: Moderately toxic by ingestion.

4,5-dihydro-((6-chloro-3-pyridinyl)methyl)-1h-imidazol-2-amine.
CAS: 115970-17-7. $C_9H_{11}ClN_4$.
Hazard: A poison.
Use: Agricultural chemical.

4,5-dihydro-1-((6-chloro-3-pyridinyl) methyl)-2-(nitroamino)-1h-imidazole-4,5-diol.
CAS: 155802-65-6. $C_9H_{10}ClN_5O_4$.
Hazard: A poison.
Use: Agricultural chemical.

dihydrocoumarin.   See benzodihydropyrone.

n-((3s)-2,3-dihydro-6-(2,6-difluorophenyl) methoxy)-3-benzofuranyl)-n-hydrourea, xy-.
CAS: 162750-10-9. $C_{16}H_{14}F_2N_2O_4$.
Hazard: A reproductive hazard.

20,22-dihydrodigitoxin.
CAS: 3786-76-3. $C_{41}H_{66}O_{13}$.
Hazard: Low toxicity.
Source: Natural product.

2,3-dihydro-2,2-dimethyl-7-benzofuranyl (butoxysulfinyl)methylcarbamate.
CAS: 77267-60-8. $C_{16}H_{23}NO_5S$.
Hazard: A poison by ingestion.
Use: Agricultural chemical.

2,3-dihydro-2,2-dimethyl-7-benzofuranyl (butylthio)methylcarbamate.
CAS: 50673-08-0. $C_{16}H_{23}NO_3S$.
Hazard: A poison by ingestion.
Use: Agricultural chemical.

2,3-dihydro-2,2-dimethyl-7-benzofuranyl (dodecylthio)methylcarbamate.
CAS: 50673-09-1. $C_{24}H_{39}NO_3S$.
Hazard: A poison by ingestion.
Use: Agricultural chemical.

2,3-dihydro-2,2-dimethyl-7-benzofuranyl((4-fluorophenyl)thio)methylcarbamate.
CAS: 50802-69-2. $C_{18}H_{18}FNO_3S$.
Hazard: A poison by ingestion.
Use: Agricultural chemical.

2,3-dihydro-2,2-dimethylbenzofuranyl-7-(methyl)(t-butoxyfulfinyl)carbamate.
CAS: 77267-59-5. $C_{16}H_{23}NO_5S$.
Hazard: A poison by ingestion.
Use: Agricultural chemical.

2,3-dihydro-2,2-dimethyl-7-benzofuranylmethyl((4-nitrophenyl)thio)carbamate.
CAS: 50802-70-5. $C_{18}H_{18}N_2O_5S$.
Hazard: A poison by ingestion.
Use: Agricultural chemical.

2,3-dihydro-2,2-dimethyl-7-benzofuranyl methyl(phenoxysulfinyl)carbamate.
CAS: 77267-47-1. $C_{18}H_{19}NO_5S$.
Hazard: A poison by ingestion.
Use: Agricultural chemical.

2,3-dihydro-2,2-dimethyl-7-benzofuranylmethyl(1,1,2,2-tetrachloroethyl)thiocarbamate.
CAS: 50673-06-8. $C_{14}H_{15}Cl_4NO_3S$.
Hazard: A poison by ingestion.
Use: Agricultural chemical.

2,3-dihydro-2,2-dimethyl-7-benzofuranylmethyl((trichloromethyl)thio) carbamate.
CAS: 37430-50-5. $C_{13}H_{14}Cl_3NO_3S$.
Hazard: A poison by ingestion.
Use: Agricultural chemical.

2,5-dihydro-1,2-dimethyl-3-(2-naphthalenyl)-1h-pyrrole, rel-(2r,3r)-2,3-dihydroxybutanedioate (1:1).
CAS: 326800-80-0. $C_{16}H_{17}N \cdot C_4H_6O_6$.
Hazard: A poison.

6-(2,5-dihydro-1,2-dimethyl-1h-pyrrol-3-yl)-2-naphthalenol hydrochloride.
CAS: 326800-75-3. $C_{16}H_{17}NO \cdot ClH$.
Hazard: A poison.

9,10-dihydro-9,10-dioxoanthracene-1,8-disulfonic acid, disodium salt.
CAS: 903-46-8. $C_{14}H_6O_8S_2 \cdot 2Na$.
Hazard: Moderately toxic.

6,7-dihydrodipyrido(1,2-a:2′,1′-c) pyrazidinium salt.   See diquat.

9,10-dihydroergocornine.
CAS: 25447-65-8. $C_{31}H_{41}N_5O_5$.
Hazard: A reproductive hazard.

dihydroetorphine hydrochloride.
CAS: 155536-45-1. $C_{25}H_{35}NO_4·ClH$.
Hazard: A reproductive hazard.

dihydrofolic reductase.   An enzyme acting as an essential catalyst of DNA synthesis.

dihydrogen ferrous EDTA.   See ethylenediaminetetraacetic acid (note).

6,15-dihydrohydroxy-5,9,14,18-anthrazinetetrone.
CAS: 1324-28-3. $C_{28}H_{14}N_2O_5$.
Hazard: Moderately toxic.

dihydro-2(3)-imidazolone.   See ethylene urea.

2,3-dihydroindene.   See indan.

1,2-dihydro-4-iodo-1,5-dimethyl-2-phenyl-3h-pyrazol-3-one.   See iodophenazone.

1-((1-(4-(3,4-dihydro-2(1h)-isoquinolinyl)butyl)-1,2,3,4-tetrahydro-7-methyl-2,4-dioxo-3-phenylpyrido(2,3-d)pyrimidin-5-yl)carbonyl)piperidine.
CAS: 272774-96-6. $C_{33}H_{37}N_5O_3$.
Hazard: A poison.

2,5-dihydro-3-(6-methoxy-2-naphthalenyl)-1,2-dimethyl-1h-pyrrole,   rel-(2r,3r)-2,3-dihydroxybutanedioate (1:1).
CAS: 326800-79-7. $C_{17}H_{19}NO·C_4H_6O_6$.
Hazard: A poison.

4,5-dihydro-*n*-(1-methylethyl)-3-(phenylamino)-2h-benz(g)indazole-2-acetamide.
CAS: 301644-18-8. $C_{22}H_{24}N_4O$.
Hazard: A poison by ingestion.

4,5-dihydro-6-methyl-4-((3-pyridinylmethylene)amino)-1,2,4-triazin-3(2h)-one, (e)-.
CAS: 123312-89-0. $C_{10}H_{11}N_5O$.
Hazard: A poison.
Use: Agricultural chemical.

4,5-dihydro-n-nitro-1-(3-pyridinylmethyl)-1h-imidazol-2-amine.
CAS: 105828-05-5. $C_9H_{11}N_5O_2$.
Hazard: A poison.
Use: Agricultural chemical.

dihydronivalenol.
CAS: 24393-94-0. $C_{15}H_{22}O_7$.
Hazard: A poison.
Source: Natural product.

9,20-dihydro-9-oxoanthracene.   See anthrone.

2,3-dihydroperfluoropentane.
CAS: 138495-42-8. $C_5H_2F_{10}$.
Hazard: Low toxicity by ingestion, inhalation, and skin contact.

2,2-dihydroperfluoropropane.   See 1,1,1,3,3,3-hexafluoropropane.

4-((4,5-dihydro-3-(phenylamino)-2h-benz(g)indazol-2-yl)acetyl)morpholine.
CAS: 301644-21-3. $C_{23}H_{24}N_4O_2$.
Hazard: A poison by ingestion.

4,5-dihydro-3-(phenylamino)-*n*-(phenylmethyl)-2h-benz(g)indazole-2-acetamide.
CAS: 301644-20-2. $C_{26}H_{24}N_4O$.
Hazard: A poison by ingestion.

dihydropicrotoxinin.
CAS: 17617-46-8. $C_{15}H_{18}O_6$.
Hazard: A poison.

3,4-dihydro-1-propyl-4,4,6-trimethyl-2(1h)-pyrimidinethione.
CAS: 18957-53-4. $C_{10}H_{18}N_2S$.
Hazard: A poison.

2,3-dihydropyran.
CAS: 25512-65-6. C_5H_8O.
Properties: Colorless, mobile liquid; etherlike odor. Bp 84.3C, fp −70C, d 0.927 (20/4C), refr index 1.4180 (25C), flash p 0F (−17.7C). Soluble in water and alcohol.
Hazard: Highly flammable, dangerous fire risk. Toxic by ingestion and inhalation.
Use: Intermediate.

1,2-dihydro-3,6-pyridazinedione.   See maleic hydrazide.

dihydrostreptomycin.   (DHS).
CAS: 128-46-1. $C_{21}H_{41}N_7O_{12}$.
A derivative of streptomycin in which the carbonyl group of the streptose portion has been reduced by the addition of two hydrogen atoms. It has antibiotic properties similar to streptomycin, but its toxicity has limited its usefulness for the treatment of tuberculosis.

4-dihydrotestosterone.   (anaboleen; anabolex; anaprotin; andractim; androlone; androstanolone; 5-α-androstan-17-β-ol-3-one;   cristerona   mb;

7β-hydroxy-5α-androstan-3-one; dht; dihydrotestosterone; 4,5-α-dihydrotestosterone; 17-β-hydroxy-5-α-androstan-3-one; (5-α,17-β)-hydroxy-androstan-3-one; neodrol; proteina; protona; stanaprol; stanolone; (5S,8R,9S,10S,13S,14S,17S)-17-hydroxy-10,13-dimethyl-1,2,3,4,5,6,7,8,9,11,12,13,14,15,16,17-tetradecahydrocyclopenta[a]phenanthren-3-one).
CAS: 521-18-6. $C_{19}H_{30}O_2$.
A potent adrogenic metabolite of testosterone. It is more potent than testosterone and simulates the development of most secondary sexual characteristics in male as well as the endocrine-based adult male sexual functions. It may also induce virilization during embryogenesis.
Derivation: Formed in a reaction catalyzed by 5α-reductase in many androgen-responsive tissues.
Hazard: Teratogen; mutagen.

2,5-dihydrothiophene-1,1-dioxide. (sulfolene).

O$_2$SCH$_2$CH:CHCH$_2$.

Properties: Solid. Bp (decomposes), d 1.314 (70C), flash p 235F (112C). Partially soluble in water, acetone, and toluene. Combustible.

1,2-dihydro-2,2,4-trimethylquinoline.
(Agerite Resin D).
CAS: 147-47-7. $C_{12}H_{15}N$.
Properties: Brown pellets or flakes. D 1.03 (25C), melting range 75–100C (initial melt).
Use: Rubber antioxidant.

dihydroxyacetone. (DHA; dihydroxypropanone).
CAS: 96-26-4. HOCH$_2$COCH$_2$OH.
Properties: Colorless, crystalline solid. Mp 80C. Hygroscopic; soluble in water and alcohol.
Derivation: Action of sorbose bacterium on glycerol.
Use: Intermediate, emulsifier, humectant, plasticizers, fungicides, cosmetics (creates synthetic suntan).

2,4-dihydroxyacetophenone. See 4-acetylresorcinol.

dihydroxy alcohol. (dimethyl ethyl carbinol; *tert*-pentyl alcohol).
Properties: Alcohol, contains two functional hydroxyl groups.

dihydroxyaluminum sodium carbonate.
CAS: 16482-55-6. Al(OH)$_2$OOCONa.
White powder.
Derivation: Reaction of aluminum isopropoxide and an aqueous soluble of sodium bicarbonate.
Use: Medicine (antacid).

1,3-dihydroxy-2-amino-4-octadecene. See sphingosine.

1,8-dihydroxyanthranol. See anthralin.

1,2-dihydroxyanthraquinone. See alizarin.

1,4-dihydroxyanthraquinone. See quinizarin.

1,5-dihydroxyanthraquinone. (anthrarufin).
CAS: 117-12-4. $C_{14}H_6O_2(OH)_2$.
Properties: Yellow crystals. Mp 280C. Soluble in alcohol; sparingly soluble in water.
Derivation: By heating anthraquinone with boric acid and sulfuric anhydride.
Use: Intermediate for alizarin and indanthrene dyes.

1,8-dihydroxyanthraquinone. (chrysazin; danthron).
CAS: 117-10-2. $C_{14}H_6O_2(OH)_2$.
Properties: Orange powder or reddish-brown needles. Mp 191C. Soluble in alcohol; sparingly soluble in water.
Derivation: From 1,8-anthraquinone potassium disulfonate.
Grade: Technical, NF.
Use: Dyes, medicine.

1,8-dihydroxy-anthrone.
CAS: 1143-38-0. $C_{14}H_{10}O_3$.
Hazard: A mild skin irritant.

5,5-dihydroxy barbituric acid.
CAS: 3237-50-1. $C_4H_4N_2O_5$.
Hazard: Moderately toxic by ingestion.

m-dihydroxybenzene. See resorcinol.

o-dihydroxybenzene. See pyrocatechol.

p-dihydroxybenzene. See hydroquinone.

2,4-dihydroxybenzenecarboxylic acid. See β-resorcylic acid.

2,4-dihydroxybenzoic acid. See β-resorcylic acid.

2,5-dihydroxybenzoic acid. See gentisic acid.

3,5-dihydroxybenzoic acid. See α-resorcylic acid.

2,4-dihydroxybenzophenone.
CAS: 131-56-6. $C_6H_5COC_6H_3(OH)_2$.
Properties: Light-yellow, crystalline solid. Mp 142C, bp 194C (1 mm Hg), d 5.8 lb/gal (20C). Insoluble in water; soluble in ethanol, methanol, methyl ethyl ketone, and ethyl acetate.
Use: UV absorber in polymers.

2,5-dihydroxybenzoquinone.
CAS: 615-94-1. $C_6H_2(OH)_2O_2$.
Properties: Yellow–orange solid. Mp 216C (decomposes). Soluble in concentrated sulfuric acid; slightly soluble in ethanol, acetone, water, and benzene.
Derivation: From hydroquinone.
Hazard: Irritant to eyes and skin.
Use: Metal chelating, insecticides, polymerization inhibitor, tanning agent, dyestuff manufacture.

2,3-dihydroxybutane.   See 2,3-butylene glycol.

2,5-dihydroxycholorobenzene.   See chlorohydroquinone.

3-(3,4-dihydroxycinnamoyl)quinic acid.
See chlorogenic acid.

dihydroxydiaminomercurobenzene.
$OHNH_2C_6H_3HgC_6H_3OHNH_2$.
A mercury compound analogous to arsphenamine.
Use: In medicine as a source of mercury.
Hazard: See mercury.

2,2′-dihydroxy-5,5′-dichlorophenylmethane.
See dichlorophene.

dihydroxydiethyl ether.   See diethylene glycol.

2,2′-dihydroxy-5,5′-difluorodiphenyl sulfide.
$FC_6H_3(OH)S(OH)C_6H_3F$.
Properties: White, amorphous solid. Mp 119–121C. Soluble in acetone, ether, chloroform, ethanol, ethyl acetate, and glacial acetic acid; moderately soluble in benzene; insoluble in water.
Use: Fungicide (textile, agricultural chemical).

2,4-dihydroxy-3,3-dimethylbutyric acid.
See pantolactone.

n-(2,4-dihydroxy-3,3-dimethylbutyryl)-β-alanine.   See pantothenic acid.

5,7-dihydroxydimethylcoumarin.
$C_{11}H_{10}O_4$.
Properties and uses closely resemble those of 5,7-dihydroxy-4-methylcoumarin.

dihydroxydiphenylsulfone.   (sulfonyl bisphenol).
CAS: 5397-34-2. $(C_6H_4OH)_2SO_2$.
The commercial product is a mixture of isomers.
Properties: White, free-flowing crystals; odorless. Mp 215–240C. Soluble in alcohol and acetone; insoluble in water.
Grade: Technical.
Use: Electroplating, phenolic resins, polyvinyl chloride, intermediate.

5,5′-dihydroxy-7,7′-disulfonic-2,2′-dinaphthylurea.   (6,6′-ureylenebis-1-naphthol-3-sulfonic acid).
CAS: 854812-04-7. $HOC_{10}H_5(SO_3H)NHCONH$
$C_{10}H_5(SO_3H)OH$.
Properties: (Crude) Light-gray paste. Soluble in water; very soluble in alkaline solutions.
Derivation: Phosgenation of J acid.
Use: Dye intermediate.

(1r,2s,3s,4r)-3,4-dihydroxy-1,2-epoxy-1,2,3,4-tetrahydrodibenz(c,h)acridine.
CAS: 93780-95-1. $C_{21}H_{15}NO_3$.
Hazard: Experimental carcinogenic data reported.

p-di-(2-hydroxyethoxy)benzene.   See hydroquinone di(β-hydroxyethyl) ether.

di(2-hydroxyethyl)amine.   See diethanolamine.

N,N-dihydroxyethyl ethylenediamine.
CAS: 4439-20-7. $(CH_2NHC_2H_4OH)_2$.
Properties: Solid crystals. Mp 98C, bp 196C (10 mm Hg), flash p 355F (179C). Combustible.
Use: Manufacture of textile-finishing assistants.

dihydroxyethyl sulfide.   See thiodiglycol.

N,N-dihydroxyethyl-m-toluidine.
CAS: 91-99-6. $CH_3C_6H_4N(C_2H_4OH)_2$.
Properties: Light-gray solid. Mp 62C, distillation range 175–185C (2 mm Hg).

2,2′-dihydroxy-3,5,6,3′,5′,6′-hexachlorodiphenylmethane.   See hexachlorophene.

1,3-dihydroxy-4-hexylbenzene.   See hexylresorcinol.

3′,4′-dihydroxy-2-(isopropylamino) acetophenone hydrochloride.
CAS: 16899-81-3. $C_6H_3(OH)_2COCH_2NH(C_3H_7)\cdot$
HCl.
Light-colored, crystalline powder with a faint odor.
Hazard: Moderately toxic.
Use: Intermediate.

5,8-dihydroxy-3-methoxyxanthone-1-o-glucoside.   See swertianolin.

3,4-dihydroxy-α-(methylaminomethyl) benzyl alcohol.   See epinephrine.

5,7-dihydroxy-4-methylcoumarin.
CAS: 2107-76-8. $C_{10}H_8O_4\cdot H_2O$.
Properties: Yellow to white solid, fluoresces blue, absorbs UV light. Melting range 270–285C. Insoluble in water, benzene, and ether; soluble in alcohol and sodium hydroxide.
Derivation: From phloroglucinol.

Use: In suntan oils as a sunscreen, in wall paints as a whitening agent.

1,1′-dihydroxymethylferrocene.
$[C_5H_4(CH_2OH)]_2Fe$.
Properties: Yellow, crystalline solid. Mp 85–86C.
Use: Intermediate.
See ferrocene.

1,3-dihydroxynaphthalene. (naphthoresorcinol).
CAS: 132-86-5. $C_{10}H_6(OH)_2$.
Properties: Transparent, crystalline plates. Mp 124–125C. Soluble in alcohol, ether, and water. Combustible.
Derivation: By heating naphthalene-1,3-disulfonic acid with alkali at 230C under pressure.
Grade: Technical, reagent.
Use: Dyes; pharmaceuticals; analytical reagent for sugars, oils, and glucuronic acid.
Note: There are several other isomeric forms of dihydroxynaphthalene, (1,5-; 1,6-; 1,7-; 1,8-; 2,3-; 2,6-; 2,7-). They are derived by heating a naphthalene disulfonic acid isomer with caustic soda and are soluble in alcohol and ether, and sparingly soluble in water; mp ranges from 136C (1,6-) to 260C (2,6-).
Use: Organic dyes.

2,8-dihydroxy-3-naphthoic acid.
CAS: 89-35-0. $(C_{10}H_5(OH)_2COOH)$.
Properties: Light-green powder. Mp 235–240C. Slightly soluble in hot water; soluble in alcohol and acetone.
Use: Intermediate.

1,8-dihydroxy-10-(1-oxopentyl)-9(10h)-anthracenone.
CAS: 75464-12-9. $C_{19}H_{18}O_4$.
Hazard: A skin irritant.

L-dihydroxyphenylalanine. (L-dopa).
An amino acid used in treating Parkinson's disease, manganese poisoning, and muscular dystonia. Derived from several types of beans including vanilla; also made synthetically.

17-α,21-dihydroxy-14-α-pregn-4-ene-3,20-dione 21-iodoacetate.
CAS: 27953-64-6. $C_{23}H_{31}IO_5$.
Hazard: A poison.

1,2-dihydroxypropane. See 1,2-propylene glycol.

dihydroxypropanone. See dihydroxyacetone.

n-(1,3-dihydroxy-2-propyl)valiolamine.
CAS: 83480-29-9. $C_{10}H_{21}NO_7$.
Hazard: Low toxicity by ingestion. A reproductive hazard.

dihydroxystearic acid.
$C_{17}H_{33}(OH)_2COOH$.
Properties: White crystals; odorless; tasteless. Mp 135C. Soluble in alcohol and ether; insoluble in water.
Derivation: By heating dibromide of isooleic acid with silver oxide.
Use: Cosmetics, lotions.

dihydroxysuccinic acid. See tartaric acid.

3,5-dihydroxytoluene. See orcin.

DII. Abbreviation for diesel ignition improver.

diindiol. The reaction product of diacetylene and acetone or another ketone.

diiodacetylene. (diiodethyne).
CAS: 624-74-8. IC≡CI.
Properties: White crystals; unpleasant odor. Mp 78.5C (decomposes). Light acts upon it, causing a gradual change in color to red and a separation of iodine. Soluble in alcohol, ether, and benzene; insoluble in water.
Derivation: By dissolving iodine in liquid ammonia and passing acetylene into the solution.
Hazard: Highly volatile. Toxic by inhalation; vapors irritating to eyes and mucous membranes.
Use: Organic synthesis, military poison.

2,2′-diiododiacetamide.
CAS: 117900-35-3. $C_4H_5I_2NO_2$.
Hazard: Moderately toxic by ingestion and skin contact. A mild skin irritant.

***sym*-diiododibromoethylene.**
BrIC:CIBr.
Properties: Crystals. Mp 95–96C.
Derivation: Reaction of iodine and dibromoacetylene.
Use: Organic synthesis.

diiododiethyl sulfide.
$(ICH_2CH_2)_2S$.
Properties: Bright-yellow prisms. Mp 62C. Slowly decomposes, the rate being accelerated by light and by heat. Hydrolyzed by alkali solutions. Soluble in alcohol, benzene, and ether; insoluble in water.
Derivation: Interaction of dichlorodiethyl sulfide with an acetic acid solution of sodium iodide.
Hazard: Toxic by ingestion and inhalation; strong irritant.

4′,5′-diiodofluorescein. (CI 45425A, Solvent #73).
$C_{20}H_{10}I_2O_5$.
Properties: Orange powder. Soluble in alcohol; slightly soluble in water.
Use: Dyeing textiles, etc., analytical reagent.

diiodomethane.   See methylene iodide.

3,5-diiodosalicylic acid.
CAS: 133-91-5. $I_2C_6H_2(OH)COOH$.
Properties: White to pale-pink, crystalline powder. Slightly soluble in water.
Use: Source of iodine for animal nutrition.

diiodothyronine.   (3,5-diiodothyronine).
$HOC_6H_4OC_6H_2I_2CH_2CH(NH_2)COOH$.
A thyronine derivative that is an intermediate obtained in the manufacture of synthetic thyroxine, also probably an intermediate in the synthesis of thyroxine by the thyroid gland.

diisobutyl adipate.   (DIBA).
CAS: 141-04-8. $[C_2H_4COOCH_2CH(CH_3)_2]_2$.
Properties: Colorless liquid; odorless. D 0.950 (25C), bp 278–280C, fp −20C, wt/gal 7.95 lb, acidity (as adipic acid) <0.05%. Compatible with most natural and synthetic polymers. Soluble in most organic solvents; insoluble in water. Combustible.
Use: Plasticizer.

diisobutylaluminum chloride.   (DIBAC).
CAS: 1779-25-5. $[(CH_3)_2CHCH_2]_2AlCl$.
Properties: Colorless liquid. Bulk d 0.905, fp −39.5C.
Derivation: Reaction of isobutylene and hydrogen with aluminum chloride.
Hazard: Strong irritant to tissue.
Use: Polyolefin catalyst.

diisobutylaluminum hydride.   (DIBAL-H).
CAS: 1191-15-7. $[(CH_3)_2CHCH_2]_2AlH$.
Properties: Colorless liquid. Pyrophoric. Fp −80C, bulk d 0.798, bp 105C (0.2 mm Hg). Miscible with hydrocarbon solvents; dilute solutions nonpyrophoric.
Derivation: Reaction of isobutylene and hydrogen with aluminum.
Hazard: Ignites spontaneously in air, dangerous fire risk.
Use: Reducing agent in pharmaceuticals.

diisobutylamine.
CAS: 110-96-3. $[(CH_3)_2CHCH_2]_2NH$.
Properties: Water-white liquid. D 0.745 (20C), boiling range 136–140C, amine odor, flash p 85F (29.4C).
Hazard: Flammable, moderate fire risk. Toxic by ingestion.
Use: Intermediate.

diisobutylcarbinol.   See 2,6-dimethyl-4-heptanol.

diisobutyl carbinyl acetate.   See *n*-nonyl acetate.

diisobutylene.   A group of isomers of the formula C_8H_{16}, of which 2,4,4-trimethylpentene-1 and 2,4,4-trimethylpentene-2 are the most important, since they are formed in appreciable amounts when isobutene (isobutylene) is polymerized.
Properties: Colorless liquids. D 0.7227 (15.5C), boiling range 101–104C, flash p 20F (−6.6C).
Hazard: Fire risk. Narcotic in high concentrations.
Use: Alkylation, intermediate, antioxidants, surfactants, lubricant additive, plasticizers, rubber chemicals.

α-diisobutylene.   See 2,4,4-trimethylpentene-1.

β-diisobutylene.   See 2,4,4-trimethylpentene-2.

diisobutyl ketone.   (2,6-dimethyl-4-heptanone).
CAS: 108-83-8. $(CH_3)_2CHCH_2COCH_2CH(CH_3)_2$.
Properties: Colorless liquid; stable; mild odor. D 0.8089 (20/20C), bp 168.1C, vap press 1.7 mm Hg (20C), flash p 140F (60C), bulk density 6.7 lb/gal (20C), fp −41.5C, coefficient of expansion 0.00101 K^{-1} (20C). Miscible with most organic liquids; immiscible with water. Combustible.
Grade: Technical.
Hazard: Toxic. Eye and upper respiratory tract irritant.
Use: Solvent for nitrocellulose, rubber, and synthetic resins; lacquers; coating compositions; organic synthesis; roll-coating inks; stains.

diisobutyloxostannane.
CAS: 61947-30-6. $C_8H_{18}OSn$.
Hazard: A poison by ingestion. An eye and severe skin irritant.

diisobutyl phenol.   See octyl phenol.

diisobutyl phthalate.
CAS: 84-69-5. $C_6H_4[COOCH_2CH(CH_3)_2]_2$.
Properties: Liquid. Refr index 1.4900 (25C), d 1.040 (20/20C), flash p 385F (196C), bp 327C. Combustible.
Use: Plasticizer.

diisobutyl sodium sulfosuccinate.
CAS: 127-39-9. $C_{12}H_{21}NaO_7S$.
Properties: The commercial product is a mixture of three esters in the form of a fine, white powder; quite soluble in water and glycerol; insoluble in acetone, benzene, and carbon tetrachloride; surface tension of 1% water solution is 50 dynes/cm.
Hazard: Irritant to eyes, skin, and mucous membranes.
Use: Surfactant, emulsifier.

diisobutyltin oxide.   See diisobutyloxostannane.

diisocyanate.   An organic compound with two isocyanate groups (-NCO), formed by treating diamines, (e.g., toluene-2,4-diamine, hexamethylenediamine, *p*,*p*′-diaminodiphenylmethane) with phosgene. Combustible.

Use: Production of polyurethane foams and elastomers, in phenol-formaldehyde resins to improve water and alkali resistance; bonding rubber to rayon or nylon.
See polyurethane.

1,3-diisocyanatomethylbenzene polymer with Niax E 488.
CAS: 87177-09-1.
Hazard: A Moderately toxic by skin contact.

diisodecyl adipate.   (DIDA).
CAS: 27178-16-1. $C_{10}H_{21}OOC(CH_2)_4COOC_{10}H_{21}$.
Properties: Light-colored, oily liquid; mild odor. D 0.918 (20/20C), fp −71C, boiling range 239–246C (4 mm Hg), refr index 1.450 (25C), bulk d 7.5 lb/gal, flash p 225F (107.2C). Insoluble in glycerol, glycols, and some amines; soluble in most other organics. Combustible.
Use: Primary plasticizer for polymers.

diisodecyl-4,5-epoxytetrahydrophthalate.
(Flexol PEP).
$OC_6H_8(COOC_{10}H_{21})_2$.
Properties: Liquid. D 0.9867 (20/20C), bulk density 8.2 lb/gal, pour p 38C, oxirane oxygen 3%. Combustible.
Use: Plasticizer–stabilizer resistant to fungi, in vinyl plastics for outdoor use.

diisodecyl phthalate.   (DIDP).
CAS: 26761-40-0. $C_6H_4(COOC_{10}H_{21})_2$.
Properties: Clear liquid; mild odor. D 0.966 (20/20C), fp −50C, bp 250–257C (4 mm Hg), refr index 1.483 (25C), viscosity 108 cP (20C), bulk density 8 lb/gal, flash p 450F (232C). Insoluble in glycerol, glycols, and some amines; soluble in most other organics. Combustible.
Grade: Technical.
Hazard: An irritant.
Use: Plasticizer.
See: di-(C9–C11 alkyl) phthalate.

diisononyl adipate.
CAS: 33703-08-01. $[C_9H_{19}OOC(CH_2)_4COOC_9H_{19}]$.
A low-volatility plasticizer based on isononyl alcohol. Combustible.

diisononyl phthalate.
CAS: 68515-48-0.
Hazard: A reproductive hazard.

diisononyltin dichloride.   See bis(trimethylhexyl)tin dichloride.

diisooctyl acid phosphate.
CAS: 27215-10-7. $(C_8H_{17})_2HPO_4$.
Hazard: Irritant to skin and eyes.

diisooctyl adipate.   (DIOA).
$C_8H_{17}OOC(CH_2)_4COOC_8H_{17}$.

Properties: A light-straw–colored liquid; mild odor. D 0.924 (25C), bp 214–226C (4 mm Hg), fp >75C, flash p 370F (187C). Combustible.
Use: Plasticizer, especially at low temperatures.

diisooctyl azelate.   (DIOZ).
$C_8H_{17}OOC(CH_2)_7COOC_8H_{17}$.
Liquid diester of azelaic acid, bp 237C (5 mm Hg), pour p −85F, acid number 1.0, viscosity (min) 3.2 cs (210F), d 0.92 (20C), flash p 415F (212C). Combustible.
Use: Plasticizer for vinyl resins, base for synthetic lubricants.

diisooctyl phthalate.   (DIOP).
CAS: 27554-26-3. $(C_8H_{17}COO)_2C_6H_4$.
Isomeric esters obtained from phthalic anhydride and the mixed octyl alcohols made by the Oxo process.
Properties: Nearly colorless, viscous liquid; mild odor. Bp 370C, d 0.980–0.983 (20/20C), bulk d 8.20 lb/gal (20C), flash p 450F (232C). Insoluble in water; compatible with vinyl chloride resins and some cellulosic resins. Combustible.
Grade: Technical.
Use: Plasticizer for vinyl, cellulosic and acrylate resins, and synthetic rubber.
See isooctyl alcohol.

diisooctyl sebacate.   (DIOS).
$C_8H_{17}OOC(CH_2)_8COOC_8H_{17}$.
Properties: Liquid. D 0.915 (25/25C), flash p 440F (226C), pour p −40C, viscosity 24 cP (20C), bulk d 7.65 lb/gal. Combustible.
Use: Plasticizer.

diisopentyloxostannane.
CAS: 63979-62-4. $C_{10}H_{22}OSn$.
Hazard: A poison by ingestion. An eye and severe skin irritant.

diisopentyltin oxide.   See diisopentyloxostannane.

diisopropyl.   See 2,3-dimethylbutane.

diisopropanolamine.   (DIPA).
$(CH_3CH_2OCH_2)_2NH$.
Properties: White, crystalline solid. D 0.9890 (45/20C), bp 248.7C, bulk d 8.2 lb/gal (45C), vap press 0.02 mm Hg (42C), mp 42C, viscosity 1.98 cP (45C), flash p 260F (126C). Miscible with water. Combustible.
Use: Emulsifying agents for polishes, textile specialties, leather compounds, insecticides, cutting oils, and water paints.

diisopropylamine.
CAS: 108-18-9. $[(CH_3)_2CH]_2NH$.
Properties: Colorless, volatile liquid; amine odor. Bp 84.1C, fp −96.3C, d 0.7178 (20/20C), bulk d 6.0 lb/gal (20C), refr index 1.3924 (20C), flash p 30F (−1.11C) (OC). Slightly soluble in water; soluble in most organic solvents.
Derivation: From isopropyl chloride and ammonia.
Grade: Technical.

Hazard: Flammable, dangerous fire risk. Toxic by ingestion, inhalation, and skin absorption. Eye and upper respiratory tract irritant.
Use: Intermediate, catalyst.

diisopropylaminoethanol. See *N,N*-diisopropylethanolamine.

2-(diisopropylamino)ethyl chloride hydrochloride. See 2-chloroethyldiisopropylamine hydrochloride.

β-diisopropylaminoethyl chloride hydrochloride. (DIC).
[$(CH_3)_2CH]_2NCH_2CH_2Cl \cdot HCl$.
Use: Organic synthesis especially for introduction of the β-diisopropylaminoethyl radical.

m-diisopropylbenzene.
CAS: 25321-09-9. $C_{12}H_{18}$
Properties: Colorless liquid. D 0.8559 (20/4C), fp −63C, bp 231C, refr index 1.4883 (20C), flash p 170F (76.6C), autoign temp 840F. Insoluble in water; miscible with alcohol, ether, acetone, benzene, and carbon tetrachloride. Combustible.
Use: Solvent, intermediate

p-diisopropylbenzene.

Properties: White solid. Mp 64C, d 0.8568 (20/4C), bp 210C, refr index 1.4898 (20C). Insoluble in water; miscible with alcohol, ether, acetone, benzene, and carbon tetrachloride. Combustible.
Use: Solvent, intermediate.

diisopropylbenzene hydroperoxide. 1,3-Diisopropylbenzolhydroperoxid; 3,5-Diisopropyl benzene hydroperoxide; Bis(1-methylethyl)phenyl-hydroperoxide; HYDROPEROXIDE, BIS(1-METHYLETHYL)PHENYL; 1,4-phenylene-dipropane-2,2-diyl dihydroperoxide; Diisopropyl-benzene hydroperoxide, isomer mixture $C_{12}H_{18}O_2$.
Available forms: A 52% solution in a nonvolatile solvent. Colorless to pale-yellow liquid, strong oxidizing agent.
Hazard: Dangerous fire risk in contact with organic materials.

N,N-diisopropylbenzothiazyl-2-sulfenamide.
$C_6H_4NC[SN(C_3H_7)_2]S$.
Rubber accelerator.

diisopropyl carbinol. (2,4-dimethylpentanol-3).
CAS: 600-36-2. [$(CH_3)_2CH]_2CH_2O$.
Properties: Colorless liquid. Bp 140C, fp approximately −70C, bulk density 6.9 lb/gal, flash p 120F (48.9C). Combustible.
Hazard: Moderate fire risk.
Use: Solvent, organic synthesis (intermediate), denaturant.

diisopropyl cresol. Antioxidant or stabilizer in medicine (external).
See isopropyl cresol.

N,N'-diisopropyldiamidophosphoryl fluoride. (mipafox).
$(CH_3)_2CHNHPO(F)NHCH(CH_3)_2$.
Hazard: Cholinesterase inhibitor.
Use: Insecticide.

diisopropyl dixanthogen.
CAS: 105-65-7. $(C_3H_7OCS_2)_2$.
Properties: Yellow to greenish pellets. D 1.28, mp 52C (min), purity 98% (min). Insoluble in water; soluble in ethyl alcohol, acetone, benzene, and gasoline.
Hazard: Toxic by ingestion and inhalation, strong irritant.
Use: Modifier in polymerization reactions, additive for lubricants, flotation reagent, fungicide, weed killer.

N,N-diisopropylethanolamine. (diisopropylaminoethanol).
CAS: 96-80-0. [$(CH_3)_2CH]_2NCH_2CH_2OH$.
Properties: Colorless liquid. D 0.8742 (20C), vap press 0.08 mm Hg (20C), fp −39.3C, bp 191C, flash p 175F (79.4C). Slightly soluble in water. Combustible.
Hazard: Strong irritant to tissue.
Use: Organic synthesis.

diisopropyl ether. See isopropyl ether.

diisopropyl fluorophosphate. (DFP; isoflurophate).
CAS: 55-91-4. [$(CH_3)_2CHO]_2POF$.
Oily liquid, forms hydrogen fluoride in the presence of moisture. One member of a series of compounds, the fluorophosphate alkyl esters, characterized by extremely high toxicity, marked mitotic effects noted even in concentrations that are chemically undetectable.
Hazard: Cholinesterase inhibitor.
Use: Insecticide.

3,5-diisopropyl-4-hydroxyphenyl methylcarbamate. See methylcarbamic acid 4-hydroxy-3,5-diisopropylphenyl ester.

diisopropyl ketone.   (2,4-dimethylpentanone-3).
CAS: 565-80-0.   [(CH$_3$)$_2$CH]$_2$CO.
Properties: Colorless liquid. Bp 123.7C, bulk d 6.9 lb/gal.
Use: Solvent.

2,6-diisopropylnaphthalene.
CAS: 24157-81-1.   C$_{16}$H$_{20}$.
Properties: Clear, yellowish-brown liquid; faint, sweet odor. Boiling range 290–295C; flash p 284F (140C) (COC); d 0.95 (30C); bulk density 7.9 lb/gal; insoluble in water. Combustible.
Hazard: Avoid inhalation of vapors and prolonged skin contact.
Use: Organic intermediate.

diisopropyl oxide.   See isopropyl ether.

diisopropyloxostannane.
CAS: 23668-76-0.   C$_6$H$_{14}$OSn.
Properties: Solid. Insol in water.
Hazard: A poison by ingestion. An eye and severe skin irritant.

diisopropyl peroxydicarbonate.   (isopropyl percarbonate; isopropyl peroxydicarbonate; IPP).
CAS: 105-64-6. (CH$_3$)$_2$CHOC(O)OOC(O)OCH(CH$_3$)$_2$.
Properties: Colorless, crystalline solid. D 1.080 (15.5/4C), mp 8–10C, refr index 1.4034 (20C). Almost insoluble in water; miscible with aliphatic and aromatic hydrocarbons, esters, ethers, and chlorinated hydrocarbons.
Derivation: By reaction of sodium peroxide with isopropyl chloroformate.
Grade: Stabilized and unstabilized.
Hazard: Spontaneous decomposition at room temperature releases flammable and corrosive products; dangerous fire hazard; explodes on heating. Store in open containers at low temperature with adequate ventilation.
Use: Low-temperature polymerization catalyst.

2,6-diisopropylphenol.
CAS: 2078-54-8.   C$_6$H$_3$OH[CH(CH$_3$)$_2$]$_2$.
Properties: Light-straw-colored liquid. Fp 18C, d 0.955 (20C), bp 242C, flash p 240F (115.5C). Soluble in toluene and alcohol; insoluble in water. Combustible.
Use: Intermediate for synthetic polymers, plasticizers, surface-active agents.

diisopropyl-*p*-phenylenediamine.
H$_7$C$_3$NHC$_6$H$_4$NHC$_3$H$_7$.
Properties: Dark-red liquid. D 0.88. Soluble in alcohol.
Use: Gasoline antioxidant and sweetener (permissible for aviation gasoline, ASTM D910-64T).

o,o-diisopropylthiolphosphoric acid.   See isopropyl phosphorothioate.

N,N'-diisopropylthiourea.
CAS: 2986-17-6.   CH$_3$($_2$CHNHCSNHCH(CH$_3$)$_2$.
Properties: Grayish-white solid. Mp 138.5–142.5C. Slightly soluble in water; soluble in methanol, acetone, and ethyl acetate; insoluble in ether, benzene, and gasoline.
Use: Corrosion inhibitor; metal pickling with hydrochloric acid or sulfuric acid for reducing corrosion of ferrous metals and aluminum alloys in brine; intermediate.

diisopropyltin dichloride.
CAS: 38802-82-3.   C$_6$H$_{14}$Cl$_2$Sn.
Properties: Colorless crystals. Sol in water. Mp: 84C
Hazard: A poison.

diisopropyltin oxide.   See diisopropyloxostannane.

o,o-diisopropyl-s-tricyclohexyltin phosphorodithioate.
CAS: 49538-98-9.   C$_{24}$H$_{47}$O$_2$PS$_2$Sn.
Hazard: Moderately toxic by ingestion.

dikaryotic.   Having two different and distinct nuclei per cell; found in the fungi. A dikaryotic individual is called a dikaryon.

diketene.   (acetyl ketene).
CAS: 674-82-8

CH$_2$:CCH$_2$C(O)O.

Properties: Colorless, nonhygroscopic liquid; pungent odor. Readily polymerizes on standing. D 1.096 (20/20C), fp −7.5C, bp 127.4C, flash p 93F (33.9C) (TOC). Soluble in common organic solvents
Derivation: By spontaneous polymerization of ketene obtained by thermal decomposition of acetone or from bromoacetylbromide and zinc.
Hazard: Flammable, moderate fire risk. An irritant.
Use: Production of acetoarylamides, pigments and toners, pesticides, food preservatives, pharmaceutical intermediates.

diketobutane.   See diacetyl.

2,5-diketohexane.   See acetonylacetone.

2,5-diketopyrrolidone.   See succinimide.

2,5-diketotetrahydrofurane.   See succinic anhydride.

Dilan.   A 1:2 mix of Prolan and Bulan used as an insecticide.
Hazard: Toxic.

dilatancy. A system is said to be dilatant if its rate of increase of strain decreases with increased shear. Among the better-known systems exhibiting dilatant behavior are pastry doughs, highly pigmented paints, and many other industrially important materials. Dilatancy is usually associated with suspension, especially those containing a high concentration of suspended matter, which is often of colloidal dimensions.

"Dilaudid" [Purdue]. Proprietary name for the drug dihydromorphine hydrochloride.

dilaurylamine. (didodecylamine).
CAS: 3007-31-6. $(C_{12}H_{25})_2NH$.
Properties: Liquid. Mp 45C, d 0.89. Almost insoluble in water.
Use: Chemical intermediate.

dilauryl ether. (didodecyl ether).
CAS: 4542-57-8. $(C_{12}H_{25})_2O$.
Properties: Liquid. Mp 33C, bp 190–195C (1 mm Hg), d 0.8147 (33/4C). Combustible.
Grade: 95% (min) purity.
Use: Electrical insulators, water repellents, lubricants for plastic molding and processing, antistatic substances, chemical intermediates.

dilauryl phosphite.
CAS: 21302-09-0. $(C_{12}H_{25}O)_2PHO$.
Properties: Water-white liquid. Combustible.
Use: Synthesis of organophosphorus compounds for extreme pressure lubricants, adhesives, textile finishing agents, pesticides; catalyst in polymerization of unsaturated compounds.

dilauryl sulfide. (didodecyl thioether).
CAS: 2469-45-6. $(C_{12}H_{25})_2$.
Properties: Liquid. Mp 40–40.5C, bp 260–263C (4 mm Hg), d 0.8275 (40/4C). Combustible.
Grade: 95% (min) purity.
Use: Organic synthesis (formation of sulfonium compounds).

dilauryl thiodipropionate. (didodecyl-3,3'-thiodipropionate; thiodipropionic acid, dilauryl ester).
CAS: 123-28-4. $(C_{12}H_{25}OOCCH_2CH_2)_2S$.
Properties: White flakes having sweetish odor. Mp 40C, d 0.975 (solid 25C). Insoluble in water; soluble in most organic solvents; extremely resistant to heat and hydrolysis. Combustible.
Grade: FCC.
Use: Antioxidant, additive for high-pressure lubricants and greases, plasticizer and softening agent, antioxidant for edible fats and oils (up to 0.02% oil content).

dilinoleic acid.
CAS: 6144-28-1. $C_{34}H_{62}(COOH)_2$.

Properties: Light-yellow, viscous liquid; slight odor. D 0.921 (100C), refr index 1.4851 (40C), iodine value 80. Combustible.
Use: Modifier in alkyd and polyamide resins, polyester or metallic soap for petroleum additive, emulsifying agent, adhesives, shellac substitute, to upgrade drying oils.

dilithium *n*-acetyl-L-aspartate.
CAS: 32093-26-8. $C_6H_7NO_5 \cdot 2Li$.
Hazard: A poison.

dilithium sodium phosphate.
CAS: 37726-18-4. Li_2NaPO_4.
A commercial source of lithium found in Searles Lake brine.

dilituric acid. See 5-nitrobarbituric acid.

dill oil. See dill seed oil, American type.

dill oil, Indian type. See dill seed oil, Indian type.

dill seed oil, American type.
Properties: From steam distillation of the salks, leaves, and seeds of *Anethum graveolens* L. Yellowish liquid. D: 0.884–0.900, refr index: 1.480 at 20C. Sol in propylene glycol; insol in glycerin.
Use: Food additive.

dill seed oil, Indian type.
Properties: From steam distillation of the dried ripe fruit of *Anethum sowa* DC (Fam. *Umbelliferae* (FCC III). Yellowish liquid; harsh caraway odor and taste. D: 0.925–0.980, refr index: 1.486 at 20C. Sol in fixed oils and mineral oil; slightly sol in propylene glycol; insol in glycerin.
Use: Food additive.

diluent. (1) An ingredient used to reduce the concentration of an active material to achieve a desirable and beneficial effect. Examples are combination of diatomaceous earth with nitroglycerin to form the much less shock-sensitive dynamite; addition of sand to cement mixes to improve workability with no serious loss of strength; addition of an organic liquid having no solvent power to a paint or lacquer to reduce viscosity and achieve suitable application properties.
See thinner.
(2) Low-gravity materials used primarily to reduce cost, e.g., blown asphalt, wood floc, etc., in rubber and plastic mixes. In this sense, there is no clear distinction between a diluent and an extender.
(3) An ingredient of rocket fuels, such as helium, hydrazine, or hydrogen.

dilute acetic acid. (vinegar; acetic acid).
CAS: 64-19-7. $C_2H_4O_2$.
Properties: Aqueous solution containing 6% acetic acid by weight.

Derivation: Occurs naturally in foods such as cheese, coffee, grapes, peaches, raspberries, strawberries; product of the oxidation of ethanol and the destructive distillation of wood.

Use: A food additive; a counterirritant; a reagent.

dilution ratio.　　(hydrocarbon tolerance).

The maximum number of unit volumes of hydrocarbon that can be added per unit volume of active solvent to cause the first trace of gelation to occur, when the concentration of nitrocellulose in the solution is 8 g/100 mL. This may be used to evaluate the solvent power of active solvents by comparing them with a standard hydrocarbon, or to evaluate hydrocarbon solvents by comparing them with an active solvent.

See solvent; lacquer.

diluted alcohol.　　(diluted ethanol).

Alcohol that contains 42% of ethanol by weight.

Use: Solvent.

dimagnesium orthophosphate.　　See magnesium phosphate, dibasic.

dimagnesium phosphate.　　See magnesium phosphate, dibasic

dimedone.　　(1,1-dimethyl-3,5-diketocyclohexane).

CAS: 126-81-8.　$(CH_3)_2C_6H_6O_2$.

Properties: Greenish-yellow needles or prisms. Mp 148–149C. Slightly soluble in cold water and naphtha; soluble in alcohol, chloroform, and benzene.

Use: Reagent for the detection of ethyl alcohol and the identification of aldehydes.

dimefox.　　(bis(dimethylamino) fluorophosphate; tetramethyldiamidophosphoric fluoride; BFPO).

CAS: 115-26-4.　$[(CH_3)_2N]_2POF$.

Properties: Liquid; fishy odor. D 1.1151 (20/4C), bp 67C (4.0 mm Hg) 86C (15 mm Hg), refr index 1.4267 (20C). Soluble in water, ether, and benzene.

Hazard: Cholinesterase inhibitor, use may be restricted.

Use: A systemic pesticide primarily for ornamental and non–food plants.

dimenhydrinate.　　(2-(benzohydryloxy)-*N,N*-dimethylethylamine-8-chlorotheophyllinate; "Dramamine").

CAS: 523-87-5.　$C_{17}H_{22}NO \cdot C_7H_6ClN_4O_2$.

Properties: White, crystalline powder; odorless. Mp 102–107C, pH (saturated solution) 6.8–7.3. Freely soluble in alcohol and chloroform; soluble in benzene; sparingly soluble in ether; slightly soluble in water.

Grade: USP.

Hazard: Can cause drowsiness. Driving and using machinery may be hazardous.

Use: For prevention and treatment of nausea and vomiting in relation to motion sickness. It can also be used for some balance disorders of the inner ear, e.g., Merniere's disease.

Dimension.　　A Nylon 6 alloy.

Use: Injection and blow molding and extrusion.

dimer.　　An oligomer whose molecule is composed of two molecules of the same chemical composition. For example, 1,4-dioxane is a dimer of ethylene oxide.

A dimer acid is a high-molecular-weight dibasic acid that is liquid (viscous), stable, and resistant to high temperatures, and polymerizes with alcohols and polyols to yield a variety of products such as plasticizers, lubricating oils, and hydraulic fluids. It is produced by dimerization of unsaturated fatty acids at midmolecule and usually contains 36 carbons. Trimer acid, which contains three carboxyl groups and 54 carbons, is similar.

See polymer.

dimercaprol.　　USP name for 2,3-dimercaptopropanol.

2,3-dimercaptopropanol.　　(BAL; British Anti-Lewisite; dimercaprol; 1,2-diethioglycerol).

CAS: 59-52-9.　$CH_2(SH)CH(SH)CH_2OH$.

Properties: Colorless, oily, viscous liquid; strong odor of mercaptans. Bp 80C (1.9 mm Hg) 140C (40 mm Hg), mp 77C, d 1.2385 (25/4C), refr index 1.5720 (25C). Soluble in vegetable oils; moderately soluble in water with decomposition; soluble in alcohol.

Derivation: Bromination of allyl alcohol, followed by reaction with NaSH.

Grade: USP, as dimercaprol.

Use: Antidote to Lewisite, organic arsenicals, and heavy metals.

dimerized fat acids, residual.　　The acids obtained by heat polymerization of methyl esters of semidrying oils, removal of unpolymerized monomers, and saponification of residual esters. It consists chiefly of mixture of polymers of linoleic acid.

dimetan.　　(5,5-dimethyldihydroresorcinol dimethylcarbamate).

$C_{11}H_{17}NO_3$.

Properties: Yellow crystals. Mp 43–45C. Slightly soluble in water and oils; readily soluble in organic solvents.

Hazard: Toxic by ingestion and inhalation.

Use: Insecticide, especially for aphids.

dimethazone.　　See 2-(2-chlorobenzyl)-4,4-dimethyl-1,2-oxazolidin-3-one.

dimethicone.

CAS: 9006-65-9.　$CH_3[Si(CH_3)_2O]Si(CH_3)_3$.

Properties: Colorless silicone oil consisting of dimethylsiloxane polymers (range in viscosity from

0.65 to 1,000,000 centistokes at room temperature). Viscosity grades higher than 50 centistokes are immiscible with water; miscible with chloroform and ether.

Use: Ointments and topical drug ingredient, skin protectant.

dimethisoquin   hydrochloride.   (3-butyl-1-(2-dimethylaminoethoxy)isoquinoline hydrochloride).
CAS: 2773-92-4. $CH_3(CH_2)_2CH_2C_9H_5NOCH_2$ $CH_2N(CH_3)_2 \cdot HCl$.
Properties: White powder; odorless; bitter numbing taste. Mp 144–147C. Freely soluble in alcohol; soluble in water; pH (1% solution) 3.5–5.0.
Grade: NF.
Use: Medicine (topical anesthetic).

dimethoate.   (O,O-dimethyl-S-(N-methylcarbamoylmethyl) phosphorodithioate).
CAS: 60-51-5. $(CH_3O)_2PSSCH_2CONHCH_3$.
Properties: White solid. Mp 51–52C. Moderately soluble in water; soluble in most organic solvents except hydrocarbons.
Hazard: A cholinesterase inhibitor; use has been restricted.
Use: Insecticide (dry formulations not permitted).

dimethoxane.   (6-acetoxy-2,4-dimethyl-m-dioxane).
CAS: 828-00-2. $CH_3COOC_4H_5O_2(CH_3)_2$.
Properties: Clear-yellow to light-amber liquid. D 1.069–1.076 (25/25C), refr index 1.431–1.438 (20C), bp 66–68C (3 mm Hg), fp <−25C. Soluble in or miscible with water and organic solvents. Combustible.
Hazard: Questionable carcinogen.
Use: Preservative for cosmetics, inks, emulsions; gasoline additive.

1,2-dimethoxy-4-allylbenzene.   See   methyl eugenol.

2,5-dimethoxyaniline.
CAS: 102-56-7. $NH_2C_6H_3(OCH_3)_2$.
Properties: Gray flakes. Mp 69–73C. Insoluble in cold water; soluble in organic solvents and hot water.
Hazard: See aniline.
Use: Intermediate for dyes, pharmaceuticals, and insecticides; antioxidant.

1,2-dimethoxybenzene.   See veratrole.

1,3-dimethoxybenzene.   See resorcinol dimethyl ether.

1,4-dimethoxybenzene.   See hydroquinone dimethyl ether.

2,3-dimethoxybenzaldehyde.
CAS: 86-51-1. $C_9H_{10}O_3$.
Hazard: A poison by ingestion.

2,5-dimethoxybenzaldehyde.
CAS: 93-02-7. $(CH_3O)_2C_6H_3CHO$.
Properties: Flaked solid. Mp 46–49C. Soluble in organic solvents; insoluble in water. Combustible.
Use: Organic synthesis.

dimethoxybenzidine.   See dianisidine.

3,3′-dimethoxybenzidine-4,4′-diisocyanate.
See dianisidine diisocyanate.

3,4-dimethoxybenzyl alcohol.
CAS: 93-03-8. $C_6H_3(OCH_3)_2CH_2OH$.
Properties: Viscous, brown liquid or low-melting solid. Combustible.
Use: Organic synthesis.

2,4-dimethoxy-5-chloroaniline.
CAS: 6358-64-1. $NH_2(Cl)C_6H_2(OCH_3)_2$.
Properties: Violet-gray crystals. Mp 70–71C. Insoluble in water; soluble in alcohol, benzene, and other organic solvents.
Grade: 98.5%.
Hazard: See aniline.
Use: Intermediate for dyes and other organics.

dimethoxydimethylsilane.
CAS: 1112-39-6. $C_4H_{12}O_2Si$.
Hazard: Low toxicity by inhalation. A moderate eye irritant.

$p,p′$-dimethoxydiphenylamine.
CAS: 101-70-2. $(CH_3OC_6H_4)_2NH$.
A rubber antioxidant.

1,2-dimethoxyethane.   See   ethylene   glycol dimethyl ether.

(2-dimethoxyethyl) adipate.
$CH_3OC_2H_4OOC(CH_2)_4COOC_2H_4OCH_3$.
Properties: Liquid. D 1.075 (25C), refr index 1.439 (25C), bp 185–190C (11 mm Hg), fp −16C. Slightly soluble in water. Combustible.
Use: Plasticizer.

di(2-methoxyethyl) phthalate.   (DMEP).
CAS: 117-82-8. $C_6H_4(COOCH_2CH_2OCH_3)_2$.
Properties: Oily liquid; mild odor. D 1.172 (20/20C), bp 340C, fp −45C, flash p 381F (194C) (OC). Combustible.
Use: Plasticizer, especially for cellulose acetate; solvent.

1-(3,5-dimethoxy-4-hydroxycinnamoyl)-4-hexahydroazepinylcarbonylmethylpiperazine.
CAS: 57061-77-5. $C_{23}H_{33}N_3O_5$.
Hazard: Moderately toxic by ingestion.

4-(3,5-dimethoxy-4-hydroxycinnamoyl)-n-propyl-1-piperazineacetamide.

CAS: 57061-73-1. $C_{20}H_{29}N_3O_5$.
Hazard: Moderately toxic by ingestion.

7,8-dimethoxyisoquinoline.
CAS: 16503-95-0. $C_{11}H_{11}NO_2$.
Hazard: A poison by ingestion.

dimethoxymethane.   See methylal.

2,5-dimethoxy-4-methylamphetamine.
(STP; DOM).
A hallucinogenic, habit-forming drug, used in medicine; manufacture and use controlled by law in the U.S.
See amphetamine.

3,4-dimethoxy-3′,4′-methylenedioxystilbene.
CAS: 76306-39-3. $C_{17}H_{16}O_4$.
Hazard: Moderately toxic by ingestion.

2,5-dimethoxymethylamphetamine.   See STP (hallucinogen).

dimethoxymethylsilane.   See methyldimethoxysilane.

3,4-dimethoxyphenethylamine.   See homoveratrylamine.

3,4-dimethoxyphenylacetic acid.   See homoveratric acid.

3-((2-((2-(3,4-dimethoxyphenyl)ethyl)amino)-2-oxoethyl)amino)-*n*-methylbenzamide.
CAS: 104775-36-2. $C_{20}H_{25}N_3O_4$.
Hazard: Low toxicity by ingestion.

2,6-dimethoxyphenyllithium.
$(CH_3O)_2C_6H_3Li$.
Properties: White to tan, free-flowing, pyrophoric powder. Moderately soluble in toluene and benzene; slightly soluble in ethyl ether; stable indefinitely in sealed containers.
Hazard: Ignites spontaneously in air.

1-(3,4-dimethoxyphenyl)-2-nitro-1-propene.
$(CH_3O)_2C_6H_3CH:C(NO_2)CH_3$.
Properties: Yellow crystals. Mp 68–75C.
Use: Intermediate.

3-(dimethoxyphosphinyloxy)-*N,N*-dimethyl-*cis*-crotonamide.
CAS: 141-66-2. $(CH_3O)_2P(O)(OC)CH_3:CHC(O)N(CH_3)_2$.
Properties: Brown liquid. Bp 400C. Miscible with water, ethanol, and xylene; very slightly soluble in kerosene.
Hazard: Cholinesterase inhibitor, use may be restricted.
Use: Insecticide.

dimethoxystrychnine.   See brucine.

dimethoxytetraglycol.   (tetraethylene glycol dimethyl ether; bis(2-methoxyethoxyethyl ether; tetraglyme).
CAS: 143-24-8. $CH_3(OCH_2CH_2)_4OCH_3$.
Properties: Water-white, liquid; practically odorless. D 1.0132 (20/20C), bp 275.8C, 189C (100 mm Hg), vap press <0.01 mm Hg (20C), flash p 285F (141C), bulk density 8.4 lb/gal (20C), fp −29.7C, viscosity 0.0405 cP (20C), coefficient of expansion 0.00091 (20C). Stable; soluble in hydrocarbons and water. Combustible.
Grade: Technical.
Use: Solvent.

dimethrin.   Generic name for 2,4-dimethylbenzyl-2,2-dimethyl-3-(2-methylpropenyl)cyclopropane carboxylate (2,4-dimethylbenzyl chrysanthemumate).
CAS: 70-38-2. $C_{19}H_{26}O_2$.
Properties: Amber liquid. D 0.986 (20C), bp 175C (3.8 mm Hg). Insoluble in water; soluble in petroleum hydrocarbons, aromatic petroleum derivatives, alcohols, and methylene chloride; decomposed by strong alkali.
Hazard: Toxic by ingestion.
Use: Insecticide.

dimethyl.   See ethane.

dimethylacetal.   (ethylidenedimethyl ether).
CAS: 534-15-6. $CH_3CH(OCH_3)_2$.
Properties: Colorless liquid; strongly aromatic odor. Soluble in water, alcohol, ether, and chloroform. D 0.848 (25C), bp 62–63C, flash p approximately 80F (26.6C).
Derivation: By heating acetaldehyde with methyl alcohol and glacial acetic acid and distilling.
Hazard: Flammable, dangerous fire risk. Toxic by ingestion and inhalation.
Use: Medicine, organic synthesis.

***N,N*-dimethyl acetamide.**   (DMAC).
CAS: 127-19-5. $CH_3CON(CH_3)_2$.
Properties: Colorless liquid. Bp 166C, d 0.9366 (25C), refr index 1.4351 (25C), flash p 171F (77C). Miscible with water, aromatics, esters, ketones, and ethers. Combustible.
Derivation: From acetic anhydride and dimethylformamide.
Grade: Technical, high-purity certified.
Hazard: Toxic by inhalation, absorbed by skin, strong irritant. Liver and embryo/fetal damage. Questionable carcinogen.
Use: Solvent for plastics, resins, gums, and electrolytes; intermediate; catalyst; paint remover; high-purity solvent for crystallization and purification.

***N,N*-dimethylacetoacetamide.**
CAS: 2044-64-6. $CH_3COCH_2CON(CH_3)_2$.

Properties: Liquid. Bp 220C, d 1.048–1.053 (20/20C), refr index 1.4379 (20C), flash p 252F (122C) (COC). Miscible with water and organic solvents. Combustible.
Use: Chemical intermediate.

2,4-dimethylacetophenone.
CAS: 89-74-7. $CH_3COC_6H_3(CH_3)_2$.
Properties: Colorless liquid; odor of mimosa. D 0.994–0.997, refr index 1.532–1.534. Soluble in four volumes of 60% alcohol. Combustible.
Use: Perfumery, flavoring.

dimethylacetyl chloride.   See isobutyric acid chloride.

dimethylacetylene.   See crotonylene.

2,5-dimethyl-3-acetylfuran.   See 3-acetyl-2,5-dimethylfuran.

dimethyallyl.   See 2,5-dimethylhexadiene-1,5.

dimethylamine. (DMA).
CAS: 124-40-3. $(CH_3)_2NH$.
Properties: (Anhydrous) Gas with ammoniacal odor. (25% water solution) d 0.6865 (−6C), bp 6.88C, fp −92.2C, flash p 0F (−17.7C), bulk density approximately 7.8 lb/gal; soluble in alcohol and ether, autoign temp (anhydrous) 806F (430C).
Derivation: Interaction of methanol and ammonia over a catalyst at high temperatures. The mono-, di-, and trimethylamines are all produced.
Method of purification: Azeotropic or extractive distillation.
Grade: Technical (anhydrous, 25%, and 40% aqueous solutions), 99%.
Hazard: Flammable, dangerous fire risk, explosive limits in air 2.8–14%. Upper respiratory tract and gastrointestinal irritant. Questionable carcinogen.
Use: Acid–gas absorbent, solvent antioxidants, manufacture of dimethylformamide and dimethylacetamide, dyes, flotation agent, gasoline stabilizers, pharmaceuticals, textile chemicals, rubber accelerators, electroplating, dehairing agent, missile fuels, pesticide propellant, rocket propellants, surfactants, reagent for magnesium.

dimethylamine-epichlorohydrin copolymer.
CAS: 25988-97-0.
Use: Food additive.

n′-dimethylaminoacetylpartricin a dimethylaminoethylamide diaspartat.
CAS: 143563-20-6. $C_7H_7NO_4 \cdot 1/_2C_{67}H_{103}N_5O_{19}$.
Hazard: A poison.

dimethylaminoaniline.   See p-aminodimethylaniline.

dimethylaminoazobenzene.   (methyl yellow; butter yellow).

CAS: 60-11-7. $C_6H_5NNC_6H_4N(CH_3)_2$.
Properties: Yellow, crystalline leaflets. Mp 116C. Soluble in alcohol, ether, strong mineral acids, and oils; insoluble in water.
Derivation: Action of benzenediazonium chloride on dimethyl aniline.
Hazard: Toxic by inhalation and skin absorption. A carcinogen. May not be used in foods or beverages (FDA). Possible carcinogen.
Use: Organic research, indicator, dyes.

p-dimethylaminobenzaldehyde.
CAS: 100-10-7. $C_6H_4[N(CH_3)_2]CHO$.
Properties: Yellow, crystalline plates. Mp 73C, bp 176–177C (17 mm Hg). Soluble in alcohol and ether. Combustible.
Derivation: By mixing dimethylaniline, anhydrous chloral, and phenol, and allowing the mixture to stand. The phenol is removed by shaking with dilute caustic soda and the residue dissolved in water and hydrochloric acid and crystallized.
Grade: Technical, reagent.
Hazard: Toxic by ingestion.
Use: Dyes, medicine, reagent.

dimethylaminobenzene.   See xylidine.

p-dimethylaminobenzene diazonium chloride, zinc chloride double salt.   See p-diazodimethylaniline; zinc chloride double salt.

p-dimethylaminobenzenediazo sodium sulfonate.   ("Dexon" [Covidien]).
CAS: 140-56-7. $(CH_3)_2NC_6H_4NNSO_3Na$.
Properties: Solid. Melts with decomposition above 200C. Soluble in water.
Hazard: Toxic by ingestion. Questionable carcinogen.
Use: Fungicide for protection of germinating seed and seedlings.

3-dimethylaminobenzoic acid.
$(CH_3)_2NC_6H_4COOH$.
Properties: Pale-yellow crystals. Mp 147–153C.
Use: Intermediate.

8-((dimethylamino)carbonyl)-5-oxo-2,4,9-trimethyl-6,11-dioxa-3-thia-2,4,7,10-tetraazadodeca-7,9-dienoic acid, 2,3-dihydro-2,2-dimethyl-7-benzofuranyl ester.
CAS: 90293-54-2. $C_{21}H_{29}N_5O_7S$.
Hazard: A poison by ingestion.
Use: Agricultural chemical.

8-((dimethylamino)carbonyl)-5-oxo-2,4,9-trimethyl-6,11-dioxa-3-thia-2,4,7,10-tetraazadodeca-7,9-dienoic acid, 1-naphthalenyl ester.
CAS: 90293-53-1. $C_{21}H_{25}N_5O_6S$.
Hazard: A poison by ingestion.
Use: Agricultural chemical.

8-((dimethylamino)carbonyl)-5-oxo-2,4,9-trimethyl-6,11-dioxa-3-thia-2,4,7,10-tetraazadodeca-7,9-dienoic acid, 4-nonylphenyl ester.
CAS: 90293-55-3. $C_{26}H_{41}N_5O_6S$.
Hazard: Moderately toxic by ingestion.
Use: Agricultural chemical.

***n*-((8-((dimethylamino)carbonyl)-2,4,9-trimethyl-1,5-dioxo-6,11-dioxa-3-thi a-2,4,7,10-tetraazadodeca-7,9-dien-1-yl)oxy)-ethanimidothioic acid, methyl ester.**
CAS: 90293-52-0. $C_{14}H_{24}N_6O_6S_2$.
Hazard: A poison by ingestion.
Use: Agricultural chemical.

***dl*-dimethylamino-4,4-diphenyl-3-heptanone hydrochloride.**   See methadone hydrochloride.

3-(dimethylamino)-1-(2,2-diphosphonoethyl) pyrazinium inner salt.
CAS: 203264-12-4. $C_8H_{15}N_3O_6P_2$.
Hazard: A reproductive hazard.

2-dimethylaminoethanol.   (deanol; dimethylethanolamine).
CAS: 108-01-0. $(CH_3)_2NCH_2CH_2OH$.
Properties: Colorless liquid; amine odor. Bp 134.6C, fp −59.0C, d 0.8879 (20/20C), bulk density 7.4 lb/gal (20C), refr index 1.4300 (20C), flash p 105F (40.5C) (OC). Miscible with water, acetone, ether, and benzene. Combustible.
Derivation: Prepared from ethylene oxide and dimethylamine.
Grade: Anhydrous and 70% aqueous solution.
Hazard: Moderate fire risk.
Use: Intermediate in the synthesis of dyestuffs, textile auxiliaries, pharmaceuticals, and corrosion inhibitors; curing epoxy, amine, and polyamide resins; emulsifier in paints and coatings.

β-dimethylaminoethyl chloride hydrochloride.   (DMC).
$(CH_3)_2NCH_2CH_2Cl:HCl$.
Use: Manufacture of antihistamines and other pharmaceuticals. Organic intermediate for introduction of β-dimethylaminoethyl radical.

2-(dimethylamino)ethylhydrazine.   See 2-hydrazino-*N,N*-dimethylethanamine.

dimethylaminoethyl methacrylate.
CAS: 2867-47-2. $CH_2:C(CH_3)COOCH_2CH_2-N(CH_3)_2$.
Properties: Liquid. D 0.933 (25C), bp 182–190C, refr index 1.4376 (25C), flash p 165F (73.9C) (TOC). Combustible.
Hazard: Irritant to skin, eyes, and mucous membranes; strong lachrymator.
Use: Binders for coatings, textile chemicals, dispersing agents for nonaqueous systems, antistatic

agents, stabilizers for chlorinated polymers, ion exchange resins, emulsifying agents, cationic precipitating agents.

2-(2-(dimethylamino)ethyl)pyridine hydrochloride.
CAS: 25877-27-4. $C_9H_{14}N_2 \cdot ClH$.
Hazard: A poison.

5-(2′-(*N,N*-dimethylamino)-2′-methyl)ethyl-10,11-dihydro-5h-dibenz(b,f)azepine.
CAS: 2064-23-5. $C_{19}H_{24}N_2$.
Hazard: A poison.

dimethylaminomethyl phenol.
CAS: 25338-55-0. $C_6H_4OHCH_2N(CH_3)_2$.
Exists as *o*-, *m*-, and *p*-isomers; the commercial material is a mixture of *o*- and *p*-.
Properties: Dark-red liquid; phenolic odor, free of methylamine. D 1.010 (25/25C), refr index 1.530 (25C), distillation range 80–130C (2 mm Hg), water content (Karl Fischer) 0.5%. Soluble in organic solvents; moderately soluble in water.
Hazard: See phenol.
Use: Antioxidants, stabilizers, catalysts, intermediates.

4-dimethylamino-3-methylphenolmethyl carbamate (ester).
CAS: 2032-59-9. $(CH_3)_2NC_6H_3(CH_3)OOCNHCH_3$.
Properties: Tan, crystalline solid. Mp 93–94C. Slightly soluble in water; moderately soluble in aromatic solvents; unstable in highly alkaline media.
Hazard: Toxic by ingestion, inhalation, and skin absorption.

2-(dimethylamino)-*n*-(((methyl(((2-phenyl-1,3-dioxan-5-yl)methoxy) sulfinyl)amino)carbonyl)oxy)-2-oxo-ethanimidothioic acid, methyl ester.
CAS: 81862-00-2. $C_{18}H_{25}N_3O_7S_2$.
Hazard: A poison by ingestion.
Use: Agricultural chemical.

dimethlaminonaphthalenesulfonyl chloride.   See dansyl chloride.

***L*-dimethylaminonapthalene-5-sulfonic acid.**
(DANS; *N,N*-dimethyl-4-[(E)-2-(4-nitrophenyl) ethenyl]aniline).
$C_{16}H_{16}N_2O_2$.
A green fluorescent compound.
Use: In histochemistry to detect antigens.

5-dimethylamino-1-napthalene sulfonyl chloride.   See dansyl chloride.

2-(dimethylamino)-2-oxoethyl-*n*-(((methylamino)carbonyl)oxy)ethanimidothioate.
CAS: 92065-83-3. $C_8H_{15}N_3O_3S$.

Hazard: Low toxicity by ingestion.
Use: Agricultural chemical.

2-dimethylamino-1-phenylpropane.   See
N,N,α-trimethylphenethylamine.

5-dimethylamino-3-piperidinoacetylindole.
CAS: 24955-83-7. $C_{13}H_{16}N_2O$.
Hazard: A poison.

1-dimethylamino-2-propanol.
CAS: 108-16-7. $(CH_3)_2NCH_2CH_2OCH_3$.
Properties: Water-white liquid; amine odor. Bp
125.6C, d 0.850 (20/20C), refr index 1.421 (20C),
flash p 90F (32C). Soluble in water and most organic
solvents.
Hazard: Flammable, moderate fire risk.
Use: Organic synthesis.

3-dimethylaminopropylamine.
CAS: 109-55-7. $(CH_3)_2NCH_2CH_2CH_2NH_2$.
Properties: Colorless liquid. Bp 123C, fp −70C (sets
to a glass below this temperature), d 0.8100 (30C),
refr index 1.4328 (25C), flash p 95F (35C) (TCC).
Soluble in water and organic solvents.
Hazard: Flammable, moderate fire risk. Toxic by
ingestion and inhalation, strong irritant.
Use: Curing agent for epoxy resins, organic interme-
diate.

1-dimethylamino-2-propyl chloride.
(β-dimethylaminoisopropyl chloride).
$(CH_3)_2NCH_2CHClCH_3$.
Properties: Yellow liquid that darkens with age. Distillation range 113–120C, refr index 1.422–1.423
(25C).
Use: Intermediate.

1-dimethylamino-3-propyl chloride.
(DMPC).
$(CH_3)_2NCH_2CH_2CH_2Cl$.
Use: Intermediate for pharmaceutical and organic
synthesis.

**α-(3-(dimethylamino)propyl)-5-(1,1-
dimethylethyl)-1-phenyl-1h-pyrazole-4-
methanol.**
CAS: 296269-50-6. $C_{19}H_{29}N_3O$.
Hazard: A poison by ingestion.

**α-(3-(dimethylamino)propyl)-1,5-diphenyl-
1h-pyrazole-4-methanol.**
CAS: 296269-51-7. $C_{21}H_{25}N_3O$.
Hazard: A poison by ingestion.

**1-(3-dimethylaminopropyl)-3-
ethylcarbodiimide hydrochloride.**   See
EDC (reagent).

**α-(3-(dimethylamino)propyl)-5-ethyl-1-
phenyl-1h-pyrazole-4-methanol.**
CAS: 296269-47-1. $C_{17}H_{25}N_3O$.
Hazard: A poison by ingestion.

dimethylaminopropylmethacrylamide.
(DMAPMA).
CAS: 5205-93-6. $H_2C=C(CH_2)COOCH_2N(CH_3)_3$.
Properties: Pale-yellow liquid. Flash p 140C (CC),
refr index 1.4763, d 0.94, bulk density 7.8 lb/gal.
Use: Reactive cationic monomer, incorporation of
amine group into polymers.

**α-(3-(dimethylamino)propyl)-5-(1-
methylethyl)-1-phenyl-1h-pyrazole-4-
methanol.**
CAS: 296269-49-3. $C_{18}H_{27}N_3O$.
Hazard: A poison by ingestion.

**α-(3-(dimethylamino)propyl)-5-methyl-1-
phenyl-1h-pyrazole-4-methanol.**
CAS: 296269-46-0. $C_{16}H_{23}N_3O$.
Hazard: A poison by ingestion.

**α-(3-(dimethylamino)propyl)-1-phenyl-5-
propyl-1h-pyrazole-4-methanol.**
CAS: 296269-48-2. $C_{18}H_{27}N_3O$.
Hazard: A poison by ingestion.

**α-(3-(dimethylamino)propyl)-1-phenyl-1h-
pyrazole-4-methanol.**
CAS: 296269-45-9. $C_{15}H_{21}N_3O$.
Hazard: A poison by ingestion.

2-dimethylaminopyridine.
CAS: 5683-33-0. $C_7H_{10}N_2$.
Hazard: A poison.

**4-dimethylamino-3,5-xylyl-N-methyl   carba-
mate.**
$(CH_3)_2N(CH_3)_2C_6H_2OOCNHCH_3$.
Properties: White, crystalline solid. Mp 85C.
Almost insoluble in water; soluble in benzene, alco-
hol, and chloroform.
Hazard: Toxic by ingestion.
Use: Systemic insecticide.

N,N-dimethylaniline.   (aniline, N,N-dimethyl).
CAS: 121-69-7. $C_6H_5N(CH_3)_2$.
Properties: Yellowish to brownish, oily liquid. D
0.954, mp 2.5C, bp 192.5–193.5C, flash p 145F
(62.7C) (CC), refr index 1.5582, autoign temp 700F
(370C). Soluble in alcohol and ether; insoluble in
water. Combustible.
Derivation: By heating a mixture of aniline, aniline
hydrochloride, and methyl alcohol (free from acetone) in an autoclave and distilling.
Grade: Technical, reagent, 99.9% pure.
Hazard: Toxic material absorbed by skin. Methemoglobinemia. Questionable carcinogen.
Use: Dyes, intermediates, solvent, manufacture of
vanillin, stabilizer (acid acceptor), reagent.

2-(3,4-dimethylanilino)-2-oxazoline.   See
3,4-xylidino-2-oxazoline.

dimethyl anthranilate.   (*N*-methyl methyl anthranilate).
CAS: 85-91-6. $CH_3OOCC_6H_4NHCH_3$.
Properties: Colorless or pale-yellow liquid; grape-like odor. D 1.132–1.138 (15C), refr index 1.578–1.581 (20C). Soluble in three volumes or more of 80% alcohol, and in benzyl benzoate, diethyl phthalate, fixed oils, mineral oils, and volatile oils; insoluble in glycerol; slightly soluble in propylene glycol. Congealing p 18C (4% impurity) to 10C (20% impurity). Combustible.
Derivation: Methylation of methyl anthranilate or esterification of *N*-methylanthranilic acid.
Use: Perfumes, flavorings, and drugs.

dimethylarsinic acid.   See cacodylic acid.

2,2′-dimethyl-2,2′-azodibutyronitrile.   See 2,2′-azobisisovaleronitrile.

9,10-dimethyl-1,2-benzanthracene.   (Dba; 9,10-dimethylbenzanthracene; 9,10-benz[a]anthracene; dimethylbenzanthrene; 7,12-dimethylbenzo[a]anthracene; 1,4-dimethyl-2,3-benzphenanthrene; dmba; 7,12-dmba).
CAS: 57-97-6. $C_{20}H_{16}$.
Properties: Polycyclic aromatic hydrocarbon comprising water-insoluble plates or leaflets, slightly soluble in ethanol, soluble in benzene.
Derivation: Found in tobacco smoke.
Hazard: Toxic; teratogenic; mutagenic; tumorigenic; questionable carcinogen; neoplastigenic.

1,2-dimethylbenzene.   See *o*-xylene.

1,3-dimethylbenzene.   See *m*-xylene.

1,4-dimethylbenzene.   See *p*-xylene.

3,3-dimethylbenzidine.   See *o*-tolidine.

dimethylbenzylcarbinyl acetate.   See α,α-dimethylphenethyl acetate.

2,5-dimethylbenzyl chloride.   (2,5-dimethyl-α-chlorotoluene; α-chloro-*p*-xylene).
$C_6H_3(CH_3)_2CH_2Cl$.
Properties: Colorless to pale-yellow liquid; sharp pungent odor. D 1.035–1.045 (25/25C), bp 221–226C, refr index 1.5350–1.5360 (25C). Soluble in alcohols and ethers; insoluble in water. Combustible.
Hazard: Irritant to eyes and mucous membranes, a lachrymator.
Use: Intermediate for pharmaceuticals, dyes, perfumes, plasticizers, resins, wetting agents, germicides, etc.

2,4-dimethylbenzyl   chrysanthemumate.
See dimethrin.

α,β-dimethylbenzyl hydroperoxide.   See cumene hydroperoxide.

dimethylbenzyl phosphonate.
CAS: 773-47-7. $(CH_3O)_2P(O)CH_2C_6H_5$.
Properties: Water-white liquid. Bp 114C (0.75 mm Hg).
Use: Extraction of mineral salts from special solutions, fire retardant, low-temperature plasticizer.

3,3′-dimethyl-4,4′-biphenylene diisocyanate.   (4,4′-bi-*o*-tolylene diisocyanate).
Properties: Flaked, white, crystalline solid. Mp 69C min. Assay 98% min.
Use: High-strength elastomers, coatings, rigid plastics.

N,N′-dimethyl-4,4′-bipyridinium dibromide.   See paraquat dibromide.

1,1′-dimethyl-4,4′-bipyridinium salt.   See paraquat.

dimethylbis(phenylthio)stannane.   See bis(phenylthio)dimethyltin.

dimethyl brassylate.
CAS:1472-87-3. $CH_3OOC(CH_2)_{11}COOCH_3$.
An ester of a 13-carbon saturated aliphatic dibasic acid. Waxy, low-melting solid, made by an ozone oxidation process.
Use: Preparation of ethylene brassylate, a synthetic musk; chemical intermediate.

2,3-dimethyl-1,3-butadiene.
CAS: 513-81-5. C_6H_{10}.
Properties: Dark liquid. Bp 70C, fp −76C, d 0.727, refr index 1.43.
Derivation: Shale oil fractions, by distillation of pinacol.
Use: Manufacture of elastomers.

2,2-dimethylbutane.   See neohexane.

2,3-dimethylbutane.   (diisopropyl).
CAS: 79-29-8. $(CH_3)_2CHCH(CH_3)_2$

$$CH_3-CH-CH-CH_3$$
$$\quad\ \ CH_3\ CH_3$$

Properties: Colorless liquid. Bp 57.9C, d 0.66164 (20C), fp −128.41C, refr index 1.37495 (20C), flash p −20F (−28.9C), autoign temp 788F (420C).
Derivation: Alkylation of ethylene with isobutane using aluminum chloride catalyst.
Grade: Technical 95%, 99%, 99.8 mole %.
Hazard: Flammable, dangerous fire and explosion risk.
Use: High-octane fuel, organic synthesis.

2,2-dimethyl-1,3-butanediol.
CAS: 76-35-7. $CH_3CH(OH)C(CH_3)CH_2OH$.
Properties: Liquid. D 0.9700, bp 202.4C, fp −12.8C, bulk density 8.1 lb/gal. Very soluble in water. Combustible.

1,3-dimethyl butanol.   See amyl methyl alcohol.

2,3-dimethylbutene-1.
CAS: 563-78-0. $CH_3CH(CH_3)C(CH_3):CH_2$.
Properties: Liquid. Fp −157.27C, bp 55.6C, d 0.678 (20/4C); bulk density 5.68 lb/gal (60F), refr index 1.390 (20C), flash p −20F (−28.9C).
Hazard: Flammable, dangerous fire and explosion risk.
Use: Perfume synthesis, isomerization reactions.

2,3-dimethylbutene-2.
CAS: 563-79-1. $CH_3C(CH_3):C(CH_3)CH_3$.
Properties: Liquid. Fp −74.3C, bp 73.2C, d 0.708 (20.4C), bulk density 5.94 lb/gal (60F), refr index 1.412 (20C), flash p 0F (−17.7C).
Hazard: Flammable, dangerous fire and explosion risk.
Use: Isomerization reactions.

***N,N*-dimethylbutylamine.**   See butyldimethylamine.

2,4-dimethyl-6-*tert*-butyl phenol.
$(CH_3)_2C_6H_2OH[C(CH_3)_3]$.
Properties: A low-viscosity, straw-colored liquid. D 0.961. Readily soluble in gasoline and oils. Combustible.
Use: Antiskinner; antioxidant, mainly for gasoline (ASTM D91064T).

dimethylcadmium.
CAS: 506-82-1. C_2H_6Cd.
Properties: Liquid; unpleasant odor. Fp −4.0C, bp 105C, d 1.98, refr index 1.54. Soluble in hydrocarbon solvents.
Hazard: Explodes on heating to 150C, pyrophoric.
Use: Polymerization catalyst, organic synthesis.

dimethylcarbamoyl chloride.   (dimethylcarbamyl chloride).
CAS: 79-44-7. $(CH_3)_2NCOCl$.
Properties: A colorless liquid that hydrolyzes rapidly. Bp 165C, fp −33C, d 1.67.
Hazard: A probable carcinogen and a strong lachrymator. Nasal cancer. Upper respiratory tract irritant.
Use: Chemical intermediate in production of dyes, pharmaceuticals, and pesticides.

dimethylcarbate.   (*cis*-dimethyl ester-5-norbornene-2,3-dicarboxylic acid).
CAS: 5826-73-3. $C_{11}H_{14}O_4$.
Generic name for bicyclo(2,2,1)-5-heptene-2,3-dicarboxylic acid dimethyl ester.

Properties: Clear, oily liquid or crystalline solid. D 1.165 (35/4C). Insoluble in water.
Derivation: Esterification of Diels-Alder condensation product of maleic anhydride and cyclopentadiene.
Hazard: Toxic by ingestion and inhalation.
Use: Insect repellent.

dimethylcarbinol.   See isopropyl alcohol.

dimethylcarbonate.   Legal label name for methyl carbonate.

dimethylchloroacetal.   (chloroacetaldehydedimethylacetal).
CAS: 97-97-2. $ClCH_2CH(OCH_3)_2$.
Properties: Colorless liquid with a pleasant odor. Boiling range 126–132C, flash p 110F (43.3C), d 1.082–1.092 (25/4C), refr index 1.4110–1.4130 (25C), purity 97% (min), bulk density 9.07 lb/gal. Combustible.
Grade: Technical.
Hazard: Moderate fire risk.
Use: Organic synthesis, pharmaceuticals, solvent.

dimethylchloroacetoacetamide.   See 2-chloro-*N,N*-dimethyl-3-oxobutanamide.

2,5-dimethyl-α-chlorotoluene.   See 2,5-dimethylbenzyl chloride.

dimethylchlortetracycline hydrochloride.
$C_{21}H_{21}ClN_2O_8 \cdot HCl$.
Properties: Yellow, crystalline powder; odorless; bitter taste. Partially soluble in water; slightly soluble in alcohol.
Grade: NF.
Use: Medicine (antibiotic).

dimethyl L-curine dimethiodide.   See *N,N',o,o'*-tetramethylcurinium diiodide.

dimethyl cyanamide.
CAS: 1467-79-4. $(CH_3)_2NCN$.
Properties: Colorless, mobile liquid. Fp −41C, d 0.876, bp 160C, flash p 160F (71.1C) (TCC). Combustible.
Hazard: Toxic by ingestion and inhalation.
Use: Chemical intermediate, solvent.

dimethylcyclohexane.   (hexahydroxylene).
CAS: 589-90-2.   Mix of *o*-, *m*-, and *p*-isomers. $C_6H_{10}(CH_3)_2$.
Properties: Water-white liquid; mild odor. D 0.776 (15/15C), boiling range 120–129C, fp <−65C, flash p approximately 50F (10C) (CC). Soluble in most common solvents; almost insoluble in water.
Hazard: Flammable, dangerous fire risk.
Use: Synthesis, special solvent.
Note: There are several isomers of dimethylcyclohexane, i.e., 1,3-; 1,4-; *cis*-1,2-; *trans*-1,2-. They

have properties and uses closely similar to those given above. All are flammable.

dimethyl-1,4-cyclohexane dicarboxylate.
CAS: 94-60-0. $CH_3OOCC_6H_{10}COOCH_3$.
Properties: Partially crystalline solid. D 1.102 (35/4C), consists of approximately 60% *cis* and 40% *trans* isomers, bp (mixed isomer) 265C, bulk density 9.18 lb/gal (20C). Soluble in all proportions in most organic solvents.
Use: Plasticizers, polymers.

N,N-dimethylcyclohexylamine.
CAS: 98-94-2. $(CH_3)_2NC_6H_{11}$.
Properties: Water-white liquid. Distilling range 157–160C, d 0.8490 (20/20C), fp approximately −77C, flash p 110F (43.3C) (COC). Partly soluble in water; miscible with alcohol, benzene, and acetone. Combustible.
Hazard: Moderate fire risk.
Use: Catalyst for polyurethane foams, intermediate for rubber accelerators, treatment of textiles.

1,2-dimethylcyclopentane.
CAS: 2452-99-5. $C_5H_8(CH_3)_2$.
Properties: Colorless liquid. *cis*: bp 99.5C, d 0.772 (20C); *trans*: bp 91.8C, d 0.751 (20C). Combustible.
Grade: Technical.
Use: Organic synthesis.

2,2′-dimethyl-1,1′-dianthraquinone.
$C_{30}H_{18}O_4$.
Properties: Yellow crystals. Mp 365–367C. Soluble in hot nitrobenzene, aniline, and chlorobenzene.
Use: Dye intermediate.

2,5-dimethyl-2,5-di(*tert*-butylperoxy)hexane.
CAS: 78-63-7. $C_4H_9OOC(CH_3)_2CH_2CH_2C$ $(CH_3)_2OOC_4H_9$.
Properties: Stable, colorless liquid. Bp 50–52C (0.1 mm Hg), active oxygen 10.5% min, fp 8C, flash p 185F (85C). Soluble in alcohol; insoluble in water. Combustible.
Hazard: Strong oxidizer, may ignite organic materials. Irritant to eyes and skin.
Use: Catalyst in polyethylene cross-linking, styrene polymerization, polyester resins.

dimethyl dicarbonate.
CAS: 4525-33-1. $(CH_3OCO)_2O$.
Properties: Bp: 44–47C at 5 mm.
Use: Food additive; fungicide.

dimethyldichlorosilane.   (dichlorodimethylsilane).
CAS: 75-78-5. $(CH_3)_2SiCl_2$.
Properties: Colorless liquid. Bp 70C, fp −86C, d 1.062 (20C), refr index 1.4023 (25C), flash p 16F

(−8.9C) (COC). Reacts with water to form complex mixture of dimethylsiloxanes and liberates hydrogen chloride. Soluble in benzene and ether.
Derivation: Action of silicon on methyl chloride in the presence of a copper catalyst or by Grignard reaction from methyl chloride and silicon tetrachloride.
Grade: Technical.
Hazard: Flammable, dangerous fire and explosion risk.
Use: Intermediate for silicone products.

dimethyldichlorovinyl phosphate.   See dichlorovos.

1,2-dimethyl-3-diethylaminopropyl *p*-isobutoxybenzoate.   See ganglefene.

2,2-dimethyl-2,3-dihydrobenzofuran-7-yl-*n*-(4-bromophenylthio)-*n*-methylcarbamate.
CAS: 50539-74-7. $C_{18}H_{18}BrNO_3S$.
Hazard: A poison by ingestion.
Use: Agricultural chemical.

5,5-dimethyldihydroresorcinol dimethylcarbamate.   See dimetan.

dimethyl diisopropyl pyrophosphate.
CAS: 63919-09-5. $C_8H_{20}O_7P_2$.
Hazard: A poison.

1,1-dimethyl-3,5-diketocyclohexane.   See dimedone.

dimethyldiketone.   See diacetyl.

n,5-dimethyl-4-((dimethylamino)carbonyl)-*n*-((4-(1,1-dimethylethyl)phenyl)thio)-2,7-dioxa-3,6-diazaocta-3,5-dienamide.
CAS: 90293-48-4. $C_{19}H_{28}N_4O_4S$.
Hazard: A poison by ingestion.
Use: Agricultural chemical.

N,N′-dimethyl-*N,N′*-di(1-methylpropyl)-*p*-phenylene diamine.   Forms a continuous protective film.
Properties: Volatile, reddish-brown liquid.
Use: Antiozonant in rubber.

N,N-dimethyl-*N,N*-dinitrosterephthalamide.
$C_6H_4[CON(CH_3)NO]_2$.
Use: Blowing agent liberating nitrogen at 100C, with a residue of dimethyl terephthalate.

dimethyldioxane.
CAS: 25136-55-4

$$\overline{OCH(CH_3)CH_2OCH_2CH(CH_3)}.$$

Properties: Water-white liquid. D 0.9268, bp 117.5C, flash p 75F (23.9), vap press 154.4 mm Hg at 20C. Soluble in water.
Hazard: Flammable, dangerous fire risk.

2,2-dimethyl-1,3-dioxolane-4-methanol.
CAS: 100-79-8. $C_6H_{12}O_3$.
Properties: Colorless, liquid; odorless. Bp 82C (10 mm Hg), d 1.065, refr index 1.43, flash p 194F (90C). Soluble in water and most organic solvents. Optically active. Combustible.
Use: General solvent, plasticizer.

N,N-dimethyl-2,2-diphenyl acetamide. See diphenamid.

3,3-dimethyl-1,4-dithian-2-one-o-((methyl ((trichloromethyl)thio)amino)carbonyl) oxime.
CAS: 55391-31-6. $C_9H_{13}Cl_3N_2O_2S_3$.
Hazard: A poison by ingestion.
Use: Agricultural chemical.

dimethylenemethane. See allene.

3,4,3′,4′-dimethylenedioxystilbene.
CAS: 76306-40-6. $C_{16}H_{12}O_4$.
Hazard: Moderately toxic by ingestion.

2-dimethylethanolamine. See 2-dimethylaminoethanol.

dimethyl ether. (methyl ether; methyl oxide; wood ether).
CAS: 115-10-6. CH_3OCH_3.
Properties: Colorless compressed gas or liquid. D 0.661, bp −24.5C, fp −141.4C, flash p −42F (−41C), autoign temp 662F (350C). Soluble in water, alcohol, and organic solvents.
Derivation: By dehydration of methanol.
Grade: Technical, 99.5%.
Hazard: Highly flammable, dangerous fire and explosion hazard.
Use: Refrigerant, solvent, extraction agent, propellant for sprays, chemical reaction medium, catalyst and stabilizer in polymerization.

dimethylethoxysilane. See ethoxydimethylsilane.

2-((1,1-dimethylethyl)amino)ethanol. See t-butylethanolamine.

(e)-1-((1,1-dimethylethyl)azo)cyclopentanol.
See (e)-1-t-butylazo-1-hydroxycyclopentane.

dimethylethyl carbinol. See tert-amyl alcohol.

1,1-dimethylethyl chloroacetate. See tert-butyl chloroacetate.

dimethylethylene. See butene-2.

n′,n″-dimethylethylenedihydrazine. See 1,1′-(1,2-ethanediyl)bis(1-methylhydrazine).

sym-dimethylethylene glycol. See 2,3-butylene glycol.

5,5-dimethyl-2-(ethylimino)-1,3-dithiolan-4-one-o-((methylamino)carbonyl)oxime.
CAS: 71108-04-8. $C_9H_{15}N_3O_2S_2$.
Hazard: A poison by ingestion.
Use: Agricultural chemical.

(4-(1,1-dimethylethyl)phenyl)methyl ethyl 3-pyridinylcarbonimidodithioate.
CAS: 51308-53-3. $C_{19}H_{24}N_2S_2$.
Hazard: Moderately toxic by ingestion.
Use: Agricultural chemical.

s-((4-(1,1-dimethylethyl)phenyl)methyl) o-ethyl-3-pyridinylcarbonimidothioate.
CAS: 51308-61-3. $C_{19}H_{24}N_2OS$.
Hazard: Moderately toxic by ingestion.
Use: Agricultural chemical.

(4-(1,1-dimethylethyl)phenyl)methyl heptyl-3-pyridinylcarbonimidodithioate.
CAS: 51308-60-2. $C_{24}H_{34}N_2S_2$.
Hazard: Moderately toxic by ingestion.
Use: Agricultural chemical.

(4-(1,1-dimethylethyl)phenyl)methyl hexyl 3-pyridinylcarbonimidodithioate.
CAS: 51308-58-8. $C_{23}H_{32}N_2S_2$.
Hazard: Moderately toxic by ingestion.
Use: Agricultural chemical.

(4-(1,1-dimethylethyl)phenyl)methyl methyl-3-pyridinylcarbonimidodithioate.
CAS: 51308-52-2. $C_{18}H_{22}N_2S_2$.
Hazard: Moderately toxic by ingestion.
Use: Agricultural chemical.

s-((4-(1,1-dimethylethyl)phenyl)methyl) o-octyl-3-pyridinylcarbonimidothioate.
CAS: 51308-71-5. $C_{25}H_{36}N_2OS$.
Hazard: Moderately toxic by ingestion.
Use: Agricultural chemical.

(4-(1,1-dimethylethyl)phenyl)methyl pentyl-3-pyridinylcarbonimidodithioate.
CAS: 51379-04-5. $C_{22}H_{30}N_2S_2$.
Hazard: Moderately toxic by ingestion.
Use: Agricultural chemical.

s-((4-(1,1-dimethylethyl)phenyl)methyl) o-pentyl-3-pyridinylcarbonimidothioate.
CAS: 51308-67-9. $C_{22}H_{30}N_2OS$.
Hazard: Moderately toxic by ingestion.
Use: Agricultural chemical.

s-((4-(1,1-dimethylethyl)phenyl)methyl) o-propyl-3-pyridinylcarbonimidothioate.
CAS: 51308-62-4. $C_{20}H_{26}N_2OS$.
Hazard: Moderately toxic by ingestion.
Use: Agricultural chemical.

α-(5-(1,1-dimethylethyl)-1-phenyl-1h-pyrazol-4-yl)-1-piperidinebutanol.
CAS: 296269-57-3. $C_{22}H_{33}N_3O$.
Hazard: A poison by ingestion.

2-(1,1-dimethylethyl)pyrimidine.
CAS: 61319-99-1. $C_8H_{12}N_2$.
Hazard: Moderately toxic by ingestion and skin contact. A mild skin irritant.

***O,O*-dimethyl-*S*-2-(ethylsulfinyl)ethyl phosphorothioate.**   (oxydemetonmethyl).
CAS: 301-12-2. $(CH_3O)_2P(O)SC_2H_4SOC_2H_5$.
Properties: Amber liquid. Bp 106C (0.01 mm Hg), d 1.28 (20/4C). Miscible with water.
Hazard: Cholinesterase inhibitor, use may be restricted.
Use: Systemic insecticide.

dimethyl ferrocenoate.
(1,1'-ferrocenedicarboxylic acid dimethyl ester).
$(C_5H_4COOCH_3)_2Fe$.
Properties: Orange, crystalline solid. Mp 114–115C.
See ferrocene.

***N,N*-dimethyl formamide.**   (DMF).
CAS: 68-12-2. $HCON(CH_3)_2$.
Properties: Water-white liquid. A dipolar aprotic solvent. Bp 152.8C, fp −61C, refr index 1.4269 (25C), d 0.953-0.954 (15.6/15.6C), flash p 136F (57.7C), autoign temp 833F (445C). Miscible with water and most organic solvents (except halogenated hydrocarbons). Combustible.
Derivation: Reaction of methyl formate with dimethylamine.
Hazard: Moderate fire risk. Toxic by skin absorption. Strong irritant to skin and tissue. Liver damage. Questionable carcinogen.
Use: Solvent for vinyl resins and acetylene, butadiene, and acid gases; polyacrylic fibers; catalyst in carboxylation reactions; organic synthesis; carrier for gases.

dimethylfuran.

$$\overline{OC(CH_3)CHCHC(CH_3)}.$$

Properties: Colorless liquid. D 0.8900, bp 94C, flash p 45F (7.2C). Insoluble in water.
Hazard: Flammable, dangerous fire risk.

dimethyl glycol phthalate.
CAS: 117-82-8. $C_6H_4(COOCH_2CH_2OCH_3)_2$.

Properties: Colorless liquid. D 1.17, bp 230C. Combustible.
Use: Solvent mix for cellulose esters, plasticizing mix for cellulose esters.

dimethylglyoxal.   See diacetyl.

dimethylglyoxime.   (butane dioxime).
CAS: 95-45-4. $CH_3C(NOH)C(NOH)CH_3$.
Properties: White crystals or powder. Mp 242C. Soluble in alcohol and ether, insoluble in water.
Use: Analytical chemistry, especially as a reagent for nickel, biochemical research.

2,6-dimethyl-4-heptanol.   (diisobutyl-carbinol).
CAS: 108-82-7. $[(CH_3)_2CHCH_2]_2CH_2O$.
Properties: Colorless liquid. Refr index 1.423 (21C), d 0.8121 (20C), fp (sets to a glass at approximately −65C), bp 178C (750 mm Hg); insoluble in water, soluble in alcohol and ether; flash p 162F (72.2C) (TOC). Combustible.
Use: Surface-active agents, lubricant additives, rubber chemicals, flotation agents, antifoam agent.

2,6-dimethyl-4-heptanone.   See diisobutyl ketone.

2,6-dimethyl-5-hepten-1-al.
CAS: 106-72-9. $(CH_3)_2C{:}CH(CH_2)_2CH(CH_3)CHO$.
Properties: Yellow liquid. Moderately stable but not likely to cause discoloration. D 0.845–0.855 (25/25C), refr index 1.441–1.447 (20C), flash p 144F (62.2C) (TCC). Soluble in two parts of 70% alcohol. Combustible.
Use: Perfumery.

2,6-dimethylheptene-3.
$(CH_3)_2CHCH{:}CHCH_2CH(CH_3)_2$.
Mixed *cis* and *trans* isomers.
Properties: Liquid. Distillation range 128–129C, d 0.722 (60/60F), refr index 1.412 (20C), flash p 70F (21.1C) (TOC).
Grade: 95%.
Hazard: Flammable, dangerous fire risk.

2,5-dimethylhexadiene-1,5.   (dimethyallyl).
$CH_2{:}C(CH_3)CH_2CH_2C(CH_3){:}CH_2$.
Properties: Water-white liquid; hydrocarbon odor. D 0.740–0.760 (25/25C), refr index 1.426–1.429 (25C), ASTM distillation 90% between 114C and 123C, flash p 56F (13.3C). Soluble in hydrocarbons; insoluble in water.
Hazard: Flammable, dangerous fire risk.
Use: Solvent.

2,5-dimethylhexane-2,5-dihydroperoxide.
CAS: 3025-88-5. $(CH_3)_2C(OOH)CH_2CH_2C(OOH)(CH_3)_2$.
Properties: Fine powder, 90% peroxide. Mp 102–104C. Insoluble in hydrocarbons; slightly soluble in water, soluble in alcohols.

Hazard: Dangerous fire risk, strong oxidizer. Store away from organic materials.
Use: High-temperature catalyst for polyester premix compounds and silicone resins.

dimethylhexanediol. (2,5-dimethylhexane-2,5-diol).
CAS: 110-03-2. $(CH_3)_2COH(CH_2)_2COH(CH_3)_2$.
Properties: White crystals. Mp 88.5–89C, bp 204–215C, d 0.898 (20/20C). Soluble in water, acetone, and alcohol; insoluble in benzene, carbon tetrachloride, and kerosene. Combustible.
Use: Chemical intermediate.

2,5-dimethylhexane-2,5-diperoxybenzoate.
$C_{22}H_{26}O_6$.
Properties: Fine, white granules. Mp 114C. Insoluble in alcohols and hydrocarbons; soluble in acetone and chlorinated hydrocarbons.
Hazard: Strong oxidant, fire risk in contact with organic materials.
Use: Oxidizing agent, polymerization agent.

dimethylhexynediol. (2,5-dimethyl-3-hexyne-2,5-diol).
CAS: 142-30-3. $(CH_3)_2CHOC\equiv CCOH(CH_3)_2$.
Properties: White crystals. Mp 94–95C, bp 205–206C, d 0.949 (20/20C). Soluble in water; slightly soluble in benzene, carbon tetrachloride, and naphtha; very soluble in acetone, alcohol, and ethyl acetate.
Use: Wiredrawing lubricant, antifoaming agent, coupling agent in resin coatings, chemical intermediate.

dimethylhexynol. (3,5-dimethyl-1-hexyne-3-ol).
CAS: 107-54-0. $HC\equiv CCOH(CH_3)CH_2CH(CH_3)_2$.
Properties: Colorless liquid; camphorlike odor. Bp 150–151C, sets to a glass approximately −68C, d 0.8545 (20/20C), slightly soluble in water, flash p 134F (56.6C) (TOC). Combustible.
Hazard: Moderate fire risk.
Use: Stabilizer for chlorinated organic compounds, surface active agent, intermediate, solvent lubricant.

5,5-dimethylhydantoin. (DMH).
CAS: 77-71-4. $HNCONHCOC(CH_3)_2$.
Properties: White, crystalline solid. Mp 178C. Soluble in water, alcohol, and ether.
Derivation: (1) From acetone, urea, and ammonium carbonate; (2) from acetone, potassium cyanate, and hydrogen cyanide.
Hazard: Central nervous system depressant.
Use: Synthesis, preparation of water-soluble polymers.

dimethylhydantoin-formaldehyde polymer.
Properties: Light-colored, brittle resin containing 0.3% max of formaldehyde. D 1.30, softening point 59–80C. Dissolves readily in cold and hot water,

methanol, ethylacetate, methyl ethyl ketone, chloroform, methylene chloride, and hot glycerol; insoluble in benzene, xylene, diethyl ether, trichloroethylene, and carbon tetrachloride.
Use: Sizing, adhesives, blending agent, aerosol hair sprays.

1,1-dimethylhydrazine.
CAS: 57-14-7. $(CH_3)_2NNH_2$.
An unsymmetrical compound.
Properties: Colorless, hygroscopic liquid; ammonia-like odor. Fumes in air; evolves heat on contact with water. Fp −58C, bp 63C, d 0.782 (25C), flash p approximately 5F (−15C), autoign temp 480F (249C). Soluble in water and alcohol.
Derivation: (1) Reaction of dimethylamine and chloramine; (2) catalytic oxidation of dimethylamine and ammonia.
Hazard: Flammable, dangerous fire risk. Toxic by skin absorption. Corrosive to skin. Possible carcinogen. Nasal cancer. Upper respiratory tract irritant.
Use: Component of jet and rocket fuels, chemical synthesis, stabilizer for organic peroxide fuel additives, absorbent for acid gases, photography, plant growth control agent.

1,2-dimethylhydrazine. (*N,N'*-dimethylhydrazine; *sym*-dimethylhydrazine; DMH; RCRA Waste Number U099; SDMH; symetrycznadwumethylohydrazyna; hydrazomethane).
CAS: 540-73-8. $C_2H_8N_2$.
A DNA alkylating agent.
Properties: Clear, colorless (gradually turning yellow), flammable, mobile, fuming (in air), hygroscopic liquid, fishy ammoniacal odor; miscible with water, ethanol, ethyl ether, dimethylformamide, and hydrocarbons.
Hazard: Corrosive; moderately toxic; probable carcinogen; mutagen; neoplastigen; tumorigen; teratogen; poison.
Use: High-energy liquid rocket propellant; to induce colon tumors in experimental animals.

1,2-dimethylhydrazine dihydrochloride.
(*n,n'*-dimethylhydrazine hihydrochloride; *sym*-dimethylhydrazine dihydrochloride; DMH).
$C_2H_{10}Cl_2N_2$.
Properties: A white to grayish white, slightly water-soluble, crystalline powder.
Hazard: Carcinogen; neoplastigen; tumorigen; poison.
Use: A reagent for arabinose and lactose.

9-(2,2-dimethylhydrazino)acridine.
CAS: 28846-35-7. $C_{15}H_{15}N_3$.
Hazard: A poison by ingestion.
Use: Agricultural chemical.

9-(2,2-dimethylhydrazino)acridine monohydrochloride.
CAS: 1086-34-6. $C_{15}H_{15}N_3\cdot ClH$.

Hazard: A poison by ingestion.
Use: Agricultural chemical.

9-(2,2-dimethylhydrazino)acridine mono(methyl sulfate).
CAS: 28846-38-0. $C_{15}H_{15}N_3 \cdot CH_4O_4S$.
Hazard: A poison by ingestion.
Use: Agricultural chemical.

dimethylhydroquinone.   See hydroquinone dimethyl ether.

dimethylhydroxybenzene.   See xylenol.

3,7-dimethyl-7-hydroxyoctenal.   See hydroxycitronellal.

dimethyl isophthalate.   See "Morflex 1129" *[Vertellus]*.

dimethylisopropanolamine.
CAS: 996-35-0. $(CH_3)_2NCH_2CH(OH)CH_3$.
Properties: Colorless liquid. D 0.8645 (25/20C), bulk d 7.4 lb/gal (20C), bp 125.8C, miscible with water, viscosity 1.51 cP (20C), vap press 9 mm Hg (20C), fp (sets to a glass at approximately −85C), refr index 1.4189 (20C), flash p 95F (35C) (OC). Solubility of water in compound complete at 20C.
Hazard: Flammable, moderate fire risk.
Use: Synthesis of methadone, other chemical syntheses. Combining the properties of tertiary amine and secondary alcohol.

4,4-dimethyl-1-isopropyl-2-nonyl-2-imidazoline.
CAS: 74038-78-1. $C_{17}H_{34}N_2$.
Hazard: A poison.

dimethyl itaconate.
$CH_2C(COOCH_3)CH_2(COOCH_3)$.
Properties: White crystals; slight odor. Mp 36C, bp 91.5C (10 mm Hg), d 1.27 (24C), refr index 1.441 (20C). Slightly soluble in water. Combustible.
Use: Polymers and copolymers, plasticizers, intermediate.

dimethylketol.   See acetylmethylcarbinol.

dimethylketone.   See acetone.

o,o-**dimethyl malathion.**   See carbomethoxy malathion.

dimethyl maleate.
$CH_3OOCCH:CHCOOCH_3$.
Properties: Colorless liquid. D 1.153, bulk density 9.62 lb/gal, bp 200.4C, flash p 235F (112C). Combustible.

dimethyl malonate.
CAS: 108-59-8. $CH_2(COOCH_3)_2$.

Properties: Colorless liquid. Bp 180–181C, fp −62C, refr index 1.4140, flash p 194F (90C). Very slightly soluble in water; soluble in alcohol and ether. Combustible.
Use: Chemical intermediate.

dimethylmethane.   See propane.

5,5-dimethyl-2-((1-methylethyl)imino)-1,3-dithiolan-4-one-*o*-((methylamino) carbonyl)oxime.
CAS: 71108-05-9. $C_{10}H_{17}N_3O_2S_2$.
Hazard: A poison by ingestion.
Use: Agricultural chemical.

5,5-dimethyl-2-((1-methylethyl)imino)1,3-dithiolan-4-one, *o*-((methyl((trichloromethyl)thio)amino)carbonyl)oxime.
CAS: 71108-06-0. $C_{11}H_{16}Cl_3N_3O_2S_3$.
Hazard: A poison by ingestion.
Use: Agricultural chemical.

3,5-dimethyl-*n*-(2-methylphenyl)-4-nitro-1h-pyrazole-1-acetamide.
CAS: 302542-44-5. $C_{14}H_{16}N_4O_3$.
Hazard: Moderately toxic by ingestion.

3,5-dimethyl-*n*-(4-methylphenyl)-1h-pyrazole-1-acetamide.
CAS: 302542-60-5. $C_{14}H_{17}N_3O$.
Hazard: Moderately toxic by ingestion.

3,5-dimethyl-*n*-(2-methylphenyl)-1h-pyrazole-1-acetamide.
CAS: 173381-90-3. $C_{14}H_{17}N_3O$.
Hazard: Moderately toxic by ingestion.

3,5-dimethyl-*n*-(3-methylphenyl)-1h-pyrazole-1-acetamide.
CAS: 302542-50-3. $C_{14}H_{17}N_3O$.
Hazard: Moderately toxic by ingestion.

o,o-**dimethyl s-((5-(methylthio)-1,3,4-thiadiazol-2-yl)methyl) phosphorodithioate.**
CAS: 38090-92-5. $C_6H_{11}N_2O_2PS_4$.
Hazard: A poison by ingestion.

(4,4-dimethyl-5-(((((methyl((trichloromethyl) thio)amino)carbonyl)oxy)imino)-1,3-dithiolan-2-ylidene)propanedinitrile.
CAS: 71108-21-9. $C_{11}H_9Cl_3N_4O_2S_3$.
Hazard: Moderately toxic by ingestion.
Use: Agricultural chemical.

2,6-dimethylmorpholine.
CAS: 141-91-3

CAS: 141-91-3.

OCH(CH$_3$)CH$_2$NHCH$_2$CH(CH$_3$).

Properties: Liquid. D 0.99346, bp 146.6C, fp −85C, bulk d 7.8 lb/gal, flash p 112F (44.4C). Soluble in water. Combustible.

Hazard: Moderate fire risk.

Use: Corrosive inhibitors, stabilizers for chlorinated solvents, rubless floor polishes, rubber accelerators, germicides, and textile finishing agents.

2-(2,6-dimethyl-4-morpholinothio) benzothiazole.

$C_{13}H_{16}N_2OS_2$.

Properties: Cream to light-yellow powder. Mp 88C, d 1.26 (25C).

Use: Delayed-action vulcanization accelerator.

1,2-dimethyl-3-(2-naphthalenyl)(2r,3s)-rel-3-pyrrolidinol drochloride.

CAS: 302959-32-6. $C_{16}H_{19}NO·ClH$.

Hazard: A poison.

2,3-dimethyl-1,4-naphthoquinone.

CAS: 2197-57-1. $C_{12}H_{10}O_2$.

Hazard: Moderately toxic by ingestion.

dimethyl-α-naphthylamine.

$C_{10}H_7N(CH_3)_2$.

Properties: Colorless liquid. D 1.045, bp 275C. Soluble in alcohol and ether; insoluble in water.

Derivation: Action of methyl sulfate on α-naphthylamine.

Grade: CP, analytical.

Use: Determination of nitrates.

dimethyl-β-naphthylamine.

$C_{10}H_7N(CH_3)_2$.

Properties: Crystalline solid. D 1.039 (70/70C), mp 46C, bp 305C. Soluble in alcohol and ether; insoluble in water.

Derivation: Interaction of dimethylamine and β-naphthol.

dimethylnitrobenzene.   See nitroxylene.

O,O-dimethyl-O-p-nitrophenyl phosphorothioate.   See methyl parathion.

dimethylnitrosamine.   See N-nitrosodimethylamine.

N,N-dimethyl-p-nitrosoaniline.

(p-nitrosodimethylaniline).

CAS: 138-89-6. $(CH_3)_2NC_6H_4NO$.

Properties: Green leaflets. Mp 93C. Soluble in alcohol and ether; insoluble in water.

Derivation: Nitrous acid and N-dimethylaniline.

Hazard: Flammable. Toxic by ingestion.

Use: Production of methylene blue, vulcanization accelerator.

3,6-dimethyl-3,6-octanediol.

CAS: 78-65-9. $C_2H_5(CH_3)COH(CH_2)_2COH(CH_3)$ C_2H_5.

Properties: White, waxy solid. Mp 44C, bp 241–242C, d 0.919 (20/20C). Soluble in water, acetone, alcohol, benzene, and carbon tetrachloride.

Use: Nonfoaming, surface-active agent; chemical intermediate.

dimethyloctanoic acid.   See isodecanoic acid.

3,6-dimethyl-3-octanol.

CAS: 151-19-9. $(C_2H_5)CHCH_3(CH_2)_2COHCH_3$ (C_2H_5).

Properties: Colorless liquid; sweet odor. D 0.8366 (20/20C), refr index 1.4370 (20C), bp 202–203C, fp −67.5C. Combustible.

Use: Perfumery (floral odors), flavoring.

3,7-dimethyl-1-octanol.   (tetrahydrogeraniol).

CAS: 151-19-9. $(CH_3)_2CH(CH_2)_3CH(CH_3)CH_2$ CH_2OH.

Properties: Colorless liquid; sweet odor. Refr index 1.4350–1.4450 (20C), d 0.826–0.842 (25C). Soluble in mineral oil and in propylene glycol; insoluble in glycerol. Combustible.

Grade: FCC.

Use: Flavoring agent, perfumery.

3,7-dimethyl-3-octanol.   See tetrahydrolinalool.

2,6-dimethyl-1,5,7-octatriene.   See ocimene.

2,6-dimethyl-2,4,6-octatriene.   See alloocimene.

3,7-dimethyl-6-octenal.   See citronellal.

3,7-dimethyl-6(or 7)-octen-1-ol.   See citronellol.

N,N-dimethyl-n-octyl-1-octanaminium chloride.

CAS: 5538-94-3. $C_{18}H_{40}N·Cl$.

Hazard: A poison by ingestion.

dimethyloctynediol.   (3,6-dimethyl-4-octyne-3,6-diol).

CAS: 78-66-0. $C_2H_5(CH_3)COHC≡CCOH(CH_3)$ C_2H_5.

Properties: White crystals. Mp 55–56C, bp 222C, d 0.923 (solid 20C), 0.908 (liquid 60C). Moderately soluble in water; slightly soluble in kerosene; very soluble in acetone, alcohol, benzene, and carbon tetrachloride. Combustible.

Use: Surface-active agent, intermediate.

dimethylolethylene urea.

$$\overline{OCN(CH_2OH)CH_2CH_2N(CH_2OH)}.$$

Use: Wrinkle-resistant textile finishes.

dimethylolpropionic acid. (DMPA; 2,2-bis(hydroxymethyl)propionic acid).
CAS: 4767-03-7. $CH_3C(CH_2OH)_2COOH$.
Properties: Off-white, crystalline solid. Mp 192–194C. Soluble in water and methanol; slightly soluble in acetone; insoluble in benzene.
Use: Water-soluble alkyd resins, textile finishing, cosmetics, plasticizers.

dimethylolurea. (DMU; 1,3-bishydroxymethylurea).

$$\underset{NHCH_2OH}{\overset{NHCH_2OH}{C\!\!=\!\!O}}$$

Properties: Colorless crystals. Mp 126C (technical 85–90C), d 1.34 (20C), slight formaldehyde odor. Soluble in water and methanol; insoluble in ether. Capable of polymerization.
Derivation: Combination of urea and formaldehyde in the presence of salts or alkaline catalysts.
Hazard: Irritant to skin.
Use: First stage of urea formaldehyde resins; impregnating wood to increase hardness and fire resistance and to form self-binding laminations for plywood manufacture; permanent-press fabrics.

2,2-dimethyl-3-(3-oxo-3-(2,2,2-trifluoro-1-(trifluoromethyl) ethoxy)-1-propenyl) cyclopropanecarboxylic acid, cyano(3-phenoxyphenyl)methyl ester, (1r-(1-α(s∗),3-α(z)))-.
CAS: 101007-06-1.
Hazard: Moderately toxic by skin contact. Low toxicity by ingestion.
Use: Agricultural chemical.

2,3-dimethylpentaldehyde.
CAS: 32749-94-3. $CH_3CH_2CH(CH_3)CH(CH_3)CHO$.
A branched-chain heptanal.
Properties: Liquid. D 0.8293, bp 140.5C, fp −110C, bulk density 6.91 lb/gal (20C), flash p 94F (34.4C). Slightly soluble in water.
Hazard: Flammable, moderate fire risk.
Use: Intermediate.

2,4-dimethylpentane.
CAS: 108-08-7. $(CH_3)_2CHCH_2CH(CH_3)_2$.
Properties: Colorless liquid. D 0.6684 (25C), bp 80.5C, refr index 1.382 (20C), flash p 10F (−12.1C), fp −119C. Soluble in alcohol; insoluble in water.
Grade: 95%, 99%, research, 99.7 mole %.
Hazard: Flammable, dangerous fire hazard.
Use: Organic synthesis.
Note: Other isomers (2,2-; 2,3-; 3,3-) of closely similar properties are available.

2,4-dimethylpentanol-3. See diisopropylcarbinol.

2,4-dimethylpentanone-3. See diisopropyl ketone.

α,α-dimethylphenethyl acetate. (dimethylbenzylcarbinyl acetate).
$C_6H_5CH_2C(CH_3)_2OOCCH_3$.
Properties: Colorless liquid; floral, fruity odor. Refr index 1.4910–1.4950 (20C) in supercooled liquid form, d 0.995–1.002 in supercooled liquid form, mp 29–30C. Solidifies at room temperature. Soluble in mineral oil; insoluble in glycerol. Combustible.
Grade: FCC.
Use: Flavoring agent.

dimethylphenol. See xylenol.

dimethyl-*p*-phenylenediamine. See *p*-aminodimethylaniline.

n-(2,3-dimethylphenyl)maleimide. See *n*-2,3-xylylmaleimide.

n-(2,6-dimethylphenyl)-*n*-(methoxyacetyl)-alanine methyl ester. See metalaxyl.

n-(2,6-dimethylphenyl)-*n*-methoxyalanine methyl ester.
CAS: 123298-28-2. $C_{13}H_{19}NO_3$.
Hazard: Moderately toxic by ingestion and inhalation.
Use: Agricultural chemical.

1-(2,4-dimethylphenyl)-3-(4-(2-methoxyphenyl)-1-piperazinyl)-1-propanone.
CAS: 302561-65-5. $C_{22}H_{28}N_2O_2 \cdot ClH$.
Hazard: A poison.

2,3-dimethyl-1-phenyl-3-pyrazolin-5-one.
See antipyrine.

dimethyl phosphite.
CAS: 868-85-9. $(CH_3O)_2P(O)H$.
Properties: Mobile, colorless liquid; mild odor. D 1.200 (20/4C), refr index 1.400 (25C), bp 72–73C (25 m/H/), flash p 205F (96C). Soluble in water; miscible with most common organic solvents. Combustible.
Use: Lubricant additive, intermediate, adhesive.

***O*,*O*-dimethyl phosphorochloridothioate.** (methyl PCT).
CAS: 2524-03-0. $(CH_3O)_2P(S)Cl$.
Properties: Colorless to light-amber liquid. Bp 66–67C (16 mm Hg) d 1.320 (25C), refr index 1.4795 (25C). Soluble in alcohol, benzene, acetone, carbon tetrachloride, chloroform, and ethyl acetate; slightly soluble in hexane; insoluble in water.
Grade: 96–100% purity.
Hazard: Strong irritant to eyes, skin, and mucous membranes; cholinesterase inhibitor; use may be restricted.

Use: Intermediate for insecticides, pesticides, fungicides, oil and gasoline additives, plasticizers, corrosive inhibitors, flame retardants, and flotation agents.

dimethyl phthalate.
CAS: 131-11-3. $C_6H_4(COOCH_3)_2$.
Properties: Colorless, oily liquid. Refr index 1.5138 (25C), heat of combustion 5769 cal/g, d 1.189 (25/25C), bp 282C, flash p 300F (149C), bulk density 9.93 lb/gal (68F), vap press <0.1 mm Hg (20C), autoign temp 1032F (555C). Miscible with alcohol and ether; insoluble in water and paraffinic hydrocarbons; slightly soluble in mineral oil. Combustible.
Grade: Technical.
Hazard: Irritant to eyes, mucous membranes, and upper respiratory tract, not absorbed by skin.
Use: Plasticizer for nitrocellulose and cellulose acetate, resins, rubber, and in solid rocket propellants; lacquers; plastics; rubber; coating agents; safety glass; molding powders; insect repellent.

N,N'-dimethylpiperazine. (1,4-dimethylpiperazine).
CAS: 106-58-1. $(CH_3)_2C_4H_8N_2$.
Properties: Colorless, mobile liquid. D 0.8565 (20/4C), bp 131C, fp −1C, flash p 176F (80C) (TOC). Combustible.
Use: Curing agent for polyurethane foams, intermediate for cationic surface-active agents.
Note: Two other isomers are available, *cis*-2,5- and *trans*-2,5-; neither is flammable, but both are combustible.

2,6-dimethylpiperidine. (2,6-lupetidine).
CAS: 504-03-0. $(CH_3)_2C_5H_8NH$.
Properties: Liquid. Bp 127.9C, d 0.8199 (20/20C), refr index 1.4383 (20C). Miscible with water at 20C. Combustible.
Use: Intermediate.

1,1-dimethylpiperidinium chloride. See mepiquat chloride.

dimethylpolysiloxane. A liquid defoaming agent, refr index 1.40, viscosity 300 centistokes.
Hazard: Use in foods limited to 10 ppm (0 ppm in milk).

dimethyl-POPOP. See 1,4-bis[2-(4-methyl-5-phenyloxazolyl)]benzene.

2,2-dimethylpropane. See neopentane.

2,2-dimethyl-1,3-propanediol. See neopentyl glycol.

2,10-dimethyl-6-(2-propenyloxy)-4,8-dioxa-3,9-dithia-2,10-diazaundecanedioic acid, di-1-naphthalenyl ester, 3,9-dioxide.

CAS: 81862-18-2. $C_{30}H_{30}N_2O_9S_2$.
Hazard: Moderately toxic by ingestion.
Use: Agricultural chemical.

1,1-dimethylpropynylamine. See 2-methyl-3-butyn-2-amine.

3,6-dimethylpyrazine-2-thiol.
CAS: 5788-49-8.
Hazard: A reproductive hazard.

dimethylpyridine. See 2,6-lutidine.

2,6-dimethyl-3,5-pyridine dicarboxylic acid, dimethylester.
CAS: 182620-63-9. $C_{34}H_{40}N_4O_5$.
Hazard: A poison.

2,5-dimethylpyrrole.
CAS: 625-84-3. C_6H_9N.
Properties: Colorless to yellow oily liquid. D: 0.935–0.945 at 20 degrees/4 degrees, refr index: 1.503–1.506, bp: 165C at 760 mm. Very sol in alc and ether; very slightly sol in water.
Use: Food additive.

N,N-dimethyl-3-(pyrrolidin-1-yl) propionamide.
CAS: 22041-39-0. $C_9H_{18}N_2O$.
Hazard: A poison.

2,9-dimethylquinacridone.
CAS: 980-26-7. $C_{22}H_{16}N_2O_2$.
Hazard: Moderately toxic by skin contact. Low toxicity by ingestion.

2,7-dimethylquinoline.
CAS: 93-37-8. $(CH_3)_2C_9H_5N$.
Properties: Liquid. Fp approximately −40C, distillation range 140–150C (20 mm Hg). Soluble in benzene and diethyl ether. Combustible.
Use: Organic synthesis, dye intermediate.

o,o'-dimethyl s,s'-2,3-quinoxalinediyl thiocarbonate. See 2,3-bis(carbomethoxymercapto) quinoxaline.

dimethyl resorcinol. See resorcinol dimethyl ether.

dimethyl sebacate.
CAS: 106-79-6. $[(CH_2)_4COOCH_3]_2$.
Properties: Liquid, water-white. D 0.9896 (20/20C), mp 24.5C, flash p 293F (145C), bp approximately 294C, refr index 1.4376 (20C). Combustible.
Grade: Technical.
Use: Solvent or plasticizer for nitrocellulose, vinyl resins; intermediate.

dimethylsilicone. General term for a family of silicones of composition $[(CH_3)_2SiO]_x$, the

more volatile materials formed on hydrolysis of dimethyldichlorosilane. Colorless oils with boiling point ranging from 134C (for $x = 3$) to 188C (20 mm Hg) (for $x = 9$).
Use: Transformer liquid, brake fluids.
See polydimethylsiloxane.

dimethyl sulfate.   (methyl sulfate).
CAS: 77-78-1. $(CH_3)_2SO_4$.
Properties: Colorless liquid. Soluble in alcohol, ether, and water. D 1.3516, fp −26.8C, bp 188C (decomposes), flash p 182F (83.3C) (CC). Combustible.
Derivation: By adding fuming sulfuric acid to methanol and distilling in vacuo.
Grade: Technical.
Hazard: Strong eye and skin irritant, absorbed by skin, a carcinogen, induces tumors in animals, protective clothing required. Probable carcinogen.
Use: Methylating agent for amines and phenols, polyurethane-based adhesives.

dimethyl sulfide.   (methyl sulfide).
CAS: 75-18-3. $(CH_3)_2S$.
Properties: Colorless, volatile liquid; disagreeable odor. D 0.845 (20C), fp −83C, bp 37.5C, evolves sulfur dioxide when heated, autoign temp 403F (206C), flash p 0F (−17.7C). Soluble in alcohol and ether; insoluble in water.
Derivation: (1) From kraft pulping black liquor by heating with inorganic sulfur compounds; (2) by interaction of a solution of potassium sulfide and methyl chloride in methanol.
Hazard: Flammable, dangerous fire risk, moderate explosion risk. Flammable limits in air 2.2–19.7%. Upper respiratory tract irritant.
Use: Gas odorant, solvent for many inorganic substances, catalyst impregnator.

2,4-dimethylsulfolane.
CAS: 1003-78-7. $C_6H_{12}O_2S$.
Properties: Slightly yellow liquid. Bp 280C, d 1.13, flash p 290F (143C), refr index 1.47. Slightly soluble in water; partially soluble in alkanes and alkenes; miscible with most aromatics.
Hazard: Toxic by ingestion.
Use: Solvent extraction (liquid–liquid and vapor–liquid).

dimethylsulfone.   (methylsulfone; methylsulfonylmethane).
CAS: 67-71-0. $C_2H_6O_2S$.
Properties: Colorless crystals. Mp 110C, bp 237C. Soluble in water, alcohol, and acetone.
Derivation: Oxidation of dimethyl sulfide.
Use: Solvent.

dimethyl   sulfoxide.   (methyl   sulfoxide; DMSO).
CAS: 67-68-5. $(CH_3)_2SO$.
Properties: Colorless, hygroscopic liquid; nearly odorless; slightly bitter taste. Bp 189C, mp 18.5C,

d 1.10 (20/20C), specific heat 0.7 cal/g K, flash p 203F (95C) (OC), dielectric constant 48.9 (20C). Extremely powerful aprotic solvent. Soluble in water, alcohol, benzene, acetone, and chloroform. Combustible.
Derivation: Oxidation of dimethyl sulfide with nitrogen tetroxide under anhydrous conditions; sulfide waste liquors.
Hazard: Readily penetrates skin and other tissues; approved by FDA for humans but must comply with FDA regulations.
Use: Solvent for polymerization and cyanide reactions; analytical reagent; spinning polyacrylonitrile and other synthetic fibers; industrial cleaners, pesticides, paint stripping; hydraulic fluids; preservation of cells at low temperatures; diffusion of drugs, etc., into bloodstream by topical application; medicine (anti-inflammatory); veterinary medicine; plant pathology and nutrition; pharmaceutical products; metal-complexing agent.

dimethyl terephthalate.   (DMT).
CAS: 120-61-6. $C_6H_4(COOCH_3)_2$.
Properties: Colorless crystals. Mp 140C, sublimes above 300C. Insoluble in water; soluble in ether and hot alcohol.
Derivation: Oxidation of p-xylene or mixed xylene isomers with concurrent esterification.
Use: Polyester resins for film and fiber production, especially polyethylene terephthalate; intermediate.

dimethyl 2,3,5,6-tetrachloroterephthalate.
$C_6Cl_4(COOCH_3)_2$.
Properties: Crystals. Mp 156C. Insoluble in water; slightly soluble in acetone and benzene.
Use: Herbicide.

1,3-dimethyl-2-tetradecyl-2-thiopseudourea hydriodide.
CAS: 5339-43-5. $C_{17}H_{36}N_2S·HI$.
Hazard: Moderately toxic.

2,2-dimethyl-3-(2,3,3,3-tetrafluoro-1-propenyl)cyclopropanecarboxylic acid, (3-phenoxyphenyl)methyl ester, (1-α,3-β(e))-(+−)-.
CAS: 73176-66-6. $C_{22}H_{20}F_4O_3$.
Hazard: A poison by ingestion.
Use: Agricultural chemical.

3,5-dimethyltetrahydro-1,3,5(2H) thiadiazine-2-thione.
CAS: 533-74-4. $C_5H_{10}N_2S_2$.
Properties: Crystals. Mp 100C, d 1.30 (20C). Slightly soluble in water and alcohol; soluble in acetone.
Hazard: Toxic by ingestion and inhalation.
Use: Herbicide, nematocide, preservative for adhesives and proteinaceous additives, fungicide.
See Crag.

3,4-dimethyl-2,5-thiomorpholinedione, 2-(o-((methyl((trichloromethyl)thio)amino) carbonyl)oxime).
CAS: 66637-35-2. $C_9H_{12}Cl_3N_3O_3S_2$.
Hazard: A poison by ingestion.
Use: Agricultural chemical.

dimethyltin bis(dibutyldithiocarbamate).
See bis(dibutyldithiocarbamato)dimethylstannane.

dimethyl-tin bis(isooctylthioglycollate).
See bis(isooctyloxycarbonylmethylthio)dimethyl-stannane.

dimethyltin dibromide.
CAS: 2767-47-7. $C_2H_6Br_2Sn$.
Properties: Colorless or white crystals. Mp 76°, bp: 208–213C. Sol in water and org solvs.
Hazard: A poison.

dimethyltin dichloride.
CAS: 753-73-1. $(CH_3)_2SnCl_2$.
Properties: White crystals. Mp 106–108C. Soluble in water, alcohol, and hydrocarbons.
Hazard: Toxic material by skin absorption.
Use: Electroluminescence, PVC stabilizer, catalyst.

dimethyltin difluoride.   See difluorodimethyl-stannane.

dimethyltin oxide.
CAS: 2273-45-2. $(CH_3)_2SnO$.
Properties: White powder. 98.5% min purity.
Hazard: Toxic material by skin absorption.
Use: Intermediate, PVC stabilizer.

dimethyl-o-toluidine.   See N,N-dimethyl-o-toluidine.

dimethyl-p-toluidine.   See N,N,4-trimethylaniline.

N,N-dimethyl-o-toluidine.
CAS: 609-72-3. $C_9H_{13}N$.
Properties: A liquid. Bp: 184.8°.
Hazard: Moderately toxic by ingestion.

o,o-dimethyl (2,2,2-trichloro-1-(chloroacetoxy)ethyl)phosphonate.
CAS: 74940-61-7. $C_6H_9Cl_4O_5P$.
Hazard: A poison by ingestion.
Use: Agricultural chemical.

N,N′-dimethylurea.   (DMU; sym-dimethylurea; 1,3-dimethylurea).
CAS: 96-31-1. $(CH_3NH)_2CO$.
Properties: Colorless prisms. D 1.14, mp 106C, bp 270C. Soluble in water and alcohol; insoluble in ether.
Use: Intermediate in synthesis of drugs.

1,3-dimethylxanthine.   See theophylline.

3,7-dimethylxanthine.   See theobromine.

dimethylzinc.   (zinc dimethyl; zinc carbanide).
CAS: 544-97-8. C_2H_6Zn.
Properties: A white, crystalline, solid, organozinc, melting point of −40°C, boiling point of 46°C, soluble in ethyl ether, miscible with hydrocarbons, volatile mobile liquid at room temperature, undergoes self-ignition in air, reacts violently with water.
Hazard: Toxic; poison.
Use: An analeptic agent.

dimetilan.   Generic name for 1-dimethylcarbamoyl-5-methyl-3-pyrazolyl dimethylcarbamate.
Properties: Colorless to red–brown solid. Mp 68–71C, bp 200C (13 mm Hg). Soluble in water and most organic solvents.
Hazard: Toxic by ingestion and inhalation.
Use: Insecticide.

Dimezone S.   A line of photographic chemicals.

dimolybdenum trioxide.   See molybdenum sesquioxide.

dimpylate.   See Diazinon.

Dimroth rearrangement.   Rearrangement of N-alkylated or arylated iminoheterocycles to the corresponding alkylamino or arylamino heterocycles.

dimyristylamine.   (ditetradecylamine).
$(C_{14}H_{29})_2NH$.
Properties: Solid. Mp 52C, d 0.89. Almost insoluble in water.
Use: Intermediate.

dimyristyl ether.   (ditetradecyl ether).
CAS: 5412-98-6. $(C_{14}H_{29})_2O$.
Properties: Liquid. Mp 38–40C, bp 238–248C (4 mm Hg) d 0.8127 (45/4C). Combustible.
Grade: 95% (min) purity.
Use: Electrical insulators, water repellents, lubricants in plastic molding, antistatic substances, chemical intermediates.

dimyristyl sulfide.   (ditetradecyl sulfide, dimyristyl thioether).
$(C_{14}H_{29})_2S$.
Properties: Solid. Mp 49–50C, bp (decomposes), d 0.8258 (50/4C).
Grade: 95% (min) purity.
Use: Organic synthesis (formation of sulfonium compounds).

N,N′-di-β-2-naphthyl-m-phenylenediamine.
$C_6H_4(NHC_{10}H_7)_2$.

Properties: Colorless needles. Mp 191C. Sparingly soluble in alcohol; insoluble in water and ether.
Derivation: By heating *m*-phenylenediamine with β-naphthol and extraction with alcohol.
Use: Organic synthesis.

N,N′-di-β-naphthyl-*p*-phenylenediamine.
(*sym*-di-β-naphthyl-*p*-phenylamine; DNPD).
CAS: 93-46-9. $C_6H_4(NHC_{10}H_7)_2$.
Properties: Gray powder. D 1.25, set point 225C (min), purity 98% (min). Insoluble in water; slightly soluble in acetone and chlorobenzene.
Use: Antioxidant, stabilizer, polymerization inhibitor, intermediate in organic synthesis.

dineric.   Solution of two immiscible solvents with a single solute soluble in each.

dineric interface.   (liquid–liquid interface).
Boundary surface between two immiscible liquids.

dinitolmide.   See zoalene.

dinitraniline orange.   (permanent orange).
A pigment made from dinitroaniline and β-naphthol. It is a reddish shade of orange that has excellent lightfastness.
Hazard: Toxic by ingestion.

dinitroaminophenol.   See picramic acid.

2,4-dinitroaniline.   (2,4-dinitraniline).
CAS: 97-02-9. $C_6H_3NH_2(NO_2)_2$.
Properties: Yellow crystals. Slightly soluble in alcohol; insoluble in water. D 1.615, mp 188F, flash p 435F (223C). Combustible.
Derivation: Nitration of *p*-nitroaniline with hot mixed acid.
Grade: Technical, pure.
Hazard: Toxic by ingestion and inhalation, strong irritant.
Use: Intermediate for azo pigments, toner pigment in printing inks, corrosion inhibitor.

2,4-dinitroanisole.   (2,4-dintriophenyl methyl ether).
$CH_3OC_6H_3(NO)_2$.
Properties: Colorless to yellow, monoclinic needles from water or alcohol. Mp 88C, d 1.341 (20/4C), sublimes. Slightly soluble in hot water; soluble in alcohol and ether.
Hazard: Toxic by ingestion.
Use: Ovicide, effective against moths, furniture and carpet beetles, cockroaches, and body lice.
3,5-dinitrobenzamide. Use: Drug (veterinary); food additive.

dinitrobenzene.
CAS: *o*- 528-29-0; *m*- 99-65-0; *p*- 100-25-4.
   $C_6H_4(NO_2)_2$.
m-, *o*-, and *p*- isomers.

Properties: Yellow crystals. D *m*- 1.546, *o*- 1.565, *p*- 1.6; mp *m*- 89.9C, *o*- 117.9C, *p*- 172–173C; bp *m*- 302.8C, *o*- 319C, *p*- 299C. Soluble in chloroform and ethyl acetate; sparingly soluble in benzene; slightly soluble in water.
Derivation: Nitration of nitrobenzene with hot mixed acid.
Hazard: *o*-isomer explodes when shocked or heated. Toxic by inhalation and ingestion, absorbed by skin. Eye damage. Methemoglobinemia.
Use: Organic synthesis, dyes, camphor substitute in cellulose nitrate.

m-dinitrobenzene.   (1,3-dinitrobenzene; 2,4-dinitrobenzene).
CAS: 99-65-0. $C_6H_4N_2O_4$.
Properties: Yellowish crystals from alcohol. Mw 168.12, mp 89C, bp 291C.
Hazard: A suspected carcinogen. A human poison by ingestion. Mixture with nitric acid is a high explosive. Mixture with tetranitromethane is a high explosive very sensitive to sparks.
See *o*-dinitrobenzene and *p*-dinitrobenzene.

o-dinitrobenzene.   (1,2-dinitrobenzene).
CAS: 528-29-0. $C_6H_4N_2O_4$.
Properties: Colorless needles or plates from alcohol. Mw 168.12, mp 118C, bp 319C, flash p 302F (CC), d 1.571 (0C/4C), vap d 5.79. Soluble in EtOH and $CHCl_3$; slightly soluble in water.
Hazard: A poison by inhalation and ingestion. Combustible. A severe explosion hazard when shocked or exposed to heat or flame.
Use: In bursting charges and to fill artillery shells.
See *n*-dinitrobenzene and *p*-dinitrobenzene.

p-dinitrobenzene.
CAS: 100-25-4. $C_6H_4N_2O_4$.
Properties: White crystals, needles, or prisms from alcohol. Mw 168.12, mp 173C, bp 299C. Volatile with steam.
Hazard: A poison by ingestion. Mixture with nitric acid is a high explosive.
See *n*-dinitrobenzene and *o*-dinitrobenzene.

1,2-dinitrobenzene.   See *o*-dinitrobenzene.

1,3-dinitrobenzene.   See *m*-dinitrobenzene.

2,4-dinitrobenzene.   See *m*-dinitrobenzene.

2,4-dinitrobenzenethiol.
CAS: 2218-96-4. $C_6H_4N_2O_4S$.
Hazard: A poison by ingestion.
Use: Agricultural chemical.

5,7-dinitro-1,2,3-benzoxadiazole.   See diazodinitrophenol.

3,5-dinitrobenzoyl chloride.
   $2NO_2(_2C_6H_3COCl)$.

COCl

NO$_2$ NO$_2$

Properties: Yellow crystals. Mp 66–68C, bp 196C (12 mm Hg). Decomposed by water and alcohol.
Hazard: An irritant.
Use: Reagent for amino acids and presence of alcohols in acetals and ketals.

2-(2,4-dinitrobenzyl)pyridine.
CAS: 1151-97-9. $C_5H_4NCH_2C_6H_3(NO_2)_2$.
Behaves photochromically.
Use: To make plastics sensitive to light.
See photochromism.

2,4-dinitro-6-*sec*-butylphenol.   (2-*sec*-butyl-4,6-dinitrophenol; dinoseb; DNBP).
CAS: 88-85-7. $CH_3(C_2H_5)CHC_6H_2(NO_2)_2OH$.
A plant growth regulator.
Properties: Reddish-brown liquid. Slightly soluble in water; soluble in alcohol and other organic solvents. Forms salts with metals and organic bases.
Hazard: Possible fire risk. Absorbed by skin, strong irritant.
Use: Insecticide and ovicide, but must be used in the dormant growth season or as a salt to reduce toxicity. Herbicide for pre-emergence treatment, increases yield of corn 5–10%.

dinitrochlorbenzene.   See 1-chloro-2,4-dinitrobenzene.

4,6-dinitro-*o*-cresol.   (dinitro-*o*-cresol; 4,6-dinitro-2-methylphenol).
CAS: 534-52-1. $CH_3C_6H_2(NO_2)_2OH$.
Properties: Yellow solid. Mp 85.8C. Slightly soluble in water; soluble in alcohol, acetone, and ether.
Hazard: Toxic, absorbed by skin, use may be restricted. Basal metabolism.
Use: Dormant ovicidal spray for fruit trees (highly phytotoxic and cannot be used successfully on actively growing plants), herbicide, insecticide.

2,6-dinitro-*p*-cresol.   (DNPC).
CAS: 609-93-8. $(NO_2)_2CH_3C_6H_2OH$.
Properties: Light-yellow, crystalline solid.
Grade: Presscake (36–43% active 2,6-DNCP).
Use: Parent compound for intermediates, dyes, and pharmaceuticals.

4,6-dinitro-*o*-cresol barium derivative.
CAS: 63989-83-3.
Hazard: A poison by ingestion.
Use: Agricultural chemical.

dinitrocyclohexylphenol.   (2-cyclohexyl-4,6-dinitrophenol;   dinitro-*o*-cyclohexylphenol; DNOCHP).

CAS: 131-89-5. $C_6H_{11}C_6H_2(NO_2)_2OH$.
Properties: Crystalline solid.
Use: Control of mites on citrus fruits.

3,5-dinitro-n^4,n^4-dipropylsulfanilamide.
See oryzalin.

2,4-dinitrofluorobenzene.   (1-fluoro-2,4-dinitrobenzene; DNFB).
CAS: 70-34-8. $C_6H_3F(NO_2)_2$.
Properties: Crystals. Mp 26C; bp 137C. Soluble in ether, benzene, and propylene glycol.
Hazard: Toxic by ingestion, mutagenic, and carcinogenic.
Use: Alkylating agent, reagent in elucidating amino acid sequence in proteins.

dinitrogen tetroxide.   See nitrogen dioxide; nitrogen oxides; NO_x.

dinitrogen trioxide.   See nitrogen trioxide; nitrogen oxides; NO_x.

dinitroglycol.   See diethylene glycol dinitrate.

2,4-dinitro-4-hydroxydiphenylamine.
CAS: 119-15-3. $(NO_2)_2C_6H_3NHC_6H_4OH$.
Properties: Yellow solid. Mp 190C. Insoluble in water.
Derivation: Condensation of 2,4-dinitro-1-chlorobenzene and *p*-aminophenol.

dinitro(1-methylheptyl)phenyl   crotonate.
See dinocap.

4,6-dinitro-2-methylphenol.   See 4,6-dinitro-*o*-cresol.

dinitronaphthalene.
CAS: 27478-34-8. $C_{10}H_6(NO_2)_2$.
Isomers: (1) 1,5-; (2) 1,8-.
Properties: (1) Yellowish-white needles; (2) yellowish-white, thick, crystalline tablets; mp (1) 217C; (2) 172C); (1) sparingly soluble in pyridine; (2) soluble in pyridine; bp (1) sublimes; (2) decomposes.
Derivation: By dissolving α-nitronaphthalene in sulfuric acid and adding nitric acid.
Method of purification: Crystallization.
Hazard: Moderate fire and explosion risk.
Use: Dyes, especially sulfur colors; intermediates.

2,4-dinitro-1-naphthol-7-sulfonic acid.   (flavianic acid).
CAS: 483-84-1. $C_{10}H_6O_8N_2S$.
Properties: Yellow needles. Mp 151C. Very soluble in water.
Use: Intermediate, precipitant for organic bases, reagent for amino acids.

dinitrophenol.
$C_6H_3OH(NO_2)_2$.

CAS: 25550-58-7.

Commercial product is usually a mixture of 2,3-, 2,4-, and 2,6-isomers.

Properties: Yellow crystals, (2,3) d 1.681, mp 144C; (2,4) d 1.683, mp 114–115C; (2,6) mp 63C. Soluble in alcohol, ether, benzene, and chloroform; slightly soluble in water.

Derivation: (1) By heating phenol with dilute sulfuric acid, cooling the product, and then nitrating, keeping the temperature approximately 50C; (2) by nitration with mixed acid with careful temperature control.

Method of purification: Crystallization.

Hazard: Severe explosion hazard when dry. Absorbed by skin; dust inhalation may be fatal.

Use: Dyes, especially sulfur colors; picric acid, picramic acid, preservation of lumber, manufacture of the photographic developer diaminophenol hydrochloride, explosives manufacture; indicator, reagent for potassium and ammonium ions.

dinitrophenylhydrazine.
CAS: 119-26-6. $(NO_2)_2C_6H_3NHNH_2$

Properties: Red, crystalline powder. Mp about 200C. Slightly soluble in water and alcohol; soluble in moderately dilute inorganic acids; readily soluble in diglyme.

Hazard: Severe explosion and fire risk.

Use: Explosive; reagent for aldehydes and ketones.

dinitrophenylmethane.   See dinitrotoluene.

2,4-dinitrophenyl methyl ether.   See 2,4-dinitroanisole.

3,5-dinitrosalicylic acid.
CAS: 609-99-4. $C_6H_2(OH)(NO_2)_2COOH$.

Properties: Yellow crystals. Mp 174C. Slightly soluble in water; soluble in alcohol and benzene.

Derivation: Nitration of salicylic acid.

Use: Determination of glucose.

dinitrosopentamethylenetetramine. (DNPT).
CAS: 101-25-7. $(NO)_2C_5H_{10}N_4$.

A bicyclic compound.

Properties: Light-cream-colored powder. Decomposes in air at 190–200C. Soluble in dimethyl formamide; somewhat soluble in pyridine, methyl ethyl ketone, and acetonitrile. Combustible.

Hazard: May explode at 200C (390F). Questionable carcinogen.

Use: Blowing agent for rubber and plastics.

2,4-dinitrosoresorcinol.
$C_6H_2(OH)_2(NO)_2 \cdot H_2O$.

Properties: Light-brown powder. Mp 162–163C. Decomposes, sometimes violently; soluble in water and most organic solvents.

Grade: Technical (13.7% N).

Hazard: Severe explosion risk when shocked or heated. An irritant.

Use: Chelation of heavy metals, cross-linking agent, blasting caps, and explosive primers.

3,5-dinitro-*o*-toluamide.   See zoalene.

dinitrotoluenes.   (dinitrophenylmethane; methyldinitrobenzene).
CAS: 25321-14-6. $C_7H_6N_2O_4$.

Properties: Mw: 182.15

Hazard: Possible carcinogen and poison. Flammable. Cardiac impairment and reproductive effects.

2,4-dinitrotoluene.   (DNT).
CAS: 121-14-2. $C_6H_3CH_3(NO_2)_2$.
(1) 2,4-; (2) 3,4-; (3) 3,5-.

Properties: Yellow crystals. D (1) 1.3208; (2) 1.32; (3) 1.277; mp (1) 70.5C; (2) 61C; (3) 92.3C. Soluble in alcohol and ether; very slightly soluble in water. A commercial grade, consisting of a mixture of the three isomers, is an oily liquid. Combustible.

Derivation: Nitration of nitrotoluene with mixed acid.

Method of purification: Crystallization.

Hazard: Absorbed by skin. Possible carcinogen.

Use: Organic synthesis, toluidines, dyes, explosives.

dinker.   A machine for cutting forms from flat sheets of plastic, rubber, metal, paper, etc., by impact of a metal die.

dinocap.   Generic name for 2-(2-methylheptyl)-4,6-dinitrophenyl crotonate.
$CH_3CH:CHCOOC_6H_2(NO_2)_2CH(CH_3)C_6H_{13}$.

Properties: Brown liquid. Bp 138–140C (0.05 mm Hg). Insoluble in water; soluble in most organic solvents.

Hazard: Toxic by ingestion and inhalation, strong irritant.

Use: Acaricide, fungicide.

dinonyl adipate.   Ester of nonyl alcohol.

Properties: Colorless liquid. Bp 201–210C (1 mm Hg), d 0.926 (25C), refr index 1.4523 (20C), viscosity 14.9 centistokes (100F). Combustible.

Use: Plasticizer where special low-temperature properties are desired.

dinonyl carbonate.
$(C_9H_{19})_2CO_3$.

Properties: Ester of nonyl alcohol, colorless liquid. Bp 135–140C (0.3 mm Hg), d 0.894 (25C), refr index 1.4427 (20C).

Hazard: Combustible.

dinonyl ether.
CAS: 2456-27-1. $C_9H_{19}OC_9H_{19}$.
Properties: Colorless liquid. Bp 148–153C (5 mm Hg), d 0.817 (25C), refr index 1.4405 (20C). Can be made from nonyl alcohol plus nonyl halide by the Williamson synthesis. Combustible.

dinonyl maleate.
CAS: 2787-64-6. $C_9H_{19}OOCCH:CHCOOC_9H_{19}$. Ester of nonyl alcohol.
Properties: Colorless liquid, bp 157–167C (0.1 mm Hg), d 0.941 (25C), refr index 1.4586 (20C), viscosity 6900 centistokes (−40F), 17.47 centistokes (100F), 3.50 centistokes (210F). Combustible.

dinonylphenol.
CAS: 137-99-5. $(C_9H_{19})_2C_6H_3OH$.
Properties: Colorless liquid. Insoluble in water; soluble in common organic solvents. Combustible.
Hazard: See phenol.
Use: Solvent.

dinonyl phthalate.   (DNP).
CAS: 84-76-4. $C_6H_4(COOC_9H_{19})$.
Ester of nonyl alcohol.
Properties: Colorless liquid. Bp 205–220C (1 mm Hg), d 0.979 (25C), refr index 1.4871 (20C), viscosity 55.3 centistokes (37.7C), flash p 420F (215C). Combustible.
Use: General-purpose, low-volatility plasticizer for vinyl resins; pure grade as stationary liquid phase in chromatography.

dinoseb.   Legal label name (air) for 2,4-dinitro-6-*sec*-butylphenol.

DIOA.   Abbreviation for diisooctyl adipate.

dioctadecylamine.   See distearylamine.

2,6-dioctadecyl-*p*-cresol.   See 2,6-distearyl-*p*-cresol.

dioctadecyl ether.   See distearyl ether.

dioctadecyl sulfide.   See distearyl sulfide.

3,3′-dioctadecyl thiodipropionate.   See distearyl thiodipropionate.

dioctyl adipate.   See di(2-ethylhexyl) adipate.

dioctylamine.   See di-(2-ethylhexyl)amine.

dioctylaminoethanol.   See di(2-ethylhexyl) ethanolamine.

dioctyl azelate.   See di(2-ethylhexyl) azelate.

dioctyl chlorophosphate.   (dioctyl phosphorochloridate).

$(C_8H_{17}O)_2P(O)Cl$.
Properties: Water-white liquid. D 0.991 (25C), refr index 1.445 (25C). Decomposes on distillation. Soluble in common inert organic solvents; insoluble in water. Combustible.
Use: Intermediate, insecticide.

di(*n*-octyl-*n*-decyl) adipate.   (DNODA).
Properties: Clear, oily liquid. D 0.912-0.920 (25/25C); refr index 1.443–1.447 (25C). Combustible.
Use: Low-temperature plasticizer.

di(*n*-octyl-*n*-decyl) phthalate.   (DNODP).
Properties: Clear, oily liquid; slight odor. Acidity (as phthalic acid) 0.01% max, d 0.968–0.977 (25/25C), crystallizing point −30C, bp 232–267C (4 mm Hg), flash p 426F (219C). Combustible.
Use: Plasticizer for polyvinyl chloride and other vinyls.

dioctyldi(lauroyloxy)stannane.
CAS: 3648-18-8. $C_{40}H_{80}O_4Sn$.
Hazard: Mildly toxic by ingestion.

di-*n*-octyldiphenylamine.
CAS: 101-67-7. $C_8H_{17}C_6H_4NHC_6H_4C_8H_{17}$.
Properties: Light-tan powder. D 0.99, mp 80–90C. Soluble in benzene, gasoline, acetone, and ethylene dichloride; insoluble in water.
Use: Antioxidant for petroleum-based and synthetic lubricants and plastics.

dioctyldiphenyltin.
CAS: 103270-64-0. $C_{28}H_{44}Sn$.
Hazard: A severe skin and eye irritant.

dioctyl ether.
CAS: 10143-60-9. $(C_8H_{17})_2O$.
Properties: Liquid. Fp −7C, bp 291.7C, d 0.805 (17/4C) refr index 1.4329 (24C). Combustible.
Grade: 95% (min) purity.
Use: Electrical insulator, water repellent, mold lubricant, antistatic agent, intermediate.

dioctyl(ethylenedioxybis(carbonylmethylthio))stannane.   See di-*n*-octyltin ethyleneglycol dithioglycolate.

dioctyl fumarate.   See di(2-ethylhexyl) fumarate.

dioctyl hexahydrophthalate.   See di(2-ethylhexyl) hexahydrophthalate.

dioctylmethylamine.   See *n*-methyl-*n*-octyl-1-octanamine.

dioctyloxostannane.
CAS: 870-08-6. $C_{16}H_{34}OSn$.
Hazard: Moderately toxic by ingestion.

N,N'-di-*n*-octyl-*p*-phenylenediamine.
$C_6H_4(NHC_8H_{17})_2$.
Properties: Colorless liquid. Bp approximately 390C, d 0.912 (15C), pour p −4C, flash p (Pensky–Martin) 395F (201C) refr index 1.5129 (20C). Completely miscible with methanol, pentane, and benzene; vap press (absolute) 0.33 mm Hg (150C). Combustible.
Use: Antioxidant, antiozonant for gasoline, mercaptans, synthetic rubber.

dioctyl phosphite.
(dioctyl phosphonate).
CAS: 3658-48-8. $(C_8H_{17}O)_2P(O)H$.
Properties: Water-white liquid. Bp 150–155C (2-3 mm Hg), d 0.929 (25C), refr index 1.4418 (25C). Soluble in common organic solvents. Combustible.
Use: Solvent, antioxidant, intermediate.

dioctylphosphoric acid.   See di(2-ethylhexyl) phosphoric acid.

dioctyl phosphorochloridate.   See dioctyl chlorophosphate.

dioctyl phthalate.   See di(2-ethylhexyl) phthalate.

di(2-octyl) phthalate.   See dicapryl phthalate.

dioctyl sebacate.   See di(2-ethylhexyl) sebacate.

dioctylsodium   sulfosuccinate.   (di(2-ethylhexyl)sodium sulfosuccinate; sodium dioctyl sulfosuccinate).
CAS: 577-11-7. $C_8H_{17}OOCCH_2CH(SO_3Na)COO$ C_8H_{17}.
An anionic surface-active agent.
Properties: White, waxlike solid; characteristic odor. Soluble in water; freely soluble in alcohol, glycerol, carbon tetrachloride, acetone, and xylene. Saponification value 240–253; stable in acid and neutral solutions; hydrolyzes in alkaline solutions.
Derivation: By esterification of maleic anhydride with 2-ethylhexyl alcohol, followed by addition of sodium bisulfite.
Hazard: Use in food products restricted.
Grade: NF, FCC.
Use: Food additive (processing aid in sugar industry, stabilizer for hydrophilic colloids), wetting agent, dispersant, emulsifier.

dioctylstannylene   maleate.   See dioctyltin maleate.

dioctyl succinate.   See di(2-ethylhexyl) succinate.

dioctyl sulfide.   (dioctyl thioether).
CAS: 2690-08-6. $(C_8H_{17})_2S$.

Properties: Liquid. Mp 0.5C, bp 180C (10 mm Hg), d 0.8419 (17/17C), refr index 1.4606 (20C). Combustible.
Grade: 95% (min) purity.
Use: Organic synthesis (formation of sulfonium compounds).

dioctylthioacetoxystannane.
CAS: 15535-79-2. $C_{18}H_{36}O_2SSn$.
Hazard: Moderately toxic by ingestion.

dioctyl thioether.   See dioctyl sulfide.

dioctyl thiopropionate.   See 3,3'-(2-ethylhexyl) thiodipropionate.

dioctylthioxostannane.
CAS: 3572-47-2. $C_{16}H_{34}SSn$.
Hazard: A poison.

di-*n*-octyltin bis(butyl maleate).   See bis(butoxymaleoyloxy)dioctylstannane.

di-*n*-octyltin bis(2-ethylhexyl maleate).
CAS: 10039-33-5. $C_{40}H_{72}O_8Sn$.
Hazard: Moderately toxic by ingestion.

di-*n*-octyltin   bis(2-ethylhexyl)   mercaptoacetate.
CAS: 15571-58-1. $C_{36}H_{72}O_4S_2Sn$.
Hazard: Moderately toxic by ingestion.

dioctyltinbis(isooctyl maleate).   See bis(isooctyloxymaleoyloxy)dioctylstannane.

di(*n*-octyl)tin-*S,S'*-bis-(isooctylmercaptoacetate).
CAS: 26401-97-8.
A heat stabilizer for PVC food-packaging materials, especially for clear plastic bottles. Approved by FDA for all foods except malt beverages, carbonated soft drinks, milk, and other dairy products.

di-*n*-octyltin bis(laurylthioglycolate).
CAS: 69226-43-3. $C_{44}H_{88}O_4S_2Sn$.
Hazard: Moderately toxic by ingestion.

dioctyltin bis(monolauryl maleate).   See di-(*n*-octyl)tin bis-*o,o'*-(monolauryl maleate).

di-(*n*-octyl)tin bis-*o,o'*-(monolauryl maleate).
CAS: 7324-77-8. $C_{48}H_{88}O_8Sn$.
Hazard: A severe eye irritant.

di-*n*-octyltindichloride.
CAS: 3542-36-7. $C_{16}H_{34}Cl_2Sn$.
Hazard: A poison by ingestion.

di-*n*-octyltin dilaurate.   See dioctyldi(lauryloxy)stannane.

di-*n*-octyltin dimonobutylmaleate.   See
bis(butoxymaleoyloxy)dioctylstannane.

di-*n*-octyltin ethyleneglycol dithioglycolate.
CAS: 69226-44-4. $C_{22}H_{42}O_4S_2Sn$.
Hazard: Moderately toxic by ingestion.

dioctyltin maleate.
CAS: 16091-18-2. $C_{20}H_{36}O_4Sn$.
Hazard: Moderately toxic by ingestion.

di-*n*-octyltin mercaptide.
CAS: 58229-88-2.
Hazard: Moderately toxic by ingestion.

dioctyltin oxide.   See dioctyloxostannane.

di-*n*-octyltin sulfide.   See dioctylthioxostannane.

dioctyltin thioglycolate.   See dioctylthioacetoxystannane.

"Diofan" *[Solvin].*   TM for dispersions of vinylidene copolymers for paper coating.

diol.   Synonym for glycol or dihydric alcohol.

diolefin.   (diene; alkadiene).
An aliphatic compound (olefin) containing two double bonds, e.g., butadiene.

DIOP.   Abbreviation for diisooctyl phthalate.

DIOS.   Abbreviation for diisooctyl sebacate.

dioxacarb.   (*O*-(1,3-dioxolan-2-yl)-phenyl-*N*-
methylcarbamate).
CAS: 6988-21-2.
Properties: Mp 114-15C, vap press 0.3 mm Hg
(20C). Soluble in water to 0.6%.
Use: Control cockroaches and stored product pests.

1,4-dioxacyclotridecane-5,13-dione.   See
ethylene azelate.

1,5,2,4-dioxadithiepane-2,2,4,4-tetraoxide.
CAS: 99591-73-8. $C_3H_6O_6S_2$.
Hazard: A poison by ingestion.

1,4-dioxane.   (diethylene ether; 1,4-diethylene
dioxide; diethylene oxide; dioxyethylene ether;
dioxane).
CAS: 123-91-1

$$H_2C \overset{O}{\underset{O}{<}} CH_2$$
$$H_2C \qquad CH_2$$

Properties: Colorless liquid; ethereal odor. Bp
101.3C, fp 10–12C, d 1.0356 (20/20C), 8.61 lb/gal
(20C), refr index 1.4221 (20C), flash p 65F (18.3C)
(ASTM OC), autoign temp 356F (180C). Stable.
Miscible with water and most organic solvents.
Derivation: (1) From ethylene glycol by treatment
with acid; (2) from β,β-dichloroethyl ether by treatment with alkali.
Grade: Reagent, technical, spectrophotometric,
scintillation.
Hazard: Flammable, dangerous fire risk, may form
explosive peroxides. Toxic by inhalation, absorbed
by skin. Liver damage. Possible carcinogen.
Use: Solvent for cellulosics and wide range of
organic products; lacquers; paints; varnishes; paint
and varnish removers; wetting and dispersing agent
in textile processing, dye baths, and stain and printing compositions; cleaning and detergent preparations; cements; cosmetics; deodorants; fumigants;
emulsions; polishing compositions; stabilizer for
chlorinated solvents; scintillation counters.

4,9-dioxadodecane-1,12-diamine.   See   1,4-
bis(3-aminopropoxy)butane.

dioxathion.   (Delnav;   *p*-dioxane-2,3-diylethyl
phosphorodithioate).
CAS: 78-34-2. $C_4H_6O_2[SPS(OC_2H_5)_2]_2$.
Properties: Viscous, brown liquid. Fp −20C, bulk d
1.257 (26C). Insoluble in water; soluble in hexane.
The technical material is a mixture of *cis* and *trans*
isomers.
Hazard: Toxic by inhalation, ingestion, and skin
absorption; cholinesterase inhibitor; use may be
restricted. Questionable carcinogen.
Use: Insecticide, miticide.

2,2-dioxide.   See propane sultone.

dioxin.
CAS: 1746-01-6.
The commonly accepted, though chemically
imprecise, name for the compound 2,3,7,8-
tetrachlorodibenzo-*p*-dioxin (TCDD), which is
only one of >70 members of the family of chlorinated dioxins. It was found to be a contaminant
of the herbicide 2,4,5-T (trichlorophenoxyacetic acid)
some 10 years after the latter was approved for
use; it was then banned by FDA for most purposes.
Synthesized in 1957, dioxin is a white, crystalline
solid in pure form. It was present as a contaminant in defoliants used in Vietnam (Agent Orange),
and its toxicity was widely publicized. Though it
is undoubtedly harmful to humans, no deaths have
occurred. It is a carcinogen, teratogen, and mutagen. Its toxicity to laboratory animals varies widely
with the species; it is lethally toxic to guinea pigs,
but hamsters appear relatively unaffected. Where
soil contamination occurs, the concentration is no
greater than several parts per million at most. Wastes
contaminated with dioxin must be disposed of in

officially approved landfills. Human toxicology is under continuing investigation.
Hazard: Confirmed carcinogen.

Dioxitol.   Diethylene glycol monoethyl ether, bp 202C, d 0.990 (20/20C).

3,5-dioxo-1,2-diphenyl-4-*n*-butylpyrazolidine.   See phenylbutazone.

1,3-dioxolane.   (ethylene glycol formal).
CAS: 646-06-0

OCH₂CH₂OCH₂.

A cyclic acetal.
Properties: Water-white liquid. D 1.065, bp 74C, flash p 35F (1.67C) (OC), vap press 70 mm Hg (20C) bulk density 8.2 lb/gal (20C). Soluble in water; stable under neutral or slightly alkaline conditions.
Derivation: Reaction of formaldehyde with ethylene glycol.
Hazard: Flammable, dangerous fire risk. Toxic by inhalation and ingestion. Hematologic effects.
Use: Low-boiling solvent and extractant for oils, fats, waxes, dyes, and cellulose derivatives.

dioxolone-2.   See ethylene carbonate.

dioxopurine.   See xanthine.

dioxyanthraquinone.   See   dihydroxyanthraquinone.

dioxybenzene.   See dihydroxybenzene.

1,1′-(dioxy)dimethylcyclohexanol.
CAS: 49796-88-5.   mf: C₁₄H₂₆O₄.
Hazard: Moderately toxic by ingestion.

dioxyethylene ether.   See 1,4-dioxane.

dioxygenase.   Any enzyme that catalyzes the introduction of two oxygen atoms derived from free molecular oxygen into a substrate.

DIOZ.   Abbreviation for diisoctyl azelate.

DIPA.   Abbreviation for diisopropanolamine.

dipalmitylamine.   (dihexadecylamine).
(C₁₆H₃₃)₂NH.
Properties: Solid. Mp 65C, d 0.83. Slightly soluble in water.
Use: Intermediate.

dipentaerythritol.
CAS: 126-58-9. (CH₂OH)₃CCH₂OCH₂C(CH₂OH)₃.

Occurs in technical pentaerythritol, an off-white, free-flowing powder. The molecule contains six primary hydroxyl groups, all esterifiable. Mp 212–220C, d 1.33 (25/4C).
Use: Paints and coatings.

dipentamethylenethiuram tetrasulfide.

[CH₂(CH₂)₄NCS]₂S₄.

Properties: Light-gray powder. D 1.53, mp 110C min. Soluble in chloroform, benzene, and acetone; insoluble in water.
Use: Ultraaccelerator for rubber.

dipentene.   (cinene; limonene, inactive; *dl-p*-mentha-1,8-diene; cajeputene).
CAS: 138-86-3.   C₁₀H₁₆.
Commercial form is high in dipentene content but also contains other terpenes and related compounds in varying amounts.
Properties: Colorless liquid; lemonlike odor. D 0.847 (15.5/15.5C), bp 175–176C, refr index 1.473 (20C), flash p 113F (45C) (CC), fp −97C, wt/gal 7.15 lb (15.5C), autoign temp 458F (236C). Miscible with alcohol; insoluble in water.
Derivation: (1) From various essential oils; (2) by close fractionation of wood turpentine; (3) by-product in making synthetic camphor.
Grade: Steam-distilled, destructively distilled.
Hazard: Moderate fire risk. Combustible.
Use: Solvent for oleoresinous products, rosin, ester gum, alkyd resins, waxes, metallic soap driers, rubber, etc.; rubber compounding and reclaiming; dispersing agent for oils, resins, resin-oil combinations, pigments, and driers; paints, enamels, lacquers, and varnishes; general wetting and dispersing agent; printing inks; perfumes; flavors; floor waxes and furniture polishes; synthetic resins, polyterpenes.

dipentene dioxide.   (limonene dioxide).
CAS: 96-08-2.   C₁₀H₁₆O₂.
Properties: Liquid. D 1.0287 (20C), bp 242C, fp −100C. Soluble in water.
Use: Intermediate for plasticizers, epoxy resins; pharmaceuticals.

dipentene glycol.   See terpin hydrate.

dipentene monoxide.   (limonene monoxide).
CAS: 106168-39-2.   C₁₀H₁₆O.
Properties: Liquid. D 0.929 (20C), fp −6C, bp 75C, flash p 152F (66C). Combustible.
Use: Organic intermediate, epoxy resins.

Dipentite.   Diphenylpentaerythritol diphosphite.

di-*n*-pentylamine.   See di-*n*-amylamine.

2,5-di(*tert*-pentyl)hydroquinone. See 2,5-di(*tert*-amyl)hydroquinone.

dipentyloxostannane.
CAS: 2273-46-3. $C_{10}H_{22}OSn$.
Hazard: A poison by ingestion. An eye and severe skin irritant.

dipentyltin dichloride.
CAS: 1118-42-9. $C_{10}H_{22}Cl_2Sn$.
Hazard: A poison.

dipentyltin oxide. See dipentyloxostannane.

diphacinone. Generic name for 2-diphenylacetyl-1,3-indandione.
Use: Rodenticide.
See diphenadione.

diphemanil methyl sulfate.
(4-diphenylmethylene-1,1-dimethylpiperidinium methyl sulfate).

$(C_6H_5)_2CCCH_2CH_2N(CH_3)_2CH_2CH_2\bullet CH_3SO_4$.

Properties: White, crystalline solid; faint characteristic odor; bitter taste. Mp 189–196C. Very slightly soluble in ether; slightly soluble in alcohol, chloroform, and water; stable to heat and light; somewhat hygroscopic; pH (1% solution) 4.0–6.0.
Grade: NF.
Use: Medicine (anticholinergic).

diphenadione. (2-diphenylacetyl-1,3-indandione; diphacinone).
CAS: 82-66-6. $C_{23}H_{16}O_3$.
Properties: Yellow, odorless crystals or crystalline powder. Mp 145–147C. Practically insoluble in water; soluble in acetone and acetic acid.
Hazard: Prevents blood clotting.
Use: Medicine (anticoagulant), rodenticide.

diphenamid.
CAS: 957-51-7.
Generic name for *N,N*-dimethyl-2,2-diphenylacetamide. $(C_6H_5)_2CHCON(CH_3)_2$.
Properties: White solid. Mp 134.5–135.5C. Soluble in water, acetone, dimethyl formamide, and phenyl "Cellosolve" *[Union]*.
Hazard: Toxic by ingestion.
Use: Herbicide, plant growth regulator.

diphenatrile. See diphenylacetonitrile.

diphenhydramine hydrochloride.
[(2-benzhydryloxy)-*N,N*-dimethylethylamine hydrochloride].
CAS: 147-24-0. $(C_6H_5)_2CHOCH_2CH_2N(CH_3)_2\cdot$ HCl.
Properties: White, crystalline powder; odorless; darkens slowly on exposure to light. Mp 166–170C.

Solutions practically neutral to litmus paper; soluble in water, alcohol, and chloroform; very slightly soluble in benzene and ether.
Grade: USP.
Hazard: Toxic. Prescription only; do not use with alcohol or other central nervous system depressants.
Use: Medicine (antihistamine).

diphenic acid. (2,2'-biphenyldicarboxylic acid).
CAS: 482-05-3. $HOOCC_6H_4C_6H_4COOH$.
Properties: White needles. Mp 228–229C. Soluble in hot water.
Use: Synthesis of dyes, detergents, pharmaceuticals.

diphenolic acid. See 4,4-bis(4-hydroxyphenyl) pentanoic acid.

diphenprofos. See *o*-(4-((4-chlorophenyl)thio) phenyl) *o*-ethyl s-propyl.

diphenyl. (biphenyl).
CAS: 92-52-4. $C_6H_5C_6H_5$.
Several crystalline forms are known

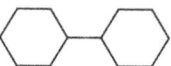

Properties: White scales; pleasant odor. D approximately 1, mp 70C, bp 256C, flash p 235F (112.7C). Soluble in alcohol and ether; insoluble in water.
Derivation: (1) By slowly passing benzene through a red-hot iron tube; (2) by heating bromobenzene and sodium with subsequent distillation.
Hazard: Toxic. Pulmonary function inhibitor.
Use: Organic synthesis, heat-transfer agent, fungistat in packaging of citrus fruit, plant disease control, manufacture of benzidine, dyeing assistant for polyesters.
See chloridiphenyl.

diphenylacetic acid.
CAS: 117-34-0. $(C_6H_5)_2CHCOOH$.
Properties: Colorless, crystals; odorless. Bp (sublimes), mp 147.8–148.2C. Soluble in hot water, alcohol, ether, and chloroform.

diphenylacetonitrile. (diphenatrile).
$(C_6H_5)_2CHCN$.
Properties: Yellow, crystalline powder. Mp 73–73.5C. Insoluble in water; very soluble in alcohol.
Use: Preparation of diphenylacetic acid, synthesis of antispasmodics, herbicide.

diphenylacetylene. See tolan.

2-diphenylacetyl-1,3-indandione. See diphenadione.

diphenylamine. (DPA; *N*-phenylaniline).
CAS: 122-39-4. $(C_6H_5)_2NH$.

Properties: Colorless to grayish crystals. D 1.159, mp 52.85C, bp 302C, flash p 307F (152.7C), autoign temp 1173F (633C). Soluble in carbon disulfide, benzene, alcohol, and ether; insoluble in water. Combustible.

Derivation: By heating equal formula weights of aniline and aniline hydrochloride in an autoclave. The product is boiled with dilute hydrochloric acid to remove the unaltered aniline, and the residue is distilled.

Grade: Technical, refined, flake, and fused.

Hazard: Toxic by ingestion. Liver and kidney damage; hematologic effects. Questionable carcinogen.

Use: Rubber antioxidants and accelerators, solid rocket propellants, pesticides, dyes, pharmaceuticals, veterinary medicine, storage preservation of apples, stabilizer for nitrocellulose, analytical chemistry.

diphenylamine chloroarsine. (adamsite; phenarsazine chloride; DM).
CAS: 578-94-9. $C_6H_4(AsCl)(NH)C_6H_4$.
Properties: Canary-yellow crystals. Sublimes readily, d 1.65, mp 195C, bp 410C (decomposes). Insoluble in water; soluble in benzene, xylene, and carbon tetrachloride.
Derivation: By heating diphenylamine with arsenic trichloride.
Hazard: Toxic by inhalation and ingestion, strong irritant.
Use: Military poison gas, wood treating.

diphenylaminochloroarsine.
CAS: 16758-26-2. $C_{12}H_{10}AsClN$.
Hazard: Low toxicity by inhalation. Human systemic effects. A mild eye irritant.

9,10-diphenylanthracene.
CAS: 1499-10-1. $C_{14}H_8(C_6H_5)_2$.
Properties: Crystals. Mp 248–250C. Insoluble in water and alcohol; slightly soluble in toluene.
Grade: Purified.
Use: Primary fluor or wavelength shifter in soluble scintillators.

1,2-diphenylbenzene. See o-terphenyl.

1,4-diphenylbenzene. See p-terphenyl.

m-diphenylbenzene. See m-terphenyl.

diphenylbenzidine.
CAS: 531-91-9. $C_6H_5HNC_6H_4C_6H_4NHC_6H_5$.
Properties: White powder. Mp 242C. Insoluble in water; slightly soluble in alcohol and acetone; soluble in boiling toluene, sensitive to light.
Derivation: Diphenylamine and fuming sulfuric acid.
Hazard: May be carcinogenic; see benzidine.
Use: Determination of zinc and nitrites.

2,7-diphenylbenzo(lmn)(3,8)phenanthroline-1,3,6,8(2h,7h)-tetrone.
CAS: 24259-89-0. $C_{26}H_{14}N_2O_4$.
Hazard: A poison.

2,5-diphenyl-p-benzoquinone.
CAS: 844-51-9. $C_6H_5C_6H_2O_2C_6H_5$.
Properties: Greenish-yellow solid. Mp 210–214C. Slightly soluble in styrene, benzene, acetone, and ethyl acetate.
Use: Polymerization inhibitor.

diphenylbis(phenylthio)stannane. See diphenylbis(phenylthio)tin.

diphenylbis(phenylthio)tin.
CAS: 1103-05-5. $C_{24}H_{20}S_2Sn$.
Properties: White solid from EtOH. Mp: 65–66C. Sol in org solvs.
Hazard: A poison.

diphenylbromoarsine.
$(C_6H_5)_2AsBr$.
Properties: White crystals. Mp 54–56C.
Derivation: (1) By heating hydrobromic acid and diphenylarsenious oxide together for approximately 4 hours at 115–120C; (2) by action of arsenic tribromide on triphenyl arsine at 300–350C.
Hazard: Strong irritant. A poison.

1,3-diphenyl-2-buten-1-one. See dypnone.

diphenylcarbazide.
CAS: 140-22-7. $(C_6H_5NHNH)_2CO$.
Properties: White crystals or flakes. Mp 173C, decomposes in light. Insoluble in water; soluble in alcohol and acetone.
Derivation: Phenylhydrazine and urea.
Use: Determination of chromium and other metals, indicator for iron.

diphenylcarbinol. See benzhydrol.

diphenyl carbonate.
CAS: 102-09-0. $(C_6H_5O)_2CO_3$.
Properties: White, crystalline solid. Can be halogenated and nitrated in characteristic manner. Readily undergoes hydrolysis and ammonolysis. Bp 302C, mp 78C, d 1.1215 (87/4C). Soluble in acetone, hot alcohol, benzene, carbon tetrachloride, ether, glacial acetic acid, and other organic solvents; insoluble in water.
Grade: Technical.
Use: Plasticizer and solvent, synthesis of polycarbonate resins.

diphenyl, chlorinated. See chlorodiphenyl.

diphenylchloroarsine.
CAS: 712-48-1. $(C_6H_5)_2AsCl$.
Properties: Colorless crystals or dark-brown liquid that slowly becomes semisolid. D 1.363 (40C)

(solid), or 1.356 (45C) (liquid), bp 333C (in CO_2 atm), mp 41C. Decomposed by water (slowly). Soluble in carbon tetrachloride, chloropicrin, and phenyldichloroarsine; almost insoluble in water.
Derivation: Benzene and arsenic trichloride are heated in the presence of aluminum chloride.
Hazard: A poison. Toxic by inhalation, strong irritant to tissue.
Use: Military poison gas.

diphenyldecyl phosphite.
$(C_6H_5O)_2POC_{10}H_{21}$.
Properties: Nearly water-white liquid. D 1.023 (25/15.5C), mp 18C, refr index 1.5160 (25C). Combustible.
Use: Chemical intermediate, stabilizer for polyvinyl and polyolefin resins.

2,4′-diphenyldiamine.   (2,4′-biphenyldiamine; 2,4′-diaminodiphenyl).
CAS: 492-17-1. $C_{12}H_{12}N_2$.
Properties: Crystalline needles. Mp 45C, bp 360C. Slightly soluble in alcohol and ether.
Derivation: Reduction of azobenzene with tin and hydrochloric acid.
Hazard: Questionable carcinogen.
Use: Azo-dye manufacture, tungsten determination.

diphenyldichlorosilane.
CAS: 80-10-4. $(C_6H_5)_2SiCl_2$.
Properties: Colorless liquid. Bp 305C, fp −22C, d 1.19 (20C), refr index 1.5773 (25C), flash p 288F (142C)(COC). Readily hydrolyzed by moisture with liberation of hydrogen chloride. Combustible.
Derivation: (1) Reaction of powdered silicon and chlorobenzene in the presence of copper powder as catalyst; (2) reaction of phenylmagnesium chloride with silicon tetrachloride.
Grade: Technical.
Hazard: Strong irritant to tissue.
Use: Intermediate for silicone lubricants.

diphenyldi-*n*-dodecylsilane.
$(C_6H_5)_2Si(C_{12}H_{25})_2$.
Properties: Colorless oil.
Derivation: Reaction of didodecyldichlorosilane with phenyllithium.
Use: High-temperature lubricant.

diphenyldiimide.   See azobenzene.

diphenyldimethoxysilane.
CAS: 6843-66-9. $(C_6H_5)_2Si(OCH_3)_2$.
Properties: Liquid. D 1.080 (25C), bp 191C (53 mm Hg), refr index 1.5404 (25C). Soluble in acetone, benzene, and methyl alcohol. Combustible.
Use: Treatment of powders, glass, paper, and fabrics.

diphenyleneimine.   See carbazole.

α-diphenylenemethane.   See fluorene.

diphenylene oxide.   (dibenzofuran).
CAS: 132-64-9. $C_{12}H_8O$ (tricyclic).
Properties: Crystalline solid. Mp 87C, bp 288C. Insoluble in water; slightly soluble in alcohol, ether, and benzene.
Derivation: From coal tar.
Use: Insecticide.

uns-diphenylethane.   (1,1-diphenylethane).
$(C_6H_5)_2CHCH_3$.
Properties: Colorless liquid. Bp 286C, d 1.004 (20C), fp −21.5C, flash p 264F (129C). Soluble in chloroform, ether, and carbon disulfide. Combustible.
Derivation: Action of acetaldehyde on benzene in the presence of concentrated sulfuric acid.
Use: Solvent for nitrocellulose, organic synthesis.

sym-diphenylethane.   (bibenzyl; dibenzyl; 1,2-diphenylethane).
CAS: 103-29-7. $C_6H_5CH_2CH_2C_6H_5$.
Properties: White, crystalline needles or small plates. D 0.9782, bp 284C, mp 52C. Soluble in alcohol, chloroform, ether, and carbon disulfide; insoluble in water.
Derivation: (1) By treating benzyl chloride with metallic sodium; (2) by action of benzyl chloride on benzylmagnesium chloride.
Use: Organic synthesis.

1,2-diphenylethanedione monoxime.   See α-benzil monoxime.

diphenyl ether.   See diphenyl oxide.

diphenylethylene.   See stilbene.

N,N-diphenylethylenediamine.
(ethyldiphenyldiamine).
CAS: 1140-29-0. $C_6H_5NHCH_2CH_2NHC_6H_5$.
Properties: Cream-colored solid. D 1.14, softening point 64C. Insoluble in water; soluble in acetone, ethylene dichloride, benzene, and gasoline.
Use: Antioxidant in rubber compounding.

diphenylglycolic acid.   See benzilic acid.

N,N′-diphenylguanidine.   (DPG; melaniline).
CAS: 102-06-7. $HN:C(NHC_6H_5)_2$.
Properties: White powder; bitter taste; slight odor. D 1.13, mp 147C, decomposes above 170C. Soluble in ethanol, carbon tetrachloride, chloroform, hot benzene, and toluene; slightly soluble in water.
Derivation: Treatment of aniline with cyanogen chloride.
Hazard: Toxic by ingestion.
Use: Basic rubber accelerator, primary standard for acids.

1,6-diphenylhexatriene. (DPH).
CAS: 1720-32-7. C_6H_5HC:CHCH:CHCCH:CHC$_6H_5$.
Use: Wavelength shifter in scintillation-counting solutions.

diphenyl isodecyl phosphite. See isodecyl diphenyl phosphite.

diphenyl isophthalate. (DPIP).
CAS: 744-45-6. $C_6H_5OOCC_6H_4COOC_6H_5$.
Use: Manufacture of polybenzimidazoles, high-temperature–resistant polymers.

diphenyl ketone. See benzophenone.

diphenylmethane. (benzylbenzene).
$(C_6H_5)_2CH_2$

Properties: Long, colorless needles. D 1.0056, mp 26.5C, bp 264.7C, flash p 266F (130C). Soluble in alcohol, chloroform, hexane, benzene, and ether; insoluble in liquid ammonia. Combustible.
Derivation: Condensation of benzyl chloride or methylene chloride and benzene in the presence of aluminum chloride.
Hazard: Toxic by ingestion.
Use: Organic synthesis, dyes, perfumery.

diphenylmethane-4,4′-diisocyanate. (MDI; methylene-di-p-phenylene isocyanate; methylene(bisphenyl isocyanate)).
CAS: 101-68-8. $CH_2(C_6H_4NCO)_2$.
Properties: Light-yellow, fused solid. Solidification point 37C, d 1.197 (70C). Soluble in acetone, benzene, kerosene, and nitrobenzene. Combustible.
Derivation: p,p′-Diaminodiphenylmethane and phosgene.
Hazard: Toxic by inhalation of fumes; strong irritant. Respiratory sensitization. Questionable carcinogen.
Use: Preparation of polyurethane resin and spandex fibers; bonding rubber to rayon and nylon.

diphenylmethane dye.
Properties: Dye in which the central carbon atom links two phenyl groups.

diphenyl methanephosphonate. See methanephosphonic acid, diphenyl ester.

diphenylmethanol. See benzhydrol.

diphenylmethyl bromide. (benzhydryl bromide).
CAS: 776-74-9. $BrCH(C_6H_5)_2$.

Properties: Solid. Mp 45C, bp 193C (26 mm Hg). Decomposes in hot water; soluble in alcohol; very soluble in benzene.
Hazard: Strong irritant to eyes and skin.
Use: Organic synthesis.

diphenylmethylchlorosilane.
CAS: 144-79-6. $(C_6H_5)_2(CH_3)SiCl$.
Properties: Colorless liquid. D 1.107 (25C), bp 295C, flash p 135F (57.2C). Combustible.
Derivation: Grignard reaction of diphenyldichlorosilane with methylmagnesium chloride.
Hazard: Moderate fire risk.
Use: Intermediate, end stopper for silicone oils.

diphenylnaphthylenediamine.
$C_{10}H_6(NHC_6H_5)_2$.
Properties: Silvery, crystalline plates. Mp 164C. Slightly soluble in alcohol; insoluble in water.
Derivation: By heating 2,7-dihydroxynaphthalene with aniline and aniline hydrochloride.
Use: Organic synthesis.

diphenylnitrosamine. See p-nitrosodiphenylamine.

diphenylolpropane disodium salt. See Bisphenol A disodium salt.

2,5-diphenyloxazole. (DPO).

OOC_6H_5:CC$_6H_5$.

Properties: White, fluffy solid. Mp 70–72C.
Grade: Scintillation.
Use: Primary fluor used in scintillation counters or as wavelength shifters.

diphenyl oxide. (phenyl ether; diphenyl ether).
CAS: 101-84-8. $(C_6H_5)_2O$.
Properties: Colorless crystals or liquid, geranium-like odor. D 1.072–1.075, mp 27C, bp 259C, flash p 239F (115C), autoign temp 1144F (618C). Soluble in alcohol and ether; insoluble in water. Combustible.
Derivation: Reaction of bromobenzene and sodium phenate heated under pressure.
Grade: Technical, perfume, industrial.
Hazard: Toxic by inhalation of vapor. Eye and upper respiratory tract irritant; nausea.
Use: Perfumery, particularly soaps; heat-transfer medium resins for laminated electrical insulation; chemical intermediate for such reactions as halogenation, acylation, alkylation, etc.

diphenylpentaerythritol diphosphite.
$C_6H_5OP(OCH_2)_2C(CH_2O)_2POC_6H_5$ (spiro).
Properties: White powder. Bp 190–200C (0.1 mm Hg).
Use: Stabilizer for resins.

N,N'-diphenyl-*m*-phenylenediamine.
CAS: 27137-31-1. $C_6H_4(NHC_6H_5)_2$.
Properties: Crystalline needles. Mp 95C. Soluble in hot alcohol; insoluble in water. Combustible.
Derivation: By heating resorcinol with aniline in the presence of calcium chloride and zinc chloride.
Use: Organic synthesis.

N,N'-diphenyl-*p*-phenylenediamine.
(DPPD).
CAS: 74-31-7. $(C_6H_5NH)_2C_6H_4$.
Properties: Gray powder. D 1.28, mp 145–152C, purity 9.2% (min). Insoluble in water; soluble in acetone, benzene, monochlorobenzene, and isopropyl acetate. Combustible.
Use: Flex-resistant antioxidant in rubbers; stabilizer; polymerization inhibitor; retards copper degradation; intermediate for dyes, drugs, plastics, and detergents.

diphenyl phosphite.
CAS: 4712-55-4. $(C_6H_5O)_2PHO$.
Properties: Clear, straw-colored liquid. Mp 12C, refr index 1.557 (25C), d 1.221 (25/15C), flash p 350F (176C). Combustible.
Use: Synthesis of organophosphorus compounds.

diphenyl phthalate.
CAS: 84-62-8. $C_6H_4(COOC_6H_5)_2$.
Properties: Yellow-white powder. Mp 68–70C, d 1.28 (20C), flash p 435F (224C), bp 405C, bulk d 10.68 lb/gal, refr index 1.572 (74C). Combustible.
Use: Plasticizer for ethylcellulose, nitrocellulose, and various polymers.

diphenyl-4-piperidylmethane.
$(C_6H_5)_2(C_5H_{10}H)CH$.
Properties: White solid. Fp 99.7C min. Difficultly soluble in water; readily soluble in dilute acids; moderately soluble in organic solvents.
Use: Intermediate.

1,3-diphenyl-1,3-propanedione.
See dibenzoylmethane.

N,N'-diphenylpropylenediamine.
$C_6H_5NHCH_2CH(CH_3)NHC_6H_5$.
Properties: Clear, deep-reddish-brown, thick liquid. D 1.07, stable in storage. Insoluble in water; soluble in acetone, ethylene dichloride, benzene, and gasoline; readily disperses. Combustible.
Use: Antioxidant for rubber latexes.

diphenylpropylphosphine.
See propyldiphenylphosphine.

diphenyl-4-pyridylcarbinol.
$(C_6H_5)_2(C_5H_4N)COH$.
Properties: White solid. Mp 236–241C. Very weak base. Slightly soluble in most organic solvents; soluble in hot glacial acetic acid.
Use: Intermediate.

diphenyl-4-pyridylmethane.
CAS: 3678-72-6. $(C_6H_5)_2(C_5H_4N)CH$.
Properties: White to pale-yellow, crystalline solid. Bp 234C, fp 123C min. Moderately soluble in common organic solvents.
Use: Intermediate.

diphenylstannane.
CAS: 1011-95-6. $C_{12}H_{12}Sn$.
Properties: Yellow powder; air and light-sensitive crystals from pet ether/CH_2Cl_2. Mp: 226C. Insol in water.
Hazard: A poison. Ignites on contact with fuming nitric acid.
Use: Agricultural chemical.

diphenylsilanediol.
CAS: 947-42-2. $(C_6H_5)_2Si(OH)_2$.
Properties: White solid. Mp 130–150.
Derivation: Hydrolysis of diphenyldichlorosilane.
Use: Silicone chemical.

p,p'-diphenylstilbene.
CAS: $C_6H_5C_6H_4CH:CHC_6H_4C_6H_5$.
Properties: Crystals. Mp 308–310C.
Use: In purified form as fluor in plastic scintillators.

diphenylsulfone.
CAS: 127-63-9. $C_{12}H_{10}SO_2$.
Properties: White, crystalline solid. Mp 128C, bp 376C. Soluble in benzene and hot alcohol; insoluble in water.
Derivation: Reaction of benzene with sulfuric acid.
Hazard: Toxic by ingestion.
Use: Larvicide and ovicide.

N,N-diphenylthiourea.
See thiocarbanilide.

diphenyltin.
See diphenylstannane.

diphenyltin dibromide.
See dibromodiphenylstannane.

diphenyltin dichloride.
CAS: 1135-99-5. $C_{12}H_{10}Cl_2Sn$.
Properties: Colorless crystals from pet ether. Decomp by water. Mp: 42C, bp: 333–337C (decomp).
Hazard: Moderately toxic by ingestion.
Use: Agricultural chemical; drug.

diphenyltin oxide polymer.
CAS: 31671-16-6. $(C_{12}H_{10}OSn)_x$.
Hazard: A poison.

1,3-diphenyltriazene.
See diazoaminobenzene.

diphenylurea.
(carbanilide).
CAS: 102-7-8. $(NHC_6H_5)CO(NHC_6H_5)$.

Properties: Colorless prisms. Soluble in alcohol and ether; very slightly soluble in water. D 1.239, mp 235C, bp 260C.
Use: Organic synthesis.

diphenyl-*o*-xenyl phosphate.
$(C_6H_5O)_2(C_6H_5C_6H_4O)PO$.
Properties: D 1.20 (20C), refr index 1.582–1.590 (60C), boiling range 250–285C (5 mm Hg), flash p 225F. Insoluble in water. Combustible.
Use: Plasticizer.

diphosgene. Legal label name (air) for trichloromethyl chloroformate.

diphosphopyridine nucleotide. See nicotinamide adenine dinucleotide phosphate.

diphosphoric acid. See pyrophosphoric acid.

dipicrylamine. See hexanitrodiphenylamine.

dipicryl sulfide. Legal label name (air) for hexanitrodiphenyl sulfide.

1,3-di-4-piperidylpropane. (4-di-pip).
CAS: 16898-52-5. $(HNC_5H_9)CH_2CH_2CH_2C_5H_9NH$.
Properties: Solid. Fp 67C, bp 329C. Soluble in water. A stable, high-boiling, strong organic base.
Use: Intermediate.

diploid. A full set of genetic material consisting of paired chromosomes, one from each parental set. Most animal cells, except the gametes, have a diploid set of chromosomes. The diploid human genome has 46 chromosomes.
See haploid.

diploid life cycle. Occurs when the only multicellular stage in an organism's life cycle is diploid.

dip oil. Tar acid oil with about 25% tar acids.
Use: An insecticide for animals.

dipole, electric. An assemblage of atoms or subatomic particles having equal electric charges of opposite sign separated by a finite distance; for instance, the hydrogen and chlorine atoms of a hydrogen chloride molecule.
See polar.

dipole moment, electric. In many molecules, the atoms and their electrons and nuclei are so arranged that one part of the molecule has a positive electrical charge while the other part is negatively charged. Such a molecule, therefore, becomes a small electric dipole. Electric fields cause the molecule to turn in one direction or another, depending on the orientation of the field. The dipole

moment (μ) is the distance between the charge centers multiplied by the quantity of charge in electrostatic units.
See polar.

dipotassium dimagnesium trisulfate. See magnesium potassium sulfate.

dipotassium orthophosphate. See potassium phosphate, dibasic.

Dippel's oil. (bone oil).
Dark-brown oil obtained from distillation of bones. Contains pyridine and substituted pyridines.
Use: To denature alcohol.

dipropargyl. (bipropargyl; 1,5-hexadiyne).
CAS: 628-16-0. $HC:CCH_2CH_2C:CH$.
Properties: Colorless liquid. D 0.805, bp 85C, fp −6.0C. Soluble in alcohol; insoluble in water.
Hazard: Moderate fire and explosion risk when exposed to heat.

di-2-propenylamine. See diallylamine.

di-*n*-propylamine.
CAS: 142-84-7. $(C_3H_7)_2NH$.
Properties: Water-white liquid; amine odor. Fp −63C, d 0.741 (20C), bp 109C, bulk density 6.1 lb/gal, flash p 63F (17.2C) (TOC). Soluble in water.
Hazard: Flammable, dangerous fire risk. Skin irritant.
Use: Intermediate.

2-(r,s)-(di-*n*-propylamino)-6-(4-methoxyphenylsulfonylmethyl)-1,2,3,4-tetrahydronaphthalene.
CAS: 175442-95-2. $C_{24}H_{33}NO_3S$.
Hazard: A poison.

dipropyl cadmium. (cadmium dipropyl).
$C_6H_{14}Cd$.
Properties: Flammable, oily liquid, melting point of 83C and boiling point of 84C, reacts with water, decomposes above 150C.

dipropylene glycol. (2,2′-dihydroxydopropyl ether).
CAS: 25265-71-8.
Properties: Colorless, slightly viscous liquid. D 1.0252 (20/20C), bp 233C, vap press 0.01 mm Hg (20C), flash p 280F (137.7C) (OC), bulk d 8.5 lb/gal (20C), coefficient of expansion 0.00073 K^{-1} (20C), viscosity 1.07 cP (20C). Soluble in toluene and water. Combustible.
Grade: Technical.
Use: Polyester and alkyd resins, reinforced plastics, plasticizers, solvent.

dipropylene glycol dibenzoate. (Benzoflex; 3,3′-oxydyl-1-propanol dibenzoate).
CAS: 94-51-9. $C_{20}H_{22}O_5$.

Properties: Light-colored liquid. D 1.1271 (20/20C), bulk d 9.4 lb/gal (20C), bp 250C (10 mm Hg), mp 200C, viscosity 227 cP (20C). Insoluble in water. Combustible.
Use: Plasticizer.

dipropylene glycol dimethyl ether.
CAS: 111109-77-4. C$_8$H$_{18}$O$_3$.
Hazard: A reproductive hazard.

dipropylene glycol dipelargonate.
[CH$_3$(CH$_2$)$_7$COOCH(CH$_3$)CH$_2$]$_2$O.
A synthetic lubricant.

dipropylene glycol monomethyl ether.
(dipropylene glycol methyl ether).
CAS: 34590-94-8. CH$_3$OC$_3$H$_6$OC$_3$H$_6$OH.
Properties: Colorless liquid. D 0.950 (25/4C), bp 189C, 74.5C (10 mm Hg), fp −80C, viscosity 3.5 cP (25C), refr index 1.419 (25C), flash p 185F (85C) (OC). Completely miscible with water, VM&P naphtha, acetone, ethanol, benzene, carbon tetrachloride, ether, methanol, monochlorobenzene, and petroleum ether. Combustible.
Hazard: Toxic. Absorbed by skin.
Use: Solvent, hydraulic brake fluids.

dipropylene glycol monosalicylate.
(dipropylene glycol monoester; salicylic acid dipropylene glycol monoester).
C$_3$H$_6$(COOC$_6$H$_4$OH)OC$_3$H$_6$OH.
Properties: Light-colored oil; fragrant odor. D 1.16 (40C), refr index approximately 1.52. Soluble in alcohol, insoluble in water.
Use: UV light—screening agents, protective coatings, plasticizers.

dipropylenetriamine. See 3,3'-iminobispropylamine.

dipropyl ketone. (butyrone; 4-heptanone).
CAS: 123-19-3. (CH$_3$CH$_2$CH$_2$)$_2$CO.
Properties: Stable, colorless liquid; pleasant odor. Bp 143.7C, fp −32.1C, d 0.8162 (20/20C), bulk d 6.79 lb/gal (20C), refr index 1.4068 (20C), surface tension 25.2 dynes/cm (25C) viscosity 0.0074 cP (20C), vap press 5.2 mm Hg (20C), flash p 120F (40C) (CC). Insoluble in water; miscible with many organic solvents.
Grade: Technical.
Hazard: Moderate fire risk. Toxic by inhalation, skin and upper respiratory tract irritant.
Use: Solvent for nitrocellulose, raw and blown oils, resins, and polymers; lacquers; flavoring.

dipropylmethane. See *n*-heptane.

dipropyloxostannane.
CAS: 7664-98-4. C$_6$H$_{14}$OSn.
Properties: Polymeric powder.
Hazard: A poison by ingestion. An eye and severe skin irritant.

dipropyl phthalate.
CAS: 131-16-8. C$_6$H$_4$(COOC$_3$H$_7$)$_2$.
Properties: Colorless liquid. D 1.071 (25C), refr index 1.494 (25C), bp 129–132C (1 mm Hg). Solubility in water 0.015% by weight. Combustible.
Use: Plasticizer.

di-*n*-propyltin bismethanesulfonate.
CAS: 73927-87-4. C$_8$H$_{20}$O$_6$S$_2$Sn.
Hazard: A poison.

dipropyltin dichloride. See dichlorodipropylstannane.

diprotic acid. An acid having two dissociable protons, e.g., H$_2$SO$_4$ or oxalic acid.

"Dipterex" [Bayer]. TM for trichlorfon.

α,α'-dipyridyl. (2,2'-bipyridine).
CAS: 366-18-7.
(C$_5$H$_4$N)$_2$.
Properties: White crystals. Mp 69–70C, bp 272–273C. Slightly soluble in water; soluble in alcohol, ether, benzene, chloroform, and petroleum ether.
Grade: Reagent.
Use: Reagent for iron determination.

2,2-dipyridylamine.
CAS: 1202-34-2. (C$_5$H$_4$N)$_2$NH.
Properties: Solid. Fp 92.3C (min), bp 222C (50 mm Hg). Very slightly soluble in water. Combustible.
Derivation: From 2-aminopyridine.
Use: Intermediate.

dipyridylethyl sulfide.
[C$_5$H$_4$N(CH$_2$)$_2$]$_2$S.
Properties: Liquid. D 1.113 (25C), refr index 1.5841 (20C), mp 1.5C. Soluble in water and common organic solvents. Combustible.
Grade: Technical (95% purity).
Use: Synthesis of pharmaceuticals, dyestuffs, rubber chemicals, flotation agents, insecticides, fungicides, plasticizers, textile assistants, herbicides, oil additives, rust preventives, and pickling inhibitors.

2,2'-dipyridyl ketone hydrazone.
CAS: 2215-33-0. C$_{11}$H$_{10}$N$_4$.
Hazard: Moderately toxic by ingestion and skin contact. A mild eye irritant.

diquat. (6,7-dihydrodipyrido(1,2-a:2',1'-c)pyrazidinium salt; 1,1'-ethylene-2,2'-dipyridinium dibromide).
CAS: 2764-72-9. (C$_5$H$_4$NCH$_2$$^-$)$_2Br_2$.
Properties: Yellow crystals. Mp 335C. Soluble in water.
Hazard: Toxic by ingestion. Lower respiratory tract irritant. Questionable carcinogen.
Use: Herbicide and plant growth regulator, sugarcane flowering suppressant.

direct bordeaux. See C.I. direct red 13, disodium salt.

direct black 32. See C.I. direct black 32, trisodium salt.

direct brown 2. See C.I. direct brown 2.

direct dye. See dye, direct.

directed evolution. A laboratory process used on isolated molecules or microbes to cause mutations and identify subsequent adaptations to novel environments.

directed mutagenesis. Alteration of DNA at a specific site and its reinsertion into an organism to study any effects of the change.

directed sequencing. Successively sequencing DNA from adjacent stretches of chromosome.

direct scarlet. See C.I. direct red 23.

diresorcinol. (tetrahydroxydiphenyl). $(OH)_2C_6H_3C_6H_3(OH)_2$.
Properties: White to slightly yellowish, crystalline powder. Mp 310C. Slightly soluble in cold water.
Derivation: By fusing resorcinol and phenol with caustic soda.
Use: Organic synthesis.

diresorcinolphthalein. See fluorescein.

dirty bomb. A device designed to disperse radioactive material by means of conventional explosives. When the initial blast occurs, the explosion disperses radioactive contamination over a large area, potentially killing a larger number of people over a longer period of time. Depending on the type and size of radioactive material released and the force of the blast. Such bombs could range in size from a miniature device to a large truck bomb and are simpler and much easier to produce than an actual nuclear weapon.

disaccharide. A carbohydrate consisting of two covalently joined monosaccharide units. Sucrose, e.g., is composed of glucose and fructose.

N,N'-disalicylidene-1,2-diaminopropane.
(N,N'-disalicylidenepropylenediamine; disalicylalaminopropane; disalicylalpropylenediimine). $HOC_6H_4CH:NCH_2CH(CH_3)N:CHC_6H_4OH$.
Properties: (80% active compound) Liquid. D 1.08 (15.5/15.5C), bulk d 9.0 lb/gal, pour p 0F (−17.7C), flash p approximately 70F (21C) (TOC), viscosity 25 cs (37.7C), insoluble in water; miscible with benzene and xylene.
Hazard: Flammable, dangerous fire risk.
Use: Metal deactivator in motor fuels.

Discaloy. An austenitic iron-based alloy containing nickel, chromium, and relatively small proportions of molybdenum, titanium, silicon, and manganese. This alloy is precipitation-hardened and was developed primarily to meet the need for improved gas-turbine disks, one of the most critical components of jet engines.

discharging agent. A substance capable of destroying a dye or mordant present within the fibers of a fabric. There are various methods of utilizing this property so that it is possible to produce a colorless figure upon a colored ground, or a colored figure upon a different-colored ground. Some examples are titanous sulfate, sodium hydrosulfide, zinc formaldehyde sulfoxylate.
See stripping (2).

Dische reaction. The reaction between the Dische reagent (diphenylamine, acetic acid, and sulfuric acid) and 2-deoxypentoses, resulting in the development of a characteristic blue color.

disease-associated genes. Alleles carrying particular DNA sequences associated with the presence of disease.

dishwashing detergent.
Properties: Contains chlorine in a dry form that is released when wet.
Use: To wash items in a dishwasher.

disilane. (disilicoethane; disilicane; disilicon; hexahydride; silicoethane).
CAS: 1590-87-0. H_6Si2.
Properties: Gas with repulsive odor; combusts spontaneously in air; reacts violently with carbon tetrachloride, chloroform, oxygen, sulfur hexafluoride, and trichloromethane; decomposes in water; soluble in benzene, carbon disulfide, ethanol, and ethyl silicate.
Hazard: Powerful irritant; toxic; poisonous.

disilanyl. See silane.

disiloxane. See siloxane.

disinfectant. A substance used on inanimate objects that destroy harmful microorganisms or inhibit their activity. Disinfectants are either complete or incomplete. Complete disinfectants destroy spores as well as vegetative forms of microorganisms; incomplete disinfectants destroy vegetative forms of the organism but do not injure spores.
Some representative disinfectants are (1) mercury compounds (mercuric chloride, phenylmercuric borate); (2) halogens and halogen compounds (chlorine, iodine, fluorine, bromine, calcium and sodium hypochlorite); (3) phenols, including cresol

from coal tar, *o*-phenylphenol; (4) synthetic detergents (anionic, such as sodium alkylbenzene sulfonates, and cationic, such as quaternary ammonium compounds); (5) alcohols of low molecular weight, except methanol; (6) natural products (pine oil); and (7) gases (sulfur dioxide, formaldehyde, ethylene oxide). Heat and electromagnetic waves are used as disinfectants.

A number of compounds (mercurous and mercuric chlorides, copper sulfate and carbonate, and a mixture of zinc oxide and zinc hydroxide) have been employed as seed disinfectants.

Effectiveness of disinfectants is rated by the phenol coefficient.

See antiseptic.

disk.   (disc).
A small thin circular section or platelet of a material, especially of a biological specimen. *Disk* is the spelling preferred by scientists.

dislocation.   Any variation from perfect order and symmetry in a crystalline lattice. In some cases, the imperfection is due to missing atoms, resulting in "holes" or vacancies in the lattice, or one or more atoms of another element may be present, effecting important changes in the conductivity, hardness, and other properties of the crystals. Disordered arrangement of the planes of atoms (dislocations) may also occur.

See impurity; semiconductor; hole.

dismutase.
CAS: 9054-89-1.
Any of a class of enzymes that catalyzes the reaction of two molecules of a single compound to yield two new compounds that have different oxidation states; one is oxidized and the other is reduced.

disodium acetarsenate.   (aricyl).
$NaOOCCH_2As(OH)O(ONa) \cdot 2H_2O$.
Properties: White, crystalline powder. Soluble in water.
Derivation: By reacting sodium arsenite with sodium monochloracetate.

disodium bisethylphenyltriaminotriazine stilbenedisulfonate.
CAS: 24565-13-7. $C_{36}H_{34}N_{12}O_6S_2 \cdot 2Na$.
Hazard: Moderately toxic by skin contact. Low toxicity by ingestion.

2,7-disodiumdibromo-4-hydroxymercurifluorescein.   See merbromin.

disodium dibutyl-*o*-phenylphenoldisulfonate.
Properties: Light-brown paste. Soluble in alcohol, acetone, dibutyl tartrate, and ethylene glycol.
Use: Wetting, penetrating, and spreading agent used in kier boiling; scouring and dyeing textiles; industrial cleaners; deodorant preparations; insecticidal

formulations; metal cleaning; stabilizer and wetting agent for latex used to treat cord or other fabrics.

disodium dihydrogen pyrophosphate.   See sodium pyrophosphate, acid.

disodium   1,2-dihydroxybenzene-3,5-disulfonate.   (4,5-dihydroxy-*m*-benzenedisulfonic acid disodium salt, sodium catechol disulfonate).
$C_6H_2(OH)_2(SO_3Na)_2$.
Properties: Nonhygroscopic crystals. Freely soluble in water. Produces water-soluble colored compounds with metal salts.
Use: Colorimetric reagent for iron, manganese, titanium, molybdenum.

disodium   S,S'-(2-dimethylamino-1,3-propanediyl)bis(thiosulfate).
CAS: 52207-48-4. $C_5H_{11}NO_6S_4 \cdot 2Na$.
Hazard: A poison by ingestion.

disodium diphosphate.   See sodium pyrophosphate, acid.

disodium EDTA.   (ethylenediaminetetraacetic acid disodium salt.)
CAS: 139-33-3. $C_{10}H_{14}N_2Na_2O_8 \cdot 2H_2O$.
Properties: White, crystalline powder. Freely soluble in water. Bulk density 6.5 lb/gal; pH (5% solution) between 4 and 6.
Grade: USP, FCC.
Use: Food preservative, chelating and sequestering agent.

disodium endothal.   See endothal.

disodium ethylenebisdithiocarbamate.   See nabam.

disodium guanylate.   See sodium guanylate.

disodium hydrogen phosphate.   See sodium phosphate, dibasic.

disodium inosinate.   See sodium inosinate.

disodium methylarsonate.   (DMA; disodium methanearsonate; methanearsonic acid disodium salt).
CAS: 144-21-8. $CH_3AsO(ONa)_2$, sometimes with $6H_2O$.
Properties: Colorless, crystalline solid. Hygroscopic. Mp >355C (hexahydrate mp 132–139C). Soluble in water and methanol.
Derivation: Reaction of methyl chloride with sodium arsenate.
Grade: 55–65% powder concentration; 31.5% blend.
Hazard: Toxic by ingestion and inhalation.
Use: Pharmaceuticals, herbicide (crabgrass killer).

disodium monomethanearsonate. (DMSA; 2,3-bis(sulfanyl)butanedioic acid).
CAS: 304-55-2. $C_4H_6O_4S_2$.
The disodium salt of methanearsonic acid and a mercaptodicarboxylic acid.
Hazard: Moderately toxic.
Use: Selective postemergence herbicide used to control grass and weeds in cotton and rubber plantations; an antidote to heavy metal poisoning.

disodium orthophosphate. See sodium phosphate, dibasic.

disodiumphenyl phosphate.
$C_6H_5Na_2PO_4$.
Properties: White powder. Soluble in water; insoluble in acetone and ether.
Use: Reagent for milk pasteurization.

disodium phosphate. See sodium phosphate, dibasic.

disodium pyrophosphate. See sodium pyrophosphate, acid.

disodium tartrate. See sodium tartrate.

dispersal. The scattering of organisms of a species, often following a major reproductive event. Spores and larvae are commonly dispersed into the environment. Pollen or gametes may also be dispersed, but in this case, the intent is to target another individual so that reproduction may occur.

disperse dye. See dye, disperse.

disperse phase. See phase (2); colloid chemistry.

dispersing agent. A surface-active agent added to a suspending medium to promote uniform and maximum separation of extremely fine solid particles, often of colloidal size. True dispersing agents are polymeric electrolytes (condensed sodium silicates, polyphosphates, lignin derivatives); in nonaqueous media sterols, lecithin and fatty acids are effective.
Use: Wet-grinding of pigments and sulfur; preparation of ceramic glazes, oil-well drilling muds, insecticidal mixtures, carbon black in rubber, and water-insoluble dyes.
See emulsion; detergent.

dispersion. (1) A two-phase system where one phase consists of finely divided particles (often in the colloidal size range) distributed throughout a bulk substance, the particles being the disperse or internal phase, and the bulk substance, the continuous or external phase. Under natural conditions, the distribution is seldom uniform, but under controlled conditions, the uniformity can be increased

by addition of wetting or dispersing agents (surfactants) such as a fatty acid. The various possible systems are gas–liquid (foam), solid–gas (aerosol), gas–solid (foamed plastic), liquid–gas (fog), liquid–liquid (emulsions), solid–liquid (paint), and solid–solid (carbon black in rubber). Some types, such as milk and rubber latex, are stabilized by a protective colloid that prevents agglomeration of the dispersed particles by an abherent coating. Solid-in-liquid colloidal dispersions (loosely called solutions) can be precipitated by adding electrolytes that neutralize the electrical charges on the particles. Larger particles will gradually coalesce and either rise to the top or settle out, depending upon their specific gravity.
See suspension; colloid chemistry.
(2) In the field of optics, dispersion denotes the retardation of a light ray, usually resulting in a change of direction as it passes into or out of a substance, to an extent depending on the frequency. Dispersion is a critically important property of optical glass.
See refraction.

Dispersite. Water dispersions of natural, synthetic, and reclaimed rubbers and resins.
Use: Adhesives for textiles, paper, shoes, leather, and tapes; coatings for metal, paper, fabrics, and carpets; protective (strippable) for saturating paper, felt, book covers, tape, and jute pads; for dipping tire cords. Can be applied by spraying, spreading, impregnation, and saturation.

Disperson. Wettable grades of zinc, calcium, and other metallic stearates.
Use: Where easy dispersion in water is desired.

Disperson OS. An oil-soluble emulsifying agent composed of an 8% solution of a polyethenoxy compound in isopropanol. Designed especially for dispersion of oil spills in seawater. Claimed to be biodegradable and to have low toxicity for fish and other marine organisms. Amount needed said to be from 20 to 25% of the oil volume.

displacement. Chemical change in which one element enters a compound in place of another, the latter being set free.

displacement series. See activity series.

disposal, waste. See waste control; chemical waste; radioactive waste.

disproportionation. A chemical reaction in which a single compound serves as both oxidizing and reducing agent and is thereby converted into a more oxidized and a more reduced derivative. Thus, a hypochlorite upon appropriate heating yields a chlorate and a chloride, and an ethyl radical formed as an intermediate is converted into ethane and ethylene.
See transalkylation.

dissociation. The process by which a chemical combination breaks up into simpler constituents as a result of either (1) added energy, as in the case of gaseous molecules dissociated by heat, or (2) the effect of a solvent on a dissolved polar compound (electrolytic dissociation), e.g., water on hydrogen chloride. It may occur in the gaseous, solid, or liquid state, or in solution. All electrolytes dissociate to a greater or less extent in polar solvents. The degree of dissociation can be used to determine the equilibrium constant for dissociation, an important factor in ascertaining the extent of a chemical process. See ionization.

dissociation constant. An equilibrium constant for the dissociation of one species into two. Examples would be the dissociation of a complex of two or more biomolecules into its components, for instance, dissociation of a substrate from an enzyme, described as a Kd; an acid dissociating into its conjugate base and a proton, described as a Ka; the dissociation of water into protons and hydroxide ions, described as a Kw; or a base dissociating into its conjugate acid and a hydroxide ion, described as a Kb.

dissolution. Molecular dispersion of a solid in a liquid.

dissolved oxygen (DO). One of the most important indicators of the condition of a water supply for biological, chemical, and sanitary investigations. Adequate dissolved oxygen is necessary for the life of fish and other aquatic organisms, and is an indicator of corrosivity of water, photosynthetic activity, septicity, etc.
See biochemical oxygen demand.

dissolved solids. (DS).
Decomposed material contained in water. Such materials may make the water toxic or otherwise unfit to drink or use in industrial processes. Excessive amounts reduce the level of dissolved oxygen and otherwise make the water unfit to drink or to use industrially.

dissymmetry. Lack of symmetry.

distearylamine. (dioctadecylamine).
CAS: 112-99-2. $(C_{18}H_{37})_2NH$.
Properties: Solid. D 0.85, mp 69C. Almost insoluble in water.
Use: Intermediate.

2,6-distearyl-*p*-cresol. (2,6-dioctadecyl-*p*-cresol).
$(C_{18}H_{37})_2CH_3C_6H_2OH$.
Properties: A viscous, pale-yellow liquid. Refr index 1.4825–1.4855 (25C). Soluble in most nonpolar solvents. Combustible.
Use: Antioxidant, heat stabilizer for polypropylene.

distearyldimethylammonium chloride. See "Arosurf TA100" *[Evonik]*.

distearyl ether. (dioctadecyl ether).
CAS: 6297-03-6. $(C_{19}H_{37})_2O$.
Properties: Solid. Mp 58–60C, bp (decomposes).
Grade: 95% (min) purity.
Use: Electrical insulators, water repellents, lubricants in plastic molding and processing, antistatic agent, intermediate.

distearyl sulfide. (dioctadecyl sulfide; distearyl thioether).
CAS: 1844-09-3. $(C_{18}H_{37})_2S$.
Properties: Solid. Mp 68–69C, bp (decomposes), d 0.8148 (70/4C).
Grade: 95% (min) purity.
Use: Organic synthesis (formation of sulfonium compounds).

distearyl thiodipropionate. (*e,e'*-dioctadecyl thiodipropionate; thiodipropionic acid distearyl ester).
CAS: 693-36-7. $(C_{18}H_{37}OOCCH_2CH_2)_2S$.
Properties: White flakes. Mp 58–62C. Insoluble in water; very soluble in benzene and olefin polymers. Resistant to heat and hydrolysis.
Use: Antioxidant, plasticizer, softening agent.

distearyl thioether. See distearyl sulfide.

distillate. Any product of distillation, especially of petroleum.

distillation. A separation process in which a liquid is converted to vapor and the vapor then condensed to a liquid. The latter is referred to as the distillate, while the liquid material being vaporized is the charge or distilland. Distillation is thus a combination of evaporation, or vaporization, and condensation. A simple example is the condensation of steam from a tea kettle on a cold surface.
The usual purpose of distillation is purification or separation of the components of a mixture. This is possible because the composition of the vapor is usually different from that of the liquid mixture from which it is obtained. Alcohol has been so purified for generations to separate it from water, fusel oil, and aldehydes produced in the fermentation process. Gasoline, kerosene, fuel oil, and lubricating oil are produced from petroleum by distillation. It is the key operation in removing salt from seawater.
Use: Chemical analysis, in laboratory research, and for manufacture of many chemical products.
See destructive distillation; batch distillation; extractive distillation; rectification; dephlegmation; flash distillation; continuous distillation; simple distillation; reflux; fractional distillation; azeotropic distillation; vacuum distillation; molecular distillation; hydrodistillation.

distilled water.
CAS: 7732-18-5. H_2O.

Water that has been purified by boiling the water, separating off the steam, and condensing it back to a liquid in a clean container. This not only leaves the impurities behind but it is also sterile.

disulfide bridge. A covalent cross-link between two polypeptide chains formed by a cystine residue (two Cys residues oxidized together).

disulfiram. See tetraethylthiuram disulfide.

3,5-disulfobenzoic acid.
CAS: 121-48-2. $C_6H_3(HSO_3)_2COOH$.
Properties: White powder. Soluble in water.
Grade: CP.
Use: Intermediate for detergents, dyes, and pharmaceuticals.

disulfoton. (Di-Syston; O,O-diethyl S-[2-(ethylthio)ethyl]phosphorodithioate).
CAS: 298-04-4. $(C_2H_5O)_2P(S)SCH_2CH_2SCH_2CH_3$.
Properties: Pure compound: yellow liquid; technical compound: brown liquid. Bp 108C (0.01 mm Hg), d 1.144 (20C). Insoluble in water.
Hazard: Toxic by ingestion and inhalation, cholinesterase inhibitor, absorbed by skin, use may be restricted. Questionable carcinogen.
Use: Systemic insecticide, acaricide.

disulfur decafluoride. See sulfur pentachloride.

disulfuryl chloride. See pyrosulfuryl chloride.

"Di-Syston" [Bayer]. TM for disulfoton. See disulfoton.

diterpenoid. Compound of four 2-methylbutane units, arranged according to the isoprene rule, with additional functionality such as ethylenic unsaturation and carboxyl, phenol, and alcohol groups.

ditetradecylamine. See dimyristylamine.

ditetradecyl ether. See dimyristyl ether.

ditetradecyl sulfide. See dimyristyl sulfide.

"Dithane" [Dow Agrosciences]. TM for agricultural fungicides based on salts of ethylene bis-dithiocarbamate. Supplied in zinc, manganese, and sodium forms as wettable powder or liquid concentration.

1,4-dithiacycloheptyliden-6-iminyl n-methyl-carbamate.
CAS: 24031-96-7. $C_7H_{12}N_2O_2S_2$.
Hazard: A poison by ingestion.
Use: Agricultural chemical.

dithiadenoxide.
CAS: 153049-45-7. $C_{17}H_{19}NOS_2 \cdot C_4H_4O_4$.
Hazard: Moderately toxic by ingestion.

dithiadenoxid hydrogen maleate. See dithiadenoxide.

1,4-dithiane. (diethylene disulfide).
CAS: 505-29-3

$$\overline{SCH_2CH_2SCH_2CH_2.}$$

Properties: White crystals. Bp 115.6C (60 mm Hg), mp 108C. Volatile in steam. Soluble in alcohol and ether; slightly soluble in water.
Derivation: Interaction of dichloroethyl sulfide with sodium or potassium sulfide.
Use: Organic synthesis.

dithiane methiodide.
$C_4H_8S_2 \cdot CH_3I$.
Properties: Crystals. Mp 174C. Soluble in hot water; slightly soluble in alcohol; insoluble in ether.
Derivation: Interaction of dichloroethyl sulfide and methyliodide.

dithianone.
CAS: 3347-22-6. $C_{14}H_4N_2O_2S_2$.
Properties: Brownish solid. Mp 220C. Insoluble in water; soluble in acetone, chloroform, and dioxane.
Hazard: Toxic by ingestion.
Use: Fungicide.

(((((1,4-dithian-2-ylideneamino)oxy)carbonyl)methylamino)thio)methylcarbamic acid, (1-methylethylidene)di-4,1-phenylene ester.
CAS: 87767-48-4. $C_{31}H_{38}N_6O_8S_6$.
Hazard: Moderately toxic by ingestion.
Use: Agricultural chemical.

β-β′-dithiobisalanine. See cystine.

2,2′-dithiobis(benzothiazole). (benzothiazolyl disulfide; benzothiazyl disulfide; 2-mercaptobenzothiazyl disulfide; MBTS).
CAS: 120-78-5. $(C_6H_4SCN)_2S_2$.
Properties: Pale-yellow, free-flowing; odorless powder. D 1.54, mp 168C. Insoluble in water; sparingly soluble in organic solvents.
Use: Primary accelerator in natural and nitrile rubber and SBR, plasticizer and vulcanization retarder in neoprene type G, cure modifier in neoprene type W, oxidation cure activator in butyl. For extruded and molded goods, tires and tubes, wire and cable, and sponge.

2,4-dithiobiuret.
CAS: 541-53-7. $C_2H_5N_3S_2$.
Properties: Colorless crystals. Bulk d 1.52, decomposes 181C. Partially soluble in boiling water, acetone, and "Cellosolve" [Union].
Derivation: Reaction of dicyandiamide and hydrogen sulfide.

Hazard: Toxic by ingestion and inhalation, causes respiratory paralysis.
Use: Intermediate in manufacture of insecticides, synthetic resins; plasticizer, rubber accelerator.

dithiocarbamic   acid.   (aminodithioformic acid).
CAS: 594-07-0. NH_2CS_2H.
Properties: Colorless needles. Soluble in alcohol.
Use: The metal salts of the acid are important as strong (ultra) rubber accelerators, as are the thiuram disulfide derivatives. Seed disinfectant.
See thiuram; selenium diethyldithiocarbamate; zinc dibutyldithiocarbamate; zinc diethyldithiocarbamate; ziram.

2,2'-dithiodibenzoic acid.
$(C_6H_4COOH)_2S_2$.
Properties: Tan to gray powder. Mp 280C (min).
Use: Intermediate for pharmaceuticals.

4,4'-dithiodimorpholine.
CAS: 103-34-4. $C_4H_8ONSSNOC_4H_8$.
Properties: Gray to tan powder. Mp 122C min, d 1.36 (25C).
Hazard: Toxic by ingestion, inhalation, and skin absorption.
Use: Rubber accelerator, fungicide.

1,2-dithioglycerol.   See 2,3-dimercaptopropanol.

1,3-dithiolane-2,4-dione, 5,5-dimethyl-, 2-(dimethyhydrazone), 4-(o-((methylethyl ((trichloromethyl)thio)amino)carbonyl) oxime).
CAS: 71108-14-0.   mf: $C_{10}H_{15}Cl_3N_4O_2S_3$.
Hazard: Moderately toxic by ingestion.
Use: Agricultural chemical.

1,3-dithiolan-4-one, 5,5-dimethyl-2-(1,1-(dimethylethyl)imino)-, o-((methylamino) carbonyl) oxime.
CAS: 71108-08-2. $C_{11}H_{19}N_3O_2S_2$.
Hazard: A poison by ingestion.
Use: Agricultural chemical.

o-(1,3-dithiolan-2-yl)phenyl   dimethylcarbamate.
CAS: 21709-44-4. $C_{12}H_{15}NO_2S_2$.
Hazard: A poison by ingestion.
Use: Agricultural chemical.

6,8-dithiooctanoic acid.   See dl-α-lipoic acid.

dithiooxamide.   (rubeanic acid).
CAS: 79-40-3. $SC(NH_2)C(NH_2)S$.
Properties: Stable, orange-red powder. Decomposes at 140C. Insoluble in water; soluble in acetone and chloroform. Forms highly colored stable complexes with many metal ions; can form a series of N,N'-derivatives.
Use: Chemical intermediate.

dithiothreitol.   (Cleland's reagent).
Properties: Solid. Mp 42–43.5C, 99% pure.
Available forms: 1-, 5-, 25-g quantities.
Use: Reducing agent for proteins and enzymes, biochemical research.

dithizone.   (phenyldiazinecarbothioic acid 2-phenylhydrazide;   (phenylazo)thioformic acid 2-phenylhydrazide;   diphenylthiocarbazone; 1-(aniline)-3-phenyliminothiourea).
CAS: 60-10-6. $C_{13}H_{12}N_4S$.
Chelating agent.
Properties: reagent.
Use: In the analysis of the heavy metals Co, Cu, Pb, and Hg.
Hazard: Toxic; causes diabetes.

dithymol diiodide.   See thymol iodide.

ditiocarb sodium.   (diethylcarbamodithioic acid sodium salt; diethyldithiocarbamic acid sodium salt; diethyldithiocarbamate; dithiocarb, sodium; Ddc; ddtc; dedc; dedt; dtc).
CAS: 148-18-5. $C_5H_{11}NNaS_2$.
Properties: Water-soluble, chelating agent; strong affinity for copper, mercury, and zinc.
Use: Antidote for nickel and cadmium poisoning.

1,4-di-p-toluidinoanthraquinone.   (FD&C Green No. 6).
CAS: 128-80-3. $C_{14}H_6O_2(NHC_6H_4CH_3)_2$.
Properties: Solid. Mp 213C.
Use: Dye, approved for restricted use in drugs and cosmetics.

di-o-tolylcarbodiimide.
CAS: 1215-57-2. $(CH_3)C_6H_4N:C:NC_6H_4(CH_3)$.
Properties: Yellow to red-brown liquid; faint acrid odor. D 1.063 (25/4C), bp 140C (0.9 mm Hg), refr index 1.6248 (20C). Soluble in organic solvents such as chloroform, carbon tetrachloride, benzene, and dioxane. Combustible.
Hazard: Toxic by inhalation and ingestion, irritant to skin and eyes.
Use: Surface coatings, textile processing, stabilizers of polyesters and urethanes, scavenger compounds for materials sensitive to active hydrogen.

N,N-di-o-tolylethylenediamine.
CAS: 94-92-8. $CH_3C_6H_4NHCH_2CH_2NHC_6H_4CH_3$.
Properties: Light-brown to purple granular solid. D 1.13, softens at approximately 57C. Insoluble in water; soluble in acetone, ethylene dichloride, benzene, and gasoline.
Use: Antioxidant for light-colored rubber goods.

di-o-tolylguanidine.   (DOTG accelerator).
CAS: 97-39-2. $(CH_3C_6H_4NH)_2CNH$.

Properties: White powder. D 1.10, mp 179C. Non-hygroscopic; very slightly soluble in water; soluble in warm alcohol, from which it crystallizes on cooling.
Derivation: Desulfurization of di-*o*-tolylthiourea with a lead compound in the presence of ammonia.
Hazard: Toxic by ingestion.
Use: Basic rubber accelerator.

2,7-di-*p*-tolylnaphthylenediamine.
$C_{10}H_6(NHC_6H_4CH_3)_2$.
Properties: Fine needles. Mp 237C. Sparingly soluble in alcohol; insoluble in water.
Derivation: By heating 2,7-dihydroxynaphthalene with *p*-toluidine and *p*-toluidine hydrochloride.

1,3-di-*p*-tolylphenylenediamine.
$C_6H_4(NHC_6H_4CH_3)_2$.
Properties: Long needles. Mp 137C. Soluble in alcohol and ether; insoluble in water.
Derivation: By heating resorcinol and *p*-toluidine in the presence of zinc chloride.

di-*o*-tolythiourea. (DOTT).
$SC(NHC_6H_4CH_3)_2$.
Properties: Colorless, crystalline leaflets; pungent odor. Mp 144–148C. Not hygroscopic; soluble in alcohol, ether, and benzene; insoluble in water.
Derivation: By the interaction of *o*-toluidine and carbon disulfide.
Hazard: Toxic by ingestion.
Use: Metal-pickling inhibitor.

ditridecyl phthalate. (DTDP; Jayflex).
CAS: 119-06-2. $C_6H_4(COOC_{13}H_{27})_2$.
Use: Plasticizer.

ditridecyl thiodipropionate. (3,3′-
tetramethylnonyl thiodipropionate; thiodipropionic acid ditridecyl ester).
$(C_{13}H_{27}OOCCH_2CH_2)_2S$.
Properties: Colorless liquid. D 0.932 (25C). Insoluble in water; soluble in most organic solvents. Combustible.
Use: Stabilizer, plasticizer, and softening agent for plastics; lubricant additive.

diuretic.
A drug that promotes water elimination from the body via kidney function.

diuron. (3-(3,4-dichlorophenyl)-1,1-
dimethylurea).
CAS: 330-54-1. $C_6H_3Cl_2NHCON(CH_3)_2$.
Properties: White, crystalline solid. Mp 159C, vap press 2×10^{-7} mm Hg (30C). Very low solubility in hydrocarbon solvents; approximately 42 ppm at 25C in distilled water. Stable toward oxidation and moisture; decomposes at 180C.
Hazard: Toxic. Upper respiratory tract irritant. Questionable carcinogen.

Use: Pre-emergence herbicide, sugarcane flowering suppressant.

divalent carbon. See carbene.

divanadyl tetrachloride. See vanadyl chloride.

divinylacetylene.
CAS: 821-08-9. $H_2C:CHC:CCH:CH_2$.
Trimer of acetylene formed by passing it into a hydrochloric acid solution containing a metallic catalyst.
Use: Intermediate in manufacture of neoprene.

divinylbenzene. (DVB; vinylstyrene).
CAS: 1321-74-0. $C_6H_4(CH:CH_2)_2$.
Exists as *o*-, *m*- and *p*-isomers. The commercial form contains the three isomeric forms together with ethylvinylbenzene and diethylbenzene.
Properties: (Pure *m*-isomer) Water-white liquid, easily polymerized. Bp 199.5C, fp −66.90C, d 0.9289 (20C), viscosity 1.09 cP (20C), refr index 1.5772 (20C), flash p 165F (73.9C). (Divinylbenzene 55%) Pale-straw–colored liquid. Fp −87C, bp 195C, d 0.918 (25/25C). Insoluble in water; soluble in methanol and ether. Combustible.
Grade: 50–60%, 20–25%.
Hazard: Velocity of polymerization involves an explosion risk. If uninhibited, store at <90F (32.3C). Toxic by inhalation. Upper respiratory tract irritant.
Use: Polymerization monomer for special synthetic rubbers, drying oils, ion exchange resins, casting resins, and polyesters.
Note: Should contain inhibitor when stored or shipped.

divinyl ether. Legal label name for vinyl ether.

divinyl oxide. See vinyl ether.

3,9-divinylspirobi-*m*-dioxane. (3,9-divinyl-
2,4,8,10-tetraoxaspiroundecane).
CAS: 78-19-3. $[CH_2:CHCH(OCH_2)_2]_2C$.
Bicyclic.
Properties: Liquid. Mp 42C, 120C (2 mm Hg), d 1.251 (20/20C), flash p 290F (143C) (COC). Slightly soluble in water. Combustible.
Use: Intermediate and monomer.

divinyl sulfide.
$(CH_2:CH)_2S$.
Properties: Mobile liquid; unpleasant odor. Polymerizes readily. D 0.9174 (15C), bp 85–86C. Combustible.
Derivation: Interaction of dichlorodiethyl sulfide and an alcoholic solution of potassium hydroxide.
Grade: Technical.

divinyl sulfone.
CAS: 77-77-0. $CH_2:CHSO_2CH:CH_2$.

Properties: Liquid. D 1.1788 (20/20C), bp 234C, fp −26C, soluble in water, flash p 255F (124C). Combustible.
Use: Monomer used in manufacture of polymers with diols, urea, and malonic esters; shrinkage control agent (textiles).

divinyltin dichloride.
CAS: 7532-85-6. $C_4H_6Cl_2Sn$.
Hazard: Moderately toxic by ingestion. Low toxicity by inhalation. A severe eye irritant.

dixanthogen.
CAS: 502-55-6. $(C_2H_5OC=S)_2S_2$.
Properties: Yellow crystals; strong odor. Mp 30C. Insoluble in water; soluble in benzene and ether.
Use: Herbicide, insecticide, parasiticide.

di-*o*-xenylphenyl phosphate.
$(C_6H_5C_6H_4O)_2(C_6H_5O)PO$.
Properties: Liquid. D 1.20 (60C), refr index 1.603–5 (60C), boiling range 285–330C (5 mm Hg), flash p 250F (121C). Insoluble in water. Combustible.
Use: Plasticizer.

di-*p*-xylylene.
CAS: 1633-22-3. $(-CH_2C_6H_4CH_2-)_2$.
Properties: Stable, white crystals. Mp 280C.
Derivation: Pyrolysis of *p*-xylene at 398C in the presence of steam. An organic quench (benzene or toluene) gives the dimer in yields of 10–15%.
Use: Parylene resins.

dixylylethane.   (DXE).
$[C_6H_3(CH_3)_2]_2HCCH_3$.
Properties: Colorless liquid. Bp 315C, pour p −34C, flash p 323F (162C), d 0.97.
Use: Dielectric fluid, heat-transfer medium, solvent, chemical intermediate.

DKP.   Abbreviation for dipotassium phosphate.
See potassium phosphate, dibasic.

DL (DL).
A prefix indicating that a compound contains equal parts of D and L stereoisomers. Small capitals are often used to indicate the right- and left-handed structure of such molecules; they do not indicate rotation.
See *dl*; racemic substance.

dl.   (dl).
A prefix denoting a crystal that rotates the plane of polarized light equally to both left and right, resulting in optical inactivity. The symbol ñ is preferably used.

"DLC" *[Natrochem]*.   TM for dry liquid concentrate. Created by uniformly applying any of hundreds of oils and resins over the surface area of a dry carrier powder to yield a free-flowing powder.

Use: Activators, plasticizers, polymers, waxes, resins, tackifiers, processing aids.

DM.   See diphenylamine chloroarsine.

D2M.   Diethylene glycol dimethyl ether.
Use: Anhydrous reaction medium for organometallic syntheses.

DMA.   Abbreviation for dimethylamine or disodium methyl arsonate.

DMAC.   Abbreviation for dimethyl acetamide.

DMAPMA.   Abbreviation for dimethylaminopropylmethacryl amide.

DMB.   Abbreviation for dimethoxybenzene.
See hydroquinone dimethyl ether.

DMC.   (1) Abbreviation for dichlorophenylmethylcarbinol.
See di(*p*-chlorophenyl)ethanol.
(2) Abbreviation for β-dimethylaminoethyl chloride hydrochloride.

DMDT.   Abbreviation for dimethoxydiphenyltrichloroethane.
See methoxychlor.

DMF.   Abbreviation for dimethyl formamide.

DMH.   Abbreviation for dimethylhydantoin.

DMHF.   Abbreviation for dimethylhydantoin-formaldehyde.
See methyloldimethylhydantoin.

DMN.   See *N*-nitrosodimethylamine.

DMNA.   See *N*-nitrosodimethylamine.

DMP.   Dimethylaminomethyl-substituted phenols.
Use: Chemical intermediate, curing agents for epoxy resins, antioxidants.

DMPA.   See *O*-(2,4-dichlorophenyl)-*O*-methyl isopropylphosphoramidothioate,   dimethylol propionic acid.

DMPC.   Abbreviation for 1-dimethylamino-3-propyl chloride.

DMSO.   Abbreviation for dimethyl sulfoxide.

DMT.   Abbreviation for dimethyl terephthalate.

DMU.   Abbreviation for dimethylurea and dimethylolurea.

DNA.   Abbreviation for deoxyribonucleic acid.

DNA bank.   A service that stores DNA extracted from blood samples or other human tissue.

DNA chimera.   A DNA containing genetic information derived from two different DNA molecules.

DNA cloning.   The production of large numbers of identical DNA molecules or cells from a single ancestral DNA molecule, nucleus, or cell.

DNA library.   A random collection of cloned DNA fragments designed to include all or most of the genome of a given organism; also called a genomic library.

DNA ligase.   An enzyme that creates a phosphodiester bond between the 3′ end of one DNA segment and the 5′ end of another.

DNA polymerase.   An enzyme that catalyzes template-dependent synthesis of DNA from its dNTP precursors.

DNA primase.   An enzyme that catalyzes synthesis of the short strands of RNA that initiate the synthesis of DNA strands.

DNA probe.   See probe.

DNA repair genes.   Genes encoding proteins that correct errors in DNA sequencing.

DNA replicase system.   The entire complex of enzymes and specialized proteins required in biological DNA replication.

DNA replication.   The use of existing DNA as a template for the synthesis of new DNA strands. In humans and other eukaryotes, replication occurs in the cell nucleus.

DNase.   Deoxyribonuclease, a class of enzymes that digest DNA. The most common is DNase I, an endonuclease that digests both single- and double-stranded DNA.

DNA sequence.   The relative order of base pairs, whether in a DNA fragment, gene, chromosome, or an entire genome.
See base sequence analysis.

DNA supercoiling.   The coiling of DNA upon itself into a more condensed structure.

DNA transposition.   The movement of a gene or set of genes from one site in the genome to another.

DNBP.   Abbreviation for dinitro-*o-sec*-butylphenol.

DNC.   Abbreviation for dinitrocresol.

DNFB.   See 2,4-dinitrofluorobenzene.

DNOC.   Abbreviation for 4,6-dinitro-*o*-cresol.

DNOCHP.   Abbreviation for dinitro-*o*-cyclohexylphenol.

DNODA.   Abbreviation for di(*n*-octyl,*n*-decyl) adipate.

DNODP.   Abbreviation for di(*n*-octyl,*n*-decyl) phthalate.

DNP.   Abbreviation for dinonyl phthalate.

DNPC.   See 2,6-dinitro-*p*-cresol.

DNPD.   Abbreviation for *N,N′-di-β-naphthyl*-p-phenylenediamine.

DNPT.   Abbreviation for dinitrosopentamethylenetetramine.

DNT.   Abbreviation for dinitrotoluene.

DO.   Abbreviation for dissolved oxygen.

DOA.   Abbreviation for dioctyl adipate.
See di(2-ethylhexyl) adipate.

dob.   (bromo-dma; 1-(2,5-dimethoxy-4-bromophenyl)-2-aminopropane; 2,4-dihydroxoybenzoic acid).
CAS: 89-86-1. $C_7H_6O_4$.
A powerful serotonin 5-$^{ht}_2$ agonist.
Hazard: Dangerous; potentially lethal; extremely toxic; hallucinations; loss of touch with reality; cold and painful extremities; peripheral arterial spasm; gangrene.
Use: Synthetic hallucinogenic agent.

DOC.   Abbreviation for dichromate oxygen consumed.
See oxygen consumed.

***n*-docosane.**
$C_{22}H_{46}$.
Properties: Solid. Mp 45.7C, bp 230C (15 mm Hg), d 0.778 (45/5C), refr index 1.4400 (45C). Combustible.
Grade: 95% (min) purity.
Use: Organic synthesis, calibration, temperature-sensing devices.

docosanoic acid.   See behenic acid.

1-docosanol.   See behenyl alcohol.

***cis*-13-docosenoic acid.**   See erucic acid.

doctor knife. A metal straightedge (or equivalent) that detaches adhering product from a rotating drum or roll or removes excess coating from a web. No cutting is involved.

Use: To scrape dried films, such as soap flakes, from drum dryers; to remove ink from printing rolls; clean residual paper stock from rolls and to provide a crimped paper for toweling, etc.; and to remove soft or sticky rubber and adhesive compositions from mixing rolls. An interesting application is the so-called air doctor (air knife); it is actually a stream of air delivered by a turbine blower through a tube with a discharge slot. Located at the takeoff end of a paper-coating machine, it efficiently levels off the coating suspension across the full width of the web.

doctor treatment. A method of improving or "sweetening" the odor of gasoline, petroleum solvents, or kerosene. A doctor solution of sodium plumbite, Na_2PbO_2, is made by dissolving litharge in sodium hydroxide solution; the feed to be sweetened is passed through the doctor solution. The action of the sodium plumbite and the lead sulfide formed from it, in conjunction with free sulfur (either naturally present or added), converts the malodorous mercaptans to the pleasanter disulfides.

dodecacarbonium chloride.
$C_{23}H_{41}ClN_2O$.
Properties: Crystalline solid; acidic taste. Mp 145C. Insoluble in benzene and acetone; soluble in water and alcohol.
Hazard: Toxic by ingestion.
Use: Biocide, disinfectant.

dodecahydrosqualane. See squalane.

dodecamethylpentasiloxane.
CAS: 141-63-9. $C_{12}H_{36}O_4S_5$.
Properties: Inert, colorless liquid. Fp about −75C, bp 230C, refr index 1.40. Soluble in benzene and low-molecular-weight hydrocarbons. Little change in viscosity with temperature.
Use: Silicone fluids, antifoam agent in lubricating oils.

dodecanal. See lauryl aldehyde.

n-**dodecane.** (dihexyl).
CAS: 112-40-3. $CH_3(CH_2)_{10}CH_3$.
Properties: Colorless liquid. D 0.749 (20/4C), fp −10C, bp 213C, refr index 1.4221 (20C), flash p 160F (71.1C), autoign temp 400F. Soluble in alcohol, acetone, and ether; insoluble in water. Combustible.
Grade: 95%, 99%, 99.7 mole %.
Use: Solvent, organic synthesis, distillation chaser, jet fuel research.

1,12-dodecanedioic acid.
CAS: 693-23-2. $HOOCC_{10}H_{20}COOH$.

A 12-carbon, straight-chain, saturated aliphatic dibasic acid.
Properties: White, crystalline powder. Mp 130–132C. Soluble in hot toluene, alcohols, and hot acetic acid; slightly soluble in hot water.
Use: Intermediate for plasticizers, lubricants, adhesives, polyesters, etc.

dodecanoic acid. See lauric acid.

n-**dodecanol.** See lauryl alcohol.

trans-**2-dodecen-1-al.**
CAS: 20407-84-5. $C_{12}H_{22}O$.
Properties: Slightly yellow liquid; fatty, citruslike odor. D: 0.839–0.049, refr index: 1.462. Sol in alc, fixed oils; insol in water.
Use: Food additive.

dodecene.
CAS: 6842-15-5. $C_{12}H_{24}$.
Many possible isomers.
See 1-dodecene; tetrapropylene; and sodium dodecylbenzenesulfonate.

1-dodecene. (α-dodecylene; tetrapropylene).
CAS: 6842-15-5. $H_2C:CH(CH_2)_9CH_3$.
Properties: Colorless liquid. D 0.7600 (20/4C), fp −33.6C, bp 213C, refr index 1.4327 (20C). Soluble in alcohol, acetone, ether, petroleum, coal tar solvents; insoluble in water. Combustible.
Hazard: Irritant and narcotic in high concentration.
Use: Flavors, perfumes, medicine, oils, dyes, resins.

(e)-9-dodecenol acetate.
CAS: 35148-19-7. $C_{14}H_{26}O_2$.
Hazard: Low toxicity by ingestion and skin contact.
Source: Natural product.

dodecenylsuccinic acid.
CAS: 11059-31-7. $HOOCCH(C_{12}H_{23})CH_2COOH$.
Properties: Extremely viscous liquid. Practically insoluble in water; miscible with oil. Combustible.
Use: Synthesis, corrosion inhibitor in oils, waterproofing.

dodecenylsuccinic anhydride. (DDS).
CAS: 25377-73-5.

$$\overset{\displaystyle\lceil\qquad\qquad\rceil}{C_{12}H_{23}CHCOOOCCH_2}.$$

The normal and at least two branched-chain dodecenyls are used commercially. The following properties are those of a branched-chain compound.
Properties: Light-yellow, clear, viscous oil. Bp 180–182C (5 mm Hg), d 1.002 (25C), flash p 352F (177C) (COC), viscosity 400 cP (20C), 15.5 cP (70C). Combustible.

Use: Alkyd, epoxy and other resins; anticorrosive agents; plasticizers; wetting agents for bituminous compounds.

dodecyl acetate. (acetate C-12; lauryl acetate).
CAS: 112-66-3. $C_{12}H_{25}OOCCH_3$.
Properties: Colorless liquid; fruity odor. D 0.860–0.862, refr index 1.430–1.433, bp 150.5–151.5C. Soluble in three volumes of 80% alcohol. Combustible.
Use: Perfumery, flavoring.

n-dodecyl alcohol. See lauryl alcohol.

dodecyl aldehyde. See lauryl aldehyde.

dodecylaniline.
CAS: 68411-48-3. $C_{12}H_{25}C_6H_4NH_2$ (probably the p-isomer).
Properties: Colorless liquid. D 0.907–0.912 (25/25C), bp 340–350C. Oil-soluble aromatic amine; insoluble in water; soluble in most organic solvents. Combustible.
Hazard: See aniline.
Use: Intermediate.

dodecylbenzene. (detergent alkylate).
CAS: 29986-57-0. $C_{12}H_{25}C_6H_5$.
A commercial blend of isomeric, predominantly monoalkyl benzenes. The side chains are saturated, averaging 12 carbon atoms. Flash p 285F (140C). Combustible.
Derivation: Alkylation of benzene with isomeric dodecenes, obtained usually by polymerization of propylene.
Hazard: Toxic by ingestion.
Use: Detergents of ABS or LAS type.
Linear (normal) dodecylbenzene is called sodium dodecylbenzenesulfonate.
See tetrapropylene; sodium dodecylbenzenesulfonate.

dodecylbenzenesulfonic acid. (DDBSA).
See sodium dodecylbenzenesulfonate.

dodecylbenzenesulfonic acid triethanolamine salt. See triethanolamine dodecylbenzene sulfonate.

dodecylbenzyl mercaptan.
$C_{12}H_{25}C_6H_4CH_2SH$.
Offered as branched-chain isomers.
Properties: Light-yellow oil; unpleasant odor. Bp approximately 150C (0.5 mm Hg). Soluble in acetone, benzene, and heptane; insoluble in water and alcohol. Combustible.
Hazard: Irritant.
Use: Polymerization modifier, intermediate, metal-cleaning and polishing compounds.
See -thiol.

n-dodecyl bromide. See lauryl bromide.

6-dodecyl-1,2-dihydro-2,2,4-trimethylquinoline.
$C_{12}H_{25}C_9H_5N(CH_3)_3$.
Properties: Dark viscous liquid. D 0.93 (45C). Combustible.
Use: Rubber antioxidant (flex-cracking).

dodecyldimethyl(2-phenoxyethyl)ammonium bromide. See domiphen bromide.

n-dodecylguanidine acetate. See dodine.

n-dodecyl mercaptan. (DDM).
CAS: 112-55-0. $C_{12}H_{25}SH$.
Many isomers of dodecyl mercaptan are possible, and a variety of these occur in the technical material known as tert-dodecyl mercaptan, or lauryl mercaptan, or simply dodecyl mercaptan.
See lauryl mercaptan.
Properties: (n-isomer) Colorless liquid. Bp 143C, refr index 1.4589, flash p 262F (127C). Soluble in ether and alcohol; insoluble in water. Combustible.
See -thiol.

4-dodecyloxy-2-hydroxybenzophenone.
CAS: 2985-59-3. $C_{12}H_{25}OC_6H_3(OH)C(O)C_6H_5$.
Properties: Pale-yellow flakes. Setting point 43C. Soluble in polar and nonpolar organic solvents.
Use: UV light inhibitor in plastics.

dodecylphenol.
CAS: 27193-86-8. $C_{12}H_{25}C_6H_4OH$.
A mix of isomers.
Properties: Straw-colored liquid; phenolic odor. D 0.94 (20/20C), flash p 325F (162.7C), boiling range 310–335C. Soluble in water. Combustible.
Hazard: See phenol.
Use: Solvent, intermediate for surface-active agents, oil additives, resins, fungicides, bactericides, dyes, pharmaceuticals, adhesives, rubber chemicals.

dodecyltrichlorosilane.
CAS: 4484-72-4. $C_{12}H_{25}SiCl_3$.
Properties: Colorless to yellow liquid. Bp 288C, d 1.026 (25/25C), refr index 1.4521 (25C). Readily hydrolyzed with liberation of hydrogen chloride.
Derivation: By Grignard reaction of silicon tetrachloride and dodecylmagnesium chloride.
Hazard: Strong irritant to tissue.
Use: Intermediate for silicones.

dodecyltrimethylammonium chloride.
CAS: 112-00-5. $C_{12}H_{25}N(CH_3)_3Cl$.
Properties: White liquid. Surface tension 0.1% in water, 33 dynes/cm (25C). Soluble in water and alcohol.
Use: Germicides, fungicides, textile fiber softeners, cationic emulsifiers, flotation reagents, mildew proofing.

dodine. (N-dodecylguanidine acetate).
CAS: 2439-10-3. $C_{12}H_{25}NHC(:NH)NH_2 \cdot CH_3COOH$.

Properties: Crystals. Mp 136C. Soluble in hot water and alcohol.
Hazard: Strong irritant to eyes and skin at >50% concentration.
Use: Fungicide.

DOE. Abbreviation for the U.S. Department of Energy.

Doebner-Miller reaction (Beyer method for quinolines). Synthesis of quinolines from primary aromatic amines and α, β-unsaturated carbonyl compounds in acid catalysis. The carbonyl compounds may be prepared in situ from two molecules of aldehyde or an aldehyde and methyl ketone. The latter is known as the Beyer method.

Doebner reaction. Formation of substituted cinchoninic acids from aromatic amines on heating with aldehydes and pyruvic acid.

Doering-LaFlamme allene synthesis. Treatment of an olefin with bromoform in the presence of an alkoxide and reaction of the 1,1-dibromocyclopropane with an active metal to produce an allene in which a carbon atom is inserted between the carbon atoms of the original olefinic double bond.

DOF. See di(2-ethylhexyl) fumarate.

β-dolabrin.
CAS: 4570-11-0.
Properties: A minor component of Thujopsis dolabrata SIEB. et ZUCC. var. hondai MAKINO mf: $C_{10}H_{12}O_2$.
Hazard: A poison.

dolichol. ((6E,10E,14E,18E,22Z,26E,30E,34E, 38E,42E,46E,50E,54E,58Z,62E,66E,70E,74E)-3,7,11,15,19,23,27,31,35,39,43,47,51,55,59,63,67, 71,75,79-icosamethyloctaconta-6,10,14,18,22,26, 30,34,38,42,46,50,54,58,62,66,70,74,78-nonadecaen-1-ol).
$C_{100}H_{164}O$.
A polyisoprenoid lipid carrier that participates in the assembly of N-glycans and gpi anchors. It is found in animal tissues that contains about 20 isoprene residues, the one carrying the alcohol group being saturated.

Dollo's law. The rule that any substantial evolutionary change is virtually irreversible because genetic changes are not likely to be reversed in an order exactly opposite to the order in which they originally developed.

dolomite.
CAS: 16389-88-1. $CaMg(CO_3)_2$.
A carbonate of calcium and magnesium.
Properties: Color gray, pink, white; vitreous luster. D 2.85, Mohs hardness 3.5-4; good cleavage in three directions. Similar to calcite but less soluble in acids (reacts with acid when powdered or with hot acid). Noncombustible.
Use: Refractory for furnaces; manufacture of magnesium compounds and magnesium metals; as building material; in fertilizers; stock feeds; papermaking; ceramics; mineral wool; removal of sulfur dioxide from stack gases.

dolphin oil. See porpoise oil.

DOM. A hallucinogenic drug.
Use: Restricted by FDA.
See di(2-ethylhexyl) maleate.

domain. A distinct structural unit of a polypeptide. Domains may have separate functions and some are known to fold as independent, compact units, such as the Fc and Fab regions of an antibody molecule.

domic acid.
An amino acid produced by certain species of algae. It binds to a receptor involved in the regulation and transport of ions across the cell membranes of nerve cells. It alsters a receptor such that an uncontrolled flux of ions is permitted with consequent damage to and eventual death of the nerve cell.

dominant. An allele that is almost always expressed, even if only one copy is present.
See gene; genome.

domiphen bromide. (dodecyldimethyl[2-phenoxyethyl]-ammonium bromide).
CAS: 538-71-6. $C_6H_5OC_2H_4N(C_{12}H_{25})(CH_3)_2Br$.
A quaternary ammonium salt.
Properties: Crystals. Mp 112C. Soluble in water and organic solvents.
Use: Medicine (anti-infective).

Donnan hydrolysis. When a neutral salt in water exchanges its cation for hydrogen ions from the water to make the latter alkaline.

donor. An atom that furnishes a pair of electrons to form a covalent bond or linkage with another atom, called the acceptor.
See bond, chemical; Lewis electron theory.

DOP. Abbreviation for dioctyl phthalate.
See di(2-ethylhexyl) phthalate.

dopa. (levodopa; 3-hydroxytyrosine; 3-(3,4-dihydroxyphenalalanine; β-(3,4-(dihydroxy-phenyl)-α-alanine; dihydroxyphenylalanine; (2S)-2-amino-3-(3,4-dihydroxyphenyl)propanoic acid).
CAS: 59-92-7. $C_9H_{11}NO_4$.
An oxidation product of tyrosine. The naturally occurring form of dihydroxypheylalanine and the immediate precursor of dopamine. It is rapidly

taken up by dopaminergic neurons and converted to dopamine.
Properties: A crystalline amino acid.
Derivation: Occurs in the posterior salivary glands of octopod mollusks (class cephalopoda); various fruits and vegetables; and in the seedlings, pods, and beans of *Vicia faba* (broad beans) and *stizolobium deeringianum* (velvet beans).
Use: In the treatment of parkinsonian disorders.

L-dopa.   See L-dihydroxyphenylalanine.

dopamine.   (4-(2-aminoethyl)-1,2-benzenediol; 4-(2-aminoethyl)pyrocatechol; 3-hydroxytyramine; 3,4-dihydroxyphenethylamine; α-(3,4-dihydroxyphenyl)-β-aminoethane; 4-(2-aminoethyl)benzene-1,2-diol).
CAS: 51-61-6. $C_8H_{11}NO_2$.
The decarboxylation product of Dapa. An endogenous monoamine transmitter with α- and β-adrenergic activity and is an intermediate in the biosynthesis of adrenalin (epinephrine) and noradrenalin (norepinephrine). A major transmitter in the extrapyramidal system of the brain and important in regulating movement. Dopamine is an inhibitory transmitter in many animals. Dopamine receptors in higher vertebrates reside on pre- and postsynaptic sites within the central nervous system, in the hypophysis, the area posterma (chemoreceptor trigger zone), and in the peripheral vasculature.
Derivation: From tyrosine.
Hazard: Highly toxic; poison; teratogen; mutagen; causes nausea, vomiting, nervousness, irritability, tachycardia, cardiac arrhythmias, dilated pupils, blurred vision, chills, pallor or cyanosis, fever, suicidal behavior, spasms, and convulsions.

dopamine hydroxylase.   (dbh; 1,2,3,4,5,6-hexachlorocyclohexane).
CAS: 9013-38-1. $C_6H_6Cl_6$.
A copper-containing enzyme that catalyzes the synthesis of norepinephrine from dopamine.

dope.   (1) Sizing formulation consisting of solutions of nitrocellulose, cellulose acetate, or other cellulose derivations applied to crepe yard to set the twist and assist creping, and to leather to form a high-gloss finish. (2) A combustible, such as wood pulp, starch, sulfur, etc., used in "straight" dynamites. (3) A trace impurity introduced into ultrapure crystals to obtain desired physical properties, especially electrical properties. Examples: erbium oxide doped with thulium for use as laser crystals; germanium or silicon doped with boron or arsenic for use as semiconductors.

Doppler effect.   A shift toward longer or shorter wavelengths for waves reaching an observer when the source of the waves is moving away from or toward the observer.

dormancy.   A period of suspended growth and metabolic activity. Many plants, seeds, spores, and some invertebrates become dormant during unfavorable conditions.

dormant oil.   (refined mineral oil).
Properties: Viscosity 90–150 s (Saybolt 100F). Unsolfonable residue of 50–70%.
Use: As insecticide.

dormant spray.   Agricultural insecticidal spray applied during winter or early spring when plant is dormant.

dosimetry, radiation.   Measurement of the amount of radiation delivered to the body of an individual. The permissible dose is the quantity of radiation that may be received by an individual over a given period with no detectable harmful effects. For X- or γ-ray exposure, the permissible dose is 0.3 roentgen/week, measured in air. All workers with radioactive materials are expected to wear some device for detecting incident radiation. A dosimeter based on fiber optics has been developed for possible application in radiation therapy.
See rad; rem.

DOT.   Abbreviation for Department of Transportation, the agency responsible for the shipping regulations for hazardous products in the U.S.

dot blot.   A technique for measuring the amount of one specific DNA or RNA in a complex mixture. The samples are spotted onto a hybridization membrane (such as nitrocellulose or activated nylon, etc.), fixed and hybridized with a radioactive probe. The extent of labeling (as determined by autoradiography and densitometry) is proportional to the concentration of the target molecule in the sample. Standards provide a means of calibrating the results.

DOTC.   See di-*n*-octyltindichloride.

DOTG.   Abbreviation for di-*o*-tolylguanidine.

dotriacontane.   (dicetyl).
CAS: 544-85-4. $CH_3(CH_2)_{30}CH_3$.
Properties: Crystals. D 0.823, bp 310C, mp 70C.
Use: Research.

DOTT.   Abbreviation for di-*o*-tolylthiourea.

double bond.   See unsaturation.

double decomposition.   (double displacement, double replacement, metathesis).
Reaction of two compounds to form two new compounds.

double displacement.   See double decomposition.

double helix. The natural coiled conformation of two complementary, antiparallel DNA chains by the formation of A-T and G-C base pairs.

double layer, electric. See electric double layer.

double membrane. In mitochondria and plastids, there is a two-layered membrane that surrounds the organelle. This is believed to be the result of endosymbiosis, with the outer membrane coming from the eukaryotic cell and the inner membrane belonging to the original prokaryote, which was "swallowed."

double replacement. See double decomposition.

double salt. A hydrated compound resulting from crystallization of a mixture of ions in aqueous solution. Common examples are the alums, made by crystallizing from solution either potassium or ammonium sulfate and aluminum sulfate; Rochelle salt (potassium sodium tartrate), made from a water solution of potassium acid tartrate treated with sodium carbonate; and Mohr's salt (ferrous ammonium sulfate), crystallized from mixed solutions of ferrous sulfate and ammonium sulfate. See nickel ammonium sulfate.

doublet. A quantum state of a system with a spin of 1/2.

"Doverphos" *[Dover].* TM for tris(nonylphenyl)phosphite.
CAS: 58968-53-9.
Available forms: Liquid.
Use: In rubber, vinyl stabilizers, and various polymers as an antioxidant.

"Dowanol" *[Dow].* TM for a series of glycol monoethers.
Use: Solvents, intermediates for plasticizers, bactericidal agents, and fixatives for soap and perfumes.

"Dowclene" *[Dow].* TM for a series of solvents for specialized cleaning. A stabilized emulsion of caustic soda, a detergent, and a sequestering agent. EC: A colorless liquid, fp −56.6, bp 77–122C, d 1.381. WR: Inhibited 1,1,1-trichloroethane.

"Dow Corning" *[Dow].* TM for a wide range of silicone and polysiloxane products including emulsions, lubricants, greases, mold-release agents, laminating polymers, electrical varnishes, and heat-resistant coatings.

Dowetch. Magnesium photoengraving sheet, plate, and extruded tube. Also applied to chemicals used in the one-step engraving process.

Dowex. A series of synthetic ion exchange resins made from styrene-divinylbenzene copolymers having a large number of ionizable or functional groups attached to this hydrocarbon matrix. These functional groups determine the chemical behavior and types of ion exchange resin. The strong acid cation resins are capable of exchanging cations, e.g., sodium for calcium and magnesium, as in softening water. The strong base anion resins are capable of exchanging anions.

"Dowfax 9N" *[Dow].* TM for a series of nonylphenolethylene oxide adducts.
Use: Surfactants; textiles, pulp and paper, leather, latex paint.

"Dowflake" *[Dow].* TM for calcium chloride, 77–80%, in a special flake form.

"Dowfrost" *[Dow].* TM for a heat-transfer medium consisting of inhibited propylene glycol.
Use: Immersion freezing of poultry and other foods.

Dowfume. A series of proprietary products used as fumigants and pesticides.

"Dowlex" *[Dow].* TM for a hybrid polyethylene of both low and high density, i.e., it has both linear and branched molecular structure. It can be used for both containers and bags. Forms available are film, injection molding, and extrusion.

Dowmetal. Magnesium alloys containing >85% magnesium.

downcomer. A pipe or flue that conveys gases, vapors, or condensate downward in blast furnaces, distillation towers, refineries, etc.

downstream. See upstream/downstream.

"Dowtherm" *[Dow].* TM for a group of liquid heat transfer media.

DOZ. Abbreviation for dioctyl azelate. See di(2-ethylhexyl) azelate.

DP. Abbreviation for degree of polymerization.

DPA. Abbreviation for diphenylamine, also for diphenolic acid.

DPG. Abbreviation for diphenylguanidine.

DPH. Abbreviation for 1,6-diphenylhexatriene.

DPIP. See diphenyl isophthalate.

DPN. Abbreviation for diphosphopyridine dinucleotide.
See nicotinamide adenine dinucleotide phosphate.

DPO.   Abbreviation for 2,5-diphenyloxazole.

DPPD.   Abbreviation   for   *N,N'-diphenyl*-p-phenylenediamine.

Dragendorf's reagent.   (Kraut's reagent; potassium iodobismuthate).
Use: Testing alkaloids.

Dragonic acid.   See anisic acid.

dragon's blood.
Properties: Deep-red, amorphous lumps. Mp 120C. Soluble in alcohol, ether, and volatile and fixed oils; insoluble in water.
Chief constituents: dracoalban, dracoresene, draconine, and esters.
Derivation: The resin from the surface of the fruit of several species of *Daemonoraps*; habitat is Indonesia and Borneo.
Use: Pigment for coloring paints, polishes, lacquers, etc.; photoengraving, to protect zinc plates from acid.

draft sequence.   The sequence generated by the HGP as of June 2000 that, while incomplete, offers a virtual road map to an estimated 95% of all human genes. Draft sequence data are mostly in the form of 10,000 base pair-sized fragments whose approximate chromosomal locations are known.
See sequencing; finished DNA sequence; working draft DNA sequence.

drain cleaner.
A product used to chemically clean drains, usually of sinks. Lye, an extremely corrosive substance, is the chief component of such cleaners. Most drain cleaners, if mixed with ammonia, toilet bowl cleaners, household cleaners, or even other drain cleaners, release toxic fumes and may cause a violent eruption from the drain cleaner.
Hazard: Corrosive; can cause skin damage; if swallowed can erode tissues of the gastrointestinal tract; fatal.

dram.   Unit of weight used in pharmacy, equivalent to 0.125 oz, 60 grains, or 3.888 g.

Dramamine.   See dimenhydrinate.

"Drapex" [Galata].
CAS: 8013-07-8.
TM for an epoxidized soybean–oil vinyl plasticizer.
Use: For food packaging.

draw.   In metalworking technology, to gradually reduce the diameter of a metal rod or plastic cylinder by pulling it through perforations of successively diminishing size in a series of plates. Wire and various specialty filaments are made in this way; they may be either cold-drawn (without preheating) or hot-drawn (heated to softening point). Dies made

of low-grade diamonds are often used for precision work. An analogous method is used in sheet-metal forming.

drazoxolon.   (3-methyl-4-(*o*-chlorophenylhydrazono)-5-isoxazolone).
CAS: 5707-69-7. $C_{10}H_8ClN_3O_2$.
Properties: Yellowish crystals. Mp 167C. Insoluble in water, acids, and straight-chain hydrocarbons; partially soluble in chloroform, alkalies, ketones, and aromatic solvents.
Hazard: Toxic by ingestion.
Use: Antifungal agent.

Dresinate.   Liquid, paste, and powder forms of sodium and potassium soaps of rosins and modified rosins used as emulsifiers, detergents, and dispersants in soluble oils, cleaning compounds, and other compositions.

drevogenin a.
CAS: 10163-83-4. $C_{28}H_{42}O_7$.
Hazard: A poison.
Source: Natural product.

"Drewmulse" [Stepan].   TM for a series of glycerol and glycol esters, and sorbitan and polyoxyethylene sorbitan esters of fatty acids.
Use: Emulsifiers and opacifiers.

Drewplast.   Glyceryl and polyglyceryl fatty acid esters.
Use: Slip additives for PVC and polyolefin resin films, antistatic and antifogging agent for clear plastic wraps, and as internal lubricants.

Dricoid.   Xanthan gum blends made by galactomannan interaction.

drier.   A substance used to accelerate the drying of paints, varnishes, printing inks, and the like by catalyzing the oxidation of drying oils or synthetic resin varnishes such as alkyds. The usual driers are salts of metals with a valence of ≥ 2, and unsaturated organic acids. The approximate order of effectiveness of the more common metals is cobalt, magnesium, cerium, lead, chromium, iron, nickel, uranium, and zinc. These are usually prepared as the linoleates, naphthenates, and resinates of the metals. Paste driers are commonly the metal salts (acetates, borates, or oxalates) dispersed in a dry oil.
See soap (2).
Note: The spelling *dryer* refers to equipment used for drying (of paper, textiles, food products, etc.).

"Drierite" [W. A. Hammond.].   TM for a special form of anhydrous calcium sulfate having a highly porous granular structure and a high affinity for water. Absorbs water vapor both by hydration and by capillary action.
Grade: Regular, indicating (turns blue to red in use), Du-Cal (for drying air and gases).

Use: Drying of solids, liquids, gases.

drilling fluid. (drilling mud).
A suspension of barytes and bentonite or attapulgite clay in either water or a petroleum oil. When circulated through oil-well drilling pipes, it acts as a coolant and lubricant and keeps the hole free from bore cuttings. To be effective, it must have a specific gravity of at least 2.0 and should be thixotropic, with appreciable gel strength. Lignosulfonates are used as thinners in the water-based type. Oil-based muds require thickening additives such as blown asphalt and metallic soaps of tall oil and rosin acids.

Drinox. A series of insecticides. H-34: A liquid containing 24.5% heptachlor. PX: A planter-box seed treatment containing 25% heptachlor.
Hazard: Toxic by ingestion, inhalation, and skin absorption.
Use: Seed treatment.

Driocel. A solid, granular desiccant for drying process liquids and gases. Manufactured from a selected grade of natural bauxite, reduced to the required particle size, and thermally activated to its maximum absorbing activity. Mesh grades 4/8, 4/10, 8/14.

"Dri-Pax" [Grace]. TM for a group of silica–gel products used in packaging pharmaceuticals and similar products. Extends shelf life and deodorizes products having unpleasant odors.

Drosophila melanogaster. A species of small fruit fly.
Use: In experimental genetics and embryology and mutagenicity testing.

dross. See slag.

Drucomine 9650. A cationic fatty amino amide.
Use: Softener and finisher.

drug. (1) A substance that acts on the central nervous system, e.g., a narcotic, hallucinogen, barbiturate, or psychotropic drug; (2) a substance that kills or inactivates disease-causing infectious organisms; (3) a substance that affects the activity of a specific bodily organ or function; (4) according to the FDA, a drug is a substance "intended for use in the diagnosis, cure, mitigation, treatment or prevention of disease, or to affect the structure or function of the body." Under this definition, high-potency preparations of vitamins A and D are classified as drugs, though other vitamins are not; antiperspirants and suntan lotions are also considered to be drugs.
The Drug Enforcement Administration is the federal agency responsible for supervision and control of the manufacture and use of drugs, especially those in class (1). The term *ethical drug* is equivalent to *prescription drug*, i.e., a drug sold only on a physician's prescription.

See pharmaceutical; see Appendix II for history of the industry.

Drupene. Nonionic ethoxylated fatty alcohols, available as liquid or paste.
Use: Detergents and penetrants.

Drustate. A series of cationic fatty amino amide condensates, alkyl sulfates.
Use: Antistatic agents.

dry cell. (Lechanche cell).
A primary battery having a zinc anode, a carbon (graphite) cathode surrounded by manganese dioxide, and a paste containing ammonium chloride as electrolyte. Such batteries are not reversible and, therefore, have a limited operating life. Their chief use is in flashlights and similar devices requiring low voltage.
See battery.

dry chemical. A mixture of inorganic substances containing sodium bicarbonate (or, frequently, potassium bicarbonate) with small percentages of added ingredients to render it free-flowing and water repellent.
Use: Fire extinguisher on fires in electric equipment, oils, greases, gasoline, paints, and flammable gases.

dry deposition. Deposition of materials from the atmosphere without the aid of rain or snow, e.g., particulates in the range of 2.5 microns as well as pollutant gases (SO_2, NO_2).
See acid precipitation.

dryer. Any of numerous types of equipment used in the chemical industries to remove water from a product during processing. Space does not permit description of even a few of the great variety and multiplicity of choices available. The major types include the following:
belt fluid-bed screw
centrifugal freeze spray
convection pan tubular
conveyor rotary drum tunnel
flash rotary tray truck tray
rotary vacuum vibrating
See drying.

dry ice. See carbon dioxide.

drying. (1) Polymerization of the glycerides of unsaturated vegetable oils induced by exposure to air or oxygen.
See drying oil; drier.
(2) Removal of 90–95% of the water from a material, usually by exposure to heat. Industrial drying is performed by both continuous and batch methods. The type of equipment and the temperatures used depend on the physical state of the material,

i.e., whether liquid (solution or slurry), semiliquid (paste), solid units, or sheet. *Continuous drying*: The rotating-drum dryer is used for flaked or powdered products (soap flakes); a heated metal drum revolves slowly in contact with a solution of the material, the dry product being removed with a doctor knife. In paper manufacturing, drying is performed by a battery of staggered, steam-heated, revolving drums located at the dry end of the fourdrinier machine, the paper passing around the drums at high speed; the moisture content is thus reduced from 60% to about 5%. In spray drying, milk, egg white, and other liquid food products are passed through an atomizing device into a stream of hot air. In tunnel drying, the product travels on a conveyor belt through a heated chamber of considerable length. *Batch drying*: Steam-jacketed pans are used if the material is in paste or slurry form, or in removable trays placed in an oven, if in solid units (fruits, vegetables, meats, etc.). The revolving-tube dryer, used for granular solids and coarse powders, is a long, horizontal cylinder in which a current of warm air runs counter to the movement of the material. Freeze-drying is a specialized technique utilizing high vacuum and low temperatures.
See dehydration; evaporation; freeze-drying.

drying oil.   An organic liquid that, when applied as a thin film, readily absorbs oxygen from the air and polymerizes to form a relatively tough, elastic film. The oxidation is catalyzed by such metals as cobalt and manganese.
Drying oils are usually natural products such as linseed, tung, perilla, soybean, fish, and dehydrated castor oils but are also prepared by combination of natural oils or their fatty acids with various synthetic resins. The drying ability is due to the presence of unsaturated fatty acids, especially linoleic and linolenic, usually in the form of glycerides. The degree of unsaturation of an oil, and hence its drying ability, is expressed by its iodine number. The drying oils have the greatest capacity for iodine, and the nondrying oils the least.
See drier.

"Dryspersion" [Kenrich].   TM for dry dispersion of rubber compounding chemicals in powder form. Deagglomerated and treated with nonstaining oil.
Use: Rubber compounding.

DS2.
Properties: Contains 70% diethylene triamine, 28% ethylene glycol monomethyl ether, and 2% sodium hydroxide.
Use: Decontaminant for chemical warfare agents.

DSC.   See differential scanning calorimetry.

DSMA.   (disodium methanearsonate).
CAS: 144-21-8.
Use: Herbicide.

DSP.   Abbreviation for disodium phosphate.
See sodium phosphate dibasic.

D-stoff.   See phosgene.

DTA.   See differential thermal analysis.

DTBP.   Abbreviation for di-*tert*-butyl peroxide.

DTBT.   See bis(trifluoroacetoxy)dibutyltin.

DTDP.   Abbreviation for ditridecyl phthalate.

dubnium.
Db.
A transfermium element. Atomic number 105. Very short half-life.

Duclean.   Acids containing pickling inhibitors.
Use: Pickling iron and steel. No. 1: sulfuric acid, 60 degrees Bé, d 1.706, fp −10.8C. No. 2: hydrochloric acid technical, d 1.142, fp 40C.
Hazard: Corrosive liquids.

Duff reaction.   The *ortho*-formylation of phenols or *para*-formylation of aromatic amines with hexamethylenetetramine in the presence of an acidic catalyst.

Duhring's rule.   Relates the vapor pressure of similar substances at different temperatures. A straight or nearly straight line results if the temperature at which a liquid exerts a particular pressure is plotted on a graph against the temperature at which some similar reference liquid exerts the same vapor pressure. Water is most frequently used as a reference liquid since its vapor pressure at various temperatures is well known.

dulcin.   (4-ethoxyphenylurea).
CAS: 150-69-6. $H_2NOCNHC_6H_4OC_2H_5$.
This substance should not be confused with dulcitol.
Properties: White, needle-like crystals or powder; taste approximately 200 times as sweet as sugar. Mp 173–174C. Soluble in alcohol and ether; moderately soluble in hot water.
Derivation: From *p*-aminophenol.
Hazard: Questionable carcinogen.
Use: Artificial sweetening agent, prohibited by FDA for use in foods.

dulcitol.   (dulcite; dulcose).
CAS: 608-66-2. $C_6H_8(OH)_6$.
A sugar.
Properties: White crystalline powder; slightly sweet taste. D 1.466, mp 188.5C. Soluble in hot water; slightly soluble in cold water; very slightly soluble in alcohol.
Derivation: By hydrogenation of lactose; occurs naturally in *Melampyrum nemorosum*.
Grade: Technical, reagent.
Use: Bacteriology, medicine.

Dulong and Petit's law. The atomic heat capacity (atomic weight times specific heat) of elementary substances is a constant whose average value at room temperature is 6.2 cal/g K. A few elements, notably boron, carbon, and silicon, obey the law only at high temperatures.

Dumas method. Condensation of nitrogenous organic compounds with copper oxide in a stream of carbon dioxide, collection of the elementary nitrogen over aqueous potassium hydroxide, and volumetric estimation in an azotometer. Any nitrogen oxides formed are reduced by passage over a red-hot copper spiral.

dumortierite.
$8Al_2O_3 \cdot 6SiO_2 \cdot B_2O_3 \cdot H_2O$ (perhaps).
A natural basic aluminum silicate, bright smalt-blue to greenish-blue in color, vitreous luster, d 3.26–3.36, Mohs hardness 7.
Occurrence: U.S. (New York, Arizona, Nevada), France, Norway.
Use: Spark-plug porcelain, special refractories.

Duo 1 Cleaner. A cleaning liquid.
Use: Cleaning compound for printing presses.

"Duon" [Propex]. TM for engineered nonwoven fabric.
Use: Automotive, consumer, home furnishings, and industrial applications.

"Duo-Sol" process. A proprietary process for refining lubricating oils by extraction with a solvent consisting of liquid propane and a cresol base.

Duponol. A series of surface-active agents based on lauryl sulfate. These have detergent, emulsifying, dispersing, and wetting properties and are used in the textile, paper, leather, cosmetic, and electroplating industries; in dental and medical preparations, and in agricultural products.

"Duracool" [Duracool]. TM for metalworking coolants and corrosion preventatives.
Use: In the machining industry.

Duradene. A stereoregular copolymer of butadiene and styrene with exceptional purity, soluble in aromatics and aliphatics.
Use: Rubber products having superior abrasion resistance, good wet-skin resistance, flex-cracking resistance, and good resilience. Automotive tire treads, belt conveyors, shoe soles, and resilient products.

duralumin. A high-strength aluminum alloy, containing 4% copper, 0.5% magnesium, 0.25–1.0% manganese, and low percentages of iron and silicon. Resistant to corrosion by acids and seawater.
Use: Aircraft parts, railroad cars, boats, machinery.

"Duranickel" 301 [Huntington]. TM for an alloy of 94% nickel and 4.5% aluminum.

"Duraphos" [Rhodia]. TM for a complex sodium–calcium–aluminum polyphosphate containing 67% phosphate.
Use: Slow-dissolving glassy phosphate.

"Duraplex" [Kentucky]. TM for drying and nondrying oil-modified alkyd resins derived from phthalic anhydride, polyhydric alcohols, and vegetable oils. Air drying, baking, and nondrying grades, in solvent solution or viscous 100% resins. Produce tough, glossy, light-colored coatings with excellent durability.
Use: Primers, lacquers, and enamels; metal decorating; automotive coatings; furniture finishes; architectural enamels; inks.

"Durazone 37" [Bayer].
TM for 2,4,6-tris-(-*N*-1,4-dimethylpentyl-*p*-phenylenediamino)-1,3,5-triazine.
Use: A nonstaining and nondiscoloring antiozonant for static and dynamic applications used in natural and synthetic rubbers.

Durcupan. An embedding medium for electron microscopy including water soluble and "Araldite" embedding agents.

durene. (*sym*-1,2,4,5-tetramethylbenzene).
$C_6H_2(CH_3)_4$

Properties: Colorless crystals; camphorlike odor. D 0.838, mp 77C, bp 196–197C, flash p 165F (74C) (COC). Soluble in alcohol, ether, and benzene; insoluble in water; sublimes and is volatile with steam. Combustible.
Derivation: By heating *o*-xylene and methyl chloride in presence of aluminum chloride. Occurs in coal tar.
Use: Organic synthesis, plasticizers, polymers, fibers.

"Durethan" [Lanxess]. TM for injection molding and extrusion-grade thermoplastic polyamide resins.
Use: For automotive components, electric components, and furniture

"Durez" [Sumitomo]. TM for synthetic resins, plastics, molding compositions, phenolformaldehyde resins, furfuryl alcohol resins, polyester resins, and polyurethane resins.
Use: Industrial chemicals; curing agents, rubber accelerators, plasticizers, and binders.

"Durfax EOM" [Loders].   (ethoxylated glyceryl monostearate).
TM for emulsifier and dough conditioner.
Use: In bakery applications and in personal care products.

"Durite" [Momentive].   TM for a series of phenolformaldehyde resins used in the manufacture of grinding wheels, brake linings, clutch facings, lamp-base cements.

Durobrite.   Zinc cyanide plating brightening agents; amber liquids.

Durometer hardness.   See hardness.

Duron Microwax.   Microcrystalline wax additive.
Available forms: Prills, slab, or pellets.
Use: In nonskid floor polish, mold release, hot melt, coatings, printing ink, corrosion protection, and metal-cutting fluids.

"Dursban" [Dow Agrosciences].   (chlorpyrifos).
CAS: 2921-88-2.
TM for insecticides containing O,O-diethyl-O-3,5,6-trichloro-2-pyridyl phosphorothioate. $C_9H_{11}Cl_3NO_3PS$.
Hazard: Toxic. Use may be restricted. Toxic by skin absorption. Cholinesterase inhibitor. Questionable carcinogen.
Use: Control of chinch bugs in Gulf Coast states, tick control on cattle and sheep in Australia.

dust.
Fine dry particles or powder that is light enough to remain suspended in air for some time. Respirable dust particles such as coal dust, cotton dust, and any of a number of proliferative dusts can be quite hazardous.

"Dustac" [Georgia Pacific].   TM for lignin.
Available forms: Liquid and dry grades.
Use: As a dust suppressant, binder, emulsifier, viscosity reducer, sequestering agent, dispersant, and flotation aid.

"Dust-Ban" [Nalco].   TM for chemical program.
Use: In suppression of mining and road dust.

dust, industrial.   Finely divided solid particles that may have damaging effects on personnel by inhalation, or that constitute a fire hazard. They are air suspensions of particles ≤10 microns in diameter, though sizes up to 50 microns may be present. Such dusts include (1) metallic particles of all types, some being more harmful than others; (2) silica, mica, talc, quartz, graphite, clays, calcium carbonate, asbestos; (3) organic materials such as chemicals, pesticides, flour or other cereals, cellulose, coal, etc. The size range of mineral dust most damaging to the lungs is between 0.5 and 2 microns. Silicosis is caused by chronic exposure to uncombined silica (quartz, cristobalite) in mines and quarries. Bag houses, dust collectors, or electrostatic precipitators can be used for dust control.
Hazard: Dust suspensions in enclosed industrial areas are a serious fire hazard regardless of the chemical nature of the dust; an explosion may be initiated by static sparks or any open flame. Threshold Limit Values (TLV) for industrial dusts are given by the ACGIH.
See Threshold Limit Value; American Conference of Governmental Industrial Hygienists.

dusting agent.   A powdery solid used as an abherent and mold-release agent in the plastics and rubber industries. Typical materials in general use are talc (soapstone), mica, slate, flour, and clay. Graphite and mica have a flat, crystalline structure that causes them to act as lubricants, and thus are especially effective in preventing sheets or slabs of hot solid mix from sticking together when stacked.
See antiblock agent.

Dutch oil.   See ethylene dichloride.

Dutch process.   Process for making white lead.
See lead carbonate, basic.

Dutrex.   A series of aromatic hydrocarbon concentrates derived from petroleum.
Properties: Colors ranging from light amber to black; odor slight to none; very low volatilities. Viscosity 3–10,000 cP at 210F, d approximately 1.0.
Hazard: Flammable.
Use: Rubber processing and extending oils, plasticizer extenders for polyvinyl chloride, resin solvents, tackifier for nitrile-butadiene rubber.

Dutt-Wormall reaction.   Preparation of diazoaminosulfinates by reaction of diazonium salts with aryl- or alkylsulfonamides, followed by alkaline hydrolysis to the corresponding sulfinic acid of the sulfonamide and the azide.

Du Vigneaud, Vincent (1901–1978).   An American biochemist who won the Nobel Prize for chemistry in 1955. His work involved the study of the metabolism of biologically significant sulfur compounds that led to the finding of transmethylation in mammalian metabolism. He isolated and proved the structure of the vitamin biotin and synthesized penicillin, oxytocin, and the vasopressin hormone of the posterior pituitary. His education was at Rochester, Yale, St. Louis, and George Washington Universities.

DVB.   Abbreviation for divinylbenzene.

DXE.   See dixylylethane.

Dy. Symbol for dysprosium.

dyclonine hydrochloride. (4'-butoxy-3-piperidinopropiophenone hydrochloride).
CAS: 994-22-9. $C_{18}H_{27}NO_2 \cdot HCl$.
Properties: Crystals. Mp 175–176C. Soluble in water, alcohol, acetone, and phenol.
Grade: ND.
Use: Medicine (topical anesthetic).

Dycril. A photosensitive plastic bonded to steel, aluminum, or Cronar base supports. Exposure to a UV light source renders exposed areas insoluble to a subsequent mild alkaline washout solution, so that a relief image of the exposed pattern remains. The depth of image is dependent on the thickness of the photosensitive plastic layer. There are eight types available.
Use: Printing plates.

Dycron.
CAS: 1344-28-1.
Aluminum oxide.

dye.
Any colored material that, when in solution, is able to color substances to which it is applied. The result may or may not be permanent. Most dyes now in use today are synthetic and many are unsaturated organic compounds that contain conjugated double bonds. The bond or moiety of a dye that is responsible for the color is called a chromophore.
Hazard: Can cause allergic reactions.
Use: Color fabric, leather, cosmetics, and various pigments; in science to stain tissues and other materials, as test reagents; therapeutic agents in medicine.

dye, azo. See azo dye.

dye, azoic. See azoic dye.

dye, certified. See food color; FD&C color.

dye, direct. A water-soluble dye taken up directly by fibers from an aqueous solution containing an electrolyte, presumably due to selective adsorption. Usually applied to cellulosic fibers. Dyeing assistants such as sodium chloride or sodium sulfate are used to obtain a high concentration of dye on the fibers.
See dye, synthetic.

dye, disperse. A dye in any of three clearly defined chemical classes: (1) nitroarylamine; (2) azo; and (3) anthraquinone; almost all contain amino or substituted amino groups but no solubilizing sulfonic acid groups. They are water-insoluble dyes introduced as a dispersion or colloidal suspension in water and are absorbed by the fiber, after which they may remain untreated or be after-treated (diazotized) to produce the final color. Their use is primarily for cellulose acetate, nylon, polyester, and other synthetic fibers, and for thermoplastics.

dye, fiber-reactive. A synthetic dye containing reactive groups capable of forming covalent linkages with certain portions of the molecules of natural or synthetic fibers, e.g., covalently bound azo dye on cellulose. Such dyes were not produced until 1953, when cholortriazinyl dyes were introduced. Since then, hundreds of fiber-reactive dyes for cellulose have been patented.

dyeing assistant. Any material added to a dye bath to promote or control dyeing. The action of assistants differs with the classes of dyes, but in most cases, they aid in level deposition of the dye by either delaying its absorption, increasing its solubility, or assisting the dye solution to penetrate the material. Examples: sodium sulfate decahydrate, pine oils, alkylaryl sulfonates.

dye intermediate. See azo-dye intermediate.

dye, metal. An organic dye suitable for use with a metal such as aluminum or steel. Alizarin Cyanin RR, Alizarin Green sulfur, Nigrosine 2Y, and Naphthalene Blue RS are used for this purpose.

dye, natural. An organic colorant obtained from an animal or plant source. Few of these are now in major use. Among the best known are madder, cochineal, logwood, and indigo. The distinction between natural dyes and natural pigments is often arbitrary.
See colorant; pigment.

dye retarding agent. An additive to dye baths to prevent, by decreasing absorption of the dyes, the rapid exhaustion of the bath. Examples: sodium lauryl sulfate, sulfonated oils.

dye, solvent. Any of several organosoluble dyes used in plastics, printing inks, cosmetics, etc., as well as for polyester and other synthetic fibers. They are chemically related to disperse dyes. The solvents used are such chlorinated hydrocarbons as 1,1,1-trichloroethane, trichloroethylene, and perchloroethylene. Use of solvent dyes avoids the necessity of fine-grinding and gives ideal dispersion. (The term *solvent dye* is somewhat misleading, as it would apply equally well to dyes that are soluble in water.)

dyestuffs Synonym for dyes, the chemicals used in dyeing that impart color.
See dye entries.

dye, synthetic. An organic colorant derived from coal tar– and petroleum-based intermediates and applied by a variety of methods to impart bright, permanent colors to textile fibers. Organic dyes were first synthesized by Perkin in England (1856) and were later developed by Hofmann in Germany. Some, called *fugitive*, are unstable to sunlight, heat, and acids or bases; others, called *fast*, are not. Direct (or substantive) dyes can be used effectively without "assistants"; indirect dyes require either chemical reduction (vat type) or a third substance (mordant), usually a metal salt or tannic acid, to bind the dye to the fiber. A noteworthy development is the fiber-reactive group, wherein the dye reacts chemically with cellulose. Dyes may be either acidic or basic, and their effectiveness on a given fiber depends on this factor. Some types are soluble in water, others are not but can be made so by specific chemical treatment.

The central problem is the affinity of a dye for a fiber, and this involves both the chemical nature and the physical state of the dye, i.e., whether acidic or basic and whether colloidal, molecular, or ionic. Neither colloidal particles nor dissociated ions are accepted by a fiber; the dye must be in molecular dispersion to be effective. Further details will be found in the following entries: acetate dye, anthraquinone dye, acid dye, azo dye, alizarin, aniline, fiber-reactive dye, eosin, intermediate, resist, stilbene dye, and sulfide dye. The chemical classes of coloring matters and their arrangement according to chemical structures have been designated numerically in the *Color Index*, a British publication.
See colorant; pigment.

dye, vat.   See vat dye.

"Dyflor" [Evonik].   TM for polyvinylidene fluoride.
Use: For a barrier layer material to prevent fuel permeation.

dyfonate.   (fonofos).
CAS: 944-22-9. $C_6H_5SPS(C_2H_5O)_2$.
Hazard: Cholinesterase inhibitor. Toxic by skin absorption. Questionable carcinogen.
Use: Soil insecticide.

Dylox.   Trichlorfon.

Dylux.   A photosensitive paper free from silver compounds and coated with organic dyes that are activated by UV radiation. Gives good continuous tones and high resolution. Used in lithography, photoproofing, and information-handling systems. Permits instantaneous, direct recording without processing.

Dynacerin 660.   A liquid wax ester similar to jojoba oil.

Dynacoll.   A polyester for reactive hot melts.

Dynacool.   Cooling-water dispersants.
Properties: Liquid.
Use: For prevention of fouling, scale, and microbiological deposits.

dynad.   Group of atoms in which every atom is combined with neighboring atoms by forces stronger than those that join the atom to atoms belonging to other groups.

Dynaflex.   Flexible polyurethane security sealant.

Dynaflock.
CAS: 11138-49-1.
Sodium aluminate.

dynamite.   An industrial explosive detonated by blasting caps. The principal explosive ingredient is nitroglycerin (straight dynamite) or especially sensitized ammonium nitrate (ammonia gelatin dynamite) dispersed in carbonaceous materials. Diethylene glycol dinitrate, which is also explosive, is often added as a freezing point depressant. A dope such as wood pulp and an antacid such as calcium carbonate are also essential.
Hazard: Moderately sensitive to shock or heat, dynamite is a rather serious explosion hazard, also may ignite or explode as a result of contact with powerful oxidizing agents.
Note: Use of dynamite is decreasing; it may be entirely replaced by safer and more efficient explosives.
See explosive, high.

"Dynapol" [Evonik].   TM for hot-melt adhesives.
Use: Bonding or laminating of fabrics, films, foils, or leather.

"Dynasan" [Cremer].   TM for fatty acid esters.
See Witepsol.

dyphylline.   [7-(2,3-dihydroxypropyl)theophylline; hyphylline].
CAS: 479-18-5. $C_{10}H_{14}N_4O_4$.
Properties: Crystals; bitter taste. Mp 158C. Soluble in water, alcohol, and chloroform.
Use: Medicine.

dypnone.   (phenyl β-methylstyryl ketone; 1,3-diphenyl-2-butene-1-one).
CAS: 1322-90-3. $C_6H_5COCHC(CH_3)C_6H_5$.
Properties: Stable, light-colored liquid; mild, fruity odor. Bp 340C, d 1.093 (20/20C), fp (sets to a glass approximately −30C), flash p 350F (176C). Insoluble in water. Combustible.
Use: Softening agent, plasticizer, and perfume base. High absorption of UV light, low water solubility, low evaporation rate, and good solvent action make it useful in light-stable coatings and sunscreen preparations.

dysprosia.   See dysprosium oxide.

dysprosium.
Dy.

Rare-earth or lanthanide element having atomic number 66; aw 162.50; valence of 3; six stable isotopes.
Properties: Noncorroding metal. Does not react with moist air to form hydroxide. Mp 1407C, bp 2330C, d 8.54. Reacts slowly with water and halogen gases; soluble in dilute acids. High cross section for thermal neutrons.
Derivation: Reduction of the fluoride with calcium.
Available forms: High-purity lumps, ingots, sponge, powder.
Use: Measurement of neutron flux, reactor fuels, fluorescence activator in phosphors.
See rare earth.

dysprosium nitrate.
CAS: 10031-49-9. $Dy(NO_3)_3 \cdot 5H_2O$.
Properties: Yellow crystals. Mp 88.6C (in its water of crystallization). Soluble in water.
Derivation: Treatment of oxides, carbonates, or hydroxide with nitric acid.
Grade: Up to 99.9% pure.
Hazard: Strong oxidizing agent, may ignite organic materials.

dysprosium oxide. (dysprosia).
CAS: 1308-87-8. Dy_2O_3.
Properties: White powder. Much more magnetic than ferric oxide; slightly hygroscopic, absorbing moisture and carbon dioxide from the air. D 7.81 (27/4C), mp 2340C. Soluble in acids and alcohol.
Derivation: Ignition of hydroxides and oxyacids (carbonates, oxalates, sulfates, etc.).
Use: With nickel, in cermets used as nuclear reactor control rods that do not require watercooling.

dysprosium salts.   Salts available commercially are the chloride $DyCl_3 \cdot {}_xH_2O$, fluoride $DyF_3 \cdot 2H_2O$, arsenide, antimonide, and phosphide.
Use: The last three are used as high-purity binary semiconductors.

dysprosium sulfate.
CAS: 10031-50-2. $Dy_2(SO_4)_3 \cdot 8H_2O$.
Properties: Brilliant-yellow crystals. Stable at 110C and completely dehydrated at 360C. Soluble in water.
Derivation: Dissolving the hydroxide, carbonate, or oxide in dilute sulfuric acid.
Use: Atomic-weight determination.

dystetic mixture.   A mix of two or more substances that has the highest possible melting point of all mixtures of these substances.

**"Dytek" *[Invista].*   (2-methylpentamethylene diamine).
CAS: 15520-10-2.
TM for corrosive urethanes, epoxies, polyamide fibers, films, resins, corrosion inhibitors, wet-strength resins.

Dytol.   Aliphatic primary alcohols derived from natural fats and oils (C_{10}–C_{18}).
Use: Additives for cosmetic creams, polymerization regulators for elastomers and plastics, detergents and viscosity-index improvers for lubricating oils, finishing and softening agents for textiles, preparation of quaternary ammonium compounds, surfactants, water evaporation control, and antifoam.

E

EADC.   Abbreviation for ethylaluminum dichloride.

EAK.   Abbreviation for ethyl amyl ketone.

earth.   (1) Any siliceous or clay-like compound or mixture, e.g., fuller's earth, diatomaceous earth. (2) A natural metallic oxide, sometimes used as a pigment, e.g., red and yellow iron oxide, ocher, or umber. (3) An oxide of any of a series of chemically related metals that are difficult to separate from their oxides or other combined forms, specifically, rare earths, alkaline earths.
See specific entry.

earth wax.   General name for ozocerite, ceresin, and montan waxes.
See wax.

EASC.   Abbreviation for ethylaluminum sesquichloride.

"Easibrom" *[Nalco].*   TM for microbiological control.
Use: For cooling and water systems.

Easperol.   Sulfated fatty alcohol-vegetable-oil mixtures.
Use: Replacement additives for sulfated sperm-oil types in production of leather.

Eastar.   PETG copolyester useful in a variety of processes and applications because of its toughness, clarity, and good melt strength.
Use: For blister packaging and thermoformed containers for refrigerated/frozen foods, candies, and sundries.

East India.   A type of fossil or semi-recent resin similar to dammar. Varieties are batu, black, and pale.
Use: In spirit and oleoresinous varnishes.

"Eastobond" *[Eastman].*   TM for a series of hot-melt adhesives used in the packaging industry for bonding paper, board, film, foil, and glassine.

"Eastone" *[Eastman].*   TM for disperse dyestuffs for use with acetate and nylon fibers.

"Eastotac" *[Eastman].*   TM for hydrogenated C_5 aliphatic hydrocarbon tackifying resins.

Derivation: Produced from petroleum feedstock by polymerization followed by hydrogenation.

Easy Glide.   Ski wax.
Use: For no-wax skis for better performance and to avoid icing.

"Ebecryl 605a" *[AI Chem].*
CAS: 84593-14-6.
Hazard: A mild skin irritant.
Use: Industrial coating for wood, plastics, paper, and electronics including parquet and safety glass interlayers.

ebonite.   See rubber, hard.

ebullator.   A solid substance or surface used to prevent superheating of liquids above their boiling point.

ebullioscope.   (ebulliscope). An instrument to determine concentration by means of boiling points of liquids.

ebulliscope.   See ebullioscope.

ecabet sodium.
CAS: 86408-72-2. $C_{20}H_{27}O_5S•Na$.
Hazard: Moderately toxic by ingestion.

"Ecdel" *[Eastman].*   TM for an elastomer which imparts excellent low-temperature resistance that significantly reduces product loss due to flex cracking and shattering during cold weather distribution.
Use: Applications where low extraction of plasticizers from the product, flex crack resistance, and utility in harsh environments are required.

α-ecdysone.   (cholest-7-en-6-one; 2,3,14,22,25-pentahydroxycholest-7-en-6-one; ecdysone;5-β-cholest-7-en-6-one, 2-β,3-β,14,22,25-pentahydroxy-, (20s,22r)-).
CAS: 3604-87-3. $C_{27}H_{44}O_6$.
The main ecdysone of insects and curstraceans and the first to be isolated in 1954.
Hazard: Carcinogen; tumorigen; mutagen.

β-ecdysone.   (2,3,14,20,22,25-hexahydroxy cholest-7-en-6-one; 20-hydroxyecdysone; ecdysterone; crustecdysone; isoinokosterone). The chief ecdysone of plants.

Hawley's Condensed Chemical Dictionary, Sixteenth Edition. Michael D. Larrañaga, Richard J. Lewis, Sr., and Robert A. Lewis.
© 2016 John Wiley & Sons, Inc. Published 2016 by John Wiley & Sons, Inc.

echinomycin.
CAS: 512-64-1. $C_{51}H_{64}N_{12}O_{12}S_2$.
Properties: Colorless crystals. Slightly hygroscopic, mp 218C. Soluble in fats, chloroform, dioxane; insoluble in water.
Use: Medicine (antibiotic).

ECM. See extracellular matrix.

ecology. The study of the interactions between plant and animal organisms, and their environment; the latter is conceived to include everything that is not an intrinsic part of the organism and thus includes both living and nonliving components. Though primarily a branch of biology, ecology does involve chemistry in respect to plant and animal nutrients, metabolism, photosynthesis, etc., especially interferences that may occur in connection with these. Thus, insecticides, chemical-waste disposal, air and water pollution, oil spills, and radioactive contaminants have direct bearing on the ecology of a given area.
See environmental chemistry.

economic poison. See pesticide.

economics, chemical. See chemical economics.

economizer. A device that acts like a heat exchanger whereby the heat produced by an operation is used to warm incoming air or water. It is widely used in the paper industry, in boilers, and in chemical processing. Heat recovery up to 30% is possible.

ecosystem. The organisms of an ecologic community together with the physical environment that they occupy.

Ecozyme. A xylanase-based enzyme that reduces chemical bleaching agents up to 25% with enhanced pulp quality.
Use: Reduces chlorinated effluent discharge levels in pulp mills.

ECTEOLA-cellulose. (epichlorohydrin triethanolamine cellulose). A dry, powdered cellulose derivative containing tertiary amine groups.
Use: As an anion exchanger in chromatography. It is less basic than DEAE-cellulose and serves to separate viruses, nucleic acids, and nucleoproteins.

ectocrine. An ectohormone that is secreted in the external medium of an organism that affects the metabolism of other organisms and stimulates or inhibits plant growth.

ectoparasite. An external parasite, one that lives on the surface of the host's body.

ectoparasiticide. An agent that, when applied directly to the exterior of the body, destroys parasites that live on the exterior of its host.
Use: Pesticide.

EDAP. See EDC (reagent).

EDB. Abbreviation for ethylene dibromide.

EDC (reagent).
CAS: 25952-53-8. $C_8H_{17}N_3 \bullet ClH$.
Use: An antibiotic.
Hazard: A poison.

Edeleanu process. A solvent extraction process using liquid sulfur dioxide for the removal of undesirable aromatics from heavy lubrication oils.

edestin. A protein having a molecular weight of 310,000 obtained from hempseed. Closely similar proteins occur in seeds of pumpkin, squash, etc.

edetate. See ethylenediaminetetraacetic acid (Note).

edetate calcium disodium. ([[N,nn-1,2-eth-anediylbis[n-(carboxymethyl)glycinato]](4)-N,n,n,o,on,on,onn]calciate(2)disodium; ethylenediaminetetraacetic acid calcium disodium chelate; calcium disodium (ethylenedinitrilo)tetraacetate; calcium disodium ethylenediaminetetraacetate; Edta calcium; edathamil calcium disodium; calcium disodium edatate; edetic acid calcium disodium salt; sodium calciumedetate; calcium disodium 2-[2-bis(2-oxido-2-oxoethyl)amino]ethyl-(2-oxido-2-oxoethyl)amino]acetate).
CAS: 62-33-9. $C_{10}H_{14}CaN_2Na_2O_8$.
A chelating agent that sequesters a variety of polyvalent cations.
Properties: Water-soluble, chelating agent.
Use: To treat heavy metal poisoning; in pharmaceutical manufacturing; a food additive.

edetate disodium. (N,nn-1,2-ethanediylbis[n-(carboxymethyl)glycine] disodium salt; (ethylenedinitrilo)tetraacetic acid disodium salt; ethylene diaminetetraacetic acid disodium salt; ethylene bis(iminodiacetic acid) disodium salt; edetic acid disodium salt; edathamil disodium; disodium edathamil; edta disodium; tetracemate disodium; disodium ethylenediaminetetraacetate; disodium edetate; 2-[2-[bis(carboxymethyl)amino]ethyl-(carboxymethyl)amino]acetic acid).
CAS: 6381-92-6. $C_{10}H_{14}N_2O_8Na_2$
A chelating agent that sequesters a variety of polyvalent cations.
Hazard: Moderately toxic.
Use: In pharmaceutics as a sequestering agent and in chelation therapy.

edetate sodium. (N,nn-1,2-ethanediylbis[n-(car-boxymethyl)glycine] tetrasodium salt;

(ethylenedinitrilo)tetraacetic acid tetrasodium salt; sodium edetate; tetrasodium ethylenediamine-tetraacetate; ethylenebis(iminodiacetic acid) tetrasodium salt; tetrasodium ethylenebis-(iminodiacetate); edta tetrasodium; edetic acid tetrasodium salt; tetracemate tetrasodium; tetrasodium edetate; tetracemin; tetrasodium 2-[2-[bis(2-oxido-2-oxoethyl)amino]ethyl-(2-oxido-2-oxoethyl)amino]acetate).
CAS: 64-02-8. $C_{10}H_{12}N_2Na_4O_8$.
The tetra-sodium salt of edatic acid. It is a chelating agent that reacts with most divalent and trivalent metal ions.
Use: In pharmaceutical manufacturing and as a food additive.

edetate trisodium. (n,nn-1,2-ethanediylbis[n-(carboxymethyl)glycine] trisodium salt; (ethylene-dinitrilo)tetraacetic acid trisodium salt; edta trisodium; ethylenediaminetetraacetic acid trisodium salt; trisodium ethylenediaminetetra-acetate; trisodium edetate; edetc acid trisodium salt; trisodium 2-(2-(bis(2-oxido-2-oxoethyl) amino)ethyl-(carboxymethyl)amino)acetate).
CAS: 150-38-9. $C_{10}H_{13}N_2Na_3O_8$
The trisodium salt of edetic acid.
Use: As a chelating agent.

edetic acid. (n,nn-1,2-ethanediylbis[n-(carboxy-methyl)glycine]; 3,6-bis-(carboxymethyl)-3,5-dia-zooctanedioic acid; ethylenebisaminodi-acetic acid; ethylenediamine-n,n,nn,nn-tetraacetic acid; (ethylenedinitrilotetraacetic acid; ethylenediamine-tetraacetic acid; ethylenediamine tartrate; edathamil; edta; edta acid; 2-[2-[bis(carboxy methyl)amino]ethyl-(carboxymethyl)amino]acetic acid).
CAS: 60-00-4. $C_{10}H_{16}N_2O_8$.
A powerful chelator of +2 and +3 cations. The sodium salt binds heavy metals most strongly and can form as many as six bonds.
Properties: A colorless, flammable, crystalline sub-stance, ammoniacal odor, slightly soluble in water, insoluble in common organic solvents.
Hazard: Extremely caustic; toxic; teratogenic; mutagenic; blood coagulant.
Use: A chelator of metals in inorganic chemistry; administered by injection in the treatment of heavy metal poisoning and in calcinosis therapy; many uses in the food-processing industry; may be used to help curb the effects of certain viper venom; in pharmaceutical manufacturing.

edible oil. As commonly used, the term refers to any fatty oil obtained from the flesh or seeds of plants that is used primarily in foodstuffs (mar-garine, salad dressing, shortening, etc.). Among these are olive, safflower, cottonseed, coconut, peanut, soybean, and corn oils, some of which may be hydrogenated to solid form. They vary in degree of unsaturation, ranging from 78% for safflower to about 10% for coconut. Castor oil, though techni-cally edible, is not usually considered in this classi-fication, nor are medicinal oils derived from animal sources (cod liver, mineral oil, etc.).

Edman degradation. Sequential degradation of peptides beginning at the N-terminal residue based on the reaction of phenylisocyanate with the α-amino group of the terminal amino acid of the peptide chain.

EDTA. Abbreviation for ethylenediaminete-traacetic acid.

EDTAN. Abbreviation for ethylene diamine tetraacetonitrile.

EDTA Na$_4$. Abbreviation for ethylene diamine tetraacetic acid tetrasodium salt.
See tetrasodium EDTA.

effect. An evaporation-condensation unit.
See evaporation.

effective charge. The value of charge that, mul-tiplied by the actual distance between two atoms of a heteropolar, diatomic molecule, gives the actual dipole moment.

effervescence. The rapid escape of gas from a liquid or a mixture.

efflorescence. Loss of combined water molecules by a hydrate when exposed to air, resulting in partial decomposition indicated by presence of a powdery coating on the mate-rial. This commonly occurs with washing soda ($Na_2CO_3 \cdot 10H_2O$), which loses almost all its water constituent spontaneously.

effluent. Any gas or liquid emerging from a pipe or similar outlet; usually refers to waste products from chemical or industrial plants, as stack gases or liquid mix.

egg. (1) A large gamete without flagellae that is fertilized by a sperm cell. An egg cell is also called an ovum. (2) A complex multicellular structure in which an animal embryo develops.

egg oil. Fatty oil obtained from egg yolk by extraction with ethylene dichloride; insoluble in water, but readily forms emulsions on strong agita-tion.
Use: Ointments, cosmetic creams.

egg yolk.
Properties: Yellow, semisolid mass. D 0.95, mp 22C. High cholesterol content.
Grade: Technical, edible.
Use: Baking, dairy products, mayonnaise, pharma-ceuticals, soap, perfumery.
See albumin, egg.

EHEC. Abbreviation for ethyl hydroxyethyl cellulose.

Ehrlich, Paul. (1854–1915). A native of Silesia, Ehrlich is considered the founder of the science of chemotherapy, or the treatment of diseases by chemical agents. He did fundamental work on immunity, which earned him the Nobel Prize in medicine in 1908, and also developed the famous neoarsphenamine (salvarsan or 606) treatment for syphilis (1910), which was not improved upon until the discovery of penicillin.

Ehrlich–Sachs reaction. Formation of anils by the base-catalyzed condensation of compounds containing active methylene groups with aromatic nitroso compounds.

eicosamethyl nonasiloxane.
CAS: 2652-13-3. $C_{20}H_{60}O_8Si_9$.
Properties: Inert liquid. Bp 173C (5 mm Hg), d 0.918. Soluble in benzene and light hydrocarbons; slightly soluble in alcohol.
Use: Silicone fluids, foam suppressor in lubricating oils.

eicosane.
CAS: 112-95-8. $C_{20}H_{42}$.
Most technical eicosane is a mixture of predominantly straight-chain hydrocarbons averaging 20 carbon atoms to the molecule.
Properties: (pure *n*-eicosane): White, crystalline solid. Fp 36.7C, bp 205C (15 mm Hg), flash p 212F (100C), refr index 1.4348 (20C), d 0.778 (at melting point). Insoluble in water; soluble in ether. Can be readily chlorinated. Combustible.
Grade: Pure normal (99+%), technical.
Use: Cosmetic, lubricants, plasticizers.

eicosanoic acid. See arachidic acid.

eicosanoid. Any of a number of biochemically active compounds resulting from enzymic oxidation of arachidonic acid, e.g., prostaglandins, thromboxanes, prostacyclin, and leukotrienes. As a group, they compose what is known as the arachidonic acid cascade. They have many pharmacological and medical possibilities.
See prostaglandin; arachidonic acid.

1-eicosanol. See arachidyl alcohol.

5,8,11,14-eicostetraenoic acid. See arachidonic acid.

eigenenergy. Quantity of energy corresponding to a time-invarient atomic state.

eigenfunction. Wave function corresponding to a state with a definite value of some quantity.

Eigen, Manfred. (1927–). A German physicist who won the Nobel Prize for chemistry in 1967. His research concerned the rate of hydrogen-ion formation through disassociation of water. He also was concerned with enzyme control. He received his degree at the University of Gottingen.

Einhorn–Brunner reaction. Formation of substituted 1,2,4-triazoles by condensation of hydrazines or semicarbazides with diacylamines in the presence of acid catalysts.

einstein. The energy acquired by a gram-molecular weight of a substance when each molecule absorbs a quantum of excitation energy.

einsteinium. Es. A synthetic radioactive element with atomic number 99 and aw 253, discovered in the debris from the 1952 hydrogen-bomb explosion. Einsteinium has since been prepared in a cyclotron by bombarding uranium with accelerated nitrogen ions, in a nuclear reactor by irradiating plutonium or californium with neutrons, and by other nuclear reactions. The element is named for Albert Einstein. It has chemical properties similar to those of the rare-earth metal holmium. Isotopes are known with mass numbers ranging from 246 to 253. Einsteinium has valence of 2, and the lowest heat of vaporization of any divalent element.
See actinide series.

eka-. Prefix referring to element in next-lower position in the same group in the periodic system.

ekahafnium. One of the last-discovered transuranic elements; atomic number 104. It has two α-emitting isotopes (257 and 259) and possibly a third (258). The former, made by bombardment of californium-249 with carbon-12 nuclei, has a half-life of 5 seconds and decays into nobelium-253. The 259 isotope, made by merging a carbon-13 nucleus with californium-249, has a half-life of 3 seconds and decays to nobelium-255.

Ekonol. An engineering plastic composed of poly-*p*-oxybenzoate. Resistant to temperatures above 600C, self-lubricating surface.
Use: In pumps handling corrosive liquids, protective coating for titanium skins on supersonic transports, disk brakes, etc.

ELA. An elastomer lubricating agent, a mixture of phosphate esters. Light-amber liquid.
Use: Unvulcanized rubbers.

elaboration. A term used in biochemistry to describe chemical transformations within an organism resulting in the formation of specific types of substances; for example, plants elaborate proteins and fats, and poisonous snakes elaborate their venom. It also refers to formation of metabolic end products such as purines and uric acid.

elaidic acid. (*trans*-9-octadecenoic acid).
CAS: 112-79-8. $CH_3[CH_2]_7HC:CH[CH_2]_7COOH$.
The *trans* form of an unsaturated fatty acid of which
the *cis* form is oleic acid.

$$CH_3(CH_2)_7 \diagdown C=C \diagup H \atop H \diagup \quad \diagdown (CH_2)_7CO_2H$$

Properties: White solid. D 0.8505 (79/4C), mp
43.7C, bp 288C (100 mm Hg), 234C (15 mm Hg),
refr index 1.4358 (79C). Insoluble in water; soluble
in alcohol, ether, benzene, and chloroform. Combustible.
Derivation: Synthesized from oleic acid by elaidinization.
Grade: Purified, 99+%.
Use: Medical research, reference standard in chromatography.

elaidinization. Originally the reaction by which
oleic acid is converted into elaidic acid, but now
used in a more general sense to indicate the
conversion of any unsaturated fatty acid or related
compound from the geometric *cis* to the corresponding *trans* form. Nitrous acid and selenium
compounds are commonly used as catalysts for this
reaction. The resulting *trans* acids are more stable
to oxidation.

elargin.
CAS: 2169-75-7. $C_{23}H_{27}NO \cdot C_6H_8O_7$.
Hazard: A poison by ingestion.

elasticity. The ability of a material to recover
its original shape partially or completely after the
deforming force has been removed. The small
amount of deformation that is not recovered is called
permanent set or permanent elongation. Among
common materials, glass and some metals are virtually 100% elastic, whereas vulcanized rubber and
other elastomeric substances are around 90% elastic
after extension to rupture.
So-called perfect elasticity is a property of atoms that
show no energy loss on collision.
See modulus of elasticity; stress; strain; plasticity.

elastic modulus. See modulus of elasticity.

elastin. A scleroprotein that occurs in connective
tissue.
Properties: Yellow, fibrous mass. Insoluble in water,
dilute acids, alkalies, salt solutions, and alcohol. Is
partially digested by pepsin solution and wholly by
trypsin.

ElastoFlo. An ethylene-propylene rubber polymer.
Properties: A free-flowing granular solid.
Use: For automotive hose, tubing and extrusions,
roofing membranes and gaskets.

Elastoguard. An antimicrobial, anti-fatigue
solution.

elastomer. As originally defined by Fisher
(1940), this term referred to synthetic thermosetting high polymers having properties similar to
those of vulcanized natural rubber, namely, the ability to be stretched to at least twice their original
length and to retract very rapidly to approximately
their original length when released. Among the
better-known elastomers introduced since the 1930s
are styrene-butadiene copolymer, polychloroprene
(neoprene), nitrile rubber, butyl rubber, polysulfide
rubber ("Thiokol"), *cis*-1,4-polyisoprene, ethylene-
propylene terpolymers (EPDM rubber), silicone
rubber, and polyurethane rubber. These can be
cross-linked with sulfur, peroxides, or similar
agents. The term was later extended to include
un-cross-linked polyolefins that are thermoplastic;
these are generally known as TPO rubbers. Their
extension and retraction properties are notably different from those of thermosetting elastomers, but
they are well-adapted to such specific uses as wire
and cable coating, automobile bumpers, vibration
dampers, and specialized mechanical products.

Elbs persulfate oxidation. Hydroxylation of
phenols to *p*-diphenols by potassium persulfate in
alkaline solution.

Elbs reaction. Formation of anthracenes by
intramolecular condensation of diaryl ketones containing a methyl or methylene substituent adjacent
to the carbonyl group.

elcatonin. See carbocalcitonin.

electric double layer. A diffuse aggregation
of positive and negative electric charges surrounding a suspended colloidal particle which
aids in maintaining its stability. According to the
Gouy–Freundlich theory, advanced about 1920,
a close-packed array of charges is attached to
the surface of the particle while a diffuse layer
of charges of opposite sign extends into the
liquid. The particle is electrically neutral. There
is an electrokinetic potential gradient across the
double layer that is called the zeta potential. The
diagram is an approximation of this phenomenon.
Modifications of this theory have been introduced
in recent years, notably by Derjaguin and Landau
and by Verwey and Overbeek (DLVO theory).
See zeta potential.

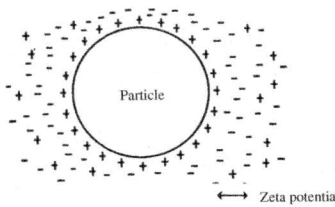

← → Zeta potential

electric furnace.   See furnace.

electric steel.   Steel made in an electric furnace.

electric vehicle propulsion.   See storage battery.

electride.   An experimental compound composed of an alkali-metal cation and an electron in which the electron functions as a chemical element (e.g., a halogen) in salt formation. Several such compounds have been made in the U.S. and abroad. The phenomenon is reported to be one that challenges accepted concepts of compound formation.

"Electrocarb" *[Electro Abrasives].*   TM for silicon carbide.
CAS: 409-21-2.
Properties: High thermal conductivity, abrasion resistant, high hardness (2550 knoop). Electrical conductor.
Grade: Grit sizes 8–1600 mesh.
Use: Wear resistant filler, thermal conductor in compounds, high temperature paints, ceramics, refractories. As a polishing and lapping abrasive.

electrochemical equivalent.   The number of grams of an element or group of elements liberated by the passage of one coulomb of electricity (one ampere for one sec).

electrochemical gradient.   The sum of the gradients of concentration and of electric charge of an ion across a membrane. Used as the driving force for processes, such as oxidative phosphorylation and photophosphorylation.

electrochemical potential.   The energy required to maintain a separation of charge and of concentration across a membrane.

Electrochemical Society.   (ECS). Established in 1902, this society was organized to promote the advancement of the science of electrochemistry and related fields. It comprises 11 divisions, each devoted to a special branch of electrochemistry, e.g., corrosion, batteries, rare metals, and electrodeposition. It publishes a journal and sponsors books relating to its major interests. Its office is at 65 South Main Street, Pennington, NJ 08534-2839. Website: http://www.electrochem.org

electrochemistry.   That branch of chemistry concerned primarily with the relationship between electrical forces and chemical reactions. This relationship is fundamental and far-reaching, as the structure of matter is based on electrical effects. Electrochemistry is directly involved in chemical bonding, ionization, electrolysis (e.g., production of aluminum), metallurgy, battery science, fuel cells, and corrosion—in short, in any situation in which a chemical change is caused by or associated with electrical phenomena. It has certain applications in biochemistry (in nerve reactions and the electric organs of fish) as well as in organic chemistry. Michael Faraday (1791–1867) is generally regarded as the founder of electrochemistry.

electrocoating.   A process of applying primer paint to household appliances, automobiles, etc., in which the metal piece to be coated becomes the anode in a tank of water-based paint. The coating deposited on the metal is uniform regardless of the shape of the article. Large-scale use of electrocoating is on automobile bodies.
See electrostatic coating.

electroconductive polymer.   See polymer, electroconductive.

electrocratic.   Descriptive of a liquid colloidal dispersion of insoluble solid particles whose stability is maintained by either positive or negative charges on the particles. As the like charges are mutually repellent, they offset the attraction of gravity and prevent the particles from sinking or coalescing. A colloidal gold suspension is a well-known example. Electrocratic dispersions can be readily precipitated by addition of an oppositely charged electrolyte, as in the purification of water with Al^{+3} ions from aluminum sulfate.

Electrocure.   A process for hardening or curing paint films on plastics by use of low-voltage electron beams. Paints so treated cure in 4 seconds or less at room temperature; the finishes produced are said to be superior in resistance to abrasion and chemicals. The process can also be applied to other substrates such as wood, glass, metals, etc.

electrode.   Either of two substances having different electromotive activity that enables an electric current to flow in the presence of an electrolyte. Electrodes are sometimes called plates or terminals. Commercial electrodes are made of a number of materials which vary widely in electrical conductivity, i.e., lead, lead dioxide, zinc, aluminum, copper, iron, manganese dioxide, nickel, cadmium, mercury, titanium, and graphite; research electrodes may be calomel (mercurous chloride), platinum, glass, or hydrogen. Electrodes are essential components of both batteries and electrolytic cells; in batteries the negative plate is the anode and the positive plate the cathode, whereas in electrolytic cells the reverse is the case.
Use: Electrodes are also used in welding devices.
See anode; cathode; battery; electrolytic cell.

electrode, glass.   A thin, glass membrane that when immersed in a suitable liquid medium develops a measurable electrical potential that can be readily related to the activity of ionic species present in the solution. By appropriate manipulation of the glass composition, careful pretreatment of the glass surface, and reproducible experimental conditions, electrodes can be devised that not only yield information about the concentration of ions in solution but also have the ability to discriminate in the sense of a selective response between a number of different ions of similar chemical characteristics. Because of their ability to give both qualitative and quantitative information about ions in solution, glass electrodes are widely used for purposes of chemical measurement, especially in electrochemical research.

electrode, hydrogen. A platinum surface coated with platinum black immersed in a solution and bathed with a stream of pure hydrogen gas. The potential developed depends on the equilibrium between the hydrogen gas and the hydrogen ions in solution.
Use: Standard reference electrode.

electrodeposition. The precipitation of a material at an electrode as the result of the passage of an electric current through a solution or suspension of the material, e.g., alkaline-earth carbonates, rubber from latex, and paint films on metal. A technique for electrodepositing refractory carbide coatings on metal has been reported. The electrode is in the shape of the desired article. An important advantage of electrodeposition is its ability to coat complex shapes having small and irregular cavities with exact thickness control.
See electrophoresis.

electrodialysis. A form of dialysis in which an electric current aids the separation of substances that ionize in solution. Seawater can be desalted by this method on a large scale by placing it in the center chamber of a three-compartment container having two semipermeable membranes and a positive electrode in one end chamber, and a negative electrode in the other. The ions migrate to their respective electrodes under a difference of potential, leaving the water salt-free.
See dialysis; desalination; demineralization.

electroforming. An electrolytic plating process for manufacturing metal parts. A mold of the object to be reproduced is made in a soft metal or in wax (by impression). The mold surface is made conducting by coating with graphite. Some suitable metal is then deposited electrolytically on the mold surface. This mold is then (in most cases) a negative of the object to be produced. Other industrial applications are phonograph records, plastic tile, ducting, tubing, etc.

electrogenic. Contributing to the generation of an electrical potential across a membrane.

electroless coating. A protective coating of copper, cobalt, nickel, gold, or palladium deposited in a bath without application of an electric current, i.e., by chemical reduction.

electroluminescence. Luminescence generated in crystals by electric fields or currents in the absence of bombardment or other means of excitation. It is a solid-state phenomenon involving p- and n-type semiconductors, and is observed in many crystalline substances, especially silicon carbide, zinc sulfide, and gallium arsenide, as well as in silicon, germanium, and diamond.
See phosphor.

electrolysis. Decomposition of water and other inorganic compounds in aqueous solution by means of an electric current, the extent being proportional to the quantity of electricity passing through the solution. The positive and negative ions formed are carried by the current to the oppositely charged electrodes, where they are collected (if wanted) or released (if unwanted). Metallic ions deposited on the electrode form a coating. A simple electrolysis is the separation of water into oxygen and hydrogen; this is one method of producing hydrogen. Somewhat more complicated is electrolysis of brine to chlorine and sodium hydroxide; this is carried out in electrolytic cells of the diaphragm or mercury type, with water taking part in the reaction. In electroplating, metal salts dissociate into their constituent ions, the positively charged metal ions coating the cathode. There are a number of variations of this process (electrodeposition, electrocoating, electroforming).
See electrolytic cell; ionization.

electrolyte. A substance that will provide ionic conductivity when dissolved in water or when in contact with it; such compounds may be either solid or liquid. Familiar types are sulfuric acid and sodium chloride, which ionize in solution. One solid electrolyte, used originally in fuel cells, is a polymer of perfluorinated sulfonic acid used as the core of a water electrolysis cell for production of hydrogen and oxygen. When saturated with water it has high conductivity. Another solid type is a ceramic mixture of sodium, aluminum, lithium, and magnesium used as a separating medium in the liquid-sodium-sulfur (β) battery under continuing development. The most common application of electrolytes is in electroplating of metals in which dissolved (ionized) metal salts are the electrolytes.
See electrolysis; electroplating.

electrolyte acid. Legal label name for battery acid.

electrolytic cell. An electrochemical device in which electrolysis occurs when an electric current is passed through it. Ionizable compounds dissociate in the aqueous solution with which the electrodes are in contact. Such cells are of two types: (1) the diaphragm cell, which has two compartments separated by a porous membrane; and (2) the mercury cell, in which mercury is the cathode. The anodes of both types have long been made of graphite; because this decomposes rapidly as electrolysis progresses, they are being replaced with dimensionally stable types consisting of titanium coated with oxides of ruthenium and other rare metals, which are also much more efficient. In electrolysis of sodium chloride, the current causes chloride ion to migrate to the anode, where it is collected as chlorine gas; sodium hydroxide and hydrogen are also formed, the hydrogen being discharged. The overall cell reaction is: $2NaCl + 2H_2O \rightarrow H_2 + Cl_2 + 2NaOH$. This principle is applied in the electroplating of metals, electrodeposition of colloids, and similar processes.

See diaphragm cell; mercury cell; electroplating.

electromagnetic separation.   Separation of isotopes, especially those of uranium, by first accelerating them by means of an electrostatic field and then passing them through a magnetic field. The effect of this is to cause all the particles to take a curved path; the heavier ones, having higher kinetic energy, describe a wider arc than the lighter ones. Thus, two isotopes of closely similar masses can be separated and collected.

See mass spectrometry; magnetic separation.

electromagnetic spectrum.   See radiation.

electrometallurgy.   Application of the principles and techniques of electrochemistry to the production of such metals as aluminum and titanium.

electromotive series.   See activity series.

electron.   Discovered by J. J. Thompson in 1896, the electron is a fundamental particle of matter that can exist either as a constituent of an atom or in the free state. It has a negative electric charge (4.8×10^{-10} esu) and a mass 1/1837 that of a proton, equivalent to 9.1×10^{-28} gram. The number of electrons in an atom of any element is the same as the number of protons in the nucleus, i.e., the atomic number. Thus, the range is from 1 electron in hydrogen to 103 in lawrencium. As the negative charge of the electrons equals the positive charge of the protons, all atoms are electrically neutral. Electrons are arranged in from one to seven shells around the nucleus; the maximum number of electrons in each shell is strictly limited by the laws of physics. The tendency of electrons to form complete outer shells accounts for the valence of an element, and they play an essential part in chemical bonding. The outer shells are not always filled: sodium has two electrons in the first shell, eight in the second, and only one in the third. A single electron in the outer shell may be attracted into an incomplete shell of

another element, leaving the original atom with a net positive charge. Each atom is then called an ion. Valence electrons are those that can be captured by or shared with another atom.

Electrons can be removed from the atoms of metals and some other elements by heat, light, electric energy, or bombardment with high-energy particles (see radiation, ionizing). In such cases, they are totally free from the atomic orbit and their energy can be utilized by means of a conductor (electricity) or a vacuum tube or semiconductor. Current is generated by detaching the electrons of a metallic conductor (silver, copper) by means of an electric or magnetic field; the electrons then flow along the conductor to a positively charged terminal. The science of electronics was made possible by the ability of a heated metal cathode to emit a continuous stream of electrons in a vacuum tube.

Free electrons, called β particles, are spontaneously emitted by decaying radioactive nuclei; they have comparatively low energy but can be accelerated to velocities approaching that of light.

The basic nature of the electron has been the subject of much research of the highest order of mathematical rigor. In simplest terms, the electron has the properties of both a particle and a wave, i.e., a standing wave is associated with an electron moving in its orbit. The energy state of any electron in an atom is described by four quantum numbers.

See shell; atom; orbital theory; Lewis electron theory.

electron acceptor.   A substance that receives electrons in a redox reaction.

electron-beam welding.   See welding.

electron carrier.   A protein or other chemical species, such as a cytochrome or coenzyme Q, that can reversibly gain and lose electrons. They function in the transfer of electrons from organic nutrients to oxygen or some other terminal acceptor as in the case of anaerobes.

electron donor.   A substance that donates electrons in a redox reaction.

electronegativity.   All atoms (except those of helium) that have fewer than eight electrons in their highest principal quantum level have low-energy orbital vacancies capable of accommodating electrons from outside the atom. The existence of these vacancies is evidence that within these regions, the nuclear charge can exert a significant attraction for such electrons, even though as a whole the atom is electrically neutral. This attraction is called "electronegativity." To the extent that the initially neutral atom may be able to acquire electrons from outside, it will acquire also their negative charge. The word "electronegativity" means "tendency to

become negatively charged." The concept of electronegativity is an extremely useful one in chemistry, since the attractive force exerted by the nuclei of atoms that have vacancies in their outer shells makes possible the formation of both covalent and ionic bonds and is thus a fundamental factor in the formation of chemical compounds. The most highly electronegative elements are the halogens, sulfur, and oxygen.
See bond, chemical.

electron microscope. A microscope in which the source of illumination is a stream of electrons emanating from a tungsten cathode in a high vacuum, and accelerated by a strong electric impulse (300 kV). The electrons are focused by a series of magnetic fields that function as lenses in the same way as glass affects waves of visible light, i.e., the electron stream curves as it passes through the magnetic field. Such lenses were developed in Germany by Busch, Knoll, and Tuska in the 1930s, and were adapted to microscopy by Zworykin and Hillier at the RCA Laboratories in the early 1940s, when the first commercial instruments were produced. The electron microscope is characterized by extremely high resolving power due to the ultrashort wavelength of electronic radiation—a small fraction of an angstrom unit (Å). Resolution of 2 Å is possible, which permits determination of the structure of macromolecules (DNA) and even observation of large atoms (uranium). Instruments of the scanning transmission type with a resolving power of 0.05 Å were developed at the University of Chicago.
Two kinds of electron microscopies are in general.
Use: the transmission type, in which the electrons penetrate the specimen, and the scanning type introduced in 1970, in which the electrons, condensed to a fine beam, repeatedly traverse the surface of the specimen, producing a three-dimensional contour effect by means of secondary electrons emanating from the specimen itself. Pictures of astonishing accuracy have been obtained, especially of surface structures, a matter of great importance in the study of catalysis and other critical phenomena in both industry and the biological sciences. A unique combination of these techniques is the scanning transmission electron microscope (STEM), by means of which colored motion pictures of uranium atoms on a thin-film carbon substrate have been obtained.
See optical microscope; resolving power; ultramicroscope; field-ion microscope.

electron octet. Group of eight valence electrons. The most stable configuration of the outermost, or valency, electron shell of the atom.

electron paramagnetic resonance. (EPR). A method of spectroscopic analysis similar to nuclear magnetic resonance except that microwave radiation is employed instead of radio frequencies. It is used for studying free radicals, crystalline centers, transition elements, and structures involving unpaired electrons.

electron transport chain. See acceptor control.

electron-volt. (eV). An extremely small unit used in measuring the energy of electrons and other atomic constituents. It is the energy developed by an electron in falling through a potential difference of 1 V, equivalent to 1.6×10^{-19} J. The rupture of a carbon-to-carbon bond has been calculated to yield approximately 5 eV.

electrophile. An electron-deficient group with a strong tendency to accept electrons from an electron-rich group (nucleophile).
See nucleophile; Lewis acid.

electroporation. A process using high-voltage current to make cell membranes permeable to allow the introduction of new DNA.
Use: In recombinant DNA technology.
See transfection.

electrophoresis. Migration of suspended or colloidal particles in a liquid such as rubber latex due to the effect of a potential difference across immersed electrodes. The migration is toward electrodes of charge opposite to that of the particles. Most solids, being negatively charged, migrate to the anode, the exception being basic dyes, hydroxide solutions, and colloids that have adsorbed positive ions, all of which are positively charged and migrate to the cathode. Migrating particles lose their charge at the electrode and generally agglomerate around it. Clay suspensions can be filtered by means of forced-flow electrophoresis.
Electrophoresis is important in the study of proteins because the molecules of such materials act like colloidal particles and their charge is positive or negative according to whether the surrounding solution is acidic or basic. Thus, the acidity of the solution can be used to control the direction in which a protein moves upon electrophoresis. It has been found that electrophoresis can be carried out more efficiently under zero-gravity conditions in outer space than on earth.
Use: Experimentally to separate mixtures of electrically charged species such as proteins or nucleic acids.
See electrodeposition.

electroplating. The deposition of a thin layer or coating of metal (chromium, nickel, copper, silver, etc.) on an object by passing an electric current through an aqueous solution of a salt containing ions of the element being deposited, e.g., Cu^{+2}. The material being plated (usually a metal but often a plastic) constitutes the cathode. The anode is often composed of the metal being deposited; ideally it dissolves as the process proceeds. The thin layer deposited is sometimes composed of two or more metals, in which case it is an alloy. The solution or plating bath contains dissolved salts of all the metals being deposited. Electrolytic cells are used for this process.

The anode must be an electrical conductor, but may or may not be of the same chemical composition as the material being deposited, and may or may not dissolve during the process. The purpose of electroplating is usually protection of the base metal from corrosion. Silver is electroplated on copper for economy reasons; plastics may be electroplated for decorative effects.

See electrophoresis; protective coating; "electroless" coating; throwing power; current density.

electropolishing. A nonmechanical method of polishing metal surfaces that is actually the reverse of electroplating. This is achieved by making the object to be polished the anode in an electrolytic circuit, the cathode usually being carbon. The electrolytes used are phosphoric, hydrofluoric, nitric, and sulfuric acids (sometimes called polishing acids).

electrostatic bond. Alternative name for an ionic bond.
See bond, chemical.

electrostatic coating. A metal-painting technique in which electrostatically charged pigment particles are sprayed onto a substrate metal, followed by baking. The electric charge attracts the particles to the metal and holds them in place until heat treatment is applied. Maintenance of the charge is thus essential; factors affecting this are relative humidity (the lower the better) and the chemical nature of the pigment, for example, phthalocyanine blue retains the charge much longer than titanium dioxide.

electrostatic precipitator. See Cottrell, Frederick G.

electrovalent bond. Alternative name for an ionic bond.
See bond, chemical; ionic bond.

electrowinning. The technique of extracting a metal from its soluble salt by an electrolytic cell. It is used in recovery of zinc, cobalt, chromium, and manganese, and has been applied to copper when in the form of a silicate ore. For any specific metal, the salt in solution is subjected to electrolysis and is electrodeposited on a cathode made of the metal being extracted.

element. One of the 112 presently known kinds of substances that compose all matter at and above the atomic level. According to a theory that has gained acceptance, the lightest elements were formed in less than half an hour from a primordial complex called ylem, a mixture of neutrons and electromagnetic radiation. The smallest unit of any element is the atom. All the atoms of a given element are identical in nuclear charge and number of electrons and protons, but they may differ in mass, e.g., hydrogen has mass numbers of 1, 2, and 3, called hydrogen, deuterium, and tritium, respectively. These are the isotopes of hydrogen; most elements have isotopic forms that are due to the presence of one or more extra neutrons in the nucleus. The atomic number of an element indicates its position in the periodic table and represents the number of protons present, which is the same as the number of electrons.

All elements heavier than lead are unstable and radioactive. About 90% of the earth's crust is made up of elements with even numbers of protons and neutrons. No stable elements heavier than nitrogen have an odd number of both protons and neutrons. Elements of even atomic number normally have several isotopes, while those of odd atomic number never have more than two stable isotopes. All elements beyond uranium (transuranic) were nonexistent in 1940. They are artificially created by bombardment of other elements with neutrons or other heavy particles. Research on new elements is actively carried on at many laboratories including the Lawrence Livermore Laboratories, which reported discovery of element 106 in 1974. Creation of element 109 was announced in 1982, and of element 112 in 1996. A single atom of element 109 was made by West German physicists by bombarding ^{209}Bi with ^{58}Fe nuclei. Controversy continues on the naming of elements 104 through 109. Many more (possibly up to 150) are theoretically possible according to Dr. G. T. Seaborg.

See periodic table; isotope; radioactivity; abundance.
Note: For origin of elements, see nucleogenesis.

elemi. A soft, balsam-like resin obtained from a tree in the Philippines, soluble in coal tar hydrocarbons, but not in petroleum solvents, alcohols, and ketones.
Use: Plasticizer, adhesion of lacquers to metals, cements and adhesives, wax compositions, printing inks, textile and paper coatings, perfumery, waterproofing, engraving.

elimination reaction. A chemical reaction involving elimination of a portion of a reactant compound, resulting in a second compound. Many reactions eliminate water.

"Elimin-Ox" [Nalco]. TM for chemicals for scavenging oxygen.
Use: In water systems.

elongation factors.   Specific proteins required in the elongation of polypeptide chains by ribosomes, or DNA strands by replisomes.

Eltekoff reaction.   Production of highly branched hydrocarbons by methylation of olefins with methyl chloride, or methyl iodide in the presence of lead oxide, or calcium oxide at high temperatures.

eluant.   The liquid used to separate or extract one material from another.
See chromatography.

eluate.   The solution that results from the elution process.
See chromatography.

elutriation.   A process of washing, decantation, and settling that separates a suspension of a finely divided solid into parts according to their weight. It is especially useful for very fine particles below the usual screen sizes, and is used for pigments, clay dressing, and ore flotation.

"Elvace" [H. B. Fuller].   TM for a series of vinyl acetate ethylene emulsions.

"Elvacite" [Lucite].   (butyl methacrylate resins). TM for polymer products.
Use: Resistance to UV weathering and degrading; as base for lacquers, paints, coatings, and inks.

"Elvanol" [Du Pont].   TM for various grades of polyvinyl alcohol.

EMA resins.   Ethylene-maleic anhydride copolymers. Water-soluble resins that serve as dispersing agents in cold-water detergents, thickeners, binders, stabilizers, and emulsifiers.

embosser.   See fiber roll.

embrittlement.   Hardening of a metal (especially steel) or of an ABS resin, resulting in loss of strength and impairment of other physical properties. In metals, the primary cause is exposure to hydrogen, though other factors such as corrosion also are involved. In copolymer plastics, such as ABS resins, embrittlement is due to formation of a vitreous matrix as well as to oxidation of the butadiene particles in the matrix. Embrittlement due to thermal shock occurs in pressurized-water reactors. This may result in rupture of reactor walls and is a constant source of trouble in reactors of this type.

embryo.   Once a zygote begins to undergo cellular divisions, it becomes an embryo.

embryonic stem (ES) cells.   An embryonic cell that can replicate indefinitely, transform into other types of cells, and serve as a continuous source of new cells.

embryotoxin.   Any naturally-occurring substance that destroy or harmfully affect the growth or development of embryos.

Emde degradation.   Modification of the Hofmann degradation method for reductive cleavage of the carbon–nitrogen bond by treatment of an alcoholic or aqueous solution of a quaternary ammonium halide with sodium amalgam. Also used as a catalytic method with palladium and platinum catalysts. The method succeeds with ring compounds not degraded by the Hofmann procedure.

emedastine difumarate.
CAS: 87233-62-3. $C_{17}H_{26}N_4O \cdot 2C_4H_4O_4$.
Hazard: A poison by ingestion.

emerald green.   A pigment consisting of copper acetoarsenite.
Hazard: Toxic by ingestion.

emeralds, synthetic.   Artificial crystals produced by a high-pressure, high-temperature process from beryllium aluminum silicate containing a small amount of chromium.
Use: Lasers, masers, semiconductors.

emery.   See corundum; abrasive.

emetine.   (cephaeline methyl ether; 6′,7′,10,11-tetramethoxyemetan).
CAS: 483-18-1. $C_{29}H_{40}O_4N_2$.
An alkaloid from ipecac.
Properties: White powder; very bitter taste. Mp 74C. Darkens on exposure to light. Soluble in alcohol and ether, slightly soluble in water.
Derivation: By extraction from root of *Cephalis ipecacuanha* (ipecac) or synthetically.
Hazard: Toxic by ingestion.
Use: Medicine (antiamebic).

emission spectroscopy.   Study of the composition of substances and identification of elements by observation of the wavelengths of radiation they emit as they return to a normal state after excitation by an external energy source. When atoms or molecules are excited by energy input from an arc, spark, or flame, they respond in a characteristic manner; their identity and composition are signaled by the wavelengths of incident light they emit. The spectra of elements are in the form of lines of distinctive color, such as the yellow sodium D line of sodium; those of molecules are groups of lines called bands. The number of lines present in an emission spectrum depends on the number and

position of the outermost electrons and the degree of excitation of the atoms. The first application of emission spectra was identification of sodium in the solar spectrum (1814).
See spectroscopy.

emivirine.
CAS: 149950-60-7. $C_{17}H_{22}N_2O_3$.
Hazard: Moderately toxic by ingestion.

emmenagogue.   A drug used to induce menstruation.

Emmert reaction.   Formation of 2-pyridyldialkylcarbinols by condensation of ketones with pyridine or its homologs in the presence of aluminum or magnesium amalgam.

emodin.   (frangula emodin; frangulic acid; 1,3,8-trihydroxy-6-methylanthraquinone).
CAS: 518-82-1. $C_{14}H_4O_2(OH)_3CH_3$.
Occurrence: Either free, or combined with a sugar in a glucoside, in rhubarb, cascara sagrada, and other plants. A synthetic product is also available.
Properties: Orange crystals. Mp 256C. Soluble in alcohol, insoluble in water.
Use: Medicine (cathartic).

emoren.
CAS: 13930-31-9. $C_{28}H_{41}N_3O_3 \cdot xClH$.
Hazard: Moderately toxic by ingestion.

empirical formula.   See formula, chemical.

EMTS.   Abbreviation for ethylmercury-*p*-toluene sulfonanilide.

"Emulmetik" *[Cargill Texturizing].*   TM for hydrogenated lecithin.

emulsifier.   A surface-active agent.
See emulsion.

emulsifying oil.   See soluble oil.

Emulsilac-S.   (sodium stearoyl lactylate) Emulsifier, dough conditioner-strengthener, and whipping agent.
Use: For baked goods, puddings, dips, cheese substitutes, sauces, whipped toppings, and fillings.

emulsion.   A stable mixture of two or more immiscible liquids held in suspension by small percentages of substances called emulsifiers. These are of two types: (1) Proteins or carbohydrate polymers which act by coating the surfaces of the dispersed fat or oil particles, thus preventing them from coalescing; these are sometimes called protective colloids. (2) Long-chain alcohols and fatty acids, which are able to reduce the surface tension at the interface of the suspended particles because of the solubility properties of their molecules. Soaps behave in this manner; they exert cleaning action by emulsifying the oily components of soils. All such substances, both natural and synthetic, are known collectively as detergents.

Polymerization reactions are often carried out in emulsion form; a wide variety of food and industrial products are emulsions of one kind or another, e.g., floor and glass waxes, drugs, paints, shortenings, textile and leather dressings, etc.

All emulsions consist of a continuous phase and a disperse phase: in an oil-in-water (o/w) emulsion, such as milk, water is the continuous phase and butterfat (oil) the disperse phase; in a water-in-oil (w/o) emulsion, such as butter, free fat (from crushed fat globules) is the continuous phase and unbroken fat globules plus water droplets are the disperse phase.
See colloid; protective; phase (2); detergent; surface-active agent; wetting agent.

emulsion breaker.   See demulsification.

emulsion paint.   See paint emulsion.

emulsion polymerization.   Polymerization reaction carried out with the reactants in emulsified form. Performed at normal pressure and −20 to +60C. Many copolymers (synthetic rubbers) are made in this way.

"Emulsiphos" *[Symrise].*   TM for sodium phosphate.
CAS: 7601-54-9.
Grade: Powders.
Use: Food-grade phosphate-emulsifying agent used in processed cheese and cheese food.

Emulvis.   Polyoxyethylene stearate.
Use: Viscosity builder and solubility retarder for cosmetics, soaps, and shampoos.

en.   Abbreviation for ethylenediamine, used in formulas for coordination compounds, e.g., the cobalt complex $Co[en]_3(NO_3)_3$.
See dien; *pn*; *py*.

enamel.   (1) A type of paint consisting of an intimate dispersion of pigments in a varnish or resin vehicle. The vehicle may be an oil-resin mix or entirely synthetic resin. Those containing drying oils are converted to films by oxidation; those composed wholly of synthetic resins may be converted by either heat or oxidation, or both.
See baking finish.
   (2) Porcelain enamel.

enamine.   A group of amino olefins; the name refers especially to unsaturated tertiary amines of the general formula

$$R_2N-\overset{|}{C}=\overset{|}{C}-$$

where R is any alkyl group. Though of little use as end products, enamines are valuable intermediates for many organic syntheses.

enanthaldehyde.   See heptanal.

enanthic acid.   See *n*-heptanoic acid.

enanthyl alcohol.   See heptyl alcohol.

enantiomer.   (enantiomorph). One of a pair of optical isomers containing one or more asymmetric carbon atoms C* whose molecular configurations have left- and right-hand (chiral) forms. These forms are conventionally designated dextro (D) and levo (L) because they compare to each other structurally as do the right and left hands when the carbon atoms are lined up vertically. This is apparent in the enantiomorphic forms of glyceraldehyde; the two structures are mirror images of each other and cannot be made to coincide.

$$\begin{array}{ccc} \text{CHO} & & \text{CHO} \\ | & & | \\ \text{HO}-\overset{*}{\text{C}}-\text{H} \quad (\text{L}) & & \text{H}-\overset{*}{\text{C}}-\text{OH} \quad (\text{D}) \\ | & & | \\ \text{CH}_2\text{OH} & & \text{CH}_2\text{OH} \end{array}$$

Several pairs of enantiomers are possible, depending on the number of asymmetric carbon atoms in the molecule. Compounds in which an asymmetric carbon is present display optical rotation.
See asymmetry; optical isomer; optical rotation.

enantiomorph.   See enantiomer.

encapsulation.   The process in which a material or an assembly of small, discrete units is coated with or imbedded in a molten film, sheath, or foam, usually of an elastomer. A foam-forming plastic may be used to fill the spaces between various electrical or electronic components so that they are imbedded in and supported by the foam. Plastics and other materials used for this purpose are often called potting compounds. A specialized use of this technique is in growing crystals for semiconductors, in which a coating of liquid boric oxide is the encapsulating agent. Use of a glassy silicate coating to encapsulate nuclear waste for permanent disposal is under investigation.
See microencapsulation.

3′ end.   The end of a nucleic acid that lacks a nucleotide bound at the 3′ position of the terminal residue.

endergonic reaction.   A chemical reaction that consumes energy (that is, for which DG is positive). This is a nonspontaneous process.

Endic anhydride.   *Endo-cis*-bicyclo(2.21)-5-heptene-2,3-dicarboxylic anhydride. ($C_9H_8O_3$).

Properties: White crystals. Mp 163C. Soluble in aromatic hydrocarbons, acetone, ethanol.
Use: Elastomers, plasticizers, fire retardant chemicals, resins, and epoxy curing systems.

endo-.   A prefix used in chemical names to indicate an inner position, specifically (1) in a ring rather than a side chain or (2) attached as a bridge within a ring.
See *exo-*.

endocrine.   A hormonal substance that is secreted by glands or other organs directly into the blood or lymph that affects specific tissues or organs.

endocrine glands.   Groups of cells specialized to synthesize hormones and secrete them into the blood to regulate other types of cells. The pancreas, for example.

endocytosis.   The uptake of extracellular material by its inclusion within a vesicle (endosome) formed by an invagination of the plasma membrane.

endoergic.   See Endothermic.

endogenous ammonia.   Ammonia that has been detoxified by conversion to urea, which is temporarily stored in the urinary bladder.

endomycin.   An antifungal antibiotic complex produced by streptomyces.

endonuclease.   An enzyme that digests nucleic acids starting in the middle of the strand (as opposed to an exonuclease, which must start at an end). Examples include the restriction enzymes, DNase I and RNase A.
See restriction enzyme.

endopeptidase.   Any proteolytic enzyme that catalyzes the breakage of a polypeptide chain at an internal peptide linkage.

endoplasm.   The innermost, more fluid portion of the cytoplasm of a cell.

endoplasmic reticulum.   An extensive system of double membranes in the cytoplasm of eukaryotic cells. The two major types are smooth and rough endoplasmic reticulum. Functions in calcium storage as well as in protein and lipid synthesis.

Endor.   A rubber peptizing agent containing activated zinc salt of pentachlorothiophenol. $(C_6Cl_5S)_2Zn$ and 80% inert filler.
Properties: Grayish-green powder. D 2.39.

endorphin.   Any of a group of polypeptides formed in the brain tissue and pituitary gland of higher animals that are thought to control the transfer of signals at nerve junctions, thus ensuring that behavior patterns in the individual remain normal. Imbalance or malfunction of these polypeptides has been reported to be a factor in irrational and violent actions and other emotional disorders, as well as in epilepsy and memory processes. This belongs to a developing field of medicinal chemistry called neuropharmacology.

β-endorphin.
CAS: 60617-12-1.
Hazard: A reproductive hazard.

endosmosis.   The passing of a fluid inward through a porous partition toward another fluid of different character.
See exosmosis; Perrin rule.

endosulfan.   (6,7,8,9,10,10-hexachloro-1,5,5a,6,9,9a-hexahydro-6,9-methano-2,4,3-benzodioxathiepin-3-oxide).
CAS: 115-29-7. $C_6H_6Cl_6O_3S$.
Properties: (commercial product): Brown crystals. Mp 70–100C (pure: mp 106C). Mixture of two isomers. Mp 108–109C and 206–208C.
Hazard: Toxic by ingestion, inhalation, and skin absorption; use may be restricted. Lower respiratory tract irritant; liver and kidney damage. Questionable carcinogen.
Use: Insecticide.

endosymbiosis.   When one organism takes up permanent residence within another, so that the two become a single functional organism. Mitochondria and plastids are believed to have resulted from endosymbiosis.

endothall.   (7-oxalobicyclo-[2.2.1]-heptane-2,3-dicarboxylic acid disodium salt).
CAS: 145-73-3. $C_8H_8Na_2O_5$.
Hazard: Strong irritant to eyes and skin.
Use: Defoliant, herbicide.

endothermic.   A process or change that takes place with absorption of heat and requires high temperature for initiation and maintenance. An example is production of carbon monoxide and hydrogen by passing steam over hot coke.

endothion.   [S-(5-methoxy-4-oxo-4H-pyran-2-yl)-methyl] O,O-dimethylphosphorothioate).
CAS: 2778-04-3.

$(CH_3O)_2P(O)SCH_2C:CHC(O)C(OCH_3):CHO.$

Properties: Crystals. Mp 90–91C. Very soluble in water; soluble in chloroform and ethanol.
Hazard: Toxic by ingestion, cholinesterase inhibitor; use may be restricted.
Use: Insecticide.

end point.   (1) In chemical analysis, the point during a titration at which a marked color change is observed, indicating that no more titrating solution is to be added.
See indicator.
(2) The highest temperature reached during an assay distillation of hydrocarbon liquids, indicating the overall volatility of the liquid (ASTM).

end-product inhibition.   Inhibition of an allosteric enzyme at the beginning of a metabolic sequence by the end product of the sequence.

endrin.   (1,2,3,4,10,10-hexachloro-6,7-epoxy-1,4,4a,5,6,7,8,8a-octahydro-1,4-*endo,endo*-5,8-dimethanonaphthalene).
CAS: 72-20-8. $C_{12}H_8OCl_6$.
A stereoisomer of dieldrin which is the *endo, exo* isomer.
Properties: White, crystalline powder. Mp approximately 200C (rearranges above this point). Insoluble in water and methanol; moderately soluble in other common organic solvents; compatible with nonacidic fertilizers, herbicides, fungicides, and insecticides.
Hazard: Toxic by inhalation and skin absorption, use may be restricted. Headache, liver damage, and central nervous system impairment. Questionable carcinogen.
Use: Insecticide.
See aldrin.

-ene.   (1) Suffix denoting open-chain unsaturated hydrocarbons having one double bond, as in propene. (2) Suffix denoting an aromatic cyclic hydrocarbon, as in benzene.

enercology.   A coined word defined as the balanced relationship between energy and ecology. A foundation devoted to practical applications of this relationship has been established at Alma College, Michigan.

Ene reaction.   Addition of an olefin with an allylic hydrogen (ene) to a compound with a double bond (enophile), involving the allylic shift of one double bond, transfer of the allylic hydrogen to the enophile, and bonding between the two unsaturated termini.

energy.   The fundamental active entity in the universe, defined as the capacity for doing work. Two famous equations relate energy to mass as either $E = mc^2$ (Einstein) or $E = h\beta n$ (Planck). These equations show that energy cannot be completely divorced from mass, as the two are, to some extent interconvertible.
The law of conservation of energy, simply stated, is that the sum total of energy in the universe is constant; therefore, energy cannot be either created or destroyed, but only converted from one form to another.

Radiant energy (light) comprises the electromagnetic spectrum; all wavelengths of light are composed of photons or packets of energy traveling at the speed of light. Theoretically, they have no mass except that associated with their speed. Protons, electrons, and neutrons are forms of highly condensed energy that possess determinable mass but move at lower speeds.

Energy is directly related to chemical phenomena in the formation and decomposition of compounds, in the many important reactions that occur in electrochemistry, and in the release of energy in nuclear fission and fusion. Free energy is a thermodynamic function; in chemical reactions, it is a measure of the extent to which a substance can react. Kinetic energy (the energy of motion) is most clearly exhibited in gases, in which molecules have much greater freedom of motion than in liquids and solids.

See radiation; matter; thermodynamics; free energy.

Note: "One of the most difficult challenges we face is to find ways to ensure that all peoples of the world share more equitably the vast human benefits that energy can bring. The foundation of worldwide energy policy must be based on energy conservation and the development of additional sources through a judicious application of science and technology." (Glenn T. Seaborg, ACS meeting, April 1976).

energy, activation.   The minimum energy which a molecule must acquire before it can be regarded as being activated. The difference between the energy of an activated molecule and the mean energy of all the molecules.

energy balance.   (1) Energy required to carry on an operation or to maintain the desired operating conditions. (2) Calculation of energy required to balance a reaction.

energy band.   Energy spectrum of valence electrons in a polyatomic material.

energy charge.   The fractional degree to which the ATP/ADP/AMP system is filled with high-energy phosphate groups.

energy converter.   Any element or compound having the ability to convert the radiant energy of sunlight into electrical, thermal, or chemical energy. Prominent among them are silicon, selenium, and tellurium, as well as the chlorophyll of plants in photosynthesis.

See solar cell; magnetohydrodynamics.

energy coupling.   The transfer of energy from one process to another. For example, the coupling of ATP hydrolysis with glucose phosphorylation.

energy gap.   Forbidden part of energy spectrum of valence electrons.

energy level.   See quantum state.

Energy Smart.   A nutritive carbo sweetener made from fruit juice and dextrin.

Use: Nutritive sweetener that is stable at room temperatures.

energy sources.   Nonrenewable energy sources are materials of geologic origin, i.e., petroleum, natural gas, coal, shale oil, and uranium, which cannot be replaced once their supply is exhausted. Renewable sources, on the other hand, are those that can be replenished on a predictable time basis and are known collectively as biomass; these include such cellulosic products as wood, bagasse, agricultural waste, and residue from the forest products industries (shavings, bark, sawdust, etc.), which yield the same heating value per dry ton as one barrel of crude oil. Several of these sources have been in use for some years as on-site fuels in the sugarcane, plywood, and paper pulp industries. Another instance is methane obtained from animal manures (biogas), which is being developed on a large scale in the west and southwest. Interest in the mass cultivation of algae and hydrocarbon-producing plants such as guayule and copaiba for fuel has been reported.

There are also a number of mechanical energy sources whose development involves engineering rather than chemistry, namely, solar radiation, wind, water flow, tides, and thermal gradients in ocean water. Several of these are already in limited operation, but in most cases, their development will be slow and costly. Hydroelectric power has long been an accomplished fact, though it accounts for only about 1% of electrical needs. Solar energy is being actively researched, but is unlikely to be a factor of consequence for at least another decade.

The following entries discuss the chemical sources of energy more specifically:

battery	hydrocarbon (note)
biogas	hydrogen
biomass	methane
breeder	natural gas
carbohydrate	nuclear energy
cellulose	ocean thermal energy
coal	oil sands
conversion	peat
copaiba	petroleum
ethyl alcohol (note)	plutonium
fission	radiation
fuel cell	shale oil
fuel oil	solar cell
fusion	solar energy
gasification	solar pond
gasohol	storage (4)
gasoline	thermoelectricity
geothermal	energy uranium
guayule	wood

energy storage. See storage (4).

enfleurage. Extraction of odoriferous components of flowers by means of fats or mixtures of fat and tallow, the process being carried out at room temperature to avoid decomposition of the desired perfumes. The latter are separated from the fat by washing with alcohol.
See essential oil; perfume.

enflurane. (2-choro-1,1,2-trifluoroethyl difluoromethyl ether).
CAS: 13838-16-9. $C_3H_2ClF_5O$.
Properties: Clear, colorless liquid; mild, sweet odor. Mw 184.50, d 1.5167 (25C), bp 56.5C, vap press 174.5 mm Hg (20C). Soluble in organic solvents; slightly soluble in water.
Hazard: Volatile with anesthetic properties, but nonflammable. Cardiac and central nervous system impairment. Questionable carcinogen.
Use: Clinical anesthetic.

engineering material. A metal, alloy, plastic, or ceramic used in the fabrication of machinery and its components, structural shapes, chemical process equipment, castings, tools, instruments, drums, tanks, piping, ductwork, and auxiliary items (excluded are materials used chiefly as protective coatings or as components of alloys). Engineering materials are characterized by hardness, strength, machinability, dimensional stability, nonflammability, and resistance to corrosion, most acids, solvents, and heat. The more important of these are listed below; for specific uses, see individual entries.

Metals and alloys	Plastics
aluminum	ABS resin
beryllium	acetal resin
brass	acrylic resin
bronze	fluorocarbon polymer
cast iron	nylon
copper	phenolformaldehyde resin
lead	polybutylene terephthalate
magnesium	polycarbonate
steel	polyethylene
tantalum	polyphenylene oxide
titanium	polypropylene
zinc	polystyrene
zirconium	polyvinyl chloride
various trademarked	reinforced plastics (FRP)
alloys	ureaformaldehyde resin
Ceramics	
glass	
porcelain	

Enhance. An ultra-high solid floor polish.

enhanced oil recovery. Any of several methods for increasing the productivity of oil wells (after pumping is no longer effective) by emulsifying as much as possible of the oil trapped in the rock structure. Techniques that have been researched and used to some extent involve the pumping into the formations of pressurized carbon dioxide, water, steam, detergent solutions, brine, and a mixture of various high polymers, guar gum, xanthan biopolymer, bauxite, or sand. Considerable success has been achieved, but such methods have proved to be both slow and inefficient.
See chemical flooding; hydraulic fracturing.

enhancer. (1) A food additive that brings out the taste of a food product without contributing any taste of its own. Sodium glutamate is the most widely used substance of this class; its effective concentration is in parts per thousand. (2) a nucleotide sequence to which transcription factor(s) bind, and which increases the transcription of a gene. It is NOT part of a promoter; the basic difference being that an enhancer can be moved around anywhere in the general vicinity of the gene (within several thousand nucleotides on either side or even within an intron), and it will still function. It can even be clipped out and spliced back in backwards, and will still operate. A promoter, on the other hand, is position- and orientation-dependent. Some enhancers are "conditional" in that they enhance transcription only under certain conditions.
See potentiator.

enocianina. See grape skin extract.

enol. A chemical grouping containing both a double bond (ene) and a hydroxyl group (OH), forming an intermediate and reversible product. Enols are characteristic of racemic compounds.

enolase. An enzyme active in glycolysis that catalyzes the conversion of 2-phosphoglyceric acid to the phosphorylated enol of pyruvic acid.

enology. (oenology). The art and science of winemaking.

Enovid. Norethynodrel with mestranol. Oral contraceptive approved by FDA.

enprostil.
CAS: 73121-56-9. $C_{23}H_{28}O_6$.
Hazard: Moderately toxic by ingestion.

enriched uranium. Uranium in which the proportion of uranium-235 has been increased by the mechanical removal of uranium-238.

enrichment. (1) In food technology, the addition to a foodstuff of various nutrient substances during manufacture to increase the dietary value of the food, e.g., addition to wheat flour of vitamins B_1,

B_2, niacin, and iron. In this way, the food is brought up to a specific nutritional standard.

(2) Increase in the abundance of certain isotopes of an element by any of several methods: (a) By a chemical reaction accompanied by irradiation from a laser beam; enrichment of boron, chlorine, and sulfur isotopes has been achieved in this way in a number of research laboratories. Uranium enrichment is also possible, either by adding previously prepared ^{235}U to natural uranium or by the laser technique. (b) Uranium can also be enriched by gas centrifugation and gaseous diffusion. The latter is the usual procedure.

(3) Addition of oxygen to air to increase its combustion-supporting ability.

enrobe. To coat candy centers, fruits, nuts, etc., with a liquid confection such as chocolate or sugar solution by mechanical dipping.

ensilage. A feed for livestock prepared by long storage of corn husks, stalks, etc., in an air-tight vertical cylindrical structure (silo) in which the material undergoes anaerobic fermentation. It is a component of biomass.

entegenen. (E). A descriptor for substances that have achiral elements due to the presence, on opposite sides of the vertical reference plane, of the highest priority groups.

enterbotactin. (enterochelin). A siderophore found in enteric bacteria such as *Escherichia coli*.

enterokinase. An enzyme found in the small intestine, which converts trypsinogen into trypsin.

enterotoxin b, staphylococcal.
CAS: 11100-45-1.
Hazard: A poison.
Source: Natural product.

enthalpy. The increase in heat content of a substance or system that accompanies its change from one state to another under constant pressure. The internal energy plus the product of the volume and pressure of a working substance.

entrainer. An additive for liquid mixtures difficult to separate by ordinary distillation. The entrainer usually forms an azeotrope with one of the compounds of the mixture and thereby aids in the separation of such a compound from the remainder of the mixture.

entrainment. The presence of minute drops of water or other liquid in the vapor produced by evaporation or distillation; also small bubbles of air or other gas in a liquid as a result of turbulence induced by agitation. Entrained water or gas is often undesirable, and care needs to be taken to eliminate or prevent it; in certain cases it is beneficial.
See venturi; entrainer.

entropy. (*S*). Thermodynamic concept referring to heat content. If, during a reversible change at temperature *T*, an amount dQ of heat enters the substance, its entropy is augmented by

$$dS = \frac{dQ}{T}$$

environmental chemistry. That aspect of chemistry concerned with air and water pollution, pesticides, and chemical and radioactive waste disposal. A random selection of specific areas of research includes lead and other toxic chemicals in the air, effects of increased burning of coal, biological modification of wastes, detoxification methods, pesticide content of fish, environmental analytical and monitoring techniques, utilization of biomass, drinking water quality, organic contaminants in lakes and rivers, and effect of deforestation on carbon dioxide and oxygen content of air.

Environmental Protection Agency. (EPA). A federal agency established in 1970. Under the Toxic Substances Control Act of 1976, it is required to ensure the safe manufacture, use, and transportation of hazardous chemicals. It includes an Office of Hazardous Materials Control, which administers Congressional legislation pertaining to this field. The EPA may require manufacturers to conduct tests on materials or products that adversely affect the environment or public health and safety. One of the most important aspects of its activities is the establishment and supervision of automotive emission standards. It is also concerned with pesticides, fungicides, and other potentially detrimental materials, as well as industrial waste disposal. It operates in close conjunction with the U.S. Department of Agriculture and the Food and Drug Administration. The construction of new plants for manufacturing products must conform to EPA standards, especially as regards effluents that contribute to water pollution. The headquarters are located at 1200 Pennsylvania Avenue, N.W., Washington, DC 20460. Website: http://www.epa.gov
See toxic substances; environmental chemistry.

enzyme. Any of a unique class of proteins that catalyze a broad spectrum of biochemical reactions. Enzymes are formed in living cells; they are comprised one or more polypeptide chains with molecular weight from 10,000 to a million or more. An important characteristic of enzymes is their specificity, i.e., a given enzyme can catalyze one particular reaction and no others. Six types are recognized, which catalyze the following reactions: (1) redox (oxidoreductases), (2) transfer of specific radicals or groups (transferases), (3) hydrolysis (proteolytic), (4) removal from or addition to the substrate of specific chemical groups (lyases), (5) isomerizaton (isomerases), (6) combination or binding together of substrate units (ligases). The names of enzymes invariably terminate in either *-ase* or *-in*. The following partial list indicates some of the more important functions performed by enzymes; among these are the ability to cleave the peptide bonds of proteins (hydrolysis) with simultaneous formation of water; and to decompose sugars and starches to ethyl alcohol and carbon dioxide (fermentation). Enzymes are essential to many biochemical processes, especially in the food, beverage, and pharmaceutical industries.

amylase	starch hydrolysis
carboxylase	decomposes pyruvic acid
cellulase	converts cellulose to glucose
cholinesterase	inactivates acetylcholine
chymotrypsin	hydrolysis of proteins
invertase	converts sucrose to glucose and fructose
lipase	hydrolysis of fats
maltase	converts maltose to glucose
pepsin	hydrolysis of proteins
protease	hydrolysis of peptide linkages
rennin	hydrolysis of proteins
ribonuclease	decomposes RNA
trypsin	splits proteins to amino acids
urease	decomposes urea to NH_4 and CO_2
zymase	converts sugars to alcohol and CO_2 (fermentation)

Recent research in biomimetic chemistry has succeeded in creating synthetic enzymes that imitate the behavior of natural enzymes, e.g., chymotrypsin, and are almost as effective.
See catalyst; biomimetic chemistry; fermentation; hydrolysis.

eolotropic. See aeolotropic.

eosin. (bromeosin; CI 45380; tetrabromofluorescein).
CAS: 15086-94-9. $C_{20}H_8Br_4O_5$.
Properties: Red, crystalline powder. Soluble in alcohol and acetic acid; insoluble in water. The potassium and sodium salts are soluble in water.
Derivation: Bromination of fluorescein.
Hazard: Questionable carcinogen.
Use: Dyeing silk, cotton, and wool; red writing ink; cosmetic products; biological stain; coloring motor fuel.

EP. (1) Abbreviation for extreme pressure as applied to lubricants. (2) Abbreviation for ethylene-propylene.

EPA. Abbreviation for Environmental Protection Agency.

EPAL. Linear primary alcohols.

EPC black. Abbreviation for easy-processing channel black.
See carbon black.

EPDM. Abbreviation for a terpolymer elastomer made from ethylene-propylene diene monomer.
See ethylene-propylene terpolymer.

ephedrine. (1-phenyl-2-methylaminopropanol).
CAS: 299-42-3. $C_6H_5CH(OH)CH(NHCH_3)CH_3$.
Optically active (levorotatory) form.
Properties: White to colorless granules, pieces, or crystals; unctuous to touch. Mp 33–40C, bp 255C (decomposes). Hygroscopic, gradually decomposes on exposure to light. Soluble in water, alcohol, ether, chloroform, and oils.
Derivation: Isolation from stems or leaves of *Ephedra*, especially Ma huang (China and India).
Grade: Technical, NF.
Hazard: Toxic by ingestion.
Use: Medicine (bronchodilator).
See racephedrine.

epi. (1) A prefix denoting a bridge or intramolecular connection, e.g., epoxide. (2) An abbreviation for epichlorohydrin.

EPI. A cleaning agent.
Use: For cleaning thermoplastic processing equipment.

(+−)-epibatidine dihydrochloride.
CAS: 162885-01-0. $C_{11}H_{13}ClN_2$•2ClH.
Hazard: A poison.
Use: Agricultural chemical.

epichlorohydrin. (chloropropylene oxide).
CAS: 106-89-8. An epoxide.

$\overline{CH_2OCHCH_2Cl}$.

$$\begin{array}{l} CH_2 \\ | \diagdown \\ CH O \\ | \diagup \\ CH_2Cl \end{array}$$

Properties: Highly volatile, unstable liquid; chloroform-like odor. D 1.1761 (20/20C); bp 115.2C; bulk d 9.78 lb/gal; vap press 12.5 mm Hg (20C); fp −25C; viscosity 1.12 cP (20C) refr index 1.4358 (25C); flash p 93F (33.9C) (TOC). Miscible with most organic solvents; slightly soluble in water.
Derivation: By removing hydrogen chloride from dichlorohydrin.
Hazard: Toxic by inhalation, ingestion, and skin absorption; strong irritant, a carcinogen. Flammable, moderate fire risk. TLV: 0.5 ppm; animal carcinogen.

Use: Major raw material for epoxy and phenoxy resins, manufacture of glycerol, curing propylene-based rubbers, solvent for cellulose esters and ethers, high-wet-strength resins for paper industry.

epichlorohydrin-bis(3-aminopropyl)methylamine copolymer.
CAS: 41941-50-8. $(C_7H_{19}N_3 \cdot C_3H_5ClO)_x$.
Hazard: Low toxicity by ingestion.

epichlorohydrin-dimethylamine copolymer.
See dimethylamine-epichlorohydrin copolymer.

epichlorohydrin triethanolamine cellulose.
See ECTEOLA-cellulose.

epigallocatechin 3-gallate.
CAS: 989-51-5. $C_{22}H_{18}O_{11}$.
Hazard: Moderately toxic by ingestion.

epigenetic carcinogen. (nongenotoxic carcinogen). A carcinogen that does not interact directly with DNA. Such agents may alter methylation partners or the tertiary structure of DNA.

epimer. An isomer that differs from the compound with which it is being compared only in the relative positions of an attached hydrogen and hydroxyl. The isomerism may be represented as – HCOH– and –HOCH–. It is common in sugars. See diastereoisomer.

epimerases. Enzymes that catalyze the reversible interconversion of two epimers.

epimers. Two stereoisomers differing in configuration at one asymmetric center in a compound having two or more asymmetric centers.

epinephrine. (l-methylaminoethanolcatechol; Adrenaline).
CAS: 51-43-4.

OH
OH
CHOH—CH₂—NH—CH₃

A hormone of the adrenal glands.
Properties: (*l*-Form) Light brown or nearly white, crystalline powder; odorless. Affected by light. Mp 211–212C, specific rotation −50 to −53.5 (25C). Sparingly soluble in water; insoluble in alcohol, chloroform, ether, acetone, oils; readily soluble in aqueous solutions of mineral acids, sodium hydroxide, and potassium hydroxide.
Derivation: From the adrenal glands of sheep and cattle or synthetically from pyrocatechol.

Grade: USP.
Use: Medicine (vasoconstrictor).

Epiphen. An epoxy resin in liquid form. Epiphen ER-823 is used in adhesives for rubber, steel, aluminum, or glass. Catalyst is supplied for specific end uses.

3-episarmentogenin.
CAS: 465-12-3. $C_{23}H_{34}O_5$.
Hazard: A poison.
Source: Natural product.

epistasis. One gene interferes with or prevents the expression of another gene located at a different locus.

epitaxy. An oriented crystalline growth between two crystalline solid surfaces of different chemical composition in which the surface of one crystal offers suitable positions for deposition of a second crystal. This behavior is characteristic of some types of high polymers.

EPI-TDPA. Plastic film products.
Use: For films that will degrade in landfills, photodegrade and biodegrade.

epithelial cell. Any cell that forms part of the outer covering of an organism or organ.

epitope. A portion of an antigenic macromolecule recognized and bound by a specific antibody.

EPN. (*o*-ethyl-*o,p*-nitrophenyl phenylphosphorothioate).
CAS: 2104-64-5. $C_6H_5P(C_2H_5O)(S)OC_6H_4NO_2$.
Properties: Light-yellow crystals. Mp 36C, d 1.5978 (30C). Insoluble in water; soluble in most organic solvents. Decomposes in alkaline solutions.
Grade: Wettable powders and dusts.
Hazard: A cholinesterase inhibitor, absorbed by skin, use may be restricted. Questionable carcinogen.
Use: Cotton insect pest control, acaricide.

epolamine. See 1-pyrrolidineethanol.

"Eponol" Resins [Momentive]. TM for high molecular weight linear copolymers of bisphenol A and epichlorohydrin; produce outstanding surface coatings by solvent evaporation alone.

"Epon" resins [Momentive]. TM for a series of condensation products of epichlorohydrin and bisphenol A having excellent adhesion, strength, chemical resistance, and electrical properties when formulated into protective coatings, adhesives, and structural plastics.

"Epotuf" *[Reichhold].* TM for epoxy resins, epoxy hardeners, and epoxy esters used as coating vehicles.

epoxidation. Reaction in which olefinic unsaturation is converted to a cyclic, three-membered ether by active oxygen agents.

epoxide. An organic compound containing a reactive group resulting from the union of an oxygen atom with two other atoms (usually carbon) that are joined in some other way as indicated:

$$=\overset{O}{\overset{\triangle}{C-C}}=$$

This group, commonly called "epoxy," characterizes the epoxy resins. Epichlorohydrin and ethylene oxide are well-known epoxides. The compounds are also used in certain types of cellulose derivatives and fluorocarbons.

epoxidized linseed oil. See "Drapex" *[Galata].*

epoxidized soybean oil.
Properties: Iodine number maximum of 6, oxirane oxygen minimum of 6.0%.
Use: Food additive.

1,2-epoxybutane. See 1,2-butylene oxide.

3,4-epoxycyclohexane carbonitrile.
$O(C_6H_9)CN$.
Properties: Liquid. D 1.0929 (20/20C), bp 244.5C, fp −33C. Soluble in water.
Hazard: Toxic by skin absorption, ingestion, and inhalation.
Use: Intermediate, stabilizer.

epoxyethane. See ethylene oxide.

2,3-epoxy-2-ethylhexanol.

$C_3H_7CHOC(C_2H_5)CH_2OH$.

Properties: Liquid. D 0.9517 (20/20C), bp (decomposes), fp −65C. Slightly soluble in water. Combustible.
Hazard: Skin irritant.
Use: Stabilizer, intermediate.

epoxyheptachlor. (HCE; heptachlor epoxide).
CAS: 1024-57-3. $C_{10}H_5Cl_7O$.
Hazard: Confirmed carcinogen and poison by ingestion.

11,13-(epoxymethano)-13h-cyclopenta(a) phenanthrene, pregn-4-ene-3,20-deriv.
CAS: 6251-69-0. $C_{21}H_{28}O_5$.
Hazard: A reproductive hazard.
Use: Hormone.

epoxy novolak. Epoxy resin made by the reaction of epichlorohydrin with a novolak resin (phenol-formaldehyde; see novolak). These have a repeating epoxide structure that offers better resistance to high temperatures than the epichlorohydrin—bisphenol A type, and are especially useful as adhesives.

2,3-epoxy-1-propanol. See glycidol.

epoxy resin. A thermosetting resin based on the reactivity of the epoxide group. One type is made from epichlorohydrin and bisphenol A. Aliphatic polyols such as glycerol may be used instead of the aromatic bisphenol A. Molecules of this type have glycidyl ether structures, $-OCH_2CHOCH_2$, in the terminal positions, have many hydroxyl groups, and cure readily with amines.
Another type is made from polyolefins oxidized with peracetic acid. These have more epoxide groups, within the molecule as well as in terminal positions, and can be cured with anhydrides, but require high temperatures. Many modifications of both types are made commercially. Halogenated bisphenols can be used to add flame-retardant properties.
See epoxy novolak.
The reactive epoxies form a tight, cross-linked polymer network and are characterized by toughness, good adhesion, corrosive-chemical resistance, and good dielectric properties.
Most epoxy resins are the two-part type, which harden when blended. A one-component liquid type for filament winding and a pelletized type for injection molding are available under the TM "Arnox."
Hazard: Strong skin irritant in uncured state.
Use: Surface coatings, as on household appliances and gas storage vessels; adhesive for composites and for metals, glass, and ceramics; casting metal-forming tools and dies; encapsulation of electrical parts; filament-wound pipe and pressure vessels; floor surfacing and wall panels; neutron-shielding materials; cements and mortars; non-skid road surfacing; rigid foams; oil wells (to solidify sandy formations); matrix for stained-glass windows; low-temperature mortars.

(3-α)-12,13-epoxytrichothec-9-ene-3,4,15-triol 15-acetate.
CAS: 2623-22-5. $C_{17}H_{24}O_6$.
Hazard: A poison by ingestion. A mild skin irritant.
Source: Natural product.

EPR. Abbreviation for ethylene propylene rubber, also for electron paramagnetic resonance.

epsilon acid. (1-naphthylamine-3,8-disulfonic acid).
CAS: 117-43-1. $C_{10}H_5(NH_2)(SO_3H)_2$.
Properties: White, crystalline scales. Soluble in hot water.

Derivation: Naphthalene-1,5- and 1,6-sulfonic acids are nitrated and reduced, giving 1-naphthylamine-3,8- and 4,8-disulfonic acids. Separation is effected by crystallizing out the acid sodium salts of 1-naphthylamine-3,8-disulfonic acid.
Use: Azo-dye intermediate.

Epsom salts.   See magnesium sulfate.

EPT.   Abbreviation for ethylene-propylene terpolymer.

Eptac No. 1.   Zinc dimethyldithiocarbamate, an ultraaccelerator for rubber.

EPTC.   (*S*-ethyl di-*N*,*N*-propylthiocarbamate). CAS: 759-94-4. $C_2H_5SC(O)N(C_3H_7)_2$.
Available forms: Liquid and granular formulations.
Use: A preemergence herbicide.

eq.   Abbreviation for gram equivalent weight, i.e., the equivalent weight in grams. Recommended as an international unit.

Equanil.   Meprobamate.
Use: Sedative.

equation of state.   The mathematical formula that expresses the relationships between pressure, volume, and temperature of a substance in any state of aggregation.

equilibrium.   (1) Chemical equilibrium is a condition in which a reaction and its opposite or reverse reaction occur at the same rate, resulting in a constant concentration of reactants; for example, ammonia synthesis is at equilibrium when ammonia molecules form and decomposes at equal velocities $(N_2 + 3H_2 \leftrightarrow 2NH_3)$.
(2) Physical equilibrium is exhibited when two or more phases of a system are changing at the same rate so that the net change in the system is zero. An example is the liquid-to-vapor-vapor-to-liquid interchange in an enclosed system, which reaches equilibrium when the number of molecules leaving the liquid is equal to the number entering it.

equilibrium constant.   A number that relates the concentrations of starting materials and products of a reversible chemical reaction to one another. For example, for a chemical reaction represented by the equation aAB + bCD $\leftrightarrow$ cAD + dBC, the equilibrium constant would be $K = [(AD)c(BD)d]/[(AB)a(CD)b]$, where a, b, c, and d are the numbers of molecules of AB, CD, AD, and BC that occur in the balanced equation, and (AD), (BC), (AB), and (CD) are the molecular concentrations of AD, BC, AB, and CD in any mixture that is at equilibrium. At any one temperature, K is usually at least approximately constant, regardless of the relative quantities of the several substances, so that when K is known, it is often possible to predict the concentrations of the products when those of the starting materials are known. The constant changes markedly with temperature. The constant can often be calculated from the relations of thermodynamics if the free energy for the chemical reaction is known, or by measuring all concentrations in one or more carefully conducted experiments.

equilibrium   diagram.   (constitutional diagram).
(1) A simplified boiling-point diagram, showing for a liquid mixture, the composition of the vapor in equilibrium with the liquid. (2) A chart showing the relation between a solution and the solids that may be crystallized from it. (3) A diagram showing the limits of composition and temperature in which the various phases or constituents of an alloy are stable.

equipartition, law of.   Every particle, heavy or light, gaseous or liquid, and independent of its chemical nature or form, always possesses the same mean energy of translation at a given temperature.

equipotential energy.   The energy existing at a constant potential throughout a system.

equivalent electrons.   Electrons of equal azimuthal quantum numbers and principal quantum numbers. They have identical orbital properties but may have a difference in sign of their orbital moments.

equivalent weight.   (combining weight). The weight of an element that combines chemically with 8 g of oxygen or its equivalent. Since 8 g of oxygen combines with 1.008 g of hydrogen, the latter is considered equivalent to 8 g of oxygen. When 8 g is selected for the combining weight of oxygen, no element has a combining-weight value greater than one. The equivalent weight of an acid is the weight that contains one atomic weight of acidic hydrogen, i.e., hydrogen that reacts during neutralization of acid with base. The equivalent weight of a base or hydroxide is the weight that will react with an equivalent weight of acid. Equivalent weights of other substances are defined in a similar manner.

Er.   Symbol for erbium.

ER.   See endoplasmic reticulum.

ERAD.   Endoplasmic   reticulum-associated degradation (of protein). Incorrectly formed proteins are destroyed in the ER instead of being secreted.

erbia.   See erbium oxide.

erbium.   Er. Element with atomic number 68, aw 167.26, valence of 3; one of the rare-earth elements of the yttrium subgroup.
See rare earth.

Properties: Soft, malleable solid with metallic luster. Salts are pink to red. D 9.16 (15C), mp 1522C, bp approximately 2500C, high electrical resistivity. Insoluble in water; soluble in acids.
Derivation: Reduction of the fluoride with calcium, electrolysis of the fused chloride.
Source: Rare-earth minerals.
Available forms: Lumps, ingots of high purity, sponge, powder.
Hazard: Flammable in finely divided form.
Use: Nuclear controls, special alloys, room-temperature laser.

erbium nitrate. $Er(NO_3)_3 \cdot 5H_2O$.
Properties: Large, reddish crystals. Loses $4H_2O$ at 130C. Soluble in water, alcohol, ether, and acetone.
Derivation: Treatment of oxides, carbonates, or hydroxides with nitric acid.
Grade: 99%.
Hazard: May explode if shocked or heated.

erbium oxalate. $Er_2(C_2O_4)_3 \cdot 10H_2O$.
Properties: Reddish, microcrystalline powder. Decomposes at 575C, d 2.64. Soluble in water and dilute acids.
Hazard: Corrosive.
Use: Oxalates of the rare-earth metals are used to separate the latter from common metals.

erbium oxide. (erbia). Er_2O_3.
Properties: Pink powder that readily absorbs moisture and carbon dioxide from the atmosphere. D 8.64, specific heat 0.065, infusible. Insoluble in water; slightly soluble in mineral acids.
Derivation: By heating the oxalate or other oxy-acid salts.
Grade: 98–99%.
Use: Phosphor activator, infrared-absorbing glass. See rare earth.

erbium sulfate. $Er_2(SO_4)_3 \cdot 8H_2O$.
Properties: Pink, monoclinic crystals. D 3.68, dehydrated at 400C. Soluble in water.
Derivation: Dissolving hydroxides, carbonates, or oxides in dilute sulfuric acid.
Grade: 99.9%.
Use: To determine atomic weight of a rare-earth element.

ERE. (estrogen response element). A binding site in a promoter to which the activated estrogen receptor can bind. The estrogen receptor is essentially a transcription factor which is activated only in the presence of estrogens. The activated receptor will bind to an ERE, and transcription of the adjacent gene will be altered. See also "Response element".

erbon. (generic name for 2-(2,4,5-trichloro phenoxy)ethyl-2,2-dichloropropionate).
CAS: 136-25-4. $(Cl)_3C_6H_2OC_2H_4OC(O)C(Cl)_2CH_3$.
Properties: Solid. Mp 49–50C (technical), bp 161C (0.5 mm Hg). Insoluble in water; soluble in most organic solvents.

Hazard: Toxic by inhalation and ingestion, strong irritant to eyes and skin.
Use: Herbicide.

erepsin. A mixture of peptidase enzymes formerly thought to be a single enzyme that catalyzes the hydrolysis of peptides in the small intestine.

ergamine. See histamine.

ergocalciferol. (calciferol; vitamin D_2).
CAS: 50-14-6. $C_{28}H_{44}O$.

Properties: White, odorless crystals. Affected by air and light. Mp 115–118C, specific rotation +103 to +106 degrees. Insoluble in water; soluble in alcohol, chloroform, ether, and fatty oils.
Derivation: From ergosterol by irradiation with UV light.
Grade: USP, FCC, as vitamin D_2.
Use: Nutrition, dietary supplement.

ergocornine, dihydro-. See 9,10-dihydroergocornine.

ergonovine. ([ββ(s)]-9,10-didehydro-n-(2-hydroxy-1-methylethyl)-6-methylergoline-8-carboxamide; n-[α-(hydroxymethyl)ethyl]-d-lysergamide; d-lysergic acid;-2-propanol-amide; erometrine; ergobasine; ergostetrine; (6aR,9R)-N-[(2S)-1-hydroxypropan-2-yl]-7-methyl-6,6a,8,9-tetrahydro-4H-indlo[4,3-fg]quinolone-9-carboxamide).
CAS: 60-79-7. $C_{19}H_{23}N_3O_2$.
An ergot alkaloid with uterine and vascular smooth muscle contractile properties.
Properties: Crystalline ergot alkaloid. Contains a lysergic acid moiety linked to a single amine-bearing group.
Hazard: Extremely toxic; vaso-constrictor; central nervous system stimulant; adrenergic blocking compound.
Use: Medically.

ergonovine maleate. (ergometrine maleate; metricalvin, ergomet).
CAS: 129-51-1. $C_{23}H_{27}N_3O_6$.
An ergot alkaloid with uterine and vascular smooth muscle contractile properties.
Use: medically as an oxytocic agent.

ergosterol. (provitamine D_2). $C_{28}H_{43}OH$.
CAS: 50-14-6. A plant sterol widely distributed in nature.

Properties: Colorless crystals. Mp 166C (with 1.5 H_2O), bp 250C (0.01 mm Hg), d 1.04, specific rotation −135 (in chloroform). Insoluble in water; soluble in alcohol, benzene, ether.
Derivation: Synthesized by yeast from simple sugars, obtained from fungus ergot.
Hazard: Due to its ability to catalyze calcium deposition in the bony structure (thus preventing rickets), overdosage of vitamin D may be harmful.
Use: Antirachitic vitamin; when irradiated with UV light, it has vitamin D activity; source of estradiol.

ergot. (secale cornutum; rye ergot). A fungus growth, *Claviceps purpurea*, on rye.
Occurrence: Europe, cultivated in Spain and Russia.
Grade: Spanish, Russian.
Hazard: Toxic by ingestion.
Use: Source of many alkaloids, medicine (vasoconstrictor).

ergot alkaloid. Any of a large group of alkaloids that are constituents of Claviceps (ergot). The four main types are clavinet alkaloids, lysergic acids, lysergic acid amides, and ergot peptide alkaloids.
Properties: Contain lysergic acid combine with various amine moieties.
Hazard: Very toxic.

ergotamine.
CAS: 113-15-5. Alkaloid present in rye ergot.
Properties: Mw 581.73.

ergot peptide alkaloid. Any of a type of alkaloid isolated from Claviceps (ergot). They are: ergotamine, ergosine, ergocristine, ergocryptine, ergocornine, and the isomers ergotaminine, ergosinine, ergocristinine, ergocryptinine, and ergocortinine.
Properties: Composed of lysergic acid, dimethyl pyruvic acid, proline and phenylalanine moieties linked by amide bridges.

ergovaline mesylate. See ergovaline monomethanesulfonate.

ergovaline monomethanesulfonate.
CAS: 3398-46-7. $C_{29}H_{35}N_5O_5 \cdot CH_4O_3S$.
Hazard: A reproductive hazard.
erionite.
CAS: 66733-21-9. $Al_2O_{18}Si_7 \cdot 1/2Ca \cdot 7H_2O \cdot 1/2Na$.
Hazard: Confirmed carcinogen.
Source: Natural product.

"Erionyl" *[Huntsman].* TM for a group of condensation products of aromatic sulfonic acids used as fixative/reserving agents in dyeing nylon and nylon-cellulosic blends.

Erlenmeyer flask. A useful type of laboratory glassware, it is an open container whose dimensions are, for example, about 8 inches tall with a relatively narrow neck section about 1.5 inches in diameter and 2 inches long, below which the contour becomes cone shaped. The bottom is flat. It was named after its inventor, a German chemist.
Use: Numerous experiments involving liquids, especially titrations and extractive testing.

Erlenmeyer–Plöchl azlactone and amino acid synthesis. Formation of azlactones by intramolecular condensation of acylglycines in the presence of acetic anhydride. The reaction of azlactones with carbonyl compounds followed by hydrolysis to the unsaturated α-acylamino acid and by reduction yields the amino acid; drastic hydrolysis gives the α-oxo acid.

Ernst, Richard R. (1933–). A native of Switzerland who won the Nobel Prize in chemistry in 1991 for important methodological developments in NMR spectroscopy. He invented Fourier-transform NMR (FT-NMR), which multiplied sensitivity 10 to 100 times compared to dispersive instruments. He also devised two-dimensional NMR techniques, increasing resolution and enabling structure determinations of biologically important macromolecules. Ernst received his Ph.D. from the Federal Technical Institute (ETH) in Zurich, Switzerland.

Ertl, Gerhard. (1936–). A German chemist, who was the recipient of 2007 Nobel Prize. Ertl's research centered on surface chemistry, and he developed groundbreaking methods to analyze reactions between gases and solid surfaces. Ertl carried out advanced work in the Haber–Bosch process (production of artificial fertilizers form atmospheric nitrogen). His pioneering work on surface chemistry provided essential understanding of automobile catalytic converters and hydrogen fuel cells. He earned his Ph.D. from the Technical University of Munich. Ertl was the director of physical chemistry departments at the Technical University of Hannover, the University of Munich, and the Fritz Haber Institute. He was named Professor Emeritus (2004) at the Fritz Haber Institute.

erucamide. (erucylamide).
CAS: 112-84-5. $C_{21}H_{41}CONH_2$.
Properties: Solid. D 0.888, mp 75–80C, iodine value 70–80. Soluble in isopropanol; slightly soluble in alcohol and acetone. Combustible.
Use: Foam stabilizer, solvent for waxes and resins, emulsions, antiblock agent for polyethylene.
See "Crodamide" *[Croda]*

erucic acid.   (*cis*-13-docosenoic acid).
CAS: 112-86-7. $C_8H_{17}CH:CH(CH_2)_{11}COOH$.
A C_{22} (solid) fatty acid with one double bond; a homolog of oleic acid with four more carbons.
Properties: Mp 33–34C, bp 264C (15 mm Hg), iodine value 75. Combustible.
Derivation: Fats and oils from mustard seed, rapeseed, and crambe seed.
Use: Preparation of dibasic acids and other chemicals, polyethylene-film additive, water-resistant nylon.

"Erucical H-102" *[Vantage].*   TM for a high erucic acid rapeseed oil referred to as HEAR oil.
Use: For lubricity in metal working fluids, as a base oil in hydraulic fluid applications.

erucyl alcohol.
CAS: 629-98-1. $C_{22}H_{43}OH$.
A C_{22} (solid) fatty alcohol having one double bond.
Properties: White, soft solid; almost odorless. D 0.8486, cloud point 27.2C, boiling range 334–376C, iodine value 83, flash p 395F (201C). Soluble in alcohol and most organic solvents. Combustible.
Derivation: Sodium reduction of erucic acid.
Use: Lubricants, surfactants, petrochemicals, plastics, textiles, rubber.

erythorbic acid.   (*d*-erythroascorbic acid; isoascorbic acid).
CAS: 89-65-6. $C_6H_8O_6$.
Properties: Shiny, granular crystals. Decomposes 164–169C. Soluble in water, alcohol, pyridine; moderately soluble in acetone; slightly soluble in glycerol.
Grade: FCC.
Use: Antioxidant (industrial and food), especially in brewing industry, reducing agent in photography.

erythorbic acid sodium salt.
CAS: 6381-77-7. $C_6H_8O_6 \cdot Na$.
Hazard: Low toxicity by ingestion. A mild eye irritant.

erythrite.   (1) Synonym for erythritol. (2) Cobalt bloom. $Co_3(As)_4)_2 \cdot 8H_2O$. A natural hydrated cobalt arsenate.
Properties: Crimson, peach, red, pink, or pearl gray. Contains 37.5% cobalt oxide. D 2.91–2.95, Mohs hardness 1.5–2.5, mp (decomposes). Soluble in hydrochloric acid.
Occurrence: U.S. (California, Colorado, Idaho, Nevada), Ontario.
Hazard: Toxic by ingestion and inhalation.
Use: Coloring glass and ceramics.

erythritol.   (tetrahydroxybutane).
CAS: 149-32-6. $CH_2OHCH_2OCH_2OCH_2OH$.
A tetrahydric alcohol found in *Protococcus vulgaris* and other lichens of *Rocella* species. Can be made synthetically.

$$
\begin{array}{c}
CH_2OH \\
| \\
HCOH \\
| \\
HCOH \\
| \\
CH_2OH
\end{array}
$$

Properties: White, sweet crystals. Mp 121–122C, bp 329–331C, d 1.45. Soluble in water; slightly soluble in alcohol; insoluble in ether.
Use: Manufacture of erythrityl tetranitrate.

erythritol anhydride.   (butadiene dioxide).
CAS: 564-00-1. $C_4H_6O_2$.
Properties: Colorless liquid. Bp 138C, fp −18C, d 1.11. Hydrolyzed to erythritol when dissolved in water.
Hazard: Quite toxic.
Use: Cross-linking agent, biocide, bacteriostat.

erythrityl tetranitrate.   (erythrol tetranitrate).
$CH_2ONO_2(CHONO_2)_2CH_2ONO_2$.
Properties: Crystals. Mp 61C. Soluble in alcohol, ether, and glycerol; insoluble in water.
Derivation: By nitration of erythritol.
Hazard: Severe explosion risk when shocked or heated.
Use: Medicine (diluted with lactose in nonexplosive tablets).

erythroascorbic acid.   See erythorbic acid.

erythrocyte.   A red blood cell, containing hemoglobin, the iron-carrying protein of the blood. See leucocyte.

erythrogenic acid.   See isanic acid.

erythromycin.
CAS: 114-07-8. $C_{37}H_{67}NO_{13}$.
An antibiotic produced by growth of *Streptomyces erythreus* Waksman. It is effective against infections caused by Gram-positive bacteria, including some β-hemolytic streptococci, pneumococci, and staphylococci.
Properties: White or slightly yellow crystalline powder; odorless; bitter. Mp 133–138C. Freely soluble in alcohol, chloroform, and ether; very slightly soluble in water; slightly hygroscopic; pH (saturated solution) 8–10.5; pH <4 is destructive. Alcoholic solution is levorotatory.
Grade: USP.
Use: Medicine (antibiotic); various salts are available.
Note: The erythromycin molecule was synthesized by the Harvard chemist Robert Woodward. The synthesis was almost completed at the time of his death in 1979, and was finished by his associates in 1981. It is an extremely complex structure containing a lactone ring of 14 members with 10 asymmetric centers; it also has two specialized

sugar molecules, L-cladinose and D-desosamine. The reported molecular configuration is:

erythromycin thiocyanate.
Properties: White or almost white powder.
Use: Food additive; drug (veterinary).

erythrosine. (Sodium or potassium salt of iodeosin; CI 45430; FD&C Red No. 3). CAS: 16423-68-0. $C_{20}H_6I_4Na_2O_5$.
Properties: Brown powder, forms cherry-red solution in water. Soluble in alcohol.
Use: Biological stain, certified food color.

Es. See einsteinium.

ESCA. Electron spectroscopy for chemical analysis.

eschka mixture. Consists of two parts magnesium oxide and one part anhydrous sodium carbonate.
Use: As a fusion mixture for determining sulfur in coal.

Eschweiler–Clarke reaction. Reductive methylation of primary or secondary amines with formaldehyde and formic acid (special form of the Leuckart–Wallach reaction).

ESD. Electron-stimulated desorption.

eserine. See physostigmine.

-esis. Suffix denoting state, condition, or process.

esmolol. See methyl 4-(2-hydroxy-3-((1.

esparto. The leaves of a desert plant (*Stipa tenacissima* and *Lygeum spartum*) of the Mediterranean area. Relatively high content of cellulose and α-cellulose make it usable for high-quality paper for intaglio color printing. A wax derived from this plant is used as a substitute for carnauba, especially in carbon paper.

ESR. See spin resonance.

essential. (1) Containing the characteristic odor or flavor, (i.e., the essence) of the original flower or fruit: an essential oil, usually obtained by steam distillation of the flowers or leaves or cold-pressing of the skin.
(2) As applied to certain amino acids, fatty acids, and vitamins, this term is used by biochemists to mean that the compound in question is a necessary nutritional factor that is not synthesized within the body of the animal and thus must be obtained from external sources. Eight amino acids are classified as essential on this basis.
See amino acid.

essential amino acids. Amino acids that cannot be synthesized by humans (or other vertebrates) and must be obtained from the diet.

essential fatty acids. The group of polyunsaturated fatty acids produced by plants, but not humans, and thus required in the human diet: linoleic, linolenic and arachidonic acids.

essential oil. A volatile oil derived from the leaves, stem, flower, or twigs of plants, and usually carrying the odor or flavor of the plant. Chemically, they are often principally terpenes (hydrocarbons), but many other types also occur. Essential oils (except for those containing esters) are unsaponifiable. Some are nearly pure single compounds, as oil of wintergreen, which is methyl salicylate. Others are mixtures, as turpentine oil (pinene, dipentene) and oil of bitter almond (benzaldehyde, hydrocyanic acid). Some contain resins in solution and are called oleoresins or balsams.
Properties: Pungent taste and odor; usually nearly colorless when fresh but becoming darker and thick on exposure to the air. Optically active, d 0.850–1.100. Soluble in alcohol, carbon disulfide, carbon tetrachloride, chloroform, petroleum ether, and fatty oils; insoluble in water except for individual constituents of some oils which may be partially water-soluble, resulting in a loss of these constituents during steam distillation.
Derivation: (1) By steam distillation; (2) by pressing (fruit rinds); (3) by solvent extraction; (4) by maceration of the flowers and leaves in fat and treating the fat with a solvent; (5) by enfleurage.
Use: Perfumery, flavors, thinning precious-metal preparations used in decorating ceramic ware.
See terpeneless oil.
Further information can be obtained from the Essential Oil Association of U.S.
Note: Many essential oils are now made synthetically for a wide variety of fragrances and flavoring agents. Use of these synthetics is increasing because of a shortage of natural products.

EST. See expressed sequence tag.

ester. An organic compound corresponding in structure to a salt in inorganic chemistry. Esters are

considered as derived from acids by the exchange of the replaceable hydrogen of the latter for an organic radical. The usual reaction is that of an acid (organic or inorganic) with an alcohol or other organic compound rich in OH groups. Esters of acetic acid are called acetates, and esters of carbonic acid carbonates.
See fatty ester.

esterase. A hydrolase that catalyzes the hydrolysis of an ester into, or synthesis from, its constituent alcohol and acid.

esterase-lipase.
Properties: Derived from *Mucor miehei*.
Use: Food additive.

ester gum. Hard, semisynthetic resin produced by esterification of natural resins (especially rosin) with polyhydric alcohols (principally glycerol but also pentaerythritol). Flash p 375F (190C). Combustible.
Grade: By color, also as gum rosin or wood rosin.
Use: Paints, varnishes, and cellulosic lacquers.

esterification. The process of producing an ester by reaction of an alcohol with an acid.

ester interchange. (transesterification; interification). The reaction between an ester and another compound with exchange of alkoxy or acyl groups to form a different ester.

ester number. (1) The number of milligrams of alkali necessary for saponification of the glyceryl esters in a fat or oil. (2) The difference between acid number and saponification number.

esterogenic. (estrogenous). Having an action similar to that of an estrogen. Estrogenic compounds include estradiol, chlordecone, coumestrol, diethylstilbestrol, genistein, and zearalinone.

"Esteron" [Dow Agrosciences]. TM for a series of weed- and brush-control products; they are formulated esters of 2,4-D and 2,4,5-T.

estradiol.
CAS: 50-28-2. $C_{18}H_{24}O_2$.
A female sex hormone. It occurs in two isomeric forms, α and β. β-estradiol has the greatest physiological activity of any naturally occurring estrogen. The α form is relatively inactive. Commonly used preparations are the benzoate, dipropionate, and valerate, as well as ethinylestradiol.

Properties: (β-form) White or slightly yellow, small crystals or crystalline powder; odorless. Mp 173–179C, stable in air. Almost insoluble in water; soluble in alcohol, acetone, dioxane, and solutions of alkali hydroxides; sparingly soluble in vegetable oils.
Derivation: Isolated from human and mare pregnancy urine, commercial synthesis from cholesterol or ergosterol.
Grade: NF (β-form).
Hazard: A carcinogen (OSHA).
Use: Medicine (estrogenic hormone).

estradiol monopamitate.
Use: Food additive; drug (veterinary).

estragole. (chavicol methyl ether; methyl chavicol).
CAS: 140-67-0. $C_6H_4(C_3H_5)(OCH_3)$.
Properties: Colorless liquid; anise odor. D 0.965–0.975 (20/4C), 1.5230 (17.5C), bp 216C. Soluble in alcohol and chloroform.
Occurrence: In tarragon oil, basil oils, anise bark oil, and others.
Use: Perfumes, flavors.

4,9,11,-estratrien-17β-ol-3-one. See trenbolone.

estriol.
CAS: 50-27-1. $C_{18}H_{24}O_3$.
Properties: White, odorless, microcrystalline powder. Mp 282C. Exhibits reddish fluorescence under filtered UV light. Undergoes phase change at 270–275C. Practically insoluble in water; soluble in alcohol, dioxane, and oils.
Derivation: Isolation from pregnant human urine, isolated from human placenta, organic synthesis.
Hazard: A carcinogen (OSHA).
Use: Medicine (estrogenic hormone).

estrogen. A general term for female sex hormones. They are responsible for the development of the female secondary sex characteristics, such as the deposition of fat and the development of the breasts. The naturally occurring estrogens, such as estradiol, estrone, and estriol, are steroids. Estrogens are produced by the ovary and, to a lesser degree, by the adrenal cortex and testis. Some synthetic nonsteroid compounds such as diethylstilbestrol and hexestrol have estrogenic activity.
Hazard: Carcinogenic, damaging side effects (thromboembolism).

Use: Medicine, biochemical research, oral contraceptives.

See antifertility agent.

estrogen response element. See ERE.

estrone.
CAS: 53-16-7. $C_{18}H_{22}O_2$.
A steroid with some estrogenic activity.

Properties: Small, white crystals or white, crystalline powder; odorless. Mp 258–262C, stable in air. Insoluble in water; soluble in alcohol, acetone, dioxane, and solutions of fixed alkali hydroxides.
Derivation: Isolated from pregnant human urine, synthesis from ergosterol.
Hazard: A carcinogen (OSHA).

estrone 17-methoxime.
CAS: 3342-64-1. $C_{19}H_{25}NO_2$.
Hazard: A reproductive hazard.

Et. Informal abbreviation for ethyl. Example: EtOH, ethyl alcohol.

etamycin.
CAS: 299-20-7. $C_{44}H_{62}N_8O_{11}$.
Hazard: Moderately toxic by ingestion.

Etard reaction. Oxidation of an arylmethyl group to an aldehyde by means of chromyl chloride.

"Ethacure" *[Albemarle].* TM for curing agents for thermosetting plastics.

"Ethafoam" *[Sealed].* TM for a light-weight, low-density polyethylene foam.

ethanal. See acetaldehyde.

ethanamide. See acetamide.

ethane. (dimethyl; methylmethane).
CAS: 74-84-0. C_2H_6.
Properties: Colorless gas; odorless.Bp −88.63C, fp −183.23C (triple point), d of liquid 0.446 (0C), d of vapor (air = 1) 1.04 (0C), critical temp 32.1C, critical press (absolute) 718 psi, specific heat at constant press 0.897, specific heat at constant vol 0.325, ratio of specific heats (*cp/cv*) 1.224, heat of combustion approximately 22,300 Btu/lb or 1800 Btu/cu ft, flash p −211F (−135C), autoign temp 959F (515C). Insoluble in water; soluble in alcohol. Relatively inactive chemically.
Derivation: Fractionation of natural gas.
Grade: 95%, 99%, research, 99.98%.
Hazard: Severe fire risk if exposed to sparks or open flame. Flammable limits in air 3–12%. An asphyxiant gas.
Use: Petrochemicals (source of ethylene, halogenated ethanes), refrigerant, fuel.

n,n′-(1,2-ethanedioxysulfinyl)bis(s-methyl-n-methylcarbamoyloxythioacetimidate).
CAS: 81861-89-4. $C_{12}H_{22}N_4O_8S_4$.
Hazard: A poison by ingestion.
Use: Agricultural chemical.

ethanedioyl chloride. See oxalyl chloride.

1,2-ethanedithiol. (dithioethyleneglycol; ethylenedimercaptan).
CAS: 540-63-6. $HSCH_2CH_2SH$.
Properties: Liquid. Mw 94.20, bp 144–146C, d 1.123. Soluble in alcohol and alkalies.
Hazard: Vapors cause severe headache and nausea.
Use: Metal-complexing agent. Reverses the inhibition by α-keto aldehydes on mitosis in *E. coli.*

n,n′-(1,2-ethanedithiosulfinyl)bis(2,3-dihydro-2,2-dimethylbenzofuranyl-7) methylcarbamate.
CAS: 81861-95-2. $C_{26}H_{32}N_2O_8S_4$.
Hazard: A poison by ingestion.
Use: Agricultural chemical.

n,n′-(1,2-ethanedithiosulfinyl)bis(s-methyl-n-methylcarbamoyloxythioacetimidate).
CAS: 81861-94-1. $C_{12}H_{22}N_4O_6S_6$.
Hazard: A poison by ingestion.
Use: Agricultural chemical.

1,1′-(1,2-ethanediyl)bis(1-methylhydrazine).
CAS: 20247-50-1. $C_4H_{14}N_4$.
Hazard: Moderately toxic by ingestion and inhalation.

1,1′-(1,2-ethanediylbis(oxy))bis(2,3,4,5,6-pentabromobenzene).
CAS: 61262-53-1. $C_{14}H_4Br_{10}O_2$.
Hazard: Moderately toxic by inhalation and skin contact. Low toxicity by ingestion.

ethane hydrate. See gas hydrate.

ethanethiol. (ethyl sulfhydrate; ethyl mercaptan).
CAS: 75-08-1. C_2H_5SH.
Properties: Colorless liquid, has one of the most penetrating and persistent odors known (skunk). D 0.83907 (20/4C), bp 36C, fp −121C, refr index 1.4305 (20C), flammable limits in air 2.8–18.2%

by volume, flash p approximately 80F (26.6C) (CC), autoign temp 570F (298C). Slightly soluble in water; soluble in alcohol, ether, petroleum naphtha.

Derivation: By saturating potassium hydroxide solution with hydrogen sulfide, mixing with calcium ethylsulfate solution, and distilling in a water bath.

Hazard: Toxic by ingestion and inhalation. Flammable, dangerous fire risk.

Use: LPG odorant, adhesive, stabilizer, chemical intermediate.

Note: Tomato juice is reported to deodorize materials contaminated with this compound.

ethanethiolic acid. See thioacetic acid.

ethanoic acid. See acetic acid.

ethanol. See ethyl alcohol.

ethanolamine. (MEA; monoethanolamine; colamine; 2-aminoethanol; 2-hydroxyethylamine).
CAS: 141-43-5. $HOCH_2CH_2NH_2$.
Properties: Colorless; ammoniacal odor. D 1.0179 (20/20C), bp 170.5C, mp 10.5C, vap press 0.48 mm Hg (20C), flash p 200F (93.3C) (OC), bulk d 8.5 lb/gal (20C). Hygroscopic, viscous liquid. Strong base. Miscible with water, methanol, acetone. Combustible.
Derivation: Reaction of ethylene oxide and ammonia gives a mix of mono-, di-, and triethanolamines.
Grade: Technical, NF.
Hazard: Skin irritant.
Use: Scrubbing acid gases (H_2S, CO_2), especially in synthesis of ammonia, from gas streams; nonionic detergents used in dry cleaning, wool treatment, emulsion paints, polishes, agricultural sprays; chemical intermediates, pharmaceuticals, corrosion inhibitor, rubber accelerator.

ethanol formamide. $HOCH_2NHOCH$.
Properties: Somewhat viscous liquid. Bp 143C (2.5 mm Hg), fp approximately −72C, d 1.170 (25/4C), flash p 347F (175C). Miscible with water, alcohol, and glycerol. Compatible with polyvinyl alcohol, many cellulosic and natural resins. Combustible.

ethanol hydrazine. See β-hydroxyethylhydrazine.

ethanolurea.
CAS: 625-52-5. $NH_2CONHCH_2CH_2OH$.
Properties: Liquid. Solidification point 71–74C. Formaldehyde condensation products are permanently thermoplastic and water-soluble. As increasing amounts of simple urea are mixed with ethanolurea, the condensation products gradually change from pliable film-forming resins into the brittle types. Thus, almost any degree of water solubility and flexibility may be obtained in the final resin. The modified resins formed with ethanolurea are compatible with polyvinyl alcohol, methyl cellulose, cooked starch, and other water-dispersible materials.

ethanone, 1-(5,6,7,8-tetrahydro-3,5,5,6,8,8-hexamethyl-2-naphthalenyl)-.
CAS: 21145-77-7. $C_{18}H_{26}O$.
Hazard: A poison by ingestion. Low toxicity by skin contact.

"Ethanox" [Albemarle]. TM for hindered phenolic antioxidants.

ethaphos.
CAS: 38527-91-2. $C_{11}H_{15}Cl_2O_3PS$.
Hazard: A poison by ingestion.
Use: Agricultural chemical.

ethaverine. (1-[(3,4-diethoxyphenyl)methyl]-6,7-diethoxyisoquinoline; isoquinoline).
CAS: 486-47-5. $C_{24}H_{29}NO_4$.
Properties: Crystals. Mw 395.48 (99–101C). Insoluble in water; very soluble in hot alcohol; slightly soluble in ether and chloroform.
Use: Antispasmodic drug.

Ethazcate. Zinc diethyldithiocarbamate. See zinc diethyldithiocarbamate.

ethchlorvynol. (1-chloro-3-ethyl-1-penten-4-yn-3-ol; β-chlorovinylethylethynylcarbinol).
CAS: 113-18-8. $HC≡CCOH(C_2H_5)CH=CHCl$.
Properties: Colorless to yellow liquid; pungent aromatic odor. Darkens on exposure to light and air. D 1.068–1.071, refr index 1.4765–1.4800 (25C), bp 173–181C. Immiscible with water; miscible with most organic solvents.
Grade: NF.
Hazard: Abuse may cause addiction.
Use: Medicine (sedative).

ethene. See ethylene.

ethenol. See vinyl alcohol.

ethenylbenzene tribromo deriv. homopolymer.
CAS: 57137-10-7. $(C_8H_5Br_3)_x$.
Hazard: Moderately toxic by skin contact. Low toxicity by ingestion. A mild eye irritant.
Use: Fragrance chemical.

ethenyl 2-propenoate.
CAS: 2177-18-6. $C_5H_6O_2$.
Hazard: A poison by ingestion. Low toxicity by inhalation. A severe skin and eye irritant.

ethephon. See 2-chloroethylphosphonic acid.

ether. A class of organic compounds in which an oxygen atom is interposed between two carbon atoms (organic groups) in the molecular structure, giving the generic formula ROR. They may be derived from alcohols by elimination of water, but the major method is catalytic hydration of olefins. Only the lowest member of the series, methyl ether, is gaseous; most are liquid, and the highest members are solid (cellulose ethers). The term *ether* is often used synonymously with *ethyl ether* and is the legal label name for it.

Hazard: The lower molecular weight ethers are dangerous fire and explosion hazards; when containing peroxides they can detonate on heating.

Use: See ethyl ether; polymer, water-soluble; ethylene oxide; propylene oxide; diethylene glycol; ethylcellulose; polyether.

Note: An illogical and archaic use of the term *ether* survives in such names as petroleum ether. See crown ether.

ethereal. Descriptive of a liquid characterized by high volatility, often a mixture of ethyl ether and an essential oil.

ethereal oil.
Properties: Aromatic substances comprised of terpenes.
Hazard: Volatile.

ethical drug. A prescription drug. See drug.

ethinylestradiol. (ethynylestradiol; 19-nor-17-α-pregna-1,3,5(1)-trien-20-yne-3,17-diol).
CAS: 57-63-6. $C_{20}H_{24}O_2$.
Properties: Fine, white to creamy white crystalline powder; odorless. Sensitive to light. Mp 142–146C. May also exist in a polymorphic modification with mp 180–186C. Soluble in acetone, alcohol, chloroform, dioxane, ether, and vegetable oils; practically insoluble in water; soluble in solutions of sodium hydroxide or potassium hydroxide; slightly dextrorotatory in dioxane solution.
Derivation: Preparation from estrone.
Grade: USP.
Use: Medicine (estrogenic hormone).

ethion. (generic name for *O,O,O',O'*-tetraethyl-*S,S'*-methylenediphosphorodithioate).
CAS: 563-12-2. $[(C_2H_5O)_2P(S)S]_2CH_2$.
Properties: Liquid. D 1.220 (20C), fp −13C. Slightly soluble in water; soluble in acetone, xylene, chloroform, and methylated naphthalene.
Hazard: Cholinesterase inhibitor, use may be restricted. Questionable carcinogen.
Use: Insecticide and miticide.

ethisterone. (preneninolone; anhydrohydroxyprogesterone; ethynyltestosterone).
CAS: 434-03-7. $C_{21}H_{28}O_2$.

Properties: White or slightly yellow crystals or crystalline powder; odorless. Stable in air. Affected by light. Mp 267–275 (decomposes). Almost insoluble in water; slightly soluble in alcohol, chloroform, ether, and vegetable oils.
Derivation: From progesterone and other steroids.
Grade: NF.
Use: Medicine (estrogenic hormone).

"Ethocel" [Dow]. TM for ethylcellulose resins are able to withstand shock and maintain toughness over a temperature range of +93 to −40C. Available in transparent, translucent, and opaque colors. Insoluble in water, soluble in most organic solvents.
Use: Household articles, automotive parts, tools for aircraft industry.

ethodin. (6,9-diamino-2-ethoxyacridine lactate monohydrate).
CAS: 1837-57-6. $C_{15}H_{15}N_3O \cdot C_3H_6O_3 \cdot H_2O$.
Properties: Pale-yellow crystals. Darkens at 200C, mp 235C. Slowly soluble in 15 parts water; soluble in 9 parts boiling water; soluble in 110 parts alcohol (22C). Solutions are yellow, fluorescent, and stable to boiling.
Purity: 97% (dry basis).
Use: Bactericide, surgical antisepsis, preparation of pure γ-globulin.

ethofumesate. (2-ethoxy-2,3-dihydro-3,3-dimethyl-5-benzofuranol methanesulfonate).
CAS: 26225-79-6. $C_{13}H_{18}O_5S$.
Properties: Whitish crystals. Mp 71C. Does not hydrolyze with water at neutral pH.
Use: Herbicide, silvicide.

ethoheptazine. (ethyl heptazine; 1-methyl-4-carbethoxy-4-phenylhexamethyleneimine).
CAS: 77-15-6. $C_{16}H_{23}NO_2$.
Properties: Liquid. D 1.038 (26/4C), bp 133–134C (1.0 mm Hg), refr index 1.5210 (26C).
Use: Medicine (analgesic).

ethohexadiol. USP name for 2-ethylhexanediol-1,3.

ETHONIC. Ethoxylated alcohol surfactants.

ETHONIC 11. Ethoxylated alcohols.
Use: As wetting agents and emulsifiers in detergents, personal care, textiles, and paper.

ethopabate.
CAS: 59-06-3. $C_{12}H_{15}NO_4$.
Properties: Odorless white to pink crystals from MeOH (aq). Mp 148–149C. Soluble in methanol, ethanol, acetone, acetonitrile, isopropanol, *p*-dioxane, ethyl acetate, and methylene chloride.
Use: Drug (veterinary); food additive.

ethoxazene. See *p*-ethoxychrysoidine.

p-ethoxyacetanilide.   See acetophenetidin.

4-ethoxyamphetamine hydrochloride.
CAS: 135014-87-8. $C_{11}H_{17}NO \cdot ClH$.
Hazard: A reproductive hazard.

ethoxycarbonyl isothiocyanate.   (ethyl isothiocyanateformate).
CAS: 16182-04-0. $C_2H_5O_2CNCS$.
Properties: Moisture-sensitive solid. Mw 131.15, bp 56C (18 mm Hg), d 1.112.
Hazard: Lachrymator.
Use: Versatile reagent for organic synthesis.

o-(4-(1-(((ethoxycarbonyl)oxy)imino)ethyl) phenyl) o,o-diethyl phosphorothioate.
CAS: 22936-34-1. $C_{15}H_{23}NO_6PS$.
Hazard: Moderately toxic by ingestion.
Use: Agricultural chemical.

p-ethoxychrysoidine.
CAS: 94-10-0. $C_{14}H_{16}N_4O$.
Hazard: A poison.

2-ethoxy-3,4-dihydro-2H-pyran.

$OCH:CHCH_2CH_2CHOC_2H_5$.

Properties: Liquid. D 0.970 (20/20C), bp 143C, fp −100C, flash p 111F (OC). Sets to glass below this temperature. Very slightly soluble in water. Combustible.
Hazard: Moderate fire risk.
Use: Stabilizer, intermediate.

6-ethoxy-1,2-dihydro-2,2,4-trimethylquinoline.   (ethoxyquin).
CAS: 91-53-2. $C_{14}H_{19}NO$.
Properties: Yellow liquid. Bp 125C (2 mm Hg), mp approximately 0C, refr index 1.569–1.572 (25C), d 1.029–1.031 (25C). Discolors and stains badly.
Hazard: Toxic by ingestion.
Use: Insecticide, antioxidant, flex-cracking inhibitor, postharvest preservation of apples (scald inhibitor).

ethoxydimethylsilane.   (dimethylethoxysilane).
CAS: 14857-34-2. $C_4H_{12}OSi$.
Hazard: An inhalation hazard. Headache, eye and upper respiratory tract irritant.

2-ethoxyethanol.   See ethylene glycol monoethyl ether.

2-ethoxyethyl acetate.   See ethylene glycol monoethyl ether acetate.

n-(2-ethoxyethyl)-4,5-dihydro-3-(phenylamino)-2h-benz(g)indazole-2-acetamide.
CAS: 301644-19-9. $C_{23}H_{26}N_4O_2$.
Hazard: A poison by ingestion.

2-ethoxyethyl-p-methoxycinnamate.
$CH_3OC_6H_4CH:CHCOOC_2H_4OC_2H_5$.
Properties: Slightly yellow, viscous liquid; practically odorless. D 1.1000–1.1035 (25/25C), refr index 1.5650–1.5675 (20C), flash p >212F (>100C) (TCC). Miscible with alcohol and isopropyl alcohol; almost insoluble in water. Combustible.
Use: UV absorber in suntan preparations.

3-ethoxy-4-hydroxybenzaldehyde.   See ethyl vanillin.

1-ethoxy-2-hydroxy-4-propenylbenzene.
See propenyl guaethol.
2-[1-(ethoxyimino)butyl]-5-[2-(ethylthio)propyl]-3-hydroxy-2-cyclohexene-1-one.
Use: Food additive; herbicide.

"Ethoxylan" *[Cognis].*   TM for additive lanolin.
Available forms: Paste, waxy solid.
Use: Emollient conditioner; lubricant emulsifier for creams, lotions, makeup, lip products, and nail and suntan products.

ethoxylated glyceryl monostearate.   See "Durfax EOM" *[Loders].*

ethoxylated mono- and diglycerides.
Properties: Mix of stearate, palmitate, and lesser amounts of myristate partial esters of glycerin condensed with approx. 20 moles of ethylene oxide per mole of α-monoglyceride reaction mixtures. (FCC III) Pale, slightly yellow, oily liquid; mildly bitter taste. Soluble in water, alcohol, xylene; slightly soluble in mineral oil, vegetable oil.
Use: Food additive.

4-ethoxy-3-methoxybenzaldehyde.
CAS: 120-25-2. $C_6H_3(OC_2H_5)(OCH_3)CHO$.
Properties: White to light-brown crystals; slight vanillin odor. Mp 62–64C. Combustible.
Use: Intermediate.

4-ethoxy-3-methoxyphenylacetic acid.
CAS: 120-13-8. $C_6H_3(OC_2H_5)(OCH_3)CH_2COOH$.
Properties: An off-white powder. Mp 119–122C.
Use: Intermediate.

1-ethoxy-2-methoxy-4-propenylbenzene.
See isoeugenyl ethyl ether.

2-ethoxy-4-methyl-2,3-dihydro-4h-pyran.
CAS: 10138-44-0. $C_8H_{14}O_2$.
Hazard: Moderately toxic by ingestion. Mild skin and eye irritant.

ethoxymethylenemalononitrile.
CAS: 123-06-8. $C_2H_5OCH:C(CN)_2$.

Properties: Colorless liquid. Bp 160C, mp 64C, flash p 311F (155C). Combustible.
Use: Chemical intermediate.

4-ethoxyphenol.   See hydroquinone monoethyl ether.

r-(-)-5-(2-((2-(2-ethoxyphenoxy)ethyl)amino) propyl)-2-methoxybenzenesulfonamide hydrochloride.
CAS: 106463-17-6. $C_{20}H_{28}N_2O_5S\cdot ClH$.
Hazard: Moderately toxic by ingestion.

2-(*m*-ethoxyphenyl)imidazo(2,1-a)isoquinoline.
CAS: 61001-14-7. $C_{19}H_{16}N_2O$.
Hazard: A reproductive hazard.

ethoxyquin.   Coined name for 6-ethoxy-1,2-dihydro-2,2,4-trimethylquinoline.

ethoxytriglycol.
CAS: 112-50-5. $C_2H_5O(C_2H_4O)_3H$.
Properties: Colorless liquid. D 1.0208 (20/20C), bulk d 8.5 lb/gal (20C), bp 255.4C, vap press <0.01 (20C), fp −18.7C, viscosity 7.80 cP (20C). Completely soluble in water. Flash p 275F (135C) (CC). Combustible.
Use: Chemical intermediate.

ethy abietate. $C_{19}H_{29}COOC_2H_5$.
Properties: Amber-colored, viscous liquid; hardens upon oxidation. Bp 350C, flash p 352F (177C), mp 45C, refr index 1.4980, d 1.02. Soluble in ether and most varnish solvents, insoluble in water. Combustible.
Derivation: (1) By heating together ethyl chloride and an alcoholic solution of rosin and caustic soda. (2) By reacting ethyl iodide with silver abietate.
Hazard: Irritant.
Use: Varnishes, lacquers, and coating compositions.

n-ethylacetamide.
CAS: 625-50-3 $CH_3CONHC_2H_5$.
Properties: Colorless liquid; faint odor. D 0.920 (20/20C), boiling range 206–208.5C, flash p 230F (110C). Combustible.

ethyl acetamidocyanoacetate.   (acetamido-cyanoacetic ester; ethyl-*n*-acetyl-α-cyanoglycine).
CAS: 4977-62-2 $NCCH(NHCOCH_3)COOC_2H_5$.
Properties: Solid. Mp 129C.
Use: Synthesis of amino acids and related compounds.

n-ethylacetanilide.   (ethylphenylacetamide).
CAS: 529-65-7. $C_6H_5NC_2H_5COCH_3$.
Properties: White, crystalline solid; faint odor. D 0.994, bp 258C, flash p 126F (52.2C), mp 54C. Soluble in most organic solvents; insoluble in water. Combustible.
Grade: Technical.

Hazard: Toxic by ingestion, moderate fire risk.
Use: Substitute for camphor in nitrocellulose.

ethyl acetate.   (acetic ether; acetic ester; vinegar naphtha).
CAS: 141-78-6. $CH_3COOC_2H_5$.
Properties: Colorless, fragrant liquid. Bp 77C, vap press 73 mm Hg (20C), fp −83.6C, bulk d 0.8945 g/ml (25C), flash p 24F (−4.4C), autoign temp 800F (426C). Soluble in chloroform, alcohol, and ether; slightly soluble in water.
Derivation: By heating acetic acid and ethyl alcohol in the presence of sulfuric acid and distilling.
Grade: Commercial 85–88%, 95–98%, 99%, NF (99%), FCC.
Hazard: Toxic by inhalation and skin absorption; irritant to eyes and skin. Flammable, dangerous fire and explosion risk, flammable limits in air 2.2–9%. Questionable carcinogen.
Use: General solvent in coatings and plastics, organic synthesis, smokeless powders, pharmaceuticals, synthetic fruit essences.

ethyl-*o*-acetate.
CAS: 78-39-7. $CH_3C(OC_2H_5)_3$.
Properties: Colorless liquid. Bp 144-148C, refr index 1.395 (25C). Insoluble in water; soluble in alcohol and ether, flash p 131F (55C). Combustible.
Hazard: Moderate fire risk.
Use: Intermediate.

ethyl acetate, anhydrous.   ethyl acetate, grade 99%.

ethyl acetic acid.   See butyric acid.

ethyl acetoacetate.   (diacetic ester; acetoacetic ester).
CAS: 141-97-9. $CH_3COCH_2COOC_2H_5$ (keto form), $CH_3C(OH){:}CHCOOC_2H_5$ (enol form).
This compound is a tautomer at room temperature consisting of about 93% keto form and 7% enol form.
Properties: Colorless liquid; fruity odor. D 1.0250 (20/4C), fp (enol) −80C; (keto) −39C, bp 180–181C, bulk d 8.5 lb/gal, vap press 0.8 mm Hg (20C), flash p 185F (85C) (COC), coefficient of expansion 0.00101/C. Soluble in water and common organic solvents. Combustible.
Derivation: Action of metallic sodium on ethyl acetate with subsequent distillation.
Grade: Technical, 98%.
Hazard: Toxic by ingestion and inhalation; irritant to skin and eyes.
Use: Organic synthesis; antipyrine, lacquers, dopes, plastics; manufacture of dyes, pharmaceuticals anti-malarials, vitamin B; flavoring.

ethyl acetone.   See methyl propyl ketone.

ethyl-*n*-acetyl-α-cyanoglycine.   See ethyl acetamidocyanoacetate.

ethylacetylene. (1-butyne).
CAS: 107-00-6. $C_2H_5C{\equiv}CH$.
Properties: Available as liquefied gas. Bp 8.3C, d 0.669 (0/0C), fp −130C, flash p <20F (−6.6C) (TOC), sv 7.2 cu ft/lb (21.2C). Insoluble in water.
Hazard: Flammable, dangerous fire risk.
Use: Specialty fuel, chemical intermediate.

α-ethylacrolein.
CAS: 922-63-4. C_5H_8O.
Hazard: Moderately toxic by inhalation.

α-ethylacrylaldehyde. See α-ethylacrolein.

ethyl acrylate.
CAS: 140-88-5. $CH_2{:}CHCOOC_2H_5$.
Properties: Colorless liquid. Bp 99.4C, fp −72.0C, d 0.9230 (20/20C), refr index 1.4037 (25C), bulk d 7.6 lb/gal (20C), flash p 60F (15.5C) (OC). Soluble in alcohol and ether. Readily polymerized.
Derivation: (1) Ethylene cyanohydrin, ethyl alcohol, and dilute sulfuric acid; (2) Oxo reaction of acetylene, carbon monoxide, and ethyl alcohol in the presence of nickel or cobalt catalyst.
Grade: Technical (inhibited, usually with hydroquinone or its monomethyl ether), pure uninhibited.
Hazard: Toxic by ingestion, inhalation, skin absorption; irritant to skin and eyes. Flammable, dangerous fire and explosion hazard. Possible carcinogen.
Use: Monomer for acrylic resins.
See acrylate; acrylic resin.

ethyl alcohol. (alcohol; grain alcohol; ethanol; EtOH).
CAS: 64-17-5. C_2H_5OH.
Properties: (Pure 100% absolute alcohol, dehydrated) Colorless, limpid, volatile liquid; ethereal vinous odor; pungent taste. Bp 78.3C, fp −117.3C, refr index 1.3651 (15C), surface tension 22.3 dynes/cm (20C), viscosity 0.0141 cP (20C), vap press 43 mm Hg (20C), specific heat 0.618 cal/g K (23C), flash p 55F (12.7C), d 0.816 (15.56C), bp 78C, fp −114C, autoign temp 793F (422C). Miscible with water, methanol, ether, chloroform, and acetone. (95% alcohol).
Derivation: (1) From ethylene by direct catalytic hydration or with ethyl sulfate as intermediate; (2) fermentation of biomass, especially agricultural wastes; (3) enzymatic hydrolysis of cellulose.
See cellulase.
Grade: USP (95% by volume), absolute, pure, completely denatured, specially denatured, industrial, various proofs (one-half the proof number is the percentage of alcohol by volume).
Hazard: Classified as a depressant drug. Though it is rapidly oxidized in the body and is therefore noncumulative, ingestion of even moderate amounts causes lowering of inhibitions, often succeeded by dizziness, headache, or nausea. Larger intake causes loss of motor nerve control, shallow respiration, and in extreme cases unconsciousness and even death.

Degree of intoxication is determined by concentration of alcohol in the brain. Of primary importance is the fact that intake of moderate amounts together with barbiturates or similar drugs is extremely dangerous and may even be fatal. Flammable, dangerous fire risk; flammable limits in air 3.3–19%. Possible carcinogen.
Use: Solvent for resins, fats, fatty acids, oils, hydrocarbons; extraction medium; manufacture of acetaldehyde, acetic acid, ethylene, butadiene, 2-ethyl hexanol, dyes, pharmaceuticals, elastomers, detergents, cleaning preparations, surface coatings, cosmetics, explosives, antifreeze, beverages, antisepsis, gasohol, yeast-growth medium; octane booster in gasoline.
See alcohol, denatured; alcohol, industrial; biomass.
Note: Ethanol from fermentation of biomass and hydrolysis of cellulose is a significant alternate energy source, especially as an automotive fuel. Its use in gasoline will continue to increase.

ethyl α-allylacetoacetate.
$CH_3COCH(CH_2CH{:}CH_2)COOC_2H_5$.
Properties: Water-white liquid. D 0.989 (20C), bulk d 8.24 lb/gal (20C). Combustible.
Use: Intermediate for pharmaceuticals, perfumes, fungicides, insecticides, fine chemicals.

ethylaluminum dichloride. (EADC).
CAS: 563-43-9. $C_2H_5AlCl_2$.
Properties: Clear, yellow, pyrophoric liquid. Bp (extrapolated) 194C, fp 32C, d 1.222, bulk d 10.28 lb/gal (25C).
Derivation: Reaction of aluminum chloride with ethylaluminum sesquichloride.
Hazard: Ignites on contact with air, dangerous fire risk, reacts violently with water. Skin irritant.
Use: Catalyst for olefin polymerization, aromatic hydrogenation; intermediate.

ethylaluminum sesquichloride. (EASC).
CAS: 12075-68-2. $(C_2H_5)_3Al_2Cl_3$.
Properties: Clear, yellow, pyrophoric liquid. Bp 204C, fp −50C, d 1.08.
Derivation: Reaction of ethyl chloride and aluminum.
Grade: Commercial.
Hazard: Ignites on contact with air, dangerous fire risk, reacts violently with water.
Use: Catalyst for olefin polymerization, aromatic hydrogenation; intermediate.

ethylamine. (monoethylamine; aminoethane).
CAS: 75-04-7. $CH_3CH_2NH_2$.
Properties: Colorless, volatile liquid (or gas). Ammonia odor, strong alkaline reaction, bp 16.6C, fp −81.2C, d 0.689 (liquid 15/15C), bulk d 5.7 lb/gal (20C), flash p approximately 0F (−17.7C) (OC), autoign temp 723F (383C). Miscible with water, alcohol, and ether.
Derivation: From ethyl chloride and alcoholic ammonia under heat and pressure.

Grade: Technical (anhydrous and 70% aqueous solution), pure 98.5% min.

Hazard: Strong irritant. Flammable, dangerous fire risk, flammable limits in air 3.5–14%.

Use: Dye intermediate, solvent extraction, petroleum refining, stabilizer for rubber latex, detergents, organic synthesis.

ethylamine hydrobromide.
CAS: 593-55-5. $C_2H_5NH_2 \cdot HBr$.

Properties: White, almost odorless granules. Mp 158–161C. Very soluble in water.

Use: Intermediate (where liquid ethylamine or liquid hydrobromic acid cannot be used).

ethyl-o-aminobenzoate.
See ethyl anthranilate.

ethyl-p-aminobenzoate hydrochloride.
(anesthesol; benzocaine; procaine hydrochloride).

CAS: 51-05-8. $C_6H_4NH_2CO_2C_2H_5 \cdot HCl$.

Properties: White, crystalline; odorless; tasteless powder. Stable in air. Mp 88–92C. Soluble in dilute acids; less soluble in chloroform, ether, and alcohol; very slightly soluble in water.

Derivation: Ethylation of p-nitrobenzoic acid followed by reduction.

Grade: Technical, pure, NF (as benzocaine).

Hazard: Toxic by ingestion.

Use: Medicine (local anesthetic), suntan preparations.

ethylaminoethanol.
See N-ethylethanolamine. Mixed ethylaminoethanols; may also contain diethylaminoethanol.

2-ethylamino-4-isopropylamino-6-methylthio-s-triazine.
CAS: 834-12-8. $C_2H_5HNC_3N_3(SCH_3)NHCH(CH_3)_2$.

Properties: White, crystalline powder. Mp 84–85C. Slightly soluble in water; soluble in organic solvents.

Hazard: Toxic by ingestion.

Use: Weed-killing agent in pineapple and sugarcane.

ethyl-1-(p-aminophenyl)-4-phenylisonipecotate.
See anileridine.

ethyl amyl ketone.
(EAK; 5-methyl-3-heptanone).

CAS: 541-85-5. $CH_3CH_2CO(CH_2)_4CH_3$.

Properties: Colorless liquid; pungent odor. Bp 157C, bulk d 83 lb/gal, d 0.819–0.824, refr index 1.416, flash p 138F (58C). Insoluble in water; soluble in four volumes of 60% alcohol. Combustible.

Hazard: Narcotic in high concentration. Moderate fire risk.

Use: Perfumery, solvent for nitrocellulose and vinyl resins.

n-ethylaniline.
CAS: 103-69-5. $C_2H_5NHC_6H_5$.

Properties: Colorless liquid, becoming brown on exposure to light. Soluble in alcohol; insoluble in water. D 0.9631, fp −63.5C, bp 206C, refr index 1.5559 (20C), flash p 185F (85C) (OC). Combustible.

Derivation: By heating aniline and ethyl alcohol in the presence of sulfuric acid, with subsequent distillation.

Hazard: Toxic by ingestion, inhalation, and skin absorption.

Use: Organic synthesis.

o-ethylaniline.
CAS: 578-54-1. $C_6H_4(NH_2)C_2H_5$.

Properties: Brown liquid. Fp −44C, d 0.982 (20C), bp 214C, flash p 185F (85C) (OC). Soluble in alcohol and toluene; insoluble in water. Combustible.

Use: Intermediate for pharmaceuticals, dyestuffs, pesticides, and other products.

ethyl anthranilate.
(ethyl-o-aminobenzoate).

CAS: 87-25-2. $C_6H_4(NH_2)COOCH_2CH_3$.

Properties: Colorless liquid; fruity odor. D 1.117, refr index 1.564, bp 260C. Soluble in alcohol and propylene glycol. Combustible.

Grade: Technical, FCC.

Use: Perfumery and flavors, similar to methyl anthranilate.

2-ethylanthraquinone.
CAS: 84-51-5. $C_{14}H_8O_2C_2H_5$.

Properties: Buff to light-yellow paste. Mp 108C.

Use: Synthesis, especially of hydrogen peroxide.

Ethyl Antiknock Compounds.
A series of fuel additives containing various percentages of tetraethyl lead, ethylene dibromide, ethylene dichloride, dye, kerosene, and antioxidant. All are used to improve the octane rating of motor fuels.

ethylarsenious oxide.
C_2H_5AsO.

Properties: Colorless oil; garlic-like, nauseating odor. D 1.802 (11C); bp 158C (10 mm Hg). Oxidizes in air and forms colorless crystals. Soluble in acetone, benzene, ether.

Hazard: Highly toxic.

Use: Organic synthesis.

2-ethylaziridine.
See ethylethyleneimine.

ethyl p-benzamidobenzoate.
CAS: 736-40-3. $C_{16}H_{15}NO_3$.

Hazard: Moderately toxic by ingestion.

Use: Agricultural chemical.

ethylbenzene.
(phenylethane).

CAS: 100-41-4. $C_6H_5C_2H_5$.

Properties: Colorless liquid; aromatic odor. Vapor heavier than air, bp 136.187C, refr index 1.49594 (20C), d 0.867 (20C), fp −95C, bulk d 7.21 lb/gal (25C), flash p 59F (15C), autoign temp 810F (432C), specific heat 0.41 cal/gal/K, viscosity 0.64 cP (25C). Soluble in alcohol, benzene, carbon tetrachloride, and ether; almost insoluble in water.
Derivation: (1) By heating benzene and ethylene in the presence of aluminum chloride, with subsequent distillation; (2) by fractionation directly from the mixed xylene stream in petroleum refining.
Grade: Technical, pure, research.
Hazard: Toxic by ingestion, inhalation, and skin absorption; irritant to skin and eyes. Flammable, dangerous fire risk. Possible carcinogen.
Use: Intermediate in production of styrene, solvent.

_n_-ethylbenzenesulfonamide.
CAS: 5339-67-3. $C_8H_{11}NO_2S$.
Hazard: Moderately toxic by ingestion.
Use: Agricultural chemical.

4-ethylbenzenesulfonic
   acid.   (*p*-ethylbenzene-
   sulfonic acid).
CAS: 98-69-1. $C_2H_5C_6H_4SO_3H$.
Properties: Solid. Mw 186.23, d 1.229, fp >110C.
Hazard: Corrosive.

ethyl benzoate.
CAS: 93-89-0. $C_6H_5CO_2C_2H_5$.
Properties: Colorless aromatic liquid. D 1.043–1.046, fp −32.7C, bp 212.9C, refr index 1.505, flash p 200F (93C). Soluble in alcohol and ether; insoluble in water. Combustible.
Derivation: By heating ethanol and benzoic acid in the presence of sulfuric acid.
Grade: Technical, FCC.
Use: Flavoring, perfumery, solvent mixture, lacquers, solvent for many cellulose derivatives and natural and synthetic resins.

ethyl benzoylacetate.
CAS: 94-02-0. $C_6H_5COCH_2COOC_2H_5$.
Properties: Light-yellow oil. Bp 265C (decomposes), d 1.111–1.117 (20C), flash p 275F (135C). Soluble in most organic solvents; insoluble in water. Combustible.
Derivation: Reaction of ethyl acetate and ethyl benzoate with metallic sodium.
Method of purification: Vacuum distillation.
Grade: 95% pure.
Use: Dye and pharmaceutical intermediate.

ethyl-*o*-benzoyl benzoate.
CAS: 604-61-5. $C_6H_5COC_6H_4COOC_2H_5$.
Properties: Yellowish-white solid; odorless. Mp 56–58C; bp 325C. Insoluble in water; soluble in alcohol, acetone, ethyl acetate, and benzene. Combustible.
Use: Plasticizer for nitrocellulose and synthetic resins.

1-ethylbenzyl alcohol.   (1-phenyl-1-propanol; ethylphenylcarbinol).
CAS: 93-54-9. $C_9H_{11}OH$.
Properties: Oily liquid. Bp 220C, d 0.99, refr index 1.51. Soluble in ethanol and methanol alcohols, benzene, and toluene.
Derivation: Benzaldehyde or phenyl ethyl ketone.
Use: Heat-transfer fluid.

ethylbenzylaniline.
CAS: 92-59-1. $C_6H_5N(C_2H_5)CH_2C_6H_5$.
Properties: Light-yellow liquid. D 1.034, bp 286C. Soluble in alcohol and ether; insoluble in water. Combustible.
Derivation: By heating ethylaniline, benzyl chloride, and aqueous caustic soda, with subsequent distillation.
Hazard: Toxic by ingestion and inhalation.
Use: Dyestuffs, organic synthesis.

ethylbenzyl chloride.
(1-chloromethylethylbenzene).
CAS: 26968-58-1. $ClCH_2C_6H_4C_2H_5$.
Consists of 70% *p*- and 30% *o*-ethylbenzyl chloride.
Properties: Colorless liquid. D 1.0460–1.0475 (25/25C), refr index 1.5290–1.5305 (25C). Soluble in alcohols; insoluble in water.
Hazard: Irritant to eyes, a lachrymator.
Use: Intermediate.

ethyl biscoumacetate.   (ethyl bis(4-hydroxycoumarinyl)acetate).
CAS: 548-00-5. $C_{22}H_{16}O_8$.
A synthetic derivative of bishydroxycoumarin.
Properties: White, crystalline solid; odorless; bitter taste. Mp 177–182C. Another form melts at 154–157C. Soluble in acetone and benzene; slightly soluble in alcohol and ether; insoluble in water.
Grade: NF.
Use: Medicine (anticoagulant).

ethyl borate.   Legal label name for triethyl borate.

ethyl bromide.   (bromoethane).
CAS: 74-96-4. C_2H_5Br.
Properties: Colorless liquid. D 1.431 (20/4C), bp 38.4C, bulk d 12–12.1 lb/gal, vap press 386 mm Hg (20C), autoign temp 952F (511C), fp −119C, flash p approximately 80F (26C). Soluble in alcohol and ether; sparingly soluble in water.
Derivation: From ethanol or ethylene and hydrobromic acid. One process uses γ-radiation to initiate the combination.
Grade: Technical (98%).
Hazard: Toxic by ingestion, inhalation, and skin absorption; strong irritant. Questionable carcinogen. Flammable, dangerous fire hazard, explosion limits in air 6–11%.
Use: Organic synthesis, medicine (anesthetic), refrigerant, solvent, grain and fruit fumigant.

ethyl bromoacetate.
CAS: 105-36-2. $CH_2BrCOOC_2H_5$.
Properties: Clear, colorless liquid. Partially decomposed by water. D 1.53 (4C), bp 168C, fp −13.8C, vap d 5.8. Soluble in alcohol, benzene, ether; insoluble in water.
Derivation: Interaction of bromine and acetic acid in the presence of red phosphorus.
Grade: Technical.
Hazard: Toxic by ingestion, inhalation, and skin absorption; strong irritant.

ethylbromopyruvate.
CAS: 70-23-5. $BrCH_2COCO_2C_2H_5$.
Properties: Light-yellow liquid. Mw 195.02, bp 90–100C (10 mm Hg), d 1.554, fp 98C.
Hazard: Severe poison, lachrymator, irritant.
Use: Pharmaceutical intermediate.

2-ethylbutadiene.
CAS: 3404-63-5. C_6H_{10}.
Hazard: A poison.

ethyl butanoate.   See ethyl butyrate.

2-ethylbutanol.   See 2-ethylbutyl alcohol.

2-ethyl-1-butene.   (*uns*-diethylethylene).
CAS: 760-21-4. $CH_3CH_2(C_2H_5)C{:}CH_2$.
Properties: Colorless liquid. D 0.6894 (20/4C), bp 64.95C, refr index 1.3969 (20C). Soluble in alcohol, acetone, ether, and benzene; insoluble in water. Combustible.
Grade: 95% pure.
Use: Organic synthesis of flavors, perfumes, medicines, dyes, resins.

3-(2-ethylbutoxy)propionic acid.
CAS: 10213-74-8. $CH_3CH_2CH(C_2H_5)$
$CH_2OCH_2CH_2COOH$.
Properties: Water-white liquid. D 0.9600 (20/20C), bp 200C (100 mm Hg), vap press <0.1 mm Hg (20C), fp (glass at approximately −90C), flash p 280F (137C). Insoluble in water. Combustible.
Use: Preparation of metallic salts for paint driers and gelling agents.

2-ethylbutyl acetate.
CAS: 123-66-0. $C_2H_5CH(C_2H_5)CH_2OOCCH_3$.
Properties: Colorless liquid; mild odor. D 0.875–0.881 (20/20C), boiling range 155–164C, purity not less than 90% ethylbutyl acetate, bulk d 7.33 lb/gal (20C), flash p 130F (54.4C) (OC).
Hazard: Moderate fire risk.
Use: Solvent for nitrocellulose lacquers, flavoring.

2-ethylbutyl alcohol.   (2-ethylbutanol; pseudohexyl alcohol).
CAS: 97-95-0. $CH_3CH_2CH(C_2H_5)CH_2OH$.
Properties: Colorless liquid. Bp 148.9C, fp −114C, d 0.8328 (20/20C), bulk d 6.93 lb/gal (20C), refr index 1.4229 (20C), flash p 137F (58.3C) (ASTM

OC), vap press 0.9 mm Hg (20C). Stable; miscible with most organic solvents; slightly soluble in water. Combustible.
Hazard: Moderate fire risk.
Use: Solvent for oils, resins, waxes, dyes; diluent; synthesis of perfumes, drugs; flavoring.

n-ethylbutylamine.
CAS: 617-79-8. $C_2H_5NHCH_2CH_2CH_2CH_3$.
Properties: Water-white liquid. Bp 108C, fp −78C, d 0.7401 (20/20C), refr index 1.407 (20C), flash p 65F (18.3C). Partly soluble in water.
Hazard: Flammable, dangerous fire risk. Toxic by ingestion.
Use: Intermediate.

ethylbutyl carbonate. $C_2H_5CO_3C_4H_9$.
Properties: Colorless liquid. D 0.92–0.93 (20C), bp 135–175C, flash p 122F (50C) (CC). Combustible.
Hazard: Moderate fire risk.
Use: Solvent for many natural and synthetic resins, in mixtures for nitrocellulose.

ethyl-n-butyl ether.   (*n*-butyl ethyl ether).
CAS: 628-81-9. $C_2H_5OC_4H_9$.
Properties: Liquid. D 0.7528 (20C), fp −103C, bp 92.2C, flash p 40F (4.4C), vap press 43 mm Hg (20C). Slightly soluble in water.
Hazard: Flammable, dangerous fire risk.
Use: Extraction solvent, inert reaction medium.

ethyl butyl ketone.   (3-heptanone).
CAS: 106-35-4. $CH_3CH_2CH_2CH_2COCH_2CH_3$.
Properties: Clear liquid. D 0.8191 (20/20C), fp −39C, boiling range 142.8–147.8C, 95% purity, bulk d 6.8 lb/gal, flash p 115F (46C) (OC). Insoluble in water; soluble in alcohol. Combustible.
Hazard: Moderate fire risk.
Use: Solvent mix for air-dried and baked finishes, for polyvinyl and nitrocellulose resins.

2-ethyl-2-butylpropanediol-1,3.   (2-butyl-2-ethylpropanediol-1,3).
CAS: 115-84-4. $HOCH_2C(C_2H_5)(C_4H_9)CH_2OH$.
Properties: White crystals or liquid. D 0.931 (50/20C), bp 178C (50 mm Hg), fp 41.4C. Solubility in water 0.8% by wt (20C), flash p 280F (137C). Combustible.
Use: Synthesis of lubricants, emulsifying agents, insect repellents, plastics.

2-ethylbutyl silicate.
$[CH_3CH_2CH(C_2H_5)CH_2O]_4Si$.
Properties: Colorless liquid. Bp 164C (1 mm Hg).
Derivation: Reaction of silicon tetrachloride with 2-ethylbutanol.
Use: Hydraulic fluid, heat-transfer liquid.

2-ethylbutyraldehyde.   (diethylacetaldehyde).
CAS: 97-96-1. $(C_2H_5)_2CHCHO$.
Properties: Colorless liquid. D 0.8164 (20/20C), bp 116.8C, vap press 13.7 mm Hg (20C), flash p 70F

(21.1C) (OC), bulk d 6.8 lb/gal (20C), fp −89C. Insoluble in water.

Grade: Technical.

Hazard: Irritant to eyes and skin. Flammable, dangerous fire risk.

Use: Organic synthesis, pharmaceuticals, rubber accelerators, synthetic resins.

ethyl butyrate.   (ethyl butanoate).
CAS: 105-54-4. $C_3H_7CO_2C_2H_5$.

Properties: Colorless liquid; pineapple-like odor. D 0.8788, fp −93.3C, bp 120.6C, refr index 1.400 (20C), flash p 78F (25.5C) (CC), autoign temp 865F (462C). Soluble in alcohol and ether; almost insoluble in water and glycerol.

Derivation: Ethyl alcohol and butyric acid heated in the presence of sulfuric acid, with subsequent distillation.

Grade: Technical, FCC.

Hazard: Irritant to eyes and mucous membranes, narcotic in high concentration. Flammable, dangerous fire risk.

Use: Flavoring extracts, perfumery, solvent mixture for cellulose esters and ethers.

2-ethylbutyric acid.   (diethylacetic acid).
CAS: 88-09-5. $(C_2H_5)_2CHCOOH$.

Properties: Water-white liquid, resembles butyric acid in most properties except that its odor is less pronounced and its water solubility limited. D 0.9225 (20/20C), bp 190C, vap press 0.08 mm Hg (20C), flash p 210F (98C), bulk d 7.7 lb/gal, fp −15C. Combustible.

Use: Ester formation; intermediate for drugs, dyestuffs, chemicals; flavoring.

ethyl caffeate.   (ethyl-3,4-dihydroxycinnamate).
CAS: 102-37-4. $C_6H_3(OH)_2CH:CHCOOC_2H_5$.

Properties: Yellow to tan crystals; characteristic aromatic odor. Insoluble in water; very soluble in alcohol.

Grade: CP.

Use: Food antioxidant.

ethyl caprate.   (ethyl decanoate).
CAS: 110-38-3. $C_9H_{19}COOC_2H_5$.

Properties: Colorless liquid; fragrant odor. D 0.862, bp 243C. Soluble in alcohol and ether; insoluble in water. Combustible.

Derivation: By heating capric acid, absolute alcohol, and sulfuric acid, with subsequent distillation.

Use: Organic synthesis, flavoring agent.

ethyl caproate.   (ethyl hexoate; ethyl hexanoate).
CAS: 123-66-0. $C_5H_{11}COOC_2H_5$.

Properties: Colorless to yellowish liquid; pleasant odor. D 0.873, bp 167. Soluble in alcohol and ether; insoluble in water and glycerol. Combustible.

Derivation: Heating absolute alcohol and n-caproic acid in the presence of sulfuric acid, with subsequent distillation.

Grade: Technical, FCC.

Use: Organic synthesis, artificial fruit essences.

2-ethylcaproic acid sodium salt.   See sodium 2-ethylhexanoate.

ethyl caprylate.   (ethyl octoate; ethyl octanoate).
CAS: 106-32-1. $CH_3(CH_2)_6COOC_2H_5$.

Properties: Colorless liquid; pineapple odor. D 0.865–0.869 (25C), fp −48C, bp 207–209C. Soluble in alcohol and ether; insoluble in water and glycerol. Combustible.

Derivation: Heating caprylic acid, alcohol, and sulfuric acid, with subsequent distillation.

Grade: Technical, FCC.

Use: Flavoring, fruit essences.

ethyl carbamate.   See urethane.

ethyl carbazate.
CAS: 4114-31-2. $H_2NNHCO_2C_2H_5$.

Properties: Off-white crystals. Mw 104.11, mp 44–47C, bp 108–110C (22 mm Hg), fp 86C.

Hazard: Irritant.

Use: Synthetic intermediate.

N-ethylcarbazole.   (9-ethylcarbazole).
CAS: 86-28-2. $(C_6H_4)_2NC_2H_5$ (tricyclic).

Properties: Leaflets. Mp 69–70C, bp 175C (5 mm Hg). Soluble in ether and hot alcohol.

Derivation: Action of ethyl chloride on the potassium salt of carbazole.

Use: Intermediate for dyes, pharmaceuticals; agricultural chemicals.

ethylcarbitol acrylate.
CAS: 7328-17-8.
Properties: Liquid.

Hazard: Moderately toxic. A severe skin irritant.

ethyl carbonate.   See diethyl carbonate.

ethylcellulose.   An ethyl ether of cellulose.

Properties: White, granular thermoplastic solid, lowest density of the commercial cellulose plastics; properties vary with extent to which hydroxyl radicals of cellulose have been replaced by ethoxy groups. Standard commercial product has 47–48% ethoxy content, d 1.07–1.18, refr index 1.47, high dielectric strength, softening point 100–130C. Soluble in most organic liquids; compatible with resins, waxes, oils, and plasticizers; inert to alkalies and dilute acids; insoluble in water and glycerol. Combustible.

Derivation: From alkali cellulose and ethyl chloride or sulfate; from cellulose and ethanol in the presence of dehydrating agents.

Grade: Technical, NF, FCC.

Use: Hot-melt adhesives and coatings for cables, paper, textiles, etc.; extrusion-wire insulation; protective coatings; pigment-grinding bases; toughening agent for plastics; printing inks; molding powders; proximity fuses; vitamin preparations; casing for rocket propellants; food and feed additive.

ethyl centralite.   See *sym*-diethyldiphenylurea.

Ethylch *[Albemarle].*   TM for antiknock fuel additive, tetraethyl lead.

ethyl chloride.   (chloroethane).
CAS: 75-00-3. C_2H_5Cl.
Properties: Gas at room temperature; when compressed, a colorless volatile liquid; ether-like odor; burning taste. D 0.9214, fp −140.85C, bp 12.5C, critical point 187.2C (52 atm, d 0.33), vap press 1000 mm Hg (20C), flash p −58F (−50C) (CC) autoign temp 966F (518C). Stable and noncorrosive when dry, but will hydrolyze in the presence of water or alkalies. Miscible with most of the commonly used solvents; slightly soluble in water.
Derivation: (1) From ethylene and hydrogen chloride; (2) by passing hydrogen chloride into a solution of zinc chloride and ethanol.
Grade: Technical, USP.
Hazard: Highly flammable, severe fire and explosion risk; flammable limits in air 3.8–15.4%. Irritant to eyes. Questionable carcinogen.
Use: Manufacture of tetraethyl lead and ethylcellulose; anesthetic; organic synthesis; alkylating agent; refrigeration; analytical reagent; solvent for phosphorus, sulfur, fats, oils, resins, and waxes; insecticides.

ethyl chloroacetal. $ClCH_2CH(OC_2H_5)_2$.
Properties: Water-white liquid; pleasant odor. D 1.022 (20C), boiling range 54–61C (20 mm Hg), 149–153C, fp −32C, flash p 117F (47.2C), refr index 1.418 (20C). Soluble in alcohol and ethyl ether; insoluble in water. Combustible.
Hazard: Moderate fire risk.

ethyl chloroacetate.
CAS: 105-39-5. $CCH_2ClCO_2C_2H_5$.
Properties: Water-white, mobile liquid; pungent fruity odor. Decomposed by hot water and alkalies. D 1.1585 (20C), bp 144.2C, vap d 4.23–4.46, flash p 131F (55C), refr index 1.4227 (20C). Soluble in alcohol, benzene, and ether; insoluble in water. Combustible.
Derivation: (1) Action of chloroacetyl chloride on alcohol; (2) by treating chloroacetic acid with alcohol and sulfuric acid.
Hazard: Strong irritant to eyes.
Use: Solvent, organic synthesis, military poison, vat dyestuffs.

ethyl chlorobenzoxazoline-2-thione-3-carboxylate.
CAS: 33388-24-8. $C_{10}H_8ClNO_3S$.

Hazard: Moderately toxic by ingestion.
Use: Agricultural chemical.

ethyl chlorocarbonate.   (ethyl chloroformate).
CAS: 541-41-3. $ClCOOC_2H_5$.
Properties: Water-white liquid; irritating odor. D 1.135–1.139 (20/20C), bp 93–95C, refr index 1.3974 (20C), flash p 61F (16.1C) (CC). Decomposes in water and alcohol; soluble in benzene, chloroform, and ether.
Derivation: Reaction of carbon monoxide with gaseous chlorine, producing phosgene ($COCl_2$), which is then treated with anhydrous ethanol, giving ethyl chlorocarbonate and splitting off hydrogen chloride.
Grade: Technical.
Hazard: Flammable, dangerous fire risk. Strong irritant to eyes and skin.
Use: Organic synthesis, intermediate in making diethyl carbonate, flotation agents, polymers, isocyanates.

ethyl chloroformate.   See ethyl chlorocarbonate.

ethyl 4-chloro-2-oxo-3(2h)-benzothiazoleacetate.
CAS: 25059-80-7. $C_{11}H_{10}ClNO_3S$.
Hazard: Moderately toxic by ingestion and skin contact.
Use: Agricultural chemical.

o-**ethyl s-4-chlorophenyl ethylphosphonodithioate.**
CAS: 2984-64-7. $C_{10}H_{14}ClOPS_2$.
Hazard: A poison by skin contact. Moderately toxic by ingestion and inhalation.
Use: Agricultural chemical.

ethyl chlorosulfonate.
CAS: 625-01-4. $C_2H_5OClSO_2$.
Properties: Colorless, oily liquid; pungent odor. D 1.379 (0C); bp 153C; vap d 5 (air = 1.29); volatility 18,000 mg/m^3 (20C). Fumes in moist air. Decomposed by water. Attacks lead and tin but copper only mildly; iron and steel not affected. Soluble in chloroform and ether; insoluble in water.
Derivation: (1) Action of fuming sulfuric acid on ethylchloroformate; (2) interaction of ethylene and chlorosulfonic acid.
Grade: Technical.
Hazard: Strong irritant to eyes and skin, evolves phosgene when heated.
Use: Organic synthesis, military poison.

3-ethylcinchoninic acid ethyl ester.
CAS: 21233-74-9. $C_{14}H_{15}NO_2$.
Hazard: Moderately toxic by ingestion.
Use: Agricultural chemical.

ethyl cinnamate.   (ethyl phenylacrylate; cinnamylic ether).
CAS: 103-36-6. $C_6H_5CH:CHCOOC_2H_5$.

Properties: Limpid, oily liquid; strawberry-like odor. D 1.045–1.048, refr index 1.560 (20C), congealing point 7C (min), bp 271C. Soluble in alcohol and ether. Insoluble in water. Combustible.
Derivation: Heating ethyl alcohol and cinnamic acid in the presence of sulfuric acid.
Use: Perfumery, flavoring extracts.

ethyl citrate. See triethyl citrate.

ethyl crotonate.
CAS: 623-70-1. $CH_3CH:CHCOOC_2H_5$.
Properties: Water-white solid or liquid; characteristic, pungent, persistent odor. D 0.9207 (20/20C), mp (solid) 45C, bp (solid) 209C, (liquid) 139C, flash p 36F (2.2C), refr index 1.4242 (20C), bulk d 7.65 lb/gal (20C). Soluble in alcohol and ether, insoluble in water.
Hazard: Flammable, dangerous fire risk. Strong irritant.
Use: Solvent and softening agent, lacquers, organic synthesis.

ethyl cyanide. (propionitrile; propanenitrile).
CAS: 107-12-0. C_2H_5CN.
Properties: Mobile, colorless liquid; ethereal odor. D 0.7829 (20/20C), refr index 1.3664 (20C), bp 97.4C, fp −92.9C, flash p 61F (16.1C) (OC). Soluble in alcohol and water.
Derivation: Heating barium ethyl sulfate and KCN with subsequent distillation.
Hazard: Toxic by ingestion and inhalation. Flammable, dangerous fire risk.
Use: Solvent, dielectric fluid, intermediate.

ethyl cyanoacetate. (malonic ethyl ester nitrile).
CAS: 105-56-6. $CNCH_2COOC_2H_5$.
Properties: Colorless liquid. Bp 206–208C, fp −22.5C, refr index 1.41751 (20C). Soluble in alcohol and ether; soluble in alkaline solutions. Flash p 230F (110C). Combustible.
Derivation: Esterification of cyanoacetic acid with ethanol; reaction of an alkali cyanide and chloroacetic ethyl ester.
Method of purification: Vacuum distillation.
Grade: Reagent, technical.
Hazard: Toxic by ingestion and inhalation.
Use: Organic synthesis, pharmaceuticals, dyes.

ethyl cyanoacrylate. (ethyl α-cyanoacrylate; ethyl 2-cyanoacrylate; ethyl 2-cyano-2-propenoate).
CAS: 7085-85-0. $C_6H_7NO_2$.
Hazard: An inhalation hazard.
Use: An adhesive.

ethyl α-cyanoacrylate. See ethyl cyanoacrylate.

ethyl 2-cyano-2-propenoate. See ethyl cyanoacrylate.

5-ethyl-5-cycloheptenylbarbituric acid.
See heptabarbital.

ethylcyclohexane.
CAS: 1678-91-7. $C_2H_5C_6H_{11}$.
Properties: Colorless liquid. D 0.787, bp 131.8C, refr index 1.4330 (20C), flash p 95F (35C), autoign temp 504F (262C).
Hazard: Flammable, moderate fire risk; flammable limits in air 0.9–6.6%.
Use: Organic synthesis.

N-ethylcyclohexylamine.
CAS: 5459-93-8. $C_6H_{11}NHC_2H_5$.
Properties: Liquid. Mw 127.23, bp 165C, d 0.844, fp 43C.
Hazard: Corrosive and toxic.
Use: Intermediate in production of herbicides and pharmaceuticals.

ethyl 2-cyclohexylpropionate.
CAS: 2511-00-4. $C_{11}H_{20}O_2$.
Hazard: Low toxicity by inhalation.

ethylcyclopentane.
CAS: 1640-89-7. $C_2H_5C_5H_9$.
Properties: Colorless liquid. D 0.766, bp 103.5C, refr index 1.4198 (20C), autoign temp 504F (262C).
Hazard: Flammable, moderate fire risk; flammable limits in air 1.1–6.7%.

ethyl cyclopentanone-2-carboxylate. See 2-carbethoxycyclopentanone.

ethyl decanoate. See ethyl caprate.

ethyl diazoacetate.
CAS: 623-73-4. $N_2CHCO_2C_2H_5$.
Properties: Liquid. Mw 114.10, mp 22C, bp 140–141C (720 mm Hg), d 1.085.
Available forms: Research quantities.
Hazard: Flammable liquid that explodes when heated.

ethyl 2,3-dibromopropanoate.
CAS: 3674-13-3. $C_5H_8Br_2O_2$.
Hazard: A poison by ingestion and skin contact. A severe skin and eye irritant.

ethyldichloroarsine. (dichloroethylarsine).
CAS: 598-14-1. $C_2H_5AsCl_2$.
Properties: Colorless, mobile liquid; becomes yellowish under the action of light and air. Fruit-like odor (high dilution). Decomposed by water. Attacks brass but not iron (dry). D 1.742 (14C), bp 156C (decomposes), fp −65C, coefficient of thermal expansion 0.0011, vap d 6 (air = 1.29),

volatility 20,000 mg/m^3 (20C), vap press 2.29 mm Hg (21.5C). Soluble in alcohol, benzene, ether, and water.
Derivation: Chlorination of ethyl arsenious oxide.
Hazard: Toxic by ingestion, inhalation, and skin absorption; strong irritant.
Use: Military poison.

ethyl-4,4′-dichlorobenzilate.   See "Chlorobenzilate" [Novartis].

ethyl dichlorophenoxyacetate.   See 2,4-D.

ethyldichlorosilane.
CAS: 1789-58-8.
$C_2H_5SiHCl_2$.
Properties: Colorless liquid. Bp 75.5C, d 1.088 (25/25C), flash p 30F (−1.1C) (COC). Readily hydrolyzed by moisture, with the liberation of hydrogen and hydrogen chloride.
Derivation: By Grignard reaction of trichlorosilane and ethylmagnesium chloride.
Hazard: Flammable, dangerous fire risk. Strong irritant to eyes and skin.
Use: Intermediate for silicones.

ethyldiethanolamine.
CAS: 139-87-7. $C_2H_5N(CH_2CH_2OH)_2$.
Properties: Water-white liquid; amine odor. D 1.015 (20C), boiling range 246–252C, flash p 255F (123C). Combustible.
Use: Solvent, detergents.

ethyl-2-(diethoxyphosphinyl)-3-oxobutanoate.
CAS: 3730-54-9. $C_{10}H_{19}O_6P$.
Hazard: A poison by ingestion.
Use: Agricultural chemical.

ethyl 3-((diethylphosphinothioyl)oxy)-2-butenoate.
CAS: 73263-81-7. $C_{10}H_{19}O_5PS$.
Hazard: A poison by ingestion.

ethyl diglyme.   See diethylene glycol diethyl ether.

20-ethyl-6-β,8-dihydroxy-1-α-methoxy-4-methylheteratisan-14-one.
CAS: 3328-84-5. $C_{22}H_{33}NO_5$.
Hazard: A poison.
Source: Natural product.

ethyldimethylmethane.   See isopentane.

ethyldimethyl-9-octadecenylammonium bromide.
CAS: 6458-13-5. $C_{22}H_{46}NBr$.
Cationic surfactant and algicide, soluble in propylene glycol and isopropyl alcohol.

n-ethyl-1,2-dimethylpropylamine.
CAS: 2738-06-9. $C_7H_{17}N$.
Hazard: Low toxicity by ingestion. A severe skin and moderate eye irritant.

ethyldiphenyltin acetate.
CAS: 77405-29-4. $C_{16}H_{18}O_2Sn$.
Hazard: A poison by ingestion. Questionable carcinogen.

ethyldipropylmethane.   See 4-ethylheptane.

S-ethyl di-N,N-propylthiocarbamate.   See EPTC.

ethyl diselenide.
CAS: 628-39-7. $C_4H_{10}Se_2$.
Hazard: A poison by ingestion and skin contact. Low toxicity by inhalation. A moderate eye irritant.

ethyl enanthate.   (ethyl heptanoate; cognac oil).
CAS: 106-30-9. $CH_3(CH_2)_5COOC_2H_5$.
Properties: Clear, colorless oil; fruity odor and taste. D 0.87; bp 187C. Soluble in alcohol, chloroform, and ether; insoluble in water. Combustible.
Derivation: By heating enanthic acid and ethanol in the presence of sulfuric acid and subsequent recovery by distillation.
Grade: Technical.
Use: Artificial cognac flavor, flavor for liquors and fruity-type soft drinks.

ethylene.   (bicarburetted hydrogen; ethene).
CAS: 74-85-1. $H_2C:CH_2$.
Properties: Colorless gas; sweet odor and taste. Fp −169C, bp −103.9C, flash p −213F (−135C), d (liquid) 0.610 (0C), vap d 0.975 (air = 1.29), critical temp 9.5C, autoign temp 1009F (543C), critical press (absolute) 744 psi. Purity not less than 96% ethylene by gas volume, not more than 0.5% acetylene, not more than 4% methane and ethane. Sp vol 13.4 cu ft/lb (15.6C). Slightly soluble in water, alcohol, and ethyl ether.
Derivation: (1) Thermal cracking of hydrocarbon gases (800–900C), (2) dehydration of ethanol, (3) from synthesis gas with Ru as catalyst.
Grade: Technical (95% min), 99.5% min, 99.9 mol%.
Hazard: Highly flammable, dangerous fire and explosion risk; explosive limits in air 3–36% by volume. Simple asphyxiant; questionable carcinogen.
Use: Manufacture of polyethylene, polypropylene, ethylene oxide, ethylene dichloride, ethylene glycols, aluminum alkyls, vinyl chloride, vinyl acetate, ethyl chloride, ethylene chlorohydrin, acetaldehyde, ethyl alcohol, polystyrene, styrene, polyvinyl chloride, SBR, polyester resins, trichloroethylene, etc.; as a refrigerant, in welding and cutting of metals, an anesthetic, and in orchard sprays to accelerate fruit ripening.

ethylene azelate.
CAS: 4471-27-6. $C_{11}H_{18}O_4$.
Hazard: Low toxicity by ingestion and skin contact. A mild skin and eye irritant.

ethylene bis(dithiocarbamate).   (Ebdc; etho-xyethene).
CAS: 109-92-2. C_4H_8O.
Hazard: Carcinogen.
Use: Dithiocarbamate fungicides used on many fruits and vegetables.

ethylene bis(iminodiacetic acid).   See ethy-lenediaminetetraacetic acid.

ethylene   bis(oxyethylenenitrilo)tetraacetic acid.   See ethylene glycol-bis(β-aminoethyl ether)-*N,N*-tetraacetic acid.

ethylene bis(pentabromophenoxide).   See 1,1′-(1,2-ethanediylbis(oxy))bis(2,3,4,5,6-pentabromobenzene).

ethylene bis(tetrabromophthalimide).
CAS: 32588-76-4. $C_{18}H_4Br_8N_2O_4$.
Hazard: Low toxicity by ingestion, inhalation, and skin contact. A mild eye irritant.

ethylene bromide.   See ethylene dibromide.

ethylene bromohydrin.   (glycol bromohydrin; 2-bromoethyl alcohol).
CAS: 540-51-2. $BrCH_2CH_2OH$.
Properties: Hygroscopic liquid. D 1.7629 (20C), bp 149–150C (750 mm Hg), refr index 1.4915 (20C). Soluble in most organic solvents; completely miscible with water. Aqueous solutions have a sweet, burning taste. Hydrolysis of aqueous solutions is accelerated by heat, acids, and alkalies. Combustible.
Derivation: Action of hydrogen bromide on ethylene oxide.
Hazard: Irritant to eyes and mucous membranes.
Use: Organic synthesis.

ethylene   carbonate.   (glycol   carbonate; dioxolone-2).
CAS: 96-49-1. $(-CH_2O)_2CO$.
Properties: Colorless, solid or liquid; odorless. Mp 36.4C, bp 248C, d 1.3218 (39/4C), refr index 1.4158 (50C), flash p 290F (143C) (OC). Miscible (40%) with water, alcohol, ethyl acetate, benzene, and chloroform; soluble in ether, *n*-butanol, and carbon tetrachloride. Combustible.
Derivation: Interaction of ethylene glycol and phosgene.
Use: Solvent for many polymers and resins, plasticizer, intermediate for pharmaceuticals, rubber chemicals, textile finishing agents, hydroxyethylation reactions.

ethylene chloride.   See ethylene dichloride.

ethylene chlorohydrin.   (2-chloroethyl alcohol; glycol chlorohydrin).
CAS: 107-07-3. $ClCH_2CH_2OH$.
Properties: Colorless liquid; faint ethereal odor. D 1.2045 (20/20C), bp 128.7C, refr index 1.4419 (20C), vap press 4.9 mm Hg (20C), flash p 140F (60C) (OC), bulk d 10.0 lb/gal (20C), coefficient of expansion 0.00089 (20C), fp −62.6C, viscosity 0.0343 cP (20C), autoign temp 797F (425C). Soluble in most organic liquids and completely miscible with water. Combustible.
Derivation: Action of hypochlorous acid on ethylene.
Grade: Anhydrous, 38%.
Hazard: Deadly via ingestion, inhalation, and skin absorption; strong irritant, penetrates ordinary rubber gloves and protective clothing. Moderate fire hazard. Questionable carcinogen.
Use: Solvent for cellulose acetate, ethylcellulose; introduction of hydroxyethyl group in organic synthesis; to activate sprouting of dormant potatoes; manufacture of ethylene oxide and ethylene glycol; insecticides.

ethylene chlorothioarsenate(III).
CAS: 3741-32-0. $C_2H_4AsClS_2$.
Hazard: A poison by ingestion.
Use: Agricultural chemical.

ethylene cyanide.   (ethylene dicyanide; succi-nonitrile).
CAS: 110-61-2. $C_2H_4(CN)_2$.
Properties: Colorless, waxy solid. Mp 57–57.5C; bp 265.7C; flash p 270F (132C). Soluble in alcohol, water, and chloroform. Combustible.
Derivation: Interaction of ethylene dibromide and potassium cyanide in the presence of alcohol.

ethylene cyanohydrin.
(β-hydroxypropionitrile).
CAS: 109-78-4. $HOCH_2CH_2CN$.
Properties: Straw-colored liquid. Fp −46C, bp 227–228C (decomposes) d 1.0404 (25/4C), vap press 0.08 mm Hg (25C), 20 mm Hg (117C). Miscible with water, acetone, methyl ethyl ketone, ethanol, chloroform, and diethyl ether; insoluble in benzene, carbon tetrachloride, and naphtha. Combustible.
Derivation: Ethylene oxide and hydrogen cyanide.
Hazard: Toxic by ingestion.
Use: Solvent for certain cellulose esters and inorganic salts; organic intermediate for acrylates.

ethylenediamine.   (1,2-diaminoethane).
CAS: 107-15-3. $NH_2CH_2CH_2NH_2$.
Properties: Colorless, alkaline liquid; ammonia odor. D 0.8995 (20/20C), bulk d 7.50 lb/gal (20C), bp 116–117C, vap press 10.7 mm Hg (20C), mp 8.5C, viscosity 0.0154 cP (25C), flash p 93F (33.9C) (CC), refr index 1.4540 (26C), pH of 25% solution 11.9 (25C). Strong base, soluble in water and alcohol; slightly soluble in ether; insoluble in benzene. Readily absorbs carbon dioxide from air.

Derivation: Catalytic reaction of ethylene glycol or ethylene dichloride and ammonia (nickel or copper catalysts are used).

Method of purification: Redistillation.

Grade: Technical, USP 97%, solutions of various strengths.

Hazard: Toxic by inhalation and skin absorption, strong irritant to skin and eyes. Flammable, moderate fire risk. Questionable carcinogen.

Use: Fungicide, manufacture of chelating agents (EDTA), dimethylolethylene-urea resins, chemical intermediate, solvent, emulsifying agent, textile lubricants, antifreeze inhibitor.

ethylenediamine carbamate. See "Diak" *[Vanderbilt].*

ethylenediamine dihydroiodide.
$(-CH_2NH_2)_2 \cdot 2HI$.
Use: Feed additive.

ethylenediamine-di-*o*-hydroxyphenylacetic acid. $[-CH_2NHCH(COOH)C_6H_4OH]_2$. A phenolic analog of ethylenediaminetetraacetic acid.

Properties: Mp 218C (decomposes). Insoluble in water and most organic solvents; soluble in sulfuric acid; also forms water-soluble alkali-metal and ammonium salts.

Use: Chelating iron in mildly alkaline solutions.

ethylenediamine tartrate.
Use: To make piezoelectric crystals for control of electric frequencies, etc., as in television.

ethylenediaminetetraacetic acid. (EDTA; ethylenebisiminodiacetic acid; ethylenedinitrilotetraacetic acid).
CAS: 60-00-4. $(HOOCCH_2)_2NCH_2CH_2N(CH_2COOH)_2$.
An organic chelating agent.

NaO—C—H₂C,   CH₂—C--ONa
       N—CH₂—CH₂—N
NaO—C—H₂C   CH₂—C--ONa

Properties: Colorless crystals. Decomposes at 240C. Slightly soluble in water; insoluble in common organic solvents; neutralized by alkali metal hydroxides to form a series of water-soluble salts containing from one to four alkali metal cations.

Derivation: (a) Addition of sodium cyanide and formaldehyde to a basic solution of ethylenediamine (forms the tetrasodium salt); (b) heating tetrahydroxyethylethylenediamine with sodium hydroxide or potassium hydroxide with cadmium oxide catalyst.

Use: Detergents; liquid soaps; shampoos; agricultural chemical sprays; metal cleaning and plating; metal chelating agent; treatment of chlorosis; decontamination of radioactive surfaces; metal deactivator in vegetable oils, oil emulsions, pharmaceutical products, etc.; anticoagulant of blood; eluting agent in ion exchange; to remove insoluble deposits of calcium and magnesium soaps; in textiles to improve dyeing, scouring, and detergent operations; antioxidant; clarification of liquids; analytical chemistry, spectrophotometric titration; aid in reducing blood cholesterol; in medicine to treat lead poisoning and calcinosis; food additive (preservative).

Note: A number of salts of EDTA are available with uses identical or similar to the acid. The USP salts are called edetates (calcium disodium, disodium edetates); others are usually abbreviated to EDTA (tetrasodium, trisodium EDTA). Other salts, known chiefly under trademark names, are the sodium ferric, dihydrogen ferrous and a range of disodium salts with magnesium, divalent cobalt, manganese, copper, zinc, and nickel.

ethylenediaminetetraacetonitrile.
(EDTAN).
CAS: 5766-67-6. $[-CH_2NCH_2CN_2]_2$.
Properties: White, crystalline solid. Melting range 126–132C, bulk d 48.4 lb/cu ft. Slightly soluble in water; soluble in acetone.
Hazard: Toxic by ingestion and inhalation.
Use: Chelating agent and intermediate.

ethylene dibromide. (EDB; 1,2-dibromoethane; ethylene bromide).
CAS: 106-93-4. $BrCH_2CH_2Br$.
Properties: Colorless, liquid; sweetish odor. D 2.17–2.18 (20C), bulk d 18.1 lb/gal, bp 131C, vap press 17.4 mm Hg (30C), fp 9C, refr index 1.5337 (25C), flash p none. Emulsifiable; miscible with most solvents and thinners; slightly soluble in water. Nonflammable.
Derivation: Action of bromine on ethylene.
Hazard: Probable carcinogen. Toxic by inhalation, ingestion, and skin absorption; strong irritant to eyes and skin.
Use: Scavenger for lead in gasoline, grain fumigant, general solvent, waterproofing preparations, organic synthesis, fumigant for tree crops.
Note: May poison platinum catalysts.

ethylene dichloride. (*sym*-dichloroethane; 1,2-dichloroethane; ethylene chloride; Dutch oil).
CAS:107-06-2. $ClCH_2CH_2Cl$.
Properties: Colorless, oily liquid; chloroform-like odor; sweet taste. Bp 83.5C, fp −35.5C, d 1.2554 (20/4C) (10.4 lb/gal), refr index 1.444, flash p 56F (13.3C). Stable to water, alkalies, acids, or active chemicals; resistant to oxidation, will not corrode metals. Miscible with most common solvents; slightly soluble in water.

Derivation: Action of chlorine on ethylene, with subsequent distillation with metallic catalyst; also by reaction of acetylene and hydrochloric acid.

Grade: Technical, spectrophotometric.

Hazard: Toxic by ingestion, inhalation, and skin absorption; strong irritant to eyes and skin; a carcinogen. Flammable, dangerous fire risk, explosive limits in air 6–16%. Possible carcinogen.

Use: Production of vinyl chloride, trichloroethylene, vinylidene chloride, and trichloroethane; lead scavenger in antiknock gasoline; paint, varnish, and finish removers; metal degreasing; soaps and scouring compounds; wetting and penetrating agents; organic synthesis; ore flotation; solvent, fumigant.

ethylene dicyanide. See ethylene cyanide.

3,3'-(ethylenediiminodimethylene)bis(5,5-diphenylhydantoin).
CAS: 21322-39-4. $C_{34}H_{32}N_6O_4$.
Hazard: A poison.

ethylenedinitrilotetraacetic acid. See ethylenediaminetetraacetic acid.

ethylenedinitrilotetra-2-propanol. See N,N,N',N'-tetrakis(2-hydroxypropyl) ethylenediamine.

ethylenediphenyldiamine. See N,N-diphenylethylenediamine.

1,1'-ethylene-2,2'-dipyridinium dibromide. See diquat.

ethylenedithiocyanate.
CAS: 629-17-4. $C_4H_4N_2S_2$.
Hazard: A poison by ingestion and skin contact.

ethylene glycol. (ethylene alcohol; glycol; 1,2-ethanediol).
CAS: 107-21-1. CH_2OHCH_2OH.
The simplest glycol.
Properties: Clear, colorless, syrupy liquid; odorless; sweet taste. D 1.1155 (20C), bp 197.2C, fp −13.5C, bulk d 9.31 lb/gal (15/15C), refr index 1.430 (25C), flash p 240.8F (116C), autoign temp 775F (412C). Hygroscopic, lowers fp of water, relatively nonvolatile. Soluble in water, alcohol, and acetone. Combustible.
Derivation: (1) Air oxidation of ethylene followed by hydration of the ethylene oxide formed; (2) acetoxylation; (3) from carbon monoxide and hydrogen (synthesis gas) from coal gasification; (4) Oxirane process.
Grade: Technical.
Hazard: Questionable carcinogen. Toxic by ingestion and inhalation. Lethal dose reported to be 100 cc.
Use: Coolant and antifreeze; asphalt-emulsion paints; heat-transfer agent; low-pressure laminates, brake fluids; glycol diacetate; polyester fibers and films; low-freezing dynamite; solvent; extractant for various purposes; solvent mixture for cellulose esters and ethers, especially cellophane; cosmetics (up to 5%); lacquers; alkyd resins; printing inks; wood stains; adhesives; leather dyeing; textile processing; tobacco; ingredient of deicing fluid for airport runways; humectant; ballpoint pen inks; foam stabilizer.

ethylene glycol-bis(β-aminoethyl ether)-N,N-tetraacetic acid. (ethylene bis(oxyethylenenitrilo)tetraacetic acid). [−$CH_2OC_2H_4N(CH_2COOH)_2]_4$.
Properties: Crystals. Mp 241C (decomposes). Soluble in water.
Use: Chelating agent.

ethylene glycol bis(chloromethyl)ether. See bis-1,2-(chloromethoxy)ethane.

ethylene glycol bis(mercaptopropionate). See glycol dimercaptopropionate.

ethylene glycol bisthioglycolate. See glycol dimercaptoacetate.

ethylene glycol diacetate. (glycol diacetate).
CAS: 111-55-7. $CH_3COOCH_2CH_2OOCCH_3$.
Properties: Colorless liquid; faint odor. D 1.1063 (20/20C), bp 190.5C, vap press 0.3 mm Hg (20C), flash p 205F (96C) (OC), bulk d 9.2 lb/gal (20C), fp −31C, refr index 1.415 (20C). Soluble in alcohol, ether, benzene; slightly soluble in water (10%). Combustible.
Derivation: (a) Ethylene glycol and acetic acid; (b) ethylene dichloride and sodium acetate.
Use: Solvent for cellulose esters and ethers, resins, lacquers, printing inks; perfume fixative; nondiscoloring plasticizer for ethyl and benzyl cellulose.

ethylene glycol dibutyl ether.
CAS: 112-48-1. $C_4H_9OC_2H_4OC_4H_9$.
Properties: Almost colorless liquid; slight odor. D 0.8374 (20/20C), bulk d 7.0 lb/gal (20C), bp 203.1C, vap press 0.09 mm Hg (20C), fp −69.1C, flash p 185F (85C). Slightly soluble in water. Combustible.
Use: High-boiling inert solvent; specialized solvent and extraction applications.

ethylene glycol dibutyrate. (glycol dibutyrate).
CAS: 105-72-6. $(-CH_2OCOC_3H_7)_2$.
Properties: Colorless liquid. D 1.024 (0C), refr index 1.424 (25C), bp 240C, fp <−80C. Solubility in water 0.050% by weight. Combustible.
Use: Plasticizer.

ethylene glycol diethyl ether. (ethyl glyme).
CAS: 16484-86-9. $C_2H_5OCH_2CH_2OC_2H_5$.
Properties: Colorless liquid; slight odor. Stable, d 0.8417 (20/20C), bp 121.4C, vap press 9.4 mm Hg

(20C), flash p 95F (35C), bulk d 7 lb/gal (20C), fp −74C. Immiscible with water.

Grade: Technical.

Hazard: Flammable, moderate fire risk.

Use: Organic synthesis (reaction medium), solvent and diluent for detergents.

ethylene glycol diformate.   (glycol diformate).
CAS: 629-15-2. $HCOOCH_2CH_2OOCH$.

Properties: Water-white liquid. D 1.2277 (20/20C), bulk d 10.2 lb/gal (20C), bp 177.1C, flash p 200F (93C), vap press 0.5 mm Hg (20C), fp −10C; hydrolyzes slowly, liberating formic acid. Soluble in water, alcohol and ether. Combustible.

Hazard: Toxic by ingestion.

Use: Embalming fluids.

ethylene glycol dimethyl ether.   (GDME; glycol dimethyl ether; 1,2-dimethoxyethane; monoglyme).
CAS: 110-71-4. $CH_3OCH_2CH_2OCH_3$.

Properties: Water-white liquid; mild odor. D 0.8683 (20C), bp 85.2C, fp −69C, refr index 1.3792 (20C), flash p 104F (40C). Soluble in water and hydrocarbons, pH 8.2.

Hazard: Moderate fire risk.

Use: Solvent.

ethylene glycol dinitrate.
CAS: 628-96-6.   A freezing-point depressant for nitroglycerine.

Use: Low-freezing dynamites.

Hazard: Toxic by skin absorption.

ethylene glycol dipropionate.   (glycol propionate; glycol dipropionate).
CAS: 123-80-8. $(-CH_2OCOC_2H_5)_2$.

Properties: Liquid. D 1.054 (15C), refr index 1.419 (25C), bp 211C, fp <−80C. Soluble in water 0.16% by weight. Combustible.

Use: Plasticizer.

ethylene glycol distearate.   See "MAPEG" *[BASF]*.

ethylene glycol monoacetate.   (glycol monoacetate).
CAS: 542-59-6. $HOCH_2CH_2OOCCH_3$.

Properties: Colorless liquid; almost odorless. Bp 181–182C, d 1.108, flash p 215F (101C). Soluble in water, alcohol, ether, benzene, and toluene. Combustible.

Derivation: (a) Heating ethylene glycol with acetic acid (glacial) or acetic anhydride; (b) passing ethylene oxide into hot acetic acid containing sodium acetate or sulfuric acid.

Use: Solvent for nitrocellulose, cellulose acetate, camphor.

ethylene glycol monobenzyl ether.   (benzyl "Cellosolve" *[Union]*).
CAS: 622-08-2. $C_6H_5CH_2OC_2H_4OH$.

Properties: Water-white liquid; faint rose-like odor. D 1.070 (20/20C), bp 255.9C, vap press 0.02 mm Hg (20C), flash p 265F (129C), bulk d 8.9 lb/gal (20C), autoign temp 665F (351C). Combustible.

Hazard: Toxic by ingestion.

Use: Solvent for cellulose acetate, dyes, inks, resins, perfume fixative; organic synthesis (selective hydroxyethylating agent); coating compositions for leather, paper, and cloth; lacquers.

ethylene glycol monobutyl ether.   (2-butoxyethanol; butyl "Cellosolve" *[Union]*).
CAS: 111-76-2. $HOCH_2CH_2OC_4H_9$.

Properties: Colorless liquid; mild odor. Bp 171.2C, d 0.9019 (20/20C), bulk d 7.51 lb/gal (20C), refr index 1.4190 (25C), vap press 0.76 mm Hg (20C), flash p 142F (61C), autoign temp 472F (244C). High dilution ratio with petroleum hydrocarbons. Soluble in alcohol and water. Combustible.

Grade: Technical.

Hazard: A toxic material. Eye and upper respiratory tract irritant. Questionable carcinogen.

Use: Solvent for nitrocellulose resins, spray lacquers, quick-drying lacquers, varnishes, enamels, dry-cleaning compounds, varnish removers, textile (preventing spotting in printing or dyeing); mutual solvent for "soluble" mineral oils to hold soap in solution and to improve the emulsifying properties.

ethylene glycol monobutyl ether acetate. (butyl "Cellosolve" acetate *[Union]*).
CAS: 112-07-2. $C_4H_9OCH_2CH_2OOCCH_3$.

Properties: Colorless liquid; fruity odor. Bp 192.3C, d 0.9424 (20/20C), fp −63.5C, flash p 190F (87.7C). Soluble in hydrocarbons and organic solvents; insoluble in water. Combustible.

Grade: Technical.

Use: High-boiling solvent for nitrocellulose lacquers, epoxy resins, multicolor lacquers; film-coalescing aid for polyvinyl acetate latex.

ethylene glycol monobutyl ether laurate. (butoxyethyl laurate).
CAS: 109-37-5. $C_{11}H_{23}COO(CH_2)_2OC_4H_9$.

Properties: Liquid. D 0.985 (25C), fp −10 to −15C. Insoluble in water. Combustible.

Use: Plasticizer.

ethylene glycol monobutyl ether oleate. (butoxyethyl oleate).
CAS: 109-39-7. $C_{17}H_{33}COOCH_2CH_2OC_4H_9$.

Properties: Liquid. D 0.892 (25C), fp <−45C. Insoluble in water. Combustible.

Use: Plasticizer.

ethylene glycol monobutyl ether stearate. (butoxyethyl stearate).
CAS: 109-38-6. $C_{17}H_{35}COOC_2H_4OC_4H_9$.

Properties: Colorless liquid. D 0.882 (20C), vap press <0.01 mm Hg (20C), bp 210–233C (4 mm Hg), mp 16.5C. Insoluble in water. Combustible.

Use: Plasticizer and solvent.

ethylene glycol mono-dicyclopentenyl ether.
CAS: 68586-17-4. $C_{12}H_{18}O_2$.
Hazard: Moderately toxic by ingestion and skin contact. A severe eye and moderate skin irritant.

ethylene glycol monoethyl ether. (2-ethoxyethanol; "Cellosolve" solvent *[Union]*).
CAS: 110-80-5. $HOCH_2CH_2OC_2H_5$.
Properties: Colorless liquid; practically odorless. Bp 135.6C, d 0.9311 (20/20C), bulk d 7.74 lb/gal (20C), refr index 1.4060 (25C), flash p 120F (48.9C), pour point <100F (37.7C), autoign temp 460F (237C). Miscible with hydrocarbons and water. Combustible.
Grade: Technical.
Hazard: Toxic by skin absorption. Moderate fire risk.
Use: Solvent for nitrocellulose, natural and synthetic resins; mutual solvent for formulation of soluble oils; lacquers and lacquer thinners; dyeing and printing textiles; varnish removers; cleaning solutions; leather; antiicing additive for aviation fuels.

ethylene glycol monoethyl ether acetate.
("Cellosolve" acetate *[Union]*; 2-ethoxyethyl acetate).
CAS: 111-15-9. $CH_3COOCH_2CH_2OC_2H_5$.
Properties: Colorless liquid; pleasant ester-like odor. Bp 156.3C, d 0.9748 (20/20C), bulk d 8.1 lb/gal (20C), refr index 1.4030 (25C), viscosity 1.32 cP (20C), flash p 120F (48.9C), fp −61.7C, vap press 2 mm Hg (20C). Miscible with aromatic hydrocarbons; slightly miscible with water. Combustible.
Grade: Technical.
Hazard: Toxic by ingestion and skin absorption. Toxic by skin absorption.
Use: Solvent for nitrocellulose, oils, and resins; retards "blushing" in lacquers, varnish removers, wood stains, textiles, and leather.

ethylene glycol monoethyl ether laurate.
$C_{11}H_{23}COO(CH_2)_2OC_2H_5$.
Properties: Liquid. D 0.89 (25C), fp −7 to −11C. Insoluble in water. Combustible.
Use: Plasticizer.

ethylene glycol monoethyl ether ricinoleate.
$C_{17}H_{32}(OH)COO(CH_2)_2OC_2H_5$.
Properties: Liquid. D 0.929 (25C), fp <−10C. Insoluble in water. Combustible.
Use: Plasticizer.

ethylene glycol monohexyl ether. (*n*-hexyl "Cellosolve" *[Union]*). $C_6H_{13}OCH_2CH_2OH$.
Properties: Water-white liquid. D 0.8887 (20/20C), bulk d 7.4 lb/gal (20C), bp 208.1C, vap press 0.05 mm Hg (20C), fp −50.1C, flash p 195F (90.5C). Combustible.
Use: High-boiling solvent.

ethylene glycol monomethyl ether. (2-methoxyethanol; methyl "Cellosolve" *[Union]*).
CAS: 109-86-4. $CH_3OCH_2CH_2OH$.
Properties: Colorless liquid; mild agreeable odor; stable. Bp 124.5C, d 0.9663 (20/20C), bulk d 8.0 lb/gal (20C), refr index 1.4021 (20C), flash p 110F (43.3C), fp −85.1C, autoign temp 551F (288C). Miscible with hydrocarbons, alcohols, ketones, glycols, and water. Combustible.
Derivation: From ethylene oxide.
Grade: Technical.
Hazard: Toxic by ingestion and inhalation. Moderate fire risk. Toxic by skin absorption. Questionable carcinogen.
Use: Solvent for nitrocellulose, cellulose acetate, alcohol-soluble dyes, natural and synthetic resins, solvent mixtures, lacquers, enamels, varnishes, leather; perfume fixative; wood stains; sealing moisture-proof cellophane; jet fuel deicing additive.

ethylene glycol monomethyl ether acetate.
(methyl "Cellosolve" acetate *[Union]*).
CAS: 110-49-6. $CH_3COOCH_2CH_2OCH_3$.
Properties: Colorless liquid; pleasant odor. D 1.0067 (20/20C), bp 145C, vap press 2 mm Hg (20C), flash p 120F (48.9C), bulk d 8.4 lb/gal (20C), fp −65.1C. Stable, miscible with common organic solvents; soluble in water. Combustible.
Hazard: Toxic by ingestion, inhalation, and skin absorption. Moderate fire risk. Toxic by skin absorption.
Use: Solvent for nitrocellulose, cellulose acetate, various gums, resins, waxes, oils; textile printing; photographic film; lacquers; dopes.

ethylene glycol monomethyl ether acetyl ricinoleate.
CAS: 140-05-6. $C_{17}H_{32}(OCOCH_3)COOCH_2CH_2OCH_3$.
Properties: Liquid. D 0.966, refr index 1.460, boiling range 220–260C, fp <−60C, flash p 425F (218C). Insoluble in water. Combustible.
Use: Plasticizer.

ethylene glycol monomethyl ether ricinoleate.
$C_{17}H_{32}(OH)COOCH_2CH_2OCH_3$.
Properties: Liquid. D 0.935 (25C), fp <−60C. Insoluble in water. Combustible.
Use: Plasticizer.

ethylene glycol monomethyl ether stearate.
$C_{17}H_{35}COOCH_2CH_2OCH_3$.
Properties: Liquid. D 0.890, mp 21C. Insoluble in water. Combustible.
Use: Plasticizer.

ethylene glycol mononitrate vinyl ether.
CAS: 53987-27-2. $C_4H_7NO_4$.
Hazard: Moderately toxic by ingestion. Low toxicity by inhalation.

ethylene glycol monooctyl ether.
CAS: 10020-43-6. $C_4H_9CHC_2H_5CH_{23}$
OCH_2CH_2OH.
Properties: Colorless liquid; odorless. Bp 228.3C, d 0.8859, flash p 230F (110C), vap press 0.02 mm Hg (20C). Combustible.
Use: Solvent for cellulose esters, plasticizers.

ethylene glycol monophenyl ether. (2-phenoxyethanol; phenyl "Cellosolve" [Union]).
CAS: 122-99-6. $C_6H_5OCH_2CH_2OH$.
Properties: Colorless liquid; faint aromatic odor. Stable in presence of acids and alkalies, partially soluble in water. D 1.1094 (20/20C), bp 244.9C, fp 14C, vap press <0.01 mm Hg (20C), flash p 250F (121C), phenol 0.3% (max), bulk d 9.2 lb/gal. Combustible.
Grade: Technical.
Use: Solvent for cellulose acetate, dyes, inks, resins; perfume fixative; bactericidal agent; organic synthesis of plasticizers, germicides, pharmaceuticals; insect repellent.

ethylene glycol monoricinoleate.
CAS: 106-17-2. $C_{17}H_{32}(OH)COO(CH_2)_2OH$.
Properties: Clear, moderately viscous, pale-yellow liquid; mild odor. Miscible with most organic solvents. D 0.965 (25/25C), saponification number 170, hydroxyl value 270, solidifies at −20C, insoluble in water. Combustible.
Derivation: Castor oil and ethylene glycol.
Grade: Technical.
Use: Plasticizer, greases, urethane polymers.

ethylene glycol monostearate. (glycol stearate).
CAS: 111-60-4. $C_{17}H_{35}COO(CH_2)_2OH$.
Properties: Yellow, waxy solid. Mp 57–60C; d 0.96 (25C). Soluble in alcohol, hot ether, and acetone; insoluble in water. Combustible.

ethylene glycol silicate.
CAS: 17622-94-5. $(HOCH_2CH_2)_4SiO_4$.
Properties: Colorless liquid. Slowly hydrolyzed by acids; miscible with water.
Use: Nonvolatile bonding agent for pigments; weatherproofing paints for protecting concrete, stone, brick, and plastic surfaces.

ethylene hexene-1-copolymer. Product developed especially for use in blow-molded items. Other possible uses are coatings for pipe, wire, and cables; sheeting; and monofilament.

ethylene hydrate. See gas hydrate.

ethyleneimine. (aziridine; ethylenimine).
CAS: 151-56-4.

$$CH_2-CH_2$$
$$\diagdown\diagup$$
$$NH$$

Properties: Clear, colorless liquid; amine odor. Bp 57C, d 0.832 (20/4C), refr index 1.4123 (25C), fp −78C, flash p 12F (−11.1C), autoign temp 612F (322C). Miscible with water and most organic solvents.
Derivation: From ethylene dichloride and ammonia by use of an acid acceptor.
Hazard: Corrosive, absorbed by skin, causes tumors; exposure should be minimized; a carcinogen. Dangerous fire and explosion hazard, flammable limits in air 3.6–46%. Toxic by skin absorption; possible carcinogen.
Use: Intermediate and monomer for fuel-oil and lubricant refining, ion exchange, protective coatings, pharmaceuticals, adhesives, polymer stabilizers, surfactants. Alkyl-substituted forms, called alkyl azranes, are used as intermediates and for microbial control; aziridinyl compounds are also used as polymers and intermediates.

ethylene-maleic anhydride copolymer.
CAS: 9006-26-2.
Properties: Fine, water-soluble powder available as both straight-chain and cross-linked polymers in a variety of molecular weights. May be in form of free acid, sodium, or amide-ammonium salts. Reacts readily with alcohols and amines.
Hazard: Low toxicity by ingestion and skin contact. A mild eye irritant.
Use: Oil-well drilling muds; stabilizers and thickeners in liquid detergents, cosmetics, paints; textile sizes; printing inks; suspending agents; ceramic binders.

ethylenenaphthalene. See acenaphthene.

ethylene oxide. (epoxyethane; oxirane).
CAS: 75-21-8.

$$CH_2-CH_2$$
$$\diagdown\diagup$$
$$O$$

Properties: Colorless gas at room temperature, liquid at approximately 12C. Fp −111.3C, bp 10.73C, d 0.8711 (20/20C), bulk d 7.25 lb/gal (20C), viscosity 0.32 cP (0C), flash p approximately 0F (−17.7C) (TOC), autoign temp 805F (429C). Soluble in organic solvents; miscible with water.
Derivation: (1) Oxidation of ethylene in air or oxygen with silver catalyst; (2) action of an alkali on ethylene chlorohydrin.
Grade: Technical, pure (99.7%).
Hazard: Irritant to eyes and skin. Confirmed carcinogen. Highly flammable, dangerous fire and explosion risk, flammable limits in air 3–100%.
Use: Manufacture of ethylene glycol and higher glycols, surfactants, acrylonitrile, ethanolamines; petroleum demulsifier; fumigant; rocket propellant; industrial sterilant (e.g., medical plastic tubing); fungicide.

ethylene oxide and propylene oxide block polymer. See poloxamer 331.

ethylene oxide polymer.
Use: Food additive.

ethylene-propylene-diene monomer. See ethylene propylene terpolymer.

ethylene-propylene rubber. (EPR). An elastomer made by the stereospecific copolymerization of ethylene and propylene. Has no unsaturation; cannot be vulcanized with sulfur but can be cured with peroxides.

ethylene-propylene terpolymer. (EPT; EPDM). An elastomer based on stereospecific linear terpolymers of ethylene, propylene, and small amounts of a nonconjugated diene, e.g., a cyclic or aliphatic diene (hexadiene, dicyclopentadiene, or ethylidene norbornene). The unsaturated part of the polymer molecule is pendant from the main chain, which is completely saturated. Can be vulcanized with sulfur.
Properties: (Vulcanizate) Light-cream to white; excellent resistance to ozone, to high and low temperatures (from −51 to +148C), and to acids and alkalies; good electrical resistance, susceptible to attack by oils; pelletized forms available.
Use: Automotive parts, gaskets, cable coating, mechanical rubber products, cover strips for tire sidewalls, tire tubes, safety bumpers, coated fabrics, footwear, wire and cable coating, thermoplastic resin modifier.
See Nordel.

ethylene thiourea.
CAS: 96-45-7.

NHCH$_2$CH$_2$NHCS (2-imidazolidinethione).

Properties: White to pale-green crystals; faint amine odor. Mp 199–204C. Slightly soluble in cold water; very soluble in hot water; slightly soluble at room temperature in methanol, ethanol, acetic acid, naphtha.
Hazard: Questionable carcinogen.
Use: Electroplating baths; intermediate for antioxidants, insecticides, fungicides, vulcanization accelerators, dyes, pharmaceuticals, synthetic resins.

ethylene urea. (2-imidazolidinone; dihydro-2(3)-imidazolone; 2-imidazolidone).
CAS: 120-93-4.

CH$_2$CH$_2$NHCONH.

Properties: White, lumpy powder; odorless. Mp 125–128C. Soluble in water.
Derivation: From ethylenediamine and urea.
Use: Drip-dry textile products, ingredient of plasticizers and adhesives, insecticide.

ethylene-vinyl acetate copolymer. (EVA). An elastomer used to improve adhesion properties of hot-melt and pressure-sensitive adhesives, as well as for conversion coatings and thermoplastics. See Ultrathene.

ethylenimine. See ethyleneimine.

N-ethylethanolamine. (ethylaminoethanol).
CAS: 110-73-6. C$_2$H$_5$NHCH$_2$CH$_2$OH.
Properties: Colorless liquid; amine odor. D 0.914 (20C), boiling range 167–169C flash p 160F (71.1C) (OC). Soluble in water, alcohol and ether. Combustible.
Use: Solvent, intermediate.

ethyl ether. (ether; diethyl ether; sulfuric ether; ethyl oxide; diethyl oxide).
CAS: 60-29-7. (C$_2$H$_5$)$_2$O.
Properties: Colorless, volatile, mobile liquid. Hygroscopic, aromatic odor, burning and sweet taste. Bp 34.5C, fp −116.2C, d 0.7147 (20/20C), surface tension 17.0 dynes/cm (20C), refr index 1.3526 (20C), viscosity 0.00233 cP (20C), vap press 442 mm Hg (20C), specific heat 0.5476 cal/g (30C), flash p −49F (−45C), autoign temp 356F (180C), latent heat of evaporation 83.96 cal/g at bp, electric conductivity 4×10^{-3} mho/cm (25C), bulk d 6 lb/gal (20C). Soluble in alcohol, chloroform, benzene, solvent naphtha, and oils; slightly soluble in water.
Derivation: By the action of sulfuric acid on ethanol or ethylene followed by distillation.
Method of purification: Rectification, dehydration, treatment with alkali and charcoal.
Grade: USP (for anesthesia), ACS Reagent, ACS Absolute, CP, concentrated, USP 1880, washed, motor, electronic.
Hazard: CNS depressant by inhalation and skin absorption. Very flammable, severe fire and explosion hazard when exposed to heat or flame. Forms explosive peroxides. Explosive limits in air 1.85–48%.
Use: Organic synthesis; smokeless powder; industrial solvent (nitrocellulose, alkaloids, fats, waxes, etc.); analytical chemistry; anesthetic, extractant.

ethyl-3-ethoxypropionate.
CAS: 763-69-9. C$_2$H$_5$OOCCH$_2$CH$_2$OC$_2$H$_5$.
Properties: Liquid. D 0.9496 (20C), bp 170.1C, vap press 0.9 mm Hg (20C), sets to glass at −100C, flash p 180F (82.2C) (OC). Slightly soluble in water. Combustible.
Use: Intermediate for vitamin B$_1$, other chemicals.

ethylethylene. See butene-1.

ethylethyleneimine. (2-ethylaziridine).

C$_2$H$_5$HCNHCH$_2$.

Properties: Water-white liquid. Bp 87–90C, d 0.812–0.816 (25/25C).

Use: Organic intermediate whose derivatives are used in the textile, paper, rubber, and pharmaceutical industries.

ethyl(5-ethylmercuri-3-(1,2,4-thiadiazolyl)thio)mercury(ii).
CAS: 73928-12-8. $C_6H_{10}Hg_2N_2S_2$.
Hazard: A poison.

ethyl 3-ethyl-4-quinolinecarboxylate.   See 3-ethylcinchoninic acid ethyl ester.

2-ethyl fenchol.
CAS: 137255-07-3. $C_{12}H_{22}O$.
Properties: Pale-yellow liquid; camphor, earthy odor. D 0.946–0.967, refr index 1.470–1.491. Soluble in alc, propylene glycol, fixed oils; insoluble in water.
Use: Food additive.

ethyl ferrocenoate.   (ferrocenecarboxylic acid ethyl ester). $(C_5H_4COOC_2H_5)_2Fe$.
Properties: Orange, crystalline solid. Mp 63–64C.
Use: Intermediate.

Ethylflo.   Synthetic hydrocarbon fluids.

ethylfluoroformate. $FCOOC_2H_5$.
Properties: Liquid. D 1.11 (33C), bp 57C.
Derivation: Interaction of ethylchloroformate and a reactive fluoride.
Hazard: Strong irritant.

ethylfluorosulfonate. $C_2H_5OFSO_2$.
Properties: Liquid; ethereal odor.
Hazard: Strong irritant.

ethyl formate.
CAS: 109-94-4. $HCOOC_2H_5$.
Properties: Water-white, unstable liquid; pleasant aromatic odor. Flash p −4F (−20C) (CC); explosive limits in air 2.8–13.5%; autoign temp 851F (455C), d 0.9236 (20/20C); fp −80.5C, bp 54.3C, vap press 200 mm Hg (20.6C), 300 mm Hg (30.2C); wt/gal 7.61 lb (68F); refr index 1.35975 (20C). Miscible with benzene, ether, alcohol; slightly soluble in water; gradual decomposition in water.
Derivation: Heating ethanol with formic acid in the presence of sulfuric acid.
Grade: Technical, FCC.
Hazard: Narcotic and irritant to skin and eyes; use in foods restricted to 0.0015%. Highly flammable, dangerous fire and explosion risk. Questionable carcinogen.
Use: Solvent for cellulose nitrate and acetate, acetone substitute, fumigant, larvicide, synthetic flavors, organic synthesis.

ethyl-o-formate.
CAS: 122-51-0. $CH(OC_2H_5)_3$.

Properties: Colorless liquid. Bp 145.9C, fp <−18C, refr index 1.392 (20C), flash p 85F (29.4C). Slightly soluble in water; soluble in alcohol and ethers.
Hazard: Flammable, moderate fire risk.
Use: Intermediate.

ethyl-3-formylpropionate.
   $C_2H_5OOCC_2H_4C(O)H$.
Properties: Liquid. D 1.0625 (20/20C), bp 190.9C, fp <−80C, bulk d 8.9 lb/gal, flash p 200F (93.3C). Somewhat soluble in water. Combustible.
Hazard: Toxic by ingestion and inhalation.
Use: Solvent for lacquers, antibiotic extraction, acetic acid separation, coalescing acids for emulsion paints.

2-ethylfuran.
Properties: Liquid. Mw 96.13, bp 92–93C (768 mm Hg), d 0.912.
Grade: Food.
Hazard: Flammable liquid.

ethyl furoate.
CAS: 1335-40-6. $C_4H_3OCO_2C_2H_5$.
Properties: White leaflets or prisms. D 1.1174 (20.8/4C), mp 34C, bp 195C (706 mm Hg). Insoluble in water; soluble in alcohol and ether.

ethyl glycolate isocyanate.   See ethyl isocyanatoacetate.

ethyl glyme.   See ethylene glycol diethyl ether.

4-ethylheptane.
(ethyldipropylmethane).
CAS: 2216-32-2. $CH_3(CH_2)_2CHC_2H_5(CH_2)_2CH_3$.
Properties: Colorless liquid. D 0.730, bp 141.2C, refr index 1.4109 (20C). Combustible.
Grade: Technical.
Use: Organic synthesis.

ethyl heptanoate.   See ethyl enanthate.

ethyl heptazine.   See ethoheptazine.

2-ethylhexaldehyde.
(butylethylacetaldehyde; octylaldehyde; 2-ethylhexanal).
CAS: 123-05-7. $C_4H_9CH(C_2H_5)CHO$.
Properties: Colorless high-boiling liquid; mild odor. Miscible with most organic solvents; slightly soluble in water. D 0.8205 (20C), bp 163.4C, vap press 1.8 mm Hg (20C), flash p 125F (51.6C) (OC), bulk d 6.8 lb/gal. Combustible.
Grade: Technical.
Hazard: Ignites in air.
Use: Organic synthesis, perfumes.

2,2′-(2-ethylhexamido)diethyl di(2-ethylhexoate).
(Flexol 8N8). $(C_7H_{15}OCOC_2H_4)_2NCOC_7H_{15}$.

Properties: Light-colored liquid. D 0.9564 (20/20C), bulk d 8.0 lb/gal (20C), bp 256C (5 mm Hg), flash p 420F (215C), vap press 0.60 mm Hg (200C), viscosity 139.2 cP (20C). Insoluble in water. Combustible.
Use: Plasticizer.

2-ethylhexanal. See 2-ethylhexaldehyde.

2-ethylhexanediol-1,3. (ethohexadiol; 2-ethyl-3-propyl-1,3-propanediol).
CAS: 94-96-2. $C_3H_7CH(OH)CH(C_2H_5)CH_2OH$.

$$CH_3CH_2CH_2\overset{\underset{|}{OH}}{C}HCHCH_2OH$$
$$\underset{CH_2CH_3}{|}$$

Properties: Colorless, slightly viscous, liquid; odorless. Hygroscopic. D 0.9422 (20/20C), bulk d 7.8 lb/gal (20C), bp 244C, vap press <0.01 mm Hg (20C), flash p 260F (126C), fp approximately −40C, refr index 1.4465–1.4515, viscosity 323 cP (20C). Soluble in alcohol and ether, partially soluble in water. Combustible.
Grade: USP (as ethohexadiol), industrial.
Use: Insect repellent, cosmetics, vehicle and solvent in printing inks, medicine, chelating agent for boric acid.

(r)-2-ethylhexanoic acid sodium salt.
CAS: 139906-72-2. $C_8H_{15}O_2$•Na.
Hazard: A reproductive hazard.

2-ethylhexanol. See 2-ethylhexyl alcohol.

((2-ethylhexanoyl)oxy)tributylstannane.
See tributyltin-2-ethylhexanoate.

2-ethylhexenal. See 2-ethyl-3-propylacrolein.

2-ethyl-1-hexene.
CAS: 1632-16-2. $CH_3(CH_2)_3(C_2H_5)C:CH_2$.
Properties: Colorless liquid. D 0.7270 (20/4C), bp 120C, refr index 1.4157 (20C). Soluble in alcohol, acetone, ether, petroleum, coal tar solvents; insoluble in water. Combustible.
Grade: 95% min purity.
Hazard: Toxic by ingestion and inhalation.
Use: Organic synthesis of flavors, perfumes, medicines, dyes, resins.

ethyl hexoate. See ethyl caproate.

2-ethylhexoic acid. (butylethylacetic acid).
CAS: 78-42-2. $C_4H_9CH(C_2H_5)COOH$.
Properties: Mild-odored liquid. D 0.9077 (20/20C), bulk d 7.6 lb/gal (20C), bp 226.9C, vap press 0.03 mm Hg (20C), fp −83C, viscosity 7.73 cP (20C), acid number 370, flash p 260F (126C). Slightly soluble in water. Combustible.
Grade: 99%.

Use: Paint and varnish driers (metallic salts). Ethylhexoates of light metals are used to convert some mineral oils to greases. Its esters are used as plasticizers.

2-ethylhexyl.
CAS: 103-09-3. $CH_3(CH_2)_3CH(C_2H_5)CH_2$-.
An 8-carbon radical. Many of its compounds were formerly called *octyl*.

2-ethylhexyl acetate.
CAS: 103-09-3. $CH_3COOCH_2CHC_2H_5C_4H_9$.
Properties: Water-white stable liquid. D 0.8733 (20C), bp 198.6C, fp −93C, vap press 0.4 mm Hg (20C), flash p 180F (82.2C), bulk d 7.3 lb/gal (20C). Very slightly soluble in water; miscible with alcohol. Combustible.
Grade: Technical (approximately 95%).
Use: Solvent for nitrocellulose, resins, lacquers, baking finishes.

2-ethylhexyl acrylate.
CAS: 103-11-7. $CH_2:CHCOOCH_2CH(C_2H_5)C_4H_9$.
Properties: Liquid; pleasant odor. D 0.8869, bp 214–218C, vap press 0.1 mm Hg (20C), sets to glass at −90C, flash p 180F (82.2C) (OC). Insoluble in water. Combustible.
Hazard: Questionable carcinogen.
Use: Monomer for plastics, protective coatings, paper treatment, water-based paints.

2-ethylhexyl acrylate-hydroxyethyl acrylate-methyl acrylate polymer.
CAS: 50601-74-6. $(C_{11}H_{20}O_2$•$C_5H_8O_3$•$C_4H_6O_2)_x$.
Hazard: Low toxicity by ingestion and skin contact. A moderate eye irritant.

2-ethylhexyl alcohol. (2-ethylhexanol; octyl alcohol).
CAS: 104-76-7. $CH_3(CH_2)_3CHC_2H_5CH_2OH$.
Properties: Colorless liquid. D 0.83 (20C), bp 183.5C, fp −76C, vap press 0.36 (20C), refr index 1.4300 (20C), bulk d 6.9 lb/gal (20C), flash p 178F (81.1C). Miscible with most organic solvents, slightly soluble in water. Combustible.
Derivation: (a) Oxo process from propylene and synthesis gas; (b) aldolization of acetaldehyde or butyraldehyde, followed by hydrogenation; (c) from fermentation alcohol.
Grade: Technical.
Use: Plasticizer for PVC resins, defoaming agent, wetting agent; organic synthesis; solvent mix for nitrocellulose, paints, lacquers, baking finishes; penetrant for mercerizing cotton; textile finishing compounds; plasticizers; inks; rubber; paper; lubricants; photography; dry cleaning.

2-ethylhexylamine.
CAS: 104-75-6. $C_4H_9CH(C_2H_5)CH_2NH_2$.
Properties: Colorless liquid. D 0.7894 (20/20C), bulk d 6.56 lb/gal (20C), bp 169.2C, vap press 1.2 (20C), viscosity 1.11 cP (20C), flash p 140F

(60C) (OC). Solubility in water 25.3% (20C). Combustible.

Hazard: Moderate fire risk. Toxic by ingestion and inhalation.

Use: Synthesis of detergents, rubber chemicals, oil additives, and insecticides.

n-2-ethylhexylaniline.
CAS: 10137-80-1. $C_6H_5NHCH_2CH(C_2H_5)C_4H_9$.

Properties: Light-yellow liquid; mild odor. D 0.9119 (20/20C), bp 194C (50 mm Hg), vap press <0.01 mm Hg (20C), fp (sets to a glass at approximately −70C), viscosity 7.4 cP (20C), flash p 325F (162C) (COC). Solubility in water <0.01% (20C). Combustible.

Hazard: Chronic exposure is toxic.

Use: Solvent, organic synthesis.

2-ethylhexyl bromide.
CAS: 18908-66-2. $C_4H_9CH(C_2H_5)CH_2Br$.

Properties: Water-white liquid. Bp 56–58C (6 mm Hg). Insoluble in water.

Use: Introduction of the 2-ethylhexyl group in organic synthesis; preparation of disinfectants, pharmaceuticals.

2-ethylhexyl chloride.
CAS: 123-04-6. $C_4H_9CH(C_2H_5)CH_2Cl$.

Properties: Colorless liquid. D 0.8833 (20C), bp 172.9C, refr index 1.4310, bulk d 7.33 lb/gal, flash p 140F (60C) (OC), fp −135C. Insoluble in water. Combustible.

Hazard: Moderate fire risk.

Use: Synthesis of cellulose derivatives, dyestuffs, pharmaceuticals, textile auxiliaries, insecticides, resins.

2-ethylhexyl cyanoacetate.
CAS: 13361-34-7. $CNCH_2COOCH_2CH$ $(C_2H_5)C_4H_9$.

Properties: Liquid. Bp 150C, refr index 1.4389 (20C). Insoluble in water; soluble in alcohol, benzene, and ether. Combustible.

Use: Organic intermediate.

2-ethylhexyl-2-cyano-3′,3′-diphenyl acrylate.
See "Uvinul" *[BASF]*

2-ethylhexyl 2-cyano-3-phenylcinnamate.
See octocrilene.

2-ethylhexyl (3-isocyanatomethylphenyl) carbamate.
CAS: 58240-57-6. $C_{17}H_{24}N_2O_3$.

Hazard: A poison by ingestion and skin contact. A severe eye irritant.

2-ethylhexylisodecyl phthalate.
$C_{26}H_{43}O_4$.

Properties: Colorless, high-boiling liquid. Good dielectric properties, refr index 1.478–1.488 (25C),

d 0.969–0.977 (25/25C). Miscible with most common solvents, thinners, and oils. Combustible.

Use: Plasticizer.

2-ethylhexylmagnesium chloride.
$C_4H_9CH(C_2H_5)CH_2MgCl$.

Grignard reagent available commercially in tetrahydrofuran solution.

2-ethylhexyloctylphenyl phosphite.
$(C_8H_{17}O)_2(C_8H_{17}C_6H_{17}C_6H_4O)P$.

Properties: Colorless to light-yellow liquid; characteristic odor. D 0.935–0.950 (20/4C), flash p 385F (196C). Insoluble in water. Combustible.

Use: Antioxidant, plasticizer, flame-retardant, lubricating-oil additive.

3,3′-(2-ethylhexyl) thiodipropionate.
(dioctyl thiopropionate). $(C_8H_{17}OOCCH_2CH_2)_2S$.

Properties: Colorless liquid. D 0.952 (25C). Insoluble in water; soluble in most organic solvents. Combustible.

Use: Antioxidant, stabilizer, and lubricant.

ethyl(hydrogen *p*-mercaptobenzenesulfonato)mercury sodium salt.
See sodium thimerfonate.

ethyl-2-hydroxy-2,2-bis(4-chlorophenyl) acetate.
See "Chlorobenzilate" *[Novartis]*.

ethyl-3-hydroxybutyrate.
CAS: 5405-41-4. $CH_3CH(OH)CH_2CO_2C_2H_5$.

Properties: A solid. Mw 132.16, bp 170C, d 1.017, fp 64C.

Grade: Food.

ethyl hydroxyethyl cellulose.
(EHEC). A cellulose ether.

Properties: White, granular solid. Soluble in mixtures of aliphatic hydrocarbons containing alcohol; soluble in water. Combustible.

Available forms: Extra low- and high-viscosity types.

Use: Stabilizer, thickener, binder, formation of film in silk-screen and gravure printing inks; protective coatings; aqueous, aqueous-organic, and organic solvent systems.

n-ethyl-*n*-(2-hydroxyethyl)-3-methyl-4-nitrosoaniline.
CAS: 58066-96-9. $C_{11}H_{16}N_2O_2$.

Hazard: A poison by ingestion.

ethyl-α-hydroxyisobutyrate.
$(CH_3)_2COHCOOC_2H_5$.

Properties: Water-white liquid. D 0.978–0.986 (20C), bp 149–150C. Soluble in water, alcohol, and ether. Combustible.

Use: Solvent for nitrocellulose and cellulose acetate; solvent mixture for cellulose ethers, organic synthesis, pharmaceuticals.

4-ethyl-7-hydroxy-3-(*p*-methoxyphenyl) coumarin.
CAS: 5219-17-0. $C_{18}H_{16}O_4$.
Hazard: A reproductive hazard.

ethyl 4-hydroxy-2-methyl-2h-1,2-benzothiazine-3-carboxylate 1,1-dioxide.
CAS: 24683-26-9. $C_{12}H_{13}NO_5S$.
Hazard: Low toxicity by ingestion and inhalation.

2-ethyl-2-(hydroxymethyl)-1,3-propanediol triacetoacetate.
CAS: 22208-25-9. $C_{18}H_{26}O_9$.
Hazard: A poison by ingestion and skin contact.

3-((10-ethyl-11-(*p*-hydroxyphenyl)dibenz (b,f)oxepin-3-yl)oxy)-1,2-propanediol hydrate (4:1).
CAS: 85850-93-7. $C_{25}H_{24}O_5 \cdot 1/_4 H_2O$.
Hazard: A reproductive hazard.

ethylidene chloride.   (1,1-dichloroethane).
CAS: 75-34-3. CH_3CHCl_2.
Properties: Colorless, neutral, mobile liquid; aromatic ethereal odor; saccharin taste. D 1.174 (17C), bp 57–59C, fp −98C, refr index 1.4166 (20C). Soluble in alcohol, ether, fixed and volatile oils; very sparingly soluble in water. Combustible.
Hazard: Toxic. Eye and upper respiratory tract irritant; kidney and liver damage. Questionable carcinogen.
Use: Extraction solvent, fumigant.

ethylidenediethyl ether.   See acetal.

ethylidenedimethyl ether.   See dimethylacetal.

ethylidene fluoride.   See 1,1-difluoroethane.

5-ethylidene-2-norbornene.   (ENB).
CAS: 16219-75-3.   A diene used as the third monomer in EPDM elastomers.
Hazard: Toxic.

4,4′,4″-ethylidynetrisphenol.
CAS: 27955-94-8. $C_{20}H_{18}O_3$.
Hazard: Low toxicity by ingestion and skin contact. A mild eye irritant.

ethyl iodide.   (iodoethane).
CAS: 75-03-6. C_2H_5I.
Properties: Colorless liquid, turning brown on exposure to light. D 1.90–1.93 (25/25C), fp −108C, bp 72C, refr index 1.5168 (15C). Soluble in alcohol and ether; slightly soluble in water. Combustible.

Derivation: By digesting red phosphorus with absolute ethanol, after which iodine is added and the mixture distilled.
Hazard: Toxic by inhalation and skin absorption, narcotic in high concentration.
Use: Medicine, organic synthesis.

ethyl iodoacetate.
CAS: 623-48-3. $CH_2ICOOC_2H_5$.
Properties: Dense, colorless liquid. Decomposed by light and air, also (very slowly) by alkaline solutions and water. D 1.8, bp 179C, vapor d 7.4, vap press 0.54 mm Hg (20C).
Derivation: Interaction of potassium iodide with ethyl bromo- or chloroacetate.
Hazard: Strong irritant to eyes and skin.

5-ethyl-5-isoamylbarbituric acid.   See amobarbital.

ethylisobutrazine hydrochloride.
CAS: 3737-33-5. $C_{20}H_{26}N_2S \cdot ClH$.
Hazard: A poison.

ethylisobutylmethane.   See 2-methylhexane.

ethyl isobutyrate.
CAS: 97-62-1. $(CH_3)_2CHCOOC_2H_5$.
Properties: Colorless, volatile liquid. D 0.870, bp 110–111C, fp −88C, refr index 1.3903 (20C). Soluble in alcohol and ether; slightly soluble in water. Combustible.
Derivation: Heating isobutyric acid and ethyl alcohol, with subsequent distillation.
Use: Organic synthesis, flavoring extracts.

ethyl isocyanate.
CAS: 109-90-0. C_2H_5NCO.
Properties: Liquid. D 0.898, bp 60C. Soluble in chlorinated and aromatic hydrocarbons.
Hazard: Strong irritant to tissue.
Use: Pharmaceutical and pesticide intermediate.

ethyl isocyanatoacetate.
CAS: 2949-22-6. $C_5H_7NO_3$.
Hazard: Moderately toxic by ingestion and inhalation.

ethyl 4-isocyanobenzoate.
CAS: 1983-99-9. $C_{10}H_9NO_2$.
Hazard: Low toxicity by ingestion.
Use: Agricultural chemical.

2-ethylisohexanol.
CAS: 111767-90-9. $(CH_3)_2CHCH_2CH(C_2H_5) CH_2OH$.
Properties: Liquid. D 0.825–8.035 (20/20C), boiling range 173.0–181.0C, refr index 1.4235 (20C), bulk d 6.89 lb/gal (20C), flash p 170F (76.6C) (COC). Combustible.
Use: Chemical intermediate.

ethyl isopropenyl ketone. See isopropenyl ethyl ketone.

ethyl isothiocyanate. See ethyl thiocarbimide.

ethyl 4-isothiocyanatobutanoate.
CAS: 17126-65-7. $C_7H_{11}NO_2S$.
Hazard: A poison.

ethyl isovalerate. (ethyl valerate; ethyl 2-methylbutyrate).
CAS: 108-64-5. $(CH_3)_2CHCH_2COOC_2H_5$.
Properties: Colorless, oily liquid; fruity odor. Bp 135C, fp −99C, d 0.864 (20/20C), refr index 1.3950–1.3990 (20C). Slightly soluble in water; miscible with alcohol, ether, and benzene. Combustible.
Derivation: Heating sodium valerate and ethanol in the presence of sulfuric acid or hydrochloric acid, with subsequent distillation.
Grade: Technical, FCC.
Use: Essential oils, perfumery, artificial fruit essences, flavoring.

ethyl lactate.
CAS: 97-64-3. $CH_3CH_2OCOOC_2H_5$.
Properties: Colorless liquid; mild odor. D 1.020–1.036 (20/20C), bp 154C, flash p 115F (46.1C) (CC); bulk d 8.55 lb/gal (20C), autoign temp 752F (400C). Miscible with water, alcohols, ketones, esters, hydrocarbons, oil. Combustible.
Derivation: (a) By the esterification of lactic acid with ethanol; (b) by combining acetaldehyde with hydrogen cyanide to form acetaldehyde cyanohydrin, which is converted into ethyl lactate by treatment with ethanol and an inorganic acid.
Grade: Technical (96%).
Hazard: Moderate fire risk.
Use: Solvent for nitrocellulose, cellulose acetate, many cellulose ethers, resins; lacquers; paints; enamels; varnishes; stencil sheets; safety glass; flavoring.

ethyl levulinate.
CAS: 539-88-8. $CH_3CO(CH_2)_2COOC_2H_5$.
Properties: Colorless liquid. D 1.012, bp 205–206C, refr index 1.4229 (20C). Soluble in water; miscible with alcohol. Combustible.
Use: Solvent for cellulose acetate and starch ethers, flavoring.

ethyllithium. (lithium ethyl).
Properties: Transparent crystals. Decomposed by water. The commercial product is a 2 M suspension of C_2H_5Li in benzene.
Hazard: Ignites in air, reacts with oxidizing materials.
Use: Grignard-type reactions.

ethylmagnesium bromide.
CAS: 925-90-6. C_2H_5MgBr.
Properties: Liquid. D 1.01.

Available forms: Dissolved in ether.
Hazard: Flammable, dangerous fire risk.
Use: Grignard-type reactions.

ethylmagnesium chloride.
CAS: 2386-64-3. C_2H_5MgCl.
Available forms: Dissolved in ether (also offered commercially in tetrahydrofuran).
Properties: Liquid. D 0.85.
Hazard: Flammable, dangerous fire risk.
Use: Grignard-type reactions.

N-ethylmaleimide. (1-ethyl-1-H-pyrrole-2,5-dione).
CAS: 128-53-0. $C_6H_7NO_2$.
Properties: Crystals. Mw 125.13, mp 45C.
Hazard: Lachrymator when liquid, a strong irritant.
Use: Cancer research.

ethyl malonate. (malonic ester; diethyl malonate).
CAS: 105-53-3. $CH_2(COOC_2H_5)_2$.
Properties: Colorless liquid; sweet ester odor. Bp 198C, fp −50C, d 1.055 (25/25C), flash p 200F (93.3C). Insoluble in water; soluble in alcohol, ether, chloroform, and benzene. Combustible.
Derivation: By passing hydrogen chloride into cyanoacetic acid dissolved in absolute alcohol with subsequent distillation.
Use: Intermediate for barbiturates and certain pigments, flavoring.

ethylmemazine monohydrochloride. See ethylisobutrazine hydrochloride.

ethyl mercaptan. Legal label name for ethanethiol.

ethylmercuric acetate.
CAS: 109-62-6. $C_2H_5HgOOCCH_3$.
Properties: White, crystalline powder. Mp 178C. Slightly soluble in water; soluble in many organic solvents. May be steam distilled.
Hazard: Strong irritant; see mercury.
Use: Seed fungicide as dust or slurry with water. See mercury.

ethylmercuric chloride.
CAS: 107-27-7. C_2H_5HgCl.
Properties: Crystals. D 3.482, mp 193C, sublimes readily. Insoluble in water; slightly soluble in ether; soluble in hot alcohol.
Derivation: Reaction of zinc diethyl and mercuric chloride.
Hazard: Strong irritant.
Use: Fungicide for seed or bulb treatment, either alone or with other organic mercury compounds. See mercury.

ethylmercuric dicyandiamide.
CAS: 63869-03-4. $C_4H_8HgN_4$.
Hazard: A poison.

ethylmercuric phosphate.
CAS: 2440-45-1. $(C_2H_5HgO)_3PO$.
Properties: White powder; garlic-like odor. Soluble in water.
Derivation: Reaction of ethylmercuric acetate with phosphoric acid.
Hazard: Strong irritant.
Use: Seed fungicide, timber preservative.
See mercury.

ethylmercurithiosalicylic acid, sodium salt.
See thimerosal.

n-ethylmercuri-p-toluenesulfonanilide.
CAS: 517-16-8. $C_{15}H_{17}SO_2NHg$.
Properties: Crystalline solid; strong, garlic-like odor. Insoluble in water.
Hazard: Ingestion may be fatal.
Use: Agricultural chemical (grain smut control).

ethylmercury-2,3-dihydroxypropyl mercaptide.
CAS: 2597-92-4. $C_2H_5HgSCH_2CH_2OCH_2OH$.
Organic mercurial.
Hazard: Very toxic; see mercury.
Use: Fungicidal dust, or in slurry treatment, for control of seed borne disease and to reduce losses from seed decay and damping off of wheat, oats, rye, etc.
See mercury.

ethylmercury-p-toluenesulfonanilide.
(EMTS; Ceresan M). $C_6H_5N(HgC_2H_5)$
$SO_2C_6H_4CH_3$.
Properties: Crystals; pungent odor. Mp 154–157C. Nearly insoluble in water; soluble in acetone and chloroform.
Hazard: Very toxic.
Use: Dust or slurry for control of seed-borne diseases of fungi by treatment of seeds or bulbs.
See mercury.

ethylmercury toluenesulfonate.
CAS: 2654-47-9. $C_8H_{10}HgO_3S$.
Hazard: A poison.

ethyl methacrylate.
CAS: 97-63-2. $H_2C:CCH_3COOC_2H_5$.
Properties: Colorless liquid. Bp 119C, fp approximately −75C, d 0.911, refr index 1.4116 (25C), flash p 70F (21.1C) (OC). Insoluble in water; readily polymerized.
Derivation: Reaction of methacrylic acid or methyl methacrylate with ethanol.
Grade: Technical (inhibited).
Hazard: Flammable, dangerous fire and explosion hazard. An irritant.
Use: Polymers, chemical intermediates.
See acrylic resin.

n-(2-(5-ethyl-2-methoxyphenyl)ethyl)propionamide.
CAS: 156482-87-0. $C_{14}H_{21}NO_2$.
Hazard: Moderately toxic by ingestion.

N-ethyl-3-methylaniline.
See n-ethyl-m-toluidine.

n-ethyl-2(or 4)-methylbenzenesulfonamide.
CAS: 8047-99-2. $C_9H_{13}NO_2S$.
Hazard: Moderately toxic by ingestion. Low toxicity by skin contact. A mild eye irritant.

ethyl-2-methylbutyrate. See ethyl isovalerate.

ethyl methyl cellulose. A water-soluble cellulose ether used for thickening, sizing, emulsifying, and dispersing.
See polymer, water-soluble.

o-ethyl-s-(3-methyl-4-chlorophenyl)ethyl phosphonodithioate.
CAS: 1942-78-5. $C_{11}H_{16}ClOPS_2$.
Hazard: A poison by ingestion and skin contact.
Use: Agricultural chemical.

n-ethyl-3,4-methylenedioxyamphetamine.
CAS: 82801-81-8. $C_{12}H_{17}NO_2$.
Hazard: Human systemic effects.

ethyl methyl ether.
CAS: 540-67-0. $C_2H_5OCH_3$.
Properties: Colorless liquid. D 0.725, bp 10.8C, flash p −35F (−37.2C), autoign temp 374F (190C). Soluble in water; miscible with alcohol and ether.
Hazard: Highly flammable, dangerous fire and explosion risk.
Use: Medicine (anesthetic).

ethyl (4-(1-methylethyl)phenyl)methyl 3-pyridinylcarbonimidodithioate.
CAS: 51308-73-7. $C_{18}H_{22}N_2S_2$.
Hazard: Moderately toxic by ingestion.
Use: Agricultural chemical.

2-ethyl-4-methylimidazole.
CAS: 931-36-2. $C_2H_5C_3N_2H_2CH_3$.
Properties: A supercooled amber liquid that if crystallized is a low-melting solid. Mp 45C, bp 154C (10 mm Hg).
Use: Curing epoxy-resin systems.

ethyl methyl ketone. See methyl ethyl ketone.

ethyl 3-methyl-4-(methylthio)phenyl(1-methylethyl) phosphoramide. See fenamiphos.

ethyl methylphenylglycidate. (so-called aldehyde C-16; "strawberry aldehyde").
CAS: 77-83-8.

$$CH_3(C_6H_5)C\overset{\displaystyle\overline{\quad\quad}}{O}CHCOOC_2H_5.$$

Properties: Colorless to yellowish liquid; strong odor suggestive of strawberry. D 1.104–1.123, refr index 1.509–1.511. Soluble in three volumes of 60% alcohol. Combustible.
Grade: Technical, FCC (as aldehyde C-16).
Use: Perfumery, flavors.

o-ethyl s-1-methylpropyl s-1,1-dimethylethyl phosphorodithioate.
CAS: 86073-23-6. $C_{10}H_{23}O_2PS_2$.
Hazard: A poison by ocular route.

2-ethyl-3-methylpyrazine.
CAS: 15707-23-0. $C_7H_{10}N_2$.
Properties: Colorless to slightly yellow liquid; strong, raw potato odor. D 0.980–0.999 @ 20C, refr index 1.502. Soluble in water, organic solvents.
Hazard: Moderately toxic by ingestion.
Use: Food additive.

5-ethyl-2-methylpyridine.   See   2-methyl-5-ethylpyridine.

O-ethyl-O-[4-(methylthio)phenyl]   S-propyl phosphorodithioate.   See sulprofos.

7-ethyl-2-methyl-4-undecanol.   (tetradecanol).
CAS: 103-20-8. $C_4H_9CH(C_2H_5)C_2H_4CH(OH)$
$CH_2CH(CH_3)_2$.
Properties: Liquid. D 0.8355 (20/20C), bp 264C, flash p 285F (140C). Insoluble in water. Combustible.
Use: Intermediate for synthetic lubricants, defoamers, and surfactants.

ethylmorphine hydrochloride.
CAS: 125-30-4. $C_{19}H_{23}NO_3 \cdot HCl \cdot 2H_2O$.
Properties: White, crystalline powder; odorless. Mp approximately 123C (decomposes). Soluble in water and alcohol; slightly soluble in ether and chloroform.
Derivation: Action of hydrochloric acid on ethylmorphine, which is made by action of ethyl iodide on morphine in alkaline solution.
Grade: Technical, NF.
Hazard: Toxic by ingestion; habit-forming narcotic.
Use: Medicine (analgesic).

N-ethylmorpholine.
CAS: 100-74-3.

$CH_2CH_2OCH_2CH_2NCH_2CH_3$.

Properties: Colorless liquid; ammoniacal odor. D 0.916 (20/20C), fp −63C, bp 138C, bulk d 7.6 lb/gal (20C), flash p 90F (32.3C) (OC). Miscible with water.
Grade: Technical.
Hazard: Irritant to skin and eyes, absorbed by skin. Flammable, moderate fire risk. Toxic by skin absorption.

Use: Intermediate for dyestuffs, pharmaceuticals; rubber accelerators and emulsifying agents; solvent for dyes, resins, oils; catalyst in making polyurethane foams.

ethyl mustard oil.   See ethyl thiocarbimide.

ethyl myristate.   (ethyl tetradecanoate).
CAS: 124-06-1. $CH_3(CH_2)_{12}COOC_2H_5$.
Properties: Liquid. D 0.856, mp 12C, bp 295C. Insoluble in water, soluble in alcohol; slightly soluble in ether. Combustible.
Use: Flavoring.

n-ethyl-α-naphthylamine.   (n-ethyl-1-naphthylamine).
CAS: 118-44-5. $C_{10}H_7NCH_2H_5$.
Properties: Colorless liquid. Refr index 1.6475, bp 305C. Insoluble in water; soluble in alcohol and ether. Combustible.
Use: Intermediate.

ethyl nitrate.
CAS: 625-58-1. $C_2H_5NO_3$.
Properties: Colorless liquid; pleasant odor; sweet taste. D 1.116, fp −112C, bp 87.6C, vapor three times heavier than air, flash p 50F (10C) (CC), lower explosive limit 3.8%. Soluble in alcohol and ether; insoluble in water.
Derivation: By heating alcohol, urea nitrate, and nitric acid, with subsequent distillation.
Hazard: Flammable, dangerous, fire and explosion risk.
Use: Organic synthesis, drugs, perfumes, dyes, rocket propellant.

ethyl nitrite.
CAS: 109-95-5. $C_2H_5NO_2$.
Properties: Yellowish volatile liquid. D 0.90, bp 16.4C, flash p −31F (−35C), decomposes spontaneously at 194F (90C), explosive limits in air 3–50%. Soluble in alcohol and ether; decomposes in water. Narcotic in high concentration.
Hazard: Highly flammable, dangerous, explodes.
Use: Organic reactions, synthetic flavoring.

ethyl 2-(m-nitrobenzylidene)acetoacetate.
CAS: 39562-16-8. $C_{13}H_{13}NO_5$
Hazard: Moderately toxic by ingestion, inhalation, and skin contact.

ethyl 2-nitrophenyl sulfide.
CAS: 3058-46-6. $C_8H_9NO_2S$.
Hazard: Moderately toxic by ingestion and skin contact. A mild skin and moderate irritant.

n-ethyl-n-nitrosourea.   (aenh; aethylnitrosoharnstoff; enu; n-ethyl-n-nitrosocarbamide; ethylnitrosourea; 1-ethyl-1-nitro-sourea; ethylnitrosourea; 1-ethyl-1-nitrosourea; neu; nitrosoethylurea; NSC-45403; RCRA waste number U176).
CAS: 759-73-9. $C_3H_7N_3O_2$.

Properties: Pale yellow, crystalline alkylating agent.
Hazard: Tumorigen, carcinogen, neoplastigen; teratogen; poison; toxic.

ethyl nonanoate.   See ethyl pelargonate.

ethyl octanoate.   See ethyl caprylate.

ethyl octoate.   See ethyl caprylate.

ethyl oenanthate.   See ethyl enanthate.

ethyl oleate.
CAS: 111-62-6. $C_{17}H_{33}COOC_2H_5$.
Properties: Light-colored, yellowish, oily liquid. Bulk d 7.27 lb/gal (20C), refr index 1.45189 (20C), bp 205C, fp approximately −32C, d 0.867, flash p 348F (175C). Insoluble in water, soluble in alcohol and ether; solubility of water in ethyl oleate 1.0 cc/100 cc (202C). Combustible.
Use: Solvent, plasticizer, lubricant, water-resisting agent, flavoring.

ethyl orthopropionate.
CAS: 115-80-0. $CH_3CH_2C(OC_2H_5)_3$.
Properties: Colorless liquid. Bp 155–160C, refr index 1.401 (20C), flash p 140F (60C). Insoluble in water; soluble in alcohol and ether. Combustible.
Hazard: Moderate fire risk.
Use: Intermediate.

ethyl oxalate.   (diethyl oxalate).
CAS: 95-92-1. $(COOC_2H_5)_2$.
Properties: Colorless, unstable, oily, aromatic liquid. D 1.09 (20/20C), bp 186C, fp −40.6C, bulk d approximately 8.96 lb/gal (20C), flash p 168F (75.5C) (CC). Miscible with alcohol, ether, ethyl acetate, and other common organic solvents; slightly soluble in water and gradually decomposed by it. Combustible.
Derivation: By standard esterification procedure using ethanol and oxalic acid. The final purification calls for unusual technique and equipment. The last traces of water are most difficult to remove, and this is accomplished by a special step in the rectification.
Hazard: Toxic by ingestion, strong irritant to skin and mucous membranes.
Use: Solvent for cellulose esters and ethers, many natural and synthetic resins; radio-tube cathode fixing lacquers; dye intermediate; pharmaceuticals, perfume preparations; organic synthesis.

ethyl-2-oxocyclopentanecarboxylate.   See 2-carbethoxycyclopentanone.

ethyl oxyhydrate.
Properties: Colorless liquid; sharp rum-like odor. Miscible in alcohol, glycerin, propylene glycol.
Use: Food additive.

ethyl pabate.   See ethopabate.

ethylparaben.   (ethyl-*p*-hydroxybenzoate).
CAS: 120-47-8. $C_9H_{10}O_3$.
Properties: Colorless crystals. Mp 115C, decomposes at 297C. Soluble in alcohol and ether; almost insoluble in water.
Derivation: Esterification of *p*-hydroxybenzoic acid.
Use: Pharmaceutical preservative.
See Parabens.

ethyl parathion.   See parathion.

ethyl PCT.   See *O,O*-diethyl phosphorochloridothioate.

ethyl pelargonate.   (ethyl nonanoate; wine ether).
CAS: 123-29-5. $CH_3(CH_2)_7COOC_2H_5$.
Properties: Colorless liquid; fruity odor. Refr index 1.4220 (20C), d 0.866 (18/4C), bp approximately 220C, fp −44C. Insoluble in water; soluble in alcohol and ether. Combustible.
Grade: Technical, FCC.
Use: Flavoring alcoholic beverages; perfumes; chemical intermediate.

3-ethylpentane.   (triethylmethane).
CAS: 617-78-7. $(C_2H_5)_3CH$.
Properties: Colorless liquid. Bp 93.5C, fp −118.6C, d 0.69818 (20C), refr index 1.3934 (20C). Soluble in alcohol; insoluble in water. Combustible.
Grade: Technical.
Use: Organic synthesis.

***m*-ethylphenol.**   (3-ethylphenol).
CAS: 620-17-7. $HOC_6H_4C_2H_5$.
Properties: Colorless liquid. Fp −4C, bp 214C, d 1.001. Very slightly soluble in water; miscible with alcohol and ether. Combustible.
Hazard: Toxic material.
See phenol.

***p*-ethylphenol.**   (4-ethylphenol).
CAS: 123-07-9. $HOC_6H_4C_2H_5$.
Properties: Colorless needles. Mp 46C, bp 219C, flash p 219F. Soluble in alcohol or ether; slightly soluble in water. Combustible.
Hazard: Toxic material.
See phenol.

3-(2-ethylphenoxy)-1-((1s)-1,2,3,4-tetrahydronaphth-1-ylamino)-(2s)-2-propanol oxalate.
CAS: 174689-39-5. $C_{21}H_{27}NO_2 \cdot C_2H_2O_4$.
Hazard: A poison.

ethylphenylacetamide.   See *n*-ethylacetanilide.

ethyl phenylacetate.
CAS: 101-97-3. $C_6H_5CH_2COOC_2H_5$.
Properties: Colorless liquid; honey-like odor. D 1.027–1.032, refr index 1.498, bp 226C. Soluble

in 8 parts of 60% alcohol; insoluble in water. Combustible.
Grade: Technical, FCC.
Use: Perfumery; flavors; intermediate, especially for phenobarbital; herbicide.

ethyl phenylacrylate.   See ethyl cinnamate.

5-ethyl-5-phenylbarbituric acid.   See phenobarbital.

ethyl phenylcarbamate.   (phenylurethane; ethyl phenylurethane).
CAS: 101-99-5. $C_6H_5NHCOOC_2H_5$.
Properties: White, crystalline solid; aromatic odor; clove-like taste. Mp 51C. Soluble in alcohol, ether, and boiling water; insoluble in cold water.
Derivation: Action of ethanol on phenyl isocyanate.

ethylphenyldichlorosilane.
CAS: 1125-27-5. $C_2H_5(C_6H_5)SiCl_2$.
Properties: Colorless liquid that fumes strongly in moist air.
Hazard: Toxic by inhalation and ingestion, strong irritant to eyes and skin.

N,N-ethylphenylethanolamine.
$C_6H_5NC_2H_5CH_2CH_2OH$.
Properties: Liquid. D 1.04 (20/20C), bp 268C (740 mm Hg), bulk d 8.7 lb/gal. Combustible.
Use: Organic synthesis, dyestuffs.

2-ethyl-2-phenylglutarimide.   (glutethimide).
CAS: 77-21-4. $C_{13}H_{15}NO_2$.
Properties: White, crystalline powder. Melting range 85–87C. Saturated solution is slightly acid. Freely soluble in acetone, ethyl acetate, and chloroform; soluble in ethanol and methanol; practically insoluble in water.
Grade: NF (as glutethimide).
Hazard: Manufacture and use controlled by law.
Use: Medicine (sedative).

ethyl phenyl ketone.   See propiophenone.

4-ethyl-n-(phenylmethylene)benzenamine n-oxide.
CAS: 42790-35-2. $C_{15}H_{15}NO$.
Hazard: Moderately toxic by ingestion and skin contact. A mild eye irritant.

α-(5-ethyl-1-phenyl-1h-pyrazol-4-yl)-1-piperidinebutanol.
CAS: 296269-54-0. $C_{20}H_{29}N_3O$.
Hazard: A poison by ingestion.

ethyl phenylurethane.   See ethyl phenylcarbamate.

ethylphosphoric acid. $C_2H_5H_2PO_4$.
Properties: Pale-straw-colored liquid. D 1.33 (25C). Can be neutralized with alkalies or amines to give water-soluble salts. Combustible.
Grade: Purity: 97%, remainder being orthophosphoric acid and ethanol.
Hazard: An irritant.
Use: Catalyst, rust remover, soldering flux, intermediate.

ethyl phthalate.   See diethyl phthalate.

ethyl phthalyl ethyl glycolate.
CAS: 84-72-0. $C_2H_5OCOC_6H_4COOCH_2COOC_2H_5$.
Properties: Liquid with slight odor. D 1.180 (25C), refr index 1.498 (25C), bp 190C (5 mm Hg). Solubility in water 0.018% by weight; miscible with most organic solvents. Combustible.
Use: Plasticizer; food additive.

5-ethyl-2-picoline.   See 2-methyl-5-ethylpyridine.

2-ethylpiperidine.
Properties: Liquid. Mw 113.20, bp 143C, d 0.850.
Hazard: Flammable; irritant.

1-ethyl-1-propanol.   See 3-pentanol.

ethyl propiolate.
CAS: 623-47-2. $HC\equiv CCO_2C_2H_5$.
Properties: Liquid. Mw 98.10, bp 120C, d 0.968.
Hazard: Flammable; lachrymator.

ethyl propionate.
CAS: 105-37-3. $C_2H_5COOC_2H_5$.
Properties: Water-white liquid; pineapple odor. D 0.888 (25C), bp 99C, fp −73C, flash p 54F (12.2C) (CC), autoign temp 890F (476C), refr index 1.3844 (20C). Soluble in alcohol and ether; solubility in water 2.5% 15C.
Derivation: Treating ethanol with propionic acid.
Grade: Commercial 85–90% ester content, FCC.
Hazard: Flammable, dangerous fire risk.
Use: Solvent for cellulose ethers and esters, various natural and synthetic resins; flavoring agent; fruit syrups; cutting agent for pyroxylin.

ethylpropionyl.   See diethyl ketone.

2-ethyl-3-propylacrolein.   (2-ethylhexenal).
CAS: 645-62-5. $C_3H_7CH:C(C_2H_5)CHO$.
Properties: Yellow liquid; powerful odor. D 0.8518 (20/20C), bp 175.0C, vap press 10 mm Hg (20C), flash p 155F (68.3) (OC), bulk d 7.1 lb/gal (20C), viscosity 0.113 cP (20C). Combustible.
Hazard: Toxic by inhalation and ingestion; strong irritant.
Use: Insecticide, organic synthesis (intermediate), warning agents, and leak detectors.

α-ethyl-β-propylacrylanilide.
$C_6H_5NHCOC(C_2H_5):CH(CH_2)_2CH_3$.
Use: Rubber accelerator.

2-ethyl-3-propylacrylic acid.
$C_3H_7CH:C(C_2H_5)COOH$.
Properties: Liquid. D 0.9484 (20C), fp −7.8C, bp 232.1C, vap press <0.1 mm Hg (20C), flash p 330F (165C). Insoluble in water. Combustible.
Use: Pharmaceuticals, resins and plastics, lubricants.

ethyl propyl ketone.   See 3-hexanone.

ethyl propyl phosphorothioate.
CAS: 20195-06-6. $C_7H_{17}O_3PS$.
Hazard: A poison.

3-ethyl-3-propyl-1,3-propanediol.   See 2-ethylhexanediol-1,3.

2-(5-ethyl-2-pyridyl)ethyl acrylate.
CAS: 144809-27-8. $CH_2:CHCOOC_2H_4C_5$
$H_3NC_2H_5$.
Properties: Liquid. D 1.0458 (20C), bp 181C (50 mm Hg), fp −75C. Very slightly soluble in water. Combustible.
Use: Manufacture of plastics and fibers, adhesives, textile finishes and sizes.

ethyl pyrophosphate.   See tepp.

ethyl salicylate.
CAS: 118-61-6. $C_6H_4(OH)COOC_2H_5$.
Properties: Colorless liquid; faint odor of methyl salicylate. D 1.127–1.130, refr index 1.523, bp 231–234C. Soluble in ether and alcohol; slightly soluble in water. Combustible.
Use: Perfumery, flavors.

ethyl silicate.   (tetraethyl orthosilicate).
CAS: 78-10-4. $(C_2H_5)_4SiO_4$.
Properties: Colorless liquid; faint odor. D 0.9356 (20/20C), bp 168.1C, mp 110C (sublimes), vap press 1.0 mm Hg (20C), flash p 125F (51.6C) (OC), bulk d 7.8 lb/gal (20C), fp −77C, viscosity 0.0179 cP (20C). Miscible with alcohol; hydrolyzed to an adhesive form of silica. Combustible.
Grade: 29% silicon, 40% silicon.
Hazard: Moderate fire risk. Strong irritant to eyes, nose, throat.
Use: Weatherproof and acid-proof mortar and cements, refractory bricks, other molded objects, heat-resistant paints, chemical-resistant paints, protective coatings for industrial buildings and castings, lacquers, bonding agent, intermediate.

ethyl silicate, condensed.   Light-yellow liquid with mild odor consisting of 85% by weight tetraethyl orthosilicate and 15% polyethoxysiloxanes.
Properties: Yields high-purity, refractory silica on hydrolysis or ignition.

Hazard: Toxic.
Use: Intermediate for siloxane compounds; for precision casting of high-melting alloys; pigment binder for paints; surface hardener for sandstones.
See ethyl silicate.

ethyl sodium oxalacetate.
CAS: 52980-17-3. $C_2H_5OOCC(ONa)$:
$CHCOOC_2H_5$.
Properties: Light-yellow powder, 92% pure.
Derivation: Reaction of pure ethyl acetate and diethyl oxalate with metallic sodium.
Use: Dyes, synthesis.

ethyl sulfate.   See diethyl sulfate.

ethyl sulfhydrate.   See ethanethiol.

ethyl sulfide.   (diethyl sulfide).
CAS: 352-93-2. $(C_2H_5)_2S$.
Properties: Colorless, oily liquid; garlic-like odor. D 0.837, fp −102C, bp 92–93C, refr index 1.4423 (20C). Soluble in alcohol and ether, slightly soluble in water. Combustible.
Derivation: Heating sodium ethyl sulfate and potassium sulfide, with subsequent distillation.
Use: Organic synthesis, solvent for mineral salts, electroplating baths.

ethylsulfonylethyl alcohol.   (2-(ethylsulfonyl) ethanol).
CAS: 513-12-2. $CH_3CH_2SO_2CH_2CH_2OH$.
Properties: Colorless crystals. Mp 45C, bp 153C (2.5 mm Hg). Soluble in water and common organic solvents; hygroscopic.
Derivation: Reaction of 2-ethylthioethanol with hydrogen peroxide.
Use: Solvent and organic intermediate, antistatic agent, humectant.

ethylsulfuric acid.   (acid ethylsulfate).
CAS: 507-09-5. $C_2H_5HSO_4$.
Properties: Colorless, oily liquid. D 1.316; bp 280C (decomposes). Soluble in water, alcohol, and ether. Combustible.
Derivation: Action of sulfuric acid on ethanol.
Hazard: Toxic by ingestion and inhalation; strong irritant to eyes and skin.
Use: Precipitant for casein, organic synthesis.

4-ethyl-1,4-thiazane. $C_4H_8NSC_2H_5$.
Properties: Colorless, mobile liquid. D 0.9929 (15C); bp 184C (763 mm Hg). Soluble in water. Combustible.
Derivation: Interaction of dichlorodiethyl sulfide and an aliphatic amine in the presence of alcohol and sodium carbonate.

3-(ethylthio)butyraldehyde.
CAS: 27205-24-9.
$C_6H_{12}OS$.

Hazard: Moderately toxic by ingestion and skin contact. Low toxicity by inhalation. An eye irritant.

ethyl thiocarbimide.   (ethyl mustard oil; ethyl isothiocyanate). C_2H_5NCS.
Properties: Colorless liquid; pungent odor. D 1.004 (15/4C), fp −5.9C, bp 131–132C. Soluble in alcohol; insoluble in water.
Hazard: Toxic by inhalation; strong irritant to skin and mucous membranes.
Use: Military poison gas.

ethyl thioethanol.
CAS: 110-77-0. $C_2H_5SC_2H_4OH$.
Properties: Pale-straw-colored liquid. D 1.015–1.025 (20/20C), distillation range 180–184C, refr index 1.486 (20C). Combustible.
Grade: 95% min.
Use: Organic intermediate (pesticides, plasticizers, etc.).

ethylthiopyrophosphate.   See sulfotepp.

(ethylthio)trioctylstannane.
CAS: 70303-46-7. $C_{26}H_{56}SSn$.
Hazard: Moderately toxic.
Use: Drug.

ethyltin trichloride.
CAS: 1066-57-5. $C_2H_5Cl_3Sn$.
Properties: Colorless liquid. Mp −10C, bp 196–198C.
Hazard: A poison by ingestion.

m-ethyltoluene.
CAS: 620-14-4. C_9H_{12}.
Hazard: Moderately toxic. Low toxicity by inhalation.

n-ethyl-p-toluenesulfonamide.
CAS: 80-39-7. $C_2H_5NHSO_2C_6H_4CH_3$.
Properties: Colorless crystals. Mp 64C, flash p 260F (126C). Soluble in alcohol. Combustible.
Grade: A mixture of the o- and p-isomers is available commercially.
Use: Plasticizer.

ethyl-p-toluenesulfonate.
CAS: 80-40-0. $CH_3C_6H_4SO_3C_2H_5$.
Properties: Unstable solid. Mp 33C, bp 221.3C, d 1.17, flash p 316F (157C). Soluble in many organic solvents; insoluble in water. Combustible.
Use: Plasticizer for cellulose acetate, ethylating agent.

N-ethyl-m-toluidine.   (N-ethyl-3-methylaniline).
CAS: 102-27-2. $CH_3C_6H_4NHC_2H_5$.
Properties: Light-amber liquid. Refr index 1.5451 (20C), bp 221C. Combustible.
Use: Chemical intermediate.

N-ethyl-o-toluidine.
CAS: 94-68-8. $CH_3C_6H_4NHC_2H_5$.
Properties: Colorless to yellowish oil. D 0.9534; bp 214C. Soluble in alcohol, ether, and hydrochloric acid; insoluble in water. Combustible.
Derivation: Heating ethanol with o-toluidine and hydrochloric acid.

ethyl triacetylgallate.
$C_2H_5OOCC_6H_2(OOCCH_3)_3$.
Properties: Colorless crystals or white crystalline powder; insipid taste; odorless. Mp 134–136C. Soluble in warm alcohol and acetone; slightly soluble in ether and alcohol; insoluble in water.

ethyl-3,5,6-tri-o-benzyl-d-glucofuranoside.
CAS: 10310-32-4. $C_{29}H_{34}O_6$.
Hazard: Low toxicity by ingestion.

ethyltrichlorosilane.
CAS: 115-21-9. $C_2H_5SiCl_3$.
Properties: Colorless liquid. Bp 99.5C, d 1.236 (25/25C), refr index 1.4257 (25C), flash p 72F (22.2C) (OC). Soluble in benzene, ether, heptane, perchloroethylene; readily hydrolyzed with liberation of hydrogen chloride.
Derivation: By reaction of ethylene and trichlorosilane in the presence of a peroxide catalyst.
Hazard: Flammable, dangerous fire risk, may form explosive mixture with air. A strong irritant.
Use: Intermediate for silicones.

Ethyl Tuex.   Tetramethylthiuran disulfide.

ethyl urethane.   See urethane.

ethyl valerate.   See ethyl isovalerate.

ethyl vanillate.
CAS: 617-05-0. $(CH_3O)HOC_6H_3COOC_2H_5$.
Properties: Solid. Mp 44C, bp 291–293C. Insoluble in water; soluble in alcohol and ether.
Use: Food preservative, medicine, sunburn preventive.

ethyl vanillin.   (3-ethoxy-4-hydroxybenzaldehyde).
CAS: 121-32-4. $OHC_6H_3(OC_2H_5)CHO$.
Properties: Fine, white crystals; intense odor of vanillin. Affected by light, mp 76.5C. Soluble in alcohol, chloroform, and ether; slightly soluble in water. Combustible.
Grade: NF, FCC.
Use: Flavors, to replace or fortify vanillin.

ethyl vinyl ether.   See vinyl ethyl ether.

ethyl vinyl ether polymer.
CAS: 25104-37-4.
Hazard: Moderately toxic.

ethylxanthic acid.   See xanthic acid.

ethylzinc.   See diethylzinc.

ethyne.   See acetylene.

ethynylation.   Condensation of acetylene with a reagent such as an aldehyde to yield an acetylenic derivative. The best example is the union of formaldehyde and acetylene to produce butynediol.

1-ethynylcyclohexanol.
CAS: 78-27-3. $HC≡CC_6H_{10}OH$.
Properties: Colorless, low-melting solid; sweet odor. Mp 30–31C, bp 180C, d 0.967 (20/20C). Slightly soluble in water; soluble in most organic solvents.
Use: Stabilization of chlorinated organic compounds, intermediate, corrosion inhibitor for mineral acids, medicine (sedative).

ethynylenebis(carbonyloxy)bis(triphenyl-stannane).   See bis(triphenyltin) acetylenedicarboxylate.

ethynylestradiol.   See ethinylestradiol.

ethythrin.   Ethyl analog of allethrin.
Use: Insecticide with applications similar to allethrin.
See barthrin; cyclethrin; furethrin.

etrimfos.   (o-(6-ethoxy-2-ethyl-4-pyrimidinyl)-phosphorothioic acid o,o-dimethyl ester; o,o-dimethyl   o-(2-ethyl-4-ethoxypyrimidinyl)-6-thinophosphate;   (6-ethoxy-2-ethylpyrimidin-4-yl)oxy-dimethoxy-sulfanylidene-5-phosphane).
$C_{10}H_{17}N_2O_4PS$.
Use: Organophosphorus insecticide used against pests of fruit and vegetable crops.
Hazard: Toxic.

ETU.   See ethylene thiourea.

Eu.   Symbol for europium.

eucalyptol.   (cineol; cajeputol).
CAS: 470-82-6. $C_{10}H_{18}O$.
Properties: Colorless essential oil; a terpene ether having a camphor-like odor and pungent, cooling, spicy taste. D 0.921–0.923 (25C); bp 174–177C; congealing point not less than about 0C, refr index 1.4550–1.4600 (20C). Slightly soluble in water; miscible with alcohol, chloroform, ether, glacial acetic acid, and fixed or volatile oils. Combustible.
Derivation: By fractionally distilling eucalyptus oil followed by freezing. The oil is imported from Spain, Portugal, and Australia.
Grade: Technical, NF.
Use: Pharmaceuticals (cough syrups, expectorants), flavoring, perfumery.

euchromatin.   Chromatin, which is in an uncondensed form, is more accessible and contains the majority of actively expressed genes. See also heterochromatin.

eucryptite.   A lithium silicate mineral $(LiAlSiO_4)$ containing up to 4.8% lithia.
Use: Glass manufacture; source of lithium.

eugenic acid.   See eugenol.

eugenics.   The study of improving a species by artificial selection; usually refers to the selective breeding of humans.

eugenol.   (4-allyl-2-methoxyphenol; caryophyllic acid; eugenic acid).
CAS: 97-53-0. $C_3H_5C_6H_3(OH)OCH_3$.
Properties: Colorless or yellowish liquid, oily; becomes brown in air; spicy odor and taste. D 1.064–1.070, bp 253.5C, fp −9C, refr index 1.5400–1.5420 (20C), optically inactive. Soluble in alcohol, chloroform, ether, and volatile oils; very slightly soluble in water. Combustible.
Derivation: By extraction of clove oil with aqueous potash, liberation with acid, and rectification in a stream of carbon dioxide.
Grade: Technical, USP, FCC.
Hazard: Questionable carcinogen.
Use: Perfumes, essential oils, medicine (analgesic), production of isoeugenol for the manufacture of vanillin, flavoring.

eugenyl methyl ether.   See methyl eugenol.

eukaryote.   A unicellular or multicellular organism with cells having a membrane-bounded nucleus. All possess, in addition, multiple chromosomes and internal organelles.

Eureka.   A soldering flux crystalline composition based on zinc chloride and ammonium chloride.

europia.   See europium oxide.

europium.   Eu. Atomic number 63, one of the lanthanide or rare-earth elements of the cerium subgroup, aw 151.96, valences of 2, 3; two stable isotopes.
Properties: Steel-gray metal, difficult to prepare, quite soft and malleable (DPH = 20). Mp 826C, bp approximately 1489C, d 5.24, oxidizes rapidly in air. May burn spontaneously. Most reactive of the rare-earth metals. Liberates hydrogen from water. Reduces metallic oxides.
Derivation: Reduction of the oxide with lanthanum or misch metal.
Source: Rare-earth minerals.
Grade: High purity (ingots, lumps).
Hazard: Highly reactive, may ignite spontaneously in powder form.

Use: Neutron absorber in nuclear control, color-TV phosphors to activate yttrium, phosphors in postage-stamp glues to permit electronic recognition of first-class mail.

europium chloride.
CAS: 10025-76-0. $EuCl_3 \cdot 6H_2O$.
Properties: Yellow needles. D 4.89 (20C), mp 850C. Soluble in water.
Derivation: Obtained by treating the oxide with hydrochloric acid.

europium fluoride. $EuF_3 \cdot 3H_2O$.
Properties: Crystals. Mp 1390C, bp 2280C. Insoluble in water and dilute acids.

europium nitrate. $Eu(NO_3)_3 \cdot 6H_2O$.
Colorless to pale-pink crystals, mp 85C (sealed tube), soluble in water, obtained by treating the oxide with nitric acid.

europium oxalate. $Eu_2(C_2O_4)_3 \cdot 10H_2O$.
White powder, insoluble in water, slightly soluble in acids.
Grade: 25–50% and 99.8% Eu salt. Impure grade may be colored.

europium oxide.   (europia). Eu_2O_3.
Properties: Pale-rose powder. D 7.42. Insoluble in water; soluble in acids to give the corresponding salt.
Derivation: Calcination of the oxalate, solvent extraction, or liquid ion-exchange processes.
Grade: 25–50%, 99.9%.
Use: Nuclear-reactor control rods, especially in red- and infrared-sensitive phosphors.

europium sulfate. $Eu_2(SO_4)_3$ anhydrous and $8H_2O$.
Properties: Colorless to pale-pink crystals. D (anhydrous) 4.99 (20C). Hydrate loses $8H_2O$ at 375C; slightly soluble in water.

Eurovanillan.   Aromatic ethyl vanillan.
Use: An aroma additive.

eutactic.   Characterized by precise molecular order, like that of a perfect crystal, the interior of a protein molecule, or a machine-phase system; contrasted to the disorder of bulk materials, solution environments, or biological structures on a cellular scale.

eutectic.   The lowest melting point of an alloy or solution of two or more substances (usually metals) that is obtainable by varying the proportion of the components. Eutectic alloys have definite and minimum melting points in contrast to other combinations of the same metals.
See alloy, fusible.

eutectic point.   Melting point of eutectic mixture. The lowest temperature at which the eutectic mixture can exist in liquid phase.

eutrophication.   The unintentional enrichment or "fertilization" of either fresh or saltwater by chemical elements or compounds present in various types of industrial wastes. Phosphates and nitrogenous compounds in detergent and chemical-processing wastes are particularly effective eutrophying agents. They supply nutrients to algae, which proliferate so abundantly that a large proportion die for lack of light; their decomposition products deplete the water of its dissolved oxygen and thus cause the death of fish and other marine life. One process for removing phosphates involves addition of a metal-ion source to waste effluents so as to insolubilize dissolved phosphates, after which the particles are agglomerated by anionic polymers.
See algae; nitrilotriacetic acid.

eutropic series.   Series of substances in which the physical properties and crystal form show regularity in their variation.

eV.   Abbreviation for electron-volt.

EVA.   Abbreviation for ethylene-vinyl acetate copolymer.

evaporation.   The change of a substance from the solid or liquid phase to the gaseous or vapor phase. In some cases (e.g., ice, snow, dry ice), the substances do not go through a liquid phase; the phenomenon is called sublimation. The rate of evaporation of liquids varies with their chemical nature and the temperature; in general, organic liquids (benzene, gasoline) evaporate at lower temperatures and higher rates than water. The thermal energy required to vaporize a given volume of a liquid is known as its latent heat of vaporization; it remains in the vapor (steam, in the case of water heated to its boiling point) and is released when the vapor condenses. For water, this latent-heat value is 540 cal/g. In chemical processing installations requiring a series of evaporations and condensations, the units are set up in series and the latent heat of vaporization from one unit is utilized to supply energy for the next. Such units are called "effects" in engineering parlance, as, e.g., a triple-effect evaporator.
See distillation.

evaporite.   A mineral or rock formed by precipitation of crystals from evaporating water.

EVE.   Abbreviation for ethyl vinyl ether.
See vinyl ethyl ether.

eve carbamate.
CAS: 204442-82-0. $C_9H_4F_{13}NO_4$.
Hazard: Low toxicity by ingestion. A mild skin and moderate eye irritant.

"Everflex" *[Covidien].*   TM for vinyl acetate polymer emulsions.
Use: Paints, paper coating, and acoustic tile coatings.

evolution.   Emission of a gas as a result of a chemical reaction, as hydrogen from acids react with a metal or carbon dioxide from the action of an acid on a carbonate. In formulas it may be indicated by an upward vertical arrow.

$$\uparrow$$

evolutionarily conserved.   See conserved sequence.

evolutionary rootprinting.   One can infer which portions of a gene are important by comparing the sequence of that gene with its cognates from other species. A plot showing the regions of high conservation indicates the regions that are functional in all the test species.

exchange reaction.   A chemical reaction in which atoms of the same element in two different molecules or in two different positions in the same molecule change places. Exchange reactions are usually studied with the aid of a tracer or tagged atom.

excimer.   A transient excited dimer formed between a solute and a solvent molecule.

excipient.   A natural, inert, and somewhat tacky material used in the pharmaceutical industry as a binder in tablets, etc. Commonly used materials are gum arabic, honey, and beet pulp.

excitation.   Addition of energy to a system whereby it is transferred from its ground state to a state of higher energy.

excitatory amino acid.   An amino acid that induces firing or increases the rate of firing of neurons that have receptors for the particular amino acid. Some can induce sustained repetitive firing that can damage or kill the affected neuron.

excited state.   A higher than normal energy level of the electrons of an atom, radical, or molecule, typically resulting from absorption of photons (quanta) from a radiation source (arc, flame, spark, etc.) in any wavelength of the electromagnetic spectrum. X-ray, UV, visible, infrared, microwave, and radio frequencies are used for excitation in various types of spectroscopy. When the energizing source is removed or discontinued, the atom or molecule returns to its normal or stable state either by emitting the absorbed photons or by transferring the energy to other atoms or molecules. The emission by the atom or molecule yields line or band spectra characteristic of its structure, thus permitting identification. Photochemical reactions are induced by excited chemical entities, which are also responsible for the phenomena of luminescence (phosphorescence and fluorescence).
See spectroscopy; photochemistry; absorption (2).

exciton.   An energetic entity induced in semiconductor crystals by incident radiation and occurring in the field area between a hole and its displaced electron. It is conceived as behaving like an uncharged particle having quantum-mechanical properties.

excitotoxin.   A substance that can evoke a cytotoxic effect.

excluded volume.   The presence of one molecule reduces the volume available for other molecules; resulting reductions in their entropy are termed excluded volume effects.

exell.
CAS: 104559-06-0.
Hazard: Moderately toxic by ingestion.

exemestane.   See 6-methylenandrosta-1,4-diene-3,17-dione.

exergonic reaction.   A chemical reaction that proceeds with the release of free energy (that is, for which DG is negative and the reaction spontaneous).

exhaust emission control.   See air pollution.

Exkin 1,2,3.   A series of anti-skinning agents of the volatile oxime type.
Use: Paints.

exo-.   A prefix used in chemical names to indicate attachment to a side chain rather than to a ring.
See *endo-.*

exocarbon.   Cyclic compound in which all or most carbon atoms are in the form of carbonyl groups or their equivalents.

exocytosis.   The fusion of an intracellular vesicle with the plasma membrane, releasing the vesicle contents to the extracellular space.

exoergic.   See exothermic.

exogenous DNA.   DNA originating outside an organism that has been introduced into the organism.

exon.   Those portions of a genomic DNA sequence which will be represented in the final, mature mRNA. The term "exon" can also be used for the equivalent segments in the final RNA. Exons

may include coding sequences, the 5′ untranslated region, or the 3′ untranslated region.
See: Intron.

exonuclease.   An enzyme that digests nucleic acids starting at one end. An example is Exonuclease III, which digests only double-stranded DNA starting from the 3′ end.

exosmosis.   Osmosis in an outward direction. See endosmosis.

exothermic.   Describes a process or chemical reaction that is accompanied by evolution of heat, e.g., combustion reactions.

exotic.   A term applied to materials of various functional types that have extra power and that are often derived from unusual sources. Examples are rocket propellants derived from boron hydrides and certain special-purpose solvents.

exotoxin.   (ectotoxin;   2-[5-[[5-(6-aminopurin-9-yl)-3,4-dihydroxyoxolan-2-yl]methoxy]-3,4-dihydroxy-6-(hydroxymethyl)oxan-2-yl]oxy-3,5-dihydroxy-4-phosphonooxyhexanedioic   acid). $C_{22}H_{32}N_5O_{19}P$. Any proteinaceious toxin produced by bacteria that is found outside the bacteria shell or is released into the surrounding medium. They are detoxified with treatment by agents such as formaldehyde and include cardiotoxins, hemotoxins, neurotoxins, botulinum toxin, diphtheria toxin, plague toxin, and various disrupting enzymes such as lacithinase and collagenase.
Properties: Heat liable.
Hazard: Extremely Toxic; Among the most toxic substances known.

expanded polymer.   Cellular polymer in which the material has been produced by allowing a gas to expand within a polymer melt followed by cooling. The gas is thus trapped as bubbles within the final solid polymer. The expansion of the gas is either due to injection of the gas under pressure or by chemical decomposition, usually induced by use of a blowing agent or by the high temperatures of processing.

expander.   (1) A mixture of lampblack, barium sulfate, and an organic material usually derived from lignin that increases the capacity of storage batteries, presumably by coating the anode and thus preventing the deposit of lead sulfate on the underlying lead metal.
(2) A substance used in medicine as a substitute for blood plasma, e.g., dextran, gelatin, or polyvinylpyrollidone.

expeller.   See extractor; expression.

explosive, high.   A chemical compound, usually containing nitrogen, that detonates as a result of shock or heat (see detonation). Dynamite was the most widely used explosive for blasting and other industrial purposes until 1955, when it was largely replaced by prills-and-oil and slurry types. The former consists of 94% ammonium nitrate prills and 6% fuel oil (ANFO). Slurry blasting agents (SBA) are based on thickened or gelatinized ammonium nitrate slurries sensitized with TNT, other solid explosives, or aluminum. An unusual type of explosive is represented by acetylides of copper and silver, which are examples of commercially used explosives that contain neither oxygen nor nitrogen. High explosives vary greatly in their shock sensitivity; most sensitive are mercury fulminate and nitroglycerin, whereas TNT and ammonium nitrate are comparatively difficult to detonate, requiring the use of blasting caps or similar activating device.

explosive, initiating.   An explosive composition used as a component of blasting caps, detonators, and primers. They are highly sensitive to flame, heat, impact, or friction. Examples are lead azide, silver acetylide, mercury fulminate, diazodinitrophenol, nitrosoguanidine, lead styphnate, and pentaerythritol tetranitrate.

explosive limits.   The range of concentration of a flammable gas or vapor (% by volume in air) in which explosion can occur upon ignition in a confined area. Explosive limits for some common substances are

Substance	Lower (%)	Upper (%)
carbon disulfide	1	50
benzene	1.5	8
methane	5	15
butadiene	2	11.5
butane	1.9	8.5
propane	2.4	9.5
natural gas	3.8	17
hydrogen	4	75
acetylene	2.5	80

explosive, low.   An explosive, such as black powder, that deflagrates rather than detonates.
See deflagration.

explosive, permissible.   Explosives approved for use in coal mines. Usually they are modified dynamites.

exposure testing.   Determination of the degradation of a material by exposing samples to an environment selected for its adverse effect. Materials most frequently tested are paints, metals and alloys, rubber, and plastics. An area frequently chosen is the coast of southern Florida, where the combination of high temperature, strong sunlight, salt air, and moisture is particularly severe, especially as

regards metal corrosion. Burial of metals in acid soils for long periods and immersion of impregnated wood samples in seawater are other exposure-testing techniques.
See testing.

expressed sequence tag.   (EST). A short strand of DNA that is a part of a cDNA molecule and can act as identifier of a gene. Used in locating and mapping genes.
See cDNA; sequence tagged site.

expression.   (1) Removal of a liquid from a solid by hydraulic pressure, as in manufacture of vegetable oils from seeds, rinds, or meal cake. Worm devices similar to extrusion machines are also used for this purpose; they are called *expellers* or *extractors*. (2) To "express" a gene is to cause it to function. A gene which encodes a protein will, when expressed, be transcribed and translated to produce that protein. A gene which encodes an RNA rather than a protein (e.g., a rRNA gene) will produce that RNA when expressed.

expression clone.   This is a clone designed to produce a protein from the DNA insert. Since mammalian genes do not function in bacteria, to get bacterial expression from mammalian cDNA. It is required to put its coding region (i.e. no introns) immediately adjacent to bacterial transcription/translation control sequences.

Ext. D&C.   Product certified for external use only in drugs and cosmetics.

extender.   A low-gravity material used in paint, ink, plastic, and rubber formulations chiefly to reduce cost per unit volume by increasing bulk. Extenders include diatomaceous earth, wood flock, mineral rubber, liquid asphalt, etc. Microscopic droplets of water fixed permanently in a plastic matrix are an efficient extender for polyester resins. In the food industry, the term refers to certain extruded proteins, especially those derived from soybeans, which are used in meat products to provide equivalent nutrient values at lower cost. Made from defatted soy flour, they are often called *textured proteins*.
See diluent (2); filler.

extinction.   When all the members of a clade or taxon die, the group is said to be extinct.

extinguishing agent.   See fire extinguishment.

extracellular matrix.   (ECM). Region outside of metazoan cells which includes compounds attached to the plasma membrane, as well as dissolved substances attracted to the surface charge of the cells. The ECM functions both to keep animal cells adhered together as well as to buffer them from their environment.

extraction, solvent.   See solvent extraction.

extractive distillation.   A variety of distillation that always involves use of a fractionating column and is characterized by use of a purposely added substance that modifies the vaporization characteristics of the materials undergoing separation to make them easier to separate. The additive is often called a solvent and is usually chosen to be much less volatile than any of the substances being separated. It is added to the down flowing liquid-reflux stream near the top of the column and is removed from the still pot or reboiler at the base. The addition of furfural to mixtures of butadiene and butene hydrocarbons to separate the butadiene more easily is an example of extractive distillation.

extractive metallurgy.   That portion of metallurgical science devoted to the technology of mining and processing of metals and their ores.

extractor.   A machine designed for expression of the oil from seeds such as cottonseed, flaxseed, etc., or for extracting the juice from fruits. It is similar in principle to an extrusion machine having a tapering screw rotating in a cylindrical barrel; the oil or juice is delivered to a container while the residual pulp and fiber are extruded as waste material. Fruit is fed into the machine after being crushed and preheated. Juices can also be separated centrifugally. Machines of this type are also called *expellers*.

"Extrema" *[JJISCO].*   (trichloroethane). TM for a chlorine source.
Use: Thermal oxidation of silicon and furnace-tube cleaning.

extreme-pressure additive.   (EP additive). (1) Material added to cutting fluids to impart high film strength. They are mainly sulfur, chlorine, and occasionally phosphorus compounds. Actual conditions, amounts, etc. are proprietary. (2) Lubricating oil and grease additives that prevent metal-to-metal contact in highly loaded gears. Some react with the metal gears to form a protective coating. Saponified lead salts are often used.

Extru Seal.   A butyl glazing tape for construction.

extrusion.   A fundamental processing operation in many industries in which a material is forced through a metal forming die, followed by cooling or chemical hardening. The material may be liquid (molten glass or a polymer dispersion); a viscous polymer, as in injection molding; a semisolid mass, such as a rubber or plastic mix; or a hot metal billet. High-viscosity materials are fed into a rotating screw of variable pitch that forces the materials through the die with considerable pressure; a ram is used for metals at temperatures from 537 to 1093C. Film is made by passing a low-viscosity mixture

through a narrow slit. Molten glass and polymer suspensions are forced through a nozzle having a tiny orifice (spinnerette); the latter are hardened after extrusion by immersion in a bath of formaldehyde or similar agent. Food items (spaghetti, etc.) are also extruded. Extrusion involves rheological principles of some complexity, critical factors being viscosity, temperature, flow rate, and die design.
See injection molding.

eyespot. Light-sensitive organelle found in many groups of protists, and in some metazoans.

E-Z Mulse. An emulsifier.
  Use: To combine *d*-limonene and water to make stable formulations; emulsifies solvents or oil with water.

Ezn-Chek. Heat stabilizers.

F

F. Symbol for fluorine; the molecular formula is F_2.

F1. See filial generation.

F2. See filial generation.

FAB. Abbreviation of fast atom bombardment.

fabric. A textile structure composed of mechanically interlocked fibers or filaments. It may be randomly integrated (nonwoven) or closely oriented by warp and filler strands at right angles to each other (woven). While the word usually refers to wool, cotton, or synthetic fibers, fabrics can also be made of glass fiber and graphite.

fabrication. The molding, forming, machining, assembly, and finishing of metals, rubber, and plastics into end-use products. In the paper industry, the term *converting* is used in this sense.

fabric dye. Any dye used on fabric.
Hazard: Very toxic.

Fabrikoid. Pyroxylin-coated fabrics that are water-resistant and soap and water washable.
Use: Book binding, luggage.

Fabrilite. Vinyl-coated fabrics and selected vinyl compounds without fabric backing.
Use: Pocketbooks, bags, upholstery, etc.

Fabrisoil. Engineered nonwoven fabric.
Use: Landfill daily cover.

face-centered cubic structure. An internal crystal structure, determined by X-rays, in which the equivalent points are at the corners of the unit cell and at the centers of the six faces of a cube.

F acid. See Casella's acid.

facilitated diffusion. Diffusion of a molecule across a biological membrane via a protein transporter down a concentration gradient. Also called passive diffusion.

factice. (vulcanized oil). A soft, mealy material made by reaction of sulfur or sulfur chloride with a vegetable oil.

Use: Erasers, rubber goods (bath spray tubing, etc.) to give soft "hand."

factor. A term used chiefly by biochemists to indicate any member of a biologically active complex, especially if its exact chemical nature is unknown or if its function in cellular metabolism has not been elucidated. Several of the B-complex vitamins were originally referred to as factors until their identity had been established by research. There are a number of blood-coagulation factors.
See Rh factor.

facultative. See bacteria.

facultative cells. Cells that can live in the presence or absence of oxygen.

FAD. (flavin adenine dinucleotide). The coenzyme of some redox enzymes. It contains riboflavin.

Fahrenheit. The scale of temperature in which 212 degrees is the boiling point of water at 760 mm Hg and 32 degrees is the freezing point of water. The scale was invented by a German physicist, G. D. Fahrenheit (1686–1736), who introduced the use of mercury instead of alcohol in thermometers. The entry for centigrade contains a method of converting from Fahrenheit to centigrade. The entry for absolute temperature contains a method for converting Fahrenheit to absolute Rankine.
See centigrade.

falecalcitriol. See hexafluorocalcitriol.

fallout. Deposition on the earth of the radioactive particles resulting from a nuclear explosion, e.g., strontium-90.

Falone. Tris(2,4-dichlorophenoxyethyl) phosphite.
Properties: Viscous, amber liquid. D 1.434, mp 70–72C. Soluble in benzene, xylene, and aromatic hydrocarbons; insoluble in water; available as an emulsifiable concentration and a granular solid.
Use: A preemergence herbicide.

famphur. (famophos; (generic name for O,O-dimethyl-O-[p-(dimethylsulfamoyl) phenyl] phosphorothioate).
CAS: 52-85-7. $(CH_3O)_2P(S)OC_6H_4SO_2N(CH_3)_2$.

Hawley's Condensed Chemical Dictionary, Sixteenth Edition. Michael D. Larrañaga, Richard J. Lewis, Sr., and Robert A. Lewis.
© 2016 John Wiley & Sons, Inc. Published 2016 by John Wiley & Sons, Inc.

Properties: Crystalline powder. Mp 55C. Very soluble in chloroform and carbon tetrachloride; slightly soluble in water.
Hazard: Cholinesterase inhibitor, use may be restricted.
Use: Insecticide.

"Fanal" *[BASF].* TM for phosphotungstic lakes. Characterized by brilliancy of shade and good fastness to light.
Use: Printing inks.

faraday. The quantity of electricity that can deposit (or dissolve) one gram-equivalent weight of a substance during electrolysis (approximately 96,500 coulombs).

Faraday, Michael. (1791–1867). A native of England, Faraday did more to advance the science of electrochemistry than any other scientist. A profound thinker and accurate experimentalist and observer, he was the first to propound correct ideas as to the nature of electrical phenomena, not only in chemistry but also in other fields. His contributions to chemistry include the basic laws of electrolysis, electrochemical decomposition (the basis of corrosion of metals) of battery science, and electrometallurgy. His work in physics led to the invention of the dynamo. Faraday was in many respects the exemplar of a true scientist, combining meticulous effort and interpretive genius.

farnesol. (generic name for 3,7,11-trimethyl-2,6,10-dodecatrienol).
CAS: 4602-84-0. $C_{15}H_{25}OH$.
Properties: Colorless liquid; delicate floral odor. D 0.885 (15C), bp 145–146C (3 mm Hg). Soluble in three volumes of 70% alcohol. Combustible.
Derivation: Found in nature in many flowers and essential oils such as cassia, neroli, cananga, rose, balsams, ambrette seed.
Use: Perfumery, flavoring, insect hormone.

fast. (1) Descriptive of a dye or pigment whose color is not impaired by prolonged exposure to light, steam, high temperature, or other environmental conditions. Inorganic pigments are normally superior in this respect to organic dyes. (2) In nuclear technology, the term refers to neutrons moving at the speed at which they emerge from a ruptured nucleus as opposed to "slow" or thermal neutrons whose speed has been reduced by impinging on a neutral substance called a moderator. Fast neutrons are used in breeder reactors.

fast atom bombardment. (FAB). One of several techniques for ionizing solids from solutions. In FAB, a thin film of the dissolved solid to be analyzed is bombarded with fast atoms. These dislocate ions by impact; the ions are then analyzed by mass spectroscopy. Peptide ions with molecular weight of approximately 6000 have been produced and analyzed by this method.

fastogen super magneta r. See 2,9-dimethyl-quinacridone.

fat. A glyceryl ester of higher fatty acids such as stearic and palmitic. Such esters and their mixtures are solids at room temperature and exhibit crystalline structure. Lard and tallow are examples. There is no chemical difference between a fat and an oil, the only distinction being that fats are solid at room temperature and oils are liquid. The term fat usually refers to triglycerides specifically, whereas lipid is all-inclusive.
See lipid.

fat dyes. Oil-soluble dyes for candles, wax, etc.

fatigue. Incremental weakening of a material as a result of repeated cycles of stresses that are far lower than its breaking load, ending in failure. For metals, to which the term usually refers, the number of low-stress cycles may be of the order of 10^7. Failure is due to development of cumulative imperfections in the crystal structure, with consequent minute interior cracks. Gear failure is often caused by fatigue. It has been reported in experimental windmills for power generation in which steel blades have failed after a few hundred hours of operation due to centrifugal stress. In elastomeric materials, fatigue involves complete dissipation of their resilient energy by repeated cycles of low-order stresses.

fat liquoring agent. An oil-in-water emulsion usually made from raw oils such as neatsfoot, cod, etc., made soluble by dispersing agents such as sulfonated oils.
Use: Leather processing to replace natural oils removed from hides by tanning operations.
See neatsfoot oil; emulsion.

fat splitting. See hydrolysis.

fatty acid. A carboxylic acid derived from or contained in an animal or vegetable fat or oil. All fatty acids are composed of a chain of alkyl groups containing from 4 to 22 carbon atoms (usually an even number) and characterized by a terminal carboxyl group –COOH. The generic formula for mentioned acetic is $CH_3(CH_2)_xCOOH$ (the carbon atom count includes the carboxyl group). Fatty acids may be saturated or unsaturated (olefinic), and solid, semisolid, or liquid. They are classed among the lipids, together with soap and waxes.
Saturated: A fatty acid in which the carbon atoms of the alkyl chain are connected by single bonds. The most important of these are butyric (C_4), lauric (C_{12}), palmitic (C_{16}), and stearic (C_{18}). They have a variety of special uses (see specific entry). Stearic acid leads all other fatty acids in industrial

use, primarily as a dispersing agent and accelerator activator in rubber products and soaps.

Unsaturated: A fatty acid in which there are one or more double bonds between the carbon atoms in the alkyl chain. These acids are usually vegetable derived and consist of alkyl chains containing 18 or more carbon atoms with the characteristic end group –COOH. Most vegetable oils are mixtures of several fatty acids or their glycerides; the unsaturation accounts for the broad chemical utility of these substances, especially of drying oils. The most common unsaturated acids are oleic, linoleic, and linolenic (all C_{18}). Safflower oil is high in linoleic acid, peanut oil contains 21% linoleic acid, olive oil is 38% oleic acid, palmitoleic acid is abundant in fish oils. Aromatic fatty acids are now available.
See phenylstearic acid.

Note: Linoleic, linolenic, and arachidonic acids are called essential fatty acids by biochemists because such acids are necessary nutrients that are not synthesized in the animal body.
Use: Special soaps, heavy-metal soap, lubricants, paints and lacquers (drying oils), candles, salad oil, shortening, synthetic detergents, cosmetics, emulsifiers.

fatty acid enol ester.   A fatty acid reacted with enolic form of acetone for the purpose of increasing the chemical reactivity of the acid. Stearic acid (18-carbon) combined with acetone (3-carbon) gives isopropenyl stearate (21-carbon). This is effective in making the fatty stearoyl group available for synthesis of polymers, medicinals, and the like.
See fatty ester.

fatty acid pitch.   A by-product residue from (1) soap stock and candle stock manufacture; (2) refining of vegetable oils; (3) refining of refuse greases; and (4) refining of wool grease.
Properties: Dark brown to black. Properties analogous to complex hydrocarbons, contains fixed carbon (5–35%), soluble in naphtha and carbon disulfide.
Use: Manufacture of black paints and varnishes, tarred papers, printers' rolls, rubber filing agent, impregnating agent, electrical insulations, marine caulking, waterproofing, sealant.

fatty alcohol.   A primary alcohol (from C_8 to C_{20}), usually straight chain. High molecular weight alcohols are produced synthetically by the Oxo and Ziegler processes. Those from C_8 to C_{11} are oily liquids; those greater than C_{11} are solids. Other methods of production are (1) reduction of vegetable seed oils and their fatty acids with sodium; (2) catalytic hydrogenation at elevated temperatures and pressures; and (3) hydrolysis of spermaceti and sperm oil by saponification and vacuum fractional distillation. The more important commercial saturated alcohols are octyl, decyl, lauryl, myristyl, cetyl, and stearyl. The commercially important unsaturated alcohols, such as oleyl, linoleyl, and linolenyl, are also normally included in this group. The odor tends to disappear as the chain length increases.
Use: Solvent for fats, waxes, gums, and resins; pharmaceutical salves and lotions; lubricating-oil additives; detergents and emulsifiers; textile antistatic and finishing agents; plasticizers; nonionic surfactants; cosmetics.

fatty amine.   A normal aliphatic amine derived from fats and oils. May be saturated or unsaturated, and primary, secondary, or tertiary, but the alkyl groups are straight chain and have an even number of carbons in each. The length varies from 8 to 22 carbon atoms.
Derivation: Fatty acids are treated with ammonia and heated to form fatty acid amides, which are converted to nitriles and reduced to the amine.
Use: Organic bases, soaps, plasticizers, tire cords, fabric softeners, water-resistant asphalt, hair conditioners, cosmetics, medicinals.

fatty ester.   A fatty acid with the active hydrogen replaced by the alkyl group of a monohydric alcohol. The esterification of a fatty acid, RCOOH, by an alcohol, R'OH, yields the fatty ester RCOOR'. The most common alcohol used is methanol, yielding the methyl ester $RCOOCH_3$. The methyl esters of fatty acids have higher vapor pressures than the corresponding acids and are distilled more easily.

fatty nitrile.   (RCN). An organic cyanide derived from a fatty acid.
Derivation: Fatty acids are treated with ammonia and heated to form fatty acid amides, which are converted to nitriles.
Use: Intermediates for fatty amines, lubricating oil additives, plasticizers.

faujasite. $Na_2CaO \cdot Al_2O_3 \cdot 5SiO_2 \cdot 10H_2O$. A mineral.
Use: Zeolite; molecular sieve.

Favorskii–Babayan synthesis.   Synthesis of acetylenic alcohols from ketones and terminal acetylenes in the presence of anhydrous alkali.

Favorskii rearrangement.   Base-catalyzed rearrangement of α-haloketones to acids or esters. The rearrangement of α,α'-dibromocyclohexanones to 1-hydroxycyclopentanecarboxylic acids, followed by oxidation to the ketones, is known as the Wallach degradation.

FBR.   Abbreviation for fast breeder reactor.
See breeder.

FCC.   (1) Abbreviation for Food Chemicals Codex, a publication giving specifications and test methods for chemicals used in foods. (2)

Abbreviation for fluid-cracking catalyst as used in the petroleum refining industry. Examples are powdered silica alumina, in which alumina is impregnated with dry synthetic silica gel, and various natural clays impregnated with alumina.

FDA.   Abbreviation for Food and Drug Administration.

FD&C color.   A series of colorants permitted in food products, marking inks, etc., certified by the FDA. Among the more important are the following:
Blue No. 1: disodium salt of 4-((4-(*N*-ethyl-*p*-sulfobenzylamino)-phenyl)-(2-sulfoniumphenyl)-methylene)-(1-(*N*-ethyl-*N*-*p*-sulfobenzyl)-$\Delta^{2,5}$-cyclohexadienimine).
Blue No. 2: disodium salt of 5,5'-indigotin disulfonic acid.
Green No. 3: disodium salt of 4-((4-(*N*-ethyl-*p*-sulfobenzylamino)-phenyl-(4-hydroxy-2-sulfonium   phenyl)-methylene)-(1-(*N*-ethyl-*N*-*p*-sulfobenzyl))-$\Delta^{2,3}$-cyclohexadienimine).
Green No. 6: 1,4-di-*p*-toluidinoanthraquinone.
Red No. 2: trisodium salt of 1-(4-sulfo-1-naphthylazo)-2-naphthol-3,6-disulfonic acid. Formerly the largest volume food color in commercial use. A carcinogen. Use prohibited by FDA. Red No. 40 is currently a permissible substitute.
Red No. 3: disodium salt of 9-*o*-carboxyphenyl-6-hydroxy-2,4,5,7-tetraiodo-3-isoxanthone (erythrosin).
Red No. 4: disodium salt of 2-(5-sulfo-2,4-xylylazo)-naphthyl-4-sulfonic acid. Use in foods prohibited by FDA.
Violet No. 1: monosodium salt of 4-((*N*-ethyl-*p*-sulfobenzylamino)-phenyl)-(4(*N*-ethyl-*p*-sulfoniumbenzylamino)-phenyl)-methylene)-(*N*,*N*-dimethyl-$\Delta^{2,5}$-cyclohexadienimine). Use prohibited by FDA in 1973 and by USDA in 1976.
Yellow No. 5: trisodium salt of 3-carboxy-5-hydroxy-1-*p*-sulfophenyl-4-sulfophenylazopyrazole.
Yellow No. 6: disodium salt of 1-*p*-sulfophenylazo-2-naphthol-6-sulfonic acid.
See food color; food additive.

Fe.   Symbol for iron.

feathers.   See keratin.

Federal Trade Commission.   (FTC). The Federal Trade Commission enforces a variety of federal antitrust and consumer protection laws. The Commission seeks to ensure that the nation's markets function competitively, and are vigorous, efficient, and free of undue restrictions. The Commission also works to enhance the smooth operation of the marketplace by eliminating acts or practices that are unfair or deceptive. Its main office is located at 600 Pennsylvania Avenue, N.W., Washington, DC 20580.
Website: http://www.ftc.gov

feedback inhibition.   Inhibition of an allosteric enzyme at the beginning of a metabolic sequence by the end product of the sequence.

feeder.   An accessory equipment unit that provides controlled flow of materials of a wide range of particulate sizes to or from processing operations. Major types include the following: (1) Vibratory: an enclosed bowl or open trough, activated electromagnetically, that vibrates at a constant rate of 3600 oscillations a minute (electromechanical, hydraulic, and pneumatic types are also used). Capacities are up to 2000 lb/hr. The bowl type is applicable to large-size units of materials up to several inches in diameter (wood, plastics, ceramics, etc.). (2) Volumetric: an enclosed device that meters a particulate by volume; there are a number of types, including the rotary lock, the helix, and the roll. Bulk density, particle size, and moisture content of the material handled are important factors. (3) Gravimetric: a belt conveyor provided with a scale that continuously weighs the material passing over it. These are used in operations that are not suitable for volumetric feed.

feedstock.   Gaseous or liquid petroleum-derived hydrocarbons or mixture of hydrocarbons from which gasoline, fuel oil, and petrochemicals are produced by thermal or catalytic cracking. It is also called charging stock. Feedstocks commonly used include ethane, propane, butane, butene, benzene, toluene, xylene, naphtha, and gas oils.

FEFO.   See formaldehyde bis(2-fluoro-2,2-dinitroethyl) acetal.

Fehling's solution.   A reagent used as a test for sugars, aldehydes, etc. It consists of two solutions, copper sulfate and alkaline tartrate, which are mixed just before use. Benedict's modification is a one-solution preparation.
Additional details are available in the Book of Methods, Association of Official Analytical Chemists.

Feist–Benary synthesis.   Formation of furans from α-halogeno ketones or ethers and 1,3-dicarbonyl compounds in the presence of pyridine. With ammonia as condensing agent, pyrrole derivatives are always formed as secondary products.

felbamate.
CAS: 25451-15-4. $C_{11}H_{14}N_2O_4$.
Hazard: Low toxicity by ingestion. Human systemic effects.

feldspar.   (potassium aluminosilicate). General name for a group of sodium, potassium, calcium, and barium aluminum silicates. Commercially, feldspar usually refers to the potassium feldspars with the formula $KAlSi_3O_8$, usually with a little sodium. Noncombustible.

Occurrence: North Carolina, Colorado, New Hampshire, South Dakota, California, Arizona, Wyoming, Virginia, Texas.

Grade: Usually based on silicon dioxide content, potassium–sodium ratio, iron content, and fineness of grinding.

Hazard: Toxic as fine-ground powder.

Use: Pottery, enamel, and ceramic ware; glass; soaps; abrasive; bond for abrasive wheels; cements and concretes; insulating compositions; fertilizer; poultry grit; tarred roofing materials.

felsic rock. A silicon-rich igneous rock that contains only a small percentage of iron and magnesium. Granite is the most abundant example. Felsic rocks dominate the crusts of continents.

felt. A compressed, porous, nonwoven fabric usually made of wool and used as a vibration damper and caulking agent. Its moisture-absorbing property is utilized in the drying section of fourdrinier machines.

FEMA. See Flavor Extract Manufacturers Association.

femto-. Prefix meaning one-quadrillionth (10^{-15}). Laser pulses as short as 30 femtoseconds have been produced.

fenac. (2,3,6-trichlorophenylacetic acid). CAS: 85-34-7.
Use: Herbicide.

fenamidone.
CAS: 161326-34-7. $C_{17}H_{17}N_3OS$.
Hazard: Moderately toxic by ingestion and skin contact. Low toxicity by inhalation.
Use: Agricultural chemical.

fenamiphos. (ethyl-3-methyl-4-(methylthio) phenyl(1-methylethyl) phosphoramide).
CAS: 22224-92-6. $C_{13}H_{22}NO_3PS$.
Properties: Tan, waxy solid. Mw 303.4, mp 49.2C (pure) 40C (tech). Soluble in organic solvents; slightly soluble in water.
Hazard: Toxic by inhalation and skin contact. Questionable carcinogen.
Use: Nematocide and insecticide.

fenchone.
CAS: 1195-79-5. $C_{10}H_{16}O$.
Properties: Oil with camphor-like odor. D 0.9465 (19C), bp 193C. Soluble in ether; insoluble in water. Combustible.
Derivation: A ketone found (1) as dextrofenchone in oil of fennel; (2) as levofenchone in oil of thuja.
Use: Flavoring.

fenchyl alcohol. (fenchol; 2-fenchanol; 1-hydroxyfenchane).

CAS: 512-13-0. $C_{10}H_{18}O$.

Properties: Colorless, oily liquid (*d* and *l* forms) or solid (*dl* form). D approximately 0.96, bp 201C, mp 39C, refr index 1.473.
Derivation: Pine oil, fennel oil, also made synthetically.
Use: Solvent, organic intermediate, odorant, flavoring.

fenhexamid.
CAS: 126833-17-8. $C_{14}H_{17}Cl_2NO_2$.
Hazard: Low toxicity by ingestion, inhalation, and skin contact.
Use: Agricultural chemical.

fenitrothion. [*O,O*-diemthyl-*O*-(3-methyl-4-nitrophenyl)phosphorothioate].
CAS: 122-14-5. $C_9H_{12}NO_5PS$.
Properties: Yellow, oily liquid. Bp 118C (0.05 mm Hg), d 1.322, refr index 1.552. Insoluble in water; soluble in most organic solvents except aliphatics.
Hazard: Cholinesterase inhibitor, use may be restricted.
Use: Insecticide.

fenitrothion–malathion mixture.
CAS: 8067-98-9. $C_{10}H_{19}O_6PS_2 \cdot C_9H_{12}NO_5PS$.
Hazard: Human systemic effects.

Fenn, John Bennett. (1917–2010). An American born in New York City who won the Nobel Prize for chemistry in 2002 for his pioneering work concerning for the development of methods for identification and structure analyses of biological macromolecules. Awarded an undergraduate degree from Berea College and Allied Schools, and a Ph.D. from Yale University. He was a long-time professor at Princeton University.

fenoprop. See silvex.

Fenox. A flowable system fungicide.
Use: Controls disease caused by Oomycete class of fungi.

fenpropidine.
CAS: 67306-00-7. $C_{19}H_{31}N$.
Hazard: Moderately toxic by ingestion, inhalation, and skin contact.

fenprostalene.
CAS: 69381-94-8. $C_{23}H_{30}O_6$.
Use: Drug (veterinary); food additive.

fenpyrate.
CAS: 55512-33-9. $C_{19}H_{23}ClN_2O_2S$.
Hazard: Moderately toxic by ingestion and skin contact.
Use: Agricultural chemical.

fenson.   (*p*-chlorophenyl benzene sulfonate; murvesco).
CAS: 80-38-6. $ClC_6H_4OSO_2C_6H_5$.
Properties: Colorless crystals. Mp 61–62C. Soluble in organic solvents; insoluble in water.
Use: Acaricide.

fensulfothion.   (*O,O*-diethyl-*O*-[*p*-(methylsulfinyl)phenyl]phosphorothioate).
CAS: 115-90-2. $C_{11}H_{17}O_4PS_2$.
Properties: Liquid. Bp 138C (0.01 mm Hg).
Hazard: Cholinesterase inhibitor. Questionable carcinogen.
Use: Insecticide, especially for nematocide control.

fentanyl.   (phentanyl; *n*-phenyl-*n*-[1-(2-phenylethyl)-4-piperidinyl]propanamide; *n*-(1-phenylethyl-4-piperidyl)propionanilide; *n*-(1-phenylethyl-4-piperidinyl)-*n*-phenylpropionamide; *n*-phenyl-*n*-{1-(2-phenylethyl)piperidin-4-yl} propanamide).
CAS: 437-38-7. $C_{22}H_{28}N_2O$.
A potent narcotic analgesic. It is primarily a mu-opioid agonist.
Hazard: Toxic.
Use: Adjunct to general anesthetics, an anesthetic for induction and maintenance.

fenthion.   (generic name for *O,O*-diethyl-*O*-[4-(methylthio)-*m*-tolyl]phosphorothioate (generic)).
CAS: 55-38-9. $(CH_3O)_2P(S)OC_6H_3(CH_3)SCH_3$.
Properties: Brown liquid. Bp 105C (0.01 mm Hg). Insoluble in water; soluble in most organic solvents.
Hazard: Toxic by ingestion, inhalation, and skin absorption; use may be restricted, cholinesterase inhibitor. Toxic by skin absorption; questionable carcinogen.
Use: Insecticide, acaricide.

fenticlor.   (2,2′-thiobis(4-chlorophenol); novex). $C_{12}H_8Cl_2O_2S$.
Properties: Acicular crystals. Mp 170C. Soluble in alcohol, hot benzene, and sodium hydroxide solution.
Derivation: Cholorination of bis(2-hydroxyphenyl) sulfide.
Hazard: Toxic by ingestion.
Use: Fungicide.

Fenton reaction.   Oxidation of α-hydroxy acids with hydrogen peroxide and ferrous salts (Fenton's reagent) to α-keto acids or of 1,2-glycols to hydroxy aldehydes.

Fenton's reagent.   A solution of sulfuric acid and a ferrous salt.
Use: Oxidation of polyhydric alcohols.

fenuron.   (generic name for 3-phenyl-1,1-dimethylurea).
CAS: 101-42-8. $C_6H_5NHCON(CH_3)_2$.
Properties: White, crystalline solid. Mp 127–129C. Almost insoluble in water (0.3% at 25C), sparingly soluble in hydrocarbon solvents, stable toward oxidation and moisture.
Use: Weed and brush killer.

fenvalerate.   (cyano(3-phenoxyphenyl)methyl-4-chloro-α-(1-methylethyl)phenylacetate).
CAS: 51630-58-1.
Hazard: Questionable carcinogen.
Use: Insecticide.

fenylfosfin.   See phenylphosphine.

FEP resin.   Abbreviation for fluorinated ethylenepropylene resin.

ferbam.   (generic name for ferric dimethyldithiocarbamate).
CAS: 14484-64-1. $[(CH_3)_2NCSS]_3Fe$.
Properties: Black or dark-colored, fluffy powder. Decomposes above 180C. Usually readily dispersible but very slightly soluble in water; pH of saturated solution 5.0.
Derivation: By addition of carbon disulfide to an alcoholic solution of dimethylamine and precipitation with a ferric salt.
Grade: 76% wettable powder, 87% technical powder.
Hazard: Irritant to eyes and mucous membranes. Questionable carcinogen.
Use: Fungicide.

Fermate.   (ferric diethyl dithiocarbamate). A wettable powder containing 76% ferbam.
Use: Fruit fungicide.

fermentation.   A chemical change induced by a living organism or enzyme, specifically bacteria or the microorganisms occurring in unicellular plants such as yeast, molds, or fungi. The reaction usually involves the decomposition of sugars and starches to ethanol and carbon dioxide, the acidulation of milk, or the oxidation of nitrogenous organic compounds. The basic reaction is catalyst $C_6H_{12}O_6 \rightarrow 2C_2H_5OH + 2CO_2$ Enzymes are usually involved in such reactions; with yeast, the effective enzyme is zymase. Fermentation is essential in the preparation of breads and other food products and in the manufacture of beer, wine, and other alcoholic beverages, as well as of citric acid, gluconic acid, sodium gluconate, and synthetic biopolymers. Much of the industrial alcohol used in the U.S. is made by fermentation of blackstrap molasses, a by-product of sugar manufacture. Antibiotics are produced by various forms of microorganisms active in molds, especially bacteria and *actinomycetes*. The activated sludge process for sewage digestion is a form of fermentation. A continuous fermentation process for deriving edible protein from petroleum has been introduced. Fermentation is also used in making synthetic amino acids. Research in this field is being directed toward conversion of agricultural, urban, and animal wastes to fuels by fermentation processes.

See yeast; sewage sludge; antibiotics; bacteria; biotechnology.

fermentation alcohol.   See ethanol.

Fermi, Enrico.   (1901–1954). An Italian physicist who later became a U.S. citizen. He developed a statistical approach to fundamental problems of physical chemistry based on Pauli's exclusion principle. He discovered induced or artificial radioactivity resulting from neutron impingement, as well as slow or thermal neutrons. He was professor of physics at Columbia (1939) and awarded the Nobel Prize in physics in 1938. He was the first to achieve a controlled nuclear chain reaction, directed the construction of the first nuclear reactor at the University of Chicago (1942), and worked on the atomic bomb at Los Alamos. He also carried on fundamental research on subatomic particles using sophisticated statistical techniques. Element 100 (fermium) is named after him.

fermion.   A type of elementary particle with a half-integral spin, that displays wavefunction antisymmetry and obeys Fermi-dirac statistics. Fermions are restricted by the pauli exclusion principle such that two fermions may not occupy the same quantum state. Fermions generally form atomic and nuclear structure.

fermium.   (Fm). Element 100, aw 254, valence of 3, half-life 3 h. A synthetic radioactive element with atomic number 100 discovered in 1952. Fermium has since been prepared in a nuclear reactor by irradiating californium, plutonium, or einsteinium with neutrons in a cyclotron; by bombarding uranium with accelerated oxygen ions; and by other nuclear reactions. The element is named for Enrico Fermi. It has chemical properties similar to those of the rare earth erbium. Isotopes are known with mass numbers 254, 255, and 256.

Use: Tracer studies.

See actinide element.

Ferrario reaction.   Formation of phenoxathiins by cyclization of diphenyl ethers with sulfur in the presence of aluminum chloride.

ferrate.   See ferrite (2).

ferredoxin.   An iron-containing protein thought to be involved in photosynthesis as an acceptor of energy-rich electrons from chlorophyll. It occurs in green plants and in bacteria that metabolize elemental hydrogen.

ferric acetate, basic.   (iron acetate, basic).
CAS: 10450-55-2. $Fe(C_2H_3O_2)_2OH$.
Properties: Red powder. Soluble in alcohol and acids; insoluble in water. Combustible.
Derivation: Action of acetic acid on ferric hydroxide, with subsequent crystallization.
Use: Mordant in dyeing textiles, wood preservative, medicine.

ferric acetylacetonate.
CAS: 14024-18-1. $Fe[OC(CH_3):CHC(O)CH_3]_3$.
Properties: Crystalline powder. Mp 179–182C. Slightly soluble in water; soluble in most organic solvents, resistant to hydrolysis, a chelating nonionizing compound. Combustible.
Use: Moderating and combustion catalyst, solid fuel catalyst, bonding agent, curing accelerator, intermediate.

ferric ammonium alum.   See ferric ammonium sulfate.

ferric ammonium citrate.   (iron ammonium citrate).
Properties: Thin, transparent, garnet-red scales or granules or a brownish-yellow powder; odorless (or slight ammonia odor); saline, mildly ferruginous taste. Deliquescent, affected by light. Soluble in water; insoluble in alcohol. Combustible.
Derivation: Addition of citric acid to ferric hydroxide, addition of ammonium hydroxide, followed by filtration.
Grade: Technical.
Use: Medicine, blueprint photography, feed additive.

ferric ammonium ferrocyanide.
CAS: 25869-00-5. $C_6FeN_6 \bullet Fe \bullet H_4N$.
Hazard: Low toxicity.

ferric ammonium oxalate.   (iron ammonium oxalate; ammonioferric oxalate).
CAS: 14221-47-7. $(NH_4)_3Fe(C_2O_4)_3 \bullet 3H_2O$.
Properties: Green crystals. Soluble in water and alcohol; sensitive to light.
Derivation: Interaction of ammonium binoxalate and ferric hydroxide.
Hazard: Irritant to skin and mucous membranes.
Use: Blueprint photography.

ferric ammonium sulfate.   (iron ammonium sulfate; ferric ammonium alum; ammonio ferric sulfate).
CAS: 10138-04-2. $FeNH_4(SO_4)_2 \bullet 12H_2O$.
Properties: Lilac to violet, efflorescent crystals. D 1.71, mp 39–41C, bp loses $12H_2O$ at 230C. Soluble in water; insoluble in alcohol.
Derivation: By mixing solutions of ferric sulfate and ammonium sulfate followed by evaporation and crystallization.

Use: Medicine, analytical chemistry, textile dyeing (mordant).

ferric arsenate.
CAS: 10102-49-5. $FeAsO_4 \cdot 2H_2O$.
Properties: Green or brown powder. D 3.18, decomposes on heating. Insoluble in water; soluble in dilute mineral acids. Nonflammable.
Hazard: Toxic by ingestion and inhalation, strong irritant.
Use: Insecticide.

ferric arsenite.
CAS: 63989-69-5. $2FeAsO_3 \cdot Fe_2O_3 \cdot 5H_2O$.   A basic salt of variable composition.
Properties: Brownish-yellow powder. Soluble in acids; insoluble in water. Nonflammable.
Hazard: Toxic by ingestion and inhalation, strong irritant.
Use: Combined with ammonium citrate (ferric ammonium citrate) and used in medicine.

ferric bromide.   (ferric tribromide; iron bromide).
CAS: 10031-26-2. $FeBr_3$.
Properties: Dark red, deliquescent crystals. Mp (sublimes). Soluble in water, alcohol, and ether.
Derivation: By the action of bromine on iron filings.
Use: Bromination catalyst.

ferric chloride, anhydrous.   (ferric trichloride; ferric perchloride; iron chloride, iron trichloride, iron perchloride).
CAS: 7705-08-0. $FeCl_3$.
Properties: Black-brown solid. D 2.898 (25C), mp 306C (partly decomposes), bp 319C. Soluble in water, alcohol, glycerol, methanol, and ether. Noncombustible.
Derivation: Action of chlorine on ferrous sulfate or chloride.
Grade: Anhydrous 96%; 42 degrees Bé solution, photographic and sewage grades.
Hazard: Toxic by ingestion, strong irritant to skin and tissue.
Use: Treatment of sewage and industrial wastes; etching agent for engraving, photography, and printed circuitry; condensation catalyst in Friedel-Crafts reactions; mordant; oxidizing, chlorinating, and condensing agent; disinfectant; pigment; feed additive; water purification.

ferric chromate.   (iron chromate).
CAS: 10294-52-7. $Fe_2(CrO_4)_3$.
Properties: Yellow powder. Soluble in acids (especially hydrochloric) insoluble in water and alcohol.
Derivation: By adding sodium chromate to a solution of a ferric salt.
Hazard: Carcinogenic. Toxic by ingestion and inhalation. Strong irritant. Moderate fire risk by reaction with reducing agents.
Use: Metallurgy, ceramics (color), paint pigment.

ferric citrate.   (iron citrate).
CAS: 2338-05-8. $FeC_6H_5O_7 \cdot 5H_2O$.
Properties: Reddish-brown scales. Keep away from light. Soluble in water; insoluble in alcohol.
Derivation: By the action of citric acid on ferric hydroxide and crystallization.
Use: Medicine, blueprint paper.

ferric dichromate.   (iron dichromate; ferric bichromate). $Fe_2(Cr_2O_7)_3$.
Properties: Reddish-brown granules. Soluble in water and acids.
Derivation: By heating aqueous chromic acid and moist ferric hydroxide.
Hazard: Toxic by inhalation and ingestion, strong irritant. Moderate fire risk by reaction with reducing agents.
Use: Preparation of pigments.

ferric   dimethyldithiocarbamate.   See ferbam.

ferric ferrocyanide.   (iron ferrocyanide; Prussian blue). Blue pigment described under iron blue.

ferric fluoride.   (iron fluoride).
CAS: 7783-50-8. FeF_3.
Properties: Green crystals. D 3.52. Soluble in dilute hydrogen fluoride; insoluble in alcohol and ether.
Hazard: Strong irritant.
Use: Ceramics (porcelain, pottery), catalyst.

ferric fluoroborate.
Use: Rebuilding of worn iron parts, such as cylinders, stereotypes, and electrotypes; plating of solder iron tips.

ferric glycerophosphate.   (iron glycerophosphate). $Fe_2[C_3H_5(OH)_2PO_4]_3 \cdot xH_2O$.
Properties: Yellowish scales; odorless; nearly tasteless. Soluble in water; insoluble in alcohol.
Use: Pharmaceuticals.

ferrichrome.
CAS: 15258-80-7. $C_{27}H_{42}FeN_9O_{12}$. A cyclic iron chelate compound.
Properties: Yellow needles. Soluble in water and methanol; slightly soluble in alcohol, acetone, and chloroform.
Derivation: Isolated from rust fungus in 1952, synthesized in 1969.
Use: Growth-promoting factor in medicine.

ferric hydroxide.   (ferric hydrate; iron hydroxide; iron hydrate; iron oxide, hydrated; ferric oxide, hydrated).
CAS: 20344-49-4. $Fe(OH)_3$.
Properties: Brown flocculant precipitate that dries as the oxide. D 3.4–3.9, mp loses water at approximately 500C. Soluble in acids; insoluble in water, alcohol, and ether. Noncombustible.

Derivation: Addition of ferrous sulfate solution to ammonia solution.

Use: Water purification, manufacturing pigments, rubber pigment, catalyst.

ferric hypophosphite.   (iron hypophosphite).
CAS: 7783-84-8.   Fe(H$_2$PO$_2$)$_3$.
Properties: White or grayish-white powder; odorless; tasteless. Slightly soluble in water; more soluble in boiling water.
Hazard: Explosion may occur if triturated or heated with nitrates, chlorates, or other oxidizing agents.

ferric naphthenate.
Properties: A heavy-metal soap. Combustible.
Derivation: Fusion method by heating naphthenic acids with the metallic oxide.
Use: Conditioning and waterproofing agent, sludge preventive, fungicide, and paint drier.
See soap (2).

ferric nitrate.   (iron nitrate).
CAS: 10421-48-4.   Fe(NO$_3$)$_3$•9H$_2$O.
Properties: Violet crystals. D 1.684, mp 47.2C, decomposes at 125C. Soluble in water and alcohol.
Derivation: Action of concentrated nitric acid on scrap iron or iron oxide and crystallization.
Hazard: Dangerous fire risk in contact with organic materials. Strong oxidant and irritant.
Use: Dyeing (mordant for buffs and blacks), tanning, analytical chemistry.

ferric octoate.   See soap (2).

ferric oleate.   (iron oleate).
CAS: 1120-45-2.   Fe(C$_{18}$H$_{33}$O$_2$)$_3$.
Properties: Brownish-red lumps. Soluble in alcohol, ether, and acids; insoluble in water. Combustible.
See soap (2).

ferric oxalate.
CAS: 2944-66-3.   Fe$_2$(C$_2$O$_4$)$_3$.
Properties: Pale-yellow amorphous scales or powder; odorless. Decomposes on heating to 100C. Soluble in water and acids; insoluble in alkali. Combustible.
Hazard: Toxic by ingestion and inhalation.
Use: Catalyst in making O$_2$, silvertone photographic printing papers.

ferric oxide.   (ferric oxide, red; iron oxide; red iron trioxide; ferric trioxide).
CAS: 1309-37-1.   Fe$_2$O$_3$.
Properties: Dense, dark red powder or lumps. D 5.12–5.24, mp 1565C. Soluble in acids; insoluble in water.
Grade: Technical, 99.5% pure, electronic.
Use: Metallurgy, gas purification, paint and rubber pigment, component of thermite, polishing compounds, mordant, laboratory reagent, catalyst (*p*-hydrogen), feed additive, electronic pigments for

TV, permanent magnets, memory cores for computers, magnetic tapes.
See iron oxide reds.

ferric perchloride.   See ferric chloride.

ferric phosphate.   (iron phosphate).
CAS: 10045-86-0.   FePO$_4$•2H$_2$O.
Properties: Yellowish-white powder. D 2.87. Insoluble in water; soluble in acids.
Derivation: By adding a solution of sodium phosphate to a solution of ferric chloride. The product is filtered and then dried.
Use: Fertilizers, feed and food additive.

ferric pyrophosphate.   (iron pyrophosphate).
CAS: 10058-44-3.   Fe$_4$(P$_2$O$_7$)$_3$•xH$_2$O.
Properties: Yellowish-white powder. Insoluble in water; soluble in dilute acid. Contains 24% iron min, not to be confused with ferric pyrophosphate, soluble.
Use: Source of nutritional iron, catalyst, pigments, flame-retardant.

ferric pyrophosphate, soluble.   A combination of ferric pyrophosphate and sodium citrate.
Properties: Apple-green crystals. Very soluble in water; insoluble in alcohol. Protect from light, 11% iron.
Use: Feed additive.

ferric resinate.   (iron resinate).
Properties: Reddish-brown powder. Soluble in ligroin, carbon disulfide, ether, oil of turpentine; slightly soluble in alcohol; insoluble in water.
Use: Drier (paints, varnish).
See soap (2).

ferric sodium oxalate.   (iron sodium oxalate).
Na$_3$Fe(C$_2$O$_4$)$_3$•5.5H$_2$O.
Properties: Emerald-green crystals, decomposed by heat or light. D 1.973 (18C), decomposes at 300C, protect from light. Soluble in water and alcohol.
Derivation: By the interaction of sodium acid oxalate and ferric hydroxide.
Use: Photography, blueprinting.

ferric sodium pyrophosphate.   See sodium ferric pyrophosphate.

ferric stearate.   (iron stearate).
CAS: 555-36-2.   Fe(C$_{18}$H$_{35}$O$_2$)$_3$.
Properties: Light brown powder. Soluble in alcohol and ether; insoluble in water. Combustible.
Derivation: Interaction of solutions of ferric sulfate and sodium stearate.
Use: Varnish driers, photocopying.
See soap (2).

ferric sulfate.   (iron sulfate; ferric trisulfate; iron tersulfate; iron persulfate).

CAS: 10028-22-5.   (1) $Fe_2(SO_4)_3$, (2) $Fe_2(SO_4)_3 \cdot 9H_2O$.
Properties: Yellow crystals or grayish-white powder. D (1) 3.097, (2) 2.0–2.1, mp decomposes at 480C, (1) slightly soluble in water, (2) very soluble in water. Keep well closed and protected from light. Noncombustible.
Derivation: By adding sulfuric acid to ferric hydroxide.
Grade: Technical, CP, partly hydrated.
Use: Pigments, reagent, etching aluminum, disinfectant, textiles (dyeing and calico printing), flocculant in water and sewage purification, soil conditioner, polymerization catalyst, metal pickling, chelated iron products, intermediate.

ferric tallate.   See soap (2).

ferric tannate.   (iron tannate; iron gallotannate). $Fe_2(C_{14}H_7O_9)(OH)_3$.
Properties: Dark brown or bluish-black powder. Variable composition. Soluble in alkalies and dilute acids; insoluble in water, alcohol, and ether. Combustible.
Derivation: Interaction of ferric acetate and tannic acid solutions.
Use: Medicine.

ferric tribromide.   See ferric bromide.

ferric trichloride.   See ferric chloride.

ferric trioxide.   See ferric oxide.

ferric trisulfate.   See ferric sulfate.

ferric vanadate.   (iron metavanadate). $Fe(VO_3)_3$.
Properties: Grayish-brown powder. Soluble in acids; insoluble in water and alcohol. Noncombustible.
Derivation: By adding a solution of a ferric salt to the liquor obtained by leaching vanadium ores with caustic soda solution or by lixivating the slags obtained when vanadium ores are fused with soda ash, etc.
Grade: Technical.
Use: Metallurgy.

ferric versenate.   See iron(III)-edta complex.

ferrite.   (1) Iron in the body-centered cubic form; commonly occurs in steels, cast iron, and pig iron at approximately 910C. α and β iron are the common varieties of ferrite, and the name is also applied to Δ iron. (2) A compound, a multiple oxide, of ferric oxide with another oxide, as sodium ferrite, $NaFeO_2$, but more commonly a multiple oxide crystal. Ferrites are made by dissolving hydrated ferric oxide in concentrated alkali solution; by fusing ferric oxide with alkali metal chloride, carbonate, or hydroxide; or by simply heating metal oxides with

ferric oxide. Ceramic ferrites are made by press-forming powdered ingredients (with a binder) into a sheet, then sintering or firing.
Use: The oxide ferrites in rectifiers on memory and record tapes, for permanent magnets, semiconductors, insulating materials, dielectrics, high-frequency components, and various related uses in radio, television, radar, computers, and automatic control systems.

ferritin.   An iron–protein complex in which iron is stored in the intestinal mucosa, spleen, and liver. It regulates iron transport from the intestinal lumen to plasma in circulating blood.
Derivation: Prepared from rat liver protein by precipitation with a cadmium salt.
Hazard: Questionable carcinogen; neoplastigen; mutagen.

ferro-alloy.   An alloy of iron with some element other than carbon used as a vehicle for introducing such an element into steel during its manufacture. The element may alloy with the steel by solution or as the carbide, neutralize the harmful impurities by combining with them, and separating from the steel as flux or slag before solidification.

ferroboron.   A ferro-alloy used as a hardening agent in special steels. It also is an efficient deoxidizer. Boron steel is used in controlling the operating rate of nuclear reactors. Two grades are available, 10 and 17% boron.

Ferrocarbo.   Briquetted or granular silicon carbide.
Use: Cupola addition in the production of gray iron or as a ladle addition to steel. Decomposes into its component elements and acts as a powerful deoxidizer and graphitizer. Machinability and strength of the iron or steel are increased with no loss of hardness.

ferrocene.   (dicyclopentadienyliron).
CAS: 102-54-5. $(C_5H_5)_2Fe$. A coordination compound of ferrous iron and two molecules of cyclopentadiene in which the organic portions have typically aromatic chemical properties. Its activity is intermediate between phenol and anisole. The first compound shown to have the "sandwich" structure found in certain types of metallocene molecules. Two structures of ferrocene are shown below:

Properties: Orange, crystalline solid; camphor-like odor. Mp 173–174C; resists pyrolysis at 400C. Insoluble in water; soluble in benzene, ether, and alcohol. Iron content 29.4–30.6%.

Derivation: From ferrous chloride and cyclopentadiene sodium.

Hazard: Moderate fire risk. Evolves toxic products on decomposition and heating.

Use: Additive to fuel oils to improve efficiency of combustion and eliminate smoke, antiknock agent, catalyst, coating for missiles and satellites, high-temperature lubricant, intermediate for high-temperature polymers, UV absorber.
See metallocene.

ferrocenecarboxylic acid ethyl ester. See ethyl ferrocenoate.

1,1′-ferrocenedicarboxylic acid diethyl ester. See 1,1′-diethyl ferrocenoate.

1,1′-ferrocenedicarboxylic acid dimethyl ester. See dimethyl ferrocenoate.

1,1′-ferrocenediyl dichlorosilane.
$(C_5H_5)_2FeSiCl_2$. An experimental ferrocene derivative that prevents oxidative deterioration of the surfaces of photoelectrodes with which it is in contact. It increases the stability of light-sensitive electrodes in energy-conversion reactions occurring in liquid media.

ferrocenoyl chloride. See chlorocarbonyl ferrocene.

ferrocenoyl dichloride. See 1,1′-dichlorocarbonyl ferrocene.

ferrocenylborane polymer.
Properties: Long-term heat resistance at 315C, short-term stability approximately 815C, good resistance to oxidation and hydrolysis, contains up to 30% iron directly combined.
Use: Specialty plastics, coatings, fibers; ablative material for space vehicles.

ferrocenyl methyl ketone. See acetylferrocene.

ferrocerium. A pyrophoric alloy of iron and misch metal.

ferrochromium. (ferrochrome). An alloy, composed principally of iron and chromium, used as a means of adding chromium to steels (low, medium, and high-carbon) and cast iron. Available in several classifications and grades, generally containing between 60 and 70% chromium, in crushed sizes and lumps up to 75 pounds that readily dissolve in molten steel.

ferroconcrete. See concrete.

ferroelectric. A crystalline material such as barium titanate, monobasic potassium phosphate, or potassium-sodium tartrate (Rochelle salts) that, over certain limited temperature ranges, has a natural or inherent deformation (polarization) of the electrical fields or electrons associated with the atoms and groups in the crystal lattice. This results in the development of positive and negative poles and a consequent "direction" of polarization, which can be reversed when the crystal is exposed to an external electric field. Ferroelectric crystals are internally strained and, as a consequence, show unusual piezoelectric and elastic properties.
Use: Capacitors, transducers, computer technology.
See ceramic, ferroelectric.

ferroin chelation group. A functional group characteristic of heterocyclic ring nitrogen compounds:

$$=N-\overset{\|}{C}-\overset{\|}{C}-N=$$

Among such compounds are 2,2′-bipyridine; 1,10-phenanthroline; and the 2-pyridyl triazines. These provide a large number of terminal ($\equiv$C–H) groups in which the hydrogen can be replaced by many chemical groupings (carboxyl, hydroxyl, halogen, etc.). Thus, synthesis of an almost endless number of substituted ferroin reactants is possible. About 200 such chelation reagents have been synthesized. Ferroin chelation chemicals in general form complex undissociated cations with divalent metal ions, e.g., $[(C_{12}H_8N_2)_3Fe]^{2+}$.

ferromagnesite. An iron-bearing variety of magnesite.
Use: Refractory owing to its ability to bond under heat.

ferromagnetic oxide. See ferrite (2).

ferromanganese. An alloy consisting of manganese (approximately 48%), plus iron and carbon.
Available forms: Standard, low-carbon, and medium-carbon grades in ground, crushed, and lump sizes ranging from 80 mesh to 75-lb lumps, suitable for ladle or furnace addition.
Use: Vehicle for adding manganese to steel.
See nodules.

ferromolybdenum. An alloy, composed largely of iron and molybdenum, used as a means of adding molybdenum to steel. Engineering steels rarely contain more than 1% molybdenum, stainless steels may contain 3%, and tool steels as much as 10%. Ferromolybdenum is available in several grades in which molybdenum ranges from 55 to 75% and the maximum carbon content is 1.10%, 0.60%, or

2.50%. It is generally added in the furnace since it does not oxidize under steelmaking conditions. Mp approximately 1630C. Available in crushed sizes up to one inch.

ferroniobium. An alloy of iron and niobium made by reducing the ore columbite with silicon.
Use: Stainless steels and other alloys for welding rods.

ferrophosphorus. An alloy of iron and phosphorus used in the steel industry for adjustments of phosphorus content of special steels.
Grade: (1) 18% phosphorus, (2) 25% phosphorus.
Use: In preventing thin sheets from sticking together when rolled and annealed in bundles.

ferrosilicon. An alloy of iron and silicon used to add silicon to steel and iron, d 5.4, insoluble in water. Small quantities of silicon deoxidize the iron, and larger amounts impart special properties.
Available forms: Six grades containing from 20 to 95% silicon. The 20% grade is made in a blast furnace, but grades of higher silicon content are made in electric furnaces.
Hazard: Ferrosilicon containing from 30 to 90% silicon is flammable and evolves gases in the presence of moisture.
Use: Pidgeon process for producing metallic magnesium.

ferrosoferric oxide. See iron oxide, black.

ferrotitanium. An alloy composed principally of iron and titanium, used to add titanium to steel. It is often made from titanium scrap. Three classifications are available: low, high, and medium carbon content. Furnished in various lump, crushed, and ground sizes.

ferrotungsten. An alloy of iron and tungsten used as a means of adding tungsten to steel. Contains 70 to 80% tungsten and no more than 0.6% carbon. Melting range 1648–2750C, dissolves readily in molten steel. Furnished in ground and crushed sizes up to one inch.
See tungsten steels.

ferrous acetate. (iron acetate).
$Fe(C_2H_3O_2)_2 \cdot 4H_2O$.
Properties: Greenish crystals when pure and unexposed to air; usually partly brown from action of air. Soluble in water and alcohol. Oxidizes to basic ferric acetate in air. Combustible.
Derivation: Action of acetic acid or pyroligneous acid on iron with subsequent crystallization.
Use: Textile dyeing, medicine, dyeing leather, wood preservative.

ferrous ammonium sulfate. (Mohr's salt; iron ammonium sulfate). $Fe(SO_4) \cdot (NH_4)_2SO_4 \cdot 6H_2O$.

Properties: Light green crystals. Soluble in water; insoluble in alcohol. D 1.865, decomposes at 100–110C, deliquescent. Affected by light.
Derivation: By mixing solutions of ferrous sulfate and ammonium sulfate, followed by evaporation and subsequent crystallization.
Use: Analytical chemistry, metallurgy.

ferrous arsenate. (iron arsenate).
CAS: 10102-50-8. $Fe(AsO_4)_2 \cdot 6H_2O$.
Properties: Green, amorphous powder. Insoluble in water; soluble in acids.
Derivation: Interaction of solutions of sodium arsenate and ferrous sulfate.
Use: Insecticide.

ferrous ascorbate.
CAS: 14536-17-5.
Properties: Blue-violet solid.
Hazard: A nuisance dust.
Use: Food additive.

ferrous bromide. (iron bromide).
$FeBr_2 \cdot 6H_2O$.
Properties: Green, crystalline powder; very deliquescent. D 4.636, mp 27C. Readily oxidized in moist air; soluble in water and alcohol.
Derivation: Action of bromine on iron filings.
Use: Polymerization catalyst.

ferrous carbonate.
CAS: 563-71-3. $CFeO_3$.
Properties: White solid; odorless; gray solid. Decomposes yielding CO_2 + FeO at 2C. Insol in H_2O; sol acids to give CO_2; sol in H_2O saturated with CO_2 to give $Fe(HCO_3)_2$ which then oxidizes.
Hazard: A nuisance dust.
Use: Food additive.

ferrous chloride. (iron chloride; iron dichloride; iron protochloride).
CAS: 7758-94-3. (1) $FeCl_2$, (2) $FeCl_2 \cdot 4H_2O$.
Properties: Greenish-white crystals. D (1) 3.16 (25C), (2) 1.93, mp (1) 670–674C, deliquescent, readily oxidized. Soluble in alcohol and water.
Derivation: Action of hydrochloric acid on an excess of iron, with subsequent crystallization.
Use: Mordant in dyeing, metallurgy, pharmaceutical preparations, manufacture of ferric chloride, sewage treatment.

ferrous citrate.
CAS: 23383-11-1. $C_6H_6FeO_7$.
Properties: White crystals or sltly colored powder.
Use: Food additive.

ferrous-2-ethylhexoate. A paint drier.
See soap (2).

ferrous fluoride. (iron fluoride).
CAS: 7789-28-8. FeF_2.

Properties: Green crystals. D 4.09. Soluble in acids; slightly soluble in water; insoluble in alcohol and ether.
Hazard: Strong irritant.
Use: Ceramics, catalyst.

ferrous fumarate.
CAS: 141-01-5. $FeC_4H_2O_4$.
Anhydrous salt of a combination of ferrous iron and fumaric acid, stable, odorless, substantially tasteless. Reddish-brown, anhydrous powder, contains 33% iron by weight, does not melt at temperatures up to 280C, insoluble in alcohol, very slightly soluble in water. Combustible.
Grade: USP.
Use: Dietary supplement.

ferrous gluconate. (iron gluconate).
CAS: 299-29-6. $Fe(C_6H_{11}O_7)_2 \cdot 2H_2O$.
Properties: Yellowish-gray or pale-greenish-yellow fine powder or granules with slight odor, solution (1 in 20) is acid to litmus. Soluble in water and glycerol; insoluble in alcohol. Combustible.
Grade: Pharmaceutical, NF.
Use: Feed and food additive, vitamin tablets.

ferrous iodide. (iron iodide; iron protoiodide).
$FeI_2 \cdot 4H_2O$.
Properties: Dark violet to black hygroscopic leaflets. Soluble in water and alcohol. D 2.873, decomposes at 90–98C, mp (anhydrous) 177C, deliquescent, affected by light.
Derivation: By the action of iodine on iron filings.
Use: Manufacture of alkali metal iodides, pharmaceutical preparations, catalyst.

ferrous lactate. (iron lactate).
CAS: 5905-52-2. $Fe(C_3H_5O_3) \cdot 3H_2O$.
Properties: Greenish-white crystals; slight peculiar odor. Soluble in water; insoluble in alcohol. Deliquescent, affected by light. Combustible.
Derivation: By interaction of calcium lactate with ferrous sulfate or direct action of lactic acid on iron filings.
Use: Food additive and dietary supplement.

ferrous metal. A magnetic metal derived from iron or steel.

ferrous naphthenate. A soap based on mixed naphthenic acids. Available commercially as a liquid containing 6% iron.
See soap (2).

ferrous octoate. A paint drier.
See soap (2).

ferrous oxalate. (iron oxalate).
$FeC_2O_4 \cdot 2H_2O$.
Properties: Pale-yellow, crystalline powder; odorless. Soluble in acids; insoluble in water. D 2.28, decomposes at 160C releasing carbon monoxide.

Derivation: By the interaction of solutions of ferrous sulfate and sodium oxalate.
Hazard: Toxic. Evolves carbon monoxide on heating.
Use: Photographic developer, pigment in glass, plastics, paints.

ferrous oxide. (iron monoxide).
CAS: 1345-25-1. FeO.
Properties: Black powder. D 5.7, mp 1420C. Insoluble in water; soluble in acid.
Derivation: Prepared from the oxalate by heating, but the product contains some ferric oxide.
Use: Catalyst, glass colorant, steel manufacture.

ferrous phosphate. $Fe_3(PO_4)_2 \cdot 8H_2O$.
Properties: Bluish-gray powder. D 2.58. Soluble in inorganic acids; insoluble in water. Hygroscopic.
Use: Catalyst, ceramics.

ferrous phosphide. (iron phosphide).
CAS: 1310-43-6. Fe_2P.
Properties: Bluish-gray powder. D 6.56, mp 1290C. Ferromagnetic. Insoluble in water.
Grade: 24–25% phosphorus.
Hazard: Evolves toxic and flammable products on exposure to moisture or acids.
Use: Iron and steel manufacture.

ferrous selenide.
CAS: 1310-32-3. FeSe.
Properties: Black, shiny solid. D 6.8. Almost insoluble in water; soluble in hydrochloric acid, evolving selenium hydride.
Use: Semiconductor technology.

ferrous sulfate. (iron sulfate; iron vitriol; copperas; green vitriol; sal chalybis).
CAS: 7720-78-7. $FeSO_4 \cdot 7H_2O$.
Properties: Greenish or yellow-brown crystals or granules; odorless. Soluble in water with saline taste; insoluble in alcohol. D 1.89, mp 64C, loses $7H_2O$ by 300C, pH 3.7 (10% solution), hygroscopic.
Derivation: (1) By-product from the pickling of steel and many chemical operations, (2) by action of dilute sulfuric acid and iron, (3) oxidation of pyrites in air followed by leaching and treatment with scrap iron.
Method of purification: Recrystallization.
Grade: Technical, anhydrous, CP, USP.
Hazard: Ingestion causes intestinal disorders.
Use: Iron oxide pigment; other iron salts; ferrites; water and sewage treatment; catalyst, especially for synthetic ammonia; fertilizer; feed additive; flour enrichment; reducing agent; herbicide; wood preservative; process engraving.

ferrous sulfide. (iron sulfide; iron protosulfide).
CAS: 1317-37-9. FeS.

Properties: Dark brown or black metallic pieces, sticks, or granules. D 4.75, mp 1195, bp (decomposes). Soluble in acids; insoluble in water.
Derivation: By fusing iron and sulfur.
Use: Generating hydrogen sulfide, ceramics, other sulfides, pigment.
See pyrite.

ferrovanadium.
CAS: 12604-58-9.
An iron–vanadium alloy used to add vanadium to steel. Vanadium is used in engineering steels to the extent of 0.1–0.25% and in high-speed steels to the extent of 1–2.5% or higher. Melting range 1482–1521C. Furnished in a variety of lump, crushed, and ground sizes.
Derivation: By reduction of the vanadium oxide with aluminum or silicon in the presence of iron in an electric furnace.
Grade: Available containing from 50 to 80% vanadium.
Hazard: Moderate fire risk.

ferrozirconium. Alloys used in the manufacture of steel. (1) 12–15% zirconium alloy. Approximate analysis: zirconium 12–15%, silicon 39–43%, iron 40–45%; application: steel of high silicon content. (2) 35–40% zirconium alloy. Approximate analysis: zirconium 35–40%, silicon 47–52%, iron 8–12%; application: steel of low silicon content.

ferrum. Latin name for iron, hence the symbol Fe.

fertile material. In nuclear technology, any substance not capable of fission but that can be converted into a fissionable material in a nuclear reactor. ^{238}Uranium (converted to ^{239}plutonium) and ^{232}thorium (converted to ^{233}uranium) are the most important fertile materials.

fertile nuclide. A nuclide that can be transformed into one that undergoes spontaneous fission.

fertilizer. A substance or mixture that contains one or more of the primary plant nutrients and sometimes also secondary and/or trace nutrients. The primary nutrients are nitrogen (supplied as anhydrous ammonia or solutions containing nitrogen derived from ammonia, ammonium nitrate, or urea), phosphorus (as superphosphates derived from phosphate rock), and potassium (in the form of KCl from sylvite ore or natural brines). Secondary nutrients are calcium, magnesium, and sulfur. Trace elements (iron, copper, boron, manganese, zinc, and molybdenum) are also among the 12 elements considered essential for plant growth. Nitrogen solutions and anhydrous ammonia are used both in fertilizer manufacture and for direct application to the soil. Substantial amounts of both separate materials and mixtures are used in liquid form. Controlled-release fertilizers are those whose particles are coated with

polymeric sulfur by a proprietary process. Their advantages include more uniform supply of nutrient, lower labor costs, and reduced leaching losses in areas of irrigation and high rainfall.
See superphosphate; nutrient solution.

fertilization. The process by which an egg is made capable of generating offspring. It is often synonymous with syngamy.

ferulic acid. (3-(4-hydroxy-3-methoxyphenyl)-2-propenoic acid; 4-hydroxy-3-methoxycinnamic acid; 3-methoxy-4-hydroxycinnamic acid; caffeic acid 3-methyl ether; (E)-3-(4-hydroxy-3-methoxyphenyl)prop-2-enoic acid).
CAS: 1135-24-6. $C_{10}H_{10}O_4$. A plant growth inhibitor.
Properties: The cis-form is a yellow oil, the trans-form precipitates from water solutions as orthorhombic crystals.
Derivation: From cinnamic acid and found in low concentrations in many species of plants.

festucine.
CAS: 25161-91-5. $C_8H_{14}N_2O$.
Hazard: A poison.
Source: Natural product.

FFA. Abbreviation for free fatty acid.
Use: Describing specifications for fatty esters, glycerides, oils, etc.

FFPA. Abbreviation for "free from prussic acid."

FGAN. Fertilizer-grade ammonium nitrate.
Use: In blasting agents, as well as fertilizers, because its coating of kieselguhr and its prilled form, making it safer to handle than the usual grades.

fiber. A fundamental form of solid (usually crystalline) characterized by relatively high tenacity and an extremely high ratio of length to diameter (several hundred to one). Natural fibers are animal, e.g., wool and silk (proteins), vegetable, e.g., cotton (cellulose), and mineral (asbestos). Cotton fiber is called staple and rarely exceeds 2 inches in length. Semisynthetic fibers include rayon and inorganic substances extruded in fibrous form, such as glass, boron, boron carbide, boron nitride, carbon, graphite, aluminum silicate, fused silica, and some metals (steel). Synthetic fibers are made from high polymers (polyamides, polyesters, acrylics, and polyolefins) by extruding from spinnerettes (nylon, "Orlon," etc.). Some are being used in specialty papers, though the primary use is in textile fabrics.
See "Fiberfax."
Metal fibers are used in several ways: (1) As whiskers, which are single-crystal fibers up to 2 inches long having extremely high tensile strength; they are made from tungsten, cobalt, tantalum, and

other metals and are used largely in composite structures for specialized functions. (2) As filaments, which are alloys drawn through diamond dies to diameters as small as 0.002 cm; steel for tire cord and antistatic devices has been developed for such applications. (3) In biconstituent structures composed of a metal and a polymeric material; for example, aluminum filament covered with cellulose acetate butyrate.

Hollow fibers of cellulose acetate and nylon are used as membranes in the reverse osmosis method of water purification.

See filament; denier; whiskers; glass fiber.

fiber, biconstituent. A composite fiber comprising a dispersion of fibrils of one synthetic material within, and parallel to, the axis of another; also a fiber made up of polymeric material and a metal or alloy filament.

fiberfill. A fiber designed specifically for use as a filling material in such products as pillows, comforters, quilted linings, and furniture battings, e.g., sisal, jute.

Fiberfrax. Ceramic fiber made from alumina and silica. Available in bulk as blown, chopped and washed, long staple, paper, rope, roving, blocks.
Properties: Retains properties to 1260C and under some conditions used to 1648C, light weight, inert to most acids and unaffected by hydrogen atmosphere, resilient.
Use: High-temperature insulation of kilns and furnaces, packing expansion joints, heating elements, burner blocks; rolls for roller hearth furnaces and piping; fine filtration; insulating electrical wire and motors; insulating jet motors; sound deadening.

fiber gear. A driver gear made of a material of somewhat lower strength than the driven gear (cast iron); for example, a composite such as fiberglass-reinforced plastic or an engineering plastic (e.g., nylon). It is intended to fail under overload, thus protecting the driven master gear from destructive stress.

Fiberglas. A variety of products made of or with glass fibers or glass flakes, including insulating wools, mats and rovings, coarse fibers, acoustical products, yarns, electrical insulation, and reinforced plastics.
See glass fiber; reinforced plastic.

fiber glass. See glass fiber.

fiber, graphite. See graphite fiber.

fiber, optical. A fine-drawn silica (glass) fiber or filament of exceptionally high purity and specific optical properties (refractive index) that transmits laser light impulses almost instantaneously with high fidelity. Such fibers are made from quartz coated with germanium-doped silica by vapor deposition; 100 or more filaments are assembled into a cable that has extremely high data-carrying capacity. These are applicable not only to telephonic communication systems, for which they are now being used, but also to remote-sensing devices that permit analysis of samples at widely separated locations. Thus, one of the most important developing uses of optical fibers is in analytical instrumentation. Because, as they are nonelectrical and noncorrosive, optical fiber cables are safe to use in highly toxic or explosive environments, e.g., radioactive separations and hazardous waste analyses. The laser beam is coupled to the end of the cable (which may be up to 1000 m long) by a device called an optrode; the light traverses the cable and interacts with the sample, eliciting a signal that is reflected back through the same cable to a spectrometer. Fiber optics are also used in other forms of instrumentation, e.g., radiation dosimeters and high-temperature thermometers. In the latter case, the fibers are made from single crystals of alumina.
See glass, optical; laser; thermometer (5).

fiber-reactive dye. See dye, fiber-reactive.

fiber roll. A calender roll constructed of specially prepared papers or fabrics on a steel base. The fibrous material is cut into circular sheets with a hole at the center; these are stacked on a steel core and then compressed under high pressure, producing a dense, hard material with a smooth surface. As such, it is used in supercalenders for paper finishing. An intaglio design can be impressed on it by an engraved steel roll; this operation requires several days and is facilitated by application of water, soap, or other softener. So prepared, it is used in embossing calenders for applying decorative patterns on special paper or plastic products.
See supercalender.

fibrid. Generic name for a fibrous form of synthetic polymeric material used for example as a binder material in the manufacture of textryl.

fibril. (1) Extremely fine fiber or cell-like mass formed during first stages of gel formation; (2) protein filament of hide fiber.

fibrin. An insoluble blood protein resulting from the hydrolysis of fibrinogen by the action of thrombin; it polymerizes to form blood clots. Recent research has found that it forms a protective coating on tumors that inhibits antigenic activity, thus protecting the tumor and neutralizing the immune system of the organism.

fibrinogen. A sterile fraction of normal human blood plasma, dried from the frozen state. In solution, it has the property of being converted into insoluble fibrin when thrombin is added. It is an essential factor in the blood-clotting mechanism.

Properties: White or grayish amorphous substance.
Grade: USP.
Use: Medicine (coagulant).

fibrinolysin. (plasmin). A proteolytic enzyme
that dissolves fibrin and hastens the solution of clots
that may form in the bloodstream. It is prepared by
activating a fraction of normal human plasma with
highly purified streptokinase.

fibroblast. A cell of connective tissue that
secretes connective tissue proteins, such as collagen.

fibroin. The fibrous material in silk; a scleroprotein containing glycine and alanine; light yellow,
silk-like mass; insoluble in water; soluble in concentrated alkalies and concentrated acids.

fibrolite. See sillimanite.

fibrous glass. See glass fiber.

fibrous proteins. Water-insoluble proteins that
serve in a protective or structural role. The ratio of
length to width generally exceeds 10.

ficin. A proteolytic enzyme hydrolyzing casein,
collagen, edestin, fibrin, liver, and other protein-
like material.
  Properties: Buff to cream-colored powder; acrid
  odor. Very hygroscopic. Partially soluble in water;
  insoluble in organic solvents.
  Source: Fig latex or sap, commercially prepared by
  filtering and drying the latex.
  Use: Food industry, bating leather, tenderizing meat,
  shrink proofing wool, coagulation of milk, chill
  proofing beer, Rh factor determination.

Fi-Con. A fiber additive.
  Use: Concrete additive.

fictile. Descriptive of certain molecules that have
no permanent structure but are constantly changing
their shapes and arrangements. An example is the
metal carbonyl $Fe_3(CO)_{12}$, in which, according to
Dr. F. Albert Cotton, originator of the term, "carbonyl groups readily move from one iron atom to
another through the rapid formation and dissolution
of carbonyl bridges between iron atoms."

field-ion microscope. A type of microscope
whose unique feature is that it has no lens system. Invented by Muller in 1951, it is capable
of resolving metal atoms 2–3 Å in diameter. Its
essential components are an evacuated glass chamber through which runs a wire carrying an electric impulse of 30,000 volts that establishes a field
strength of 500 million volts/cm. A specimen of
the metal to be observed, which is machined to an
extremely fine tip and is positively charged, is connected to the wire. An inert gas such as helium or
neon is then admitted. The positively charged tip
of the specimen attracts electrons from the helium
atoms, creating positive helium ions. These are
strongly repelled by the metal atoms and stream to
the negatively charged fluorescent screen, producing an image of the individual atoms of the metal.
Magnifications of one million times or more have
been obtained of atoms of indium, tungsten, and
others.
See electron microscope.

Fieser, Louis F. (1899–1977). A distinguished
American chemist, Fieser became professor of
organic chemistry at Harvard in 1930 after teaching
for several years at Bryn Mawr. He achieved the synthesis of vitamin K_1 and did fundamental research
on cortisone, the chemistry of steroids, and aromatic carcinogens. His achievements as a chemist
and educator are recognized throughout the world.
Unique in his facility in laboratory demonstration
and as a lecturer and author, he exemplified that
rare combination of a great teacher and a profound
scholar.

filament. A continuous fiber usually made by
extrusion from a spinnerette (nylon, rayon, glass,
polyethylene). It also may be a drawn metal (tungsten, gold) or a metal carbide.
See fiber.

filament winding. The process of winding fibers
under tension onto a prepared core. Before or during
the winding operation, the assembly is impregnated
with a thermosetting resin. Structures of considerable size and strength can be made in this way.
The fibers used are chiefly glass, boron, or silicon
carbide.
See filament.

filial generation. (F1; F2). Each generation of
offspring in a breeding program, designated F1, F2,
etc.

filler.
(1) An inert mineral powder of rather high specific
gravity (2.00–4.50) used in plastic products and rubber mix to provide a certain degree of stiffness and
hardness and to decrease cost. Examples are calcium carbonate (whiting), barytes, blanc fixe, silicates, glass spheres and bubbles, slate flour, soft
clays, etc. Fillers have neither reinforcing nor coloring properties, and the term should not be applied
to materials that do, i.e., reinforcing agents or pigments. Fillers are similar to extenders and diluents
in their cost-reducing function; exact lines of distinction between these terms are difficult, if not
impossible, to draw. Use of fillers and extenders in
plastics has increased in recent years due to short-
ages of basic materials.

(2) The cross or transverse thread in a fabric or other textile structure.

(3) A metal or alloy used in brazing and soldering to effect union of the metals being joined.

See diluent; extender; reinforcing agent.

film. An extremely thin continuous sheet of a substance that may or may not be in contact with a substrate. There is no precise upper limit of thickness, but a reasonable assumption is 0.010 inch. The protective value of any film depends on its being 100% continuous, i.e., without holes or cracks, since it must form an efficient barrier to molecules of atmospheric water vapor, oxygen, etc. A long-chain fatty acid or alcohol on water produces a film whose thickness is the length of one molecule (approximately 200 Å). The fatty acid molecules are oriented with the radical end in the water. Such films are good evaporation barriers and have been successfully imposed on glass. Soap bubbles are elastic films about one micron thick and have considerable strength.

Film-forming agents (drying oils) are essential in paints and lacquers. Oxide films formed automatically on the surface of aluminum protect it from corrosion. Thin metallic oxide films are widely used in electronic and semiconducting devices. Electrodeposited metals (chromium, copper, nickel) are conventionally (and perhaps illogically) called coatings.

The term film is also applied to sheets of cellophane, polyethylene, polyvinylidene chloride, etc., used for wrapping and packaging of food products, meats, and poultry (especially shrink films that are stretched before application). These function as a moisture vapor barrier. Plastic films are also used as slip surfaces in concrete structures such as airstrips, ice rinks, and highways. Photographic film is made from cellulose acetate.

filter. See filtration; leaf, filter; baghouse.

filter aid. See filter media; filtration.

filter alum. See aluminum sulfate.

filter medium. Almost any water-insoluble, porous material having a reasonable degree of rigidity can serve as a filter. Sand is used in simple large-scale water filtration, the voids between the grains providing the porosity. In industrial operations, cotton duck, woven wire cloth, nylon cloth, and glass cloth are used. For laboratory work, Whatman filter paper, diatomaceous earth, and closely packed glass fibers are standard materials. Plastics membranes containing more than a million pores per square inch are used in bacteriological filtration.

See filtration; screen.

filter sand. Sand used to separate sediment and suspended matter from water.

filtration. The operation of separating suspended solids from a liquid (or gas) by forcing the mixture through a porous barrier (see filter media). The construction and operation of the many kinds of industrial filtration equipment are too detailed to permit description. The most widely used types may be classified as follows: (1) gravity filters, used largely for water purification and consisting of thick beds of sand and gravel that retain the flocculated impurities as the water passes through; (2) pressure filters of plate-and-frame or shell-and-leaf construction that utilize filter cloths of coarse fabric as a separating medium; (3) vacuum or suction filters of the rotating drum or disk type, used on thick sludges and slurries; (4) edge filters; (5) clarification filters; (6) bag filters (dust collectors). Gel filtration is a chromatographic technique involving separation at the molecular level. For bacteriological filtration, membranes having more than a million pores per square inch are used, e.g., collodion or synthetic film. Some types of viruses will pass through such membranes and are thus known as filterable viruses.

See baghouse.

filtration barrier. A barrier that separates the blood in the glomerular capillaries of the kidney from that of the capsular space of the renal corpuscle.

fine chemical. A chemical produced in comparatively small quantities and in a relatively pure state. Examples are pharmaceutical and biological products, perfumes, photographic chemicals, and reagent chemicals. They are sold on the basis of specifications (on the basis of what they are, not what they do).

fines. The portion of a powder composed of particles that are smaller than a specified size (MPA definition, MPA Standard 9-50T).

fine structure. See ultrastructure.

fingerprinting. The characteristic two-dimensional pattern (on paper or gel) formed by the separation of a mixture of peptides resulting from partial hydrolysis of a protein; also known as peptide mapping.

finished DNA sequence. High-quality, low error, gap-free DNA sequence of the human genome. Achieving this ultimate 2003 HGP goal requires additional sequencing to close gaps, reduce ambiguities, and allow for only a single error every 10,000 bases—the agreed-upon standard for HGP finished sequence.

See sequencing; draft sequence.

finishing compounds.   Materials that impart softness, flexibility, stiffness, color, water and fire resistance, etc.
Use: In the final or finishing stages of manufacture of a product, usually textiles and leather, to make them suitable for specific purposes.

Finkelstein reaction.   Reaction of alkyl halides with sodium iodide in acetone.

Firebrake ZB.   (zinc borate). A synergist with antimony compounds and alumina trihydrate.
Use: Flame and smoke suppressant.

fireclay.   See refractory.

firedamp.   Methane that seeps into coal mines that accumulates forming an explosive mixture. The degraded air remaining in the mine following such an explosion is called blackdamp, and the carbon monoxide that is generated by such an explosion is called afterdamp.

fire extinguishment.   Fires are divided into 4 classes, each requiring special treatment. The essential point in extinguishing all types is exclusion of air from the fire by an effective means.
Class A includes fires in combustible materials, such as wood, paper, and cloth, where the quenching and cooling effect of quantities of water or of solutions containing a high percentage of water is of primary importance. Fire extinguishers utilizing the pressure of carbon dioxide to throw a stream of water onto the fire are the most widely used for this class. In the soda-acid extinguisher, the carbon dioxide is generated within the cylinder at the time of use. In another type, carbon dioxide gas is stored in the cylinder under pressure and is released by opening a valve.
Class B includes fires in flammable liquids where a blanketing or smothering effect is essential. Carbon dioxide gas, dry chemical, or foam are suitable. Water should not be used.
Class C includes fires in electrical equipment. The use of carbon dioxide gas or dry chemical extinguishers is recommended. Water should not be used.
Class D fires are burning metals. A powder formulation such as "Met-L-X" powdered graphite or trimethoxyboroxine will extinguish a metal fire. Water should not be used.
In general, for small fires, salt (sodium chloride) and sodium bicarbonate, either dry or in concentrated solution are effective. Carbon tetrachloride and methyl bromide should be avoided as extinguishing agents because of the toxicity of their decomposition products, e.g., phosgene.
See foam, fire-extinguishing.

"Firefrax" *[Saint-Gobail].*   TM for a group of refractory cements made from kaolin or fireclay base materials for applications where aluminum silicate cements are best suited.

Use: Laying and repairing fireclay and silica brickwork, bond for crushed firebrick or ganister for patching furnace linings and for making rammed-up or monolithic linings, patching materials for by-product coke ovens, and as a wash for small pouring ladles in nonferrous foundry.

fire point.   The lowest temperature at which a liquid evolves vapors fast enough to support continuous combustion. It is usually close to the flash point. See autoignition temperature.

fire-retarding   agent.   See   flame-retarding agent.

fire sand.   See furnace sand.

fir-needle oil.   (fir oil). An essential oil obtained by the steam distillation of needles and twigs of several varieties of coniferous trees (Abies) native to both Canada and Siberia.
Use: Odorant in perfumery, flavoring agent.

fir needle oil, Canadian type.
Properties: Found in the needles and twigs of *Abies balsamea* L. Mill (family Pinacea). Colorless to faintly yellow liquid; pleasant odor. Sol in fixed oils, mineral oil; sltly sol in propylene glycol; insol in glycerin.
Use: Food additive.

first law of thermodynamics.   Energy can be neither created nor destroyed, only converted to other forms.

Fischer, Ernst Otto.   (1918–2007). A German inorganic chemist who won the Nobel Prize for chemistry in 1973 with Wilkinson for their independent work on the chemistry of organometallic "sandwich compounds." He was the contributor to many publications on organometallic chemistry. His education and work were primarily in Munich.

Fischer, Hans.   (1881–1945). A German biochemist who studied under Hermann Fischer. He was awarded the Nobel Prize in chemistry in 1930 for his synthesis of the blood pigment hemin. He also did important fundamental research on chlorophyll, the porphyrins and carotene.

Fischer-Hepp   rearrangement.   Rearrangement of secondary aromatic nitrosamines to *p*-nitrosoarylamines.

Fischer, Hermann Emil Louis.   (1852–1919). A German organic chemist, recipient of the Nobel Prize in chemistry (1902) for his original research in the chemistry of purines and sugars. He was professor of chemistry at the University of Berlin (1882), succeeding Hofmann. He synthesized fructose and glucose and elucidated their sterochemical configurations; he also established the nature of uric acid and its derivatives. Additional work included enzyme chemistry, proteins, synthetic nitric acid, and ammonia production.

Fischer indole synthesis.   Formation of indoles on heating aryl hydrazones of aldehydes or ketones in the presence of catalysts such as zinc chloride, or other Lewis acids, or proton acids.

Fischer oxazole synthesis.   Condensation of equimolar amounts of aldehyde cyanohydrins and aromatic aldehydes in dry ether in the presence of dry hydrochloric acid.

Fischer peptide synthesis.   Formation of polypeptides by treatment of an α-chloro or α-bromo acyl chloride with an amino acid ester, hydrolysis to the acid, and conversion to a new acid chloride that is again condensed with a second amino acid ester, and so on. The terminal chloride is finally converted to an amino group with ammonia.

Fischer phenylhydrazine synthesis.   Formation of arylhydrazines by reduction of diazo compounds with excess sodium sulfite and hydrolysis of the substituted hydrazine sulfonic acid salt with hydrochloric acid. The process is a standard industrial method for production of arylhydrazines.

Fischer phenylhydrazone and osazone reaction.   Formation of phenylhydrazones and osazones by heating sugars with phenylhydrazine in dilute acid.

Fischer projection formulas.   Formulas which, by convention, show bonds coming out of the plane drawn vertically, and bonds that project behind the plane drawn horizontally.

Fischer–Speier esterification method. Esterification of acids by refluxing with excess alcohol in the presence of hydrochloric acid or other acid catalysts.

Fischer's reagent.   A reagent used as a test for sugars.
 Derivation: Three parts of sodium acetate and two parts of phenylhydrazine hydrochloride in 20 parts of water.
Note: Do not confuse with Karl Fischer reagent.

Fischer's salt.   See cobalt potassium nitrite.

Fischer–Tropsch process.   Synthesis of liquid or gaseous hydrocarbons or their oxygenated derivatives from the carbon monoxide and hydrogen mixture (synthesis gas) obtained by passing steam over hot coal. The synthesis is carried out with metallic catalysts such as iron, cobalt, or nickel at high temperature and pressure. The process was developed in Germany in 1923 by F. Fischer and H. Tropsch and was used there for making synthetic

fuels before and during World War II. It has never been used for this purpose in the U.S.; the only coal-to-gasoline conversion plant using this process is Sasol in South Africa, though the closely related Lurgi process is being used rather extensively in a number of locations. Easing of the petroleum crisis has tended to diminish conversion activity in the U.S.

Fischer–Tropsch synthesis.   (Synthol process; Oxo synthesis).
 Synthesis of hydrocarbons, aliphatic alcohols, aldehydes, and ketones by the catalytic hydrogenation of carbon monoxide using enriched synthesis gas from passage of steam over heated coke. The ratio of products varies with conditions. The high-pressure Synthol process gives mainly oxygenated products and addition of olefins in the presence of cobalt catalyst (Oxo synthesis) produces aldehydes. Normal-pressure synthesis leads mainly to petroleum-like hydrocarbons.

fisetin.   (3,7,3'-tetrahydroxyflavone).
 CAS: 528-48-3. $C_{15}H_{10}O_6$.
 See flavanol.

FISH.   (Florescent *in situ* hybridization). A technique for uniquely identifying whole chromosomes or parts of chromosomes using florescent tagged DNA.

Fisher's solution.   See physiological salt solution.

fish glue.   An adhesive derived from the skins of commercial fish (chiefly cod). A ton of skins yields about 50 gal liquid glue. Bond strength on wood is approximately 2500 psi, pH approximately 6.5–7.2. Compatible with animal glues, some dextrins, some polyvinyl acetate emulsions, and rubber latex. Chief applications are in gummed tape, cartons, blueprint paper, and letterpress printing plates. Fish glue can be made light sensitive by adding ammonium bichromate and water insoluble by UV radiation, hence its usefulness in the photoengraving process.
 See adhesive.

fish-liver oil.   An oil containing a high percentage of vitamin A. High-potency livers, as from cod, shark, and halibut, contain from 100,000 to 1,500,000 A units/g. The oil is extracted by cooking the livers under low-pressure steam and removing the oil, which floats on the condensate. Livers of low oil content are processed with a weak solution of sodium hydroxide or sodium carbonate, which extracts the oil in emulsified form.
 Use: Medicine and dietary supplement.
 See fish oil.

fish meal.   A fishery by-product consisting essentially of processed scrap from the filleting operation

or from whole fish. In the dry process, the waste from cod, halibut, and haddock heads is disintegrated and dried. The oil and proteins are largely retained. In the wet process, the whole fish (chiefly menhaden and pilchard) are used. These are steam-cooked and run through a screw press to remove the oil. The resulting meal is then dried and packed. Its chief use is now for animal feeds and as a raw material for fish protein concentrate.
Hazard: Flammable, strong tendency for spontaneous heating.

fish oil.   A drying oil obtained chiefly from menhaden, pilchard, sardine, and herring. Extracted from the entire body of the fish by cooking and compressing. Should not be confused with fish-liver oil. It contains approximately 20% polyunsaturated fatty acids, which enables it to lower cholesterol content of the human diet. Chemically modified fish oil is used in soaps, detergents, protective coatings, and alkyd resins. The hydrogenated product is used as a base for margarines and shortenings and as an industrial dispersing agent.
Hazard: Subject to spontaneous heating.

fish protein concentrate.   (FPC). A flour or paste-like product prepared from whole fish including bones and viscera, of a size and type not acceptable for sale as such. Both biological (enzymatic) and chemical (solvent extraction) methods are used to obtain the proteins.

fissile.   Synonymous with fissionable.
See fission.

fissiochemistry.   The process by which a chemical change or reaction is brought about by nuclear energy, e.g., the production of anhydrous hydrazine from liquid ammonia in a nuclear reactor.

fission.   The splitting of an atomic nucleus induced by bombardment with neutrons from an external source and propagated by the neutrons so released. When a fissionable (unstable) nucleus, such as ^{235}U or ^{239}Pu, is struck by a neutron in a critical area, the following events occur: (1) the nucleus disintegrates to form several other elements, called fission products or fragments, all of which are radioactive and have high kinetic energy; (2) the disrupted nucleus emits an average of 2.5 neutrons (^{235}U) or three neutrons (^{239}Pu), which in turn split other nuclei of the fissionable material in a chain reaction that is self-perpetuating; (3) it also emits the energy equivalent of the mass defect of the nucleus, usually approximately 200 MeV per nucleus, some of which is in the form of γ rays.
A nuclear explosion will not occur until a critical mass of fissionable material is attained, i.e., the smallest amount capable of sustaining a chain reaction. Similarly, a nuclear reactor will not produce power until the assembly achieves a critical activity. This occurs as follows: The neutrons introduced to

the system are continually escaping—some are lost through the walls, some are captured by structural materials, and some are absorbed by the fissionable atoms themselves without fission taking place. When the neutrons entering the system are very slightly in excess of those lost to it, the assembly is said to be critical and measurable power generation takes place. The ratio is carefully controlled, the rate of energy production rising exponentially. Control rods made of cadmium absorb neutrons so readily that the reactor can function at precisely predetermined levels of activity. Nuclear fission is used for electric power generation and for making radioactive isotopes.
See nuclear reactor.

Fittig's synthesis.   The preparation of aromatic hydrocarbons by condensation of aryl halides with alkyl halides in the presence of metallic sodium.

fix.   (1) To cause an unreactive element, e.g., nitrogen to combine into a chemical compound, as in ammonia synthesis. (2) To hold a dye permanently on a fiber or fabric by chemical or mechanical action or a combination of both. (3) To retard the evaporation rate of the volatile components of essential oils, as in perfumes.
See nitrogen fixation.

fixation.   See nitrogen fixation.

fixative.   (1) See fixing agent, perfume; (2) a substance applied as a spray or solution to harden and preserve objects for microscopic examination, or to pencil and ink drawings to prevent blurring, e.g., a sodium silicate solution.

fixed oil.   A nonvolatile, fatty oil characteristic of vegetables as opposed to the volatile essential oils of flowers.

fixing agent, chemical.   (1) A substance that aids fixation of mordants on textiles by uniting chemically with them and holding them on the fiber until the dyes can react with them. (2) A substance that causes actual precipitation of mordant on the fiber by double decomposition.

fixing agent, mechanical.   (1) A substance (e.g., albumin) capable of holding pigments permanently on textile fibers. (2) Certain gums and starches that hold dyes and other substances on textile fibers long enough to permit a desirable reaction to take place.

fixing agent, perfume.   (fixative). A substance that prevents too rapid volatilization of the components of a perfume and tends to equalize their rates of volatilization. It thus increases the odor life of a perfume and keeps the odor unchanged. For

many years, the chief fixatives were animal products (ambergris, civet, musk, castoreum), but these have been largely replaced by synthetics.
See perfume.

flagellum. Hair-like structure attached to a cell, used for locomotion in many protists and prokaryotes. The prokaryotic flagellum differs from the eukaryotic flagellum in that the prokaryotic flagellum is a solid unit composed primarily of the protein flagellin, while the eukaryotic flagellum is composed of several protein strands bound by a membrane and does not contain flagelli.

"Flagyl" Tablets *[Searle].* TM for metronidazole.
Use: Drug.

flake lead. See lead carbonate, basic.

flame cracking. See ethylene (note).

flame photometry. See emission spectroscopy.

flame-retarding agent. A substance applied to or incorporated in a combustible material to reduce or eliminate its tendency to ignite when exposed to a low-energy flame such as a match or cigarette. There are three methods of application: (1) as a coating or surface finish (nondurable, readily removed); (2) in solution form to penetrate the fibers (semidurable, reasonably stable); and (3) as an integral part of the polymer structure of a synthetic fiber (durable, not removable). The latter method provides permanent protection because it not only makes the material self-extinguishing but cannot be leached out by laundering or dry cleaning. Substances commonly used in methods (1) and (2) include such inorganic salts as ammonium sulfamate, zinc borate, and antimony oxychloride; chlorinated organic compounds such as chlorendic anhydride; alumina trihydrate: and certain organic phosphates and phosphonates. Method (3) is exemplified by a polyester fiber, Trevira 271, composed of polyethylene terephthalate and an undisclosed flame-retardant chemically linked to the polymer molecule. A copolymer of styrene and phosphazene has also been researched. Certain types of fibers (polyamides and aramids) are inherently flame-retardant, e.g., nylon, "Nomex" *[Du Pont]*, "Kevlar" *[Du Pont]*.
Use: Carpets, rugs, upholstery, plastics used in construction and miscellaneous wearing apparel.

flammability. The ease with which a material (gas, liquid, or solid) will ignite either spontaneously (pyrophoric) or from exposure to a high-temperature environment (autoignition) or a spark or open flame. It also involves the rate of spreading of a flame once it has started. The more readily ignition occurs, the more flammable the material; less easily ignited materials are said to be combustible,

but the line of demarcation is often indefinite and depends on the state of subdivision of the material, as well as its chemical nature. The Flammable Fabrics Act establishes standards of flammability to which all textile manufacturers must conform.
See flammable material; combustible material.

flammable material. Any solid, liquid, vapor, or gas that will ignite easily and burn rapidly. Flammable solids are of several types: (1) dusts or fine powders (metals or organic substances such as cellulose, flour, etc.); (2) those that ignite spontaneously at low temperatures (white phosphorus); (3) those in which internal heat is built up by microbial or other degradation activity (fish meal, wet cellulosic materials); (4) films, fibers, and fabrics of low-ignition-point materials.
Flammable liquids are defined by the National Fire Protection Association and the Department of Transportation (DOT) as those having a flash point (flash p) less than 100F (37.7C) (CC) and a vapor pressure of not more than 40 psi(a) at 100F.
Flammable gases are ignited very easily; the flame and heat propagation rate is so great as to resemble an explosion, especially if the gas is confined. The most common flammable gases are hydrogen, carbon monoxide, acetylene, and other hydrocarbon gases. Oxygen, though essential for the occurrence of combustion, is not itself either flammable or combustible; neither are the halogen gases, sulfur dioxide, or nitrogen. Flammable gases are extremely dangerous fire hazards and require precisely regulated storage conditions.
Note: The terms *flammable, nonflammable,* and *combustible* are difficult to delimit. Since any material that will burn at any temperature is combustible by definition, it follows that this word covers all such materials, irrespective of their ease of ignition. Thus, the term *flammable* actually applies to a special group of combustible materials that ignite easily and burn rapidly. Some materials (usually gases) classified in shipping and safety regulations as nonflammable are actually noncombustible. The distinction between these terms should not be overlooked. For example, sodium chloride, carbon tetrachloride, and carbon dioxide are noncombustible; sugar, cellulose, and ammonia are combustible but nonflammable.
See combustible material.

flamprop-*m*-isopropyl.
CAS: 63782-90-1. $C_{19}H_{19}ClFNO_3$.
Hazard: Moderately toxic by ingestion and skin contact.
Use: Agricultural chemical.

flash. The overflow of rubber or plastic at the parting line of a mold when subjected to full pressure. It is removed in the finishing operation.

flash distillation. Distillation in which an appreciable proportion of a liquid is quickly converted to vapor in such a way that the final vapor is in equilibrium with the final liquid. This method is now widely used for desalination of seawater.

flash photolysis. A method of investigating the mechanism of extremely rapid photochemical reactions involving the formation of free radicals (both inorganic and organic) by irradiating a given reaction mixture with a flash of high-intensity light, thus producing the short-lived radicals that activate photochemical reactions. These products are instantaneously analyzed spectroscopically, which permits identification of the radical species from the spectra obtained. The time lapses involved in this technique are approximately 1/100,000 second. It has also been applied to study of the exceedingly fast reactions occurring in flames and explosions.
See photochemistry; free radical; photolysis.

flash point. The temperature at which a liquid or volatile solid gives off vapor sufficient to form an ignitable mixture with the air near the surface of the liquid or within the test vessel (NFPA). For the purposes of the official shipping regulations, the flash point is determined by the Tagliabue open cup method (ASTM D1310-63), usually abbreviated TOC. (IATA also permits the Abel or Abel-Pensky closed cup tester.) Other methods used, generally for the higher flash points, are the Tag closed cup (Tagliabue closed cup, TCC) and Cleveland open cup (COC). The open cup method more nearly approximates actual conditions.
See flammable material.

flatting agent. A substance ground into minute particles of irregular shape and used in paints and varnishes to disperse incident light rays so that a dull or flat effect is produced. Standard flatting agents are heavy-metal soaps, finely divided silica, and diatomaceous earth.

flavanol. (3-hydroxyflavone). A derivative of flavanone; yellow needles melting at 169C, has violet fluorescence in concentrated sulfuric acid. It is a flavonoid pigment. Dyes cotton a bright yellow when mordanted with aluminum hydroxide. Other hydroxyflavones are chrysin, fisetin, and quercitin. Eleven different flavonols are known. Not identical to flavonol.

flavanone. (2,3-dihydroflavone). A group of colorless derivatives of flavone distributed in higher plants either in free form or as glucosides. About 25 different types have been isolated. It comprises one of the major groups of flavonoids. Examples are hesperidin and naringin.

flavanthrene. (indanthrene yellow; chloranthrene yellow).
CAS: 475-71-8. $C_{28}H_{12}O_2N_2$.
Properties: Brownish-yellow needles. Soluble in dilute alkaline solutions.

Derivation: Action of antimony pentachloride on β-aminoanthraquinone in boiling nitrobenzene.
Use: Vat dye for textiles, etc.

flavianic acid. See 2,4-dinitro-1-naphthol-7-sulfonic acid.

flavin.
(1) Isoalloxazine. $C_{10}H_6N_4O_2$. The nucleus of various natural yellow pigments.
See riboflavin; flavin enzymes.
(2) Tetrahydroxyflavanol. $C_{15}H_{10}O_7 \cdot 2H_2O$. A yellow dye derived from oak bark.

flavine. See acriflavine.

flavin adenine dinucleotide. (FAD; [(2R,3E,4R,5R)-5-(6-aminopurin-9-yl)-3,4-dihydroxyoxolan-2-yl]methyl phosphono hydrogen phosphate).
CAS: 146-14-5. $C_{27}H_{22}N_2O_{15}P_2$. An adenine nucleotide containing two phosphate groups esterified to the sugar moiety at the 5'-position.

flavin enzyme. (flavoprotein). An enzyme composed of protein linked to coenzymes that are mono- or dinucleotides containing riboflavin. Because of their distinctive color, they are also called yellow enzymes. The flavin enzymes function in tissue respiration as dehydrogenases, the hydrogen atoms being taken up by the riboflavin group.

flavin-linked dehydrogenases. Dehydrogenases requiring one of the riboflavin coenzymes, either FMN or FAD.

flavin mononucleotide. See riboflavin phosphate.

flavin nucleotides. Nucleotide coenzymes containing riboflavin, either FMN or FAD.

flavone. (2-phenylchromone). One of a group of flavonoid plant pigments existing as colorless needles, that are insoluble in water and melting at 100C. It fluoresces violet in concentrated sulfuric acid. It can be synthesized. Treatment with alcoholic alkali yields flavanone. The flavones produce ivory and yellow colors in plants and flowers.
See flavonoid.

flavonoid. A group of aromatic, oxygen-containing, heterocyclic pigments widely distributed among higher plants. They constitute most of the yellow, red, and blue colors in flowers and fruits. Exceptions are the carotenoids. The flavonoids include the following subgroups: (1) catechins; (2) leucoanthocyanidins and flavanones; (3) flavanins, flavones, and anthocyanins; and (4) flavonols.
For details consult specific entries.

flavonol. (flavon-3-ol). A flavonoid plant pigment giving ivory and yellow colors to flowers. Not identical with flavanol.

flavoprotein. An enzyme containing a FMN or FAD as a tightly bound prosthetic group.
See flavin enzyme.

flavor. (1) The simultaneous physiological and psychological response obtained from a substance in the mouth that includes the senses of taste (salty, sour, bitter, sweet), smell (fruity, pungent), and feel. The sense of feel as related to flavor encompasses only the effect of chemical action on the mouth membranes such as heat from pepper, coolness from peppermint, and the like (Institute of Food Technologists). No reliable correlation of taste with chemical structure has yet been possible. Flavor is a critical factor in the acceptability of foods, medicines, confectionery, and beverages. Flavors are used in insect and animal baits to induce ingestion of the bait, also to prevent rodent attack on organic materials, e.g., tributyl tin in cable covers. Substances that affect flavor often have a synergistic effect (e.g., monosodium glutamate and certain nucleotides). Sodium chloride is classed as a seasoning agent.
See potentiator; enhancer.
(2) Any substance or mixture of substances that contributes a positive taste to a food product, such as vanillin, cacao, and fruit extracts among natural products, together with numerous synthetic compounds that imitate or duplicate these tastes. Undesirable or off flavors occur in milk, meat, and other food products as a result of improper preparation, oxidation, and incipient rancidity. There are over 1500 flavoring materials listed as food additives under provisions of the Food, Drug, and Cosmetics Act.
See odor.

Flavor Extract Manufacturers Association. (FEMA). The FEMA makes recommendations to the FDA on safety aspects of flavoring materials. Membership is composed of firms engaged in the manufacture and sale of flavoring extracts. It is located at 1101 17th Street NW, Suite 700, Washington, DC 20036. Website: http://www.femaflavor.org

flavoskyrin. (2H-dibenzo(*c,mn*)naphtha(2,3-*g*) xanthene-6,13,18(3H)-trione). $C_{20}H_{21}O_{10}$. An anthraquinoid mycotoxin that is extremely toxic to cultured Helas, the protozoan *Tetrahymena pyriformis*, and *Escherichia coli* mutants.

flax. Bast fibers, approximately 20 inches long, obtained from the stems of the linseed plant, *Linum usitatissimum*. Stronger and more durable than cotton. Combustible.
Use: Apparel fabrics (linens), thread, rope, twine, cigarette paper, duplicating papers.

flaxseed oil. See linseed oil.

flecainide.
CAS: 54143-55-4. $C_{17}H_{20}F_6N_2O_3$.
Hazard: Human systemic effects.

Flectol. Polymerized 1,2-dihydro-2,2,4-trimethylquinoline.
Use: Rubber antioxidant to retard oxidation, deterioration, and normal aging in belts, tires, hoses, retread, rubber, and general mechanics.

Fleming, Sir Alexander. (1881–1955). A Scottish biochemist and bacteriologist who discovered (1928) the bactericidal properties of molds produced from the plant *Penicillium notatum*. A broad spectrum of antibiotics has been developed from this discovery.
See antibiotic.

"Flexamine G" [Chemtura]. TM for a mixture of *N,N'*-diphenyl-*p*-phenylenediamine and a complex diarylamine-ketone reaction product.
Properties: Brownish-gray granules. D 1.20, melting range 75–90C. "Flexamine S" is soluble in acetone, benzene, and ethylene dichloride and insoluble in water and gasoline.
Use: Antioxidant used in tires, camelback, wire insulation, neoprene belting, and soles.

Flexol. A series of plasticizers and stabilizers including phthalates, adipates, polyalkylene glycol derivatives, polymeric epoxies, decanoates, octoates, hexoates, tri(2-ethylhexyl)phosphate, and dibutyltin dilaurate.
Use: Film and sheeting; flooring; coated fabrics; wire, cable and other extrusions; organosols; plastisols and plastigels; lacquers; and rubbers.

"Flexomer" [Union]. TM for very low density, extremely flexible, olefin copolymers that bridge the gap between plastics and rubber.
Use: For hose and tubing, ice and frozen-food bags, food packaging and stretch wrap, and impact modifiers when blended with other polymers.

"Flexsil" [Adfast]. TM for a high silica refractory cloth filter designed to extract dross, slag, refractory particles, and nonmetallic inclusions from molten metals. An important property is its chemically active surface.
Use: In-mold filtration of gray, malleable, or white, compacted graphite, and ductile cast irons, as well as nonferrous aluminum- and copper-based alloys.

Flexzone. A series of antiozonants, antioxidants, and stabilizers based on *p*-phenylenediamine.

flint. A crystalline form of native silica or quartz.
Properties: Smoky-gray, brownish, blackish or dull yellowish in color; waxy to greasy luster. Mohs

hardness 6.5–7, d 2.60–2.65. More easily soluble in hot caustic alkali than is crystallized quartz.
Occurrence: Europe, U.S.
Use: Abrasive; balls for ball mills; paint extender; filler for fertilizer, insecticides, rubber, plastics, and road asphalt; ceramics; chemical tower packing.

"Flintflex" *[Axalta].*   TM for air- or force-dried organic coating system.
Use: Linings in interiors of containers that haul dry bulk ladings of edibles or chemicals. Complies with FDA regulations. Accepted by the Meat Inspection Division of the USDA for interiors of freight cars, motor trucks, and trailers and in federally inspected meat-processing plants.

flint glass.   See glass, optical.

FLIT.   An insecticide containing coal tar oil and refined petroleum.

floatation.   Purification and/or classification of finely divided solids, e.g., clays, by passing them through an air blast. Do not confuse with *flotation.*

flocalcitriol.   See hexafluorocalcitriol.

flocculant.   A substance that induces flocculation. Flocculants are used in water purification, liquid waste treatment, and other special applications. Inorganic flocculants are lime, alum, and ferric chloride; polyelectrolytes are examples of organic flocculants.

flocculation.   The combination or aggregation of suspended colloidal particles in such a way that they form small clumps or tufts. The word is derived from this appearance. Carbon black displays a tendency to flocculate in rubber when improperly dispersed, and some clays have the same property. Oil-well drilling muds are made alkaline to prevent flocculation of their components. Flocculation can often be reversed by agitation, because the cohesive forces are relatively weak. This is not true of other forms of aggregation (coalescence and coagulation), which are irreversible.
See agglomeration; aggregation.

flocculation value.   See coagulation value.

flock.   A light powder, composed of ground wood or cotton fibers, used as an extender or filler in plastics, low-grade rubber, and flooring compositions.

"Flo-Fre" *[Oil-Dri].*   (bentonite).
CAS: 1302-78-9.
TM for flowability aid for soybean meal and other feeds.
Use: Drying agent in fertilizer impregnations.

"Flo-Gard" *[PPG].*   TM for amorphous calcium polysilicate used as an anticaking agent for salt.

Flomet-Z.   A fine, white, grit-free powder containing 12.5–14.0% zinc oxide.
Use: Lubricant in powdered iron metallurgy.

flooding, chemical.   See chemical flooding.

Flood reaction.   Formation of trialkylsilyl halides from hexaalkyldisiloxanes using concentrated sulfuric acid in the presence of ammonium chloride or fluoride, or by treatment of the intermediate silane sulfates with hydrochloric acid in the presence of ammonium sulfate.

Floropryl.   Diisopropyl fluorophosphate.

Flory, Paul John.   (1910–1985). An American chemist who won the Nobel Prize in 1974 for his work in polymer chemistry. He published extensive work on the physical chemistry of polymers and macromolecules. He held many medals and awards. Flory received his doctorate from Ohio State University in 1934. He was the C. H. Wood Professor of Chemistry at Stanford University.

flotation.   A method of separating minerals from waste rock or solids of different kinds by agitating the pulverized mixture of solids with water, oil, and special chemicals that cause preferential wetting of solid particles of certain types by the oil, while other kinds are not wet. The unwetted particles are carried to the surface by the air bubbles and thus separated from the wetted particles. A frothing agent is also used to stabilize the bubbles in the form of a froth that can be easily separated from the body of the liquid (froth flotation). Do not confuse with floatation.

flow cytometry.   Analysis of biological material by detection of the light-absorbing or fluorescing properties of cells or subcellular fractions (i.e., chromosomes) passing in a narrow stream through a laser beam. An absorbance or fluorescence profile of the sample is produced. Automated sorting devices, used to fractionate samples, sort successive droplets of the analyzed stream into different fractions depending on the fluorescence emitted by each droplet.

flow diagram.   (flow sheet). A chart or line drawing used by chemical engineers to indicate successive steps in the production of a chemical, materials input and output, by-products, waste, and other relevant data.

flow karyotyping.   Use of flow cytometry to analyze and/or separate chromosomes on the basis of their DNA content.

flowers.   A fine powder usually resulting from sublimation of a substance, e.g., flowers of sulfur. The term is now obsolete.

flowers of Benjamin.   See benzoic acid.

flox. A mixture of liquid fluorine (30%) and liquid oxygen (70%), designed for use as a space-vehicle propellant.
Hazard: Explosively flammable.

fluazifop-butyl.
CAS: 69806-50-4. $C_{19}H_{20}F_3NO_4$.
Properties: Pale straw-colored, odorless liquid. Bp 165C, mp 13C, d 1.21 @ 20C.
Hazard: Moderately toxic by ingestion.
Use: Food additive; herbicide; agricultural chemical.

fluazinam. See 3-chloro-n-(5-chloro-2,6-dinitro-4-trifluoromethylphenyl)-5.

flucarbazone-sodium.
CAS: 181274-17-9. $C_{12}H_{10}F_3N_4O_6S\cdot Na$.
Hazard: Moderately toxic.

flucythrinate. See (+)cyano(3-phenoxyphenyl)methyl(ñ)-1-(difluoromethoxy).

fluid. Any material or substance that changes shape or direction uniformly in response to an external force imposed on it. The term applies not only to liquids but also gases and finely divided solids. Fluids are broadly classified as Newtonian and non-Newtonian depending on their obedience to the laws of classical mechanics.
See liquid, Newtonian; rheology; fluidization; hydraulic fluid.

fluid bed. See fluidization.

fluidization. A technique in which a finely divided solid is caused to behave like a fluid by suspending it in a moving gas or liquid. The solids so treated are frequently catalysts, hence the term *fluid catalysis*. The fluidized catalyst, e.g., alumina-silica gel, is brought into intimate contact with the suspending liquid or gas mix, usually a petroleum fraction. Local overheating of the catalyst is greatly reduced, and portions of catalyst can be easily removed for regeneration without shutting down the unit. There are also noncatalytic applications of fluidization, e.g., reduction of iron ore. Important uses of the fluidized bed process are (1) cracking of petroleum fractions, (2) gasification of coal, (3) application of organic coatings to metals (fusion bond method), (4) coal combustion, in which sulfur-bearing coal (1.33 inch diameter) is fed into a fluidized bed of limestone. Combustion occurs at 1600F, at which temperature the limestone is reduced to lime, which reacts with the sulfur in the coal to form gypsum. This technique makes possible the use of high-sulfur coal without necessity of scrubbers. The bed material is approximately 5% coal and 95% limestone products.

fluid mosaic model. A model proposed by Singer and Nicholson, describing biological membranes as a fluid lipid bilayer with embedded proteins that create a mosaic. Both protein and lipid components of the bilayer are able to move but exhibit both structural and functional asymmetry.

fluid, supercritical. See supercritical fluid.

fluoboric acid. (fluoroboric acid; hydrogen tetrafluoroborate).
CAS: 16872-11-0. HBF_4.
Properties: Colorless. Strongly acidic liquid. D approximately 1.84; bp 130C (decomposes). Miscible with water and alcohol.
Derivation: Action of boric and sulfuric acids on fluorspar.
Grade: Technical (approximately 48%), pure.
Hazard: Highly toxic, corrosive, irritant.
Use: Production of fluoborates, electrolytic brightening of aluminum, throwing power aid in electrolytic plating baths, esterification catalyst, metal cleaning, making stabilized diazo salt.

fluometuron. (N-(3-trifluoromethylphenyl)-N', N'-dimethylurea).
CAS: 2164-17-2. $C_{10}H_{11}F_3N_2O$.
Properties: Crystalline solid. Mp 163C. Partially soluble in water; soluble in alcohol and acetone.
Derivation: Reaction of dimethylamine with 3-trifluoromethylphenyl isocyanate.
Hazard: Toxic by ingestion. Questionable carcinogen.
Use: Herbicide.

fluophosphate alkyl ester. See diisopropyl fluorophosphate.

fluor. See phosphor.

fluoranthene. (idryl).
CAS: 206-44-0. $C_{16}H_{10}$.
A tetracyclic hydrocarbon.
Properties: Colored needles. Fp 107C, bp 250C (60 mm Hg). Insoluble in water; soluble in ether and benzene. Combustible.
Hazard: Questionable carcinogen.
Derivation: From coal tar.

fluorapatite. See apatite.

fluorbenside.
CAS: 405-30-1. $ClC_6H_4CH_2SC_6H_4F$.
(generic name accepted for p-chlorobenzyl-p-fluorophenyl sulfide).
Properties: Crystals, mp 36C, insoluble in water, soluble in acetone and oils.
Use: Acaricide.

"Fluorel" [3M]. TM for a fully saturated fluorinated polymer containing more than 60% fluorine by weight.
Use: O-rings, gaskets, hoses, wire and fabric coatings, diaphragms, fuel cells, expellent bladders, sealants, insulation, containers.

fluorene.   (α-diphenylenemethane).
CAS: 86-73-7.

Properties: Small, white, crystalline plates; fluorescent when impure. Mp 116C, bp 295C (decomposes). Soluble in alcohol, ether, benzene, and carbon disulfide; insoluble in water. Combustible.
Derivation: By reduction of diphenylene ketone with zinc, from coal tar.
Grade: Technical, 98% pure.
Hazard: Questionable carcinogen.
Use: Resinous products, dyestuffs.

9-fluorenone.
CAS: 486-25-9. $C_{13}H_8O$.
Properties: Solid. Mw 180.21, mp 82–85C, bp 342C.
Use: Intermediate.

fluorescein.   (resorcinolphthalein;   diresorcinolphthalein; CI 45350).
CAS: 2321-07-5. $C_{20}H_{12}O_5$.
Properties: Orange-red, crystalline powder. Very dilute alkaline solutions exhibit intense greenish-yellow fluorescence by reflected light, while the solution is reddish-orange by transmitted light. Mp decomposes at 290C; soluble in dilute alkalies, boiling alcohol, ether, dilute acids, and glacial acetic acid; insoluble in water, benzene, chloroform. Combustible.
Derivation: By heating phthalic anhydride and resorcinol.
Grade: The sodium salt (uranine) and potassium salt are marketed.
Use: Dyeing seawater for spotting purposes, tracer to locate impurities in wells, dyeing silk and wool, diagnostic aid in ophthalmology, indicator and reagent for bromine.
See uranine.

fluorescence.   A type of luminescence in which an atom or molecule emits visible radiation in passing from a higher to a lower electronic state. The term is restricted to phenomena in which the time interval between absorption and emission of energy is extremely short (10^{-8} to 10^{-3} second). This distinguishes fluorescence from phosphorescence, in which the time interval may extend to several hours. Fluorescent materials may be liquid or solid, organic or inorganic. Fluorescent crystals such as zinc or cadmium sulfide are used in lamp tubes, television screens, scintillation counters, and similar devices. Fluorescent dyes are used for labeling molecules in biochemical research.
See phosphorescence; phosphor; resonance (2).

fluorescent   antibody.   An   immunoglobulin coupled with a fluorescent dye. A fluorescence microscope is used to identify antigen on cells or microorganisms.

fluorescent brightener 85.   See C.I. fluorescent brightener 85.

fluorescent brightener 220.   See C.I. fluorescent brightener 220.

fluoridation.   Addition to public drinking water supplies of 1 ppm of a fluoride salt for the purpose of reducing the incidence of dental caries. The chemicals most commonly used for city fluoridation programs are fluosilicic acid, sodium silicofluoride, and sodium fluoride. The concentration used has been established to be far below the permissible level of toxicity of fluorine-containing compounds in the human body. The program was successfully tested for over 20 years on local populations and since then has been widely adopted in large cities in the U.S. The protection is especially effective for children, whose teeth are usually more susceptible to caries than those of adults. Fluorine is a bone-seeking element; tooth protection is due to the ability of fluoride ion to replace other ions in hydroxyapatite, the chief mineral component of bones and teeth. Fluorides are used in toothpastes and other dentifrices.

fluoride.   (fluoride(1-); fluoride ion; fluoride ion (1-); perfluoride).
CAS: 16984-48-8.   F.
Any inorganic salt of hydrofluoric acid in which fluorine has a valence of −1. Fluorides are major environmental pollutants released into the atmosphere from aluminum reduction, steel manufacturing, and coal-fired power plants. Fluoride pollution is associated with extensive damage to livestock, agricultural crops, and timber. Fluoride has a strong tendency to bioaccumulate and can be hazardous at very low environmental concentrations.
Hazard: Highly toxic; poison; mutagen; can cause convulsions, changes in the respiratory system, liver and kidneys.

fluorinated ethylene-propylene resin.   (FEP resin). A copolymer of tetrafluoroethylene and hexafluoropropylene with properties similar to polytetrafluoroethylene resin. The repeating structure of the molecule is $[–CF_2–CF_2–CF_2CF(CF_3)–]_n$.
See Teflon.
Properties: Similar to polytetrafluoroethylene but has a higher coefficient of friction.
Available forms: Extrusion and molding powder, aqueous dispersion, film, monofilament fiber, and nonsticking finish.
Use: Wire and cable insulation, pipe linings, lining for processing equipment. Fibers are used for filtration screening and mist separators.

fluorine.
CAS: 7782-41-4. F.
Nonmetallic halogen element in group 17 of the periodic table. An 9, aw 18.99840, valence of 1, no other stable isotopes, the most electronegative element and most powerful oxidizing agent known.
Properties: Pale-yellow diatomic gas or liquid; pungent odor. Bp −188C, fp −219C, d (gas) 1.695 (air = 1.29), d (liquid) 1.108 (−188C), sp vol 10.2 cu ft/lb (21C). Reacts vigorously with most oxidizable substances at room temperature, frequently with ignition; forms fluorides with all elements except helium, neon, and argon.
Occurrence: Widely distributed to the extent of 0.03% of the earth's crust. The chief minerals are fluorapatite, cryolite, and fluorspar (Spain, Mexico, South Africa).
Derivation: Electrolysis of molten anhydrous hydrofluoric acid-potassium fluoride melts with special copper-bearing carbon anodes, steel cathodes and containers, and Monel screens.
Hazard: Powerful oxidizing agent; though nonflammable, it reacts violently with a wide range of both organic and inorganic compounds and thus is a dangerous fire and explosion risk in contact with such materials. Toxic by inhalation, extremely strong irritant to tissue.
Use: Production of metallic and other fluorides; production of fluorocarbons, active constituent of fluoridating compounds used in drinking water, toothpastes, etc.

fluorine cyanide. See cyanogen fluoride.

fluorine nitrate.
CAS: 7789-26-6. FNO_3.
Properties: Gas or liquid. Ignites in contact with alcohol, aniline, and ether.
Hazard: Strong oxidizing agent, liquid explodes on shock or friction.
Use: Rocket propellants.

fluorine perchlorate. (chlorine tetroxyfluoride; perchloryl hypofluorite). $ClFO_4$.
Properties: Colorless, explosive gas with a pungent, acrid odor.
Hazard: Extremely poisonous.

fluorite. (calcium difluoride).
CAS: 14542-23-5. CaF_2.
A pure naturally occurring form of calcium fluoride.
Use: Was widely used as a flux in metallurgical industries.

fluoroacetamide. (AFL 1081; compound 1081; FAA; fluorakil 100; 2-fluoroacetamide; fluoroacetic acid amide; fussol; megatox; monofluoroacetamide; navron; RCRA waste number P057; rodex; yanock). C_2H_4FNO.
Properties: Crystalline solid, freely soluble in water and Mg_2CO; slightly soluble in $ClCH_4$.
Hazard: Extremely toxic; poison; mutagen.
Use: Rodenticide, insecticide.

fluoroacetate. (2-fluoroacetic acid).
$C_2H_3FO_2$. Any salt of fluoroacetic acid in which carbon-bound hydrogen atoms are replaced by fluorine atoms.
Hazard: Extremely poisonous; may cause increase respiratory rate, vomiting, facial paresthesias, central nervous system stimulation, convulsions, cardiac arrhythmias, death from ventricular fibrillation.
Use: Rodenticide.

fluoroacetic acid.
CAS: 144-49-0. CH_2FCOOH.
Properties: Colorless crystal. Mp 33C, bp 165C. Soluble in water and alcohol.
Hazard: Toxic by ingestion.
Use: Rodenticide.

fluoroacetophenone. (phenacyl fluoride; phenyl fluoromethylketone).
$C_6H_5COCH_2F$.
Properties: Brown liquid; pungent odor. Bp 98C (8 mm Hg).
Derivation: By Friedel-Crafts synthesis.
Hazard: Irritant.

p-**fluoroaniline.**
CAS: 371-40-4. $FC_6H_4NH_2$.
Properties: Liquid. D 1.1524 (25C), bp 187.4C, fp −2C, refr index 1.5395 (20C).
Hazard: Toxic material.
Use: Intermediate (herbicides), preparation of *p*-fluorophenol.
See aniline.

fluorobenzene. (phenyl fluoride).
CAS: 462-06-6. C_6H_5F.
Properties: Colorless liquid; benzene odor. D 1.0252 (20C), refr index 1.4646 (25C), bp 84.9C, fp −40C. Insoluble in water, miscible with alcohol and ether.
Hazard: Flammable, dangerous fire risk. Irritant.
Use: Insecticide and larvicide intermediate, identification reagent for plastic or resin polymers.

1-(2-(4-(6-fluoro-1,2-benzisoxazol-3-yl)-1-piperidinyl)ethyl)-3-phenyl-2-imidazolidinone.
CAS: 200398-40-9. $C_{23}H_{25}FN_4O_2$.
Hazard: A poison by ingestion.

p-**fluorobenzoic acid 2-phenylhydrazide.**
CAS: 1496-02-2. $C_{13}H_{11}FN_2O$.
Hazard: A poison by ingestion.
Use: Agricultural chemical.

fluoroboric acid. Legal label name (Air) for fluoboric acid.

fluorocarbon. Any of a number of organic compounds analogous to hydrocarbons in which the

hydrogen atoms have been replaced by fluorine. The term is loosely used to include fluorocarbons that contain chlorine; these should properly be called chlorofluorocarbons or fluorocarbon chloride, since they deplete the ozone layer of the upper atmosphere.
See chlorofluorocarbon.

Properties: Fluorocarbons are chemically inert, nonflammable, and stable to heat up to 260–315C. They are denser and more volatile than the corresponding hydrocarbons and have low refractive indices, low dielectric constants, low solubilities, low surface tensions, and viscosities comparable to hydrocarbons. Some are compressed gases and others are liquids.

Hazard: Nonflammable; reacts violently with reactive substances, e.g., barium, sodium, and potassium.

Use: Refrigerants, solvents, blowing agents, fire extinguishment, lubricants and hydraulic fluids, flotation and damping fluids, dielectric, plastics, electrical insulation, wax coatings for alkali cleaning tanks, air- conditioning.

Note: Many of these compounds are designated by a number system preceded by the word *refrigerant*, *propellant*, *fluorocarbon* or by a TM ("Genetron" *[Honeywell]* and "UCon" *[Union]*).
They are cross-referenced in this book as follows:

11	See trichlorofluoromethane.
12	See dichlorodifluoromethane.
13	See chlorotrifluoromethane.
14	See tetrafluoromethane.
21	See dichlorofluoromethane.
22	See chlorofluoromethane.
23	See fluoroform.
113	See 1,1,2-trichloro-1,2,2-trifluoroethane.
114	See 1,2-dichloro-1,1,2,2-tetrafluoroethane.
115	See chloropentafluoroethane.
116	See hexafluoroethane.

fluorocarbon polymer. This term includes polytetrafluoroethylene, polymers of chlorotrifluoroethylene, fluorinated ethylene-propylene polymers, polyvinylidene fluoride, hexafluoropropylene, etc.

Properties: Thermoplastic; resistant to chemicals and oxidation; broad useful temperature range (up to 287C); high dielectric constant; resistant to moisture, weathering, ozone, and UV radiation. Their structure comprises a straight backbone of carbon atoms symmetrically surrounded by fluorine atoms. Noncombustible.

Available forms: Powders, dispersions, film, sheet, tubes, rods, tapes, and fibers.

Use: High-temperature wire and cable insulation, electrical equipment, drug and chemical equipment, coating of cooking utensils, piping gaskets, continuous sheet, bonding industrial diamonds to metal (grinding wheels).
See fluoroelastomer.

fluorochemical. Organic compounds, not necessarily hydrocarbons, in which a large percentage of the hydrogen directly attached to carbon has been replaced by fluorine. The presence of two or more fluorine atoms on a carbon atom usually imparts stability and inertness to the compound, and fluorine usually increases the acidity of organic acids.

Derivation: (1) Electrolysis of solutions in hydrogen fluoride (Simons process); (2) replacement of chlorine or bromine by fluorine with hydrogen fluoride in the presence of a catalyst (antimony trifluoride or pentafluoride); (3) addition of hydrogen fluoride to olefins or acetylene.

Use: Dielectric and heat-transfer liquids, pump sealants, surfactants, metering devices, special solvents.
See fluorocarbon; fluoroelastomer.

fluorochrome. ((2',7'-dibromo-3',6'-dihydroxy-3-oxospiro[2-benzofuran-1,9'-xanthene]-4'-yl) mercury; sodium; hydrate). $C_{20}H_{11}Br_2HgNa_2O_6$. Any fluorescent dye.

Use: In staining tissues and cells for examination by fluorescence microscopy.

5-fluorocytosine. (flucytosine).
CAS: 2022-85-7. $C_4H_4FN_3O$.
Properties: Crystalline solid. Mw 129.09, mp 295C (decomposes), light sensitive.

fluorodichloromethane. See dichlorofluoromethane.

fluoroelastomer. Any elastomeric high polymer containing fluorine; they may be homopolymers or copolymers. Fluorocarbon polymers include a large group of fluoroelastomers, including a copolymer in which the molecular skeleton is a -P=N- chain containing approximately equal numbers of tri- and heptafluoroethoxy side groups. Such polymers are amorphous, thermally stable, noncombustible. Generally resistant to attack by solvents and chemicals, have low glass transition temperature (−77C), and are generally resistant to attack by solvents and chemicals.

fluoroform. (trifluoromethane; propellant 23; refrigerant 23).
CAS: 75-46-7. CHF_3.
Properties: Colorless gas. Bp −84C, mp −160C. Nonflammable.
Grade: 98% min purity.
Use: Refrigerant, intermediate in organic synthesis, direct coolant for infrared detector cells, blowing agent for urethane foams.

fluoroformyl fluoride. See carbonyl fluoride.

Fluorographite. Fluorinated graphite cathode material.
Use: Lithium anode batteries, lubricant chromatographic materials.

Fluoroinert. A series of perfluorinated liquids used for cleaning electronic components after testing, bp range 31–173C, high dielectric strength, colorless, nonflammable.

fluorol. See sodium fluoride.

"Fluorolubes" *[Gabriel].* TM for polymers of trifluorovinyl chloride $(-CF_2-CFCl-)_x$ containing 49% fluorine and 31% chlorine. Products are light oils, heavy oils, and grease-like materials.
Use: Lubricant and sealant for plug cocks, valves, and vacuum pumps; impregnant for gaskets and packings; fluid for hydraulic equipment, heat exchange, and instrument damping.

fluoromethane. (methyl fluoride).
CAS: 593-53-3. CH_3F.
Properties: Colorless gas. Bp −78.2C, fp −142C, d 1.19 (air = 1.29). Soluble in alcohol and ether.
Hazard: Flammable. Narcotic in high concentrations.

fluorometholone. (9-fluoro-11,17-dihydroxy-6-methylpregna-1,4-diene-3,20-dione; fluormetholone).
CAS: 426-13-1. $C_{22}H_{29}FO_4$.
Properties: Crystalline solid. Mp 292–303C.
Use: A steroid, glucocorticoid, antiinflammatory.

3-(6-fluoro-2-naphthalenyl)-2,5-dihydro-1,2-dimethyl-1h-pyrrole, el-(2r,3r)-2,3-dihydroxybutanedioate (1:1).
CAS: 326800-76-4. $C_{16}H_{16}FN•C_4H_6O_6$.
Hazard: A poison.

3-(6-fluoro-2-naphthalenyl)-1,2-dimethyl (2r,3s)-rel-3-pyrrolidinol hydrochloride.
CAS: 302959-28-0. $C_{16}H_{18}FNO•ClH$.
Hazard: A poison.

fluoronitrofen.
CAS: 13738-63-1. $C_{12}H_6Cl_2FNO_3$.
Hazard: Moderately toxic by ingestion.
Use: Agricultural chemical.

p-**fluorophenol.** (4-fluorophenol).
CAS: 371-41-5. FC_6H_4OH.
Properties: White, crystalline solid. D 1.1889 (56C), mp 48.2C (stable form), 28.5C (unstable form), bp 185.6C. Soluble in water.
Hazard: Irritant.
Use: Fungicide, intermediate for pharmaceuticals.

2-(4-fluorophenyl)imidazo(2,1-a)isoquinoline.
CAS: 61001-09-0. $C_{17}H_{11}FN_2$.
Hazard: A reproductive hazard.

3-(4-fluorophenyl)-n-(4-propylcyclohexyl)-2-propenamide.

CAS: 315706-75-3. $C_{18}H_{24}FNO$.
Hazard: A poison by ingestion.

8-fluoro-*n*-(2-(4-phenyl-2-thiazolyl)ethyl)-4-quinolinamine.
CAS: 124533-50-2. $C_{20}H_{16}FN_3S$.
Hazard: Moderately toxic by ingestion.

fluorophosphoric acid, anhydrous.
CAS: 13537-32-1. H_2PO_3F.
Properties: Colorless, viscous liquid. D 1.1818 (25C). Miscible with water.
Hazard: Strong irritant to tissue.
Use: Metal cleaners, electrolytic or chemical polishing agents, formation of protective coatings for metal surfaces, catalyst.
See difluorophosphoric acid; hexafluorophosphoric acid.

4-fluoro-*n*-(4-propylcyclohexyl)benzamide.

CAS: 315706-67-3. $C_{16}H_{22}FNO$.
Hazard: A poison by ingestion.

6-fluoro-7-(1-pyrrolyl)-1-ethyl-1,4-dihydro-4-oxo-3-quinolinecarboxylic acid.
CAS: 91524-15-1. $C_{16}H_{13}FN_2O_3$.
Hazard: Low toxicity by ingestion.

fluorosilicate. (fluosilicate; fluorosilicate salt; hexafluorosilicon(2-)). F_2Si_2. Any salt of fluorosilic acid.
Properties: More soluble than the corresponding fluoride salt.
Hazard: Very toxic.

fluorosilicic acid. See fluosilicic acid.

fluorosulfonic acid. See fluosulfonic acid.

fluorosulfuric acid. See fluosulfonic acid.

fluorothane. (ethyl fluoride). CH_3CH_2F.
Use: Replaces ether in surgery.

n-**(2-fluoro-1,1,2,2-tetrachloroethylthio)-methanesulfoanilide.**
CAS: 22729-75-5. $C_9H_8Cl_4FNO_2S_2$.
Hazard: Low toxicity by ingestion.
Use: Agricultural chemical.

5-fluoro-3-(tetrahydro-2-furyl)uracil.
CAS: 63901-83-7. $C_8H_9FN_2O_3$.
Hazard: Moderately toxic by ingestion.

fluorothene. See chlorotrifluoroethylene polymer.

8-fluoro-*n*-(2-(2-thienyl)ethyl)-4-quinolinamine.
CAS: 124533-68-2. $C_{15}H_{13}FN_2S$.
Hazard: Moderately toxic by ingestion.

fluorotrichloromethane. See trichlorofluoromethane.

5-fluorouracil. [5-fluoro-2,4(1H,3H)-pyrimidinedione].
CAS: 51-21-8. $C_4H_3FN_2O_2$.
Properties: Crystalline solid. Mw 130.08, mp 282C (decomposes). Soluble in water or methanol-water mixtures.
Hazards: Questionable carcinogen.
Use: Antineoplastic agent.

fluorspar. (fluorite; florspar).
CAS: 7789-75-5. CaF_2. Natural calcium fluoride; yellow, green, or purple crystals. Mohs hardness 4, d 3.2, mp 1350C.
Grade: Metallurgical; ceramic and acid, containing more than 85 and 98% CaF_2, respectively.
Occurrence: U.S., Canada, Europe, Mexico.
Use: Principal source of fluorine and its compounds by way of hydrogen fluoride, flux in open-hearth steel furnaces and in metal smelting, in ceramics, for synthetic cryolite, in carbon electrodes, emery wheels, electric arc welders, certain cements, dentifrices, phosphors, paint pigment, catalyst in wood preservatives, optical equipment.

fluosilicate. A salt of fluosilicic acid, H_2SiF_6.

fluosilicic acid. (hydrofluosilicic acid; fluorosilicic acid; hexafluorosilicic acid; hydrogen hexafluorosilicate; hydrosilicofluoric acid; hydrofluorosilicic acid).
CAS: 16961-83-4. H_2SiF_6.
Properties: Aqueous solution: Colorless fuming liquid. Attacks glass and stoneware.
Derivation: By-product of the action of sulfuric acid on phosphate rock containing fluorides and silica or silicates. The hydrogen fluoride acts on the silica to produce silicon tetrafluoride, SiF_4, which reacts with water to form fluosilicic acid, H_2SiF_6.
Grade: Technical, CP.
Hazard: Extremely corrosive by skin contact and inhalation.
Use: Water fluoridation, ceramics (to increase hardness), disinfecting copper and brass vessels, hardening cement, etc., wood preservative and impregnating compounds, electroplating, manufacture of aluminum fluoride, synthetic cryolite and hydrogen fluoride, sterilizing bottling and brewing equipment (1–2% solution).

fluosulfonic acid. (fluorosulfuric acid; fluorosulfonic acid).
CAS: 7789-21-1. HSO_3F.
Properties: Colorless, fuming liquid. D 1.745 (15C); fp −87C; bp 165C. Soluble in nitrobenzene. Reacts violently with water, does not attack glass.
Derivation: Reaction of anhydrous hydrogen fluoride with sulfuric acid or sulfuric acid anhydride.
Hazard: Extremely irritating to eyes and tissue.

Use: Catalyst in organic synthesis, electropolishing, fluorinating agent.

flushed color. A pigment dispersed in oil, varnish, etc., the transfer from the water phase to the oil phase having been effected without the usual drying and subsequent grinding of the dry pigment. It is claimed that flushed colors are ready for use without grinding.

fluvoxamine.
CAS: 54739-18-3. $C_{15}H_{21}F_3N_2O_2$.
Hazard: A poison.

fluvoxamine maleate.
CAS: 61718-82-9. $C_{15}H_{21}F_3N_2O_2 \cdot C_4H_4O_4$.
Hazard: A poison by ingestion. Human systemic effects.

flux. (1) A substance that promotes the fusing of minerals or metals or prevents the formation of oxides. For example, in metal refining lime is added to the furnace charge to absorb mineral impurities in the metal. A slag is formed that floats on the bath and is run off. (2) A substance applied to metals that are to be united. On application of heat, it aids the flow of solder and prevents formation of oxides. (3) Any readily fusible glass or enamel used as a base or ground in ceramic processing. (4) The rate of flow or transfer of electricity, magnetism, water, heat, energy, etc., the term being used to denote the quantity that crosses a unit area of a given surface in a unit of time. (5) The intensity of neutron radiation, expressed as the number of neutrons passing through one square centimeter in one second. (6) A mixture of sodium nitrate and sodium nitrite; oxidizing agent used as a low explosive.

fluxing lime. See calcium oxide.

fly ash. The very fine ash produced by combustion of powdered coal with forced draft and often carried off with the flue gases. Special equipment is required for effective recovery, e.g., electrostatic precipitators. Fly ash is a mixture of alumina, silica, unburned carbon, and various metallic oxides. It is reported to have mutagenic properties after passing through stack precipitators. The alumina is recoverable by calsintering, which makes it a possible alternative source of aluminum.
Use: Cement additive for oil-well casings, absorbent for oil spills (silicone-coated), to replace lime in scrubbing sulfur dioxide from flue gas, as a filler in plastics, source of germanium (England), proposed as catalyst for coal liquefaction, removal of heavy metals from industrial wastewaters, separation of oil-sand tailings.
See calsintering.

Fm. Symbol for fermium.

FMN. Abbreviation for flavin mononucleotide. See riboflavin phosphate.

foam. A dispersion of a gas in a liquid or solid. The gas globules may be of any size, from colloidal to macroscopic, as in soap bubbles. Bakers' bread and sponge rubber are examples of solid foams. Typical liquid foams are those used in fire-fighting, shaving creams, etc. In such foams, the liquid must have sufficient cohesion to form an elastic film, e.g., soap, oil, protein, fatty acids, etc. Surfactant-induced foams have been developed to increase the efficiency of fuel cells.

Foams made by mechanical incorporation of air are widely used in the food industry, e.g., whipped cream, egg white, ice cream, etc. Useful foams for automobile seats, mattresses, and similar uses are made from natural and synthetic latexes, e.g., polystyrene, polyurethane.

A glass foam is based on sodium silicate and rock wool, and vitreous ceramic foams are also available. Metals can be caused to foam. Concrete foams are also in general use.

Foams designed for fire extinguishment are agglomerations of small bubbles of gas produced by two methods: (1) by chemical reaction between aluminum sulfate and sodium bicarbonate to generate carbon dioxide (chemical foams), and (2) by mixing or agitation of air with water containing the foaming ingredients (mechanical foams). The two types are equally efficient in fire-extinguishing ability.

Besides the foaming ingredients, the foams contain stabilizing agents to assure permanence; there are many of these, e.g., soaps, proteins, extract of licorice root, fatty acids, and sulfite liquors. The ingredients of chemical foams are assembled in two separate units, which generate the foam on blending.

Fire foams are used primarily on fires in hydrocarbon liquids (class B fires). There are many special types tailored for specific uses. Some fire-protection systems include an instant-generating foam for cabin interiors, using a 2.5% aqueous solution of alkyl sulfonate and rigid polyurethane foams for use in fuel tanks.

See fire extinguishment.

An unusually stable foam that remains intact much longer than fire-fighting foams has been developed at Sandia National Laboratories. It results from a synergistic action caused when a water-soluble polymer and a fatty alcohol are blended with a solvent and a surfactant. A possible agricultural use would be in insecticide application.

foam, metal. A cellular metallic structure, usually of aluminum or zinc alloys, made by incorporating titanium or zirconium hydride in the base metal. This subsequently evolves hydrogen to produce a uniform, foam-like material. Its density is approximately that of seawater so that it is weightless when submerged. The principal use of foamed metals is in absorption of shock impact without elastic rebound. Fiber-reinforced light-metal foams have potential application in reducing the weight of automobile bodies.

foam, plastic. A cellular plastic that may be either flexible or rigid. Flexible foams may be polyurethane, rubber latex, polyethylene, or vinyl polymers; rigid foams are chiefly polystyrene, polyurethane, epoxy, and polyvinyl chloride. The blowing agents used are sodium bicarbonate, halocarbons such as CCl_3F, and hydrazine. Flexible polystyrene foam is available in extruded sheets and also in the form of beads made by treating a polystyrene suspension with pentane; these expand from 30 to 50 times on heating and are used as automobile radiator sealants. Rigid foams are widely used for boat construction, filtration, fillers in packing cases, absorption of oil spills, and building insulation. The latter application involves a fire risk described below.

Hazard: The most widely used types of organic foam plastics (polystyrene, polyurethane, polyisocyanurate) are combustible; even when fire-retardant agents are incorporated, such foams will burn. The extent of burning or fire severity will vary with surface treatment, end-use location, recipe, and degree of protection. Thin coatings of fire-retardant paint, metal, or automatic sprinkler systems may not adequately protect against rapid fire spread. Organic foamed plastic surfaces should not be left exposed. Multiple adjacent surfaces such as walls and ceiling create a most severe hazard because of the chemical kinetics associated with radiative, conductive, and convective currents developed during a building fire.

New methods of making such plastic foams as polyurethane that are reported to reduce their combustibility have been developed, e.g., use of trichlorobutylene oxide instead of propylene oxide.

Use of urea-formaldehyde foams for building insulation has been restricted in some areas due to potential toxic effect of the formaldehyde.

fob. Abbreviation for "freight on board," a designation used in shipping a material to indicate that freight charges are to be paid by the purchaser. This is in contrast to "freight prepaid and allowed," indicating that freight charges are paid by the manufacturer.

fog. A suspension of liquid droplets in air; an aerosol. The size of the droplets ranges from colloidal to macroscopic. "Synthetic" fogs can be produced on a laboratory scale by ultrasonic vibrations, and natural fogs can be precipitated by the same means. Mists or fogs composed of atomized particles of oil are used as military concealment screens and for insecticidal purposes in orchards and truck gardens.

See smog; chemical smoke; aerosol.

folacin. See folic acid.

folic acid.   (pteroylglutamic acid; folacin; PGA).
CAS: 59-30-3. $C_{19}H_{19}N_7O_6$.
Considered a member of the vitamin B-complex. At least three substances with folic acid activity occur in nature, one of which, pteroylglutamic acid, is made synthetically.
Properties: Pteroylglutamic acid: Orange-yellow needles or platelets; tasteless; odorless. Slightly soluble in methanol and sparingly soluble in water; insoluble in acetone, ether, benzene; moderately soluble in dilute alkali hydroxide and carbonate solutions; stable in heat in neutral and alkaline solution; destroyed by heating with acid; inactivated by light.
Source: Green plant tissue, fresh fruit, liver, and yeast. Synthetic pteroylglutamic acid made by the reaction of 2,3-di-bromopropanol, 2,4,5-triamino-6-hydroxypyrimidine, and p-aminobenzoyl glutamic acid.
Grade: 10% feed grade, USP.
Use: Medicine, nutrition, food additive (maximum daily ingestion not to exceed 0.01 mg).
See folinic acid.

folinic acid.   (5-formyl-5,6,7,8-tetrahydrop-teroyl-l-glutamic acid; citrovorum factor; leucoverin).
CAS: 58-05-9. $C_{20}H_{23}N_7O_7$.
A member of the folic acid group of vitamins and a growth factor for the bacterium *Leuconostoc citrovorum*. Folinic acid is an important metabolite of folic acid and may be the active form in cellular metabolism. It is an effective hematopoietic factor. Ascorbic acid and vitamin B_{12} are essential for the conversion of folic acid to folinic acid.
Properties: (*dl* form): Crystals. Decompose 240–250C. Sparingly soluble in water.
Derivation: (1) Prepared by catalytic reduction of folic acid, (2) produced microbially.
Use: Medicine, nutrition, biochemical research.

follicle-stimulating hormone.   (anthrogon; fertinorm; follitropin; FSH; FSH-P; hebin; luteoantine; menotrophin; menotropins; prolan b; thylaentrin; urinary hebin; urofolitrophin).
CAS: 9002-68-0.
A gonadotropic glycoprotein secreted by the anterior pituitary gland. It regulates the metabolic activity of the granulosa of the mammalian ovary; stimulates maturation of ovarian follicles, and production of estrogens. Spermatogenesis is partially under the control of this hormone.
Properties: Solid; soluble in water.

folpet.   (phaltan; (*N*-(trichloromethylthio)-phthalimide)).
CAS: 133-07-3. $C_6H_4(CO)_2NSCCl_3$.
Properties: Light-colored powder. Insoluble in water, slightly soluble in organic solvents.
Use: Fungicide-bactericide for vinyls, paints, and enamels.

Fomerez.   A series of polyester and polyether resins, stannous octoate catalysts, and coupling agents used in the manufacture of urethane foams.

fonofos.   See dyfonate.

"Fonoline" [Sonneborn].   TM for petrolatum of soft consistency and low melting point with color range of white to yellow and meeting USP or NF purity requirements for petrolatum.

fonophos.   (ethylphosphonodithioic acid-*O*-ethyl-*S*-phenyl ester).
CAS: 944-22-9. $C_{10}H_{15}OPS_2$.
Properties: Yellow liquid. Insoluble in water; miscible with organic solvents.
Hazard: Cholinesterase inhibitor.
Use: Insecticide, soil fumigant.

food.   Any substance or mixture that, when ingested by humans or animals, contributes to the maintenance of vital processes. With the exception of sodium chloride and water, all foods are derived from plants, either by direct consumption or by ingestion of animal tissue or such animal products as eggs, milk, etc., which are derived metabolically from vegetable sources. Basic foods are composed of proteins, fats (lipids), and carbohydrates, together with vitamins and minerals. Ancillary items that are associated with foods, though with little or no nutritive value, are collectively called food additives, e.g., flavorings, spices, preservatives, and colorants. Many of these are also plant-derived, though some are now made synthetically.
See plant (1); nutrient.

food additive.   (1) Intentional: The Food Protection Committee of the National Research Council states that a food additive is "a substance or mixture other than a basic foodstuff that is present in food as a result of any aspect of production, processing, storage, or packaging." (2) Unintentional: Substances that may become part of a food product as a result of chance contamination, such as insecticide residues, fertilizers, and the like. The permissible content of insecticide residues has been established by the FDA.
The Food Additives Amendment to the Food, Drug, and Cosmetic Act empowers the FDA to disapprove or discontinue any food additive that it determines to be unsafe at the level of intended use, based on data supplied by the manufacturer.
Note: Unintentional additives should more logically be called contaminants.

Food and Drug Administration.   (FDA). The federal agency responsible for administering and enforcing the Food, Drug and Cosmetics Act, including the Food Additives Amendment, which went into effect in March 1960 and has been extended and updated regularly. It has the authority to require proof of the efficacy and safety of drugs,

foods, and pharmaceuticals; to conduct and evaluate screening tests; and to compel withdrawal from the market of any such product that it finds ineffective or hazardous. It establishes tolerances on food and animal feed additives of all types, including pesticides, as well as on cosmetic products, flammable fabrics, and packaging and labeling materials. It also requires specific statements on labels of the components or ingredients of a product, as well as precautionary warnings. Headquarters are located at 10903 New Hampshire Havenue, Silver Spring, MD 20993. Website: http://www.fda.gov
See Environmental Protection Agency; food additive.

food chain. The sequence of nutritional steps in an ecosystem, with producers at the bottom and consumers at the top.

food color. (certified color). A colorant that may be either dye (soluble) or a lake (insoluble) permissible for use in foods, drugs, or cosmetics by the FDA. The dyes color by solution and the lakes by dispersion. All must satisfy strict regulations as to toxicity.
See FD&C colors.

food engineering. Application of engineering principles to the design of equipment for large-scale food processing, e.g., automatic harvesting devices, dryers of various types, crystallizers, ovens and heat exchangers, comminuting and mixing equipment, distillation units, packaging machines. Food engineering requires an understanding of thermodynamics, conditions of state and equilibrium, rate processes, and transport phenomena. The unit operations of chemical engineering and basic physics and mechanics are also involved.

food starch, modified.
Properties: White powders; tasteless and odorless. Insol in water, alc, ether, chloroform.
Use: Food additive.

food technology. Practice of the techniques used in the preparation of foods for large-scale human use. Among others these include harvesting, postharvest treatment, all forms of cooking, tenderizing, preservation by chemicals, heating, dehydration, drying and freezing, distillation and solvent extraction, milling, refining, hydrogenation, emulsification, packaging materials and storage, labeling, and transportation. Other aspects of food technology are bacteriology, sanitation, quality control, and formulation of ingredients for a wide variety of end products. A recent development of importance is the growth of convenience and quick-service foods.

food web. The nutritional structure of an ecosystem in which more than one species occupies

each level. Thus, there are usually several producer species and several consumer species in a food web.

footprinting. A technique for identifying the nucleic acid sequence bound either by a DNA- or by an RNA-binding protein. This sequence is protected from digestion by the protein.

foots. (soapstock). The mixture of soap, oil, and impurities that precipitates when natural fatty oils are refined by treatment with caustic soda or soda ash. Usually contains 30–50% free and combined fatty acids. A related meaning is the suspended solid matter in crude oils.
Use: Manufacture of relatively low-grade soaps, as a source of free fatty acids.

foots oil. The oil sweated out of slack wax. It takes its name from the fact that it goes to the bottom, or foot, of the pan when sweated.
See foots; slack; wax.

Foray. A monoammonium phosphate-based formulation used to extinguish fires in flammable liquids (Class B fires) and in combustible materials such as wood and paper (Class A fires).

forensic chemistry. See legal chemistry.

forensics (DNA). The use of DNA for identification. Some examples of DNA use are to establish paternity in child support cases; establish the presence of a suspect at a crime scene; and identify accident victims.

forge. A furnace used for heating and softening a metal in preparation for hot working, e.g., wrought iron.

forging. Shaping a heated metal by means of repeated impact, thus improving its strength.

formal. See methylal.

formaldehyde. (oxymethylene; formic aldehyde; methanal).
CAS: 50-00-0. HCHO.
A readily polymerizable gas. Formaldehyde is commercially offered as a 37–50% aqueous solution that may contain up to 15% methanol to inhibit polymerization. These commercial grades are called *formalin*. It is one of the few organic compounds known to exist in outer space.

$$H-C-H$$
$$\overset{\|}{O}$$

Properties: (Gas) Strong, pungent odor. Vap d 1.067 (air = 1.000), vap d 0.815, (−20/4C), bp −19C, fp −118C, autoign temp 806F (430C), soluble in water and alcohol. (Aqueous 37% solution with 15%

methanol) Bp 96C, flash p 122F (50C). (Methanol free) Bp 101C, flash p 185F (85C).

Derivation: Oxidation of synthetic methanol or low-boiling petroleum gases such as propane and butane. Silver, copper, or iron-molybdenum oxide are the most common catalysts.

Grade: Aqueous solutions: 37, 44, 50% inhibited (with varying percentages of methanol) or stabilized or unstabilized (methanol-free); also available in solution in *n*-butanol, ethanol, or urea; USP (37% aqueous solution containing methanol).

See paraformaldehyde.

Hazard: Moderate fire risk. Explosive limits in air 7–73%. Toxic by inhalation, strong irritant, a carcinogen. (Solution) Avoid breathing vapor and avoid skin contact. Confirmed carcinogen.

Use: Urea and melamine resins, polyacetal resins, phenolic resins, ethylene glycol, pentaerythritol, hexamethylenetetramine, fertilizer, disinfectant, biocide, embalming fluids, preservative, reducing agent as in recovery of gold and silver, corrosion inhibitor in oil wells, durable-press treatment of textile fabrics, industrial sterilant, treatment of grain smut, foam insulation, particle board, plywood, a versatile chemical intermediate.

formaldehyde aniline. (formaniline). $C_6H_5NCH_2$.
Properties: Colorless to yellowish crystals. Initial mp 133C, bp 271C, d 1.14 (but these vary somewhat from sample to sample). Soluble in water, ether, and alcohol.
Derivation: Condensation of formaldehyde and aniline.
Hazard: Toxic by ingestion.
Use: Rubber accelerator, intermediate.

formaldehyde bis(2-fluoro-2,2-dinitroethyl) acetal.
CAS: 17003-79-1. $C_5H_6F_2N_4O_{10}$.
Hazard: Moderately toxic by ingestion.

formaldehyde cyanohydrin. See glycolonitrile.

formaldehyde, polymer with (chloromethyl) oxirane and phenol.
CAS: 9003-36-5. $(C_6H_6O \cdot C_3H_5ClO \cdot CH_2O)_x$.
Hazard: Moderately toxic by ingestion and skin contact. A mild skin irritant.

formaldehyde, polymer with 4-nonylphenol and oxirane.
CAS: 30846-35-6. $(C_{15}H_{24}O \cdot C_2H_4O \cdot CH_2O)_x$.
Hazard: A mild eye irritant.

formaldehyde-*p*-toluidine. (methylene-*p*-toluidine). $(CH_3C_6H_4NCH_2)_x$.
Properties: White powder with grayish-yellow cast; aromatic odor. D 1.11. Soluble in acetone.
Derivation: Reaction between formaldehyde and *p*-toluidine.
Use: Rubber accelerator, dyes.

formalin. An aqueous 37–50% solution of formaldehyde that may contain 15% methyl alcohol.
See formaldehyde.

formamidine disulfide. $C_2H_6N_4S_2$. An organic compound and its derivatives formed by substitution of nitrogen.

formamide. (methanamide).
CAS: 75-12-7. $HCONH_2$.
Properties: Colorless, hygroscopic, oily liquid. D 1.146, bp 200–212C with partial decomposition beginning at approximately 180C, mp 2.5C, flash p 310F (154C). Soluble in water and alcohol. Combustible.
Derivation: Interaction of ethyl formate and ammonia, with subsequent distillation.
Hazard: Toxic material. Toxic by skin absorption.
Use: Solvent, softener, intermediate in organic synthesis.

formaniline. See formaldehyde aniline.

"Formcel" *[Celanese].* TM for a series of water-free formaldehyde solutions in alcohols.
Use: Alcoholated urea and melamine resins, embalming fluids.

formetanate. (3-dimethylaminoethyleneimino-phenyl-*N*-methylcarbamate hydrochloride).
CAS: 23422-53-9.
Properties: Water soluble.
Use: Acaricide for deciduous fruits.

Formica Brand Laminate. High-pressure laminated sheets of melamine and phenolic plastics for decorative applications such as surfacing, adhesives for bonding laminated plastic to other surfaces.

formic acid. (hydrogen carboxylic acid; methanoic acid).
CAS: 64-18-6. HCOOH.
Properties: Colorless, fuming liquid; penetrating odor. D 1.2201 (20/4C), mp 8.3C, bp 100.8C, flash p 156F (69C) (OC), bulk d 10.16 lb/gal (20C), refr index 1.3719 (20C), autoign temp 1114F (600C), strong reducing agent. Soluble in water, alcohol, and ether. Combustible.
Derivation: (1) By treatment of sodium formate and sodium acid formate with sulfuric acid at low temperatures and distilling in vacuo; (2) by acid hydrolysis of methyl formate; (3) as a by-product in the manufacture of acetaldehyde and formaldehyde.
Method of purification: Rectification.
Grade: Technical, 85%, 90%, CP, FCC.
Hazard: Corrosive to skin and tissue.
Use: Dyeing and finishing of textile; leather treatment; chemicals (formates, oxalic acid, organic esters); manufacture of fumigants, insecticides, refrigerants; solvents for perfumes, lacquers;

electroplating; brewing (antiseptic); silvering glass; cellulose formate; natural latex coagulant; ore flotation; vinyl resin plasticizers.

formic aldehyde.   See formaldehyde.

formicidae.   A family of polymorphic colonial hymenopterous insects (ants), 2 to 25 mm long. They are usually black, brown, or reddish in color. Most species are scavengers or predators. Most ants are venomous. The venom is used in defense and to kill insects and other small animals on which some species prey. Envenomation by ants may cause painful reactions, but is rarely fatal to humans.

formonitrile.   See hydrocyanic acid.

Formopon.   Sodium formaldehyde hydrosulfite. Formopon Extra is the basic zinc salt.

formothion.   [S-(N-formyl-N-methylcarbamoyl-methyl)-dimethyl phosphorodithioate]. CAS: 2540-82-1. $C_6H_{12}NO_4PS_2$.
Properties: Yellow liquid. Fp 25C. Insoluble in water; miscible with common organic solvents.
Hazard: Cholinesterase inhibitor.
Use: Systemic insecticide.

Form Release Gold.   A general use premium type form release product.
Use: For steel, aluminum, plywood and composition forms.

Formrez.   Specialty oxylates or polyethers.
Use: In polyurethane foams, prepolymers, coatings, elastomers, adhesives, caulks, and sealants.

formula, chemical.   A written representation using symbols of a chemical entity or relationship. There are several kinds of formulas: (1) Empirical. Expresses in simplest form the *relative* number and the kind of atoms in a molecule of one or more compounds; it indicates composition only, not structure. Example: CH is the empirical formula for both acetylene and benzene. (2) Molecular. Shows the actual number and kind of atoms in a chemical entity (i.e., a molecule, group, or ion). Examples: H_2 (one molecule of hydrogen), $2H_2SO_4$ (two molecules of sulfuric acid), CH_3 (a methyl group), $Co(NH_3)_6^{2+}$ (an ion). (3) Structural. Indicates the location of the atoms, groups, or ions relative to one another in a molecule, as well as the number and location of chemical bonds. Examples:

$$CH_2{=}\overset{\overset{\textstyle CH_3}{|}}{C}{-}CH{=}CH_2 \text{ (isoprene)}$$

benzene

Since all molecules are three-dimensional, they cannot properly be shown in the plane of the paper. This third dimensionality is sometimes indicated by extra-heavy lines or three-dimensional artwork (configurational formula) as in this representation of an ethanol molecule:

(4) Generic. Expresses a generalized type of organic compound in which the variables stand for the number of atoms or for the kind of radical in a homologous series. Examples:

C_nH_{2n+2} C_nH_{2n}
a paraffin   an olefin

ROR   ROH
an ether   an alcohol
                                  (R = a hydrocarbon radical)

(5) Electronic. A structural formula in which the bonds are replaced by dots indicating electron pairs, a single bond being equivalent to one pair of electrons shared by two atoms. Example: the electronic formula for methane is

$$\begin{matrix} & H & \\ H{:}&\!\!\ddot{C}\!\!&{:}H \\ & \ddot{H} & \end{matrix}$$

formula, product.   A list of the ingredients and their amounts or percentages required in an industrial product. Such formulas (or recipes) are mixtures, not compounds; they are generally used in such industries as adhesives, food, paint, rubber, and plastics.
See formulation.

formulation.   Selection of components of a product formula or mixture to provide optimum specific properties for the end use desired. Formulation by experienced technologists is essential for products intended to meet specifications or special service conditions.

formula weight.   The sum of the atomic weights represented in a chemical formula. Thus, since the atomic weight of hydrogen is 1 and that of oxygen is 16, the formula weight of water (H_2O) is 18 (approximate atomic weights used).

2-formyl-3,4-dihydro-2H-pyran.   See acrolein dimer.

formyl fluoride.   HCOF.
Properties: Gas at normal temperature and pressure. Bp −26C, fp −142C. Decomposes slowly with formation of hydrogen fluoride and carbon monoxide. Soluble in water (decomposes).
Derivation: Interaction of benzoyl chloride and a formic acid solution of potassium fluoride.
Grade: Technical.
Hazard: Toxic by inhalation, strong irritant to tissue.
Use: Organic synthesis (acetylating agent).

2-formyl-2′-methyl-1,1′-(oxydimethylene) dipyridinium, dichloride oxime.
CAS: 70441-84-8. $C_{14}H_{17}N_3O_2 \cdot 2Cl$.
Hazard: A poison.

formyloxytribenzylstannane.
CAS: 17977-68-3. $C_{23}H_{24}O_2Sn$.
Hazard: A poison by ingestion. A skin and eye irritant.

1-formylpiperidine.

$C_5H_5NCH{=}O$.

Properties: Colorless liquid. Liquid from −30 to 222C, aprotic, low volatility. Miscible with alcohols, esters, ketones, amines, amides, inorganic acids, organometallics; soluble in water and hexane.
Use: Solvent for polar and nonpolar compounds, as well as many high polymers; gas absorption; plastics modifiers.
See N,N-dimethylformamide.

3-formyltetrahydrofuran.
CAS: 79710-86-4. $C_5H_8O_2$.
Hazard: A mild eye irritant.

Forster reaction.   Formation of secondary amines by condensation of a primary amine with an aldehyde, addition of alkyl halide to the Schiff base, and subsequent hydrolysis.

"FORTEX" *[Johns Manville].*   TM for an oxidized hydrocarbon wax.

Forticel.   A cellulosic thermoplastic for use in injection molding, extrusion, rotational casting, and blow molding.
Properties: Pellets (crystals, translucent, metallic, and opaque colors). D 1.20, highest use temperature 80C. Soluble in organic solvents; insoluble in mineral oils. Combustible.

Use: Pen and pencil barrels, telephone bases, eyeglass frames, tool handles, sheeting, steering wheels, etc.

fortification.   In food technology, addition to a food ingredient or product of nutrients that are not normally present, e.g., addition of vitamin D to milk or of vitamin C to cake fillings. Nutritionists apply this term to foods especially designed for schoolchildren and older persons.
See nutrification; enrichment.

"Fortiflex" *[VIRBAC].*   TM for a high-density polyethylene consisting mainly of long molecules with occasional short side branches. Thermoplastic.
Properties: Milk-white, translucent pellets (colors are also available). D 0.95; melt index 0.2–8; tensile strength 3100–3700 psi; highest use temperature 225F. Combustible.
See polyethylene.

Fortisan.   A cellulosic fiber manufactured by partial saponification of stretched cellulose acetate. A semisynthetic product, it resembles cellulose (cotton) in many respects. The high-tenacity product has a dry strength of 5–7 lb/denier (100,000–130,000 psi); wet strength is 85% of dry strength. It has relatively low elongation under stress, elastic recovery is approximately 70% after extension to break, immediate elastic recovery is 46%, and delayed recovery is 30% at 5% strain. Young's modulus 1650; can be dyed in the same way as cotton. The monofilament can be produced to a fineness of one denier. D 1.50. Resists stretching both dry and wet. Combustible.

fortovase.
CAS: 149845-06-7. $C_{38}H_{50}N_6O_5 \cdot CH_4O_3S$.
Hazard: Moderately toxic by ingestion.

Fortrel.   A polyester-type synthetic fiber.

fosetyl aluminum.   (phosphonic acid monoethyl ester aluminum salt; fractionation; alliette; aluminum tris-(o-ethylphosphonate; aluminum ethoxy-oxido-oxophosphanium). $C_6H_{15}AlO_3P_5I_5$. The physical separation and identification of the components of a mixture often by distillation.

Fosfodril.   A glassy phosphate of high molecular weight (sodium hexametaphosphate).
Use: Thickener for drilling muds, water treatment in oil-well flooding operations.

fosfomycin.
CAS: 23155-02-4. $C_3H_7O_4P$.
Properties: Water-soluble crystals. Mp 95C.
Derivation: Produced by *Streptomyces*, also made synthetically.
Use: General antibiotic.

fossil. Any material that results from an animal or vegetable source in past geologic ages and has been buried (compressed) in the earth. Examples are fossil fuels (petroleum, natural gas, coal, lignite), fossil waxes (ozocerite, montan), fossil resins (amber), and fossil woods partially preserved (petrified) by the action of silica.

Fosterite. A family of resins. Largest application is as solventless varnishes for electric insulation, also as a photoelastic resin and as a bond for impregnating and laminating asbestos sheets. Rods made of this plastic will carry a beam of light without the dispersion that occurs in air, making it possible to bend the beam.

fostion. (*O,O*-diethyl-*S*-(*N*-isopropyl carbamoyl-methyl) phosphorodithioate).
CAS: 2275-18-5. $(C_2H_5O)_2SPSCH_2SCH(CH_3)_2$.
Properties: White, crystalline solid. Mp 24C. Soluble in most organic solvents.
Hazard: Toxic by ingestion and skin absorption, cholinesterase inhibitor.
Use: Fungicide, ovicide.

Fotoceram. A photosensitive, crystalline ceramic.
Use: Electronics and industrial arts.

Foundrez. A group of water-soluble phenol-formaldehyde and urea-formaldehyde resins for foundry applications.

foundry sand. (greensand; molding sand). Sand containing zirconium, titanium, and other metals and mixed with suitable binders used in making molds for casting metals. The sand/binder mixture is either rammed into place around the mold or baked into a core at 204–260C (dry-sand molding). The binders used are resins of various types, casein, etc.

fourdrinier. The machine most widely used for papermaking, named for its English inventors who introduced it in the mid-19th century. It provides a wide range of papers from heavy board to light tissue and is of impressive size and complexity. Its unique feature is the traveling mesh belt onto which the slurry of fiber and water is run from the head-box. The wire was formerly a screen made of specially annealed bronze and brass in 55–85 mesh size range. This has largely been replaced by single and multilayer woven polyester-forming fabrics, which are plastic screens woven with approximately 0.20 mm polyester monofilament. The plastic-forming fabric is longer lasting and more adaptable.
The sheet is formed on this wire almost instantly, most of the water draining through the interstices of the wire. After leaving the wire, the sheet (called the web) passes through the press section of the machine where a number of rollers express enough of the remaining water to enable the sheet to hold

together. It then moves into the multiroller drying section. The dried sheet (4–6% moisture content) is then fed to a high-speed calender for compaction and finishing. The entire process is continuous and rapid, the machine often operating for several days without shutdown.
See calender; drying (2); paper; couch roll; dandy roll; supercalender.

Fourier-transform NMR. (FT-NMR). A type of spectroscopy, developed by Ernst, which multiplies sensitivity 10 to 100 times compared to dispersive NMR instruments.
See spectroscopy.

FP Acids. A series of fluorophosphoric acid.

FPC. (1) Abbreviation for fish protein concentrate. (2) Abbreviation for Federal Power Commission.

Fr. Symbol for francium.

fraction. Any portion of a mixture characterized by very similar properties. The most important fractions of petroleum are naphtha, gasoline, fuel oil, kerosene, and tarry or waxy residues. These are obtained by fractional distillation.
See separation.

fractional distillation. Distillation in which rectification is used to obtain product as nearly pure as possible. A part of the vapor is condensed and the resulting liquid contacted with more vapor, usually in a column with plates or packing. The term is also applied to any distillation in which the product is collected in a series of separate components of similar boiling range.
See reflux.

fractionation. In general, the separation or isolation of components of a mixture or a micromolecular complex. In distillation, this is done by means of a tower or column in which rising vapor and descending liquid are brought into contact (countercurrent flow). Macromolecular components (proteins and other high polymers) can be separated by a number of methods, including electrophoresis, gel filtration, chromatography, centrifugation, foam fractionation, and partition.
See reflux.

fracturing, hydraulic. See hydraulic fracturing.

fragile sites. A non-staining gap of variable width that usually involves both chromatids and is always at exactly the same point on a specific chromosome derived from an individual or kindred.

fragrance. An odorant used to impart a pleasant smell to lotions, toothpastes, cosmetics, etc.; balsamic and piney odors are typical.

fraissite. (benzyl iodide). $C_6H_5CH_2I$.
Use: A tear gas.

frame shift. A mutation caused by insertion or deletion of one or more paired nucleotides, changing the reading frame of codons and therefore the amino acids incorporated into proteins.

Franchimont reaction. Carboxylic acid dimerization to 1,2-dicarboxylic acids by treating α-bromocarboxylic acids with potassium cyanide followed by hydrolysis and decarboxylation.

francium. Fr. Element of atomic number 87, group IA of the periodic table system, aw 223, valence of 1; it appears to exist only as radioactive isotopes. One isotope is actinium K (^{223}Fr). Other isotopes have been made artificially: ^{223}Fr is the longest-lived isotope, having a half-life of 21 minutes, and is the only natural isotope. Francium is the heaviest of the alkali-metal family.

frankincense. (olibanum). A gum resin.

frankincense oil. See olibanum oil.

Frankland–Duppa reaction. Formation of α-hydroxycarboxylic esters by reaction of dialkyl oxalates with alkyl halides in the presence of zinc, or amalgamated zinc, and acid.

Frankland synthesis. Synthesis of zinc dialkyls from alkyl halides and zinc.

franklinite. (iron, manganese, zinc). $(FeMn)_2O_4$. Black mineral resembling magnetite.

Frary metal. A lead-based bearing metal containing 97–98% lead alloyed with 1–2% each of barium and calcium; excellent for low-pressure bearings at moderate temperatures.

Frasch process. A process by which much of the world's sulfur is obtained. Developed about 1900 by Herman Frasch, the process involves melting sulfur underground by introducing superheated water through a pipe under pressure and forcing the molten sulfur to the surface by compressed air.

Fraunhofer lines. See spectroscopy.

free electron. Electron not attached to any one atom and not restricted by potential gradients.

free energy. (ΔG; Gibbs free energy). An exact thermodynamic quantity used to predict the maximum work obtainable from the spontaneous transformation of a given system. It also provides a criterion for the spontaneity of a transformation or reaction and predicts the greatest extent to which the reaction can occur, i.e., its maximum yield. Transformation of a system can be brought about by either heat or mechanical work. Free energy is derived from the internal energy and entropy of a system in accordance with the laws of thermodynamics.
See also enthalpy.

free radical. A molecular fragment having one or more unpaired electrons, usually short-lived and highly reactive. In formulas, a free radical is conventionally indicated by a dot, as in Cl and $^-$, $(C_2H_5)^\bullet$. In spite of their transitory existence, they are capable of initiating many kinds of chemical reactions by means of a chain mechanism. Free radicals are formed only by the splitting of a molecular bond. A chain can result only if (1) radicals attack the substrate and (2) the radicals lost by this reaction are regenerated. Chain mechanisms for the thermal decomposition of many substances have been established. Free radicals are known to be formed by ionizing radiation and thus play a part in deleterious degradation effects that occur in irradiated tissue. They also act as initiators or intermediates in such basic phenomena as oxidation, combustion, photolysis, and polymerization.
See carbonium ion.

free sulfur. Sulfur that is left chemically uncombined after vulcanization of a rubber compound. When this exceeds 1%, the upper limit of solubility of sulfur in rubber, blooming will occur. Most rubber products are vulcanized with as low a sulfur content as possible so that the free sulfur content of the product is seldom over 0.5%.
See bloom; vulcanization.

freeze-drying. (lyophilization). A method of dehydration or of separating water from biological materials. The material is first frozen and then placed in a high vacuum so that the water (ice) vaporizes in the vacuum (sublimes) without melting and the nonwater components are left behind in an undamaged state.
Use: Blood plasma, certain antibiotics, vaccines, hormone preparations, food products such as coffee and vegetables. One technique prepares freeze-dried ceramic pellets from water solutions of metal salts.

"Freezene" [Crompton]. TM for a series of refrigeration white mineral oils.
Use: Low-temperature lubrication.

freezing point. See melting point.

Freon. A series of fluorocarbon products used in refrigeration and air-conditioning equipment, as blowing agents, fire-extinguishing agents, and cleaning fluids and solvents.
Properties: Clear, water-white liquids; vapors have a mild, somewhat ethereal, odor and are not irritating; essentially stable and inert. Nonflammable, nonexplosive, noncorrosive.
See fluorocarbon.
Note: Many types contain chlorine, as well as fluorine, and should be called chlorofluorocarbons.

Freon 22. See 1,1,1-chlorodifluoroethane.

Freon 112. *sym*-Tetrachlorodifluoroethane.

Freon C-51-12. See perfluorodimethylcyclobutane.

Freon 225. See 3,3-dichloro-1,1,1,2,2-pentafluoropropane.

Freon E. A series of hydrogen endcapped tetrafluoroethylene epoxide polymers having a DP up to 10, boiling range 39–49 C, high dielectric constant.
Use: Coolants in electronic devices.

Fresh Notes. Flavor and fragrance fractions.
Use: A flavor and fragrance additive.

Freundlich isotherm. The relationship between the amount of a substance adsorbed and the concentration of the solute:

$$\frac{x}{m} = Kc^{1/n}$$

where x is the amount adsorbed, m the weight of adsorbent, c the concentration or pressure, and K and n constants.

Freund's acid. 1-Naphthylamine 3,6-disulfonic acid.

Freund synthesis. Formation of alicyclic hydrocarbons by the action of sodium (Freund) or zinc (Gustavson) on open-chain dihalo compounds; 1,3-dichloropropane derived from the chlorination of propane, obtained from natural gas, is cyclized in the Hass process by treating with zinc dust in aqueous alcohol in the presence of sodium iodide as catalyst.

friction. A soft and extremely tacky mixture of rubber and softener applied to a fabric by means of a three-roll calender. The differential speed of the calender rolls drives the material into the interstices of the fabric, forming a strongly adherent coating. Uncured friction on a light sheeting is used for electrical insulating or friction tape. Heavy-weave fabric-coated with high-grade friction (rubber plus softener and curing agents) is used as piles in tire carcasses, transmission belts, and other laminated products that are vulcanized.
See calender.

friction welding. See welding.

Friedel-Crafts reaction. A type of reaction involving anhydrous aluminum chloride and similar metallic halides as catalysts, discovered in 1877 by Charles Friedel, a French chemist (1832–1899), and James Mason Crafts, an American chemist (1830–1917), during joint research in France; it has been developed since then for many important industrial uses, exemplified by the condensation of ethyl chloride and benzene to form ethylbenzene and the manufacture of acetophenone from acetyl chloride and benzene. The name is now applied to a wide variety of acid-catalyzed organic reactions.
Use: Alkylation and acylation in general. Some examples are the production of high-octane gasoline, cumene, detergent alkylate, and various plastics and elastomers.

Friedlaender synthesis. Base-catalyzed condensation of 2-aminobenzaldehydes with ketones to form quinoline derivatives.

Fries rearrangement. Rearrangement of phenolic esters to *o*- and/or *p*-phenolic ketones on heating with aluminum chloride or other Lewis acid catalysts.

Fries rule. The most stable form of a polynuclear hydrocarbon is that in which the maximum number of rings has the benzenoid arrangement of three double bonds. A benzenoid electronic configuration is energetically favored and, therefore, particularly stable.

frit. A ground glass used in making glazes and enamels and also for making so-called frit seals. Finely powdered glass may be called a frit. The term is also used for finely ground inorganic minerals, mixed with fluxes and coloring agents that become a glass or enamel on heating.

Fritsch–Buttenberg–Wiechell rearrangement. The rearrangement of 1,1-diaryl-2-haloethylenes to diaryl acetylenes with strong bases.

FR-N. A series of butadieneacrylonitrile elastomeric polymers and latexes.
Properties: Oil and solvent resistance combined with good flexibility and resistance to low temperatures, water absorption, and permanent set.
Use: Oil-resistant seals, shoe soles, gasoline hose, belt conveyors, plasticizers, paper saturation, adhesives, leather finishes, and carpet backing.

froth flotation. See flotation.

FRP. Abbreviation for glass fiber-reinforced plastic.
See reinforced plastic.

FR-S. General-purpose rubbers and latexes, composed of copolymers of butadiene and styrene.
Use: (Rubber): Tires, hose, belting, and packing; molded and extruded automotive and industrial products; soles and heels; hard rubber. (Latex): Adhesives, foamed rubber, textile and rug backing, paper coating and impregnation, modification of plastics to produce high-impact strength, asphalt additive.

fructose. (fruit sugar; D(−)-fructose; levulose). $C_6H_{12}O_6$. A sugar occurring naturally in a large number of fruits and in honey. It is the sweetest of the common sugars.
Properties: White crystals. Mp 103–105C (decomposes); specific rotation −89 to −91 degrees. Soluble in water, alcohol, and ether. Combustible.
Derivation: Hydrolysis of inulin; hydrolysis of beet sugar followed by lime separation, from cornstarch by enzymic or microbial action.
Grade: Technical, NF, food, parenteral.
Use: Foodstuffs, medicine, preservative.

fructose-1,6-diphosphate. (FDP; fructosediphosphoric acid; Harden-Young ester). H_2PO_4 $(C_6H_{10}O_4)H_2PO_4$.
Can be prepared from fructose and certain other sugars by the use of yeasts. It is known to take part in cell metabolism; an intermediate in carbohydrate metabolism. Usually handled in the form of its barium or calcium salts, white amorphous powders, soluble in ice water and dilute acid solutions, insoluble in hot water and alcohol.
Use: Organic synthesis, research in cell metabolism.

fruit acids. See α hydroxy acids.

frustule. The mineral "skeleton" of a diatom or other unicellular organism.

3-FT. See 5-fluoro-3-(tetrahydro-2-furyl)uracil.

FT black. Abbreviation for fine thermal black.
See thermal black.

FTC. Abbreviation for Federal Trade Commission.

fuchsin. (basic fuchsin; magenta).
CAS: 632-99-5.
A synthetic rosaniline dyestuff, a mixture of rosaniline and p-rosaniline hydrochlorides.
Properties: Dark green powder or greenish crystals with a bronze luster; faint odor. Soluble in water and alcohol.
Grade: NF.
Hazard: Possible carcinogen.

Use: Textiles and leather industries, red dye, pharmaceutical.

fuel. Any substance that evolves energy in a controlled chemical or nuclear reaction. The most common type of chemical reaction is combustion, the type of oxidation occurring with petroleum products, natural gas, coal, and wood; more-rapid oxidation takes place in rocket fuels (hydrogen, hydrogen peroxide, hydrazine) and approaches the rate of an explosion. The nuclear fuels used for power generation release their energy by fission of the atomic nucleus (uranium, plutonium, thorium).
See combustion; fission.

fuel cell. (1) An electrochemical device for continuously converting chemicals—a fuel and an oxidant—into direct-current electricity. It consists of two electronic-conductor electrodes separated by an ionic-conducting electrolyte with provision for the continuous movement of fuel, oxidant, and reaction product into and out of the cell. The fuel can be gaseous, liquid, or solid; the electrolyte liquid or solid; the oxidant gaseous or liquid. The electrodes are solid but may be porous and may contain a catalyst. Fuel cells differ from batteries in that electricity is produced from chemical fuels fed to them as needed, so that their operating life is theoretically unlimited. The cell products can be regenerated externally into fuel for return to the cell, e.g., carbon dioxide from the cell can be reacted with coal to form carbon monoxide for feed to the cell. Fuel is oxidized at the anode (negative electrode), giving electrons to an external circuit; the oxidant accepts electrons from the anode and is reduced at the cathode. Simultaneously with the electron transfer, an ionic current in the electrolyte completes the circuit. One type of electrolyte is a solid polymer of perfluorinated sulfonic acid. The fuels range from hydrogen, carbon monoxide, and carbonaceous materials to redox compounds, alkali metals, and biochemical materials. Fuel cells based on hydrogen and oxygen have a significant future as a primary energy source. Cells of this type are under development for use as a power source for electric automobiles, the hydrogen being derived from methanol. Large-scale development of fuel cells for on-site power generation for housing units is well advanced, and research has been completed for construction of a 26-megawatt cell capable of serving the needs of a small community.
(2) An aircraft fuel tank or container made of or lined with an oil-resistant synthetic rubber.

fuel element. A fabricated rod, form, or other shape that consists of or contains the fissionable fuel for a nuclear reactor. The term does not refer to a chemical element but rather to a device from which power is derived.

fuel oil. Any liquid petroleum product that is burned in a furnace for the generation of heat or used in an engine for the generation of power, except oils having a flash point of approximately 100F (37.7C) and oils burned in cotton or wool-wick burners. The oil may be a distillated fraction of petroleum, a residuum from refinery operations, a crude petroleum, or a blend of two or more of these.

Because fuel oils are used with burners of various types and capacities, different grades are required. ASTM has developed specifications for six grades of fuel oil. No. 1 is a straight-run distillate, a little heavier than kerosene, used almost exclusively for domestic heating. No. 2 (diesel oil) is a straight-run or cracked distillate used as a general purpose domestic or commercial fuel in atomizing-type burners. No. 4 is made up of heavier straight-run or cracked distillates and is used in commercial or industrial burner installations not equipped with preheating facilities. The viscous residuum fuel oils, Nos. 5 and 6, sometimes referred to as bunker fuels, usually must be preheated before being burned. ASTM specifications list two grades of No. 5 oil, one of which is lighter and under some climatic conditions may be handled and burned without pre-heating. These fuels are used in furnaces and boilers of utility power plants, ships, locomotives, metal-lurgical operations, and industrial power plants.

Hazard: Questionable carcinogen.

Use: Domestic and industrial heating, power for heavy units (ships, trucks, trains), source of syn-thesis gas.

See diesel oil.

Fujimoto-Belleau reaction. Synthesis of cyclic α-substituted α,β-unsaturated ketones from enol lactones and Grignard reagents prepared from primary halides.

Fukui, Kenichi. (1918–1998). A Japanese pro-fessor who was corecipient of the Nobel Prize for chemistry along with Hoffmann in 1981. His work involved quantum mechanical studies of chemical reactivity. Fukui's entire career has been at Kyoto University.

fullerines. General name for closed spheroidal aromatic molecules with even numbers of carbons. See buckminsterfullerene.

fuller's earth. A porous colloidal aluminum sil-icate (clay) that has high natural adsorptive power. Gray to yellow color, noncombustible.

Occurrence: Florida, England, Canada.

Use: Decolorizing of oils and other liquids, oil-well drilling muds, insecticide carrier, floor-sweeping compounds, cosmetics, rubber filler, carrier for cat-alysts, filtering medium.

See bentonite; diatomite.

full gene sequence. The complete order of bases in a gene. This order determines which protein a gene will produce.

fulminates. Materials with carbon-nitrogen-oxygen groups.

Use: Sensitive explosives.

fumagillin.
CAS: 23110-15-8. $C_{27}H_{36}O_7$.
An antibiotic substance produced by *Aspergillus fumigatus*.
Properties: Light yellow crystals from dilute methanol. Mp 189–194C. Insoluble in water, dilute acids, saturated hydrocarbons; soluble in most other organic solvents.
Use: Medicine.

fumarase. (fumarate hydratase).
CAS: 9032-88-6.
An enzyme that catalyzes the interconversion of fumaric acid and malic acid. It is an important reac-tant in the tricarboxylic acid cycle.

fumaric acid. (boletic acid; lichenic acid; allo-maleic acid; *trans*-butenedioic acid).
CAS: 110-17-8.
The trans isomer of maleic acid.

$$HO-\overset{\overset{O}{\|}}{C}-\overset{}{C}-H$$
$$H-\overset{}{C}-\overset{\underset{\|}{O}}{C}-OH$$

Properties: Colorless, odorless crystals; fruit acid taste. Stable in air, d 1.635, sublimes at 290C, mp 287C (sealed tube), soluble in water 0.63g/100 g (25C), soluble in alcohol 5.76 g/100 g (30C). Insol-uble in chloroform and benzene. Combustible.

Derivation: (1) Isomerization of maleic acid, (2) cat-alytic oxidation of benzene.

Grade: Technical, crystals, FCC.

Use: Modifier for polyester, alkyd, and phenolic resins; paper-size resins; plasticizers; rosin esters and adducts; alkyd resin coatings; upgrading natu-ral drying oils (especially tall oil) to improve dry-ing characteristics; in foods as substitute for tar-taric acid, as acidulant and flavoring agent (FDA approved); mordant; organic synthesis, printing inks.

fumaryl chloride.
CAS: 627-63-4. ClCOCH:CHCOCl.
Properties: Clear, straw-colored liquid. Bp 158–160C, 62–64C (13 mm Hg), d 1.408 (20C).
Hazard: Corrosive to eyes and skin.
Use: Chemical intermediate for pharmaceuticals, dyestuffs, and insecticides.

fume. The particulate, smoke-like emanation from the surface of heated metals. Also the vapor evolved from concentrated acids (sulfuric, nitric), from evaporating solvents, or as a result of com-bustion or other decomposition reactions (exhaust fume). Many of these fumes are toxic.

fumigant. A toxic agent in vapor form that destroys rodents, insects, and infectious organisms; a type of pesticide. The most effective temperature for their use is approximately 70F. They are used chiefly in enclosed or limited areas (barns, greenhouses, ships' holds, and the like) and also are applied locally to soils, grains, fruits, and garments. Some commonly used fumigants are formaldehyde, sulfur dioxide and other sulfur compounds, hydrogen cyanide, methyl bromide, carbon tetrachloride, p-dichlorobenzene, and ethylene oxide. Care must be used when handling and applying fumigants because of their toxicity.
See repellent; pesticide.

Fumigatoxin. (fumigacin, fumigation, fumitermorgins; (2Z)-2-[(4S,5S,6S,8S,9S,10R,13R,14S,16S)-6,16-diacetyloxy-4,8,10,14-tetramethyl-3,7-dioxo-5,6,9,11,12,13,15,16-octahydro-4H-cyclopenta[a]phenanthren-17-ylidene]-6-methylhept-5-enoic acid). $C_{20}H_{11}O_6$. A mycotoxin.
Derivation: Produced by *Aspergillus fumagatus*.
Hazard: Very toxic.

fuming. A characteristic of some highly active liquids that evolve visible smoke-like emanations on contact with air. Most familiar are the forms of nitric and sulfuric acids designated as fuming. These are not pure, concentrated acids; low percentages of nitrogen dioxide and water are present in fuming nitric acid, and fuming sulfuric acid contains sulfur trioxide. Hydrofluoric acid (a mixture of hydrogen fluoride and water) also fumes. Pure compounds in which fuming occurs are fluosilicic acid and hydrazine.

functional gene tests. Biochemical assays for a specific protein, indicating a specific gene is not merely present but active.

functional genomics. The study of genes, their resulting proteins, and the role played by the proteins in the body's biochemical processes.

fundamental particle. (elementary particle). One of the many constituents of atoms in either the normal or the excited state, i.e., protons, electrons, and neutrons, that are directly involved in chemical reactions. Other subatomic species are mesons, positrons, neutrinos, hyperons, etc. Photons are light quanta, i.e., "particles" of energy that have no rest mass.
See particle; photon.

fundamental research. (basic research). Scientific investigations undertaken primarily to increase knowledge of a given field on a long-range basis. It is free from the time factor usually present in applied research and is comparatively unlimited by economic restrictions. In general, it seeks basic causes for phenomena rather than immediate results. It has no predetermined goal or purpose.

Nonetheless, tremendous achievements in chemistry and other sciences have resulted, and it will always remain the essential cornerstone of science. "Fundamental research is essentially a matter of inquiring into nature. The motivation for this activity is only imperfectly understood but it is primarily an intellectual pursuit. Unlike applied research, the reward being sought is attained through the ability to understand and explain natural phenomena. The interest centers on elucidating the laws of nature, not on manipulating or exploiting them" (Howard McMahon).

fungal protease enzyme. See acid fungal protease.

fungicide. Any substance that kills or inhibits the growth of fungi. Older types include a mixture of lime and sulfur, copper oxychloride, and Bordeaux mixture. Copper naphthenate has been used to impregnate textile fabrics such as tenting and military clothing. Dithiocarbamate and quinone types were introduced about 1940. Mercury compounds are also effective but have been discontinued because of their toxicity to humans. Hypochlorite solutions are used in swimming pools and water-cooled heat exchangers. Some types of fungi that infect the human body are extremely hard to eradicate and require highly specific medical treatment.

fungizone intravenous.
CAS: 58501-21-6. $C_{47}H_{73}NO_{17} \cdot C_{24}H_{40}O_4 \cdot Na$.
Hazard: A poison. Human systemic effects.

fungocin. See bacillus subtilis BPN.

fungus. Any of a plant-like group of organisms that does not produce chlorophyll; they derive their food either by decomposing organic matter from dead plants and animals or by parasitic attachment to living organisms, thus often causing infections and disease. Examples of fungi are molds, mildews, mushrooms, and the rusts and smuts that infect grain and other plants. They grow best in a moist environment at temperatures of about 25C, little or no light being required.
See mycotoxin.

Funk, Kazimierz. (1884–1967). Born in Poland and later becoming an American Citizen, Funk in 1911 isolated a food factor, extracted from rice hulls, that he found to be a cure for a disease caused by malnutrition (beri-beri). Believing this to be an amine compound essential to life, he coined the name *vitamine*, from which the final *e* was later dropped. The various types and functions of vitamins were not differentiated until some years later as a result of the work of McCollum, Szent-Gyorgi, R. J. Williams and others.

furacrylic acid. See furylacrylic acid.

"Furadan" *[FMC].* (carbofuran; 2,3-dihydro-2,2-dimethyl-7-benzofuranylmethylcarbamate). CAS: 1563-66-2.
TM for a pesticide designed to combat corn rootworm and rice water weevil. Approved by USDA. Also effective on alfalfa, sugarcane, rice, peanuts, and potatoes.
Hazard: Toxic material. Cholinesterase inhibitor. Questionable carcinogen.

furamide. See furoamide.

furan. (furfuran; tetrol).

CAS: 110-00-9. HC:CHCH:CHO.
A heterocyclic compound. Its basic structure is

$$\begin{array}{ccc} C & \!\!\!\!—\!\!\!\! & C \\ \| & & \| \\ C & & C \\ & \diagdown_O\diagup & \end{array}$$

Properties: Colorless liquid turning brown on standing, this color change is retarded if a small amount of water is added. D 0.938 (20/4C), fp −86C, bp 31.4C, flash p <32F (0C) (TOC), refr index 1.4216 (20C). Insoluble in water; soluble in alcohol and ether.
Derivation: Dry distillation of furoic acid from furfural.
Hazard: Flammable, dangerous fire risk, flammable limits 2–24%, forms peroxides on exposure to air. Absorbed by skin. Possible carcinogen.
Use: Organic synthesis, especially for pyrrole, tetrahydrofuran, thiophene.
See furan polymer.

furancarboxylic acid. See furoic acid.

2,5-furandione. See maleic anhydride.

2-furanmethanethiol.

OCH:CHCH:CCH$_2$SH.
Properties: Liquid. D 1.1319 (20/4C), bp 155C, refr index 1.5324 (20C). Insoluble in water.
Use: Ingredient for synthetic coffee compositions and fortifier for natural coffee blends and flavor adjunct. Also inhibits the corrosive power of nitric acid.

furanodiene.
CAS: 57566-47-9. C$_{15}$H$_{20}$O.
Hazard: A poison by ingestion.

furanose. A simple sugar structurally analogous to the five-membered furan ring.

furan polymer. A plastic derived (1) from furfuryl alcohol or (2) from furfural or reaction products of furfural and a ketone. The materials are dark colored and resistant to solvents, most nonoxidizing acids, alkalies, and specific corrosives such as dinitrogen pentoxide.
Properties: Physical properties of a typical asbestos-reinforced furan polymer are d 1.7, tensile strength 5000 psi, flexural strength 7500 psi, impact strength 0.5 ft lb/in. notch, water absorption 0.2%, and coefficient of thermal expansion 3.0–10^{-5} in/in/F.
Use: Coating asphaltic pavements, foundry sand cores, shell molding and corrosion-resistant materials of construction. Since furfural is readily obtainable by heating pentosan-containing products such as corncobs with mineral acid, these resins are inexpensive and have great potential use where products with their characteristics are required.

3-((2-furanylmethylene)amino)-2-oxazolidone.
CAS: 6270-33-3. C$_8$H$_8$N$_2$O$_3$.
Hazard: Moderately toxic by ingestion.

furazolidone. (*N*-(5-nitro-2-furfurylidene-3-amino)-2-oxazolidinone).
CAS: 67-45-8. C$_8$H$_7$N$_3$O$_5$.
Properties: Yellow powder; odorless. Mp 255C. Slightly soluble in polyethylene glycol; insoluble in water, alcohol, and peanut oil.
Derivation: Synthetically from furfural, hydroxyethylhydrazine, and diethyl carbonate.
Grade: NF.
Hazard: A questionable carcinogen, use has been restricted.
See nitrofuran.

furcellaran.
Properties: Seaweed gum (a natural phycocolloid) available as an odorless white powder, soluble in warm water. It form gels at low concentrations. Reputed to be more stable to heat and acids than other vegetable gums. Available in the form of salts.
Derivation: From the seaweed *Furcellaria fastigiata.*
Use: Gelling-agent, viscosity-control agent, puddings, jams, toothpastes, bacterial cultural media, pharmaceuticals.

furethrin. (generic name for 3-furfuryl-2-methyl-4-oxo-2-cyclopenten-1-ylchrysanthemumate (generic)).
C$_{21}$H$_{27}$O$_4$.
A synthetic analog of allethrin substituting the 2-furfuryl for allyl in the side chain.
Properties: Yellow liquid. Bp 187–188C (0.4 mm Hg.) Insoluble in water; soluble in light oils.
Use: Insecticide, use like allethrin.

furfural. (ant oil, artificial; pyromucic aldehyde; furfuraldehyde; bran oil).

CAS: 98-01-1. C_4H_3OCHO.

$$
\begin{array}{cc}
CH\!-\!\!-\!CH \\
\| \quad \| \\
CH \quad CCHO \\
\diagdown O \diagup
\end{array}
$$

Properties: Colorless liquid when very pure, becomes reddish-brown on exposure to light and air; odor somewhat similar to benzaldehyde. Forms condensation products with many types of compounds, phenol, amines, urea, etc. D 1.1598 (20/4C), fp −36.5C, bp 161.7C, heat of vaporization 107.5 cal, refr index 1.5260 (20C), flash p 140F (60C) (CC), autoign temp 392C (797F). Soluble in alcohol, ether, and benzene; 8.3% soluble in water at 20C. Combustible.
Derivation: From oat hulls, rice hulls, corncobs, bagasse, and other cellulosic waste materials by steam-acid digestion.
Grade: Technical, refined.
Hazard: Absorbed by skin; irritant to eyes, skin, and mucous membranes. Toxic by skin absorption; questionable carcinogen.
Use: Solvent refining of lubricating oils, butadiene, rosin, and other organic materials; solvent for nitrocellulose, cellulose acetate, shoe dyes; intermediate for tetrahydrofuran and furfuryl alcohol, phenolic and furan polymers; wetting agent in manufacture of abrasive wheels and brake linings; weed killer; fungicide; adipic acid and adiponitrile; road construction; production of lysine; refining of rare earths and metals; flavoring; analytical reagent.

furfuralacetic acid.   See furylacrylic acid.

furfuraldehyde.   See furfural.

furfuramide.   See hydrofuramide.

furfuran.   See furan.

furfuryl acetate.
CAS: 623-17-6. $C_4H_3OCH_2OOCCH_3$.
Properties: Colorless liquid turning brown on exposure to light and air; pungent odor. D 1.1175 (20/4C), bp 175–177C, refr index 1.4627. Insoluble in water; soluble in alcohol and ether. Combustible.
Derivation: By treatment of furfuryl alcohol with acetic anhydride.
Grade: Refined.
Use: Flavor.

furfuryl alcohol.   (furyl carbinol).
CAS: 98-00-0. $C_4H_3OCH_2OH$.

$$
\begin{array}{cc}
CH\!-\!\!-\!CH \\
\| \quad \| \\
CH \quad CCH_2OH \\
\diagdown O \diagup
\end{array}
$$

Properties: Colorless, mobile liquid becoming brown to dark red on exposure to light and air. It autopolymerizes with acid catalysts, often with explosive violence, to form a thermosetting resin that cures to an insoluble, infusible solid, highly resistant to chemical attack. D 1.1285 (20/4C), bp 170C, refr index 1.4850 (25C), flash p 167F (75C) (OC), autoign temp 915F (490C). Soluble in alcohol, chloroform, benzene, and water; insoluble in paraffin hydrocarbons. Combustible.
Derivation: Continuous vapor-phase hydrogenation of furfural.
Grade: Technical, refined.
Hazard: May react explosively with mineral acids and some organic acids. Toxic by inhalation and skin absorption. Approved for food products. Toxic by skin absorption.
Use: Wetting agent, furan polymers, corrosion-resistant sealants and cements, foundry cores, modified urea-formaldehyde polymers, penetrant, solvent for dyes and resins, flavoring. The polymer is used as a mortar for bonding acid-proof brick and chemical masonry.
See furan polymer.

α-furfuryl amine.
CAS: 617-89-0. $C_4H_3OCH_2NH_2$.
Properties: Colorless liquid. D 1.0550 (17C), bp 145C (757 mm Hg), refr index 1.4900 (17C), flash p 99F (37.2C) (OC). Soluble in water, alcohol, and ether.
Derivation: Furfural and ammonia.
Hazard: Flammable, moderate fire risk.
Use: Corrosion inhibitor, component of soldering flux, chemical intermediate.

6-furfurylaminopurine.   See kinetin.

2-(2-furfuryl)furan.
CAS: 1197-40-6. $C_9H_8O_2$.
Hazard: A poison by skin contact. Low toxicity by inhalation. A moderate eye irritant.

furfuryl mercaptan.   See 2-furanmethanethiol.

furilazole.
CAS: 121776-33-8. $C_{11}H_{13}Cl_2NO_3$.
Hazard: Moderately toxic by ingestion, inhalation, and skin contact.
Use: Agricultural chemical.

α-furildioxime.   (di-2-furanylethanedione dioxime).
CAS: 23789-33-5. $C_{10}H_8N_2O_4$.
Properties: Needle-like crystals. Mp 167C, mw 220.18. Very soluble in alcohol, ether; slightly soluble in benzene, petroleum ether.
Use: Reagent in alcohol solution for nickel, gives an orange-red compound.

furisyl.
CAS: 72239-53-3. $C_{16}H_{20}N_6O_4$.
Hazard: A poison.

furnace.   An enclosed chamber or structure lined with firebrick or similar refractory and containing a heat source (coal, coke, gas, or electric elements). It may have various designs depending on its function. Furnaces are used for steel production (open-hearth and basic oxygen types), for smelting iron and other ores (blast furnace) and for manufacture of furnace carbon black, etc. Electric furnaces are used for special types of steel, as well as for high-temperature reactions such as the manufacture of pyrolytic graphite, synthetic diamonds, silicon, silicon carbide, and salt cake. Temperatures obtained range up to 3000C. Laboratory electric furnaces are used for high-temperature experiments and product testing.
See muffle furnace; reverberatory furnace; Mannheim furnace; kiln; forge; cupola.

furnace black.   See carbon black.

furnace oil.   Usually No. 1 fuel oil. See fuel oil.

furnace sand.   (fire sand). Sand used to line furnace bottoms or walls, particularly in open-hearth steel furnaces.

furnish.   Term used by papermakers for the mixture containing the constituents of paper as supplied to the fourdrinier wire on which the sheet is formed.

furoamide.   (pyromucamide; furamide). C_4H_3OCON.
Properties: Crystals. Sublimes partly at 100C, mp 142C.
Derivation: Treatment of furoyl chloride with ammonia.

furocoumarin.   (furanocoumarin; furo[3,2-g]chromen-7-one).
CAS: 66-97-7. $C_{11}H_6O_3$.
Any of a group of photoactive antifungal dyestuffs found in plants of the families *apiaceae* and *rutaceae* that cause primary photosensitization in domestic fowl and livestock.
Derivation: Occur naturally as derivatives of psoralen or its isomer, angelicin.

furoic acid.   (pyromucic acid; furancarboxylic acid).
CAS: 88-14-2.

C_4H_3OCOOH or $OCH:CHCH:CCOOH$.

Properties: Colorless crystals. Mp 133–134C, sublimes at 130C (50–60 mm Hg). Slightly soluble in cold water; very soluble in hot water, alcohol, and ether; insoluble in paraffin hydrocarbons. Combustible.

Derivation: Cannizzaro reaction from furfural, oxidation of furfural.
Method of purification: Sublimation, fractional crystals from hot water.
Grade: Technical.
Use: Preservative, bactericide, furoates for perfume and flavoring, fumigant, textile processing, chemical intermediate.

Furol viscosity.   The efflux time in seconds (SFS) of 60 mL of sample flowing through a calibrated Furol orifice in a Saybolt viscometer under specified conditions. Furol viscosity is approximately 1/10 of Saybolt Universal viscosity and is used for fuel oil and residual materials of relatively high viscosity. Furol is derived from the words *fuel* and *road oils*.

furosemide.   (5-(aminosulfonyl)-4-chloro-2[(2-furanylmethyl)amino]benzoic acid; 4-chloro-N-furfuryl-5-sulfamoylanthranilic acid; 4-chloro-n-(2-furylmethyl)-5-sulfamoylanthranilic acid; frusemide; fursemide; 4-chloro-2-(furan-2-ylmethylamino)-5-sulfamoylbenzoic acid).
CAS: 54-31-9. $C_{12}H_{11}ClN_2O_5S$.
A benzoic-sulfonamide-furan. It is a diuretic with fast onset and short duration and anti-hypertensive agent.
Hazard: Moderately toxic.
Use: To treat cardiovascular and renal diseases.

furoyl chloride.
CAS: 527-69-5. C_4H_3OCOCl.
Properties: Colorless liquid. Powerful lachrymator. Decomposes in water, mp −2C, bp 176C. Soluble in ether. Combustible.
Derivation: Treatment of furoic acid with phosphorus pentachloride.
Hazard: Strong irritant to eyes and skin.
Use: Substitute for chloropicrin in disinfecting grain elevators.

furylacrylic acid.   (furfural acetic acid; furacrylic acid). $C_4H_3OCH:CHCOOH$.
Properties: White powder. Mp 141C, bp 117C (8 mm Hg), 286C. Slightly soluble in cold water; easily soluble in hot water; soluble in alcohol, ether, and glacial acetic acid.
Derivation: From furfural.
Use: Derivatives used in perfumes.

furyl carbinol.   See furfuryl alcohol.

2-furyl-1-nitroethene.
CAS: 699-18-3. $C_6H_5NO_3$.
Hazard: A poison by ingestion. Moderately toxic by inhalation.
Use: Agricultural chemical.

furyltriazine.
CAS: 4685-18-1. $C_7H_7N_5O$.
Hazard: Moderately toxic.

"Fusabond" *[DuPont].* TM for resin used for coupling in mineral-filled polypropylene systems. Works best with long glass fiber systems. The resins act to bond the filler into the polymer matrix. Also improves surface wetting and dispersion of the filler. It enhances tensile and impact strength of the composite.

fusaric acid. (5-butyl-2-pyridinecarboxylic acid; 5-butylpicolinic acid; 5-butylpyridine-2-carboxylic acid). $C_{10}H_{13}NO_2$. A mycotoxin and picolinic acid, that is an antibiotic and wilting agent that causes yellowing of infected plants.
Derivation: Produced by fungi of the genus *Fusarium*, especially *Fusarium moniliforme*.
Hazard: Very toxic.
Use: A medical research tool.

Fusarium. A genus of fungi commonly found in soil and water affected buildings. Some are allergens and important plant pathogens that may infect animals and humans, especially the immunocompromised host. Some species produce Type B Tricothecene mycotoxins, T-2 toxin, and zearalenone (F-2 toxin), vomitoxin, deoxynivalenol, and fumonisin.
Hazard: Toxic, pathogenic.

fuse. Of a solid, to melt, e.g., fused salt. An electric fuse acts as a circuit breaker by the melting of a thin strip of metal. The term has the connotation of uniting or joining, as in welding. The union of hydrogen nuclei to yield energy is called fusion.

fused ring. A ring having one or more of its sides in common with another ring, as shown:

fused salt. A salt (i.e., ionic compound) in the molten state. Halides and nitrates are the salts most commonly used. High temperatures (500–1000C for the alkali halides) are usually required. Most fused salts are liquids with viscosities, diffusion coefficients, thermal conductivities, and surface tensions in the same range as water. They conduct electricity exceptionally well.
Use: Production of sodium by electrolysis, heat transfer agents, reaction medium in chemical synthesis, heat-treatment of metals (from 350 to 2400F), solvents for the metals corresponding to their cations, nuclear power reactors.
See salt bath.

fused silica. See silica.

fusel oil. (amyl alcohol; fermentation; grain oil; potato oil). A volatile, oily mixture consisting largely of amyl alcohols. Isoamyl alcohol (isobutyl carbinol) and active amyl alcohol (2-methyl-1-butanol) are chief constituents. Ethyl, propyl, butyl, hexyl, and heptyl alcohols, as well as other alcohols, have been separated. Acids, esters, and aldehydes are also present. Normal primary amyl alcohol (1-pentanol) is not found in fusel oil. Combustible.
Hazard: Moderate fire risk. Toxic by ingestion and inhalation.
Use: Chemicals (amyl ether, amyl acetate, pure amyl alcohols, nitrous ether, various esters), explosives (gelatinizing agent), solvent for fats and oils, intermediate, pharmaceuticals, nitrocellulose plastics, synthetic rubber, varnishes, lacquers, solvent for resins and waxes, perfumery.

fusible alloy. See alloy, fusible.

fusidate sodium. (sodium fusidate; (2Z)-2-[(3R,4S,5S,8S,9S,10S,11R,13R,14S,16S)-16-acetyloxy-3,11-dihydroxy-4,8,10,14-tetramethyl-2,3,4,5,6,7,9,11,12,13,15,16-dodecahydro-1H-cyclopenta[a]phenanthren-17-ylidene]-6-methylhept-5-enoate).
$C_{31}H_{47}NaO_6$.
The sodium salt of fusidic acid.
Use: Bactericide.

fusidic acid. (3α,4α,8α,9ß,11α,13α,14ß,16ß,17z)-16-(acetyloxy)-3,11-dihydroxy-29-nordammara-17(20),24-dien-21-oic acid; 3α, 11α,16ß-trihydroxy-29-nor-8α,9ß,13α, 14ß-dammara-17(20),24-dien-21-oic acid 16-acetate; 3α,11α,16ß-trihydroxy-4α,8,14-trimethyl-18-nor-5α,8α,9ß,13α, 14ß-cholesta-17(20),24-dien-21-oic acid 16-acetate; 3,11, 16-trihydroxy-4,8,10,14-tetramethyl-17-(1n-carboxyisohept-4n-enylidene)cyclopentanoperhydrophenanthrene 16-acetate; ramycin; (2Z)-2-[(3R,4S,5S,8S,9S,10S,11R,13R,14S,16S)-16-acetyloxy-3,11-dihydroxy-4,8,10,14-tetramethyl-2,3,4,5,6,7,9,11,12,13,15,16-dodecahydro-1H-cyclopenta[a]phenanthren-17-ylidene]-6-methylhept-5-enoic acid).
CAS: 6990-06-3. $C_{31}H_{48}O_6$.
Properties: Terpene.
Derivation: Isolated from the fermentation broth of fusidium coccideum.
Hazard: Moderately toxic inhibitor of translocation during protein synthesis.
Use: Antibiotic, very active against staphylococci.

fusion. (thermonuclear reaction). An endothermic nuclear reaction yielding large amounts of energy in which the nuclei of light atoms (chiefly the hydrogen isotopes D [deuterium] and T [tritium]) unite or fuse to form helium. Uncontrolled fusion was achieved some years ago in the hydrogen bomb, in which the initiating temperature was supplied by a fission reaction. Research efforts are now being devoted to developing a controlled and sustained fusion reaction that would utilize the deuterium and tritium in water. Several reactions are possible, but the most efficient is, for each fusion event, $D + T + e \rightarrow {}^4He + n + 17.5$ MeV. An energy input (e) equivalent to at least 44 million degrees C is necessary.

One approach utilizes powerful laser beams impinging on a mixture of deuterium and tritium in glass microspheres coated with "Teflon" and beryllium, which have an ablative effect. Fusion reactions have been successfully attained by this method at the Lawrence Livermore Laboratory. The other approach is a magnetic fusion device called a tokamak, located in Princeton, NJ. The Joint European Torus (JET) is located in Cambridge, England.

Fusion has two great advantages over fission as an energy source: (1) it utilizes water and readily available lithium as its raw materials instead of scarce and costly uranium; (2) it produces only tritium as a radioactive by-product. As indicated, the $D + T$ reaction yields ${}^4He$ nuclei, as well as 24 MeV neutrons, which carry off 80% of the energy.
See JET; tokamak.

fusion protein. Depending on the context, means one of two different things. Most commonly, it refers to the protein product of a gene created by the fusion of two distinct genes, or portions of genes. Less commonly it refers to one of a family of proteins that participate in membrane fusion.

futile cycle. A set of enzyme-catalyzed cyclic reactions that result only in the net release of thermal energy by the hydrolysis of ATP.

Fybrene. Petrolatum, USP, of medium melting point and medium consistency.
Use: Paper industry.

G

γ.   See gamma.

G.   See guanine.

g.   (1) Abbreviation of gram, (2) acceleration due to gravity.

μg.   Abbreviation of microgram.

"G-30" [Haynes].   TM for a high-strength, nickel-based alloy with outstanding corrosion resistance to strong oxidizing media, and excellent resistance to localized corrosion and stress corrosion cracking while maintaining ease of welding and fabrication.
Available forms: Sheet, plate, bars, rods, welding electrodes, and wire.
Use: Fabrication into all types of process equipment.

G-942.   An aqueous solution of the partial sodium salt of a polymeric carboxylic acid.
Use: Tanning agent.

Ga.   Symbol for gallium.

gaa.   ((2R,3R,4S,5R,6R)-2-(hyrdroxymethyl)-6-(3-nitrophenoxy)oxane-3,4,5-triol; 3-nitrophenyl-alpha-d-galactopyranoside).
$C_{12}H_{15}NO_3$.
A codon of glutamate that directs the placement of glutamate into a polypeptide.

GABA.   Abbreviation for γ-aminobutyric acid. See aminobutyric acid.

Gabriel–Colman rearrangement.   Formation of isoquinoline derivatives by the action of sodium ethoxide on phthalimidoacetic esters.

Gabriel ethylenimine method.   Formation of ethylenimines (aziridines) by elimination of hydrogen halides from aliphatic vicinal haloamines with alkali. The method can be extended to the preparation of five- and six-membered ring amines.

Gabriel synthesis.   Conversion of alkyl halides to primary amines by treatment with potassium phthalimide and hydrolysis of the N-alkylphthalimides formed.

gac.   ([(2R,3S,4R,5R,6R)-6-[(2R,3S,4R,5R,6R)-4,5-dihydroxy-2-(hydroxymethyl)-6-[(2R,3S,4R,

5R,6S)-4,5,6-trihydroxy-2-(hydroxymethyl)oxan-3-yl]oxyoxan-3-yl]oxy-4,5-dihydroxy-2-methyloxan-3-yl]-[(1S,2S,3S,4R,5R)-2,3,4-trhydroxy-5-(hydroxymethyl)cyclohexyl]azanium.
$C_{25}H_{16}NO_{16}I$.
A codon of aspartate that directs the placement of aspartate into a polypeptide.

G acid.   (2-naphthol-6,8-disulfonic acid).

Properties: White needles. Soluble in water.
Derivation: Sulfonation of β-naphthol.
See Schaeffer acid.
Use: Azo dye intermediate.

gadodiamide hydrate.
CAS: 122795-43-1. $C_{16}H_{28}GdN_5O_9 \cdot xH_2O$.
Hazard: Low toxicity. A reproductive hazard.

gadolinium.
CAS: 7440-54-2.   Gd.
A rare-earth element of the lanthanide series, atomic number 64, group IIIB of the periodic table, aw 157.25, valence of 3; seven natural isotopes.
Properties: Lustrous metal. D 7.87, mp 1312C, bp 3000C; reacts slowly with water. Soluble in dilute acid; insoluble in water. Exhibits a high degree of magnetism, especially at low temperatures; salts are colorless; has highest neutron absorption cross section of any known element; has superconductive properties, burns in air to form the oxide. Combustible.
Derivation: (1) Reduction of the fluoride with calcium, (2) electrolysis of the chloride with sodium chloride or potassium chloride in an iron pot that serves as an anode and graphite cathode.

Grade: Ingots, lumps, turnings, powder up to 99.9+% pure.
Use: Neutron shielding, garnets in microwave filters, phosphor activator, catalyst, scavenger for oxygen in titanium production.

gadolinium chloride.
$GdCl_3 \cdot xH_2O$.
Colorless crystals, soluble in water, purities up to 99.9%.

Hawley's Condensed Chemical Dictionary, Sixteenth Edition. Michael D. Larrañaga, Richard J. Lewis, Sr., and Robert A. Lewis.
© 2016 John Wiley & Sons, Inc. Published 2016 by John Wiley & Sons, Inc.

Use: Gadolinium sponge metal, by contact with a reducing metal vapor; source of gadolinium.

gadolinium fluoride.
CAS: 13765-26-9.　$GdF_3 \cdot 2H_2O$.
Available up to 99.9% purity.
Properties: Semisolid mass.
Hazard: Toxic material.
Use: Source of gadolinium.

gadolinium nitrate.
CAS: 94219-55-3.　$Gd(NO_3)_3 \cdot xH_2O$.
Colorless crystals, soluble in water, decomposes at 110C, purities up to 99.9% gadolinium salt.
Hazard: Fire risk in contact with organic materials.

gadolinium oxalate.
CAS: 867-64-1.　$Gd_2(C_2O_4)_3 \cdot 10H_2O$.
Properties: White powder. Insoluble in water; slightly soluble in acids; loses $6H_2O$ at 110C.
Available forms: Purities up to 99.9% gadolinium salt.

gadolinium oxide.
CAS: 12064-62-9.　Gd_2O_3.
Properties: White- to cream-colored powder, d 7.41, mp 2330C, insoluble in water, soluble in acids to form the corresponding salts, hygroscopic, absorbs carbon dioxide from the air, purities up to 99.8% gadolinium oxide.
Derivation: Calcination of gadolinium salts, liquid–liquid ion-exchange separation.
Use: Neutron shields; catalysts; dielectric ceramics; filament coatings; special glasses; TV phosphor activator; lasers, masers, and telecommunication; laboratory reagent.

gadolinium sulfate.
$Gd_2(SO_4)_3 \cdot 8H_2O$.
Properties: Colorless crystals. D 3.01 (15C), purities up to 99.9% gadolinium salt. Slightly soluble in hot water; more soluble in cold water.
Use: Cryogenic research; the selenide is used in thermoelectric generating devices.

gafamide coa.　　See coconut amine oil condensate.

GAF Carbonyl Iron Powders.　Microscopic spheres of extremely pure iron. Produced in 11 carefully controlled grades ranging in particle size from 3 to 20 microns in diameter. The iron content of some types is as high as 99.5%.
Use: High-frequency cores for radio, telephone, television, short-wave transmitters, radar receivers, direction finders; alloying agents; catalysts; powder metallurgy; magnetic fluids.

gag.　(3-oxo-1,2-oxazole-4-carboxylic acid; glycosaminoglycan).
$C_2H_3NO_2$.

A codon of glutamate that directs the placement of glutamate into a polypeptide.

g agent.　(g-series nerve agent).
A military designation for any of a series of potent, relatively nonpersistent inhibitors of acetylcholinesterase. The number and severity of the symptoms depend on the quantity and route of entry of the nerve agent into the body. Some g-agents may be thickened with certain substances that increase their persistence, thus increasing the amount that penetrates the skin.
Hazard: Difficulty breathing, drooling, excessive sweating, nausea, vomiting, cramps, twitching and jerking, staggering, headache, confusion, drowsiness, coma, and convulsions.

Gal.　Abbreviation for galactose.

d-galactosamine hydrochloride.　(2-amino-2-dioxy-d-galactose).
CAS: 1772-03-8.　$C_6H_{14}ClNO_5$.
Properties: Crystalline solid. Mw 215.64, mp 182–185C (decomposes).
Derivation: Amino sugar isolated from chondroitin sulfate.

galactose.
CAS: 59-23-4.　$C_6H_{12}O_6$.
A monosaccharide commonly occurring in milk sugar or lactose.
Properties: White crystals. Mp 165–168C. Soluble in hot water and pyridine; slightly soluble in glycerol.
Derivation: By acid hydrolysis of lactose.
Use: Organic synthesis, medicine (diagnostic aid).

α-galactosidase.
Properties: Derived from *Mortierella vinaceae* var. *raffinoseutilizer*.
Use: Food additive.

D(+)-galacturonic acid.
CAS: 685-73-4.　$COOH(CH_2O)_4CHO$.
A major constituent of plant pectins. It exhibits mutarotation, having both an α and β form.
Properties: The α form melts with decomposition at 159–160C. Soluble in water, slightly soluble in hot alcohol; insoluble in ether.
Derivation: Hydrolysis of pectins.
Use: Biochemical research.

galena.　(galenite; lead glance).
CAS: 1314-87-0.　PbS.
Natural lead sulfide.
Properties: Lead gray in color, lead-gray streak, metallic luster, good cubic cleavage. D 7.4–7.6, Mohs hardness 2.5. Soluble in strong nitric acid, in excess of hot hydrochloric acid.
Occurrence: Western U.S., Canada, Africa, South America.

Use: Chief ore of lead, frequently recovered for the silver it sometimes contains.

"Galex" *[Natrochem].* **TM for disproportioned gum rosin or wood rosin.**
Properties: Stable. Softening point 64–72C, acid no. 150-170 degrees.
Use: Adhesives, extender in rubber formulations.

gallic acid. (3,4,5-trihydroxybenzoic acid).
CAS: 149-91-7. $C_6H_2(OH)_3CO_2H$.

Properties: Colorless or slightly yellow crystalline needles or prisms. D 1.694, mp 222–240C. Soluble in alcohol and glycerol; sparingly soluble in water and ether.
Derivation: Action of mold on solutions of tannin or by boiling the latter with strong acid or caustic soda.
Use: Photography, writing ink, dyeing, manufacture of pyrogallol, tanning agent and manufacture of tannins, paper manufacture, pharmaceuticals, engraving and lithography, analytical reagent.

gallium.
CAS: 7440-55-3. Ga.
Metallic element of atomic number 31, group IIIA of the periodic table, aw 69.72, valences of 2, 3; two stable isotopes.
Properties: Silvery-white liquid at room temperature. Mp 29.7C, bp 2403C; may be undercooled to almost 0C without solidifying, d 5.9 (25C). More dense as a liquid than as a solid. Soluble in acid, alkali; slightly soluble in mercury. Reacts with most metals at high temperatures.
Grade: Up to 99.9999% purity.
Occurrence: Prepared commercially from zinc ores and bauxite.
Derivation: Extraction of gallium as gallium chloride by ethyl ether or isopropyl ether and subsequent electrodeposition from an alkaline gallium oxide solution.
Use: The metal has no significant commercial uses. Its compounds are used as semiconductors.

gallium antimonide.
CAS: 12064-03-8. GaSb.
Available in an electronic grade.
Use: Semiconducting devices.

gallium arsenide.
CAS: 1303-00-0. GaAs.
Properties: Crystals. Mp 1238C. Electroluminescent in infrared light.
Grade: Ingots, polycrystalline form in high-purity electronic grade, single crystals. Often alloyed with gallium phosphide or indium arsenide.
Hazard: Toxic metal. Questionable carcinogen.
Use: Semiconductor in light-emitting diodes for telephone dials, injection lasers, solar cells, magnetoresistance devices, thermistors, microwave generation.
See arsenic.

gallium oxides.
The sesquioxide, Ga_2O_3, and suboxide, Ga_2O, are known. Both are stable at ordinary temperatures.
See gallium sesquioxide.

gallium phosphide.
CAS: 12063-98-8. GaP.
Properties: Pale-orange, transparent crystals or whiskers up to 2 cm long, made by vapor phase reaction at relatively low temperatures between phosphorus and gallium suboxide. These crystals are intermediate between normal semiconductors and insulators or phosphors. They operate over a temperature range of −55 to 500C. Gallium phosphide is electroluminescent in visible light.
Grade: Polycrystalline form in high-purity electronic grade, single crystals, whiskers.
Use: Semiconducting devices.

gallium sesquioxide.
CAS: 12024-21-4. Ga_2O_3.
Properties: (α form) White crystals. D 6.44, mp 1900C, changes to β form at 600C. (β form) white crystals. D 5.88. Both forms are insoluble in hot acid.
Grade: High-purity electronic.
Use: Spectroscopic analysis.

gallocyanine. (CI 51030).
CAS: 1562-85-2. $C_{15}H_{13}ClN_2O_5$.
Properties: Greenish solid. Insoluble in water; soluble in alcohol, glacial acetic acid, alkali carbonate, and concentrated hydrochloric acid.
Derivation: Reaction of gallic acid with nitrosodimethylaniline.
Use: Analytical reagent for lead, dye, and biological stain.

gallotannic acid. See tannic acid.

galls. (nutgalls). Excrescences on various kinds of oak trees resulting from the deposition of insect eggs.
Grade: The best grades (55–60% tannic acid) come from Iran, Syria, Turkey, and Tripoli
Use: Source of gallic and gallotanic acids, leather tanning, writing inks, medicine, textile printing, pharmaceuticals.

"Galvan" *[Cabot].* TM for battery and electrically conductive carbon black.

galvanized steel. Steel whose surface has been coated with a layer of zinc to prevent corrosion.

galvanizing. Coating of a ferrous metal by passing it through a bath of molten zinc or by electrodeposition of zinc. In the former process, the iron and zinc combine to form an intermetallic compound at the interface, the outer surface being relatively pure zinc, which crystallizes as it cools to form the characteristic spangle. The electrodeposition method gives a uniform surface that may be either dull or bright. Duration of corrosion protection is directly related to the thickness of the zinc coating.
See sacrificial protection.

"Galvoline" [Timminco]. TM for a cored magnesium ribbon used as a continuous anode for the cathode protection of buried pipelines and other metal structures. Combustible.

Galvomag. A magnesium alloy composition used in anodes in cathodic protection.

"Galvorod" [Timminco]. TM for a cored magnesium rod used as an anode in the cathodic protection of water-heater tanks. Combustible.

gamete. A sex cell (egg or sperm) that carries half the normal complement of chromosomes and combines with another sex cell to produce a new individual possessing the normal complement.

gametes. Reproductive cells (sperm or egg) that are haploid.

gametocide. A substance that can control pollination of plants by selectively killing plant sex cells (gametes). Some suggested that gametocides are maleic hydrazide and sodium-α,β-dichloroisobutyrate.

gametophyte. The haploid stage in the life cycle of an organism undergoing alternations of generations. The gametophyte is multicellular and mitotically produces gametes. In plants, the gametophyte nourishes the zygote and young sporophyte.

gamma. (γ).
A prefix having meanings analogous to those of α, namely, to designate locations of substituents in a compound or a particular form or modification of an organic substance (γ-globulin) or a metal crystal. It also identifies the most intense form of short-wave radiation.
See gamma ray.

gamma-acid. See γ-acid.

gamma globulin. γ globulin.
See globulin.

gamma iron. An allotrope of nonmagnetic iron that is stable between 910 and 1400°C.

gamma ray. (γ).
Electromagnetic radiation of extremely short wavelength and intensely high energy. γ rays originate in the atomic nucleus; they usually accompany α and β emission, as in the decay of radium, and always accompany fission. γ rays are extremely penetrating and are absorbed by dense materials like lead and depleted uranium.
Hazard: Exposure to γ radiation may be lethal. Complete protection is essential.
Use: To initiate chemical syntheses (ethyl bromide) and cross-linking of polyethylene and other polymers; biochemical research; food preservation; analytical chemistry.
See X-ray; radiation.

gamma-ray spectroscopy. (γ-ray spectroscopy).
An analytical technique involving the use of γ radiation, which is emitted from radioactive nuclei in discrete energies. The spectrum of energies and the relative intensities of the γ rays often characterize the radionuclide that emits them; these are determined by γ-ray spectroscopy. It is possible to identify quantitatively the elements present by their characteristic γ-ray spectra, as well as to determine radioactive decay rates.

Ganex. A series of modified polyvinylpyrrolidone-based products. Available in a wide range of physical forms, from liquid to waxy solid to granular solid. Used as emollients and lubricity additives, dispersing and suspending agents, pour-point depressants, sizing additives.

ganglefene.
CAS: 299-61-6. $C_{20}H_{33}NO_3$.
Hazard: Moderately toxic.

gangliosides. Sphingolipids containing complex oligosaccharides as head groups.

gangue. The minerals and rock mined with a metallic ore but valueless in themselves or used only as a by-product. They are separated from the ore in the milling and extraction processes, often as slag. Common gangue materials are quartz, calcite, limonite, feldspar, pyrite, etc.

"Gantrez" AN [ISP]. TM for an interpolymer of methyl vinyl ether and maleic anhydride.
Properties: Soluble in water and many organic solvents. Compatible with a wide range of gums, resins, and plasticizers and with most metallic salts. Stable in acid and alkaline solutions. Available in a range of molecular weights.
Use: Ammonium nitrate slurry explosives, for its suspending action and cross-linking ability; rust-preventive films; antistatic and finishing agent in

natural and synthetic textiles and glass fibers; adhesives and coatings; thickening agent and protective colloid; flocculant and foam stabilizer in papermaking; photoreproduction; pharmaceutical preparations; cosmetics; etc.

"Gantrez" M [ISP].   TM for polyvinyl methyl ether.

gardenia yellow.
  CAS: 94238-00-3.
  Hazard: Moderately toxic by ingestion.
  Source: Natural product.

gardrin.   See enprostil.

garlic oil.
  CAS: 8000-78-0.
  Hazard: Moderately toxic by ingestion.
  Source: Natural product.

"Garlon" [Dow Agrosciences].   TM for herbicides containing water-soluble salts or oil-soluble esters of 3,5,6-trichloro-2-pyridyloxyacetic acid.

garnet, natural.   A group of silicate minerals. Garnets in nature are usually composed of a mix of various garnet subspecies.
  Use: Abrasive, blast cleaning of buildings.

garnet, synthetic.   Single-crystal garnets designed, by the introduction of bismuth, calcium, and vanadium, for microwave devices such as low-frequency ferrimagnetic resonators. Special yttrium, iron, and aluminum garnets are used in lasers and electronic devices.

garnierite. $(Ni, Mg)_6(OH)_6Si_4O_{11} \cdot H_2O$. A natural nickel–magnesium silicate, occurring as a natural alteration of magnesium silicate rocks. A nickel ore.

gas.   (↑). A state of matter characterized by very low density and viscosity (relative to liquids and solids), comparatively great expansion and contraction with changes in pressure and temperature, ability to diffuse readily into other gases, and ability to occupy with almost complete uniformity the whole of any container. Gases may be either elements (argon) or compounds (carbon dioxide); elemental gases may be monatomic (helium), diatomic (chlorine), or triatomic (ozone). All exist in the gaseous state at standard temperature and pressure but can be liquefied by pressure. The most abundant gases on earth are oxygen, hydrogen, nitrogen (diatomic), and carbon dioxide. Gases are used for fundamental research on the behavior of matter largely because the low concentration permits isolation of the phenomena far better than is possible in liquids or solids.

A "perfect" or ideal gas closely conforms to the simple gas laws (Boyle's law, Charles' law) for expansion and contraction.
Note: Use of the word *gas* for gasoline, natural gas, or the anesthetic nitrous oxide is acceptable only in informal communication.
See kinetic theory; compressed gas; noxious gas; noble gas; vapor.

gas, asphyxiant.   See asphyxiant gas.

gas black.   See carbon black.

gas chromatography.   (GC; gas–liquid chromatography; GLC; vapor–phase chromatography; VPC).
The process in which the components of a mix are separated from one another by volatilizing the sample into a carrier gas stream that is passing through and over a bed of packing consisting of a 20–200 mesh solid support. The surface of the latter is usually coated with a relatively nonvolatile liquid (the stationary phase). This gives rise to the term *gas–liquid chromatography*. If the liquid is not present, the process is *gas–solid chromatography*, which is also widely used for analysis. Different components move through the bed of packing at different rates and so appear separately, at the effluent end, where they are detected and measured by thermal conductivity changes, density differences, or ionization detectors.
Gas chromatography is advantageous as a means of analysis of minute quantities of complex mixtures from industrial, biological, and chemical sources and is also of potential value in actually preparing moderate quantities of highly purified compounds otherwise difficult to separate from the mixture in which they occur.

gas, compressed.   See compressed gas; liquefied petroleum gas.

gas constant (R).   Pressure times volume divided by the temperature of one gram-molecule of any ideal gas in the general gas law.

$$R = \frac{p_0 v_0}{273C}$$

gas constant, molar. R in general gas law, $PV = nRT$, equal to 0.082060 L-atm mole^{-1} degree^{-1}.

gaseous diffusion.   See diffusion, gaseous.

gas hydrate.   A clathrate compound formed by a gas (either noble or reactive) and water. The compounds are crystalline solids and are insoluble in water. They usually form (only at relatively low temperatures and high pressures) directly by contact of gas and liquid water. From 6 to 18 molecules of water may combine with each molecule of gas, depending on the nature of the gas.

The best-known gas hydrates are those of ethane, ethylene, propane, and isobutane. Others include methane and 1-butene, most of the fluorocarbon refrigerant gases, nitrous oxide, acetylene, vinyl chloride, carbon dioxide, methyl and ethyl chloride, methyl and ethyl bromide, cyclopropane, hydrogen sulfide, methyl mercaptan, and sulfur dioxide.

Interest in the gas hydrates originated mainly because of the nuisance of such compound formation in gas pipelines. In recent years, propane has been used successfully to precipitate water from salt solution (or seawater), thus yielding potable water.

gasification. Production of gaseous or liquid hydrocarbon fuels from coal (1) by direct addition of hydrogen to form methane (hydrogasification); (2) by reacting steam with hot coal (800C) in the presence of air or oxygen to form carbon monoxide and hydrogen (synthesis gas), followed by a methanation reaction (Fischer–Tropsch method); and (3) by underground aeration. In methods (1) and (2), the coal enters the reaction sequence in finely divided form; hydrocarbon gases, naphtha, and fuel oils of various grades can be produced by several modifications. In method (3), air is pumped down to previously ignited coal seams and the combustible products evolved are collected.

Many variations of methods (1) and (2) have been researched in the U.S., but none is economically competitive with crude oil. Several have advanced to the pilot stage, and projections of cost, investment, and installation for large-scale operation have been made. Environmental and geographic considerations have also been investigated. The controlling factor for future development is the price of crude.

All three methods mentioned are technically feasible. During World War II, hydrocarbon fuels were produced from coal on a large scale in Germany by both hydrogasification and Fischer–Tropsch methods. Large-scale coal conversion has been successfully achieved by the Lurgi and Sasol methods; the latter, in South Africa, is a large-scale installation utilizing the Fischer–Tropsch technology. The Lurgi technique is in production at a number of locations.

Catalytic gasification of peat for production of methanol is now utilized on a commercial scale; use of lignite and wood for this purpose is under development.

See Fischer–Tropsch process; synthesis gas.

gas, inert. See noble gas, inert.

gas laws. See Boyle's law; Charles' law; Gay-Lussac's law; ideal gas.

gas, liquefied petroleum. See liquefied petroleum gas.

gas liquid. See light hydrocarbon.

gas, natural. See natural gas.

gas, noble. See noble.

gasohol. Gasoline blended with alcohol. Typically 10% methanol or ethanol is blended with the gasoline.

gas oil. A liquid petroleum distillate with viscosity and boiling range between those of kerosene and lubricating oil. Boiling range 232–426C, flash p 150F (65.5C), autoign temp 640F (337C). Combustible.
Use: Absorption oil, manufacture of ethylene.

gasoline.
CAS: 8006-61-9.
A mixture of volatile hydrocarbons suitable for use in a spark-ignited internal-combustion engine and having an octane number of at least 60. The major components are branched-chain paraffins, cycloparaffins, and aromatics. There are several methods of production: distillation or fractionation, which yields straight-run product of relatively low octane number, used primarily for blending; thermal and catalytic cracking; reforming; polymerization; isomerization; and dehydrocyclodimerization. All but the first are various means of converting hydrocarbon gases into motor fuels by modifications of chemical structure, usually involving catalysis. The present commercial source of gasoline is petroleum, but it may also be produced from shale oil and Athabasca tar sands, as well as by hydrogenation or gasification of coal.
anti-knock gasoline. A gasoline to which a low percentage of methyl-*tert*-butyl ether (MBTE) has been added to eliminate knocking and increase octane number. This compound has almost completely replaced tetraethyl lead. Gasolines of octane number 100 or more are used chiefly as aviation fuel; those having a research octane number of approximately 90 are in general automotive use.
See antiknock agent; octane number.
casinghead gasoline. See natural gasoline (below).
cracked gasoline. Gasolines produced by the catalytic decomposition of high-boiling components of petroleum. In general, such gasolines have much higher octane ratings (80–100) than that produced by fractional distillation. The difference is due to the prevalence of unsaturated, aromatic, and branched-chain hydrocarbons in the cracked gasoline. The actual properties vary widely with the nature of the starting material and the temperature, time, pressure, and catalyst used in cracking.
high-octane gasoline. A gasoline with an octane number of 90–100.
See antiknock gasoline; octane number.
lead-free gasoline. An automotive fuel containing no more than 0.05-g lead per gallon, designed for use in engines equipped with catalytic converters.

natural gasoline. A gasoline obtained by recovering the butane, pentane, and hexane hydrocarbons present in small proportion in certain natural gases. Used in blending to produce a finished gasoline with adjusted volatility but low octane number. Do not confuse with natural gas.

polymer gasoline. A gasoline produced by polymerization of low-molecular-weight hydrocarbons such as ethylene, propene, and butenes. Used in small amounts for blending with other gasolines to improve their octane numbers.

pyrolysis gasoline. Gasoline produced by thermal cracking as a by-product of ethylene manufacture. It is used as a source of benzene by the hydrodealkylation process.

reformed gasoline. A high-octane gasoline obtained from low-octane gasoline by heating the vapors to a high temperature or by passing the vapors through a suitable catalyst.

straight-run gasoline. Gasoline produced from petroleum by distillation without the use of cracking or other chemical conversion processes. Its octane number is low.

white gasoline. An unleaded gasoline especially designed for use in motorboats; it is uncracked and strongly inhibited against oxidation to avoid gum formation and is usually not colored to distinguish it from other grades. It also serves as a fuel for camp lanterns and portable stoves.

Hazard: Highly flammable, dangerous fire, and explosion risk. Eye and upper respiratory tract irritant, and central nervous system impairment. Possible carcinogen.

gas, perfect.   See ideal gas.

Gastaldi synthesis.   Formation of dicyanopyrazines by cyclization of two molecules of an aminocyanomethyl ketone, produced by treatment of an isonitrosomethyl ketone bisulfite compound with potassium cyanide, heating in hydrochloric acid, and oxidation.

gastric juice.   A mixture of hydrochloric acid and pepsin secreted by glands in the stomach in response to a conditioned nerve reflex. Its pH is about 2.0. Its action in the metabolic breakdown of food components is essential to the digestive process, though carbohydrate decomposition is initiated by the saliva.
See digestion (1).

gastric mucosal barrier.   A physiologic barrier that limits the diffusion of nonionic materials through gastric mucosa. Certain agents impair the effectiveness of this barrier.

gastrin tetrapeptide amide.
CAS: 1947-37-1. $C_{37}H_{42}N_6O_8S$.
Hazard: Moderately toxic by ingestion.

gate.   The opening in an injection mold leading from the sprue or runner to the mold cavity; the term also refers to molded material removed from the aperture when the product is ejected.

Gattermann aldehyde synthesis.   Preparation of aldehydes of phenols, phenol ethers, or heterocyclic compounds by treatment of the aromatic substrate with hydrogen cyanide and hydrochloric acid in the presence of Lewis acid catalysts.

Gattermann–Koch reaction.   Formulation of benzene, alkylbenzenes, or polycyclic aromatic hydrocarbons with carbon monoxide and hydrochloric acid in the presence of aluminum chloride at high pressure. Addition of cuprous chloride allows the reaction to proceed at atmospheric pressure.

gau.   ((4S)-4-amino-5-hydroxypentanoic acid). $C_5H_{11}NO_3$.
A codon of aspartate that directs the placement of glutamate into a polypeptide.

gauge.   An instrument for measuring and indicating such process variables as pressure (hydraulic, steam, air), liquid level, thickness, vacuum, etc. The many types of gauges are activated by mechanical, ultrasonic, electronic, magnetic, and pneumatic means. Some operate on the principle of automatic control. In materials technology, the term *gauge* is often synonymous with thickness, especially in the metals, rubber, and plastics fields. Light gauge refers to thicknesses from about 0.005 to 0.05 inches, and heavy gauge to thicknesses from about 0.05 to 0.150 inches.
See mil; meter (2).

Gay-Lussac, Joseph Louis.   (1778–1850). A French chemist and physicist noted for the brilliance and accuracy of his reasoning and experimental work. He contributed greatly to the knowledge of gases in his discovery (1808) of the law of combining volumes and his independent discovery (1802) of the law of Charles, the relationship of temperature to the volume of gases. He graduated from and taught at the Ecole Polytechnique, becoming a full professor in 1810. His work in chemistry was extensive, resulting in the discovery of boron, which he named, with Louis–Jacques Thenard, and a variety of compounds such as boron trifluoride, chloric acid, and dithionic acid ($H_2S_2O_6$). He identified iodine as an element, named it, and studied its properties. He investigated the relationship of acids and bases and introduced many analytical techniques (such as the use of litmus as an indicator). Among his many contributions to industrial chemistry were improvements in the production of sulfuric acid. Much of the progress of chemistry in the early 19th century is associated with his career.

Gay–Lussac's law.   A modification of Charles' law to state the following: At constant pressure the volume of a confined gas is proportional to its absolute temperature. The volumes of gases involved in a chemical change can always be represented by the ratio of small whole numbers.

GC.   Abbreviation for gas chromatography.

gca.   (6-[(3,5-dimethylphenyl)methyl]-1-(ethoxy-mthyl)-5-propan-2-ylpyrimidine-2,4-dione. $C_{19}H_{26}N_2O_3$.
A codon of alanine that directs the placement of alanine into a polypeptide.

gcg.   ((2S)-2-amino-5-[[(2R)-1-[[2-[4-[3-[[2-[[(2R)-2-[[(4S)-4-amino-4-carboxybutanoy]amino]-3-sulfanylpropanoyl]amino]acetyl]amino]propylamino]butylamino]-2-oxoethyl]amino]-1-oxo-3-sulfanylpropan-2-yl]amino]-5-oxopentanoic acid). $C_{27}H_{49}N_9O_{10}S_2$.
A codon of alanine that directs the placement of alanine into a polypeptide.

GC-rich area.   Many DNA sequences carry long stretches of repeated G and C, which often indicate a gene-rich region.

gcu.   ((2S,3S,4S,5R,6R)-3,4,5,6-tetrahydroxyoxane-2-carboxylic acid). $C_6H_{10}O_7$.
A codon of alanine that directs the placement of alanine into a polypeptide.

Gd.   Symbol for gadolinium.

GDME.   Abbreviation for glycol dimethyl ether. See ethylene glycol dimethyl ether.

GDNF.   Abbreviation for glial cell line-derived neurotrophic factor.

GDP.   Abbreviation for guanosine diphosphate. See guanosine phosphates.

GDUE.   See 1h,3h,5h-oxazolo(3,4-c)oxazole-7a(7h)-methanol.

Ge.   Symbol for germanium.

gel.   A colloid in which the disperse phase has combined with the continuous phase to produce a viscous jellylike product. Only 2% gelatin in water forms a stiff gel. A gel is made by cooling a solution, whereupon certain kinds of solutes (gelatin) form submicroscopic crystalline particle groups that retain much solvent in the interstices (so-called "brush-heap" structure). Gels are usually transparent but may become opalescent.
See pectin.

gelatin.   A mixture of proteins obtained by hydrolysis of collagen by boiling skin, ligaments, tendons, etc. Its production differs from that of animal glue in that the raw materials are selected, cleaned, and treated with special care so that the product is cleaner and purer than glue. Type A gelatin is obtained from acid-treated raw materials, and type B from alkali-treated raw materials. Gelatin is strongly hydrophilic, absorbing up to 10 times its weight of water and forming reversible gels of high strength and viscosity. It can be chemically modified to make it insoluble in water for such special applications as microencapsulation of fish nutrients for fish culture.
Properties: Flakes or powder; odorless; tasteless. Soluble in warm water and glycerol; insoluble in organic solvents.
Grade: Edible, photographic, technical, USP.
Use: Photographic film; sizing; textile and paper adhesives; cements; capsules for medicinals; matches; light filters; clarifying agent; desserts, jellies, etc.; culture medium for bacteria; blood plasma volume expander; microencapsulation; printing inks; nutrient; protective colloid in ice cream.

gelatin dynamite.   A high explosive that contains nitrocellulose in addition to nitroglycerin. The product is a gelatinized mass, less sensitive to shock and friction than straight dynamite.

gel electrophoresis.   A procedure for separating a mixture of molecules through a stationary material (gel) in an electrical field.

gel filtration.   A type of fractionation procedure in which molecules are separated from each other according to differences in size and shape; the action is similar to that of molecular sieves. Dextran gels (three-dimensional networks of polysaccharide chains) are usually used in this method known as gel-filtration chromatography.
See fractionation; molecular sieve.

"Gelflex Blue" *[Versa Flex].*   TM for a reusable molding rubber.

"Gelgard" *[GelTech].*   TM for a synthetic polymeric water-gelling material.
Use: Fire control.

gelled hydrogen.   Liquid hydrogen thickened with silica powder.
Use: Rocket fuel.

gel paint.   (thixotropic paint).
A paint formulation that has a semisolid or gel consistency when undisturbed, but that flows readily under the brush or when stirred or shaken. After removal of the stress, it becomes stiff again and has little tendency to spill, drip, or run. The thixotropic

quality is obtained by the carefully controlled reaction of a relatively small proportion of a polyamide resin with an alkyd resin vehicle.
See thixotropy.

gel-permeation chromatography. (GPC).
The form of liquid chromatography that sorts polymer molecules in a gel-packed column according to their size in solution.

gelsemine.
CAS: 509-15-9. $C_{20}H_{22}N_2O_2$.
Hazard: A toxic plant alkaloid.
Use: Medicine, as a CNS stimulant.

gel shift assay. gel mobility shift assay (GMSA); band shift assay (BSA); electrophoretic mobility shift assay (EMSA).
A method by which one can determine whether a particular protein preparation contains factors that bind to a particular DNA fragment. When a radiolabeled DNA fragment is run on a gel, it shows a characteristic mobility. If it is first incubated with a cellular extract of proteins (or with purified protein), any protein-DNA complexes will migrate slower than the naked DNA—a shifted band.

"Gelva" [Henkel]. TM for a family of acrylic multipolymer products designed for high-performance, pressure-sensitive adhesive applications where skin adhesion, high temperature, and/or difficult-to-adhere-to-surface are involved. Provided for both solvent and water-based grades.
Use: Medical, automotive, solar film, graphics, and specialty tape products.

gem-. Prefix. Abbreviation of geminate, meaning two identical groups attached to the same carbon atom.

gemcitabine.
CAS: 95058-81-4. $C_9H_{11}F_2N_3O_4$.
Hazard: Human systemic effects

gene. A complex of nucleoproteins (chiefly DNA) that is the active transmitter of genetic information. Each of the body's 50,000–100,000 genes contains the code for a specific product, typically, a protein such as an enzyme.
Genes occur on the chromosomes of every living cell, where they are arranged in a linear order. Genetic mechanism is the same in all organisms, ranging from the lowest forms of life, both plant and animal, to humans. Every organism has a large number of different genes; there are over 10,000 in each cell of the human body. These control the intricate and well-balanced system of biochemical reactions in the cell. The first synthesis of a gene was reported in 1970. In 1976, it was announced that a synthetic gene was successfully introduced into a microorganism in which it functioned in the same manner as would a normal gene. Because of the

ability of DNA to store and transfer coded genetic instructions, genes determine the sequence of amino acids in specific polypeptides (see proteins), thus they prescribe the structure of the proteins synthesized. In viruses, the genes consist of ribonucleic acid (RNA). Mutation of genes is a change in the basic sequence of amino acids in DNA and may be induced by ionizing radiation. Mutations can occur in any living cell.
See genetic code; radiation, ionizing; recombinant DNA; molecular biology.

gene amplification. Any process by which specific DNA sequences are replicated disproportionately larger than their representation in the parent molecules; during development, some genes become amplified in specific tissues.
See gene; oncogene.

gene chip technology. Development of cDNA microarrays from a large number of genes. Used to monitor and measure changes in gene expression for each gene represented on the chip.

gene deletion. The total loss or absence of a gene.

gene expression. The process by which a gene's coded information is translated into the structures present and operating in the cell (either proteins or RNAs).

gene families. Groups of closely related genes that make similar products.

gene library. See genomic library.

gene mapping. Determination of the relative positions of genes on a DNA molecule (chromosome or plasmid) and of the distance, in linkage units or physical units, between them.

gene pool. All the variations of genes in a species.
See allele; gene; polymorphism.

gene prediction. Predictions of possible genes made by a computer program based on how well a stretch of DNA sequence matches known gene sequences

gene product. The biochemical material, either RNA or protein, resulting from expression of a gene. The amount of gene product is used to measure how active a gene is; abnormal amounts can be correlated with disease-causing alleles.

gene testing. See genetic testing; genetic screening.

gene therapy. An experimental procedure aimed at replacing, manipulating, or supplementing

nonfunctional or misfunctioning genes with healthy genes.
See gene; inherit; somatic cell gene therapy; germ line gene therapy.

gene transfer.   Incorporation of new DNA into and organism's cells, usually by a vector such as a modified virus. Used in gene therapy.
See mutation; gene therapy; vector.

general acid–base catalysis.   Catalysis involving a proton transfer from/to a molecule other than water. Compare to specific acid–base catalysis.

generic formula.   See formula, chemical.

"Genesolv" [Honeywell].   TM for ultrapure halogenated hydrocarbon solvents of the methane and ethane series.
Properties: High densities and low viscosities, combined residue (soluble plus insoluble matter) is less than 1 ppm.
Grade: Standard and electronic, Genesolv D, 200, and 2004.
Hazard: Nonflammable, nonexplosive, thermally stable, and relatively nonhydrolyzable.
Use: Removal of greases and oils from metal, plastic, elastomer, and paint or varnish surfaces. Used with all cleaning techniques on assembled motors and parts, electronic devices, precision components, motion-picture film, refrigeration systems, etc. Also used for isolation of viruses, for fire extinguishing, and as dielectric coolants.

gene-splicing.   The enzymatic attachment of one exon to another.
See genetic engineering.

gene testing.   Examining a sample of blood or other body fluid or tissue for biochemical, chromosomal, or genetic markers that indicate the presence or absence of genetic disease.

gene therapy.   Treatment that alters genes (the basic units of heredity found in all cells in the body). In early studies of gene therapy for cancer, researchers are trying to improve the body's natural ability to fight the disease or to make the tumor more sensitive to other kinds of therapy.

genetic code.   Information stored in the genes that program the linear sequence of amino acids within the protein polypeptide chain synthesized during cell development. The agencies involved are DNA (deoxyribonucleic acid) and RNA (ribonucleic acid). It is this information code (analogous to a computer language) that determines (1) the nature and type of an organism, i.e., its heredity, and (2) the kind of cell structure to be formed (muscle, bone, organ, etc.).
See protein; amino acid; ribonuclease.

genetic counseling.   Provides patients and their families with education and information about genetic-related conditions and helps them make informed decisions.

genetic discrimination.   Prejudice against those who have or are likely to develop an inherited disorder.

genetic engineering.   Transference of genetic material from the genes of one species to those of another by uniting a portion of the DNA of one organism with extranuclear sections of DNA from another organism (also called gene-splicing). Such recombinant molecules will replicate in the same way as in normal DNA behavior. This technique is being utilized in basic genetic research sponsored by the National Institutes of Health.
There have been a number of outstanding achievements in genetic engineering in which bacteria are made to form chains of amino acids programmed by nucleic acids supplied by the experimenters: (1) Production of human insulin from *E. coli* by implanting synthetic genes in the bacterium; commercial production of such synthetic insulin has been going on for some time, and the product has been approved by the FDA. (2) Formation and secretion by *E. coli* of the protein ovalbumin. (3) Synthesis of rat growth hormone in bacteria. (4) Synthesis of human growth hormone in bacteria; it will probably be developed on a commercial scale. (5) Production of a biologically active protein closely similar to interferon by exposing to the action of viruses bacteria that have been programmed to form it by introduction of the appropriate genetic code. (6) Creation of a wholly synthetic bacterium characterized by unusual ability to consume crude oil.
Various patents are pending on some of these techniques for commercial exploitation. The Supreme Court has ruled that these and other results of genetic engineering are patentable inventions.
See biotechnology.

genetic engineering technology.   See recombinant DNA technology.

genetic illness.   Sickness, physical disability, or other disorder resulting from the inheritance of one or more deleterious alleles.

genetic information.   The hereditary information contained in a sequence of nucleotide bases.

genetic linkage map.   A chromosome map showing the relative positions of the known genes on the chromosomes of a given species.

genetic map.   A diagram showing the position of specific genes along a chromosome relative to markers.

genetic marker. A gene or other identifiable portion of DNA whose inheritance can be followed. See chromosome; DNA; gene; inherit.

genetic material. See genome

genetic mosaic. An organism in which different cells contain different genetic sequence. This can be the result of a mutation during development or fusion of embryos at an early developmental stage.

genetic polymorphism. Difference in DNA sequence among individuals, groups, or populations (e.g., genes for blue eyes vs. brown eyes).

genetic predisposition. Susceptibility to a genetic disease. May or may not result in actual development of the disease.

genetics. The scientific study of heredity—how particular qualities or traits are transmitted from parents to offspring.

genetic screening. Testing groups of individuals to identify defective genes capable of causing hereditary conditions.

genetic testing. Analyzing an individual's genetic material to determine predisposition to a particular health condition or to confirm a diagnosis of genetic disease.

genetic variation. A phenotypic variance of a trait in a population attributed to genetic heterogeneity.

"Genetron" [Honeywell]. TM for a group of fluorinated hydrocarbons of the methane and ethane series.
Properties: Low-moisture content, noncorrosive, nonflammable, stable, low power requirement per unit of refrigeration, high dielectric strength.
Use: Refrigerant and foam expansion agent.

Geneva System. A system of nomenclature for organic compounds recommended in 1892. It is based on compounds derived from hydrocarbons as a starting point, the names corresponding to the longest straight carbon chain present. The position is indicated by numbers applied to the carbons of the straight chain, beginning with the end carbon nearest the substituent element or group.
Normal compounds, in which the carbons are linked in a straight chain, are named ethane, propane, pentane, etc. Normal propane is

$$
\begin{array}{c}
H \quad H \quad H \\
| \quad\; | \quad\; | \\
H-C-C-C-H \\
| \quad\; | \quad\; | \\
H \quad H \quad H
\end{array}
$$

Branched compounds are so named to show the proper deviation, e.g., 2-methyl propane

$$
\overset{1}{H_3C}-\overset{2}{\underset{\underset{CH_3}{|}}{CH}}-\overset{3}{CH_3}
$$

is a derivation of propane, not of methane. Similarly, isopentane

$$
H_3C-\underset{\underset{CH_3}{|}}{\overset{\overset{CH_3}{|}}{C}}-CH_3
$$

is named 2-methylbutane. The four carbons in a straight line give the name *butane*, and the side group is attached to the second carbon. Another pentane has the structure

$$
\overset{1}{H_3C}-\overset{2}{CH}-\overset{3}{\underset{\underset{CH_3}{|}}{CH}}-\overset{4}{CH}
$$

and was once called tetramethyl pentane. Under the Geneva System, it is named in terms of two methyl groups joined to carbon number two in a propane chain (e.g., 2,2-dimethyl propane).
Some special recommendations of the Geneva System define the use of certain suffixes: open-chain hydrocarbons with one double bond end in *-ene*; those with two double bonds end in *-diene* (*alkenes* and *alkadienes* rather than *olefins* and *diolefins*). Triple-bonded compounds end in *-yne*.
Alcohols are named by use of the hydrocarbon name with *-ol* as the characteristic suffix, i.e., methanol, ethanol, etc. In order to extend this plan to polyhydric alcohols, a syllable such as *di-, tri-,* or *tetra-* is inserted between the name of the parent hydrocarbon and the suffix *-ol*.

CH_2OHCH_2OH 1,2-ethane*di*ol

$\overset{1}{CH_2}OH\overset{2}{CH}OH\overset{3}{CH_2}OH$ 1,2,3-propane*tri*ol

Sulfides, disulfides, sulfoxides, and sulfones are named like ethers, the *oxy* term being replaced with *thio-, dithio-, sulfinyl,* and *sulfonyl*. The acids are also named in accordance with the Geneva System. See chemical nomenclature.

genin. The steroid portion linked to a sugar residue in certain glycosides. Important genins are found in the digitalis glycosides used as heart stimulants.

genome. The total DNA contained in each cell of an organism. Mammalian genomic DNA (including that of humans) contains 6×10^9 base pairs of DNA per diploid cell. There are somewhere in the order of a hundred thousand genes, including coding regions, 5′ and 3′ untranslated regions, introns, and 5′ and 3′ flanking DNA. Also present in the genome are structural segments, such as telomeric and centromeric DNAs and replication origins, and intergenic DNA.

genome maps. Charts that indicate the ordered arrangement of the genes or other DNA markers within the chromosomes.

genome project. Research and technology-development effort aimed at mapping and sequencing the genome of human beings and certain model organisms.
See Human Genome Initiative.

genomic blot. A type of Southern blot specifically used to analyze a mixture of DNA fragments derived from total genomic DNA. Because genomic DNA is very complicated, when it has been digested with restriction enzymes, it produces a complex set of fragments ranging from tens of base pairs to tens of thousands of base pairs. However, any specific gene will be reproducibly found on only one or a few specific fragments. A million identical cells will produce a million identical restriction fragments for any given gene, so probing a genomic Southern with a gene-specific probe will produce a pattern of perhaps one or just a few bands.

genomic clone. A piece of DNA taken from the genome of a cell or animal, and spliced into a bacteriophage or other cloning vector. A genomic clone may contain coding regions, exons, introns, 5′ flanking regions, 5′ untranslated regions, 3′ flanking regions, and 3′ untranslated regions, or it may contain none of these.

genomic library. A collection of clones made from a set of randomly generated overlapping DNA fragments representing the entire genome of an organism.

genomics. The study of genes and their function.

genomic sequence. See DNA.

genotype. The genetic constitution of an organism, as distinguished from its physical appearance (its phenotype).

gentamicin. (gentamycin).
CAS: 1403-66-3. $C_{21}H_{43}N_5O_7$.
Properties: Amorphous solid. Mp 102–108C. Freely soluble in water, pyridine, acid solutions; moderately soluble in methanol, ethanol, and acetone; practically insoluble in benzene and halogenated hydrocarbons.
Use: Antibacterial.

gentian violet. USP name for methyl violet.

gentisic acid. (2,5-dihydroxybenzoic acid).
CAS: 490-79-9. $C_6H_3(OH)_2COOH$.
Properties: Crystals. Mp 199–200C. Soluble in water, alcohol, and ether; insoluble in carbon disulfide, chloroform, and benzene.
Use: Medicine, as sodium gentisate (analgesic).

geochemistry. The study of the chemical composition of the earth in terms of the physicochemical and geological processes and principles that produce and modify minerals and rocks. Of practical importance in discovering and establishing the limits of ore deposits, petroleum, tar sands, salt, sulfur, and other valuable resources.

geometric isomer. A type of stereoisomer in which a chemical group or an atom occupies different spatial positions in relation to the double bonds. If the double bond is between two carbon atoms, the isomers are called *cis* and *trans*, as in crotonic acid, where the H and COOH reverse locations:

$$CH_3 \diagdown \quad \diagup COOH \qquad CH_3 \diagdown \quad \diagup H$$
$$\qquad C=C \qquad\qquad\qquad C=C$$
$$H \diagup \quad \diagdown H \qquad\qquad H \diagup \quad \diagdown COOH$$
$$\qquad cis \qquad\qquad\qquad\qquad trans$$

If the double bond is between a nitrogen and a carbon atom, the isomers are named *anti* and *syn*, as in benzaldoxime, where the OH group shifts locations.

$$C_6H_5-C-H \qquad C_6H_5-C-H$$
$$\qquad\quad \| \qquad\qquad\qquad\quad \|$$
$$\quad HO-N \qquad\qquad\qquad N-OH$$
$$\qquad anti \qquad\qquad\qquad\quad syn$$

This phenomenon also occurs in saturated ring compounds having three or more members in the ring, as well as in certain coordination compounds.
See stereochemistry.

Geon.
CAS: 9002-86-2.
A polyvinyl chloride formulated for moldings, extrusions, binders, and coatings.
Use: For appliance applications, electronic parts, medical devices, construction applications, bottles, wiring, and cable.

geothermal energy. Superheated water and steam trapped in rock strata in areas characterized by volcanic activity or by intrusions of molten magma. Associated temperatures range from 150 to 300C. It escapes from either natural surface vents (geysers, fumaroles, hot springs) or bore holes drilled through the strata. A contributing source of heat is the natural radioactivity of rocks in the earth's upper mantle. Power was produced in Italy from geothermal sources as long ago as 1913; since then geothermal power plants have been installed in Iceland, New Zealand, France, Hungary, Japan, Mexico, and El Salvador. Many nonelectrical uses have been developed for home heating and industrial purposes, especially in Iceland.

The chief source of geothermal energy in the continental U.S. is the California–Nevada area. The geysers in central California have been generating electric power from steam in substantial amounts for some years. The same is true of Hawaii. Geological formations appropriate for geothermal heat are so few that this form of energy will always have limited potential.

geranial.   See citral.

geranialdehyde.   See citral.

geraniol.   (*trans*-3,7-diemthyl-2,6-octadien-1-ol).
CAS: 106-24-1. $(CH_3)_2C:CH(CH_2)_2C(CH_3):$
$CHCH_2OH.$
A terpene alcohol.
Properties: Colorless to pale-yellow, liquid oil; pleasant geranium-like odor. D 0.870–0.890 (15C), fp −15C, bp 230C, refr index 1.4710–1.4780 (20C), optical rotation −2 to +2 degrees. Soluble in alcohol, ether, mineral oil, fixed oils; insoluble in water and glycerol. Combustible.
Derivation: From citronella oil (Java), citronellol-free grades from palmerosa oil, and (synthetically) from pinene. These are of higher quality.
Grade: Standard, soap, synthetic, FCC, EOA.
Use: Perfumery, constituent of synthetic fragrances and synthetic linalool.

geranyl acetate.   (geraniol acetate).
CAS: 105-87-3. $CH_3COOC_{10}H_{17}.$
Properties: Clear, colorless liquid; odor of lavender. D 0.907–0.918 (15C), bp 242C, optical rotation −2 to +2 degrees, refr index 1.4580–1.4640 (20C). Soluble in alcohol and ether; insoluble in water and glycerol. Combustible.
Derivation: (1) Constituent of several essential oils. (2) By heating geraniol and sodium acetate with acetic anhydride.
Grade: Technical, FCC.
Use: Perfumery, flavoring.

geranyl butyrate.   (geraniol butyrate).
CAS: 106-29-6. $C_3H_7COOC_{10}H_{17}.$
Properties: Colorless liquid; rose-like odor. Bp 151C (18 mm Hg), d 0.9008 (17/4C). Insoluble in water and glycerol; soluble in alcohol, ether. Occurs in several essential oils. Combustible.
Grade: FCC.
Use: Perfumes and soaps, flavoring, synthetic attar of rose.

geranyl formate.   (geraniol formate).
CAS: 105-86-2. $HCOOC_{10}H_{17}.$
Properties: Colorless liquid. Bp 113 (15 mm Hg), d 0.927 (20/4C), rose-like odor. Insoluble in alcohol and ether. Occurs in several essential oils. Combustible.

Grade: FCC.
Use: Perfumes and soaps, flavoring, synthetic neroli oil.

geranyl propionate.   (geraniol propionate).
CAS: 105-90-8. $C_2H_5COOC_{10}H_{17}.$
Properties: Colorless liquid; roselike odor. D 0.896–0.913 (25C), refr index 1.4570–1.4650 (20C). Soluble in most oils; insoluble in glycerol. Combustible.
Use: Perfumery, flavoring.

germacrone.
CAS: 6902-91-6. $C_{15}H_{22}O.$
Hazard: A poison by ingestion.

"Germall" [ISP].   TM for a broad spectrum highly effective preservative.

"Germall 115" [ISP].
CAS: 39236-46-9.
TM for imidazolidinyl urea.
Use: Preservative in cosmetics.

germane.   A germanium hydride of the general formula $Ge_nH_{2n+2}.$
See germanium tetrahydride.

germanium.
Ge.
Nonmetallic element of atomic number 32, aw 72.59, valences of 2, 4; group IVa of the periodic table.
Properties: Grayish-white solid. A *p*-type semiconductor, conductivity depends largely on added impurities, d 5.323, mp 937.4C, bp 2830C, oxidizes readily at 600–700C, does not volatilize to approximately 1350C, Mohs hardness 6. Attacked by nitric acid and aqua regia; stable to water, acids, and alkalies in the absence of dissolved oxygen.
Derivation: Recovered from residues from refining of zinc and other sources, by heating in the presence of air and chlorine. It is also present in some coals and can be recovered from their ash.
Occurrence: Missouri, Kansas, Oklahoma.
Method of purification: The chloride is distilled and then hydrolyzed to the oxide, which is reduced by hydrogen to the metal. Zone melting is used for final purification. Single crystals are made by vaporization of germanium diiodide. The impurities in germanium are of controlling importance in its use in transistors. These are added to high-purity germanium in trace amounts during growth of single crystals.
Grade: Transistor, i.e., impurities 1 part in 10^{10}.
Available forms: Ingots, single crystals, pure or doped, powder.
Use: Solid-state electronic devices (transistors, diodes), semiconducting applications, brazing alloys, phosphors, gold and beryllium alloys, infrared-transmitting glass.

germanium dichloride.
$GeCl_2.$

White powder, mp (decomposes), decomposes in water, soluble in germanium tetrachloride, insoluble in alcohol and chloroform.

germanium dioxide. (germanium oxide).
CAS: 1310-53-8. GeO_2.
Properties: White powder, hexagonal, tetragonal, and amorphous.
Grade: Technical, semiconductor, 99.999% pure.
Use: Phosphors, transistors and diodes, infrared-transmitting glass.

germanium monoxide.
CAS: 20619-16-3. GeO.
Properties: Black solid. Mp 710C (sublimes); insoluble in water; soluble in oxidizing agents. Available commercially 99.999% pure.

germanium potassium fluoride. (potassium germanium fluoride).
K_2GeF_6.
Properties: White crystals. Soluble in water (hot); insoluble in alcohol.
Grade: Technical.
Hazard: Toxic material.

germanium telluride.
CAS: 12025-39-7. GeTe.
Properties: An efficient semiconductor. Mp 725C.

germanium tetrachloride.
CAS: 10038-98-9. $GeCl_4$.
Properties: Colorless liquid. D 1.874 (25/25C), fp −49.5C (1 atm), bp 83.1C, refr index 1.464, decomposes in water. Insoluble in concentrated hydrochloric acid; soluble in carbon disulfide, chloroform, benzene, alcohol, and ether.
Derivation: Reaction of chlorine with elemental germanium.

germanium tetrahydride.
CAS: 7782-65-2. GeH_4.
Properties: Colorless gas, d 3.43 g/L, fp −165C, bp −88C, decomposes at 350C, insoluble in water, soluble in liquid ammonia, slightly soluble in hot hydrochloric acid.
Hazard: Toxic material.

germ cell. The reproductive cells, either egg or sperm cells. Germ cells are haploid and have only one set of chromosomes (23 in all), while all other cells have two copies (46 in all).

germicide. See bactericide.

germ line. The continuation of a set of genetic information from one generation to the next.
See inherit.

germination. The process by which a seedling emerges and develops from a seed, or by which a sporeling emerges and develops from a spore.

germ-line cell. A class of animal cells formed early in embryogenesis that are set aside for reproductive function (give rise to gametes). All other cells are somatic cells.

germ line gene therapy. An experimental process of inserting genes into germ cells or fertilized eggs to cause a genetic change that can be passed on to offspring.
Use: To alleviate effects associated with a genetic disease.
See genomics; somatic cell gene therapy.

germ line genetic mutation. See mutation.

getter. See scavenger.

gga. ((5E,9E,13E)-6,10,14,18-tetramethylnonadeca-5,9,13,17-tetraen-2-one).
$C_{22}H_{39}O$.
A codon of glycine that directs the placement of glycine into a polypeptide.

ggc. ((2S)-2-amino-5-[[(2R)-1-(carboxymethyl-amino)-3-(naphthalene-1-ylmethylsulfanyl)-1-oxopropan-2-yl]amino]-5-oxopentanoic acid).
$C_{21}H_{25}N_3O_6S$.
A codon of glycine that directs the placement of glycine into a polypeptide.

ggg. (3-(2-methoxyphenoxy)propane-1,2-diol).
$C_{19}H_{14}O_4$.
A codon of glycine that directs the placement of glycine into a polypeptide.

ggu. A codon of glycine that directs the placement of glycine into a polypeptide.

ghatti gum. Exudation from the stem of *Anogeissus latifolia*.
Properties: Colorless to pale-yellow tears, rounded or vermiform. Almost tasteless and odorless, partially soluble in water. Can be solubilized by autoclaving.
Use: Thickener and protective colloid; emulsifier for oils, fats, waxes.

Ghost Remover. An ink haze and stencil residue remover.
Use: Effective on most inks and emulsions.

Giauque, William F. (1895–1982). An American chemist who achieved distinction for his studies of the properties of matter at temperatures approaching absolute zero (−273C). This research established the science of cryogenics. Giauque received the Nobel Prize in chemistry in 1949. He was professor and research director at the University of California at Berkeley. One of his most significant contributions was the invention of a magnetic cooling device that made it possible to attain cryogenic temperatures. An important property of

matter discovered as a result of his work is super-conductivity.
See cryogenics; superconductivity.

gibberellic acid.
CAS: 77-06-5.　$C_{19}H_{22}O_6$.
A plant-growth-promoting hormone. It is a tetra-cyclic dihydroxylaectonic acid. It was synthesized in 1978.
Properties: Crystals. Mp 233–235C. Slightly soluble in water; soluble in methanol, ethanol, acetone; soluble in aqueous solutions of sodium bicarbonate and sodium acetate.
Use: Agriculture and horticulture, malting of barley with improved enzymatic characteristics.

gibberellin.
A group of plant-growth regulators (hormones), isolated in 1938 and widely distributed in flowering plants that promotes elongation of shoots and coleoptiles. They differ from auxins in not stimulating the growth of roots and in various other properties. The presence of auxin appears to be necessary for gibberellins to function. All gibberellins have closely related structures, being weak acids with a ring system containing double bonds and eight asymmetric carbons.

Gibberellin A$_3$

See gibberellic acid; plant-growth regulator.

Gibbs–Duhem equation.
(GDE).
An exact thermodynamic relation that permits computation of the changes of chemical potential for one component of a uniform mixture over a range of compositions, provided the changes of potential for each of the other components have been measured over the same range.

Gibbs, Josiah Willard.
(1839–1903). The father of modern thermodynamics. During his life-long post as professor of mathematical physics at Yale, he stated the fundamental concepts embraced by the three laws of thermodynamics, especially the nature of entropy. A theorist rather than an experimenter, Gibbs was the first to expound with mathematical rigor the "relation between chemical, electrical, and thermal energy and capacity for work." It has been said that throughout his adult life, Gibbs did nothing but think. The results established him as a great creative scientist.
See thermodynamics.

Gibb's free energy.
See free energy.

Gibbs phase rule.
See phase rule.

Gibbs phthalic anhydride process.
Oxidation of naphthalene to phthalic anhydride with air at 360C over vanadium pentoxide and other catalysts.

giga-.
Prefix meaning 10^9 units (symbol = G). 1 Gg = 1 gigagram = 10^9 grams.

Gilbert, Walter.
(1932–). An American molecular biochemist who won the Nobel Prize for chemistry in 1980 along with Berg and Sanger for their studies of the chemical structure of nucleic acid. Author of many papers on theoretical physics and molecular biology, he has been at Harvard since 1972.

gilsonite.
An asphaltic material or solidified hydrocarbon found only in Utah and Colorado. One of the purest (9.9%) natural bitumens. Said to be the first solid hydrocarbon to be converted to gasoline.
Hazard: Irritant, skin sensitizer.
Use: Acid, alkali, and waterproof coatings; black varnishes, lacquers, baking enamels, and japans; wire-insulation compounds; linoleum and floor tile; paving; insulation; diluent in low-grade rubber compounds; possible source of gasoline, fuel oil, and metallurgical coke.
See asphalt; bitumen.

gin.
An alcoholic beverage made by distilling alcohol through a mixture of herbs and berries (juniper, coriander, etc.) and adjusting to 80–100 proof.
Properties: Flash p 90F (32.3C).
Hazard: Flammable, moderate fire risk. Slight irritant, intoxicant.

ginkgo biloba l., root extract.
CAS: 90045-36-6.
Hazard: Low toxicity by ingestion. Human systemic effects.
Source: Natural product.

ginsenoside.
CAS: 74749-74-9.
Hazard: A poison.
Source: Natural product.

Girard's reagent.
(Girard's "P": carboxymethylpyridinium chloride hydrazide; acethydrazide-pyridinium chloride).
$C_5H_5NClCH_2CONHNH_2$.
(Girard's "T": carboxymethyl trimethyl ammonium chloride hydrazide; trimethylacethydrazide ammonium chloride).
$(CH_3)_3NClCH_2CONHNH_2$.
Properties: White to faintly pinkish crystals; little or no odor. Mp 190–200C. Soluble in water; insoluble in oils. T is hygroscopic.
Use: Separation of aldehydes and ketones from natural oily or fatty materials; extraction of hormones.

Girbotol absorption.
(amine absorption).

A process for the removal of hydrogen sulfide or carbon dioxide from a gaseous mixture. An organic amine (ethanolamine or diethanolamine, which are basic) is allowed to flow down a tortuous path through a tower where it is contacted by and absorbs (acidic) hydrogen sulfide or carbon dioxide from the gas to be purified as it moves up the tower. The amine, contaminated with these products, is then sent from the bottom of the tower to a steam stripper where it flows countercurrent to steam, which strips the hydrogen sulfide or carbon dioxide from it. The amine is then returned to the top of the tower. The process is widely used in the petroleum industry for purifying refinery and natural gases and for recovery of hydrogen sulfide for sulfur manufacture. Removal of carbon dioxide from gases is usually done with monoethanolamine.

glacial.   A term applied to a number of acids, e.g., acetic and phosphoric, that have a freezing point slightly below room temperature when in a highly pure state. For example, glacial acetic acid is 99.8% pure and crystallizes at 16.6C.

glacial acetic acid.   (acetic acid)
CAS: 64-19-7. $C_2H_4O_2$.
Properties: Pure acetic acid; density of 1.05 kg/l; clear, colorless, vesicant liquid.
Derivation: Produced by destructive distillation of wood, by the reaction of acetylene and water; by the reaction of methanol with carbon monoxide, oxidation of acetaldehyde; liquid- and vapor-phase oxidation of petroleum gases; the fermentative oxidation of ethanol or lactic acid.
Hazard: Corrosive; exposure of small amounts can severely erode the lining of the gastrointestinal tract; may cause vomiting, diarrhea, bloody feces and urine; cardiovascular failure and death.
Use: Locally, occasionally internally, as a counterirritant and a reagent.

"Glacier" [NuMark].   TM for a new class of metal-working fluids. It is biodegradable and nontoxic and generates no oil mist while in use. Provides low maintenance, environmental compatibility, and worker safety in metal operations.
Use: Desirable when aggressive metal removal and extreme requirements are desired. For grinding, milling, and turning in automotive, aviation, tooling, and bearings.

glance.   A mineralogical term meaning brilliant or lustrous; used to describe hard, brittle materials that exhibit a bright reflecting surface when fractured. Examples of such materials are hard asphalts (glance pitch) and ores of certain metals such as lead glance (galena).

Glaser coupling.   Coupling of terminal acetylenes by shaking an aqueous solution of cuprous chloride-ammonium chloride and the alkyne in an atmosphere of air or oxygen.

glass.   A ceramic material consisting of a uniformly dispersed mixture of silica (sand) (75%), soda ash (20%), and lime (5%), often combined with such metallic oxides as those of calcium, lead, lithium, cerium, etc., depending on the specific properties desired. The blend (or "melt") is heated to fusion temperature (approximately 700–800C) and then gradually cooled (annealed) to a rigid, friable state, often referred to as vitreous. Technically, glass is an amorphous, undercooled liquid of extremely high viscosity that has all the appearances of a solid. It has almost 100% elastic recovery.
See glass, optical.
Properties: (Soda-lime glass.) Lowest electrical conductivity of any common material (below 10^{-6} mho/cm). Low thermal conductivity. High tensile and structural strength. Relatively impermeable to gases. Inert to all chemicals except hydrofluoric, fluosilicic, and phosphoric acids and hot, strong alkaline solutions. Continuous highest-use temperature about 121C but may be higher, depending on composition. Good thermal insulator in fibrous form. Molten glass is extrudable into extremely fine filaments. Glass is almost opaque to UV radiation; in the absence of added colorant it transmits 95–98% of light to which it is exposed. Noncombustible.
Occurrence: Natural glass is rare but exists in the form of obsidian in areas of volcanic activity and meteor strikes. Excellent sand for glassmaking occurs in Virginia (James River), Pennsylvania, Massachusetts, New Jersey, West Virginia, Illinois, and Maryland; also in southern Germany and the Czech Republic.
Available forms: Plate, sheet, fiber, filament, fabric, rods, tubing, pipe, powder, beads, flakes, hollow spheres.
See sodium silicate.
Use: Windows, structural building blocks, chemical reaction equipment, pumps and piping, vacuum tubes, light bulbs, glass fibers, yarns and fabrics, containers, optical equipment. Minute glass spheres with partial vacuum interior and treated exterior are available for compounding with resins for use in deep-sea floats, potting compounds, and other composites.

glass, borosilicate.   See glass, heat-resistant.

glass ceramic.   A devitrified or crystallized form of glass whose properties can be made to vary over a wide range.
Properties: Rupture modulus up to 50,000 psi, d 2.5, thermal shock resistance 900C, highest continuous-use temperature 700C. Glass ceramics lie between borosilicate glasses and fused silica in high-temperature capability.
Derivation: A standard glass formula to which a nucleating agent, such as titania, has been added, is melted, rolled into sheet, and cooled. It is then heated to a temperature at which nucleation occurs, causing formation of crystals.

Use: Range and stove tops, laboratory bench tops, architectural panels, restaurant heating and warming equipment, telescope mirrors.
See nucleation; "Pyroceram" *[Corning]*; Cer-vit.

"Glassclad 6C" *[Corning]*.   TM for a chlorine terminated polydi methylsiloxane telomer.
Use: To form a siliconized surface.

glass cleaner.
Properties: Mixture of ammonia with water and a colorant.
Hazard: Highly irritant, moderately toxic.
Use: To clean glass.

glass electrode.   See electrode, glass.

glass enamel.   A finely ground flux, basically lead borosilicate, intimately blended with colored ceramic pigments. Different grades give characteristics of acid resistance, alkali resistance, sulfide resistance, or low lead release to meet requirements for various uses. Firing range 540–760C.
Use: For fired-on labels and decorations on glassware, tumblers, milk bottles, beverage bottles, glass containers, illuminating ware, architectural glass, and signs.
See porcelain enamel.

glass fiber.   (fibrous glass; fiberglass).
Generic name for a manufactured fiber in which the fiber-forming substance is glass (Federal Trade Commission). Noncombustible.
Properties: Tensile strength 15 g/denier, elongation 3–4%, d 2.54, no moisture regain, loses strength at >315C, softens at approximately 815C.
See glass.
Derivation: Molten glass is extruded at high speed through extremely small orifices.
Hazard: A possible carcinogen.
Use: Thermal, acoustic, and electrical insulation (coarse fibers in bats or sheets); decorative and utility fabrics such as drapes, curtains, table linen, carpet backing, tenting, etc.; tire cord as belt between tread and carcass; filter medium; reinforced plastics; light transmission for communication signals; reinforcement of cement products for construction use.
See fiber, optical.

glass, heat-resistant.   (1) A soda-lime glass containing approximately 5% boric oxide, which lowers the viscosity of the silica without increasing its thermal expansion. Such glasses (known as borosilicates) have a very low expansion coefficient and high softening point (about 593C). Tensile strength is approximately 10,000 psi. Continuous-use temperature 482C. Transmits UV light in higher wavelengths and is used in sunlight lamps and similar equipment for this reason.
See "Pyrex" *[Corning]*.

Note: Use of borosilicate glass as a storage-disposal medium for high-level radioactive wastes has been under research for some time. Tentative conclusions based on high-temperature and high-pressure autoclave tests indicate that this method would be suitable for geologic storage of such wastes.
See radioactive waste.
(2) A pure silica glass trademarked "Vycor" *[Grace]* that softens at about 1482C.
See silica, fused.

glassine.   A thin, transparent, and very flexible paper obtained by excessive beating of the pulp. It may contain an admixture of urea formaldehyde to improve strength.
Use: Packaging, dust covers for books; general household purposes.

glassmaker's soap.   Term for manganese dioxide (MnO_2).

glass, metallic.   Metal alloys having an amorphous atomic structure similar to that of silica glass; achieved by cooling of the molten alloy so rapidly that no crystalline structure is formed. Such alloys are said to be harder than their crystalline counterparts and are more resistant to corrosion. Those containing iron have unusual ferromagnetic properties that make them suitable for use as transformer coils.
See "Metglas" *[Metglas]*.

glass, optical.   Glasses intended for vision correcting and such applications as lenses for cameras, microscopes, and other instruments; must be of extremely high quality and uniformity to meet requirements for refractive index and light dispersion. Optical glass may be either crown (lime) or flint (lead). Lead oxide is a major ingredient of flint glass, imparting high refractive index and dispersion, as well as surface brilliance. Flint glasses are also used in vacuum tubes and electrical equipment. Many special ingredients are used in both crown and flint glasses for specific refractivity and dispersion properties.

glass, photochromic.   A glass that changes color on exposure to light and returns to original color when the light has been removed. One type is a silicate glass containing dispersed crystals of colloidal silver halide that is precipitated within the melt during cooling. Alkali borosilicates are the most suitable types of glasses for this purpose.
Use: Variable-tint prescription lenses that darken in sunlight and return to original clearness indoors (85% light transmission when clear, 45% in sunlight).

glass, photosensitive.   A glass containing a small amount of a photosensitive substance such as a gold, silver, or copper compound. When UV light is passed through a photographic negative onto the surface of this glass, a latent image formed within the glass is converted to a visible image made up of tiny metal particles when the glass is heated. In a special type of photosensitive glass (photosensitive opal), the metal particles of the photographic image within the glass serve as nuclei for the growth of nonmetallic crystals; crystalline growth is confined to the area of the image. These crystalline areas are dissolved much more rapidly than the adjacent glass by hydrogen fluoride. Thus, the glass can be formed into intricate shapes without the use of mechanical tools.

glass, plate. Plate glass has the same composition as window glass (soda-lime silica), differing from it only in method of manufacture. These differences are primarily (1) the longer time of annealing (3 or 4 days), which eliminates the distortion and strain effects of rapid cooling; and (2) intensive grinding and polishing, which removes local imperfections and produces a bright, highly reflecting finish.

glass, ruby. A deep-red glass made by incorporating colloidal gold into the silicate mixture. It is used chiefly in the decorative arts.

glass, safety. See safety glass.

glass transition temperature. (T_g). The temperature at which an amorphous material (such as glass or a high polymer) changes from a brittle vitreous state to a plastic state. Many high polymers, such as acrylics, and their derivatives have this transition point, which is related to the number of carbon atoms in the ester group. T_g depends on its composition and extent of annealing.

glass, water. See sodium silicate.

Glauber's salt. See sodium sulfate decahydrate.

glauconite.
$K_2(Mg, Fe)_2Al_6(Si_4O_{10})_3(OH)_{12}$.
A natural silicate of potassium, aluminum, iron, and magnesium found in greensands and other sedimentary rocks.
Properties: Green in color, earthy luster. D 2.3.
Occurrence: New Jersey, Virginia.
Use: Water softener, foundry molds, fertilizer.

allo-glaucotoxigenin.
CAS: 14155-65-8. $C_{23}H_{32}O_6$.
Hazard: A poison.
Source: Natural product.

glaze. A mixture similar to porcelain enamel, applied to a ceramic substrate. It may refer to (1) a vitreous coating on pottery or enamelware; (2) the mixed dry powders of the batch to be used for the coating; or (3) a water suspension of these materials (wet glaze). Glazes must be low in sodium and are usually mixtures of silicates and flint, lead compounds, boric acid, calcium carbonate, etc. See frit; porcelain enamel.

glaze stain. Finely ground calcined oxide of cobalt, copper, iron, and manganese.
Use: Coloring ceramic glazes.

GLC. Abbreviation for gas–liquid chromatography.
See gas chromatography.

gliadin. A prolamin occurring in gluten, the protein of wheat, rye, and other grains. Wheat gliadin has the following composition: 52.7% carbon, 17.7% nitrogen, 21.7% oxygen, 6.9% hydrogen, 1.0% sulfur. It is composed of 18 amino acids, 40% being glutamic acid. Insoluble in water, soluble in 70–90% alcohol, soluble in dilute acid and in alkali.
Use: Chemical synthesis of spinal anesthetics, pharmaceutical preparations.

gliftor.
CAS: 8065-71-2. $C_3H_6ClFO·C_3H_6F_2O$.
Hazard: A poison by ingestion, inhalation, and skin contact. An eye irritant.

gliquidone.
CAS: 33342-05-1. $C_{27}H_{33}N_3O_6S$.
Hazard: Low toxicity by ingestion. A reproductive hazard.

Globar. Silicon carbide heating elements and resistors and accessories.
Properties: Elements have a working temperature up to 1510C, which can be extended to 1648C for short periods; low coefficient of expansion; structure not affected by rapid heating and quick cooling; resistance remains practically constant above 482C.
Use: Electric resistors and heating elements, terminals, and other accessories for electric heating elements, electric heating appliances, electric furnaces.

globular proteins. Soluble proteins with a globular (somewhat rounded) shape. Their axial to diameter ratio is less that 10:1 and usually closer to 1:1.

globulin. (1) Any of a group of proteins synthesized by the body when invaded by infective organisms. They are coagulated by heat; insoluble in water; soluble in dilute solutions of salts, strong acids, and strong alkalies. Enzymes and acids cause hydrolysis to amino acids as the only products. Examples are immune serum or γ-globulins in blood and myosin in muscle. The blood globulins are used in immunizing against specific diseases and in medical research. The γ-globulin molecule is reported to consist of 19,996 atoms associated in 1320 amino acid units.

(2) A protein occurring naturally in wheat and other cereal grains.
See immunochemistry.

γ-globulin.   See immune endoglobulin.

GLP.   (good laboratory practices).
Refers to the body of regulations that govern the collection of laboratory data. As two examples, data collected must be traceable to the exact person who collected the data, and computer files must have their name changed if the file contents change.

Glu.   Abbreviation for glutamic acid.

glucagon.   (hyperglycemic–glycogenolytic factor; HG-factor; HGF).
Produced by the α cells of the islands of Langerhans and also by the gastric mucosa. It is opposite in effect to insulin. It appears to be a straight-chain polypeptide with a molecular weight of approximately 3500. Small amounts have been detected in commercial insulin preparations.
Grade: USP.
Use: Medicine, biochemical research.

"Glucam E-10" *[Lubrizol].*   TM for humectant from corn.
Use: To reduce overall irritancy from other ingredients in hair- and skin-care preparations.

"Glucam E-20" *[Lubrizol].*   TM for humectant from corn.
Use: For a highly active, mild skin moisturizer that reduces water loss from cream and lotion formulations.

"Glucam P-20" *[Lubrizol].*   TM for a fragrance fixative.

4-α-*d*-glucanotransferase.   (dextrin transglycosylase or glycosyltransferase; *d*-enzyme; disproportionating enzyme; amylomaltase).
A 4-glycosyltransferase that coverts maltodextrins into amylose and glucose by transferring portions of 1,4-glucan chains to new 4-positions on glucose or other 1,4-glucans.

glucase.   See maltase.

glucinium.   Former name for the element beryllium, named because the salts of beryllium are sweet tasting.

glucocorticoid response element.   See GRE.

glucogenic amino acids.   Amino acids with carbon chains that can be metabolically converted into glucose via gluconeogenesis. Contrast with ketogenic amino acids.

gluconeogenesis.   The biosynthesis of a carbohydrate from simpler, noncarbohydrate precursors such as alanine or pyruvate.

gluconic acid.   (glyconic acid; glycogenic acid).
CAS: 299-27-4. $CH_2OH(CHO)_4COOH$.
Properties: Pure product is crystalline; light-brown color. Mp 131C, d 1.24. Commercial grade is 50% aqueous solution. Insoluble in organic solvents.
Derivation: Bacterial, chemical, or electrochemical oxidation of glucose. It is the chief acid in honey.
Grade: Technical, 50% solution.
Use: Pharmaceutical and food products, cleaning and pickling metals, sequestrant, cleansers for bottle washing, paint strippers, alkaline derusters, catalyst in textile printing (ammonium salt).

glucono-Δ-lactone.   (D-gluconic acid, Δ-lactone).
CAS: 90-80-2. $CH_2OHCHCH(CHO)_3C(O)O$.
Properties: White crystals. Mp 155C, bp (decomposes). Readily soluble in water; slightly soluble in alcohol.
Derivation: Oxidation of glucose.
See gluconic acid.

8-β-*d*-glucopyranosyl-apigenin.
CAS: 3681-93-4. $C_{21}H_{20}O_{10}$.
Hazard: A poison.

D(+)-glucosamine.
CAS: 3416-24-8. $CH_2OH(CHO)_3CHNH_2CHO$.
Properties: (β form.) Colorless needles. Mp 110C (decomposes). Very soluble in water; slightly soluble in methanol and ethanol; insoluble in ether and chloroform.
Use: Biochemical research.

glucose.   (dextrose; grape sugar; corn sugar).
CAS: 50-99-7.

$$
\begin{array}{c}
CH{=}O \\
H{-}C{-}OH \\
HO{-}C{-}H \\
H{-}C{-}OH \\
H{-}C{-}OH \\
CH_2OH
\end{array}
$$

Properties: Colorless crystals or white granular powder; odorless; sweet taste. D 1.544, mp 146C. Soluble in water; slightly soluble in alcohol. It has the D (right-handed) configuration and is dextrorotatory. Combustible.
Occurrence: Formed in plants by photosynthesis, also in the blood.
Derivation: Hydrolysis of cornstarch with acids or enzymes, hydrolysis of cellulosic wastes.
Grade: Technical, USP, anhydrous, hydrated.
Use: Confectionery, infant foods, medicine, brewing and wine making, intermediate, caramel coloring,

baking and canning, source of methane by anaerobic fermentation, source of certain amino acids (e.g., lysine) by fermentation.
See glucose syrup; invert sugar.

glucose isomerase enzyme preparations, insoluble.
Use: Food additive.

glucose oxidase.   An enzyme commercially available under various trademarks; catalyzes oxidation of glucose to gluconic acid. Water-soluble amorphous powder. Removes excess oxygen from canned foods, beer, etc., and from stored food.
Use: Food preservative, analytical reagent for glucose, stabilizer for vitamins C and B_{12}.

glucose-6-phosphatase.   An enzyme produced by the kidney and liver of vertebrates. It catalyzes the conversion of glucose-6-phosphate into glucose, enabling release into the circulating blood.

glucose-6-phosphate.   ((3,4,5,6-tetrahydroxyxan-2-yl)methyl dihydrogen phosphate; d-glucose-6-(dihydrogen)phosphate; glucose-6-phosphoric acid; g6p).
CAS: 56-73-5. $C_6H_{13}O_9P$.
A constituent of resting muscle and an important intermediate in carbohydrate metabolism.

glucose syrup.   (corn syrup).
A mixture of D-glucose, maltose, and maltodextrins made by hydrolysis of cornstarch by the action of acids or enzymes. The degree of conversion of the starch varies with consequent effect on the dextrose equivalent (DE) or reducing power of the syrup.
Use: Food industry as a sweetener (high DE), thickener, or bodying agent (low DE) in soft drinks.
See glucose.

α-glucosidase.   See maltase.

β-glucosidase.   See emulsion.

glucoside.   See glycoside.

"Glucquat 100" [Lubrizol].   TM for humectant from corn.
Use: For conditioning, plasticizing, and foam retention.

D(+)-glucuronic acid.
CAS: 6556-12-3. $COOH(CH_2O)_4CHO$.
A widely distributed substance in both plants and animals, usually occurs as part of a larger molecule as in various gums or combined with phenols or alcohols.
Properties: Needle-like crystals. Exhibits mutarotation; the β form has mp 165C. Soluble in water and alcohol.
Derivation: From gum acacia.
Use: Biochemical research, medicine.

β-D-glucuronidase.   (glusulase; glycuronidase).
An enzyme that catalyzes the hydrolysis of various glucuronides with the liberation of glucuronic acid.

glucuronide.   ((2S,3S,4S,5R,6S)-6-[(2E,4E,6E,8E0-3,7-dimethyl-9-(2,6,6-trimethylcyclohexen-1-yl)nona-2,4,6,8-tetraenoyl]oxy-3,4,5-trihydroxyoxane-2-carboxylic acid).
$C_{26}H_{36}O_8$.
Any glucoside of glucuronic acid. They react with xenobiotics through a nucleophilic displacement (sn_2 reaction) of the functional group of the substrate, catalyzed by a glucuronosyltransferase. Many xenobiotic chemicals, their intermediates, and certain other products of catabolism in vertebrates are conjugated in the liver with glucuronide and excreted in the urine.
Derivation: Result from the reaction of uridine diphosphate glucuronic acid with an aglycone.

D-glucuronolactone.
CAS: 32449-92-6. $C_6H_8O_6$.
The γ-lactone of glucuronic acid. Found in plant gums and animal connective tissues.
Properties: Colorless or white powder; odorless. D 1.76 (30/4C), mp 172–178C. Soluble in water.
Derivation: From glucuronic acid or by synthesis.
Use: Growth factor, medicine, pharmaceutical intermediate.

glue.   A colloidal suspension of various proteinaceous materials in water. Most familiar are those derived by boiling animal hides, tendons, or bones, which are high in collagen. Chief sources are slaughterhouse wastes and fish scraps. Other animal-derived glues are made from casein (milk) and blood. The most important vegetable glue is made from soybean protein. Combustible in the solid form.
See adhesive; fish glue; collagen; gelatin.

Glueglis.   A composition containing chiefly glue and glycerol used in printing-press rollers. Has soft, rubbery consistency but is easily decomposed by heat.

glufosinate.
CAS: 53369-07-6. $C_5H_{12}NO_4P$.
Hazard: Moderately toxic by ingestion.
Use: Agricultural chemical.

gluside.   See saccharin.

glutamate.   (2-aminopentanedioic acid).
CAS: 56-86-0. $C_5H_9NO_4$.
A salt or ester of glutamic acid that is a common acceptor of amino acids in mammals. It is a nonessential amino acid naturally occurring in the L-form and is the most common excitatory neurotransmitter in the central nervous system.

glutamic acid.   (α-aminoglutaric acid; 2-amino-pentanedioic acid).
CAS: 56-86-0. $COOH(CH_2)_2CH(NH_2)COOH$.
A nonessential amino acid. The naturally occurring form is L(+)-glutamic acid.
Properties: DL-glutamic acid (synthetic racemic mix): crystals. Mp 224–227C (decomposes). Slightly soluble in ether, alcohol, and petroleum ether. D 1.4601 (20/4C). Moderately soluble in water L(-)-glutamic acid. Mp 247–249C (decomposes), d 1.538 (20/4C) L(+)-glutamic acid: crystals. Sublimes at 200C, decomposes at 247–249C. Slightly soluble in ether, acetone, cold glacial acetic acid; insoluble in alcohol. D 1.538 (20/4C), specific rotation +37 to +38.9 degrees (25C). Available commercially.
Derivation: Hydrolysis of vegetable protein (e.g., beet sugar waste, wheat gluten), organic synthesis based on acrylonitrile. It comprises 40% of the gliadin in wheat gluten.
Grade: FCC (L-form).
Use: Medicine, biochemical research, salt substitute, flavor enhancer (L-form only).
See sodium glutamate.

glutamine.   (2-amino-4-carbamoylbutanoic acid).
CAS: 56-85-9. $H_2NC(O)(CH_2)_2CH(NH_2)COOH$.
A nonessential amino acid. Both the L- and DL-forms are available.
Properties: White, crystalline powder. Soluble in water; insoluble in most organic solvents. Should be kept dry and refrigerated. Mp (L-form) 184–185C (decomposes) (DL-form) 176C. Its presence in wheat gluten contributes to the elastic properties of flour by hydrogen bonding and disulfide cross-linking.
Derivation: Action of enzymes on gluten, from beet roots, constituent of many proteins.
Use: Medicine, culture media, biochemical research, feed additive.

glutamine synthetase.   An enzyme that catalyzes the lamination of glutamic acid to glutamine with concomitant hydrolysis of ATP to ADP and Pi.

7-n-((2-((2-(glutamylamino)ethyl)dithio)ethyl))mitomycin c.
CAS: 118359-59-4. $C_{24}H_{34}N_6O_8S_2$.
Hazard: A poison.

γ-glutamylcysteinylglycine.   See glutathion.

glutaraldehyde.
CAS: 111-30-8. $OCH(CH_2)_3CHO$.
Properties: Liquid. D 0.72, bp 188C (decomposes), fp −14C, vap press 17 mm Hg (20C). Soluble in water and alcohol. No flash point. Nonflammable.
Grade: 99%, 50% biological solution, 25% solution.
Hazard: Irritant. Questionable carcinogen.

Use: Intermediate; fixative for tissues; cross-linking protein and polyhydroxy materials, tanning of soft leathers.

glutaric acid.   (n-pyrotartaric acid; pentanedioic acid).
CAS: 110-94-1. $HOOC(CH_2)_3COOH$.
Properties: Colorless crystals. Mp 97C; refr index 1.419 (106C). Soluble in water, alcohol, ether, benzene, and chloroform.
Derivation: From cyclopentanone.
Use: Organic synthesis, biochemical research.

glutaric anhydride.   (pentanedioic acid anhydride).
CAS: 108-55-4. $H_2C(CH_2CO)_2O$.
Properties: Solid. Mp 56.5C, bp 303C. Soluble in benzene and toluene; soluble in water on complete hydrolysis.
Hazard: Irritant.
Use: Plasticizers, resin, lubricant, adhesive synthesis, dyes, and pharmaceuticals.

glutaronitrile.   (trimethylenedicyanide; pentanedinitrile).
CAS: 544-13-8. $NC(CH_2)_3CN$.
Properties: Colorless to straw-colored, viscous liquid. Bp 286C, d 0.989. Soluble in water and alcohol; insoluble in ether and carbon disulfide. Combustible.
Use: Chemical intermediate.

glutathione.   (γ-glutamylcysteinylglycine).
CAS: 70-18-8. $C_{10}H_{17}O_6N_3S$.
A universal component of the living cell. Contains glutamic acid, cysteine, and glycine. These are chemically bound but can be separated by hydrolysis.
Properties: White, crystalline powder; odorless; mild, sour taste. Mp 190–192C. Soluble in water and dilute alcohol.
Use: Nutritional and metabolic research.

gluten.   A mixture of many proteins in which gliadin, glutenin, globulin, and albumin predominate; it occurs in highest percentage in wheat (Manitoba wheat contains approximately 12%) and also to some extent in other cereal grains, usually associated with starch. It comprises 18 amino acids. Gluten is insoluble in water and hydrophilic. Its specific adaptability to bread making is due to its elastic, cohesive nature that enables it to retain the bubbles of carbon dioxide evolved by leavening agents; this also imparts to doughs their characteristic dilatant properties. This behavior is due to disulfide cross-links and hydrogen bonding between the proteins or their constituent amino acids.
Use: Special breakfast foods and other cereals and foods, cattle food, adhesives, production of certain amino acids.

glutenin. One of the proteins present in wheat flour in substantial percentage. It is composed of 18 amino acids.

Gluten Intolerance Group. (GIG). A nonprofit group that provides support to persons with gluten intolerances, including celiac disease, dermatitis herpetiformis, and other gluten sensitivities. Its mailing address is 31214, 124th Ave SE, Auburn, WA 98092. Website: http://www/gluten.net

gluten sensitive enteropathy. (GSE). See: celiac disease.

glutethimide. See 2-ethyl-2-phenylglutarimide.

Gly. Abbreviation for glycine.

glycan. A general term for a polymer of monosaccharide units joined by glycosidic bonds, may or may not have other components, e.g., peptidoglycan.

glycarbylamide. (4,5-imidazoledeicarboxamide). $C_3H_2N_2(CONH_2)_2$.
Properties: White powder. Melts above 360C. Insoluble in water.
Use: A coccidiostat for chickens.

glyceraldehyde. (glyceric aldehyde). CAS: 56-82-6. $HOCH_2CH_2OCHO$. Isomeric with dihydroxyacetone. It is produced by the oxidation of sugars in the body. As the simplest aldose, the conformation of D- and L-glyceraldehydes has been designated the reference standard for D- and L-carbohydrates and derivatives.

In these isomers, the central carbon atom (C*) is asymmetric.
Properties: (DL-glyceraldehyde.) Tasteless crystals from alcohol–ether mixture. Mp 145C. Insoluble in benzene, petroleum ether, pentane.
Grade: 40% aqueous solution.
Use: Biochemical research; intermediate; nutrition; preparation of polyesters, adhesives; cellulose modifier; leather tanning.

glyceride. An ester of glycerol and fatty acids in which one or more of the hydroxyl groups of the glycerol have been replaced by acid radicals. The latter may be identical or different so that the glyceride may contain up to three different acid groups. Glycerides can be made synthetically. The most common are based on fatty acids that occur naturally in oils and fats. See monoglyceride; triglyceride.

glycerin. See glycerol.

glycerin carbonate. (hydroxymethylethylene carbonate). $CH_2O(CO)OCHCH_2OH$.
Properties: Pale-yellow hygroscopic liquid; odorless. Boiling range 125–130C (0.1–0.2 mm Hg), fp supercools to a glass, d 1.4000 (20/4C), refr index 1.4580 (20C), flash p 415F (212C). Miscible with water, alcohol, ether; soluble in ethylene dichloride; insoluble in carbon tetrachloride, benzene, and aliphatic hydrocarbons. Combustible.
Grade: Technical.
Use: Solvent, intermediate.

glycerin 1-isopropyl ether. CAS: 17226-43-6. $C_6H_{14}O_3$.
Hazard: Low toxicity by ingestion and inhalation. A mild skin and eye irritant.

glycerol. (glycerin; glycyl alcohol; 1,2,3-propanetriol). CAS: 56-81-5.

A trihydric (polyhydric) alcohol.
Properties: Clear, colorless, syrupy liquid; odorless; sweet taste; hygroscopic. D anhydrous 1.2653, USP greater than 1.249 (25/25C), dynamite 1.2620, mp 18C, bp 290C, flash p 320F (160C), autoign temp 739F (392C). Soluble in water and alcohol (aqueous solutions are neutral); insoluble in ether, benzene, and chloroform and in fixed and volatile oils. Combustible.
Derivation: (1) By-product of soap manufacture; (2) from propylene and chlorine to form allyl chloride, which is converted to the dichlorohydrin with hypochlorous acid; this is then saponified to glycerol with caustic solution; (3) isomerization of propylene oxide to allyl alcohol, which is reacted with peracetic acid, (the resulting glycidol is hydrolyzed to glycerol); (4) hydrogenation of carbohydrates with nickel catalyst; and (5) from acrolein and hydrogen peroxide.
Method of purification: Redistillation, ion-exchange techniques.
Grade: USP, CP (pharmaceutical and commercial, where the highest grade is required), saponification soap lye, crude yellow distilled (for

commercial purposes where color and extreme purity are not factors), high gravity or dynamite (dehydrated to 99.8–99.9% purity), natural, synthetic, FCC.
Use: Alkyd resins, dynamite, ester gums, pharmaceuticals, perfumery, plasticizer for regenerated cellulose, cosmetics, foodstuffs, conditioning tobacco, liquors, solvent, printer's ink rolls, polyurethane polyols, emulsifying agent, rubber stamp and copying inks, binder for cements and mixes, special soaps, lubricant and softener, bacteriostat, penetrant, hydraulic fluid, humectant, fermentation nutrients, antifreeze mixtures.

glycerol boriborate.
Properties: Pale-yellow liquid obtained by heating glycerol, sodium borate, and boric acid; composition varies, soluble in cold water, absolute alcohol, other alcohols, glycerol.
Use: Adhesive, binder, fabric softener, fire-retardant on fabrics.

glycerol dichlorohydrin.   See α-dichlorohydrin.

glycerol-1,3-distearate.   (glyceryl-1,3-distearate).
$C_{39}H_{76}O_5$.
Properties: Solid. Mp 29.1. Very slightly soluble in cold alcohol and ether; soluble in hot organic solvents.

glycerol ester of partially dimerized rosin.
Properties: Hard, pale amber-colored resin. Sol in acetone, benzene; insol in water.
Use: Food additive.

glycerol ester of partially hydrogenated wood rosin.
Properties: Medium hard, pale amber resin. Sol in acetone, benzene; insol in water, alc.
Use: Food additive.

glycerol ester of polymerized rosin.
Properties: Hard, pale amber resin. Sol in acetone, benzene; insol in water, alc.
Use: Food additive.

glycerol ester of tall oil rosin.
Properties: Pale amber resin. Sol in acetone, benzene; insol in water.
Use: Food additive.

glycerol ester of wood rosin.
Properties: Hard, pale amber resin. Sol in acetone, benzene; insol in water.
Use: Food additive.

glycerol-lacto oleate.
Use: Food additive.

glycerol-lacto palmitate.
Use: Food additive.

glycerol-lacto stearate.
Use: Food additive.

glycerol monolaurate.   (glyceryl monolaurate).
$C_{11}H_{23}COOCH_2CH_2OCH_2OH$.
Properties: Cream-colored paste; faint odor. Mp 23–27C, d 0.98, FFA <2.5%, iodine value 5–8, pH 8.0–8.6 (25C) (5% aqueous dispersion). Dispersible in water; soluble in methanol and ethanol, toluene, naphtha, mineral oil, cottonseed oil, ethyl acetate. Combustible.
See monoglyceride.
Grade: Edible, technical.
Use: Emulsifying and dispersing agent for food products, oils, waxes, and solvents; antifoaming agent; dry-cleaning soap base.

glycerol monooleate.
CAS: 25496-72-4.
Properties: Clear amber or pale yellow liquid. Insol in water.
Use: Food additive.
See Aldo MO.

glycerol monooleate.   (glyceryl monooleate).
$C_{17}H_{33}COOCH_2CH_2OCH_2OH$.
Properties: Yellow oil or soft solid. D 0.95, mp 14–19C depending on purity, iodine value 65–80. Insoluble in water; somewhat soluble in alcohol and most organic solvents. Combustible.
See monoglyceride.
Grade: Edible, technical.
Use: Foods, pharmaceuticals, and cosmetics; rust-preventive oils; textile finishing; vinyl light stabilizers; odorless base paints, flavoring.

glycerol monoricinoleate.   (glyceryl monoricinoleate).
$C_6H_{13}CH_2OC_{10}H_{18}COOCH_2CH_2OCH_2OH$.
Properties: Yellow liquid. D 1.10, mp <−5C, iodine value 65–70. Dispersible in water; soluble in most organic solvents. Combustible.
See monoglyceride.
Use: Nondrying emulsifying agent; solvent; plasticizer; in polishes; in cosmetics; in textile, paper, and leather processing; low-temperature lubricant. Stabilizes latex paints against breakdown due to repeated freeze–thaws.

glycerol monostearate.   (GMS; glyceryl monostearate; monostearin).
$(C_{17}H_{35})COOCH_2CH_2OCH_2OH$.
Properties: Pure white or cream-colored, waxlike solid; faint odor; fatty agreeable taste. Affected by light, mp 58–59C (capillary tube), d 0.97, FFA <5%, iodine value 3–4, dispersible in hot water,

soluble (hot) in alcohol, oils, and hydrocarbons. Combustible.
See monoglyceride.
Grade: Edible, cosmetic, NF.
Use: Thickening and emulsifying agent for margarine, shortenings, and other food products; flavoring; emulsifying agent for oils, waxes, and solvents; protective coating for hygroscopic powders; cosmetics; pharmaceuticals; opacifier; detackifier; resin lubricant.
See Atmul.

glycerol phthalate. See glyceryl phthalate.

glycerol tributyrate. See glyceryl tributyrate.

glycerol tripropionate. See glyceryl tripropionate.

glycerol tristearate. See stearin; "Neustrene 064" *[ACH]*.

α-glycerophosphate dehydrogenase. (GDP; ((2R,3S,4R,5R)-5-(2-amino-6-oxo-3H-purin-9-yl)-3,4-dihydroxyoxolan-2-yl)methyl phosphono hydrogen phosphate). $C_{10}H_{15}N_5O_{11}P_2$.
A guanine nucleotide enzyme that contains two phosphate groups esterified to the sugar moiety and occurs in the cytoplasm of muscle cells that contributes to the synthesis of adenosine triphosphate.

glycerophospholipid. An amphipathic lipid with a glycerol backbone. A (usually saturated) fatty acid is esterified to C-1 and a (usually unsaturated) fatty acid is esterified to C-2 of glycerol. A polar alcohol is attached through a phosphodiester linkage to C-3.

glycerophosphoric acid.
CAS: 57-03-4. $C_3H_5(OH)_2H_2PO_4$.
Properties: Colorless liquid; odorless. D 1.60; fp −25C. Soluble in water and alcohol. Combustible.
Derivation: Interaction of glycerol and phosphoric acid.
Use: Manufacture of glycerophosphates.

glyceryl abietate. An ester gum.
Use: Additive in citrus-flavored beverages.

glycerylaminophenaquine.
CAS: 3820-67-5. $C_{19}H_{17}ClN_2O_4$.
Hazard: Moderately toxic by ingestion. Human systemic effects.

glyceryl behenate.
Properties: Off-white powder. Tasteless.
Use: Food additive.

glyceryl α-chlorohydrin. See chlorohydrin.

glyceryl diacetate. See diacetin.

glyceryl-1,3-distearate. See glycerol-1,3-distearate.

glyceryl ditolyl ether. See dicresyl glyceryl ether.

glyceryl monoacetate. See acetin.

glyceryl monolaurate. See glycerol monolaurate.

glyceryl monooleate. See glycerol monooleate.

glyceryl monoricinoleate. See glycerol monoricinoleate.

glyceryl monostearate. See glycerol monostearate.

glyceryl monothioglycolate.
CAS: 30618-84-9. $C_5H_{10}O_4S$.
Hazard: A mild skin and eye irritant.

glyceryl phthalate. (glycerol phthalate).
Properties: Water-white, solid resin. D 1.29, saponifiction value 605–615, acid value 300–315, softening point approximately 67C. Insoluble in water; soluble (hot) in methanol and ethanol, acetone, ethyl acetate; partly soluble in toluene, naphtha.

Grade: Technical.
Use: Varnishes, lacquers, etc.
See alkyd resin.

glyceryl ricinoleate. See glyceryl triricinoleate.

glyceryl triacetate. See triacetin.

glyceryl tri-(12-acetoxystearate). (castor oil, acetylated and hydrogenated).
CAS: 139-43-5. $C_3H_5(OOCC_{17}H_{34}OCOCH_3)_3$.
Properties: Clear, pale-yellow, oily liquid; mild odor. D 0.955 (25/25C), saponification value 298, iodine value 3, solidifies at 4C. Soluble in most organic solvents; insoluble in water. Combustible.
Derivation: Hydrogenation of acetylated castor oil.
Grade: Technical.
Use: Plasticizer for nitrocellulose, ethylcellulose, and polyvinyl chloride; lubricants; protective coatings; food additive.

glyceryl tri-(12-acetylricinoleate). (castor oil, acetylated).
$C_3H_5(OOCC_{17}H_{32}OCOCH_3)_3$.
Properties: Clear, pale-yellow, oily liquid; mild odor. D 0.967 (25/25C), saponification value 300, iodine value 76, solidifies at −40C. Soluble in most organic liquids; insoluble in water. Combustible.
Grade: Technical.
Derivation: Acetylation of castor oil.

Use: Plasticizer for nitrocellulose, ethylcellulose, and polyvinyl chloride; lubricants; protective coatings.

glyceryl tributyrate. (tributyrin; butyrin; glyceryl tributyrate).
CAS: 60-01-5. $C_3H_5(OCOC_3H_7)_3$.
Properties: Colorless, oily liquid. D 1.035 (20C), refr index 1.4359 (20C), bp 315C, soluble in water 0.010%. Soluble in alcohol and ether. Combustible.
Grade: Technical, FCC.
Use: Plasticizer, flavoring.

glyceryl tri(12-hydroxystearate). (castor oil, hydrogenated).
$C_3H_5(OOCC_{17}H_{34}OH)_3$.
Glyceryl triricinoleate in which hydrogen has saturated the ricinoleic groups.
Properties: Hard, brittle, waxlike solid; yellowish to milky-white in color. Mp 86–88C, d 0.899 (100/25C).
Use: Lubricants, heavy-metal soaps, waxes, plasticizers, cosmetics, chemical intermediate. The lithium compound is used in high-temperature greases.

glyceryl trinitrate. See nitroglycerin.

glyceryl trioleate. See olein.

glyceryl tripalmitate. See tripalmitin.

glyceryl tripropionate. (glycerol tripropionate; tripropionin).
CAS: 139-45-7. $C_3H_5(OCOC_2H_5)_3$.
Properties: Solid. D 1.078 (20C), refr index 1.431 (20C), bp 177–182C (20 mm Hg), fp < 50C. Solubility in water 0.313% of weight; soluble in alcohol.
Use: Plasticizer.

glyceryl triricinoleate. (glyceryl ricinoleate).
$C_3H_5(OOCC_{17}H_{32}OH)_3$.
The triglyceride of ricinoleic acid. It constitutes approximately 80% of castor oil.
Properties: Light-amber, oily liquid.
Use: Emulsifying agent.

glyceryl tristearate. See stearin.

glycidol. (2,3-epoxy-1-propanol).
CAS: 556-52-5.

$$\overset{\displaystyle O}{\overset{\displaystyle /\backslash}{CH_2OHCHCH_2}}$$

An epoxide.
Properties: Colorless liquid. Bp 162C, d 1.12. Soluble in water, alcohol, and ether. Combustible.
Derivation: Treatment of monochlorohydrin with bases; reaction product of allyl alcohol and perbenzoic acid.

Hazard: Toxic material. Probable carcinogen.
Use: Stabilizer for natural oils, demulsifier, dye-leveling agent, stabilizer for vinyl polymers.

γ-glycidoxypropyltrimethoxysilane.

$\overline{}$
$OCH_2CHCH_2O(CH_2)_3Si(OCH_3)_3$

Properties: Liquid. D 1.070 (25C), bp approximately 120C (2 mm Hg), refr index 1.4280 (25C). Soluble in acetone, benzene, ether; reacts with water.
Derivation: Addition of hypochlorite to allyl alcohol and reaction with soda lime.
Use: Coupling agent for glass- and mineral-filled plastics.

glycidyl acrylate.
CAS: 106-90-1.
$H_2C:CHCOOCH_2\overline{CHCH_2O}$.
Properties: Liquid. D 1.1074 (20/20C), bp 57C (2 mm Hg) with polymerization, fp −41.5C, flash p 141F (60.5C) (TOC). Insoluble in water. Combustible.
Hazard: Irritant to skin and eyes.
See acrylate.

glycidyl ether.
$\overline{}$
OCH_2CHOCH_2.
Appears as the terminal group of epoxy resin structures resulting from reaction of epichlorohydrin and bisphenol A with alkaline catalyst. Also reacts with novolac resins.
Properties: Combustible liquid.

glycidyl isocyanurate. (1,3,5-triglycidyl-*S*-triazinetrione; TGT).
CAS: 2451-62-9. $C_{12}H_{15}N_3O_6$.
Properties: Mw 297.30.

glycine.
(1) (aminoacetic acid).
CAS: 56-40-6. NH_2CH_2COOH.
A nonessential amino acid.
Properties: White, sweet, crystals; odorless. Mp 232–236C with decomposition, d 1.1607. Combines with hydrochloric acid to form the hydrochloride; soluble in water; insoluble in alcohol and ether.
Derivation: Action of ammonia on chloroacetic acid; occurs in many proteins and is especially abundant in silk fibroin, gelatin, and sugarcane.
Grade: Technical, NF, FCC
Hazard: Use in fats restricted to 0.01%.
Use: (1) Organic synthesis, nutrient, biochemical research, buffering agent, chicken-feed additive, reduces bitter taste of saccharin, retards rancidity in animal and vegetable fats.

(2) The extreme dilution of methylacetophenone gives a perfume resembling the odor of the climbing plant glycine (*Wisteria sinensis*), native to China and cultivated elsewhere. The name is also given to bouquets made from violet, lilac, and jasmin ottos.
(3) *p*-Hydroxyphenylglycine. A photographic developer.

glycocholic acid.   (cholylglycine). $C_{26}H_{43}NO_6$.
The sodium salt occurs in bile, where it is formed by the combination of glycine with cholic acid. It aids in the digestion and absorption of fats.
Properties: Crystallizes from water with 1.5 moles H_2O, becomes anhydrous at 100C, anhydrous form decomposes at 165C. Practically insoluble in water. The sodium salt is soluble in water and alcohol.
Derivation: Precipitation from bile.
Use: Biochemical research, food emulsifying agent (up to 0.1%).

glycocoll.   See glycine (1).

glycocoll-*p*-phenetidine hydrochloride.   See phenocoll hydrochloride.

glycogen.   (animal starch; liver starch).
CAS: 9005-79-2. $(C_6H_{10}O_5)$.
A glycose polysaccharide, the storage carbohydrate of the animal organism, found especially in the liver and rested muscle.
Properties: White powder; sweet taste. Forms a dextrorotatory colloidal solution; insoluble in alcohol; soluble in water.
Derivation: Isolated from liver by treatment with 30% sodium hydroxide solution.
Use: Biochemical research.

glycogenic acid.   See gluconic acid.

glycol.   A general term for dihydric alcohols that are physically and chemically similar to glycerol. See ethylene glycol.

glycol bromohydrin.   See ethylene bromohydrin.

glycol carbonate.   See ethylene carbonate.

glycol chlorohydrin.   See ethylene chlorohydrin.

glycol diacetate.   See ethylene glycol diacetate.

glycol dibutyrate.   See ethylene glycol dibutyrate.

glycol diformate.   See ethylene glycol diformate.

glycol dimercaptoacetate.   (ethylene glycol bisthioglycolate).
CAS: 123-81-9. $HSCH_2COOCH_2CH_2OOCCH_2SH$.

Properties: Liquid. D 1.313, bp 137–139C (2 mm Hg), refr index 1.519 (25C). Insoluble in water; soluble in alcohol, acetone, and benzene. Combustible.
Use: Crosslinking agent for rubbers, accelerator in curing epoxy resins.

glycol   dimercaptopropionate.   [ethylene glycol bi-(mercaptopropionate)].
$(HSCH_2CH_2COOCH_2)_2$.
Properties: Liquid. D 1.219 (25C), bp 175–195C, refr index 1.5150 (25C). Insoluble in water and hexane; soluble in alcohol, acetone, and benzene. Combustible.
Use: Cross-linking agent for polymers, especially epoxy resins; chemical intermediate.

glycol dimethyl ether.   See ethylene glycol dimethyl ether.

glycol dipropionate.   See ethylene glycol dipropionate.

glycolic acid.   See hydroxyacetic acid.

glycolic aldehyde.
CAS: 141-46-8. $C_2H_4O_2$.
Hazard: Moderately toxic by ingestion.

glycolipid.   A lipid containing a small amount of carbohydrate.

glycol monoacetate.   See ethylene glycol monoacetate.

glycolonitrile.   (glyconitrile; formaldehyde cyanohydrin).
CAS: 107-16-4. $HOCH_2CN$.
Properties: Mobile oil; colorless; odorless. Supplied commercially as a 70% aqueous solution stabilized with phosphoric acid. Bp 183C (slight decomposition), mp does not solidify when cooled to $-72C$, d 1.1039 (19C), refr index 1.4090 (25C), electrolytic dissociation constant $K = 0.843 \times 10^{-5}$ (25C).
Derivation: Formaldehyde and hydrogen cyanide.
Hazard: Toxic by ingestion, inhalation, and skin absorption.
Use: Solvent and organic intermediate.

(glycoloyloxy)tributylstannane.
CAS: 5847-48-3. $C_{14}H_{30}O_3Sn$.
Hazard: A poison. Moderately toxic by ingestion. Tributyl tin compounds are extremely toxic to marine life.

glycol propionate.   See ethylene glycol dipropionate.

glycol stearate.   See ethylene glycol monostearate.

glycolysis.   Enzymatic (anaerobic) decomposition of sugars, starches, and other carbohydrates with release of energy, a type of reaction occurring in yeast fermentation and in certain metabolic processes. Lactic acid is one of the products formed.

"Glycomul" *[Lonza].*   TM for PEG, glyceryl, sorbitan, polysorbate esters, and fatty acid esters. See Pegosperse Aldo.

glyconic acid.   See gluconic acid.

glyconitrile.   See glycolonitrile.

glycoprotease.
  CAS: 120720-15-2.
  Hazard: A poison.
  Source: Natural product.

glycoprotein.   A composite molecule made up of a carbohydrate group and a simple protein. An example is the taste-modifying sweetener occurring in the so-called "miracle fruit."

glycosaminoglycan.   A molecule composed of alternating monosaccharide units. One is either Glc-NAc or GalNac, and the other is a usually glucuronic acid. Formerly called mucopolysaccharide.

glycosidase.   An enzyme that catalyzes the hydrolysis of a glucosidic linkage between two sugar molecules.

glycoside.   One of a group of organic compounds, of abundant occurrence in plants that can be resolved by hydrolysis into sugars and other organic substances known as aglycones. Specifically glycosides are acetals derived from a combination of various hydroxy compounds with various sugars. They are designated individually as glucosides, mannosides, galactosides, etc. Glycosides were formerly called glucosides, but the latter term now refers to any glycoside having glucose as its sugar constituent.

glycosidic bonds.   Bonds between a sugar and another molecule (typically an alcohol, purine, pyrimidine, or a second sugar). Divided into *N*-glycosidic or *O*-glycosidic linkages, depending on the point of attachment.

"Glycosperse" *[Lonza].*   TM for PEG, glyceryl, sorbitan, polysorbate esters, and fatty acid esters. See Pegosperse Aldo.

5-glycosyl-1,4,5-trihydroxy maphthalene.
  A labile glycoside that releases trihydroxymaphthalene that is readily oxidized to juglone.

glycothiourea.   See 2-thiohydantoin.

glycoylurea.   See hydantoin.

glycyl alcohol.   See glycerol.

glycyrrhizin.   A glycoside of the triterpene group; the active principle of licorice root from which it is extracted. It has an intensely sweet taste and is used as a humectant in tobacco and a flavoring in confectionery and pharmaceutical products. The ammoniated derivative, which is 50 times as sweet as sucrose, is used as a foaming agent in root beer and mouthwashes; as a sweetener in chocolate, cocoa, and chewing gum; and as a taste-masking agent in pharmaceuticals such as aspirin. Its ability to exert strong synergistic action with sucrose makes it useful in low-calorie foods (from 30 to 100 ppm is effective).
See sweetener, nonnutritive.

glyme.   Trivial name for a series of glycol ethers used as aprotic solvents. The group includes monoglyme (bp 85C), ethyl glyme (bp 121C), diglyme (bp 162C), ethyl diglyme (bp 190C), triglyme (bp 216C), butyl diglyme (bp 256C), and tetraglyme (bp 276C). Each is separately listed and referred to its conventional name.

glyodin.   (generic name for 2-heptadecyl-2-imidazoline acetate; 2-heptadecylglyoxalidine acetate).
  CAS: 556-22-9. $C_{17}H_{35}C_3H_5N_2 \cdot CH_3COOH$.
  Properties: Light-orange crystals. Mp 62C, d 1.03. Insoluble in water.
  Derivation: Ethylenediamine and stearic acid.
  Use: Fungicide (fruits and vegetables).

glyoxal.
  CAS: 107-22-2.   OHCCHO.
  Properties: Yellow crystals or light-yellow liquid; mild odor. Mp 15C, bp 51C, d 1.14 (20/20C), bulk d 10.0 lb/gal (20C), vapor has a green color and burns with a violet flame, refr index 1.3826 (20C). Polymerizes on standing or in presence of a trace of water. An aqueous solution contains monomolecular glyoxal and reacts weakly to acid. Undergoes many addition and condensation reactions with amines, amides, aldehydes, and hydroxyl-containing materials. Glyoxal VP resists discoloration.
  Derivation: Oxidation of acetaldehyde.
  Grade: 40% solution; pure, solid; VP.
  Hazard: Mixture of vapor and air may explode. Questionable carcinogen.
  Use: Permanent-press fabrics; dimensional stabilization of rayon and other fibers. Insolubilizing agent for compounds containing polyhydroxyl groups (polyvinyl alcohol, starch, and cellulosic materials); insolubilizing of proteins (casein, gelatin, and animal glue); embalming fluids; leather tanning; paper coatings with hydroxyethylcellulose; reducing agent in dyeing textiles.

glyoxaline.   See imidazole.

glyoxaline-5-alanine   monohydrochloride.
  See histidine monohydrochloride.

glyoxylate cycle. A variant of the citric acid cycle present in bacteria and some plant cells that is used for the net conversion of lipid into carbohydrate.

glyoxyldiureide. See allantoin.

glyoxysome. A specialized peroxisome containing the enzymes of the glyoxylate cycle. Particularly prevalent and important in germinating seeds.

glyphosate. (N-(phosphonomethyl)glycine).
CAS: 1071-83-6.
Use: Herbicide.

glyphosine.
$C_4H_{11}NO_8P_2$.
A ripening agent for sugarcane.

"Glyphosphate" *[Monsanto].* (2-(phosphonomethylamino)acetic acid; n-(phosphonomethyl) glycine; *roundup*^{TM}; *kleen-up*^{TM}).
CAS: 1071-83-6. $C_3H_8NO_5P$.
Hazard: Low toxicity.
Use: Herbicide.

G-M-F.
$HONC_6H_4NOH$
p-quinonedioxime. A rubber accelerator.

GMP. Abbreviation for guanosine monophosphate.
See guanosine phosphates; guanylic acid; sodium guanylate.

GMS. Abbreviation for glycerol monostearate.

Gogte synthesis. Formation of α-pyrone derivatives by rearrangement of acyl-substituted glutaconic anhydrides.

gold.
CAS: 7440-57-5. Au.
Metallic element of atomic number 79, Group IB of the periodic table, aw 196.9665, valences of 1, 3; no stable isotopes.
Properties: Yellow, ductile metal; relatively soft. D 19.3, mp 1063C, bp 2800C. Does not corrode in air but is tarnished by sulfur. Chemically nonreactive; attacked by chlorine and cyanide solutions in the presence of oxygen. Soluble in aqua regia. Insoluble in acids. Excellent reflector of infrared and heat; electrical resistivity (0C) 2.06 microhm-cm; extremely high light reflectivity.
Occurrence: South Africa, the former U.S.S.R., northwest Canada, U.S. (South Dakota, Nevada, Utah, Alaska, California), Australia. Oceans are estimated to contain 70 million tons in solution, with 10 billion additional tons on the ocean floor. There is no present economical method of exploiting these resources.

Derivation: Ore is treated with cyanide solution, and the dissolved gold cyanate is recovered by precipitation with zinc dust or aluminum or by hydrolysis. Placer methods are also used.
Available forms: Ingots; sheet; wire; tubing; powder; leaf; alloys with copper or other metals, the gold content being expressed in carats (the number of parts of gold in 24 parts of alloy). Single crystals and aqueous colloidal suspensions. Leaf may be made in near-colloidal thickness; one troy ounce covers 68 sq ft at 0.0001-inch thickness.
Use: Infrared reflectors; electrical contact alloys; brazing alloys; polarographic electrodes; spinnerettes; laboratory ware; decorative arts (ceramics); in electronics for bonding transistors and diodes to wires, for metallizing ceramic and mica capacitors, and for printed circuits; space vehicle instruments; dental alloys; jewelry. Colloidal dispersions are used in coloring glass, as a nucleating agent in photosensitive glasses, and for specialized medical treatments; gold leaf is used in surgery. Further details can be obtained from Gold Information Center, Box 934, Madison Sq. Station, New York, NY 10159.

^{198}gold. Radioactive gold of mass number 198.
Properties: Half-life, 2.7 days.
Derivation: Neutron irradiation of the element.
Available forms: Gold metal, colloidal gold. The NF solution is colloidal ^{198}Au.
Hazard: Toxic from β- and γ radiation.
Use: Internal radiation therapy, to detect leaks in bacterial filters, to locate solidification boundary in continuously cast aluminum, to determine metallic silver in photographic materials.
See radio-gold; gold sodium thiosulfate.

gold, artificial. See stannic sulfide.

gold bromide. (aurous bromide).
AuBr.
Properties: Yellowish-gray mass. Decomposes at approximately 165C. Insoluble in water.
Grade: Technical.
See gold tribromide.

gold bronze. See aluminum bronze.

gold chloride. See gold trichloride.

gold–cobalt alloys. See cobalt–gold alloys.

gold cyanide. (auric cyanide; cyanoauric acid).
$Au(CN)_3 \cdot 3H_2O$ or $HAu(CN)_4 \cdot 3H_2O$.
Properties: Colorless, hygroscopic crystals. Mp 50C (decomposes). Very soluble in water; soluble in alcohol and ether.
Hazard: Toxic material.
Use: Electrolyte in the electroplating industry.

gold, filled. A thin gold alloy bonded to a base metal, also called clad stock.
Use: Inexpensive jewelry.

gold hydroxide.   (auric hydroxide).
Au(OH)$_3$.
  Properties: Brown powder. Sensitive to light, keep in amber bottle. Probably a hydrated gold oxide (Au$_2$O$_3$) and loses water easily. Soluble in hydrochloric acid, solutions of sodium cyanide, and alkali hydroxides; insoluble in water.
  Use: Gilding liquids, decorating porcelain, gold plating.

gold, liquid bright.   See Liquid Bright Gold.

gold oxide.   (auric oxide; auric trioxide; gold trioxide).
Au$_2$O$_3$.
  Properties: Brownish-black powder, decomposed by heat. Keep in dark bottle. Soluble in hydrochloric acid; insoluble in water.
  Use: Gold plating.

gold potassium chloride.   See potassium gold chloride.

gold potassium cyanide.   See potassium gold cyanide.

gold salts.   See sodium gold chloride.

Goldschmidt, Hans.   See thermite.

Goldschmidt law.   The structure of a crystal is determined by the ratio of the numbers, sizes, and properties of polarization of its atoms or ions.

Goldschmidt process.   Formation of sodium formate by absorption of carbon monoxide in caustic soda at increased pressures and temperatures around 200C. At temperatures above 400C, alkali formate liberates hydrogen and yields alkali oxalate.

gold–silicon alloy.   (silicon–gold alloy).
Formed in amorphous foils 10 microns thick by cooling molten gold and silicon almost instantaneously by spreading on a moving wheel. The atoms are "frozen" before crystals can form.
  Use: Electronics.

gold sodium chloride.   See sodium gold chloride.

gold sodium cyanide.   See sodium gold cyanide.

gold sodium thiomalate.
  CAS: 12244-57-4. NaOOCCH(SAu)CH$_2$COONa·H$_2$O.
  Properties: White to yellowish-white powder; odorless; metallic taste. Affected by light. Very soluble in water; practically insoluble in alcohol and ether; aqueous solutions are colorless to pale yellow; pH (5% solution) 5.8–6.5.

Derivation: Reaction of sodium thiomalate with a gold halide.
  Grade: USP.
  Use: Medicine (antirheumatic).

gold solder.   A solder usually composed of gold, silver, copper, zinc, or brass.
  Use: Principally by jewelers.

gold tin purple.   (purple of Cassius; gold stannate; aurous stannate; gold–tin precipitate; CI 77482).
  Properties: Brown powder. Insoluble in water; soluble in ammonia.
  Derivation: By the reaction of a neutral solution of gold trichloride with stannous and stannic chlorides, yielding a mixture of colloidal gold and tin oxide in varying proportions.
  Grade: Technical.
  Use: Manufacture of ruby glass, coloring enamels and porcelain.

gold tribromide.   (auric bromide; gold bromide).
AuBr$_3$.
  Properties: Brownish-black powder. Mp 160C with decomposition. Soluble in water and alcohol.
  Use: Analysis (testing for alkaloids, spermatic fluid), medicine.

gold trichloride.   (1) (auric chloride; gold chloride).
CAS: 13453-07-1.   AuCl$_3$.
   (2) AuCl$_3$·2H$_2$O.
   (3) (chlorauric acid; chloroauric acid; gold trichloride, acid).
AuCl$_2$·HCl·4H$_2$O or HAuCl$_4$·4H$_2$O.
  Properties: Yellow to red crystals, decomposed by heat. Soluble in water, alcohol, and ether.
  Derivation: Action of aqua regia on gold.
  Grade: Technical, CP, usually as chlorauric acid.
  Use: (3): Photography, gold plating, special inks, medicine, ceramics (enamels, gilding, and painting porcelain), glass (gilding, ruby glass), manufacture of finely divided gold and purple of Cassius.

gold trioxide.   See gold oxide.

gold, white.   A jeweler's alloy consisting of approximately 58% gold, 17% nickel, 7% zinc, and 17% copper.

Golgi apparatus.   Eukaryotic organelle that package cell products, such as enzymes and hormones, and coordinate their transport to the outside of the cell.

Golgi complex.   A complex and dynamic membranous organelle of eukaryotic cells. Golgi bodies function in the posttranslational modification (glycosylation) and secretion of proteins or insertion into membranes. Composed of *cis*, medial and *trans* compartments, as well as a less well defined "trans-golgi network."

Gomberg–Bachmann reaction. Formation of diaryl compounds from aryl diazonium salts and aromatic compounds in the presence of alkali.

Gomberg free radical reaction. Formation of free radicals by abstraction of the halogen from triarylmethyl halides with metals.

gonad. Either of the two primary sex organs (ovary and testis in higher animals). These produce sex hormones and gametes (eggs and sperm). The reproductive functions of these organs can be seriously impaired or destroyed by many toxicants.

Gooch. See crucible.

good laboratory practices. See GLP.

Goodyear, Charles. (1800–1860). Born in Woburn, MA, Goodyear was the first to realize the potentialities of natural rubber. Frustrated by its lack of stability to temperature and other weaknesses in the uncured state, he experimented with additives such as magnesium and sulfur. The discovery of vulcanization was not accidental, as is often stated, but the result of intelligent trials and correct evaluation of their results. Though Goodyear's patents were contested by Hancock in England, he well merits the credit for making rubber usable in countless ways and helping to make the automobile possible. See vulcanization.

Gordon Research Conferences. See American Association for Advancement of Science.

"Gorilla Glue" *[Gorilla Glue].* TM for liquid glue.
Use: To bond and repair interior and exterior wood, stone, metal ceramic, foam board, and fiberglass.

"Gorilla Tape" *[Gorilla Glue].* TM for tape.
Use: Bonds to brush stucco, wood, and more—a double thick adhesive.

gossyplure. (7,11-hexadecadien-1-ol acetate).
CAS: 50933-33-0. $C_{18}H_{32}O_2$.
Properties: A yellow liquid. Mw 280.46, bp 130–132C. Soluble in most organic solvents.
Hazard: Extremely flammable.
Use: Sex attractant for pink bollworm.

gossypol.
CAS: 303-45-7. $C_{30}H_{30}O_8$.
A natural polyphenol.
Properties: Yellow, crystalline pigment having three modifications. Insoluble in water; soluble in alcohol.
Occurrence: Cottonseed kernels.

Hazard: Toxic by ingestion but is inactivated by heat, 0.04% max allowed in foods.
Use: Stabilizer for vinyl polymers, has possibilities as a biodegradable insecticide and male contraceptive, actively investigated in China.

Gould–Jacobs reaction. Synthesis of 4-hydroxyquinolines from anilines and diethyl ethoxymalonate via cyclization of the intermediate anilinomethylenemalonate followed by hydrolysis and decarboxylation.

GPC. See gel permeation chromatography.

GPCR. G-protein coupled receptor. Usually a serpentine receptor.

GPF black. Abbreviation for general purpose furnace black.
See carbon black.

gracilin.
CAS: 23522-05-6. $C_{15}H_{20}O_3$.
Hazard: A reproductive hazard.

grade. Any of a number of purity standards for chemicals and chemical products established by various specifications. Some of these grades are as follows:
ACS (American Chemical Society specifications) reagent (analytical reagent quality)
CP (chemically pure)
USP (conforms to U.S. Pharmacopeia specifications)
NF (conforms to National Formulary specifications)
FCC (Food Chemicals Codex specifications)
Other common grade designations are chemical, commercial, feed, food, injectable, nitration, purified, radio, research, semiconductor, spectro, technical (industrial chemicals).

graduate. A cylindrical glass container with etched volumetric gradations usually ranging from 5 to 100 or more milliliters.
Use: Measuring liquids in chemical and biological laboratories.

Graebe–Ullmann synthesis. Formation of carbazoles by the action of nitrous acid on 2-aminodiphenylamines, followed by decomposition of the resulting benzotriazoles.

graft copolymer. Polymer having branches of varying length made up of different monomeric units on a common truck chain.

grafting. A deposition technique whereby organic polymers can be bonded to a wide variety of other materials, both organic and inorganic, in the form of fibers, films, chips, particles, or other shapes. Grafting occurs at specific catalyst sites on the host materials, which must have some capacity for ion exchange, metathesis, or complex formation. Ionizable groups may be added artificially.

One proprietary application is polymerization of acrylonitrile with wood pulp fibers to make synthetic soil blocks; the polymer imparts high water-holding capacity to the pulp. Plant nutrient materials are added and the mixture pressed into blocks to be used for starting seedlings.

graft polymer. A copolymer molecule comprised of a main backbone chain to which side chains containing different atomic constituents are attached at various points. The main chain may be either a homopolymer or a copolymer. This process may be applied to the union of cellulosic molecules (cotton, rayon) with synthetic polymers (except polyesters, acrylics, and polypropylene) to form modified fibers having improved flame-resistance, dimensional stability, resilience, and bacterial resistance. An intermediate called cellulose thiocarbonate is formed in this proprietary process.

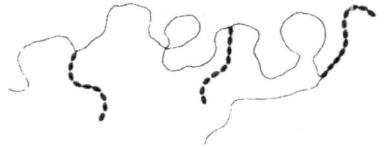

See polyorganosilicate graft polymer.

Graham's salt.   See sodium metaphosphate.

Graham, Thomas.   (1805–1869). Born in Scotland, Graham is famous for his basic studies in diffusion that led to the development of colloid chemistry. He was the first to observe a marked difference in the rate of passage of certain types of substances through a parchment membrane. Those that readily crystallize, like sugar, pass rapidly through the membrane, but gelatinous types are "slow in the extreme." Graham designated the latter, which comprise albumin, starch, gums, etc., as colloids and their solutions as colloidal solutions. The former, which he called crystalloids, form "true" or molecularly dispersed solutions.

See colloid chemistry.

grain.   (1) The smallest unit of mass in the avoirdupois system; 1 grain = 0.0648 grams; one ounce contains 437.5 grains.

(2) Any cereal plant, such as wheat, corn, barley, etc.

(3) Crystalline particles of metals.

(4) The dehaired side of a skin or hide.

grain alcohol.   See ethanol.

grain dust.   Small particles of broken grain dust generated when grain is stored.
  Hazard: Highly explosive.

grain oil.   See fusel oil.

gram.   (g).
One one-thousandth kilogram. It is the mass of 1 mL (approximately 1 cubic centimeter) of water at 4C. One pound contains 453.5 grams.

gram atomic weight.   The atomic weight of an element in grams; e.g., the gram atomic weight of oxygen is 15.994 grams.
See mole.

gramicidin.   An antibiotic produced by the metabolic processes of the bacteria *Bacillus brevis*. It is a polypeptide that is active against most Gram-positive pathogenic bacteria. It is one of the two antibiotic components of tyrothricin but has been isolated and used alone.
  Properties: White, crystalline platelets. Mp 229–230C. Soluble in lower alcohols, acetic acid, and pyridine; moderately soluble in dry acetone and dioxane; almost insoluble in water, ether, and hydrocarbons. Depresses surface tension, forms a fairly stable colloidal emulsion in distilled water.
  Derivation: From tyrothricin by extraction with a mixture of equal volumes of acetone and ether, followed by concentration in vacuo and dissolving in hot acetone.
  Grade: NF.
  Use: Medicine (antibacterial).

gramicidin S.
A cyclic peptide that acts on gram-positive bacteria.
  Derivation: Produced by certain species such as *B. brevis*.

gramine.   [3-(dimethylaminomethyl)indole].
CAS: 87-52-5. $C_{11}H_{14}N_2$.
  Properties: Shiny, flat needles. Mw 174.24, mp 138–139C. Soluble in alcohol, ether, chloroform; slightly soluble in cold acetone; practically insoluble in petroleum ether, water.

gram molecular weight.   The molecular weight of a compound in grams, e.g., the gram molecular weight of carbon dioxide is 44.01 grams.
See mole.

gram-positive, -negative.   A characteristic property of bacteria in reacting to a staining method developed by Gram around 1880. The bacteria are stained with crystalline violet, treated with Gram's solution, and again stained with safranin. If the dye is retained, the bacteria are called gram-positive, and vice versa.
See bacteria.

grana.   Stacks of thylakoids in chloroplasts.

granule.   A piece of gravel of small size (between 2 and 4 mm).

grape color extract.
  Use: Food additive.

grapefruit oil.   See citrus peel oil.

grape skin extract.
  Properties: Red to purple powder or liquid concentrate.
  Use: Food additive.

grape sugar.   See glucose.

graphite.   (black lead; plumbago).
  CAS: 7440-44-0.
  The crystalline allotropic form of carbon.
  Occurrence: Naturally in Madagascar, Ceylon, Mexico, Korea, Austria, the former U.S.S.R., and China. Also produced synthetically by heating petroleum coke to approximately 3000C in an electric resistance furnace. Approximately 70% used in U.S. is synthetic.
  Properties: Relatively soft, greasy feel; steel-gray to black color with a metallic sheen. D 2.0–2.25 depending on origin. Apparent d of artificial graphite 1.5–1.8. High electrical and thermal conductivity, specific heat 0.17 at room temperature, 0.48 at 1500C, tensile strength 400–2000 psi, compressive strength usually approximately 2000–8000 psi. Coefficient of friction 0.1 microns. Resistant to oxidation and thermal shock. Sublimes at 3650C.
  Grade: Powdered, flake, crystals, rods, plates, fibers.

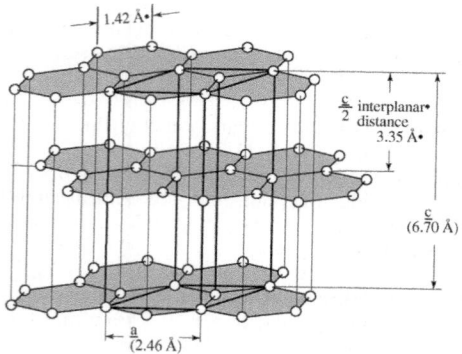

Hexagonal form of graphite.

Hazard: (Powder, natural) Fire risk.
Use: Crucibles, retorts, foundry facings, molds, lubricants, paints and coatings, boiler compounds, powder glazing, electrotyping, monochromator in X-ray diffraction analysis, fluorinated graphite polymers with fluorine-to-carbon ratios of 0.1–1.25, electrodes, bricks, chemical equipment, motor and generator brushes, seal rings, rocket nozzles, moderator in nuclear reactors, cathodes in electrolytic cells, pencils, fibers, self-lubricating bearings, intercalation compounds.
See carbon; graphite fiber; carbon, industrial.

graphite fiber.   High-tensile fibers or whiskers made from (1) rayon, (2) polyacrylonitrile, or (3) petroleum pitch.

Properties: (1) Amorphous structure, oxidizes readily, requiring silicon carbide coating. Resistant to acids and bases, including hydrogen fluoride; tensile strength 50,000–150,000 psi, elastic modulus 4–9 million psi, temperature limit in oxidizing atm 500C, self-lubricating, resistant to electricity, lightweight. (2) Polycrystalline structure, stensile strength up to 350,000 psi, elastic modulus up to 70 million psi, smooth surface, lightweight. (3) Properties similar to (2).
Derivation: (1) Heating rayon in air at 1400–1700C. (2) Heating polyacrylonitrile in air at 220C (20 hours) to oxygenate, then in hydrogen to 1000C (24 hours) to carbonize, finally in argon at 2500C (2 hours). (3) Heating pitch materials (petroleum residues, asphalt, etc.) to carbonization temperature.
Available forms: Filament, yarn, fabric, whiskers. Fibers may be 7–8 microns in diameter.
Use: Heating pads (combined with glass fiber), protection clothing, polyester and epoxy composites for jet engine components, spacecraft, compressor blades, airframe structure, electrodes for spark-hardening metals, flame-proof textile products, engineering thermoplastics.
See fiber; whiskers; composite; carbon, industrial.

graphite, pyrolytic.   (pyrographite).
  A dense, nonporous graphite stronger and more resistant to heat and corrosion than ordinary graphite, intended for rocket nozzles, missiles in general, and nuclear reactors; exhibits tensile strengths 5–10 times higher than commercial graphite and maintains its strength above 3300C. It is ultrapure, impermeable to all fluids, and an anisotropic thermal and electrical conductor.

graphitic oxide.   (GO).
  $C_7O_2(H)_2$.
  Properties: Light-yellow flakes or plates that deposit in layers to form a membrane, usually supported on a cellulose ester.
  Derivation: Slow oxidation of graphite with potassium nitrate in nitric and sulfuric acids.
  Use: Reverse-osmosis membrane for desalination of seawater.

GRAS.   Abbreviation for "generally recognized as safe," applied to food additives approved by the FDA.

gravel.   Sediment particles larger than sand (larger than 2 mm).

gravimetric analysis.   A type of quantitative analysis involving precipitation of a compound that can be weighed and analyzed after drying. It is also used in determining specific gravity.

grayanol A.
  CAS: 52557-31-0. $C_{20}H_{32}O_6$.
  Hazard: A poison.
  Source: Natural product.

grayanol B.
CAS: 52611-78-6. $C_{20}H_{32}O_6$.
Hazard: A poison.
Source: Natural product.

grayanotoxin II.
CAS: 4678-44-8. $C_{20}H_{32}O_5$.
Hazard: A poison.

grayanotoxin III.
CAS: 4678-45-9. $C_{20}H_{34}O_6$.
Hazard: A poison.
Source: Natural product.

grayanotoxin III 6,14-diacetate.
CAS: 30460-34-5. $C_{24}H_{38}O_8$.
Hazard: A poison.
Source: Natural product.

grayanotoxin VI.
CAS: 30460-36-7. $C_{20}H_{32}O_5$.
Hazard: A poison.

GRE. (glucocorticoid response element).
The glucocorticoid receptor is essentially a transA binding site in a promoter to which the activated glucocorticoid receptor can bind to a transcription factor that is activated only in the presence of glucocorticoids. The activated receptor will bind to a GRE, and transcription of the adjacent gene will be altered.
See response element.

Greek fire. Any of a number of deadly combustible mixtures used in ancient Greece and medieval warfare. The precise compositions of these material remain largely unknown, but pitch and sulfur were probably contributors. A petroleum-based mixture invented by the Byzantine Greeks in the 7th century was a true, extremely deadly Greek fire. It apparently combusted spontaneously and could not be quenched with water. During an attack on Constantinople in 673 by Arabs, their fleet was decimated by Greek fire launched from ship-mounted tubes.

Greek letters. See alpha; beta; gamma.

green.
Use: In the process industries in a number of senses in addition to that of a dye or pigment. For example, it often is applied to any material that is uncured or untreated, e.g., leather, rubber, cement, etc.
See green liquor; greensand; greensalt; green soap; Paris green.

green cross. (superpalite).
See trichloromethyl chloroformate.

greenhouse effect. Gradual rise in average global temperature due to absorption of infrared radiation by increasing amounts of carbon dioxide in the air, which retards dissipation of heat from the earth's surface. This phenomenon, which has been well documented, is ascribed to the burning of fossil fuels, especially coal, aided by aerosols and other contaminants. Definitive studies by NRC (1979) and NASA (1981) project a substantial increase in average temperature during the next century, perhaps as much as 3–4C. The increase of 1C reported to have occurred from 1900 to 1950 indicates that this has been going on for the last 100 years or more, coinciding with the industrial revolution.
Since the CO_2 concentration tends to be greatest at the poles (perhaps due to the absence of vegetation), it is probable that the ice caps there will deteriorate, with a concomitant rise in sea levels throughout the world. Other deleterious effects are also predicted, e.g., droughts in North America and central Asia caused by basic changes in rainfall distribution. As combustion of organic fuels is the underlying cause of this situation, reduction in their use is desirable wherever possible, combined with accelerated development of energy sources that do not produce carbon dioxide.

greenhouse gas. Any gas in the atmosphere that interferes with the radiation of heat from the surface of the earth by absorbing infrared rays and thereby trapping heat near the surface of the earth. A wide variety of chemicals can be classed as greenhouse gases with the major ones being water vapor, carbon dioxide, methane, nitrous oxide, ozone, and chlorofluorocarbons.

Greenland spar. See cryolite, natural.

green liquor. See liquor (b).

greenockite.
CdS.
A native cadmium sulfide containing 77.7% cadmium. Ore of cadmium.

greensalt. A wood preservative containing chromated copper arsenate.
Hazard: Highly toxic.

greensand. See foundry sand.

green soap. A liquid soap made with potassium hydroxide and a vegetable oil (except coconut and palm kernel oil).
See soap.

greenstone.
Available forms: Mineral granules, various screen gradings and colors.
Use: In roofing materials where outdoor weathering resistance is required, as a filler for asbestos shingles, tennis court surfaces.

green-verditer. A paint pigment consisting of the hydroxycarbonate of copper.
See copper carbonate; malachite.

grex. See denier.

Griess diazo reaction. Formation of aromatic diazonium salts from primary aromatic amines and nitrous acid or other nitrosating agents.

Grignard degradation. Stepwise dehalogenation of a polyhalo compound through its Grignard reagent, which on treatment with water yields a product containing one fewer halogen atom.

Grignard reaction. Addition of organomagnesium compounds (Grignard reagents) to carbonyl groups or other unsaturated groups to give alcohols or ketones.

Grignard reagent. An important class of reagents used in synthetic organic chemistry, made by union of metallic magnesium with an organic chloride, bromide, or iodide, usually in the presence of an ether and in the complete absence of water. General formula RMgX, where R is an alkyl, aryl, or other organic group and X a halogen. The value of the reagents lies in their ease of reaction with water, carbon dioxide, alcohols, aldehydes, ketones, amines, etc., to produce a great variety of organic compounds, usually with good yields. Examples are ethyl magnesium chloride (C_2H_5MgCl), methyl magnesium bromide (CH_3MgBr), etc.

Hazard: Because the heat of decomposition of Grignard reagents with water is great and the ether in which they are dissolved is highly volatile and flammable, they must be handled with extreme care. Some, especially the solution of MeMgBr in ethyl ether, may ignite spontaneously on contact with water or even damp floors; nearly all will ignite on a wet rag or similar material. Since all Grignard reagents react rapidly with both water and oxygen, contact must be avoided. Ordinary materials of construction are satisfactory for use with Grignard reagents.

Grignard, Victor. (1871–1935). A French chemist who worked with Sabatier to win the Nobel Prize in 1912. He authored "Theses sur les combinaisons organomagnesiennes mixtes et leurs applications a des syntheses." As a student of Berbier, he went from Lyons to Besancon to Nancy, France. He discovered organomagnesium compounds that are used for organic synthesis by the Grignard reaction.

"Grilbond" *[EMS-CHEMIE].* TM for solutions for bonding coating and sealing components in auto production.

"Grilonit" *[EMS-CHEMIE].* TM for solutions for bonding coating and sealing components in auto production.

grinding media. Balls ranging in diameter from 1/4 to 2 inches (in special cases even larger) made of porcelain, flint, bronze-alumina, or alloy steels of various compositions. These perform an efficient size-reduction action on solid materials that are softer than the balls themselves. Steel rods are also used.
See ball mill; rod mill.

Grinsted. Locust bean gum.
Available forms: Powder.
Use: A gelling, thickening, stabilizing agent for ice cream, cream cheese, sauces, and desserts.

griseofulvin.
CAS: 126-07-8. $C_{17}H_{17}ClO_6$.

Properties: White to creamy white, odorless powder. Insoluble in water; slightly soluble in alcohol, benzene, acetone, and chloroform.
Derivation: By growth of *Penicillium griseofulvum* or synthetically.
Grade: USP.
Hazard: Possible carcinogen.
Use: Medicine (antifungal), feed additive.

grizzly. A coarse size-grading device used for ores, coal, and mineral aggregate. It consists of a grid of parallel steel bars separated by a distance determined by the size separation desired. It is set at an angle of approximately 45 degrees so that lumps that are too large to pass through it can roll down the bars while the smaller sizes fall into a container beneath.

Grob fragmentation reaction. Carbon–carbon bond cleavage involving a system of four atoms.

grog. Crushed refractory materials added to ceramic mixes to reduce lamination in plastic clays and also to reduce shrinkage on drying. Materials crushed for this purpose are pottery, firebrick, quartz, quartzite, burned ware, and saggers.

ground-nut oil. See peanut oil.

ground state. Condition in which the orbital electrons of an atom are in there lowest energy state.

groundwood. A type of wood pulp produced by direct friction of a rotating stone on the end of a log. In the paper trade this is called mechanical pulp.
Use: Newsprint and other low-grade papers.
See pulp, paper.

group. (1) One of the major classes or divisions into which elements are arranged in the periodic table (vertical columns). The classification is made according to the properties of the elements, those whose properties are similar occupying one group. Groups I through VII are divided into subgroups (A and B), but group VIII, which contains the nobel gases, is not so divided.
See periodic table.
Note: Though often loosely so used, the word group should not be applied to a number of elements of similar properties that are not actual groups in the periodic table; the proper term for these is series, e.g., lanthanide series, rare-earth series, etc.
(2) A combination of two or more closely associated elements that tend to remain together in reactions, usually behaving chemically as if they were individual entities, i.e., in respect to valence, ionization, and related properties. Among the more familiar are OH (hydroxyl), COOH (carboxyl), CO_3 (carbonate), NH_4 (ammonium), SO_4 (sulfate), CH_3 and homologs (methyl, etc.), SH (sulfhydryl), and CO (carbonyl). When a group acquires an electric charge, it is called a radical.
(3) Any combination of elements that has a specific functional property, i.e., a chromophore group in dyes.

group transfer potential. A measure of the ability of a compound to donate an activated group (such as a phosphate group). Computed and expressed from the standard free energy of hydrolysis.

growth. (1) In biochemistry, the continuous process of cell division and reproduction characteristic of all living organisms. The basic phenomenon is considered to be osmosis, by which nutrients are transferred through cell walls and tissue structures; it is thus essential to the metabolic functioning of the organism.
(2) In crystallography, the process of crystalline formation and development by nucleation and accretion. Crystals of many kinds are artificially produced for a variety of uses (e.g., lasers) by vapor condensation, electrodeposition, or rapid cooling of a saturated solution.

growth hormone. See somatotropic hormone.

growth regulator. See plant growth regulator.

Grubbs, Robert H. (1942–). An American born in Calvert City, Kentucky, who won the Nobel Prize for chemistry in 2005 for his pioneering work concerning the development of the metathesis method in organic synthesis. Grubbs was awarded a B.S. and an M.S. in chemistry from the University of Florida, and obtained his Ph.D. at Columbia University. He is the Victor and Elizabeth Atkins Professor of Chemistry at the California Institute of Technology and is a member of the National Academy of Sciences.

Grundmann aldehyde synthesis. Transformation of an acid into an aldehyde of the same chain length by conversion of the acid chloride via the diazo ketone to the acetoxy ketone, reduction with aluminum isopropoxide and hydrolysis to the glycol, and cleavage with lead tetraacetate.

G salt. The sodium or potassium salt of 2-naphthol-6,8-disulfonic acid (G acid).
Use: Dye intermediate.

GTP. Abbreviation for guanosine triphosphate. See guanosine phosphates.

gua. (2-amino-3,7-dihydropurin-6-one). $C_5H_5N_5O$.
A codon of valine that directs the placement of valine into the polypeptide.

guaiac. (guaiac gum; guaiac resin).
A resin from certain Mexican and West Indian trees, especially *Guaiacum santum* and *G. officinale.*
Properties: Brownish lumps. Mp 85C. Insoluble in water; soluble in alcohol, ether, acetone, chloroform, and caustic soda.
Use: Flavoring agent in foods; medicine (diagnostic aid for blood or hemoglobin).

guaiacol. (methylcatechol; pyrocatechol methyl ether; pyrocatechol methyl ester; *o*-methoxyphenol; *o*-hydroxyanisole).
CAS: 90-05-1. $OHC_6H_4OCH_3$.

Properties: Faintly yellowish, limpid, oily liquid, or yellow crystals; aromatic odor. D 1.1395, mp 27.9C, bp 205C, flash p 180F (82.2C) (OC). Constitutes 60–90% of beechwood creosote. Soluble in alcohol, ether, chloroform, and glacial acetic acid; moderately soluble in water. Combustible.
Derivation: (1) By extracting beechwood creosote with alcoholic potash, washing with ether, crystallizing the potash compound from alcohol, and decomposing it with dilute sulfuric acid. (2) Also from *o*-anisidine by diazotization and subsequent action of dilute sulfuric acid.
Hazard: Toxic by ingestion and skin absorption.
Use: Synthetic flavors, medicine (expectorant).

guaiacwood oil.
Properties: Yellow to amber semisolid mass; floral odor. D 0.965-0.975, optical rotation −6 to −7 degrees. Soluble in alcohol, ether, and chloroform; insoluble in water.

Derivation: Steam distillation of guaiacwood (Paraguay).
Chief constituent: Guaiol.
Use: Perfume fixative and modifier, soap odorant, fragrances, production of guaiacwood acetate.

guaiol.
CAS: 489-86-1. $C_{15}H_{26}O$.
A bicyclic sesquiterpene alcohol found in guaiac wood oil.
Properties: Crystalline solid. Mp 91C. Insoluble in water; soluble in alcohol.

guanethidine sulfate. ([2-(hexahydro-1(2H)-azocinyl)ethyl]guanidine sulfate).
CAS: 60-02-6. $(C_{10}H_{22}N_4)_2 \cdot H_2SO_4$.
Properties: White crystalline powder; strong odor. Very soluble in water; slightly soluble in alcohol; practically insoluble in chloroform.
Grade: USP.
Use: Medicine (antihypertensive agent).

guanidine. (carbamidine; iminourea).
CAS: 113-00-8. $HN=C(NH_2)_2$.
Properties: Colorless crystals. Mp 50C, decomposes at 160C. Soluble in water and alcohol. Combustible.
Derivation: (1) By heating calcium cyanamide with ammonium iodide. (2) By treating urea with ammonia under pressure.
Use: Organic synthesis.

guanidine-aminovaleric acid. See arginine.

guanidine carbonate.
CAS: 593-85-1. $(H_2NCNHNH_2)_2 \cdot H_2CO_3$.
Properties: White granules. Decomposes without melting at 197–199C, d 1.25. Soluble in water; slightly soluble in alcohol and acetone. Combustible.
Derivation: From dicyandiamide.
Grade: Technical, over 95% pure.
Hazard: Toxic by ingestion.
Use: As a strong organic alkali, organic intermediate, soap and cosmetic products.

guanidine hydrochloride.
CAS: 50-01-1. $HNC(NH_2)_2 \cdot HCl$.
Properties: White powder. Mp approximately 183C. Soluble in water and alcohol, pH (aqueous solution) 6.2 for 10% solution.
Grade: 88% and 95% pure.
Hazard: Toxic by ingestion, evolves hydrogen chloride fumes on heating.

guanidine nitrate.
CAS: 506-93-4. $H_2NC(NH)NH_2 \cdot HNO_3$.
Properties: White granules. Melting range 206–212C. Soluble in water and alcohol; slightly soluble in acetone.
Derivation: From cyanamide or dicyandiamide.
Grade: Technical, greater than 95% pure.

Hazard: Strong oxidant, may ignite organic materials on contact, may explode by shock or heat.
Use: Manufacture of explosives, disinfectants, photographic chemicals.

guanidine thiocyanate.
CAS: 593-84-0. $H_2NC(=NH)NH_2 \cdot HSCN$.
Properties: Solid. Mw 118.16, mp 114–119C. Contains 2% hydrochloric acid.
Hazard: Irritant.
Use: Potent protein denaturant used in isolation of intact DNA, RNA.

guanine. (G; 2-amino-6-oxypurine).
CAS: 73-40-5. $C_5H_5N_5O$.
A purine constituent of ribonucleic acid and deoxyribonucleic acid. Usual sources are guano, sugar beets, yeast, clover seed, and fish scales.
Properties: Colorless, rhombic crystals. Mp 360C (decomposes). Insoluble in water; sparingly soluble in alcohol and ether; freely soluble in ammonium hydroxide, alkali hydroxides, and dilute acids. Available as hydrochloride or hemisulfate.
Derivation: Isolation following hydrolysis of nucleic acids (usually from yeast), organic synthesis.
Use: Biochemical research, cosmetics.
See: Base pair; pearl essence; nucleotide.

guano. Excrement of sea birds found chiefly on islands off the coasts of Peru, Chile. Contains 10% nitrogen, 6% phosphorus, 2% potassium. Formerly the main source of nitrates.

guanosine. (guanine riboside).
CAS: 118-00-3. $C_{10}H_{13}N_5O_5$.
The nucleoside containing guanine and D-ribose.
Properties: White, crystalline powder; odorless; mild saline taste. Mp 237–240C (decomposes). Very slightly soluble in cold water; soluble in boiling water, dilute mineral acids, hot acetic acid, and dilute bases; insoluble in alcohol, ether, chloroform, and benzene.
Derivation: Found in pancreas, clover, coffee plant, and pollen of pines; prepared from yeast nucleic acid.
Use: Biochemical research.

guanosine monophosphate. See guanylic acid.

guanosine phosphates. Nucleotides used by the body in growth processes; important in biochemical and physiological research. Those isolated are the monophosphate (GMP), the diphosphate (GDP), and the triphosphate (GTP).

guanosine phosphoric acid. See guanylic acid.

guanoxan sulfate. ((1,4-benzodioxan-2-ylmethyl)g-anidine sulfate; [N'-(2,3-dihydro-1,4-benzodioxin-2-ylmethyl)carbamimidoyl]azanium sulfate).
$C_{20}H_{28}N_6O_8S$.
An antihypertensive agent.

guanylic acid. (GMP; guanosine monophosphate; guanosine phosphoric acid).
CAS: (3') 117-68-0; (5') 85-32-5. $C_{10}H_{14}N_5O_8P$.
The monophosphoric ester of guanine, i.e., the nucleotide containing guanine, D-ribose, and phosphoric acid. The phosphate may be esterified to either the 2, 3, or 5 carbon of ribose, yielding guanosine-2'-phosphate, guanosine-3'-phosphate, and guanosine-5'-phosphate, respectively. It is important in growth processes of the body.
Derivation: (Commercial product): Isolation from nucleic acid of yeast or pancreas; also made synthetically.
Use: Biochemical research, flavor potentiator (disodium salt).

guanyl nitrosaminoguanylidene hydrazine.
A high explosive.

guanyl nitrosaminoguanyl tetrazene. See tetrazene.

guanylurea sulfate. (carbamylguanidine sulfate).
CAS: 5338-16-9. $(C_2H_6ON_4)_2 \cdot H_2SO_4 \cdot 2H_2O$.
Properties: White powder, greater than 97% pure. Soluble in water and alcohol.
Derivation: From cyanamide or dicyandiamide.
Use: Analytical reagent for nickel, manufacture of dyes, organic synthesis.

"Guardkote" [Shell]. TM for fast-setting two-component liquid systems based on "Epon" resins and designed specifically for highway resurfacing and repair. They are used in combination with sharp aggregates to waterproof and deslick Portland cement or bituminous concrete pavements.

Guareschi–Thorpe condensation. Synthesis of pyridine derivatives by condensation of cyanoacetic ester with acetoacetic ester in the presence of ammonia. In a second type of synthesis, a mixture of cyanoacetic ester and a ketone is treated with alcoholic ammonia.

guar gum. (guar flour).
CAS: 9000-30-0.
A water-soluble plant mucilage obtained from the ground endosperms of *Cyanopsis tetragonoloba*, cultivated in India and Pakistan as livestock feed, as well as in southwestern U.S. The water-soluble portion of the flour (85%) is called guaran and consists of 35% galactose, 63% mannose, probably combined in a polysaccharide, and 5–7% protein.

Properties: Yellowish-white free-flowing powder. Completely soluble in hot or cold water. Practically insoluble in oils, greases, hydrocarbons, ketones, esters. Water solutions are tasteless, odorless, non-toxic. Has 5-8 times the thickening power of starch. Reduces the friction drag of water on metals.
Grade: Industrial, technical, FCC.
Use: Paper coating, cosmetics, pharmaceuticals, binder in tablet mixtures, interior coating of fire hose nozzles, fracturing aid in oil wells, textiles, printing, polishing, thickener and emulsifier in food products (e.g., cheese spreads, ice cream, and frozen desserts).

guayule. A rubberlike hydrocarbon, almost identical with *cis*-polyisoprene, obtained from a shrub that is a member of the *Compositae* family. It is grown extensively in Mexico and the southwestern U.S. It contains approximately 15% rubber hydrocarbon in the form of latex in cells in the roots and stem. Conventional practice involves crushing and parboiling the entire plant; this coagulates the latex, which is removed by milling. An improved process utilizes steam from 200 to 240C for 6 minutes instead of milling. Resins, which comprise 10–15% of the plant, are then removed with acetone followed by treatment with hexane to extract the hydrocarbon. This method can increase the yield of hydrocarbon by up to 50%. The yield is 1 lb rubber/6 lb comminuted shrub.

guc. A codon of valine that directs the placement of valine into the polypeptide.

Guerbet reaction. Condensation of alcohols at high temperature and pressure in the presence of sodium alkoxide or copper by a dehydrogenation, aldol condensation, and hydrogenation sequence.

gug. A codon of valine that directs the placement of valine into the polypeptide.

Guignet's green. A chrome-green pigment made by fusing potassium chromate and boric acid.

Guldberg and Waage law. See Mass action law.

gum arabic. See arabic, gum.

gum benzoin. See benzoin resin.

gum camphor. See camphor.

gum, gasoline. A viscous oxidation product occurring after long standing in gasoline that is not stabilized with an antioxidant.

gum, natural. A carbohydrate high polymer that is insoluble in alcohol and other organic solvents, but generally soluble or dispersible in water. Natural gums are hydrophilic polysaccharides composed of monosaccharide units joined by glycosidic bonds.

They occur as exudations from various trees and shrubs in tropical areas or as phycocolloids (algae) and differ from natural resins in both chemical composition and solubility properties. Some contain acidic components, and others are neutral.

Use: Protective colloids and emulsifying agents in food products and pharmaceuticals, as sizing for textiles, and in electrolytic deposition of metals.

See arabic; tragacanth; guar; karaya.

Note: The terminology of natural gums and resins is inconsistent and often confusing. The word *gum*, often used as an adjective, seems to acquire a different meaning from the noun. For example, the resinous products obtained from pine pitch (produced by the parenchyma cells of softwoods) are conventionally called *gum turpentine*, and *gum rosin*. There are also such *gum resins* as *gum benzoin*, gum camphor, and others. The so-called ester gum is a semisynthetic reaction product of rosin and a polyhydric alcohol. All these are actually resinous products having properties quite different from those of natural gums. Furthermore, resins are complex *mixtures*, whereas gums are *compounds* that can be represented by a formula. Still further complicating the matter is the common application of the word *gum*, to such plant lattices as chicle and natural rubber, which are different from both carbohydrate gums and resins. It is probable that the confusion originated in the casual use of *gum* to refer to any soft, sticky product derived from trees. In view of this situation, any specific definition of these terms is likely to be controversial.

See resin; natural; polymer, water soluble.

gum rosin.   See rosin; gum, natural (note).

gum sugar.   See arabinose.

gum, synthetic.   See ester gum.

gum tragacanth.   See tragacanth gum.

gum turpentine.   See turpentine; gum, natural (note).

gun blueing.
Properties: A solution consisting of acetic acid, selenious acid, and cupric salt.
Use: To develop friction ridge detail on metal surfaces.

guncotton.   See nitrocellulose.

gun metal.   An alloy of copper with 10% tin.

gunpowder.   See black powder.

"Guthion" *[Makhteshim].*   TM for *O,O*-dimethyl-*S*-4-oxo-1,2,3-benzotriazin-3(4*H*)-ylmethyl phosphorodithioate.
See azinphos methyl.
Hazard: A cholinesterase inhibitor.

Gutknecht pyrazine synthesis.   Cyclization of α-amino ketones produced by reduction of isonitroso ketones; the dihydropyrazines formed are dehydrogenated with Hg_2O or $CuSO_4$, or sometimes with atmospheric oxygen.

gutta percha.   (*trans*-polyisoprene).
A geometric isomer of natural rubber obtained from trees native to Malaya. It is stiff, hard, and inelastic when cold; softens at 60C to moldable condition and melts at 100C, insoluble in water, essentially soluble in carbon disulfide, petroleum ether, and chloroform. It can be vulcanized with sulfur.
Use: Formerly used widely for golf-ball covers, its present uses are in dentistry, surgical accessories, and as an insulating medium in electrical devices.

Gutzeit test.   A test for arsenic. Zinc and dilute sulfuric acid are added to the substance, which is then covered with a filter paper moistened with mercuric chloride solution. A yellow spot forms on the paper if arsenic is in the sample.
See Bettendorf's reagent.

guu.   A codon of valine that directs the placement of valine into the polypeptide.

gynandromorph.   Organisms that have both male and female cells and, therefore, express both male and female characteristics.

glyocardia oil.   See chaulmoogra oil.

Glyp.   See Glyptal.

Glyptal.   (Glyp).
Synthetic resin made by heating phthalic anhydride and glycerin together.
Use: Shellac, lacquers, paints, or varnishes.

gyplure.   (generic name for *cis*-9-octadecen-1,12-diol-12-acetate).
It is a synthetic product used as a sex attractant for the male gypsy moth. The natural product, found in female moths, is said to be *d*-10-acetoxy-1-hydroxy-*cis*-7-hexadecene.

gypsum **(calcium(II)** **sulfate** **dihydrate (1:1:2)).**
CAS: 10101-41-4. $CaSO_4 \cdot 2H_2O$.
A mineral that consists of calcium sulfate with water molecules attached, or the rock that consists primarily of this mineral.
See calcium sulfate.

gypsum cement.   (plaster of Paris; Keene's cement; Parian cement; Martin's cement; Mack's cement).
A group of cements that consist essentially of calcium sulfate and are produced by the partial dehydration of gypsum to the hemihydrate $CaSO_4 \cdot 1/2H_2O$. They usually contain additions of

various sorts. For example, Keene's cement contains alum or aluminum sulfate; Mack's cement contains sodium or potassium sulfate; Martin's cement contains potassium carbonate; and Parian cement contains borax.

gyratory crusher.   A machine for large-scale size reduction of rocks, ore, or mineral aggregate. Essentially, it comprises a vertical cone-shaped crushing head or mantle that moves in a shell shaped like an inverted cone that is open at the top where the feed enters. The space between the mantle and the shell decreases gradually, being smallest at the bottom where the crushed material is discharged. The lower third of the mantle and shell have concave surfaces to minimize packing of the crushed product. The crushing head functions by executing a gyratory or circular motion, i.e., a rolling action in which the upper portion describes a circle while the lower portion is almost stationary. From 300 to 500 gyrations per minute are possible. The mechanism is activated by a shaft attached to the crushing head, the bottom of which rests in a gear-driven eccentric. Heavy-duty gyratories have feed openings up to 72 inches; fine-reduction sizes are as small as 12 inches.

H

H.   Symbol for hydrogen, the molecular formula is H_2.

h.   See Planck's constant.

Haber, Fritz.   (1868–1934). Born in Breslau, Germany, Haber's great contribution to chemistry, for which he was awarded the Nobel Prize in 1918, was his development (with Bosch) of a workable method for synthesizing ammonia by the water–gas reaction from hot coke, air, and steam; the gas mixture obtained includes nitrogen from the air, as well as hydrogen from the steam. It was the first successful attempt to "fix" atmospheric nitrogen in an industrial process. This discovery was developed to production scale in approximately 1912; it enabled Germany to manufacture an independent supply of explosives for World War I.

habit.   The type of geometric structure that a given crystalline material invariably forms, e.g., cubic, orthorhombic, monoclinic, tetragonal, hexagonal, etc. Each of these types has several subclasses. Thus, crystals may have the form of thin sheets or plates, cubes, rhomboids, and even more complicated geometric structures. For example, the crystalline habit of mica is monoclinic, with formation of extremely thin sheets.
See crystal.

H acid.   (1-amino-8-napthol-3,6-disulfonic acid).
Properties: Gray powder. Soluble in water, alcohol, and ether.

Derivation: Fusion of 1-naphthylamine-3,6,8-trisulfonic acid (Koch acid) with 30% sodium hydroxide solution at 180C.
Use: Azo dye intermediate.

haemo-.   Prefix indicating relationship to blood.

haemosiderin.   A protein that serves as the source of iron in the synthesis of hemoglobin and other synthetic pathways that involve the incorporation of iron.

hafnia.   See hafnium oxide.

hafnium.
CAS: 7440-58-6.   Hf.
Metallic element of atomic number 72, Group IVB of the periodic table, aw 178.49, valences of 2, 3, 4; 6 stable isotopes.
Properties: Generally similar to zirconium. Gray crystals. D 13.1, mp approximately 2150C, bp above 5400C, high thermal neutron cross section (115 barns). Good corrosive resistance and high strength.
See zirconium.
Occurrence: Zirconium ores.
Derivation: Extremely difficult to separate from zirconium. Most important methods are (1) solvent extraction of the thiocyanates by hexone, (2) solvent extraction of the nitrates by tributyl phosphate, and (3) fractional crystallization of the double fluorides.
Available forms: Powder, rods, single crystals.
Hazard: Toxic by inhalation; compounds are also toxic. Powder form is explosive in air, either dry or wet, with <25% water.
Use: Control rods in water-cooled nuclear reactors; lightbulb filaments, electrodes, special glasses, getter in vacuum tubes.

hafnium boride.
CAS: HfB_2.
Crystalline solid, mp 3100C.
Use: Refractory material.

hafnium carbide.
CAS: 12069-85-1.   HfC.
Properties: High thermal neutron absorption cross section. Very high mp 3890C (7030F). Most refractory binary substance known.
Use: Control rods in nuclear reactors.

hafnium disulfide.
CAS: 18855-94-2. HfS_2.
Available in a particle size of 40 microns as a solid lubricant. The diselenide and ditelluride are also available.

hafnium nitride.
CAS: 25817-87-2.   HfN.
Yellow-brown crystals, mp 3305C, the most refractory of known metal nitrides.

hafnium oxide.   (hafnia).
CAS: 71243-80-6. HfO_2.
Properties: White solid. D 9.68 (20C), mp 2812C, bp approximately 5400C. Insoluble in water.
Use: Refractory metal oxide.

Hahn, Otto. (1879–1968). A German physical chemist who won the Nobel Prize for chemistry in 1944 for atom splitting and the principle of the chain reaction. Well-known for work on nuclear fission, he discovered protactinium and transuranium elements with atomic numbers 94, 95, and 96. After receiving his doctorate at the University of Munich, he worked in Canada before returning to Europe.

hair. See keratin.

halazone. (*p-N,* *N*-dichlorosulfamylbenzoic acid; *p*-sulfondichloraminobenzoic acid). CAS: 80-13-7. $HOOCC_6H_4SO_2NCl_2$.
Properties: White crystalline powder; strong chlorine odor. Mp 195C with decomposition. Affected by light. Soluble in glacial acetic acid, benzene; slightly soluble in water, chloroform; insoluble in petroleum ether.
Grade: NF.
Use: Water disinfectant.

half-life. The time required for an unstable element or nuclide to lose one-half of its radioactive intensity in the form of α, β, and γ radiation. It is a constant for each radioactive element or nuclide. Half-lives vary from fractions of a second for some artificially produced radioactive elements to millions of years. The half-life of ^{235}U is 7.1 × 10^8 years; that of ^{239}Pu is 24,360 years.

halibut liver oil. (haliver oil).
Properties: Pale-yellow to dark-red liquid; fishy odor and taste. D 0.920–0.930, saponification number 160–180, iodine number 120–136, refr index about 1.47. Soluble in alcohol, ether, chloroform, and carbon disulfide; insoluble in water.
Derivation: By expressing and boiling halibut livers.
Grade: Crude, refined.
Use: Source of vitamins A and D, leather dressing.

halide. Any salt, ester, or negative ion of the group VIIA elements in the periodic table.

halides. Binary compounds of the halogens.

halite. The mineral that consists of sodium chloride (NaCl), popularly known as rock salt, or the rock that consists primarily of this mineral.

Hall, Charles Martin. (1863–1914). A native of Ohio, Hall invented a method of reducing aluminum oxide in molten cryolite by electrochemical means. This discovery made possible the large-scale production of metallic aluminum and resulted in formation of the Aluminum Company of America. The process requires high electric power input. Hall is generally considered the founder of the aluminum industry.
See Hall process.

"Hallco" [Hallstar]. TM for specialty esters based on acetic, lauric, and sebacic acids.

Hallcomid. A series of *N,N'*-dimethyl amides and esters of fatty acids.

"Hallcote" [Hallstar]. Clay containing antiblocking coatings (slab dips) used in rubber processing.

Haller-Bauer reaction. Cleavage of nonenolizable ketones with sodium amide, most often applied to ketones $ArCOCR_3$ to yield trisubstituted acetic acids.

halloysite.
$Al_2O_3·3SiO_2·2H_2O$.
A clay used in refractories and as a catalyst support.

Hall process. The electrolytic recovery of aluminum from bauxite or, more specifically, from the alumina extracted from it (see Bayer process). A typical cell for this process consists of a rectangular steel shell lined with insulating brick and block carbon. The cell holds a molten cryolite-alumina electrolyte, commonly called the "bath." The carbon bottom is covered by a pad of molten aluminum and serves as the cathode. The anodes are prebacked carbon blocks suspended in the electrolyte. The cathodic current is collected from the carbon bottom by embedded steel bars that protrude through the shell to connect with the cathode bus. During electrolysis, aluminum is deposited on the metal pad, and the oxygen, liberated at the anode, reacts with the carbon to form carbon dioxide, some of which is reduced to carbon monoxide by secondary reactions. At 24- to 48-hour intervals, aluminum is tapped from the cell by a siphon. The process requires large amounts of electric power (from 4 to 5% of total U.S. production). Disposition of the toxic fluoride waste is a problem.
See Toth process.

hallucinogen. Any of a number of drugs acting on the central nervous system in such a way as to cause mental disturbance, imaginary experiences, coma, and even death. Many of these are narcotics and/or alkaloids; some are derived from plants and others are made synthetically. They differ in degree of addiction and hallucinatory effect. Their sale and possession (other than by physicians) is illegal in the U.S. The most common hallucinogens are cannabis (marijuana, hashish), lysergic acid (LSD), amphetamine, and numerous morphine derivatives.

halocarbon. A compound containing carbon, one or more halogens, and sometimes hydrogen. The lower members of the various homologous series are used as refrigerants, propellant gases, fire-extinguishing agents, and blowing agents for urethane foams. When polymerized, they yield plastics characterized by extreme chemical resistance, high electrical resistivity, and good heat resistance. See fluorocarbon.

halocarbon 112a. See 1,1,2-tetrachloro-2,2-difluoroethane.

halofuginone hydrobromide.
mf: $C_{16}H_{18}Br_2ClN_3O_3$.
Properties: Crystals.
Use: Drug (veterinary); food additive.

halogen. One of the electronegative elements of group VII A of the periodic table (fluorine, chlorine, bromine, iodine, astatine, listed in order of their activity, fluorine being the most active of all chemical elements).

halogenation. Incorporation of one of the halogen elements, usually chlorine or bromine, into a chemical compound. Thus, benzene (C_6H_6) is treated with chlorine to form chlorobenzene (C_6H_5Cl), and ethylene (C_2H_4) is treated with bromine to form ethylene dibromide ($C_2H_4Br_2$). Compounds of chlorine and bromine are sometimes used as the source of the halogen, e.g., phosphorus pentachloride.

halogen azide. Any azide that contains a halogen.
Properties: Colorless compounds with hexagonal crystals, extremely reactive, can explode spontaneously.
Hazard: Irritant; poisonous; toxic.

halohydrin. Any alcohol that contains a halogen atom substituted at a saturated carbon atom that otherwise bears only hydrogen or hydrocarbyl groups.

haloid salt. A salt comprising a base and a halogen.

halomethane. Any compound comprising one or more halogen atoms and one or more methane groups.
Hazard: Carcinogen.

halon. (1-chloro-1,2,2-trifluoroethane).
C_2ClF_3.
A halogenated compound such as carbon tetrachloride in which one or more halogen atoms are attached to carbon atoms.

Halon. Tetrafluoroethylene polymer.
Properties: Inertness to almost all chemicals, resistance to high and low temperatures, zero moisture absorption, high impact strength, excellent dielectric properties, nonstick surface with low coefficient of friction, self-extinguishing (ASTM D-635).
Grade: Varies with particle size as 600-, 300-, 20-micron average size.
See polytetrafluoroethylene; fluorocarbon polymer.

Halon 1211. (bromochlorodifluoromethane).
CAS: 353-59-3.
A compressed gas.
Use: Fire-extinguisher compound.

Halon 1301. See trifluorobromomethane.

haloperidol. (4-[4-(4-chlorophenyl)-4-hydroxy-1-piperidinyl]-1-(4-fluorophenyl)-1-butanone; Haldol; pernox).
CAS: 52-86-8. $C_{21}H_{23}ClFNO_2$.
A phenyl-piperidinyl-butyrophenone that is a potent antiemetic.
Hazard: Moderately toxic, cumulative central nervous system depressant; acute poisoning can cause immediate death.
Use: A highly potent antipsychotic and a major tranquilizer.

halophyte. A plant of seashores and estuaries or mashes that thrive on soils of high salt content.

halosafen. See 5-(2-chloro-6-fluoro-4-(trifluoromethyl)phenoxy)-n.

halothane. (2-bromo-2-chloro-1,1,1-trifluoroethane).
CAS: 151-67-7. $CF_3CHBrCl$.
Properties: Colorless, volatile liquid; sweetish odor. D 1.872–1.877 (20/4C), bp 50.2C, 20C (243 mm Hg). Light sensitive. May be stabilized with 0.01% thymol. Slightly soluble in water; miscible with many organic solvents.
Grade: USP.
Hazard: Questionable carcinogen.
Use: Medicine (anesthetic).

Halowax 1051. See octachloronaphthalene.

Hamilton, Alice. (1869–1970). The first American physician to devote her life to the practice of industrial medicine. In studying the lead industries in Illinois, she discovered and ameliorated lead poisoning among bathtub enamelers in Chicago. She wrote about phossy jaw, which occurred among American matchmakers who used white or yellow phosphorus. She studied the effects of carbon monoxide among steelworkers, the toxicity of nitroglycerin among munitions makers during World War I, the symptoms of hatters exposed to mercury in Danbury, Connecticut, and the "dead fingers" syndrome of workers utilizing the early jackhammers. She also described the toxic effects to the blood-forming cells from benzol, and the neurologic and psychological responses of workers in the viscose rayon industry. In 1919, Dr. Hamilton was appointed assistant professor of industrial medicine at Harvard Medical School. The first woman on the Harvard faculty, she gave occupational medicine respectability as an academic pursuit.

Hamiltonian function. An expression of the sum of potential and kinetic energy in a system in terms of Cartesian coordinates and momenta.

hammer mill. A crushing or shredding device consisting of four or more rectangular metal hammers or sledges mounted on a rotating shaft, the hammers being free to swing on a pin. As the shaft rotates, the hammers impact the material introduced from above, crushing it against a stationary breaker plate. The fragments are carried downward and sorted by a grid located beneath the shaft; those too large to pass through it are returned for another cycle. A complete mill is made up of several such units working on a single shaft, with suitable housing. Products such as coal, limestone, and other mineral aggregates, as well as wood, sugarcane, and similar materials, can be effectively disintegrated with equipment of this type.

Hammick reaction. Decarboxylation of α-picolinic or related acids in the presence of carbonyl compounds accompanied by the formation of a new carbon–carbon bond.

Hammond-Leffler postulate. States that the structure of the transition state more closely resembles the product or the starting material, depending on which is higher in enthalpy.

Hampamide. Mixtures of mono-, di-, and triamides of nitrailotriacetic acid in solution of dry salt form.

Hamp-ene. Chemical compositions comprising amino acid or carbohydrate chelating or sequestering agents or compositions containing such sequestering agents.
Available forms: Bulk and package.

Hamp-Ex. Chemical compositions comprising amino acid or carbohydrate chelating or sequestering agents or compositions containing such sequestering agents.
Available forms: Bulk and package form.

Hampirex. Chelated micronutrients.

Hamp-Iron. Iron chelates for agricultural purposes.

Hamp-Ol. Chemical compositions comprising amino acid or carbohydrate chelating or sequestering agents or compositions containing such sequestering agents.
Available forms: Bulk and package form.

"Hamposyl" *[Elcat].* TM for chemically modified fatty acids.

Hampshire. Chemical products, namely, amino acids, chelating agents, metal chelates, surfactants, sequestering agents, and resins.

Hamp-Tex. Chelating agent for textile applications.

hand. (handle).
A term used chiefly in the textile industry to describe a fabric in a qualitative manner, as determined by the sensory perception of its feeling. It is also used to some extent in the leather industry. Such pragmatic physical perception tests are of great practical value, although the properties being ascertained are often not objectively definable.
See texture.

handedness. See asymmetry; enantiomer; chiral.

Hansa. A group of yellow to orange, insoluble azo pigments based on toluidine and β-naphthol. Have good light fastness in deep shades but tend to fade in pastels. Notably poor resistance to bleeding, good weather resistance, and relatively unaffected by acids and alkalies. Used chiefly in emulsion paints, toy enamels, and other applications.

Hantach pyrrole synthesis. Formation of pyrrole derivatives from α-chloromethyl ketone, β-keto esters, and ammonia or amines.

Hantzch pyridine synthesis. Synthesis of alkylpyridines by condensation of two moles of a β-dicarbonyl compound with one mole of an aldehyde in the presence of ammonia. The resulting dihydropyridine is dehydrogenated with an oxidizing agent.

Hanus reagent. Reagent consisting of 20% iodine monobromide in glacial acetic acid.
Use: Determination of iodine values.

haploid. A single set of chromosomes (half the full set of genetic material), present in the egg and sperm cells of animals and in the egg and pollen cells of plants. Human beings have 23 chromosomes in their reproductive cells.
See: Diploid.

haploid-diploid life cycle. Occurs when a multicellular diploid phase, or sporophyte, alternates with a multicellular haploid phase, or gametophyte. Only plants and certain algae possess this kind of life cycle, which is also called "alternation of generations."

haploid life cycle. Occurs when the only multicellular stage in an organism's life cycle is haploid.

haplotype.   A way of denoting the collective genotype of a number of closely linked loci on a chromosome.

hapten.   (incomplete antigen).
A molecule bound to an antigenic detergent used to elicit an immune response.
Properties: Low-molecular-weight organic molecule.

haptoglobin.   A serum protein that binds free hemoglobin for transport to the liver for destruction.

Harborlite.   Perlite minerals and mineral substances.
Use: Filtration for industry, agriculture, horticulture, and forestry applications.

hard.   A nontechnical word, used by chemists, that has a variety of meanings. It describes the following: (1) Water in which calcium carbonate or calcium sulfate is present.
See water, hard.
(2) An acid having high positive oxidation and whose valence electrons are not readily excited.
See acid.
(3) Extremely short-wave radiation, e.g., hard X-rays.
(4) The resistance of a pesticide or detergent to biodegradation.
(5) Rubber cured with 30% or more of sulfur.
See rubber, hard.
(6) Wood from deciduous trees.
See hardwood; hardness.

Harden, Sir Arthur.   (1865–1940). An English chemist who won the Nobel Prize in chemistry in 1929 along with Hans von-Euler-Chelpin. He discovered fermentation enzymes and demonstrated the structure of zymase. His fermentation work proved how inorganic phosphates speeded the process. Born in England, he received his doctorate in Germany.

hardness.   The resistance of a material to deformation of an indenter of specific size and shape under a known load. This definition applies to all types of hardness scales except the Mohs scale, which is based on the concept of scratch hardness and is used chiefly for minerals. The most generally used hardness scales are Brinell (for cast iron), Rockwell (for sheet metal and heat-treated steel), diamond pyramid, Knoop, and scleroscope (for metals). Durometer hardness is used for materials such as rubber and plastics.

hardwood.   In papermaking terminology, the wood from deciduous trees (maple, oak, birch), regardless of whether it is actually hard or soft.
Use: Components of paper pulp, but in much lower volume than softwoods.

Hardy-Weinberg Law.   The concept that both gene frequencies and genotype frequencies will remain constant from generation to generation in an infinitely large, interbreeding population in which mating is at random and there is no selection, migration, or mutation.

hard water/acid tolerant synthetic lube.   A conveyor lubricant.

Hargreaves process.   The manufacture of sodium sulfate (salt cake) from sodium chloride and sulfur dioxide. A mixture of sulfur dioxide and air is passed over briquettes of sodium chloride in a countercurrent manner to produce sodium sulfate and hydrogen chloride. This process accounts for only a small amount of the salt cake produced in the U.S.

Harries ozonide reaction.   Treatment of olefins with ozone as a method of cleaving olefinic linkages. On hydrolysis or catalytic hydrogenation, the initially formed ozonide yields two molecules of carbonyl compounds.

"Hartex" *[Firestone].*   TM for natural rubber latex.
Use: Dipped goods and adhesives.

hartshorn.   See ammonium carbonate.

hartshorn oil.   See bone oil.

HAS.   Abbreviation for hydroxylamine acid sulfate.

hashish.   Extract of cannabis (marijuana), more concentrated and thus more powerful than the base drug.
See cannabis.

Hass chlorination rules.   A series of rules pertaining to the chlorination of saturated hydrocarbons.

Hassel, Odd.   (1897–1981). A Norwegian chemist who won the Nobel Prize for chemistry in 1969 with Barton. A great deal of his work was concerned with using X-ray and electron differentiation methods of crystal and molecular structures. He also researched stereochemistry and conformational analysis. His education and teaching career were in his homeland.

hassium.   Hs.
A transfermium element. Atomic number 108. Very short half-life.

Hass vapor-phase nitrogen.   Production of nitroparaffins by vapor phase reaction of aliphatic hydrocarbons with gaseous nitric acid at 420C; a series of 13 rules governing these reactions.

Hastelloy. A family of high-strength, nickel-based alloys with a high resistance to uniform corrosion, outstanding resistance to localized attack, and excellent stress corrosion cracking resistance while maintaining ease of welding and fabrication.
Available forms: Sheet, plate, bars, rods, welding electrodes, and wire.
Use: Fabrication into all types of process equipment.

Hauptman, Herbert A. (1917–2011).
An American biophysicist who won the Nobel Prize for chemistry in 1985 along with Karle. Work involved developing equations that allow determination of phase information from X-ray crystallography intensity patterns. The use of computers permitted use of the equations to determine the conformation of thousands of chemicals. Hauptman was director of research and vice president of the Medical Foundation of Buffalo and a professor of biophysics in Buffalo at The State University of New York.

"Haveg" *[Ametek].* TM for a series of molding compounds fabricated into corrosive-resistant chemical process equipment. Composites of various synthetic polymers with silicate fillers.
Use: Tanks, towers, scrubbers, agitators, piping, etc. in the handling of sulfuric, hydrochloric, and phosphoric acids and chlorinated solvents. Also used as ablative products to protect metal surfaces from heat of launching missiles.

Havelast. An elastomeric binder or impregnant for various reinforcing materials such as "Siltemp," as fabric or rovings, asbestos, glass, or graphite.
Use: Rocket and missile industry when resiliency is desired.

Haworth methylation. Formation of methylated methyl glycosides from monosaccharides with dimethyl sulfate and 30% sodium hydroxide. The glycosidic methyl group is hydrolyzed with acid to yield the free methylated sugar.

Haworth perspective formulas. A method for representing cyclic chemical structures (sugars) in order to define the configuration around each chiral center.

Haworth phenanthrene synthesis. Acylation of aromatic compounds with aliphatic dibasic acid anhydrides to β-aroylpropionic acids, reduction of the carbonyl group according to Clemmensen or Wolff-Kishner procedures, cyclization of the γ-arylbutyric acid with 85% sulfuric acid, and conversion of the cyclic ketone to polycyclic hydroaromatic and subsequently to aromatic compounds.

Haworth, Sir Walter N. (1883–1950). An English chemist who received the Nobel Prize in chemistry in 1937 along with Paul Karrer. He recommended the name *ascorbic acid* and synthesized vitamin C. He accomplished much work on carbohydrate structure and developed a substitute for blood plasma using carbohydrates. During World War II, he developed gaseous diffusion separation on uranium isotopes. He received his Ph.D. in Manchester.

Hayashi rearrangement. Rearrangement of α-benzoylbenzoic acids in the presence of sulfuric acid or phosphorus pentoxide.

Haynes. A series of high-temperature nickel and cobalt-based alloys designed to resist high-temperature corrosion processes up to 2200F. Engineered to provide high-temperature strength, resistance to creep, oxidation, and sulfidation resistance while maintaining ease of welding and fabrication.
Available forms: Sheet, plate, bars, rods, welding electrodes, and wire.
Use: Fabrication into all types of process equipment.

Haystellite. Tungsten carbide products, principally in the form of hard-surfacing rods for protecting parts from severe abrasion.

hazardous air pollutant. Any air pollutant that is not bound by ambient air quality standards but which, under the Clean Air Act, may reasonably be expected to cause or contribute to irreversible illness or death of humans.

hazardous material. Any material or substance that in normal use can be damaging to the health and well-being of humans. Such materials cover a broad range of types that may be classified as follows:
(1) Toxic agents (see poison; toxicity) including drugs, chemicals, and natural or synthetic products that in normal use are in any way harmful, ranging from poisons to skin irritants and allergens. When improperly used, all materials can be hazardous to humans.
(2) Corrosive chemicals such as sodium hydroxide or sulfuric acid that destroy or otherwise damage the skin and mucous membranes on external contact or inhalation.
(3) Flammable materials including (a) organic solvents, (b) finely divided metals or powders, (c) some classes of fibers, textiles, or plastics, and (d) chemicals that either evolve or absorb oxygen during storage, thus constituting a fire risk in contact with organic materials.
(4) Explosives and strong oxidizers such as peroxides and nitrates.
(5) Materials in which dangerous heat buildup occurs on storage, either by oxidation or by microbiological action. Examples are fish meal, wet cellulosics, and other organic waste materials.
(6) Radioactive chemicals that emit ionizing radiation. Packaging, labeling, and shipping of hazardous materials by rail, highway, water, and air are regulated in the U.S. by the Department of Transportation (DOT). In general, its regulations

for air shipment follow those of the International Air Transport Association (IATA), though there are numerous points of difference.

An authoritative guide to the safety labeling of hazardous materials and products is issued by the American National Standards Institute.

See toxicity; flammable material; label.

hazardous waste. See chemical waste; radioactive waste.

Hb. Abbreviation for hemoglobin.

HCCH. Abbreviation for hexachlorocyclohexane.

hc color. Any color approved only for coloring hair. Colors thus classified include aniline, azo, and peroxide dyes.
Hazard: Skin rash, eczema, bronchial asthma, gastritis, occasional complications, and death.

HCE. See epoxyheptachlor.

HCG. Abbreviation for human chorionic gonadotropin.

HCR. An ion exchange resin used in water-treating and chemical process applications; strong acid cation exchange resin, 8% divinylbenzene cross-linked.

HDPE. Abbreviation for high-density polyethylene.

He. Symbol for helium.

HEA. Abbreviation for 2-hydroxyethyl acrylate.

head box. A large container, holding the prepared slurry of paper pulp and additives, located at the head of the fourdrinier machine; the slurry is fed continuously from slots in the bottom of the box to the wire on which the sheet is formed.

head space. A term used in the canning industry to refer to the space intentionally left between the top of the filled liquid and the cover of the can to allow for expansion of the contents during heat processing.

"Healthya Water" [Kao]. TM for first sport drink approved as a food for specific health use.

heat. A form of energy associated with and proportional to molecular motion. It can be transferred from one body to another by radiation, conduction, or convection; unlike latent heat, sensible heat is accompanied by a change in temperature.
—*of combustion*: The heat evolved when a definite quantity of a substance is completely oxidized (burned).

—*of crystallization*: The heat evolved or absorbed when a crystal forms from a saturated solution of a substance;
—*of dilution*: The heat evolved per mole of solute when a solution is diluted from one specific concentration to another.
—*of formation*: The heat evolved or absorbed when a compound is formed in its standard state from elements in their standard states at a specified temperature and pressure.
—*of fusion*: The heat required to convert a substance from the solid to the liquid state with no temperature change (also called latent heat of fusion or melting).
—*of hydration*: The heat associated with the hydration or solvation of ions in solution; also, the heat evolved or absorbed when a hydrate of a compound is formed.
—*of reaction*: The heat evolved or absorbed when a chemical reaction occurs in which the final state of the system is brought to the same temperature and pressure as that of the initial state of the reacting system.
—*of solution*: The heat evolved or absorbed when a substance is dissolved in a solvent.
—*of sublimation*: The heat required to convert a unit mass of a substance from the solid to the vapor state (sublimation) at a specified temperature and pressure without the appearance of the liquid state.
—*of transition*: The heat evolved or absorbed when a unit mass of a given substance is converted from one crystalline form to another.
—*of vaporization*: The heat required to convert a substance from the liquid to the gaseous state with no temperature change (also called latent heat of vaporization).

heat exchanger. A vessel in which an outgoing hot liquid or vapor transfers a large part of its heat to an incoming cool liquid; in the case of vapors, the latent heat of condensation is thus utilized to heat the entering liquid. The shell-and-tube type is widely used; here the hot liquid or vapor is contained in the shell while the cool liquid passes through the tubes, which are usually arranged in coils for maximum contact with the heat source. Heat exchangers are used in many chemical operations, e.g., evaporation and pulp manufacture, as well as to produce steam from the heat developed in nuclear reactors for power generation.
See evaporation; heat transfer.

heat transfer. Transmission of thermal energy from one location to another by means of a temperature gradient existing between the two locations. It may take place by conduction, convection, or radiation. Heat transfer is involved in many types of industrial operations, including distillation, evaporation, canning of foods, baking, curing, etc. In some cases, a heat exchanger is utilized. Fluids of high heat capacity are widely used to remove unwanted heat or to transfer it from one place to

another within a system; examples are air, water, ethylene glycol.
See coolant.

heavy.   A nontechnical word used in a number of scientific senses: (1) referring to atomic weight (heavy water, heavy metal); (2) referring to production volume (heavy chemical); (3) referring to physical weight (heavy spar); (4) referring to thickness (heavy-gauge wire); and (5) referring to distillation range (heavy oil).

heavy chemicals.   A chemical produced in tonnage quantities, often in a relatively impure state. Examples are sodium chloride, sulfuric acid, soda ash, salt cake, sodium hydroxide, etc.
See basic chemicals; fine chemical.

heavy hydrogen.   See deuterium; tritium.

heavy metal.   A metal of atomic weight greater than sodium (23) that forms soaps on reaction with fatty acids, e.g., aluminum, lead, cobalt.
See soap (2).

heavy oil.   An oil distilled from coal tar between 230C and 330C, the exact range not at all definite.
See coal tar.

heavy oxygen.   See oxygen-18.

heavy spar.   See barite.

heavy water.   (deuterium oxide; DOD; tritium oxide; TOT).
Water composed of two atoms of deuterium and one atom of oxygen; or one or two atoms of tritium and one of oxygen. In lower states of purity, the proportion of heavy water molecules decreases. Deuterium oxide is present to the extent of approximately one part in 6500 of ordinary water. Deuterium oxide freezes at 3.8C, boils at 101.4C, and has density of 1.1056 at (25C).
Derivation: There are several methods of separating or concentrating DOD: (1) fractional distillation, (2) Girdler–Spevack process, (3) hydrogen–sulfide exchange process, (4) electrolysis, and (5) cryogenic methane distillation.
Available forms: DOD 99.75% pure, in up to 5000 g lots.
Use: Moderator in some types of nuclear reactors.
Tritium (hydrogen of atomic weight 3) combines with oxygen to give another variety of heavy water, TOT, i.e., tritium oxide.
See tritium.

HEBP.   See 2-hydroxyethyl dibutyl phosphate.

Heck, Richard F.   (1931–2015). An American chemist who was awarded the 2010 Nobel Prize jointly with Ei-ichi Negishi and Akira Suzuki. The scientists discovered the use of palladium as a catalyst to produce organic molecules. The technique of palladium catalysis found extensive use in the pharmaceutical, agricultural, and electronics industries. Heck began the process in the late 1960s, and Negishi and Suzuki refined palladium catalysis for widespread use. Heck was educated at the University of California, Los Angles and was awarded Ph.D. (1954). For his remarkable body of work, he was awarded the Wallace H. Carothers Award (2005) and Herbert C. Brown Award (2006).

hecto-.   (h).
Prefix meaning 10^2 units, e.g., 1 hg = 1 hectogram = 100 grams.

hectorite.   (hector clay).
$Na_{0.67}(Mg,Li)_6Si_8O_{20}(OH,F)_4$.
One of the montmorillonite group of minerals that are principal constituents of bentonite.
Hazard: Respiratory irritant.
Use: Chill-proofing of beer.

hedeoma oil.   See pennyroyal oil.

hedonal.   (methylpropylcarbinolurethane).
CAS: 120-36-5. $CH_3CH_2CH(CH_3)OCONH_2$.
Properties: White, crystalline powder; faint aromatic odor and taste. Fusing p 76C, bp 215C. Soluble in alcohol, ether, organic solvents; sparingly soluble in cold water; more soluble in hot water.
Use: Medicine (sedative).

α-hederine.
CAS: 27013-91-8. $C_{41}H_{66}O_{12}$.
Hazard: Low toxicity by ingestion.

HEDTA.   See hydroxyethylethylenediaminetriacetic acid.

HEED.   Abbreviation for high-energy electron diffraction.

Heeger, Alan Jay.   (1936–).
Awarded the Nobel Prize in chemistry in 2000 jointly with Alan G. MacDiarmid and Hideki Shirakawa for the discovery and development of conductive polymers. He performs his research at Texas Instruments, Dallas, Texas.

Hehner number.   The percentage of weight of water-insoluble fatty acids in oils and fats.

Heisenberg, Werner Karl.   (1901–1976).
A native of Germany, Heisenberg received his doctorate from the University of Munich in 1923, after which he was closely associated for several years with Niels Bohr in Copenhagen. He was awarded the Nobel Prize in physics in 1932 for his brilliant work in quantum mechanics. In 1946, he became director of the Max Planck Institute. His notable contributions to theoretical physics, best known of which was the uncertainty principle, imparted new

impetus to nuclear physics and made possible a better understanding of atomic structure and chemical bonding.
See uncertainty principle.

helenine.   A nucleoprotein derived from the mold *Penicillium funiculosum.*
Use: An antiviral drug.

Helferich method.   Glycosidation of an acetylated sugar by heating with a phenol in the presence of zinc chloride or *p*-toluenesulfonic acid as catalyst.

Helianthine B.   See methyl orange.

helicase.   An enzyme that catalyzes the unwinding of strands in a DNA molecule before replication.

Helindon.   Vat dyestuffs used for dyeing wool.

Heliogen.   Phthalocyanine dyestuffs.
Use: Paint, lacquers, printing inks, wallpaper, coated paper, rubber, and plastics.

Helisorb.   UV stabilizer.
Properties: Free-flowing nonhazardous powder.
Use: An effective UV stabilizer for UVA and UVB protection in low concentrations for cosmetics and plastics.

He-Li-Ox 500 Series.   A series of PVC heat stabilizers for extruding, injection molding, calendering, and blow molding. All ingredients conform to FDA requirements for food packaging.

Heliozone.   A blend of waxy material.
Use: Retards sun-checking and cracking of rubbers.

helium.
CAS: 7440-59-7.   He.
Noble element of atomic number 2, first element in the noble gas group of the periodic table, aw 4.00260, valence of 0. Helium nuclei are alpha particles. Most important isotope is helium-3.
Properties: Colorless, noncombustible gas; odorless and tasteless. Liquefies at 4.2K to form helium I; at 2.2K, there is a transition (λ) point at which helium II is formed. See note below.
Properties: Bp −268.9C (1 atm), fp −272.2C (25 atm) lowest of any substance, bulk d 0.1785 g/L (0C). Very slightly soluble in water; insoluble in alcohol; rate of diffusion through solids is three times that of air; an asphyxiant gas.
See noble gas.
Occurrence: Texas, Oklahoma, Kansas, New Mexico, Arizona, and Canada. Originally discovered in the sun's atmosphere (1868) and confirmed in the atmosphere of Jupiter.
Derivation: From natural gas, by liquefaction of all other components, followed by purification over activated charcoal.
Grade: USP, technical, 99.995% pure.

Use: To pressurize rocket fuels, welding, inert atmosphere (growing germanium and silicon crystals), inflation of weather and research balloons, heat-transfer medium, leak detection, chromatography, cryogenic research, magnetohydrodynamics, luminous signs, geological dating, aerodynamic research, lasers, diving, and space-vehicle breathing equipment. Possible future uses include coolant for nuclear fusion power plants and in superconducting electric systems.
Note: Liquid helium has unique thermodynamic properties too complex to be adequately described here. Liquid He I has refr index 1.026, d 0.125, and is called a "quantum fluid" because it exhibits atomic properties on a macroscopic scale. Its bp is near absolute zero and viscosity is 25 micropoises (water = 10,000). He II, formed on cooling He I below its transition point, has the unusual property of superfluidity, extremely high thermal conductivity, and viscosity approaching zero.
See superconductivity.

α-helix.   A helical conformation of a predominantly right-handed polypeptide chain with maximal intrachain hydrogen bonding of the peptide bonds.

helix-loop-helix.   A protein structural motif characteristic of certain DNA-binding proteins.
hellebrigenin 3-acetate.
CAS: 4064-09-9. $C_{26}H_{34}O_7$.
Hazard: A poison.
Source: Natural product.

Hell, Stefan W.   (1962–). A Romanian-born German chemist who won the 2014 Nobel Prize jointly with Eric Betzig and William E. Moerner. The scientists developed a groundbreaking method to bypass the natural resolution limit in optical microscopy by using fluorescent molecules. This method enables to track process of active viruses and molecules in living cells. Hell used a modified form of fluorescence microscopy called stimulated emission depletion microscopy, where one laser beam illuminates the fluorescent molecules and another causes a small area to become dark. He created images of *E. coli* bacteria and yeast cells by moving laser beams over the specimens. His Ph.D. was awarded by the University of Heidelberg. He is also an honorary professor of experimental physics at the University of Gottingen.

Hell-Volhard-Zelinsky reaction.   α-Halogenation of carboxylic acids with halogen and phosphorus, presumably involving the enol form of the intermediate acyl halide.

17-α-helveticoside.
CAS: 6869-17-6. $C_{29}H_{42}O_9$.
Hazard: A poison.
Source: Natural product.

hem-.   Prefix indicating relationship to blood.

HEMA.   See hydroxyethyl methacrylate.

hema-.   Prefix indicating relationship to blood.

hemagglutinin.   (haemagglutinin).
A type of antibody that agglutinates erythrocytes.

hemagogue.   (emmenagogue).
A substance that elevates the flow of blood in the peripheral circulation.

hemalum.   A solution of hematoxylin and alum used as a nuclear stain in histology, especially with eosin as a counterstain.
Properties: Solution of hematoxylin and alum.
Use: A nuclear stain in histology.

hematein.
CAS: 475-25-2. $C_{16}H_{12}O_6$.
A tetracyclic compound. An oxidation product of hematoxylin, the coloring principle of logwood. Not to be confused with hematin.
Properties: Dark-purple solid. Mp 200+C. Almost insoluble in water, slightly soluble in alcohol and ether; soluble in dilute sodium hydroxide, giving a bright-red color; soluble in ammonia with brownish-violet color.
Derivation: By adding ammonia to logwood extract and exposing to air.
Use: Indicator, biological stain.

hematin.
CAS: 15489-90-4. $C_{34}H_{32}N_4O_4FeOH$.
The hydroxide of heme. Not to be confused with hematein.
Properties: Blue to brown-black powder. Decomposes at 200C without melting. Soluble in alkalies, hot alcohol, or ammonia; slightly soluble in hot pyridine; insoluble in water, ether, and chloroform.
Derivation: By dissolving hemin in dilute potassium hydroxide, precipitating with acetic acid, and recrystallizing from pyridine.
Use: Biochemical research.

hematite.   (iron (3+); oxygen(2-); hydrate).
CAS: 1317-60-8. $Fe_2H_2O_4$.
The naturally occurring α-form of ferric oxide.
Properties: Steel gray to black crystals or as ready earthy material.
Hazard: Benign pneumoconiosis.

hematite, red.   (red iron ore; bloodstone; iron oxide).
CAS: 1317-60-8. Fe_2O_3 with impurities.
Properties: Brilliant black to blackish-red or brick-red mineral with brown to cherry-red streak and metallic to dull luster. D 4.9–5.3, Mohs hardness 6. Noncombustible.
Hazard: A questionable carcinogen.
Use: The most important ore of iron. Also, certain varieties are used as paint pigments and for rouge.
See iron oxide reds; ferric oxide.

hemato-.   Prefix indicating relationship to blood.

hematoporphyrin.
CAS: 14459-29-1. $C_{34}H_{38}O_6N_4$.
Deep-red crystals; soluble in alcohol, sparingly soluble in ether; insoluble in water.
Derivation: Obtained from hemin or hematin by the action of strong acids.
Hazard: Toxic. Reported to be preferentially absorbed by cancerous tissues, making them fluoresce under UV light.
Use: Medicine (antidepressant).

hematoxylin.
CAS: 517-28-2. $C_{16}H_{14}O_6 \cdot 3H_2O$.
Properties: Slightly yellow crystals, turning red in light. Mp 100–120C. Soluble in hot water and alcohol, in glycerol, and in alkali hydroxides.
Hazard: May be carcinogenic.
Use: Colorant in inks, biological stains.

heme.   (hem).
$C_{34}H_{32}FeN_4O_4$.
The nonprotein portion of hemoglobin and myoglobin, consisting of reduced (ferrous) iron bound to protoporphyrin.
See porphyrin hemin.
Use: Medical and biochemical research.

hemel.   (generic name for hexamethylmelamine; HMM).
$C_3N_3[N(CH_3)_2]_3$ (cyclic).
Properties: Solid. Insoluble in water; soluble in acetone.
Use: Chemosterilant for insects.

hemellitic acid.   (2,3-xylic acid).
CAS: 603-79-2. $C_6H_3(CH_3)_2COOH$.
Properties: Mp 144C.

heme protein.   A protein containing a heme as a prosthetic group.

hemi-.   Prefix meaning half.

hemicellulose.   Cellulose having a degree of polymerization of 150 or less. A collective term for α and γ cellulose. It is that portion of holocellulose that is soluble in mild caustic solution. The pure form is obtained from corn grain hulls. It is not an important component of cellulosic products and is of chiefly theoretical interest. Hemicellulose obtained by treating a mixture of hard and softwoods with steam has been used as an animal feed supplement.

hemicholinium.
CAS: 16478-59-4. $C_{24}H_{34}N_2O_4$.
Hazard: A poison by ingestion.

hemiketal.   Organic compound formed by the reaction of one molecule of ketone (RR'CO) with one molecule of alcohol (R''CO). Hemiketals are the rings of ketose sugars.
See ketals.

hemimellitene. (1,2,3-trimethylbenzene).
CAS: 526-73-8. $C_6H_3(CH_3)_3$.
Properties: Liquid. D 0.8944 (20/4C), fp −25.5C, bp 176C. Insoluble in water; soluble in alcohol. Occurs in some petroleums.
Hazard: Combustible. Central nervous system impairment, asthma, and hematologic effects.
Use: Raw material for chemical synthesis.

hemin. (Teichmann's crystals; the chloride of heme).
$C_{34}H_{32}N_4O_4FeCl$.

Properties: Crystals that are brown by transmitted light and steel blue by reflected light. Sinters at 240C. Freely soluble in ammonia water; soluble in strong organic bases; insoluble in carbonate solutions, dilute acid solutions; insoluble but stable in water.
Derivation: By heating hemoglobin with acetic acid and sodium chloride.
Use: Identification of blood stains, biochemical research, complexing agent.
See chelate.

hemizygous. Having only one copy of a particular gene. For example, in humans, males are hemizygous for genes found on the Y chromosome.

hemo-. (haemo-, hem-, hema-, hemato-).
Prefix indicating relationship to blood.

hemocyanin. A dioxygen-carrying protein, an analog of heme.
Derivation: Found in certain invertebrate animals including a few insects and molluscs.

hemoglobin. (Hb).
The respiratory protein of the red blood cells, it transfers oxygen from the lungs to the tissues and carbon dioxide from the tissues to the lungs. Its affinity for carbon monoxide is >200 times that for oxygen.
Hemoglobin is a conjugated protein of molecular weight 65,000, consisting of approximately 94% globin (protein portion) and 6% heme. Each molecule can combine with one molecule of oxygen to form oxyhemoglobin (HbO_2). The iron (in the heme portion) must be in the reduced (ferrous)

state to enable the hemoglobin to combine with oxygen.
Oxyhemoglobin is available commercially as a brownish-red powder or crystals, soluble in water.
Use: Medicine, usually called hemoglobin.

hemolysin. (erythrocytolysin; erythrolysin).
A sensitizing, complement-fixing antibody produced by a living system that can lyse erythrocytes with the release of hemoglobin.

hemolytic gas. A poisonous gas.
Hazard: Causes hemoglobinuria, jaundice, gastroenteritis, nephritis, and hemolysis.

hemosiderin. An iron–protein molecule that serves as a source of iron in the synthesis of hemoglobin and as a source of iron in certain other syntheses.

hemotoxin. (hematotoxin; hematoxin).
A substance that destroys erythrocytes.

hemp. Soft, white fibers 3–6 ft long. It is coarser than flax but stronger, more glossy, and more durable than cotton. Obtained from the stems of Cannabis sativa.
Source: Central Asia, Italy, the former U.S.S.R., India, U.S.
Hazard: Combustible. May ignite spontaneously when wet.
Use: Blended with cotton or flax in toweling and heavy fabrics, twine, cordage, packing.
See cannabis.

hempa. (generic name for hexamethylphosphoric triamide; hexamethylphosphoramide; HMPA; hexametapol).
CAS: 680-31-9. $[N(CH_3)_2]_3PO$.
Properties: Water-white liquid; mild amine odor. Bp 230–232C (739.4 mm Hg), d 1.021 (15.5/15.5C). Soluble in water and polar and nonpolar solvents.
Hazard: Possible carcinogen. Toxic by skin contact. Combustible.
Use: UV inhibitor in polyvinyl chloride, chemosterilant for insects, promoting stereospecific reactions, specialty solvent.

hempseed oil. An oil obtained by pressing hemp seeds and manufactured from varieties of Cannabis sativa that do not contain significant amounts of tetrahydrocannabinol (THC). Approximately 30-35% of the weight of the hempseed is an edible oil that contains about 80% of essential fatty acids.
Properties: A drying oil similar in properties and uses to linseed. Edible, iodine value approximately 160, d 0.923, refr index 1.470-1.472. Contains approximately 10% saturated fatty acids (palmitic and stearic), unsaturated acids present are linoleic, linolenic, and oleic. Saponification value 190-193.
Source: Produced chiefly in the former U.S.S.R., southern Europe, Japan, Chile, little in U.S. With

the expected legalization in cannabis in the U.S., increased production is possible.
Hazard: Combustible.
Use: Limited use in foodstuffs due to the high concentration of essential fatty acids. In retail, as an ingredient in body care products, soaps, shampoos, and detergents. Industrial hempseed oil is used in lubricants, paints, inks, fuels, and plastics.

henda- compounds.   See unde-.

Henderson-Hasselbalch equation.   An equation relating the pH, the pK_a, and the ratio of the concentrations of the proton-acceptor (A-) and proton-donor (HA) species in a solution.

n-heneicosanoic acid.
CAS: 2363-71-5. $CH_3(CH_2)_{19}COOH$.
A saturated fatty acid not normally found in natural fats or waxes.
Properties: White, crystalline solid. Mp 74.3C. Synthetic product available for organic synthesis, 99% purity.
Hazard: Combustible.

(z)-6-1-heneicosen-11-one.
CAS: 54844-65-4. $C_{21}H_{40}O$.
Hazard: Low toxicity by ingestion and skin contact.
Source: Natural product.

henequen.
Properties: Hard, strong, reddish fibers obtained from the leaves of *Agave fourcroydes*. It is similar to sisal but coarser and stiffer.
Available forms: Denier ranges from 300 to 500.
Source: Mexico, Cuba.
Hazard: Combustible.
Use: Binder twine, cordage.

Henkel reaction.   Industrial-scale thermal rearrangement or disproportionation of alkaline salts of aromatic acids to symmetrical diacids in the presence of cadmium or other metallic salts.

henna.   A coloring principle obtained from dried leaves of certain tropical plants (North Africa, India).
Use: Commercial hair-dyeing preparations to give a yellow-red color; medicine as an antifungal agent.

Hennig purifier.   A preparation having a soda–ash base and other materials. Produced as walnut-sized briquettes. Packed in 100-lb paper bags.
Use: Ladle addition to produce cleaner steel by aiding in removal of dissolved oxides and silicates and fluxing nonmetallic inclusions to slag.

Henry reaction.   Formation of nitroalcohols by an aldol-type condensation of nitroparaffins with aldehydes in the presence of base (Henry) or by the condensation of sodium salts of acinitroparaffins

with the sodium bisulfite addition products of aldehydes in the presence of a trace of alkali or weak acid (Kamlet). Widely used in sugar chemistry.

Henry's law.   When a liquid and a gas are in contact, the mass of the gas that dissolves in a given quantity of liquid is proportional to the pressure of the gas above the liquid. Thus, if air is kept in contact with water at standard atmospheric pressure, each kg of water dissolves 0.017 g oxygen at 20C; if this pressure is halved (by doing the experiment at high altitude where the pressure is only 0.5 atm), the water dissolves only 0.0085 g oxygen. The law holds true only for equilibrium conditions, i.e., when enough time has elapsed so that the quantity of gas dissolved is no longer changing.

hentriacontane.
$C_{31}H_{64}$ or $CH_3(CH_2)_{29}CH_3$.
Properties: Crystals. D 0.781 (68C), bp 302C (15 mm Hg), mp 68C.
Hazard: Combustible.

HEOD.   See dieldrin.

heparin.
CAS: 9005-49-6.
A complex organic acid (mucopolysaccharide) present in mammalian tissues; a strong inhibitor of blood coagulation; a dextrorotatory polysaccharide built up from hexosamine and hexuronic acid units containing sulfuric acid ester groups. Precise chemical formula and structure uncertain; a formula of $(C_{12}H_{16}NS_2Na_3)_{20}$ and molecular weight of 12,000 have been suggested for sodium heparinate.
Properties: White or pale-colored amorphous powder; nearly odorless; hygroscopic. Soluble in water; insoluble in alcohol, benzene, acetone, chloroform, and ether; pH in 17% solution between 5.0 and 7.5.
Derivation: Animal livers or lungs.
Grade: USP.
Hazard: May cause internal bleeding.
Use: Medicine (anticoagulant), biochemical research, rodenticides.

hepasynthyl.
CAS: 5634-42-4. $C_{19}H_{26}O_4 \cdot C_4H_{11}NO_2$.
Hazard: Moderately toxic by ingestion.

hepatocyte.   The major cell type of liver tissue, which is parenchymal tissue. Liver also contains fixed macrophages called Kupfer cells and other cells.

hepatoxane.   See hepasynthyl.

"Hepsera Tabs" [Gilead].   TM for tablets.
Use: For antiviral treatment of chronic hepatitis B.

heptabarbital.   (5-[1-cyclohepten-1-yl]-5-ethylbarbituric acid; 5-ethyl-5-cycloheptenylbarbituric acid).
CAS: 509-86-4. $C_{13}H_{18}N_2O_3$.

Properties: White, crystalline powder; odorless; slightly bitter taste. Mp 174C. Very sparingly soluble in water; slightly soluble in alcohol; soluble in alkaline solutions. Forms water-soluble sodium, magnesium, and calcium salts.
Use: Medicine (sedative).
See barbiturate.

heptachlor. (generic name for 1, 4, 5, 6, 7, 8, 8-heptachloro-3a,4,7,7a-tetrahydro-4,7-methanoindene. Generic).
CAS: 76-44-8. $C_{10}H_7Cl_7$.
Properties: White to light-tan, waxy solid. Mp 95–96C, d 1.57–1.59. Insoluble in water; soluble in xylene and alcohol.
Hazard: Toxic by ingestion, inhalation, and skin absorption; use has been restricted and discontinued except for termite control. Possible carcinogen.
Use: Insecticide.

heptachlorepoxide.
CAS: 1024-57-3. $C_{10}H_9Cl_7O$.
A degradation product of heptachlor that also acts as an insecticide.
Hazard: Possible carcinogen.
See epoxyheptachlor.

heptachlorotetrahydromethanoindene.
See heptachlor.

***n*-heptacosane.**
CAS: 593-49-7. $CH_3(CH_2)_{25}CH_3$.
Properties: Crystals. D 0.804, bp 270C (15 mm Hg), mp 59.5C. Soluble in alcohol; insoluble in water.
Hazard: Combustible.

***n*-heptadecane.**
CAS: 629-78-7. $C_{17}H_{36}$ or $CH_3(CH_2)_{15}CH_3$.
Properties: Leaflets, soluble in alcohol. D 0.778, bp 303C, mp 22.5C. Insoluble in water.
Hazard: Combustible.

***n*-heptadecanoic acid.** (margaric acid).
CAS: 506-12-7. $CH_3(CH_2)_{15}COOH$.
A saturated fatty acid not normally found in natural fats or waxes.
Properties: Colorless crystals. Mp 61C, d 0.8355 (90.6/4C), bp 363.8C, 230.7C (16 mm Hg), refr index 1.4324 (70C). Soluble in alcohol and ether; insoluble in water.
Available forms: 99% pure synthetic product.
Use: Organic synthesis.

heptadecanol. Any saturated C_{17} alcohol.
Hazard: Combustible.
See *n*-heptadecanol; 3,9-diethyl-6-tridecanol.

***n*-heptadecanol.**
CAS: 1454-85-9. $C_{17}H_{35}OH$.
Properties: Colorless liquid. D 0.8475 (20/20C), bp 308.5C, vap press <0.01 mm Hg (20C), flash p 310F (154C), bulk d 7.1 lb/gal (20C). Slightly soluble in water.
Grade: Technical.

Hazard: Combustible.
Use: Organic synthesis, plasticizer, intermediates, perfume fixatives, soaps and cosmetics, manufacture of wetting agents and detergents.

2-(8-heptadecenyl)-2-imidazoline-1-ethanol.
See amine 220.

2-heptadecylglyoxalidine. (2-heptadecylimidazoline).
$C_{20}H_{40}N_2$ or $C_{17}H_{35}C_3H_5N_2$.
Properties: Waxy solid. Mp 85C, bp 200C (2 mm Hg). Slightly soluble in water; soluble in alcohol, benzene; hydrolyzes on standing to form *N*-2-aminoethyl stearamide.
Derivation: By reacting stearic acid with ethylene diamine.
Hazard: Combustible.
Use: Fungicide.

2-heptadecylglyoxalidine acetate. See glyodin.

2-heptadecylimidazoline. See 2-heptadecylglyoxalidine.

2-heptadecyl-2-imidazoline acetate. See glyodin.

heptafluorobutyric acid. (perfluorobutyric acid).
CAS: 375-22-4. C_3F_7COOH.
Properties: Colorless, hygroscopic liquid with sharp odor. Bp 120C (735 mm Hg), fp −17.5C, d 1.641 (25C), refr index 1.290 (25C), surface tension 15.8 dynes/cm (30C). Miscible with water, acetone, ether, and petroleum ether; soluble in benzene and carbon tetrachloride; insoluble in carbon disulfide and mineral oil.
Derivation: By electrolysis of a solution of butyric acid in hydrogen fluoride.
Hazard: Irritant to tissue.
Use: Intermediate, surfactant, acidulant.

γ-heptalactone.
CAS: 105-21-5. $C_7H_{12}O_2$.
Properties: Colorless, slightly oily liquid; coconut, sweet, malty, caramel odor. D: 0.997–1.004 at 20C, refr index: 1.439–1.445. Misc in alc, fixed oils; very slightly sol in water.
Hazard: A skin irritant.
Use: Food additive.

heptaldehyde. See heptanal.

heptalin acetate. See methylcyclohexanol acetate.

heptamethylene. See cycloheptane.

heptamethylnonane.
$C_{16}H_{34}$.
Isomer of cetane (hexadecane). In 1964, it replaced α-methylnaphthalene as ignition standard for diesel fuels.

Properties: Ignition value 15 on cetane-α-methylnaphthalene scale.
Hazard: Flammable, moderate fire risk.

heptanal. (heptaldehyde; enanthaldehyde; aldehyde C-7).
CAS: 111-71-7. $C_6H_{13}CHO$.
Properties: Oily, colorless liquid; penetrating, fruity odor; hygroscopic. D 0.814–0.819, refr index 1.42, mp 43C, bp 153C. Soluble in three volumes of 60% alcohol, slightly soluble in water. Soluble in ether.
Derivation: Castor oil, from decomposition of the ricinoleic acid glyceride.
Hazard: Combustible.
Use: Manufacture of 1-heptanol, organic synthesis, perfumery, pharmaceuticals, flavoring.

n-heptane. (dipropylmethane).
CAS: 142-82-5. $CH_3(CH_2)_5CH_3$.
Properties: Volatile, colorless liquid. Fp –90.595C, bp 98.428C, refr index 1.38764 (20C), d 0.68368 (20C), flash p 25F (–3.89C) (CC). Soluble in alcohol, ether, and chloroform; insoluble in water; distillation range 93.3–98.9C; vap press 2.0 psi (a)(37.7C) (max). Color Saybolt +30 (min), maximum sulfur content 0.01 wt %, corrosive passes ASTM D 130-30 test, autoign temp 433F (222C).
Derivation: Fractional distillation of petroleum, purified by rectification.
Grade: Commercial, 99%, spectro, ASTM reference fuel, research, 99.92 mole %.
Hazard: Toxic by inhalation. Flammable, dangerous fire risk.
Use: Standard for octane-rating determinations (pure normal heptane has zero octane number), anesthetic, solvent, organic synthesis, preparation of laboratory reagents.
See octane number.

1,7-heptanedicarboxylic acid. See azelaic acid.

1,7-heptanedioic acid. See pimelic acid.

n-heptanoic acid. (enanthic acid; n-heptylic acid; heptoic acid).
CAS: 111-14-8. $CH_3(CH_2)_5COOH$.
Properties: Clear, oily liquid; unpleasant odor. D 0.9181 (20/4C), fp 8C, bp 221.9C, refr index 1.4229. Soluble in alcohol and ether, insoluble in water.
Derivation: By oxidizing heptanal with potassium permanganate in dilute sulfuric acid.
Hazard: Combustible.
Use: Organic synthesis, production of special lubricants for aircraft and brake fluids.

1-heptanol. See heptyl alcohol.

2-heptanol. See methyl amyl carbinol.

3-heptanol.
CAS: 589-82-2. $CH_3CH_2CH(OH)C_4H_9$.

Properties: Liquid. D 0.8224 (20C), fp –70C, bp 156.2C, flash p 140F (60C) (CC). Slightly soluble in water.
Hazard: Toxic by ingestion. Moderate fire risk.
Use: Flotation frother, solvent and diluent in organic coatings, intermediates.

2-heptanone. See methyl-n-amyl ketone.

3-heptanone. See ethyl butyl ketone.

4-heptanone. See dipropyl ketone.

heptanoyl chloride.
CAS: 2528-61-2. $CH_3(CH_2)_5COCl$.
Properties: Solid. Mw 148.63, bp 173C, d 0.960, fp 58C.
Hazard: Corrosive and a lachrymator.

1-heptene. (1-heptylene).
CAS: 592-76-7. $CH_3(CH_2)_4CH:CH_2$.
Properties: Colorless liquid. D 0.6968 (20/4C), bp 93.3C, fp –10C, flash p 32F (0C), refr index 1.3994 (20C). Soluble in alcohol, acetone, ether, petroleum, and coal tar solvents; insoluble in water.
Hazard: Flammable, dangerous fire risk.
Use: Organic synthesis.

2-heptene. (2-heptylene).
CAS: 592-77-8. $CH_3(CH_2)_3CH:CHCH_3$ (cis and trans isomers).
Properties: Colorless liquid. D (cis) 0.708, (trans) 0.704, (commercial) 0.7010–0.7050 (20/4C), bp (trans) 98C, (cis) 98.5C, (commercial) 97–99C, refr index 1.406 (20C); flash p (commercial) 28F (–2.2C). Soluble in alcohol, acetone, ether, petroleum, and coal tar solvents; insoluble in water.
Hazard: Flammable, dangerous fire risk.
Use: Suggested as plant growth retardant.

3-heptene. (3-heptylene).
CAS: 14686-14-7. $C_3H_7CH:CHC_2H_5$.
Properties: (Mixed cis and trans isomers) Colorless liquid. Bp 95C, d 0.705 (15.5/15.5C), refr index 1.405 (20C), flash p 21F (–6.1C).
Hazard: Flammable, dangerous fire risk.
Use: Suggested as plant-growth retardant.

6-heptenoic acid, 7-(2-cyclopropyl-4-(4-fluorophenyl)-3-quinolinyl)-3,5-dihydroxy-,calcium salt (2:1), (3r,5s,6e)-. See NK 104 (acid).

heptenophos. (phosphoric acid 7-chlorobicyclo-[3.2.0]hepta-2,6-dien-6-yl dimethyl ester; 7-chlorobicyclo-[3.2.0]hepta-2,6-dien-6-yl dimethylphosphate; (6-chloro-7-bicyclo[3.2.0]hepta-3,6-dienyl) dimethyl phosphate; o,o-dimethyl-o-(6-chlorobicyclo(3.2.0)heptadien-1,5-phoshate); 5-(o,o-dimethylphosphoryl)-6-chlorobicyclo(3.2.0)-hepta-1,5-dien; HOE 2982; HOE 2982 OJ; hosta-quick; hostavik; ragadan; XOE 2982).
$C_9H_{12}ClO_4P$.

A nonpersistent contact and systematic phosphate insecticide.
Properties: Pale amber liquid; miscible in most organic solvents; soluble in xylene, acetone, and methanol.
Hazard: Poison; moderately toxic.
Use: An ectoparasiticide.

heterocytotropic antibody.   A cytophilic antibody with an activity similar to homocytotropic antibody that has an affinity for cells of a different species that is not closely related.

heptoic acid.   See heptanoic acid.

1-heptyl acetate.
CAS: 112-06-1. $C_7H_{15}OOCCH_3$.
Liquid with fruity odor.
Use: Artificial fruit essences.

heptyl alcohol.   (1-heptanol; enanthyl alcohol).
CAS: 111-70-6. $C_7H_{15}OH$.
Properties: Colorless, fragrant liquid. Fp −34.6C, bp 175C, d 0.824 (20/4C), refr index 1.4233 (20C), flash p 170F (76.6C). Slightly soluble in water; miscible with alcohol and ether.
Derivation: From heptaldehyde by reduction.
Hazard: Combustible.
Use: Organic intermediate, solvent, cosmetic formulations.

heptylamine.
CAS: 111-68-2. $C_7H_{15}NH_2$.
Properties: Colorless liquid. D 0.777 (20/4C), fp −23C, bp 155C, flash p 140F (60C) (OC). Slightly soluble in water; soluble in alcohol and ether.
Hazard: Combustible.

heptyl formate.
CAS: 112-23-2. $HCOOC_7H_{15}$.
Properties: Colorless liquid; fruity odor. Bp 176.7C, d 0.894 (0C).
Hazard: Combustible.
Use: Artificial fruit essences.

heptyl heptoate.
$C_7H_5OOCC_6H_{13}$.
Properties: Colorless liquid; fruity odor. D 0.865 (19C), bp 273–274C (754 mm Hg).
Hazard: Combustible.
Use: Artificial fruit essences.

n-heptyl *p*-hydroxybenzoate.   See heptylparaben.

n-heptylic acid.   See heptanoic acid.

heptyl (4-(1-methylethyl)phenyl)methyl 3-pyridinylcarbonimidodithioate.
CAS: 51308-74-8. $C_{23}H_{32}N_2S_2$.
Hazard: Moderately toxic by ingestion.
Use: Agricultural chemical.

heptylparaben.
CAS: 1085-12-7. $C_{14}H_{20}O_3$.
Properties: Small colorless crystals or white crystalline powder; odorless, burning taste. Mp: 48–51C. Sol in alc, ether; very slightly sol in water.
Use: Food additive.

heptyl pelargonate.
CAS: 71605-85-1. $C_7H_{15}OOCC_8H_{17}$.
Properties: Liquid with pleasant odor. D 0.866 (15.5/15.5C), bp 300C, refr index 1.4360.
Hazard: Combustible.
Use: Flavors and perfumes.

α-heptyl-3,4,5-trimethoxyphenethylamine.
CAS: 67293-51-0. $C_{18}H_{31}NO_3$.
Hazard: A poison.

herbicide.   (weed killer).
A pesticide, either organic or inorganic, used to destroy unwanted vegetation, especially various types of weeds, grasses, and woody plants. Until 1924, inorganics such as sodium chlorate, sodium chloride, ammonium sulfamate, arsenic, and boron compounds were used. At that time, the more specific organics were introduced, typified by 2,4-dichlorophenoxyacetic acid (2,4-D). Herbicides may be of two major types: (1) selective, such as 2,4-D, 2,4,5-T, phenols, carbamates, and urea derivatives, permitting elimination of weeds without injury to the crop, and (2) nonselective, comprising soil sterilants (sodium compounds) and silvicides (ammonium sulfamate). The latter kill woody plants and trees. Some types act as overstimulating growth hormones. Many herbicides are highly toxic and should be handled and applied with care; use of chlorinated types may be restricted.
See defoliant.

herbifert.
CAS: 39324-65-7. $C_9H_{17}N_5S \cdot C_8H_{14}ClN_5$.
Hazard: Moderately toxic by ingestion. Low toxicity by skin contact. SO_x and Cl^-.

Herclor.   A group of specialty elastomers based on epichlorohydrin, claimed to have unique service performance properties.
Use: Automotive and aircraft parts, wire and cable coating, seals and gaskets, packings, hose, belting, and coated fabrics. Herclor H is a homopolymer, and Herclor C a copolymer with ethylene oxide; both have high resistance to ozone, heat, solvents, and chemical attack.

"Hercoflex" [Hercules].   TM for a series of plasticizers.
600 High-boiling ester of pentaerythritol and a saturated aliphatic acid.
707 High-molecular-weight polyol ester.
707A High-molecular-weight polyol ester.
Use: High-temperature vinyl electrical insulation.
900 High-molecular-weight polyester.
Use: Plasticizer for polyvinyl acetate.

"Hercolube" [Hercules]. TM for synthetic lubricant base stocks.
J15 Saturated aliphatic ester of pentaerythritol for plasticizing vinylidene chloride.
Derivation: Derived from pentaerythritol esters of saturated fatty acids.

Hercolyn D. A pale, viscous liquid; the hydrogenated methyl ester of rosin.
Use: Plasticizing resin.

Hercules trap. Water-measuring liquid trap used in aquametry when the material collected is heavier than water.

Herculoid. Nitrocellulose containing 10.9–11.2% nitrogen.
Hazard: See nitrocellulose.
Use: Pyroxylin plastics.

"Herculon" [PPG]. TM for polypropylene olefin fibers. Available in bulked continuous and continuous multifilament yarns, staple, and uncut tow.
Use: Apparel, home furnishings, and industrial applications.

hereditary cancer. Cancer that occurs due to the inheritance of an altered gene within a family.
See sporadic cancer.

hereditary mutation. A gene change in the body's reproductive cells (egg or sperm) that becomes incorporated in the DNA of every cell in the body; also called germline mutation.

heroin. See diacetylmorphine.

herring oil. See fish oil.

Herschbach, Dudley R. (1932–). Awarded the Nobel Prize in chemistry in 1986 for work reporting that the energies of reactions of crossed molecular beams of isolated alkali metal atoms and alkyl halide molecules appeared mostly as vibrational excited states of products. This method of studying all types of chemical reactions led to a more detailed knowledge of reaction processes. Doctorate awarded from Harvard in 1958.

Hershko, Avram. (1937–). Born in Karcag, Hungary, Hershko won the Nobel Prize for chemistry in 2004 for his pioneering work concerning the discovery of ubiquitin-mediated protein degradation. He received his M.D. in 1965 and his Ph.D. in 1969 from the Hadassah Faculty of Medicine of The Hebrew University of Jerusalem. He is currently a distinguished professor at the Rappaport Family Institute for Research in Medical Sciences at the Technion (Israel Institute of Technology) in Haifa and adjunct professor of Pathology at New York University. In 2000, he received the Albert Lasker Award for Basic Medical Research.

Herzberg, Gerhard. (1904–1999). A German-born physicist who won the Nobel Prize for chemistry in 1971 for his work on the composition of molecules. His research involved the spectroscopy of atoms and molecules and their excitation behavior. He became a Canadian citizen and was the director of the Division of Pure Physics of the National Research Council of Canada.

Herzig-Meyer determination of *N*-alkyl groups. *N*-alkylamines are refluxed with hydriodic acid, and the quaternary alkyl ammonium iodides are pyrolyzed to split off alkyl iodide, which is determined gravimetrically by conversion to silver iodide or titrated as iodate.

Herz reaction; Herz compounds. Formation of *o*-aminothiophenols by heating aromatic amines with excess sulfur monochloride. The first products formed are thiazothionium halides known as Herz compounds. If the position next to the amino group is unoccupied, chloride is substituted at this position during the reaction.

hesperidin.
CAS: 520-26-3. $C_{28}H_{34}O_{15}$.
A natural bioflavonoid of the flavanone group.
Properties: Fine needles. Mp 258–262C. Soluble in dilute alkalies and pyridine.
Derivation: Extraction from citrus fruit peel.
Use: Synthetic sweetener research.

Hess's law. The heat evolved or absorbed in a chemical process is the same whether the process takes place in one or in several steps; also known as the law of constant heat summation.

hetastarch. A starch derivative containing 90% amylopectin.
Use: Blood plasma volume expander.

heteratisine. See 20-ethyl-6-β,8-dihydroxy-1-α-methoxy-4.

heteratisine 6-benzoate.
CAS: 99759-48-5. $C_{29}H_{37}NO_6$.
Hazard: A poison.
Source: Natural product.

heteroaromatic. See heterocyclic.

heteroazeotrope. Azeotropic mixture having more than one liquid phase in equilibrium with the vapor phase at the boiling points.

heterochromatin. Chromatin which is in a condensed, less accessible state and thus frequently transcriptionally silent. However, some DNA is constitutive heterochromatin (e.g., the second X chromosome is inactivated to a Barr body for dosage compensation) whereas other DNA is facultative heterochromatin. An example might be genes required under time of stress only, or during a stage of the cell cycle.

heterocyclic. Designating a closed-ring structure, usually of either five or six members, in which one or more of the atoms in the ring is an element other than carbon, e.g., sulfur, nitrogen, etc. Examples are pyridine, pyrole, furan, thiophene, and purine.

thiophene   pyridine

heteroduplex DNA. Duplex DNA containing complementary strands derived from two different DNA molecules with similar or partially identical sequences. Sometimes from two unmanipulated species, at other times a gene has been modified and reintroduced.

heterogeneity. The production of identical or similar phenotypes by different genetic mechanisms.

heterogeneous. (Latin "different kinds").
Any mixture or solution comprising two or more substances regardless of whether they are uniformly dispersed. Common examples are such diverse materials as air (a mixture of 20% oxygen and 80% nitrogen), milk, marble, paint, gasoline, blood, and mayonnaise. In all such cases, the mixtures can be separated mechanically into their components. "Homogenized" milk is as heterogeneous as regular milk and the term is, strictly speaking, a misnomer. See homogeneous; mixture.

heterogeneous catalysis. See catalysis, heterogeneous.

heterologous hemagglutinin. An antibody that agglutinates erythrocytes of species other than the producing species.

heteromolybdates. (heteropolymolybdates).
A large group of complex molybdenum salts and acids in which the anion contains oxygen atoms and from 2 to 18 hexavalent molybdenum atoms, as well as one or more other metal or nonmetal atoms (phosphorus, arsenic, iron, and tellurium). The latter are referred to as hetero atoms, and any of approximately 35 elements may be present in this manner. Example: $Na_3PMo_{12}O_{40}$, sodium phospho-12-molybdate. The molecular weights of these compounds range up to 3000. The acids and most of the salts are very soluble in water, and the acids and some salts are soluble in organic solvents.
Use: Phosphomolybdates and phosphotungstates are used as precipitants for basic dyes to form lakes and toners. The phospho- and silicomolybdate groups are of key importance in the functioning of certain enzymes. There are many uses in analytical chemistry.

heterophos.
CAS: 40626-35-5. $C_{11}H_{17}O_3PS$.
Hazard: A poison by ingestion.
Use: Agricultural chemical.

heteropolysaccharide. A polysaccharide containing two or more types of sugars.

heterotroph. An organism that requires complex nutrient molecules, such as glucose, as an energy source.

heterotropic enzyme. An allosteric enzyme requiring a molecule other than its substrate as a modulator.

heterozygosity. The presence of different alleles at one or more loci on homologous chromosomes.

heterozygote. A heterozygous organism that carries two or more different alleles of a given gene. See heterozygosity.

HETP. (1) Abbreviation for hexaethyl tetraphosphate; (2) abbreviation for height equivalent to a theoretical plate.
See theoretical plate.

"Hetron" [Ashland]. TM for an unsaturated polyester resin.
Use: Corrosion and chemical resistance, fire-retardant, castings, marine gel coats, laminating, and molding.

Heumann-Pfleger indigo synthesis. Cyclization of phenylglycine to indoxyl, followed by oxidation by air or oxidizing agents, such as ferric chloride, to yield indigo.

heuristic. See computational chemistry.

Hevesy, Georg de. (1885–1966). A Hungarian chemist who won the Nobel Prize in chemistry in 1943. He discovered the element hafnium in 1923. One of his interesting projects involved the calculation of the percentages of chemical elements in the universe. He was also involved in research using radioactive lead and phosphorus traces. His work included the separation of isotopes by physical means. His Ph.D. was granted at Freiburg in 1908.

hexa-. Prefix signifying six.

hexabromobenzene.
CAS: 87-82-1. C_6Br_6.
Hazard: A poison.

hexabromoethane.
C_2Br_6.
Properties: Yellowish-white, rhombic needles. Slightly soluble in water and alcohol. Mp 149C (decomposes with separation of bromine).
Derivation: Action of bromine on diiodoacetylene.
Use: Organic synthesis.

1,2,3,4,6,7-hexabromonaphthalene.
CAS: 75625-24-0. $C_{10}H_2Br_6$.
Hazard: A poison by ingestion.

1,1,1,3,3,3-hexabutyldistannthiane.
CAS: 4808-30-4. $C_{24}H_{54}SSn_2$.
Properties: Colorless oil. Bp: 208C (decomp).
Hazard: Moderately toxic by ingestion.
Use: Drug.

hexacalcium phytate.   See calcium phytate.

hexachloroacetone.   (hexachloro-2-propa-
none).
CAS: 116-16-5. $Cl_3CCOCCl_3$.
Properties: Yellow liquid. Bp 204C, fp −3C, d 1.744
(12/12C). Slightly soluble in water; soluble in acetone.
Hazard: Toxic by ingestion and inhalation, strong
irritant. Combustible. Evolves phosgene when
heated.
Use: Desiccant, herbicide.

hexachlorobenzene.   (perchlorobenzene).
CAS: 118-74-1. C_6Cl_6.
Properties: White needles. D 2.04, mp 229C, bp
326C, flash p 468F (242C). Soluble in benzene and
boiling alcohol; insoluble in water.
Hazard: Possible carcinogen. Toxic by ingestion.
Combustible.
Use: Organic synthesis, fungicide for seeds, wood
preservative.

2,2′,4,4′,6,6′-hexachlorobiphenyl.
CAS: 33979-03-2. $C_{12}H_4Cl_6$.
Hazard: A poison.

2,3′,4,4′,5′,6-hexachlorobiphenyl.
CAS: 59291-65-5. $C_{12}H_4Cl_6$.
Hazard: A poison.

hexachlorobutadiene.
CAS: 87-68-3. $Cl_2C:CClCCl:CCl_2$.
Properties: Clear, colorless liquid; mild odor. Freezing range −19 to −22C, boiling range 210–220C,
refr index 1.552 (20C), flash p none, d 1.675
(15.5/15.5C), bulk d 13.97 lb/gal (15.5C), purity
98% (min). Vap press 22 mm Hg (100C), 500 mm
Hg (200C). Viscosity (37.7C) 2.447 cP, 1.479 centistokes; (98.9C) 1.131 cP, 0.724 centistokes. Insoluble in water, compatible with numerous resins,
soluble in alcohol and ether. Nonflammable.
Hazard: Toxic by ingestion and inhalation, a questionable carcinogen.
Use: Solvent for elastomers, heat-transfer liquid,
transformer and hydraulic fluid, wash liquor for
removing C_4 and higher hydrocarbons.

1,2,3,4,5,6-hexachlorocyclohexane.   (BHC;
HCCH; HCH; TBH; benzene hexachloride).
CAS: 608-73-1. $C_6H_6Cl_6$.
A systemic insecticide. The γ isomer is known as
lindane.

Properties: White or yellowish powder or flakes;
musty odor. Melting point varies with isomeric
composition. D 1.87, vap press approximately 0.5
mm Hg (60C). Stable toward moderate heat but
decomposed by alkaline substances. Mp of the pure
isomers are (α-*trans*) 157–158C, (β-*cis*) 297C (sublimes), (γ) 112.5C, (Δ) 138–139C, (ε) 217–219C.
Insoluble in water; soluble in 100% alcohol, chloroform, and ether.
Derivation: Chlorination of benzene in actinic light.
Method of purification: Fractional crystallization.
The technical grade may run 10–15% γ isomer but
can be brought up to 99% (lindane).
Grade: Technical (mixture of isomers), 25% γ isomer, and 99% γ isomer (lindane).
Hazard: Toxic by ingestion and inhalation, absorbed
by skin, strong irritant to skin and eyes. CNS depressant. Use may be restricted.
Use: Component of insecticides; toxic to flies, cockroaches, aphids, grasshoppers, wireworms, and boll
weevils.

τ-hexachlorocyclohexane.
(1α,2α,3β,4α,5α,6β-hexachlorocyclohexane;
τ-hch; γ-benzene hexachloride; γ-hexachlor;
γ-hexane).
CAS: 608-73-1. $C_6H_6Cl_6$.
An environmentally persistent constituent of a wide
variety of agricultural, medical, and veterinary
products.
Use: Insecticide.
Hazard: Highly toxic; questionable carcinogen.

hexachlorocyclopentadiene.   (perchlorocyclopentadiene).
CAS: 77-47-4. C_5Cl_6.
Properties: Pale-yellow liquid; pungent odor. Bp
239C, fp 9.6C, d 1.717 (15/15C), bulk d 14.30 lb/gal
(15.5C), refr index 1.563 (25C), flash p none. Nonflammable.
Grade: Technical.
Hazard: Toxic by ingestion, inhalation, and skin
absorption. Questionable carcinogen.
Use: Intermediate for resins, dyes, pesticides, fungicides, pharmaceuticals.

hexachlorocyclopentadiene.   (C-56; graph-
lox; HCCPD; hexachlorcyklopentadien; hexa-
chloro-1,3-cyclopentadiene; 1,2,3,4,5,5-hexach-
lorocyclopenta-1,3-diene; hexachlorocyclopentadiene; HRS 1655; NCI C55607; PCL; perchlorocyclopentadiene; RCRA waste number 130).
CAS: 77-47-4. C_5Cl_5.
Properties: highly reactive, cyclic alkenyl halide
with two double bonds, readily undergoes substitution and addition reactions, freezing point of
11C, boiling point of 239C, density of 1.7 g/cm³;
greenish-yellow to amber-colored liquid with a pungent odor.
Hazard: Deadly poison; moderately toxic; teratogen.
Use: originally as an agricultural fumigant and as an
intermediate in the manufacture of insecticides.

2,3,4,4,5,5-hexachloro-2-cyclopenten-1-one.
CAS: 2514-52-5. C_5Cl_6O.
Hazard: A poison by ingestion. Moderately toxic by inhalation.

2,2',4,4',5,5'-hexachlorodiphenyl ether.
CAS: 71859-30-8. $C_{12}H_4Cl_6O$.
Hazard: A reproductive hazard.

2,2',4,4',5,6'-hexachlorodiphenyl ether.
CAS: 106220-81-9. $C_{12}H_4Cl_6O$.
Hazard: A reproductive hazard.

hexachlorodiphenyl oxide.   (chlorinated diphenyl oxide).
CAS: 31242-93-0. $C_{12}H_4Cl_6O$.
Properties: Light-yellow, very viscous liquid. Bp 230–260C (8 mm Hg), d 1.60 (20/20C), bulk d 13.12 lb/gal (25C), refr index 1.621 (25C), flash p none. Soluble in methanol and ether; very slightly soluble in water. Nonflammable.
Hazard: Toxic by ingestion.
Use: Solvent, intermediate.

1,1,1,4,4,4-hexachloro-1,4-disilabutane.
See bis(trichlorosilyl)ethane.

hexachloroendomethylene   tetrahydrophthalic acid.   See chlorendic acid.

hexachloroendomethylene   tetrahydrophthalic anhydride.   See chlorendic anhydride.

hexachloroethane.   (perchloroethane; carbon trichloride; carbon hexachloride).
CAS: 67-72-1. Cl_3CCCl_3.
Properties: Colorless crystals: camphorlike odor. D 2.091, mp 185C, bp sublimes at 185C. Soluble in alcohol and ether; insoluble in water.
Hazard: Toxic by ingestion and inhalation, strong irritant, absorbed by skin. Possible carcinogen.
Use: Organic synthesis, retarding agent in fermentation, camphor substitute in nitrocellulose, pyrotechnics and smoke devices, solvent, explosives.

hexachlorohexane.   Several isomers that are environmentally persistent with varying pharmacological properties.

hexachloromethylcarbonate.   (triphosgene).
$(OCCl_3)_2CO$.
Properties: White crystals; odor similar to that of phosgene. D approximately 2, bp 205–206C (partial decomposition), mp 78–79C. Decomposed by hot water and alkali hydroxides. Only slowly acted upon by cold water. Soluble in alcohol, benzene, and ether.
Derivation: Chlorination of dimethyl carbonate exposed to direct sunlight.
Hazard: Strong irritant to eyes and skin.
Use: Lachrymator.

hexachloromethyl ether.
$O(CCl_3)_2$.
Properties: Liquid; phosgenelike odor. D 1.538 (18C), bp 98C (partial decomposition).
Derivation: Chlorination of dichloromethyl ether.
Hazard: Strong irritant to eyes and skin.

hexachloronaphthalene.
CAS: 1335-87-1. $C_{10}H_2Cl_6$.
Properties: White solid.
Hazard: Toxic by inhalation, strong irritant, absorbed by skin.

1,2,3,4,6,7-hexachloronaphthalene.
CAS: 103426-96-6. $C_{10}H_2Cl_6$.
Hazard: A reproductive hazard.

1,2,3,5,6,7-hexachloronaphthalene.
CAS: 103426-97-7. $C_{10}H_2Br_6$.
Hazard: A poison by ingestion.

hexachlorophene.   (2,2'-methylene-bis-(3,4,6-trichlorophenol); bis-(3,5,6-trichloro-2-hydroxyphenyl)methane).
CAS: 70-30-4. $(C_6HCl_3OH)_2CH_2$.
Properties: White, free-flowing powder; odorless. Mp 161–167C. Soluble in acetone, alcohol, ether, and chloroform; insoluble in water.
Derivation: Condensation of 3,4,5-trichlorophenol with formaldehyde in the presence of sulfuric acid.
Hazard: FDA prohibits use unless prescribed by a physician. Questionable carcinogen.
Use: Topical anti-infective (restricted), germicidal soaps, veterinary medicine.

hexachloro-2-propane.   See hexachloroacetone.

hexachloropropylene.   (hexachloropropene; perchloropropylene).
CAS: 1888-71-7. $CCl_3CCl:CCl_2$.
Properties: Water-white liquid. Bp 210C. Insoluble in water; miscible with alcohol, ether, and chlorinated compounds.
Use: Solvent, plasticizer, hydraulic fluid.

hexacontane.
CAS: 7667-80-3. $C_{60}H_{122}$.
High-molecular-weight hydrocarbon.
Properties: Waxy solid. Mp 101C.
Hazard: Combustible.

hexacosanoic acid.   See cerotic acid.

***n*-hexadecane.**   (cetane).
CAS: 544-76-3. $C_{16}H_{34}$.
Properties: Colorless liquid. D 0.77335 (20/4C), flash p 200F (93C), bp 286.5C, mp 18.14C, refr index 1.43435 (20C). Soluble in alcohol, acetone, and ether; insoluble in water, autoign temp 401F (205C).
Grade: Technical, ASTM.
Hazard: Combustible.

Use: Solvent, organic intermediate, ignition standard for diesel fuels.
See cetane number.

hexadecanoic acid.   See palmitic acid.

1-hexadecanol.   See cetyl alcohol.

hexadecanoyl chloride.   See palmitoyl chloride.

(z)-7-hexadecenal.
CAS: 56797-40-1. $C_{16}H_{30}O$.
Hazard: Low toxicity by ingestion.
Source: Natural product.

(z)-9-hexadecenal.
CAS: 56219-04-6. $C_{16}H_{30}O$.
Hazard: Low toxicity by ingestion.
Source: Natural product.

1-hexadecene.   (cetene; α-hexadecylene).
CAS: 629-73-2. $CH_3(CH_2)_{13}CH:CH_2$.
Properties: Colorless liquid. Mp 4C, bp 274C, flash p 200F (93C), d 0.784 (15/4C), refr index 1.441 (20C). Insoluble in water; soluble in alcohol, ether, petroleum, and coal tar solvents.
Derivation: Treatment of cetyl alcohol with phosphorus pentoxide.
Grade: 95% purity.
Hazard: Combustible.
Use: Organic synthesis.

cis-9-hexadecenoic acid.   See palmitoleic acid.

(z)-11-hexadecenol.
CAS: 56683-54-6. $C_{16}H_{32}O$.
Hazard: Low toxicity by ingestion.
Source: Natural product.

(z)-7-hexadecen-1-ol acetate.
CAS: 23192-42-9. $C_{18}H_{34}O_2$.
Hazard: Moderately toxic by ingestion and skin contact. Low toxicity by inhalation.
Use: Agricultural chemical.

6-hexadecenolide.   See ambrettolide.

hexadecyl mercaptan.   See cetyl mercaptan.

tert-hexadecyl mercaptan.
CAS: 25360-09-2. $C_{16}H_{33}SH$.
Properties: Colorless liquid; unpleasant odor. Boiling range 121–149C (5 mm Hg), d 0.874 (60/60F), refr index 1.474 (20C), flash p 265F (129.4C).
Hazard: Combustible.
Use: Polymer modification.
See thiol.

hexadecyltrichlorosilane.
CAS: 5894-60-0. $C_{16}H_{33}SiCl_3$.
Properties: Colorless to yellow liquid. Bp 269C, d 0.996 (25/25C), refr index 1.4568 (25C), flash p 295F (146C).

Derivation: By Grignard reaction of silicon tetrachloride and hexadecylmagnesium chloride.
Grade: Technical.
Hazard: Strong irritant. Combustible. Evolves hydrogen chloride in the presence of moisture.
Use: Intermediate for silicones.

hexadecyltrimethylammonium   bromide.
See cetyl trimethylammonium bromide.

hexadecyltrimethylammonium chloride.
CAS: 112-02-7. $C_{19}H_{42}N \cdot Cl$.
Hazard: A poison by ingestion and skin contact.

1,4-hexadiene.
CAS: 42296-74-2. $H_2C:CHCH_2HC:CHCH_3$.
Properties: Colorless liquid. D 0.6996 (20/4C), bp 64C (745 mm Hg), refr index 1.4162 (20C), flash p −6F (−21.1C). Insoluble in water.
Derivation: Reaction between ethylene and butadiene with a special catalyst.
Hazard: Highly flammable, explosive limits in air 2–6.1%.
Use: As third monomer in EPDM synthetic elastomers.

2,4-hexadienedial, (e,e)-.
CAS: 18409-46-6. $C_6H_6O_2$.
Hazard: A poison. Possible carcinogen.

2,4-hexadienoic acid.   See sorbic acid.

1,5-hexadiyne.   See dipropargyl.

hexa-2-ethylbutoxydisiloxane.
$[(CH_3CH(C_2H_5)CH_2CH_2O)_3Si]_2O$.
Properties: Colorless oil. Bp 195C (0.2 mm Hg).
Derivation: Reaction of silicon tetrachloride, 2-ethylbutanol, and water.
Hazard: Combustible.
Use: Aircraft hydraulic fluid.

hexaethyldistannoxane.
CAS: 1112-63-6. $C_{12}H_{30}OSn_2$.
Properties: Air-sensitive liquid. D: 1.377 at 20C, bp: 272C.
Hazard: A poison.
Use: Drug.

hexaethyldistannthiane.
CAS: 994-50-3. $C_{12}H_{30}SSn_2$.
Properties: A liquid. D: 1.431 at 20C, bp: 187–188C at 20 mm.
Hazard: A poison.
Use: Drug.

hexaethyl tetraphosphate.   (HETP).
CAS: 757-58-4.
A mixture of ethyl phosphates and ethyl pyrophosphates (TEPP).

Properties: Yellow liquid. D 1.26–1.28 (25/4C), fp −90C, refr index 1.427, decomposes at high temperatures. Soluble or miscible with water and many organic solvents except kerosene; hydrolyzes in low concentration; hygroscopic.
Hazard: Toxic by ingestion, inhalation, and skin absorption; cholinesterase inhibitor.
Use: Contact insecticide.
Note: Hexaethyl tetraphosphate and compressed gas mixture not accepted by air or by passenger rail. For details, consult regulations.

hexafluoroacetone.
CAS: 684-16-2.   CF$_3$COCF$_3$.
Properties: Colorless, hygroscopic, highly reactive gas. Bp −27C, fp −122C, liquid density 1.33 (25C), minimum purity 95%.
Hazard: Toxic by inhalation and skin absorption. Reacts vigorously with water and other substances, releasing considerable heat. Nonflammable.
Use: Intermediate in organic synthesis.

hexafluoroacetone bisphenol a.
CAS: 1478-61-1.   C$_{15}$H$_{10}$F$_6$O$_2$.
Hazard: Moderately toxic by ingestion.

hexafluorobenzene.
CAS: 392-56-3.   C$_6$F$_6$.
Properties: Liquid. Bp 80.26C, mp 5.2C, d 1.613.
Hazard: Toxic by inhalation. Combustible.
Use: Chemical intermediate, solvent in NMR spectroscopy.

hexafluorocalcitriol.
CAS: 83805-11-2.   C$_{27}$H$_{38}$F$_6$O$_3$.
Hazard: A poison by ingestion.

hexafluorodiphenylolpropane.   See hexafluoroacetone bisphenol a.

hexafluoroethane.   (fluorocarbon 116).
CAS: 76-16-4.   CF$_3$CH$_3$.
Properties: A gas. Bp −78.2C, d 1.59. Insoluble in water; slightly soluble in alcohol. One of the most stable of all organic compounds.
Grade: 99.6% pure.
Use: Dielectric and coolant, aerosol propellant, refrigerant.

2-(1,1,2,3,3,3-hexafluoro-2-(heptafluoropropoxy)propoxy)-2,3,3,3-tetrafluoropropanoic acid.
CAS: 13252-14-7.   C$_9$HF$_{17}$O$_4$.
Hazard: Moderately toxic by inhalation and skin contact. A moderate eye irritant.

hexafluorophosphoric acid.
CAS: 16940-81-1.   HPF$_6$.
Properties: (65% solution) Colorless, fuming liquid. D 1.81, mp 31C (6H$_2$O). Stable in neutral and alkaline solutions.
Hazard: Strong irritant to tissue.

Use: Metal cleaners, electrolytic or chemical polishing agents for the formation of protective coatings for metal surfaces, and as a catalyst.

1,1,1,2,3,3-hexafluoropropane.
CAS: 431-63-0.   C$_3$H$_2$F$_6$.
Hazard: Low toxicity by inhalation. A reproductive hazard.

1,1,1,3,3,3-hexafluoropropane.
CAS: 690-39-1.   C$_3$H$_2$F$_6$.
Hazard: Moderately toxic by inhalation route.

hexafluoropropylene.   (perfluoropropene).
CAS: 116-15-4.   CF$_3$CF:CF$_2$.
Properties: Gas. Fp −156C, bp −29C, d 1.583 (−40/4C).

hexafluoropropylene epoxide.
(HFPO).
CF$_2$CF$_2$CF$_2$O.
Derivation: Oxidation of hexafluoropropylene with alkaline hydrogen peroxide at approximately −30C.
Use: Monomer for HFPO polymers that are heat resistant to 410C. Noncombustible.
See Freon E; "Krytox" *[Du Pont]*.

hexafluorosilicic acid.   See fluosilicic acid.

hexaglycerol.   See trimethylolpropane, polyglycerol.

hexahydric alcohol.   See mannitol, sorbitol, and dulcitol.

hexahydroaniline.   See cyclohexylamine.

hexahydrobenzene.   See cyclohexane.

hexahydrobenzoic   acid.   (cyclohexanecarboxylic acid [a naphthenic acid]).
CAS: 98-89-5.   C$_6$H$_{11}$COOH.
Properties: Colorless monoclinic prisms. Mp 31C, bp 233C, d 1.048 (15/4C), refr index 1.4561 (33.8C). Slightly soluble in water; soluble in alcohol and ether.
Use: Paint and varnish driers, dry-cleaning soaps, lubricating oils, stabilizer for rubber.

hexahydrocresol.   See methylcyclohexanol.

hexahydro-4,7-methanoindandimethanol.
See tricyclodecanedimethanol.

hexahydromethylphenol.   See methylcyclohexanol.

hexahydrophenol.   See cyclohexanol.

hexahydrophthalic   anhydride.   (1,2-cyclohexanedicarboxylic anhydride).
C$_6$H$_{10}$(CO)$_2$O.

Properties: Clear, colorless, viscous liquid that becomes a glassy solid at 35–36C. Bp 158C (17 mm Hg), d 1.19 (40C). Miscible with benzene, toluene, acetone, carbon tetrachloride, chloroform, ethanol, and ethyl acetate; slightly soluble in petroleum ether.
Hazard: Toxic by inhalation, strong irritant to eyes and skin.
Use: Intermediate for alkyds, plasticizers, insect repellents, and rust inhibitors; hardener in epoxy resins.

hexahydropyridine.   See piperidine.

hexahydrotoluene.   See methylcyclohexane.

hexahydro-1,3,5-trinitro-*sym*-triazine.   See cyclonite.

hexahydroxycyclohexane.   See inositol.

hexahydroxylene.   See dimethylcyclohexane.

hexaisobutylditin.
CAS: 3750-18-3. $C_{24}H_{54}Sn_2$.
Hazard: A poison.

hexakis(methoxymethyl)melamine.

NC(NR)NC(NR)NC(NR) where R = $(CH_2OCH_3)_2$.

Use: Cross-linking agent for alkyds, epoxies, cellulosics, and vinyls.

***n*-hexaldehyde.**   (caproic aldehyde).
CAS: 66-25-1. $CH_3(CH_2)_4CHO$.
Properties: Colorless liquid; sharp aldehyde odor. D 0.8156 (20/20C), bp 128.6C, vap press 10.5 mm Hg (20C), flash p 90F (32.2C) (OC), bulk d 6.9 wt/gal (20C), fp −56.3C. Immiscible with water.
Grade: Technical.
Hazard: Flammable, moderate fire risk.
Use: Organic synthesis of plasticizers, rubber chemicals, dyes, synthetic resins, insecticides.

Hexalin.   Cyclohexanol (usually shipped with 2.25% methanol as antifreeze).
Hazard: Toxic by ingestion.

hexametapol.   See hempa.

hexamethonium   chloride.   (hexamethylenebis(trimethylammonium) chloride).
CAS: 60-25-3.
$(CH_3)_3NCl(CH_2)_6NCl(CH_3)_3$.
Properties: White, crystalline, hygroscopic powder; faint odor. Mp 289–292C (decomposes). Very soluble in water; soluble in alcohol, methanol, and *n*-propanol; insoluble in chloroform and ether. Available commercially as unhydrated form or as dihydrate.
Use: Medicine (antihypertensive).

hexamethylbenzene.
$C_{12}H_{18}$ or $C_6(CH_3)_6$.
Properties: Colorless plates. Bp 265C, mp 165.5C. Soluble in alcohol; insoluble in water.
Hazard: Combustible.

hexamethylcyclotrisilazane.
CAS: 1009-93-4. $C_6H_{21}N_3Si_3$.
Hazard: Moderately toxic by ingestion.

hexamethyldiaminoisopropanol   diiodide.
See propiodal.

hexamethyldisilazane.   (HMDS).
CAS: 999-97-3. $(CH_3)_3SiNHSi(CH_3)_3$.
Properties: Liquid. D 0.77 (25C), refr index 1.4057 (25C), bp 125C, flash p 77F (25C). Soluble in acetone, benzene, ethyl ether, heptane, and perchloroethylene; reactive with methanol and water.
Grade: 99% min.
Hazard: Flammable, moderate fire risk.
Use: Chemical intermediate, chromatographic packings.

hexamethyldistannane.   See hexamethylditin.

hexamethylditin.
CAS: 661-69-8. $C_6H_{18}Sn_2$.
Properties: Crystals from pet ether. Mp: 28C, bp: 182C.
Hazard: A poison by ingestion.
Use: Agricultural chemical.

hexamethylene.   See cyclohexane.

hexamethylenediamine.   (1,6-diaminohexane; 1,6-hexanediamine).
CAS: 124-09-4. $H_2N(CH_2)_6NH_2$.
Properties: Colorless leaflets. Mp 39–42C, bp 205C. Soluble in water; slightly soluble in alcohol and benzene.
Derivation: (1) Reaction of adipic acid and ammonia (catalytic vapor phase) to yield adiponitrile, followed by liquid-phase catalytic hydrogenation. (2) Chlorination of butadiene, followed by reaction with sodium cyanide (cuprous chloride catalyst) to 1,4-dicyanobutylene and hydrogenation.
Hazard: Toxic by ingestion, strong irritant to tissue. Combustible.
Use: Formation of high polymers, e.g., nylon 66.

hexamethylenediamine   carbamate.   See "Diak" [*Vanderbilt*].

hexamethylene diisocyanate.
CAS: 822-06-0. $OCN(CH_2)_6NCO$.
Properties: Liquid. D 1.04 (25/15.5C), flash p 284F (140C).
Hazard: Combustible.
Use: Chemical intermediate.

hexamethylene glycol. (1,6-hexanediol).
CAS: 629-11-8. $CH_2OH(CH_2)_4CH_2OH$.
Properties: Crystalline needles. Mp 42C, bp 210C, refr index 1.457, d 0.953 (50C), flash p 130C (266F).
Derivation: Reduction of adipic acid ester with copper chromite catalyst.
Hazard: Toxic by ingestion.
Use: Solvent, intermediate for high polymers (nylon, polyesters), coupling agent, coil coating.

hexamethyleneimine.
CAS: 111-49-9. $C_6H_{12}NH$ (cyclic).
Properties: Clear, colorless liquid; ammonia-like odor. Bp 138C, fp −37C, d 0.8799 (20/4C).
Hazard: Toxic by ingestion, strong irritant to tissue.
Use: Intermediate for pharmaceutical, agricultural, and rubber chemicals.

hexamethylenetetramine. (methenamine; HMTA; aminoform; hexamine, erroneously hexamethyleneamine).
CAS: 100-97-0. $(CH_2)_6N_4$.
A heterocyclic fused ring structure.

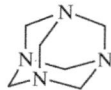

Properties: White, crystalline powder or colorless lustrous crystals; practically odorless. D 1.27 (25C); soluble in water, alcohol, and chloroform; insoluble in ether; sublimes approximately 200C; partly decomposes.
Derivation: Action of ammonia on formaldehyde.
Grade: Technical, NF (as methenamine).
Hazard: Skin irritant. Flammable, dangerous fire risk.
Use: Curing of phenolformaldehyde and resorcinol-formaldehyde resins, rubber-to-textile adhesives, protein modifier, organic synthesis, pharmaceuticals, ingredient of highly explosive cyclonite, fuel tablets, rubber accelerator, fungicide, corrosion inhibitor, shrink-proofing textiles, antibacterial.

hexamethylmelamine. See hemel.

hexamethylpararosaniline chloride. See methyl violet.

hexamethylphosphoramide. See hempa.

hexamethylphosphoric triamide. See hempa.

hexamethyltetracosahexaene. See squalene.

hexamethyltetracosane. See squalane.

hexamine. See hexamethylenetetramine.

hexanaphthene. See cyclohexane.

n-hexane.
CAS: 110-54-3. $CH_3(CH_2)_4CH_3$.
Properties: Colorless, volatile liquid; faint odor. D 0.65937 (20/4), bp 68.742C, fp −95C, refr index 1.37486 (20C), flash p −9F (−22.7C), autoign temp 500F (260C). Soluble in alcohol, acetone, and ether; insoluble in water.
Derivation: By fractional distillation from petroleum (molecular sieve process).
Grade: 85%, 95%, 99%, spectro, research, and nanograde.
Hazard: Flammable, dangerous fire risk.
Use: Solvent, especially for vegetable oils; low-temperature thermometers; calibrations; polymerization reaction medium; paint diluent; alcohol denaturant.

2-hexanecarboxylic acid.
CAS: 4536-23-6. $C_7H_{14}O_2$.
Hazard: A reproductive hazard.

hexanedioic acid. See adipic acid.

1,6-hexanediol. See hexamethylene glycol.

hexanedione-2,5. See acetonyl acetone.

1,2,3,4,5,6-hexanehexol. (manna sugar; mannite; mannitol; d-mannitol; NCI-C50362; osmitrol).
CAS: 69-65-8. $C_6H_{14}O_6$.
A sugar alcohol with six hydroxyl groups. It is an osmotic diuretic that is metabolically inert in humans and occurs naturally, as a sugar or sugar alcohol, in fruits and vegetables. It elevates blood plasma osmolality, resulting in enhanced flow of water from tissues, including the brain and cerebrospinal fluid, into interstitial fluid and plasma.
Properties: White, crystalline powder or needles; odorless; soluble in water; slightly soluble in lower alcohols and amines; almost insoluble in organic solvents.
Hazard: Mildly toxic; mutagen.
Use: Medically to reduce cerebral edema, elevated intracranial pressure, and cerebrospinal fluid volume and pressure; used for the promotion of diuresis before irreversible renal failure becomes established; the promotion of urinary excretion of toxic substances; as an antiglaucoma agent; as a renal function diagnostic aid.

1,2,6-hexanetriol.
CAS: 106-69-4. $HOCH_2CH(OH)CH_2CH_2CH_2CH_2OH$.
Properties: Water-white liquid. D 1.1063, sets to glass at approximately −20C (fp under controlled conditions 32.8C), bp (178C) 5 mm Hg, flash p 380F (193C). Miscible with water.
Hazard: Combustible.
Use: Alkyd and polyester resin intermediate, softener, moistening agent, and solvent.

hexanitrodiphenylamine.   (hexil; hexyl; hexite; dipicrylamine).
$(NO_2)_3C_6H_2NHC_6H_2(NO_2)_3$.
Properties: Yellow solid. Mp 238–244C, decomposes violently at higher temperatures. Insoluble in water and alcohol; soluble in alkalies and warm acetic or nitric acid.
Derivation: Nitration of diphenylamine, also from dinitrochlorobenzene.
Hazard: Explodes on shock or exposure to heat, dangerous.
Use: Booster explosive, analysis for potassium.

hexanitrodiphenyl sulfide.   (dipicryl sulfide).
$[(NO_2)_3C_6H_2]_2S$.
Properties: Golden-yellow leaflets. Mp 234C. Sparingly soluble in alcohol and ether; more soluble in glacial acetic acid and acetone.
Derivation: Interaction of picryl chloride and sodium thiosulfate in alcohol solution in the presence of magnesium carbonate.
Hazard: Explodes on shock or exposure to heat; dangerous.
Use: High explosive.

hexanitromannite.   See mannitol hexanitrate.

hexaoctyldistannoxane.   See 1,1,1,3,3,3-hexaoctyldistannoxane.

1,1,1,3,3,3-hexaoctyldistannoxane.
CAS: 2787-93-1. $C_{48}H_{102}OSn_2$.
Hazard: Moderately toxic.
Use: Drug.

hexaoctyldistannthiane.
CAS: 13413-18-8. $C_{48}H_{102}SSn_2$.
Hazard: Moderately toxic.
Use: Drug.

hexanoic acid.   Legal label name for caproic acid.

1-hexanol.   See hexyl alcohol.

2-hexanone.   See methyl-*n*-butyl ketone.

3-hexanone.   (ethyl propyl ketone).
CAS: 589-38-8. $C_2H_5CO(CH_2)_2CH_3$.
Properties: Colorless liquid. Bp 124C, d 0.813 (22C), flash p 95F (35C) (OC).
Hazard: Toxic by ingestion and inhalation, strong irritant. Flammable, moderate fire risk.
Use: Solvent.

_n_-(_n_-hexanoyl)aniline.
CAS: 621-15-8. $C_{12}H_{17}NO$.
Hazard: Moderately toxic by ingestion.

hexanoyl chloride.
CAS: 142-61-0. $CH_3(CH_2)_4COCl$.

Properties: Colorless liquid. Bp 151–153C, refr index 1.4867 (20C). Decomposed by water and alcohol; soluble in ether and chloroform.
Hazard: Combustible.
Use: Chemical intermediate.

hexaphenyldisilane.
$[(C_6H_5)_3Si]_2$.
Properties: White powder, mp 352C.
Derivation: Sodium condensation of triphenylchlorosilane.
Use: High-temperature applications.

"Hexaphos" [ICL].   TM for a glassy phosphate of high molecular weight having superior water-softening properties.
Use: Water-softening, boiler-scale control, component of cleansers, laundry mixes, dishwashing compounds, pitch control in pulp industry, prevention of lime soap deposits in textile operations.

hexapropyldistannthiane.
CAS: 7328-05-4. $C_{18}H_{42}SSn_2$.
Hazard: A poison.
Use: Drug.

hexatriacontane.
CAS: 630-06-8. $C_{36}H_{74}$.
Properties: Waxy solid. D 0.797, mp 75C.
Hazard: Combustible.
See paraffin wax.

1-hexene.   (hexylene).
CAS: 592-41-6. $CH_3CH_2CH_2CH_2CH:CH_2$.
Properties: Colorless liquid. D 0.6734 (20/4C), bp 63.55C, fp −139.8C, refr index 1.3876 (20C), flash p −15F (−26.1C). Insoluble in water; soluble in alcohol.
Grade: 95%, 99%, research.
Hazard: Irritant. Highly flammable, dangerous fire risk.
Use: Synthesis of flavors, perfumes, dyes, resins; polymer modifier.

2-hexene.
CAS: 592-43-8. $CH_3CH_2CH_2CH:CHCH_3$.
Properties: (Mixed *cis* and *trans* isomers.) Colorless liquid. Bp 68C, fp −146C, refr index 1.3948 (20C), d 0.686 (15.5/15.5C), flash p -5F (20.5C). Insoluble in water; soluble in alcohol.
Grade: 95%, 99%.
Hazard: Highly flammable, dangerous fire risk.
Use: Chemical intermediate.

5-hexene-2-one.   See allylacetone.

hexenol.   (3-hexen-1-ol; leaf alcohol).
CAS: 111-28-4. $C_6H_{11}OH$.
Properties: Liquid; odor of green leaves. Bp 156C, refr index 1.438, d 0.85.
Occurrence: Grasses, leaves, herbs, tea, etc.
Hazard: Combustible.
Use: Odorant in perfumery.

hexestrol. (*p,p′*-(1,2-diethylethylene)diphenol).
CAS: 84-16-2. $HOC_6H_4CH(C_2H_5)CH(C_2H_5)C_6H_4OH$.
A nonsteroid, synthetic estrogen.
Properties: Odorless, white, crystalline powder; mp 185–188C; soluble in ether, acetone, alcohol, and methanol; practically insoluble in water; sensitive to light.
Derivation: From anethole, by reaction of diacetyl peroxide on *p*-methoxy-*n*-propylbenzene.
Use: Medicine (estrogenic hormone).

hexetidine. (amino-1,3-bis[β-ethylhexyl]-5-methylhexahydropyrimidine).
CAS: 141-94-6. $C_{21}H_{45}N_3$.
Properties: Liquid. D 0.860-0.875 (25/25C), bp 172–176C (1 mm Hg), refr index 1.460–1.466 (25C). Soluble in methanol, benzene, and petroleum ether; insoluble in water.
Grade: Technical, NF.
Hazard: Combustible.
Use: Fungicide, bactericide, algicide, antistatic agent for synthetics, insect repellent, medicine (antifungal agent).

hexil.   See hexanitrodiphenyl amine.

hexite.   See hexanitrodiphenyl amine.

hexobarbital. (*n*-methyl-5-cyclohexenyl-5-methylbarbituric acid).
CAS: 56-29-1. $C_{12}H_{16}N_2O_3$.
Properties: White crystals. Mp 145–147C.
Use: Medicine (sedative).
See barbiturate.

Hexogen Octoate.   A series of paint driers made with odorless solvents, essentially solutions of metallic salts of 3-ethylhexoic acid.
Available forms: Supplied in a variety of high metal concentrations including calcium 4%, calcium 5%, cobalt 6%, lead 24%, manganese 6%, iron 6%, and zinc 8%.

hexoic acid.   See caproic acid.

hexokinase.   An enzyme that catalyzes the formation of adenosine diphosphate and hexose-6-phosphate from adenosine triphosphate and glucose or fructose.
Use: Biochemical research.

hexone.   See methyl isobutyl ketone.

hexosamine.   A derivative of hexose in which the hydroxyl group is replaced by an amino acid at the C-2 position.

hexose.   Aldehyde or ketone sugar containing six carbons.

hexyl.   (1) The straight-chain group C_6H_{13}; (2) Hexanitrodiphenylamine.

hexyl acetate.
CAS: 142-92-7. $CH_3COOC_6H_{13}$.
Properties: Colorless liquid; sweet ester odor. Bp 169.2C, d 0.890. Insoluble in water; very soluble in alcohol and ether.
Derivation: From primary and *sec*-hexyl alcohols.
Hazard: Combustible.
Use: Solvent for cellulose esters and other resins, spray base.

sec-**hexyl acetate.**   See methyl amyl acetate.

hexyl alcohol.   (1-hexanol; amyl carbinol).
CAS: 111-27-3. $CH_3(CH_2)_4CH_2OH$.
Properties: Colorless liquid. D 0.8186, fp −51.6C, bp 157.2C, bulk d 6.8 lb/gal (20C), refr index 1.1469 (25C), flash p 149F (65C) (TOC), autoign temp 559F (292C). Slightly soluble in water; soluble in alcohol and ether.
Derivation: (1) By reduction of ethyl caproate, (2) from olefins by the Oxo process.
Grade: Technical (90–99%), purified (99.8%).
Hazard: Combustible.
Use: Pharmaceuticals (introduction of hexyl group into hyponotics, antiseptics, perfume esters, etc.), solvent, plasticizer, intermediate for textile and leather finishing agents.

n-**hexylamine.**
CAS: 111-26-2. $CH_3(CH_2)_5NH_2$.
Properties: Water-white liquid; amine odor. Boiling range 126–132C, fp −21C, d 0.767 (20/20C), refr index 1.419 (20C), flash p 85F (29.4C) (OC). Slightly soluble in water.
Hazard: Toxic by ingestion, inhalation, and skin absorption. Flammable, moderate fire risk.

n-**hexyl bromide.**   (1-bromohexane).
CAS: 111-25-1. $CH_3(CH_2)_5Br$.
Properties: Colorless to slightly yellow liquid. D 1.165 (20/20C), bp 155.5C. Soluble in alcohol, esters, and ethers; insoluble in water.
Grade: 96–98% pure.
Use: Intermediate, for introduction of hexyl group.

hexyl-2-butenoate.
CAS: 19089-92-0. $C_{10}H_{18}O_2$.
Properties: Colorless liquid; fruity odor. D: 0.880, refr index: 1.428–1.449. Sol in alc, fixed oils; insol in water, propylene glycol.
Hazard: Low toxicity by ingestion and skin contact. A skin irritant.
Use: Food additive.

n-**hexyl 2-butenoate.**   See hexyl-2-butenoate.

n-**hexyl "Carbitol" [Union].**   TM for diethylene glycol monohexyl ether.

n*-hexyl "Cellosolve" *[Union].   TM for ethylene glycol monohexyl ether.

hexyl cinnamaldehyde.
CAS: 101-86-0. $C_6H_{13}C(CHO):CHC_6H_5$.
Properties: Pale-yellow liquid; jasminelike odor, particularly on dilution. D 0.953–0.959 (25C), refr index 1.5480–1.5520 (20C). Soluble in most fixed oils and in mineral oil; insoluble in glycerol and in propylene glycol.
Grade: FCC.
Hazard: Combustible.
Use: Flavoring agent.

hexyl crotonate.   See hexyl-2-butenoate.

hexylene.   See 1-hexene.

hexylene glycol.   (4-methyl-2,4-pentanediol).
CAS: 107-41-5. $(CH_3)_2COHCH_2CH_2OCH_3$.
Properties: Colorless liquid; nearly odorless. D 0.9216 (20/4C), bp 198.3C, refr index 1.4276 (20C), flash p 200F (OC) (93C); bulk d 7.69 lb/gal. Miscible with water, hydrocarbons, and fatty acids.
Hazard: Toxic by ingestion and inhalation; irritant to skin, eyes, and mucous membranes. Combustible.
Use: Hydraulic brake fluids, printing inks, coupling agent and penetrant for textiles, fuel and lubricant additive, emulsifying agent, inhibitor of ice formation in carburetors, cosmetics.
See 1,6-hexanediol.

***n*-hexyl ether.**
CAS: 112-58-3. $C_6H_{13}OC_6H_{13}$.
Properties: Colorless liquid; characteristic odor. D 0.7942 (20/20C), bulk d 6.6 lb/gal (20C), fp −43.0C, viscosity 1.68 cP (20C), flash p 170F (76.6C), autoign temp 369F (187C). Very slightly soluble in water.
Hazard: Combustible.
Use: Extraction processes, manufacture of collodion, photographic film, and smokeless powder.

hexylic acid.   See caproic acid.

hexyl isovalerate.
CAS: 10632-13-0. $C_{11}H_{22}O_2$.
Properties: Colorless liquid; pungent, fruity odor. D: 0.853, refr index: 1.417. Sol in alc, fixed oils; insol in water.
Use: Food additive.

hexyl mercaptan.
CAS: 111-31-9. $C_6H_{13}SH$.
Properties: Colorless liquid; unpleasant odor. Bp 149–150C (768 mm Hg), d 0.8450 (20/4C), refr index 1.4492 (20C).
Grade: 95% min purity.
Hazard: Combustible.
Use: Intermediate, synthetic rubber processing.
See thiol.

hexyl methacrylate.
CAS: 142-09-6. $C_6H_{13}OOCC(CH_3):CH_2$.
Properties: Liquid. D 0.88, bp 67–85C (8 mm Hg).
Hazard: Combustible.
Use: Monomer for plastics, molding powder, solvent coatings, adhesives, oil additives; emulsions for textile, leather, and paper finishing.

2-(2-(2-(hexyloxy)ethoxy)ethoxy)ethanol.
CAS: 25961-89-1. $C_{12}H_{26}O_4$.
Hazard: A poison by ingestion.
Use: Agricultural chemical.

((hexyloxy)sulfinyl)methylcarbamic   acid-2,3-dihydro-2,2-dimethyl-7-benzofuranyl ester.
CAS: 77248-43-2. $C_{18}H_{27}NO_5S$.
Hazard: A poison by ingestion.
Use: Agricultural chemical.

***p-tert*-hexylphenol.**
$C_6H_{13}C_6H_4OH$.
Properties: Water-white liquid; faint phenol odor. D 0.986 (20/20C), boiling range 155–165C, refr index 1.520 (20C), flash p 285F (140C).
Hazard: Combustible.
Use: Organic synthesis, preparation of resinous condensation products.

hexylresorcinol.   (1,3-dihydroxy-4-hexylbenzene).
CAS: 136-77-6. $C_6H_{13}C_6H_{13}(OH)_2$.

Properties: Yellow, viscous liquid that solidifies on standing, or needle-shaped crystals; faint fatty odor; astringent taste. Mp 62–67C, bp 333C. Slightly soluble in water; freely soluble in alcohol, glycerol, and vegetable oils.
Grade: NF.
Hazard: Irritant to respiratory tract and skin, concentrated solutions are vesicant.
Use: Medicine (topical antiseptic).

hexyltrichlorosilane.
CAS: 928-65-4. $C_6H_{13}SiCl_3$.
Properties: Colorless liquid; sharp penetrating odor. Fumes strongly in moist air.
Hazard: Toxic by ingestion and inhalation, strong irritant. Combustible.
Use: Chemical intermediate.

1-hexyne.   (butyl acetylene).
CAS: 693-02-7. $C_4H_9C{\equiv}CH$.
Properties: Water-white liquid; characteristic odor. D 0.7152 (20/4C), refr index 1.3990 (20C), bp 71.4C, fp −132C.
Hazard: Probably flammable.

hexynol.   (1-hexyn-3-ol).
CAS: 105-31-7. $CH_3(CH_2)_2CH_2OC{\equiv}CH$.
Properties: Light-yellow liquid; strong odor. Bp 142C, d 0.882 (20/20C). Slightly soluble in water; miscible with most hydrocarbons, chlorinated solvents, ketones, alcohols, and glycols.
Hazard: Toxic by ingestion and inhalation, absorbed by skin. Probably flammable.
Use: Corrosion inhibitor against mineral acids, high-temperature oil-well-acidizing inhibitor.

Heyrovsky, Jaroslav.   (1890–1967). A Czechoslovakian physiochemist who won the Nobel Prize for chemistry in 1959. He is known for work in electrochemistry. He developed polarographic and oscillo-polarographic methods. Although his Ph.D. was from the University of Prague, he later studied in London.

Hf.   Symbol for hafnium.

HFPO.   See hexafluoropropylene epoxide.

HFG.   See glucagon.

Hg.   Symbol for mercury (Latin: hydrargyrum).

HGP.   Human Genome Project.

"HGR" [Calgon].   TM for an ion exchange resin used in water-treating and chemical process applications, strong acid cation exchange resin, 10% divinylbenzene cross-linked.

HHDN.   Abbreviation for hexachlorohexahydrodimethanonaphthalene.
See aldrin.

HHMI.   Howard Hughes Medical Institute.

hidden maximum system.   See peritectic system.

hiding power.   See opacity.

hi-flash naphtha.   See naphtha (2a).

high-energy compound.   A compound that on hydrolysis undergoes a large decrease in free energy under standard conditions.

highly enriched uranium.   (heu). Uranium that contains a high percentage (>80%) of the U235 isotope. It is ordinarily considered to be weapons-grade material.

high-fructose corn syrup.   A common sweetener in foodstuffs made from corn.
Properties: Water-white to light yellow viscous liquid; sweet taste. Misc with water.
Use: Food additive.

highly conserved sequence.   DNA sequence that is very similar across several different types of organisms.
See gene; mutation.

high-performance liquid chromatography.   (HPLC).
A type of chromatography using relatively high pressures and small diameter column packings to achieve sharp and highly reproducible elution profiles. Used to be called high-pressure liquid chromatography.

high polymer.   See polymer, high.

high-throughput sequencing.   A fast method of determining the order of bases in DNA.
See sequencing.

high vacuum distillation.   See molecular distillation.

Hilbert-Johnson reaction.   Reaction of 2,4-dialkoxypyrimidines with halogenoses to yield pyrimidine nucleosides.

Hildebrand, Joel Henry.   (1881–1983).
One of the most distinguished American chemists and teachers. Born in New Jersey, he obtained his doctorate in chemistry and physics from the University of Pennsylvania. After studying abroad under Nernst and van't Hoff, he became professor of Chemistry at the University of California, Berkeley, in 1913 where he remained until retirement. He made many important contributions to physical chemistry, particularly in the area of nonelectrolyte solutions; his treatise on the subject is a recognized classic and his textbook on the principles of chemistry established a new standard of excellence. He also made important contributions to the thermodynamics of vaporization of liquids. He proposed the use of helium in deep-sea diving equipment, which has become accepted practice. A gifted teacher and lecturer, he continued his constructive research to the end of his long life. He was unusually active in outdoor sports such as swimming, skiing, and hiking. Among his numerous awards were the Nichols and William Gibbs medal and the Priestley medal.

hindered isocyanate.   See isocyanate generator.

hindrance.   See steric hindrance.

Hinsberg oxindole synthesis.   Formation of oxindoles from secondary aryl amines and sodium bisulfite addition compound of glyoxal; primary aryl amines give glycine or glycinamide derivatives.

Hinsberg reaction.   Reaction of primary and secondary amines with sulfonyl halides to give sulfonamides; because the products from primary amines are soluble in alkali and those from secondary amines are not, and since tertiary amines do not react, this method is useful for the separation and identification of amines.

Hinsberg sulfone synthesis. Formation of sulfonylquinol derivatives by addition of quinones to cold, dilute aqueous solutions of sulfinic acids.

Hinsberg synthesis of thiophene derivatives. Formation of thiophene carboxylic acids from α-diketones and dialkyl thiodiacetate.

Hinshelwood, Sir Cyril Norman. (1897–1967).
An English chemist who won the Nobel Prize for chemistry in 1956 along with Semenov, a Russian. He authored "The Kinetics of Chemical Change," "The Structure of Physical Chemistry," and many other journal articles. His work clarified inorganic and organic reactions. He was educated at Oxford before he began lecturing and research.

Hippuran. A brand of iodohippurate sodium, a water-soluble X-ray contrast medium.

hippuric acid. (benzaminoacetic acid; benzoylaminoacetic acid; benzoylglycocoll; benzoylglycin).
CAS: 495-69-2. $C_6H_5CONHCH_2COOH$.
Properties: Colorless crystals. D 1.371 (20C), mp 188C, decomposes on further heating. Soluble in hot water, alcohol, and ether.
Use: Organic synthesis and medicine.

His. Abbreviation for histidine.

"Hi-Sil" [PPG]. TM for a group of hydrated, amorphous silicas used as reinforcing pigments in elastomers, as fillers and brightening agents in paper and paints, and as flow conditioners.
Use: In adhesives, caulks, and sealants.

histaminase. An enzyme occurring in the animal digestive system; it converts histidine to histamine.

histamine. (4-aminoethylglyoxaline; 4-(2-aminoethyl)imidazole; 4-imidazole ethylamine).
CAS: 51-45-6. $NH_2CH_2CH_2C_3H_3N_2$.

$$HC\!=\!\!=\!\!CCH_2CH_2NH_2$$
$$\underset{N\diagdown\underset{\underset{H}{|}}{C}\diagup NH}{|\qquad\qquad|}$$

Properties: White crystals. Mp 83–84C, bp 209–210C (18 mm Hg). Soluble in water; slightly soluble in alcohol. A product of the degradation of histidine, histamine occurs in animal and human body tissues and is liberated by injury to the tissue or whenever a protein is decomposed by putrefactive bacteria.

Use: (As hydrochloride or phosphate) Medicine (diagnostic aid).
See antihistamine.

histidine. (α-amino-β-imidazolepropionic acid).
CAS: 71-00-1. $HOOCCH(NH_2)CH_2C_3H_3N_2$.
An amino acid essential for rats. It is found naturally in the L(-) form.

$$H\!-\!C\!=\!\!=\!CCH_2CH\!-\!COOH$$
$$\underset{CH}{\underset{\diagdown\diagup}{\underset{N}{|}\qquad\underset{NH}{|}\quad\underset{NH_2}{|}}}$$

Properties: Colorless crystals. DL-histidine, mp 285–286C with decomposition; D(+)-histidine, mp 287–288C; L(−)-histidine, mp 277C with decomposition. Soluble in water; insoluble in alcohol and ether. Shows optical activity. Available commercially as L(+)-histidine hydrochloride and as the free base.
Derivation: From blood corpuscles, organic synthesis.
Use: Medicine, feed additive, biochemical research, dietary supplement.

histidine monohydrochloride.
mf: $C_6H_9N_3O_2\cdot HCl\cdot H_2O$.
Properties: White needles, plates, or crystalline powder; slightly bitter taste. Decomp 250C. Sol in water; insol in alc, ether.
Use: Food additive.

histochemistry. A branch of biochemistry devoted to the study of the chemical composition and structure of animal and plant tissues. It involves the use of microscopic, X-ray diffraction, and radioactive tracer techniques in examining cellular composition and structure of bones, blood, muscle, and other animal and vegetable tissues. It is also applied to the study of the action of herbicides, defoliants, etc.
See cytochemistry.

histone. The family of five basic proteins (H1, H2A, H2B, H3, and H4) that associate tightly with DNA in the chromosomes of all eukaryotic cells. They contain varying amounts of the amino acids lysine, arginine, cysteine, and glycine.

Histoplasma. Fungi that are commonly found in soil usually associated with bird or bat droppings. Inhalation of *Histoplasma capsulatum* spores may result in histoplasmosis, a lung infection similar to pneumonia.
Hazard: Pathogen.

history, chemistry. See Appendix II A–E.

Hitec. A eutectic mixture composed of sodium nitrite, sodium nitrate, and potassium nitrate.

Use: Heat-transfer medium for both heating and cooling operations in the range of 149–537C, such as maintaining reactor temperature, high-temperature distillation, and preheating of reactants.

"HiTEC" Antioxidants *[Albemarle].* TM for a series of gasoline antioxidants based on phenols (chiefly di-*tert*-butyl phenol). They inhibit formation of gum and peroxides in gasoline and the formation of decomposed products of jet fuels in storage. Also used for steam turbine and industrial oils and to retard decomposition of antiknock compounds in gasoline.

HiTEC Antioxidants. A series of gasoline antioxidants based on phenols (chiefly di-*tert*-butyl phenol). They inhibit formation of gum and peroxides in gasoline and the formation of decomposed products of jet fuels in storage. Also used for steam turbine and industrial oils and to retard decomposition of antiknock compounds in gasoline.

HMAF. See 6-(hydroxymethyl)acylfulvene.

HMDS. Abbreviation for hexamethyldisilazane.

HMF black. Abbreviation for high-modulus furnace black.
See carbon black.

HMM. Abbreviation for hexamethylmelamine.
See hemel.

HMPA. Abbreviation for hexamethylphosphoramide.
See hempa.

HMTA. Abbreviation for hexamethylenetetramine.

HNM. Abbreviation for hexanitromannite.
See mannitol hexanitrate.

hn1. (2-chloro-n-(2-chloroethyl)-*n*-ethylethanamine; 2,2′-dichlorotriethylamine; bis(2-chloroethyl)ethylamine; ethylbis(2-chloroethyl)-amine).
$C_6H_{13}Cl_2N$.
Properties: Volatile, water-insoluble, liquid nitrogen mustard, slightly fishy odor.
Hazard: Extreme irritant; deadly vesicant.

hnRNA. Heterogeneous nuclear RNA; refers collectively to the variety of RNAs found in the nucleus, including primary transcripts, partially processed RNAs, and snRNA. The term hnRNA is often used just for the unprocessed primary transcripts, however.

Ho. Symbol for holmium.

hob. A hardened steel master die used to make multiple mold cavities by forcing it into soft steel or beryllium-copper blanks.

Hoch-Campbell aziridine synthesis. Formation of aziridines by treatment of ketoximes with Grignard reagents and subsequent hydrolysis of the organometallic complex.

Hodgkin, Dorothy C. (1910–1994). An Egyptian-born chemist who was recipient of the Nobel Prize for chemistry in 1964. Her work involved determining the structure of vitamin B_{12}, cholesterol iodide, and the antibiotic penicillin by using X-ray crystallographic analysis. She was educated at Oxford and Cambridge.

Hoffman, Roald. (1937–). A Polish-born chemist who won the Nobel Prize for chemistry with Fukui in 1981. His work involved applying the theories of quantum mechanics to predict the course of chemical reactions.

Hofmann, August Whilhelm. (1818–1892). A German organic chemist who studied under Liebig. While professor of chemistry at the Royal College of Chemistry in London, he did original research on coal tar derivatives that later led him into a study of organic dyes. Perkin, who first synthesized the dye mauveine in England, was a student of Hofmann. When the latter returned to Germany, he continued his work in the field of dyes, which became the basis of German leadership in synthetic dye manufacture that continued until World War I.

Hofmann degradation. Formation of an olefin and a tertiary amine by pyrolysis of a quaternary ammonium hydroxide; useful for the preparation of some cyclic olefins and for opening nitrogen-containing ring compounds.

Hofmann isonitrile synthesis. Formation of isonitriles by the reaction of primary amines with chloroform in the presence of an alkali; the odor of the isocyanide is a test for a primary amine.

Hofmann-Loffler-Freytag reaction. Formation of pyrrolidines or piperidines by thermal or photochemical decomposition of protonated *N*-haloamines.

Hofmann-Martius rearrangement. Thermal conversion of *N*-alkylaniline hydrohalides to *o*- and *p*-alkylanilines.

Hofmann rule. When a quaternary ammonium hydroxide containing different primary alkyl radicals is decomposed, the least substituted olefin is formed preferentially.

Hofmann-Sand reaction. Olefin mercuration with mercuric salts (halides, acetates, nitrates, or sulfates) in aqueous solution. In alcoholic solutions, the accelerated reaction produces alkoxylalkyl compounds.

Hofmann's reaction.   Reaction used for preparation of a primary amine from an amide by treatment with a halogen (bromine usually) and caustic soda. The resulting amine has one fewer carbon atom than the amide used.

Hofmann's Violet.   (triethylrosaniline hydrochloride; CI 42530).
$C_{26}H_{32}N_3HCl$.
Water-soluble green powder.
Use: Dye for inks and textiles, biological stain.

hog.   A large enclosed chamber equipped either with rotating knives or heavy hammers by which wood is disintegrated to a uniform degree of fineness.

hole.   In semiconductor terminology, a hole is an energy deficit in a crystalline lattice due to (1) electrons ejected from unsatisfied covalent bonds at sites where an atom is missing, i.e., a vacancy, or (2) electrons supplied by atoms of impurities in the crystal, e.g., arsenic or boron. The free electrons from these sources move through the crystal, leaving positively charged energy deficits that are considered to move as they become alternately filled and vacated by electrons, creating a flow of positive electricity.
See semiconductor.

Holliday intermediate.   An intermediate in genetic recombination in which two double-stranded DNA molecules are joined by virtue of a reciprocal crossover involving one strand of each molecule.

holmium.
Ho.
Metallic element of atomic number 67, group IIIB of the periodic table, one of the rare-earth elements of the yttrium subgroup, aw 164.9303, valence of 3, no stable isotopes.
See rare-earth metals.
Properties: Crystalline solid with metallic luster. D 8.803, mp 1470C, bp 2720C. Reacts slowly with water; soluble in dilute acids. Has one of the highest nuclear moments of any rare earth. Important magnetic and electrical properties.
Occurrence: In gadolinite and monazite.
Derivation: Reduction of the fluoride by calcium.
Grade: Lumps, ingots, bulk sponge, powder. Highest purity is nuclear grade 99.9+%.
Use: Getter in vacuum tubes, research in electrochemistry, spectroscopy.

holmium chloride.
CAS: 10138-62-2. $HoCl_3$.

Properties: Bright-yellow solid. Mp 718C, bp 1500C. Soluble in water.

holmium fluoride.
CAS: 13760-78-6. HoF_3.
Properties: Bright-yellow solid. Mp 1143C, bp above 2200C. Insoluble in water.

holmium oxide.   (holmia).
CAS: 12055-62-8. Ho_2O_3.
Properties: Light-yellow solid. Slightly hygroscopic; soluble in inorganic acids.
Grade: 98–99%.
Use: Refractories, special catalyst.

holocellulose.   The entire water-insoluble carbohydrate fraction of wood (60–80%). It is composed of α-cellulose and hemicellulose; it contains hexosan and pentosan polymers and varies widely in degree of crystallinity.

holoenzyme.   A catalytically active enzyme including all necessary subunits, prosthetic groups, and cofactors.

holography.   Production of a unique three-dimensional image on photographic film by means of an interference pattern created by a laser beam that is split by a mirrorlike device. The beam is divided in such a way that one portion is reflected from the subject while the other forms a direct image; the resulting superimposition results in an unusual three-dimensional effect.
Use: The technique has a number of practical applications in airplane flight control and missile guidance systems; other uses will undoubtedly develop.

holopulping.   A method for making paper pulp without use of sulfur compounds that has been proposed as an eventual replacement for the kraft process (sodium sulfate). More selective delignification of the wood fibers is obtained by alkaline oxidation of extremely thin (0.03 inch) wood chips at low temperature and pressure, followed by solubilization of the lignin fraction. Holopulping has other advantages over the kraft process: (1) A 65–80% carbohydrate yield compared to 45–50% for kraft. (2) Holopulp may be used for a dense paper such as glassine or for a bulky board. Its use in tissue and printing grades offers improved strength. (3) Low temperatures and atmospheric pressure. (Kraft pulping is carried out at 170C and under pressure.) Readily adaptable to continuous operation and automatic control. (4) Air pollution is greatly reduced because the organic materials are burned and few odorous compounds are formed. Stream pollution is minimized by countercurrent washing. The remaining calcium sludge waste is easily disposed of without harmful effects.

holothurin. Steroid glucoside (saponin) having antibiotic properties, extracted from the sea cucumber. It is reported to have suppressed growth of tumors in mice.

holozoic. Pertaining to or denoting organisms that feed entirely on living organisms or on other complex, organic matter, following which the material is digested, absorbed, and assimilated.

homoaromoline.
CAS: 17132-74-0.   mf: $C_{37}H_{40}N_2O_6$.
Hazard: A poison.

homatropine.
CAS: 87-00-3. $C_{16}H_{21}NO_3$.
An alkaloid.
Properties: White crystals. Mp 95.5C. Slightly soluble in water.
Derivation: Condensation of tropine and mandelic acid.
Hazard: Toxic by ingestion and inhalation.
Use: Medicine (usually in the form of its salts).

homeobox. A short stretch of nucleotides whose base sequence is virtually identical in all the genes that contain it. It has been found in many organisms from fruit flies to human beings. In the fruit fly, a homeobox appears to determine when particular groups of genes are expressed during development.

homeobox geme. A short stretch of nucleotides whose bas sequence is virtually identical in all the genes that contain it. Found in many organisms from fruit flies to human beings.

homeodomain protein. A protein with a DNA-binding motif of 60 amino acids that is encoded by a homeobox.

homocytotropic antibody. (reaginic antibody).
An antibody that has an affinity for cells of the same or closely related species of animal. It combines with specific antigen, thereby triggering the release of pharmacological mediators of anaphylaxis from the cells to which it attaches.

homeostasis. The maintenance of a dynamic steady state by regulatory mechanisms that compensate for changes in circumstances. For example, maintaining body temperature whether in a 20C or a 4C environment.

homeotic genes. Genes originally defined as regulating the development of the pattern of segments in the Drosophila body plan. Similar genes are now known to occur in most vertebrates.

homo-. A prefix meaning the same or similar; usually designating a homolog of a compound, differing in formula from the latter by an increase of CH_2.
See homologous series.

homoaromoline.
CAS: 17132-74-0. $C_{37}H_{40}N_2O_6$.
Hazard: A poison.

homocyclic. A ring compound containing only one kind of atom in the ring structure, e.g., benzene.
See heterocyclic.

homogeneous. (Latin, "the same kind"). This term, in its strict sense, describes the chemical constitution of a compound or element. A compound is homogeneous since it is composed of one and only one group of atoms represented by a formula. For example, pure water is homogeneous because it contains no other substance than is indicated by its formula, H_2O. Homogeneity is a characteristic property of compounds and elements (collectively called substances) as opposed to mixtures. The term is often loosely used to describe a mixture or solution composed of two or more compounds or elements that are uniformly dispersed in each other. Actually, no solution or mixture can be homogeneous; the situation is more accurately described by the phrase "uniformly dispersed." Thus, so-called homogenized milk is not truly homogeneous; it is a mixture in which the fat particles have been mechanically reduced to a size that permits uniform dispersion and consequent stability.
See mixture; compound; heterogeneous; substance.

homogeneous catalysis. See catalysis, homogeneous.

homogeneous reaction. A chemical reaction in which the reacting substances are in the same phase of matter, i.e., solid, liquid, or gaseous.
See catalysis, homogeneous.

homogenization. A mechanical process for reducing the size of the fat particles of an emulsion (usually milk) to uniform size, thus creating a colloidal system that is unaffected by gravity. The original diameter of the fat particles (6–10 microns) is reduced to 1–2 microns, with an increase in total surface area of 4–6 times. This is done by passing the milk through a homogenizer (or colloid mill), a machine having small channels, under a pressure of 2000–2500 psi at a speed of approximately 700 ft/sec. This operation not only brings about a permanently stable system but also changes the properties of the milk in respect to taste, color, and the chemical nature of the protective coating on the fat particles. It also increases its sensitivity to light and its tendency to foam. The forces involved are shear, impingement, distention, and cavitation.
See homogeneous; colloid mill.

homolog. A member of a chromosome pair in diploid organisms, or a gene that has the same origin and functions in two or more species.

homologies. Similarities in DNA or protein sequences between individuals of the same species or among different species.

homologous chromosomes. A pair of chromosomes containing the same linear gene sequences, each derived from one parent.

homologous genetic recombination. Recombination between two DNA molecules of similar sequence, occurring in all cells; occurs during meiosis and mitosis in eukaryotes.

homologous proteins. Proteins from different species having similar sequences and similar functions in each. For example, the many species that have hemoglobin, which transports oxygen in all.

homologous recombination. Swapping of DNA fragments between paired chromosomes.

homologous series. A series of organic compounds in which each successive member has one more CH_2 group in its molecule than the preceding member. For instance, CH_3OH (methanol), C_2H_5OH (ethanol), C_3H_7OH (propanol), etc., form a homologous series.

homomenthyl salicylate. (3,3,5-trimethylcyclohexyl salicylate). $(CH_3)_3C_6H_8OOCC_6H_4OH$. A homolog of menthyl salicylate.
Properties: Light-yellow oil; odorless. Neutral and nonirritating to the skin. Absorbs UV radiation in sunlight (2940–3200 Å). Insoluble in water; soluble in alcohol, chloroform, and ether.
Use: UV filter for antisunburn creams.

homomorphs. Molecules similar in size and shape. They need to have no other characteristics in common. Many properties of several homomorphs can be predicted by knowing properties of one.

homophthalic acid.
CAS: 89-51-0. $C_6H_4(CH_2COOH)COOH$.
Properties: Light-tan powder.
Use: Intermediate.

homopolar adsorption. See apolar adsorption.

homopolymer. A natural or synthetic high polymer derived from a single monomer. An example of a natural homopolymer is rubber hydrocarbon, whose monomer is isoprene; a synthetic homopolymer is typified by polychloroprene or polystyrene, whose monomers are, respectively, chloroprene and styrene.
See polyblend.

homopolysaccharide. A polysaccharide made up of only one type of monosaccharide unit.

homosalate.
CAS: 118-56-9. $C_{16}H_{22}O_3$.
Properties: Liquid. Bp 162C (4 mm Hg), d 1.05, refr index 1.51.
Use: Sunscreening agent.

o-**homosalicylic acid.** See cresotic acid.

homotropic enzyme. An allosteric enzyme that is modulated by its substrate.

homotropine.
CAS: 87-00-3. $C_{16}H_{21}NO_3$.
Hazard: Moderately toxic by ingestion and subcutaneous routes.

homoveratric acid. (3,4-dimethoxyphenylacetic acid).
CAS: 93-40-3. $(CH_3O)_2C_6H_3CH_2COOH$.
Properties: Crystals. Mp 94–101C. Very slightly soluble in water; soluble in most organic solvents.

homoveratrylamine. (3,4-dimethoxyphenylethylamine).
CAS: 93-40-3. $(CH_3O)_2C_6H_3(CH_2)_2NH_2$.
Properties: Colorless to pale-yellow liquid; slight vanilla odor. D 1.09 (25/25C), solidifies 15C, bp 295C (decomposes), refr index 1.5442–1.5452 (25C).

homozygote. Having identical alleles at one or more loci in homologous chromosome segments.

honey. A unique mixture of a number of low-molecular-weight sugars (except sucrose) but including invert sugar. It is considerably sweeter than glucose.
Use: A food and sweetener since the beginning of civilization, also has applications in medicine and tobacco processing.

"Honeywell Bulk Etch" *[Honeywell].* TM for liquid, wafer-thinning materials.
Use: For rapid and uniform removal of all remnants of individual back grind damage.

"Honeywell Polish Etch I & II" *[Honeywell].* TM for a revealing liquid for wafer-thinning materials.
Use: Used as a contour etch to delineate cracks and scratches.

"Honeywell Texture Etch" *[Honeywell].* TM for a liquid texture material.
Use: On wafer-thinning materials to increase surface area with uniform dimples to promote improved back metal tensile strength.

Hooker reaction. Oxidation of 2-hydroxy-3-alkyl-1,4-quinones with dilute alkaline permanganate with shortening of the alkyl side chain by a methylene group and simultaneous exchange of hydroxyl and alkyl or alkenyl group positions.

Hooke's law. When a load is applied to any elastic body so that the body is deformed or strained, then the resulting stress (the tendency of the body to resume its normal condition) is proportional to the strain. Stress is measured in units of force per unit area; strain is the extent of the deformation. For example, when a bar of metal is subjected to a stretching load, the extent of the increase in length of the bar is directly proportional to the force per unit area, i.e., to the stretching load or stress. In general, Hooke's law applies only up to a certain stress called the yield strength.

hoolamite.
Properties: Contains fuming sulfuric acid, iodine pentoside, and powdered pumice.
Use: A detector whose color changes from light gray to green in the presence of carbon monoxide.

hopcalite. A mixture of oxides of copper, cobalt, manganese, and silver.
Use: In gas masks, as a catalyst converting carbon monoxide to carbon dioxide.
Hazard: Not safe for a respirator when nitroparaffin vapors are present.

hop extract, modified.
Properties: An extract of hops by a variety of organic solvent extractions.
Use: Food additive.

hops. A plant of the genus *Humulus* widely grown in temperate climates.
Use: Beer making.
See brewing.

hops oil. A brownish yellow aromatic essential oil obtained from hops and used chiefly in flavoring cereal beverages.
Properties: From steam distillation of cones from female *Humulus lupulus* L. or *Humulus americanus* Nutt. (Fam. *Moraceae*). Yellow liquid; aromatic odor. D: 0.825–0.926, refr index: 1.470–1.494 at 20C. Sol in fixed oils and mineral oil; insol in glycerin and propylene glycol.
Use: Food additive.

horizon. See soil.

"Hormodin" [OHP]. TM for a formulation of indolebutyric acid.

17-hormoforin. See DHEA.

hormone. An organic compound (peptide, steroid) synthesized in small amounts by an endocrine tissue and carried in the blood to another tissue, where it acts as a messenger to regulate the function of the target tissue.
Hormones regulate such physiological processes as metabolism, growth, reproduction, molting, pigmentation, and osmotic balance. They are sometimes called "chemical messengers." Hormones produced by one species usually show similar action in other species. They vary widely in chemical nature, some are steroids—estrogen, progesterone, cortisone—while others are amino acids (e.g., thyroxine), polypeptides (vasopressin), low-molecular-weight proteins (insulin), and conjugated proteins. Amino acid and steroid hormones have been isolated and many (including insulin) have been synthesized and are manufactured for medical purposes. Other types are made directly from the endocrine organs of animals.

hormone, plant. See plant growth regulator.

hormone receptor. A protein in, or on, the surface of the target cells of a tissue that functions as a sensor for the hormone by binding the hormone and initiating a cellular response.

hormone therapy. Treatment that prevents certain cancer cells from getting the hormones they need to grow.

Hornstone. A zinc fluosilicate concrete hardener.

host-guest chemistry. Study of molecules that can enclose ions or atoms within their structure without the normal chemical bonding.
See cavitands; buckminsterfullerene.

host strain (bacterial). The bacterium used to harbor a plasmid. Typical host strains include HB101 (general purpose *E. coli* strain), JM101 and JM109 (suitable for growing M13 phages), XL1-Blue. Many host strains are available commercially.

hot. Slang for highly radioactive, e.g., hot laboratory.

hot-melt composition. See adhesive, hot-melt; sealant; asphalt; bitumen.

hot-short. See short.

Houben-Fischer synthesis. Formation of aromatic nitriles by basic hydrolysis of trichloromethyl aryl ketimines obtained by Hoesch synthesis (but not from dichloro- or monochloromethyl aryl ketimines). Acidic hydrolysis yields ketones.

Houben-Hoesch reaction. Synthesis of acylphenols from phenols or phenolic ethers by the action of organic nitriles in the presence of hydrochloric acid and aluminum chloride as catalyst.

Houdriflow process.   A moving-bed type of catalytic cracking in which the catalyst pellets move downward through a reactor concurrently with the feed. The catalyst is then separated and regenerated for further use.
See Thermofor process.

Houdry cracking process.   Decomposition of petroleum or heavy petroleum fractions into more useful lower-boiling materials by heating at 500C and 30 psi over a silica-alumina-manganese oxide catalyst.

housekeeping genes.   Those genes expressed in all cells because they provide functions needed for sustenance of all cell types.

household insecticide.   Any insecticide produced for use in the home.

household pesticide.   Any pesticide used in the home. In the United States alone, about 2.5 million human are thought to be affected annually by such substances; 70% of those affected are children under the age of 5 years.
  Hazard: Nausea, coughing, dyspnea, eye irritation, dizziness, weakness, blurred vision, muscle spasms, and convulsions; lipid soluble cumulative poisons.

"Howmet Castings" [Howmet].   TM for a wide variety of aluminum cast products.

HPA.   Abbreviation for hydroxypropyl acrylate.

HPC black.   Abbreviation for hard-processing channel black.
See carbon black.

HPLC.   Abbreviation of high-performance liquid chromatography.
See high-performance liquid chromatography.

HPMPC.   See (s)-1-(3-hydroxy-2-phosphonyl-methoxypropyl)cytosine.

HPRT locus.   (hypoxanthine guanine phosphoribosyltransferase locus).
  A gene locus that supports the incorporation of purines from the medium into cultured mammalian cells. These purines can then be converted into nucleic acids. A mutation at this locus prevents such uptake of toxic as well as normal purines. Consequently, the cultured can produce purines de novo and will grow.
  Use: The basis for mutagenicity tests in which cultured mammalians are exposed to toxic purines together with a test mutagen; if these grow, one may infer that a mutation in the hgprt locus has occurred.

HS.   (1) Symbol for hassium. (2) Abbreviation for hydroxylamine sulfate.

"HTH" [Haiyan].   TM for a high-test calcium hypochlorite product commercially available as a stable, water-soluble material in both granular and tablet forms, containing a minimum of 70% available chlorine as calcium hypochlorite.
  Use: Bleaching, sterilizing, oxidizing.

"HTH-15" [Haiyan].   TM for an all-purpose germicide, disinfectant, and stain remover. Contains 15% of available chlorine and yields sodium hypochlorite solutions directly when added to water.
  Use: Dairy and poultry farm sanitation, for sterilizing glasses and food utensils and general sanitation.

"HTH Sock-It" [Haiyan].   TM for a shock system for swimming pools.
  Use: To kill bacteria and control algae.

HTST.   Abbreviation for high-temperature short-time, refers to processes such as pasteurization, sterilization, etc.

HTU.   Abbreviation for height of a transfer unit: the height of a distillation column or fractionating tower in which unit separation is achieved by transfer from liquid to vapor or vice versa, of the materials being separated. Unit separation is defined by the differential equation that takes into account the varying concentrations along the column. HTU is also applied to extraction and other countercurrent separation processes.

"Huberfill 93 and 96" [Huber].   TM for sodium magnesium alumino silicate.
  Grade: Dry powder, slurry.
  Use: Print quality enhancements, paper fillers, carbonless copy intensifier, and titanium dioxide extension.

Huber, Robert.   (1937–). Awarded Nobel Prize for chemistry in 1988, along with Deisenhofer and Michel, for work that revealed the three-dimensional structure of closely linked proteins that are essential to photosynthesis. Doctorate awarded in 1963 by Technical University of Munich, Germany.

Huber's reagent.   An aqueous solution of ammonium molybdate and potassium ferrocyanide used for detecting free mineral acid. With the exception of boric acid and arsenic trioxide, free mineral acids produce a reddish-brown precipitate or a turbidity with the reagent.

Hubl's reagent. (1) 50-g iodine dissolved in 1 L of 95% alcohol; (2) 60-g mercuric chloride dissolved in 1 L of alcohol; (3) Make up an iodine monochloride solution from (1) and (2). Add an excess to a known weight of the fat or oil dissolved in chloroform. The excess of iodine chloride can be estimated by the potassium iodide and thiosulfate method. By running a blank test, the amount of iodide absorbed can be estimated.
Use: Determination of iodine values of oils and fats.

Hudson isorotation rules. For anomeric (α and β) sugars, Hudson's isorotation rule states that (1) the rotation of carbon 1 in many sugar derivatives is affected in only a minor degree by changes in the structure of the rest of the molecule and (2) changes in the structure of carbon 1 affect in only a minor degree the rotation of the remainder of the molecule. Another way of stating the rule is to say that the rotation of any aldose derivative is the algebraic sum of A and B, where A is the contribution of the anomeric center and B is the contribution of the rest of the molecule.

Hudson lactone rule. The value of the rotation of aldonic acid lactones is decisively affected by the configuration of that carbon atom whose hydroxyl group is engaged in the cyclization. If, in the normal Fischer projection formula, the lactone ring is written on the right, the lactone is dextrorotatory; if it is written on the left, the lactone is levorotatory.

human artificial chromosome (HAC). A vector used to hold large DNA fragments.

human gene therapy. Insertion of normal DNA directly into cells to correct a genetic defect.

human genome. The full collection of genes needed to produce a human being.

Human Genome Initiative. Collective name for several projects begun in 1986 to (1) create an ordered set of DNA segments from known chromosomal locations; (2) develop new computational methods for analyzing genetic map and DNA sequence data; and (3) develop new techniques and instruments for detecting and analyzing DNA. This DOE initiative is now known as the Human Genome Program. The national effort, led by DOE and NIH (now including the National Human Genome Research Institute), is known as the Human Genome Project.

Human Genome Project. An international research effort (led in the United States by the National Institutes of Health and the Department of Energy) aimed at identifying and ordering every base in the human genome.

humectant. A substance having affinity for water with stabilizing action on the water content of a material. A humectant keeps within a narrow range the moisture content caused by humidity fluctuations. Example, glycerol.
Use: Tobacco, baked products, dentifrices.

humic acid. A brown, polymeric constituent of soils, lignite, and peat; it contains the brownish-black pigment melanin. It is soluble in bases, but insoluble in mineral acids and alcohols. It is not a well-defined compound but a mixture of polymers containing aromatic and heterocyclic structures, carboxyl groups, and nitrogen. An excellent chelating agent, important in the exchange of cations in soils. It is a natural stream pollutant and is thought to be capable of triggering the "red tide" phenomenon due to microorganisms in seawater. Detectable to 0.1 ppm in water.
Use: Drilling fluids, printing inks, plant growth.

humidity, absolute. The pounds of water vapor per pound of dry air in an air-water-vapor mixture.

humidity indicator. A cobalt salt (e.g., cobaltous chloride) that changes color as the humidity of the environment changes. Cobaltous compounds are pink when hydrated and greenish-blue when anhydrous.

humidity, relative. The percentage relation between the actual amount of water vapor in a given volume of air at a definite temperature and the maximum amount of water vapor that would be present if the air were saturated with water vapor at that temperature.

humoral antibody. A free molecule that is liberated into blood and tissue fluids where they may coat bacteria or combine with bacterial toxins, thus promoting phagocytosis.

"Humulin" [Lilly]. TM for synthetic insulin. It is the first recombinant DNA product to be made commercially; approved by FDA.

humus. The organic component of soils containing humic acid, fulvic acid, and humin. It is formed by the decay of leaves, wood, and other vegetable matter.
Use: Top soil additive in horticulture, golf courses, and truck gardens; available in 100-lb paper bags and bulk shipments.
See soil (1).

hungarian red. A red protein stain.
Use: To visualize bloody friction ridge detail.

Hunsdiecker reaction. Synthesis of organic halides by thermal decarboxylation of silver salts of carboxylic acids in the presence of halogens.

Hunter process. Production of titanium by reduction with sodium in an atmosphere of argon or helium.

Huskey. Specialty lubricants with multiple applications.

HVP. See acid hydrolyzed proteins.

Hyalure.
CAS: 9067-32-7.
Hyaluronic acid.
Use: In orthopedic, ophthalmic, wound care, drug delivery, and surgical applications.

hyaluronic acid. A polymer of acetylglycosamine, $C_8H_{15}NO_6$, and glucuronic acid occurring as alternate units with a high molecular weight. Found in vitreous humor (of the eye), synovial fluid, pathologic joints, group A and C hemolytic streptococci, and skin. It appears to bind water in the interstitial spaces, forming a gellike substance that holds the cells together. Its solutions are highly viscous. The polymeric structure is broken down by the enzyme hyaluronidase.
See "Actimoist" [Active].

hyaluronidase. (hyaluronate glycanohydrolase; hyaluronic lyase; hyaluronoglucuronidase; hyaluronate lyase).
CAS: 9001-54-1.
Any of a class of soluble enzymes produced in mammalian testes that catalyze cleavage of glycosidic bonds of oartucular mucopolysaccharides. As a component of venoms, it reduces the viscosity of connective tissues and promotes tissue penetration of the venom.

hyaluronoglucosaminidase.
CAS: 37326-33-0.
A hyaluronidase that attacks the 1,4-linkages in hyaluronates.

"Hyamine" [Lonza]. TM for quaternary-ammonium-type bactericides, algicides, and fungicides, supplied as water-soluble crystals or aqueous solutions.

Hyatt, John Wesley. (1837–1920). Hyatt is generally credited as being the father of the plastics industry. In 1869, he and his brother patented a mixture of cellulose nitrate and camphor, which could be molded and hardened. Its first commercial use was for billiard balls. The TM "Celluloid" was the first ever applied to a synthetic plastic product; its flammability hazard limits its use.

hybrid. The offspring of genetically different parents.
See heterozygote.

hybridization. The reaction by which the pairing of complementary strands of nucleic acid occurs. DNA is usually double-stranded, and when the strands are separated they will rehybridize under the appropriate conditions. Hybrids can form between DNA–DNA, DNA—RNA, or RNA–RNA. They can form between a short strand and a long strand containing a region complementary to the short one. Imperfect hybrids can also form, but the more imperfect they are, the less stable they will be (and the less likely to form). To "anneal" two strands is the same as to "hybridize" them.

Hydan. Methionine hydroxy analog c. 90%.
Use: A source of methionine (an essential amino acid) for poultry, dog, and livestock feeds.

hydantoin. (glycolylurea).
CAS: 461-72-3. $NHCONHCOCH_2$.
Properties: White, odorless solid; crystallizing in needles. Mp 220C. Slightly soluble in water and ether; soluble in alcohols and solutions of alkali hydroxides.
Use: Intermediate in the synthesis of pharmaceuticals, textile lubricants, and certain high polymers, including epoxy resins.

hydnocarpic acid.
CAS: 459-67-6. $C_{16}H_{28}O_2$.
A component of chaulmoogra oil.

hydrabamine pencillin V. (hydrabamine phenoxymethylpenicillin).
Properties: A water-insoluble mixture of crystalline phenoxymethylpenicillin salts consisting chiefly of the salt of N,N'-bis(dehydroabietyl)ethylenediamine with smaller amounts of the salts of the dihydro and tetrahydro derivatives.
Use: Medicine (antibacterial).

hydracrylic acid, acrylate.
CAS: 24615-84-7. $C_6H_8O_4$.
Hazard: A severe skin irritant.

"Hydraid" [International Nutrition]. TM for a family of water-soluble organic polymers, some cationic, anionic, and nonionic of various molecular weights and coagulation properties.
Use: Paper and pulp mill retention, drainage, and clarification aids.

hydralazine hydrochloride. (1-hydrazinophthalazine hydrochloride).
CAS: 304-20-1. $C_8H_5N_2NHNH_2HCl$.
Properties: White, crystalline powder; odorless. Mp 270–280C (decomposes). Very slightly soluble in ether and alcohol; soluble in water, pH (2% solution) 3.5–4.5.
Grade: NF.
Use: Medicine (antihypertensive agent).

"Hydral"700 Series *[Huber Specialty].*
$Al_2O_3 \cdot 3H_2O$ or $Al(OH)_3$, of extremely fine, uniform particle size.
TM for several grades of hydrated aluminum oxides.
Properties: Fluffy, snow-white powders.
Use: As fillers in rubber, paper, plastics, adhesives, polishes, inks, paints, and cosmetics, and as a flame retarder in plastics.

Hydraphthal. A combination solvent and detergent for textile scouring.

hydrargaphen. (phenylmercury methylenedinaphthalenesulfonate).
CAS: 14235-86-0. $C_{33}H_{24}Hg_2O_6S_2$.
Properties: Extremely fine powder. Insoluble in water. Forms colloidal dispersions with strong adsorptive power in sodium or potassium dinaphthylmethane disulfonates.
Hazard: A poison.
Use: Biocide for protection of wool, leather, paints, and wood products.

hydrase. See hydrolase.

hydrate. See hydration.

hydrated aluminum oxide. See alumina trihydrate.

hydrated silica. See silicic acid.

hydration. (1) The reaction of molecules of water with a substance in which the H-OH bond is not split. The products of hydration are called hydrates, e.g.,
$CuSO_4 + 5H_2O \rightarrow CuSO_4 \cdot 5H_2O$.
A given compound often forms more than one hydrate; the hydration of sodium sulfate can give $Na_2SO_4 \cdot 10H_2O$ (decahydrate), $Na_2SO_4 \cdot 7H_2O$ (heptahydrate), and $Na_2SO_4 \cdot H_2O$ (monohydrate). In formulas of hydrates, the addition of the water molecules is conventionally indicated by a centered dot. The water is usually split off by heat, yielding the anhydrous compound.
See water of crystallization; gas hydrate.
(2) The strong affinity of water molecules for particles of dissolved or suspended substances that is the fundamental cause of electrolytic dissociation. Ions and other charged particles thus acquire a tightly held film of water, an effect that is important in the stabilization of colloidal solutions. The phenomenon is also called solvation. The term "hydration" is used in the paper industry to describe the combination of water with wood pulp in the beater, as a result of which fiber-to-fiber adhesion is increased by hydrogen bonding.
See solvation.

hydratropic aldehyde dimethyl acetal. See 2-phenylpropionaldehyde dimethyl acetal.

hydraulic. (1) Descriptive of a machine or operation in which a liquid is used to exert or transfer pressure, e.g., hydraulic press, hydraulic fracturing. The liquid is usually water, but it may also be of higher viscosity such as a heavy oil or glycol-type lubricant, as in brake fluid. (2) Descriptive of a material that hardens on addition of water, e.g., hydraulic cement.

hydraulic barking. Removal of bark from logs by impingement of a stream of water delivered from one or more nozzles at a pressure of 1200–1400 psi. Several types of machines are used, the best known being the Hansel barker.

hydraulic cement. See cement, hydraulic.

hydraulic fluid. A liquid or mixture of liquids designed to transfer pressure from one point to another in a system on the basis of Pascal's law, i.e., pressure on a confined liquid is transmitted equally in all directions. For industrial use, such fluids are based on paraffinic and cycloparaffinic petroleum fractions, usually with added antioxidant and viscosity index improvers. Flame-resistant types include additives such as phosphate esters or emulsions of water and ethylene glycol. The brake fluids used in autos are composed of (1) a lubricant (polypropylene glycol of 1000–2000 mw, a castor oil derivative, or a synthetic polymeric mixture of monobutyl ethers of oxyethylene and oxypropylene glycols); (2) a solvent blend (mixture of glycol ethers); and (3) additives for corrosive resistance, buffering, etc.; bp 375–550F. The composition and performance characteristics are specified by the Society of Automotive Engineers.

hydraulic fracturing. A method of enhanced recovery of natural gas and petroleum. An aqueous solution of a water-soluble gum (e.g., guar), in which coarse sand or sintered bauxite is suspended, is introduced through a well bore under extremely high pressure into the rock structure in which the gas or oil is entrained. This creates minute fissures (fractures) in the rock, which are held open by the suspended particles after the liquid has drained off. The hydrocarbon flows through these fissures to the well bore and is evacuated to a pipeline. The sand and bauxite are called "proppants" by petroleum engineers as they prevent the fissures from closing. Sand is used in shallower wells and bauxite in formations over 10,000 ft deep.
See chemical flooding.

hydraulic lime. See lime, hydraulic.

hydraulic press. A simple machine (the only one discovered since prehistoric times) that operates on Pascal's principle (1650): pressure applied to a unit area of a confined liquid is transmitted equally in all directions throughout the liquid. A hydraulic press comprises a large piston in an enclosed chamber; its top is attached to a platen that rests on the members of a metal frame when the press is open. Water (or oil) is pumped into the chamber through a valve; once it has been filled, whatever pressure per square inch is applied at the valve will be transmitted to *every* square inch of the piston and of the walls of the chamber as well. Thus, for a piston whose cross-sectional area is 100 sq inch, 10 psi at the valve will exert 1000 lb pressure on the bottom of the piston, causing it to rise and the press to close. The pressure on the object being pressed varies inversely with its area. Hydraulic presses exerting pressures up to 15 tons are used for shaping steel products. Less dramatic are those for molding rubber and plastics, compressing laminates, de-watering solids, and expressing vegetable oils. Some have up to a dozen platens (decks) for multiple product work. The same principle is used to activate plungers on injection-molding presses.

hydrazine.   (hydrazine base; hydrazine, anhydrous; diamine).
CAS: 302-01-2. H_2NNH_2.
Properties: Colorless, fuming, hygroscopic liquid. Mp 2.0C, bp 113.5C, d 1.004 (25/4C), bulk density 8.38 lb/gal, flash p 126F (52.2C) (OC), autoign temp 518F (270C). Miscible with water and alcohol; insoluble in chloroform and ether. Strong reducing agent and diacidic but weak base. Combustion of hydrazine is highly exothermic, yielding 148.6 kcal/mole; nitrogen and water are products.
Derivation: The preferred method is a two-step process: (1) reaction of sodium hypochlorite and ammonia to yield chloramine (NH_2Cl) and sodium hydroxide; (2) reaction of chloramine, ammonia, and sodium hydroxide to yield hydrazine, sodium chloride, and water. Noteworthy is the need to carry out the reactions in the presence of such colloidal materials as gelatin, glue, or starch to prevent unwanted side reactions that would reduce the yield of hydrazine. An older method utilized the reaction of sodium hypochlorite or calcium hypochlorite with urea.
Grade: To 99% pure.
Hazard: Severe explosion hazard when exposed to heat or by reaction with oxidizers. Toxic by ingestion, inhalation, and skin absorption; strong irritant to skin and eyes; a confirmed carcinogen.
Use: Reducing agent for many transition metals and some nonmetals (arsenic, selenium, tellurium), as well as uranium and plutonium; corrosion inhibitor in boiler feedwater and reactor cooling water; waste water treatment; electrolytic plating of metals on glass and plastics; nuclear fuel reprocessing; redox reactions; polymerization catalyst; shortstopping agent; fuel cells; blowing agent; scavenger for gases; drugs and agricultural chemicals (maleic hydrazide); component of high-energy fuels; rocket propellant.

hydrazine acid tartrate.   (hydrazine tartrate).
$N_2H_4 \cdot C_4H_6O_6$.
Properties: Colorless crystals. Mp 182–183C. Soluble in water.
Hazard: See hydrazine.

hydrazine, compd. with borane (1:1).
CAS: 14931-40-9. BH_7N_2.
Hazard: A poison by ingestion and inhalation. May be unstable at room temperature.

hydrazine dihydrochloride.
CAS: 5341-61-7. $N_2H_4 \cdot 2HCl$.
Properties: Colorless crystals. D 1.42, mp 198C (loses hydrogen chloride), bp 200C (decomposes). Soluble in water; slightly soluble in alcohol.

hydrazine hydrate.   (diamide hydrate).
CAS: 7803-57-8. $H_2NNH_2 \cdot H_2O$.
Properties: Colorless fuming liquid. Fp −51.7C, bp 119.4C, d 1.032, bulk d 8.61 lb/gal, flash p (OC) 163F. Miscible with water and alcohol; insoluble in chloroform and ether. Strong reducing agent; weak base. Combustible.
Hazard: See hydrazine.
Use: Chemical intermediate, catalyst, solvent for inorganic materials.

hydrazine monobromide.
$N_2H_4 \cdot HBr$.
Properties: White, crystalline flakes. Mp 81–87C, decomposes at approximately 190C. Soluble in water and lower alcohols; insoluble in most organic solvents.
Grade: 95%.
Use: Soldering flux.

hydrazine monochloride.
$N_2H_4 \cdot HCl$.
Properties: White, crystalline flakes. Mp 87–92C, decomposes at approximately 240C. Soluble in water (37g/100g H_2O at 20C); somewhat soluble in lower alcohols; insoluble in most organic solvents.

hydrazine nitrate.
CAS: 13464-97-6. $N_2H_4NO_3$.

Hazard.   Severe   explosion   risk.   Poison.
hydrazine perchlorate
$N_2H_4 \cdot HClO_4 \cdot 1/2H_2O$.
Properties: Solid. D 1.939, mp 137C, bp 145C. Decomposes in water; soluble in alcohol; insoluble in ether, benzene, chloroform, and carbon disulfide.
Hazard: Severe explosion risk.
Use: Rocket propellant.

hydrazine sulfate.   (diamine sulfate; diamidogen sulfate).
CAS: 10034-93-2. $NH_2NH_2 \cdot H_2SO_4$.
Properties: White, crystalline powder. D 1.37, mp 85C. Very soluble in hot water, solubility ratio in

cold water is 1:33, insoluble in alcohol, stable in storage, strong reducing agent.

Hazard: A carcinogen (OSHA).

Use: Manufacture of chemicals, condensation reactions, catalyst in making acetate fibers. Analysis of minerals, slags, and fluxes; determination of arsenic in metals; separation of polonium from tellurium; fungicide, germicide.

hydrazine tartrate.
$C_4H_{10}N_2O_6$.
Properties: Colorless crystals. Mp 182C.
Use: Deposition of metals, as in silvering mirrors.

2-hydrazino-*n,n*-dimethylethanamine.
CAS: 1754-57-0. $C_4H_{13}N_3$.
Hazard: A poison. Moderately toxic by ingestion and inhalation.

hydrazobenzene.   (1,2-diphenylhydrazine).
CAS: 122-66-7. $C_6H_5NHNHC_6H_5$.
Properties: Mw 184.24, mp 123–126C, d 1.158(16/4C).

hydrazoic acid.   (hydrogen azide).
CAS: 7782-79-8. HN_3.
Properties: Colorless, volatile liquid; obnoxious odor. Fp −80C, bp 37C. Soluble in water.
Derivation: Reaction of hydrazine and nitrous acid, or of nitrous oxide and sodium amide (with heat).
Hazard: Dangerous explosion risk when shocked or heated. Strong irritant to eyes and mucous membranes.

"Hydrex" *[Enegenx].*   (sodium magnesium aluminosilicate).
TM for titanium dioxide extenders.
Use: Print enhancement; imparting high brightness and opacity; as a paper filer; and carbonless copy intensifier.

hydride.   An inorganic compound of hydrogen with another element. Some are ionic and others are covalent. Hydrides may be either binary or complex; the latter are transition metal complexes, e.g., carbonyl hydrides and cyclopentadienyl hydrides. Most common are hydrides of sodium, lithium, aluminum, boron, etc.
Hazard: Flammable, dangerous fire risk, react violently with water and oxidizing agents. Irritant.
See lithium aluminum hydride; sodium borohydride.

hydriodic acid.
CAS: 10034-85-2.
Properties: Colorless or pale-yellow liquid (an aqueous solution of hydrogen iodide, which is a gas at room temperature), a constant boiling solution is formed (bp 127, d 1.7) containing 57% hydrogen iodide; strong acid and an active reducing agent.
Derivation: (1) By passing hydrogen with iodine vapor over warm platinum sponge and absorption

in water. (2) By the action of iodine on a solution of hydrogen sulfide.
Grade: Technical, 47%; NF, diluted, 10%.
Hazard: Strong irritant to eyes and skin.
Use: Preparation of iodine salts, organic preparations, analytical reagent, disinfectant, pharmaceuticals.
See hydrogen iodide.

hydroabietyl alcohol.   See dihydroabietyl alcohol.

hydrobiotite.   A natural ore of magnesium, iron, and aluminum silicate occurring in Montana. Source of verxite.

hydroboration.   The reaction of diboranes either with alkenes (olefins) to form trialkylboron compounds or with acetylene to yield alkenylboranes. Much research has been devoted to developing these reactions, the products of which are called organoboranes. They are useful in many complex organic syntheses, including prostaglandins and insect pheromones.
See borane; organoborane; carborane.

hydrobromic acid.
CAS: 10035-10-6.
Hydrogen bromide in aqueous solution.
Properties: Colorless or faintly yellow liquid consisting of an aqueous solution of hydrogen bromide, which is a gas at room temperature. Soluble in water and alcohol, a constant boiling solution is formed of d 1.49, containing 48% hydrogen bromide; bp at 700 mm Hg (122C), saturated solution contains 68.8% hydrogen bromide at 0C. Hydrobromic acid is a strong acid and sensitive to light. Noncombustible.
Derivation: By dissolving hydrogen bromide in water or by distilling from a mixture of sodium bromide and 50% sulfuric acid.
Grade: Technical 40%; medicinal 48%, 62%.
Hazard: Strong irritant to eyes and skin.
Use: Analytical chemistry, solvent for ore minerals, manufacture of inorganic and some alkyl bromides, alkylation catalyst.
See hydrogen bromide.

hydrocarbon.   An organic compound consisting exclusively of the elements carbon and hydrogen. Derived principally from petroleum, coal tar, and plant sources. Following is a resume of the principal types.

I. Aliphatic (straight-chain)
   (1) Paraffins (alkanes): generic formula C_nH_{2n+2}. Saturated, single bonds only.
   (2) Olefins: generic formula C_nH_{2n}.
       (a) Alkenes: unsaturated (one double bond).
       (b) Alkadienes: unsaturated (two double bonds) (butadiene).

(3) Acetylenes: generic formula C_nH_{2n} 2. Unsaturated (triple bond).

(4) Acyclic terpenes. Unsaturated (polymers of isoprene, C_5H_8).

Note: Some aliphatic compounds have branched chains in which the subchain also contains carbon atoms (isobutane); both chains are essentially straight.

II. Cyclic (closed ring)

(1) Alicyclic: Three or more carbon atoms in a ring structure with properties similar to those of aliphatics.

  (a) Cycloparaffins (naphthenes): Saturated compounds often having a boat or chair structure, e.g., cyclohexane, cyclopentane.

  (b) Cycloolefins: Unsaturated, two or more double bonds, e.g., cyclopentadiene (2), cyclooctatetraene (4).

  (c) Cycloacetylenes (cyclynes): Unsaturated (triple bond).

(2) Aromatic: Unsaturated, hexagonal ring structure (three double bonds), single rings and double or triple fused rings.

  (a) Benzene group (1 ring).

  (b) Naphthalene group (2 rings).

  (c) Anthracene group (3 rings).

(3) Cyclic terpenes: Monocyclic (dipentene); dicyclic (pinene).

Note: Olefinic (isoprenoid) hydrocarbons are produced by a number of plants, notably *Hevea braziliensis* (rubber), guayule, and various members of the Euphorbiaceae family. Current research on the latter group indicates that they could be used as a source of liquid fuels and chemical feedstocks by genetic modification of the plants and control of their molecular constitution. It is estimated that oil obtained by large-scale cultivation of such plants, which grow well in semiarid environments, could become economically competitive with petroleum within a few years.

See guayule; biomass; copaiba.

hydrocarbon gas streams. A hydrocarbon such as methane is contacted with a catalyst under moderate conditions of temperature and pressure and decomposed into carbon, which remains on the catalyst, and hydrogen, which is mechanically removed. The hydrogen produced is about 94% pure when the charge stock is methane.

hydrocarbon, halogenated. A hydrocarbon in which one or more of the hydrogen atoms has been replaced by fluorine, chlorine, bromine, or iodine. Examples: Carbon tetrachloride, chlorobenzene, chloroform, trifluoromethane. This greatly increases the anesthetic and narcotic action of aliphatic hydrocarbons. Many halogenated hydrocarbons are highly toxic; some may detonate on contact with barium. A number of the chlorinated types are used as insecticides.

See fluorocarbon; chlorofluorocarbon.

hydrocellulose.   See cellulose, hydrated.

hydrochlorbenzethylamine.
CAS: 17692-34-1. $C_{23}H_{31}ClN_2O_3$.
Hazard: Moderately toxic by ingestion.

hydrochloric acid.   (HCl).
CAS: 7647-01-0.
Hydrogen chloride in aqueous solution.
Properties: Colorless or slightly yellow, fuming, pungent liquid. Flash p none. A constant boiling acid containing 20% hydrochloric acid is formed. Hydrochloric acid is a strong, highly corrosive acid. The commercial "concentrated" or fuming acid contains 38% hydrochloric acid and has a d 1.19. Soluble in water, alcohol, and benzene. Noncombustible.
Derivation: Dissolving hydrogen chloride in water at various concentrations.
Grade: USP (35–38%), NF dilution (10%), technical (usually 18, 20, 22, 23 degrees Bé, corresponding to 28, 31, 35, 37% hydrogen chloride), FCC.
Hazard: Toxic by ingestion and inhalation, strong irritant to eyes and skin.
Use: Acidizing (activation) of petroleum wells, boiler scale removal, chemical intermediate, ore reduction, food processing (corn syrup, sodium glutamate), pickling and metal cleaning, industrial acidizing, general cleaning, e.g., of membrane in desalination plants, alcohol denaturant, laboratory reagent.
See hydrogen chloride.

hydrocinnamic acid.   (3-phenylpropionic acid).
CAS: 501-52-0. $C_6H_5CH_2CH_2COOH$.
Properties: Crystals with hyacinth–rose odor. Mp 46C, bp 280C. Soluble in hot water, alcohol, benzene, and ether.
Derivation: Reduction of cinnamic acid with sodium amalgam.
Use: Fixative for perfumes, flavoring.

hydrocinnamic alcohol.   See phenylpropyl alcohol.

hydrocinnamic aldehyde.   See phenylpropyl aldehyde.

hydrocinnamyl acetate.   See phenylpropyl acetate.

hydrocolloid. A hydrophilic colloidal material used largely in food products as emulsifying, thickening, and gelling agents. They readily absorb water, thus increasing viscosity and imparting smoothness and body texture to the product, even in concentrations of <1%. Natural types are plant exudates (gum arabic), seaweed extracts (agar), plant seed gums or mucilages (guar gum), cereal gums (starches), fermentation gums (dextran), and animal products (gelatin). Semisynthetic types are modified celluloses and modified starches. Completely synthetic types are also available, e.g., polyvinylpyrolidone. Most are carbohydrate polymers but a few such as gelatin and casein are proteins.

hydrocortisone. (17-hydroxycorticosterone; cortisol; hydrocortisone alcohol).
CAS: 50-23-7. $C_{21}H_{30}O_5$.
An adrenal cortical steroid hormone.
Properties: White, crystalline powder; odorless; bitter taste. Sensitive to light. Mp 212–220C with some decomposition. Soluble in sulfuric acid; partially soluble in alcohol and propylene glycol.
Derivation: Isolation from extracts of adrenal glands, synthesis from other steroids.
Grade: USP.
Use: Medicine (anti-inflammatory agent), also used as the acetate and sodium succinate salts.
See cortisone.

hydrocracking. The cracking of petroleum or its products in the presence of hydrogen. Special catalysts are used, e.g., platinum on a solid base of mixed silica and alumina or zinc chloride.
See hydrogenation; hydrogenolysis; hydroforming.

hydrocyanic acid. (prussic acid; hydrogen cyanide; formonitrile).
CAS: 74-90-8. HCN.
Properties: Water-white liquid at temperatures below 26.5C; faint odor of bitter almonds. Usual commercial material is 96–99% pure. D (Liquid) 0.688 (20/4C), (gas) 0.938 g/L, bp 26.5C, fp −13.3C, flash p 0F (−17.7C). Soluble in water. The solution is weakly acidic, sensitive to light. When not absolutely pure or stabilized, hydrogen cyanide polymerizes spontaneously with explosive violence. Miscible with water, alcohol, soluble in ether, autoign temp 1000F (537C).
Derivation: (1) By catalytically reacting ammonia and air with methane or natural gas. (2) By recovery from coke oven gases. (3) From bituminous coal and ammonia at 1250C. Hydrogen cyanide occurs naturally in some plants (almond, oleander).
Grade: Technical (96–98%), 2, 5, and 10% solutions. All grades usually contain a stabilizer, usually 0.05% phosphoric acid.
Hazard: Flammable, dangerous fire risk, explosive limits in air 6–41%. Toxic by ingestion, inhalation, and skin absorption. TLV: ceiling 4.7 ppm.
Use: Manufacture of acrylonitrile, acrylates, adiponitrile, cyanide salts, dyes, chelates, rodenticides, pesticides.

"Hydro-Darco" *[Norit].* TM for activated carbon-based adsorbent.
Use: For purification of water, fine chemicals, wastewater, air, and pharmaceuticals. Used for recovery of valuable products as in gold and organic solvents.

hydrodealkylation.
(HDA).
A type of hydrogenation used in petroleum refining in which heat and pressure in the presence of hydrogen are used to remove methyl or larger alkyl groups from hydrocarbon molecules, or to change the position of such groups. The process is used to upgrade products of low value, such as heavy reformate fractions, naphthenic crudes, or recycle stocks from catalytic cracking. Also, toluene and pyrolysis gasoline are converted to benzene and methyl naphthalenes to naphthalene by this process.
See transalkylation.

hydrodistillation. (steam distillation).
Removal of essential oils from plant components (flowers, leaves, bark, etc.) by the use of high-temperature steam. The process is used chiefly in the perfume and fragrance industry.

hydrofining. A petroleum-refining process in which a limited amount of hydrogenation converts the sulfur and nitrogen in a petroleum fraction to forms in which they can be easily removed. Hydrofining is generally a separate treatment prior to more extensive hydrogenation. The usual catalysts are oxides of cobalt and molybdenum. Desulfurization, ultrafining, and catfining have a similar meaning.

hydroflumethiazide. (trifluoromethylhydrothiazide).
CAS: 135-09-1. $C_8H_8F_3N_3O_4S_2$.
Properties: White, crystalline solid; odorless. Mp 260–275C. Insoluble in water and acid; soluble in dilute alkali but unstable in alkaline solutions.
Grade: NF.
Use: Medicine (antihypertensive).

hydrofluoric acid.
CAS: 7664-39-3.
Hydrogen fluoride in aqueous solution.
Properties: Colorless, fuming, mobile liquid. Bp (38% solution) 112C. Will attack glass and any silicon-containing material.
Derivation: Dissolving hydrogen fluoride in water to various concentrations.
Grade: CP, technical, 38%, 47%, 53%, 70%.
Hazard: Toxic by ingestion and inhalation, highly corrosive to skin and mucous membranes.
Use: Aluminum production, fluorocarbons, pickling stainless steel, etching glass, acidizing oil wells, fluorides, gasoline production (alkylation), processing uranium.
See hydrogen fluoride.

hydrofluorosilicic acid. Legal label name (Rail) for fluosilicic acid.

hydrofluosilicic acid. See fluosilicic acid.

Hydrofol.   (stearic acid).
CAS: 57-11-4.
An additive.
Use: In esters, greases, candles, crayons, cosmetics, metallic salts, mono- and diglyceride, shaving creams, textile auxiliaries, white stearates, waxes, and rubber compounding.

hydroforming.   The use of hydrogen in the presence of heat, pressure, and catalysts (usually platinum) to convert olefinic hydrocarbons to branched chain paraffins (isomerization) to yield high-octane gasoline. Catforming and similar terms are often used in the same sense.

hydrofuramide.   (furfuramide).
CAS: 494-47-3. $OC_4H_3CH(NCHC_4H_3O)_2$.
Properties: Light-brown to white powder. Mp 117C, boils about 250C with decomposition. Insoluble in cold water; soluble in alcohol and ether.
Derivation: Treatment of furfural with ammonia.
Use: Rubber accelerator, hardening agent for resins, rodenticides, fungicides.

hydrogasification.   Production of gaseous or liquid fuels by direct addition of hydrogen to coal.
See gasification.

hydrogen.
CAS: 1333-74-0. H_2.
Nonmetallic element of atomic number 1, group IA of periodic table, atomic weight 1.0079, valence of 1. Molecular formula is H_2. Isotopes: deuterium (^2D), tritium (^3T). Hydrogen discovered by Cavendish in 1766, named by Lavoisier in 1783.
Properties: A diatomic gas. D 0.08999 g/L, d 0.0694 (air = 1.0), specific volume 193 cu ft/lb (21.1C), fp −259C, bp −252C, autoign temp 1075F (580C). Very slightly soluble in water, alcohol, and ether; noncorrosive; can exist in crystalline state at from 4-1 K; classed as an asphyxiant gas; rate of permeation through solids is approximately four times that of air.
Occurrence: Chiefly in combined form (water, hydrocarbons, and other organic compounds), traces in earth's atmosphere. Unlimited quantities in sun and stars. It is the most abundant element in the universe.
Derivation: (1) Reaction of steam with natural gas (steam reforming) and subsequent purification; (2) partial oxidation of hydrocarbons to carbon monoxide and interaction of carbon monoxide and steam; (3) gasification of coal (see Note 1); (4) dissociation of ammonia; (5) thermal or catalytic decomposition of hydrocarbon gases; (6) catalytic reforming of naphtha; (7) reaction of iron and steam; (8) catalytic reaction of methanol and steam; (9) electrolysis of water (see Note 2). In view of the importance of hydrogen as a major energy source of the future, development of the most promising of these methods may be expected.
See gasification.

Note 1: The projected cost of producing hydrogen from coal by proven gasification techniques has been estimated to be competitive with gasoline.
Note 2: More efficient methods than electrolysis for obtaining hydrogen from water are under investigation. One of these is thermochemical decomposition. Another is photochemical decomposition by solar radiation, either directly or via a solar power generator. Photolytic decomposition of water with platinum catalyst has been achieved. Hydrogen can also be obtained by photolytic decomposition of hydrogen sulfide with cadmium sulfide catalyst.
See photolysis; thermochemistry.
Method of purification: By scrubbing with various solutions (see especially the Girbitol absorption process). For very pure hydrogen, by diffusion through palladium.
Grade: Technical, pure, from an electrolytic grade of 99.8% to ultrapure with <10 ppm impurities.
See para-hydrogen.
Hazard: Highly flammable and explosive, dangerous when exposed to heat or flame, explosive limits in air 4–75% by volume.
Use: Production of ammonia, ethanol, and aniline; hydrocracking, hydroforming, and hydrofining of petroleum; hydrogenation of vegetable oils; hydrogenolysis of coal; reducing agent for organic synthesis and metallic ores; reducing atmosphere to prevent oxidation; as oxyhydrogen flame for high temperatures; atomic-hydrogen welding; instrument-carrying balloons; making hydrogen chloride and hydrogen bromide; production of high-purity metals; fuel for nuclear rocket engines for hypersonic transport; missile fuel; cryogenic research. A treatise on the physical properties of hydrogen compiled by the National Bureau of Standards is available from the U.S. Government Printing Office.
Note: A safe storage method for hydrogen for possible use as automotive fuel involves the use of metal hydrides from which the hydrogen is released at specified temperatures. Iron titanium hydride has been found the most satisfactory.

hydrogenase.   Any of a class of enzymes; dihydrogen acceptor oxidoreductases that catalyzes the reversible dissociation of molecular hydrogen into hydrogen ions and electrons.

ortho-hydrogen.   See para-hydrogen.

para-hydrogen.   Type of molecular hydrogen preferred for rocket fuels. Molecular hydrogen (H_2) exists in two varieties, ortho- and para-, named according to their nuclear spin types. Ortho-hydrogen molecules have a parallel spin, para- an antiparallel spin. By cooling to liquid air temperature and use of a ferric oxide gel catalyst, the normal equilibrium of 3 ortho- to 1 para- is displaced and para-hydrogen may be isolated. It is being produced with <5 ppm impurities.

hydrogenated fish oil.
CAS: 8016-14-6.
Properties: Oil. Mp: >32C.
Use: Food additive.

hydrogenated sperm oil.
A waxy liquid with a faint odor composed of wax esters with small proportions of triglycerides. Prior to the discovery of kerosene and the widespread use of electricity, it was used as a primary fuel for illumination due to its bright and odorless flame.
Use: Food additive.

hydrogenated terphenyls.
CAS: 61788-32-7.
Complex mixtures of ortho- meta-, and para-terphenyls in various stages of hydrogenation.
Use: A heat transfer medium and plasticizer.

hydrogenated vegetable oil.
See "Neobee" [Stepan].

hydrogenation.
Any reaction of hydrogen with an organic compound. It may occur either as direct addition of hydrogen to the double bonds of unsaturated molecules, resulting in a saturated product, or it may cause rupture of the bonds of organic compounds, with subsequent reaction of hydrogen with the molecular fragments. Examples of the first type (called addition hydrogenation) are the conversion of aromatics to cycloparaffins and the hydrogenation of unsaturated vegetable oils to solid fats by addition of hydrogen to their double bonds. Examples of the second type (called hydrogenolysis) are hydrocracking of petroleum and hydrogenolysis of coal to hydrocarbon fuels.
See hydrogenolysis; hydrocracking; hydroforming.

hydrogen azide.
See hydrazoic acid.

hydrogen bond.
An attractive force, or bridge, occurring in polar compounds such as water in which a hydrogen atom of one molecule is attracted to two unshared electrons of another. The hydrogen atom is the positive end of one polar molecule and forms a linkage with the electronegative end of another such molecule. In the formula below, the hydrogen atom in the center is the "bridge."

$$H:\ddot{O}:H:\ddot{O}:H:\ddot{O}:$$
$$\ddot{H}\qquad\ddot{H}\qquad\ddot{H}$$

Hydrogen bonds are only one-tenth to one-thirteenth as strong as covalent bonds but they have pronounced effects on the properties of substances in which they occur, especially as regards melting point, boiling point, and crystalline structure. They are found in compounds containing such strongly electronegative atoms as nitrogen, oxygen, and fluorine. They play an important part in the bonding of cellulosic compounds, e.g., in the paper industry, and occur also in many complex structures of biochemical importance, e.g., adenine-uracil linkage in DNA.

hydrogen bromide.
CAS: 10035-10-6. HBr.
Properties: Colorless gas. D 2.71 (air = 1.00), fp −86C, bp −66.4C, specific volume 4.8 cu ft/lb (70F, 1 atm). Soluble in water and alcohol. Nonflammable.
Derivation: (1) By passing hydrogen with bromine vapor over warm platinum sponge which acts as a catalyst; (2) As a by-product in the bromination of organic compounds.
Grade: Up to 99.8% min purity.
Hazard: Toxic by inhalation, strong irritant to eyes and skin.
Use: Organic synthesis, makes bromides by direct reaction with alcohols, pharmaceutical intermediate; alkylation and oxidation catalyst, reducing agent.
See hydrobromic acid.

hydrogen chloride.
CAS: 7647-01-0. HCl.
Properties: Colorless gas or fuming liquid; suffocating odor. D 1.268 (air = 1.00), fp −114C, bp −85C, specific volume 10.9 cu ft/lb (21.1C, 1 atm). Very soluble in water; soluble in alcohol and ether. Nonflammable.
Derivation: (1) By-product of organic chlorination reactions (approximately 90%); (2) reaction of sodium chloride and sulfuric acid; (3) burning hydrogen in an atmosphere of chlorine in absence of air.
Hazard: Toxic by inhalation, strong irritant to eyes and skin. Questionable carcinogen.
Use: Production of vinyl chloride from acetylene and alkyl chlorides from olefins, hydrochlorination (see rubber hydrochloride), polymerization, isomerization, alkylation, and nitration reactions.
See hydrochloric acid.

hydrogen cyanide.
See hydrocyanic acid.

hydrogen dioxide.
See hydrogen peroxide.

hydrogen electrode.
See electrode, hydrogen.

hydrogen fluoride.
CAS: 7664-39-3. HF.
Properties: Colorless, fuming gas or liquid. Very soluble in water. The liquid and gas consist of associated molecules. The vapor density corresponds to hydrogen fluoride only at high temperatures. Fp −83C, bp 19.5C, d (liquid) 0.988 (14C), sp vol 17 cu ft/lb (21.1C, 1 atm). Nonflammable.
Derivation: Distillation from the reaction product of calcium fluoride and sulfuric acid, also from fluosilicic acid.
Grade: To 99.9% min purity.

Hazard: Toxic by ingestion and inhalation, strong irritant to eyes, skin, and mucous membranes.
Use: Catalyst in alkylation, isomerization, condensation, dehydration, and polymerization reactions; fluorinating agent in organic and inorganic reactions; production of fluorine and aluminum fluoride; additive in liquid rocket propellants; refining of uranium.
See hydrofluoric acid.

hydrogen hexafluorosilicate.   See fluosilicic acid.

hydrogen iodide.
CAS: 10034-85-2.   HI.
Properties: Colorless gas. Bp −35C, fp −51C, fumes in moist air, d 5.2 (25C). Freely soluble in water. Nonflammable.
Hazard: Strong irritant. Poison.
Use: Making hydriodic acid.

hydrogen ion concentration.   See pH.

hydrogenolysis.   (destructive hydrogenation).
A type of hydrogenation reaction in which molecular cleavage of an organic compound occurs with addition of hydrogen to each portion. An important application is hydrocracking (hydrogenative splitting) of large organic molecules, with formation of fragments that react with hydrogen by use of catalysts and high temperatures. Hydrogenolysis of coal to gaseous and liquid fuels was used in Germany in the 1940s; a similar method (Oil/Gas Process) is under development in the U.S. The German process used pulverized coal made into a paste with heavy oil and a metallic catalyst. The mixture plus the necessary hydrogen was subjected to 300–700 atm at approximately 500C. The coal was converted into heavy oil, distillable oil, gasoline, and hydrocarbon gases. Large quantities of hydrogen are necessary.
See gasification; hydrogenation.

hydrogen overvoltage.   The difference between actual cathode potential for hydrogen evolution and the equilibrium (theoretical) potential of hydrogen in the same electrolyte.

hydrogen peroxide.
CAS: 7722-84-1. H_2O_2 (molecular formula); H−O−O−H (structural formula).
Properties: (pure anhydrous) density of solid, 1.71 g/cc, density of liquid 1.450 g/cc at 20C, viscosity, liquid 1.245 cP, surface tension 80.4 dynes/cm at 20C, fp −0.41C, bp 150.2C. Soluble in water and alcohol. (Solutions): pure hydrogen peroxide solutions, completely free from contamination, are highly stable; a low percentage of an inhibitor such as acetanilide or sodium stannate is usually added to counteract the catalytic effect of traces of impurities such as iron, copper, and other heavy metals. A relatively stable sample of hydrogen peroxide

typically, decomposes at the rate of approximately 0.5% per year at room temperature.
Derivation: (1) Autoxidation of an alkyl anthrahydroquinone such as the 2-ethyl derivative in a cyclic continuous process in which the quinone formed in the oxidation step is reduced to the starting material by hydrogen in the presence of a supported palladium catalyst; (2) by electrolytic processes in which aqueous sulfuric acid or acidic ammonium bisulfate is converted electrolytically to the peroxydisulfate, which is then hydrolyzed to form hydrogen peroxide; (3) by autoxidation of isopropyl alcohol. Method (1) is most widely used.
Grade: USP (3%), technical (3, 6, 27.5, 30, 35, 50, and 90%), FCC. Most common commercial strengths are 27.5, 35, 50, and 70%.
Hazard: Dangerous fire and explosion risk, strong oxidizing agent. Concentrated solutions are highly toxic and strongly irritating. Questionable carcinogen.
Use: Bleaching and deodorizing of textiles, wood pulp, hair, fur, etc.; source of organic and inorganic peroxides; pulp and paper industry; plasticizers; rocket fuel; foam rubber; manufacture of glycerol; antichlor; dyeing; electroplating; antiseptic; laboratory reagent; epoxidation, hydroxylation, oxidation, and reduction; viscosity control for starch and cellulose derivatives; refining and cleaning metals; bleaching and oxidizing agent in foods; neutralizing agent in wine distillation; seed disinfectant; substitute for chlorine in water and sewage treatment.
See "Kastone" *[Kasten]*.

hydrogen phosphide.   See phosphine.

hydrogen, phosphoretted.   See phosphine.

hydrogen selenide.
CAS: 7783-07-5. H_2Se.
Properties: Colorless gas. Bp −42C; fp −64C; d 2.00 (air = 1). Soluble in water, carbon disulfide, phosgene.
Grade: 98% pure.
Hazard: Dangerous fire and explosion risk; reacts violently with oxidizing materials. Toxic by inhalation, strong irritant to skin, damaging to lungs and liver.
Use: Preparation of metallic selenides and organoselenium compounds; in doping as mix for preparation of semiconductor materials containing controlled amounts of significant impurities.

hydrogen slush.   A mixture of solid and liquid hydrogen at the hydrogen triple point 13.8K and 1.02 psia. It is denser and less hazardous than liquid hydrogen.

hydrogen sodium selenite.   See selenious acid, monosodium salt.

hydrogen sulfide.   (sulfuretted hydrogen).
CAS: 7783-06-4. H_2S.

Properties: Colorless gas; offensive odor. D 1.189 (air = 1.00), fp −83.8C, bp −60.2C, sp vol 11.23 cu ft/lb (21.1C, 1 atm), autoign temp 500F (260C). Soluble in water and alcohol.

Derivation: (1) By the action of dilute sulfuric acid on a sulfide, usually iron sulfide; (2) by direct union of hydrogen and sulfur vapor at a definite temperature and pressure; (3) as a by-product of petroleum refining.

Grade: Technical 98.5%, purified 99.5% min, CP.

Hazard: Highly flammable, dangerous fire risk, explosive limits in air 4.3–46%. Toxic by inhalation, strong irritant to eyes and mucous membranes.

Use: Purification of hydrochloric acid and sulfuric acid, precipitating sulfides of metals, analytical reagent, source of sulfur and hydrogen.

hydrogen tellurate.   See telluric acid.

hydrogen telluride.
H_2Te.
Properties: Colorless gas. D (liquid): 2.57 (−20C), fp −49C, bp −2C. Soluble in water but unstable; soluble in alcohol and alkalies.
Hazard: See hydrogen selenide.

Hydroholac.   Plasticized nitrocellulose lacquer emulsions, including clear finishes, binders, and colors. Produce flexible, lacquer-type, cleanable leather finishes from aqueous systems.
Use: Finishes on glove, garment, handbag, and shoe leather.

α-hydro-ω-hydroxypoly(oxy(methyl-1,2-eth-anediyl)) ether with bis((2-(hydroxye-thyl)amino)methyl)phenol (3:1).
CAS: 68909-26-2.
Hazard: A moderate eye irritant.

hydrol.   See tetramethyldiaminobenzhydrol.

hydrolase.   (hydrase).
An enzyme that catalyzes the removal of water from the substrate.
See enzyme.

"Hydrolin" *[Olin].*   TM for ammonium nitrate preparations.
Hazard: See ammonium nitrate.
Use: Bleaching agent for pulp, clay, and mineral processing.

hydroliquefaction.   Production of liquid hydro-carbon fuels by hydrogenation of coal.
See gasification; Oil/Gas Process.

hydrolube.   A water–glycol base noncombustible hydraulic fluid.

hydrolysis.   A chemical reaction in which water reacts with another substance to form two or more new substances. This involves ionization of the water molecule as well as splitting of the compound hydrolyzed, e.g., $CH_3COOC_2H_5 + H_2O \rightarrow CH_3COOH + C_2H_5OH$. Examples are conversion of starch to glucose by water in the presence of suitable catalysts; conversion of sucrose (cane sugar) to glucose and fructose by reaction with water in the presence of an enzyme or acid catalyst; conversion of natural fats into fatty acids and glycerol by reaction with water in one process of soap manufacture; and reaction of the ions of a dissolved salt to form various products, such as acids, complex ions, etc.

hydrolyzed vegetable proteins.   See Hydro-triticum.

hydrometer.   Device for measuring the density of liquids.
See Baumé.

hydronium ion.   An ion (H_3O^+) formed by the transfer of a proton (hydrogen nucleus) from one molecule of water to another; a companion ion (OH^-) is also formed, the reaction is $2H_2O \rightarrow H_3O^+ + OH^-$. Formation of such ions is statistically rare, resulting from the interaction of water molecules in a ratio of 1:556 million.

hydropathy index.   A scale that expresses the relative hydrophobic and hydrophilic tendencies of amino acid R groups. Used to predict membrane-spanning regions.

hydroperoxide.   An organic peroxide having the generalized formula ROOH. An example is ethyl hydroperoxide (C_2H_5OOH). Methyl and ethyl hydroperoxides are unstable and thus are strong oxidizing agents and explosion hazards; those of higher molecular weight are more stable. Hydroperoxides can be derived by oxidation of saturated hydro-carbons or by alkylating hydrogen peroxide in a strongly acidic environment. They are used as poly-merization initiators.

hydrophilic.   Having a strong tendency to bind or absorb water, which results in swelling and forma-tion of reversible gels. This property is characteris-tic of carbohydrates, such as algin, vegetable gums, pectins, and starches, and of complex proteins such as gelatin and collagen.

hydrophilic amino acid.   Any acid or basic amino acid.

hydrophobic amino acid.   Any amino acid that is insoluble or nearly insoluble in water.

hydrophobic.   (1) Antagonistic to water; inca-pable of dissolving in water. This property is char-acteristic of all oils, fats, waxes, and many resins, as well as of finely divided powders like carbon black and magnesium carbonate.

(2) The association of nonpolar groups or compounds with each other in aqueous systems due to their insolubility in water.

hydroponics.   See nutrient solution.

hydroquinol.   See hydroquinone.

hydroquinone.   (quinol; hydroquinol; *p*-dihydroxybenzene).
CAS: 123-31-9. $C_6H_4(OH)_2$.
Properties: White crystals. D 1.330, mp 170C, bp 285C, flash p 329F (165C), autoign temp 960F (515.5C). Soluble in water, alcohol, and ether. Combustible.
Derivation: Aniline is oxidized to quinone by manganese dioxide and is then reduced to hydroquinone.
Grade: Technical, photographic.
Hazard: Toxic by ingestion and inhalation, irritant. Questionable carcinogen.
Use: Photographic developer (except color film); dye intermediate; inhibitor; stabilizer in paints and varnishes, motor fuels, and oils; antioxidant for fats and oils; inhibitor of polymerization; skin hyperpigmentation.

hydroquinone benzyl ether.   See *p*-benzyloxyphenol.

hydroquinone dibenzyl ether.
CAS: 103-16-2. $C_6H_5CH_2OC_6H_4OCH_2C_6H_5$.
Properties: Tan powder. Mp 119C (min). Purity 90% (min). Insoluble in water; soluble in acetone, benzene, and chlorobenzene. Combustible.
Use: Solvent; perfumes, soap, plastics, and pharmaceuticals.

hydroquinone di-*n*-butyl ether.   (1,4-dibutoxybenzene).
$C_6H_4[O(CH_2)_3CH_3]_2$.
Properties: White flakes; no appreciable odor. Mp 45–46C; bp 124C (1.3 mm Hg), 158C (15.0 mm Hg). Insoluble in water; soluble in benzene, acetone, ethyl acetate, and alcohol. Combustible.

hydroquinone diethyl ether.   (1,4-diethoxybenzene).
CAS: 122-95-2. $C_6H_4(OC_2H_5)_2$.
Properties: White granular solid with anise-like odor. Mp 71–72C, bp 246C. Neither boiling caustic nor acid solution causes any hydrolysis. Absorbs UV light. Insoluble in water; soluble in benzene, acetone, ethyl acetate, and alcohol. Combustible.

hydroquinone di(β-hydroxyethyl) ether.   (*p*-di-[2-hydroxyethoxy]benzene).
$C_6H_4(OC_2H_4OH)_2$.
Properties: White solid. Mp 99C, bp 185–200C (0.3 mm Hg). Slightly soluble in water and most organic solvents; miscible with water at 80C. Combustible.

Use: Preparation of polyester, polyolefins, polyurethanes, and hard waxy resins, organic synthesis.

hydroquinone dimethyl ether.   (1,4-dimethoxybenzene; DMB; dimethyl hydroquinone).
CAS: 654-42-2. $C_6H_4(OCH_3)_2$.
Properties: White flakes with sweet clover odor. Bp 213C, mp 56C, d 1.0293 (65C), viscosity 1.04 cP (65C), dielectric constant 2.8, absorbs UV light in range 2800–3000Å. Soluble in benzene and alcohol; insoluble in water. Combustible.
Use: Weathering agent in paints and plastics, fixative in perfumes, dyes, resin intermediate, cosmetics, especially suntan preparations, flavoring.

hydroquinine mono-*n*-butyl ether.
$CH_3(CH_2)_3OC_6H_4OH$.
Properties: White flakes. Mp 64–65C; bp 115C (1.4 mm Hg). Insoluble in water; soluble in benzene, acetone, ethyl acetate, and alcohol. Combustible.

hydroquinone monoethyl ether.   (4-ethoxyphenol).
CAS: 622-62-8. $C_2H_5OC_6H_4OH$.
Properties: White solid. Mp 63–65C, bp 246–247C. Slightly soluble in water; soluble in benzene, acetone, ethyl acetate, and alcohol. Combustible.
Use: See hydroquinone monomethyl ether.

hydroquinone monomethyl ether.   (4-methoxyphenol; *p*-hydroxyanisole).
CAS: 150-76-5. $CH_3OC_6H_4OH$.
Properties: White, waxy solid. Mp 52.5C, bp 243C, d 1.55 (20/20C). Slightly soluble in water; readily soluble in benzene, acetone, ethanol, and ethyl acetate. Combustible.
Hazard: Eye irritant and skin damage.
Use: Manufacture of antioxidants, pharmaceuticals, plasticizers, dyestuffs; stabilizer for chlorinated hydrocarbons and ethyl cellulose, inhibitor for acrylic monomers and acrylonitriles, UV inhibitor.

hydrosilicofluoric acid.   See fluosilicic acid.

hydrosolvation.   Solvent extraction of coal (containing up to 5% sulfur) under hydrogen pressure with the use of a catalyst such as zinc chloride; pressures from 1000 to 2000 psi are necessary for suitable conversion. This process offers a means of deriving fuel oil and petrochemical feedstocks directly from coal.

hydrosulfite.   See sodium hydrosulfite.

hydrosulfite-formaldehyde.   One of several mixtures of sodium formaldehyde hydrosulfite and sodium formaldehyde bisulfite used as discharges and stripping or reducing agents in dyeing and other textile operations. In some cases, the zinc derivatives are used. Derivation is by the action

of formaldehyde on aqueous sodium hydrosulfite or from zinc, formaldehyde, sulfur dioxide, and sodium hydroxide.
Hazard: Toxic by ingestion.

hydrothermal energy. See geothermal energy.

Hydrotriticum. (hydrolyzed vegetable proteins).
Wheat protein.
Use: For hair waving systems, shampoos, conditioners, and skin care products.

hydrotrope. A chemical that has the property of increasing the aqueous solubility of various slightly soluble organic chemicals.
Use: Formulation of liquid detergents.

hydroxocobalamin. (USAN) (α[5,6-dimethylbenzimidazolyl]hydroxocobaltamide).
CAS: 13422-51-0. $C_{62}H_{89}CoN_{13}O_{15}P$.
A form of vitamin B_{12}.
Use: Medicine.
See vitamin B_{12}.

hydroxyacetal. See hydroxycitronellal dimethyl acetal.

p-**hydroxyacetanilide.** See *p*-acetylaminophenol.

hydroxyacetic acid. (glycolic acid).
CAS: 79-14-1. $CH_2OHCOOH$.
Properties: Colorless crystals, deliquescent. Mp 78–79C, bp decomposes. Soluble in water, alcohol, and ether; available commercially as a 70% solution, light straw-colored liquid, odor like burnt sugar, d 1.27, mp 10C. Combustible.
Derivation: From chloroacetic acid by reaction with sodium hydroxide, or by reduction of oxalic acid. Occurs naturally in sugar cane syrup.
Grade: Technical 70% solution, pure crystals.
Use: Leather dyeing and tanning; textile dyeing; cleaning, polishing, and soldering compounds; copper pickling; adhesives; electroplating; breaking of petroleum emulsions; chelating agent for iron; chemical milling; pH control.

hydroxyacetone. See hydroxy-2-propanone.

o-**hydroxyacetophenone.**
$C_6H_4(OH)COCH_3$.
Properties: Greenish-yellow liquid with minty odor. D 1.1307 (20.8C), bp 213C (717 min), refr index 1.5580 (20C). Slightly soluble in water. Combustible.

α-**hydroxy acids.** (fruit acids; alpha-hydroxy acids).
Said to diminish fine skin lines and pigmentation spots and to stimulate collagen to allow the skin to repair itself.

2-hydroxyadipaldehyde.
$OHCCH_2CH_2CH_2CH_2OCHO$.
Properties: (25% aqueous solution) d 1.066 (20C), bp 37C (50 mm Hg), vap press 17 mm Hg (20C), fp −3.5C, pH approximately 3.0. Combustible.
Hazard: Toxic by inhalation.
Use: Intermediate, insolubilizing agent for proteins and polyhydroxy materials, cross-linking agent for polyvinyl compounds, shrinkage control agent (textiles).

β-**hydroxyalanine.** See serine.

5-hydroxy-3-(β-aminoethyl)indole. See serotonin.

n-**hydroxy *p*-aminooctanoylphenone.**
CAS: 253883-44-2. $C_{14}H_{21}NO_2$.
Hazard: A poison.

p-**hydroxyamphetamine hydrochloride.**
CAS: 6078-07-5. $C_9H_{13}NO \cdot ClH$.
Hazard: A reproductive hazard.

3-β-hydroxyandrosten-17-one acetate.
CAS: 1239-31-2. $C_{21}H_{32}O_3$.
Hazard: Low toxicity by ingestion. A reproductive hazard.
Use: Hormone.

hydroxyaniline. See aminophenol.

o-**hydroxyanisole.** See guaiacol.

p-**hydroxyanisole.** See hydroquinone monomethyl ether.

9-hydroxyanthracene. See anthranol.

hydroxyapatite.
$Ca_{10}(PO_4)_6(OH)_2$.
The major constituent of bone and tooth mineral. It is finely divided, crystalline, nonstoichiometric material rich in surface ions (carbonate, magnesium, citrate), which are readily replaced by fluoride ion, thus affording protection to the teeth.
See fluoridation; calcium phosphate tribasic.

p-**hydroxyazobenzene-*p*-sulfonic acid.**
$HOC_6N_4NNC_6H_4SO_3H$.
Properties: Orange-red crystals. Very soluble in water.
Use: Analytical reagent, precipitant for numerous organic bases.

m-**hydroxybenzaldehyde.**
CAS: 100-83-4. HOC_6H_4CHO.
Properties: Orange-pink crystals. Mp 101.5C. Slightly soluble in cold water; very soluble in hot water and aromatic hydrocarbons.
Hazard: Toxic by ingestion.

Use: Intermediate for dyes, plastics, pharmaceuticals, and bactericides; color reagent for Schiff's reagent, sensitizing agent in photographic emulsions.

o-hydroxybenzaldehyde.   See salicylaldehyde.

p-hydroxybenzaldehyde.
CAS: 123-08-0. HOC_6H_4CHO.

Properties: Colorless needles. D 1.129, mp 116C (sublimes). Soluble in alcohol, ether, or hot water.
Use: Pharmaceuticals.

o-hydroxybenzamide.   See salicylamide.

hydroxybenzene.   See phenol.

2-hydroxy-1',2'-benzocarbazole-3-carboxylic acid.
$C_{17}H_{11}NO_3$.
Properties: A four-ring structure, light-green powder. Mp 315–320C. Soluble in ethanol and acetone; insoluble in water.
Use: Manufacture of dye intermediates and other organic chemicals.

m-hydroxybenzoic acid.
CAS: 99-06-9. $C_6H_4(OH)COOH$.
Properties: White powder. Mp 200C. Soluble in water and hot alcohol.
Use: Intermediate for plasticizers, resins, light stabilizers, petroleum additives, pharmaceuticals.

o-hydroxybenzoic acid.   See salicylic acid.

p-hydroxybenzoic acid.
CAS: 99-96-7. $C_6H_4(OH)COOH·H_2O$.
Properties: Colorless crystals. D 1.46, mp 210C. Soluble in alcohol and ether; partially soluble in water.
Derivation: Interaction of p-aminobenzoic acid and nitrous acid.
Use: Intermediate, synthetic drugs, food preservative (up to 0.1%) (approved by FDA). Its methyl, propyl, and butyl esters are preservatives for cosmetics and pharmaceuticals.

2-hydroxybenzophenone.
CAS: 117-99-7. $C_6H_5COC_6H_4OH$.
Properties: A solid. Mp 41C, bp 210C (27 mm Hg). Insoluble in water; soluble in alcohol.
Use: UV absorber in plastics.

4-hydroxybenzothiophene.
CAS: 3610-02-4. C_8H_6OS.
Hazard: A poison.

o-hydroxybenzyl alcohol.   See salicyl alcohol.

β-hydroxybutyraldehyde.   See aldol.

p-hydroxybutyranilide.   (4'-hydroxybutyranilide).
$C_{10}H_{13}NO_2$.
Properties: Acicular crystals. Mp 138C. Soluble in alcohol; partially soluble in hot water.
Use: Antioxidant for petroleum products.

β-hydroxybutyric acid.
CAS: 502-85-2.
$CH_3CH(OH)CH_2COOH$.
Properties: Viscid yellow mass. Mp 48–50C, bp 130C (12 mm Hg). Very soluble in water, alcohol, and ether.
Derivation: Interaction of acetoacetic acid and sodium amalgam.
Grade: Technical, reagent.
Use: Intermediate.

2-hydroxycamphane.   See borneol.

hydroxy-β-carotene.   See cryptoxanthin.

3-β-hydroxycholestane.   See dihydrocholesterol.

hydroxycitronellal.   (citronellal hydrate; 3,7-dimethyl-7-hydroxyoctenal).
CAS: 107-75-5.
$(CH_3)_2C(OH)(CH_2)_3CH(CH_3)CH_2CHO$.
Properties: Viscous, colorless, or faintly yellow liquid; sweet lily-type odor. D 0.925–0.930 (15C), refr index 1.448–1.450 (20C), optical rotation (Java type) +9 to +10.5 degrees, boiling range 94–96C (1 mm Hg). Soluble in alcohol (50%), fixed oils; slightly soluble in water, glycerol, and mineral oil. Combustible.
Derivation: From citronellal (*Java citronella* or *Eucalyptus citriodora*).
Grade: Perfume, FCC.
Use: Perfumery (fixative, muguet odor), flavoring, soap and cosmetic fragrances.

hydroxycitronellal dimethyl acetal.   (hydroxyacetal).
$(CH_3)_2C(OH)(CH_2)CH(CH_3)CH_2CH(OCH_3)_2$.
Properties: Colorless liquid; light floral odor. D 0.925–0.930 (25/25C), refr index 1.4410–1.4440 (20C). Soluble in most fixed oils, mineral oil, and propylene glycol; insoluble in glycerol. Combustible.
Grade: Perfume, FCC.
Use: Flavoring agent in foods, perfumery.

hydroxycitronellal-methyl anthranilate Schiff base.
$C_{18}H_{27}O_3N$.
Properties: Linden-orange-flower odor; yellow, honeylike, viscous liquid. Stable, refr index 1.5350–1.5460 (20C), flash p 206F (96.6C) (TCC).

Soluble in two parts of 70% alcohol, one part of 80% alcohol. Combustible.
Use: Perfumery.

17-hydroxycorticosterone.   See   hydrocortisone.

trans-1-(4-hydroxycyclohexyl)-4-(4-fluorophenyl)-5-(2-methoxypyrimidin-4-yl)-imidazole.
CAS: 193551-21-2. $C_{20}H_{21}FN_4O_2$.
Hazard: A poison by ingestion.

3-hydroxy-_p_-cymene.   See thymol.

1-hydroxy-2,4-diamylbenzene.   See   diamyl phenol.

6-hydroxy-3,7-dimethyloctanoic   acid   lactone.
$C_{10}H_{18}O_2$.
Properties: Colorless solid; maple syrup odor. D: 0.966, refr index: 1.457–1.461. Sol in alc; very slightly sol in water.
Use: Food additive.

hydroxydiphenyl.   See phenylphenol.

_p_-hydroxydiphenylamine.   (anilinophenol).
CAS: 122-37-2. $C_6H_5NHC_6H_4OH$.
Gray solid leaflets, mp approximately 70C, purity 98% (min), bp 330C. Insoluble in water; soluble in alcohol, ether, acetone, chloroform, alkali, and benzene.
Hazard: Irritant.

7-hydroxy-2-(dipropylamino)tetraline.
CAS: 74938-11-7. $C_{16}H_{25}NO$.
Hazard: A poison.

8-hydroxy-2-(di-_n_-propylamino)tetraline.
CAS: 78950-78-4. $C_{16}H_{25}NO$.
Hazard: A poison.

1-hydroxyestradiol.
CAS: 4147-05-1. $C_{18}H_{24}O_3$.
Hazard: A reproductive hazard.
Use: Hormone.

4-hydroxyestrone.
CAS: 3131-23-5. $C_{18}H_{22}O_3$.
Hazard: A reproductive hazard.
Use: Hormone.

2-hydroxyethanesulfonic acid.   See isethionic acid.

hydroxyethylacetamide.   See _n_-acetylethanolamine.

2-hydroxyethyl acrylate.   (HEA).
A functional monomer for the manufacture of thermosetting acrylic resins.

2-hydroxyethylamine.   See ethanolamine.

_n_-(2-hydroxyethyl)ammonium benzothiazole-2-thiolate.
CAS: 5902-85-2. $C_7H_5NS_2 \cdot C_2H_7NO$.
Hazard: Moderately toxic by ingestion. A severe eye irritant.

2-hydroxyethyl carbamate.
CAS: 589-41-3. $H_2NCOOCH_2CH_2OH$.
Properties: Crystalline, deliquescent solid. Mp 43C, bp 130–135C (1 mm Hg), d 1.2852 g/cc (20C), flash p 370F (187.7C). Miscible with water; soluble in alcohol and acetone; and insoluble in benzene and chloroform. Combustible.
Use: Chemical intermediate, especially for wash-and-wear cotton finishing agents.

2-hydroxyethyl dibutyl phosphate.
CAS: 130525-77-8. $C_{10}H_{23}O_5P$.
Hazard: Moderately toxic by ingestion and skin contact. A mild skin and eye irritant.

hydroxyethylcellulose.   ("Cellosize" _[Union]_).
Properties: Nonionic, water-soluble cellulose ether; white, free-flowing powder; insoluble in organic solvents; soluble in hot or cold water; stable in concentrated salt solutions; grease and oil resistant. Combustible.
Use: Thickening and suspending agent; stabilizer for vinyl polymerization; retards evaporation of water in cements, mortars, and concrete; binder in ceramic glazes, films, and sheeting; protective colloid; paper and textile sizing; secondary petroleum recovery.

_n_-(2-hydroxyethyl)-5,8,11,14-eicosatetraenamide (all-z)-.
CAS: 94421-68-8. $C_{22}H_{37}NO_2$.
Hazard: A reproductive hazard.

hydroxyethylethylenediamine.
(aminoethylethanol amine).
CAS: 111-41-1. $NH_2CH_2CH_2NHCH_2CH_2OH$.
Properties: Hygroscopic liquid; mild ammoniacal odor. D 1.0304 (20/20C), bp 243.7C, vap press 0.01 mm Hg (20C), flash p 275F (135C), bulk d 8.6 lb/gal (20C). Soluble in water. Combustible.
Use: Textile finishing compounds (antifuming agents, dyestuffs, cationic surfactants), resins, rubber products, insecticides, and certain medicinals.

hydroxyethylethylenediaminetriacetic   acid.
(HEDTA).
CAS: 150-39-0. $C_{10}H_{18}O_7N_2$.
Properties: Solid. Mp 159C. Soluble in water and methanol.
Use: Chelating compound.

n-(2-hydroxyethyl)ethyleneimine.   See   1-
aziridineethanol.

**1-(2-hydroxyethyl)-2-*n*-heptadecenyl-2-imi-
dazoline.**
$C_{22}H_{42}N_2O$.
A liquid cationic detergent.
Use: Corrosion inhibitor, acid-stable emulsifier.

β-hydroxyethylhydrazine.   (ethanolhy-
drazine).
CAS: 109-84-2. $HOCH_2CH_2NHNH_2$.
Properties: Colorless, slightly viscous liquid. D 1.11
(20C), fp −70C, boiling range 145–153C (25 mm
Hg), flash p 224F (106.6C). Miscible with water;
soluble in lower alcohols; and slightly soluble in
ether. Combustible.
Grade: 70%.
Hazard: Moderate fire risk. Irritant.
Use: Intermediate, plant growth regulator, flowering
inducer for pineapples.

**2-hydroxy-2-ethyl-3-isobutyl-9,10-dime-
thoxy-1,2,3,4,6,7-hexahydrobenzo(a)
chinolizin.**
CAS: 303-75-3. $C_{21}H_{33}NO_3$.
Hazard: A poison.

2-hydroxyethyl methacrylate.   (HEMA).
CAS: 868-77-9. $C_6H_{10}O_3$.
Properties: Clear, mobile liquid. D 1.064 (77/60F),
fp −12C, refr index 1.4505 (25C). Miscible with
water; soluble in common organic solvents. An
inhibitor is usually added to solutions to prolong
shelf life. The 30% solution is made with xylene.
Hazard: 30% grade (with xylene): Flammable, moderate fire risk.
Grade: 30%, 96%.
Use: Acrylic resins, binder for nonwoven fabrics,
enamels.

**n-(2-hydroxyethyl)-*n*-methyl-2-nitro-1h-imi-
dazole-1-acetamide.**
CAS: 197004-63-0.
Properties: 2-Nitroimidazole hypoxic cell radiosen-
sitizer mf: $C_8H_{12}N_4O_4$.
Hazard: A poison.

N-hydroxyethylmorpholine.   See   *n*-mor-
pholine ethanol.

N-hydroxyethyl piperazine.
CAS: 103-76-4. $HOCH_2CH_2$

$HOCH_2CH_2NCH_2CH_2NHCH_2CH_2$.

Properties: Liquid. D 1.0614 (20/20C), bp 246.3C,
fp −10C, flash p 255F (124C). Miscible with water.
Combustible.

Use: Intermediate   for   pharmaceuticals,
anthelmintics, surface-active agents, and synthetic fibers.

N-2-hydroxyethylpiperidine.   See   2-piperi-
dinoethanol.

(2-hydroxyethyl)-2-propenenitrile.
CAS: 69521-64-8. C_5H_7NO.
Hazard: A poison by ingestion and skin contact. A
severe eye irritant.

hydroxyethylpyrrolidine.   See   1-pyrrolidi-
neethanol.

**hydroxyethyltrimethylammonium bicarbon-
ate.**   (choline bicarbonate).
$(CH_3)_3NCH_2CH_2OH \cdot HCO_3$.
Properties: Colorless liquid. D 1.0965 (25/4C), fp
−21.3C, refr index 1.3967 (25C). Miscible with
water.
Grade: Industrial grade is a 44–46% aqueous solution of a mixture of carbonate and bicarbonate.
Use: Alkaline catalyst, intermediate for choline salts
and surfactants.

**β-hydroxyethyltrimethylammonium hydrox-
ide.**   See choline.

1-hydroxyfenchane.   See fenchyl alcohol.

13-hydroxygermacrone.
CAS: 213908-53-3. $C_{15}H_{22}O_2$.
Hazard: A poison by ingestion.

5-hydroxy-1h-indole-3-ethanol.
CAS: 154-02-9. $C_{10}H_{11}NO_2$.
Hazard: A reproductive hazard.

**m-hydroxyisopropylphenyl-*n*-
methylcarbamate.**
CAS: 17710-63-3. $C_{11}H_{15}NO_3$.
Hazard: A poison.
Use: Agricultural chemical.

4-hydroxy-2-keto-4-methylpentane.   See
diacetone alcohol.

hydroxylamine.   (oxammonium).
CAS: 7803-49-8.
NH_2OH.
Properties: Colorless crystals. D 1.227, mp 33C, bp
70C (60 mm Hg), the free base is unstable. Soluble
in alcohol, acids, and cold water.
Derivation: By decomposing hydroxylamine
hydrochloride or sulfate with a base and distilling
in vacuo.
Hazard: Decomposes rapidly at room temperature,
violently when heated, detonates in flame-heated
test tube. Irritant to tissue.
Use: Reducing agent, organic synthesis.

hydroxylamine acid sulfate. (hydroxylam-monium acid sulfate; hydroxylamine hemisulfate; HAS).

$NH_2OH \cdot H_2SO_4$.

Properties: White to brown crystalline solid. Bulk d 15–16 lb/gal (20C), mp indefinite (decomposes), pH of 0.1 molar aqueous solution 1.6. Very hygroscopic; soluble in water and methanol; slightly soluble in alcohol.

Use: Reducing agent, photographic developer, purification agent for aldehydes and ketone, synthesis of dyes and pharmaceuticals; rubber chemicals, reagent.

hydroxylamine hydrochloride. (hydroxylammonium chloride).

CAS: 5470-11-1. $NH_2OH \cdot HCl$.

Properties: Colorless, hygroscopic crystals. D 1.67 (17C). Soluble in water, glycerol, and alcohol; insoluble in ether; mp 152C (decomposes); pH of 0.1 molar aqueous solution 3.4.

Derivation: Reduction of ammonium chloride, frequently by electrolysis.

Hazard: Toxic by ingestion, strong irritant to tissue.

Use: Organic synthesis, photographic developer, medicine, controlled reduction reactions, nondiscoloring short-stopper for synthetic rubbers, antioxidant for fatty acids.

hydroxylamine sulfate. (HS; hydroxylammonium sulfate).

CAS: 10039-54-0. $(NH_2OH)_2 \cdot H_2SO_4$.

Properties: Colorless crystals. Mp 177C (decomposes). Solution has a corrosive action on the skin. Soluble in water, slightly soluble in alcohol.

Hazard: Irritant to tissue.

Use: Reducing agent, photographic developer, purification agent for aldehydes and ketones, chemical synthesis, textile chemical, oxidation inhibitor for fatty acids, catalyst, biological and biochemical research, making oximes for paints and varnishes, rustproofing, nondiscoloring short-stopper for synthetic rubbers, unhairing hides.

hydroxylammonium acid sulfate. See hydroxylamine acid sulfate.

hydroxylammonium chloride. See hydroxylamine hydrochloride.

hydroxylammonium sulfate. See hydroxylamine sulfate.

hydroxylated lecithin. Modified lecithin by treating it with hydrogen peroxide and an organic acid.

Properties: Light yellow liquid to paste; characteristic odor. Moderately sol in water.

Use: Food additive.

hydroxyl group. The univalent group -OH, occurring in many inorganic compounds that ionize in solution to yield OH^- radicals. Also the characteristic group of alcohols.

See -ol.

o-hydroxymercuriobenzoic acid.

CAS: 14066-61-6. $C_7H_6HgO_3$.

Hazard: A poison.

p-hydroxymercuribenzoic acid.

CAS: 1126-48-3. $C_7H_6HgO_3$.

Hazard: A poison.

N-(7-hydroxyl-1-naphthyl)acetamide. (1-acetylamino-7-naphthol).

$C_{10}H_6(OH)NHCOOH_3$.

A granular paste.

Use: Chemical intermediate.

hydroxymercurichlorophenol. (2-chloro-4-[hydroxymercuri]phenol).

$C_6H_3Cl(HgOH)(OH)$.

Properties: Insoluble in water and common organic solvents; soluble in solutions of acids and alkalies with the formation of salts.

Hazard: Toxic by ingestion and inhalation, strong irritant to eyes and skin, use may be restricted.

Use: Seed disinfectant and fungicide.

hydroxymercuricresol.

$C_6H_3CH_3(HgOH)(OH)$.

Hazard: Toxic by ingestion and inhalation, strong irritant to eyes and skin, use may be restricted.

Use: Pesticide.

hydroxymercurinitrophenol.

$C_6H_3NO_2(HgOH)(OH)$.

Hazard: Toxic by ingestion and inhalation, strong irritant to eyes and skin, use may be restricted.

Use: Pesticide.

2-hydroxymesitylene. See mesitol.

2-hydroxymestranol.

CAS: 26011-40-5. $C_{21}H_{26}O_3$.

Hazard: A reproductive hazard.

Use: Hormone.

6-(hydroxymethyl)acylfulvene.

CAS: 158440-71-2. $C_{15}H_{18}O_3$.

Hazard: A poison.

2-hydroxy-3-methylbenzoic acid. See cresotic acid.

3-hydroxy-3-methylbutan-2-one.

$CH_3COH(CH_3)C(O)CH_3$.

Colorless liquid, d 0.95 (25/25C). Combustible.

Use: Intermediate.

7-hydroxy-4-methylcoumarin. See β-methylumbelliferone.

4-hydroxymethyl-2,6-di-*tert*-butylphenol.
$[(CH_3)_3C]_2C_6H_2(OH)(CH_2OH)$.
Properties: White, crystalline powder. Mp 140–141C, bp 162C (2.6 mm Hg), flash p 375F (190.5C), vap press 0.03 mm Hg (100C), bulk d 26 lb/ft^3. No odor. Partially soluble in methanol, ethanol, and acetone; insoluble in water. Combustible.
Hazard: Toxic by ingestion.
Use: Antioxidant for gasoline and other hydrocarbons.

hydroxymethylethylene   carbonate.   See glycerin carbonate.

5-hydroxymethyl-2-furaldehyde.   (5-hydroxymethyl-2-formylfuran).
$C_6H_6O_3$.
Properties: Crystalline solid (needles). D 1.20, refr index 1.56, UV absorption max: 283 nm, soluble in water and common organic solvents, protect from light and oxygen.
Derivation: From glucose, cornstarch, sucrose, molasses.
Use: Organic synthesis for glycols, acetals, aldehydes, etc.

***dl*-α-hydroxy-γ-methylmercaptobutyric acid, calcium salt.**   See methionine hydroxy analog calcium.

2-hydroxymethyl-5-norbornene.   See 5-norbornene-2-methanol.

4-hydroxy-4-methylpentan-2-one.   See diacetone alcohol.

2-(2′-hydroxy-5′-methylphenyl)benzotriazole.
$HOC_6H_3(CH_3)N_3C_6H_4$.
Properties: Off-white, crystalline powder. Mp 129–130C, bp 225C (10 mm Hg). Insoluble in water; soluble in methylethyl ketone and methyl methacrylate.
Use: Protector against UV radiation, thermal stabilizer, antioxidant, chelation of metals.

α-hydroxy-α-methylpropiophenone.
CAS: 7473-98-5. $C_{10}H_{12}O_2$.
Hazard: Moderately toxic by ingestion.

3-hydroxy-2-methyl-1,4-pyrone.   See maltol.

2-hydroxy-4-(methylthio)butanenitrile.
CAS: 17773-41-0. C_5H_9NOS.
Hazard: A poison by ingestion and skin contact. A moderate skin irritant.

hydroxynaphthalene.   See naphthol.

3-(6-hydroxy-2-naphthalenyl)-1,2-dimethyl (2r,3s)-rel-3-pyrrolidinol ydrochloride.
CAS: 210828-80-1. $C_{16}H_{19}NO_2 \cdot ClH$.
Hazard: A poison.

3-hydroxy-2-naphthoic acid.   (β-hydroxynaphthoic acid; 3-naphthol-2-carboxylic acid; β-oxynaphthoic acid).
CAS: 92-70-6. $C_{10}H_6OHCOOH$.
Properties: Yellow, rhombic leaflets. Mp 217–219C. Soluble in alcohol and ether; insoluble in water.
Derivation: By treating sodium-2-naphtholate with carbon dioxide under pressure.
Hazard: Irritant.
Use: Dyes and pigments.

β-hydroxynaphthoic anilide.   (naphthol AS; azoic coupling component 2).
$C_{10}H_6OHCONHC_6H_5$.
Properties: Cream-colored crystals. Mp 246.0C. Sodium salt is soluble in water.
Derivation: Condensation of β-hydroxynaphthoic acid and aniline.
Use: Dyes.

2-hydroxy-1,4-naphthoquinone.
CAS: 83-72-7. $C_{10}H_5O_2(OH)$.
Properties: Yellow to orange-yellow needles or powder. Mp 192–195C (decomposes). Redox potential 0.362 volt. Soluble in cold water, benzene, carbon tetrachloride, and petroleum ether.
Use: Intermediate for pharmaceuticals, henna hair and wool dyes, bactericides; seed disinfectant.

5-hydroxy-1,4-naphthoquinone.   (juglone).
CAS: 481-39-0.
This compound derived from walnuts is the starting point in the 16-step synthesis of oxytetracycline. It has also been found to be a natural herbicide with allelopathic properties.

4-hydroxy-3-nitrobenzenearsonic acid.   (3-nitro-4-hydroxyphenylarsonic acid).
$HOC_6H_3(NO_2)AsO(OH)_2$.
Properties: Pale-yellow crystals.
Derivation: Heating phenol with arsenic and treating the *p*-hydroxyphenyl arsonate with nitric and sulfuric acids.
Hazard: Toxic by ingestion.

4-hydroxy-3-nitrophenylarsonic acid.   (Roxarsone).
$C_6H_6AsNO_6$.
An arsenic derivative that has anticoccidial action and promotes growth in animals.
Properties: Pentavalent organic arsenical.
Hazard: Highly toxic.
Use: A growth promoter in animal feed and antibacterial for domestic fowl.

4-hydroxy-nonachlorodiphenyl ether.
CAS: 21567-21-5. $C_{12}HCl_9O_2$.
Hazard: A poison.

4-hydroxynonanoic acid, γ-lactone. See γ-nonyl lactone.

***cis*-12-hydroxyoctadec-9-enoic acid.** See ricinoleic acid.

12-hydroxyoleic acid. See ricinoleic acid.

15-hydroxypentadecanoic acid lactone. See pentadecanolide.

3-hydroxyphenol. See resorcinol.

2-hydroxy-2-phenylacetophenone. See benzoin.

β-*p*-hydroxyphenylalanine. See tyrosine.

2-(2′-hydroxyphenyl)benzotriazole.
CAS: 3147-75-9. $HOC_6H_4N_3C_6H_4$.
An UV light absorber.
Use: Plastics.

***p*-hydroxyphenyl benzyl ether.** See *p*-benzyloxyphenol.

4-(*p*-hydroxyphenyl)-2-butanone acetate.
See Cuelure.

(+-)-4-hydroxy-α1-(((6-(4-phenylbutoxy)-hexyl)amino)methyl)-1,3-benzenedime-thanol.
CAS: 89365-50-4. $C_{25}H_{37}NO_4$.
Hazard: A poison by inhalation.

***N*-(*p*-hydroxyphenyl)glycine.**
(glycine[photographic]; photo-glycin).
$HOC_6H_4NHCH_2COOH$.
Properties: White to buff crystals or powder. Mp 240C (with decomposition). Slightly soluble in water; soluble in alkaline solutions.
Derivation: By condensation of *p*-aminophenol with chloroacetic acid.
Use: Photographic developer, cellulose and nitro-cellulose acetate lacquers and varnishes, analytical reagent.

2-hydroxyphenylmercuric chloride. (chloromercuriphenol).
CAS: 90-03-9. HOC_6H_4HgCl.
Properties: White to faint-pink fine crystals. Mp 152C. 0.1 Part in 100 soluble in water (25C); soluble in hot water, alkali, and alcohol.
Hazard: Skin irritant.
Use: Antiseptic, fungicide.

3-hydroxy-*n*-phenyl-2-naphthalenecarbox-amide. See C.I. azoic coupling component 2.

hydroxyphenylstearic acid. A derivative of phenylstearic acid potentially useful as oxidation and corrosion inhibitor.

(s)-1-(3-hydroxy-2-phosphonylmethoxypro-pyl)cytosine.
CAS: 113852-37-2. $C_8H_{14}N_3O_6P$.
Hazard: A severe skin irritant.

1-hydroxypiperidine.
$C_5H_{10}NOH$.
Properties: Soluble in water, organic solvents, aqueous acids, and bases.
Use: Intermediate, reducing agent, polymerization inhibitor for vinyl monomers, antioxidant, metal ion reduction.

11-α-hydroxyprogesterone.
$C_{21}H_{30}O_3$.
Properties: White, crystalline powder. Mp 163, specific rotation +179 degrees. Insoluble in water; soluble in alcohol.
Derivation: From progesterone by microbiological oxidation.
Use: A steroid intermediate, biochemical research.

4-hydroxyproline. (Hyp; 4-hydroxy-2-pyrro-lidinecarboxylic acid).
HOC_4H_7NCOOH.

Properties: Colorless crystals. Very soluble in water; slightly soluble in alcohol; insoluble in ether. Optically active. *dl*-hydroxyproline, mp 261–262C with decomposition. *l*-hydroxyproline, mp 270C (natural). *d*-hydroxyproline, mp 274C.
Derivation: Hydrolysis of protein (gelatin), organic synthesis.
Use: Biochemical research. Available commercially as *l*-hydroxyproline.

3-hydroxy-1-propanesulfonic acid sultone.
See propane sultone.

2-hydroxy-1,2,3-propanetricarboxylic acid.
See citric acid.

hydroxy-2-propanone. (acetol; acetonyl alcohol; acetylcarbinol; hydroxyacetone; pyruvic alcohol).
CH_3COCH_2OH.
Properties: Colorless liquid. D 1.0824 at 20/20C, bp 146C, fp −17C. Soluble in water, alcohol, and ether. Combustible.
Derivation: (1) By action of potassium acetate or potassium formate on a solution of bromo- or chloroacetone in dry methanol; (2) by bacterial fermentation of propylene glycol.

Grade: Technical.
Use: Solvent for nitrocellulose.

α-hydroxypropionic acid. See lactic acid.

α-hydroxypropionitrile. See lactonitrile.

β-hydroxypropionitrile. See ethylene cyanohydrin.

2-hydroxypropyl acrylate. (HPA).
CAS: 999-61-1. $CH_2CHCOOCH_2CH_2OCH_3$.
A functional monomer used in manufacture of thermosetting acrylic resins for surface coatings.
Properties: A liquid. Mw 130.14, bp 77C at 5 mm Hg.
Hazard: TLV: 0.5 ppm. Corrosive to skin and eyes. Toxic by skin absorption.
Use: In manufacture of thermosetting resins for surface coatings.

2-hydroxypropylamine. See isopropanolamine.

hydroxypropyl cellulose.
CAS: 9004-64-2.
A cellulose ether with hydroxypropyl substitution.
Properties: White powder. Soluble in water, methyl and ethyl alcohols, and other organic solvents. Thermoplastic; can be extruded and molded. Insoluble in water >37.7C. Combustible.
Grade: FCC.
Use: Emulsifier, film former, protective colloid, stabilizer, suspending agent, thickener, food additive.

s-(3-hydroxypropyl)-*l*-cysteine.
CAS: 13189-98-5. $C_6H_{13}NO_3S$.
Hazard: A reproductive hazard.

hydroxypropylglycerin.
Properties: Pale straw-colored liquid. D 1.084 (25/25C), refr index 1.459 (25C), flash p 380F (193C), pour p −23C. Soluble in water and methanol. Combustible.
Use: Intermediate for alkyd resins and polyesters, plasticizer for cellulosics, glue, starch, etc.

hydroxypropyl methacrylate.
$CH_3CH_2OCH_2OOCC(CH_3):CH_2$.
Properties: Clear, mobile liquid. D 1.066 (25/16C), refr index 1.446 (25C), flash p 206F (96.6C). Limited solubility in water; soluble in common organic solvents. Combustible.
Use: Monomer for acrylic resins, nonwoven fabric binders, detergent lubricating-oil additives.

hydroxypropyl methylcellulose. (methylcellulose; propylene glycol ether).
Properties: White powder. Swells in water producing clear to opalescent, viscous, colloidal solution. Insoluble in anhydrous alcohol, ether, and chloroform. Combustible.

Grade: NF, FCC.
Use: Food products (except confectionery), as thickening agent, stabilizer, emulsifier; thickener in paint-stripping preparations.
N-β-hydroxypropyl-*o*-toluidine.
$CH_3C_6H_4NHCH_2CH(OH)CH_3$.
Properties: Amber color. Distillation range 170–180C (20 mm Hg), d 1.035–1.045 (20/20C), refr index 1.540–1.550 (20C).
Use: Dye intermediate.

4-hydroxy-2H-pyran-3,3,5,5(4H,6H)tetramethanol. See anhydroenneaheptitol.

2-hydroxypyridine-*N*-oxide. Bactericidal agent related to aspergillic acid, made from pyridine-N-oxide.

1-hydroxy-2-pyridine thione. (2-pyridine-thiol-1-oxide).
$C_5H_4NOH(S)$.
Apparently exists in equilibrium with the -SH form. Forms chelates with iron, manganese, zinc, etc.
Use: Fungicide, bactericide.

4-hydroxy-2-pyrrolidinecarboxylic acid. See hydroxyproline.

8-hydroxyquinoline. (8-quinolinol; oxyquinoline; oxine).
CAS: 148-24-3. C_9H_6NOH.
Properties: White crystals or powder; phenolic odor. Mp 73–75C; bp 267C. Darkens when exposed to light. Technical grade usually tan. Almost insoluble in water; soluble in alcohol, acetone, chloroform, benzene, and in formic, acetic, hydrochloric, and sulfuric acids, and alkalies
Grade: CP, technical.
Hazard: Toxic by ingestion. Questionable carcinogen.
Use: Precipitating and separating metals, preparation of fungicides, chelating agent, disinfectant.

8-hydroxyquinoline benzoate.
CAS: 86-75-9. $C_9H_6NOH:C_6H_5COOH$.
Properties: Yellowish-white crystals with a saffron odor. Mp 56–61C. Almost insoluble in water; soluble in alcohol and glycerol.
Use: Antiseptics, fungicide, recommended against Dutch elm disease.

8-hydroxyquinoline sulfate.
CAS: 134-31-6. $(C_6H_7NO)_2 \cdot H_2SO_4$.
Properties: Pale-yellow powder; slight saffron odor; burning taste. Melting range 167–182C. Soluble in water; slightly soluble in alcohol; and insoluble in ether.
Use: Antiseptic, antiperspirant, deodorant, fungicide.

4-hydroxysalicylic acid. See β-resorcylic acid.

12-hydroxystearic acid.
CAS: 106-14-9. $CH_3(CH_2)_5(CH_2O)(CH_2)_{10}COOH$.
A C_{18} straight-chain fatty acid with an -OH group attached to the carbon chain, mp 79–82C. It is produced by hydrogenation of ricinoleic acid. Combustible.
Use: Lithium greases, chemical intermediates.

1,12-hydroxystearyl alcohol.   (1,12-octade-canediol).
A long-chain fatty alcohol made by reduction of 12-hydroxystearic acid by replacing the -COOH group with a -CH_2OH. Combustible.
Properties: Boiling range 315–335C, mp 69C.
Use: Chemical intermediate, synthetic fibers, organic synthesis, pharmaceuticals, surface-active agents, plastics and resins, protective coatings.

hydroxysuccinic acid.   See malic acid.

2-hydroxy-1,2,3,3-tetrahydro-3h-pyrano (3,2-f)quinoline-8(7h)-one.
CAS: 128202-33-5. $C_{12}H_{11}NO_3$.
Hazard: Moderately toxic by ingestion.

4-hydroxy-3-(1,2,3,4-tetrahydro-3-(4-(4-(tri-fluoromethyl)phenoxy) phenyl)-1-naphtha-lenyl)2h-1-benzopyran-2-one.
CAS: 90035-14-6. $C_{32}H_{23}F_3O_4$.
Hazard: A poison by ingestion.
Use: Agricultural chemical.

4-hydroxy-3-(1,2,3,4-tetrahydro-3-(4-(trifluo-romethyl)phenyl)-1-naphthalenyl)2h-ben-zopyran-2-one.
CAS: 90034-99-4. $C_{26}H_{19}F_3O_3$.
Hazard: A poison by ingestion.
Use: Agricultural chemical.

hydroxythiospasmin.
CAS: 3569-58-2. $C_{18}H_{27}O_3S \cdot I$.
Hazard: A poison by ingestion.

hydroxytoluene.   See cresol.

α-hydroxytoluene.   See benzyl alcohol.

hydroxytoluic acid.   See cresotic acid.

1-hydroxytriacontane.   See 1-triacontanol.

(hydroxy)tributylstannane.   See tributyltin hydroxide.

hydroxytributylstannane-4,4-dimethylocta-noate.   See tributyltin neodecanoate.

2-hydroxy-3,5,6-trichloropyridine.
CAS: 6515-38-4. $C_5H_2Cl_3NO$.
Hazard: Moderately toxic by ingestion and skin contact.

hydroxytrimethylstannane.   See trimethyltin hydroxide.

5-hydroxytryptamine.   See serotonin.

4-hydroxyundecanoic acid, γ-lactone.   See γ-undecalactone.

hydrozincite.   (zinc bloom).
CAS: 12122-17-7. $Zn_5(OH)_6(CO_3)_2$.
A natural basic carbonate of zinc found in the upper zones of zinc deposits.
Properties: White to gray or yellowish color, dull to silky luster. Fluorescent in UV light. D 3.5–4.0, Mohs hardness 2.0–2.5.
Occurrence: Missouri, Pennsylvania, Utah, California, Nevada, Europe.
Use: An ore of zinc.

Hydrozon.   A system based on ozone.
Use: To treat drinking water that incorporates oxidation, coagulation, flocculation, filtration, and disinfection.

hydyne.   Mixture of 60% (by weight) of *uns*-dimethylhydrazine and 50% diethylenetriamine.
Use: High-energy fuel.

Hyform.   Water emulsions of pure paraffin wax, microcrystalline wax, or a modification of one of these waxes.
Use: Binders for pressed ceramic pieces, lubricants for die or mold release, and plasticizers during mold forming.

hygromycin.
CAS: 31282-04-9. $C_{23}H_{29}NO_{12}$.
Properties: White powder. Weakly acidic. Freely soluble in water and alcohol. Produced by *Streptomyces hygroscopicus*.
Use: Medicine (broad-spectrum antibiotic).

hygroscopic.   Descriptive of a substance that has the property of adsorbing moisture from the air, such as silica gel, calcium chloride, or zinc chloride. The water vapor molecules are held by the molecules of the agent, which is called a desiccant when used primarily for this purpose. Paper and cotton fabrics are hygroscopic, normally containing 5–8% water after standing in an atmosphere of normal humidity; they are usually kept in constant humidity rooms before use. Many dry chemicals are hygroscopic and should be kept in well-stoppered bottles or tightly closed containers.
See deliquescent.

"Hylene" [Parker].   TM for a series of organic isocyanates.
M-50, 50% methylenebis(4-phenyl isocyanate) $(C_6H_4NCO)_2$, in monochlorobenzene. Used in adhesives.
See diphenylmethane diisocyanate.
MP Bisphenol adduct of methylenebis(4-phenyl) isocyanate. Used as bonding agent for adhering "Dacron" *[Du Pont]* fiber to rubber compositions.

T Toluene-2,4-diisocyanate, used for urethane products.
TM 80% toluene-2,4-diisocyanate and 20% toluene-2,6-diisocyanate.
TM-65 65% 2,4-compound and 35% 2,6-compound.

hyoscyamine. (atropine; daturine; α-(hydroxymethyl)benzeneacetic acid; hyoscyamine; 1-hyoscyamine; 8-methyl -8-azobicyclo[3.2.1]-oct-3-yl ester; 1-hyoscyamine; 1αh,5αh-tropan-3α-ol(-)-tropate; 3α-tropanyl s-(-)tropate; 1-tropic acid ester with tropine; 1-tropine tropate; [(1R,5S)-8-methyl-8-azabicyclo[3.2.1]octan-3-yl]3-hydroxy-2-phenylpropanoate).
CAS: 101-31-5. $C_{17}H_{23}NO_3$.
An alkaloid that is the 3(s)-endo isomer of atropine.
Properties: White, crystalline, anticholinergic alkaloid; freely soluble in ethanol and dilute acids.
Derivation: From *hyoscyamus niger, Atropa belladonna, Datura stramonium, mandragora*, and other plants of the family Solanaceae.
Hazard: Extremely toxic, poison, paralyzes the parasympathetic nervous system by blocking the action of acetylcholine at nerve endings.
Use: Medically as preanesthetics, antispasmodics, analgesics, sedatives, antidotes to cholinesterase inhibitors.

hyoscyamus extract. A preparation of hyoscyamine by extraction from solanaceous plants of the genus *Hyoscyamus*.

Hyp. Abbreviation for hydroxyproline.

"Hypalon" *[Du Pont Performance].* TM for chlorosulfonated polyethylene, a synthetic rubber.
Properties: White chips. D 1.10–1.28. Resistant to ozone as well as the weather, oil, solvents, chemicals, and abrasion.
Use: Insulation for wire and cable, shoe soles and heels, automotive components, building products, coatings, flexible tubes and hoses, seals, gaskets, diaphragms. "Hypalon" 45 can accept large amounts of filler and is used as a binder for powdered metal to produce magnetic gaskets for doors and sheet goods for X-ray barriers.

hyperchromic effect. The large increase in light absorption at 260 nm occurring as a double-helical DNA is melted.

hyperfine structure. Structure of spectral line making each ordinary line a multiplet.

hyperglycemic-glycogenolytic factor. See glucagon.

hypergolic fuel. A liquid rocket fuel or propellant that consists of combinations of fuels and oxidizers, which ignite spontaneously on contact.

hyperon. Short-lived nuclear particle with mass greater than that of the neutron.

hypersorption. Process in which activated carbon selectively adsorbs the less-volatile components from a gaseous mix, while the more volatile components pass on unaffected. Particularly applicable to separations of low-boiling mixtures such as hydrogen and methane, ethane from natural gas, ethylene from refinery gas, etc.

hypertensin. See angiotensin.

hypertonic. (1) Possessing greater osmotic pressure than a standard; (2) having a higher than isotonic concentration.

hypnone. See acetophenone.

hypo-. (1) A prefix used in chemical terminology to indicate a compound (usually an acid) in its lowest oxidation state or containing the lowest proportion of oxygen in a series of compounds, e.g., nitric acid (HNO_3), nitrous acid (HNO_2), hyponitrous acid ($H_2N_2O_2$). (2) Common term for a photographic chemical.
See sodium thiosulfate.

hypoallergenic. A term used in the cosmetics industry and defined by the FDA as describing a cosmetic product that is less likely to cause adverse allergenic reactions than are competing products. Claims of hypoallergenicity must be substantiated by specific dermatological tests made by the manufacturer of the product.

hypobromous acid.
HBrO.
An unstable compound resulting from hydrolysis of bromine chloride.
Use: Bactericide and wastewater disinfectant.

hypochlorite. (hypochlorous acid).
CAS: 14380-61-1. ClO.
An oxyacid of chlorine containing monovalent chlorine that acts as an oxidizing or reducing agent.
Properties: Oxidizing, bleachng agents.
Hazard: Highly toxic, caustic, dangerous farm and household poisons.
Use: Dry or in solution as bleaches.

hypochlorite salt.
CAS: 7681-52-9. ClHONa.
A metallic salt of hypochlorous acid.

hypochlorite solution. An aqueous solution of a metallic salt of hypochlorous acid. Strong oxidizing agent.
Hazard: Irritant to skin and eyes.
Use: Bleaching of textiles, antiseptic agent.

hypochlorous acid.
CAS: 7790-92-3. HOCl.

Properties: Greenish-yellow aqueous solution. Highly unstable, weak acid. Decomposes to hydrogen chloride and oxygen. Can exist only in dilute solutions.

Derivation: Water solution of chloride of lime (bleaching powder).

Hazard: Irritant to skin and eyes.

Use: Textile and fiber bleaching, water purification, antiseptic.

See hypochlorite solution; calcium hypochlorite; sodium hypochlorite.

hypoglycine. Either of two closely related potent hypoglycemic agents isolated from the seeds and fruit of *Blighia sapida* (the akee), a poisonous tree indigenous to tropical West Africa, but cultivated in Florida and tropical America.

hypoglycine A. (α-amino-2-methylenecyclopropanepropanoic acid; 2-methylenecyclopropanealanine; 2-amino-4,5-methylenehex-5-enoic acid; α-amino-β-(2-methylenecyclopropyl)porpanoic acid; 2-amino-3-(2-methylidenecyclopropyl)propano acid).
$C_7H_{11}NO_2$.
A more potent substance than hypoglycine B that is heat liable and the ripe aril, if cooked, is considered a delicacy.

hypoglycine B. (alanine; *n*-l-®-glutamyl-3-(2-methylenecyclopropyl)alanine; γ-l-glutamyl-α-amino-β-(2-methylenecyclopropyl)propionic acid dipeptide; ®-l-glutamylhypoglycine; 2-amino-5-[[1-hydroxy-3-(2-methylidenecyclopropyl)-1-oxopropan-2-yl]amino]-5-oxopentanoic acid; hypoglycin b; glutamine, N-(1-carboxy-2-(methylenecyclopropyl)ethyl)-(7CI)).
CAS: 502-37-4. $C_{12}H_{18}N_2O_5$.
A hypoglycemic agent.

Properties: Needles from Me_2CO.

Hazard: Poison; teratogen.

"Hypol" [Dow]. TM for a family of hydrophilic polyurethane polymers derived from toluene diisocyanate (TDI). Foams with addition of water. Good fire retardancy, high additive loading.

Use: Consumer, medical, biomedical, and industrial applications.

α-hypophamine. See oxytocin.

β-hypophamine. See vasopressin.

hypophosphoric acid.
CAS: 6303-21-5. H_3O_6P (CHANGED FORMULA).

Properties: Platelike crystals. Commercially available as water solution. Readily forms hydrates. Mp 70C, mp (hydrated) 55C, the solution decomposes when concentrated.

Use: Baking powder (sodium salt).

hypophosphorous acid.
CAS: 6303-21-5. H_3PO_2.

Properties: Colorless, oily liquid or deliquescent crystals; sour odor. D 1.439, mp 26.5C. Soluble in water. A strong monobasic acid and reducing agent, sold in solution.

Derivation: Heating concentrated baryta water with white phosphorus and decomposing the barium hypophosphite with sulfuric acid, filtering the liquid, and concentrating under reduced pressure.

Grade: Technical, NF (30–32% solution, d 1.13), 50% purified.

Hazard: Fire and explosion risk in contact with oxidizing agents.

Use: Preparation of hypophosphites, electroplating baths.

hypoxanthine.
CAS: 68-94-0. $C_5H_4N_4O$.
An intermediate in the metabolism of animal purines, also widely distributed in the vegetable kingdom.

Properties: White-to-cream powder. Decomposes at 150C. Almost insoluble in cold water; slightly soluble in boiling water; and soluble in dilute acids and alkalies.

Derivation: Deamination of adenine, reduction of uric acid.

Use: Biochemical research.

hypoxanthine riboside. See inosine.

hypoxanthine riboside-5-phosphoric acid. See inosinic acid.

hysteresis. (Derived from the Greek word meaning "to lag behind.")
A retardation of the effect, as if from viscosity, when the forces acting upon the body are changing. A common illustration is the retentivity of induction in ferromagnetic materials such as iron and its alloys when the magnetizing force is changed. When such a substance is placed in a magnetizing coil and the magnetizing field is gradually increased to a given value, and then decreased, the magnetic induction, in decreasing does not follow the same relation to the magnetizing field that it did when the field was increasing, but lags behind the decreasing field. Hysteresis is analogous to mechanical inertia, and the energy lost is analogous to that lost in mechanical friction. It presents a major problem in the design of electrical machines with iron cores such as transformers and rotating armatures. In instruments designed for very high frequencies, the retardation and losses are so great as to render iron cores useless.
The stress–strain curves of vulcanized rubber also display hysteresis, in that strain (elongation, crystallization) persists when the deforming stress is removed, thus producing a hysteresis *loop* instead of a reversible pathway of the curves. This loop indicates a *loss* of resilient energy (Norman E. Gilbert).

Note: The simple diagram shown below is a generalized representation of a hysteresis loop.

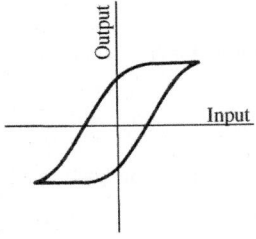

"Hystrene" [PMC].   (behenic acid).
TM for fatty acids.
Properties: Light colors and stable.
Grade: Food grade products.
Use: Waxes, textiles, pharmaceuticals, emulsifiers, and personal care products, lubricants, esters, chemical synthesis, and specialties.

"Hytrel" [Du Pont].   TM for a series of polyester elastomers.
Use: In sports equipment, automobile parts, hoses, tubes, belts, couplings, gears, seals, and electronic parts.

I

¹³¹**I.** See iodine-131.

IAA. Abbreviation for 3-indoleacetic acid.

IATA. Abbreviation for International Air Transport Association (Geneva), which publishes annual regulations for air shipment of hazardous materials.

IBIB. Abbreviation for isobutyl isobutyrate.

ibotenic acid. (α-amino-2,3-dihydro-3-oxo-5-isoxazoleacetic acid; α-amino-3-hydroxy-5-isoazoleacetic acid; amino-(3-hydroxy-5-isoxazolyl) acetic acid; azoleacetic acid; 2-amino-2-(3-oxo-1,2-oxazol-5-yl)acetic acid). $C_5H_6N_2O_4$. A isoxazole which, together with muscimol, is largely responsible for the toxicities of a number of mushrooms including those of the genus *Amanita*. It is a potent excitatory amino acid.
Derivation: Found in Amanita mushrooms.
Hazard: Neurotoxic; causes motor depression, ataxia, changes in mood, perceptions and feelings.
Use: Insecticide.

ibuprofen. (*p*-isobutyl-hydratropic acid; 2-(4-isobutylphenyl)propionic acid).
CAS: 15687-27-1. $C_{13}H_{18}O_2$.
Properties: Mw 206.31.
Use: Ingredient in over-the-counter pain relievers.

-ic. A suffix, used in naming inorganic compounds, that indicates that the central element is present in its highest oxidation state. Thus in ferric chloride ($FeCl_3$) the iron has an oxidation number of +3, equivalent to its valence: in an ionized state it would have three positive charges (Fe^{3+}). (A recommended change in this system of nomenclature is to use the common name of the element (iron) together with a Roman numeral showing the oxidation number; thus, ferric chloride would be iron(III) chloride.)

ICC. Abbreviation for ignition control compound; also Interstate Commerce Commission.

ice. (H_2O).
Properties: An allotropic, crystalline form of water. Mp 0C (32F), latent heat of melting 80 cal/g, d 0.91. Its property of melting under pressure accounts for slipperiness. Occurs in nature as ice I, but several other forms are known.

Use: Preservation of fish at sea, medicine (reduction of swelling).
See water.

Iceland moss.
Properties: A water-soluble gum that gels on cooling.
Derivation: A lichen growing in Scandinavia and Iceland.
Use: Flavoring for alcoholic beverages, food additive, cosmetics.

Iceland spar. A form of calcite having unique optical properties.
Use: Polarizing light (Nicol prism).

ichthammol. (ammonium ichthosulfonate).
Properties: Brownish-black, syrupy liquid. Burning taste. Tarry odor. Incompatible with acids, alkaloids, carbonate, hydroxides, mercuric chloride; soluble in water, alcohol-ether, or alcohol-ether-water mixture; partially soluble in alcohol and ether; miscible with glycerol.
Derivation: Aqueous solution of sulfonated ammonium compounds derived from the action of sulfuric acid on distillates from bituminous shales.
Grade: NF.
Use: Pharmaceutical products such as skin ointments, cosmetic preparations, special dermatological soaps.

Ichthymall. Ichthammol.

"Ichthyol" *[Ichthyol-Gesellschaft]*. TM for ichthammol.

ICP. See inductively coupled plasma emission spectroscopy.

-ide. A suffix used in naming compounds composed of two elements; in such names the first (electropositive) element retains its name without change, while the second (electronegative) bears the suffix *-ide* as a modification of the elemental name. Examples: sodium hydroxide, magnesium chloride, hydrogen sulfide. Similarly, oxygen is modified to oxide, fluorine to fluoride, phosphorus to phosphide, and carbon to carbide.

ideal black body. (perfect black body). A hypothetical object that absorbs all incident radiant energy emitted in the form of thermal radiation.

ideal crystal.   A crystal in which there are no defects or impurities.

ideal gas.   (perfect gas). A gas in which there is complete absence of cohesive forces between the component molecules; the behavior of such a gas can be predicted accurately by the ideal gas equation through all ranges of temperature and pressure. The concept is theoretical, since no actual gas meets the ideal requirement; carbon dioxide especially lacks conformity. The generalized ideal gas law is derived from a combination of the laws of Boyle and Charles, namely $pv = RT$, where p is pressure, v is volume, T is absolute temperature, and R is the gas constant.

ideal solution.   A solution that exhibits no change of internal energy on mixing and complete uniformity of cohesive forces. Its behavior is described by Raoult's law over all ranges of temperature and concentration.

identification limit.   Minimum quantity of a material that can be revealed by a given reaction or test.

identity period.   The repeating unit or monomer that occurs n times in a natural or synthetic polymer molecule; for example, the anhydroglucose unit in cellulose is enclosed in brackets:

IFT.   Abbreviation for the Institute of Food Technologists.

Ig.   Abbreviation for immunoglobulin.

Igenal.   A series of dyestuffs for chrome-tanned leather. Characterized by unusual tinctorial power.

ignition control compound.   A substance, such as methyl diphenyl phosphate or trimethyl phosphate, added to gasoline to control spark plug fouling, surface ignition, and engine rumble.

ignition point.   See autoignition temperature.

ignotine.   See carnosine.

IILE.   Ion-induced light emission.

"Illium" *[Stainless Foundry].*   TM for a series of stainless steel and nickel-base alloys with a high corrosion resistance.

illoxan.
CAS: 51338-27-3. $C_{16}H_{14}Cl_2O_4$.
Hazard: Moderately toxic by ingestion and skin contact. Low toxicity by inhalation.
Use: Agricultural chemical.

illuminating gas.   A flammable, poisonous mixture of gases that can be used to provide illumination.

ilmenite.   (titanic iron ore). $FeO \cdot TiO_2$. Iron-black mineral, black to brownish-red streak, submetallic luster, resembles magnetite in appearance but is readily distinguished by feeble magnetic character, d 4.5–5, Mohs hardness 5–6.
Occurrence: Widely in U.S., Canada, Sweden, the former U.S.S.R., India; also made synthetically.
Use: Titanium paints and enamel, source of titanium metal, welding rods, titanium alloys, ceramics.

Imag.   A potassium iodide mixture containing 90% potassium iodide and made free-flowing with 8% magnesium carbonate and 2% potassium hydroxide.

imaging contrast agent.   Added to a sample to increases the intensity of the signal detected by an imaging technique such as MRI and ultrasound. For example, gadolinium fixed to a targeting antibody is used in MRI.

imazaquin.
CAS: 81335-37-7. $C_{17}H_{17}N_3O_3$.
Hazard: Moderately toxic by ingestion and skin contact. Low toxicity by inhalation.
Use: Agricultural chemical.

Imhoff tank.   A reinforced concrete structure of considerable size (approximately 35 ft high) designed especially for sewage clarification. Its principal features are (1) an upper or sedimentation compartment in which in-flowing sewage deposits its suspended solids by gravity (residence time 2–3 hours), the free water being drawn off through an outlet, and (2) a separate lower compartment in which digestion of the accumulated sediment (sludge) takes place. The sludge is passed from the upper compartment to the digestion chamber through an inclined slot or channel. The gases generated by digestion are released through suitably located vents. The digested sludge is removed through outlet pipes at intervals of about 6 months. The dried sludge contains 2–3% ammonia and 1% phosphoric acid, which make it suitable as a soil conditioner.
See sewage sludge.

imidazo(2,1-a)isoquinoline,   5,6-dihydro-2-(*m*-methoxyphenyl)-.
CAS: 61001-21-6. $C_{18}H_{16}N_2O$.
Hazard: A reproductive hazard.

imidazole.   (glyoxalin).
CAS: 288-32-4.

HNCHNCHCH.   A dinitrogen ring compound. An
antimetabolite and inhibitor of histamine. Colorless crystals; mp 90C; bp 257C; soluble in water,
alcohol, and ether.
Use: Biological control of pests, especially fabric-
feeding insects, often in combination with *dl-p*-
fluorophenylalanine, an amino-acid inhibitor; also
as a contact insecticide in an oil spray. The mechanism is that of structural antagonism rather than
active toxicity.
See antihistamine; antagonist, structural.

4,5-imidazoledicarboxamide.   See glycarbylamide.

4-imidazole ethylamine.   See histamine.

2-imidazolidinone.   See ethylene urea.

2-imidazolidone.   See ethylene urea.

**2-(1h-imidazol-4-ylmethyl)-8h-indeno(1,2-
d)thiazole monofumarate.**
CAS:159081-23-9. $C_{14}H_{11}N_3S•C_4H_4O_4$.
Hazard: A poison by ingestion.

(1-imidazolyl)tributylplumbane.
CAS: 16128-42-0. $C_{15}H_{30}N_2Pb$.
Hazard: A poison.

imidazo(4,5-d)pyrimidine.   See purine.

imide.   A nitrogen-containing acid having two
double bonds.
See succinimide; phthalimide.

imidine.   Derivative of an acid amide. R–C:NH–
NH_2.

imine.   A nitrogen-containing organic substance
having a carbon-to-nitrogen double bond.

$$R-CH$$
$$\|$$
$$NH$$

Such compounds are highly reactive, even more so
than the carbon-nitrogen triple bond characteristic
of nitriles.

3,3′-iminobispropylamine.   (dipropylene tri-
amine; 3,3′-diaminodipropylamine).
CAS: 56-18-8. $H_2NC_3H_6NHC_3H_6NH_2$.
Properties: Colorless liquid. D 0.9307 (20/20C),
bp 240.6C, fp −6.1C, flash p 175F (79.4C) (CC).
Soluble in water and polar organic solvents. Com-
bustible.
Hazard: Toxic by ingestion and inhalation; irritant.

Use: Intermediate for soaps, dyestuffs, rubber chemicals, emulsifying agents, petroleum specialties,
insecticides, and pharmaceuticals.

iminodiacetic acid disodium salt hydrate.
(iminodiethanoic acid disodium salt hydrate).
CAS: 142-73-4. $HN(CH_2CO_2NA)_2•xH_2O$.
Properties: Crystalline solid.
Hazard: Irritant.
Use: Intermediate for surface-active agents, complex
salts, chelating agents, and aminocarboxylic acid
synthesis.

iminodiacetonitrile.
CAS: 628-87-5. $HN(CH_2CN)_2$.
Properties: Light-tan, crystalline solid. Mp 77–78C.
Soluble in water and acetone.
Use: Chemical intermediate.

l-N^6)-(1-iminoethyl)lysine hydrochloride.
CAS: 150403-89-7. $C_{18}H_{17}N_3O_2•ClH$.
Hazard: A poison.

iminourea.   See guanidine.

Imlar.   A vinyl resin-base finish used where
extreme resistance to abnormal chemical exposure
is required.

Immedial.   A series of sulfur dyestuffs. Characterized by very good fastness to light and good fastness to washing and perspiration.
Use: Dyeing of cotton and rayon.

immiscible.   Descriptive of substances of the
same phase or state of matter that cannot be uniformly mixed or blended. Though usually applied
to liquids such as oil and water, the term also may
refer to powders that differ widely in some physical
property, e.g., specific gravity, such as magnesium
carbonate and barium sulfate.
See miscibility.

immune endoglobulin.
CAS: 9007-83-4.
Hazard: Low toxicity by ingestion. Human systemic
effects.
Source: Human natural product.

immune response.   The generation of antibodies
to an antigen by a vertebrate organism.

immune serum.   Blood serum of an actively
immunized animal that contains specific antibodies. Such serum, when introduced into the body,
produces passive immunization by virtue of the
antibodies that it contains.

immune serum globulin.   A sterile solution of
globulins that contains those antibodies normally

present in adult blood. Over 90% of the total protein is globulin. It is a transparent, nearly colorless, nearly odorless liquid. Must be kept refrigerated.
Derivation: From a plasma or serum pool of venous or placental blood from 1000 or more individuals. Grade: USP.
Use: Medicine (immunology).
See antigen; globulin.

immune system. The complex group of cells and organs that defends the body against infection and disease.

immunizing agent. A substance that, when introduced into the body of an animal, will increase the concentration of antibodies in the blood to an extent that the immune system will be able to resist or overcome infection by a specific pathogen or group of closely related pathogens.

immunochemistry. That branch of chemistry concerned with the various defense mechanisms of the animal organism against infective agents, particularly the response between the body and foreign macromolecules (antigens) and the interaction between the products of the response (antibodies) and the agents that have elicited them. This involves study of the many proteins (serum globulins, enzymes, bacteria, and viruses) involved in these responses. It developed from the original work of Jenner (1775) and Pasteur (1880).
See antigen-antibody; complement.

immunogen. An antigen that is able to produce an immune response.

immunoglobulin. See globulin; immune serum globulin.

immunotherapy. Using the immune system to treat disease, for example, in the development of vaccines. May also refer to the therapy of diseases caused by the immune system.
See cancer.

immunotoxin. Any semi-synthetic conjugate of a toxic molecule and a specific immune molecule. The toxic moiety is transported by the conjugate to a site where it can exert a toxic effect.

IMP. (1) Abbreviation for inosine monophosphate.
See inosinic acid; sodium inosinate.
(2) Abbreviation for insoluble metaphosphate (Maddrell's salt).
See sodium metaphosphate.

impact strength. The ability of a material to accept a sudden blow or shock without fracture or other substantial damage, measured by standard impact-testing equipment (Izod, Charpy). It is a property of hard, friable materials such as metals,

hard rubber, engineering plastics, Portland cement, glass, etc.

impalpable. Descriptive of a state of subdivision of particles so fine that the individual particles cannot be distinguished as such by pressing a powder between the thumb and index finger.

impeller. A type of agitator used in mixing or blending fluids of low viscosity, usually in a cylindrical chamber, either open or closed. The motion induced by an impeller is a combination of flow and turbulence, the proportion of each depending on the size, speed, and position of the impeller. In common use are the marine propeller, the turbine, and the helical ribbon types. Propeller and turbine impellers are attached to a power-driven rotating shaft that enters the container either vertically (top entering) or at an angle (side entering), they may be centered in a liquid or placed off-center, depending on the flow pattern desired. The propeller type has from two to four elliptical blades, whereas the turbine has a number of rectangular blades set vertically or at an angle. A wide range of flow-turbulence patterns can be obtained with either type. Helical ribbon impellers are used for viscous liquids and dry powders. Many variations of impellers are available for liquids up to medium viscosity for a multitude of special mixing techniques.
See agitator; mixing.

imperial green. See copper acetoarsenite.

Implex. Thermoplastic, high-impact acrylic molding powder, supplied in natural and colored forms. Maximum toughness, gloss, stain- and heat-resistant grades.
Use: Shoe heels, business-machine and musical instrument keys, housings, automotive parts, knobs, metalized parts, etc.

imprinting. A biochemical phenomenon that determines, for certain genes, which one of the pair of alleles, the mother's or the father's, will be active in that individual.

impurity. The presence of one substance in another, often in such low concentration that it cannot be measured quantitatively by ordinary analytical methods. It is impossible to prepare an ideally pure substance. In certain metal crystal lattices, foreign substances can exist in as low a concentration as one-millionth of an atomic percent. For example, arsenic atoms are present in germanium crystals in this percentage; this fact is largely responsible for the semiconducting properties of germanium. Here the impurity is beneficial, but often it is detrimental, for example, in graphite used as a moderator in nuclear reactors, and in many metallic catalysts. In the air, trace amounts of sulfur dioxide and carbon monoxide are potentially

dangerous impurities in concentration of 5 ppm of sulfur dioxide and 50 ppm of carbon monoxide.
See purity; chemical; trace element; air pollution; semiconductor; purification.

inborn errors of metabolism. Inherited diseases resulting from alterations in genes that code for enzymes.

incendiary gel. (1) Mixture of thermite suspended in oil set to a jelly with a small amount of soap; it undergoes spontaneous ignition on contact with air. Another type may contain magnesium in jellied oil. (2) Jellied gasoline combined with thickening agents such as napalm or finely divided magnesium.

incineration. Disposal of solid and liquid organic waste materials by burning at temperatures from 1200 to 1500C. This method is approved by the EPA for use on very toxic organic chemicals and chemical wastes. Use of specially equipped incinerator ships for burning chemical wastes at sea has become a common practice.
See waste control.

inclusion complex. (adduct). An unbonded association of two molecules in which a molecule of one component is either wholly or partly locked within the crystal lattice of the other. There are several types of such complexes, the most familiar being the so-called clathrates (from Latin, "crossbars of a grating"). The clathrate compound $3C_6H_4(OH)_2 \cdot SO_2$ may be depicted as

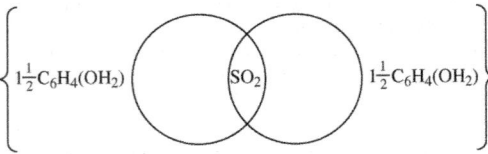

where the interlocked rings denote mutual enclosure of two identical cages. The formula for any clathrate compound is determined by the ratio of available cavities to the amount of cage material. Inclusion compounds can be used to separate molecules of different shapes, e.g., straight-chain hydrocarbons from those containing side chains, as well as structural isomers. They can also be used as templates for directing chemical reactions.
See clathrate compound; gas hydrate.

"Incoloy" [Huntington]. TM for a series of corrosion-resistant alloys of nickel, iron, and chromium.

incomplete penetrance. The gene for a condition is present, but not obviously expressed in all individuals in a family with the gene.

Inconel. A series of corrosion-resistant alloys of nickel and chromium.

IND. Abbreviation for investigational new drug.

Indalone. n-butyl mesityl oxide oxalate.
Use: Insect repellents that can be applied directly on the skin.
See butopyronoxyl.

indan. (hydrindene; 2,3-dihydroindene).
CAS: 496-11-7.

$\overline{CHCHCHCHCCCH_2CH_2CH_2}$.

Bicyclic.
Properties: Colorless liquid. Bp 176.5C, fp −51.4C, refr index 1.5388 (16.4C), d 0.965 (20/4C). Insoluble in water; soluble in alcohol and ether. Combustible.
Derivation: From coal tar.
Hazard: Irritant to skin and eyes.
Use: Organic synthesis.

indanthren blue 3g. See 6,15-dihydrohydroxy-5,9,14,18-anthrazinetetrone.

indanthrene. See indanthrone.

indanthrene yellow. See flavanthrene.

indanthrone. (Indanthrene Blue R; 6,15-dihydro-5,14,18-anthrazinetetrone).
CAS: 81-77-6. $C_{28}H_{14}N_2O_4$.
A blue vat dye or pigment. The molecule consists of two anthraquinone nuclei linked through two −NH groups.
Properties: Heat- and light-stable blue powder. Decomposes at 470C. Soluble in concentrated sulfuric acid and dilute alkaline solutions.
Derivation: By fusion of β-aminoanthroquinone with caustic potash in the presence of potassium nitrate.
Use: Dyeing unmordanted cotton, pigment in quality paints and enamels.

indazol blue r. See C.I. basic blue 16.

indene.
CAS: 95-13-6.

$\overline{CHCHCHCHCCCHCHCH_2}$.

Properties: Colorless liquid. D 0.996 (20/4C), fp −3.5C, bp 182C, refr index 1.5726 (25C), flash p 173F (78.3C). Insoluble in water; soluble in most organic solvents; oxidizes readily in air. Forms polymers on exposure to air and sunlight. Combustible.
Derivation: Contained in the fraction of crude coal tar distillates that boils from 176–182C.

Hazard: Toxic by inhalation.
Use: Preparation of coumarone-indene resins, intermediate.
See coumarone.

indenolol hydrochloride.
CAS: 68906-88-7. $C_{15}H_{21}NO_2 \cdot ClH$.
Hazard: Moderately toxic by ingestion.

indeno(1,2,3-cd)pyrene.
CAS: 193-39-5. $C_{22}H_{12}$.
Properties: Yellow crystals from cyclohexane; bright-yellow plates from pet ether.
Hazard: Possible carcinogen.

independent assortment.   During meiosis each of the two copies of a gene is distributed to the germ cells independently of the distribution of other genes.
See linkage.

indeterminate error.   An error that cannot be identified.

indian bean.   (Indian licorice; mienie-mienie). *Abrus precatorius*.

indian ox.   *Bos indicus*.

Indian red.   (iron saffron). A red (maroon) pigment made by calcining copperas to obtain red ferric oxide. Fine particle size.
Use: Pigment in paint, rubber, and plastics; polishing agent.
See iron oxide red; rouge.

Indian yellow.   (1) (Aureolin) A yellow pigment distinguished by being unaffected by hydrogen sulfide. It is durable, without action on other pigments, and is permanent in oils and water-color. It consists of a double nitrite of cobalt and potassium and is prepared by adding excess of potassium nitrite solution to a solution of cobalt nitrate acidified with acetic acid. (2) Sometimes made of the yellow die primuline.
See cobalt potassium nitrite.

indicator.   An organic substance (usually a dye or intermediate) that indicates by a change in its color the presence, absence, or concentration of some other substance, or the degree of reaction between two or more other substances. The most common example is the use of acid-base indicators such as litmus, phenolphthalein, and methyl orange to indicate the presence or absence of acids and bases, or the approximate concentration of hydrogen ion in a solution.
Use: Analytical chemistry.

The pH ranges of several typical indicators are as follows:

alizarin yellow R	10.1–12.0	yellow to red
methyl orange	3.1–4.4	red to yellow
phenolphthalein	8.3–10.0	colorless to red
phenol red	6.8–8.4	yellow to red
litmus	4.4–8.3	red to blue
Congo red	3.0–5.2	blue to red
bromothymol blue	6.0–7.6	yellow to blue
chlorophenol red	5.2–6.8	yellow to red
cresol purple	7.4–9.0	yellow to purple

See titration; pH.

indicator   yellow.   The   chromophore   of rhodopsin.

indigo.   (indigotin; synthetic indigo blue; CI 73000).
CAS: 482-89-3. $C_{16}H_{10}N_2O_2$.
A double indole derivative.

Properties: Dark blue, crystalline powder; bronze luster. D 1.35, sublimes at 300C (decomposes). Soluble in aniline, nitrobenzene, chloroform, glacial acetic acid, and concentrated sulfuric acid; insoluble in water, ether, and alcohol.
Derivation: From aniline and chloroacetic acid, and fusing the resulting phenylglycine with alkali and sodium amide. Formerly from plants of genus *Indigofera*.
Grade: Technical, pure.
Use: Textile dyeing and printing inks, manufacture of indigo derivatives, paints, analytical reagents.

indinavir sulfate.
CAS: 157810-81-6. $C_{36}H_{47}N_5O_4 \cdot H_2O_4S$.
Hazard: Moderately toxic by ingestion.

indirect dye.   A mordant dye.

indium.
CAS: 7440-74-6.   In.
Metallic element atomic number 49, of group IIIA of the periodic table, aw 114.82, valences of 1, 3; 2 stable isotopes.
Properties: Ductile shiny silver-white metal. D 7.31 (20C), mp 156C, bp 2075C, Mohs hardness 1.2. Softer than lead. Soluble in acids; insoluble in alkalies. Corrosion-resistant at room temperature. Oxidizes readily at higher temperatures.

Occurrence: Not found native but in a variety of zinc and other ores. The indium content is generally very low, rarely exceeding 0.001%. Indium-bearing ores occur in the western U.S., Canada, Peru, Japan, Europe, the former U.S.S.R.

Available forms: Small ingots or bars, shot, pencils, wire, sheets, powder, single crystals.

Purity: Technical, high purity (below 10 ppm impurities).

Hazard: Metal and its compounds are toxic by inhalation.

Use: Automobile bearings, electronic and semiconductor devices, low-melting brazing and soldering alloys, reactor control rods, electroplated coatings on silver-plated steel aircraft bearings that are tarnish resistant, radiation detector.

indium acetylacetonate. $In(C_5H_7O_2)_3$.
Properties: Mp 186C.
Hazard: See indium.
Use: Catalyst.

indium antimonide.
CAS: 1312-41-0.   InSb.
Properties: Crystalline solid. Mp 535C.
Hazard: See indium; antimony.
Use: Electronic grade used for semiconductor devices and infrared detector, computer technology (Hall effect).

indium arsenide.
CAS: 1303-11-3.   InAs.
Properties: Crystals. Mp 943C. Insoluble in acids.
Hazard: See indium; arsenic.
Use: Electronic grade for semiconductor devices, in injection lasers.

indium chloride.   (indium trichloride).
CAS: 10025-82-8. $InCl_3$.
Properties: White powder. Deliquescent. D 3.46 (25C), mp 586C, sublimes at 300C. Soluble in alcohol and water.
Derivation: Direct union of the elements or by the action of hydrochloric acid on the metal.
Hazard: See indium.

indium cyanide.
CAS: 13074-68-5. InC_3N_3.
Properties: Colorless, water-soluble, powder.
Hazard: Poisonous.

indium oxide.   (indium sesquioxide; indium trioxide).
CAS: 1312-43-2. In_2O_3.
Properties: White to light-yellow powder in both amorphous and crystalline forms, depending on temperature. D 7.179. Soluble in hot acid (amorphous); insoluble (crystals).
Derivation: By burning the metal in air or heating the hydroxide, nitrate, or carbonate.
Hazard: See indium.
Use: Manufacture of special glasses.

indium phosphide.
CAS: 22398-80-7.   InP.
A brittle metallic mass, mp 1070C, slightly soluble in mineral acids.
Hazard: Probable carcinogen. See indium.
Use: Electronic grade in semiconductor devices, injection lasers, and experimental solar cells.

indium sulfate.
CAS: 13464-82-9. $In_2(SO_4)_3$.
Properties: Grayish powder. Deliquescent. D 3.438. Soluble in water. Decomposed by heat.
Hazard: See indium; tellurium.

indium telluride.
CAS: 1312-45-4. In_2Te_3.
Properties: Black, friable crystals; d 5.8, mp 665C.
Hazard: See indium.
Use: Semiconductor technology.

indium trichloride.
CAS: 10025-82-8. $InCl_3$.
Properties: Tan to yellow deliquescent crystals. D 4.0, mp 585C (sublimes). Soluble in water. Keep in closed containers.
Hazard: See indium.
Use: Electroplating baths.

Indo Carbon.   Sulfur dyestuffs.
Use: Dyeing and printing of cotton and rayon.

Indocin.   The antiinflammatory drug indomethacin.
Use: Treatment of arthritis.

indole.   (2,3-benzopyrrole).
CAS: 120-72-9.

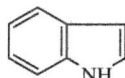

Properties: White to yellowish scales, turning red on exposure to light and air; unpleasant odor in high concentration but pleasant in dilute solutions. Mp 52C, bp 254C. Soluble in alcohol, ether, hot water, and fixed oils; insoluble in mineral oil and glycerol.
Derivation: From indigo and by numerous syntheses. Also can be produced from the 220–260C fraction of coal tar.
Grade: Technical, CP, FCC.
Hazard: A carcinogen.
Use: Chemical reagent, perfumery.

3-indoleacetic acid.   (IA; IAA; β-indoleacetic acid).
CAS: 87-51-4. $C_8H_6NCH_2COOH$.
A plant hormone.
Properties: Crystals. Mp 168–170C. The natural material is levorotatory, specific rotation −3.8 degrees in alcohol (20C). Insoluble in water and chloroform; soluble in alcohol and ether.

Use: Agriculture and horticulture, growth-promoting hormone, plant cell enlarger.
See auxin; plant growth regulator.

indole acetonitrile. A naturally occurring auxin.

indole alkaloid. $C_{21}H_{24}N_2O_3$. Any of a group of alkaloids that contain an indole nucleus.
Derivation: From tryptophan or phenylalanine.

indole-α-aminopropionic acid. See tryptophan.

3-indolebutyric acid. (hormodin).
CAS: 133-32-4. $C_8H_6N(CH_2)_3COOH$.
Properties: White or off-white powder; essentially odorless. Mp 123C. Insoluble in water; soluble in alcohols and ketones.
Use: Plant hormone, especially used in rooting plants.
See auxin; plant growth regulator.

indole-2,3-dione. See isatin.

indomethacin. (USAN name for 1-(p-chlorobenzoyl)-5-methoxy-2-methylindole-3-acetic acid).
CAS: 53-86-1. $C_{19}H_{16}ClNO_4$.
Use: Medicine for treatment of arthritis.

indomethacin sodium trihydrate.
CAS: 74252-25-8. $C_{19}H_{15}ClNO_4 \cdot 3H_2O \cdot Na$.
Hazard: A poison by ingestion.

indoor air. Air in a confined area such as a home or office building typically treated by mechanical ventilation.

indophenine reaction. Intense blue coloration given by thiophene or its derivatives when treated with concentrated sulfuric acid with a trace of isatin.

"Indopol" [INEOS]. TM for viscous polybutenes.
Available forms: Liquid.
Grade: 10 viscosity.
Use: In lubricating-oil additives for ashless dispersant; processing, rolling, and compressor oils; caulks, sealants, adhesives; as an elastomeric process aid; and cling improver in cling films.

indospicine monohydrochloride monohydrate.
CAS: 16377-01-8. $C_7H_{15}N_3O_2 \cdot ClH \cdot H_2O$.
Hazard: A reproductive hazard.
Source: Natural product.

indoxacarb.
CAS: 173584-44-6. $C_{22}H_{17}ClF_3N_3O_7$.
Hazard: A poison by ingestion. Low toxicity by inhalation and skin contact.
Use: Agricultural chemical.

induced enzyme. (adaptive enzyme; inducible enzyme). An enzyme whose production is induced or greatly accelerated by a specific small molecule which is the substrate of the enzyme or is very similar in structure to the substrate.

induced fit. A change in the conformation of an enzyme in response to substrate binding that renders the enzyme catalytically active. It is one of the models used to explain substrate specificity.

inducer. A signal molecule that, when bound to a regulatory protein, produces an increase in the expression of a given gene.

inducible protein. A protein that is synthesized in differing amounts depending on cellular signals.

inducing agent. (inductor; enzyme inducing agent). A substance, usually a small molecule, that promotes the production of a specific enzyme in an organism.

induction. An increase in the expression of a gene in response to a change in the activity of a regulatory protein.

inductively coupled plasma emission spectroscopy. (ICP). An analysis system in which the energy of a plasma excites the atoms of an injected sample causing the excited atoms to emit light at signature wavelengths. Three-dimensional computer-generated images are used to interpret the results.

indulines. Blue azine dyestuffs obtained by the interaction of amino-azo benzene and aniline hydrochloride.

indurite. Explosive containing 40% guncotton and 60% nitrobenzene.

"Industrene" [Crompton]. TM for fatty acids.
Use: Rubber compounding, foam dispersants, lubricants, water repellents, polishes, metallic soaps, crayons, alkyd resins, mineral flotation adjuvants, and emulsifiers.

industrial alcohol. See alcohol, industrial.

industrial carbon. See carbon, industrial.

industrial chemistry. See chemical technology; chemical process industry.

industrial diamonds. See diamonds, industrial.

industrial dust. See dust, industrial.

industrial waste. See waste control; chemical waste.

inert. A term used to indicate chemical inactivity in an element or compound. Helium, neon, and argon are practically inert gaseous elements; carbon dioxide is a gaseous compound of low activity. Ingredients added to mixtures chiefly for bulk and weight purposes are said to be inert.
See noble; extender.

inert gas. Gaseous element of group 18 of the periodic table, such as helium or argon, which is nonreactive under ordinary conditions. These gases are not completely unreactive and inert gas compounds have been synthesized.

inert ingredient. A substance that is not an active ingredient. When discussing pesticides, it is a component of the formulation such as solvent, carrier or surfactant that does not attack a target pest, but can be chemically or biologically active causing health and environmental problems.

infinite dilution. Point of maximum dissociation of an electrolyte at which point the greatest amount of conductivity has been reached.

informatics. The study of the application of computer and statistical techniques to the management of information. In genome projects, informatics includes the development of methods to search databases quickly, to analyze DNA sequence information, and to predict protein sequence and structure from DNA sequence data.
See bioinformatics.

informational macromolecules. Biomolecules containing information in the form of specific sequences of different monomers; for example, many proteins, lipids, polysaccharides, and nucleic acids.

informed consent. An individual willingly agrees to participate in an activity after first being advised of the risks and benefits.
See privacy.

infrared. The region of the electromagnetic spectrum including wavelengths from 0.78 micron to approximately 300 microns (i.e., longer than visible light and shorter than microwave).
Use: Spectroscopic analysis, medicine, baking of enamels, drying, photography.
See radiation.

infrared spectroscopy. An analytical technique that may measure either (1) the range of wavelengths in the infrared that are absorbed by a specimen, which characterize its molecular constitution (absorption spectroscopy), or (2) the infrared waves emitted by excited atoms or molecules (emission spectroscopy). Extremely hot bodies (stars) emit spectra in which the atomic composition can be determined by characteristic lines such as the sodium D line in the sun's spectrum. Infrared absorption bands identify molecular components and structures, some of which are

Absorption band (μm)	Structure indicated
2.3–3.2	OH and NH groups; H$_2$
3.2–3.3	aromatics, olefins
3.33–3.55	aliphatics
5.7–6.1	aldehydes, ketones, acids, amides

See microwave spectroscopy; absorption (2).

"Infrax" [Carborundum]. TM for a refractory insulation used as primary linings of fuel-fired and electric furnaces only when protected by a cement facing. Available in brick form.

infusion. An aqueous solution obtained by treating drugs with hot or cold water, without boiling. Generally prepared by pouring boiling water upon the vegetable substance and macerating the mixture in a tightly closed vessel until the liquid cools. When not otherwise specified, they are of 5% strength by weight.

infusorial earth. See diatomaceous earth.

ingot iron. Highly refined steel with a maximum of 0.15% impurity. Due to high purity it has excellent ductility and resistance to rusting.

ingrain dye. An insoluble dye developed by impregnating a fabric with one or more intermediates and then producing the dye by reaction with a different intermediate.

inherit. In genetics, to receive genetic material from parents through biological processes.

inherited. See inherit.

"Inhibisil" [PPG]. TM for a non-toxic corrosion inhibiting silica pigment for metal finishes.
Use: Paints and coatings.

inhibitor. (1) A compound (usually organic) that retards or stops an undesired chemical reaction, such as corrosion, oxidation or polymerization. Examples are acetanilide which retards decomposition of hydrogen peroxide and salicylic acid, used to prevent prevulcanization of rubber. Such substances are sometimes called negative catalysts. (2) A biological antagonist used to retard growth of pests and insects and in medicine.
See antagonist, structural; antioxidant.

Inhibitor NPH. A synthetic organic chemical that provides an effective means of controlling hard polymer formation in synthetic rubber production.

Properties: Fine white to yellow-white platelets; ammoniacal odor. Bulk d 5 lb/gal, mp 160–164C (with decomposition).

inhibitory enzyme. An enzyme that blocks a chemical reaction.

initiating explosive. See explosive, initiating.

initiation codon. (AUG). Codes for the first amino acid in all polypeptide sequences, which is N-formyl-methionine in prokaryotes and methionine in eukaryotes.

initiation complex. A complex of all ribosomal components, a mRNA, and the initiating Met-tRNA^{Met} (eukaryotes) or fMet-tRNAMet (prokaryotes).

initiator. An agent used to start the polymerization of a monomer. Its action is similar to that of a catalyst, except that it is usually consumed in the reaction. Organic peroxides and similar compounds are often used and shortwave radiation has a similar initiating effect. Free radicals usually play a part. See activator; free radical.

injection molding. A plastic-molding operation, introduced about 1935, performed in a single machine capable of producing both small articles of complex geometry (combs), and large units that could not be made economically in any other way (auto body parts, tubs, etc.). Though used primarily for thermoplastics, the injection method can be modified to handle thermosets. A simplified description is as follows: (1) A molding powder is fed into the heating chamber of the machine, which holds several times as much material as is necessary to fill the mold. The powder is heated to a viscous liquid. (2) An amount of molding powder that is just sufficient to fill the mold cavity is then forced into the rear of the heating chamber by a plunger, thus injecting an equal amount of liquid plastic from the front of the heating chamber into the mold. (3) The material remains in the mold under high pressure until it cools, and is then ejected. The rheological properties of the fluid plastic are of critical importance, as it must flow readily through the sprue and mold gate and fill the mold uniformly. The amount of material injected into a mold can range from less than 1 ounce to 25 pounds or more. Modern injection-molding machines have many specialized mechanical features and are of impressive dimensions.

Of comparatively recent development is the technique of reaction injection molding, in which a two-part semiliquid resin blend is flowed through a nozzle into the mold cavity, where it polymerizes as a result of chemical reaction. One of the components contains the activating agent or curative; the two parts must be mixed with the greatest care for satisfactory results. Isocyanate resins (urethanes) and epoxy resins are well adapted to this fast, low-energy process.

ink. See printing ink; writing ink.

Inoc. Bioaugmentation products.
Use: In the treatment of wastewater for removal of organics.

Inol k65.
CAS: 188364-50-3.
Hazard: Moderately toxic by ingestion and skin contact. A severe skin irritant. An antioxidant.
Use: In a wide range of petroleum products including lubricants, industrial oils, and refinery streams.

inorganic anhydride. An anhydride of an inorganic acid that is usually an oxide of a nonmetallic element.

inorganic carcinogen. Any inorganic substance that may not act directly on DNA but may alter DNA indirectly and is carcinogenic.

inorganic catalyst. An inorganic substance, such as a finely divided metal that acts as a catalyst.

inorganic chemistry. A major branch of chemistry that is generally considered to embrace all substances except hydrocarbons and their derivatives, or all substances that are not compounds of carbon, with the exception of carbon oxides and carbon disulfide. It covers a broad range of subjects, among which are atomic structure, crystallography, chemical bonding, coordination compounds, acid-base reactions, ceramics, and the various subdivisions of electrochemistry (electrolysis, battery science, corrosion, semiconduction, etc.). It is important to state that inorganic and organic chemistry often overlap. For example, chemical bonding applies to both disciplines, electrochemistry and acid-base reactions have their organic counterparts; catalysts and coordination compounds may be either organic or inorganic.

Regarding the importance of inorganic chemistry, R. T. Sanderson has written: "All chemistry is the science of atoms, involving an understanding of why they possess certain characteristic qualities and why these qualities dictate the behavior of atoms when they come together. All properties of material substances are the inevitable result of the kind of atoms and the manner in which they are attached and assembled. All chemical change involves a rearrangement of atoms. Inorganic chemistry [is] the only discipline within chemistry that examines specifically the differences among all the different kinds of atoms."

inosine. (hypoxanthine riboside).
CAS: 58-63-9. $C_{10}H_{12}N_4O_5$.

An important intermediate in animal purine metabolism. Also available as its barium salt.

Properties: (Dihydrate.) Crystallizes in needles from water. Mp 90C (anhydrous), 218C (decomposes). Levorotatory in solution. Slightly soluble in water; soluble in alcohol.

Derivation: By deamination of adenosine.

Use: Biochemical research.

inosine dialdehyde.
CAS: 23590-99-0. $C_{10}H_{10}N_4O_5$.
Hazard: Moderately toxic.

inosine 5-diphosphate. ([(2R,3S,4R,5R)-3,4-dihydroxy-5-(6-oxo-3H-purin-9-yl)oxolan-2-yl]methyl phosphono hydrogen phosphate).
CAS: 81012-88-6. $C_{10}H_{14}N_4O_{11}P_2$.
An inosine nucleotide containing a pyrophosphate group esterified to C5 of the sugar moiety.

inosinic acid. (IMP; hypoxanthine riboside-5-phosphoric acid).
CAS: 131-99-7. $C_{10}H_{13}N_4O_8P$.
An important intermediate in the synthesis and metabolism of animal purines.

Properties: Syrup; agreeable sour taste. Freely soluble in water and formic acid; very sparingly soluble in alcohol and ether.

Derivation: From meat extract or by enzymatic deamination of muscle adenylic acid.

Use: Biochemical research, flavor enhancer.
See sodium inosinate.

inositol. (hexahydroxycyclohexane).
CAS: 87-89-8. $C_6H_6(OH)_6 \cdot 2H_2O$.
A constituent of body tissue. There are nine isomeric forms of inositol (*myo*-inositol or *meso*-inositol or specifically the *cis*-1,2,3,5-*trans*-4,6-hexahydroxycyclohexane has vitamin activity).

Properties: (I-inositol.) White crystals; odorless; sweet taste. Mp 224–227C, d 1.524 (dihydrate), 1.752 (anhydrous), dihydrate melts at 215–216C. Soluble in water; insoluble in absolute alcohol and ether. Stable to heat, strong acid, and alkali. Amounts are expressed in milligrams of inositol.

Source: Food Source: vegetables, citrus fruits, cereal grains, liver, kidney, heart, and other meats. Commercial Source: corn steep liquor by precipitation and hydrolysis of crude phytate.

Grade: NF, FCC.

Use: Medicine, nutrition, intermediate.

inositol hexaphosphoric acid. See phytic acid.

inositol hexaphosphoric acid ester, sodium salt. See sodium phytate.

INPC. Abbreviation for isopropyl-*N*-phenyl-carbamate.
See IPC.

INS. Ion neutralization spectroscopy.

insecticide. A type of pesticide designed to control insect life that is harmful to humans, either directly as disease vectors or indirectly as destroyers of crops, food products, or textile fabrics. General types are as follows: (1) Inorganic: arsenic, lead, and copper (inorganic compounds and mixtures); the use of these has diminished sharply in recent years because of the development of more effective types less toxic to humans. (2) Natural organic compounds, such as rotenone and pyrethrins (relatively harmless to humans since they quickly decompose to nontoxic compounds), nicotine, copper naphthenate, and petroleum derivatives. (3) Synthetic organic compounds: (a) chlorinated hydrocarbons such as DDT, dieldrin, endrin, chlordane, lindane, *p*-dichlorobenzene; (b) the organic esters of phosphorus (the parathions and related substances). (4) Of comparatively recent development are pyrethroids, or insect growth regulators, which act as neurotoxins, preventing larvae from becoming adult forms (juvenile hormones); and metabolic inhibitors, e.g., imidazole, which function as structural antagonists.

Hazard: Insecticides are toxic to humans in varying degrees. Among the safest are the pyrethrins, rotenone, and methoxychlor. Most of the organophosphorus types (parathion and related compounds) are highly toxic but are reasonably biodegradable. The chlorinated hydrocarbons resist biodegradation; their use has been restricted and in some cases (DDT) banned for agricultural application, due to their harmful ecological effects. EPA is constantly monitoring new insecticides and issues warnings or restrictions on their use; its approval is required before they can be registered.
See pesticide; fumigant; herbicide; rodenticide; repellent.

insect wax. See shellac; Chinese insect wax.

insert. In a complete plasmid clone, there are two types of DNA—the "vector" sequences and the "insert". The vector sequences are those regions necessary for propagation, antibiotic resistance, and all those functions necessary for useful cloning. The insert is the piece of DNA of interest.

insertion. A chromosome abnormality in which a piece of DNA is incorporated into a gene and thereby disrupts the gene's normal function.
See chromosome; DNA; gene; mutation.

insertion mutation. A mutation caused by insertion of one or more extra bases between bases in DNA, or a mutagen which itself is inserted between bases.
See insertion.

insertion sequence. The specific base sequences at either end of a transposon which allow for insertion into DNA.

in situ hybridization. Use of a DNA or RNA probe to detect the presence of the complementary DNA sequence in cloned bacterial or cultured eukaryotic cells.

inspired gas. Any gas that is inhaled and undergoes humidification at body temperature.

instantizing. See agglomeration (2).

Institute of Food Technologists. (IFT). Founded in 1939, the Institute of Food Technologists is a nonprofit scientific society with members working in food science, food technology, and related professions in industry, academia and government. The Institute advances the science and technology of food through the exchange of knowledge. Its world headquarters are at 525 W. Van Buren, Ste 1000, Chicago, IL 60607. Website: http://www.ift.org

instrument. Any of a wide variety of devices used for one of the following purposes: (1) observation (microscope); (2) measurement (thermometer, thermocouple, flowmeter, balance); (3) chemical analysis (spectrometer).
See analytical chemistry; instrumentation.

instrumentation. (1) *Plant.* An inclusive term for sensing devices that measure, record, and control temperature, flow rate, thickness, pH, liquid level, and other process variables on a continuous basis. Some types, e.g., thickness gauges, utilize radioisotopes. Particularly sophisticated computerized instrumentation is required in petroleum refining and nuclear reactor control. (2) *Laboratory.* A broad range of analytical techniques and devices utilized in many types of chromatography and spectroscopy. Significant advances in analytical instrumentation have been made in recent years, e.g., in liquid chromatography (HPLC). A notable event in this field is the annual Pittsburgh Conference on Analytical Chemistry and Applied Spectroscopy. (3) *General.* Of special significance for both laboratory and process control is the introduction of fiber optical devices that can transmit signals from many remote locations. They are particularly useful in analyzing samples of radioactive chemicals and other hazardous materials. They are also used in high-temperature thermometry and dosimetry.
See fiber, optical.

insulating oil. See transformer oil.

insulator. Any substance or mixture that has an extremely low dielectric constant, low thermal conductivity, or both. Electrical insulators are either solid or liquid, the latter being used in transformers (askarel, mineral oils, silicone oils). A wide variety of solid types includes porcelains, glass, mica, alumina, various high polymers (epoxies, polyethylene, polystyrene, phenolics), cellulosic materials, nylon, and silicone resins. All these may be used alone or combined with other insulators as composites.
See dielectric; transformer oil.
Thermal insulators comprise an equally broad range of materials. Such inorganics as mineral fibers, magnesia, aluminum silicate, cellulose, and glass fibers are widely used for steam and hot-water pipes, furnaces, and blown-in home insulation. Organic products that are effective include plastic foams (polyurethane, polyvinyl chloride, polystyrene) and cellular rubber. There are a number of materials that may be called double insulators, since they have both electrical and thermal insulating properties, e.g., polystyrene, PVC, cellulose, glass, magnesia, and aluminum silicate.
Air is a unique case, as it is the only gaseous material in actual use as an insulator. Its dielectric constant is 1.0058, far less than that of any other dielectric material, and it has low thermal conductivity as well. It is particularly effective when trapped within a solid network, as in wool, cellular plastics, or glass fibers, or as an interlayer between wall panels.

insulin. A polypeptide hormone having a molecular weight of 5733. It is formed in the islets of Langerhans located in the pancreas and was so named for this reason. Insulin is composed of 16 amino acids arranged in a coiled chain and cross-linked in several places by the disulfide bonds of cystine residues. The sequence of amino acids has been elucidated. The insulin molecule was synthesized in 1963. In 1977, rat insulin was produced in the bacterium *E. coli* by recombinant DNA techniques. A year later, human insulin was generated after chemically synthesized genes were added to *E. coli*. This synthetic insulin is now in commercial production and has been approved by FDA. Insulin regulates carbohydrate metabolism in the body by decreasing the blood glucose level. A systemic deficiency leads to diabetes.
Properties: White powder or hexagonal crystals. Readily soluble in dilute acids; soluble in water.
Derivation: Extraction of minced pancreas with acidified dilute alcohol, followed by precipitation with absolute alcohol; also by gene-splicing methods.
Grade: USP, in various solutions or suspensions that include insulin injection; isophane insulin suspension; protamine zinc insulin suspension; NF, as globin zinc insulin injection.
Hazard: Overdosage can be fatal.
Use: Medicine (for diabetes control).
See Banting; recombinant DNA.

integral membrane proteins. Proteins inserted into a membrane by hydrophobic interactions. Contrast with peripheral proteins.

intercalating agent. (4-*N*-(6-chloro-2-methoxyacridin-9-yl)-1-*N*,1-*N*-diethylpentane-1,4-diamine dihydrochloride).

CAS: 83-89-6. $C_{23}H_{30}ClN_3O$.

A chemical, usually an aromatic compound, that insets itself between adjacent base pairs in double-stranded DNA. This leads to an adjustment in the secondary structure of the DNA, since adjacent base pairs are usually closely packed.

Use: As an anthelmintic, in the treatment of giardiasis and malignant effusions; in cell biological experiments as an inhibitor of phospholipase A2.

intercalating mutagen. A mutagen that inserts itself between two successive bases in a nucleic acid causing a frame-shift mutation.

intercalation compound. A compound composed of a crystalline lattice that acts as an electron donor, and "foreign" electron acceptor atoms interspersed or diffused between the planes of the lattice. An important group of intercalated compounds is composed of graphite, where bromine, for example, can act as electron acceptor. Graphite is particularly susceptible to this phenomenon because of its orderly stacked layers of crystals. Anhydrous metal nitrates such as copper and zinc nitrates also form intercalated compounds with graphite. A further example is trilithium nitride, whose structure consists of a series of layers of dilithium nitride, between which is a layer of lithium atoms. This markedly increases the conductivity, so that the material becomes an effective solid electrolyte in batteries. Other substances having this property are sodium β-alumina, titanium disulfide, and some metal dioxides. The phenomenon does not impair the crystalline structure and is reversible. Intercalated compounds are used for superconductors, synthetic lubricants, catalysts, and storage batteries. They are used in biochemical research; an acridine-based compound that can intercalate between stacked pairs of bases in a DNA helix is used in cancer research.

Intercar. A series of water-dispersible paint driers.

Available forms: Supplied in high metal concentration (calcium 4%, cobalt 6%, lead 24%, manganese 6%).

"Intercide" [Akros]. TM for a series of fungicides and mildewcides.

Use: Wood and fabric preservatives.

interface. The area of contact between two immiscible phases of a dispersion which may involve either the same or different states of matter. Five types are possible: (1) solid–solid (carbon black–rubber), (2) liquid–liquid (water–oil), (3) solid–gas (smoke–air), (4) solid–liquid (clay–water), (5) liquid–gas (water–air). At a fresh surface of either liquid or solid, the molecular attraction exerts a net inward pull. Hence, the characteristic property of a liquid is surface tension and that of a solid surface is adsorption. Both have the same cause, namely, the inward cohesive forces acting on the molecules at the surface. These phenomena provide to some degree the fundamental mechanism for many industrially important processes (catalysis, emulsification, mixing, alloying) and products (detergents, adhesives, lubricants, paints). Such properties as wettability of solid powders, spreading coefficients of liquids, and protective action of colloidal substances are intimately associated with interfacial behavior.

See surface; surface tension; catalysis; emulsion; detergent; wetting agent.

interfacial film. Skin-like structure that develops at the interface between the two emulsion phases, consisting of adsorbed and oriented molecules of the emulsifying agent in a state of loose aggregation. It surrounds the droplets of the internal phase and keeps them from flowing together.

interfacial surface energy. The tension measured in dynes per centimeter.

interference. One crossover event inhibits the chances of another crossover event. Also known as positive interference. Negative interference increases the chance of a second crossover.

See crossing over.

interferon. An antiviral protein produced by vertebrate cells in response to virus infection. Discovered in 1957, it is a product of the infected cell, rather than of the disease-inducing virus. It can be formed by any type of infected cell and is effective against many viruses, but only in the cells of the organism that produced it. It is quite different from an antibody which is produced only by specialized cells and acts by combining directly with a specific virus. Interferon does not inactivate viruses directly, but reacts with susceptible cells which then resist virus multiplication. Prostaglandins are believed to be effective in maintaining normal interferon production in the body. Four types of interferon are presently known. As a result of advanced recombinant DNA techniques, a biologically active protein almost identical with interferon has been made outside the body by exposing virus activated bacteria to the genetic code controlling interferon formation. The "synthetic" interferon is available in quantities large enough to permit its use in treatment of various virus diseases. The interferon molecule itself has not yet been synthesized.

See recombinant DNA.

intergenic. Between two genes, for example, intergenic DNA is the DNA found between two genes. The term is often used to mean nonfunctional DNA (or at least DNA with no known importance to the two genes flanking it). Alternatively, one might speak of the "intergenic distance" between two genes as the number of base pairs from the polyA site of the first gene to the cap site of the second. This usage might therefore include the promoter region of the second gene.

interification.   See ester interchange.

interleukin-2.   A type of biological response modifier (a substance that can improve the body's natural response to disease). It stimulates the growth of certain disease-fighting blood cells in the immune system. Also called IL-2.

intermediary metabolism.   The enzyme-catalyzed reactions that extract chemical energy from nutrient molecules and utilize it to synthesize molecules and perform cell functions.

intermediate.   An organic compound, either cyclic (derived from coal tar or petroleum products such as benzene, toluene, naphthalene, etc.) or acyclic (e.g., ethyl and methyl alcohol). These compounds may be considered as chemical stepping stones between the parent substance and the final product. The cyclic type (e.g., aniline, β-naphthol, and benzoylbenzoic acid) still predominate as intermediates for synthetic dyes and has few other uses; the acyclic type in general has many independent uses. Exceptions are hexamethylenetetramine, an acyclic intermediate for phenolformaldehyde resin and butadiene for synthetic elastomers. Intermediates are the foundation of the modern approach to organic technology. The distinction between an intermediate and an end product is not always precise.
See azo dye intermediate.

intermetallic compound.   (1) A compound or alloy formed by two metals that have been placed in intimate contact during the process of brazing or coating, the compound occurring at the interface between the metal surfaces. In some cases, as in galvanizing, the metals form bimetallic compounds; in others they form alloys or solid solutions of varying composition. (2) A two- or three-component metal system prepared with special metals having semiconducting properties, e.g., gallium, for use in lasers, diodes, transistors, etc.
See galvanizing; semiconductor.

intermolecular.   Describes an interaction (e.g., a chemical reaction) between different molecules.

internal compensation.   (intramolecular compensation; *meso* form).
The optical neutralization produced in a molecule when two asymmetric centers in the molecule produce equal, opposite amount of rotation of the plane of polarization.
See *meso-*.

internal conversion.   The effect of an atom produced by a γ-ray photon emerging from the nucleus, and giving up its energy on meeting with an extra nuclear electron.

internal energy.   The sum of the kinetic and potential energies (including electromagnetic field energies) of the particles that make up a system.

International Union of Pure and Applied Chemistry.   (IUPAC). A voluntary nonprofit association of national organizations representing chemists in 77 member countries. It was formed in 1919 with the objective of facilitating international agreement and uniform practice in both academic and industrial aspects of chemistry. Examples are nomenclature, atomic weights, symbols and terminology, physiochemical constants, and certain methods of analysis and assay. In addition to standardization, IUPAC carries on several hundred research projects relating to food technology, water and air quality, single-cell proteins, etc. Its secretariat can be reached at P.O. Box 13757, Research Triangle Park, NC 27709. Website: http://www.iupac.org

interphase.   The period in the cell cycle when DNA is replicated in the nucleus. It is followed by mitosis.

interstellar space.   See astrochemistry.

interstitial.   (1) Descriptive of a nonstoichiometric compound of a metal and a nonmetal whose structure conforms to a simple chemical formula, but exists over a limited range of chemical composition. Interstitial compounds are represented by borides, nitrides, and carbides of the transition metals. (2) Descriptive of an atom of an impurity that causes a defect or dislocation in a crystalline lattice, e.g., an atom of carbon or nitrogen in an iron crystal, or of arsenic in a semiconductor. (3) In a biological sense, the term describes cells located between or within layers of tissue.

Interwax.   A series of wax modifiers.
"P" = High molecular weight polybutene and paraffin wax.
"M" = High molecular weight polybutene and amber-colored, microcrystalline wax.
"280" = A hard, high-melting-point synthetic wax.
Use: In lacquers and varnishes, as a plastic lubricant and antiblocking agent, as an extender or substitute for carnauba wax.

"Interwet" [AKZO].   TM for a series of wetting agents for emulsion paints, latex paints, and vinyl plastisols.

intestinal anthrax.   A form of anthrax in humans that is acquired by eating infected meat.

Hazard: Chills, high fever, vomiting, headache, back pain, pain in the extremities, bloody diarrhea, bleeding from the mucous membranes and skin, prostration, and death.

intracellular enzyme. (endoenzyme). An enzyme that acts within the cell that produces it.

"Intracron" [Yorkshire]. TM for a group of fiber-reactive dyes.
Available forms: Liquid and powder.
Use: For cellulosic fibers and blends.

intramolecular compensation. See internal compensation; *meso-*.

intramolecular reaction. Change in position of an atom or group of atoms in a molecule giving rise to a different, though isomeric compound.

"Intrasil" [Sensient]. TM for a group of disperse dyes.
Available forms: Liquid and powder.
Use: Acetate and polyester apparel and polyester automotive fabrics.

intrinsic antigen. An antigen that is a cellular constituent.

intrinsically disordered proteins. Any of a distinct class of proteins that are marked by similar functional properties. Included are molecular recognition, molecular assembly, protein modification, and entropic chains.

intrinsically safe. Equipment and wiring that is incapable of releasing sufficient electrical or thermal energy under normal or abnormal conditions to cause ignition of a specified hazardous atmospheric mixture of flammable materials.

introfier. Substance used to lower interfacial tension between two phases, usually a liquid soluble in both phases; a wetting accelerator.

introns. The DNA base sequences interrupting the protein-coding sequences of a gene; these sequences are transcribed into RNA but are cut out of the message before it is translated into protein. See Exons.

intumescence. The foaming and swelling of a plastic or other material when exposed to high surface temperatures or flames.
Use: Polyurethane base coating materials for rocket reentry.

inulin. (alantin; alant atarch).
Properties: Spherical crystals. Mp 158–165C. Hygroscopic in moist air. Soluble in hot water.
Source: From chicory root.
Use: Diagnostic aid for kidney function.

Invar. An iron–nickel alloy containing 40–50% nickel and characterized by an extremely low coefficient of thermal expansion.
Use: Precision instruments, measuring tapes, weights, etc.

invariant point. State of system where the number of degrees of freedom is zero.

invention. The chief requirement of an invention is that it be unobvious to a person having ordinary skill in the art to which the claim pertains and knowing everything that has gone before, as shown by publication anywhere or public use in the U.S. When reliance for patentability is placed on a new mixture of components that have been used separately, it must be shown that there is some unexpected coaction between the ingredients and not just the additive effects of the several materials. The inventor is the one who contributes the inventive concept in workable detail, not the one who demonstrates it or tests it.

inversion. Splitting up of higher sugars into molecules of a lower sugar.

inversion point. Temperature at which one polymorphic form of a substance changes into another under variant conditions.

invertase. (sucrase; invertin). Enzyme produced by yeast and by the lining of the intestines. It is a white powder, soluble in water. It catalyzes the conversion of sucrose (ordinary sugar) to glucose and levulose (fructose) during fermentation of sugars.
Use: Production of invert sugar for syrups and candy; analytical reagent for sucrose.

invert soaps. Cationic detergents with disinfectant properties.

invert sugar. A mixture of 50% glucose and 50% fructose obtained by the hydrolysis of sucrose. It absorbs water readily and is usually handled only as a syrup. Because of its fructose content, invert sugar is levorotatory in solution and sweeter than sucrose. Invert sugar is often incorporated in products where loss of water must be avoided. Commercially it is obtained from the inversion of a 96% cane sugar solution.
Use: Food industry, brewing industry, confectionery, humectant.

investment casting. A ceramic or metal casting method originally used to make reproductions of sculptured pieces (lost wax process) and adapted to industry for manufacture of precision metal parts. It is generally known as precision investment casting. The sequence of operations is (1) a wax prototype is made in a metal mold; (2) several of these are attached to a central member, also made of wax to form a "tree"; (3) the tree is dipped and dried eight times in a ceramic glaze, thus building up a coating or "investment"; (4) the assembly is baked in an oven, thus melting out the wax, leaving a cavity which is then used as a mold for liquid metal. Great accuracy can be obtained with this method. The exact nature of the waxes and coatings is not disclosed.

in vitro.   A condition in which a reaction is carried out in a laboratory experiment (i.e., a glass container, test tube, or beaker), as opposed to a reaction occurring in a living organism (in vivo).

iocetamic acid.
CAS: 16034-77-8. $C_{12}H_{13}I_3N_2O_3$.
Hazard: Low toxicity by ingestion.

iodeosin.   (tetraiodofluorescein). $C_{20}H_8I_4O_5$.
Properties: Red powder. Soluble in dilute alkalies; slightly soluble in alcohol and ether; insoluble in water.
Derivation: Interaction of fluorescein and iodine in presence of iodic acid.
Use: Indicator in analytical chemistry.

iodic acid.
CAS: 7782-68-5. HIO_3.
Properties: Colorless, rhombic crystals or white, crystalline powder. D 4.629, mp 110C (decomposes). A moderately strong acid. Soluble in cold and hot water.
Derivation: By adding sulfuric acid to a solution of barium iodate and subsequent filtration and crystallization.
Hazard: Toxic by ingestion, strong irritant to eyes and skin.
Use: Analytical chemistry, medicine (1–3% solution).

iodic acid anhydride.   (iodine pentoxide).
CAS: 7790-35-4. I_2O_5.
Properties: White, crystalline powder. D 4.799, mp 300C (decomposes). Soluble in water and nitric acid; insoluble in absolute alcohol, chloroform, ether, carbon disulfide.
Hazard: Toxic by ingestion, strong irritant to eyes and skin, strong oxidizing agent.
Use: Oxidizing agent, organic synthesis.

iodide.
CAS: 20461-54-5. I^-.
A compound that contains pentavalent iodine, which is usually ionically bound to electropositive atoms.

iodide peroxidase.   (iodotyrosine deiodase; iodinase). An oxidoreductase that catalyzes reactions of iodine and water that yield iodide and peroxide. It also catalyzes the iodination and deiodination of tyrosine compounds.

iodinated casein.
Use: Drug (veterinary); food additive.

iodine.
CAS: 7553-56-2.   I.
Nonmetallic halogen element of atomic number 53; group VIIA of the periodic table; the least reactive of the halogens, aw 126.9045; valences = 1,3,5,7; no stable isotopes but many artificial radioactive isotopes.
Properties: Heavy, grayish-black plates or granules having a metallic luster; characteristic odor. Readily sublimed having a violet vapor. D 4.98, mp 113.5C, bp 184C. Soluble in alcohol, carbon disulfide, chloroform, ether, carbon tetrachloride, glycerol, and alkaline iodide solutions; insoluble in water. A semiconductor.
Derivation: (a) From brine wells in Michigan, Oklahoma, Japan, Indonesia; (b) from mother liquors of Chilean nitrate. It can also be extracted from kelp.
Method of purification: Sublimation.
Grade: Crude, CP, USP.
Hazard: Toxic by ingestion and inhalation, strong irritant to eyes and skin. Questionable carcinogen.
Use: Dyes (aniline dyes, phthalein dyes), alkylation and condensation catalyst, iodides, iodates, antiseptics and germicides, X-ray contrast media, food and feed additive, stabilizers, photographic film, water treatment, pharmaceuticals, medicinal soaps, unsaturation indicator.

iodine 131.   (^{131}I). Radioactive iodine of mass number 131.
Properties: Half-life 8 days. Radiation: β and γ.
Derivation: By pile irradiation of tellurium and from the fission products of nuclear reactor fuels.
Available forms: As elemental iodine and in a weakly basic solution of sodium iodide in sodium sulfite; iodine 131 is also available in tagged compounds such as dithymol diiodide, potassium iodate, diiodofluorescein, insulin, ACTH, etc.
Grade: USP lists iodinated ^{131}I serum albumin, rose bengal sodium ^{131}I injection, sodium iodide ^{131}I capsules and solution, sodium iodohippurate ^{131}I injection.
Use: Diagnosis and treatment of goiter, hyperthyroidism, and other thyroid disorders; internal radiation therapy; in film gauges to measure film thicknesses of the order of one micron; for detecting leaks in water lines; as a source of radiation in oil field tests; as a tracer in chemical analysis; as a tracer in studying diet iodine for cattle, the functions of the thyroid gland, the efficiency of mixing pulp fibers, the thermal stability of potassium iodate in bread dough, chemical reaction mechanisms, etc.

iodine bromide.   See iodine monobromide.

iodine chloride.   See iodine monochloride; iodine trichloride.

iodine cyanide.   See cyanogen iodide.

iodine monobromide.   (bromine iodide).
CAS: 7789-33-5.   IBr.

Properties: Crystals, purplish-black mass. Mp 42C, bp 116C (decomposes), d 4.41. Soluble in water with decomposition, alcohol, and ether.
Derivation: By the interaction of iodine and bromine.
Hazard: Toxic by ingestion and inhalation, vapors corrosive to tissue.
Use: Organic synthesis.

iodine monochloride.
CAS: 7790-99-0.   ICl.
Properties: Black crystals (α and β forms) or reddish-brown, oily liquid. Mp (α) 27C, (β) 14C, bp 101C (decomposes), d (α) 3.18 (0C), (β) 3.24 (liquid at 34C). Soluble in alcohol, water (with decomposition), and dilute hydrochloric acid.
Derivation: By the action of dry chlorine on iodine.
Hazard: Toxic by ingestion and inhalation, strong irritant to eyes and skin.
Use: Analytical chemistry, organic synthesis.

iodine number.   (iodine value). The percentage of iodine that will be absorbed by a chemically unsaturated substance (vegetable oils, rubber, etc.) in a given time under arbitrary conditions. A measure of unsaturation.

iodine pentafluoride.
CAS: 7783-66-6. IF_5.
Properties: Fuming liquid. Bp 98C, mp 9.4C, d 3.189 (25C). Attacks glass.
Derivation: Passing fluorine over iodine. Available in cylinders at 98.0% min purity.
Hazard: Dangerous fire risk, reacts violently with water. Toxic by ingestion and inhalation, corrosive to skin and mucous membranes.
Use: Fluorinating and incendiary agent.

iodine pentoxide.   See iodic acid anhydride.

iodine test.   Placing a few drops of potassium iodide solution on a sample to detect the presence of starch. The test is positive if sample turns blue.

iodine tincture.   A solution of iodine and potassium iodide or sodium iodide in alcohol, a reddish-brown liquid having the odors of iodine and alcohol, contains 44–50% by volume alcohol, 2 g iodine and 2.4 g sodium iodide per 100 cc.
Grade: USP.
Hazard: Toxic by ingestion, avoid using on open cuts.
Use: Antiseptic (use on skin surface only).

iodine trichloride.
CAS: 865-44-1. ICl_3.
Properties: Orange-yellow, deliquescent, crystalline powder; pungent irritating odor. Mp 33C, d 3.11. Soluble in water (with decomposition), alcohol, and benzene.
Derivation: By interaction of iodine and chlorine.

Hazard: Toxic by ingestion and inhalation, corrosive to tissue.
Use: Agent for introducing iodine and chlorine in organic synthesis; topical antiseptic.

iodine value.   See iodine number.

iodipamide.
CAS: 606-17-7. $(CH_2)_4(CONHC_6HI_3COOH)_2$.
Properties: White, crystalline powder; nearly odorless. Very slightly soluble in alcohol, chloroform, ether; insoluble in water; pH of saturated solution is 3.5–3.9.
Hazard: Toxic by ingestion.
Use: Medicine (X-ray contrast medium).

iodisan.   See propiodal.

iodized oil.   An iodine addition product of vegetable oil or oils containing 38–42% organically combined iodine.
Properties: Thick, viscous, oily liquid; affected by air and light. Soluble in solvent naphtha.
Hazard: Toxic by ingestion.
Use: Medicine (radiopaque medium).

iodoacetic acid, sodium salt.   (sodium iodoacetate).
CAS: 305-53-3. ICH_2CO_2Na.
Properties: Colorless or white crystals. Mw 207.93, mp 210C. Soluble in water, alcohol, and very slightly soluble in ether. Hygroscopic in moist air.
Hazard: Toxic.
Use: Analytical reagent.

(iodoacetoxy)tributylstannane.   See tributyltin iodoacetate.

(iodoacetoxy)tripropylstannane.   See tripropyltin iodoacetate.

o-iodobenzoic acid tributylstannyl ester.   See tributyltin-o-iodobenzoate.

(o-iodobenzoyloxy)tripropylstannane.
CAS: 73927-94-3. $C_{16}H_{25}IO_2Sn$.
Hazard: A poison.

4-iodo-3,5-dimethyl-n-(2-methylphenyl)-1h-pyrazole-1-acetamide.
CAS: 302542-42-3. $C_{14}H_{16}IN_3O$.
Hazard: Moderately toxic by ingestion.

4-iodo-3,5-dimethyl-n-(3-methylphenyl)-1h-pyrazole-1-acetamide.
CAS: 302542-51-4. $C_{14}H_{16}IN_3O$.
Hazard: Moderately toxic by ingestion.

4-iodo-3,5-dimethyl-n-(4-methylphenyl)-1h-pyrazole-1-acetamide.
CAS: 302542-63-8. $C_{14}H_{16}IN_3O$.
Hazard: Moderately toxic by ingestion.

iodoethane.   See ethyl iodide.

iodoethylene.   See tetraiodoethylene.

iodoform.   (triiodomethane).
CAS: 75-47-8. CHI_3.
Properties: Small greenish-yellow or lustrous crystals or powder; penetrating odor. D 4.08, mp 115C. Soluble in benzene and acetone; partially soluble in alcohol, glycerol, chloroform, carbon disulfide, and ether; insoluble in water.
Derivation: (a) By heating acetone or methanol with iodine in presence of an alkali or alkaline carbonates. (b) Electrolytically, by passing a current through a solution containing potassium iodide, alcohol, and sodium carbonate.
Grade: Technical, NF.
Hazard: Irritant. Decomposes violently at 400F (204C).
Use: Medicine (antiseptic for external use only).

iodomethane.   See methyl iodide.

1-iodooctadecane.   (octadecyl iodide).
CAS: 629-93-6. $CH_3(CH_2)_{17}I$.
Properties: Solid. Mw 380.40, mp 33–35C, bp 194–197C (2 mm Hg), fp 110C. Light sensitive.
Hazard: Irritant.
Use: Reagent for introduction of C_{18} chain.

iodopanoic acid.   See iopanoic acid.

iodophenazone.
CAS: 129-81-7. $C_{11}H_{11}IN_2O$.
Hazard: Moderately toxic by ingestion.

iodophor.   ("tamed iodine"). (1) A complex of iodine with certain types of surface-active agents that have detergent properties. (2) More generally, any carrier of iodine.

iodophosphonium.   See phosphonium iodide.

2-iodopropane.   See isopropyl iodide.

(iodopropionyloxy)tributylstannane.   See tributyltin-β-iodopropionate.

iodoprotein.   A protein that contains iodine bound to tyrosine.

iodopyracet meglumine.
CAS: 3736-90-1. $C_7H_5I_2NO_3 \cdot C_7H_{17}NO_5$.
Hazard: Low toxicity.

N-iodosuccinimide.   (succiniodimide). ($-CH_2$ $CO)_2NI$).
Properties: Colorless crystals. Mp 200–201C. Soluble in acetone and methanol; insoluble in carbon tetrachloride and ether; decomposes in water.
Hazard: Skin irritant.
Use: Iodinizing agent in synthetic organic chemistry.

iodotributylstannane.
CAS: 7342-47-4. $C_{12}H_{27}ISn$.
Hazard: A poison by ingestion and skin contact. Moderately toxic by inhalation. Tributyl tin compounds are extremely toxic to marine life.

iodotrimethyltin.
CAS: 811-73-4. C_3H_9ISn.
Properties: White powder, insoluble in water and organic solvents.
Hazard: A poison.

iodotriphenylstannane.
CAS: 894-09-7. $C_{18}H_{15}ISn$.
Hazard: A poison.

iodotripropylstannane.   See tripropyltin iodide.

iomeprol.
CAS: 78649-41-9. $C_{17}H_{22}I_3N_3O_8$.
Hazard: Low toxicity.

ion.   An atom or radical that has lost or gained one or more electrons and has thus acquired an electric charge. Positively charged ions are cations and those having a negative charge are anions. An ion often has entirely different properties from the element (atom) from which it was formed.
In sodium chloride solution, sodium exists as sodium ion (Na^+), i.e., sodium atoms that have lost one electron. The chlorine is present as chloride ion (Cl^-), i.e., chlorine atoms that have gained one electron. Copper sulfate solution contains copper ion (Cu^{2+}), i.e., copper atoms that have lost two electrons, and sulfate ion (SO_4^{2-}), i.e., sulfate radicals that have gained two electrons.
Ions occur in water solution or in the fused state (except in the case of gases). Compounds that form ions are called electrolytes because they enable the solution to conduct electricity. Ion formation causes an abnormal increase in the boiling point of water and also lowers the freezing point, the extent depending on the concentration of the solution. Ions are also formed in gases as a result of electrical discharge.
See ionization; electrolysis; ion exchange.

ion channel.   An integral membrane protein that provides for the regulated transport of a specific ion or ions across a membrane.

ion electrode.   Electrode that develops and measures an electrical potential in response to the activity of an ion in solution to which it is selective.

ion exchange.   A reversible chemical reaction between a solid (ion exchanger) and a fluid (usually a water solution) by means of which ions may be interchanged from one substance to another. The superficial physical structure of the solid is not affected. The customary procedure is to pass the

fluid through a bed of the solid, which is granular and porous and has only a limited capacity for exchange. The process is essentially a batch type in which the ion exchanger, upon nearing depletion, is regenerated by inexpensive brines, carbonate solutions, etc. Ion exchange occurs extensively in soils.
Ion exchange resins are synthetic resins containing active groups (usually sulfonic, carboxylic, phenol, or substituted amino groups) that give the resin the property of combining with or exchanging ions between the resin and a solution. Thus, a resin with active sulfonic groups can be converted to the sodium form and will then exchange its sodium ions with the calcium ions present in hard water.
Some specific applications of ion exchange: water softening, milk softening (substitution of sodium ions for calcium ions in milk), removal of iron from wine (substitution of hydrogen ions), recovery of chromate from plating solutions, uranium from acid solutions, streptomycin from broths, removal of formic acid from formaldehyde solutions, demineralization of sugar solutions, recovery of valuable metals from wastes, recovery of nicotine from tobacco-dryer gases, catalysis of reaction between butyl alcohol and fatty acids, recovery and separation of radioactive isotopes from atomic fission, chromatography, establishment of mass micro standards, and in cigarette filters to remove polonium from smoke.
See zeolite.

ion-exchange chromatography.   A chromatographic method based on the ability of polymers to sorb ionized solutes reversibly, e.g., cross-linked resins with exchangeable hydrogen or hydroxyl ions. It can be carried out both in columns and on sheets.

ion-exchange resin.   See ion exchange.

ion exclusion.   The process in which a synthetic resin of the ion exchange type absorbs nonionized solutes such as glycerine or sugar while it does not absorb ionized solutes that are also present in a solution in contact with the resin. Thus, sodium chloride and glycerine can be separated by passage of their aqueous solution through a bed of particles of an ion exclusion resin.

ionic bond.   (electrovalent bond). Refers to the formation of ions by transfer of one or more electrons from one atom to another.
See bond, chemical.

ionic detergent.   See detergent.

ionization.   A chemical change by which ions are formed from a neutral molecule of an inorganic solid, liquid, or gas. The most common type of ionization occurs when an ionically bonded inorganic compound such as sodium chloride or sulfuric acid is dissolved in water (or other solvent); the molecule

separates or dissociates into two ions, the metallic ion being positively charged by loss of an electron and the nonmetallic ion being negatively charged by gaining an electron. The degree of dissociation varies with the type of compound, the solvent, and the temperature. Molecules or atoms of gases are ionized by passage of an electric current through the gas; this removes electrons and leaves a positive charge.
Compounds that ionize in solution greatly increase the conductivity of the solvent. Ionization is most effective in water because its high dielectric constant lowers the ionic bonding forces in the solute molecules enough to cause separation of their constituent atoms. Ion formation produces a notable rise in the boiling point and a depression of the freezing point of water. An electric current passed through a solution containing ions causes them to move to the oppositely charged electrode; this effect is the basis of many industrial electrochemical operations, such as electroplating and the manufacture of sodium hydroxide and chlorine.
See ion; electroplating; electrolysis; electrolyte; dissociation.

ionizing radiation.   See radiation, ionizing.

ionomer resin.   A copolymer of ethylene and a vinyl monomer with an acid group, such as methacrylic acid. They are cross-linked polymers in that the linkages are ionic as well as covalent bonds.
There are positively and negatively charged groups which are not associated with each other and this polar character makes these resins unique.
Properties: Transparent, electrically conductive, resilient, and thermoplastic. Cannot be completely dissolved in any commercial solvents. High resistance to abrasion, cracking, corona attack; high tensile strength. Working temperature approximately -108 to $+65.5$C.
Use: Break-resistant transparent bottles, packaging films, mercury flasks, protective equipment, pipe and tubing, electric distribution elements. As foam, insulation of fresh concrete.
See bond, chemical.

ionone.   (α- or β-cyclocitrylidenacetone).
CAS: 8013-90-9. $C_{13}H_{20}O$.
Properties: Light-yellow to colorless liquid; violet odor. D 0.927–0.933 (25C), bp 126–128C (12 mm Hg), refr index 1.4970–1.5020 (20C). Soluble in alcohol, ether, mineral oil, and propylene glycol; insoluble in water and glycerol.
Grade: 95%, 99%, mixed isomers, FCC.
Derivation: (1) Condensing citral with acetone followed by ring closure with an acid, (2) reaction of acetylene with acetone followed by hydrogenation, condensation with diketone, and further reaction with acetylene.
Use: Perfumery, chemical synthesis, flavoring, vitamin A production (β isomer).

ionophore.   A compound that binds metal ions and diffuses across membranes carrying the bound ion. An example is valinomycin, which transports potassium.

ion retardation.   A process based on amphoteric (bifunctional) ion-exchange resins containing both anion and cation adsorption sites. These sites will associate with mobile anions and cations in solution and thus remove both kinds of ions from solutions. These ions may be eluted by rinsing with water. Process can make clean separations of ionic-nonionic mixtures. Has also been suggested for demineralization of salt solutions.
See ion.

iopanoic acid.   (iodopanoic acid; 3-amino-α-ethyl-2,4,6-triiodohydrocinnamic acid).
CAS: 96-83-3. $C_6HI_3NH_2CH_2CH(COOH)C_2H_5$.
Properties: Cream-colored powder; tasteless; faintly aromatic odor. Mp 152–158C (decomposes), darkens on exposure to light. Soluble in acetone, ether, alcohol, chloroform, and dilute alkalies; insoluble in water.
Grade: USP.
Use: Medicine (radiopaque medium).

ioxynil.   (3,5-diiodo-4-hydroxybenzonitrile; 3,5-diiodo-4-hydroxyphenyl cyanide; 2,6-diiodo-4-cyanophenol; 4-hydroxy-3,5-diiodobenzonitrile). $C_7H_3I_2NO$. A post-emergence contact herbicide with little residual activity.
Use: Herbicide to control broad-leaved seedling weeds in cereal crops and sports turf.

IPA.   (1) Abbreviation for isophthalic acid. (2) Abbreviation for isopropyl alcohol.

IPAE.   See isopropylaminoethanol.

Ipatieff, Vladimir N.   (1890–1952). Born in Russia, Ipatieff was an army officer as well as a chemist. He was a member of the Academy of Science and carried out organic research at the Institute of Chemistry in Leningrad. He left the former U.S.S.R. under the Stalin regime and at the invitation of Gustav Egloff joined the Universal Oil Products Co. He and his close associate, Herman Pines, did basic development on catalytic alkylation and isomerization of hydrocarbons of the greatest importance for high-octane aviation gasoline.

IPC.   (INPC; isopropyl-*N*-phenylcarbamate).
CAS: 122-42-9. $C_6H_5NHCOOCH(CH_3)_2$.
Properties: White to gray crystalline needles; odorless when pure. Mp 84C (technical grade). Soluble in alcohol, acetone, isopropyl alcohol; insoluble in water.
Hazard: Toxic by ingestion.
Use: Herbicide.

IPC, chloro-.   See chloro-IPC.

ipecac.   Dried root of *Cephaelis ipecacuanha*.
Occurrence: Brazil and Bolivia, cultivated in India.
Grade: Technical, USP.
Hazard: Contains emetine, a toxic alkaloid.
Use: Medicine (emetic), production of emetine.

IPM.   See *m*-phthalodinitrile.

ipronidazole.   (2-isopropyl-1-methyl-5-nitroimidazole;   1-methyl-5-nitro-2-propan-2-ylimidazole). $C_7H_{11}N_3O_2$.   An antihistomonal agent that promotes growth and feed utilization in poultry.
Hazard: Low toxicity.

Ir.   Symbol for iridium.

"Ircogel" [Lubrizol].   TM for a variety of calcium sulfonates, thixotropic.
Grade: 900 (paste), 903, 904, 905, 906, 907 (gel), 2064 (pourable liquid), 2050B (polymeric organic phosphate).
Available forms: Gel, pourable liquids.
Use: A thixotrope for PVC plastisols, organosol, and high solids solvent-based paints; 2064: corrosion inhibitor for latex paint and other water-based coatings; 2085B: rust-preventative coatings.

"Irganox" [BASF].   TM for a series of complex, high molecular weight stabilizers that inhibit oxidation and thermal degradation of many organic materials. They contain multifunctional chemical groupings. Several are hindered polyphenols. All are white, crystalline, free-flowing powders; nonstaining, nonvolatile, and odorless.

"Irgasan" [BASF].   TM for broad-spectrum antibacterial agent.
Use: Deodorant soaps and underarm deodorants.

iridic chloride.   (iridium chloride; iridium tetrachloride).
CAS: 10025-97-5. $IrCl_4$.
Properties: Brownish-black mass, hygroscopic. Soluble in water, alcohol, and dilute hydrochloric acid.
Derivation: Action of chlorine or aqua regia on the ammonium salt $(NH_4)_2IrCl_6$.
Use: Analysis (testing for nitric acid in the presence of nitrous acid), microscopy, plating solution.

iridium.
CAS: 7439-88-5.   Ir.
Metallic element of atomic number 77, one of the platinum metals, group VIII of the periodic table, aw 192.22, valences = 1,2,3,4,6, two stable isotopes.
Properties: Silver-white, low ductility. Does not tarnish in air. On heating strongly, a slightly volatile oxide is formed. Bulk d 22.65 (20C calculated) The most dense element known. Mp 2443C, bp approximately 4500C, most corrosion-resistant element, modulus of elasticity is one of highest (75,000,000

psi), Brinell hardness (cast) 218, highly resistant to chemical attack. Insoluble in acids; slowly soluble in aqua regia and in fused alkalies.

Occurrence: Canada, South Africa, the former U.S.S.R., Alaska.

Derivation: Occurs with platinum, remains insoluble when the crude platinum is treated with aqua regia, occurs as iridosmine. The powder is obtained by hydrogen reduction of ammonium chloroiridate.

Available forms: Powder, single crystals.

Use: Alloy with platinum for ammonia fuel-cell catalyst, electric contacts and thermocouples, commercial electrodes and resistance wires, laboratory ware, extrusion dies for glass fibers, jewelry. Primary standards of weight and length.

^{192}iridium.
Radioactive iridium of mass number 192.

Properties: Half-life, 74 days. Radiation: β and γ.

Derivation: Pile irradiation of iridium.

Available forms: Iridium metal or potassium or sodium chloroiridate in hydrochloric acid solution.

Use: Radiography of light castings, treatment of cancer.

iridium bromide. See iridium tribromide.

iridium chloride. See iridic chloride; iridium trichloride.

iridium potassium chloride. (potassium chloroiridate; potassium iridium chloride). K_2IrCl_6.

Properties: Dark-red crystals. D 3.549. Soluble in water (hot).

Use: Black pigment (porcelain decoration). The iridium (III) salt is also known: K_3IrCl_6, greenish-yellow.

iridium sesquioxide.
CAS: 16920-56-2. Ir_2O_3.

Properties: Black powder. Slightly soluble in hydrochloric acid (concentrated); insoluble in water. Decomposes at 400C.

Derivation: By heating the chloroiridate K_2IrCl_6 with sodium carbonate.

Use: Ceramics (porcelain decoration).

iridium tetrachloride. See iridic chloride.

iridium tribromide. (iridium bromide). $IrBr_3 \cdot 4H_2O$.

Properties: Olive-green, brown, or black crystals. Soluble in water; insoluble in alcohol. Prepared by action of hydrobromic acid on iridium trihydroxide.

iridium trichloride. (iridium chloride). CAS: 12645-45-3. $IrCl_3$.

Properties: Dull green to blue-black particles. Mp 763C (decomposes). Insoluble in water and alcohol.

Prepared by action of chlorine on iridium powder at 600C.

iridomyrmecin.
CAS: 485-43-8. $C_{10}H_{16}O_2$.

Properties: Colorless crystals. Mp 60C, bp 104C (1.5 mm Hg), characteristic odor, refr index 1.46, optically active. Soluble in ether and fatty alcohols; almost insoluble in water.

Use: Insecticide, biocide.

iridosmine.
(osmiridium). A natural alloy of iridium and osmium containing some platinum, rhenium, ruthenium, iron, copper, and palladium.

Properties: Tin-white to light steel gray in color; streak, same; metallic luster. D l8.8–21.12, Mohs hardness 6–7. Composition is variable ranging from 10.0–77.2% iridium, 17.2–80.0% osmium, 0–10.1% platinum, 0–17.2% rhenium, 0–8.9% ruthenium, 0–1.5% iron, 0–0.9% copper, trace, palladium. Unattacked by aqua regia.

Occurrence: Alaska, South Africa.

Use: Fountain-pen point tips, surgical needles, watch pivots, compass bearings, hardening platinum (standard weights, jewelry), source of iridium and osmium.

irinotectan hydrochloride hydrate.
CAS: 136572-09-3. $C_{33}H_{38}N_4O_6 \cdot ClH \cdot 3H_2O$.

Hazard: Moderately toxic by ingestion.

Irish moss. See carrageenan.

Irish Moss. A beer clarifier.

Use: Clarifier added in last 15 minutes of boil.

irloxacin. See 6-fluoro-7-(1-pyrrolyl)-1-ethyl-1,4-dihydro-4-oxo-3.

IRN. A group of magnetic iron oxides, black ferroso-ferric oxides (Fe_3O_4) and brown γ-ferric oxides (Fe_2O_3), with application in ferric cores and electronic parts, magnetic inks, data-handling accounting systems, recording and instrumentation tapes.

iron.
CAS: 7439-89-6. Fe.

Metallic element of atomic number 26, group VIII of the periodic table, aw 55.847, valences = 2,3; four stable isotopes, 4 artificially radioactive isotopes.

Properties: Silver-white malleable metal. Tensile strength 30,000 psi, Brinell hardness 60, mp 1536C, bp 3000C, d 7.87 (20C), magnetic permeability 88,400 gauss (25C). The only metal that can be tempered. Mechanical properties are altered by impurities, especially carbon.

Iron is highly reactive chemically, a strong reducing agent, oxidizes readily in moist air, reacts with steam when hot, to yield hydrogen and iron oxides. Dissolves in nonoxidizing acids (sulfuric and hydrochloric acid) and in cold dilute nitric acid. For biochemical properties, see below.

Source: Hematite, limonite, magnetite, siderite, also taconite (low-grade 25–30% iron).

Occurrence: Minnesota (Mesabi), Alabama, Labrador, Yukon, Europe, South America.

Available forms: (1) Molten (or pig) iron. Derived (a) by smelting ore with limestone and coke in blast furnaces (purity 91–92%), (b) by continuous direct reduction in which iron ore and limestone are preheated in a fluidized bed, followed by heating to 926C, by melting at 1926C, and reduction to iron at 1648C with powdered coal (purity 99%—proprietary process).

Use: Steels of various types, other alloys (cast and wrought iron), source of hydrogen by reaction with steam.

See steel.

Available forms: (2) Powdered iron. Derived (a) by treatment of ore or scrap with hydrochloric acid to give ferrous chloride solution, which is then purified by filtration, vacuum crystallized, and dehydrated to ferrous chloride dehydrate powder; this is reduced at 800C to metallic iron (briquettes or powder) of 99.5% purity; (b) by thermal decomposition of iron carbonyl [$Fe(CO_5)$] at 250C (99.6–99.9% pure); (c) by hydrogen reduction of high-purity ferric oxide or oxalate; (d) by electrolytic deposition from solutions of a ferrous salt (99.9% pure).

Use: Powder metallurgy products, magnets, high-frequency cores, auto parts, catalyst in ammonia synthesis.

Available forms: (3) Cast iron and wrought iron are mixtures of iron and other materials.

Available forms: (4) Single crystals and whiskers are also available.

Available forms: (5) Iron sponge.

Hazard: Dust and fine particles suspended in air are flammable and an explosion risk. Biochemistry: Iron is a constituent of hemoglobin and is essential to plant and animal life, an important factor in cellular oxidation mechanism.

Use: Medicine and dietary supplements.

[55]iron. Radioactive iron, mass number 55.

Properties: Half-life 2.91 years, decays through K-capture.

See iron 59.

Hazard: Toxic material.

[59]iron. Radioactive iron of mass number 59.

Properties: Half-life 46.3 days. Radiation: β and γ.

Derivation: Pile irradiation of iron metal, giving a product that contains iron 55 impurity. Both iron 55 and iron 59 are produced pure in the cyclotron. Enriched samples of each are available.

Hazard: Toxic material.

Use: Medicine, tracer element in biochemical and metallurgical research.

iron acetate liquor.   (iron liquor; black liquor; black mordant; iron pyrolignite).

Properties: Intensely black liquor, sometimes containing copperas or tannin. Absorbs oxygen from the air. D 1.09–1.115, containing 5–5.5% iron.

Derivation: (1) By the action of pyroligneous acid on iron filings, (2) double decomposition of ferrous sulfate with calcium pyrolignite (calcium acetate).

Use: Mordant, especially for alizarine and nitroso dyes.

iron alum.   See ferric potassium sulfate.

iron(II) ascorbate.   See ferrous ascorbate.

iron black.   (contains no iron).

Properties: Fine black powder.

Derivation: Action of zinc upon an acid solution of an antimony salt, a black antimony being precipitated as a fine powder.

Use: Imparting the appearance of polished steel to papier mâché and plaster of Paris.

iron blue.   A pigment (of which there are several varieties) prepared by precipitating ferrous ferrocyanide from a solution of ferrocyanide and ferrous sulfate. Subsequent oxidation produces a complex ferriferrocyanide whose shade and pigment properties are dependent upon the oxidizing agent, reactant concentrations, pH, temperature, size of batch, and other conditions of manufacture. Common oxidants are nitric acid, sulfuric acid, potassium dichromate with sulfuric acid, perchlorates, and peroxides.

Properties: Insoluble in water, oils, alcohol, hot paraffin, organic solvents; unaffected by dilute acids. Unstable to alkalies of all concentrations or reducing media. Resistant to light and ordinary baking temperatures.

Use: Paints, printing inks, plastics, cosmetics (eye shadow), artist colors, laundry blue, paper dyeing, fertilizer ingredient, baked enamel finishes for autos and appliances, industrial finishes.

iron caprylate.
Use: Food additive.

iron(II) carbonate.   See ferrous carbonate.

iron carbonate, precipitated.   See iron oxide, brown.

iron carbonyl.   See iron pentacarbonyl.

iron, cast.   See cast iron.

iron(II) citrate.   See ferrous citrate.

iron compounds.   See ferric-; ferrous-.

iron(III) diethyldithiocarbamate.
CAS: 13963-59-2. $C_{15}H_{30}FeN_3S_6$.
Hazard: Moderately toxic.

irone. (6-methylionone).
CAS: (α) 79-69-6; (β) 79-70-9; (γ) 79-68-5. $C_{14}H_{22}O$.
Properties: Yellowish liquid; Mw 206.32, mp 89–90C, refractive index 1.5017 (20C), d 0.9434 (21/4C). Soluble in alcohol. A mixture of three isomers (α-, β- and γ-irone).
Use: Perfumery, violet odor. The α isomer is also used as a flavoring agent.

iron(III)-edta complex.
CAS: 15275-07-7. $C_{10}H_{12}FeN_2O_8$.
Hazard: A poison.

iron formation. Complex sedimentary or weakly metamorphosed rock that usually consists of oxides, sulfides, or carbonates of iron, often in association with chert.

iron linoleate.
Use: Food additive.

iron mass. See iron sponge.

iron naphthenate.
Use: Food additive.

iron octoate.
Use: Catalyst for curing silicone resins and rubbers. See soap (2).

iron-ore cement. Cements in which ferric oxide replaces a large part of the alumina. There must be some alumina present, however. Iron ore cement is rather slow-setting and hardening, but is more resistant to sea water than is Portland cement. It is light to chocolate brown in color, d about 3.31, higher than Portland cement.

iron ore, chrome. See chromite.

iron ore, magnetic. See magnetite.

iron ore, red. See hematite, red.

iron oxide, black. (ferrosoferric oxide; ferroferric oxide; iron oxide, magnetic; black rouge).
CAS: 1309-37-1. $FeO \cdot Fe_2O_3$ or Fe_3O_4.
See magnetite.
Properties: Reddish or bluish-black amorphous powder. D 5.18, mp 1538C (decomposes). Soluble in acids; insoluble in water, alcohol, and ether.
Derivation: (1) Action of air, steam, or carbon dioxide on iron; (2) specially pure grade by precipitating hydrated ferric oxide from a solution of iron salts, dehydrating, and reducing with hydrogen, (3) occurs in nature as the mineral magnetite.
Grade: Technical, pure (96% min).
Hazard: Pneumoconiosis. Questionable carcinogen.
Use: Pigment, polishing compound, metallurgy, magnetic inks, and in ferrites for electronic industry, coatings for magnetic tape, catalyst.

iron oxide, brown. (iron subcarbonate; iron carbonate, precipitated).
Properties: Reddish-brown powder containing ferric carbonate with ferric hydroxide $Fe(OH)_3$ and ferrous hydroxide $Fe(OH)_2$ in varying quantities; not a true oxide. Soluble in acids; insoluble in water and alcohol.
Derivation: By the interaction of solution of ferrous sulfate and sodium carbonate.
Grade: Technical.
Use: Paint pigment.

iron oxide, hydrated. See ferric hydroxide.

iron oxide, metallic brown. A naturally occurring earth, principally ferric oxide to which extenders have been added.
Use: Paint pigment.

iron oxide process. A process for the removal of sulfides from a gas by passing the gas through a mixture of iron oxide, Fe_2O_3 (called iron sponge or iron mass), and wood shavings. The iron oxide is converted to iron sulfide and can be regenerated by allowing the iron sulfide to contact air.
See iron sponge, spent.

iron oxide red. (burnt sienna; Indian red; red iron oxide; red oxide; rouge (1); Turkey red).
CAS: 1332-37-2. Pigments composed mainly of ferric oxide (Fe_2O_3).
Use: Marine paints, metal primers, polishing compounds, pigment in rubber and plastic products, theatrical rouge, grease paints.
See ferric oxide.

iron oxide yellow. (hydrated ferric oxide).
CAS: 12259-21-1. $Fe_2O_3 \cdot H_2O$.
A precipitated pigment of finer particle size and greater tinctorial strength than the naturally occurring oxides such as ocher; excellent lightfastness and resistance to alkali.
Use: Paints, rubber products, plastics.

iron pentacarbonyl. (iron carbonyl).
CAS: 13463-40-6. $Fe(CO)_5$.
Properties: Mobile, yellow liquid. D 1.466 (18C), bp 102.8C (749 mm Hg), decomposes at 200C, mp −21C, flash p 5F (−15C). Evolves carbon monoxide on exposure to air or light. Soluble in nickel tetracarbonyl and most organic solvents; soluble with decomposition in acids and alkalies; insoluble in water.
Derivation: Finely divided iron is treated with carbon monoxide, in the presence of a catalyst such as ammonia.
Hazard: Flammable, dangerous fire risk. Toxic by ingestion, inhalation, and skin absorption.
Use: Catalyst in organic reactions, carbonyl iron for high-frequency coils.

iron pigment.　　Pigments that contain iron. Many play critical biological roles.
Hazard: Toxic.

iron powder.　　See iron.

iron protochloride.　　See ferrous chloride.

iron protoiodide.　　See ferrous iodide.

iron protosulfide.　　See ferrous sulfide.

iron pyrite.　　See pyrite.

iron pyrolignite.　　See iron acetate liquor.

iron pyrophosphate.　　See ferric pyrophosphate.

iron red.　　Red varieties of ferric oxide that are used as pigments.
See iron oxide red.

iron sponge.　　(iron mass, iron oxide). Finely divided iron oxide, distributed on a support so as to give a large surface area. One form is a mixture of wood shavings covered with a hydrated iron oxide. This may be made by mixing wet wood shavings with iron borings or similar material and allowing rusting to occur to produce finely divided iron oxide. In another method, wood shavings are mixed with a slurry of the hydrated ferric oxide produced in purifying alum and then dried. The iron sponge or iron mass is used for removing sulfur from coal gas or similar materials.
Use: For precipitating copper or lead from solutions of their salts, and in electric furnace steel operations.
See iron oxide process; iron sponge, spent.

iron, sponge.　　See Sponge iron.

iron sponge, spent.　　(iron mass, spent; spent oxide). Iron sponge after saturation with sulfur. It is liable to spontaneous heating.
See iron oxide process.

iron, stainless.　　Alloys containing 3–38% chromium, with or without traces of nickel, essentially magnetic and ferritic in character. High chromium irons are brittle after welding. Most popular composition for fabrication is 15–18% chromium, 0.1% C (max).
See steel, stainless.

iron-sulfur center.　　A prosthetic group of some redox proteins involved in respiration and other electron transfers. Fe^{2+} or Fe^{3+} is with Cys groups in the protein.

iron tallate.
Use: Food additive.

irradiation.　　Exposure to radiation of wavelengths shorter than those of visible light (γ, X-ray, or UV) either for medical purposes (cancer therapy, removal of skin blemishes), for destruction of bacteria in milk and other foodstuffs or for inducing polymerization of monomers or vulcanization of rubber. UV irradiation was formerly used to induce activation of vitamin D in milk and has been used for some time to sterilize the air in operating rooms, etc.

irreversible.　　A permanent chemical or physicochemical change. Examples are: (1) a chemical reaction that can proceed only to the right, giving a product that is stable and that cannot revert to the original constituents; most reactions are of this type; (2) a colloidal system that cannot be restored to its original form after coagulation or precipitation, e.g., the hardening of egg white or milk protein by heat, the formation of butter from milk (mechanical action), and the coagulation of rubber latex by acid (chemical action).
See reversible.

irreversible colloid.　　A colloidal substance that when coagulated cannot be brought back to the colloidal state.

Irron.　　8-quinolinol-7-iodo-5-sulfonic acid.
Use: Colorimetric determination of iron.

Irugalan.　　1:2 metal complex dyes for wool and polyamide fibers.

ISAF black.　　Abbreviation for intermediate super abrasion furnace black.

isanic acid.　　(erythrogenic acid).
CAS: 506-25-2.　$C_{18}H_{26}O_2$.
A fatty acid with acetylenic triple bonds at the 9 and 11 positions.
Properties: Crystals. Mp 39C, d 0.93, refr index 1.49. Readily polymerizes and may explode on rapid heating to 250C. Soluble in acetone and alcohol. Turns red on exposure to air and heat.
Derivation: Vegetable oil from isano nuts grown in Zaire.

isatin.　　(indole-2,3-dione).
CAS: 91-56-5.　$C_6H_4COC(OH)N$.

Properties: Yellowish-red or orange crystals; bitter taste. Mp 200–203C. Soluble in water, alcohol, and ether.
Derivation: From indigo by oxidation.
Use: Dyestuffs, pharmaceuticals, analytical reagent.

isatoic anhydride.
CAS: 118-48-9. $C_8NO_3H_5$.
A bicyclic molecule composed of a benzene ring attached at the *o*- and *m*- positions to a heterocyclic ring. It forms useful anthranilic acid derivatives by reaction with hydrogen.
Properties: Tan powder.
Grade: Technical, 96% min.
Use: Intermediate for polymer curing agents, anthranilic esters, heterocyclic compounds, and benzyne.

isepamicin disulfate.
CAS: 68000-78-2. $C_{22}H_{43}N_5O_{12} \cdot 2H_2O_4S$.
Hazard: Low toxicity by ingestion.

isethionic acid.   (2-hydroxyethanesulfonic acid).
CAS: 107-36-8. $HOCH_2CH_2SO_3H$.
Properties: Liquid. Bp 100C (decomposes). Very soluble in water; insoluble in alcohol.
Use: Detergents, surfactants, synthesis.

iso-.   A prefix meaning "the same" as in such terms as *isomer* (the same part), *isotope* (the same place), *isometric* (the same measure), *isobar* (the same pressure), etc. In organic chemistry, it denotes an isomer of an alkyl hydrocarbon, alcohol, etc., having a subordinate chain of one or more carbon atoms attached to a carbon of the straight chain.
See isobutane; isooctane; branched chain.

isoactivity line.   Locus of all points that represent a certain chemical activity of a component in an isothermal section of a ternary diagram.

isosafrole.
CAS: 120-58-1. $C_{10}H_{10}O_2$. Bicyclic.
Properties: Colorless, fragrant liquid; odor of anise. D 1.117–1.120, ref index 1.576, bp 253C. Soluble in alcohol, ether, and benzene. Combustible.
Derivation: Treatment of safrole with alcoholic potash.
Hazard: Questionable carcinogen.
Use: Manufacture of heliotropin, perfumes, flavors, pesticide synergists.

isoalloxazine.   (flavin).
CAS: 490-59-5. $C_{10}H_6N_4O_2$.
A derivative of isoalloxazine, widely distributed in plants and animals, usually as yellow pigments.
See riboflavin.

isoamyl acetate.
CAS: 123-92-2. $CH_3COOCH_2CH_2CH(CH_3)_2$.
Properties: Colorless liquid; banana-like odor. Bp 142C, fp −78.5C, d 0.876 (15/4C), bulk d 7.30 lb/gal, (15C), flash p 80F (26.6C). Slightly soluble in water; miscible with alcohol and ether.
Derivation: Rectification of commercial amyl acetate.
Grade: Reagent, technical.

Hazard: Flammable, moderate fire risk. Irritant. Explosive limits in air 1–7.5%.
Use: Flavoring, perfumes, solvent for nitrocellulose, masking undesirable odors.
See amyl acetate.

***sec*-isoamyl alcohol.**   (diethyl carbine; diethylcarbinol; 3-pentanol; pentanol-3; pentan-3-ol; 3-methylbutan-2-ol).
CAS: 584-02-1. $C_5H_{12}O$.
A branched chain secondary alcohol and isomer of amyl alcohol.
Properties: Liquid; acetone-like odor; soluble in alcohol and ether; slightly soluble in water.
Hazard: Moderately toxic.
Use: A flotation agent, a solvent, and in organic synthesis.

isoamyl alcohol, primary.   (3-methyl-1-butanol; isopentyl alcohol; isobutyl carbinol).
CAS: 123-51-3. $(CH_3)_2CHCH_2CH_2OH$.
Properties: Colorless liquid; pungent taste; disagreeable odor. Bp 132.0, fp −117.2C, d 0.813 (15/4C), bulk d 6.79 lb/gal, refr index 1.4075 (20C), flash p 109F (42.7C) (CC), autoign temp 657F (347C). Slightly soluble in water; miscible with alcohol and ether. Combustible.
Derivation: Distillation of fusel oil or the mixed alcohols resulting from the chlorination and hydrolysis of pentane.
Grade: Technical.
Hazard: Moderate fire risk. Vapor is toxic and irritant. Explosive limits in air 1.2–9%.
Use: Photographic chemicals, organic synthesis, pharmaceutical products, solvent, determination of fat in milk, microscopy.
See fusel oil.

isoamyl alcohol, secondary.   Similar to the primary form except bp is 113C, flash p 103F (39.4C), d 0.819. Combustible.

isoamyl benzoate.   (amyl benzoate).
CAS: 94-46-2. $C_6H_5COOC_5H_{11}$.
Properties: Colorless liquid; fruity odor. D 0.986–0.989, refr index 1.493, bp 260C. Insoluble in water; soluble in alcohol. Combustible.
Use: Perfumery, flavors, cosmetics.

isoamyl benzyl ether.   (benzyl isoamyl ether).
CAS: 122-73-6. $C_5H_{11}OCH_2C_6H_5$.
Properties: Colorless liquid; fruity odor. D 0.904–0.908, refr index 1.481–1.485. Soluble in four parts of 80% alcohol. Combustible.
Grade: Technical.
Use: Soap perfumes.

isoamyl butyrate.
CAS: 106-27-4. $C_5H_{11}OOCC_3H_7$.
Properties: Practically water-white. D 0.866–0.868 (15.5C), bp 189C, flash p 138F (58C). Soluble in

alcohol and ether; slightly soluble in water. Combustible.

Derivation: By treating isoamyl alcohol with butyric acid.

Method of purification: Distillation.

Grade: Commercial 95–100% ester content, FCC (as amyl butyrate).

Use: Flavoring extracts, solvent and plasticizer for cellulose acetate.

isoamyl chloride.
CAS: 107-84-6. $(CH_3)_2CHCH_2CH_2Cl$.
Any of several compounds or mixtures thereof may be referred to by this name, since numerous isomers are possible, the most common of which is 1-chloro-3-methylbutane. Combustible.

Properties: Colorless or slightly yellow liquid. Bp 99.7C (758 mm Hg), d 0.893, refr index 1.410. Slightly soluble in water; soluble in alcohol and ether.

Derivation: Isoamyl alcohol and hydrochloric acid or chlorination of isopentane.

Use: Mixture, usually also containing normal amyl chloride. Solvent (nitrocellulose, varnishes, lacquers, neoprene), rotogravure inks, soil fumigation, organic compounds.

isoamyldichloroarsine.
CAS: 64049-23-6. $C_5H_{11}AsCl_2$.
Properties: Oily liquid; sweetish odor. Bp 88.5–91.5C (15 mm Hg). Decomposed by water. Combustible.

Derivation: Interaction of phosphorus trichloride and isoamylarsenic acid.

Hazard: Strong irritant.

α-isoamylene.   See 3-methyl-1-butene.

β-isoamylene.   See 3-methyl-2-butene.

isoamyl ether.
CAS: 544-01-4. $[(CH_3)_2CHCH_2CH_2]_2O$.
Properties: Colorless liquid; pleasant odor. Bp 172C, d 0.783 (12/4C), refr index 1.40. Insoluble in water; soluble in alcohol and ether. Combustible.

Use: Grignard reaction solvent, lacquer solvent.

isoamyl formate.
CAS: 110-45-2. $HCOOCH_2CH_2CH(CH_3)_2$.
Properties: Colorless liquid; fruity odor. Bp 122–124C, d 0.877. Partially soluble in water; soluble in alcohol and ether. Combustible.

Hazard: Strong irritant.

Use: Artificial fruit essences.

isoamyl furoate. $C_4H_3OCO_2C_5H_{11}$.
Properties: Colorless liquid, becoming brown in light. D 1.0335 (20/4C), bp 232–234C, 135–137C (25 mm Hg), refr index 1.4720. Insoluble in water; soluble in alcohol and ether.

isoamyl isovalerate.   See isoamyl valerate.

sec-isoamyl mercaptan.   (2-methylbutyl-3-thiol). $(CH_3)_2CHCH(SH)CH_3$.
Properties: Colorless liquid; offensive odor. Distillation range 101–127C, d 0.841 (20/4C), refr index 1.445 (20C), flash p 46F (7.7C). Insoluble in water; soluble in organic solvents.

Hazard: Flammable, dangerous fire risk.

Use: Polymerization modifier, insecticide intermediate, vulcanization accelerator intermediate, nonionic surface-active agent.

isoamyl nitrite.   See amyl nitrite.

isoamyl pelargonate.
CAS: 7779-70-6. $(CH_3)_2CHCH_2CH_2OOCC_8H_{17}$.
Properties: Liquid; fruity odor. D 0.860 (15.5/15.5C), bp 260C, refr index 1.4300 (20C). Combustible.

Use: Flavors and perfumes, chemical intermediate.

isoamyl phthalate.
CAS: 605-50-5. $C_{18}H_{26}O_4$.
Properties: Colorless liquid; no odor. Bp 225C, d 1.02. Insoluble in water; soluble in organic solvents.

Use: Plasticizer for nitrocellulose, rubber cements, foam suppressant.

isoamyl propionate.   See amyl propionate.

isoamyl salicylate.   (amyl salicylate).
CAS: 87-20-7. $C_6H_4OHCOOC_5H_{11}$.
Properties: Water-white liquid; sometimes having a faint yellow tinge which should not be pink or red; orchid-like odor. D 1.053 (15C), refr index 1.5050–1.5080 (20C), optical rotation 0 to +2.30 degrees, bp 280C, flash p 132C. Soluble in alcohol, ether; insoluble in water and glycerol. Combustible.

Derivation: By esterifying salicylic acid with amyl alcohol. The usual article of commerce is the isoamyl ester.

Method of purification: Distillation.

Grade: A pure grade of at least 99% ester content which should not exceed 100% on analysis (indicating lower esters).

Use: Soap perfumes.

isoamyl valerate.   ("apple essence," "apple oil," isoamyl isovalerate, amyl valerianate, amyl valerate).
CAS: 2050-09-1. $C_4H_9CO_2C_5H_{11}$.
Properties: Clear liquid; odor of apple when diluted with alcohol. D 0.8812, bp 203.7C. Soluble in alcohol and ether; slightly soluble in water. Combustible.

Derivation: By adding sulfuric acid to a mixture of amyl alcohol and valeric acid. Subsequent recovery by distillation.

Grade: Technical, FCC (as isoamyl isovalerate).

Use: Fruit essences, flavoring agent.

isoascorbic acid.   See erthorbic acid.

isobars. Nuclides having the same mass number but different atomic numbers, in contrast to isotopes, which have the same atomic number but different mass number. C-14 and N-14 are isobars.

isobenzan. (generic name for 1,3,4,5,6,7,8,8-octachloro-1,3,3a,4,7,7a-hexahydro-4,7-methanoisobenzofuran).
CAS: 297-78-9. $C_9H_6OCl_8$.
Properties: Solid. Mp 248–257C. Soluble in acetone and ether; slightly soluble in alcohols and kerosene; insoluble in water.
Use: Insecticide.

isobestic points. Points of equal absorption for solutions of the same total concentration.

isoborneol.
CAS: 124-76-5. $C_{10}H_{17}OH$.
A geometrical isomer of borneol.
Properties: White solid; camphor odor. Mp 216C (sublimes). More soluble in most solvents than borneol.
Derivation: By reduction of camphor.
Use: Perfumery, chemical esters.

isobornyl acetate.
CAS: 125-12-2. $C_{10}H_{17}OOCCH_3$.
Properties: Colorless liquid; pine-needle odor. D 0.978 (20C), bp 220–224C. Soluble in most fixed oils and in mineral oil; insoluble in glycerol and water. Combustible.
Derivation: By heating camphene (50–60C) with glacial acetic acid and sulfuric acid and separating by adding water.
Grade: Technical, FCC.
Use: Compounding pine-needle odors, toilet waters, bath preparations, antiseptics, theater sprays, soaps, making synthetic camphor, flavoring agent.

isobornyl acrylate.
CAS: 5888-33-5. $C_{13}H_{20}O_2$.
Hazard: Moderately toxic by ingestion. Low toxicity by skin contact. A moderate skin and mild eye irritant.

p-(isobornyloxy)aniline.
CAS: 1740-15-4. $C_{16}H_{23}NO$.
Hazard: Moderately toxic by ingestion. Low toxicity by skin contact. A severe eye irritant.

isobornyl salicylate.
Properties: Viscous, colorless oil; sweet odor. Ester content 96%. Combustible.
Grade: Technical.
Use: Perfumery (fixative), cosmetics (filter for suntan preparations).

isobornyl thiocyanoacetate.
CAS: 115-31-1. $C_{10}H_{17}OOCCH_2SCN$.

The technical grade contains 82% or more of isobornyl thiocyanoacetate, also other terpenes and derivatives.
Properties: Yellow oily liquid; terpene-like odor. D 1.1465 (25/4C), refr index 1.512 (25C), acid number 1.19. Very soluble in alcohol, benzene, chloroform, and ether; practically insoluble in water. Combustible.
Derivation: By treating isoborneol with chloroacetyl chloride and potassium thiocyanate.
Hazard: Toxic by ingestion, strong irritant.
Use: Insecticide chiefly in cattle spray, medicine.

isobornyl valerate. $(CH_3)_2CHCH_2COOC_{10}H_{17}$
Properties: Colorless, neutral liquid; oily taste; peculiar, aromatic odor. Bp 132–138C (12 mm Hg), d 0.954. Does not irritate the stomach. Soluble in alcohol and ether; sparingly soluble in water.

isobutane. (2-methylpropane; trimethylmethane).
CAS: 75-28-5. $(CH_3)_2CHCH_3$.

A liquefied petroleum gas.
Properties: Colorless gas; slight odor. Stable. Bp −11.73C, fp −159C, d 0.5572 (20C at saturation pressure), d (air = 1) 2.01, soluble in water, slightly soluble in alcohol, flash p −117F (−83C), autoign temp 864F (462C). Does not react with water. Has no corrosive action on metals. Soluble in ether.
Derivation: An important component of natural gasoline, refinery gases, wet natural gas; also obtained by isomerization of butane.
Grade: Technical, 99 mol% (pure grade), 99.96 mol% (research grade), and other high-purity grades.
Hazard: Highly flammable, dangerous fire and explosive risk; explosive limits in air 1.9–8.5%.
Use: Organic synthesis, refrigerant, motor fuels, aerosol propellant, synthetic rubber, instrument calibration fluid.

isobutane hydrate. See gas hydrates.

isobutanol. See isobutyl alcohol.

isobutanolamine. See 2-amino-2-methyl-1-propanol.

isobutanyl chloride. (3-chloro-2-methylpropene; 3-chloro-2-methylprop-1-ene). C_4H_7Cl.

isobutene. (2-methylpropene; isobutylene).
CAS: 115-11-7. $(CH_3)_2C{:}CH_2$.

A liquefied petroleum gas.

$$CH_3$$
$$CH_3-C=CH_2$$

Properties: Colorless, volatile liquid or easily lique-fied gas; coal gas odor. Bp $-6.9C$, fp $-139C$, flash p $-105F$ ($-76C$), d 0.6 (20C), autoign temp 869F (465C). Soluble in organic solvents. Polymerizes easily and also reacts easily with numerous materials.
Derivation: Fractionation of refinery gases, catalytic cracking of MTBE.
Hazard: Highly flammable, dangerous fire and explosion risk, explosive limits in air 1.8–8.8%.
Use: Production of isooctane, high-octane aviation gasoline, butyl rubber, polyisobutene resins, *tert*-butyl chloride, *tert*-butanol methacrylates; copolymer resins with butadiene, acrylonitrile, etc.; methyl-*tert*-butyl ether.

3-isobutoxy-2-pyrrolidino-*n*-phenyl-*n*-benzylpropylamine hydrochloride hydrate.
CAS: 74764-40-2. $C_{24}H_{34}N_2O\bullet ClH\bullet H_2O$.
Hazard: Moderately toxic by ingestion.

isobutyl acetate.
CAS: 110-19-0. $C_4H_9OOCCH_3$.
Properties: Colorless, neutral liquid; fruitlike odor. Bp 116–117C, flash p 64F (17.7C) (CC), d 0.8685 (15C), refr index approximately 1.392, bulk d 7.23 lb/gal, fp $-99C$, autoign temp 793F (422C). Soluble in alcohols, ether, and hydrocarbons; partially soluble in water.
Derivation: Treating isobutanol with acetic acid in the presence of catalysts.
Grade: Technical, solvent, perfume, FCC.
Hazard: Flammable, dangerous fire risk.
Use: Solvent for nitrocellulose; in thinners, sealants, and topcoat lacquers; perfumery; flavoring agent.

isobutyl acrylate.
CAS: 106-63-8. $(CH_3)_2CHCH_2OOCCH:CH_2$.
Properties: Liquid. Bp 61–63C (51 mm Hg), d 0.884 (25C), refr index 1.4124 (25C), flash p 86F (30C) (TOC). Contains 100 ppm monomethyl ether hydroquinone as inhibitor.
Hazard: Flammable, moderate fire risk.
Use: Monomer for acrylate resins.

isobutyl alcohol.   (isopropylcarbinol; 2-methyl-1-propanol).
CAS: 78-83-1. $(CH_3)_2CHCH_2OH$.
Properties: Colorless liquid. D 0.806 (15C), bp 107C, flash p 100F (37.7C) (OC), fp $-108C$, refr index 1.397 (15C), autoign temp 800F (426C). Partially soluble in water; soluble in alcohol and ether.
Derivation: By-product of synthetic methanol production, purified by rectification.

Hazard: Flammable, moderate fire risk. Strong irritant.
Use: Organic synthesis, latent solvent in paints and lacquers, intermediate for amino coating resins, substitute for *n*-butanol. Paint removers, fluorometric determinations, liquid chromatography, fruit flavor concentrates.

$$CH_2OH$$
$$CH_3-CH$$
$$CH_3$$

isobutyl aldehyde.   See isobutyraldehyde.

isobutylamine.
CAS: 78-81-9. $(CH_3)_2CHCH_2NH_2$.
Properties: Colorless liquid; amine odor. D 0.731 (20C), boiling range 66–69C, fp $-85C$, flash p 15F ($-9.4C$), autoign temp 712F (377C). Strongly caustic. Soluble in water, alcohol, ether, and hydrocarbons.
Hazard: Flammable, dangerous fire risk. Strong irritant to skin and mucous membranes.
Use: Organic synthesis, insecticides.

isobutyl-*p*-aminobenzoate.
CAS: 94-14-4. $NH_2C_6H_4COOCH_2CH(CH_3)_2$.
Properties: White, crystalline scales. Mp 64–65C. Almost insoluble in water; soluble in alcohol, benzene and acetone.
Use: Medicine (topical anesthetic), sunscreen preparations.

2-isobutylaminoethanol.
CAS: 17091-40-6. $C_6H_{15}NO$.
Hazard: A poison.

isobutylated urea formaldehyde.
CAS: 68002-18-6.
Hazard: Low toxicity by ingestion and skin contact. A severe eye irritant.

isobutylbenzene.
CAS: 538-93-2. $(CH_3)_2CHCH_2C_6H_5$.
Properties: Liquid. D 0.8532 (20/4C), fp $-51.6C$, bp 171.1, refr index 1.486 (20C), flash p 140F (60C), autoign temp 802F (427C). Combustible.
Hazard: Moderate fire risk. Toxic in high concentration, a skin and eye irritant.

isobutyl benzoate.   (eglantine).
$C_6H_5CO_2CH_2CH(CH_3)_2$.
Properties: Colorless liquid; characteristic odor. D 1.002, bp 237C. Insoluble in water; miscible with alcohol and ether. Combustible.
Use: Perfumes and flavors.

isobutyl-2-butenoate. $C_8H_{14}O_2$.
Properties: Colorless liquid; powerful fruity odor. D: 0.880, refr index: 1.426–1.430. Soluble in

alcohol, propylene glycol, fixed oils; slightly soluble in water.
Use: Food additive.

isobutyl carbinol. See isoamyl alcohol, primary.

isobutyl chloroformate.
ClCO$_2$CH$_2$CH(CH$_3$)$_2$.
Properties: Liquid. Mw 136.58, bp 128.8C, d 1.053, flash p 27C.
Hazard: Flammable. Corrosive.
Use: Peptide reagent.

isobutyl cinnamate.
CAS: 122-67-8. C$_4$H$_9$OOCCH:CHC$_6$H$_5$.
Properties: Colorless oil; amber fragrance. D 1.001–1.004, refr index 1.541. Soluble in two volumes of 70% alcohol. Combustible.
Use: Perfumery.

isobutyl cyanoacrylate. A tissue adhesive effective in surgery and medicine to retard bleeding from internal organs. Also applicable to the mounting of pearls and other jewels. Available as an aerosol spray.

isobutylene. See isobutene.

isobutylene-isopreme rubber. See butyl rubber.

isobutylene-isoprene copolymer.
CAS: 9010-85-9.
Properties: Viscosity controlling agent.
Use: Food additive.
See butyl rubber.

isobutyl furoate. C$_4$H$_3$OCO$_3$C$_4$H$_9$.
Properties: Colorless liquid becoming brown in light. D 1.0383 (26.5/4C), bp 221–223C (corrosive), refr index 1.4676 (26.5C). Insoluble in water; soluble in alcohol and ether. Combustible.

N-isobutylhendecenamide. See N-isobutylundecylenamide.

isobutyl heptyl ketone. See 2,6,8-trimethyl-4-nonanone.

isobutyl isobutyrate. (IBIB).
CAS: 97-85-8. (CH$_3$)$_2$CHCOOCH$_2$CH(CH$_3$)$_2$.
Properties: Colorless liquid; fruity odor. D 0.853–0.857 (20/20C), fp −80.7C, bp 148.7C, refr index 1.3999 (20C). Insoluble in water; soluble in alcohol and ether. Combustible.
Hazard: Toxic by inhalation.
Use: Flavoring, insect repellent, nitrocellulose lacquers and thinners, substitute for methyl amyl acetate.

isobutyl mercaptan. See 2-methyl-1-propanethiol.

isobutyl methacrylate.
CAS: 97-86-9. (CH$_3$)$_2$CHCH$_2$OOCC(CH$_3$):CH$_2$.
Properties: Liquid. Bulk d 0.882 g/mL (25/25C), boiling range 155C, refr index 1.4172 (25C), flash p 120F (49C) (TOC). Contains 25 ppm hydroquinone monomethyl ether as inhibitor. Combustible.
Hazard: Moderate fire risk.
Use: Monomer for acrylic resins.

isobutyl phenylacetate.
(CH$_3$)$_2$CHCH$_2$OOCCH$_2$C$_6$H$_5$.
Properties: Colorless liquid; honeylike odor. D 0.984–0.988 (25C), refr index 1.4860–1.4880 (20C). Soluble in most fixed oils; insoluble in glycerol, mineral oil, and propyene glycol. Combustible.
Grade: Technical, FCC.
Use: Flavoring agent, perfumes.

isobutyl propionate.
CAS: 540-42-1. CH$_3$CH$_2$COOCH$_2$CH(CH$_3$)$_2$.
Properties: Water-white liquid. D 0.86–0.8635 (20/20C), bp 138C, fp −71.4C, refr index 1.3975 (20C). Insoluble in water; very soluble in alcohol and ether. Combustible.
Use: Solvent, fruit flavor concentration.

isobutyl salicylate.
CAS: 87-19-4. HOC$_6$H$_4$COOCH$_2$CH(CH$_3$)$_2$.
Properties: Colorless liquid; may have slightly yellowish tinge. D 1.064–1.065 (25C), bp 259C. Soluble in alcohol and mineral oil, insoluble in water and glycerol. Combustible.
Grade: Technical, FCC.
Use: Perfumery, flavoring.

isobutyl stearate.
CAS: 646-13-9. C$_{22}$H$_{44}$O$_2$.
Properties: Waxy, crystalline solid. Mp 20C.
Use: Cosmetics, inks, coatings, polishes.

N-isobutylundecylenamide.
(N-isobutylhendecenamide).
CH$_3$(CH$_2$)$_7$CH:CHCONHC$_4$H$_9$. A synergist for pyrethrum.
Use: Insecticides.

isobutyl valerate.
CAS: 10588-10-0. C$_9$H$_{19}$O$_2$.
Properties: Colorless liquid; pleasant odor. D 0.85, bp 170C, refr index 1.40. Insoluble in water; soluble in ether and alcohol.
Use: Food industry as flavoring, fruit extracts, and similar products.

isobutyl vinyl ether. See vinyl isobutyl ether.

isobutyraldehyde. (isobutyl aldehyde).
CAS: 78-84-2. (CH$_3$)$_2$CHCHO.

Properties: Transparent, highly refractive liquid; colorless; pungent odor. D 0.794 (20/4C), bp 65C, fp −66C, refr index 1.3730 (20C), flash p (CC) −40F (−40C), (OC) −11F (−23.9C), explosive limits in air 1.6–10.6%, autoign temp 490F (254C). Soluble in alcohol; insoluble in water.
Derivation: (1) Oxo process reaction of propylene with carbon monoxide and hydrogen, (2) dehydrogenation of isobutanol.
Hazard: Highly flammable, dangerous fire and explosion risk. Irritant to skin and eyes.
Use: Intermediate for rubber antioxidants and accelerators, for neopentyl glycol; synthesis of amino acids, cellulose esters, flavors, etc.

isobutyric acid. (2-methylpropanoic acid). CAS: 79-31-2. $(CH_3)_2CHCOOH$.

$$CH_3CH-\underset{\underset{O}{\|}}{C}-OH$$
$$CH_3$$

Properties: Colorless liquid. D 0.946–0.950 (20/20C), bp 154.4C, fp −47C, refr index 1.3930 (20C), flash p 170F (76.6C) (TOC), autoign temp 935F (501C). Soluble in water. Alcohol and ether. Combustible.
Grade: Technical.
Hazard: Toxic by ingestion, strong irritant to tissue.
Use: Manufacture of esters for solvents, flavors and perfume bases, disinfecting agent, varnish, deliming hides, tanning agent.

isobutyric acid chloride. CAS: 79-30-1. C_4H_7ClO.
Hazard: Low toxicity by inhalation. A severe eye irritant.

isobutyric anhydride. CAS: 97-72-3. $[(CH_3)_2CHCO]_2O$.
Properties: Liquid. Boiling range 180–187C, d 0.951–0.956 (20/20C), flash p 139F (59.4C), autoign temp 665F (351C). Combustible.
Hazard: Strong irritant to tissue.
Use: Chemical intermediate.

isobutyronitrile. (2-methylpropanenitrile; isopropyl cyanide). CAS: 78-82-0. $(CH_3)_2CHCN$.
Properties: Colorless liquid. D 0.773 (20/20C), bp 107C, fp −75C. Slightly soluble in water; very soluble in alcohol and ether. Combustible.
Hazard: Toxic by ingestion, inhalation, and skin absorption.
Use: Intermediate for insecticides, etc.

isobutyryl chloride. (isobutyryl chloride; 2-methylpropanoyl chloride). CAS: 79-30-1. $(CH_3)_2CHCOCl$.
Properties: Colorless liquid. Refr index 1/4079, d 1.017 (20/4C), fp −90C, bp 92C. Soluble in ether; reacts with water and alcohol.
Use: Chemical intermediate.

isocetyl laurate. CAS: 89527-28-6. $C_{11}H_{23}COOC_{16}H_{33}$.
Properties: Oily liquid; almost odorless. D 0.858, fp approximately −65C, viscosity 19.6 cP at 25C. Insoluble in water; soluble in most organic solvents. Combustible.
Use: Cosmetics and pharmaceuticals (lubricant, fixative and solvent), plasticizer, mold release agent, textile softener.

isocetyl myristate. CAS: 83708-66-1. $C_{13}H_{27}COOC_{16}H_{33}$.
Properties: Oily liquid; practically no odor. D 0.857, fp −39C, viscosity 25.6 cP at 25C. Insoluble in water; soluble in most organic solvents. Combustible.
See isocetyl laurate.

isocetyl oleate. $C_{17}H_{33}COOC_{16}H_{33}$
Properties: Oily liquid; practically no odor. D 0.862, fp −57C, viscosity 29.0 cP at 25C. Insoluble in water; soluble in most organic solvents. Combustible.
See isocetyl laurate.

isocetyl stearate. CAS: 25339-09-7. $C_{17}H_{35}COOC_{16}H_{33}$.
Properties: Oily liquid; practically no odor. D 0.862, fp 275C, viscosity 29.0 cP at 25C. Insoluble in water; soluble in most organic solvents. Combustible.
See isocetyl laurate.

isochromosome. A metacentric chromosome produced during mitosis or meiosis when the centromere splits transversely instead of longitudinally; the arms of such chromosome are equal in length and genetically identical, however, the loci are positioned in reverse sequence in the two arms.

isocil. (generic name for 5-bromo-3-isopropyl-6-methyluracil). CAS: 314-42-1.

$OCCBrC(CH_3)NHC(O)NCH(CH_3)_2$.

Properties: Crystals. Mp 158C. Soluble in absolute alcohol.
Hazard: Toxic by ingestion.
Use: Herbicide.

isocinchomeronic acid. (2,5-pyridinedicarboxylic acid). CAS: 100-26-5. $HOOC(C_5H_3N)COOH$.
Properties: Light-tan powder, leaflets, or prisoms; no odor. Mp 254C, sublimes as nicotinic acid above this temperature. Insoluble in cold water, alcohol, ether, benzene; slightly soluble in boiling water, boiling alcohol; soluble in hot dilute mineral acids.
Use: Intermediate for nicotinic acid, insecticides, polymers, and dyestuffs.

isocrotonic acid. See crotonic acid.

isocrotyl chloride. (α-chloroisobutylene; 1-chloro-2-methylpropene; 1-chloro-2-methylprop-1-ene; β,β-dimethylvinyl chloride; NCI-C54819). C_4H_7Cl.
Properties: Liquid.
Hazard: Carcinogen; neoplastigen; moderately toxic; local irritant, and narcotic.

isocurcumenol.
CAS: 24063-71-6. $C_{15}H_{22}O_2$.
Hazard: A poison by ingestion.

isocyanate. A compound containing the isocyanate radical –NCO. Monoisocyanates are in use, as in the treatment of cellulose to obtain a cellulose tricarbamate, but the term isocyanate usually refers to a diisocyanate.

isocyanate generator. (hindered isocyanate). An isocyanate derivative that will decompose to an isocyanate upon heating. In one type, phenol is combined with an isocyanate, and the resulting urethane is stable at room temperature but dissociates at 160C to the original phenol and isocyannate. These generators are used commercially in a mixture with a polyester, which can be stored indefinitely but upon heating produces a polyurethane resin.

isocyanate resin. See polyurethane resin.

isocyanatoacetic acid ethyl ester. See theyl isocyanatoacetate.

1-isocyanato-2-methylbenzene.
CAS: 614-68-6. C_8H_7NO.
Hazard: A poison by inhalation. Moderately toxic by ingestion.

1-isocyanato-4-methylbenzene.
CAS: 622-58-2. C_8H_7NO.
Hazard: A Poison by inhalation. Moderately toxic by ingestion.

3-isocyanatopropyltriethoxysilane.
CAS: 24801-88-5. $C_{10}H_{21}NO_4Si$.
Hazard: A poison by skin contact. Moderately toxic by inhalation. A mild skin and severe eye irritant.

isocyanic acid. (cyanic acid). HN=C=O. A gas resulting from depolymerization of cyanuric acid at 300–400C in a stream of carbon dioxide. An intermediate product in the formation of urethanes and allophanates. Direct synthesis from nitric oxide, carbon monoxide, and hydrogen with iridium or palladium catalyst at 300C was reported by Bell Laboratories in 1978.
Hazard: Severe explosion risk. Strong irritant to eyes, skin and mucous membranes.

isocyanide. See carbylamines.

1-isocyano-2-methoxy-4-nitrobenzene.
CAS: 2008-62-0. $C_8H_6N_2O_3$.
Hazard: Moderately toxic by ingestion.
Use: Agricultural chemical.

2-isocyano-1-methoxy-4-nitrobenzene.
CAS: 1983-95-5. $C_8H_6N_2O_3$.
Hazard: Moderately toxic by ingestion.
Use: Agricultural chemical.

1-isocyano-4-methyl-3-nitrobenzene.
CAS: 1930-92-3. $C_8H_6N_2O_2$.
Hazard: Moderately toxic by ingestion.
Use: Agricultural chemical.

1-isocyanonaphthalene.
CAS: 1984-04-9. $C_{11}H_7N$.
Hazard: A poison by ingestion.
Use: Agricultural chemical.

1-isocyano-4-nitrobenzene.
CAS: 1984-23-2. $C_7H_4N_2O_2$.
Hazard: Low toxicity by ingestion.
Use: Agricultural chemical.

isocyanurate. A compound closely related to isocyanate but containing three NCO groups. Its products are similar to polyurethane resins and are particularly useful as rigid foams for insulation in the building and construction industry. For combustibility, see foam, plastic.
See isocyanuric acid.

isocyanuric acid. (s-triazine-2,4,6-trione).
CAS: 108-80-5.

OCNHCONHCONH.

Properties: White, crystalline powder. The ketone isomer of cyanuric acid. Derivatives of isocyanuric acid, such as dichloro- and trichloroisocyanuric acid, and potassium and sodium dichloroisocyanurate, are bleaches and sanitizers.

isodecaldehyde.
CAS: 3085-26-5. $C_9H_{19}CHO$.
Mixed isomers.
Properties: Liquid. D 0.8290, bp 197.0C, fp −80C, bulk d 6.9 lb/gal, flash p 185F (85C). Insoluble in water. Combustible.
Use: Intermediate for pharmaceuticals, dyes, and resins.

isodecane. See 2-methylnonane.

isodecanoic acid.
CAS: 5963-14-4. $C_{10}H_{20}O_2$.
Mixture of branched-chain acids, primarily trimethylheptanoic and dimethyloctanoic.
Properties: Liquid. Bp 254C, 137C (10 mm Hg), d 0.9019 (20/20C), fp glass at approximately −60C,

very slightly soluble in water, viscosity 12.9 cP at 20C, refr index 1.4358 (20C). Combustible.
Use: Intermediate for metal salts, ester type lubricants, plasticizers.

isodecanol.
CAS: 68526-85-2. $C_{10}H_{21}OH$.
Mixed isomers.
Properties: Colorless liquid. D 0.8395, flash p 220F (104C), bp 220C. Insoluble in water. Combustible.
Use: Antifoaming agent in textile processing.

isodecyl chloride. $C_{10}H_{21}Cl$. Mixed isomers.
Properties: Colorless liquid. D 0.8767, bp 210.6C, fp glass at −180C, flash p 200F (93.3C). Insoluble in water. Combustible.
Use: Solvent for oils, fats, greases, resins, gums; extractants, cleaning compounds; intermediate for insecticides, pharmaceuticals, plasticizers, polysulfide rubbers, resins, and cationic surfactants.

isodecyl diphenyl phosphite.
CAS: 26544-23-0. $C_{22}H_{31}O_3P$.
Hazard: A mild eye irritant.

isodecyl octyl adipate.
Properties: Light-colored oily liquid. D 0.924 (20/20C), mid-bp 227C (4 mm Hg), refr index 1.448 (25C), viscosity 20 cP at 20C, flash p 400F (204C). Combustible.
Use: Plasticizer.

isodecyl pelargonate. $(CH_3)_2CH(CH_2)_6CH_2$ $OOC(CH_2)CH_3$. A synthetic lubricant.

isodecyl phosphite.
CAS: 25448-25-3. $C_{30}H_{63}O_3P$.
Hazard: Low toxicity by ingestion, inhalation, and skin contact. A skin irritant.

isodiazine.
CAS: 1912-25-0. $C_{10}H_{18}ClN_5$.
Hazard: Moderately toxic by ingestion.
Use: Agricultural chemical.

isodiphenylbenzene. See *m*-terphenyl.

isodrin. (generic name for 1,2,3,4,10,10-hexachloro-1,4,4a,5,8,8a-hexahydro-1,4-*endo,endo*-5,8-dimethanonaphthalene).
CAS: 465-73-6. $C_{12}H_8Cl_6$.
An isomer of aldrin.
Properties: Crystals. Decomposes above 100C.
Hazard: Toxic, use may be restricted.
Use: Insecticide.

isodurene. (1,2,3,5-tetramethylbenzene).
CAS: 527-53-7. $(CH_3)_4C_6H_2$.
Properties: Liquid. D 0.896, bp 197C, fp −24C. Soluble in alcohol and ether; insoluble in water. Combustible.
Derivation: From coal tar.

Grade: Technical.
Use: Organic synthesis.

isoelectric focusing. An electrophoretic method for separating macromolecules on the basis of their pI.

isoelectric point. The pH at which the net charge on a molecule in solution is zero. At this pH, amino acids exist almost entirely in the zwitterion state, that is, the positive and negative groups are equally ionized. A solution of proteins or amino acids at the isoelectric point exhibits minimum conductivity, osmotic pressure, and viscosity. Proteins coagulate best at this point. Typical isolectric points (pH) are glycine 6.6, gelatin 4.7, and serum albumin 5.4.

isoelectronic. Two molecules are described as isolectronic if they have the same number of valence electrons in similar orbitals, although they may differ in their distribution of nuclear charges (e.g., H–CN and H–N⁺C⁻).

isoenanthic acid.
CAS: 628-46-6. $C_7H_{14}O_2$.
Hazard: A reproductive hazard.

isoenzyme. An enzyme performing the same function as another enzyme but having a different set of amino acids. The two enzymes may function at different speeds.

isoeugenol. (1-hydroxy-2-methoxy-4-propenyl-benzene).
CAS: 97-54-1. $(CH_3CHCH)C_6H_3OHOCH_3$.
Properties: Pale-yellow oil; spice-clove odor. D 1.081–1.084, mp 19C, bp 268C, refr index 1.5739 (19C). Soluble in alcohol, ether and other organic solvents; slightly soluble in water. Combustible.
Derivation: From eugenol by isomerization with caustic potash.
Grade: Perfumer's grade, FCC.
Use: Perfumes, vanillin, flavoring agent.

isoeugenol acetate. See acetylisoeugenol.

isoeugenol ethyl ether. (1-ethoxy-2-methoxy-4-propenyl-benzene). $C_3H_5(CH_3O)C_6H_3OC_2H_5$.
Properties: Synthetic, white, crystalline powder. Mp 64C. Insoluble in water; soluble in alcohol, ether, and benzene. Combustible.
Use: Sweetening agent and odorant fixative.

isoflurophate. See diisopropyl fluorophosphates.

isoforming. A process for fixed-bed hydroisomerization, requiring a non-noble-metal catalyst. Claimed to give high yields of C_8 (xylene) isomers with low hydrogen consumption and minimal catalyst regeneration.

isoheptane.   See 2-methylhexane.

isohexane.   (1,2-dimethylbutane;   2-methylpentane).
CAS: 107-83-5. C_6H_{14}.
A mixture of branched-chain isomers.
Properties: Colorless liquid. Boiling range 54–61C, d 0.671 (15.5/15.5C), flash p −26F (−32C) (CC).
Grade: Commercial.
Hazard: Highly flammable, dangerous fire and explosion risk, explosive limits in air 1–7%.
Use: Solvent, freezing-point depressant.

isohexenyl cyclohexenyl carboxaldehyde.
CAS: 37677-14-8. $C_{13}H_{20}O$.
Hazard: Low toxicity by ingestion. A moderate skin and mild eye irritant.

isohexyl alcohol.   See amyl methyl alcohol.

isolac.   A whey protein powder.
Use: For dietary, health, sport food and beverage systems.

isolan.   See   1-isopropyl-3-methyl-5-pyrazolyl dimethylcarbamate.

isolated double bond.   Double bond separated by more than one single bond linkage from the next double bond.

isolation.   Identification and separation of a pure substance that is present in trace amounts in a complex mixture. A famous instance of this was the isolation of polonium (1898) and radium (1912) from pitchblende by the Curies by coprecipitation techniques followed by repeated fractional crystallization.

isoleucine.   (2-amino-3-methylpentanoic   acid; Ile).
CAS: 73-32-5. $CH_3CH_2CH(CH_3)CH(NH_2)COOH$.
An essential amino acid, found naturally in the L(+) form.
Properties: Crystals. Slightly soluble in water; nearly insoluble in alcohol; insoluble in ether.
Derivation: Hydrolysis of protein (zein, edestin), amination of α-bromo-β-methylvaleric acid.
Use: Medicine, nutrition, biochemical research.

***dl*-isoleucine.**
CAS: 443-79-8. $C_6H_{13}NO_2$.
Properties: White crystalline powder from EtOH; slightly bitter taste. Mp 290C (decomp). Soluble in water; insoluble in alcohol and ether.

isolysin.   An antibody that lyses the cells of animals of the same species that produced it.

"isomate" *[Washington].*   TM for isocyanate foam systems. Available as nonburning, pour-in-place froth, or spray foams.

isomer.   (1) One of two or more molecules having the same number and kind of atoms and hence the same molecular weight, but differing in respect to the arrangement or configuration of the atoms. Butanol (C_4H_9OH or $C_4H_{10}O$) and ethyl ether ($C_2H_5OC_2H_5$ or $C_4H_{10}O$) have the same empirical formulas but are entirely different kinds of substances; normal butanol ($CH_3CH_2CH_2CH_2OH$) and isobutanol ($[CH_3]_2CHCH_2OH$) are the same kinds of substances, differing chiefly in the shape of the molecules; *sec*-butanol ($CH_3CH_2OCH_2CH_3$) exists in two forms, one a mirror image of the other (enantiomer). Isomers often result from location of an atom or group of a compound at various positions on a benzene ring; e.g., xylene, dichlorobenzene. (2) Nuclides (i.e., kinds of atomic nuclei) having the same atomic and mass numbers, but existing in different energy states. One is always unstable with respect to the other, or both may be unstable with respect to a third. In the latter instance the energy of transformation in the two cases will differ.
See geometic isomer; optical isomer.

isomerases.   Enzymes that catalyze the transformation of compounds into their positional isomers.

isomerization.   A method used in petroleum refining to convert straight-chain to branched-chain hydrocarbons, or alicyclic to aromatic hydrocarbons, to increase their suitability for high-octane motor fuels. For example, butane (a gaseous paraffin hydrocarbon, $CH_3CH_2CH_2CH_3$) can be slightly modified in structure by catalytic reactions to give the isomeric isobutene ($CH_3CH_3CHCH_3$) used as a component of aviation fuel. Similarly, methylcyclopentane can be isomerized to cyclohexane, which is then dehydrogenated to benzene. Isomerization techniques were introduced on a large scale during World War II.
See isomer; chain.

α-isomethylionone.   (γ-methylionone).
$C_{14}H_{22}O$.
Properties: Slightly yellow liquid. D 0.925–0.929 (25/25C), refr index 1.5000–1.5010 (20C), flash p 217F (102.7C) (TCC). Soluble in 5 parts of 70% alcohol. A synthetic product. Combustible.
Use: Floral perfumes, particularly of a violet character; flavoring.

isomorphism.   The state in which two or more compounds that form crystals of similar shape have similar chemicals properties and can usually be prepresented by analogous formulas, e,g., Ag_2S and Cu_2S.

isonipecaine hydrochloride.   See meperidine hydrochloride.

isonitrile.   See carbylamines.

isonol C100.
CAS: 89750-17-4. $C_6H_5N[CH_2CH(CH_3)OH]_2$.
An aromatic reinforcing polyol.
Properties: Amber liquid. Viscosity (50C) 1000 cP (max), d 1.055 (23C), water content 0.05%. Combustible.
Use: Ingredient of polyurenthane foams, coatings, sealants, and elastomers; intermediate in organic synthesis.

isononyl alcohol.
CAS: 27458-94-2. $C_6H_{17}CH_2OH$.
A Higher alcohol developed in early 1968. Combustible.
Use: Basis of plasticizers such as diisononyl adipate.

isooctane.   (2,2,4-trimethylpentane).
CAS: 540-84-1. $(CH_3)_3CCH_2CH(CH_3)_2$.

$$CH_3-\underset{\underset{CH_3}{|}}{\overset{\overset{CH_3}{|}}{C}}-CH_2-\underset{\underset{CH_3}{|}}{CH}-CH_3$$

A branched chain hydrocarbon.
Properties: Colorless liquid. D 0.6919 (20/4C), fp −107.4C, bp 99.2C, refr index 1.3914 (20C), flash p 10F (−12.2.C), autoign temp 784F (417C). Insoluble in water; slightly soluble in alcohol and ether.
Grade: Technical, pure, research, spectrophotometric.
Hazard: Flammable, dangerous fire risk, explosive limits in air 1.1–6%. Toxic by ingestion and inhalation.
Use: Organic synthesis, solvent, motor fuel; used with normal heptane to prepare standard mix to determine antiknock property of gasoline.
See octane number.

isooctene.
CAS: 107-39-1. C_8H_{16}.
Mixture of isomers.
Proeprties: Colorless liquid. Boiling range 87.7–93.3C bromine number 137, d, 0.726 (60/60F), flash p approximately 20F (−6.6C).
Hazard: Flammable, dangerous fire risk.

isooctyl adipate.
CAS: 1330-86-5. $(C_8H_{17}OOCCH_2CH_2)_2$.
Plasticizer providing low-temperature stability.
Use: In calendaring film, sheeting, vinyl dispersions, extrusions.

isooctyl alcohol.   (isooctanol).
CAS: 26952-21-6.   General term applied to any isomer of the formula $C_7H_{15}CH_2OH$ in which the eight carbon atoms form a branched chain. Usually refers to a mixture of isomers made by the Oxo process. A selected C_7 hydrocarbon fraction is reacted with hydrogen and carbon monoxide in the presence of a catalyst at pressures up to 3000 psi. The crude alcohol is recovered and purified.

Properties: Clear liquid. Distillation range 182–195C, bulk d 6.95 lb/gal, d 0.832 (20/20C), flash p 180F (82.2C) (TOC). Combustible.
Hazard: Toxic by skin absorption.
Use: Ingredient of plasticizers; intermediate for nonionic detergents and surfactants, synthetic drying oils, cutting and lubricating oils, hydraulic fluids; resin solvent; emulsifier; antifoaming agent; intermediate for introducing the isooctyl group into other compounds.

isooctyl isodecyl phthalate.
$C_8H_{17}OOCC_6H_4COOC_{10}H_{21}$.
Properties: Clear liquid; mild odor. D 0.976 (20/20C), flash p 445F (229C). Combustible.
Grade: Technical.
Use: Plasticizer.

isooctylmercapto acetate.   See isooctyl thioglycolate.

isooctyl palmitate.
CAS: 1341-38-4. $C_8H_{17}OOCC_{15}H_{31}$.
Properties: Clear liquid. D 0.863 (20C), acidity 0.2% max (palmitic), moisture 0.05% max, mp 6–9C, bp 228C (5 mm Hg). Soluble in most organic solvents. Combustible.
Use: Secondary plasticizer for synthetic resins, extrusion aid and plasticizer.

isooctylphenoxypolyoxyethylene ethanol.
(isooctylphenylpolyethylene   glycol   ether).
$C_{32}H_{55}O_{10}$.
Properties: Slightly viscous, pale-amber-colored liquid; oily, musty odor. Mp 2–5C, bp 150C (initial) at 1 µm, bulk d 1.06 (20C), flash p 227F (108C). Combustible.
Use: Surface-active agent.

isooctyl   thioglycolate.   (isooctylmercaptoacetate).
CAS: 25103-09-7. $HSCH_2COOCH_2C_7H_{15}$.
Properties: Water-white liquid; faint fruity odor. Bp 125C (17 mm Hg) D 0.9736 (25C), refr index 1.4606 (21C), acid number less than 1. Combustible.
Grade: 99% (minimum purity).
Hazard: Toxic by ingestion.
Use: Antioxidants, fungicides, oil additives, plasticizers, insecticides, stabilizers, polymerization modifiers, stabilizer in tin-sulfur compounds, stripping agent for polysulfide rubber.

isoparaffinic petroleum hydrocarbons, synthetic.
CAS: 64742-88-7.
Use: Coating agent; float; food additive; froth-flotation cleaning; insecticide formulations component.

isopentaldehyde.
CAS: 590-86-3. C_4H_9CHO.
A mix of isomeric five-carbon aldehydes.

Properties: Water-white liquid; sharp odor. D 0.8089 (20/20C), bp 98.6C, fp −95.4C. Water dissolves 0.85% aldehyde at 20C, water soluble to 2.2% in the aldehyde. Combustible.

Use: Possible intermediate for bis-phenols, epoxy and polycarbonate resins, and modified formaldehyde resins.

isopentane.   (2-methylbutane;   ethyldimethylmethane).
CAS: 78-78-4. $(CH_3)_2CHCH_2CH_3$.
Properties: Colorless liquid; pleasant odor. Fp −159.890C, bp 27.854C, d 0.61967 (20C), flash p −70F (−57C), autoign temp 788F (420C). Soluble in hydrocarbons, oils, ether; very slightly soluble in alcohol, insoluble in water.
Derivation: Fractional distillation from petroleum, purified by rectification.
Grade: Research (99.99%), pure (99%), technical (95%), commercial.
Hazard: Highly flammable, dangerous fire risk.
Use: Solvent, manufacture of chlorinated derivatives, blowing agent for polystyrene.

isopentanoic acid.
CAS: 503-74-2. C_4H_9COOH.
A mixture of isomeric five-carbon acids.
Properties: Water-white liquid; penetrating odor. D 0.9388 (20/20C), bp 183.2C, vap press 0.14 mm Hg (20C), fp −44C. Water dissolves 3.24 wt% of acid at 20C; acid dissolves 10.4% water at 20C. Combustible.
Hazard: Strong irritant to tissue.
Use: Intermediate for plasticizers, synthetic lubricants, pharmaceuticals, metallic salts, vinyl stabilizers; extractant for mercaptans from hydrocarbons.

isopentyl alcohol.   See isoamyl alcohol, primary.

isophane insulin suspension.   See insulin.

isophorone.   (3,5,5-trimethyl-2-cyclohexen-1-one).
CAS: 78-59-1.

C(O)CHC(CH₃)CH₂C(CH₃)₂CH₂.

Properties: Water-white liquid. D 0.9229 (20/20C), bulk d 7.7 lb.gal (20C), bp 215.2C, vap press 0.2 mm Gh (20C), fp −8.1C, viscosity 2.62 cP (20C), flash p 205F (96C), autoign temp 864F (462C). Has high solvent power for vinyl resins, cellulose esters, ether, and many substances soluble with difficulty in other solvents; slightly soluble in water. Combustible.
Hazard: Irritant to skin and eyes. Possible carcinogen.
Use: In solvent mixtures for finishes, for polyvinyl and nitrocellulose resins, pesticides, stoving lacquers.

isophorone diisocyanate.   (3-isocyanateomethyl-3,5,5-trimethylcyclohexyl-isocyanate; IPDI).
CAS: 4098-71-9. $C_{12}H_{18}N_2O_2$.
Properties: Colorless to slightly yellow liquid. Mw 222.3, mp about −60C, bp 158C (10 mm Hg), vap press 0.0003 mm Hg (20C), d 1.056. Completely miscible with esters, ketones, ethers and aromatic and aliphatic hydrocarbons.
Hazard: A severe irritant, toxic by skin absorption.
Use: Yields polyurethanes with high stability, resistance to light discoloration, and chemical resistance.

isophoronenitrile.
CAS: 7027-11-4. $C_{11}H_{15}NO$.
Hazard: A poison by ingestion. Low toxicity by skin contact. A mild eye irritant.

isophthalic acid.   (m-phthalic acid; IPA).
CAS: 121-91-5. $C_6H_4(COOH)_2$.
Properties: Colorless crystals. Mp 345–348C, sublimes. Slightly soluble in water; soluble in alcohol and acetic acid; insoluble in benzene and petroleum ether. Combustible.
Derivation: (1) Oxidation of m-xylene; (2) liquid phase oxidation of mixed xylenes; (3) direct oxidation of mixed alkyl aromatics with heavy metal salts and bromine as catalysts.
Grade: Technical.
Use: Polyester, alkyd; polyurethane and other high polymers, plasticizers.

isophthalodinitrile.   See m-phthalodinitrile.

isophthaloyl chloride.   (m-phthalyl dichloride).
CAS: 99-63-8. $C_6H_4(COCl)_2$.

Properties: Crystalline solid. Mp 41C, bp 276C, flash fp 356F (180C) ©C). Soluble in ether and other organic solvents; reactive with water and alcohol. Combustible.
Use: Intermediate, dyes, synthetic fibers, resins, films, protective coatings, laboratory reagent.

isopimpinellin.
CAS: 482-27-9. $C_{13}H_{10}O_5$.
Properties: Coumarin derivative isolated from Cnidii Monnieri Fructus extract.
Hazard: A poison by ingestion.

"Isoplast" *[Lubrizol Advanced].*   TM for an engineering-grade thermoplastic polyurethane for extrusion and injection molding. It is an impact-resistant glass based on methylenediphenyl isocyanate.

isopolyester. A polyester resin based on isophthalic acid.

isoprene. (3-methyl-1,3-butadiene; 2-methyl-1,3-butadiene).
CAS: 78-79-5. CH_2=$C(CH_3)CH$=CH_2.
The molecular unit of natural rubber.
Properties: Colorless, volatile liquid. Fp --146C, bp 34.08C, refr index 1.4216 (20C), d 0.6808 (20/4C), flash p −55F (−48C), autoign temp 802F (427C). Insoluble in water; soluble in alcohol, ether, and hydrocarbon solvents.
Derivation: (1) From cracked products of heavy petroleum oils; (2) dehydrogenation of isopentene; (3) pyrolysis of methyl pentene or of isobutylene-formaldehyde condensation products; (4) dehydration of methyl butenol.
Grade: Polymerization (min purity 99%), research (99.99%).
Hazard: Highly flammable, dangerous fire and explosion risk. Irritant. Possible carcinogen.
Use: Monomer for manufacture of polyisoprene, chemical intermediate, component of butyl rubber. See polyisoprene; rubber, natural.

isoprenoid. A compound based on the isoprene structure. These include many naturally occurring materials such as terpenes, rubber, cholesterol, and other steroids.

isopropanol. Legal label name for isopropyl alcohol.

isopropanolamine. (MIPA; 2-hydroxypropylamine; 1-amino-2-propanol).
CAS: 78-96-6. $CH_3CH(OH)CH_2NH_2$.
Properties: Liquid; slight ammonia odor. D 0.9619, mp 1.4C, bp 159.9C, refr index 1.4462 (20C), flash p 170F (76.6C). Soluble in water. Combustible.
Use: Emulsifying agent, drycleaning soaps, soluble textile oils, wax removers, metal-cutting oils, cosmetics, emulsion paints, plasticizers, insecticides.

isopropenyl acetate.
CAS: 108-22-5. $CH_3COOC(CH_3)$:CH_2.
Properties: Water-white liquid. D 0.9226, bp 97.4C, fp −92.9C, solubility in water 3.25% by wt (20C), refr index 1.4020 (20C), flash p 60F (15.5C) (OC).
Hazard: Flammable, dangerous fire risk.
Use: Acylation reagent.

ispropenylacetylene. (2-methyl-1-buten-3-yne). H_2C=$C(CH_3)C$≡CH.
Properties: Colorless liquid. Bp 33–34C, fp −113C, d 0.695 (20/20C), refr index 1.4168 (20C), flash p below 20F (−6.6C) (TOC). Very slightly soluble in water; miscible with acetone, alcohol, benzene, carbon tetrachloride, and kerosene.
Hazard: Flammable, dangerous fire risk.

isopropenyl chloride. See chloropropene.

isopropenylchloroformate.
CAS: 57933-83-2. $ClCOOC(CH_3)$:CH_2.
Properties: Liquid. D 1.103 (20C), bp 93C (746 mm Hg).
Derivation: Distillation of the reaction products of acetone and phosgene.
Hazard: Strong irritant to eyes and skin.

isopropenyl ethyl ketone.
CAS: 25044-01-3. $C_6H_{10}O$.
Hazard: Moderately toxic by ingestion and skin contact. A severe skin and mild eye irritant.

isoprophenamine hydrochloride.
CAS: 6933-90-0. $C_{11}H_{16}ClNO$•ClH.
Hazard: Moderately toxic by ingestion.

***p*-isopropoxydiphenylamine.**
CAS: 101-73-5. $C_6H_5NHC_6H_4OCH(CH_3)_2$.
Properties: Dark-gray flakes. D 1.10, set point 80–86C, purity 92% (min), ash 0.10% (max). Insoluble in water; soluble in ethanol, acetone, benzene, and gasoline.
Use: Rubber antioxidant.

2-isopropoxyethanol. (IPE; isopropyl glycol; isopropyl "Cellosolve" *[Union]*).
CAS: 109-59-1. $(CH_3)_2CHOCH_2CH_2OH$.
Properties: A mobile liquid. Mw 104.15, d 0.91, bp 139.5–144.5, vap press 2.6 torr (20C).
Hazard: Combustible liquid with flash p 49C. Toxic by skin absorption.
Use: As a component of lacquers and other coatings, and as a solvent.

2-isopropoxy-4-hydroxyphenyl methylcarbamate.
CAS: 17595-59-4. $C_{11}H_{15}NO_4$.
Hazard: A poison.
Use: Agricultural chemical.

1-isopropoxypentachlorobutadiene.
CAS: 68334-67-8. $C_7H_7CL_5O$.
Hazard: Moderately toxic by ingestion and skin contact. Low toxicity by inhalation. A severe skin irritant.

2-isopropoxyphenyl (methyl) (*t*-butoxysulfinyl)carbamate.
CAS: 77276-08-5. $C_{15}H_{23}NO_5S$.
Hazard: A poison by ingestion.
Use: Agricultural chemical.

***o*-isopropoxyphenyl-*N*-methylcarbamate.** (propoxur).
CAS: 114-26-1. $(CH_3)_2CHOC_6H_4OOCNHCH_3$.
Properties: White, crystalline powder; odorless. Mp 91C. Soluble in most polar solvents; very slightly soluble in water; unstable in highly alkaline media; stable under normal use conditions.

Hazard: Toxic by ingestion and inhalation. Cholinesterase inhibitor. Possible carcinogen.
Use: Insecticide, molluscicide.

2-isopropoxyphenyl　　(methyl)(n-hexoxysulfinyl)carbamate.
CAS: 77267-49-3.　$C_{17}H_{27}NO_5S$.
Hazard: Moderately toxic by ingestion.
Use: Agricultural chemical.

n,n'-(3-isopropoxy-1,2-propanedioxysulfinyl)bis(s-methyl-n-methylcarbamoyloxythioacetamidate).
CAS: 81861-97-4.　$C_{16}H_{30}N_4O_9S_4$.
Hazard: A poison by ingestion.
Use: Agricultural chemical.

n,n'-(3-isopropoxy-1,2-propanedioxysulfinyl)bis(3-methylphenylmethylcarbamate).
CAS: 81862-11-5.　$C_{24}H_{32}N_2O_9S_2$.
Hazard: Moderately toxic by ingestion.
Use: Agricultural chemical.

n,n'-(3-isopropoxy-1,2-propanedioxysulfinyl)bis(1-naphthylmethylcarbamate).
CAS: 81826-19-3.　$C_{30}H_{32}N_2O_9S_2$.
Hazard: Moderately toxic by ingestion.
Use: Agricultural chemical.

β-isopropoxypropionitrile.
$(CH_3)_2CHO(CH_2)_2CN$.
Properties: Colorless to straw-colored liquid. Combines the chemical and physical properties of ethers and nitriles. Fp −67C, bp 82–86C (25 mm Hg) 65–65.5C (10 mm Hg), d 0.9058 (25C), flash p 155F (68.3C). Slightly soluble in water; soluble in organic solvents. Combustible.

isopropyl acetate.
CAS: 108-21-4.　$CH_3COOCH(CH_3)_2$.
Properties: Colorless liquid; aromatic odor. Bp 89.4C, d 0.8690 (25/4C), refr index 1.378 (20C), specific heat 0.46 cal/g, fp −73.4C, heat of vaporization 135 Btu/lb, viscosity 0.49 c (25C), solubility in water 2.9 wt%, flash p 40F (4.4C), bulk d 7.17 lb/gal (20C), autoign temp 860F (460C). Miscible with most organic solvents.
Derivation: By reacting isopropyl alcohol with acetic acid in the presence of catalysts.
Grade: 95%, 85–88%.
Hazard: Flammable, dangerous fire risk.
Use: Solvent for nitrocellulose, resin gums, etc.; paints, lacquers, and printing inks; organic synthesis, perfumery.

isopropylacetone.　　See methyl isobutyl ketone.

n-isopropylacrylamide.　　(NIPAM).

Properties: Crystalline solid. Homopolymers and copolymers prepared with this material show inverse solubility in water.
Use: Binders in textiles, paper, adhesives, detergents, cosmetics.
See acrylic resin.

isopropyl alcohol.　　(IPA; dimethylcarbinol; sec-propyl alcohol; isopropanol; 2-propanol).
CAS: 67-63-0.　$(CH_3)_2CH_2O$.
Properties: Colorless liquid; pleasant odor. Bp 82.4C, d 0.7863 (20.20C), refr index 1.3756 (20C), specific heat 0.65 cal/g, fp −86C, critical temperature 235C, critical pressure 53 atm, vap press 33 mm Hg (20C), fash p 53F (11.7C) (TOC), heat of combustion 14.346 Bt/lb, heat of vaporization 288 Btu/lb, viscosity 2.1 cP (25C), autoign temp 850F (453C). Soluble in water, alcohol and ether.
Derivation: By treatment of propylene with sulfuric acid and hydrolyzing.
Method of purification: Rectification.
Grade: 91%, 95%, 99%, NF (99%), nanograde.
Hazard: Flammable, dangerous fire risk, explosive limits in air 2–12%. Toxic by ingestion and inhalation. Eye and upper respiratory tract irritant, central nervous system impairment. Questionable carcinogen.
Use: Manufacture of acetone and its derivatives, manufacture of glycerol and isopropyl acetate, solvent for essential and other oils, alkaloids, gums, resins, etc.; latent solvent for cellulose derivatives, coatings solvent, deicing agent for liquid fuels, lacquers, extraction processes, dehydrating agent, preservative, lotions, denaturant.

isopropylamine.　　(2-aminopropane).
CAS: 75-31-0.　$(CH_3)_2CHNH_2$.
Properties: Colorless, volatile liquid; amine odor. Strong alkaline reaction, bp 32.4C, fp −101C, d 0.6881 (20/20C), bulk d 5.7 lb/gal (20C), refr index 1.3770 (15C), flash p −35F (−37.2C) (OC), autoign temp 756F (402C). Miscible with water, alcohol and ether.
Derivation: From acetone and ammonia under pressure.
Hazard: Highly flammable, dangerous fire risk. Strong irritant to tissue.
Use: Solvent, intermediate in synthesis of rubber accelerators, pharmaceuticals, dyes, insecticides, bactericides, textile specialties, and surface-active agents, dehairing agent, solubilizer for 2,4-D acid.

isopropyl 3-aminocrotonate.
CAS: 14205-46-0.　$C_7H_{13}NO_2$.
Hazard: Moderately toxic by ingestion. Low toxicity by inhalation. A mild skin and eye irritant.

p-isopropylaminodiphenylamine.　　See N-isopropyl-N'-phenyl-p-phenylenediamine.

isopropylaminoethanol. (IPAE).
CAS: 109-56-8. A commercial mixture of approximately 60% isopropylethanolamine ($(CH_3)_2CHNHCH_2CH_2OH$) and 40% isopropyldiethanolamine ($(CH_3)_2CHN(CH_2CH_2OH)$).
Properties: Amber to straw-colored liquid. Distillation range 110–165C, fp approximately −50C, d 0.91–0.94 (20/20C), flash p 145–155F (62.7–68.3C) (OC). Combustible.
Use: Synthesis of emulsifiers.

n-isopropylaniline.
CAS: 768-52-5. $C_6H_5NHCH(CH_3)_2$.
Properties: Yellowish liquid. Bp 206C, pour p below −67C, refr index 1.5365 (20C), flash p 190F (87.7C) (COC). Combustible.
Hazard: Toxic by inhalation and skin absorption.
Use: Dyeing acrylic fibers, chemical intermediate.

p-isopropylaniline. See cumidine.

isopropyl antimonite. $[(CH_3)_2CHO]_3Sb$.
Properties: Colorless liquid. Bp 82C (7 mm Hg).
Derivation: Reaction of antimony trichloride with isopropanol.
Use: Cross-linking agent, flameproofing agent.

p-isprorylbenzaldehyde. See cuminic aldehyde.

isopropylbenzene. See cumene.

isopropyl n-benzoyl-n-(3-chloro-4-fluorophenyl)-d-alaninate. See flamprop-m-isopropyl.

p-isopropylbenzyl alcohol. See cuminic alcohol.

isopropyl 6-benzyloxy-4-methoxymethyl-β-carboline-3-carboxylate.
CAS: 111841-85-1. $C_{24}H_{24}N_2O_4$.
Hazard: Human systemic effects.

isopropylbiphenyl. See Tanacol CG.

isopropyl bromide.
CAS: 75-26-3. $CH_3CHBrCH_3$.
Properties: Colorless liquid. D 1.304 (25/25C), bp 58.5–60.5C, fp −90C, refr index 1.422 (25C), flash p none. Slightly soluble in water; soluble in ethanol and ether. Nonflammable.
Use: Synthesis of pharmaceuticals, dyes, other organics.

isopropyl butyrate.
CAS: 638-11-9. $(CH_3)_2CHOOCC_3H_7$.
Properties: Colorless liquid. D 0.8652 (13C), bp 128C.
Use: Solvent for cellulose ethers, flavoring.

isopropylcarbinol. See isobutyl alcohol.

isopropyl carbitol.
CAS: 5412-01-1. $C_7H_{16}O_3$.
Hazard: A poison by ingestion and skin contact. A moderate eye irritant.

isopropyl chloride.
CAS: 75-29-6. $CH_3CHClCH_3$.
Properties: Colorless liquid. D 0.858 (25/25C), bp 34.8C, fp −117.6C, refr index 1.374 (25C), flash p −26F (−32.3C), autoign temp 1100F (593C). Slightly soluble in water; soluble in ethanol and ether.
Hazard: Highly flammable, fire and explosion risk, explosive limits in air 2.8–10.7%.
Use: Solvent, intermediate.

n-isoprpyl-α-chloroacetanilide. See 2-chloro-N-isopropylacetanilide.

isopropyl-3-chlorocarbanilate. See chloro-IPC.

isopropyl chloroformate.
CAS: 108-23-6. $(CH_3)_2CHOOCCl$.
Properties: Colorless liquid. A phosgene derivative.
Hazard: Toxic by inhalation.
Use: Chemical intermediate for free-radical polymerization initiators, organic synthesis.

isopropyl-N-(3-chlorophenyl)carbamate. See chloro-IPC.

isopropyl citrate.
Properties: Viscous colorless syrup. Crystallizes upon standing.
Use: Food additive.

isopropyl cresol. A mix of di- and monoisopropyl cresols.
Use: Antioxidant.
See thymol; carvacrol.

isopropyl-m-cresol. See thymol.

isopropyl-o-cresol. See carvacrol.

isopropyl cyanide. See isobutyronitrile.

isopropyl dichlorophenoxyacetate. See 2,4-D.

isopropyldiethanolamine. See isopropylaminoethanol.

isopropyl 3,4-diethoxycarbanilate.
CAS: 87130-20-9. $C_{14}H_{21}NO_4$.
Hazard: Low toxicity by ingestion, inhalation, and skin contact.

**2-isopropyl-4-dimethylamino-
5-methyl-phenyl-1-
piperidinecarboxylate methyl chloride.**
$C_3H_7C_6H_2(CH_3)N(CH_3)_2OOCNC_5H_{10 \cdot c}H_3Cl$.
Properties: White solid. Mp 151–153C. Insoluble in ether; soluble in methanol.
Use: A plant tranquilzer or antigibberellin, which causes some plants to become dwarfs without otherwise affecting their growth or health.

isopropyl dimethyl carbinol.   See amyl methyl alcohol.

isopropylethanolamine.   See isopropylaminoethanol.

isopropyl ether.   (diisopropyl ether; diisopropyl oxide).
CAS: 108-20-3. $(CH_3)_2CHOCH(CH_3)_2$.
Properties: Colorless, volatile liquid; ethereal odor. Somewhat similar to ethyl ether in properties but does tend to form peroxides more readily than ethyl ether. Consequently, the presence or absence of peroxides should be determined; if present, they should be destroyed with sodium sulfite before distillation. Bp 67.5C, d 0.723 (15.5/4C), refr index 1.368, fp −60C, flash p approx. 0F (−17.1C), autoign temp 830F (443C), bulk d 6.05 lb/gal (15.5C). Soluble in water 0.65% wt (25C); miscible with most organic solvents.
Grade: Technical.
See ether.
Hazard: Flammable, dangerous fire risk, explosive limits in air 1.4–21%. Toxic by inhalation, strong irritant.
Use: Solvent for animal, vegetable and mineral oils, waxes, and resins; extraction of acetic acid from aqueous solutions; solvent for dyes in presence of small amount of alcohol; paint and varnish removers; spotting compositions; rubber cements.

isoproylethylene.   See 3-methyl-1-butene.

isopropyl furoate.
CAS: 6270-34-4. $C_4H_3OCO_2C_3H_7$.
Properties: Colorless liquid becoming brown in light. D 1.0655 (23.7/4C), bp 198.6C (corrosive), refr index 1.4682 (23.7C). Insoluble in water; soluble in alcohol and ether.

α-isopropyl glycerol ether.   See glycerin 1-isopropyl ether.

isopropyl glycidyl ether.   (IGE).
CAS: 4016-14-2. $C_6H_{12}O_2$.
Properties: A liquid. Mw 116.16, d 0.9186, bp 127C, vap press about 9.4 torr (25C).
Hazard: Flammable liquid with flash p 92F (33.33C). Skin, eye and respiratory tract irritant.
Use: Stabilizer of chlorinated solvents, and viscosity reducer of epoxy resins.

3-isopropyl-4-hydroxyphenyl methylcarbamate.
CAS: 19189-02-7. $C_{11}H_{15}NO_3$.
Hazard: A poison.
Use: Agricultural chemical.

isopropylideneacetone.   See meistyl oxide.

p,p-isopropylidenediphenol.   See bisphenol A.

4,4′-isopropylidenediphenol alkyl (C12-C15) phosphite.
CAS: 93356-94-6.
Hazard: A severe skin irritant.

isopropyl iodide.   (2-iodopropane).
CAS: 75-30-9. CH_3CHICH_3.
Properties: Colorless liquid that discolors in air and light; miscible with chloroform, ether, alcohol and benzene; slightly soluble in water. D 1.703, fp −90C, bp about 90C, refr index 1.5026 (20C).
Use: Organic synthesis, pharmaceuticals.

isopropyl mercaptan.
CAS: 75-33-2. $(CH_3)_2CH(HS)$.
Properties: Liquid; extremely unpleasant odor. D 0.814 (15.5/15.5C), boiling range 51–55C, flash p −30F (−34.4C).
Derivation: Propylene and hydrogen sulfide.
Hazard: Highly flammable, dangerous fire hazard.
Use: Standard for petroleum analysis, intermediate.

***n*-isopropylmethoxamine.**
CAS: 550-53-8. $C_{14}H_{23}NO_3$.
Hazard: Moderately toxic by ingestion.

2-isopropyl-5-methylbenzoquinone.   See *p*-thymoquinone.

5-isopropyl-2-methyl-1,3-cyclohexadiene.
See α-phellandrene.

3-isopropyl-6-methylene-1-cyclohexene.
See β-phellandrene.

1-isopropyl-2-methylethylene.   See 4-methyl-2-pentene.

***n*-isopropyl-α-methylphenethylamine hydrochloride.**
CAS: 26640-60-8. $C_{12}H_{19}N \cdot ClH$.
Hazard: A poison. Human systemic effects.

1-isopropyl-3-methyl-5-pyrazolyl dimethylcarbamate.   (isolan).
CAS: 119-38-0. $C_{10}H_{17}N_3O_2$.
Properties: Liquid. D 1.07 (20C), bp 103C (0.7 mm Hg). Miscible with water.
Derivation: By treating 1-isopropyl-3-methyl-5-pyrazolone with dimethylcarbamoyl chloride.
Hazard: Cholinesterase inhibitor.
Use: Insecticide.

isopropylmyristate.
CAS: 110-27-0. $CH_3(CH_2)_{12}CO_2CH(CH_3)_2$.
Properties: Colorless oil; practically odorless. D 0.850–0.860, fp 3C, refr index 1.435–1.438 (20C). Soluble in most organic solvents, vegetable oils; dissolves waxes, lanolin, and similar products; insoluble in water. Combustible.
Grade: Double-distilled from coconut oil.
Use: Cosmetic creams, topical medicinals.

2-isopropylnaphthalene.
CAS: 2027-17-0. $C_{13}H_{14}$.
Properties: Clear, yellowish-brown liquid; faint sweet odor. Bp 268C, flash p 252F (122C) (COC), autoign temp 475C, bulk d 8.1 lb/gal, d 0.973. Insoluble in water. Combustible.
Hazard: Avoid inhalation of vapors and prolonged skin contact.
Use: Organic intermediate.

isopropyl nitrate.   (2-propanol nitrate).
CAS: 1712-64-7. $(CH_3)_2CHNO_3$.
Properties: Colorless liquid. Bp 102C.
Hazard: Oxidizing material, fire risk in contact with organic materials.

isopropyl palmitate.
CAS: 142-91-6. $(CH_3)_2CHOOCC_{15}H_{31}$.
Properties: Colorless liquid. D 0.850–0.855 (25/25C), refr index 1.4350–1.4390 (20C), mp 14C. Soluble in 4 parts 90% alcohol; soluble in mineral and fixed oils; insoluble in water. Combustible.
Use: Emollient and emulsifier in lotions, creams, and similar cosmetic products.

isopropyl percarbonate.   Legal label name for diisopropyl peroxydicarbonate.

isopropyl peroxydicarbonate.   See diisopropyl peroxydicarbonate.

m,p-isopropylphenol. $(CH_3)_2CHC_6H_4OH$.
Properties: A solid mixture of the m- and p-isomers. Fp (m-) 25.9C, (p-) 63.2C, bp (m-) 228.6C, (p-) 228.5C. Completely soluble in 10% sodium hydroxide. Combustible.

o-isopropylphenol.
CAS: 88-69-7. $(CH_3)_2CHC_6H_4OH$.
Properties: Light yellow liquid. Bp 214C, fp 17C, d 0.995 (20C), flash p 220F (104C) (OC). Insoluble in water; soluble in isopentane, toluene, ethyl alcohol, 10% sodium hydroxide. Combustible.
Use: Intermediate for synthetic resins, plasticizers, surface-active agents, perfumes.

isopropyl-N-phenylcarbamate.   See IPC.

n-(4-isopropylphenyl)-n',n'-dimethylurea.
CAS: 34123-59-6. $C_{12}H_{18}N_2O$.

Hazard: Moderately toxic by ingestion, inhalation, and skin contact.
Use: Agricultural chemical.

3-isopropylphenyl n-isobutyryl-n-methylcarbamate.
CAS: 5748-26-5. $C_{15}H_{23}NO_3$.
Hazard: A poison by ingestion.

3-isopropylphenyl (methyl)(n-hexoxysulfinyl) carbamate.
CAS: 77267-48-2. $C_{17}H_{27}NO_4S$.
Hazard: Moderately toxic by ingestion.
Use: Agricultural chemical.

n-isopropyl-N'-phenyl-p-phenylenediamine.
(p-isopropylaminodiphenylamine).
CAS: 101-72-4. $C_3H_7NHC_6H_4NHC_6H_5$.
Properties: Dark gray to black flakes. Fp range 72–76C, d 1.04 (25C). Soluble in benzene and gasoline; insoluble in water.
Use: Protection of rubbers against oxidation, ozone, flex-cracking, and poisoning by copper and manganese.

isopropyl phosphorothioate.
CAS: 4486-44-6.
$C_6H_{15}O_3PS$.
Hazard: Moderately toxic by ingestion.

4-isopropylpyridine.
CAS: 696-30-0.
$C_5NH_4C_3H_7$.
Properties: Liquid. Bp 182.2, d 0.9282 (20C), refr index 1.4960 (20C). Solubility in 100 g, 19.4 g at 20C.

isopropyl rubbing alcohol.
Properties: Aqueous solution, contains 68–72% isopropyl alcohol by volume; flammable.
Hazard: Toxic.
Use: Rubefacient.

isopropyl titanate.   See tetraisopropyl titanate.

isopropyltoluene.   See cymene.

isopropyl-2,4,5-trichlorophenoxyacetate.
See 2,4,5-trichlorophenoxyacetic acid.

isopropyltrimethylmethane.   See 2,2,3-trimethylbutane.

isopropyl unoprostone.
CAS: 120373-24-2. $C_{25}H_{44}O_5$.
Hazard: A poison.

isopropyl xanthogen disulfide.
CAS: 105-65-7. $C_8H_{14}O_2S_4$.
Hazard: Moderately toxic by ingestion.

isoprothiolane.
CAS: 50512-35-1. $C_{12}H_{18}O_4S_2$.
Hazard: Moderately toxic by ingestion.
Use: Agricultural chemical.

isopulegol.   (1-methyl-4-
isopropenylcyclohexan-3-ol).
CAS: 7786-67-6. $C_{10}H_{17}OH$.
A terpene derivative.
Properties: Water-white liquid; mint-like odor. D
0.904–0.911, refr index 1.471–1.474. Combustible.
Available forms: The acetate.
Use: Perfumery (geranium and rose compounds), fla-
voring.

isoquinoline.
CAS: 119-65-3.

CHCHCHCHCCHCHCNCH

Properties: Colorless plates or liquid. D 1.099
(20C), mp 26.48C, bp 243C, refr index 1.6223
(25C). Insoluble in water; soluble in dilute mineral
acids and most organic solvents. Combustible.
Derivation: From coal tar, also synthetic.
Hazard: Toxic by ingestion.
Use: Manufacture of pharmaceuticals (such as nicotinic acid), dyes, insecticides, rubber accelerators,
and in organic synthesis.

isoquinoline   alkaloid.   A   neuromuscular
blocker and any of a group of alkaloids derived
from tyrosine and phenylalanine that contain an isoquinoline nucleus. Any alkaloid that possesses an
isoquinoline skeleton is known as an isoquinoline
alkaloid (e.g., papaverine and morphine).
Derivation: Occur in plants of the family *papaver-
aceae* (poppies) and related plants.
Use: Medication, including as pain killers, anesthetics, antifungal agents, analgesics, and to treat anti-
hypertension; narcotic.

1,3-isoquinolinediol. $C_9H_5N(OH)_2$.
Properties: Cream-colored paste. Solids approx.
80%.
Use: Intermediate.

isosorbide.   (1,4,3,6-dianhydrosorbitol).
CAS: 625-67-5.

$OCH_2CH_2OCHCHCH_2OCH_2O$.

A polyol with a hydroxyl group attached to each of
two cis-oriented saturated furan rings. Intermediate
for pharmaceuticals. Combustible.
Use: Medicine (diuretic).

isostearic acid.   A coined name for a C_{18} saturated fatty acid of the formula $C_{17}H_{35}COOH$. It
is a complex mixture of isomers, primarily of the

methyl-branched series that are mutually soluble
and virtually inseparable.
Use: Similar to stearic or oleic acids.

isostilbene.   See stilbene.

isotatic polymer.   See polymer, isotactic.

isoterism.   Similarity in physical properties of
elements, ions, or compounds, due to similar or
identical outer shell electron arrangements.

isotherm.   Constant temperature line used on climatic maps or in graphs of thermodynamic relations, particularly the graph of pressure-volume
relations at constant temperature.

isothermal.   Occurring at constant temperature.

isothiazolinone chloride.
CAS: 55965-84-9. $C_4H_5NOS \cdot C_4H_4ClNOS$.
Hazard: A poison by ingestion.

isothiocyanate.   (sulfocarbamide). The radical
of isothiocyanic acid.

isothiocyanatotrimethyltin.
CAS: 15597-43-0. C_4H_9NSSn.
Properties: White needles from C_6H_6. Mp 108.5C.
Hazard: A poison.

(isothiocyanato)tripropylstannane.   See
tripropyltin isothiocyanate.

isothiocyanic acid allyl ether.   (allyl isothiocyanate; thiourea; 3-isothiocyanatoprop-1-ene).
CAS: 57-06-7. C_4H_5NS.

isotone.   A nuclide that has the same excess of
neutrons over protons as another nuclide.

isotonic.   A solution having the same osmotic
pressure as another solution, for example, human
blood and physiological salt solution.

isotope.   One of two or more forms or species of
an element that have the same atomic number, i.e.,
the same position in the periodic table, but different
masses. The difference in mass is due to the presence of one or more extra neutrons in the nucleus.
Thus, "regular" hydrogen, with atomic number 1
and a mass of 1 (proton), is one of the three isotopes of hydrogen. The other two are, the naturally occurring deuterium, which has a neutron in
its nucleus as well as a proton, giving a mass of
2; and the artificially produced tritium (1 proton
and 2 neutrons) with a mass of 3 (approximately).
The atomic weight of an element is the average
weight percent of all its natural isotopes. The heavier isotopes usually occur very rarely in the atomic
population (1 part in 4500 for ^2H, and 1 part in 140

for U-235; in the exceptional case of chlorine, the ratio of isotopes 35 and 37 is about 3 to 1).

The occurrence of isotopes among the 83 most abundant elements is widespread, but separation methods are complicated and costly. Twenty-one elements have no isotopes, each consisting of only one kind of atom (see note below). The remaining 62 natural elements have from 2 to 10 isotopes each. There are 287 different isotopic species in nature; noteworthy among them are oxygen-17, carbon-14, uranium-235, cobalt-60, and strontium-90, all but the first being radioactive.

There are three kinds of isotopes: (1) natural nonradioactive, (2) natural radioactive, and (3) artificially radioactive (made by neutron bombardment).

Use: (Nonradioactive.) Preparation of heavy water to moderate nuclear reactors. (Radioactive.) Tracers in biochemical, metallurgical, and medical research; in geochemical and archeological research (^{14}C); irradiation source for polymerization, sterilization, etc.; therapeutic agents in various diseases (iodine, sodium, gold, etc.); electric power generation.

See Aston; chemical dating; radioactivity; heavy water; decay; tracer; nuclide.

Note: According to this definition, it is strictly improper to refer to elements that exist in only one atomic form as having "one isotope"; actually such elements as beryllium, aluminum, arsenic, iodine, and others have no isotopes, that is, they have no other atomic form that is like them in all respects except mass. The term isotope requires the existence of at least two elemental forms, in the same sense that the word twin requires the existence of a pair.

isotoxin.   A toxin that is poisonous to other animals of the same species.

isotropic.   Descriptive of the property of transmitting light equally in all directions, cubic (isometric) crystals have this property, as well as liquids, gases and most glasses.
See anisotropic.

isotype.   Any class of immunoglobulins that all have similar structures but differing molecular weights and concentrations in circulating blood.

isourea.   (urea).
CAS: 57-13-6. CH_4N_2O.
A compound formed in the liver from ammonia produced by the deamination of amino acids. It is the principal end-product of protein catabolism and constitutes about one half of the total urinary solids. It is the imidic acid tautomer of urea and its hydrocaryl derivatives.

isovaleral.   See isovaleraldehyde.

isovaleraldehyde.   (isovaleral; isovaleric aldehyde; 3-mehylbutryaldehyde).
CAS: 590-86-3. $(CH_3)_2CHCH_2CHO$.

Occurs in orange, lemon, peppermint, and other essential oils.
Properties: Colorless liquid; apple-like odor. D 0.785, fp −51C, bp 92C, refr index 1.390 (20C). Soluble in alcohol and ether; slightly soluble in water. Combustible.
Derivation: Oxidation of isoamyl alcohol, also by Oxo process from petroleum.
Use: Flavoring, perfumes, pharmaceuticals, synthetic resins.

isovaleric acid.   (isopropylacetic acid).
CAS: 503-74-2. $(CH_3)_2CHCH_2COOH$.
Occurs in valerian, hop oil, tobacco, and other plants.
Properties: Colorless liquid; disagreeable taste and odor. D 0.931 (20/20C), bp 176C, refr index 1.4043 (20C), fp −37C. Slightly soluble in water; soluble in alcohol and ether. Combustible.
Derivation: With other valeric acids, by distillation from valerian, by oxidation of isoamyl alcohol.
Use: Medicine, flavors, perfumes.

isovaleric aldehyde.   See isovaleraldehyde.

isovaeryl chloride.   (3-methylbuanoyl chloride). $(CH_3)_2CHCH_2COCl$.
Properties: Colorless liquid. Refr index 1.4136 (24C), d 0.9854 (24/4C), bp 113C. Soluble in ether; decomposed by water and alcohols.
Use: Intermediate in synthesis.

isovaleryl-*p*-phenetidine.
$C_2H_5OC_6H_4NHCOCH_2CH(CH_3)_2$.
Properties: White, glistening needles. Almost insoluble in water and ether; soluble in alcohol and chloroform.
Derivation: By heating isovaleric acid with *p*-phenetidine.

isovaltrate.
CAS: 31078-10-1. $C_{22}H_{30}O_8$.
Hazard: A poison.

isoxazole.   (1,2-oxazole). C_3H_3NO. Any of a type of chemical produced by poisonous mushrooms such as *Amanita muscaria*, *Amanita pantherina*, and *Tricholoma muscarium*. It is an azole with an oxygen and a nitrogen next to each other at the 1,2 positions.
Hazard: Narcotic and psychomimetic effects.
Use: Natural insecticide.

isozymes.   Two or more forms of an enzyme that catalyze the same reaction but differ from each other in their physical properties, such as amino acid sequence, substrate affinity, V_{max}, tissue expression, or regulatory properties.

itaconic acid.   (methylene succinic acid).
CAS: 97-65-4. $CH_2:C(COOH)CH_2COOH$.

Properties: White, odorless, hygroscopic crystals. Mp 167–168C. Soluble in water, alcohols, and acetone; sparingly soluble in other organic solvents.
Derivation: Submerged fermentation by mold of various carbohydrates.
Use: Copolymerizations, resins, plasticizers, lube-oil additive, intermediate.

-ite. A suffix indicating an intermediate oxidation state of a metallic salt, analogous to "-ous" for acids, e.g., sodium sulfite ($NaSO_3$), containing one fewer oxygen atom than the sulfate.

IUPAC. Abbreviation for International Union of Pure and Applied Chemistry.

Ivanov reagent. Reaction product of aryl acetic acids and excess Grignard reagent. The reagent may be used in condensations with carbonyl compounds and other Grignard-type reactions.

IVE. Abbreviation of isobutyl vinyl ether. See vinyl isobutyl ether.

Izod. An impact-testing device of the notched-bar type.

J

J acid.   (2-amino-5-naphthol-7-sulfonic acid).

Properties: Gray needles, white when pure. Soluble in hot water; sparingly soluble in cold water.
Derivation: β-naphthylamine sulfate is sulfonated with 66% oleum, the filtrate yields 2-naphthylamine-1,5,7-trisulfonic acid. On reaction with dilute sulfuric acid at 125C, this yields 2-naphthylamine-5,7-disulfonic acid; from this J acid is obtained on reaction with 58% sodium hydroxide solution at 200C and 210 psi.
Use: Azo dye intermediate.

J acid urea.   See 5,5′-dihydroxy-7,7′-disulfonic-2,2′-disulfonic-2,2′-dinaphthylurea.

Jacobsen rearrangement.   Reaction of polymethylbenzenes with concentrated sulfuric acid to give rearranged polymethylbenzenesulfonic acids. Under identical conditions, halogenated polymethylbenzenes undergo disproportionation.

Jacquemart's reagent.   Analytical reagent used to test for ethyl alcohol. Consists of an aqueous solution of mercuric nitrate and nitric acid.

jalapin.   A natural glucoside classified as a derivative of ethylene.

Janovsky reaction.   Reaction of aldehydes and ketones containing α-methylene groups with m-dinitrobenzenes in the presence of a strong base, resulting in the formation of an intense purple coloration, used for the detection of carbonyl compounds. The color is due to the formation of a Meisenheimer complex.

Japan black.   A varnish yielding a hard, glossy, dark-colored film. Japans are usually dried by baking at relatively high temperatures (ASTM D 16–52). True Japan varnishes contain a strongly irritating chemical, more recent types contain kauri or copal resin, linseed oil, lead oxide, pigments, and solvents such as kerosene or turpentine.
Hazard: Flammable, irritant to eyes and skin.

Use: Coatings for miscellaneous wood and metal products.

Japan wax.   (Japan tallow; sumac wax).
Properties: Pale-yellow solid; tallowlike rancid odor. Contains 10–15% palmatin and other glycerides. Soluble in benzene and naphtha; insoluble in water and in cold alcohol. D 0.970–0.980, mp 53C, saponification number 220, iodine number 10–15. Combustible.
Derivation: From a species of Rhus by boiling the fruit in water.
Use: Candles, floor waxes, polishes, substitute for beeswax, food packaging.

Japp–Klingemann reaction.   Formation of hydrazones by coupling of aryldiazonium salts with active methylene compounds in which at least one of the activating groups is acyl or carboxyl. This group usually cleaves during the process.

Jaracol.   A line of hair dye intermediates.

Jarcoal.   A potassium acetate based fluid.
Use: For spraying interior of coal barge, rail cars, and trailers to prevent freezeup.

Jargrip.   A potassium acetate based fluids.
Use: For deicing runways.

jar mill.   An assembly of small ceramic jars containing porcelain or flint pebbles; the jars are often arranged in parallel tiers, each containing four or more jars, each tier being mounted on a rotating shaft. Such mills are used for production of small quantities of pulverized material.
See ball mill; pebble mill.

jasmine absolute.   See oil of jasmine.

jasmine aldehyde.   See α-amyl cinnamic aldehyde.

jasmine oil.   An essential oil in perfumery and flavoring. It is dextrorotatory.

jamolin. $C_{21}H_{30}O_3$. Insecticidal principle of pyrethrum.

jasmone.   (3-methyl-2-(2-pentenyl)-2-cyclopenten-1-one).
CAS: 488-10-8. $C_{11}H_{16}O$. A ketone found in jasmine oil and other flower oils.

Properties: Odor of jasmine. D 0.944 (22/0C).
Use: Perfumery.

jatrophone. A diterpenoid growth inhibitor isolated from an alcohol extract of the plant *Jatropha gossypiifolia*. Its unique structure includes a 12-membered ring. It is readily attacked by nucleophiles. Useful in study of tumor growth inhibition and other biochemical research.

Javel water. NaHClO.
Use: A bleach.

"Jayflex" [Exxon]. TM for a plasticizer.

"Jeffamine" [Chase]. TM for polyoxypropyleneamines.
Grade: 3 diamine, 1 triamine.
Available forms: Liquid.
Hazard: Corrosive.
Use: Curing agent for epoxy resin systems in adhesives, elastomers, and foam formulations, and as an intermediate for textile and paper-treating chemicals.

jelly. A modified form of the word *gel* widely used in popular language but also used in chemical literature to refer to the mechanical strength of the gel structures occurring with pectins, gelatin, and various natural gums. "Jelly strength" is frequently specified in the food industry. Other uses of the word are found in "petroleum jelly" obtained as a distillation product of petroleum residues (petrolatum) and in the so-called *royal jelly*, a natural nutrient mixture of proteins and carbohydrates produced by bees as food for the queen bee.
See gel.

Jel-O-Mer. A thixotropic additive used in coatings.

Jenner, Edward. (1749–1823). An English physician, Jenner studied medicine in London and established his practice in the rural area of Gloucestershire. Here he discovered the technique of vaccination as a preventive of smallpox (1776). The idea of utilizing cowpox, a disease of cattle, as a protective medium was suggested by his observation that personnel working in dairies developed immunity to smallpox after contracting the much milder cowpox. Jenner's work not only led to almost complete elimination of smallpox in Europe, but also anticipated the development of immune reactions by Pasteur a century later. His success was no accident, but rather the result of detailed observations from which he drew correct conclusions. He was a scientist of the highest caliber and a noteworthy benefactor of mankind.

jequirity bean. (love bean; lucky bean). *Abrus precatorius*.
Properties: Plant belonging to the family Leguminosae.
Hazard: Poison; moderately toxic.

JET. Abbreviation for Joint European Torus, an experimental nuclear fusion device in England. It is a project jointly undertaken by several European countries.
See tokamak; fusion.

jet fuel. A fuel for jet (turbine) engines, usually a petroleum distillate similar to kerosine. A number of types with somewhat different compositions and properties have been used.

Jetset. A fast-setting cement developed by the Portland Cement Association. Reported to harden in 20 minutes after pouring. Accelerating agent has not been disclosed.

JH. (methyl-*cis*-10,11-epoxy-7-ethyl-3,11-dimethyl-*trans,trans*-2,6-tridecadienoate). A synthetic hormone containing a 13-carbon chain; said to have possibilities as an insecticide. It acts by preventing insects from maturing. Its future depends on the possibility of large-scale production.
See juvenile hormone.

JHR Compound. A thermoplastic compound impervious to mineral acid; does not decompose hydrogen peroxide.
Use: To coat interiors of tanks and containers for shipment of hydrogen peroxide, acids, etc.

Jiffix. An acid-hardening, ammonium thiosulfate fixing bath. Ready-mixed and rapid-acting.

JinBest Paste. A semi discharge paste.

JIN Print. A synthetic thickener for pigment printing.

Joint European Torus. See JET.

jojoba oil.
Properties: Colorless, odorless, waxy liquid; chemically similar to sperm oil.
Derivation: By crushing seeds of an evergreen desert shrub found in southwestern U.S. and northern Mexico. Experimental cultivation in California and Israel. Yield of oil from seeds approaches 50%.
Use: Substitute for sperm oil, especially in transmission lubricants, high-pressure lubricant, antifoam agent (antibiotic fermentation), substitute for carnauba wax and beeswax, cosmetic preparations.

Joliot-Curie, Frederick. (1900–1958). A French physicist who, along with his wife Irene Joliot-Curie, won the Nobel Prize in chemistry in 1935. His important discoveries included artificial radioactivity. He did much work on atom structure, dematerialization of electrons, and inverse transformation. Work on hormone synthesis and thyroid substances containing radioactively labeled elements was significant. Sc.D. from the University of Paris was followed by a distinguished career filled with honors and appointments.

Joliot-Curie, Irene. (1897–1956). A French nuclear scientist who won the Nobel Prize for chemistry with her husband Frederic Joliet-Curie. Their joint work involved production of artificial radioactive elements by using α-rays to bombard boron. They discovered that hydrogen-containing material when exposed to what they considered γ-rays would emit protons. They were involved in many firsts: they gave the first chemical proof of artificial transmutation and of capture of alpha particles, and were the first to prepare positron emitters. Her career started with a Sc.D., at the University of Paris, and included scores of honors and awards.

Jones oxidation. The oxidation of primary and secondary alcohols to acids and ketones by the addition of the calculated amount of chromic anhydride in dilute sulfuric acid to a solution of the alcohol in acetone. This procedure does not attack triple bonds or shift double bonds into conjugation with the ketone formed in the oxidation.

jonkmari, extract.
CAS: 84929-59-9.
Hazard: A poison.
Source: Natural product.

jordan. See beater.

Joule–Thomas coefficient. The change in temperature per atmosphere change of pressure on a gas or other fluid when the enthalpy remains constant. It is found by measuring the temperature change from T' to T when the pressure p' of a gas on one side of a porous plug changes to Phosphorus on the other side. The change of temperature $(T' - T)$ and of $(p' - p)$ are measured under conditions such that no heat is gained or lost, and the pressure of the plug is great enough to insure a nearly constant pressure in the incoming and outgoing gas. The ratio of $(T' - T)/(p' - p)$ at several pressure ranges is extrapolated to the limiting case as $(p' - p)$ approaches zero. This limiting value is the Joule–Thomson coefficient, μ. Thus,

$$\mu = \left(\frac{\alpha T}{\alpha P}\right)_H$$

The subscript H indicates that the enthalpy, H, remains constant during the expansion ($\Delta H = 0$).

Jourdan–Ullmann–Goldberg synthesis. Synthesis of substituted diphenylamines. The reaction products can be used as intermediates in the synthesis of acridones.

juglone. See 5-hydroxy-1,4-naphthoquinone.

juniper tar oil. (oil of cade, cade oil, oleum cadium). Chief constituent is cadinene, a sesquiterpene.
Properties: Yellow oil. D 0.980–1.055. Soluble in alcohol.
Derivation: By product of distillation of *Juniperus oxycedrus*.

junk DNA. (non-coding DNA). Stretches of DNA that do not code for genes; most of the genome consists of so-called junk DNA which may have regulatory and other functions.

justicidin. Any of several lignans isolated from various species of fish (genus *Justicia*; family Acanthaceae).

justicidin B. (dehydrocollinusin).
Use: Piscicidal agent.

jute. Bast fibers, 4–10 ft long, obtained from the stems of several species of *Corchorus*, especially *C. capularis*. Contains a higher proportion of lignin and less cellulose than any other commercial vegetable fiber and has relatively poor strength and durability. The fibers are soft and lustrous but lose strength when wet. Combustible. Not self-extinguishing.
Hazard: Flammable in form of dust, may ignite spontaneously when wet.
Use: Burlap, sacking, linoleum, twine, carpet backing, packing, coarse paper.

juvenile hormone. One of several hormones that retard the development of insects in the larval stage; so called because they prevent the insect from maturing by maintaining its juvenile characteristics. Obtained naturally from silk moths; various syntheses indicate possible use as insecticides, especially for fire ants. Composition of one type is $C_{18}H_{30}O_3$.
See JH.

K

K.   (1) Symbol for potassium (from Latin kalium).
(2) Symbol for Kelvin scale.

k.   Abbreviation for kilo-, as in kcal.

K acid.   (1-amino-8-naphthol-4,6-disulfonic
acid). $C_{10}H_4NH_2OH(SO_3H)_2$.
Derivation: Fusion of a naphthylamine trisulfonic
acid with sodium hydroxide.
Use: Azo dye intermediate.

kainic acid.   ([2S-(2a,3b,4b)]-2-carboxy-4-(1-
methylethenyl)-3-pyrrolidineacetic acid).
CAS: 487-79-6. $C_{10}H_{15}NO_4$.
Properties: White crystalline powder. Soluble in
water or dilute aqueous base.
Use: To simulate brain degeneration in a laboratory
environment.

kainite. $MgSO_4 \cdot KCl \cdot 3H_2O$.
Properties: A natural, hydrated, double salt of potassium and magnesium. White, gray, reddish, or colorless solid; streak, colorless; vitreous luster. D
2.05–2.13, hardness 2.5–3. Contains 30% potassium chloride.
Occurrence: Germany; one of the Stassfurt minerals.
Use: Chemicals (potassium salts), fertilizer (as such).
See potash.

Kairomone.   A biologically active chemical produced and disseminated by living organisms of one
species that is advantageous to recipient species.

"Kalrez" [DuPont].   TM for engineering resins
and high-end specialty polymers.

kalsomine.   See calcimine.

Kamicryl 2089.   A water-based overprint varnish hard nonfilm-forming liquid.
Use: In high gloss varnish and water-based inks.

Kamicryl Paint CP651.   A vinyl acetate acrylic
copolymer.
Use: In a superior quality emulsion paint.

Kamicryl Thickner 5360.   A cross-linked
acrylic copolymer emulsifier.
Use: Keeps pigment and filler in suspension and prevents sedimentation.

kanamycin sulfate.
CAS: 25389-94-0. $C_{18}H_{36}N_4O_{11} \cdot H_2SO_4$. A broad-
spectrum antibiotic.
Properties: White, crystalline powder; odorless.
Decomposes over a wide range above 250C. Solu-
ble in water; practically insoluble in methanol and
ethanol.
Grade: USP.
Use: Medicine (antibacterial).

Kansil.   Potassium silicate.
Grade: Liquids, solids, and powders in $SiO_2:K_2O$
weight ratio of 1.60 to 2.5.
Use: Ceramics, welding rod coatings, detergents,
corrosion inhibitor, electronics, and inorganic coatings.

kaolin.   (China clay).
CAS: 1332-58-7. A white-burning aluminum silicate
that, due to its great purity, has a high fusion point
and is the most refractory of all clays.
Composition: Mainly kaolinite (50% alumina, 55%
silica, plus impurities and water).
Properties: White to yellowish or grayish fine powder. D 1.8–2.6. Darkens and develops clay odor
when moistened. Insoluble in water, dilute acids,
and alkali hydroxides; has high lubricity (slipperiness). Noncombustible.
Occurrence: Southeastern U.S., England, France.
Grade: Technical, NF, also graded on basis of color
and particle size.
Hazard: Respirable fraction; questionable carcinogen.
Use: Filler and coatings for paper and rubber, refractories, ceramics, cements, fertilizers, chemicals
(especially aluminum sulfate), catalyst carrier, anticaking preparations, cosmetics, insecticides, paint,
source of alumina, adsorbent for clarification of liquids, electrical insulators.
See Toth process; kaolinite; aluminum silicate; clay;
pharmolin.

kaolinite. $Al_2O_3 \cdot 2SiO_2 \cdot 2H_2O$. A clay mineral
rarely found pure, the main constituent of kaolin
and some other clays.

kapok.   Cotton-like fibers obtained from the seed-
pods of various species of *Ceiba* and *Bombax*.
Extremely light and resilient but too brittle for spin-
ning. Combustible; not self-extinguishing.
Source: Indonesia, Philippines, Ecuador.
Use: Life jackets, insulation, pillows, upholstery.

Hawley's Condensed Chemical Dictionary, Sixteenth Edition. Michael D. Larrañaga, Richard J. Lewis, Sr., and Robert A. Lewis.
© 2016 John Wiley & Sons, Inc. Published 2016 by John Wiley & Sons, Inc.

karatavic acid.
CAS: 21800-49-7. $C_{24}H_{28}O_5$.
Hazard: A poison.

"Karathane" [Dow].   TM for an agricultural fungicide-miticide based on dinitro(1-methylheptyl)phenyl crotonate, supplied as a wettable powder or liquid concentration. May be combined with most other insecticides and fungicides, except oil-based products.
Use: Controls powdery mildew and various species of mites on plants.
See dinocap.

karaya gum.   (sterculia gum; India tragacanth; kadaya gum). A hydrophilic polysaccharide which exudes from certain Indian trees of the genus *Sterculia*. Color varies from white to dark brown or black.
Properties: A carbohydrate polymer of varying chemical composition. The properties depend on freshness and time of storage. Viscosity greatly decreases over 6 months storage. Forms a translucent colloidal gel in water.
Grade: Technical, FCC.
Use: Pharmaceuticals, textile coatings, ice cream and other food products, adhesives, protective colloids, stabilizers, thickeners, emulsifiers.
See gum, natural.

Karle, Jerome.   (1918–2013). An American physical chemist who won the Nobel Prize for chemistry along with Hauptman in 1985. He developed a series of mathematical equations that allow determination of phase information from X-ray crystallography intensity patterns. The advent of computers allowed the use of the equations to determine the conformation of thousands of chemicals. The work was done at the Naval Research Laboratory in Washington, D.C. where Karle headed the Laboratory for the Structure of Matter.

Karl Fischer reagent.   A solution of iodine, sulfur dioxide, and pyridine in methanol or methyl "Cellosolve" [Union].
Use: Determination of water.
Note: Do not confuse with Fischer's reagent.

Karplus, Martin.   (1930–). An American-Austrian chemist who was awarded the 2013 Nobel Prize jointly with Michael Levitt and Arieh Warshel. The scientist developed accurate computer models to understand and predict chemical processes. As the computer simulations could produce the exact output, this created an efficient method to study and test the effect of drugs on living molecules, leading to drug design for cancer therapy. One of Karplus' main interests was simulating how the shape of retina in the eye changes when exposed to light, and in 1974, he successfully developed a computer model of the retinal change.

He was awarded his Ph.D. by the California Institute of Technology followed by a renowned career, which included awards such as the Irving Langmuir Award (1987), Foreign Member of the Royal Society (2000), and Linus Pauling Award (2004).

Karrer, Paul.   (1889–1971). A recipient of the Nobel Prize for chemistry in 1937 with Haworth. Although born in Moscow, he attended European universities and received his doctorate in Zurich. He initiated work on flavins, carotenoids, and vitamins A and B, and accomplished work on structure and synthesis of vitamin B2 as well as vitamins A and E.

karyogamy.   A process of fusion of the nuclei of two cells; the second step in syngamy.

karyotype.   A photomicrograph of an individual's chromosomes arranged in a standard format showing the number, size, and shape of each chromosome type; used in low-resolution physical mapping to correlate gross chromosomal abnormalities with the characteristics of specific diseases.

"Kastone" [Kasten].   (hydrogen peroxide). CAS: 7722-84-1. TM for oxidizer and bleaching agent.
Use: For textile and pulp bleaching, cyanide destruction, wastewater and sewage treatment, food processing, electronics industry.

Katadyne process.   A method of sterilizing water and other potable liquids with a specially prepared form of silver.

katapol vp-532.
CAS: 63091-06-5.
Hazard: A severe skin and eye irritant.

katharometer.   A gas and air detector instrument. Its multiple uses include the comparison of two gases via comparison of the rate of heat loss from heating coils surrounded by the gases. Also can be used in gas chromatography as a detector and to detect impurities in the air.

kauri.   A fossil (hard) copal resin, derived from the kauri pine (*Agathis australis*) of New Zealand. Soluble in alcohols and ketones, acid value 60–80, must be heat-treated (cracked) before use in varnishes. Combustible.
Use: Varnishes and lacquers, paints, organic cements, to evaluate the solvent power of hydrocarbons.

kauri-butanol value.   A measure of the aromatic content and hence the solvent power of a hydrocarbon liquid. Kauri gum is readily soluble in butanol but insoluble in hydrocarbons. The kb value is the measure of the volume of solvent required to produce turbidity in a standard solution containing kauri gum dissolved in butanol. Naphtha fractions have a kb value of about 30, and toluene about 105.

kb. Abbreviation for kilobase—one thousand bases.

kcal. Abbreviation for kilocalorie; Cal has the same meaning.

K-capture. (K-radiation). A type of radioactive decay in which an electron is captured by an atomic nucleus and immediately combines with a proton to form a neutron. The product of this radioactivity has the same mass number as the parent but the atomic number is one unit less. Thus, ^{55}iron with atomic number 26 decays by K-capture to form ^{55}manganese, with atomic number 25. Terms synonymous with *K-capture* are *K-electron capture* and *orbital* electron capture.

KC Finings, Super-Kleer. A two pouch system containing liquid Kieselsol and chitosan finings.
Use: In brewing.

KE-106A Natural Insecticide. A limonene-based (non-toxic) insecticide.
Use: Safe for use around children.

KE Heavy Duty. A degreaser.
Use: Environmentally safe degreaser.

Keene's cement. See gypsum cement.

Kekule, August. (1829–1896). Born in Darmstadt, Germany, Kekule laid the basis for the ensuing development of aromatic chemistry. His idea of a hexagonal structure for benzene in 1865 was a monumental contribution to theoretical organic chemistry. "This had been preceded in 1858 by the remarkable notion that carbon was tetravalent and that carbon atoms could be joined to each other in molecules." The theory of the benzene ring has been called the "most brilliant piece of scientific prediction to be found in the whole field of organic chemistry, for besides promulgating the idea, he had predicted the number and types of isomers which might be expected in various substitutions on the ring" (L. B. Clapp).

Kelacid. Alginic acid.

Kelate CU.
CAS: 14025-15-1.
Copper disodium EDTA.
Use: Sequestering agent in cosmetics.

"Kelcoloid HVF" *[ISP].* TM for xanthan gum.

"Kel-F" *[3M].* TM for a series of fluorocarbon products including polymers of chlorotrifluoroethylene and certain copolymers available as extrusion and molding powders, resins, dispersions, gums, oils, waxes, and greases that are characterized by high thermal stability, resistance to chemical corrosion, high dielectric strength, and high impact, tensile, and compressive strength.
Use: Corrosion control, contamination prevention, insulation, electrical equipment, molded and fabricated industrial equipment, lubricants, gyro and damping fluids. Especially useful under extreme conditions, including jet and space technology.
See fluorocarbon polymer.

"Kelgin" *[Monsanto].* TM for sodium alginates.

Kelmar. Potassium alginates.

kelp. A large coarse seaweed occurring chiefly off the coast of California. It is a type of algae and is mechanically harvested by specially equipped barges. Dried kelp contains 2–4% ammonia, 1–2% phosphoric acid, 15–20% potash, and traces of iodine.
Use: Fertilizers, plastics and conversion to methane by microorganisms; permissible chewing-gum base.

"Keltex" *[ISP].* TM for alginate.

"Keltose" *[FMC].* TM for self-gelling alginates.

Kelvin scale. See absolute temperature.

Kelzan. Xanthan gum.

Kelzan XC. Xanthan gum.

Kemamide.
Properties: High-melting solids. Fatty amides and bisamides.
Use: For slip agents and for friction reduction in polyolefin films. As process lubricants for styrenics and PVC, mold release for phenolics, as antiblock agents for polyolefins.

Kemester EGDS. (ethylene glycol distearate). Pearlizing agents.
Use: For shampoos, liquid soaps, and detergents.

Kemstrene. Glycerin.
Use: For a solvent and humectant in cosmetics, liquid soaps, confections, inks, lubricants, in polyester and polyurethane formulations.

Kenamine. A series of straight-chain amines from primary through quaternary; available in chain lengths from C_{12} through C_{22}. They are strongly hydrophobic and are biodegradable.
Use: Cationic intermediate, removal of moisture from surfaces.

Ken-Cel. A series of formulated blowing agent masterbatches.

Kencolor.   A series of dispersions based on silicone binders.
Use: Pigmenting of silicone elastomers.

Ken-Cure.   A series of specialty curative chemicals.

Kendall–Mattox reaction.   Formation of a conjugated ketone from an α-bromoketone via a phenylhydrazone or semicarbozone.

Kendrew, John C.   (1917–1997). An English molecular biologist who won the Nobel Prize for chemistry in 1962 with Perutz. His work verified Pauling's earlier thesis concerning the α-helix structure of the polypeptide chain. After receiving his Ph.D. from Cambridge, he was science advisor to the allied air commander-in-chief during World War II. He was also editor of the journal *Molecular Biology*.

"Kenflex" *[Kenrich]*.   TM for a series of aromatic oligomers of dimethyl naphthalene and blends thereof.
Use: Processing and compounding aid for neoprene, "Hypalon," SBR, vinyl compounds and other plastics; potting compounds; protective coatings; paper and textile coatings; insecticides; ink; chemical synthesis.

"Kenite" *[Imerys]*.   TM for diatomaceous earth and related products.

"Ken-Kem" *[Kenrich]*.   TM for a series of specialty chemicals.

"Kenlastic" *[Kenrich]*.   TM for a series of elastomer-based dispersions of rubber chemicals.
Use: Dispersion for elastomers.

"Ken-Mag" *[Kenrich]*.   TM for a bar form of dispersed magnesium oxide.
Use: Curing of elastomers.

"Kenmix" *[Kenrich]*.   TM for a series of paste dispersions of rubber chemicals.
Use: Dispersions for elastomers.

"Kenplast" *[Kenrich]*.   TM for a series of plasticizers based on aromatic hydrocarbons or derivatives of cumyl phenol.
Use: Plasticizer for nitrile elastomers and reactive diluents for epoxy.

"Ken-React" *[Kenrich]*.   TM for a series of organometallic coupling agents based on titanium, zirconium, and aluminum.
Use: Bonding agents, polymer process aids, catalysts, blowing agent activators, corrosion inhibitors, dispersing agents.

"Ken-React KR TTS" *[Kenrich]*.   TM for a series of alkyl titanates.
Available forms: Liquid, powder, and pellets.
Use: Catalyst, intermediates for ceramic coatings.

"Ken-Stat" *[Kenrich]*.   TM for a series of antistats based on combined neoalkoxy titanates or combined neoalkoxy zirconates.

"Ken-Zinc" *[Kenrich]*.   TM for a bar form of dispersed zinc oxide.
Use: Curing of elastomers.

kepone.   (chlordecone; decachlorooctahydro-1,3, 4-metheno-2H-cyclobuta[cd]pentalene-2-one).
CAS: 143-50-0. $C_{10}Cl_{10}O$.
Properties: Crystalline solid. Mp 350C (decomposes). Soluble in acetic acid, alcohols, ketones, acetone; slightly soluble in water.
Hazard: A carcinogen. Toxic by ingestion, inhalation, and skin absorption. Manufacture and use have been prohibited.
Use: Insecticide, fungicide.

keratin.   A class of natural fibrous proteins occurring in vertebrate animals and humans, they are characterized by their high content of several amino acids, especially cystine, arginine, and serine. They are generally harder than the fibrous collagen group of proteins. The softer keratins are components of the external layers of skin, wool, hair, and feathers, while the harder types predominate in such structures as nails, claws, and hoofs. The hardness is largely due to the extent of cross-linking by the disulfide bonds of cystine by the mechanism shown below:

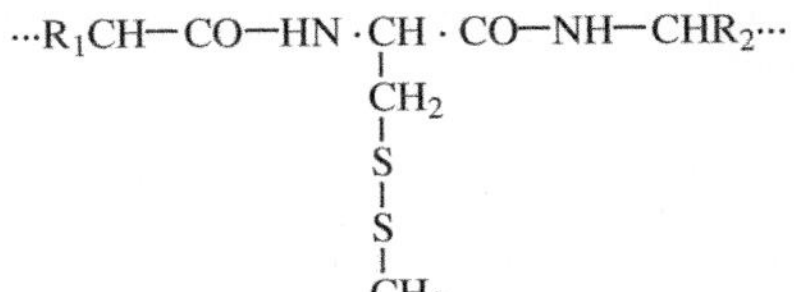

$$\cdots R_1CH-CO-HN \cdot CH \cdot CO-NH-CHR_2 \cdots$$
$$| $$
$$CH_2$$
$$|$$
$$S$$
$$|$$
$$S$$
$$|$$
$$CH_2$$
$$|$$
$$\cdots R_3CH \cdot OC-HN \cdot CH \cdot CO-NH-CHR_4 \cdots$$

Keratins are insoluble in organic solvents but do absorb and hold water. The molecules contain both acidic and basic groups and are thus amphoteric.
Use: Tablet coatings that dissolve only in the intestines, foam extinguishers, protein hydrolyzates.

keratinase.   A water-soluble, proteolytic enzyme having the ability to digest the keratin in wool and other forms of hair, converting a portion of it to a water-soluble form. It thus acts as a depilatory and is used in removing hair from pelts and hides, as well as from human skin. It is inactivated by heating to 100C.

"Kerlone" *[Sanofi].* TM for betaxolol hydrochloride.
Use: Drug.

kernite. $Na_2B_4O_7 \cdot 4H_2O$. A natural sodium borate found in Kern County, California.
Properties: Colorless to white, two good cleavages, luster vitreous to pearly. Mohs hardness 3, d 1.95. Noncombustible.
Use: Major source of borax and boron compounds.

kerogen. The organic component of oil shale, it is a bitumen-like solid whose approximate composition is 75–80% carbon, 10% hydrogen, 2.5% nitrogen, 1% sulfur, and the balance oxygen. It is a mixture of aliphatic and aromatic compounds of humic and algal origin and comprises a substantial proportion of the shale; after fractionating and refining, the oil is reported to yield 18% gasoline, 30% kerosene, 27% gas oil, 15% light lube oil, and 10% heavy lube oil.

kerosene. (kerosine).
CAS: 8008-20-6.
Properties: Water-white, oily liquid; strong odor. D 0.81, boiling range 180–300C, flash p 100–150F (37.7–65.5C), autoign temp 444F (228C). Combustion properties can be greatly improved by a proprietary hydrotreating process involving a selective catalyst.
Derivation: Distilled from petroleum.
Hazard: Moderate fire risk, explosive limits in air 0.7–5.0%. Toxic by inhalation. Questionable carcinogen.
Use: Rocket and jet engine fuel, domestic heating, solvent, insecticidal sprays, diesel and tractor fuels.

ketal. Organic compound produced by addition of an alcohol to a ketone. Analogous to acetal.
See hemiketal.

ketene.
CAS: 463-51-4. $H_2C=C=O$.
Properties: Colorless gas; disagreeable odor. Readily polymerizes; cannot be shipped or stored in a gaseous state. Mp −151C, bp −56C.
Derivation: Pyrolysis of acetone or acetic acid by passing its vapor through a tube at 500–600C.
Hazard: Toxic by inhalation, strong irritant to skin and mucous membranes.
Use: Acetylating agent, generally reacting with compounds having an active hydrogen atom; reacts with ammonia to give acetamide. Starting point for making various commercially important products, especially acetic anhydride and acetate esters.

ketimine. A type or class of curing agent for epoxy resins that makes it possible to use very-high-solids content coatings in spray equipment. Reacts with epoxies very slowly and thus delays curing time, which prevents setting up of the resin during spraying operation. In presence of water or water vapor, ketimine breaks down to a polyamine and a ketone. Epoxy coatings cured with ketimine should not exceed a thickness of 10 mils.

4-ketobenzotriazine. (benzazimide; 4-keto-(3H)-1,2,3-benzotriazine).
CAS: 90-16-4. $C_7H_5N_3O$.
Bicyclic.
Properties: Tan powder. Mp 210C (decomposes). Soluble in alkaline solutions and organic bases.
Use: Organic synthesis.

1-keto-2,3-dihydrocyclopentindole oxime.
CAS: 22942-83-2. $C_{11}H_{10}N_2O$.
Hazard: A poison.

keto-enol tauterism. A compound with isomers in equilibrium between the keto form –CH2–CO– and the enol form –CH=C(OH)–. It occurs by migration of a hydrogen atom between a carbon atom and the oxygen on an adjacent carbon.
See isomerization.

ketogenic amino acids. Amino acids with carbon skeletons that are metabolized to ketone bodies. Contrast with the glucogenic amino acids.

α-ketoglutaric acid. (2-oxopentanedioic acid). $HOOCCH_2CH_2COCOOH$.
Properties: Mp 113.5C. Soluble in water and alcohol. Important in amino-acid metabolism.

β-ketoglutaric acid. (ADA, acetonedicarboxylic acid). $HOOCCH_2CH_2COCOOH$.
Properties: Colorless needles. Mp 135C (decomposes). Soluble in water and alcohol; insoluble in benzene and chloroform.
Derivation: By heating dehydrated citric acid and concentrated sulfuric acid together.
Use: Organic synthesis.

α-ketoglutaric acid, disodium salt.
CAS: 305-72-6. $C_5H_4O_5 \cdot 2Na$.
Hazard: Moderately toxic by ingestion.

ketohexamethylene. See cyclohexanone.

ketone. A class of liquid organic compounds in which the carbonyl group, C=O, is attached to two alkyl groups; they are derived by oxidation of secondary alcohols. The simplest member of the series is acetone, $CH_3C(O)CH_3$, but many more complex ketones are known.
Use: Solvents, especially for cellulose derivatives, in lacquers, paints, explosives, and textile processing.
See acetone; diethyl ketone; methyl ethyl ketone.

ketone bodies. A group of molecules normally produced in very low amounts, but increased amounts are found during fasting or in untreated diabetes mellitus.

ketone, Michler's. See tetramethyldiaminobenzophenone.

ketonimine dye. A dye whose molecules contain the –NH=C= chromophore group. There are only two members in the class: auramine and a closely related homolog, methyl aurin, in which a methyl group replaces one of the hydrogen atoms of aurin. These are basic dyes used on cotton with tannin or tartar emetic as mordant.

4-ketopentenal.
CAS: 5729-47-5. $C_5H_6O_2$.
Hazard: A poison.

α-ketopropionic acid. See pyruvic acid.

ketose. A simple monosaccharide in which the carbonyl group is a ketone.

ketosis. A metabolic condition in which the concentration of ketone bodies in the blood, tissues, and urine is abnormally high.

γ-ketovaleric acid. See levulinic acid.

"Kevlar" [Du Pont]. TM for an aromatic polyamide fiber of extremely high tensile strength and greater resistance of elongation than steel. Its high energy-absorption property makes it particularly suitable for use in belting radial tires, for which it was specifically developed; it is also used as a reinforcing material for plastic composites in bullet-proof vests and in cordage products.
See aramid.

Keyes process. A distillation process involving the addition of benzene to a constant-boiling 95% alcohol-water solution to obtain absolute (100%) alcohol. On distillation, a ternary azeotropic mixture containing all three components leaves the top of the column while anhydrous alcohol leaves the bottom. The azeotrope (which separates into two layers) is redistilled separately for recovery and reuse of the benzene and alcohol.

kg. Abbreviation for kilogram. Equals 1000 grams.

Kick's law. The amount of energy required to crush a given quantity of material to a specified fraction of its original size is the same, regardless of the original size.

kier. A large metal tank or vessel in which wool or cotton fibers or fabrics are scoured, bleached, or dyed, usually in an alkaline solution (kier boiling).

kieselguhr. See diatomaceous earth.

kieserite. $MgSO_4 \cdot H_2O$. A natural magnesium sulfate occurring in enormous quantities in the Stassfurt salt beds (Germany), Austria, and India. See magnesium sulfate.

Kiliani–Fischer synthesis. Extension of the carbon atom chain of aldoses by treatment with cyanide. Hydrolysis of the cyanohydrins followed by reduction of the lactone yields the homologous aldose.

"killed" steel. Steel deoxidized by the addition of aluminum, ferrosilicon, etc., while the mixture is maintained at melting temperature until all bubbling ceases. The steel is quiet and begins to solidify at once without any evolution of gas when poured into the ingot molds.

kiln. (1) A refractory-lined cylinder, either stationary or rotary.
Use: Calcination of lime, magnesia, cement, ores, etc., and for incinerating gaseous, liquid, and solid wastes.
(2) A furnace for firing ceramic products.

kilo-. Prefix meaning 10^3 units (symbol k), e.g., 1 kg = 1 kilogram = 1000 grams.

kilogram. (1) A mass identical with that of the international kilogram at the International Bureau of Weights and Measures in France. It is the mass of a liter of water at 4C.
(2) A force equal to the weight of one kilogram mass, measured at sea level.

kinase. A kinase is in general an enzyme that catalyzes the transfer of a phosphate group from ATP to something else. In molecular biology, it has acquired the more specific verbal usage for the transfer onto DNA of a radiolabeled phosphate group. This would be done in order to use the resultant "hot" DNA as a probe.

kinematic viscosity. See viscosity.

kinetic chain length. The average number of molecules of a monomer converted to a polymer for each active center formed in an initiation reaction.

kinetics, chemical. Chemical phenomena can be studied from two fundamental approaches: (1) *thermodynamics*, a rigorous and exact method concerned with equilibrium conditions of initial and final states of chemical changes; (2) *kinetics*, which is less rigorous and deals with the rate of change from initial to final states under nonequilibrium conditions. The two methods are related. Thermodynamics, which yields the driving potential—a measure of the tendency of a system to change from one state to another—is the foundation upon which kinetics is built. The rate at which a change will occur depends upon two factors: (1) directly with driving force or potential and (2) inversely with a resistance. A measure of the tendency of a system to resist chemical change is the so-called activation energy, which is independent of the driving force or so-called free energy of the reaction.

The diagram is a mechanical analogy illustrating the difference between activation energy and driving potential. The chemical system is represented by a sphere resting in a valley. The initial equilibrium state, A, is at a higher elevation than the final state, B. The difference in elevation between A and B is a measure of the free energy change of the reaction, that is, the driving force that will take the system from A to B. This quantity, ΔG, is determined by the classical methods of thermodynamics. Now A and B are equilibrium states represented by the valleys. For the system to go from A to B, it must first overcome the hill separating the valleys. The elevation of this hill from the valley of the initial state is a measure of the resistance to change in the system in going from A to B. The quantity ΔG^*, known as free energy of activation, is determined by the methods of kinetics.

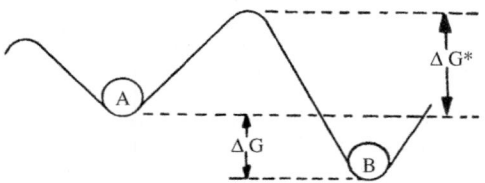

The system of molecules that is undergoing reaction consists of these molecules in different energy states. If the temperature of a gas is raised, there is an increase in the energy of these states and hence an increase in the collisions of molecules which have the necessary activation energy; as a result, the rate of the reaction will increase. Also, if by means of a catalyst, the activation energy is decreased, more colliding molecules will react and again the rate of reaction will increase.

Note: Adapted from an article by Roger Gilmont in "Encyclopedia of Chemistry," Hampel and Hawley, editors, 3rd ed., 1973.

See thermodynamics, chemical; thermodynamics.

kinetic theory.   A theory of matter based on the mathematical description of the relationship between the volumes, temperatures, and pressures of gases (P-V-T phenomena). This relationship is summarized in the so-called gas laws as follows: (1) Boyle's law (at constant temperature the volume of a gas is inversely proportional to its pressure); (2) Charles's law (at constant volume the pressure exerted by a gas is proportional to its absolute temperature); (3) Avogadro's law (equal volumes of the same or different gases under the same conditions of temperature and pressure contain the same number of molecules).

The theory involves the basic concept of matter as composed atoms and/or molecules that move

more rapidly (gases) or vibrate more energetically (solids) as temperature increases. Thus, crystals melt at a point where the heat or energy input exceeds the bond energy of the solid state.

See kinetics, chemical; gas; thermodynamics.

kinetin.   (6-furfurylaminopurine).
CAS: 525-79-1. $C_{10}H_9N_5O$.
A plant growth regulator used to promote growth in bacterial cultures and as dormancy breaker.

king's green.   See copper acetoarsenite.

kinin.   (cytokinin). One of a group of plant growth regulators that promote cell division and differentiation.
See kinetin.

Kipp's apparatus.   An apparatus for producing a gas by the action of a liquid on a solid that consists of three globes. The top one has a funnel-shaped extension in to the bottom globe which stores the liquid. The middle globe contains the solid.

Kishner cyclopropane synthesis.   Formation of cyclopropane derivatives by decomposition of pyrazolines formed by reacting α,β-unsaturated ketones or aldehydes with hydrazine.

Kistiakowsky, George B.   (1900–1982). Born in Kiev, Russia, where he fought in the White Russian Army, he studied in Germany under Maxwell Bodenheim, where he obtained his doctorate in chemistry. In 1926, he came to the U.S. and became an American citizen in 1933. For 41 years, he was Professor of Chemistry at Harvard; in addition to many chemical awards, he was the recipient of the Priestley Medal as well as distinguished Medals of Honor from three Presidents of the U.S. President Eisenhower appointed him as his assistant for science and technology, and he was chairman of the Science Advisory Committee from 1957 to 1963. Among his many achievements in both chemistry and physics, he was a world-famous authority on explosives. A key member of the Manhattan Project, he devised the detonating mechanism for the first experimental atomic bomb in New Mexico, at which time he was head of the Los Alamos Laboratory. Though he ranks high among those who developed the bomb, he perceived the awesome destructive potential of nuclear weapons and became an ardent opponent of their future use. Resigning from the Pentagon in 1967, he returned to teaching at Harvard. Among other distinguished organizations, he was a member of the Royal Society of London, the AAAS, and the ACS.

Kjeldahl flask.   A round bottom flask with a long wide neck that is used in the determination of nitrogen by Kjeldahl's method.

Kjeldahl test.   An analytical method for determination of nitrogen in certain organic compounds. It involves addition of a small amount of anhydrous potassium sulfate to the test compound, followed by heating the mixture with concentrated sulfuric acid, often with a catalyst such as copper sulfate. As a result ammonia is formed. After alkalyzing the mixture with sodium hydroxide, the ammonia is separated by distillation, collected in standard acid, and the nitrogen determined by back-titration.

Kjelgest.   Anhydrous granular magnesium perchlorate for use as a desiccant.

Kleanrol.   A soldering flux crystal based on zinc chloride and ammonium chloride.

"Klearol" [Sonneborn].   TM for a white mineral oil, technical grade, d 0.828–0.838.
Use: Cosmetic preparations, eggshell preservation.

Klein's reagent.   A saturated solution of cadmium borotungstate, formula variously given, possibly $2CdO \cdot B_2O_3 \cdot 18H_2O$, d 3.28.
Use: Separation of minerals by specific gravity.

"Klucel" [Hercules].   TM for hydroxypropyl cellulose.
Use: A thickener for paint removers, tablet binders, aqueous film-coating polymers, sustained-release matrix polymer for pharmaceuticals.

Klug, Aaron S.   (1926–). A South African-born chemist who won the Nobel Prize for chemistry in 1982 for his work with the electron microscope and research into the structure of nucleic and protein complexes. He used crystallographic electron microscopy to analyze the structures of biologically important complex chemicals was noteworthy. He was cited in particular for his establishment of Fourier microscopy.

kneading.   Blending of soft, plastic, semisolid materials into a uniform mixture by subjecting them to a rolling pressure exerted by agitators of specific shape rotating in trough-like containers. The action is a combination of turning, folding, and pressing. The operation is used in processing bakery doughs, printing inks, clays, and pastes of various types.
See blend; sigma blade.

knife.   See doctor knife.

knock.   Ignition of a portion of the gasoline in the cylinder head due to spontaneous oxidation reactions rather than to the spark. It causes serious power loss, especially in high-compression engines.
See octane number.

knock-out experiment.   A technique for deleting, mutating or otherwise inactivating a gene in a mouse. This laborious method involves transfecting a crippled gene into cultured embryonic stem cells, searching through the thousands of resulting clones for one in which the crippled gene exactly replaced the normal one (by homologous recombination), and inserting that cell back into a mouse blastocyst. The resulting mouse will be chimaeric but its germ cells will carry the deleted gene. A few rounds of careful breeding can then produce progeny in which both copies of the gene are inactivated.
See gene; locus; model organisms.

Knoevenagel condensation; Doebner modification.   Condensation of aldehydes or ketones with active methylene compounds (specifically malonic ester) in the presence of ammonia or amines; the use of malonic acid and pyridine is known as the Doebner modification.

Knoop hardness.   See hardness.

Knoop–Oesterlin amino acid synthesis.   Preparation of α-amino acids by catalytic hydrogenation of α-oxo acids in aqueous ammonia in the presence of platinum, palladium, or Raney nickel catalysts, probably via an unstable iminocarboxylate ion intermediate.

Knoop scale.   Comparative hardness scale, ranges from glass (300–600) to diamond (6000–6500).

Knorr pyrazole synthesis.   Formation of pyrazole derivatives from hydrazines, hydrazides, semicarbazides, and aminoguanidines by condensation with 1,3-dicarbonyl compounds; with substituted hydrazines, two structurally isomeric pyrazoles are formed.

Knorr pyrrole synthesis.   Formation of pyrrole derivatives by condensation of α-amino ketones as such or generated *in situ* from isonitrosoketones with carbonyl compounds containing active α-methylene groups.

Knorr quinoline synthesis.   Formation of α-hydroxyquinolines from β-ketoesters and arylamines above 100C. The intermediate anilide undergoes cyclization by dehydration with concentrated sulfuric acid.

Knowles, William Standish.   (1917–2012). An American born in Taunton, Massachusetts who won the Nobel Prize for chemistry in 2001 for his pioneering work concerning chirally catalyzed hydrogenation reactions. He was awarded an undergraduate degree from Harvard and worked many years for the Monsanto Corporation. Knowles received the ACS Award for Creative Invention 1982.

KOALA-TY.   Laboratory apparatus.

Kobilka, Brian K. (1955–). As American physician and molecular biologist who was awarded the 2012 Nobel Prize jointly with Robert J. Lefkowitz. The scientists' groundbreaking discovery of the G-protein-coupled receptors in living cell has revolutionized the understanding of how cells respond to stimuli such as hormones and how certain types of drugs have their effect, this remarkable discovery lead to major advances in cell biology and medicine. Kobilka has dedicated most of his career to research structure and workings of G-protein-coupled receptors. He was educated at the University of Minnesota and received his M.D. from Yale University. He serves as a professor and chair of molecular and cellular physiology at Stanford University School of Medicine; he is also the Helene Irwin Fagan Chair in cardiology.

Koch acid. (1-naphthylamine-3,6,8,-trisulfonic acid).

Properties: White solid. Slightly soluble in water.
Derivation: Naphthalene-β-sulfonic acid is sulfonated with 60% oleum at 165C, the resulting naphthalene-1,3,6-trisulfonic acid is nitrated and the product reduced with iron.
Use: Azo dye intermediate.

Koch–Haaf carboxylations. Formation of tertiary carboxylic acids by treating alcohols with carbon monoxide in strong acid.

Kochi reaction. Synthesis of organic chlorides by decarboxylation of carboxylic acids in the presence of lead tetraacetate and lithium chloride.

Kodo-cytochalasin-1.
Hazard: Extremely toxic.

Koenigs–Knorr synthesis. Formation of glycosides from acetylated glycosyl halides and alcohols or phenols in the presence of silver carbonate or silver oxide. The reaction proceeds with inversion of configuration.

Kohlrausch's law. Ions have independent migrations, and the conductance of a solution is the sum of the conductances of the anions and cations.

KOH number. In the electrometric titration of latex with KOH, the number of grams of KOH per 100 grams of rubber necessary to obtain a pH of 10.7–11.

Kohn, Walter. (1923–). Awarded Nobel Prize in chemistry in 1998 jointly with John A. Pople for their pioneering contributions in developing methods that can be used for theoretical studies of the properties of molecules and the chemical processes in which they are involved. He performs his research at the University of California at Santa Barbara.

kojic acid. [5-hydroxy-2-(hydroxymethyl)-4-pyrone].
CAS: 501-30-4. $C_6H_6O_4$.
Properties: Crystals. Mp 152–154C. Soluble in water, acetone, alcohol; slightly soluble in ether; insoluble in benzene. Mildly antibiotic.
Derivation: Fermentation of starches and sugars by certain molds.
Hazard: Questionable carcinogen.
Use: Chemical intermediate, metal chelates, insecticide, antifungal and antimicrobial agent.

kola. See cola.

Kolbe electrolytic synthesis. Formation of hydrocarbons by the electrolysis of alkali salts of carboxylic acids (decarboxylative dimerization).

Kolbe–Schmidt reaction. The preparation of salicylic acid or its derivatives from carbon dioxide and sodium or potassium phenolate.

"Kolene" DGS Salt [Kolene]. TM for an anhydrous molten oxidizing salt bath using a sodium hydroxide base with additives necessary to provide controlled chemical oxidizing and dissolving properties.
Use: Descaling of heat-treated and hot-work oxides and scales; deglassing (removal of glass-drawing lubricants), investment, and silica removal; removal of burned-in carbon deposits; cleaning of oils, greases, and organic materials from the surface of metals.

Komarowsky reaction. The reaction between certain alcohols and p-hydroxybenzaldehyde in dilute sulfuric acid solutions to give soluble colored complexes. 1,2-Propylene glycol gives a colored product while ethylene glycol does not. The reaction has also been employed to determine cyclohexanol in cyclohexanone.

Kondakov rule. Olefins that add mineral acids readily react with chlorine or bromine to give unsaturated monohalides; those that do not add mineral acids readily form dihalides.

Konowaloff rule. The vapor over a liquid is relatively rich in the component whose addition to the liquid mixture results in an increase of the total vapor pressure.

Kopp's law. The molecular heat of a solid compound is an additive function of the atomic heat capacities of its individual atoms. The molecular volume of a liquid is equal to the sum of the atomic volumes of its constituent atoms.

korax.
CAS: 600-25-9.
Generic name for 1-chloro-1-nitropropane. $ClCH_2CH(CH_3)NO_2$.
Properties: Liquid. Bp 170.6C (745 mm Hg), flash p 144F (62.2C). Miscible with most organic solvents; slightly soluble in water. Combustible.
Hazard: Moderate fire risk. Strong irritant. Eye irritant and pulmonary edema.
Use: Fungicide.

Kornberg, Roger D. (1947–). An American chemist who received the Nobel Prize in 2006. Kornberg discovered how transcription (process where DNA is converted to RNA) occurs and how it is controlled at a molecular level in a group of organisms called eukaryotes (organisms with well-defined nucleus). He was the first to create an actual picture of how DNA's genetic blueprint is read and used to direct the process for protein manufacturing. Kornberg's discovery explains that complications in transcription process can lead to birth defects, cancer, and other diseases. Kornberg received his Ph.D. from Stanford University. He was a faculty of Harvard Medical School and later became a professor at Stanford University.

Kostanecki acylation. Formation of chromones or coumarines by acylation of o-hydroxyaryl ketones with aliphatic acid anhydrides followed by cyclization.

KR. A prefix for a series of monoalkoxy, chelate, and coordinate type titanates and zirconates.

Kr. Symbol for krypton.

Krafft degradation. Conversion of carboxylic acids, especially of high molecular weight, into the next-lower homolog by dry distillation of the alkaline earth salt with the corresponding acetate, followed by chromic acid oxidation of the methyl ketone.

Krafft point. Temperature (°C) at which cogel or crystal formation takes place in a soap solution and produces opacity.

kraft paper. A strong and relatively cheap paper made chiefly from pine by digestion with a mixture of caustic soda, sodium sulfate, sodium carbonate, and sodium sulfide. It is by far the largest volume paper made in the U.S.
See holopulping process.

"Kraton" 101 *[Kraton].* TM for a styrene-butadiene elastomer that requires no vulcanization, while displaying most of the properties of conventional vulcanized polymers. White, free-flowing crumb, readily soluble in a large number of commercially used solvents.

Use: In pressure-sensitive wet lay-up and hot-melt adhesives, dip coating, spraying, and spreading applications.

"Kraton Liquid l-207" *[Kraton].*
CAS: 103599-19-5.
TM for a basic polymer liquid.
Hazard: Moderately toxic by skin contact. Low toxicity by ingestion. A mild skin irritant.

Krebs cycle. See TCA cycle.

"Krenite" 10 *[Albaugh].* TM for a solution of urea and ammonium nitrate in aqueous ammonia; contains 43.5% nitrogen.
Use: Manufacture of mixed fertilizers.

"K-Resin" *[Chevron].* TM for a styrene-butadiene copolymer.
Use: Clear packaging, film, medical applications, and toys.

Krohnke aldehyde synthesis. Transformation of benzyl halides into aldehydes via their pyridinium salts which, on treatment with p-nitrosodimethylaniline, give nitrones. Hydrolysis of the nitrones yields aldehydes.

Kroll process. A widely used process for obtaining titanium metal. Titanium tetrachloride is reduced with magnesium metal at red heat and atmospheric pressure, in the presence of an inert gas blanket of helium or argon. Magnesium chloride and titanium metal are produced. The reaction is $TiCl_4 + 2Mg \rightarrow Ti + 2MgCl_2$. Essentially the same process is also used for obtaining zirconium.

"Kromatherm" *[Stollberg].* TM for high-temperature pigments designed for silicone- and fluorocarbon-resin-based paint vehicles.

"Kronitex" *[Great Lakes].* TM for a series of synthetic phosphate esters to replace such natural products as tricresyl and cresyl diphenyl phosphates.
Use: Flame-retardant plasticizers for vinyls, dust filter medium, gas additives, wood-treating chemical, foam control.

krotiline.
CAS: 2971-38-2. $C_{12}H_{11}Cl_3O_3$.
Hazard: Moderately toxic by ingestion and inhalation.
Use: Agricultural chemical.

Kroto, Sir Harold W. (1939–). A British chemist who won the Nobel Prize for chemistry along with Robert F. Curl, Jr. and Richard E. Smalley in 1996, the 100th anniversary of Alfred Nobel's death. The trio won for the discovery of the C_{60} compound called buckminsterfullerene. He received a Ph.D. from the University of Sheffield.

See buckminsterfullerene; Curl, Robert F., Jr.; Smalley, Richard E.

Krum.   Horticultural perlite.

krypton.
CAS: 7439-90-9.   Kr.
Element of atomic number 36, noble-gas group of the periodic table, aw 83.80, valence = 2 (possibly others), has six stable isotopes and a number of artificially radioactive forms.
Properties: Colorless gas; odorless. Bp −152.9C (1 atm), fp −157.1C, d 2.818 (air = 1), sp vol 4.61 cu ft/lb (21C, 1 atm). Only slightly soluble in water. Known to combine with fluorine at liquid nitrogen temperature by means of electric discharges or ionizing radiation to make KrF_2 and KrF_4. These compounds decompose at room temperature. Noncombustible.
See noble gas.
Derivation: By fractional distillation of liquid air. Air contains 0.000108% of krypton by volume.
Use: Incandescent bulbs and fluorescent light tubes, lasers, high-speed photography.
Note: Solid krypton exists at cryogenic temperatures as a white, crystalline substance; mp 116 K.

krypton-86.   Radioactive krypton of mass number 85.
Properties: Half-life 10.3 years. Radiations: β with a small component of γ. Low radiotoxicity.
Derivation: A fission product extracted from irradiated nuclear fuel.
Available forms: Gas of high chemical purity, but mixed with other isotopes of krypton in sealed glass flasks.
See kryptonates.
Use: Activation of phosphors for self-luminous markers, detecting leaks, medicine to trace blood flow, and in measurement of standard meter.

kryptonates.   Materials impregnated with krypton-85 in such a way that the radioactive atoms are held within the crystalline lattice structure. Elements, alloys, glasses, inorganic compounds, rubbers, and plastics have been so impregnated with tracer atoms.

"Krytox" [Du Pont].   TM for a series of hexafluroropropylene epoxide polymers of medium molecular weight, used as lubricating oils and greases; in high-temperature or corrosive conditions, good chemical inertness, even with boiling sulfuric acid; low solubility in most solvents; good lubricity under load; nonflammable, have thermal stability up to 260C.

K-Selectride.   Potassium tri-*sec*-butylborohydride, 1.0 molar solution in tetrahydrofuran. $KB[CH(CH_3)C_2H_5]_3H$.
Properties: Moisture-sensitive liquid. Mw 222.27, d 0.913, fp −17C.

Hazard: Flammable liquid, handle under nitrogen.
Use: Stereoselective reduction of ketones; conjugate reduction and alkylation of α,β-unsaturated ketones.

"K-Tea Algacide" [SePro]. TM for an algacide.
Use: To kill most forms of algae in lakes and fish hatcheries, rivers, streams, and golf course waters.

KTPP.   Abbreviation for potassium tripolyphosphate.

Kucherov reaction.   Hydration of acetylenic hydrocarbons with dilute sulfuric acid in the presence of mercuric sulfate or boron trifluoride as catalyst.

Kuhn, Richard.   (1900–1967). A German chemist who won the Nobel Prize in 1938. He worked on cartinoids and synthetic vitamins and discovered the chemical formula for vitamin B6. He also discovered a method for dissolving symplexes from plants using invert soaps. He received his Ph.D. in Munich, and went on to teach in Switzerland.

Kuhn–Roth method for C-methyl determination.   Oxidation of organic compounds with chromic and sulfuric acids in such a manner that the C-methyl groups are converted to acetic acid which can be assayed volumetrically. The method has been modified and extended (1) to saturated fatty acids and alcohols containing up to about 20 carbon atoms, and (2) to aliphatic long-chain compounds of very high molecular weight.

Kuhn–Winterstein reaction.   Conversion of 1,2-glycols into *trans* olefins by reaction with diphosphotetraiodide (P_2I_4) or other halogenated reagents. This reaction is useful in the preparation of polyenes.

K-Universal Line Klisters.   A ski wax.
Use: Versatile wax products utilizing newest raw materials.

Kuron.   A hormone-type weed and brush killer.

Kurroll's salt. $NaPO_3(IV)$. A high-temperature crystalline form of sodium metaphosphate.

Kuzo.   A beeswax formulation.
Use: For dreadlocks and braid ends.

kyanite.   (cyanite; disthene; rhoetizite). Al_2O_3. SiO_2. A mineral.

"Kydex" [Kydex].   TM for a thermoformed acrylic polyvinyl chloride alloy plastic sheet.

Use: Housings, trays, covers, containers, protective guards, and decorative parts exposed to severe service conditions.

"Kynol" *[Gun EI].* TM for a flame-resistant fiber available as 1.5 inch staple, 1.7 denier. It is a cross-linked, amorphous phenolic polymer, inert to all solvents and with fair resistance to oxidizing acids and strong alkalies. Will not ignite up to 2500C, but will char slowly at 260C. Potential uses are as ablative agent in spacecraft, flameproof apparel, protective clothing, etc.
See ablation.

kynurenine. $C_{10}H_{12}N_2O_3$. An amino acid that is a metabolic product of tryptophan.
Use: Biochemical and nutritional research, especially in connection with B-complex deficiencies.

L

L.   (l). Abbreviation for liter.

L-.   (L-). Prefix indicating the left-handed enantiomer of an optical isomer. Often displayed as a small capital letter.
See D-.

l-.   (l-). Prefix indicating that a compound is levorotatory. A minus sign (−) is now preferred.
See *d*-.

La.   Symbol for lanthanum.

label.   (1) A warning notice required by DOT or IATA to be placed on the shipping container of a hazardous material transported by air, highway, rail, or water. Names of some labels are as follows:

corrosive	irritant
dangerous when wet	nonflammable gas
explosive	organic peroxide
flammable gas	oxidizer
flammable liquid	poison
flammable solid	radioactive

(2) A notice required to appear on a food product indicating its composition and nutritional value (RDA), or on a pharmaceutical or household product stating its hazardous properties.
(3) A radioactive isotope or fluorescent dye added to a chemical compound to trace its course and behavior through a series of reactions, usually in living organisms. This technique has also been used to measure frictional wear of moving parts of automotive engines.
See tracer; tagged compound.

labeled compound.   Refers to material containing radioactive atoms.
Use: As a tracer in chemical reactions and medical diagnosis.

labeled water.   A naturally occurring form of water which contains ^{18}O, a rare isotope of oxygen.

labeling.   (labelling). Replacement of a stable atom in a compound with a radioisotope of the same element. This facilitates tracing the path of the chemical through a biological or mechanical system by detection of the radiation. Tritiated compounds have a hydrogen atom replaced by a tritium atom.

labile.   Descriptive of a substance that is inactivated by high temperature or radiation, e.g., a heat-labile vitamin, unstable.

laboratory alcohol.   Usually a mixture of 70% ethanol and 30% water.

laboratory conditions.   An ideal set of conditions in which all variant factors except the one under test can be held constant, as for example, rooms provided with constant temperature and humidity control, clean rooms, and the like. Less specifically, the term refers to experimental or small-batch conditions as opposed to large-scale production.

laboratory machinery.   Small-scale working models of basic types of equipment used for experimental purposes in the laboratories of many chemical process industries. Commercially available are such items as calenders, mills of various kinds, mixers, autoclaves, extruders, electric furnaces, distillation columns, etc. A machine as complex as a fourdrinier can be duplicated in laboratory size (about 10 ft by 18 in).

laccase.   An enzyme that oxidizes phenols or *o*- and *p*-quinones.
Occurrence: It occurs in the latex of the lac tree, in potatoes, sugar beets, apples, cabbages, and other plants.

lac dye.   (CI Natural Red 25; lacchaid acid). CAS: 60687-93-6. A brilliant red dye obtained by maceration of crude lac.
See shellac.

lachrymator.   (lacrimator). A material (gas) that is strongly irritant to the eyes; tear gas.

L acid.   (1-naphthol-5-sulfonic acid). $C_{10}H_6(OH)SO_3H$.
Properties: White solid. Soluble in water.
Derivation: From naphthalene-1,5-disulfonic acid by fusion with sodium hydroxide.
Use: Azo dye intermediate.

lacmoid.   (resorcinol blue). $(HO)_2C_6H_3N[C_6H_2(OH)_3]_2$.
Properties: Lustrous, dark-violet, crystalline scales. Soluble in alcohol, ether, acetone, phenol, and glacial acetic acid; slightly soluble in water; pH 4.4–6.2.

Hawley's Condensed Chemical Dictionary, Sixteenth Edition. Michael D. Larrañaga, Richard J. Lewis, Sr., and Robert A. Lewis.
© 2016 John Wiley & Sons, Inc. Published 2016 by John Wiley & Sons, Inc.

Derivation: From resorcinol by treatment with sodium nitrite.
Use: Indicator in analytical chemistry (more sensitive than litmus).

lacquer.   A protective or decorative coating that dries primarily by evaporation of solvent, rather than by oxidation or polymerization. Lacquers were originally comprised of high-viscosity nitrocellulose, a plasticizer (dibutyl phthalate or blown castor oil), and a solvent. Later, low-viscosity nitrocellulose became available; this was frequently modified with resins, such as ester gum or rosin. The solvents used are ethanol, toluene, xylene, and butyl acetate. Together with nitrocellulose, alkyd resins are used to improve durability. The nitrocellulose used for lacquers has a nitrogen content of 11–13.5% and is available in a wide range of viscosities, compatibilities, and solvencies. Chief uses of nitrocellulose-alkyd lacquers are for coatings for metal, paper products, textiles, plastics, furniture, and nail polish. Various types of modified cellulose are also used as lacquer bases, combined with resins, and plasticizers. Many noncellulosic materials such as vinyl and acrylic resins are also used, as are bitumens, with or without drying oils, resins, etc.
See nitrocellulose.
Hazard: Flammable, dangerous fire and explosion risk.

lactalbumin.   See albumin, milk.

lactam.   A cyclic amide produced from amino acids by the removal of one molecule of water. An example is caprolactam, $CH_2(CH_2)_4CONH$, derived from epsilon-aminocaproic acid, $NH_2(CH_2)_5COOH$.

lactase.   An enzyme present in intestinal juices and mucosa that catalyzes the production of glucose and galactose from lactose.
Use: Biochemical research.

lactase enzyme preparations from *Kluyveromyces lactis*.
Properties: Derived from *Kluyveromyces lactis*.
Use: Food additive.

lactate.   An ester or salt of lactic acid, e.g., ethyl lactate.

lactated mono-diglycerides.
Properties: Soft to hard waxy solid. Dispersible in hot water; moderately soluble in hot isopropanol, xylene, cottonseed oil.
Use: Food additive.

lactic acid.   (α-hydroxypropionic acid; milk acid).

CAS: 50-21-5. CH_3CH_2OCOOH.

$$CH_3-\overset{\overset{\displaystyle H}{|}}{\underset{\underset{\displaystyle OH}{|}}{C}}-COOH$$

Properties: Colorless or yellowish syrupy liquid; odorless; hygroscopic. Bp 122C (15 mm Hg), mp 18C, d 1.2. Miscible with water, alcohol, glycerol, and furfural; insoluble in chloroform, petroleum ether, carbon disulfide. Cannot be distilled at atmospheric pressure without decomposition; when concentrated above 50% it is partially converted to lactic anhydride. It has one asymmetric carbon and two enantiomorphic isomers. The commercial form is a racemic mixture.
Derivation: (1) By fermenting starch, milk whey, molasses, potatoes, etc. and neutralizing the acid as soon as formed with calcium or zinc carbonate. The solution of lactates is concentrated and decomposed with sulfuric acid. (2) Synthetically by hydrolysis of lactonitrile.
Grade: Technical 22% and 44%, food 50–80%, plastic 50–80%, USP (85–90%), CP, FCC.
Use: Cultured dairy products, as acidulant, chemicals (salts, plasticizers, adhesives, pharmaceuticals), mordant in dyeing wool, general-purpose food additive, manufacture of lactates.

lactic acid dehydrogenase.   An enzyme found in animal tissues and yeast which takes part in controlling carbohydrate metabolism in the cell.
Use: Biochemical research.

lactoferrin.   An iron-binding protein in milk that is structurally similar to a transferrin.

lactoflavin.   (7,8-dimethyl-10-[(2S,3S,4R)-2,3,4,5-tetrahydroxypentyl]benzo[g]pteridine-2,4-dione; vitamin B2; Riboflavin).
CAS: 83-88-5. $C_{17}H_{20}N_4O_6$.
Nutritional factor found in milk, eggs, malted barley, liver, kidney, heart, and leafy vegetables. The richest source of yeast. It occurs in the free form only in the retina of the eye, in whey, and in urine. Its principal forms in tissues and cells are as flavin mononucleotide and flavin-adenine dinucleotide.

lactogenic hormone.   See luteotropin.

lactoglobulin.   A protein found in milk. It comprises from 7 to 12% of the skim milk protein and is closely associated with casein.

lactol.   A cyclic hemiacetal produced by the intramolecular addition of a hydroxyl group to an aldehydic or ketonic carbonyl group.

lactone.   An inner ester of a carboxylic acid formed by intramolecular reaction of hydroxylated or halogenated carboxylic acids with elimination of water. The resulting ring compound contains the group –CO•O–.

Derivation: In nature as odor-bearing components of various plant products; also made synthetically. See butyrolactone; propiolactone.

lactonitrile. (α-hydroxypropionitrile; acetaldehyde cyanohydrin).
CAS: 78-97-7. CH_3CH_2OCN.
Properties: Straw-colored liquid. Acid to methyl red, fp −40C, bp 182–184C (slight decomposition), d 0.9919 (18.4C), refr index 1.4058 (18.4C), flash p 170F (76.6C). Soluble in water and alcohol; insoluble in petroleum ether and carbon disulfide. Combustible.
Derivation: Acetaldehyde and hydrocyanic acid.
Grade: Technical, 95–97% purity.
Hazard: Toxic by inhalation, ingestion, and skin absorption; evolves hydrocyanic acid in presence of alkali.
Use: Solvent, intermediate in production of ethyl lactate and lactic acid.

lactoprotein. Any protein that normally occurs in milk.

lactose. (milk sugar; saccharum lactis).
CAS: 63-42-3. $C_{12}H_{22}O_{11}•H_2O$.

Properties: White, hard, crystalline mass or white powder; sweet taste; odorless; stable in air. D 1.525, mp decomposes at 203.5C. Soluble in water; insoluble in ether and chloroform; very slightly soluble in alcohol.
Derivation: From whey, by concentration and crystallization. Cows' milk contains about 5% lactose.
Grade: Crude, fermentation, spray-dried, edible, USP.
Use: Pharmacy; infant foods; bacteriology; baking and confectionery; margarine and butter manufacture; manufacture of penicillin, yeast, edible protein, and riboflavin; culture media; adsorbent in chromatography.

lactylic esters of fatty acids.
Properties: Hard, waxy solid to liquid. Dispersible in hot water; soluble in organic solvents, vegetable oil.
Use: Food additive.

lac wax. A wax obtained from lac consisting of myricyl and ceryl alcohols, free and combined with various fatty acids. Combustible.
See shellac.

LAD. Abbreviation for lithium aluminum deuteride.

ladder polymer. An ordered molecular network of double-stranded chains connected by hydrogen or chemical bonds located at regular intervals along the chains. Many complex proteins, including DNA, are of this nature.

Ladenburg rearrangement. Thermal rearrangement of an alkyl- or benzylpyridinium halide to an alkyl- or benzylpyridine.

ladle. A refractory-lined pot or vessel equipped a spout and with lugs for handling by a crane used to transport molten metal from the furnace to ingot molds, into which the metal is poured from the ladle.

lagging strand. The DNA strand that, during replication, must be synthesized in the direction opposite to that in which the replication fork moves. It is synthesized via a series of fragments called Okazaki fragments.

lagoon. A scientifically constructed pond 3–5 ft deep in which sewage and other organic wastes are decomposed by the action of algae, sunlight, and oxygen, thus restoring water to a purity equal to that obtained with other types of treatment. These so-called oxidation ponds are often used following activated sludge treatment. The waste may be retained in the lagoon for as long as 30 days and is then chlorinated and passed through a trickle filter.

LAH. Abbreviation for lithium aluminum hydride.

laid (papermaking). See dandy roll.

lake. An organic pigment produced by the interaction of an oil-soluble organic dye, a precipitant, and an absorptive inorganic substrate. Insoluble in water; poor lightfastness makes lakes unsuitable for use in exterior paints.
Use: Printing inks, wallpaper inks, metal decorative coatings, coated fabrics, rubber, plastics, food colorants.
See toner; alizarin; madder lake.

Lake Red carbon. Red pigments made by coupling 2-chloro-5-aminotoluene-4-sulfonic acid with β-naphthol and forming various metal salts.
Properties: Good resistance to bleeding, reasonable light resistance, good transparency. Produces inks with good flow.
Grade: Resinated and nonresinated.
Use: General purpose colors for letterpress, gravure, flexographic, moisture-set, heat-set inks, especially for offset printing inks.

lambda-cyhalothrin.
CAS: 91465-08-6. $C_{23}H_{19}ClF_3NO_3$.
Hazard: A poison by ingestion. Moderately toxic by skin contact and inhalation.
Use: Agricultural chemical.

lamb shift.   A small energy difference between two levels ($^2S_{1/2}$ and $^2P_{1/2}$) in the hydrogen spectrum. The shift results from the quantum interaction between the electromagnetic radiation and the atomic electron. It was first explained by Willis Eugene Lamb.

"Lamchem PE-130" [Vantage].   TM for a phosphate ester.
Use: As a PE additive lubricant and metal-working fluid.

lamepon.   An acetylated peptide used as a surface-active agent.

laminate.   A composite made of any one of several types of thermosetting plastic (phenolic, polyester, epoxy, or silicone) bonded to paper, cloth, asbestos, wood, or glass fiber. High tensile and dielectric strength and low moisture absorption are characteristic of these products. Available as sheet, rod, or tubing in mechanical, electrical, and general-purpose grades (National Electrical Manufacturers Association). Plywood is composed of a veneer with grain oriented at a 90-degree angle on successive layers and bonded with a thermosetting adhesive of the urea or phenol-formaldehyde type to give a high strength, dimensionally stable, weather-resistant construction material. It can be made nonflammable by treatment with salt solution. Polyvinyl butyral sheet is used in safety glass. See reinforced plastic.

L-amino acid.   An isomeric form of amino acid that occurs consistently among naturally occurring peptides.

"Lamitex" [Franklin].   TM for laminated fiber material.

lampblack.   A black or gray pigment made by burning low-grade heavy oils or similar carbonaceous materials with insufficient air and in a closed system such that the soot can be collected in settling chambers.
Properties: Markedly different from carbon black. Strongly hydrophobic. Nonflammable.
Use: Black pigment for cements, ceramic ware, mortar, inks, linoleum, surface coatings, crayons, polishes, carbon paper, soap, etc.; ingredient of insulating compositions, liquid-air explosives, matches, fertilizer, furnace lutes, lubricating compositions, carbon brushes; reagent in cementation of steel. See carbon black.

"Lanaset" [Huntsman].   TM for wool dyes with optimal fastness and minimal fiber damage.

land.   (1) The portion of a mold that provides the separation or cutoff of the flash from the molded article. (2) The surface along the top of the flights or ribs of the screw of an extruder. (3) In an extrusion die, the surface parallel to the flow of material.

"Lanette O" [BASF Care].   (cetearyl alcohol). TM for viscosity builder.
Use: For lotions, creams, and other cosmetic emulsions.

"Lanfrax" [Cognis].   TM for lanolin wax emulsifier, stabilizer, emollient, crystallization inhibitor, and film former.
Properties: Waxy solid.
Use: In creams lotions, make-up, lip products, and suntan preparations.

langbeinite. $K_2Mg_2(SO_4)_3$. A natural sulfate of potassium and magnesium found in salt deposits.
Properties: Colorless to yellowish, reddish greenish; luster vitreous. Mohs hardness 3.5–4, d 2.83.
Occurrence: New Mexico, Germany, India.
Use: Source of potash.

Langmuir, I.   (1881–1957). An American physical chemist who was awarded the Nobel Prize in 1932 for his fundamental research in surface chemistry, especially monomolecular films. This led to development of modern knowledge of emulsification and detergency. Langmuir also investigated electrical discharges in gases and did pioneer work on cloud-seeding techniques.

"Lannate" [Du Pont].   TM for methomyl insecticides.
See methomyl.

"Lanogene" [Lubrizol].   TM for liquid lanolin consisting of low-melting point esters. A crystal-clear, amber, viscous liquid; anhydrous. Soluble in mineral and vegetable oils.
Use: Nonionic without emulsifier; pigment dispersant; plasticizer; emollient.

lanolin.
Properties: (Hydrous): Yellowish to gray semisolid containing 25–30% water. Slight odor. (Anhydrous): Brownish-yellow semisolid containing no more than 0.25% water but can be mixed with about twice its weight of water without separation.
Derivation: Purification of degras, a crude grease obtained by solvent-treatment of wool. Contains cholesterol esters of higher fatty acids. Hydrogenated, ethoxylated, and acetylated derivatives are available.
Grade: Technical, cosmetic, USP.

Use: Ointments, leather finishing, soaps, face creams, facial tissues, hair-set, and suntan preparations.
See degras.

lanolin, anhydrous.
Properties: Yellow-white semisolid. Insoluble in water; soluble in chloroform, ether.
Use: Food additive.

lanosterol.
(isocholesterol). $C_{30}H_{50}O$. An unsaturated sterol closely related to cholesterol.
Properties: Optically active crystals. Mp 139–140C.

lanthana.
See lanthanum oxide.

lanthanide series.
(lanthanoid series). The rare earth series of elements, atomic numbers 58 through 71.
See rare earth.

lanthanum.
CAS: 7439-91-0. La.
Metallic element of atomic number 57, group IIIB of the periodic table, a rare earth of the cerium group, aw 138.9055, valence = 3; two stable isotopes.
Properties: White, malleable ductile metal. Oxidizes rapidly in air. D 6.18–6.19, mp 920C, bp 3454C. Corrodes in moist air. Soluble in acids; decomposes water to lanthanum hydroxide and hydrogen. Superconducting at approximately 6K.
Derivation: By cracking of monazite or bastnasite ores with concentrated sulfuric acid and subsequent separation.
Available forms: Ingots, rods, 20-mil sheets, powdered, 99.9% pure.
Hazard: Ignites spontaneously in powdered form. May delay blood clotting and cause liver injury upon ingestion.
Use: Lanthanum salts, electronic devices, pyrophoric alloys, rocket propellants, reducing agent catalyst for conversion of nitrogen oxides to nitrogen in exhaust gases (usually in combination with cobalt, lead, or other metals), phosphors in X-ray screens.
See rare-earth metal.

lanthanum acetate.
$La(C_2H_4O_2)_3 \cdot xH_2O$.
Properties: White powder. Soluble in water and acids.
Grade: Purities up to 99.9+%.

lanthanum ammonium nitrate.
Properties: Colorless crystals. Soluble in water.
Grade: Purities to 99.9+%.
Hazard: Oxidizer; explosion and fire risk.

lanthanum antimonide.
LaSb. A binary semiconductor.

lanthanum arsenide.
LaAs. Made in high purity for use as a binary semiconductor.
Hazard: Highly toxic.

lanthanum carbonate.
$La_2(CO_3)_3 \cdot 8H_2O$.
Properties: White powder. D 2.6. Insoluble in water; soluble in acids.
Grade: Up to 99.9+% lanthanum salts.

lanthanum chloranilate.
$La_2(O:C_6Cl_2O_2:O)_3 \cdot xH_2O$.
Use: Reagent for fluoride determination.

lanthanum chloride.
CAS: 10099-58-8. $LaCl_3 \cdot 7H_2O$.
Properties: White crystals, transparent, hygroscopic. Mp (hydrate) 91C (decomposes); (anhydrous) d 3.842 (25C), mp 87.C; soluble in alcohol, water, acids.
Derivation: Treatment of lanthanum carbonates or oxides with hydrochloric acid in an atmosphere of dry hydrogen chloride.
Grade: Purities to 99.9+%, single crystals available.
Use: Anhydrous trichloride of rare-earth metal is often used to prepare the metal.

lanthanum fluoride.
LaF_3.
Properties: White powder. Insoluble in water, acids.
Grade: Purities up to 99.9+%, single crystals available.
Hazard: Toxic by ingestion.
Use: Phosphor lamp coating (gallium arsenide solid-state lamp), carbon arc electrodes, lasers.

lanthanum nitrate.
CAS: 10099-59-9. $La(NO_3)_3 \cdot 6H_2O$.
Properties: White crystals, hygroscopic. Bp 126C, mp 40C. Soluble in alcohol, water, acids.
Grade: Purities to 99.9+%.
Hazard: Explosion and fire risk.
Use: Antiseptic, gas mantles.

lanthanum oxalate.
$La_2(C_2O_4)_3 \cdot 9H_2O$.
Properties: White powder. Insoluble in water; soluble in acids.
Grade: Purities to 99.9+%.

lanthanum oxide.
(lanthana; lanthanum trioxide; lanthanum sesquioxide). La_2O_3.
Properties: White or buff amorphous powder. D 6.51 (15C), mp 2315C, bp 4200C. Soluble in acids; insoluble in water.

Derivation: Ignition of hydroxide or oxyacid (oxalate, sulfate nitrate, etc.), direct combustion of free metal (burns with brilliant, white light).
Grade: Purities to 99.9+%.
Use: Calcium lights, optical glass, technical ceramics, cores for carbon-arc electrodes, fluorescent phosphors, refractories.

lanthanum phosphide.   LaP. Made in high purity for use as binary semiconductor.

lanthanum sesquioxide.   See lanthanum oxide.

lanthanum sulfate. $La_2(SO_4)_3 \cdot 9H_2O$.
Properties: White crystals. D 2.821, refr index 1.564 (20C). Insoluble in alcohol; slightly soluble in water, acids.
Derivation: By dissolving hydroxide, carbonate, or oxide in dilute sulfuric acid.
Grade: Purities to 99.9+%.
Use: The sulfates of the rare-earth elements are often used for atomic weight determination of the element.

lanthanum trioxide.   See lanthanum oxide.

lanthionine. $S(CH_2CHNH_2COOH)_2$. A nonessential amino acid first obtained from deaminated wool.

"Lantrol" [Cognis].   TM for lanolin oil emollient and moisturizer.
Properties: Liquid.
Use: In make-up as a pigment dispersant and cream, lotions, hair care products, bath oils, and medicinal preparations.

lard.   Purified internal fat of the hog.
Properties: Soft, white, unctuous mass; faint odor; bland taste. Mp 36–42C. High in saturated fats. Soluble in ether, chloroform, light petroleum hydrocarbons, carbon disulfide; insoluble in water. Combustible.
Chief constituents: Stearin, palmitin, olein.
Use: Cooking, pharmacy (ointments, cerates), perfumery (pomades), source of lard oil.

lard oil.
Properties: Colorless or yellowish liquid; peculiar odor; bland taste. Mp −2C, refr index 1.470 (20C), d 0.915, saponification value 195–196, iodine value 56–74, subject to spontaneous heating, flash p 420F (215C), autoign temp 883F (472C). Soluble in benzene, ether, chloroform, and carbon disulfide; slightly soluble in alcohol. Combustible.
Chief constituents: Olein with a small percentage of the glycerides of solid fatty acids.
Derivation: By cold-pressing lard.
Grade: Prime winter edible, prime winter inedible, antibiotic, off prime, extra #1, #1, #2.
Use: Lubricant, metal cutting compounds, oiling wool, soap manufacture, antibiotic fermentation.

lard (unhydrogenated).
Properties: Whitish fat rendered from pork fat. Mp 42°C.
Use: Food additive.

larva.   Among invertebrates, an immature stage in the life cycle which usually is much smaller than, and morphologically different from, the adult. In insects with metamorphosis, the larva must become a pupa before reaching adulthood.

larvicide.   A chemical agent that kills the eggs of insects. Examples are p-dichlorobenzene, chloropicrin, and copper acetoarsenite.

LAS.   See alkyl sulfonate, linear.

laser.   A device that produces a beam of coherent or monochromatic light as a result of photon-stimulated emission. Such beams have extremely high energy, because they consist of a single wavelength and frequency. Laser is an acronym for light amplification by stimulated emission of radiation.
Materials capable of producing this effect are certain high-purity crystals (ruby, yttrium garnet, and metallic tungstates or molybdates doped with rare-earth ions); semiconductors such as gallium arsenide, neodymium-doped glass; various gases, including carbon dioxide, helium, argon, and neon; and plasmas. A chemical laser is one in which the excitation energy is furnished by a chemical reaction, for example, $H + Cl_2 \rightarrow HCl$ (active) + Cl; or combustion of carbon monoxide to form excited carbon dioxide.
Hazard: Laser radiation can irreparably damage the eyes. Proper shielding is essential at all times.
Use: Laser beams are used in industry for cutting diamonds that are used for wire-drawing dies, and in flash photolysis, spectroscopy, and photography. They also have developing applications in medicine and surgery. They are being used for controlled fusion reactions, for biomedical investigations, for organic chemical research, for sophisticated analytical techniques, and in three-dimensional photography (holography). It is possible to increase the abundance of certain isotopes of such elements as uranium, chlorine, and boron by use of laser irradiation. Research on uranium enrichment by this method has been under way for several years.
See fusion (2); enrichment; holography.

laser-induced nuclear polarization.   A technique for making the spin vectors of an assembly of nuclei point preferentially in one direction by means of an optical pumping process using either circularly or linearly polarized laser light.

laser spectroscopy.   A branch of spectroscopy in which a laser is used as an intense

monochromatic light source. In particular it includes saturation spectroscopy as well as the application of laser sources to Raman spectroscopy and other techniques.
See spectroscopy.

LATB. See lithium aluminum tri-*tert*-butoxy-hydride.

latent heat. The quantity of energy in calories per gram absorbed or given off as a substance undergoes a change of state, that is, as it changes from liquid to solid (freezes), from solid to liquid (melts), from liquid to vapor (boils), or from vapor to liquid (condenses). No change in temperature occurs. Water has unusually high latent heat values; the latent heat of fusion (melting) of ice is 80 cal/g and the latent heat of condensation of steam (latent heat of vaporization of water) is 540 cal/g. The considerable energy delivered by steam condensation is utilized for power generation and for heating a variety of chemical plant equipment (dryers, evaporators, reactors, and distillation columns).
See evaporation; heat.

latent solvent. See solvent, latent.

laterite. A low-grade ore similar to bauxite, but containing only half as much aluminum oxide. Possible substitute for bauxite.

latex. A white, tacky, aqueous suspension of a hydrocarbon polymer occurring naturally in some species of trees, shrubs, or plants, or made synthetically. The most important natural latex is that of the tropical tree *Hevea brasiliensis*, which was the only source of rubber up to 1945. It is composed of globules of rubber hydrocarbon coated with protein; the particles are of irregular shape, varying from 0.5 to 3 microns in diameter; the suspension is stabilized by electric charges. The composition is about 60% water, 35% hydrocarbon, 2% protein, and low percentages of sugars and inorganic salts. For commercial purposes, rubber latex can be concentrated by evaporation or centrifugation. Ammonia is added as a preservative. Coagulation is induced by addition of acetic or formic acid. A vulcanized form is available. Natural latex is used in the manufacture of thin articles (surgeons' gloves and other medical equipment), as an adhesive, in foamed products, and for coating various products such as tire cord. Conversion of latex to gasoline via zeolite catalysts has been reported.
Other sources of rubber-containing latex are guayule, a shrub grown in Mexico and the southwestern U.S., and several types of dandelions and related species. The botanical function of latex in the plant is unknown.
Synthetic latexes are made by emulsion polymerization techniques from styrene-butadiene copolymer,

acrylate resins, polyvinyl acetate, and similar materials. Their particle size is much smaller than in natural latex, ranging from 0.05 to 0.15 micron; thus, they are truly colloidal suspensions. Their chief use is as a binder in exterior and interior paints, replacing drying oils; they are also used for foams and coatings.
See guayule; gutta percha; electrophoresis; paint; emulsion; "Vultex."

latex paint. See paint; emulsion.

"Latron AG-98" *[J. R. Simplot].* TM for alkylaryl polyoxyethylene glycols.
Use: As a spray adjuvant for agricultural pesticides.

lattice. (1) The structural arrangement of atoms in a crystal. Accurate information is obtained by X-rays, which are diffracted by the lattice at various angles. As the atoms are from 1.5 to 3 Å apart in most crystals, the lattice acts as a diffraction grating.
See crystals; dislocation.
(2) The array of nuclear fuel elements and moderator in a nuclear reactor.

Latyl. A group of disperse dyes developed particularly for coloration of "Dacron" polyester fiber, on which they have exceptionally good light- and wetfastness properties.

laudanidine. (*l*-laudanine; tritopine). $C_{20}H_{25}O_4N$. An alkaloid.
Properties: White crystals. Mp 182–185C. Insoluble in water; soluble in alcohol, benzene, chloroform, and slightly soluble in ether.
Derivation: From opium.
Hazard: Toxic by ingestion.
Use: Medicine (analgesic).

laudanum. Tincture of opium (tinctura opii BP).

laughing gas. See nitrous oxide.

laundry sour. See sour.

lauraldehyde. See lauryl aldehyde.

lauramidopropyl betaine.
CAS: 86438-78-0.
Hazard: An eye irritant.

Laurent's acid. (1-naphthylamine-5-sulfonic acid).

Properties: White or pink needles, gives greenish fluorescence in dilute aqueous solution.

Derivation: From naphthalene-α-sulfonic acid by nitration, reduction with iron, and separation from the 1-naphthylamine-8-sulfonic acid also formed.
Use: Azo dye intermediate.

Laurent's α-acid.   (1-nitronaphthalene-5-sulfonic acid). $C_{10}H_6(NO)_2(SO_3H)$.
Properties: Pale-yellow needles. Soluble in water, alcohol, and ether. Combustible.
Derivation: By sulfonating nitronaphthalene with a mixture of chlorohydrin and sulfuric acid.
Use: Azo dye intermediate.

"Laurex" *[Huntsman].*   TM for the zinc salts of a mixture of fatty acids in which lauric acid predominates.
Properties: Yellowish-white granulated waxy powder. D 1.15, mp 95–105C. Soluble in benzene; insoluble in acetone, gasoline, ethylene dichloride, and water. Combustible.
Use: Accelerator, activator, and plasticizer for rubber.

lauric acid.   (dodecanoic acid).
CAS: 143-07-7. $CH_3(CH_2)_{10}COOH$.
A fatty acid occurring in many vegetable fats as the glyceride, especially in coconut oil and laurel oil. Combustible.
Properties: Colorless needles. D 0.833, mp 44C, bp 225C (100 mm Hg), refr index 1.4323 (45C). Insoluble in water; soluble in benzene and ether.
Derivation: Fractional distillation of mixed coconut or other acids.
Grade: 99.8% pure, technical, FCC.
Use: Alkyd resins, wetting agents, soaps, detergents, cosmetics, insecticides, food additives.

lauric aldehyde.   See lauryl aldehyde.

laurone.   An aliphatic ketone, insoluble in water, stable to high temperatures, acids, alkalies. Compatible with high-melting vegetable waxes, fatty acids, paraffins, etc. Incompatible with resins, polymers, and organic solvents at room temperature, but compatible with them at high temperature.
Use: As antiblock agent.

N-lauroyl-p-aminophenol.
$HO(C_6H_4)NHCOCH_2(CH_2)_9CH_3$.
Properties: White to off-white powder. Mp 123–126C. Insoluble in water; soluble in polar organic solvents (especially when heated), including alcohol, acetone, and dimethylformamide.
Use: Rubber antioxidant.

lauroyl chloride. $C_{11}H_{23}COCl$.
Properties: Water-white liquid. Refr index 1.445 (20C), fp −17C, bp 145C (18 mm Hg). Decomposes in water and alcohol, soluble in ether.

Use: Surfactant, polymerization initiator, antienzyme agent, foamer; synthesis of lauroyl peroxide, sodium *N*-lauroyl sarcosinate, and other sarcosinates.

(lauroyloxy)tributylstannane.   See tributyltin laurate.

lauroyl peroxide.   (dodecanoyl peroxide).
CAS: 105-74-8. $(C_{11}H_{23}CO)O_2$.
Properties: White, coarse powder; tasteless; faint odor. Mp 49C. Soluble in oils and in most organic solvents. Slightly soluble in alcohols. Insoluble in water.
Grade: Technical (about 95%).
Hazard: Dangerous fire and explosion risk, will ignite organic materials, strong oxidizer. Toxic by ingestion and inhalation, strong irritant to skin. Questionable carcinogen.
Use: Bleaching agent, intermediate and drying agent for fats, oils, and waxes; polymerization catalyst.

N-lauroylsarcosine. $CH_3(CH_2)_{10}CON(CH_3)$ CH_2COOH.
Properties: White solid. Mp 31–35C, d 0.970.
Use: Surfactant, antienzyme, in cosmetics and pharmaceuticals. Also used in form of sodium *N*-lauroyl sarcosinate.

lauryl acetate.   See dodecyl acetate.

lauryl alcohol.   (alcohol C12; *n*-dodecanol; dodecyl alcohol).
CAS: 112-53-8. $CH_3(CH_2)_{10}CH_2OH$.
Properties: Colorless solid; fatty repulsive odor. D 0.830–0.836, refr index 1.428, mp 24C, bp 259C, flash p above 212F (100C) (CC). Insoluble in water; soluble in two parts of 70% alcohol. Combustible.
Derivation: Reduction of coconut-oil fatty acids.
Grade: Technical, FCC.
Use: Synthetic detergents, lube additives, pharmaceuticals, rubber, textiles, perfumes; flavoring agent.

lauryl aldehyde.   (lauric aldehyde; dodecyl aldehyde; aldehyde C-12 lauric; dodecanal; lauraldehyde).
CAS: 112-54-9. $CH_3(CH_2)_{10}CHO$.
Properties: Colorless solid or liquid; strong fatty floral odor. D 0.828–0.836, refr index 1.433–1.440, mp 44C. Soluble in 90% alcohol; insoluble in water. Combustible.
Grade: Technical, FCC.
Use: Perfumery, flavoring agent.

lauryl bromide.   (*n*-dodecyl bromide; 1-bromododecane). $C_{12}H_{25}Br$.
Properties: Amber liquid; coconut odor. Low volatility, d 1.026 (25/25C), bp 175C (45 mm Hg), fp −15.5C, flash p 291F (144C). Insoluble in water; soluble in alcohol, ether. Combustible.

Grade: Technical, approximately 60% pure.
Derivation: Coconut oil.
Use: Intermediate for quaternary ammonium compounds, organometallics, and vinyl stabilizers.

lauryl chloride.
Properties: Commercially a mixture of *n*-alkyl chlorides, with $C_{12}H_{25}Cl$ dominant. A clear, water-white, oily liquid; faint fatty odor. Crystallization point −19C, d 0.863 (15.5/15.5C), distillation range 112–160C (5 mm Hg), flash p 235F (112C). Completely miscible with most organic solvents; slightly miscible with alcohol; immiscible with water. Combustible.
Grade: Refined, technical.
Use: Synthesis of esters, sulfides, lauryl mercaptan (used in styrene-butadiene polymerization), other organics.

lauryldimethylamine.
CAS: 112-18-5. $CH_3(CH_2)_{11}N(CH_3)_2$.
A liquid cationic detergent.
Use: Corrosion inhibitor, acid-stable emulsifier.

lauryldimethylamine oxide.
CAS: 1643-20-5. $CH_3(CH_2)_{11}N(CH_3)_2 \cdot H_2O$.
A nonionic detergent.
Use: As a foam stabilizer; stable at high concentration of electrolytes and over a wide pH range.
See Admox.

lauryl lactate.   See "Ceraphyl" *[ISP]*.

lauryl mercaptan.   (*n*-dodecyl mercaptan).
CAS: 112-55-0. $C_{12}H_{25}SH$ (approximately).
Properties: (Technical material, mixture of isomers): Water-white or pale-yellow liquid; mild odor. D 0.85 (20/20C), fp −7.5C, distillation range 200–235C, refr index 1.45–1.47, flash p 210F (99C). Insoluble in water; soluble in methanol, ether, acetone, benzene, gasoline, and ethyl acetate. Combustible.
Grade: 95% min.
Hazard: May be injurious to eyes.
Use: Manufacture of synthetic rubber and plastics, also in the synthesis of pharmaceuticals and in insecticides and fungicides; nonionic detergent.
See thiol.

lauryl methacrylate.
CAS: 142-90-5. $CH_2{:}C(CH_3)COO(CH_2)_{11}CH_3$.
The commercial material is a mixture containing both lower and higher fatty derivatives.
Properties: Boiling range 272–344C, bulk d 0.868 g/mL, flash p 270F (132C) (COC). Combustible.
Use: Polymerizable monomer for plastics, molding powders, solvent coatings, adhesives, oil additives; emulsions for textile, leather, and paper finishing.
See acrylic resin.

lauryl pyridinium chloride. $C_5H_5NClC_{12}H_{25}$.
Properties: Mottled-tan semisolid. Soluble in water and organic solvents. Flash p 347F (175C). Combustible.
Grade: Technical, contains higher and lower fatty acid derivatives.
Use: Cationic detergent, dispersing and wetting agent, ingredient of fungicides and bactericides.

lauryltrimethylammonium chloride.
CAS: 112-00-5. $C_{15}H_{34}N \cdot Cl$.
Hazard: A poison by ingestion.

lautal.   A hard aluminum alloy containing 4–5% copper, 1.5–2% silicon, and fractional percentages of other metals such as iron, manganese, or magnesium.

lava.   Molten rock (magma) that has reached the surface of the earth.

lavandin oil.   See lavender oil.

lavender oil.
CAS: 8000-28-0. An essential oil used in perfumery, 35% ester content as linalyl acetate required. Terpeneless grade has about twice the concentration of the natural oil.

Lavoisier, Antoine Laurent.   (1743–1794). A French chemist generally regarded as the "father" of chemistry. His "Traite elementaire de chimie" (1789) listed 30 elements, clarified the nomenclature of acids, bases, and salts, and described the composition of numerous organic substances. He erroneously believed that oxygen is the characteristic element of acids. However, his fundamental work on combustion, as a result of which he identified and named nitrogen (azote), and on the separation of hydrogen from water by a unique reduction experiment carried out in a heated gun barrel, earned him a leading position among early chemists.
See Mendeleyev.

law of mass action.   The law stating that the rate of any given chemical reaction is proportional to the product of the concentrations of the reactants. At extremely low concentrations, activities are used.

Lawrence, Ernest O.   (1901–1958). An American physicist who invented the cyclotron in 1929. Both the element lawrencium and the Lawrence Livermore Research Laboratory at the University of California were named after him.
See cyclotron; bombardment.

lawrencium.   Lr. A synthetic radioactive element with atomic number 103, discovered in 1961, aw 257, only one other isotope is known (256); the 257 isotope has a half-life of 8 sec. It has been made by bombarding californium with boron ions. It exhibits α radiation.
See actinide series.

lay-up.   In the reinforced-plastics industry, a term used to refer to placement of the reinforcing material in the mold.

L-64 Barrier Foaming Quat.   A drain and floor protection system.
Use: Floor cleaner for the food industry.

LC$_{50}$.   (lethal concentration 50%). That quantity of a substance administered by inhalation that is necessary to kill 50% of test animals exposed to it within a specified time. The test applies not only to gases and vapors but to fumes, dusts, and other particulates suspended in air.

LCL.   Abbreviation for "less than carload lot," used by shippers, traffic managers, railroads, etc.

LD$_{50}$.   (lethal dose 50%). That quantity of a substance necessary to kill 50% of exposed animals in laboratory tests within a specified time. A substance having an oral LD$_{50}$ of less than 400 mg/kg of body weight is considered to be highly toxic.

LDPE.   Abbreviation for low-density polyethylene.

leaching.   See solvent extraction.

lead.   (from Latin *plumbum*).
CAS: 7439-92-1.   Pb.
Metallic element of atomic number 82, group IVA of the periodic table, aw 207.2, valences = 2,4, four stable isotopes. The isotopes are the end products of the disintegration of three series of natural radioactive elements uranium (206), thorium (208), and actinium (207).
Properties: Heavy, ductile, soft, gray solid. D 11.35, mp 327.4C, bp 1755C. Soluble in dilute nitric acid; insoluble in water but dissolves slowly in water containing a weak acid. Resists corrosion; relatively impenetrable to radiation. Poor electrical conductor, good sound and vibration absorber. Noncombustible.
Occurrence: U.S., Mexico, Canada, South America, Australia, Africa, Europe.
Derivation: Roasting and reduction of galena (lead sulfide), anglesite (lead sulfate), and cerussite (lead carbonate). Also from scrap.
Method of purification: Desilvering (Parkes process), electrolytic refining (Betts process), pyrometallurgical refining (Harris process). Bismuth is removed by Betterton–Kroll process.
Grade: High purity (less than 10 ppm impurity), pure (99.9+%), powdered (99% pure), pig lead, paste.
Available forms: Ingots, sheet, pipe, shot, buckles or straps, grids, rod, wire, etc.; paste; powder; single crystals.

Hazard: Toxic by ingestion and inhalation of dust or fume. For ambient air the EPA standard is 1.5 μg/m^3. A cumulative poison. FDA regulations require zero lead content in foods and less than 0.05% in house paints. Probable carcinogen.
Use: Storage batteries, tetraethyllead (gasoline additive), radiation shielding, cable covering, ammunition, chemical reaction equipment (piping, tank linings, etc.), solder and fusible alloys, type metal, vibration damping in heavy construction, foil, Babbitt and other bearing alloys.

lead acetate.   (sugar of lead).
CAS: 301-04-2.   Pb(C$_2$H$_3$O$_2$)$_2$•3H$_2$O.
Properties: White crystals or flakes (commercial grades are frequently brown or gray lumps); sweetish taste. D 2.50, mp loses water at 75C, at 200C decomposes, bp (anhydrous) 280C. Absorbs carbon dioxide when exposed to air, becoming insoluble in water. Soluble in water; slightly soluble in alcohol; freely soluble in glycerol. Combustible.
Derivation: By the action of acetic acid on litharge or thin lead plates.
Grade: Powdered, granular, crystals, flakes, CP.
Hazard: Toxic by ingestion, inhalation, and skin absorption; use may be restricted.
Use: Dyeing of textiles, waterproofing, varnishes, lead driers, chrome pigments, gold cyanidation process, insecticide, antifouling paints, analytical reagent, hair dye.

lead alkyl, mixed.   A mixture containing various methyl and ethyl derivatives of tetraethyl lead and tetramethyl lead. Thus, methyl triethyl lead, dimethyl diethyl lead, and ethyl trimethyl lead may all be present with or without tetraethyl and tetramethyl lead.
Hazard: Toxic by ingestion and skin absorption.
Use: Antiknock agents in aviation gasoline.

lead antimonate.   (Naples yellow; antimony yellow). Pb$_3$(SbO$_4$)$_2$.
Properties: Orange-yellow powder; insoluble in water. D 6.58 (20C). Noncombustible.
Derivation: Interaction of solutions of lead nitrate and potassium antimonate, concentration, and crystallization.
Hazard: Toxic by inhalation.
Use: Staining glass, crockery, and porcelain.

lead arsenate.   (lead orthoarsenate).
CAS: 7784-40-9.   Pb$_3$(AsO$_4$)$_2$.
Properties: White crystals. D 5.8, mp 1042C (decomposes). Soluble in nitric acid; insoluble in water.
Derivation: By the action of a soluble lead salt on a solution of sodium arsenate, concentration, and crystallization.
Hazard: Highly toxic. A confirmed carcinogen.
Use: Insecticide, herbicide.

lead arsenite.
CAS: 10031-13-7. $Pb(AsO_2)_2$.
Properties: White powder. D 5.85. Soluble in nitric acid; insoluble in water.
Hazard: Highly toxic.
Use: Insecticide.

lead azide.
CAS: 13424-46-9. $Pb(N_3)_2$.
Properties: Colorless, very sensitive needles. An initiating explosive. Should always be handled submerged in water.
Derivation: Reaction of sodium azide with a lead salt.
Hazard: Severe explosion risk, detonates at 350C (660F). Highly toxic.
Use: Primary detonating compound for high explosives.
Note: Explosions have occurred in cases where azide compounds have reacted with the lead in plumbing after being washed down sinks.

lead-base Babbitt.   See Babbitt metal.

lead biorthophosphate.   See lead phosphate, dibasic.

lead, blue.   A term applied to galena to distinguish it from white lead ore. It is also applied to blue basic lead sulfate.

lead borate. $Pb(BO_2)_2 \cdot H_2O$.
Properties: White powder. D 5.598, mp 160C (loses water). Soluble in dilute nitric acid; insoluble in water. Noncombustible.
Derivation: Interaction of solutions of lead hydroxide and boric acid with subsequent crystallization.
Hazard: Toxic by inhalation.
Use: Varnish and paint drier, waterproofing paints, lead glass, electrically conductive ceramic coatings.

lead borosilicate.   A constituent of optical glass, composed of a mixture of the borate and silicate of lead.

lead bromate.
CAS: 34018-28-5. $Pb(BrO_3)_2 \cdot H_2O$.
Properties: Colorless crystals. D 5.53, decomposes at about 180C. Soluble in hot water.
Hazard: Toxic by inhalation or ingestion.

lead bromide. $PbBr_2$.
Properties: White powder. D 6.66, bp 916C, mp 373C. Slightly soluble in hot water; insoluble in alcohol.
Hazard: Toxic by inhalation.

lead carbolate.   See lead phenate.

lead carbonate. $PbCO_3$.
See lead carbonate, basic.

lead carbonate, basic.   (lead subcarbonate; white lead; lead flake). $2PbCO_3 \cdot Pb(OH)_2$.
Properties: White amorphous powder. D 6.86. Soluble in acids; insoluble in water, decomposes at 400C. Noncombustible.
Derivation: (1) Dutch process. By the corrosion of lead buckles in pots by acetic acid and carbon dioxide generated by the fermentation of waste tanbark. (2) Carter process. By treating very finely divided lead in revolving wooden cylinders with dilute acetic acid and carbon dioxide.
Hazard: Toxic by inhalation.
Use: Exterior paint pigment, ceramic glazes.

lead chloride.
CAS: 7758-95-4. $PbCl_2$.
Properties: White crystals. D 5.88, mp 498C, bp 950C. Slightly soluble in hot water; insoluble in alcohol and cold water. Noncombustible.
Derivation: By the addition of hydrochloric acid or sodium chloride to a solution of a lead salt with subsequent crystallization.
Hazard: Toxic by inhalation.
Use: Preparation of lead salts, lead chromate pigments, analytical reagent.

lead chromate.   (chrome yellow).
CAS: 7758-97-6. $PbCrO_4$.
Properties: Yellow crystals. D 6.123, mp 844C. Soluble in strong acids and alkalies; insoluble in water.
Derivation: Reaction of sodium chromate and lead nitrate in solution.
Hazard: Toxic by ingestion and inhalation; probable carcinogen.
Use: Pigment in industrial paints, rubber, plastics, ceramic coatings; organic analysis.
See chrome pigment.

lead coating.   Coatings of lead or lead-rich alloys are (1) deposited by dipping into the molten metal after applying a layer of tin to secure good adhesion of the lead coating, (2) by electroplating from a fluosilicate or fluoborate bath, or (3) by spraying.

lead cyanide.
CAS: 592-05-2. $Pb(CN)_2$.
Properties: White to yellowish powder. Slightly soluble in water; decomposes in acid.
Derivation: Interaction of solutions of potassium cyanide and lead acetate.
Hazard: Toxic by ingestion, inhalation, and skin absorption.
Use: Metallurgy.

lead dimethyldithiocarbamate.
CAS: 19010-66-3. $Pb[SCSN(CH_3)_2]_2$.
Properties: White powder. D 2.43, melting range 310C. Insoluble in all common organic solvents; slightly soluble in cyclohexanone.
Use: Vulcanization accelerator with litharge.

lead dioxide.   (lead oxide, brown; plumbic acid, anhydrous; lead peroxide; lead superoxide).
CAS: 1309-60-0. PbO_2.
Properties: Brown, hexagonal crystals. D 9.375, mp 290C (decomposition), an oxidizing agent. Soluble in glacial acetic acid; insoluble in water and alcohol.
Derivation: By adding bleaching powder to an alkaline solution of lead hydroxide.
Hazard: Dangerous fire risk in contact with organic materials.
Use: Oxidizing agent, electrodes, lead-acid storage batteries, curing agent for polysulfide elastomers, textiles (mordant, discharge in dyeing with indigo), matches, explosives, analytical reagent.
Use: Oxidizing agent, electrodes, lead-acid storage batteries, curing agent for polysulfide elastomers, textiles (mordant, discharge in dyeing with indigo), matches, explosives, analytical reagent.

lead dross.   (lead scrap). Consists of the scrap, dross, or waste from sulfuric acid tanks; a mixture of metallic lead, lead sulfate, and free sulfuric acid.

lead, electrolytic.   Pure lead obtained by electrolytic deposition.
See Betts process.

leader.   A short sequence near the amino terminus of a protein or the 5a′ end of an RNA molecule that has a specialized targeting or regulatory function. Often the leader is cleaved after the targeting event.

lead flake.   See lead carbonate, basic.

lead fluoborate.
CAS: 13814-96-5. $B_2F_8 \cdot Pb$.
Properties: Liquid. Mw 380.81.
Grade: Technical 51%.
Use: Salt for electroplating lead; can be mixed with stannous fluoborate to electroplate any composition of tin and lead as an alloy.

lead fluoride.
CAS: 7783-46-2. PbF_2.
Properties: Colorless crystals. D 8.2 g/cc, mp (approximately) 824C. Very slightly soluble in water. Noncombustible.
Grade: Crystals, 99.93%.
Hazard: Toxic. Strong irritant.
Use: Electronic and optical applications, starting materials for growing single crystal solid-state lasers, high-temperature dry film lubricants in the form of ceramic-bonded coatings.

lead fluosilicate.   (lead silicofluoride).
CAS: 25808-74-6. $PbSiF_6 \cdot 2H_2O$.
Properties: Colorless crystals. Soluble in water; decomposes when heated.
Hazard: Toxic. Strong irritant.
Use: Solution for electrorefining lead.

lead formate. $Pb(CHO_2)_2$.
Properties: Brownish-white lustrous needles. D 4.63. Soluble in water; decomposes at 190C. Noncombustible.
Hazard: Toxic by ingestion.
Use: Reagent in analytical determinations.

lead glass.   See glass.

lead hydroxide.   (lead hydrate; hydrated lead oxide). $Pb(OH)_2$.
Properties: White, bulky powder. D 7.592, mp decomposes at 145C, absorbs carbon dioxide from air. Soluble in alkalies; slightly soluble in water; soluble in nitric and acetic acids. Noncombustible.
Derivation: By the addition of sodium or ammonium hydroxide to a solution of a lead salt with subsequent filtration and drying.
Hazard: Toxic material.
Use: Lead salts, lead dioxide.

lead hyposulfite.   See lead thiosulfate.

leading strand.   The DNA strand that, during replication, is synthesized in the same direction in which the replication fork moves. Synthesized in a continuous fashion.

lead iodide. PbI_2.
Properties: Golden-yellow crystals or powder; odorless. D 6.16, mp 402C, bp 954C. Soluble in potassium iodide and concentrated sodium acetate solutions; soluble in boiling water. Noncombustible.
Derivation: Interaction of lead acetate and potassium iodide.
Hazard: Toxic material.
Use: Bronzing, printing, photography, cloud seeding.

lead linoleate.   (lead plaster). $Pb(C_{18}H_{31}O_2)_2$.
Properties: Yellowish-white paste. Soluble in oils; insoluble in water. Combustible.
Derivation: By heating a solution of lead nitrate with sodium linoleate.
Grade: Technical, fused (contains 26.5% lead).
Hazard: Toxic material. Absorbed by skin.
Use: Medicine, drier in paints and varnishes.

lead maleate, tribasic. $C_4H_6O_5 \cdot Pb$.
Properties: Soft, yellowish-white, crystalline powder. D 6.3, refr index 2.08.
Hazard: Toxic material. Absorbed by skin.
Use: Vulcanizing agent for chlorosulfonated polyethylene. Highly basic stabilizer with high heat stability in vinyls.

lead metavanadate.   See lead vanadate.

lead molybdate. $PbMoO_4$.
Properties: Yellow powder. D 5.9, mp 1060–1070C. Soluble in nitric acid; insoluble in water and alcohol. Noncombustible.

Derivation: By adding a solution of lead nitrate to a solution of ammonium molybdate, concentration, and crystallization.
Hazard: Toxic material.
Use: Analytical chemistry, pigments.
Single crystals are available for electronic and optical uses.
See molybdate oranges.

lead monohydrogen phosphate.   See lead phosphate, dibasic.

lead mononitroresorcinate.
CAS: 51317-24-9. $PbO_2C_6H_3NO_2$.
Hazard: An initiating explosive, dangerous. Forbidden for transport.

lead monoxide.   See litharge, massicot.

lead β-naphthalenesulfonate. $Pb(C_{10}H_7SO_3)_2$.
Properties: White, crystalline powder. Soluble in alcohol; insoluble in water. Combustible.
Derivation: By the action of lead acetate on β-naphthalenesulfonic acid.
Hazard: Toxic material. Absorbed by skin.
Use: Organic preparations.

lead naphthenate.
CAS: 61790-14-5. $C_7H_{12}O_2 \cdot xPb$.
Properties: Soft, yellow, resinous, semitransparent. Mp approximately 0C. Gives deposits in highly acid oils but not when mixed with suitable quantities of cobalt or manganese; soluble in alcohol. Combustible.
Derivation: Addition of lead salt to aqueous sodium naphthenate solution.
Grade: liquid 16%, 24% lead, solid 37% lead.
Hazard: Toxic material. A known carcinogen (OSHA), absorbed by skin.
Use: Paint and varnish drier, wood preservative, insecticide, catalyst for reaction between unsaturated fatty acids and sulfates in the presence of air, lube oil additive.
See soap (2).

lead nitrate.
CAS: 10099-74-8. $Pb(NO_3)_2$.
Properties: White crystals. D 4.53, decomposes at 470C. Soluble in water and alcohol.
Derivation: By the action of nitric acid on lead.
Hazard: Strong oxidizing material, dangerous fire risk in contact with organic materials.
Use: Lead salts, mordant in dyeing and printing calico, matches, mordant for staining mother-of-pearl, oxidizer in the dye industry, sensitizer in photography, explosives, tanning, process engraving, and lithography.

lead nitrite, basic.   (basic lead nitrite; lead subnitrite).

Properties: Light-yellow powder, variable composition, essentially $3PbO \cdot N_2O \cdot H_2O$. Soluble in dilute nitric acid; easily decomposed.
Hazard: Toxic material.

lead ocher.   See massicot (1).

lead octoate.   See soap (2).

lead oleate.
CAS: 1120-46-3. $[CH_3(CH_2)_7CH:CH(CH_2)_7COO]_2Pb$.
Properties: White powder or ointment-like granules or mass. Soluble in alcohol, ether, turpentine, and benzene; insoluble in water. Combustible.
Derivation: Reaction of oleic acid with lead hydrate or carbonate, or of lead acetate and sodium oleate.
Hazard: Toxic material. Absorbed by skin.
Use: Varnishes, lacquers, paint drier, high-pressure lubricants.
See soap (2).

lead orthoarsenate.   See lead arsenate.

lead orthophosphate, normal.   See lead phosphate.

lead orthosilicate.   See lead silicate.

lead oxide, black.   See lead suboxide.

lead oxide, brown.   See lead dioxide.

lead oxide, hydrated.   See lead hydroxide.

lead oxide, red.   (red lead; minium; lead tetroxide).
CAS: 1314-41-6. Pb_3O_4.
Properties: Bright red powder. D reported variously 8.32–9.16, decomposes between 500 and 530C. Partly soluble in acids; insoluble in water. An oxidizing agent may react with reducing agents.
Derivation: By carefully heating litharge in a furnace in a current of air.
Grade: Technical, 95%, 97%, 98%.
Hazard: Toxic as dust.
Use: Storage batteries, glass, pottery and enameling, varnish, purification of alcohol, packing pipe joints, metal-protective paints, fluxes, and ceramic glazes.

lead oxide, yellow.   See litharge.

lead perchlorate.
CAS: 13637-76-8. $Pb(ClO_4)_2 \cdot 3H_2O$.
Properties: White crystals. D 2.6, mp 100C (decomposes). Very soluble in cold water; soluble in alcohol.
Hazard: Dangerous in contact with organic materials, strong oxidizing agent. Very toxic material.

lead peroxide.   See lead dioxide.

lead phenate.   (lead phenolate; lead carbolate). $Pb(OH)OC_6H_5$.
Properties: Yellowish to grayish-white powder. Soluble in nitric acid; insoluble in water and alcohol.
Derivation: By boiling phenol with litharge.
Hazard: Toxic material. Absorbed by skin.

lead phenolsulfonate.   (lead sulfocarbolate). $Pb(C_6H_4OHSO_3)_2 \cdot 5H_2O$.
Properties: White crystals or powder. Soluble in water and alcohol.
Hazard: Toxic material. Absorbed by skin.

lead phosphate.   (normal lead orthophosphate). CAS: 7446-27-7. $Pb_3(PO_4)_2$.
Properties: White powder. D 6.9–7.3, mp 1014C. Insoluble in water; soluble in acids and alkalies.
Hazard: Toxic material. Probable carcinogen.
Use: Stabilizing agent in plastics.

lead phosphate, dibasic.   (lead monohydrogen phosphate; lead biorthophosphate). $PbHPO_4$.
Properties: Soft, white powder or fine, plate-like crystals. D 5.66 (15C), mp decomposes.
Hazard: Toxic material.
Use: Imparting heat resistance and pearlescence to polystyrene and casein plastics.

lead phosphite, dibasic.
CAS: 1344-40-7. $2PbO \cdot PbHPO_3 \cdot 1/2H_2O$.
Properties: Fine, white, acicular crystals. D 6.94, refr index 2.25. Insoluble in water.
Hazard: Toxic material. Store in closed containers away from open flame or sparks and at temperatures not to exceed 400F.
Use: Heat and light stabilizer for vinyl plastics and chlorinated paraffins. As an UV screening and antioxidizing stabilizer for vinyl and other chlorinated resins in paints and plastics.

lead phthalate, dibasic. $C_6H_4(COO)_2Pb \cdot PbO$.
Properties: Fluffy white crystalline powder. D 4.5, refr index 1.99 (avg). Insoluble in water.
Derivation: By boiling litharge with phthalic acid.
Hazard: Toxic by inhalation and skin absorption.
Use: Heat and light stabilizer for general vinyl use.

lead pigment.   Any pigment that contains lead compounds.
Hazard: potentially toxic.

lead plaster.   See lead linoleate.

lead protoxide.   See litharge.

lead, red.   See lead oxide, red.

lead resinate. $Pb(C_{20}H_{29}O_2)_2$
Properties: Brown, lustrous, translucent lumps or yellow-white powder, or yellowish-white paste. Insoluble in most solvents. Combustible.
Derivation: By heating a solution of lead acetate and rosin oil.
Grade: Precipitated 23% lead.
Hazard: Toxic material. Absorbed by skin.
Use: Paint and varnish drier, textile waterproofing agent.

lead salicylate. $Pb(OOCC_6H_4OH)_2 \cdot H_2O$
Properties: Soft, creamy-white, crystalline powder. D 2.3, refr index 1.78. Soluble in hot water and alcohol. Combustible.
Use: Stabilizer or costabilizer for flooring and other vinyl compounds requiring good light stability.

lead selenide.   PbSe
Properties: Gray crystals. D 8.10 (15C), mp 1065C. Insoluble in water; soluble in nitric acid.
Hazard: Moderate fire hazard as dust or in presence of moisture.
Use: As a semiconductor in infrared detectors and thermoelectric devices.

lead sesquioxide. Pb_2O_3
Properties: Reddish-yellow powder. Soluble in alkalies and acids; insoluble in water; decomposes at 370C.
Derivation: By gently heating metallic lead.
Hazard: Toxic material.
Use: Ceramics, ceramic cements, metallurgy, varnishes.

lead silicate.   (lead metasilicate).
CAS: 10099-76-0. $PbSiO_3$.
Properties: White, crystalline powder. Insoluble in most solvents. Noncombustible.
Derivation: Interaction of lead acetate and sodium silicate.
Hazard: Toxic material.
Use: Ceramics, fireproofing fabrics.

lead silicate, basic.   (white lead silicate; lead silicate sulfate). A pigment made up of an adherent surface layer of basic lead silicate and basic lead sulfate cemented to silica.
Properties: Excellent film-forming properties with drying oils combined with low density.
Derivation: Fine silica is mixed with litharge and sulfuric acid. The mixture is then furnaced in a rotary kiln and ground to break up agglomerates.
Hazard: Toxic material.
Use: Pigment in industrial paints.

lead silicochromate.   A yellow, lead-silicon pigment. Normal lead silicon chromate is used as a yellow prime pigment for traffic marking paints. Basic lead silicon chromate is used as a corrosive

inhibitive pigment for metal protective coatings, primers and finishers. Also for industrial enamels requiring a high gloss.
Hazard: Toxic material.

lead silicofluoride.   See lead fluosilicate.

lead-soap lubricants.   Lead salts saponified with fats, hard at low temperatures, viscous at medium temperatures, but become somewhat fluid on heating by friction.
Use: "Extreme-pressure lubricants," but are not suited for high speeds.
See lead naphthenate; lead oleate; lead stearate; soap (2).

lead sodium hyposulfite.   See lead sodium thiosulfite.

lead sodium thiosulfate.   (lead sodium hyposulfite; sodium lead hyposulfite; sodium lead thiosulfate). $PbS_2O_3 \cdot 2Na_2S_2O_3$.
Properties: Heavy, small, white crystals. Soluble in solutions of thiosulfates.
Hazard: Toxic material.
Use: Matches.

lead stannate. $PbSnO_3 \cdot 2H_2O$.
Properties: Light-colored powder. Insoluble in water. Approximate temperature of dehydration 170C.
Hazard: Toxic material.
Use: Additive in ceramic capacitors, pyrotechnics.

lead stearate.
CAS: 7428-48-0. $Pb(C_{18}H_{35}O_2)_2$.
Properties: White powder. Mp 100–115C, d 1.4. Soluble in hot alcohol; insoluble in water. Combustible.
Derivation: By heating a solution of lead acetate with sodium stearate.
Hazard: Toxic material. Absorbed by skin.
Use: Varnish and lacquer drier, high-pressure lubricants, lubricant in extrusion processes, stabilizer for vinyl polymers, corrosion inhibitor for petroleum, component of greases, waxes, and paints.

lead styphnate.   Legal label name for lead trinitroresorcinate.

lead subacetate.
CAS: 1335-32-6. $(2Pb(OH)_2Pb(C_2H_3O_2)_2$.
Properties: White, heavy powder. Mw 807.75. Partially soluble in cold water; more soluble in hot water; absorbs atmospheric carbon dioxide becoming almost insoluble.
Hazard: Suspected carcinogen and poisonous. Toxic by ingestion.
Use: Decolorizing agent (sugar solutions, etc.).

lead subcarbonate.   See lead carbonate, basic.

lead subnitrite.   See lead nitrite.

lead suboxide.   (lead oxide, black; litharge, leaded). Pb_2O.
Properties: Black amorphous material. D 8.342, decomposes on heating. Insoluble in water; soluble in acids and bases.
Hazard: Toxic material.
Use: In storage batteries.

lead, sugar of.   See lead acetate.

lead sulfate.
CAS: 7446-14-2. $PbSO_4$.
Properties: White, rhombic crystals. D 6.12–6.39, mp 1170C. Slightly soluble in hot water; insoluble in alcohol; soluble in sodium hydroxide solution and concentrated hydriodic acid. Noncombustible.
Derivation: Interaction of solutions of lead nitrate and sodium sulfate.
Hazard: Strong irritant to tissue.
Use: Storage batteries, paint pigments.

lead sulfate, basic.   (white lead, sublimed; white lead sulfate). Approximate formula $PbSO_4 \cdot PbO$.
Properties: White, monoclinic crystals. D 6.92, mp 977C. Slightly soluble in hot water or acids. Noncombustible.
Grade: Vary from 72–85% lead sulfate and remainder lead oxide. Sold dry or ground in oil.
Derivation: Three methods are used: (1) lead sulfide ore (galena) is subjected to high temperatures in an oxidizing atmosphere; (2) molten lead is sprayed into a jet of ignited fuel gas and air in the presence of sulfur dioxide gas; (3) atomized metallic lead is mixed with water, and sulfuric acid is added under controlled conditions.
Hazard: Toxic material.
Use: Paints, ceramics, pigments.

lead sulfate, blue basic.   (sublimed blue lead; blue lead). Composition: Lead sulfate (min) 45%, lead oxide (min) 30%, lead sulfide (max) 12%, lead sulfite (max) 5%, zinc oxide 5%, carbon and undetermined matter (max) 5%.
Properties: Blue-gray, corrosion-inhibiting pigment. D 6.2. Insoluble in water or alcohol. Noncombustible.
Derivation: By heating lead ores in special furnaces.
Hazard: Toxic material.
Use: Component of structural-metal priming coat paints, rust-inhibitor in paints, lubricants, vinyl plastics, and rubber products.

lead sulfate, tribasic. $3PbO \cdot PbSO_4 \cdot H_2O$.
Properties: Fine, white powder. D 6.4, refr index 2.1.
Hazard: Toxic material.
Use: Electrical and other vinyl compounds requiring high heat stability.

lead sulfide. (plumbous sulfide).
CAS: 1314-87-0. PbS.
Properties: Silvery, metallic crystals or black powder. D 7.13–7.7, mp 1114C, sublimes at 1281C. Soluble in acids; insoluble in water and alkalies.
Derivation: (1) Found in nature as the mineral galena, (2) by passing hydrogen sulfide gas into an acid solution of lead nitrate.
Grade: Technical, CP, electronic.
Hazard: Toxic by ingestion and inhalation.
Use: Ceramics, infrared radiation detector, semiconductor, ceramic glaze, source of lead.

lead sulfite. PbSO$_3$.
Properties: White powder. Decomposes on heating. Insoluble in water; soluble in nitric acid.
Hazard: Toxic by ingestion and inhalation.

lead sulfocarbolate. See lead phenolsulfonate.

lead sulfocyanide. See lead thiocyanate.

lead superoxide. See lead dioxide.

lead tallate. A mixture of lead and tall oil. Combustible.
Grade: Liquid: 16% lead, 24% lead: solid: 30% lead.
Derivation: By the fusion process.
Hazard: Toxic material. Absorbed by skin.
See soap (2).

lead telluride. PbTe.
Properties: Crystalline solid. D 8.2, mp 905C. Insoluble in water and most acids.
Hazard: Toxic by ingestion and inhalation.
Use: Single crystals used as photoconductor and semiconductor in thermocouples.

lead tetraacetate.
CAS: 546-67-8. Pb(CH$_3$COO)$_4$.
Properties: Colorless or faintly pink crystals, sometimes moist with glacial acetic acid. Mp 175C, d 2.228 (17C). Soluble in benzene, chloroform, nitrobenzene, hot glacial acetic acid. Combustible.
Derivation: From red lead (Pb$_3$O$_4$) and glacial acetic acid in the presence of acetic anhydride.
Hazard: Toxic material. Absorbed by skin.
Use: Oxidizing agent in organic synthesis, laboratory reagent.

lead tetraethyl. See tetraethyl lead.

lead tetroxide. See lead oxide, red.

lead thiocyanate. (lead sulfocyanate).
CAS: 592-87-0. Pb(SCN)$_2$.
Properties: White or light-yellow crystalline powder. D about 3.8. Soluble in potassium thiocyanate, nitric acid; slightly soluble in cold water; decomposes in hot water.
Hazard: Toxic by ingestion and inhalation.

Use: Ingredient of priming mix for small-arms cartridges, safety matches, dyeing.

lead thiosulfate. (lead hyposulfite).
CAS: 502-87-4. PbS$_2$O$_3$.
Properties: White crystals. D 5.18, mp decomposes. Soluble in acids and sodium thiosulfate solution; insoluble in water.
Derivation: By the interaction of solutions of lead nitrate and sodium thiosulfate, concentration, and crystallization.
Hazard: Toxic by ingestion and inhalation.

lead titanate.
CAS: 12060-00-3. PbTiO$_3$.
Properties: Pale-yellow solid. D 7.52. Insoluble in water.
Derivation: Interaction of oxides of lead and titanium at a high temperature. Contains lead sulfate and lead oxide as impurities.
Hazard: Toxic by ingestion and inhalation.
Use: Industrial paint pigment.

lead trinitroresorcinate. (lead styphnate).
CAS: 63918-97-8. C$_6$H(NO$_2$)$_3$(O$_2$Pb).
Properties: Monohydrate is monoclinic orange-yellow crystals. D 3.1 (monohydrate), 2.9 (anhydrous). Practically insoluble in water.
Derivation: Prepared by adding a solution of magnesium styphnate (from magnesium oxide and styphnic acid) to a lead salt solution.
Hazard: Detonates at 500F (260C), dangerous explosion risk, an initiating explosive.

lead, tris(lauroyloxy)phenyl-. See phenyllead trilaurate.

lead tungstate. (lead wolframate). PbWO$_4$.
Properties: White powder. D 8.235, mp 1130C. Soluble in sodium hydroxide solutions; insoluble in water.
Derivation: By mixing solutions of lead nitrate and sodium tungstate, concentrating, and crystallizing.
Hazard: Toxic material.
Use: Pigment.

lead vanadate. (lead metavanadate).
CAS: 10099-79-3. Pb(VO$_3$)$_2$.
Properties: Yellow powder. Insoluble in water; decomposes in nitric acid.
Hazard: Toxic material.
Use: Preparation of other vanadium compounds, pigment.

lead water. A 1% solution of basic lead acetate.
Hazard: Toxic. See lead acetate.

lead, white. See lead carbonate, basic; lead silicate, basic; and lead sulfate, basic.

lead wolframate. See lead tungstate.

lead wool. Fine filaments or threads of metallic lead, prepared and used for packing pipe joints.

lead zirconate titanate. (LZT).
CAS: 12626-81-2. $PbTiZrO_3$.
Available forms: Piezoelectric crystals.
Use: Element in high fidelity stereo units and as a transducer for ultrasonic cleaners, ferroelectric materials in computer memory units.

leaf, filter. A unit of a shell-and-leaf filter press on which the cake is formed. In general, a leaf consists of a circular or rectangular metal frame in which is fastened a coarse wire screen. This is covered on both sides with a fine-mesh wire cloth, over which is placed the filter medium proper, e.g., nylon fabric. The filtrate passes through the fabric and into an escape pipe to the discharge port. Each shell may contain as few as six or as many as 50 leaves of varying dimensions; the entire assembly can be pulled out of the shell for cake removal. In some models the leaves rotate.

leafseal. A formulation of decenylsuccinic acid and its esters.
Use: Direct application to plants to enable them to resist frost and drought.

leather. An animal skin or hide that has been permanently combined with a tanning agent that causes a physicochemical change in the protein components of the skin. This change renders it resistant to putrefactive bacteria, enzymes, and hot water, increases its strength and abrasion resistance, and makes it serviceable for long periods of time. Tanning agents are vegetable, mineral, or synthetic. Hides from cows or steers are chiefly used for men's shoes, transmission belting, and other heavy-duty service. These are usually vegetable-tanned. Lighter grades made from the skins of sheep, calves, or reptiles are used for shoe uppers, luggage, gloves, and similar end products (chrome-tanned).
Leather is a naturally poromeric material that retains the microporosity of the original skin; this property makes it uniquely applicable to footwear; to a limited extent it is able to conform to the contour of the individual foot. Leather is made in many colors, weights, and finishes. However, it has been replaced to an increasing extent by plastics for many minor uses, and by synthetics for shoe uppers and soling. See poromeric; tanning.

leavening agent. See yeast; baking powder.

Lebedev process. Formation of butadiene from ethanol by catalytic pyrolysis. The catalysts used are mixtures of silicates and aluminum and zinc oxides.

Le Blanc. (1742–1806). A French inventor of the first successful process for making soda ash. His patent was confiscated by the Revolutionist government, and the process was used widely for years without either acknowledgment or remuneration. His original formula was 100 parts salt cake, 100 parts limestone, and 50 parts coal.

Le Chatelier. (1850–1936). A French physical chemist, famous chiefly for his statement of the equilibrium principle (often known as Le Chatelier's law). His work included investigations of cements, alloys, and gaseous combustion. The principle may be stated: every system in equilibrium is conservative and tends to resist changes upon it by reacting in such a way as to help nullify the imposed change.

Lecisun. A skin-friendly mineral sunscreen.

lecithin. $C_8H_{17}O_5NRR'$, R and R' being fatty acid groups. Pure lecithin is a phosphatidyl choline. The lecithins are mixtures of diglycerides of fatty acids linked to the choline ester of phosphoric acid. The lecithins are classed as phosphoglycerides or phosphatides (phospholipids). Commercial lecithin is a mixture of acetone-insoluble phosphatides. FCC specifies not less than 50% acetone-insoluble matter (phosphatides).
Properties: Light-brown to brown, viscous semiliquid with a characteristic odor. Partly soluble in water and acetone; soluble in chloroform and benzene.
Derivation: Usually from soybean oil, also from corn, other vegetable seeds, egg yolk, and other animal sources.
Grade: Technical, unbleached, bleached; fluid, plastic, edible, FCC, 96+% for biochemical or chromatographic standards.
Use: Emulsifying, dispersing, wetting, penetrating agent, and antioxidant; in margarine, mayonnaise, chocolate and candies, baked goods, animal feeds, paints, petroleum industry (drilling, leaded gasoline), printing inks, soaps and cosmetics, mold release for plastics, blending agent in oils and resins, rubber processing, lubricant for textile fibers.

Leclanche cell. See dry cell.

lectin. A type of protein occurring in the seeds of certain plants, especially legumes, characterized by unusual binding specificity; their precise function in the plant is being researched. Studies have been made on the molecular structure and carbohydrate content of the lectin found in the European herb sainfoin.

Ledu's rule. States that the volume occupied by a gas mixture is equal to the sum of the volumes occupied separately by each constituent at the same temperature and pressure as the mixture.

LEED. Low-energy electron diffraction.

lees. The sediment at bottom of wine storage tank.

Leeuwenhoek, van. See van Leeuwenhoek, Anton.

Lee, Yuan T. (1936–). Awarded Nobel Prize in chemistry in 1986 jointly with Polanyi and Herschbach. A former student of Herschbach, Lee refined molecular-beam and laser techniques, combining them with theory to perform definitive studies of reactions of individual complex molecules. Doctorate from University of California in 1965.

Lefkowitz, Robert J. (1943–). An American physician and molecular biologist who was awarded the 2012 Nobel Prize jointly with Brian K. Kobilka. The scientists' groundbreaking discovery of the G-protein-coupled receptors in living cell has revolutionized the understanding of how cells respond to stimuli such as hormones and how certain types of drugs illicit their effect. This remarkable discovery led to major advances in development of pharmaceuticals. Lefkowitz received his M.D. form Columbia University followed by a renowned career which comprises more than 60 awards, including the Shaw Prize in Life Science and Medicine (2007) and the National Medal of Science (2007). He serves as James B. Duke Professor of Medicine and professor of biochemistry and chemistry at Duke University Medical Center.

legal chemistry. (forensic chemistry). The application of chemical knowledge and procedures to matters involving civil or criminal law and to all questions where control of chemical compounds, products, or processes is vested in agencies of Federal or state governments. Legal chemistry applies to the following areas:
(1) Crime detection: primarily identification of poisons, of bloodstains, writing and typewriter inks, and a host of miscible materials such as textile fibers from clothing, hair, skin, etc. A variety of analytical methods are used in police laboratories, including microscopes, spot tests, color reactions, and spectrophotometry.
(2) Food, drugs, and cosmetics are under the control of the U.S. Food and Drug Administration. New products and proposed additives must be submitted by the manufacturer and approved before being placed on public sale. Control of the manufacture of illicit drugs is an important phase of legal chemistry.
(3) Pesticides are subject to Federal regulation. New products must be registered, labeling must be specific as to chemical composition, active and inert ingredients, and directions for use.
(4) Marketing and competitive pricing of chemical products' fair trade agreements and discriminatory practices are also under Federal supervision (Robinson–Patman Act). This includes mergers, tie-in sales, and other merchandising practices.
(5) Interstate shipment and labeling of hazardous chemicals is regulated by the Department of Transportation and the Federal Aviation Agency as well as by state and local laws.
See labeling, toxicology.
(6) Patent law comprises a vast body of legal practice and court decisions. The patent system is designed to protect inventions and new discoveries, and most chemical companies retain legal counsel in this field.
(7) Water pollution is subject to federal regulations (Federal Water Pollution Control Act, 1956). This covers the discharge of contaminating industrial waste, sewage, oil, etc., into navigable streams and their tributaries as well as into coastal waters.
(8) Flammability of fabrics.
(9) Use of volatile, toxic solvents on an industrial scale.
(10) Air pollution including gases and particulates from industrial stacks and auto exhaust emissions.

Lehmstedt–Tanasescu reaction. Preparation of acridones (and 10-hydroxyacridones) from *o*-nitrobenzaldehyde and a halobenzene in the presence of concentrated sulfuric acid containing nitrous acid as catalyst.

Lehn, Jean-Marie Pierre. (1939–). Awarded Nobel Prize for chemistry, together with Cram, in 1987 for work in elucidating mechanisms of molecular recognition, which are fundamental to the enzymic catalysis, regulation, and transport. He also studied three-dimensional cyclic compounds that maintained a rigid structure, accepting substrates in a structurally preorganized cavity. Lehn named these compounds cryptands, while Cram called them cavitands. Awarded Doctorate by University of Strasbourg, France in 1963.

lehr. A long oven designed for controlled slow cooling of glass (annealing). The hot glass is carried through the lehr on a conveyor at a predetermined rate, traversing areas of gradually decreasing temperature. The entire process may require several days.
See annealing.

Leloir, Luis F. (1906–1987). A French-born biochemist who won the Nobel Prize for chemistry in 1970 for work in biosynthesis of carbohydrates. He discovered chemical compounds that affect the storage of chemical energy in humans and animals. He headed the Department of Biochemistry at the University of Buenos Aires for many years.

Lemol. A series of polyvinyl alcohols in partially and fully hydrolyzed form at various molecular weights. Supplied as nondusting, white granules with d 1.2–1.3.
Use: Adhesives, emulsions, polymerization, film coatings, polyester release agents, textile printing, finishing, and sizing.

lemon chrome. See barium chromate.

lemongrass oil. (verbena oil, Indian). An essential oil obtained by steam distillation of a grass (*Cymbopogon*).
Properties: Dark-yellow to light-brown-red; pronounced heavy lemon-like odor. D 0.900–0.910 (15/15C), optical rotation −3 to +1 degrees, refr

index 1.4830–1.4890 (20C). Soluble in alcohol; slightly soluble in glycerol. Combustible.
Source: India, East and West Indies, Guatemala.
Chief constituents: Citral (75–85%), geraniol, ethylheptenone.
Use: Perfumes, flavoring, isolates and ionones, source of citral.

lemon oil.   See citrus peel oils.

lenacil.
CAS: 2164-08-1. $C_{13}H_{18}N_2O_2$.
Properties: Colorless solid. Mp 315C, d 1.32. Almost insoluble in water; soluble in pyridine.
Use: Herbicide.

lentagran.   See fenpyrate.

lepicidin A.
CAS: 131929-60-7. $C_{41}H_{65}NO_{10}$.
Hazard: Moderately toxic by ingestion and inhalation.

lepidine.   (γ-methylquinoline; cincholepidine). CAS: 491-35-0. $C_9H_6NCH_3$.
An alkaloid.
Properties: Oily liquid; quinoline-like odor. Turns red-brown on exposure to light. D 1.086, bp 266C, solidifies at about 0C, soluble in alcohol, ether, and benzene. Slightly soluble in water.
Derivation: From cinchonine.
Use: Organic preparations, medicine.

lepidolite.   (lithia mica). $K_2Li_3Al_4Si_7O_{21}(OH, F)_3$. A fluosilicate of potassium, lithium, and aluminum, found in pegmatites. Rubidium occurs as an impurity. A variety of mica.
Properties: Pink and lilac to gray color; pearly luster; perfect micaceous cleavage. Mohs hardness 2.5–4, d 2.8–3.0.
Occurrence: California, South Dakota, New Mexico, South Africa.
Use: Source of lithium and rubidium, flux in glass and ceramics production.

lepton.   (5,5-diethyl-1,3-oxazinane-2,4-dione). $C_8H_{13}NO_3$. Any of six types of fermions, the electron and its antiparticle, the muon and its antiparticle, the tau and its antiparticle. Included also are the three types of neutrino and antineutrino associated with each of these particles, the electron neutrino, the muon neutron and the tau neutrino. The latter participate in electromagnetic and weak interactions and have a half-integral spin. Leptons are the lightest class of particles that have nonzero rest mass. They are weakly interacting fermions and can result from the slow decay of nuclear particles such as the neutron but do not exhibit a strong attraction toward these particles. Leptons can be described by Fermi-dirac statistics since they are restricted by the pauli exclusion principle such that two identical leptons cannot occupy the same quantum state.

leptophos.   (abar;   o-(4-bromo-2,5-dichlorophenyl)-o-methyl   phenylthiophosphonate;   o-(2,5-dichloro-4-bromophenyl-o-methyl phenylthiophosphonate; fosvel; K62-105; MBCP; o-methyl-o-(4-bromo-2,5-dichlorophenyl)phenyl thiophosphonate; NK 711; phenylphosphonothioic acid o-(-4-bromo-2,5-bromo-2,5-dichlorophenyl) o-methyl ester; phosvel; velsicol 506; veliscol vcs 506; 4-bromo-2,5-dichlorophenoxy)-methoxy-phenyl-sulfanylidene-(5)-phosphane). $C_{15}H_{10}BrCl_2O_2PS$. An organothiophosphate.
Hazard: Extremely toxic; neurotoxic; poison; teratogen; mutagen; causes muscular weakness, staggering gait, tremors, paresthesia, tingling sensations, blurred vision, disorientation, temporary memory loss, auditory hallucinations, paralysis of the legs; teratogenic and mutagenic effects.
Use: Was used in the U.S. during the early to mid-1970s as an insecticide and pesticide.

lethal gene.   A gene that leads to the death of an organism before reproduction is possible or one that precludes reproduction.

lethal mutation.   A mutation that inactivates a biological function essential to the life of the cell or organism.

"Lethane" *[Dow]*.   TM for a group of thiocyanate insecticides. 60:2-thiocyanoethyl laurate. 384:β-butoxy-β'-thiocyanodiethyl ether. A-70 diethylene glycol dithiocyanate.
Hazard: Toxic by ingestion.

Letts nitrile synthesis.   Formation of nitriles by heating aromatic carboxylic acids with metal thiocyanates.

Leu.   Abbreviation for leucine.

leucine.   (α-amino-γ-methylvaleric acid; α-aminoisocaproic acid).
CAS: 61-90-5. $(CH_3)_2CHCH_2CH(NH_2)COOH$. An essential amino acid. Found naturally in the L(−) form.
Properties: White crystals. Soluble in water; slightly soluble in alcohol; insoluble in ether. Optically active (natural form). DL-leucine: mp 332C with decomposition; L(−)-leucine: mp 295C, d 1.239 (18/4C).
Derivation: Hydrolysis of protein (edestin, hemoglobin, zein), organic synthesis from the α-bromo acid.
Grade: Commercial (DL), FCC (L-).
Use: Nutrient and dietary supplement, biochemical research.

1-leucin.   (α-aminoisocaproic acid; 2-amino-4-methylpentanoic acid; α-amino-γ-methylvaleric acid; 2-amino-4-methylvaleric acid; 1,2-amino-4-methylvaleric acid; leucin; leucine; norvaline, 4-methyl-; valeric acid, 2-amino-4-methyl-).

CAS: 61-90-5. $C_6H_{13}NO_2$. An essential branched-chain amino acid important for hemoglobin formation.
Properties: White crystals; glistening, hexagonal plates frol aqueous alcohol; soluble in water; slightly soluble in alcohol; insoluble in ether.
Derivation: Isolated from gluten in 1826.
Hazard: Moderately toxic; teratogen.

leucine zipper. A motif found in certain proteins in which Leu residues are evenly spaced through an α-helical region, such that they would end up on the same face of the helix. Dimers can form between two such proteins. The Leu zipper is important in the function of transcription factors such as Fos and Jun and related proteins.

Leuckart thiophenol reaction. Decomposition of diazoxanthates by warming gently in faintly acidic cuprous media to the corresponding aryl xanthates which produce aryl thiols on alkaline hydrolysis and aryl thioethers on warming.

Leuckart–Wallach reaction. Reductive alkylation of ammonia or of primary or secondary amines with carbonyl compounds and formic acid or formamides as reducing agents.

leuco-compound. See vat dye.

leucocrystal violet. (LCV; 4-[bis-[4-(dimethylamino)penyl]methyl]-*N,N*-dimethylaniline). $C_{25}H_{31}N_3$. Reagent.
Use: To detect bloody friction ridge detail by either fluorescent or nonfluorescent staining.

leucomalachite green. (4-[[4-9dimethylamino)phenyl]-phenylmethyl]-*N,N*-dimethylaniline). $C_{23}H_{26}N_2$. Reagent.
Use: To detect bloody friction ridge detail.

leucothol B.
CAS: 38302-26-0. $C_{20}H_{32}O_5$.
Hazard: A poison.
Source: Natural product.

leucovorin. Preferred name for folinic acid.

Leukanol. Synthetic tanning assistants of the sulfonic type supplied in liquid and solid grades. Powerful dispersants for vegetable tannins, and bleaches for chrome-tanned leather.

leukocyte. A white blood cell.

leukomaine. Any of a large group of basic metabolic products that resemble alkaloids that commonly occur in living tissues.
Hazard: Toxic.

leukotoxic serum. A serum that destroys leukocytes.

leukotriene. One of a group of physiologically active compounds derived directly from arachidonic acid. They are chemically related to the prostaglandins and occur in white blood cells (leukocytes). They cause muscle contractions that constrict air passages in the lungs and are involved in asthma.

leutenizing hormone-releasing factor (pig), 6-*d*-leucine-9-(*n*-ethyl-*l*-prolinamide)-10-deglycinamide-, monoacetate (salt).
CAS: 74381-53-6. $C_{59}H_{84}N_{16}O_{12} \cdot C_2H_4O_2$.
Hazard: A poison.

leveling. (1) A term used in the paint industry to describe the application properties of a paint, i.e., its ability to cover a dry surface easily and to hold its level without sagging or running. (2) The ability of a nickel-plated coating to cover surface irregularities of the substrate, achieved by the incorporation of one or more brighteners in the plating formulation. (3) Aiding the uniform dispersion of a dye in a dye bath or solution by addition of a suitable material, e.g., lignin.

Levelume. A bright high-leveling nickel process. Prepared from nickel sulfate, nickel chloride, boric acid, and organic addition agents.
Use: Electrical appliances, automotive trim, plumbing fixtures.

Levene–Hudson phenylhydrazide rule. The direction of rotation of the phenylhydrazides of the sugar acids indicates the configuration of the hydroxyl on the α-carbon atom. If the phenylhydrazide rotates to the right, the hydroxyl on the α-carbon is to the right, and vice-versa. The rule was shown to be valid for salts, amides, and corresponding acylated nitriles. In connection with the rule, Hudson mentioned that, "the sugar benzylphenylhydrazones rotate to the left when the asymmetric α-carbon atom of the configuration has its hydroxyl to the right, and vice-versa."

Levitt, Michael. (1947–). An American-British-Israeli chemist who was awarded the 2013 Nobel Prize jointly with Martin Karplus and Arieh Warshel. The scientist developed accurate computer models to understand and predict chemical processes. As the computer simulations could produce the exact output, this created an efficient method to study and test effect of drugs on living molecules, leading to drug design for cancer therapy and, Alzheimer's disease. Levitt's study concentrates on analysis of protein, DNA, and RNA molecules using computer models. Levitt and Warshel developed the first generalized computer model of enzymatic reactions, and this model proved significant as it could be used to model other molecules. He was awarded Ph.D. jointly by the medical Research Council Laboratory of Molecular Biology in Cambridge and the

University of Cambridge. He serves as a professor and chair in Computational Structural Biology at Stanford University School of Medicine.

Levnlite.   Sodium aluminum phosphate.
Grade: Regular and high calcium powder.
Use: Imparts volume, lightness, tenderness, tolerance in use, and good crumb resiliency to cake products.

levocabastine hydrochloride.
CAS: 79547-78-7. $C_{26}H_{29}FN_2O_2 \cdot ClH$.
Hazard: Moderately toxic by ingestion.

levo form.   An optical isomer that produces levorotation in a beam of plane polarized light.
See levorotatory.

levorotatory.   Having the property when in solution of rotating the plane of polarized light to the left or counterclockwise. Levorotatory compounds may have the prefix l- to distinguish them from their dextrorotatory or d- isomers, but the minus sign (−) is preferred.

levulinic acid.   (γ-ketovaleric acid; acetylpropionic acid; 4-oxopentanoic acid; levulic acid).
CAS: 123-76-2. $CH_3CO(CH_2)_2COOH$.
Properties: Crystals. Bp 245–246C, mp 33–35C, d 1.1447 (25/4C), refr index 1.442 (16C). Miscible with water, alcohol, esters, ethers, ketones, aromatic hydrocarbons; insoluble in aliphatic hydrocarbons. Combustible.
Use: Intermediate for plasticizers, solvents, resins, flavors, pharmaceuticals, acidulant and preservative; chrome plating; solder flux; stabilizer for calcium greases; control of lime deposits.

levulose.   See fructose.

Lewis acid.   Any molecule or ion (called an electrophile) that can combine with another molecule or ion by forming a covalent bond with two electrons from the second molecule or ion. An acid is thus an electron acceptor. Hydrogen ion (proton) is the simplest substance that will do this, but many compounds such as boron trifluoride, BF_3, and aluminum chloride, $AlCl_3$, exhibit the same behavior and are therefore properly called acids. Such substances show acid effects on indicator colors and when dissolved in the proper solvents.

Lewis base.   A substance that forms a covalent bond by donating a pair of electrons, neutralization resulting from reaction between the base and the acid with formation of a coordinate covalent bond. It is also called a nucleophile.
See Lewis electron theory.

Lewis electron theory.   A theory involving acid and base formation, neutralization, and related phenomena on the basis of exchange of electrons between substances and the formation of coordinate bonds. It represented an important advance in chemical theory, largely replacing earlier concepts. Advanced in 1923 by Gilbert N. Lewis, it contributed much to the development of coordination chemistry in which the base is represented by the ligand and the acid by the metal ion.

Lewis, Gilbert N.   (1875–1946). An American chemist, native of Massachusetts, professor of chemistry at MIT from 1905 to 1912 after which he became dean of chemistry at University of California at Berkeley. His most creative contribution was the electron-pair theory of acids and bases which laid the groundwork for coordination chemistry. He was also a leading authority on thermodynamics.

lewisite.   Legal label name for β-chlorovinyldichloroarsine.

Lewis metal.   Alloy of one part tin and one part bismuth.
Properties: Expands when cooling. Mp 138C.
Use: Sealing and holding die parts.

"Lewisol" 28 *[Pinova].*   TM for a pale, hard resin, a maleic-modified glycerol ester of rosin. Acid number 36, softening point 141C, USDA color WG.

Lewis, Warren Kendall.   (1882–1975). Born in Laurel, Maryland, graduated from MIT in 1905, Ph. D. from University of Breslau, Germany in 1908. He became professor of chemical engineering at MIT in 1910. He is often regarded as the father of chemical engineering in the U. S., as his outstanding books and other publications did much to establish the fundamental principles of this field.

Leyden temperature scale.   A low-temperature thermometer scale based on a boiling point of hydrogen equal to −252.74C and of oxygen equal to −182.95C.

L-310 Fatty Acid.   A fatty acid derived from linseed oil.
Properties: The major component acids are oleic, linoleic, and linolenic. Light yellow liquid at ambient temperature, obtained from naturally occurring triglycerides.
Use: Chemical intermediate, paint and varnishes, alkyd resins and soaps.

Li.   Symbol for lithium.

Libby, Willard F.   (1908–1980). An American chemist who won the Nobel Prize for chemistry in 1960 and in 1980. His first Nobel work involved radiochemistry and isotope tracer work in the space program. He is known for the "atomic time clock" which is a way of estimating the age of ancient materials by measuring the amount of radioactive ^{14}C in organic or carbon-containing objects. His Doctorate was awarded at Wesleyan University. He also worked at Syracuse, Carnegie Institute of Technology, and Georgetown institutions, among others.

library. Biology: An unordered collection of clones (i.e., cloned DNA from a particular organism), whose relationship to each other can be established by physical mapping.
See genomic library.

Librium Hydrochloride. Chlordiazepoxide hydrochloride. Manufacture and use restricted.

Lica. A prefix for a series of neoalkoxy zirconates.

licanic acid. (4-keto-9,11,13-octadecatrienoic acid). $CH_3(CH_2)_3(CH:CH)_3(CH_2)_4COCH_2COOH$.
Properties: White crystals. α-licanic acid (naturally occurring isomer) melts at 74–75C, readily isomerizes to the β-form, mp 99.5C. Soluble in organic solvents.
Derivation: Occurs in oiticica and other oils as glycerides.

lichenic acid. See fumaric acid.

licorice. See glycyrrhizin.

lidocaine. (α-diethylaminoaceto-2,6-xylidide). CAS: 137-58-6. $C_6H_3(CH_3)_2NHCOCH_2N(C_2H_5)_2$.
Properties: White or slightly yellow crystalline powder; characteristic odor. Mp 66–69C, bp 180–182C (at 4 mm Hg). Soluble in alcohol, ether, or chloroform; insoluble in water.
Derivation: By action of diethylamine on chloroacetylxylidide.
Grade: USP.
Use: Medicine (local anesthetic).

Lieben iodoform reaction. Cleavage of methyl ketones with halogens (mostly iodine) and base to carboxylic acids and haloform.

Liebig, Justus von. (1803–1873). A German chemist who founded the *Annalen*, a world-famous chemical journal. He was a great teacher of chemistry, training such men as Hoffmann, who did basic work on organic dyes. Liebig contributed original research in the fields of human physiology, plant life, soil chemistry, and was the discoverer of chloroform, chloral, and cyanogen compounds. He was the first to recommend addition of nutrients to soils and thus may be considered the originator of the fertilizer industry.

life, origin. (biogenesis). The succession of chemical events that led up to the appearance of living organisms on earth about 3.3 billion years ago. According to one theory, substantiated by experimental evidence, this occurred as follows. The inorganic compounds originally present were carbides, water, ammonia, and carbon dioxide. The carbides reacted with water to form methane, which in turn reacted with ammonia and water vapor as a result of an electric impulse to form amino acids, porphyrins, and nucleotides (or their precursors). All these compounds have been created artificially in the laboratory. It has further been shown that amino acids and nucleotides can be concentrated into proteins (and probably nucleic acids) by the action of zinc-bearing clays, which were present along the shores of the primeval oceans. Little or no free oxygen existed in the primordial atmosphere, that consisted chiefly of reducing gases. The complex chemical reactions which eventually resulted in the formation of DNA took place in an anaerobic aqueous environment, and the earliest living organisms developed in a nutrient solution in which free oxygen finally appeared as the result of photosynthesis by blue-green bacteria. Another theory advances the idea that essential life chemicals such as purines and amino acids were formed under primitive conditions from aqueous solutions of hydrogen cyanide.

ligand. A molecule, ion, or atom that is attached to the central atom of a coordination compound, a chelate, or other complex. Thus, the ammonia molecules in $[Co(NH3)_6]^{3+}$ and the chlorine atoms in $[PtCl_6]^{2-}$ are ligands. Ligands are also called complexing agents, e.g., EDTA, ammonia.
See chelate; coordination compound.

ligase. An enzyme, T4 DNA ligase, which can link pieces of DNA together. The pieces must have compatible ends (both of them blunt, or else mutually compatible sticky ends), and the ligation reaction requires ATP.

ligation. The process of splicing two pieces of DNA together. In practice, a pool of DNA fragments are treated with ligase (see "ligase") in the presence of ATP, and all possible splicing products are produced, including circularized forms and end-to-end ligation of 2, 3 or more pieces.

light hydrocarbon. One of a group of hydrocarbon products derived from natural gas or petroleum; ethane, propane, iso- and normal butane, and natural gasoline (C_5 and heavier). Produced largely in southwest Texas and Louisiana, these are used as feedstocks for a wide variety of organics.
See liquefied petroleum gas.

Light-Lead. Lead-lined polyethylene shielding bricks. Weight of one brick is 7 pounds.

light metal. In engineering terminology, a metal of specific gravity less than three that is strong enough for construction use (aluminum, magnesium, beryllium).

light microscope. See optical microscope.

light oil. (coal tar light oil). A fractional distillate from coal tar with bp range from 110–210C, consisting of a mixture of benzene, pyridine, toluene, phenol, and cresols. The term is also sometimes used for oils of about the same bp range, but from other sources.
Grade: Technical.
Hazard: Highly flammable, dangerous fire risk.
Use: Source of benzene, solvent naphthas, toluene, phenol, and cresols.

light reactions. The reactions of photosynthesis occurring only in the light.

light water. (1) A fire-fighting agent consisting of a water solution of perfluorocarbon compounds mixed with a water-soluble thickener of the polyoxyethylene compound type. It can be used simultaneously with a dry chemical to smother gasoline or similar fires. (2) Ordinary water (as distinct from heavy water) used to both cool and moderate nuclear reactors.

lignaloe oil. See bois de rose oil.

lignin. A phenylpropane polymer of amorphous structure comprising 17–30% wood. It is so closely associated with the holocellulose which makes up the balance of woody material that it can be separated from it only by chemical reaction at high temperature. It is believed to function as a plastic binder for the holocellulose fibers. It is recovered from wood-processing wastes in limited amounts.
Use: Stabilization of asphalt emulsions, ceramic binder and deflocculant, dye leveler and dispersant, drilling-fluid additive, precipitation of proteins, extender for phenolic plastics, special molded products, source of vanillin, phenol, and of a component of battery expanders.

lignin sulfonate. (lignosulfonate). A metallic sulfonate salt made from the lignin of sulfite pulp-mill liquors, mw range 1000–20,000.
Properties: Light-tan to dark-brown powder; no pronounced odor. Stable in dry form and relatively stable in aqueous solution. Nonhygroscopic, no definite mp, decomposes above 200C, d about 1.5. Forms colloidal solutions or dispersions in water; practically insoluble in all organic solvents.
Use: Dispersing agent in concrete and carbon black-rubber mixes, extender for tanning agents, oil-well drilling mud additives, ore flotation agents, production of vanillin, industrial cleaners, gypsum slurried, dyestuffs, pesticide formulations. Commercially available as the salts of most metals and of ammonium.
See Raykrome.

lignite. (brown coal). A low rank of coal between peat and subbituminous; it contains 35–40% water. It occurs in the continental U.S., Alaska, Germany, and the Netherlands. Its Btu value is low. Drying,

crushing, and pelletizing lignite with an asphaltic binder for direct use as fuel has been successfully demonstrated. Polymer resins (polyesters and polyamides) can be derived from lignite by oxidation with nitric acid, followed by extraction of the nitrocoal acids, which are the basis of the polymer molecules. Peat can also be used. A process for gasification of lignite to produce methanol alcohol is approaching commercial development in Sweden.
See peat; gasification.

lignoceric acid. (n-tetracosanoic acid). $CH_3(CH_2)_{22}COOH$. A long-chain saturated fatty acid found in minor quantities in most natural fats.
Properties: Crystals. Mp 84.2C, bp 272C (10 mm Hg), d 0.8207, refr index 1.4287 (100C). Nearly insoluble in ethanol.
Source: Lignite and beechwood tar, peanut oil, sphingomyelin.
Grade: Technical, 99%.
Use: Biochemical research.

lignosulfonate. See lignin sulfonate.

ligroin. A saturated volatile fraction of petroleum boiling in the range 60–110C. There is a special grade of ligroin known as petroleum benzin.
Hazard: Highly flammable, dangerous fire risk. Toxic by ingestion and inhalation.
Use: Solvent for resins, paints, varnishes, etc.

lime. Specifically, calcium oxide (CaO); more generally, any of the various chemical and physical forms of quicklime, hydrated lime, and hydraulic lime (adapted from ASTM definition C41–47). Noncombustible.
Hazard: Unslaked lime (quicklime) yields heat on mixing with water and is a caustic irritant.
Use: See calcium oxide, calcium hydroxide.
See calcium oxide.

lime acetate. See calcium acetate.

lime, agricultural. Lime slaked with a minimum amount of water to form calcium hydroxide.

lime, air-slaked. Lime which has absorbed carbon dioxide and moisture from the atmosphere. It consists of a powder composed of calcium carbonate and calcium hydroxide.

lime, chlorinated. (chloride of lime; bleaching powder). $CaCl(ClO) \cdot 4H_2O$.
Properties: White powder; chlorine odor. Mp (decomposes). Decomposes in water, acids.
Derivation: By conducting chlorine into a box-like structure containing slaked lime spread upon perforated shelves.
Grade: 35–37% active chlorine, technical.

Hazard: Evolves chlorine and at higher temperatures oxygen. With acids or moisture evolves chlorine freely at ordinary temperatures.
Use: Textile and other bleaching applications, organic synthesis, deodorizer, disinfectant.
See calcium hypochlorite; bleach.

lime citrate.   See calcium citrate.

lime, fat.   A pure lime that combines readily with water to form a fine, white powder, free from grit, and makes a smooth, stiff paste with excess of water. Must not be loaded hot.
See lime, lean.

lime, hydrated.   See calcium hydroxide.

lime, hydraulic.   A variety of calcined limestone that when pulverized absorbs water without swelling or heating and gives a cement that hardens under water. The limestone burned for this purpose usually contains 10–17% silica, alumina, and iron, and 40–45% lime, magnesia sometimes replacing lime. Must not be loaded hot.

lime hypophosphite.   See calcium hypophosphite.

lime, lean.   A lime which does not lake freely with water because it has been prepared from limestone containing a high percentage of impurities, e.g., silica, iron, alumina. Must not be loaded hot.

lime-nitrogen.   See calcium cyanamide.

lime oil, distilled.
CAS: 8008-26-2.   Colorless to greenish-yellow, volatile oil obtained by distillation from the juice or whole crushed fruit of *Citrus aurantifolia* Swingle.
Properties: Refr index 1.4745–1.4770 (20C), d 0.855–0.863 (25C), angular rotation +34 to +47 degrees. Soluble in most fixed oils and mineral oil; insoluble in glycerol and propylene glycol. Combustible.
Chief constituents: Terpineol, citral.
Grade: FCC (contains between 0.5 and 2.5% of aldehydes, calculated as citral).
Use: Extracts, flavoring, perfumery, toilet soaps, cosmetics.

lime oil, expressed.   See citrus peel oil.

lime saltpeter.   See calcium nitrate.

lime, slaked.   See calcium hydroxide.

limestone.
CAS: 1317-65-3.   CaCO$_3$. A noncombustible solid characteristic of sedimentary rocks and composed mainly of calcium carbonate in the form of the mineral calcite, Mohs hardness about 3. Limestones are sometimes classed according to the impurities contained. For example, dolomitic limestone: usually a limestone containing more than 5% magnesium carbonate; magnesium limestone: dolomitic limestone, used as a solid diluent and carrier in pesticides; argillaceous limestone: contains clays, used in cement manufacture as "cement rock"; siliceous limestone: a limestone containing sand or quartz. Limestones are also named according to the formation in which they occur.
See marble; dolomite.
Use: Building stone, metallurgy (flux), manufacture of lime, source of carbon dioxide, agriculture, road ballast, cement (Portland and natural), alkali manufacture, removal of sulfur dioxide from stack gases and sulfur from coal.

lime, sulfurated.   (calcium sulfide, crude). A mix of calcium sulfide and calcium sulfate.
Properties: Yellowish-gray or grayish-white powder; odor of hydrogen sulfide. Soluble in acids; insoluble in water and alcohol. Noncombustible.
Derivation: By roasting calcium sulfate with coke.
Use: Medicine, depilatory, luminous paint.

lime-sulfur solution.   A solution made by boiling together lime (50 lb), sulfur (100 lb), and water (100 gal) and diluting to one-tenth strength. Contains calcium polysulfide, free sulfur, and calcium thiosulfate.
Use: Fungicidal spray on fruit trees, sheep dip.

lime, unslaked.   See calcium oxide.

lime water.   (calcium hydroxide solution).
CAS: 1305-62-0.
Properties: Clear, colorless, odorless, alkaline aqueous solution of calcium hydroxide containing more than 0.14 g of Ca(OH)$_2$ in each 100 mL at 25C (the strength varies with the temperature at which the solution is stored), d about 1.00 (25C), absorbs carbon dioxide from air.
Grade: USP (as calcium hydroxide solution).
Use: Medicine (external).

limonene.
CAS: 138-86-3.   C$_{10}$H$_{16}$.
A widely distributed, optically active terpene, closely related to isoprene. It occurs naturally in both *d*- and *l*-forms. The racemic mixture of the two isomers is known as dipentene.

Properties: Colorless liquid. (1) D 0.8411 (20C), bp 176–176.4C; (2) d 0.8422 (20C), bp 176–176.4C. Oxidizes to a film in air, oxidation behavior similar to that of rubber or drying oils.

Derivation: (1) Lemon, bergamot, caraway, orange, and other oils, (2) peppermint and spearmint oils.

Use: Flavoring, fragrance and perfume materials, solvent, wetting agent, resin manufacture.

limonene dioxide.   See dipentene dioxide.

limonene, inactive.   (or racemic or dl). See dipentene.

limonene monoxide.   See dipentene monoxide.

linalool.   (linalol; 3,7-dimethyl-1,6-octadien-3-ol).
CAS: 78-70-6. $(CH_3)_2C:CHCH_2CH_2C(CH_3)OH-CH:CH_2$.
Linalool is the *l*-isomer, coridandrol is the *d*-isomer.
Properties: Colorless liquid; odor similar to that of bergamot oil and French lavender. D 0.858–0.868 (25C), bp 195–199C, angular rotation −2 to +2 degrees. Soluble in alcohol, ether, fixed oils. Combustible.
Derivation: Citrus peel oils, especially from oranges. Made synthetically from geraniol.
Method of purification: Rectification.
Grade: Ex bois de rose oil, synthetic, FCC.
Use: perfumery, flavoring agent.

linalool, dehydro-.
CAS: 29171-20-8. $C_{10}H_{16}O$.
Hazard: A poison by ingestion. A mild skin and moderate eye irritant.

linalool oxide.   (tetrahydro-α,α-5-trimethyl-5-vinylfurfuryl alcohol).
CAS: 1365-19-1. $C_{10}H_{17}O_2$.
Properties: Liquid. Refr index 1.4523 (20C).
Derivation: Synthetically from acetone.
Use: Perfuming and flavoring agent.

linalyl acetate.
CAS: 115-95-7. $C_{10}H_{17}C_2H_3O_2$.
Properties: Clear, colorless, oily liquid; odor of bergamot. Bp 108–110C, d 0.908–0.920, refr index 1.450–1.458 (20C), angular rotation −1 to +1 degrees. Soluble in alcohol, ether, diethyl phthalate, benzyl benzoate, mineral oil, fixed oils, alcohol; slightly soluble in propylene glycol; insoluble in water, glycerol. Combustible.
Derivation: Action of acetic anhydride on linalool in presence of sulfuric acid. May also be obtained from bergamot and other oils.
Method of purification: Rectification.
Grade: Ex bois de rose oil 92%, 96–98%, FCC (natural and synthetic).
Use: Extracts, perfumery, flavoring agent, substitute for petitgrain oil.

linalyl formate.
CAS: 115-99-1. $C_{10}H_{17}OOCH$.

Properties: D 0.915 (25/4C), bp 100–103C (10C), refr index 1.456–1.457 (20C). Insoluble in water; soluble in alcohol. Combustible.
Derivation: Synthetically from acetone.
Use: Perfume, flavoring.

linalyl isobutyrate.
CAS: 78-35-3. $C_{10}H_{17}OOCCH(CH_3)_2$.
Properties: liquid. D 0.890, refr index 1.4490. Combustible.
Derivation: Synthetically from acetone.
Use: Perfume, flavoring.

linalyl propionate.
CAS: 144-39-8. $C_{10}H_{17}OOCC_2H_5$.
Properties: Colorless liquid; floral odor similar to bergamot oil. D 0.895–0.902 (25C), refr index 1.4500–1.4550 (20C). Soluble in most fixed oils and in mineral oil; slightly soluble in propylene glycol; insoluble in glycerol. Combustible.
Derivation: Synthetically, starting with acetone.
Grade: FCC.
Use: Perfumery, flavoring agent.

lincomycin.
CAS: 154-21-2. $C_{18}H_{34}N_2O_6S$.
An antibacterial substance produced by *Streptomyces lincolnensis*. The hydrochloride is stable in the dry state and in aqueous solution for at least 24 months, soluble in water at room temperature in concentrations up to 500 mg/mL.

Lindahl, Thomas.   (1938–). A Swedish biochemist that received the 2015 Nobel Prize jointly with Paul Modrich and Aziz Sancar. The scientists mapped how cells repair damaged DNA while safeguarding the genetic information at a genetic level. This work has provided fundamental knowledge of how a living cell functions and is used for the development of new cancer treatments. His work defied orthodoxy about DNA stability and he discovered a molecular system that constantly counteracts DNA collapse. His Ph.D. was awarded by the Karolinska Institutet, followed by an illustrious career that included awards such as the Royal Society's Royal Medal (2007), Copley Medal (2010), and the INSERM Prix Etranger (France, 2009).

lindane.   (γ-benzene hexachloride).
CAS: 58-89-9. $C_6H_6Cl_6$.
Legal label name for γ isomer of 1,2,3,4,5,6-hexachlorocyclohexane.

Hazard: Toxic by inhalation, ingestion, and skin absorption. Possible carcinogen.
Use: Pesticide. Use may be restricted.

linear alkylbenzene A-315.
CAS: 68648-87-3.
Hazard: Low toxicity by ingestion and skin contact. A mild eye and severe skin irritant.

linear molecule. See alkyl sulfonate, linear; polyethylene, high-density.

Lineweaver–Burk equation. An algebraic transformation of the Michaelis–Menten equation (plot of 1/V vs 1/[S]), allowing determination of Vmax and Km by extrapolation of [S] to infinity.

linkage. (a) The greater association in inheritance of two or more nonallelic genes than is to be expected from independent assortment; genes are linked because they reside on the same chromosome. (b) Analysis of pedigree; the tracking of a gene through a family by following the inheritance of a (closely associated) gene or trait and a DNA marker.

linkage analysis. A gene-hunting technique that traces patterns of heredity in large, high-risk families in an attempt to locate a disease-causing gene mutation by identifying traits that are co-inherited with it.

linkage disequilibrium. Where alleles occur together more often than can be accounted for by chance. Indicates that the two alleles are physically close on the DNA strand.
See Mendelian inheritance.

linkage map. A map of the relative positions of genetic loci on a chromosome, determined on the basis of how often the loci are inherited together. Distance is measured in centimorgans (cM).

linker. A small piece of synthetic double-stranded DNA which contains something useful, such as a restriction site. A linker might be ligated onto the end of another piece of DNA to provide a desired restriction site.

linoleamide.
CAS: 3999-01-7.
Properties: An antistatic agent, opacifier, viscosity controlling agent.
Use: Food additive.

linoleic acid. (linolic acid).
CAS: 60-33-3.

$$CH_3(CH_2)_4 \underset{H}{\overset{}{C}}=\underset{H}{\overset{}{C}} \overset{CH_2}{} \underset{H}{\overset{}{C}}=\underset{H}{\overset{}{C}} (CH_2)_7CO_2H$$

A polyunsaturated fatty acid (2 double bonds) existing in both conjugated and unconjugated forms. A plant glyceride essential to human diet.

Properties: Colorless to straw-colored liquid. D 0.905 (15/4C), fp −5C, bp 228C (14 mm Hg), refr index 1.4710 (15C). Insoluble in water; soluble in alcohol and ether. Combustible.
Source: Linseed oil, safflower oil, tall oil.
Grade: Technical, purified (99+%), edible.
Use: Soaps, special driers for protective coatings, emulsifying agents, feeds, biochemical research, dietary supplement, margarine.
Note: Do not confuse with linolenic acid.

linoleic acid amide. See linoleamide.

linolein. A glyceride of linoleic acid. It is one of the constituents of linseed oil which induces drying.

linolenic acid. (9,12,15-octadecatrienoic acid).
CAS: 463-40-1.

$$CH_3CH_2 \underset{H}{\overset{}{C}}=\underset{H}{\overset{}{C}} \overset{CH_2}{} \underset{H}{\overset{}{C}}=\underset{H}{\overset{}{C}} \overset{CH_2}{} \underset{H}{\overset{}{C}}=\underset{H}{\overset{}{C}} (CH_2)_7CO_2H$$

A polyunsaturated fatty acid (3 double bonds) which occurs as the glyceride in many seed fats. It is an essential fatty acid in the diet. It is also a constituent of drying oils.
Properties: Colorless liquid. D 0.916 (20/4C), fp −11C, bp 230C (17 mm Hg). Soluble in most organic solvents; insoluble in water. Combustible.
Grade: Purified 99+%.
Use: Nutrient, biochemical research, drying oils.
Note: Do not confuse with linoleic acid.

linolenin. A glyceride of linolenic acid. Like linolein it is a constituent of drying oils.
See linseed oil.

linolenyl alcohol. (octadecatrienol).
CAS: 24149-05-1. $C_{18}H_{32}O$.
The fatty alcohol derived from linolenic acid.
Properties: Colorless solid. Iodine value 190, cloud point 50.0F (10C), d 0.864. Combustible.
Available forms: Available commercially as 50% pure.
Derivation: Reduction of acid made from linseed oil.
Impurities: Oleyl and linoleyl alcohols with some saturated alcohols.
Use: Paints, flotation, lubricants; surface-active agents, resins, synthetic fibers.

(linoleoyloxy)tributylstannane.
CAS: 24124-25-2. $C_{30}H_{58}O_2Sn$.
Properties: Antifoulant, disinfectant. Bp <−45°C, d 1.17–1.18.
Hazard: A poison. Tributyl tin compounds are extremely toxic to marine life.

linoleyl alcohol.
CAS: 506-43-4. $CH_3(CH_2)_4CH:CHCH_2CH:CH(CH_2)_7CH_2OH$.
The fatty alcohol derived from linoleic acid.

Properties: Colorless solid. Iodine value 137, d 0.855, cloud point 59F (15C). Combustible.
Derivation: Reduction of linoleic acid.
Available forms: Available commercially as 50–60% pure alcohol.
Impurities: Mostly oleyl alcohol with some linolenyl and saturated alcohols.
Use: Paints, flotation, paper, surface-active agents, resins, leather.

linoleyltrimethylammonium bromide.   An amorphous solid from yellow to very light-brown in color. Soluble in water, alcohol.
Use: Germicide, deodorant, algicide, slime control.

linolic acid.   See linoleic acid.

linseed cake.   The press cake formed when the seeds are crushed and the oil is extracted.
See linseed-oil meal.

linseed oil.   (flaxseed oil).
CAS: 8001-26-1.
Properties: Golden-yellow, amber, or brown drying oil; peculiar odor; bland taste. D 0.921–0.936, iodine value 177, saponification value 189–195, acid number (max) 4 (ASTM D 234–48), flash p 432F (222C), autoign temp approximately 650C (343C), polymerizes on exposure to air. Soluble in ether, chloroform, carbon disulfide, and turpentine; slightly soluble in alcohol; spontaneous heating. Combustible.
Chief constituents: Glycerides of linolenic, oleic, linoleic, and saturated fatty acids. The drying property is due to the linoleic and linolenic groups.
Derivation: From seeds of the flax plant *Linum usitatissimum* by expression or solvent extraction. Various refining and bleaching methods are used.
Grade: Raw, boiled, doubled-boiled, blown, varnish maker's, refined.
See linseed oil, boiled.
Use: Paints, varnishes, oilcloth, putty, printing inks, core oils, linings, and packings, alkyd resins, soap, pharmaceuticals.
Note: Use of linseed oil in paints has decreased sharply since the introduction of emulsion paints.

linseed oil, blown.   Linseed oil whose viscosity is increased by air bubbled through it at 93C. The reaction is mainly oxidation followed by polymerization. The resulting product dries to a harder film than heat-bodied oils and is used in interior paints and enamels.

linseed oil, boiled.   The term is a misnomer since the oil does not boil. Small amounts of driers (e.g., oxides of manganese, lead, or cobalt or their naphthenates, resinates, or linoleates) are added to hot linseed oil to accelerate drying. The "boiled oil" becomes thicker and darker.

linseed oil, heat-bodied.   Linseed oil that has been polymerized by heating at 287–315C. This increases viscosity and acid content and reduces iodine value. Bodied oil dries much faster than unprocessed oil.

linseed-oil meal.   (linseed cake). The crushed and extracted residue from flaxseed (linseed), generally prepared by crushing the seeds, cooking with steam, and hydraulic expression of the oil. The resulting cake is sold by its protein content.

linters, cotton.   See cotton linters.

LINUP.   See laser-induced nuclear polarization.

linuron.   (generic name for 3-(3,4-dichlorophenyl)-1-methoxy-1-methylurea).
CAS: 330-55-2. $C_6H_3Cl_2NHC(O)N(OCH_3)CH_3$.
Properties: Solid. Mp 93–94C. Slightly soluble in water; partially soluble in acetone and alcohol.
Use: Herbicide.

Lipal.   A series of polyoxyethylene esters and ethers of fatty acids.
Use: Detergents, emulsifiers, solubilizers, wetting agents, and coupling agents.

lipase.
CAS: 9001-62-1. Any of a class of enzymes that hydrolyze fats to glycerol and fatty acids. Lipase is abundant in the pancreas but also occurs in gastric mucosa, in the small intestine, and in fatty tissue. It is found in milk, wheat germ, and various fungi. Commercial pancreatin and most trypsin preparations contain lipase.
Derivation: From pork pancreas or calf glands.
Use: Manufacture of cheese and similar foods, for removal of fat spots in dry cleaning or grease accumulations, in analytical chemistry of fats because it selectively hydrolyzes only fatty acids on the ends of triglycerides.

lipases.   Enzymes that catalyze the hydrolysis of triacylglycerols.

lipid.   (lipide). An inclusive term for fats and fat-derived materials. It includes all substances which are (1) relatively insoluble in water but soluble in organic solvents (benzene, chloroform, acetone, ether, etc.); (2) related either actually or potentially to fatty acid esters, fatty alcohols, sterols, waxes, etc.; and (3) utilizable by the animal organism. One of the chief structural components of living cells.
See phospholipid; fat; fatty acid.

lipid bilayer.   The inferred structure of a cell membrane based on its interaction with the hydrophobic regions of phospholipids.

lipid metabolism.   (lipometabolism). Complex catabolic and anabolic processes that involve lipids.

Disruption of lipid metabolism as by a toxicant can result in pathological accumulations of lipids in the liver.

Lipmann, Fritz.   (1899–1986). A German-born biochemist who won the Nobel Prize in 1953 for the discovery of coenzyme A (CoA). He earned doctorates at the University of Berlin in both chemistry and medicine. He worked at Cornell and Harvard Universities. He founded the biochemistry department of Brandeis University and later joined the faculty of Rockfeller University.

lipoate.   (lipoic acid). A coenzyme involved as a carrier of acyl groups.

Lipo GMC Glyceryl Cocoate.   An emulsifier, and emulsion stabilizer.

_dl_-α-lipoic acid.   (6,8-dithiooctanoic acid; thioctic acid; POF). CAS: 62-46-4.

SSCH$_2$CH$_2$CH(CH$_2$)$_4$COOH. A pyruvate oxidation factor. Pyruvate is a normal intermediate in carbohydrate metabolism.
Properties: Crystals. Mp 60–61C, bp 160–165C. Practically insoluble in water; soluble in fat solvents; forms a water-soluble sodium salt.
Source: Yeast and liver.
Use: Nutrition, biochemical research.
See lipoate.

"Liponate CL" _[Lipo]_.   TM for a cetyl lactate formulation.
Use: Adds a creamy feel in skin products.

lipopolysaccharide, -esc.
CAS: 93572-42-0.
Hazard: A poison by ingestion.
Source: Natural product.

lipoprotein.   A protein-lipid complex that functions in the transport of lipids in the blood. In cyclical fashion, present as chylomicrons, VLDL, LDL, IDL, HDL or chylomicron remnants. The protein component alone is an apolipoprotein.

liposome.   A nanoparticle consisting of lipids, or fat molecules, surrounding a water core. They were the first type of nanoparticle used to create therapeutic agents.

lipostat.
CAS: 81131-70-6.   C$_{23}$H$_{35}$•Na.
Hazard: Low toxicity by ingestion. Human systemic effects.

lipotropic agent.   An agent that because of its affinity for fats and oils, helps to regulate the

metabolism of fat and cholesterol in the animal body. Inositol is an example.

Lipovol MOS-70.   Blends of specialty esters that exhibit tactile properties of mineral oil.

Lipowitz's metal.   A fusible alloy. See alloy, fusible.

lipoxidase.   An enzyme that catalyzes the addition of oxygen to the double bonds of unsaturated fatty acids of plant origin.
Use: Biochemical research, whitening bread.

Lipscomb, William Nunn Jr.   (1919–2011). An American chemist who won the Nobel Prize for chemistry in 1976 for his studies on the structure and bonding mechanisms of boranes. Much of the research concerned structure and function of enzymes and natural products in organic and theoretical chemistry. He studied at the Universities of Kentucky, California, and Minnesota.

liquation.   The separation of two or more components of a mixture by heating to a temperature at which one component melts, leaving the others as solids.
Use: Separation of alloy components.

liquefaction.   (coal).
See gasification; Fischer–Tropsch process.

liquefied natural gas.   See natural gas.

liquefied petroleum gas.   (compressed petroleum gas; liquefied hydrocarbon gas; LPG). CAS: 68476-85-7. A compressed or liquefied gas obtained as a by-product in petroleum refining or natural gasoline manufacture, e.g., butane, isobutane, propane, propylene, butylenes and their mixtures.
Properties: Colorless. Noncorrosive. Flash p −100F (−74C), autoign temp 800–1000F (426–537C).
Hazard: Highly flammable, dangerous fire and explosion risk.
Use: Domestic and industrial fuel, automotive fuel, welding, brazing, and metal cutting.
See compressed gas; natural gas.

liqueur.   (cordial). Distilled spirits treated with fruit, herb, pulp, juice, and natural flavorings that are sugar sweetened.

liquid.   An amorphous (noncrystalline) form of matter intermediate between gases and solids in which the molecules are much more highly concentrated than in gases but much less concentrated than in solids. The molecules of liquids are free to move within the limits set by intermolecular attractive forces. At the air-liquid interface the vibration of the molecules causes some to be ejected from the liquid at a rate depending on the surface tension. The tendency of molecules to escape from a liquid surface is called fugacity and is largely responsible for evaporation, which occurs when the air space above the liquid is unrestricted. In a closed system, where the air space is restricted, the escaping molecules eventually saturate the air, and thus the number of molecules leaving the liquid will be equal to those returning to it as a result of molecular attraction. In these circumstances the liquid-air system is said to be in equilibrium.

Liquids vary greatly in viscosity, melting point, vapor pressure, and surface tension. Mercury has a density of 13.6 and the highest surface tension of all liquids. Glass has the highest viscosity. Polar liquids are those whose molecules have opposite electrical charges on their terminal atoms or groups, that impart a force called dipole moment. Water is a polar liquid with high dielectric constant. Pure hydrocarbon liquids are generally nonpolar.

See liquid, Newtonian; glass; amorphous; solid; liquid crystals; kinetic theory.

liquid aerosol. A colloidal suspension of liquid droplets in a gas, usually air.

liquid air. Air cooled below −189C.
Properties: A cloudy or blue-grey liquid.
Use: As a source of oxygen and nitrogen.

liquid chromatography. An analytical method based on separation of the components of a mixture in solution by selective adsorption. All systems include a moving solvent, a means of producing solvent motion (such as gravity or a pump), a means of sample introduction, a fractionating column, and a detector. Innovations in functional systems provide the analytical capability for operating in three separation modes: (1) liquid–liquid: partition in which separations depend on relative solubilities of sample components in two immiscible solvents (one of which is usually water); (2) liquid–solid: adsorption where the differences in polarities of sample components and their relative adsorption on an active surface determine the degree of separation; (3) molecular size separations which depend on the effective molecular size of sample components in solution. Solvents, often referred to as carriers, include isooctane, methyl ethyl ketone, acetone/chloroform, tetrahydrofuran, hexane, and toluene. Packing materials in columns of various lengths include silica gel, alumina, glass beads, polystyrene gel, and ion exchange resins.
High-performance liquid chromatography (HPLC) is the term applied to new and more effective instrumental techniques developed in recent years that have greatly increased the scope of this analytical method. It can now be applied to biological as well as chemical research. Among the separations possible are peptides (by reverse phase chromatography), proteins and enzymes (hydrophobic and size exclusion modes of chromatography), amino acids, and inorganic and organometallic compounds (neutral species, including clusters, by adsorption and size exclusion, and ionic species including coordination compounds). A comparatively recent development is the use of supercritical fluids as solvents, e.g., carbon dioxide and sulfur hexafluoride.
See supercritical fluid; gas chromatography; paper chromatography; thin-layer chromatography; instrumentation.

liquid crystal. An organic compound in an intermediate or mesomorphic state between solid and liquid. This phenomenon was first noted in 1888 in cholesteryl benzoate, a crystalline solid. It becomes a turbid liquid, or liquid crystals, when heated to 145C; on further heating to 179C the liquid becomes isotropic. This sequence is reversed when the substance is cooled. Color changes occur on both heating and cooling. Many organic compounds, e.g., sodium benzoate, exhibiting this behavior are known and used extensively in electric and electronic displays, thermometers, color TV tubes, electronic clocks and calculators, and similar devices dependent on temperature determination. Liquid crystals have several varieties of molecular order: nematic, smectic (nine types), and cholesteric. They indicate small temperature differences by changing color when applied to the skin and are used in medicine for this purpose. They can align with dichroic dye molecules in a thin-layer cell to produce color changes.
Available forms: Microencapsulated.
See nematic; smectic; cholesteric.

liquid dioxide. See nitrogen dioxide.

liquid-drox.
Properties: Fluorescent yellow solution.
Use: To develop friction ridge detail on the adhesive and non-adhesive sides of dark colored tape.

liquid-liquid extraction. The transfer of components between two liquid phases.

liquid-liquid interface. See dineric interface.

liquid, Newtonian. Characteristic of liquids is their ability to flow, a property depending largely on their viscosity and sometimes also on the rate of shear. A Newtonian liquid is one that flows immediately on application of force and for which the rate of flow is directly proportional to the force applied. Water, gasoline, and motor oils at high temperatures are examples. Some liquids have abnormal flow response when force is applied, that is, their viscosity is dependent on the rate of shear. Such liquids are said to exhibit non-Newtonian flow properties. Some will not flow until the force exerted is greater than a definite value called the yield point. (W. A. Gruse).

liquid pitch oil. See creosote; coal tar.

liquid rosin.   See tall oil.

liquid rubber.   See rubber, liquid.

liquid nitrogen.   Elemental nitrogen used in its liquid state (T <−196C).
Properties: A very cold liquid that boils at temperatures above −196C.
Use: Separation of adhesive surfaces, to enhance the fluorescence of Zinc Chloride and Zinc Nitrate treated prints for visualization and photography, to store biological samples (blood, reproductive cells, sperm, and egg) at very low temperatures and other uses.
Hazard: Severe freezing of tissue, a simple asphyxiant that displaces oxygen in confined spaces, explosive pressures generated if confined in a vessel.

Liquigreen.   A liquid with clear, self-flowing, low salt index, stable at all temperatures.
Use: As a turf fertilizer.

"Liquinox" [Alconox].   TM for an anionic liquid detergent.
Use: For hard and soft water cleaning in laboratories.

liquor.   In chemical technology, any aqueous solution of one or more chemical compounds. In sugar manufacturing, it refers to the syrups obtained from various refining steps (mother liquors). The paper industry uses this term extensively as follows: (1) black liquor is liquid digester waste (also called spent sulfate liquor) containing sulfonated lignin, rosin acids, and other waste wood components from which tall oil is made; (2) green liquor is a solution made by dissolving chemicals recovered in the alkaline pulping process in water; (3) white liquor is made by adding caustic soda to sodium sulfide solution. In dyeing technology, red liquor is an alternate name for mordant rouge.

lisinopril.
CAS: 76547-98-3. $C_{21}H_{31}N_3O_5$.
Hazard: Low toxicity by ingestion. Human systemic effects.

liter.   (L, l). The volume of one kilogram of water at its temperature of maximum density (4C) at standard atmospheric pressure. A liter is 1.05 quart, or 0.26 gallon.

Lithafrax.   A ceramic material made from β-spodumene.

litharge.   (lead oxide, yellow; plumbous oxide; lead monoxide).
CAS: 1317-36-8.   PbO.
An oxide of lead made by controlled heating of metallic lead.
Properties: Yellow crystals. D 9.53, mp 888C. Insoluble in water; soluble in acids and alkalies. A strong base. Commercial grades are yellow to reddish, depending on treatment and purity.
See massicot.
Derivation: By oxidizing metallic lead in air. Various forms of lead and various temperatures from 500–1000C are used.
Grade: CP, fused, powdered.
Hazard: Toxic by ingestion and inhalation.
Use: Storage batteries, ceramic cements and fluxes, pottery and glazes, glass, chromium pigments, oil refining, varnishes, paints, enamels; assay of precious metal ores, manufacture of red lead, cement (with glycerol), acid-resisting compositions, matchhead compositions, other lead compounds, rubber accelerator.

litharge-glycerol cement.   Made by mixing glycerol with $1/6$ to $1/2$ portion of water and mixing with enough litharge to give a paste of desired consistency. Must be used as soon as mixed. Fillers retard the setting and avoid cracking. The product is somewhat resistant to acids.
Hazard: Toxic by ingestion.

litharge, leaded.   See lead suboxide.

lithia.   See lithium oxide.

lithic acid.   See uric acid.

lithium.
CAS: 7439-93-2.   Li.
Metallic element of atomic number 3, group IA of periodic table, aw 6.941, valence = 1; two isotopes. It is the lightest and least reactive of the alkali metals and the lightest solid element.
Properties: Very soft, silvery metal. D 0.534 (20C), mp 179C, bp 1317C, Mohs hardness 0.6, viscosity of liquid lithium less than water, heat capacity about the same as water. Reacts exothermally with nitrogen in moist air at high temperatures; high electrical conductivity; soluble in liquid ammonia.
Source: Spodumene, lepidolite, ambylgonite occurring in U.S., Canada, central Africa, Brazil, Argentina, Australia, Europe. Also desert lake brines.
Derivation: By electrolysis of a mixture of lithium chloride and potassium chloride, high-temperature extraction from spodumene by sodium carbonate, solar evaporation of lake brines.
Grade: 99.86 to 99.9999%.
Available forms: Ingots, rods, wire, ribbon, and pellets.
Hazard: Ignites in air near its melting point; dangerous fire and explosion risk when exposed to water, acids, or oxidizing agents. Extinguish lithium fires only with chemicals designed for this purpose. Lithium in solution is toxic to the central nervous system.

Use: Production of tritium, reducing and hydrogenating agents, alloy hardeners, pharmaceuticals, Grignard reagents. Scavenger and degasifier for stainless and mild steels in molten state, modular iron, soaps and greases, deoxidizer in copper and copper alloys, catalyst, heat-transfer liquid, storage batteries (with sulfur, selenium, tellurium, and chlorine). Rocket propellants, Vitamin A synthesis, silver solders, underwater buoyancy devices, nuclear reactor coolant.
See "Lith-Ex."

lithium acetate, dihydrate.   (acetic acid, lithium salt; Quilonum).
CAS: 546-89-4. $CH_3CO_2Li \cdot 2H_2O$.
Properties: White, crystalline powder. Mw 102.02, mp 53–56C. Soluble in water and alcohol.

lithium aluminate.
CAS: 12003-67-7. $LiAlO_2$.
Properties: White powder. Mp 1900–2000C, d 2.55 (25C). Insoluble in water.
Grade: Ceramic.
Use: As a flux in high-refractory porcelain enamels.

lithium aluminum deuteride.   (LAD).
$LiAlD_4$.
Properties: White to gray crystals. D 1.02. Stable in dry air at room temperature, but very sensitive to moisture. Decomposes above 140C liberating deuterium. Soluble in diethyl ether, tetrahydrofuran; slightly soluble in other low molecular weight ethers. Preparation: By reacting aluminum chloride with lithium deuteride.
Hazard: Flammable, dangerous fire risk, requires special handling; ignites in air.
Use: Introduction of deuterium atoms into molecules by reduction of same groups attacked by lithium aluminum hydride.

lithium aluminum hydride.   (LAH).
CAS: 16853-85-3. $LiAlH_4$.
Properties: White powder. D 0.917. Sometimes turns gray on standing. Stable in dry air at room temperature, but highly sensitive to moisture, including atmospheric. Decomposes to lithium hydride, aluminum metal, and hydrogen above 125C without melting. Soluble in diethyl ether, tetrahydrofuran, dimethyl "Cellosolve"; slightly soluble in dibutyl ether; insoluble or very slightly soluble in hydrocarbons and dioxane.
Preparation: Reaction of aluminum chloride with lithium hydride.
Hazard: Flammable, dangerous fire risk; may ignite spontaneously on grinding or rubbing, or from static sparks. Reacts violently with air, water, and many organic materials. Fires must be extinguished with powdered limestone or dry chemical.
Use: Reducing agent for over 60 different functional groups, especially for pharmaceutical, perfume, and fine organic chemicals; converts esters, aldehydes, and ketones to alcohols and nitriles to

amines; source of hydrogen; propellant; catalyst in polymerizations.

lithium aluminum hydride, ethereal.
$LiAlH_4$, plus ether.
Properties: Colorless solution in ether. Very reactive to water.
Derivation: From lithium hydride and ether solution of aluminum chloride.
Hazard: Flammable, dangerous fire risk.
Use: See lithium aluminum hydride.

lithium aluminum-tri-*tert*-butoxyhydride.
(LATB; lithium-tri-*tert*-butoxyaluminohydride).
CAS: 17476-04-9. $LiAl[OC(CH_3)_3]_3H$.
Properties: White powder. D 1.03, stable in dry air but sensitive to moisture. Soluble in the diemthyl ether of diethylene glycol, tetrahydrofuran, diethyl ether; slightly soluble in other ethers.
Hazard: Evolves flammable hydrogen at 400C. Dangerous.
Use: For stereospecific reductions of steroid ketones and for reductions of acid chlorides to aldehydes.

lithium amide.
CAS: 7782-89-0. $LiNH_2$.
Properties: White, crystalline solid; ammonia-like odor. D 1.18, mp 380–400C. Decomposes in water to form ammonia.
Derivation: Reaction of lithium hydride with ammonia.
Grade: 92–95% lithium amide.
Hazard: Flammable, dangerous fire risk. Do not use water to extinguish.
Use: Organic synthesis, including antihistamines and other pharmaceuticals.
See Lith-Ex.

lithium arsenate. Li_3AsO_4.
Properties: White powder. D 3.07 (15C). Slightly soluble in water; soluble in dilute acetic acid.
Hazard: A poison.

lithium benzoate.
CAS: 553-54-8. $LiC_7H_5O_2$.
Properties: White crystals or powder. Soluble in water and alcohol.
Derivation: Reaction of benzoic acid with lithium carbonate.

lithium bicarbonate.
CAS: 5006-97-3. $LiHCO_3$.
A lithium salt formed by dissolving lithium carbonate in water with excess carbon dioxide. The solution, called Lithia Water, is used in medicine and prepared mineral waters.

lithium borate.   See lithium tetraborate.

lithium borohydride.
CAS: 16949-15-8. $LiBH_4$.

Properties: White to gray crystalline powder. D 0.66; extremely hygroscopic. Decomposes in vacuum above 200C, in air at 275C. Soluble in water, lower primary amines, and ethers.
Derivation: Reaction of sodium borohydride and lithium chloride.
Hazard: Flammable, dangerous fire and explosion risk.
Use: Source of hydrogen and borohydrides; reducing agent for aldehydes, ketones, and esters.

lithium bromide.
CAS: 7550-35-8. LiBr.
Properties: White, cubic, deliquescent crystals, or as a white to pinkish-white, granular powder; odorless; sharp bitter taste. D 3.464, mp 547C, bp 1265C. Very soluble in water, alcohol, and ether; slightly soluble in pyridine; soluble in methanol, acetone, glycol; a hot concentrated solution dissolves cellulose; forms addition compounds with ammonia and amines; forms double salts with $CuBr_2$, $HgBr_2$, HgI_2, $Hg(CN)_2$, and $SrBr_2$; greatly depresses vapor pressure over its solutions.
Derivation: Reaction of hydrobromic acid with lithium carbonate.
Grade: 53% (min) LiBr brine; solid, single crystals.
Use: Pharmaceuticals, air conditioning, humectant, drying agent, batteries, low-temperature heat-exchange medium, medicine (sedative).

lithium *tert*-butanolate.
CAS: 1907-33-1. $C_4H_9O\bullet Li$.
Hazard: Moderately toxic by ingestion.

lithium butyl. See butyllithium.

lithium carbide.
CAS: 1070-75-3. Li_2C_2.
Properties: Crystalline, white powder. D 1.65 (18C). Decomposes in water; soluble in acid with evolution of acetylene.
Hazard: Fire risk in contact with acids.

lithium carbonate.
CAS: 554-13-2. Li_2CO_3.
Properties: White powder. D 2.111, mp 735C, bp (decomposes) at 1200C. Slightly soluble in water; insoluble in alcohol; soluble in dilute acid.
Derivation: (1) Finely ground ore is roasted with sulfuric acid at 250C, lithium sulfate is leached from the mass and converted to the carbonate by precipitation with soda ash. (2) Reaction of lithium oxide with carbon dioxide or ammonium carbonate solution.
Grade: Technical, CP.
Hazard: Water solution is strong irritant.
Use: Ceramics and porcelain glazes, pharmaceuticals, catalyst, other lithium compounds, coating of arc-welding electrodes, nucleonics, luminescent paints, varnishes and dyes, glass ceramics, aluminum production.

lithium chlorate.
CAS: 13453-71-9. $LiClO_3$.
Properties: Needle-like crystals. Deliquescent, d 1.119 (18C), mp 128C, decomposes at 270C. More soluble in water than any other inorganic salt (313 g per 100 mL water at 18C); very soluble in alcohol.
Hazard: Dangerous explosion hazard when shocked or combined with organic materials. Strong oxidant.
Use: Air conditioning, inorganic and organic chemicals, propellants.
See "Lith-X" *[Tyco]*.

lithium chloride.
CAS: 7447-41-8. LiCl.
Properties: White, deliquescent crystals. D 2.068, mp 614C, bp 1360C. Very soluble in water, alcohols, ether, pyridine, nitrobenzene. One of the most hygroscopic salts known. Not to be used as dietary salt substitute.
Derivation: Reaction of lithium ores with chlorides, natural brines.
Grade: Technical 99% minimum assay, 35–40% brine, inhibited; single crystals.
Use: Air conditioning, welding and soldering flux, dry batteries, heat-exchange media, salt baths, desiccant, production of lithium metal, soft drinks and mineral water to reduce escape of carbon dioxide.

lithium chromate.
CAS: 7789-01-7. $Li_2CrO_4\bullet 2H_2O$.
Properties: Yellow, crystalline, deliquescent powder. Soluble in water and forms a eutectic at −60C; soluble in alcohols.
Hazard: Toxic by ingestion.
Use: Corrosion inhibitor in alcohol-base antifreezes and water-cooled reactors. Oxidizing agent for organic material, especially in the presence of light; heat-transfer medium.

lithium citrate.
CAS: 919-16-4. $Li_3C_6H_5O_7\bullet 4H_2O$.
Properties: White powder or granules. Loses four waters at 105C; mp (decomposes). Soluble in water; slightly soluble in alcohol.
Derivation: Reaction of citric acid with lithium carbonate.
Use: Beverages, pharmaceuticals, dispersion stabilizer (clay deflocculant).

lithium cobaltite.
CAS: 57035-78-6. $LiCoO_2$.
Properties: Dark-blue powder. Insoluble in water. The compound exhibits both the fluxing property of lithium oxide and the adherence-promoting property of cobalt oxide.
Use: Ceramics.

lithium deuteride. LiD.
Properties: Gray crystals. D 0.906. Reacts slowly with moist air. Thermally stable to its mp of 680C.
Use: Thermonuclear fusion.
See deuterium.

lithium dichromate.
CAS: 10022-48-7. $Li_2Cr_2O_7 \cdot 2H_2O$.
Properties: Yellowish-red, crystalline powder. D 2.34 (30C), mp 130C, deliquescent. Soluble in water; forms eutectic at $-70C$.
Use: Dehumidifying, refrigeration.

lithium ferrosilicon.
CAS: 64082-35-5.
Properties: Dark, crystalline, brittle, metallic lumps or powder. Evolves flammable gas in contact with moisture. Must be kept cool and dry.
Hazard: Flammable, dangerous fire risk.

lithium fluoride.
CAS: 7789-24-4. LiF.
Properties: Fine white powder. D 2.635 (20C), mp 842C, bp 1670C. Slightly soluble in water; does not react with water at red heat. Soluble in acids; insoluble in alcohol.
Derivation: Reaction of hydrogen fluoride with lithium carbonate.
Grade: Guaranteed 98% (min) lithium fluoride, CP, single pure crystals.
Hazard: Strong irritant to eyes and skin.
Use: Welding and soldering flux, ceramics, heat-exchange media, synthetic crystals in infrared and ultraviolet instruments, rocket fuel component, radiation dosimetry. Component of fuel for molten salt reactors, X-ray diffraction.

lithium fluorophosphate. $LiF \cdot Li_3PO_4 \cdot H_2O$.
Properties: White crystals.
Derivation: Interaction of lithium fluoride and lithium phosphate.
Use: Ceramics.

lithium grease. A grease using a lithium soap of the higher fatty acids as a base. Water-resistant, stable when heated above its melting point and cooled again. Lithium hydroxy-stearate from hydrogenated castor oil is also widely used.
Use: Aircraft and low-temperature service.
See lithium stearate; lubricating grease.

lithium hydride.
CAS: 7580-67-8. LiH.
Properties: White, translucent, crystalline mass or powder. Commercial product is light bluish-gray due to minute amount of colloidally dispersed lithium. D 0.82 (20C), mp 680C, decomposition pressure nil at 25C, 0.7 mm Hg (500C), 760 mm Hg (approximately 850C). Decomposed by water-forming hydrogen and lithium hydroxide. Insoluble in benzene and toluene; soluble in ether.
Derivation: Reaction of molten lithium with hydrogen.
Grade: 93–95%, based on hydrogen evolution.
Hazard: Flammable, dangerous fire risk, ignites spontaneously in moist air, use dry chemical to extinguish.

Use: Desiccant, source of hydrogen, condensing agent in organic synthesis, preparation of lithium amide and double hydrides, nuclear shielding material, reducing agent.
See "Lith-X" *[Tyco]*.

lithium hydroxide.
CAS: 1310-65-2. LiOH.
Properties: Colorless crystals. D 2.54, mp 470C, bp 924C (decomposes). Slightly soluble in alcohol; soluble in water; absorbs carbon dioxide and water from air.
Derivation: Causticizing of lithium carbonate, action of water on metallic lithium, or by addition of Li_2O to water.
Hazard: Water solutions are strongly irritant.
Use: Storage battery electrolyte, carbon dioxide absorbent in space vehicles, lubricating greases, ceramics, catalyst, photographic developers, lithium soaps.

lithium hydroxystearate. $LiOOC(CH_2)_{10}$ $CH_2O(CH_2)_5CH_3$.
Properties: White powder. Mp 205C. Dissolves in hot petroleum oil to form greases. Combustible.
Derivation: From hydrogenated castor oil.
Use: Lubricating greases.

lithium hypochlorite.
CAS: 13840-33-0. LiOCl.
Hazard: Strong oxidant may ignite organic materials.
Use: Bleach, sanitizing agent.

lithium iodate.
CAS: 13765-03-2. $LiIO_3$.
Properties: White powder. D 4.487 (25C), mp 50–60C (transition point from α to β form). Soluble in water; insoluble in alcohol. An oxidizing material.

lithium iodide trihydrate. $LiI \cdot 3H_2O$.
Properties: White crystals. D 3.48 (25C), mp 72C, loses water (anhydrous) 450C. bp 1171C, extremely hygroscopic. Soluble in water and in alcohol.
Derivation: Action of hydriodic acid on lithium hydroxide with subsequent crystallization.
Use: Air conditioning, catalyst in acetal formation, solubilizes lithium in propylamine, nucleonics.

lithium lactate.
CAS: 16891-53-5. $LiC_3H_5O_3$.
Properties: White, odorless powder. Very soluble in water with practically neutral reaction. Lithium lactate is nonhygroscopic and stable, whereas sodium lactate can be prepared in solution only.
Use: Wherever a dry alkali lactate is required.

lithium manganite.
CAS: 12163-00-7. Li_2MnO_3.
Properties: Reddish-brown powder. Insoluble in water; extremely stable.
Use: Smelter addition in the manufacturing of frit, as a mill addition, ceramic-bonded grinding wheels.

lithium metaborate dihydrate.
CAS: 15293-74-0. $LiBO_2 \cdot 2H_2O$.
Properties: White, crystalline powder. Mp 840C (anhydrous). Soluble in water.
Use: Ceramics (flux in enamel cover coats, increases resistance to torsion), welding and brazing (anhydrous form), nucleonics.

lithium metasilicate.   (lithium silicate).
CAS: 10102-24-6. Li_2SiO_3.
Properties: White powder. Mp 1201C, d 2.52 (25C). Insoluble in water.
Use: Flux in glazes and ceramic enamels, welding-rod coating.

lithium methoxide.   (lithium methylate).
CAS: 865-34-9. $LiOCH_3$.
Properties: White, free-flowing powder. Soluble in methanol. Must be protected from moisture.
Available forms: A 10% solution in methanol.
Hazard: Solution very toxic by ingestion.
Use: Strong base and chemical intermediate where water is undesirable.

lithium methylate.   See lithium methoxide.

lithium molybdate.
CAS: 13568-40-6. Li_2MoO_4.
Properties: A white, crystalline compound. Mp 705C, d 2.66. Soluble in water.
Use: Steel coating, petroleum cracking catalyst.

lithium niobate.   (lithium metaniobate).
CAS: 12031-63-9. $LiNbO_3$.
Properties: Mp 1250C.
Available forms: Single crystals available in sizes up to 6 inches long and 0.5 inch in diameter.
Use: Infrared detectors, transducer in laser technology.

lithium nitrate.
CAS: 7790-69-4. $LiNO_3$.
Properties: Colorless powder. D 2.38, mp 261C. Soluble in water and alcohol.
Derivation: Reaction of nitric acid with lithium carbonate.
Grade: Technical, commercially pure, reagent.
Hazard: Dangerous explosion risk when shocked or heated, strong oxidizing agent.
Use: Ceramics, pyrotechnics, salt baths, heat-exchange media, refrigeration systems, rocket propellant.

lithium nitride.
CAS: 26134-62-3. Li_3N.
Properties: Brownish-red crystals of hexagonal structure or fine, free-flowing red powder. D approximately 1.3 (25C), mp 845C, decomposition pressure not measurable below 1250C. Water vapor in moist air causes slow decomposition. Reacts with water giving LiOH and ammonia; insoluble in polyethers.

Hazard: Ignites in air, use dry chemicals to extinguish (Lith-X).
Use: Nitriding agent in metallurgy, reducing and nucleophilic reagent in organic reactions, source of bound nitrogen in organic reactions.

lithium octadecanoate.   See lithium stearate.

lithium orthophosphate.   See lithium phosphate.

lithium oxide.   (lithia).
CAS: 12057-24-8. Li_2O.
Properties: White powder. Mp 1427C, d 2.023 (25C). A strong alkali. Absorbs carbon dioxide and water from air.
Hazard: Water solutions of high concentration are strongly irritant.
Use: Ceramics and special glass formulations, carbon dioxide absorbent (mineral water).

lithium perchlorate.
CAS: 7791-03-9. $LiClO_4$.
Properties: Colorless, deliquescent crystals. D 2.429, mp 236C, decomposes at 430C, has more available oxygen than liquid oxygen (on a volume basis), reacts with $4NH_3$ to form an ammoniate. Forms $LiClO_4 \cdot 3H_2O$, colorless crystals with d 1.84, mp 75C. Soluble in water and alcohol.
Hazard: Oxidizing agent, dangerous fire and explosion risk in contact with organic materials. Irritant to skin and mucous membranes.
Use: Solid rocket propellants.

lithium peroxide.
CAS: 12031-80-0. Li_2O_2.
Properties: Fine white powder. Mp (decomposes), d 2.14 (20C), in closed container no detectable loss of available oxygen. Soluble in water, 8% (20C); anhydrous acetic acid 5.6% (20C); insoluble in absolute alcohol (20C).
Derivation: Addition of hydrogen peroxide to lithium hydroxide.
Hazard: Dangerous fire and explosion risk in contact with organic materials. Strong oxidizing agent.
Use: As supplier of active oxygen, commercial samples have 32.5–34% available oxygen content.

lithium phosphate.   (lithium orthophosphate).
CAS: 10377-52-3. $2Li_3PO_4 \cdot H_2O$.
Properties: White, crystalline powder. D 2.41. Soluble in dilute acids; slightly soluble in water.

lithium ricinoleate.
CAS: 15467-06-8. $LiOOC_{17}H_{32}OH$.
Properties: White powder. Mp 174C. Insoluble or limited solubility in most organic solvents. Combustible.
Derivation: Castor oil.
Use: Alcoholysis and ester interchange catalyst.

lithium silicate.   See lithium metasilicate.

lithium silicon.
CAS: 68848-64-6. An alloy of lithium and silicon.
Properties: Black, shiny lumps or powder; sharp irritant odor. Keep cool and dry.
Hazard: Dangerous fire risk, reacts with water to form flammable gases.

lithium stearate. (lithium octadecanoate; octadecanoic acid, lithium salt).
CAS: 4485-12-5. $LiC_{18}H_{35}O_2$.
Properties: White crystals. D 1.025, mp 220C. Insoluble in cold and hot water, alcohol and ethyl acetate; forms gels and mineral oils.
Derivation: Reaction of stearic acid with lithium carbonate.
Grade: Grease, cosmetic.
Hazard: Low toxicity by ingestion. This substance is spontaneously combustible.
Use: Cosmetics, plastics, waxes, greases, lubricant in powder metallurgy, corrosive inhibitor in petroleum, flatting agent in varnishes and lacquers, high-temperature lubricant.

lithium sulfate.
CAS: 10377-48-7. $Li_2SO_4 \cdot H_2O$.
Properties: Colorless crystals. D 2.06, mp 130C. Soluble in water; insoluble in 80% alcohol. Does not form alums.
Derivation: Reaction of sulfuric acid with lithium carbonate or with spodumene ore.
Grade: Technical and pharmaceutical.
Use: Pharmaceutical products, ceramics.

lithium tetraborate.
CAS: 12007-60-2. $Li_2B_4O_7 \cdot 5H_2O$.
Properties: White, crystalline powder. Mp loses water at 200C. Very soluble in water; insoluble in alcohol.
Derivation: Reaction of boric acid with lithium carbonate.
Use: Ceramics, vacuum spectroscopy, metal refining and degassing.

lithium titanate.
CAS: 12031-82-2. Li_2TiO_3.
Properties: White powder. Insoluble in water. Has strong fluxing properties in small percentage in titanium-bearing enamels. The insolubility permits its use in vitreous and semivitreous glazes.

lithium tri-*tert*-butoxyaluminohydride. See lithium aluminum tri-*tert*-butoxyhydride.

lithium tungstate.
CAS: 13568-45-1. Li_2WO_4.
Properties: White crystals. D 3.71. Soluble in water.

lithium vanadate. (lithium metavanadate).
CAS: 15060-59-0. $LiVO_3 \cdot 2H_2O$.
Properties: Yellowish powder. Soluble in water.

lithium zirconate.
CAS: 12031-83-3. Li_2ZrO_3.
Properties: White powder. Insoluble in water. Efficient flux in glasses containing zirconium dioxide. Recommended as a flux in zirconium-opacified enamels, glazes, and porcelains.

lithium-zirconium silicate.
CAS: 12027-82-6. $Li_2O \cdot ZrO_2 \cdot SiO_2$.
Properties: White powder. A strong flux in enamels, glazes, and porcelains. It can be used in place of lithium zirconate.

lithocholic acid.
CAS: 434-13-9. $C_{24}H_{39}O_2OH$.
A bile acid.
Properties: Crystals in leaflets from alcohol. Mp 184–189C. Not precipitated by digitonin. Freely soluble in hot alcohol; soluble in ethyl acetate; slightly soluble in glacial acetic acid; insoluble in water, ligroin.
Derivation: From bile and gallstones, from deoxycholic acid or cholic acid.
Use: Biochemical research.

"Lithol" Red *[BASF]*. TM for one of a group of pigments made by combining Tobias acid and β-naphthol.
Properties: Poor resistance to sunlight and weathering, generally good resistance to bleeding and to chemicals.
Available forms: Sodium, barium, and calcium lithols.
Use: Industrial enamels, toys and dipping enamels, rubber, plastics, etc.

"Lithol" Rubine *[BASF]*.
$OOCC_{10}H_5(OH)N{:}NC_6H_3(CH_3)SO_2OCa$. TM for the calcium salt of an azo pigment made by diazotizing *p*-toluidine-*m*-sulfonic acid and coupling with 3-hydroxy-2-naphthoic acid. Has poor hiding power, good resistance to bleeding and baking, fair lightfastness and alkali resistance.
Use: Paints, plastics, printing inks, cosmetics.

lithopone.
CAS: 1345-05-7. A white pigment consisting of zinc sulfide, barium sulfate, and zinc oxide; formerly used widely in paints, white rubber goods, paper, white leather etc. It has been largely replaced by titanium dioxide.

"Lith-X" *[Tyco]*. TM for a graphite-base, dry chemical extinguishing agent suitable for fires of lithium, titanium, and zirconium.

litmus. (lichen blue).
Properties: A blue, amorphous powder frequently compressed into small cakes or strips (paper). Soluble in water; changes color with acidity of solution, red at pH 4.5, blue at pH 8.3.

Derivation: By treating lichens (particularly *Variolaria*, *Lecanora*, and *Roccella*) with ammonia and potash and then fermenting the mass.
Use: Indicator in analytical chemistry where precision is not required, soil testing.

litmus milk agar.   Agar plus milk tinted with litmus to detect acidity.

Little, Arthur D.   (1863–1935). Born in Boston, Little was a pioneer in the field of industrial research and chemical consulting. Originally an authority on paper technology, he established a consulting industrial chemical laboratory in 1886, which has since become a large institution of worldwide reputation, located in Cambridge, MA. It has served as a prototype of many industrially oriented consulting firms that have become a significant factor in the growth of research in the last half century. It has made significant contributions in such fields as flavors, food chemistry and acceptability, paper chemistry, and rubber chemistry, as well as in corporate management.

liver.   A term formerly used to characterize compounds containing sulfides or sulfates, possibly because of the dark-brown color, e.g., liver of sulfur, liver of lime. The word is outmoded and should be avoided.
See livering.

livering.   A pronounced increase in the viscosity of a paint or printing ink as a result of which it becomes a coagulated and unusable mass. This may be due to oxidation or to partial polymerization of the components.

lixiviation.   The separation of mixtures by dissolving soluble constituents in water.
See solvent extraction.

LNG.   Abbreviation for liquefied natural gas.
See natural gas.

loam.   A soil comprising 25–50% sand, 30–50% silt, and 5–25% clay.
See soil.

Lo-Bax.   A hypochlorate product.
Use: Bactericide, fungicide, germicide, disinfectant, and deodorizer.

Lobry de Bruyn–van Ekenstein transformation.   Isomerization of carbohydrates in alkaline media, considered to embrace both epimerization of aldoses and ketoses and aldose–ketose interconversion.

localize.   Determination of the original position (locus) of a gene or other marker on a chromosome.

locus.   The position on a chromosome of a gene or other chromosome marker; also, the DNA at that position. The use of locus is sometimes restricted to mean regions of DNA that are expressed. Plural is loci.
See: gene expression.

Lochschmidt's constant.   (Lochschmidt number). The number of particles per unit volume of an ideal gas at STP. It has the value 2.686763×10^{25} m^{-3}. Named after its discoverer Joseph Lochschmidt (1821–1895).

"Loctite" [Henkel].   TM for anaerobic polymers that retain liquid while exposed to air and that automatically harden without heat or catalysts when confined between closely fitted metal parts. Applied to mating surfaces before, during, and after assembly.
Available forms: A range of strengths and viscosities for a wide variety of applications.

locust-bean gum.   See carob-seed gum.

lod score.   Logarithm of the odd score; a measure of the likelihood of two loci being within a measurable distance of each other.

Lo Foam.   Concentrated iodophor.
Use: For formulation of sanitizers, disinfectants, and handsoaps.

logwood.   (hematoxylin). $C_{10}H_{14}O_6 \cdot 3H_2O$.
See hematin; dye, natural.

lolinine.   See *n*-acetylloline.

"Lomotil" [Searle].   TM for dipenoxylate hydrochloride with atropine sulfate.
Use: Antidiarrhea drug.

London dispersion force.   See van der Waals forces.

London purple.
CAS: 8012-74-6.   An insecticide containing arsenic trioxide, aniline, lime, and ferrous oxide; insoluble in water.
Hazard: Toxic by ingestion and inhalation.

lone pair.   Two valence electrons of an atom that share an orbital but do not participate in a bond.

long-range restriction mapping.   Restriction enzymes are proteins that cut DNA at precise locations. Restriction maps depict the chromosomal positions of restriction-enzyme cutting sites. These are used as biochemical "signposts," or markers of specific areas along the chromosomes. The map will detail the positions where the DNA molecule is cut by particular restriction enzymes.

lopressor.   ((2R,3R)-2,3-dihydroxybutanedioic acid; 1-[4-(2-methoxyethyl)phenoxy]-3-(propan-2-ylamino)propan-2-ol; 1-[4-(2-methoxyethyl)phenoxy]-3-(propan-2-ylamino)propan-2-ol). $C_{34}H_{56}N_2O_{10}$. A beta-adremergic blocking agent.

lorica.   A vase-shaped or cup-shaped outer covering. Found in many protists, including some flagellates, ciliates, chrysophytes, and choanoflagellates, as well as in some animal cells.

"Lorox" [Tessenderlo].   TM for either a powder or solution containing linuron.

Lossen rearrangement.   Conversion of a hydroxamic acid, via a Hofmann-like arrangement, to an amine containing one less carbon than the original hydroxamic acid.

"Lo-Vel" [PPG].   TM for flatting silicas to reduce the gloss of paints, varnishes, lacquers, and industrial finishes.
Use: Automotive and specialty coating finishes.

"Lovibond" [Tintometer].   TM for color measurement and water testing equipment used in standard methods specified by ASTM, AOCS, APHA, European and U.S. Pharmacopoeia, ISO, BA, and IP.

low-alloy steel.   See alloy steel.

low-energy phosphate compound.   A phosphorylated compound with a relatively small standard free energy of hydrolysis.

low explosive.   See explosive, low.

low-melting alloy.   See alloy, fusible.

low-pressure resin.   See contact resin.

low-soda alumina.   Aluminum oxide (Al_2O_3) with <0.15% sodium oxide content.
Use: High-grade electric insulator and other ceramic bodies.

LOX.   Abbreviation for liquid oxygen, especially when used as a rocket fuel.

loxiglumide.
CAS: 107097-80-3. $C_{21}H_{30}Cl_2N_2O_5$.
Hazard: Moderately toxic by ingestion.

LPG.   Abbreviation for liquefied petroleum gas.

Lr.   Symbol for element lawrencium.

LSD.   Abbreviation for lysergic acid diethylamide.

L-Selectride.   Lithium tri-*sec*-butylborohydride, 1.0 M solution in tetrahydrofuran. LiB[CH-$(CH_3)C_2H_5]_3$H.

Properties: Liquid. Mw 190.11, d 0.890, freezing p −17C. Moisture sensitive. Packaged under nitrogen.
Hazard: Pyrophoric, must be handled under inert atmosphere.
Use: Reagent for the stereoselective reduction of ketones. Has been used in prostaglandin synthesis.

LTH.   Abbreviation for luteotropic hormone. See luteotropin.

Lu.   Symbol for lutetium.

lube-oil additive.   A chemical added in small amounts to lubricating oils to impart special qualities, such as low pour point when chlorinated hydrocarbons are added. Other special properties include the following:

low viscosity index	butene polymers
detergent and suspensoid properties	metallic stearate soaps
oxidation stability	calcium stearate
reduced foaming tendency	silicone compounds
resistance to high operating temperatures	phosphorus pentasulfide, zinc dithiophosphate

lubricant, solid.   A material having a characteristic crystalline habit that causes it to shear into thin, flat plates, which readily slide over one another and thus produce an antifriction or lubricating effect, e.g., mica, graphite, molybdenum disulfide, talc, boron nitride.

lubricant, synthetic.   Any of a number of organic fluids having specialized and effective properties that are required in cases where petroleum-derived lubricants are inadequate. Each type has at least one property not found in conventional lubricants. Though their cost is much higher, they can be used over a wide range of temperatures and are stable to heat and oxidation. The major types are polyglycols (hydraulic and brake fluids), phosphate esters (fire-resistant), dibasic acid esters (aircraft turbine engines), chlorofluorocarbons (aerospace), silicone oils and greases (electric motors, antifriction bearings), silicate esters (heat-transfer agents and hydraulic fluids), neopentyl polyol esters (turbine engines), and polyphenyl ethers (excellent heat and oxidation resistance, but poor performance at low temperatures). An unusual property of synthetic lubricants is their exceptional resistance to ionizing radiation.

lubricating grease.   A mixture of mineral oil or oils with one or more soaps. The most common soaps are those of sodium, calcium, barium, aluminum, lead, lithium, potassium, and zinc. Oils thickened with residuum, petrolatum, or wax may

be called greases. Some form of graphite may be added. Greases range in consistency from thin liquids to solid blocks, and in color from transparent to black. The specifications for a grease are determined by the speed, load, temperature, environment, and metals in the desired application. Texture of grease may be smooth, buttery, ropy or stringy, fibrous, spongy, or rubbery. The texture does not necessarily indicate the viscosity of the grease, but is related to the formulation and methods of manufacture.
See lubricating oil.

lubricating oil. (lube oil). A selected fraction of refined mineral oil used for lubrication of moving surfaces, usually metallic and ranging from small precision machinery (watches) to the heaviest equipment. Lubricating oils usually have small amounts of additives to impart special properties such as viscosity index and detergency. They range in consistency from thin liquids to grease-like substances. In contrast to lubricating greases, lube oils do not contain solid or fibrous materials.
See porpoise oil; lubricant, synthetic; extreme-pressure additive (2); lube oil additive.

lubrication. The introduction of a substance of low viscosity between two adjacent solid surfaces, one of which is in motion (bearing). From an engineering point of view, the chemical nature of the substance is not of critical importance. Thus, materials as diverse as air, water, and molasses could theoretically be used as lubricants under appropriate conditions. Air and water have been used, as well as some solids such as graphite, but in general oils, fats, and waxes are utilized. The ability of a substance to act as a lubricant is sometimes called lubricity.

lubricity. See lubrication.

"Lubriseal" [Thomas]. TM for a low-vapor-pressure, stable grease.
Use: Lubricating ground glass and metal joints and stopcocks, and for sealing desiccators and similar vessels.

luciculine.
CAS: 5008-52-6. $C_{22}H_{33}NO_3$.
Hazard: A poison.
Source: Natural product.

luciferase. An enzyme occurring in fireflies.
See luciferin.

luciferin. An albumin present in some life forms (notably fireflies) that under the influence of the enzyme luciferase, exhibits bioluminescence. When luciferin (along with luciferase) comes in contact with adenosine triphosphate, present in all living cells, a chemical reaction occurs that causes luminescence. By correlating its intensity with the amount of adenosine triphosphate present, a means

of measuring bacterial growth is obtained. Applications in biomedical technology are envisaged for this so-called luminescence biometer.

Lucifer Yellow CH. $C_{13}H_9Li_2N_5O_9S$. The dye spreads rapidly through the injected cell and binds effectively to tissue by a variety of fixatives. The movement from cell to cell is termed dye-coupling.
Properties: Fluffy-orange, hygroscopic powder. Mw 457.25.
Use: Highly fluorescent dye for marking nerve cells.

"Lucite" [Lucite]. TM for acrylic resins consisting of a series of polymeric esters of methacrylic acid, $CH_2C(CH_3)COOR$, in which R is methyl, ethyl, n-butyl, isobutyl, or combinations of these alkyl groups. These are water-white, transparent, thermoplastic resins in granular or solution forms, used in lacquers, coatings, adhesives, modifiers for other resins. "Lucite" is also a TM for an acrylic monomer, methyl methacrylate, for acrylic resins in the form of injection molding and extrusion powders; for acrylic resin dental materials; and for acrylic syrup, a liquid methyl methacrylate for use in translucent and decorative panels.
See acrylic resin.

"Ludox" [Grace]. TM for series of aqueous colloidal silica solutions.
Use: Floor polishes, paints, adhesives, paper coatings, catalyst supports, latex products, textile cleaning treatments, photosensitized paper, binder for inorganic fibrous materials, reactant for synthetic silicates.

Ludwig–Soret effect. The difference in concentration of one of the components of a mixed crystal due to temperature.

Lugatol. Anionic dyes having good tinctorial and fastness properties for all kinds of leather.

Luggin probe. A device that transmits a significant current density on the surface of an electrode to measure its potential.

Lugol's solution. An aqueous solution containing 5 g iodine and 10 g potassium iodide per 100 mL water. Grade USP.
Use: Medicine and pharmaceuticals.

Lumatex. A series of mineral and organic pigment dyestuffs that can be fixed on all types of textile fibers with suitable binders.

lumen. (lm). The SI unit of luminous flux equal to the flux emitted by a uniform point source of 1 candela in a solid angle of 1 steradian.

luminescence. The emission of visible or invisible radiation unaccompanied by high temperature

by any substance as a result of absorption of exciting energy in the form of photons, charged particles, or chemical change. It is a general term which includes both fluorescence and phosphorescence. Special types are chemiluminescence, bioluminescence, electroluminescence, photoluminescence, and triboluminescence. Common examples are light from the firefly, fluorescent lamp tubes, and television screens.
See fluorescence; phosphorescence; phosphor.

LUMO. Acronym for lowest unoccupied molecular orbital.

"Lumulse EST-300" [Vantage]. TM for an emulsifier and lubricant, a unique high HLB ester easy to formulate.
Use: Additive to many base fluids.

2,6-lupetidine. See 2,6-dimethylpiperidine.

"Luran S" [BASF]. (ASA polymer). TM for acrylic ester-modified styrene-acrylonitrile terpolymer. A thermoplastic material claimed to be excellent for outdoor use. Highly resistant to sunlight and weathering.

Lurgi process. See gasification.

luster. The appearance of the surface of a substance in reflected light. The term is used particularly in describing minerals. Types of luster include (1) metallic, like metals or the mineral pyrite; (2) vitreous, like glass or quartz; (3) adamantine, exceedingly brilliant, like diamond; (4) resinous, like resin, or sphalerite; (5) dull, not bright or shiny, like chalk.
See delustrant.

"Lustralite" [Onyx]. TM for sulfonamide formaldehyde resins used as coating vehicles.

"Lustran" [Ineos]. TM for ABS and SAN molding and extruding compounds.

Lustrasol. Acrylic and acrylic-modified alkyd solutions.

Lustrex. Styrene molding compound and extruding and calendering material.

Lustrone. Plasticized nitrocellulose lacquers tinted with dyes.
Use: Transparent color effects on leather.

lutein. A yellow pigment isolated from the corpus luteum and found in body fats and egg yolks. It is a carotenoid and is similar to or identical with xanthophyll.

luteolin-7-o-rutinoside.
CAS: 20633-84-5. $C_{27}H_{30}O_{15}$.

Properties: Flavonoid glycoside isolated from the aerial part of *Mentha piperita* L.
Hazard: A poison by ingestion.

luteotropin. (adenohypophyseal luteotropin; prolactin; lactogenic hormone; LTH).
One of the hormones secreted by the anterior lobe of the pituitary gland. It aids in causing growth of the mammary gland and initiates milk secretion by the mammary gland, it also influences the activity of the corpus luteum, including the secretion of progesterone.
Properties: Crystalline protein. Mw approximately 33,300. Almost insoluble in water; soluble in dilute acids and acidified methanol and ethanol.

lutes. Cements and adhesives composed of oxides, clays, and silicas.

lutetia. See lutetium oxide; rare earths.

lutetium. Lu. Metallic element, atomic number 71, group IIIB of the periodic table, a lanthanide rare earth, aw 174.97, valence = 3. It has one natural radioactive isotope (^{176}Lu) with half-life of 2.2 × 10^{10} years.
See rare-earth metals.
Properties: Metallic luster, soft and ductile. D 0.849, mp 1652C, bp 3327C. Reacts slowly with water; soluble in dilute acids; difficult to isolate.
Occurrence: Monazite (approximately 0.003%) (India, Brazil, Africa, Australia, U.S.).
Derivation: Reduction of the fluoride or chloride with calcium.
Grade: Regular, high purity (ingots, lumps).
Use: Nuclear technology.

lutetium chloride. $LuCl_3 \cdot xH_2O$ (anhydrous).
Properties: D 3.98, mp 905C. Water-soluble.
Grade: Up to 99.9% lutetium salts.

lutetium fluoride.
CAS: 13760-81-1. $LuF_3 \cdot 2H_2O$ (anhydrous).
Properties: Mp 1182C, bp 2200C. Insoluble in water.
Grade: Up to 99.9% lutetium salts.
Hazard: Irritant.

lutetium nitrate.
CAS: 100641-16-5. $Lu(NO_3)_3 \cdot 6H_2O$.
Grade: Up to 99.9% lutetium salts.
Hazard: Fire and explosion risk.

lutetium oxide. (lutetia).
CAS: 12032-20-1. Lu_2O_3.
Properties: White solid. Slightly hygroscopic. Mp approximately 2500C, d 9.42. Absorbs water and carbon dioxide.
Derivation: Monazite sand.
Grade: Up to 99.9% purity.

lutetium sulfate. $Lu_2(SO_4)_3 \cdot 8H_2O$.
Properties: Soluble in water.
Grade: Up to 99.9% lutetium salts.

2,6-lutidine. (2,6-dimethylpyridine).

$\overline{NC(CH_3)CHCHCHCCH_3}$.

Properties: Colorless, oily liquid; peppermint odor. D 0.932, bp 143C, fp −6.6C, refr index 1.4973 (20C). Derived from coal tar.
Available forms: 2,3-, 2,4-, 2,5-, 3,4-, 3,5-isomers.
Hazard: Flammable.
Use: Pharmaceuticals, resins, dyestuffs, rubber accelerators, insecticides.

"Lutonal" [BASF]. TM for vinyl ether polymers, solid and in solution.

lutrelin acetate. See luteinizing hormone-releasing factor (pig), 6-d-tryptophan.

lux. (lx). The SI unit of illuminance equal to the illumination produced by a luminous flux of one lumen distributed uniformly over an area of one square meter.

"Luxol" [Dow]. TM for a series of spirit and lacquer-soluble dyes.
Use: Lacquers, wood stains, stamp inks, ball-point pen inks.

"LX-685" [Neville]. TM for a heat-reactive resin used in the manufacture of ready-mixed aluminum paints, grease- and gasoline-resistant coatings, floor and deck enamels, and concrete curing compounds.

lyases. Enzymes that catalyze the removal of a group from a molecule to form a double bond, or the addition of a group to a double bond.

lycopene. (carotene). $C_{40}H_{56}$. A long-chain hydrocarbon with 13 double bonds of which 11 are conjugated. Combustible.
Properties: Red crystals. Mp 175C. Soluble in chloroform and carbon disulfide; slightly soluble in alcohol, insoluble in water.
Source: The main pigment of tomato, paprika, rose hips, etc.

lycopodium. Spores from *Lycopodium clavatum*, a fungus native to North America and Europe.
Properties: A flammable yellowish powder.
Hazard: Flammable, protect from open flame.
Use: In the pharmaceutical industry as a coating, in flashlight powders and pyrotechnics.

"Lycra" [Invista]. TM for a spandex fiber in the form of continuous monofilaments. Available in yarn with denier from 40 to 2240.

Properties: Tensile strength 0.8 gpd, d 1.21, break elongation 550%, moisture regain 1.3%, mp 230C, tensile recovery 95% from 50% elongation. Soluble in dimethyl formamide (hot).
Available forms: Yarn with denier from 40–2240.
Hazard: Combustible, but self-extinguishing.
Use: Foundation garments, swim wear, surgical hose, and other elastic products.

lye. See sodium hydroxide; potassium hydroxide.

Lykopon. Sodium hydrosulfite.

lymph. An almost colorless fluid that travels through the lymphatic system and carries cells that help fight infection and disease.

lymphatic system. The tissues and organs, including the bone marrow, spleen, thymus, and lymph nodes, that produce and store cells that fight infection and disease. This system also has channels that carry lymph.

lymphatolytic serum. A serum that is destructive to lymphatic tissues.

lymph nodes. Small, bean-shaped organs located along the channels of the lymphatic system. Bacteria or cancer cells that enter the lymphatic system may be found in the nodes. Also called lymph glands.

lymphocytes. A subclass of leukocytes characterized as mononuclear, with a small amount of cytoplasm. Involved in the immune response and divided into two types: B lymphocytes (synthesize and secrete antibodies) and T lymphocytes (participate in cell-mediated immunity).

lymphocytotoxic antibody. An antibody that is specified for antibody histocompatibility antigens of lymphocytes. The antigens to which it attaches are damaged or killed.

lyocratic. See lyophilic.

lyophilic. Characterizing a material that readily goes into colloidal suspension in a liquid; if into water, it is called hydrophilic. The colloid is stabilized by the formation of an adsorbed layer of molecules of the dispersing medium about the suspended particles. Systems of this type are said to be lyocratic. Examples: glue, gelatin, milk-fat particles.

lyophilic colloid. See reversible colloid.

lyophilization. See freeze-drying.

lyophobic. Characterizing a material that exists in the colloidal state but with a tendency to repel liquids; if the liquid is water, the material is called hydrophobic. Such colloids are generally stabilized by the adsorption of ions and coagulate when the charge is neutralized. Examples: colloidal gold, colloidal arsenic sulfide.

lyophobic colloid.   Colloid that is readily precipitated form solution and cannot be redispersed by an addition of the solvent.

lyotropic series.   Ions, radicals, or salts placed in order of their effect on various catalytic, colloidal, and physiological phenomena.

Lys.   Abbreviation for lysine.

lysergic acid. $C_{16}H_{16}N_2O_2$. Structure based on a condensed four-ring nucleus.
  Properties: Crystals (with 1–2 molecules of water) in plates from water. Mp 240C (decomposes). Amphoteric; moderately soluble in pyridine; slightly soluble in water and neutral organic solvents; soluble in alkaline and acid solutions.
  Derivation: Alkaline hydrolysis of ergot alkaloids, organic synthesis.
  Use: Medical research, synthesis of ergonovine.

D-lysergic acid diethylamide.   (LSD).
  CAS: 50-37-3. $C_{20}H_{25}N_3O$.
  Synthetic derivative of lysergic acid. Only the D-form is active. Tasteless, colorless, odorless.
  Hazard: A strong hallucinatory and habit-forming drug with possible mutagenic effects. Manufacture and sale are under legal restraint.

lysine.   (α,ε-diaminocaproic acid; Lys).
  CAS: 56-87-1. $NH_2(CH_2)_4CH(NH_2)COOH$.
  An essential amino acid.
  Properties: Colorless crystals. Optically active: D(−)-lysine, mp 224C with decomposition; L(+)-lysine: mp 224C with decomposition. Soluble in water; slightly soluble in alcohol; insoluble in ether.
  Derivation: Extraction of natural proteins, synthetically by fermentation of glucose or other carbohydrates, and by synthesis from caprolactam.

Use: Biochemical and nutritional research, pharmaceuticals, culture media, fortification of foods and feeds (wheat flour), nutrient and dietary supplement, animal feed additive.
Available forms: As DL- and L-lysine monohydrochloride.

lysine.   (aminutrin; α,ε-diaminocaproic acid; 2,6-diaminohexanoic acid; 1-lysine; 1-(+)-lysine; lysine acid).
  CAS: 56-87-1. $C_6H_{14}N_2O_2$.
  A type of antibody destroys cells by lysis of the cell membrane (plasmalemma) or the cell wall in certain cases.
  Properties: Needles from water, hexagonal plates from dilute alcohol; soluble in water; slightly soluble in alcohol; insoluble in either.
  Hazard: Teratogen.

lysergic acid α-hydroxyethylamide.
  CAS: 3343-15-5. $C_{18}H_{21}N_3O_2$.
  Hazard: A reproductive hazard.

lysosome.   Eukaryotic organelle which carries digestive enzymes. The lyzosome fuses with a vacuolar membrane containing ingested particles, which are then acted upon by the enzymes.

lysozyme.   An antibiotic enzyme found in egg white. By hydrolyzing certain sugar linkages in glycoproteins, it can dissolve the mucopolysaccharides found in the walls of certain bacteria and hence acts as a mild antiseptic. Its three-dimensional structure has been determined by X-ray crystallography. It is a molecule containing approximately 2200 atoms comprising 129 amino acid units strung together and intricately folded (mw 14,500).

lyxoflavine. $C_{17}H_{20}N_4O_6$.
  Properties: Yellowish crystals. Mp 284C (decomposes). Optically active. Almost insoluble in water and alcohol; soluble in alkaline solutions. It is a chemical analog of riboflavin.
  Use: Growth promoter for agricultural products.

M

M. Abbreviation for molar, used to characterize the concentration of a solution. A molar solution contains one mole of a substance in one liter of solution.

m. Abbreviation for meter.

m-. (m-). Abbreviation for meta-.

M13. A bacteriophage which infects certain strains of *Escherichia coli*. The salient feature of this phage is that it packages only a single strand of DNA into its capsid. If the investigator has inserted some heterologous DNA into the M13 genome, copious quantities of single-stranded DNA can subsequently be isolated from the phage capsids. M13 is often used to generate templates for DNA sequencing.

μm. Abbreviation for micrometer.

mμ. Abbreviation for millimicron.

MAA. See methylarsenic acid.

MAC. Abbreviation for methyl allyl chloride.

MacDiarmid, Alan Graham. (1927–2007). Awarded Nobel Prize in chemistry in 2000 jointly with Alan J. Heeger and Hideki Shirakawa for the discovery and development of conductive polymers. He performs his research at the University of Pennsylvania, Philadelphia.

Mace. (Chemical Mace). A riot-control gas dispersed as an aerosol. See chloroacetophenone.

macerate. To soften or break up a fibrous substance by soaking long in water at or near room temperature, often accompanied by mechanical action, as in the preparation of paper stock in the beater. In the plastics industry, to comminute a fabric so that it can be used as a filler in a plastics composition. The term is also used in pharmacy to describe a method of preparing medicinal compositions.

"Machadet EZ-154 High Foam Powder" [Rhodia]. TM for shampoo blend.

M acid. (1-amino-5-naphthol-7-sulfonic acid). $C_{10}H_5NH_2OHSO_3H$.

Properties: Gray needles. Slightly soluble in cold water; soluble in hot water and alcohol.
Use: Azo dye intermediate.

"Mackadet EQ 112 P" [Rhodia]. TM for pearlized baby shampoo concentrate.

"Mackalene 110" [Rhodia]. TM for cocamidopropyl dimethylamine lactate.
Grade: Cosmetic and industrial.
Use: Hair conditioning agent for shampoos and conditioners.

"Mackamide CMA" [Rhodia]. TM for cocamide monoethanol amine (MEA).
Properties: Low solubility in water.
Grade: Flaked and cosmetic.
Use: Viscosity builder and foam enhancement.

"Mackine 301" [Rhodia]. TM for stearamidopropyl dimethylamine.
Grade: Flaked, cosmetic, and industrial.
Use: Hair conditioning agent for shampoos and conditioners.

MacKinnon, Roderick. (1956–). An American born in Burlington, MA, who won the Nobel Prize for chemistry in 2003 for his pioneering work discovering channels in cell membranes, in particular for the structural and mechanistic studies of ion channels. He received a B.A. in biochemistry from Brandeis University and an M.D. from Tufts University School of Medicine. MacKinnon is a member of the National Academy of Sciences and was awarded the 1999 Albert Lasker Basic Medical Research Award.

Macpearl DR-140V. A glycol disterate pearl concentrate.

macroanalysis. Analysis of chemicals in gram quantities.
See microanalysis.

macrocyclic. An organic molecule with a large ring structure containing over 15 carbon atoms.

macromolecule. A molecule, usually organic, composed an aggregation of hundreds or thousands of atoms. Such giant molecules are generally of two types. (1) Individual entities (compounds) that

cannot be subdivided without losing their chemical identity. Typically these are proteins, many of which have molecular weights running into the millions. (2) Combinations of repeating chemical units (monomers) linked together into chain or network structures called polymers; each monomer has the same chemical constitution as the polymer, e.g., isoprene (C_5H_8) and polyisoprene ($C_5H_8)_n$. Synthetic elastomers (plastics) are typical of this kind of macromolecule; cellulose is the most common example found in nature. Most macromolecules are in the colloidal size range.
See polymer, high; protein; colloid chemistry.

macrorestriction map.   Map depicting the order of and distance between sites at which restriction enzymes cleave chromosomes.

macrose.   See dextran.

madder.   A natural dyestuff.
See alizarin; lake; dye, natural.

Maddrell's salts.   (IMP). Insoluble sodium metaphosphate, $NaPO_3$-II and $NaPO_3$-III.
See sodium metaphosphate.

Madelung synthesis.   Formation of indole derivatives by intramolecular cyclization of an *N*-(2-alkylphenyl)alkanamide by a strong base at high temperature.

"MagClean HSAIO" [Produits].   TM for a micronized high activity MgO.
 Use: A scorch retarder in chloroprene and fluorelastomers, thickening agents for SMC/BMC acid acceptors in halogenated and chloroprene adhesives.

magenta.   See fuchsin.

magic acid.   A superacid containing equal molar concentrations of antimony pentafluoride (SbF_6) and fluorosulfonic acid (FSO_2OH).

magic numbers.   Atoms whose nuclei have the number of protons, neutrons, or both with any of the values, 8, 20, 28, 50, 82, 126. These elements have a stability and binding energy that are greater than average and have other special properties.

"Maglite" [MAG].   (magnesium oxide).
CAS: 1309-48-4. TM for acid acceptor and stabilizer.
 Use: In rubber, plastic thickener filler, and smoke suppressant.

mag-lith.   A magnesium-lithium alloy used as a structural metal in space vehicles.

magma.   (1) In medicine, a class of preparations in which finely divided, freshly precipitated, insoluble, inorganic hydroxides are suspended in water to form a viscous, opaque mixture that may settle out on standing. Magmas of bismuth, magnesium, and iron are used, commonly called milk of bismuth, milk of magnesia, etc.
(2) In geology, a molten mass within the earth's crust (e.g., lava). The source of igneous rock.

Magnacide H Herbicide.   A liquid herbicide.
 Use: To destroy submersed aquatic weeds in irrigation systems.

magnalium.   An alloy of aluminum and magnesium.

magnesia.   Magnesium oxide that has been specially processed.
See magnesium oxide.

magnesia alba.   See magnesium carbonate; magnesium carbonate, basic.

magnesia-alumina. $MgO\bullet Al_2O_3$. A synthetic spinel.

magnesia, burnt.   See magnesite, dead-burned.

magnesia, calcined.   See magnesite, caustic-calcined.

magnesia, caustic-calcined.   See magnesite, caustic-calcined.

magnesia-chromia. $MgO\bullet Cr_2O_3$. A synthetic spinel.

magnesia, dead-burned.   See magnesite, dead-burned.

magnesia, fused.
 Use: As a refractory and to handle electricity at high temperatures.

magnesia, lightburned.   A special high-purity magnesium oxide.

magnesite.   (natural magnesium carbonate).
CAS: 546-93-0.   ($MgCO_3$). The term magnesite is loosely used as a synonym for magnesia as are also the terms *caustic-calcined magnesite, dead-burned magnesite,* and *synthetic magnesite.*
 Hazard: A nuisance particulate.
 Properties: White, yellowish, grayish-white, or brown crystalline solid. D 3–3.12, Mohs hardness 3.5–4.5.
 Occurrence: U.S. (California, Washington, Nevada), Austria, Greece.
 Use: To make the various grades of magnesium oxide, to produce carbon dioxide, refractory.
See magnesium carbonate.

magnesite, burnt.   See magnesite, dead-burned.

magnesite, caustic-calcined. (caustic-calcined magnesia; calcined magnesite; calcined magnesia. Principally magnesia (magnesium oxide) MgO). The product obtained by firing magnesite or other substances convertible to magnesia upon heating at some temperature below 1450C so that some carbon dioxide is retained (2–10%) and the magnesium oxide displays adsorptive capacity or activity.
Grade: Technical, chemical, synthetic rubber, USP (light, medium light, heavy).
Use: Magnesium oxychloride and oxysulfate cements, 85% magnesia insulation, rubber (reinforcing agent, accelerator), uranium processing, chemical processing, rayon, refractories, paper pulp, acid-neutralizing fertilizers, welding-rod coatings, fillers, glass constituents, abrasives.
See magnesium oxide.

magnesite, dead-burned. (burnt magnesia; dead-burned magnesia; refractory magnesia; burnt magnesite; magnesium oxide). MgO. The granular product obtained by burning (firing) magnesite or other substances convertible to magnesia upon heating above 1450C long enough to form granules suitable for use as a refractory (ASTM). Synthetic magnesium hydroxide or chloride is sometimes used instead of magnesite as a source.
Grade: 85–87% (from magnesite ores); 97–99% (from seawater and brines).
Use: Refractories, as grains or basic brick, the latter especially in open hearth furnaces for steel, furnaces for nonferrous metal smelting, and in cement and other kilns.
See magnesium oxide.

magnesite, synthetic. Magnesium oxide, MgO, as obtained from seawater, seawater bitterns, or well brines. The preliminary product is usually magnesium hydroxide or chloride, which is then heated, or sometimes treated with steam and heated in the case of the chloride, to obtain the oxide. Synthetic magnesite constitutes the purer grades of dead-burned magnesite.

magnesium.
CAS: 7439-95-4. Mg. Metallic element of atomic number 12, group IIA of the periodic table, aw 24.305, valence = 2; three isotopes. Magnesium is the central element of the chlorophyll molecule; it is also an important component of red blood corpuscles.
Properties: Silvery, moderately hard, alkaline-earth metal; readily fabricated by all standard methods. Lightest of the structural metals; strong reducing agent; electrical conductivity similar to aluminum. D 1.74, mp 650C, bp 1107C. Soluble in acids; insoluble in water.
Source: Magnesite and dolomite; seawater and brines.
Derivation: (1) Electrolysis of fused magnesium chloride (Dow seawater process); (2) reduction of magnesium oxide with ferrosilicon (Pidgeon process).
Available forms: Ingots, bars, fine powder (up to 99.6% pure), sheet and plate, rods, tubing, ribbon, flakes.
Hazard: (Solid metal) Combustible at 650C. (Powder, flakes, etc.) Flammable, dangerous fire hazard. Use dry sand or talc to extinguish.
Use: Aluminum alloys for structural parts, die-cast auto parts, missiles, space vehicles; powder for pyrotechnics and flash photography, production of iron, nickel, zinc, titanium, zirconium; anti-knock gasoline additives; magnesium compounds and Grignard syntheses; cathodic protection; reducing agent; desulfurizing iron in steel manufacture; precision instruments; optical mirrors; dry and wet batteries.

magnesium acetate.
CAS: 142-72-3. (1) $Mg(OOCCH_3)_2$ or (2) $Mg(OOCCH_3)_2 \cdot 4H_2O$.
Properties: Colorless, crystalline aggregate or monoclinic crystals; acetic acid odor. (1) Mp 323C, d 1.42; (2) mp 80C, d 1.45. Soluble in water and dilute alcohol.
Derivation: Interaction of magnesium carbonate and acetic acid.
Use: Dye fixative in textile printing, deodorant, disinfectant, and antiseptic.

magnesium acetylacetonate. $Mg(C_5H_7O_2)_2$. Crystalline powder, slightly soluble in water, resistant to hydrolysis, a chelating nonionizing compound.

magnesium amide.
CAS: 7803-54-5. $Mg(NH_2)_2$.
Properties: Whitish to gray crystals. D 1.40. Decomposes when heated.
Derivation: Reaction of magnesium with ammonia under elevated pressure.
Hazard: A pyrophoric material igniting in air at room temperature. Evolves ammonia on vigorous reaction with water.
Use: Catalyst for polymerization.

magnesium ammonium arsenate dihydrate.
CAS: 14644-70-3. $AsO_4 \cdot H_3N \cdot Mg \cdot 2H_2O$.
Hazard: A poison by ingestion and skin contact.

magnesium ammonium orthophosphate. See magnesium ammonium phosphate.

magnesium ammonium phosphate. (magnesium ammonium orthophosphate).
CAS: 7785-21-9. $MgNH_4PO_4 \cdot 6H_2O$.
Properties: White powder. D 1.71, mp (decomposes to magnesium pyrophosphate, $Mg_2P_2O_7$). Soluble in acids; insoluble in alcohol and water.

Derivation: By the interaction of solutions of a magnesium salt and ammonium phosphate.
Use: Fire retardant for fabrics, fertilizer.

magnesium arsenate.
(arsenic acid, magnesium salt).
CAS: 10103-50-1. $Mg_3(AsO_4)_2 \cdot xH_2O$.
Properties: White powder. When pure it is insoluble in water. Technical material is highly hydrated and made from magnesium carbonate and arsenic acid.
Hazard: Toxic by ingestion and inhalation.
Use: Insecticide.

magnesium benzoate.
CAS: 553-70-8. $Mg(C_7H_5O_2)_2 \cdot 3H_2O$.
Properties: White, crystalline powder. Loses $3H_2O$ at 110C, mp approximately 200C. Soluble in water and alcohol.

magnesium biphosphate.
See magnesium phosphate, monobasic.

magnesium borate.
CAS: 13703-82-7. $3MgO \cdot B_2O_3$ (orthoborate) or $Mg(BO_2)_2 \cdot 8H_2O$ (metaborate).
Properties: Transparent, colorless crystals or white powder. Soluble in alcohol, acetic acid, and inorganic acids; slightly soluble in water.
Derivation: By heating magnesium oxide, boric anhydride.
Use: Preservative, antiseptic, fungicide.

magnesium borocitrate.
$Mg(BO_2)_2 \cdot Mg_3(C_6H_5O_7)_2 \cdot 14H_2O$.
Properties: White powder or small, white, lustrous scales. Soluble in water.
Derivation: By mixing magnesium borate and magnesium citrate.

magnesium boron fluoride.
Grade: Technical.
Hazard: Strong irritant.
Use: Metal flux.

magnesium bromate.
CAS: 14519-17-6. $Mg(BrO_3)_2 \cdot 6H_2O$.
Properties: White crystals or crystalline powder. D 2.29, mp loses $6H_2O$ at 200C, bp (decomposes). Soluble in water; insoluble in alcohol.
Derivation: By adding magnesium sulfate to a solution of barium bromate.
Hazard: Dangerous fire risk in contact with organic materials.
Use: Analytical reagent, oxidizing agent.

magnesium bromide.
CAS: 7789-48-2. $MgBr_2 \cdot 6H_2O$.
Properties: Colorless, very deliquescent crystals; bitter taste. D 2.00, mp 172C, mp (anhydrous) 700C. Soluble in water; slightly soluble in alcohol.

Derivation: Reaction of hydrobromic acid with magnesium oxide and subsequent crystallization.
Use: Organic syntheses, medicine (sedative).

magnesium carbonate.
CAS: 13717-00-5. $MgCO_3$. The term *magnesium carbonate* is generally reserved for the synthetic, pure variety. The naturally occurring material is called magnesite.
Properties: Light, bulky, white powder. Bulk d approximately 4 lb/ft^3, d 3.0, decomposes 350C, refr index about 1.52. Soluble in acids; very slightly soluble in water; insoluble in alcohol. Noncombustible.
Derivation: (1) Mined as natural material; (2) carbonation of magnesium oxide or $Mg(OH)_2$ with CO_2; (3) reaction of a soluble magnesium salt solution with sodium carbonate or bicarbonate.
Grade: Technical, NF, FCC.
Use: Magnesium salts, heat insulation and refractory, rubber reinforcing agent, inks, glass, pharmaceuticals, dentifrice and cosmetics, free-running table salts, antacid, making magnesium citrate, filtering medium. In foods as drying agent, color retention agent, anticaking agent, carrier.

magnesium carbonate, basic.
(magnesia alba). Various formulas are given and may all be possible because of the method of derivation. A typical formula is $Mg(OH)_2 \cdot 3MgCO_3 \cdot 3H_2O$. Properties and uses are almost identical with those listed for magnesium carbonate.
Derivation: Precipitation from magnesium salt solution.
See magnesium carbonate.

magnesium chlorate.
CAS: 10326-21-3. $Mg(ClO_3)_2 \cdot 6H_2O$.
Properties: White powder; bitter taste. Very hygroscopic. D 1.8, mp 35C (decomposes at 120C). Soluble in water; slightly soluble in alcohol.
Hazard: Dangerous fire risk in contact with organic materials, strong oxidizing agent.
Use: Defoliant, desiccant.

magnesium chloride.
CAS: 7786-30-3. (1) $MgCl_2$, (2) $MgCl_2 \cdot 6H_2O$.
Properties: Colorless or white crystals. Deliquescent, d (1) 2.32, (2) 1.56; mp (1) 708C, (2) loses $2H_2O$ at 100C. if heated rapidly melts at 116–118C. Bp (1) 1412C, (2) decomposes to oxychloride. Soluble in water and alcohol.
Derivation: Action of hydrochloric acid on magnesium oxide or hydroxide, especially the latter when precipitated from seawater or brines (Great Salt Lake).
Method of purification: Recrystallization.
Grade: Technical (crystals, fused, flakes, granulated), CP.
Hazard: Toxic by ingestion.

Use: Source of magnesium metal, disinfectants, fire extinguishers, fireproofing wood, magnesium oxychloride cement, refrigerating brines, ceramics, cooling drilling tools, textiles (size, dressing and filling of cotton and woolen fabrics, thread lubricant, carbonization of wool), paper manufacture, road dust-laying compounds, floor sweeping compounds, flocculating agent, catalyst.

magnesium chromate.
CAS: 16569-85-0. $MgCrO_4 \cdot 5H_2O$.
Properties: Small, readily soluble, yellow crystals.
Use: Since it does not produce a fusible alkaline residue when thermally decomposed, it is used as a corrosion inhibitor in the water coolant of gas turbine engines. Insoluble basic magnesium chromates also are available. Their potential applications are in the treatment of light metal surfaces.
Hazard: Toxic by ingestion.

magnesium citrate, dibasic.   (acid magnesium citrate).
CAS: 7779-25-1. $MgHC_6H_5O_7 \cdot 5H_2O$.
Properties: White or slightly yellow granules or powder; odorless. Soluble in water; insoluble in alcohol.
Derivation: Reaction of citric acid and magnesium hydroxide or carbonate.
Use: Laxative, dietary supplement.

magnesium dichromate.
CAS: 13423-61-5. $MgCr_2 \cdot 6H_2O$.
Properties: Characterized by high solubility in water. It is an orange-red, deliquescent, crystalline hydrate.
Use: Potential applications are in formulations for corrosion prevention and metal treatment. Noncombustible.

magnesium dioxide.   See magnesium peroxide.

magnesium fluoride.   (magnesium flux).
CAS: 7783-40-6. MgF_2.
Properties: White crystals. D 3.15, mp 1263C, bp 2239C. Exhibits fluorescence by electric light. Soluble in nitric acid; insoluble in alcohol and water. Noncombustible.
Derivation: By adding sodium fluoride or hydrofluoric acid to a solution of magnesium salt.
Grade: Technical, CP, single crystals.
Hazard: Strong irritant. TLV: 2.5 mg(F)/m³.
Use: Ceramics, glass, single crystals for polarizing prisms, lenses, and windows.

magnesium fluosilicate.   (magnesium silicofluoride).
CAS: 18972-56-0. $MgSiF_6 \cdot 6H_2O$.
Properties: White, efflorescent, crystalline powder. D 1.788, decomposes 120C. Soluble in water.
Derivation: By treating magnesium hydroxide or carbonate with hydrofluosilicic acid.

Grade: Technical (crystals, solution).
Hazard: Strong irritant.
Use: Ceramics, concrete hardeners, waterproofing, mothproofing, laundry sour, magnesium casting.

magnesium flux.   See magnesium fluoride.

magnesium formate. $Mg(CHO_2)_2 \cdot 2H_2O$.
Properties: Colorless crystals. Soluble in water; insoluble in alcohol and ether. Combustible.
Derivation: Action of formic acid on magnesium oxide.
Use: Analytical chemistry.

magnesium gluconate.
CAS: 3632-91-5. $Mg(C_6H_{11}O_7)_2 \cdot 2H_2O$.
Properties: White powder or fine needles; odorless; almost tasteless. Soluble in water. Combustible.
Derivation: Magnesia or magnesium carbonate dissolved in gluconic acid.
Grade: Pharmaceutical.
Use: Medicine, vitamin tablets.

magnesium glycerophosphate.   (magnesium glycerinophosphate).
CAS: 927-20-8. $CH_2(OH)CH(OH)CH_2OP(O)O_2Mg$.
Properties: Colorless powder. Soluble in water; insoluble in alcohol. Combustible.
Derivation: Action of glycerophosphoric acid on magnesium hydroxide.
Use: Food additive; plasticizer.

magnesium hydrogen phosphate.   See magnesium phosphate, dibasic.

magnesium hydroxide.   (magnesium hydrate in aqueous suspension; milk of magnesia; magnesia magma).
CAS: 1309-42-8. $Mg(OH)_2$.
Properties: White powder; odorless. D 2.36, mp decomposes at 350C. Soluble in solution of ammonium salts and dilute acids; almost insoluble in water and alcohol. Noncombustible.
Derivation: Precipitation from a solution of a magnesium salt by sodium hydroxide, precipitation from seawater with lime. It occurs naturally as brucite.
Grade: Technical, NF, FCC.
Use: Intermediate for obtaining magnesium metal, sugar refining, medicine (antacid, laxative), residual fuel oil additive, sulfite pulp, uranium processing, dentifrices, in foods as drying agent, color retention agent, frozen desserts.

magnesium lauryl sulfate.
$Mg(OSO_3C_{12}H_{25})_2$.
Properties: Pale-yellow liquid; mild odor. Soluble in methanol, acetone, and water; insoluble in kerosene. Combustible.

Derivation: Sulfonation of lauryl alcohol and interaction with a magnesium salt.

Use: Surfactant and anionic detergent, foaming, wetting, and emulsifying agent.

magnesium lime. Same as magnesium limestone.
See limestone.

magnesium limestone. See limestone.

magnesium methoxide. (magnesium methylate).
CAS: 27428-49-5. $(CH_3O)_2Mg$.
Properties: Colorless, crystalline solid. Decomposes on warming.
Derivation: Reaction of magnesium and methanol.
Use: Dielectric coating, cross-linking agent to form stable gels, catalyst.

magnesium methylate. See magnesium methoxide.

magnesium molybdate.
CAS: 12013-21-7. $MgMoO_4$.
Properties: Crystalline powder. Absolute d 2.8, mp approximately 1060C. Soluble in water.
Use: Electronic and optical applications.

magnesium nitrate.
CAS: 10377-60-3. $Mg(NO_3)_2•2H_2O$.
Properties: White crystals. D 1.45, mp 95–100C, decomposes at 330C. Soluble in water and alcohol; deliquescent.
Derivation: Action of nitric acid on magnesium oxide with subsequent crystallization.
Hazard: Dangerous fire and explosion risk in contact with organic materials, strong oxidizing agent.
Use: Pyrotechnics.

magnesium oleate.
CAS: 1555-53-9. $Mg(C_{18}H_{33}O_2)_2$.
Properties: Yellowish mass. Soluble in linseed oil, hydrocarbons, alcohol, and ether; insoluble in water. Combustible.
Derivation: Interaction of magnesium chloride and sodium oleate.
Use: Varnish driers, in dry-cleaning solvents (to prevent spontaneous ignition), emulsifying agent, lubricant for plasticizers.

magnesium orthophosphate. See magnesium phosphate.

magnesium oxide. (magnesia).
CAS: 1309-48-4. MgO. Two forms are produced, one a light, fluffy material prepared by a relatively low-temperature dehydration of the hydroxide, the other a dense material made by high-temperature furnacing of the oxide after it has been formed from the carbonate or hydroxide.
See periclase.

Properties: White powder, either light or heavy depending upon whether it is prepared by heating magnesium carbonate or the basic magnesium carbonate. D approximately 0.36 (varies), mp 2800C, bp 3600C. Slightly soluble in water; soluble in acids and ammonium salt solutions. Noncombustible.
Derivation: (1) By calcining magnesium carbonate or magnesium hydroxide, (2) by treating magnesium chloride with lime and heating or by heating it in air, (3) from seawater via the hydroxide.
Grade: Technical, CP, USS, FCC, 99.5%, fused, low boron, rubber, semiconductor, single crystals.
Hazard: Toxic by inhalation of fume. Upper respiratory tract irritant, and metal fume fever. Questionable carcinogen.
Use: Refractories, especially for steel furnace linings, polycrystalline ceramic for aircraft windshields, electrical insulation, pharmaceuticals and cosmetics, inorganic rubber accelerator, oxychloride and oxysulfate cements, paper manufacture, fertilizers, removal of sulfur dioxide from stack gases, adsorption and catalysis, semiconductors, pharmaceuticals, food and feed additive.
See "Maglite" *[MAG]*.

magnesium oxychloride cement. (Sorel cement). A mixture of magnesium chloride and magnesium oxide that reacts with water to form a solid mass, presumed to be magnesium oxychloride. Fillers such as wood flour, sawdust, sand, powdered stone, talc, etc., are usually present. A variety of proprietary mixtures are available. Strength ranges from 7000 to 10,000 psi. Copper powder minimizes water solubility.

magnesium palmitate. $Mg(C_{16}H_{31}O_2)_2$.
Properties: Crystalline needles or white lumps. Mp 121.5C; insoluble in water and alcohol. Combustible.
Use: Varnish drier, lubricant for plastics.

magnesium perborate.
CAS: 17097-11-9. $Mg(BO_3)_2•7H_2O$.
Properties: White powder. Sparingly soluble in water; decomposes with evolution of oxygen.
Derivation: Action of peroxide or electrolytic oxidation of borate solutions.
Hazard: Moderate fire risk in contact with organic materials.
Use: Driers, bleaching, antiseptic (tooth powders).

magnesium perchlorate.
CAS: 10034-81-8. (1) $Mg(ClO_4)_2$; (2) $Mg(ClO_4)_2•6H_2O$.
Properties: White crystals. Deliquescent; very soluble in water and alcohol. (1) D 2.21 (18C), decomposes at 251C (2) d 1.98, decomposes at 185–190C.
Derivation: Reaction of magnesium hydroxide and perchloric acid.

Hazard: Dangerous fire and explosion risk in contact with organic materials.
Use: (1) As a regenerable drying agent for gases and (2) oxidizing agent.

magnesium permanganate.
$Mg(MnO_4)_2 \cdot 6H_2O$.
Properties: Bluish-black, friable, deliquescent crystals. D 2.18, mp (decomposes). Soluble in water.
Hazard: Fire hazard in contact with organic materials. Powerful oxidizer.
Use: Polymerization catalyst, antiseptic.

magnesium peroxide. (magnesium dioxide).
CAS: 14452-57-4. MgO_2.
Properties: White, powder; tasteless; odorless. Decomposes above 100C. Insoluble in water; soluble in dilute acids with formation of hydrogen peroxide. Available oxygen 28.4%. Keep cool and dry. A powerful oxidizing material.
Derivation: From sodium or barium peroxide with magnesium sulfate in a concentrated solution.
Grade: Technical, 15, 25, and 50%.
Hazard: Powerful oxidizer and dangerous fire risk, reacts with acidic materials and moisture.
Use: Bleaching and oxidizing agent, medicine (antacid).

magnesium phosphate. (magnesium orthophosphate).
See magnesium phosphate, dibasic; magnesium phosphate, monobasic; magnesium phosphate, tribasic.

magnesium phosphate, dibasic. (dimagnesium orthophosphate; dimagnesium phosphate; magnesium phosphate, secondary; magnesium hydrogen phosphate).
CAS: 7782-75-4. $MgHPO_4 \cdot 3H_2O$.
Properties: White, crystalline powder. D 2.13, loses water at 205C, decomposes at 550–650C, decomposes to pyrophosphate on heating. Soluble in dilute acids; slightly soluble in water. Nonflammable.
Derivation: Action of orthophosphoric acid on magnesium oxide.
Grade: Technical, FCC.
Use: Stabilizer for plastics, food additive, medicine (laxative).

magnesium phosphate, monobasic. (magnesium biphosphate; acid magnesium phosphate; magnesium tetrahydrogen phosphate).
$MgH_4(PO_4)_2 \cdot 2H_2O$.
Properties: White, hygroscopic, crystalline powder. Decomposes to metaphosphate on heating. Soluble in water and acids; insoluble in alcohol. Nonflammable.
Derivation: Action of orthophosphoric acid on magnesium hydroxide.
Use: Fireproofing wood, stabilizer for plastics.

magnesium phosphate, neutral. See magnesium phosphate, tribasic.

magnesium phosphate, secondary. See magnesium phosphate, dibasic.

magnesium phosphate, tribasic. (Magnesium phosphate, neutral; trimagnesium phosphate).
CAS: 7757-87-1. $Mg_3(PO_4)_2 \cdot 8H_2O$ or $4H_2O$.
Properties: Soft, bulky, white powder; odorless; tasteless. Loses all water at 400C. Soluble in acids; insoluble in water. Nonflammable.
Derivation: Reaction of magnesium oxide and phosphoric acid at high temperatures.
Grade: Technical, reagent, NF (5 H_2O variety), FCC (4, 5, or 8 H_2O).
Use: Dentifrice polishing agent, pharmaceutical antacid, adsorbent, stabilizer for plastics, food additive and dietary supplement.

magnesium potassium sulfate.
CAS: 13826-56-7. $O_{12}S_3 \cdot 2K \cdot 2Mg$.
Hazard: Moderately toxic by inhalation. Low toxicity by ingestion.

magnesium pyrophosphate.
CAS: 10102-34-8. $Mg_2P_2O_7 \cdot 3H_2O$.
Properties: White powder. D 2.56, loses water at 100C, mp (anhydrous) 1383C. Soluble in acids; insoluble in alcohol and water.

magnesium ricinoleate.
CAS: 22677-47-0. $Mg(OOCC_{17}H_{32}OH)_2$.
Properties: Coarse, yellow granules; faint fatty acid odor. Mp 98C, d 1.03 (25/25C). Combustible.
Use: Cosmetics.

magnesium salicylate.
CAS: 18917-89-0. $Mg(C_7H_5O_3)_2 \cdot 4H_2O$.
Properties: Colorless, efflorescent, crystalline powder. Soluble in water and alcohol.
Derivation: Action of salicylic acid on magnesium hydroxide.
Use: Medicine (anti-infective).

magnesium silicate.
CAS: 1343-88-0. $3MgSiO_3 \cdot 5H_2O$ (variable). The FCC specifies a ratio of $2MgO:5SiO_2$.
See magnesium trisilicate; serpentine; talc.
Properties: Fine, white powder. Insoluble in water or alcohol. An absorbent. Noncombustible.
Derivation: Interaction of a magnesium salt and a soluble silicate.
Grade: Technical, FCC.
Hazard: Toxic by inhalation, use in foods restricted to 2%.
Use: Rubber filler; ceramics, glass, refractories; absorbent for crude oil spills; manufacture of permanently dry resins and resinous compositions; paints, varnishes, and paper (filler); animal and vegetable oils (bleaching agent); odor absorbent; filter

medium; catalyst and catalyst carrier; anticaking agent in foods.
See asbestos.

magnesium silicide.
CAS: 22831-39-6. Mg_2Si.
Properties: Bluish crystals. Mp 1085C, d 1.9, decomposes on heating above 500C, also by water and hydrochloric acid.
Derivation: By heating magnesium powder with silicon in ratio of 20:6.
Use: Semiconductor technology, electrical equipment.

magnesium silicofluoride. See magnesium fluosilicate.

magnesium stannate.
CAS: 12032-29-0. $MgSnO_3 \cdot 3H_2O$.
Properties: White, crystalline powder. Decomposes at 340C. Soluble in water.
Hazard: Toxic by inhalation.
Use: Additive in ceramic capacitors.

magnesium stannide.
CAS: 1313-08-2. Mg_2Sn.
Properties: Blue-white crystals. Mp 775C. Soluble in water and dilute hydrochloric acid. Has electrical and magnetic properties.
Use: Semiconductor technology, magnetochemistry, thermoelectric research.

magnesium stearate. (octadecanoic acid, magnesium salt).
CAS: 557-04-0. $Mg(C_{18}H_{35}O_2)_2$ or with one H_2O. Technical grade contains small amounts of the oleate and 7% magnesium oxide, MgO.
Properties: Soft, white, light powder; tasteless; odorless. D 1.028, mp 88.5C (pure), 132C (technical). Insoluble in water and alcohol. Nonflammable.
Grade: Technical, USP FCC.
Use: Dusting powder, lubricant in making tablets, drier in paints and varnishes, a flatting agent, stabilizer and lubricant for plastics, an emulsifying agent in cosmetics, a dietary supplement, and in medicines.

magnesium sulfate.
CAS: 7587-88-9. (1) $MgSO_4$, (2) (epsom salts) $MgSO_4 \cdot 7H_2O$.
Properties: Colorless crystals; saline bitter taste. Neutral to litmus. D (1) 2.65, (2) 1.678. (1) Decomposes at 1124C, (2) loses 6 H_2O at 150C, loses water at 200C. Soluble in glycerol; very soluble in water; sparingly soluble in alcohol. Noncombustible.
Derivation: (1) Action of sulfuric acid on magnesium oxide, hydroxide, or carbonate, (2) mined in a high degree of purity.
Grade: Technical, CP, USP, FCC.
Use: Fireproofing, textiles (warp-sizing and loading cotton goods, weighting silk, dyeing and calico printing), mineral waters, catalyst carrier, ceramics,

fertilizers, paper (sizing), cosmetic lotions, dietary supplement.

magnesium sulfide.
CAS: 12032-36-9. MgS.
Properties: Red-brown, crystalline solid. D 2.84. Decomposes above 2000C; decomposes in water.
Use: Source of hydrogen sulfide, laboratory reagent.

magnesium sulfite.
CAS: 7757-88-2. $MgSO_3 \cdot 6H_2O$.
Properties: White, crystalline powder. D 1.725, mp loses 6 H_2O at 200C, bp (decomposes). Slightly soluble in water; insoluble in alcohol.
Derivation: Action of sulfurous acid on magnesium hydroxide.
Use: Manufacture of paper pulp (as bisulfite).

magnesium tetrahydrogen phosphate. See magnesium phosphate, monobasic.

magnesium trisilicate. USP specifies not less than 20% magnesium oxide and 45% SiO_2, similar to the FCC requirements for magnesium silicate. See talc.
Properties: Fine, white; odorless; tasteless powder; free from grittiness. Insoluble in water and alcohol; readily decomposed by mineral acids. Noncombustible.
Derivation: By reaction of soluble magnesium salts with soluble silicates.
Grade: Technical, USP.
Use: Industrial odor absorbent, decolorizing agent, antioxidant, medicine (antacid).

magnesium tungstate. (magnesium wolframate).
CAS: 13573-11-0. $MgWO_4$.
Properties: White crystals. D 5.66. Soluble in acids; insoluble in water and alcohol. Noncombustible.
Derivation: Interaction of solutions of magnesium sulfate and ammonium tungstate.
Use: Fluorescent screens for X-rays, luminescent paint.

magnesium zirconate.
CAS: 12032-31-4. MgO,ZrO_2.
Properties: Powder. D 4.23, mp 2060C.
Use: Electronics.

magnesium zirconium silicate.
CAS: 52110-05-1. $MgZrSiO_5$, or $MgO \cdot ZrO_2 \cdot SiO_2$.
Properties: White solid. Mp 1760C, d 80 lb/ft^3. Insoluble in water and alkalies; slightly soluble in acids. Noncombustible.
Use: Electrical resistor, ceramics, glaze opacifier.

"Magnesol" *[Dallas].* TM for a synthetic, amorphous, adsorptive magnesium silicate.
Use: Solvent purification, clarification, and recovery; oil refining and recovery; deoderizing and

decolorizing oils and fats; chill-proofing fruit and vegetable-based beverages; anticaking; flavor and fragrance carrier.

magnetic double refraction.   See Cotton–Mouton effect.

magnetic separation.   Use of a magnetic field to remove unwanted magnetic particulates from solid or liquid mixtures of nonmagnetic materials, e.g., removal of impurities from clays, bauxite, glass sands, and mineral processing. Low-gradient fields are suitable for separation of strongly magnetic materials, whereas high-gradient fields can separate particles of materials that are weakly magnetic, such as coliform bacteria from municipal wastes and sulfur from coal. Removal of magnetic impurities from industrial wastewater is called magnetic filtration, e.g., reconditioning of boiler water and regeneration of condensate in power plants.
See electromagnetic separation; mass spectrometry.

magnetite.   (lodestone; iron ore, magnetic).
CAS: 1309-38-2. Fe_3O_4 often with titanium or magnesium. A component of taconite.
Properties: Black mineral, black streak, submetallic or dull to metallic luster. Contains 72.4% iron. Readily recognized by strong attraction by magnet. Soluble in powder form in hydrochloric acid. Decomposes at 1538C to ferric oxide Fe_2O_3, d 4.9–5.2, hardness 5.5–6.5.
See iron oxide, black.

magnetochemistry.   A subdivision of chemistry concerned with the effect of magnetic fields on chemical compounds; analysis and measurement of these effects, (e.g., magnetic moment and magnetic susceptibility) are important tools in crystallographic research and determination of molecular structures. Substances that are repelled by a magnetic field are diamagnetic (water, benzene); those that are attracted are paramagnetic (oxygen, transition element compounds). Diamagnetic materials have only induced magnetic moment; paramagnetic materials have permanent magnetic moment. Magnetochemistry has been useful in detection of free radicals, elucidation of molecular configurations of highly complex compounds, and in its application to catalytic and chemisorption phenomena.
See nuclear magnetic resonance.

magnetohydrodynamics.   (MHD). The behavior of high-temperature ionized gases passed through a magnetic field. A power-generating method using MHD involves an open cycle in which hot combustible gases from coal, seeded with cerium or potassium to increase electrical conductivity, constitute the working fluid. These are sent through a nozzle surrounded by a magnet; the electricity induced by movement of the ionized gas through the magnetic field is passed to electrodes and the gas sent to a steam generator. Efficiency is

rated at 50–60% compared with 40% for conventional fossil fuel plants and 33% for plants using nuclear fuels. Two-phase liquid-metal systems are being studied as auxiliary units for a number of energy converters. MHD is an important field of expansion of research activity on new sources of energy; its high efficiency and low pollution factor indicate that it may have a significant future in electric power supply.
See plasma (2).

Magnifloc.   Water soluble polymers for flocculation, coagulation, dispersion, and viscosification.
Available forms: Emulsion, dry and liquid forms, cationic, anionic, and nonionic charges, and a broad range of molecular weights and chemistries.
Use: Liquid and solid separation applications.

Maillard reaction.   See browning reaction.

Maintain.   A line of products for cleaning, inhibiting corrosion, and coating architectural copper, brass, and bronze.

maitotoxin.
CAS: 59392-53-9. $C_{164}H_{256}O_{68}S_2 \cdot 2Na$.
Hazard: A poison.
Source: Natural product.

Malachite Green.   (Benzaldehyde Green; CI 42,000; Victoria Green).
CAS: 569-64-2. $C_{23}H_{25}ClN_2$.
Properties: Green crystals. Soluble in water; ethyl, methyl, and amyl alcohol.
Derivation: Condensation of benzaldehyde with N,N-dimethylaniline, oxidation of the phenylmethane product, and reaction with hydrochloric acid. It may be formed as a double salt of zinc chloride.
Hazard: Toxic by ingestion.
Use: Dyeing textiles, either directly or with mordant; plant fungicide, staining bacteria, antiseptic.

Malachite Green G.   See Brilliant Green.

Malaprade reaction.   Compounds containing two hydroxyl groups or a hydroxyl and an amino group attached to adjacent carbon atoms undergo cleavage of the carbon-to-carbon bond when treated with periodic acid.

malathion.   (S-[1,2-bis(ethoxy carbonyl)ethyl]-O,O-dimethyl phosphorodithioate).
CAS: 121-75-5. $(CH_3O)_2P(S)SCH(COOC_2H_5)$ $CH_2COOC_2H_5$.
Properties: Yellow, high-boiling liquid. Bp 156–157C under 0.7 mm Hg with slight decomposition), mp 3.0C, refr index 1.4985 (25C), d 1.2315 (25C), vap press (20C) approximately 0.00004 mm Hg. Miscible with most polar organic solvents; slightly soluble in water. Combustible.

Derivation: From diethyl maleate and dimethyldithiophosphoric acid.
Grade: Technical grade is ≥95% pure.
Hazard: Absorbed by skin, cholinesterase inhibitor. Questionable carcinogen.
Use: Insecticide; has been used effectively on the Mediterranean fruit fly.

malathion-fenitrothion mixture.   See fenitrothion-malathion mixture.

malaxate.   To soften and mix dry materials in the presence of water or other liquid by rubbing, kneading, or rolling, thus producing a soft plastic mass. This term is used by one manufacturer, the Fitzpatrick Co., 832 Industrial Drive, Elmhurst, IL 60126, to describe a machine designed for this purpose, it provides continuous mixing of dry solids with one or more liquids by means of single or double screw-type agitators rotating in a channel.

maleic acid.   (maleinic acid; *cis*-butenedioic acid).
CAS: 110-16-7.

$$\begin{array}{c} O \\ \| \\ HC-COH \\ \| \\ HC-COH \\ \| \\ O \end{array}$$

Properties: Colorless crystals; repulsive astringent taste; faint odor. D 1.59, mp 130–131C. Soluble in water, alcohol and acetone; very slightly soluble in benzene. At temperatures slightly higher than its melting point, it is converted partly to fumaric acid. Combustible.
Derivation: Same as for maleic anhydride with special recovery conditions.
Grade: Technical, reagent.
Hazard: Toxic by ingestion.
Use: Organic synthesis (malic, succinic, aspartic, tartaric, propionic, lactic, malonic and acrylic acids); dyeing and finishing of cotton, wool and silk; preservative for oils and fats.

maleic anhydride.   (2,5-furandione).
CAS: 108-31-6.

$$\begin{array}{c} O \\ \| \\ HC-C \\ \| \quad O \\ HC-C \\ \| \\ O \end{array}$$

Properties: Colorless needles. D 0.934 (20/4C), mp 53C, bp 200C, flash p 218F (103C), autoign temp 890F (476C). Soluble in water, acetone, alcohol, and dioxane; partially soluble in chloroform and benzene.
Derivation: (1) Vapor-phase oxidation of benzene with atmospheric oxygen with V_2O_5 catalyst at 400C. (2) Under development is a fixed-bed process involving oxidation of butane with undisclosed catalyst.
Grade: Technical; rods, flakes, lumps, briquettes, and molten.
Hazard: Irritant to tissue. Dermal and respiratory sensitization. Questionable carcinogen.
Use: Polyester resins, alkyd coating resins, fumaric and tartaric acid manufacture, pesticides, preservative for oils and fats, permanent-press resins (textiles), Diels–Alder reactions.

maleic hydrazide.   (1,2-dihydro-3,6-pyridazinedione).
CAS: 123-33-1.

HC:CHC(O)NHNHC(O).

A plant growth regulator.
Properties: Crystals. Mp 297C. Slightly soluble in hot alcohol; more soluble in hot water.
Derivation: By treating maleic anhydride with hydrazine hydrate.
Hazard: Toxic by ingestion. Questionable carcinogen.
Use: Systemic herbicide, treatment of tobacco plants, postharvest sprouting inhibitor, weed control, sugar content stabilizer in beets.

maleimide.   (pyrrole-2,5-dione).
CAS: 541-59-3. $C_4H_3NO_2$.
Properties: Plates. Mw: 97.08, mp: 93C.
Hazard: A poison.

maleinic acid.   See maleic acid.

maleo-pimaric acid.   Reaction product of maleic anhydride and *l*-pimaric acid; derived from pine gum.
Properties: Crystalline solid. Mp approximately 225C. Soluble in most organic solvents; insoluble in water or aliphatic hydrocarbons.
Use: Resins.

malformin A.   (4-butan-2-yl-7-(2-methylpropyl)-10-propan-2-yl-15,16-dithia-2,5,8,11,19-pentazabicyclo[11.4.2]nonadeca-3,6,9,12,18-pentone). $C_{23}H_{39}N_5O_5S_2$. A mycotoxin.
Derivation: Produced by *Aspergillus niger*.

malic acid.   (hydroxysuccinic acid; apple acid).
CAS: 6915-15-7. $COOHCH_2CH(OH)COOH$.
Note: Do not confuse with maleic acid.
Properties: Colorless crystals; sour taste. D (*dl*-form) 1.601, (*d*- or *l*- form) 1.595 (20/4C); mp (*dl*) 128C, (*d*- or *l*-) 100C; bp (*dl*) 150C (decomposes), (*d*- or *l*-) 140C (decomposes). Very soluble in water and alcohol; slightly soluble in ether. Combustible.
Derivation: Occurs naturally in unripe apples and other fruits. Made synthetically by catalytic oxidation of benzene to maleic acid, which is converted to malic acid by heating with steam under pressure.

Grade: Technical, active, and inactive; FCC. The natural material is levorotatory but the synthetic material is inactive.

Use: Manufacture of various esters and salts, wine manufacture, chelating agent, food acidulant, flavoring.

malignant anthrax. A fatal form of the disease in cattle and sheep that also occurs in humans. It is marked by the development of hard ulcers at the point of inoculation and eventually extreme prostration and death.

malloside.
CAS: 17489-40-6. $C_{29}H_{44}O_9$.
Hazard: A poison.
Source: Natural product.

malonamide. (carboxamidoacetamide; malonic acid diamide).
CAS: 108-13-4. $C_3H_6N_2O_2$.
Properties: Dimorphous, tetragonal or monoclinic. Mw: 102.11, mp: 170C. Soluble in water @ 8C; insoluble in EtOH and Et_2O; insoluble in ether.
Hazard: Mildly toxic by ingestion.

malonamide nitrile. See cyanoacetamide.

malonic acid. (methanedicarbonic acid).
CAS: 141-82-2. $CH_2(COOH)_2$.
Properties: White crystals. Mp 132–134C, bp (decomposes), d 1.63. Soluble in water, alcohol and ether.
Derivation: From monochloroacetic acid by reaction with potassium cyanide followed by hydrolysis.
Hazard: Strong irritant.
Use: Intermediate for barbiturates and pharmaceuticals.

malonic acid diamide. See malonamide.

malonic dinitrile. (malononitrile).
CAS: 109-77-3. $CH_2(CN)_2$.
Properties: Colored crystals. Mp 32.1C, bp 220C.
Hazard: Toxic by ingestion and inhalation.
Use: Organic synthesis, leaching agent for gold.

malonic ester. See ethyl malonate.

malonic ester synthesis. Syntheses based on the strongly activated methylene group of malonic esters which on reaction with sodium ethoxide form a resonance-stabilized ion that can be alkylated and acylated. After hydrolysis, the free alkylmalonic acids readily decarboxylate to mono- or disubstituted monocarboxylic acids.

malonic ethyl ester nitrile. See ethyl cyanoacetate.

malonic methyl ester nitrile. See methyl cyanoacetate.

malonic mononitrile. See cyanoacetic acid.

malononitrile. See malonic dinitrile.

malonylurea. See barbituric acid.

malt. Yellowish or amber-colored grains of barley that have been partially germinated by artificial means. It contains dextrin, maltose, and amylase; characteristic odor and taste. Black malt is grain that has been scorched in the drying process.
Use: Brewing, malted milk and similar food products, extract of malt (with 10% glycerol).

maltase. (glucase; α-glucosidase). An enzyme that hydrolyzes maltose to glucose. Occurs in the small intestine, in yeast, molds, and malt; usually associated with the enzyme amylase.

malt extract. (maltine).
Properties: Light-brown, sweet, viscous liquid; contains dextrin, maltose, a little glucose, and an amylolytic enzyme. It is capable of converting not less than five times its weight of starch into water-soluble sugars; soluble in cold water but more readily soluble in warm water, d greater than 1.350 and less than 1.430 (25C).
Derivation: By infusing malt with water at 60C, concentrating the expressed liquid below 60C, and adding 10% by weight of glycerol.
Use: Nutrient, emulsifying agent.

maltha. A black, viscous, natural bitumen consisting of a complex mixture of hydrocarbons. Its viscosity and rheological properties lie between those of crude oil and semisolid asphalt. It is the chief component of Athabasca oil sands.

malthenes. See petrolenes.

maltodextrin.
CAS: 9050-36-6. $(C_6H_{10}O_5)_n$.
Properties: White powder or solution from partial hydrolysis of wheat or corn starch.
Use: Food additive.

maltol. (3-hydroxy-2-methyl-4-pyrone).
CAS: 118-71-8. $CH_3C_5H_2O(O)(OH)$.
Properties: White, crystalline powder; characteristic caramel-butterscotch odor and suggestive of a fruity-strawberry aroma in dilute solution. Melting range 160–164C. Slightly soluble in water; more soluble in alcohol and propylene glycol.
Grade: FCC.
Use: Flavoring agent in bakery products.

maltose. (malt sugar; maltobiose).
CAS: 69-79-4. $C_{12}H_{22}O_{11} \cdot H_2O$. The most common reducing disaccharide, composed of two

molecules of glucose. Found in starch and glycogen.

Properties: Colorless crystals. Mp 102–103C. Soluble in water; insoluble in ether; slightly soluble in alcohol. Combustible.

Derivation: By the enzymatic action of diastase (usually obtained from malt extract) on starch.

Use: Nutrient, sweetener, culture media, stabilizer for polysulfides, brewing.

malt syrup.
Properties: Derived from barley (*Hordeum vulgare* L.). Brown liquid; sweet taste. Sol in water.
Use: Food additive.

Man.   Abbreviation for mannose, also for methyacrylonitrile.

mancude.   (mancude-ring system). A ring in a molecule that has the maximum number of non-cumulative double bonds.

mandarin oil.   See citrus peel oils.

mandarin oil, coldpressed.
Properties: From expression of peel of *Citrus reticulata* Blanco var. *mandarin*. Clear orange to brown-orange liquid; orange odor. D: 0.846. Sol in fixed oils, mineral oil; slt sol in propylene glycol; insol in glycerin.
Use: Food additive.

mandelic acid.   (phenylglycolic acid; α-phenylhydroxyacetic acid; benzoglycolic acid, known also as amygdalic acid).
CAS: 90-64-2. $C_6H_5CH_2OCOOH$. Exists in stereoisomeric forms. The properties are those of the *dl*-form.

$$C_6H_5-\underset{\underset{OH}{|}}{\overset{\overset{H}{|}}{C^*}}-\underset{\underset{O}{||}}{C}-OH$$

Properties: Large, white crystals or powder; faint odor. D 1.30, mp 117–119C, darkens on exposure to light. Soluble in ether; slightly soluble in water and alcohol.
Derivation: Hydrolysis of the cyanohydrin formed from benzaldehyde, sodium bisulfite, and sodium cyanide. Can be obtained from amygdalin.
Hazard: Toxic by ingestion.
Use: Organic synthesis, medicine (urinary antiseptic).

mandelonitrile.   (benzaldehyde cyanohydrin; Laetrile).
CAS: 532-28-5. $C_6H_5CH(OH)CN$.
Properties: Oily, yellow liquid. D 1.1165 (20/4C), fp −10C, bp 170C (decomposes). Soluble in alcohol, chloroform, ether; nearly insoluble in water.

mandelyltropeine.   See homotropine.

maneb.   (generic name for manganese ethylenebisdithiocarbamate).
CAS: 12427-38-2. $(SSCNCH_2CH_2NHCSS)Mn$.
Properties: Brown powder. Decomposes on heating. Partially soluble in water; soluble in chloroform.
Derivation: Reaction of disodium ethylenebisdithiocarbamate and a manganese salt.
Use: Fungicide for foliage.
See zineb.

manganese.
CAS: 7439-96-5.   Mn. Metallic element of atomic number 25, group VIIB of periodic table, aw 64.9380, valences = 2, 3, 4, 6, 7; no stable isotopes; four artificial radioisotopes.
Properties: There are four allotropic forms of which α is most important. Brittle silvery metal, d 7.44, Mohs hardness 5, mp 1245C, bp 2097C, decomposes water. Readily dissolves in dilute mineral acids. Pure manganese cannot be fabricated. Manganese is considered essential for plant and animal life.
Occurrence: Usually associated with iron ores in submarginal concentration. Important ores of manganese are pyrolusite, manganite, psilomelane, rhodochrosite. An important source of manganese is open-hearth slags. Ores occur chiefly in Brazil, India, South Africa, Gabon, Ghana, Zaire, Montana; 90% of U.S. consumption is imported. So-called nodules rich in manganese and containing also cobalt, nickel, and copper have been found in huge quantities (estimated at 1.5 trillion tons) on the floor of the Pacific south of Hawaii. Such nodules have been located in other areas such as in Lake Michigan.
Derivation: Reduction of the oxide with aluminum or carbon. Pure manganese is obtained electrolytically from sulfate or chloride solution.
Grade: Technical, pure or electrolytic, powdered.
Hazard: Dust or powder is flammable. Use dry chemical to extinguish. Toxic. Central nervous system impairment. Questionable carcinogen.
Use: Ferroalloys (steel manufacture), nonferrous alloys (improved corrosion resistance and hardness), high-purity salt for various chemical uses, purifying and scavenging agent in metal production, manufacture of aluminum by Toth process.

manganese acetate.   (acetic acid manganese(2+) salt).
CAS: 638-38-0. $Mn(C_2H_3O_2)_2 \cdot 4H_2O$.
Properties: Pale-red crystals. D 1.59, mp 80C. Soluble in water and alcohol. Combustible.
Derivation: Action of acetic acid on manganese hydroxide.
Use: Textile dyeing, oxidation catalyst, paint and varnish (drier, boiled oil manufacture), fertilizers, food packaging, feed additive.

manganese ammonium sulfate.   (manganous ammonium sulfate).
CAS: 13566-22-8. $MnSO_4 \cdot (NH_4)_2SO_4 \cdot 6H_2O$.
Properties: Light-red crystals. D 1.83. Soluble in water.

manganese arsenate.   See manganous arsenate.

manganese binoxide.   See manganese dioxide.

manganese, bis(2-benzoylbenzoato)bis(3-(1-methyl-2-pyrrolidinyl)pyridine)-, tri-hydrate.   See nicotine, compounded with manganese(ii) o-benzoyl benzoate.

manganese black.   See manganese dioxide.

manganese borate.
CAS: 12228-91-0. MnB_4O_7.
Properties: Reddish-white powder. Insoluble in water.
Derivation: By the action of boric acid on manganese hydroxide.
Grade: Technical.
Use: Varnish and oil drier.

manganese-boron.   An alloy of manganese and boron used in making brass, bronze, and other alloys.

manganese bromide.   See manganous bromide.

manganese bronze.   Alloy of 55–60% copper, 38–42% zinc, up to 3.5% manganese with or without small amounts of iron, aluminum, tin, or lead.

manganese caprylate.
Use: Food additive.

manganese carbonate.   (manganous carbonate; rhodochrosite).
CAS: 598-62-9. $MnCO_3$.
Properties: Rose-colored crystals, almost white when precipitated. D 3.125, mp (decomposes). Soluble in dilute inorganic acids; almost insoluble in common organic acids; both concentrated and dilute, insoluble in water.
Derivation: (1) A precipitate from the addition of sodium carbonate to a solution of a manganese salt; (2) hydrometallurgical treatment of manganiferous iron ore.
Grade: Chemical (46% manganese).

manganese carbonyl.
CAS: 10170-69-1. $Mn_2(CO)_{10}$.
Properties: Yellow crystals. Mp 154C. Decomposition begins at 110C in absence of CO, d 1.75. Insoluble in water; soluble in most organic solvents.
Hazard: Toxic material.
Use: Antiknock gasoline, catalyst.

manganese chloride.   See manganous chloride.

manganese chromate.   See manganous chromate.

manganese citrate.   (manganous citrate).
CAS: 10024-66-5. $Mn_3(C_6H_5O_7)_2$.
Properties: White powder. Soluble in water in the presence of sodium citrate. Combustible.
Derivation: Action of citric acid on manganese hydroxide.
Use: Feed additive, food additive, and dietary supplement.

manganese cyclopentadienyl tricarbonyl.
CAS: 12079-65-1. $C_5H_4Mn(CO)_3$.
Hazard: Toxic material absorbed by skin. Central nervous system impairment and skin irritant.
Use: Antiknock agent.

manganese dioxide.   (manganese binoxide; manganese black; battery manganese; manganese peroxide).
CAS: 1313-13-9. MnO_2.
Properties: Black crystals or powder. D 5.026, mp (decomposes). Soluble in hydrochloric acid; insoluble in water.
Derivation: (1) Natural as pyrolusite and as a special African ore of different atomic structure used exclusively for the battery grade; (2) by electrolysis; (3) by heating manganese oxide in presence of oxygen; (4) by decomposition of manganese nitrate.
Grade: Technical, CP, Battery.
Hazard: Oxidizing agent, may ignite organic materials.
Use: Oxidizing agent, depolarizer in dry cell batteries (African and synthetic types only), pyrotechnics, matches, etc., catalyst, laboratory reagent, scavenger and decolorizer, textile dyeing, source of metallic manganese (as pyrolusite).

manganese dithionate. MnS_2O_6.
Properties: Crystals. D 1.76. Soluble in water.

manganese ethylenebisdithiocarbamate.
See maneb.

manganese ethylhexoate.   See manganese octoate.

manganese fluoride.   See manganous fluoride.

manganese gluconate.
CAS: 6485-39-8. $Mn(C_6H_{11}O_7)_2 \cdot 2H_2O$.
Properties: Light-pinkish powder or coarse, pink granules. Soluble in water; insoluble in alcohol and benzene. Combustible.
Grade: Pharmaceutical, FCC.
Use: Feed additive, food additive and dietary supplement, vitamin tablets.

manganese glycerophosphate.
CAS: 1320-46-3. $CH_2OHCH_2OCH_2OP(O)O_2Mn$.
Properties: Yellow-white or pinkish powder; odorless. Nearly tasteless. Soluble in water in the presence of citric acid; insoluble in alcohol.
Derivation: Action of glycerophosphoric acid on manganese hydroxide.
Grade: Technical, FCC.
Use: Food additive and dietary supplement.

manganese green.   See barium manganate.

manganese hydrate.   See manganic hydroxide, manganous hydroxide.

manganese hydrogen phosphate.   See manganous phosphate, acid.

manganese hydroxide.   See manganic hydroxide, manganous hydroxide.

manganese hypophosphite.
CAS: 10043-84-2. $Mn(H_2PO_2)_2 \cdot H_2O$.
Properties: Pink crystals or powder; odorless; tasteless. Soluble in water; insoluble in alcohol.
Derivation: Interaction of manganese sulfate and calcium hypophosphite.
Grade: Technical, FCC.
Hazard: Dangerous fire and explosion risk when heated (evolves phosphine) or in presence of strong oxidizing agents.
Use: Food additive and dietary supplement.

manganese iodide.   See manganous iodide.

manganese linoleate.
CAS: 6904-78-5. $Mn(C_{18}H_{31}O_2)_2$.
Properties: Dark-brown, plaster-like mass. Soluble in linseed oil. Combustible.
Derivation: By boiling a manganese salt, sodium linoleate, and water.
Use: Paint and varnish drier, pharmaceutical preparations.

manganese monoxide.   See manganous oxide.

manganese naphthenate.
Properties: Hard, brown, resinous mass. It is a pale buff in color when precipitated in the cold, but darkens immediately in solution. Mp approximately 130–140C, commercial solution contains 6% manganese. Soluble in mineral spirits. Hardens on exposure to air. Combustible.
Derivation: Precipitation from a mixture of soluble manganese salts and aqueous sodium naphthenate solution.
Hazard: Solution is flammable.
Use: Paint and varnish drier.

manganese nitrate.   See manganous nitrate.

manganese octoate.   (manganese ethylhexoate).
CAS: 15956-58-8. $Mn(OOCC_6H_{13}[C_2H_5])_2$.
Commercially formed from 2-ethylhexoic acid and manganous hydroxide. Sold as a clear brown solution in a petroleum solvent containing 6% manganese.
Hazard: Solution is flammable.
Use: Primarily as drier for paints, enamels, varnishes, and printing inks.

manganese oleate.
CAS: 23250-73-9. $Mn(C_{18}H_{33}O_2)_2$.
Properties: Brown, granular mass. Soluble in oleic acid and ether; insoluble in water.
Derivation: By boiling manganese chloride, sodium oleate, and water.
Hazard: Solution is flammable.
Use: Paint and varnish drier.

manganese oxalate.
CAS: 640-67-5. $MnC_2O_4 \cdot 2H_2O$.
Properties: White, crystalline powder. D 2.453, loses $2H_2O$ at 100C. Soluble in dilute acids; very slightly soluble in water. Combustible.
Derivation: By adding sodium oxalate to manganese chloride.
Use: Paint and varnish drier.

manganese peroxide.   See manganese dioxide.

manganese phosphate.   See manganous phosphate; manganous phosphate, acid.

manganese phosphate, dibasic.   See manganous phosphate, acid.

manganese protoxide.   See manganese oxide.

manganese pyrophosphate.   See manganous pyrophosphate.

manganese resinate.
CAS: 9008-34-8. $Mn(C_{20}H_{29}O_2)_2$.
Properties: Dark, brownish-black mass or flesh-colored powder. Soluble in hot linseed oil; insoluble in water.
Derivation: By boiling manganese hydroxide, rosin oil, and water.
Hazard: Flammable, dangerous fire risk.
Use: Varnish and oil drier.

manganese sesquioxide.   See manganic oxide.

manganese silicate.   See manganous silicate.

manganese sulfate.   See manganous sulfate.

manganese sulfide.   See manganous sulfide.

manganese sulfite.   See manganous sulfite.

manganese tallate.   Manganese salts of tall oil fatty acids. Marketed as a solution containing 6% manganese. Combustible.
Use: Drier.

manganese tetroxide.   (manganese oxide; trimanganese tetroxide).
CAS: 1317-35-7. Mn_3O_4.
Properties: A brownish powder. Mw 228.79, d 4.876, mp 1564C. Insoluble in water; soluble in hydrochloric acid with evolution of chlorine.
Occurrence: Generated in the pouring and casting of molten ferromanganese.
Hazard: Chronic manganese poisoning and pulmonary effects.

manganese-titanium.   An alloy containing manganese, titanium, aluminum, iron, silicon.
Properties: (Regular) Mp 1454C, (special) mp 1332C.
Use: (Regular) Deoxidizer in high grade steel; (special) nonferrous alloys deoxidizer.

manganic acetylacetonate.
CAS: 14284-89-0. $Mn[OC(CH_3):CHCO(CH_3)]_3$.
Properties: Brown, crystalline solid. Mp 172C. Combustible.
Derivation: Reaction of a manganese salt with acetylacetone and sodium carbonate.

manganic fluoride.
CAS: 7783-53-1. MnF_3.
Properties: Red, crystalline solid. D 3.54. Decomposed by water and by heat. Soluble in acids. Attacks glass when hot.
Hazard: Toxic material.
Use: Fluorinating agent.

manganic hydroxide.   (manganese hydroxide; hydrated manganic oxide).
CAS: 1332-62-3. $Mn(OH)_3$; rapidly loses water to form MnO(OH).
Properties: Brown powder. D 4.2–4.4, mp (decomposes). Decomposes in acids. Insoluble in water.
Derivation: By the action of oxygen on precipitated manganous hydroxide.
Use: Pigment for fabrics, ceramics.

manganic oxide.   (manganese sesquioxide).
CAS: 1317-34-6. Mn_2O_3. In nature as manganite, a manganese ore.
Properties: Black, lustrous powder, sometimes tinged brown. Very hard. D 4.5, decomposes at 1080C. Soluble in cold hydrochloric acid; not soluble in nitric acid (decomposes), hot sulfuric acid; insoluble in water. Noncombustible.
Hazard: Flammable in finely divided form. Toxic by inhalation of dust.

manganic oxide, hydrated.   See manganic hydroxide.

manganous ammonium sulfate.   See manganese ammonium sulfate.

manganous arsenate.   (manganese arsenate; manganous arsenate, acid). $MnHAsO_4$.
Properties: Reddish-white powder. Hygroscopic. Soluble in acids; slightly soluble in water.
Hazard: Highly toxic.

manganous bromide.   (manganese bromide).
CAS: 13446-03-2. $MnBr_2 \cdot 4H_2O$.
Properties: Red crystals. Loses water at 64C. Very soluble in water; deliquescent. Noncombustible.
Derivation: Action of hydrobromic acid with manganous carbonate or manganous hydroxide.
Hazard: Irritant.

manganous carbonate.   See manganese carbonate.

manganous chloride.   (manganese chloride).
CAS: 7773-01-5.   (1) $MnCl_2$, (2) $MnCl_2 \cdot 4H_2O$.
Properties: Rose-colored crystals; deliquescent. D (1) 2.98, (2) 1.913; mp (1) 650C, (2) 87.5C; bp (1) 1190C. Very soluble in water; soluble in alcohol; insoluble in ether. Noncombustible.
Grade: CP, anhydrous.
Use: Catalyst in the chlorination of organic compounds, paint drier, dyeing, pharmaceutical preparations, fertilizer compositions, feed additive, dietary supplement.

manganous chromate.   (manganese chromate; manganous chromate, basic). $2MnO \cdot CrO_3 \cdot 2H_2O$.
Properties: Brown powder. Slightly soluble in water with hydrolysis.
Hazard: Toxic by inhalation.

manganous citrate.   See manganese citrate.

manganous fluoride.   (manganese fluoride).
CAS: 7782-64-1. MnF_2.
Properties: Reddish powder. D 3.98, mp 856C. Soluble in acids; insoluble in water, alcohol, and ether. Noncombustible.
Derivation: Action of hydrogen fluoride on manganous hydroxide.
Grade: Technical.

manganous hydroxide.   (manganese hydroxide). $Mn(OH)_2$. Occurs naturally as pyrochroite.
Properties: White to pink crystals. D 3.258, Mohs hardness 2.5, decomposes with heat. Insoluble in water and alkali; soluble in acids and ammonium salts.

manganous iodide.   (manganese iodide). (1) MnI_2, (2) $MnI_2 \cdot 4H_2O$.
Properties: (1) White, deliquescent, crystalline mass; (2) rose crystals. D (1) 5.01, mp (1) 638C

(in vacuum), bp (1) 1061C. Soluble in water with gradual decomposition; soluble in alcohol.
Derivation: Action of hydriodic acid on manganous hydroxide.

manganous nitrate. (manganese nitrate).
CAS: 10377-66-9. $Mn(NO_3)_2 \cdot 6H_2O$.
Properties: Pink crystals. D 1.82, bp 129C, mp 26C. Very soluble in water; deliquescent; soluble in alcohol.
Hazard: Fire and explosion risk in contact with organic materials.
Use: Ceramics, intermediates, catalyst, manganese dioxide.

manganous orthophosphate. See manganous phosphate.

manganous oxide. (manganese protoxide; manganese monoxide).
CAS: 1317-35-7. MnO.
Properties: Grass-green powder. D 5.45, mp 1650C, but converted to Mn_3O_4 if heated in air. Soluble in acids; insoluble in water. Noncombustible.
Derivation: (1) By reduction of the dioxide in hydrogen, (2) by heating the carbonate with exclusion of air.
Grade: Technical.
Use: Textile printing, analytical chemistry, catalyst in manufacture of allyl alcohol, ceramics, paints, colored glass, bleaching tallow, animal feeds, fertilizers, food additive and dietary supplement.

manganous phosphate. (manganese phosphate; manganous orthophosphate).
$Mn_3(PO_4)_2 \cdot 7H_2O$.
Properties: Reddish-white powder. Soluble in mineral acids; insoluble in water.
Derivation: By the action of orthophosphoric acid on manganous hydroxide.
Use: Conversion coating of steels, aluminum, and other metals.

manganous phosphate, acid. (manganese hydrogen phosphate; manganous phosphate, secondary; manganese phosphate, dibasic).
$MnHPO_4 \cdot 3H_2O$.
Properties: Pink powder, contains some tribasic phosphate. Soluble in acids; slightly soluble in water.
Use: Feed additive.

manganous pyrophosphate. (manganese pyrophosphate). (1) $Mn_2P_2O_7$, (2) $Mn_2P_2O_7 \cdot 3H_2O$.
Properties: White powder. D (1) 3.71, mp 1196C. Soluble in solutions of potassium or sodium pyrophosphate; insoluble in water.

manganous silicate. (manganese silicate).
$MnSiO_3$. Occurs naturally as rhodonite.

Properties: Red crystals or yellowish-red powder. D 3.72, mp 1323C. Insoluble in water. Noncombustible.
Derivation: Interaction of manganous salts with sodium silicate.
Use: Colorant for glass and ceramic glazes.

manganous sulfate. (manganese sulfate).
CAS: 7785-87-7. $MnSO_4 \cdot 4H_2O$.
Properties: Translucent, pale-rose-red, efflorescent prisms. D 2.107, mp 30C, anhydrous mp 700C, decomposes at 850C. Soluble in water; insoluble in alcohol.
Derivation: By-product of production of hydroquinone, or by the action of sulfuric acid on manganous hydroxide or carbonate.
Grade: Technical, CP, fertilizer, feed.
Use: Fertilizers, feed additive, paints and varnishes, ceramics, textile dyes, medicines, fungicides, ore flotation, catalyst in viscose process, synthetic manganese dioxide.

manganous sulfide. (manganese sulfide).
CAS: 18820-29-6. MnS. Green crystals, d 3.99, decomposes on melting, almost insoluble in water.
Use: Additive in steel making.

manganous sulfite. (manganese sulfite; manganous sulfite, normal). $MnSO_3$.
Properties: Grayish-black or brownish-red powder. Soluble in solution of sulfur dioxide; insoluble in water.

Manila fiber. See abaca.

Manila resin. A type of copal resin similar to Congo and kauri.
Properties: D 1.062, mp 230–250C. Soluble in ether, methanol, and ethanol; partially soluble in amyl alcohol; insoluble in water. Combustible.
Occurrence: Philippine Islands and East Indies.
Use: Varnishes, paints, lacquers, printing inks.

manna. The water-soluble exudate of a plant occurring in the Mediterranean area, used in medicine as a laxative. It has a high carbohydrate content, especially of mannitol.

Mannheim furnace. A muffle furnace used for the manufacture of salt cake and hydrogen chloride. It consists of a firebrick compartment in which is a circular pan some 10–15 ft in diameter, usually made of cast iron. The charge is fed through a hopper at the top and plows maintain continuous movement of the materials being heated. The temperature of the hydrogen chloride gas evolved is approximately 145C. Salt cake is removed through an opening at the side. The furnace is run continuously. The Mannheim furnace is no longer widely used because most salt cake is now obtained from natural sources.

Mannich reaction. Reaction of active methylene compounds with formaldehyde and ammonia or primary or secondary amines to give β-aminocarbonyl compounds.

mannitol. (manna sugar; mannite).
CAS: 69-65-8. $C_6H_8(OH)_6$. A straight-chain hexahydric alcohol.

$$
\begin{array}{c}
CH_2OH \\
| \\
HCOH \\
| \\
HCOH \\
| \\
HOCH \\
| \\
HOCH \\
| \\
CH_2OH
\end{array}
$$

Properties: White, crystalline powder or granules; odorless; faint sweet taste. D 1.52, mp 165–167C, specific rotation (20C) between +23 and +24 degrees, bp 290–295C (3–3.5 mm Hg). Soluble in water; slightly soluble in lower alcohols and amines; almost insoluble in other organic solvents. Combustible.
Derivation: (1) A natural plant exudate; (2) by hydrogenation of corn sugar or glucose.
Grade: Reagent, commercial, NF, FCC.
Use: Base for dietetic foods, diluent, determination of boron, pharmaceutical products, medicine, thickener and stabilizer in food products.

mannitol hexanitrate. (hexanitromannite; HNM; nitromannite; nitromannitol).
CAS: 15825-70-4. $C_6H_8(ONO_2)_6$.
Properties: Colorless crystals. Mp 112–113C. Soluble in alcohol, acetone, ether; insoluble in water.
Derivation: By nitrating mannitol with mixed acid, purifying the precipitate from organic solvents and stabilizing.
Grade: Technical, pharmaceutical.
Hazard: Dangerous fire and explosion risk; an initiating explosive.
Use: Explosive cap ingredient, medicine (admixed with at least 10 parts of carbohydrate).

D(+)-mannose.
CAS: 3458-28-4. $C_6H_{12}O_6$. A carbohydrate occurring in some plant polysaccharides.
Properties: Crystals from alcohol or acetic acid; sweet taste with bitter aftertaste. Mp 132C (decomposes).
Use: Biochemical research.

Manugel C. Specialty algin blends.

manure salts. Potash salts that have a high proportion of chloride and 20–30% potash.
Use: in fertilizers.

"MAPEG" [BASF]. (ethylene glycol distearate). TM for emulsifier and pearlizing opacifier.
Use: For cosmetics and pharmaceutical creams, lotions, and ointments.

MAPP. See methylacetylene-propadiene, stabilized.

mapping population. The group of related organisms used in constructing a genetic map.

maraging steel. An alloy steel containing nickel, cobalt, and titanium.
Use: Solid rocket cases.

maranta. See arrowroot.

Marathon-Howard process. A treatment of waste sulfite liquor from sulfite pulp manufacture to recover chemicals and reduce stream pollution. The waste sulfite is treated with lime and precipitates (1) calcium sulfite for use in preparing fresh cooking acid for the sulfite pulp process; and (2) a basic calcium salt of lignin sulfonic acid (lignin sulfonates) that can be pressed and used as a fuel or used as raw material for vanillin, lignin plastics, and other chemicals. The remaining liquor with its BOD reduced 80% is the effluent.

marble. A metamorphic form of calcium carbonate, usually containing admixtures of iron and other minerals that impart variegated color patterns. Marble chips are often used as source of carbon dioxide in laboratory experiments.

Marcus, Rudolph A. (1923–). An American who won the Nobel Prize for chemistry in 1992 for his theories of electron transfer. The processes Marcus has studied, the transfer of electrons between molecules in solution, underlie a number of exceptionally important chemical phenomena, and the practical consequences of his theory extend over all areas of chemistry. The Marcus theory describes, and makes predictions concerning, such widely differing phenomena as the fixation of light energy by green plants, photochemical production of fuel, chemiluminescence ("cold light"), the conductivity of electrically conducting polymers, corrosion, the methodology of electrochemical synthesis and analysis, and more.

margaric acid. See n-heptadecanoic acid.

marijuana. See cannabis.

marjoram oil. A yellowish essential oil, optically active.
Use: Soap perfume and in toilet preparations. The Spanish grade is used as a flavoring ingredient.

marker. (1) Molecular weight size marker is a piece of DNA of known size, or a mixture of pieces with known size, used on electrophoresis gels to determine. by comparison, the size of unknown DNAs. (2) Genetic marker is a known site on the chromosome. It may, for example, be the site of a locus with some recognizable phenotype, or it may be the site of a polymorphism that can be experimentally discerned.

Mark–Houwink equation. Defines the relationship between the intrinsic viscosity and molecular weight for homogeneous linear polymers.

Markownikoff rule. When a halogen acid adds to an asymmetrical ethylenic compound, the halogen usually appears on the carbon atom carrying the smaller number of hydrogen atoms; this order of addition is frequently reversed with hydrogen bromide if peroxides are present (peroxide effect).

Marlate. Technical grade methoxychlor insecticides, supplied in 50% wettable powder and an emulsifiable formulation containing 2 lb/gal.

"Marlex" [Chevron]. TM for high-density polyethylene.
Use: Packaging, consumer and industrial goods, pipe, film, transportation, building, and construction.

Marquis' reagent. Formaldehyde-sulfuric acid reagent for alkaloids.

marsh gas. See methane.

martensite. The chief constituent of hardened carbon tool steels. It is a solution of carbon or Fe_3C in β-iron, or an exceedingly fine-grained α-iron with carbon or Fe_3C in atomic or molecular dispersion. Carbon content up to 1%; easily obtained by quenching small bodies of hypereutectoid steel in cold water; more difficult to obtain in low-carbon steels.

Martin, Archer John Porter. (1910–2002). An English chemist and engineer who won the Nobel Prize for chemistry in 1952 along with Synge. His work involved partition chromatography in analysis which led to development of new antibiotics and amino acids. He received his doctorate from Cambridge University.

Martinet dioxindole synthesis. Formation of derivatives of dioxindole from esters of mesoxalic acid and aromatic amines or amino quinolines.

martonite. A lachrymator gas containing 80% bromoacetone and 20% chloroacetone.

Marvess. Olefin (polypropylene) fiber.
Use: Consumer, home furnishings, and industrial applications.

mash. A mixture of malted barley (or other grain) and water used for preparing wort in brewing operations. Also a mixture of grain, etc., prepared for fermentation in distilling, e.g., "sour mash whiskey."

Masonex. A hemicellulose extract.
Properties: Liquid containing 65% solids, 55% carbohydrates, pH 5.5. Water soluble; the spray-dried product is also water soluble with 97% solids and 84% carbohydrates (pH 3.7).
Derivation: From wood fibers after mild acid hydrolysis in steam.
Use: (Liquid) Intermediate in furfuryl production, animal feeds; (dried) binder in refractory bricks, tackifier in adhesives.

Masonite. A composition hardboard made by treating wood chips with steam at high pressure and compressing the resulting fibers into mats from which rigid panels are made by hot-pressing. The fiber is waterproofed with a water repellent sizing agent having a paraffin base.

masonry cement. A group of special cements more workable and plastic than Portland cement. Some are similar to waterproofed Portland cement while others are Portland cement mixed with hydrated lime, crushed limestone, diatomaceous earth, or granulated slag.

mass. The quantity of matter contained in a particle or body regardless of its location in the universe. Mass is constant, whereas weight is affected by the distance of a body from the center of the earth (or of any other planet or satellite, e.g., the moon). At extremely high temperatures (e.g., the sun's interior), mass is converted into energy. According to the Einstein equation $E = mc^2$, all forms of energy, such as radiant energy and energy of motion, possess a mass equivalent, even though they have no independent rest mass (photons); thus there is no absolute distinction between mass and energy.
See energy; matter.

mass action law. The rate of a chemical reaction for a uniform system of constant temperature is proportional to the concentrations of the substances reacting.
See chemical laws (1).

mass–action ratio. The ratio which describes the kinetics of a reaction, given as products over reactants. For a hypothetical reaction aA + bB cC + dD, the ratio would be [A]a [B]b/[C]c [D]d.

mass conservation law. See chemical laws (4).

mass defect. The difference between the total mass of the constituents of an atomic nucleus

(protons and neutrons) taken independently, and the actual mass of the nucleus as a whole. The latter is always slightly less than the sum of the masses of the constituents, the difference being the mass equivalent of the energy of formation (binding energy) of the nucleus. This accounts for the high energy release obtained from nuclear fission.
See fission, nuclear; mass number.

massecuite.   Term used in the sugar industry for the mixture of sugar and molasses prior to removal of the molasses.

massicot.   PbO. (1) (Lead ocher) Natural lead monoxide, PbO. Contains 92.8% lead; found in U.S. (Colorado, Idaho, Nevada, and Virginia). (2) An oxide of lead corresponding to the same formula as litharge (PbO), but having a different physical state.
Properties: It is a yellow powder formed by the oxidation of a bath of metallic lead at approximately 345C so that the oxide formed is not melted. D 9.3, mp 600C. If the oxide is melted, it is converted into litharge.

mass number.   The number of neutrons and protons in the nucleus of an atom. Thus, the mass number of normal helium is 4; of carbon, 12; of oxygen, 16; and of uranium 238. A given atom is characterized by its atomic number, equivalent to the number of protons, which give it its charge and thus determine the kind of element, and by its mass number, which includes the neutrons that make up the remainder of its mass. Helium has 2 protons and 2 neutrons (mass number 4 and atomic number 2). Protons and neutrons each have very close to unit mass, and since the mass change associated with binding the particles together in the nucleus is very small, the mass number is always within 1/10th unit of the atomic weight of the nuclide.
See mass; mass defect; atomic weight.

mass spectrometry.   A method of chemical analysis in which the substance to be analyzed is heated and placed in a vacuum. The resulting vapor is exposed to a beam of electrons that causes ionization to occur, either of the molecules or their fragments. The ions thus produced are accelerated by an electric impulse and then passed through a magnetic field, where they describe curved paths whose directions depend on the speed and mass-to-charge ratio of the ions. This has the effect of separating the ions according to their mass (electromagnetic separation). Because of their greater kinetic energy, the heavier ions describe a wider arc than the lighter ones and can be identified on this basis. The ions are collected in appropriate devices as they emerge from the magnetic field.

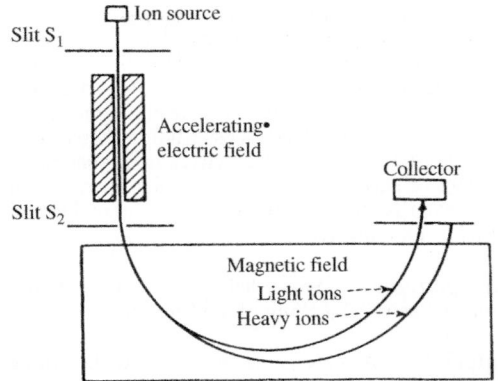

mass susceptibility.   Magnetic susceptibility of a compound per gram.

mast cell degranulating peptide.   (*Vespa basalis*).
CAS: 137354-65-5. $C_{78}H_{138}N_{20}O_{16}$.
Properties: Isolated from the venom of the Taiwan hornet *Vespa basalis*.
Hazard: A poison.
Source: Natural product.

master batch.   A previously prepared mixture composed of a base material and a high percentage of an ingredient (usually a dry powder) that is critical to the product being manufactured. Aliquot parts of this mixture are added to production-size quantities (batches) during the mixing operation. This method permits uniform dispersion of very small amounts (less than 1%) of such additives as dry curing agents in rubber and colorants in plastics and paints. Modifying elements may be incorporated in alloys in this way. Master batches of organic dyes dispersed in rubber or plastic are prepared by manufacturers of colorants for direct use. Master-batched accelerators (mixtures of rubber, zinc oxide and accelerator) are commonly used in rubber mixes.

mastic.   (1) A solid, resinous exudation of the tree *Pistacia lentiscus*, of recent geologic origin. Found in the Mediterranean area. Soluble in alcohol and ether, balsamic odor, turpentine taste. Used in chewing gum, varnishes, and to some extent in adhesives and dentistry. (2) A soft, putty-like sealant; often packed in cartridges and applied by a gun with a nozzle of appropriate size or in bulk for application by a knife or other spreading device. Such mastics often contain bituminous ingredients or polymerized rosin acids, used for laying floor and wall tiles and similar applications.

mastication.   Permanent softening of crude natural rubber and certain other elastomers by application of mechanical energy, as on a roll mill or in a Banbury mixer. The softening is said to be due to formation of free radicals resulting from the rupture of the polymer chain and the addition of oxygen at these active points. The study of this phenomenon has been called mechanochemistry. This "breaking down" of a high polymer substance, first practiced by Hancock in England, is essential in preparing it for the incorporation of curatives and other modifying substances and is an essential operation in the manufacture of useful products from rubber and rubber-like substances.

mastigoneme. Small hair-like filaments found on the "hairy" flagellum of the Chromista.

material. A nonspecific term used with various shades of meaning in the technical literature. It should not be used as a synonym for substance, but is generally used in the collective expressions "raw materials" and "materials handling." The term *material balance*. in chemical engineering denotes the sum of all the substances entering a reaction and all those that leave it in a given time period. *Material* also loosely refers to closely associated mixtures, either of natural origin (petroleum, wood, ores) or man-made (glass, cement, composites). See substance; engineering material; raw material.

material balance. See material.

materials handling. A general term that includes the methods used for in-plant transportation, distribution, and storage of raw materials and semiprocessed products, as by forklift trucks, elevators, conveyor systems, pipelines, etc., as well as safe practices for storage and movement of toxic and flammable substances.

matte. A product containing a metal sulfide as obtained after roasting and fusion of sulfide minerals. Oxides or metals may also be present. Common examples are copper matte and nickel matte.

matter. Anything that has mass or occupies space.
See mass. (1) *States*: There are three generally accepted states (phases) in which substances can exist, i.e., solid, liquid, and gas (vapor). From time to time it has been proposed that specialized forms of matter be regarded as states, such as the vitreous (glassy) state, the colloidal state, and the plasma state, but none of these suggestions has gained substantial acceptance.
See phase (1); solid; liquid; gas. (2) *Levels*: Matter is basically composed of particles in the following levels of size and complexity: (a) subatomic (protons, neutrons, electrons); (b) atomic and molecular (below 10 Å); (c) colloidal (from 10 Å to 1 micron); (d) microscopic; (e) macroscopic; (f) space or celestial. Level (a) is invisible by any means; levels (b) and (c) can be resolved in field-ion or electron microscopes; level (d) lies in the range of the optical microscope; (e) is visible to the naked eye; (f) requires telescopes for detailed observation. See particle; electron microscope; field-ion microscope.

Maxaquin. Lomefloxacin hydrochloride.
Use: Drug.

"Maxarome" *[DSM].* TM additive for UV coatings.

Maxatase. Subtilisins.

maxilon golden yellow gl. See C.I. basic yellow 28.

maxilon yellow 3gl. See C.I. basic yellow 11.

MaxiThermo. Primary grade natural gas standard now available for measuring the BTU content of natural gas. It offers direct traceability for natural gas.

Mayer's reagent. See mercuric potassium iodide.

"Mazawax" *[BASF].* TM for all-purpose emulsifying waxes.
Available forms: Solid and flaked.
Use: In creams, lotions, hair relaxant, and hair depilatory formulas.

MBMC. Abbreviation for monobutyl-*m*-cresol. See *tert*-butyl-*m*-cresol.

MBT. Abbreviation for mercaptobenzothiazole.

MBTE. Abbreviation for methyl-*tert*-butyl ether.

MBTS. Abbreviation for 2-mercaptobenzothiazyl disulfide.
See 2,2'-dithiobisbenzothiazole.

MCA. Abbreviation for monochloroacetic acid. See chloroactic acid.

mCi. Abbreviation for milliCurie.
See curie.

McFadyen–Stevens reaction. Base-catalyzed thermal decomposition of aroylbenzenesulfonylhydrazines to aldehydes.

McLafferty rearrangement. Electron-impact-induced cleavage of carbonyl compounds having a hydrogen in the γ-position, to an enolic fragment and an olefin.

McMillan, Edwin M. (1907–1991). An American physicist who won the Nobel Prize for chemistry in 1951 along with Seaborg. His work included research in nuclear physics and particle accelerator development as well as microwave radar and sonar. He and his colleagues discovered neptunium and plutonium. He was the recipient of the Atoms for Peace prize in 1963. His Ph.D. in physics was awarded from Princeton University.

MCPA. (2-methyl-4-chlorophenoxyacetic acid; 4-chloro-2-methylphenoxyacetic acid).
CAS: 94-74-6. $CH_3ClC_6H_3OCH_2COOH$.
Properties: White, crystalline solid. Mp 118–119C. Free acid insoluble in water but sodium and amine salts are soluble.
Grade: Emulsifiable concentration.
Use: Selective herbicide.

MCPB. (4-(2-methyl-4-chlorophenoxy)butyric acid).
CAS: 94-81-5.
Hazard: Toxic by ingestion.
Use: Herbicide.

MCPP. (2-(2-methyl-4-chlorophenoxy)-propionic acid).
See mecoprop.

Md. Symbol for mendelevium.

MDA. (1) Abbreviation for metal deactivator. (2) Abbreviation for p,p'-methylenedianiline; p,p'-diaminodiphenylmethane.

MDAC. 4-methyl-7-diethylaminocoumarin. See 7-diethylamino-4-methylcoumarin.

MDEA. See n-ethyl-3,4-methylenedioxy-amphetamine.

MDI. methylene di-p-phenylene isocyanate. See diphenylmethane-4,4'-diisocyanate.

Me. Symbol for methyl.

MEA. Abbreviation for monoethanolamine. See ethanolamine.

mean free path. In a given system, the average distance that particles travel between collisions with other particles. The particles may be molecules of a gas, free electrons, neutrons, etc. The value is determined by statistical methods.

mecamylamine hydrochloride. (methylaminoisocamphane hydrochloride).
CAS: 826-39-1. $C_{11}H_{21}N•HCl$.
Properties: White, crystalline powder; almost odorless. Mp 245C with some decomposition. Soluble in water, alcohol, chloroform; somewhat soluble in benzene, isopropyl alcohol; insoluble in ether.

Grade: USP.
Use: Medicine (antihypertensive).

mecarbam. (S-[(ethoxycarbonyl)methylcarbamoyl]methyl-O,O-diethyl phosphorodithioate).
CAS: 2595-54-2. $(C_2H_5O)_2P(S)SCH_2CON(CH_3)COOC_2H_5$.
Properties: Yellow oil. Bp 144C (0.02 mm Hg).
Hazard: Highly toxic, cholinesterase inhibitor. Use may be restricted.
Use: Insecticide.

mechanochemistry. See mastication.

mechlorethamine hydrochloride. [methyl-bis-(2-chloroethyl)amine hydrochloride].
CAS: 55-86-7. $CH_3N(CH_2CH_2Cl)_2•HCl$. A nitrogen mustard.
Properties: White, crystalline, hygroscopic powder; nasal irritant and a vesicant. Mp 108–111C. Soluble in water.
Grade: USP.
Hazard: Highly toxic, vesicant, and strongly irritant to mucous membranes.
Use: Medicine (antineoplastic).

meclizine hydrochloride. (1-p-chlorobenzhydryl-4-methylbenzylpiperazine dihydrochloride).
CAS: 36236-67-6. $C_{25}H_{27}ClN_2•2HCl•H_2O$.
Properties: White or yellowish powder or crystals; slight odor. Insoluble in water and ether; very soluble in chloroform, pyridine, and acid-alcohol-water mixture; slightly soluble in dilute acids and alcohol.
Grade: USP.
Use: Medicine (antihistamine).

meclofenoxate hydrochloride.
CAS: 3685-84-5. $C_{12}H_{17}Cl_2NO_3$.
Properties: Colorless crystals. Mp 135C. Soluble in water, insoluble in benzene.
Use: Plant growth hormone.

"Mecopex" *[Morton].* TM for a broad-leaf herbicide containing potassium salt of mecoprop (31.5%).
Hazard: See mecoprop.

mecoprop. (2-(4-chloro-2-methylphenoxy)propionic acid; MCPP; CMPP).
CAS: 93-65-2. $ClC_6H_3(CH_3)OCH(CH_3)COOH$.
Properties: Solid. Mp 93–94C. Insoluble in water; soluble in alcohol, acetone, and ether.
Hazard: Toxic by ingestion and inhalation, irritant to skin and eyes.
Use: Herbicide.

medical waste. (infectious waste; red bag waste). Any waste from a hospital, clinic, or other health care facility that contain or have come into contact with diseased tissues or infectious microorganisms.

medicinal chemistry.   A subdivision of chemistry that deals with the effects of drugs and pharmaceuticals on the human body and on various infective organisms, and with the synthesis of compounds specifically for certain diseases, such as antimalarials and antihypertensive agents. It also is concerned with immunology, hormone activity, etc.
See clinical chemistry.

medlure.   (*sec*-butyl-4-(or 5)-chloro-2-methylcyclohexanecarboxylate). $C_{12}H_{21}ClO_2$.
Properties: Liquid. Bp 78–79C (0.25 mm Hg). Insoluble in water; soluble in most organic solvents.
Use: Insect attractant.

2,2-MEDP.   See 2-formyl-2′-methyl-1,1′-(oxydimethylene)dipyridinium.

medroxyprogesterone acetate.   (17-hydroxy-6a-methyl-preg-4-ene-3,20-dione-17-acetate).
CAS: 71-58-9. $C_{24}H_{34}O_4$. A hormone derivative.
Properties: White to off-white, crystalline powder; odorless. Melting range 200–208C. Stable in air. Insoluble in water; freely soluble in chloroform; sparingly soluble in alcohol.
Hazard: Possible carcinogen.
Use: Medicine (injectable contraceptive).

medulla.   Soft inner part as of a living organ.

Meerwein–Ponndorf–Verley reduction.
Reduction of aldehydes or ketones to the corresponding alcohols with aluminum alkoxides.

mega-.   Prefix meaning 10^6 units (symbol M), e.g., 1 megaton = 1,000,000 tons.

megabase (Mb).   Unit of length for DNA fragments equal to 1 million nucleotides and roughly equal to 1 cm.

megacin.   Any of a group of bacteriocins isolated strains of *Bacillus megaterium*.

megatomoic acid.   (*trans*-3,*cis*-5-tetradecadienoic acid). The active ingredient of the sex attractant in the female carpet beetle. Research is underway on synthesis of this substance to be used as a lure in control of this pest.

megaton.   One million tons, usually used in defining the blast effect of a nuclear explosion. A 1-megaton bomb is equivalent in destructive potential to 1 million tons of TNT.

megestrol   acetate.   (17-hydroxy-6-methylpregna-4,6-diene-3,20-dione acetate).
CAS: 595-33-5. $C_{24}H_{32}O_4$.
Properties: Crystals. Mw 384.50, mp 214–216C. Soluble in water.
Use: Antineoplastic and oral contraceptive.

meglumine   diatrizoate.   (methylglucamine diatrizoate; diatrizoate methylglucamine).
CAS: 131-49-7. $(CH_3CONH)_2C_6I_3COOH \cdot CH_3 NHCH_2(CH_2O)_4 \cdot CH_2OH$.
Properties: Available in solution for injection, pH between 6.0–7.6.
Grade: USP (as injection).
Use: Medicine (radiopaque medium).
Note: The iodipamide and iothalamate are also available.

meiosis.   A two-stage type of cell division in sexually reproducing organisms. In meiosis, a diploid cell divides to produce four haploid cells, each with half the original chromosome content. For this reason, meiosis is often called "reduction division." In organisms with a diploid life cycle, the products of meiosis are usually called gametes. In organisms with alternation of generations, the products of meiosis are called spores.

Meisenheimer complexes.   Sigma-complexes obtained as brightly colored solutions on interaction of 1,3,5-trinitrobenzene to similar compounds.

Meisenheimer rearrangement.   Rearrangement of tertiary amine oxide to *O,N,N*-trisubstituted hydroxylamines.

meitnerium.   Mt. A transfermium element. Atomic number 109. Very short half-life.

MEK.   Abbreviation for methyl ethyl ketone.

"MEKON" [Baker Hughes].   TM for a hard grade of petroleum microcrystalline wax.

melamine.   (cyanurtriamide; 2,4,6-triamino *s*-triazine).
CAS: 108-78-1.

Properties: White, monoclinic crystals. D 1.573 (14C), mp 354C. Sparingly soluble in water, glycol, glycerol, pyridine; very slightly soluble in ethanol; insoluble in ether, benzene, carbon tetrachloride. Nonflammable.
Derivation: (1) By heating urea and ammonia. The resulting mixture of isocyanic acid and ammonia reacts over a solid catalyst at approximately 400C to form melamine. (2) From cyanamide, dicyanamide, or cyanuric chloride.
Method of purification: Recrystallization from water.
Grade: 99% min.

Hazard: Toxic by ingestion, skin, and eye irritant. Questionable carcinogen.
Use: Melamine resins, organic syntheses, leather tanning.

melamine, formaldehyde, toluenesulfonamide polymer, butylated.
CAS: 68891-01-0.
Hazard: Low toxicity by ingestion and skin contact. A moderate eye irritant.

melamine resin. A type of amino resin made from melamine and formaldehyde. The first step in resin formation is the production of trimethylol melamine, $C_3N_3(NHCH_2OH)_3$, the molecules of which contain a ring with three carbon and three nitrogen atoms, the –NHCH$_2$OH groups being attached to the carbon atoms. This molecule can combine further with others of the same kind by a condensation reaction. Excess formaldehyde or melamine can also react with trimethylol melamine or its polymers, providing many possibilities of chain growth and cross-linking. The nature and degree of polymerization depend upon pH, but heat is always needed for curing. Melamine resins are more water- and heat-resistant than urea resins. They may be water-soluble syrups (low molecular weight) or insoluble powders (high molecular weight) dispersible in water. Widely used as molding compounds with α-cellulose, wood flour, or mineral powders as fillers and with coloring materials; also for laminating, boil-proof adhesives, increasing wet strength of paper, textile treatment, leather processing, and for dinnerware and decorative plastic items. Butylated melamine resins are formed by incorporating butyl or other alcohols during resin formation, whereupon the –NHCH$_2$OH groups convert to –NHCH$_2$OC$_4$H$_9$. These resins are soluble in paint and enamel solvents and in surface coatings, often in combination with alkyds. They give exceptional curing speed, hardness, wear resistance, and resistance to solvents, soaps, and foods. Melamine-acrylic resins are water soluble and used for formation of water-base industrial and automotive finishes.
See urea-formaldehyde resin.

melaniline. See diphenylguanidine.

melanin. A brownish-black pigment that occurs normally in the retina, skin, and hair of higher animals with the exception of albinos. Formed from tyrosine by the action of tyrosinase.

melanocyte. Cells in the epidermis of the skin that produce the pigment melanin.

melanoma. Cancer of the cells that produce pigment in the skin. Melanoma usually begins in a mole.

melinamide. See *dl-n-*(α-methylbenzyl) linoleamide.

melissic acid. (triacontanoic acid).
CAS: 506-50-3. $CH_3(CH_2)_{28}COOH$. A long-chain fatty acid.
Properties: Crystalline solid. Mp 94C. Soluble in benzene and hot alcohol; insoluble in water. Combustible.
Derivation: By oxidation of 1-triacontanol, occurs in minor amounts in many plant and insect waxes and in montan wax.
Use: Biochemical research.

melissyl alcohol. See 1-triacontanol.

melittin.
CAS: 37231-28-0. $C_{131}H_{229}N_{39}O_{31}$. A polypeptide derived from bee venom that has strong antibacterial activity, especially against *Staphylococcus aureus* 80 which is resistant to penicillin. It inhibits growth of many Gram-positive and Gram-negative bacteria.
Properties: White powder. Water soluble.
Use: Antirheumatic drug.

mellitate. An ester or salt of mellitic acid.

"Melmac" [Allnex]. TM for products molded from melamine-formaldehyde resins.

"Melonite" [Durferrit]. TM for an anhydrous molten salt bath used to nitride ferrous work pieces. The bath operates at a subcritical temperature and produces a continuous ε-iron nitride layer on carbon steels, and alloy nitride surfaces on alloy steels.
Use: For many components to enhance fatigue strength, wear and corrosion resistance. It is also used for tooling and dies to extend service life.

melphalan. (*p*-di(2-chloroethyl) aminophenylalanine; formerly called sarcolysin).
CAS: 148-82-3. $(ClCH_2CH_2)_2NC_6H_4CH_2CH(NH_2)COOH$. Melphalan is both the USAN name for the acid and the generic name for the hydrochloride.
Properties: A nitrogen mustard, crystals. Mp 180C.
Grade: ND (in medicine, for the acid).
Hazard: Strong irritant to eyes and mucous membranes. Confirmed carcinogen.
Use: Medicine, insect chemosterilant.

melt index. The viscosity of a thermoplastic polymer at a specified temperature and pressure, it is a function of the molecular weight. Specifically, the number of grams of such a polymer that can be forced through a 0.0825-inch orifice in 10 minutes at 190C by a weight of 2160 g.

melting point. (mp). The melting point or freezing point of a pure substance is the temperature at which its crystals are in equilibrium with the liquid phase at atmospheric pressure. The term *melting point* is used when the equilibrium temperature is approached by heating the solid. Ordinarily, mp refers to temperatures above 0C, the melting point of ice. The terms *melting point* and *freezing point* are often used interchangeably, depending on whether the substance is being heated or cooled. The number of calories required to convert one mole of pure crystals to the liquid state is called the molar heat of fusion.

"Melurac" [Cytec]. TM for urea-melamine-formaldehyde condensation products used mainly as adhesives for bonding of veneers for the production of exterior-grade plywood.

membrane. Semi-fluid structure which bounds all cells and partitions the interior of eukaryotic cells. It consists primarily of two lipid layers, with proteins "dissolved" in the lipids.

membrane cell. See diaphragm cell.

membrane equilibrium. Equilibrium between the ions of a salt solution and the giant and normal ions of a colloidal electrolyte separated by a membrane.

membrane hydrolysis. Hydrolysis that occurs when a colloidal electrolyte, consisting of giant ions and normal ions, is separated from pure water by a membrane. The diffusing ions are substituted by the OH^- or H^+ ions of the dissociating water.

membrane, semipermeable. A microporous structure, either natural or synthetic, that acts as a highly efficient filter in the range of molecular dimensions, allowing passage of ions, water, and other solvents and very small molecules, but is almost impermeable to macromolecules (proteins) and colloidal particles. The thickness is about 100 Å, the pore diameter is from 8 Å for the walls of tissue cells to 100 Å or more for manufactured membranes. Plant cell wall membranes are proteinaceous substances that function in natural osmosis. Membranes of cellophane, collodion, asbestos fiber, etc., are used in such industrial operations as waste-liquor recovery, desalination, and electrolysis; processing whey proteins, biomedical research; gas separations, e.g., adjusting carbon monoxide/hydrogen ratios for ammonia production from synthesis gas; metal extraction and recovery. See osmosis; dialysis.

membrane transport. Movement of a polar molecule across a membrane via a specific membrane protein (a transporter).

memtetrahydrophthalic anhydride. (methyl norbornene dicarboxylic anhydride). CAS: 85-43-8. $C_8H_8O_3$.

Properties: Clear, transparent, slightly viscous liquid; colorless to light yellow.
Hazard: Strong irritant to eyes and skin.
Use: Curing epoxy resins, electrical laminating and filament winding, intermediate for polyesters, alkyd resins, and plasticizers.

MENA. Abbreviation for the methyl ester of naphthaleneacetic acid.
See α-naphthaleneacetic acid methyl ester.

menadiol diphosphate tetrasodium salt. See 1:4-naphthaquinol-bisdisodium phosphate.

menadione. (2-methyl-1,4-naphthoquinone; menaphthone; vitamin K_3).
CAS: 58-27-5. $C_{10}H_5CH_3O_2$.
Properties: Yellow, crystalline powder; nearly odorless. Mp 105–107C. Affected by sunlight. Soluble in alcohol, benzene, and vegetable oils; insoluble in water.
Derivation: Oxidation of β-methylnaphthalene.
Grade: USP.
Hazard: Irritant to skin and mucous membranes, especially the alcoholic solution.
Use: Medicine, fungicides, animal feed additives.

menazon. (generic name for S-[(4,6-diamino-s-triazin-2-yl)methyl]-O,O-dimethyl phosphorodithioate).
CAS: 78-57-9.

$(CH_3O)_2P(S)SCH_2C:NC(NH_2):NC(NH_2):N.$

Properties: Off-white solid. Mp 160–162C. Slightly soluble in water and organic solvents.
Hazard: Highly toxic, cholinesterase inhibitor.
Use: Acaricide, insecticide.

mendelevium. (Md). Synthetic radioactive element produced in a cyclotron by bombarding einsteinium with alpha particles; atomic number 101, aw 256, 4 isotopes, valence = 3. Mendelevium decays by spontaneous fission with a half-life of 1.5 hr. The heaviest isotope, Md-258, has half-life of 60 days. Mendelevium is thought to have chemical properties similar to those of the rare earth thulium. It is made in research quantities only and no uses are reported.
See actinide series.

Mendeleev, Dmitri Ivanovich. (1834–1907). Born in Siberia, Mendeleev made a fundamental contribution to chemistry in 1869 by establishing the principle of periodicity of the elements. His first periodic table was compiled on the basis of arranging the elements in ascending order of atomic weight and grouping them by similarity of properties. So accurate was Mendeleev's thinking that he predicted the existence and atomic weights of several elements that were not actually discovered until years later. The original table has been modified and corrected several times, notably by Moseley, but it has accommodated the discovery of isotopes, rare gases, etc. Its importance in the development of chemical theory can hardly be overestimated.

Mendelian inheritance. One method in which genetic traits are passed from parents to offspring. Named for Gregor Mendel, who first studied and recognized the existence of genes and this method of inheritance.
See autosomal dominant; recessive gene; sex-linked.

menhaden oil.
Properties: Yellowish-brown or reddish-brown, drying oil; characteristic odor. D 0.927–0.933, saponification value 191–196, iodine value 139–180, refr index 1.480. Soluble in ether, benzene, naphtha, and carbon disulfide. Combustible.
Derivation: By cooking or pressing the body of the menhaden fish.
Method of purification: Filtration and bleaching with fuller's earth.
Grade: Prime crude, brown strained, strained, bleached, winter oil, bleached winter white oil.
Hazard: Subject to spontaneous heating.
Use: Hydrogenated fats for cooking and industrial use (soap, rubber compounding), printing inks, animal feed, leather dressing lubricants, paint drier, cleansers, lipstick.

meniscus. The concave curve of a liquid surface in a graduate or narrow tube. Caused by surface tension. In reading a value (e.g., 5 cc), it is conventional to ignore the higher liquid around the perimeter. In the case of mercury, which does not wet the tube because of its extremely high surface tension, the meniscus is convex.

Menschutkin reaction. Reaction of tertiary amines with alkyl halides to form quaternary salts.

menthanediamine. (*p*-menthane-1,8-diamine).
CAS: 80-52-4.

$(CH_3)_2C(NH_2)CHC_2H_4C(CH_3)(NH_2)CH_2CH_2.$

A primary alicyclic diamine, also a *tert*-alkylamine.
Properties: Clear liquid; terpene odor. Boiling range 107–126C (10 mm Hg), fp −45C, refr index 1.4794 (25C). Miscible with water and most organic solvents.
Hazard: Strong irritant to eyes and skin, calls for eye protection.
Use: Curing agent for epoxy resins, chemical intermediate.

p-menthane-8-hydroperoxide.
CAS: 80-47-7. $C_{10}H_{20}O_2.$
Properties: Clear, pale-yellow liquid. D 0.910–0.925 (15.5/4C), refr index 1.460–1.475 (20C).

Hazard: Strong oxidizing agent, dangerous in contact with organic materials. Strong irritant to skin and eyes.
Use: Catalyst for rubber and polymerization reactions, coatings.

p-menthan-3-one. See menthone.

menthol. (hexahydrothymol; methylhydroxyisopropyl-cyclohexane; *p*-menthan-3-ol; peppermint camphor).
CAS: 89-78-1. $CH_3C_6H_9(C_3H_7)OH$. It may be *l*- (from natural sources) or *dl*- (natural or synthetic).
Properties: White crystals; cooling odor and taste. Mp 41–43C (*l*-form), congealing temperature 27–28C (*dl*-form), specific rotation −45 to −51 degrees (25C) (*l*-menthol), −2 to +2 degrees (*dl*-menthol). Soluble in alcohol, light petroleum solvents, glacial acetic acid, and fixed or volatile oils; slightly soluble in water. Combustible.
Occurrence: Brazil (natural product), Japan.
Derivation: By freezing from peppermint oil, by hydrogenation of thymol.
Grade: Technical, USP, FCC.
Hazard: Irritant to mucous membranes on inhalation.
Use: Perfumery, cigarettes, liqueurs, flavoring agent, chewing gum, chest rubs, cough drops.

menthol acetic ester. See menthyl acetate.

menthol valerate. (menthyl isovalerate). $(CH_3)_2CHCH_2COOC_{10}H_{19}.$
Properties: Colorless liquid; mild odor; cooling, faintly bitter taste. D 0.907 (15.4C). Insoluble in water; soluble in alcohol, chloroform, ether, and oils.
Derivation: By action of valeric acid on menthol.
Use: Medicine, flavoring.

menthone. (*p*-menthan-3-one).
CAS: 1074-95-9. $C_{10}H_{18}O.$
Properties: Colorless, oily, mobile liquid; slight peppermint odor. D 0.897 (15C), bp 207C. Slightly soluble in water; soluble in organic solvents. Combustible.
Derivation: A ketone found in oil of peppermint.
Use: Flavoring.

menthyl acetate. (menthol acetic ester).
CAS: 16409-45-3. $C_{10}H_{19}OOCCH_3.$
Properties: Colorless liquid; menthol-like odor. Bp 227–228C, d 0.922–0.927, optical rotation −72 degrees 47 minutes to −73 degrees 18 minutes, refr index 1.447. Slightly soluble in water; miscible with alcohol and ether. Combustible.
Derivation: (1) By boiling menthol with acetic anhydride in the presence of sodium acetate; (2) peppermint oil.
Use: Perfumery, flavoring.

menthyl isovalerate. See menthyl 3-menthylbutyrate.

menthyl 3-menthylbutyrate.
CAS: 16409-46-4. $C_{15}H_{28}O_2$.
Hazard: Low toxicity by ingestion and skin contact. A moderate skin contact irritant.
Source: Natural product.

menthyl salicylate.
CAS: 89-46-3. $C_6H_4(OH)COOC_{10}H_{19}$.
Properties: Colorless liquid. Miscible with alcohol, ether, chloroform, and fatty oils; insoluble in water; soluble in organic solvents. Combustible.
Use: Sunscreen preparations.
See homomenthyl salicylate.

MEP. Abbreviation for methyl ethyl pyridine.

meperidine hydrochloride. (Demerol).
CAS: 50-13-5. $C_{15}H_{21}NO_2HCl$. An addictive drug, use by prescription only.
Use: Analgesic.

mepiquat chloride.
CAS: 24307-26-4. $C_7H_{16}N\cdot Cl$.
Hazard: Moderately toxic by ingestion. Low toxicity by inhalation and skin contact.
Use: Agricultural chemical.

mepivacaine. (Carbocaine; 1-methyl-2-(2,6-xylylcarbamoyl)piperidine).
Use: Local anesthetic.

meprobamate. (2-methyl-2-n-propyl-1,3-propanediol dicarbamate).
CAS: 57-53-4. $H_2NCOOCH_2C(CH_3)(C_3H_7)$ CH_2OOCNH_2.
Properties: White powder; characteristic odor; bitter taste. Mp 103–107C. Slightly soluble in water and ether; soluble in alcohol and acetone.
Grade: NF.
Hazard: Central nervous system depressant, use restricted by law.
Use: Medicine (antianxiety agent).

meq. Abbreviation for milliequivalent.
See eq.

merbromin. (dibromohydroxymercurifluorescein disodium salt; 2,7-disodiumdibromo-4-hydroxymercurifluorescein; mercurochrome).
CAS: 129-16-8. $C_{20}H_8Br_2HgNa_2O_6$.
Properties: Iridescent green scales or granules; odorless. Soluble in water; insoluble in alcohol, acetone, chloroform, or ether; stable in air.
Derivation: From dibromofluorescein and mercuric acetate.
Grade: Technical, NF.
Hazard: Toxic by ingestion.
Use: Medicine (antiseptic).

mercapsol process. Process of removing mercaptans from gasoline by counter-current liquid–liquid extraction with various liquids, as with a water solution of caustic soda and tar acids.

mercaptamine. See 2-aminoethanethiol.

mercaptan. See thiol.

mercaptoacetic acid. See thioglycolic acid.

2-mercaptobenzoic acid. See thiosalicylic acid.

2-mercaptobenzothiazole. (MBT).
CAS: 149-30-4.

CHCHCHCHCHCCSC(SN)N.

Properties: Yellowish powder; distinctive odor (depends on degree of purification). D 1.52, melting range 164–175C. Soluble in dilute caustic soda, alcohol, acetone, benzene, chloroform; insoluble in water and gasoline. Combustible.
Grade: Technical, 97%.
Use: Vulcanization accelerator for rubber (requires use of stearic acid for full activation), tire treads and carcasses, mechanical specialties, etc., fungicide, corrosion inhibitor in cutting oils and petroleum products, extreme-pressure additive in greases.

2-mercaptobenzothiazole monoethanolamine salt. See n-(2-hydroxyethyl)ammonium benzothiazole-2-thiolate.

(2-mercaptobenzothiazolyl)-2-(2-aminothiazol-4-yl)-2-methoxyiminoacetate (syn).
CAS: 80756-85-0. $C_{13}H_{10}N_4O_2S_3$.
Hazard: Low toxicity by ingestion.

2-mercaptobenzothiazyl disulfide. See 2,2'-dithiobisbenzothiazole.

2-mercaptoethanol. $HSCH_2CH_2OH$.
CAS: 60-24-2.
Properties: Water-white, mobile liquid; disagreeable odor. D 1.1168 (20/20C), 9.29 lb/gal (20C), bp 157.1C, vap press 1.0 mm Hg (20C), viscosity 3.43 cP (20C), flash p 165F (74C), fp sets to a glass approximately −100C, refr index 1.5011 (20C). Miscible with water, and most organic solvents. Combustible.
Hazard: Toxic by inhalation and ingestion.
Use: Solvent for dyestuffs, intermediate for producing dyestuffs, pharmaceuticals, rubber chemicals, flotation agents, insecticides, plasticizers, water-soluble reducing agent, biochemical reagent, PVC stabilizers, agricultural chemicals, textile auxiliary.

β-mercaptoethylamine hydrochloride.
CAS: 156-57-0.   HS(CH$_2$)$_2$NH$_2$•HCl.
Properties: Hygroscopic, white powder. Mp 71C.
Very soluble in water and ethyl alcohol.
Hazard: Toxic by inhalation and ingestion.

N-(2-mercaptoethyl)benzenesulfonamide.
(S-(O,O-diisopropyl phosphorodithioate; N-β-
diisopropyldithiophosphorylethyl)benzenesulfona-
mide).
CAS: 741-58-2.   C$_6$H$_5$SO$_2$NH(CH$_2$)$_2$SP(S)
[OCH(CH$_3$)$_2$]$_2$.
Properties: Colorless liquid; slight odor. D 1.25
(22C). Insoluble in water; soluble in kerosene and
xylene.
Hazard: Highly toxic, cholinesterase inhibitor. Use
may be restricted.
Use: Herbicide.

2-mercapto-4-hydroxypyrimidine.   See
thiouracil.

2-mercaptoimidazoline.   See NA-22.

3-mercaptolvaline.   See l-penicillamine.

mercaptomerin.   See thiomerin.

3-mercapto-n-methylpropanamide.   See n-
methyl 3-mercaptopropionamide.

β-mercaptopropionic acid.
HSCH$_2$CH$_2$COOH.
Properties: Clear liquid. D 1.218 (21C), mp 16.8C,
bp 111C (15 mm Hg). Soluble in water and alcohol.
Use: Stabilizer, antioxidant, catalyst, chemical inter-
mediate.

6-mercaptopurine.   (6-MP; 6-purinethiol).
CAS: 50-44-2.   C$_5$H$_4$N$_4$S•H$_2$O. A sulfur-
containing purine base not found in animal
nucleoproteins.
Properties: Yellow, crystalline powder; nearly odorless. Mp 308C (decomposes). Insoluble in water;
soluble in hot alcohol and dilute alkali solutions;
slightly soluble in dilute sulfuric acid.
Grade: USP.
Hazard: Questionable carcinogen.
Use: Medicine (to prevent rejection of kidney transplants).

mercaptosuccinic acid.   See thiomalic acid.

2-mercaptothiazoline.   C$_2$H$_4$NSSH.   White
crystals.
Use: Synthesis of pharmaceuticals.

d-3-mercaptovaline.   See penicillamine.

mercerized cotton.   See cotton, mercerized.

mercerizing assistant.   A compound used to
increase the penetration of mercerizing baths. Cresylic acid derivatives, special sulfonated oils, and
other wetting agents are typical.

Mercer's liquor.   A solution containing potassium ferricyanide.
Use: For etching.

merchant acid.   A crude phosphoric acid (51–
54% P$_2$O$_5$).

**mercurate(1-),   (l-cysteinato(2-)-s)methyl-,
hydrogen.**
CAS: 32754-35-1.   C$_4$H$_8$HgNO$_2$S•H.
Hazard: A reproductive hazard.

mercuric acetate.   (mercury acetate).
CAS: 1600-27-7.   Hg(C$_2$H$_3$O$_2$)$_2$.
Properties: White, crystalline powder. D 3.2544.
Soluble in alcohol and water; sensitive to light.
Hazard: Toxic by ingestion, inhalation, and skin
absorption; strong irritant.
Use: Catalyst in organic synthesis, pharmaceuticals.

mercuric arsenate.   (mercury arsenate).
CAS: 7784-37-4.   HgHAsO$_4$.
Properties: Yellow powder. Soluble in hydrochlo-
ric acid; slightly soluble in nitric acid; insoluble in
water.
Hazard: Toxic by ingestion, inhalation, and skin
absorption; strong irritant.
Use: Waterproof paints, antifouling paints.

mercuric barium bromide.   (barium mercury
bromide; mercury barium bromide). HgBr$_2$•BaBr$_2$.
Properties: Colorless, crystalline mass. Very hygroscopic; soluble in water.
Hazard: Toxic by ingestion, inhalation, and skin
absorption; strong irritant.

mercuric barium iodide.   (barium mercury
iodide; mercury barium iodide). HgI$_2$•BaI$_2$•5H$_2$O.
Properties: Reddish or yellow crystalline mass.
Unstable; deliquescent; soluble in alcohol and
water.
Hazard: Toxic by ingestion, inhalation, and skin
absorption; strong irritant.
Use: Microanalysis (testing for alkaloids), preparing
Rohrbach solution.

mercuric benzoate.   (mercury benzoate).
CAS: 583-15-3.   Hg(C$_7$H$_5$O$_2$)$_2$•H$_2$O.
Properties: White crystals. Mp 165C. Sensitive to
light. Soluble in solutions of sodium chloride and
ammonium benzoate; slightly soluble in alcohol
and water.
Derivation: By the interaction of a mercuric salt and
sodium benzoate.
Hazard: Toxic by ingestion, inhalation, and skin
absorption; strong irritant.
Use: Medicine (antisyphilitic).

mercuric bromide. (mercury bromide).
CAS: 7789-47-1. $HgBr_2$.
Properties: White, rhombic crystals. Sensitive to light. D 6.11, mp 235C, bp 322C. Soluble in alcohol and ether; sparingly soluble in water.
Derivation: By adding potassium bromide to a solution of a mercuric salt and crystallizing.
Hazard: Toxic by inhalation, ingestion, and skin absorption; strong irritant.
Use: Medicine.

mercuric chloride. (mercury bichloride; mercury chloride; corrosive sublimate).
CAS: 7487-94-7. $HgCl_2$.
Properties: White crystals or powder; odorless. D 5.44 (25C), bp 303C, mp 276C. Soluble in water, alcohol, ether, pyridine, glycerol, and acetic acid.
Derivation: (1) Direct combination of chlorine with mercury heated to volatilizing point, (2) by subliming mercuric sulfate with common salt.
Method of purification: Recrystallization and sublimation.
Impurities: Mercurous chloride.
Grade: Technical, crystals, granular, powder, CP, NF.
Hazard: Toxic by ingestion, inhalation, and skin absorption; a poison.
Use: Manufacture of calomel and other mercury compounds, disinfectant, organic synthesis, analytical reagent, metallurgy, tanning, catalyst for vinyl chloride, sterilant for seed potatoes; fungicide, insecticide, and wood preservative; embalming fluids, textile printing, dry batteries, photography, process engraving and lithography.

mercuric chloride, ammoniated. See mercury, ammoniated.

mercuric cuprous iodide. (copper mercury iodide; mercury copper iodide).
CAS: 13876-85-2. $HgI_2•2CuI$.
Properties: Dark-red, crystalline powder. D 6.12. Insoluble in alcohol and water.
Hazard: Highly toxic.
Use: Thermoscopy (detecting overheating of machine bearings) by reversible color change.

mercuric cyanate. See mercury fulminate.

mercuric cyanide. (mercury cyanide).
CAS: 592-04-1. $Hg(CN)_2$.
Properties: Colorless, transparent prisms darkened by light. D 4.018, mp (decomposes). Soluble in water and alcohol.
Derivation: Interaction of mercuric oxide and an aqueous solution of hydrogen cyanide.
Grade: Technical, CP.
Hazard: Toxic by ingestion, inhalation, and skin absorption.
Use: Medicine (antiseptic), germicidal soaps, manufacturing cyanogen gas, photography.

mercuric dichromate. See mercury dichromate.

mercuric dimethyldithiocarbamate.
CAS: 15415-64-2. $C_6H_{12}N_2S_4Hg$.
Hazard: Highly toxic by ingestion, inhalation, and skin absorption.
Use: Turf fungicides.

mercuric fluoride. (mercury fluoride).
CAS: 7783-39-3. HgF_2.
Properties: Transparent crystals. D 8.95, mp 645C, decomposes. Moderately soluble in water and alcohol.
Derivation: Mercuric oxide and hydrogen fluoride.
Hazard: Highly toxic by ingestion, inhalation, and skin absorption; strong irritant.
Use: Synthesis of organic fluorine compounds.

mercuric iodide. (mercury iodide, red).
CAS: 7774-29-0. HgI_2.
Properties: Red, tetragonal crystals; turn yellow when heated to 130C, returning to red on cooling. D 6.28, mp 259C, bp 349C. Soluble in boiling alcohol and in solutions of sodium thiosulfate or potassium iodide or other hot alkali chloride solutions, almost completely insoluble in water.
Derivation: By the direct union of mercury and iodine.
Grade: Technical, reagent.
Hazard: Highly toxic by ingestion, inhalation, and skin absorption; strong irritant.
Use: Medicine (antiseptic), analytical reagents (Nessler's reagent, Mayer's reagent).
See mercuric potassium iodide.

mercuric lactate.
CAS: 814-82-4. $Hg(C_3H_5O_3)_2$.
Properties: White, crystalline powder. Soluble in water; decomposed by heat.
Hazard: Toxic by ingestion, inhalation, and skin absorption.
Use: Fungicide.

mercuric naphthenate. (mercury naphthenate).
Properties: Dark-amber liquid. Bulk d 10.4 lb/gal. Soluble in lubricating oils and mineral spirits.
Grade: 25% mercury.
Hazard: Poison by ingestion, inhalation, and skin absorption.
Use: Gasoline antiknock compounds and as a paint antimildew promoter.

mercuric nitrate. (mercury nitrate; mercury pernitrate).
CAS: 10045-94-0. $Hg(NO_3)_2$.
Properties: Colorless crystals or white, deliquescent powder. D 4.3, mp 79C. Decomposed by heat; soluble in water and nitric acid; insoluble in alcohol.
Derivation: By action of hot nitric acid on mercury.

Hazard: Dangerous fire risk in contact with organic materials. Very toxic.

Use: Nitration of aromatic organic compounds, felt manufacture, mercury fulminate manufacturing.

mercuric oleate.   (mercury oleate).
CAS: 1191-80-6. $C_{36}H_{66}HgO_4$.
Properties: Yellowish to red semisolid or solid mass. Soluble in fixed oils; insoluble in water; slightly soluble in alcohol or ether.
Derivation: By mixing yellow mercuric oxide with oleic acid.
Grade: Technical, pharmaceutical.
Hazard: Highly toxic.
Use: Antiseptic, antifouling paints.

mercuric oxide, red.   (red precipitate; mercury oxide, red).
CAS: 21908-53-2.   HgO.
Properties: Heavy, bright orange-red powder. D 11.00–11.29, mp (decomposes). Soluble in dilute hydrochloric acid and nitric acid;, insoluble in water, alcohol, and ether.
Derivation: By heating mercurous nitrate.
Grade: Technical, paint, ACS reagent.
Hazard: Fire risk in contact with organic materials. Highly toxic.
Use: Chemicals, paint pigment, perfumery and cosmetics, pharmaceuticals for topical disinfection, ceramics (pigment), dry batteries, especially for miniaturized equipment, polishing compounds, analytical reagent, antifouling paints, fungicide, antiseptic.

mercuric oxide, yellow.   (mercury oxide, yellow; yellow precipitate).
CAS: 21908-53-2.   HgO.
Properties: Light, orange-yellow powder; odorless. Stable in air but turns dark on exposure to light; finer powder than the red form. D 11.03 (27.5C), mp decomposes. Slightly soluble in cold water, more so after boiling; soluble in dilute hydrochloric acid and nitric acid, potassium iodide solution, concentrated solutions of alkaline-earth chloride, magnesium chloride; insoluble in alcohol.
Derivation: (1) By the action of either potassium hydroxide or sodium hydroxide on mercuric chloride. (2) By the action of sodium carbonate upon mercuric nitrate solution.
Grade: CP, technical, NF.
Hazard: Fire risk in contact with organic materials. Highly toxic.
Use: Antiseptic, mercury compounds.
See mercuric oxide, red.

mercuric oxycyanide.
CAS: 1335-31-5. $HgO \cdot Hg(CN)_2$.
Properties: White, crystalline powder. D 4.44. Moderately soluble in water.
Hazard: Detonates on heating, dangerous explosion risk. Highly toxic.
Use: Medicine (topical antiseptic).

mercuric phosphate.   (normal mercuric phosphate; neutral mercuric phosphate; trimercuric orthophosphate; mercuric phosphate, tertiary; mercury phosphate).
CAS: 10451-12-4. $Hg_3(PO_4)_2$.
Properties: Heavy, white or yellowish powder. Soluble in acids; insoluble in alcohol, water.
Hazard: Highly toxic.

mercuric potassium cyanide.   (mercury potassium cyanide).
CAS: 591-89-9. $Hg(CN)_2 \cdot 2KCN$.
Properties: Colorless crystals. Soluble in water and alcohol.
Derivation: By mixing mercuric and potassium cyanides and crystallizing.
Hazard: Highly toxic by ingestion, inhalation, and skin absorption.
Use: Silvering glass in mirror manufacture.

mercuric potassium iodide.   (Mayer's reagent; potassium mercuric iodide). K_2HgI_4 or $2KI \cdot HgI_2$.
Properties: Odorless, yellow crystals; deliquescent in air; crystallizes with either 1, 2, or 3 molecules of water. The commercial product is the anhydrous form containing approximately 25.5% mercury; d 4.29; neutral or alkaline to litmus; very soluble in water; soluble in alcohol, ether, and acetone.
Derivation: (1) Action of hydrochloric acid and potassium iodide on mercuric cyanide or mercuric chloride; (2) action of potassium iodide on mercuric oxide.
Hazard: Highly toxic by ingestion, inhalation, and skin absorption.
Use: Analytical chemistry.
See Nessler's reagent.

mercuric salicylate.   (salicylated mercury).
CAS: 5970-32-1.
Properties: White powder, yellow or pink tinge; odorless; tasteless. A compound of mercury and salicylic acid of somewhat varying composition, mercury replacing both phenolic and carboxylic hydrogen; contains more than 54% but less than 59.5% mercury. Soluble in solutions of the fixed alkalies or their carbonates, and in warm solutions of the alkali halides; insoluble in water and alcohol.
Derivation: By gently heating freshly precipitated yellow mercuric oxide and salicylic acid in the presence of water.
Hazard: Highly toxic by ingestion, inhalation, and skin absorption.
Use: Medicine (topical antiseptic).

mercuric silver iodide.   (mercury silver iodide; silver mercury iodide).
CAS: 7784-03-4. $HgI_2 \cdot 2AgI$.
Properties: Yellow powder, becomes red at 40–50C. D 6.08. Soluble in solutions of potassium cyanide or potassium iodide; insoluble in dilute acids and water.

Hazard: Highly toxic by ingestion, inhalation, and skin absorption.

Use: Thermoscopy (detecting overheating in journal bearings).

mercuric sodium phenolsulfonate.
CAS: 535-55-7. $C_{12}H_8HgNa_2O_8S_2$.
Properties: Colorless, water-soluble powder.
Derivation: Reaction of sodium salt of *p*-phenolsulfonic acid and mercuric oxide.
Hazard: Toxic by ingestion and skin absorption.
Use: Antiseptic and germicide, in soaps, ointments, etc. (1% concentration).

mercuric stearate. (mercury stearate).
CAS: 645-99-8. $(C_{17}H_{35}CO_2)_2Hg$.
Properties: Yellow, granular powder. Soluble in fatty oils; insoluble in water and alcohol.
Hazard: Highly toxic by ingestion, inhalation, and skin absorption.
Use: Germicide.

mercuric sulfate. (mercury persulfate; mercury sulfate).
CAS: 7783-35-9. $HgSO_4$.
Properties: White, crystalline powder. D 6.466, decomposes at red heat. Soluble in acids; insoluble in alcohol; decomposes in water.
Derivation: By the action of sulfuric acid on mercury with subsequent crystallization.
Hazard: Highly toxic by ingestion, inhalation, and skin absorption.
Use: Calomel and corrosive sublimate, catalyst in the conversion of acetylene to acetaldehyde, extracting gold and silver from roasted pyrites, battery electrolyte.

mercuric sulfide, black. (mercury sulfide, black).
CAS: 1344-48-5. HgS.
Properties: Black powder. D 7.55–7.70, mp sublimes at 446C, mp 583C. Soluble in sodium sulfide solution; insoluble in water, alcohol and nitric acid.
Derivation: By passing hydrogen sulfide into a solution of mercury salt, reaction of mercury with sulfur.
Hazard: Highly toxic by ingestion, inhalation, and skin absorption.
Use: Pigment.

mercuric sulfide, red. (vermilion; quicksilver vermilion; chinese vermilion; red mercury sulfide; artificial cinnabar; red mercury sulfuret).
CAS: 1344-48-5. HgS.
Properties: Fine, bright-scarlet powder. D 8.06–8.12, mp sublimes at 583C. Insoluble in water and alcohol.
Derivation: By heating mercury and sulfur with subsequent recovery by sublimation. A precipitated form is known as English vermilion.
Hazard: Highly toxic by ingestion, inhalation, and skin absorption.
Use: Pigment.

mercuric thiocyanate. (mercuric sulfocyanate; mercuric sulfocyanide; mercury sulfocyanate; mercury thiocyanate).
CAS: 592-85-8. $Hg(SCN)_2$.
Properties: White powder. Mp decomposes. Slightly soluble in alcohol; insoluble in water.
Derivation: By precipitation of mercuric nitrate with ammonium sulfocyanate and subsequent solution in a large amount of hot water and crystallizing.
Hazard: Highly toxic by ingestion, inhalation, and skin absorption.
Use: Photography, pyrotechnics.

"Mercurochrome" [BD]. TM for merbromin.

mercurous acetate. (mercury proto-acetate; mercury acetate).
CAS: 631-60-7. $C_4H_6Hg_2O_4$.
Properties: Colorless scales or plates, decomposed by boiling water and by light into mercury and mercuric acetate. Slightly soluble in water; insoluble in alcohol and ether; soluble in dilute nitric acid.
Derivation: Reaction of sodium acetate with mercurous nitrate solution acidified with nitric acid.
Grade: Technical.
Hazard: Highly toxic by ingestion, inhalation, and skin absorption.
Use: Medicine (antibacterial).

mercurous acetylide. Hg_2C_2.
Properties: White powder. A salt of acetylene.
Derivation: Reaction of acetylene with aqueous solution of mercurous salts.
Hazard: Severe explosion risk when shocked or heated. Highly toxic.

mercurous bromide. (mercury bromide).
CAS: 10031-18-2. HgBr or Hg_2Br_2.
Properties: White powder or colorless crystals; odorless; tasteless. Becomes yellow on heating, returning to white on cooling. Darkens on exposure to light. D 7.307, sublimes at 340–350C, mp 405C. Soluble in fuming nitric acid (prolonged heating), hot concentrated sulfuric acid, hot ammonium carbonate or ammonium succinate solutions; sparingly soluble in water; insoluble in alcohol and ether.
Derivation: (1) Action of potassium bromide on solution of mercurous nitrate in dilute nitric acid; (2) sublimation from mixture of mercury and mercuric bromide.
Hazard: Highly toxic by ingestion, inhalation, and skin absorption.

mercurous chlorate. (mercury chlorate).
$Hg_2(ClO_3)_2$.
Properties: White crystals. D 6.409, mp 250C (decomposes). Soluble in alcohol, water, and acetic acid.
Hazard: Explodes in contact with organic or combustible materials. Keep away from light. Highly toxic.

mercurous chloride. (mercury monochloride; mercury protochloride; mercury chloride, mild; calomel).

CAS: 10112-91-1. Hg_2Cl_2.

Properties: White, rhombic crystals or crystalline powder; odorless. Stable in air but darkens on exposure to light. D 6.993, mp 302C, bp 384C; decomposed by alkalies. Insoluble in water, ether, alcohol, and cold dilute acids.

Derivation: By heating mercuric chloride and mercury with subsequent sublimation.

Grade: Technical, CP, NF.

Hazard: Toxic dose is uncertain.

Use: Fungicide, electrodes, pharmaceuticals, pyrotechnics, ceramic painting, maggot control in agriculture.

mercurous chromate. (mercury chromate). Hg_2CrO_4.

Properties: Brick-red powder, variable composition. Decomposes on heating. Soluble in nitric acid (concentrated); insoluble in alcohol and water.

Hazard: Moderate fire hazard in contact with organic materials. Highly toxic.

Use: Ceramics (coloring green).

mercurous iodide. (mercury protoiodide).

CAS: 7783-30-4. HgI or Hg_2I_2.

Properties: Bright-yellow powder, becoming greenish on exposure to light due to decomposition into metallic mercury and mercuric iodide. Becomes dark yellow, orange, and orange-red on heating. Undergoes same color change in opposite order on cooling. Odorless and tasteless. D 7.6445–7.75, sublimes at 140C, mp 290C (with partial decomposition). Soluble in castor oil, liquid ammonia, aqua ammonia; insoluble in water, alcohol, and ether.

Derivation: (1) Action of potassium iodide on a mercurous salt. (2) Boiling a solution of mercurous nitrate containing nitric acid with excess of iodine.

Grade: Technical.

Hazard: Toxic by ingestion, inhalation, and skin absorption.

Use: Medicine (topical antibacterial).

mercurous nitrate, hydrated.

CAS: 10415-75-5. $HgNO_3 \cdot 2H_2O$.

Properties: Short prismatic crystals, effloresces and becomes anhydrous in dry air, sensitive to light. D 4.785 (3.9C), mp 70C (decomposes). Soluble in small quantities of warm water (hydrolyzes in larger quantities)and water acidified with nitric acid.

Derivation: Action of cold dilute nitric acid upon an excess of mercury and warming slightly.

Hazard: May be explosive if shocked or heated. Highly toxic.

Use: Analytical agent.

mercurous oxide.

CAS: 15829-53-5. Hg_2O.

Properties: Black powder. D 9.8, decomposes at 100C. Soluble in acids; insoluble in water.

Derivation: Action of sodium hydroxide on mercurous nitrate.

Hazard: Highly toxic.

mercurous sulfate.

CAS: 7783-36-0. Hg_2SO_4.

Properties: White to yellow, crystalline powder. D 7.56, decomposes on heating. Soluble in hot sulfuric acid, dilute nitric acid; slightly soluble in water.

Derivation: (1) Dissolving mercury in sulfuric acid and heating gently; (2) adding sulfuric acid to mercurous nitrate solution.

Hazard: Highly toxic.

Use: Chemical (admixed with sulfuric acid as a catalyst, in oxidation of naphthalene to phthalic acid), batteries (Clark cell, Weston cell).

mercury. (quicksilver; hydrargyrum).

CAS: 7439-97-6. Hg. Metallic element of atomic number 80, group IIB of the periodic table, aw 200.59, valences = 1,2; 4 stable isotopes and 12 artificially radioactive isotopes.

Properties: Silvery, extremely heavy liquid, sometimes found native. D 13.59, fp −38.85C, bp 356.6. Insoluble in hydrochloric acid; soluble in sulfuric acid upon boiling; readily soluble in nitric acid; insoluble in water, alcohol, and ether; soluble in lipids;, extremely high surface tension (480 dynes/cm) giving it unique rheological behavior; high electric conductivity. Noncombustible.

Source: Cinnabar.

Occurrence: Spain, Yugoslavia, Mexico, Canada, Algeria.

Derivation: By heating cinnabar in air or with lime and condensing the vapor.

Method of purification: Distillation; an important proportion of used mercury is recovered by redistillation.

Grade: Technical, virgin, redistilled, ACS.

Hazard: Central nervous system impairment, peripheral nervous system impairment, and kidney damage. (1) Mercury, metallic: Highly toxic by skin absorption and inhalation of fume or vapor, absorbed by respiratory and intestinal tract. FDA permits zero addition to the 20 micrograms of mercury contained in average daily diet. Questionable carcinogen. (2) All inorganic compounds of mercury are highly toxic by ingestion, inhalation, and skin absorption. Questionable carcinogen. (3) Most organic compounds of mercury are highly toxic; inorganic mercury can be converted to methylmercury by bacteria in water.

Note: Spillage may be a toxic hazard due to droplet proliferation. Clean-up requires special care.

Use: Amalgams, catalyst, electrical apparatus, cathodes for production of chlorine and caustic soda, instruments (thermometers, barometers, etc.), mercury vapor lamps, extractive metallurgy, mirror coating, arc lamps, boilers, coolant, and neutron absorber in nuclear power plants.

mercury, ammoniated. (mercuric chloride, ammoniated; ammonobasic mercuric chloride; ammoniated mercury chloride, white precipitate; fusible; aminomercuric chloride; mercury cosmetic). $HgNH_2Cl$.
Properties: White, pulverulent lumps or powder; earthy, metallic taste; odorless. Stable in air, darkens on exposure to light. Soluble in ammonium carbonate and sodium thiosulfate solutions and in warm acids; insoluble in water and alcohol.
Derivation: By precipitating mercuric chloride with ammonium hydroxide in excess.
Grade: USP, technical.
Hazard: Highly toxic.
Use: Medicine (local anti-infective), pharmaceuticals.

mercury bichloride. Legal label name for mercuric chloride.

mercury cell. An electrolytic cell for the production of caustic soda and chlorine from sodium chloride brine. Continuously fed brine is decomposed in one compartment between graphite anodes, where chlorine is liberated, and a mercury cathode, where a sodium amalgam is formed. The amalgam flows continuously or intermittently to a second compartment where it is decomposed with water, forming a caustic solution. The decomposition is usually performed electrolytically by making the amalgam anodic with respect to an iron or graphite cathode. Pure water is supplied to the decomposition compartment at such a rate as to maintain a constant concentration of caustic in the product. With respect to the diaphragm cell, the mercury cathode cell has generally a more concentrated solution (50–70%); it has the disadvantages of higher operating voltage and lower efficiency (52–55%) and a high capital investment in mercury. Examples of mercury-cathode cells are Castner cell and DeNora cell.

mercury compounds. See mercury chlorate; mercurous chlorate; mercury oxide; mercuric oxide.

mercury dichromate. (mercuric dichromate; mercury bichromate). $HgCr_2O_7$.
Properties: Heavy, red, crystalline powder. Soluble in acids; insoluble in water.
Hazard: Highly toxic.

mercury fulminate. (mercuric cyanate).
CAS: 628-86-4. $Hg(CNO)_2$.
Properties: Gray, crystalline powder. D 4.42, mp (explodes). Soluble in alcohol, ammonium hydroxide, and hot water; slightly soluble in cold water.
Derivation: By treating mercury with strong nitric acid and alcohol.
Grade: Technical.
Hazard: Explodes readily when dry, keep moist till use, an initiating explosive. Highly toxic.

Use: Manufacture of caps and detonators for producing explosions of military, industrial, and sporting purposes.

mercury nitrilotriacetate.
CAS: 73128-65-1. $C_{12}H_{12}HgN_2O_{12}$.
Hazard: A poison by ingestion.

mercury selenide.
CAS: 20601-83-6. HgSe. Sublimes in a vacuum, d 8.266, insoluble in water.
Hazard: Highly toxic.
Use: Semiconductor in solar cells, thin-film transistors, infrared detectors, ultrasonic amplifiers.

mercury telluride.
CAS: 12068-90-5. HgTe.
Grade: 99.99%.
Use: Semiconductor in solar cells, thin-film transistors, infrared detectors, ultrasonic amplifiers.

Merlon. A polycarbonate resin. Rods, tubes, pipes, sheets, film extrusions, and special types may be produced by extrusion, thermoforming, injection, and blow moldings.
Properties: Transparent, slightly straw-colored resin. D 1.2, refr index 1.587.
Use: Electrical and electronic component industries, protective equipment, graphic arts and photographic film, cable wrapping and protective overlays for other thermoplastic sheeting, missile applications.

merocyanine. See cyanine dye.

Merpentine. A surface-active compounded anionic wetting, penetrating and leveling agent used in leather and textile fields.

"Merpol" [Stepan]. TM for fatty alcohol-ethylene oxide condensates.
Use: As nonionic dispersing and emulsifying agents for paper, leather, and textile industries.

Merrifield, R. Bruce. (1921–2006). An American chemist who won the Nobel Prize for chemistry in 1984. Merrifield was cited for work on the use of a solid matrix as an aid to chemical synthesis of complex peptides and proteins. His synthesis techniques have been used in the development of solid matrix-bound inorganic and organic agents. He was awarded a doctorate from U.C.L.A.

Merrifield solid-phase peptide synthesis. Synthesis of large peptides by stepwise chain elongation on a polymeric support.

"Mertect" [Syngenta]. TM for a post harvest fungicide.

Merthiolate. A 0.1% solution of thiomersal in alcohol.

Hazard: Highly toxic by ingestion, for external use only.
Use: General antiseptic.

mescaline. (3,4,5-trimethoxymethoxy-phenethylamine).
CAS: 54-04-6. $(CH_3O)_3C_6H_2CH_2CH_2NH_2$. An alkaloid derived from a Mexican cactus.
Properties: Crystals. Mp 35–36C, bp 180C. Soluble in water, alcohol, chloroform, benzene; nearly insoluble in ether.
Hazard: Highly toxic by ingestion.
Use: Biochemical and medical research, hallucinogenic drug.

mesh. The number of apertures per square inch of a screen or sieve, it is the square of the number of strands of metal or plastic per linear inch.
See screen.

mesitol. (2,4,6-trimethyl phenol; 2-hydroxy mesitylene).
Properties: Needles. Mp 69C, bp 220C. Soluble in water and alcohol.

mesitylene. (1,3,5-trimethylbenzene; *sym-*trimethylbenzene).
CAS: 108-67-8. $C_6H_3(CH_3)_3$.
Properties: Liquid. D 0.863, fp -52.7C, bp 164.6C. Insoluble in water; soluble in alcohol and ether. Derived from coal tar. Combustible.

Hazard: Moderate fire hazard. Toxic by inhalation. Central nervous system impairment, asthma, and hematologic effects.
Use: Intermediate, including anthraquinone vat dyes, UV oxidation stabilizers for plastics.

2-mesitylenesulfonyl chloride. 1,3,5-$(CH_3)_3$ $C_6H(SO_2Cl)_2$.
Properties: White crystalline solid. Mw 317.21, mp 123–125C, moisture sensitive.
Hazard: Corrosive.
Use: Pharmaceutical intermediate, protective reagent for the guanidino group in peptide synthesis, selective sulfonating agent.

mesityl oxide. (isopropylideneacetone; methyl isobutenyl ketone; 4-methyl-3-penten-2-one).
CAS: 141-79-7. $(CH_3)_2$:CHCOCH$_3$.
Properties: Oily, colorless liquid; honey-like odor. D 0.8569 (20/20C), bp 130–131C, vap press 8.7 mm Hg (20C), flash p 90F (32.3C), bulk d 7.1 lb/gal (20C), fp −46.4C, viscosity 0.0060 cP (20C),

autoign temp 652F (344C). Soluble in water, alcohols, and ethers.
Derivation: Dehydration of acetone or diacetone alcohol.
Grade: Technical.
Hazard: Flammable, moderate fire risk. Toxic by ingestion, inhalation, and skin absorption. Eye and upper respiratory tract irritant, central nervous system impairment.
Use: Solvent for cellulose esters and ethers, vinyl resins, lacquers, roll-coating inks, stains, ore flotation, paint and varnish-removers, insect repellent, manufacture of methyl isobutyl ketone.

meso-. A prefix meaning middle or intermediate, specifically (1) an optically inactive stereoisomeric form resulting from the presence of an even number of dextro and levorotatory isomers in a natural substance. This cancels the optical activity and causes formation of an intermediate structure which is its own mirror image. This effect is called internal compensation. Tartaric acid is an example.
See racemization; tartaric acid. (2) An intermediate hydrated form of an inorganic acid. (3) Designating a middle position in certain cyclic organic compounds. (4) A ring system characterized by a middle position of certain rings.

mesokaryotic. Nuclear condition unique to the dinoflagellates in which the chromosomes remain permanently condensed.

mesomerism. See resonance (1).

mesomorphic. A molecular arrangement intermediate between crystals (solid) and amorphous (liquid) that characterizes liquid crystals.

meson. A type of strongly interacting elementary subatomic particle of medium size between that of a proton and electron that actually consist of quark–antiquark pairs.

messenger RNA. (mRNA). RNA that serves as a template for protein synthesis.
See genetic code.

mesyl chloride. See methanesulfonyl chloride.

Met. Symbol for methionine.

meta-. A prefix. For definition of *meta-*compounds.
See *ortho-.*

metabolism. The chemical transformations occurring in an organism from the time a nutrient substance enters it until it has been utilized and the waste products eliminated. In animals and humans, digestion and absorption are primary steps, followed by a complicated series of degradations, syntheses, hydrolyses, and oxidations, in

which agents such as enzymes, bile acids, and hydrochloric acid take part. These transformations are often localized with respect to organs, tissues, and types of cells involved. Basal metabolism is the rate of total heat production of an individual who is awake but in complete mental and physical repose, at comfortable temperature and without having had food for at least 12 hours. Under these conditions, oxidation of stored nutrients provides the sole source of energy expended and heat is measurable by calorimetry.
See digestion (1).

metabolite. An intermediate substance produced and used in the processes of a living cell or organism. Metabolites effect replacement and growth in living tissue and are also a source of energy in the body. Examples are nucleic acids, enzymes, glucose, cholesterol, and many similar substances.

metaboric acid. HBO_2.
Properties: White crystals. D 2.486, mp 236C. Slightly soluble in water.

"Metacaulk 350i" *[Rectorseal].* TM for a general purpose fire rated sealant.
Use: Sealant for construction joints.

metacetone. See diethylketone.

Metachrome Yellow. (alizarine yellow GG; sodium-*m*-nitrobenzeneazosalicylate; CI 14025).
CAS: 584-42-9. $C_{13}H_8N_3NaO_5$.
Properties: Finely divided yellow crystals. Soluble in hot water.
Derivation: Diazotizing *m*-nitroaniline and coupling with salicylic acid.
Use: Acid–base indicator, biological stain.

metaformaldehyde. See *sym*-trioxane.

metal. An element that forms positive ions when its compounds are in solution and whose oxides form hydroxides rather than acids with water. Approximately 75% of the elements are metals that occur in every group of the periodic table except VIIA and the noble gas groups. Most are crystalline solids with metallic luster, conductors of electricity, and have rather high chemical reactivity; many are quite hard and have high physical strength. They also readily form solutions (alloys) with other metals. The presence of very low percentage of other elements (not necessarily metals) profoundly affects the properties of many metals, e.g., carbon in iron. Mercury, cesium, and gallium are liquid at room temperature. Geologically, metals usually occur in the form of compounds that must be physically or chemically processed to yield the pure metal; common methods are application of heat (smelting), carbon reduction, electrolysis,

and reduction with aluminum or magnesium. Metals fall into the following classifications, which are not mutually exclusive:

alkali metals	rare metals
alkaline-earth metals	rare-earth metals
transition metals	actinide metals
noble metals	light metals
platinum metals	heavy metals

The chemistry of metals, i.e., their behavior as atoms or ions, is a fundamental factor in electrochemical reactions, as well as in the metabolism of plants and animals, where many have essential nutrient and other biochemical functions. Among these are iron, copper, cobalt, potassium, and sodium, often in traces. Some metals are quite toxic, especially cadmium, mercury, lead, barium, chromium, and beryllium, both in elemental form and as compounds.
See alloy; trace element; electroplating.

metal acetylide. An alkyne in which the terminal hydrogen atom has been replaced by a metal.

metal alkyl. One of the family of organometallic compounds. It is a combination of an alkyl organic radical with a metal atom or atoms.
See Ziegler catalyst.

metalaxyl.
CAS: 57837-19-1. $C_{15}H_{21}NO_4$.
Properties: White crystals.
Hazard: Moderately toxic by ingestion.
Use: Food additive; fungicide; agricultural chemical.

metal cleaner. (metal polish). Any paste or liquid used to clean and/or polish metal surfaces. They usually contain ammonium hydroxide and a variety of unknown petroleum distillates of varying toxicity.

metal deactivator. A compound added to gasoline to neutralize the catalytic effect of copper in promoting fuel oxidation.

metaldehyde.
CAS: 9002-91-9. $(CH_3CHO)_n$. A polymer of acetaldehyde in which *n* usually is 4–6.
Properties: White prisms, decomposes with partial regeneration of acetaldehyde, when heated above 80C. Sublimes at 112–115C; mp 246C in sealed tube. Soluble in benzene, chloroform; slightly soluble in alcohol, ether; insoluble in water.
Hazard: Flammable, dangerous fire risk. Strong irritant to skin and mucous membranes.
Use: Fuel to replace alcohol, to destroy snails and slugs.

metal dye. See dye, metal.

metal fiber. See fiber.

metal, foamed.   A light metal such as aluminum or zinc to which titanium hydride has been added. The latter evolves hydrogen to give a blown structure with about the same specific gravity as water.

metal glass.   See glass, metallic.

metal indicator. (complexometric indicator; chelatometric indicator).   A chelating dye which changes color with an increase or decrease in metal ion concentration by forming a metal–dye complex.

metallic bond.   See bond, chemical.

metallic soap.   See soap (2).

metallized dye.   A soluble dye including any one of a variety of metals chemically combined, applied to wool in an acid bath, by use of sodium chloride to salt out the dye onto the fiber.

metallizing.   Coating a plastic or similar material with a deposit of metal (usually aluminum) by means of vacuum deposition. The thickness of such films may vary from 0.01 to as much as 3 mils. Metallized plastic films are used for yarns, packaging, stamping foil, labels, etc.

metallocene.   An organometallic coordination compound obtained as a cyclopentadienyl derivative of a transition metal or metal halide. The metal is bonded to the cyclopentadienyl ring by electrons moving in orbitals extending above and below the plane of the ring (pi bond). There are three types of metallocenes: (1) dicyclopentadienyl-metals with the general formula $(C_5H_5)_2M$, (2) dicyclopentadienyl-metal halides with the general formula $(C_5H_5)_2MX_{1-3}$, (3) monocyclopentadienyl-metal compounds with the general formula $C_5H_5MR_{1-3}$, where R is CO, NO, halide group, alkyl group, etc. Types (1) and (2) are known as molecular sandwiches since the two cyclopentadiene rings lie above and below the plane on which the metal atom is situated.
Most metallocenes are crystalline and those belonging to the first transition series of the periodic table, from vanadium to nickel, have the same melting point, 173C. The metal complexes are soluble in many organic solvents while the halides are soluble primarily in polar solvents. When the central metal atom is in a stable oxidation state the metallocene is not decomposed by high temperature, air, water, dilute acids, or bases.
Some metallocenes, such as ferrocene, undergo a wide variety of aromatic ring substitution reactions, including Friedel–Crafts acylation, arylation, and sulfonation; a few such as nickelocene and cobaltocene are too unstable to be directly substituted.

Derivation: The most important industrial method is the reaction of a cyclopentadienide (e.g., the sodium salt) with a transition metal halide in an organic solvent.
Use: Catalysts, polymers, UV absorbers, reducing agents, free radical scavengers, antiknock agents.
See ferrocene; cobaltocene; nickelocene; titanocene dichloride; zirconocene dichloride; uranocene.

metallochromic indicator.   A complexing agent for the metal ion being assessed which changes color at the end point of a titration.

metalloendoprotease.   See glycoprotease.

metalloenzyme.   An enzyme that functions enzymatically only when one or two metal atoms are incorporated into is structure.

metalloid.   See nonmetal; semiconductor.

metallophyte.   A plant that tolerates substrates that have very high levels of heavy metals.

metalloprotein.   A protein having a metal ion as its prosthetic group.

metallothionein.   (Mt). Inducible protein that is involved in the detoxification of heavy metals.
Properties: Intracellular, cytosolic, low-molecular weight, metal-bind; consists of two similar low-molecular-weight moieties; high cysteine content.
Derivation: Synthesis is induced by sublethal dose of heavy metals and by stressors such as cold, exercise, and food restrictions.

metallurgical coke.   See coke.

metal, powdered.   Metals are produced in powdered form for a variety of uses in several industries. In this form they are the raw materials for powder metallurgy in which the powders are pressed in molds and heated (sintered) at high temperature. Metal powders range in size from −325 mesh (0.045–0.060 mm "diameter") to +100 mesh and are available in practically all industrial metals. They are produced by machining, milling, shotting, granulation, atomizing, condensation, reduction, chemical precipitation, or electrodeposition. Their properties and purities vary with the method of preparation.
Hazard: Dangerous fire risk, especially as dust. Toxic by inhalation.
Use: Electric, automotive, machinery, tool and refractory-metal industries; paint pigments, flares and incendiary bombs, brazing materials, calorizing, metal-spraying, metallurgical agents; heat-generating agents, catalysts, etc.

"Metalyn" [Eastman].   TM for a series of distilled methyl esters of tall-oil fatty acids.

metamorphosis.   A process of developmental change whereby a larva reaches adulthood only after a drastic change in morphology; occurs in most amphibians and insects; for some insects, this change may include another stage (pupa) before the adult stage.

"Metamucil" *[Procter & Gamble].*   Proprietary preparation of psyllium mucilloid, dextrose, benzyl benzoate, $NaHCO_3$, KH_2PO_4, and citric acid.
Use: Laxative.

metanilic   acid.   (*m*-sulfanilic   acid;   *m*-aminobenzenesulfonic acid).
CAS: 121-47-1. $C_6H_4(NH_2)SO_3H$.
Properties: Small, colorless needles. Soluble in water, alcohol, and ether.
Derivation: By the reduction of metal-nitrobenzenesulfonic acid. Nitrobenzene is sulfonated until the product is soluble in water. The mixture is then poured into water and reduced with iron, made alkaline with lime, and the lime salt dissociated with sodium carbonate.
Hazard: See aniline.
Use: Azo dye manufacture (sodium salt), sulfa drug synthesis.

metaphase.   A stage in mitosis or meiosis during which the chromosomes are aligned along the equatorial plane of the cell.

***meta*-phosphoric acid.**   See phosphoric acid, *meta-*.

metapon.   (methyldihydromorphine). A narcotic drug.
See narcotic.

"Metasol" *[QOL].*   TM for a series of chemicals used as fungicides, mildewicides, bactericides, slime control agents for pulp- and paper-mill systems, and as preservatives for all types of systems. Type (J-26) is claimed to be a broad-spectrum antimicrobial agent with a minimum ecological hazard.

metastable.   A system is metastable if it is above its minimum-energy state, but requires an energy input before it can reach a lower-energy state.

metastasis.   The spread of cancer from one part of the body to another. Cells in the metastatic (secondary) tumor are like those in the original (primary) tumor.

Meta-Systox R.   *O,O*-dimethyl-*S*-2-(ethylsulfinyl)ethyl phosphorothioate.

metathesis.   See double decomposition.

metazine.
CAS: 67704-68-1. $C_{11}H_{19}N_7$.
Hazard: Moderately toxic by ingestion.
Use: Agricultural chemical.

metazoa.   A subkingdom of multicellular animals whose bodies in most cases are composed of many specialized cells that are organized into tissues and organs. They possess a coordinating nervous system. Included are all animals except the *protozoa* and *parazoa* (sponges).

metepa.   (generic   name   for   tris(methyl-1-aziridinyl)phosphine oxide).
CAS: 57-39-6. $(C_3H_6N)_3PO$.
Properties: Amber liquid; amine odor. Bp 188C (1 mm Hg), d 1.079. Miscible with water and organic solvents.
Hazard: Toxic by ingestion and skin absorption, strong irritant to skin.
Use: Insect chemosterilant, addition products for textile treatments, adhesives, paper and rubber processing, cross-linking agent in polymer systems which contain active hydrogens, monomer.

meter.   (1) The basic unit of length of the metric system (39.37 inches). Originally defined as one ten-millionth of the distance from the equator to the North Pole. Now defined as 1,650,763.73 wavelengths of the orange-red line of the isotope krypton 86. (2) A device for measuring the flow rate of liquids, gases, or particulate solids, e.g., flowmeters, rotameters, proportioning equipment.
See gauge.

"Metglas" *[Metglas].*   TM for an amorphous metal alloy developed for use as transformer coils.

Methac.   A series of blends of methyl acetate with methanol in varying proportions.
Hazard: Toxic by ingestion.
Use: Lacquer solvents, paint removers, organic synthesis.

methacrolein.   (methacrylaldehyde).
CAS: 78-85-3. $CH_2:C(CH_3)CHO$.
Properties: Liquid. D 0.8474 (20/20C), bp 68.0C, flash p 35F (1.7C) (OC). Soluble in water 5.9% by wt (20C), shipped with 0.1% hydroquinone as polymerization inhibitor.
Hazard: Flammable, dangerous fire risk. Strong irritant.
Use: Copolymers, resins.

methacrylaldehyde.   See methacrolein.

methacrylamide.   (methacrylic acid amide).
CAS: 79-39-0. $CH_2:C(CH_3)CONH_2$.
Properties: Solid. Mp 110C.
Use: A monomer for acrylic resins.

methacrylamidopropyltrimethylammonium chloride. (MAPTAC).

$$\underset{\substack{| \\ O}}{\overset{\substack{CH_3 \\ |}}{CH_2=CCNHCH_2CH_2CH_2N(CH_3)_3}} \; Cl$$

Properties: (50% water solution): Amber liquid. D 1.059, refr index 1.427, flash p none (CC), bulk d 8.66 lb/gal.
Use: Reactive cationic monomer for wide range of industrially useful polymers.

methacrylate ester. CH_2:$C(CH_3)COOR$ where R is usually methyl, ethyl, isobutyl, or *n*-butyl-isobutyl (50:50).
Esters of methacrylic acid; supplied commercially as the polymers.
See acrylic resin.

methacrylate resin.   See acrylic resin.

methacrylatochromic chloride.
H_2C:$C(CH_3)$

$$H_2C:C(CH_3)C:OCrCl_2OHCrCl_2O$$

Properties: Water-soluble solid.
Derivation: Reaction of methacrylic acid with basic chromic chloride.
Use: Water repellent, nonadhesive, insolubilizer for vinyl polymer.
See "Volan" *[Zaclon].*

methyacrylic acid. (α-methacrylic acid) (monomer).
CAS: 79-41-4. H_2C:$C(CH_3)COOH$.

$$\underset{H_2C=C-C-OH}{\overset{\substack{CH_3 \quad O \\ | \qquad ||}}{}}$$

Properties: Colorless liquid. Mp 15–16C, bp 161–162C, d 1.015 (20C), flash p 170F (76.6C). Soluble in water, alcohol, ether, most organic solvents. Polymerizes readily to give water-soluble polymers. Combustible.
Derivation: Reaction of acetone cyanohydrin and dilute sulfuric acid; oxidation of isobutylene.
Grade: 40% aqueous solution, bp 76–78C (25 mm Hg); crude monomer 85% pure, glacial (99% assay).
Hazard: Toxic material. Strong irritant to eyes and skin. Questionable carcinogen.
Use: Monomer for large-volume resins and polymers, organic synthesis. Many of the polymers are based on esters of the acid, as the methyl, butyl, or isobutyl esters.
See acrylic resin.

α-methacrylic acid.   See methacrylic acid.

β-methacrylic acid.   See crotonic acid.

methacrylic acid-divinylbenzene copolymer.
Use: Food additive.

methacrylonitrile. (2-cyanopropene-1; isopropenenitrile).
CAS: 126-98-7. $CH_2=C(CH_3)C\equiv N$.
Properties: Clear, colorless liquid. Bp 90.3C, fp −38.8C, flash p 55F (12.7C) (TOC), d 0.789. Slightly soluble in water; soluble in acetone. Thermoplastic; resistant to acids and alkalies.
Hazard: Flammable. Toxic by ingestion, inhalation, and skin absorption.
Use: Vinyl nitrile monomer, copolymer with styrene, butadiene, etc., elastomers, coatings, plastics.

γ-methacryloxypropyltrimethoxysilane.
CH_2:$C(CH_3)COOCH_2CH_2CH_2Si(OCH_3)_3$.
Properties: Liquid. D 1.045 (25C), bp approximately 80C (1 mm Hg), refr index 1.4285 (25C), flash p 135F (57.2C). Soluble in acetone, benzene, ether, methanol, and hydrocarbons. Combustible.
Grade: 97% min purity.
Hazard: Moderate fire risk.
Use: Coupling agent for promotion of resin–glass, resin–metal, and resin–resin bonds, for formulation of adhesives having "built in" primer systems.

methacryloyl chloride.
CAS: 920-46-7. $H_2C=C(CH_3)COCl$.
Properties: Liquid. Mw 104.54, bp 95–96C, d 1.070, fp 2C.
Available forms: Technical 90% stabilized with phenothiazine.
Hazard: Flammable and corrosive liquid.

methadone hydrochloride.
(*dl*-6-dimethylamino-4,4-diphenyl-3-heptanone hydrochloride).
CAS: 1095-90-5. $(C_6H_5)_2C(COC_2H_5)$
$CH_2CH(CH_3)N(CH_3)_2 \cdot HCl$. A synthetic narcotic.
Properties: Crystalline substance with a bitter taste; no odor. Mp 232–235C. Soluble in water, alcohol, and chloroform; practically insoluble in ether and glycerol; pH (1% aqueous solution) 4.5–6.5.
Grade: USP.
Hazard: Toxic. Addictive narcotic. Use restricted.
Use: Medicine (sedative, treating heroin addiction).

methallenestril. (β-ethyl-6-methoxy-α,α-dimethyl-2-naphthalenepropionic acid).
CAS: 517-18-0. $CH_3OC_{10}H_6CH(C_2H_5)C(CH_3)_2$
COOH.
Properties: Crystals. Mp 132.5C. Soluble in ether, vegetable oils.
Use: Medicine (estrogen).

methallyl acetate.   See methylallyl acetate.

methallyl alcohol.   See methylallyl alcohol.

β-methallyl chloride.   See β-methylallyl chloride.

methallylidene diacetate.
CAS: 10476-95-6. CH_2:$C(CH_3)CH(OCOCH_3)_2$.
Properties: Liquid. D 1.510 (20/20C), bp 191.0C, fp −15.4C, flash p 215F (101C) (COC). Slightly soluble in water. Combustible.
Use: Chemical intermediate, can provide controlled release of methacrolein in acid solution.

methamidophos.   (*O,S*-dimethyl phosphoramidothioate).
CAS: 10265-92-6.
Properties: Mp 39–41C. Water-miscible.
Use: Insecticide for cotton, cole crops, lettuce, potatoes.

d-**methamphetamine.**   See *d*-1-phenyl-2-methylaminopropan.

methamphetamine hydrochloride.   See amphetamine.

methanal.   See formaldehyde.

methanamide.   See formamide.

methanation.   A reaction by which methane is formed from the hydrogen and carbon monoxide derived from coal gasification. It requires a catalyst, e.g., nickel, and temperatures in the range of 500C. In one process the reaction is performed in an adiabatic fixed-bed reactor. The reaction is $3H_2 + CO \rightarrow CH_4 +$ water.
See gasification.

methane.   (marsh gas; methyl hydride).
CAS: 74-82-8. CH_4. The first member of the paraffin (alkane) hydrocarbon series.
Properties: Colorless gas; odorless; tasteless. Lighter than air. Practically inert toward sulfuric acid, nitric acid, alkalies, and salts but reacts with chlorine and bromine in light (explosively in direct sunlight). Flash p −306F (−188C), bp −161.6C, fp 182.5C, autoign temp 1000F (537C), vap d 0.554 (0C), critical temperature −82.1C, critical pressure 672 psia, heating value 1009 Btu/cu ft. Soluble in alcohol, ether; slightly soluble in water.
Occurrence: Natural gas and coal gas, from decaying vegetation and other organic matter in swamps and marshes.
Derivation: (1) From natural gas by absorption or adsorption. (2) From coal mines for use as fuel gas. (3) From a mixture of carbon monoxide and hydrogen (synthesis gas) obtained by reaction of hot coal with steam; the mixed gas is passed over a nickel-based catalyst at high temperature. See methanation. Methane can also be obtained by a nickel-catalyzed reaction of carbon dioxide and hydrogen. (4) Anaerobic decomposition of manures and other agricultural wastes. (5) By horizontal drilling of coal seams.
Grade: Research 99.99%; CP 99%; technical 95%; Btu grade, which must have heating value of 1000 Btu/cu ft at 15.5C and a pressure of 30 inches of mercury.
Hazard: Severe fire and explosion hazard, forms explosive mixture with air (5–15% by volume). An asphyxiant gas.
Use: Source of petrochemicals by conversion to hydrogen and carbon monoxide by steam cracking or partial oxidation. Important products are methanol, acetylene, hydrogen cyanide, and ammonia. Chlorination gives carbon tetrachloride, chloroform, methylene chloride, and methyl chloride. In the form of natural gas, methane is used as a fuel, as a source of carbon black, and as the starting material for manufacture of synthetic proteins.
See natural gas; synthetic natural gas; biogas.

methanecarboxylic acid.   See acetic acid.

methanedicarbonic acid.   See malonic acid.

methanephosphonic acid, diphenyl ester.
CAS: 7526-26-3. $C_{13}H_{13}O_3P$.
Hazard: A poison by ingestion.

methanesulfonic acid.
CAS: 75-75-2. CH_3SO_3H.
Properties: Liquid at room temperature. D 1.4812 (18/4C), mp 17–20C, bp 200C, refr index 1.4317 (16C), flash p none. Soluble in water, alcohol, ether.
Grade: 70%.
Hazard: Corrosive to tissue (eyes, skin, and mucous membranes).
Use: Catalyst in esterification, alkylation, olefin polymerization, peroxidation reactions.

methanesulfonic acid, nonamethylene ester.
CAS: 4248-77-5. $C_{11}H_{24}O_6S_2$.
Hazard: A reproductive hazard.

methanesulfonyl chloride.   (mesyl chloride).
CAS: 124-63-0. CH_3SO_2Cl.
Properties: Pale-yellow liquid. D 1.485 (20/20C), bp 164C, fp −32C. Soluble in most organic solvents; insoluble in water (hydrolyzes slowly).
Grade: 98%, 99+%.
Use: Intermediate, flame-resistant products, stabilizer for liquid sulfur trioxide, biological chemicals.

n,n′-**methanetetrayl biscyclohexanamine.**
CAS: 538-75-0. $C_{13}H_{22}N_2$.
Hazard: A poison by skin contact. Moderately toxic by ingestion and inhalation.

n,n′-**methanetetraylbis(2,6-bis(1-methylethyl)benzenamine).**   See bis(2,6-diisopropylphenyl)carbodiimide.

methanethiol.   (methyl mercaptan).
CAS: 74-93-1. CH_3SH.
Properties: Water-white liquid when below its boiling point, or colorless gas; powerful, unpleasant odor. Fp −121C, d 0.87 (20C), flash p 0F (−17C), bp 5.96C. Slightly soluble in water; soluble in alcohol, ether, and petroleum naphtha.
Derivation: Methanol and hydrogen sulfide.
Grade: 98.0% purity.
Hazard: Flammable, dangerous fire risk. Explosive limits in air 3.9–21.8%. Strong irritant. Liver damage.
Use: Synthesis, especially of methionine, jet fuel additives, fungicides; also as catalyst.

methanethiomethane.   See dimethyl sulfide.

methanoic acid.   See formic acid.

methanol.   See methyl alcohol.

methapyrilene.
CAS: 91-80-5. $C_{14}H_{19}N_3S$.
Properties: Colorless liquid. Bp 173C (3 mm Hg). Soluble in water and alcohol; insoluble in benzene.
Hazard: Toxic by ingestion, a carcinogen.
Use: Medicine (antihistamine).

"Methazate" *[Chemtura].*   TM for zinc dimethyldithiocarbamate.
See ziram.

methemoglobin. (Methb; ferrihemoglobin; methaemoglobin; ferrihemoglobin; methemoglobinferrihemoglobin).   A form of hemoglobin that does not bind oxygen reversibly. The iron of methb has been oxidized to the ferric state, with essentially ionic bonds. Methemoglobin is normally present in circulating blood of vertebrates but its concentration is increased under certain circumstances such as poisoning by nitrates and certain other chemicals.

methenamine.   See hexamethylenetetramine.

Methendic Anhydride.   A mix of bicyclic unsaturated dibasic anhydrides as a relatively non-volatile liquid at room temperature.
Properties: Pale-amber liquid, color Gardner 3–6. D 1.2–1.3 (25C), 1.5052 (27C), flash p 275–285F (135–140C) (COC). Miscible with acetone, aromatic and aliphatic hydrocarbons at room temperature. Combustible.
Use: Cross-linking or curing agent for epoxy-type resin systems.

methenyl tribromide.   See bromoform.

methicillin.
CAS: 61-32-5. $C_6H_3(OCH_3)C=O$. A semisynthetic antibiotic.
Use: Effective against resistant staphylococci.

methidathion.   (an ester of dithiophosphoric acid).
CAS: 950-37-8. $C_6H_{11}N_2O_4PS_3$.
Properties: Crystalline solid. Almost insoluble in water; soluble in common organic solvents.
Hazard: Toxic by ingestion, a cholinesterase inhibitor.
Use: Insecticide.

methiocarb.   (4-methylthio-3,5-xylyl-*N*-methylcarbamate).
CAS: 2032-65-7.
Properties: Mp 121C.
Use: Insecticide for vegetables and fruits.

methiodal sodium.   See sodium methiodal.

methiomeprazine hydrochloride.
CAS: 14056-64-5. $C_{19}H_{24}N_2S_2 \cdot ClH$.
Hazard: A poison.

methionamine.   See *n*-acetyl-*l*-methionine.

methionine.   (2-amino-4-(methylthio)butyric acid).
CAS: 59-51-8. $CH_3SCH_2CH_2CH(NH_2)COOH$.
An optically active, essential sulfur-containing amino acid important in biological *trans*-methylation processes. The levo form is biologically active.
Properties: (*dl*-racemic mix) White, crystalline platelets or powder; faint odor. Soluble in water, dilute acids and alkalies, very slightly soluble in alcohol, practically insoluble in ether, pH (1% aqueous solution) 5.6–6.1.
Derivation: Hydrolysis of protein, synthesized from hydrogen cyanide, acrolein, and methyl mercaptan.
Grade: NF, feed 98%.
Use: Pharmaceuticals, feed additive, vegetable-oil enrichment, single-cell protein.

methionine hydantoin.
CAS: 13253-44-6. $C_6H_{10}N_2O_2S$.
Hazard: Low toxicity by ingestion and skin contact.

methionine hydroxy-analog calcium.   (*dl*-α-hydroxy-γ-methylmercaptobutyric acid, calcium salt; 2-hydroxy-4-methylthiobutyric acid, calcium salt). $(CH_3SCH_2CH_2CH_2OCOO)_2Ca$. Free methionine hydroxy analog is a metabolite in methionine utilization.
Properties: Free-flowing light tan powder. Soluble in water; insoluble in common organic solvents.
Use: Animal feed, synthesis of pharmaceuticals.

methiotepa.   (generic name for tris(2-methyl-1-aziridinyl)-phosphine sulfide).
See metepa.

"Methocel" *[Dow].*   TM for methylcellulose.

methomyl. (*S*-methyl-*N*-(methylcarbamoy-loxy)-thioacetimidate).
CAS: 16752-77-5. $C_5H_{10}N_2O_2S$.
Properties: Crystalline solid. Mp 78C, d 1.30. Partially soluble in water, alcohol, and acetone; soluble in methanol.
Hazard: Toxic by ingestion. Acetyl cholinesterase inhibitor, male reproductive damage, and hematologic effects. Questionable carcinogen.
Use: Insecticide, nematocide.

methoprene.
CAS: 40596-69-8. $C_{19}H_{34}O_3$. An insecticidal preparation said to act in the manner of a juvenile hormone, which arrests development of insects in the larval stage.

methotrexate. (amethopterin; 4-amino-10-methylfolic acid).
CAS: 59-05-2.
Properties: Orange-brown, crystalline powder. Insoluble in water, alcohol, chloroform, ether; slightly soluble in dilute hydrochloric acid; soluble in dilute solutions of alkali hydroxides and carbonates; folic acid antagonist.
Grade: USP.
Hazard: Very toxic. Questionable carcinogen.
Use: Chemosterilant, cancer treatment.

methoxsalen. (8-methoxypsoralen).
CAS: 298-81-7. $C_{12}H_8O_4$, tricyclic.
Properties: White to cream-colored, crystalline solid; odorless. Slightly soluble in alcohol; practically insoluble in water. Combustible.
Use: Suntan accelerator, sunburn protector.

methoxyacetaldehyde. A highly reactive aldehyde derivative available in 77% aqueous solution. Clear colorless liquid with odor characteristic of lower aldehydes, miscible with water and many organic solvents. Resembles butyraldehyde in structure and some properties. Claimed as possible antimicrobial agent, preservative, and polymer modifier.

methoxyacetic acid.
CAS: 625-45-6. CH_3OCH_2COOH.
Properties: Liquid. Mp (min) 7.7C, boiling range 197–198C (min at 733 mm Hg), d 1.1738 (25/4C), refr index 1.415 (25C), flash p 260F (126C), acid number 612 (min). Combustible.
Use: Synthesis.

***p*-methoxyacetophenone.** (*p*-acetoanisole; acetanisole; *p*-acetylanisole).
CAS: 100-06-1. $CH_3OC_6H_4COCH_3$.
Properties: Crystalline solid; pleasant odor. Bp 258C, congealing p 36.5C. Soluble in alcohol, ether, fixed oils. Combustible.
Derivation: Interaction of anisole and acetyl chloride in the presence of aluminum chloride and carbon disulfide.

Grade: Technical, FCC.
Use: Perfumery (for floral odors), flavoring.

methoxyamine. (hydroxylamine methyl ether).
CAS: 67-62-9. CH_3ONH_2.
Properties: Colorless liquid; unpleasant odor. Bp 50C. Soluble in water and alcohol.
Derivation: From hydroxylamine disulfonic acid and methyl sulfate.
Hazard: Toxic by ingestion, strong skin irritant.
Use: Analytical reagent, mainly for ketones and aldehydes.

***o*-methoxyaniline.** See *o*-anisidine.

***p*-methoxyaniline.** See *p*-anisidine.

methoxybenzaldehyde. See anisaldehyde.

methoxybenzene. See anisole.

***p*-methoxybenzoic acid.** See anisic acid.

***p*-methoxybenzoic acid 2-phenyl-hydrazide.**
CAS: 15089-03-9. $C_{14}H_{14}N_2O_2$.
Hazard: Moderately toxic by ingestion.
Use: Agricultural chemical.

***p*-methoxybenzyl acetate.** See anisyl acetate.

***p*-methoxybenzyl alcohol.** See anisic alcohol.

***p*-methoxybenzyl formate.** See anisyl formate.

2-methoxy-4,6-bis(isopropylamino)-*s*-triazine. (2,4-bis(isopropylamino)-6-methoxy-*s*-triazine).
$CH_3OC_3N_3[NHCH(CH_3)_2]_2$.
Properties: White solid. Almost insoluble in water.
Use: Herbicide.

3-methoxybutanol.
CAS: 2517-43-3. $CH_3CH(OCH_3)CH_2CH_2OH$.
Properties: Liquid. D 0.9229, bp 161.1C, flash p 165F (74C), vap press 0.9 mm Hg (20C), sets to glass at −85C. Soluble in water. Combustible.
Hazard: Toxic by ingestion.
Use: High-boiling lacquer solvent, coupling agent for brake fluids, intermediate for plasticizers, herbicides, film-forming additive in PVA emulsions, solvent for pharmaceuticals.

1-(methoxycarbonyl)aziridine. See methyl 1-aziridineacetate.

3-methoxycarbonyl-6-*n*-butyl-7-benzyloxy-4-oxoquinoline. See nequinate.

1-methoxycarbonyl-1-propen-1-yl dimethylphosphate. See mevinphos.

methoxychlor.   (methoxy DDT; DMDT; 2,2-bis(*p*-methoxyphenol)-1,1,1-trichloroethane).
CAS: 72-43-5. $Cl_3CCH(C_6H_4OCH_3)_2$.
Properties: White, crystalline solid. Mp 89C. Insoluble in water; soluble in alcohol. Not compatible with alkaline materials.
Derivation: Reaction of methyl phenyl ether and chloral hydrate.
Hazard: Toxic material. Liver damage and central nervous system impairment. Questionable carcinogen.
Use: Insecticide effective against mosquito larvae and house flies; recommended for use in dairy barns.

2-methoxy-3,6-dichlorobenzoic acid.   See dicamba.

***n*-methoxy-3-(3,5-di-*tert*-butyl-4-hydroxybenzylidene)-2-pyrrolidone.**
CAS: 107746-52-1. $C_{20}H_{29}NO_3$.
Hazard: A poison by ingestion.

**2-methoxy-4-(*o*-(*o,o*-diethylphosphorothioyl))
benzaldoximino-*n*-methylcarbamate.**
CAS: 22936-03-4. $C_{14}H_{21}N_2O_6PS$.
Hazard: Moderately toxic by ingestion.
Use: Agricultural chemical.

9-methoxyellipticine lactate.
CAS: 26691-08-7. $C_{18}H_{16}N_2O•C_3H_6O_3$.
Hazard: A poison.

2-(β-methoxyethoxy)ethanol.   See diethylene glycol monomethyl ether.

**α-((2-methoxyethoxy)methyl)-ω-hydroxypoly(oxy-1,2-ethanediyl)coco
alkyl ethers.**
CAS: 73507-40-1.
Hazard: A severe eye irritant.

methoxyethylene.   See vinyl methyl ether.

2-methoxyethylmercury acetate.
CAS: 151-38-2. $CH_3OCH_2CH_2HgOOCCH_3$. A fungicide and disinfectant used in treating seeds.
Hazard: Highly toxic.

**2-methoxyethyl 2-(*m*-nitrobenzylidene)
acetoacetate.**
CAS: 39562-22-6. $C_{14}H_{15}NO_6$.
Hazard: Low toxicity by ingestion and inhalation. A moderate eye irritant.

methoxyethyl oleate.
CAS: 111-10-4. $CH_3OCH_2CH_2OOCC_{17}H_{33}$.
Properties: Oily liquid; mild odor. Fp approximately −18C, d 0.898 (25C), boiling range 180–206C (4 mm Hg), flash p 385F (196C) (OC), viscosity 8 cP (25C). Combustible.
Use: Plasticizer and solvent.

methoxyethyl stearate.
$CH_3OCH_2CH_2OOCC_{17}H_{35}$.
Properties: Oily liquid; mild odor. Fp 19–24C, boiling range 186–205C (4 mm Hg), flash p 378F (192C) (OC), viscosity 9 cP at 25C. Combustible.
Use: Plasticizer and solvent.

methoxyfenozide.
CAS: 161050-58-4. $C_{22}H_{28}N_2O_3$.
Hazard: Low toxicity by ingestion, inhalation, and skin contact.

methoxyflurane.   See   2,2-dichloro-1,1-difluoroethyl methyl ether.

3-methoxy-4-hydroxybenzaldehyde.   See vanillin.

**methoxyhydroxymercuripropylsuccinyl
urea.**   3-hydroxymercuri-2-methoxypropylcarbamoylsuccinamic   acid. $C_9H_{16}HgN_2O_6$.
Properties: Bitter crystals. Mp 198.5C.
Derivation: Made by the mercuration of allylsuccinylurea.
Hazard: Highly toxic.

9-(1-(methoxyimino)ethyl)-6-oxo-n,n,2,2,5-pentamethyl-7-oxa-3,4-dithia-5,8-diazadec-8-en-10-amide.
CAS: 90293-50-8. $C_{13}H_{24}N_4O_4S_2$.
Hazard: A poison by ingestion.
Use: Agricultural chemical.

***n*-(2-(5-methoxy-4-indolyl)ethyl)acetamide.**
CAS: 190650-04-5. $C_{13}H_{16}N_2O_2$.
Hazard: Low toxicity by ingestion.

**4′-methoxy-2-(*p*-methoxyphenyl)
acetophenone.**   See deoxyanisoin.

2-methoxy-5-methylaniline.   See 5-methyl-*o*-anisidine.

6-methoxy-2-methyl-2h-1-benzopyran-3-carbonitrile.
CAS: 57543-75-6. $C_{12}H_{11}NO_2$.
Hazard: Moderately toxic.

**2-(((((methoxy(1-methylethoxy)phosphinothi
oyl)thio)methyl)thio)ethylethylcarbamate.**
CAS: 3853-74-5. $C_{10}H_{22}NO_4PS_3$.
Hazard: A poison by ingestion.
Use: Agricultural chemical.

4-methoxy-4-methylpentanol-2.
$CH_3C(CH_3)(OCH_3)CH_2CH_2OCH_3$.
Properties: Liquid. Boiling range 163.8–167C. Combustible.
Use: Solvent for resin-coating formulation.

4-methoxy-4-methylpentan-2-one.
$CH_3C(CH_3)(OCH_3)CH_2COCH_3$.
Properties: Water-white liquid. Boiling range 147–163C, flash p 141F (60.5C). Combustible.
Derivation: Diacetone alcohol.
Hazard: Moderate fire risk. Irritant to skin and eyes.
Use: Solvent for a variety of resin coatings.

n-(2-(2-methoxy-5-methylphenyl)ethyl)propionamide.
CAS: 156482-84-7. $C_{13}H_{19}NO_2$.
Hazard: Moderately toxic by ingestion.

n-(2-(2-methoxy-5-methylphenyl)ethyl)trifluoroacetamide.
CAS: 156482-85-8. $C_{12}H_{14}F_3NO_2$.
Hazard: Low toxicity by ingestion.

2-methoxy-2-methylpropane. See methyl *t*-butyl ether.

2-(methoxy(methylthio)phosphinylimino)-3-methyl-1,3-thiazoline.
CAS: 33918-12-6. $C_6H_{13}N_2O_2PS_2$.
Hazard: A poison by ingestion and skin contact.
Use: Agricultural chemical.

2-methoxynaphthalene. See β-naphthyl methyl ether.

3-(6-methoxy-2-naphthalenyl)-1,2-dimethyl (2r,3s)-rel-3-pyrrolidinol hydrochloride.
CAS: 302959-30-4. $C_{17}H_{21}NO_2 \cdot ClH$.
Hazard: A poison.

n-(2-(2-methoxynaphthyl)ethyl)butyramide.
CAS: 156482-71-2. $C_{17}H_{21}NO_2$.
Hazard: Moderately toxic by ingestion.

n-(2-(2-methoxy-1-naphthyl)ethyl) cyclopropylcarboxamide.
CAS: 156482-69-8. $C_{17}H_{19}NO_2$.
Hazard: Moderately toxic by ingestion.

n-(2-(2-methoxy-1-naphthyl)ethyl)pentanamide.
CAS: 156482-72-3. $C_{18}H_{23}NO_2$.
Hazard: Moderately toxic by ingestion.

1-methoxy-4-nitrobenzene. See *p*-nitroanisole.

o-methoxyphenol. See guaiacol.

4-methoxyphenol. (*p*-methoxyphenol).
See hydroquinone monomethyl ether.

p-methoxyphenylacetic acid. (*p*-methoxy-α-toluic acid). $CH_3OC_6H_4CH_2COOH$.
Properties: Off-white to pale-yellow flakes. Mp 85C.
Use: Preparation of pharmaceuticals, other organic compounds.

p-methoxyphenylbutanone. See anisylacetone.

4-(4-methoxyphenyl)-6h-1,3,5-oxathiazine.
CAS: 58955-83-2. $C_{10}H_{11}NO_2S$.
Hazard: Moderately toxic by ingestion.
Use: Agricultural chemical.

methoxypolyethylene glycol. One of a series of compounds with properties similar to the polyethylene glycols of comparable molecular weight; slightly viscous liquids to soft, wax-like solids.
Use: Manufacture of detergents and emulsifying and dispersing agents through the preparation of the mono derivatives of fatty acids.

methoxypropanol. (monopropylene glycol methyl ether).
CAS: 107-98-2. $CH_3OCH_2CH(OH)CH_3$.
Properties: Colorless liquid. Bp 120C, flash p 102F (39C), d 0.92 (25C). Combustible.
Hazard: Moderate fire risk.
Use: Antifreeze and coolant for diesel engines.

2-methoxy-1-propanol.
CAS: 1589-47-5. $C_4H_{10}O_2$.
Hazard: A reproductive hazard.

p-methoxypropenylbenzene. See anethole.

p-methoxypropiophenone.
$C_2H_5COC_6H_4OCH_3$.
Properties: Clear colorless liquid. Distillation range 110–140C (3 mm Hg), refr index 1.543–1.545 (25C).

3-methoxypropylamine. (3-MPA).
CAS: 5332-73-0. $CH_3OCH_2CH_2CH_2NH_2$.
Properties: Colorless liquid. Bp 119C, d 0.873 (20/20C), refr index 1.4153 (25C), flash p 80F (26.6C) (TCC). Miscible with water, ethanol, toluene, acetone, carbon tetrachloride, hexane, and ether.
Hazard: Flammable, moderate fire risk. Toxic by ingestion and inhalation.
Use: Organic intermediate, emulsifier in anionic coatings and wax formulations.

6-methoxy-8-(5-propylaminoamylamino) quinoline phosphate.
CAS: 3818-70-0. $C_{18}H_{27}N_3O \cdot xH_3O_4P$.
Hazard: A poison by ingestion.

8-methoxypsoralen.   See methoxsalen.

p-**methoxytoluene.**   See *p*-methylanisole.

p-**methoxy-α-toluic   acid.**   See   *p*-methoxyphenylacetic acid.

methoxytriethylene   glycol   acetate.   See methoxytriglycol acetate.

methoxytriglycol.
CAS: 112-35-6. $CH_3O[C_2H_4O]_3H$.
Properties: Colorless liquid. D 1.0494, bp 249C, fp −44C, flash p 245F (118C). Soluble in water. Combustible.
Use: Plasticizer intermediate.

methoxytriglycol   acetate.   (methoxytriethyleneglycol acetate).
CAS: 3610-27-3. $CH_3COO(C_2H_4O)_3CH_3$.
Properties: Colorless liquid; fruity odor. D 1.0940 (20/20C), bulk d 9.2 lb/gal (20C), bp 244.0C, flash p 260F (126C). Water soluble. Low volatility. Combustible.
Use: Antidusting agent for finely powdered materials, especially for certain dyestuffs.

methyl abietate.
CAS: 127-25-3. $C_{19}H_{29}COOCH_3$.
Properties: Colorless to yellow liquid. D 1.033–1.043 (20C), refr index 1.525–1.535, flash p 360F (182C), bp 365C. Miscible with most organic solvents. Combustible.
Use: Solvent and plasticizer, lacquers, varnishes, coating compositions, adhesives.

methyl acetate.
CAS: 79-20-9. $CH_3CO_2CH_3$.
Properties: Colorless, volatile liquid; fragrant odor. Miscible with the common hydrocarbon solvents, soluble in water, d 0.92438, fp −98.05C, bp 54.05C, flash p 15F (−9.4C), refr index 1.3619 (20C), bulk d 7.76 lb/gal (20C), autoign temp 935F (501C).
Derivation: By heating methanol and acetic acid in the presence of sulfuric acid and distilling.
Grade: Technical, CP.
Hazard: Flammable, dangerous fire and explosion risk, explosive limits in air 3–16%. Irritant to respiratory tract. Headache, dizziness, nausea, eye damage (degeneration of ganglion cells in the retina).
Use: Paint remover compounds, lacquer solvent, intermediate, synthetic flavoring.

methylacetic acid.   See propionic acid.

methyl acetoacetate.
CAS: 105-45-3. $CH_3COCH_2CO_2CH_3$.

Properties: Colorless liquid. D 1.0785 (20/20C), bp 171.7C, vap press 0.7 mm Hg (20C), flash p 170F (78C), bulk d 9.0 lb/gal (20C), fp −80C. Soluble in alcohol; slightly soluble in water. Combustible.
Hazard: Toxic by ingestion and inhalation.
Use: Solvent for cellulose ethers, ingredient of solvent mixture for cellulose esters, organic synthesis.

methyl acetone.
Properties: Water-white, anhydrous liquid, consisting of various mixtures of acetone, ethyl acetate, and methanol; miscible with hydrocarbons, oils, and water. Flash p near 0F (−17C).
Derivation: A by-product in the wood-distillation industry, also synthetic.
Hazard: Flammable, dangerous fire risk. Toxic by ingestion.
Use: Solvent for nitrocellulose, cellulose acetate, rubber, gum, resins; lacquers, paint and varnish removers, extracting perfumes, dewaxing natural gums.

methylacetophenone.   (methyl-*p*-tolyl ketone).
$CH_3C_6H_4COCH_3$.
Properties: Colorless or pale-yellow liquid; fragrant coumarin odor. D 1.001–1.004, refr index 1.533–1.535. Soluble in seven parts of 50% alcohol and in most fixed oils. Combustible.
Derivation: Action of acetic anhydride on toluene.
Grade: Technical, FCC.
Use: Perfumery, flavoring.
See glycine (2).

methylacetopyranone.   See dehydroacetic acid.

methyl acetylene.   (allylene; propyne).
CAS: 74-99-7. $CH_2C \equiv CH$.
Properties: Colorless, liquefied gas. Bp −23.1C, fp −101.5C, sp vol 9.7 cu ft/lb (70F).
Hazard: Flammable, dangerous fire risk. Toxic by inhalation. Central nervous system impairment.
Use: Specialty fuel, chemical intermediate.

methylacetylene-propadiene, stabilized.
(MAPP).
CAS: 59355-75-8.
Properties: Colorless, liquefied gas. D (liquid): 0.576 (15/15C), boiling range −39 to −20C, flame temperature (in oxygen) 2925C. A mixture containing 60–66.5% methylacetylene and propadiene, with the balance being propane and butane.
Hazard: Flammable gas; dangerous fire risk. Toxic by inhalation. Central nervous system impairment.
Use: Industrial fuel gas for cutting, welding, brazing, heat treating, and metallizing.

methyl acetylricinoleate.
CAS: 140-03-4. $C_{17}H_{32}(OCOCH_3)COOCH_3$.
Properties: Pale-yellow, low viscosity, oily liquid. Mild odor, d 0.938 (25/25C), solidifies at −26C.

Soluble in most organic liquids; insoluble in water. Combustible.

Derivation: Castor oil, methanol and acetic anhydride.

Use: Plasticizer, lubricant, protective coatings, synthetic rubbers, vinyl compounds.

methyl acetylsalicylate.
CAS: 580-02-9. $CH_3COOC_6H_4COOCH_3$.
Properties: White crystals. Mp 52C, 134–136C (9 mm Hg).
Derivation: By heating methyl salicylate with a slight excess of acetic anhydride, adding alcohol, then water, and separating the resulting precipitate.
Use: Perfumery (fixative).

methyl acid phosphate.   See methylphosphoric acid.

β-methylacrolein.   See crotonaldehyde.

methyl acrylate.
CAS: 96-33-3. $CH_2:CHCOOCH_3$.
Properties: Colorless, volatile liquid. Bp 80.5C, fp −76.5C, vap press 65 mm Hg (20C), d 0.9574 (20/20C), bulk d 8.0 lb/gal, flash p 25F (−3.8C) (TOC). Slightly soluble in water. Readily polymerized.
Derivation: (1) Ethylene cyanohydrin, methanol, and dilute sulfuric acid; (2) Oxo reaction of acetylene, carbon monoxide, and methanol in the presence of nickel or cobalt catalyst; (3) from β-propiolactone.
Grade: Technical (inhibited).
Hazard: Flammable, dangerous fire and explosion risk. Toxic by inhalation, ingestion, and skin absorption; irritant to skin, eyes and upper respiratory tract irritant; eye damage. Questionable carcinogen.
Use: Acrylic polymers, amphoteric surfactants, vitamin B_1, chemical intermediate.
See acrylate.

β-methylacrylic acid.   See crotonic acid.

methylal.   (dimethoxymethane; formal).
CAS: 109-87-5. $CH_3OCH_2OCH_3$.
Properties: Colorless, volatile liquid; chloroform-like odor; pungent taste. Fp −105C, d 0.86 (20/4C), bp 42.3C, flash p approximately 0F (−17.7C) (OC)(toc, autoign temp 459F (237C). Soluble in water at 20C to extent of 32 wt%; miscible with alcohol, ether, and hydrocarbons.
Hazard: Flammable, dangerous fire and explosion risk. Toxic by ingestion and inhalation. Eye irritant and central nervous system impairment.
Use: Solvent, organic synthesis, perfumes, adhesives, and protective coatings; special fuel.

2-methylalanine.   See aminoisobutyric acid.

methyl alcohol.   (methanol; wood alcohol).
CAS: 67-56-1. CH_3OH.

Properties: Clear, colorless, mobile, highly polar liquid. D 0.7924, fp −97.8C, bp 64.5C, bulk d 6.59 lb/gal (20C), refr index 1.329 (20C), surface tension 22.6 dynes/dm (20C), viscosity 0.00593 cP (20C), vap press 92 mm Hg (20C), flash p 54F (12.2C) (OC), autoign temp 867F (464C). Miscible with water, alcohols, and ether.
Derivation: (1) By high-pressure catalytic synthesis from carbon monoxide and hydrogen; (2) partial oxidation of natural gas hydrocarbons; (3) several processes for making methanol by gasification of wood, peat, and lignite have been developed but have not yet proved out commercially; (4) from methane with molybdenum catalyst (experimental).
Method of purification: Rectification.
Grade: Technical, CP (99.85%), electronic (used to cleanse and dry components), fuel.
Hazard: Flammable, dangerous fire risk. Explosive limits in air 6–36.5% by volume. Toxic by ingestion (causes blindness). Headache, eye damage, dizziness, and nausea.
Use: Manufacture of formaldehyde, acetic acid, and dimethyl terephthalate; chemical synthesis (methyl amines, methyl chloride, methyl methacrylate); antifreeze; solvent for nitrocellulose, ethylcellulose, polyvinyl butyral, shellac, rosin, manila resin, dyes; denaturant for ethanol; dehydrator for natural gas; fuel for utility plants (methyl fuel); feedstock for manufacture of synthetic proteins by continuous fermentation; source of hydrogen for fuel cells; home-heating-oil extender.

methylallyl   acetate.   (methyallyl   acetate).
$CH_2:C(CH_3)CH_2OOCCH_3$.
Properties: Colorless liquid. D 0.9162 (20C), bulk d 7.6 lb/gal.
Hazard: Probably flammable.
Use: Monomer, preparation of methallyl derivatives.

methylallyl   alcohol.   (methallyl alcohol; 2-methyl-2-propen-1-ol). $H_2C:C(CH_3)CH_2OH$.
Properties: Colorless liquid; pungent odor. D 0.8515 (20/4C), bp 115C, refr index 1.4255 (25C), flash p 92F (33.3C). Soluble in water, alcohols, esters, ketones, and hydrocarbons.
Grade: 98.5% min purity.
Hazard: Flammable, moderate fire risk. Irritant to eyes and skin.
Use: Intermediate.

β-methylallyl chloride.   (methallyl chloride; 3-chloro-2-methyl-1-propene; MAC).
CAS: 563-7-3. $CH_2:C(CH_3)CH_2Cl$.

$$H_2C{=}\underset{\underset{CH_3}{|}}{C}{-}\underset{\underset{CH_2}{|}}{C}{-}Cl$$

Properties: Colorless to straw-colored, volatile liquid; sharp penetrating odor. D 0.925 (20C), bp 73C, refr index 1.427 (25C), flash p −3F (−19.4C) (TOC).

Hazard: Flammable, dangerous fire risk. Explosive limits in air 3.2–8.1%. Toxic by ingestion, irritant to eyes and skin. Questionable carcinogen.
Use: Intermediate for production of insecticides, plastics, pharmaceuticals, and other organic chemicals; fumigant for grains, tobacco, and soil.

methylaluminum sesquibromide.
CAS: 12263-85-3. $(CH_3)_3Al_2Br_3$.
Properties: Cloudy, yellow liquid at 25C. Fp −4C, bp (extrapolated) 166C, d 1.514 (25C).
Hazard: Ignites spontaneously in air, reacts violently with water, keep out of contact with air and moisture.
Use: Catalyst for polymerization of olefins and hydrogenation of aromatics, chemical intermediate.

methylaluminum sesquichloride.
CAS: 12542-85-7. $(CH_3)_3Al_2Cl_3$.
Properties: Colorless liquid at 25C. Fp 22.8C, bp (extrapolated) 143.7C, d 1.1629 (25C), 9.705 lb/gal (25C).
Hazard: Ignites spontaneously in air, reacts violently with water, keep out of contact with air and moisture.
Use: Catalyst for polymerization of olefins, catalyst for hydrogenation of aromatics.

methylamine.   (monomethylamine; aminomethane).
CAS: 74-89-5. CH_3NH_2.
Properties: Colorless gas; strong ammoniacal odor. Bp −6.79C, fp −92.5C, flash p (gas) 14F (−10C), (30% solution) 34F (1.1C) (TOC), autoign temp 806F (430C). Soluble in water, alcohol, ether.
Derivation: Interaction of methanol and ammonia over a catalyst at high temperature. The mono-, di-, and trimethylamines are all produced and yields are regulated by conditions. They are separated by azeotropic or extractive distillation.
Grade: Technical (anhydrous, 30–40% solutions).
Hazard: (Gas and liquid) Dangerous fire risk. Explosive limits in air 5–21%. Strong irritant to tissue. Eye, skin and upper respiratory tract irritant.
Use: Intermediate for accelerators, dyes, pharmaceuticals, insecticides, fungicides, surface active agents, tanning, dyeing of acetate textiles, fuel additive, polymerization inhibitor, component of paint removers, solvent, photographic developer, rocket propellant.

methylaminoacetic acid.   See sarcosine.

methyl-*o*-aminobenzoate.   See methyl anthranilate.

2-(4-methylaminobutoxy)diphenylmethane hydrochloride.
CAS: 62232-46-6. $C_{18}H_{23}NO•ClH$.
Hazard: Moderately toxic by ingestion.

methylaminodimethylacetal.
CAS: 122-07-6. $(CH_3O)_2CHCH_2NHCH_3$.
Properties: Water-white to slightly yellow, clear liquid; sharp ammoniacal odor. Refr index 1.406–1.409 (20C), d 0.924 (25/25C). Combustible.

***l*-methylaminoethanolcatechol.**   See epinephrine.

2-(methylamino)glucose.   See *N*-methyl-glucosamine.

2-methylaminoheptane.
$CH_3(CH_2)_4CH(NHCH_3)CH_3$.
Properties: Oily liquid; slight amine odor. Bp 155C. Somewhat soluble in water.
Use: Medicine, usually as the hydrochloride.

***N*-methyl-*p*-aminophenol.** $CH_3NHC_6H_4OH$.
Properties: Colorless needles. Mp 87C. Soluble in water, alcohol, and ether. Combustible.
Derivation: (1) Interaction of hydroquinone and methylamine; (2) methylation of *p*-aminophenol hydrochloride.
Hazard: Eye and skin irritant.
Use: Organic synthesis, photographic developer.

***N*-methyl-*p*-aminophenol sulfate.**
$HOC_6NHCH_3•1/2H_2SO_4$.
Properties: Colorless needles. Mp 250–260C with decomposition. Soluble in water and alcohol; insoluble in ether; discolors in air. Combustible.
Derivation: By methylation of *p*-aminophenol and conversion of the resulting methylated base by neutralization with sulfuric acid.
Grade: CP, photographic.
Use: Photographic developer.

methylamyl acetate.   (methylisobutyl carbinol acetate; *sec*-hexyl acetate; 4-methyl-2-pentanol acetate).
CAS: 108-84-9. $CH_3COOCH(CH_3)CH_2CH(CH_3)_2$.
Properties: Colorless liquid; mild odor. D 0.8598 (20/20C), bp 146.3C, fp −64C, vap press 3 mm Hg (20C), flash p 110F (43.3C) (OC), bulk d 7.1 lb/gal. Insoluble in water; soluble in alcohol. Combustible.
Grade: Technical.
Hazard: Moderate fire risk. Toxic by inhalation.
Use: Solvent for nitrocellulose and other lacquers.

methylamyl alcohol.   (methylisobutyl carbinol; MIBC; 4-methylpentanol-2).
CAS: 108-11-2. $(CH_3)_2CHCH_2CH(CH_3)OH$.
Properties: Colorless, stable liquid. Bp 131.8C, d 0.8079 (20/20C), bulk d 6.72 lb/gal (20C), sets to a glass at approximately −90C, refr index 1.4089 (25C), vap press 2.8 mm Hg (20C), flash p 105F (40.5C) (OC). Miscible with most common organic solvents; slightly soluble in water. Combustible.
Derivation: From methyl isobutyl ketone.
Grade: Technical.

Hazard: Moderate fire risk, explosive limits in air 1–5.5%. Eye and upper respiratory tract irritant, and central nervous system impairment.

Use: Solvent for dyestuffs, oils, gums, resins, waxes, nitrocellulose, and ethylcellulose; organic synthesis; froth flotation; brake fluids.

methyl-*n*-amyl carbinol. (heptanol-2,2-methyl-1-pentanol).
CAS: 543-49-7. $CH_3(CH_2)_4CH_2OCH_3$.
Properties: Stable, colorless liquid; mild odor. D 0.8187 (20/20C), bp 160.4C, vap press 1.0 mm Hg (20C), flash p 130F (54.4C), bulk (OC) d 6.8 lb/gal (20C). Miscible with common organic liquids. Combustible.
Grade: Technical.
Hazard: Moderate fire risk.
Use: Solvent for synthetic resins, frothing agent in ore flotation.

methyl-*n*-amyl ketone. (2-heptanone).
CAS: 110-43-0. $CH_3CH_2CH_2CH_2CH_2COCH_3$.
Properties: Water-white liquid. D 0.8166 (20/20C), bp 150.6C, vap press 2.6 mm Hg (20C), flash p 102F (38.9C), autoign temp 991F (532C), refr index 1.4110 (20C), bulk d 6.8 lb/gal (20C), nitrocellulose–toluene dilution ratio 3.9, fp −35C. Almost insoluble in water; miscible with organic solvents. Combustible.
Grade: Technical.
Hazard: Moderate fire risk. Toxic by inhalation, skin and eye irritant, narcotic in high concentration.
Use: Solvent for nitrocellulose lacquers, synthetic flavoring, perfumery.

N-methylaniline.
CAS: 100-61-8. $C_6H_5NH(CH_3)$.
Properties: Colorless to reddish-brown oily liquid, discolors on standing. D 0.991, fp −57C, bp 190–191C. Soluble in alcohol and ether. Combustible.
Derivation: By heating aniline chloride and methanol and subsequent distillation.
Grade: Technical.
Hazard: Toxic by ingestion, inhalation, and skin absorption. Methemoglobinemia and central nervous system impairment.
Use: Organic synthesis, solvent, acid acceptor.

α-methylanisalacetone. (1-[*p*-methoxy-phenyl]-1-penten-3-one).
CAS: 104-27-8. $CH_3OC_6H_4CH:CHCOCH_2CH_3$.
Properties: White to pale-yellow solid; sharp odor. Stable, mp 60C min, 1 g is soluble in 5 mL 95% alcohol. Combustible.
Grade: 99% pure min.
Use: Flavoring.

5-methyl-*o*-anisidine. (*p*-cresidine; 2-methoxy-5-methylaniline).
CAS: 120-71-8. $CH_3C_6H_3(NH_2)OCH_3$.
Properties: White crystals. Mp 51.5C, bp 235C. Insoluble in water; soluble in organic solvents.

Derivation: 2-Nitro-*p*-cresol, obtained by the action of nitrous and excess nitric acids upon *p*-toluidine, is methylated and reduced.
Grade: Technical.
Hazard: A possible carcinogen.
Use: Dyes.

p-methylanisole. (*p*-cresyl methyl ether; *p*-methoxytoluene; methyl-*p*-cresol).
CAS: 104-93-8. $CH_3C_6H_4OCH_3$.
Properties: Colorless liquid; strong floral odor. D 0.966–0.970, refr index 1.5100–1.5130 (20C); one part dissolves in three parts of 80% alcohol. Combustible.
Grade: FCC, technical.
Use: Perfumery, flavoring.

α-methylanthracene. (1-methylanthracene).
$C_{15}H_{12}$ or $C_6H_4(CH)_2C_6H_3CH_3$ (a tricyclic aromatic).
Properties: Colorless leaflets. D 1.101, bp 200C, mp 86C. Soluble in alcohol; insoluble in water. Combustible.
Grade: Technical.
Use: Organic synthesis.

1-methylanthracene. See α-methylanthracene.

methyl anthranilate. (methyl-*o*-amino-benzoate; neroli oil, artificial).
CAS: 134-20-3. $H_2NC_6H_4CO_2CH_3$.
Properties: Crystals or pale-yellow liquid with bluish fluorescence; orange blossom odor. D 1.167–1.175 (15C), refr index 1.5820–1.5840 (20C), bp 135C, mp 23.8C (min). Soluble in five volumes or more of 60% alcohol; soluble in fixed oils, propylene glycol, volatile oils; slightly soluble in water, mineral oil; insoluble in glycerol. Combustible.
Derivation: By heating anthranilic acid and methanol in the presence of hydrochloric acid with subsequent distillation. Occurs in many flower oils.
Grade: Technical, FCC.
Use: Flavoring, perfume (cosmetics and pomades).

methylanthraquinone.
CAS: 84-54-8. $CH_3C_6H_3(CO)_2C_6H_4$ (tricyclic).
Properties: White needles. Mp 177C, bp sublimes. Soluble in organic solvents; insoluble in water. Combustible.
Derivation: By heating anthraquinone and methanol in the presence of sulfuric acid.
Use: Organic synthesis.

methyl apholate. 2,2,4,4,6,6-hexahydro-2,2,4,4,6,6-hexakis(2-methyl-1-aziridinyl)-1,3,5,2,4,6-triazatriphosphorine.

$N_3P_3(N\overline{CH_2CHCH_3})_6$.

Use: Insect chemosterilant.

methyl arachidate. (methyl eicosanoate).
CAS: 1120-28-1. $CH_3(CH_2)_{18}COOCH_3$. The methyl ester of arachidic acid.
Properties: Wax-like solid. Mp 45.8C, bp 284C (100 mm Hg), 216C (10 mm Hg), refr index 1.4352 (50C). Insoluble in water; soluble in alcohol and ether.
Derivation: Esterification of arachidic acid with methanol and vacuum distillation.
Grade: Purified (99.8%+).
Use: Special synthesis, intermediate for pure arachidic acid, reference standard for gas chromatography, medical research.

methylarsenic acid.
CAS: 124-58-3. CH_5AsO_3.
Properties: Plates from EtOH or Me_2CO. Mp: 159.8°C. Sol in H_2O and EtOH.
Hazard: Moderately toxic by ingestion.
Use: Agricultural chemical.

methylated cytosine. (mc). The methylated form of cytosine. It significantly affects gene activity.

methyl azinphos. See azinphos methyl.

methyl 1-aziridineacetate.
CAS: 14745-52-9. $C_5H_9NO_2$.
Hazard: A poison.

methyl behenate. (methyl docosanoate).
CAS: 929-77-1. $CH_3(CH_2)_{20}COOCH_3$. The methyl ester of behenic acid.
Properties: Wax-like solid. Mp 53.2C, bp 215.5C (3.75 mm Hg), refr index 1.4262 (80C). Insoluble in water; soluble in alcohol and ether. Combustible.
Derivation: Esterification of behenic acid with methanol followed by fractional distillation.
Grade: Purified (99.8%+).
Use: Special synthesis, intermediate for pure behenic acid, biochemical and medical research, reference standard in gas chromatography.

methylbenzaldehydes. See tolyl aldehydes.

methylbenzene. See toluene.

4-methylbenzenesulfonyl isocyanate.
CAS: 4083-64-1. $C_8H_7NO_3S$.
Hazard: Moderately toxic by ingestion. Low toxicity by inhalation. A mild skin and moderate eye irritant.

methylbenzethonium chloride.
(benzyldimethyl(2-[2-(p-1,1,3,3-tetramethylbutylcresoxy)ethoxy]ethyl) ammonium chloride).
CAS: 25155-18-4. A quaternary ammonium compound.

Properties: Colorless crystals; odorless; bitter taste. Mp 161–163C. Readily soluble in alcohol, hot benzene, "Cellosolve" *[Union]*, chloroform, and water; insoluble in carbon tetrachloride and ether.
Grade: NF.
Use: Medicine (bactericide).

methyl benzoate. (benzoic acid, methyl ester; niobe oil).
CAS: 93-58-3. $C_6H_5COOCH_3$.

$$CH_3O-C=O$$

Properties: Liquid of fragrant odor, colorless, oily. D 1.085–1.088, refr index 1.514, fp −12.3C, bp 198.6C, flash p 181F (82.7C). Soluble in three parts of 60% alcohol, in most fixed oils, in ether; insoluble in water. Combustible.
Derivation: (1) By heating methanol and benzoic acid in presence of sulfuric acid; (2) passing dry hydrogen chloride through a solution of benzoic acid and methanol; (3) occurs naturally in oils of clove, ylang-ylang, tuberose.
Grade: Technical, FCC.
Hazard: Toxic by ingestion.
Use: Perfumery; solvent for cellulose esters and ethers, resins, rubber; flavoring.

methylbenzoic acid. See o-toluic acid; m-toluic acid; p-toluic acid.

methylbenzophenone. See phenyl tolyl ketone.

methyl-o-benzoylbenzoate.
$C_6H_5COC_6H_4COOCH_3$.
Properties: Colorless liquid. D 1.190 (25C), refr index 1.587 (25C), vap press 4.0 mm Hg (175C), bp 351C, mp 40C, flash p 350F (176C). Very slightly soluble in water. Combustible.
Use: Plasticizer.

α-methylbenzyl acetate. (methylphenylcarbinyl acetate; styrallyl acetate; *sec*-phenylethyl acetate; phenylmethylcarbinyl acetate). $C_6H_5CH(CH_3)OOCCH_3$.
Properties: Colorless liquid; strong floral odor. D 1.023–1.026 (25/25C), refr index 1.4935–1.4960 (20C), flash p 178F (81.1C) (TCC). Soluble in 70% alcohol, glycerol, and mineral oil; insoluble in water. Combustible.
Grade: 98% min, FCC.
Use: Perfumery, flavoring.

α-methylbenzyl alcohol. (styrallyl alcohol; phenylmethylcarbinol; 1-phenylethan-1-*sec*-phenethyl alcohol; methylphenylcarbinol).
CAS: 589-18-4. $C_6H_5CH(CH_3)OH$.

Properties: Colorless liquid; mild floral odor; congeals below room temperature. D 1.009–1.014 (25C), mp 20.7C, bp 204C, refr index 1.525–1.529 (20C), flash p 205F (96.1C) (COC). Soluble in alcohol, glycerol, mineral oil; slightly soluble in water. Combustible.
Grade: FCC.
Use: Perfumery, flavoring dyes, laboratory reagent.

α-methylbenzylamine.
CAS: 98-84-0. $C_6H_5CH(CH3)NH_2$.
Properties: Water-white liquid; mild ammoniacal odor. D 0.9535 (20/20C), refr index 1.5366 (20C), bp 80C (18 mm Hg), vap press 0.5 mm Hg (20C), fp sets to a glass approximately −65C, flash p 175F (79.4C) (COC). Soluble in most organic solvents; somewhat soluble in water. Combustible.
Hazard: Toxic by ingestion.
Use: Synthesis, emulsifying agent.

α-methylbenzyldiethanolamine.
$C_6H_5CH(CH_3)N(C_2H_4OH)_2$.
Properties: Dark-amber liquid; ammonia-like odor. D 1.0812 (20C), bp 244C (50 mm Hg), flash p 370F (187C) (OC), vap press less than 0.01 mm Hg (20C), sets to glass at −7C. Moderately soluble in water. Combustible.
Use: Emulsifying agents, textile specialties, quaternaries.

α-methylbenzyldimethylamine.
$C_6H_5CH(CH_3)N(CH_3)_2$.
Properties: Colorless liquid. D 0.9044 (20/20C), bp 195.6C, vap press 0.6 mm Hg (20C), fp sets to a glass approximately −70C, refr index 1.5024 (20C), flash p 175F (79.4C) (OC), viscosity 1.85 cP (20C). Slightly soluble in water. Combustible.
Use: Polymerization catalyst.

α-methylbenzyl ether.
CAS: 93-96-9. $C_6H_5CH(CH_3)OCH(CH_3)C_6H_5$.
Properties: Straw-yellow, mobile liquid; faint odor. D 1.0017 (20/20C), bp 286.3C, vap press less than 0.01 mm Hg (20C), fp sets to a glass at approximately −30C, very slightly soluble in water, flash p 275F (135C) (COC). Soluble in most organic solvents. Combustible.
Use: Solvent, styrenating agent, softener for synthetic rubbers.

dl-n-(α-methylbenzyl)linoleamide.
CAS: 14417-88-0. $C_{26}H_{41}NO$.
Hazard: Low toxicity by ingestion.

methyl blue. (sodium triphenyl-*p*-rosaniline sulfonate; CI 42780).
CAS: 28983-56-4. $C_{37}H_{27}N_3O_3S•2NaSO_3$. A dark-blue powder or dye.
Use: Medicine, as an antiseptic, and in biological and bacteriological stains.
Note: Do not confuse with methylene blue.

methyl borate. See trimethyl borate.

methyl bromide. (bromomethane).
CAS: 74-83-9. CH_3Br.
Properties: Colorless, transparent, easily liquefied gas or volatile liquid; burning taste; chloroform-like odor. D 1.732 (0C), bp 3.46C, vap press 1250 mm Hg (20C), fp −94C, flash p none, nonflammable in air; burns in oxygen. Miscible with most organic solvents; forms a voluminous crystalline hydrate with cold water.
Derivation: Action of bromine on methanol in presence of phosphorus with subsequent distillation.
Grade: Technical, pure (99.5% min).
Hazard: Toxic by ingestion, inhalation, and skin absorption; strong irritant to skin and upper respiratory tract. Questionable carcinogen.
Use: Soil and space fumigant; disinfestation of potatoes, tomatoes, and other crops; organic synthesis; extraction solvent for vegetable oils.

methyl bromoacetate.
CAS: 96-32-2. $BrCH_2COOCH_3$.
Properties: Colorless to straw-colored liquid. Fp approximately −50C, bp 145.0–146.7C, d 1.655 (25/25C), ref index 1.456 (25C). Very slightly soluble in water; soluble in methanol, ether.
Hazard: Vapor is strong irritant to eyes.
Use: Synthesis of weed killers, dyes, vitamins, pharmaceuticals; lachrymator.

2-methyl-1,3-butadiene. See isoprene.

2-methylbutanal. See 2-methylbutyraldehyde.

2-methylbutane. See isopentane.

2-methyl-2-butanethiol. (*tert*-butyl mercaptan). $(CH_3)_2CSH(C_2H_5)$.
Properties: Boiling range 95–119C, d 0.828 (15.5/15.5C), refr index 1.438 (20C), flash p 30F (−1.1C); strong offensive odor.
Grade: 95%.
Hazard: Flammable, dangerous fire risk.
Use: Odorant, intermediate, bacterial nutrient.

2-methyl-1-butanol. (amyl alcohol, primary, active; *sec*-butyl carbinol).
CAS: 137-32-6. $CH_3CH_2CH(CH_3)CH_2OH$. The active alcohol from fusel oil. The synthetic product is a racemic mixture of both dextro- and levorotatory compounds and therefore not optically active.
Properties: Colorless liquid. D 0.81–0.82 (20C), fp below −70C, bp 128C, refr index 1.41 (20C), flash p (OC) 115F (46.1C). Slightly soluble in water; miscible with alcohol and ether. Combustible.
Derivation: Occurs in fusel oil; is made synthetically by fractional distillation of the mixed alcohols resulting from the chlorination and alkaline hydrolysis of pentane.
Hazard: Moderate fire and explosion risk. Toxic by ingestion, inhalation, and skin absorption.

Use: Solvent, organic synthesis (introduction of active amyl group), lubricants, plasticizers, additives for oils and paints.

2-methyl-2-butanol. See *tert*-amyl alcohol.

3-methyl-1-butanol. See isoamyl alcohol, primary.

methylbutanolbutanol. (3-methyl-1-butanol-butanol). An alcohol composed of four carbon atoms bonded linked to each other by single bonds and to one alcohol or one hydroxyl group.

3-methyl-2-butanone. See methyl isopropyl ketone.

2-methyl-1-butene.
CAS: 563-46-2. C_5H_{10} or $H_2C{:}C(CH_3)CH_2CH_3$.
Properties: Colorless, volatile liquid; disagreeable odor. Bp 31.11C, refr index 1.378 (20C), d 0.650 (20/20C), fp −137.52C, flash p approximately −20F (−28C). Soluble in alcohol; insoluble in water.
Derivation: Refinery gas.
Grade: 95%, 99%, and research.
Hazard: Highly flammable, dangerous fire and explosion risk.
Use: Organic synthesis, pesticide formulations.

2-methyl-2-butene. See 3-methyl-2-butene.

3-methyl-1-butene. (isopropylethylene; α-isoamylene).
CAS: 563-45-1. C_5H_{10} or $H_2C{:}CHCH(CH_3)_2$.
Properties: Colorless, extremely volatile liquid or gas; disagreeable odor. Bp 20.1C, refr index 1.3643 (20C), d 0.6272 (20C), fp −168.5C, flash p −70F (−57C). Soluble in alcohol; insoluble in water.
Derivation: Cracking of petroleum, a component of refinery gas.
Grade: Research, 99% min, technical 95% min.
Hazard: Highly flammable, dangerous fire and explosion risk, explosive limits 1.6–9.1%.
Use: Organic synthesis, high-octane fuel manufacture.

3-methyl-2-butene. (2-methyl-2-butene; trimethylethylene; β-isoamylene).
CAS: 513-35-9. C_5H_{10} or $H_3CCH{:}C(CH_3)_2$.
Properties: Colorless, volatile liquid, disagreeable odor. Bp 38.51C, refr index 1.387 (20C), d 0.6623 (20/4C), fp −133.83C, flash p −50F (−45C). Soluble in alcohol; insoluble in water.
Derivation: Cracking of petroleum, a component of refinery gas.
Grade: 90%, 95% (technical), 99% (pure), and research.
Hazard: Highly flammable, dangerous fire and explosion risk.
Use: Organic synthesis, hydrogenation, halogenation, alkylation, condensation reactions.

cis-**2-methyl-2-butenoic acid.** See angelic acid.

trans-**2-methyl-2-butenoic acid.** See tiglic acid.

2-methyl-1-buten-3-yne. See isopropenylacetylene.

1-methylbutyl alcohol. See 2-pentanol.

N-methylbutylamine.
CAS: 110-68-9. $CH_3CH_2CH_2CH_2NHCH_3$.
Properties: Liquid. D 0.7335, bp 91.1C, fp −75.0C, flash p 35F (1.6C) (TOC). Miscible with water.
Hazard: Flammable, dangerous fire risk.
Use: Intermediate.

methyl-*tert*-butyl ether. (MTBE).
CAS: 1634-04-4. $(CH_3)_3COCH_3$.
Properties: Colorless liquid. Mw 88.17, bp 55C, fp −110C, d 0.74, bulk d 6.18 lb/gal, heat of vaporization 145 Btu/lb (7 kcal/mole) (55C), heat of combustion 101,000 Btu/gal (804 kcal/mole) (25C). Solubility in water 4 wt%; solution of water in 1.3 wt%. Octane blending value: 115–125 (Research), 98–110 (Motor).
Derivation: Catalytic reaction of methanol and isobutene (38–93C at 100–200 psi). There are several variations of the process.
Hazard: Slightly toxic by ingestion and inhalation. Flammable when exposed to heat or flame. Upper respiratory tract irritant and kidney damage. Questionable carcinogen.
Use: Octane booster for unleaded gasoline (up to 7% by volume), manufacture of isobutene. Approved by EPA.
See octane number.

2-methylbutyl isovalerate.
CAS: 2445-77-4. $C_{10}H_{20}O_2$.
Properties: Colorless liquid; herbaceous, fruity odor. D: 0.852, refr index: 1.413. Sol in alc, fixed oils; insol in water.
Use: Food additive.

methyl butyl ketone. (propylacetone; 2-hexanone).
CAS: 591-78-6. $CH_3COC_4H_9$.
Properties: Colorless liquid. Bp 127.2C, d 0.830 (20/20C), refr index 1.4024 (20C), vap press 10 mm Hg (20C), flash p 95F (35C) (OC). Soluble in alcohol and ether.
Grade: Technical.
Hazard: Flammable, moderate fire risk, explosive limits 1.2–8% in air. Irritant to eyes and mucous membranes, narcotic in high concentration, absorbed by skin. Causes peripheral neuropathy and testicular damage.
Use: Solvent.

2-methylbutyl-3-methylbutanoate. See 2-methylbutyl isovalerate.

2-methyl-6-*tert*-butylphenol.
$C_6H_3(OH)(CH_3)tert$-C_4H_9).
Properties: Crystalline solid, light straw color. Mp 28C, d 0.9618 (30C), bp 230C, flash p 220F (104C) (OC). Soluble in methyl ethyl ketone, ethanol, benzene, and isooctane; insoluble in water. Combustible.
Use: Chemical intermediate.

2-methylbutyl-3-thiol. See *sec*-isoamyl mercaptan.

2-methyl-4-*tert*-butylthiophenol. (4-*tert*-butyl-*o*-thiocresol). $(CH_3)_3CC_6H_3(CH_3)SH$.
Properties: Water-white liquid, no mercaptan odor. D 0.983 (25C), refr index 1.546 (25C), fp −4C, bp 177C (100 mm Hg). Soluble in aliphatic and aromatic hydrocarbons; insoluble in water. Combustible.
Use: Chemical intermediate.

2-((*o*-(*n*-methyl-*n*-(*tert*-butylthiosulfenyl) carbamoyl)oximino))-1,3-dithiolane.
CAS: 76858-53-2. $C_9H_{16}N_2O_2S_4$.
Hazard: A poison by ingestion.
Use: Agricultural chemical.

2-methyl-3-butyn-2-amine.
CAS: 2978-58-7. C_5H_9N.
Hazard: Moderately toxic by ingestion. A moderate eye irritant.

methyl butynol. (2-methyl-3-butyn-2-ol).
CAS: 115-19-5. $HC\equiv CCOH(CH_3)_2$.
Properties: Colorless liquid; fragrant odor. Bp 104–105C, mp 2.6C, d 0.8672 (20/20C), refr index 1.4211 (20C), flash p 77F (25C) (TOC). Miscible with water; soluble in most organic solvents.
Grade: Technical, 95% min.
Hazard: Flammable, dangerous fire risk.
Use: Stabilizer in chlorinated solvents, viscosity reducer and stabilizer, electroplating brightener, intermediate.

2-methylbutyraldehyde. (2-methylbutanal).
CAS: 96-17-3. $CH_3CH_2CH(CH_3)CHO$.
Properties: Liquid. D 0.8029 (20/4C), bp 92–93C, refr index 1.3869 (20C). Soluble in alcohol and ether; insoluble in water. Combustible.
Use: Flavoring.

3-methylbutyraldehyde. See isovaleraldehyde.

methyl butyrate.
CAS: 623-42-7. $CH_3CH_2CH_2COOCH_3$.
Properties: Colorless liquid. D 0.898 (20C), bp 102C, fp −92C, refr index 1.3875 (20C), flash p

57F (14C) (CC). Slightly soluble in water; soluble in alcohol.
Grade: Technical.
Hazard: Flammable, dangerous fire risk.
Use: Solvent for ethylcellulose, solvent mixture for nitrocellulose, flavoring.

2-methylbutyric acid. See isopentanoic acid.

3-methylbutyric acid. See isopentanoic acid.

methyl caprate. (methyl decanoate).
CAS: 110-42-9. $CH_3(CH_2)_8COOCH_3$.
Properties: Colorless liquid. D 0.8733 (20/4C), fp −13.3C, bp 224C, 130.6C (30 mm Hg), refr index 1.4237 (25C). Insoluble in water; soluble in alcohol and ether. Combustible.
Derivation: Esterification of capric acid with methanol or alcoholysis of coconut oil, purified by fractional vacuum distillation.
Grade: Technical, 99.8% pure.
Use: Intermediate for detergents, emulsifiers, wetting agents, stabilizers, resins, lubricants, plasticizers.

methyl caproate. (methyl hexanoate).
CAS: 106-70-7. $CH_3(CH_2)_4COOCH_3$. The methyl ester of caproic acid.
Properties: Colorless liquid. D 0.8850 (20/4C), fp −71C, bp 151.2C, 63.0C (30 mm Hg), refr index 1.4054 (20C). Insoluble in water; soluble in alcohol and ether. Combustible.
Derivation: Esterification of caproic acid with methanol or alcoholysis of coconut oil.
Grade: Technical, 99.8+%.
Use: Intermediate for caproic acid detergents, emulsifiers, wetting agents, stabilizers, resins, lubricants, plasticizers, flavoring.

2-methylcaproic acid. See 2-hexanecarboxylic acid.

5-methylcaproic acid. See isoenanthic acid.

methyl caprylate. (methyl octanoate).
CAS: 111-11-5. $CH_3(CH_2)_6COOCH_3$. The methyl ester of caprylic acid.
Properties: Colorless liquid. D 0.8784 (20/4C), fp −37.3C, bp 192C (759 mm Hg), 98.3 (30 mm Hg), refr index 1.4152 (25C). Insoluble in water; soluble in alcohol and ether. Combustible.
Derivation: (1) Esterification of caprylic acid with methanol, (2) alcoholysis of coconut oil.
Grade: Technical, 99.8%.
Use: Intermediate for caprylic acid detergents, emulsifiers, wetting agents, stabilizers, resins, lubricants, plasticizers, flavoring.

methylcarbamic acid 4-hydroxy-*m*-cumenyl ester. See 3-isopropyl-4-hydroxyphenyl methylcarbamate.

methylcarbamic acid 4-hydroxy-3,5-diisopropylphenyl ester.
CAS: 17710-64-4. $C_{14}H_{21}NO_3$.
Hazard: A poison.
Use: Agricultural chemical.

2-(o-(methylcarbamoyl)oximino)-3,3-dimethyltetrahydro-1,4-thiazin-5-one.
CAS: 66637-25-0. $C_8H_{13}N_3O_3S$.
Hazard: A poison by ingestion.
Use: Agricultural chemical.

2-(o-(methylcarbamoyl)oximino)-1,4-dithiane.
CAS: 55391-34-9. $C_6H_{10}N_2O_2S_2$.
Hazard: A poison by ingestion.
Use: Agricultural chemical.

methyl "Carbitol" [Union].   TM for diethylene glycol monomethyl ether.

methyl "Carbitol" acetate. [Union].   TM for diethylene glycol monomethyl ether acetate.

methyl carbonate.   (dimethyl carbonate).
CAS: 616-38-6. $CO(OCH_3)_2$.
Properties: Colorless liquid; pleasant odor. D 1.0718 (20C), bp 90.6C, mp 0C. Miscible with acids and alkalies; stable in the presence of water; soluble in most organic solvents; insoluble in water.
Derivation: Interaction of phosgene and methanol.
Grade: Technical.
Hazard: Flammable, dangerous fire risk. Toxic by inhalation, strong irritant.
Use: Organic synthesis, specialty solvent.

methyl "Cellosolve" [Union].   TM for ethylene glycol monomethyl ether.

methylcellulose.   (cellulose   methyl   ether; "Methocel" [Dow]).
CAS: 9004-67-5.
Properties: Grayish-white, fibrous powder; aqueous suspensions neutral to litmus. Swells in water to a viscous colloidal solution. Insoluble in alcohol, ether, chloroform, and in water warmer than 50.5C; soluble in glacial acetic acid; unaffected by oils and greases; stable up to approximately 300C; stable to light. Combustible. Molecular weights vary from 40,000 to 180,000. Specifications call for methoxy group content of narrow or wide ranges within 25–33%.
Derivation: From cellulose by conversion to alkali cellulose and then reacting this with methyl chloride, dimethyl sulfate, or methanol and dehydrating agents. The proportions of the reacting materials are varied to control the properties of the product, such as water solubility and viscosity of water solutions.
Grade: USP, technical, FCC.
Use: Protective colloid in water-based paints to prevent flocculation of pigment; film and sheeting; binder in ceramic glazes; leather tanning; dispersing, thickening, and sizing agent; adhesive; food additive.
See cellulose, modified; carboxymethylcellulose; hydroxyethylcellulose.

methylcellulose, propylene glycol ether.   See hydroxypropyl methylcellulose.

methyl cerotate.   (methyl hexacosanoate).
CAS: 5802-82-4. $CH_3(CH_2)_{24}COOCH_3$.   The methyl ester of cerotic acid.
Properties: Wax-like solid. Mp 62.9C, bp 237C (1.95 mm Hg), refr index 1.4301 (80C). Insoluble in water; soluble in alcohol and ether. Combustible.
Derivation: Esterification of cerotic acid with methanol.
Grade: Purified (99+%).
Use: Intermediate in special synthesis, medical research, reference standard for gas chromatography.

methyl chloride.   (chloromethane; monochloromethane).
CAS: 74-87-3. CH_3Cl.
Properties: Colorless compressed gas or liquid; faintly sweet; ethereal odor. D 0.92 (20C), bp −23.7C, fp −97.6C, flash p approximately 32F (0C), refr index 1.3712 (−23.7C), critical temperature 143C, critical pressure 970 psi absolute, autoign temp 1170F (632C), bulk d 7.68 lb/gal (20C). Slightly soluble in water, by which it is decomposed; soluble in alcohol, chloroform, benzene, carbon tetrachloride, glacial acetic acid; attacks aluminum, magnesium, and zinc.
Derivation: (1) Chlorination of methane; (2) action of hydrochloric acid on methanol either in vapor or liquid phase.
Grade: Pure (99.5% min), technical, and two refrigerator grades.
Hazard: Flammable, dangerous fire risk, explosive limits in air 10.7–17%. Narcotic. Psychic effects. Central nervous system impairment; liver, kidney and testicular damage, and teratogenic effects. Questionable carcinogen.
Use: Catalyst carrier in low-temperature polymerization (butyl rubber), tetramethyl lead, silicones, refrigerant, fluid for thermometric and thermostatic equipment, methylating agent in organic synthesis, such as methylcellulose, extractant and low-temperature solvent, herbicide, topical anesthetic.

methyl chloroacetate.
CAS: 96-34-4. $ClCH_2COOCH_3$.
Properties: Colorless liquid; sweet pungent odor. D 1.236 (20/4C), fp −32.7C, bp 131C, refr index 1.419–1.420 (25C). Insoluble in water; miscible with alcohol and ether. Combustible.
Hazard: Toxic by ingestion and inhalation.
Use: Solvent, intermediate.

methyl (chlorocarbonyl)(4-(trifluoromethoxy)phenyl)carbamate.
CAS: 173903-15-6. $C_{10}H_7ClF_3NO_4$.
Hazard: Moderately toxic by ingestion. A moderate eye irritant.

methyl chloroform.   See 1,1,1-trichloroethane.

methyl chloroformate.   (methyl chlorocarbonate).
CAS: 79-22-1. $ClCOOCH_3$.
Properties: Colorless liquid. Decomposed by hot water; stable to cold water. D 1.23 (15C), bp 71.4C, vapor d 3.9 (air = 1), flash p 54F (12.2C). Soluble in methanol alcohol, ether, and benzene.
Derivation: Reaction between methanol and carbonyl chloride.
Grade: Technical (95% min).
Hazard: Flammable, dangerous fire risk. Highly corrosive and irritant to skin and eyes.
Use: Military poison (lachrymator), organic synthesis, insecticides.

methylchloromethyl   ether.   (chloromethyl methyl ether).
CAS: 107-30-2. $ClCH_2OCH_3$.
Properties: Colorless liquid. D 1.0625 (10/4C), fp −103.5C, bp 59.5C. Decomposes in water; soluble in alcohol and ether.
Hazard: Flammable, dangerous fire and explosion risk. Toxic by ingestion and inhalation. TLV: Suspected human carcinogen.

2-methyl-4-chlorophenoxyacetic acid.   See MCPA.

4-(2-methyl-4-chlorophenoxy)butyric   acid.
See 4-MCPB.

2-(2-methyl-4-chlorophenoxy)propionic acid.
See mecoprop.

2-methyl-3-(2-chlorophenyl)chinazolon-4.
CAS: 340-57-8. $C_{15}H_{11}ClN_2O$.
Hazard: Moderately toxic by ingestion.

methyl chlorosilane.
CAS: 993-00-0. CH_5ClSi. One of several intermediates in the formation of silicones or siloxanes, they react with hydroxyl groups on many types of surfaces to produce a permanent, thin-surface film of silicone that imparts water repellency. Examples are methyltrichlorosilane, dimethyldichlorosilane, and trimethylchlorosilane.
Hazard: Toxic by ingestion and inhalation, strong irritant to skin and eyes.

methyl chlorosulfonate.
CAS: 812-01-1. CH_3OSO_2Cl.
Properties: Colorless liquid; pungent odor. D 1.492 (10C), bp 133–135C (decomposes), fp −70C, vap

d 4.5 (air = 1). Decomposed by water. Soluble in alcohol, carbon tetrachloride, chloroform; insoluble in water.
Derivation: Interaction of sulfuryl chloride and methanol.
Grade: Technical.
Hazard: Highly toxic by ingestion and inhalation, strong irritant to skin and eyes.
Use: Organic synthesis, military poison.

methyl-3-((5-(2-chloro-4-(trifluoromethyl)phenoxy)-2-nitrophenyl)amino)butyrate.
CAS: 116929-00-1. $C_{18}H_{16}ClF_3N_2O_5$.
Hazard: A reproductive hazard.
Use: Agricultural chemical.

methylcholanthrene.
CAS: 56-49-5. $C_{21}H_{16}$. A polynuclear hydrocarbon.

Properties: Yellow crystals. Mp 180C. Soluble in benzene; insoluble in water.
Derivation: From bile acids via 1,2-benzanthracene.
Hazard: Powerful carcinogen.
Use: Biochemical research.

methyl cinnamate.
CAS: 103-26-4. $C_6H_5CH:CHCOOCH_3$.
Properties: White crystals; strawberry-like odor. D 1.0415, mp 34C, bp 259.6C. Soluble in alcohol and ether, in glycerol, most fixed oils, and mineral oil; insoluble in water. Combustible.
Derivation: By heating methanol, cinnamic acid, and sulfuric acid with subsequent distillation.
Grade: Technical, FCC.
Use: Perfumes, flavoring.

methylcobalamin.   (carbanide;   cobalt(3+); [5-(5,6-dimethylbenzimidazol-1-yl)-4-hydroxy-2-(hydroxymethyl)oxolan-3-yl]   1-[3-[(4Z,9Z,14Z)-2,13,18-tris(2-amino-2-oxoethyl)-7,12,17-tris(3-amino-3-oxopropyl)-3,5,8,8,13,15,18,19-octamethyl-2,7,12,17-tetrahydro-1H-corrin-21-id-3-yl]propanoylamino]propan-2-yl phosphate)
CAS: 13422-55-4. $C_{63}H_{91}CoN_{13}O_{14}P$. A vitamin B_{12} analog used as an intermediate in the synthesis of methane. It is responsible for the methylation of inorganic mercury by anaerobic bacteria in bottom sediments. Through the action of methylcobalamin in an anaerobic bacteria in bottom sediments of aquatic systems, arsenic(III) is methylated to methanearsonic acid then to cacodylic acid.

methylcoumarin.
CAS: 92-48-8. $C_{10}H_8O_2$.
Properties: White crystals with vanilla flavor. Exists as α and β forms, mp (α) 90C, (β) 82C, both forms are soluble in alcohol. Combustible.
Use: Perfumes, flavoring.

methyl-*p*-cresol.   See *p*-methylanisole.

cis-**α-methylcrotonic acid.**   See angelic acid.

trans-**α-methylcrotonic acid.**   See tiglic acid.

methyl cyanide.   See acetonitrile.

methyl cyanoacetate.   (malonic methyl ester nitrile).
CAS: 105-34-0. $CNCH_2COOCH_3$.
Properties: Colorless liquid. Bp 203C (115C) at 6 mm Hg, fp −22.5C, d 1.1225 (15/4C), refr index 1.419–1.420 (20C). Soluble in water, alcohol, and ether. Combustible.
Derivation: Esterification of cyanoacetic acid with methanol, reaction of an alkali cyanide and chloroacetic methyl ester.
Use: Organic synthesis, pharmaceuticals, dyes.

methyl-2-cyanoacrylate.
CAS: 137-05-3. $CH_2{:}C(CN)COOCH_3$.
Properties: Colorless liquid. Bp 48–49C (2.5–2.7 mm Hg), d 1.1044 (27/4C), viscosity 2.2 cP (25C).
Hazard: Toxic by inhalation. Eye and upper respiratory tract irritant.
Use: Adhesive, dentistry.
See cyanoacrylate adhesives.

methyl cyanoethanoate.   See cyanomethyl acetate.

methyl cyanoformate.
CAS: 17640-15-2. $CNCOOCH_3$.
Properties: Colorless liquid; ethereal odor. D approximately 1.00 (20C), bp 100C. Decomposed by alkalies and water. Soluble in alcohol, benzene, ether.
Derivation: Methylchloroformate is dissolved in methanol and subjected to the action of (hot) sodium cyanide or potassium cyanide.
Use: Organic synthesis.

methylcyclohexane.   (hexahydrotoluene).
CAS: 108-87-2. $CH_3C_6H_{11}$.
Properties: Colorless liquid. D 0.769, bp 100.8C, fp −126.9C, refr index 1.42312, flash p 25F (−3.89C) (CC), autoign temp 545F (285C).
Source: Petroleum.
Grade: Technical (95%), 99%, and research.
Hazard: Flammable, dangerous fire risk. Lower explosive limit 1.2% in air. Upper respiratory tract irritant, central nervous system impairment, liver and kidney damage.
Use: Solvent for cellulose ethers, organic synthesis.

methylcyclohexanol.   (hexahydromethyl phenol; hexahydrocresol).
CAS: 25639-42-3. $CH_3C_6H_{10}OH$. *o*-, *m*-, and *p*- forms.
Properties: Colorless, viscous liquid; aromatic menthol-like odor. Bp 155–180C, d 0.924, flash p 154F (67.7C) (CC). Combustible.
Derivation: (1) A mixture of three isomeric (*o*-, *m*-, and *p*-) cyclic secondary alcohols made by the hydrogenation of cresol, (2) catalytic oxidation of methylcyclohexane.
Grade: Technical.
Hazard: Toxic by ingestion. Eye and upper respiratory tract irritant.
Use: Solvent for cellulose esters and ethers and for lacquers, antioxidant for lubricants, blending agent for special textile soaps and detergents.

methylcyclohexanol acetate.   (heptalin acetate; methyl cyclohexyl acetate).
$C_7H_{13}OOCCH_3$.
Properties: Colorless liquid, ester-like odor. Slower rate of evaporation than amyl acetate. Bp 176–193C, d 0.941, flash p 147F (64C) (CC), toluene dilution ratio 2.5. Combustible.
Derivation: Catalytic hydrogenation and esterification of cresols by means of acetic acid.
Use: Solvent.

o-**methylcyclohexanone.**
CAS: 583-60-8. $CH_3C_5H_9CO$.
Properties: Water-white to pale-yellow liquid; acetone-like odor. A mixture of cyclic ketones. Closely resembles cyclohexanone in physical properties. Bp 160–170C, d 0.925, flash p 138F (58.9C). Miscibility, tolerance for nonsolvent and solvent action. Combustible.
Derivation: By high-temperature, catalytic hydrogenation of cresols or by the dehydrogenation of methylcyclohexanol.
Hazard: Moderate fire risk. Toxic by ingestion, inhalation, and skin absorption. Eye and upper respiratory tract irritant, and central nervous system impairment.
Use: Solvent, lacquers.

methylcyclohexanone glyceryl acetal.
$CH_3C_6H_9I_2C_3H_5OH$. Spiro rings.
Properties: Colorless liquid. D 1.074 (20C), refr index 1.474 (20C), bp 130–140C (20 mm Hg), flash p 200F (93.3C). Insoluble in water. Combustible.
Use: Plasticizer.

methylcyclohexanyl oxalate.
$(CH_3C_6H_{10}OOC)_2$.
Properties: Colorless liquid; odorless; neutral; stable. Comprising a mixture of isomers. Miscible with most lacquer solvents and diluents.

4-methylcyclohex-1-ene. $CH_3C_6H_9$.
Properties: Colorless liquid. Fp −121.1C, distillation range 110–117C, d 0.818 (60/60F), refr index

1.450 (20C), flash p 30F (−1.1C). Insoluble in water; soluble in alcohol.

Grade: Pure, 99.0 mol%; technical, 95.0 mol%.

Hazard: Flammable, dangerous fire risk. May be irritant to skin and eyes.

Use: Intermediate.

6-methyl-3-cyclohexene carboxaldehyde.

CH$_3$CHCH$_2$CH:CHCH$_2$CHCHO.

Properties: Colorless liquid. D 0.9484, bp 176.4C, fp −39.0C. Soluble in water 0.3% by wt at 20C.

Use: Intermediate.

N-methyl-5-cyclohexenyl-5-methylbarbituric acid.   See hexobarbital.

N-methylcyclohexylamine.

CAS: 100-60-7.   C$_6$H$_{11}$NHCH$_3$.

Properties: Water-white liquid. D 0.86 (20C). Soluble in alcohol and ether; slightly soluble in water. Purity 99%, distillation range 5–95 cc within 2C, including 149C, corrected to 760 mm Hg. Combustible.

Hazard: Toxic. Strong irritant to tissue.

Use: Intermediate, solvent, acid acceptor.

methylcyclohexylcarbinol.   See 1-cyclohexylethanol.

4-methylcyclohexylmethylcarbinol.

CAS: 34884-20-3.   C$_9$H$_{18}$O.

Hazard: Moderately toxic by ingestion and skin contact. A severe skin and eye irritant.

2-methyl-1,3-cyclopentadiene.

CAS: 3727-31-9.   C$_6$H$_8$.

Hazard: Moderately toxic by ingestion.

methylcyclopentadiene dimer.   (methyl-1,3-cyclopentadiene).

CAS: 26472-00-4.   C$_{12}$H$_{16}$.

Properties: Colorless liquid. D 0.9341 (20/4C), bp 78–183C, flash p 140F (60C) (TOC). Insoluble in water; very soluble in alcohol, benzene, and ether. Combustible.

Hazard: Moderate fire risk.

Use: High-energy fuels, curing agents, plasticizers, resins, surface coatings, pharmaceuticals, dyes.

methylcyclopentadienyl manganese tricarbonyl.

CAS: 12108-13-3.   CH$_3$C$_5$H$_4$Mn(CO)$_3$.

Derivation: Reaction of methylcyclopentadiene with manganese carbonyl.

Hazard: Toxic by ingestion, inhalation, and skin absorption. Central nervous system impairment; lung, liver and kidney damage.

Use: Antiknock agent for unleaded gasoline. Its use has been prohibited by EPA because of its deleterious effect on catalytic converters.

methylcyclopentane.

CAS: 96-37-7.   C$_5$H$_9$CH$_3$.

Properties: Colorless liquid. D 0.750 (20/4C), fp −142.5C, refr index 1.40983 (20C), bp 72C (742 mm Hg), flash p approximately 20F (−6.6C). Immiscible with water.

Grade: Technical (95%), 99%, and research.

Hazard: Flammable, dangerous fire and explosion risk. May be irritant and narcotic.

Use: Organic synthesis, extractive solvent, azeotropic distillation agent.

methylcyclopentenolone.

Use: In soap, cosmetic perfumery, and flavor compounding.

5-methylcytosine.   (5-methyl-2-oxy-4-aminopyrimidine).

CAS: 554-01-8.   C$_5$H$_7$N$_3$O. A pyrimidine found in deoxyribonucleic acids, nucleotides, and nucleosides.

Properties: Crystals in prisms from water. Mp 270C (decomposes).

methyl decanoate.   See methyl caprate.

methyl demeton.   (generic name for O,O-dimethyl-O(and S)-2-ethylthio)ethyl phosphorothioates).

CAS: 8022-00-2.

Properties: Slightly soluble in water. Soluble in most organic solvents.

Hazard: Cholinesterase inhibito. Use may be restricted, absorbed by skin.

Use: Systemic insecticide.

methyl dichloracetate.

CAS: 116-54-1.   Cl$_2$CHCOOCH$_3$.

Properties: Liquid. D 1.3759–1.3839 (20/20C), refr index 1.4374–1.4474 (20C), bp 143C, flash p 176F (80C). Slightly soluble in water; soluble in ether and alcohol. Combustible.

Grade: 99.0% pure.

Hazard: Strong irritant to tissue. Forms corrosive products on hydrolysis, keep dry.

Use: Organic intermediate.

methyldichloroarsine.

CAS: 593-89-5.   CH$_3$AsCl$_2$.

Properties: A colorless, mobile liquid; agreeable odor. Decomposed by water. Bp 136C, mp −59C, d 1.838.

Hazard: Toxic. Highly irritant.

Use: Military poison, intermediate.

methyl-3,4-dichlorocarbanilate.   See swep.

methyl-N-3,4-dichlorophenylcarbamate.
See swep.

methyldichlorosilane.
CAS: 75-54-7. CH_3SiHCl_2.
Properties: Colorless liquid. Bp 41C, d 1.10 (27C), flash p −26F (−32.2C). Soluble in benzene, ether, heptane.
Derivation: Reaction of methyl chloride with silicon and copper.
Hazard: Flammable, dangerous fire risk. Very toxic.
Use: Manufacture of siloxanes.

methyl dichlorostearate.
CAS: 27986-38-5. $C_{17}H_{33}Cl_2COOCH_3$.
Properties: Light-yellow, oily liquid; slight odor. Freezing range +7 to −5C, bp decomposes 250C, d 0.997 (15.5/15.5C), refr index 1.4599 (25C), flash p 358F (181C). Completely miscible with most organic solvents. Combustible.
Use: Intermediate, plasticizer extender.

methyldiethanolamine.
CAS: 105-59-9. $CH_3N(C_2H_4OH)_2$.
Properties: Colorless liquid; amine-like odor. Miscible with benzene and water, d 1.0418 (20C), bp 247.2C, bulk d 8.7 lb/gal, vap press less than 0.01 mm Hg (20C), fp −21.0C, refr index 1.4699, flash p 260F (126.6C) (COC). Combustible.
Grade: Technical.
Use: Intermediate, absorption of acidic gases, catalyst for polyurethane foams, pH control agent.

methyldiethylamine.
CAS: 616-39-7. $CH_3N(C_2H_5)_2$.
Properties: Water-white to straw-colored liquid. Bp 62.5C, d 0.724 (20C). Has inverse water solubility. Combustible.
Use: Desalination of brackish water, chemical intermediate, acid neutralizer.

4-methyl-7-(diethylamino)coumarin. See 7-diethyl-amino-4-methylcoumarin.

3-methyl-2,5-dihydrothiophene-1,1-dioxide.
(3-methylsulfolene).
$CH_3C_4H_5SO_2$ (cyclic).
Properties: Solid. Mp 63C, bp decomposes. Slightly soluble in water, acetone, and toluene. Combustible.
Use: Chemical intermediate, catalyst.

methyl 3,3-dimethoxy-2-oxa-7,10-diaza-3-silatridecan-13-oate.
CAS: 1067-66-9. $C_{12}H_{28}N_2O_5Si$.
Hazard: A moderate skin irritant.

methyl-3-(dimethoxyphosphinyloxy)crotonate.
See mevinphos.

methyldimethoxysilane.
CAS: 16881-77-9. $C_3H_{10}O_2Si$.
Hazard: A poison by ingestion. A moderate eye irritant.

methyl 1,1-dimethylethyl ether. See methyl t-butyl ether.

methyl-N,3,7-dimethyl-7-hydroxyoctyliden anthranilate. See hydroxycitronellal-methyl anthranilate Schiff base.

methyldinitrobenzene. See dinitrotoluene.

Methyl Dioxitol. Diethylene glycol monomethyl ether having an ASTM distillation range 192–196C.

methyldioxolane. (2-methyl-1,3-dioxolane).
CAS: 497-26-7.

$OCH(CH_3)OCH_2CH_2$.
Properties: Water-white liquid. D 0.982 (20/20C), bp 81C. Soluble in water. Combustible.
Hazard: Irritant.
Use: Extractant and solvent for oils, fats, waxes, dyestuffs, and cellulose derivatives.

methyldiphenylamine. (diphenylmethylamine).
CAS: 552-82-9. $(C_6H_5)_2NCH_3$.
Properties: Colorless liquid. D 1.048, fp −7.5C, bp 295C, refr index 1.62. Insoluble in water; soluble in alcohol.
Derivation: Heating diphenylamine with methanol and hydrochloric acid.
Hazard: Toxic by ingestion.
Use: Analytical reagent, dye synthesis.

methyl diphenyl phosphate. (methyl phenyl phosphate).
CAS: 115-89-9. $CH_3OP(O)(OC_6H_5)_2$.
Properties: Clear, oily liquid. D 1.225–1.235 (25C), refr index 1.5370 (20C), pour p −85F (−65C). Nonflammable.
Use: Ignition control compound.

methyldipropylmethane. See 4-methylheptane.

2-methyl-m-dithiane-2-carboxaldehyde o-(methylcarbamoyl)oxime.
CAS: 26419-72-7. $C_8H_{14}N_2O_2S_2$.
Hazard: A poison by ingestion.
Use: Agricultural chemical.

methyl docosanoate. See methyl behenate.

methyl dodecanoate. See methyl laurate.

methyl eicosanoate. See methyl arachidate.

methyl elaidate.
CAS: 1937-62-8. $(CH_3(CH_2)_7CH:CH(CH_2)_7$
$COOCH_3$. The methyl ester of elaidic acid (*trans*-octadec-9-enoic acid).
Properties: Colorless liquid. D 0.8702 (25C), mp less than 15C, bp 213.5C (15 mm Hg), refr index 1.4462 (25C). Insoluble in water; soluble in most organic solvents. Combustible.
Derivation: Prepared from oleic acid by elaidinization and esterification.
Use: Pure grade (99+%) used in biochemical research.

6-methylenandrosta-1,4-diene-3,17-dione.
CAS: 107868-30-4. $C_{20}H_{24}O_2$.
Hazard: A reproductive hazard.

methylene. (1) The divalent hydrocarbon group $^-CH_2^-$, in which the carbon atom has its normal valence of 4; it is derived from methane by dropping two hydrogen atoms. It occurs in many organic compounds, e.g., methylene chloride (dichloromethane), CH_2Cl_2. (2) See carbene.

N,N'-methylenebisacrylamide.
CAS: 110-26-9. $CH_2(NHCOCH:CH_2)_2$.
Properties: Colorless, crystalline powder. Mp 185C.
Hazard: Toxic by ingestion.
Use: Organic intermediate, cross-linking agent.

4,4'-methylenebis(2-chloroaniline). (3,3'-dichloro-4,4'-diaminodiphenylmethane; *p,p'*-methylene-bis-*o*-chloroaniline; MOCA).
CAS: 101-14-4. $CH_2(C_6H_4ClNH_2)_2$.
Properties: Tan-colored pellets. D 1.44, melting range 99–107C. Soluble in hot methyl ethyl ketone, acetone, esters, and aromatic hydrocarbons.
Hazard: Toxic. A confirmed carcinogen, absorbed by skin. Causes bladder cancer and methemoglobinemia.
Use: Curing agent for polyurethanes and epoxy resins.

p,p'-methylenebis(o-chloroaniline). See 4,4'-methylenebis(2-chloroaniline).

2,2'-methylenebis(4-chlorophenol). See dichlorophene.

methylene bis(4-cyclohexylisocyanate).
CAS: 5124-30-1. $C_{15}H_{22}N_2O_2$.
Properties: Liquid. Mw 262.35, sp g 1.07, fp $<-10C$, vp 0.4 torr at 150C. Soluble in acetone; reacts with water or ethanol.
Hazard: Strong skin and eye irritant. Respiratory sensitization and lower respiratory tract irritant.
Use: To produce non-yellowing urethane products.

4,4'-methylenebis(2,6-di-*tert*-butylphenol).
$[(C_4H_9)_2C_6H_2(OH)]_2CH_2$. A sterically hindered bisphenol.
Properties: Light-yellow powder. Mp 155C, bp 217C (1 mm Hg), d 0.99 (20C). Insoluble in water and 1.0 N sodium hydroxide. Combustible.
Use: Oxidation inhibitor and antiwear agent for motor oils, aviation piston engine oils, industrial oils, antioxidant for rubbers, resins, adhesives.

2,2'-methylenebisfuran. See 2-(2-furfuryl) furan.

3,3'-methylenebis(4-hydroxycoumarin). See bishydroxycoumarin.

methylenebis(phenylisocyanate). See diphenylmethane diisocyanate.

methylene bisthiocyanate. See "Organiclear" *[Standard]*.

2,2'-methylenebis(3,4,6-trichlorophenol). See hexachlorophene.

methylene blue. (CI 52015; methylthionine chloride).
CAS: 61-73-4. $C_{16}H_{18}N_3SCl\cdot3H_2O$ (medicinal); $(C_{16}H_{18}N_3SCl)_2\cdot ZnCl_2\cdot H_2O$ (dye).
Properties: Dark green crystals or powder with bronze-like luster; odorless or slight odor. Stable in air. Soluble in water, alcohol, chloroform. Water solutions are deep blue.
Derivation: By oxidation of *p*-aminodimethylaniline with ferric chloride in the presence of hydrogen sulfide. The dye is the zinc chloride double salt of the chloride.
Grade: USP, technical.
Hazard: Toxic by ingestion.
Use: Dyeing cotton and wool, biological and bacteriological stains, reagent in oxidation–reduction titrations in volumetric analysis, indicator.
Note: Do not confuse with methyl blue.

methylene bromide. (dibromomethane).
CAS: 74-95-3. CH_2Br_2.
Properties: Clear, colorless liquid. D 2.47, solidifies $-52C$, bp 97C. Slightly soluble in water; miscible with alcohol, ether, chloroform, and acetone. Nonflammable.
Use: Organic synthesis, solvent.

methylene chloride. (methylene dichloride; dichloromethane).
CAS: 75-09-2. CH_2Cl_2.
Properties: Colorless, volatile liquid; penetrating ether-like odor. D 1.335 (15/4C), fp $-97C$, bp 40.1C, bulk d 11.07 lb/gal (20C), refr index 1.4244 (20C), viscosity 0.430 cP (20C), autoign temp 1224F (662C). Soluble in alcohol and ether; slightly

soluble in water. Nonflammable and nonexplosive in air.

Derivation: Chlorination of methyl chloride and subsequent distillation.

Hazard: Toxic. A narcotic. Central nervous system impairment and carboxyhemoglobinemia. Possible carcinogen.

Use: Paint removers, solvent degreasing, plastics processing, blowing agent in foams, solvent extraction, solvent for cellulose acetate, aerosol propellant.

methylene   chlorobromide.   See   bromochloromethane.

4,4′-methylenedianiline.   See *p,p,′*-diaminodiphenylmethane.

methylene dichloride.   See methylene chloride.

3,4-methylenedioxybenzaldehyde.   See piperonal.

methylenedioxyethamphetamine.   See *n*-ethyl-3,4-methylenedioxyamphetamine.

7,8-methylenedioxyisoquinoline.
CAS: 234-17-3. $C_{10}H_7NO_2$.
Hazard: A poison by ingestion.

3,4-methylenedioxypropylbenzene.   (dihydrosafrole).
CAS: 94-58-6. $C_3H_7C_6H_3O_2CH_2$.
Properties: Colorless liquid. D 1.065 (25/25C). Somewhat soluble in alcohol. Combustible.
Use: Essential oil compositions.

methylene-di-*p*-phenylene isocyanate.   See diphenylmethane-4,4′-diisocyanate.

5,5′-methylenedisalicylic acid.
CAS: 27496-82-8. $CH_2[C_6H_3(OH)COOH]_2$.
Properties: Nonhygroscopic, light-tan, coarse powder. Stable in air; darkens in light. Tends to decarboxylate at very high temperatures. Decomposes at 238C. Soluble in alcohol, ether, and acetone; insoluble in benzene, chloroform. Combustible.
Use: Alkyd resins and modified phenolic compositions for paints and varnishes, intermediate for dyestuffs and printing ink.

methylene glutaronitrile.   (acrylonitrile dimer; 2,4-dicyanobutene-1).
CAS: 1572-52-7. $NCC_2H_4C(:CH_2)CN$.
Properties: Colorless liquid. Bp 103C (5 mm Hg), fp −9.6C, refr index 1.4504 (25C), d 0.9756 (25/4C). Slightly soluble in water; insoluble in aliphatic and alicyclic hydrocarbons, soluble in aromatic hydrocarbons and polar organic solvents.
Derivation: From acrylonitrile by catalytic dimerization.

Use: Vinyl monomer for polymerization and copolymerization, intermediate in making fibers and pharmaceuticals.

methylene iodide.   (diiodomethane).
CAS: 75-11-6. CH_2I_2.
Properties: Yellow liquid. D 3.33, mp 6C, bp 180C (decomposes). Soluble in alcohol and ether; insoluble in water.
Hazard: May be irritating and narcotic.
Use: Separating mixture of minerals, organic synthesis, X-ray contrast media.

5-methylene-2-norbornene.   A copolymer in EPDM elastomers.

methylene succinic acid.   See itaconic acid.

methylenesuccinyloxybis(tributylstannane).   See bis(tributyltin) itaconate.

methylene-*p*-toluidine.   See formaldehyde-*p*-toluidine.

6-*o*-methylerythromycin.
CAS: 81103-11-9. $C_{38}H_{69}NO_{13}$.
Hazard: Moderately toxic by ingestion. Human systemic effects.

methyl esters.   Any of a group of fatty esters derived from coconut and other vegetable oils, tallow, etc.; alkyl groups range from C_8 to C_{18} in varying percentages.
Use: Lubricants for metal-cutting fluids, high-temperature grinding, cold-rolling of steel.

N-methylethanolamine.
CAS: 109-83-1. $CH_3NHC_2H_4OH$.
Properties: Liquid. D 0.9414, bp 159.5C, vap press 0.7 mm Hg (20C), fp −4.5C, flash p 165F (73.9C). Soluble in water. Combustible.
Use: Textile chemicals, pharmaceuticals.

methyl ether.   See dimethyl ether.

4-methyl-7-ethoxycoumarin.

$\overline{C_2H_5OC_6H_3C(CH_3):CHC(O)O}$.

Properties: White solid; walnut odor. Mp 113–114C. Slightly soluble in alcohol.
Use: Perfumery, flavoring.

2-(1-methylethoxy)phenol methylcarbamate.
$(CH_3)_2CHOC_6H_4OOCNHCH_3$.
Properties: White, crystalline powder; odorless. Mp 91C. Soluble in most alcohols; very slightly soluble in water; unstable in highly alkaline media; stable under normal conditions.
Hazard: Toxic by ingestion and inhalation.
Use: Insecticide.

1-methylethyl 3-amino-2-butenoate. See isopropyl 3-aminocrotonate.

methylethylcarbinol. See *sec*-butyl alcohol.

methyl ethyl cellulose. The methyl ether of ethylcellulose in which both methyl and ethyl groups are attached to anhydroglucose units by ether linkages.
Properties: White to pale-cream-colored fibrous solid or powder; practically odorless. Disperses in cold water to form aqueous solutions which undergo a reversible transformation from solution to gel upon heating and cooling, respectively. Combustible.
Grade: Technical, FCC.
Use: Emulsifier, stabilizer, foaming agent. See cellulose, modified.

methyl ethyl diketone. See acetyl propionyl.

2-methyl-2-ethyl-1,3-dioxolane.

$(CH_3)(C_2H_5)COCH_2CH_2O.$

Properties: Colorless liquid. D 0.9392, bp 117.6C, fp −81.96C, flash p 74F (23.3C) (OC). Soluble in water 2.2% by wt.
Hazard: Flammable, dangerous fire risk.
Use: Solvent.

sym-**methylethylethylene.** See 2-pentene.

methylethylglyoxal. See acetyl propionyl.

methylethylhydantoin formaldehyde resin. A reaction product of 5,5-methylethylhydantoin,

$NHCONHCOC(CH_3)C_2H_5$, and formaldehyde.
Properties: Clear, pale, solid resin. Softening point (B&R) 85C, d 1.30, pH (10% aqueous solution) 6.5–7.5.

4,4′-(1-methylethylidene)bis(2,6-dibromophenol).
CAS: 79-94-7. $C_{15}H_{12}Br_4O_2$.
Hazard: Moderately toxic by inhalation and skin contact. An eye irritant.
Use: Reproductive effector.

9-(3′-methyl-4′-ethylidene-thiosemicarbazido)acridine.
CAS: 28846-43-7. $C_{17}H_{16}N_4S$.
Hazard: A poison by ingestion.
Use: Agricultural chemical.

methyl ethyl ketone. (ethyl methyl ketone; 2-butanone; MEK).
CAS: 78-93-3. $CH_3COCH_2CH_3$.
Properties: Colorless liquid; acetone-like odor. Bp 79.6C, d 0.8255 (0/4C), 0.805 (20/4C) and 0.7997 (25/4C), refr index 1.379 (20C), sp heat 0.549 cal/g, fp −86.4C, viscosity 0.40 cP (25C), soluble in water 22.6 wt%, solubility of water 9.9 wt%, flash p (TOC) 24F (−4.4C), bulk d 6.71 lb/gal (20C), autoign temp 960F (515C). Soluble in benzene, alcohol, and ether; miscible with oils.
Derivation: (1) From mixed *n*-butylenes and sulfuric acid to cause hydrolysis followed by distillation to separate *sec*-butyl alcohol which is dehydrogenated; (2) by controlled oxidation of butane; (3) by fermentation.
Grade: Technical.
Use: Solvent in nitrocellulose coatings and vinyl films, "Glyptal" resins, paint removers, cements and adhesives, organic synthesis, manufacture of smokeless powder, cleaning fluids, printing, catalyst carrier, acrylic coatings.
Note: Does not dissolve cellulose acetate and most waxes.
Hazard: Flammable, dangerous fire risk, explosive limits in air 2–10%. Toxic by inhalation. Upper respiratory tract irritant, central nervous system and peripheral nervous system impairment.

methyl ethyl ketone peroxide. (ethyl methyl ketone peroxide).
CAS: 1338-23-4. $C_8H_{16}O_4$.
Properties: Colorless liquid, strong oxidizing agent.
Hazard: Fire risk in contact with organic materials. Strong irritant to skin and tissue. Liver and kidney damage.
Use: Manufacture of acrylic resins, hardening agent for fiberglass-reinforced plastics.

methyl 3-((ethyl(propylamino)phosphino thioyl)oxy)-2-butenoate.
CAS: 37902-85-5. $C_{10}H_{20}NO_3PS$.
Hazard: A poison by ingestion and skin contact.

2-methyl-5-ethylpyridine. (MEP; aldehydine; aldehyde collidine; 5-ethyl-2-picoline).
CAS: 104-90-5. $CH_2C_5H_3NC_2H_5$.
Properties: Colorless liquid; sharp penetrating odor. D 0.921 (20/20), bp 178.3C, fp −70.3C, flash p (COC) 165F (73.9C), refr index 1.4970 (20C). Almost insoluble in water; soluble in alcohol, ether, benzene, concentrated sulfuric acid.
Derivation: Paraldehyde is treated with ammonia under high pressure and in the presence of ammonium acetate as a catalyst. Picolines and other substituted pyridines are by-products.
Grade: Technical.
Hazard: Toxic. Corrosive and strong irritant to tissue.
Use: Nicotinic acid and nicotinamide, vinyl pyridines for copolymers, intermediates for germicides and textile finishes, corrosion inhibitor for chlorinated solvents.

methyl eugenol. (Generic name for 4-allyl-1,2-dimethyoxybenzene; 4-allyl veratrole; 1,2-dimethoxy-4-allylbenzene; eugenyl methyl ether). CAS: 93-15-2. $(CH_3O)_2C_6H_3CH_2CH:CH_2$.
Properties: Colorless to pale-yellow liquid. Bp 91–95C (0.3 mm Hg), d 1.032–1.036 (25C). Insoluble in water; soluble in most organic solvents. Combustible.
Grade: Technical, FCC.
Hazard: Possible carcinogen.
Use: Insect attractant, flavoring.

methyl fluoride. Legal label name (Air) for fluoromethane.

methyl fluorosulfonate.
CAS: 421-20-5. CH_3OSO_2F.
Properties: Colorless liquid; ethereal odor. Bp 92C, d 1.42.
Hazard: Toxic. Strong irritant to tissue, inhalation of fume must be avoided. Reacts with water, steam, and acids, evolving corrosive vapor.
Use: Methylating agent, organic synthesis.

N-methylformanilide.
CAS: 93-61-8. $C_6H_5N(CH_3)CHO$.
Properties: Colorless to light-yellow liquid. Refr index 1.5570–1.5600 (25C), distillation range 127–131C (16 mm Hg).
Grade: 95% min.
Use: Organic synthesis.

methyl formate.
CAS: 107-31-3. $HCOOCH_3$.
Properties: Colorless liquid; agreeable odor. Saponified by water or alkaline solutions. D 0.950–0.980 (20/20C), fp −99.8C, bp 31.8C, flash p −2F (−18.9C), bulk d 8.03 lb/gal (68F), refr index 1.3431 (20C), autoign temp 853F (456C). Soluble in water, alcohol, and ether.
Derivation: By heating methanol with sodium formate and hydrochloric acid with subsequent distillation.
Grade: Technical, refined, FCC.
Hazard: Flammable, dangerous fire and explosion risk, explosive limits in air 5.9–20%. Eye, upper and lower respiratory tract irritant.
Use: Organic synthesis, cellulose acetate solvent, fumigant, larvicides.

methyl fuel. Auxiliary fuel for automotive equipment, electric power production, fuel cells, etc. A mixture of methanol and alcohols of up to four carbon atoms, it is made by catalytic treatment of synthesis gas. It can be blended with gasoline in low percentage. Its combustion products contain less polluting components than No. 5 fuel oil.
See methanol.

2-methylfuran.
CAS: 534-22-5. $C_4H_3OCH_3$.

Properties: Colorless liquid; ether-like odor. Fp −88.68C, bp 63.2–65.6C, d 0.913 (20/4C), refr index 1.4320 (20C), flash p −22F (−30C). Insoluble in water 0.3 g/100 g water. Miscible with most organic solvents. Forms a binary azeotrope with methanol, a ternary azeotrope with methanol-water.
Hazard: Highly flammable, dangerous fire and explosion risk. Irritant.
Use: Chemical intermediate.

N-methylfurfurylamine.
CAS: 4753-75-7. $C_4H_3OCH_2NHCH_3$.
Properties: Colorless to light yellow. Refr index 1.4700–1.4720 (25C), distilling range 144–153C. Combustible.
Use: Intermediate.

methyl-2-furoate. (methyl pyromucate).
CAS: 611-13-2. $C_4H_3OCO_2CH_3$.
Properties: Colorless liquid turning yellow in light; pleasant odor. D 1.1739 (15/15C), bp 181.3C (corrosive), refr index 1.4860 (20C). Insoluble in water; soluble in alcohol and ether. Combustible.
Derivation: By esterification of furoic acid.
Use: Solvent, organic synthesis.

methyl gallate. (methyl-3,4,5-trihydroxybenzoate).
CAS: 99-24-1. $C_6H_2(OH)_3COOCH_3$.
Properties: White, crystalline powder.
Use: Industrial antioxidant.

N-methylglucamine.
CAS: 6284-40-8. $CH_2OH(CH_2O)_4CH_2NHCH_3$.
Properties: White crystals. Mp 128C. Soluble in water; slightly soluble in alcohol; complexes with metals.
Derivation: From glucose and methylamine.
Use: Detergents, pharmaceuticals, dyes.

methyl-α-d-glucopyranoside. See methyl glucoside.

α-methyl glucoside. (methyl-α-d-glycopyranoside).

$$\overline{CH_2OHCH(CH_2O)_3CHOOCH_3}.$$

Properties: Odorless, white crystals. Mp 168C, bp 200C (0.2 mm Hg). Specific optical rotation (aqueous solution) +158.9C (20C), d (30/4C) 1.46. Soluble in water; slightly soluble in 80% alcohol and methanol, insoluble in ether. Combustible.
Derivation: (1) By treating dextrose with methanol in the presence of hydrochloric acid or cation exchange resin, (2) enzymatic synthesis from yeast.
Grade: Technical.

Use: Plasticizer for phenolic, amine, and alkyd resins; nonionic surfactants; tall-oil varnishes; reclaiming drying oils; polyurethane foams.

methyl glycocoll. See sarcosine.

methyl glycol. See propylene glycol.

methyl group. The simplest alkyl group, CH_3, formed by dropping a hydrogen atom from methane (CH_4). It occurs at both ends of paraffinic molecules having two or more carbon atoms in the chain, as well as in many other organic compounds. See alkyl.

N-methyl-N-guanylglycine. See creatine.

methyl heneicosanoate.
CAS: 6064-90-0. $CH_3(CH_2)_{19}COOCH_3$. The methyl ester of heneicosanoic acid.
Properties: White wax-like solid. Mp 48–9C, bp 207C (3.75 mm Hg). Insoluble in water; soluble in alcohol and ether. Combustible.
Grade: Purified 96%, 99.5%.
Use: Intermediate in organic synthesis.

methyl heptadecanoate. (methyl margarate).
CAS: 1731-92-6. $CH_3(CH_2)_{15}COOCH_3$. The methyl ester of heptadecanoic acid (margaric acid).
Properties: White, wax-like solid. Mp 29C, bp 184–187C, 130C (1 mm Hg). Insoluble in water; soluble in alcohol and ether. Combustible.
Grade: Purified 96%, 99.5%.
Use: Intermediate in organic synthesis.

2-methylheptane.
CAS: 592-27-8. $(CH_3)_2CH(CH_2)_4CH_3$.
See isooctane.

3-methylheptane.
CAS: 589-81-1. $C_2H_5CH(CH_3)(CH_2)_3CH_3$.
Properties: Colorless liquid. Fp −120.5C, bp 118.9C, d 0.70582 (20/4C), refr index 1.39849 (20C).
Grade: 99%, 95%.
Hazard: Flammable, dangerous fire risk.
Use: Calibration, organic synthesis.

4-methylheptane. (methyldipropylmethane).
CAS: 589-53-7. C_8H_{18} or $CH_3(CH_2)_2CHCH_3 (CH_2)_2CH_3$.
Properties: Colorless liquid. D 0.7161, bp 122.2C. Soluble in alcohol and ether; insoluble in water.
Grade: Technical.
Hazard: Flammable, dangerous fire risk.
Use: Organic synthesis.

methylheptenone. (6-methyl-5-heptene-2-one).
CAS: 110-93-0. $(CH_3)_2C:CH(CH_2)_2COCH_3$.
Constituent of lemongrass oil and many other essential oils.

Properties: Colorless liquid. D 0.860 (20C), fp −67.1C, bp 173–174C. Insoluble in water; miscible with alcohol or ether. Combustible.
Derivation: From oil of lemon grass or by controlled oxidation of corresponding secondary alcohol.
Use: Organic synthesis, inexpensive perfumes, flavoring.

methyl heptyne carbonate. (methyl-2-octynoate).
CAS: 111-12-6. $CH_3(CH_2)_4C\equiv CCOOCH_3$.
Properties: Colorless liquid; strong violet-type odor. D 0.919–0.923, refr index 1.446–1.450 (20C). Soluble in most fixed oils and mineral oil; soluble in five parts of 70% alcohol. Combustible.
Derivation: From heptaldehyde.
Grade: Technical, FCC.
Use: Perfumery, flavoring.

methyl hexacosanoate. See methyl cerotate.

methyl hexadecanoate. See methyl palmitate.

2-methylhexane. (ethylisobutylmethane; isoheptane).
CAS: 591-76-4. $(CH_3)_2CH(CH_2)_2CH_2CH_3$.
Properties: Colorless liquid. D 0.6789, bp 90.0C, fp −118.5C, refr index 1.38498 (20C), flash p approximately 0F (−17.7C). Soluble in alcohol; insoluble in water.
Grade: Technical.
Hazard: Flammable, dangerous fire risk, explosive limits in air 1–6%.
Use: Organic synthesis.

3-methylhexane.
CAS: 589-34-4. $H_3CCH_2CH(CH_3)CH_2CH_2CH_3$.
Properties: Colorless liquid. Bp 92C, d 0.692 (15.5/15.5C), refr index 1.388 (20C), flash p 25F (−3.9C).
Grade: Technical 95%.
Hazard: Flammable, dangerous fire risk.
Use: Organic synthesis, oil extender solvent.

methyl hexanoate. See methyl caproate.

5-methyl-2-hexanone. See methyl isoamyl ketone.

methyl hexyl ketone. (2-octanone).
CAS: 111-13-7. $CH_3COC_6H_{13}$.
Properties: Colorless liquid; pleasant odor; camphor taste. D 0.82 (20/4C), mp 20.9C, bp 173.5C, distillation range 166–173C, flash p 160F (71.1C), refr index 1.416 (20C). Insoluble in water; soluble in alcohol, hydrocarbons, ether, esters, etc. Combustible.
Derivation: By distilling sodium ricinoleate with caustic soda.

Use: Perfumes, high-boiling solvent, especially for epoxy resin coatings, leather finishes, flavoring, odorant, antiblushing agent for nitrocellulose lacquers.

methyl histidine. (3-methyl-L-histidine histidine). A variant of 3-methyl-histidine and an amino acid constituent of actin that contains three atoms of deuterium in the methyl group. It is released into the urine during the degradative metabolism of actin and myosin. Its output in urine is a reliable indicator of the breakdown of myofibrillar protein in muscle.
Use: As a tracer to follow the pathway of muscle protein degradation in the body.

methyl hydrazine. (monomethylhydrazine; MMH).
CAS: 60-34-4. CH_3NHNH_2.
Properties: Colorless, hygroscopic liquid; ammonia-like odor. D 0.874 (25C), fp −52.4C, bp 87.5C, flash p approximately 80F (26.6C). Soluble in water, hydrazine, hydrocarbons, and monohydric alcohols.
Hazard: Flammable, dangerous fire risk, vapors may explode, may self-ignite in air and on contact with oxidizing agents. Toxic by ingestion and inhalation. Eye and upper respiratory tract irritant, lung cancer and liver damage. Possible carcinogen.
Use: Missile propellant, intermediate, solvent.

methyl hydride. See methane.

methylhydrogen sulfate. See methylsulfuric acid.

methyl-*p*-hydroxybenzoate. See methyl paraben.

methylhydroxybutanone. (3-methyl-3-hydroxybutan-2-one). $(CH_3)_2COHCOCH_3$.
Properties: Clear, colorless liquid; sweet camphor-like odor. Bp 140.3C, fp −86.5C, d 0.9553 (20/20C), refr index 1.4153 (20C). Miscible with water, acetone, benzene, mineral spirits. Combustible.
Use: Specialty solvent, chemical intermediate, flavor formulations.

4-methyl-7-hydroxycoumarin diethoxythiophsphate. See potasan.

methyl-3-hydroxy-α-crotonate dimethyl phosphate. See mevinphos.

methylhydroxyisopropylcyclohexane. See menthol.

methyl 4-(2-hydroxy-3-((1-methylethyl) amino)propoxy)benzenepropanoate.
CAS: 81147-92-4. $C_{16}H_{25}NO_4$.
Hazard: A reproductive hazard.
Use: Agricultural chemical.

4-methyl-4-hydroxy-1-octyne.
CAS: 22128-43-4. $C_9H_{16}O$.
Hazard: Moderately toxic by ingestion. A moderate eye irritant.

methyl-12-hydroxystearate.
CAS: 141-23-1. $C_{17}H_{34}OHCOOCH_3$.
Properties: White, waxy solid in the form of short flat rods. Mp 48C, acid value 4, saponification value 177, iodine value 5. Insoluble in water; limited solubility in organic solvents. Combustible.
Use: Adhesives, inks, cosmetics greases.

2-methylimidazole. (2MZ).
CAS: 693-98-1.

CHCHNC(CH₃)NH.

Properties: Solid. Mp 142–143C.
Hazard: Possible carcinogen.
Use: Dyeing auxiliary for acrylic fibers, plastic foams.

3-methylindole. See skatole.

methyl iodide. (iodomethane).
CAS: 74-88-4. CH_3I.
Properties: Colorless liquid, turns brown on exposure to light. D 2.24–2.27 (25/25C), fp −66.1C, bp 42C, refr index 1.526–1.527 (25C). Soluble in alcohol and ether; partially soluble in water. Nonflammable.
Derivation: Interaction of methanol, sodium iodide, and sulfuric acid with subsequent distillation.
Hazard: Toxic by ingestion, inhalation, and skin absorption; narcotic, irritant to skin. Eye damage and central nervous system impairment. Questionable carcinogen.
Use: Organic synthesis, microscopy, testing for pyridine.

methylionone. (irone).
CAS: 1335-46-2. $C_{14}H_{22}O$.
Properties: Colorless to amber-yellow liquid; floral odor.
Grade: Several isomers are available as α, β, γ, Δ, and mixtures. The constants are approximately: d 0.926–0.939, refr index 1.501–1.504, bp 144C (16 mm Hg). Soluble in alcohol, insoluble in water.
Derivation: Oil of orris.
Use: Perfumery, flavoring.
See α-isomethylionone.

methyl isoamyl ketone. (5-methyl-2-hexanone; MIAK).
CAS: 110-12-3. $CH_3COC_2H_4CH(CH_3)_2$.
Properties: Colorless, stable liquid; pleasant odor. D 0.8132 (20/20C), refr index 1.4062 (20C), bp 144C, fp −73.9C, bulk d 6.77 lb/gal, flash p (OC) 110F (43.3C). Slightly soluble in water; miscible with most organic solvents. Combustible.
Grade: 97.5%.

Hazard: Moderate fire risk. Central nervous system impairment and upper respiratory tract irritant.
Use: Solvent for nitrocellulose, cellulose acetate butyrate, acrylics, and vinyl copolymers.

methyl isobutenyl ketone.   See mesityl oxide.

methylisobutyl carbinol.   See methylamyl alcohol.

methylisobutyl carbinol acetate.   See methylamyl acetate.

methyl isobutyl ketone.   (hexone; 4-methyl-2-pentanone; isopropylacetone).
CAS: 108-10-1. $(CH_3)_2CHCH_2COCH_3$.
Properties: Colorless, stable liquid; pleasant odor. D 0.8042 (20/20C), bp 115.8C, fp −85C, bulk d 6.68 lb/gal (20C), vap press 15.7 mm Hg (20C), refr index 1.3959 (20C), flash p 73F (22.7C), autoign temp 860F (460C). Slightly soluble in water; miscible with most organic solvents.
Derivation: Mild hydrogenation of mesityl oxide.
Grade: Technical, 98.5%.
Hazard: Flammable, dangerous fire risk, explosive limits in air 1.4–7.5%. Avoid ingestion and inhalation. Upper respiratory tract irritant, dizziness, and headache. Possible carcinogen.
Use: Solvent for paints, varnishes, nitrocellulose, lacquers; manufacture of methyl amyl alcohol; extraction processes including extraction of uranium from fission products; organic synthesis; denaturant for alcohol.

methyl isocyanate.
CAS: 624-83-9. CH_3NCO.
Properties: Colorless liquid. D 0.9599 (20/20C), bp 39.1C, reacts with water, flash p less than 20F (−6.6C).
Hazard: Flammable, dangerous fire risk. Toxic by skin absorption and a strong eye and upper respiratory tract irritant.
Use: Intermediate.

methylisoeugenol.   (propenyl guaiacol).
CAS: 93-16-3. $CH_3CH:CHC_6H_3(OCH_3)_2$.
Properties: Colorless to light-yellowish liquid; spicy odor. D 1.050–1.053, bp 262–264C, refr index 1.566–1.569, soluble in two parts of 70% alcohol. Almost insoluble in mineral oil; insoluble in glycerol. Combustible.
Grade: Technical, FCC.
Use: Perfumery, flavoring agent.

methyl isonicotinate. $C_5NH_4COOCH_3$.
Properties: Clear amber to red liquid; mild odor. D 1.15 (20/20C).
Use: Intermediate for synthesis of isonicotinic acid hydrazide.

1-methyl-4-isopropenylcyclohexan-3-ol.
See isopulegol.

methyl isopropenyl ketone.
CAS: 814-78-8. $CH_3COC(CH_3):CH_2$.
Properties: Colorless liquid; pleasant odor; sweet taste. D 0.854 (20C). Polymerizes readily.
Hazard: Flammable, dangerous fire risk, explosive limits in air 1.8–9.0%.
Use: Plastics.

5-methyl-2-isopropyl-2-hexenal. $C_{10}H_{18}O$.
Properties: Sltly yellow liquid; herbaceous, woody, fruity, chocolate odor. D: 0.845–0.860, refr index: 1.448. Sol in alc, fixed oils; insol in water, propylene glycol.
Use: Food additive.

methyl isopropyl ketone.   (3-methyl-2-butanone).
CAS: 563-80-4. $CH_3COCH(CH_3)_2$.
Properties: Colorless liquid. Bp 93C, fp −92C, refr index 1.38788 (16C), d 0.815 (15/4C). Very slightly soluble in water; soluble in alcohol and ether.
Derivation: Synthetic, also by fermentation.
Hazard: Toxic material. Embryo/fetal damage and neonatal toxicity.
Use: Solvent for nitrocellulose lacquers.
See ketones.

2-methyl-5-isopropylphenol.   See carvacrol.

5-methyl-2-isopropylphenol.   See thymol.

methyl-*p*-isopropylphenyl propyl aldehyde.
See cyclamen aldehyde.

methyl isothiocyanate.   (methyl mustard oil).
CAS: 556-61-6. $CH_3N:CS$.
Properties: Colorless crystals. Mp 35C, bp 120C. Soluble in alcohol and ether; partially soluble in water.
Derivation: Reaction of methylamine and carbon disulfide.
Hazard: Toxic by ingestion, strong irritant to eyes and skin.
Use: Insecticide, a possible military poison.

methyl lactate.
CAS: 547-64-8. $CH_3CH_2OCOOCH_3$.
Properties: Colorless liquid. Decomposed by water, bp 144.8C, fp approximately −66C, refr index 1.4156 (20C), flash p (CC) 125F (51.6C), autoign temp 725F (385C), bulk d 9 lb/gal (68F). Soluble in alcohol and ether. Combustible.
Hazard: Moderate fire risk.
Use: Solvent for cellulose acetate, nitrocellulose, cellulose acetobutyrate, cellulose acetopropionate, lacquers, stains.

methyllactonitrile.   See acetone cyanohydrin.

methyl laurate.   (methyl dodecanoate).
CAS: 29972-79-0. $CH_3(CH_2)_{10}COOCH_3$. The methyl ester of lauric acid.

Properties: Water-white liquid. D 0.8702 (20/4C), mp 4.8C, bp 262C (766 mm Hg), 160C (30 mm Hg), refr index 1.4301 (25C). Insoluble in water. Noncorrosive. Combustible.
Derivation: From coconut oil.
Method of purification: Vacuum fractional distillation.
Grade: 69%, 74%, 90%, 96%, 99.8%.
Use: Intermediate for detergents, emulsifiers, wetting agents, stabilizers, lubricants, plasticizers, textiles, flavoring.

methyl lauroleate.
$CH_3CH_2CH:CH(CH_2)_7COOCH_3$. The methyl ester of lauroleic acid.
Properties: Colorless liquid. Insoluble in water; soluble in common organic solvents. Combustible.
Grade: Purified product, 99.5%.
Use: Organic synthesis, reference standard for gas chromatography, biochemical research.

methyl lignocerate.   (methyl tetracosanoate).
CAS: 2442-49-1. $CH_3(CH_2)_{22}COOCH_3$.   The methyl ester of lignoceric acid.
Properties: Wax-like solid. Mp 57.8C, bp 232C (3.75 mm Hg), refr index 1.4283 (80C). Insoluble in water; soluble in alcohol and ether. Combustible.
Derivation: Esterification of lignoceric acid with methanol followed by vacuum distillation.
Grade: Purified (99.8%+).
Use: Intermediate in special synthesis, biochemical research, reference standard in gas chromatography.

methyl linoleate.
CAS: 68605-14-1. $C_{19}H_{32}O_2$. The methyl ester of linoleic acid (*cis,cis,cis*-octadec-9,12,15-dienoic acid).
Properties: Colorless oil. D 0.8886 (18/4C), fp −35C, bp 212C (16 mm Hg), refr index 1.4593 (25C). Insoluble in water; soluble in alcohol and ether. Combustible.
Derivation: Urea fractionation and vacuum distillation of methyl esters of safflower oil.
Grade: Technical, purified (99+%).
Use: Intermediate for detergents, emulsifiers, wetting agents, stabilizers, resins, lubricants, plasticizers, textiles, reference standard in gas chromatography, biochemical research.

methyllithium.
CAS: 917-54-4. CH_3Li. Commercially available as a 5% solution in ether.
Hazard: Flammable, dangerous fire and explosion risk, self-ignites in air.
Use: In Grignard-type reactions.

n-methylloline.
CAS: 22143-50-6. $C_9H_{16}N_2O$.
Hazard: A poison.
Source: Natural product.

methylmagnesium bromide.
CAS: 75-16-1. CH_3MgBr. Available in solution in ether.
Derivation: Reaction of magnesium and methylbromide.
Hazard: Flammable, dangerous fire and explosion risk.
Use: Alkylating agent in organic synthesis, Grignard reagent.

methylmagnesium chloride.
CAS: 676-58-4. CH_3MgCl. Available as a solution in tetrahydrofuran.
Hazard: Flammable, dangerous fire and explosion risk.
Use: Alkylating agent in organic synthesis, Grignard reagent.

methylmagnesium iodide.
CAS: 917-64-6. CH_3MgI. Available as solution in ether.
Derivation: Reaction of magnesium and methyl iodide.
Hazard: Flammable, dangerous fire and explosion risk.
Use: Alkylating agent in organic synthesis, Grignard reagent.

methylmaleic anhydride.   See citraconic anhydride.

methyl margarate.   See methyl heptadecanoate.

methyl mercaptan.   Legal label name for methanethiol.

n-methyl 3-mercaptopropionamide.
CAS: 52334-99-3. C_4H_9NOS.
Hazard: Moderately toxic by ingestion and skin contact. A mild skin and severe eye irritant.

methylmercury.   (methyl-mercury(1+); methylmercury(I) cation; methylmercury ion; methylmercury ion(1+))
CAS: 22967-92-6. CH_3Hg. Either of two compounds that contain the methyl group and common salts of monomethylmercury.
Properties: Water soluble, lipid soluble, high vapor pressure.
Hazard: Mutagen; possible carcinogen; teratogen; central nervous system poisons that easily pass through the blood–brain barrier.

methylmercury acetate. $CH_3HgOOCCH_3$.
Hazard: Toxic by ingestion, absorbed by skin.
Use: Seed disinfectant.

methylmercury   cyanide.   (methylmercury nitrile).
CAS: 2597-97-9. CH_3HgCN.
Properties: Crystals. Mp 95C. Soluble in water and organic solvents.

Hazard: Toxic by ingestion, absorbed by skin.
Use: Fungicide, seed disinfectant.

methylmercury cysteine. See mercurate(1-), (l-cysteinato(2-)-s)methyl-, hydrogen.

methylmercury dicyandiamide. See cyano(methylmercury)guanidine.

methylmercury-2,3-dihydroxy-propylmercaptide.
$CH_3HgSCH_2CH_2OCH_2OH$.
Hazard: Toxic by ingestion, absorbed by skin.
Use: Seed disinfectant.

methylmercury-8-hydroxyquinolate. See methylmercury quinolinolate.

methylmercury quinolinolate.
(methylmercury-8-hydroxyquinolate; methylmercury oxyquinolinate).
CAS: 86-85-1. $C_9H_6NOHgCH_3$.
Properties: Yellow crystals. Mp 133–137C.
Hazard: Toxic material.
Use: Seed disinfectant.

methylmercury toluenesulfonate.
CAS: 63869-06-7.
Hazard: A poison.

methyl methacrylate.
CAS: 80-62-6. $CH_2:C(CH_3)COOCH_3$. Acrylic resin monomer.
Properties: Colorless, volatile liquid. Bp 101C, fp −48.2C, d 0.940 (25/25C), flash p (OC) 50F (10C), autoign temp 790F (421C). Slightly soluble in water; soluble in most organic solvents. Readily polymerized by light, heat, ionizing radiation and catalysts. Can be copolymerized with other methacrylate esters and many other monomers.
Derivation: (1) Acetone cyanohydrin, methanol, and dilute sulfuric acid; (2) oxidation of *tert*-butyl alcohol to methacrolein and then to methacrylic acid, followed by reaction with methanol.
Grade: Technical (inhibited).
Hazard: Flammable, dangerous fire risk, explosive limits in air 2.1–12.5%. Eye and upper respiratory tract irritant, body weight effects, and pulmonary edema. Questionable carcinogen.
Use: Monomer for polymethacrylate resins, impregnation of concrete.
See acrylic resin.

methylmethane. See ethane.

methyl (E)-2-methoxyimino-(2-(o-tolyloxymethyl)phenyl)acetate.
CAS: 143390-89-0. $C_{18}H_{19}NO_4$.
Hazard: Moderately toxic by skin contact. Low toxicity by ingestion and inhalation.
Use: Agricultural chemical.

N-methyl methyl anthranilate. See dimethyl anthranilate.

7-methyl-3-methylene-1,6-octadiene. See myrcene.

2-(o-(n-methyl-n-(n'-methyl-n'-ethoxycarbonylaminosulfenyl)carbamoyl) oximino)-1,4-dithiane.
CAS: 64029-07-8. $C_{10}H_{17}N_3O_4S_3$.
Hazard: A poison by ingestion.
Use: Agricultural chemical.

4-methyl-α-(1-methylethyl)benzeneacetic acid, (5-(2-furanylmethyl)-2-thienyl) methyl ester.
CAS: 51629-74-4. $C_{22}H_{24}O_3S$.
Hazard: Moderately toxic by ingestion.
Use: Agricultural chemical.

4-methyl-α-(1-methylethyl)benzeneacetic acid, (4,5,6,7-tetrahydrobenzo(b)thien-2-yl) methyl ester.
CAS: 51629-79-9. $C_{21}H_{26}O_2S$.
Hazard: Moderately toxic by ingestion.
Use: Agricultural chemical.

methyl (4-(1-methylethyl)phenyl)methyl 3-pyridinylcarbonimidodithioate.
CAS: 51308-72-6. $C_{17}H_{20}N_2S_2$.
Hazard: Moderately toxic by ingestion.
Use: Agricultural chemical.

methyl((methyl(((((5-methyl-1,3-oxathiolan-4-ylidene)amino)oxy)carbonyl)amino)thio) carbamic acid, ethyl ester.
CAS: 64029-08-9. $C_{10}H_{17}N_3O_5S_2$.
Hazard: A poison by ingestion.
Use: Agricultural chemical.

2-methyl-n-(n'-methyl-n'-(4-morpholino-sulfenyl)carbamoyloxy)thioacetimidate.
CAS: 62382-21-2. $C_9H_{17}N_3O_3S_2$.
Hazard: A poison by ingestion.
Use: Agricultural chemical.

α-(3-methyl-1-(2-methylpropyl)butyl)-ω-hydroxypoly(oxy-1,2-ethanediyl).
CAS: 132299-20-8. $(C_2H_4O)_nC_9H_{20}O$.
Hazard: Moderately toxic by ingestion. Low toxicity by skin contact. A mild skin and moderate eye irritant.

2-methyl-2-(methylthio)propanol-o-((n-methyl-n-morpholinosulfenyl) carbamoyl)oxime.
CAS: 62382-23-4. $C_{11}H_{21}N_3O_3S_2$.
Hazard: A poison by ingestion.
Use: Agricultural chemical.

2-methyl-2-(methylthio)propionaldehyde-*o*-((methyl)(decoxysulfinyl)carbamoyl)oxime.
CAS: 77248-45-4. $C_{17}H_{34}N_2O_4S_2$.
Hazard: A poison by ingestion.
Use: Agricultural chemical.

methylmorphine.   See codeine.

***N*-methyl morpholine.**
CAS: 109-02-4.

$\overline{CH_2CH_2OCH_2CH_2NCH_3}$.

Properties: Water-white liquid; ammonia odor. D
0.921 (20/20C, bp 115.4C, fp −66C, flash p 75F
(23.9C) (TOC). Forms constant-boiling mixture
with 25% water and boiling at 97C. Miscible with
benzene, water.
Grade: Technical.
Hazard: Flammable, dangerous fire risk. Skin irritant.
Use: Catalyst in polyurethane foams, extraction solvent, stabilizing agent for chlorinated hydrocarbons, self-polishing waxes, oil emulsions, corrosion
inhibitors, pharmaceuticals.

methyl myristate.   (methyl tetradecanoate).
CAS: 124-10-7. $CH_3(CH_2)_{12}COOCH_3$.   The
methyl ester of myristic acid.
Properties: Colorless liquid. Mp 17.8C, bp 186.8C
(30 mm Hg), 157.5C (1 mm Hg), refr index 1.4351
(25C). Insoluble in water. Combustible.
Derivation: (1) Esterification of myristic acid with
methanol, (2) alcoholysis of coconut oil with
methanol.
Method of purification: Vacuum fractional distillation.
Grade: Technical (93%), purified (99.8+%).
Use: Intermediate for myristic acid detergents, emulsifiers, wetting agents, stabilizers, resins, lubricants,
plasticizers, textiles, animal feeds, standard for gas
chromatography, flavoring.

methyl myristoleate.
CAS: 56219-06-8. $CH_3(CH_2)_3CH:CH(CH_2)_7$
$COOCH_3$. The methyl ester of myristoleic acid (*cis*-
tetradec-9-enoic acid).
Properties: Colorless liquid. Bp 108.9C (1 mm Hg).
Insoluble in water; soluble in alcohol and ether.
Combustible.
Use: Purified product used in medical research and
organic synthesis.

α-methylnaphthalene.
CAS: 90-12-0. $C_{10}H_7CH_3$.
Properties: Colorless liquid. D 1.025, fp −32C, bp
240–243C, refr index 1.6140 (25C), autoign temp
984F (529C). Insoluble in water; soluble in alcohol
and ether. Combustible.
Derivation: From coal tar.
Hazard: Moderate fire risk. Lower respiratory tract
irritant and lung damage. Questionable carcinogen.
Use: Organic synthesis.

β-methylnaphthalene.
CAS: 91-57-6. $C_{10}H_7CH_3$.
Properties: Solid. D 0.994 (40/4C), bp 241–242C,
mp 34C, refr index 1.6015 (25C). Insoluble in
water; soluble in alcohol and ether. Combustible.
Derivation: From coal tar.
Grade: Technical, 95% min.
Hazard: Lower respiratory tract irritant and lung
damage. Questionable carcinogen.
Use: Organic synthesis, insecticides.

2-methyl-1,4-naphthoquinone.   See menadione.

**methyl naphthyl dodecyl dimethylammonium
chloride.** $C_{10}H_7HC_2C_{12}H_{25}(CH_3)_2NCl$. Quaternary ammonium salt.
Properties: White to slightly yellow, crystalline
powder; mild odor and taste. Mp 159–160C, bulk
d 5 lb/gal. Soluble in cold water, lower alcohols,
glycerin, and acetone; pH of 5% solution 8.9.
Grade: Min purity 97%.
Use: Germicide.

methyl naphthyl ether.   See β-naphthyl methyl
ether.

methyl nitrate.
CAS: 598-58-3. CH_3NO_3.
Properties: Colorless liquid. Bp 66C (explodes), d
1.217 (15C). Slightly soluble in water; soluble in
alcohol and ether.
Derivation: By reaction of nitric acid and methanol
in the presence of urea.
Hazard: Explodes when heated, severe hazard when
exposed to heat or shock. Narcotic, strong irritant
to tissue.
Use: Rocket propellant.

methyl nitrite.
CAS: 624-91-9. CH_3NO_2.
Properties: A gas. Bp −12C, fp −17C, d 0.991
(15C).
Hazard: Severe explosion risk when shocked or
heated. Toxic by inhalation, narcotic.
Use: Synthesis of nitrile and nitroso esters.

methylnitrobenzene.   See nitrotoluene.

4-methyl-2-nitrophenol.   See 2-nitro-*p*-cresol.

***n*-methyl-4-nitrophthalimide.**   See 4-nitro-*n*-
methylphthalimide.

***n*-methyl-*n*-nitroso-4-aminopyridine.**   See
n-nitroso-*n*-methyl-4-aminopyridine.

methyl-4-nitrosoperfluorobutyrate.
A monomer for a nitroso ester terpolymer, prepared by reacting methyl nitrite with perfluorosuc-
cinic and perfluoroglutaric anhydrides. The reaction
is exothermic and gives high conversion to liquid

nitrile esters, which yield nitroso esters on pyrolysis or photolysis.

N-methyl-N-nitroso-p-toluenesulfonamide.
$CH_3C_6H_4SO_2N(NO)CH_3$.
Properties: Fine, yellow crystals. Mp 56–59C. Soluble in ether and most organic solvents; insoluble in water.
Use: Reagent for the preparation of diazomethane.

methyl nonadecanoate.
CAS: 1731-94-8. $CH_3(CH_2)_{17}COOCH_3$. The methyl ester of nonadecanoic acid.
Properties: White, waxy solid. Mp 39.5C, bp 190.5C (3.75 mm Hg). Insoluble in water; soluble in alcohol and ether. Combustible.
Grade: Purified 96%, 99.5%.
Use: Intermediate in organic synthesis, medical research.

2-methylnonane. (isodecane).
CAS: 871-83-0. $(CH_3)_2CH(CH_2)_6CH_3$.
Properties: Colorless liquid. D 0.728, fp −74.7C, bp 167C. Combustible.

methyl nonanoate. (methyl pelargonate).
CAS: 1731-84-6. $CH_3(CH_2)_7COOCH_3$. The methyl ester of pelargonic acid.
Properties: Colorless liquid; fruity odor. Fp −35C, bp 213.5C, d 0.877 (18C), refr index 1.4302 (25C). Insoluble in water; soluble in alcohol and ether. Combustible.
Derivation: Esterification of nonanoic (pelargonic) acid with methanol followed by fractional distillation.
Grade: Purified (96+%).
Use: Perfumes, flavors, reference standard for gas chromatography, intermediate in organic synthesis, medical research.

methyl-2-nonenoate.
CAS: 111-79-5. $CH_3(CH_2)_5HC:CHCOOCH_3$.
Properties: Colorless to slightly yellow liquid; strong violet-leaf odor. D 0.893–0.898 (25/25C), refr index 1.4400–1.4440. Stable; soluble in alcohol. Combustible.
Use: Perfumes.

methylnonylacetaldehyde. (aldehyde C-12; MNA; methylundecanal).
CAS: 110-41-8. $CH_3(CH_2)_8CH(CH_3)CHO$.
Properties: Colorless liquid; fruity odor. D 0.824–0.828, refr index 1.432–1.435 (20C). Soluble in three volumes of 80% alcohol, in fixed oils, mineral oil. Combustible.
Grade: Technical, FCC (as methylundecanal).
Use: Perfumery, flavoring.

methyl nonyl ketone. (2-undecanone).
CAS: 112-12-9. $CH_3COC_9H_{19}$.
Properties: Oily liquid; strong odor. D 0.822–0.826, bp 225C, refr index 1.429–1.433, flash p 192F

(88.9C). Soluble in two parts of 70% alcohol. Combustible.
Derivation: Oil of rue, also made synthetically.
Use: Perfumery, flavoring, animal repellent.

methyl norbornene dicarboxylic anhydride.
Legal label name for memtetrahydrophthalic anhydride.

methyl octadecanoate. See methyl stearate.

4-methyloctin-4-ol. See 4-methyl-4-hydroxy-1-octyne.

n-methyl-n-octyl-1-octanamine.
CAS: 4455-26-9. $C_{17}H_{37}N$.
Hazard: A severe skin irritant.

methyl-2-octynoate. See methyl heptyne carbonate.

methylolacrylamide.
CAS: 90456-67-0. Available in 60% aqueous solution.
Hazard: Questionable carcinogen.
Use: Intermediate for copolymerization of vinyl acetate and acrylic acid, polymers for coatings, varnishes, adhesives, crease-proof and wrinkle-resistant fabrics, permanent-press textiles by irradiation bonding.

methylol dimethylhydantoin. (dimethylhydantoin formaldehyde; DMHF).

$(CH_3)_2CN(CH_2OH)CONHCO$.

Properties: White, odorless, crystalline solid. Mp 110–117C. Soluble in water, methanol, acetone; insoluble in hydrocarbons.
Use: Textile and paper finishing, preservative for cosmetics; source of formaldehyde.
See dimethylhydantoinformaldehyde resin.

methyl oleate.
CAS: 112-62-9. $CH_3(CH_2)_7COOCH_3$. The methyl ester of oleic acid (*cis*-octadec-9-enoic acid).
Properties: Colorless to amber clear liquid; faint fatty odor. D 0.8739 (20C), fp −19.9C, bp 218.5C (20 mm Hg), refr index 1.4521 (20C). Soluble in alcohols and most organic solvents; insoluble in water. Combustible.
Derivation: Esterification of oleic acid, vacuum fractional distillation, solvent crystallization.
Grade: Technical, purified 99+%.
Use: Intermediate for detergents, emulsifiers, wetting agents, stabilizers, textile treatment, plasticizers for duplicating inks, rubbers, waxes, etc.; biochemical research, chromatographic reference standard.

methylol formaldehyde. See glycolic aldehyde.

methylol imidazolidone. $C_3H_4N_2(O)CH_2OH$.
Properties: Straw-colored to water-white, clear liquid; mild odor. Bulk d 10 lb/gal (15.5C).
Use: Wash-and-wear fabrics.

methylol riboflavin. A mixture of methylol derivatives of riboflavin exhibiting the same activity.
Properties: Orange to yellow, hygroscopic powder. Nearly odorless. Soluble in water; nearly insoluble in alcohol, benzene, chloroform and ether; dextrorotatory. Dry powder is unstable and on standing loses biological activity by liberation of formaldehyde.
Derivation: Action of formaldehyde on riboflavin in weakly alkaline solutions.
Use: Nutrition, vitamin source.

methylolurea.
CAS: 1000-82-4. $H_2NCONHCH_2OH$.
Properties: Colorless crystals. Mp 11C. Soluble in water and methanol; insoluble in ether; capable of polymerization.
Derivation: Combination of urea and formaldehyde, in the presence of salts or alkaline catalysts.
Use: Urea-formaldehyde resins, molding adhesives, treating textiles and wood.
See A-stage resin; urea-formaldehyde resin.

methyl orange. (*p*-(*p*-dimethylamino phenylazo)-benzene sulfonate of sodium; Helianthine B; orange III; gold orange; tropaeolin D; CI 13025).
CAS: 547-58-0. $(CH_3)_2NC_6H_4NNC_6H_4SO_3Na$.
Properties: Orange-yellow powder. Soluble in hot water; insoluble in alcohol.
Use: Acid–base indicator; red in acid, yellow-orange in alkali, pH range 3.1–4.4.
See indicator.

methyl orthophosphoric acid. See methylphosphoric acid.

***n*-methyloxazolidine.** See 3-methyl-1,3-oxazolidine.

3-methyl-1,3-oxazolidine.
CAS: 27970-32-7. C_4H_9NO.
Hazard: A poison by ingestion and skin contact. A severe eye irritant.

methyl oxide. See dimethyl ether.

methyl Oxitol. Ethylene glycol monomethyl ether whose ASTM distillation range is 123.5–125.5C.

methyl palmitate. (methyl hexadecanoate).
CAS: 112-39-0. $CH_3(CH_2)_{14}COOCH_3$. The methyl ester of palmitic acid.

Properties: Colorless liquid. Mp 29.5C, bp 211.5C (30 mm Hg), 180.5C (10 mm Hg), refr index 1.4310 (45C). Insoluble in water; soluble in alcohol and ether. Combustible.
Derivation: Esterification of palmitic acid with methanol or alcoholysis of palm oil plus vacuum distillation.
Grade: 80%, pure (99.8%).
Use: Intermediate for detergents, emulsifiers, wetting agents, stabilizers, resins, lubricants, plasticizers, animal feeds, medical research.

methylparaben. (methyl-*p*-hydroxybenzoate).
CAS: 99-76-3. $CH_3OOCC_6H_4OH$.
Properties: Colorless crystals or white, crystalline powder; odorless or faint characteristic odor; slight burning taste. Mp 125–128C. Soluble in alcohol, ether; slightly soluble in water, benzene, and carbon tetrachloride.
Grade: USP, FCC.
Hazard: Toxic. Use in foods restricted to 0.1%.
Use: Food additive (preservative), antimicrobial agent.
See Parabens.

methylparafynol. See 3-methyl-1-pentyn-3-ol.

methyl parathion. (*O,O*-dimethyl-*O-p*-nitrophenylphosphorothioate).
CAS: 298-00-0. $(CH_3O)_2P(SO)OC_6H_4NO_2$. The methyl homolog of parathion.
Properties: White, crystalline solid. Mp 35–36C. D 1.358 (20/4C), refr index 1.5515 (35C). Slightly soluble in water; miscible in all proportions with acids and alcohols, esters and ketones. Slightly decomposed by acid solutions; rapidly in dilute alkalies. The commercial product is a tan liquid (xylene solution) with pungent odor, decomposes violently at 248F. It is not classed as a "hard" insecticide, but is relatively biodegradable.
Available forms: Emulsifiable concentrate, wettable powder, dusts, technical (80%), liquid solution, encapsulated.
Hazard: Explosion risk when heated. Toxic by skin absorption, inhalation, and ingestion; cholinesterase inhibitor. Use has been restricted. Questionable carcinogen.
Use: Insecticide, especially for cotton.

methyl PCT. See *O,O*-dimethyl phosphorochloridothioate.

methyl pelargonate. See methyl nonanoate.

methyl pentadecanoate.
CAS: 7132-64-1. $CH_3(CH_2)_{13}COOCH_3$. The methyl ester of pentadecanoic acid.
Properties: Colorless liquid. D 0.8618 (25/4C), mp 18.5C, bp 199C (30 mm Hg), refractive index 1.4374 (25C). Insoluble in water; soluble in alcohol and ether. Combustible.

Grade: Reagent, 96%, and 99.5%.
Use: Intermediate in organic synthesis, reagent medical research.

methylpentadiene.
CAS: 54363-49-4. C_6H_{10}. Numerous isomers are possible. Commercially available mixture contains 2- and 4-methyl-1,3-pentadiene.
Properties: Colorless liquid. D 0.7184 (20/4C), bp 75–77C, flash p −30F (−34.4C). Reactive with halogens, hydrohalogens, sulfur dioxide, and maleic anhydride.
Hazard: Flammable, dangerous fire risk.
Use: Organic synthesis, alkyd and other polymers.

2-methylpentaldehyde.
CAS: 123-15-9. $C_3H_7CH(CH_3)CHO$.
Properties: Colorless liquid. D 0.8092, bp 118.3C, fp −100C. Soluble in water 0.42% by wt, flash p 68F (20C) (OC).
Hazard: Flammable, dangerous fire risk. Strong irritant to skin and mucous membranes.
Use: Intermediates for dyes, resins, pharmaceuticals.

2-methylpentamethylenediamine. See "Dytek" *[Invista]*.

2-methylpentane. (dimethylpropylmethane).
CAS: 107-83-5. $CH_3(CH_2)_2CH(CH_3)_2$.
Properties: Colorless liquid. Fp −153C, bp 60C, refr index 1.372 (20C), d 0.658 (60/60F), flash p −10F (−23.3C), autoign temp 583F (306C).
Grade: 95%, 99%, research.
Hazard: Flammable, dangerous fire risk, reacts vigorously with oxidizing materials.
Use: Organic synthesis, solvent.

3-methylpentane. (diethylmethylmethane).
CAS: 96-14-0. $CH_3CH_2CH(CH_3)CH_2CH_3$.
Properties: Colorless liquid. D 0.6645 (20/4C), bp 64.0C, refr index 1.37662 (20C), flash p approximately 20F (−6.6C). Soluble in alcohol; insoluble in water; slightly soluble in ether.
Grade: Technical (95%), 99%, research.
Hazard: Flammable, dangerous fire risk.
Use: Organic synthesis, solvent.

2-methyl-1,3-pentanediol.
$C_2H_5CH(OH)CH(CH_3)CH_2OH$.
Properties: Colorless liquid. D 0.9745, bp 220.3C, fp −30C. Freely soluble in water. Combustible.
Use: Solvent, coupling agent.
See hexylene glycol.

2-methyl-2,4-pentanediol. (4-methyl-2,4-pentanediol).
CAS: 107-41-5. $CH_3CH_2OCH_2COH(CH_3)CH_3$.
Properties: Colorless liquid. D 0.9235 (20/20C), bp 197.1C, viscosity 34 cP (20C), vap press 10.8 mm Hg (95.2C), 334 mm Hg (169.7C), flash p (OC) 201F (93.9C), 7.59 lb/gal (20C). Miscible with water and most organic solvents, including lower aliphatic hydrocarbons. Combustible.
Use: Coupling agent, chemical synthesis.
See hexylene glycol.

4-methyl-2,4-pentanediol. See 2-methyl-2,4-pentanediol.

2-methylpentanoic acid.
CAS: 97-61-0. $CH_3CH_2CH_2CH(CH_3)COOH$.
Properties: Water-white liquid. D 0.9242 (20/20C), bp 196.4C, vap press 0.02 mm Hg (20C), fp sets to glass approximately −85C, flash p 225F (107C). Soluble in water 1.3% by wt (20C). Combustible.
Use: Plasticizers, vinyl stabilizers, metallic salts, alkyd resins.

4-methylpentanoic acid.
CAS: 646-07-1. $(CH_3)_2CH(CH_2)_2COOH$.
Properties: Colorless liquid. Bp 197C, d 0.921 (20/4C), bulk d 7.66 lb/gal (20C). Miscible with alcohol, benzene, and acetone; low solubility in water. Combustible.
Use: Intermediate for plasticizers, pharmaceuticals and perfumes.

2-methyl-1-pentanol.
CAS: 105-30-6. $C_3H_7CH(CH_3)CH_2OH$.
Properties: Colorless liquid. D 0.8252, bp 148.0C, vap press 1.1 mm Hg (20C), flash p 135F (57.2C) (OC). Solubility in water 0.31% by wt. Combustible.
Hazard: Moderate fire risk.
Use: Solvent, intermediate.

4-methyl-2-pentanol. See methylamyl alcohol.

4-methyl-2-pentanol acetate. See methylamyl acetate.

4-methyl-2-pentanone. See methyl isobutyl ketone.

2-methyl-1-pentene. (1-methyl-1-propylethylene).
CAS: 763-29-1. $H_2C:C(CH_3)CH_2CH_2CH_3$.
Properties: Colorless liquid. D 0.6820 (20/4C), bp 62.6C, fp −135C, refr index 1.3925 (20C), flash p −15F (26.1C). Soluble in alcohol, acetone, ether, petroleum, coal tar solvents; insoluble in water.
Grade: 95%, 99%, research.
Hazard: Flammable, dangerous fire risk.
Use: Organic synthesis, flavors, perfumes, medicines, dyes, oils, resins.

4-methyl-1-pentene.
CAS: 691-37-2. $H_2C:CHCH_2CH(CH_3)CH_3$.
Properties: Colorless liquid. Fp −153C, bp 53.5C, d 0.6640 (20/4C), refr index 1.3826 (20C), flash p −25F (−31.6C).
Grade: 95%, 99%, research.
Hazard: Same as for 2-methyl-1-pentene.

Use: Organic synthesis, monomer for plastics used in automobiles, electronic components, and laboratory ware.

4-methyl-2-pentene.   (1-isopropyl-2-methylethylene (*cis-trans* mixture)).
CAS: 674-76-0. $CH_3CH:CHCH(CH_3)_2$.
Properties: Colorless liquid. D 0.670 (20/4C), bp (mixture) 55C, (*cis*) 56.1C, (*trans*) 58.3C, refr index 1.388 (20C), flash p −20F (−28.9C). Soluble in alcohol, acetone, ether, petroleum and coal tar solvents; insoluble in water.
Hazard: Flammable, dangerous fire risk.
Use: Organic synthesis.

2-methyl-1-penten-3-one.   See   isopropenyl ethyl ketone.

4-methyl-3-penten-2-one.   See mesityl oxide.

3-methyl-1-pentyn-3-ol.   (meparfynol; methylparafynol; methyl pentynol).
$HC{\equiv}CCOH(CH_3)CH_2CH_3$.
Properties: Colorless liquid. Bp 121.4C, fp −30.6C, d 0.8721 (20/20C), refr index 1.4318 (20C), flash p 101F (38.3C) (TOC). Moderately soluble in water; miscible with acetone, benzene, carbon tetrachloride, ethyl acetate. Combustible.
Grade: High purity 98.5%, technical 95.0% min, pharmaceutical.
Hazard: Moderate fire risk. Toxic.
Use: Stabilizer in chlorinated solvents, viscous reducer, electroplating brightener, intermediate in syntheses of hypnotics and isoprenoid chemicals, solvent for polyamide resins, acid inhibitor, prevention of hydrogen embrittlement, medicine (soporific and anesthetic).

methylphenethylamine.   See amphetamine.

methyl phenylacetate.
CAS: 101-41-7. $C_6H_5CH_2COOCH_3$.
Properties: Colorless liquid; honey-like odor. D 1.062–1.066 (25C), refr index 1.506–1.509 (20C), bp 218C. Soluble in five parts of 60% alcohol, in fixed oils; insoluble in water. Combustible.
Grade: Technical, FCC.
Use: Perfumery, flavors for tobacco, flavoring.

methylphenylcarbinol.   See α-methylbenzyl alcohol.

methylphenylcarbinyl acetate.   See α-methylbenzyl acetate.

methylphenyldichlorosilane.
CAS: 149-74-6. $CH_3(C_6H_5)SiCl_2$.
Properties: Colorless liquid. Bp 82C (13 mm Hg), 205C, refr index 1.5199 (25C), d 1.19, flash p 83F (28.3C). Soluble in benzene, ether, methanol.

Derivation: From chlorobenzene Grignard reagent and methyltrichlorosilane or from benzene and methyldichlorosilane.
Hazard: Flammable, moderate fire risk, reacts strongly with oxidizing materials. Irritant.
Use: Manufacture of silicones.

methyl phenyl ether.   See anisole.

2-methylphenyl isocyanate.
See 1-isocyanato-2-methylbenzene.

4-methylphenyl isocyanate.
See 1-isocyanato-4-methylbenzene.

3-methylphenylmethyl(((2-phenyl-1,3-dioxan-5-yl)methoxy)sulfinyl) carbamate.
CAS: 81862-06-8. $C_{20}H_{23}NO_6S$.
Hazard: Moderately toxic by ingestion.
Use: Agricultural chemical.

2-methyl-4-phenyl-6h-1,3,5-oxathiazine.
CAS: 58955-85-4. $C_{10}H_{11}NOS$.
Hazard: Moderately toxic by ingestion.
Use: Agricultural chemical.

4-(4-methylphenyl)-6h-1,3,5-oxathiazine.
CAS: 58955-82-1. $C_{10}H_{11}NOS$.
Hazard: Moderately toxic by ingestion.
Use: Agricultural chemical.

4-methyl-1-phenyl-2-pentanone.   (benzyl isobutyl ketone). $C_6H_5CH_2C(O)CH_2CH(CH_3)_2$.
Properties: Combustible.
Use: Flavoring.

methyl phenyl phosphate.   See methyl diphenyl phosphate.

9-methyl-2-(3-(4-phenyl-1-piperazinylpropyl))-1,2,3,4-tetrahydro-β-carbolin-1-one 2hcl.
CAS: 124824-14-2. $C_{25}H_{30}N_4O{\cdot}2ClH$.
Hazard: Moderately toxic by ingestion.

2-methyl-2-phenylpropane.   See *tert*-butylbenzene.

3-(4-methylphenyl)-*n*-(4-propylcyclohexyl)-2-propenamide.
CAS: 315706-79-7. $C_{19}H_{27}NO$.
Hazard: A poison by ingestion.

***n*-(2-methylphenyl)-1h-pyrazole-1-acetamide.**
CAS: 302542-40-1. $C_{12}H_{13}N_3O$.
Hazard: Moderately toxic by ingestion.

***n*-(3-methylphenyl)-1h-pyrazole-1-acetamide.**
CAS: 302542-49-0. $C_{12}H_{13}N_3O$.
Hazard: Moderately toxic by ingestion.

n-(4-methylphenyl)-1h-pyrazole-1-acetamide.
CAS: 302542-57-0. $C_{12}H_{13}N_3O$.
Hazard: Moderately toxic by ingestion.

4-methyl-1-phenyl-3-pyrazolidone. See Phenidone B.

3-methyl-1-phenyl-2-pyrazolin-5-one. (3-methyl-1-phenyl-5-pyrazolone; 1-phenyl-3-methyl-5-pyrazolone). $C_6H_5NN:C(CH_3)CH_2CO$.
Properties: White powder or crystals. Bp 287C (205 mm Hg), mp 127C, vap press less than 0.01 mm Hg (20C). Soluble in water; slightly soluble in alcohol or benzene; insoluble in ether.
Derivation: By condensation of phenylhydrazine with ethylacetoacetate.
Hazard: Toxic by ingestion.
Use: Intermediate for dyes and drugs, sensitive reagent for detection of cyanide.

α-(5-methyl-1-phenyl-1h-pyrazol-4-yl)-1-piperidinebutanol.
CAS: 296269-53-9. $C_{19}H_{27}N_3O$.
Hazard: A poison by ingestion.

3-((4-methylphenyl)sulfonyl)-2-propenenitrile. See 3-(*p*-tolylsulfonyl) acrylonitrile.

1-methyl-4-phenyl-1,2,3,6-tetrahydropyridine. See 1,2,3,6-tetrahydro-1-methyl-4-phenylpyridine.

methylphloroglucinol. (2,4,6-trihydroxytoluene).
CAS: 88-03-9. $C_6H_2(OH)_3CH_3$.
Properties: Cream to light tan fine crystals; odorless. Mp 210–214C. Soluble in water, alcohol and ether; insoluble in benzene. Combustible.
Use: Reactive coupling agent, dye and plastic intermediate.

methylphosphonic acid.
CAS: 993-13-5. $CH_3PO(OH)_2$.
Properties: White solid. Mp 103–104C.
Use: Organic synthesis.

methylphosphoric acid. (methyl orthophosphoric acid; methyl acid phosphate).
$CH_3H_2PO_4$.
Properties: Pale-straw-colored liquid. D 1.42 (25C). Can be neutralized with alkalies or amines to give water-soluble salts. Combustible.
Grade: 97% (remainder orthophosphoric acid and methanol).
Use: Textile and paper processing compounds, catalysts in urea-resin formation, polymerizing agent for resins and oils, rust remover, soldering flux, chemical intermediate.

methyl phoxim.
CAS: 14816-16-1. $C_{10}H_{11}N_2O_3PS$.
Hazard: Moderately toxic by ingestion.
Use: Agricultural chemical.

3′-methylphthalanilic acid. (*N-m*-tolyl-phthalamic acid).
CAS: 85-72-3. $C_{15}H_{13}NO_3$.
Properties: Crystalline solid. Mp 150C. Soluble in alcohol and other polar solvents with decomposition; slightly soluble in water.
Hazard: Toxic. Avoid ingestion.
Use: In agriculture as an antiscission agent for fruits and vegetables.

methyl phthalyl ethyl glycolate.
$C_2H_5OOCCH_2OOCC_6H_4COOCH_3$.
Properties: Colorless liquid; slight characteristic odor, d 1.217–1.227, flash p 375F (190C). Miscible with most organic solvents; very slightly soluble in water. Combustible.
Use: Plasticizer, solvent.

methyl picrate. See trinitroanisole.

N-methylpiperazine.
CAS: 109-01-3.

$\overline{CH_3NCH_2CH_2NHCH_2CH_2}$.

Properties: Colorless liquid; amine-like odor. D 0.9038, bp 138.0C, fp −6.4C, hygroscopic, flash p 108F (42.2C). Combustible.
Hazard: Moderate fire risk.
Use: Intermediate for pharmaceuticals, surface agents, synthetic fibers.

2-methylpiperidine. (2-pipecoline).
CAS: 109-05-7. $C_5H_{10}NCH_3$.
Properties: Colorless liquid. Bp 118.2C, fp −4.2C, d 0.8401 (20/20C), refr index 1.4457 (20C). Miscible with water at 20C. Combustible.

1-methyl-4-piperidinol.
CAS: 106-52-5. $CH_3C_5H_9NOH$.
Properties: Colorless to light amber, oily liquid; amine-like odor. Crystals may be present, or material may be solid, but will melt on slight warming. Refr index 1.4757–1.4777 (25C), congealing temperature 27.5C min. Combustible.
Use: Organic synthesis.

3-(2-methylpiperidino)propyl-3,4-dichlorobenzoate. See piperalin.

methylprednisolone.
CAS: 83-43-2. $C_{22}H_{30}O_5$. A steroid.
Properties: White, crystalline powder; odorless. Mp 240C with decomposition. Sparingly soluble in alcohol; slightly soluble in acetone; insoluble in water.

Grade: NF.
Use: Medicine (hormone).

2-methylpropane.   See isobutane.

2-methylpropanenitrile.   See isobutyronitrile.

2-methyl-1-propanethiol.   (isobutyl mercaptan).
CAS: 513-44-0. $(CH_3)_2CHCH_2SH$.
Properties: Liquid; unpleasant odor. Bp 85–95C d 0.8363 (15.5C), flash p 15F (−9.4C).
Grade: 95%.
Hazard: Flammable, dangerous fire risk.

2-methyl-2-propanethiol.   (*tert*-butyl mercaptan).
CAS: 75-66-1. $(CH_3)_3CSH$.
Properties: Colorless liquid; strong skunk odor. D 0.79–0.82 (60/60F), distillation range 62–67C, refr index 1.422 (20C), flash p −15F (−26.1C), bulk d 6.71lb/gal.
Hazard: Flammable, dangerous fire risk. Very toxic.
Use: Intermediate, gas odorant for detecting leaks.

2-methylpropanoic acid.   See isobutyric acid.

2-methyl-1-propanol.   See isobutyl alcohol.

2-methyl-2-propanol.   See *tert*-butyl alcohol.

2-methylpropene.   See isobutene.

2-methyl-2-propen-1-ol.   See methylallyl alcohol.

methyl propionate.
CAS: 554-12-1. $CH_3CH_2COOCH_3$.
Properties: Colorless liquid. D 0.937 (4C), refr index 1.3769 (20C), boiling range 78.0–79.5C, autoign temp 876F (468C), flash p 28F (−2.2C) (CC), 7.58 lb/gal. Soluble in most organic solvents; somewhat soluble in water.
Grade: Technical.
Hazard: Flammable, dangerous fire risk, explosive limits in air 2.5–13%.
Use: Solvent for cellulose nitrate, solvent mixture for cellulose derivative, lacquers, paints, varnishes, coating compositions, flavoring.

1-(1-methylpropoxycarbonyl)-2-(2-hydroxyethyl)piperidine.
CAS: 119515-38-7. $C_{12}H_{23}NO_3$.
Hazard: A reproductive hazard.
Use: Agricultural chemical.

methylpropylbenzene.   See cymene.

methyl propyl carbinol.   See 2-pentanol.

methyl propyl carbinol urethane.   See hedonal.

4-methyl-*n*-(4-propylcyclohexyl)benzamide.
CAS: 315706-71-9. $C_{17}H_{25}NO$.
Hazard: A poison by ingestion.

methyl propyl ketone.   (ethyl acetone; 2-pentanone; MPK).
CAS: 107-87-9. $CH_3(CH_2)_2COCH_3$.
Properties: Water-white liquid. The commercial material consists of a mixture of methyl propyl and diethyl ketones in the approximate ratio of 3:1 and contains at least 97% of these ketones, the balance being *sec*-amyl alcohol. D 0.809 (20/20C), fp −77.5C, bp 101.7C, refr index 1.3895 (20C), viscosity 0.473 cP (25C), flash p 45F (7.22C), autoign temp 941F (505C). Soluble in alcohol and ether; slightly soluble in water.
Grade: Technical.
Hazard: Flammable, dangerous fire risk, explosive limits in air 1.6–8.2%. Eye irritant and affects pulmonary function.
Use: Solvent, substitute for diethyl ketone, flavoring.

2-methyl-2-*n*-propyl-1,3-propanediol dicarbamate.   See meprobamate.

methylprylon.   (3,3-diethyl-5-methyl-2,4-piperidinedione).
CAS: 125-64-4. $C_5NH_4(O)_2(C_2H_5)_2(CH_3)$.
Properties: Nearly white, crystalline powder; slight characteristic odor; bitter taste. Melting range 74–77C. Soluble in water; very soluble in alcohol, chloroform, ether, and benzene.
Grade: NF.
Hazard: Abuse may cause addiction.
Use: Medicine (sedative, hypnotic).

methylpyridine.   See picoline.

methyl pyridyl ketone.   (1-(3-pyridinyl) ethanone; 3-acetylpridine; β-acetylpyridine; methyl 3-pyridyl ketone; methyl β-pyridyl ketone; 1-pyridin-3-ylethanone)
CAS: 350-03-8. C_7H_7NO. It is a nicotinic acid antagonist.
Properties: A liquid that dissolves freely in acids.

methyl pyromucate.   See methyl-2-furoate.

***N*-methylpyrrole.**
CAS: 96-54-8. $C_4H_4NCH_3$.
Properties: Colorless liquid. Bp 112C, fp −57C, d 0.914 (20C) refr index 1.4898 (17D), flash p 61F (16.1C). Insoluble in water; soluble in alcohol.
Grade: 98% min purity.
Hazard: Flammable, dangerous fire risk.
Use: Organic synthesis.

***N*-methylpyrrolidine.**

$$CH_3NCH_2CH_2CH_2CH_2.$$

Properties: Colorless liquid; ammonia-like odor. Refr index 1.4200–1.4230 (25C), bp 80.5C, fp −90C, d 0.805, flash p 7F (−13.9C).
Hazard: Flammable, dangerous fire risk. Irritant to skin and eyes.

N-methyl-2-pyrrolidone.
CAS: 872-50-4.

CH₃NCH₂CH₂CH₂CO.

Properties: Colorless liquid; mild amine odor. Mw 99.1, d 1.027, fp −24C, bp 202C, flash p 204F (95.5C). Miscible with water, various organic solvents, castor oil. Combustible.
Derivation: High-pressure synthesis from acetylene and formaldehyde.
Hazard: Severe skin and eye irritant. Explosive limits 2.2–12.2%.
Use: Solvent for resins, acetylene, etc., pigment dispersant, petroleum processing, spinning agent for polyvinyl chloride, microelectronics industry plastic solvent applications, intermediate.
Note: A proprietary adaptation of this solvent to clean-up of vinyl chloride reaction vessels is available under trademark of "M-Pyrol" *[GAF].*
See "M-Pyrol" *[GAF].*

methyl 2-pyrrolyl ketone. See 2-acetylpyrrole.

α-methylquinoline. See quinaldine.

γ-methylquinoline. See lepidine.

2-methylquinone. See toluquinone.

6-methyl-2,3-quinoxalinedithiol cyclic carbonate. (6-methyl-2-oxo-1,3-dithio-(4,5-b)quinoxaline).
CAS: 2439-01-2. C₁₀H₆N₂S₂O.
Properties: Yellow, crystalline powder. Mp 172C. Insoluble in water; slightly soluble in acetone and alcohol; soluble in organic solvents.
Hazard: Toxic by ingestion and inhalation.
Use: Acaricide, insecticide, fungicide.

methyl red. (*p*-dimethylaminoazobenzenecarboxylic acid; CI 13020).
CAS: 493-52-7. (CH₃)₂NC₆H₄NNC₆H₄COOH.
Properties: Dark-red powder or violet crystals. Mp 180C. Insoluble in water; soluble in alcohol, ether, glacial acetic acid. Color fades quickly due to reduction.
Hazard: Questionable carcinogen.
Use: Acid–base indicator in the range pH 4.2–6.2 (red to yellow).
Note: No longer widely used because of instability.

methylresorcinol. See orcin.

methyl ricinoleate.
CAS: 41989-07-5. C₁₉H₃₆O₃. The methyl ester of ricinoleic acid.
Properties: Colorless liquid. D 0.9236 (22/4C), fp −4.5C, bp 245C (10 mm Hg), refr index 1.4628. Insoluble in water; soluble in alcohol and ether. Combustible.
Derivation: Esterification of ricinoleic acid or alcoholysis of castor oil, purification by vacuum distillation.
Grade: Technical, purified (99+%).
Use: Plasticizer, lubricant, cutting oil additive, wetting agent.

methylrosaniline chloride. See methyl violet.

methyl salicylate. (gaultheria oil; wintergreen oil; betula oil; sweet-birch oil).
CAS: 119-36-8. C₆H₄OHCOOCH₃.
Properties: Yellow to red liquid; odor of wintergreen. Refr index 1.535–1.538, d 1.180–1.185, fp −8.3C, flash p 214F (101C), autoign temp 850F (454C), bp 222.2C; natural oil optically inactive, synthetic oil, angular rotation not more than −1.5 degrees, soluble in 7 parts of 70% alcohol. Soluble in ether and in glacial acetic acid; sparingly soluble in water. Combustible.
Derivation: By heating methanol and salicylic acid in presence of sulfuric acid or by distillation from leaves of Gaultheria procumbens or bark of *Betula lenta.*
Method of purification: Rectification.
Grade: Technical, USP, FCC.
Hazard: Toxic by ingestion; use in foods restricted by FDA, lethal dose 30 cc in adults, 10 cc in children.
Use: Flavor in foods, beverages, pharmaceuticals, odorant, perfumery, UV absorber in sunburn lotions.

4-methylsalinomycin. See narasin.

methyl silicate. (tetramethoxy silane).
CAS: 681-84-5. (CH₃O)₄Si.
Properties: Colorless needles. Mw 147.18, d 1.0232 at 20C, bp 121C. Soluble in alcohol.
Hazard: Eye damage and upper respiratory tract irritant.
Use: Coating inside of TV picture tubes.

methyl silicone. See dimethyl silicone; silicone; siloxane.

methylsilylidynetris(2-ethylhexanoate).
See tris((2-ethylhexanoyl)oxy)methylsilane.

methyl stearate. (methyl octadecanoate).
CAS: 27234-05-5. CH₃(CH₂)₁₆COOCH₃. The methyl ester of stearic acid.
Properties: Semisolid. Mp 37.8C, bp 234.5C (30 mm Hg), 204.5C (10 mm Hg), flash p 307F (152C), refr index 1.4328 (50C). Insoluble in water; soluble

in ether and alcohol. Combustible. Most technical methyl stearate is 55% stearate and 45% methyl palmitate.

Derivation: Esterification of stearic acid with methanol or alcoholysis of stearin with methanol.

Method of purification: Vacuum fraction distillation.

Grade: Distilled, pressed, technical, pure (99.8+%).

Use: Intermediate for stearic acid detergents, emulsifiers, wetting agents, stabilizers, resins, lubricants, plasticizers.

4-α-methylstrylphenol.   See Prodox 121.

α-methylstyrene.
CAS: 98-83-9. $C_6H_5C(CH_3):CH_2$.

Properties: Colorless liquid. Subject to polymerization by heat or catalysts. Bp 165.38C, fp −23.21C, d 0.9062 (25/25C), viscosity 0.940 cP (20C), flash p 129F (53.9C), autoign temp 1065F (573C), refr index 1.5359 (25/25C), insoluble in water. A polymerization inhibitor such as *tert*-butyl catechol is usually present in commercial quantities. Combustible.

Derivation: From benzene and propylene by use of aluminum chloride and hydrochloric acid to yield cumene, which is then dehydrogenated.

Hazard: Moderate fire risk. Explosive limits in air 1.9–6.1%. Avoid inhalation and skin contact. Upper respiratory tract irritant, kidney and female reproductive damage. Possible carcinogen.

Use: Polymerization monomer, especially for polyesters.

methyl styryl ketone.   See benzylidene acetone.

methylsuccinic acid.   See pyrotartaric acid.

methyl sulfate.   Legal label name (Rail) for dimethyl sulfate.

methyl sulfide.   Legal label name (Rail) for dimethyl sulfide.

3-methylsulfolane.   See 3-methyltetrahydrothiophene-1,1-dioxide.

3-methylsulfolene.   See 3-methyl-2,5-dihydrothiophene-1,1-dioxide.

methylsulfonic acid.   (methanesulfonic acid).
CAS: 75-75-2. CH_3SO_2OH.

Properties: Colorless solid. D 1.48, mp 20C, bp 167C (10 mm Hg). Slightly soluble in benzene; almost insoluble in other organic solvents. Attacks iron, copper, and steel.

Derivation: Reaction of methane and sulfur trioxide.

Hazard: Toxic by ingestion, skin irritant, corrosive to tissue.

Use: Polymerization catalyst.

((methylsulfonyl)oxy)tributylstannane.
See tri-*n*-butyltin methanesulfonate.

((methylsulfonyl)oxy)triphenylstannane.
See triphenyltin methanesulfonate.

methyl sulfoxide.   See dimethyl sulfoxide.

methylsulfuric acid.   (acid methyl sulfate; methyl hydrogen sulfate). CH_3OSO_2OH or CH_3HSO_4.

Properties: Oily liquid. Bp 188C, d 1.352, fp −27C. Soluble in anhydrous ether; slightly soluble in alcohol and water.

Derivation: Interaction of methanol and chlorosulfonic acid.

Use: Sulfonating agent, specialty solvent.

N-methyltaurine.   (sodium-*N*-methyltaurate).
$CH_3NHCH_2CH_2SO_3Na$. Available in commercial quantities as an aqueous solution of the sodium salt.

Properties: (Solution) Clear, light-colored liquid, approximately 34–36% sodium salt. D 1.21 (25/4C), at fp (−28C average) becomes a suspension of white crystals.

Use: Intermediate for detergents, dyestuffs, pharmaceuticals and other organics.

17-methyltestosterone.
CAS: 58-18-4. $C_{20}H_{30}O_2$. A synthetic androgenic steroid.

Properties: White or creamy-white crystals or crystalline powder; odorless. Stable in air, slightly hygroscopic, affected by light. Mp 163–168C. Soluble in ethanol; methanol, ether and other organic solvents; insoluble in water.

Derivation: By organic synthesis.

Grade: NF.

Use: Medicine (hormone).

methyl tetracosanoate.   See methyl lignocerate.

methyl tetradecanoate.   See methyl myristate.

2-methyltetrahydrofuran.
CAS: 96-47-9. $C_4H_7CH_3O$.

Properties: Colorless liquid; ether-like odor. Bp 80.2C, fp −136C, d 0.854 (20/4C), refr index 1.4025 (25C), flash p 12F (−11.1C) (TOC). Soluble in water 15.1 g/100 g water (25C); solubility in water increases with a decrease in temperature. Freely soluble in most organic solvents.

Hazard: Flammable, dangerous fire risk.

Use: Chemical intermediate, reaction solvent.

3-methyltetrahydrothiophene-1,1-dioxide.
(3-methylsulfolane).
$CH_3C_4H_7SO_2$.

Properties: Colorless liquid. D 1.194 (20/4C), mp 0C, bp 276C, flash p 325F (162C). Completely soluble in water, acetone, and toluene. Combustible.
Use: Solvent and extractive medium.

methyl-2-thienyl ketone.   See 2-acetyl thiophene.

***m*-(methylthio)aniline.**
CAS: 1783-81-9. $H_2NC_6H_4SCH_3$.
Properties: Pale-yellow oil. D 1.140 (25C), bp 163–165C (16 mm Hg), fp −3.0C. Insoluble in water; soluble in alcohol, benzene, and acetic acid. Combustible.
Hazard: See aniline.
Use: Pharmaceutical intermediate.

6-methyl-2-thio-2h-1,3-benzoxazine-2,4(3h)-dione.
CAS: 23611-64-5. $C_9H_7NO_2S$.
Hazard: A poison by ingestion.
Use: Agricultural chemical.

4-(methylthio)-3,5-dimethylphenyl-*n*-methylcarbamate.   [(4-methylthio)-3,5-xylyl   methylcarbamate].
$CH_3S(CH_3)_2C_6H_2OOCNHCH_3$.
Properties: Solid. Mp 121C. Insoluble in water; soluble in alcohol and acetone.
Use: Pesticide.

3-methyl-2,5-thiomorpholinedione   2-(*o*-((methylamino)carbonyl)oxime).
CAS: 66637-32-9. $C_7H_{11}N_3O_3S$.
Hazard: A poison by ingestion.
Use: Agricultural chemical.

methylthionine chloride.   See methylene blue.

methylthiopropionic cyanohydrin.   See 2-hydroxy-4-(methylthio)butanenitrile.

methyltin.   See organotin compounds.

methyl-*p*-toluate.
CAS: 99-75-2. $CH_3C_6H_4COOCH_3$.
Properties: White, crystalline solid. Fp 34C, d 1.058 (25/25.6C).
Use: Organic synthesis.

methyl-*p*-toluenesulfonate.
CAS: 80-48-8. $CH_3C_6H_4SO_3CH_3$.
Properties: White, damp crystals. Solidification point 24C; bp 157C (8 mm Hg), decomposes 262C. A vesicant material. Insoluble in water; soluble in alcohol and benzene.
Grade: 97% min.
Hazard: Toxic by ingestion and inhalation, strong irritant to skin and eyes.
Use: Accelerator, methylating agent, catalyst for alkyd resins.

methyl-*p*-tolyl ketone.   See methyl acetophenone.

methyltransferase.
CAS: 9012-25-3. Any of a class of enzymes that catalyze the transfer of a methyl group to an organic molecule. The methyl donor is s-adenosylmethionine, formed from methionine and ATP.

2-((*n*-methyl-*n*-trichloromethanesulfenyl)carbamoyloximino)-1,4-dithiane.
CAS: 55391-23-6. $C_7H_9Cl_3N_2O_2S_3$.
Hazard: A poison by ingestion.
Use: Agricultural chemical.

4-((*n*-methyl-*n*-trichloromethanesulfenyl)carbamoyloximino)-1,3-dithiolane.
CAS: 55391-24-7. $C_6H_7Cl_3N_2O_2S_3$.
Hazard: A poison by ingestion.
Use: Agricultural chemical.

methyltrichlorosilane.
CAS: 75-79-6. CH_3SiCl_3.
Properties: Colorless liquid. Bp 66.4C, d 1.270 (25/25C), refr index 1.4085 (25C), flash p 8F (−13.3C). Readily hydrolyzed by moisture with the liberation of hydrogen chloride.
Derivation: By Grignard reaction of silicon tetrachloride and methylmagnesium chloride.
Hazard: Flammable, dangerous fire risk, may form explosive mixture with air. Strong irritant.
Use: Intermediate for silicones.

methyltrichlorostannane.
CAS: 993-16-8. CH_3Cl_3Sn.
Properties: Crystals from pet ether or C_6H_6 or by sublimation. Mp: 53°C.
Hazard: Moderately toxic. Experimental reproductive effects.

methyl tricosanoate.
CAS: 2433-97-8. $CH_3(CH_2)_{21}COOCH_3$. The methyl ester of tricosanoic acid.
Properties: White, wax-like solid. Mp 55–56C. Insoluble in water; soluble in alcohol and ether. Combustible.
Grade: Purified 96% and 99.5%.
Use: Intermediate in organic synthesis, biochemical research.

methyl tridecanoate.
CAS: 1731-88-0. $CH_3(CH_2)_{11}COOCH_3$. The methylester of tridecanoic acid.
Properties: Colorless liquid. Mp 5.5C, bp 130–132C (4 mm Hg) refr index 1.4327 (25C). Insoluble in water; soluble in alcohol and ether. Combustible.
Derivation: Esterification of tridecanoic acid with methanol followed by fractional distillation.
Grade: Purified 96% and 99.5%.
Use: Intermediate in organic synthesis, biochemical research, reference standard in gas chromatography.

methyl trimethylolmethane.   See trimethylol ethane.

methyltris(2-ethylhexyloxycarbonylmethyl thio)stannane.
CAS: 57583-34-3. $C_{31}H_{60}O_6S_3Sn$.
Hazard: Moderately toxic by ingestion.

β—methylumbelliferone.   (7-hydroxy-4-methylcoumarin; BMU). $C_{10}H_8O_3$.
Properties: White to light-tan powder. Mp 186–188C. Soluble in concentrated sulfuric acid; partly soluble in ethanol, isopropanol, 5% aqueous sodium carbonate solution. Very slightly soluble in water. Very dilute aqueous alkaline solutions give a bright blue-white fluorescence in daylight or UV light.
Grade: Technical.
Use: Optical bleach on soaps, starches and laundry products, suntan lotions.

2-methylundecanal.   See methylnonylacetaldehyde.

methylundecanoate.
CAS: 1731-86-8. $(CH_3(CH_2)_9COOCH_3$.   The methyl ester of undecanoic acid.
Properties: Colorless liquid. Bp 123C (10 mm Hg), refr index 1.4270 (25/4C). Insoluble in water; soluble in alcohol and ether. Combustible.
Derivation: Esterification of undecanoic acid with methanol followed by fractional distillation.
Grade: Purified 96%, 99.5%.
Use: Organic intermediate for synthesis, flavoring, biochemical research.

5-methyluracil.   See thymine.

n-methylurethanebenzenesulfohydrazine.
CAS: 1879-26-1. $C_8H_{11}N_3O_4S$.
Hazard: A poison by ingestion and inhalation.

methylvinyldichlorosilane.
CAS: 124-70-9. $(CH_3)(C_2H_3)SiCl_2$.
Properties: Colorless liquid. Bp 92C, d 1.08 (25C), refr index 1.4270 (25C), flash p 40F (4.4C). Soluble in benzene and ether. Reacts with methanol and water.
Derivation: From methyldichlorosilane and acetylene or vinyl chloride.
Hazard: Flammable, dangerous fire risk. Irritant.
Use: Manufacture of silicones.

methyl vinyl ether.   See vinyl methyl ether.

methyl vinyl ketone.   Legal label name for vinyl methyl ketone.

methyl vinyl oximino silane.
CAS: 73160-32-4. $C_{11}H_{22}N_2O_2Si$.
Hazard: A poison by ingestion and skin contact.

2-methyl-5-vinylpyridine.
$CH_3C_5H_3NCH:CH_2$.
Properties: Clear to faintly opalescent liquid. D 0.978–0.982 (20/20C), bp 181C, refr index 1.5400–1.5454 (20C), fp (anhydrous) −14.3C, flash p (TOC) 165F (73.9C). Combustible.
Use: Monomer for resins, oil additive, ore flotation agent, dye acceptor.

methyl   violet.   (Gentian   Violet,   USP; hexamethyl-*p*-rosaniline chloride; CI 42555).
CAS: 8004-87-3. $C_{25}H_{30}N_3Cl$.
Properties: Green powder. Soluble in water and chloroform; partially soluble in alcohol and glycerol.
Use: Medicine (topical antibacterial and antiallergen), acid–base indicator, alcohol denaturant, biological stain, textile dye.

methylviologen   dibromide.   See   paraquat dibromide.

methyl yellow.   See dimethylaminoazobenzene.

metiram.
CAS: 12122-67-7. Mixture of (ethylenebis (dithiocarbamato))zinc ammoniates with ethylenebis(dithiocarbamic acid)anhydrosulfides.
Use: Fungicide.

"Met-L-KYL" [Tyco].   TM for a dry chemical that extinguishes fires caused by pyrophoric liquids such as triethylaluminum and adsorbs the spilled metal alkyl to prevent reignition.

"Met-L-X" [Tyco].   TM for a dry chemical based on sodium chloride, approved for use on sodium, potassium, sodium-potassium alloy, and magnesium fires.

metobromuron.
CAS: 3060-89-7. $C_9H_{11}BrN_2O_2$.
Properties: Colorless crystals. Mp 95C. Soluble in alcohols; slightly soluble in water.
Use: Preemergence herbicide.

metofos.
CAS: 67481-15-6. $C_{16}H_{15}Cl_3O_2 \bullet C_{12}H_{14}Cl_3O_4P$.
Hazard: A poison by ingestion. Moderately toxic by skin contact.
Use: Agricultural chemical.

metolachlor.   (2-chloro-*N*-(2-ethyl-6-methyl-phenyl)-*N*-(2-methoxy-1-methylethyl)acetamide).
CAS: 51218-45-2.
Use: A herbicide.

metominostrobin.
CAS: 133408-50-1. $C_{16}H_{16}N_2O_3$.
Hazard: Moderately toxic by ingestion.
Use: Agricultural chemical.

metopon hydrochloride. (6-methyldihydro-morphinone hydrochloride). $C_{18}H_{21}O_3N \cdot HCl$. A morphine derivative.
Properties: White, odorless, crystalline powder. Very soluble in water; sparingly soluble in alcohol; insoluble in benzene.
Hazard: Addictive narcotic, prescription only.
Use: Medicine (analgesic).

metoquinone.
CAS: 622-91-3. $C_{20}H_{24}N_2O_4$.
Properties: Colorless crystals. Mp 130C (decomposes). Partially soluble in water; slightly soluble in alcohol and acetone; almost insoluble in benzene.
Use: Photographic developer.

metribuzin.
CAS: 21087-64-9. $C_8H_{14}N_4OS$.
Properties: Colorless crystals. Mp 125C. Soluble in alcohols; slightly soluble in water.
Hazard: Toxic material. Liver damage and hematologic effects. Questionable carcinogen.
Use: Herbicide.

MeV. Abbreviation for million electron-volts.

mevalonic acid. (3,5-dihydroxy-3-methyl-pentanoic acid; 3,5-dihydroxy-3-methlvaleric acid; β,α-dihydroxy-β-methylvaleric acid; (3R)-3,5-dihydroxy-3-methylpentanoic acid). $C_6H_{12}O_4$. An organic acid, it is an intermediate in the biosynthesis of squalene, cholesterol, and coenzyme q in plants and animals.

mevinphos. (2-carbomethoxy-1-methylvinyl dimethyl phosphate; methyl-3-hydroxy-α-crotonate dimethyl phosphate).

CAS: 7786-34-7. $(CH_3O)_2P(O)OC(CH_3:CHCOOCH_3$.
Properties: Yellow liquid. Bp 99–103C (0.03 mm Hg). Slightly soluble in oils; miscible with water and benzene.
Hazard: Toxic by ingestion, inhalation, and skin absorption; use may be restricted; cholinesterase inhibitor. Questionable carcinogen.
Use: Insecticide and acaricide.

mexacarbate.
CAS: 315-18-4. $C_{12}H_{18}N_2O_2$.
Properties: Colorless crystals. Mp 85C. Soluble in acetone, alcohol, and benzene; insoluble in water.
Hazard: Toxic by ingestion.
Use: Insecticide.

mexicain. Proteolytic enzyme with properties and uses similar to papain.

Meyer reaction. Preparation of alkylstannonic acids by reacting alkali stannite with an alkyl iodide. When applied to alkali arsenites or plumbites, the reaction yields alkylarsonic and alkylplumbonic acids, respectively.

Meyer–Schuster rearrangement. Acid catalyzed rearrangement of secondary and tertiary α-acetylenic alcohols to α,β-unsaturated carbonyl compounds: aldehydes when the acetylenic group is terminal, ketones when it is internal.

Meyers aldehyde synthesis. Synthesis of aldehydes $R-CH_2CHO$ from alkylhalides $R-X$ and 2-lithiomethyl-tetrahydro-3-oxazine.

Meyer synthesis. Formation of aliphatic nitrites and nitro derivatives by the reaction of aliphatic halides with metal nitrites.

Mg. Symbol for magnesium.

mg. Abbreviation for milligram.

MH. See maleic hydrazide.

"MH-30" [Chemtura]. TM for a 30% solution of maleic hydrazide.

MHD. See magnetohydrodynamics.

MIAK. See methyl isoamyl ketone.

miazine. See pyrimidine.

MIBC. Abbreviation for methylisobutyl carbinol.
See methylamyl alcohol.

MIBK. Abbreviation for methyl isobutyl ketone.

mica.
CAS: 12001-26-2. Any of several silicates of varying chemical composition but with similar physical properties and crystalline structure. All characteristically cleave into thin sheets that are flexible and elastic. Synthetic mica is available. It has electrical and mechanical properties superior to those of natural mica; it is also water free.
Properties: Soft, translucent solid; colorless to slight red (ruby), brown to greenish yellow (amber). D 2.6–3.2. Mohs hardness 2.8–3.2, refr index 1.56–1.60, dielectric constant 6.5–8.7. Noncombustible. Heat resistant to 600C.
Derivation: From muscovite (ruby mica), phlogopite (amber mica), and pegmatite. Synthetic single crystals are "grown" electrothermally.
Occurrence: U.S., Canada, India, Brazil, Malagasy Republic.
Available forms: Block, sheet, powder, single crystals.
Grade: Dry-ground, wet-ground.
Hazard: (Dust) Irritant by inhalation, may be damaging to lungs. Pneumoconiosis.
Use: Electrical equipment, vacuum tubes, incandescent lamps, dusting agent, lubricant, windows in high-temperature equipment, filler in exterior paints, cosmetics, glass and ceramic flux, roofing,

rubber, mold-release agent, specialty paper for insulation and filtration, wallpaper and wallboard joint cement, oil-well drilling muds.

Micarta. A group of laminated plastics composed of paper or fabric made from cellulose, glass, asbestos or synthetic fibers bonded with phenolic or melamine resins and cured at elevated temperature and pressure.
Use: Plating barrels, rayon-manufacturing equipment, pickling tanks, electrical and thermal insulation, oil-handling equipment, steel rolling-mill bearings, chemical-handling valve bodies, paper-mill suction box covers and equipment.

micellar flooding. See chemical flooding.

micelle. An electrically charged colloidal particle, usually organic in nature, composed of aggregates of large molecules, e.g., in soaps and surfactants. The term is also applied to the casein complex in milk.
See colloid chemistry.

Michaelis–Arbuzov reaction. Formation of monoalkylphosphonic esters from alkyl halides and trialkyl phosphites, via the intermediate phosphonium salt.

Michaelis–Menten constant. The substrate concentration at which an enzyme-catalyzed reaction proceeds at one-half of its maximum velocity.

Michaelis–Menten equation. The equation describing the hyperbolic dependence of the initial reaction velocity, Vo, on substrate concentration, [S].

Michaelis–Menten kinetics. A kinetic pattern observed in enzyme-catalyzed reactions when the initial rate (as a function of substrate concentration) exhibits a hyperbolic shape.

Michael reaction. Base catalyzed addition of carbanions to activated unsaturated systems.

Michel, Hartmut. (1948–). Awarded Nobel Prize for chemistry in 1988, along with Deisenhofer and Huber, for work that revealed the three-dimensional structure of closely-linked proteins which are essential to photosynthesis. Doctorate awarded in 1977 by University of Wurtzburg, Germany.

Michler's hydrol. See tetramethyldiaminobenzhydrol.

Michler's ketone. See tetramethyldiaminobenzophenone.

"mickey finn". *[Babco].* An alcoholic beverage doctored with a powerful purgative or a stupefying drug, usually used on an unsuspecting victim (e.g., "slipping someone a mickey") to ease the commission of a crime such as robbery. In practice, any of a variety of substances have been used for such a purpose including chloral hydrate.
Hazards: Extremely hazardous.

micro-. Prefix meaning 10^{-6} unit (symbol μ), e.g., 1 microgram = 0.000001 g or 1 μg.
See micron.

microanalysis. Analysis of chemicals using miniaturized equipment and microscopic techniques. The sample is in microgram quantities.
See macroanalysis; microchemistry.

microarray. Sets of miniaturized chemical reaction areas that may also be used to test DNA fragments, antibodies, or proteins.

microballoon. Hollow, finely divided, hole-free, low-density particles of synthetic resins or similar film-forming materials. Glass is one of the materials used.
Use: To form a protecting layer of the tiny spheres over liquid surfaces, such as oils in big tanks, to reduce evaporation; to separate helium from natural gas because of the wide difference in relative rates of diffusion through the spheres; as an extender in plastics to achieve low density.

microbial genetics. The study of genes and gene function in bacteria, archaea, and other microorganisms. Often used in research in the fields of bioremediation, alternative energy, and disease prevention.
See model organisms; biotechnology; bioremediation.

microbody. Refers to structures less than 1 micron in diameter, delimited by a single membrane and containing a moderately dense matrix in which a crystal-like core and other inclusions are sometimes present.

"Microcel" *[Blanver].* TM for finely divided calcium or magnesium silicate.
Use: For use as a filler in rubber compounding and other purposes.

"Micro-Chek" *[Ferro].* TM for antimicrobials.

microchemistry. A branch of analytical chemistry that involves procedures that require handling of very small quantities of materials. Specifically, it refers to carrying out various chemical operations (weighing, purification, quantitative and qualitative analysis) on samples ranging from 0.1 to 10 mg; this often involves use of a microscope, and still more often chromatography.
See microscopy, chemical.

microcrystalline. A form in which a number of high-polymeric substances have been prepared. They include cellulose, chrysotile asbestos, amylose (starch), collagen, nylon, and certain mineral waxes. On the microscopic level, these substances are composed of colloidal microcrystals connected by molecular chains. The process involves breaking up the network of microcrystals (by acid hydrolysis in the case of cellulose) and separating them by mechanical agitation. The size range of the microcrystals is from 2.5 to 500 nanometers (millimicrons). The products form extremely stable gels that have a number of commercial use possibilities. Petroleum-derived waxes of high molecular weight have been available in microcrystalline form for many years. Chlorophyll has a natural microcrystalline structure.
See cellulose; wax, microcrystalline; "Avicel" [FMC].

microcrystalline wax. (amorphous or petrolatum wax). A hydrocarbon wax obtained from petroleum that consists of extremely fine crystals. Stickier than paraffin wax and has higher viscosity than paraffin wax when melted.
Properties: Mp 145–190F.

microCurie. See Curie.

Microcystis. A species of cyanobacteria.
Properties: Cyclic polypeptide that contains aspartic acid, glutamic acid, d-serine, valine, ornithine, alanine, and leucine.

microcystis aeruginosa toxin. See 5-l-argininecyanoginosin la.

microencapsulation. Enclosure of a material in hollow spheres or capsules (microspheres) in the micron size range (20–150μ), they can be made of glass, silica, various high polymers or proteins (gelatin, albumen). The silica type can be incorporated in plastics, elastomers, and metals for weight-saving purposes; they can also be bonded to one another to give extremely thin sheets of silica. Microspheres coated with layers of "Teflon" [Du Pont] and beryllium are used to contain the deuterium and tritium used in laser fusion experimentation.
Polymeric or proteinaceous microspheres are used to introduce drugs to specific locations in the body. The coating material acts as a semipermeable membrane, permitting slow release and high concentration of a drug at the desired site. Enzymes, hormones, and other biochemical substances can be temporarily immobilized by this technique.

microfilaments. Thin filaments composed of actin found throughout the cytoplasm of eukaryotic cells. Their function is to give the cell shape and movement.

microfluidics. Study of the behavior of fluids at volumes thousands of times smaller than a common droplet. It is a multidisciplinary field that includes physics, chemistry, engineering, and biotechnology. Microfluidic components form the basis of so-called "lab-on-a-chip" devices that can process microliter and nanoliter volumes and conduct highly sensitive analytical measurements.

microgram (μg). One millionth (10^{-6}) gram.

microinjection. A technique for introducing a solution of DNA into a cell using a fine microcapillary pipette.

"Microlith" *[Precision].* TM for organic pigment stir-in dispersions compatible with a broad range of organic solvents and polymers.

micrometer. (μm). One millionth (10^{-6}) meter, or 1 micron (10,000 Å units).

micron. See micrometer.

micronized sulfur. Sulfur that consists of extremely fine particles.
Use: Dusts or sprays in the control of certain plant diseases.

micronuclei. Chromosome fragments that are not incorporated into the nucleus at cell division.

micronutrient. See trace element.

microorganism. An organism of microscopic size generally considered to include bacteria, molds (actinomyces), and fungi, but excluding viruses.
See bacteria.

microprobe. An instrument for chemical microanalysis of a sample, in which a beam of electrons is focused on an area less than a micrometer in diameter. The characteristic X-rays emitted as a result are dispersed and analyzed in a crystal spectrometer to provide a qualitative and quantitative evaluation of chemical composition.
See microanalysis.

microsatellite. A microsatellite is a simple sequence repeat (SSR). It might be a homopolymer ("... TTTTTTT ... "), a dinucleotide repeat ("....CACACACACACACA ....."), a trinucleotide repeat ("....AGTAGTAGTAG-TAGT ... "), etc. Due to polymerase slip (a.k.a. polymerase chatter), during DNA replication there is a slight chance these repeat sequences may become altered; copies of the repeat unit can be created or removed. Consequently, the exact number of repeat units may differ between unrelated individuals. Considering all the known microsatellite markers, no two individuals are identical. This is the basis for forensic DNA identification and for testing of familial relationships (e.g., paternity testing).

microscope. See optical microscope; electron microscope; field-ion microscope; ultramicroscope.

microscopy, chemical. Use of a microscope primarily for study of physical structure and identification of materials. This is especially useful in forensic chemistry and police laboratories. Many types of microscopes are used in industry; most important are the optical, ultra-, polarizing, stereoscopic, electron, and X-ray microscopes. Organic dyes of various types are used to stain samples for precise identification.

"MicroSelect" [Enpath]. TM for ultra pure basic reagents for use in biochemistry and the life sciences where relatively large amounts of reagents are required.

"Microsol" [TIFA]. TM for aqueous pigment dispersions for spin-coloring of regenerated cellulose fibers.

microsomes. Membranous vesicles formed by fragmentation of the endoplasmic reticulum (and probably Golgi) of eukaryotic cells. Recovered by differential centrifugation.

microsphere. See microencapsulation.

microtubule organizing center. (MTOC). Bundles of protein tubes which may be found at the base of a eukaryotic flagellum. In animals, they also function in creating the arrays of microtubules that pull the chromosomes apart during mitosis.

microtubules. Type of filament in eukaryotic cells composed of units of the protein tubulin. Among other functions, it is the primary structural component of the eukaryotic flagellum.

microvilli. Thin finger-like protrusions from the surface of a cell, often used to increase absorptive capacity or to trap food particles. The "collar" of choanoflagellates is actually composed of closely spaced microvilli.

microwave spectroscopy. A type of absorption spectroscopy used in instrumental chemical analysis that involves use of that portion of the electromagnetic spectrum having wavelengths in the range between the far infrared and the radiofrequencies, i.e., between 1 mm and 30 cm. Substances to be analyzed are usually in the gaseous state. Klystron tubes are used as microwave source.

middle oil. A fraction distilled from coal tar. See coal tar.

middlings. The granular part of the interior of the wheat berry obtained in the process of milling. This product, when reduced by grinding to the desired fineness, produces the finest quality of flour.

Midgley, Thomas, Jr. (1889–1944). An American chemist and inventor. One of the most creative and brilliant chemists of his era, Midgley's early work was in the field of rubber chemistry and technology, especially in the development of synthetic and substitute rubbers that were being introduced in the 1930s. He worked with Kettering at General Motors and then became vice president of Ethyl Corporation, as well as of the Ohio State University Research Foundation. His innovative genius was responsible for the development of organic lead compounds for antiknock gasoline and later for the discovery of fluorocarbon refrigerants for which he did the basic research. He was recipient of many of chemistry's highest honors including the Nichols medal, the Perkin medal, and the Priestly medal.

Miescher degradation. Adaptation of the Barbier–Wieland carboxylic acid degradation to permit simultaneous elimination of three carbon atoms, as in degradation of the bile acid side chain to the methyl ketone stage. Conversion of the methyl ester of the bile acid to the tertiary alcohol, followed by dehydration, bromination, dehydrohalogenation, and oxidation of the diene yields the required degraded ketone.

"Miglyol" [Cremer]. TM for fatty acid esters. See "Witepsol" [Cremer].

Mignonac reaction. Formation of amines by catalytic hydrogenation of aldehydes and ketones in liquid ammonia and absolute ethanol in the presence of a nickel catalyst.

migration. Movement of a substance from one material to another with which it is in intimate contact, e.g., from the container into packaged produce. Similarly, a portion of the sulfur in a rubber mixture may migrate into a material to which it is laminated.

migration area. A term used in nuclear technology as a measure of the moderation or slowing down of neutrons. It is one-sixth of the mean square distance a neutron travels before thermal capture.

mike. A term adopted by the American Standards Association for a microinch or 10^{-6} inch.

mil. One-thousandth (0.001) inch.
Use: In reference to surface coatings, metal sheet, films, cable covers, friction tape, etc.

Milad. Clarifying agents used in polyproylene products.
Use: To prevent haziness and discoloring.

Milas hydroxylation of olefins.   Formation of *cis*-glycols by reaction of alkenes with hydrogen peroxide and either a catalytic amount of osmium, vanadium, or chromium oxide or UV light.

mildew preventive.   A compound used to prevent the growth of parasitic fungi, usually stain-producing, on such organic materials as textiles, leather, paper, farinaceous products, etc. Compounds used widely include cresols, phenols, benzoic acid, formaldehyde, and organic derivatives or salts of copper, zinc, and mercury.

Milestone.   A multi-component fouling control solution.

military gas.   A chemical warfare agent delivered as a gas or dispersed as an aerosol.

military poison.   A chemical used as or developed for use as a warfare agent.
Hazard: Toxic.

milk.   A heterogeneous liquid secreted by the mammary gland and composed (for cows' milk) of approximately 87% water, 3.8% emulsified particles of fat and fatty acids, 3% casein, 5% sugar (lactose), serum proteins, calcium, phosphorus, potassium, iron, magnesium, copper, and several vitamins. The fat particles are from 6 to 10 micrometers in diameter, much larger than colloidal size; they are coated with an adsorbed layer of protein (protective colloid), that maintains the emulsion. The casein is closely associated with calcium, forming micelles of calcium caseinate that are of colloidal dimensions. The white color of milk is largely due to light scattering by these particles, and to some extent by the fat particles, rather than to the presence of a pigment. The casein complex coagulates (1) when high temperature or bacteria convert the lactose to lactic acid, as in souring, or (2) when acid or certain enzymes (rennet) are intentionally added. The serum proteins lactalbumin and lactoglobulin are also colloidally dispersed. The lactose and mineral salts are in true molecular solution. Thus milk is a complex system exhibiting several levels of dispersion, from the molecular through colloidal and into the microscopic size range. Important processes applied to milk on an industrial scale include pasteurization, homogenization, coagulation, dehydration, and condensation. The milks of animals other than cows show considerable variations in composition, especially fat and protein content.
See colloid chemistry; emulsion; casein.

milk-clotting   enzyme   from   *Bacillus cereus*.
Properties: Derived from *Bacillus cereus* (fam. *Bacillaceae*).
Use: Food additive.

milk-clotting enzyme from *Endothia parasitica*.
Properties: Derived from *Endothia parasitica* (fam. *Diaporthacessae*).
Use: Food additive.

milk-clotting enzyme from *Mucor miehei*.
Properties: Derived from *Mucor miehei Cooney et Emerson* (fam. *Mucoraceae*).
Use: Food additive.

milk-clotting enzyme from *Mucor pusillus*.
Properties: Derived from *Mucor pusillus Lindt* (fam. *Mucoraceae*).
Use: Food additive.

milk of lime.   (lime water). Calcium hydroxide suspended in water.

milk of magnesia.   (magnesia magma). A white, opaque, more or less viscous suspension of magnesium hydroxide in water from which varying proportions of water usually separate on standing.
Grade: USP.
Use: Medicine (laxative).

milk sugar.   See lactose.

milk thistle extract.   See *Silybum marianum* (linn.) gaertn., extract.

mill.   See ball mill; pebble mill.

Millester.   Hydroxyl terminated polyesters.
Use: For building blocks for water-based, solvent-based, and 100% active polyurethanes.

milli-.   Prefix meaning 10^{-3} unit, or 1/1000th part.

milliCurie.   See Curie.

milliequivalent.   (meq). One-thousandth of the equivalent weight of a substance.

milligram.   (mg). One-thousandth gram (10^{-3} gram).

milliliter.   (mL). One thousandth liter, the volume occupied by one gram of pure water at 4C and 760 mm Hg pressure. One milliliter (mL) equals 1.000027 cubic centimeters (cc).

millimeter.   (mm). One-thousandth meter, about 0.03937 inch.

millimicron.   (mμ; mu). One-thousandth micron, 10 Å, 1 billionth meter, 1 nanometer.

millirem.   One-thousandth rem.
See rem.

Millon's reagent.   Solution produced by dissolving mercury in equal weight of nitric acid (d: 1.41), and solution diluted to twice the volume and decanted from the precipitate.
Use: Test for albumin.

Milloxane.   Thermoplastic polyurethanes.
Use: For coated fabrics such as ski clothing, rainwear, inflatables, tents; adhesives.

Mills–Nixon effect.   Selective reactivity of certain substituted aromatic systems suggestive of certain degree of preponderance of one of the tautomeric Kekule forms in the benzenoid nucleus.

"Milorganite" *[Milwaukee].*   TM for blended fertilizer containing approximately 20% of an activated sludge marketed in dry granular form by the Milwaukee sewage disposal plant. Contains 5–10% moisture, 6.5–7.5% ammonia, 2.5–3.5% available phosphoric acid, 3–4% total phosphoric acid. Source of vitamin B_{12}.

Milori blue.   Any of a number of the varieties of iron blue pigments.
See iron blues.

mineral.   A widely used general term referring to the nonliving constituents of the earth's crust that includes naturally occurring elements, compounds, and mixtures that have a definite range of chemical composition and properties. Usually inorganic, but sometimes including fossil fuels (e.g., coal), minerals are the raw materials for a wide variety of elements (chiefly metals) and chemical compounds. Minerals can be and many are synthesized to achieve purity greater than that found in natural products. The term *mineral industry* statistically comprehends the mining and production of metals (ores) fossil fuels, clay, gemstones, cement, glass, rocks, sulfur, sand, etc. Mineralogy is the study and classification of minerals by source, chemical composition, and properties, chiefly physical, such as color, hardness, and crystalline structure. This term was used by early chemists to describe a variety of substances; many of these uses are obsolescent, but a few persist including the following:

mineral black: inorganic black pigments
mineral blue: varieties of blue pigments
mineral dust: industrial dust, nuisance dust
mineral green: copper carbonate
mineral oil: a liquid petroleum derivative
mineral pitch: asphalt
mineral red: iron oxide red
mineral rubber: blown asphalt
mineral spirits: a grade of naphtha
mineral water: natural spring water containing sulfur, iron, etc.

mineral wax: a wax found in the earth (ozokerite), or derived from petroleum
mineral wool: fibers made by blowing air or steam through slag

As used by nutritionists the term refers to such components of foods as iron, copper, phosphorus, calcium, iodine, selenium, fluorine, and trace micronutrients.

mineral alkali.   An inorganic alkali.

mineral oil.   (mineral oil, white; molol; nujol; oil mist, mineral; paraffin oil).
CAS: 8012-95-1.
Properties: Colorless, oily liquid; practically tasteless and odorless. D 0.83–0.86 (light), 0.875–0.905 (heavy), flash p 444F (OC), ULC: 10–20. Insoluble in water and alcohol Soluble in benzene, chloroform, and ether. A mixture of liquid hydrocarbons from petroleum.
Hazard: Inhalation of vapor or particulates can cause aspiration pneumonia. Combustible liquid. Upper respiratory tract irritant. Questionable carcinogen.
Use: As a binder, defoaming agent, fermentation aid, lubricant, coating (protective), and release agent.

mineral spirits.   (AMSCO 140; mineral turps; petroleum spirits; soltrol; soltrol 50; soltrol 100; soltrol 180)
CAS: 64475-85-0. A medium petroleum distillate.
Properties: Flash point of 100F.
Hazard: Low toxicity.
Use: As a solvent in insecticides and certain household products.

minim.   In the U.S., a unit of volume equal to approximately 0.06 mL.
Use: Pharmacy.

minium.
CAS: 1314-41-6. Pb_3O_4. Natural red oxide of lead.
Occurrence: Found in Colorado, Idaho, Utah, Wisconsin.
See lead oxide, red.

MIPA.   Abbreviation for monoisopropanolamine.
See isopropanolamine.

mipafox.   See *N,N'*-diisopropyldiamidophosphoryl fluoride.

miracle fruit.   (miraculin).
See sweetener, nonnutritive; glycoprotein.

mirbane oil.   See nitrobenzene.

mirex. (bichlorendo; CG-1283; dechlorane 4070; dodecachlorooctahydro-1,3,4-metheno-2H-cyclobuta(c,d) pentalene; 1,1a,2,2,3,3a,4,5,5,5a,5b,6-dodecachlorooctahydro-1,3,4-metheno-1H-cyclobuta(c,d)pentalene; dodecachloropentacyclodecane; dodecachloropentacyclo(3,2,2,O^{26}, O^{36}, O$^{5\cdots}$)decane; ENT 25,719; ferriamicide;-hexachlorocyclopentadienedimer; 1,3,4,5,5-hexachloro-1,3-cyclopentadiene dimer; HRS 1276; NCI-C06428; perchlorodihomocubane; perchloropentacyclodecane; perchloropentacyclo(5.2.1.O^{26}, O^{36}, O^{53})decane)
CAS: 2385-85-5. C$_{10}$Cl$_2$. A chlorinated hydrocarbon.
Properties: Intensely white, odorless, crystalline substance, insoluble in water, soluble in benzene and dioxane.
Use: Insecticide.
Hazard: Nephrotoxic, highly toxic, possible carcinogen, teratogen, tumorigen, mutagen, poisonous, environmentally persistent, and bioaccumulates.

mirror-image molecules. See optical isomerism; enantiomer; chiral.

mirror nuclei. A pair of atomic nuclei, each of which would be transformed into the other by changing all its neutrons into protons and vice versa.

misch metal. The primary commercial form of mixed rare-earth metals (95%) prepared by the electrolysis of fused rare earth chloride mixture, d approximately 6.67, mp approximately 648C. Form: waffle-like plates weighing 40–60 lb packed in oiled paper, immersed in oil, or coated with vinyl paint.
Hazard: Flammable, dangerous fire risk.
Use: Lighter flints, ferrous and nonferrous alloys, cast iron, aluminum, nickel, magnesium and copper alloys, getter in vacuum tubes, magnetic alloys.

miscibility. The ability of a liquid or gas to dissolve uniformly in another liquid or gas. Gases mix with one another in all proportions. This may or may not be true of liquids, whose miscibility properties depend on their chemical nature. Alcohol and water are completely miscible because of their chemical similarity, but some liquids are only partially miscible with others because of their chemical difference, e.g., benzene and water. Many gases are miscible with liquids to a greater or lesser extent, e.g., formaldehyde mixes readily with water; CO$_2$ is partially miscible with water, and oxygen only very slightly. Liquids that do not mix at all are said to be immiscible, as oil and water. The term *solubility* is often used synonymously with *miscibility* in reference to liquids, but it more properly applies to solids.

Miscible G-125. A complex solvent blend cleaning fluid.

missense mutation. A change in the base sequence of a gene that alters or eliminates a protein.

Mitchell, Peter. D. (1920–1992). A British biochemist who was the recipient of the Nobel Prize for chemistry in 1978 for his work on studies of cellular energy transfer. A graduate of Cambridge and recipient of many awards, he was the Director of Research, Glynn Research Institute.

miticide. A pesticide that kills mites, small animals of the spider class, among them the European red mite and the common red spider which infest fruit trees.

mitochondria. (mitochondrion). Particles of cytoplasm found in most respiring cells. They synthesize most of the cell's adenosine triphosphate and are the chief energy sources of living cells. They are highly plastic, mobile structures that may fragment or fuse together at random. Many enzymes, especially those involved in converting food-derived energy into a form usable by the cell, are located in the mitochondria, and DNA molecules have also been found there. Yeast is a particularly rich source of mitochondria for research purposes. The complex organelle found in most eukaryotes is believed to be descended from free-living bacteria that established a symbiotic relationship with a primitive eukaryote.
See double membrane.

mitochondrial DNA. The genetic material found in mitochondria, the organelles that generate energy for the cell. Not inherited in the same fashion as nucleic DNA.
See cell; DNA; genome; nucleus.

mitochondrial gene. Any gene located in the mitochondria. Human mitochondrial DNA is only 16,569 base pairs in length. It codes for ribosomal RNAs and transfer RNAs used in the mitochondrion. Only 13 recognizable genes are present that code for polypeptides. The mitochondrial genome is directly transmitted through the maternal line and is therefore ideal for tracing family lineages.

mitochondrial matrix. The aqueous portion of a mitochondrion bounded by the inner mitochondrial membrane.

mitomycin C.
CAS: 50-07-7. C$_{15}$H$_{18}$N$_4$O$_5$. Antibiotic derived from *Streptomyces,* stated to be effective against tumors.
Hazard: Possible carcinogen.

mitosis. The division of a cell nucleus to produce two new cells, each having the same chemical and genetic constitution as the parent cell. The deoxyribose (nucleic acid) component of the chromosomes

is present in duplicate in the original nucleus. The amount of nucleic acid is doubled just before cell division begins; subsequent events (called phases) permit separation of the products of replication to form the new nuclei. Each half-chromosome carries the identical nucleic acids of the original chromosome.
See cell (1).

Mitsunobu reaction. Intermolecular dehydration reaction occurring between alcohols and acidic components on treatment with diethyl azodicarboxylate and triphenyl phosphine under mild neutral conditions. The reaction exhibits streospecificity and regional and functional selectivity.

mixed acid. (nitrating acid). A mixture of sulfuric and nitric acids used for nitrating, e.g., in the manufacture of explosives, plastics, etc. Consists of 36% nitric acid and 61% sulfuric acid.
Hazard: Spillage may cause fire or liberate dangerous gas. Causes severe burns, irritant by ingestion and inhalation, may cause NO_x poisoning.

mixed expired gas. The gas that results from one or more deep exhalations. It consists of thoroughly mixed gas from the dead space of the respiratory tree and the alveoli.

mixed-function oxidases. (oxygenases). Enzymes that use O_2 to simultaneously oxidize a substrate and a cosubstrate (commonly NADH or NADPH).

mixed glyceride. A diglyceride or triglyceride that contains more than one type of fatty acid connected to glycerol via an ester linkage.

mixing. Effecting a uniform dispersion of liquid, semisolid, or solid ingredients of a mixture by means of mechanical agitation. Low-viscosity liquids and suspensions are mixed with impellers of the turbine or propellor type. The mixing action results both from direct contact of the impeller blades with the liquid and from the turbulence induced by the impeller in the outer portions of the liquid. For this reason the diameter of the impeller need be only from one-fourth to one-half that of the container. For liquids of medium viscosity, revolving paddles of various shapes are used. Thicker mixtures involving volatile solvents are mixed in closed containers (churns) equipped with fin-like members mounted on a rotating shaft. For liquids of very high viscosity, helical rotors, sigma blades, and similar devices are necessary. Because turbulence cannot be initiated in such fluids, the blades must fit closely within the walls of the container so as to make contact with every part of the material being mixed. While most industrial mixing of such pasty materials is done batchwise in kneaders, Banbury mixers, etc., continuous mixing of these can also be effectively carried out in horizontal compartments equipped with rotating screws whose pitch

and flight are contoured in such a way as to provide both rotary and axial motion. There are a number of ingeniously engineered types of these for mixing plastics, rubber, food products, and similar products. Dry, solid particulates are mixed in rotating cylinders or tumbling barrels.
See impeller; agitator; kneading; muller.

mixo-. Prefix indicating mixtures of isomers having the same functional group.

mixture. (mix). A heterogeneous association of substances that cannot be represented by a chemical formula. Its components may or may not be uniformly dispersed and can usually be separated by mechanical means. Liquids that are uniformly dispersed are called solutions. Mixtures may be natural or artificial, as indicated by the following:

Natural	Artificial
air	glass
petroleum	paint
milk	cement
blood	perfumes
marble	plastics
wood	cermets
latex	alloys
vegetable oils	seawater

See compound; blend; solution; mixing.

MKP. Abbreviation for monopotassium phosphate.
See potassium phosphate, monobasic.

mL. Abbreviation for milliliter.

MLA. Abbreviation for mixed lead alkyls.

mm. Abbreviation for millimeter.

MMH. Abbreviation for monomethylhydrazine.

Mn. Symbol for manganese.

Mo. Symbol for molybdenum.

mobile liquid. See mobility.

mobility. The ease with which a liquid moves or flows. Hydrocarbon liquids (nonpolar) that have low viscosity, surface tension, and density respond more readily to an applied force than does water (a polar liquid). For this reason, fires involving hydrocarbon liquids should be extinguished with foam rather than with a direct stream of water.

MOCA. See 4,4'-methylenebis(2-chloroaniline).

modacrylic fiber. A manufactured fiber in which the fiber-forming substance is any long-chain synthetic polymer composed of less than 85% but

at least 35% by weight of acrylonitrile units, – $CH_2CH(CN)$– (Federal Trade Commission). Other chemicals, such as vinyl chloride, are incorporated as modifiers. Characterized by moderate tenacity, low water absorption, and resistance to combustion; self-extinguishing.
Hazard: Questionable carcinogen.
Use: Deep pile and fleece fabrics, industrial filters, carpets, underwear, blends with other fibers.
See acrylonitrile; acrylic fiber.

"Modaflow" *[AI Chem].* TM for ethyl acrylate.
Available forms: Viscous clear liquid.
Use: It improves flow and leveling in products; eliminates lumpiness.

model. A representation, either abstract or physical, of a system, arrangement, or structure that cannot be perceived objectively. (1) A mathematical model is one in which all or most of the parameters of a complex system such as an ocean are assigned symbolic values that can be utilized to give a theoretical approximation of actuality. Such models are useful in physical chemical analyses. (2) A space-lattice model is a three-dimensional duplication of the shape and structure of a crystal in which the atoms composing the lattice are plastic spheres or balls connected by rods to represent bonds. (3) A molecular model is similar, except that it represents an individual chemical compound rather than a crystal. The spheres are made to scale based on the known diameter of the atoms represented; they are often colored to suggest the nature of the element (black for carbon, white for hydrogen, red for halogens, etc.). In one type, both single and double bonds are plastic rods that join the spheres at appropriate angles; in another the spheres are fused in clusters. The two types are illustrated by the models of isobutane shown; a clustered model of the DNA molecule is shown in the entry on deoxyribonucleic acid. Both space-lattice and molecular models are useful for classroom demonstration.

modeling. The use of statistical analysis, computer analysis, or model organisms to predict outcomes of research.

model organisms. A laboratory animal or other organism useful for research.

moderator. A substance of low atomic weight, such as beryllium, carbon (graphite), deuterium (in heavy water), or ordinary water, which is capable of reducing the speed of neutrons but that has little tendency toward neutron absorption. The neutrons lose speed when they collide with the atomic nuclei of the moderator. Moderators are used in nuclear reactors, because slow neutrons are most likely to produce fission. A typical graphite-moderated reactor may contain 50 tons of uranium for 472 tons of graphite. Reactors in the U.S. are cooled and moderated with light water.

modification. A chemical reaction in which some or all of the substituent radicals of a high polymer are replaced by other chemical entities, resulting in a marked change in one or more properties of the polymer without destroying its structural identity. Cellulose, e.g., can be modified by substitution of its hydroxyl groups by carboxyl or alkyl radicals along the carbon chain. These reactions are usually not stoichiometric. Their products have many properties foreign to the original cellulose, e.g., water solubility, high viscosity, gel- and film-forming ability. Other polymeric substances that can undergo modification are rubber, starches, polyacrylonitrile, and some other synthetic resins.
See cellulose, modified.

modified polyacrylamide resins.
Properties: Produced by copolymerization of acrylamide with not more than 5 mol% of β-methacrylyloxyethyl trimethylammonium methyl sulfate.
Use: Flocculent; food additive.

modifier gene. A nonallelic gene that controls or changes the manifestation of a gene by interfering with transcription.

Modrich, Paul. (1946–). An American biochemist that received the 2015 Nobel Prize jointly with Thomas Lindahl and Aziz Sancar. The scientists mapped how cells repair damaged DNA while safeguarding the genetic information at a genetic level. This work has provided fundamental knowledge of how a living cell functions and is used for the development of new cancer treatments. His research showed how cellular machinery fixes errors that arise during DNA replication. His Ph.D. was awarded by Stanford University, and he is currently an investigator at the Howard Hughes Medical Institute and James B. Duke Professor at the Duke University Medical Center. His interest in nature and science was sparked by the natural landscape of his hometown of Raton, New Mexico.

modulator.   A metabolite that, when bound to the allosteric site of an enzyme, alters its kinetic characteristics; can be homotropic or heterotropic, inhibitor or activator.

modulus of elasticity.   (elastic modulus). A coefficient of elasticity representing the ratio of stress to strain as a material is deformed under dynamic load. It is a measure of the softness or stiffness of the material (Young's modulus).

moellon degras.   See degras.

Moerner, William E.   (1953–). An American chemist who won the 2014 Nobel Prize jointly with Eric Betzig and Stefan W. Hell. The scientists developed a groundbreaking method to bypass the natural resolution limit in optical microscopy by using fluorescent molecules. Moerner and Bitzig's method used light to excite or deactivate fluorescent molecules of proteins and to create an image by combining images of different activated molecules of the protein. This method enables tracking of active viruses and molecules in living cells. He was received his M.S. and Ph.D. from Cornell University.

moexipril hydrochloride.
CAS: 82586-52-5. $C_{27}H_{34}N_2O_7$•ClH.
Hazard: Moderately toxic by ingestion.

mohair.   A natural fiber, similar to wool, obtained from angora goats.
Properties: Tenacity 14 g/denier. Combustible.
Use: Fabrics for outer clothing, draperies, upholstery.

Mohr's salt.   See ferrous-ammonium sulfate.

Mohs scale.   An empirical scale of the hardness of mineral or mineral-like materials originally consisting of 10 values, ranging from talc, with a rating of 1, to diamond, with a rating of 10—the rating based on the ability of each material to scratch the one directly below it in the series. The number of materials has been expanded from 10 to 15 with the addition of several synthetically produced substances (e.g., silicon carbide) between the original 9 and 10 positions. The scale is named after the German mineralogist, Friedrich Mohs (1773–1839). See hardness (1).

moiety.   An indefinite portion of a sample.

Moissan, Henri.   (1852–1907). A Native of Paris, Moissan was a professor at the School of Pharmacy from 1886 to 1900 and at the Sorbonne from 1900 to 1907. At the former institution, he first isolated and liquefied fluorine in 1886 by the electrolysis of potassium acid fluoride in anhydrous hydrogen fluoride. His work with fluorine undoubtedly shortened his life as it did that of many other early experimenters in the field of fluorine chemistry. He won great fame by his development of the electric furnace and pioneered its use in the production of calcium carbide, making acetylene production and use commercially feasible in the preparation of pure metals, such as magnesium, chromium, uranium, tungsten etc. and in the production of many new compounds, e.g., silicides, carbides, and refractories. In 1906, he was awarded the Nobel Prize in chemistry.

"Moisture-Lok" [Grayson].   TM for a powder composed of dextrose, dextrin, and fiber.
Use: Binds moisture to meat products for longer than other ingredients.

molal.   A concentration in which the amount of solute is stated in moles and the amount of solvent in kilograms. The unit of molality is moles of solute per kilogram of solvent and is designated by a small *m*, 1 mole of NaCl in 1 kg of solvent is a 1 molal concentration.
Note: Do not confuse with molar.

molar.   A concentration in which one molecular weight in grams (1 mole) of a substance is dissolved in enough solvent to make one liter of solution. Molarity is indicated by an italic capital *M*. Molar quantities are proportional to the molecular weight of the substances.

molassess.   The thick liquid left after sucrose has been removed from the mother liquor in sugar manufacture. Blackstrap molasses is the syrup from which no more sugar can be obtained economically. It contains approximately sucrose 20%, reducing sugars 20%, ash 10%, organic nonsugars 20%, water 20%. Combustible.
Use: Feed, food, raw material for various alcohols, acetone, citric acid, and yeast propagation. Sodium glutamate is made from Steffens molasses, a waste liquor from beet sugar manufacture.
See fermentation.

mold.   See fungus.

mold and mildew cleaner.   Any cleaner used to control mold and mildew. They often contain phenol, kerosene, or pentachlorophenol.
Hazard: Toxic, death may result from ingestion.

molding.   Forming a plastic or rubber article in a desired shape by application of heat and pressure, either in a negative cavity, usually of metal, or in contact with a contoured metal or phenolic resin surface.
See injection molding; blow molding; compression molding.

molding powder. A mixture in a granular or pelleted form of a plastic base material together with necessary modifying ingredients (filler, plasticizer, pigment, etc.). Such mixtures are normally prepared by resin manufacturers and sold as such to processors ready for use in injection molding or extrusion operations.

molding sand. See foundry sand.

mold preventive. See mildew preventive.

mold release agent. See abherent.

"Moldwiz" *[Axel].* TM for external mold release agents that are advanced polymeric formulations.
Use: Permit multiple release from a single application leaving parts clean for secondary process.

mole. The amount of pure substance containing the same number of chemical units as there are atoms in exactly 12 g of carbon-12 (i.e., 6.023×10^{23}). This involves the acceptance of two dictates—the scale of atomic masses and the magnitude of the gram. Both have been established by international agreement. Formerly, the connotation of "mole" was "gram molecular weight." Current usage tends to apply the term "mole" to an amount containing Avogadro's number of whatever units are being considered. Thus, it is possible to have a mole of atoms, ions, radicals, electrons, or quanta. This usage makes such terms as *gram-atom*, *gram-formula* weight unnecessary. All stoichiometry essentially is based on the evaluation of the number of moles of substance. The most common involves the measurement of mass. Thus 25.000 g of water will contain 25.000/18.015 moles of water, 25.000 g of sodium will contain 25.000/22.990 moles of sodium. The convenient measurements on gases are pressure, volume, and temperature. Use of the ideal gas law constant R allows direct calculation of the number of moles: $n = PV/RT$. T is the absolute temperature, R must be chosen in units appropriate for P, V, and T. The acceptance of Avogadro's law is inherent in this calculation; so too are approximations of the ideal gas.
See Avogadro's law.

molecular biology. A subdivision of biology that approaches the subject of life at the molecular level. This applies to phenomena occurring within the cell nucleus, where the chromosomes and genes are located. These structures, which determine heredity, are in turn composed of nucleic acids, which direct the selection and assembly of amino acids in the dividing chromosomes. Much of the essential mechanism of life can be understood by study of specific protein molecules (DNA and RNA) and their determination of the amino acid composition of the genes.
See genetic code; deoxyribonucleic acid; recombinant DNA.

molecular distillation. (high-vacuum distillation). Distillation at low pressures of the order of 0.001 mm Hg. A molecular distillation is distinguished by the fact that the distance from the surface of the liquid being vaporized to the condenser is less than the mean free path (the average distance traveled by a molecule between collisions) of the vapor at the operating pressure and temperature. This distance is usually of the order of magnitude of a few inches. This process is useful in separation of extremely high boiling and heat-sensitive materials such as glycerides and some vitamins.

molecular farming. The development of transgenic animals to produce human proteins for medical use.

molecular formula. See formula, chemical.

molecular gas. A gas composed of a single species, such as oxygen, chlorine, or neon.

molecular genetics. The study of macromolecules important in biological inheritance.

molecular manufacturing. The production of complex structures via nonbiological mechanosynthesis (and subsequent assembly operations).

molecular medicine. The treatment of injury or disease at the molecular level. Examples include the use of DNA-based diagnostic tests or medicine derived from DNA sequence information.

molecular orbital. The wave function of an electron as it moves in the field of all other electrons and the nuclei constituting the molecule. It is usually expressed as a linear combination of atomic orbitals.

molecular oxygen. (free oxygen; dioxygen).
CAS: 7782-44-7. O_2. The most abundant element on earth, making up about 47% of the earth's mass, and essential for respiration.
Properties: Diatomic molecule, colorless, odorless, tasteless, can be broken down into oxygen atoms by electromagnetic radiation of wavelengths below 242 nm.

molecular rearrangement. See rearrangement.

molecular sandwich. (sandwich molecule).
See metallocene.

molecular sieve. A microporous structure composed of either crystalline aluminosilicates, chemically similar to clays and feldspars and belonging to a class of materials known as zeolites, or crystalline aluminophosphates derived from

mixtures containing an organic amine or quaternary ammonium salt. Pore sizes range from 5 to 10Å. The outstanding characteristic of these materials is their ability to undergo dehydration with little or no change in crystalline structure. The dehydrated crystals are interlaced with regularly spaced channels of molecular dimensions, that compose almost 50% of the total volume of the crystals. The empty cavities in activated "molecular sieve" crystals have a strong tendency to recapture the water molecules that have been driven off. This tendency is so strong that if no water is present they will accept any material that can get into the cavity. However, only those molecules that are small enough to pass through the pores of the crystals can enter the cavities and be adsorbed on the interior surface. This sieving or screening action, which makes it possible to separate smaller molecules from larger ones, is the most unusual characteristic of molecular sieves. They are used in many fields of technology; to dry gases and liquids; for selective molecular separations based on size and polar properties; as ion-exchangers; as catalysts; as chemical carriers; in gas chromatography; and in the petroleum industry to remove normal paraffins from distillates.
See zeolite; gel filtration; pore.

molecular weight. The sum of the atomic weights of the atoms in a molecule. That of methane (CH_4) is 16.043, the atomic weights being carbon = 12.011, hydrogen = 1.008. The chemical formula used in such a calculation must be the true molecular formula of the substance designated. For example, the molecular formula of oxygen is O_2 and its molecular weight is 31.998 (atomic weight of oxygen = 15.999). For ozone the molecular formula is O_3 and the molecular weight is 47.997. The true molecular weight of a gas or vapor is found by measuring the volume of a given weight and then calculating the weight of 22.4 L at 0C and 760 mm Hg. The molecular weight of many complex organic molecules runs as high as a million or more (proteins and high polymers).
See Avogadro; atomic weight; mole.

molecule. A chemical unit composed of one or more atoms. The simplest molecules contain only one atom, for example, helium molecules (1 atom/molecule). Oxygen molecules (O_2) are composed of two atoms, and ozone (O_3) of three. Molecules may contain several different sorts of atoms. Water contains two different kinds, hydrogen and oxygen, and dimethylamine [$(CH_3)_2NH$] has three kinds. Molecules of many common gases [hydrogen (H_2), oxygen (O_2), nitrogen (N_2), and chlorine (Cl_2)] consist of two atoms each. The atoms of a molecule are held together by chemical bonds. Molecules vary in size from less than 1 to more than 500 millimicrons and in weight from 4 (He) to 40 million for tobacco mosaic virus.
See macromolecule; bond, chemical; atom.

Molina, Mario, J. (1943–). Mexican who won the Nobel Prize for chemistry along with Paul Crutzen and Frank Sherwood Rowland in 1995 for their work in atmospheric chemistry, particularly concerning the formation and decomposition of ozone.
See Crutzen, Paul; Rowland, Frank Sherwood.

molinate sulfoxide.
CAS: 52236-29-0. $C_9H_{17}NO_2S$.
Hazard: A reproductive hazard.

molten salt.   See fused salt.

molybdate chrome orange.   See molybdate orange.

molybdate orange. (molybdenum orange; molybdate chrome orange).
CAS: 12656-85-8. A solid solution of lead chromate, lead molybdate and lead sulfate.
Properties: Fine, dark-orange or light-red powder.
Derivation: By adding solutions of sodium chromate, sodium molybdate, and sodium sulfate to a lead nitrate solution under carefully controlled conditions and filtering off the precipitates.
Hazard: Toxic by ingestion.
Use: Pigment in printing inks, paints, plastics.

molybdenite. (molybdenum glance).
CAS: 1317-33-5. MoS_2. Natural molybdenum sulfide found in igneous rocks and metallic veins.
Properties: Bluish-gray color, gray-black streak, metallic luster. One perfect cleavage, greasy feel. Mohs hardness 1–1.5, d 4.6–4.8. Similar in appearance to graphite. Soluble in sulfuric and concentrated nitric acids.
Occurrence: Colorado, Utah, New Mexico, Chile.
Use: Principal ore of molybdenum.

molybdenite concentrate. Commercial molybdenite ore after the first processing operations. Contains approximately 90% molybdenum disulfide along with quartz, feldspar, water, and processing oil.

molybdenum.
CAS: 7439-98-7.   Mo. Metallic element of atomic number 42, group VIB of the periodic table, aw 95.94, valences = 2, 3, 4, 5, 6. Seven stable isotopes.
Properties: Gray metal or black powder. Does not occur free in nature. D 10.2, mp 2610C, bp 5560C, high strength at very high temperatures, oxidizes rapidly above 1000F in air at sea level, but is stable in upper atmosphere. Insoluble in hydrochloric acid and hydrogen fluoride, ammonia, sodium hydroxide, or dilute sulfuric acid; soluble in hot concentrated sulfuric or nitric acids; insoluble in water. A necessary trace element in plant nutrition.
Derivation: By aluminothermic, hydrogen, or electric furnace reduction of molybdenum trioxide or ammonium molybdate.

Available forms: Ingots, rods, wire, powder, ingots (from powder), high ductility sheets, also as large single crystal.

Grade: Rods and wire 99.9%, powder 99.9%.

Hazard: Flammable in form of dust or powder. Lower respiratory tract irritant. Questionable carcinogen.

Use: Alloying agent in steels and cast iron; high-temperature alloys, tool steels; pigments for printing inks, paints, and ceramics; catalyst; solid lubricants; missile and aircraft parts; reactor vessels; cermets; die-casting copperbase alloys; special batteries.

See ferromolybdenum; heteromolybdates.

molybdenum acetylacetonate.
CAS: 14284-90-3. $Mo(C_5H_7O_2)$.
Properties: Crystalline powder. Slightly soluble in water; resistant to hydrolysis. A chelating nonionizing compound.
Use: Catalyst for polymerization of ethylene and formation of polyurethane foam.

molybdenum aluminide.   A cermet that can be flame-sprayed.

molybdenum anhydride.   See molybdenum trioxide.

molybdenum boride.
CAS: 12007-97-5. Several borides are known: Mo_2B, mp 2000C; MO_3B_2, mp 2070C; MoB (ordinary and β-forms), mp 2180C; Mo_2B_5, mp 1600C (transforms to MoB_2).
Derivation: By heating molybdenum powder and boron to 1500–1600C in hydrogen.
Use: Brazes to join molybdenum, tungsten, tantalum, and niobium parts, especially electronic components, corrosion and abrasion-resistant parts, cutting tools, refractory cermets.

molybdenum carbonyl.   See molybdenum hexacarbonyl.

molybdenum dioxide.
CAS: 18868-43-4. MoO_2.
Properties: Lead-gray, nonvolatile powder. D approximately 6.4. Insoluble in hydrochloric acid and hydrogen fluoride and alkalies, sparingly soluble in sulfuric acid.
Derivation: Reduction of molybdenum trioxide or molybdates by hydrogen, partial oxidation of metallic molybdenum.
Hazard: Toxic material.

molybdenum diselenide.
CAS: 12058-18-3. $MoSe_2$. Available as a 40-micron powder.
Use: Solid lubricant.

molybdenum disilicide.
CAS:12136-78-6. $MoSi_2$. A cermet.

Properties: Dark-gray, crystalline powder. D 6.31 (20C), mp 1870–2030C. Not affected by air up to 1648C. Not attacked by most inorganic acids including aqua regia, but very soluble in hydrofluoric and nitric acids. Has high stress-rupture strength.
Derivation: By fusion of hydrogen-reduced molybdenum with silicon.
Available forms: Cylinders, lumps, granules, powder, whiskers. May be coated on materials by vapor deposition and by flame-spraying.
Grade: 98%, 99.5%, mesh size 200 and 325.
Hazard: Toxic material.
Use: Electrical resistors, protective coatings at high temperatures, engine parts in space vehicles (molybdenum coated with molybdenum disilicide).

molybdenum disulfide.   (molybdic sulfide; molybdenum sulfide).
CAS: 1317-33-5. MoS_2.
Properties: Black, crystalline powder. D 4.80, mp 1185C Mohs hardness 1, coefficient of friction 0.02–0.06. Soluble in aqua regia, sulfuric acid (concentrated); insoluble in water.
Hazard: Toxic material.
Derivation: Purification of molybdenite, reaction of sulfur or hydrogen sulfide on molybdenum trioxide.
Use: Lubricants in greases, oil dispersions, resin-bonded films, dry powders, etc., especially at extreme pressures and high vacua; hydrogenation catalyst.
See molybdenite.

molybdenum ditelluride.
CAS: 12058-20-7. $MoTe_2$. Available as a 40-micron powder.
Use: Solid lubricant.

molybdenum glance.   See molybdenite.

molybdenum hexacarbonyl.   (molybdenum carbonyl).
CAS: 13939-06-5. $Mo(CO)_6$.
Properties: White, shiny crystals. Decomposes at 150C (sublimes), d 1.96, bp approximately 155C, vap press approximately 0.1 mm Hg (20C), approximately 43 mm Hg (101C). Insoluble in water; soluble in ceresin, paraffin oil, benzene, aminoanthraquinone; slightly soluble in ether and other organic solvents.
Derivation: From molybdenum pentachloride by reaction with zinc dust and carbon monoxide in ether at high pressures.
Hazard: Decomposes above 150C to evolve carbon monoxide.
Use: Plating molybdenum, i.e., molybdenum mirrors; intermediate.

molybdenum hexafluoride.
CAS: 7783-77-9. MoF_6.
Properties: White, crystalline solid; hygroscopic. Mp 17.5C, bp 35C, d (liquid) approximately 2.5. Readily hydrolyzed.

Derivation: Action of fluorine on molybdenum powder.
Hazard: Strong irritant.
Use: Separation of molybdenum isotopes.

molybdenum lake.   See phosphomolybdic pigment.

molybdenum metaphosphate.
CAS: 13520-60-0.   Mo(PO$_3$)$_6$.
Properties: Yellow powder. D 3.28 (0C). Insoluble in water and in most acids; slightly soluble in hot aqua regia.

molybdenum naphthalene.
Properties: Dark purple, viscous liquid. Soluble in most hydrocarbons.
Hazard: Toxic material.
Use: Catalyst for commercial production of propylene oxide using hydroperoxides.

molybdenum orange.   See molybdate orange.

molybdenum (III) oxide.   See molybdenum sesquioxide.

molybdenum oxides.   See molybdenum sesquioxide, molybdenum dioxide, molybdenum trioxide.

molybdenum pentachloride.
CAS: 10241-05-1.   MoCl$_5$.
Properties: Green-black solid, dark red as liquid or vapor. Mp 194C, bp 268C, d 2.9. Hygroscopic, reacting with water and air. Soluble in dry ether, dry alcohol, and other organic solvents.
Derivation: Direct action of chlorine on finely divided molybdenum metal.
Hazard: Irritant.
Use: Chlorination catalyst, vapor-deposited molybdenum coatings, component of fire-retardant resins, brazing and soldering flux, intermediate for organometallic compounds, e.g., molybdenum hexacarbonyl.

molybdenum sesquioxide.   (dimolybdenum trioxide; molybdenum (III) oxide). Mo$_2$O$_3$. Known only in the hydrated form, Mo(OH)$_3$, although commonly assigned the formula Mo$_2$O$_3$. A compound formed by a dry reaction of molybdenum and oxygen which approximates the composition of the sesquioxide is probably a mixture of molybdenum and molybdenum dioxide.
Properties: Gray-black powder. Slightly soluble in acids; insoluble in alkalies and water.
Derivation: Zinc reduction of acid solutions of molybdic acids and molybdates, electrolytic deposition from acid solutions of molybdates.
Use: Catalyst in organic synthesis, decoration and protection for metal articles, feed additive.

molybdenum silicide.   Alloy of 60% molybdenum, 30% silicon, and 10% iron used as means of introducing molybdenum into steel.

molybdenum sulfide.   See molybdenum disulfide.

molybdenum trioxide.   (molybdenum anhydride; molybdic oxide; molybdic acid hydride).
CAS: 1313-27-5.   MoO$_3$.
Properties: White powder at room temperature, yellow at elevated temperatures. D 4.69, mp 795C, sublimes starting at 700C, bp 1150C. Sparingly soluble in water; very soluble in excess alkali with formation of molybdates; soluble in concentrated mixture of nitric acid and hydrochloric acid or nitric and sulfuric acids. Two hydrates are known: MoO$_3$•H$_2$O and MoO$_3$•2H$_2$O. Readily combines with acids and bases to form a series of polymeric compounds.
Derivation: Roasting of molybdenite; by ignition of the metal, sulfides, lower oxides, and molybdic acids.
Method of purification: Sublimation.
Grade: Technical, pure, reagent, ACS.
Hazard: Toxic material.
Use: Source of molybdenum compounds, agriculture, analytical chemistry, manufacture of metallic molybdenum, introduction of molybdenum in alloys, corrosion inhibitor, ceramic glazes, enamels, pigments, catalyst.

molybdic acid.   Molybdic acid of commerce is either ammonium molybdate (molybdic acid 85%) or molybdenum trioxide. The use of the term interchangeably for these compounds has caused confusion. Solutions of molybdic acid are very complex chemically since they show a great tendency to polymerize.

molybdic acid, anhydride.   See molybdenum trioxide.

molybdic oxide.   See molybdenum trioxide.

molybdic sulfide.   See molybdenum disulfide.

molybdophosphates.   See heteromolybdates.

12-molybdophosphoric acid.   See phosphomolybdic acid.

molybdosilicates.   See heteromolybdates.

12-molybdosilicic acid.   (silicomolybdic acid). H$_4$SiMo$_{12}$O$_{40}$•xH$_2$O where x is usually 6–8.
Properties: Yellow, crystalline powder. D 2.82. Soluble in water, ethanol, acetone; insoluble in benzene and cyclohexane; decomposes in strongly basic solutions; thermally stable.
Grade: Reagent.
Hazard: Strong oxidizing agent in aqueous solution.

Use: Catalysts; reagents; photography; precipitants and ion exchangers in atomic energy; additives in plating processes; imparting water resistance to plastics, adhesives, and cement.
See heteromolybdates.

MON. Abbreviation for Motor Octane Number.

monacetin. See acetin.

Monacide. A series of insecticides containing 5% DDVP.

Monafax 1214. Deceth-4 phosphate.
Properties: Hypochloride stable surfactant.
Grade: 100% active (free acid) liquid.
Hazard: Neutralize with caustic before contacting hypochloride.
Use: Wetting agent, surface-tension reduction, detergent.

monalide. (N-(4-chlorophenyl)-2,2-dimethylpentanamide). $C_{13}H_{18}ClNO$.
Use: Post-emergent analide herbicide used to destroy weeds of crops such as carrots, celery, and parsley.

Monamate LNT-40. Ammonium lauryl sulfosuccinate.
Hazard: Extremely mild to skin and eyes.
Use: High-foaming anionic surfactant for use in non-irritating shampoos, soaps, and bubble bath.

"Monamids" [Mona]. TM for a group of dialkylolamides that includes various grades of coconut fatty acid monoethanolamide, coconut fatty acid monopropanolamide, lauric acid monoethanolamide, lauric acid monoisopropanolamide, and stearic acid monoethanolamide.

Monamines. A group of dialkylolamides used as detergent, detergent additives, foam boosters, wetters, emulsifiers, dispersing agents, thickeners, and conditioners.

Monastrip. A solvent stripper for uncured and cured epoxy, polyester, and silicone rubber casting and encapsulating compounds.

Monateric CAB. Cocamidopropyl betaine.
Properties: Low color, low pH.
Grade: Liquid.
Use: High-foaming surfactant for viscosity control and detergent action in cosmetic and industrial cleansers.

Monaterics. A special group of substituted imidazolines classified as amphoteric surfactants. Excellent wetting, emulsifying, penetrating, and spreading properties in systems requiring broad pH ranges.

monatomic. See diatomic.

"Monawets" [Mona]. TM for a group of surfactants of di-octyl, di-hexyl, di-isobutyl, and di-tridecyl sulfosuccinates known for their wetting, spreading, penetrating, and emulsifying power.

monazite. A natural phosphate of the rare-earth metals, principally the cerium and lanthanide metals, usually with some thorium. Yttrium, calcium, iron, and silica are frequently present. Monazite sand is the crude natural material and is usually purified from other minerals before entering commerce.
Properties: Yellowish to reddish-brown color, vitreous to resinous luster, white streak. Mohs hardness 5–5.5, d 4.9–5.3.
Occurrence: North Carolina, South Carolina, Idaho, Colorado, Montana, Florida, Brazil, India, Australia, Canada.
Use: Source of thorium, cerium, and other rare-earth metals and compounds.

Monazolines. A series of cationic imidazolines useful as emulsifiers, antistatic agents, water displacers, and corrosion inhibitors in agricultural sprays, acid and solvent cleaners, cosmetics, and water–oil systems.

Mond process. Mixed ores, obtained from roasting crude ores, are heated from 50 to 80C in a stream of producer gas. Oxides other than nickel are reduced to the metallic state, whereas nickel forms nickel carbonyl [$Ni(CO)_4$], which passes off as a vapor. The vapor is subsequently resolved into carbon monoxide and free nickel.

"Mondur" [Bayer Material]. TM for a series of isocyanates.
Use: Surface coatings; adhesives; chemical intermediate; hydrophobic agent to increase water repellency of textiles, leather, and paper products.

Monel. A series of corrosion-resistant alloys of nickel and copper.

"Monex" [Chemtura]. TM for tetramethylthiuram monosulfide.

monhydrin. See propadrine hydrochloride.

mono-. Prefix denoting one, e.g., monochloroacetic acid.
See chloroacetic acid.

monoacetoxyscirpenol. See (3-α)-12,13-epoxytrichothec-9-ene-3,4,15-triol 15.

mono-o-acetylsolanoside.
CAS: 4420-65-9. $C_{32}H_{48}O_9$.
Hazard: A poison.
Source: Natural product.

monoamine oxidase. (mao; amine oxidase; adrenalin oxidase; tyraminase).
CAS: 9001-66-5. An enzyme that is distributed widely among animals. It inactivates a number of biogenic amines by catalyzing the oxidative deamination of these compounds.

monoamine oxidase inhibitor. (mao inhibitor; mao-i). Any of a group of psychiatric drugs that act chiefly as central nervous system stimulants. They presumably change the chemical composition of the blood reportedly responsible for the depression. The antidepressant action of this type of drug is not fully explained by the inhibition of monoamine oxidase. Moa inhibitors potentiate the action of barbiturates, antihistamines, and other antidepressants.
Hazard: Toxic; may cause trembling, nausea, vomiting, lethargy, dry mouth, ataxia, stupor, rise or fall in blood pressure, fever, tachycardia, acidosis, convulsions, liver damage, jaundice, death from cardiac or respiratory failure.
Use: In medicine to control depression, help control anxiety, to treat migraine headaches and hypertension.

monoammonium glycyrrhizinate. See ammoniated glycyrrhizin.

monoanhydrosorbitol. See sorbitan.

monoazo dye. See azo dye.

monobasic. Descriptive of acids having one displaceable hydrogen atom per molecule. Acids having two, three, or more displaceable hydrogen atoms are called dibasic, tribasic, and polybasic, respectively.
See acid.

monobasic acid. An acid that contains one ionizable hydrogen atom per molecule.

Monobed. Intimate mixture of "Amberlite" cation and anion exchange resins.
Use: Complete removal of ionizable impurities from water and other solutions in a one-step treatment.

monocalcium phosphate. See calcium phosphate, monobasic.

monochloroacetic acid. Legal label name (Rail) for solid chloroacetic acid.

monochloroacetone. Legal label name (Rail) for chloroacetone.

monochlorobenzene. See chlorobenzene.

monochlorodifluoromethane. Legal label name (Rail) for chlorodifluoromethane.

monochloroethane. See ethyl chloride.

monochloromethane. See methyl chloride.

monochloropentafluoroethane. Legal label name (Rail) for chloropentafluoroethane.

monochlorophenol. See chlorophenol.

1-monochloropinacoline.
CAS: 13547-70-1. $C_6H_{11}ClO$.
Hazard: A poison by ingestion. Low toxicity by inhalation.

monochlorotetrafluoroethane. Legal label name (Rail) for chlorotetrafluoroethane.

monochlorotriazinyl dye. A fiber-reactive dye for cellulose fibers.
See dye, fiber-reactive.

monochlorotrifluoromethane. Legal label name (Rail) for chlorotrifluoromethane.

Monochrome. (USA). A series of mordant dyestuffs. Characterized by good fastness properties.
Use: Dyeing of wool.

monoclonal antibodies. Substances that can locate and bind to cancer cells wherever they are in the body. They can be used alone, or they can be used to deliver drugs, toxins, or radioactive material directly to the tumor cells.

monocrotalic acid.
CAS: 26543-09-9. $C_8H_{12}O_5$.
Hazard: Moderately toxic.
Source: Natural product.

monocrotophos. See Azodrin.

mono- and diglycerides.
Properties: Yellow liquids to ivory-colored plastics to hard solids; bland odor and taste. Sol in alc, ethyl acetate, chloroform, other chlorinated hydrocarbons; insol in water.
Use: Food additive.

mono and diglycerides, sodium sulfoacetate derivatives. See sodium sulfoacetate derivatives of mono and diglycerides.

mono-, di-, and tripotassium citrate.
Use: Food additive.

mono-, di-, and trisodium citrate.
Use: Curing accelerator; food additive; stabilizer.

mono-, di-, and tristearyl citrate.
Use: Food additive.

monoethanolamine. See ethanolamine.

monoethylamine.   Legal label name (Rail) for ethylamine.

monoethyl butylphosphonate anhydride with diethyl phosphate.
CAS: 63886-51-1. $C_{10}H_{24}O_6P_2$.
Hazard: A poison.

monofilament.   A single, continuous strand of glass or synthetic fiber as extruded from a spinnerette.
See filament.

monogenic disorder.   A disorder caused by mutation of a single gene.
See mutation; polygenic disorder.

monogenic inheritance.   See monogenic disorder.

monoglyceride.   A glycerol ester of fatty acids in which only one acid group is attached to the glycerol group. A typical formula is $RCOOCH_2CH_2OCH_2OH$. Small amounts of monoglycerides occur naturally.
Derivation: Produced synthetically by the alcoholysis of fats with glycerol, yielding a mixture of mono-, di-, and triglycerides that is predominantly monoglycerides.
Use: Emulsifiers, cosmetics, lubricants.
See glycerol monostearate; glycerol monolaurate.

monoglyceride citrate.
Properties: Soft, white-colored, lard-like, waxy solid; bland odor and taste. Sol in fat solvents, alc; insol in water.
Use: Food additive.

monoglyme.   See ethylene glycol dimethyl ether.

monohydrate salt.   (dried cupric sulfate).
Properties: Hygroscopic, water-soluble, off-white powder, nearly insoluble in ethanol.

monohydroxy alcohol.   An alcohol in which a hydroxyl group (–OH) has replaced one of the hydrogen atoms of a hydrocarbon, for example:

C_2H_5H C_2H_5OH
ethane   ethanol

RH   ROH
alkane   alkyl alcohol

There are a number of classifications analogous to those of hydrocarbons: (1) paraffinic or simple alcohols, whose formula may be represented as C_nH_{2n+1}; (2) olefinic or fatty alcohols that contain one or more double bonds; (3) alicyclic alcohols, closed-ring structures that may or may not contain a double bond, e.g., cyclohexanol; (4) aromatic alcohols in which the hydroxyl group is attached to a benzene nucleus as in phenol; (5) heterocyclic alcohols, based on the pentagonal furan ring; and (6) polycyclic alcohols of high molecular weight, known collectively as sterols. Any of these types that contain 12 or more carbon atoms are semisolid to solid and have a wax-like consistency; the others are colorless liquids. Monohydric alcohols are also classified as primary, secondary, or tertiary on the basis of the number of alkyl (methyl) groups substituted for the hydrogen atoms on the central or methanol carbon atom.
See primary.

monoisopropyl citrate.
Use: Food additive.

monolayer.   A single layer of oriented lipid molecules or cells.

monomer.   (momer). A molecule or compound, usually containing carbon and of relatively low molecular weight and simple structure, that is capable of conversion to polymers, synthetic resins, or elastomers by combination with itself or other similar molecules or compounds. Thus, styrene is the monomer from which polystyrene resins are produced and vinyl chloride is the monomer of polyvinyl chloride. Other common monomers are methyl methacrylate, adipic acid, and hexamethylenediamine.

monomethylamine.   Legal label name (Rail) for methylamine.

monomethylarsinic acid.   See methylarsenic acid.

monomethylolformaldehyde.   See glycolic aldehyde.

monomethylmercury.   (methylmercury ion; methylmercury ion (1+); methylmercury (II) cation; methyl mercury). CH_3Hg. An ionic compound that rarely occurs free in the environment as it readily ionizes and usually occurs as a component of another compound. It is often found in association with dimethyl mercury in contaminated fish and aquatic birds. It is synthesized in sediments from inorganic mercury by anaerobic bacteria in neutral or slightly acid sediments through the action of methylcobalamin.
Hazard: Highly reactive, poisonous, mutagenic.
Use: A fungicide.

monomolecular film.   See film.

mono-*n*-octyltin   trichloride.   See octyltrichlorostannane.

monophyletic.   Term applied to a group of organisms which includes the most recent common ancestor of all of its members and all of the descendants of that most recent common ancestor. A monophyletic group is called a clade.

"Monoplex" *[Brandt].*   TM for monomeric liquid plasticizers for polyvinyl chloride and other high polymers. Primarily esters but also some epoxides that impart heat and light stability.
Use: Plasticizers, stabilizers, process aids.

monopropellant.   A propellant that combines fuel and oxidizer in one compound or mixture. Gunpowder is an example of a solid monopropellant. Liquid monopropellants, for rockets, include: methyl nitrate, nitromethane, a mixture of hydrocarbons with tetranitromethane, a mixture of methyl nitrate and methanol.
See rocket fuel.

monoprotic acid.   An acid which has only one proton to lose to a base, HCL, HSO_4.

monosaccharide.   Any of several simple sugars having the formula $C_6H_{12}O_6$; the best-known are glucose, fructose, and galactose. Monosaccharides combine to form more-complex sugars known as oligo- and polysaccharides.

monosodium acid methanearsonate.

monosodium glutamate.   See sodium glutamate.

monosodium methanearsonate.   (ANSAR 170; arsonate liquid; asazol; bueno; daconate 6; dal-e-rad; herb-all; herban m; merge; mesamate; mesamate concentrate; methylarsenic acid, sodium salt; monate; monosodium acid methanearsonate; monosodium acid metharsonate; monosodium methanearsonate; monosodium methanearsonic acid; msma; NCI-C60071; phyban; silvisar 550; sodium acid methanearsonate; sodium hydroxyl(methyl)arsenate; sodium methanearsonate; target msma; trans-vert; weed 108; weed-e-rad; weed-hoe).
CAS: 2163-80-6. CH_5AsNaO_3. The monosodium salt of methylarsonic acid.
Properties: Crystals from water; insoluble in most organic solvents.
Hazard: Toxic; poison.
Use: A herbicide.

monosodium phosphate derivatives of mono- and diglycerides.   See mono- and diglycerides, monosodium phosphate derivatives.

monosomy.   Possessing only one copy of a particular chromosome instead of the normal two copies. See cell; chromosome; gene expression, trisomy.

monostearin.   See glycerol monostearate.

monostearyl citrate.   See "Morflex MSC" *[Vertellus].*

monotonic function.   Function that steadily increases or decreases with x.

monotrichloro-tetra(monopotassium dichloro)-penta-s-triazinetrione.
CAS: 64474-06-2. $C_3HCl_2N_3O_3$•1/$_4C_3Cl_3N_3O_3$•K.
Hazard: Moderately toxic by ingestion. Low toxicity by skin contact.

monovalent serum.   Antiserum that contains antibody to only one strain or species of microorganism or to one type of antigen.

monoxychlor.   Trichloroethane analog of DDT; less toxic to warm-blooded animals than DDT.

Monsell's salt.   Basic ferric sulfate.

montan wax.   (lignite wax).
Properties: White, hard-earth wax; crude product is dark brown. Mp 80–90C. Soluble in carbon tetrachloride, benzene, and chloroform; insoluble in water. Combustible.
Derivation: By countercurrent extraction of lignite. American and German lignite are usual sources.
Method of purification: Distillation with superheated steam.
Grade: Crude, refined.
Use: Substitute for carnauba and beeswax, shoe and furniture polishes, phonograph records, roofing paints, rendering paints waterproof, adhesive pastes, electric insulating compositions, paper-sizing compositions, carbon papers, wire coating, sun-crack preventive in rubber products.

montmorillonite. Al_2O_3•$4SiO_2$•H_2O (approximately). A type of clay. One of the major components of bentonite and fuller's earth.

monuron.   (3-(p-chlorophenyl)-1,1-dimethylurea; CMU).
CAS: 150-68-5. $ClC_6H_4NHCON(CH_3)_2$. A plant growth regulator.
Properties: White, crystalline solid; odorless. Mp 175C. Very low solubility in water and hydrocarbon solvents; slightly soluble in oils; partially soluble in alcohols; stable toward oxidation and moisture.
Hazard: Questionable carcinogen.
Use: Herbicide, sugarcane-flowering suppressant.

Moore, Stanford.   (1913–1982). An American biochemist who won the Nobel Prize for chemistry in 1972, with Anfinsen and Stein, for enzyme studies. He was involved with the analysis of the action of the complex enzyme deoxyribonuclease. His Ph.D. was granted from the University of Wisconsin.

"Mor-Ad" *[Dow].*   TM for a series of adhesives consisting of solvent-free-one component urethane, cross-linked water-borne epoxy, and polypropylene bonding adhesives.

Use: Bonding structural panel components such as metal, wood, and plastics to themselves or to polystyrene foam and untreated polypropylene.

morantel tartrate.
CAS: 26155-31-7. $C_{12}H_{16}N_2S \cdot C_4H_6O_6$.
Use: Drug (veterinary); food additive.

morbid map. A diagram showing the chromosomal location of genes associated with disease.

mordant. A substance capable of binding a dye to a textile fiber. The mordant forms an insoluble lake in the fiber, the color depending on the metal of the mordant. The most important mordants are trivalent chromium complexes, metallic hydroxides, tannic acid, etc. Mordants are used with acid dyes, basic dyes, direct dyes, and sulfur dyes. Premetalized dyes contain chromium in the dye molecule. A mordant dye is a dye requiring use of a mordant to be effective.
See dye, fiber-reactive.

Mordantine. Liquid antimony lactate containing 11% available antimony oxide. Recommended as a replacement for technical tartar emetic.

mordanting assistant. A chemical such as lactic, oxalic, and sulfuric acids, tartrates, etc.
Use: In conjunction with mordants to bring about a gradual decomposition of the latter and to assist in producing a uniform deposition of the actual mordant on and within textile materials.

mordant rouge. See aluminum diacetate.

Morestan. 6-methyl-2,3-quinoxalinedithiol cyclic carbonate.

"Morflex" *[Vertellus].* TM for specialty plasticizers (including phthalates, adipates, azelates, sebacates, trimellitates, triacetin, and polymerics) for vinyl compounds.

"Morflex 150" *[Vertellus].* (dicyclohexyl phthalate).
CAS: 84-61-7. TM for adhesives and nitrocellulose lacquers.

"Morflex 190" *[Vertellus].* (*n*-butyl phthalyl-*n*-butyl glycolate).
CAS: 85-70-1. TM for a plasticizer.
Use: Vinyl.

"Morflex 530" *[Vertellus].* (tridiisodecyl trimellitate).
CAS: 36631-30-8. TM for a plasticizer.
Use: Vinyl.

"Morflex 560" *[Vertellus].* (tri-*n*-hexyl trimellitate). TM for a plasticizer.
Use: Vinyl.

"Morflex 1129" *[Vertellus].* (dimethyl isophthalate).
CAS: 1459-93-4. TM for chemical intermediate.

"Morflex MSC" *[Vertellus].* (monostearyl citrate). TM for chelating agent, surface lubricant.

"Morflex P-51A" *[Vertellus].* TM for an acetylated polymeric of mw 2000 with good solvent resistance.

morin. (2′,3,4,5,7-pentahydroxyflavone).
CAS: 480-16-0. $C_{15}H_{10}O_7 \cdot 2H_2O$. One of the two coloring principles of yellow brazilwood.
Properties: Colorless needles. Mp 285C (decomposes). Soluble in alcohol, alkaline solutions; slightly soluble in boiling water. Combustible.
Use: Mordant dye, spot-test reagent for metal salts, luminescence indicator.
See brasilin.

"Morlex" *[Union].* TM for corrosion inhibitors.
Use: Steam boilers and steam-eating systems, for example, Corrosion Inhibitor A. A mixture of 91% morpholine in water.

Morosodren. A fungicide concentration containing methylmercury dicyandiamide (2.2%).
Hazard: Toxic by ingestion.

morphactin. (methyl 2-chloro-9-hydroxyfluorene-9-carboxylate).
$C_{15}H_{11}ClO_3$
Any of a class of derivatives of fluorine-9-carboxylic acids that powerfully regulate plant growth and development. They inhibit shoot elongation and cause morphological abnormalities in intact plants. They also promote the visions that give rise to lateral primordial in the root pericycle.

morphine.
CAS: 57-27-2. $C_{17}H_{19}NO_3 \cdot H_2O$.
Properties: White, crystalline alkaloid. Bp 254C (decomposes), d 1.31. Slightly soluble in water, alcohol, and ether.
Derivation: From opium by extraction and crystallization. Opium contains approximately 10% morphine.
Hazard: Narcotic, habit-forming drug, sale restricted by law in the U.S.
Use: Analgesic (in form of acetate, hydrochloride, tartrate, and other soluble salts).

p-**morphine.** See thebaine.

morpholine. (tetrahydro-1,4-oxazine).
CAS: 110-91-8.

$OCH_2CH_2NHCH_2CH_2$ or C_4H_8ONH.
Properties: Colorless, hygroscopic liquid; amine-like odor. Bp 128.9C, fp −4.9C, d 1.002 (20/20C),

bulk d 8.34 lb/gal (20C), vap press 6.6 mm Hg (20C), viscosity 2.23 cP (20C), flash p (100F) (37.7C OC), autoign temp 590F (310C). Soluble in water and organic solvents.

Derivation: Dehydration of diethanolamine.

Grade: Technical, 98%.

Hazard: Flammable, moderate fire risk. Toxic by ingestion and inhalation, irritant to skin, absorbed by skin. Eye damage and upper respiratory tract irritant. Questionable carcinogen.

Use: Rubber accelerator, solvent, additive to boiler water, waxes and polishes, optical brightener for detergents, corrosion inhibitor, preservation of book paper, organic intermediate (catalyst, antioxidants, pharmaceuticals, bactericides, etc.).

morpholine borane.
CAS: 4856-95-5. $C_4H_8ONH \cdot BH_3$.
Properties: White, needle-shaped, crystalline compound. Mp 93C. Soluble in hot water and alcohol; insoluble in carbon tetrachloride.
Use: Reducing agent for aldehydes and ketones. Useful in acid media where sodium borohydride is ineffectual because of its instability in acid.

morpholine ethanol. (*N*-hydroxyethylmorpholine).
CAS: 622-40-2. $C_4H_8ONCH_2CH_2OH$.
Properties: Colorless liquid. D 1.0724, bp 225.5C, flash p 210F (98.9C). Miscible with water. Combustible.

9-(morpholinoamino)acridine.
CAS: 28846-41-5. $C_{17}H_{17}N_3O$.
Hazard: A poison by ingestion.
Use: Agricultural chemical.

9-(morpholinoamino)acridine mono(methyl sulfate).
CAS: 28846-42-6. $C_{17}H_{17}N_3O \cdot CH_4O_4S$.
Hazard: A poison by ingestion.
Use: Agricultural chemical.

2-(morpholinothio)benzothiazole. (*N*-oxydiethylene-2-benzothiazolesulfenamide).
$C_{11}H_{12}N_2OS_2$.
Properties: Buff to brown flakes; sweet odor. Mp 80C min, d 1.34 (25C). Insoluble in water; soluble in benzene, acetone, methanol.
Use: Delayed-action vulcanization accelerator.

morphology. A term borrowed from the biological sciences by physical chemists to denote the shape, structure, or form of such substances as high polymers, crystals, reinforcing agents, and the like, e.g., the morphology of carbon black in rubber.

morphothion. (generic for O,O-dimethyl-*S*-(morpholinocarbonylmethyl)phosphorodithioate).

$(CH_3O)_2P(S)SCH_2C(O)NCH_2CH_2OCH_2CH_2$.

Properties: Colorless solid. Mp 65C. Soluble in acetone, dioxane, and acetonitrile.
Hazard: Cholinesterase inhibitor, use may be restricted.
Use: Insecticide.

Morse equation. An equation according to which the potential energy of a diatomic molecule in a given electronic state is given by a Morse potential.

mortar. (1) A type of adhesive or bonding agent that may be either inorganic or organic, soft, and workable when fresh but sets to a hard, infusible solid on standing, either by hydraulic action or by chemical cross-linking. The chief ingredients of inorganic mortars are cement, lime, silica, sulfur, and sodium or potassium silicate. Organic mortars are based on various synthetic resins (epoxy, phenolic, polyester, and furan). All types are resistant to acids. Some (potassium silicate) are useful up to 1600F. Others are used for bonding acid-proof brick, tile, etc.; for masonry construction; and for lining chemical reaction equipment.
See sealant; adhesive; cement. (2) A ceramic receptacle used by pharmacists for preparing mixtures of medicinals and for hand-pulverizing soft solids.

mortar, metallic. A mixture of powdered metal and other ingredients that have been mixed with water. Lead, tungsten, and depleted uranium have been used as the metal component. These mortars resist weathering, mild acids and alkalies, intense radiation, and extreme temperature variation.
Use: Space technology.

Mortrim. A series of one- and two-component solvent-based and water-based polyurethane adhesives.
Use: Manufacture of automotive headliners, door panels, and instrument panels.

morzid. (generic for bis(1-aziridinyl)-morpholinophosphine sulfide). $CH_2CH_2OCH_2CH_2$ $NP(NCH_2CH_2)_2 : S$.
Use: Insect chemosterilant.

mosaic protein. Any proteins that have many, often repeated, domains. While the domains are interconnected, each behaves independently with respect to structure, function, and folding behavior.

mosapride.
CAS: 112885-41-3. $C_{21}H_{25}ClFN_3O_3$.
Hazard: A poison by ingestion.

Moseley curve. See Moseley diagram.

Moseley diagram. (Mosely curve). A graph showing the relationship between atomic numbers of a series of elements and the wavelengths of their corresponding spectral lines.

Moseley, Henry. (1887–1915). A British chemist who studied under Rutherford and brilliantly developed the application of X-ray spectra to the study of atomic structure; his discoveries resulted in a more accurate positioning of elements in the periodic table by closer determination of atomic numbers. Tragically for the development of science, Moseley was killed in action at Gallipoli in 1915.

Moseley's law. The square root of the frequency of a given line of an element in the X-ray spectrum is directly proportional to the atomic number of the element.

Mossbauer effect. A nuclear phenomenon discovered in 1957. Defined as the elastic (recoil-free) emission of a γ-particle by the nucleus of a radioactive isotope and the subsequent absorption (resonance scattering) of the particle by another atomic nucleus. Occurs in crystalline solids and glasses but not in liquids. Examples of γ-emitting isotopes are: iron-57, nickel-61, zinc-67, tin-119. The Mossbauer effect is used to obtain information on isomer shift, on vibrational properties and atomic motions in a solid, and on location of atoms within a complex molecule.

mossy zinc. Zinc powder formed by pouring molten zinc into water.

mother. (1) A mold of bacterial complex containing enzymes that promote fermentation, as in manufacture of vinegar from cider or of cultured dairy products from milk. (2) A substance secreted by epithelial cells of the oyster. (3) A mother liquor is a concentrated solution from which the product is obtained by evaporation and/or crystallization, e.g., in sugar manufacture.
See nacre.

motile. Able to move oneself about, capable of self-locomotion.

Motor Octane Number. (MON).
See octane number.

mountain blue. (copper blue).
Derivation: The mineral azurite in ground form.
Use: Paint pigment.

mouse model. See model organisms.

6-MP. Abbreviation for 6-mercaptopurine.

MPA. Abbreviation for multipurpose additive.

3-MPA. Abbreviation for 3-methoxypropylamine.

MPC black. Abbreviation for medium processing channel black.
See carbon black.

MPK. Abbreviation for methyl propyl ketone.

MPP+. See cyperquat.

MPS-500. A stabilized chlorinated ester of a fatty acid. A viscous, light-yellow liquid recommended as a low-cost plasticizer for polyvinyl-chloride formulations.

M-Pyrol. An aprotic solvent, N-methyl-2-pyrrolidone. Claimed to be an effective cleaner of vinyl chloride reaction equipment.
See N-methyl-2-pyrrolidone.

MRI. A procedure using a magnet linked to a computer to create pictures of areas inside the body. Also called magnetic resonance imaging.

MRM-10.
Combination of basic yellow 40, Rhodamine 6G and MBD dyes which produce fluorescence when exposed to selected wavelengths of light.
Use: To visualize cyanoacrylate fumed friction ridge detail.

mRNA. (messenger RNA). An RNA which contains sequences coding for a protein. The term mRNA is used only for a mature transcript with poly(A) tail and with all introns removed, rather than the primary transcript in the nucleus. As such, an mRNA will have a 5′ untranslated region, a coding region, a 3′ untranslated region, and (almost always) a poly(A) tail. Typically about 2% of the total cellular RNA is mRNA.

"MS" [Grace]. TM for microspherical silica alumina.
Use: Cracking catalyst.

MSG. Abbreviation for monosodium glutamate.
See sodium glutamate.

MSP. Abbreviation for monosodium phosphate.
See sodium phosphate, monobasic.

Mt. Symbol for meitnerium.

MTBE. See methyl t-butyl ether.

MT black. Abbreviation for medium thermal black.

MTD. Abbreviation for m-tolylenediamine.
See toluene-2,4-diamine.

MTOC. See microtubule organizing center.

mucic acid. (saccharolactic acid; galactaric acid; tetrahydroxyadipic acid). $HOOC(CH_2O)_4COOH$.
Properties: White, crystalline powder. Mp approximately 210C (decomposes). Soluble in water; insoluble in alcohol. Combustible.

Derivation: Oxidation of lactose or similar carbohydrates with nitric acid.

Use: Substitute for tartaric acid, sequestrant for metal ions (calcium, iron), retards hardening of concrete intermediate for synthesis of heterocyclic compounds (pyrroles).

mucilage. A plant product obtained from seeds, roots, or other parts of plants by extraction with either hot or cold water. Mucilages give slippery or gelatinous solutions, e.g., those from guar bean, linseed, locust bean, and related leguminous plant seeds. Generally plant mucilates are insoluble in alcohol, but some are partly soluble in water and partly soluble in alcohol. From various types of saltwater algae the so-called seaweed mucilages, such as agar, algin, and carrageenin (sometimes referred to as *algal polysaccharides*) may be obtained by extraction with hot water. Mucilages are closely related to gums, and the distinction between them is not always clear.
See adhesives; gum, natural.

mucleoside p. Enzymes that catalyze the phosphorolysis of a nuclotide, forming free purine of pyrimidine plus ribose (or deoxyribose 1-phosphate).

mucopolysaccharide. A polysaccharide composed of alternate units of uronic acids and amino sugars (in which a hydroxyl group is replaced with an amino group, which in turn may be *N*-substituted by other groups). The mucopolysaccharides act as structural support for connective tissue and mucous membranes of animal organisms.

mucronatine. See *trans*-retrorsine.

mud, drilling. See drilling fluid.

muffle furnace. A furnace or kiln in which the materials being heated are kept out of direct contact with the heat source, the combustion being effected by heat reflected from the walls of the furnace.
See Mannheim furnace; reverberatory furnace.

muller. A device for uniform mixing of dry and wetted solids by a combined rubbing and smearing action analogous to that of a mortar and pestle. It consists of a stationary circular pan within which two heavy wheel-like members, together with scrapers (plows), revolve. The mulling wheels have flat, wide surfaces (outer rims) that ride on the material and effect the mixing action. Because the inner edges of the wheels travel less distance than the outer edges, a smearing effect is provided across the surface of the wheels. The plows continually rake the material into the path of the wheels as the unit revolves. Continuous mulling is obtained with two such machines arranged in tandem, space being provided for constant recirculation of the material

between them. Mullers are used for fine dispersion and blending of a wide range of products that are dense enough to support the wheels and fluid enough to provide traction, (e.g., putty, explosives, and heavy pastes).

"Mullfrax" *[Saint Gobain].* TM for refractory products made from mullite produced in electric furnaces.
Use: Construction materials for furnaces and kilns.

Mulliken, Robert S. (1896–1986). An American chemist, physicist, and educator who won the Nobel Prize for chemistry in 1961. He did research on isotope separation and on spectroscopy of electrons in molecule formation. After M.I.T. granted his Ph.D., he did postgraduate work before working in industry, government, and academia.

Mullis, Kary Banks. (1944–). An American who won the Nobel Prize for chemistry in 1993 for the invention of the polymerase chain reaction (PCR). In PCR two short oligonucleotides are synthesized so that they are bound correctly to opposite strands of the DNA segment it is wished to replicate. At the points of contact an added enzyme (DNA polymerase) can start to read off the genetic code and link code words through which two new double strands of DNA are formed. The sample is then heated, which makes the strands separate so that they can be read off again. The procedure is then repeated time after time, doubling at each step the number of copies of the desired DNA segment. Through such repetitive cycles it is possible to obtain millions of copies of the desired DNA segment within a few hours.
See Smith, Michael.

mullite. $3Al_2O_3 \cdot 2SiO_2$. A stable form of aluminum silicate formed by heating other aluminum silicates (such as cyanite, sillimanite, and andalusite) to high temperatures; also found in nature.
Properties: Colorless crystals. D 3.15, mp 1810C. Insoluble in water.
Use: Refractories, glass.
See aluminum silicate.

multicellular. Any organism which is composed of many cells is termed multicellular.

multienzyme system. A group of related enzymes participating in a given metabolic pathway and often channeling substrates from one to another.

multifactorial. A characteristic influenced in its expression by many factors, both genetic and environmental.
See polygenic disorder.

multigenic disorder. See polygenic disorder.

multiple-effect evaporator. A series of evaporator bodies so connected that the vapor from one is the heating medium of the next.

multiple proportions law. See chemical laws (3).

multiplexing. A laboratory approach that performs multiple sets of reactions in parallel (simultaneously), greatly increasing speed and throughput.

"Multisorb" A.R. *[Baker Hughes].* TM for a grade of manganese dioxide used as a solid absorbent for SO_2 and NO_2 in analytical chemistry.

"Multiwax" *[Sonneborn].* TM for refined microcrystalline wax obtained from crude petroleum. Composed primarily of alkylated naphthenes and isoparaffins, with small amounts of normal paraffins.

Multranil 176. A polymer that in the solid form is similar in appearance to SBR. A base resin of a two-component system that when mixed with the proper curing agents, forms a versatile adhesive.

"Multrathane" *[Bayer].* TM for a series of compounds used mainly in formulations for solid urethane elastomers. Some are also used in the formulations of spandex fibers, urethane coatings, and adhesives.

municipal waste. See sewage sludge.

Muntz metal. An alloy containing approximately 60% copper and 40% zinc; a low percentage of lead is sometimes added for free-cutting. It is classified as brass and used primarily for condenser tube plates and other electrical applications. It is formed by hot-working and is not amenable to cold-working.
See brass.

murexide.
CAS: 3051-09-0. $C_8H_8N_6O_6$.
Properties: Dark-red crystals. Partially soluble in hot water; insoluble in alcohol.
Use: Indicator.

muriatic acid. Obsolete name for hydrochloric acid. The related term, *muriate*, indicating presence of chlorine in an inorganic compound, is also obsolete.

murine. Organism in the genus *Mus*. A rat or mouse.

murotox.
CAS: 116397-83-2. $C_{15}H_{16}O \cdot C_4H_{10}O_3 \cdot B_4Na_2O_7 \cdot 10H_2O$.
Hazard: Low toxicity by ingestion, inhalation, and skin contact.
Use: Agricultural chemical.

murvesco. See fenson.

muscone. See musk.

muscovite. (white mica; potassium mica; isinglass). $KAl_2(AlSi_3O_{10})(OH)_2$. A natural hydrous potassium aluminum silicate of the mica group.

musk.
CAS: 300-54-9.
Properties: An unctuous, brownish, semiliquid when fresh; dried, in grains or lumps with color resembling dried blood. Strong characteristic odor. The odor-bearing constituent is muscone, $CH_3C_{15}H_{27}O$, a 15-carbon ring with ketone oxygen.
Derivation: (Natural) Secretion from preputial follicles of the musk deer. Synthetic: (1) Ketones and lactones with 15- or 16-carbon rings, structurally resembling the odoriferous principles of natural musk, civet, and musky-type plants. Among these are ambrettolide, civetone, muscone, exaltolide. (2) Nitrated compounds, usually nitrated *tert*-butyltoluenes or xylenes or related compounds. The three most commonly used in perfumery are musk ambrette, musk ketone, and musk xylene.
Use: Cosmetics and perfumery (fixative), fragrances, mothproofing agent.

musk tonalid. See ethanone, 1-(5,6,7,8-tetrahydro-3,5,5,6,8,8-hexamethyl-2-naphthalenyl)-.

mustard gas. Legal label name (Air) for dichlorodiethyl sulfide.

mustard oil. Any of several organic compounds having the formula R–N=C=S, in which R is an alkyl or aryl radical and –NCS an isothiocyanate group. Its best-known member is allyl isothiocyanate, the characteristic ingredient of mustard oils.
See nitrogen mustard.
Hazard: Irritant to mucous membranes.

mustard oil, artificial. See allyl isothiocyanate.

mutagen. See mutagenic agent.

mutagenic agent. (1) Any of a number of chemical compounds able to induce mutations in DNA and in living cells. The alkyl mustards, as well as dimethyl sulfate, diethyl sulfate, and ethylmethane sulfonate, comprise a group of so-called alkylating agents, reacting with the nitrogen atoms of guanine, a constituent of both RNA and DNA. This reaction affects the guanine molecule in such a way as to ultimately induce a mutation in DNA by depurination. Nitrous oxide can deaminate both guanine and cytosine. If DNA having transforming activity is exposed to such deamination conditions, it is slowly deactivated. Nitrous oxide also produces mutants in whole cells, whole bacteriophage, some viruses, and DNA having transforming ability. (2) Ionizing radiation.

mutagenicity.   The capacity of a chemical or physical agent to cause permanent genetic alterations.
See somatic cell genetic mutation.

mutant gene.   A gene that has been changed from an ancestral type.

mutarotation.   The change in specific rotation of a sugar, as equilibrium between its alpha- and beta-anomeric and open chain forms occurs.

mutase.   An enzyme that catalyzes the transposition of functional groups within a molecule or between molecules.

mutases.   Enzymes that catalyze the transposition of functional groups.

mutation.   Any heritable change in DNA sequence.
See polymorphism.

mutuality of phases.   The rule that if two phases, with respect to a reaction, are in equilibrium with a third phase at a certain temperature, then they are in equilibrium with respect to each other at that temperature.

MVE.   Abbreviation for methyl vinyl ether.
See vinyl methyl ether.

mw.   Abbreviation for molecular weight.

"MX" [Carborundum].   TM for fiber-bonded abrasives.
Properties: High tensile strength and resistance to impact and heat shock, unusually resilient.
Use: Finishing and polishing flutes of taps, drill end mills, reamers, etc.; removing burrs from milling and drilling operations; breaking edges of cast aluminum parts, etc.; cleaning cast iron molds; removing flash from molded plastics.

MXDA.   See m-xylene-α,α'-diamine.

myclobutanil.
CAS: 88671-89-0. $C_{15}H_{17}ClN_4$.
Hazard: Moderately toxic by ingestion, inhalation, and skin contact.
Use: Food additive; fungicide.

"Mycoban" [Cultor].   TM for sodium and calcium propionates. These salts inhibit the growth of many fungi and of some microorganisms, particularly *Bacillus mesentericus*, for commercially significant periods of time.
Use: Inhibit mold and rope in bakery products.

"Mycostatin" [T-Rex].   TM for nystatin.

mycotoxin.   A highly toxic principle produced by molds or fungi. One type, the aflatoxins, is produced by the *Aspergillus flavus* fungus; another is a member of the trichothecene group produced by the fusarium fungus. This has been identified in samples of the so-called "yellow rain" in Southeast Asia, where it is said to have been the cause of many deaths among war refugees. Its presence there is subject to some conjecture, since the fusarium fungus cannot germinate in the humid environment of that area. There is substantial evidence (blood tests, autopsies, and contaminated gas masks) that the former U.S.S.R. have used such lethal agents in Afghanistan also.

mydatoxin.   (mydatoxine). $C_6H_{13}NO_2$.   A ptomaine.
Derivation: From putrefying viscera and flesh or from human intestines maintained at low temperatures for a long time.

mydriatic alkaloid.   An alkaloid such as atropine and cocaine that dilates the pupil of the eye.

myelin.   A unique, sheath-like structure that encloses major nerve trunks, somewhat like insulation around a wire. It is composed of approximately 80% lipid, the balance being made up of proteins, polysaccharides, salts, and water. The lipid fraction is composed of sphingolipids and glycerophosphates, which in turn contain long-chain fatty acids. It has a low concentration of polyunsaturated lipids and high concentration of long-chain sphingolipids. Its composition is essentially constant in different species of animals and also between adults and infants. The breakdown of the lipid structure of myelin is a characteristic of multiple sclerosis.

Mylar.   A polyester film. Seven available types used for electrical, industrial, and packaging purposes.
Available forms: Roll and sheet.

Mylone.   A fungicide.
See "Crag" [Immuno-Mycologics].

myofibril.   A unit of thick and thin filaments of muscle fibers.

myoglobin.   A protein-iron-porphyrin molecule similar to hemoglobin. The chief difference is that myoglobin complexes one heme group per molecule, whereas hemoglobin complexes four heme groups.
See heme.

myo-inositol.   See inositol.

myokinase. An enzyme found in muscle and other tissues that catalyzes the reaction 2ADP ↔ ATP + AMP.

myosin. A protein, of molecular weight above 500,000, that is an essential component of muscular tissue and strongly affects its contractile properties.

myrac aldehyde. See isohexenyl cyclohexenyl carboxaldehyde.

myrcene. (7-methyl-3-methylene-1,6-octadiene). $C_{10}H_{16}$. A triply unsaturated aliphatic hydrocarbon found in oil of bay, verbena, hops, and others.
Properties: Yellow, oily liquid; pleasant odor. Bp 167C, d (80% myrcene) 0.806 (15.5/15.5C), refr index (81% myrcene) 1.471 (20C). Insoluble in water; soluble in alcohol, chloroform, ether, glacial acetic acid. Combustible.
Use: Preparation of perfume chemicals, flavoring.

myrcia oil. (bay oil; bayleaf oil). A yellow essential oil, slightly levorotatory.
Use: Bay rum, fragrances, and flavors.

myricyl alcohol. See 1-triacontanol and 1-hentriacontanol.

myricyl palmitate. $C_{30}H_{61} \cdot C_{16}H_{31}O_2$ (approximately). A wax ester found in beeswax.

myristic acid. (tetradecanoic acid).
CAS: 544-63-8. $CH_3(CH_2)_{12}COOH$.
Properties: Oily, white, crystalline solid. D 0.8739 (80C), bp 326.2C, 204.3C (20 mm Hg), mp 54.4C, refr index 1.4310 (60C). Soluble in alcohol and ether, soluble in water. Combustible.
Derivation: Fractional distillation of coconut acid and other vegetable oils, occurs in sperm oil.
Grade: Technical, 99.8%, FCC.
Use: Soaps, cosmetics, synthesis of esters for flavors and perfumes, component of food-grade additives.

myristicin. (5-allyl-1-methoxy-2,3-(methylenedioxy)benzene; 4-methoxy-6-prop-2-enyl-1,3-benzodioxole). $C_{11}H_{12}O_3$. The chief substance in the ripe seeds of nutmeg (*Myristica fragrans*). It reduces the levels of cytochrome p-450 and inhibits monooxygenations catalyzed by this enzyme.
Properties: Crystalline phenolic ether with a strong odor.
Derivation: Occurs in a number of essential oils.
Hazard: Poison; moderately toxic; mutagen.

myristin. (glyceryl trimyristate).
CAS: 555-45-3. $C_3H_5(OOCC_{13}H_{27})_3$. A triglyceride occurring, usually to a small extent, in natural fatty oils.

myristoleic acid. (*cis*-tetradec-9-enoic acid).
CAS: 544-64-9. $CH_3(CH_2)_3CH:CH(CH_2)_7COOH$.
Properties: Colorless liquid. Mp −4C. Found in fat of some seeds and in fish oil.

myristoyl peroxide.
CAS: 3530-28-7. $(C_{13}H_{27}CO)_2O_2$.
Properties: Soft granules, 90% peroxide.
Hazard: Oxidizing materials, dangerous fire and explosion risk.
Use: Catalyst for vinyl type monomers.

myristyl alcohol. (1-tetradecanol).
CAS: 112-72-1. $C_{14}H_{29}OH$.
Properties: White solid. D 0.8355 (20/20C), bp 264.1C (20 mm Hg) 171.5C, mp 38C, flash p 285F (140.5C), bulk d 7.0 lb/gal (20C). Insoluble in water; soluble in ether; partially soluble in ethanol. Combustible.
Grade: Technical.
Use: Organic synthesis, plasticizers, antifoam agent, intermediate, perfume fixative for soaps and cosmetics, wetting agents and detergents, ointments and suppositories, shampoos, toothpaste cold creams, specialty cleaning preparations.

myristyl chloride. See tetradecyl chloride.

myristyldimethylamine.
CAS: 112-75-4. $CH_3(CH_2)_{13}N(CH_3)_2$. A liquid cationic detergent; acid stable.
Use: Corrosion inhibitor.

myristyldimethylbenzylammonium chloride.
CAS: 139-08-2. $C_{14}H_{29}(CH_3)_2C_6H_5CH_2NCl$. A quaternary ammonium compound. Free-flowing powder.
Use: Surfactant and detergent.

myristyl lactate. See "Ceraphyl" [ISP].

myristyl mercaptan. See tetradecyl thiol.

myrrh. Gum resin obtained from various species of *Balsamodendron* and *Commiphora*.
Use: Perfumery, incense, and toiletries.

Mytab.
CAS: 1119-97-7. Myristyltrimethylammonium bromide.
Use: Emulsifier; antibacterial.
See myristyltrimethylammonium bromide.

myxin. (6-methoxy-1-phenazinol-5,10-dioxide).
CAS: 13925-12-7. $C_{13}H_{10}N_2O_4$.
Properties: Reddish, acicular crystals. Mp 120C. Evolves heat near 150C and can decompose with explosive violence at this temperature, soluble in acetone.
Use: Bacteriostat and antifungal agent, antibiotic.

N

N. (1) Symbol for nitrogen. The names of certain compounds (such as *N,N*-dibutyl urea) contain this symbol as an indication that the group or groups appearing next in the name (i.e., the butyl groups in the example cited) are joined to the nitrogen atoms in the molecule. The molecular formula is N_2. (2) Mathematical symbol for Avogadro's number. (3) Abbreviation for normal solution. See normal (2).

n. Symbol for refractive index: n20/D is refractive index under standard conditions of temperature and wavelength (sodium D line).

n-. Abbreviation for normal. See normal (1).

Na. Symbol for sodium.

NA-22. 2-mercaptoimidazoline.

$(CH_2CH_2NC(SH)NH)$.

Properties: A white powder. D 1.42, mp above 195C.
Use: To accelerate vulcanization of neoprene.

nabam. (disodium ethylenebisdithiocarbamate). CAS: 142-59-6. $NaSSCNHCH_2CH_2NHCSSNa$.
Properties: Colorless crystals when pure. Easily soluble in water.
Derivation: (a) Addition of carbon disulfide to an alcoholic solution of ethylenediamine followed by neutralization with sodium hydroxide, or (b) by reaction of ethylenediamine with carbon disulfide in aqueous sodium hydroxide.
Grade: 19% aqueous solution.
Hazard: Irritant to skin and mucous membranes, narcotic in high concentrations, use may be restricted.
Use: Plant fungicide, starting material for derivatives that are also pesticides.

NAC. Abbreviation for National Agricultural Chemicals Association.

nacre. (mother of pearl). A form of calcium carbonate secreted by the epithelial cells in the mantle of the oyster. The crystals are bonded by conchiolin $(C_{32}H_{98}N_2O_{11})$; the layers built up by excretion form pearls.

nacreous pigment. A pigment, containing guanine crystals obtained from fish scales or skin, that produces a pearly luster. May be applied as surface coatings, as in simulated pearls, or incorporated into plastics. The pigment particle is generally a very thin platelet of high index of refraction. The crystals are readily oriented into parallel layers because of their shape. Being transparent, each crystal reflects only part of the incident light reaching it and transmits the remainder to the crystal below. The nacreous effect is obtained from the simultaneous reflection of light from the many parallel microscopic layers.

NAD. See nicotinamide adenine dinucleotide phosphate.

nad+. ([[(2R,3S,4R,5R)-5-(6-aminopurin-9-yl)-3,4-dihydroxyoxolan-2-yl]methoxy-hydroxyphosphoryl][(2R,3S,4R,5R)[5[(3-carbamoylpyridin-1-ium-1-yl)-3,4-dihydroxyoxolan-2-yl]methyl phosphate).
CAS: 53-84-9. $C_{21}H_{27}N_7O_{14}P_2$.
The oxidized form of nicotinamide adenine dinucleotide. It exists as an anion under normal physiologic conditions.

NADH. [(2R,3S,4R,5R)-5-(6-aminopurin-9-yl)-3,4-dihydroxyoxolan-2-yl]methyl [[(2R,3S,4R,5R)-5-(3-carbamoyl-4H-pyridin-1-yl)-3,4-dihydroxyoxolan-2-yl]methoxy-hydroxyphosphoryl] hydrogen phosphate).
CAS: 606-68-8. $C_{21}H_{27}N_7O_{14}P_2Na_2$.
A dehydrogenase complex that is the reduced form of NAD.

nadide. (adenosine 5N-(trihydrogen diphosphate) 5n65n-ester with 3-(aminocarbonyl)-1-β-d-ribofuranosylpyridinium, hydroxide, inner salt; 3-carbamoyl-1-β-d-ribofuranosylpyridinium hydroxide 5n-ester with adenosine 5n-pyrophosphate inner salt; diphosphopyridine nucleotide; nicotinamide-adenine dinucleotide; codehydogenase; coenzyme I, co I, cozymase; dpn; adenine-d-ribophosphate-phosphate-d-ribose-nicotinamide; [[(2R,3S,4R,5R)-5-(6-aminopurin-9-yl)-3,4-dihydroxyoxolan-2-yl]methoxy-hydroxyphosphoryl][(2R,3E,4R,5R)-5-(3-carbamoylpyridin-1-ium-1-yl)-3,4-dihydroxyoxolan-2-yl]methyl phosphate).

Hawley's Condensed Chemical Dictionary, Sixteenth Edition. Michael D. Larrañaga, Richard J. Lewis, Sr., and Robert A. Lewis. © 2016 John Wiley & Sons, Inc. Published 2016 by John Wiley & Sons, Inc.

CAS: 53-84-9. $C_{21}H_{27}N_7O_{14}P_2$.
A component of fresh baker's yeast. It is the coenzyme of apozymase, composed of ribosylnicotinamide 5'-diphosphate coupled to adenosine 5'-phosphate by pyrophosphate linkage, which is the essential catalyst of the alcoholic fermentation of glucose.
Properties: Hygroscopic white powder.
Derivation: Naturally occurring form of nicotinamide-adenine dinucleotide.
Use: Therapeutically as a narcotic and alcohol antagonist.

nad-nucleotidase. An enzyme known from at least nine species of venomous snakes that catalyzes the hydrolysis of nicotinamide n-ribosidic linkage of nad (nicotinamide adenine dinucleotide).

NADP. See nicotinamide adenine dinucleotide phosphate.

NADP+. [(2R,3R,4R,5R)-5-(6-aminopurin-9-yl)-3-hydroxy-4-phosphonooxyoxolan-2-yl]methyl[[(2R,3S,4R,5R)-5-(3-carbamoylpyridin-1-ium-1-yl)-3,4-dihydroxyoxolan-2-yl]methoxy-hydroxyphosphoryl] hydrogen phosphate).
CAS: 53-59-8. $C_{21}H_{28}N_7O_{17}P_3$.
The oxidized form of nicotinamide adenine dinucleotide phosphate (NADP) that receives electrons from photosystem I during photosynthesis. It exists as an anion under normal physiologic conditions.

NADPh. ([(2R,3R,4R,5R)-5-(6-aminopurin-9-yl)-3-hydroxy-4-phosphonooxyoxolan-2-yl]methyl[[(2R,3S,4R,5R)-5-(3-carbamoyl-4H-pyridin-1-yl)-3,4-dihydroxyoxolan-2-yl]methoxy-hydroxyphosphoryl] hydrogen phosphate).
CAS: 53-57-6. $C_{21}H_{30}N_7O_{17}P_3$.
A coenzyme composed of ribosylnicotinamide 5'-phosphate coupled by pyrophosphate linkage to the 5'-phosphate adenosine 2',5'-bisphosphate. The reduced from of NADP. It is an energy-storage form that can be transferred to the Calvin cycle where it participates in the production of carbohydrate.

nad synthetase. A key enzyme in the biosynthesis of Nad.

nafidimide.
CAS: 69408-81-7. $C_{16}H_{17}N_3O_2$.
Hazard: A poison.

Naflon. A perfluorosulfonic acid membrane.
Use: Manufacture of chlorine and caustic soda. It is a chemically stable ion-exchange resin.

naja mossambica mossambica α-neurotoxin i.
CAS: 115722-23-1. $C_{48}H_{70}N_{10}O_7$.
Hazard: A poison.
Source: Natural product.

NaK. (sodium–potassium alloy).
Properties: Soft, silvery solid or liquid. (1) 78% potassium, 22% sodium: mp −11C, bp 784C, d 0.847 (100C); (2) 56% potassium, 44% sodium: mp 19C, bp 825C, d 0.886 (100C). Must be kept away from air and moisture. The liquid forms come under the class name potassium (or sodium) metallic liquid alloy.
Hazard: Ignites in air; explodes in the presence of moisture, oxygen, halogens, acids. Store under kerosene. Use *dry* salt or soda ash to extinguish, *not* water or foam.
Use: Heat-exchange fluid, electric conductor, organic synthesis and catalysis.

Nalan. Durable water repellents used in the textile industry.

Nalclean. Cleaning composition.
Use: Removing water and process-formed deposits from industrial equipment.

"Nalclear" *[Nalco].* TM for flocculating chemicals.
Use: Treating wastewater.

"Nalco" *[Nalco].* TM for a broad class of chemicals, organic or otherwise, employed in the treatment of water and hydrocarbons; paper-making chemicals; cleaning compounds; combustion aids; weed and brush controls; lubricating and antilubricating compositions; apparatuses; pumps, and mechanisms for proportioning chemicals.

Nalcoag. A colloidal silica available in particle sizes from 4 to 100 mμ.
Use: Reinforcing agent, antiblock agent, and dispersing agent.

"Nalcolyte" 671 *[Nalco].* TM for a synthetic high polymer used for clarifying industrial plant water and municipal water supplies. A coagulant behaving as a polyelectrolyte. Effective at concentrations of less than 1 ppm; also used in still lower concentrations as a filter aid.
Use: Treating water and wastewater.

naled. (1,2-dibromo-2,2-dichloroethyl dimethyl phosphate).
CAS: 300-76-5. $(CH_3O)_2P(O)OC(Br)HCBr(Cl)_2$.
Properties: Pure compound is a solid. Mp 26C. Technical compound is a moderately volatile liquid. Bp 110C (0.5 mm Hg). Insoluble in water; slightly soluble in aliphatic solvents; very soluble in aromatic solvents; hydrolyzes in water.
Hazard: Cholinesterase inhibitor; use may be restricted. Toxic by skin absorption. Questionable carcinogen.
Use: Insecticide, acaricide.

"Nalflote" *[Nalco].* TM for flotation reagent.
Use: Handling, processing, and manufacturing of coal.

Nalgene. Plastic laboratory ware for industry, research, and education. Made of polypropylene, polyethylene, Teflon, FEP, polyallomer, and polycarbonate.

nalidixic acid. (USAN; 1-ethyl-7-methyl-1,8-naphthyridin-4-one-3-carboxylic acid). CAS: 389-08-2. $C_{12}H_{12}N_2O_3$. An antibacterial compound used in medicine.

"Nalkat" *[Nalco].* TM for cationic polymers. Use: Manufacture of pulp and paper.

"Nalmet" *[Nalco].* TM for services in the wastewater-treatment chemical industry. Use: Metals recovery and reclamation.

nalorphine. (*N*-allylnormorphine). CAS: 62-67-9. $C_{19}H_{21}NO_3$. The allyl ($-CH_2-CH=CH_2$) derivative of morphine. It is able to "antagonize" or neutralize most of the effects of narcotic drugs (morphine, codeine) but not those of other types of depressants. Use: Biochemical research tool for studying the mechanism of narcotic action; also as an antidote for acute morphine poisoning. See narcotic.

"Nalprep" *[Nalco].* TM for corrosion inhibitor. Use: To protect cooling-water systems.

NaMBT. See sodium MBT.

nameplate. The officially rated capacity of a chemical plant, as opposed to effective or actual maximum; the latter is usually 85–95% of nameplate.

name reaction. A chemical reaction, usually organic, that is commonly identified by the name of its discoverer(s), e.g., Friedel–Crafts, Fischer–Tropsch, Claisen, Clemmensen, Willegerodt, Diels–Alder. Many have important industrial applications.

nano-. Prefix meaning 10^{-9} unit (symbol n); 1 ng = 1 nanogram = 0.000000001 gram; 1 nanometer = 1 millimicron.

nanoarray. An ultra-sensitive, ultra-miniaturized array for biomolecular analysis.

nanobalance. A nanoscale balance for determining mass; small enough to weigh viruses and other sub-micron scale particles.

nanobeads. (nanodots; nanocrystals; quantum beads). Polymer beads with diameters from 0.1 to 10 micrometers.

nanobiotechnology. Applying the tools and processes of nanotechnology to build devices for studying biosystems.

nanobubbles. Nano-sized air bubbles on colloid surfaces.

nanocantilever. The simplest micro-electromechanical system. They are easily machined and mass-produced by the same techniques used to make computer chips. The ability to detect extremely small displacements make nanocantilever beams an ideal device for detecting extremely small forces, stresses, and masses.

nanochondria. Nanomachines existing inside living cells and participating in their biochemistry or assembling various structures.

nanocomposite. See nanophase carbon materials.

nanocomputer. A computer made from components (mechanical, electronic, or otherwise) built at the nanometer scale.

nanocontainers. Nanoscale polymeric containers that could be used to selectively deliver hydrophobic drugs to specific sites within individual cells.

nanocrystals. (nanoscale semiconductor crystals). Aggregates of a few hundred to tens of thousands of atoms combining into a crystalline form of matter known as a cluster. Typically around ten nanometers in diameter, nanocrystals are larger than molecules but smaller than bulk solids.

nanodiamond. See nanophase carbon materials.

nanoelectromechanical systems. (NEMS). A generic term to describe nanoscale electrical or mechanical devices.

nanoelectronics. Electronics on a nanometer scale.

nanofabrication. Construction of items using assemblers and stock molecules.

nanofluidics. Controlling nanoscale amounts of fluids.

nanogate. A device that precisely meters the flow of tiny amounts of fluid.

"Nanograde" *[Mallinckrodt].* TM for a grade of chemical purity. Impurities guaranteed to be less than ten parts per trillion.

nanoimprinting. (soft lithography). Use of lithographic masters with nanoscale features.

nanolithography. Writing on the nanoscale. Used for etching computer circuits.

nanomachining. Involves mechanically changing the structure of nanoscale materials or molecules.

nanomanipulation. Manipulating at an atomic or molecular scale in order to produce precise structures.

nanometer. (nm). One-billionth (10^{-9}) of a meter, equal to 1 millimicron or 10 Å units.

nanoparticle. A nanoscale spherical or capsule-shaped structure. Many are hollow, which provides a central space that can be filled with substances. Various substances can also be attached to the surface of a nanoparticle for transport and reaction purposes. Most nanoparticles are small enough to pass through blood capillaries and enter cells.

nanopharmaceuticals. Nanoscale particles used to modulate drug transport for drug uptake and delivery applications.

nanophase carbon materials. (carbon nanotubes, nanodiamond, nanocomposite). Form of matter in which small clusters of atoms form the building blocks of a larger structure.

nanorods. Multi-wall carbon nanotubes.

nanoropes. Formed by connecting and stringing together nanotubes.

nanoshell. A nanoparticle composed of a metallic shell surrounding a semiconductor. These nanoscale metal spheres can absorb or scatter light at virtually any wavelength. When attached to a target cancer cell, they can be irradiated to cause the nanoshell to become hot, killing the cell.

nanosystem. A eutactic set of nanoscale components working together to serve a set of purposes; complex nanosystems can be of macroscopic size.

nanotechnology. The interactions of cellular and molecular components and engineered materials typically clusters of atoms, molecules, and molecular fragments at the most elemental level of biology. Such nanoscale objects—with dimensions smaller than 100 nanometers—typically, to not exclusively, can be useful by themselves or as part of larger devices containing multiple nanoscale objects.

nanotubes. Carbon nanotubes are fullerene-related structures that consist of graphene cylinders closed at either end with caps containing pentagonal rings. They are in the size range of cylinders 10–20 nm in diameter.
See: buckminsterfullerene, fullerines.

nanowire. A nanometer-scale wire made of metal atoms, silicon, or other materials that conduct electricity. Nanowires are built atom by atom on a solid surface, often as part of a microfluidic device. They can be coated with molecules, such as antibodies, that will bind to proteins and other substances of interest to researchers and clinicians.

napalm. An aluminum soap of a mixture of oleic, naphthenic, and coconut fatty acids.
Properties: Granular powder. Mixed with gasoline it forms a sticky gel that is stable from −40 to 100C.
Hazard: Flammable, dangerous fire risk.
Use: Incendiary agent.

naphite. See trinitronaphthalene.

naphtha. (benzin).
CAS: 8030-30-6. (1) *Petroleum* (petroleum ether). A general term applied to refined, partly refined, or unrefined petroleum products and liquid products of natural gas, not less than 10% of which distill below 347F (175C) and not less than 95% of which distill below 464F (240C) when subjected to distillation in accordance with the Standard Method of Test for Distillation of Gasoline, Naphtha, Kerosene, and Similar Petroleum Products (ASTM D86); fp −73C, bp 30–60C, flash p −57F (−50C), autoign temp 550F (287C), d 0.6.
Hazard: Flammable, dangerous fire risk, explosive limits in air 1–6%.
Use: Source (by various cracking processes) of gasoline, special naphthas, and petroleum chemicals, especially ethylene. Cracking for ethylene also produces propylene, butadiene, pyrolysis gasoline, and fuel oil; source of synthetic natural gas.
(a) *VM&P (Varnish Makers and Painters)* (petroleum spirits, petroleum thinner).
CAS: 8032-32-4. Any of a number of narrow-boiling-range fractions of petroleum with bp of approximately 93–204C, according to the specific use.
Properties: Distillation range 119–143C, d 07543, bulk d 6.280 lb/gal, pour p approximately −70F (−56C), flash p 20F (−6.6C) (TCC), autoign temp 450F (232C).
Hazard: Flammable, dangerous fire risk.
Use: Thinners in paints and varnish.
(b) *Blending.* A petroleum fraction with volatility similar to the higher boiling fractions of gasoline. It is used primarily in blending with natural gasoline to produce a finished gasoline of specified volatility.
(c) *Cleaners'.* A dry-cleaning fluid derived from

petroleum and similar to Stoddard solvent but not necessarily meeting all its specifications; flash p 100F (37.7C).

(2) *Coal tar.*

(a) *Heavy* (high-flash naphtha).

Properties: Deep-amber to dark-red liquid. A mixture of xylene and higher homologs. D 0.885–0.970, bp 160–220C (approximately 90% at 200C), flash p 100F (37.7C), evaporation 303 min.

Derivation: From coal tar by fractional distillation.

Hazard: Moderate fire risk. Toxic by ingestion, inhalation, and skin absorption.

Use: Coumarone resins; solvent for asphalts, road tars, pitches, etc.; cleansing compositions; process engraving and lithography; rubber cements (solvent); naphtha soaps; manufacture of ethylene and acetic acid.

(b) *Solvent* (160-degree benzol).

Properties: A mixture of a small percentage of benzene, toluene, xylene, and higher homologs from coal tar. (1) Crude: dark straw-colored liquid, (2) refined: water-white liquid; d (1) 0.862–0.892, (2) 0.862–0.872; bp (1) approximately 160C (80%), (2) approximately 160C (90%); flash p (1) and (2) approximately 78F (25.5C).

Derivation: From coal tar by fractional distillation.

Grade: Dark-straw, water-white.

Hazard: Flammable, dangerous fire risk.

Use: Solvent; xylene; cumene; nitrated, for incorporation in dynamite.

naphthacene.   (tetracene; rubene).

CAS: 92-24-0. $C_{18}H_{12}$.

The molecule consists of four fused benzene rings.

Properties: Orange solid; slight green fluorescence in daylight. D 1.35, mp approximately 350C. Not easily soluble.

Occurrence: In commercial anthracene and coal tar.

Hazard: Explodes when shocked, reacts with oxidizing materials.

Use: Organic synthesis.

naphthalene.   (tar camphor).

CAS: 91-20-3. $C_{10}H_8$.

Properties: White, crystalline, volatile flakes; strong coal tar odor. D 1.145 (20/4C), mp 80.2C, bp 217.96C, flash p 176F (80C), sublimes at room temperature, autoign temp 979F (526C). Soluble in benzene, absolute alcohol, and ether; insoluble in water. Combustible.

Derivation: (1) From coal tar oils boiling between 200 and 250C (middle oil) by crystallization and distillation. (2) From petroleum fractions after various catalytic processing operations.

Available forms: Flakes, cubes, spheres, powder.

Grade: By melting point, 74C min (crude) to <79C (refined); scintillation (80–81C).

Hazard: Toxic by inhalation. Upper respiratory tract irritant, cataracts and hemolytic anemia. Possible carcinogen.

Use: Intermediate (phthalic anhydride, naphthol, Tertralin, "Decalin" *[Sigma-Aldrich Biotechnology]*, chlorinated naphthalenes, naphthyl and naphthol derivatives, dyes), moth repellent, fungicide, smokeless powder, cutting fluid, lubricant, synthetic resins, synthetic tanning, preservative, textile chemicals, emulsion breakers, scintillation counters, antiseptic.

α-naphthaleneacetic acid.   (1-naphthylacetic acid).

CAS: 86-87-3. $C_{10}H_7CH_2COOH$.

A plant growth regulator.

Properties: White crystals; odorless. Mp 132–135C. Soluble in acetone, ether, and chloroform; slightly soluble in water and alcohol.

Grade: Usually supplied in dilute form, either as a powder or liquid solution ready for use.

Hazard: Skin irritant.

Use: Inducing rooting of plant cuttings, spraying apple trees to prevent early drop, fruit thinner.

α-naphthaleneactic acid, methyl ester.
(MENA). $C_{10}H_7CH_2COOCH_3$. A plant growth regulator.

Use: Delaying sprouting of potatoes, weed control, thinning of peaches, etc.

naphthalene, chlorinated.   See chlorinated naphthalene.

naphthalenediamine.   See naphthylenediamine.

1,5-naphthalene diisocyanate.

CAS: 3173-72-6. $C_{10}H_6(NCO)_2$.

White to light-yellow, crystalline solid; mp 127–131C.

Hazard: Irritant. Questionable carcinogen.

Use: Manufacture of polyurethane solid elastomers.

naphthalene-1,5-disulfonic acid.   See Armstrong's acid.

naphthalene-2,7-disulfonic acid.

CAS: 92-41-1. $C_{10}H_6(SO_3H)_2$.

Properties: White, crystalline solid. Soluble in water. Combustible.

Derivation: Sulfonation of naphthalene at high temperature and separation from 2,6-isomer.

Use: Intermediate for dyes.

α-naphthalenesulfonic acid.   (1-naphthalenesulfonic acid). $C_{10}H_7SO_3H \cdot H_2O$.

Properties: Deliquescent crystals. Mp 90C. Soluble in water, alcohol, and ether. Combustible.

Derivation: Interaction of naphthalene and sulfuric acid.

Use: Starting point in the manufacture of α-naphthol, α-naphthalene sulfonic acid, α-naphthylaminesulfonic acid; solvent (Na salt) for phenol in the manufacture of disinfectant soaps.

β-naphthalenesulfonic   acid.   (2-naphthalenesulfonic acid).
CAS: 120-18-3. $C_{10}H_7SO_3H$ or $C_{10}H_7SO_3H \cdot H_2O$.
Properties: Nondeliquescent, white plates. Mp 124–125C. Soluble in water, alcohol, and ether. Combustible.
Derivation: Sulfonation of naphthalene.
Use: Starting point in the manufacture of β-naphthol, β-naphtholsulfonic acid, β-naphthylaminesulfonic acid, etc.

1,3,6-naphthalenetrisulfonic acid, trisodium salt. $C_{10}H_5(SO_3Na)_3$. Fine buff crystals.
Use: Diazo-type stabilizer.

1-naphthalenyl   ((hexyloxy)sulfinyl)methylcarbamate.
CAS: 77267-52-8. $C_{18}H_{23}NO_4S$.
Hazard: Moderately toxic by ingestion.
Use: Agricultural chemical.

(naphthalenyl)methanone.
CAS: 131543-22-1. $C_{27}H_{26}N_2O_3$.
Hazard: A poison.

1,8-naphthalic acid anhydride.
CAS: 81-84-5. $C_{12}H_6O_3$.
Properties: Light-tan powder. Mp 268–270C.
Use: Dyestuffs, organic synthesis in general.

Naphthanil.   A series of dye bases. Before coupling the bases must first be diazotized to form the diazo salt. Also represents a series of diazo pigments.
Use: Widely used on cotton and rayon textiles.

naphtha, petroleum.   See naphtha (1).

1:4-naphthaquinol-bisdisodium   phosphate.

CAS: 131-13-5. $C_{11}H_{12}O_8P_2 \cdot 4Na$.
Hazard: A poison.

naphtha, solvent.   See naphtha (2b).

naphtha, VM&P.   See naphtha (1a).

naphthene.   The term *naphthene* is misleading and obsolete.
See cycloparaffin.

naphthenic acid.   Any of a group of saturated higher fatty acids derived from the gas-oil fraction of petroleum by extraction with caustic soda solution and subsequent acidification. Gulf and West coast crudes are relatively high in these acids. The commercial grade is a mixture, usually of dark color and unpleasant odor, corrosive to metals. The chief use of naphthenic acids is in the production of metallic naphthenates for paint driers and cellulose preservatives. Other uses are as solvents, detergents, rubber reclaiming agent, etc.

naphthenic oils.
CAS: 67254-74-4.
Hazard: Questionable carcinogen.

naphthionic acid.   (1-naphthylamine-4-sulfonic acid; 1-aminonaphthalene-4-sulfonic acid; 4-amino-1-naphthalenesulfonic acid).

Properties: White crystals or powder. Soluble in water; slightly soluble in alcohol and ether.
Derivation: Heating equimolar amounts of α-naphthylamine and sulfuric acid at 10–15 mm Hg (several hours).
Use: Intermediate for azo dyes, e.g., Congo red.

α-naphthol.   (1-naphthol;   1-hydroxynaphthalene).
CAS: 90-15-3. $C_{10}H_7OH$.

Properties: Colorless or yellow prisms or powder; disagreeable taste. D 1.224 (4C), 1.0954 (95/4C), mp 96C, bp 278C, volatile in steam, sublimes, refr index 1.6206 (98.7C). Soluble in benzene, alcohol, and ether; insoluble in water. Combustible.
Derivation: By fusing sodium α-naphthalene sulfonate and caustic soda. The melt is decomposed with hydrochloric acid and distilled.
Hazard: Toxic by ingestion and skin absorption.
Use: Dyes, organic synthesis, synthetic perfumes.

β-naphthol.   (2-naphthol;   2-hydroxynaphthalene).
CAS: 135-19-3. $C_{10}H_7OH$.

Properties: White, lustrous, bulky leaflets or white powder. Darkens with age; faint phenol-like odor. Stable in air but darkens on exposure to sunlight. D

1.217, mp 121.6C, bp 285C, flash p 307F (152.7C). Soluble in alcohol, ether, chloroform, glycerol, oils, and alkaline solutions; almost insoluble in water. Combustible.
Derivation: By fusing sodium β-naphthalene sulfonate with caustic soda. The product is distilled in vacuo and then sublimed.
Grade: Technical, sublimed, resublimed.
Hazard: See α-naphthol.
Use: Dyes; pigments; antioxidants for rubber, fats, oils; insecticides; synthesis of fungicides; pharmaceuticals; perfumes; antiseptic.

naphthol AS.   See β-hydroxynaphthoic anilide.

3-naphthol-2-carboxylic acid.   See 3-hydroxy-2-naphthoic acid.

Naphthol Green B.   A dye used in crystallizing solar salt, it increases the evaporation rate by added absorption of energy; 5 ppm is said to increase salt production 15–20%.

β-naphthol methyl ether.   See β-naphthyl methyl ether.

naphtholsulfonic acid.   Any of several sulfonated aromatic acids derived from α- or β-naphthol or naphthalene and used as azo dye intermediates.

naphthol yellow S.   (C.I. 103161; 8-hydroxy-5,7-dinitro-7-naphathlene sulfonic acid; sodium 5,7-dinitro-8-oxidonaphthalene-2-sulfonate). $C_{10}H_4N_2Na_2O_8S$. An acid dye.
Use: To stain basic proteins in microspectroscopy-photometry.

1,2-naphthoquinone.   (β-naphthoquinone).
CAS: 524-42-5. $C_{10}H_6O_2$.
Properties: Yellow crystals. Mp 120C (decomposes). Soluble in ether, benzene, alcohol.
Hazard: Irritant.
Use: Chemical reagent and intermediate.

1,4-naphthoquinone.   (α-naphthoquinone).
CAS: 130-15-4. $C_{10}H_6O_2$.
Properties: Yellow powder; odor like benzoquinone. Mp 123–126C, sublimes at 100C. Slightly soluble in water; soluble in ethanol, ethyl ether, chloroform, benzene, and acetic acid. Combustible.
Hazard: Irritant.
Use: Polymerization regulator for rubber and polyester resins, synthesis of dyes and pharmaceuticals, fungicide, algicide.

naphthoquinone oxime.   See 1-nitroso-2-naphthol; 2-nitroso-1-naphthol.

1,2-naphthoquinone-4-sulfonic acid.   (β-naphthoquinone-4-sulfonic acid).
CAS: 521-24-4. $C_{10}H_5(O)_2SO_3H$.

Derivation: Oxidation with nitric acid of 2-amino-1-naphthol-4-sulfonic acid or 1-amino-2-naphthol-4-sulfonic acid.
Use: Dye intermediate, identification of sulfonamide derivatives.

naphthoresorcinol.   See 1,3-dihydroxynaphthalene.

β-naphthoxyacetic acid.   (2-naphthoxyacetic acid).
CAS: 120-23-0. $C_{10}H_7OCH_2COOH$.
A plant growth regulator.
Properties: Crystals. Mp 156C. Soluble in water, alcohol, acetic acid.
Use: Rooting clippings, inhibits early fall of fruit, growth promoter.

1-naphthylacetic acid.   See naphthaleneacetic acid.

α-naphthylamine.   (1-naphthylamine).
CAS: 134-32-7. $C_{10}H_7NH_2$.
Properties: White crystals becoming red on exposure to air. Flash p 157C, d 1.13, mp 50C, bp 301C. Soluble in alcohol and ether; slightly soluble in water. Combustible.
Derivation: Reduction of α-nitro-naphthalene with iron and hydrochloric acid. The mass is then mixed with milk of lime and distilled.
Method of purification: Crystallization.
Hazard: Toxic, especially if containing the β isomer; a questionable carcinogen.
Use: Dyes and dye intermediates, agricultural chemicals.

β-naphthylamine.   (2-naphthylamine).
CAS: 91-59-8. $C_{10}H_7NH_2$.
Properties: White to reddish, lustrous leaflets. Soluble in hot water, alcohol, ether. Commercial: mp 109.5C, d 1.061 (98.4C), bp 306C. Combustible.
Derivation: From β-naphthol by heating in an autoclave with ammonium sulfite and ammonia (Bucherer reaction).
Method of purification: Distillation.
Hazard: Toxic by ingestion, inhalation, skin absorption; a confirmed carcinogen. Causes bladder cancer.

α-naphthylamine hydrochloride. $C_{10}H_7$ NH_2•HCl.
Properties: White to gray, crystalline powder. Soluble in water, alcohol, and ether.
Derivation: By the action of hydrochloric acid on α-naphthylamine.
Use: Dyes, organic synthesis.

naphthylaminesulfonic acid.   Any of several sulfonated aromatic acids derived from α- or β-naphthylamine and used as azo dye intermediates.

o-2-naphthyl-*m*-*N*-dimethylthiocarbanilate.
See tolnaftate.

1,5-naphthylenediamine. (1,5-diaminonaphthalene).
CAS: 2243-62-1. $C_{10}H_6(NH_2)_2$.
Properties: Colorless crystals. Mp 190C, bp (sublimes). Soluble in alcohol and hot water; very sparingly soluble in cold water. Combustible.
Derivation: (1) By the reduction of α-dinitronaphthalene; (2) by heating dihydroxynaphthalene with aqueous ammonia.
Hazard: Questionable carcinogen.
Use: Organic synthesis.

1,8-naphthylenediamine. (1,8-diaminonaphthalene).
CAS: 479-27-6. $C_{10}H_6(NH_2)_2$.
Properties: Colorless crystals. Mp 66C, bp 205C (12 mm Hg), d 1.12, refr index 1.68. Soluble in alcohol; slightly soluble in water.
Derivation: Reduction of 1,8-dinitronaphthalene with PI_3.
Use: Lubricating-oil antioxidant, analytical reagent.

N-α-naphthylethylenediamine dihydrochloride.
CAS: 1465-25-4. $C_{10}N_7NHCH_2CH_2NH_2 \cdot 2HCl$.
Properties: Colorless crystals. Soluble in water.
Use: Reagent for the quantitative determination of sulfa drugs, for the detection of nitrogen dioxide in air.

β-naphthyl ethyl ether. (nerolin). $C_{10}H_7OC_2H_5$.
Properties: White crystals; orange-blossom odor. Congealing p 35C. Soluble in 5 parts of 95% alcohol. Combustible.
Derivation: Interaction of β-naphthol and ethanol in the presence of sulfuric acid.
Use: Perfumes, soaps, flavoring.

1-naphthyl isocyanide. See 1-isocyanonaphthalene.

1-naphthyl-*n*-(2,3,4,5-di-*o*-isopropylidene-1-α-*o*-fructopyranosylsulfinyl)-*n*-methylcarbamate.
CAS: 81862-21-7. $C_{24}H_{29}NO_9S$.
Hazard: Moderately toxic by ingestion.
Use: Agricultural chemical.

1-naphthyl-*N*-methylcarbamate. See carbaryl.

β-naphthyl methyl ether. (β-naphthol methyl ether; 2-methoxynaphthalene; methyl naphthyl ether). $C_{10}H_7OCH_3$.
Properties: White, crystalline scales. Mp 72C, bp 274C. Soluble in alcohol and ether; insoluble in water. Combustible.

Derivation: (1) By heating β-naphthol and methanol in the presence of sulfuric acid. (2) By methylating β-naphthol with dimethyl sulfate.
Use: Perfumery (soaps).

α-naphthylphenyloxazole. (NPO; ANPO; 2-(1-naphthyl)-5-phenyloxazole). $C_{19}H_{13}NO$.
Properties: Fluorescent yellow needles. Mp 104–106C.
Grade: Scintillation.
Use: Scintillation counter or wavelength shifter in solution scintillators.

N-1-naphthylphthalamic acid. $C_{10}H_7NHCOC_6H_4COOH$.
Properties: Crystalline solid. Mp 185C. Almost insoluble in water; slightly soluble in acetone, benzene, and ethanol. Not stable in solutions above pH 9.5 nor at temperatures above 180C. Noncorrosive. Do not store near seeds or fertilizers. Combustible.
Use: Selective preemergence herbicide.

α-naphthylthiourea. (ANTU).
CAS: 86-88-4. $C_{10}H_7NHCSNH_2$.
Properties: Odorless, gray powder. Mp 198C. Insoluble in water and only very slightly soluble in most organic solvents.
Derivation: From α-naphthylthiocarbamide and alkali or ammonium thiocyanate.
Hazard: Toxic by ingestion.
Use: Rodenticide.

Naples yellow. See lead antimonate.

napropamide. (n,n-diethyl-2-(1-naphthalenyloxy)propanamide; n,n-diethyl-2-(1-maphthyloxy)propionamide; 2-(maphthyloxy)propionamide; N,N-diethyl-2-naphthalen-1-yloxypropanamide).
CAS: 15299-99-7. $C_{17}H_{21}NO_2$.
Use: Herbicide.
Hazard: Low toxicity.

naptalam. (α-naphthylphthalamic acid).
CAS: 132-67-2. $C_{18}H_{13}NO_3$.
Properties: Colorless crystals. Mp 203C, d 1.40. Soluble in alkaline solutions; slightly soluble in alcohol, benzene, and acetone.
Use: Analytical reagent (thorium, zirconium); herbicide.

narasin.
CAS: 55134-13-9. $C_{43}H_{72}O_{11}$.
Properties: Crystals from Me_2CO (aq). Mp: 98–100C. Soluble in alcohols, acetone, chloroform, ethyl acetate. Insoluble in water.
Use: Drug (veterinary); food additive.

narcotic. (1) A natural, semisynthetic, or synthetic nitrogen-containing heterocyclic drug that characteristically effects sleep (coma) and relief of pain but also may result in addiction, i.e., a biochemical situation in which the body tissues become so adapted to the drug that they can no longer function normally without it. Natural narcotics are the plant products morphine and codeine (constituents of opium), both of which are alkaloids. Opium is obtained from the seed of the oriental poppy, *Papaver somniferens*. Semisynthetic narcotics are modifications of the morphine molecule, e.g., diacetylmorphine (heroin), ethylmorphine ("Dionin"), and methyldihydromorphine (metopon).

Synthetic narcotics are meperidine, ethadone, and phenazocine (there are other addictive agents that are not narcotics). The sale of narcotics is strictly controlled by law in the U.S. (2) Inducing sleep or coma. Many chemicals that are not narcotics in sense (1) have this property (chloroform, barbiturates, benzene, etc.).

l-α-narcotine. See noscapine.

naringin. (naringenin-7-rhamnoglucoside; naringenin-7-rutinoside; aurantiin). CAS: 10236-47-2. $C_{27}H_{32}O_{14}$.
Properties: A flavanone glycoside (bioflavonoid), crystals, bitter taste. Mp 171C. Soluble in acetone, alcohol, warm acetic acid, and warm water.
Source: Extracted from flowers and rind of grapefruit and immature fruit.
Use: Beverages, sweetener research.

narrow-spectrum antibiotic. An antibiotic that is effective against either gram-positive or gram-negative bacteria.

NAS. See the National Academy of Sciences-National Research Council.

nascent. Descriptive of the abnormally active condition of an element, for example, the atomic oxygen released from hydrogen peroxide and sulfur atoms evolved from thiuramsulfide accelerators. The term is now obsolete.

"NA-SUL" [King]. TM for dinonyl naphthalene sulfonates.
Available forms: Liquid.
Use: Corrosion inhibitors for industrial lubricants, rust preventatives, and metalworking specialties.

National Academy of Sciences – National Research Council. (NAS). A private, nonprofit organization of scientists devoted to the expansion of science and its use for the general welfare. The Academy was established in 1863, in part to act as adviser to the federal government on scientific matters; the council was established in 1916, its members being appointed by the president of the academy. Its headquarters are located at 500 Fifth Street, NW, Washington, DC 20001. Website: http://www.nasonline.org

National Fire Protection Association. (NFPA). An organization devoted to promoting knowledge of fire-protection methods. For many years its publication (the NFPA Handbook) has been the accepted standard for all matters relating to combustion and flammable materials, firefighting methods, safety, and protection of property. Its headquarters are at 1 Batterymarch Park, Quincy, MA 02169. Website: http://www.nfpa.org

National Institute for Occupational Safety and Health. (NIOSH). A federal agency under the Department of Health and Human Services, Public Health Service. It is responsible for investigating the toxicity of workroom environments and all other matters relating to safe industrial practice. Its research laboratories are located at 395 E. Street, S.W., Suite 9200, Patriots Plaza Building, Washington, DC 20201. Website: http://www.cdc.gov/niosh

National Institute of Standards and Technology. (NIST). Founded in 1901, NIST is a non-regulatory federal agency within the U.S. Commerce Department's Technology Administration. NIST's mission is to promote U.S. innovation and industrial competitiveness by advancing measurement science, standards, and technology in ways that enhance economic security and improve our quality of life. Its main laboratories are located at 100 Bureau Drive, Stop 1070, Gaithersburg, MD 20899-1070. Website: http://www.nist.gov/public_affairs/general2.htm

National Nanotechnology Coordination Office. (NNCO). The NNCO assists in the preparation of multi-agency planning, budget, and assessment documents. The NNCO is the point of contact on Federal nanotechnology activities for regional, state, and local initiatives, as well as, government organizations, academia, industry, professional societies, foreign organizations, and others to exchange technical and programmatic information. In addition, the NNCO develops and makes available printed and other materials as directed by the NSET Subcommittee. Contact address is 4201 Wilson Blvd., Stafford II Rm 405, Arlington, VA 22230. Website: www.nano.gov

National Nanotechnology Initiative. (NNI). The goals of the NNI are to: (1) maintain a world-class research and development program aimed at realizing the full potential of nanotechnology; (2) facilitate transfer of new technologies into products for economic growth, jobs, and other public benefit; (3) develop educational resources, a skilled workforce, and the supporting infrastructure and tools to advance nanotechnology; and, (4) Support responsible development of nanotechnology. Website: http://www.nano.gov

National Research Council. See National Academy of Sciences – National Research Council.

National Science and Technology Council. (NSTC). A primary objective of the NSTC is the establishment of clear national goals for Federal science and technology investments in a broad array of areas spanning virtually all the mission areas of the executive branch. The Council prepares research and development strategies that are coordinated across Federal agencies to form investment packages aimed at accomplishing multiple national goals. The work of the NSTC is organized under for primary committees: Science, Technology, Environment and Natural Resources, and Homeland and National Security. Each of these committees oversees subcommittees and working groups focused on different aspects of science and technology and the coordination of research and development strategies working to coordinate across the federal government. Contact address is Eisenhower Executive Office Building, 1650 Pennsylvania Avenue, Washington, DC 20504. Website: http://www.whitehose.gov/administration/eop/ostp/nstc

National Science Foundation. (NSF). The National Science Foundation (NSF) is an independent federal agency created by Congress in 1950 "to promote the progress of science; to advance the national health, prosperity, and welfare; to secure the national defense." With an annual budget of $7.2 billion, NSF is the funding source for approximately 24% of all federally supported basic research conducted by America's colleges and universities. In many fields such as mathematics, computer science, and the social sciences, NSF is the major source of federal backing. Contact address is 4201 Wilson Boulevard, Arlington, VA 22230. Website: http://www.nsf.gov

native conformation. The biologically active conformation of a protein.

native ore. A metal that occurs free or uncombined in the environment.

Natralith. A line of functional additives for ink and coatings.
Use: To improve, add lubricity and wetting/dispersing in both metallic and non-metallic pigments.

Natralube. A line of rust preventatives, lubricity, and anti-wear components.
Use: For metal-working, industrial lubricant, and grease formulations.

natrium. The Latin name for sodium, hence the symbol Na in chemical nomenclature.

natrolite. $Na_2Al_2Si_3O_{10} \cdot 2H_2O$. A mineral of the zeolite group.
Properties: Colorless or white to gray, yellow, greenish, or red. D 2.2–2.25, Mohs hardness 5–5.5.
See zeolite.

natron. A complex salt found in dry lake beds of Egypt. Originally an ingredient of ceramic glazes. It has the following percentage composition: sodium carbonate 4.9, sodium bicarbonate 12.6, sodium chloride 30.6, sodium sulfate 20.6, silica 10, calcium carbonate 2, magnesium carbonate 1.9, alumina 0.7, iron oxide 0.3, water 4.7, and organic matter 11.7.

Natrorez. Coumarone-indene resins plasticized to various softening points or viscosities.
Use: Tackifier, processing aid.

"Natrosol" [Hercules]. TM for hydroxyethylcellulose.
CAS: 9004-62-0.
Grade: 10 viscosity, 4 biostable, 7 cosmetic, 5 NF, 3 European Pharmacopeia.
Available forms: Solid.
Use: Thickening agent.

"Natrusol" [Hercules, Ltd.]. TM for a hydroxyethyl cellulose.

Natta catalyst. A stereospecific catalyst made from titanium chloride and aluminum alkyl or similar materials by a special process that includes grinding the materials together to produce an active catalyst surface.
See Ziegler catalyst.

Natta, Giulio. (1903–1979). An Italian chemist born in Imperia on the Riviera, corecipient (with Karl Ziegler) of the Nobel Prize in 1963 for his fundamental work on catalytic polymerization. In 1954, he developed isotactic polypropylene at his laboratory at the Polytechnic Institute of Milan, which led to wide application of various stereospecific polymers with organometallic catalysts such as triethylaluminum. He was for many years consultant for the Montecatini chemical firm. The researchers of Natta, together with those of Ziegler, made possible the chemical manipulation of monomers to form specifically ordered three-dimensional polymers having predetermined properties, to which the term *tailor-made* is often applied.

natural. Descriptive of a substance or mixture that occurs in nature; the opposite of synthetic or manufactured. Elements 1–92 are natural substances that may occur in either the free or combined state. The transuranium elements and all artificial isotopes of other elements are synthetic. Many mixtures occur naturally, e.g., petroleum, shale oil, wood, metallic ores, natural gas; others are synthetic modifications of natural compounds, e.g., glass, cement, paper, gasoline. Such materials may be considered as semisynthetic. The term *synthetic natural* is often applied to synthesized compounds that are identical with the natural substance; e.g., synthetic natural gas and synthetic natural rubber. The

term *natural product* is defined as any organic compound formed by living organisms; *natural gasoline* has a specialized meaning.
See synthesis; natural gas; biomass; gasoline.

natural dye. Any organic colorant that occurs in plant or animal materials, few of which are in use today.

natural gas.
CAS: 8006-14-2. A mixture of low molecular weight hydrocarbons obtained in petroleum-bearing regions throughout the world. Its composition is 85% methane 10% ethane, the balance being made up of propane, butane, and nitrogen. In the U.S., it occurs chiefly in the southwestern states and Alaska. An as-yet-unexploited source of natural gas under extremely high pressure (so-called geopressurized gas) exists in Texas and Louisiana at depths of 15,000–20,000 ft. The tremendous pressures involved present formidable engineering problems. Natural gas is classed as a simple asphyxiant. It should not be confused with natural gasoline. About 3% of the natural gas consumed in the U.S. is used as feedstocks by the chemical industries.
Properties: Colorless, flammable gas or liquid; almost odorless. Autoign temp 900–1100F, heating value 1000 Btu/ft^3. A warning odor is added to household fuel gas as a safety precaution.
Hazard: Flammable, dangerous fire and explosion risk; explosive limits in air 3.8–17%.
Use: Fuel and cooking gas, ammonia synthesis, formaldehyde and other petrochemical feedstocks, source of synthesis gas and methanol. Asphixiant.
See liquefied petroleum gas; synthetic natural gas.

natural gasoline. See gasoline.

natural insecticide. Any insecticide traditionally derived from plants. These substances have traditionally been regarded as safe for use on animals.
Hazard: Poisonous.

natural product. See natural.

Natural Resources Defense Council. (NRDC). A private environmental advisory group, founded in 1970, whose function is to point out serious environmental hazards and to oversee the enforcement of regulations pertaining to them. It was influential in postponement of the breeder reactor and in recognition of the chlorofluorocarbon-ozone problem. It operates primarily through the courts. Its main office is at 40 West 20th Street, New York, NY 10011. Website: http://www.nrdc.org

"Naugard PANA" [Chemtura]. TM for phenyl-α-naphthylamine.
Available forms: Solid.
Grade: Crushed and fused.

Use: Aromatic amine antioxidant for synthetic lubricants, lubricating greases, and industrial oils.
See N-phenyl-α-naphthylamine.

"Naugard Q" [Chemtura]. TM for polymerized 1,2-dihydro-2,2,4-trimethylquinoline.
Use: As an antioxidant that gives good protection against degradation due to heat and oxygen.

"Naugard SP" [Chemtura]. TM for styrenated phenol (mixture of mono, di, tri).
Use: Antioxidant for nonstaining and nondiscoloring applications.

"Naugawhite" [Chemtura]. TM for major component 2,2'-methylene bis(4-methyl-6-nonylphenol).
Use: General-purpose antioxidant for nonstaining and nondiscoloring for rubber and latex in foam sponge, tire carcasses, refrigerator gaskets, footwear, proofing, wire insulation, and sundries.

naval stores. Historically, the pitch and rosin used on wooden ships. The term now includes all products derived from pinewood and stumps, including rosin, turpentine, pine oils, and tall oil and its derivatives.

"Navane" [Pfizer]. TM for the *cis* isomer of N,N-dimethyl-9[3-4-methyl-1-piperazinyl-propylidene]-thioxanthene-2-sulfonamide. An antipsychotic drug (thiothixene) said to be as effective as most of the phenothiazines. Approved by the FDA.

"NAXONATE" Hydrotropes [NEASC]. TM for products used to obtain uniform, clear, and fluid synthetic detergent formulations. Some examples of the chemicals are sodium, potassium, and ammonium salts of xylene, cumene, and toluene.
Available forms: Powder and liquid.
Use: Synthetic detergents.

Nazarov cyclization reaction. Synthesis of cyclopentenones by the acid-catalyzed electrocyclic ring closure of divinyl or allylvinyl ketones available by hydration of divinylacetylenes.

Nb. Symbol for niobium.

NBA. Abbreviation for N-bromoacetamide.

NBR. Abbreviation for nitrile-butadiene rubber.

NBS. Abbreviation for N-bromosuccinimide.

NC. Abbreviation for nitrocellulose.

NCI. Abbreviation for National Cancer Institute.

NCS. Abbreviation for N-chlorosuccinimide.

Nd. Symbol for neodymium.

ND. Abbreviation for new drug.

NDGA. Abbreviation for nordihydroguaiaretic acid.

Ne. Symbol for neon.

neatsfoot oil.
  Properties: A fixed pale-yellow oil with a peculiar odor. D 0.916, saponification value 194–199, flash p 470F (243C), autoign temp 828F (442C), iodine value 70. Soluble in alcohol, ether, chloroform, and kerosene. Combustible.
  Derivation: By boiling in water the shinbones and feet (without hooves) of cattle and separating the oil from the fat obtained.
  Grade: 15, 20, 30, 40F cold test (the temperature in degrees Fahrenheit at which stearin separates).
  Hazard: Subject to spontaneous heating.
  Use: Leather industry for fat liquoring and softening, lubricant, oiling wool.

Neber rearrangement. Formation of α-amino ketones by treatment of sulfonic esters of ketoximes with potassium ethoxide, followed by hydrolysis.

neburon. (generic name for 1-*n*-butyl-3-(3,4-dichlorophenyl)-1-methylurea).

nectar. The fabled drink of the gods. The honey-like secretion by the nectar gland of flowers. This material is gathered by bees to make honey and used as food by other insects.

neem extract.
  CAS: 116580-64-4.
  Hazard: Moderately toxic by skin contact. Low toxicity by ingestion and skin contact.
  Use: Agricultural chemical.

Nef reaction. Formation of aldehydes and ketones from primary and secondary nitroparaffins, respectively, by treatment of their salts with sulfuric acid.

Nef synthesis. Addition of sodium acetylides to aldehydes and ketones to yield acetylenic carbinols; occasionally and erroneously referred to as the Nef reaction.

negative cooperativity. A phenomenon of some multisubunit enzymes, or proteins, in which binding of a ligand or substrate to one subunit impairs binding of a second ligand molecule to another subunit.

negative feedback. Regulation of a biochemical pathway achieved when a reaction product inhibits an earlier step in the pathway.

negatol. A condensation product of *m*-cresolsulfonic acid with formaldehyde. A polymerized dihydroxydimethyldiphenylmethanedisulfonic acid. It is dispersible in water, forming very acidic colloidal solutions. The pH of a 5% dispersion is approximately 1.0.
  Use: Medicine (antiinfective).

Negishi, Ei-ichi. (1935–). A Japanese chemist who was awarded the 2010 Nobel Prize jointly with Richard F. Heck and Suzuki Akira. The scientists discovered the use of palladium as a catalyst to produce organic molecules. Negishi's research with the palladium catalyst reaction enabled synthesis of discodermolide, which may have great potential as a cancer treatment. Negishi earned his Ph.D. from the University of Pennsylvania. His illustrious career includes awards from The Chemical Society of Japan Award (1996), Humboldt Senior Researcher Award (1998), Royal Society of Chemistry's Sir Edward Frankland Prize Lectureship (2010); he has received several Honorary Doctor of Science degrees from universities around the world.

nematic. A linear molecular structure occurring in some liquid crystals and characterized by a thread-like appearance under a polarizing microscope.

nematocide. An agent that is destructive to soil nematodes (roundworms or threadworms).

NEMS. See nanoelectromechanical systems.

Nencki reaction. The ring acylation of phenols with acids in the presence of zinc chloride, or the modification of the Friedel–Crafts alkylation-acylation procedure by substitution of ferric chloride for aluminum chloride.

Nenitzescu indole synthesis. Hydrogenative acylation of cycloolefins with acid chlorides in the presence of aluminum chloride; with five- and six-membered rings, no change in ring size occurs, but with seven-membered rings, rearrangement takes place with formation of a cyclohexane derivative.

neo-. (1) A prefix meaning new and designating a compound related in some way to an older one, e.g., neoprene. (2) A prefix indicating a hydrocarbon in which at least one carbon atom is connected directly to four other carbon atoms; as in neopentane, neohexane.

"Neobee" [Stepan]. (hydrogenated vegetable oil). TM for oil.
  Use: For a flavor carrier, vitamin solubilizer, antibiotic vehicle, nutritional fluid, lubricant, emollient, and medical diluent.

Neochel. A wetting agent used in copper plating. A liquid replacement for Rochelle salt with proprietary additives.

neocurdione.
CAS: 108944-67-8. $C_{15}H_{24}O_2$.
Hazard: A poison by ingestion.

neodecanoic acid.
CAS: 26896-20-8. $C_9H_{19}COOH$.
Clear, colorless liquid in 97% purity; available commercially. Its derivatives are especially effective as paint driers and are being widely used. Applications as plasticizers and lubricants are also possible.

"Neodol 25" [Shell].   TM for a C_2–C_{15} linear, primary alcohol.
Use: Manufacture of biodegradable surfactants, dispersants, solvents, emulsifiers, and chemical intermediates.

neodymia.   See neodymium oxide.

neodymium.
CAS: 7440-00-8.   Nd.
Metallic element having atomic number 60, group IIIB of the periodic table, aw 144.24, valence of 3. A rare-earth element of the lanthanide (cerium) group. There are seven isotopes.
Properties: Soft, malleable, yellowish metal; tarnishes easily. D 7.0, mp 1024C, bp approximately 3030C, ignites to oxide (200–400C). Liberates hydrogen from water; soluble in dilute acids. High electrical resistivity, paramagnetic. Readily cut and machined. Store under mineral oil or inert gas to prevent tarnish and corrosion.
Derivation: Monazite, bastnasite, allanite. Ores are cracked by heating with sulfuric acid.
Available forms: Ingots, rods, sheet, powder to 99.9+% purity.
Hazard: (Salts) Irritant to eyes and abraded skin.
Use: Neodymium salts, electronics, alloys, colored glass, (especially in astronomical lenses and lasers), to increase heat resistance of magnesium, metallurgical research, yttrium-garnet laser dope, gas scavenger in iron and steel manufacture.
See didymium.

neodymium acetate.
CAS: 16648-22-9. $Nd(C_2H_3O_2)_3 \cdot xH_2O$.
Properties: Pink powder. Soluble in water.

neodymium ammonium nitrate. $Nd(NO_3)_3 \cdot 2NH_4NO_3 \cdot 4H_2O$.
Properties: Pink crystals. Soluble in water. Technical grade contains 75% neodymium salt, principal impurities praseodymium and samarium compounds.

neodymium carbonate.
CAS: 5895-46-5. $Nd(CO_3)_3 \cdot xH_2O$.

Properties: Pink powder. Insoluble in water; soluble in acids.
Grade: 75, 95, and 99% neodymium salt.

neodymium chloride.   (1) $NdCl_3$, (2) $NdCl_3 \cdot 6H_2O$.
Properties: (1) Violet crystals. D 4.134 (25C), mp 784C, bp 1600C. Very soluble in water; soluble in alcohol; insoluble in ether and chloroform. (2) Red crystals. Mp 124C, loses $6H_2O$ at 160C. Very soluble in water and alcohol.
Grade: 75, 95, 99, and 99.9%.

neodymium fluoride.
CAS: 13709-42-7. NdF_3.
Properties: Pink powder. Mp 1410C, bp 2300C. Insoluble in water.
Grade: 65, 75, 99, and 99.9%.
Hazard: Irritant.

neodymium nitrate.
CAS: 10045-95-1. $Nd(NO_3)_3 \cdot 6H_2O$.
Properties: Pink crystals. Very soluble in water; soluble in alcohol and in acetone.
Grade: 75, 95, 99, and 99.9%.
Hazard: Possible explosion risk.

neodymium oxalate.
CAS: 14551-74-7. $Nd_2(C_2O_4)_3 \cdot 10H_2O$.
Properties: Pink powder. Insoluble in water; slightly soluble in acids.
Grade: 75, 95, and 99%.

neodymium oxide.   (neodymia).
CAS: 1313-97-9. $Nd_2(SO_4)_3$.
Properties: Pure product is a blue-gray powder; technical grade is a brown powder. D 7.24, mp 2270C. Insoluble in water; soluble in acids, hygroscopic. Absorbs carbon dioxide from the air.
Grade: 65, 75, 85, 95, 99, and 99.9% oxide.
Use: (65%) To counteract color of iron in glass. (Purified grade) Ceramic capacitors, coloring glass, refractories, carbon arc-light electrodes, color TV tubes, dehydrogenation catalyst.

neodymium sulfate.
CAS: 13477-91-3. $Nd(SO_4)_3 \cdot 8H_2O$.
Properties: Pink crystals. D 2.85, mp 800C (decomposes). Soluble in cold water; sparingly soluble in hot water.
Grade: 75, 99, and 99.9%.

"Neoflex" [Neoperl].   TM for a series of linear primary alcohols, from C_6 through C_{11}, including blends.
Properties: (C_7) Colorless liquid. D 0.8217 (25/25C), distillation range 174–182C.
Hazard: Combustible, moderate fire risk.
Use: Solvent, reaction intermediate, esterifying agent.

neohexane. (2,2-dimethylbutane).
CAS: 75-83-2. C_6H_{14}.

$$CH_3-\underset{\underset{CH_3}{|}}{\overset{\overset{CH_3}{|}}{C}}-CH_2-CH_3$$

Properties: Colorless, volatile liquid. Bp 49.7C, refr index 1.3659 (25C), d 0.6570 (25C), fp −99.7C, flash p −54F (−47C), octane rating 100+, autoign temp 797F (425C).
Derivation: By the thermal or catalytic union (alkylation) of ethylene and isobutane, both recovered from refinery gases.
Grade: 95%, 99%, research.
Hazard: Highly flammable, dangerous fire and explosion risk, explosion limits 1.2–7%.
Use: Component of high-octane motor and aviation fuels, intermediate for agricultural chemicals.

"Neolyn" *[Hercules]*. TM for a series of soft-to-medium hard modifying rosin-derived alkyd resins. Available in solution and/or solid forms of various grades.
Use: Adhesives, lacquers, organosols, plastisols, and floor tile.

neomycin.
CAS: 1404-04-2. An antibiotic complex obtained from *Streptomyces fradiae*; it is soluble in water and methanol but insoluble in most organic solvents. It consists of three component substances, all of which function as antiinfective agents; some derivatives have fungicidal properties. The three types are A (also called neamine): $C_{12}H_{26}N_4O_6$; B: $C_{23}H_{46}N_6O_{13}$ (also available as hydrochloride and sulfate); and C: $C_{23}H_{46}O_{13}$.

neon.
CAS: 7440-01-9. Ne.
Inert element of atomic number 10, noble gas group of the periodic table, aw 20.179. Three stable isotopes.
Properties: Colorless, odorless; tasteless gas. Does not form compounds, but ionizes in electric discharge tubes. Liquefies −245.92C, fp −248.6C, d 0.6964 (air = 1), sp vol 11.96 cu ft/lb (21C, 1 atm). Slightly soluble in water. An asphyxiant gas. Noncombustible.
Derivation: By fractional distillation of liquid air. It constitutes 0.0012% of normal air.
Grade: Technical, highest purity.
Hazard: Simple asphyxiant.
Use: (Gas) Luminescent electric tubes and photoelectric bulbs, electronic industry, high-voltage indicators, lasers. (Liquid) cryogenic research.

neonicotine. See anabasine.

neopentane. (2,2-dimethylpropane; tetramethylmethane).

CAS: 463-82-1. C_5H_{12} or $C(CH_3)_4$.
Present in small amounts in natural gas.

$$CH_3-\underset{\underset{CH_3}{|}}{\overset{\overset{CH_3}{|}}{C}}-CH_3$$

Properties: Colorless gas or very volatile liquid. Bp 9.5C, d 0.591 (20/4C), fp −19.5C, flash p −85F (−65C), autoign temp 842F (450C). Soluble in alcohol; insoluble in water.
Grade: Technical 95%, pure 99%, research 99.9%.
Hazard: Highly flammable, dangerous fire risk, explosive limits in air 1.4–7.5%.
Use: Research, butyl rubber.

neopentanoic acid. See trimethylacetic acid.

neopentyl glycol. (2,2-dimethyl-1,3-propane-diol).
CAS: 126-30-7. $HOCH_2C(CH_3)_2CH_2OH$.
Properties: White, crystalline solid. Boiling range 95% between 204–208C, mp 120–130C, d 1.066 (25/4C). Partially soluble in water; miscible with alcohol and ether.
Use: Polyester foams, insect repellent, alkyd modifier, plasticizers, urethanes, synthetic lubricants.

neopinamine.
CAS: 66525-27-7. $C_{19}H_{25}NO_4$.
Hazard: Moderately toxic by ingestion and inhalation. Low toxicity by skin contact.
Use: Agricultural chemical.

neoplasm. An abnormal growth of tissue in the body that may or may not be malignant.
See antineoplastic.

neoprene. (polychloroprene).
CAS: 9010-98-4. $(CH_3ClC:CHCH_3)_n$.
A synthetic elastomer available in solid form, as a latex or as a flexible foam.
Properties: Vulcanized with metallic oxides rather than sulfur. D 1.23; resistant to oils, oxygen, ozone, corona discharge, and electric current. An isocyanate-modified form has high flame resistance. Combustible but less so than natural rubber.
Hazard: Questionable carcinogen.
Use: (Solid) Mechanical rubber products, lining oil-loading hose and reaction equipment, adhesive cement, binder for rocket fuels, coatings for electric wiring, gaskets and seals. (Liquid) Specialty items made by dipping or electrophoresis from the latex. (Foam) Adhesive tape to replace metal fasteners for automotive accessories, seat cushions, carpet backing, sealant.

neopyrithiamine. See pyrithiamine.

Neosalvarsan. (sodium 3:3′-diamino-4:4′-dihydroxyarsenobenzene formaldehyde sulfoxylate).

CAS: 457-60-3. $C_{13}H_{14}O_4N_2SAs_2Na$.
Use: Treatment against syphilis.

"Neosol" *[Hangsterfer's].*   TM for ethanol based on a formulation approved by the Bureau of Internal Revenue.

"Neosporin" *[Johnson & Johnson].*   Proprietary formulation of polymyxin B sulfate, neomycin sulfate, and gramicidin.
Use: Ocular antibiotic.

neotridecanoic acid.
CAS: 26403-14-5. $C_{12}H_{25}COOH$.
Colorless liquid of 97% purity. Suggested for plasticizers, lubricants, paint driers, fungicides, cosmetics, alkyd resins.

Neozone.   A group of rubber antioxidants. An *N*-phenyl-α-naphthylamine. "D" is *N*-phenyl-β-naphtylamine. "C" is fusion of "Neozone" A and toluene-2,4-diamine.

nephelite.   (nepheline). $(Na,K)(Al,Si)_2O_4$.
Essentially a silicate of sodium, found in silica-poor igneous rocks.
Properties: Colorless, white, or yellowish; vitreous to greasy luster. Mohs hardness 5.5–6, d 2.55–2.65.
Occurrence: The former U.S.S.R., Ontario, Norway, South Africa, Maine, Arkansas, New Jersey.
Use: Ceramic and glass manufacture, enamels, source of potash and aluminum (the former U.S.S.R.).

nephelometry.   Photometric analytical techniques for measuring the light scattered by finely divided particles of a substance in suspension. It is used to estimate the extent of turbidity in such products as beer and wine, in which colloidally dispersed particles are present.

nephrolytic serum.   (nephrotoxic serum). A serum that is specifically toxic to the kidney.

nephrotoxic agent.   A poison that is harmful or damaging to the kidney of vertebrates. There are numerous such agents including many antibiotics, biological, botanicals, herbicides, oxalosis-inducing agents, pesticides, solvents, all nonsteroidal anti-inflammatory drugs and most heavy metals.

nephrotoxin.   A cytotoxin that specifically damages or destroys kidneys.

neptunium.   Np. A radioactive transuranic element having atomic number 93, first formed by bombarding uranium with high-speed deuterons aw 237.0482, valences of 3, 4, 5, 6; d 20.45. Neptunium-237, the longest-lived of the 11 isotopes, has been found naturally in extremely small amounts in uranium ores. It is produced in

weighable amounts as a by-product in the production of plutonium.
Metallic neptunium is obtained by first preparing neptunium trifluoride, which is reduced with barium vapor at 1200C. It is a silvery white metal; mp 640C. Neptunium is similar chemically to uranium; it forms such intermetallic compounds as $NpAl_2$ and $NpBe_{13}$, as well as NpC, $NpSi_2$, NpN, NpF_3, NpF_6, NpF_4, $NpNO_2$, Np_3O_8, etc.; soluble in hydrochloric acid, strong reducing agents.
Use: Np-237 is used in neutron detection instruments.
Hazard: A radioactive poison.

neptunium dioxide.
CAS: 12035-79-9. NpO_2.
Properties: Dark-olive, free-flowing powder.
Derivation: Neptunium oxalate is precipitated from solutions containing nitric acid and neptunium. The neptunium oxalate is calcined at 500–550C, producing neptunium dioxide.
Hazard: Toxic. A radioactive poison.
Use: Fabrication by powder metallurgy into target elements to be irradiated to produce Pu-238.

nequinate.
CAS: 13997-19-8. $C_{22}H_{23}NO_4$.
Properties: Crystals. Mp 287–288C.
Use: Drug (veterinary); food additive.

Nernst potential.   (thermodynamic potential).
(1) The potential that exists between the interiors of the phases of a two-phase system. (2) Potential (E) of a single electrode in a galvanic cell.

Nernst, Walther.   (1864–1941). A German chemist who won the Nobel Prize in 1920. He was educated at Zurich and Berlin and received his Ph.D. at Wurzburg. He wrote many works concerning theory of electric potential and conduction of electrolytic solutions. He developed the third law of thermodynamics, which states that at absolute zero, the entropy of every material in perfect equilibrium is zero, and therefore volume, pressure, and surface tension all become independent of temperature. He also invented Nernst's lamp, which required no vacuum and little current.

"Neroflex" *[Millenium].*   TM for floral composition additive.
Use: To insure consistent odor in sweet, floral compositions.

nerol.   (*cis*-3,7-dimethyl-2,6-octadien-1-ol).
CAS: 106-25-2. $(CH_3)_2C:CH(CH_2)_2C(CH_3):CH\cdot CH_2OH$.
The *cis* isomer of geraniol.
Properties: Colorless liquid, rose-neroli odor. Soluble in absolute alcohol. Combustible.
Derivation: Iodization of geraniol with hydriodic acid, followed by treatment with alcoholic soda.
Use: Perfumery, flavoring.

nerolidol. (3,7,11-trimethyl-1,6,10-dodecatrien-3-ol).
CAS: 7212-44-4. $(CH_3)_2C:CH(CH_2)_2C(CH_3):$ $CH(CH_2)_2(CH_3) \bullet (OH)CH:CH_2$.
A sesquiterpene alcohol.
Properties: Straw-colored liquid; odor similar to rose and apple mixtures. D 0.878, refr index 1.480–1.482, angular rotation (natural) +11 to +14 degrees; (synthetic) optically inactive, stable in air. Soluble in alcohol and most fixed oils; insoluble in glycerol. Combustible.
Derivation: Occurs naturally in Peru balsam and oils of orange flower, sweet orange, and ylang-ylang. Also made synthetically.
Grade: FCC.
Use: Perfumery, flavoring.

nerolin. See β-naphthyl ethyl ether.

neroli oil. An essential oil used in perfumery and flavoring.

nerve agent. (2-[fluoro(methyl)phosphory]oxypropane). $C_2H_{10}FO_2P$. Any of a number of organophosphate chemical warfare agents that attack the nervous system. Their effects are due chiefly to their ability to inhibit acetylcholinesterase activity.
Properties: In the pure state they are colorless, mobile liquids; in an impure state they may have a yellowish to brown color; a faint fruity odor.
Hazard: Extremely toxic; excessive amounts of acetylcholine accumulate at synapses and neuro-muscular junctions of neurons to the smooth muscles of the iris, ciliary body, bronchial tree, gastrointestinal tract, bladder and blood vessels and to the salivary glands and secretory glands of the gastrointestinal and respiratory tracts; cardiac muscles and sympathetic innervation of nerves to the sweat glands are also affected; extensive systematic effects.

nerve gas. (nerve poison). One of several toxic chemical warfare agents developed in Germany during World War II. They are organic derivatives of phosphoric acid (principally alkyl phosphates, fluorophosphates, and thiophosphates). They inhibit the enzyme cholinesterase and cause acetylcholine poisoning and cessation of nerve transmission. They are colorless, odorless, tasteless liquids of low volatility and are absorbed rapidly through the eyes, lungs, or skin; they are lethally toxic to higher animals and humans. Many insecticides have the same structure and properties. Antidotes are atropine sulfate and pralidoxime iodide.
See parathion; cholinesterase inhibitor.

Nessler's reagent
CAS: 7783-33-7. Solution of mercuric iodide in potassium iodide.

Hazard: High toxicity.
Use: Detecting the presence of ammonia, particularly in very small amounts.

Nessler tubes. Standardized glass tubes for filling with standard solution colors for visual color comparison with similar tubes filled with solution samples.
See Nessler's reagent.

neticonazole hydrochloride.
CAS: 130773-02-3. $C_{17}H_{22}N_2OS \bullet ClH$.
Hazard: Moderately toxic by ingestion.

netobimin.
CAS: 88255-01-0. $C_{14}H_{20}N_4O_7S_2$.
Hazard: A reproductive hazard.

Neuberg blue. A mixture of copper blue (powdered azurite) and an iron blue (Prussian blue). It can be more easily ground in oil than pure copper blue.

"Neulon" [Ashland]. TM for polyvinyl acetate low-profile additives.
Use: For polyester modifiers, fast cure processes, and pigmentable systems.

Neumann's law. Molecular heat in compounds of analogous constitution is always the same.

Neumann triangle. Graphical representation of equilibrium of three surface tensions at point of contact of two immiscible liquids with air.

neurine. (trimethylvinylammonium hydroxide).
CAS: 463-88-7. $CH_2:CHN(CH_3)_3OH$.
A poisonous ptomaine formed during putrefaction by the dehydration of choline.
Properties: Syrupy liquid; fishy odor. Absorbs carbon dioxide from the air; soluble in water and alcohol.
Hazard: Highly toxic.
Use: Biochemical research.

neurolysin. (neurotoxin). An exotoxin that has a marked affinity for nerve tissue, causing fatty degeneration of the myelin sheath of peripheral nerves, of the white matter of the brain and spinal cord, and of certain other tissues.
Hazard: Inhibition of nerve conduction or transmission at the synapse by linkage to a voltage-gated sodium channel protein.

neurolytic serum. (neurotoxic serum). A serum that is selectively toxic to the brain and spinal cord.

neuron. A cell of nervous tissue specialized for transmission of a nerve impulse.

neurotoxic agent. A substance that impairs neural or nervous system function.

Hazard: Toxic, produce focal lesions, may act preferentially only on a particular class.

neurotoxic esterase. (nte; neuropathy target enzyme).
CAS: 9032-73-9. Any of a class of membrane-bound hydrolases situated in the brain and spinal cord. They are generally accepted as the target enzyme of those organophosphates that elicit delayed neurotoxicity.

neurotoxin A (naja naja reduced).
CAS: 11080-14-1. $C_{331}H_{526}N_{98}O_{103}S_{10}$.
Hazard: A poison.
Source: Natural product.

neurotransmitter. A low molecular weight compound (usually containing nitrogen) secreted from the terminal of one neuron and bound by a specific receptor in the next neuron in order to transmit a nerve impulse.

"Neustrene" [Chemtura]. TM for hydrogenated refined glyceride products from selected animal, vegetable, and marine raw material sources.
Use: Lubricating greases, pharmaceutical intermediates, synthetic waxes, textile lubricants, mold-release agents, buffering compounds, emulsifying agents, adhesives, textile softeners, and lubricants.

"Neustrene 064" [Chemtura]. (glyceryl tristearate).
CAS: 555-43-1. TM for plastic lubricant and mold-release agent.
Use: Synthetic wax for cosmetic sticks and pencils.

neutral. (1) Of particles, without electric charge. See neutron; atom.
(2) Of solutions, neither acidic nor basic. See pH.

neutral flame. Gas flame produced by a mixture of fuel and oxygen so as to be neither oxidizing nor reducing.

neutralization. A chemical reaction in which water is formed by mutual interaction of the ions that characterize acids and bases when both are present in an aqueous solution, i.e., $H^+ + OH^- \rightarrow H_2O$, the remaining product being a salt. R. T. Sanderson states: "An aqueous solution containing an excess of hydronium ions is called acidic. It readily releases protons to electron-donating substances. An aqueous solution containing an excess of hydroxyl ions is called basic. It readily accepts protons from substances that can release them, and is in general an excellent donor. No aqueous solution can contain an excess of both hydronium and hydroxyl ions, because when these ions collide, a proton is immediately transferred from the hydronium to the hydroxyl ion, and both become water molecules."

Neutralization occurs with both (1) inorganic and (2) organic compounds:
(1) $Ca(OH)_2 + H_2SO_4 \rightarrow CaSO_4 + 2H_2O$; (2) $HCOOH + NaHCO_3 \rightarrow HCOONa + CO_2 + H_2O$. It should be noted that neutralization can occur without formation of water, as in the reaction $CaO + CO_2 \rightarrow CaCO_3$. Neutralization does not mean the attaining of pH 7.0; rather it means the equivalence point for an acid-base reaction. When a strong acid reacts with a weak base, the pH will be less than 7.0, and when a strong base reacts with a weak acid, the pH will be greater than 7.0.

neutral oil. A lubricating oil of medium or low viscosity obtained by distillation and dewaxing of crude petroleum or its cracking products.

neutral red. (toluylene red; 3-amino-7-(dimethylamino)-2-methylphenazine monohydrochloride; CI 50040). $(CH_3)_2NC_6H_3N_2C_6H_2CH_3 NH_2 \cdot HCl$ (tricyclic).
Properties: Green powder. Dissolves in water or alcohol to give red color.
Use: Acid-base indicator in the pH range 6.8–8.0 (red in acid, yellow-brown in alkali); biological stain.

neutron. Discovered by Chadwick in 1932, the neutron is a fundamental particle of matter having a mass of 1.009 but no electric charge. It is a constituent of the nucleus of all elements except hydrogen, the number of neutrons present being the difference between the mass number and the atomic number of the element. Neutrons may be liberated from the nucleus by fission of uranium-235, plutonium, and a few other elements, each nucleus yielding an average of 2.5 neutrons; they can also be produced by bombardment of other elements, e.g., beryllium with positively charged particles.
Because free neutrons are uncharged they have tremendous penetrating power as a result of their electrical neutrality; hence, they have a highly damaging effect on living tissue, requiring the use of shielding of all equipment in which they are produced. Neutrons directly emitted from atomic nuclei are termed fast; these bring about the chain reaction in the atomic bomb. In a nuclear power reactor, where a less rapid reaction is desired, the energy of fast neutrons is partially absorbed by the moderator, and the neutrons so retarded are called slow or thermal.
See electron; proton; fission.
Use: Nuclear fission; manufacture of plutonium and radioactive isotopes; activation analysis.

neutron activation analysis. See activation analysis.

neutron diffraction. See diffraction, neutron.

Neville-Winter acid. (1-naphthol-4-sulfonic acid; α-naphtholsulfonic acid; NW acid).

Properties: Transparent plates. Mp 170C. Soluble in water.

Derivation: From sodium salt of naphthionic acid by hydrolysis of the amino group.

Use: Azo dye intermediate, e.g., Congo Corinth.

Nevindene. High-melting coumarone-indene resins of extreme hardness.

Use: Dental compounds, fast-drying varnishes, rotogravure inks, aluminum paints, and insulating compounds.

Nevinol. A series of (alkyl) hydroxy resins.

Use: Adhesives, lacquers, paper coatings, special inks, and varnishes.

Newtonian flow. Flow characterized by a rate of shear that is directly proportional to the shearing force.

See non-Newtonian behavior.

Newtonian liquid. A liquid conforming to the law that the homogeneous shearing stress is the product of the coefficient of viscosity and rate of shear.

NF. Abbreviation for National Formulary, a compendium of pharmaceutical formulations widely used as a standard reference.

NFPA. Abbreviation for National Fire Protection Association.

Ni. Symbol for nickel.

Niacide. Fungicidal products containing dimethyl dithiocarbamates used mainly for scab control.

niacin. (nicotinic acid; pyridine-3-carboxylic acid).

CAS: 59-67-6.

The antipellagra vitamin, essential to many animals for growth and health. In humans, niacin is believed necessary, along with other vitamins, for the prevention and cure of pellagra. It functions in protein and carbohydrate metabolism. As a component of two important enzymes, coenzymes I and II, it functions in glycolysis and tissue respiration.

Properties: Colorless needles; odorless; sour taste. Mp 236C, sublimes above melting point, d 1.473. Soluble in water and alcohol; insoluble in most lipid solvents, quite stable to heat and oxidation. A vasodilator in high concentration. Amounts of niacin are expressed in milligrams.

Source: Food sources: meat, fish, milk, whole grains, yeast. Commercial sources: synthetic niacin is made by oxidation of nicotine, quinoline, or 2-methyl-5-ethylpyridine (from ammonia and formaldehyde or acetaldehyde).

Grade: NF, FCC, blended with soy flour (animal feeds).

Use: Medicine (cholesterol-lowering agent), nutrition, feeds, enriched flours, dietary supplement. See niacinamide.

niacinamide. (nicotinamide; nicotinic acid amide).

CAS: 98-92-0. $C_5N_4NCONH_2$.

Same biological function as niacin.

Properties: Colorless needles; bitter taste. Mp 129C, d 1.40. Soluble in water, ethanol, and glycerol.

Source: Synthetic made by conversion of niacin to the amide.

Grade: USP, FCC, Also commercially available as the hydrochloride.

Use: Medicine, dietary supplement.

niacinamide ascorbate. A complex of ascorbic acid and niacinamide.

Properties: Lemon-yellow powder; odorless or with a very slight odor. May gradually darken upon exposure to air. Soluble in water and alcohol, sparingly soluble in glycerol, practically insoluble in benzene.

Grade: FCC.

Use: Dietary supplement.

nialamide. (1-(2-benzylcarbamyl)ethyl-2-isonicotinoylhydrazine).

CAS: 51-12-7. $C_5H_4NCO(NH)_2(CH_2)_2CONHC-H_2C_6H_5$.

Properties: White, crystalline powder. Low solubility in water; good solubility in slightly acid solution. It is stable in crystalline form, suspension, and solution.

Use: Medicine (as an antidepressant).

Hazard: Toxic in overdose.

Nialk. Chlorine, caustic soda, caustic potash, carbonate of potash, paradichlorobenzene, and trichloroethylene.

nicarbazin. Equimolar complex of 4,4′-dinitrocarbanilide and 2-hydroxy-4,6-dimethylpyrimidine.

Properties: Forms crystals. Decomposes at 265–275C. Insoluble in water.

Use: Coccidiostat.

niccolite. (arsenical nickel). NiAs. An ore of nickel.

Properties: Pale-copper-red mineral with dark tarnish, metallic luster. Contains 43.9% nickel.

D 7.3–7.67, Mohs hardness 5–5.5. Soluble in concentrated nitric acid.
Hazard: Toxic by inhalation of dust.

Nichrome. An alloy containing 60% nickel, 24% iron, 16% chromium, 0.1% carbon.
Use: It is used principally for electric resistance purposes. It also offers good resistance to mine and seawaters and moist sulfurous atmospheres.

nickel.
CAS: 7440-02-0. Ni.
Metallic element of atomic number 28, group VIII of the periodic table, aw 58.70, valences of 2, 4. Five stable isotopes.
Properties: Malleable, silvery metal. Readily fabricated by hot- and cold-working, takes high polish, excellent resistance to corrosion. D 8.908, mp 1455C, bp 2900C, electrical resistivity (20C) 6.844 $\mu\Omega$-cm. Attacked slightly by hydrochloric acid and sulfuric acid, somewhat more by nitric acid, highly resistant to strong alkalies.
Occurrence: Ontario, Cuba, Norway, Dominican Republic.
Derivation: Nickel ores are of two types, sulfide and oxide, the former accounting for two-thirds of the world's consumption. Sulfide ores are refined by flotation and roasting to sintered nickel oxide, and either sold as such or reduced to metal, which is cast into anodes and refined electrolytically or by the carbonyl process (see Mond process). Oxide ores are treated by hydrometallurgical refining, e.g., leaching with ammonia. Much/most of secondary nickel is recovered from scrap.
Grade: Electrolytic, ingot, pellets, shot, sponge, powder, high-purity strip, single crystals.
Hazard: Flammable and toxic as dust or fume. Dermatitis and pneumoconiosis. A confirmed carcinogen.
Use: Alloys (low-alloy steels, stainless steel, copper and brass, permanent magnets, electrical resistance alloys), electroplated protective coatings, electroformed coatings, alkaline storage battery, fuel cell electrodes, catalyst for methanation of fuel gases and hydrogenation of vegetable oils.
See Raney's nickel.

nickel acetate.
CAS: 373-02-4. $Ni(OOCCH_3)_2 \cdot 4H_2O$.
Properties: Green, monoclinic crystals; effloresces somewhat in air. D 1.74, decomposes on heating to 250C. Soluble in water and alcohol.
Derivation: By heating nickel hydroxide with acetic acid in the presence of metallic nickel.
Hazard: Toxic by ingestion, a carcinogen (OSHA).
Use: Textiles (mordant), catalyst.

nickel acetylacetonate.
CAS: 3264-82-2. $(CH_3COCHCOCH_3)_2Ni$.
Properties: Green, crystalline solid. Mp 230C, bp 220C (11 mm Hg), d 1.45. Soluble in water, alcohol, and benzene; insoluble in ether.
Use: Catalyst for organic reactions.

nickel aluminide. A cermet that can be flame-sprayed.

nickel ammonium chloride. (ammonium nickel chloride). (1) $NiCl_2 \cdot NH_4Cl$, (2) $NiCl_2 \cdot NH_4Cl \cdot 6H_2O$.
Properties: (1) Yellow powder, (2) green crystals, d 1.65, soluble in water, deliquescent.
Hazard: See nickel.
Use: Electroplating, dyeing (mordant).

nickel ammonium sulfate. (nickel salts, double; ammonium nickel sulfate).
CAS: 15699-18-0. $NiSO_4 \cdot (NH_4)_2SO_4 \cdot 6H_2O$.
Properties: Green crystals. D 1.929. Decomposed by heat. Soluble in water; less in ammonium sulfate solution; insoluble in alcohol.
Derivation: An aqueous solution of nickel sulfate is acidified with sulfuric acid, then an aqueous solution of ammonium sulfate is added. On concentrating, crystals of the double sulfate separate out.
Hazard: See nickel.
Use: Nickel electrolyte for electroplating.

nickel arsenate. (nickelous arsenate). $Ni_3(AsO_4)_2 \cdot 8H_2O$.
Properties: Yellow-green powder. D 4.98. Soluble in acids; insoluble in water.
Hazard: Highly toxic.
Use: Catalyst (hardening fats used in soap).

nickel bromide. (nickelous bromide). (1) $NiBr_2$, (2) $NiBr_2 \cdot 3H_2O$.
Properties: (1) Brownish-yellow solid or yellow, lustrous scales. D 5.098, mp 963C. (2) Deliquescent, greenish scales; loses $3H_2O$ at 300C. Soluble in water, alcohol, ether, and ammonium hydroxide.
Derivation: Bromination of nickel powder or nickel carbonyl.
Hazard: See nickel.

nickel carbonate, basic.
CAS: 3333-67-3. $NiCO_3 \cdot 2Ni(OH)_2 \cdot 4H_2O$.
Properties: Light-green crystals or brown powder. D 2.6. Insoluble in water; soluble in ammonia and dilute acids.
Derivation: (1) In nature as the mineral zaratite. (2) Synthetically, by addition of soda ash to a solution of nickel sulfate.
Hazard: Confirmed carcinogen.
Use: Electroplating, preparation of nickel catalysts, ceramic colors and glazes.

nickel carbonyl. (nickel tetracarbonyl).
CAS: 13463-39-3. $Ni(CO)_4$.
A zero-valent compound. The four carbonyl groups form a tetrahedral arrangement and are linked covalently through the metal through the carbons.
Properties: Colorless liquid. D 1.3185, fp −19C, bp 43C, vap press 400 mm Hg (25.8C). Soluble in alcohol and many organic solvents; soluble in concentrated nitric acid; insoluble in water.

Derivation: By passing carbon monoxide gas over finely divided nickel.

Grade: Technical.

Hazard: Flammable, dangerous fire risk, explodes at 60C (140F). A lung irritant and confirmed carcinogen.

Use: Production of high-purity nickel powder by Mond process, continuous nickel coatings on steel and other metals.

nickel chloride.   (nickelous chloride).

CAS: 7718-54-9.   (1) $NiCl_2$, (2) $NiCl_2 \cdot 6H_2O$.

Properties: (1) Brown scales, deliquescent, (2) green scales, deliquescent. Soluble in water, alcohol, and ammonium hydroxide. (1) D 3.55, mp 1001C. Nonflammable.

Derivation: Action of hydrochloric acid on nickel.

Hazard: Confirmed carcinogen.

Use: Electroplated nickel coatings, reagent chemical.

nickel cyanide.

CAS: 557-19-7. $Ni(CN)_2 \cdot 4H_2O$.

Properties: Apple-green plates or powder. Mp loses $4H_2O$ at 200C, bp decomposes. Soluble in ammonium hydroxide and potassium cyanide solution; insoluble in water and acids.

Derivation: By adding potassium cyanide to a solution of a nickel salt.

Hazard: Highly toxic.

Use: Metallurgy, electroplating.

nickel dibutyldithiocarbamate.

CAS: 13927-77-0. $Ni[SC(S)N(C_4H_9)_2]_2$.

Properties: Dark-green flakes. D 1.26, mp 86C min.

Use: Antioxidant for synthetic rubbers.

nickel dimethylglyoxime.

CAS: 13478-93-8. $C_8H_{14}N_4NiO_4$.

Properties: Bright-red powder. Sublimes at 250C. Insoluble in water; soluble in mineral acids.

Use: Pigment in paints, cosmetics, cellulosics.

nickel formate.

CAS: 3349-06-2. $(HCOO)_2Ni \cdot 2H_2O$.

Properties: Green crystals. D 2.15. Partially soluble in water; insoluble in alcohol.

Derivation: (1) Reaction of sodium formate and nickel sulfate, (2) dissolving nickel hydroxide in formic acid.

Hazard: See nickel.

Use: Production of nickel catalysts.

nickel hydroxide.   See nickelous hydroxide; nickelic hydroxide.

nickelic hydroxide.   (nickel hydroxide).

CAS: 12054-48-7. $Ni(OH)_3$.

Properties: Black powder. Mp decomposes.

Derivation: By adding a hypochlorite to a solution of a nickel salt.

Hazard: Confirmed carcinogen.

Use: Nickel salts.

nickelic oxide.   (nickel peroxide; nickel(III) oxide; nickel sesquioxide; black nickel oxide).

CAS: 1314-06-3. Ni_2O_3.

Properties: Gray-black powder. D 4.84, mp (reduced to nickel oxide) 600C. Soluble in acids; insoluble in water.

Derivation: By gentle heating of the nitrate or chlorate.

Hazard: Confirmed carcinogen.

Use: Storage batteries.

nickel iodide.   (nickelous iodide). NiI_2 or $NiI_2 \cdot 6H_2O$.

Properties: Black, crystalline powder or blue-green crystals; hygroscopic. D 5.834, sublimes at 797C without melting. Soluble in alcohol and water.

Derivation: Direct combination of nickel and iodine.

Hazard: Toxic by ingestion.

See nickel.

nickel–iron alloy.   See iron-nickel alloy.

Nickel-Lume.   A bright nickel electroplating process; materials used are nickel sulfate, nickel chloride, boric acid, and organic addition agents.

nickel matte.   See matte.

nickel nitrate.   (nickelous nitrate).

CAS: 13138-45-9. $Ni(NO_3)_2 \cdot 6H_2O$.

Properties: Green, deliquescent crystals. D 2.065, mp 55C, bp 136.7C. Soluble in water, ammonium hydroxide, and alcohol.

Derivation: By the action of nitric acid on nickel, or on nickel oxide.

Grade: Technical, reagent.

Hazard: Dangerous fire risk, strong oxidizing agent.

Use: Nickel plating, preparation of nickel catalysts, manufacture of brown ceramic colors.

nickel nitrate, ammoniated.   (nickel nitrate tetrammine). $Ni(NO_3)_2 \cdot 4NH_3 \cdot 2H_2O$.

Properties: Green crystals. Soluble in water; insoluble in alcohol; decomposes in air.

Derivation: By adding ammonium hydroxide to a nitric acid solution of nickel oxide with subsequent crystallization.

Hazard: See nickel nitrate.

Use: Nickel plating.

nickelocene.   (dicyclopentadienylnickel).

CAS: 1271-28-9. $(C_5H_5)_2Ni$.

Properties: Dark-green crystals. Mp 171–173C. Soluble in most organic solvents; insoluble in water. Decomposes in acetone, alcohol, and ether; highly reactive compound that decomposes rapidly in air.

Hazard: Toxic by inhalation and skin contact, a carcinogen (OSHA).

Use: Catalyst, complexing agent.

See metallocene.

nickelous arsenate.   See nickel arsenate.

nickelous bromide. See nickel bromide.

nickelous chloride. See nickel chloride.

nickelous hydroxide. (nickel hydroxide).
CAS: 12125-56-3. Ni(OH)$_2$.
Properties: Fine green powder. D 4.15, mp (decomposes) 230C. Very slightly soluble in water; soluble in acids and ammonium hydroxide.
Derivation: By adding caustic soda to a solution of nickelous salt.
Hazard: See nickel.
Use: Nickel salts.

nickelous iodide. See nickel iodide.

nickelous nitrate. See nickel nitrate.

nickelous phosphate. See nickel phosphate.

nickel oxide. (nickelous oxide; nickel(II) oxide; nickel protoxide; green nickel oxide).
CAS: 1313-99-1. NiO.
Properties: Green powder that becomes yellow. D 6.6–6.8, absorbs oxygen at 400C forming nickelic oxide, which is reduced to nickel oxide at 600C. Soluble in acids and ammonium hydroxide. Insoluble in water and caustic solutions.
Derivation: By heating nickel above 400C in the presence of oxygen.
Hazard: Confirmed carcinogen.
Use: Nickel salts, porcelain painting, fuel cell electrodes.

nickel oxide, black. See nickelic oxide.

nickel oxide, green. See nickel oxide.

nickel peroxide. See nickelic oxide.

nickel phosphate. (nickelous phosphate; trinickelous-orthophosphate).
CAS: 14396-43-1. Ni$_3$(PO$_4$)$_2$•7H$_2$O.
Properties: Light-green powder. Soluble in acids and ammonium hydroxide; insoluble in water.
Use: Electroplating.

nickel potassium sulfate. (potassium nickel sulfate).
CAS: 13842-46-1. NiSO$_4$•K$_2$SO$_4$•6H$_2$O.
Properties: Blue-green crystals. D 2.124. Soluble in water.

nickel protoxide. See nickel oxide.

nickel–rhodium. Alloys containing nickel and 25–80% rhodium, but sometimes also some platinum, iridium, palladium, molybdenum, tungsten, copper, iron, or cobalt.
Use: Electrodes, chemical apparatus, reflectors.

nickel salt, double. See nickel ammonium sulfate.

nickel salt, single. See nickel sulfate.

nickel sesquioxide. See nickelic oxide.

nickel-silver. Nonferrous alloy of nickel, copper, and zinc having a silver appearance.
Use: Etching, enameling, silver plating, and chromium plating.

nickel stannate.
CAS: 12035-38-0. NiSnO$_3$•2H$_2$O.
Properties: Light-colored, crystalline powder. Approximate temperature of dehydration 120C.
Hazard: Toxic by ingestion and inhalation. TLV: 0.1 mg(Ni)/m^3.
Use: Additive in ceramic capacitors.

nickel sulfate. (nickel salts, single; blue salt).
CAS: 7786-81-4. (1) NiSO$_4$, (2) NiSO$_4$•6H$_2$O, (3) NiSO$_4$•7H$_2$O.
Properties: (1) Yellow-green crystals, (2) blue or emerald green crystals, (3) green crystals. D (1) 3.68, (2) 2.031, (3) 1.98; mp (1) 840C (loses SO$_3$), (2) and (3) loses 6H$_2$O at 103C. All the sulfates are soluble in water; (2) and (3) are soluble in alcohol; (1) is insoluble in alcohol and ether.
Derivation: Action of sulfuric acid on nickel.
Grade: Technical, CP, single crystals.
Hazard: Toxic material. Questionable carcinogen.
Use: Manufacture of nickel ammonium sulfate, nickel catalysts, nickel plating, mordant in dyeing and printing textiles, coatings, ceramics.

nickel tetracarbonyl. See nickel carbonyl.

nick translation. A method for incorporating radioactive isotopes (typically 32P) into a piece of DNA. The DNA is randomly nicked by DNase I, and then starting from those nicks DNA polymerase I digests and then replaces a stretch of DNA. Radiolabeled precursor nucleotide triphosphates can thus be incorporated.

nicol. An optical material (Iceland spar) that functions as a prism, separating light rays that pass through it into two portions, one of which is reflected away and the other transmitted. The transmitted portion is called plane-polarized light.
See calcite; optical isomerism.

"Nicomo-12" [Grace]. TM for a hydrodesulfurization catalyst of nickel, cobalt, and molybdenum on alumina.

"Nicon" [Nicon]. Brand name for diethyldithiocarbamate used in the colorimetric determination of nickel.

nicotinamide. See niacinamide.

nicotinamide adenine dinucleotide phosphate. (NADP; diphosphopyridine nucleotide;

DPN; coenzyme I; Co I; codehydrogenase I). $C_{22}H_{27}O_{14}N_7P_2$. Entry name is recommended by the International Union of Biochemistry and IUPAC. A coenzyme necessary for the alcoholic fermentation of glucose and the oxidative dehydrogenation of other substrates. It occurs widely in living tissue, especially in liver.
Use: Biochemical research, chromatography.

nicotinamide mononucleotide.   (Nmn; [(2R, 3S,4R,5R)-5-(3-carbamoylpyridin-1-ium-1-yl)-3, 4-dihydroxyoxolan-2-yl]methyl hydrogen phosphate). $C_{11}H_{15}N2O_8P$. A condensation product of nicotinamide and ribose 5-phosphate, in which the nitrogen of nicotinamide is linked to the (β) c-1 of the ribose.

nicotinate.   (pyridine-3-carboxylic acid).
CAS: 63-75-2. $C_8H_{13}NO_2$.
A salt or ester of nicotinic acid.

nicotine.   (β-pyridyl-α-*n*-methylpyrrolidine).
CAS: 54-11-5. $C_5H_4NC_4H_7NCH_3$.

Properties: Alkaloid from tobacco; thick, water-white, levorotatory oil turning brown on exposure to air, also in form of dust or powder. Hygroscopic. Bp 247C, d 1.00924, autoign temp 471F (243C). Soluble in alcohol, chloroform, ether, kerosene, water, and oils. Combustible.
Derivation: By distilling tobacco with milk of lime and extracting with ether.
Hazard: Toxic by ingestion, inhalation, and skin absorption. Gastrointestinal damage, central nervous system impairment, and cardiac impairment.
Use: Insecticide, fumigant; use as insecticide may be restricted. Available as the dihydrochloride, salicylate, sulfate, and bitartrate.

nicotine, compd. with manganese(ii) *o*-benzoyl benzoate, trihydrate (2:1).
CAS: 64092-22-4. $C_{48}H_{46}MnN_4O_6 \cdot 3H_2O$.
Hazard: A poison by ingestion.
Use: Agricultural chemical.

nicotine-*n'*-oxide.
CAS: 491-26-9. $C_{10}H_{14}N_2O$.
Hazard: A poison.

nicotine salts.   (1) Hydrochloride, $C_{10}H_{14}N_2 \cdot$ HCl; (2) salicylate, $C_{10}H_{14}N_2 \cdot C_7H_6O_3$; (3) sulfate, $(C_{10}H_{14}N_2)_2 \cdot H_2SO_4$ (40% solution = "Black Leaf Forty"); (4) tartrate, $C_{10}H_{14}N_2 \cdot 2C_4O_6H_6 \cdot H_2O$.
Properties: (1) Colorless oil; (2) white crystals. Mp 117.4C; (3) white crystals; (4) white plates. Mp 89C. All the salts are soluble in water, alcohol, and ether.

Derivation: By the action of the respective acid on the alkaloid.

nicotine sulfate.   (ENT 2,453; 1,1-methyl-2-(3-pyridyl)-pyrrolidine sulfate; (s)-3-(1-methyl-2-pyrrolidinyl)pyridine sulfate (2:1); 1, 3-(1-methyl-2-pyrrolidyl)pyridine sulfate; 1,10-methyl-2-(3-pyridyl)-pyrrolidine sulfate; (S)-3-(1-methyl-2-pyrrolidinyl)pyridine sulfate (2:1); 1,3-(1-methyl-2-pyrrolidinyl)pyridine sulfate; 3-(1-methylpyrrolidin-2-yl)pyridine; nikotinsulfat; pyridine, 3-(1-methyl-2-pyrrolidinyl)-, (s)-, Sulfate (2L1); pyrrolidine, 1-methl-2-(3-pyridyl)-, sulfate; sulfate de nicotine; sulfuric acid). $C_{20}H_{30}N_4O_4S$.
Hazard: Poisonous; causes tremors, incoordination, nausea, disturbed respiration, dark bloody hemorrhages in the heart and in the lungs, congestion of the brain.
Use: Insecticide.

nicotinic acid.   See niacin.

nicotinic acid, aluminum salt.   See aluminum nicotinate.

nicotinic acid amide.   See niacinamide.

nidultoxin.   A mycotoxin.
Derivation: Produced by *Aspergillus nidulans*.

Niementowski quinazoline synthesis.   Formation of 4-oxo-3,4-dihydroquinazolines by cyclization of the reaction products of anthranilic acid and amides.

Niementowski quinoline synthesis.   Formation of γ-hydroxyquinoline derivatives from anthranilic acids and carbonyl compounds.

Nierenstein reaction.   Formation of omega-chloroacetophenones by reaction of diazomethane in dry ether with aroyl chlorides. Coumaranones are obtained if an ortho-hydroxy group is present.

Nieuland, Father J. A.   (1878–1936). A Jesuit whose research on polymers of acetylene formed the basis for the development of polychloroprene (neoprene) in 1931.
See Carothers, Wallace H.

(-)-nigaldipine hydrochloride.
CAS: 189624-85-9. $C_{41}H_{42}N_4O_6 \cdot 2ClH$.
Hazard: A poison by ingestion.

nigre.   The dark-colored layer containing some soap, as well as salts and impurities, formed in soap manufacture between the layers of soap proper and caustic solution.

nigrescigenin.
CAS: 6785-70-2. $C_{23}H_{32}O_7$.
Hazard: A poison.
Source: Natural product.

nigrosine.   A class of dark-blue or black dyes, some soluble in water, some in alcohol, and some in oil.
Use: Manufacture of ink and shoe polish dyeing leather, wood, textiles, etc. It is also used as a shark repellent.

NIH.   Abbreviation for National Institutes of Health (Bethesda, MD).

"Nikanol" [Mitsubishi].   TM for a liquid resin.

ninhydrin.   (triketohydrindene hydrate).
CAS: 485-47-2. $C_9H_4O_3 \cdot H_2O$.
Properties: White crystals or powder. Mp 240–245C, becomes red when heated above 100C. Soluble in water and alcohol; slightly soluble in ether and chloroform.
Hazard: Irritant.
Use: Chemical intermediate; reagent for determination of amines, amino acids, ascorbic acid.

ninhydrin reaction.   A color reaction given by free amino groups of amino acids and peptides on heating with ninhydrin; widely used for their detection and estimation.

"Ninol" [Stepan].   TM for an emulsifier and lubricant.
Use: Anti-static properties for textiles.

niobe oil.   See methyl benzoate.

niobic acid.   Any hydrated form of Nb_2O_5. It forms as a white, insoluble precipitate when a potassium hydrogen sulfate fusion of a niobium compound is leached with hot water or when niobium fluoride solutions are treated with ammonium hydroxide. Soluble in concentrated sulfuric acid, concentrated hydrochloric acid, hydrogen fluoride, and bases. Important in analytical determination of niobium.

niobite.   See columbite.

niobium.   (columbium).
CAS: 7440-03-1.   Nb.
The name *niobium* is officially approved by chemical authorities, but columbium is still used chiefly by metallurgists. Metallic element, atomic number 41, group VB of the periodic table, aw 92.9064, valences of 2, 3, 4, 5; no stable isotopes.
Properties: Gray or silvery, ductile metal. Does not tarnish or oxidize at room temperature. D 8.57, mp 2468C. Reacts with oxygen and halogens only when heated, less corrosion resistant than tantalum at high temperature. Not attacked by nitric acid up to 100C, but vigorously attacked by a mixture of nitric and hydrofluoric acids. Hot concentrated hydrochloric acid, sulfuric acid, and phosphoric acid attack it, but hot concentrated nitric acid does not. Unaffected at room temperature by most acids and by aqua regia.

It is attacked by alkaline solutions to some extent at all temperatures.
Occurrence: Found in two major ores, columbite and pyrochlore (a carbonate-silicate rock). Chief sources are Brazil, Nigeria, Canada.
Derivation: Niobium is so closely associated with tantalum that they must be separated by fractional crystallization or by solvent extraction, with subsequent purification.
Grade: Plates, rods, powder, single crystals.
Use: Superconducting and magnetic alloys (with tin and titanium), cermets, missiles and rockets, cryogenic equipment, ferroniobium for alloy steels.

niobium carbide.   NbC.
Properties: Lavender-gray powder. Mp approximately 3500C, d 7.82. Insoluble in water and in all acids except a mixture of nitric acid and hydrogen fluoride.
Derivation: By direct combination of niobium with carbon or by the reduction of niobium oxide with lampblack.
Use: Cemented carbide tipped tools, special steels, preparation of niobium metal, coating graphite for nuclear reactors.

niobium chloride.   (niobium pentachloride).
CAS: 10026-12-7. $NbCl_5$.
Properties: Yellow, crystalline solid. Mp 205C, bp 254C, d 2.75. Deliquescent; decomposes in moist air with evolution of hydrogen chloride fumes. Soluble in carbon tetrachloride, hydrochloric acid, concentrated sulfuric acid.
Derivation: Direct combination of niobium and chlorine, chlorination of niobium oxide in the presence of carbon.
Hazard: May evolve fumes of hydrogen chloride. Keep dry.
Use: Preparation of pure niobium; intermediate.

niobium diselenide.
CAS: 12034-77-4. $NbSe_2$.
Properties: Gray-black solid. Mp >1316C, vacuum stable from −430 to 2400F (−170 to 1315C). Has higher electrical conductivity than graphite.
Use: Lubricant and conductor at high temperatures and high vacuum.

niobium oxalate.
CAS: 21348-59-4. $NbO(HC_2O_4)_3 \cdot 4H_2O$.
Properties: White crystals, 99.99% pure. Very soluble in water.
Use: Intermediate, special catalysts.

niobium oxide.   (niobium pentoxide).
CAS: 1313-96-8. Nb_2O_5.
Properties: White powder. D 4.5–5.0, mp 1520C. Insoluble in acids (except hydrofluoric and hot sulfuric acids); insoluble in water; soluble in fused potassium hydrogen sulfate, or carbonates or hydroxides of the alkali metals.

Derivation: Strong ignition of niobic acid.
Use: Intermediate, electronics.

niobium pentachloride.   See niobium chloride.

niobium pentoxide.   See niobium oxide.

niobium potassium oxyfluoride.   (potassium niobium oxypentafluoride; potassium oxyfluoniobate).
CAS: 17523-77-2. $K_2NbOF_5 \cdot H_2O$.
Properties: White, lustrous leaflets; greasy to touch. Soluble in hot water.
Hazard: Toxic by ingestion, strong irritant. TLV: 2.5 mg(F)/m^3.
Use: Separation of niobium from tantalum, electrolytic preparation of niobium metal.

niobium silicide.
CAS: 12034-80-9. $NbSi_2$.
Properties: Crystalline solid. Mp 1950C.
Use: Refractory material.

niobium-tin. Nb_3Sn. Alloy used for special wire for superconducting magnets to obtain high magnetic fields for use in communication and containment of thermonuclear fusion plasmas.

niobium–titanium.   Alloy used for magnetic devices with fields up to 100,000 gauss.

niobium–uranium.   Niobium alloyed with 20% uranium yields a nuclear fuel that maintains tensile strength and hardness at 871C.

NIOSH.   Abbreviation for National Institute for Occupational Safety and Health.

NiPar S-10.   1-nitropropane.
CAS: 108-03-2.
Grade: Industrial.
Available forms: Liquid.
Use: Solvent for inks; chemical intermediate; diesel fuel additive.
See 1-nitropropane.

NiPar S-20.   (2-nitropropane).
CAS: 79-46-9. A solvent.
Use: For inks and coatings based on resins of vinyl, epoxy, acrylic, polyurethane, polyamide, nitrocellulose, chlorinated rubber. Also used in automotive finishes, extractions, crystallization, diesel fuel additives, and intermediates.

"Nirez" [Galvo].   TM for polyterpene, terpene, phenol, and resins used in adhesives, coatings, chewing gums, and printing inks.

NIRMS.   Noble gas-ion reflection mass spectroscopy.

nisin.   Antibiotic containing 34 amino acid residues, produced by *Streptomyces lactis*.

Use: Food preservative, especially in canned products.

nisin preparation.
CAS: 1414-45-5. $C_{143}H_{230}N_{42}O_{37}S_7$.
Properties: Crystals from ethanol. Derived from *Streptococcus lactis* Lancefield Group N.
Use: Food additive.

NIST.   See National Institute of Standards and Technology.

nital.   Solution of nitric acid in methyl or ethyl alcohol; 1.5% by volume.
Use: Etching agent in ferrous metallography.

niter.   (nitre; saltpeter).
CAS: 7757-79-1. KNO_3.
A natural potassium nitrate.

niter cake.   A common name for sodium bisulfate ($NaHSO_4$) because it was a product of the reaction by which nitric acid was first made: $NaNO_3 + H_2SO_4 \rightarrow NaHSO_4 + HNO_3$.
See sodium bisulfate.

niter, Chile.   See sodium nitrate.

"Niterox" [SQM].
CAS: 7631-99-4.   TM for sodium nitrate.
Use: Charcoal, glass, explosives, treating metal, antifreeze, chemical compounding, wastewater treatment, cleaning compounds, and pharmaceuticals.

nitralin.   A herbicide used largely to control weeds in cotton and soybeans. Research indicates possible use in plant breeding; treatment of corn roots induces abnormal chromosome formation and cell-wall deterioration.

nitralloy.   See nitriding.

nitramine.   See tetryl.

Nitrane.   Industrial chemicals, mainly nitroparaffin solvents.

nitranilic acid.   (2,5-dihydroxy-3,6-dinitroquinone).
CAS: 479-22-1. $C_6O_2(NO_2)_2(OH)_2$.
Properties: Flat, yellow crystals. Loses H_2O at 100C, decomposes explosively at 170C. Soluble in water and alcohol; insoluble in ether.
Hazard: Fire risk when heated. Evolves toxic fumes on decomposition.

nitraniline.   See nitroaniline.

nitrapyrin.   (2-chloro-6-(trichloromethyl)pyridine; "N-Serve" [Dow Agrosciences]).
CAS: 1929-82-4. $C_6H_3Cl_4N$.

Properties: Crystals. Mp 62.5–62.9, bp 136–137.5 (11 torr).
Hazard: Liver damage. Questionable carcinogen.
Use: A fertilizer additive.

nitrate.
CAS: 14797-55-8. NO_3^-.
A salt or ester of nitric acid. They exist in gases in the atmosphere and as dissolved gases in water. They occur in ground water chiefly as a result of agricultural practices and are common air pollutant.
Properties: A colorless liquid.
Hazard: Moderately toxic.
Use: In the treatment of angina pectoris; in the manufacture of inorganic and organic nitrates and nitro compounds for fertilizers, dye intermediates, explosives, and many different organic chemicals.

nitrate-nitrogen. The expression of nitrate concentration in terms of total nitrogen present in a specimen or sample.

nitrate reductase. A metalloenzyme that contains molybdenum. It catalyzes the reduction of nitrate to nitrite.

nitrating acid. Legal label name for mixed acid.

nitration. A reaction in which a nitro group ($–NO_2$) replaces a hydrogen on a carbon atom by the use of nitric acid or mixed acid. An example is the nitration of cellulose to nitrocellulose. It is widely used in aromatic reactions to form such compounds as nitrobenzene, trinitrotoluene, nitroglycerin, and other explosives. Aromatic nitrations are usually effected with mixed acid, a mixture of nitric and sulfuric acids, at 0–120C. Aliphatic nitration is less common than aromatic, but propane can be nitrated under pressure to yield nitroparaffins.

nitrene. Electron-deficient, uncharged monovalent nitrogen species with either singlet or triplet electronic configurations; analog of carbene.

nitric acid. (aqua fortis; engraver's acid; azotic acid).
CAS: 7697-37-2. HNO_3.
Properties: Transparent, colorless, or yellowish fuming, suffocating, hygroscopic, corrosive liquid. Will attack almost all metals. The yellow color is due to release of nitrogen dioxide on exposure to light; strong oxidizing agent. Bp (decomposes) 78C, fp −42C, d 1.504 (25/4C), vap press 62 mm Hg (25C), refr index 1.3970 (24C), viscosity 0.761 cP (25C). Miscible with water, decomposes in alcohol.
Derivation: (1) Oxidation of ammonia by air or oxygen with platinum catalyst. (*Note:* A pelleted catalyst not containing platinum or other noble metals is available.) Air oxidation yields 60% acid; concentration is achieved by (a) distillation with sulfuric acid, (b) extractive distillation with magnesium nitrate, or (c) neutralizing the weak acid with soda

ash, evaporating to dryness, and treating with sulfuric acid. Method (c) yields synthetic niter cake ($NaHSO_4$) as a by-product. (2) High-pressure oxidation of nitrogen tetroxide (yields 98% acid). (3) Reaction of nitrogen and oxygen in nuclear reactors; two tons of nitric acid are said to be produced from one gram of enriched uranium. Not in commercial use.
Grade: 36, 38, 40, 42 degrees Bé; 58–63.5%; 95%.
Hazard: Dangerous fire risk in contact with organic materials. Highly toxic by inhalation, corrosive to skin and mucous membranes, strong oxidizing agent. Eye and upper respiratory tract irritant and dental erosion.
Use: Manufacture of ammonium nitrate for fertilizer and explosives, organic synthesis (dyes, drugs, explosives, cellulose nitrate, nitrate salts), metallurgy, photoengraving, etching steel, ore flotation, urethanes, rubber chemicals, reprocessing spent nuclear fuel.

nitric acid, fuming. (1) White fuming nitric acid (WFNA) contains more than 97.5% nitric acid, less than 2% water, and less than 0.5% NO_x. It is a colorless or pale-yellow liquid that fumes strongly. It is decomposed by light or elevated temperatures, becoming red in color from nitrogen dioxide. (2) Red fuming nitric acid (RFNA) contains more than 85% nitric acid, approximately 6–15% NO_x (as nitrogen dioxide), and less than 5% water.
Derivation: From dilute nitric acid, nitrogen dioxide, and oxygen.
Hazard: Toxic by inhalation, corrosive to skin and mucous membranes. Strong oxidizing agent, may explode in contact with strong reducing agents. Dangerous fire risk.
Use: Preparation of nitro compounds, rocket fuels, laboratory reagent.

nitric acid, magnesium salt, hexahydrate.

CAS: 13446-18-9. $N_2O_6 \cdot Mg \cdot 6H_2O$.
Hazard: Low toxicity by ingestion. A mild skin and eye irritant.

nitric oxide.
CAS: 10102-43-9. NO.
Properties: Colorless gas (readily reacts with oxygen at room temperature to form nitrogen dioxide, NO_2, a reddish brown gas). Bp −152C, fp −164C, d at bp 1.27. Slightly soluble in water. Noncombustible.
Derivation: Oxidation of ammonia above 500C, decomposition of nitrous acid (aqueous solution). Also from atmospheric oxygen and nitrogen in the electric-arc process for fixation of nitrogen.
Grade: Pure (99%).
Hazard: Supports combustion. Toxic by inhalation, strong irritant to skin and mucous membranes. Hypoxia/cyanosis, nitrosyl-hemoglobin formation, and upper respiratory tract irritant.

Use: Intermediate in production of nitric acid from ammonia, preparation of nitrosyl carbonyls, bleaching rayon.
See nitrogen dioxide.

nitride.   A compound of metal and nitrogen. As in aluminum nitride.

nitriding.   A process of case hardening in which a ferrous alloy, usually of special composition, is heated in an atmosphere of ammonia or in contact with nitrogenous material to produce surface hardening by absorption of the nitrogen without quenching. The alloys used for nitriding are known as nitroalloys. Several types are available, with ranges of composition as follows: aluminum 0.85–11.2%, carbon 0.20–0.45%, chromium 0–8%, molybdenum 0.15–1.00%, manganese 0.4–0.7%, silicon 0.2–0.4%.

nitrile.   An organic compound containing the $-C{\equiv}N$ grouping, e.g., acrylonitrile $H_2C{=}CHC{\equiv}N$ and acetonitrile ($CH_3C{\equiv}N$).
Hazard: Some organic cyanide compounds are flammable (acetonitrile, acrylonitrile). Most organic cyanide compounds are toxic (acetonitrile, acrylonitrile).

nitrile rubber.   (acrylonitrile rubber; acrylonitrilebutadiene rubber; nitrile-butadiene rubber; NBR). A synthetic rubber made by random polymerization of acrylonitrile with butadiene by free radical catalysis. Alternating copolymers using Natta–Ziegler catalyst have been developed. Approximately 20% of the total is used as latex; also available in powder form. Its repeating structure may be represented as $-CH_2CH{=}CHCH_2CH_2$ $CH(CN)-$.
Properties: (Medium acrylonitrile) D (polymer) 0.98, tensile strength (psi) 1000–3000, elongation (%) 100–700, maximum service temperature 121–148C. Combustible.
Use: (High acrylonitrile) Oil well parts, fuel-tank liners, fuel hoses, gaskets, packing oil seals, hydraulic equipment.
(Medium acrylonitrile) General-purpose oil-resistant applications, shoe soles, kitchen mats, sink topping, and printing rolls. (Low acrylonitrile) Gaskets, grommets, and O-rings (flexible at a very low temperature), adhesives. Fuel binder in solid rocket propellants.

nitrile-silicone rubber.   (NSR). Combines the characteristic properties of silicones with the oil resistance of nitrile rubber. Resistant to jet fuels, solvents, and hot oils.

nitrilotriacetic acid.   (NTA; triglycine; TGA; triglycolamic acid).
CAS: 139-13-9. $N(CH_2COOH)_3$.
Properties: White, crystalline powder. Mp 240C (with decomposition). Insoluble in water and most organic solvents; forms mono-, di-, and tribasic salts that are water soluble. Combustible. 70% biodegradable.
Available forms: Di- and trisodium salts.
Hazard: Possible carcinogen.
Use: Synthesis, chelating agent, eluting agent in purification of rare-earth elements, detergent builder.

nitrilotriacetonitrile.   (NTAN). $N(CH_2CH)_3$.
Properties: White, crystalline solid. Mp 130–134C. Insoluble in water; soluble in acetone.
Hazard: See nitriles.
Use: Intermediate and chelating agent.

2,2′,2″-nitrilotris(ethylamine).
CAS: 4097-89-6. $C_6H_{18}N_4$.
Hazard: A poison by ingestion and skin contact.

"Nitrix" [Glanbia].   TM for synthetic rubber latices of the butadiene-acrylonitrile type.
Use: Paper saturation, leather finishing, plasticizer for resin latices.

p-nitroacetanilide. $NO_2C_6H_4NHCOCH_3$.
Properties: White crystals. Mp 214–216C. Soluble in alcohol and ether; very slightly soluble in cold water; soluble in hot water and in potassium hydroxide solution. Combustible.
Derivation: By acetylating aniline, then nitrating.
Use: Manufacture of nitraniline.

p-nitro-o-aminophenol.
CAS: 99-57-0. $C_6H_3OHNH_2NO_2$.
Properties: Yellow-brown leaflets containing water of crystallization. Melting at 80–90C, anhydrous melts at 154C. Soluble in acid.
Derivation: From dinitrophenol.
Use: Dyes.

m-nitroaniline.   (m-nitraniline).
CAS: 99-09-2. $NO_2C_6H_4NH_2$.
Properties: Yellow needles. D 1.43, mp 111.8C, bp 306C. Soluble in alcohol and ether; slightly soluble in water.
Derivation: From aniline by nitration after acetylation, with subsequent removal of the acetyl group by hydrolysis; from m-nitrobenzoic acid.
Hazard: Moderate fire risk. Toxic when absorbed by skin.
Use: Dry intermediate.

o-nitroaniline.   (o-nitraniline).
CAS: 88-74-4. $NO_2C_6H_4NH_2$.
Properties: Orange-red needles. D 1.443, mp 69.7C. Not lightfast, flash p 335F (168C), autoign temp 970F (521C). Soluble in alcohol and ether; slightly soluble in water.
Derivation: From aniline by nitration after acetylation, with subsequent removal of the acetyl group by hydrolysis; from o-dinitrosobenzene.

Hazard: Explosion risk. Toxic when absorbed by skin.

Use: Dye intermediate, synthesis of photographic antifogging agent, *o*-phenylenediamine, coccidiostats.

p-nitroaniline.   (*p*-nitraniline).
CAS: 100-01-6. $NO_2C_6H_4NH_2$.

O_2N—⟨benzene ring⟩—$\ddot{N}H_2$

Properties: Yellow needles. D 1.437, mp 148C, flash p 390F (198C). Soluble in alcohol and ether; insoluble in water. Combustible.
Derivation: (1) From *p*-chloronitrobenzene, (2) from aniline by nitration after acetylation, (3) from acetanilide.
Hazard: Explosion risk. Toxic when absorbed by skin. Methemoglobinemia, liver damage and eye irritant. Questionable carcinogen.
Use: Dye intermediate, especially for *p*-nitraniline red; intermediate for antioxidants; gasoline gum inhibitors; corrosion inhibitors.

o-nitroanisole.   (2-nitroanisole).
CAS: 91-23-6. $C_6H_4OCH_3NO_2$.
Properties: Light-reddish or amber liquid. D 1.255 (20/20C), fp 9.6C, boiling range 268–271C, refr index 1.5602 (20C). Soluble in alcohol and ether; insoluble in water. Combustible.
Derivation: From *o*-nitrophenol by methylation or from *o*-nitrochlorobenzene by action of methanol and caustic soda.
Grade: Technical.
Hazard: Questionable carcinogen.
Use: Organic synthesis, intermediate for dyes and pharmaceuticals.

p-nitroanisole.   (1-methoxy-4-nitrobenzene).
CAS: 100-17-4. $NO_2C_6H_4OCH_3$.
Properties: Colorless crystals. Mp 54C, bp 260C. Insoluble in water; soluble in alcohol and ether. Combustible.
Grade: Technical.
Use: Intermediate for dyes.

1-nitroanthraquinone. $C_{14}H_7NO_4$.
Properties: Yellow needles from AcOH; yellow plates by sublimation; soluble in EtOH and Et_2O; slightly soluble in EtOAc, benzene and $CHCl_3$; insoluble in water.
Hazard: Hepatotoxic; moderately toxic; may cause hepatocellular neoplasms, subcutaneous fibromas, subcutaneous hemangiosarcomas, and type II pneumocyte proliferation.

5-nitrobarbituric acid.   (dilituric acid).
CAS: 480-68-2.

$\overline{O_2NHCCONHCONHCO}$.

Properties: Prisms and leaflets from water. Mp 176C (decomposes). Slightly soluble in water; soluble in alcohol and sodium hydroxide solution; insoluble in ether.
Use: Microreagent for potassium.

3-nitrobenzaldehyde.   (*m*-nitrobenzaldehyde).
CAS: 99-61-6. $NO_2C_6H_4CHO$.
Properties: Yellowish, crystalline powder. Mp 58C, bp 164C (23 mm Hg). Almost insoluble in water; soluble in alcohol, chloroform, ether.
Use: Synthesis of dyes, pharmaceuticals, surface-active agents, vapor-phase corrosion inhibitor, antioxidant for chlorophyll, mosquito repellent.

2-nitrobenzaldehyde dimethylacetal.
CAS: 20627-73-0. $C_9H_{11}NO_4$.
Hazard: Moderately toxic by ingestion and inhalation.

nitrobenzene.   (oil of mirbane).
CAS: 98-95-3. $C_6H_5NO_2$.

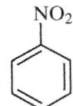

Properties: Greenish-yellow crystals or yellow, oily liquid. D 1.19867, mp 5.70C, bp 210.85C, flash p 190F (87.7C), autoign temp 900F (482C). Soluble in alcohol, benzene, and ether; slightly soluble in water. Combustible.
Derivation: From benzene by nitrating with nitric acid-sulfuric acid mix.
Method of purification: By washing and distilling with steam, then redistilling.
Grade: Technical, redistilled, 97%.
Hazard: Toxic by ingestion, inhalation, and skin absorption. Methemoglobinemia. Possible carcinogen.
Use: Manufacture of aniline; solvent for cellulose ethers; modifying esterification of cellulose acetate; ingredient of metal polishes and shoe polishes; manufacture of benzidine, quinoline, azobenzene, etc.

p-nitrobenzeneazoresorcinol.
$NO_2C_6H_4N_2C_6H_3(OH)_2$.
Properties: Red crystals. Mp 198C. Slightly soluble in water; soluble in nitrobenzene. Combustible.
Derivation: Diazotized *p*-nitroaniline is coupled with resorcinol.
Grade: Analytical.
Use: Determination of magnesium.

m-nitrobenzenesulfonic acid.   (3-nitrobenzenesulfonic acid).
CAS: 98-47-5. $C_6H_4NO_2SO_3H$.
Properties: Crystals. Mp 70C. Soluble in water and alcohol.
Use: Organic synthesis. The sodium salt is a protective antireduction agent.

6-nitrobenzimidazole.
CAS: 94-52-0.

$\overline{O_2NC_6H_4NHCH{:}N}$

Properties: Solid. Mp 203C. Very soluble in alcohol; slightly soluble in water, ether, and benzene.
Use: Antifogging agent in photographic developers.

5-nitrobenzimidazole nitrate.
CAS: 27896-84-0. $C_7H_5N_3O_2 \cdot HNO_3$.
Hazard: Moderately toxic by ingestion.

nitrobenzoic acid.   (1) *m*-, (2) *o*-, (3) *p*-(nitrodracylic acid).
CAS: (*m*-)   121-92-6;   (*p*-)   62-23-7.
$C_6H_4(NO_2)COOH$.
Properties: Yellowish-white crystals. (1) d 1.494, mp 140–141C; (2) d 1.575, mp 147.7C; (3) d 1.5497, mp 238C. (1) Soluble in alcohol and ether; slightly soluble in water; (2) soluble in water, alcohol, and ether; (3) soluble in alcohol, sparingly soluble in water. Combustible.
Derivation: (1) Nitration of benzoic acid; (2) oxidation of *o*-nitrotoluene with MnO_2 and sulfuric acid, (3) oxidation of *p*-nitrotoluene by hot chromic acid mixture.
Use: (1) Dye intermediate, reagent for alkaloids; (2) and (3) organic synthesis.

m-nitrobenzotrifluoride.   (3-nitrobenzotrifluoride; *m*-nitrotrifluoromethylbenzene).
CAS: 98-46-4. $NO_2C_6H_4CF_2$.
Properties: Pale-straw, thin, oily liquid; aromatic odor. Distillation range 200.5–208.5C, fp −5.0C, d 1.437 (15.5C), bp 203C, flash p 214F (101C) (OC), bulk d 11.98 lb/gal (15.5C), viscosity 2.35 cP (37.7C). Soluble in organic solvents; insoluble in water. Combustible.
Hazard: Toxic material.

m-nitrobenzoyl chloride.
CAS: 121-90-4. $NO_2C_6H_4COCl$.
Properties: Yellow to brown liquid. Partially crystallized at room temperature. Mp approximately 34C, bp 278C. Soluble in ether; decomposes in water and alcohol. Combustible.
Hazard: Toxic by ingestion.
Use: Manufacture of dyes for fabrics and color photography, intermediate in preparation of pharmaceuticals.

p-nitrobenzoyl chloride.
CAS: 122-04-3. $NO_2C_6H_4COCl$.

Properties: Yellow, crystalline solid. Mp 72C, bp 154C (15 mm Hg). Decomposes in water and alcohol; soluble in ether. Combustible.
Use: Intermediate for procaine hydrochloride, dyestuffs.

1-(p-nitrobenzoyl)-2-thiobiuret.
CAS: 127019-55-0. $C_9H_8N_4O_4S$.
Hazard: A poison.

p-nitrobenzyl cyanide.   (*p*-nitro-α-tolunitrile).
CAS: 555-21-5. $NO_2C_6H_4CH_2CN$.
Properties: Crystals. Mp 116–118C. Insoluble in water; soluble in alcohol and ether.
Derivation: Action of concentrated nitric acid on benzyl cyanide.
Hazard: Toxic material.
Use: Intermediate for dyestuffs and pharmaceuticals; preparation of *p*-nitrophenylacetic acid.

3-nitrobenzylidene ethyl ester.   See ethyl 2-(*m*-nitrobenzylidene)acetoacetate.

o-nitrobiphenyl.   (ONB; *o*-nitrodiphenyl).
CAS: 92-93-3. $C_6H_5C_6H_4NO_2$.
Properties: Light-yellow to reddish solid or liquid. D 1.203 (25/25C), bulk d 10 lb/gal, crystallizing p 34.5C (min), refr index approximately 1.613 (25C), bp approximately 330C, flash p 290F (143C), autoign temp 356F (180C). Soluble in carbon tetrachloride, mineral spirits, pine oil, turpentine, benzene, acetone, glacial acetic acid, and perchloroethylene; insoluble in water. Combustible.
Derivation: By controlled nitration of biphenyl.
Hazard: Toxic by ingestion and skin contact. Confirmed carcinogen.
Use: Dye intermediate, fungicide, plasticizer for cellulosics, wood preservative.

p-nitro blue tetrazolium chloride.
CAS: 298-83-9. $C_{40}H_{30}N_{10}O_6 \cdot 2Cl$.
Hazard: Moderately toxic by ingestion.

nitrobromoform.   See bromopicrin.

2-nitro-1-butanol.
CAS: 609-31-4. $CH_3CH_2CHNO_2CH_2OH$.
Properties: Colorless liquid. Soluble in water (20 g/100 cc at 20C), d 1.133 (20/20C), bp 105C (10 mm Hg), fp −48 to −47C, bulk d 9.44 lb/gal (20C), refr index 1.4390 (20C), pH of 0.1M solution 4.51. Combustible.
Use: Organic synthesis.

nitro carbon nitrate.   A blasting agent consisting of ammonium nitrate sensitized with diesel oil. Will burn with explosive violence.
Hazard: Dangerous explosion risk, strong oxidizing agent.

nitrocellulose. (cellulose nitrate; nitrocotton; guncotton; pyroxylin).
CAS: 9004-70-0.

Properties: Pulpy, cotton-like, amorphous solid (dry); colorless liquid to semisolid (solution). Contains from 10 to 14% nitrogen. D 1.66, flash p 55F (12.7C), autoign temp 338F (170C). High-nitrogen form (explosives) is soluble in acetone; insoluble in ether–alcohol mixture. Low-nitrogen form (pyroxylin) is soluble in ether-alcohol mix and acetone (collodion and lacquers).
Derivation: Treatment of cellulose (as cotton linters, wood pulp) with mixture of nitric and sulfuric acids. By varying strength of acids, temperature and time of reaction, and acid/cellulose ratio, widely different products are obtained.
Available forms: Colloided block, colloided flake or granular, flakes, powder, solutions of several viscosities (from 1/4 to 1000 sec). May be dry or wet with alcohol or water.
Hazard: Flammable, dangerous fire and explosion risk. Somewhat less flammable when wet.
Use: Fast-drying automobile lacquers, high explosives, collodion, rocket propellant, printing ink base, flashless propellant powder, coating bookbinding cloth, leather finishing, manufacture of "Celluloid."
See Hyatt, John Wesley.

nitrocellulose lacquer. See lacquer.

nitrochlorobenzene. Legal label name (Rail) for chloronitrobenzene.

nitrochloroform. See chloropicrin.

p-nitro-o-chlorophenyl dimethyl thionophosphate. See dicapthon.

nitrocobalamin. $C_{62}H_{90}N_{14}O_{16}PCo$. One of the active forms of vitamin B_{12} in which a nitro group is attached to the central cobalt atom.

nitrocotton. See nitrocellulose.

2-nitro-p-cresol. (4-methyl-2-nitrophenol). $NO_2(CH_3)C_6H_3OH$.
Properties: Yellow crystals. D 1.24 (38/4C), mp approximately 35C, bp 234C. Slightly soluble in water; soluble in alcohol and ether. Combustible.
Hazard: Toxic by ingestion, inhalation, and skin absorption.
Use: Intermediate.

nitrodichloro derivative. See the corresponding dichloronitro derivative.

8-nitro-dihydro-1,3-benzoxazine-2-thione-4-one.
CAS: 23611-69-0. $C_8H_4N_2O_4S$.
Hazard: A poison by ingestion.
Use: Agricultural chemical.

o-nitrodiphenyl. See o-nitrobiphenyl.

o-nitrodiphenylamine. $C_6H_5NHC_6H_4NO_2$.
Properties: Red-brown, crystalline powder. Mp 75–76C. Combustible.
Use: Stabilizer for nitroglycerin; chemical intermediate.

Nitrodisc. Nitroglycerin transdermal patch system.
Use: Drug administered by transdermal patch.

nitrodracylic acid. See p-nitrobenzoic acid.

nitro dye. A dye whose molecules contain the NO_2 chromophore group.

nitroethane.
CAS: 79-24-3. $CH_3CH_2NO_2$.
A nitroparaffin.
Properties: Colorless liquid. D 1.052 (20/20C), fp −50C, bp 114C, vap press 15.6 mm Hg (20C), flash p 106F (41C), autoign temp 779F (415C), bulk d 8.75 lb/gal (20C), refr index 1.3917 (20C). Solubility in water 4.5 cc/100 cc (20C); solubility of water in nitroethane 0.9 cc/100 (20C).
Derivation: By reaction of propane with nitric acid under pressure.
Hazard: Moderate fire risk. Upper respiratory tract irritant, central nervous system impairment, and liver damage.
Use: Solvent for nitrocellulose, cellulose acetate, cellulose acetopropionate, cellulose acetobutyrate, vinyl, alkyd and many other resins, waxes, fats, and dyestuffs; Friedel–Crafts synthesis; propellant research; fuel additive.

2-nitro-2-ethyl-1,3-propanediol.
CAS: 597-09-1. $CH_2OHC(C_2H_5)NO_2CH_2OH$.
Properties: White, crystalline solid. Mp 56–75C, bp (decomposes) (10 mm Hg), pH 0.1M aqueous solution 5.48. Soluble in organic solvents; very soluble in water.
Use: Organic synthesis.

nitrofuran. Any of several synthetic antibacterial drugs used to treat mammary gland infections in cows and to inhibit disease in swine, chickens, etc. Among them are nitrofurazone, furazolidone, nihydrazone, and furaltadone. All of the latter have been found to cause cancer in laboratory animals, and their use has been discontinued.
See furazolidone.

nitrofurantoin. (*N*-(5-nitro-2-furfurylidene)-1-aminohydantoin).
CAS: 67-20-9. $C_8H_6N_4O_5$.
Properties: Yellow, bitter powder; slight odor. Mp (decomposes) 270–272C. Very slightly soluble in alcohol; practically insoluble in ether and water.
Grade: USP.
Hazard: Questionable carcinogen.
Use: Medicine (antibacterial).

nitrogen.
CAS: 7727-37-9. N.
Gaseous element of atomic number 7 of group VA of the periodic table, aw 14.0067, valences of 1, 2, 3, 4, 5. There are two stable and four radioactive isotopes; the molecular formula is N_2.
Properties: Colorless, diatomic gas; odorless; tasteless; constituting approximately four-fifths of the air. Colorless liquid, chemically unreactive. D 1.251 g/L (0C, 1 atm), d (gas) 0.96737 (air = 1.00), (liquid) 0.804, (solid) 1.0265, fp −210C, bp −195.5C. Slightly soluble in water and alcohol. Noncombustible, an asphyxiant gas.
Derivation: From liquid air by fractional distillation, by reducing ammonia.
Grade: USP, prepurified 99.966% min, extra dry 99.7% min, water pumped 99.6% min.
Hazard: Asphyxiant.
Use: Production of ammonia, acrylonitrile, nitrates, cyanamide, cyanides, nitrides; manufacture of explosives; inert gas for purging, blanketing, and exerting pressure; electric and electronic industries; in-transit food refrigeration and freeze drying; pressurizing liquid propellants; quick-freezing foods; chilling in aluminum foundries; bright annealing of steel; cryogenic preservation; food antioxidant; source of pressure in oil wells; inflating tires; component of fertilizer mixtures.

nitrogen-15. A stable isotope, with an atomic mass of 15.00011, present in naturally occurring nitrogen to the extent of 0.37%. Many N-15 compounds are commercially available.

nitrogenase. Enzyme that fixes nitrogen and can be isolated from soil bacteria. It is possible to synthesize ammonia from nitrogen and hydrogen without high temperatures and pressures by means of nitrogenase. Pyruvic acid is an adjunct of the reaction.

nitrogenase complex. A system of enzymes capable of anaerobic reduction of atmospheric nitrogen to ammonia in the presence of ATP.

nitrogen, available. Water-soluble nitrogen compounds plus that which is rendered soluble or converted to free ammonia.

nitrogen base. Compound, as in an amine, that could be considered a substitution product of ammonia. A compound containing trivalent nitrogen, as ammonia, capable of combining with acids and forming salts containing pentavalent nitrogen.

nitrogen chloride. See nitrogen trichloride.

nitrogen cycle. The cycling of various forms of biologically available nitrogen (N2, nitrate, nitrite, ammonia, etc.) through the plant, animal, and microbial worlds and through the atmosphere and geosphere.

nitrogen dioxide.
CAS: 10102-44-0. NO_2.
Properties: Red to brown gas above 21.1C, brown liquid below 21.1C; colorless solid approximately −11C. The pressurized liquid is nitrogen tetroxide (dinitrogen tetroxide) because of admixture of N_2O_4 with NO_2, fp −9.3C, bp 21.1C. Noncombustible but supports combustion.
Derivation: By oxidation of nitric acid; an intermediate stage in the oxidation of ammonia to nitric acid.
Grade: Pure, 99.5% min.
Hazard: Inhalation may be fatal. Can react strongly with reducing materials. Lower respiratory tract irritant. Questionable carcinogen.
Use: Production of nitric acid, nitrating agent, oxidizing agent, catalyst, oxidizer for rocket fuels, polymerization inhibitor for acrylates.

nitrogen fixation. Utilization of atmospheric nitrogen to form chemical compounds. In nature, this function is performed by bacteria located on the root hairs of plants; as a result, plants are able to synthesize proteins. Industrial nitrogen fixation is exemplified by the synthesis of ammonia from a gaseous mixture containing carbon monoxide, hydrogen, and nitrogen. Nitrogen can also be fixed by the electric arc process and the cyanamide process. The latter involves reaction of water and calcium cyanamide: $CaCN_2 + 3H_2O \rightarrow CaCO_3 + 2NH_3$.
See synthesis gas.

nitrogen fluoride. (nitrogen trifluoride; trifluoroamine).
CAS: 7783-54-2. F_3N.
Properties: A colorless gas.
Hazard: Mottling of the teeth and skeletal changes; liver and kidney damage; methemoglobininemia.
Use: An oxidizer of high-energy fuels.

nitrogen monoxide. See nitrous oxide.

nitrogen mustard.
CAS: 51-75-2. A class of compounds with fishy odor and lachrymatory properties. They are named from their similarity in structure to mustard gas (dichlorodiethyl sulfide). The sulfur of the mustard gas is replaced by an amino nitrogen.

Typical nitrogen mustards are halogenated alkylamines, such as methyl bis(2-chloroethyl)amine: $(ClCH_2CH_2)_2NCH_3$. Other examples are triethylene melamine, triethylene thiophosphoramide, and triethylene phosphoramide.

Hazard: Strong irritant to tissues, lachrymatory. Probable carcinogen.

Use: Medicine, military poison gas.

See mechlorethamine hydrochloride.

nitrogenous base.   A nitrogen-containing molecule having the chemical properties of a base. DNA contains the nitrogenous bases adenine (A), guanine (G), cytosine (C), and thymine (T).

See DNA.

nitrogen oxide red. N_2O_4. A nitrogen dioxide. It is often produced together with nitrogen dioxide during the fermentation of fodder to form silage and may contribute to silo-filler's disease.

nitrogen oxides. NO_x. The nitrogen oxides provide an example of the law of multiple proportions and the +1 to +5 oxidation states of nitrogen. They are as follows: N_2O (nitrous oxide), NO (nitric oxide), N_2O_3 (nitrogen trioxide or nitrogen sesquioxide), N_2O_4 (dinitrogen tetroxide or nitrogen peroxide), NO_2 (nitrogen dioxide), N_2O_5 (dinitrogen pentoxide), N_3O_4 (trinitrogen tetroxide), and NO_3, which is unstable.

Hazard: Toxic by inhalation, especially NO_2.

nitrogen solutions.   Aqueous solutions of ammonium nitrate, ammonia, and/or urea. They are graded according to total nitrogen content and composition.

Use: Direct application to soil as fertilizer, neutralizing superphosphate fertilizers.

nitrogen tetroxide.   See nitrogen dioxide.

nitrogen trichloride.   (nitrogen chloride).
CAS: 10025-85-1. NCl_3.

Properties: Yellow oil or rhombic crystals. D 1.653, bp below 71C, fp below −40C. Insoluble in cold water; decomposes in hot water; soluble in chloroform, phosphorus trichloride, and carbon disulfide.

Hazard: Explodes when heated to approximately 200F (93C) or when exposed to direct sunlight. Toxic by ingestion and inhalation, strong irritant.

Use: Flour bleach (no longer permitted in U.S.).

nitrogen trifluoride.   (nitrogen fluoride).
CAS: 7783-54-2. NF_3.

Properties: Colorless, unstable gas; moldy odor. D 1.537 (−129C), gas d 0.1864 lb/cu ft (70F), fp −206.6C, bp −128.8C. Very slightly soluble in water. Nonflammable.

Derivation: Electrolysis of ammonium acid fluoride.

Hazard: Severe explosion hazard. Corrosive to tissue. Methemoglobinemia, liver and kidney damage.

Use: Oxidizer for high-energy fuels, chemical synthesis.

nitrogen triiodide.   (nitrogen iodide).
CAS: 13444-85-4. NI_3.

Properties: Black, unstable crystals.

Hazard: Explodes at slightest touch when dry; when handled it should be kept wet with ether. Too sensitive to be used as explosive, because it cannot be stored, handled, or transported safely.

nitrogen trioxide.   (dinitrogen trioxide; nitrogen sesquioxide). N_2O_3.

Properties: Blue liquid. D 1.447 (2C), bp 3.5C (1 atm), fp −102C (1 atm). Partially dissociated into nitric oxide (NO) and nitrogen dioxide (NO_2).

Derivation: Prepared by passing nitric oxide into nitrogen tetroxide at approximately 0C until the stoichiometric amount has been absorbed.

Hazard: Toxic by inhalation, strong irritant.

Use: Oxidant in special fuel systems, identification of terpenes, preparation of pure alkali nitrites.

nitroglycerin.   (glyceryl trinitrate; trinitroglycerin).
CAS: 55-63-0. $CH_2NO_3CHNO_3CH_2NO_3$.

Properties: Pale-yellow, viscous liquid. D 1.6009, fp 13.1C, explosion p 424F (218C). Soluble in alcohol and ether; slightly soluble in water.

Derivation: By dropping glycerol through cooled, mixed acid and stirring, followed by repeated washing with water.

Grade: Technical.

Hazard: Severe explosion risk, highly sensitive to shock and heat. Toxic by ingestion, inhalation, and skin absorption. Toxic by skin absorption. Vasodilator.

Use: High explosive, production of dynamite and other explosives, medicine (vasodilator) combating fires in oil wells, rocket propellants.

nitroguanidine.
CAS: 556-88-7. $H_2NC(NH)NHNO_2$.

Exists in two forms, α and β.

Properties: (α) Long, thin, flat, flexible lustrous needles; (β) small, thin, elongated plates. For both, melting ranges are from 220 to 250C. The β form appears to be more soluble in water.

Derivation: (α) Results when guanidine nitrate is dissolved in concentrated sulfuric acid and the solution is poured into water. (β) Nitration of a mixture of guanidine sulfate and ammonium sulfate that results from the hydrolysis of dicyandiamide by sulfuric acid.

Hazard: May explode when shocked or heated.

Use: High explosives, especially flashless propellant powder (with nitrocellulose), chemical intermediate.

nitrohydrochloric acid.   Legal label name for aqua regia.

3-nitro-2-hydroxybenzoic acid. See *m*-nitrosalicylic acid.

4-nitro-3-hydroxymercuri-*o*-cresol anhydride. See nitromersol.

3-nitro-4-hydroxyphenylarsonic acid. See 4-hydroxy-3-nitrobenzenearsonic acid.

nitrolarginine.
CAS: 2149-70-4. $C_6H_{13}N_5O_4$.
Hazard: A reproductive hazard.

nitromannite. See mannitol hexanitrate.

nitromersol. (4-nitro-3-hydroxymercuri-*o*-cresol anhydride). $C_6H_2(CH_3)(NO_2)(OHg)$.
Properties: Brownish-yellow or yellow granules or powder; odorless; tasteless. Insoluble in water; almost insoluble in alcohol, acetone, ether; soluble in solutions of alkalies, ammonia, by opening the anhydride ring and salt formation.
Grade: NF.
Hazard: Toxic by ingestion and inhalation.
Use: Disinfectant for skin in extremely dilute solution (1 part in 1000 is maximum strength).

nitrometer. (azotomer). A glass apparatus used to measure and collect nitrogen and other gases evolved by a chemical reaction.

nitromethane.
CAS: 75-52-5. CH_3NO_2.
A nitroparaffin.
Properties: Colorless liquid. D 1.139 (20/20C), bp 101C, vap press 27.8 mm Hg (20C), flash p 95F (35C), bulk d 9.5 lb/gal, refr index 1.3817 (20C), fp −29C, autoign temp 785F (418C). Soluble in water and alcohol; solubility of water in nitromethane 2.2 cc/100 cc (20C).
Derivation: By reaction of methane or propane with nitric acid under pressure.
Hazard: Dangerous fire and explosion risk, lower explosion limit 7.3% in air. Toxic by ingestion and inhalation. Thyroid effects, upper respiratory tract irritant, and lung damage. Possible carcinogen.
Use: Solvent for cellulosic compounds, polymers, waxes, fats, etc.; chemical synthesis; rocket fuel; gasoline additive.

2-nitro-4-methoxyaniline.
$NO_2C_6H_3(OCH_3)NH_2$.
Properties: Orange-red powder, mp 118–120C, sparingly soluble in cold water and alcohol, soluble in hot water and dioxane.
Use: Chemical intermediate.

2-(nitromethylene)-1-(phenylmethyl)imidazolidine.
CAS: 61532-76-1. $C_{11}H_{13}N_3O_2$.
Hazard: A poison.
Use: Agricultural chemical.

4-nitro-*n*-methylphthalimide.
CAS: 41663-84-7. $C_9H_6N_2O_4$.
Hazard: Moderately toxic by ingestion and skin contact. A mild skin irritant.

2-nitro-2-methyl-1,3-propanediol.
$CH_2OCH(CH_3)NO_2CH_2OH$.
Properties: White, crystalline solid. Mp 147–149C, bp (decomposes) (10 mm Hg), pH 0.1M solution 5.42. Solubility in water 80 g/100 cc (20C).
Hazard: Toxic by ingestion.
Use: Intermediate.

nitromide and sulfanitran.
Use: Drug (veterinary); food additive.

nitron. (1,4-diphenyl-3,5-*endo*-anilino-4,5-dihydro-1,2,4-triazole). $C_{20}H_{16}N_4$.
Properties: Lemon-yellow, fine, crystalline needles. Soluble in chloroform, acetone, and acetic acid ester; slightly soluble in ether and alcohol; insoluble in water.
Use: Reagent for detection of nitrate, perchlorate, boron, rhodium.

α-nitronaphthalene. (1-nitroanphthalene).
CAS: 86-57-7. $C_{10}H_7NO_2$.
Properties: Yellow crystals. D 1.331, mp 55–56C, bp 304C, flash p 327F (164C). Soluble in alcohol and ether; insoluble in water. Combustible.
Derivation: Action of a mixture of nitric and sulfuric acids on finely ground naphthalene.
Hazard: Irritant. The β-isomer is highly toxic by ingestion. Questionable carcinogen.
Use: Dyes, naphthylamine, added to mineral oils to mask fluorescence.

1-nitronaphthalene-5-sulfonic acid. See Laurent's α-acid.

nitronitramide ammonium salt.
CAS: 140456-78-6. $H_4N_4O_4$.
Hazard: Moderately toxic by ingestion and inhalation.

nitronium perchlorate. NO_2ClO_4.
Properties: White, crystalline solid. Mp 120–140C, hygroscopic, noncorrosive. Soluble in water to form nitric and perchloric acids. Highly reactive.
Derivation: From ozone, nitrogen dioxide, chlorine dioxide.
Hazard: Strong oxidizing agent, may explode in contact with organic materials. Irritant to skin and mucous membranes.
Use: Suggested as propellant oxidizer.

nitroparaffin. (nitroalkane). Any of a homologous series of compounds whose generic formula is $C_nH_{2n+1}NO_2$, the nitro groups being attached to a carbon atom via the nitrogen.
Properties: Colorless, mobile liquids; pleasant odor. Bp range 101–131C, fp range −18 to −104C, d

range 0.983–1.131, dielectric constant range 23–35, flash p range 75–100F (24–38C). Slightly soluble in water.
Hazard: Toxic by inhalation.
Derivation: Nitration of propane and other paraffin hydrocarbons under pressure.
Use: Solvents for dyes, waxes, resins, gums, and various polymeric substances; gravure and flexographic inks; propellants and fuel additives; chemical intermediates.
See nitroethane; nitromethane; nitropropane.

p-nitrophenetole.
CAS: 100-29-8. $NO_2C_6H_4OC_2H_5$.
Properties: Crystals in prisms. Mp 58C, bp 283C. Soluble in alcohol and ether.
Derivation: By ethylation of *p*-nitrophenol with ethyl chloride.
Use: Dyes and other intermediates.

nitrophenide.   [bis(3-nitrophenyl)disulfide]. $(NO_2C_6H_4)_2S$.
Properties: Yellow, rhomboid crystals. Mp 83C. Insoluble in water, soluble in ether; slightly soluble in alcohol.
Derivation: Reduction of *m*-nitrobenzenesulfonyl chloride with hydriodic acid.
Use: Pharmaceutical intermediate.

m-nitrophenol.
CAS: 554-84-7. $NO_2C_6H_4OH$.

Properties: Pale-yellow crystals. D 1.485 (20C), mp 96–97C, bp 194C (70 mm Hg). Slightly soluble in water; soluble in alcohol. Combustible.
Derivation: Diazotized *m*-nitroaniline is boiled with water and sulfuric acid.
Hazard: Toxic by ingestion.
Use: Indicator.

o-nitrophenol.
CAS: 88-75-5. $NO_2C_6H_4OH$.

Properties: Yellow crystals. D 1.295 (45C), 1.657 (20C), mp 44–45C. Soluble in hot water, alcohol, ether.
Derivation: Action of dilute nitric acid on phenol at low temperature; *p*-nitrophenol formed at same time. They are separated by steam distillation.
Hazard: Toxic by ingestion.
Use: Intermediate in organic synthesis, indicator, reagent for glucose.

p-nitrophenol.
CAS: 100-02-7. $NO_2C_6H_4OH$.

Properties: Yellowish, monoclinic, prismatic crystals. D 1.479–1.495 (20C), mp 111.4–114C (sublimes), bp 279C (decomposes). Soluble in hot water, alcohol, ether.
Derivation: (1) From *p*-chloronitrobenzene; (2) as in *o*-nitrophenol.
Hazard: Toxic by ingestion.
Use: Intermediate in organic synthesis, production of parathion, fungicide for leather, indicator.
See *o*-nitrophenol.

nitrophenol herbicide.
Properties: Organic chemical.
Use: Selective herbicide.

p-nitrophenol, sodium salt.   (sodium-*p*-nitrophenolate; *p*-nitro sodium phenolate). $NO_2C_6H_4ONa$. Yellow, crystalline solid; soluble in water. Intermediate.
Hazard: Toxic by ingestion.

p-nitrophenoxytributyltin.
CAS: 3644-32-4. $C_{18}H_{31}NO_3Sn$.
Hazard: A poison.

p-nitrophenylacetic acid.   (*p*-nitro-α-toluic acid).
CAS: 104-03-0. $NO_2C_6H_4CH_2COOH$.
Properties: Yellow needles. Mp 152.3C. Slightly soluble in cold water; soluble in alcohol and chloroform. Combustible.
Derivation: Hydrolysis of *p*-nitrobenzyl cyanide with 50% sulfuric acid.
Use: Intermediate for dyestuffs, pharmaceuticals, penicillin precursors, local anesthetics.

p-nitrophenylazosalicylate sodium.   See alizarin yellow R.

p-nitrophenylhydrazine.
CAS: 100-16-3. $C_6H_7N_3O_2$.
Properties: Reddish to orange crystals. Mp 155C (decomposes). Soluble in hot water, alcohol, and chloroform.
Use: Analytical reagent (aldehydes, ketones).

2-(4-nitrophenyl)imidazo(2,1-a)isoquinoline.
CAS: 61001-13-6. $C_{17}H_{11}N_3O_2$.
Hazard: A reproductive hazard.

4-nitrophenyl isocyanide.   See 1-isocyano-4-nitrobenzene.

m-nitrophenyl isocyanide.
CAS: 2008-59-5. $C_7H_4N_2O_2$.
Hazard: Low toxicity by ingestion.
Use: Agricultural chemical.

_o_-nitrophenylpropiolic acid.
CAS: 530-85-8. $C_9H_5NO_4$.
Properties: Yellow to brown crystals. Mp 157C (decomposes). Partially soluble in hot water and alcohol; insoluble in carbon disulfide.
Hazard: Decomposition may occur with explosive violence.
Use: Analytical reagent (alkaloids).

3-(4-nitrophenyl)-_n_-(4-propylcyclohexyl)-2-propenamide.
CAS: 315706-80-0. $C_{18}H_{24}N_2O_3$.
Hazard: A poison by ingestion.

_n_-(_p_-nitrophenyl)sulfanilamide. See sulfanitran.

_p_-nitrophenylthiol acetate.
CAS: 15119-62-7. $C_8H_7NO_3S$.
Hazard: Moderately toxic by ingestion.
Use: Agricultural chemical.

"Nitrophoska" *[Euro-Chem].* TM for a group of complete (N-P-K) fertilizers for agricultural and horticultural crops.

nitrophosphate. A nitrogen-phosphorus fertilizer produced by the action of nitric acid or a mixture of nitric and sulfuric or phosphoric acids on phosphate rock. Potassium salts usually are added to produce complete fertilizers. Typical analysis: Available nitrogen 15%, available P_2O_5 15%, available K_2O 15%.
Use: Fertilizer.
See superphosphate; triple superphosphate.

1-nitropropane.
CAS: 108-03-2. $CH_3CH_2CH_2NO_2$. A nitroparaffin.
Properties: Colorless liquid. D 1.003 (20/20C), miscible with organic solvents, bp 132C, vap press 7.5 mm Hg (20C), flash p 93F (34C) (TOC), bulk d 8.4 lb/gal (20C), refr index 1.4015 (20C), fp −108C, autoign temp 789F (420C). Soluble in water 1.4 mL/100 mL (20C); solubility of water in 1-nitropropane 0.5 cc/100 cc (20C).
Derivation: By reaction of propane with nitric acid under pressure.
Hazard: Flammable, moderate fire risk, moderate explosion hazard when shocked or heated. Liver damage, eye and upper respiratory tract irritant. Questionable carcinogen.
Use: Solvent, chemical synthesis, rocket propellant, gasoline additive.

2-nitropropane.
CAS: 79-46-9. $CH_3CHNO_2CH_3$. A nitroparaffin.
Properties: Colorless liquid. D 0.992 (20/20C), bp 120C, vap press 12.9 mm Hg (20C), flash p 75F (24C) (TOC), bulk d 8.3 lb/gal (20C), refr index 1.3941 (20C), fp −93C, autoign temp 802F (427C). Soluble in water 1.7 mL/100 mL (20C); solubility

of water in 2-nitropropane 0.6 cc/100 cc (20C); miscible with organic solvents.
Derivation: By reaction of propane with nitric acid under pressure.
Hazard: Flammable, dangerous fire risk, moderate explosion hazard when shocked or heated. Liver damage and liver cancer. Possible carcinogen.
Use: Solvent (especially for vinyl and epoxy coatings), chemical synthesis, rocket propellant, gasoline additive.
See "NiPar S-20" *[Angus].*

β-nitropropionic acid. (3-nitropropionic acid; BNP; bovinocidin; hiptagenic acid; NCI-C03076). $C_3H_5NO_4$.
Properties: A mycotoxin very soluble in water, EtOH, Et_2O, and in warm $CHCl_3$; insoluble in ligroin.
Derivation: Produced by *Aspergillus flavus.*
Hazard: Poison; neoplastigen; mutagen; questionable carcinogen.

5-nitro-2-_n_-propoxyaniline.
CAS: 553-79-7. $C_9H_{12}N_2O_3$.
Properties: Artificial sweetener.
Use: Food additive (prohibited).

4-nitro-_n_-(4-propylcyclohexyl)benzamide.

CAS: 315706-72-0. $C_{16}H_{22}N_2O_3$.
Hazard: A poison by ingestion.

Nitrorace. Liquid fuel for internal combustion engines.

_m_-nitrosalicylic acid. (3-nitro-2-hydroxybenzoic acid). $C_6H_3COOH(OH)NO_2$.
Properties: Yellowish crystals. Mp 144C. Soluble in benzene and alcohol.
Derivation: Nitration of salicylic acid.
Use: Intermediate, azo dyes.

nitrosamine. Any of a series of organic compounds in which $=N–N=O$ is attached to alkyl or aryl group, e.g., diphenylnitrosamine $(C_6H_5)_2NNO$ or dimethylnitrosamine $(CH_3)_2NNO$. Such compounds are formed by reaction between an amine and NO_x or nitrites. They occur in many food products, whiskey, herbicides, and cosmetics, as well as in industrial environments such as tanneries, rubber factories, and iron foundries. They are also formed within the body by reaction of amine-containing drugs with the nitrites resulting from bacterial conversion of nitrates. Nitrosamines have been found to be strong carcinogens in laboratory animals.

s-nitrosocaptopril.
CAS: 122130-63-6. $C_9H_{14}N_2O_4S$.
Hazard: Moderately toxic by ingestion.

n-nitroso compound. (noc). Any of a class of compounds that contain the nitroso radical.
Hazard: carcinogen.

n-nitrosodiethylamine. (dana; diethylnitro-samin; *n*-ethyl-*n*-nitrosoethanamine; diethylni-trosamine; den; dena; ndea; *N,N*-diethylnitrous amide; nitrosodiethylamine RCRA waste number U174).
CAS: 55-18-5. $C_2H_{10}N_2O$.
Properties: Slightly yellow oil; soluble in water, alcohol and ether, antioxidant, stabilizer.
Hazard: Possible carcinogen; mutagen; neoplasti-gen; tumorigen; poison; teratogen.
Use: Additive to gasoline and lubricants.

p-nitrosodiphenylamine. (diphenylnitrosa-mine; nitrous diphenylamide).
CAS: 156-10-5. $(C_6H_5)_2NNO$.
Properties: Greenish crystals. Mp 145C. Soluble in alcohol, benzene, ether, sulfuric acid.
Hazard: Questionable carcinogen.
Use: Rubber accelerator.

p-nitro sodium phenolate. See *p*-nitrophenol, sodium salt.

nitroso dyes. (quinone oxime dyes). Dyes whose molecules contain the –N=O or =N–OH chro-mophore group.

nitroso ester terpolymer. See nitroso polymer.

nitrosoguanidine.
CAS: 674-81-7. $ONNHC(NH)NH_2$.
Pale-yellow, crystalline powder.
Derivation: Cooling the reaction products of nitroguanidine, ammonium chloride, and zinc dust.
Hazard: Severe explosion risk.
Use: An initiating explosive.

nitrosohemoglobin. A form of hemoglobin in which the nitroso radical binds to the heme moi-ety of hemoglobin. It is toxicologically similar to carbon monoxide hemoglobin.

n-nitroso-n-methyl-4-aminopyridine.
CAS: 16219-99-1. $C_6H_7N_3O$.
Hazard: A poison by ingestion.

1-nitroso-2-naphthol. (naphthoquinone oxime; [one of several isomers] α-nitroso-β-naphthol).
CAS: 131-91-9. $C_{10}H_7NO_2$.
Properties: Yellow needles. Mp 110C. Soluble in alcohol and ether; insoluble in water. Combustible.
Derivation: Action of nitrous acid on β-naphthol.
Use: Organic synthesis, prevention of gum formation in gasoline, dye intermediate, analytical reagent.

p-nitrosophenol.
CAS: 104-91-6. C_6H_4OHNO.
Properties: Crystals of light-brown leaflets. Mp 140C (decomposes). Soluble in alcohol, ether, and acetone; moderately soluble in water.

Derivation: From phenol by action of nitrous acid in the cold.
Hazard: Dangerous fire and explosion risk, reacts violently with acids and alkalies, may explode spon-taneously, intraplant transport must be in tightly covered steel barrels.
Use: Dyes.

nitroso polymer. (nitroso rubber). A flame and heat-resistant copolymer derived from triflu-oronitrosomethane (CF_3NO) and tetrafluoroethy-lene ($F_2C{=}CF_2$). A third monomer (methyl-4-nitrosoperfluorobutyrate) provides more easily cross-linked functional groups than heretofore available. This terpolymer, called nitroso ester ter-polymer (NET), is a notable improvement on the earlier copolymer and is especially effective for fire-proof interiors of aerospace vehicles and airplanes. Cross-linking is by a peroxide mechanism. Proper-ties of the terpolymer are a glass transition tem-perature of approximately −50C, decomposition temperature of 275C, tensile strength 750 psi, elon-gation 500%, Shore hardness 78. Nonflammable.

nitrostarch. (starch nitrate).
CAS: 9056-38-6. $C_{12}H_{12}(NO_2)_8O_{10}$.
Properties: Orange-colored powder. Contains 16.5% nitrogen. Soluble in ether-alcohol.
Hazard: Flammable, dangerous fire risk, severe explosion risk when dry.
Use: Explosives.

trans-4-nitrostilbene.
CAS: 1694-20-8. $C_{14}H_{11}NO_2$.
Hazard: A poison.

β-nitrostyrene. $C_6H_5CH{:}CHNO_2$.
Properties: Liquid. Mp 58C. Available as a 30% solution in styrene.
Hazard: Moderate fire and explosion risk.
Use: Chain stopper in styrene-type polymerization.

p-nitrosulfathiazole.
CAS: 473-42-7. $NO_2C_6H_4SO_2NHC_3H_2NS$.
Properties: Pale-yellow powder; odorless; slightly bitter taste. Mp 258–266C. Slightly soluble in alco-hol; very slightly soluble in chloroform, ether and water; practically insoluble in benzene.
Use: Medicine (antibacterial).

nitrosyl chloride.
CAS: 2696-92-6. NOCl.
One of the oxidizing agents in aqua regia.
Properties: Yellow-red liquid or yellow gas. Bp −5.5C, fp −61.5C, d (gas) 2.3 (0C, air = 1), d (liquid) 1.273 (20C). Decomposed by water; dis-sociates into nitric oxide and chlorine on heating; soluble in fuming sulfuric acid. Nonflammable.
Derivation: By action of chlorine on sodium nitrate, also from nitrosylsulfuric acid and hydrochloric acid.
Grade: Pure, 93% min.

Hazard: Strong irritant, especially to lungs and mucous membranes.

Use: Synthetic detergents, catalyst, intermediate.

nitrosyl fluoride.
CAS: 7789-25-5. NOF.
Properties: Colorless gas. Fp −132C, bp −60C, d (liquid): 1.32, (solid) 1.72. Attacks glass severely and corrodes quartz.
Hazard: Toxic by inhalation, irritant to skin and mucous membranes.
Use: Fluorinating agent, rocket fuel oxidizer.

nitrosylsulfuric acid.
CAS: 7782-78-7. HNO_5S.
Properties: Straw-colored, oily liquid; furnished as a 40% solution in 87% sulfuric acid. Stable at room temperature.
Hazard: Strong irritant to skin and mucous membranes.
Use: Diazotizing agent for dyes, chemical intermediate, drugs and pharmaceuticals.

p-**nitrotetrazolium blue.**
CAS: 298-83-9. $C_{40}H_{30}N_{10}O_6 \cdot 2Cl$.
Hazard: Moderately toxic by ingestion.

m-**nitrotoluene.** (*m*-methylnitrobenzene).
CAS: 99-08-1. $NO_2C_6H_4CH_3$.
Properties: Yellow liquid. D 1.1571 (20/4C), fp 15C, bp 232.6C, refr index 1.5466 (20C), flash p 223F (106). Insoluble in water; soluble in alcohol and ether. Combustible.
Hazard: Toxic by inhalation, ingestion, skin absorption. Methemoglobinemia. Probable carcinogen.
Use: Organic synthesis.

o-**nitrotoluene.** (*o*-methylnitrobenzene).
CAS: 88-72-2. $NO_2C_6H_4CH_3$.
Properties: Yellow liquid. D 1.1629 (20C), fp −9.3C, bp 220.4C, refr index 1.544 (25C), flash p 223F (106C). Insoluble in water; miscible with alcohol and benzene. Combustible.
Derivation: From toluene by nitration and separation by fractional distillation.
Hazard: Toxic by inhalation, ingestion, skin absorption. Methemoglobinemia. Probable carcinogen.
Use: For production of toluidine, tolidine, fuchsin, and various synthetic dyes.

p-**nitrotoluene.** (*p*-methylnitrobenzene).
CAS: 99-99-0. $NO_2C_6H_4CH_3$.
Properties: Yellow crystals. D 1.299 (0/4C), mp 51.7C, bp 238.3C, refr index 1.5382 (15C), flash p 223F (106C). Insoluble in water; soluble in alcohol, ether, and benzene. Combustible.
Derivation: From toluene by nitration and separation by fractional distillation.
Hazard: Toxic by inhalation, ingestion, skin absorption. Methemoglobinemia. Questionable carcinogen.

Use: For production of toluidine, fuchsin, and various synthetic dyes.

p-**nitro-α-toluic acid.** See *p*-nitrophenylacetic acid.

m-**nitro-*p*-toluidine.** (3-nitro-4-toluidine).
CAS: 119-32-4. $NO_2C_6H_3(CH_3)NH_2$.
Properties: Orange-red crystals. Mp 117C, flash p 315F (157C). Soluble in alcohol and concentrated sulfuric acid. Combustible.
Derivation: From acetyl-*p*-toluidine by nitration.
Hazard: Toxic by ingestion.
Use: Intermediate for dyes and pigments.

p-**nitro-*o*-toluidine.** (4-nitro-2-toluidine).
CAS: 99-52-5. $NO_2C_6H_3(CH_3)NH_2$.
Properties: Yellow, crystalline solid. Mp 104C, flash p 315F (157C). Soluble in alcohol and ether. Combustible.
Derivation: From *o*-toluidine by nitration.
Use: Intermediate for dyes and pigments.

p-**nitro-α-tolunitrile.** See *p*-nitrobenzyl cyanide.

nitrotrichloromethane. See chloropicrin.

m-**nitrotrifluoromethylbenzene.** See *m*-nitrobenzotrifluoride.

2-nitro-4-trifluoromethylbenzonitrile.
CAS: 778-94-9. $NO_2(CF_3)C_6H_3CN$.
Properties: Liquid. Bp 156–158C (18–19 mm Hg), mp 47–48C.
Derivation: From a cyclic halogen compound by heating with copper cyanide in the presence of amines.
Use: Dyes.

2-nitro-4-trifluoromethyl chlorobenzene.
$NO_2C_6H_3(CF_3)Cl$.
Properties: Pale-yellow oil. Fp −6F. Soluble in acetone, alcohol, and other solvents; insoluble in water.
Use: Intermediate for dyes and organic chemicals.

nitrourea.
CAS: 556-89-8. $NH_2CONHNO_2$.
Properties: White, crystalline powder. Mp 158–159C. Slightly soluble in water; soluble in alcohol, acetone, and acetic acid.
Derivation: Reaction of urea nitrate and concentrated sulfuric acid.
Hazard: Severe explosion risk.
Use: Explosives.

nitrous acid.
CAS: 7782-77-6. HNO_2.
Properties: A weak acid occurring only in the form of a light-blue solution.

Derivation: Reaction of strong inorganic acids with nitrites, e.g., HCl + NaNO$_2$.

Use: Formation of diazotizing compounds by reaction with primary aromatic amines, source of nitric oxide.

nitrous bacteria.   Soil bacteria that converts ammonia to nitrates.

nitrous diphenylamide.   See *N*-nitrosodiphenylamine.

nitrous oxide.   (nitrogen monoxide; laughing gas).
CAS: 10024-97-2.   N$_2$O.

Properties: Colorless, sweet-tasting gas. D 1.52 (air). Liquid D 1.22 (−89C), fp −90.8C, bp −88.5C. Soluble in alcohol, ether, and concentrated sulfuric acid; slightly soluble in water; an asphyxiant, anesthetic gas. Noncombustible.

Derivation: Thermal decomposition of ammonium nitrate, controlled reduction of nitriles or nitrates.

Grade: Pure, 98.0% min, USP (97% min).

Hazard: Supports combustion, can form explosive mixture with air. Narcotic in high concentration. Central nervous system impairment, hematologic effects, and embryo/fetal damage. Questionable carcinogen.

Use: Anesthetic in dentistry and surgery, propellant gas in food aerosols, leak detection.

β-nitrovinylfuran.   See 2-furyl-1-nitroethene.

nitroxanthic acid.   See picric acid.

nitroxylene.   (dimethylnitrobenzene).
CAS: 25168-04-1.   C$_6$H$_3$(CH$_3$)$_2$NO$_2$.
There are three isomers (2,4-; 3,4-; and 2,5-).

Properties: Yellow liquid (2,4- and 2,5-), or crystalline needles (3,4-). D 1.135, fp 2C, bp 246C. Soluble in alcohol and ether; insoluble in water.

Derivation: By nitrating xylene and separating from the resulting mixture by rectification.

Hazard: Toxic by ingestion, inhalation, and skin absorption.

Use: Organic synthesis, gelatinizing accelerators for pyroxylin.

nitryl chloride.
CAS: 13444-90-1.   ClNO$_2$.

Properties: Colorless gas; odor of chlorine; liquid and solutions have yellow tinge. Bp −14C, fp −145C, d (liquid) 1.33.

Derivation: Reaction of chlorosulfonic acid with nitric acid.

Hazard: May explode on contact with organic materials. Corrosive to tissue.

Use: Organic synthesis (nitrating and chlorinating agent).

nitryl fluoride.
CAS: 10022-50-1.   FNO$_2$.

Properties: Colorless gas. Fp −165C, bp −72C, d (liquid) 1.80. Strong oxidizing agent; hydrolyzes to form nitric and hydrogen fluoride acids.

Derivation: Reaction of fluorine with nitrogen dioxide.

Hazard: Ignites on contact with selenium, iodine, phosphorus, arsenic, antimony, boron, silicon, molybdenum. Corrosive to tissue.

Use: Rocket propellants, fluorinating agent.

Nivco.   A precipitation-hardening cobalt-base alloy with good strength-to-weight ratio for use at temperatures up to 1200F. It contains somewhat greater than 20% nickel and small amounts of titanium, zirconium, and iron. Damping material for steam-turbine blade applications.

nivonorm.
CAS: 90-26-6.   C$_{10}$H$_{13}$NO.
Hazard: Moderately toxic by ingestion.

"Nix-Stix L0-014" *[HA].*   TM for a formulated rust inhibitor film for stored materials.

"Nix-Stix L-806" *[HA].*   TM for a fluorocarbon resin emulsifier in water.
Use: Making a surface paintable and non-flammable.

"Nix-Stix M-301" *[HA].*   TM for a suspension aerosol solvent.
Use: For fast dry, multiple releases.

NK 104 (acid).
CAS: 147526-32-7.   C$_{50}$H$_{46}$F$_2$N$_2$O$_8$•Ca.
Hazard: A poison by ingestion.

nm.   Abbreviation for nanometer.

NMR.   Abbreviation for nuclear magnetic resonance.

NNCO.   See: National Nanotechnology Coordination Office.

NNI.   See: National Nanotechnology Initiative.

No.   Symbol for the element nobelium.

Nobel, Alfred B.   (1833–1896). A native of Sweden, Nobel devoted most of his life to a study of explosives and was the inventor of a mixture of nitroglycerin and diatomaceous earth, which he called dynamite. He also invented blasting gelatin and smokeless powder. With the fortune he accumulated from his work, Nobel established the foundation that bears his name, which annually recognizes outstanding work in physics, chemistry, medicine, literature, and human relations. The Nobel Prize is still the world's most valued scientific award.

Nobel Foundation.   A private institution established in 1900 based on the will of Alfred Nobel.

The foundation manages the assets made available through the will for the awarding of the Nobel Prize in Physics, Chemistry, Physiology or Medicine, Literature and Peace. It represents the Nobel institutions externally and administers informational activities and arrangements surrounding the presentation of the Nobel Prize. The Foundation also administers Nobel symposia in the different prize areas. Its address is P. O. Box 5232, SE-102 45 Stockholm, Sweden. Website: http://www.nobelprize.org

nobelium. No. Synthetic radioactive element number 102, aw 254, one of the actinide series of elements; its discovery has been claimed by research groups in Sweden, the former U.S.S.R., and California. It can be produced in a cyclotron by bombarding copper with nuclei of carbon-13 accelerated to high energies. The name nobelium has been accepted by the IUPAC Commission on Nomenclature of Inorganic Chemistry. It has nine isotopes (251–259) that are so short-lived that their chemical properties have not been determined. It has no known uses or compounds.

noble. In chemical terminology, this term describes an element that either is completely unreactive or reacts only to a limited extent with other elements. Six noble gases constitute group VIIIA in the periodic table that is variously called the zero group (as the first three of its members have a valence of 0), the inert gas group, and the noble gas group. The last is preferable, as three of the gases, though unreactive, are not inert. The gases of this group are helium, neon, argon, krypton, xenon, and radon. The noble metals are generally considered to be gold, silver, platinum, palladium, iridium, rhenium, mercury, ruthenium, and osmium. The term has no reference to their commercial value. See inert.

noble gas. Inert gas such as helium, argon, or neon.
See noble.

noble metals. Refers to the unreactive metals (examples are gold and platinum) that are not readily dissolved by acids and not oxidized by heating in air.

NODA. Abbreviation for *n*-octyl-*n*-decyl adipate.

nodular iron. A ductile form of cast iron.

nodules. Aggregates of ferromanganese occurring in huge quantities on the ocean floors. It has been estimated that they cover approximately 10% of the bottom of the Pacific. They are also present in other oceans and even in large lakes. The size averages approximately 4 cm, varying from small pellets to masses several meters in diameter. Their composition is approximately 55% manganese and 35%

iron, the balance being cobalt, copper, and nickel. Their origin has not been definitely established. Proposals for commercial exploitation have been made, but no economically satisfactory method has been developed. Considerations of international law and *ownership* of the sea floor are also involved.

NOLA. See omega-nitrolarginine.

nomenclature, chemical. See chemical nomenclature.

"Nomex" [Du Pont]. TM for an aramid fiber or fabric.
Use: Bag filters for industrial dust control and insulation of spacecraft.
See aramid.

nomograph. A chart in which a straight line, either drawn or indicated by a ruler, intersects three scales at points that represent values that satisfy an equation or a given set of physical conditions. It thus enables one to determine the value of a dependent variable when that of the independent variable is known.
Use: By chemical engineers to determine relationships between the properties of materials.

non-. (nona). Prefix meaning nine.

nona. See non-.

nonachloro-4-phenoxyphenol. See 4-hydroxy-nonachlorodiphenyl ether.

nonadecane.
CAS: 629-92-5. $C_{19}H_{40}$ or $CH_3(CH_2)_{17}CH_3$.
Properties: Leaflets. Soluble in alcohol and ether, insoluble in water, d 0.777, mp 32C, bp 330C. Combustible.
Use: Organic synthesis.

***n*-nonadecanoic acid.**
CAS: 646-30-0. $CH_3(CH_2)_{17}COOH$.
A saturated fatty acid normally not found in natural vegetable fats or waxes. Combustible.
Properties: Colorless crystals. Mp 68.7C, bp 297C (100 mm Hg). Soluble in alcohol and ether; insoluble in water. Synthetic product available 99% pure for organic synthesis.

***trans,trans*-2,4-nonadienal.**
CAS: 5910-87-2. $C_9H_{14}O$.
Properties: Slightly yellow liquid; strong, fatty, floral odor. D 0.850–0.870, bp: 97–98C @ 10 mm Hg, refr index 1.522. Soluble in alcohol, fixed oils; insoluble in water.
Use: Food additive.

***trans,cis*-2,6-nonadienol.**
CAS: 28069-72-9. $C_9H_{16}O$.

Properties: White to yellow liquid; powerful, vegetable odor. D 0.860–0.880, bp: 98–100C @ 11 mm Hg, refr index 1.464. Insoluble in water.
Hazard: Low toxicity by ingestion and skin contact. A skin irritant.
Use: Food additive/natural product.

γ-nonalactone.   See γ-nonyl lactone.

δ-nonalactone. $C_9H_{16}O_2$.
Properties: Colorless to pale-yellow liquid; coconut odor. D 0.980, refr index 1.452.
Use: Food additive.

nonallyl chloride.   See pelargonyl chloride.

nonanal.   (pelargonic aldehyde; *n*-nonyl aldehyde; aldehyde C-9).
CAS: 124-19-6. $C_8H_{17}CHO$.
Properties: Colorless liquid with an orange-rose odor. D 0.822–0.830, refr index 1.424–1.429. Soluble in three volumes of 70% alcohol, in mineral oil; insoluble in glycerol. Combustible.
Grade: Technical, FCC.
Use: Perfumery, flavoring agent.

nonane.   (nonyl hydride).
CAS: 111-84-2. C_9H_{20} or $CH_3(CH_2)_7CH_3$.
Properties: Colorless liquid. D 0.722, bp 150.7C, fp −54C, refr index 1.40561 (20C), flash p 86F (30C) (CC), autoign temp 403F (206C).
Grade: Technical (95%), 99%, research. Soluble in alcohol; insoluble in water.
Hazard: Flammable, moderate fire risk. Irritant, narcotic in high concentration. Central nervous system impairment.
Use: Organic synthesis, biodegradable detergents, distillation chaser.

nonane-1,9-dimethanesulfonate.   See methanesulfonic acid, nonamethylene ester.

nonanedioic acid.   See azelaic acid.

nonane-1,3-diol monoacetate. $C_9H_{18}OOC-CH_3$.
Properties: A mixture of isomers, colorless to slightly yellow liquid. Stable; soluble in alcohol. Combustible.
Use: Perfumery, flavoring.

n-nonanoic acid.   See pelargonic acid.

nonanol.   See nonyl alcohol.

(nonanoyloxy)tributylstannane.   See tributyltin nonanoate.

non-coding DNA.   See junk DNA.

non-coding strand.   An anti-sense strand.
See: Sense strand.

noncombustible material.   A solid, liquid, or gas that will not ignite or burn, regardless of how high a temperature it is exposed to, e.g., silicon dioxide, water, carbon dioxide.
See nonflammable material.

noncompetitive inhibition.   A type of enzyme inhibition not reversed by increasing the substrate concentration. The inhibitor binds to a site on the enzyme other than the active site.

non-cyano (type I) pyrethroid.   A pyrethroid that lacks the alpha-cyano group.
Hazard: Cause hyperexcitability, repetitive nerve firing, incoordination, whole-body tremors.

noncyclic electron flow.   The light-induced flow of electrons from water to NADP in oxygen-evolving photosynthesis involving both photosystems I and II.
See: Cyclic photophosphorylation.

nondestructive testing.   A test that does not involve destruction of the sample or material tested. Such tests are usually carried out by radiographic methods on large, finished metal products (castings and machine components) to determine the presence of internal defects likely to cause operational failure. An infrared camera-scanning device for detection of internal weaknesses in tires is a recent development in nondestructive testing. X-radiation is used to determine the authenticity of paintings and other objects of art.
See testing.

nondrying oil.   See drying oil.

nonelectrolyte.   A compound that resists passage of electricity (nonconductor) both in liquid form and in solution. Included in this classification are most organic compounds, with the exception of acids and amides, and such inorganic compounds as nonmetal halides. Nonelectrolytes are covalently bonded; some inorganic compounds having covalent bonds form electrolytic solutions in aqueous solution.
See electrolyte.

nonene.   See nonylene.

***cis*-6-nonen-1-ol.** $C_9H_{18}O$.
Properties: White to slightly yellow liquid; powerful, melon-like odor. D 0.850–0.870, refr index 1.448. Insoluble in water.
Use: Food additive.

***trans*-2-nonen-1-ol.** $C_9H_{18}O$.
Properties: White liquid; fatty, violet odor. D 0.830–0.850, refr index 1.444–1.448. Insoluble in water.
Use: Food additive.

nonessential amino acids. Amino acids that can be made by humans and other vertebrates from simpler precursors, and thus, are not required in the diet.

nonheme iron proteins. Proteins, most often acting in redox reactions, which contain iron but no porphyrin groups.

nonionic detergent. Any detergent in which the surface active moiety is nonionic at neutral pH. This type of detergent acts by means of hydrogen-bonding.

nonmetal. (1) Any of a number of elements whose electronic structure, bonding characteristics, and consequent physical and chemical properties differ markedly from those of metals, particularly in respect to electronegativity and thermal and electrical conductivity. In general, nonmetals have very low to moderate conductivity and high electronegativity. The 25 elements classified as nonmetals may be considered in two groups: (a) those having moderate electrical conductivity (semiconductors), all of which are solids, and (b) those having very low conductivity, many of which are gases. The semiconductors of group (a) were formerly called metalloids since they more nearly resemble metals than those of the group (b), but this term is no longer used by chemists. The nonmetals are given below based on this subgrouping, though any such list is open to challenge:

(a)	(b)
antimony	halogens
arsenic	hydrogen
boron	nitrogen
carbon	oxygen
germanium	noble gases
phosphorus	
polonium	
selenium	
silicon	
sulfur	
tellurium	

(2) Loosely, any material that is not a metal, e.g., petroleum, plastics, waxes. The term is widely used in this sense by engineers and specification writers.

non-Newtonian behavior. A property possessed by some fluids and many plastic solids, including lubricating grease, of having a variable relationship between shear stress and rate of shear.

nonoic acid. An acid of the formula $C_8H_{17}COOH$ of which there are many possible isomers. Pelargonic acid is the normal or straight-chain acid. A mixture of various branched-chain nonoic acids is recovered from products of the Oxo process.

***n*-nonoic acid.** See pelargonic acid.

nonoxynol. (nonoxinol; α-(4-nonylphenyl)-α-hydroxypoly-(oxy-1,2-ethanediyl); polyethyleneglycols mono(nonylphenyl) ether; macrogol nonylphenyl ether; nonoxinol; polyoxyethylene-(*n*)nonylphenyl ether; nonylphenyl polyethyleneglycol ether; nonylphenoxypolyethoxyethanol; 2-(4-nonylphenoxy)ethanol; nonylphenoxypolyethoxyethanol). $C_{17}H_{28}O_2$. An alkyl phenoxy polyethoxy ethanol.
Hazard: Moderately toxic.
Use: Spermicidal agents, pharmaceutic aids, nonionic detergents, emulsifiers, wetting agents, defoaming agents dispersants, and as intermediates in the synthesis of anionic surfactants.

nonpolar. A substance whose molecules possess no permanent electric moments. One whose atoms are linked by sharing of an electron pair and that are not electrolytes in the liquid state. They do not ionize, or ionize weakly, in solution.

nonpolar molecule. A molecule with polar bonds that does not have an overall dipole. The vector sum of its dipoles is zero.

nonselective herbicide. A herbicide that acts on a wide range of plant species.

nonsense codon. A codon that does not specify an amino acid, but instead signals the termination of a polypeptide chain (UAA, UAG, UGA). See stop codons.

nonsense mutation. A mutation that results in the premature termination of a polypeptide chain.

nonsense suppressor. A mutation, usually in the gene for a tRNA, that causes an amino acid to be inserted into a polypeptide in response to a termination codon.

nonsoap cleaner. Any synthetic detergent.
Hazard: Slightly to very toxic.

nonviscous neutral oil. A neutral oil of viscosity less than 135 SUS at 37.7C.

nonwoven fabric. A fabric made from staple lengths of cotton, rayon, glass, or thermoplastic synthetic fiber mechanically positioned in a random manner and usually bonded with a synthetic adhesive or rubber latex. The sheets thus formed can be pressed together to form porous mats of high absorptivity and good elastic recovery on deformation. Permanent bonds are formed where the fibers touch each other as a result of heat treatment when the fibers are thermoplastic or by use of a high-polymer binder. Disposable filters of polyethylene and polyester can be made without a binder.

A specialty nonwoven (so-called "melded") fabric trademarked "Cambrelle" is a composite of two different polymers; heating the exterior layer to its melting point causes the fibers to fuse into a fabric.

Use: Applications are as a backing for plastic film; padding for surgical dressings, diapers, sanitary napkins; drapes and other decorative textile products; filtration; shoe liners; industrial wiping and polishing fabrics; disposable clothing; carpet backing.

n-nonyl acetate.   (acetate C-9).
CAS: 143-13-5. $CH_3COO(CH_2)_8CH_3$.
Properties: Colorless liquid; strong pungent odor. D 0.864–0.868, refr index 1.422–1.426. Soluble in four volumes of 70% alcohol, flash p 155F (68.3C), several isomers exist. Combustible.
Use: Perfumery, flavoring.

n-nonyl alcohol.   (nonanol; alcohol C-9; octyl carbinol; pelargonic alcohol).
CAS: 143-08-8. $CH_3(CH_2)_7CH_2OH$.
Properties: Colorless liquid; floral odor. D 0.826–0.829, refr index 1.431–1.435, fp –5C, bp 215C, flash p 165F (73.9C). Soluble in seven volumes of 50% alcohol; insoluble in water. Several isomers exist. Combustible.
Use: Perfumery, flavoring (lemon oil).

n-nonylamine.
CAS: 112-20-9. $C_9H_{19}NH$.
Properties: Colorless liquid. Bp 75–85C (20 mm Hg), d 0.798 (25C), refr index 1.4366 (20C). May be prepared from nonyl halides by conventional techniques. Combustible.
Hazard: Moderate fire risk. Skin irritant.

tert-nonylamine.   Principally $tert$-$C_9H_{19}NH_2$
and $tert$-$C_{10}H_{21}NH_2$.
Properties: Colorless liquid. Boiling range 160–174C, d 0.789 (25C), refr index 1.428 (25C), flash p 120F (48.9C). Insoluble in water; soluble in common organic solvents; especially in petroleum hydrocarbons. Combustible.
Hazard: Moderate fire risk. Skin irritant.
Use: Intermediate for rubber accelerators, insecticides, fungicides, dyestuffs, pharmaceuticals.

nonylbenzene.   (1-phenylnonane).
CAS: 1081-77-2. $C_9H_{19}C_6H_5$.
Properties: Light-straw liquid; aromatic odor. Boiling range 245–252C, d 0.864 (20/20C), refr index 1.488 (20C), viscosity 41.9 cP (20C), flash p (CC) 210F (98.9C). Combustible.
Hazard: Irritant; narcotic in high concentration.
Use: Manufacture of surface-active agents.

nonyl bromide.
CAS: 693-58-3. $C_9H_{19}Br$.
Properties: Liquid. Bp 81–85C (10 mm Hg), d 1.101 (25C), refr index 1.4583 (20C). Nonflammable.
Derivation: High yields of nonyl bromide are obtained by passing hydrogen bromide into the alcohol while heating or by refluxing with aqueous hydrogen bromide in the presence of an acid catalyst.

nonyl chloride.
CAS: 2473-01-0. $C_9H_{19}Cl$.
Properties: Liquid. Bp 58–63C (8 mm Hg), d 0.878 (25C), refr index 1.4379 (20C). Nonflammable.
Derivation: Nonyl alcohol reacts with hydrochloric acid at elevated temperatures and pressures to give nonyl chloride. It can also be made by refluxing a mixture of concentrated hydrochloric acid with alcohol in the presence of zinc chloride.

1-nonylene.   (1-nonene). C_9H_{18} or $CH_3(CH_2)_6$
$CH:CH_2$.
Properties: Colorless liquid. D 0.7433, bp 149.9C. Soluble in alcohol; insoluble in water. Combustible.
Derivation: From propylene.

nonyl hydride.   See nonane.

n-nonylic acid.   See pelargonic acid.

γ-nonyl lactone.   (aldehyde C-18; 4-hydroxynonanoic acid, γ-lactone; γ-nonalactone). CH_3 $(CH_2)_4CHCH_2CH_2C(O)O$.
Properties: Yellowish to almost colorless liquid; coconut-like odor. D 0.956–0.963, refr index 1.447. Soluble in five volumes of 50% alcohol; soluble in most fixed oils and mineral oil; practically insoluble in glycerol. Combustible.
Grade: Technical, FCC.
Use: Perfumery, flavors.

nonyl nonanoate.   (nonyl pelargonate).
CAS: 83852-51-1. $C_9H_{19}OOCC_8H_{17}$.
Properties: Liquid; floral odor. D 0.863 (25C), bp 315C, refr index 1.4419 (20C). Combustible.
Use: Flavors, perfumes, organic synthesis.

nonylphenol.
CAS: 25154-52-3. $C_9H_{19}C_6H_4OH$.
A mixture of isomeric monoalkyl phenols, predominantly p-substituted.
Properties: Pale-yellow, viscous liquid; slight phenolic odor. D 0.950 (20/20C), bp 293C, fp –10C (sets to glass below this temperature), viscosity 563 cP (20C), flash p 285F (140.5C). Insoluble in water; soluble in most organic solvents. Combustible.
Grade: Technical.
Use: Nonionic surfactant (nonbiodegradable), lubricating-oil additives, stabilizers, petroleum demulsifiers, fungicides, antioxidants for plastics and rubber.

nonylphenol polyethylene glycol ether.
CAS: 127087-87-0. $C_{23}H_{40}O_5$.
Hazard: A reproductive hazard.

4-nonylphenol polymer with formaldehyde and oxirane. See formaldehyde, polymer with 4-nonylphenol and oxirane.

nonylphenoxyacetic acid.
CAS: 28065-23-8. $C_9H_{19}C_6H_4OCH_2COOH$.
Properties: Light-amber liquid. Miscible with organic solvents. D 1.02 (20C), flow p 5C, viscosity 6500 cP (25C). Insoluble in water; soluble in alkali. Combustible.
Use: Corrosion inhibitor for turbine oils, lubricants, fuels, greases; antifoaming agent in gasoline, hydraulic fluids, cutting oils.

1-(4-nonylphenyl)ethanone oxime.
CAS: 64128-32-1. $C_{17}H_{27}NO$.
Hazard: A severe skin irritant.

α-(4-nonylphenyl)-o-hydroxypoly(oxy-1,2-ethanediyl)-branched.
CAS: 127087-87-0.
Hazard: A reproductive hazard.

nonyl thiocyanate. $C_9H_{19}SCN$.
Properties: Bp 84–86.5C (1 mm Hg), d 0.919 (25C), refr index 1.4696 (20C). Nonyl thiocyanate can be made from nonyl chloride by refluxing with alcoholic sodium thiocyanate solution using conventional techniques.

nonyltrichlorosilane. (trichlorononylsilane).
CAS: 5283-67-0. $C_9H_{19}SiCl_3$.
Properties: Water-white liquid; pungent irritating odor. Fumes readily in moist air.
Hazard: Strong irritant to skin and mucous membranes.

nopinene. See β-pinene.

NOPON. See p-bis[2-(5-α-naphthyloxazolyl)-benzene].

(+-)-noradrenaline bitartrate.
CAS: 3414-63-9. $C_8H_{11}NO_3 \cdot C_4H_6O_6$.
Hazard: A poison.

norbormide. (generic name for 5-(α-hydroxy-α-2-pyridylbenzyl)-1-(α-2-pyridylbenzylidene)-5-norbornene-2,3-dicarboximide).
CAS: 991-42-4.
Properties: White solid. Mp 180–190C. Insoluble in water; soluble in dilute acids.
Hazard: Toxic by ingestion.
Use: Rodenticide.

5-norbornene-2-methanol. (2-hydroxymethyl-5-norbornene; Cyclol). $C_7H_9CH_2OH$.

Properties: Stable, colorless liquid. D 1.022–1.024 (25/25C), refr index 1.4985–1.4995 (25C), a high-boiling solvent, bp 192–198C, flash p (TOC) 183F (83.9C). Miscible with most common organic solvents; slightly soluble in hot water. Combustible.
Use: Monomer for the modification of condensation and addition polymers for coatings.

5-norbornene-2-methyl acrylate.
CAS: 95-39-6. $(C_3H_4O_2)_4$.
Properties: Colorless liquid. Bp 103–105C, d 1.027–1.031 (25/25C). Soluble in most organic solvents; inhibitor 0.02% MEHQ.
Use: A difunctional monomer particularly suited for in situ cross-linking of vinyl acetate emulsions and polymer systems.

Nordel. An elastomer based on an ethylene-propylene-hexadiene terpolymer, sulfur curable.
Use: Automotive and appliance components; wire insulation for electrical accessories, belts, hose, mechanical products.

nordhausen acid.
CAS: 7664-93-9. Fuming sulfuric acid of d 1.86–1.90.
Hazard: Corrosive to skin and tissue.
See sulfuric acid, fuming.

nordihydroguaiaretic acid. (NDGA; 4,4'-(2,3-dimethyltetramethylene)dipyrocatechol).
CAS: 500-38-9. $[C_6H_3(OH)_2CH_2CH(CH_3)]_2$.
Properties: Crystals from acetic acid. Mp 184–185C. Soluble in methanol, ethanol, ether; slightly soluble in hot water, chloroform; nearly insoluble in benzene, petroleum ether.
Grade: Technical, FCC.
Derivation: Extraction from guaiac, also synthetically.
Hazard: Use in foods prohibited by FDA.
Use: Antioxidant to retard rancidity of fats and oils.

norea. (3-(hexahydro-4,7-methanoindan-5-yl)-1,1-dimethylurea).
CAS: 18530-56-8. $C_{10}H_{14}NHCON(CH_3)_2$.
Properties: White solid. Mp 168–169C. Very slightly soluble in water; soluble in polar solvents such as cyclohexanone, acetone, alcohol; less soluble in nonpolar solvents such as benzene, xylene, hexane, and kerosene.
Use: Selective herbicide.

dl-norephedrine hydrochloride. See propadrine hydrochloride.

norepinephrine. (adrenor; aktamin; 1,2-amino-1-(3-,4-dihydroxyphenyl)ethanol; (R)-4-(2-amino-1-hydroxyethyl)-1,2-benzenediol; 1-α-(aminomethyl)-3,4-dihydroxybenzyl alcohol; 2-amino-1-(3,4-dihydroxyphenyl)ethanol; 1-(3,4-dihydroxyphenyl)-2-aminoethanol; noradrenaline; 4-[(1R)-2-amino-1-hydroxyethyl]benzene-1,2-diol;

arterenol; 1-arterenol; 1,1-(3,4-dihydroxyphenyl)-2-aminoethanol; 1,3,4-dihydroxyphenylethanolamine; levarterenol; levoarterenol; levonoradrenaline; levonoreinphrine; levophed; (−)-noradrec; noradrenalin; noradrenalina; noradrenaline; (−)-noradrenaline; d-(−)-noradrenaline; 1-noradrenaline; noradriniline; norartrinal; 1-norepinephrine; (−)-norepinephrine; norepirenamine; sympathin E).
CAS: 51-41-2. $C_8H_{11}NO_3$.
An adrenergic hormone produced by the chromaffin cells of the adrenal medulla of vertebrates. It is a vasoconstrictor and mediates transmission of synthetic nerve impulses. A precursor of epinephrine that is secreted by the adrenal medulla. It is the principal transmitter of most postganglionic sympathetic fibers and of the diffuse projection system in the brain arising from the locus ceruleus.
Hazard: May cause local tissue necrosis, headache, bradycardia, hypertension; poison; teratogen; mutagen.
Use: Therapeutically to maintain blood pressure in acute hypotension resulting from trauma, heart disease, or cardiovascular collapse.

18-norestrone methyl ether.
CAS: 4147-10-8. $C_{18}H_{22}O_2$.
Hazard: A reproductive hazard.
Use: Hormone.

norethisterone oxime.
CAS: 23965-86-8. $C_{20}H_{27}NO_2$.
Hazard: A reproductive hazard.
Use: Hormone.

norflurazon.
Use: Food additive; herbicide.

norgestrel.
CAS: 797-63-7. A synthetic steroid hormone used as an ingredient of oral contraceptives. Approved by FDA.
See "Ovral" [Wyeth].

"Norkool" [Union]. TM for inhibited glycol-based coolants, corrosion inhibitors, and cleaners.
Use: Industrial applications, such as compressor stations, generators, and high-speed, stationary engines.

norleucine. (α-aminocaproic acid).
CAS: 327-57-1. $CH_3(CH_2)_3CH(NH_2)COOH$.
A nonessential amino acid found naturally in the L(+) form.
Properties: Leaflets, crystals from water, dl-norleucine. Soluble in water and acids; slightly soluble in alcohol. Decomposes at 327C.
L(+)-norleucine: slightly sweet taste. Sublimes at 275–280C, mp 301C (decomposes).
D(−)-norleucine: bitter taste. Partially sublimes at 275–280C, mp 301C (decomposes).

Derivation: Found in traces in proteins, also synthetically.
Use: Biochemical research.

normal. (1) Of hydrocarbon molecules: containing a single, unbranched chain of carbon atoms, usually indicated by the prefix n-, as n-butane, n-propane, etc.
(2) Of solutions: containing one equivalent weight of a dissolved substance per liter of solution; a standard measure of concentration, indicated by N, e.g., 2N, 0.5N.
See molar.
(3) Perpendicular or at right angles to, for example, incident light rays normal to a surface.

normal antibody. (natural antibody; neutralizing antibody). Any of a type of antibody in the serum or plasma of various persons or animals that is not known to have been stimulated by a specific antigen. Such antibodies inhibit the act of, or destroy, the targeted infectious agents.

normalize. To temper steels and other alloys by heating to a predetermined temperature and cooling at a controlled rate to relieve internal stresses and improve strength (analogous to annealing of glass).

Normant reagents. Vinylmagnesium halides that were previously unavailable are now prepared by reacting the parent vinyl halides with magnesium in tetrahydrofuran. These compounds are called Normant reagents and behave like typical Grignard reagents.

nornicotine. (3-(2-pyrrolidinyl)pyridine).
CAS: 479-97-3. $C_9H_{12}N_2$.
Properties: Moderately thick, hygroscopic liquid. Bp 270C, d 1.07, optical rotation −89 degrees, refr index 1.53. Soluble in water, alcohol, and kerosene.
Hazard: Toxicity about one-third that of nicotine.
Use: Insecticide.

"Norpace" Capsules [Searle]. TM for disopyramide phosphate.
Use: Drug.

norpentazocine.
CAS: 25144-78-9. $C_{14}H_{19}NO \cdot ClH$.
Hazard: A poison.

norphytane. See pristane.

Norrish, Ronald G. W. (1897–1978). An English physical chemist who was recipient of the Nobel Prize in 1967 with Eigen and Porter. His analysis of reactions of one ten-billionth of a second were made possible by disturbing the chemical equilibrium with short energy pulses. After receiving a doctorate, he went to the Sorbonne before returning to Cambridge to teach. His career was long and distinguished by many awards.

Norrish-type cleavage.   Homolytic cleavage of aldehydes and ketones originating from their excited $n\pi^*$ state. Synthetically useful for the ring cleavage of cyclic ketones.

"Norsolene IR-140" *[Cray].* TM for ink additive.
Properties: High softening point for fast solvent release. Soluble and dilutable in aliphatic solvents. High viscosity with infinite dilutability.
Use: Offset inks.

Northern blot.   A gel-based laboratory procedure that locates on a gel mRNA sequences that are complementary to a piece of DNA used as a probe. Named after Dr. E. M. Southern, who invented the Southern blot, it was adapted to RNA and named the "Northern" blot.
See DNA.

Northrup, John H.   (1891–1987). An American chemist who won a Nobel Prize in chemistry in 1946 along with Sumner and Stanley. His work was primarily concerned with isolation and crystallization of enzymes. Many firsts included the production of the enzyme trypsin in the laboratory and isolation of the first bacterial virus. He was also responsible for producing diptheria antitoxin in crystalline form. His education was at eastern schools including Harvard, Yale, and Princeton.

noruron.
CAS: 18530-56-8. $C_{13}H_{22}N_2O$.
Properties: Colorless crystals. Mp 175C. Insoluble in water; soluble in acetone; slightly soluble in benzene.
Hazard: Highly toxic.
Use: Herbicide.

Norway saltpeter.   See ammonium nitrate.

Norwegian saltpeter.   See calcium nitrate.

NOS.   Abbreviation for "not otherwise specified."
Use: In shipping regulations, for classes of substances to which a restriction applies, individual members of which are not listed in the regulations.

noscapine.   (l-α-narcotine; narcosine).
CAS: 128-62-1. $C_{22}H_{23}NO_7$.
An isoquinoline alkaloid of opium.
Properties: Fine, white, crystalline powder. Mp 176C, sublimes at 150–160C. Insoluble in water; practically insoluble in vegetable oils; slightly soluble in hot solutions of potassium hydroxide and sodium hydroxide; soluble in most organic solvents. Salts formed with acids are dextrorotatory and unstable in water.
Derivation: An opium alkaloid from the seed capsules of *Papaver somniferum.*
Grade: USP.

Hazard: Narcotic, use legally restricted.
Use: Medicine (cough control).

nose irritant gas.   (diphenylchloroarsine).
Properties: Smoke.
Hazard: Irritant; can cause intense pain in the nose, mouth and upper respiratory tree; sneezing, headache, aching teeth and jaws, vomiting, and acute mental depression.

Nouryset 200.
CAS: 142-22-3.   A colorless monomer of diethylene glycol bis(allyl carbonate).
Use: Transparent organic glass in the manufacture of optical corrective lenses.

novobiocin.
CAS: 303-81-1. $C_{31}H_{36}H_2O_{11}$.
Properties: A light-yellow to white antibiotic produced by *Streptomyces niveus.* Available as calcium and sodium salts.
Hazard: May have damaging side effects.

"Novogel" *[Organovo].*
CAS: 1309-42-8.   TM for magnesium hydroxide.
Grade: Technical, food, USP, directly compressible powder.
Use: Flame-retardant, smoke suppressant, acid scavenger. USP grade used for tablet and liquid antacids.

"Novol" *[Croda].*   TM for fatty alcohols.
Use: Thickeners, opacifiers, structural agents for antiperspirant sticks, creams, and lotions.

novolak.   (novolac). A thermoplastic phenol–formaldehyde–type resin obtained primarily by the use of acid catalysts and excess phenol. Generally alcohol soluble and requires reaction with hexamethylenetetramine, *p*-formaldehyde, etc. for conversion to cured, cross-linked structures by heating at 200–400F.
Use: Molding materials, bonding materials, bonding agent in brake linings, abrasive grinding wheels, electrical insulation, clutch facings, air-drying varnishes, reinforcing agent and modifier for nitrile rubber.

novoldiamine.   (1-diethylamino-4-aminopentane).
CAS: 140-80-7. $C_9H_{22}N_2$.
Properties: Colorless liquid. D 0.82, bp 200C, refr index 1.44. Soluble in water and alcohol.
Derivation: From 2-diethylaminoethanol and ethyl acetoacetate.
Use: Manufacturing antimalarials (quinacrine and related compounds).

noxious gas.   Any natural or by-product gas or vapor that has specific toxic effects on humans or animals (military poison gases are not included in this group). Examples of noxious gases are ammonia, carbon monoxide, nitrogen oxides, hydrogen sulfide, sulfur dioxide, ozone, fluorine, and vapors evolved by benzene, carbon tetrachloride, and a number of chlorinated hydrocarbons. Gases that act as simple asphyxiants are not classified as noxious. See air pollution.

Noyori, Ryoji. (1938–). Born in Kobe (now Ashiya), Japan, Noyori won the Nobel Prize for chemistry in 2001 for his pioneering work concerning chirally catalysed hydrogenation reactions. Awarded a doctorate from Kyoto University in 1967. Noyori is a professor at Nagoya and Harvard Universities. Among many other awards, he received the Wolf Prize in Chemistry (Wolf Foundation, Israel) 2001, and the Roger Adams Award in Organic Chemistry (ACS), 2001.

Np. Symbol for neptunium.

NPG. The glycol of choice for premium-quality gel coats. It reacts rapidly with isophthalic acid to produce low-color gel-coat resins.

NPK mixtures. Fertilizers containing nitrogen, phosphorus, and potassium. These are usually characterized by numbers such as 5-10-10, meaning 5% nitrogen, 10% phosphorus as P_2O_5, and 10% potassium as K_2O. The percentages refer to the amount of nitrogen, phosphorus, or potassium present in available form rather than to the amount of the compounds.

NPN. Abbreviation for *n*-propyl nitrate.

NPO. See α-naphthylphenyloxazole.

NQR spectroscopy. See nuclear quadrupole resonance.

NR. Abbreviation for natural rubber.

NRC. Abbreviation of Nuclear Regulatory Commission.

NRDC. Abbreviation for National Resources Defense Council.

"N-Serve" *[Dow Agrosciences].* TM for nitrapyrin.

NSF. See National Science Foundation.

NSR. Abbreviation for nitrile silicone rubber.

NSTC. See National Science and Technology Council.

NT. Abbreviation for nucleotide, i.e., the monomeric unit from which DNA or RNA are built. One can express the size of a nucleic acid strand in terms of the number of nucleotides in its chain; hence "NT" can be a measure of chain length.

NTA. Abbreviation for nitrilotriacetic acid.

NTAN. Abbreviation for nitrilotriacetonitrile.

nuclear chemistry. The division of chemistry dealing with changes in or transformations of the atomic nucleus. It includes spontaneous and induced radioactivity, the fission or splitting of nuclei, and their fusion or union; also the properties and behavior of the reaction products and their separation and analysis. The reactions involving nuclei are usually accompanied by large energy changes, far greater than those of chemical reactions; they are carried out in nuclear reactor for electric power production and manufacture of radioactive isotopes for medical use, also (in research work) in cyclotrons.
See fission; fusion; radiochemistry; nucleus (1).

nuclear energy. The energy liberated by (1) the splitting or fission of an atomic nucleus, (2) the union or fusion of two atomic nuclei, and (3) the radioactive decay of a nucleus (transmutation). For details refer to the following entries: nuclear reaction, fission, fusion, tokamak, radioactivity, ionizing radiation, nuclear chemistry, transmutation, breeder, uranium-235, plutonium, nuclear reactor.
See acceptable risk.

nuclear fuel. A fissionable material that is the source of energy in a nuclear reactor, specifically uranium-235, thorium, and plutonium-239. For thermonuclear (fusion) reactions the "fuels" are the hydrogen isotopes deuterium and tritium.
See breeder; fission; fusion.

nuclear fuel element. A rod, tube, plate, or other mechanical shape or form into which a nuclear fuel is fabricated for use in a reactor. The word *element* is used here in its electrical engineering sense rather than its chemical sense. Cladding materials are usually ceramics or zirconium alloys.

nuclear grade graphite. A form of carbon.
Use: A neutron moderator and reflector in some nuclear reactors.

nuclear magnetic resonance. (NMR). A type of radio-frequency spectroscopy based on the magnetic field generated by the spinning of electrically charged atomic nuclei. This nuclear magnetic field is caused to interact with a very large (10,000–50,000 gauss) magnetic field of the instrument magnet. The magnetic properties of atomic nuclei are the spin number and the magnetic moment. Hydrogen nuclei, fluorine, phosphorus, boron, nitrogen, carbon-13, and oxygen-17 have distinctive magnetic properties. The molecular or chemical environment of the nucleus produces characteristic shifts and fine structure in the NMR spectra. Because of its dependence on molecular structure, NMR has become a fundamental research tool for structure determinations in organic chemistry. Studies of hydrogen locations in crystals have been useful, as these cannot be determined directly by means of X-rays. NMR techniques have been applied to studies of electron densities and chemical bonding, the composition of mixtures, percentage purity determinations, and elemental hydrogen analyses.

nuclear membrane. The double membrane that surrounds the eukaryotic nucleus. It has many pores in its surface which regulate the flow of large compounds into and out of the nucleus.

nuclear quadrupole resonance. (NQR). A spectroscopic technique related to nuclear magnetic resonance that utilizes the electric fields naturally present in crystals instead of a magnet. It is useful for studies of electrical field gradients around nuclei, chemical bonding, space groupings in crystals, and molecular structure.

nuclear reaction. A reaction that involves a change in the nucleus of an atom, as opposed to a chemical reaction in which only the electrons take part. Nuclear reactions usually result in release of tremendous amounts of energy, while the energy obtained from chemical reactions is slight; for example, rupture of a chemical bond evolves approximately 5 eV compared with 200 MeV resulting from the fission of an atomic nucleus. There are three types of nuclear reactions: (1) transmutation (radioactive decay), (2) fission, and (3) fusion (thermonuclear).

nuclear reactor. An assembly of fissionable material (uranium-235 or plutonium-239) designed to produce a sustained and controllable chain reaction for the generation of electric power. The first reactor (then called a "pile") was constructed at the University of Chicago in 1942 under the leadership of Enrico Fermi. The essential components of a modern nuclear reactor are (1) The core, composed of metal- or ceramic-clad rods containing enough fissionable material to maintain a chain reaction at the necessary power level; as much as 50 tons of uranium may be required (d 19). (2) A source of neutrons to initiate the reaction, such as a mixture of polonium and beryllium. (3) A moderator to reduce the energy of fast neutrons for more efficient fission (called slow or thermal neutrons); materials such as graphite, beryllium, heavy water, and light water are used. (4) A coolant to remove the fission-generated heat; water is generally used converted to steam in heat exchangers and used to drive turbines. Sodium, helium, and nitrogen may also be used. The heating of streams and estuaries by reactor effluent is a serious environmental problem, which can be surmounted by construction of special cooling towers. (5) A control system such as rods of boron or cadmium that have high capture cross sections, to absorb neutrons rapidly when their concentration becomes too high. (6) Adequate shielding, remote-control equipment, and appropriate instrumentation are essential for personnel safety and efficient operation.

The primary use of nuclear reactors is for electric power generation. This has been a reasonably successful endeavor; nuclear energy has made a notable contribution to the overall energy supply situation in the U.S. But the efficiency of reactors is often reduced far below their potential by failures resulting from the highly corrosive operating conditions to which the materials of construction of the reactor and the associated hardware are exposed. Frequent and extended shutdowns are commonplace. This, together with the environmental radiation hazard and escalating construction costs, has restricted the development of nuclear energy as a power source. Plans for construction of many plants have been delayed or cancelled. A further negative fact is that the estimated safety factor has changed radically since 1975, at which time the risk of serious accident that might result in core meltdown was reported to be 1 in 20,000 years of operation. A study conducted by ORNL based on accident data in the decade 1969–1979 concluded that this risk has risen to 1 in 1000 operating years—a 20-fold decrease in the safety factor.

See fission; breeder; acceptable risk.

Nuclear Regulatory Commission. (NRC). A federal agency established in 1975 to regulate all commercial uses of atomic energy, including construction and operation of nuclear power plants, nuclear fuel reprocessing plants, and research applications of radioactive materials. It is also responsible for safety and environmental protection. Its mailing address is U.S. Nuclear Regulatory Commission, Washington, DC 20555. Website: http://www.nrc.gov

nuclear run-on. A method used to estimate the relative rate of transcription of a given gene, as opposed to the steady-state level of the mRNA transcript (which is influenced not just by transcription rates, but by the stability of the RNA). This technique is based on the assumption that a highly-transcribed gene should have more molecules of RNA polymerase bound to it than will the same gene in a less-active state. If properly prepared, isolated nuclei will continue to transcribe genes and incorporate 32P into RNA, but only in those transcripts that were in progress at the time the nuclei were isolated. Once the polymerase molecules complete the transcript they have in progress, they should not be able to re-initiate transcription. If that is true, then the amount of radiolabel incorporated into a specific type of mRNA is theoretically proportional to the number of RNA polymerase complexes present on that gene at the time of isolation.

nuclear transfer. A laboratory procedure in which a cell's nucleus is removed and placed into an oocyte with its own nucleus removed, so the genetic information from the donor nucleus controls the resulting cell. Such cells can be induced to form embryos. This process was used to create the cloned sheep "Dolly."
See cloning.

nuclear waste. See radioactive waste.

nuclease. An enzyme that degrades nucleic acids. A nuclease can be DNA-specific (a DNase), RNA-specific (RNase), or non-specific. It may act only on single stranded nucleic acids, or only on double-stranded nucleic acids, or it may be non-specific with respect to strandedness. A nuclease may degrade only from an end (an exonuclease), or may be able to start in the middle of a strand (an endonuclease). To further complicate matters, many enzymes have multiple functions; for example, Bal31 has a 3′-exonuclease activity on double-stranded DNA, and an endonuclease activity specific for single-stranded DNA or RNA.

nuclease protection assay. See "RNase protection assay."

nucleases. Enzymes that hydrolyze the phosphodiester linkages of nucleic acids.

nucleation. The process by which crystals are formed from liquids, supersaturated solutions, or saturated vapors (clouds). Crystals originate on a minute trace of a foreign substance acting as a nucleus. These are provided by impurities or by container walls in laboratory apparatus. Crystals form initially in tiny regions of the parent phase and then propagate into it by accretion. Rain and snow are formed in this way in moisture-laden air, the moisture condensing on minute particles of dust or ice. Cloud seeding with silver iodide or carbon dioxide is based on this principle. The former has a crystalline structure similar to that of ice, and the latter causes rapid formation of ice nuclei by intense local lowering of temperature.
See crystals; whiskers.
There are various methods for growing crystals, including (1) evaporation of a solution; (2) cooling of a saturated solution or melt; (3) condensation of a vapor (plasmas are sometimes used to generate the vapor); (4) electrodeposition; (5) growth in gel media (in order to slow growth). Methods (2) through (5) are particularly suited for the growth of large crystals, which are widely used in modern technology.

nucleation rate. The number of crystal nuclei formed per unit volume of reaction phase.

nucleic acid. Any of several complex compounds occurring in living cells, usually chemically bound to proteins to form nucleoproteins. Nucleic acids are of high molecular weight are easily changed by many mild chemical reagents. They contain carbon, hydrogen, oxygen, nitrogen (15–16%), and phosphorus (9–10%).
The fundamental units of nucleic acid, are nucleotides, nucleic acids, and polynucleotides in which the nucleotides are linked by phosphate bridges. Upon extensive heating in the presence of water (hydrolysis), nucleic acids yield a mixture of purines and pyrimidines, d-ribose or d-deoxyribose, and phosphoric acid. Nucleic acids are subdivided into two types: ribonucleic acid (RNA), containing the sugar d-ribose, and deoxyribonucleic acid (DNA), containing the sugar d-deoxyribose.
See deoxyribonucleic acid; ribonucleic acid; nucleoside; nucleotide; nucleoprotein.
Good sources of nucleic acids are salmon, thymus, yeast, and wheat kernel embryo.
Use: Biochemical and medical research in genetics, virus diseases, and cancer.

nucleogenesis. The original synthesis of the chemical elements resulting from a huge explosion of undifferentiated energy that astrophysicists believe initiated the universe approximately 20 billion years ago, at least 7 billion years before the formation of the Milky Way galaxy (the so-called big bang theory). Information obtained with radiotelescopes indicates that temperatures were so inconceivably high in this continuum of radiation that the elementary particles that were undoubtedly present were unable to combine. As the continuum expanded rapidly after the original eruption, it cooled within a few minutes to the point where protons and electrons could unite to form hydrogen. After further cooling to 10^7C, thermonuclear fusion of hydrogen to helium occurred in a matter of seconds. But no more complex combinations took place in the young universe for several million years, until the shrinking cores of giant stars, many times the mass of the sun, raised their internal temperatures to 10^8C, high enough to fuse helium nuclei to form carbon. The heavier elements were eventually synthesized during explosions of supernovae, which produced temperatures in the range of 600×10^6C. Thus, the sequence of creation of the elements is believed to be hydrogen and helium, then after a few million years came carbon, oxygen, magnesium, silicon, sulfur, and iron; after yet another interval, the heavy metals were formed in another round of stellar explosions. All this occurred over a period of 5–7 billion years.
See fusion; astrochemistry; life; origin.

nucleohistone. A complex of DNA and histone.

nucleoid. Region in prokaryotes where the DNA is concentrated. Unlike a nucleus, it is not bound by a membrane.

nucleolar organizing region. A part of the chromosome containing rRNA genes.

nucleolus. A densely staining structure in the nucleus of eukaryotic cells in interphase. Contains the machinery involved in rRNA synthesis and ribosome formation.

nucleon. General name applied to neutrons and protons, the essential constituents of atomic nuclei; also used as a class name for fundamental particles of that mass. The study of subatomic particles is often called nucleonics.

nucleophile. An ion or molecule that donates a pair of electrons to an atomic nucleus to form a covalent bond; the nucleus that accepts the electrons is called an electrophile. This occurs, for example, in the formation of acids and bases according to the Lewis concept, as well as in covalent carbon bonding in organic compounds.
See Lewis base; donor.

nucleoplasm. The aqueous portion of a cell's contents enclosed by the nuclear membrane; also called the nuclear matrix.

nucleoprotein. A type of protein universally present in the nuclei and the surrounding cytoplasm of living cells. A nucleoprotein is composed of a protein, which is rich in basic amino acids, and a nucleic acid. The nucleic acid portions can be isolated and used in medical and biochemical research.
See deoxyribonucleic acid; chromatin; virus.

nucleoside. A compound of importance in physiological and medical research, obtained during partial decomposition (hydrolysis) of nucleic acids and containing a purine or pyrimidine base linked to either *d*-ribose, forming ribosides, or *d*-deoxyribose, forming deoxyribosides. They are nucleotides minus the phosphorus group.
See adenosine; cytidine; guanosine; uridine.

nucleoside diphosphate kinase. An enzyme that catalyzes the transfer of the terminal phosphate of a nucleoside 5'-triphosphate (most often ATP) to a nucleoside 5'-diphosphate (GDP, for example).

nucleoside diphosphate sugar. An activated form of a sugar molecule used in the enzymatic synthesis of polysaccharides and sugar derivatives. The base depends on the sugar; for example, UDP-glucose but GDP-fucose.

nucleoside monophosphate kinase. An enzyme that catalyzes the transfer of the terminal phosphate of ATP to a nucleoside 5'-monophosphate.

nucleosome. Structural unit for packaging DNA into chromatin. A nucleosome consists of a DNA strand wound around an octameric histone core.

nucleotide. A fundamental unit of nucleic acids; some are important coenzymes. The four nucleotides found in nucleic acids are phosphate mono-esters of nucleosides, adenylic acid, guanylic acid, uridylic acid, and cytidylic acid. Great progress has been made in determining the nucleotide sequence in fundamental materials, such as yeast genes.
The term is also applied to compounds not found in nucleic acids and that contain substances other than the usual purines and pyrimidines. Such compounds are modified vitamins and function as coenzymes; examples are riboflavin phosphate (flavin mononucleotide), flavin adenine nucleotide, nicotinamide adenine dinucleotides, nicotine adenine dinucleotide phosphate, and coenzyme A. The nucleotides inosine-5'-monophosphate and guanosine-5'-monophosphate are used as flavor potentiators.

nucleus. (1) The positively charged central mass of an atom, it contains essentially the total mass in the form of protons and neutrons. The nucleus of the hydrogen atom consists of one proton, while that of uranium is composed of 92 protons and 146 neutrons. (2) The central portion of a living cell, consisting primarily of nucleoplasm in which chromatin is dispersed. It is enclosed by a membrane that separates it from the surrounding cytoplasm. All the most important functions of the cell, including the mechanics of division (mitosis) and the programming of the genetic code, take place in the nucleus.
See gene; chromosome.
(3) The characteristic structure of a group of chemical compounds, e.g., the benzene nucleus. (4) Any small particle that can serve as the basis for crystal growth (see nucleation).
Note: The multiple meanings of *nucleus* and *resonance* can be a source of confusion, especially when these terms are closely associated, as in nuclear magnetic resonance and resonance of a molecular nucleus. In the first of these expressions, *nucleus* is used in sense (1) under nucleus and *resonance* in sense (2) under resonance. In the second expression, *nucleus* is used in sense (3) under nucleus and *resonance* in sense (1) under resonance.

nuclide. A particular species of atom, characterized by the mass, the charge (number of protons), and the energy content of its nucleus. A radionuclide is a radioactive nuclide. Example: carbon-14 is a radionuclide of carbon.
See isotope.

Nuclon.   A polycarbonate thermoplastic resin.
  Properties: High impact resistance not appreciably
  reduced by temperature fluctuations, good dimensional stability, or good electrical resistance. Naturally transparent, of light-straw color.
  Use: Engineering plastic for *hard-service* parts and
  components.

"Nucrel" *[DuPont].*   TM for an ethylene-
  methacrylic acid copolymer resin available for use
  in conventional extrusion coating, coextrusion coating, and extrusion laminating equipment designed
  to process polyethylene resins. Higher melt flow
  index results in fewer resin leaks.
  Use: Packaging foil containing shampoos, toothpaste, towelettes, and miscellaneous food pouches.

nuisance particulate.   Fine particles (dusts) that
  are not very toxic in low concentrations. Among
  them are clay, calcium carbonate, emery, glass fiber,
  silicon carbide, gypsum, starch, Portland cement,
  marble, and titanium dioxide.
  See dust, industrial.

Nujol.   Former TM for a mineral oil used to prepare mulls for infrared analysis.

numerals.   For their use and meaning in chemical names see Geneva System; benzene; chemical
  nomenclature.

"Nuroz" *[Reichhold].*   TM for a polymerized
  wood rosin.
  Use: Adhesives, gloss oils, paper label coatings, oleoresinous varnishes, solder flux, spirit varnishes,
  waxed paper and hot melt compounds, synthetic
  resins.

Nusselt number.   A value used in heat-transfer
  studies and calculations to compare heat losses by
  conduction from various shaped objects under various conditions. It is combined into a single number
  that is the function of the actual heat loss (Q), the
  temperature difference (δT) between the body and
  its surroundings, the size (d) and shape of the body,
  and the thermal conductivity (k) of the fluid surrounding the object, in the equation Nu = $Qd/\delta Tk$.

"Nutrasweet" *[Nutrasweet].*   TM for "Aspartame."

"NuTriene" *[Eastman].*   TM for a nutraceuti-
  cal supplement. A mixture of naturally occurring
  tocotrienols and tocopherols are extracted and concentrated from rice bran oil, a common cooking oil
  in Asia.
  Use: Suitable for incorporation into soft gelatin capsules or liquid emulsions.

nutrient.   Any element or compound that is essential to the life and growth of plants or animals, either

as such or as transformed by chemical or enzymatic reactions. In plants, nutrients include numerous mineral elements, as well as nitrogen, carbon
dioxide, and water. In animals and humans, the
primary nutrients are the proteins, carbohydrates,
and fats obtained from plants, either directly or
indirectly, supplemented by vitamins and minerals.
Water and oxygen are included in this definition.
All told, there are 43 basic nutrients.
See food.

nutrient solution.   A water solution of minerals and their salts necessary for plant growth that
  is used instead of soil, the plants being supported
  by mechanical means. Such solutions contain combined nitrogen, potassium, phosphorus, calcium,
  sulfur, and magnesium, together with traces of iron,
  boron, zinc, and copper. They are extensively used
  for commercial growing of flowers and vegetables,
  particularly on islands, and also to some extent for
  house plants.

nutrification.   Addition of nutrients to a food
  to replace those lost in processing (restoration),
  to provide nutrients that are not normally present
  in the food (fortification), or to bring the food
  into conformity with a specific standard for that
  food.

"Nutrifos" *[TCL].*   (sodium phosphate).
  CAS: 7601-54-9.   TM for food-grade phosphate.
  Use: Curing meat products; washing fruits and veg-
  etables.

"Nutriphos STP" *[TCL].*   TM for sodium
  phosphate.
  CAS: 7601-54-9.
  Grade: Powders.
  Use: To cure ham and related meat products and in
  washing fruits and vegetables.

nutrition.   The effects of nutrients on living organisms and the biochemical mechanisms
  involved in bringing them about; also, the subdivision of biochemistry that deals specifically with
  these effects. In plant nutrition the essential requirements are carbon dioxide and water, from which the
  plant forms carbohydrates by photosynthesis; nitrogen, which is essential for the synthesis of proteins
  by the plant, with the aid of nitrogen-fixing bacteria;
  as well as phosphorus, calcium, potassium, and a
  number of trace elements (micronutrients). Besides
  proteins and carbohydrates, plants also synthesize
  vitamins and various fats and oils. Thus they provide
  a basis for human nutrition, both directly (grain and
  other vegetables) and indirectly (meats and dairy
  products), though the conversion to protein values
  for human nutrition is only approximately 10% for
  meats.
  Human diet requires proteins (milk, eggs, fish, and
  some vegetables), carbohydrates (plants), fats (oils)

from both plants and animals, minerals from milk and meats, salt (chloride), vitamins from green vegetables and citrus fruits, and water. Micronutrients are furnished by seafood, cereals, vegetables, and fruit.

Human digestive processes involve primarily the hydrolysis of complex carbohydrates to simple sugars, of proteins to a mixture of amino acids, and of fats to glycerol and higher fatty acids. Hydrolysis is catalyzed by various enzymes in the saliva and digestive tract. The end products of digestion are absorbed across a semipermeable membrane in the intestine and thus enter the bloodstream, unusable products being eliminated. The efficiency of digestion plus absorption is approximately 92% for protein, 95% for fat, and 98% for carbohydrates.

See metabolism; digestion (1); plant (1); nutrient; RDA.

"Nuts & Bolts 434" [Hernon].   TM for a medium-strength thread-locking adhesive, offering 80 lbs breakaway torque.

nut shells.   In a fine-ground state, the shells of coconuts and other nuts are a source of decolorizing carbon; the pits of peaches and similar fruits have been used for gas-adsorbent carbon.

nux vomica extract.   ((4aR,5aS,8aR,15bR)-4a,5,5a,7,8,13a,15,15a,15b,16-decahydro-2H-4,6-methanoindolo[3,2,1-ij]oxepino[2,3,4-de]pyrrolo-[2,3-h]quinolone-14-one). $C_{21}H_{22}N_2O_2$. An alkaloid found in the seeds of *Strychnos Nux-Vomica*.

Properties: Powder form, contains 7–7.5 grams of strychnine per 100 grams of strychnine.

Use: As an analeptic; in the treatment of nonketotic hyperglycinemia and sleep apnea; as a rat poison.

Hazard: Convulsant.

NW acid.   Abbreviation for Neville-Winter acid.

Nylander's reagent.   Basic solution of bismuth subnitrate and Rochelle salt.

Use: To detect glucose in urine.

nylidrin hydrochloride.   (*p*-hydroxy-α-[1-(1-methyl-3-phenyl-propylamino)-ethyl]benzyl alcohol hydrochloride).

CAS: 849-55-8. $C_{19}H_{25}O_2N \cdot HCl$.

Properties: White, crystals or powder, odorless; tasteless. Slightly soluble in water and alcohol; very slightly soluble in chloroform, ether; pH of 1% solution is between 4.5 and 6.5.

Grade: NF.

Use: Medicine (treatment of heart disease).

nylon.

CAS: 63428-83-1. $(C_6H_{11}NO)_n$.

Generic name for a family of polyamide polymers characterized by the presence of the amide group –CONH. By far the most important are nylon 66 (75% of U.S. consumption) and nylon 6 (25% of U.S. consumption). Except for slight difference in melting points, the properties of the two forms are almost identical, though their chemical derivations are quite different. Other types are nylons 4, 9, 11, and 12 (see Grade).

Properties: Crystals, thermoplastic polymers. May be extruded as monofilaments over a wide dimensional range. Filaments are oriented by cold-drawing. Tensile strength (high tenacity) up to 8 g/denier (approximately 100,000 psi); d 1.14; mp (66) 264C, (6) 223C; low water absorption. Good electrical resistance but accumulates static charges. Highly elastic, with rather high percentage of delayed recovery at low strain values; low permanent elongation; moisture absorption 4% at 65% R.H. Wet strength approximately 90% of dry strength. Can be dyed with ionic and nonionic dyestuffs. Attacked by mineral acids but resistant to alkalies and cold abrasion; soluble in hot phenols, cresols, and formic acid; insoluble in most organic solvents; difficult to ignite, self-extinguishing, melts forming beads; resistant to attack by moths, carpet beetles, etc.; compatible with wool and cotton, increases wear and crease resistance in 30% blends with natural fibers; rods and blanks are machinable.

Available forms: Monofilaments, yarns, bristles, molding powders, rods, bars, sheets. Microcrystalline nylon is now available.

Grade: Nylon 66 is a condensation product of adipic acid and hexamethylenediamine developed by Carothers in 1935. Adipic acid is obtained by catalytic oxidation of cyclohexane. Nylon 6 (CAS: 25038-54-4) is a polymer of caprolactam, originated by I. G. Farbenindustrie in 1940. It is a questionable carcinogen. Nylon 4 is based on butyrolactam (2-pyrrolidone); its tenacity, abrasion resistance, and melting point are said to be about the same as for the 6 and 66 grades. It has excellent dyeability.

Nylon 610 (TM "Tynex" [Du Pont]) is obtained by condensation of sebacic acid and hexamethylene-diamine, and nylon 11 (TM "Rilsan" [Arkema]) from castor bean oil (developed in France). Nylon 12 (also called "Rilsan" 12 [Arkema]) is made from butadiene, also by a French process involving photonitrosation of cyclododecane by actinic light from mercury lamps. Its properties are similar to those of nylon 11. Nylon 9 can be made from 9-aminononanoic acid, present in soybean oil. It has properties specifically desired in metal coatings and electrical parts; higher electrical resistance than 6 and 66 absorbs less moisture and has better distortion resistance.

Use: Tire cord; hosiery; wearing apparel component; bristles for toothbrushes, hairbrushes, paint brushes (nylon 610); cordage and towlines for gliders; fish nets and lines; tennis rackets; rugs and carpets; molded products; turf for athletic fields; parachutes; composites; sails; automotive upholstery; film; gears and bearings; wire insulation; surgical sutures; artificial blood vessels; metal coating;

pen tips; osmotic membranes; fuel tanks for automobiles.

See polyamide; aramid.

Note: Not all nylons are polyamide resins, nor are all polyamide resins nylons, e.g., Versamide. One class of polyamide resins distinct from nylons is derived from ethylenediamine; they may be liquids or low-melting solids and have lower molecular weight than nylons. Another class, called aramids, is aromatic in nature.

nymph.   In aquatic insects, the larval stage.

"Nytal 2K" *[Vanderbilt].*   TM for an epoxy heavy duty tank lining.

nytril.   Generic name for a manufactured fiber containing at least 85% of a long-chain polymer of vinylidene dinitrile, $-CH_2C(CH)_2-$, where the vinylidene dinitrile content is no less than every other unit in the polymer chain (Federal Trade Commission).

Properties: Soft, resilient fabric is obtained; easy to clean, does not pill, resists wrinkling and retains shape after pressing.

Use: Fur-like pile fabrics, sweaters, yarns, blended fabrics for coats and suits.

NZ.   A prefix for a series of neoalkoxy zirconates.

O

O.   Symbol for oxygen; the molecular formula is O_2.

o-.   (o-). Abbreviation for ortho-.

oakum.   Hemp fiber impregnated with tar or pitch.
Use: Caulking.

Obermayer's reagent.   Ferric chloride in concentrated hydrochloric acid (4 g $FeCl_3$ in 1 L concentrated HCl).
Use: Determining indoxyl in urine.

OC.   Abbreviation of oxygen consumed.

Occupational Safety and Health Administration.   (OSHA). A federal agency responsible for establishing and enforcing standards for exposure of workers to harmful materials in industrial atmospheres, and other matters affecting the health and well-being of industrial personnel. A number of OSHA regulations and proposals have been controversial. Its main office is at 200 Constitution Avenue, N.W., Washington, DC 20210. Website: http://www.osha.gov

ocean thermal energy conversion.   (OTEC). Utilization of ocean temperature differentials between solar-heated surface water and cold deep water as a source of electric power. In tropical areas such differences amount to 35–40F. A pilot installation now operating near Hawaii utilizes a closed ammonia cycle as a working fluid, highly efficient titanium heat exchangers, and a polyethylene pipe 2000 ft long and 22 inches inside diameter to handle the huge volume of cold water required. Alternate uses for such a system, such as electrolysis of water, ammonia production, and desalination, are envisaged. There has been active interest in the possibilities of this energy source in France from the time of d'Arsonval (1885) that continues especially in Japan and Hawaii. Ongoing research indicates that OTEC may be harder to commercialize than once projected.

ocean water.   (seawater). A uniform solution of essentially constant composition containing 96.5% water and 3.5% ionized salts, associated compounds, elements, and ionic complexes. Na^+ and Cl^- are completely ionized, but $MgSO_4$ and AuCl remain in bound form. Dissolved gases, e.g., oxygen and nitrogen are also present.

Composition: More than 60 elements, by far the most important of which are (in tons/cubic mile): chlorine 89,500,000; sodium 49,500,000; magnesium 6,400,000; sulfur 4,200,000; calcium 1,900,000; potassium 1,800,000; bromine 306,000; carbon 132,000. Other elements present include (tons/cubic mile): copper 14, indium 47, silver 1, lead 0.1, gold 0.02. Total solids content/cubic mile is 166 million tons.
Properties: Colorless liquid; bitter taste; faint odor (depending on organic impurities). D 1.02, pH 7.8–8.2, fp −2.78C, bp 101.1C. Average temperature about 5C.
Hazard: Ingestion of substantial amounts will create bodily chloride imbalance with harmful effects.
Use: Source of magnesium, bromine, sodium chloride, and fresh water; source of hydrogen.

Ocenol.   Technical grades of oleyl alcohol rich in cetyl and unsaturated alcohols, used as antifoam agents.

ocher.   Any of various colored earthy powders consisting essentially of hydrated ferric oxides mixed with clay, sand, etc. Some grades are calcined (burnt ocher). The colors are yellow, brown, or red. Noncombustible.
Use: Paint pigments, cosmetics, theatrical makeup. See umber; sienna.

ochratoxin.   Hepatoxic mycotoxins produced by *Aspergillus ochraceus* and *A. niger* and some *Penicillium* species, mainly *P. verrocosum* and *P. carbonarius* that contaminate cottonseed, breads, and raw foods such as stored grains, fruits, such as corn and peanuts.
Hazard: Poison; extremely toxic; teratogen.

ochratoxin A.   ((2S)-2-[[(3R)-5-chloro-8-hydroxy-3-methyl-1-oxo-3,4-dihydroisochromene-7-carbonyl]amino]-3-phenylpropanoic acid). CAS: 303-47-9. $C_{20}H_{18}ClNO_6$.
One of the most abundant food-contaminating mycotoxins.
Derivation: Produced by any of a number of fungi.
Hazard: Hepatotoxic, nephrotoxic, extremely toxic; possible carcinogen.

ochratoxin B.   ((2S)-2-[[(3R)-8-hydroxy-3-methyl-1-oxo-3,4-dihydroisochromene-7-carbonyl]amino]-3-phenylpropanoic acid). $C_{20}H_{19}NO_6$. A mycotoxin.
Derivation: Produced by *Aspergillus ochraceus*.

ocimene.   (2,6-dimethyl-1,5,7-octatriene).
CAS: 29714-87-2. $CH_2:C(CH_3)(CH_2)_2CH:C$
$(CH_3)CH:CH_2$.
Mixture of isomers. A terpene obtained from sweet basil oil, bp 81C (30 mm Hg), insoluble in water, soluble in alcohol, d 0.8031 (15C). Combustible.
Use: Flavors and perfumes.

ocotea oil.
CAS: 68917-09-9. A volatile oil derived from *Ocotea cymbarum* used for its safrole content for the manufacture of heliotropin.

octabenzone.
CAS: 1843-05-6. $C_{21}H_{26}O_3$.
Properties: Colorless crystals. Mp 45C.
Use: UV absorber, especially for polyethylene.

octabromodiphenyl ether.
CAS: 32536-52-0. $C_{12}H_2Br_8O$.
Hazard: Moderately toxic by inhalation and skin contact. Low toxicity by ingestion.

octachloronaphthalene.   (Halowax 1051).
CAS: 2234-13-1. $C_{10}Cl_8$.
Properties: Pale-yellow, waxy solid. Mw 403.74, specific gravity 2.00, mp 192C, bp 440C. Insoluble in water; slightly soluble in alcohol; soluble in benzene and chloroform.
Hazard: Toxic by inhalation and skin contact. Toxic by skin absorption. Liver damage.
Use: Fireproofing and waterproofing in cable insulation and lubricant additive.

(z,z)-3,13-octadecadien-1-ol acetate.
CAS: 53120-27-7. $C_{20}H_{36}O_2$.
Hazard: Low toxicity by ingestion.
Source: Natural product.

***n*-octadecane.** $C_{18}H_{48}$ or $CH_3(CH_2)_{16}CH_3$.
Properties: Colorless liquid. D 0.7767 (28/4C), bp 318C, mp 28.0C, refr index 1.4369 (28C), flash p 200F (93C). Soluble in alcohol, acetone, ether, petroleum, and coal tar hydrocarbons; insoluble in water. Combustible.
Use: Solvents, organic synthesis, calibration.

octadecadienoic acid.   See linoleic acid.

1,12-octadecanediol.   See 1,12-hydroxystearyl alcohol.

octadecaneuropeptide diazepam-binding inhibitor (33–50).
CAS: 95237-86-8. $C_{81}H_{138}N_{24}O_{29}$.
Hazard: A poison.

***n*-octadecanoic acid.**   See stearic acid.

octadecanoic acid, ammonium salt.   See ammonium stearate.

octadecanoic acid, calcium salt.   See calcium stearate.

octadecanoic acid, lithium salt.   See lithium stearate.

1-octadecanol.   See stearyl alcohol.

***n*-octadecanoyl chloride.**   See stearoyl chloride.

octadecatrienol.   See linolenyl alcohol.

9-octadecen-1,12-diol.   See ricinoleyl alcohol.

α,α′-((9-octadecenylimino)di-2,1-ethanediyl) bis(w-hydroxy-poly(oxy-1,2-ethanediyl).
CAS: 58253-49-9. $(C_2H_4O)_n(C_2H_4O)_nC_{22}H_{45}$
NO_2.
Hazard: A severe eye irritant.

1-octadecene. $C_{18}H_{36}$ or $CH_3(CH_2)_{15}CH:CH_2$.
Properties: Colorless liquid. D 0.7884 (20/4C), refr index 1.4456 (20C), bp 180C (15 mm Hg), flash p 200F (93C). Soluble in alcohol, acetone, ether, petroleum, and coal tar solvents; insoluble in water. Combustible.
Grade: 95% min purity.
Use: Organic synthesis, surfactants.

octadecene-octadecadieneamine.   See oleyllinoleylamine.

***cis*-9-octadecenoic acid.**   See oleic acid.

***trans*-9-octadecenoic acid.**   See elaidic acid.

octadecenol.   See oleyl alcohol.

***cis*-9-octadecenoyl chloride.**   See oleoyl chloride.

octadecenyl aldehyde.   (oleyl aldehyde).
$C_{17}H_{35}CHO$.
Properties: Liquid. Bp 167C (20 mm Hg), refr index 1.4620 (25C), d 0.847 (25C).
Hazard: Flammable, moderate fire risk. Irritant to skin and mucous membranes.
Use: Intermediate for vulcanization accelerators, rubber antioxidants, synthetic drying oils, and pesticides.

octadecyl alcohol.   See stearyl alcohol.

octadecyldimethylbenzylammonium chloride. $C_{18}H_{37}(CH_3)_2(C_6H_5CH_2)NCl$. A quaternary ammonium salt.
Properties: White, crystalline powder. Soluble in water, chloroform, benzene, acetone, xylene.

octadecyl isocyanate. A straight-chain, saturated monoisocyanate consisting principally of a mixture of C_{18} and C_{16} alkyls, $CH_3(CH_2)_{16}CH_2NCO$, and $CH_3(CH_2)_{14}CH_2NCO$.
Properties: Colorless, slightly cloudy liquid at room temperature. Fp 10–20C (2 mm Hg), flash p 355F (179C) (OC). Combustible.
Hazard: By inhalation, irritant to skin.
Use: Intermediate in synthesis, water-repellent textiles, paper, and other surfaces.

octadecyl mercaptan. See stearyl mercaptan; thiol.

octadecyltrichlorosilane. (trichlorooctadecyl-silane). $C_{18}H_{37}SiCl_3$.
Properties: Water-white liquid; pungent odor. D 0.984 (25C), refr index 1.4580 (25C), bp 380C, flash p 193F (89.4C). Soluble in benzene, ethyl ether, heptane, and perchloroethylene. Combustible.
Hazard: Strong irritant to skin and mucous membranes.
Use: Intermediate for silicones.

octafluoro-2-butene. (perfluoro-2-butene).
CAS: 360-89-4. $CF_3CF{:}CFCF_3$.
Properties: Colorless, gas or liquid. D (liquid) (at bp) 1.5297 g/cc, bp 1.2C, fp −136 to −134C, sp vol 2.7 cu ft/lb (70F, 1 atm). Nonflammable.
Grade: 95%.
Use: Organic intermediate, fluorocarbon polymers.

octafluorocyclobutane.
CAS: 115-25-3. C_4F_8.
Properties: Colorless gas. Fp −41C, bp −6C. Nonflammable.
Use: Refrigerant, heat-transfer agent.

octafluoroisobutene. See octafluoro-*sec*-butene.

octafluoroisobutylene. See octafluoro-*sec*-butene.

octafluoropropane. (perfluoropropane). C_3F_8.
Properties: Colorless, gas. D 1.29, bp −36.7C, fp −160C, sp vol 2.02 cu ft/lb (21C), 1 atm.
Derivation: Electrofluorination of various organic compounds.
Grade: 98% min purity. Nonflammable.
Use: Refrigerant (when combined with chlorofluorohydrocarbons); gaseous insulator, especially for radar wave guides.

octafluoro-*sec*-butene. (octafluoroisobutene; octafluoroisobutylene; perfluoroisobutylene).
CAS: 382-21-8. C_4F_8.
Properties: A gas at room temp. Mw 200.04, d 1.592 (0C), bp 5–6C (740 mm).
Hazard: A poison by inhalation. A skin, eye, and mucous membrane irritant. Upper respiratory tract irritant and hematologic effects.

octahydro-(1,2,4,5)tetrazino(1,2-a)(1,2,4,5) tetrazine.
CAS: 1743-13-1. $C_4H_{12}N_6$.
Hazard: A poison by ingestion and inhalation.

octakis(2-hydroxypropyl)sucrose.
Properties: Viscous, straw-colored liquid. D 1.170 (70/20C), refr index 1.485 (25C), pour p 38C, flash p 485F (251C). Soluble in water, methanol, and ether. Combustible.
Use: Crosslinking agent for urethane foams; plasticizer for cellulosics, glue, starch, and many resins.

octamethylcyclotetrasiloxane.
CAS: 556-67-2. $C_8H_{24}O_4Si_4$.
Properties: Smooth, viscous liquid. Bp 175C, mp 17C, d 0.95, refr index 1.4.
Derivation: Hydrolysis of dimethyldichlorosilane.
Use: Silicone oils and related products.

octamethyl pyrophosphoramide. See schradan.

octamethylsilanetetramine.
CAS: 1624-01-7. $C_8H_{24}N_4Si$.
Hazard: A poison by ingestion and skin contact. A severe eye irritant.

octamethyltrisiloxane. $C_8H_{24}O_2Si_3$.
Properties: Colorless liquid. Bp 151C, d 0.820, fp −80C. Soluble in light hydrocarbons.
Use: Silicone oils, antifoam agent in lubricating oils.

"Octamine" [*Chemtura*]. TM for a reaction product of diphenylamine and diisobutylene.
Properties: Light-brown, granular, waxy solid. D 0.99, mp 75–85C, good storage stability. Soluble in gasoline, benzene, ethylene dichloride, and acetone; insoluble in water.
Use: Rubber antioxidant.

1-octanal. (*n*-octyl aldehyde; aldehyde C-8; caprylic aldehyde).
CAS: 124-13-0. $(CH_3CH_2)_6CHO$.
Properties: Colorless liquid; strong fruity odor. D 0.820–0.830, refr index 1.418–1.425, bp 163C, flash p 125F (51.6C) (CC). Soluble in 70% alcohol and mineral oil; insoluble in glycerol. Combustible.
Grade: Technical, FCC.
Hazard: Moderate fire risk.
Use: Perfumery, flavors.

n-octane. CAS: 111-65-9. $CH_3(CH_2)_6CH_3$.
Properties: Colorless liquid. D 0.7026 (20/4C), refr index 1.39745 (20C), bp 125.6C, fp −56.798C, flash p 56F (13.3), autoign temp 428F (220C). Soluble in alcohol and acetone; insoluble in water.
Grade: 95%, 99% research.
Hazard: Flammable, dangerous fire risk.
Use: Solvent, organic synthesis, calibrations, azeotropic distillations.

1,8-octanedicarboxylic acid. See sebacic acid.

octanedioic acid. See suberic acid.

octane number. A number indicating the anti-knock properties of an automotive fuel mixture under standard test conditions. Pure normal heptane (a very high-knocking fuel) is arbitrarily assigned an octane number of zero, while 2,2,4-trimethylpentane, or isooctane (a branched-chain paraffin), is assigned 100. Thus, a rating of 80 for a given fuel indicates that its degree of knocking in a standard test engine is that of a mixture of 80 parts isooctane and 20 parts n-heptane. Octane ratings as high as 115 have been obtained by addition of tetraethyllead to isooctane. Premium leaded gasolines have a research octane rating of about 100, but this value drops to 85–90 for unleaded gasolines, though it may be improved by addition of methyl-*tert*-butyl ether. The octane rating scale ends at 125, and any higher figure is meaningless. Research octane numbers are those obtained under test or *laboratory* conditions; they generally run about ten points higher than the so-called motor octane numbers, which represent actual road operating conditions.

1-octanesulfonyl fluoride.
CAS: 40630-63-5. $C_8H_{17}FO_2S$.
Hazard: A mild skin and eye irritant.

octanoic acid. (octylic acid; octoic acid; caprylic acid).
CAS: 124-07-2. $CH_3(CH_2)_6COOH$.
Properties: Colorless, oily liquid; slight odor; rancid taste. D 0.9105 (20/4C), mp 16C, bp 237.9C, 147.9 (30 mm Hg), refr index 1.4278 (20C), flash p 270F (132C). Slightly soluble in water; soluble in alcohol and ether. Combustible.
Derivation: By saponification and subsequent distillation of coconut oil.
Method of purification: Crystallization or rectification.
Grade: Technical, 99%, FCC.
Use: Synthesis of various dyes, drugs, perfumes, antiseptics, and fungicides; ore separations; synthetic flavors.

octanol. See octyl alcohol.

2-octanone. See methyl hexyl ketone.

octanoyl chloride. (capryloyl chloride; sometimes called caprylyl chloride). $CH_3(CH_2)_6COCl$.
Properties: Water-white to straw-colored liquid; pungent odor. Fp −70C, distillation range 183–212C, d 0.9576 (15.5/15.5C), refr index 1.4357 (20C), flash p 180F (82.2C). Miscible with most common solvents; reacts with alcohol and water. Combustible.

Hazard: Irritant to eyes and skin.
Use: Organic synthesis.

***trans*-2-octen-1-al.** $C_8H_{14}O$.
Properties: Slightly yellow liquid; green odor. D: 0.830–0.850, refr index: 1.421–1.424. Soluble in alcohol, fixed oils; slightly soluble in water.
Use: Food additive.

1-octene. (1-octylene; 1-caprylene).
CAS: 111-66-0. C_8H_{16} or $CH_3(CH_2)_5CH{:}CH_2$.
Properties: Colorless liquid. D 0.7160 (20/4C), bp 121.27C, fp −102.4C, refr index 1.4088 (20C), flash p 70F (21.1C) (TOC). Soluble in alcohol, acetone, ether, petroleum, and coal tar solvents; insoluble in water.
Grade: 95%, 99% research.
Hazard: Flammable, dangerous fire risk.
Use: Organic synthesis, plasticizer, surfactants.

2-octene. C_8H_{16} or $CH_3(CH_2)_4CH{:}CHCH_3$. Both *cis* and *trans* forms exist.
Properties: Colorless liquid. D (*cis*) 0.7243, (*trans*) 0.7199, (commercial) 0.7185–0.7200 (20.4C); bp (*cis*) 125.6C, (*trans*) 125.0C, (commercial) 124–127C; fp −94.04C; refr index (*cis*) 1.4150, (*trans*) 1.4132, (commercial) 1.4120–1.4145 (20C); flash p (mixed isomers) 70F (21.2C) (TOC). Soluble in alcohol, acetone, ether, petroleum, and coal tar solvents; insoluble in water.
Grade: 95, 99 mole %.
Hazard: Flammable, dangerous fire risk.
Use: Organic synthesis, lubricants.

***cis*-3-octen-1-ol.** $C_8H_{16}O$.
Properties: White to yellowish liquid; musty, mushroom odor. D: 0.830–0.850, refr index: 1.440. Insoluble in water.
Use: Food additive.

1-octen-3-yl acetate. $C_{10}H_{18}O_2$.
Properties: Colorless liquid; metallic, mushroom odor. D: 0.865–0.886, refr index: 1.414–1.434 @ 25C. Soluble in fixed oils; insoluble in water, propylene glycol.
Use: Food additive.

1-octen-3-yl butyrate. $C_{12}H_{22}O_2$.
Properties: Colorless liquid; metallic, mushroom odor. D: 0.859–0.880, refr index: 1.416–1.437 @ 25C. Soluble in alcohol, fixed oils; slightly soluble in propylene glycol; insoluble in water.
Use: Food additive.

octocrilene.
CAS: 6197-30-4. $C_{24}H_{27}NO_2$.
Hazard: Low toxicity by ingestion.

octoic acid. An eight-carbon acid, usually designates caprylic acid.

octopirox. See piroctone olamine.

octyl. The general name describing all eight-carbon radicals having the formula C_8H_{17}, often used interchangeably for the 2-ethylhexyl isomer.

n-octyl acetate. (acetate C-8; caprylyl acetate). CAS: 103-09-3. $CH_3COO(CH_2)_7CH_3$.
Properties: Colorless liquid, floral-fruity odor, slightly soluble in water, soluble in alcohol and most other organic liquids, d 0.865–0.869, flash p 180F (82.2C) (OC), refr index 1.419–1.422, bp 199C. Combustible.
Use: Perfumery, flavors, solvent.

sec-n-octyl alcohol. (2-n-octanol; methyl hexyl carbinol). $CH_3(CH_2)_5CH_2OCH_3$. Frequently called capryl alcohol.
Properties: Colorless, oily liquid; refractive; aromatic odor. D 0.825 (15C), bp 178–179C, fp −38C, refr index 1.437 (20C), flash p 140F (60C). Miscible with alcohol, ether; slightly soluble in water. Combustible.
Derivation: By distilling sodium ricinoleate with an excess of sodium hydroxide.
Grade: Technical 92–99%, pure.
Use: Solvent, manufacture of plasticizers, wetting agents, foam-control agents, hydraulic oils, petroleum additives, perfume intermediates, masking of industrial odors.

n-octyl alcohol, primary. (1-n-octanol; alcohol C-8; heptyl carbinol).
CAS: 111-87-5. $CH_3(CH_2)_6CH_2OH$. In industrial practice, the term *octyl alcohol* has been used for both 1-octanol and 2-ethylhexanol. The latter is also sometimes called isooctanol. The term *capryl alcohol* has been used for both 1-octanol and 2-octanol. It therefore seems preferable to designate the normal primary alcohol as 1-n-octanol.
Properties: Colorless liquid; penetrating aromatic odor. D 0.826 (20C), bp 194–195C, 108.7C (30 mm Hg), refr index 1.430 (20C), fp −16C, flash p 178F (81.1C). Miscible with alcohol, chloroform, mineral oil; immiscible with water and glycerol. Combustible.
Derivation: By reduction of caprylic acid.
Grade: Technical, CP, pure, perfume, FCC.
Use: Perfumery, cosmetics, organic synthesis, solvent manufacture of high-boiling esters, antifoaming agent, flavoring agent.

octyl aldehyde. See octanal and 2-ethylhexaldehyde.

octylamine. $CH_3(CH_2)_7NH_2$.
Properties: Water-white liquid; amine odor. Boiling range 170–179C, d 0.779 (20/20C), refr index 1.431 (20C), flash p 140F (60C). Combustible.

tert-octylamine. $(CH_3)_3CCH_2C(CH_3)_2NH_2$.
Properties: Colorless liquid; amine odor. Bp 137–143C, d 0.771 (25C), refr index 1.423 (25C), flash p 92F (33.3C) (OC). Insoluble in water; soluble in common organic solvents, especially petroleum hydrocarbons.
Hazard: Flammable, moderate fire risk. Skin irritant.
Use: Intermediate for rubber accelerators, insecticides, fungicides, dyestuffs, pharmaceuticals.

octylammonium methylarsonate.
CAS: 6379-37-9. $C_8H_{19}N \cdot CH_5AsO_3$.
Hazard: Moderately toxic by ingestion.
Use: Agricultural chemical.

n-octylbicycloheptene dicarboximide.
(n-(2-ethylhexyl)-bicyclo(2,2,1)-5-heptene-2,3-dicarboximide). $C_8H_{17}NC_9H_8O_2$.
Properties: Liquid. Bp 158C (2 mm Hg), d 1.05 (18C), refr index 1.505 (20C). Miscible with most organic solvents and oils. Combustible.
Derivation: From maleic anhydride, cyclopentadiene, and 2-ethylhexylamine.
Hazard: By ingestion and skin absorption.
Use: Insecticide and pesticide synergist.

n-octyl bromide. (capryl bromide; caprylic bromide).
CAS: 111-83-1. $CH_3(CH_2)_6CH_2Br$.
Properties: Colorless liquid. D 1.118 (15C), bp 202C, fp −55C, refr index 1.4503 (25C). Miscible with alcohol and ether; immiscible with water.
Grade: Technical.
Use: Synthesis of quaternary ammonium compounds, organometallics, vinyl stabilizers.

octyl carbinol. See nonyl alcohol.

n-octyl chloride. $CH_3(CH_2)_6CH_2Cl$.
Properties: Colorless liquid. D 0.8697 (25/15.5C), refr index 1.4288 (25C), fp −62C, bp 181.6C, flash p 158F (70C). Soluble in most organic solvents. Combustible.
Use: Chemical intermediate, manufacture of organometallics.

n-octyl-n-decyl adipate. (NODA).
Properties: Liquid; mild odor. D 0.92–0.98 (20/20C), fp −50C, boiling range 220–254C (4 mm Hg), refr index 1.447. Combustible.
Use: Low-temperature plasticizer.
See Staflex Noda.

n-octyl-decyl alcohol. A blend of alcohols; available in tank cars and trucks. Combustible.
Use: Intermediate for plasticizers.

n-octyl-n-decyl phthalate.
CAS: 119-07-3. $C_{26}H_{42}O_4$.
Properties: Clear liquid; mild characteristic odor. D 0.972–0.976 (20/20C), fp −40C, boiling range 232–267C (4 mm Hg), refr index 1.482 (25C), flash p 455F (235C). Combustible.
Use: Plasticizer for vinyl resins.

octyl 2,4-dichlorophenoxyacetate.
CAS: 1928-44-5. $C_{16}H_{22}Cl_2O_3$.
Hazard: Moderately toxic by ingestion and skin contact.
Use: Agricultural chemical.

octylene.   See octene.

octylene glycol titanate.
Properties: Light-yellow solid.
Derivation: Reaction of butyl titanate with octylene glycol.
Use: Cross-linking agent, surface-active agent. See titanium chelate.

octylene oxide.

Mixed $CH_3(CH_2)_5CHCH_2O$

and $CH_3(CH_2)_4CHCH(CH_3)O$. D (liquid) 0.830 (25C).

Combustible.
Use: Organic intermediate, epoxy resins.

octyl formate.
CAS: 112-32-3. $C_8H_{17}OOCH$.
Properties: Colorless liquid; fruity odor. D 0.869–0.872 (25C), refr index 1.4180–1.4200 (20C). Soluble in mineral oil; practically insoluble in glycerol, 1 mL dissolves in 5 mL of 70% alcohol. Combustible.
Grade: FCC.
Use: Flavoring agent.

octylic acid.   See caprylic acid.

2-octyl iodide.   (caprylic iodide; secondary capryl iodide). $CH_3(CH_2)_5CHICH_3$.
Properties: Oily liquid. D 1.318 (18C), bp 210C (decomposes).
Hazard: By ingestion and inhalation. Keep away from light and air.
Use: Organic synthesis.

octylmagnesium chloride. $C_8H_{17}MgCl$. A Grignard reagent available in tetrahydrofuran solution.

n-octyl mercaptan.
CAS: 111-88-6. $C_8H_{17}SH$.
Properties: Water-white liquid; mild odor. Bp 199C, d 0.8395 (25/4C), refr index 1.4497 (25C), flash p 115F (46.1C) (OC). Combustible.
Hazard: Moderate fire risk.
Use: Polymerization conditioner, synthesis.

tert-octyl mercaptan. $C_8H_{17}SH$.
Properties: Colorless liquid. Boiling range 154–166C, d 0.848 (15.5C), refr index 1.454 (20C), flash p 105F (40.5C). Combustible.
Grade: 95%.

Hazard: Moderate fire risk.
Use: Polymer modification, lubricant additive.

n-octyl methacrylate.
$H_2C:C(CH_3)COOC_8H_{17}$.
Properties: Water-insoluble, colorless liquid. Polymerizes to a resin if unstabilized.

octyl peroxide.   (caprylyl peroxide).
Properties: Straw-colored liquid; sharp odor. Immiscible with water.
Hazard: Dangerous fire risk, strong oxidizing agent.

octyl phenol.   (diisobutyl phenol).
$C_8H_{17}C_6H_4OH$. Probably a mixture of isomers.
Properties: White flakes. Mp 72–74C, d 0.89 (90C), bp 280–302C, hydroxyl coefficient 259–275. Insoluble in hot and cold water; limited solubility in alkalies; soluble in 1:1 mixture of methanol and 50% aqueous potassium hydroxide, also in alcohol, acetone, fixed oils. Combustible.
Derivation: (p-tert-isomer) Catalytic alkylation of phenol with olefins.
Use: Nonionic surfactants, plasticizers, antioxidants, fuel oil stabilizer, intermediate for resins, fungicides, bactericides, dyestuffs, adhesives, rubber chemicals.

4-octylphenol.
CAS: 1806-26-4. $C_{14}H_{22}O$.
Hazard: A reproductive hazard.

p-tert-octylphenoxy polyethoxyethanol.
CAS: 9002-93-1. $(CH_3)_3CCH_2C(CH_3)_2C_6H_4O$ $(CH_2CH_2O)xH$. Anhydrous liquid mixture of mono-p-(1,1,3,3-tetramethylbutyl)phenyl esters of polyethylene glycols in which x varies from 5 to 15.
Properties: Yellow, viscous liquid; faint odor; bitter taste. D 1.060, refr index 1.489 (25C). Soluble in water, alcohol, acetone, benzene, toluene; insoluble in hexane; pH 7–9. Combustible.
Grade: NF.
Use: Food packaging, probably as a plasticizer for films.

p-octylphenyl salicylate.
$C_6H_4OHCOOC_6H_4C_8H_{17}$.
Properties: White solid. Mp 72–74C, bulk d 5.6 lb/gal (20C). Soluble in hexane, benzene, acetone, and ethanol; insoluble in water.
Use: Prevents photooxidation in polyethylene and polypropylene.

octyl phosphate.   See trioctyl phosphate.

n-octyl sulfoxide isosafrole.   See sulfoxide.

octyl trichlorosilane.   (trichlorooctylsilane).
CAS: 5283-66-9. $C_8H_{17}SiCl_3$.

Properties: Water-white liquid; pungent irritating odor. Fumes readily in moist air to evolve corrosive vapors.
Hazard: Moderate fire risk in contact with oxidizing materials. Toxic by ingestion and inhalation, strong irritant to skin and mucous membranes.
Use: Intermediate for silicones.

octyltrichlorostannane.
CAS: 3091-25-6. $C_8H_{17}Cl_3Sn$.
Hazard: Mildly toxic by ingestion.

octyltris(2-ethylhexyloxycarbonylmethylthio) stannane.
CAS: 27107-89-7. $C_{38}H_{74}O_6S_3Sn$.
Hazard: Moderately toxic by ingestion.

odor. An important property of many substances, manifested by a physiological sensation caused by contact of their molecules with the olfactory nervous system. Odor and flavor are closely related, and both are profoundly affected by submicrogram amounts of volatile compounds. Attempts to correlate odor with chemical structure have produced no definitive results. Objective measurement techniques involving chromatography are under development. Even potent odors must be present in a concentration of 1.7×10^7 molecules/cc to be detected. It has been authentically stated that the nose is 100 times as sensitive in detection of threshold odor values as the best analytical apparatus. Many compounds have a characteristic odor that is an effective means of identification. Toxic and noxious gases have distinctive odors often utilized for warning purposes. An important exception is carbon monoxide, which is almost odorless. The penetrating, banana-like odor of amyl acetate has been used in mine rescue work. Among the most powerful unpleasant odors are those of organic sulfur compounds, especially ethyl mercaptan (skunk). Organic substances having a pleasant odor are broadly designated as aromatic, regardless of chemical nature. The cyclic aromatic (benzene) series of hydrocarbons was so named for this reason. Most essential oils have a pleasant odor and are the basis of perfumes and fragrances. Odor research, including evaluation by test panels, is conducted at the Olfactronics and Odor Sciences Center at Illinois Institute of Technology, Chicago.

odorant. A substance having a distinctive, sometimes unpleasant odor that is deliberately added to essentially odorless materials to provide warning of their presence. For example, mercaptan derivatives may be added to natural gas for this purpose. In a broad sense, perfumes are odorants that are added to cosmetics, toilet goods, etc., largely for consumer appeal.
See odor; deodorant; perfume; fragrance.

odorless light petroleum hydrocarbons.
Properties: Liquid; faint odor. Bp: 300–650F.

Use: Coating agent; defoamer; float; food additive; froth-flotation cleaning; insecticide formulations component.

odor masking. Addition of a substance with strong odor to obtain a less-offensive effect without changing the composition of the original odorous substance.

ODPN. Abbreviation for β,β′-oxydipropionitrile.

Oenanthic acid. See n-heptanoic acid.

oenology. See enology.

OFA. TM for chemical foaming agent for additives.
Use: To create foam or tiny bubbles in plastic to make products lighter and use less resin.

oil. The word *oil* is applied to a wide range of substances that are quite different in chemical nature. Oils derived from animals or from plant seeds or nuts are chemically identical with fats, the only difference being one of consistency at room temperature. They are composed largely of glycerides of the fatty acids, chiefly oleic, palmitic, stearic, and linolenic. As a rule the more hydrogen the molecule contains, the thicker the oil becomes. Petroleum (rock oil) is a hydrocarbon mixture comprising hundreds of chemical compounds. It is thought to be derived from the remains of tiny sea animals laid down in past geologic ages. Following is a classification of oils by type and function.

I. Mineral
  (1) Petroleum
    (a) Aliphatic or wax base (Pennsylvania)
    (b) Aromatic or asphalt base (California)
    (c) Mixed-base (midcontinent of U.S.)
  (2) Petroleum derived
    (a) Lubricants: engine oil, machine oil, cutting oil
    (b) Medicinal: refined paraffin oil
II. Vegetable (chiefly from seeds or nuts)
  (1) Drying (linseed, tung, oiticica)
  (2) Semidrying (soybean, cottonseed)
  (3) Nondrying (castor, coconut)
  (4) Inedible soap stocks (palm, coconut)
III. Animal
  These usually occur as fats (tallow lard, stearic acid). The liquid types include fish oils, fish-liver oils, oleic acid, sperm oil, etc. They usually have a high fatty acid content.
IV. Essential
  Complex volatile liquids derived from flowers, stems, and leaves, and often the entire plant. They contain terpenes (pinene, dipentene, etc.) and are used chiefly for perfumery and flavorings. Usually resinous products are admixed with them. Turpentine is a highly resinous essential oil.

V. Edible
Derived from fruits or seeds of plants and used chiefly in foodstuffs (margarine, etc.). Most common are corn, coconut, soybean, olive, cottonseed, and safflower. They have varying degrees of saturation.

oil black. A carbon black made from oil, usually an aromatic-type petroleum oil.
See furnace black.

oil blue. Violet-blue copper sulfide pigment used in varnishes.

oil cake. The residue obtained after the expression of vegetable oils from oil-bearing seeds, used as cattle feed and fertilizer. When ground they are known as meal.
See cottonseed cake and meal; peanut oil meal.

oil gas. A gas made by the reaction of steam at high temperature on gas oil or similar fractions of petroleum, or by high-temperature cracking of gas oil. One typical analysis is heating value 554 Btu/ft^3, illuminants 4.2%, carbon monoxide 10.4%, hydrogen 47.6%, methane 27.0%, carbon dioxide 4.6%, oxygen 0.4%, nitrogen 5.8%, autoign temp 637F (336C).
Hazard: Flammable, dangerous fire and explosion risk. Toxic by inhalation.

oiliness. That property of a lubricant that causes a difference in coefficient of friction when all the known factors except the lubricant itself are the same. This concept is also expressed by the term *lubricity*.

oil of bitter almond. See almond oil.

oil of cinnamon, ceylon. See cinnamon leaf oil.

oil of cubeb. See cubeb oil.

oil of garlic. See garlic oil.

oil of jasmine.
CAS: 8022-96-6.
Hazard: Low toxicity by ingestion.
Source: Natural product.

oil of lime oil, coldpressed.
Properties: Expressed from the peel of *Citrus aurantifolia Swingle* (Mexican type) or *Citrus latifolia* (Tahitian type). Yellow to brown-green liquid. Soluble in fixed oils, mineral oil; insoluble glycerin, propylene glycol.
Use: Food additive.

oil of mirbane. See nitrobenzene.

oil of muscatel. See clary oil.

oil of vitriol. See sulfuric acid.

oil of wintergreen. See methyl salicylate.

oil sands. (tar sands). Porous sandstone structures occurring on the surface and to depths of 100 m or more in certain localities; they contain a high proportion of bitumen composed chiefly of asphaltenes and maltha, together with substantial percentage of sulfur and heavy metals. Its viscosity is about midway between that of crude oil and soft asphalt. The largest deposit in North America is in the Athabasca region of Alberta; there are smaller ones in the western U.S. Venezuela and Trinidad have large deposits. The Athabasca sands have been successfully mined and have made a substantial contribution to Canadian energy resources over the past decade.

oil shale. Extensive sedimentary rock deposits in the mountains of Colorado, Utah, and Wyoming contain a high percentage of kerogen, which can be separated from the shale either by heating in retorts (surface mining) or by direct combustion in situ in interior excavations. The deposits range in thickness from 10 to 800 ft and yield from 25 to 30 gal oil/ton shale. Only 33% of the oil content is recoverable by present techniques.
See shale oil; kerogen.

oil varnish. See varnish.

oil, vulcanized. See factice.

oil white. One of several mixtures of lithopone and white lead or zinc white. It may also contain gypsum, magnesia, whiting, or silica.
Use: White-lead substitute.

ointment. (salve). A semisolid pharmaceutical preparation based on a fatty material such as lanolin and often containing petrolatum or zinc oxide together with specific medication for relief of rashes and other forms of dermatitis.

oiticica oil.
Derivation: By expression from the seeds of the Brazilian oiticica tree, *Licania rigida*.
Chief constituents: Glycerides of α-licanic acid (4-keto-9,11,13-octadecatrienoic acid).
Use: Drying oil in paints, varnishes, etc.

okazaki fragment.
A short segment of single-stranded DNA that is an intermediate in DNA synthesis. In bacteria, such fragments are 1000–2000 bases in length. In eukaryotes, the fragments are 100–200 bases in length.

"Okerin" *[International].* TM for rubber waxes and paraffin products.
Available forms: Flake, prill, or slab.
Use: To provide controlled migration for ozone protection of rubber.

-ol.   A suffix indicating that one or more hydroxyl groups (OH) are present in an organic compound, e.g., alcohol, phenol, menthol. Thiol is an exception, the oxygen of the OH group being replaced by sulfur. There are a few other exceptions among the essential oils, for example, eucalyptol.

Olah, George A.   (1927–). Born in Hungary, now an American citizen, he won the Nobel Prize for Chemistry in 1994 for his work with carbocations. These are positively charged hydrocarbons with lifetimes on the order of microseconds. Olah developed methods of studying carbocations with different physical techniques, changing the direction of this field. He received a Ph.D. from the Technical University of Budapest in 1949.

olanzapine.   See zyprexa.

olealkonium chloride.
CAS: 37139-99-4. $C_{27}H_{48}N \cdot Cl$.
Hazard: A severe skin and eye irritant.

oleamide. cis-$CH_3(CH_2)_7CH:CH(CH_2)_7CONH_2$.
Properties: Ivory-colored powder. Mp 72C, d 0.94. Combustible.
Grade: Refined.
Use: Slip agent for extrusion of polyethylene, wax additive, ink additive.

oleandomycin.
CAS: 3922-90-5. $C_{35}H_{61}NO_{12}$.
Hazard: Low toxicity by ingestion.

oleate.   Salt made up of a metal or alkaloid with oleic acid. It is used for external medications and in soaps and paints.

olefin.   (alkene). A class of unsaturated aliphatic hydrocarbons having one or more double bonds, obtained by cracking naphtha or other petroleum fractions at high temperatures (1500–1700F). Those containing one double bond are called alkenes, and those with two are called alkadienes, or diolefins. They are named after the corresponding paraffins by adding -ene or -ylene to the stem. α-olefins are particularly reactive because the double bond is on the first carbon. Examples are 1-octene and 1-octadecene, which are used as the starting point for medium-biodegradable surfactants. Other olefins (ethylene, propylene, etc.) are starting points for certain manufactured fibers.
See diolefin.

olefin fiber.   Synthetic long-chain polymer fiber composed of at least 85% by weight of ethylene, propylene, or other crystalline polyolefins.

olefinic alcohol.   (allyl alcohol; prop-2-en-1-ol).
CAS: 107-18-6. C_3H_6O. An alcohol in which the hydroxyl group is attached to an unsaturated hydrocarbon skeleton.

oleic acid.   (cis-9-octadecenoic acid; red oil).
CAS: 112-80-1. $CH_3(CH_2)_7CH:CH(CH_2)_7COOH$.
A monounsaturated fatty acid, it is a component of almost all natural fats, as well as tall oil. Most oleic acid is derived from animal tallow or vegetable oils.
Properties: (Commercial grades) Yellow to red oily liquid; lard-like odor, darkens on exposure to air. Insoluble in water; soluble in alcohol, ether, and most organic solvents, fixed and volatile oils. Solvent for other oils, fatty acids and oil-soluble materials. (Purified grades) Water-white liquid. D 0.895 (20/4C), fp 4C, bp 286C (100 mm Hg), 225C (10 mm Hg), refr index 1.4599 (20C), acid value 196–204, iodine value 83–103, saponification value 196–206, flash p 372 (189C), Combustible.
Derivation: The free fatty acid is obtained from the glyceride by hydrolysis, steam distillation, and separation by crystallization or solvent extraction. Filtration from the press cake results in the oleic acid of commerce (red oil), which is purified and bleached for specific uses.
Grade: Variety of technical grades, grade free from chick edema factor, USP, FCC, 99+%. A purified technical oleic acid containing 90% or more oleic, 4% maximum linoleic, and 6% maximum saturated acids is available.
Use: Soap base, manufacture of oleates, ointments, cosmetics, polishing compounds, lubricants, ore flotation, intermediate, surface coatings, food-grade additives.

oleic acid diethanolamide.
CAS: 93-83-4. $C_{22}H_{43}NO_3$.
Hazard: A poison by ingestion. A moderate skin irritant.

olein.   (triolein; glyceryl trioleate).
$C_{17}H_{33}COO)_3C_3H_5$. The triglyceride of oleic acid, occurring in most fats and oils. It constitutes 70–80% of olive oil.
Properties: Yellow, oily liquid. D 0.915, mp −4 to −5C. Soluble in chloroform, ether, carbon tetrachloride; slightly soluble in alcohol. Combustible.
Impurities: Stearin, linolein.
Derivation: Refined natural oils.
Use: Textile lubricants.

oleoresin.   Any of a number of mixtures of essential oils and resins characteristic of the tree or plant from which they are derived. Most types are semisolid and tacky at room temperature, becoming soft and sticky at high temperatures. They have various distinctive odors.
See balsam; rosin.

oleoyl chloride.   (cis-9-octadecenoyl chloride).
$CH_3(CH_2)_7CH:CH(CH_2)_7COCl$.
Properties: Liquid. Bp 175–180C (3 mm Hg). Soluble in hydrocarbons and ethers; reacts slowly with water. Combustible.
Use: Chemical intermediate.

n-oleoylsarcosine.
$C_{17}H_{33}C(O)N(CH_3)CH_2COOH$.
Properties: Amber liquid. D 0.955 (20/20C), refr index 1.4703 (20C), 95% pure. Combustible.
Use: Surfactants.

oleum. The Latin word for oil, applied to fuming sulfuric acid. (Sulfuric acid was originally called oil of vitriol.)

oleyl alcohol. (octadecenol).
CAS: 143-28-2. $CH_3(CH_2)_7CH:CH(CH_2)_7CH_2OH$. The unsaturated alcohol derived from oleic acid.
Properties: Clear, viscous liquid at room temperature. Iodine value 88, cloud p −6.6C, bp 333C, fp −75C, d 0.84. Combustible.
Impurities: Linoleyl, myristyl, and cetyl alcohols.
Derivation: Reduction of oleic acid, occurs in fish and marine mammal oils.
Grade: Technical, commercial (80–90% pure).
Use: Surfactants, metal cutting oils, printing inks, textile finishing, antifoam agent, plasticizer.

oleyl aldehyde. See octadecenyl aldehyde.

oleylhydroxamic acid. $C_{17}H_{33}CONH_2O$.
Properties: Waxy solid, off-white color. D 0.897 (70/25C). Insoluble in water; soluble in aqueous potassium hydroxide and organic solvents.

oleyl-linoleylamine. (octadecene-octadecadieneamine).
Properties: Highly unsaturated primary amine. D 0.83, mp 19C, bp 198–209C, amine no. 200–210, iodine value 90 min. Soluble in many organic solvents; insoluble in water.
Use: Organic intermediate.

oleyl methyl tauride. See sodium-N-methyl-N-oleoyl taurate.

olibanum oil.
CAS: 8050-07-5.
Properties: Distilled from a gum from the trees *Boswellia carterii* and other *Boswellia* species (Fam. Burseraceae). Pale liquid; pleasant balsamic odor. D: 0.862–0.889, refr index: 1.465–1.482 @ 20C. Soluble in fixed oils, mineral oil; insoluble in glycerin, propylene glycol.
Hazard: A skin irritant.
Use: Food additive.

Oligga-Fiber. Inulin recognized for its nutritional and functional benefits in food.
Use: For softer, long shelf life.

oligo-. A prefix meaning "a few" or "very little."

oligodynamic. Literally, active in small amounts. In technical literature, the term describes the sterilizing or purifying action of a substance, for example, silver.

oligogenic. A phenotypic trait produced by two or more genes working together.
See polygenic disorder.

oligomer. A polymer molecule consisting of only a few monomer units (dimer, trimer, tetramer).

oligomeric protein. A multisubunit protein having two or more polypeptide chains.

oligomycin C.
CAS: 11052-72-5. $C_{28}H_{46}O_6$.
Hazard: A poison.
Source: Natural product.

oligonucleotide. A molecule usually composed of 25 or fewer nucleotides; used as a DNA synthesis primer.
See nucleotide.

oligopeptide. A peptide made up of not more than 10 amino acids.

oligosaccharide. A carbohydrate containing from two to ten simple sugars linked together (e.g., sucrose, composed of dextrose and fructose). Beyond ten they are called polysaccharides.

olive oil.
CAS: 8001-25-0.
Properties: Pale yellow or greenish-yellow liquid; slight odor and taste. D 0.910–0.918, saponification value 188-196, iodine value 77–88, flash p 437F (225C), cloud p −6.6 to −1.1C. A nondrying oil. Soluble in ether, chloroform, and carbon disulfide; sparingly soluble in alcohol. Combustible.
Use: Salad dressings and other foods; ointments, liniments, etc.; Castile soap, special textile soaps; lubricant; sulfonated oils; cosmetics.

olivetol. (5-pentylresorcinol). $CH_3(CH_2)_4C_6H_3$-1,3-$(OH)_2$. Store under nitrogen.
Properties: Off-white solid. Mw 180.25, mp 42–44, flash p above 110C.
Hazard: Skin irritant; combustible.
Use: Pharmaceutical intermediate.

olivine. (chrysolite). $(Mg,Fe)_2SiO_4$. Natural magnesium-iron silicate, found in igneous and metamorphic rocks, meteorites, and blast furnace slags. A complete series exists from Fe_2SiO_4 to Mg_2SiO.
Grade: Crude, 20 mesh, 100 mesh.
Use: Refractories, cements.

"Omadine" [Arch]. TM for a series of derivatives of pyridinethione (such as 1-hydroxy-2-pyridinethione, $C_5H_4NOH(S)$) having bactericide-fungicide properties.
Use: Cosmetics, textiles, cutting oils, coolant systems, vinyl films, and rubber products.

OMC.   Abbreviation of oxidized microcrystalline waxes.

Omega Gold.   An omega 3 rich ground flax seed, premix supplement.
Use: To improve swine health, gestation, lactation, and breeding.

Omega Horseshine.   An omega 3 stabilized flax supplement.
Use: To maintain healthy coat, solid hooves, and long life.

Omega Ultra Egg.   An omega 3 rich stabilized flax seed premix.
Use: Food suppement for laying hens.

OMPA.   Abbreviation for octamethyl pyrophosphoramide.
See schradan.

ONB.   Abbreviation for *o*-nitrobiphenyl.

oncogen.   Any substance that will cause either benign or malignant tumors in test animals. EPA pesticide regulations use this term instead of *tumorogenic* and *carcinogenic*.

oncogene.   A gene of cellular or viral origin that causes cells to exhibit rapid, uncontrolled proliferation.
See also proto-oncogene.

oncogene protein.   Any protein that is coded by an oncogene or by an oncogene fused to another gene.

one-color indicator.   An indicator that loses its color at the end point of a titration.

one-step resin.   See A-stage resin.

Onsager, Lars.   (1903–1976). A Norwegian chemist who won the Nobel Prize for chemistry in 1968. He studied and wrote on the theory of electrolytic conduction and theory of dielectrics. He also worked with superfluids and crystal statistics and reciprocal relations in irreversible processes. After receiving his doctorate in Norway, he came to the U.S. and became a citizen.

Onyx Classica. $Al_2O_3 \cdot 3H_2O$. A filler; flame-retardant in the premium products of the synthetic marble industry.
Properties: Exceptionally high purity, whiteness, and particle size consistency.

opacity.   The optical density of material, usually a pigment; the opposite of transparency. A colorant or paint of high opacity is said to have good hiding power or covering power, by which is meant its ability to conceal another tint or shade over which it is applied. Apparatus for measuring opacity is available.

OPDN.   Abbreviation for β,β-oxydipropionitrile.

open reading frame.   (ORF). The sequence of DNA or RNA located between the start-code sequence (initiation codon) and the stop-code sequence (termination codon).

open system.   A system that exchanges matter and energy with its surroundings.

operation.   See unit operation.

operations research.   The use of statistical and other mathematical methods for studying and evaluating the best procedures to be used for carrying out a particular operation and all functions associated with it.

operator.   Biology — A region of DNA that interacts with a regulatory protein to control the expression of a gene or group of genes.

operon.   A unit of genetic expression in prokaryotes consisting of one or more related genes, and the operator and promoter sequences that regulate their transcription.

OPG.   Abbreviation for oxypolygelatin.

opium.   A mixture of alkaloids.
Derivation: The air-dried, milky exudate obtained from the unripe capsules of *Papaver somniferum*.
Available forms: Deodorized, granulated, powdered.
Hazard: A habituating narcotic; importation and sale restricted by law in U.S.
Use: Source of morphine.

"Oppanol" *[BASF]*.   TM for a series of polyisobutylenes, varying from oily liquids through highly viscous materials to rubberlike solids according to degree of polymerization.

Oppenauer oxidation.   The aluminum alkoxide-catalyzed oxidation of a secondary alcohol to the corresponding ketone (the reverse of the Meerwein-Ponndorf-Verley reduction).

Optanol drops.   A liquid drug.
Use: For treatment of symptoms of allergic conjunctivitis.

optical activity.   See optical rotation.

optical bleach.   See brightening agent.

optical brightener.   (optical bleach; colorless dye; fluorescent brightener). A colorless, fluorescent, organic compound that absorbs UV light and emits it as visible blue light. The blue light masks the undesirable yellow of textiles, paper, detergents, and plastics. Some examples are derivatives of 4,4'-diaminostilbene-2,2'-disulfonic acid, coumarin derivatives such as 4-methyl-7-diethylaminocoumarin.

optical crystal.   A comparatively large crystal, either natural or synthetic, used for infrared and ultraviolet optics, piezoelectric effects, and short-wave radiation detection. Examples are sodium chloride, potassium iodide, silver chloride, calcium fluoride, and (for scintillation counters) such organic materials as anthracene, naphthalene, stilbene, and terphenyl.

optical fiber.   See fiber, optical.

optical glass.   See glass, optical.

optical isomer.   Either of two kinds of optically active three-dimensional isomers (stereoisomers). One kind is represented by mirror-image structures called enantiomers that result from the presence of one or more asymmetric carbon atoms in the compound (glyceraldehyde, lactic acid, sugars, tartaric acid, amino acids). The other kind is exemplified by diastereoisomers, which are not mirror images. These occur in compounds having two or more asymmetric carbon atoms; thus, such compounds have 2_n optical isomers, where n is the number of asymmetric carbon atoms.
See enantiomer; diastereoisomer; optical rotation; asymmetry.

optical microscope.   (light microscope). A magnifying lens system that utilizes light in the visible wavelength range of the electromagnetic spectrum (5000 Å). A convex glass lens bends or focuses light waves because of the difference in density between glass and air. Invented in 1590 by the Janssen brothers and later improved by van Leeuwenhoek, the compound microscope has three lenses: a condenser lens, which concentrates the incident light; an objective lens, which gives an enlarged reverse image of the specimen; and a projector lens, which further enlarges the image and returns it to normal position. Its maximum resolving power is 0.5 micron, compared with 100 microns for the human eye. The compound microscope is particularly useful in studying bacteria and other microorganisms in their natural state without interfering with their behavior. It has been of untold benefit to biologists and bacteriologists and also has innumerable uses in chemical and metallurgical research, as well as in forensic chemistry.
See resolving power; electron microscope; ultramicroscope; field-ion microscope; van Leeuwenhoek, Anton.

optical rotation.   The change of direction of the plane of polarized light to either the right or the left as it passes through a molecule containing one or more asymmetric carbon atoms, for example, sugars. The direction of rotation, if to the right, is indicated by either a plus sign (+) or a d-; if to the left, by a minus sign (−) or an l-. Molecules having a right-handed configuration (D) usually are dextrorotatory, D(+), though they may be levorotatory, D(−); those having a left-handed configuration (L) are usually levorotatory, L(−), but may be dextrorotatory, D(+). Compounds having this property are said to be optically active and are isomeric. The amount of rotation varies with the compound but is the same for any two isomers, though in opposite directions.
See optical isomer; nicol.

optical spectroscopy.   See spectroscopy; absorption spectroscopy; emission spectroscopy.

"Optidose 2000" [Catalent].   TM for a carboxylic sulfonated copolymer.
Use: For water treatment and traceable scale inhibitor for phosphates and zinc.

optimax.
CAS: 62602-94-2. $C_{11}H_{12}N_2O_2 \cdot C_8H_{11}NO_3 \cdot C_6H_8O_6 \cdot ClH$.
Hazard: Human systemic effects.

Optimer.   Flocculating chemicals.
Use: Dewatering of sludge.

Optimum.   Guanidine hydrochloride and guanidine thiocyanate.
Available forms: Solid.
Grade: Bio-tech.

optimum pH.   The characteristic pH at which an enzyme has maximal catalytic activity.

optrode.   A component of fiber-optical analytical systems that is analogous to an electrode. Its function is to couple the laser beam to the sample solution being analyzed. There are three types: cuvette, sapphire ball, and membrane.

Oracle, 4E Dicamba Herbicide.   A broad spectrum control for turf weeds.

oral contraceptive.   See antifertility agent.

orange B.
CAS: 15139-76-1. $C_{22}H_{16}N_4Na_2O_9S_2$.
Properties: Dull orange crystals.
Use: Food additive.

orange cadmium.   See cadmium sulfide.

orange III.   See methyl orange.

orange IV.   See tropaeolin 00.

orange mineral.   A red lead oxide pigment made in a furnace by roasting lead carbonate or sublimed litharge; it is a very bright orange but has low tinting strength.
Properties: Fine powder, −325 mesh. D 9.0, contains 95.5% red lead (Pb_3O_4).
Use: Pigment in printing inks and primers.

orange oil, bitter, coldpressed.
Properties: Oil expressed from the peel of *Citrus aurantium* L. Osbeck (Fam. Rutaceae). Pale yellow to yellow-brown liquid; characteristic orange odor and bitter taste. D: 0.845–0.851, refr index: 1.472 @ 20C. Misc in abs alc, in 1 vol glacial acetic acid; soluble in fixed oils, mineral oil; insoluble in glycerin.
Use: Food additive.

orange oil, distilled.
Properties: From steam distillation of fresh peel of *Citrus sinensis* L. Osbeck (Fam. Rutaceae). Colorless to pale yellow liquid; odor of fresh orange peel. Soluble in fixed oils, mineral oil, alcohol; insoluble in glycerin, propylene glycol.
Use: Food additive.

orange peel.   A term used in the paint industry to refer to a roughened film surface caused by too rapid drying.

orange peel oil.   See citrus peel oils.

orange seed oil.   See citrus seed oils.

orange toner.   A diazo dyestuff coupled to diacetoacetic acid anhydride. It contains no sulfonic or carboxylic groups.
Use: Printing inks.

"Orasol" [BASF].   TM for solvent dyes for transparent coatings, inks, and plastics.

orbital.   The area in space about an atom or molecule in which the probability of finding an electron is greatest.

orbital electron.   Electron remaining with high degree of probability in the immediate neighborhood of a nucleus where it occupies a quantized orbital.

orbital theory.   The quantum theory of matter applied to the nature and behavior of the electron either in a single atom (atomic orbital) or combined atoms (molecular orbital). A combination of Schrodinger's wave mechanics and Heisenberg's uncertainty principle, the orbital theory was formulated in 1926. It has yielded a better understanding of the electron and its critical part in chemical bonding than is possible with Newtonian mechanics. In simple language, the orbital theory considers the electron not as a particle but as a three-dimensional wave that can exist at several energy levels; its exact location and position in the "shell" (which in most elements is a group of orbitals) cannot be precisely determined but only predicted by the laws of mathematical probability. The orbital levels and the movement of electrons within them are expressed by wave functions and quantum numbers. The probability that an electron will be found in a given volume (i.e., the square of the one-electron wave function) is called the orbital of that electron, and the shape of the orbital is defined by surfaces of constant probability (i.e., spheres and elliptically shaped doughnuts). The electron orbital, described in terms of probability, is like a cloud, with indefinite boundaries. The energy state of each electron is given by four quantum numbers that describe its principal level, its angular momentum, its magnetic moment, and its spin. This concept has exerted a profound effect on modern ideas about chemical bonding, transition metal complexes, semiconductors, and solid-state physics.

orcin.   (dihydroxytoluene; methylresorcinol; orcinol).
CAS: 504-15-4. $CH_3C_6H_3(OH)_2$.

Properties: White, crystalline prisms becoming red in air; intensely sweet, unpleasant taste. D 1.2895, mp (anhydrous) 107C, (hydrated) 56C; bp 287–290C. Soluble in water, alcohol, and ether.
Derivation: By fermentation of various species of lichens (*Roccella*) and extraction.
Use: Reagent for certain carbohydrates (beet sugar, lignin, pentoses, etc.).

order of magnitude.   A range of values applied to numbers, distances, dimensions, etc., that begins at any value and extends to 10 times that value; for example, 2 is of the same order of magnitude as any number between itself and 20; and 5 miles is of the same order of magnitude as any distance between 5 and 50 miles. The expression usually applies to extremely large or extremely small units, that is, the size ranges of atoms, molecules, colloidal particles, etc., or astronomical distances.

ore.   An aggregate of valuable minerals and gangue from which one or more metals can be extracted at a profit.

ore flotation.   See flotation.

organelle.   A portion of a cell having specific functions, distinctive chemical constituents, and characteristic morphology; it is a unit subsystem of a cell. Examples are mitochondria and chromosomes. Organelles are often closely associated with enzymes. The lysosome (an enzyme-bearing organelle) has been synthesized.

organic acid. An acid that is characterized by a carbon-atom backbone and one or more ionizable carboxyl groups. Major classes of organic acids include carboxylic acids, dicarboylic acids, aliphatic and aromatic fatty acids, and amino acids.

organic anhydride. (acyl anhydride). An acid anhydride of an organic compound that contains the carbonyl group.
Properties: Highly reactive; yield two carboxy groups on hydrolysis.
Derivation: Produced by dehydrating acids and by reacting an acyl halide with the sodium salt of the acid.
Use: In acylation reactions.

organic chemistry. A major branch of chemistry that embraces all compounds of carbon except such binary compounds as the carbon oxides, carbides, carbon disulfide, etc.; such ternary compounds as the metallic cyanides, metallic carbonyls, phosgene ($COCl_2$), carbonyl sulfide (COS), etc.; and the metallic carbonates, such as calcium carbonate and sodium carbonate. The total number of organic compounds is indeterminate, but some 6,000,000 have been identified and named. These fall into several structural groups as follows:

I. Aliphatic (straight chain)
  (1) Hydrocarbons (petroleum and coal derived)
    (a) Paraffins or alkanes (saturated) (C_nH_{2n+2}) methane and homologs; halogen-substituted derivatives, e.g., fluorocarbons
    (b) Olefins (unsaturated)
      1. Alkenes (one double bond) (C_nH_{2n}) ethylene and homologs
      2. Alkadienes (two double bonds) (C_nH_{2n-1}) butadiene, allene
    (c) Acetylenes or alkynes (triple bond)
  (2) Alcohols (ROH): methanol and homologs
  (3) Ethers (ROR): methyl ether and homologs
  (4) Aldehydes (RCHO): formaldehyde, acetaldehyde
  (5) Ketones (RCOR): acetone, methylethylketone
  (6) Carboxylic acids (RCOOH)
  (7) Carbohydrates ($C_nH_{2n}O_n$)
    (a) Sugars: glucose, fructose, sucrose, gums
    (b) Starches: wheat, corn, potato
    (c) Cellulose: cotton, plant fibers
II. Cyclic (closed ring)
  (1) Alicyclic hydrocarbons (properties similar to aliphatics)
    (a) Cycloparaffins (naphthenes) (saturated): cyclohexane, cyclopentane, etc.
    (b) Cycloolefins (unsaturated): cyclopentadiene, cyclooctatetraene
    (c) Cycloacetylenes (triple bond)

  (2) Aromatic hydrocarbons (arenes): unsaturated compounds; hexagonal ring structure; single and multiple fused rings
    (a) Benzene group (1 ring)
    (b) Naphthalene group (2 rings)
    (c) Anthracene group (3 rings)
    (d) Polycyclic group (steroids, sterols)
  (3) Heterocyclic: unsaturated; usually pentagonal rings containing at least one other element besides carbon
    (a) Pyrroles
    (b) Furans
    (c) Thiazoles
    (d) Porphyrins
III. Combinations of aliphatic and cyclic structures
  (1) Terpene hydrocarbons
  (2) Amino acids (some are aliphatic and others combinations)
  (3) Proteins and nucleic acids (coiled or helical formations)
IV. Organometallic compounds
V. Synthetic high polymers, including silicones

Important areas of organic chemistry include polymerization, hydrogenation, isomerization, fermentation, photochemistry, and stereochemistry. There is no sharp dividing line between organic and inorganic chemistry; the two often tend to overlap. See inorganic chemistry.

organic disulfide.
Hazard: Toxic; may produce a contact dermatitis, hemolytic anemia.

"Organiclear" [Gilman]. (methylene bisthiocyanate).
CAS: 6317-18-6. TM for wood preservative.
Use: To stop wood rot, termite attack, mildew, and decay.

organo-arsenical herbicide. Any organoarsenical that is a principle ingredient in an herbicide.
Use: Herbicide.

organoborane. A compound composed of an unsaturated organic group and a borane obtained by the hydroboration reaction. Such compounds are useful catalytic reagents in organic syntheses of some complexity, e.g., *cis*- or *trans*-olefins, optically pure alcohols, alkanes, and ketones. Prostaglandins and insect pheromones have been synthesized by this means. A particularly versatile example is triphenylboron, $B(C_6H_5)_3$.
See hydroboration; carborane; borane.

organochlorine insecticide. (chlorinated organic insecticide). Any chlorinated hydrocarbon that is a principle ingredient in an insecticide.
Hazard: Cumulative poison; very toxic; carcinogenic; may cause vomiting; paresthesia and itching of lips, tongue, and face, headache, sore throat,

fatigue, tremors, ataxia, weakness, confusion, convulsions, coma, and death from respiratory failure.
Use: Insecticide.

organoclay. (organopolysilicate). A clay such as kaolin or montmorillonite, to which organic structures have been chemically bonded; since the surfaces of the clay particles, which have a lattice-like arrangement, are negatively charged, they are capable of binding organic radicals. When this type of structure is in turn reacted with a monomer such as styrene, a complex known as a polyorganosilicate graft polymer results.

organoleptic. A term widely used to describe consumer testing procedures for food products, perfumes, wines, and the like in which samples of various products, flavors, etc. are submitted to groups or panels. Such tests are a valuable aid in determining the acceptance of the products and thus may be viewed as a marketing technique. They also serve psychological purposes and are an important means of evaluating the subjective aspects of taste, odor, color, and related factors. The physical and chemical characteristics of foods are stimuli for the eye, ear, skin, nose, and mouth, whose receptors initiate impulses that travel to the brain, where perception occurs.

organomercurial pesticide. Any of a number of alkylated mercury compounds, aromatic mercury compounds, alkoxy-alkylmercury compounds that are prominent in the formulation of an pesticide.
Use: Pesticide.

organometallic compound. An organic compound composed of a metal attached directly to carbon (RM); such compounds have been prepared of practically all the metals, as well as with such nonmetals as silicon and phosphorus. Metallic salts (soaps) of organic acids are excluded. Examples are diethylzinc (the first known organometallic), Grignard compounds such as methyl magnesium iodide (CH_3MgI), and metallic alkyls such as butyllithium (C_4H_9Li), tetraethyllead, triethyl aluminum, tetrabutyl titanate, sodium methylate, copper phthalocyanine, and metallocenes. Some are highly toxic or flammable; others are coordination compounds.
Reactive and moderately reactive organometallic compounds will react with all functional groups; two major types of reaction in which they are involved are oxidation and cleavage by acids. Probably the most important organometallic reactions are those involving addition to an unsaturated linkage. Many of them are powerful catalysts and form useful coordination complexes.
See catalysis; metallocene; coordination compound.

organophosphate insecticide. (proganophosphorus insecticide). Organophosphate esters form a very large class of neurotoxic insecticides.

Hazard: Neurotoxic; causes restricted breathing, vomiting, diarrhea, twitching, cramps, restlessness, anxiety, emotional instability, headache, insomnia, respiratory and circulatory depression, convulsions, coma and death due to respiratory paralysis.
Use: Insecticide.

organophosphate warfare agent. (organophosphate military poison; organophosphorus warfare agent). Any of a number of organophosphate compounds developed for use as chemical warfare agents. They are potent inhibitors of acetylcholinesterase.
Hazard: Extremely toxic; deadly.

organophosphorus compound. Any organic compound containing phosphorus as a constituent. These fall into several groups, chief of which are the following: (1) phospholipids, or phosphatides, which are widely distributed in nature in the form of lecithin, certain proteins, and nucleic acids; (2) esters of phosphinic and phosphonic acids, used as plasticizers, insecticides, resin modifiers, and flame-retardants; (3) pyrophosphates, e.g., tetraethyl pyrophosphate, which are the basis for a broad group of cholinesterase inhibitors used as insecticides; (4) phosphoric esters of glycerol, glycol, sorbitol, etc., which are components of fertilizers. While many of these compounds play an important part in animal metabolism, those in group (3) are toxic and should be handled with extreme care.

organopolysilicate. See organoclay.

organosilane. See organosilicon.

organosilicon. An organic compound in which silicon is bonded to carbon (organosilane). Such compounds were first made by Friedel and Crafts in 1863. Silicon was found to have a remarkable chemical similarity to carbon, which it can replace in organic compounds. The silicon-carbon bond is about as strong as the carbon-carbon bond, and compounds containing them are similar in properties to all-carbon compounds. Organosilicon oxides (organosiloxanes or silicones) were discovered by F. S. Kipping in England in 1900; he found that Grignard reagents would react with silicon tetrachloride to form silicon-carbon-bonded polymers of both ring and chain types. These were named silicones because of the similarity of their empirical formula (R_2SiO) to that of ketones (R_2CO). An organosilicon compound (tetramesityldisilene) containing a silicon to silicon double bond has been synthesized. It is a crystalline solid, mp 176C, and has reactive properties similar to olefins. Compounds of this type are called silylenes.
See silicone.

organosol. Colloidal dispersion of any insoluble material in an organic liquid; specifically the finely divided or colloidal dispersion of a synthetic resin in

plasticizer in which dispersion the volatile content exceeds 5% of the total.
See plastisol.

organotin compounds. A family of alkyl tin compounds widely used as stabilizers for plastics, especially rigid vinyl polymers used as piping, construction aids, and cellular structures. Some have catalytic properties. They include butyl tin trichloride, dibutyltin oxide, etc., and various methyltin compounds. They are both liquids and solids.
Hazard: All are highly toxic. TLV: 0.1 mg/m^3.
See dibutyltin-.

origanum oil. An essential oil used in pharmacy and as a flavoring.

origin. Biology: The nucleotide sequence or site in DNA where DNA replication is initiated.

Orlon. A copolymer containing at least 85% acrylonitrile. Available in various types of staple and tow.
Properties: Tensile strength (psi) 32,000–39,000, d 1.14–1.17, break elongation 20–28%, moisture regain 1.5% (21.2C, 65% RH), softens at 235C, soluble in butyrolactone (hot), dimethyl formamide (hot), ethylene carbonate (hot), resistant to mineral acids, fair to good resistance to weak alkalies. Insoluble in alcohol, acetone, benzene, carbon tetrachloride, and petroleum ether; soluble in dimethyl sulfoxide, maleic anhydride, ethylene carbonate, nitriles, and nitrophenols.
Hazard: Combustible, burns freely and rapidly.
Use: In apparel, usually blended with wool or other fibers.

ormetoprim.
Properties: White, tasteless, odorless powder.
Use: Drug (veterinary); food additive.

Orn. Abbreviation for ornithine.

ornithine. (2,5-diaminovaleric acid). NH_2 $(CH_2)_3CH(NH_2)COOH$. A nonessential amino acid produced by the body and important in protein metabolism.
Properties: (L(+)-ornithine.) Crystals from alcohol-ether. Mp 140C. Soluble in water and alcohol. (DL-ornithine) Crystals from water. Slightly soluble in alcohol.
Derivation: Isolated from proteins after hydrolysis with alkali.
Use: Biochemical research; medicine.

ornithine carbamyl transferase. (ornithine transcarbamoylase). An enzyme of the urea cycle that catalyzes the transfer of carbamyl groups from carbamyl phosphate to ornithine with the formation of citrulline. It is a normal constituent of mitochondria in livers, but does not normally occur in significant amounts in serum. Its presence in serum is a specific indicator of damage to livers.

ORNL. Abbreviation for Oak Ridge National Laboratory.

orosomucoid. An old term for a1-acid glycoprotein.

"Orotan" TV *[Dow].* TM for a synthetic tanning agent with attributes of vegetable tannins. Dark-red, viscous solution; 31% tannin. Imparts high degree of tannage, strength, fullness, and solidity to leather. Solubilizing, penetrating, and bleaching agent.

orotic acid. (uracil-6-carboxylic acid; 6-carboxyuracil).
CAS: 65-86-1. $C_4N_2H_3(O)_2COOH$. Occurs in cow's milk and has also been isolated from certain strains of molds (*Neurospora*). A growth factor for certain microorganisms.
Properties: Crystals. Mp 345–346C.
Use: Biochemical research, especially the biosynthesis of nucleic acids.

orphan protein. Any hormone receptors that has not been matched with the hormones which activate them.

orpiment. Obsolete name for arsenic trisulfide.

orris root oil.
Properties: From steam distillation of peeled, dried, aged rhizomes of *Iris pallida* L. (Fam. Iridaceae). Light yellow to brown solid at room temp. Mp: 38–50C. Soluble in fixed oils, mineral oil, propylene glycol; insoluble in glycerin.
Use: Food additive.

orthamine. See *o*-phenylenediamine.

ortho-. (*o*-). A prefix meaning "straight ahead." Compare with meta- (*m*-) meaning "beyond," para- (*p*-) meaning "opposite." These prefixes are used in organic chemistry in naming disubstitution products derived from benzene in which the substituent atoms or radicals are located in certain definite positions on the benzene ring. This is illustrated in the diagram, where A and B represent the substituent atoms or groups. When attached to adjoining carbon atoms, B is in the *o*- position in respect to A (also called the 1,2-position). If B is located on the third carbon atom in respect to A, it is in the *m*-position (also called 1,3-); when B is attached to the opposite carbon atom, it is in the *p*- position (1,4).

ortho　　　　meta　　　　para

In organic compounds, these prefixes usually appear in italics (often abbreviated *o*-, *m*-, and *p*-) and are ignored in alphabetizing.

In inorganic chemistry, the prefix *ortho* designates the most highly hydrated acid, or its salt, to contrast with the *meta*, or less hydrated acid or salt. For example, H_3PO_4 is orthophosphoric acid and HPO_3 is metaphosphoric acid. These are considered in alphabetizing.

orthoarsenic acid.　　See arsenic acid.

orthoboric acid.　　See boric acid.

Orthochrom.　　Pigmented plasticized nitrocellulose lacquers and thinners. They produce durable, washable, flexible, colored lacquer finishes of good lightfastness.
Use: Finishing of belt, garment, upholstery, and other leathers.

Orthoclear.　　Permanently plasticized nitrocellulose binders and lacquers in various solvents. They produce clear, durable, flexible finishes.
Use: Top-coat finishes for glazing or high-gloss leather coatings.

orthoform.　　See methyl-*m*-amino-*p*-hydroxybenzoate.

"Ortholite" [O2].　　TM for clear and pigmented vinyl lacquers, binders, and solvents. They produce finishes of outstanding abrasion resistance and low-temperature flexibility.
Use: Finishes on upholstery, automotive, luggage, and case leathers.

orthosilicic acid.　　See silicic acid (ortho).

oryzalin.
CAS: 19044-88-3.　$C_{12}H_{18}N_4O_6S$.
Hazard: Low toxicity by ingestion and skin contact.
Use: Food additive; herbicide; agricultural chemical.

Os.　　Symbol for osmium.

osamine.　　A compound derived from a sugar.

oscillometry.　　Electrode measurement of oscillation-frequency changes to detect the progress of a titration of electrolyte solutions.

-ose.　　A suffix indicating a carbohydrate compound or polymer, usually a simple or complex sugar, e.g., sucrose, fructose, glucose, maltose, etc., and also cellulose, cellobiose, amylose (starch).

OSHA.　　Abbreviation for Occupational Safety and Health Administration.

osmic acid.　　(osmium tetroxide; perosmic acid anhydride; perosmic oxide).
CAS: 20816-12-0.　OsO_4.
Properties: A colorless dimorphic compound with both crystalline and amorphous forms; pungent, disagreeable odor. D 4.90, mp 40C, bp 130C. Soluble in water, alcohol, and ether.
Derivation: By heating powdered osmium in air, or by treating it with nitric acid, aqua regia, or chlorine.
Hazard: Toxic by inhalation; strong irritant to eyes and mucous membranes. Upper respiratory tract and skin irritant.
Use: Microscopic staining, photography, oxidation catalyst in organic synthesis.

osmiridium.　　See iridosmine.

osmium.
CAS: 7440-04-2.　Os. Metallic element having atomic number 76, in group VIII of the periodic table, aw 190.2, valences of 2, 3, 4, 6, 8; seven stable isotopes.
Properties: Hard white metal of the platinum group. D 22.5, mp 3000C, bp 5500C, it has the highest specific gravity and melting point of the platinum metals, metallurgically unworkable. On heating in air gives off poisonous fumes of osmium tetroxide. Insoluble in acids and aqua regia; attacked by fused alkalies.
Occurrence: Tasmania, South Africa, the former U.S.S.R., Canada.
Derivation: Occurs with platinum, from which it is recovered during the purification process. Also occurs with iridium as a natural alloy, iridosmine.
Hazard: Highly toxic; irritant to skin.
Use: Hardener for iridium and platinum, pen points, instrument pivots, catalyst.

osmium ammonium chloride.　　See ammonium hexachloroosmate.

osmium chloride.　　(osmium dichloride; osmous chloride).
$OsCl_2$.
Properties: Dark-green needles. Hygroscopic. Keep away from air. Soluble in alcohol, ether; insoluble in water.

osmium dichloride.　　See osmium chloride.

osmium sodium chloride. See sodium hexachloroosmate.

osmium tetroxide. See osmic acid.

osmocaine.
CAS: 532-77-4. $C_{16}H_{23}NO_2$.
Hazard: Human systemic effects.

osmocene. (dicyclopentadienylosmium).
$(C_5H_5)_2Os$.
Properties: Stable, white solid. Mp 229–230C.
Use: Intermediate, high-temperature applications, derivatives used as UV-radiation absorber.
See metallocene.

osmometry. The measurement of osmotic pressure.

osmosis. Passage of a pure liquid (usually water) into a solution (e.g., of sugar and water) through a membrane that is permeable to the pure water but not to the sugar in the solution. This passage can also occur when the two phases consist of solutions of different concentration. The membrane is called semipermeable when the molecules of the solvent, but not those of the solute, can penetrate it. This pushing of water through a membrane into a solution results from the greater tendency of water molecules to escape from water than from a solution. The term *osmosis* is usually restricted to movement through a solid or liquid barrier that prevents the phases from mixing rapidly. In test apparatuses parchment or collodion membranes are used; in plants and animals the cell wall acts as a diffusion barrier. The pressure exerted by osmosis is substantial and accounts for the elevation of sap from root systems to the tops of trees. Osmosis is considered an essential characteristic of growth. Reverse osmosis is used as a method of desalting seawater, recovering wastewater from paper mill operations, pollution control, industrial water treatment, chemical separations, and food processing. This method involves application of pressure to the surface of a saline solution, thus forcing pure water to pass from the solution through a membrane that is too dense to permit passage of sodium and chlorine ions. Hollow fibers of cellulose acetate or nylon are used as membranes, since their large surface area offers more efficient separation.
See dialysis; membrane; diffusion; desalination.

osmous chloride. See osmium chloride.

Ostromyslenskii (Ostromisslenskii) reaction.
Dehydrogenation of ethanol over copper-containing catalysts and conversion of the acetaldehyde ethanol mixture to butadiene by passage at high temperature over silica gel containing a small amount of tantalum oxide.

Ostwald ripening. Refers to the process of solution of smaller crystals and growth of larger crystals.

Ostwald, Wilhelm. (1853–1932). A German chemist who won the Nobel Prize for chemistry in 1909. He was considered to be a founder of modern physical chemistry. His work involved research in catalysis, the rates of chemical reactions, equilibrium, and conductivity of organic acids. He was an admirer of Mach and did not readily accept the atomic theory. He was educated at the University of Dorpat.

osutidine.
CAS: 140695-21-2. $C_{19}H_{28}N_4O_5S_2$.
Hazard: Low toxicity by ingestion.

OTEC. See Ocean Thermal Energy Conversion.

otto. See attar.

Otto Fuel II.
CAS: 106602-80-6. A torpedo fuel that is easy to handle, has a non-explosive classification and a low fire hazard. Developed by Otto Reitlinger at the Indian Head Division Naval Surface Warfare Center during the 1960s.

ouabain. (acocantherin astrobain; gratibain; gratus strophanthin; g-strophanthin; ouabagenin-1-rhamnsid; ouabagenin-1-rhamnoside; ouabaine; oubain; purostrophan; strophanthin G; strophoperm; 3β-[(6-deoxy-α-1-mannopyranosyl)oxy]-1β, 5β,11α,19-pentahydroxycard-20,22-enolide octahydrate; g-strophanthin; gratus strophanthin; strophanthin-g; acocantherin; 3-[(1R,3S,5S,10R,11R, 13R,14S,17R)-1,5,11,14-tetrahydroxy-10-(hydroxymethyl)-13-methyl-3-[(2R,3R,4R,5R,6S)-3,4,5-trihydroxy-6-methyloxan-2-yl]oxy-2,3,4,6,7,8,9, 11,12,15,16,17-dodecahydro-1H-cyclopenta[a] phenanthren-17-yl]-2H-furan-5-one).
CAS: 630-60-4. $C_{29}H_{44}O_{12}$. Digitalis glycoside that is a specific inhibitor of na$^{1/k}$ ATPase.
Properties: White crystalline substance or white powder, water- and ethanol-soluble, melts with decomposition at 190C.
Derivation: From certain plants of the family Apocynaceae: *Strophanthus* spp. and the *Acokanthera* group of trees.
Hazard: Extremely toxic; cardiotonic; poison; mutagen.
Use: Therapeutically to treat acute congestive heart failure, nondal paroxysmal tachycardia, and atrial flutter.

ouricury wax. A vegetable wax exuded by the leaves of *Cocos coronapa* (South America).
Properties: Brown. Acid value 10, saponification value 80, d 0.970 (15C), mp 83C, foreign matter (dirt, etc.) sometimes 18%. Combustible.

Grade: Crude, refined.
Use: Substitute for carnauba wax.

-ous. A suffix, used in naming inorganic compounds, that indicates that the central element is present in its lower oxidation state. For example, in ferrous chloride ($FeCl_2$), the iron atom is in its lower oxidation state of +2, equivalent to its valence; in an ionized state it would have two positive charges (Fe^{2+}). (A recommended change in this system of nomenclature is to use the common name of the element [iron] together with a Roman numeral showing the oxidation number; thus, ferrous chloride would be iron(II) chloride.)

outdoor air.
Atmospheric air at a given location. Usually taken to be a mixture of gases that contain about 78% nitrogen, 20.95% free oxygen, 0.93% argon, 0.03% carbon dioxide, and trace amounts of numerous other components.

outgassing. The removal of gas from a metal by heating at a temperature somewhat below melting, while maintaining a vacuum in the space around the metal. Usually done before melting but sometimes afterward.
Use: Manufacture of tubes and other vacuum devices.

outgroup. In a cladistic analysis, any taxon used to help resolve the polarity of characters, and which is hypothesized to be less closely related to each of the taxa under consideration than any are to each other.

ovalbumin. See albumin, egg.

oven. A heated chamber of varying sizes used for removing moisture from industrial products before or during processing, for example, finely divided solids, food products, tobacco, textiles, wood, etc. Laboratory sizes are used for testing the effect of heat exposure on such materials as rubber, plastics, fibers, paints, etc., as well as for sterilization, drying electronic components, and curing encapsulation compounds.

oven cleaner. Any of a variety of formulations. They contain a variety of substances, including ammonia, detergents, synthetic fragrances, and aerosol propellants.
Hazard: Toxic; power irritant; hazardous.
Use: To clean baking ovens.

overlapping clones. See genomic library.

ovex. (generic name for *p*-chlorophenyl-*p*-chlorobenzenesulfonate).
CAS: 80-33-1. $ClC_6H_4OSO_2C_6H_4Cl$.

Properties: White, crystalline solid. Mp 86.5C. Insoluble in water; soluble in acetone and aromatic solvents.
Use: Insecticide and acaricide.

ovicide. A type of pesticide that kills the eggs of insects.

Ovral. An oral contraceptive containing norgestrel, a synthetic progestogen. Contains 0.05 mg norgestrel and 0.05 mg ethinyl estradiol.

oxa-. Prefix indicating the presence of oxygen in a heterocyclic ring.

6-oxa-5,7-disilaundecane-1,11-diamine, 5,5,7,7-tetramethyl-.
CAS: 3663-42-1. $C_{12}H_{32}NOSi_2$.
Hazard: A severe eye irritant.

"OXAF" [Chemtura]. TM for the zinc salt of 2-mercaptobenzothiazole ($Zn(SCNSC_6H_4)_2$).
Properties: White to pale-yellow powder. D 1.63, melting range decomposes without melting when heated to 200C or over, excellent storage stability. Slightly soluble in ethylene dichloride and acetone; insoluble in water, benzol, and gasoline. Available in pelletized form.
Use: Rubber accelerator, especially latex foam sponge, wire insulation, air-cured footwear, druggist sundries, and specialties.

oxalate. $C_2H_2O_4$. Any salt of oxalic acid and a strong dicarboxylic acid. It is produced in the body by metabolism of glyoxylic acid or ascorbic acid. It is not metabolized but excreted in the urine.
Derivation: Found in numerous green plants and some fungi.
Hazard: Toxic.
Use: An analytical reagent and general reducing agent.

oxalatoplatinum.
CAS: 61825-94-3. $C_8H_{14}N_2O_4Pt$.
Hazard: A poison.

oxalic acid.
CAS: 144-62-7. $HOOCCOOH \cdot 2H_2O$.
Properties: Transparent, colorless crystals. Mp 187C for anhydrous form, 101.5C for dihydrate.
Derivation: Occurs naturally in many plants (wood sorrel, rhubarb, spinach) and can be made by alkali extraction of sawdust. Now manufactured by reaction of carbon monoxide and sodium hydroxide or of sodium formate with sodium hydroxide, followed by distillation of the resulting dihydrate crystals.
Grade: Technical (crystals and powder), CP.
Hazard: Toxic by inhalation and ingestion, strong irritant. Eye, skin, and upper respiratory tract irritant.

Use: Automobile radiator cleanser, general metal and equipment cleaning, purifying agent and intermediate for many compounds, leather tanning, catalyst, laboratory reagent, stripping agent for permanent press resins, bleaching of textiles, rare-earth processing, printing and dyeing auxiliary.

oxalomolybdic acid. $[MoO_3(C_3O_4)]H_2$. A water-soluble, crystalline compound used in invisible inks.

oxalonitrile. See cyanogen.

oxalyl chloride. (ethanedioyl chloride).
CAS: 79-37-8. $(COCl)_2$.
Properties: Colorless liquid; if cooled to −12C, solidifies to a white, crystalline mass; gives off carbon monoxide on heating. Bp 64C, fp −12C, d 1.43. Decomposed by water and alkaline solutions. Soluble in ether, benzene, chloroform.
Derivation: Interaction of oxalic acid and phosphorus pentachloride.
Hazard: Toxic by inhalation and ingestion.
Use: Military poison gas, chlorinating agent in organic synthesis.

oxamic acid. (aminooxoacetic acid).
H_2NCOCO_2H.
Properties: Crystalline powder. Mw 89.05, mp 210C. Sparingly soluble in water.
Use: In medical diagnostic manufacture.

oxamide.
CAS: 471-46-5. $NH_2COCONH_2$.
Properties: White, odorless powder. Mp 419C (decomposes). Probably the highest melting organic compound; slightly soluble in water; very slightly soluble in alcohol and ether; not hygroscopic; decomposes to ammonia and carbonic acid.
Use: Stabilizer for nitrocellulose preparations, possible substitute for urea as fertilizer.

oxammonium. See hydroxylamine.

Oxanal. Dyes for coloring anodized aluminum.

1,2-oxathrolane. See propane sultone.

oxazin dye. Any of a class of dyes that are structurally similar to azin dyes. One of the connecting nitrogen atoms is replaced by oxygen.

oxazole.

OCH:NCH:CH.

A five-membered heterocyclic compound valuable for its derivatives. The dihydro forms, 2- and 4-oxazoline, are the parents of increasingly useful

commercial compounds, e.g., surface-active agents, detergents, etc.; 2-oxazoline is

OCH₂NHCH:CH.

oxazoline wax. A series of synthetic waxes having the oxazoline structure. They can be made to fairly exact specifications and are miscible with most natural and synthetic waxes (and can be applied to the same uses).
See oxazole.

1h,3h,5h-oxazolo(3,4-c)oxazole-7a(7h)-methanol.
CAS: 6542-37-6. $C_6H_{11}NO_3$.
Hazard: A poison by ingestion.

oxetane. (trimethylene oxide).
CAS: 503-30-0.

CH₂OCH₂CH₂.

An oxetane group ($=COCH_2C=$) is one kind of epoxy group.
See Penton.

oxethazaine hydrochloride. See emoren.

oxidase. An enzyme whose activity results in the transfer of electrons on the substrate; an oxidizing enzyme.

oxidation. The term *oxidation* originally meant a reaction in which oxygen combines chemically with another substance, but its usage has long been broadened to include any reaction in which electrons are transferred. Oxidation and reduction always occur simultaneously (redox reactions), and the substance that gains electrons is termed the oxidizing agent. For example, cupric ion is the oxidizing agent in the following reaction: Fe (metal) + $Cu^{2+} \rightarrow Fe^{2+}$ + Cu (metal). Here, two electrons (negative charges) are transferred from the iron atom to the copper atom; thus, the iron becomes positively charged (is oxidized) by loss of two electrons, while the copper receives the two electrons and becomes neutral (is reduced). Electrons may also be displaced within the molecule without being completely transferred from it. Such partial loss of electrons likewise constitutes oxidation in its broader sense and leads to the application of the term to a large number of processes that at first sight might not be considered oxidations. Reaction of a hydrocarbon with a halogen, e.g., CH_4 + 2Cl → CH_3Cl + HCl, involves partial oxidation of the methane; halogen addition to a double bond is regarded as an oxidation. Dehydrogenation is also a form of oxidation, when two hydrogen atoms, each

having one electron, are removed from a hydrogen-containing organic compound by a catalytic reaction with air or oxygen, as in oxidation of alcohols to aldehydes.
See dehydrogenation.

oxidation number. The number of electrons that must be added to or subtracted from an atom in a combined state to convert it to the elemental form; that is, in barium chloride ($BaCl_2$), the oxidation number of barium is +2 and of chlorine is −1. Many elements can exist in more than one oxidation state.
See valence.

oxidation-reduction indicator. A substance that has a color in the oxidized form different from that of the reduced form and can be reversibly oxidized and reduced. Thus, if diphenylamine is present in a ferrous sulfate solution to which potassium dichromate is being added, a violet color appears with the first drop of excess dichromate.
See indicator.

oxidative coupling. A polymerization technique for certain types of linear high polymers. Oxidation of 2,6-dimethylphenol with an amine complex of a copper salt as catalyst forms a polyether, with splitting off of water. The product is soluble in aromatic and chlorinated hydrocarbons; insoluble in alcohols, ketones, and aliphatics. It is thermoplastic and unaffected by acids, bases, and detergents. It has a very broad useful temperature range (from −170 to + 190C). It is also dimensionally stable and has good electrical resistance. Oxidative coupling of diacetylenes and dithiols also yields promising polymers.

oxidative phosphorylation. The enzymatic phosphorylation of ADP coupled to electron transfer from a substrate to molecular oxygen.

oxide. A mineral in which metallic atoms are bonded to oxygen atoms.

3-oxidido 17-α-ethynyl 17-β-hydroxy estra-4,9,11-triene.
CAS: 19636-23-8. $C_{20}H_{23}NO_2$.
Hazard: A poison by ingestion.
Use: Hormone.

oxidimethiin.
CAS: 55290-64-7. $C_6H_{10}O_4S_2$.
Hazard: Moderately toxic by ingestion. Low toxicity by inhalation and skin contact.
Use: Agricultural chemical.

oxidizing material. Any compound that spontaneously evolves oxygen either at room temperature or under slight heating. The term includes such chemicals as peroxides, chlorates, perchlorates, nitrates, and permanganates. These can react

vigorously at ambient temperatures when stored near or in contact with reducing materials such as cellulosic and other organic compounds. Storage areas should be well ventilated and kept as cool as possible.

oxine. See 8-hydroxyquinoline.

oxirane.
CAS: 75-21-8.

$$H_2COC_2.$$

A synonym for ethylene oxide. An oxirane group is one having the structure.

$$=COC=$$

and is one kind of epoxy group.
See ethylene oxide.

Oxirane process. A method of making ethylene glycol by catalytic oxidation of ethylene to the diacetate, which is then hydrolyzed to ethylene glycol.

oxirene. (oxacyclopropene). An organic intermediate containing four pelectrons, reported to result from oxidation of acetylene.

oxochlorpromazine.
CAS: 4337-86-4. $C_{17}H_{19}ClN_2O_2S$.
Hazard: A poison.

8-oxocoptisine.
CAS: 19716-61-1. $C_{19}H_{13}NO_5$.
Hazard: A poison by ingestion.

2-oxohexamethylenimine. See caprolactam.

"Oxone" [Du Pont]. TM for an acidic, white, granular, free-flowing solid containing the active ingredient potassium peroxymonosulfate; readily soluble in water; 1% solution has pH of 2–3; minimum active oxygen content 4.5%; strong oxidizing agent.
Hazard: Fire risk in contact with organic materials.
Use: Manufacture of dry laundry bleaches, detergent-bleach washing compound, scouring powders, plastic dishware cleaners, and metal cleaners; hair-wave neutralizers, pharmaceuticals; general oxidizing reactions.

oxonium ion. See hydronium ion.

2-oxopentanedioic acid. See α-ketoglutaric acid.

4-oxopentanoic acid. See levulinic acid.

2-oxo-2-(phenylamino)ethyl selenocyanate.
CAS: 63981-21-5. $C_9H_8N_2OSe$.
Hazard: A poison by ingestion.

Oxo process. Production of alcohols, aldehydes, and other oxygenated organic compounds by passage of olefin hydrocarbon vapors over cobalt catalysts in the presence of carbon monoxide and hydrogen. Aldehydes are formed as products, but in most cases these are hydrogenated at once to the corresponding alcohol. Propylene produces normal and isobutyraldehyde; higher olefins produce a mixture of aldehydes containing one more carbon atom than the olefins; *n*-butyl, isobutyl, amyl, isooctyl, decyl, and tridecyl alcohols are produced in large quantities.

**2-((1-oxo-2-propenyl)oxy)-*n*,*n*,*n*-
trimethylethanaminium chloride.**
CAS: 44992-01-0. $C_8H_{16}NO_2 \cdot Cl$.
Hazard: A severe eye irritant.

**2-oxo-2-(1-pyrrolidinyl)ethyl-*n*-
(((methylamino)carbonyl)oxy)
ethanimidothioate.**
CAS: 92065-85-5. $C_{10}H_{17}N_3O_3S$.
Hazard: A poison by ingestion.
Use: Agricultural chemical.

oxosilane. See siloxane.

oxotremorine fumarate (2:3). See oxotremorine sesquifumarate.

oxotremorine sesquifumarate.
CAS: 17360-35-9. $C_{12}H_{18}N_2O \cdot 3/_2C_4H_4O_4$.
Hazard: A poison.

**5-oxo-2,4,8-trimethyl-6-oxa-3,9-dithia-2,4,7-
triazadec-7-enoic acid, 2-(2-(2-
methoxyethoxy) ethoxy)ethyl ester.**
CAS: 64029-10-3. $C_{14}H_{27}N_3O_7S_2$.
Hazard: A poison by ingestion.
Use: Agricultural chemical.

**5-oxo-2,4,8-trimethyl-6-oxa-3,9-dithia-2,4,7-
triazadec-7-enoic acid, (1-
methylethylidene)di-4,1-phenylene ester.**
CAS: 79006-76-1. $C_{29}H_{38}N_6O_8S_4$.
Hazard: Moderately toxic by ingestion.
Use: Agricultural chemical.

**1-oxo-2-(2,4,6-trimethylphenyl)-1h-inden-3-
yl dodecanoate.**
CAS: 53083-27-5. $C_{30}H_{38}O_3$.
Hazard: Low toxicity by ingestion.

(4-oxovaleryloxy)triphenylstannane. See triphenyltin levulinate.

"Oxsol 100" *[Emerald]*. TM for a fluorinated toluene that is exempt from regulations as a VOC.

Not an air toxic, ozone depleter, or suspected carcinogen.
Use: To clean metal, plates, electronics and glass. Dissolves resins in paint, coatings, inks, adhesives and other resin applications.

oxyacid. An acid in which acidic hydrogen atoms are attached to an oxygen atom.

oxyanion. A negatively charged ion that is formed by the loss of one or more hydrogen ions from an oxyacid.

oxybenzoic acid. See hydroxybenzoic acid.

oxybenzone. (4-methoxy-2-
hydroxybenzophenone).
CAS: 131-57-7. $C_{14}H_{12}O_3$.
Properties: Colorless crystals. Mp 65C. Soluble in common organic solvents.
Use: Sunscreen lotions.

***p*,*p*′-oxybis(benzenesulfonylhydrazide).**
[(4,4′-oxybis(benzenesulfonyl)hydrazide)].
CAS: 80-51-3. $H_2NNHSO_2C_6H_4OC_6H_4SO_2NHNH_2$.
Properties: Fine, white, crystalline powder; odorless. D 1.52, mp decomposes at 150–160C. Soluble in acetone; moderately soluble in ethanol and polyethylene glycols; insoluble in gasoline and water. Combustible.
Hazard: Teratogenic effects.
Use: Blowing agent for sponge rubber and expanded plastics.

oxybis(dibutyl(2,4,5-trichlorophenoxy)tin).
CAS: 74007-80-0. $C_{28}H_{40}Cl_6O_3Sn_2$.
Hazard: A poison.

**oxybis(2,1-ethanediyloxy-2,1-ethanediyl) 3-
(dodecylthio)propanoate.**
CAS: 64253-30-1. $C_{38}H_{74}O_7S_2$.
Hazard: A reproductive hazard.

**1,1′-(oxybis(methylenesulfonyl))bis(2-
chloroethane).**
CAS: 53061-10-2. $C_6H_{12}Cl_2O_5S_2$.
Hazard: A poison by ingestion.

1,1′-oxybis(2,4,5-trichlorobenzene). See 2,2′,4,4′,5,5′-hexachlorodiphenyl ether.

Oxy-Complete. A high potency anti-oxidant composed of amino acids, enzymes, and nutrients.
Use: Food supplement.

oxydemetonmethyl. (generic for *S*-[2-(ethylsulfinyl)ethyl]-*O*,*O*-dimethylphosphorothioate).
See *O*,*O*-dimethyl-*S*-2-(ethylsulfinyl)ethyl phosphorothioate.

***n*-oxydiethylene-2-benzothiazolesulfenamide.**
See 2-(morpholinothio)benzothiazole.

1,1'-(oxydiethylene)bis(4-formylpyridinium bromide), dioxime.
CAS: 3852-72-0. $C_{16}H_{20}N_4O_3 \cdot 2Br$.
Hazard: A poison.

β,β'-oxydipropionitrile.   (ODPN).
CAS: 1656-48-0. $C_6H_8N_2O$.
Properties: Colorless liquid. Fp −26.3C, bp 120C (1 mm Hg), bp 155C (5 mm Hg), d 1.0405 (30C), viscosity 8.00 cP (30C), refr index 1.4392 (25C), flash p 180F (82.2C) (TOC); soluble in water. It is thermally unstable, yielding acrylonitrile and water at above 175C. Hydrolyzed by strong acids and bases, quite immiscible with paraffin hydrocarbons, but dissolves aromatics. Combustible.
Derivation: From acrylonitrile.
Hazard: Toxic by inhalation and ingestion.
Use: Solvent in fractional extraction.

oxyfluorfen.
CAS: 42874-03-3. $C_{15}H_{11}ClF_3NO_4$.
Properties: Orange crystal solid. Soluble in water and most solids.
Use: Food additive; herbicide.

oxygen.
CAS: 7782-44-7.   O. Nonmetallic gaseous element of atomic number 8; group VIA of the periodic table, aw 15.9994, valence of 2, isotopes 16, 17, 18; molecular oxygen is O_2 and ozone O_3. Atmospheric oxygen is the result of photosynthesis. Oxygen was discovered by Priestley in England in 1774.
Properties: Colorless, diatomic gas; odorless; tasteless. Liquefiable at −183C to slightly bluish liquid, solidifiable at −218C. It constitutes 20% by volume of air at sea level, d (gas) 1.429 g/L (0C, 1 atm) (air = 1.29), d (gas): 1.10535 (air = 1.00), (liquid) d 1.14 (−183C). Soluble in water and alcohol. Oxygen is noncombustible, but actively supports combustion.
Derivation: Before 1971 the only commercial method for large-scale oxygen production was fractionation of liquefied air; this has largely been replaced by a process that utilizes ambient temperature separation by means of a pressure cycle in which molecular sieves of synthetic zeolite preferentially adsorb nitrogen from air, giving 95% oxygen and 5% argon. Electrolysis of water is used for small amounts and laboratory demonstration.
Grade: Low purity, high purity, USP.
Hazard: (Gaseous) Moderate fire risk as oxidizing agent; therapeutic overdoses can cause convulsions. (Liquid) May explode on contact with heat or oxidizable materials. Irritant to skin and tissue.
Use: Blast furnaces; copper smelting; steel production (basic oxygen converter process); manufacture of synthesis gas for production of ammonia, methyl alcohol, acetylene, etc.; oxidizer for liquid rocket propellants; resuscitation, heart stimulant; decompression chambers; spacecraft; chemical intermediate; to replace air in oxidation of municipal and industrial organic wastes; to counteract effect of eutrophication in lakes and reservoirs; coal gasification.
See ozone.

oxygen 18.   (heavy oxygen).
Oxygen isotope of aw 18. Occurs in proportion of 8 parts to 10,000 of ordinary oxygen in water, air, rocks, etc. The proportion may be increased by passing carbon dioxide gas repeatedly through a packed column down which water is passed. The carbon dioxide leaving the top of the tower is enriched in heavy oxygen, and the water leaving the bottom is depleted.
Use: Tracer experimentation.
See heavy water.

oxygenases.
Enzymes that catalyze reactions in which oxygen is introduced into an acceptor molecule.
See mixed-function oxidases.

Oxygenated Hydrocarbons.
A series of petroleum-derived oxidates composed primarily of organic acids and esters. Designated by TC or TX followed by a four-digit number, e.g., TC-5416, TC-6664.
Use: Corrosion inhibitors, surface-active components in wax emulsions, emulsifiable lubricants, plasticizer, intermediate, leather and cordage oils, lubricity agents and solubilizers.

oxygen balance.
Oxygen content relative to the total oxygen required for oxidation of all carbon, hydrogen, and other easily oxidizable elements to carbon dioxide, water, etc.

oxygen cell.
An electrolytic cell whose emf is due to a difference in oxygen concentration at one electrode compared with that at another electrode of the same material.

oxygen consumed.   (OC; COD; DOC).
A measure of the quantity of oxidizable components present in water. Since the carbon and hydrogen, but not the nitrogen, in organic matter are oxidized by chemical oxidants, the oxygen consumed is a measure only of the chemically oxidizable components and is dependent on the oxidant, structure of the organic compound, and manipulative procedure. Since this value does not differentiate stable from unstable organic matter, it does not necessarily correlate with the biochemical oxygen demand value. It is also known as chemical oxygen demand (COD) and dichromate oxygen consumed (DOC).
See biochemical oxygen demand; dissolved oxygen.

oxygen debt.
The extra oxygen (above the normal resting level) consumed in the recovery period after strenuous physical exertion due to the re-oxidation of lactate.

oxygen fluoride.   (oxygen difluoride; fluorine monoxide).
CAS: 7783-41-7. OF_2.
Properties: An unstable, colorless gas. Fp −224C, bp −145C. Slightly soluble in water and alcohol. Suggested as oxidizer for rocket propellants.
Hazard: Explodes on contact with water, air, and reducing agents. Corrosive to tissue. Headache, pulmonary edema and upper respiratory tract irritant.

oxygen sink.   A reservoir consisting of a chemical element or compound that combines readily with oxygen and thus removes it from the atmosphere. During the early part of Precambrian time, sulfur, iron, and other elements and compounds served as important oxygen sinks, preventing oxygen from accumulating in the atmosphere.

oxyhemoglobin.   See hemoglobin.

oxyl process.   A method for directly producing higher alcohols by catalytically reducing carbon monoxide with hydrogen.

oxyluminescence.   See chemiluminescence.

oxymethurea.   (1,3-bis(hydroxymethyl)urea).
CAS: 140-95-4. $(HOCH_2NH)_2CO$.
Properties: Crystalline solid. Mp 137C. Soluble in water, ethanol and methanol.
Use: Textile auxiliary (crease- and shrinkproofing agent for cotton), photographic developers, antiseptic.

oxymethylene.   See formaldehyde.

β-oxynaphthoic acid.   See 3-hydroxy-2-naphthoic acid.

oxyneurine.   See betaine.

n-**oxynicotine.**   See nicotine-*n'*-oxide.

oxypendyl hydrochloride.
CAS: 17297-82-4. $C_{20}H_{26}N_4OS•2ClH$.
Hazard: Moderately toxic by ingestion.

oxyphosphorane.   One of a class of compounds derived from trialkyl phosphites and *o*-quinones. Their molecules have a five-atom ring
$\overline{}$
$OCCOP(OR)_3$ in which the two carbon atoms are part of an aromatic ring. They react by liberating a phosphate ester.

oxyphyte.   A plant that grows under acidic conditions.

oxypolygelatin.   (OPG). A purified gelatin treated with glyoxal, followed by oxidation with hydrogen peroxide. A possible plasma substitute.

oxyquinoline.   See 8-hydroxyquinoline.

oxysonium iodide.   See hydroxythiospasmin.

oxystearin.
CAS: 8028-45-3.
Properties: Mixture of the glycerides of partially oxidized stearic and other fatty acids. Tan to light brown waxy solid; bland taste. Refr index: 1.465. Soluble in ether, solvent hexane, chloroform.
Hazard: Low toxicity.
Use: Food additive.

oxytetracycline.
CAS: 79-57-2. $C_{22}H_{24}N_2O_9•2H_2O$. An antibiotic obtained by fermentation from *Streptomyces rimosus*, an actinomycete. Its chemical structure is that of a modified naphthacene molecule having six asymmetrical centers. It has been synthesized.
Properties: Dull-yellow crystalline powder; odorless; slightly bitter. Mp 179–182C (decomposes). Soluble in acids and alkalies; very slightly soluble in acetone, alcohol, chloroform, and water; practically insoluble in ether; stable in air; affected by sunlight; deteriorates in solutions with pH less than 2; destroyed rapidly by alkali hydroxide solutions; pH (saturated solution) 6.115.
Grade: NF.
Use: Medicine (antibiotic). Inhibitor of lethal yellowing in coconut palm trees; feed additive; the hydrochloride is Terramycin.
Use: Medicine (antibiotic). Inhibitor of lethal yellowing in coconut palm trees; feed additive; the hydrochloride is TM "Terramycin" *[Pfizer].*

oxythioquinox.   (6-methyl-2,3-quinoxalinedithiol cyclic carbonate).
CAS: 2439-01-2.
Properties: Mp 172C. Insoluble in water.
Use: Acaricide for tree fruits.

oxytocin.   (α-hypophamine).
CAS: 50-56-6. $C_{43}H_{66}N_{12}O_{12}S_2$.   A hormone secreted by the posterior lobe of the pituitary gland. Its chief action is stimulation of the contraction of the smooth muscle of the uterus. It contains eight different amino acids. In 1955, du Vigneaud elucidated its amino acid sequence, the first such determination ever made; it may be represented:

$$
\begin{array}{ccc}
\text{Cys} & \text{Tyr} & \text{Ileu} \\
| & & | \\
\text{Cys} & \text{Asp} & \text{Glu} \\
| & & | \\
\text{Pro} & \text{Leu} & \text{Gly}
\end{array}
$$

It is available as a solution for injection (oxytocin injection, USP).

oyster shells.   Shells of *Ostrea virginica*, taken from the Gulf of Mexico coast in Texas and Louisiana and from Chesapeake Bay. Average analysis: $CaCO_3$ 93–97%, $MgCO_3$ 1%, silica 0.5–2.0%,

SO$_4$ (as CaSO$_4$) 0.3–0.4%, also miscellaneous substances.
Use: Source of lime, drilling muds, road beds, poultry and cattle feeds.

ozalid. Copying process that gives positive prints (dark on white).

ozocerite. (mineral wax; fossil wax; ozokerite).
Properties: Wax-like, hydrocarbon mixture; yellow-brown to black or green, translucent when pure and having a greasy feel. D 0.85–0.95, mp 55–110C (usually 70C). Soluble in light petroleum hydrocarbons, benzene, turpentine, kerosene, ether, carbon disulfide; slightly soluble in alcohol; insoluble in water. Combustible.
Occurrence: Utah, Australia, near the Caspian Sea.
Method of purification: Filtration.
Grade: Technical.
Use: Electric insulation, rubber products, paints, leather polish, lithographic and printing inks, electrotypers' wax, carbon paper, source of ceresin, floor polishes, impregnating furniture and parquet floor lumber, lubricating compositions, grease crayons, sizing and glossing paper, waxed paper, cosmetics, ointments, matrices for galvanoplastic work, textile sizings, waxed cloth, substitute for carnauba and beeswax.

ozone.
CAS: 10028-15-6. O$_3$. An allotropic form of oxygen.
Properties: Unstable blue gas; pungent odor. Liquefiable at −12C, more active oxidizing agent than oxygen. Contributes to formation of photochemical smog; deterioration of rubber is accelerated by traces of ozone. Bp −112C, fp −192C, d (liquid) 1.6, more soluble in water than oxygen. Inhibits penetration of UV rays through the earth's atmosphere.
Occurrence: Formed locally in air from lightning, in stratosphere by UV radiation. Also occurs in automobile engines and by electrolysis of alkaline perchlorate solutions. Commercial mixtures containing up to 2% ozone are produced by electronic irradiation of air. It is usually manufactured on the spot because it is too expensive to ship. Tonnage quantities are used.
Hazard: Dangerous fire and explosion risk in contact with organic materials. Toxic by inhalation, strong irritant. Pulmonary function effects. Questionable carcinogen.
Use: Purification of drinking water; industrial waste treatment; deodorization of air and sewage gases; bleaching waxes, oils, wet paper, and textiles; production of peroxides, bactericide. Oxidizing agent in several chemical processes (acids, aldehydes, ketones from unsaturated fatty acids), steroid hormones, removal of chlorine from nitric acid, oxidation of phenols and cyanides.
Note: Depletion of the ozone layer in the stratosphere, which acts as a shield against penetration of UV light in the sun's rays, is believed to be caused by light-induced chlorofluorocarbon decomposition resulting from increased use of halocarbon aerosol propellants. Their manufacture and use were prohibited in 1979, except for a few specialized items.

ozonide. A product of ozonolysis.

ozonolysis. (1) Oxidation of an organic material, that is, tall oil, oleic acid, safflower oil, cyclic olefins, carbon treatment, peracetic acid production by means of ozone. (2) The use of ozone as a tool in analytical chemistry to locate double bonds in organic compounds and a similar use in synthetic organic chemistry for preparing new compounds. Under proper conditions, ozone attaches itself at the double bond of an unsaturated compound to form an ozonide. Since many ozonides are explosive, it is customary to decompose them in solution and deal with the final product.
See ozone.

P

P.   Symbol for phosphorus.

p-.   (p-). Abbreviation for para-.

PA.   Abbreviation for phthalic anhydride and for polyamide.

Pa.   Symbol for protactinium.

Paal–Knorr pyrrole synthesis.   Formation of pyrroles by heating 1,4-dicarbonyl compounds with ammonia or primary amines in a sealed tube.

Paar turbidimeter.   A visual-extinction device for measurement of solution turbidity. The length of the column of liquid suspension is adjusted until the light filament can no longer be seen.
See nephelometry.

PABA.   Abbreviation for *p*-aminobenzoic acid.

PABA sodium.   See sodium-*p*-aminobenzoate.

PAC.   See P1-derived artificial chromosome.

Pace.   Water-treatment chemicals and particularly chlorinated organic compounds.
Use: Swimming-pool chlorination.

packaging.   The operation of placing materials in suitable containers or protective covering for purposes of storage, distribution, and sale. Some packages act merely as containers, but others protect perishable materials (especially foodstuffs) from environmental damage, contamination, and biological deterioration; in this respect the critical factor is exclusion of moisture vapor, bacteria, and oxygen. Some packages perform both functions simultaneously. Common packaging materials are:

(for nonperishable products)	*(for perishable products)*
wooden boxes, kegs, barrels	glass bottles
fiber drums	"tin" cans
glass bottles (perfumes, pharmaceuticals)	plastic film
polyvinyl chloride bottles (detergents)	cellophane (tobacco)
aluminum tubes (toothpaste)	polypropylene
polyethylene paperboard cartons	polyvinylidene chloride paraffin-coated paper and board
polyethylene film (textiles)	aluminum cans (beer, soft drinks)
ceramic jars (cosmetic creams)	
steel cylinders (gases)	

packing.   (1) A collar or gasket used to seal mechanical devices to prevent leakage of oil or water; often made of specially compounded rubber or a flexible plastic. (2) The operation of placing solid materials or objects in shipping containers in such a way as to secure maximum space economy and freedom from damage by vibration or impact. Barriers of paperboard, foamed plastic, or glass fiber are widely used. (3) An inert material used in distillation columns to baffle the downward flow of countercurrent liquid; it may be glass fiber or beads, metal tubes called Raschig rings, metal chains, or specially shaped devices of various kinds (saddles, helices, rings, etc.).
See tower, distillation.

paclitaxel.   (taxol;   5-β,20-epoxy-1,2-α,4,7-β, 10-β,13-α-hexahydroxy-tax-11-en-9-one   4,10-diacetate 2-benzoate   13-ester   with   (2r,3s)-n-benzoyl-3-phenylisoserine). $C_{47}H_{51}NO_{14}$.
CAS: 33069-62-4.

Properties: Needles from aq methanol. Mw 853.99, mp 213–216C (decomp). Extracted from the bark of Pacific yew tree, *Taxus brevifolia*.

Hawley's Condensed Chemical Dictionary, Sixteenth Edition. Michael D. Larrañaga, Richard J. Lewis, Sr., and Robert A. Lewis.
© 2016 John Wiley & Sons, Inc. Published 2016 by John Wiley & Sons, Inc.

Use: Treatment of ovarian cancer. See taxol.

paclobutrazol. See trimmit.

padan. [S,S'(2-dimethylaminotrimethylene-bisthiocarbamate)].
CAS: 15263-52-2. $C_7H_{15}O_2N_2S_2Cl$.
Hazard: Toxic by ingestion.
Use: Insecticide.

PAHA. See p-aminohippuric acid.

paint. A uniformly dispersed mixture having a viscosity ranging from a thin liquid to a semisolid paste and consisting of (1) a drying oil, synthetic resin, or other film-forming component, called the binder; (2) a solvent or thinner; and (3) an organic or inorganic pigment. The binder and the solvent are collectively called the vehicle. Paints are used (1) to protect a surface from corrosion, oxidation, or other type of deterioration, and (2) to provide decorative effects.
Hazard: Flammable, dangerous fire risk (except water based). Toxic if vapors are inhaled over a long period. The lead content of household paints is limited to 0.5%.
See paint; emulsion; vehicle; protective coating; antifouling paint.

paint, emulsion. (latex paint). A paint composed of two dispersions: (1) dry powders (colorants, fillers, extenders) and (2) resin. The former is obtained by milling the dry ingredients into water. The resin dispersion is either a latex formed by emulsion polymerization or a resin in emulsion form. The two dispersions are blended to produce an emulsion paint. Surfactants and protective colloids are necessary to stabilize the product. Emulsion paints are characterized by the fact that the binder is in a water-dispersed form, whereas in a solvent paint it is in soluble form. The principal latex paints are styrene-butadiene, polyvinyl acetate, and acrylic resins. Percentage composition may be 25–30% dry ingredients, 40% latex, and 20–30% water, plus stabilizers. The unique properties of emulsion paints are ease of application, absence of disagreeable odor, and nonflammability. They can be used on both interior and exterior surfaces.

paint, inorganic. A potassium silicate-based, corrosion-resistant coating designed for use on bridges and other metalwork subject to marine environments.

paint, metallic. A paint in which the primary pigment is a finely divided metal dispersed in the vehicle. Most common is aluminum paint, but other metals are also used.

paint remover. (varnish remover). A mixture in liquid or paste form containing volatile solvents

and nonvolatile components that retard evaporation of the solvent, thereby prolonging its action. Typical solvents are methanol, denatured ethanol, methylene chloride, toluene, benzene, and ethyl acetate. Paraffin is often used as the retarder. Caustic removers contain sodium phosphate, sodium silicate, caustic soda, or the like.

paint, water-based. See paint, emulsion.

Palacet. A series of organic pigments used for dyeing and printing on acetate, nylon, and polyester fibers.

Palatin. Metallized acid dyestuffs approaching the fastness of chrome colors.

palindrome. Biology: A segment of duplex DNA in which the base sequences of the two strands is the same from each end of the strands.

palconic acid. An alkali-soluble extract of redwood bark dust. A mixture of partially methylated phenolic acids.

palladium.
CAS: 7440-05-3. Pd. Metallic element of atomic number 46, aw 106.4, valences of 2, 3, 4, group VIII of the periodic table; there are 6 stable isotopes.
Properties: Silver-white, ductile metal that does not tarnish in air. It is the least noble (most reactive) of the platinum group. Absorbs up to 800 times its own volume of hydrogen. Attacked by hot, concentrated nitric acid and boiling sulfuric acid, soluble in aqua regia and fused alkalies, insoluble in organic acids, good electrical conductor. D 12.0, mp 1554C, bp 2800C, Mohs hardness 4.8, Brinell 61, Vickers (annealed) 41. Noncombustible, except as dust.
Occurrence: Siberia, Ural Mountains (the former U.S.S.R.), Ontario, South Africa.
Derivation: In ores with platinum, gold, copper, etc. Concentrated ores are dissolved in aqua regia; after gold and platinum are removed by chemical treatment, palladium is precipitated by ammonia, followed by hydrochloric acid. After further purification treatment, ignition yields palladium metal.
Available forms: Wire, leaf, powder, single crystals.
Grade: CP (99.99%), technical (99.0%).
Use: Alloys for electrical relays and switching systems in telecommunication equipment, catalyst for reforming cracked petroleum fractions and hydrogenation, metallizing ceramics, "white" gold in jewelry, resistance wires, hydrogen valves (in hydrogen separation equipment), aircraft spark plugs, protective coatings.

palladium chloride. (palladous chloride; palladium dichloride).
CAS: 7647-10-1. (1) $PdCl_2$. (2) $PdCl_2 \cdot 2H_2O$.
Properties: Dark brown, deliquescent powder or crystals. (1) D 4.0 (18C), mp 675C (decomposes).

Soluble in water, hydrochloric acid, alcohol, and acetone.
Derivation: By solution of palladium in aqua regia and evaporation.
Grade: Technical, reagent.
Use: Analytical chemistry, "electroless" coatings for metals, photography, leak detection in gas lines, indelible inks, catalyst.

palladium diacetate.
CAS: 3375-31-3. $(CH_3COO)_2Pd$.
Properties: Reddish-brown, crystalline solid. Decomposes at 200C. Insoluble in water and alcohols; soluble in acetone, chloroform, acetonitrile.
Derivation: Reaction of palladium nitrate or palladium sponge with glacial acetic acid.
Use: Catalyst for organic reactions.

palladium iodide.   (palladous iodide). PdI_2.
Properties: Black powder. D 6.003 (18C). Soluble in a solution of potassium iodide; insoluble in alcohol, water, and ether; decomposes at 350C.

palladium monoxide.   See palladium oxide.

palladium   nitrate.   (palladous   nitrate). $Pd(NO_3)_2$.
Properties: Brown salt, deliquescent, decomposed by heat. Soluble in water with turbidity; soluble in dilute nitric acid.
Hazard: Oxidizing agent, may react with organic materials.
Use: Analytical reagent, catalyst.

palladium oxide.   (palladium monoxide). PdO.
Properties: Black-green or amber solid. D 8.70 (20C), mp 750C (decomposes). Soluble in dilute acids.
Derivation: Careful ignition of the nitrate or prolonged heating of the finely divided metal at 800C.
Use: Reduction catalyst in organic synthesis.

palladium potassium chloride.   (palladous potassium chloride; potassium palladium chloride). $PdCl_2 \cdot 2KCl$.
Properties: Reddish-brown crystals. D 2.67, mp 524C. Soluble in water; slightly soluble in hot alcohol.
Use: Reagent for carbon monoxide determination.

palladium   sodium   chloride.   (palladous sodium chloride; sodium palladium chloride). $NaPdCl_2 \cdot 3H_2O$.
Properties: Brown salt. Hygroscopic, soluble in alcohol and water.
Use: Analysis (testing for carbon monoxide, ethylene, illuminating gas, iodine).

Palladon.   *p*-nitrosodimethylaniline used in the colorimetric determination of palladium and platinum.

palladous chloride.   See palladium chloride.

palladous iodide.   See palladium iodide.

palladous nitrate.   See palladium nitrate.

palladous potassium chloride.   See palladium potassium chloride.

palladous sodium chloride.   See palladium sodium chloride.

pallet.   A low platform of wood or metal used for transportation or temporary storage of materials or semifinished products, it stands on supports that are high enough to permit handling by forklift trucks.

palmarosa oil.   (geranium oil, Turkish).
CAS: 8014-19-5. A light-yellow essential oil consisting chiefly of geraniol; optically active.
Use: Source of geraniol; perfumes and flavors.

palm butter.   See palm oil.

Palmetto.   Agricultural dusting sulfur.
Use: Insecticide and fungicide.

palmitamide.
Properties: Semisynthetic compound derived from palm oil.
Use: Food additive; packaging material release agent.

palmitic acid.   (hexadecanoic acid; cetylic acid).
CAS: 57-10-3. $CH_3(CH_2)_{14}COOH$. A saturated fatty acid, it occurs in natural fats and oils, tall oil, and most commercial-grade stearic acid.
Properties: White crystals. D 0.8414 (80/4C), mp 62.9, bp 351.5C, 271.5C (100 mm Hg), 139.0C (1 mm Hg), refr index 1.4309 (70C). Soluble in hot alcohol and ether; insoluble in water. Combustible.
Derivation: From spermaceti by saponification and from palm oil, hydrolysis of natural fats.
Method of purification: Crystallization.
Grade: Technical, 99.8%, FCC.
Use: Manufacture of metallic palmitates, soaps, lubricating oils, waterproofing, food-grade additives.

palmitic acid amide.   See palmitamide.

palmitic acid cetyl ester.   See cetin.

palmitin.   See tripalmitin.

palmitoleic acid.   (*cis*-9-hexadecenoic acid).
$CH_3(CH_2)_5CH:CH(CH_2)_7COOH$. An unsaturated fatty acid found in nearly every fat, especially in marine oils (15–20%).
Properties: Colorless liquid. Mp 1.0C, bp 140–141C (5 mm Hg). Insoluble in water; soluble in alcohol and ether. Combustible.

Grade: Purified product 99%.
Use: Organic synthesis, chromatographic standard.

palmitoyl chloride. (hexadecanoyl chloride; palmitoyl chloride, so-called). $CH_3(CH_2)_{14}COCl$.
Properties: Colorless liquid. Mp 11–12C, bp 194.5C (17 mm Hg). Soluble in ether, decomposes in water or alcohol.

palmityl alcohol. See cetyl alcohol.

palmityltrimethylammonium chloride. See hexadecyltrimethylammonium chloride.

palm kernel oil (unhydrogenated).
Properties: From the kernel of the fruit of the oil palm *Elaeis guineensis*. A fatty solid; characteristic sweet nutty flavor.
Use: Food additive.

palm nut cake. (palm cake). The cakes formed in the press when the palm nut kernels are expressed to obtain the oil. Contains various useful constituents, such as unexpressed oil, carbohydrates, proteins, and salts. Typical analysis: proteins 30.4%, fats 8.4%, fiber 41.0%, water 9.5%, ash 10.6%.
Use: Cattle-food, fertilizer ingredient.

palm oil.
Properties: Yellow-brown, buttery, edible solid at room temperature. D 0.952, mp 30C, iodine number 13.5, saponification number 247.6. Soluble in alcohol, ether, chloroform, carbon disulfide. Combustible.
Occurrence: Oil palms are native to several countries in central Africa and are extensively cultivated in Malaysia, which is its chief commercial source. It is also produced in Indonesia.
Use: Soap manufacture, pharmacy, food shortening, cutting-tool lubricant, hot-dipped tin coating, terne plating, cosmetics, softener in rubber processing, cotton-goods finishing, substitute for tallow as mold-release agent.

palm oil (unhydrogenated).
Properties: From the pulp of the fruit of the oil palm *Elaeis guineensis*. A deep orange-red fatty semisolid @ 21–27C; characteristic sweet nutty flavor.
Use: Food additive.

2-PAM. Abbreviation for 2-pyridine aldoxime methiodide.

"Pamak" [Mead Westvaco]. TM for various tall oil products including a series of tall oil fatty acids and distilled tall oils containing varying percentages of rosin acids. "Pamak" TP and WTP are residues from fractionation of crude tall oil in the manufacture of tall oil fatty acids.

pamaquine naphthoate. $C_{42}H_{45}N_3O_7$.
Properties: Yellow to orange-yellow; odorless; almost tasteless powder. Insoluble in water; soluble in alcohol and acetone.
Use: Medicine (antimalarial).

"Pamolyn" [Eastman]. TM for a low odor oleic acid.

pancreas. A gland in the abdominal cavity of vertebrates that secretes digestive enzymes into the small intestine. It also secretes insulin and glucagon into the circulating blood where they control blood glucose levels. It also secretes somatostatin, a hormone that inhibits the release of somatotropin, a growth hormone.

pancreatic juice. The external secretion of the pancreas.
Properties: Clear alkaline fluid contains digestive enzymes: a-amylase, chymotrypsinogen, nucleases, trypsinogen, and triacylglycerol lipase.

pancreatin. A mixture of enzymes, principally pancreatic amylase, trypsin, and pancreatic lipase. Obtained from the pancreas of hog or ox.
Properties: Cream-colored amorphous powder; characteristic odor. Acts upon starch and proteins. Soluble in water; insoluble in alcohol. It changes protein into proteoses and derived substances, and starch into dextrins and sugars. Its greatest activity is in neutral or slightly alkaline media.
Derivation: Pancreas gland is extracted by macerating with chloroform, water, dilute boric acid, glycerol, or alcohol; filtered; and evaporated.
Grade: NF.
Use: Preparation of so-called predigested protein nutrients; in bating compounds of leather; to remove starch and protein sizings from textiles.

Paneth technique. Method demonstrating the existence of free radicals (e.g., methyl) or atoms, which is based on the removal of a metallic "mirror" by a stream of gas containing the radicals. The reaction products can be collected and assayed.

Pano-drench. A liquid soil-treatment concentrate containing 0.6% cyano(methylmercuri)-guanidine.
Hazard: Toxic by ingestion.

"Pan-o-lite" [ICL]. TM for monobasic calcium phosphate.
Use: Food chemical used in waffle, pancake, and cake mixes.

pantethine.
CAS: 16816-67-4. $C_{22}H_{42}N_4O_8S_2$. The disulfide form of *N*-pantothenylthioethanolamine. *Lactobacillus bulgaricus* growth factor (LBF). A fragment of coenzyme A, a pantothenic acid derivative.
Use: Biochemical research.

panthenol.   USAN for pantothenol.

pantocaine.   (4-butylaminobenzoic-β-dimethy-laminoethyl ester hydrochloride).
$C_4H_9NHCOOCH_2CH_2N(CH3)_2\cdot HCl$.
Use: Local anesthetic.

pantolactone.   (2,4-dihydroxy-3,3-dimethyl-butyric acid; γ-lactone).

$\overline{HC(OH)C(CH_3)_2CH_2OCO}$.

Properties: Crystals. D 1.180 (20/20C), mp 79.2C. Soluble in water.
Grade: 80% aqueous solution.
Use: Preparation of pantothenic acid.

pantothenic acid.   [*N*-(2-4-dihydroxy-3,3-dimethylbutyryl)-β-alanine].

$$\underset{\underset{OH}{|}}{HOOC-CH_2-CH_2-NH-\overset{\overset{O}{\|}}{C}-\underset{\underset{CH_3}{|}}{CH}-\overset{\overset{CH_3}{|}}{C}-CH_2-OH}$$

A member of the vitamin-B complex; it is a component of coenzyme A and may be considered a β-alanine derivative with a peptide linkage. It is involved in the release of energy from carbohydrate utilization and is necessary for synthesis and degradation of fatty acids, sterols, and steroid hormones; it also functions in the formation of porphyrins. It occurs in all living cells and tissues. The natural product is dextrorotatory [D(+)] and is the only form having vitamin activity.
Properties: Viscous, hygroscopic liquid. Soluble in water, ethyl acetate, glacial acetic acid; insoluble in benzene.
Source: (Food sources) Liver, kidney, yeast, crude molasses, milk, whole-grain cereals, rice. (Commercial sources) Produced synthetically from 2,4-dihydroxy-3,3-dimethylbutyric acid and β-alanine. See calcium pantothenate.

pantothenol.   (D(+)-pantothenyl alcohol; panthenol).
CAS: 81-13-0. $HOCH_2C(CH_3)_2CHOHCONH-(CH_2)_2CH_2OH$. The alcohol corresponding to pantothenic acid, with vitamin activity.
Properties: Viscous liquid. Specific rotation +28.36 to 30.7 degrees in water (c = 5); refr index 1.497 (20C). Soluble in water, ethanol, methanol.
Use: Biochemical research, food additive and dietary supplement.

papain.   (papayotin).
CAS: 9001-73-4.
Properties: White or gray, powder. Slightly hygroscopic; partially soluble in water and glycerol; insoluble in common organic solvents. The most thermostable enzyme known. Digests proteins.

Derivation: Obtained as dried and purified latex of *Carica papaya*.
Grade: Technical, purified. Technical grade is susceptible to decomposition in storage.
Use: Meat tenderizer, other food industries (mainly to prevent protein haze on chilling beer), tobacco, pharmaceutical, cosmetic, leather, textiles.

papaver.   Pertaining to poppies of the genus *Papaver* such as the opium poppy, *Papaver somniferum*.

papaverine.   (6,7-di-methoxy-1-veratrylisoquinoline).
CAS: 58-74-2. $(CH_3O)_2C_6H_3CH_2C_9H_4N(OCH_3)_2$. An alkaloid.
Properties: White, crystalline powder. Mp 147C. Soluble in chloroform, hot benzene, aniline, glacial acetic acid, and acetone; slightly soluble in alcohol and ether; insoluble in water.
Derivation: From opium or by synthesis.
Hazard: Toxic narcotic.
Use: A vasodilator used for treatment of hypertension (also as the hydrochloride which is soluble in water).

paper.   A semisynthetic product made by chemically processing cellulosic fibers. A wide variety of sources have been used for specialty papers (flax, bagasse, esparto, straw, papyrus, bamboo, jute, and others), but by far the largest quantity is made from softwoods (coniferous trees), such as spruce, hemlock, pine, etc.; some is also made from such hardwoods as poplar, oak, etc., as well as from synthetic fibers. Papermaking technology involves the following basic steps: (1) chipping or other subdivision of the logs (see groundwood); (2) manufacture of chemical or semichemical pulp by digestion in acidic or alkaline solutions, which separates the cellulose from the lignin (see pulp, paper); (3) beating the pulp to break down the fibers and permit proper bonding when the sheet is formed; (4) addition of starches, resins, clays, and pigments to the liquid stock (or "furnish"); (5) formation of the sheet continuously on a fourdrinier machine, where the water is screened out and the sheet dried by passing over a series of heated drums; (6) high-speed calendering for brightness and finish; (7) coating either by machine application or (for heavy finishes) by brushes. Further information can be obtained from the Technical Association of the Pulp and Paper Industry or from the Institute of Paper Chemistry, Appleton, WI.
See Appendix II for history of the industry.

paper chromatography.   (PC). A micro type of chromatography. A drop of the liquid to be investigated is placed near one end of a strip of paper. This end is immersed in solvent, which travels down the paper and selectively distributes the materials present in the original drop. Comparison with known substances makes identification possible.

paper, coated. A paper that is covered on one or both sides with a suspension of clays, starches, casein, rosin, wax, or combinations of these to serve special purposes. Machine-coated paper is required for standard book printing; the rather light coating is applied by any of several devices (air knife, trailing blade, or roll coater). Heavier coatings are applied by means of brushes or spreading devices. These are required for high-grade printing of magazines, art books, etc., where excellent photographic reproduction is essential. Special-purpose coatings as for packaging are applied in a separate operation.

paper, synthetic. Paper or paperlike material made from a polyolefin; polypropylene is usually selected. A paper made from styrene copolymer fibers has been developed to production stage in Japan. Plastic-coated cellulosic papers are available for children's books, posters, and similar applications.

"Papi" [ISACO]. TM for a series of methylene diphenyl diisocyanate urethane polymers. Average viscosity 250 cP at 25C.
Use: One-shot rigid urethane foams. "Papi" 50. 50% solution in monochlorobenzene. Adhesives (rubber to metals and synthetic fabrics), coating intermediate. "Papi" 94. Light-colored polymer. Foam seating and packaging.

paprika.
Properties: Ground dried pod of mild capsicum *Capsicum annuum* L.
Use: Food additive.

paprika oleoresin.
Properties: Derived from organic solvent extraction of ground dried pod of mild capsicum *Capsicum annuum* L.
Use: Food additive.

para-. (*p*-). A prefix.
See ortho-.

Parabens. The methyl, propyl, butyl, and ethyl esters of *p*-hydroxybenzoic acid. Antimicrobial agents for foods and pharmaceuticals. Approved by FDA as GRAS.
See GRAS.

paracasein. See casein.

paracetaldehyde. See paraldehyde.

"Paracol" [Hercules]. TM for a series of wax and wax-rosin emulsions produced from paraffin waxes, microcrystalline waxes, or combinations of these waxes with rosin.
Use: Impart water resistance to paper and allied materials.

"Paracril" [Paratec]. TM for a group of synthetic rubbers of the Buna-N or nitrile type, produced by the copolymerization of butadiene and acrylonitrile. Resist deterioration by aliphatic hydrocarbon, mineral and vegetable oils and animal fats and oils, and are particularly resistant to petroleum products. "Paracril" is also used as a plasticizer for vinyls and other thermoplastic and thermosetting resins.

"Paradene" [Neville]. TM for low-priced, dark, thermoplastic, coal tar resins (coumarone-indene) available in low to high softening-point ranges.
Use: Rubber compounding.

paradigm. In biochemistry, an experimental model held to be true.

para-dimethylaminocinnamaldehyde.
(PDMAC). A reagent that reacts with urea, amines and their salts.
Use: In forensics to develop details of friction ridges on human digits when exposed to selected wavelengths of light.

paraffin. (1) Also called alkane. A class of aliphatic hydrocarbons characterized by a straight or branched carbon chain; generic formula C_nH_{2n+2}. Their physical form varies with increasing molecular weight from gases (methane) to waxy solids. They occur principally in Pennsylvania and mid-continent petroleum. (2) Paraffin wax.

paraffin, chlorinated. A paraffin oil or wax in which some of the hydrogen atoms have been replaced by chlorine atoms. Nonflammable.
Hazard: Possible carcinogen.
Use: High-pressure lubricants, as flame retardants in plastics and textiles, as plasticizer for polyvinyl chloride in polyethylene sealants, and in detergents.

paraffin oil. An oil either pressed or dry-distilled from paraffin distillate. Liquid petrolatum is also known as paraffin oil. Combustible.
Grade: By viscosity and color.
Use: Floor treatment, lubricant.

paraffin distillate. A distilled petroleum fraction that when cooled consists of a mixture of crystalline wax and oil.

paraffin wax. (paraffin scale; paraffin).
CAS: 8002-74-2.
Properties: White, translucent solid; tasteless; odorless. Consisting of a mixture of solid hydrocarbons of high molecular weight, e.g., $C_{36}H_{74}$. D 0.880–0.915, mp 47–65C, flash p 390F (198C), autoign temp 473F (245C). Soluble in benzene, ligroin, warm alcohol, chloroform, turpentine, carbon disulfide, and olive oil; insoluble in water and acids. Combustible.

Grade: Yellow crude scale, white scale, refined wax, ASTM, NF. Also graded by melting point in F and color. The higher-melting grades are more expensive.
Hazard: Many waxes contain carcinogens. Upper respiratory tract irritant and nausea.
Use: Candles; paper coating; protective sealant for food products, beverages, etc.; glass-cleaning preparations; hot-melt carpet backing; biodegradable mulch (hot melt-coated paper); impregnating matches; lubricants; crayons; surgery; stoppers for acid bottles; electrical insulation; floor polishes; cosmetics; photography; antifrothing agent in sugar refining; packing tobacco products; protecting rubber products from sun-cracking; chewing-gum base (to ASTM specifications).

"Paraflint" *[Sasol].*   TM for a polymethylene wax. It is white, odorless, and has congealing point of 96C. Available in flaked form. Approved by FDA.

"Para-Flux" *[Hallstar].*   TM for a series of petroleum-based plasticizers for natural and synthetic rubbers.

paraformaldehyde.   (paraform).
CAS: 30525-89-4.   HO(CH$_2$O)$_n$H. A polymer of formaldehyde in which n equals 8–100. Not to be confused with the trimer, symtrioxane.
Properties: White solid; slight odor of formaldehyde. Melting range 120–170C, flash p 160F (71C) (CC), autoign temp 572F (300C). Insoluble in alcohol and ether; soluble in strong alkali solution. The higher polymers are insoluble in water Combustible.
Derivation: By evaporating an aqueous solution of formaldehyde.
Available forms: Flake, powder.
Grade: Bags, carlots.
Hazard: Toxic by ingestion.
Use: Fungicides; bactericides; disinfectants; adhesives; hardener and waterproofing agent for gelatin; contraceptive creams.

paraldehyde.   (2,4,6-trimethyl-1,3,5-trioxane).
CAS: 123-63-7. C$_6$H$_{12}$O$_3$. A cyclic polymer (trimer) of acetaldehyde. A depressant drug; may be addictive.
Properties: Colorless liquid; disagreeable taste; agreeable odor. D 0.9960 (20/20C), bp 124.5C, mp 12.6C, vap press 25.3 mm Hg (20C), flash p 96F (35.5C) (TOC), sp heat 0.434, refr index 1.40–1.42 (20C), bulk d 8.27 lb/gal (20C), autoign temp 460F (237C). Decomposes on standing; stable toward alkalies but slowly decomposed to acetaldehyde when treated with a trace of mineral acid; miscible with most organic solvents and volatile oils; soluble in water.
Derivation: Action of hydrochloric acid and sulfuric acid upon acetaldehyde.
Grade: Technical, USP.

Hazard: Flammable, moderate fire risk. Toxic by ingestion.
Use: Substitute for acetaldehyde; rubber accelerators; rubber antioxidants; synthetic organic chemicals; dyestuff intermediates; solvent for fats, oils, waxes, gums, resins; leather; solvent mixture for cellulose derivatives; sedative (hypnotic).

paralytic shellfish poisoning.   See red tide.

paramagnetic analytical methods.   Analysis of fluid mixtures by measurement of the paramagnetic susceptibilities of materials when exposed to a magnetic field.
See electron paramagnetic resonance; magnetochemistry.

paranitraniline.   See *p*-nitroaniline.

paranitraniline red.   See para red.

para-oxon.   (diethyl   para-nitrophenyl   phosphate).
CAS: 311-45-5.   (C$_2$H$_5$O)$_2$P(O)OC$_6$H$_4$NO$_2$. Generic name for the oxygen analog of parathion.
Properties: Odorless, reddish-yellow oil. Bp 148–151C (1 mm Hg), d 1.269 (25/25C), refr index 1.5060 (25C). Slightly soluble in water; soluble in most organic solvents. Decomposes rapidly in alkaline solutions.
Hazard: Poison by ingestion, inhalation, and skin absorption; cholinesterase inhibitor; use may be restricted.
Use: Insecticide.

paraphyletic.   Term applied to a group of organisms which include the most recent common ancestor of all of its members, but not all of the descendants of that most recent common ancestor.

"Paraplex" *[Standard].*   TM for polymeric plasticizers for polymers and resinous coatings. Primarily polyesters, but some are epoxidized oils that impart heat and light stability as well as plasticization. Supplied as viscous liquids in a range of molecular weights all at 100% solids. Compatible with polyvinyl chloride, polyvinyl butyral, cellulosics, and other high polymers and elastomers.
Use: Calendered sheet and film; extruded and molded items; electrical wire insulation; coatings for wood, metal, fabrics, and paper.

paraquat.   (generic name for 1,1'-dimethyl-4,4'-bipyridinium salt).
CAS: 4685-14-7.   [CH$_3$(C$_5$H$_4$N)$_2$CH$_3$]•2CH$_3$SO$_4$.
Properties: Yellow solid. Soluble in water.
Hazard: Highly toxic by ingestion, inhalation, and skin absorption; use is restricted. Lung damage.
Use: A widely used nonselective herbicide that kills all green plant tissue it contacts, especially grasses and weeds. Has been used to kill marijuana crops.

paraquat dibromide.
CAS: 3240-78-6. $C_{12}H_{14}N_2 \cdot 2Br$.
Hazard: Human systemic effects.
Use: Agricultural chemical.

paraquat dichloride.   (1,1'-dimethyl-4,4'-bipyridynium dichloride).
CAS: 1910-42-5. $C_{12}H_{14}N_2 \cdot 2Cl$.
Properties: Yellow solid. Very air-sensitive needles from MeOH/Me₂CO. Mp 300C (decomp), mw 257.18. Very soluble in water; slightly soluble MeOH, EtOH. Insoluble in nonpolar solvents.
Hazard: A poison.
Use: Herbicide.

paraquat dihydride.   (1,1'-dimethyl-4,4'-bipyridinium dihydrate; dipyridyldihydrate).
CAS: 4685-14-7. $C_{12}H_{14}N_2 \cdot 2H_2O$.
Properties: Mw 222.32.
Hazard: A poison.
Use: Herbicide.

para red.   (paranitraniline red).
$C_{10}H_6(OH)NNC_6H_4NO_2$. A pigment formed by coupling diazotized p-nitroaniline with β-naphthol. The term is also used to refer to a group of lakes based on this dye.
See para toner.

Para Resin 2457.   A petroleum-base resin.
Use: Plasticizer and softener for rubbers.

pararosaniline.   (CI 42500).
CAS: 569-61-9. $HOC(C_6H_4NH_2)_3$. A triphenyl-methane dye. Component of fuchsin.
Properties: Colorless to red crystals. Mp 205C. Soluble in alcohol; very slightly soluble in water and ether. Combustible.
Use: Dye (usually as the hydrochloride).

pararosolic acid.   See aurin.

parathion.   (generic name for O,O-diethyl-O,p-nitrophenyl phosphorothioate; ethyl parathion; O,O-diethyl-p-nitrophenyl thiophosphate; AATP).
CAS: 56-38-2. $(C_2H_5O)_2P(S)OC_6H_4NO_2$.
Properties: Deep brown to yellow liquid; usually has faint odor. Refr index 1.5367 (25C), d 1.26 (25/4C), bp 375C, fp 6C, vap press 0.003 mm Hg (24C). Very slightly soluble in water (20 ppm); completely soluble in esters, alcohols, ketones, ethers, aromatic hydrocarbons, animal and vegetable oils; insoluble in petroleum ether, kerosene, spray oils; stable in distilled water and in acid solution; hydrolyzed in the presence of alkaline materials; slow decomposition in air.
Grade: Technical grade is 95% pure. Also supplied diluted with inert carriers of various types, and in various proportions.
Derivation: From sodium ethylate, thiophosphoryl chloride, and sodium p-nitrophenate.

Hazard: Highly toxic by skin contact, inhalation, or ingestion; cholinesterase inhibitor. Repeated exposure may, without symptoms, be increasingly hazardous. Fatalities have resulted from its accidental use; use may be restricted. Toxic by skin absorption. Questionable carcinogen.
Use: Insecticide and acaricide.
See methyl parathion.

para toner.   An insoluble red pigment derived from β-naphthol and p-nitroaniline. The former is sometimes partly replaced by mono-acid F, 2-naphthol-7-sulfonic acid. By varying the conditions of temperature and acid concentration, different shades may be obtained.
Use: Paint and printing-ink pigments, making para lakes.

Parco.   Phosphoric acid and phosphate compounds for dissolving rust from the surface of metal.

"Parcolene" [Henkel].   TM for chemicals for treating metal surfaces to remove extraneous matter and/or condition the surface before other finishing operations.

paregoric.   Contains a derivative of opium that is habituating on continued use. Its use is restricted by the FDA.
Use: Medicine, especially for digestive disorders.

"Parez" [Kemira].   TM for a series of melamine-formaldehyde and urea-formaldehyde resins, designed for use in papermaking.

Paris green.   See copper acetoarsenite.

parison.   (1) An unformed mass of molten glass from which finished products are manufactured. (2) An extruded tube of plastic from which toys and similar items are made by blow molding.

Paris white.   See whiting.

Parkes process.   A standard process for the separation of silver from lead. 1–2% molten zinc is added to the lead-silver mixture, heated to above the melting point of zinc. A scum containing most of the silver and zinc forms on the surface; this is separated and the silver recovered. The separation of silver is not complete, and the process is repeated several times.

paromomycin sulfate. $C_{23}H_{47}N_5O_{18}S$. Antibiotic from a strain of *Streptomyces*.
Properties: Creamy white powder; odorless; hygroscopic. Soluble in water; insoluble in chloroform and ether.
Grade: ND.
Use: Medicine (antimicrobial).

parsimony. Refers to a rule used to choose among possible cladograms which states that the cladogram implying the least number of changes in character states is the best.

partial pressure. The pressure due to one of the several components of a gaseous or vapor mixture. In general this pressure cannot be measured directly but is obtained by analysis of the gas or vapor and calculated by use of Dalton's law.
See Raoult's law.

particle. Any discrete unit of material structure; the particulate basis of matter is a fundamental concept of science. The size ranges of particles may be summarized as follows: (1) Subatomic: protons, neutrons, electrons, deuterons, etc. These are collectively called fundamental particles. (2) Molecular: includes atoms and molecules with size ranging from a few angstroms to half a micron. (3) Colloidal: includes macromolecules, micelles, and ultrafine particles such as carbon black, resolved via electron microscope, with size ranges from 1 millimicron up to lower limit of the optical microscope (1 micron). (4) Microscopic: units that can be resolved by an optical microscope (includes bacteria). (5) Macroscopic: all particles that can be resolved by the naked eye.
See fundamental particle; particle size.

particle accelerator. A device in which the speed of charged subatomic particles (protons, electrons) and heavier particles (deuterons, alpha particles) can be greatly increased by application of electric fields of varying intensity, often in conjunction with magnetic fields. It is possible to accelerate electrons and protons to speeds approaching the speed of light if sufficiently high voltage is used. Straight-line (linear) accelerators are used for protons, and doughnut-shaped betatrons for electrons; other types are the Van de Graaff electrostatic generator, the synchrotron, and the cyclotron. Before the development of nuclear reactors, the cyclotron was used to accelerate deuterons for use in bombarding stable nuclei to produce neutrons for inducing artificial radioactivity, fission, and formation of synthetic (transuranic) elements.
See betatron; cyclotron.

particle size. This term refers chiefly to the solid particles of which industrial materials are composed (carbon black, zinc oxide, clays, pigments, and the like). The smaller the particle, the greater will be the total exposed surface area of a given mass. Activity is a direct function of surface area; i.e., the finer a substance is, the more efficiently it will react, both chemically and physically. A colloidal pigment is a more effective colorant than a coarse one because of the greater surface area of its particles. A pound of channel carbon black has a surface area of 18 acres, which largely accounts for its powerful reinforcing effect in rubber. Thus, ultrafine grinding of powders is of utmost importance in such products as paints, cement, plastics, rubber, dyes, pharmaceuticals, printing inks, and numerous others.
See particle; surface chemistry; colloid chemistry; sedimentation.

particulate matter. Solid or liquid matter that is dispersed in a gas, or insoluble solid matter dispersed in a liquid, that gives a heterogeneous mixture.

parting agent. See abherent.

partition chromatography. The method of chromatography in which equilibrium is established between two liquid phases, one of which is held in the form of a gel.
See liquid chromatography.

partition coefficient. A constant that expresses the ratio in which a given solute will be partitioned or distributed between two given immiscible liquids at equilibrium, or between a solvent and adsorbent.

partition function. An equation giving the distribution of molecules in different energy states in a system. Symmetry effects on reaction rates arise when a reaction requires overlap between two lobes of the orbitals on each of two reagents: if the algebraic signs of the wave functions in the facing lobes do not match, bond formation between those orbitals is prohibited.

parylene. Generic name for thermoplastic film polymers based on p-xylylene and made by vapor-phase polymerization.
Derivation: p-xylene, $CH_3C_6H_4CH_3$, is heated with steam at 950C to produce the cyclic dimer di-p-xylylene, a solid that can be separated in pure form. The dimer is then pyrolyzed at 550C to produce monomer vapor of p-xylylene, CH_2:C_6H_4:CH_2, which is then cooled below 50C and condenses on the desired object as a polymer having the repeating structure $-CH_2C_6H_4CH_2-)_n$, with n about 5000 and molecular weights of about 500,000. The polymer is used as a protective coating. Films as thin as 500 Å to 5 mils are obtained.
Use: Thin coatings of high purity and uniformity on almost any substrate that will resist a high vacuum, as paper, fabric, polyethylene and polystyrene film, ceramics, metals, many solid chemicals; electronic miniaturization systems; capacitors; thin film circuits.

"Parzate" *[Du Pont].* TM for a series of fungicides. "Parzate" liquid is a solution containing 22% nabam to be combined with zinc sulfate in the spray tank. "Parzate" carbon is a wettable powder containing 75% zineb. "Parzate" D is a finely divided powder containing 85% zineb.
Hazard: Irritant to skin and mucous membranes.

PAS.   Abbreviation for *p*-aminosalicylic acid. See 4-aminosalicylic acid.

Paschen series.   One of the hydrogen spectral series in the infrared region.

Passerini reaction.   Formation of 2-acyloxy amides on treatment of an isonitrile with a carboxylic acid and an aldehyde or ketone.

passive transport.   Transport of a molecule across a membrane, down a concentration gradient which does not require the input of energy.

passivity.   A property shown by iron, chromium, and related metals involving loss of their normal chemical activity in an electrochemical system or in a corrosive environment after treatment with strong oxidizing agents, like nitric acid, and when oxygen is evolved upon them during electrolysis, forming an oxide coating.

PAS sodium.   See sodium-*p*-aminosalicylate.

paste.   (1) An adhesive composition of semisolid consistency, usually water dispersible. The common pastes are based on starch, dextrin, or latex often, with the addition of gums, glues, and antioxidants. They are widely employed for the adhesion of paper and paperboard. (2) More generally, a soft, viscous mass of solids dispersed in a liquid. For example, paste resins are finely divided resins mixed with plasticizers to form fluid or semifluid mixtures, without the use of low-boiling solvents or water emulsions.

paste solder.   A paste (2) containing flux, cleaner, tinning agent, and powdered metallic solder.

pasteurization.   Heat treatment of milk, fruit juices, canned meats, egg products, etc. for the purpose of killing or inactivating disease-causing organisms. For milk, the minimum exposure is 62C for 30 min or 72C for 15 sec, the latter being called flash pasteurization. Although this treatment kills all pathogenic bacteria and also inactivates enzymes that cause deterioration of the milk, the shelf life is limited. To prolong storage life, temperatures of 80–88C for 20–40 sec must be used. Complete sterilization requires ultrahigh pasteurization at from 94C for 3 sec to 150C for 1 sec. In-can heating at 116C for 12 min and 130C for 3 min is also employed for maximum stability and long storage life. Some meat products are pasteurized by α-radiation.

Pasteur, Louis.   (1822–1895). A French chemist and bacteriologist who made three notable contributions to science: (1) As a result of extensive study of fermentation, which led him to conclude that it is caused by infective bacteria, he extended the work of Jenner on smallpox serum made from cowpox

(1775) to development of the concept of immunizing serums and the antibody–antigen relationship (1880). Pasteur was the first to inoculate for rabies and anthrax, and suggested the term *vaccination* (from Latin *vaccus* for "cow") in recognition of Jenner's achievement. (2) Initiation of the practice of heat-treating wine, and later milk and other food products, to kill or inactivate toxic microorganisms, especially the tuberculosis bacillus. (3) Discovery of the optical properties of tartaric acid, present in wine residues, which laid the basis for modern knowledge of optical isomers (right- and left-handed molecular structure), a phenomenon now often called chirality.
See pasteurization; optical isomer.

Pastewiz.   A polymer paste release agent.
Use: Hand application.

PAT.   Abbreviation for polyaminotriazole.

patchouli oil.   A yellow to brownish essential oil used in perfumery and flavoring. It is strongly levorotatory.

patentability.   The qualifications for obtaining a patent on an invention or chemical process. These are (1) the invention must not have been published in any country or in public use in the U.S., in either case for more than 1 year before the date of filing the application; (2) it must not have been known in the U.S. before date of invention by the applicant; (3) it must not be obvious to an expert in the art; (4) it must be useful for a purpose not immoral and not injurious to the public welfare; (5) it must fall within the five statutory classes on which patents may be granted, i.e., (a) composition of matter, (b) process of manufacture or treatment, (c) machine, (d) design (ornamental appearance), or (e) a plant produced asexually. Special regulations relate to atomic energy developments and subjects directly affecting national security (Robert Calvert).
Note: In 1980, the Supreme Court in a landmark decision upheld the patentability of synthetic bacteria created by recombinant DNA techniques.

patent alum.   See aluminum sulfate.

patent leather.   Fashion leather used chiefly for formal shoes, bags, etc. characterized by a high, glossy finish applied as the final step in processing. The finish greatly reduces the poromeric nature of the leather.

Paterno–Buchi reaction.   Formation of oxetanes by photochemical cycloaddition of carbonyl compounds to olefins.

pathfinder element.   An element present in small proportions less than 1%, generally metallic in nature, associated with ore deposits at the time of formation. Mapping of the concentration variation of the selected element serves to locate the main ore deposit. Examples are zinc as the pathfinder for lead, copper, and silver ores, and molybdenum associated with copper deposits.

pathway.   A sequence of reactions, usually of a biochemical nature, in which more-complex substances are converted to simple end products, as in the degradation of the components of foods to carbon dioxide and water. Its course is determined largely by preferential factors involving coenzymes and other catalysts. An example is the TCA cycle, which is the common pathway in the degradation of foodstuffs and cell constituents to carbon dioxide and water.

patina.   Variously used to refer to an ornamental and/or corrosion-resisting film on the surface of copper, copper alloys, including bronzes, and also sometimes iron and other metals. Such a film is formed by exposure to the air or by a suitable chemical treatment.

patronite.   A mixture of vanadium-bearing substances with the formula VS_4 found in Peru.

Pattinson process.   Process for the removal of silver from lead. The silver-lead mixture is melted in one of a series of pots and allowed to cool slowly. The lead, that is free from silver or poorer in silver, separates out as crystals, which are removed, leaving the silver-rich lead in the molten state. From a number of such operations in series, a lead rich in silver is obtained, collected, and the silver recovered.
See Parkes process.

Patvag.   Solutions for bonding, coating, and sealing components in auto production.

Pauli exclusion principle.   A fundamental generalization concerning the energy relationships of electrons within the atom, namely that no two electrons in the same atom have the same value for all four quantum numbers; corollary to this is the fact that only two electrons can occupy the same orbital, in which case they have opposite spins, i.e., +1/2 and −1/2. This principle has an important bearing on the sequence of elements in the periodic table and on the limiting numbers of electrons in the shells (2 in the first, 8 in the second, 18 in the third, 32 in the fourth, etc.).
See quantum number; shell; orbital theory.

Pauling, Linus.   (1901–1994). An American chemist and physicist who won the Nobel Prize for chemistry in 1954. By using X-ray diffraction analysis he determined the crystal structure of molecules. He made significant progress in the study of chemical bonds and discovered the atomic structure of many proteins including hemoglobin.

He is known to the masses for his advocacy of vitamin C. Pauling was also a recipient of the Nobel peace prize in 1962.

Paulson Tower.   See absorption tower.

Pavy's solution.   Modified Fehling's solution having enough ammonium hydroxide to redissolve any precipitated copper oxide.

Pb.   Symbol for lead.

PBAA.   Abbreviation for polybutadieneacrylic acid copolymer.

PBD.   See 1,3,4-phenylbiphenylyloxadiazole.

PBI.   See polybenzimidazole.

PBPB.   Abbreviation for pyridinium bromide perbromide.

PCB.   Abbreviation for polychlorinated biphenyl.

PCC.   See premature chromosome condensation.

PCE.   Abbreviation for pyrometric cone equivalent, a scale of melting or fusion points of refractory materials, based on comparison with the temperature at which pyrometric cones melt.

PCNB.   Abbreviation for pentachloronitrobenzene.

PCP.   (1) Abbreviation for pentachlorophenol. (2) Abbreviation for phenylcyclidene hydrochloride.

PCR.   See polymerase chain reaction.

PCTFE.   Abbreviation for polychlorotrifluoroethylene.
See chlorotrifluoroethylene resin.

pcu.   Abbreviation for pound centigrade unit, the amount of heat needed to raise 1 lb of water from 15 to 16C.
See chu.

Pd.   Symbol for palladium.

P1-derived artificial chromosome.   (PAC). One type of vector used to clone DNA fragments (100- to 300-kb insert size; average, 150 kb) in *Escherichia coli* cells. Based on bacteriophage (a virus) P1 genome.
See cloning vector.

PDB.   Abbreviation for *p*-dichlorobenzene.

PDMS.   See polydimethylsiloxane.

PE. Abbreviation for pentaerythritol and polyethylene.

Peacock. (tallow oil). Acidless tallow.
Use: Metalworking, lubricants, additives, soaps, cleaning compounds, mold releases, animal feed supplements, inks, and pigments.

peacock blue. (α,α-bis[N-ethyl-N-(4-sulfo-benzyl)aminophenyl]-α-hydroxy-o-toluenesulfonic acid sodium salt).
CAS: 2650-18-2. HSOC$_6$H$_4$COH[C$_6$H$_4$N(C$_2$H$_5$)-CH$_2$C$_6$H$_4$SO$_3$Na]$_2$. A blue organic pigment used especially in inks for multicolor printing. It is a lake of acid glaucine blue dye on alumina hydrate and is prepared from aniline, ethanol, benzyl chloride, o-chlorobenzaldehyde, sulfuric acid, and sodium bisulfite.
Note: The term *peacock blue* is sometimes applied to other pigments of similar color, such as Prussian blue, that have been treated with phosphotungstic acid.

peanut cake. The press cake resulting from the extraction of oil from the peanut.
See peanut oil meal.

peanut oil. (arachis oil; groundnut oil).
CAS: 8002-03-7. A fixed, nondrying oil.
Properties: Yellow to greenish yellow. D 0.912–0.920 (25C), solidifying p −5 to +3C, saponification value 186–194, iodine number 88–98, refr index 1.4625–1.4645 (40C), flash p 540F (282C), autoign temp 883F (472C). Soluble in ether, petroleum ether, carbon disulfide, and chloroform; insoluble in alkalies, but saponified by alkali hydroxides with formation of soaps; insoluble in water; slightly soluble in alcohol. Combustible.
Use: Substitute for olive oil and other edible oils, both hydrogenated and unhydrogenated; soaps; vehicle for medicines; salad oil; mayonnaise; margarine.

peanut oil meal. The crushed form of peanut cake resulting from the extraction of oil from the seed. Prepared with or without the shells, the oil meal of commerce contains between 39 and 45% crude protein and is sold on that basis. Typical analysis of 39% protein meal: 39.1% crude protein, 5.3% crude fiber, 34.3% nitrogen-free extract, 6.2% ether soluble (fats), 5.3% ash; total digestible nutrient 80%.
Use: Animal feeds, fertilizer ingredient.

pearl alum. See aluminum sulfate.

pearl ash. See potassium carbonate.

pearl essence. See nacreous pigment.

pearl pigment. See nacreous pigment.

pearl white. See bismuth oxychloride; bismuth subnitrate.

pear oil. See amyl acetate.

Pearson's solution. A dilute sodium arsenate solution containing 0.1% anhydrous sodium arsenate.

peat. Semicarbonized residue of plants formed in water-saturated environments (bogs and marshes). It occurs in surface layers 3–10 ft thick and has a water content of 85%. Before peat can be used for chemical or fuel purposes it must be field-dried to a water content of 30–40%. Since the dried product is susceptible to autoignition, storage conditions must minimize this risk. Peat is easily converted to hydrocarbons and is an excellent source of natural gas; when dry it can be used directly as a fuel. The U.S. has peat sources second only to those of the former U.S.S.R., located in Alaska, the north-central states, and Maine, where processing on a large scale is planned. Their total energy content is said to be equivalent to 240 billion barrels of petroleum. The peat can be gasified for production of methanol after mechanical dewatering. Experimental conversion studies have been under way for some time. Substantial quantities of oil, ammonia, and sulfur can be obtained as by-products.

pebble. A piece of gravel between 4 and 8 mm in size.

pebble mill. A jacketed steel cylinder rotating on a horizontal axis and containing flint or porcelain pebbles as the grinding medium. Its operation is similar to that of a ball mill. It is used for grinding and mixing of dry chemicals, pigments, food products, and the like. Pebble mills are usually lined with alumina, buhrstone, or similar material to protect the walls from wear.

pebulate. (propyl ethyl-n-butylthiocarbamate).
CAS: 1114-71-2. C$_{10}$H$_{21}$NOS.
Properties: Colorless liquid. Bp 142C (20 mm Hg), d 0.945, refr index 1.47. Soluble in benzene, acetone, methanol, and xylene.
Hazard: Toxic by ingestion.
Use: Herbicide.

Pechmann pyrazole synthesis. Formation of pyrazoles from acetylenes and diazomethane. The analogous addition of diazoacetic esters to the triple bond yields pyrazolecarboxylic acid derivatives.

pectic acid. An acid derived from pectin by treating it with sodium hydroxide solution, washing with isopropyl alcohol, adding alcoholic hydrochloric acid, washing again with isopropyl alcohol, and drying.
Use: Acidulant in pharmaceuticals.

pectin. A high molecular weight hydrocolloidal substance (polyuronide) related to carbohydrates and found in varying proportions in fruits and plants. Pectin consists chiefly of partially methoxylated galacturonic acids joined in long chains.
Properties: White powder or syrupy concentration. The most common characteristic of pectins is their property of jelling at room temperature, typically after addition of sugar and fruit juices in the preparation of jams or jellies. Soluble in water; insoluble in organic solvents.
Derivation: By dilute-acid extraction of the inner portion of the rind of citrus fruits, or of fruit pomaces, usually apple.
Method of purification: Following decolorization, the extracts are concentrated by evaporation or the pectins precipitated with alcohol or acetone.
Grade: Pure (NF) containing not less than 6.7% methoxy groups and not less than 74% galacturonic acid; 150-, 200-, 250-jelly grades, containing various diluents.
Use: Jellies, foods, cosmetics, drugs, protective colloids, emulsifying agents, dehydrating agents.
See gel.

pectinase. An enzyme present in most plants. It catalyzes the hydrolysis of pectin to sugar and galacturonic acid.
Use: Biochemical research, juice and jelly industry.

Pectinol. A formulated enzyme concentrate of fungal origin, with varying degrees of pectinase activity, that hydrolyzes pectic substances.
Use: Clarification of wines and fruit juices and processing of jellies.

pectin sugar. See l-arabinose.

Pedersen, Charles John. (1904–1989). Awarded Nobel Prize for chemistry in 1987 for work in elucidating mechanisms of molecular recognition, which are fundamental to the enzymic catalysis, regulation, and transport. He reported that alkali metal ions could be bound by crown ethers into a more rigid, layered structure, in which the alkali metal ion was bound into the center of the ring. This field of study is called host–guest chemistry. An American born in Korea, Pedersen received an M.S. from M.I.T. in 1927.

pedigree. A family tree diagram that shows how a particular genetic trait or disease has been inherited. See inherit.

PEG. Abbreviation for polyethylene glycol.

Pegosperse Aldo. PEG, glyceryl, sorbitan, polysorbate esters, and fatty acid esters.
Grade: Liquid, bead, flake, food, Kosher, cosmetic, and industrial grades.

pelargonic acid. (n-nonoic acid; n-nonanoic acid; n-nonylic acid).
CAS: 112-05-0. $CH_3(CH_2)_7COOH$.
Properties: Colorless or yellowish oil; slight odor. D 0.9052 (20/4C), mp 12.5C, bp 255.6C, refr index 1.4322 (20C). Soluble in alcohol, ether, and organic solvents; almost insoluble in water. Combustible.
Derivation: By the oxidation of nonyl alcohol or nonyl aldehyde; by the oxidation of oleic acid, especially by ozone.
Grade: Technical, 99%.
Hazard: Strong skin irritant.
Use: Organic synthesis, lacquers, plastics, production of hydrotropic salts, pharmaceuticals, synthetic flavors and odors, flotation agent, esters for turbojet lubricants, vinyl plasticizer, gasoline additive.

pelargonic alcohol. See nonyl alcohol.

pelargonic aldehyde. See nonanal.

pelargonyl chloride. (n-nonanoyl chloride). $CH_3(CH_2)_7COCl$.
Properties: Bp 80–85C (5 mm Hg), min assay 97%. Soluble in hydrocarbons and ethers; decomposes in water.
Hazard: Skin irritant.
Use: Intermediate in organic synthesis.

pelargonyl peroxide. $(C_8H_{17}COO–)_2$.
Properties: Water-white liquid; faint odor. D 0.926 min (25/25C), mp 10C, refr index 1.443 min (25C). Insoluble in water and glycerol; soluble in alcohol and hydrocarbons.
Hazard: Dangerous fire risk in contact with organic materials. Strong skin irritant and oxidizing agent.
Use: Initiator of polymerization reactions.

"Pelaspan" *[Storopack]*. TM for a series of expandable polystyrenes in bead or pellet form. Each bead contains its own expanding agent, which is activated by heat.

Peligot's salt. See potassium chlorochromate.

pellagra. A disease caused by deficiency of niacin in the diet.

pellet. A small unit of a light, bulky material compressed into any of several shapes and sizes, usually either spherical or rectangular. The operation is performed on a pellet mill, which consists essentially of a pair of steel rollers around which rotates a circular perforated metal die. Material is fed into the chambers above and below the inner face of the die. As the die turns, in contact with the rollers, the latter also turns thus compressing the material and forcing it through the holes in the die at the point of tangency, where the extruded segment is sheared off by knives. Pelletizing is advantageous for fluffy particulates that are difficult to handle in loose form, e.g., carbon black, clays, plastic molding powders, etc. Binding materials called excipients are often used.

pelletierine. (β-2(-piperidyl) propionaldehyde).
CAS: 2858-66-4. $C_5H_{10}N(CH_2)_2CHO$.
Properties: Liquid alkaloid from the root of the pomegranate. D 0.988 (20/4C), bp 195C. Soluble in water, alcohol, ether, chloroform, benzene.
Use: Medicine (in form of its salts, sulfate, tannate, valerate).

Pellizzari reaction. Formation of substituted 1,2,4-triazoles by the condensation of amides and acyl hydrazines. When the acyl groups of the amide and acylhydrazine are different, interchange of acyl groups may occur, with formation of a mixture of triazoles.

Pelouze synthesis. Formation of nitriles from alkali cyanides by alkylation with alkyl sulfates or alkyl phosphates.

"Penacolite" *[Indespec].* TM for resorcinol formaldehyde condensation resins.
Use: As adhesion promoters in tires and mechanical rubber products.

Pen Cote Liquid Binder. Ready-to-use liquid coating binders.
Use: Enhances optical properties and improves water holding.

penetrance. The probability of a gene or genetic trait being expressed. (a) "Complete" penetrance means that the gene or genes for a trait are expressed in all the population who have the genes. (b) "Incomplete" penetrance means the genetic trait is expressed in only part of the population. The percent penetrance also may change with the age range of the population.

penetrant. Any agent used to increase the speed and ease with which a bath or liquid permeates a material being processed by effectively reducing the interfacial tension between the solid and liquid. Penetrants are widely used in the textile, tanning, and paper industries for improving dyeing, finishing, etc. operations. Sulfonated oils, soluble pine oils, and soaps are popular among the older penetrants, and the salts of sulfated higher alcohols are typical of the synthetic organics developed for this purpose.
See wetting agent.

Penford Gum. Hydroxyethyl ether derivatized corn starches.

Penglo 65. A pale maleic modified pentaerythritol ester of a special tall oil in mineral spirits.
Use: Paint and varnish.

penicillamine. (USAN; *d*,3-mercaptovaline).
CAS: 52-67-5. $(CH_3)_2C(SH)CH(NH_2)COOH$.
Properties: Crystals. Decomposes at 178C.

Use: Medicine, as a chelate for copper, and in treatment of rheumatoid arthritis.

l-penicillamine.
CAS: 1113-41-3. $C_5H_{11}NO_2S$.
Hazard: A poison.
Use: Drug.

penicillic acid. (3-methoxy-5-methyl-4-oxo-2,5-hexadenoic acid; γ-keto-β-methoxy-δ-methylene-δ²-hexenoic acid; (2E)-3-methoxy-5-methyl-4-oxohexa-2,5-dienoic acid; pa; pencillic acid).
CAS: 90-65-3. $C_8H_{10}O_4$. A mycotoxin with antibiotic activity. It contains an α,β-unsaturated lactone moiety. It is active against both gram-positive and gram-negative bacteria.
Properties: Needles or rhombic or hexagonal plates; slightly soluble in cold water, hot ether; very soluble in hot water, alcohol, and benzene chloroform; insoluble in pentane-hexane.
Derivation: Produced by *Penicillium puberulum, Penicillium baarnese, Penicillium cyclopium, Penicillium suaveolens, Penicillium thomii, Aspergillus melleus,* and *Aspergillus ochraceus.*
Hazard: Toxic; poison; questionable carcinogen.

penicillin.
CAS: 1406-05-9. $(CH_3)_2C_5H_3NSO(COOH)NH-COR$ (bicyclic). A group of isomeric and closely related antibiotic compounds with outstanding antibacterial activity, obtained from the liquid filtrate of the molds *Penicillium notatum* and *Penicillium chrysogenum* or by a synthetic process that includes fermentation. Total synthesis of the penicillin molecule by J. C. Sheehan in 1957 was an outstanding achievement.
Derivation: The mold is grown in a nutrient solution such as corn steep liquor, lactose, or dextrose. After several days of cultivation the mold excretes penicillin into its liquid culture medium. This liquid is then filtered off, and the penicillin extracted and purified by countercurrent extraction with amyl acetate, adsorption on carbon, or other methods. Different varieties of penicillin are produced biosynthetically by adding the proper precursors to the nutrient solution.
Grade: Aluminum penicillin G, benzathine penicillin G, benzylpenicillin G, chloroprocaine penicillin O, hydrabramine penicillin V, phenoxymethylpenicillin, potassium penicillin G, potassium α-phenoxyethylpenicillin, potassium phenoxymethylpenicillin, procaine penicillin G, sodium methicillin, sodium penicillin G.

Hazard: Strong allergen, reaction may be severe in susceptible people.
Use: Medicine (antibiotic).

penicillin V.   See phenoxymethylpenicillin.

penicillinase.   An enzyme that antagonizes the antibacterial action of penicillin. Such enzymes are found in many bacteria.
Use: Pharmaceutical, biological research.
See antagonist, structural.

Penicillium.   A genus of imperfect fungi (family Moniliaceae, order Moniliales) that is commonly found in soil, spoiled foodstuffs, and water affected indoor environments. Some members of the genus Penicillium produce *antibiotics* and have had a significant positive impact on public health worldwide. *Penicillium camemberti* and *Penicillium roqueforti* are the fungi used to make camembert, roquefort, brie, and many other cheeses. Many species produce mycotoxins, which are potent poisons. Many species of Penicillium have been shown to be allergenic.
Use: Antibiotics, food production, enzyme production.
Hazard: Allergenic, pathogenic, toxic.

penitrem A.   (tremortin a). $C_{37}H_{44}ClNO_6$. An indole mycotoxin.
Properties: Amorphous solid.
Derivation: Produced by *Penicillium cyclopium, Penicillium crustosum*, and *Penicillium palitans*.
Hazard: Extremely toxic; poison; teratogen; tremorgen; neurotoxic; nephrotoxic; diuretic.

penitrem B.   (dechloro-penitrem a). $C_{37}H_{45}NO_5$. An indole mycotoxin.
Derivation: Produced by *Penicillium cyclopium.*
Hazard: Extremely toxic.

penitrem C. $C_{32}H_{44}ClNO_4$. An indole mycotoxin.
Derivation: Produced by *Penicillium crustosum.*
Hazard: Moderately to very toxic.

"Penn Drake Petrosul" *[Calumet].*   TM for a variety of sodium sulfonates.
Properties: Mw 415–619.
Grade: Available as 50, 60, and 70% in mineral oil.
Available forms: Liquid.
Use: Surface-active agent used as solubilizers, emulsifiers, rust-preventive bases, and dispersants.

pennyroyal oil.   (pulegium oil; hedeoma oil). Yellow to reddish essential oil, strongly dextrorotatory.
Use: Manufacture of pulegone, flavoring alcoholic beverages, emmenagogue.

Penros.   A polymerized wood rosin.

Use: Adhesives, gloss oils, paper label coatings, oleoresinous varnishes, solder flux, spirit varnishes, waxed paper and hot-melt compounds, synthetic resins.

pentaborane.
CAS: 19624-22-7. B_5H_9.
Properties: Colorless liquid; pungent odor. Fp −46.6C, bp 58C, d 0.61, vap press 6 mm Hg, decomposes at 150C, ignites spontaneously in air if impure, flash p 86F (30C) (OC). Hydrolyzes slowly in water.
Derivation: Hydrogenation of diborane.
Grade: Technical 95%, high purity 99%.
Hazard: Highly flammable, dangerous fire and explosion risk. Toxic by ingestion and inhalation, strong irritant. Central nervous system convulsions and impairments.
Use: Fuel for air-breathing engines, propellant.

pentabromodiphenyl ether.
CAS: 32534-81-9. $C_{12}H_5Br_5O$.
Hazard: A poison by ingestion. Moderately toxic by inhalation. Low toxicity by skin contact. A moderate skin irritant.

"Pentac"   *[Gleason].*   TM   for   miticide whose   active   ingredient   is   bis(pentachloro-2,4-cyclopentadien-1-yl).

pentacarbonyl(piperidine)chromium.
CAS: 15710-39-1. $C_{10}H_{11}CrNO_5$.
Hazard: A poison.

pentacene.   (2,3,4,7-dibenzoanthracene).
CAS: 135-48-8. $C_{22}H_{14}$. Highly reactive aromatic compound consisting of five fused benzene rings.
Properties: Deep blue-violet solid. Sublimes at 290–300C, decomposes in air above 300C. Insoluble in water; slightly soluble in organic solvents.
Use: Suggested as organic photoconductor (instead of selenium) in copying systems.

2,2′,4,5,6′-pentachlorodiphenyl ether.   See 1,2,4-trichloro-5-(2,6-dichlorophenoxy)benzene.

pentachloroethane.   (pentalin).
CAS: 76-01-7. $CHCl_2CCl_3$.
Properties: Dense, high-boiling, colorless liquid. D 1.685 (15/4C), bp 159.1C, fp −22C, refr index 1.503 (24C). Insoluble in water.
Derivation: By chlorination of trichloroethylene, obtained by a two-step process involving chlorination of acetylene to obtain tetrachloroethane, and removal of hydrogen chloride by action of alkali.
Hazard: Moderate fire and explosion risk. Toxic by inhalation and ingestion. Questionable carcinogen.
Use: As solvent for oil and grease in metal cleaning. Also used for separation of coal from impurities by density difference.
See tetrachloroethane.

pentachloronaphthalene.
CAS: 1321-64-8. $C_{10}H_3Cl_5$.
Properties: White powder.
Hazard: Action similar to chlorinated naphthalenes and chlorinated diphenyls. Liver damage and chloracne.

pentachloronitrobenzene. (PCNB).
CAS: 82-68-8. $C_6Cl_5NO_2$.
Properties: Cream crystals; musty odor. D 1.718 (25/4C), mp 142–145C, bp 328C (some decomposition). Practically insoluble in water; slightly soluble in alcohols; somewhat soluble in carbon disulfide, benzene, chloroform.
Derivation: By reacting pentachlorobenzene with fuming nitric acid.
Grade: Dust, emulsion concentrate, wettable powder.
Hazard: Skin irritant. Liver damage. Questionable carcinogen.
Use: Intermediate, soil fungicide, slime prevention in industrial waters, herbicide.

pentachlorophenol. (PCP).
CAS: 87-86-5. C_6Cl_5OH.
Properties: White powder or crystals. Mp 190C, bp 310C with decomposition, d 1.978 (22/4C). Slightly soluble in water; soluble in dilute alkali, alcohol, ether, benzene.
Derivation: Chlorination of phenol.
Hazard: Toxic by ingestion, inhalation, and skin absorption; abuse may be fatal. Toxic by skin absorption. Eye and upper respiratory tract irritant, central nervous system impairment, and cardiac impairment. Possible carcinogen.
Use: Fungicide, bactericide, algicide, herbicide; as sodium pentachlorophenate, wood preservative (telephone poles, pilings, etc.).

pentachlorophenyl butyl ether. See butoxypentachlorobenzene.

(pentachlorophenyl)glycolonitrile.
CAS: 21727-09-3. $C_8H_2Cl_5NO$.
Hazard: Moderately toxic by skin contact.
Use: Agricultural chemical.

Pentacite. Pale-colored rosin ester gums and modified rosin esters used in coating vehicles and in chewing gums and some rubber adhesives.

pentadecane. $CH_3(CH_2)_{13}CH_3$.
Properties: Colorless liquid. D 0.776, bp 270.5C, mp 10C. Soluble in alcohol; insoluble in water. Combustible.
Grade: Technical.
Use: Organic synthesis.

n-pentadecanoic acid. (pentadecylic acid). $CH_3(CH_2)_{13}COOH$. A saturated fatty acid normally not found in vegetable fats but made synthetically.

Properties: Colorless crystals. D 0.8423 (80/4C), mp 51.8–52.8C, bp 339.1C, 212C (16 mm Hg), refr index 1.4529 (60C). Insoluble in water; soluble in alcohols and ethers.
Grade: 99% pure.
Use: Organic synthesis, reference standard in gas chromatography.

pentadecanolide. (15-hydroxypentadecanoic acid lactone; pentadecalactone).

$$\overline{CH_2(CH_2)_{13}C(O)O.}$$

Properties: Colorless liquid; strong musky odor. Congeals to white crystals at room temperature. Minimum congealing p 36C, soluble in equal volume of 90% ethanol. Combustible.
Derivation: Angelica root oil.
Grade: 98% min.
Use: Perfumery.

pentadecylcatechol. (3-pentadecylcatechol; 3-pentadecyl-1,2-benzenediol; 3-pentadecylpyrocatechol; tetrahydrourushiol; hydrourushiol; dihydrohengol; 3-PDC). $C_{21}H_{36}O_2$.
Properties: A white powder; soluble in acetone, benzene, ethanol, and ethyl ether.
Use: A chemical intermediate, a diagnostic aid, and as an algicide, bactericide, and herbicide.
Hazard: Toxic.

pentadecenyl phenol. See Cardanol.

pentadecylic acid. See n-pentadecanoic acid.

1,3-pentadiene. See piperylene.

2,4-pentadienol. See 2,4-pentadien-1-ol.

2,4-pentadien-1-ol.
CAS: 4949-20-6. C_5H_8O.
Hazard: A severe skin irritant.

pentaerythrite tetranitrate. Legal label name for pentaerythritol tetranitrate.

pentaerythritol. (PE; tetramethylolmethane; monopentaerythritol).
CAS: 115-77-5. $C(CH_2OH)_4$.
Properties: White, crystalline powder; readily esterified by common organic acids. Bp 276C (30 mm Hg), mp 262C, refr index 1.54-1.56 (20C), d 1.399 (25/4C). Unaffected when boiled with dilute caustic alkali. Soluble in water; slightly soluble in alcohol; insoluble in benzene, carbon tetrachloride, ether, and petroleum ether. Combustible.
Grade: Technical, nitration, CP. Technical grade is 88% monopentaerythritol and 12% dipentaerythritol.
Derivation: Reaction of acetaldehyde with an excess of formaldehyde in an alkaline medium.
Hazard: Gastrointestinal irritant.

Use: Alkyd resins, rosin and tall oil esters, special varnishes, pharmaceuticals, plasticizers, insecticides, synthetic lubricants, explosives, paint swelling agents.

pentaerythritol ester of partially hydrogenated wood rosin.
Properties: Hard, amber-colored solid. Sol in acetone, benzene; insol in water.
Use: Food additive.

pentaerythritol ester of wood rosin.
Properties: Hard, amber-colored solid. Sol in acetone, benzene; insol in water.
Use: Food additive.

pentaerythritol tetraacetate.
CAS: 597-71-7. $C(CH_2OOCCH_3)_4$.
Properties: White, crystalline powder; extremely stable in sunlight. Mp 84C, bp 225C (30 mm Hg). Soluble in water, alcohol, and ether. Combustible.
Derivation: By the esterification of pentaerythritol with acetic acid.
Grade: Technical.

pentaerythritol tetrakis(diphenyl phosphite).
(tetra(diphenylphosphito)pentaerythritol).
$C[CH_2OP(OC_6H_5)_2]_4$. A low-melting, white, waxy solid with a slight phenolic odor, d 1.24 (25/15.5C), mp 30–60C, ref index 1.5823 (25C). Combustible.
Use: Ingredient in stabilizer systems for resins.

pentaerythritol tetra(3-mercaptopropionate).
$C(CH_2OOCCH_2CH_2SH)_4$.
Properties: Liquid. D 1.28 (25C), refr index 1.5300 (25C). Insoluble in water, alcohol, and hexane; soluble in acetone and benzene. Combustible.
Use: Curing or cross-linking agents for polymers, especially epoxy resins, intermediate for stabilizers and antioxidants.

pentaerythritol tetranitrate.   (PETN).
CAS: 78-11-5. $C(CH_2ONO_2)_4$.
Properties: White, crystalline material. D 1.75, mp 138–140C, decomposes above 150C. Very soluble in acetone, slightly soluble in alcohol and ether, insoluble in water.
Derivation: Esterification of pentaerythritol with nitric acid.
Hazard: Shock-sensitive explosive. Detonates at 210C.
Use: Demolition explosive, blasting caps, detonating compositions ("Primacord" *[Dyno]*).

pentaerythritol   tetrastearate. $C(CH_2-OOCC_{17}H_{35})_4$.
Properties: Hard, high-melting wax, ivory colored. Essentially neutral, acid number 1, softening p 67C. Combustible.
Use: Polishes, coatings, textile finishes.

pentaerythritol tetrathioglycolate.
$C(CH_2OOCCH_2SH)_4$.
Properties: Liquid. D 1.385 (25C), refr index 1.5499 (25C). Insoluble in water, alcohol, and hexane; partially soluble in benzene; soluble in acetone. Combustible.
Use: Curing or cross-linking agents for polymers, especially epoxy resins, intermediate for stabilizers and antioxidants.

pentaglycerine.   See trimethylolethane.

pentahydrate salt.   (blue copper; blue vitriol; bluestone; roman vitriol; salzburg vitriol; dicopper chloride trihydroxide). $ClCu_2H_3O_3$.
Properties: Consists of large blue triclinic crystals, blue granules or a light blue powder; soluble in water, methanol and glycerol; slightly soluble in ethanol.
Use: An agricultural algicide, bactericide, fungicide, and herbicide; in insecticidal mixtures; as a wood preservative, food and fertilizer additive, in paints and dyes, as a precursor of other copper salts, in electroplating solutions.
Hazard: Strong irritant; systemic poison; can cause pain in the mouth, esophagus, and stomach, vomiting, diarrhea, shock convulsions, paralysis, lesions, extensive capillary damage, kidney and liver damage, coma, and death.

2′,3,4′,5,7-pentahydroxyflavone.   See morin.

pentalin.   See pentachloroethane.

"Pentalyn" *[Pinova]*.   TM for a series of nonreactive and heat-reactive pentaerythritol esters of rosin. Available in solid and flake form in various grades.
Use: Varnishes, floor polishes, inks, and adhesives.

1,1,3,3,5-pentamethyl-4,6-dinitroindane.
CAS: 116-66-5. $C_{14}H_{18}N_2O_4$.
Properties: Pale-yellow crystals; musk-type odor. Mp min 132C. Slightly soluble in alcohol; soluble in diethyl phthalate.
Use: Perfumery.

pentamethylene.   See cyclopentane.

pentamethyleneamine.   See piperidine.

pentamethylene-1,1-bis(1-methylpyrrolidinium bitartrate).   See pentolinium tartrate.

pentamethylenediamine.   See cadaverine.

pentamethylene   dibromide.   (1,5-dibromopentane). $BrCH_2(CH_2)_3CH_2Br$.
Properties: Colorless, aromatic liquid. Fp −35C, bp 224C. Insoluble in water.

pentamethylene glycol. See 1,5-pentanediol.

pentamethylpararosaniline chloride. See methyl violet.

2,4,4,6,6-pentamethyl-2-phenylcyclotrisiloxane.
CAS: 17962-31-1. $C_{11}H_{20}O_3Si_3$.
Hazard: A reproductive hazard.

pentamidine. (4,4'-[1,5-pentanediylbis(oxy)]-bis-benzenecarboximidamide; 4,4'-(pentamethylenedioxy)dibenzamidine; 4,4'-diamidino-α,ω-diphenoxypentane)
CAS: 100-33-4. $C_{19}H_{24}N_4O_2$.
Properties: Colorless, crystalline liquid; slightly soluble in water, soluble in alcohol and ether.
Use: Therapeutically as an antiprotozoal against *babesia, leishmania,* and *pneumocystis.*
Hazard: May cause diabetes mellitus, central nervous system damage, and other toxic effects.

pentanal. See *n*-valeraldehyde.

***n*-pentane.** (amyl hydride).
CAS: 109-66-0. $CH_3(CH_2)_3CH_3$.
Properties: Colorless liquid; pleasant odor. Fp −129.7C, bp 36.074C, refr index 1.35748 (20C), d 0.62624, flash p −40F (−40C), autoign temp 588F (308C). Soluble in alcohol and most organic solvents; insoluble in water.
Derivation: Fractional distillation from petroleum, purified by rectification.
Grade: Pure, technical, commercial.
Hazard: Highly flammable, dangerous fire and explosion risk. Explosive limits 1.4–8% in air. Narcotic in high concentration. Narcosis and respiratory tract irritant.
Use: Artificial ice manufacture, low-temperature thermometers, solvent extraction processes, blowing agent in plastics (e.g., expandable polystyrene), pesticide.

pentanedinitrile. See glutaronitrile.

pentanedioic acid. See glutaric acid.

pentanedioic acid anhydride. See glutaric anhydride.

1,5-pentanediol. (pentamethylene glycol).
CAS: 111-29-5. $HOCH_2(CH_2)_3CH_2OH$.
Properties: Viscous liquid. Bp 240C, fp −15.6C, d 0.9921 (20/20C), bulk d 8.2 lb/gal (20C), flash p 265F (129C) (OC), autoign temp 633F (334C). Miscible with water and alcohol. Combustible.
Grade: Technical.
Use: Hydraulic fluid, lubricating-oil additive, antifreeze, plasticizer and polyester resin intermediate.

2,3-pentanedione. See acetyl propionyl.

2,4-pentanedione. See acetylacetone.

pentanethiol. (amyl mercaptan).
CAS: 110-66-7. $C_5H_{11}SH$. A mixture of isomers.
Properties: Water-white to light-yellow liquid; strong offensive odor. D 0.83–0.84 (20C), mercaptan content greater than 90.0%, initial bp above 104.0C, final bp below 130C, bulk d 6.99 lb/gal, flash p 65F (18.3C) (OC). Insoluble in water; soluble in alcohol. These properties vary with proportions of isomers.
Derivation: Mixing amyl bromide and potassium hydrosulfide in alcohol.
Hazard: Flammable, dangerous fire risk. Toxic by inhalation.
Use: Synthesis of organic sulfur compounds, chief constituent of odorant used in gas lines to locate leaks.

pentanoic acid. See *n*-valeric acid.

1-pentanol. See *n*-amyl alcohol, primary.

2-pentanol. (*sec-n*-amyl alcohol; *sec*-amyl alcohol, active; methyl propyl carbinol; 1-methylbutyl alcohol).
CAS: 6032-29-7. $CH_3CH_2CH_2CH_2OCH_3$.
Properties: (Racemic form) Colorless liquid. Fp −75C, bp 119.3C, d 0.811 (20/20C), bulk d 6.75 lb/gal (20C), refr index 1.4041 (40.5C), flash p 105F (40.5C) (OC), autoign temp 657F (347C). Soluble in water; miscible with alcohol and ether. Combustible.
Derivation: Fractional distillation of the mixed alcohols resulting from the chlorination and hydrolysis of pentanes.
Hazard: Moderate fire risk. Irritant to eyes, nose, and throat.
Use: Solvent for paints and lacquers, pharmaceutical intermediate.

3-pentanol. (*sec-n*-amyl alcohol; 1-ethyl-1-propahol; diethyl carbinol).
CAS: 584-02-1. $CH_3CH_2CHOHCH_2CH_3$.
Properties: Colorless liquid. D 0.82 (20C), fp below −75C, bp 115.6C, bulk d 6.81 lb/gal, refr index 1.41 (20C), flash p 94F (34.4C) (CC), autoign temp 650F (343C). Soluble in alcohol and ether; slightly soluble in water.
Hazard: Moderate fire risk. Irritant to eyes, nose, and throat.
Use: Solvent, flotation agent, pharmaceuticals.

2-pentanone. See methyl propyl ketone.

3-pentanone. See diethyl ketone.

penta resin. Ester gum made from rosin and pentaerythritol.

pentasodium diethylenetriaminepentaacetate. $C_{14}H_{23}N_3O_{10}$·5Na. A sodium salt of diethylenetriaminepentaacetic acid.
Use: A chelating agent.

pentasodium triphosphate. See sodium tripolyphosphate.

***n*-pentatriacontane.** $C_{35}H_{72}$ or $CH_3(CH_2)_{33}CH_3$.
Properties: Crystals. D 0.782 at 75C, bp 331C (15 mm Hg), mp 75C. Combustible.
Use: Organic synthesis.

pentazocine. (2-dimethylallyl-5,9-dimethyl-2′-hydroxy benzomorphan).
CAS: 359-83-1. $C_{19}H_{27}NO$. A synthetic drug claimed to be as effective as morphine but without its addictive properties. Has a noncumulative effect. FDA approved.

pentazocine lactate.
CAS: 17146-95-1. $C_{19}H_{27}NO$·$C_3H_6O_3$.
Hazard: A poison. A human skin irritant.

Pentecat L. A 50% aqueous solution of lithium naphthenate.
Use: Alcoholysis catalyst in alkyd varnish cooking.

1-pentene. (α-*n*-amylene; propylethylene). $CH_3CH_2CH_2CH{:}CH_2$.
Properties: Colorless liquid. Fp −165C, bp 30C, d 0.6410 (20C), flash p 0F (−17.7C) (OC), autoign temp 523F (272C). Soluble in alcohol; insoluble in water.
Derivation: Natural gasoline.
Hazard: Flammable, dangerous fire risk. Toxic by ingestion, inhalation, and skin absorption.
Use: Organic synthesis, blending agent for high-octane motor fuel, pesticide formulations.

2-pentene. (β-*n*-amylene; *sym*-methylethylethylene).
CAS: 646-04-8. $CH_3CH_2HC{:}CHCH_3$. Mixed *cis*- and *trans*-isomers are available commercially.
cis-isomer

Properties: Bp 37C, fp −180C, d 0.656 (20/4C), flash p 0F (−17.7C). Soluble in alcohol; insoluble in water.
Grade: Technical, 95.0 mole %.
trans-isomer

Properties: Bp 36.4C, fp −139C, d 0.6482 (20C), flash p 0F (−17.7C). Soluble in alcohol; insoluble in water.
Derivation: Natural gasoline.
Hazard: Flammable, dangerous fire risk.
Use: Polymerization inhibitor, organic synthesis.

"Pentex" *[Orica].* TM for tetrabutylthiuram monosulfide.
Use: Rubber accelerator. When mixed with 87.5% clay, it is used for sponge rubbers and called "Pentex Flour."

Pentite. Tetra(diphenylphosphito)pentaerythritol. See pentaerythritol tetrakis(diphenylphosphite).

pentlandite. (Fe,Ni)S.
Properties: Light-bronze-yellow mineral, metallic luster, contains 35.57% nickel. Soluble in nitric acid, d 4.6–5, Mohs hardness 3.5–4.
Occurrence: Canada (Ontario), Norway.
Use: Nickel ore.

pentobarbital. (5-ethyl-5-(1-methylbutyl)barbituric acid).
CAS: 76-74-4. $C_{11}H_{18}O_3N_2$.
See barbiturate.

pentolinium tartrate. (pentamethylene-1,1-bis(1-methylpyrrolidinium) bitartrate).
CAS: 52-62-0. $C_{23}H_{42}N_2O_{12}$.
Properties: White- to light-cream colored, crystalline powder. Slightly soluble in alcohol; insoluble in ether, chloroform; very soluble in water; pH of 1% solution in water is 3.0–4.0, decomposes 203C.
Use: Medicine (antihypertensive).

pentolite. A high explosive consisting of equal parts of pentaerythritol tetranitrate and trinitrotoluene.
Hazard: Dangerous, explodes on shock or heating.

Penton.

$(CH_2Cl)_2CCH_2OCH_2$.

A thermoplastic resin derived from 3,3-bis(chloromethyl)oxetane. A chlorinated polyether.
Properties: A linear polymer extremely resistant to chemicals and to thermal degradation at molding and extrusion temperature, d 1.4, self-extinguishing, dimensionally stable, very low water absorption, outstanding chemical resistance. Natural, black, or olive-green molding powder. Finely divided powder for coatings.
Use: Solid and lined valves, pumps, pipe, and fittings; monofilament for filter supports and column packing.

pentosan. A complex carbohydrate (hemicellulose) present with the cellulose in many woody plant tissues, particularly cereal straws and brans, characterized by hydrolysis to give five-carbon-atom sugars (pentoses). Thus the pentosan xylan yields the sugar xylose (HOH$_2$C•CHOH•CHOH•CHOH•CHO) that is dehydrated with sulfuric acid to yield furfural (C$_5$H$_4$O$_2$).

pentose. General term for sugars with five carbon atoms per molecule.

pentose phosphate pathway. A pathway involved in the oxidation of glucose, and a source of reducing equivalents (NADPH) and pentoses for biosynthetic processes; present in most organisms. Also called the phosphogluconate pathway, pentose phosphate shunt, or hexose monophosphate shunt.

"Pentothal" *[Hospira].* TM for sodium thiopental, a barbiturate.
See thiopental sodium.

pentyl. Synonym for the amyl group, C$_5$H$_{11}$–.

pentyl acetate. See amyl acetate.

***tert*-pentyl alcohol.** (*tert*-amyl alcohol; amylene hydrate; 2-methyl-2-butanol; dimethyl ethyl carbinol; ethyl dimethyl carbinol; 3-methylbutan-3-ol; *tert*-pentanol).
CAS: 75-85-4. C$_5$H$_{12}$O. A commercially available branched-chain tertiary alcohol and isomer of amyl alcohol.
Properties: Volatile liquid with a characteristic odor.
Hazard: Moderately toxic; burning taste, irritates mucous membranes, narcotic action.

pentylamine. See *n*-amylamine.

α-pentylcinnamaldehyde. See α-amylcinnamic aldehyde.

pentylenetetrazole. **(6,7,8,9-tetrahydro-5H-tetrazolo[1,5-a]azepine; metrazol).**
CAS: 54-95-5. C$_6$H$_{10}$N$_4$. It is considered a noncompetitive gamma-aminobutyric acid antagonist.
Use: A diagnostic tool in screening for latent epileptogenic foci and in basic research.
Hazard: Central nervous system stimulant, induce seizures.

***p-tert*-pentylphenol.** (*p-tert*-amylphenol).
CAS: 80-46-6. C$_{11}$H$_{15}$OH.
Properties: Crystalline solid. Mp 95C, bp 261C, d 0.962 (20/4C). Insoluble in water; soluble in organic solvents.
Derivation: Condensation of *tert*-pentanol with phenol with aluminum chloride catalyst.
Use: Pesticide intermediate, oil-soluble resin manufacture, may be useful as germicide and fumigant.

6-(pentylthio)purine.
CAS: 5443-89-0. C$_{10}$H$_{14}$N$_4$S.
Hazard: A reproductive hazard.

1-(4-pentynyloxy)-4-phenoxybenzene.
CAS: 42873-80-3. C$_{17}$H$_{16}$O$_2$.
Hazard: A reproductive hazard.
Use: Agricultural chemical.

Penzold's reagent. Solution of diazobenzosulfonic acid and potassium hydroxide.
Use: Testing for sugar in urine.

Pepha-Ctive.
Properties: Clear yellowish liquid. An algae extract.
Use: In cosmetics to protect the mitochondria and increases the ATP levels of skin cells leading to enhancement of cell turnovers.

peppermint oil. Essential oil with strong aromatic odor and taste, levorotatory, chief component is menthol.
Use: To flavor mouthwashes, chewing gum, liqueurs, toothpastes; source of menthol.

pepsin. (pepsinum).
CAS: 9001-75-6.
Properties: A digestive enzyme of gastric juice. White or yellowish-white powder or lustrous transparent or translucent scales; should have no odor; converts proteins into albumoses and peptones; soluble in water; insoluble in alcohol, chloroform, and ether.
Derivation: From the glandular layer of fresh hog stomachs.
Grade: Technical, NF.
Use: Medicine (digestive ferment); substitute for rennet in cheese making.

pepsinogen. An inactive precursor of pepsin.

peptidase. An enzyme that hydrolyzes a peptide bond.
See protease.

peptide. Any of a class of low molecular weight organic compounds composed of two or more amino acids in which the alpha carboxyl group of one bonds covalently to the alpha amino group of a neighbor. These covalent amide bonds, called peptide bonds, are formed with the loss of a molecule of water. Peptides form the component parts of proteins.
See polypeptide.

peptide bond. A planar, amide linkage between the a-amino group of one amino acid and the a-carboxyl group of another, with the elimination of a molecule of water.

peptide finger-printing. See peptide mapping.

peptide mapping. (peptide finger-printing). The characteristic pattern of fragments formed by the separation of a mixture of peptides resulting from hydrolysis of a protein or peptide.

peptidoglycan. A major component of bacterial cell walls consisting of parallel heteropolysaccharides cross-linked by short peptides.

"PeptiSelect" [Sabinsa]. TM for a series of high-quality peptides that have been specially selected.

peptization. Stabilization of hydrophobic colloidal solutions by addition of electrolytes that provide the necessary electric double layer of ionic charges around each particle. Such electrolytes are known as peptizing agents. The ions of the electrolyte are strongly adsorbed on the particle surfaces. Stable solutions of nonionizing substances acquire a charge in contact with water by preferential adsorption of the hydroxyl ions, which may be considered peptizing agents. The term is also loosely applied to the softening or liquefaction of one substance by trace quantities of another, analogous to the digestion of a protein by an enzyme (pepsin).

peptoid. A peptidomimetic substance that arises from the oligomeric construction of n-substituted glycines.

peptone.
Properties: (1) From albumin: white or pale-yellow amorphous powder. (2) From meat: light-brown amorphous powder. (3) From milk: light-brown powder. Soluble in water, insoluble in alcohol or ether.
Derivation: (1) By digestion of egg albumin by pepsin and a small quantity of dilute hydrochloric acid at 38–40C (body temperature). (2) By digestion of red meat with pancreatin at body temperature. (3) By digestion of casein.
Grade: Technical, reagent.
Use: Preparation of nutrient media in bacteriology; nutrient.

peptotoxin. A poisonous alkaloid or ptomaine that occurs in certain peptons resulting from partial or imperfect digestion in the stomach.

"Peptrix" [Synpep]. TM for a rubber peptizer.
Use: For rubber reclaiming and processing.

per-. A prefix signifying complete or extreme and specifically denoting: (1) a compound containing an element in its highest state of oxidation, as perchloric acid; (2) presence of the peroxy group, –O–O–, as peracetic and perchromic acids; (3) exhaustive substitution or addition, as perchloroethylene.

peracetic acid. (peroxyacetic acid).
CAS: 79-21-0. CH_3COOOH.
Properties: Colorless liquid; strong odor. Bp 105C, fp −30C, d 1.15 (20C), flash p 105F (40.5C) (OC). Soluble in water, alcohol, sulfuric acid; strong oxidizing agent.
Derivation: (1) Oxidation of acetaldehyde, (2) reaction of acetic acid and hydrogen peroxide with sulfuric acid catalyst.
Grade: Technical, 40% soluble in acetic acid.
Hazard: Oxidizing material, dangerous in contact with organic materials, explodes at 110C. Strong irritant. Eye, skin, and upper respiratory tract irritant. Questionable carcinogen.
Use: Bleaching textiles, paper, oils, waxes, starch; polymerization catalyst; bactericide and fungicide, especially in food processing; epoxidation of fatty acid esters and epoxy resin precursors; reagent in making caprolactam, synthetic glycerol.

per-acids. Derivatives of hydrogen peroxide, the molecules of which contain one or more directly linked pairs of oxygen atoms, –O–O–. Examples are persulfuric, perchromic, peracetic acids. Permanganic, perchloric, and periodic acids are not per-acids in this sense.
See per-(2); per-(1).

"Perapret" [BASF]. TM for finishing agents and binders (polymer dispersions), additives for wash-and-wear finishing.

perbenzoic acid. See benzoyl hydroperoxide.

perbromobiphenyl.
CAS: 13654-09-6. $C_{12}Br_{10}$.
Hazard: Confirmed human carcinogen.

percarbamide. See urea peroxide.

perchloric acid.
CAS: 7601-90-3. $HClO_4$.
Properties: Colorless, fuming, hygroscopic liquid, unstable in concentrated form. D 1.764, bp 19C (11 mm Hg), fp −112C, evolves heat on combination with water. Commercial aqueous solutions contain 65–70% $HClO_4$.
Derivation: By distilling potassium perchlorate with strong sulfuric acid (96%) under reduced pressure in an oil bath at 140–190C.
Method of purification: Rectification.
Grade: Technical, CP.
Hazard: Strong oxidizing agent, will ignite vigorously in contact with organic materials, or detonate by shock or heat. Toxic by ingestion and inhalation, strong irritant.
Use: Analytical chemistry, catalyst, manufacture of various esters, ingredient of electrolytic bath in deposition of lead, electropolishing, explosives.

perchlorobenzene. See hexachlorobenzene.

perchlorocyclopentadiene. See hexachloro-cyclopentadiene.

perchloro-2-cyclopentenone. See 2,3,4,4,5,5-hexachloro-2-cyclopenten-1-one.

perchloroethane. See hexachloroethane.

perchloroethylene. (tetrachloroethylene).
CAS: 127-18-4. $Cl_2C:CCl_2$.
Properties: Colorless liquid; etherlike odor. Extremely stable, resists hydrolysis. D 1.625 (20/20C), bp 121C, fp −22.4C, bulk d 13.46 lb/gal (26C), refr index 1.5029 (25C), flash p none. Miscible with alcohol, ether, and oils; insoluble in water. Nonflammable.
Derivation: (1) By chlorination of hydrocarbons and pyrolysis of the carbon tetrachloride also formed, (2) from acetylene and chlorine via trichloroethylene.
Method of purification: Distillation.
Grade: Purified, technical, USP, as tetrachloroethylene, spectrophotometric.
Hazard: Irritant to eyes and skin. Central nervous system impairment. Probable carcinogen.
Use: Dry-cleaning solvent, vapor-degreasing solvent, drying agent for metals and certain other solids, vermifuge, heat-transfer medium, manufacture of fluorocarbons.

perchloromethane. See carbon tetrachloride.

perchloromethyl mercaptan. Legal label name for trichloromethylsulfenyl chloride.

perchloropentacyclodecane.
CAS: 2385-85-5. $C_{10}Cl_{12}$.
Properties: White, crystalline solid. Mp (sealed tube) 485C. Nonflammable.
Use: Fire-retardant additive in elastomeric resin systems.

perchloropropylene. See hexachloropropylene.

perchloryl fluoride.
CAS: 7616-94-6. $ClFO_3$.
Properties: Colorless, noncorrosive gas or liquid; sweet odor. Fp −146C, bp −46.8C, d (liquid) 1.434 (20C). Nonflammable but supports combustion.
Hazard: Dangerous in contact with organic materials, strong oxidizing agent. Toxic by inhalation. Lower and upper respiratory tract irritant, methemoglobinemia, fluorosis.
Use: Oxidant in rocket fuels, oxidizing and fluorinating agent in chemical reactions.

perchromic acid. Probably $(HO)_4Cr(OOH)_3$.
Properties: Unstable acid formed when a solution of chromic acid is added to hydrogen peroxide; forms deep-blue crystals below −15C; the blue color can be extracted from solutions by ether; decomposes in acid solution to form chromic salts and in alkaline solution to form chromates; the blue color can be used as a test for chloride or chromate.
Hazard: Irritant.

Peregal. A series of textile chemicals. One of which is ST, a polymeric dye complexing agent.

perfect gas. See gas.

perfluidone.
CAS: 37924-13-3. $C_{14}H_{12}F_3NO_4S_2$. A preemergence herbicide.

perfluoro-2-butene. Legal label name for octafluoro-2-butene.

perfluoro(2-butoxypropyl vinyl ether).
CAS: 115659-47-7. $C_9H_{18}O_2$.
Hazard: Moderately toxic by ingestion. A severe eye irritant.

perfluorobutyric acid. See heptafluorobutyric acid.

perfluorocarbon compound. A fluorocarbon in which the hydrogen directly attached to the carbon atoms is completely replaced by fluorine.

perfluorocyclobutane. See octafluorocyclobutane.

perfluorodimethylcyclobutane. An inert fluorocarbon liquid.
Properties: Bp 45C, stable to 260C.
Use: Evaporative coolant for electronic equipment.

perfluoroethylene. See tetrafluoroethylene.

perfluoroisobutylene. See octafluoro-*sec*-butene.

perfluorooctanesulfonic acid.
CAS: 1763-23-1. $C_8HF_{17}O_3S$.
Hazard: A poison by ingestion. A reproductive hazard.
Use: Agricultural chemical.

***n*-perfluorooctanesulfonyl-*n*-methylcarbamoyl ((nonadecaethoxy)butoxy)butyl ether.**
CAS: 52032-20-9. $(C_2H_4O)_n$–$C_{14}H_{12}F_{17}NO_4S$.
Hazard: Moderately toxic by ingestion.

perfluoropropane. Legal label name (Air) for octafluoropropane.

perfluoropropene. See hexafluoropropylene.

performic acid.   (peroxyformic acid).
CAS: 107-32-4.   HCOOOH.
Properties: Colorless liquid. Miscible with water, alcohol, ether; soluble in benzene, chloroform; solutions are unstable.
Derivation: Mixture of formic acid, peroxide, and sulfuric acid is allowed to interact for 2 hours and then distilled.
Grade: 90% solution.
Hazard: Explodes when shocked or heated or in contact with reducing materials, metals, and metallic oxides. Strong irritant and oxidizing agent.
Use: Oxidation, epoxidation, and hydroxylation reactions.

perfume.   A blend of pleasantly odorous substances (usually liquids) obtained from the essential oils of flowers, leaves, fruit, roots, or wood of a wide variety of plants, either by steam distillation or solvent extraction. Flower oils (rose, jasmine) are extracted with a nonpolar solvent to give a waxy mixture called concrete; the wax is then removed by a second solvent (an alcohol), which is then in turn removed to form an absolute. It is necessary that all the solvent be eliminated to obtain the finest perfumes. The center of this industry has long been in Grasse, France. Perfume materials are also derived from animal sources (musk, ambergris) and from resinous extracts (terpenes and balsams); they are also made synthetically. Cologne and toilet water are weak alcoholic solutions (5% or less) of perfumes. Fine perfumes may contain as many as 30 ingredients, and their blending is an art rather than a science. The largest volume use of perfumes is in soaps, lotions, shaving creams, and cosmetics.
See fragrance; odorant.

perhydrosqualene.   See squalane.

peri acid.   (1-naphthylamine-8-sulfonic acid).

Properties: White needles. Slightly soluble in water.
Derivation: Nitration of naphthalene-α-sulfonic acid followed by reduction with iron and crystallization of the sodium salt.
Use: Azo dye intermediate, production of Chicago acid.

periclase.   MgO. A natural or calcined magnesium oxide used as a lining and maintenance material for basic oxygen steelmaking furnaces and other refractories.

perilla oil.
Properties: Light-yellow drying oil. D 0.932–0.945, saponification value 191–193, iodine value 187–202, refr index 1.4841. Soluble in alcohol, ether, chloroform, and carbon disulfide. Combustible. Subject to spontaneous heating.
Derivation: From the seeds of *Perilla ocymoides*, grown commonly in Japan and Korea.
Chief constituents: Linoleic and linolenic acids.
Use: Substitute for linseed oil, edible oil in Asia, manufacture of varnishes.

perindopril erbumine.   See perindopril *tert*-butylamine.

perindopril *tert*-butylamine.
CAS: 107133-36-8. $C_{19}H_{32}N_2O_5 \cdot C_4H_{11}N$.
Hazard: Moderately toxic by ingestion.

periodic acid.
CAS: 13444-71-8. $HIO_4 \cdot 2H_2O$.
Properties: White crystals. Mp 122C, decomposes at 130C, loses $2H_2O$ at 100C. Soluble in water, alcohol; slightly soluble in ether.
Derivation: By the interaction of iodine and concentrated perchloric acid, by low-temperature electrolytic oxidation of concentrated iodic acid.
Method of purification: Crystallization.
Hazard: Dangerous in contact with organic materials. Irritant and oxidizing material.
Use: Oxidizing agent, increasing wet strength of paper, photographic paper.

periodic law.   Originally stated in recognition of an empirical periodic variation of physical and chemical properties of the elements with atomic *weight*, this law is now understood to be based fundamentally on atomic *number* and atomic *structure*. A modern statement is: The electronic configurations of the atoms of the elements vary periodically with their atomic number. Consequently, all properties of the elements that depend on their atomic structure (electronic configuration) tend also to change with increasing atomic number in a periodic manner.

periodic table.   An arrangement of the chemical elements by symbol in a geometric pattern designed to represent the periodic law by aligning the elements in periods so that the corresponding parts of the several periods are adjacent. When the elements are aligned in order of increasing atomic number, they constitute a succession of period, each beginning with an alkali metal (one electron in the outermost principal quantum level) and ending with an element of the helium family, helium, neon, argon, krypton, xenon, and radon (each having eight electrons in the outermost principal quantum level, except for helium, which is limited to two). Each helium family element is followed directly in atomic number by an alkali metal, which begins a new period.

The advantage of placing each successive period so that it is adjacent to the preceding period in its corresponding parts is that similar elements are thus brought together in groups. For example, the alkali metals that begin each period have similar properties that correspond to their atomic structure, each having one outermost shell electron. In a periodic table, the elements exhibit a steady trend in properties from the beginning of each period, where they are metals, to the end, where they are nonmetals. Within each group the elements are quite similar. A given element can be predicted, on the basis of its position in a periodic table, to resemble the other elements of its group and be intermediate in properties between its adjacent neighbors within its period. The chief function of a periodic table is to serve as a fundamental framework for the systematic organization of chemistry (R. T. Sanderson).

See Mendeleyev, D. I.

Note: Dr. Glenn T. Seaborg predicted the existence of synthetic superheavy elements 107–168 on the basis of theoretical calculations. Numbers 122–153 he called "superactinides."

peripheral proteins.   Proteins that are loosely or reversibly bound to a membrane by hydrogen bonds or electrostatic forces which can be easily released from the membrane. Contrast with integral proteins.

peritectic system.   (transition type system; unstable compound system; hidden maximum system). Constitution–temperature system or diagram in which there is a binary compound unstable in contact with a melt of its own composition.

peritectic temperature.   Temperature in a peritectic system at which there is equilibrium between the solid and the remaining melt, the composition of which conforms to the peritectic point, at which the temperature line meets the liquids curve.

"Perkalink" *[Lanxess].*

CAS: 101-37-1. TM for triallyl cyanurate coagent.
Use: Coagent for peroxide cross-linking of PE, EVA, and EPDM.

Perkin alicyclic synthesis.   Synthesis of alicyclic compounds from α,W-dihaloalkanes and compounds containing active methylene groups in the presence of sodium ethoxide.

Perkin reaction.   Formation of α,β-unsaturated carboxylic acids by aldol condensation of aromatic

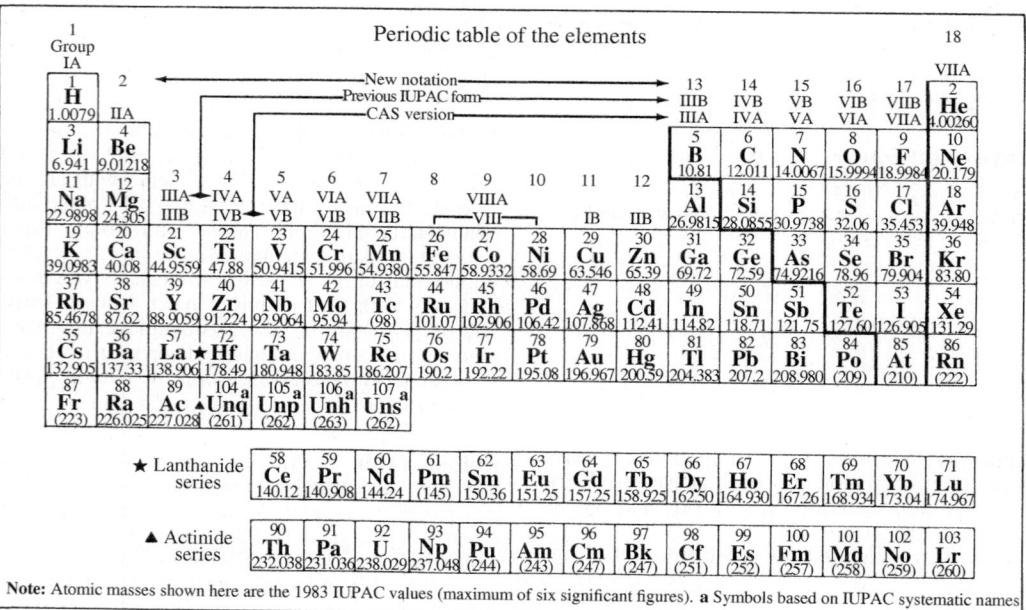

Note: Atomic masses shown here are the 1983 IUPAC values (maximum of six significant figures). a Symbols based on IUPAC systematic names

aldehydes and acid anhydrides in the presence of an alkali salt of the acid.

Perkin rearrangement. Formation of benzofuran-2-carboxylic acids and benzofurans by heating 3-halocoumarins with alkali.

Perkin, Sir William Henry. (1838–1907). An English chemist who was the first to make a synthetic dyestuff (1856). He studied under Hofmann at the Royal College of London. Perkin's first dye was called mauveine, but he proceeded to synthesize alizarin and coumarin, the first synthetic perfume. In 1907 he was awarded the first Perkin Medal, which has ever since been awarded by the American Division of the Society of Chemical Industry for distinguished work in chemistry. Notwithstanding the fact that Perkin patented and manufactured mauve dye in England, the center of the synthetic dye industry shifted to Germany, where it remained until 1914.
See Hofmann, August Wilhelm.

Perkow reaction. Formation of enol phosphates on treatment of α-halocarbonyl compounds with trialkyl phosphites.

perlite.
CAS: 93763-70-3. Eutectic between ferrite and cementite (steelmaking).
Properties: Average density of 0.13. Expands when finely ground and heated. Natural glass, amorphous mineral consisting of fused sodium potassium aluminum silicate, containing <1% quartz.
Hazard: Slightly toxic by ingestion. A nuisance dust. Questionable carcinogen.
Use: Filter aid, food additive.

"Permacore" [Rockland]. TM for high-density polyethylene pipe.
Use: Underground sewer systems, manholes, industrial, and municipal.

permafil. A mixture in which the liquid undergoes complete polymerization and hardens without the necessity of any evaporation. Anaerobic permafils harden out of contact with air.

"Permalloy" [Marigold]. TM for an alloy containing 78.5% nickel and 21.5% iron. It has high magnetic permeability and electrical resistivity.
Use: Transocean submarine cables.

"Permalume G" [Boston]. TM for a semibright nickel electroplating process. The plating bath contains nickel sulfate, nickel chloride, boric acid, and organic addition agents.

permalux. See dicatechol borate 1,3-di(o-tolyl)guanidine salt.

"Permalux" [Krokeide]. TM for a di-o-tolylguanidine salt of dicatechol borate. $(HOC_6H_4O)_2B \cdot HNC(NHC_6H_4CH_3)_2$.
Properties: Light-grayish-brown powder. D 1.27, fp>165C.
Use: Rubber antioxidant, vulcanization of neoprene.

permanent magnet. A magnet whose magnetic properties are not affected by removing it from an external magnetic field.

Permanent Orange. See dinitraniline orange.

permanent-press resin. (durable-press resin). A thermosetting resin used as a textile impregnant or fiber coating to impart crease resistance and permanent hot-creasing to suitings, dress fabrics, etc. Chemicals such as formaldehyde and maleic anhydride are the basis of these products. The resin is "cured" after the fabric has been tailored into a garment. A permanent-press fabric that requires no resin has been developed (a blend of polyester with cotton or rayon).

Permanent Red 2B Amine. See 4-amino-2-chlorotoluene-5-sulfonic acid.

permanent set. (permanent elongation). The extent to which vulcanized rubber and other elastomeric materials are permanently deformed after extension to break. In general it varies in the range of 5–10% of the original length, depending on the formulation and state of cure.
See elasticity.

permanent white. $BaSO_4$. Precipitated barium sulfate.

Permasep. A method of desalinating water by reverse osmosis using membranes composed of hollow nylon fibers that provide maximum membrane surface.
See osmosis.

permenorm. Nickel-iron alloy produced by magnetic annealing and drastic cold reduction and used for mechanical rectifiers and low-frequency amplifiers. This alloy has a rectangular hysteresis loop that eliminates arcing at the contacts of mechanical rectifiers, as well as other desirable properties.

permethrin. [(3-phenoxyphenyl)methyl-3-(2,2-dichlorovinyl)-2,2-dimethylcyclopropane-carboxylate].
CAS: 52645-53-1.
Hazard: Questionable carcinogen.
Use: Insecticide, nematocide, acaricide.

permissible explosive. See explosive, permissible.

Perone. HOOH, 30, 35, and 50% solutions by weight.
See hydrogen peroxide.

perovskite. A natural or synthetic crystalline mineral composed of calcium dioxide and titanium dioxide. The natural material was discovered in Russia in about 1850; the synthetic type can be so made as to incorporate particles of catalysts (platinum or palladium), that are protected by the crystalline structure from contamination by lead and other poisons, thus permitting use of leaded gasoline in cars equipped with emission-control devices.

peroxidase. An enzyme found in most plant cells and some animal cells, that promotes the oxidation of various substrates, such as phenols, aromatic amines, etc., by means of hydrogen peroxide.

peroxide. (1) Any compound containing a bivalent O–O group, i.e., the oxygen atoms are univalent. Such compounds release atomic (nascent) oxygen readily. Thus they are strong oxidizing agents, and fire hazards when in contact with combustible materials, especially under high-temperature conditions. The chief industrial uses of peroxides are as oxidizing agents, bleaching agents, and initiators of polymerization. (2) Hydrogen peroxide.

peroxisome. Membrane-bounded organelle in the cytoplasm of eukaryotic cells functioning in the destruction of proteins and characterized by peroxide-forming enzymes.

peroxisomicin A2.
CAS: 156768-16-0. $C_{30}H_{26}O_8$.
Hazard: A poison by ingestion.
Source: Natural product.

peroxisomicine A2. See peroxisomicin Aa2.

peroxyacetic acid. See peracetic acid.

peroxybenzoyl nitrate. A component of photochemical smog.
Hazard: More than 200 times as irritating to the eyes as formaldehyde. A concentration of 0.02 ppm causes moderate to severe conjunctival irritation.
See smog.

peroxyformic acid. See performic acid.

peroxysulfuric acid. See Caro's acid.

perphenazine.
CAS: 58-39-9. $C_{21}H_{26}ClN_3OS$.
Properties: Crystals, sensitive to light. Mp 97–100C. Insoluble in water; soluble in ethanol and acetone.
Grade: ND.
Use: Medicine.

perrhenic acid. $HReO_4$. Exists only in solution, commercially available as aqueous syrup. Strong, very stable, monobasic acid; extremely soluble in water and organic solvents.

Perrin rule. Ions of charge opposite to that of a diaphragm have by far the greatest effect, on endosmosis. The higher their valence (of opposite sign) the greater the reduction of electroosmotic flow.

Persistol. Agents for the wash-proof water-repellent finishing of textiles of natural and synthetic fibrous materials and fiber mixes.

perstoff. See trichloromethylchloroformate.

persulfuric acid. See Caro's acid.

Perthane. An agricultural insecticide based on 1,1-dichloro-2,2-bis(p-ethylphenyl)ethane. Supplied as a wettable powder or emulsifiable concentration.

Peru balsam. See balsam.

Perutz, Max F. (1914–2002). An Austrian molecular biologist who was a recipient of the Nobel Prize for chemistry in 1962 along with Kendrew. His work was concerned with crystalline protein structure, particularly the molecular structure of hemoglobin and myoglobin, nicotinamide adenine dinucleotide. He was educated in England and Austria.

PES. Abbreviation for photoelectron spectroscopy.

pesticide. Any substance, organic or inorganic, used to destroy or inhibit the action of plant or animal pests; the term thus includes insecticides, herbicides, rodenticides, miticides, etc.
See methyl parathion. Virtually all pesticides are toxic to humans to some degree. They vary in biodegradability. The use of more toxic types, especially DDT, has been restricted. Microencapsulated controlled-release forms are available.
See insecticide; biocide.

petalite. (lithium aluminum silicate). $LiAl(Si_2O_5)_2$.
Properties: Colorless, white, gray, or occasionally pink mineral, white streak, vitreous luster. Resembles spodumene in appearance. Contains up to 4.9% lithia, sometimes with partial replacement by sodium or less often by potassium. Insoluble in acids, d 2.39–2.46, Mohs hardness 6–6.5.
Occurrence: U.S. (Massachusetts, Maine); Sweden.
Use: Source of lithium salts, in ceramics and glass.

Peterson reaction (olefination). Reaction of α-silylated carbanions with carbonyl compounds, yielding β-hydroxyl-alkyl silanes, which undergo instantaneous elimination to afford olefins.

petitgrain oil. See neroli oil.

petitgrain oil, Paraguay type.
  Properties: From steam distillation of the leaves of *Citrus aurantium* L. subspecies *amara*. Yellow to brown liquid; harsh bitter odor. D: 0.878–0.889, refr index: 1.455 @ 20C. Sol in fixed oils, mineral oil, propylene glycol; insol in glycerin.
  Use: Food additive.

PETN. Abbreviation for pentaerythritol tetranitrate.

Petra. A PET (polyethylene terephthalate) resin.
  Use: Injection molding.

Petrac Crude Glyceride. Glycerol.

Petrac Vyn-Eze. Stearamide of commercial stearic acid.

Petrac Wax. Paraffin and polyethylene wax.

Petrenko-Kritschenko piperidone synthesis. Formation of piperidones by cyclization from two moles of aldehyde and one mole each of acetonedicarboxylic ester and of ammonia or a primary amine. Used widely in synthesis of tropane derivatives.

petri dish. A small concave glass plate having an easily removable cover, used for the culture of bacteria. Named after its inventor, a German biologist.

petrochemical. An organic compound for which petroleum or natural gas is the ultimate raw material. Thus, cracking of petroleum produces ethylene, which is converted to ethylene glycol, the latter being a typical petrochemical. The term is also applied to substances such as ammonia, because the hydrogen used to form the ammonia is derived from natural gas. Thus, synthetic fertilizers are considered to be petrochemicals. At least 175 substances are designated as petrochemicals including many paraffin, olefin, naphthene, and aromatic hydrocarbons (methane, ethane, propane, ethylene, propylene, butenes, cyclohexane, benzene, toluene, naphthalene, etc.) and their derivatives, even though some of their commercial production is from sources other than petroleum. The percentage of total hydrocarbon consumption represented by petrochemicals is steadily increasing; some authorities maintain that petroleum is too valuable to be used for fuel and that it should be conserved for future petrochemical development.

petrolatum. A semisolid or liquid mixture of hydrocarbons derived by distillation of paraffin-base petroleum fractions. The solid form (mineral jelly) may be either water white or pale yellow. Its chief uses are in mild ointments, cosmetics, softener in rubber mixtures and food processing (release agent in bakery products, dehydrated fruits and vegetables), protective coating (raw fruits and vegetables), and defoaming agent (beet sugar, yeast). The liquid form (white mineral oil) is used as a laxative, textile lubricant, and dispersing agent. There are three grades of both solid and liquid types with various specifications (USP, NF, and FCC).
  See ointment.

petrolatum wax. A microcrystalline wax containing hydrocarbons from $C_{33}H_{70}$ to $C_{43}H_{88}$. Solidifying range 71–83C.
  See wax, microcrystalline.

petrolenes. (malthenes). Portions of bituminous substances soluble in hexane.

petroleum. (crude oil). A highly complex mixture of paraffinic, cycloparaffinic (naphthenic), and aromatic hydrocarbons, containing a low percentage of sulfur and trace amounts of nitrogen and oxygen compounds. Said to have originated from both plant and animal sources 10–20 million years ago. The most important petroleum fractions, obtained by cracking or distillation, are various hydrocarbon gases (butane, ethane, propane), naphtha of several grades, gasoline, kerosene, fuel oils, gas oil, lubricating oils, paraffin wax, and asphalt. From the hydrocarbon gases, ethylene, butylene, and propylene are obtained; these are important industrial intermediates, being the source of alcohols, ethylene glycols, and monomers for a wide range of plastics, elastomers, and pharmaceuticals. Benzene, phenol, toluene, and xylene can be made from petroleum, and hundreds of other products, including biosynthetically produced proteins, are petroleum derived. About 5% of the petroleum consumed in the U.S. is used as feedstocks by the chemical industries.
  Occurrence: At present, half of the world's proven resources are in the Middle East and North Africa, the other half being divided among the U.S. (including Alaska), Canada, Venezuela, the former U.S.S.R., the North Sea area, Indonesia, Mexico, Romania, and Australia.
  Properties: Viscous, dark-brown liquid; unpleasant odor. D 0.78–0.97, flash p 20–90F.
  Hazard: Flammable, moderate fire risk. Toxic by ingestion, local skin irritant. For further information refer to The American Petroleum Institute.
  See natural gas; petrochemical.

petroleum benzin. A special grade of ligroin.

petroleum coke. See coke.

petroleum distillate.
CAS: 8002-05-9. Any of a wide variety of materials of varying toxicity distilled from petroleum.
Use: A vehicle for pesticides.
Hazard: Poisonous; moderately toxic; questionable carcinogen.

petroleum ether. This term is used synonymously with petroleum naphtha. It is also sometimes used as a synonym for ligroin or petroleum spirits. It is technically a misnomer, because it is not an ether in the chemical sense. For details about specified distillation ranges and other distinctive properties, consult ASTM and API specifications. See naphtha (1).

petroleum gas, liquefied. See liquefied petroleum gas.

petroleum jelly. See petrolatum.

petroleum naphtha. See naphtha (1).

petroleum spirits. In Great Britain the term *petroleum spirits* refers to a volatile hydrocarbon mixture having a flash p 32F (0C).
Hazard: Highly flammable, dangerous fire risk.
See naphtha (1a); spirits; petroleum ether.

petroleum, synthetic. See pyrolysis.

petroleum thinner. See naphtha (1a).

petroleum wax. A high molecular weight solid hydrocarbon derived from petroleum. There are three types: paraffin waxes, microcrystalline waxes, and petrolatum waxes. All are made mostly by solvent dewaxing, although pressing and sweating processes are still used.

"Petrolite" *[Baker Hughes].* TM for a hard grade of petroleum microcrystalline wax; a synthetic polymer; an oxidized hydrocarbon; a modified hydrocarbon; or a dispersion.

"Petromat" *[Propex].* TM for engineered nonwoven fabric.
Use: Paving repair.

"Petronates" *[Sonneborn].* TM for salts of petroleum sulfonic acids, varying in molecular weight and color.
Use: Emulsifying agents, dispersing agents, wetting agents, corrosion preventive.

"Petronauba" *[Baker Hughes].* TM for an oxidized hydrocarbon.

"Petrotac" *[Propex].* TM for engineered nonwoven fabric.
Use: Paving repair membranes.

"Petrothene" *[Equistar].* TM for polypropylene resins for blown, cast, and water-quenched films, substrate coating, wire and cable coating, injection molding, blow molding, thermoforming, pipe extrusion, calendering.
Available forms: Solid cubes and pellets in natural and black.

Petrowet R. A surface-active agent composed of saturated hydrocarbon sodium sulfonate. A wetting and penetrating agent effective in high concentration of electrolytes and acids, suitable for use in acidizing of oil wells.

pewter. Tin alloys with 5–15% tin, 0–3% copper, and 0–15% lead. White metal and Britannia metal are also of this general composition.

peyote. Root of cactus *Lophophora williamsii* (Mexico) known as an hallucinogen.

Pfau–Plattner azulene synthesis. Formation of azulenes by ring enlargement of indanes on addition of diazoacetic ester, hydrolysis, dehydrogenation, and decarboxylation of the resulting acid.

PFGE. See pulsed field gel electrophoresis.

Pfitzinger reaction. Formation of quinoline-4-carboxylic acids by condensation of isatic acids from isatin with α-methylene carbonyl compounds; subsequent decarboxylation yields quinolines.

Pfitzner–Moffatt oxidation. Oxidation of alcohols to carbonyl derivatives with dimethyl sulfoxide and dicyclohexylcarbodiimide. The procedure is especially useful for the conversion of a primary alcohol to an aldehyde without further oxidation to the carboxylic acid.

PFOS. See perfluorooctanesulfonic acid.

PF resins. Abbreviation for phenol-formaldehyde resins.

PG. Abbreviation for polypropylene glycol.

PGA. Abbreviation for pteroylglutamic acid. See folic acid.

PGB. See podophyllotoxin 4-*o*-glucoside.

PGDN. See propylene glycol dinitrate.

p53 gene. A gene that normally regulates the cell cycle and protects cells from damage to its genome. Cancer may result from mutations in this gene.

pH. pH is a value taken to represent the acidity or alkalinity of an aqueous solution; it is defined as the logarithm of the reciprocal of the hydrogen-ion concentration of a solution:

$$pH = \log_{10} \frac{1}{[H^+]}$$

The pH scale is designed to conveniently characterize the acidity of aqueous acid–base systems. Pure water is the standard used for the pH scale. Under ordinary conditions water molecules dissociate into the ions H^+ and OH^-, with recombination at such a rate that with very pure water at 22C there is a concentration of oppositely charged ions of 1/10,000,000, or 10^{-7}, mole per liter. This is commonly expressed by saying that pure water has a pH of 7, which means that its concentration of hydrogen ions is expressed by the exponent 7, without its minus sign.

When acids or hydroxyl-containing bases are in water solution they ionize more or less completely, furnishing varying concentrations of H^+ and OH^- ions, respectively, to the solution. Strong acids and bases ionize much more completely than weak acids and bases; thus strong acids give solutions of pH 1–3, while solutions of weak acids have a pH of 6. Strong bases give solutions of pH 12 or 13, while weak bases give solutions of pH 8. Because the pH scale is logarithmic, the intervals are exponential and thus represent far greater differences in concentration than the values themselves seem to indicate (see table).

Liquid	pH value
Pure water	7
Seawater	7.8–8.2
Electroplating bath	6.5–5
0.01 N HCl	2
0.1 N HCl	1.08
0.1 N H$_2$SO$_4$	1.17
0.01 N NaOH	12
0.1 N acetic acid	3
0.1 N NH$_4$OH	11
Gastric juices	1.7
Urine	5–7
Blood	7.3–7.5
Milk	6.5–7
Soil (optimum for crops)	6–7

In acid-base titrations, changes in pH can be detected by indicators such as methyl orange, phenolphthalein, etc. Litmus paper can also be used as a rough indication of acidity or alkalinity. In carrying out titrations, the end point signaled by an indicator does not always correspond to a neutral pH (7). pH control is of critical importance in a large number of industrial operations such as water purification, chrome tanning process for leather, in preservation of food products, in electroplating baths, dyeing, agriculture, and numerous other instances. See acid; base.

phage. A virus for which the natural host is a bacterial cell. See bacteriophage.

phagemid. A type of plasmid which carries within its sequence a bacteriophage replication origin. When the host bacterium is infected with "helper" phage, the phagemid is replicated along with the phage DNA and packaged into phage capsids.

phallolysin. A heat-sensitive glycoprotein toxin of the mushroom, *Amanita phalloides*.

phallotoxin. Any of at least six related, thermostable, cyclic polypeptide toxins that have the same heptapeptide skeleton.
  Derivation: They occur in several species of wild mushrooms of the genus *Amanita*.
  Hazard: Toxic, deadly, cause hepatic and renal failure.

phaltan. See folpet.

Pharaoh's serpent eggs. (mercuric thiocyanate). $(NCS)_2Hg$. Swells when heated.

pharmaceutical. A broad term that includes not only all types of drugs and medicinal and curative products but also ancillary products such as tonics,

pH Value	Ratio of H^+ or OH^- concentration to that of pure water at 22C
Acid side (excess of H^+ ions) 1	1,000,000
2	100,000
3	10,000
4	1000
5	100
6	10
Neutrality 7	1
Alkaline side (excess of OH^- ions) 8	10
9	100
10	1000
11	10,000
12	100,000
13	1,000,000

dietary supplements, vitamins, deodorants, and the like.
See drug.

pharmacogenomics. The study of the interaction of an individual's genetic makeup and response to a drug.

Pharmolin. (kaolin clay). Aluminum silicate.
Use: For various internal and external pharmaceutical compositions used in antidiarrheal medications.

phase. (1) One of the three states or conditions in which substances can exist, i.e., solid, liquid, or gas (vapor). The condition depends primarily on the concentration of atoms of molecules; solids are the most dense, gases the least, and liquids occupy the intermediate position. Solids are normally crystals, liquids are amorphous, and gases are without structure.
See matter. (2) A physically distinct and mechanically separable portion of a dispersion or solution. Phases may be solid, liquid, or gaseous (vapor). In any mixture or solution the major component is called the continuous or external phase and the minor component the dispersed or internal phase. The latter may or may not be uniformly dispersed in the continuous phase.
See colloid chemistry; solution.

Phaseolus. (kidney beans; bush beans; lima beans). A genus of warm-climate, cyanogenic, twining, annual or perennial herbs (family Fabaceae, formerly Leguminosae). A number are grown for their edible pods and seeds.

Phaseolus limensis. (lima bean; java bean; Burma bean; sieve bean; etc.). There are numerous varieties of this bean, all of which contain the cyanogenic glycoside, linamarin. The lima beans commonly cultivated in the U.S. and other countries in the temperate zone are essentially nontoxic. They are small, light green to nearly white in color, easily recognized, and essentially nontoxic. An exception is the black lima bean of Puerto Rico, which has proven toxic to humans. Tropical varieties are often quite toxic. They are usually small, plump beans that are solid colored or sometimes spotted. Many instances of cyanide poisoning are known from New Guinea due to eating raw lima beans (cooking is preventative). Lima beans imported into Canada or the U.S. contain less than 0.01% hydrocyanic acid (HCN), whereas tropical varieties contain as much as 0.3% HCN. Plants containing 0.02% HCN are potentially hazardous to livestock.

phase rule. Propounded by J. Willard Gibbs in 1877, the phase rule is a general system of equations of the form $F = C - P + 2$ stating the boundaries of thermodynamic equilibrium in a system of chemical reactants. The number of degrees of freedom (F) allowed in a given heterogeneous system may be examined by analysis or observation and plotted on a graph by proper choice of the components (C), the phases (P), and the independently variable factors of temperature and pressure. The principles of the phase rule apply to all multicomponent systems, including solvent blends, glass, alloys, and plastics.

phase transition. Abrupt change in physical properties as temperature is changed continuously. Water freezing at 0C for instance.

Phe. Abbreviation for phenylalanine.

α-phellandrene. (4-isopropyl-1-methyl-1,5-cyclohexadiene).
CAS: 99-83-2.

$\overline{CH_3C{:}CHCH_2CH[CH(CH_3)_2]CH{:}CH}$.

A monocyclic terpene occurring as (1) *d*- and (2) *l*-optical isomers.
Properties: Colorless oil. Insoluble in water; soluble in ether. (1) D 0.8463 (25C), bp 66–68C (16 mm Hg), refr index 1.4777; (2) d 0.8324 (20C), bp 58–59C (16 mm Hg), refr index 1.4724. Combustible.
Derivation: (1) Found in ginger oil, Ceylon, and Seychelles cinnamon oil. (2) Found in eucalyptus oil.
Hazard: Toxic by ingestion and skin absorption; strong irritant.
Use: Flavoring, perfumery.

β-phellandrene. (4-isopropyl-1-methylene-2-cyclohexene).

$\overline{CH_2C{:}CH{:}CHCH[CH(CH_3)_2]CH_2CH_2}$.

A monocyclic terpene occurring as (*1*) *d*- and (2) *l*-optical isomers.
Properties: (1) Mobile oil; pleasant odor; burning taste. D 0.8520 (20C), bp 171–172C, refr index 1.4788; (2) mobile oil, d 0.8497 (15C), bp 178–179, flash p (TCC) 120F (48.9C), refr index 1.4800. Both are insoluble in water and alcohol; soluble in ether. Combustible.

phen-. A prefix indicating phenol and benzene derivatives.

phenacaine hydrochloride. (*N,N'*-bis(*p*-ethoxyphenyl)acetamidine hydrochloride).
$C_{18}H_{22}O_2N_2 \cdot HCl \cdot H_2O$.
Properties: Small, white crystals; odorless; faintly bitter taste. Incompatible with alkalies, mp 190C. Soluble in alcohol, boiling water, and chloroform; less so in cold water; insoluble in ether.
Grade: NF, technical.
Hazard: Toxic by ingestion.
Use: Medicine (topical anesthetic).

phenacemide. (*N*-(aminocarbonyl)benzeneacetamide; (phenylacetyl)urea; phenacetylurea; phenurone; *N*-carbamoyl-2-phenylacetamide).
CAS: 63-98-9. $C_9H_{12}N_2O_2$.
Properties: A white, or nearly white, crystalline substance or fine crystalline powder; nearly insoluble in water and slightly soluble in ethanol, benzene, chloroform, and ether.
Hazard: May cause gastrointestinal distress, dermatitis, behavioral effects, aplastic anemia, and lesions.
Use: As an adjunct in the management of psychomotor, grand mal, and petit mal epilepsy and in mixed seizures.

phenacetin. USP name for acetophenetidin.

phenacyl chloride. See chloroacetophenone.

phenacyl fluoride. See fluoroacetophenone.

phenanthraquinone. See phenanthrenequinone.

phenanthrene.
CAS: 85-01-8. $C_{14}H_{10}$.

Properties: Colorless, shining crystals. D 1.063, mp 100.35C, bp 340C. Soluble in alcohol, ether, benzene, carbon disulfide, and acetic acid; insoluble in water. Combustible.
Derivation: Fractional distillation of high-boiling coal tar oils with subsequent recrystallization from alcohol.
Hazard: A questionable carcinogen.
Use: Dyestuffs, explosives, synthesis of drugs, biochemical research, manufacturing phenanthrenequinone.

phenanthrenequinone. (not phenanthraquinone).
CAS: 84-11-7. $C_{14}H_8O_2$.
Properties: Yellow-orange, needlelike crystals. D 1.4045, mp 206–207C, bp sublimes above 360C. Soluble in sulfuric acid, benzene, glacial acetic acid and hot alcohol; insoluble in water.
Derivation: By oxidation of a boiling solution of phenanthrene in glacial acetic acid with chromic acid, extraction in sodium disulfite, precipitation by means of hydrochloric acid, and recrystallization.
Use: Organic synthesis, dyes.

1,10-phenanthroline. (4,5-phenanthroline; *o*-phenanthroline).
CAS: 66-71-7. $C_{12}H_8N_2 \cdot H_2O$. A heterotricyclic compound.

Properties: White, crystalline powder. Mp 93–94C, anhydrous 117C. Slightly soluble in water; soluble in alcohol, benzene, and acetone.
Derivation: Heating *o*-phenylenediamine with glycerol, nitrobenzene, and concentrated sulfuric acid; also from 8-aminoquinoline.
Use: Forms a complex compound with ferrous ions used as an indicator; drier in coatings industry.

phenarsazine chloride. See diphenylaminechloroarsine.

phenazine. (azophenylene).
CAS: 92-82-0. $C_6H_4N_2C_6H_4$. A tricyclic compound.
Properties: Yellow crystals. Mp 170–171C, bp above 360C. Very slightly soluble in water; partially soluble in alcohol and ether. Combustible.
Use: Organic synthesis, manufacturing of dyes, larvicide.

phenazopyridine. (3,9-phenylazo-2,6-pyridine-diamine; 3-phenyldiazenylpyridine-2,6-diamine).
CAS: 94-78-0. $C_{11}H_{11}N_5$. Azo dye.
Properties: Brick red.
Hazard: Possible carcinogen, may cause methemoglobinemia and Heinz body anemia.
Use: Urinary tract analgesic.

phencyclidine. (CL-395; 1-(1-phencyclohexy)-piperidine; angel dust; hog; pcp; ci-395; sernyl).
CAS: 77-10-1. $C_{17}H_{25}N$. An arylcyclohexylamine. It is a central stimulant, depressant, anesthetic, analgesic, and hallucinogenic activity. It is similar to ketamine in structure and in many of its effects. It exerts its pharmacological action through inhibition of NMDA receptors.
Properties: Colorless crystals.
Hazard: Very toxic; poison; causes hyperactivity; slurred speech; sweating; muscular rigidity; numbness in fingers and toes; visual, auditory, and tactile illusions; delusions; eyes crossing; incoordination; lack of sensation; hyperextension; ataxic gait; facial grimaces; anxiety; hostility; feelings of inebriation; disorientation; prominent body image distortions; transient amnesia; lack of pain perception; wild movements; tachycardia, convulsions; stupor or coma; high fever; seizures; high blood pressure; convulsions; grand mal seizures; suppressed or absent reflexes; renal failure; respiratory arrest; death.
Use: As an hallucinogen.

phenelzine. ((2-phenethyl)hydrazine; β-phenylethylhydrazine; phenalzine; phenethylhydrazine; sulfuric acid).
CAS: 51-78-8. $C_8H_{12}N_{12}$. A monoamine oxidase inhibitor.
Properties: White, water-soluble, powder.
Hazard: Very toxic.
Use: To treat depression, phobic disorders and panic.

phenesterine. (cholesteryl-*p*-bis(2-chloroethyl) amino phenylacetate; (*p*-(bis(2-chloroethyl)amino) phenyl)acetic acid cholesterol ester; (4-(bis(2-chloroethyl)amino)phenyl)acetic acid cholesteryl ester; 5-cholesten-3-β-ol-3-(*p*-(bis(2-chloroethyl) amino)phenyl)acetate; fenesterin; fenestrin; phenestrin; [(3S,8S,9S,10R,13R,14S,17R)-10,13-dimethyl-17-[(2R)-6-methylheptan-2-yl]-,3,4,7,8, 9,11,12,14,15,16,17-dodecahydro-1H-cyclopenta [a]phenanthren-3-yl]2-[4-[bis(2-chloroethyl) amino]phenyl]acetate). $C_{39}H_{59}Cl_2NO_2$.
Hazard: Toxic, carcinogen.

phenethicillin. See potassium-α-phenoxyethyl penicillin.

phenethyl acetate. See 2-phenylethyl acetate.

phenethyl alcohol. (phenylethyl alcohol; 2-phenylethanol; benzyl carbinol).
CAS: 60-12-8. $C_6H_5CH_2CH_2OH$.
Properties: Colorless liquid; floral odor; sharp burning taste. D 1.017–1.020 (25C), refr index 1.5310–1.5340 (20C), fp −27C, bp 219C, flash p 216F (102C). Soluble in 50% alcohol; soluble 1 part in 50 parts of water; soluble in fixed oils, alcohol, and glycerol; slightly soluble in mineral oil. Combustible.
Derivation: (1) By reduction of phenylacetic ethyl ester by sodium in absolute alcohol. (2) By the action of ethylene oxide on phenylmagnesium bromide and subsequent hydrolysis. (3) Component of many essential oils.
Grade: Technical, NF, FCC.
Use: Organic synthesis, synthetic rose oil, soaps, flavors, antibacterial, preservative.

***sec*-phenethyl alcohol.** See α-methylbenzyl alcohol.

phenethylamine. See 2-phenylethylamine.

phenethyl anthranilate. See 2-phenylethyl anthranilate.

phenethyl isobutyrate. See 2-phenylethyl isobutyrate.

phenethyl phenylacetate. See 2-phenylethyl phenylacetate.

phenethyl propionate. See 2-phenylethyl propionate.

phenethyl salicylate. See 2-phenylethyl salicylate.

***o*-phenetidine.** (2-aminophenetole).
CAS: 94-70-2. $NH_2C_6H_4OC_2H_5$.
Properties: Oily liquid, rapidly becomes brown on exposure to light or air. Solidifies at about −20C, bp

228–230C. Soluble in alcohol and ether; insoluble in water. Combustible.
Derivation: Reduction of *o*-nitrophenetole with iron filings and hydrochloric acid.
Hazard: Toxic by ingestion, inhalation, and skin absorption.
Use: Manufacture of dyes, laboratory reagent.

***p*-phenetidine.** (4-aminophenetole).
CAS: 156-43-4. $NH_2C_6H_4OC_2H_5$.
Properties: Colorless, oily liquid; becomes red to brown on exposure to air and light. D 1.0613 (15C), mp 2–4C, bp 253–255C. Insoluble in water; soluble in alcohol. Combustible.
Derivation: Ethylating *p*-nitrophenol with ethyl sulfate or chloride in the presence of sodium hydroxide, followed by reduction with iron filings and hydrochloric acid.
Hazard: Toxic by ingestion, inhalation, and skin absorption.
Use: Dyestuffs intermediate, pharmaceuticals, laboratory reagent.

phenetole. (phenyl ethyl ether).
CAS: 103-73-1. $C_6H_5OC_2H_5$.

Properties: Colorless, oily liquid. Bp 172C, fp −30C, d 0.967 (20/4C). Insoluble in water; soluble in alcohol and ether. Combustible.

phenetsal. See *p*-acetylaminophenyl salicylate.

"Phenex" *[Abbott].* TM for α-ethyl-β-propylacrylaniline.
Use: Accelerator for natural and synthetic rubber and latexes.

phenformin. (*N*-(2-phenylethyl)imidodicarbonimidic diamide; 1-phenethylbiguanide; phenethyldiguanide; *n*′-β-phenethylformamidinyliminourea; fenformin; fenormin; 1-(diaminomethylidene)-2-(2-phenylethyl)guanidine).
CAS: 114-86-3. $C_{10}H_{15}N_5$. A biguanide hypoglycemic agent with actions and use similar to those of metformin.
Hazard: Very toxic; causes fatal lactic acidosis.
Use: Was used as an antidiabetic.

phenformin hydrochloride. (1-phenylbiguanide monohydrochloride; 1-(diaminomethylidene)-2-(2-phenylethyl)guanidine). $C_{19}H_{16}ClN_5$. A substance with an action essentially similar to that of phenformin.
Hazard: Toxic.

phenic acid.   See phenol.

Phenidone B.   4-methyl-1-phenyl-3-pyrazoli-
done.
Use: Photographic developer.

phenindione.   (2-phenyl-1,3-indanedione).
CAS: 83-12-5. $C_{15}H_{10}O_2$.
Properties: Pale-yellow crystals; practically odorless. Insoluble in water; soluble in methanol, alcohol, ether, acetone, benzene; solutions in alkalies
are red; in concentrated sulfuric acid blue.
Use: Medicine (blood anticoagulant).

pheniramine   maleate.   (prophenpyri-
damine   maleate;   1-phenyl-1-(2-pyridyl)-3-di-
methylaminopropane maleate).
CAS: 132-20-7. $C_{16}H_{20}N_2 \cdot C_4H_4O_4$.
Properties: White, crystalline powder; faint amine-
like odor. Mp 104–108C. Very soluble in alcohol
and water; slightly soluble in benzene and ether,
1% solution has pH between 4.5 and 5.5.
Grade: NF.
Use: Medicine (antihistamine).

Phenlam.   A line of phenolic resins.
Use: To provide high strength, fire and heat resistance
for low weight plastics.

Phenmad.   A 10% phenylmercuric acetate aque-
ous solution.
Use: Turf fungicide.
Hazard: Highly toxic.

phenmedipham.   ((3-methylphenyl)carbamic
acid 3[(methoxycarbonyl)amino]phenyl ester; *m*-
hydroxycarbanilic acid methyl ester m-methylcar-
banilate; methyl-3-(m-tolylcarbamoyloxy)phenyl-
carbamate; [3-(methyoxycarbonylamino)phenyl]
N-(3-methylphenyl)carbamate).
CAS: 13684-63-4. $C_{16}H_{16}N_2O_4$.
Hazard: Toxic.
Use: A selective carbamate herbicide.

phenobarbital.   (phenylbarbital;   phenylethyl-
malonylurea; 5-ethyl-5-phenylbarbituric acid).
CAS: 50-06-6. $C_{12}H_{12}N_2O_3$.
Properties: White, shining, crystalline powder;
odorless; stable. Mp 174–178C. Soluble in alcohol,
ether, chloroform, alkali hydroxides, alkali carbonate solutions; sparingly soluble in water.
Derivation: Condensation of phenylethylmalonic
acid derivatives and urea.
Grade: USP.
Hazard: May have damaging side effects.
See barbiturate.
Use: Medicine (sedative), laboratory reagent. Also
available as the sodium salt, which has good water
solubility.

pheno black ep.   See apomine black gx.

phenocoll hydrochloride.   (aminoacetophenetidide   hydrochloride;   glycocoll-*p*-phenetidine
hydrochloride).
$C_2H_5OC_6H_4NHCOCH_2NH_2 \cdot HCl$.
Properties: Fine, white, crystalline powder. Mp 95C.
Soluble in water and warm alcohol; slightly soluble
in chloroform, ether, and benzene.
Derivation: By the action of aminoacetic acid upon
phenetidine and acidifying.
Use: Medicine (analgesic).

phenocopy.   A trait not caused by inheritance of
a gene which appears to be identical to a genetic
trait.

phenol.   (1) A class of aromatic organic compounds in which one or more hydroxy groups are
attached directly to the benzene ring. Examples are
phenol itself (benzophenol), the cresols, xylenols,
resorcinol, naphthols. Though technically alcohols,
their properties are quite different. (2) Phenol (carbolic acid; phenylic acid; benzophenol; hydroxybenzene).
CAS: 108-95-2. C_6H_5OH.

Properties: White, crystalline mass that turns pink
or red if not perfectly pure or if under influence
of light; absorbs water from the air and liquefies;
distinctive odor; sharp burning taste. When in very
weak solution it has a sweetish taste. D 1.07, mp
42.5–43C, bp 182C, flash p 172.4F (78C) (CC),
autoign temp 1319F (715C). Soluble in alcohol,
water, ether, chloroform, glycerol, carbon disulfide,
petrolatum, fixed or volatile oils, and alkalies. Combustible.
Derivation: Most of the phenol in the U.S. is made by
the oxidation of cumene, yielding acetone as a by-
product. The first step in the reaction yields cumene
hydroperoxide, which decomposes with dilute sulfuric acid to the primary products, plus acetophenone and phenyl dimethyl carbinol. Several other
benzene-based processes have been used in the past;
derivation from benzoic acid is also possible.
Method of purification: Rectification.
Grade: Fused, crystals, or liquid, all as technical (82,
90, and 95%, other components mostly cresols), CP,
and USP.
Hazard: Toxic by ingestion, inhalation, and skin
absorption; strong irritant to tissue and upper respiratory tract. Toxic by skin absorption. Lung damage
and central nervous system impairment. Questionable carcinogen.
Use: Phenolic resins, epoxy resins (bisphenol-A),
nylon-6 (caprolactam), 2,4-D, selective solvent for
refining lubricating oils, adipic acid, salicylic acid,

phenolphthalein, pentachlorophenol, acetophene-tidin, picric acid, germicidal paints, pharmaceuti-cals, laboratory reagent, dyes and indicators, slimi-cide, biocide, general disinfectant.

Note: High-boiling phenols are mixtures containing predominantly *m*-substituted alkyl phenols. Their boiling points range from 238 to 288C; they set to a glass at −30C.

phenolate process. A process for removing hydrogen sulfide from gas by the use of sodium phenolate, which reacts with hydrogen sulfide to give sodium hydrosulfide and phenol. This can be reversed by steam heat to regenerate the sodium phenolate.

phenol, camphorated.
CAS: 8002-06-0. $C_{10}H_{16}O \cdot C_6H_6O$.
Hazard: Moderately toxic by ingestion. Human sys-temic effects.

phenol coefficient. In determining the effective-ness of a disinfectant using phenol as a standard of comparison, the phenol coefficient is a value obtained by dividing the highest dilution of the test disinfectant by the highest dilution of phenol that sterilizes a given culture of bacteria under standard conditions of time and temperature.
See disinfectant.

phenoldisulfonic acid. $C_6H_6O_7S_2$.
Properties: Deliquescent crystals. Mp 90C. Decom-poses above 100C; soluble in water and alcohol.
Use: Manufacture of dye intermediates.

phenol-formaldehyde resin. The first syn-thetic thermosetting polymer; the reaction product of phenol with aqueous 37–50% formaldehyde at 50–100C, with basic catalyst, discovered by Baeke-land in 1907 and trademarked "Bakelite" in 1911. Polymerization is of the condensation type, pro-ceeding through three stages. With an acid catalyst novolac resins, which are thermoplastic, are pro-duced.
Properties: Gray to black, hard, infusible solid when cured, resistant to moisture, solvents, and heat up to 200C, dimensionally stable, good electrical resis-tance, sound and noise absorbent, decomposed by oxidizing acids, fair resistance to alkalies. Cannot be successfully colored. Noncombustible.
Use: Molded and cast articles, bonding powders, ion exchange, laminating and impregnating, plywood and glass-fiber composites, ablative coatings for aerospace use, binder for oil-well sands, paint and baked enamel coatings, thermal and acoustic insu-lation, brake linings, clutch facings, shell molds, chemical equipment, machine and instrument hous-ings, chemical-resistant mortars, machine parts, electrical devices.
See A-stage resin; B-stage resin; C-stage resin; novolak; Baekeland, L. H.; phenolic resin.

phenol-furfural resin. A phenolic resin that has a somewhat sharper transition than phenol-formaldehyde from the soft, thermoplastic stage to the cured, infusible state and can be fabricated by injection molding since it has little tendency to harden before curing conditions are reached.

phenolic. See phenolic molding compound.

phenolic acids. Aromatic acids having one or more hydroxyl groups on the benzene ring.

phenolic anthraquinone. Any of a class of anthraquinones that can cause adenomas and ade-nocarcinomas of the colon and cecum.

phenolic herbicide. Any of a number of pheno-lic compounds.
Use: Herbicides.
Hazard: Poison.

phenolic molding compound. (phenolic). Thermosetting compound composed essentially of phenol-formaldehyde resin alone or intimately mixed with fillers, pigments, or dyes.

phenolic resin. Any of several types of synthetic thermosetting resin obtained by the condensation of phenol or substituted phenols with aldehydes such as formaldehyde, acetaldehyde, and furfural. Phenol-formaldehyde resins are typical and consti-tute the chief class of phenolics.

phenolics. Designation for phenol-formaldehyde resins and plastics.

phenolphthalein. (3,3-bis(*p*-hydroxyphenyl) phthalide).
CAS: 77-09-8. $(C_6H_4OH)_2C_2O_2C_6H_4$ (an approx-imation).
Properties: Pale-yellow powder; forms an almost colorless solution in neutral or acid solution, pink to deep red in presence of alkali, but colorless in the presence of large amounts of alkali. D 1.2765, mp 261C. Soluble in alcohol, ether, and alkalies; insoluble in water.
Derivation: Interaction of phenol and phthalic anhy-dride in sulfuric acid.
Grade: Technical, pure reagent, NF.
Hazard: Possible carcinogen.
Use: Acid–base indicator, laboratory reagent, medicine (laxative).

phenol red. See phenolsulfonephthalein.

phenolsulfonephthalein. (phenol red).
CAS: 143-74-8. $(C_6H_4OH)_2COSO_2C_6H_4$ (an approximation).
Properties: Bright- to dark-red, crystalline powder; stable in air. Slightly soluble in water, alcohol, and acetone; almost insoluble in chloroform and ether; soluble in alkali hydroxides and carbonates.

Derivation: Reaction of phenol with *o*-sulfobenzoic acid anhydride. Differs from phenolphthalein in containing an SO_2 group in place of a CO group.

Grade: Technical, reagent, USP. The USP spelling is *phenolsulfonphthalein*.

Use: Acid–base indicator, diagnostic reagent in medicine, laboratory reagent.

phenolsulfonic acid. (sulfocarbolic acid). $HOC_6H_4SO_3H$.

Properties: Yellowish liquid, becoming brown on exposure to air. A mixture of *o*- and *p*-phenolsulfonic acids; soluble in water and alcohol.

Derivation: Action of sulfuric acid on phenol.

Grade: Technical, reagent.

Hazard: Irritant to skin and tissue.

Use: Water analysis, laboratory reagent, electroplated tin coating baths, manufacture of intermediates and dyes, pharmaceuticals.

phenol trinitrate. See picric acid.

phenosafranin. (3,7-diamino-5-phenylphenazinium chloride; CI 50200). $C_{18}H_{15}ClN_4$.

Properties: Green, acicular crystals. Soluble in water and alcohol; solutions have purple to red color.

Use: Staining tissues for microscopy.

phenothiazine. (thiodiphenylamine).
CAS: 92-84-2. $C_{12}H_9NS$.

Properties: Grayish-green to greenish-yellow powder, granules, or flakes; tasteless; slight odor. Mp 175–185C, bp 371C, sublimes 130C (1 mm Hg). Insoluble in water; soluble in benzene, ether, hot acetic acid.

Derivation: By reaction of diphenylamine and sulfur in the presence of an oxidizing catalyst.

Grade: Technical, NF.

Hazard: Toxic by ingestion, inhalation, and skin absorption. Eye photosensitization and skin irritant.

Use: Insecticide, manufacture of dyes, parent compound for chlorpromazine and related antipsychotic drugs, polymerization inhibitor, antioxidant.

phenothrin. [(3-phenoxyphenyl)methyl-2,2-dimethyl-3-(2-methyl-1-propenyl)-cyclopropane carboxylate].

Use: Insecticide.

phenotype. The physical characteristics of an organism or the presence of a disease that may or may not be genetic.
See genotype.

phenoxyacetic acid.
CAS: 122-59-8. $C_6H_5OCH_2COOH$.

Properties: Light-tan powder. Bp 285C, mp 98C. Soluble in ether, water, alcohol, carbon disulfide, glacial acetic acid. Combustible.

Use: Intermediate for dyes, pharmaceuticals, pesticides, other organics, fungicides, flavoring, laboratory reagent, precursor in antibiotic fermentations, (especially penicillin V).

phenoxybenzamine. (*N*-(2-chloroethyl)-*n*-(1-methyl-2-phenoxyethyl)benzenemethanamine; *n*-(2-chloroethyl)-*n*-(1-methyl-2-phenoxyethyl)-benzylamine; *n*-phenoxyisopropyl-*n*-benzyl-β-chlorothylamine; bensylyt; 688a; *N*-(2-chloroethyl)-1-(phenoxy)-*N*-(phenylmethyl) propan-2-amine).
CAS: 59-69-1. $C_{18}H_{22}ClNO$. An alpha-adrenergic antagonist with long duration of action.

Properties: A crystalline, benzene-soluble compound.

Hazard: Carcinogen.

Use: To treat hypertension and as peripheral vasodilator.

2-phenoxy-4h-1,3,2-benzodioxaphosphorin 2-oxide. See saligenin cyclic phenyl phosphate.

o-phenoxybenzoic acid.
CAS: 2243-42-7. $C_{13}H_{10}O_3$.

Hazard: A poison.

p-phenoxybenzoic acid.
CAS: 2215-77-2. $C_{13}H_{10}O_3$.

Hazard: A poison.

3′-phenoxybenzyl α-ethyl-4-methoxyphenyl-acetate.
CAS: 51629-13-1. $C_{24}H_{24}O_4$.

Hazard: Moderately toxic by ingestion.

Use: Agricultural chemical.

3′-phenoxybenzyl α-ethylphenylacetate.
CAS: 51628-37-6. $C_{23}H_{22}O_3$.

Hazard: Moderately toxic by ingestion.

Use: Agricultural chemical.

3′-phenoxybenzyl α-isopropenyl-4-methoxy-phenylacetate.
CAS: 51629-15-3. $C_{25}H_{24}O_4$.

Hazard: Moderately toxic by ingestion.

Use: Agricultural chemical.

3′-phenoxybenzyl α-isopropenyl-4-methylphenylacetate.
CAS: 51629-54-0. $C_{25}H_{24}O_3$.

Hazard: Moderately toxic by ingestion.

Use: Agricultural chemical.

3′-phenoxybenzyl α-isopropyl-4-chlorophenylacetate.
CAS: 51629-48-2. $C_{24}H_{21}ClO_3$.

Hazard: Moderately toxic by ingestion.

Use: Agricultural chemical.

3'-phenoxybenzyl   α-isopropyl-4-methoxy-phenylacetate.
CAS: 51628-95-6. $C_{25}H_{26}O_4$.
Hazard: Moderately toxic by ingestion.
Use: Agricultural chemical.

3'-phenoxybenzyl α-isopropyl-4-methylphenylacetate.
CAS: 51629-37-9. $C_{25}H_{26}O_3$.
Hazard: Moderately toxic by ingestion.
Use: Agricultural chemical.

phenoxydihydroxypropane.   See phenoxypropanediol.

2-phenoxyethanol.   See ethylene glycol monophenyl ether.

α-phenoxyethylpenicillin.   See penicillin.

5'-phenoxyfurfuryl α-isopropyl-4-methoxyphenylacetate.
CAS: 51628-96-7. $C_{23}H_{24}O_5$.
Hazard: Moderately toxic by ingestion.
Use: Agricultural chemical.

phenoxymethylpenicillin.   (penicillin V).
$C_{16}H_{18}N_2O_5S$.
Properties: White, odorless, crystalline powder. Very slightly soluble in water; soluble in alcohol and acetone; insoluble in fixed oils, pH of saturated solution is 2.5–4.0, decomposes at 120C.
Grade: NF.
Use: Medicine (antibiotic for oral use). Available also as potassium salt.

(3-phenoxyphenyl)methyl 4-chloro-α-(1-methylethyl)benzeneacetate.
CAS: 51630-33-2. $C_{24}H_{23}ClO_3$.
Hazard: Moderately toxic by ingestion.
Use: Agricultural chemical.

(3-phenoxyphenyl)methyl α-ethyl-4-methyl-benzeneacetate.
CAS: 51630-04-7. $C_{24}H_{24}O_3$.
Hazard: Moderately toxic by ingestion.
Use: Agricultural chemical.

phenoxypropanediol.   (1-phenoxypropanediol-2,3).
$C_6H_5OCH_2CHOHCH_2OH$.
Properties: White, crystalline solid. Mp 53C, bp 150–155C (4 mm Hg). Soluble in water, alcohol, glycerol, carbon tetrachloride, warm benzene; insoluble in gasoline. Combustible.
Derivation: Phenol and glycerol.
Use: Plasticizer, resins, lacquers.

1-phenoxy-2-propanol.
CAS: 770-35-4. $C_9H_{12}O_2$.

Hazard: Moderately toxic by ingestion and skin contact.

phenoxypropylene oxide.
CAS: 122-60-1.

$\overline{C_6H_5OCH_2CHCH_2O}$.

Properties: Practically colorless liquid; characteristic odor. D 1.1110 (20/20C), bp 244.2C, vap press less than 0.1 mm Hg (20C), fp 2.8C, viscosity 6.93 cP (20C). Very slightly soluble in water.

phenoxy resin.   A high molecular weight thermoplastic copolymer of bisphenol A and epichlorohydrin having the basic molecular structure – $[OC_6H_4C(CH_3)_2C_6H_4OCH_2CH(OH)CH_2]_n$– ($n$ is about 100). It uses the same raw materials as epoxy resins but contains no epoxy groups. It may be cured by reacting with polyisocyanates, anhydrides, or other cross-linking agent capable of reacting with hydroxyl groups. The ductility of phenoxy resins resembles that of metals. They are transparent and also characterized by low mold shrinkage, good dimensional stability, and moderately good resistance to temperature and corrosion. Phenoxy resins are soluble in methyl ethyl ketone and have been used for coatings and adhesives. A typical injection-molded specimen has a tensile strength of 9000 psi, heat distortion point 86.6C at 264 psi load, and d 1.18. They may be extruded or blow molded. Parts may be thermally formed and heat or solvent welded. Some applications are blow-molded containers, pipe, ventilating ducts, and molded parts.

phenoxytriethylstannane.   See triethyltin phenoxide.

phenyl.   The univalent C_6H_{5+} group derived from benzene and characteristic of phenol and other derivatives.

phenylacetaldehyde.   (α-toluic aldehyde).
CAS: 122-78-1. $C_6H_5CH_2CHO$.
Properties: Colorless liquid; strong hyacinth-like odor. D 1.023–1.030 (25C), fp −10C, bp 193–194C, refr index 1.520–1.530, becomes more viscous on aging. Soluble in 2 parts of 80% alcohol; soluble in ether and most fixed oils; slightly soluble in water. Combustible.
Derivation: From phenyl-α-chloroacetic acid, by action of alkalies, oxidation of phenylethyl alcohol.
Grade: Technical, 50% soluble in benzyl alcohol, FCC.
Use: Perfumes, flavoring, laboratory reagent.

phenylacetaldehyde dimethylacetal.
(α-tolyl aldehyde dimethyl acetal).
CAS: 101-48-4. $C_6H_5CH_2CH(OCH_3)_2$.
Properties: Colorless liquid; strong odor. More stable than phenylacetaldehyde; not known to cause discoloration. D 1.000–1.004 (25/25C), refr index

1.493–1.496 (20C), flash p 191F (88.3C) (TCC). Soluble in 2 parts of 70% alcohol. Combustible.
Grade: Technical, FCC.
Use: Perfumery, flavoring, laboratory reagent.

phenylacetamide.   (α-toluamide).
CAS: 103-81-1. $C_6H_5CH_2CONH_2$.
Properties: White crystals. Bp 280–290C (decomposes), mp 156–160C. Soluble in hot water and alcohol; slightly soluble in cold water and ether. Combustible.
Derivation: From acetophenone or styrene by Willgerodt reaction, dehydration of ammonium phenyl acetate.
Use: Organic synthesis, pharmaceuticals, penicillin G precursor, laboratory reagent.

***N*-phenylacetamide.**   See acetanilide.

phenyl acetate.   (acetylphenol).
CAS: 122-79-2. $C_6H_5OOCCH_3$.

Properties: Water-white liquid. D 1.073 (25/25C), bp 195–196C, flash p 176F (80C). Soluble in alcohol and ether; almost insoluble in water. Combustible.
Derivation: (1) From phenol and acetyl chloride. (2) By heating triphenyl phosphate with potassium acetate and alcohol.
Use: Solvent, organic synthesis, laboratory reagent.

phenylacetic acid.   (α-toluic acid).
CAS: 103-82-2. $C_6H_5CH_2COOH$.
Properties: Shiny, white plate crystals; floral odor. D 1.0809, fp 76–78C, bp 262C. Soluble in alcohol, ether and hot water. Combustible.
Derivation: From benzyl cyanide refluxed with dilute hydrochloric acid.
Grade: Technical, FCC.
Use: Perfume, precursor in manufacture of penicillin G, fungicide, flavoring, laboratory reagent.

phenylacetonitrile.   See benzyl cyanide.

α-phenylacetophenone.   See deoxybenzoin.

phenylacetyl chloride. $C_6H_5CH_2COCl$.
Properties: Colorless liquid. Refr index 1.5320 (20C).
Hazard: Strong irritant.
Use: Acylating agent, including manufacture of esters for flavors; laboratory reagent.

phenyl α-acid.   (phenyl-2-amino-8-naphthol-6-sulfonic acid). $HOC_{10}H_5(NHC_6H_5)(SO_3H)$.

Properties: Gray crystals. Soluble in alkali. Combustible.
Derivation: From α-acid and aniline (condensation with heat).
Use: Azo dye manufacture.

β-phenylacrylic acid.   See cinnamic acid.

phenylacrylyl chloride.   See cinnamoyl chloride.

phenylalanine.   (α-amino-β-phenylpropionic acid).
CAS: 63-91-2. $C_6H_5CH_2CH(NH_2)COOH$.   An essential amino acid.
Properties: (L(−)-phenylalanine) Plates and leaflets from concentrated aqueous solutions, hydrated needles from dilute aqueous solutions, decomposes at 283C. Soluble in water; slightly soluble in methanol and ethanol. (D(+)-phenylalanine) Leaflets from water, decomposes 285C. Soluble in water; slightly soluble in methanol. (DL-phenylalanine) Leaflets or prisms from water or alcohol; sweet tasting. Decomposes 318–320C. Soluble in water.
Source: (L(−))-Phenylalanine is isolated commercially from proteins (ovalbumin, lactalbumin, zein, and fibrin). DL-Phenylalanine is synthesized from α-acetaminocinnamic acid.
Grade: Technical, FCC.
Use: (L isomer) Medicine and nutrition, essential ingredients of Aspartame. Available commercially as DL-dihydroxyphenylalanine and as DL-phenylalanine.

d-phenylalanyl-*l*-phenylalanyl-*l*-phenylalanyl-*l*-tryptophyl-*l*-lysyl-*l*-threon yl-*l*-phenylalanyl-*l*-threoninamide.
CAS: 133073-82-2. $C_{61}H_{75}N_{11}O_{10}$.
Hazard: A poison.

d-phenylalanyl-*l*-phenylalanyl-*l*-tyrosyl-*d*-tryptophyl-*l*-lysyl-*l*-valyl-*l*-phenylalanyl-3-(2-naphthalenyl)-d-alaninamide.
CAS: 150155-61-6. $C_{71}H_{81}N_{11}O_9$.
Hazard: A poison.

phenylallylic alcohol.   See cinnamic alcohol.

phenylamine.   See aniline.

1-phenyl-3-aminobutane.
CAS: 22374-89-6. $C_{10}H_{15}N$.
Hazard: Moderately toxic by ingestion.

1-phenyl-2-aminopropane.   See amphetamine.

***N*-phenylaniline.**   See diphenylamine.

***o*-phenylaniline.**   See *o*-aminobiphenyl.

phenylarsonic acid.
CAS: 98-05-5. $C_6H_5AsO(OH)_2$.
Properties: Crystalline powder. Mp 160C with decomposition. Soluble in water and alcohol; insoluble in chloroform.
Method of purification: Reaction of the diazonium salt and sodium arsenite.
Hazard: Toxic by ingestion.
Use: Analytical reagent for tin.

phenylazoaniline. See aminoazobenzene.

1-(phenylazo)-2-naphthylamine. See yellow AB.

p-(phenylazo)phenyl isocyanide.
CAS: 22287-69-0. $C_{13}H_9N_3$.
Hazard: Moderately toxic by ingestion.
Use: Agricultural chemical.

phenylbarbital. See phenobarbital.

phenylbenzamide. See benzanilide.

1,3,4-phenylbiphenylyloxadiazole. (PBD; phenylbiphenyloxadiazole). $C_{20}H_{14}N_2O$.
Properties: Crystals. Mp 166–168C.
Grade: Purified.
Use: As primary fluors or as wavelength shifters in soluble scintillators.

3-phenyl-3-benzoborepin. A synthetic carbon-boron heterocyclic compound with aromatic properties; stable in air. Key intermediate in its synthesis is the nonaromatic carbon-tin heterocycle 3,3-dimethyl-3-benzostannepin.

2-phenyl-benzo(de)isoquinoline-1,3-dione.
CAS: 6914-98-3. $C_{18}H_{11}NO_2$.
Hazard: A poison.

phenylbis[1-(2-methyl)aziridinyl]phosphine oxide. $C_6H_5(C_3H_6N)_2PO$.
Properties: (Technical) Straw-colored liquid. Limited solubility in water; soluble in most organic solvents. The pure material is a low-melting solid. Combustible.
Use: Polymerization initiator.

phenyl bromide. See bromobenzene.

1-phenylbutane. See n-butylbenzene.

2-phenylbutane. See sec-butylbenzene.

4-phenyl-2-butanone 3-thiosemicarbazone.
CAS: 25687-87-0.
Hazard: A poison by ingestion.

Use: Agricultural chemical.

phenylbutazone. (4-n-butyl-1,2-diphenyl-3,5-pyrazolidinedione).
CAS: 50-33-9. $C_{19}H_{20}N_2O_2$. A synthetic pyrazolone derivative, also occurs naturally in certain herbs.
Properties: White or very-light-yellow powder; slightly bitter taste; very slight aromatic odor. Mp 103–106C. Freely soluble in acetone, ether, and ethyl acetate; very slightly soluble in water; stable if stored at room temperature in closed containers in absence of moisture. Also available as sodium salt.
Grade: NF.
Hazard: May have harmful side effects.
Use: Medicine, anti-inflammatory drug licensed for human use, veterinary medicine.

1-phenylbutene-2. $C_6H_5CH_2CH:CHCH_3$.
Properties: Colorless liquid. Boiling range 174–176C, d 0.888 (15.5C), refr index 1.511 (20C), flash p 160F (71.1C) (TOC). Combustible.
Grade: Technical, 95 mole %.
Use: Organic synthesis, laboratory reagent.

phenylbutynol. (3-phenyl-1-butyn-3-ol).
$HC:CC(C_6H_5)(OH)CH_3$.
Properties: Crystals; camphor odor. Mp 51–52C, bp 217–218C, d 1.0924 (20/20C). Slightly soluble in water; soluble in acetone, benzene, most organic solvents. Combustible.
Use: Acid inhibitor, organic synthesis.

2-phenylbutyramide. See nivonorm.

2-phenylbutyric acid. See phenylethylacetic acid.

n-phenylcaproamide. See n-(n-hexanoyl) aniline.

1-phenyl-3-carbethoxypyrazolone-5.
$C_{12}H_{12}N_2O_3$.
Properties: White to light-buff powder. Stable in aqueous solution, mp 182–188C. Combustible.
Use: Dyestuff intermediate, laboratory reagent.

phenyl carbimide. See phenyl isocyanate.

phenylcarbinol. See benzyl alcohol.

phenyl carbonate.
CAS: 102-09-0. $C_6H_5OCOOC_6H_5$.
Properties: Shiny crystals. Mp 80C, bp 300C. Insoluble in water; soluble in hot alcohol, benzene.
Use: Solvent for nitrocellulose.

phenylcarbylamine chloride.
CAS: 622-44-6. $C_6H_5NCCl_2$.

Properties: Pale-yellow, oily liquid; onionlike odor. D 1.30 (15C), bp 208–210C. Mildly volatile. Soluble in alcohol, benzene, ether; insoluble in water.
Derivation: Chlorination of phenylisothiocyanate.
Grade: Technical.
Hazard: Toxic by inhalation and skin absorption, strong irritant to skin and mucous membranes.
Use: Organic synthesis, military poison.

phenyl chloride.   See chlorobenzene.

n-phenylchloroacetamide.
CAS: 587-65-5. Mf: C_8H_8ClNO.
Hazard: A poison.

phenylchloroform.   See benzotrichloride.

phenyl chloromethyl ketone.   See chloroacetophenone.

2-phenyl-6-chlorophenol. $C_{12}H_9ClO$.
Properties: Yellow, oily liquid; slight odor. Mp 6C, bp 318C (decomposes), d 1.24. Insoluble in water; soluble in organic solvents.
Use: Fungicide, disinfectant.

1-phenyl-3-chloropropane.   See phenylpropyl chloride.

phenyl cyanide.   See benzonitrile.

phenylcyclidene   hydrochloride.   (PCP). $C_{17}H_{25}N \cdot HCl$.
Properties: White crystals. Mp 243C.
Hazard: Toxic by ingestion and inhalation. A psychotropic drug that may cause serious mental disorders.
Use: Veterinary anesthetic (manufacture legally restricted to 5 kg annually).

phenylcyclohexane.   (cyclohexylbenzene).
CAS: 827-52-1. $C_6H_5C_6H_{11}$.
Properties: Colorless, oily liquid; pleasant odor. D 0.938 (25/25C), mp 5C, bp 237.5C, refr index 1.523 (25C), flash p 210F (98C). Insoluble in water, glycerol; very soluble in alcohol, acetone, benzene, carbon tetrachloride, castor oil, hexane, xylene. Combustible.
Use: High-boiling solvent, penetrating agent, intermediate, laboratory reagent.

2-phenylcyclohexanol.

$HOCH(CH_2)_4CH(C_6H_5)$.

Properties: Colorless to pale-straw-colored liquid. Pour p −18C, bp 276–281C, d 1.033 (25/25C), refr index 1.536 (25C), flash p 280F (137C). Very slightly soluble in water; soluble in methanol, ether. Combustible.
Use: Solvent, intermediate.

N-phenyl-N'-cyclohexyl-p-phenylene-diamine. $C_6H_5HNC_6H_4NHC_6H_{11}$.
Properties: Gray-violet powder. D 1.16, mp 103–107C. Soluble in acetone, benzene, MEK, ethyl acetate, ethylene dichloride; insoluble in water.
Use: Antioxidant–antiozonant for elastomers.

phenyldichloroarsine.
CAS: 696-28-6. $C_6H_5AsCl_2$.
Properties: Liquid, microcrystalline mass at the fp, decomposed by water. D 1.654 (20C), bp 255–257C, fp −20C, vap tension 0.014 mm Hg (15C), volatility 404 mg/m^3 (20C), coefficient of thermal expansion 0.00073. Soluble in alcohol, benzene, and ether; insoluble in water.
Derivation: Arsenic trichloride and phenylmercuric chloride are heated together at 100C.
Grade: Technical.
Hazard: Toxic by ingestion and inhalation, strong irritant to eyes, skin, and tissue.
Use: Lachrymator poison gas, solvent for diphenylcyanoarsine.

phenyldidecyl phosphite. C_6H_5OP $(OC_{10}H_{21})_2$.
Properties: Nearly water-white liquid; odor of alcohol. D 0.940 (25/15.5C), mp 0C, refr index 1.4785 (25C), flash p 425F (218C) (COC). Combustible.
Use: Chemical intermediate, antioxidant, ingredient in stabilizer systems for resins.

phenyl diethanolamine.
CAS: 120-07-0. $C_6H_5N(C_2H_4OH)_2$.
Properties: Colorless liquid. Mp 58C, bp 190C (1 mm Hg), vap press <0.01 mm Hg (20C), bulk d 10.0 lb/gal (20C), d 1.1203 at 60/20C, viscosity 1.19 cP (20C), flash p 375F (190C) (OC). Slightly soluble in water; soluble in ethanol and acetone. Combustible.
Grade: Technical.
Use: Organic synthesis, dyestuff, laboratory reagent.

phenyl diglycol carbonate.   [diethyl glycol bis(phenyl carbonate)]. $(C_6H_5OOCOCH_2CH_2)_2O$.
Properties: Colorless solid. D 1.23 (20/4C), mp 40C, bp 225–229C, refr index 1.525 (20C), evaporation rate 0.026 mg/sq cm/hr at 100C. Insoluble in water (very stable to hydrolysis); widely soluble in organic solvents; compatible with many resins and plastics. Combustible.
Use: Plasticizer.

N-phenyl-N'-(1,3-dimethyl butyl)-p-phenyle-nediamine. $C_6H_5HNC_6H_4NHC_6H_{13}$.
Properties: Dark-violet, staining, low-melting solid. Mp 50C, d 1.07.
Use: Antiozonant, antioxidant, and polymer stabilizer.

3-phenyl-1,1-dimethylurea.   See fenuron.

1,1'-(p-phenylenebis(carbonylmethyl))di-3-picolinium dibromide.
CAS: 63868-81-5. $C_{22}H_{22}N_2O_2 \cdot 2Br$.
Hazard: A poison.

m-phenylenebis(methylamine).   See m-xylene-α,α'-diamine.

4,4'-(1,4-phenylene-bis(1-methyl ethylidene)) bisaniline.   See bisaniline-p.

m-phenylenediamine.   (m-diaminobenzene).
CAS: 108-45-2. $C_6H_4(NH_2)_2$.
Properties: Colorless needles becoming red in air, usually in the form of the stable hydrochloride. D 1.1389, mp 63C, bp 282-287C. Soluble in alcohol, ether, and water.
Derivation: Reduction of m-dinitrobenzene or nitroaniline with iron and hydrochloric acid; purified by crystallization.
Grade: Technical, 99% min purity.
Hazard: Toxic by ingestion and inhalation, strong irritant to skin. Liver damage. Questionable carcinogen.
Use: Dyestuff manufacture, detection of nitrite, textile developing agent, laboratory reagent, vulcanizing agent, ion-exchange resins, block polymers, corrosion inhibitors, photography.

o-phenylenediamine.   (orthamine; o-diaminobenzene).
CAS: 95-54-5. $C_6H_4(NH_2)_2$.
Properties: Colorless, monoclinic crystals; darkens in air. Mp range 102–104C, bp 252–258C. Soluble in alcohol, ether, and chloroform.
Derivation: Reduction of o-dinitrobenzene or nitroaniline with iron and hydrochloric acid.
Method of purification: Crystallization.
Grade: Technical, 99.0% min purity.
Hazard: Anemia. Possible carcinogen.
Use: Manufacture of dyes, photographic developing agent, organic synthesis, laboratory reagent.

p-phenylenediamine.   (p-diaminobenzene).
CAS: 106-50-3. $C_6H_4(NH_2)_2$.
Properties: White to light-purple crystals (oxidizes on standing in air to purple and black). Mp 147C, bp 267C; flash p 312F (155C). Soluble in alcohol, ether and 100 parts water; affected by light. Combustible.
Derivation: Reduction of p-dinitrobenzene or nitroaniline with iron and hydrochloric acid.
Method of purification: Crystallization.
Grade: Technical, 99% min purity.
Hazard: Toxic by ingestion and inhalation, strong irritant to skin and upper respiratory tract. Toxic by skin absorption. Questionable carcinogen.
Use: Azo dye intermediate, photographic developing agent, photochemical measurements, intermediate in manufacture of antioxidants and accelerators for rubber, synthetic fibers, laboratory reagent, dyeing hair and fur.

2,3-phenylenepyrene.   See indeno(1,2,3-cd)pyrene.

phenylethane.   See ethylbenzene.

2-phenylethanol.   See phenethyl alcohol.

phenylethanolamine.
CAS: 7568-93-6. $C_6H_5NHCH_2CH_2OH$.
Properties: Yellow crystals. Mp 56C, bp 157C (17 mm Hg). Soluble in water.
Use: Short-stopping agent for SBR, intermediate, wax hardener.

phenyl ether.   See diphenyl oxide.

2-phenylethyl acetate.   (phenethyl acetate).
CAS: 103-45-7. $C_6H_5CH_2CH_2OOCCH_3$.
(Not the same as sec-phenylethyl acetate.)
Properties: Colorless liquid; peachlike odor. D 1.030–1.034, refr index 1.497–1.501 (20C), bp 226C, flash p 230F (110C). Soluble in alcohol, ether, and most fixed oils. Combustible.
Derivation: (1) Interaction of ethyl acetate and aluminumphenyl ethylate. (2) Interaction of acetic anhydride and phenylethyl alcohol in the presence of sodium acetate.
Grade: Technical, FCC.
Use: Perfumery, laboratory reagent.

sec-phenylethyl acetate.   See α-methylbenzyl acetate.

phenylethylacetic acid.   (2-phenylbutyric acid). $C_2H_5CHC_6H_5COOH$.
Properties: White crystals; aromatic odor. Mp 41.0C (min). Insoluble in water; soluble in alcohol, ketones, and esters. Combustible.
Use: Organic synthesis, lab reagent.

2-phenylethyl alcohol.   See phenethyl alcohol.

2-phenylethylamine.   (phenethylamine; 1-amino-2-phenylethane).
CAS: 64-04-0. $C_6H_5C_2NH_2$.
Properties: Liquid; fishy odor. Absorbs carbon dioxide from the air, strong base. D 0.9640, bp 194.5C. Soluble in water, alcohol, and ether. Combustible.
Derivation: From phenylethyl alcohol and ammonia under pressure.
Grade: Technical, scintillation.
Hazard: Toxic by ingestion, skin irritant.
Use: Organic synthesis, lab reagent, scintillation counter (CO_2 absorber).

2-phenylethyl anthranilate.   (phenethyl anthranilate).
CAS: 133-18-6. $H_2NC_6H_4COOC_2H_4C_6H_5$.
Properties: Colorless liquid that yellows with age; odor of grape and orange. D 1.14 (25/25C). Combustible.
Use: Perfumes, flavoring.

phenylethyl carbinol.   See phenylpropyl alcohol.

phenylethylene.   See styrene.

phenylethylene glycol.   See styrene glycol.

N-phenylethylethanolamine.
$C_6H_5N(C_2H_5)C_2H_4OH$.
Properties: Solid. Mp 37.2C, bp 268C (740 mm Hg), d 1.04 (20/20C), flash p 270F (132C) (COC). Very slightly soluble in water; soluble in alcohol, acetone, benzene. Combustible.
Use: Solvents, chemical intermediates, preparation of dyes for acetate rayons, laboratory reagent.

phenyl ethyl ether.   See phenetole.

2-phenylethyl isobutyrate.   (phenethyl isobutyrate).
CAS: 103-48-0. $(CH_3)_2CHCOOC_2H_4C_6H_5$.
Properties: Colorless liquid; pleasant odor. D 0.988 (25/25C), refr index 1.488 (20C). Soluble in alcohol and ether. Combustible.
Use: Perfumes, flavoring.

phenylethylmalonylurea.   See phenobarbital.

2-phenylethyl mercaptan. $C_6H_5CH_2CH_2SH$.
Properties: Liquid. Boiling range 193–225C, unpleasant odor, d 1.0264 (15.5C), refr index 1.5582 (20C), flash p 160F (71.1C). Combustible.
Use: Organic synthesis, laboratory reagent.

2-phenylethyl   phenylacetate.   (phenethyl phenylacetate).
CAS: 102-20-5. $C_6H_5(CH_2)_2OOCCH_2C_6H_5$.
Properties: White crystals; hyacinth odor. D 1.080–1.082, congealing p 27C. Combustible.
Use: Perfumery, flavors.

2-phenylethyl propionate.   (phenethyl propionate).
CAS: 120-45-6. $C_2H_5COOC_2H_4C_6H_5$.
Properties: Synthetic, colorless liquid; flower-fruit odor. D 1.012 (25/25C). Miscible with alcohols and ether. Combustible.
Use: Perfumes, flavors.

2-phenylethyl salicylate.   (phenethyl salicylate). $C_6H_5C_2H_4OOCC_6H_4OH$.
Properties: White crystals; faint aromatic odor. Congealing point 41.5C. Soluble in 14 parts of 95% alcohol. Combustible.
Use: Flavors.

phenyl ferrocenyl ketone.   See benzoylferrocene.

phenyl fluoride.   See fluorobenzene.

phenyl fluoromethyl ketone.   See fluoroacetophenone.

phenylformamide.   See formanilide.

phenylformic acid.   See benzoic acid.

phenyl glycidyl ether.   (1,2-epoxy-3-phenoxypropane; PGE).
CAS: 122-60-1. $H_2COCHCH_2OC_6H_5$.
Properties: Colorless liquid. D 1.11, bp 245C, mp 3.5C.
Hazard: Toxic by skin absorption, moderate irritant to eyes and skin. Toxic by skin absorption. Testicular damage. Possible carcinogen.

phenylglycol.   See styrene glycol.

phenylglycolic acid.   See mandelic acid.

phenylhydrazine.
CAS: 100-63-0. $C_6H_5NHNH_2$.
Properties: Colorless liquid, becomes red brown on exposure to air. D 1.0978, mp 19.35C, bp 243.5C with decomposition, flash p 192F (88.9C) (CC), autoign temp 345F (173C). Soluble in alcohol, ether, chloroform, benzene, and dilute acids; slightly soluble in water. Also available as the hydrochloride. Combustible.
Derivation: Reduction of diazotized aniline, followed by reaction with sodium hydroxide.
Grade: Commercial, CP, reagent.
Hazard: Toxic by inhalation, ingestion, and skin absorption; attacks red blood cells. Toxic by skin absorption. Skin and upper respiratory tract irritant; anemia. Possible carcinogen.
Use: Analytical chemistry (reagent for detecting aldehydes, sugars, etc.), organic synthesis (intermediates, dyestuffs, pharmaceuticals). The hydrochloride is a strong reducing agent.

α-phenylhydroxyacetic acid.   See mandelic acid.

phenylic acid.   See phenol.

n-(3-phenylimino-1-propenyl)aniline hydrochloride.
CAS: 28140-60-5. $C_{15}H_{14}N_2 \cdot ClH$.
Hazard: Moderately toxic by ingestion and skin contact.

phenylindan dicarboxylic acid.   See 1,1,3-trimethyl-5-carboxy-3-(p-carboxyphenyl)indan.

phenyl isocyanate.   (phenyl carbimide; carbanil).
CAS: 103-71-9. $C_6H_5N:C:O$.
Properties: Liquid. Bp 165C, fp −30C, d 1.095 (20/4C), refr index 1.53684 (19.6C), flash p 132F (55.5C) (TOC). Decomposes in water and alcohol; very soluble in ether. Combustible.

Hazard: Moderate fire risk. Strong irritant, a lachrymator.
Use: Reagent for identifying alcohols and amines, intermediate.

phenyl isothiocyanate.   See phenyl mustard oil.

phenyl J acid.   (phenyl-2-amino-5-naphthol-7-sulfonic acid). $HOC_{10}H_5(NHC_6H_5)(SO_3H)$.
Properties: Slate-colored crystals. Soluble in alkali. Combustible.
Derivation: From hydrogen acid and aniline (condensation with heat).
Hazard: Toxic by ingestion.
Use: Azo dye manufacturing.

phenyllead trilaurate.
CAS: 3268-27-7. $C_{42}H_{74}O_6Pb$.
Hazard: A poison.

phenyllithium. C_6H_5Li. Available in a 20% by weight solution of a 70:30 volume percentage benzene-ether mixture.
Hazard: Flammable, dangerous fire risk.
Use: Organic synthesis.

phenylmagnesium bromide. C_6H_5MgBr. A Grignard reagent available as a solution in ether, d 1.14.
Derivation: From magnesium and bromobenzene.
Hazard: Dangerous fire risk, solution highly flammable.
Use: Arylating agent in organic synthesis.

phenylmagnesium   chloride. C_6H_5MgCl. Available dissolved in tetrahydrofuran.
Hazard: Dangerous fire risk, solution highly flammable.
Use: Grignard reagent, organic synthesis.

phenyl mercaptan.
CAS: 108-98-5. C_6H_5SH.
Properties: Water-white liquid; offensive odor. Oxidizes on exposure to air; supplied under nitrogen atmosphere. D 1.080 (15.5/15.5C), refr index 1.5815 (25C), fp −15C, bp 169C. Insoluble in water; very soluble in aromatic and aliphatic hydrocarbons.
Hazard: Store out of contact with air and acids. Toxic by inhalation. Eye and skin irritant, and central nervous system impairment.
Use: Chemical intermediate, mosquito larvicide.

phenylmercuric acetate.
CAS: 62-38-4. $C_6H_5HgOCOCH_3$.
Properties: White to cream prisms. Mp 148–150C. Slightly soluble in water; soluble in alcohol, benzene, and glacial acetic acid; slightly volatile at ordinary temperatures.
Derivation: Action of heat on benzene and mercuric acetate.
Grade: CP, technical, commercial.

Hazard: Toxic by ingestion, inhalation, and skin absorption; strong irritant.
Use: Fungicide; herbicide; "mildewcide" for paints; "slimicide" in paper mills.

phenylmercuric benzoate.
$C_6H_5HgOOCC_6H_5$.
Properties: Crystals. Mp 94C (min).
Hazard: Toxic by ingestion, inhalation, and skin absorption.
Use: Bactericide, fungicide, alcohol denaturant.

phenylmercuric borate.
CAS: 102-98-7. $(C_5H_5Hg)_2HBO_3$.
Properties: White, crystalline powder. Mp 120–130C. Slightly soluble in water; soluble in alcohol.
Derivation: Reaction of phenylmercuric acetate with boric acid.
Hazard: Toxic by ingestion, inhalation, and skin absorption.
Use: Fungicide, bactericide.

phenylmercuric chloride.
CAS: 100-56-1. C_6H_5HgCl.
Properties: White, satiny crystals. Mp 251C. Insoluble in water; slightly soluble in hot alcohol; soluble in benzene, ether, pyridine.
Derivation: Reaction of phenylmercuric acetate and sodium chloride.
Hazard: Toxic by ingestion, inhalation, and skin absorption.
Use: Antiseptic, fungicide, germicide.

phenylmercuric hydroxide.
CAS: 100-57-2. C_6H_5HgOH.
Properties: Fine white to cream crystals. Mp 197–205C. Slightly soluble in water; soluble in acetic acid and alcohol.
Grade: Technical, pure.
Hazard: Toxic by ingestion, inhalation, and skin absorption.
Use: Manufacture of phenylmercuric salts, fungicide, and germicide. Principal compound in manufacturing organic mercury derivatives, denaturant for alcohol.

phenylmercuric lactate.
CAS: 122-64-5. $C_6H_5HgOOCCHOHCH_3$.
Hazard: Toxic by ingestion, inhalation, and skin absorption.
Use: Bactericide, fungicide.

phenylmercuric naphthenate.   Prepared by interaction of phenylmercuric acetate and naphthenic acid, producing a colored solution.
Hazard: Toxic by ingestion, inhalation, and skin absorption.
Use: Wood preservative, mildew-proofing agent.

phenylmercuric nitrate (basic).
CAS: 8003-05-2. $C_6H_5HgNO_3 \cdot C_6H_5HgOH$.

Properties: Fine white crystals or grayish powder. Mercury content 63–65% (theoretical 63.2%), melting range 175–185C with decomposition, ash 0.1% max. Very slightly soluble in water and alcohol; insoluble in ether, moderately soluble in glycerol.
Grade: NF.
Hazard: Toxic by ingestion, inhalation, and skin absorption.
Use: Germicide, fungicide.

phenylmercuric oleate. $C_6H_5HgOOC(CH_2)_7$ $CH:CHC_8H_{17}$.
Properties: White, crystalline powder. Mp 45C. Insoluble in water; soluble in organic solvents and some oils.
Derivation: Reaction of phenylmercuric acetate with oleic acid.
Hazard: Toxic by ingestion, inhalation, and skin absorption.
Use: Mildew-proofing agent for paints, fungicide, and germicide.

phenylmercuric propionate.
CAS: 103-27-5. $C_6H_5HgOCOCH_2CH_3$.
Properties: Technical grade, white to off-white waxlike free-flowing powder. Mp 65–70C, stable to 200C for short periods, 57% min mercury content.
Hazard: Toxic by ingestion, inhalation, and skin absorption.
Use: Fungicide and bactericide for paints and industrial finishes.

phenylmercuric salicylate. $C_6H_4(OH)(COO$ $HgC_6H_5)$.
Hazard: Toxic by ingestion, inhalation, and skin absorption.
Use: Seed disinfectant.

phenylmercuriethanolammonium acetate.
$[(HOC_2H_4)NH_2(C_6H_5Hg)]OOCCH_3$.
Properties: White, crystalline solid. Soluble in water.
Derivation: Reaction of phenylmercuric acetate with monoethanolamine.
Hazard: Toxic by ingestion, inhalation, and skin absorption.
Use: Insecticide and fungicide, may not be used on food crops.

phenylmercuritriethanolammonium lactate.
[tris(2-hydroxyethyl)(phenylmercuri)ammonium lactate].
CAS: 23319-66-6. $[(HOC_2H_4)_3NHgC_6H_5]$ $OOCCHOHCH_3$.
Properties: White, crystalline solid. Soluble in water.
Derivation: Reaction of phenylmercuric acetate with triethanolamine and lactic acid.
Hazard: Toxic by ingestion, inhalation, and skin absorption.
Use: Turf fungicide and eradicant fungicide for fruit trees.

phenylmercury formamide.
$HCONHHgC_6H_5$.
Hazard: Toxic by ingestion, inhalation, and skin absorption.
Use: Seed disinfectant.

phenylmercury urea.
CAS: 2279-64-3. $C_6H_5HgNHCONH_2$.
Hazard: Toxic by ingestion, inhalation, and skin absorption.
Use: Disinfectant and fungicide for seed treatment.

phenylmethane. See toluene.

phenylmethanol. See benzyl alcohol.

MD+ITān-((phenylmethoxy)carbonyl)-*o*-(phenylmethyl-)*l*-seryl-*l*-tryptophyl-*l*-seryl-*l*-tyrosyl-2,3,4,5,6-pentamethyl-*dl*-phenylalanyl-*l*-leucyl-*l*-arginyl-*n*-ethyl-*l*-prolinamide monoacetate (salt), dihydrate.
CAS: 70601-60-4. $C_{74}H_{97}N_{13}O_{13} \cdot C_2H_4O_2 \cdot 2H_2O$.
Hazard: A reproductive hazard.
Use: Hormone.

MD+ITān-((phenylmethoxy)carbonyl)-*o*-(phenylmethyl)-*l*-seryl-*l*-tryptophyl-*l*-seryl-*l*-tyrosyl-2,3,4,5,6-pentamethyl-*d*-phenylalanyl-*l*-leucyl-*l*-arginyl-*n*-ethyl-*l*-prolinamide, monoacetate (salt), hydrate.
CAS: 70601-54-6. $C_{74}H_{97}N_{13}O_{13} \cdot C_2H_4O_2 \cdot H_2O$.
Hazard: A reproductive hazard.
Use: Hormone.

phenylmethyl acetate. See benzyl acetate.

d-1-phenyl-2-methylaminopropan.
CAS: 537-46-2. $C_{10}H_{15}N$.
Hazard: A poison. Human systemic effects reported.

phenylmethylcarbinol. See α-methylbenzyl alcohol.

phenylmethylcarbinyl acetate. See α-methylbenzyl acetate.

N-phenylmethylethanolamine.
$C_6H_5N(CH_3)C_2H_4OH$.
Properties: Liquid. Sets to a glass at −30C, bp 192C (100 mm Hg), d 1.0661 (20/20C), flash p 280F (137C) (COC). Slightly soluble in water. Combustible.
Use: Chemical intermediate, solvent for dyes for acetate fibers.

(5-(phenylmethyl)-3-furanyl)methyl 4-chloro-α-ethylbenzeneacetate.
CAS: 51630-12-7. $C_{22}H_{21}ClO_3$.
Hazard: Moderately toxic by ingestion.
Use: Agricultural chemical.

phenyl methyl ketone.   See acetophenone.

1-phenyl-3-methyl-5-pyrazolone.   See
3-methyl-1-phenyl-2-pyrazolin-5-one.

phenyl α-methylstyryl ketone.   See dypnone.

***N*-phenylmorpholine.**

$C_6H_5NCH_2CH_2OCH_2CH_2$.

Properties: White solid. Bp 268C, mp 57C, vap
press <0.1 mm Hg (20C), flash p 220F (104C),
d 1.06 (57/20C). Soluble in water. Combustible.
Grade: Technical.
Hazard: Toxic by ingestion and skin absorption.
Use: Chemical intermediate for dyestuffs, rubber
accelerators, corrosion inhibitors, and photographic
developers; insecticide.

phenyl mustard oil.   (thiocarbanil; phenyl
isothiocyanate; phenylthiocarbonimide).
C_6H_5NCS.
Properties: Pale-yellow or colorless liquid; penetrating, irritating odor. D 1.1382, fp −21C, bp 221C.
Readily volatilized with steam; soluble in alcohol
and ether; insoluble in water. Combustible.
Derivation: (1) By action of concentrated
hydrochloric acid and sulfocarbanilide, (2) by reaction of thiophosgene with aniline.
Hazard: Toxic by ingestion and inhalation, irritant
to tissue.
Use: Medicine, organic synthesis.

2-phenyl-1,8-naphthalimide.   See 2-phenyl-
benzo(de)isoquinoline-1,3-dione.

***N*-phenyl-α-naphthylamine.**
CAS: 90-30-2. $C_{10}H_7NHC_6H_5$.
Properties: Crystallizes in prisms, white to slightly
yellowish. Mp 62C, bp 335C (260 mm Hg). Soluble
in alcohol, ether, and benzene. Combustible.
Derivation: From α-naphthylamine and aniline.
Purified by distillation.
Use: Dyes and other organic chemicals, rubber
antioxidant.

***N*-phenyl-β-naphthylamine.**
(2-naphthylphenylamine).
CAS: 135-88-6. $C_{10}H_7NHC_6H_5$.
Properties: Light-gray powder. Mp 107C, bp 395C,
d 1.24. Insoluble in water; soluble in alcohol, acetone, benzene. Combustible.
Hazard: Questionable carcinogen.
Use: Rubber antioxidant, lubricant, inhibitor (butadiene).

***N*-phenyl-1-naphthylamine-8-sulfonic   acid.**
(*N*-phenyl peri acid). $C_{16}H_{13}NO_3S$.
Properties: Greenish-gray needles. Insoluble in
water; soluble in alcohol.

Derivation: Arylation of 1-naphthylamine-8-
sulfonic acid with aniline.
Grade: Technical, mostly as sodium salt.
Use: Azo dyes.

phenylneopentyl phosphite.
Properties: Water-white liquid. Mp 19C, bp 138–
140C (10 mm Hg), refr index 1.517 (25C), d 1.135
(25/15C).
Use: Chemical intermediate.

1-phenylnonane.   See nonylbenzene.

4-phenyl-6h-1,3,5-oxathiazine.
CAS: 58955-84-3. C_9H_9NOS.
Hazard: Moderately toxic by ingestion.
Use: Agricultural chemical.

1-phenylpentane.   See *n*-amylbenzene.

2-phenylpentane.   See *sec*-amylbenzene.

phenyl phenacyl ketone.
CAS: 120-46-7. $C_{15}H_{12}O_2$.
Hazard: Moderately toxic by ingestion.

***o*-phenylphenol.**   (*o*-hydroxydiphenyl; *o*-xenol).
CAS: 90-43-7. $C_6H_5C_6H_4OH$.
Properties: Nearly white or light-buff crystals. Mp
56–58C, bp 280–284C, d 1.217 (25/25C), flash p
255F (124C). Soluble in alcohol; sodium hydroxide
solution; insoluble in water. Combustible.
Derivation: From reaction of chlorobenzene and
caustic soda solution at elevated temperatures and
pressures.
Hazard: Toxic by ingestion. Questionable carcinogen.
Use: Intermediate for dyes, germicides, fungicides,
rubber chemicals, laboratory reagents, food packaging.

***p*-phenylphenol.**   (*p*-hydroxydiphenyl; *p*-xenol).
CAS: 92-69-3. $C_6H_5C_6H_4OH$.
Properties: Nearly white crystals. Mp 164–165, bp
308C, flash p 330F (165.5C). Soluble in alcohol,
alkalies, and most organic solvents; insoluble in
water. Combustible.
Derivation: From reaction of chlorobenzene and
caustic soda solution at elevated temperatures and
pressures.
Hazard: Toxic by ingestion.
Use: Intermediate for dyes and resins, rubber chemicals, laboratory reagent, fungicide.

***N*-phenyl-*p*-phenylenediamine.**   See *p*-
aminodiphenylamine.

phenylphosphine.   (fenylfosfin; phosphaniline).
CAS: 638-21-1. $C_6H_5PH_2$.
Properties: Foul-smelling liquid. Mw 110.10, bp
160–161C, d 1.001 (15 mm). Insoluble in water;
soluble in alkali; very soluble in alcohol and ether.

Hazard: Poison by inhalation. Ignites spontaneously in air. Dermatitis, hematologic effects, and testicular damage.

phenylphosphinic acid.　See benzenephosphinic acid.

phenylphosphonic acid.　See benzenephosphonic acid.

phenyl phthalate.　$C_6H_4(COOC_6H_5)_2$.
Properties: White crystals. Mp 70C, bp 255 (14 mm Hg), d 1.57. Soluble in ketones; insoluble in water.
Use: Plasticizer in nitrocellulose lacquers.

N-phenylpiperazine.
CAS: 92-54-6.

$C_6H_5NCH_2CH_2NHCH_2CH_2$.

Properties: Pale-yellow oil. D 1.0621 (20/4C), bp 286.5C, 156–157C (10 mm Hg), mp 18.8C, flash p 285F (140.5C). Insoluble in water; soluble in alcohol and ether. Combustible.
Use: Intermediate for pharmaceuticals, anthelmintics, surface-active agents, synthetic fibers.

phenylpropane.　See n-propylbenzene.

2-phenyl-1,3-propanediol　　dicarbamate.
See felbamate.

3-phenyl-1-propanol.　See phenylpropyl alcohol.

phenylpropanolamine hydrochloride.　See propadrine hydrochloride.

1-phenylpropanone-1.　See propiophenone.

3-phenylpropenal.　See cinnamic aldehyde.

3-phenylpropenoic acid.　See cinnamic acid.

3-phenylpropenol.　See cinnamic alcohol.

1-(α-phenyl)-propenylveratrole.　See benzyl isoeugenol.

3-phenylpropionaldehyde.　See phenylpropyl aldehyde.

2-phenylpropionaldehyde dimethyl acetal.
$C_{11}H_{16}O_2$.
Properties: Colorless to sltly yellow liquid; mushroom odor. D: 0.989–0.994, refr index: 1.492–1.497. Sol in alc, ether; insol in water.
Use: Food additive.

3-phenylpropionic acid.　See hydrocinnamic acid.

phenylpropyl　　acetate.　(hydrocinnamyl acetate).
CAS: 122-72-5.　$C_6H_5CH_2CH_2CH_2OOCCH_3$.
Properties: White crystals. D 1.012–1.016, refr index 1.497. Soluble in 70% alcohol. Combustible.
Use: Perfumery, flavoring, laboratory reagent.

phenylpropyl alcohol.　(hydrocinnamic alcohol; 3-phenyl-1-propanol; phenylethyl carbinol).
CAS: 122-97-4.　$C_6H_5CH_2CH_2CH_2OH$.
Properties: Colorless liquid; floral odor. Bp 219C, d 0.998–1.000, refr index 1.524–1.528 (20C). Soluble in 70% alcohol; insoluble in water. Combustible.
Grade: Technical, FCC.
Use: Perfumery, flavoring, laboratory reagent.

phenylpropyl　　aldehyde.　(3-phenylpropionaldehyde; hydrocinnamic aldehyde).
CAS: 104-53-0.　$C_6H_5CH_2CH_2CHO$.
Properties: Colorless liquid; floral odor. D 1.010–1.020, refr index 1.520–1.532. Soluble in 50% alcohol. Combustible.
Grade: Chlorine free.
Use: Perfumery, flavors.

2-phenylpropylamine.
CAS: 582-22-9.　$C_9H_{13}N$.
Hazard: Moderately toxic by ingestion.

phenylpropyl chloride.　(hydrocinnamyl chloride; 1-phenyl-3-chloropropane).
$C_6H_5CH_2CH_2CH_2Cl$.
Properties: Colorless to pale-yellow liquid. Bp 219–220C, d 1.056 (25/4C), refr index 1.5220 (20C). Combustible.
Use: Organic synthesis, laboratory reagent.

3-phenyl-n-(4-propylcyclohexyl)-2-propenamide.
CAS: 315706-78-6.　$C_{18}H_{25}NO$.
Hazard: A poison by ingestion.

α-(1-phenyl-5-propyl-1h-pyrazol-4-yl)-1-piperidinebutanol.
CAS: 296269-55-1.　$C_{21}H_{31}N_3O$.
Hazard: A poison.

4-phenylpropylpyridine.
$C_6H_5(CH_2)_3C_5H_4N$.
Properties: Colorless liquid. Bp 322C. Soluble in organic solvents.
Use: Heat-transfer agent, chemical intermediate.

1-phenyl-3-pyrazolidone.

$C_6H_5NNHC(O)CH_2CH_2$.

Properties: Crystals. Mp 121C. Soluble in water. Combustible.
Use: Photographic developer, laboratory reagent.

α-(1-phenyl-1h-pyrazol-4-yl)-1-piperidinebutanol.
CAS: 296269-52-8. $C_{18}H_{25}N_3O$.
Hazard: A poison by ingestion.

phenyl salicylate.   (salol).
CAS: 118-55-8. $C_6H_4OHCOOC_6H_5$.
Properties: White, crystalline powder; faint aromatic odor and taste. D 1.2614, mp 41.9C, bp 172–173C, absorbs light, especially at 290–330μ. Soluble in alcohol, ether, chloroform, benzene, and fixed or volatile oils; sparingly soluble in water. Combustible.
Derivation: Heating salicylic acid and phenol with phosphorus pentachloride or other dehydrating agent.
Grade: NF, granular powder.
Hazard: Toxic by ingestion.
Use: Medicine, preservative, UV absorber in plastics, waxes, polishes, laboratory reagent.

phenyl saligenin phosphate.   See saligenin cyclic phenyl phosphate.

phenylstearic acid.   An organic fatty acid having a high degree of fluidity and no definite melting point. Pour point is −26C.
Use: Lubricant stabilizer; its potential uses include corrosion inhibitor, plasticizer, and textile auxiliary. Phenylstearic acid and its quaternary and ethoxylated derivatives are used in synthetic latices, as mineral oil emulsifiers, and in invert systems.

phenylsulfanilic acid. $C_{12}H_{11}NO_3S$.
Properties: White plates turning blue in light. Soluble in water and alcohol. Decomposes above 200C.
Use: Indicator, colorimetry, determination of nitrates.

phenylsulfohydrazide. $C_6H_5SO_2NHNH_2$.
Properties: Colorless crystals. Decomposes at 100C, evolving nitrogen.
Use: Blowing agent for cellular rubber and plastics.

phenylsulfonic acid.   See benzenesulfonic acid.

4-phenyl-1,4-thiazane.

$SCH_2CH_2N(C_6H_5)CH_2CH_2.$

Properties: White powder. Mp 108–111C. Soluble in hot toluene.
Derivation: Interaction of dichlorodiethyl sulfide and an aliphatic amine in the presence of alcohol and sodium carbonate.

phenylthiocarbonimide.   See phenyl mustard oil.

phenyl thiocyanate.   See thiocyanic acid, phenyl ester.

1-phenyl-2-thiourea.   (NCI-C02017; phenylthiocarbamide; *n*-phenylthiourea; α-phenylthiourea; 1-phenyltiourea; phenylcarbamide; phenylthiourea; PTC; PTU; RCRA waste number P093; U 6324; USAF EK-1569)
CAS: 103-85-5. $C_7H_8N_2S$.
Properties: Compound with bitter taste, needlelike crystals, soluble in water, alcohol, ethanol, and aqueous ether.
Hazard: Poisonous; teratogen.
Use: In medical genetics and production of rodenticide.

phenyl tolyl ketone.   (methylbenzophenone). $C_{14}H_{12}O$.
Properties: (*o*-) Viscous liquid. Mp below −18C, bp 309C. (*p*-) Colorless crystals. Mp 59C, bp 311C. Soluble in alcohol, benzene, and common organic solvents.
Derivation: (*o*-) Benzene and *o*-toluic acid chloride (aluminum chloride catalyst); (*p*-) benzoyl peroxide + toluene (aluminum chloride catalyst).
Use: Perfume additive (fixative).

phenyltrichlorosilane.
CAS: 98-13-5. $C_6H_5SiCl_3$.
Properties: Colorless liquid. Bp 201C, d 1.321 (25/25C), refr index 1.5240 (25C), flash p 185F (85C) (COC). Soluble in benzene, ether, perchloroethylene; readily hydrolyzed by moisture with liberation of hydrogen chloride. Combustible.
Derivation: By Grignard reaction of silicon tetrachloride and phenylmagnesium chloride, reaction of benzene with trichlorosilane.
Hazard: Strong irritant to tissue.
Use: Intermediate for silicones, laboratory reagent.

1-phenyltridecane.   See tridecylbenzene.

phenyltrimethoxysilane.
CAS: 780-69-8. $C_6H_5Si(OCH_3)_3$.
Properties: Liquid. D 1.063 (25C), refr index 1.4710 (25C), bp 211C. Soluble in acetone, benzene, perchloroethylene, methanol. Combustible.
Use: In polymers to be applied to powders, glass, paper, and fabrics.

phenylurethane.   See ethyl phenylcarbamate.

phenyl valerate. $C_4H_9COOC_6H_5$.
Properties: Colorless liquid. Slightly soluble in water; soluble in alcohol and ether.
Use: Flavors and odorants.

pheromone. A group of organic compounds, produced by insects, that function as communication means and sex attractants. Synthetic pheromones have been used experimentally to control insect pests by disrupting their mating behavior, e.g., 4-methyl-3-heptanone. Pheromones have also been synthesized by use of organoborane reactions. See phoromone.

phlogiston. The hypothetical principle of fire used for the explanation of the phenomena of oxidation and reduction before Lavoisier's discoveries. See Lavoisier, Antoine Laurent.

phloroglucinol. (phloroglucine; 1,3,5-trihydroxybenzene).
CAS: 108-73-6. $C_6H_3(OH)_3 \cdot 2H_2O$.
Properties: White to yellowish crystals; odorless. Mp 212–217C if rapidly heated, 200–209C if slowly heated; bp sublimes with decomposition. Soluble in alcohol, ether, pyridine, and in 100 parts water.
Derivation: By fusion of resorcinol with caustic soda, by reduction of trinitrobenzene.
Use: Analytical chemistry (reagent for pentoses and with vanillin for determining the presence of free hydrogen chloride), decalcifying agent for bones, preparation of pharmaceuticals and dyes, resins, preservative for cut flowers, textile dyeing and printing.

phomin.
CAS: 14930-96-2. $C_{29}H_{37}NO_5$. A type of mycotoxin that is structurally and toxicologically related to the cytochalasins and chaetoglobosins.
Derivation: Produced by a wide variety of fungi.

phorate. (generic name for O,O-diethyl-S-[(ethylthio)methyl]phosphorodithioate).
CAS: 298-02-2. $(C_2H_5O)_2P(S)SCH_2SC_2H_5$.
Properties: Liquid. Bp 118–120C (0.8 mm Hg). Insoluble in water; miscible with carbon tetrachloride, dioxane, xylene.
Hazard: Toxic by skin contact, inhalation, or ingestion. Rapidly absorbed through skin. Repeated inhalation or skin contact may, without symptoms, progressively increase susceptibility. Use may be restricted. A cholinesterase inhibitor. Toxic by skin absorption.
Use: Insecticide. Questionable carcinogen.

phorbol. (4,9,12-β-13,20-pentahydroxy-1,6-tigliadiene-3-ol).
CAS: 17673-25-5. The parent alcohol of tumor-producing compounds in croton oil.
Use: Biochemical and medical research.

phoromone. (7-ethyl-5-methyl-6,8-dioxabicyclo[3.2.1]-octane). Product excreted by bark beetles that acts as sex attractant. It has been isolated and synthesized for possible use in protection of forest timber.

phorone. (diisopropylidene acetone).
CAS: 504-20-1. $(CH_3)_2CCHCOCHC(CH_3)_2$.
Properties: Yellow liquid or yellowish-green prisms. D 0.8791 (20/20C), bp 197.9C, fp 28.0C, vap press 0.38 mm Hg (20C), flash p 185F (85C), bulk d 7.3 lb/gal (20C). Slightly soluble in water; soluble in alcohol. Combustible.
Use: Solvent for nitrocellulose, coating compositions, stains, intermediate (organic synthesis).

phosalone.
CAS: 2310-17-0. $C_{12}H_{15}ClNO_4PS_2$.
Properties: Colorless crystals. Mp 48C. Soluble in alcohols and aromatic solvents; insoluble in water and aliphatic solvents.
Hazard: Toxic by ingestion and inhalation.
Use: Insecticide, molluscicide.

"Phos-chek P-30 and P-40" [ICL]. TM for ammonium polyphosphate.
Grade: Regular and fine white powder.
Use: Phosphorus-based catalyst in organic and latex-based fire-retardant intumescent paints, mastics, and polymers.

Phosdrin. A mixture containing more than 60% of the α isomer of $2-(CH_3O)_2P(O)OC(CH_3)$: $CHCOOCH_3$ (generic name mevinphos) and less than 40% of insecticidally active related compounds. It is 100% active.
See mevinphos.

Phosflake. A uniform blend of caustic soda and trisodium phosphate prepared in flake form, especially for bottle-washing use.

phosgene. (carbonyl chloride; carbon oxychloride; chloroformyl chloride).
CAS: 75-44-5. $COCl_2$.
Properties: Liquid or easily liquefied gas, colorless to light yellow; odor varies from strong and stifling when concentrated to haylike in dilute form. D 1.392 (19/4C), fp −128C, bp 8.2C, sp vol 3.9 cu ft/lb (21.1C). Slightly soluble in water and slowly hydrolyzed by it; soluble in benzene and toluene. Noncombustible.
Derivation: By passing a mixture of carbon monoxide and chlorine over activated carbon.
Hazard: Very toxic via inhalation, strong irritant to eyes and upper respiratory tract. Pulmonary edema and pulmonary emphysema.
Use: Organic synthesis, especially of isocyanates, polyurethane and polycarbonate resins, carbamates, organic carbonates, and chloroformates; pesticides; herbicides; dye manufacture.

phosgene oxime. (dichloroformoxine).
$CHCl_2NO$
Hazard: Asphyxiant.
Hazard: Poison.
Use: A war gas.

phosmet.
CAS: 732-11-6. $C_{11}H_{12}NO_4PS_2$. A dimethyl ester of phosphorodithioic acid.
Properties: Colorless crystals. Mp 72C. Partially soluble in water; decomposes on heating.
Hazard: Toxic by ingestion, may inhibit cholinesterase.
Use: Acaricide, insecticide.

phosphamidon. (2-chloro-2-diethylcarbamoyl-1-methylvinyl dimethyl phosphate).
CAS: 13171-21-6. $(CH_3O)_2P(O)OC(CH_3)$:
$C(Cl)C(O)N(C_2H_5)_2$.
Properties: Colorless liquid. Bp 162C (1.5 mm Hg). Soluble in water and organic solvents.
Hazard: Toxic by ingestion, inhalation, skin absorption; cholinesterase inhibitor; use may be restricted.
Use: Insecticide.

cis-**phosphamidon.**
CAS: 23783-98-4. Mf: $C_{10}H_{19}ClNO_5P$.
Hazard: A poison by ingestion and skin contact.

phosphatase, alkaline. An enzyme excreted into the bile by the liver and found in the blood. It is concerned with bone formation, probably being produced by osteoblasts. It hydrolyzes phosphoric acid esters at pH 7–8, liberating phosphate ions.
Use: Biochemical research.

phosphate, condensed. A phosphorus compound with two or more phosphorus atoms in the molecule. Examples are polyphosphates, pyrophosphates.
See polyphosphoric acid.

phosphate glass. A type of glass containing phosphorus pentoxide. Aluminum-metaphosphate is frequently the basic material. Such glasses have properties not attainable in silicate glasses, e.g., resistance to hydrogen fluoride.

phosphate rock. (phosphorite). A natural rock consisting largely of calcium phosphate and used as a raw material for manufacture of phosphate fertilizers, phosphoric acid, phosphorus, and animal feeds. Recovery of uranium from the manufacture of phosphoric acid and other phosphate chemicals is expected to become an important source of this metal. Phosphate rock is the primary source of superphosphate, prepared by treatment of the pulverized rock with sulfuric acid (superphosphate having 16–18% P_2O_5) or by acidifying with phosphoric acid (triple superphosphate having 40–48% P_2O_5). Nitric acid is sometimes used, i.e., nitrophosphate. Defluorinated phosphate rock is the source of phosphate used in animal feeds and feed concentrations. Important deposits are in the U.S. (Florida, North Carolina, Tennessee, California, Wyoming, Montana, Utah, Idaho), North Africa (Morocco, Libya, Algeria), the former U.S.S.R., and various islands in the Pacific.

phosphate slag. Glassy calcium silicate, by-product of electric furnace phosphorus manufacture.
Properties: Lumps, loose bulk d 85 lb/ft^3.

phosphatide. See phospholipid.

phosphatidyl choline. See lecithin.

phosphatidyl ethanolamine. See cephalin.

phosphatidyl serine. See cephalin.

phosphazene. (phosphonitrile). A ring or chain polymer that contains alternating phosphorus and nitrogen atoms with two substituents on each phosphorus atom. Characteristic structures are cyclic trimers, cyclic tetramers, and high polymers. The substituent can be any of a wide variety of organic groups, halogen, amino, etc. Most cyclic trimers are crystalline, solids, organosoluble, and stable to weather conditions; the high polymers (polyphosphazenes) are elastomeric or thermoplastic. A copolymer of phosphazene and styrene has been investigated for use as a flame retardant.

phosphinate. A derivative of the hypothetical phosphinic acid, $H_2P(O)OH$.

phosphine. (hydrogen phosphide).
CAS: 7803-51-2. PH_3.
Properties: Colorless gas; disagreeable, garliclike odor. D 1.185, fp −133.5C, bp −85C, autoign temp 100F (37.7C). Soluble in alcohol, ether, and cuprous chloride solution; slightly soluble in cold water. Insoluble in hot water.
Derivation: By action of freshly formed hydrogen or of caustic potash on phosphorus.
Hazard: Spontaneously flammable. Toxic by inhalation, strong upper respiratory tract, gastrointestinal and central nervous system irritant.
Use: Organic preparations, phosphonium halides, doping agent for n-type semiconductors, polymerization initiator, condensation catalyst.
Note: A synthetic dye, chrysaniline yellow, is sometimes called phosphine.

phosphinidynetrismethanol. See tris(hydroxymethyl)phosphine.

phosphite. Salt of phosphorous acid, as in sodium phosphite.

phosphocysteamine. See s-(2-aminoethyl) phosphorothioate.

phosphodiesterase. An enzyme that cleaves hydrolytically the carboxy- and phosphoesters of phosphatides (phospholipids).

phosphodiester linkage. A chemical grouping that contains two alcohols esterified to one molecule of phosphoric acid, which thus serves as a bridge between them.

phosphogluconate pathway. An oxidative pathway beginning with glucose-6-phosphate and leading to ribose and NADPH. Also called the pentose phosphate pathway or hexose monophosphate shunt.

2-phosphoglyceric acid. $HOCH_2CH$ $[OPO(OH)_2]COOH$. An intermediate in the metabolism of carbohydrates in biological systems. See enolase.

phosphoglyceride. See phospholipid.

phospholine. (2-diethyoxyphosphorylsulfanyl-ethyl(trimethyl)azanium; (2-mercaptoethyl)trimethylammonium-5-ester with o,o'-diethylphosphorothioate).
CAS: 6736-03-4. $C_9H_{23}NO_3PS$. A topically administered parasympathomimetic agent that is a potent, long-acting irreversible cholinesterase inhibitor.
Hazard: Poison.
Use: To treat glaucoma and for accommodative esotropia.

phospholipase A2. (PLA; formerly lecithinase A; pyridine).
CAS: 9001-84-7. C_5H_5N. A class of enzymes that catalyze the hydrolysis of one of the fatty ester linkages in diacyl phosphatides, yielding lysophospatides and fatty acids.
Derivation: Occurs in the liver, pancreas, kidney, muscle, heart, and adrenal glands; and in the venoms of most families of venomous snakes.

phospholipase B. (formerly lecithinase B). It catalyzes the hydrolysis of lysolectithin and lysocephalin to yield a fatty acid.

phospholipase C. (PLLC).
CAS: 9001-86-9. A hydrolytic enzyme that catalyzes the hydrolysis of glycerylphosphoporylcholine into a diglyceride and choline phosphate.

phospholipase D. (lecithinase; PLD; 1-dodecyl-3-hydroxy-5-(hydroxymethyl)-2-methylpyridin-1-ium-4-carbaldehyde chloride).
CAS: 9001-87-0. $C_{20}H_{34}ClNO_3$. A type of enzyme that catalyzes the hydrolysis of phosphatidylcholine into choline and phospatidate.
Derivation: Found in the toxin of *Clostridium welchii*.

phospholipid. (phosphatide). A group of lipid compounds that yield on hydrolysis phosphoric acid, an alcohol, fatty acid, and nitrogenous base. They are widely distributed in nature and include such substances as lecithin, cephalin, and sphingomyelin.

phosphomolybdate. See heteromolybdate.

phosphomolybdenic acid, sodium salt. See sodium molybdophosphate.

phosphomolybdic acid. (phospho-12-molybdic acid; PMA; 12-molybdophosphoric acid). $H_3PMo_{12}O_{40} \cdot xH_2O$.
Properties: Yellowish crystals. D 3.15 g/cc, mp 78–90C, strong oxidizing agent in aqueous solution, strong acid in free acid form. Soluble in water, alcohol, and ether.
Grade: Technical, CP, reagent.
Hazard: (Solution) May ignite combustible materials.
Use: Reagent for alkaloids; pigments; catalyst; fixing agent in photography; additive in plating processes; imparts water resistance to plastics, adhesives, and cement.

phosphomolybdic pigment. (molybdenum lake). A pigment made by precipitating a basic organic dye with phosphomolybdic acid or a mixture of phosphomolybdic and phosphotungstic acids.
See phosphotungstic pigment.

phosphonic acid, benzyl-, dibutyl ester. See benzylphosphonic acid dibutyl ester.

phosphonitrile. See phosphazene.

phosphonium iodide. (iodophosphonium). PH_4I.
Properties: Colorless or slightly yellowish crystals. D 2.86, mp 18.5C, sublimes at 61.8C, bp 80C. Decomposed by water or alcohol evolving highly toxic phosphine.
Derivation: Hydrolysis of phosphorus tetraiodide and white phosphorus.
Hazard: Rapid heating can cause detonation. Toxic by ingestion.

2-phosphonobutanetricarbonic acid pentamethyl ester.
CAS: 37971-35-0. $C_{12}H_{21}O_9P$.
Hazard: A moderate eye irritant.

3-phosphonopropanoic acid.
CAS: 5962-42-5. $C_3H_7O_5P$.
Hazard: Low toxicity by ingestion. A severe eye irritant.

phosphonosuccinic acid tetramethyl ester.
CAS: 2788-26-3. $C_8H_{15}O_7P$.
Hazard: A moderate eye irritant.

9-((2-phosphonylmethoxy)-ethyl)guanine.
CAS: 114088-58-3. $C_8H_{12}N_5O_5P$.
Hazard: A severe skin irritant.

phosphor. (fluor). A substance, either organic or inorganic, liquid or crystals, that is capable of luminescence, i.e., of absorbing energy from sources such as X-rays, cathode rays, UV radiations, and alpha particles, and emitting a portion of the energy in the UV, visible, or infrared. When the emission of the substance ceases immediately or in the order of 10^{-8} sec after excitation, the material is said to be fluorescent. Material that continues to emit light for a period after the removal of the exciting energy is said to be phosphorescent. The half-life of the afterglow varies with the substance and may range from 10^{-6} sec to days.
Use: Fluorescent light tubes; television, radar, and cathode ray tubes, instrument dials, scintillation counters.
See fluorescence; phosphorescence.

phosphor bronze. A tin bronze that has been deoxidized by the addition of up to 0.5% phosphorus. Relatively hard, strong, and corrosion resistant. Has good cold-work properties and high strength.
Grade: Grade A (5% tin), grade C (8% tin), grade D (10% tin), grade E (1.25% tin).
Use: Springs, electrical switches, contact fingers, chains, fourdrinier wire.
See brass; bronze.

phosphorescence. A type of luminescence in which the emission of radiation resulting from excitation of a crystalline or liquid material occurs *after* excitation has ceased and may last from a fraction of a second to an hour or more. This phenomenon is characteristic of some organic compounds, as in the firefly, and also of a number of inorganic solid materials, both natural and synthetic.
Use: Industrially as phosphors.
See fluorescence.

phosphoric acid. (orthophosphoric acid). CAS: 7664-38-2. H_3PO_4.
Properties: Colorless, odorless, sparkling liquid or transparent, crystalline solid, depending on concentration and temperature. At 20C the 50 and 75% strengths are mobile liquids, the 85% is of a syrupy consistency, and the 100% acid is in the form of crystals, d 1.834 (18C), mp 42.35C, loses 1/2 H_2O at 213C (to form pyrophosphoric acid), soluble in water and alcohol, corrosive to ferrous metals and alloys.
Derivation: (1) Action of sulfuric acid on pulverized phosphate rock; (2) action of hydrochloric acid on phosphate rock, with extraction by tributylphosphate; (3) by heating phosphate rock, coke, and silica in an electric furnace, burning the elemental phosphorus produced, and then hydrating the phosphoric oxide (furnace acid).
Grade: Agricultural, technical (50, 75, 85, 90, 100%), food (50, 75, 85%), NF (85–88%), FCC (75–85%). (Polyphosphoric acid is sometimes called 115% phosphoric acid.)

Hazard: Toxic by ingestion and inhalation, irritant to skin, eyes, and upper respiratory tract.
Use: Fertilizers, soaps and detergents, inorganic phosphates, pickling and rust-proofing metals, pharmaceuticals, sugar refining, gelatin manufacture, water treatment, animal feeds, electropolishing, gasoline additive, conversion coatings for metals, catalyst for ethanol manufacture, lakes in cotton dyeing, yeasts, soil stabilizer, waxes and polishes, binder for ceramics, activated carbon, in foods and carbonated beverages as acidulant and sequestrant, laboratory reagent.

phosphoric acid, anhydrous. See phosphoric anhydride.

phosphoric acid, cyclohexyl diphenyl ester. See cyclohexyl phenyl phosphate.

phosphoric acid, dibutyl phenyl ester. See dibutyl phenyl phosphate.

phosphoric acid, meta-. (metaphosphoric acid; phosphoric acid, glacial). $(HPO_3)_x$.
Properties: Transparent, highly deliquescent, glassy mass. D 2.2–2.488. Soluble in water slowly forming the ortho-acid. Also soluble in alcohol. Noncombustible.
Derivation: By heating orthophosphoric acid to redness, by treating phosphorus pentoxide with a calculated quantity of water; by heating diammonium phosphate.
Grade: Technical, CP.
Use: Phosphorylating agent, dehydrating agent, dental cements, laboratory reagent.
See polyphosphoric acid.

phosphoric acid, trisodium salt, dodeahydrate. See sodium phosphate tribasic dodecahydrate.

phosphoric anhydride. (phosphorus pentoxide; phosphoric oxide; phosphoric acid, anhydrous). CAS: 1314-56-3. P_2O_5.
Properties: Soft, white powder. Absorbs moisture from the air with avidity, forming meta-, pyro-, or ortho-phosphoric acid, depending on the amount of water absorbed and upon conditions of absorption. D 2.30, mp 340C, bp sublimes at 360C.
Derivation: By burning phosphorus in dry air.
Grade: Technical, nitrogen-complex coated (for slow solution).
Hazard: Reacts violently with water to evolve heat; dangerous fire risk. Keep tightly sealed or stoppered. Corrosive to skin and tissue.
Use: Preparation of phosphorus oxychloride and metaphosphoric acid, acrylate esters, surfactants, dehydrating agent, condensing agent in organic synthesis, sugar refining, laboratory reagent, fire extinguishing, special glasses.

phosphoric bromide. See phosphorus pentabromide.

phosphoric chloride. See phosphorus pentachloride.

phosphoric oxide. See phosphoric anhydride.

phosphoric perbromide. See phosphorus pentabromide.

phosphoric perchloride. See phosphorus pentachloride.

phosphoric sulfide. See phosphorus pentasulfide.

phosphorolysis. Cleavage of a compound with phosphate as the attacking group: analogous to hydrolysis.

phosphorus.
CAS: 7723-14-0. P. Nonmetallic element of atomic number 15, group VA of periodic table, aw 30.97376, valences of 1, 3, 4, 5; allotropes; white (or yellow), red, and black phosphorus. No stable isotopes, several artificial radioactive isotopes with mass numbers 29–34.
White phosphorus
Properties: Crystals, waxlike, transparent solid, metastable with respect to red phosphorus, an impurity present in white allotrope. Bp 280C, vap d corresponds to formula P_4, mp 44.1C, d (solid 20C), 1.82 d (liquid 44.5C) 1.745, Mohs hardness 0.5. High electrical resistivity. Insoluble in water and alcohol; soluble in carbon disulfide, exhibits phosphorescence at room temperature. An essential dietary nutrient.
Derivation: (1) Produced in an electric furnace from phosphate rock, sand, and coke. The phosphorus vapor is driven off and condensed under water. (2) By reaction of phosphate rock with sulfuric acid, the resulting $CaSO_4$ being removed by filtration and the phosphoric acid concentrated by evaporation (wet process).
Grade: Technical 99.9%, electronic grade 99.9999%.
Occurrence: Occurs in nature in phosphate rock [impure $Ca_3(PO_4)_2$], apatite [$Ca_5(PO_4)_3F$], bones, teeth, organic compounds of living tissue, and as phosphorite nodules on the ocean floor.
Hazard: Ignites spontaneously in air at 86F (30C). Store under water and away from heat, dangerous fire risk. Toxic by ingestion and inhalation, skin contact causes burns.
Use: Rodenticides, smoke screens, analytical chemistry.
Red phosphorus
Properties: Violet-red amorphous powder, obtained from white phosphorus by heating at 240C with catalyst, sublimes at 416C, d 2.34, autoign temp 500F (260C). High electrical resistivity. Much less

reactive than white phosphorus. Insoluble in most solvents.
Hazard: Large quantities ignite spontaneously and on exposure to oxidizing materials. Reacts with oxygen and water vapor to evolve phosphine. Extinguish with foam or dry chemical (not water).
Use: Manufacture of phosphoric acid and other phosphorus compounds, phosphor bronzes, metallic phosphides, additive to semiconductors, electroluminescent coatings, safety matches, fertilizers.
Black phosphorus
Properties: Black solid resembling graphite. D 2.25–2.6. Obtained by heating white phosphorus under pressure. Insoluble in most solvents. Electrically conducting.

phosphorus-32.
CAS: 14596-37-3. Radioactive phosphorus of mass number 32.
Properties: Half-life, 14.3 days; radiation β.
Derivation: Pile irradiation of potassium dihydrogen phosphate or sulfur and sulfur compounds.
Available forms: Phosphate ion in weak hydrochloric acid solution, solid potassium dihydrogen phosphate, P-32 sterile solution, in tagged compounds such as hexaethyltetraphosphate, ribonucleic acid, triphenylphosphine, etc.
Grade: USP as sodium phosphate P-32 solution.
Hazard: Confirmed carcinogen.
Use: Biochemical radioactive tracer studies; medical treatment of leukemia, skin lesions, etc.; industrial measurements, e.g., tire tread wear, thickness of ink and paint films; lead detection.

phosphorous acid, ortho-. (orthophosphorous acid; phosphonic acid).
CAS: 13598-36-2. H_3PO_3.
Properties: White or yellowish, crystalline mass; very hygroscopic. Keep tightly sealed or stoppered, absorbs oxygen very readily with formation of orthophosphoric acid. D 1.651, bp 200C (decomposes), mp 70C. Soluble in alcohol and water.
Grade: Reagent, technical, 70%.
Use: Analysis (testing for mercury), chemical reducing agent, phosphite salts.

phosphorous acid, triallyl ester.
CAS: 102-84-1. $C_9H_{15}O_3P$.
Hazard: A poison by ingestion and skin contact. A mild eye irritant.

phosphorus chloride. See phosphorus trichloride.

phosphorus halide. Any of a class of phosphorus compounds. They react violently with water to produce the corresponding hydrogen halides plus oxo phosphorus acids. They tend to form strongly acidic solutions.
Hazard: Strong irritants; very toxic.

phosphorus heptasulfide. (tetraphosphorus heptasulfide).
CAS: 12037-82-0. P_4S_7.
Properties: Light-yellow crystals. D 2.19, mp 310C, bp 523C. Slightly soluble in carbon disulfide.
Hazard: Flammable, dangerous fire risk.

phosphorus nitride. P_3N_5.
Properties: Amorphous white solid. Insoluble in cold water; decomposes in hot water; soluble in common organic solvents. Nonhygroscopic and stable in air, decomposes at 800C.
Use: Doping semiconductors.

phosphorus oxybromide.
CAS: 7789-59-5. $POBr_3$.
Properties: Colorless crystals. D 2.82, mp 56C, bp 189C, decomposed by water. Soluble in ether and benzene; reacts strongly with organic matter.
Hazard: Strong irritant to skin and tissue, store in sealed glass containers.
Use: Chemical intermediate.

phosphorus oxychloride. (phosphoryl chloride).
CAS: 10025-87-3. $POCl_3$.
Properties: Colorless, fuming liquid; pungent odor. D 1.675 (20/20C), mp 1.2C, bp 107.2C, refr index 1.460 (25C). Decomposed by water and alcohol with evolution of heat.
Derivation: From phosphorus trichloride, phosphorus pentoxide, and chlorine.
Grade: Technical, 99.999+%.
Hazard: Toxic by inhalation and ingestion, strong irritant to skin, tissue, and upper respiratory tract.
Use: Manufacture of cyclic and acyclic esters for plasticizers, gasoline additives, hydraulic fluids and organophosphorus compounds, chlorinating agent and catalyst, dopant for semiconductor-grade silicon, tricresyl phosphate, and fire-retarding agents.

phosphorus pentabromide. (phosphoric bromide; phosphoric perbromide).
CAS: 7789-69-7. PBr_5.
Properties: Yellow, crystalline mass. Keep hermetically sealed!! Soluble in water (decomposes), carbon disulfide, carbon tetrachloride, benzene. Bp 106C (decomposes).
Grade: Technical.
Hazard: Corrosive to skin and tissue.
Use: Brominating agent.

phosphorus pentachloride. (phosphoric chloride; phosphoric perchloride).
CAS: 10026-13-8. PCl_5.
Properties: Slightly yellow, crystalline mass; irritating odor; fuming in moist air. D 3.60, mp (under pressure) 148C, ordinarily sublimes at 160–165C. Soluble in carbon disulfide and carbon tetrachloride.
Derivation: By action of chlorine on phosphorus or phosphorus trichloride.

Grade: Technical, reagent.
Hazard: Flammable, reacts strongly with water; use carbon dioxide or dry chemical to extinguish. Store in tightly closed containers. Corrosive to eyes and skin. Upper respiratory tract irritant.
Use: Chlorinating and dehydrating agent, catalyst.

phosphorus pentafluoride.
CAS: 7647-19-0. PF_5.
Properties: Colorless gas. Fumes strongly in air. Fp −94C, bp −84.8C. Decomposed by water. Available in small cylinders. Nonflammable.
Grade: 99%.
Hazard: Corrosive to eyes and skin.
Use: Polymerization catalyst.

phosphorus pentasulfide. (phosphoric sulfide; phosphorus persulfide; thiophosphoric anhydride).
CAS: 1314-80-3. P_2S_5.
Properties: Light-yellow or greenish-yellow crystalline mass; odor similar to hydrogen sulfide. Keep in sealed containers. Very hygroscopic, burns in air forming P_2O_5 and SO_2, decomposed by moist air. Mp 286–290C, bp 515C, d 2.03, vap press 1 mm Hg (300C), autoign temp 287F (141.6C). Soluble in solutions of alkali hydroxides and carbon disulfide.
Derivation: By reaction of phosphorus and sulfur.
Grade: Technical, distilled.
Hazard: Dangerous fire risk; ignites by friction; contact with water or acids liberates poisonous and flammable hydrogen sulfide. Toxic by inhalation, strong upper respiratory tract irritant.
Use: Intermediate for lubricating oil additives, insecticides (chiefly parathion and malathion), flotation agents, safety matches, ignition compounds, sulfonation.

phosphorus pentoxide. See phosphoric anhydride.

phosphorus persulfide. See phosphorus pentasulfide.

phosphorus salt. See sodium ammonium phosphate.

phosphorus sesquisulfide. (tetraphosphorus trisulfide).
CAS: 1314-85-8. P_4S_3.
Properties: Yellow, crystalline mass. D 2.00, mp 172C, bp 407.8C, autoign temp 212F (100C) (solution). Soluble in carbon disulfide; insoluble in cold water, decomposed by hot water.
Derivation: By gently heating phosphorus and sulfur.
Hazard: Dangerous fire risk, ignites by friction. An irritating poison.
Use: Organic synthesis, manufacture of matches.

phosphorus sulfide. See phosphorus trisulfide.

phosphorus tribromide.
CAS: 7789-60-8. PBr_3.
Properties: Fuming, colorless liquid; penetrating odor. D 2.852 (15C), bp 175C; fp −40C. Soluble in acetone, alcohol, carbon disulfide, hydrogen sulfide, water (decomposes).
Hazard: Corrosive to skin and tissue, store in tightly closed containers.
Use: Analysis (testing for sugar and oxygen), catalyst, synthesis.

phosphorus trichloride.   (phosphorus chloride).
CAS: 7719-12-2. PCl_3.
Properties: Clear, colorless, fuming liquid. D 1.574, fp −111.8C, bp 76C. Decomposes rapidly in moist air; soluble in ether, benzene, carbon disulfide, and carbon tetrachloride.
Derivation: By passing a current of dry chlorine over gently heated phosphorus, which ignites. The trichloride admixed with some pentachloride distills over. A small amount of phosphorus is added, and the whole distilled.
Grade: Technical, 99.9%.
Hazard: Corrosive to skin and tissue, reacts with water to form hydrochloric acid, store in tightly closed containers. Eye, skin and upper respiratory tract irritant.
Use: Making phosphorus oxychloride, intermediate for organophosphorus pesticides, surfactants, phosphites (reaction with alcohols and phenols), gasoline additives, plasticizers, dyestuffs, chlorinating agent, catalyst, preparing rubber surfaces for electrodeposition of metal, ingredient of textile finishing agents.

phosphorus trifluoride.
CAS: 7719-12-2. PF_3.
Properties: Colorless gas.
Hazard: Poisonous; eye, skin and upper respiratory tract irritant.

phosphorus triiodide. PI_3.
Properties: Red crystals. Mp 61C (decomposes), d 4.18. Hygroscopic; soluble in alcohol, carbon disulfide, water (decomposes).
Grade: Technical, reagent.
Hazard: Flammable, reacts with water. Irritating to skin and eyes.
Use: Organic synthesis.

phosphorus trisulfide.   (phosphorus sulfide; tetraphosphorus hexasulfide; thiophosphorous anhydride).
CAS: 12165-69-4. P_2S_3 or P_4S_6.
Properties: Grayish-yellow mass; tasteless; odorless. Bp 490C, mp 290C. Keep well stoppered. Decomposes in moist air; soluble in alcohol, carbon disulfide, ether.
Hazard: Flammable, highly dangerous fire risk, reacts with water.
Use: Organic chemistry (reagent).

phosphorylase.   An enzyme, occurring in muscle and liver, that catalyzes the conversion of glycogen into glucose-1-phosphate.

phosphorylation.   A reaction in which phosphorus combines with an organic compound, usually in the form of the trivalent phosphoryl group. It occurs naturally in cellular metabolism and is of particular importance in vitamin activity and enzyme formation. It is also used to produce a modified cellulose (P-cellulose) for cation exchange in chromatographic separations. Most often by enzymatic transfer of a phosphate group from ATP.

phosphorylation potential.   The actual free-energy change of ATP (or other phosphate-containing molecule) hydrolysis under the nonstandard conditions within a cell.

phosphoryl chloride.   See phosphorus oxychloride.

phosphotungstic   acid.   (phospho-12-tungstic acid; phosphowolframic acid; 12-tungstophosphoric acid).
CAS: 12067-99-1. $H_3PW_{12}O_{40} \cdot xH_2O$.
Properties: Yellowish-white solid. Mp (for $24H_2O$ of hydration) 89C. Soluble in water, acetone, and diethyl ether. Relatively insoluble in nonpolar organic solvents. Strong oxidizing agent in aqueous solution; strong acid in the free acid form.
Derivation: Addition of phosphates to sodium tungstate in the presence of hydrochloric acid.
Grade: Reagent, technical.
Hazard: Strong irritant to skin and eyes.
Use: Reagent in analytical chemistry and biology; manufacture of organic pigments; additive in plating industry; imparts water resistance to plastics, adhesives, and cement; catalyst for organic reactions; photographic fixing agent; textile antistatic agent.

phosphotungstic acid, sodium salt.   See sodium-12-tungstophosphate.

phospho-tungstic-phosphomolybdic   acid.   See Bessonoff's reagent.

phosphotungstic pigment.   (tungsten lake). A green or blue pigment manufactured by precipitating basic dyestuffs such as malachite green or Victoria blue with solutions of phosphotungstic acid, phosphomolybdic acid, or a mixture of both. See phosphomolybdic pigment.
Use: Printing inks, paper, paints, and enamels.

photocatalysis.   Refers to the increase in the speed of, or facilitation of, a reaction by exposure to light or light waves of a definite frequency.

photochemical air pollutant. Any of a large number of trace compounds that are a product of the action of sunlight on certain primary air pollutants.

photochemical equivalent law. Each molecule entering into reaction has to be excited by the absorption of one quantum of radiation.

photochemical reaction center. The part of a photosynthetic complex where the energy of an absorbed photon causes charge separation, initiating electron transfer.

photochemical smog. (smog; oxidant smog). A highly irritating, usually dense, lacrimatory haze produced by photochemical reactions that occur under the influence of strong sunlight in air polluted by various sources but most importantly by automobile exhaust gases under temperature inversion conditions.

photochemical yield. Refers to quantum efficiency of a reaction.
See quantum efficiency.

photochemistry. The branch of chemistry concerned with the effect of absorption of radiant energy (light) in inducing or modifying chemical changes. Photosynthesis is the most important example of a photochemical reaction; others are the photosensitization of solids, applied in photography, photocells, photovoltaic cells, and the formation of visual pigments; photochemical decomposition (photolysis), photo-induced polymerization, oxidation, and ionization; fluorescence and phosphorescence; and the reaction of chlorine with organic compounds. Free-radical chain mechanisms are usually involved.
See free radical.

photochromism. The ability of a transparent material to darken reversibly when exposed to light. Plastics can be made light-sensitive by certain aromatic organic nitro compounds such as 2-(2,4-dinitrobenzyl)pyridine. Such chemicals are compatible with most transparent plastics and are either blended with the base resin or applied as coatings.
See glass, photochromic.

photodecomposition. See photolysis; photochemistry.

photoelectric colorimeter. An analytical device used to measure the comparative intensity of color in solutions using photoelectric detectors.

photo-glycin. See p-hydroxyphenylglycine.

photographic chemistry. In photographic films and papers the sensitive surface usually consists of microscopic grains of a silver halide, suspended in gelatin. Exposure to light renders the halide particles susceptible to reduction to metallic silver by developing agents containing a reducing agent, as well as an accelerator, preservative, and restrainer. The accelerator increases the activity of the reducing agent (principally due to ionization of the phenolic agents to their active form) and is usually an alkaline compound. The preservative, usually sodium sulfite, minimizes air oxidation. The restrainer helps to prevent "fog" (reduction of silver halide grains that have not been exposed to light) and is almost always potassium bromide.

Color sensitizers are dyes added to silver halide emulsions to broaden their response to various wavelengths. Unsensitized emulsions are most responsive in the blue region of the spectrum and thus do not correctly represent the light spectrum striking them. Widely used sensitizers include the cyanine dyes, the merocyanines, the benzooxazoles, and the benzothiazoles. Cryptocyanine sensitizes the extreme red and infrared.

In color photography diethyl-p-phenylenediamine is an important developer because its oxidation product readily couples with a large number of phenol and reactive methylene compounds to form indophenol and indoaniline dyes, which are the basis of most of the current color processes.
See holography.

photographic grade. A chemical in which impurities known to be photographically harmful are limited to safe levels and inert impurities are restricted to levels that will not reduce the required assay strength.

photoionization. (atomic photoelectric effect). Refers to the removal of one or more electrons from an atom or molecule by absorption of a photon of visible or ultraviolet light.

photolysis. Decomposition of a compound into simpler units as a result of absorbing one or more quanta of radiation; examples are splitting of hydrogen iodide by the reaction $2HI + hn \rightarrow H_2 + I_2$; and of ketene ($H_2C=CO$) into CO and carbene (methylene) ($=CH_2$). Photodecomposition may also occur with aldehydes, ketones, azo compounds, and organometallic compounds. Continuous generation of hydrogen by photolysis of water has been achieved using platinum catalyst in conjunction with ruthenium and rhodium. Similarly, hydrogen can be split from H_2S by photolysis with cadmium sulfide as catalyst, aided by ruthenium dioxide.
See flash photolysis; photochemistry.

photometric analysis. Chemical analysis by means of absorption or emission of radiation, primarily in the near UV, visible, and infrared portions of the electromagnetic spectrum. It includes such techniques as spectrophotometry, spectrochemical analysis, Raman spectroscopy, colorimetry, and fluorescence measurements.
See spectroscopy.

photon.   The unit (quantum) of electromagnetic radiation. Light waves, γ rays, X-rays, etc. consist of photons. Photons are discrete concentrations of energy that seem to have no rest mass and move at the speed of light. Their nature can be described only in mathematical terms. Photons are emitted when electrons move from one energy state to another, as in an excited atom.
See radiation.

photophor.   See calcium phosphide.

photophosphorylation.   The enzymatic formation of ATP from ADP coupled to the light-dependent transfer of electrons from water in photosynthetic cells.

photopolymer.   A polymer or plastic so made that it undergoes a change on exposure to light. Such materials can be used for printing and lithography plates, photographic prints, and microfilm copying. The light may cause further polymerization or cross-linking, or it may cause degradation. One application involves the use of esters of polyvinyl alcohol that cross-link and so become insoluble, whereas unexposed portions of the material remain soluble.

photoreduction.   The light-induced reduction of NADP in photosynthetic cells.

photosensitive glass.   Certain clear silicate glass containing ingredients capable of forming permanent photographic images when subjected to action of X-rays or ultraviolet light and subsequent heat treatment.

photosensitizing pigment.   Any pigment, usually a plant pigment, that causes photosensitization in animals when ingested and the animal is subsequently exposed to sunlight.

photosynthesis.   The utilization of sunlight by plants and bacteria to convert two inorganic substances (carbon dioxide and water) into carbohydrates. Chlorophyll acts as the energy converter in this reaction, which is perhaps the most important on earth. The generalized reaction is $6CO_2 + 6H_2O + 672$ kcal $\rightarrow C_6H_{12}O_6 + 6O_2$. The significance of this process lies in the conversion of energy from radiant to chemical form. The chemical energy that a green plant stores by photosynthesis provides the total energy requirement of the plant. Directly or indirectly plants supply the primary organic nutrient for most other living organisms. Most fossil fuels are storehouses of the radiant energy transformed by photosynthesis in earlier geologic eras.
Photosynthesis is the principal source of atmospheric oxygen. At least two-thirds of the total photosynthetic activity of the earth takes place in the oceans. Its exact chemical mechanism is extremely complex. Essential features are the reduction of carbon dioxide and the utilization of the hydrogen of water to form carbohydrates, with oxygen being liberated; the nucleotides nicotinamide and adenosine triphosphate are involved in this conversion. Sugar (sucrose) is formed in the cytoplasm surrounding the chloroplasts. Photosynthesis has been shown to be substantially inhibited by air pollution to the extent of 20% in rural locations and 33% in urban areas.
See algae.

photosystem.   In photosynthetic cells, a functional set of light-absorbing pigments and their reaction center.

phototroph.   An organism that can use the energy of light to synthesize its own fuels from simple molecules such as carbon dioxide, oxygen, and water. Contrast with chemotroph.

photovoltaic cell.   See solar cell.

phthalamide.
CAS: 88-96-0. $C_6H_4(CONH_2)_2$. The double acid amide of phthalic acid.
Properties: Colorless crystals. Mp 220C (decomposes into phthalimide and ammonia). Very slightly soluble in water and alcohol, insoluble in ether.
Derivation: By stirring phthalimide with cold concentrated ammonia solution; by the reaction of phthalyl chloride and ammonia; or from the addition of ammonia to phthalic anhydride under pressure.
Use: Intermediate in organic synthesis, laboratory reagent.

phthalate.   (phthalic acid).
CAS: 88-99-3. $C_8H_6O_4$. Any of a class of organic compounds that contains phthalic acid esters and derivatives.
Use: As plasticizers in plastics, in detergents, in cosmetics, in industrial processes and products such as defoaming agents during paper and paperboard manufacture and dielectrics in capacitors.

phthalic acid.   (o-phthalic acid; o-benzene dicarboxylic acid).
CAS: 88-99-3. $C_6H_4(COOH)_2$.
Properties: Colorless crystals. D 1.585, mp decomposes at 191C. Soluble in alcohol; sparingly soluble in water and ether.
Derivation: Catalytic oxidation of o-toluic acid and oxidation of xylene.
Grade: Technical, reagent.
Use: Dyes, phenolphthalein, phthalimide, anthranilic acid, synthetic perfumes, laboratory reagent.

p-phthalic acid.   See terephthalic acid.

phthalic acid, diisononyl ester.
CAS: 28553-12-0. Mf: $C_{26}H_{42}O_4$.
Hazard: Questionable carcinogen. A reproductive hazard.

phthalic anhydride.
CAS: 85-44-9. $C_6H_4(CO)_2O$.

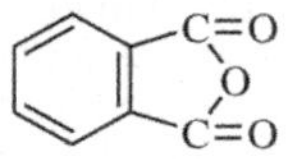

Properties: White, crystalline needles; mild odor. Sublimes below boiling point; d 1.527 (4C), mp 131.16C, bp 285C, flash p 305F (151.6C) (CC), autoign temp 1083F (583C). Soluble in alcohol, carbon disulfide, and hot water. Combustible.
Derivation: Catalytic oxidation of naphthalene.
Method of purification: Sublimation.
Grade: Pure, technical.
Hazard: Skin, eye and upper respiratory tract irritant. Questionable carcinogen.
Use: Alkyd resins, plasticizers, hardener for resins, polyesters, synthesis of phenolphthalein and other phthaleins, many other dyes, chlorinated products, pharmaceutical intermediates, insecticides, diethyl phthalate, dimethyl phthalate, laboratory reagent.

phthalimide.
CAS: 85-41-6. $C_6H_4(CO)_2NH$.

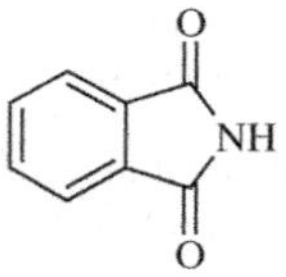

Properties: White, crystalline leaflets. Mp 233–238C, bp (sublimes). Slightly soluble in water; insoluble in benzene; soluble in boiling alcohol or acetic acid and in aqueous alkalies. Combustible.
Derivation: By dissolving phthalic anhydride in ammonium hydroxide, evaporating to dryness, and using the residue.
Use: Synthetic indigo via anthranilic acid, fungicide, organic synthesis, laboratory reagent.

phthalocyanine.
Any of a group of benzoporphyrins that have strong pigmenting power, forming a family of dyes. The basic structure of the molecule comprises four isoindole groups, $(C_6H_4)C_2N$, joined by four nitrogen atoms. Four commercially important modifications are (1) metal-free phthalocyanine, $(C_6H_4C_2N)_4N_4$, having a blue-green color (structure shown below): (2) copper phthalocyanine, in which a copper atom is held by secondary valences of the isoindole nitrogen atoms (see Pigment Blue 15); (3) chlorinated copper phthalocyanine, green, in which 15–16 hydrogen atoms are replaced by chlorine (see Pigment Green 7); and (4) sulfonated copper phthalocyanine, water soluble, green, in which two hydrogen atoms are replaced by HSO_3 groups.

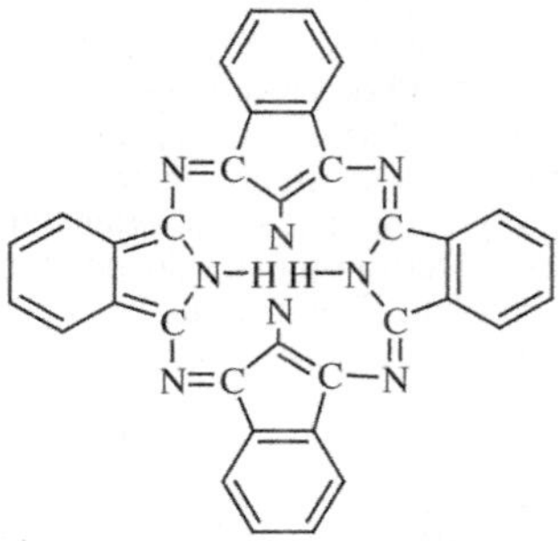

Use: Decorative enamels, automotive finishes, and similar applications where lightfastness and chemical stability are required.

phthalocyaninesulfonic acid, cobalt complexes.
See cobalt phthalocyaninesulfonate.

m-phthalodinitrile.
(isophthalodinitrile; IPM; m-dicyanobenzene; 1,3-dicyanobenzene; 1,3-benzenedicarbonitrile).
CAS: 626-17-5. $C_8H_4N_2$.
Properties: Needles. Mw 128.14, mp 141C, bp sublimes. Slightly soluble in water; insoluble in ether, acetone, hot ligroin, very soluble in alc, benzene, and chloroform.
Hazard: Eye and upper respiratory tract irritant.
Use: Intermediate.

phthalonitrile.
(o-dicyanobenzene).
CAS: 91-15-6. $C_6H_4(CN)_2$.
Properties: Buff-colored crystals. Mp 138C. Insoluble in water; soluble in acetone and benzene. Combustible.
Derivation: Vapor-phase reaction of ammonia and phthalic anhydride over alumina catalyst at high temperature.
Hazard: Toxic by ingestion.
Use: Intermediate in organic synthesis (especially pigments and dyes), base material for high-temperature lubricants and coatings, insecticide.

phthaloyl chloride.
(phthaloyl dichloride; phthalyl chloride). $C_6H_4(COCl)_2$.
Properties: Colorless, oily liquid. Mp 16C, bp 277C, refr index 1.568 (20C). Decomposed by water or alcohol; soluble in ether. Combustible.
Derivation: By the action of phosphorus pentachloride on phthalic anhydride.
Hazard: Irritating by inhalation and skin contact.
Use: Chemical intermediate, especially for plasticizers and resins; laboratory reagent.
See isophthaloyl chloride.

n-phthalyl-dl-glutamin.
CAS: 3343-29-1. $C_{13}H_{12}N_2O_5$.
Hazard: A reproductive hazard.

phycocolloid.   One of several carbohydrate polymers (polysaccharides) occurring in algae (seaweed). They are hydrophilic colloids having a tendency to absorb water, with swelling, and to form gels of varying strength and consistency. The chief types of phycocolloid are carrageenan from Irish moss, algin from brown algae, and agar from red algae. They contain complex galactose and mannose sugars and are sometimes considered seaweed mucilages.

phylanthoside.
CAS: 63166-73-4. $C_{40}H_{52}O_{17}$.
Hazard: A poison.

Phyllanthus abnormis.   (spurge). A small annual or perennial herb (family Euphorbiaceae) of western and southwestern Texas.
Hazard: Nephrotoxic; hepatotoxic.

phylloerythrin. $C_{33}H_{34}N_4O_3$. A photosensitizing porphyrin pigment that is an ordinary product of the anaerobic digestion of chlorophyll in the forestomaches of ruminants. It is normally rapidly excreted via bile.

phylogenetics.   Field of biology that deals with the relationships between organisms. It includes the discovery of these relationships, and the study of the causes behind this pattern.

phylogeny.   The evolutionary relationships among organisms; the patterns of lineage branching produced by the true evolutionary history of the organisms being considered.

physical chemistry.   Application of the concepts and laws of physics to chemical phenomena in order to describe in quantitative (mathematical) terms a vast amount of empirical (observational) information. A selection of only the most important concepts of physical chemistry would include the electron wave equation and the quantum mechanical interpretation of atomic and molecular structure, the study of the subatomic fundamental particles of matter, application of thermodynamics to heats of formation of compounds and the heats of chemical reaction, the theory of rate processes and chemical equilibria, orbital theory and chemical bonding, surface chemistry (including catalysis and finely divided particles) the principles of electrochemistry and ionization. Although physical chemistry is closely related to both inorganic and organic chemistry, it is considered a separate discipline.

physical constants.   Measured values of properties expressible in strictly physical terms; for the purpose of chemistry, they concern characteristics of chemical substances, such as atomic weight, vapor density, etc.

physical map.   A map of the locations of identifiable landmarks on DNA (e.g., restriction-enzyme cutting sites, genes), regardless of inheritance. Distance is measured in base pairs. For the human genome, the lowest-resolution physical map is the banding patterns on the 24 different chromosomes; the highest-resolution map is the complete nucleotide sequence of the chromosomes.

physiological salt solution.   A solution of sodium chloride and water (0.9%) that is identical with the concentration found in the body. Also called isotonic salt solution.
Use: Medicine to replace acute loss of water as from burns, etc.

physiologic antidote.   An agent that produces systemic effects that oppose those of the specific toxicant under treatment.

physiological saline.   (physiological saline solution; physiological salt solution). An isotonic aqueous solution of salts at a concentration that is isotonic with the tissue or body fluids of a particular type or organism.
Use: To bathe and temporarily maintain living cells in vitro, or as the carrier of drugs or other substances that are administered by injection.

physostigmine.   (eserine; calabarine).
CAS: 57-47-6. $C_{15}H_{21}O_2N_3$. An alkaloid.
Properties: Colorless or pinkish crystals. Mp 86–87C and 105–106C (unstable and stable forms), specific rotation −119 to −121 degrees. Slightly soluble in water; soluble in alcohol and dilute acids.
Derivation: By solvent extraction from the seeds of *Physostigma venenosum.*
Grade: USP.
Hazard: Toxic by ingestion.
Use: Medicine (anticholinesterase). Available as salicylate and sulfate.

phytane.   (2,6,10,14-tetramethylhexadecane).
$C_{20}H_{42}$. A hydrocarbon found in rock specimens 2.5–3 billion years old. Known to be synthesized only by living organisms (is a derivative of chlorophyll) and to withstand heat and pressure, so helps to date the existence of life on earth.
See pristane.

phytic acid.   (inositolhexaphosphoric acid).
CAS: 83-86-3. $C_6H_6[OPO(OH)_2]_6$. Occurs in nature in the seeds of many cereal grains, generally as the insoluble calcium-magnesium salt. It inhibits absorption of calcium in the intestine.
Properties: White to pale-yellow liquid; odorless; acid taste. D 1.58, bulk d 13.1 lb/gal. Soluble in water and alcohol, pH less than 1.0 (in 1% solution).
Derivation: From corn steep liquor.
Grade: Technical (as a 70% solution).

Use: Chelation of heavy metals in processing of animal fats and vegetable oils, rust inhibitor, preparation of phytate salts, metal cleaning, treatment of hard water, nutrient.

phytochemistry.   That branch of chemistry dealing with (1) plant growth and metabolism and (2) plant products. The former includes the absorption of inorganic nutrients (nitrogen, phosphorus, potassium, carbon dioxide, water, etc.) to form sugars, starches, proteins, fats, vitamins, etc. and is closely associated with photosynthesis. Plant products comprise a vast group of natural materials and chemicals; besides those used directly as foods, these include alkaloids, cellulose, lignin, dyes, glucosides, essential oils, resins, gums, tannins, rubbers, terpene hydrocarbons, and glycerides (fats and oils). Some of these are basic raw materials for industry (paper, pharmaceuticals, food, paint, perfume, flavoring, leather, rubber); there are also many miscible plant products such as drugs, poisons, and pigments. Phytochemistry also embraces the study of plant hormones or growth regulators (auxin, gibberellin, synthetic types).

phytocide.   A chemical.
Use: To destroy and control plants.

phytohormone.   (plant hormone). Any substance normally produced by a plant at a location removed from the site of action within the plant and is active in minute amounts. There are three types: auxins, cytokinins, and gibberellins.

phytol.
CAS: 150-86-7. $C_{20}H_{40}O$. An alcohol obtained by the decomposition of chlorophyll.
Properties: Odorless liquid. BP 202–204C (10 mm Hg), d 0.8478 (25/4C). Soluble in the common organic solvents; insoluble in water. Combustible.
Use: Synthesis of vitamins E and K.

phytonadione.   (2-methyl-3-phytyl-1,4-naphthoquinone; vitamin K_1).
$CH_3C_{10}H_4O_2C_{20}H_{39}$.
Properties: Clear, yellow, viscous liquid; odorless. D 0.967 (25/25C), refr index 1.5230–1.5252 (25C), stable in air. Protect from sunlight!! Insoluble in water; soluble in benzene, chloroform, and vegetable oils; slightly soluble in alcohol.
Derivation: Synthetically from 2-methyl-1,4-naphthoquinone and phytol.
Grade: USP.
Use: Food supplement.

phytosterol.   Sterol alcohols from plants.
See sterol.

pi bond.   A covalent bond formed between atoms by electrons moving in orbitals that extend above and below the plane of an organic molecule containing double bonds. A double bond consists of one pi and one sigma bond, and a triple bond consists of one sigma and two pi bonds.
See metallocene; orbital theory.

pibutidine hydrochloride.
CAS: 126463-66-9. Mf: $C_{19}H_{24}N_4O_3 \cdot ClH$.
Hazard: Moderately toxic by ingestion. A reproductive hazard.

pickle alum.   See aluminum sulfate.

pickle liquor.   A waste solution of metal salts resulting from the cleaning of metals with acids.

pickling.   (1) Removal of scale, oxides, and other impurities from metal surfaces by immersion in an inorganic acid, usually sulfuric, hydrochloric, or phosphoric. Rate of scale removal varies inversely with concentration and temperature; the usual concentration is 15% at or above 100C. The rate is also increased by electrolysis. (2) A method of food preservation involving use of salt, sugar, spices, and organic acids (acetic). (3) Preserving or preparing hides for tanning by immersion in a 6–12% salt solution, together with enough acid to maintain pH at 2.5 or less.

picloram.   (4-amino-3,5,6-trichloropicolinic acid).
CAS: 1918-02-1. $C_6H_3Cl_3N_2O_2$.
Properties: Crystalline solid. Mp 218C.
Hazard: Toxic by ingestion and inhalation. Use has been restricted. Liver and kidney damage. Questionable carcinogen.
Use: Herbicide and defoliant.

pico-.   Prefix meaning 10^{-12} unit (symbol = p); e.g., 1 pg = 1 picogram = 10^{-12} gram.

picocurie.   One trillionth (10^{-12}) of a curie. A standard measure of the intensity of radiation.

α-picoline.   (2-methylpyridine; 2-picoline).
CAS: 109-06-8.

$C_5H_4N(CH_3)$ or $\overline{NC(CH_3)CHCHCH}$.

Properties: Colorless liquid; strong unpleasant odor. D 0.952, bp 129C, fp −69.9C, refr index 1.4957 (20C), miscible with water and alcohol, flash p 102F (39C) (OC), autoign temp 1000F (537C). Combustible.
Derivation: From cyclohexylamine plus ammonia and zinc chloride, also from coal tar and bone oil.
Hazard: Moderate fire risk. Irritant.
Use: Organic intermediate for pharmaceuticals, dyes, rubber chemicals, solvent, source for vinyl pyridine, laboratory reagent.

β-picoline.　(3-methylpyridine; 3-picoline).
CAS: 108-99-6.

Properties: Colorless liquid; unpleasant odor. Bp 143.5C, fp −18.3C, d 0.9613 (15/4C), refr index 1.5060 (20C). Soluble in water, alcohol, and ether. Combustible.
Derivation: From cyclohexylamine plus ammonia and zinc chloride, also from coal tar and bone oil.
Hazard: Moderate fire risk. Irritant.
Use: Solvent in synthesis of pharmaceuticals, resins, dyestuffs, rubber accelerators, insecticides, preparation of nicotinic acid and nicotinic acid amide, waterproofing agents, laboratory reagent.

γ-picoline.　(4-methylpyridine; 4-picoline).
CAS: 108-89-4.

NCHCHC(CH$_3$)CHCH.

Properties: Colorless, moderately volatile liquid. D 0.957 (15/4C), bp 144.9C, refr index 1.5050 (20C), mp 3.8C. Soluble in water, alcohol, and ether. Flash p 134F (56.6C) (OC). Combustible.
Derivation: From cyclohexylamine plus ammonia and zinc chloride, also from coal tar and bone oil.
Hazard: Moderate fire risk. Irritant.
Use: Solvent in synthesis of pharmaceuticals, resins, dyestuffs, rubber accelerators, pesticides and waterproofing agents, laboratory reagent, making isoniazid, catalyst, curing agent.

picoline α.　(α-picoline; 2-methylpyridine). C$_6$H$_7$N.
Properties: A pyridine derivative; colorless liquid with a strongly disagreeable odor.
Derivation: Component of coal tar and bone oil.
Use: An industrial solvent and chemical intermediate in the manufacture of dyes and resins.

picoline-N-oxide.　(2-picoline-N-oxide).

N(O)C(CH$_3$)CHCHCHCH.

Properties: Crystals. Mp (2-isomer) 49.5C, (3-isomer) 40.5C, (4-isomer) 186.3C. Very soluble in water. Combustible.
Use: Organic synthesis.

4-picolylamine.　(CH$_2$NH$_2$)CHCHNCHCH. A heterocyclic compound, pyridine derivative, highly reactive, strong base. Combustible.
Use: Manufacture of polyamides, epoxy curing agents, carbinols, and amine polyols.

picramic acid.　(picraminic acid; 2-amino-4,6-dinitrophenol; dinitroaminophenol).

CAS: 96-91-3.　C$_6$H$_2$(NO$_2$)$_2$(NH$_2$)OH.

Properties: Red crystals. Mp 168C. Soluble in alcohol, benzene, glacial acetic acid, aniline, and ether; sparingly soluble in water.
Derivation: By partial reduction of picric acid.
Hazard: May explode when shocked or heated, dangerous fire risk.
Use: Azo dyes, reagent for albumin.

picramide.　Legal label name (Air) for trinitroaniline.

picraminic acid.　See picramic acid.

picric acid.　(picronitric acid; trinitrophenol; nitroxanthic acid; carbazotic acid; phenoltrinitrate). CAS: 88-89-1.　C$_6$H$_2$(NO$_2$)$_3$OH.

Properties: Yellow crystals; very bitter taste. D 1.767, mp 122C, explodes at 300C. Soluble in water, alcohol, chloroform, benzene, and ether.
Derivation: Nitration of phenolsulfonic acid, obtained by heating phenol with concentrated sulfuric acid.
Grade: Technical paste, pure paste.
Hazard: Severe explosion risk when shocked or heated, especially reactive with metals or metallic salts. Toxic by skin absorption. Skin sensitization, dermatitis and eye irritant.
Use: Explosives, matches, electric batteries, etching copper, mordant in textile dyeing, reagent, picrates.

picrolonic acid.　(3-methyl-4-nitro-1-(p-nitrophenyl)-5-pyrazolone). NO$_2$C$_6$H$_4$NNC(CH$_3$) C(NO$_2$)COH.
Properties: Yellow leaflets. Mp 116–117C, decomposes 125C. Slightly soluble in water; soluble in alcohol.
Use: Reagent for alkaloid identifications; tryptophan and phenylalanine; detection and estimation of calcium.

picronitric acid.　See picric acid.

picropodophyllotoxin.
CAS: 477-47-4. $C_{22}H_{22}O_8$.
Hazard: A poison.
Source: Natural product.

picrotoxin. (cocculin).
CAS: 124-87-8. $C_{30}H_{34}O_{13}$. A glucoside.
Properties: Flexible, shining, prismatic crystals or microcrystalline powder; odorless; very bitter taste. Stable in air; affected by light. Mp 200C. Soluble in boiling water, boiling alcohol, dilute acids, and alkalies; sparingly soluble in ether and chloroform.
Derivation: Derived from the fruit of *Anamirta paniculata* or *Cocculus indicus,* fishberries.
Hazard: Toxic in overdose.
Use: Medicine, as CNS stimulant and antidote for barbiturate poisoning.

picryl chloride. (2-chloro-1,3,5-trinitrobenzene).
CAS: 88-88-0. $C_6H_2(NO_2)_3Cl$.
Hazard: Severe explosion and fire risk. A high explosive.

Pictet–Gams isoquinoline synthesis. Formation of isoquinolines by cyclization of acylated aminomethyl phenyl carbinols or their ethers with phosphorus pentoxide in toluene or xylene.

Pictet–Hubert reaction. Phenanthridine cyclization by dehydrative ring closure of acyl-*o*-aminobiphenyls on heating with zinc chloride at 250–300C or with phosphorus oxychloride in boiling nitrobenzene.

Pictet–Spengler isoquinoline synthesis. Formation of tetrahydroisoquinoline derivatives by condensation of β-arylethylamines with carbonyl compounds and cyclization of the Schiff bases formed.

Pictol. Monomethyl-*p*-aminophenol sulfate, a photo developer.

PIDA. Abbreviation for phenylindane dicarboxylic acid.
See 1,1,3-trimethyl-5-carboxy-3-(*p*-carboxyphenyl)indane.

Pidgeon process. (ferrosilicon process; silicothermic process). Process for the production of high-purity magnesium metal from dolomite or magnesium oxide by reduction with ferrosilicon at 1150C under high vacuum.

piezochemistry. Study of reactions occurring at very high pressures, e.g., in the interior of the earth's crust.

piezoelectricity. Electric energy created by application of pressure to ceramics or plastics. Devices utilizing this phenomenon are gas flame igniters, ultrasonic welding tools, and sonar navigation aids.

pig iron. Product of blast-furnace reduction of iron oxide in the presence of limestone. About half the ore is converted to iron. Average analysis is 1% silicon, 0.03% sulfur, 0.27% phosphorus, 2.4% manganese, 4.6% carbon, balance iron. Pig iron is the basic raw material for steel and cast iron. In metal terminology a "pig" is a bar or ingot of cooled metal.
See iron.

pigment. Any substance, usually in the form of a dry powder, that imparts color to another substance or mixture. Most pigments are insoluble in organic solvents and water; exceptions are the natural organic pigments, such as chlorophyll, which are generally organosoluble. To qualify as a pigment, a material must have positive colorant value. This definition excludes whiting, barytes, clays, and talc. See fillers.
Some pigments (zinc oxide, carbon black) are also reinforcing agents, but the two terms are not synonymous; in the parlance of the paint and rubber industries these distinctions are not always observed.
Pigments may be classified as follows:

(i) Inorganic
  (a) Metallic oxides (iron, titanium, zinc, cobalt, chromium)
  (b) Metal powder suspensions (gold, aluminum)
  (c) Earth colors (siennas, ochers, umbers)
  (d) Lead chromates
  (e) Carbon black
(ii) Organic
  (a) Animal (rhodopsin, melanin)
  (b) Vegetable (chlorophyll, xanthophyll, indigo, flavone, carotene)
  See pigment, plant
  (a) Synthetic (phthalocyanine, lithos, toluidine, para red, toners, lakes, etc.)
  See dye, natural; dye, synthetic.

Pigment Blue 15. (CI 74160). $C_{32}H_{16}N_8Cu$. A bright-blue copper phthalocyanine pigment.
Derivation: By heating phthalonitrile with cuprous chloride.
Use: In paints, alkyd resin enamels, printing inks, lacquers, rubber, resins, papers, tinplate printing, colored chalks, and pencils.
See phthalocyanine.

Pigment Blue 19. (CI 42750A).
$C_{32}H_{28}N_3O_4SNa$. A bright-blue to bright-reddish-navy triphenylmethane pigment.
Use: Coloring for candles.

Pigment Blue 24.   (CI 42090).
$C_{37}H_{34}N_2O_9S_3Na_2$. A bright-greenish-blue triaryl-methane pigment.
Use: Printing inks, especially for tinplate printing; rubber; plastics; artist colors; lacquers.

Pigment E.   See barium potassium chromate.

Pigment Green 7.   (CI 74260).
$C_{32}O_{0-1}N_8Cl_{15-16}Cu$. A bright-green chlorinated copper phthalocyanine pigment.
Derivation: Heating copper phthalocyanine in sulfur dichloride under pressure.
Use: Paints, printing inks, lacquers, leather and book cloth, paper surfacing, chalks, colored pencils.

pigment, plant.   Any of a large number of organic natural colorants produced by living plants, with the exception of fungi and lichens.
They may be classified into three groups:
(1) The chlorophylls (types a, b, and c): Green color; they are magnesium-containing porphyrins and are technically considered to be microcrystalline waxes
(2) The carotenoids: yellow and orange colors
  (a) Carotene (straight-chain hydrocarbon)
  (b) Xanthophyll (straight-chain hydrocarbon containing two oxygen atoms)
(3) The flavanoids: red, yellow, blue, orange, ivory colors; they are oxygen-containing heterocyclic compounds
  (a) Catechins
  (b) Flavones, flavanols, anthocyanins
  (c) Flavanones and leucoanthocyanidins
  (d) Flavonols
Some of these pigments can be made synthetically. They have limited use as textile colorants and pharmaceutical products.

pigment, precipitated.   See lake.

pigment volume concentration.   See PVC (2).

Pigment Yellow 12.   (CI 21090).
CAS: 6358-85-6. $C_{32}H_{26}Cl_2N_6O_4$. A yellow diazo pigment.
Method of purification: Condensation of 3,3′-dichlorobenzidine di-diazotate with acetoacetanilide.
Use: Printing inks, lacquers resistant to heat and solvents, in rubber and resins, in paper coloring, textile printing.
See diazotization.

Pilate Fast Dyes.   1:1 metal complex dyes for dyeing and printing textiles of animal fibers and union materials of wool and nylon fibers.

pilchard oil.   An oil expressed from the pilchard fish, a member of the herring family.

Properties: Pale-yellow liquid, deposits stearin on long standing. D 0.931–0.933, saponification value 186–189.6, refr index 1.4751 (40C). Combustible.
Use: Potash soft soap, paints.

pill press.   A press for making small compacts in powder metallurgy.

pilot plant.   A trial assembly of small-scale reaction and processing equipment that is the intermediate stage between laboratory experiment and full-scale operation in the production of a new product. The functions of this stage are (1) to furnish chemical engineers with design data needed to construct a large-scale plant, (2) to resolve the many problems inherent in conversion from batch to continuous production, (3) to eliminate the differences that accompany change from constant laboratory conditions to a less closely controlled environment, and (4) to provide management with a basis for cost evaluation and estimation of the capital requirements of the new product. Because the size of the pilot plant varies with the nature of the product, it must be determined on an individual basis.

Piloty–Robinson synthesis.   Formation of pyrroles by heating azines of enolizable ketones with acid catalysts, usually zinc chloride or hydrochloric acid.

pilsicainide hydrochloride hemihydrate.
CAS: 88069-49-2. $C_{17}H_{24}N_2O•ClH•1/2H_2O$.
Hazard: A poison by ingestion.

l-pimaric acid.   (levopimaric acid). $C_{20}H_{30}O_2$.
Properties: Solid. Mp 150C, optical rotation ([a] 20/d) −280 degrees (c = 0.7 in alcohol). Soluble in most organic solvents; insoluble in water. Combustible.
Derivation: From pine gum.
Use: Resins.
See maleo-pimaric acid.

pimelic acid.   (1,7-heptanedioic acid).
CAS: 111-16-0. $OOC[CH_2]_5COOH$.
Properties: Crystals. Mp 105–106C. Slightly soluble in water; soluble in alcohol and ether; nearly insoluble in cold benzene. Combustible.
Use: Biochemical research, polymers, plasticizers.

pimelic ketone.   See cyclohexanone.

pinacoid.   Crystal form that consists of exactly two parallel faces.

pinacolone.   (pinacoline; methyl-t-butyl ketone; 3,3-dimethyl-2-butanone).
CAS: 75-97-8. $CH_3COC(CH3)_3$.
Properties: Bp 106C, refr index 1.3964 (20C), d 0.801, flash p 75F (23C).

PINACOL REARRANGEMENT

1086

Pinacol rearrangement. Acid-catalyzed rearrangement of vicinal glycols to aldehydes and ketones.

pindone. Coined name for 2-pivaloyl-1,3-indandione.

α-pinene. $C_{10}H_{16}$. A terpene hydrocarbon derived from sulfate wood turpentine.
Properties: Colorless, transparent liquid; terpene odor. D 0.8620–0.8645 (15.5/15.5C), refr index 1.4655–1.4670 (20C), boiling range 95% between 156 and 160C, fp −40C, flash p 90F (32.3C) (TCC), occurs in *d*-, *l*-, and racemic forms. Insoluble in water; soluble in alcohol, chloroform, and ether.
Hazard: Flammable, moderate fire risk. Skin irritant.
Use: Solvent for protective coatings, polishes, and waxes; synthesis of camphene, camphor, geraniol, terpin hydrate, terpineol, synthetic pine oil, terpene esters and ethers, lubricating oil additives, flavoring odorant.

β-pinene. (nopinene). $C_{10}H_{16}$. A terpene hydrocarbon derived from sulfate wood turpentine.
Properties: Colorless, transparent liquid; terpene odor. Insoluble in water; soluble in alcohol, chloroform, and ether. D 0.8740–0.8770 (15.5/15.5C), refr index 1.4775–1.4790 (20C), boiling range 95% between 164 and 169C, flash p 117F (47.2C), levorotatory. Combustible.
Hazard: Fire risk.
Use: Polyterpene resins, substitute for α-pinene, intermediate for perfumes and flavorings.

pinene hydrochloride. See bornyl chloride.
pine oil.
Properties: Colorless to light amber liquid; strong, piny odor. D 0.927–0.940, refr index 1.4780–1.4820 (20C), distilling range 200–225C, flash p 172F (77.7C) (CC). Miscible with alcohol. Combustible.
Chief constituents: Tertiary and secondary terpene alcohols.
Derivation: From the wood of *Pinus palustris* by extraction and fractionation or by steam distillation; also from turpentine.
Use: Odorant, disinfectant, penetrant, wetting agent, preservative (textile and paper industries), laboratory reagent, fragrances.

pine tar.
Properties: Sticky, viscous, dark-brown to black liquid or semisolid; strong odor; sharp taste. Translucent in thin layers, hardens with aging. D 1.03–1.07, boiling range 240–400C, flash p 130F (54.4C) (CC). Soluble in alcohol, acetone, fixed and volatile oils, and in sodium hydroxide solution; slightly soluble in water. Combustible.

Chief constituents: Complex phenols, turpentine, rosin, toluene, xylene, and other hydrocarbons.
Derivation: By destructive distillation of pine wood, especially *Pinus palustris*.
Grade: Kiln burnt, retort, NF.
Hazard: Fire risk, subject to spontaneous heating.
Use: Ore flotation, roofing compositions, paints and varnishes, softener in plastics and rubber processing, tar soaps, deKhotinsky cement, asphaltic compositions, marine preservative, medicine (cough syrups), laboratory reagent.

pine tar oil. See tar oil; wood.

pine tar pitch. The residue after distillation of practically all the volatile oils from pine tar. Similar to coal-tar pitch.

pinhole. (1) A small hole in electrical insulating tape caused by failure of the rubber coating to penetrate the fabric. The acceptable number of pinholes per unit area is subject to specification. (2) In paints, the presence of pimples or tiny holes in a coating.

Pinner reaction. Formation of imino esters (alkyl imidates) by addition of dry hydrogen chloride to a mixture of a nitrile and an alcohol. Treatment of alkyl imidates with ammonia or primary or secondary amines affords amidines, while treatment with alcohols yields ortho-esters.

Pinner triazine synthesis. Preparation of 2-hydroxy-4,6-diaryl-*s*-triazines by reaction of aryl amidines and phosgene.

pinocarveol. See 1-octen-3-yl acetate.

"Pinova" [Pinova]. A pale wood rosin derivative.
Use: For adhesives, flavor, and construction applications.

pintsch gas. See oil gas.

pipecoline. See 2-methylpiperidine.

piperalin. [3-(2-methylpiperidino)propyl-3,4-dichlorobenzoate].
CAS: 3478-94-2. $CH_3C_5H_9N(CH_2)_3OC(O)C_6H_3Cl_2$.
Properties: Amber liquid. Bp 156–157C (0.2 mm Hg). Slightly soluble in water; miscible with paraffin hydrocarbon, aromatic hydrocarbon, and chlorinated hydrocarbon solvents.
Use: Fungicide.

piperazine. (diethylenediamine; pyrazine hexahydride; piperazidine).
CAS: 110-85-0.

$NHCH_2CH_2NHCH_2CH_2$

Properties: Colorless, deliquescent, transparent, needlelike crystals that absorb carbon dioxide from the air. Mp 104–107C, bp 145C, flash p 190F (87.7C). Soluble in water, alcohol, glycerol, and glycols. Combustible.
Derivation: Treatment of ethylene bromide or chloride with alcoholic ammonia at 100C.
Hazard: Respiratory sensitization and asthma. Questionable carcinogen.
Use: Corrosion inhibitor, anthelmintic, insecticide, accelerator for curing polychloroprene.

piperazine dihydrochloride.
CAS: 142-64-3. $C_4H_{10}N_2 \cdot 2HCl$.
Properties: White needles. Soluble in water.
Available forms: The monochloride, $C_4H_{10}N_2 \cdot HCl$, is commercially available.
Use: Fibers, insecticides, pharmaceuticals.

piperazine hexahydrate. $C_4H_{10}N_2 \cdot 6H_2O$.
Properties: White crystals. Mp 44C, bp 125C. Soluble in water and alcohol.
Use: Fibers, insecticides, pharmaceuticals, laboratory reagent, anthelmintic.

piperidine.   (hexahydropyridine; pentamethyleneamine).
CAS: 110-89-4.

$\overline{CH_2CH_2CH_2CH_2CH_2NH}$. Completely saturated ring compound.
Properties: Colorless liquid; odor of pepper. D 0.862, bp 106C, fp −7 to −9C. Soluble in water, alcohol, and benzene; strong base. Combustible.
Derivation: By electrolytic reduction of pyridine.
Grade: 95 and 98% pure.
Hazard: Toxic by ingestion, strong irritant.
Use: Solvent and intermediate, curing agent for rubber and epoxy resins, catalyst for condensation reactions, ingredient in oils and fuels, complexing agent.

piperidine alkaloid.   Any alkaloid that contains the piperidine ring.

piperidinepentacarbonylchromium.   See pentacarbonyl(piperidine)chromium.

piperidine pentamethylene dithiocarbamate.   (Pippip). $C_{11}H_{22}N_2S_2$.
Properties: White powder.
Hazard: Toxic by ingestion, strong irritant to eyes and skin.
Use: Ultra-accelerator for rubber.

9-(piperidinoamino)acridine.
CAS: 28846-40-4. $C_{18}H_{19}N_3$.
Hazard: A poison by ingestion.
Use: Agricultural chemical.

2-piperidinoethanol.   (*n*-2-hydroxyethylpiperidine). $C_5H_{10}NCH_2OH$.

Properties: Colorless liquid. D 0.972–0.974 (20/4C), bp 115–117C (45 mm Hg), refr index 1.478–1.480 (20C). Miscible with water and most organic solvents.
Use: Intermediate.

piperocaine hydrochloride.   (3-(2-methylpiperidyl)propyl benzoate hydrochloride).
CAS: 533-28-8. $C_{16}H_{23}NO_2 \cdot HCl$.
Properties: White, crystalline powder; odorless; bitter taste. Stable in air. Mp 172–175C. Solution (1 in 10) acid to litmus; soluble in water, alcohol, and chloroform; almost insoluble in ether and fixed oils.
Use: Medicine (local anesthetic).

piperonal.   (heliotropin; piperonyl aldehyde; 3,4-methylenedioxybenzaldehyde).
CAS: 120-57-0. $C_6H_3(CH_2)OO)CHO$ (bicyclic).
Properties: White, shining crystals; turns red brown on exposure to light; floral odor. Mp 35.5–37C, bp 263C. Soluble in alcohol and ether; slightly soluble in water and glycerol. Combustible.
Derivation: By oxidation of isosafrole.
Grade: Technical, FCC.
Use: Perfumery, suntan preparations, mosquito repellent, laboratory reagent, flavoring.

piperonyl butoxide.   (generic name for α-[2-(2-butoxyethoxy)-ethoxy]-4,5-(methylenedioxy)-2-propyltoluene).
CAS: 51-03-6. $C_3H_7C_6H_2(OCH_2O)CH_2OC_2H_4$ $OC_2H_4OC_4H_9$.
Properties: Light-brown liquid; mild odor. D 1.06 (25C), refr index 1.50 (20C), bp 180C (1 mm Hg), flash p 340F (171C). Insoluble in water; soluble in alcohol, benzene, petroleum hydrocarbons. Combustible.
Hazard: Questionable carcinogen.
Use: Synergist in insecticides in combination with pyrethrins in oil solutions, emulsions, powders, or aerosols.

piperonyl cyclonene.   (generic name for a mixture of 3-alkyl-6-carbethoxy-5-(3,4-methylenedioxyphenyl)-2-cyclohexen-1-one and 3-alkyl-5-(3,4-methylenedioxyphenyl)-2-cyclohexen-1-one).

$(CH_2O_2)C_6H_3CHCH_2C(O)CH:CRCH_2$. R is usually C_6H_{13}.
Properties: Red liquid. Insoluble in water, oils, and refrigerant 12. Flash p 290F (143C). Combustible.
Use: Synergist in insecticides in combination with rotenone, pyrethrins, or rotenone-pyrethrin mixture in oil solutions, emulsions, or powders.

Pipersin.   A substitute for oleoresin of black pepper. Officially recognized by USDA Meat Inspection Division for use under its supervision.

piperylene. (1,3-pentadiene).
$CH_2:CHCH:CHCH_3$. *cis-* and *trans-* forms.
Properties: Colorless liquid. D 0.693 (60/60F), fp
(*cis*) −141C, (*trans*) −87C; bp (*cis*) 44C, (*trans*)
42C; refr index (*cis*) 1.43634 (20C), (*trans*) 1.43008
(20C); flash p (mix) −20F (−28.8C). Insoluble in
water; soluble in alcohol and ether.
Hazard: Highly flammable, dangerous fire risk.
Use: Polymers, maleic anhydride adducts, intermediate.

pipette. (pipet). A slender glass tube open at both
ends and having an expanded area at or near the center designed to contain a specific volume of liquid,
e.g., 5 ml. Liquid is drawn into the tube by oral or,
for the sake of safety, some other form of suction.
Use: Transferring measured volumes of liquid from
one container to another.

pirfloxacin. See 6-fluoro-7-(1-pyrrolyl)-1-ethyl-
1,4-dihydro-4-oxo-3.

Piria reaction. Formation of arylsulfamic acids
or sulfonation products or both by refluxing aromatic nitro compounds with a metal sulfite and boiling the mixture with dilute acid to yield the amines
and sulfamic acids.

pirimicarb. (dimethylcarbamic acid 2-
(dimethylamino)-5,6-dimethyl-4-pyrimidinyl
ester; 2-(dimethylamino)-5,6-dimethyl-4-
pyrimidinyl dimethylcarbamate; 5,6-dimethyl-2-
dimethylamino-4-dimethylcarbamoyloxypyrimid-
ine;[2-(dimethylamino)-5,6-dimethylpyrimidin-4-
yl] *N,N*-dimethylcarbamate). $C_{11}H_{18}N_4O_2$.
Use: Insecticide used as a systematic aphicide.
Hazard: Volatile, very toxic.

piroctone olamine.
CAS: 68890-66-4. $C_{14}H_{23}NO_2 \cdot C_2H_7NO$.
Hazard: Low toxicity by ingestion. A reproductive
hazard.

"Piror" *[Union].* TM for high-performance
microbiocides based on glutaraldehyde.
Use: Papermaking, industrial-process and cooling
water systems, and industrial preservatives.
See glutaraldehyde.

pitch. (1) A carbonaceous, tacky residue resulting
from distillation of coal tar, petroleum, pine tar, and
fatty acids. Some types, such as glance pitch, occur
naturally. They are used chiefly as sealants, roofing compounds, and wood preservatives. Synthetic
carbon fibers are made from petroleum pitch. (2)
In papermakers' terminology, a mixture of calcium
carbonate, calcium soaps from wood components,
and miscellaneous residues from materials used in
paper manufacture. Pitch of this type is a production
nuisance that requires close control. (3) The degree
of slope of an inclined plane as in a screw or auger,
as measured by the distance between the flights or
threads.

pitchblende. A massive variety of uraninite or
uranium oxide found in metallic veins. Contains
55–75% UO_2, up to 30% UO_3, usually a little water,
and varying amounts of other elements. Thorium
and the rare earths are generally absent.
Properties: Black color, brownish-black streak,
pitchy to dull luster. Mohs hardness 5.5, d 6.5–8.5.
Occurrence: Canada, Colorado, Europe, Zaire.
Hazard: Radioactive material.
Use: Most important ore of uranium, original source
of radium.

"Pittclor" *[Axiall].*
CAS: 7778-54-3. TM for calcium hypochlorite.
Use: Water sanitizer.

pituitrin. (coluitrin; hypophysin; pitilobin).
Use: Antidiuretic hormone.

Pitzer equation. Equation for the approximation of data for heats of vaporization for organic
and simple inorganic compounds. It is derived from
temperature and reduced temperature relationships.

pivalic acid. See trimethylacetic acid.

pivalic acid, sodium salt. See sodium 2,2-
dimethylpropanoate.

2-pivaloyl-1,3-indandione. (pivalyl-1,3-
indandione; pindone).
CAS: 83-26-1. $C_9H_5O_2C(O)C(CH_3)_3$.
Properties: Bright-yellow powder or crystals. Mp
109C. Insoluble in water; soluble in most organic
solvents.
Hazard: Toxic by inhalation and ingestion; inhibits
blood clotting.
Use: Rodenticide, insecticide, pharmaceutical intermediate.

Pizazz. A firm and uniform cheese.
Use: Minimizes fines when shredded and reduces
loss.

pK. A measure of the completeness of an incomplete chemical reaction. It is defined as the negative
logarithm (to the base 10) of the equilibrium constant, K, for the reaction in question. The pK is
most frequently used to express the extent of dissociation or the strength of weak acids (particularly
fatty acids), amino acids, and complex ions or similar species. The weaker an electrolyte, the larger is
its pK. Thus, at 25C for sulfuric acid (strong acid),
pK is about −3.0; acetic acid (weak acid), pK =
4.76; boric acid (very weak acid) pK = 9.24. In
a solution of a weak acid, if the concentration of
undissociated acid is equal to the concentration of
the anion of the acid, the pK will be equal to the
pH.

placental barrier. (placental membrane). The mammalian placenta physically separates fetal blood from maternal blood and limits the passage of large blood-borne molecules between the two. In at least some species, this barrier impedes the passage of many toxicants to some extent. Its effectiveness, probably varies considerable among placental mammals as a function of the number of layers (one of six).

plain agar. Ordinary agar with some variation dictated by the nutrient employed.
Properties: Contains 15 g agar, 10 g peptone, 5 g sodium chloride, 1000 ml bouillon stock solution, neutralized with sodium hydroxide, and filtered while hot.

Plait point. Composition conditions in which the three coexisting phases of partially soluble components of a three-phase liquid system approach each other in composition.
See triple point.

plamotoxin B_0. A mycotoxin.
Derivation: Produced by *Aspergillus flavus*.

plamotoxin G_0. A mycotoxin.
Derivation: Produced by *Aspergillus flavus*.

Planck's constant. (h). A constant that when multiplied by the frequency of radiation gives the quantity of energy contained in one quantum. Equal to $6.626176(36) \times 10^{-19}$ JHz^{-1}.

planetology, chemical. See chemical planetology.

plankton. Microscopic plant and animal life that floats in the oceans or in lake waters.

plant. (1) Any large-scale manufacturing unit including pipelines, reaction equipment, machinery, etc.
(2) A broad group of vegetable organisms composed of all types of vegetation that contain chlorophyll (algae, mosses, grasses, vegetables, trees, etc. but excluding fungi). Their metabolic processes are vital to the maintenance of life on earth and result in the following products: (1) oxygen (from respiration), (2) carbohydrates (from photosynthesis), (3) amino acids and proteins (from nitrates and nitrogen-fixing bacteria), (4) fats and oils, (5) vitamins, (6) natural fibers, (7) coal, (8) various other substances of value such as alkaloid drugs, rubber, etc.
See photosynthesis; phytochemistry.

plant agglutinin. (phytoagglutinin). A lectin of plant origin.

plant alkaloid. Any alkaloid that occurs naturally in plants.

plant growth regulator. An organic compound either natural or synthetic that modifies or controls one or more specific physiological processes within the plant. If the compound is produced by the plant it is called a plant hormone, e.g., auxin, which regulates the growth of longitudinal cells involved in bending of the stem one way or another. Substances applied externally also bring about modifications such as improved rooting of cuttings, increased rate of ripening (ethylene), and easier scission (separation of fruit from stem). A large number of chemicals tend to increase the yield of certain plants such as sugarcane, corn, etc. All these, as well as plant-produced hormones, are included in the term *plant growth regulator*.
See dinitrobutylphenol; kinin; gibberellin; abscisic acid.

plant location. Selection of a site for a new chemical or process industry plant. The problem has been compounded in recent years by the increasing number of environmental regulations and energy shortage. Among the more important considerations are (1) accessibility of essential materials, including water; (2) transportation of finished product (rail, air, truck, barge); (3) reliability of fuel and power supply; (4) liquid and solid waste-disposal restrictions; (5) commuting distance for employees in view of gasoline consumption; (6) availability of housing for employees; (7) state and local regulations (zoning, hazardous chemicals, building codes); (8) availability of qualified labor; (9) taxation; (10) weather factors (temperature range, severe storms, floods, etc.); (11) expansion possibilities.

plant pigment. (vegetable pigment). Any of a large variety of organic pigments that are produced by plants.

"Plas-Chek" [Ferro]. TM for plasticizers.

"Plasdone" [ISP]. TM for the pharmaceutical grade of polyvinylpyrrolidone.
Use: Tablet binding and coating agent, detoxicant and demulcent lubricant in ophthalmic preparations, film-forming agent in medical aerosols.

plasma. (1) The portion of the blood remaining after removal of the white and red cells and the platelets; it differs from serum in that it contains fibrinogen, which induces clotting by conversion into fibrin by activity of the enzyme thrombin. Plasma is made up of more than 40 proteins and also contains acids, lipids, and metal ions. It is an amber, opalescent solution in which the proteins are in colloidal suspension and the solutes (electrolytes and nonelectrolytes) are either emulsified or in true solution. The proteins can be separated from each other and from the other solutes by ultrafiltration, ultracentrifugation, electrophoresis, and immunochemical techniques. (2) Two kinds of plasma are recognized by physicists, namely, a particle plasma and a reactor plasma. A particle plasma is a neutral mixture of positively and negatively charged particles interacting with an electromagnetic field, which dominates their motion. Temperatures of 10,000 to 15,000C can be reached. Such plasma, formed by sudden energy releases can be utilized as an energy source, as in magnetohydrodynamics. Reactor plasmas, on the other hand, are composed of positively charged ions of hydrogen isotopes (deuterium, tritium); the electric charge is the controlling factor. These are used in nuclear fusion devices, where temperatures of 74,000,000C have been attained and still higher temperatures are expected. These plasmas also respond to electromagnetic forces that are used to confine them.

See magnetohydrodynamics; fusion; tokamak.

plasmalogen.   A phospholipid with an alkenyl ether substituent on the C-1 of glycerol.

plasma membrane.   Outer membrane of a cell, sometimes called the cell membrane.

plasma proteins.   The proteins present in blood plasma—albumin, antibodies, etc.

plasma volume expander.   A substance used to partially or wholly replace blood plasma in treatment of the injured. Most important are gelatin, polyvinylpyrrolidone, and dextran.

plasmid.   A strand or fragment of genetic material existing outside the chromosomes in certain types of bacteria. R-type plasmids, which are present in *Escherichia coli*, impart resistance to antibiotics in organisms that are exposed to them. The plasmids can be transferred from animals to humans, as well as to other, harmful bacteria that also become resistant to antibiotics. Feeding of traces of antibiotics to animals is believed to promote the growth of *E. coli* and, thus, to produce strains of pathogenic bacteria that are not amenable to antibiotic treatment. For this reason FDA has recommended elimination of certain antibiotics from animal feeds, e.g., penicillin, oxytetracycline, and chlortetracycline. Synthetic plasmids have been used successfully in recombinant DNA research.

plasmin.   See fibrinolysin.

plasminogen activator (human tissue-type protein moiety reduced), *n*-(n2-(*n*-glycyl-*l*-alanyl)-*l*-arginyl)-, glycoform.   See silteplase.

plasmogamy.   A process of fusion of the cytoplasm of two cells; the first step in syngamy.

plasmoquin.   (pamaquine; plasmochin; 8-dimethylamino-isoamyl-6-methoxyquinoline). $C_{19}H_{28}N_2O$.

Properties: Yellow powder. Mw 300.2. Insoluble in water.
Use: Antimalarial.

Plastacele.   Cellulose acetate flake, a fine white powder used for molding powders, films, sheets, rods, and tubes.

plaster of Paris.   See calcium sulfate.

"Plasthall" *[Hallstar]*.   TM for a broad range of monomeric and polymeric plasticizers used in polymers and elastomers. Types include adipates, glutarates, trimellitates, azelates, sebacates, and tallates.

plastic.   (1) Capable of being shaped or molded with or without the application of heat. Soft waxes and moist clay are good examples of this property. See plasticity. (2) A high polymer, usually synthetic, combined with other ingredients, such as curatives, fillers, reinforcing agents, colorants, plasticizers, etc.; the mixture can be formed or molded under heat and pressure in its raw state and machined to high dimensional accuracy, trimmed, and finished in its hardened state. The thermoplastic type can be resoftened to its original condition by heat; the thermosetting type cannot.

Plastics in general (including all forms) are sensitive to high temperatures, among the more resistant being fluorocarbon resins, nylon, phenolics, polyimides, and silicones, though even these soften or melt above 260C. Other types (cellulosics, polyethylene, acrylic polymers, polystyrene) are combustible when exposed to flame for a short time and still others (polyurethane) burn with evolution of toxic fumes.

Engineering plastics are those to which standard metal engineering equations can be applied; they are capable of sustaining high loads and stresses and are machinable and dimensionally stable. They are used in construction, as machine parts, automobile components, etc. Among the more important are nylon, acetals, polycarbonates, ABS resins, PPO/styrene, and polybutylene terephthalate.

Fibers, films, and bristles are examples of extruded forms. Plastics may be shaped by either compression molding (direct pressure on solid material in a hydraulic press) or injection molding (injection of a measured amount of material into a mold in liquid form). The latter process is most generally used, and articles of considerable size can be produced. Because of their dielectric properties, plastics are essential components of electrical and electronic equipment (especially for use within the human body).

Plastics can be made into flexible and rigid foams by use of a blowing agent; these foams are light and strong, and the rigid type is machinable. They are collectively called cellular plastics. Plastics can also be reinforced, usually with glass or metallic fibers, for added strength. They are laminated to paper, cloth, wood, etc. for many uses in the packaging, electrical, and furniture industries; they also can be metal plated. Plastic pipe is widely used for underground transportation of gases and liquids over long distances and intraplant.

Several natural materials (waxes, clays, and asphalts) have rheological properties similar to synthetic products, but because they are not polymeric, are not considered true plastics. Certain proteins (casein, zein) are natural high polymers from which plastics are made (buttons and other small items), but they are of decreasing importance.

Plastics have permeated industrial technology. Not only have they replaced and improved upon many materials formerly used, but they also have made possible industrial and medical applications that would have been impracticable with older technologies. Their major application areas are (1) automobile bodies and components, boat hulls; (2) building and construction (siding, piping insulation, flooring); (3) packaging (vapor-proof barriers, display cartons, bottles, drum linings); (4) textiles (carpets, cordage, suiting, hosiery, drip-dry fabrics, etc.); (5) organic coatings (paint and varnish vehicles); (6) adhesives (plywood, reinforced plastics, laminated structures); (7) pipelines; (8) electrical and electronic components; (9) surgical implants; (10) miscellaneous (luggage, toys, tableware, brushes, furniture, etc.).

The Nobel Prize was awarded in 2000 for work on developing conducting plastics.

For additional information refer to The Society of the Plastics Industry.

See polymer, high; cellular plastic; reinforced plastic; foam, plastic; plastic pipe. See Appendix III for a history of the industry.

plastic film. A thermoplastic film less than 0.022 cm (0.010 inch) in thickness.

plastic flow. A type of rheological behavior in which a given material shows no deformation until the applied stress reaches a critical value called the yield value. Most of the so-called plastics do not exhibit plastic flow. Common putty is an example of a material having plastic flow.

plastic foam. See foam, plastic; cellular plastic.

plasticity. A rheological property of solid or semisolid materials expressed as the degree to which they will flow or deform under applied stress and retain the shape so induced, either permanently or for a definite time interval. It may be considered the reverse of elasticity. Application of heat and/or special additives is usually required for optimum results.

See thermoplastic; plasticizer.

plasticizer. An organic compound added to a high polymer both to facilitate processing and to increase the flexibility and toughness of the final

product by internal modification (solvation) of the polymer molecule. The latter is held together by secondary valence bonds: the plasticizer replaces some of these with plasticizer-to-polymer bonds, thus aiding movement of the polymer chain segments. Plasticizers are classed as primary (high compatibility) and secondary (limited compatibility). Polyvinyl chloride and cellulose esters are the largest consumers of plasticizers; they are also used in rubber processing. Among the more important plasticizers are nonvolatile organic liquids and low melting solids (e.g., phthalate, adipate, and sebacate esters), polyols such as ethylene glycol and its derivatives, tricresyl phosphate, castor oil, etc. Camphor was used in the original modification of nitrocellulose to celluloid.

See plastisol; softener.

plastic pipe. Tubes, cylinders, conduits, and continuous length piping made (1) from thermoplastic polymers unreinforced (polyethylene, polyvinyl chloride, ABS polymers, polypropylene) or (2) from thermosetting polymers (polyesters, phenolics, epoxies) blended with 60–80% of such reinforcing materials as chopped asbestos or glass fibers to increase strength. The latter type is a reinforced plastic. In general the properties of plastic tubing or pipe are those of the polymers that comprise it. Most have good resistance to chemicals, corrosion, weathering, etc., combined with flexibility, light weight, and high strength. They are combustible but generally slow burning. The reinforced type is widely used as underground conduit for transportation of gases and fluids, including city water services, sewage disposal systems, etc. Its use in buildings is subject to local building codes.

plastic, reinforced. See reinforced plastic.

plastid. Any of several pigmented cytoplasmic organelles found in plant cells and other organisms having various physiological functions, such as the synthesis and storage of food.

plastisol. A dispersion of finely divided resin in a plasticizer. A typical composition is 100 parts resin and 50 parts plasticizer, forming a paste that gels when heated to 150C as a result of solvation of the resin particles by the plasticizer. If a volatile solvent is included, the plastisol is called an organosol. Plastisols are used for molding thermoplastic resins, chiefly polyvinyl chloride.

See plasticizer.

Plastisol Ink. All-purpose ink for printing on cotton and polyester fabric.

plate column. Distillation column consisting of a number of perforated, equally spaced, horizontal plates on which a layer of liquid is maintained.

plate efficiency. (1) The number of theoretical plates that are divided by the number of plates actually used in a tower. (2) Overall plate efficiency refers to the number of equilibrium states necessary for a given separation, divided by the number of actual plates required.

platelet. (thrombocyte). A proteinaceous cellular structure occurring in blood in the amount of 150–500 × 10³ units/mm³. Platelets range from 2 to 4 μm in diameter and contain no nuclei. They are rich in amine compounds, which constrict the blood vessels at the site of an injury, to which the platelets adhere; on dissolution they release thromboplastin, which initiates the coagulation mechanism.
See blood; fibrinogen; thrombin.

platen. A vertically movable plate (deck) of a compression molding press.
See hydraulic press.

platforming. The process in which octane ratings of gasoline are raised by dehydrogenating naphthenes to aromatics, cracking high-boiling paraffins, and isomerizing paraffins to form products of greater chain branching. Desulfurization also takes place in this process.

platinic. Refers to compounds containing tetravalent platinum.

platinic ammonium chloride. See ammonium hexachloroplatinate.

platinic chloride. See chloroplatinic acid, platinum chloride.

platinic oxide. See platinum dioxide.

platinic sal ammoniac. See ammonium hexachloroplatinate.

platinic sodium chloride. See sodium chloroplatinate.

platinic sulfate. See platinum sulfate.

platinous ammonium chloride. See ammonium chloroplatinate.

platinous chloride. See platinum dichloride.

platinous iodide. See platinum iodide.

platinum.
CAS: 7440-06-4. Pt. Metallic element of atomic number 78, group VIII of the periodic table, aw 195.09, valences of 2, 4. There are five stable isotopes.
Properties: Silvery, white, ductile metal. D 21.45, mp 1769C, bp 3827C. Brinell hardness 97, annealed (Vickers) 42. Insoluble in mineral and organic acids.

Soluble in aqua regia; attacked by fused alkalies. Does not corrode or tarnish; heated platinum absorbs large volumes of hydrogen. It is also a strong complexing agent. As a catalyst it is abnormally sensitive to poisons.
Occurrence: Canada (Ontario), South Africa, the former U.S.S.R., Alaska. Usually mixed with ores of copper, nickel, etc.
Derivation: By dissolving the ore concentrate in aqua regia, precipitating the platinum by ammonium chloride as ammonium hexachloroplatinate, igniting the precipitate to form platinum sponge. This is then melted in an oxyhydrogen flame or in an electric furnace.
Available forms: Powder (platinum black), single crystals, wire (2 by 0.05–0.005 inches diameter); special composition for electronics, metallizing, and decorating ceramics and metals.
Grade: Physically pure (99.99%), chemically pure (99.9%), crucible platinum (99.5%), commercial (99.0%).
Hazard: Flammable in powdered form. Soluble salts are highly toxic by inhalation. Asthma and upper respiratory tract irritant.
Use: Catalyst (nitric acid, sulfuric acid, high-octane gasoline, automobile exhaust gas converters), laboratory ware, spinnerets for rayon and glass fiber manufacture, jewelry, dentistry, electrical contacts, thermocouples, surgical wire, bushings, electroplating, electric furnace windings, chemical reaction vessels, permanent magnets.

platinum ammonium chloride. See ammonium hexachloroplatinate; ammonium chloroplatinate.

platinum barium cyanide. See barium cyanoplatinite.

platinum black. Finely divided metallic platinum.
Properties: Black powder, exhibits a metallic luster when rubbed. Soluble in aqua regia, d 15.8–17.6 (apparent).
Derivation: Reduction of solution of a platinum salt with zinc or magnesium.
Hazard: Flammable when dispersed in air.
Use: Catalyst; to absorb gases (hydrogen, oxygen, etc.) that it again liberates at red heat; gas ignition apparatus.

platinum chloride. (platinum tetrachloride; platinic chloride).
CAS: 10025-65-7. (1) $PtCl_4$; (2) $PtCl_4 \cdot 5H_2O$. The platinum chloride of commerce is usually chloroplatinic acid.
Properties: (1) Brown solid, (2) red crystals. Soluble in water and alcohol; (1) d 4.30 (25C), decomposes at 370C; (2) d 2.43, mp loses $4H_2O$ at 100C.
Derivation: By solution of platinum in aqua regia and evaporation.
Use: See chloroplatinic acid.

platinum-cobalt alloy. A 76.7 platinum/23.3 cobalt alloy forms a more powerful permanent magnet than any other known.

platinum dichloride. (platinous chloride). PtCl$_2$.
Properties: Greenish-gray powder that forms double salts with the chlorides of the alkali metals. D 5.87, mp decomposes at red heat, yielding platinum. Soluble in hydrochloric acid and ammonium hydroxide; insoluble in water.
Derivation: (1) By heating platinum sponge in the presence of dry chlorine; (2) by heating chloroplatinic acid to 200C.
Use: Platinum salts.

platinum dioxide. (platinic oxide). PtO$_2$.
Properties: Black powder. Soluble in concentrated acids and dilute solutions of potassium hydroxide.
Derivation: Reaction of platinic chloride with excess sodium hydroxide.
Use: Hydrogenation catalyst (forms platinum black when reduced by hydrogen).

platinum foil.
CAS: 7440-06-4. Pt. Thin sheets of pure platinum.
Hazard: Asthma; upper respiratory tract irritant.
Use: In various soldering procedures in dentistry, internal form in porcelain restorations during fabrication.

platinum iodide. (platinous iodide; platinum diiodide). PtI$_2$.
Properties: Heavy black powder. D 6.4, mp 300–350C (decomposes). Slightly soluble in hydriodic acid; insoluble in alkalies and water.

platinum-iridium alloy. The most important platinum alloy. Commercial alloys contain 1–30% iridium. As the iridium is increased, the hardness of the alloy increases, along with the resistance to chemical attack. The melting point of platinum is raised by the addition of iridium.
Use: Jewelry ("medium" platinum is 95% platinum, 5% iridium; and "hard" platinum is 90% platinum, 10% iridium), electrical contacts (10–25% iridium), fuse wire (10–20% iridium), hypodermic needles (20–30% iridium), and in general where high corrosion resistance is needed.
See iridium.

platinum-lithium. LiPt$_2$. Brittle solid, nonreactive with water, made by direct combination at 540C. If the lithium and platinum are combined at 200C, the product can be decomposed by water, hydrolyzing and dissolving the lithium and leaving unusually active platinum catalyst.

platinum metal. Any of a group of six metals, all members of group VIII of the periodic table: ruthenium, rhodium, palladium, osmium, iridium, and platinum. All of these are also transition metals.

platinum potassium chloride. See potassium chloroplatinate.

platinum-rhodium alloys. Alloys containing up to 40% rhodium. Such alloys are harder than platinum but not as hard as the corresponding platinum-iridium alloys. The addition of rhodium to platinum increases the resistance to attack by aqua regia. The melting points of the alloys are higher than those of platinum.
Use: Catalyst in nitric acid production, high-temperature vessels, furnace resistors, thermocouples and resistance thermometers, spinneret nozzles, components of gas-turbine aircraft engines.

platinum sodium chloride. See sodium chloroplatinate.

platinum sponge.
Properties: Grayish-black, porous mass of finely divided platinum. Soluble in aqua regia.
Derivation: By ignition of ammonium hexachloroplatinate or other salts.
Use: Catalyst.
See platinum black.

platinum sulfate. (platinic sulfate). Pt(SO$_4$)$_2$.
Properties: Greenish-black mass. Hygroscopic; soluble in acids (dilute), alcohol, ether, water.
Hazard: Toxic by inhalation.
Use: Analysis (microtesting for bromine, chlorine, iodine).

platinum tetrachloride. See platinum chloride.

platiphillin.
CAS: 480-78-4. C$_{18}$H$_{27}$NO$_5$.
Hazard: Moderately toxic by ingestion.

pleated sheet. The side-by-side, hydrogen-bonded arrangement of polypeptide chains in the extended conformation.

pleiotropy. One gene that causes many different physical traits, such as multiple disease symptoms.

Plenlizer. A line of plastic resin stabilizers.

plesiomorphy. A primitive character state for the taxa under consideration.

"Plexiglas" *[Arkema].* TM for thermoplastic poly(methyl methacrylate)-type polymers. Available in bead or granular form and sheets.
Use: Manufacture of lenses, ornaments, letters for signs, aircraft canopies and windows, light diffusers, industrial and architectural glazing, chalkboards, boat windshields, and similar products.

Plexol. Synthetic lubricants and additives for petroleum oils. Most grades are diesters of dibasic acids; some are polyesters or polyether alcohols. The ester lubricants have very low freezing points, high flash points, little change of viscosity with temperature.
Use: Aircraft-engine lubricants, hydraulic systems, instrument oils, petroleum-base lubricant formulation.

Plictran. Cyhexatin.

pliofilm. Former TM for rubber hydrochloride as transparent base.

"Pliotec" [Eliokem]. TM for a unique water borne resin.
Use: For masonry, concrete, and anti-corrosion coatings.

"Pliotone" [Eliokem]. TM for resins.
Use: In xerographic devices from PCs to high speed lasers, color copies, and latest digital multifunction machines.

"Pliovic" [Veyance]. TM for homopolymers, vinyl dispersion resin, plastisol, and organosol compounds.
Use: For coatings of rainwear, automotive filters, wall coverings, awnings, flooring, glove dipping, and laminating.

plow. A scraping device of various contours used for dislodging sediment or "mud" accumulated at the bottom of clarifying and thickening tanks. It is activated by a rotating arm that moves it circumferentially around the tank riding close to the bottom; it thus transfers the sedimented material from the periphery to the center of the tank where it is discharged into a hopper. Plows are also used in mixing equipment of the Muller type, where they serve to continuously rake the material being mixed into the path of the Muller wheels.

plumb-. (plumbo-).
Containing combined lead.

plumbago. Refers to graphite.

plumbate. A compound containing the negative group, $PbO_3{}^-$.

plumbic. The designation for lead compounds in which the lead is tetravalent.

plumbic acid, anhydrous. See lead dioxide.

plumbo-. See plumb-.

plumboplumbic oxide. See lead oxide, red.

plumbous. Refers to lead compounds in which the lead has a valence of two.

plumbous oxide. See litharge.

plumbous sulfide. See lead sulfide.

plumbum. The Latin name for lead, hence the symbol Pb and the names plumbic and plumbous.

"Pluracol" [BASF]. TM for a series of organic compounds used in hydraulic brake and other functional fluids, chemical intermediates, urethane foams, elastomers, and coatings.

Plurafac. A series of 100% active, nonionic biodegradable surfactants of straight-chain, primary aliphatic oxyethylated alcohols. Available in liquid, paste, flake, and solid form.
Use: Range from light-duty hand dishwashing formulations to heavy-duty industrial detergents, rinse aids, metal cleaners, etc.

pluripotency. The potential of a cell to develop into more than one type of mature cell, depending on environment.

"Pluronic" [BASF]. TM for a nonionic series of 28 related difunctional block-polymers terminating in primary hydroxyl groups with molecular weights ranging from 1000 to more than 15,000. They are polyoxyalkylene derivatives of propylene glycol. Available in liquid, paste, flake powder, and cast-solid forms.
Use: Defoaming agents, emulsifying and demulsifying agents, binders, stabilizers, dispersing agents, wetting agents, rinse aids, and chemical intermediates.

plutonium. Pu. Synthetic radioactive metallic element with atomic number 94, first prepared in 1941, aw 239.11, valences of 3, 4, 5, 6; there are 15 isotopes (from 232 to 246), 6 allotropic forms. Plutonium-239 (half-life 24,360 years) is produced in a nuclear reactor by neutron bombardment of the nonfissionable isotope ^{238}U. Plutonium is readily fissionable with both slow and fast neutrons and can be used for either nuclear weapons or electric power production. The critical mass of pure ^{239}Pu is 10 lb. One lb reactor-grade material contains a heat energy equivalent of 10^6 kilowatt hours. Weapons-grade plutonium contains up to 7% ^{240}Pu; though reactor-grade plutonium is somewhat less pure, it can be used for weapons. The Safeguards Manual of the International Atomic Energy Agency states: "Plutonium of any grade, in either metal, oxide, or nitrate form, can be put in a form suitable for the manufacture of explosive devices in a matter of a few days or weeks."
According to Glenn T. Seaborg, "in breeder reactors it is possible to create more new plutonium from U-238 than that consumed in sustaining the fission

chain reaction. Because of this, plutonium is the key to unlocking the enormous energy reserves in the nonfissionable U-238." Reactor fuels containing plutonium can be either liquid or solid; since plutonium forms low-melting alloys with a number of metals (gallium, bismuth, tin, iron, cobalt, and nickel), these are often used as liquid reactor fuels. Cerium also may be a component.

Hazard: The most radiotoxic of the elements and one of the most toxic substances known; dangerous ionizing radiation persists indefinitely; a confirmed carcinogen. Must be handled by remote control and with adequate shielding.

See breeder.

Plyamule.   A series of vinyl acetate homopolymers.

Plyamine.   A group of liquid water-soluble urea-formaldehyde adhesive resins.

Use: Binders in the manufacture of plywood, furniture, wood-particle products, etc.

Plyophen.   A water-soluble impregnating resin. Penetrates deeply and quickly into wood, canvas, asbestos, paper, and other laminating and molding stocks. Can be diluted as much as 8–10 parts water to 1 part resin for spraying glass fiber or rock wool.

plywood.   A composite composed of thin wood veneers (with grains placed at right angles to each other) bonded with a synthetic resin, usually phenol-formaldehyde or resorcinol-formaldehyde. It is superior to metals in strength-to-weight ratio and has low thermal expansion, high heat capacity, and low water absorption.

See laminate; composite.

Pm.   Symbol for promethium.

PMA.   Abbreviation for phosphomolybdic acid and for pyromellitic acid.

PMDA.   Abbreviation for pyromellitic dianhydride.

PMEG.   See   9-((2-phosphonylmethoxy)-ethyl)guanine.

PMHP.   Abbreviation for p-methane hydroperoxide. A polymerization catalyst.

PMP.   Abbreviation for 1-phenyl-3-methyl-5-pyrazolone.

PMTA.   Abbreviation for a mixture of phosphomolybdic and phosphotungstic acids.

Use: Making pigments.

See phosphotungstic pigment.

pn.   Abbreviation for propylenediamine, as used in formulas for coordination compounds.

See dien; en; py.

"PNF" [White].   TM for a special-purpose synthetic rubber, a phosphonitrilic fluoroelastomer, said to be flexible at −56C and serviceable up to 176C. Resistant to oils over a wide temperature range. Can be processed on standard equipment.

Po.   Symbol for polonium.

podophyllotoxin 4-o-glucoside.
CAS: 16481-54-2. $C_{28}H_{32}O_{13}$.

Hazard: Moderately toxic by ingestion.

POEMS.   Abbreviation for polyoxyethylene monostearate.

POEOP.   See polyoxyethyleneoxypropylene.

POF.   See dl-α-lipoic acid.

poise.   See centipoise.

poison.   (1) Any substance that is harmful to living tissues when applied in relatively small doses. The most important factors involved in effective dosage are (a) quantity or concentration; (b) duration of exposure; (c) particle size or physical state of the substance; (d) its affinity for living tissue; (e) its solubility in tissue fluids; and (f) the sensitivity of the tissues or organs. Sharp distinction between poisons and nonpoisons is not always possible, because many variables must be taken into consideration in each case. Poisons are divided into four classes by the shipping regulatory agencies, as follows:

Poison A: A gas or liquid so toxic that an extremely small amount of the gas or the vapor formed by the liquid is dangerous to life.

Poison B: Less toxic liquids and solids that are hazardous either by contact with the body (skin absorption) or by ingestion.

Poison C: Liquids or solids that evolve toxic or strongly irritating fumes heated or when exposed to air (excluding class A poisons).

Poison D: Radioactive materials.

See toxicity; toxic substances.

(2) In nuclear technology, any material with a high capture probability for neutrons that may divert an undesirable number of neutrons from the fission chain reaction.

(3) A substance that reduces or destroys the activity of a catalyst. Carbon monoxide and phosphorus, arsenic, or sulfur compounds have this effect on the formation of ammonia from hydrogen and nitrogen gases, and the gases, must be highly purified to avoid this. Another example is the poisoning of the platinum catalysts used in emission-control devices by organic lead compounds.

poison gas.   A toxic or irritant gas or volatile liquid designed for use in chemical warfare or riot control. They vary in toxicity from nerve gases, which

are lethal, to tear gases (lachrymators), which cause only temporary disability.
See noxious gas; chemical warfare.

poison ivy extract.   An extract from fresh leaves of poison ivy (*Rhus radicans*).
Use: For desensitization.

poison oak extract.   An extract from the fresh leaves of poison oak (*Rhus toxicodendron*).
Use: For desensitization.

Polanyi, John C.   (1929–). Awarded the Nobel Prize in chemistry in 1986 jointly with Herschbach and Lee. Herschbach reported that the energies of reactions of colliding beams of isolated alkali metal atoms and alkyl halide molecules appeared mostly as vibrational excited states of products. Polanyi characterized the excited states by the infrared light emitted by product molecules. His work also led to the development of lasers. Born in Germany, Polanyi studied in England and later became a Canadian citizen. Doctorate awarded by Manchester University, England, in 1952.

polar.   Descriptive of a molecule in which the positive and negative electrical charges are permanently separated, as opposed to nonpolar molecules in which the charges coincide. Polar molecules ionize in solution and impart electrical conductivity. Water, alcohol, and sulfuric acid are polar in nature; most hydrocarbon liquids are not. Carboxyl and hydroxyl groups often exhibit an electric charge. The formation of emulsions and the action of detergents are dependent on this behavior.
See dipole moment.

polar brilliant blue raw.   See C.I. acid blue 80.

polarimeter.   An instrument for measuring the amount by which the plane of polarization of plane-polarized light is rotated in passing through a medium (usually a liquid).

polarimetry.   Measurement of the degrees and direction of the plane of polarized light as it passes through an optically active compound. It is used in the investigation of optical isomers, especially in analysis of sugars.

polarity of characters.   The states of characters used in a cladistic analysis, either original or derived. Original characters are those acquired by an ancestor deeper in the phylogeny than the most recent common ancestor of the taxa under consideration. Derived characters are those acquired by the most recent common ancestor of the taxa under consideration.

polarization.   In electrochemistry, the increase of solution resistance due to gas accumulation at the electrode or chemical depletion in part of the solution.

polarization concentration.   The polarization in an electrolytic cell caused by changes in concentration of electrolyte surrounding the electrode.

polarization current.   The reverse current produced by polarization on removal of external electromotive force.

polarized electrode, ideal.   A system composed of a metal in contact with an electrolytic solution for which, at equilibrium, the concentration of every charged component is finite in one phase only.

polarized light.   See nicol; polarimetry.

polar liquid.   A liquid having a dipole moment, as alcohol.

polar molecule.   (1) A molecule with a positive charge on one end and a negative charge on its other end. The vector sum of its bond dipoles is not zero. (2) Molecule in which the electrons forming the valency bond are not symmetrically arranged.

polarogram.   A record on a polarograph of a variation in current or current–voltage relation.

polarograph.   A device for the automatic electroanalysis of a solution by means of the dropping mercury electrode.

polar solvent.   Solvents containing hydroxyl or carbonyl groups, having high dielectric constant and strong polarity.

Polectron.   Modified vinylpyrrolidone resins.
Properties: 40% active aqueous emulsion, stable to intense mechanical shear, freeze–thaw cycling. Compatible with other commercial latexes.
Use: Binding agents for wood, cotton, paper, glass fibers; stabilizer, opacifier and dyestuff; medium precoat for photosensitive papers. Adhesive for metal, glass, cotton, paper, and wood.

Polenske number.   A measure of the insoluble volatile fatty acids in a sample of fat or oil. The number of cubic centimeters of 0.1 N alkali required to neutralize the water-insoluble volatile acids in a 5 g sample of fat or oil.

polish.   (1) A solid powder or a liquid or semiliquid mixture that imparts smoothness, surface protection, or a decorative finish. The most widely used solid polishing agent is fine-ground red iron oxide (rouge), applied to the surface of plate glass, backs of mirrors, and optical glass. A wide variety of liquid and pastelike polishes are based on vegetable waxes (carnauba and candelilla), combined with

softeners, fillers, and pigments or emulsified in alcohol or other solvent. Furniture polishes often contain red oil, lemon oil, and petroleum solvent; most types of metal and wood polish contain organic solvents and, hence, are flammable liquids. Nail polishes are nitrocellulose lacquers, usually with amyl acetate solvent.
See electropolishing.
Hazard: May be flammable.
(2) The hard outer coating of cereal grains, especially rice, which is usually removed in processing. These coatings are rich in vitamin B_1. Their removal robs the cereal of much of its nutritive value.

Politol 711. A PVC seam sealant.
Use: For joining PVC pipe.

pollen extract. A proteinaceous liquid extract obtained from pollen.

pollucite. $Cs_4Al_4Si_9O_{26}$•H_2O. A natural cesium aluminum silicate found in pegmatites.
Properties: Colorless. Mohs hardness 6.5, d 2.9.
Use: Source of cesium, catalyst, fluxes, welding materials, ion propulsion, thermocouple units.

pollution. The introduction into any environment of substances that are not normally present therein and that are potentially toxic or otherwise objectionable. The most serious atmospheric contaminants have been (1) sulfur dioxide evolved from the fuels used in electric power production and industrial processing, and (2) automobile exhaust gases rich in carbon monoxide and tetraethyllead residues. The former is being alleviated by mandatory use of low-sulfur fuels and the latter by elimination of tetraethyllead from most gasoline and by use of catalytic converters.
Water pollution due to discharge of toxic chemical wastes is closely regulated by both the EPA and FDA. Such substances are defined in the 1972 amendment of the Federal Water Pollution Control Act as those "which will cause death, disease, cancer, or genetic malfunctions in any organisms with which they come into contact." Substances added to water for purification purposes (chlorine, aluminum sulfate, etc.) are excluded from the category of pollutants.
See Environmental Protection Agency; air pollution; water pollution.

polonium. Po. Radioactive element of atomic number 84, group VIA of the periodic table, aw 210, valences of 2, 4, 6. There are no stable isotopes. Polonium is a member of the uranium natural radioactive decay series, occurring naturally in uranium-bearing ores; it is produced artificially by bombarding bismuth with neutrons. It has been identified in cigarette smoke.
Properties: Similar to those of tellurium. Mp 254C, bp 962C, d 9.4. Dissolved by concentrated

sulfuric and nitric acids and aqua regia, and by dilute hydrochloric acid.
Hazard: Dangerous radioactive poison.
Use: Source of α-radiation and neutrons, instrument calibration, oil-well logging, moisture determination, power source.
See smoke (4).

Polonovski reaction. Demethylation of tertiary (or heterocyclic) amine N-oxides on treatment with acetyl chloride or acetic anhydride to give N-acylated secondary amines and formaldehyde, along with O-acylated aminophenols as a result of a side reaction.

poloxalene. See "Therabloat" *[Zoetis]*.

poloxalene free-choice liquid type C feed.
See Purina Bloat Block.

Poloxamer 331.
Properties: Average molecular weight 3800. Colorless liquid. D: 1.02, refr index: 1.452. Very sltly sol in water; sol in alc; insol in propylene glycol, ethylene glycol.
Engineering plastics are those to which standard metal A food additive.

poly-. A prefix signifying many. For example, a polymer is an aggregate formed by combination of a number of single molecules.
See polymer, high.

"Polyac" *[Haliburton].* TM for a butyl rubber conditioner containing 25% poly-*p*-dinitrosobenzene $[C_6H_4(NO)_2]_x$ with an inert wax. Dark-brown, waxy pellets; d 0.96.
Use: Processing aid and accelerator of vulcanization for butyl rubber.

polyacetal. See acetal resin.

polyacetylene. A linear polymer of acetylene having alternate single and double bonds, developed in 1978. It is electrically conductive, but this property can be varied in either direction by appropriate doping either with electron acceptors (arsenic pentafluoride or a halogen) or with electron donors (lithium, sodium). Thus, it can be made to have a wide range of conductivity from insulators to *n*- or *p*-type semiconductors to strongly conductive forms. Polyacetylene can be made in both *cis* and *trans* modifications in the form of fibers and thin films, the conductivity of the fibers increasing with their degree of orientation. Films can be applied on glass or metal substrates. Though still in an experimental stage, these polymers have significant possibilities for industrial applications, e.g., in batteries.
See cyclooctatetraene.

polyacrylamide. $(CH_2CHCONH_2)_n$. White solid, water-soluble high polymer.

Derivation: Polymerization of acrylamide with *N,N'*-methylene bisacrylamide.
Use: Thickening agent, suspending agent, additive to adhesives. Permissible food additive.
See acrylic resin.

polyacrylamide resins, modified.   See modified polyacrylamide resins.

polyacrylate.   See acrylic resin.

polyacrylic acid.   See acrylic acid, methacrylic acid.

polyacrylonitrile.   A polymer of acrylonitrile that is the basic material used in the manufacture of a number of synthetic fibers, e.g., "Orlon" and "Dynel." When combined with other materials it produces a hard resin having high solvent resistance

$$-CH_2-\underset{\underset{CN}{|}}{CH_2}$$

and high-temperature stability; from these are made such items as moldings, shoe soling, wall panels, and the like.
See acrylonitrile; acrylic resin; acrylic fiber.

Polyact.   A winemaking additive.
Use: For the treatment of oxidized must wine to prevent browning and pinking.

polyalcohol.   See polyol.

polyallomer.   A copolymer that has a uniform crystalline structure but a mixed chemical composition. It is prepared by anionic coordination catalysis using a Ziegler catalyst. The best-known polyallomer is a copolymer of propylene and ethylene; it has the stereoregular crystalline structure of the homopolymers of these resins but a variable chemical composition. Temperature range -40 to 99C. The physical properties of polyallomers are generally intermediate between those of the homopolymers of the component resins but give a better balance of properties than blends of the homopolymers.
Use: Vacuum-formed, injection-molded, blow-molded, and extruded products; film; sheeting; wire cables.

polyamide.   A high molecular weight polymer in which amide linkages (–CONH–) occur along the molecular chain. They may be either natural or synthetic. Important natural polyamides are casein, soybean and peanut proteins, and zein, the protein of corn, from all of which plastics, textile fibers, and adhesive compositions can be made. Synthetic polyamides are typified by the numerous varieties of nylon, though some, e.g., "Versamide", are quite different from nylon.
See nylon; polypeptide; protein; aramid.

polyamine.   A hydrocarbon compound that contains two or more amino acid groups.

polyamine-methylene resin.   A polyethylene polyamine methylene-substituted resin of diphenylol dimethylmethane and formaldehyde in basic form.
Properties: Light-amber, granular, free-flowing powder; appreciable odor. Insoluble in dilute acids and alkalies, alcohol, ether, and water.
Use: Medicine, ion-exchange resin, antacid.

polyaminotriazole.   (PAT). A synthetic polymer made from sebacic acid and hydrazine, with small amounts of acetamide. Polyoctamethylene-aminotriazole is a specific example.
See monobasic.

polyA tail.   After an mRNA is transcribed from a gene, the cell adds a stretch of A residues (typically 50–200) to its 3' end. It is thought that the presence of this "polyA tail" increases the stability of the mRNA (possibly by protecting it from nucleases). Note that not all mRNAs have a polyA tail; the histone mRNAs in particular do not.

polybenzimidazole.   (PBI). $(C_7H_6N_2)_n$. A synthetic polymer designed for high-temperature space technology applications. Reputed to withstand temperature up to 260C for 1000 hr.
Derivation: Condensation of diphenyl isophthalate and 3,3'-diaminobenzidine.
Use: Fibers, composites, adhesives (high adhesion to steel, titanium, beryllium, and aluminum alloys), coatings, ablative materials.

polyblend.   A combination in any proportion of (1) two homopolymers (natural or synthetic), (2) a homopolymer and a copolymer, or (3) two copolymers. An example of (1) is rubber-polystyrene, of (2) is rubber and butadiene-styrene, and of (3) is a mixture of butadiene-acrylonitrile and isobutylene-isoprene. A polyblend is a mixture that is made after its components have been polymerized and thus is different from a copolymer, which is made by chemical combination of two monomers.
See homopolymer; copolymer; blend.

polybrominated biphenyl.   (pbb; polybromobiphenyl; 1,2,4-tribromo-5-(2,4,5-tribromophenyl)benzene; 2,2',4,4',5,5'-hexabromobiphenyl).
CAS: 59536-65-1. $C_{12}H_4Br_4$. Any of a class of brominated hydrocarbon compounds.
Hazard: Poison; teratogen; embryotoxin; possible carcinogen.
Use: As fire retardants and components of heat resistant plastics.

polybutadiene.   A synthetic thermoplastic polymer made by polymerizing 1,3-butadiene with a stereospecific organometallic catalyst (butyl lithium), though other catalysts such as titanium

tetrachloride and aluminum iodide may be used. The *cis*-isomer, which is similar to natural rubber, is used in tire treads due to its abrasion and crack resistance and low heat buildup. Large quantities are also used as blends in SBR. The *trans*-isomer resembles gutta-percha and has limited utility. Liquid polybutadiene, which is sodium catalyzed, has specialty uses as a coating resin. It is cured with organic peroxides. Combustible.
Hazard: (Liquid) By ingestion and inhalation; skin irritant.
See polymer, stereospecific.

polybutene.   See polybutylene.

polybutylene.   (polybutene;   polyisobutylene; polyisobutene). Any of several thermoplastic isotactic (stereoregular) polymers of isobutene of varying molecular weight; also polymers of butene-1 and butene-2. Butyl rubber is a type of polyisobutene to which has been added 2% of isoprene, which provides sulfur linkage sites for vulcanization. Isobutene can be homopolymerized to various degrees in chains containing from 10 to 1000 units, the viscosity increasing with molecular weight. Combustible.
Use: Lubricating-oil additive, hot-melt adhesives, sealing tapes, special sealants, cable insulation, polymer modifier, viscosity index improvers, films, and coatings.

polybutylene terephthalate.   An engineering plastic derived from 1,4-butanediol, it is a thermoplastic polyester with a broad spectrum of uses.

polycarbonate. $(COOC_6H_5C(CH_3)_2C_6H_5O)_n$. A synthetic thermoplastic resin derived from bisphenol A and phosgene, a linear polyester of carbonic acid. Can be formed from any dihydroxy compound and any carbonate diester, or by ester interchange. Polymerization may be in either aqueous emulsion or in nonaqueous solution.
Properties: Transparent (90% light transmission), noncorrosive, weather and ozone resistant, nontoxic, stain resistant, Combustible but self-extinguishing, low water absorption, high impact strength, heat resistant, high dielectric strength, dimensionally stable, soluble in chlorinated hydrocarbons and attacked by strong alkalies and aromatic hydrocarbons, stable to mineral acids, insoluble in aliphatic alcohols. Excellent for all molding methods, extrusion, thermoforming, etc.; easily fabricated by all methods including thermoforming and fluidized-bed coating.
Use: Molded products, solution-cast or extruded film, structural parts, tubes and piping, prosthetic devices, meter face plates, nonbreakable windows, streetlight globes, household appliances.

polycarboxylic acid.   An organic acid containing two or more carboxyl (COOH) groups.

polychlor.   General name for synthetic chlorinated hydrocarbons.
Use: Pesticides.

polychlorinated biphenyl.   (PCB). CAS: 1336-36-3. One of several aromatic compounds containing two benzene nuclei with two or more substituent chlorine atoms. They are colorless liquids with d 1.4–1.5. Because of their persistence, toxicity, and ecological damage via water pollution, their manufacture was discontinued in the U.S. in 1976.
Hazard: Highly toxic. Probable carcinogen.

polychloroprene.   See neoprene.

polychlorotrifluoroethylene.   (PCTFE). See chlorotrifluoroethylene polymer.

polyclonal antibodies.   A heterogeneous pool of antibodies produced in an animal by a number of different B lymphocytes in response to an antigen. Different antibodies in the pool recognize different epitopes of the antigen.

Polyco.   A series of thermoplastic polymers in the form of water emulsions or solvent solutions, applied to vinyl acetate polymers and copolymers, butadiene-styrene copolymer latics, polystyrenes, vinyl and vinylidene chloride copolymers, acrylic copolymers, and water-soluble polyacrylates.
Use: Adhesives and coatings, in paint, leather, textiles, paper, cosmetics, and construction fields.

polycondensation.   See condensation (1); polymerization.

polycoumarone resin.   See coumarone-indene resin.

polycyclic.   An organic compound having three or more aromatic nuclei in its ring structure, which may be the same or different, e.g., anthracene, naphthacene.
See polynuclear.

poly(1-4-cyclohexylenedimethylene) terephthalate.   A linear polyester film or fiber obtained by condensation of terephthalic acid with 1,4-cyclohexanedimethanol. It has good electrical resistivity and hydrolytic stability.
Use: Carpet fibers and chemically resistant films.
See terephthalic acid.

PolyDADMAC.   Liquid cationic polymers of differing molecular weights.
Use: As primary coagulants in liquid solid separation.

polydextrose.
CAS: 68424-04-4.
Properties: Off-white to light tan solid. Sol in water.
Use: Food additive.

polydextrose solution.
  Properties: Clear, straw-colored liquid.
  Use: Food additive.

poly-1,1-dihydroperfluorobutyl acrylate.
  Properties: White, rubber-like polymer. D 1.5, begins to degrade at 148C, retains strength and elastomeric properties in contact with synthetic lubricants, solvents, hydraulic fluids, oils, etc. at temperatures in the range 148–204C. Has limited flexibility at temperatures below −17C. Nonflammable.
  Use: O-rings, seals, gaskets, diaphragms, hose, sheets, and coatings for fabrics and other surfaces.

polydimethyldiphenylsiloxane.
  CAS: 68083-14-7.
  Hazard: Low toxicity by ingestion. A mild eye irritant.

polydimethylsiloxane. (PDMS). A silicone polymer developed for use as a dielectric coolant and in solar energy installations. It also may have a number of other uses. It is stated to be highly resistant to oxidation and biodegradation by microorganisms. It is degradable when exposed to a soil environment by chemical reaction with clays and water, by which it is decomposed to silicic acid, carbon dioxide, and water.

poly-*p*-dinitrosobenzene. See "Polyac" [Haliburton].

Polydril. A synthetic water-soluble polymer.
  Use: Flocculating agent in the oil industry.

polyelectrolyte. A high-polymer substance, either natural (protein, gum arabic) or synthetic (polyethyleneimine, polyacrylic acid salts), containing ionic constituents; may be either cationic or anionic. The former type is widely used for industrial applications. Water solutions of both types are electrically conducting; some are effective in concentrations as low as 1 ppm. In a given polyelectrolyte, ions of one sign are attached to the polymer chain, while those of opposite sign are free to diffuse into the solution. Major uses are flocculation of solids (especially dissolved phosphates) in potable water, dispersion of clays in oil-well drilling muds, soil conditioning, antistatic agents, and treatment of paper-mill wastewater. Ion-exchange resins are cross-linked (stabilized) polyelectrolytes.
  See flocculant; Purifloc; "Cat-Floc" [Calgon].

polyene. Any unsaturated aliphatic or alicyclic compound containing more than four carbon atoms in the chain and having at least two double bonds. Examples are pentadiene, cyclooctatriene.

polyester fiber. Generic name for a manufactured fiber (either as staple or continuous filament) in which the fiber-forming substance is any long-chain synthetic polymer composed of at least 85% by weight of an ester of a dihydric alcohol and terephthalic acid (Federal Trade Commission).
  See "Dacron" [Invista]; polyethylene terephthalate.
  Properties: Strength (staple) 2.2–4.0 g/denier; (continuous filament) up to 9.5 g denier, mp 264C, water absorption 0.5%. Nonflammable.
  Use: Tire fabric, seat belts, reinforcement of rubber hose for seawater cooling systems, as blend in clothing fabrics, fire-hose jackets.

polyester film. Continuously extruded polyester sheet of various thicknesses, especially useful in electrical equipment because of its high resistivity. Its tensile strength of 25,000 psi is much greater than that of other plastic films. Sensitized polyester film is used in magnetic tapes, in the photocopying technique known as reprography.

polyester resin. Any of a group of synthetic resins, which are polycondensation products of dicarboxylic acids with dihydroxy alcohols. They are thus a special type of alkyd resin but, unlike other types, are not usually modified with fatty acids or drying oils. The outstanding characteristics of these resins is their ability, when catalyzed, to cure or harden at room temperature under little or no pressure. Most polyesters now produced contain ethylenic unsaturation, generally introduced by unsaturated acids. The unsaturated polyesters are usually cross-linked through their double bonds with a compatible monomer, also containing ethylenic unsaturation, and thus become thermosetting. Flame resistance is imparted by using either acid or glycol ingredients having a high content of halogens, e.g., HET acid.
  The principal unsaturated acids used are maleic and fumaric. Saturated acids, usually phthalic and adipic, may also be included. The function of these acids is to reduce the amount of unsaturation in the final resin, making it tougher and more flexible. The acid anhydrides are often used if available and applicable. The dihydroxy alcohols most generally used are ethylene, propylene, diethylene, and dipropylene glycols. Styrene and diallyl phthalate are the most common cross-linking agents. Polyesters are resistant to corrosion, chemicals, solvents, etc.
  Available forms: Sheets, powder, chips.
  Use: Reinforced plastics; automotive parts; boat hulls; foams; encapsulation of electrical equipment; protective coatings; ducts, flues, and other structural applications; low-pressure laminates; magnetic tapes; piping; bottles; nonwoven disposable filters; low-temperature mortars.
  See alkyd resin; polyester fiber.

polyethenoid. Characterizing an aliphatic compound having more than one ethene group – CH=CH–. Linoleic acid is a polyethenoid fatty acid.

polyether.   A polymer in which the repeating unit contains a C=O bond derived from aldehydes or epoxides or similar materials.

polyether, chlorinated.   A highly crystalline material that is 46% chlorine. Outstanding corrosion resistance. Good electrical resistance. Readily processed and fabricated.
Use: Fluid-bed coating, tank linings, piping, valves, laboratory equipment, chemical processing equipment.

polyether, cyclic.   See crown ether.

polyether foam.   A polyurethane foam, either rigid or flexible, made by use of a polyether as distinct from a polyester or other resin component.
Hazard: As for polyurethane.

polyether glycol.   A compound with a structural skeleton such as HO–C–C–O–C–C–O–C–C–O–C–C–OH. The length of the chain can vary widely, and the number of consecutive carbon atoms may be greater than two. Examples are polyethylene glycol and polypropylene glycol.

polyethylene.
CAS: 9002-88-4. $(-H_2C-CH_2-)_n$.
Chlorosulfonated. See "Hypalon" *[Du Pont Performance]*.
Cross-linked (XLPE).
Properties: Thermoplastic white solid, high-temperature-resistant, excellent resistance to chemicals and to creep, high impact and tensile strength, high electrical resistivity, insoluble in organic solvents, does not stress-crack. Combustible.
Derivation: (1) By irradiating linear polyethylene with electron beam or γ-radiation, cross-linking taking place through a primary valence bond, as shown.

(2) By chemical cross-linking agent such as an organic peroxide (e.g., benzoyl peroxide). All grades of polyethylene and most copolymers can be chemically cross-linked.
Use: Wire and cable coatings and insulation (low-density grades), pipe and molded fittings

(high-density grades). Special types having low electrical resistivity can be made; these can be regarded as semiconductors.
Note: In molding cross-linked polyethylene, the desired part must be formed before cross-linking is initiated, because material will not change its shape after cross-linking. The variations in composition and wide range of properties approach the ideal of a universal material more closely than most polymers.
Density
The density of polyethylene and other thermoplastic polymers is affected by the shape and spacing of the molecular chains; low-density materials have highly branched and widely spaced chains, whereas high-density materials have comparatively straight and closely aligned chains. Polymers of the latter type are called linear. The physical properties are markedly affected by increasing density.
Low density (branched chain)
Properties: Crystallinity 50–60%, d 0.915, mp 240F, tensile strength 1500 psi, impact strength above 10 ft-lb/inch/notch, thermal expansion 17×10^{-5} inch/inch/C. Soluble in organic solvents above 200F; insoluble at room temperature.
Derivation: (1) Ethylene is polymerized in a free-radical-initiated liquid-phase reaction at 1500 atm (22,000 psi) and 375F with oxygen as catalyst (usually from peroxides). (2) A much more effective and cheaper process uses pressures of only 100–300 psi at less than 212F; the catalyst is undisclosed and reaction is vapor phase.
Hazard: Questionable carcinogen.
Use: Packaging film (especially for food products), paper coating, liners for drums and other shipping containers, wire and cable coating, toys, cordage, refuse and waste bags, chewing-gum base, squeeze bottles, electrical insulation.
High density (linear)
Properties: Crystallinity 90%. D 0.95, mp 275F, tensile strength 4000 psi, impact strength 8 ft-lb/inch notch, high electrical resistivity, film is gas-permeable, hydrophobic, does not resist nitric acid.
Derivation: Ethylene polymerized by Ziegler catalysts at 1–100 atm (15–1500 psi) at from room temperature to 200F. Catalyst is a metal alkyl, e.g., triethylaluminum plus a metallic salt ($TiCl_4$) dissolved in a hydrocarbon solvent. A vapor-phase modification of this process was developed in 1965. Another method uses such metallic catalysts as Cr_2O_3 at 100–500 psi with solvents such as cyclohexane or xylene.
Use: Blow-molded products, injection-molded items, film and sheet, piping, fibers, gasoline and oil containers.
Note: Ethylene may be copolymerized with varying percentages of other materials, e.g., 2-butene or acrylic acid; a crystalline product results from copolymerization of ethylene and propylene. When butadiene is added to the copolymer blend, a vulcanizable elastomer is obtained.

Low molecular weight
Properties: Molecular weight 2000–5000. Translucent white solids, excellent electrical resistance, abrasion resistant, resistant to water and most chemicals, d 0.92. Slightly soluble in turpentine, petroleum naphtha, xylene, and toluene at room temperature; soluble in xylene, toluene, trichloroethylene, turpentine, and mineral oils at 82.2C; practically insoluble in water; slightly soluble in methyl acetate, acetone, and ethanol up to the boiling points of these solvents. Available as emulsified and nonemulsified forms. Combustible.
Use: Mold-release agent for rubber and plastics, paper and container coatings, liquid polishes, and textile finishing agents.

polyethylene glycol. (PEG; poly(oxyethylene); polyglycol; polyether glycol).
CAS: 25322-68-3. Any of several condensation polymers of ethylene glycol with the general formula $HOCH_2(CH_2OCH_2)_nCH_2OH$ or $H(OCH_2CH_2)_nOH$. Average molecular weights range from 200 to 6000. Properties vary with molecular weight.
Properties: Clear, colorless, viscous liquids to waxy solids; odorless. Soluble or miscible with water and for the most part with alcohol and other organic solvents, heat-stable; inert to many chemical agents. Do not hydrolyze or deteriorate. Have low vapor pressure. Combustible.
Derivation: By condensation of ethylene glycol or of ethylene oxide and water.
Use: Chemical intermediates (lower molecular weight varieties), plasticizers, softeners and humectants, ointments, polishes, paper coating, mold lubricants, bases for cosmetics and pharmaceuticals, solvents, binders, metal and rubber processing, permissible additives to foods and animal feed, laboratory reagent.
See "Carbowax" *[Union]*.

polyethylene glycol chloride.
$H(OCH_2CH_2)_nCl$. Any of a group of polymers, usually colorless liquids with very low vapor pressure at room temperature. Mw from 100 to 600. Miscible with water, d for a low molecular weight polymer is 1.18 (20C), for a high molecular weight polymer 1.14 (10C). The former sets to a glass at −90C, the latter sets to a waxlike solid at 20C. Combustible.
Use: Solvents for cleaning, extracting, and dewaxing.

polyethylene glycol 400, dichloride.
CAS: 27252-69-3. $(C_2H_4O)_n C_4H_8C_{l2}O$.
Hazard: A severe eye irritant.

polyethylene glycol ester. A mono- or di-ester resulting from the interaction of an organic acid with one or both of the glycol ends of the polyethylene glycol polymer. These are also called polyoxyethylene esters, polyglycol esters, or a coined generic name.

polyethylene imine.
CAS: 26913-06-4. $(CH_2CH_2NH)_n$. A synthetic polymer that is a highly viscous, hygroscopic liquid when anhydrous; completely miscible with water and lower alcohols; insoluble in benzene. Reactive toward cellulose. Combustible.
Use: Adhesive and anchoring agent for paper and cellophane, dewatering agent and wet strength improver in paper manufacture, fixative, leveling agent in textile fibers, antiblocking agent on plastic films, flocculating agent, ion-exchange resins, complexing agents, disinfectant for textiles and skins, photographic chemistry, absorbent for carbon dioxide, water purification, polyelectrolyte.

polyethylene glycol lauryl thioether.
CAS: 9014-89-5. $(C_2H_4O)_nC_{14}H_{30}OS$.
Hazard: A severe skin and eye irritant.

polyethylene glycol mono(2-(dodecylthio) ethyl)ether. See polyethylene glycol lauryl thioether.

polyethylene oxide. A plastic reported to be dimensionally stable at high and low temperatures and designed as a substitute for phenolics.

polyethylene oxide sorbitan fatty acid esters. See polysorbate.

polyethylene terephthalate.
CAS: 25038-59-9. $(C_{10}H_8O_4)_x$. A thermoplastic polyester formed from ethylene glycol by direct esterification or by catalyzed ester exchange between ethylene glycol and dimethyl terephthalate. Offered as oriented film or fiber. It melts at 265C; tenacity is 2.2–4 g/denier as staple and up to 9.0 g/denier as continuous filament; d 1.38. It has good electrical resistance and low moisture absorption. Resists combustion and is self-extinguishing.
Use: Blended with cotton, for wash-and-wear fabrics; blended with wool, for worsteds and suitings; packaging films, recording tapes, soft-drink bottles.

polyethylene thiuram sulfide.
Derivation: Oxidation of diammonium ethylene bisdithiocarbamate with calcium hypochlorite.
Grade: 50% vegetable powder, 95% technical powder.
Use: Fungicide.

polyformaldehyde. See *p*-formaldehyde.

polyformaldehyde resin. See acetal resin.

polyforming. The combined thermal reforming and polymerization processes in gasoline products.

Polyform process. The process of producing high-octane gasoline from formerly waste refinery gases and petroleum naphtha.

polyfunctional system.   A system in which one of the reactants is at least trifunctional and the others are at least bifunctional.

polyfurfuryl alcohol.   See furfuryl alcohol.

"Poly-G" *[Monument].*   TM for a series of polyethylene glycols, polypropylene glycols, and polyoxypropylene adducts of glycerol. G200, 300, 400, and 600 are liquid polyethylene glycols; G1000, 1500, GB-1530, and BG-2000 are waxy polyethylene glycols. The number indicates the molecular weight. G420P, 1020P, 2020P are propylene oxide condensation polymers of propylene glycol. G1030PG, 3030PG, 4030PG are propylene oxide condensation polymers of glycerol.

"Polygard" *[Merck].*   TM for a mixture of alkylated aryl phosphites.
Properties: Clear amber liquid. D 0.99. Soluble in acetone, alcohol, benzene, carbon tetrachloride, solvent naphtha, and ligroin; insoluble in water, but can hydrolyze.
Use: Nondiscoloring stabilizer for rubber and plastics.

polygenic disorder.   Genetic disorder resulting from the combined action of alleles of more than one gene (e.g., heart disease, diabetes, and some cancers). Although such disorders are inherited, they depend on the simultaneous presence of several alleles; thus the hereditary patterns usually are more complex than those of single-gene disorders. See single-gene disorder.

polyglycerate (60).   See ethoxylated mono- and diglycerides.

polyglycerol.   One of several mixtures of ethers of glycerol with itself, ranging from diglycerol to triacontaglycerol. Some examples are (1) diglycerol, possibly $(CH_2OHCH_2OCH_2)_2$, mw 166, a liquid with four –OH groups, viscosity of 287 cs at 65.5C; (2) hexaglycerol, mw 462, a liquid with eight –OH groups, viscosity 1671 cs at 65.5C; (3) decaglycerol, mw 758, a liquid with 12 –OH groups, viscosity 3199 cs.
Properties: Viscous liquids to solids. Soluble in water, alcohol, and other polar solvents. Act as humectants much like glycerol but have progressively higher molecular weight and boiling point. Combustible.
Derivation: Glycerol is heated with an alkaline catalyst (200–275C) at normal or reduced pressure. A stream of inert gas may be used to blanket the reaction and help remove the water of reaction.
Use: Surface-active agents, emulsifiers, plasticizers, adhesives, lubricants, and other compounds used for both edible and industrial applications.

polyglycerol ester.   One of several partial or complete esters of saturated and unsaturated fatty acids with a variety of derivatives of polyglycerols ranging from diglycerol to triacontaglycerol. Prepared by (1) direct esterification and (2) transesterification reactions. Combustible.
Some examples of (1): decaglycerol monostearate, semisolid, d 1.04, mp 51.9C; decaglycerol monooleate, viscous liquid, d 1.13, mp around 0C; decaglycerol hexaoleate, liquid, d 0.97, mp −17.7C.
Examples of (2): triglycerol monolinoleate, viscosity 322 cP (75.5C); triglycerol trilinoleate, viscosity 30.1 cP (75.5C).
Use: Lubricants, plasticizers, paint and varnish vehicles, gelling agents, urethane intermediates, adhesives, cross-linking agents, humectants, textile fiber finishes, functional fluids, surface-active agents, dispersants and emulsifiers in foods, pharmaceuticals, cosmetic preparations.

polyglycerol esters of fatty acids.
Properties: Yellow to amber oily viscous liquids; light tan to brown soft solids; tan to brown waxy solids. Dispersible in water; sol in org solvs and oils.
Use: Food additive.

polyglycol.   See polyethylene glycol.

polyglycol amine H-163.
$HO[C_2H_4O]_2C_3H_6NH_2$.
Properties: Colorless liquid. D 1.0556, bp (decomposes), fp 14.5C, bulk d 8.8 lb/gal, flash p 295F (146C). Soluble in water. Combustible.

polyglycol distearate.   (polyethylene glycol distearate).
CAS: 9005-08-7. $C_{17}H_{35}COO(CH_2CH_2O)_nOCC_{17}H_{35}$. Distearate ester of polyglycol.
Properties: A soft, off-white solid. D 1.04 (50C), mp 43C, pH of 10% dispersion 7.26, saponification number variable. Soluble in chlorinated solvents, light esters, and acetone; slightly soluble in alcohols, insoluble in glycols, hydrocarbons, and vegetable oils. Combustible.
Use: Plasticizer for various resins, component of grinding and polishing pastes to promote easy removal in water.

Polygriptex.   An adhesive especially designed for bonding polyethylene sheeting to porous surfaces, used in making polyethylene-lined bags and multiwall kraft bags. Has good adhesion to waxed surfaces. Available in viscosities from 300 to 20,000 cP.

polyhalite. $2CaSO_4 \cdot MgSO_4 \cdot K_2SO_4 \cdot 2H_2O$.   A naturally occurring potash salt found in Germany, Texas, and New Mexico.
Use: Source of potash for fertilizer.

polyhexamethyleneadipamide.   Same as nylon 66.

polyhexamethylene sebacamide. Same as nylon 610.

polyhydric alcohol. See polyol.

polyhydroxy alcohol. (polyhydric alcohol). Any alcohol that contains two or more functional hydroxyl groups.
Hazard: Toxic.
Use: Heat exchangers, in hydraulic fluids, in antifreeze.

polyhydroxy alkaloid. (indolizidine alkaloids). Any bicyclic alkaloid with fused 5- and 6-membraned ring systems.

polyimide. Any of a group of high polymers that have an imide group (–CONHCO–) in the polymer chain.
Properties: Tensile strength 13,500 psi, d 1.42, water absorption 0.3% (24 hours at 77K), heat distortion point above 260C, dielectric constant at 2000 mc 3.2, and coefficient of linear expansion 28.4×10^{-6} inch/inch/ft. High-temperature stability (up to 370C), excellent frictional characteristics, good wear resistance at high temperatures, resists radiation, exhibits low outgassing in high vacuum, resistant to organic materials at quite high temperatures, not resistant to strong alkalies and to long exposure to steam. Flame retardant.
Use: High-temperature coatings, laminates and composites for aerospace vehicles, ablative materials, oil sealants and retainers, adhesives, semiconductors, valve seats, bearings, insulation for cables, printed circuits, magnetic tapes (high- and low-temperature-resistant), flame-resistant fibers, binders in abrasive wheels.

polyindene resin. See coumarone-indene resin.

polyisobutene. See polybutylene.

polyisobutylene. See polybutylene.

polyisocyanurate. See isocyanurate.

polyisoprene. $(C_5H_8)_n$. The major component of natural rubber, also made synthetically. Forms are stereospecific cis-1,4- and trans-1,4-polyisoprene. Both can be produced synthetically by the effect of heat and pressure on isoprene in the presence of stereospecific catalysts. Natural rubber is cis-1,4-; synthetic cis-1,4- is sometimes called synthetic natural rubber. trans-1,4-polyisoprene resembles gutta-percha. Polyisoprene is thermoplastic until mixed with sulfur and vulcanized. Supports combustion.
See rubber, natural; rubber, synthetic; catalyst, stereospecific.

polyketide. Any of a diverse group of naturally occurring compounds, related to fatty acids, that are composed of alternating carbonyl and methylene groups that are biogenetically derived from repeated condensation of acetyl coenzyme A, and usually the compounds derived from them by further condensations.

polylinker. A short, usually synthetic fragment of DNA containing recognition sequences for several restriction endonucleases employed to insert DNA into plasmids.

"Polylite" [GC]. TM for a group of 100% reactive alkyd resins, dissolved in styrene and other monomers. Highly diversified applications both alone and in combination with such materials as fibrous glass. This group also includes resins for use with diisocyanate to form rigid or flexible polyurethane foams.

polymaleic acid.
CAS: 26099-09-2.
Use: Food additive.

polymaleic acid, sodium salt. **Use:** Food additive.

"POLYMEKON" [Baker Hughes]. TM for a dispersion.

polymer. A macromolecule formed by the chemical union of five or more identical combining units called monomers. In most cases the number of monomers is quite large (3500 for pure cellulose) and often is not precisely known. In synthetic polymers this number can be controlled to a predetermined extent, e.g., by short-stopping agents. (Combinations of two, three, or four monomers are called, respectively, dimers, trimers, and tetramers and are known collectively as oligomers.) A partial list of polymers by type is as follows:

I. Inorganic: siloxane, sulfur chains, black phosphorus, boron-nitrogen, silicones
II. Organic
  (1) Natural
    (a) Polysaccharides: starch, cellulose, pectin, seaweed gums (agar, etc.), vegetable gums (arabic, etc.)
    (b) Polypeptides (proteins): casein, albumin, globulin, keratin, insulin, DNA
    (c) Hydrocarbons: rubber and gutta-percha (polyisoprene)
  (2) Synthetic
    (a) Thermoplastic: elastomers (unvulcanized), nylon, polyvinyl chloride, polyethylene (linear), polystyrene, polypropylene, fluorocarbon resins, polyurethane, acrylate resins
    (b) Thermosetting: elastomers (vulcanized), polyethylene (cross-linked), phenolics, alkyds, polyesters

(3) Semisynthetic: cellulosics (rayon, methylcellulose, cellulose acetate), modified starches (starch acetate, etc.)

polymer, addition.   See addition polymer.

polymerase.   An enzyme which links individual nucleotides together into a long strand, using another strand as a template. There are two general types of polymerase: DNA polymerases (which synthesize DNA) and RNA polymerase (which make RNA).

polymerase chain reaction.   (PCR). A method for amplifying a DNA base sequence using a heat-stable polymerase and two 20-base primers, one complementary to the (+) strand at one end of the sequence to be amplified and one complementary to the (−) strand at the other end. Because the newly synthesized DNA strands can subsequently serve as additional templates for the same primer sequences, successive rounds of primer annealing, strand elongation, and dissociation produce rapid and highly specific amplification of the desired sequence. PCR also can be used to detect the existence of the defined sequence in a DNA sample.

polymerase, DNA or RNA.   Enzyme that catalyzes the synthesis of nucleic acids on preexisting nucleic acid templates, assembling RNA from ribonucleotides or DNA from deoxyribonucleotides.

polymer, atactic.   See atactic.

polymer, block.   See block polymer.

polymer, condensation.   A polymer formed by a condensation reaction.

polymer, coordination.   A polymer made by organic addition, neither free radical nor ionic, using an organometallic catalyst.

polymer, electroconductive.   A polymer or elastomer made electrically conductive by incorporation of a substantial percentage of a suitable metal powder (e.g., aluminum) or acetylene carbon black; the proportion used must be high enough to permit the particles to be in contact with one another in the mixture. Polyelectrolytes such as ion-exchange resins, salts of polyacrylic acid, and sulfonated polystyrene are electroconductive in the presence of water. Pyrolysis of polyacrylonitrile makes it electrically conductive without impairment of its structure. Polyacetylene and a few related polymers are made conductive by various doping agents such as arsenic pentafluoride and iodine.
See polyacetylene.

polymer eutactic.   A tactic polymer completely devoid of any structural disorder along its chain.

polymer, graft.   See graft polymer.

polymer, high.   An organic macromolecule composed of a large number of monomers. The molecular weight may range from 5000 into the millions (for some polypeptides). Natural high polymers are exemplified by cellulose $(C_6H_{10}O)_n$ and rubber $(C_5H_8)_n$. Proteins are natural high-polymer combinations of amino acid monomers. The dividing line between low and high polymers is considered to be in the neighborhood of 5000–6000 mw.
Synthetic high polymers (or "synthetic resins") include a wide variety of materials having properties ranging from hard and brittle to soft and elastic. Addition of such modifying agents as fillers, colorants, etc. yields an almost infinite number of products collectively called plastics. High polymers are the primary constituents of synthetic fibers, coating materials (paints and varnishes), adhesives, sealants, etc. Polymers having special elastic properties are called rubbers, or elastomers.
Synthetic polymers in general can be classified: (1) by thermal behavior, i.e., thermoplastic and thermosetting; (2) by chemical nature, i.e., amino, alkyd, acrylic, vinyl, phenolic, cellulosic, epoxy, urethane, siloxane, etc.; and (3) by molecular structure, i.e., atactic, stereospecific, linear, cross-linked, block, graft, ladder, etc. Copolymers are products made by combining two or more polymers in one reaction (styrene-butadiene).
See cross-linking; block polymer; epitaxy; homopolymer; plastics.

polymer, homo-.   A polymer consisting of only one repeated monomer.

polymer, inorganic.   A polymer in which the main chain contains no carbon atoms and in which behavior similar to that of an organic polymer can be developed, i.e., covalent bonding and cross-linking, as in silicone polymers. Here the element silicon replaces carbon in the straight chain; substituent groups are often present, forming highly useful polymers. Other inorganic high polymers are black phosphorus, boron, and sulfur, all of which can form polymeric structures under special conditions. At present these have little or no commercial significance.
Note: Some authorities consider silicone resins to be semiorganic, since their substituent groups are composed of methyl groups.

polymer, isotactic.   A type of polymer structure in which groups of atoms that are not part of the backbone structure are located either all above or all below the atoms in the backbone chain, when the latter are all in one plane.

See polymer, stereospecific.

$$-\overset{\overset{\displaystyle H}{|}}{C}-\overset{\overset{\displaystyle H}{|}}{\underset{\underset{\displaystyle R}{|}}{O}}-\overset{\overset{\displaystyle H}{|}}{C}-\overset{\overset{\displaystyle H}{|}}{\underset{\underset{\displaystyle R}{|}}{O}}-\overset{\overset{\displaystyle H}{|}}{C}-\overset{\overset{\displaystyle H}{|}}{\underset{\underset{\displaystyle R}{|}}{O}}-$$

polymerization. A chemical reaction, usually carried out with a catalyst, heat, or light, and often under high pressure, in which a large number of relatively simple molecules combine to form a chainlike macromolecule. The combining units are called monomers, e.g., styrene is the monomer for polystyrene. The linear chains can be combined (cross-linked) by addition of appropriate chemicals. The polymerization reaction occurs spontaneously in nature; industrially it is performed by subjecting unsaturated or otherwise reactive substances to conditions that will bring about combination. This may occur by *addition*, in which free radicals are the initiating agents that react with the double bond of the monomer by adding to it on one side, at the same time producing a new free electron on the other.

$$R \bullet + CH_2 = CHX \rightarrow R - CH_2 - CHX\bullet$$

By this mechanism the chain becomes self-propagating. Polymerization may also occur by condensation, involving the splitting out of water molecules by two reacting monomers, and by so-called oxidative coupling. The degree of polymerization (DP) is the number of monomer units in an average polymer unit of a given sample.

Polymerization techniques may be (1) in the gas phase at high pressures and temperatures (200C), (2) in solution at normal pressure and temperatures from -70 to $+70C$, (3) bulk or batch polymerization at normal pressure at 150C, (4) in suspension at normal pressure at 60–80C, (5) in emulsion form at normal pressure at -20 to $+60C$ (used for copolymers). Catalysts of the peroxide type are necessary with some of these methods.

Note: Polymerization, like its handmaiden, catalysis, has long been one of the most complex and productive areas of chemical research; from year to year new materials and reaction mechanisms are constantly being explored, sometimes with only marginal success. But it need only be recalled that such now commonplace materials as polyethylene, polycarbonate, nylon, neoprene, epoxies, acrylics, to mention only a few, as well as block, graft, and stereospecific polymers, have resulted from continuous and intensive research by many brilliant chemists over the last 60 years, and this research continues undiminished.

See free radical; cross-linking.

polymer, ladder. See ladder polymer.

polymer, low. A polymer composed of comparatively few monomer units and having a molecular weight from 300 to 5000.

polymer, natural. See polymer.

polymer, stereoblock. See stereoblock polymer.

polymer, stereospecific. (stereoregular). A polymer whose molecular structure has a definite spatial arrangement, i.e., a fixed position in geometrical space for the constituent atoms and atomic groups comprising the molecular chain, rather than the random and varying arrangement that characterizes an amorphous polymer. Achievement of this specific steric (three-dimensional) structure (also called tacticity) requires use of special catalysts such as those developed by Ziegler and Natta about 1950. Such polymers are wholly or partially crystalline. Synthetic natural rubber, *cis*-polyisoprene, is an example of a stereospecific polymer made possible by this means. There are five types of stereospecific (or stereoregular) structures: *cis*, *trans*, isotactic, syndiotactic, and tritactic.

See catalyst; stereospecific.

polymer, syndiotactic. See syndiotactic polymer.

polymer, synthetic. See polymer.

polymer, water-soluble. Any substance of high molecular weight that swells or dissolves in water at normal temperature. These fall into several groups, including natural, semisynthetic, and synthetic products. Their common property of water solubility makes them valuable for a wide variety of applications as thickeners, adhesives, coatings, food additives, textile sizing, etc.

See specific entries.

(1) *Natural.* This type is principally composed of gums, which are complex carbohydrates of the sugar group. They occur as exudations of hardened sap on the bark of various tropical species of trees. All are strongly hydrophilic. Examples are arabic, tragacanth, karaya.

(2) *Semisynthetic.* This group (sometimes called water-soluble resins) includes such chemically treated natural polymers as carboxymethylcellulose, methylcellulose, and other cellulose ethers, as well as various kinds of modified starches (ethers and acetates).

(3) *Synthetic.* The principal members of this class are polyvinyl alcohol, ethylene oxide polymers, polyvinyl pyrrolidone, polyethyleneimine.

polymethacrylate resin. See acrylic resin; methyl methacrylate.

polymethylbenzene. See durene; pseudocumene. Durene and pseudocumene are two members of this group with some commercial production and use.

polymethylene polyphenylisocyanate.
CAS: 9016-87-9. A polymer of diphenylmethane-4,4'-diisocyanate.
Hazard: Questionable carcinogen.

polymethylene wax.   See wax, polymethylene.

poly-4-methylpentene-1.
Properties: High resistance to all chemicals except carbon tetrachloride and cyclohexane. Excellent heat resistance, high clarity and light transmittance. Temperature limit 170C, d 0.83.
Use: Laboratory ware (beakers, graduates, etc.); electronic and hospital equipment; food packaging, especially types subject to high temperature such as trays for TV dinners, etc.; light reflectors.

poly(methyl vinyl ether).   See polyvinyl methyl ether.

polymorphic.   Describing a protein for which amino acid sequence variants exist in a population of organisms, but the variations do not destroy the protein's function. For example, antibodies which contain variable and hypervariable sites.

polymorphism.   See allotropy.

polymyxin.
CAS: 1406-11-7. Generic term for a series of antibiotic substances produced by strains of *Bacillus polymyxa*. Various polymyxins are differentiated by the letters A, B, C, D, and E. All are active against certain gram-negative bacteria. Polymyxin B is most used.
Properties: All are basic polypeptides. Soluble in water; the hydrochlorides are soluble in water and methanol, insoluble in ether, acetone, chlorinated solvents, and hydrocarbons. Permissible food additives.
Use: Medicine (antibiotic), beer production.

polymyxin E sulfate.
CAS: 1264-72-8. $C_{45}H_{85}N_{13}O_{10} \cdot H_2O_4S$.
Hazard: A poison by ingestion.

Poly-N.   (ammonium polyphosphate). A chemical used as a phosphorus-based catalyst in organic as well as latex-based products.
Use: Fire-retardant intumescent paints, mastics, and polymers.

polynuclear.   Descriptive of an aromatic compound containing three or more closed rings, usually of the benzenoid type, e.g., sterols.
See polycyclic; nucleus (3).

polynucleotide phosphorylase.   An enzyme that catalyzes the polymerization of ribonucleotide diphosphates.

polyol.   A polyhydric alcohol, i.e., one containing three or more hydroxyl groups. Those having three hydroxyl groups (trihydric) are glycerols; those with more than three are called sugar alcohols, with general formula $CH_2OH(CHOH)_nCH_2OH$, where n may be from 2 to 5. These react with aldehydes and ketones to form acetals and ketals.
See alcohol; glycerol.

polyolefin.   A class or group name for thermoplastic polymers derived from simple olefins; among the more important are polyethylene, polypropylene, polybutenes, polyisoprene, and their copolymers. Many are produced in the form of fibers. This group comprises the largest tonnage of all thermoplastics produced.

polyorganosilicate graft polymer.   An organoclay to which a monomer or an active polymer has been chemically bonded, often by the use of ionizing radiation. An example is the bonding of styrene to a polysilicate containing vinyl radicals, resulting in the growth of polystyrene chains from the surface of the silicate. Such complexes are stable to organic solvents. They have considerable use potential in the ion-exchange field, as ablative agents, reinforcing agents, and hydraulic fluids.
See organoclay; graft polymer.

"Polyox" [Bimeda].   TM for a series of water-soluble ethylene oxide polymers with molecular weights in the 100,000 to several million range.
Use: Textile warp size, paper coatings, detergents, hair spray, toothpastes, water-soluble packaging film, adhesives.

polyoxadiazole.   A polymer of oxadiazole, cyclic C_2N_2O, prepared by cyclodehydration (ring formation from a chain with subsequent loss of water) of polyisophthalic hydrazide. Because of its high temperature tolerance (above 398C), fibers made from it may be useful for space vehicles.

polyoxamide.   A nylon-type material made from oxalic acid and diamines.

polyoxetane.   See oxetane; "Penton" [Aqualon].

polyoxyethylated (C9-C10) alkyl thioether.
CAS: 9004-83-5. $(C_2H_4O)_nC_{14}H_{30}OS$.
Hazard: A severe skin and eye irritant.

polyoxyethylated (4) isodecyl alcohol.
CAS: 61827-42-7. $(C_2H_4O)_nC_{10}H_{22}O$.
Hazard: A severe skin and eye irritant.

polyoxyethylated (6) isodecyl alcohol.
CAS: 26183-52-8. $(C_2H_4O)_nC_{10}H_{22}O$.
Hazard: A moderate skin and severe eye irritant.

polyoxyethylated oleyl amine.
CAS: 58253-49-9. $(C_2H_4O)_n(C_2H_4O)_nC_{22}H_{45}$ NO_2.
Hazard: A severe eye irritant.

polyoxyethylene.   See polyethylene glycol.

polyoxyethylene(4)docyl alcohol phosphate potassium salt.
CAS: 68070-99-5.
Hazard: A severe eye irritant.

polyoxyethylene fatty acid ester.   See polysorbate.

polyoxyethylene lauryl ether phosphate.
CAS: 39464-66-9. $(C_2H_4O)_nC_{12}H_{26}O.xH_3O_4P$.
Hazard: Low toxicity by ingestion and skin contact. A severe skin and eye irritant.

polyoxyethylene (40) monostearate.
(polyethylene glycol stearate).
CAS: 9004-99-3. A mixture of the mono- and distearate esters of mixed polyoxyethylene diols and corresponding free glycols. The monostearate can be represented as $H(OCH_2CH_2)_nOCOC_{17}H_{35}$ (n is about 40).
Properties: Waxy, light-tan solid; nearly odorless. Congealing range 39–45C. Soluble in water, alcohol, ether, and acetone; insoluble in mineral and vegetable oils.
Grade: USP.
Use: Ointments, emulsifier, surfactant, food additive. See polysorbate.

polyoxyethyleneoxypropylene.   (POEOP). A polymer of ethylene and propylene glycols (ethylene oxide propylene oxide).
Use: Solvent.

polyoxyethylene (8) stearate.
CAS: 9004-99-3. A mixture of the mono- and diesters of stearic acid and mixed polyoxyethylene diols having an average polymer length of 7.5 oxyethylene units.
Properties: Cream-colored, soft, waxy, or pasty solid at 25C; faint, fatty odor; slightly bitter, fatty taste. Soluble in toluene, acetone, ether, and ethanol.
Use: Emulsifier in bakery products. See polysorbate.

polyoxymethylene.   Any of several polymers of formaldehyde and trioxane.
See acetal resin.

polyoxypropylene diamine.   (POPDA). Any of the six high molecular weight amines of low viscosity and vapor pressure, high primary amine content, and light color.
Use: As cross-linking agents in epoxy coatings, imparting high flexibility and adhesion at low temperatures. Other possible uses are in polyamide and polyurethane coatings, adhesives, elastomers and foams, as intermediates for textile and paper treatment, and viscosity-index improvers in lubricating oils.

polyoxypropylene ester.   See polypropylene glycol ester.

polyoxypropylene-glycerol adduct.   One of several condensation polymers of propylene oxide and glycerol with molecular weights in the range 1000–4000. Clear, stable, almost colorless, noncorrosive liquids.
Use: Similar to those of polypropylene glycol.

polypeptide.   (peptide). The class of compounds composed of acid units chemically bound together with amide linkages (–CONH–) with elimination of water. A polypeptide is thus a polymer of amino acids. The chain of amino acids (less than 100) are linked by peptide bonds. A segment of such a chain is as follows:

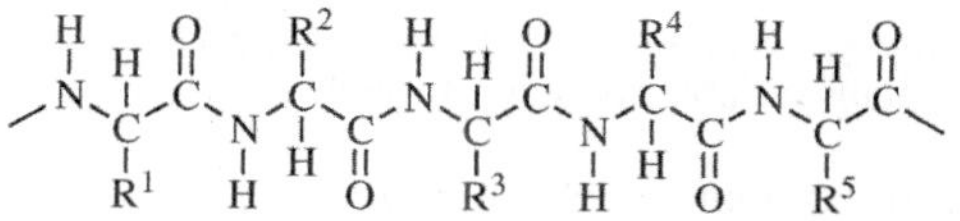

The sequence of amino acids in the chain is of critical importance in the biological functioning of the protein, and its determination is one of the most difficult problems in molecular biology. The chains may be relatively straight, or they may be coiled or helical. In the case of certain types of polypeptides, such as the keratins, they are cross-linked by the disulfide bonds of cystine. Linear polypeptides can be regarded as proteins. Synthesis of a 20-amino acid polypeptide that induces formation of antibodies for foot and mouth disease was announced in 1982. It is the first vaccine to be synthesized.
See protein; polyamide; keratin.

polyphenylene triazole.
$[-C_6H_4-C_2N_3(C_6H_5)-]_n$. A polymer stated to be serviceable up to 260C for films, coatings, adhesives, and lamination.

Polyphos.   A water-soluble glassy sodium phosphate of standardized composition, $(Na_{12}P_{10}O_{31})$, analyzing 63.5% P_2O_5 (ratio of Na_2O:P_2O_5 is 1.2:1). It is closely similar to a sodium hexametaphosphate and sodium tetraphosphate; frequently the three names are used interchangeably.
Grade: Ground, walnut-size to pea-size lumps.
Use: Boiler water compounds, detergents, textiles, leather tanning, photographic film developing, deflocculation of clays, flotation and desliming of minerals, dispersion of pigments, paper processing, industrial and municipal water treatment.

polyphosphazene.   See phosphazene.

polyphosphoric acid. $H_n 2P_n O_3 n_+ 1$, for $n > 1$. Any of a series of strong acids, from pyrophosphoric acid, $H_4 P_2 O_7$ ($n = 2$), through metaphosphoric acid (large values of n).
   Properties: Viscous, water-white liquid. Water soluble. Does not crystallize on standing; hygroscopic. The commercial acid is a mixture of orthophosphoric acid with pyrophosphoric, triphosphoric, and higher acids and is sold on the basis of its calculated content of H_3PO_4, e.g., 115%. Superphosphoric acid is a similar mixture sold at 105% H_3PO_4. These acids revert slowly to orthophosphoric acid on dilution with water.
   Hazard: Toxic by ingestion; strong irritant.
   Use: Dehydrating, catalytic, and sequestering agents, for metal treating; many applications where a concentrated monooxidizing acid is needed; laboratory reagent.
   See phosphoric acid.

polyphyletic.   Term applied to a group of organisms which does not include the most recent common ancestor of those organisms; the ancestor does not possess the character shared by members of the group.

polypropylene.
   CAS: 9003-07-0. $(C_3H_5)_n$. A synthetic, crystalline, thermoplastic polymer with molecular weight of 40,000 or more.
   Note: Low molecular weight polymers are also known that are amorphous in structure and used as gasoline additives, detergent intermediates, greases, sealants, and lubricating oil additives. Also available as a high melting wax.
   Derivation: Polymerization of propylene with a stereospecific catalyst such as aluminum alkyl.
   Properties: Translucent, white solid. D 0.90, mp 168–171C, tensile strength 5000 psi, flexural strength 7000 psi, usable up to 121C. Insoluble in cold organic solvents; softened by hot solvents. Maintains strength after repeated flexing. Degraded by heat and light unless protected by antioxidants. Readily colored; good electrical resistance; low water absorption and moisture permeability; poor impact strength below −9.4C; not attacked by fungi or bacteria; resists strong acids and alkalies up to 60C, but is attacked by chlorine, fuming nitric acid, and other strong oxidizing agents. Combustible, but slow-burning. Fair abrasion and good heat resistance if properly modified. Can be chrome-plated, injection- and blow-molded, and extruded.
   Available forms: (Molding powder) Extruded sheet, cast film (1–10 mils), textile staple and continuous filament yarn, fibers with diameters from 0.05 to 1 mm, and fiber webs down to 2 microns thick, low-density foam.
   Hazard: Questionable carcinogen.
   Use: Packaging film; molded parts for automobiles, appliances, housewares, etc.; wire and cable coating; food container closures; coated and laminated products; bottles; artificial grass and turfs; plastic pipe; wearing apparel (acid-dyed); fish nets; surgical casts; strapping; synthetic paper; reinforced plastics; nonwoven disposable filters.

polypropylenebenzene.   See dodecylbenzene.

polypropylene, chlorinated.   White, odorless, nonflammable powder. A film-forming polymer used in coatings, inks, adhesives, and paper coatings.

polypropylene glycol.   (PG).
   CAS: 25322-69-4. $HO(C_3H_6O)_nH$. One of a group of compounds comparable to polyethylene glycols but more oil soluble and substantially less water soluble. Classified by molecular weight as 425, 1025, 2025. Nonvolatile, noncorrosive liquids; lower molecular weight members are soluble in water. Solvents for vegetable oils, waxes, resins. Combustible.
   Use: Hydraulic fluids, rubber lubricants, antifoam agents, intermediates in urethane foams, adhesives, coatings, elastomers, plasticizers, paint formulations, laboratory reagent.

polypropylene glycol ester.   Exactly analogous to polyethylene glycol ester.

polypropylene glycol monobutyl ether.   See butoxy polypropylene glycol.

polypropyleneimine.   Polymeric form of propyleneimine. Available in 50% aqueous solution.
   Use: Textile, paper, and rubber industries.

polypropylene oxide. $(C_3H_6O)_n$. A derivative of propylene.
   Use: Intermediate for urethane foams.

polypyrrolidone.   Synonym for nylon 4.

"Polyrad" *[Pinova].*   TM for reaction products of "Amine D" and ethylene oxide.
   Grade: Various grades that differ in chain length of polyoxyethylene units and free amine content. Vary in viscosity at 25C from 0.5 to 24.8 cP.
   Use: Corrosion inhibitors and detergents in petroleum processing equipment, wetting and emulsifying agents, inhibiting hydrogen chloride.

"Polyram" *[BASF].*   TM for a wettable powder.
   Hazard: Toxic by ingestion and inhalation.
   Use: Fungicide, approved for many vegetables.

polyribosome.   (polysome). A complex of an mRNA molecule and many ribosomes.

polyrotaxames.   Hooplike molecules threaded as "rotors" on a linear polymer "axle." Up to 40 α-cyclodextrin (rotor) molecules have been threaded on a poly(iminooligomethylene) (hub) chain.

polysaccharide.   A combination of nine or more monosaccharides, linked together by glycosidic bonds. Examples: Starch, cellulose, glycogen.
See carbohydrate; phycocolloid.

polysiloxane.   See siloxane.

"Poly-Solv" *[Monument].*   TM for a series of glycol ether solvents for paints, varnishes, dry-cleaning soaps, cutting oils, insecticides.

polysome.   See polyribosome.

polysorbate.   (USAN for a polyoxyethylene fatty acid ester). One of a group of nonionic surfactants obtained by esterification of sorbitol with one or three molecules of a fatty acid (stearic, lauric, oleic, palmitic) under conditions that cause splitting out of water from the sorbitol, leaving sorbitan. About 20 moles of ethylene oxide per mole of sorbitol are used in the condensation to effect water solution.
Properties: Lemon to amber, oily liquids; faint odor; bitter taste. D 1.1. Most types are soluble in water, alcohol, and ethyl acetate. Combustible.
Grade: Polysorbate 20 (polyoxyethylene (20) sorbitan monolaurate). Polysorbate 60 (polyoxyethylene (20) sorbitan monostearate). Polysorbate 80 (polyoxyethylene (20) sorbitan monooleate). Polysorbate 65 (polyoxyethylene (20) sorbitan tristearate).
Use: Surfactant, emulsifying agent, dispersing agents, shortenings and baked goods, pharmaceuticals, flavoring agents, foaming and defoaming agents.
See sorbitan fatty acid ester.

polysorbate 65.
Properties: Tan, waxy solid; faint odor, bitter taste. Sol in mineral oil, vegetable oil, mineral spirits, acetone, ether, dioxane, alc, methanol; dispersible in water, carbon tetrachloride.
Use: Food additive.

polystyrene.
CAS: 9003-53-6. $(C_6H_5CHCH_2)_n$. Polymerized styrene, a thermoplastic synthetic resin of variable molecular weight depending on degree of polymerization.
Properties: Transparent, hard solid. High strength and impact resistance; excellent electrical and thermal insulator. Attacked by hydrocarbon solvents but resists organic acids, alkalies, and alcohols. Not recommended for outdoor use; unmodified polymer yellows when exposed to light, but light-stable modified grades are available. Easily colored, molded, and fabricated. Copolymerization with butadiene and acrylonitrile and blending with rubber or glass fiber increase impact strength and heat resistance; autoign temp 800F. Combustible.
Derivation: Polymerization of styrene by free radicals with peroxide initiator.
Available forms: Sheet, plates, rods, rigid foam, expandable beads or spheres.

Hazard: As for foam, plastic. Possible carcinogen.
Use: Packaging, refrigerator doors, air-conditioner cases, containers and molded household wares, machine housings, electrical equipment, toys, clock and radio cabinets. Foam: Thermal insulations, light construction (as in boats, etc.), ice buckets, water coolers, fillers in shipping containers, furniture construction. Spheres: Radiator leak stopper.
See Styron; Styrofoam.

polysulfide elastomer.   A synthetic polymer in either solid or liquid form obtained by the reaction of sodium polysulfide with organic dichlorides such as dichlorodiethyl formal, alone or mixed with ethylene dichloride. Outstanding for resistance to oils and solvents and for impermeability to gases. They have poor tensile strength and abrasion resistance but are resilient and have excellent low-temperature flexibility. Some grades have fairly strong odor, which is not objectionable in most applications. Sealant grades are furnished in two parts that cure at room temperature when blended.
Use: Gasoline and oil-loading hose, sealants and adhesive compositions, binder in solid rocket propellants, gaskets, paint spray hose.
See "Thiokol" *[Toray].*

polysulfone.   A synthetic thermoplastic polymer.
Properties: Hard, rigid, transparent solid. Tensile strength 10,000 psi, d 1.24, flexural strength 15,000 psi, good electrical resistance, minimum creep, low expansion coefficient. Soluble in aromatic hydrocarbons, ketones, and chlorinated hydrocarbons; resistant to corrosive acids and alkalies, heat, oxidation, detergents, oils, and alcohols. Dimensionally stable over the temperature range −100 to +148C; tends to absorb moisture, readily processed and fabricated. Combustible but self-extinguishing.
Derivation: Condensation of bisphenol A and dichlorophenyl sulfone.
Use: Power-tool housings, electrical equipment, extruded pipe and sheet, automotive components, electronic parts, appliances, computer components, base matrix for stereotype printing plates.

polyterpene resin.   A class of thermoplastic resins or viscous liquids of amber color, obtained by polymerization of turpentine in the presence of catalysts such as aluminum chloride or mineral acids. The resins consist essentially of polymers of α- or β-pinene and are soluble in most organic solvents.
Use: Paints, rubber plasticizers, curing concrete, impregnating paper, adhesives, hot-melt coatings, pressure-sensitive tapes.
See pinene.

polytetrafluoroethylene.   (PTFE; TFE; Teflon).
CAS: 9002-84-0. $(C_2F_4)_n$. A polymer of tetrafluoroethylene; it is essentially a straight chain of the repeating unit $[-CF_2-CF_2-]_n$. Soft and waxy with a

milk-white color; it can be molded by powder metallurgy techniques involving mixing with a diluent, that is subsequently removed, and sintering at 371C.

Properties: Highly resistant to oxidation and action of chemicals including strong acids, alkalies, oxidizing agents; resistant to nuclear radiation and UV rays, ozone, and weather; halogenated solvents at high temperatures and pressures have some adverse effect. Retains useful properties up to 287C and is strong and tough. Low coefficient of friction (0.05) and antistick properties; excellent resistance to electricity; coefficient of thermal expansion greater than other plastics and metals. Nonflammable.

Available forms: Extrusion and molding powders, aqueous dispersion, film, multifilament fiber.

Hazard: Evolves toxic fumes on heating. Questionable carcinogen.

Use: Gaskets, liners, seals, flexible hose; ablative coatings for rockets and space vehicles: chemical process equipment; coatings in aerospace; coaxial spacers; insulators; wire coating and tape in electrical and electronic fields; bearings; seals; piston rings; antistick coatings for cooking vessels, felts, packings, and bearings.

See Teflon; Halon.

polythene. (generic name for polyethylene). No longer current terminology in the U.S. but still used in England.

polythiadiazole. $[-C_6H_4-C_2N_2S-]_n$. A polymer made from polyoxathiahydrazide. Can be converted to fibers; stated to retain properties to 398C and to have resistance to thermal degradation such that it retains 60 or 70% of original tenacity after 32 hours at 398C.

polythiazyl. $(SN)_n$. An experimental polymer of sulfur nitride with covalent linkages said to have the optical and electrical properties characteristic of metals. Thin films are reported to exhibit epitaxial growth.

polytrifluorochloroethylene resin. See chlorotrifluoroethylene resin.

polyunsaturated fat. A fat or oil based at least partly on fatty acids having two or more double bonds per molecule, such as linoleic and linolenic acids. Examples are corn oil and safflower oil.

Use: Margarine and dietary foods, salad dressings, etc.

polyurethane.

CAS: 9009-54-5. A thermoplastic polymer (which can be made thermosetting) produced by the condensation reaction of a polyisocyanate and a hydroxyl-containing material, e.g., a polyol derived from propylene oxide or trichlorobutylene oxide. The basic polymer unit is formed as follows: $R_1NCO + R_2OH \rightarrow R_1NHCOOR_2$.

Fiber

Properties: High elastic modulus, good electrical resistance, highly moisture-proof, crystalline structure. Combustible.

Derivation: Reaction of hexamethylene diisocyanate and 1,4-butanediol.

Hazard: Questionable carcinogen.

Use: Chiefly in so-called spandex fibers for girdles and other textile structures requiring exceptional elasticity, bristles for brushes, etc.

Coatings

Properties: Excellent hardness, gloss, flexibility, abrasion resistance, and adhesion; resistant to impact, weathering, acids, and alkalies; attacked by aromatic and chlorinated solvents. Applied by brush, spray, or dipping. Combustible.

Derivation: Formed from "prepolymers" containing isocyanate groups (toluene and 4,4'-diphenylmethane diisocyanates) and hydroxyl-containing materials such as polyols and drying oils.

Use: Baked coatings, two-component formulations, wire coatings, tank linings, maintenance paints, masonry coating.

Elastomers

Properties: Good resistance to abrasion, weathering, and organic solvents, tend to harden and become brittle at low temperatures. Combustible.

Derivation: Reaction of polyisocyanates with linear polyesters or polyethers containing hydroxyl groups.

Use: Sealants and caulking agents; adhesives; films and linings; shoe uppers and heels; encapsulation of electronic parts; binders for rocket propellants; abrasive wheels and other mechanical items; automobile bumpers, fenders, and other components.

Foams

Properties: Both flexible and rigid foams are available. Density varies from 2 to 50 lb/ft^3, thermal conductivity as low as 0.11 Btu/hr/sq ft/F/in, excellent insulators. Combustible unless protected by effective thermal barrier.

Hazard: Evolves toxic fumes on ignition.

See foam, plastic.

Derivation: A polyether such as polypropylene glycol is treated with a diisocyanate in the presence of some water and a catalyst (amines, tin soaps, organic tin compounds). As the polymer forms, the water reacts with the isocyanate groups to cause cross-linking and also produces carbon dioxide, which causes foaming. In other cases, trifluoromethane or similar volatile material may be used as a blowing agent. Flexible foams are based on polyoxypropylenediols of 2000 molecular weight and triols up to 4000 molecular weight. Rigid foams are based on polyethers made from sorbitol, methyl glucoside, or sucrose.

Use: (Flexible) Furniture, mattresses, laminates and linings, flooring leveling, seat cushions and other automotive accessories, carpet underlays, upholstery, absorbent of crude oil spills on seawater. (Rigid) Furniture; packaging and packing; building insulation; marine flotation; automobile components; cigarette filters; light structures (as boat hulls); soundproofing; salvaging operations; ship building (for buoyancy); transportation insulation for box cars, refrigerated cars, tank and hopper cars, trucks, and trailers (claimed to be twice as effective as glass fiber); insulation for storage tanks, ships' holds, and pipelines; automobile bumpers.

polyvalent serum. Antiserum that contains antibody to more than one kind of antigen.
Derivation: Produced by mixing monovalent sera.

polyvinyl acetal. One of the family of vinyl resins resulting from the condensation of polyvinyl alcohol with an aldehyde; acetaldehyde, formaldehyde, and butyraldehyde are commonly used. The three main groups are polyvinyl acetal itself, polyvinyl formal, and polyvinyl butyral. These are all thermoplastic and can be processed by extruding, molding, coating, and casting. They are used chiefly in adhesives, paints and lacquers, and films. Polyvinyl butyral is used in sheet form as an interlayer in safety glass and shatter-resistant protection in aircraft.

$$-CH_2CH \underset{\underset{\displaystyle C_3H_7}{\overset{\displaystyle CH}{O}}{\overset{\displaystyle CH_2}{\underset{O}{}}} CH-$$

polyvinyl acetate. (PVAc).
CAS: 9003-20-7. $[-CH_2CH(OOCCH_3)-]_n$. A thermoplastic high polymer.
Properties: Colorless transparent solid; odorless. D 1.19 (15C). Insoluble in water, gasoline, oils, and fats; soluble in low molecular weight alcohols, esters, benzene, and chlorinated hydrocarbons. Resistant to weathering. Combustible.
Derivation: Polymerization of vinyl acetate with peroxide catalysts.
Hazard: Questionable carcinogen.
Use: Latex water paints; adhesives for paper, wood, glass, metals, and porcelain; intermediate for conversion to polyvinyl alcohol and acetals; sealant; shatterproof photographic bulbs; paper coating and paperboard; bookbinding; textile finishing; nonwoven fabric binder; component of lacquers, inks, and plastic wood; strengthening agent for cements.

polyvinyl alcohol. (PVA; PVOH).
CAS: 9002-89-5. $(-CH_2CHOH-)_x$. A water-soluble synthetic polymer made by alcoholysis of polyvinyl acetate.
Properties: White- to cream-colored powder. D 1.27–1.31, refr index 1.49–1.53. Properties depend on degree of polymerization and the percentage of alcoholysis, both of which are controllable in processing. Water solubility increases as molecular weight decreases; strength, elongation, tear

resistance, and flexibility improve with increasing molecular weight. Tensile strength up to 22,000 psi; decomposes at 200C. PVA has high impermeability to gases, is unaffected by oils, greases, and petroleum hydrocarbons. Attacked by acids and alkalies. It forms films by evaporation from water solution. Combustible.
Grade: Super high viscosity (mw 250,000–300,000), high viscosity (mw 170,000–220,000), medium viscosity (mw 120,000–150,000), low viscosity (mw 25,000–35,000).
Use: Textile warp and yarn size, laminating adhesives, molding powders, binder for cosmetic preparations, ceramics, leather, cloth, nonwoven fabrics and paper, paper coatings, grease-proofing paper, emulsifying agent, thickener and stabilizer, photosensitive films, cements and mortars, intermediate for other polyvinyls, imitation sponges, printing inks (glass).

polyvinylbenzyltrimethyl ammonium chloride. An electrically conductive polymer.
Use: To increase the conductivity of papers.

polyvinyl butyral. See polyvinyl acetal.

polyvinyl carbazole. A brown thermoplastic resin obtained by the reaction of acetylene with carbazole. It softens at 150C, and has excellent dielectric properties, good heat, and chemical stability but poor mechanical strength.
Use: Substitute for mica in electrical equipment and as an impregnant for paper capacitors.

polyvinyl chloride. (PVC).
CAS: 9002-86-2. $(-H_2CCHCl-)_n$. A synthetic thermoplastic polymer.
Properties: White powder or colorless granules. Resistant to weathering and moisture; dimensionally stable; good dielectric properties; resistant to most acids, fats, petroleum hydrocarbons, and fungus. Readily compounded into flexible and rigid forms by use of plasticizers, stabilizers, fillers, and other modifiers. Easily colored and processed by blow molding, extrusion, calendering, fluid-bed coating, etc. Available as film, sheet, fiber, and foam.
Derivation: Polymerization of vinyl chloride by free radicals with peroxide initiator. May be copolymerized with up to 15% of other vinyls.
Hazard: Decomposes at 148C, evolving toxic fumes of hydrogen chloride. Pneumoconiosis, lower respiratory tract irritant, and pulmonary function effects. Questionable carcinogen.
See vinyl chloride.
Use: Piping and conduits of all kinds; siding; gutters; window and door frames; officially approved for use in interior piping, plumbing, and other construction uses. Raincoats, toys, gaskets, garden hose, electrical insulation, shoes, magnetic tape, film and sheeting, containers for toiletries, cosmetics, household

chemicals, fibers for athletic supports, sealant liners for ponds and reservoirs, adhesive and bonding agent, plastisols and organosols, tennis court surfaces, flooring, coating for paper and textiles, wire and cable protection, base for synthetic turf, phonograph records, fuel in pyrotechnic devices.

Note: Use of PVC in rigid and semirigid food containers such as bottles, boxes, etc., is under restriction by FDA as well as in coatings for fresh citrus fruits. Its use in thinner items such as films and package coatings is permissible. Possibility of migration of vinyl chloride monomer into food products is the critical factor; this tends to increase with the thickness of the material.

polyvinyl chloride-acetate.
CAS: 34149-92-3. $(-C_2H_4O \cdot C_2H_3Cl-)_n$. A vinyl chloride and vinyl acetate copolymer that is more flexible than polyvinyl chloride. The copolymer usually contains 85–97% of the chloride. It generally has similar properties and uses as polyvinyl chloride.

polyvinyl dichloride. (PVDC). A chlorinated polyvinyl chloride. Has high strength and superior chemical resistance over a broad temperature range. Combustible but self-extinguishing.
Use: Pipe and fittings for hot corrosive materials up to 100C. Immune to solvation or direct attack by inorganic reagents, aliphatic hydrocarbons, and alcohols.

polyvinyl ether. See polyvinyl ethyl ether; polyvinyl isobutyl ether; polyvinyl methyl ether; polyvinyl methyl ether-maleic anhydride.

polyvinyl ethyl ether. (PVE; polyvinyl ether). $[-CH(OC_2H_5)CH_2-]_n$.
Properties: Viscous gum to rubbery solid, depending on molecular weight. Colorless when pure. D 0.97 (20C), refr index 1.45 (25C). Insoluble in water; soluble in nearly all organic solvents; stable toward dilute and concentrated alkalies and dilute acids. Compatible with a limited number of commercial resins, including rosin derivatives and some phenolics.
Derivation: Polymerization of vinyl ethyl ether.
Use: Pressure-sensitive tape, to improve adhesion to porous surfaces, cellophane, cellulose acetate, and vinyl sheet.

polyvinyl fluoride. $(-H_2CCHF-)_n$. Polymer of vinyl fluoride. In film form it is characterized by superior resistance to weather, high strength, high dielectric constant, low permeability to air and water, as well as oil, chemical solvent, and stain resistance.
Hazard: Not recommended for food packaging; evolves toxic fumes on heating.
Use: Protective material for outdoor use, packaging, electrical equipment.

polyvinyl formal. See polyvinyl acetal.

polyvinylidene chloride. (saran). A stereoregular, thermoplastic polymer.
Properties: Tasteless, odorless, abrasion resistant, low vapor transmission, impermeable to flavor, highly inert to chemical attack. Softened by chlorinated hydrocarbons, soluble in cyclohexanone and dioxane. Combustible but self-extinguishing.
Derivation: Polymerization of vinylidene chloride or copolymerization of vinylidene chloride with lesser amounts of other unsaturated compounds.
Available forms: Extruded and molded products, oriented fibers, films.
Use: Packaging of food products; specially meats and poultry, insecticide-impregnated multiwall paper bags; pipes for chemical processing equipment; seat covers; upholstery; fibers; bristles; latex coatings.
See saran fiber; Cryovac.

polyvinylidene fluoride. $(-H_2C=CF_2-)_n$. A thermoplastic fluorocarbon polymer suitable for compression and injection molding and extrusion.
Properties: Crystals. Mp 171C, thermally stable from −62 to +148C; combustible, self-extinguishing and nondripping. Tensile strength 7000 psi at 25C, yield stress 5500 psi, elongation 300%, compression strength 10,000 psi, thermal conductivity 1.05 Btu/hr/sq ft/F/in, water absorption 0.04% in 24 hours, d 1.76, refr index 1.42. Resistant to oxidative degradation, electricity, acids, alkalies, oxidizers, halogens; somewhat soluble in dimethylacetamide, attacked by hot concentrated sulfuric acid or *n*-butylamine.
Available forms: Powder, pellets, solution, and dispersion.
Use: Insulation for high-temperature wire; tank linings; chemical tanks and tubing; protective paints and coatings with exceptional resistance (30 years) to weathering and UV; valve and impeller parts; shrinkage tubing to encapsulate resistors, diodes, and soldered joints; sealant.
See fluorocarbon polymer.

polyvinyl isobutyl ether. (PVI; polyvinyl ether). $[-CH(OCH_2CH(CH_3)_2CH_2-]_n$.
Properties: White, opaque elastomer or viscous liquid depending on molecular weight; almost odorless. D 0.91–0.93 (20C), refr index 1.45–1.46 (25C). Insoluble in water, ethanol, and acetone; soluble in most other organic solvents; stable toward dilute and concentrated alkalies and dilute acids. Compatible with a limited number of commercial resins, including rosin derivatives and some phenolics. Combustible.
Derivation: Polymerization of vinyl isobutyl ether by peroxides or acid catalysts.
Grade: As 100% material in three molecular weight ranges.
Use: Adhesives, waxes, tackifiers, plasticizers, surface coatings, laminating agents, cable filling compositions, lubricating oils.

polyvinyl methyl ether.   (PVM).
$(-CH_2CHOCH_3-)_n$. A nonionic polymer of high molecular weight.
Properties: Colorless (when pure), tacky liquid. D 1.05, refr index 1.47, solidifies near 0C. Soluble in water at 32C above which it precipitates from water but redissolves on cooling. Soluble in most organic solvents except aliphatic hydrocarbons. Compatible with rubber latexes, rosin derivatives, and some phenolics. Combustible.
Derivation: Polymerization of vinyl methyl ether with peroxides or acid catalysts.
Use: Pressure-sensitive and hot-melt adhesives, bonding of paper to polyethylene, laminating adhesive, tackifier and plasticizer for coatings, heat sensitizer for rubber latex, pigment binder in textile finishing and printing inks, protective colloid in emulsions.

polyvinyl methyl ether-maleic anhydride.
(PVM/MA). $[-CH_2CHOCH_3CHCOOCOCH-]_n$.
Water-soluble copolymer of vinyl methyl ether and maleic anhydride.
Use: Protective colloid, dispersing agent, thickener, binder, adhesive and size in coatings, detergents, emulsions, paper, textiles, leather, latex, rust preventive, foam stabilizer.

polyvinylpolypyrrolidone.
Properties: White, hygroscopic powder; faint bland odor. Insol in water.
Use: Food additive.

poly-2-vinylpyridine. $(-CH(C_5H_4N)-CH_2-)_n$.
A vinyl-type polymer suggested for use as a photographic dye mordant, tablet coating material, antistatic for textiles and paper, and textile dye receptor.

polyvinylpyrrolidone.   (PVP).
CAS: 9003-39-8. $(C_6H_9NO)_n$.
Properties: White, free-flowing, amorphous powder or aqueous solution. D 1.23–1.29, bulk d 15 lb/ft^3, hygroscopic. Soluble in water and organic solvents; compatible with wide range of hydrophilic and hydrophobic resins.
Grade: Various molecular weight: 10,000, 40,000, 160,000, 360,000.
Hazard: Questionable carcinogen.
Use: Pharmaceuticals; blood plasma expander; cast films adherent to glass, metals, and plastics; complexing agent; detoxification of chemicals such as dyes, iodine, phenol, and poisonous drugs. Tableting, photographic emulsions, cosmetics (hair sprays, shampoos, hand creams, skin lotions), dentifrices, dye-stripping, textile finishes, protective colloid, detergents, adhesives, beer and wine clarification.

polyvinyl resin.   (vinyl plastic). Any of a series of polymers (resins) derived by polymerization or copolymerization of vinyl monomers

$(CH_2=CH-)$ including vinyl chloride and acetate, vinylidene chloride, methyl acrylate and methacrylate, acrylonitrile, styrene, the vinyl ethers, and numerous others. Specifically, polyvinyl chloride, acetate, alcohol, etc. and copolymers or closely related materials.
See vinyl; polyvinyl.

"POLYWAX" *[Baker Hughes].*   TM for polyethylene.

Pomeranz–Fritsch reaction.   Formation of isoquinolines by the acid-catalyzed cyclization of benzalaminoacetals prepared from aromatic aldehydes and aminoacetal.

Ponceau 3R.   (3-hydroxy-4-[2,4,5-(trimethylphenyl)azo]-2,7-naphthalenedisulfonic acid, disodium salt; CI 16155).
CAS: 3564-09-8. $(CH_3)_3C_6H_2NNC_{10}H_4(OH)(SO_3)•2Na$. A water-soluble red dye.
Properties: Dark-red powder. Soluble in water and acids to form a cherry-red solution; slightly soluble in alcohol; insoluble in alkalies.
Hazard: Prohibited by FDA for use in foods, drugs, cosmetics. Possible carcinogen.
Use: Dyeing wool; biological stain.

"Ponolith" *[Kemira].*   TM for a type of highly dispersed pigments used primarily for the coloring and tinting of paper.

pontianak.   A type of Manila resin that is semifossil, hard, and soluble in oils and hydrocarbons.
Use: Adhesives; paints, varnishes, and lacquers.
See copal resin.

pony.   A small unit of equipment used for laboratory or plant experimental work.

Ponzio reaction.   Formation of dinitrophenylmethanes from benzaldoximes by oxidation with nitrogen dioxide in ether.

POPDA.   See polyoxypropylene diamine.

Pople, John A.   (1925–2004). A British citizen awarded the Nobel Prize in chemistry in 1998, jointly with Walter Kohn, for pioneering contributions in developing methods that can be used for theoretical studies of the properties of molecules and the chemical processes in which they are involved. He performed his research at Northwestern University, Evanston, IL.

POPOP.   Abbreviation for phenyl-oxazolyl-phenyloxazolyl-phenyl.
See 1,4-bis[2-(5-phenyloxazolyl)]benzene.

population genetics.   The study of variation in genes among a group of individuals.

porcelain. (potassium aluminum silicate). ($4K_2O \cdot Al_2O_3 \cdot 3SiO_2$). A mixture of clays, quartz, and feldspar usually containing at least 25% alumina. Ball and china clays are ordinarily used. A slip or slurry is formed with water to form a plastic, moldable mass, which is then glazed and fired to a hard, smooth solid.
Properties: High impact strength, impermeable to liquids and gases, resistant to chemicals except hydrogen fluoride and hot, strong caustic solutions, usable up to 1093C but subject to heat shock. D 2.41, Mohs hardness 6–7, compression strength 100,000 psi.
Grade: Chemical and electrical.
Use: Reaction vessels, spark plugs, electrical resistors, electron tubes, corrosion-resistant equipment, ball mills and grinders, food-processing equipment, piping, valves, pumps, tower packing, laboratory ware.
See ceramics; porcelain enamel; porcelain, zircon.

porcelain enamel. A substantially vitreous inorganic coating bonded to metal by fusion above 426C (ASTM). Composed of various blends of low-sodium frit, clay, feldspar, and other silicates; ground in a ball mill; and sprayed onto a metal surface (steel, iron, or aluminum), to which it bonds firmly after firing, giving a glasslike fire-polished surface.
Properties: High hardness, abrasion, and chemical resistance except to hydrogen fluoride and hot strong caustic; low expansion coefficient. Corrosion resistant, stable to heat-shock, easy to clean and decontaminate.
Use: Chemical reaction equipment, light reflectors, laboratory bench tops, storage tanks and containers, high temperature process equipment, seawater treatment and marine engine parts, glazing frit.
See porcelain; glaze.

porcelain, zircon. ($ZrO_2 \cdot SiO_2$). A special high-temperature porcelain used for spark plugs and furnace trays because of its high mechanical strength and heat-shock resistance. Usable up to 1700C. High dielectric strength but rather lower power factor at high frequencies.

pore. (1) A minute cavity in epidermal tissue as in skin, leaves, or leather, having a capillary channel to the surface that permits transport of water vapor from within outward but not the reverse. (2) A void or interstice between particles of a solid such as sand minerals or powdered metals, that permits passage of liquids or gases through the material in either direction. In some structures, such as gaseous diffusion barriers and molecular sieves, the pores are of molecular dimensions, i.e., 4–10 Å units. Such microporous structures are useful for filtration and molecular separation purposes in various industrial operations. (3) A cell in a spongy structure made by

gas formation (foamed plastic) that absorbs water on immersion but releases it when stressed.
See membrane, semipermeable; molecular sieve.

"Porocel" [Porocel]. TM for an activated bauxite. Supplied in various standard meshes, moisture contents, and regular low iron and low silica grade.
Use: Absorbent catalysts.

poromeric. A term coined to describe the microporosity, air permeability, and water and abrasion resistance of natural and synthetic leather. The pores decrease in diameter from the inner surface to the outer and thus permit air and water vapor to leave the material while excluding water from the outside. Polyester-reinforced urethane resins have poromeric properties, and vinyl resins have been used as leather substitutes with some success, primarily for shoe uppers.

porphyrin. Any of several physiologically active nitrogenous compounds occurring widely in nature. The parent structure is composed of four pyrrole rings, shown as I, II, III, and IV in the diagram below, together with four nitrogen atoms and two replaceable hydrogens, for which various metal atoms can be readily substituted. A metal-free porphyrin molecule has the structure:

Porphyrins of this type have been made synthetically by passing an electric current through a mixture of ammonia, methane, and water vapor. Biochemists believe that this phenomenon may account for the original formation of chlorophyll and other porphyrins that have been essential factors in the origin of life.
The most important porphyrin derivatives are characterized by a central metal atom; hemin is the iron-containing porphyrin essential to mammalian blood, and chlorophyll is the magnesium-containing porphyrin that catalyzes photosynthesis. Other derivatives of importance are cytochromes, which function in cellular metabolism, and the phthalocyanine group of dyes.

porpoise oil. (dolphin oil).
Properties: Pale-yellow liquid. D 0.926–0.929, saponification value 290, iodine value 10–30. Soluble in ether, chloroform, benzene, and carbon disulfide; the refined jaw and head oil has uniquely low

pour point and high lubricity and is highly resistant to gumming, oxidation, and evaporation.
Use: Lubricant for watches and precision instruments.

Porta-Pig.   A lead-lined mobile cart.
Use: Carrying radioactive materials.

Porter, George. (1920–2002). An English chemist who won the Nobel Prize for chemistry in 1967 with Eigen and Norrish. His research concerned fast chemical reactions and the chemistry of photosynthesis. He was educated at Cambridge University and taught there before going on to other posts.

Portland cement.   See cement, Portland.

"Positek" [Nalco].   TM for retention and draining program.
Use: Papermaking.

positive cooperativity.   A phenomenon of some multisubunit enzymes or proteins in which binding of a ligand or substrate to one subunit facilitates binding of a second ligand or substrate to another subunit.

positional cloning.   A technique used to identify genes, usually those that are associated with diseases, based on their location on a chromosome.

positron.   An antiparticle whose mass and spin are the same as those of an electron but whose electric charge is positive. Positrons can be made in a linear particle accelerator for use in physical research.
See antiparticle.

post-transcriptional processing.   The enzymatic processing of an hnRNA, producing functional mRNA, tRNA, or rRNA molecules.

post-transcriptional regulation.   Any process occurring after transcription which affects the amount of protein a gene produces. Includes RNA processing efficiency, RNA stability, translation efficiency, protein stability. For example, the rapid degradation of an mRNA will reduce the amount of protein arising from it. Increasing the rate at which an mRNA is translated will increase the amount of protein product.

post-translational modification.   Enzymatic processing of a polypeptide chain after translation from its mRNA. Examples are enzymatic cleavage, phosphorylation or sulfation, and oligosaccharide addition.

post-translational processing.   The reactions which alter a protein's covalent structure, such as phosphorylation, glycosylation, or proteolytic cleavage.

potasan.   (4-methyl-7-hydroxycoumarin diethoxythiophosphate; hymecromone-O,O-diethylphosphorothioate).
CAS: 299-45-6. $C_{14}H_{17}O_5PS$.
Properties: Acicular crystals. Mp 38C, bp 210C (1 mm Hg) (decomposes), d 1.26, refr index 1.567. Soluble in common organic solvents; almost insoluble in water.
Hazard: Toxic by ingestion, cholinesterase inhibitor.
Use: Insecticide.

potash.   See potassium carbonate.

potash alum.   See aluminum potassium sulfate.

potash, caustic.   See potassium hydroxide.

potash chrome alum.   See chromium potassium sulfate.

potash feldspar.   See feldspar.

potash magnesia double salt.   A material containing potassium carbonate, magnesium sulfate, a low proportion of chloride, and 20–30% K_2O.
Use: Fertilizer.

potash, sulfurated.   (potassium, sulfurated). A mixture composed chiefly of potassium polysulfides and potassium thiosulfate, containing not less than 12.8% sulfur in combination as sulfide.
Properties: Liver brown when freshly made changing to a greenish yellow; odor of hydrogen sulfide; bitter, acrid, and alkaline taste; decomposes upon exposure to air, soluble in water leaving a residue.
Grade: USP.
Use: Medicine; decorative color effects on brass, bronze, and nickel.

potassium.
CAS: 7440-09-7.   K. Metallic element of atomic number 19, group IA of the periodic table, an alkali metal, aw 39.098, valence of 1. Potassium-40 is a naturally occurring radioactive isotope. There are also two stable isotopes. The synthetic isotope, potassium-42, is used in tracer studies, primarily in medicine. An essential element in plant growth and in animal and human nutrition; occurs in all soils.
Properties: Soft, silvery metal. Rapidly oxidized in moist air. D 0.862, mp 63C, bp 770C. Soluble in liquid ammonia, aniline, mercury, and sodium.
Derivation: Ores and deposits in Stassfurt (Germany), Carlsbad (New Mexico), Saskatchewan (Canada), Searles Lake (California), Great Salt Lake (Utah), Yorkshire (England), Ural Mountains (the former U.S.S.R.), Israel, and eastern Mediterranean area. The most important ores are carnallite, sylvite, and polyhalite. Thermochemical distillation

of potassium chloride with sodium is the chief U.S. method of production.

Grade: Technical, 99.95% pure.

Hazard: Dangerous fire risk, reacts with moisture to form potassium hydroxide and hydrogen. The reaction evolves much heat, causing the potassium to melt and spatter. It also ignites the hydrogen. Burning potassium is difficult to extinguish; dry powdered soda ash or graphite or a special mixture of dry chemicals is recommended. It can ignite spontaneously in moist air. Moderate explosion risk by chemical reaction. Potassium metal will form the peroxide and the superoxide at room temperature even when stored under mineral oil; may explode violently when handled or cut. Oxide-coated potassium should be destroyed by burning. Store in inert atmospheres, such as argon or nitrogen, under liquids that are oxygen free, such as toluene or kerosene, or in glass capsules that have been filled under vacuum or inert atmosphere.

Use: Preparation of potassium peroxide, heat-exchange alloys (see NaK); laboratory reagent; seeding of combustion gases in magnetohydrodynamic generators; component of fertilizers (as potassium chloride).

potassium abietate. $KO_2CC_{19}H_{29}$. Water-soluble soap resulting from action of rosin on potassium hydroxide.

potassium acetate.
CAS: 127-08-2. $KC_2H_3O_2$.
Properties: White, crystalline, deliquescent powder; saline taste. Mp 292, d 1.57 (25C). Soluble in water and alcohol; insoluble in ether; solutions alkaline to litmus but not to phenolphthalein. Combustible.
Derivation: By the action of acetic acid on potassium carbonate.
Grade: Pure, pure fused, CP, NF reagent.
Use: Dehydrating agent, textile conditioner, reagent in analytical chemistry, medicine, cacodylic derivatives, crystal glass, synthetic flavors.

potassium acid carbonate. See potassium bicarbonate.

potassium acid fluoride. See potassium bifluoride.

potassium acid oxalate. See potassium binoxalate.

potassium acid phosphate. See potassium phosphate, monobasic.

potassium acid saccharate.
$HOOC(CHOH)_4COOK$.
Properties: Light, off-white powder, pH of solution 3.5. Slightly soluble in cold water; soluble in hot water, acid, or alkaline solutions. Combustible.
Use: Chelating agent, rubber formulations, metal plating, soaps, and detergents.

potassium acid sulfate. See potassium bisulfate.

potassium acid sulfate, anhydrous. See potassium pyrosulfate.

potassium acid sulfite. See potassium bisulfite.

potassium acid tartrate. See potassium bitartrate.

potassium alginate. (potassium polymannuronate). $(C_6H_7O_6K)_n$. Hydrophilic colloid having a molecular weight of 32,000–250,000.
Properties: Occurs in filamentous, grainy, granular, and powdered forms. It is colorless or slightly yellow and may have a slight characteristic smell and taste. Slowly soluble in water, forming a viscous solution; insoluble in alcohol.
Grade: Technical, FCC.
Use: Thickening agent and stabilizer in dairy products, canned fruits, and sausage casings; emulsifier. See alginic acid.

potassium alum. See aluminum potassium sulfate.

potassium aluminate. $K_2Al_2O_4 \cdot 3H_2O$.
Properties: Hard crystals, lustrous. Soluble in water with hydrolysis to form strongly alkaline solution; insoluble in alcohol.
Derivation: By fusing potassium hydroxide with aluminum oxide.
Grade: Technical.
Use: Dyeing, printing (mordant); lakes, paper sizing.

potassium aluminosilicate. See feldspar.

potassium aluminum fluoride. K_3AlF_6.
Properties: White powder. Slightly soluble in water.
Derivation: Aluminum fluoride, ammonium fluoride, and potassium chloride.
Hazard: Toxic by ingestion and inhalation, strong irritant.
Use: Insecticide.

potassium aluminum sulfate. See aluminum potassium sulfate.

potassium-p-aminobenzoate. $C_7H_6KNO_2$.
Properties: Colorless crystals. Soluble in water; partially soluble in alcohol; insoluble in ether.
Use: Condensation catalyst, mainly for polyglycol ether polymers.

potassium antimonyl tartrate. See antimony potassium tartrate.

potassium argentocyanide. See silver potassium cyanide.

potassium arsenate.   (Macquer's salt).
CAS: 7784-41-0. KH_2AsO_4.
Properties: Colorless crystals. D 2.867, mp 288C. Soluble in water; insoluble in alcohol.
Hazard: Toxic by ingestion and inhalation, strong irritant.
Use: Manufacture of fly paper, insecticidal preparations, preserving hides, printing textiles.

potassium arsenite.   (potassium metaarsenite).
CAS: 10124-50-2. $KH(AsO_2)_2•H_2O$.
Properties: White powder. Hygroscopic; decomposes slowly in air. Variable composition. Keep well stoppered. Soluble in water; slightly soluble in alcohol.
Grade: Technical, reagent.
Hazard: Toxic by ingestion and inhalation, strong irritant.
Use: Reducing agent in silvering mirrors.

potassium asulam.
CAS: 14089-43-1. $C_8H_9N_2O_4S•K$.
Hazard: Low toxicity by ingestion.
Use: Agricultural chemical.

potassium aurate. $KAuO_2•3H_2O$.
Properties: Yellow crystals. Soluble in water and alcohol.
Derivation: Gold oxide dissolved in potassium hydroxide solution.
Use: To prepare other gold compounds.

potassium beryllium fluoride.   See beryllium potassium fluoride.

potassium bicarbonate.   (potassium acid carbonate; baking soda).
CAS: 298-14-6. $KHCO_3$.
Properties: Colorless, transparent crystals or white powder; odorless; slightly alkaline, salty taste. D 2.17, mp decomposes between 100 and 120C, refr index 1.482. Soluble in water and potassium carbonate solution; insoluble in alcohol.
Derivation: By passing carbon dioxide into a solution of potassium carbonate in water.
Grade: Commercial, highest purity, USP, reagent, FCC.
Use: Baking powders, soft drinks, medicine (antacid), manufacture of pure potassium carbonate, fire-extinguishing agent, low-pH liquid detergents, laboratory reagent, food additive.

potassium bichromate.   See potassium dichromate.

potassium bifluoride.   (potassium acid fluoride; potassium hydrogen fluoride).
CAS: 7789-29-9. KHF_2.
Properties: Colorless crystals. D 2.37, mp 238C. Decomposed by heat; soluble in alcohol (dilute) and water; insoluble in alcohol (absolute).
Grade: Technical.
Hazard: Corrosive to tissue.

Use: Etching glass, flux in silver solders, alkylation catalyst, electrolyte in fluorine production.

potassium   binoxalate.   (potassium acid oxalate; acid potassium oxalate; sorrel salt).
CAS: 127-95-7. $KHC_2O_4•1/2H_2O$.
Properties: White crystals; bitter, sharp taste; somewhat hygroscopic. D (anhydrous salt) 2.088. Decomposes when heated. Soluble in water; insoluble in alcohol.
Derivation: Neutral potassium oxalate and oxalic acid are dissolved in water and crystallized.
Hazard: Toxic by ingestion.
Use: Removing ink stains, scouring metals, cleaning wood, photography, laboratory reagent, mordant.

potassium biphthalate.   See potassium hydrogen phthalate.

potassium bisulfate.   (acid potassium sulfate; potassium hydrogen sulfate; potassium acid sulfate).
CAS: 7646-93-7. $KHSO_4$.
Properties: Colorless crystals. The fused salt is deliquescent. D 2.245, mp 195 (decomposes). Soluble in water yielding a solution with acid reaction, decomposes in alcohol.
Derivation: Heating potassium sulfate with sulfuric acid.
Use: Conversion of wine lees and tartrates into potassium bitartrate, flux, manufacture of mixed fertilizers, methyl acetate, ethyl acetate, laboratory reagent.

potassium bisulfide.   See potassium hydrosulfide.

potassium bisulfite.   (potassium acid sulfite; potassium hydrogen sulfite). $KHSO_3$.
Properties: White, crystalline powder; sulfur dioxide odor. Mp decomposes at 190C. Soluble in water; insoluble in alcohol.
Derivation: Sulfur dioxide is passed through a solution of potassium carbonate until no more carbon dioxide is given off; the solution is concentrated and allowed to crystallize.
Grade: Commercial, reagent, medicinal.
Use: Reduction of various organic compounds; purification of aldehydes and ketones, iodine, sodium hydrosulfite; antiseptic; source of sulfurous acid, particularly in brewing; analytical chemistry; tanning; bleaching straw and textile fibers; chemical preservative in foods (except meats and other sources of vitamin B_1).

potassium bitartrate.   (cream of tartar; potassium acid tartrate).
CAS: 868-14-4. $KHC_4H_4O_6$.
Properties: White crystals or powder; pleasant, slightly acid taste. D 1.984 (18C). Soluble in boiling water; insoluble in alcohol.

Derivation: From wine lees by extraction with water and crystallization.

Grade: Technical, NF, FCC.

Use: Baking powder, preparation of other tartrates, medicine, galvanic tinning of metals, food additive.

potassium borate.   See potassium tetraborate.

potassium borofluoride.   See potassium fluoborate.

potassium borohydride.

CAS: 13762-51-1. KBH_4.

Properties: White, crystalline powder. D 1.18, stable in moist and dry air, stable in vacuum to 500C, decomposed by acids with evolution of hydrogen. Partially soluble in water and ammonia; insoluble in ethers and hydrocarbons.

Derivation: By reaction of sodium borohydride and potassium hydroxide.

Grade: Technical, powder, pellets.

Hazard: Flammable, dangerous fire risk. Toxic by ingestion.

Use: Source of hydrogen; reducing agent for aldehydes, ketones, and acid chlorides; foaming agent for plastics.

potassium bromate.

CAS: 7758-01-2. $KBrO_3$.

Properties: White crystals or crystalline powder. D 3.27, mp (decomposes) 370C, strong oxidizing agent. Soluble in boiling water.

Derivation: By passing bromine into a solution of potassium hydroxide.

Grade: Pure, CP, FCC.

Hazard: Dangerous fire risk in contact with organic materials, strong irritant. Possible carcinogen.

Use: Laboratory reagent, oxidizing agent, permanent-wave compounds, maturing agent in flour, dough conditioner, food additive.

Note: Food additive uses restricted as to proportions used.

potassium bromide.

CAS: 7758-02-3.   KBr.

Properties: White, crystalline granules or powder; pungent, strong, bitter, saline taste; somewhat hygroscopic. D 2.749, mp 730C, bp 1435C. Soluble in water and glycerol; slightly soluble in alcohol and ether.

Derivation: Solutions of iron bromide and potassium carbonate are mixed and heated, the solution filtered and concentrated, and the bromide crystallized out.

Grade: Technical, CP, NF, reagent, single crystals.

Hazard: Toxic by ingestion and inhalation.

Use: Photography (gelatin bromide papers and plates), process engraving and lithography, special soaps, spectroscopy, infrared transmission, laboratory reagent.

potassium-*tert*-butoxide.   See potassium-*tert*-butylate.

potassium-*tert*-butylate.   (potassium-*tert*-butoxide). $(CH_3)_3COK$.   White, hygroscopic powder; a strong organic base.

Hazard: Flammable. May cause severe caustic tissue burns.

potassium carbonate.   (potash; pearl ash).

CAS: 584-08-7.   (1) K_2CO_3. (2) $K_2CO_3 \cdot 1.5H_2O$.

Properties: White, deliquescent, granular, translucent powder; alkaline reaction. (1) D 2.428 (19C), mp 891C, bp decomposes. Soluble in water; insoluble in alcohol. Noncombustible.

Derivation: (1) Engel–Precht process uses magnesium oxide, potassium chloride, and carbon dioxide, separating the Engels salt $(MgCO_3 \cdot KHCO_3 \cdot 4H_2O)$. Decomposition leaves potassium bicarbonate in solution, which can be processed to potassium carbonate. (2) Alkyl amines or ion-exchange resins can be used with potassium chloride and carbon dioxide to yield potassium bicarbonate, which is calcined to the carbonate. This is analogous to the Solvay process for sodium carbonate.

Grade: Crystals; technical; reagent; calcined 80–85%, 85–90%, 90–95%, 96–98%; hydrated 80–85%; FCC.

Hazard: Solutions irritating to tissue.

Use: Special glasses (optical and color TV tubes), potassium silicate, dehydrating agent, pigments, printing inks, laboratory reagent, soft soaps, raw wool washing, general-purpose food additive.

potassium chlorate.

CAS: 3811-04-9. $KClO_3$.

Properties: Transparent, colorless crystals or white powder; cooling, saline taste. D 2.337, mp 368C, bp decomposes at 400C giving off oxygen gas. Soluble in boiling water.

Derivation: (1) By electrolyzing a hot concentrated alkaline solution of potassium chloride. Preferably (2) by interaction of solutions of potassium chloride and sodium chlorate or calcium chlorate.

Hazard: Forms explosive mixture with combustible materials (sulfur, sugar, etc.); strong oxidizing agent.

Use: Oxidizing agent, explosives, matches, source of oxygen, textile printing, pyrotechnics, percussion caps, disinfectant, bleaching.

potassium chloraurate.   See potassium gold chloride.

potassium chloride.

CAS: 7447-40-7.   KCl.

Properties: Colorless or white crystals; strong saline taste. D 1.987, mp 772C. Soluble in water; slightly soluble in alcohol; insoluble in ether and acetone. Occurs naturally as sylvite. Noncombustible.

Derivation: (1) Mined from sylvite deposits in New Mexico and Saskatchewan, purified by fractional crystallization or flotation; (2) extracted from salt lake brines and purified by recrystallization.

Grade: Chemical (99.5 and 99.9%); agricultural grades sold as 60–62%, 48–52%, and 22% K_2O; single crystals; USP; FCC.

Use: Fertilizer, source of potassium salts, pharmaceutical preparations, photography, spectroscopy, plant nutrient, salt substitute, lab reagent, buffer solutions, food additive.

potassium chlorochromate.   (Peligot's salt). $KClCrO_3$.

Properties: Red crystals. D 2.497. Liberates chlorine on heating. Soluble in water.

Hazard: Toxic by ingestion and inhalation.

Use: Oxidizing agent.

potassium chloroiridate.   See iridium potassium chloride.

potassium chloroplatinate.   (platinum potassium chloride; potassium platinichloride).

CAS: 16921-30-5. K_2PtCl_6.

Properties: Small, orange-yellow crystals or powder. Decomposes when heated (250C), d 3.499 (24C). Insoluble in alcohol; very slightly soluble in water.

Derivation: Adding platinic chloride to a solution of potassium salt and crystallizing.

Hazard: Toxic by ingestion and inhalation.

Use: Photography, laboratory reagent.

potassium chromate.   (potassium chromate, yellow; neutral potassium chromate).

CAS: 7789-00-6. K_2CrO_4.

Properties: Yellow crystals. D 2.7319, mp 971C. Soluble in water; insoluble in alcohol.

Derivation: Roasting powdered chromite with potash and limestone, treating the cinder with hot potassium sulfate solution, and leaching.

Grade: Technical, reagent.

Hazard: Toxic by ingestion and inhalation.

Use: Reagent in analytical chemistry, aniline black, textile mordant, enamels, chromate pigments, inks.

potassium chromate, red.   See potassium dichromate.

potassium chromium sulfate.   See chromium potassium sulfate.

potassium citrate.

CAS: 866-84-2. $K_3C_6H_5O_7 \cdot H_2O$.

Properties: Colorless or white crystals or powder; cooling, saline taste; odorless; deliquescent. D 1.98, loses H_2O at 180C, decomposes at 230C. Soluble in water and glycerol; almost insoluble in alcohol.

Derivation: Action of citric acid on potassium carbonate.

Grade: Technical, CP, NF, FCC.

Use: Medicine (antacid), sequestrant, stabilizer, buffer in foods.

potassium cobaltinitrite.   See cobalt potassium nitrite.

potassium columbate.   Obsolete name for potassium niobate.

potassium copper cyanide.   (copper potassium cyanide; potassium cuprocyanide).

CAS: 13682-73-0. $C_2CuN_2 \cdot K$.

Properties: White, crystalline double salt of copper cyanide and potassium cyanide; copper content min 25.8%; free potassium cyanide 1.25–3.0%.

Hazard: Highly toxic.

Use: For preparing and maintaining cyanide copper plating baths based on potassium cyanide.

potassium copper ferrocyanide.   See copper potassium ferrocyanide.

potassium cupric ferrocyanide.   See copper potassium ferrocyanide.

potassium cuprocyanide.   Legal label name (Air) for potassium copper cyanide.

potassium cyanate.

CAS: 590-28-3.   KOCN.

Properties: Colorless crystals. D 2.05, decomposes at 700–900C. Soluble in water; insoluble in alcohol.

Derivation: Heating potassium cyanide with lead oxide.

Use: Herbicide, manufacture of organic chemicals and drugs, treatment of sickle-cell anemia.

potassium cyanide.

CAS: 151-50-8.   KCN.

Properties: White, amorphous, deliquescent lumps or crystalline mass; faint odor of bitter almonds. D 1.52 (16C), mp 634C. Soluble in water, alcohol, and glycerol.

Derivation: Absorption of hydrogen cyanide in potassium hydroxide.

Grade: Commercial, pure, solution, reagent.

Hazard: A poison as absorbed by skin.

Use: Extraction of gold and silver from ores, reagent in analytical chemistry, insecticide, fumigant, electroplating.

potassium cyanoargentate.   See silver potassium cyanide.

potassium cyanoaurite.   See potassium gold cyanide.

potassium cyclamate.   See sodium cyclamate.

potassium di-*n*-butyl dithiocarbamate.

$(C_4H_9)_2NCSSK$. A 50% aqueous solution in a straw-colored liquid, d 1.08–1.12.

Use: Ultra-accelerator for natural and synthetic latexes.

potassium dichloroisocyanurate.
(dichloroisocyanuric acid potassium salt; potassium troclosene; troclosene potassium).
CAS: 2244-21-5. $C_3HCl_2N_3O_3 \cdot K$. A cyclic compound.
Properties: White, slightly hygroscopic, crystalline powder or granules. Mw 237.07, mp 250C (decomp). Loose bulk d (approximate) (powder) 37 lb/cu ft, (granular) 64 lb/cu ft. Active ingredient 59% available chlorine; decomposes at 240C.
Hazard: Moderately toxic by ingestion. Dangerous fire risk in contact with organic materials. A strong oxidizing agent.
Use: Household dry bleaches, dishwashing compounds, scouring powders, detergent sanitizers; replacement for calcium hypochlorite; agriculture. See dichloroisocyanuric acid.

potassium 5-(2,4-dichlorophenoxy)-2-nitrobenzoate.
CAS: 53775-55-6. $C_{13}H_7Cl_2NO_5 \cdot K$.
Hazard: Low toxicity by ingestion, inhalation, and skin contact. A mild skin irritant.

potassium dichromate. (potassium bichromate; red potassium chromate).
CAS: 7778-50-9. $K_2Cr_2O_7$.
Properties: Bright, yellowish-red, transparent crystals; bitter, metallic taste. Mp 396C, d 2.676 (25C), bp decomposes at 500C. Soluble in water, insoluble in alcohol.
Derivation: Reaction of potassium chloride and sodium dichromate.
Grade: Commercial; highest purity; highest purity fused; reagent.
Hazard: Toxic by ingestion and inhalation. Dangerous fire risk in contact with organic materials. Strong oxidizing agent.
Use: Oxidizing agent (chemicals, dyes, intermediates); analytical reagent; brass pickling compositions; electroplating; pyrotechnics; explosives; safety matches; textiles; dyeing and printing; chrome glues and adhesives; chrome tanning leather; wood stains; poison fly paper; process engraving and lithography; synthetic perfumes; chrome alum manufacture; pigments; alloys, ceramic products; depolarizer in dry cell batteries; bleaching fats and waxes.

potassium dihydrogen phosphate. See potassium phosphate, monobasic.

potassium dimethyldithiocarbamate. See sodium dimethyldithiocarbamate.

potassium dinitrotetrachloroplatinate.
CAS: 53403-68-2. $Cl_4K_2N_2O_4Pt$.
Hazard: A poison.

potassium diphosphate. See potassium phosphate, monobasic.

potassium dithionate. (potassium hyposulfate). $K_2S_2O_6$.
Properties: Colorless crystals. D 2.27. Soluble in water; insoluble in alcohol.
Use: Analytical reagent.

potassium endothall. (dipotassium salt of endothall). The water solution is an amber liquid used as a contact herbicide.

potassium ethyldithiocarbonate. See potassium xanthate.

potassium ethylxanthate. See potassium xanthate.

potassium ethylxanthogenate. See potassium xanthate.

potassium feldspar. (feldspar). $(K_2O \cdot 3SiO_2) + (Al_2O_3 \cdot 3SiO_2)$. "Potassium" feldspar is usually the pink variety of feldspar.

potassium ferric oxalate.
$K_3Fe(C_2O_4)_3 \cdot 3H_2O$.
Properties: Monoclinic green crystals. Loses $3H_2O$ at 100C, decomposes at 230C. Soluble in water and acetic acid; incompatible with alkali and ammonia, since these react to precipitate ferric hydroxide.
Use: Photography and blueprinting.

potassium ferricyanide. (red prussiate of potash; red potassium prussiate).
CAS: 13746-66-2. $K_3Fe(CN)_6$.
Properties: Bright-red, lustrous crystals or powder. D 1.85 (25C). Soluble in water; slightly soluble in alcohol.
Derivation: Chlorine is passed into a solution of potassium ferrocyanide, the ferricyanide separating out.
Grade: Pure crystals, pure powder, reagent, technical.
Hazard: Decomposes on strong heating to evolve highly toxic fumes, but the compound itself has low toxicity.
Use: Tempering steel, etching liquid, production of pigments, electroplating, sensitive coatings on blueprint paper, fertilizer compositions, laboratory reagent.

potassium ferrocyanide. (yellow prussiate of potash; yellow potassium prussiate).
CAS: 13943-58-3. $K_4Fe(CN)_6 \cdot 3H_2O$.
Properties: Lemon-yellow crystals or powder; mild saline taste. D 1.853 (17C), loses its water of crystallization when heated to 70C, bp (decomposes). Effloresces on exposure to air; soluble in water; insoluble in alcohol.
Derivation: From nitrogenous waste products, iron filings, and potassium carbonate.
Grade: Technical, CP.

Hazard: It evolves highly toxic fumes on heating to red heat, but the compound itself has low toxicity.
Use: Potassium cyanide and ferricyanide, dry colors, tempering steel, dyeing, explosives, process engraving and lithography, laboratory reagent.

potassium fluoborate.   (potassium borofluoride). KBF_4.
Properties: Colorless crystals. D 2.498 (20C), decomposes at 350C. Very slightly soluble in water; insoluble in alcohol, ether, alkalies.
Derivation: By the reaction of boric acid, hydrogen fluoride, and potassium hydroxide.
Hazard: Toxic by ingestion.
Use: Sand-casting of aluminum and magnesium; grinding aid in resinoid grinding wheels; flux for soldering and brazing;, electrochemical processes; and chemical research.

potassium fluoride.
CAS: 7789-23-3.   (1) KF; (2) $KF•2H_2O$.
Properties: White, crystalline, deliquescent powder; sharp saline taste. D (1) 2.48, (2) 2.454; mp (1) 846C; bp 1505C. Soluble in water and hydrogen fluoride, insoluble in alcohol.
Derivation: Neutralizing hydrogen fluoride with potassium carbonate.
Grade: Technical, free of arsenic, CP.
Hazard: Toxic by ingestion and inhalation, strong irritant to tissue.
Use: Etching glass, preservative, insecticide, solder flux, fluorination reactions.

potassium fluosilicate.   (potassium silicofluoride).
CAS: 16871-90-2. K_2SiF_6.
Properties: White, odorless, crystalline powder. D 3.0. Slightly soluble in water; soluble in hydrochloric acid.
Hazard: Toxic by ingestion and inhalation, strong irritant to tissue.
Use: Vitreous enamel frits, synthetic mica, metallurgy of aluminum and magnesium, ceramics, insecticide, opalescent glass.

potassium fluotantalate.   See tantalum potassium fluoride.

potassium fluozirconate.   See zirconium potassium fluoride.

potassium germanium fluoride.   See germanium potassium fluoride.

potassium gibberellate. $C_{19}H_{21}KO_6$.
Properties: White crystalline powder, odorless. Deliquescent, sol in water, alc, acetone. A salt of gibberellic acid.
Use: Promote and control development of malt in grain.

potassium gluconate.
CAS: 299-27-4. $KC_6H_{11}O_7$.
Properties: Fine, white, crystalline powder; odorless; salty tasting. Mp 180C (decomposes). Soluble in water; insoluble in alcohol and benzene.
Derivation: Reaction of potassium hydroxide or carbonate with gluconic acid.
Grade: Pharmaceutical.
Use: Medicine (vitamin tablets).

potassium glutamate.   (monopotassium-l-glutamate).
CAS: 19473-49-5. $KOOC(CH_2)_2CH(NH_2)COOH•H_2O$.
Properties: White, free-flowing, hygroscopic powder; practically odorless. Freely soluble in water; slightly soluble in alcohol, pH of a 2% solution 6.9–7.1.
Grade: FCC.
Use: Flavor enhancer, salt substitute.

potassium glycerophosphate.   (potassium glycerinophosphate).
CAS: 1319-69-3. $K_2C_3H_5O_2•H_2PO_4•3H_2O$.
Properties: Pale-yellow viscous mass or liquid (75% aqueous solution); acidic taste. Soluble in alcohol; miscible with water.
Derivation: Glycerol and phosphorus pentoxide or metaphosphoric acid are mixed, warmed, and exactly neutralized with potassium carbonate, then warmed and concentrated.
Grade: Technical, 50 or 75% solution, FCC (syrup or solution).
Use: Food additive and dietary supplement.

potassium gold chloride.   (potassium aurichloride; potassium chloroaurate; gold potassium chloride). $KAuCl_4•2H_2O$.
Properties: Yellow crystals. Soluble in water, alcohol, and ether.
Derivation: Neutralizing chloroauric acid with potassium carbonate.
Use: Photography, painting porcelain and glass, medicine.

potassium gold cyanide.   (potassium cyanoaurite; gold potassium cyanide). $KAu(CN)_2$.
Properties: White, crystalline powder. Soluble in water; slightly soluble in alcohol; insoluble in ether.
Derivation: Action of hydrogen cyanide on potassium aurate.
Hazard: Highly toxic.
Use: Electrogilding.

potassium guaiacol sulfonate.
$C_6H_3OCH_3OHSO_3K•1/2H_2O$.
Properties: White powder or crystals, gradually turns pink on exposure to air and light; bitter taste, afterward becoming sweetish; odorless. Contains 60% guaiacol; soluble in water; sparingly soluble in alcohol.
Grade: NF.
Use: Medicine (expectorant), laboratory reagent.

potassium-2,4-hexadienoate. See potassium sorbate.

potassium hexafluorophosphate. KPF$_6$.
Properties: Solid. Mp 575C, bp (decomposes). Soluble in water.
Grade: 98–100%.
Hazard: Toxic by ingestion.
Use: Maintenance of fluoride atmospheres, preparation of bactericides and fungicides, laboratory reagent.

potassium hexanitrocobaltate III. See cobalt potassium nitrite.

potassium hexyl xanthane. C$_6$H$_{13}$OCSSK.
Use: Flotation agent.

potassium hydrate. See potassium hydroxide.

potassium hydride. KH. Marketed as a semidispersion of gray powder in oil.
Properties: The solid decomposes on heating or in contact with moisture.
Hazard: Dangerous fire and explosion risk, evolves toxic and flammable gases on heating and on exposure to moisture.
Use: Organic condensations and alkylations. See hydride.

potassium hydrogen fluoride. See potassium bifluoride.

potassium hydrogen phosphate. See potassium phosphate dibasic.

potassium hydrogen phthalate. (potassium biphthalate). HOOCC$_6$H$_4$COOK.
Properties: Colorless crystals. D 1.636. Soluble in water.
Derivation: Potassium hydroxide and phthalic anhydride.
Grade: CP, analytical.
Use: Alkalimetric standard, buffering agent.

potassium hydrosulfide. (potassium sulfhydrate; potassium bisulfide). KHS.
Properties: White to yellow crystals; hydrogen sulfide odor. Forms the polysulfide when exposed to air. D 1.69, mp 455C. Hygroscopic; soluble in alcohol and water.
Grade: Technical.
Use: Separation of heavy metals.

potassium hydroxide. (caustic potash; potassium hydrate; lye).
CAS: 1310-58-3. KOH.
Properties: White, deliquescent pieces, lumps, sticks, pellets, or flakes having a crystalline fracture. D 2.044, mp 405C (varies with water content). Keep well stoppered, absorbs water and carbon dioxide

from the air. Soluble in water, alcohol, glycerol; slightly soluble in ether.
Derivation: Electrolysis of concentrated potassium chloride solution.
Method of purification: Sulfur compounds are removed by the addition of potassium nitrate to the fused caustic. The purest form is obtained by solution in alcohol, filtration, and evaporation.
Grade: Commercial, ground, flake, fused (88–92%), purified by alcohol (sticks, lumps, and drops), reagent, highest purity, USP, liquid (45%), FCC.
Hazard: Toxic by ingestion and inhalation, strong caustic, handle with gloves or tongs, corrosive to tissue. Eye, skin and upper respiratory tract irritant.
Use: Soap manufacture, bleaching, manufacture of potassium carbonate and tetrapotassium pyrophosphate, electrolyte in alkaline storage batteries and some fuel cells, absorbent for carbon dioxide and hydrogen sulfide, dyestuffs, liquid fertilizers, food additive, herbicides, electroplating, mercerizing, paint removers, reagent.

potassium hypophosphite. (potassium hypophosphite, monobasic).
CAS: 7782-87-8. KH$_2$PO$_2$.
Properties: White, opaque crystals or powder; pungent saline taste; very deliquescent. Soluble in water and alcohol; decomposed by heat.
Derivation: Interaction of calcium hypophosphite and potassium carbonate.
Hazard: Moderate fire risk, may explode if ground with chlorates, nitrates, or other strong oxidizing agents.

potassium hyposulfate. See potassium dithionate.

potassium hyposulfite. See potassium thiosulfate.

potassium iodate.
CAS: 7758-05-6. KIO$_3$.
Properties: White, crystalline powder; odorless. D 3.9, mp 560C (partial decomposition). Soluble in water, sulfuric acid (dilute); insoluble in alcohol.
Grade: Technical, CP, FCC.
Use: Analysis (testing for zinc and arsenic), iodometry, reagent, feed additive, in foods as maturing agent and dough conditioner, medicine (topical antiseptic).

potassium iodide.
CAS: 7681-11-0. KI.
Properties: White crystals, granules, or powder; strong, bitter, saline taste. D 3.123, mp 686C, bp 1330C. Soluble in water, alcohol, acetone, and glycerol.
Grade: Reagent, USP, single crystals, FCC.
Use: Reagent in analytical chemistry, photographic emulsions (precipitating Ag), feed additive, spectroscopy, infrared transmission, scintillation, dietary supplement (up to 0.01% in table salt).

potassium iridium chloride. See iridium potassium chloride.

potassium laurate. $KOOCC_{11}H_{23}$.
Properties: Light-tan paste. Soluble in water and alcohol.
Use: Emulsifying agent, base for liquid soaps and shampoos.

potassium linoleate. $KOOCC_{17}H_{31}$.
Properties: Light-tan paste. Soluble in water.
Use: Emulsifying agent.

potassium magnesium sulfate.
$K_2SO_4 \cdot 2MgSO_4$.
Properties: White, tetragonal crystals. D 2.829, mp 927.
Use: Fertilizer.

potassium manganate. K_2MnO_4.
Properties: Dark-green powder or crystals. Mp 190C (decomposes). Soluble in potassium hydroxide solution and water; decomposes in acid solution.
Derivation: Fusion of pyrolusite with potassium hydroxide.
Grade: Technical.
Hazard: Dangerous fire risk in contact with organic materials, strong oxidizing agent.
Use: Bleaching skins, fibers, oils; disinfectants; mordant (wool); batteries; photography; printing; source of oxygen (dyeing); water purification; oxidizing agent.

potassium mercuric iodide. See mercuric potassium iodide.

potassium metaarsenite. See potassium arsenite.

potassium metabisulfite. (potassium pyrosulfite).
CAS: 16731-55-8. $K_2S_2O_5$.
Properties: White granules or powder; pungent sharp odor. D 2.3, decomposes at 150–190C, oxidizes in air and moisture to sulfate. Soluble in water; insoluble in alcohol.
Derivation: By heating potassium bisulfite until it loses water.
Grade: Technical, reagent, FCC.
Use: Antiseptic, reagent, source of sulfurous acid, brewing (cleaning casks and vats), winemaking (said to kill only undesirable yeasts and bacteria), food preservative, developing agent (photography), process engraving and lithography, dyeing, antioxidant, bleaching agent.

potassium, metallic, liquid alloy. A flammable mixture or alloy of potassium and another metal that is liquid at normal temperature.

potassium metaphosphate. (monopotassium metaphosphate). KPO_3 or $(KPO_3)_6$.

Properties: White powder. D 2.45. Soluble in dilute acids; slightly soluble in water.

potassium molybdate. K_2MoO_4.
Properties: White, deliquescent, microcrystalline powder. D 2.91 (18C), mp 919C. Soluble in water; insoluble in alcohol.
Use: Reagent.

potassium monophosphate. See potassium phosphate, dibasic.

potassium naphthenate.
Properties: Gray paste. Soluble in water. Combustible.
Derivation: From naphthenic acids.
Use: Driers, emulsifying agents.

potassium nickel sulfate. See nickel potassium sulfate.

potassium nicotinate.
CAS: 16518-17-5. $C_6H_4NO_2 \cdot K$.
Hazard: Low toxicity by ingestion.

potassium nitrate. (niter; nitre; saltpeter).
CAS: 7757-79-1. KNO_3.
Properties: Transparent, colorless or white, crystalline powder or crystals; slightly hygroscopic; pungent, saline taste. D 2.1062, mp 337C, bp decomposes at 400C. Soluble in water and glycerol; slightly soluble in alcohol.
Grade: Commercial, CP, FCC.
Hazard: Dangerous fire and explosion risk when shocked or heated, or in contact with organic materials, strong oxidizing agent.
Use: Pyrotechnics, explosives, matches, specialty fertilizer, reagent, to modify burning properties of tobacco, glass manufacture, tempering steel, curing foods, oxidizer in solid rocket propellants.

potassium nitrite.
CAS: 7758-09-0. KNO_2.
Properties: White or slightly yellowish prisms or sticks; deliquescent. D 1.915, bp explodes at 1000F (537C), mp 440C (decomposition starts at 350C). Soluble in water; insoluble in alcohol.
Grade: CP, technical, reagent, FCC.
Hazard: Fire and explosion risk when shocked or heated, or in contact with organic materials, strong oxidizing agent.
Use: Analysis (testing for amino acids, cobalt, iodine, urea), food additive (curing agent).

potassium oleate.
CAS: 143-18-0. $C_{17}H_{33}COOK$.
Properties: Gray-tan paste. Soluble in water and alcohol. Combustible.
Use: Textile soaps, emulsifying agent.

potassium orthophosphate. See potassium phosphate, monobasic, dibasic or tribasic.

potassium orthotungstate. See potassium tungstate.

potassium osmate. (potassium perosmate). $K_2OsO_4 \cdot 2H_2O$.
Properties: Violet crystals, hygroscopic. Soluble in water; insoluble in alcohol and ether.
Hazard: Toxic by ingestion and inhalation.
Use: Determination of nitrogenous matter in water.

potassium oxalate.
CAS: 583-52-8. $K_2C_2O_4 \cdot H_2O$.
Properties: Colorless, transparent crystals; odorless. D 2.08, decomposes when heated, efflorescent in warm dry air. Soluble in water.
Derivation: Potassium formate or carbonate mixed with a small quantity of oxalate and a slight excess of alkali is heated, the oxalate extracted with water and crystallized.
Grade: Technical, CP.
Hazard: Toxic by inhalation and ingestion.
Use: Reagent in analytical chemistry, source of oxalic acid, bleaching and cleaning, removing stains from textiles, photography.

potassium, oxalatoplatinate, dihydrate.
CAS: 14244-64-5. $C_4K_2O_8Pt \cdot 2H_2O$.
Hazard: A poison.

potassium oxide. K_2O.
Properties: Gray, crystalline mass. D 2.32 (0C), decomposes 350C. Soluble in water (forms potassium hydroxide); soluble in alcohol and ether.
Derivation: By heating potassium nitrate and metallic potassium.
Hazard: Corrosive to tissue.
Use: Reagent and intermediate.

potassium oxyfluoniobate. See niobium potassium oxyfluoride.

potassium palladium chloride. See palladium potassium chloride.

potassium penicillin G. (benzylpenicillin potassium).
CAS: 113-98-4. $C_{16}H_{17}KN_2O_4S$.
Properties: Colorless or white crystals or powder; odorless. Mp 214–217C (decomposes), moderately hygroscopic. Solutions dextrorotatory. Relatively stable to air and light. Very soluble in water, saline, and dextrose solutions; moderately soluble in alcohol; pH of solution (30 mg/mL) is 5.0–7.5.
Grade: USP.
Use: Medicine (antibiotic).

potassium penicillin V. See potassium phenoxymethylpenicillin.

potassium percarbonate. $K_2C_2O_6 \cdot H_2O$.
Properties: Granular, white mass. Mp 200–300C. Soluble in water (liberates oxygen). Keep away from light and moisture.
Grade: Technical.
Hazard: Strong irritant to tissue. Fire risk in contact with organic materials, strong oxidizing agent.
Use: Analysis (testing for cerium, chromium, vanadium, titanium), photography, textile printing.

potassium perchlorate.
CAS: 7778-74-7. $KClO_4$.
Properties: Colorless crystals or white, crystalline powder. Decomposed by concussion, organic matter, and agents subject to oxidation; more stable than potassium chlorate. D 2.524, mp 400C (decomposes) Soluble in water; insoluble in alcohol.
Grade: Technical, reagent, ordnance.
Hazard: Fire risk in contact with organic materials, strong oxidizing agent. Strong irritant.
Use: Explosives, oxidizing agent, photography, pyrotechnics and flares, reagent, oxidizer in solid rocket propellants.

potassium periodate.
CAS: 7790-21-8. KIO_4.
Properties: Small, colorless crystals or white, granular powder. D 3.168, mp 582C, explodes at 1076F (580C). Slightly soluble in water.
Grade: Technical, CP, reagent.
Hazard: Fire risk in contact with organic materials, strong oxidizing agent. Strong irritant to tissue.
Use: Analysis (oxidizing agent).

potassium permanganate.
CAS: 7722-64-7. $KMnO_4$.
Properties: Dark-purple crystals with blue metallic sheen; sweetish, astringent taste; odorless. D 2.7032, mp decomposes at 240C, oxidizing material. Soluble in water, acetone, and methanol; decomposed by alcohol.
Derivation: (1) By oxidation of the manganate in an alkaline electrolytic cell. (2) A hot solution of the manganate is treated with carbon dioxide; on cooling, the solution deposits crystals of the permanganate.
Grade: Technical, CP, USP.
Hazard: Dangerous fire and explosion risk in contact with organic materials, powerful oxidizing agent.
Use: Oxidizer, disinfectant, deodorizer, bleach, dye, tanning, radioactive decontamination of skin, reagent in analytical chemistry, medicine (antiseptic), manufacture of organic chemicals, air and water purification.

potassium perosmate. See potassium osmate.

potassium peroxide.
CAS: 17014-71-0. K_2O_2.
Properties: Yellow, amorphous mass; decomposes in water evolving oxygen. Mp 490C.
Derivation: Oxidation of potassium in oxygen.

Hazard: Dangerous fire and explosion risk in contact with organic materials, strong oxidizing agent. Irritant to skin and tissue.
Use: Oxidizing agent, bleaching agent, oxygen-generating gas masks.

potassium peroxydisulfate. See potassium persulfate.

potassium peroxymonosulfate. See Oxone.

potassium persulfate. (dipotassium persulfate; potassium peroxydisulfate).
CAS: 7727-21-1. $K_2S_2O_8$.
Properties: White crystals. D 2.477, mp decomposes below 100C. Soluble in water; insoluble in alcohol.
Derivation: By electrolysis of a saturated solution of potassium sulfate.
Hazard: Strong irritant and oxidizing agent. Fire risk in contact with organic materials.
Use: Bleaching, oxidizing agent, reducing agent in photography, antiseptic, soap manufacture, analytical reagent, polymerization promoter, pharmaceuticals, modification of starch, flour-maturing agent, desizing of textiles.

potassium α-phenoxyethylpenicillin.
(potassium penicillin 152; potassium phenethicillin; phenethicillin). $KC_{17}H_{19}N_2O_5S$. Synthetically prepared, a mixture of two stereoisomers.
Properties: White, crystalline solid; moderately hygroscopic. Decomposes above 220C. Very soluble in water; resistant to acid decomposition. Produced by N-acylation of α-phenoxypropionic acid and G-aminopenicillanic acid (produced by fermentation using *Penicillium chrysogenum*).
Grade: NF.
Use: Antibiotic.

potassium phenoxymethylpenicillin.
(potassium penicillin V).
CAS: 132-98-9. $KC_{16}H_{17}N_2O_5S$.
Properties: White, crystalline powder; odorless. Very soluble in water; slightly soluble in alcohol; insoluble in acetone.
Grade: USP.
Use: Antibiotic.

potassium phosphate, dibasic. (DKP; potassium hydrogen phosphate; potassium monophosphate; dipotassium orthophosphate). K_2HPO_4.
Properties: Hygroscopic, white crystals or powder. Very soluble in water. Converted to pyrophosphate by ignition.
Derivation: Action of phosphoric acid on potassium carbonate.
Grade: Commercial, pure, highest purity, NF, FCC.
Use: Buffer in antifreezes; ingredient of "instant" fertilizers; nutrient for penicillin culturing; humectant; in pharmaceuticals; foods as a buffer, sequestrant, and yeast food; and as a laboratory reagent.

potassium phosphate, monobasic. (MKP; potassium acid phosphate; potassium diphosphate; potassium orthophosphate; potassium dihydrogen phosphate). KH_2PO_4.
Properties: Colorless crystals. D 2.338, mp 253C. Acid in reaction; soluble in water; insoluble in alcohol.
Derivation: Action of phosphoric acid on potassium carbonate.
Grade: Technical, CP, FCC.
Use: Baking powder, nutrient solutions, yeast foods, buffer and sequestrant, lab reagent.

potassium phosphate, tribasic. (potassium phosphate, neutral; potassium phosphate normal; tripotassium orthophosphate; potassium phosphate, tertiary; tripotassium phosphate).
CAS: 7778-53-2. $K_3PO_4 \cdot H_2O$ or K_3PO_4.
Properties: Granular, white powder; deliquescent. Mp (anhydrous) 1340C, d (anhydrous) 2.564 (17C). Soluble in water giving strongly basic solution. Insoluble in alcohol.
Grade: Reagent, technical, FCC.
Use: Purification of gasoline, water softening, liquid soaps, fertilizer, in foods as an emulsifier, laboratory reagent.

potassium phosphite, monobasic. KH_2PO_3.
Properties: White powder. Hygroscopic; soluble in water; insoluble in alcohol; slowly oxidized by air to phosphate.

potassium platinichloride. See potassium chloroplatinate.

potassium polymetaphosphate. $(KPO_3)_n$.
The molecular weight may be as high as 500,000.
Properties: White powder; odorless. Insoluble in water; soluble in sodium salt solutions that may have high viscosity.
Derivation: Dehydration of monobasic potassium phosphate.
Grade: Technical, FCC.
Use: Fat emulsifier and moisture-retaining agent in foods.
See sodium metaphosphate.

potassium polysulfide. K_2S_n.
Properties: Crystals. Soluble in water and alcohol.
Hazard: Moderate fire risk. Toxic by ingestion, irritant to skin and eyes.
Use: Fungicide.

potassium prussiate, red. See potassium ferricyanide.

potassium prussiate, yellow. See potassium ferrocyanide.

potassium 3-pyridinecarboxylate. See potassium nicotinate.

potassium pyroantimonate.
$K_2H_2SbO_7 \cdot 4H_2O$.
Properties: White, crystalline powder or granules. Slightly soluble in cold water; readily soluble in hot water; insoluble in alcohol.
Grade: Reagent, technical.
Use: Starch sizes and flame-retarding compounds.

potassium pyroborate.   See potassium tetraborate.

potassium pyrophosphate.   (TKPP; tetrapotassium pyrophosphate; potassium pyrophosphate, normal). $K_4P_2O_7 \cdot 3H_2O$.
Properties: Colorless crystals or white powder. Somewhat hygroscopic in air (deliquescent at a relative humidity of above 40–45%). Similar to tetrasodium pyrophosphate except for greater solubility. D 2.33, dehydrates at about 300C, mp 1,090C. Soluble in water; insoluble in alcohol.
Grade: Technical, 99.4%, 60% solution, FCC.
Use: Soap and detergent builder, sequestering agent, peptizing and dispersing agent.

potassium pyrosulfate.   (potassium acid sulfate, anhydrous). $K_2S_2O_7$.
Properties: Colorless needles or white, crystalline powder or fused pieces. D 2.25 (25/4C), mp 325C. Soluble in water (converted to potassium bisulfate).
Use: Acid flux in analysis, laboratory reagent.

potassium pyrosulfite.   See potassium metabisulfite.

potassium rhodanide.   See potassium thiocyanate.

potassium ricinoleate. $C_{17}H_{32}OHCOOK$.
Properties: White paste. Soluble in water. Combustible.
Use: Emulsifying agent.

potassium silicate.
Properties: (Solid) Weight ratio SiO_2:K_2O varies with grade from 2.1:1 to 2.5:1; colorless anhydrous lump, shattered or granular. Soluble in water at high temperature and pressure; insoluble in alcohol. (Solution) Colorless liquid, Bé range 29–48 degrees.
Derivation: Supercooled melt of potassium carbonate and pure silica sand.
Use: (Solid) Manufacture of glass and refractory material, welding rods, high-temperature mortars, binder in carbon arc-light electrodes, detergents, catalyst, adhesives.

potassium silicofluoride.   See potassium fluosilicate.

potassium sodium carbonate.   See sodium potassium carbonate.

potassium sodium ferricyanide.
$K_2NaFe(CN)_6$.
Properties: Red crystals, over 99% pure. Mp (decomposes); nonhygroscopic and stable. Easily soluble in water.
Derivation: From ferrocyanides.
Use: Blueprint paper and photography.

potassium-sodium phosphate.   See sodium-potassium phosphate.

potassium sodium tartrate.   (Rochelle salt; sodium potassium tartrate).
CAS: 304-59-6. $KNaC_4H_4O_6 \cdot 4H_2O$. It is salt of L(+)-tartaric acid.
Properties: Colorless, transparent, efflorescent crystals or white powder; cool, saline taste. Unstable above 225C, d 1.77, mp 70–80C. Soluble in water, insoluble in alcohol, loses water of crystallization at 140C.
Derivation: Potassium acid tartrate is dissolved in water, the solution saturated with sodium carbonate, concentrated after purification, and crystallized.
Grade: Highest purity, reagent, commercial crystals or powder, NF, FCC.
Use: Baking powders, medicine (cathartic), component of Fehling's solution, silvering mirrors.

potassium sorbate.   (potassium-2,4-hexadienoate).
CAS: 590-00-1. $CH_3CH:CHCH:CHCOOK$.
Properties: White powder. Mp 270C (decomposes), d 1.36 (25/20C). Soluble in water (25C).
Grade: Technical, FCC.
Use: Bacteriostat and preservative in meats, sausage casings, wines, etc.

potassium stannate.
CAS: 12125-03-0. $K_2SnO_3 \cdot 3H_2O$.
Properties: White to light-tan crystals. D 3.197. Soluble in water; insoluble in alcohol.
Grade: Technical.
Hazard: Highly toxic.
Use: Textiles (dyeing and printing), alkaline tin-plating bath.

potassium stearate.   (stearic acid potassium salt).
CAS: 593-29-3. $C_{17}H_{35}COOK$.
Properties: White, crystalline powder; slight odor of fat. Mw 322.57. Soluble in hot water and alcohol.
Grade: Commercial, contains considerable palmitate; FCC.
Use: Anticaking agent, binder, emulsifier, stabilizer for chewing gum, base for textile softeners.

potassium strontium chlorate.   See strontium potassium chlorate.

potassium styphnate. $KC_6H_2N_3O_8 \cdot H_2O$.
Properties: Yellow prisms. Mp loses water at 120C.
Hazard: Explodes when shocked or heated.
Use: High explosive.

potassium sulfate.
CAS: 7778-80-5. K_2SO_4.
Properties: Colorless or white, hard crystals or powder; bitter saline taste. D 2.66, mp 1,072C. Soluble in water; insoluble in alcohol.
Derivation: (1) By treatment of potassium chloride either with sulfuric acid or with sulfur dioxide, air, and water (Hargreaves process); (2) by fractional crystallization of a natural sulfate ore; (3) from salt-lake brines.
Grade: Highest purity medicinal, commercial, crude, CP, agricultural, reagent, technical.
Use: Reagent in analytical chemistry, medicine (cathartic), gypsum cements, fertilizer for chloride-sensitive crops such as tobacco and citrus, alum manufacture, glass manufacture, food additive.

potassium sulfhydrate.　　See potassium hydrosulfide.

potassium sulfide.
CAS: 1312-73-8. K_2S.
Properties: Red or yellow-red crystalline mass or fused solid. D 1.75 (20/4C), mp 910C. Deliquescent in air; soluble in water, alcohol, and glycerol; insoluble in ether.
Grade: Technical.
Hazard: Flammable, dangerous fire risk, may ignite spontaneously, explosive in the form of dust or powder.
Use: Reagent in analytical chemistry, depilatory, medicine.

potassium sulfite.
CAS: 10117-38-1. $K_2SO_3 \cdot 2H_2O$.
Properties: White crystals or powder. Soluble in water; sparingly soluble in alcohol; decomposes on heating and slowly oxidizes in air.
Grade: Technical, CP, FCC.
Use: Photographic developer, medicine (cathartic), food and wine preservative.

potassium sulfocarbonate.　　(potassium trithiocarbonate). K_2CS_3.
Properties: Yellowish-red crystals. Very hygroscopic; soluble in alcohol and water.
Grade: Technical.
Hazard: Toxic by ingestion, strong irritant.
Use: Analysis (testing for cobalt, nickel), medicine, soil fumigant.

potassium sulfocyanate.　　See potassium thiocyanate.

potassium sulfocyanide.　　See potassium thiocyanate.

potassium tantalum fluoride.　　See tantalum potassium fluoride.

potassium tartrate.　　$K_2C_4H_4O_6 \cdot 1/2H_2O$.
Properties: Colorless, crystalline solid. D 1.98. Soluble in water; insoluble in alcohol; decomposed by heat (200–220C).
Grade: CP, technical.
Use: Manufacture of potassium salts, medicine (cathartic), lab reagent.

potassium tellurite.　　K_2TeO_3.
Properties: Granular, white powder. Hygroscopic; decomposes at 460–470C. Soluble in water.
Use: Analysis (testing for bacteria).

potassium tetrathiocyanodiammonochromate.　　See Reinecke salt.

potassium tetroxalate.　　$KHC_2O_4 \cdot H_2C_2O_4$.
Properties: White crystals. Soluble in water; slightly soluble in alcohol.
Use: Metal polish, spot removal, analytical chemistry.

potassium thiocyanate.　　(potassium rhodanide; potassium sulfocyanate; potassium sulfocyanide).
CAS: 333-20-0. KCNS.
Properties: Colorless, transparent, deliquescent crystals; odorless; saline cooling taste. D 1.88, mp 173C, turns brown, green, blue when fused, white again on cooling, bp decomposes at 500C. Soluble in water, alcohol, and acetone.
Derivation: By heating potassium cyanide with sulfur.
Grade: Commercial, pure, reagent, ACS.
Hazard: Toxic by ingestion.
Use: Reagent; manufacture of sulfocyanides, thioureas; printing and dyeing textiles; photographic restrainer and intensifier; synthetic dyestuffs; medicine (hypotensive).

potassium thiosulfate.　　(potassium hyposulfite).
CAS: 10294-66-3. $K_2S_2O_3$ (with varying proportions of water of crystallization).
Properties: Colorless crystals. Hygroscopic; soluble in water.
Grade: Technical, CP.
Use: Analytical reagent.

potassium titanate.　　K_2TiO_3.
Properties: White salt. Hydrolyzes in water to give a strongly alkaline solution.
Derivation: From titanic acid and potassium hydroxide.
Use: See potassium titanate fiber.

potassium titanate fiber.　　Approximate composition $K_2O \cdot (TiO_2)_n$, where n is 4–7.
Properties: High refr index. Mp 1371C, can diffuse and reflect infrared radiation.
Use: Rockets, missiles, nuclear-powered applications as an insulator, especially for the range 1300–2100F.

potassium titanium fluoride. See titanium potassium fluoride.

potassium titanium oxalate. See titanium potassium oxalate.

potassium trichlorophenate. $Cl_3C_6H_2OH \cdot K$.
Offered as a solution containing 47% potassium trichlorophenate and 3% other potassium chlorophenates, d 1.3, fp −9C.
Use: Slime control agent for pulp and paper-mill systems.

potassium tripolyphosphate. (KTPP).
$K_5P_3O_{10}$.
Properties: White, crystalline solid. Mp 620–640C, d 2.54, loose bulk d 67 lb/cu/ft. Solubility in water (26C) more than 140 g/100 mL water.
Use: Water-treating compounds, cleaners, specialty fertilizers, sequestrant.

potassium trithiocarbonate. See potassium sulfocarbonate.

potassium troclosene. See potassium dichloroisocyanurate.

potassium tungstate. (potassium ortho-tungstate; potassium wolframate). $K_2WO_4 \cdot 2H_2O$.
Properties: Heavy, crystalline powder. D 3.1, mp 921C, deliquescent, soluble in water, insoluble in alcohol.

potassium undecylenate.
$CH_2{:}CH(CH_2)_8COOK$.
Properties: Finely divided, white powder. Decomposes above 250C; limited solubility in most organic solvents; soluble in water.
Hazard: Toxic in high concentration.
Use: Bacteriostat and fungistat in cosmetics and pharmaceuticals.

potassium wolframate. See potassium tungstate.

potassium xanthate. (potassium ethyldithio-carbonate; potassium xanthogenate; potassium ethyl xanthate; potassium ethylxanthogenate).
CAS: 140-89-6. $KS_2COC_2H_5$.
Properties: Colorless or light-yellow crystals. D 1.558 (21.5C). Soluble in water and alcohol; insoluble in ether.
Derivation: Reaction of potassium ethylate and carbon disulfide.
Hazard: Toxic by ingestion.
Use: Fungicide for soil treatment, reagent in analytical chemistry.

potassium zinc iodide. (zinc potassium iodide). $ZnI_2 \cdot KI$.
Properties: Colorless crystals. Very hygroscopic.
Use: Analysis (testing for alkaloids).

potassium zinc sulfate. See zinc potassium sulfate.

potassium zirconifluoride. See zirconium potassium fluoride.

potassium zirconium chloride. See zirconium potassium chloride.

potassium zirconium sulfate. See zirconium potassium sulfate.

potential energy. The energy associated with a configuration of particles, as distinct from their motions. In macroscopic terms, potential energy can be increased (for example) by stretching a spring or by lifting a mass against a gravitational force. In molecular systems, potential energy can be increased (for example) by stretching a bond or by separating molecules against a van der Waals attraction.

potential energy surface. The potential energy of a ground-state molecular system containing N atoms is a function of its geometry, defined by 3N spatial coordinates (a configuration space). If the energy is imagined as corresponding to a height in a 3N + 1 dimensional space, the resulting landscape of hills, hollows, and valleys is the potential energy surface.

potential well. In a potential energy surface, the region surrounding a local energy minimum. Typically taken to include at least those points in configuration space such that a path of steadily declining energy can be found that leads to the minimum in question, and such that no similar path can be found to any other minimum.

potentiator. A term used in the flavor and food industries to characterize a substance that intensifies the taste of a food product to a far greater extent than does an enhancer. The most important of these are the 5′-nucleotides. They are approved by FDA. Their effective concentration is measured in parts per billion, whereas that of an enhancer such as MSG is in parts per thousand. The effect is thought to be due to synergism. Potentiators do not add any taste of their own, but intensify the taste response to substances already present in the food.
See enhancer; seasoning; flavor.

pot life. See adhesive working life.

potting compound. See encapsulation.

pour point. (1) The lowest temperature at which a liquid will flow when a test container is inverted. (2) The temperature at which an alloy is cast.

pour point depressant. An additive for lubricating and automotive oils that lowers the pour point (or increases the flow point) by 11.0C. The agents now generally used are polymerized higher esters of acrylic acid derivatives. They are most effective with low-viscosity oils.

powder. Any solid, dry material of extremely small particle size ranging down to colloidal dimensions, prepared either by comminuting larger units (mechanical grinding), combustion (carbon black, lampblack), or precipitation via a chemical reaction (calcium carbonate, etc.). Powders that are so fine that the particles cannot be detected by rubbing between thumb and forefinger are called impalpable. Typical materials used in powder form are cosmetics, inorganic pigments, metals, plastics (molding powders), dehydrated dairy products, pharmaceuticals, and explosives. Metal powders are used to make specialized equipment by sintering and pressing (powder metallurgy), as well as sprayed coatings and paint pigments (aluminum, bronze). Thermoplastic polymers in powder form are used in a technology known as powder molding, and thermosetting polymers are used in the sprayed coatings field for autos, machinery, and other industrial applications in which they have many advantages over sprayed solvent coatings.
See metal, powdered; carbon black; black powder.

powder of Algaroth. A mixture of SbOCl and Sb_2O_3.
Use: To prepare tartar emetic.

powder metallurgy. See metal, powdered; sintering.

ppb. Abbreviation for parts per billion.

ppm. Abbreviation for parts per million.

Pr. (1) Symbol for praseodymium. (2) Informal abbreviation for propyl.

pralidoxime methiodide. See 2-pyridine aldoxime methiodide.

d,d-t80-prallethrin.
CAS: 23031-36-9. $C_{19}H_{24}O_3$.
Hazard: A poison by ingestion and inhalation. Low toxicity by skin contact.

Prandtl number. For any substance, the ratio of the viscosity to the thermal conductivity. The lower the number, the higher the convection capacity of the substance. This ratio is important in heat and chemical engineering calculations.

praseodymia. See praseodymium oxide; rare earth.

praseodymium. Pr. Metallic element of atomic number 59, group IIIB of the periodic table, one of the rare earth elements of the lanthanide group, aw 140.9077, valences = 3, 4. No stable isotopes.
Properties: Yellowish metal, tarnishes easily (color of salts green). Paramagnetic, d 6.78–6.81, mp 930C, bp 3200C, ignites to oxide (200–400C), liberates hydrogen from water. Soluble in dilute acids.
Source: Monazite, cerite, and allonite; also a fission product.
Derivation: Reduction of the trifluoride with an alkaline metal or by electrolysis of the fused halides.
Grade: Ingots, rods, sheets, 98.8–99.9+% pure.
Use: Praseodymium salts, ingredient of mischmetal, core material for carbon arcs, colorant in glazes and glasses, catalyst, phosphors, lasers.
See didymium.

praseodymium oxalate. $Pr_2(C_2O_4)_3 \cdot 10H_2O$. Green powder, insoluble in water, slightly soluble in acids.
Use: Ceramics.

praseodymium oxide. (praseodymia). Pr_2O_3. Yellow-green powder, d 7.07, insoluble in water, soluble in acids, hygroscopic, absorbs carbon dioxide from air, purities to 99.8% oxide. Combustible.
Use: Glass and ceramic pigment, laboratory reagent.

prasterone. See DHEA.

PRE. See progesterone response element.

precipitant. Substance that causes precipitation of a solid from solution.

precipitate. ($\downarrow$, ppt). Small particles that have settled out of a liquid or gaseous suspension by gravity, or that result from a chemical reaction. Precipitated compounds, such as blanc fixe (barium sulfate), are prepared in this way, for example, by the reaction $BaCl_2 + Na_2SO_4 \rightarrow NaCl + BaSO_4$. In formulas, a downward vertical arrow, $\downarrow$, or "ppt" is sometimes used to indicate a precipitate. A class of organic pigments called lakes are made by precipitating an organic dye onto an inorganic substrate. Colloidal particles dispersed in a gas, as flue dust in industrial stacks, can be precipitated by introducing an electric charge opposite to that which sustains the particles.
See Cottrell; sedimentation.

precipitator, electrostatic. See Cottrell; precipitate.

precision investment casting. See investment casting.

precursor. In biochemistry, an intermediate compound or molecular complex present in a living organism; when activated physiochemically it is converted to a specific functional substance. The prefix *pro-* is usually used to indicate that the compound in question is a precursor. Examples are ergosterol (pro-vitamin D_2), which is activated by UV radiation to vitamin D; carotene (pro-vitamin A), a precursor of vitamin A; prothrombin, which forms thrombin upon activation in the bloodclotting mechanism; and phenylacetic acid, a precursor in the biosynthesis of penicillin G.

prednisolone. ($\Delta^{1,4}$-pregnadiene-11β,17α,21-triol-3,20-dione).
CAS: 50-24-8. $C_{21}H_{28}O_5$. Generic name for an analog of hydrocortisone.
Properties: White to practically white crystalline powder; odorless. Mp 235C with some decomposition. Very slightly soluble in water; soluble in alcohol, chloroform, acetone, methanol, dioxane.
Grade: USP.
Hazard: Causes sodium retention; may have side effects similar to cortisone.
Use: Medicine, also available as acetate.

prednisone. ($\Delta^{1,4}$-pregnadiene-17α,21-diol-3,11,20-trione).
CAS: 53-03-2. $C_{21}H_{26}O_5$. Generic name for an analog of cortisone.
Hazard: Questionable carcinogen.

preferential. Descriptive of the selectivity of action, either chemical or physiochemical, exhibited by a substance when in contact with two other substances; it may be either due to chemical affinity or due to surface phenomena. An example of a preferential chemical combination is that of hemoglobin with carbon monoxide, with which it unites 200 times as readily as with oxygen when exposed to a mixture of the two. Such phenomena as adsorption, corrosion, and the wetting of dry powders by liquids are other examples.

Pregl, Fritz. (1869–1930). An Austrian chemist who won the Nobel Prize in 1923. He was also a medical doctor who worked in micromechanical analysis and developed determinations for hydrogen, carbon, nitrogen, and organic groups using micromethods. He was educated at Tubingen, Leipzig, and Berlin.

pregnanediol. (5β-pregnane-3α,20α-diol).
$C_{21}H_{36}O_2$. A steroid, the metabolic product of progesterone.
Properties: Crystallizes in plates from acetone. Mp 238C, dextrorotatory in solutions. Sparingly soluble in organic solvents. Not precipitated by digitonin.
Derivation: Isolation from urine of pregnant women, cows, mares, and chimpanzees; by reduction of pregnanedione.
Use: Synthesis of progesterone, medically as a pregnancy test.

pregnenedione. See progesterone.

pregneninolone. See ethisterone.

4-pregnen-21-ol-3,20-dione. See deoxycorticosterone.

pregnenolone. (Δ^5-pregnene-3β-ol-20-one).
$C_{21}H_{32}O_2$. A steroid that is a biologically active hormone similar to progesterone and the adrenal steroid hormones. Also available as acetate salt.
Properties: Crystallizes in needles from dilute alcohol. Mp 193C. Slightly soluble in acetone, petroleum ether, benzene, and carbon tetrachloride.
Derivation: From stigmasterol or other steroids.
Use: Medicine, biochemical research.

Prelog, Vladimir. (1906–1998). A Swiss organic chemist who won the Nobel Prize for chemistry in 1975 along with Cornforth for work on chemical synthesis of organic compounds. Although educated in Yugoslavia, he spent many years in Zurich.

premature chromosome condensation. (PCC). A method of studying chromosomes in the interphase stage of the cell cycle.

premix molding. A mixture of plastic ingredients prepared in advance of the molding or extruding operation and stored in bags or bins until required. It is made by mixing the components (resin, filler, fibrous materials such as glass and necessary curatives) in a dough blender. Storage life may be from a few days to a year or more, depending on formulation. Such mixtures are then calendered or extruded after warming to suitable temperature.

prenitene. (1,2,3,4-tetramethylbenzene; prenitol). $(CH_3)_4C_6H_2$.
Properties: Colorless liquid. D 0.901, bp 204C, fp −7.7C. Soluble in alcohol; insoluble in water.

prepolymer. An adduct or reaction intermediate of a polyol and a monomeric isocyanate, in which either component is in considerable excess of the other. A polymer of medium molecular weight having reactive hydroxyl and –NCO groups.
Use: Preparation of polyurethane coatings and foams.

prepreg. A term used in the reinforced plastics field to mean the reinforcing material containing or combined with the full complement of resin before molding.

"Pre-San" *[PBI].* [N-(2-mercaptoethyl) benzenesulfonamide-S-(O,O-diisopropylphosphorodithioate)]. TM for a selective herbicide.

preservative. Any agent that prolongs the useful life of a material. Food products are preserved by (1) low temperature, (2) ionizing radiation (X- and

γ-rays), (3) antioxidants, (4) fungicides, (5) aldehydes, (6) paints, and others.
Use: Antioxidants in lubricating oils, rubber, and plastics; fungicides on textiles; aldehydes on biological specimens; paints on wood and metals.
See protective coating; antioxidant; radiation, industrial.

press, hydraulic. See hydraulic press.

Prestabit Oil V. An anionic textile chemical consisting of purified sulfated castor-oil fatty acids.
Use: Dyeing assistant for cotton and wool fiber, in viscose manufacture, clarifying agent to prevent milkiness of the yarn, antistatic agent for acetate and polyacrylonitrile fibers.

"Pretested Agaroid RS-30" *[ITC]*. TM for a powder gelatin replacement.
Use: Allows complete hydration as a gelatin replacement in confection, dairy, and low fat applications.

"Preventol" *[Lanxess]*. TM for a skin degerming agent.
Properties: White to grayish-white powder.
Use: Incorporation into bar soaps.

Prevost reaction. Hydroxylation of olefins with iodine and silver benzoate in an anhydrous solvent to give *trans*-glycols.

Priestly, Joseph. (1733–1804). Born near Leeds, England, Priestley originally planned to enter the ministry. As a youth he became interested in both physics and chemistry, and his research soon established his position as a scientist. He was elected to the Royal Society in 1766. He discovered nitrous oxide in 1772, but his greatest contribution to science was his discovery of oxygen in 1774. He emigrated from England to Northumberland, PA, where he lived from 1784 to his death. His research in America resulted in the discovery of carbon monoxide (1799).

Prigogine, Ilya. (1917–2003). A Belgian chemist who won the Nobel Prize for chemistry in 1977 for his contributions to nonequilibrium thermodynamics. He was educated at the University of Brussels. The Center for Statistical Mechanics and Thermodynamics at the University of Texas bears his name.

Prilezhaev (Prileschajew) reaction. Formation of epoxides by the reaction of alkenes with peracids.

prills. Small, round, or acicular aggregates of a material, usually a fertilizer, that are artificially prepared. In the explosives field, prills-and-oil consists of 94% coarse, porous ammonium nitrate prills and 6% fuel oil.
See explosives, high.

Primacord. A detonating composition.
See pentaerythritol tetranitrate.

primaequine phosphate. (N^4-(6-methoxy-8-quinolinyl)-1,4-pentanediamine phosphate (1:2) (9CI); Primquine diphosphate; 8-[(4-amino-1-methylbutyl)amino]-6-methoxyquinoline phosphate(1:2)). $C_{15}H_{21}N_3O \cdot 2H_3O_4P$. An agent that is especially effective against *Plasmodium vivax*.
Properties: Yellow crystals from EtOH; moderately soluble in water.
Hazard: Poison; teratogen; mutagen.
Use: Antimalarial agent.

Primafloc. A series of organic polyelectrolyte products used for flocculating suspended solids in water, waste, and process streams.
See polyelectrolyte.

Primal. Aqueous dispersions of acrylic resins, supplied in various grades that differ in hardness and flexibility and produce finishes that are water insoluble, require no plasticizer for flexibility, are unimpaired by aging, and adhere tenaciously to leather and lacquer coats.

primary. (1) In reference to monohydric alcohols, amines, and a few related compounds, this term, together with *secondary* and *tertiary*, describes the molecular structure of isomeric or chemically similar individuals. Monohydric alcohols are based on the methanol group

$$-\overset{|}{\underset{|}{C}}-OH$$

in which three of the bonds of the methanol carbon may be attached either to hydrogen atoms or alkyl groups. A primary alcohol has one alkyl group and two hydrogens,

$$CH_3-\overset{H}{\underset{H}{C}}-OH$$

except methanol, in which all three bonds are to hydrogen atoms. A secondary alcohol has two alkyl groups and one hydrogen, e.g.,

$$CH_3-\overset{CH_3}{\underset{H}{C}}-OH$$

A tertiary alcohol has three alkyl groups, e.g.,

$$CH_3-\underset{\underset{CH_3}{|}}{\overset{\overset{CH_3}{|}}{C}}-OH$$

The three types can be readily identified by the number of hydrogen atoms attached to the central (methanol) carbon atom: if it is two or more, the alcohol is primary; if one, it is secondary; and if zero, it is tertiary. For example, CH_3CH_2OH is primary; $(CH_3)_2CH_2OH$ is secondary; $(CH_3)_3COH$ is tertiary.

Primary, secondary, and tertiary amines are formed from ammonia (NH_3) when one, two, or three hydrogen atoms, respectively, are replaced by alkyl groups.

These terms are also used to name salts of orthophosphoric acid (H_3PO_4) in which one, two, or three of the hydrogen atoms have been replaced by metal or radicals: NaH_2PO_4 is primary sodium phosphate, Na_2HPO_4 is secondary sodium phosphate. The same system of names is used for salts of other acids containing three replaceable hydrogen atoms.

(2) A type of battery that is irreversible in respect to power output; a voltaic cell. A secondary battery (storage battery) is reversible and can be recharged.

(3) In the terminology of minerals, *primary* (in the case of metals) refers to direct production from the ore; in the case of petroleum it refers to production from wells by direct means. This meaning contrasts with the term *secondary* that is used to denote recovery of metal from scrap and, for petroleum, recovery by means of special techniques such as flooding and hydraulic pressure.

primary air pollutant.   A pollutant emitted into the atmosphere from an identifiable source.

primary alcohol.   An alcohol distinguished by the presence of the univalent radical.
Properties: The carbon atom bears the hydroxyl group also bonds with two hydrogen atoms.

primary amine.   An amine in which a hydrocarbon replaces one of the NH_3 group.

primary arsine.   An arsine in which one hydrogen is replaced by an alkyl radical.

primary azo dyes.   Azo dyes derived from primary amines.

primary calcium phosphate.   See calcium phosphate, monobasic.

primary carcinogen.   (direct-acting carcinogen). A carcinogen that does not require metabolic activation; one that is able to act, in its metabolically unmodified form.

primary structure.   The sequence of amino acids and any interchain and intrachain disulfide bonds of a protein.

primary transcript.   When a gene is transcribed in the nucleus, the initial product is the primary transcript—an RNA containing copies of all exons and introns. This primary transcript is then processed by the cell to remove the introns, to cleave off unwanted 3′ sequence, and to polyadenylate the 5a end. The mature message thus formed is then exported to the cytoplasm for translation.

primase.   An enzyme that synthesizes RNA oligonucleotides used as primers by DNA polymerases.

"Primene JM-T" *[Dow].*   TM for a solvent composed of branched-chain primary amines containing from 18 to 22 carbon atoms.

primer.   Short preexisting polynucleotide chain to which new deoxyribonucleotides can be added by DNA polymerase.

primer extension.   This is a method used to determine how far the start of an mRNA is upstream from a fixed site.

"Primid" *[EMS].*   TM for solutions for bonding, coating. and sealing components in auto production.

primuline dye.   See thiazole dye.

principle.   A constituent of a substance, especially one giving it some distinctive quality or effect.

Prins reaction.   Acid-catalyzed addition of olefins to formaldehyde to give 1,3-diols, allylic alcohols, or *meta*-dioxanes.

printing ink.   A viscous to semisolid suspension of finely divided pigment in a drying oil such as heat-bodied linseed oil. Alkyd, phenol-formaldehyde, or other synthetic resins are frequently used as binders; and cobalt, manganese, and lead soaps are added to catalyze the oxidative drying reaction. Some types of inks dry by evaporation of a volatile solvent rather than by oxidation and polymerization of a drying oil or resin. Use distribution is: offset 40%, gravure 23%, flexographic 18%, letterpress 9%, screen 4%, other 6%. For further information refer to National Printing Ink Institute, Lehigh University, Bethlehem, PA.
Hazard: Questionable carcinogen.

prion.   (prPon; prion protein; proteinaceous infectious particle). A small biological entity with at least one protein but no demonstrable nucleic acid. Such particles are resistant to inactivation by most procedures that modify nucleic acids, are resistant to inactivation by heat and show heterogeneity with respect to size.

pristane. (2,6,10,14-tetramethylpentadecane; norphytane).
CAS: 1921-70-6. $C_{19}H_{40}$. Found in rock specimens 2.5–3 billion years old. It is known to be synthesized only by living organisms and to withstand heat and pressure; thus it serves to date the existence of life on earth.
Properties: Colorless, transparent, stable liquid; faintly odored. Bp 290C, fp −60C, d 0.775–0.795 (20C), refr index 1.435–1.440 (20C). Soluble in most organic solvents. Combustible.
Grade: 90% purity.
Use: Precision lubricant, chromatographic oil, anticorrosive agent.
See phytane.

"Pristene" [Lyondell Basell]. TM for natural tocopherols and herbal extracts.
Use: Vegetable oils, animal fats, flavors, spices, fragrances, confections, nuts, yeast, sausage, gum feeds, beverages, cereal, desserts, dehydrated potatoes, and cosmetics.

privacy. Genetics: the right of people to restrict access to their genetic information.

probe. Single-stranded DNA or RNA molecules of specific base sequence, labeled either radioactively or immunologically, that are used to detect the complementary base sequence by hybridization.

procaine acryloyl monomer.
CAS: 25252-96-4. $C_{16}H_{22}N_2O_3$.
Hazard: A poison.

procaine hydrochloride. (procaine).
CAS: 51-05-8. $C_6H_4NH_2COOCH_2CH_2N(C_2H_5)_2 \cdot$ HCl.
Properties: Small, colorless crystals or white, crystalline powder; odorless; stable in air. Mp 153–156C. Soluble in water and in alcohol at 25C; slightly soluble in chloroform; almost insoluble in ether; solutions acid to litmus.
Derivation: (1) By heating chloroethyl-*p*-nitrobenzoic ester with diethylamine for 24 hours under pressure at 120C. The product is then reduced with tin and hydrochloric acid. (2) By condensation of ethylene chlorohydrin with diethylamine. The chloroethyldiethylamine formed is heated with sodium-*p*-aminobenzoate.
Grade: USP.
Use: Medicine (local anesthetic).

procaine penicillin G.
CAS: 54-35-3. $C_{16}H_{18}N_2O_4S \cdot C_{13}H_{20}N_2O_2 \cdot H_2O$.
Properties: White, fine crystals or powder; odorless; relatively stable to air and light; solutions dextrorotatory. Sparingly soluble in water; slightly soluble in alcohol; fairly soluble in chloroform.

Grade: USP.
Use: Antibiotic, animal feed additive.

proceroside.
CAS: 25323-74-4. $C_{29}H_{40}O_{10}$.
Hazard: A poison.
Source: Natural product.

process. See unit process.

process industry. See chemical process industry.

processing. The reactions occurring in the nucleus which convert the primary RNA transcript to a mature mRNA. Processing reactions include capping, splicing, and polyadenylation. The term can also refer to the processing of the protein product, including proteolytic cleavages, glycosylation, etc.

processivity. For any enzyme that catalyzes the synthesis of a biological polymer, the property of adding multiple subunits to the polymer without dissociating from the substrate.

prochiral molecule. A symmetric molecule that can react asymmetrically with an enzyme having an asymmetric active site, generating a chiral product. For example, fumarate converted to L-malate by fumarase.

"Prodox 121" [ProDox]. (4-α-methylstrylphenol). TM for intermediate.
Use: In fabric-finishing chemicals and antioxidants for plastics and rubber.

prodrug. A term applied in the pharmaceutical chemistry to a chemical compound that is converted into an active curative agent by metabolic processes within the body.
See precursor.

producer gas. A gas obtained by burning coal or coke with a restricted supply of air, or by passing air and steam through a bed of incandescent fuel under such conditions that the carbon dioxide formed is converted into carbon monoxide. The water vapor reacts to form carbon monoxide and hydrogen. Producer gas is cheap but has low Btu and is used where transportation is not required.
Hazard: Highly flammable and toxic. Explosive range 20–73% in air.
See water gas; synthesis gas.

proliferative dust. Any type of dust that is not readily removed from the lungs.
Hazard: May cause irreversible lung damage and pneumoconiosis.

profile. See soil.

progesterone. (Δ^4-pregnene-3,20-dione).
CAS: 57-83-0. $C_{21}H_{30}O_2$. The female sex hormone secreted in the body by the corpus luteum, adrenal cortex, or placenta during pregnancy. It is important in the preparation of the uterus for pregnancy and for the maintenance of pregnancy. It exists in two crystalline forms (α and β) of equal physiological activity. Progesterone is believed to be the precursor of the adrenal steroid hormones.

Properties: White, crystalline powder; odorless and stable in air but sensitive to light. Mp (α form) 128–133C, (β form) 121C. Practically insoluble in water; soluble in alcohol, acetone, and dioxane; sparingly soluble in vegetable oils. The international unit (IU) of progestational activity is expressed as 1 mg of progesterone.
Derivation: Isolation from corpus luteum of pregnant sows, synthesis from other steroids such as stigmasterol.
Grade: NF.
Hazard: A carcinogen (OSHA).
Use: Oral contraceptive, lab reagent.

progesterone 16,17-acetonide.
CAS: 4968-09-6. $C_{24}H_{34}O_4$.
Hazard: A reproductive hazard.
Use: Hormone.

progesterone response element. (PRE). A binding site in a promoter to which the activated progesterone receptor can bind. The progesterone receptor is essentially a transcription factor which is activated only in the presence of progesterone. The activated receptor will bind to a PRE, and transcription of the adjacent gene will be altered.
See: Response element.

proglyde dmm. See dipropylene glycol dimethyl ether.

proguanyl. (chlorguanide; 1-*p*-chlorophenyl-5-isopropyl biguanide). An antimalarial drug said to be less toxic than others.

prokaryotic. Literally "before the nucleus", the term applies to all bacteria and archaea. Prokaryotic cells have no internal membranes or cytoskeleton. Their DNA is circular, not linear.
See chromosome; eukaryote.

prolactin. See luteotropin.

prolamin. Any of a group of simple vegetable proteins, e.g., gliadin in wheat, zein in corn. When split by acids they give only amino acids.

prolan. (2-nitro-1,1-bis(*p*-chlorophenyl)-propane).
CAS: 117-27-1. $C_{15}H_{13}Cl_2NO_2$.
Properties: Thick, oily liquid. Crystalline form melts at 80C.
Hazard: Toxic by ingestion.
Use: Insecticide; when mixed with bulan, the insecticide known as dilan is formed.
See bulan; dilan.

proline. (2-pyrrolidinecarboxylic acid).
C_4H_8NCOOH.
A nonessential amino acid found naturally in the L(−) form.

Properties: Colorless crystals. Soluble in water and alcohol; insoluble in ether; optically active. DL-proline: mp 205C with decomposition. D(+)-proline: mp 215–220C with decomposition. L(−)-proline: mp 220–222C with decomposition.
Derivation: Hydrolysis of protein, also synthetically and by recombinant DNA techniques.
Use: Biochemical and nutritional research, microbiological tests, culture media, dietary supplement, lab reagent. Available commercially as the L(−)-proline.

***l*-proline.** $C_5H_9NO_2$.
Properties: White crystals or crystalline powder; odorless with sweet taste. Very sol in water, alc; insol in ether.
Use: Food additive.

proline reductase. (D-proline reductase). An oxidoreductase that catalyzes the cleavage of D-proline to 5-aminovalerate.

prolipin. A compound, sterile solution of protein obtained from nonpathogenic bacteria, various animal fats, and lipoids derived from bile.

promazine hydrochloride. (10-(3-dimethylaminopropyl)phenothiazine hydrochloride). $C_{17}H_{20}N_2S \cdot HCl$. Isomeric with promethazine hydrochloride.
Properties: White to slightly yellow, hygroscopic powder; practically odorless. Oxidizes upon prolonged exposure to air and acquires a blue or pink color. Melts within a range of three degrees between 172 and 182C, pH of 5% solution between 4.2–5.2. Soluble in water and alcohol. Insoluble in benzene.
Grade: NF.
Use: Medicine (tranquilizer).

promecarb. (3-methyl-5-isopropyl-*N*-methylcarbamate).
CAS: 2631-37-0. $C_{12}H_{17}NO_3$.
Properties: Colorless, crystalline solid. Mp 87C. Insoluble in water; soluble in alcohol.
Hazard: Toxic by ingestion.
Use: Insecticide.

promethium. Pm. Radioactive rare-earth element, a member of the lanthanide series, atomic number 61, aw 147. The 145 isotope has a half-life of 18 years, the 147 half-life is 2.64 years; the latter is the only form available.
Properties: Silvery-white metal. Mp 1,160C, density 7.2.
Derivation: The 147 isotope is recovered from spent uranium fission products, also by reduction of the chloride or fluoride with an alkali metal.
Hazard: Strong radioactive poison, use requires shielding and glove boxes.
Use: (147 isotope) Nuclear auxiliary power generators, special semiconductor battery, luminescent paint for watch dials, X-ray source, source of β-rays for thickness gauges, used with tungsten cermet for space power systems (withstands 2000C for 1000 hours).
Note: Some 30 compounds are known, but may not be commercially available.

promoted electron. Electron forced to occupy a principal quantum level higher than that of its ground-state level.

promoter. (1) A substance that, when added in relatively small quantities to a catalyst, increases its activity, e.g., aluminum and potassium oxide are added as promoters to the iron catalyst used in facilitating a combination of hydrogen and nitrogen to form ammonia. (2) In ore flotation, a substance that provides the minerals to be floated with a water-repellent surface that will adhere to air bubbles. Such reagents are generally more or less selective toward minerals of certain classes. (3) The first few hundred nucleotides of DNA "upstream" (on the 5′ side) of a gene, which control the transcription of that gene. The promoter is part of the 5′ flanking DNA, i.e., it is not transcribed into RNA, but without the promoter, the gene is not functional.

pronucleus. The nucleus of a sperm or egg prior to fertilization.
See nucleus; transgenic.

proof. The ethanol content of a liquid at 15.5C, stated as two times the percentage of ethanol by volume. One gallon of 95% alcohol is therefore equivalent to 1.9 gallons of proof alcohol. In the U.S., the alcohol tax is based on the number of proof gallons.

propachlor.
CAS: 1918-16-7. $C_{11}H_{14}NOCl$.

Properties: Tan powder. Mp 68C. Soluble in alcohol, benzene.
Use: Selective weed killer.

propadiene. See allene.

propadrine hydrochloride. (monhydrin, *dl*-norephedrine hydrochloride, phenylpropanolamine hydrochloride).
CAS: 154-41-6. Mf: $C_9H_{13}NO•ClH$.
Properties: A solid. Mw: 187.69, mp: 194C. Soluble in water.
Hazard: Poison by ingestion.
Use: Used as a raw material in cold and diet tablets.

"Propafilm" *[Innovia].* TM for coated polypropylene packaging films.
Use: A metalizable film used for packaging, barrier, and heat sealable films.

propanal. See propionaldehyde.

propane. (dimethylmethane).
CAS: 74-98-6. C_3H_8.
Properties: Colorless gas; natural-gas odor. Non-corrosive. Bp −42.5C, fp −189.9C, density of liquid 0.531 (0C), density of vapor 1.56 (0C), flash p −156F (−105C), autoign temp 874F (467C). Soluble in ether, alcohol; slightly soluble in water. An asphyxiant gas.
Grade: Technical, research (9.9%).
Derivation: From petroleum and natural gas.
Hazard: Asphyxiant. Flammable, dangerous fire risk, explosive limits in air 2.4–9.5%. For storage, see butane (note).
Use: Organic synthesis, household and industrial fuel, manufacture of ethylene, extractant, solvent, refrigerant, gas enricher, aerosol propellant, mixture for bubble chambers.
See butane (note).

1,3-propanediamine. See 1,3-diaminopropane.

1,2-propanediol. See 1,2-propylene glycol.

1,3-propanediol. See trimethylene glycol.

n,n′-(1,3-propanedioxysulfinyl)bis(2,3-dihydro-2,2-dimethylbenzofuranyl-7-methylcarbamate).
CAS: 81861-90-7. $C_{27}H_{34}N_2O_{10}S_2$.
Hazard: A poison by ingestion.
Use: Agricultural chemical.

propane hydrate. See gas hydrate.

propanenitrile. See ethyl cyanide.

propane sultone. (1,2-oxathrolane; 2,2-dioxide; 3-hydroxy-1-propanesulphonic acid sultone; 1,3-propane sultone).
CAS: 1120-71-4. $C_3H_6O_3S$.
Properties: Liquid or white crystals. Mw 122.14, mp above 31.11C (with a foul odor), bp 180C. Soluble in water and organic solvents.
Hazard: Possible carcinogen.
Use: Chemical intermediate for the production of fungicides, insecticides, cation-exchange resins, dyes, and vulcanization accelerators.

1,3-propane sultone. See propane sultone.

1-propanethiol. (*n*-propyl mercaptan).
CAS: 107-03-9. C_3H_7SH.
Properties: Liquid; offensive-smelling. Boiling range 67–73C, d 0.8408 (20/4C), refr index 1.4380 (20C), flash p −5F (−20.5C).
Grade: 95%.
Hazard: Highly flammable, dangerous fire risk.
Use: Chemical intermediate, herbicide.

n,n′,n″-(1,2,3-propanetrioxysulfinyl)tris(2-isopropoxyphenylmethylcarbamate).
CAS: 81862-23-9. $C_{36}H_{47}N_3O_{15}S_3$.
Hazard: Moderately toxic by ingestion.
Use: Agricultural chemical.

n,n′,n″-(1,2,3-propanetrioxysulfinyl)tris(1-naphthyl methylcarbamate).
CAS: 81862-20-6. $C_{39}H_{35}N_3O_{12}S_3$.
Hazard: Moderately toxic by ingestion.
Use: Agricultural chemical.

propanil. (generic name for 3,4-dichloropropionanilide).
CAS: 709-98-8. $Cl_2C_6H_3NHCOCH_2CH_3$.
Properties: (Pure) Light-brown solid. Mp 85–89C. (Technical) Liquid. Bp 91–91.5C.
Hazard: Toxic by ingestion and inhalation.
Use: Postemergence herbicide, especially for rice culture; nematocide.

propanoic acid. See propionic acid.

1-propanol. See propyl alcohol.

2-propanol. See isopropyl alcohol.

propanolamine. See 2-amino-1-propanol; 3-amino-1-propanol.

2-propanol nitrate. See isopropyl nitrate.

2-propanolpyridine. $C_5NH_4C_3H_6OH$.
Properties: Colorless liquid. Bp 260.2C, d 1.060 (25C), refr index 1.5298 (20C). Miscible with water at 20C.

4-propanolpyridine. $C_5NH_4C_3H_6OH$.
Properties: Colorless liquid. Bp 289.0C, fp 36.7C, d 1.053 (40C). Soluble in water.

2-propanone. See acetone.

2-propanone oxime. See acetoxime.

propanoyl chloride. See propionyl chloride.

propargite. (2-(*p-tert*-butylphenoxy)-cyclohexyl-2-propynylsulfite; dark oil).
CAS: 2312-35-8.
Properties: D 1.1 (25C). Insoluble in water.
Use: Acaricide for fruits, vegetables, row crops.

propargyl alcohol. (2-propyn-1-ol).
CAS: 107-19-7. $HC{\equiv}CCH_2OH$.
Properties: Colorless liquid; geranium-like odor. D 0.971, fp −48C, bp 114C, flash p 97F (36C) (OC). Soluble in water, alcohol, and ether; immiscible with aliphatic hydrocarbons.
Derivation: From acetylene by high-pressure synthesis.
Grade: Technical, 75% solution.
Hazard: Flammable, moderate fire risk. Toxic by ingestion, inhalation, and skin absorption. Eye irritant, liver and kidney damage.
Use: Chemical intermediate, corrosion inhibitor, lab reagent, solvent stabilizer, prevents hydrogen embrittlement of steel, soil fumigant.

propargyl bromide. (3-bromo-1-propyne).
CAS: 106-96-7. $HC{\equiv}CCH_2Br$.
Properties: Liquid, sharp odor. D 1.520, bp 88–90C, flash p 65F (18.3) (COC).
Derivation: From acetylene by high-pressure synthesis.
Hazard: Flammable, dangerous fire and explosion risk. Irritant.
Use: Chemical intermediate, soil fumigant.

propargyl chloride. (3-chloro-1-propyne).
$HC{\equiv}CCH_2Cl$.
Properties: Liquid. Fp −76.9C, bp 57.1, refr index 1.4310 (25C), flash p 90F (32C). Soluble in benzene, alcohol, carbon tetrachloride; insoluble in water.
Derivation: From acetylene by high-pressure synthesis.
Hazard: Flammable, moderate fire risk.
Use: Chemical intermediate, soil fumigant.

(n^{10})-propargyl-5,8-dideazafolic acid.
CAS: 76849-19-9. $C_{24}H_{23}N_5O_6$.
Hazard: A poison.

5′-propargyl-α′-ethynyl-3′-furylmethyl α-ethylphenyl acetate.
CAS: 51715-75-4. $C_{20}H_{18}O_3$.
Hazard: Moderately toxic by ingestion.
Use: Agricultural chemical.

propazine 80W.
CAS: 139-40-2. $C_9H_{16}ClN_5$. Generic name for a preemergence herbicide containing 80% 2-chloro-4,6-bis(isopropylamino)-s-triazine.
Use: Weed control in sorghum culture.

propellant. (1) A rocket fuel. (2) A compressed gas used to expel the contents of containers in the form of aerosols. Chlorofluorocarbons were once widely used because of their nonflammability. The strong possibility that they contribute to depletion of the ozone layer of the upper atmosphere has resulted in prohibition of their use for this purpose. Other propellants used are hydrocarbon gases, such as butane and propane, carbon dioxide, and nitrous oxide. The materials dispersed include insecticides, shaving cream, whipping cream, and cosmetic preparations.
See ozone (note).

2-propenal. See acrolein.

propene. See propylene.

propenenitrile. See acrylonitrile.

2-propene-nitrile, polymer with 1,3-butadiene, 3-carboxy-1-cyano-1-methylpropyl-terminated,2-hydroxy-3-((1-oxo-2-propenyl)oxy)propyl ester.
CAS: 68891-47-4. $(C_4H_6 \bullet C_3H_3N)_x$.
Hazard: A severe skin irritant.

2-propene-1-thiol. See allyl mercaptan.

propene-1,2,3-tricarboxylic acid. See aconitic acid.

propenoic acid. See acrylic acid.

2-propen-1-ol. See allyl alcohol.

propenyl alcohol. See allyl alcohol.

2-propenylamine. See allylamine.

p-propenylanisole. See anethole.

α-propenyldichlorohydrin. See α-dichlorohydrin.

propenyl guaethol. (1-ethoxy-2-hydroxy-4-propenylbenzene). $C_2H_5OC_6H_3(OH)(C_3H_5)$.
Properties: Free-flowing, white powder; odor and taste similar to vanilla, but much more powerful. Mp 85–86C. Very soluble in fats, edible solvents, and essential oils; very slightly soluble in water.
Use: Artificial vanilla flavoring, flavor enhancer.

propenyl guaiacol. See methyl isoeugenol.

2-propenyl hexanoate. See allyl caproate.

2-propenyl isothiocyanate. See allyl isothiocyanate.

propineb. (zinc-1,2-propylene-bisdithiocarbamate).
CAS: 12071-83-9. $C_5H_8N_2S_4Zn$.
Properties: Faintly yellow crystals or powder; becomes dark when heated. Decomposes in the presence of strong acids or bases. Insoluble in most common solvents.
Hazard: Toxic by ingestion.
Use: Fungicide.

γ-propiobutyrolactone. See γ-heptalactone.

propiodal. (1,3-bis(trimethylamino)-2-propanol diiodide; iodisan; hexamethyldiaminoisopropanol diiodide). $[CH_2N(CH_3)_3I]_2CHOH$.
Properties: White, crystalline solid. Mp 275C (decomposes), turns brown at 240C. Freely soluble in water; slightly soluble in alcohol; insoluble in ether and acetone.
Use: Medicine (iodine therapy).

β-propiolactone. (USAN) (BPL).
CAS: 57-57-8.

$OCH_2CH_2CO.$

Properties: Colorless liquid; pungent odor. Bp 155C with rapid decomposition, fp −33.4C, refr index 1.4131 (20C), d 1.1450 (20/4C), flash p (OC) 167F (75C). Soluble in water; miscible with ethanol, acetone, ether, and chloroform at 25C; reacts with alcohol. Stable when stored in glass at refrigeration temperature (+5 to +10C). Combustible.
Derivation: Direct combination of ketene and formaldehyde.
Hazard: Strong skin and upper respiratory tract irritant, skin cancer. Possible carcinogen. Worker exposure should be minimized.
Use: Organic synthesis, vapor sterilant, disinfectant.

propionaldehyde. (propanal; propyl aldehyde; propionic aldehyde).
CAS: 123-38-6. C_2H_5CHO.
Properties: Water-white liquid; suffocating odor. Flash p 15F (−9.4C) (OC), bp 48.8C, fp −81C, d 0.807 (20.4C), refr index 1.364 (20C), autoign temp 405F (207C). Soluble in water and alcohol.
Derivation: (1) Oxidation of propyl alcohol with dichromate, (2) by passing propyl alcohol over copper at elevated temperatures.
Hazard: Flammable, dangerous fire risk, explosive limits in air 3.0–16%. Upper respiratory tract irritant.
Use: Manufacture of propionic acid, polyvinyl, and other plastics; synthesis of rubber chemicals; disinfectant; preservative.

propionamide nitrile.   See cyanoacetamide.

propione.   See diethyl ketone.

propionic acid.   (methylacetic acid; propanoic acid).
CAS: 79-09-4. $CH_3CH_2CO_2H$.
Properties: Colorless, oily liquid; rancid odor. D 0.9942 (20/4C), fp −20.8C, refr index 1.3862 (20C), bp 140.7C, flash p 130F (54.4C), autoign temp 955F (512C). Soluble in water, alcohol, chloroform, and ether. Combustible.
Derivation: By reaction of ethanol with carbon monoxide, using a boron trifluoride catalyst; also by the reaction of carbon monoxide with hydrogen and olefins or alcohols.
Method of purification: Rectification.
Grade: Technical, 99.0%, FCC.
Hazard: Moderate fire risk. Strong eye, skin and upper respiratory tract irritant.
Use: Propionates, some of which are used as mold inhibitors in bread and fungicides in general; herbicides; preservative for grains and wood chips; emulsifying agents; solutions for electroplating nickel; perfume esters; artificial fruit flavors; pharmaceuticals; cellulose propionate plastics.

propionic aldehyde.   See propionaldehyde.

propionic anhydride.
CAS: 123-62-6. $(CH_3CH_2CO)_2O$.
Properties: Colorless liquid; pungent odor. Fp −45C, bp 167–169C, d 1.0119 (20C), vap press 1 mm Hg (20C), flash p 165F (73.9C) bulk d 8.4 lb/gal at 20C. Soluble in alcohol, ether, chloroform, and alkalies; decomposed by water. Combustible.
Hazard: Strong irritant to tissue.
Use: Esterifying agent for fats and oils; cellulose; dehydrating medium for nitrations and sulfonations; alkyd resins; dyestuffs; and pharmaceuticals.

propionitrile.   See ethyl cyanide.

propionylbenzene.   See propiophenone.

propionyl chloride.   (propanoyl chloride).
CAS: 79-03-8. CH_3CH_2COCl.
Properties: Colorless liquid; pungent odor. Fp −94C, bp 80C, d 1.065 (20/4C). Decomposes in water and alcohol.
Hazard: Strong irritant to skin.
Use: Chemical intermediate.

propionyl peroxide.
CAS: 3248-28-0. $C_2H_5C(O)OOC(O)C_2H_5$. Available as a 25% solution in a high-boiling hydrocarbon, flash p 125F (51.6C).
Hazard: Strong oxidizing agent, may explode if shocked or heated.
Use: Initiator in polymerization reactions, such as the high-pressure polymerization of ethylene.

propiophenone.   (ethyl phenyl ketone; propionylbenzene; 1-phenylpropanone-1).
CAS: 93-55-0. $C_6H_5COC_2H_5$.
Properties: Colorless to light-amber liquid or crystals; strong persistent odor. D 1.012 (20/20C), refr index 1.527 (20C), congealing temp 17.5–21C, bp 218C, flash p 210F (99C) (TOC). Insoluble in water, ethylene glycol, glycerol; miscible with alcohol, ether, benzene, and toluene. Combustible.
Use: Fixative in perfumes, starting material for synthesis of ephedrine and other important pharmaceuticals and synthetic organic chemicals, lab reagent.

Propi-Rhap.   The 2-ethylhexyl ester of 2-(2,4-dichlorophenoxy)propionic acid.
Use: Herbicide.

propoxur.   See   o-isopropoxyphenyl-N-methyl carbamate.

ProPoxy 300.   A high strength two-component epoxy adhesive anchoring gel.

propoxyphene.   (α-(+)-4-dimethylamino-1,2-diphenyl-3-methyl-2-butanol propionate ester).
CAS: 469-62-5. $C_{22}H_{29}NO_2$.   The α-diastereoisomers are optically active and are preferred for their greater pain-relieving ability. The drug has about the same analgesic effect as codeine. Abuse can cause addiction, and overdosage can be fatal. Its use has been restricted by FDA.

2-(3-propoxyphenyl)imidazo(2,1-a)isoquinoline.
CAS: 61001-15-8. $C_{20}H_{18}N_2O$.
Hazard: A reproductive hazard.

n-propoxypropanol. $C_6H_{14}O_2$.
Properties: Liquid. D 0.8865 (20/20C), bp 149.8C, fp −80C (sets to glass below this), flash p 128F (53.3C). Soluble in water.
Hazard: Moderate fire risk.
Use: Solvent for water-based enamel.

proppant.   A term coined by petroleum engineers to refer to agents such as sand, sintered bauxite, etc., that are used in hydraulic fracturing of oil wells; they are so called because they prop open the minute cracks in rock formations created by a hydraulic press.
See hydraulic fracturing.

propranolol.   (1-(isopropylamino)-3-(1-naphthyloxy)-2-propanol).
CAS: 525-66-6. $C_{16}H_{21}NO_2$.
Properties: Colorless crystals. Mp 96C. Soluble in water and alcohol; insoluble in benzene and ether.
Hazard: Highly toxic.

Use: An adrenergic blocker used in treatment of hypertension and various forms of heart disease. Approved by FDA.

"Propulse" *[Agilent].* TM for a pea protein isolate.
Use: Gives a high level of functionality and nutrition in food and beverages where protein enrichment is required.

n-propyl acetate.
CAS: 109-60-4. $C_3H_7OOCCH_3$.
Properties: Colorless liquid; pleasant odor. D 0.887, flash p 58F (14.4C), boiling range 96.0–102.0C, bulk d 7.36 lb/gal, autoign temp 842F (450C), fp −92C. Slightly soluble in water; miscible with alcohols, ketones, esters, hydrocarbons.
Derivation: Interaction of acetic acid and n-propyl alcohol in the presence of sulfuric acid.
Grade: Technical.
Hazard: Flammable, dangerous fire risk, explosive limits in air 2–8%. Eye and upper respiratory tract irritant.
Use: Flavoring agents, perfumery, solvent for nitrocellulose and other cellulose derivatives, natural and synthetic resins, lacquers, plastics, organic synthesis, lab reagent.

propyl acetone. See methyl butyl ketone.

propyl alcohol. (1-propanol; n-propyl alcohol).
CAS: 71-23-8. $CH_3CH_2CH_2OH$.
Properties: Colorless liquid; odor similar to ethanol. Bp 97.2C, fp −127C, d 0.804 (20/4C), flash p 77F (25C) (OC), autoign temp 700F (371C), refr index 1.385 (20C), viscosity 2.256 cP (20C). Soluble in water, alcohol, and ether.
Derivation: From oxidation of natural gas hydrocarbons, also from fusel oil.
Hazard: Flammable, dangerous fire risk. Explosive limits in air 2–13%. Toxic by skin absorption. Eye and upper respiratory tract irritant. Questionable carcinogen.
Use: Organic synthesis and chemical intermediate; solvent for waxes, vegetable oils, natural and synthetic resins, cellulose esters and ethers; polishing compositions; brake fluids; solvent degreasing; antiseptic.

sec-propyl alcohol. See isopropyl alcohol.

propyl aldehyde. See propionaldehyde.

n-propylamine.
CAS: 107-10-8. $C_3H_7NH_2$.
Properties: Colorless liquid; amine odor. D 0.7182 (20C), bp 47.8C, vap press 248 mm Hg (20C), fp −83C, flash p −35F (−37.2C), autoign temp 604F (317C). Soluble in water, alcohol, and ether.
Hazard: Highly flammable, dangerous fire risk, explosive limits in air 2–10%, use alcohol foam to extinguish. Strong irritant to skin and tissue.
Use: Intermediate, lab reagent.

n-propylbenzene. (1-phenylpropane).
CAS: 103-65-1. $C_6H_5CH_2CH_2CH_3$.
Properties: Colorless liquid. Bp 160C, fp −100C, d 0.862, flash p 86F (30C), refr index 1.49. Soluble in alcohol and ether; sparingly soluble in water.
Derivation: Reaction of benzylmagnesium chloride and diethyl sulfate.
Hazard: Flammable, moderate fire risk.
Use: Solvent for cellulose acetate, dyeing textiles.

n-propyl bromide. (1-bromopropane).
CAS: 106-94-5. C_3H_7Br.
Properties: Liquid. Mp −110C, bp 71C, d: 1.35 @ 20/4C, mw 123.01, autoign temp 914F, flash p 14–158F, lower explosive level 4.6%.
Hazard: A flammable liquid.
Use: Solvent, organic synthesis, pharmaceuticals, intermediate.

propyl butyrate.
CAS: 105-66-8. $C_3H_7OOCC_3H_7$.
Properties: Colorless liquid. D 0.8789 (15C), bp 142.7C, fp −95.2C. Slightly soluble in water; soluble in alcohol and ether. Combustible.
Hazard: Irritant to mucous membranes, narcotic in high concentration.
Use: Solvent mixture for cellulose ethers.

propyl chloride. (1-chloropropane).
CAS: 540-54-5. $CH_3CH_2CH_2Cl$.
Properties: Liquid. Fp −122.8C, bp 46.6C, refr index 1.3886 (20C), flash p 0F (−17.7C). Soluble in alcohol and ether; slightly soluble in water.
Hazard: Highly flammable, dangerous fire risk, explosive limits in air 2.5–11%. Irritant and narcotic.
See isopropyl chloride.

propyl chlorosulfonate.
CAS: 109-61-5. $CH_3CH_2CH_2OSO_2Cl$.
Properties: Liquid. Bp 70–72C (20 mm Hg).
Derivation: Interaction of n-propyl alcohol and sulfuryl chloride.
Hazard: Toxic by inhalation and ingestion, strong irritant to eyes.
Use: Organic synthesis, military poison gas (lachrymator).

n-propyl cyanide. See n-butyronitrile.

n-(4-propylcyclohexyl)benzamide.
CAS: 315706-70-8. $C_{16}H_{23}NO$.
Hazard: A poison by ingestion.

n-(4-propylcyclohexyl)-4-
   morpholinecarboxamide.
CAS: 315706-81-1. $C_{14}H_{26}N_2O_2$.
Hazard: A poison by ingestion.

n-(4-propylcyclohexyl)-3-(3,4,5-
   trimethoxyphenyl)-2-propenamide.
CAS: 315706-73-1. $C_{21}H_{31}NO_4$.
Hazard: A poison by ingestion.

n-propyldiethanolamine.
CAS: 6735-35-9. $C_7H_{17}NO_2$.
Hazard: A poison by ingestion and skin contact. A
severe eye irritant.

**propyl-3,5-diiodo-4-oxo-
   1(4H)pyridineacetate.**   See propyliodone.

propyldiphenylphosphine.
CAS: 7650-84-2. $C_{15}H_{17}P$.
Hazard: A poison by ingestion and skin contact. A
mild skin irritant.

propylene.   (propene).
CAS: 115-07-1. $CH_3CH:CH_2$.
Properties: Colorless gas. Bp −47.7C, fp −185.2C,
d (liquid) 0.5139 (20/4C), vap d (0C) (air = 1) 1.46,
flash p −162F (−108C), autoign temp 927F (497C).
Soluble in alcohol and ether; slightly soluble in
water.
Derivation: Catalytic and thermal cracking of ethylene with zeolite catalyst, from naphtha.
Grade: 95%, 99%, and research.
Hazard: Asphyxiant. Highly flammable, dangerous fire risk, explosive limits in air 2–11%.
Upper respiratory tract irritant. Questionable
carcinogen.
Use: Manufacture of isopropyl alcohol, polypropylene, synthetic glycerol, acrylonitrile, propylene
oxide, heptene, cumene, polymer gasoline, acrylic
acid, vinyl resins, oxo chemicals.

propylene carbonate.

$C_4H_6O_3$ or OCOCH$_2$CH(CH$_3$)O.

Properties: Colorless liquid; odorless. Fp −49.2C
(easily supercooled). Bp 241.7C, d 1.2057 (20/4C),
bulk d 10 lb/gal (20C), refr index 1.4209 (20C);
flash p 270F (132C). Miscible with acetone, benzene, chloroform, ether, ethyl acetate; moderately
soluble in water and carbon tetrachloride. Combustible.
Use: Solvent extraction, plasticizer, organic synthesis, natural gas purification, synthetic fiber spinning
solvent.

propylene chloride.   See propylene dichloride.

propylene   chlorohydrin.   (chloro-isopropyl
   alcohol; 1-chloro-2-propanol).
CAS: 78-89-7. $CH_2ClCH_2OCH_3$.
Properties: Colorless liquid; mild, nonresidual odor.
Bp 127.5C, vap press 4.9 mm Hg (20C), flash
p 125F (51.6C) (CC), bulk d 9.3 lb/gal (20C), d
1.1128 (20/20C). Soluble in water and alcohol.
Grade: Technical.
Hazard: Moderate fire risk. Toxic by ingestion and
skin absorption. Liver damage. Questionable carcinogen.
Use: Organic synthesis (introducing hydroxypropyl
group), manufacture of propylene oxide.

propylenediamine.   Legal label name for 1,2-
diaminopropane.

propylene   dichloride.   (1,2-dichloropropane;
   propylene chloride).
CAS: 78-87-5. $CH_3CHClCH_2Cl$.
Properties: Colorless, stable liquid; chloroform-like
odor. Bp 96.3C, d 1.1583 (20/20C), bulk d 9.6 lb/gal
(20C), refr index 1.4068 (20C), fp −80C, flash p
61F (16.1C), autoign temp 1035F (557C). Soluble
in water 0.26% by wt (20C); miscible with most
common solvents.
Derivation: Action of chlorine on propylene.
Grade: Refined.
Hazard: Flammable, dangerous fire risk, explosive
limits in air 3.4–14.5%. Toxic by ingestion and
inhalation. Upper respiratory tract irritant and body
weight effects. Questionable carcinogen.
Use: Intermediate for perchloroethylene and carbon
tetrachloride; lead scavenger for antiknock fluids;
solvents for fats, oils, waxes, gums, and resins; solvent mixture for cellulose esters and ethers; scouring compounds; spotting agents; metal degreasing
agents; soil fumigant for nematodes.

1,2-propylene glycol.   (1,2-dihydroxypropane;
   1,2-propanediol; methylene glycol; methyl glycol).
CAS: 57-55-6. $CH_3CHOHCH_2OH$.
Properties: Colorless, viscous, stable, hygroscopic
liquid; practically odorless and tasteless. Miscible
with water, alcohols, and many organic solvents.
Bp 187.3C, fp −60C, d 1.0381 (20/20C), bulk d
8.64 lb/gal (20C), refr index 1.4293 (27C), surface
tension 40.1 dynes/cm (25C), viscosity 0.581 cP
(20C), vap press 0.07 mm Hg (20C), specific heat
0.590 cal/g (20C), latent heat of evaporation 168.6
cal/g at bp, flash p 210F (99C) (OC), autoign temp
780F (415C), heat of combustion 431.0 kg cal/mole.
Combustible.
Derivation: By hydration of propylene oxide.
Method of purification: By distillation.
Grade: Refined, technical, USP, FCC, feed.
Use: Organic synthesis, especially polypropylene
glycol and polyester resins; cellophane; antifreeze
solution. Solvent for fats, oils, waxes, resins, flavoring extracts, perfumes, colors, soft-drink syrups,

antioxidants. Hygroscopic agent; coolant in refrigeration systems; plasticizers; hydraulic fluids; bactericide; textile conditioners. In foods as solvent, wetting agent, humectant. Emulsifier; feed additive; anticaking agent; preservative (retards molds and fungi); cleansing creams; suntan lotions; pharmaceuticals; brake fluids; deicing fluids for airport runways; tobacco.
See polypropylene glycol.

1,3-propylene glycol.   See trimethylene glycol.

propylene glycol alginate.   (hydroxypropyl alginate). ($C_9H_{14}O_7$).
 Properties: White powder; practically tasteless and odorless. Vary with degree of esterification. Soluble in water and dilute organic acids.
 Grade: FCC.
 Use: Stabilizer, thickener, emulsifier, food additive.

propylene glycol dinitrate.   (PGDN).
 CAS: 6423-43-4. $C_3H_6N_2O_6$.
 Properties: Colorless liquid; unpleasant odor. Mw 166.09, d 1.232 g/ml at 25C, fp −27.7C, bp decomposes above 121C. Slightly soluble in water.
 Hazard: Toxic by inhalation and skin absorption. Headache and central nervous system impairment.
 Use: Torpedo propellant in Otto Fuel II.

propylene glycol distearate.   See propylene glycol monostearate.

propylene glycol lactostearate.
 Properties: Soft to hard waxy solid. Dispersible in hot water; moderately sol in hot isopropanol, benzene, chloroform, soybean oil.
 Use: Food additive.

propylene glycol mono- and diesters.
 Properties: Clear liquid or white to yellow beads or flakes; bland odor and taste. Insol in water; sol in alc, ethyl acetate, chloroform.
 Use: Food additive.

propylene glycol monomethyl ether.
 (polypropylene glycol methyl ether).
 CAS: 107-98-2. $CH_3OCH_2CH_2OCH_3$.
 Properties: Colorless liquid. Fp −95C (sets to glass), bp 120.1C, d 0.9234 (20/20C) bulk d 7.65 lb/gal (25C), refr index 1.402 (25C), flash p 97F (36.1C). Soluble in water; methanol, ether.
 Hazard: Flammable, moderate fire risk. TLV: 100 ppm; STEL 150 ppm.
 Use: Solvent for celluloses, acrylics, dyes, inks, stains; solvent-sealing of cellophane.

propylene glycol monoricinoleate.
 $C_{17}H_{32}(OH)COOCH_2CH_2OCH_3$.
 Properties: Pale-yellow, moderately viscous, oily liquid; mild odor. D 0.960 (25/25C), saponification value 160, hydroxyl value 285, solidifies at −26C.

Soluble in most organic solvents; insoluble in water. Combustible.
 Derivation: Castor oil and propylene glycol.
 Grade: Technical.
 Use: Plasticizer, dye solvents, lubricant, cosmetics, urethane polymers, and hydraulic fluids.

propylene glycol monostearate.   The FCC grade is a mixture of propylene glycol mono- and diesters of stearic and palmitic acids. White beads or flakes, bland odor and taste, insoluble in water, soluble in alcohol, ethyl acetate, chloroform and other chlorinated hydrocarbons. Combustible.
 Use: Emulsifier, stabilizer.

propylene glycol phenyl ether.
 $C_6H_5OCH_2CHOHOCH_3$.
 Properties: Colorless liquid. D 1.060–1.070 (25/25C), boiling range 5.95%, 237–242C, flash p 275F (135C). Combustible.
 Use: High-boiling solvent, bactericidal agent, fixative for soaps and perfumes, intermediate for plasticizers.

propyleneimine.   (2-methylaziridine; propylenimine).
 CAS: 75-55-8.

$\overline{CH_2HCNHCH_2}$.

 Properties: Water-white liquid. Bp 66–67C, d 0.8039–0.8070 (25/25C), 1.4094–1.4109 (25C). Soluble in water and most organic solvents.
 Hazard: Flammable, dangerous fire risk. Toxic by ingestion, inhalation, and skin absorption. Upper respiratory tract irritant and kidney damage. Possible carcinogen.
 Use: Organic intermediate whose derivatives are used in the paper, textile, rubber, and pharmaceutical industries.

propylene oxide.
 CAS: 75-56-9.

$$H_2C\!-\!CH\!-\!CH_3 \atop \diagdown\!\!O\!\!\diagup$$

 Properties: Colorless liquid; ethereal odor. D 0.8304 (20/20C) bp 33.9C, vap press 445 mm Hg (20C), flash p −35F (−37.2C), bulk d 6.9 lb/gal (20C), fp −104.4C. Partially soluble in water; soluble in alcohol and ether.
 Derivation: (1) Chlorohydration of propylene followed by saponification with lime, (2) peroxidation of propylene, (3) epoxidation of propylene by a hydroperoxide complex with molybdenum catalyst.
 Hazard: Highly flammable, dangerous fire risk, explosive limits in air 2–22%. An irritant. TLV: 20 ppm; animal carcinogen.

Use: Polyols for urethane foams, propylene glycols, surfactants and detergents, isopropanol amines, fumigant, synthetic lubricants, synthetic elastomer (homopolymer), solvent.

propylene oxide and ethylene oxide block polymer.   See poloxamer 331.

propylene phenoxetol.   See 1-phenoxy-2-propanol.

propyl formate.
CAS: 110-74-7. $HCOOC_3H_7$.
Properties: Liquid. D 0.9006 (20/4C), fp −92.9C, bp 81.3C, refr index 1.3769 (20C). Slightly soluble in water, miscible with alcohol and ether. Flash p 27F (−2.8C) (COC), autoign temp 851F (455C).
Hazard: Flammable, dangerous fire risk.
Use: Flavoring.

propylformic acid.   See butyric acid.

n-propyl furoate. $C_4H_3OCO_2C_3H_7$.
Properties: Colorless, fragrant liquid; becomes yellow in light. D 1.0745 (25.9/4C), bp 210.9C (corrected), refr index 1.4737 (25.9C). Practically insoluble in water; soluble in alcohol and ether. Combustible.
Use: Flavoring.

propyl gallate.
CAS: 121-79-9. $C_3H_7OOCC_6H_2(OH)_3$.
Properties: Colorless crystals. Mp 150C. Almost insoluble in water; soluble in alcohol (50/50); somewhat soluble in oils.
Hazard: Use in foods restricted to 0.02% of fat content.
Use: Food preservative and antioxidant for animal fats and oils, flavoring, transformer oils.

propyl-p-hydroxybenzoate.   See propyl-paraben.

propyliodone.   (propyl-3,5-diiodo-4-oxo-1(4H)-pyridineacetate).
$I_2(O)C_5H_2NCH_2COOC_3H_7$.
Properties: White, crystalline powder; odorless or nearly so. Mp 187–190C. Practically insoluble in water; soluble in acetone, alcohol, and ether.
Grade: USP.
Use: Medicine (radiopaque medium).

propyl isomer.   Generic name for dipropyl-5,6,7,8-tetrahydro-7-methylnaphtho[2,3-d]-1,3-dioxole-5,6-dicarboxylate.
CAS: 83-59-0. $C_{20}H_{26}O_6$.
Properties: Orange liquid. Insoluble in water; slightly soluble in oils; soluble in most organic solvents.
Use: Insecticide synergist.

propylmagnesium bromide. C_3H_7MgBr.
Available as a solution in ether; a Grignard reagent.
Use: Alkylating agent in organic synthesis.

propylmalonic acid, diethyl ester.   (propyl diethyl malonate). $C_3H_7CH(COOC_2H_5)_2$.
Properties: Colorless liquid; fragrant odor. D 0.9860 (25C); bp 222C. Soluble in water; soluble in alcohols, ethers, esters, and ketones.
Use: Intermediate, tobacco flavoring.

n-propyl mercaptan.   Legal label name for 1-propanethiol.

2-n-propyl-4-methylpyrimidyl-6-N,N-dimethylcarbamate.
$C_3H_7C_4HN_2(CH_3)OOCN(CH_3)_2$.
Properties: A liquid. Miscible with water and most organic solvents.
Use: Insecticide.

n-propyl nitrate.   (NPN).
CAS: 627-13-4. $C_3H_7NO_3$.
Properties: White to straw-colored liquid; ethereal odor. D 1.07 (20C), bp 110C, flash p 68F (20C), autoign temp 350F (176C), fp −100C, refr index 1.3975 (20C). Insoluble in water; soluble in alcohol and ether.
Grade: 96–98% pure.
Hazard: Flammable, severe fire and explosion risk, strong oxidizing material, explosive limits in air 2–100%. Nausea and headache.
Use: Rocket fuel (monopropellant).

propylparaben.   (propyl-p-hydroxybenzoate).
CAS: 94-13-3. $C_{10}H_{12}O_3$.
Properties: Colorless crystals or white powder. Mp 95–98C. Slightly soluble in boiling water; soluble in alcohol, ether, and acetone.
Grade: USP, FCC.
Use: Food preservative, fungicide, mold control in sausage casings.
See Parabens.

propyl pelargonate. $C_3H_7OOCC_8H_{17}$.
Properties: Liquid. D 0.870 (15/15C), bp 237C. Insoluble in water; soluble in alcohol and most organic solvents. Combustible.
Use: Flavors and perfumes, bactericides and fungicides.

2-propyl-4-pentenoic acid.   See 2-allylpentanoic acid.

propylpiperidine.   See coniine.

1-propylpiperidine.
CAS: 5470-02-0. $C_8H_{17}N$.
Hazard: A poison.

6-propylpiperonyl butyl diethylene glycol ether.   See piperonyl butoxide.

n-propyl propionate.
CAS: 106-36-5. $CH_3CH_2COOCH_2CH_2CH_3$.
Properties: Colorless liquid. Boiling range 122–124C, fp −76C, flash p 174F (78.9C) (OC), bulk d 7.31 lb/gal. Soluble in most organic solvents; slightly soluble in water. Combustible.
Grade: Technical.
Use: Solvent for nitrocellulose, paints, varnishes, lacquers, coating agents.

4-*n*-propylpyridine.
$C_8H_{11}N$ or $C_3H_7C_5H_4N$.
Not to be confused with coniine (propylpiperidine).
Properties: Bp 188C.
Use: Intermediate.

5-(propylsulfonyl)-2-benzimidazolecarbamic acid methyl ester.
CAS: 75184-71-3. $C_{12}H_{15}N_3O_4S$.
Hazard: A reproductive hazard.

2-propylthiazolidine.
CAS: 24050-10-0. $C_6H_{13}NS$.
Hazard: A poison.

propylthiouracil.
(6-propyl-2-thiouracil).
CAS: 51-52-5. $C_7H_{10}N_2OS$.
Properties: White, powdery, crystalline substance; starchlike in appearance and to touch; bitter taste. Mp 218–221C, sensitive to light. Very slightly soluble in water; sparingly soluble in alcohol; soluble in ammonia and alkali hydroxides.
Derivation: Condensation of β-oxocaproate with thiourea.
Grade: USP.
Hazard: Possible carcinogen.
Use: Medicine (thyroid inhibitor).

n-propyltrichlorosilane.
CAS: 141-57-1. $C_3H_7SiCl_3$.
Properties: Colorless liquid. Bp 12.5, d 1.195 (25/25C), refr index 1.4292 (25C), flash p 100F (37.7C) (COC). Readily hydrolyzed with liberation of hydrogen chloride.
Derivation: By Grignard reaction of silicon tetrachloride and propylmagnesium chloride.
Grade: Technical.
Hazard: Flammable, moderate fire risk. Strong irritant.
Use: Intermediate for silicones.

propyl xanthate.
See xanthic acid.

propyne.
See methylacetylene.

2-propyne-1-ol.
See propargyl alcohol.

prostacycline.
See prostaglandin.

prostaglandin.
One of a group of physiologically active compounds derived from arachidonic acid, a 20-carbon fatty acid that occurs in glandular organs and the liver. It was named after the prostate gland, where it was originally found. Research on prostaglandins has been intense in recent years in view of their importance in various reproduction mechanisms and effects on blood pressure. They are believed to have significant relationships to a number of hormones; they also affect the nervous system, inhibit production of gastric juice, stimulate smooth muscles, and induce labor. They occur naturally in body tissues and biological fluids (especially semen). Both the chemical structure and metabolic functions of these compounds have been established with considerable accuracy, and several types have been synthesized; one pathway uses norbornadiene as a starting point followed by a 10-step sequence of conversions to a diol that serves as a precursor; another starts with cyclopentadiene, followed by hydroboration, yielding two intermediates from which prostaglandin can be derived. The most prolific source of prostanglandin intermediates (called syntons) is a marine organism called a gorgonian sea whip, which occurs in great numbers in coral reefs, especially in the Caribbean area. Harvesting of these has made the production of prostaglandins much less expensive. The occurrence of prostaglandin A_1 in yellow onions has been confirmed. Important derivatives of prostaglandins are prostacyclins and thromboxanes; a closely related group of compounds derived directly from arachidonic acid are the leukotrienes, which occur in white blood cells.

prosthetic group.
A chemical grouping in which a metal ion is associated with a large molecule or molecular complex, e.g., coenzymes and metal-porphyrin complexes such as chlorophyll and hemin. Such groups activate metabolic mechanisms such as phosphorylation, decarboxylation, by coordination reactions with amino acids, proteins, enzymes, and nucleic acids. The behavior of certain vitamins and other metabolites is due in part to prosthetic groups; catalysis is also involved.

protactinium.
Pa. A radioactive element of atomic number 91, a member of the actinide series, aw 231.0359, valences = 4, 5; 13 unstable isotopes, two of which occur naturally. Protactinium is a constituent of all uranium ores, 340 mg being extracted from 1 ton. Protactinium may also be produced by irradiation of thorium-230. Purification is carried out by ion-exchange and solvent-extraction techniques. The longest lived isotope, Pa-231, decays by α-emission and has a half-life of 33,000 years. Protactinium may be precipitated as the double potassium fluoride, K_2PaF_7, or the oxide, Pa_2O_5. The metal may be prepared by reducing PaF_4 with barium or by heating PaI_4 in a vacuum. It is hard and white, melting near 1600C. It is too rare for commercial use. Forms several compounds with halogens.
Hazard: Highly toxic, radioactive material.

protamines. Simplest proteins, without sulfur, molecular weights about ~3000.
Properties: Water soluble, producing basic solutions.

protease. A proteolytic enzyme that weakens or breaks the peptide linkages in proteins. They include some of the more widely known enzymes such as pepsin, trypsin, ficin, bromelain, papain, and rennin. Being water soluble they solubilize proteins and are commercially used for meat tenderizers, bread baking, and digestive aids.

protective coating. A film or thin layer of metal glass or paint applied to a substrate primarily to inhibit corrosion, and secondarily for decorative purposes. Metals such as nickel, chromium, copper, and tin are electrodeposited on the base metal; paints may be sprayed or brushed on. Vitreous enamel coatings are also used; that require baking. Zinc coatings are applied by a continuous bath process in which a strip of ferrous metal is passed through molten zinc.
See galvanizing; terne plate; electroplating; paint; corrosion; cladding.

protective colloid. See colloid, protective.

protein. A complex, high polymer containing carbon, hydrogen, oxygen, nitrogen, and usually sulfur, and composed of chains of amino acids connected by peptide linkages ($-CO\cdot NH-$). Proteins occur in the cells of all living organisms and in biological fluids (blood plasma, protoplasm). They are synthesized by plants largely through the nitrogen-fixing ability of certain soil bacteria. Their molecular weight may be as high as 40 million (tobacco mosaic virus). They have many important functional forms: enzymes, hemoglobin, hormones, viruses, genes, antibodies, and nucleic acids. They also serve as the basic component of connective tissue (collagen), hair (keratin), nails, feathers, skin, etc. Some have been synthesized in the lab.
The sequence of amino acids; in the polypeptide chain is of critical importance in genetics. Proteins can be hydrolyzed to their constituent amino acids and can be broken down into simpler forms by proteolytic enzymes. They form colloidal solutions, and behave chemically as both acids and bases simultaneously (amphoteric). They are denatured by changes in pH, and by heat, UV radiation, and many organic solvents.
Simple proteins contain only amino acids, conjugated proteins contain amino acids plus nucleic acids, carbohydrates, lipids, etc. On the basis of solubility, they can be classified as albumins (water soluble), globulins (insoluble in water but soluble in aqueous salt solutions), and prolamins (soluble in alcohol-water mixture but not in alcohol or water alone). A number of proteins have been synthesized, notably the hormone insulin. Proteins are an essential component of the diet, occurring chiefly in meat, eggs, milk, and fish. Edible proteins suitable for human food as well as cattle feed can be produced from microorganisms grown in carbonaceous or nitrogenous media to form yeastlike materials. Paraffinic hydrocarbons (methane) and petroleum-derived ethanol can be used as growth media for protein biosynthesis.
Industrial applications of proteins include plastics, adhesives, and fibers derived from casein and soybean protein, but these have been declining in recent years. Special forms in which proteins are commercially available include textured proteins for food products, and protein hydrolyzate and liquid predigested protein, both for medical use.
See ribonuclease; deoxyribonucleic acid; nutrition; amino acid; protein, textured; protein, single-cell; polypeptide.

protein hydrolyzate. Solution of protein hydrolyzed into its constituent amino acids.
Grade: USP.
Use: Medicine and surgery. Usually administered by a stomach tube or intravenous injection.

protein kinases. Enzymes that phosphorylate certain amino acid residues (most often Ser, Thr, or Tyr) in specific proteins.

protein, single-cell. (SCP). A protein nutrient derived from bacteria or yeast by hydrocarbon fermentation or from fungi by fermentation on food-plant waste. A process developed in West Germany during the 1970s utilizes the bacterium *Methylomonas clara* cultured in a mixture of methanol, ammonia, water, and air. The continuous fermentation process is followed by dewatering and spray drying. The product contains 70% protein, 10% nucleic acids, 8% fats, and 7% minerals; in this form it is suitable for animal feeds. A purified type (90% protein) is made by dissolving the product in an ammonia-methanol mixture, followed by filtration to remove the fats, and then water-washing to extract the nucleic acids. The product may prove to be satisfactory for use in human foods. The presence of nucleic acids is undesirable, because they may lead to metabolic disorders such as gout. Commercial production awaits further testing.

protein, spun. See protein, textured.

protein targeting. The process by which newly synthesized proteins are sorted and transported to their proper locations in the cell.

protein, textured. A meat extender or substitute made from defatted soybean flour or similar protein, usually by an extrusion process. Some types are used to fortify cereals and other food products. The filaments produced by extrusion are designed to simulate the fibrous structure (texture) of meats. The term *spinning* is also used, and the products are often called spun proteins.

"Protek-Sorb" *[Grace].*   TM for a group of silica gels.

proteoglycan.   A hybrid macromolecule consisting of a larger heteropolysaccharide joined to a smaller polypeptide, i.e., the polysaccharide is the major component.

proteolysis.   The structural breakdown of proteins, usually by hydrolysis, as a result of the action of an enzyme, e.g., trypsin, pepsin, papain, etc.

proteolytic enzyme.   See subtilisins.

proteome.   Proteins expressed by a cell or organ at a particular time and under specific conditions.

proteomics.   The study of the full set of proteins encoded by a genome.

prothrombin.   The precursor of the enzyme thrombin, a proteinaceous component of blood plasma that is converted into thrombin by the blood-clotting mechanism when activated at the site of an open wound.
See coagulation.

Protina.   A line of products for cleaning, inhibiting corrosion, and coating copper, brass, or bronze artwork and statuary.

protoalkaloid.   Any of a class of alkaloids derived from amino acids that lacks a heterocyclic ring.

protocatechuic aldehyde, methyl ether.   See vanillin.

Protolin.   Reducing agents based on zinc sulfoxylate and zinc formaldehyde sulfoxylate. Supplied as water-soluble, white powder.
Use: Stripping colors from fabrics, chemical synthesis.

protolytic catalysis.   Catalysis in which acids, bases, or both act as catalysts by undergoing protolytic reactions.

proton.   A fundamental unit of matter having a positive charge and a mass number of 1, equivalent to 1.67×10^{-24} g. Its mass is 1837 times that of the negatively charged electron, but is almost identical with that of the uncharged neutron. Protons are constituents of all atomic nuclei, their number in each nucleus being the atomic number of the element. An atom of normal hydrogen contains one proton and one electron. A proton is identical with a hydrogen ion (H^+).

proton acceptor.   The acceptor of a proton in an acid–base reaction: a Brønsted–Lowry base.

proton donor.   The donor of a proton in an acid-base reaction: a Brønsted acid.

proton-motive force.   The electrochemical potential inherent in a transmembrane H^+ concentration gradient. The proton gradient is used in oxidative phosphorylation and photophosphorylation to drive ATP synthesis.

proto-oncogene.   A gene of cellular or viral origin, usually encoding a regulatory protein, that can be converted into an oncogene by mutation.

"Protopet" *[Sonneborn].*   TM for petrolatum of medium consistency and ranging in color from pure white to amber, but meeting USP or NF purity requirements for petrolatum.

protoplasm.   The total contents of the living cell, including both nucleus and cytoplasm. Predominantly a mixture of proteins, protoplasm is the physical basis of life. Most of its components are in the colloidal size range. This term is falling into disuse among modern biochemists.
See cell (1).

protopolygonatoside g.
CAS: 62601-71-2.
Hazard: A poison.
Source: Natural product.

protoveratrine.   A substance isolated from the *Veratrum album* plant. It is a mixture of two alkaloids, designated protoveratrine A and protoveratrine B.
Use: Medicine to lower blood pressure.

provitamin.   The precursor of a vitamin. Examples are carotene and ergosterol, which upon activation become vitamin A and vitamin D, respectively.
See specific compounds.

proxifeine.
CAS: 65497-24-7. $C_{13}H_{21}N_5O_3 \cdot ClH$.
Hazard: Moderately toxic.

Prussian blue.   The most common and best-known name for blue iron ferrocyanide (iron blue) pigments made by a variety of procedures.
See iron blue.

prussic acid.   See hydrocyanic acid.

Pschorr reaction.   Synthesis of phenanthrene derivatives from diazotized α-aryl-omicron-aminocinnamic acids by intramolecular arylation.

pseudo-.   A prefix indicating false, similar, or closely related.

pseudoalkaloid.   A substance that is derived from a sterol, purine, terpene, aliphatic acid, or nicotinic acid.

pseudobufarenogin.
CAS: 17008-69-4. $C_{24}H_{32}O_6$.
Hazard: A poison.
Source: Natural product.

pseudobutylene glycol.   See 2,3-butylene glycol.

pseudocumene.   (1,2,4-trimethylbenzene; *uns*-trimethylbenzene).
CAS: 95-63-6. $C_6H_3(CH_3)_3$.

Properties: Liquid. Fp −43.91C, bp 168.89C, d 0.8758 (20/4C), refr index 1.5045 (20C), flash p 130F (54.4C). Insoluble in water; soluble in alcohol, benzene, and ether.
Derivation: From C_9 fraction of refinery reformate streams by fractional distillation.
Grade: 95%, 99%, and research.
Hazard: Moderate fire risk. Central nervous system depressant, irritant to mucous membranes. Asthma and hematologic effects.
Use: Manufacture of trimellitic anhydride, dyes, pharmaceuticals, and pseudocumidine.

pseudocumidine.   (2,4,5-trimethylaniline; 1,2,4-trimethyl-5-aminobenzene).
CAS: 137-17-7. $C_6H_2(CH_3)_3NH_2$.
Properties: White crystals. D 0.957, mp 62C, bp 236C. Soluble in alcohol and ether; insoluble in water. Combustible.
Derivation: From pseudocumene.
Hazard: Questionable carcinogen.
Use: Manufacture of dyes, organic synthesis.

pseudoextinction.   The apparent disappearance of a taxon. In cases of pseudoextinction, this disappearance is not due to the death of all members but to the evolution of novel features in one or more lineages, so that the new clades are not recognized as belonging to the paraphyletic ancestral group whose members have ceased to exist.

pseudogene.   A sequence of DNA similar to a gene but nonfunctional; probably the remnant of a once functional gene that accumulated mutations.

pseudohexyl alcohol.   See 2-ethylbutyl alcohol.

pseudoionone.   (6,10-dimethyl-3,5,9-undecatriene-2-one).
CAS: 141-10-6. $(CH_3)_2C:CH(CH_2)_2C(CH_3):$ $CHCH:CHCOCH_3$.
Properties: Pale-yellow liquid. D 0.8984 (20C), bp 143–145C (12 mm Hg). Soluble in alcohol and ether. Combustible.
Use: Perfumery, cosmetics.

pseudopodia.   Fingerlike extensions from an amoeboid cell; literally "false feet."

psi.   Abbreviation for pounds per square inch.

psia.   Abbreviation for pounds per square inch absolute.

psicain.   Acid tartrate of *d*-psi-cocaine.
Use: Same as cocaine hydrochloride.

psig.   Abbreviation for pounds per square inch gauge.

psilocin.
CAS: 520-53-6. $C_8H_5N(OH)C_2H_4N(CH_3)_2$. An indole derivative. An alkaloid from certain mushrooms; a hallucinogenic drug.

psilocybin.
CAS: 520-52-5. $C_8H_5N(OPO_3H_2)C_2H_4N(CH_3)_2$. An indole derivative. An alkaloid from certain mushrooms; hallucinogenic drug.

psilomelane. $BaMn_9O_{16}(OH)_4$. A natural oxide of variable composition. Calcium, nickel, cobalt, and copper frequently are present. The name sometimes refers to mixture of manganese minerals.
Properties: Black color, brownish-black streak, submetallic luster. Hardness 5–6, d 3.7–4.7.
Occurrence: The former U.S.S.R., India, South Africa, Cuba, U.S. (Arkansas, Virginia, Georgia).
Use: Important ore of manganese.

psychotropic drug.   Any of a number of therapeutic agents that affect the behavior, emotional state, or mental functioning of psychologically disturbed persons. They are widely known as *tranquilizers*, but this term is no longer accepted as clinically accurate, because the minor tranquilizers (benzodiazepine and glycerol derivatives) act quite differently from the major tranquilizers. For this reason the latter are now classified as antipsychotics and antidepressants, and the term *antianxiety agent* is applied to the minor tranquilizers. Antipsychotic agents include phenothiazines (chlorpromazine), thioxanthenes, and butyrophenones; antidepressant agents are of two major types, namely, monoamine oxidase (MAO) inhibitors and several tricyclic compounds; antianxiety agents are glycerol derivatives (meprobamate) and benzodiazepine derivatives, e.g., oxazepam.

Pt.   Symbol for platinum.

PTA.   Abbreviation for phosphotungstic acid; also for purified terephthalic acid.

pteroylglutamic acid.   See folic acid.

PTFE.   Abbreviation for polytetrafluoroethylene. A plastic tank material with excellent thermal and electrical insulation.
Use: Plastic coating, electrical insulation.

PTMA.   Abbreviation for a mixture of phosphotungstic and phosphomolybdic acids.
Use: Making pigments.
See phosphotungstic pigment.

ptomaine.   A group of highly toxic substances (derivatives of ethers of polyhydric alcohols) resulting from the putrefaction or metabolic decomposition of animal proteins. Examples that have been isolated and prepared synthetically are cadaverine (1,5-diaminopentane), muscarine (hydroxyethyltrimethylammonium hydroxide), putrescine (tetraethylenediamine), and neurine (trimethylvinylammonium hydroxide).
Note: The term *ptomaine poisoning* is usually a misnomer for other types of food poisoning.

PTSA.   Abbreviation for *p*-toluenesulfonamide.

ptyalin.   A salivary amylase that acts upon α-1,4-glycosidic linkages, converting starch to various dextrins and maltose. It can act over a pH range of 4.0–9.0; optimum pH 5.6–6.5. It requires the presence of certain negative ions for activation; chlorides and bromides are the most effective.
Use: Biochemical research.

ptychodiscus brevis toxin.   See brevetoxin.

Pu.   Symbol for plutonium.

pug mill.   A comminuting or granulating machine whose essential components are a shaft equipped with blades or arms with alloy-hardened tips rotating in a troughlike compartment.
Use: Grinding and amalgamating fertilizer ingredients, clay mixtures, cement components, and similar products.

pulegium oil.   See pennyroyal oil.

pulegone.   (1-isopropylidene-4-methyl-2-cyclohexanone).
CAS: 89-82-7. $C_{10}H_{16}O$. A ketone found in pennyroyal and hedeoma oil. Combustible.
Properties: Oily liquid; pleasant odor. D 0.9323 (20C), bp 221C, dextrorotatory, refr index 1.4894 (20C). Insoluble in water; soluble in alcohol and ether.
Use: Chemical intermediate, flavoring.

Pullulan.   A biodegradable polysaccharide made by yeast fermentation, originally developed in Japan. Its adhesive and oxygen impermeable properties enable it to be used to coat pharmaceutical products. It is water soluble, odorless, and edible; these properties have led to its use admixed with foodstuffs for special-purpose applications. The mixture can be processed into a semirigid plastic sheet or film that can serve as an emergency food source. FDA approval is pending.

pulmonary anthrax. (anthrax pneumonia; wool-sorters' pneumonia; wool sorter's disease).
A form of anthrax in humans due to inhalation of spore-bearing dust.
Hazard: May cause chills, fever, back and leg pain, dyspnea, coughing, rapid breathing, rapid pulse, and extreme prostration.

pulp, paper.   Processed cellulosic fibers derived from hardwoods, softwoods, and other plants. There are two major types of pulp: (1) Ground wood or mechanical pulp, which is merely finely divided wood without purification and is made into newsprint, cheap manila papers, and nonpermanent tissues. (2) Chemical pulp, of which there are three kinds: (a) soda process pulp obtained from the digestion of wood chips (mostly poplar) by caustic soda; (b) sulfite process pulp (mostly spruce and other coniferous woods), obtained by digestion with a solution of magnesium, ammonium, or calcium disulfite containing free sulfur dioxide; and (c) sulfate process (kraft) pulp, in which sodium sulfate is added to the caustic liquors but is reduced by the carbon present to the sulfide, which becomes a digesting agent. Sulfite and sulfate pulps (chiefly from softwoods) compose the bulk of paper pulps. Sulfate pulps are known as kraft pulps because of their strength ("Kraft" is the German word for strength), and are used for wrapping, packaging, container board, etc. A relatively new process called holopulping replaces sodium sulfate with oxidants. A synthetic pulp based on polyolefins (styrene copolymer fibers) has been developed to the production stage in Japan.
See holopulping; paper; digestion.

Pulsar 1 System.   A shock system for swimming pools.
Use: Simplifies water balance; safer than bleach or chlorine.

pulsed field gel electrophoresis.   (PFGE). A gel technique which allows size-separation of very large fragments of DNA, in the range of hundreds of kb to thousands of kb. As in other gel electrophoresis techniques, populations of molecules migrate through the gel at a speed related to their size, producing discrete bands. In normal electrophoresis,

DNA fragments greater than a certain size limit all migrate at the same rate through the gel. In PFGE, the electrophoretic voltage is applied alternately along two perpendicular axes, which forces even the larger DNA fragments to separate by size.

pultrusion. A technique for making certain products from glass-reinforced plastics, such as rods, electrical insulators, etc. It involves passage of continuous bundles of glass fiber that have been impregnated with liquid resin through an oven at the rate of 18 inches per minute at 140C (285F).

pumice. A highly porous igneous rock, usually containing 67–75% SiO_2 and 10–20% Al_2O_3, with a glassy texture. Potassium, sodium, and calcium are generally present. Insoluble in water, not attacked by acids.
Occurrence: U.S.A. (Arizona, Oregon, California, Hawaii, New Mexico), Italy, New Zealand, Greece.
Grade: Lump, powdered (coarse, medium, and fine); NF, technical.
Use: Concrete aggregate, heat and sound insulation, filtration, finishing glass and plastics, road construction, scouring preparations, paint fillers, absorbents, support for catalysts, dental abrasive, abherent for uncured rubber products, possible substitute for asbestos, polishing agent for pedicures.

Pummerer rearrangement. Rearrangement of sulfoxides to α-acyloxythioethers in the presence acyclic anhydrides.

pumpdown time. The time needed to produce a certain vacuum.

punty. A solid or hollow iron rod 4–6 ft long, usually with an insulation covering on one end.
Use: By glass workers to remove molten material from the melt preparatory to shaping finished articles.

pupa. In metamorphozing insects, a stage between the larva and adult during which the organism undergoes major developmental changes.

Purdie (Irvine–Purdie) methylation. Exhaustive methylation of a methyl glycoside by repeated treatment with methyl iodide and silver oxide, followed by hydrolysis of the pentamethyl ether with dilute acid to yield the anomeric hydroxyl group.

Purex process. See reprocessing.

purification. Removal of extraneous materials (impurities) from a substance or mixture by one or more separation techniques. A pure substance is one in which no impurity can be detected by any experimental procedure. Though absolute purity is impossible to attain, a number of standard procedures exist for approaching it to the extent of 1

ppm of impurity or less. The following fractionation techniques are widely used: crystallization, precipitation, distillation, adsorption (various types of chromatography), extraction, electrophoresis, and thermal diffusion.
See purity, chemical.

Purified Plus.
CAS: 67-56-1. Purified methanol.
Use: For chromatography, pesticide, residue analysis, spectrophotometry, and semiconductor wafer processing.

Purifloc. A polyelectrolyte.
Use: To flocculate solids in water and industrial waste treatment.

Purina Bloat Block. (poloxalene free-choice liquid type C feed).
CAS: 9003-11-6. $HO(CH_2CH_2O)_n[CH(CH_3)CH_2O]_n(CH_2CH_2O)_nH$.
Properties: Liquid nonionic surfactant polymer. A bloat preventing veterinary drug.
Use: Drug; food additive.

purine. (1) (imidazo(4,5-d)pyrimidine).
CAS: 120-73-0.

Properties: Colorless crystals. Mp 217C. Soluble in water, alcohol, toluene.
Derivation: Prepared from uric acid and regarded as the parent substance for compounds of the uric groups, many of which occur naturally in animal waste products.
Use: Organic synthesis, metabolism, and biochemical research.
(2) One of a number of basic compounds found in living matter and having a purine-type molecular structure.
See adenine; base pair; guanine; hypoxanthine; xanthine; uric acid; caffeine; theobromine.

purity, chemical. A substance is said to be pure when its physical and chemical properties coincide with those previously established and recorded in the literature, and when no change in these properties occurs after application of the most selective fractionation techniques. In other words, purity exists when no impurity can be detected by any experimental procedure. There are a number of recognized standards of purity.
See grade.

puromycin. (USAN).
CAS: 53-79-2. $C_{22}H_{29}N_7O_5$. An antibiotic that inhibits protein synthesis, prevents transfer of an

amino acid from its carrier to the growing protein. Produced by *Streptomyces alboniger*, effective against bacteria, protozoa, parasitic worms, and cancerous tumors.
Properties: Crystals. Mp 176C.
Hazard: Toxic to living cells of all kinds.

purple of Cassius. See gold-tin purple.

purpurin. (1,2,4-trihydroxyanthraquinone; CI 58205).
CAS: 81-54-9. $C_{14}H_5O_2(OH)_3$.
Properties: Reddish needles. Mp 256C. Slightly soluble in hot water; soluble in alcohol and ether.
Derivation: Occurs as a glucoside in madder root. Made synthetically by oxidation of alizarin.
Use: Dye for cotton, stain for microscopy, reagent for boron determination, manufacture of acid and chrome dyes.

purpurin red. See anthrapurpurin.

putty. A mixture of whiting (chalk) with 12–18% of linseed oil, with or without white lead or other pigment. Containers must be airtight.
Use: Sealant, glass setting, caulking agent.

putty powder. A soft abrasive composed of tin oxide.

PVA. Abbreviation for polyvinyl alcohol.

PVAc. Abbreviation for polyvinyl acetate.

PVB. Abbreviation for polyvinyl butyral.

PVC. (1) Abbreviation for polyvinyl chloride. (2) Abbreviation for pigment volume concentration, a term used in paint technology to mean pigment volume divided by the sum of the pigment volume and the vehicle solids volume, multiplied by 100.

PVC, chlorinated. See polyvinyl dichloride.

PVDC. See polyvinyl dichloride.

PVE. Abbreviation for polyvinyl ethyl ether.

PVI. Abbreviation for polyvinylisobutyl ether.

PVM. Abbreviation for polyvinyl methyl ether.

PVM/MA. See polyvinyl methyl ether/maleic anhydride.

PVOH. Abbreviation for polyvinyl alcohol.

PVP. Abbreviation for polyvinylpyrrolidone.

PVPP. See polyvinylpolypyrrolidone.

PVT. Abbreviation for pressure–volume–temperature.
Use: Chemical engineering.

py. An abbreviation for pyridine.
Use: As in formulas for coordination compounds.
See dien; *en*; *pn.*

pymetrozine. See 4,5-dihydro-6-methyl-4-((3-pyridinylcmethylene)amino)-1,2,4.

pyranose. A simple sugar structurally analogous to the six-membered pyran ring.

"Pyratex" [R. R. Street]. TM for a vinyl pyridine latex.
Properties: Total solids 40–42%, pH 10.4–11.5, d 0.96.
Use: To promote adhesion between rayon or nylon fibers and rubber, as in tire cord, belting, hose, etc.

pyrazine hexahydride. See piperazine.

pyrazole.
CAS: 288-13-1.

HNNCHCHCH.

Properties: Off-white, crystalline solid; pyridine odor. Mp 68–70C, bp 186–188C. Soluble in water and alcohol.
Hazard: Toxic by ingestion and inhalation, irritant to skin and eyes.
Use: Chemical intermediate, stabilizer for halogenated solvents and lubricating oils.

pyrazoline.

HNNCHCH₂CH₂.

Properties: Bp 144C.
Use: Organic synthesis.

pyrazolone dye. A dye whose molecules contain both the –N=N- and the =C=C= chromophore groups in their structure. These are acid dyes most used for silk and wool and to some extent for lakes. Tartrazine (CI 19140) is an important member of this group.
See dye.

Pyre-ML. Polyimide coated fabrics and laminates.
Use: Class H electrical insulating materials.

pyrene.
CAS: 129-00-0. $C_{16}H_{10}$. A condensed ring hydrocarbon.

Properties: Colorless solid (tetracene impurities give a yellow color), solutions have a slight blue fluorescence. Mp 156C, d 1.271 (23C), bp 404C. Insoluble in water; partially soluble in organic solvents.

Derivation: From coal tar.

Hazard: A questionable carcinogen, absorbed by skin.

Use: Biochemical research.

pyrethrin. (pyrethrum; pyrethrum insecticide; pyrethrum flowers; pyrethrum extract; Dalmatian insect powder; Persian insect powder; 2-cyclopentenyl-4-hydroxy-3-methyl-2-cyclopenten-1-one chrysanthemate; 3-(2-cyclopenten-1-yl)-2-methyl-4-ox-2-cyclopenten-1-yl chrysanthemumate; 3-(2-cyclopentenyl)-2-methyl-4-oxo-2-cyclopentenyl chrysanthemummonocarboxylate; cyclopentenylrethonyl chrysanthemate; ENT 22,952; (2-methyl-4-oxo-3-prop-2-enylcyclopent-2-en-1-yl) (2,2-dimethyl-3-(2-methylprop-1-enyl)cyclopropane-1-carboxylate)

CAS: 8003-34-7. $C_{21}H_{28}O_3$. A botanical composed of several esters extracted from the flowers of three species of *Chrysanthemum* (family Asteraceae) using hot carbon dioxide under high pressure. It is any of the six insecticidal allethrin analogs extracted from pyrethrum flowers (cinerins I and II, jasmolins I and II, and pyrethrins I and II).

Hazard: Highly toxic; fast-acting paralytic; hepatotoxic; nephrotoxic; questionable carcinogen; may cause gastroenteritis, central nervous system disturbance, diarrhea, nausea, vomiting, tinnitus, headache, numbness of tongue and lips, syncope, hyperexcitability, incoordination, convulsions, prostration and death due to respiratory paralysis; liver damage; lower respiratory tract irritant.

Use: Broad-spectrum contact insecticides.

pyrethrin I.
CAS: 8003-34-7. $C_{21}H_{28}O_3$. Pyrethrolone ester of chrysanthemummonocarboxylic acid. Most potent insecticidal ingredient of pyrethrum flowers.

Properties: Viscous liquid. Oxidizes readily in air. Insoluble in water; soluble in other common solvents. Incompatible with alkalies.

Hazard: Toxic by ingestion and inhalation.

Use: Household insecticide (flies, mosquitoes, garden insects, etc.), treatment of paper bags for shipping cereals, etc.

See cinerin I; cinerin II; pyrethrin II.

pyrethrin II. $C_{22}H_{28}O_5$. Pyrethrolone ester of chrysanthemumdicarboxylic acid. One of the four primary active insecticidal ingredients of pyrethrum flowers.

Properties: Similar to those of pyrethrin I.

See pyrethrin I; cinerin I; cinerin II.

pyrethroid.
CAS: 8003-34-7. Any of a group of insect growth regulators that act as neurotoxins, analogous to juvenile hormones, restricting the development of insect larvae. Thus they are especially effective against insects that are destructive in the adult stage. They are considered nontoxic to animals and humans.

See juvenile hormone.

pyrethrum.
CAS: 8003-34-7.

Properties: Viscous liquid. Bp 170 (0.1 mm) decomposes. A natural insecticide obtained by extraction of chrysanthemum flowers native to Kenya, Ecuador, and Japan. The solvent used is a hydrocarbon of the kerosene type. Pyrethrum is also made synthetically. Not compatible with alkaline material. The chief constituents are pyrethrins I and II and cinerins I and II. These compounds are nonvolatile and very slightly soluble in water.

Hazard: Toxic by ingestion and inhalation. Liver damage and lower respiratory tract irritant. Questionable carcinogen.

Use: Household insecticide (flies, mosquitoes, garden insects, etc.).

See allethrin.

"Pyrex Glass Brand No. 7740" *[Corning].* TM for a borosilicate glass.

Properties: Linear coefficient of expansion 32 × 10^{-7} in/C, elasticity coefficient 6.230 kg/sq mm, hardness (scleroscope) 120, d 2.25, specific heat 0.20, refr index 1.474 dispersion 0.00738, light and heat transmission higher than the best plate glass, dielectric constant 4.5 (25C), upper working temperature (mechanical considerations only): annealed, extreme limit 490C, normal service 230C, thermal stress resistance 53C, impact abrasion resistance 3.1, softening temperature 820C.

Use: Laboratory and pharmaceutical glassware and apparatus, electrochemical equipment, fiber manufacture, domestic ovenware apparatus, and equipment for many processes.

pyridine.
CAS: 110-86-1. $N(CH)_4CH$.

Properties: Slightly yellow or colorless liquid; nauseating odor; burning taste. Slightly alkaline in reaction. Soluble in water, alcohol, ether, benzene, ligroin, and fatty oils. D 0.987, fp −42.0C, bp 115.5C, flash p 68F (20C) (CC), autoign temp 900F (482C).

Derivation: (1) By coal carbonization and recovery both from coke-oven gases and the coal tar middle oil; (2) also synthetically from acetaldehyde and ammonia.

Grade: Technical, as 20C, 2C, etc., (meaning distillation range), medicinal, CP, spectrophotometric.

Hazard: Flammable, dangerous fire risk, explosive limits in air 1.8–12.4%. Toxic by ingestion and

inhalation. Skin irritant, liver and kidney damage. Questionable carcinogen.
Use: Synthesis of vitamins and drugs, solvent waterproofing, rubber chemicals, denaturant for alcohol and antifreeze mixtures, dyeing assistant in textiles, fungicides.

pyridine alkaloid.   Any alkaloid that contains a pyridine nucleus.

2-pyridine aldoxime methiodide.   (2-PAM). CAS: 94-63-3. $C_5NH_4CHNOH \cdot ICH_3$.
Use: Antidote for cholinesterase-inhibiting pesticides of the parathion type because of its property of reactivating the cholinesterase by removal of phosphoryl groups. Also antidote for nerve gases.

pyridine-3-carboxylic acid.   See niacin.

3-pyridine diazonium fluoborate.   Intermediate in manufacture of 3-fluoropyridine.
Properties: Explodes when dry.

2,3-pyridinedicarboxylic acid.   See quinolinic acid.

2,5-pyridinedicarboxylic acid.   See isocinchomeronic acid.

pyridine nucleotide.   A nucleotide coenzyme containing the pyridine derivative nicotinamide: NAD(H) or NADP(H).

pyridine-N-oxide.   (pyridine-1-oxide). CAS: 694-59-7. C_5H_5NO.
Properties: Fp 67.0C. Soluble in water.
Hazard: Probably flammable and toxic.
Use: Intermediate.

pyridine polymer.   A polymer or copolymer of methylvinylpyridine and vinylpyridine.
Use: Corrosion inhibitors and as intermediates for coatings and printing inks.

2-pyridinethiol-1-oxide.   See 1-hydroxy-2-pyridine thione.

pyridinium bromide perbromide.   (PBPB). $C_5H_6NBr \cdot Br_2$.
Properties: Red, prismatic crystals. Mp 135–137C (decomposes) with preliminary softening. The salt is stable in the dry state and can be used in glacial acetic acid, ethanol, and related solvents. This compound has 45–50% available bromine.
Use: Brominating phenols and addition to double bonds; mono- and polybromination of ketones, including aliphatic, alicyclic, steroid, and amino carbonyls. Micro or semimicro quantitative analysis.

C3-pyridinylarbonimidothioic acid, o-(3,4-dimethylcyclohexyl) s-((4-(1,1-dimethylethyl) phenyl)methyl) ester.
CAS: 42754-20-1. $C_{25}H_{34}N_2OS$.
Hazard: Moderately toxic by ingestion.
Use: Agricultural chemical.

3-pyridinylcarbonimidodithioic acid, (2-chlorophenyl)methyl (2,4-dichlorophenyl)methyl ester.
CAS: 34763-31-0. $C_{20}H_{15}Cl_3N_2S_2$.
Hazard: Moderately toxic by ingestion.
Use: Agricultural chemical.

3-pyridinylcarbonimidothioic acid, o-cyclohexyl s-((4-(1,1-dimethylethyl)phenyl)methyl) ester.
CAS: 42754-16-5. $C_{23}H_{30}N_2OS$.
Hazard: Moderately toxic by ingestion.
Use: Agricultural chemical.

3-pyridinylcarbonimidothioic acid, o-cyclopentyl s-((4-(1,1-dimethylethyl)phenyl)methyl) ester.
CAS: 42754-15-4. $C_{22}H_{28}N_2OS$.
Hazard: Moderately toxic by ingestion.
Use: Agricultural chemical.

3-pyridinylcarbonimidothioic acid, o-(2,5-dimethylcyclohexyl) s-((4-(1,1-dimethylethyl) phenyl)methyl) ester.
CAS: 42754-19-8. $C_{25}H_{34}N_2OS$.
Hazard: Moderately toxic by ingestion.
Use: Agricultural chemical.

3-pyridinylcarbonimidothioic acid, o-(3,5-dimethylcyclohexyl) s-((4-(1,1-dimethylethyl) phenyl)methyl) ester.
CAS: 42754-21-2. $C_{25}H_{34}N_2OS$.
Hazard: Moderately toxic by ingestion.
Use: Agricultural chemical.

3-pyridinylcarbonimidodithioic acid, 1,1-dimethylethyl (4-(1,1-dimethylethyl)phenyl)methyl ester.
CAS: 51308-57-7. $C_{21}H_{28}N_2S_2$.
Hazard: Moderately toxic by ingestion.
Use: Agricultural chemical.

3-pyridinylcarbonimidodithioic acid, (4-(1,1-dimethylethyl)phenyl)methyl 1-methylethyl ester.
CAS: 34763-49-0. $C_{20}H_{26}N_2S_2$.
Hazard: Moderately toxic by ingestion.
Use: Agricultural chemical.

3-pyridinylcarbonimidodithioic acid, (4-(1,1-dimethylethyl)phenyl)methyl1-methylpropyl ester.
CAS: 51308-56-6. $C_{21}H_{28}N_2S_2$.
Hazard: Moderately toxic by ingestion.
Use: Agricultural chemical.

3-pyridinylcarbonimidodithioic acid, (4-(1,1-dimethylethyl)phenyl)methyl2-methylpropyl ester.
CAS: 51308-55-5. $C_{21}H_{28}N_2S_2$.
Hazard: Moderately toxic by ingestion.
Use: Agricultural chemical.

3-pyridinylcarbonimidodithioic acid, (4-(1,1-dimethylethyl)phenyl)methyl 1,1-dimethylpropyl ester.
CAS: 51379-05-6. $C_{22}H_{30}N_2S_2$.
Hazard: Moderately toxic by ingestion.
Use: Agricultural chemical.

3-pyridinylcarbonimidodithioic acid, (4-(1,1-dimethylethyl)phenyl)methyl 1-ethyl-1-methylpropyl ester.
CAS: 51308-59-9. $C_{23}H_{32}N_2S_2$.
Hazard: Moderately toxic by ingestion.
Use: Agricultural chemical.

3-pyridinylcarbonimidothioic acid, s-((4-(1,1-dimethylethyl)phenyl)methyl)o-(1,2-dimethylpropyl) ester.
CAS: 51308-69-1. $C_{22}H_{30}N_2OS$.
Hazard: Moderately toxic by ingestion.
Use: Agricultural chemical.

3-pyridinylcarbonimidothioic acid, s-((4-(1,1-dimethylethyl)phenyl)methyl)-o-(1-methylbutyl) ester.
CAS: 51308-68-0. $C_{22}H_{30}N_2OS$.
Hazard: Moderately toxic by ingestion.
Use: Agricultural chemical.

3-pyridinylcarbonimidothioic acid, s-((4-(1,1-dimethylethyl)phenyl)methyl)o-(3-methylbutyl) ester.
CAS: 51308-70-4. $C_{22}H_{30}N_2OS$.
Hazard: Moderately toxic by ingestion.
Use: Agricultural chemical.

3-pyridinylcarbonimidothioic acid, s-((4-(1,1-dimethylethyl)phenyl)methyl)o-(2-methylcyclohexyl) ester.
CAS: 42754-17-6. $C_{24}H_{32}N_2OS$.
Hazard: Moderately toxic by ingestion.
Use: Agricultural chemical.

3-pyridinylcarbonimidothioic acid, s-((4-(1,1-dimethylethyl)phenyl)methyl)o-(4-methylcyclohexyl) ester.
CAS: 42754-18-7. $C_{24}H_{32}N_2OS$.
Hazard: Moderately toxic by ingestion.
Use: Agricultural chemical.

3-pyridinylcarbonimidothioic acid, s-((4-(1,1-dimethylethyl)phenyl)methyl)o-(3-methylcyclohexyl) ester.
CAS: 42723-79-5. $C_{24}H_{32}N_2OS$.
Hazard: Moderately toxic by ingestion.
Use: Agricultural chemical.

3-pyridinylcarbonimidothioic acid, s-((4-(1,1-dimethylethyl)phenyl)methyl)o-(1-methylpropyl) ester.
CAS: 51308-66-8. $C_{21}H_{28}N_2OS$.
Hazard: Moderately toxic by ingestion.
Use: Agricultural chemical.

3-pyridinylcarbonimidothioic acid, s-((4-(1,1-dimethylethyl)phenyl)methyl)o-(2-methylpropyl) ester.
CAS: 51308-65-7. $C_{21}H_{28}N_2OS$.
Hazard: Moderately toxic by ingestion.
Use: Agricultural chemical.

3-pyridinylcarbonimidothioic acid, s-((4-(1,1-dimethylethyl)phenyl)methyl) o-(3,3,5-trimethylcyclohexyl) ester.
CAS: 42754-22-3. $C_{26}H_{36}N_2OS$.
Hazard: Moderately toxic by ingestion.
Use: Agricultural chemical.

pyridoxal hydrochloride.
CAS: 58-56-0. $C_8H_9NO_3 \cdot HCl$. An aldehyde derivative of pyridoxine, with vitamin B_6 activity.
Properties: Rhombic crystals. Mp 165C (decomposes). Soluble in water and 95% ethanol.
Use: Nutrition.

pyridoxal phosphate.   (2-methyl-3-hydroxy-4-formyl-5-pyridylmethylphosphoric acid).
CAS: 54-47-7. $CH_3C_5HN(OH)(CHO)CH_2PO_4H_2$).
The coenzyme of amino acid metabolism that also is the active group of various decarboxylases and other types of enzymes. It is closely related to pyridoxine.
Derivation: Synthesized (1) through the action of adenosine triphosphate, or phosphorus oxychloride, on pyridoxal, and (2) by phosphorylation of pyridoxamine followed by oxidation with 100% H_3PO_4.
Use: Nutrition, biochemical research.

pyridoxine.   (vitamin B_6; 3-hydroxy-4,5-dimethylol-2-methylpyridine).
$CH_3C_5HN(OH)(CH_2OH)_2$.

A group name designating the naturally occurring pyridine derivatives having vitamin B_6 activity. Essential for the dehydration and desulfhydration of amino acids and for the normal metabolism of tryptophan; appears to be related to fat metabolism. Pyridoxine is required in the nutrition of all species of animals.
Source: (Food) Vegetable fats, whole grain cereals, legumes, yeast, muscle meats, liver and fish. (Commercial) Synthetic pyridoxine, pyridoxal, and

pyridoxamine are produced by a complex series of reactions from isoquinoline. Amounts are expressed in micrograms.

Use: Medicine, nutrition (available as pyridoxine hydrochloride).

pyridoxine tripalmitate.
CAS: 4372-46-7. $C_{56}H_{101}NO_6$.
Hazard: Moderately toxic.

α-pyridylamine. See 2-aminopyridine.

β-pyridylamine. See 3-aminopyridine.

n-3-pyridyl-o-butyl-s-3,4-dichlorobenzyl imidothiocarbonate.
CAS: 34763-54-7. Mf: $C_{17}H_{18}Cl_2N_2OS$.
Hazard: Moderately toxic by ingestion.
Use: Agricultural chemical.

3-pyridylcarbinol. (3-pyridine methanol).
$C_5H_4NCH_2OH$.
Properties: Colorless liquid. Bp 266C, fp −6.5C, d 1.131 g/ml (20C), refr index 1.5455 (20C). Miscible with water at 20C; hygroscopic. Combustible.
Use: Intermediate.

pyrimidine. One of a group of basic compounds found in living matter. They may be isolated following complete hydrolysis of nucleic acids. They include uracil, thymine, cytosine, and methylcytosine. Thiamine is also a pyrimidine derivative. Other pyrimidines such as alloxan and thiouracil are important in medicine and biochemical research.
See: base pair; purine.

pyrimidine dimer. A covalently joined dimer of two adjacent pyrimidine residues in DNA, induced by absorption of UV light, usually derived from two adjacent thymines (a thymine dimer).

pyrimidine herbicide. A pyrimidine that acts almost exclusively to inhibit photosynthesis and has little effect on nonphotosynthesizing organisms.

pyrimithate. $C_{11}H_{20}N_3O_3PS$. A diethyl ester of phosphoric acid.
Properties: Colorless liquid. D 1.16. Soluble in acetone, alcohol, and benzene; almost insoluble in water.
Hazard: Cholinesterase inhibitor.
Use: Acaricide, insecticide.

pyrite. (iron pyrite; fool's gold).
CAS: 12068-85-8. FeS_2. Often mixed with small amounts of copper, arsenic, nickel, cobalt, gold, selenium.
Properties: Brass-yellow or brown, tarnished mineral, greenish or brownish-black streak, metallic luster. D 4.9–5.2, hardness 6–6.5.

Use: Manufacture of sulfur, sulfuric acid and sulfur dioxide, ferrous sulfate, cheap jewelry, recovery of metals.
See ferrous sulfide.

pyrite, magnetic. See pyrrhotite.

pyrithiamine. (neopyrithiamine).
$C_{14}H_{20}Br_2N_4O$. A thiamine antagonist.
Properties: Crystallizes from acetone. Mp 219C (decomposes). Soluble in water.
Derivation: Synthetically from the condensation of 2-methyl-3-(β-hydroxyethyl)pyridine with the pyrimidine moiety of thiamine.
Use: Biochemical research.

pyro-. A prefix indicating formation by heat, specifically, an inorganic acid derived by loss of one molecule of water from two molecules of an orthoacid as pyrophosphoric acid.

pyroboric acid. (tetraboric acid). $H_2B_4P_7$. Vitreous or white powder, soluble in water and in alcohol.

Pyrobrite. A bright-leveling, pyrophosphate copper electroplating process. The materials used are copper pyrophosphate trihydrate, potassium pyrophosphate, ammonium hydroxide, and addition agents.

pyrocatechol. (1,2-benzenediol; o-dihydroxybenzene; catechol).
CAS: 120-80-9. $C_6H_4(OH)_2$.
Properties: Colorless crystals, discolors to brown on exposure to air and light, especially when moist. D 1.371, mp 104C, bp 245C, sublimes, flash p 261 (127C) (CC). Soluble in water, alcohol, ether, benzene, and chloroform, also in pyridine and aqueous alkaline solutions. Combustible.
Derivation: (1) By fusion of o-phenolsulfonic acid with caustic potash at 350C; (2) by heating guaiacol with hydriodic acid.
Grade: Technical, CP, resublimed.
Hazard: Strong irritant. Toxic by skin absorption. Eye and upper respiratory tract irritant, and dermatitis. Possible carcinogen.
Use: Antiseptic, photography, dyestuffs, electroplating, specialty inks, antioxidants and light stabilizers, organic synthesis.

pyrocatechol dimethyl ether. See veratrole.

pyrocatechol methyl ester. See guaiacol.

pyrocellulose. See guncotton.

"Pyroceram" Brand 9608 [Corning]. TM for a crystalline ceramic material made from glass by controlled nucleation.

Properties: Hard, brittle solid. D 2.5, softening temperature 1250C, specific heat 0.19 (25C), dielectric constant 6.54 (25C), flexural strength 14,000–23,000 psi, Knoop hardness 703 (100g).
Use: Telescope mirrors, special-purpose ceramic products, domestic dinnerware.

"Pyroceram" Brand Cement *[Corning].* TM for powdered glasses that are thermosetting and utilized for sealing inorganic materials. The resultant seals are crystalline and have service temperatures in excess of the sealing temperatures.

pyrochlore. $NaCaNb_2O_6F.$
Properties: A complex oxide of sodium, calcium, and niobium. Tantalum, rare-earth metals, and other elements may be present. Color brown to black, light brown streak. Hardness 5–5.5, d 4.2–6.4.
Occurrence: Canada (Quebec), Brazil, Africa.
Use: Ore of niobium.

pyrochroite.   See manganous hydroxide.

pyrogallol.   (pyrogallic acid; 1,2,3-trihydroxybenzene).
CAS: 87-66-1. $C_6H_3(OH)_3.$
Properties: White, lustrous crystals; turn gray on exposure to light. D 1.463, mp 132.5, bp 309C. Soluble in water, alcohol, and ether. A solution of pyrogallol acquires a brown color on exposure to air. This absorption of oxygen and change of color take place rapidly when the solution is made alkaline.
Derivation: By heating gallic acid with three times its weight of water in an autoclave.
Hazard: Toxic by ingestion and skin absorption.
Use: Protective colloid in preparation of metallic colloidal solutions, photography, dyes, intermediates, synthetic drugs, medicine (antibacterial), process engraving, laboratory reagent, gas analysis (an oxygen absorber), reducing agent, antioxidant in lubricating oils.

pyrolan.   (1-phenyl-3-methyl-5-pyrazolyl dimethylcarbamate).
CAS: 87-47-8. $C_{13}H_{15}N_3O_2.$
Properties: Crystalline solid. Mp 50C. Soluble in water and fats.
Hazard: Toxic by ingestion. Cholinesterase inhibitor.
Use: Insecticide.

pyroligneous acid.   (wood vinegar; pyroligneous liquor).
Properties: Crude, yellow to red liquid. A mixture of materials from wood distillation. Crude product contains methanol, acetic acid, acetone, furfural, and various tars and related products. D 1.018–1.030. Miscible with water and alcohol.
Use: Smoking meats.

pyrolusite.   (manganese dioxide, black). $MnO_2.$

Properties: Iron-black to dark steel-gray or bluish mineral; black or bluish-black streak; metallic or dull luster. D 4.73–4.86, Mohs hardness 2–2.5. Soluble in hydrochloric acid.
Occurrence: U.S. (Virginia, Georgia, Arkansas, Lake Superior region, Massachusetts, Vermont, New Mexico), Germany, Australia, India, Canada.
Use: Manganese ore. For many uses the ore and synthetic material are interchangeable, but pyrolusite is not usable for batteries.
See manganese dioxide.

pyrolysis.   Transformation of a compound into one or more other substances by heat alone, i.e., without oxidation. It is thus similar to destructive distillation. Though the term implies decomposition into smaller fragments, pyrolytic change may also involve isomerization and formation of higher molecular weight compounds. Hydrocarbons are subject to pyrolysis, e.g., formation of carbon black and hydrogen from methane at 1300C and decomposition of gaseous alkanes at 500–600C. The latter is the basis of thermal cracking (pyrolysis gasoline). One application of pyrolysis is conversion of acetone into ketenes by decomposition at about 700C; the reaction is $CH_3COCH_3 \rightarrow H_2C{=}C{=}O + CH_4.$ Pyrolysis of natural gas or methane at about 2000C and 100 mm Hg pressure produces a unique form of graphite. Synthetic crude oil can be made by pyrolysis of coal, followed by hydrogenation of the resulting tar. Large-scale pyrolysis of cellulosic wastes is being conducted for production of synthetic fuel oils and other products; the method is said to require only 30 minutes at about 537C (flash pyrolysis).
See destructive distillation.

pyrolysis gasoline.   See gasoline.

pyromellitic acid.   (PMA; 1,2,4,5-benzenetetracarboxylic acid).
CAS: 89-05-4. $C_6H_2(COOH)_4.$
Properties: White powder. D 1.79, mp 257–265C, bp (converts to dianhydride), bulk d 32 lb/cu ft. Absorbs moisture slowly if exposed to atmosphere. Soluble in alcohol, slightly soluble in water.
Grade: 99% purity.
Hazard: Skin irritant, may be carcinogenic.
Use: Intermediate for polyesters and polyamides used in electrical and nonfogging plasticizers, lubricants, and waxes.

pyromellitic dianhydride.   (PMDA).
$C_6H_2(C_2O_3)_2.$

Properties: White powder. D 1.68, mp 286C, bp 397–400C, 305–310C (30 mm Hg) bulk d 21 lb/cu ft. Soluble in some organic solvents; hydrolyzes to the acid when exposed to moisture.

Derivation: Oxidation of durene, either in wet process by nitric acid or a dichromate, or as direct air oxidation with catalyst.

Grade: 98+% purity.

Hazard: Skin irritant.

Use: Curing agent for epoxy resins used in high-temperature laminates, molds, and coatings; cross-linking agent for epoxy plasticizers in vinyls, alkyd resins; intermediate for pyromellitic acid.

pyrometer. An instrument for measuring temperatures of 1800C or higher, for example, molten steels, hot springs, volcanoes, etc. There are three kinds: (1) thermocouples of the graphite to silicon carbide type; (2) optical, in which the indications depend on the brightness at some one wavelength of the hot body whose temperature is being measured; and (3) radiation, in which the indications depend on the radiance of a source of radiant energy.
See thermocouple.

pyrometric cone. (Seger cone). A small pyramid composed of a mixture of oxides that melt at known temperatures.

Use: To measure temperatures in autoclaves, curing ovens, etc., in the 100–3700F range.

pyromucamide. See furoamide.

pyromucic acid. See furoic acid.

pyromucic aldehyde. See furfural.

pyrophoric material. Any liquid or solid that will ignite spontaneously in air at about 130F (54.4C). Titanium dichloride and phosphorus are examples of pyrophoric solids, tributylaluminum and related compounds are pyrophoric liquids. Sodium, butyllithium, and lithium hydride are spontaneously flammable in moist air because they react exothermically with water. Such materials must be stored in an atmosphere of inert gas or under kerosene. Some alloys (barium, misch metal) are called pyrophoric because they spark when slight friction is applied.

Hazard: Dangerous fire risk near combustible materials.

Use: As tips on pocket lighters and similar devices.

pyrophosphatase. An inorganic pyrophosphatase.

pyrophosphates. Are those with the anion PO^3; all are alkaline.

Hazard: Vary in toxicity.

pyrophosphoric acid. (diphosphoric acid). $H_4P_2O_7$.

Properties: A viscous, syrupy liquid that tends to solidify on long standing at room temperature. When diluted with water it is rapidly converted into orthophosphoric acid. Mp 54C. Soluble in water.

Derivation: By heating phosphoric acid at 250–260C. Further heating produces metaphosphoric acid.

Grade: Technical.

Use: Catalyst, manufacture of organic phosphate esters, metal treatment, stabilizer for organic peroxides.
See polyphosphoric acid.

pyrophyllite. (agalmatolite). $Al_2Si_4O_{10}(OH)$. A natural hydrous aluminum silicate, found in metamorphic rocks.

Properties: White, green, gray, brown color; pearly to greasy luster, good micaceous cleavage. D 2.8–2.9, Mohs hardness 1–2. Similar to talc.

Occurrence: North Carolina, California, Newfoundland, Japan.

Use: Ceramics, insecticides, slate pencils, substitute for talc, extender in paints, wallboard, buffer in high-pressure equipment, permissible animal feed additive.

pyroracemic acid. See pyruvic acid.

pyrosulfuric acid. See sulfuric acid, fuming.

pyrosulfuryl chloride. (disulfuryl chloride). CAS: 7791-27-7. $S_2O_5Cl_2$.

Properties: Colorless, mobile, very refractive, fuming liquid. D 1.819, bp 146C, fp −38C, refr index 1.449 (19C).

Hazard: Decomposes violently with water to sulfuric acid and hydrochloric acid. Corrosive to tissue.

Use: Organic synthesis.

pyrotartaric acid. (methylsuccinic acid). $HOOCCH(CH_3)CH_2COOH$.

Properties: White or yellowish crystals. D 1.4105, mp 111–112C. Soluble in water, alcohol, and ether.

Derivation: By distilling tartaric acid.

Use: Organic synthesis.

pyrotartaric acid, normal. See glutaric acid.

pyrotechnics. The formulation and manufacture of fireworks, signal flares, and military warning devices. The industry is professionally represented by the Pyrotechnic Guild International, Inc. The primary ingredients of pyrotechnic products are as follows: (1) Oxidizers: potassium nitrate, potassium chlorate, or potassium perchlorate; ammonium perchlorate; barium chlorate and nitrate; and strontium nitrate. (2) Fuels: aluminum, magnesium, antimony sulfate, dextrin, sulfur, and titanium. (3) Binders: dextrin and various polymers. Colored flames are produced by strontium compounds (red); barium

compounds (green); copper carbonate, sulfate, and oxide (blue); sodium oxalate and cryolite (yellow); and magnesium, titanium, or aluminum (white). Black powder is used as propellant.
See chemiluminescence.

pyrovanadic acid. See vanadic acid.

"Pyrovatex" CP *[Huntsman].* TM for a fiber-reactive phosphone alkyl amide.
Properties: Odorless, nontoxic, nonirritating, stable to laundering and dry-cleaning. Reduces tear and tensile strengths, no effect on "hand." Compatible with other finishing and proofing agents and pre-cured durable-press processes. From 25 to 35% by weight is required; imparts self-extinguishing properties to cotton.
Use: Flame-retardance of 100% cotton fabrics, such as those used for tenting, military uniforms, and draperies.

pyroxylin. See nitrocellulose.

pyrrhotite. (magnetic pyrites; pyrrhotine). FeS. A natural iron sulfide. Frequently has a deficiency in iron. May contain small amounts of nickel, cobalt, manganese, and copper.
Properties: Brownish-bronze color, black streak, metallic luster. Slightly magnetic. Hardness 4. D 4.6.
Occurrence: Tennessee, Pennsylvania, Europe, Canada.
Use: Ore of iron, manufacture of sulfuric acid.

pyrrobutamine phosphate. (1-[4(*p*-chloro-phenyl)-3-phenyl-2-butenyl]pyrrolidine diphosphate). $ClC_6H_4CH_2C(C_6H_5):CHCH_2NC_4H_8 \cdot 2H_3PO_4$.
Properties: Cream to off-white powder; slight odor; bitter taste. Melting range 127–131C. Soluble in water; slightly soluble in alcohol; almost insoluble in chloroform and ether.
Grade: NF.
Use: Medicine (antihistamine).

pyrrole.
CAS: 109-97-7. Pyrrole is regarded as a resonance hybrid and no one structure adequately represents it. The following is an approximation:

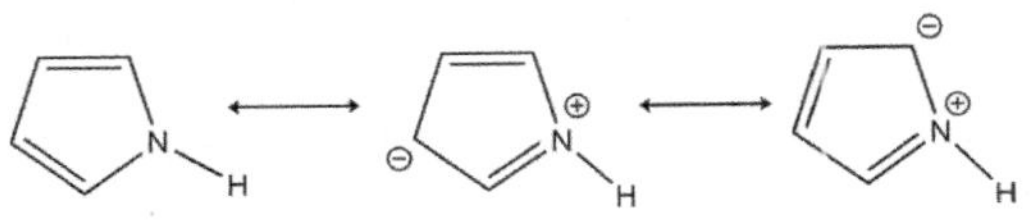

Properties: Yellowish or brown oil; burning, pungent taste; odor similar to chloroform. Readily polymerizes by the action of light and turns brown. Bp 130–131C, fp −24C, d 0.968 (20/4C), refr index 1.5091 (20C), flash p 102F (38.9C) (TCC). Soluble in alcohol, ether, dilute acids, and most organic chemicals; insoluble in water and dilute alkalies. Combustible.
Derivation: Fractional distillation of bone oil with sulfuric acid.
Method of purification: Conversion into the potassium compound (C_4H_4NK), washing with ether, and treatment with water, followed by drying and distillation.
Grade: Technical.
Hazard: Moderate fire risk. Toxic by ingestion and inhalation.
Use: Manufacture of pharmaceuticals.

pyrrole-2,5-dione. See maleimide.

pyrrolidine.
CAS: 123-75-1. C_4H_9N.
Properties: Colorless to pale-yellow liquid; penetrating aminelike odor. D 0.8660 (20/20C), fp −60C, bp 87C, refr index 1.4425 (20C), flash p 37F (2.7C) (TCC). Soluble in water and alcohol.
Grade: 95% min purity.
Use: Intermediate for pharmaceuticals, fungicides, insecticides, rubber accelerators, citrus decay control, curing agent for epoxy resins, inhibitor.
Hazard: Flammable, dangerous fire risk. Toxic by ingestion and inhalation.

2-pyrrolidinecarboxylic acid. See proline.

l-**2-pyrrolidinecarboxylic acid.** See *l*-proline.

1-pyrrolidineethanol.
CAS: 2955-88-6. $C_6H_{13}NO$.
Hazard: A reproductive hazard.

2-pyrrolidone. (2-pyrrolidinone; butyrolactam).
CAS: 616-45-5.

$CH_2CH_2CH_2C(O)NH.$

Properties: Light-yellow liquid. D 1.1, bp 245C, flash p 265F (129C). Soluble in water, ethanol, ethyl ether, chloroform, benzene, ethyl acetate, carbon disulfide. Noncorrosive. Combustible.
See polyvinylpyrrolidone.
Derivation: From acetylene and formaldehyde by high-pressure synthesis.
Use: Plasticizer and coalescing agent for acrylic latexes in floor polishes, solvent for polymers, insecticides, polyhydroxylic alcohols, sugar, iodine, specialty inks, monomer for nylon-4.

pyrrolizidine alkaloid. Any of a type of alkaloid produced by various plants such as *Senecio* and *Crotalaria.*

pyrrone. An aromatic heterocyclic polymer derived from a cyclic dianhydride and an aromatic *o*-tetramine or a derivative. Stable to 482C.
Use: Films, coatings, adhesives, laminates, and moldings.

pyruvaldehyde.   See pyruvic aldehyde.

pyruvate.   (2-oxopropanoic acid).
CAS: 127-17-3. $C_3H_4O_3$. An intermediate compound in the metabolism of carbohydrates, proteins and fats.

pyruvic acid.   (α-ketopropionic acid; acetylformic acid; pyroracemic acid).
CAS: 127-17-3. $CH_3COCOOH$. A fundamental intermediate in protein and carbohydrate metabolism in the cell.
  Properties: Liquid; odor resembling acetic acid. D 1.2272 (20/4C), mp 13.6C, bp 165C. Miscible with water, ether, and alcohol.

Derivation: Dehydration of tartaric acid by distilling with potassium acid sulfate.
Use: Biochemical research.

pyruvic alcohol.   See hydroxy-2-propanone.

pyruvic aldehyde.   (pyruvaldehyde; methyl glyoxal). CH_3COHCO.
Properties: Supplied commercially as a 30% aqueous solution. D 1.20 (20/20C), bulk d 10 lb/gal (20C).
Use: Organic synthesis, as of complex chemical compounds such as pyrethrins, tanning leather, flavoring.

Q

"Qiana" *[Du Pont].* TM for a nylon-type synthetic fiber with properties similar to silk.

Q-lure. See cue-lure.

quad. An energy unit that has come into use in recent years in predicting future energy requirements on a national basis. One quad equals 10^{15} Btu, which is the energy equivalent of 10^{12} cu ft natural gas, or 182 million barrels of oil, or 42 million tons of coal, or 293 billion kilowatt-hours of electricity.

quadruple point. The temperature at which four phases are in equilibrium, such as ice, saturated salt solution, water vapor, and salt.

quadrupole resonance. See nuclear quadrupole resonance.

qualitative analysis. See analytical chemistry.

quantitative analysis. See analytical chemistry.

quantization of energy. The assumption that the energy of a particle is not infinitely variable but can have any one of a definite set of values.

quantum dot. (Qdots). Nanometer sized semiconductor particles, made of cadmium selenide (CdSe), cadmium sulfide (CdS), or cadmium telluride (CdTe) with an inert polymer coating.

quantum efficiency. (photochemical yield). The number of electrons actually ejected per quantum of light absorbed.

quantum jump. See quantum transition.

quantum mechanics. Quantum mechanics describes a system of particles in terms of a wave function defined over the configuration space of the system. Although the concept of particles having distinct locations is implicit in the potential energy function that determines the wave function (e.g., of a ground-state system), the observable dynamics of the system cannot be described in terms of the motion of such particles from point to point. In describing the energies, distributions, and behaviors of electrons in nanometer-scale structures, quantum mechanical methods are necessary. Electron wave functions help determine the potential energy surface of a molecular system, which in turn is the basis for classical descriptions of molecular motion.

quantum number. The quantum is the basic unit of electromagnetic energy. It characterizes the wave properties of electrons, as distinct from their particulate properties. The quantum theory developed by Max Planck states that the energy associated with any quantum is proportional to the frequency of the radiation, that is, e (energy) $= hu$, where u is the frequency and h is a universal constant. An electron has four quantum numbers that define its properties. These are as follows: (1) The principal quantum number is a constant that can be any positive integer ($n = 1, 2, 3, \ldots$). It determines the principal energy level, or shell, of the electron, sometimes designated by letters such as K, L, or M, depending on the value of the principal quantum number. (2) The angular momentum constant l, also an integer, is related to n as $l = 0, 1, \ldots, n-1$. Here again, letter designations are often used. In s electrons, $l = 0$; in p electrons, $l = 1$; in d electrons, $l = 2$; and in f electrons, $l = 3$. (3) The magnetic quantum number, m, is an integer related to l as: $m = -1, \ldots, -1, 0, +1, \ldots, +l$. (4) The spin quantum number is independent of the other three and has a value of either $+1/2$ or $-1/2$, depending on the direction of rotation of the electron on its axis in the atomic frame of reference.
See orbital theory; electron; photon; radiation; Pauli exclusion principle.

quantum number, azimuthal. The quantum number specifying the angular momentum of an orbital electron.

quantum number, magnetic. The determinant of the component of angular momentum vector of an atomic electron or group of electrons along the externally applied magnetic field.

quantum number, radial. The radial motion quantum number that is an integer for any permitted stationary condition of a particle moving under the influence of a central field.

quantum state. (energy level). The energy state of an atom as determined by the frequencies of its characteristic spectral lines.

Hawley's Condensed Chemical Dictionary, Sixteenth Edition. Michael D. Larrañaga, Richard J. Lewis, Sr., and Robert A. Lewis.
© 2016 John Wiley & Sons, Inc. Published 2016 by John Wiley & Sons, Inc.

quantum transition. (quantum jump). The sharp change in an atom accompanied by emission or absorption of a quantum of radiant energy.

quantum yield. Number of photon-induced reactions of a specified type per photon absorbed.

quark. The smallest known bit of matter. Hypothetical entities carrying electrical charges of one-third or two-thirds the normal unit. Light (300 MeV) hypothetical particles used to facilitate calculation of certain observables.

quartz.
CAS: 14808-60-7. SiO_2.
Crystallized silicon dioxide (silica).
Properties: White to reddish color, luster vitreous. Mp 1713C, Mohs hardness 7, d 2.65. Insoluble in acids except hydrogen fluoride; only slightly attacked by solutions of caustic alkali. Piezoelectric and pyroelectric. Noncombustible.
Derivation: Synthetic crystals of good size and purity are grown by mass production methods under very carefully regulated conditions of temperature and concentration.
Hazard: Avoid inhalation of fine particles.
Use: Electronic components; piezoelectric control in filters, oscillators, frequency standards, wave filters, radio and TV components; barrel-finishing abrasive.
See silica.

quartz dust. (dioxosilane). SiO_2. A major component of dust in quartz mines and quarries; in metal mines where the rock between veins of ore contains free silica; and in factors that use quartz sand.
Hazard: Pulmonary fibroses, emphysema, impairment of lung function.

quartz, fused. Pure silica that has been melted to yield a glass-like material on cooling.
Use: For apparatus and equipment (such as vacuum tubes) where its high melting point, ability to withstand large and rapid temperature changes, chemical inertness and transparency (including UV radiation), and electrical resistance are valuable. Produced as fibers and fabrics for heat resistance, low expansion coefficient, and insulating value.
See glass.

quassia. (bitter ash; bitterwood).
Properties: White to bright-yellow chips or shavings; very bitter taste.
Derivation: The wood or bark of *Picrasma excelsa* or *Quassia amara*.
Use: Decoction or tincture as a fly poison, surrogate for hops, medicine (anthelminthic), hair lotion, flavoring, alcohol denaturant.

quaternary ammonium salt. A type of organic nitrogen compound in which the molecular structure includes a central nitrogen atom joined to four organic groups (the cation) and a negatively charged acid radical (the anion). The structure is indicated as

$$\left[\begin{array}{c} R \\ | \\ R-N-R \\ | \\ R \end{array} \right]^+ Z^-$$

Octadecyldimethylbenzyl ammonium chloride and hexamethonium chloride are examples. Pentavalent nitrogen ring compounds, such as lauryl pyridinium chloride, are also considered quaternary ammonium compounds. They are all cationic, surface-active coordination compounds and tend to be adsorbed on surfaces.
Use: Disinfectant, cleanser and sterilizer, cosmetics (deodorants, dandruff removers, emulsion stabilizers), fungicides, mildew control, to increase affinity of dyes for film in photography, coating of pigment particles to improve dispersibility, to increase adhesion of road dressings and paints, antistatic additive, biocide.
See detergent, synthetic; coordination compound.

***p*-quaterphenyl.** $C_6H_5C_6H_4C_6H_4C_6H_5$.
Properties: Crystals. Mp 316–318C, bp 428C (18 mm Hg).
Grade: Purified.
Use: As primary fluor or as wavelength shifter in soluble scintillators.

quebrachine. See yohimbine.

quebracho.
Properties: A wood-derived tannin, the most important tanning agent used in the American leather industry. Combustible.
Derivation: From *Aspidosperma quebracho* and *Quebracho lorentzi*, imported as logs from Argentina.
Grade: Liquid: 35–37% tannin. Solid: 65% tannin.
Use: Vegetable tanning, retanning of chrome-tanned upper leathers, dyeing, ore flotation, oil-well drilling fluids, flavoring.

Quelet reaction. Passage of dry hydrogen chloride through a solution in ligroin of a phenolic ether and an aliphatic aldehyde in the presence or absence of a dehydration catalyst to yield α-chloroalkyl derivatives by substitution in the *para* position to the ether group or in the *ortho* position in *para*-substituted phenolic ethers.

quench. In the terminology of metallurgy, quick cooling of metals or alloys by immersion in cold water or oil. This is an essential part of the tempering process, especially for steels. If the metal or alloy is in the liquid (molten) state and the quench time is extremely short (less than a second), the product will have an amorphous or glass-like structure, because no crystallization occurs.
See glass, metallic.

quercetin.
CAS: 117-39-5. $C_{15}H_{10}O_7$.
Properties: Yellow needles (dihydrate). Anhydrous form decomposes at 315C. Soluble in alcohol and glacial acetic acid; insoluble in water.
Derivation: Bark of fir trees, also synthetically.
Hazard: Questionable carcinogen.
Use: Medicine, reported formation of epoxy resins on mixing with epichlorohydrin.

quetiapine fumarate.
CAS: 111974-72-2. $C_{42}H_{50}N_6O_4S_2 \cdot C_4H_4O_4$.
Hazard: Human systemic effects.

quick-.   Prefix meaning alive or active, as in quicksilver (mercury), quicklime (unslaked lime), quicksand, quick (the flesh beneath the fingernail).

Quick-Dry Citrus Solvent.   A citrus-based solvent.
Use: To replace VOC compounds in industrial products.

quicklime.   See calcium oxide.

quicksilver.   See mercury.

"Quilon" [Du Pont].   TM for a Werner-type chromium complex in isopropanol, 30% solution of stearatochromic chloride.
Use: Water repellent and sizing treatment of cellulosic materials, treatment of negatively charged surfaces, antiblocking agent, for insolubilizing various water-soluble or swellable coatings, improving grease-resistant coatings, treatment of feathers.

quinacridone.   A lightfast pigment used in paints, printing inks, plastics, etc.

quinacridone magneta.   See 2,9-dimethylquin-acridone.

quinacrine dihydrochloride.   (3-chloro-7-methoxy-9-(1-methyl-4-diethylamino-butylamino) acridine dihydrochloride).
CAS: 69-05-6. $C_{23}H_{30}ClN_3O \cdot 2HCl \cdot 2H_2O$.
Properties: Bright yellow, crystalline powder; odorless; bitter taste. Decomposes at 248–250C. Soluble in hot water; pH of 1% water solution 4.5.
Derivation: Organic synthesis.
Grade: USP.
Use: Drug.

quinaldine.   (chinaldine; α-methylquinoline).
CAS: 91-63-4. $C_9H_6NCH_3$.
Properties: Colorless, oily liquid; odor of quinoline; darkens to reddish brown in air. Bp 246–247C, mp −2C, d 1.51. Soluble in alcohol, ether, and chloroform; insoluble in water.
Derivation: (1) By the treatment of aniline and paraldehyde with hydrochloric acid and heat; (2) from coal tar.

Hazard: Strong irritant to mucous membranes.
Use: Manufacture of dyes, pharmaceuticals, fine organic chemicals, acid-base indicators.

quinaphthol.   (quinine-β-naphthol α-sulfonate).
$C_{20}H_{24}N_2O_2 \cdot (OHC_{10}H_6 \cdot SO_3H)_2$.
Properties: Yellow crystalline powder, contains 42% quinine; bitter taste. Mp 185–186C. Moderately soluble in hot water or alcohol; insoluble in cold water.
Derivation: Interaction of quinine and β-naphtholsulfonic acid.
Use: Medicine.

quinarene.   Any mancude assembly of three carbocyclic rings.
Properties: Contains a six-membered quinonoid ring bonded at the 1,4-positions to odd-membered rings which differ in ring size by two.

quinhydrone.
CAS: 106-34-3. $C_6H_4O_2C_6H_4(OH)_2$.
Properties: Dark green crystals. Mp 171C, d 1.40. Slightly soluble in water; soluble in alcohol, ether, hot water, ammonia.
Derivation: Oxidation of hydroquinone with sodium dichromate.
Grade: Reagent, technical.
Hazard: Toxic by ingestion.
Use: Electrode for pH determination.

quinidine.   (chinidin; cin-quin; conchinin; conquinine; 6′-methoxycinchonan-9-ol; α-(6-methoxy-4-quinolyl)-5-vinyl-2-quinuclidinemethanol; 6-methoxy-α-(5-vinyl-2-quinuclidinyl)-4-quinolinemethanol; NCI-C56246; pitayine; quinicardine; quinidex; (+)-Quinidine; β-quinine; (S)-[(4S,5R,7R)-5-ethenyl-1-azabicyclo[2.2.2]octan-7-yl]-(6-methoxyquinolin-4-yl)methanol).
CAS: 56-54-2. $C_{20}H_{24}N_2O_2$.
One of the alkaloids of cinchona and a stereoisomer of quinine.
Properties: A solid.
Derivation: Extracted from the bark of the Cinchona tree and similar plant species.
Hazard: Poison.
Use: Antimalarial agent; in the treatment of atrial fibrillation and flutter, and paroxysmal ventricular tachycardia.

quinine.
CAS: 130-95-0. $C_{20}H_{24}N_2O_2 \cdot 3H_2O$.
An alkaloid.
Properties: Bulky, white, amorphous powder or crystalline alkaloid; very bitter taste; odorless. Levorotatory. Soluble in alcohol, ether, chloroform, carbon disulfide, glycerol, alkalies, and acids (with formation of salts); very slightly soluble in water.
Derivation: Finely ground cinchona bark mixed with lime is extracted with hot, high-boiling paraffin oil. The solution is filtered, shaken with dilute sulfuric acid, and the latter neutralized while still hot with

sodium carbonate; on cooling, quinine sulfate crystallizes out. The sulfate is then treated with ammonia, the alkaloid being obtained.
Source: Indonesia, Bolivia.
Hazard: Skin irritant, ingestion of pure substance adversely affects eyes.
Use: Medicine (antimalarial) as the alkaloid or as numerous salts and derivatives; flavoring in carbonated beverages.

quinine acid sulfate.   See quinine.

quinine bisulfate.   See quinine.

quinizarin.   (1,4-dihydroxyanthraquinone;   CI 58050). $C_{14}H_6O_2(OH)_2$.
Properties: Red or yellow-red crystals. Mp 194–195C. Soluble in hot water, alcohol, ether, benzene, potassium chloride, and sulfuric acid.
Use: Antioxidant in synthetic lubricants, dyes.

quinol.   See hydroquinone.

quinoline.   (chinoline).
CAS: 91-22-5. C_9H_7N.
A basic heterocyclic nitrogenous compound occurring in coal tar and obtained from it, but more frequently by synthesis.

Properties: Highly refractive, colorless liquid; darkens with age; hygroscopic; peculiar odor. D 1.0899, fp −15C, bp 238C, autoign temp 896F (480C). Soluble in water, alcohol, ether, and carbon disulfide. Combustible.
Derivation: By treatment of aniline and nitrobenzene with glycerol and sulfuric acid and heat.
Use: Preserving anatomical specimens; manufacture of quinolinol sulfate; niacin and copper-8-quinolinolate; flavoring.

quinoline dye.   See cyanine dye.

quinolinic acid.
CAS: 89-00-9. $C_7H_5NO_4$.
Hazard: A poison by skin contact. Moderately toxic by ingestion. A mild skin irritant.

8-quinolinol.   See 8-hydroxyquinoline.

(8-quinolinolato)tributylstannane.   See tributyl(8-quinolinolato)tin.

quinolizidines.   Alkaloids that have two-fused 6-membered rings that share a nitrogen atom. They are derivatives of lysine.

quinomethionate. $C_{10}H_6N_2OS_2$.
Properties: Tan to yellow crystals. Mp 170C. Soluble in benzene, toluene, and dioxane; insoluble in water.
Use: Fungicide, acaricide.

quinone.   (1,4-benzoquinone; chinone).
CAS: 106-51-4. $C_6H_4O_2$.

Properties: Yellow crystals; irritant odor. D 1.307, mp 115.7, bp sublimes. Soluble in alcohol, ether, and alkalies; slightly soluble in hot water. Volatile with steam, being in part decomposed. Combustible.
Derivation: By oxidation of aniline with chromic acid, extraction with ether and distillation.
Hazard: Toxic by inhalation, strong irritant to skin, eyes and mucous membranes. Skin damage. Questionable carcinogen.
Use: Manufacture of dyes and hydroquinone, fungicides, analytical reagent, photography, oxidizing agent.

***p*-quinonedioxime.**   See "GMF."

quinone oxime dye.   See nitroso dye.

quinoxaline.   (1,4-benzodiazine; benzo-*p*-diazine). $C_8H_6N_2$ (bicyclic). An organic base.
Properties: Colorless, crystalline powder. Mp 30C, bp 229C. Soluble in water and organic solvents.
Use: Organic synthesis.

quinoxalinedithiol   cyclic   trithiocarbonate.
See thioquinox.

(2,3-quinoxalinyldithio)dimethyltin.
CAS: 73927-90-9. $C_{10}H_{10}N_2S_2Sn$.
Hazard: A poison.

(2,3-quinoxalinyldithio)diphenyltin.
CAS: 73927-96-5. $C_{20}H_{14}N_2S_2Sn$.
Hazard: A poison.

quinoxidine.
CAS: 10103-89-6. $C_{14}H_{14}N_2O_6$.
Hazard: Moderately toxic by ingestion.

quinsol. $C_9H_6NOSO_3K \cdot H_2O$. Yellow, crystalline powder. Combustible.
Use: Fungicide.

quintozene.   See pentachloronitrobenzene.

quinuclidine.
  CAS: 100-76-5. $C_7H_{13}N$.
  Hazard: A poison by ingestion and skin contact.
  Low toxicity by inhalation. A moderate skin and
  severe eye irritant.

3-quinuclidinol.   (quinuclidinol; 1-azabicyclo
  [3.3.3]octan-3-ol;   3-hydroxyquinuclidine;   1-
  azabicyclo[2.2.2octan-8-ol). $C_7H_{13}NO$.
  Properties: Highly water-soluble compound.
  Hazard: Mutation.
  Use: Thereapeutically as a hypotensive agent and in
  glaucoma therapy.

quinuclidnyl acetate.   (3-quinuclidinyl
  acetate). The acetate ester of 3-quinuclidinol.
  Hazard: Cholinergic.
  Use: Chemical warfare agents.

quinuclidinyl benzylate.   (3-quinuclidinyl benzylate; Bz; qnb; QNB; 10azabicyclo[2.2.2]octan-8-
  yl 2-hydroxy-2,2-di)phenyl)acetate). $C_{21}H_{23}NO_3$.
  The benzylate ester of quinuclidinol and a high-
  affinity muscarinic antagonist.
  Hazard: Incapacitating.
  Use: A tool in animal and tissue studies; a chemical
  warfare agent.

Quotane.   An ointment and lotion containing
  dimethisoquin hydrochloride.

R

R. (1) Symbol used to represent an organic group in a chemical formula, e.g., CH_3, C_2H_5, C_6H_5. (2) A free radical (•, R Code: •). (3) The gas constant, equal to $p_0v_0/273C$. (4) Abbreviation of Rankine temperature scale.

Ra. Symbol for radium.

R-acid. (2-naphthol-3,6-disulfonic acid; β-naphtholdisulfonic acid).

$$HO_3S \quad \text{(naphthalene ring, positions 6 and 3)} \quad OH, SO_3H$$

Properties: Deliquescent, colorless needles. Soluble in water, alcohol, and ether.
Derivation: Sulfonation of β-naphthol. For details see Schaeffer acid.
Use: Azo dye intermediate. The disodium salt is used as a reagent in detection of nitrogen dioxide in the air.

2R acid. See RR acid.

racemate. (1) The mixture of two optically active components that neutralize the optical effect of each other. (2) Salt of racemic acid.

racemic mixture. A mixture of the D and L stereoisomers of an optically active compound.

racemic substance. A mixture of dextro- and levorotatory optically active isomers in equal amounts, the resulting mixture having no rotary power. These mixtures are prefixed by ñ or *dl*-. See *dl*-.

racemization. Conversion, by heat or by chemical reaction (e.g., enolization) of an optically active compound into an optically inactive form in which half of the optically active substance becomes its mirror image (enantiomer). This change results in a mixture of equal quantities of dextro- and levorotatory isomers, as a result of which the compound does not rotate plane-polarized light to either right or left since the two opposite rotations cancel each other. This is sometimes referred to as external compensation, as opposed to the internal compensation exhibited by *meso*-compounds.
See *meso*-(1); tartaric acid.

racemomycin E.
CAS: 3484-68-2. $C_{43}H_{82}N_{16}O_{12}$.
Hazard: A poison.
Source: Natural product.

racephedrine. (racemic ephedrine; *dl*-ephedrine). $C_{10}H_{15}NO$.
Properties: Crystals. Mp 79C. Soluble in water, alcohol, chloroform, and oils.
Derivation: Synthetic.
Use: Medicine (also as hydrochloride and sulfate). See ephedrine.

racking. Experimental cold-stretching of unvulcanized rubber, whose behavior under stress is unique among natural materials. A thin, narrow strip stretched at, for example, 500–600% at 0C will retain that extension indefinitely after release of stress as long as the low temperature persists. In this state, it loses its elasticity and has virtually 100% permanent set. It also displays a crystalline X-ray pattern similar to that of a fiber, in contrast to the amorphous structure of the unstretched state. On exposure to room temperature, it slowly retracts to its original length; higher temperature increases its rate of recovery. Tests made on racked rubber have shown that crude rubber can be exposed to any degree of low temperature for any length of time without impairment of its properties.

rad. That quantity of ionizing radiation that results in the absorption of 100 ergs of energy per gram of irradiated material, regardless of the source of the radiation. The federal radiation safety standard is 0.5 rem per person per year for nonoccupational exposure, and even this is considered too high by some authorities. Occupational exposure is set at 5 rem per year.
See rem.

radiation. Energy in the form of electromagnetic waves (also called radiant energy, or light). It is emitted from matter in the form of photons (quanta), each with an associated electromagnetic wave having frequency (ν) and wavelength (λ). The various forms of radiant energy are characterized by their wavelength, and together they compose the electromagnetic spectrum, the components of which are as follows: (1) cosmic rays (highest energy, shortest wavelength), (2) γ-rays from radioactive disintegration of atomic nuclei, (3) X-rays, (4) UV rays, (5) visible light rays, (6) infrared, (7) microwave, and (8) radio (Hertzian) and electric rays. All these are identical in every way except wavelength, those having the shortest wavelength being the most penetrating. They are not electrically charged and have no mass, their velocity of propagation is the same, all display the properties characteristic of light and have a dual nature (wave-like and corpuscular). In a looser sense the term *radiation* also includes energy emitted in the form of particles that possess mass and may or may not be electrically charged, (i.e., α [positive] and β [negative]) and also neutrons. Beams of such particles may be considered as "rays." The charged particles may all be accelerated and the high energy imparted to "beams" in particle accelerators such as cyclotrons, betatrons, synchrotrons, and linear accelerators.

Radiation is used in medicine in the form of X-rays and radioactive isotopes; it is used in industry in many ways, e.g., as vitamin activator, sterilizing agent, and polymerization initiator; it is also the basis of all types of spectroscopic analysis.

radiation barrier.   Any barrier or device of sufficient density and thickness to protect an organism from a particular type of radiation.

radiation biochemistry.   The study of substances having the ability to protect cells and body tissue against the deleterious effects of ionizing radiation. Because one of these effects is to deprive proteins of sulfhydryl (–SH) groups necessary for cell division, the injection of compounds rich in this radical (notably cysteine) has been successfully tried with laboratory animals. Thiourea has been found to protect DNA from depolymerization by X-rays; enzymes containing –SH groups inactivated by radiation are reactivated by addition of glutathione. Some of the other radiochemically induced reactions that adversely affect biochemical activity are (1) formation of hydrogen peroxide (a biological poison) by free radical mechanism; (2) denaturation of proteins; (3) change in substituent groups of amino acids; (4) oxidation of hemoglobin; (5) depolymerization of DNA.
See radiation, ionizing.

radiation curing.   See radiation, industrial (6).

radiation hybrid.   A hybrid cell containing small fragments of irradiated human chromosomes. Maps of irradiation sites on chromosomes for the human, rat, mouse, and other genomes provide important markers, allowing the construction of very precise STS maps indispensable to studying multifactorial diseases.
See sequence tagged site.

radiation, industrial.   Chemical or physiochemical changes induced by exposure to various types and intensities of radiation. (1) Synthesis of ethyl bromide from hydrogen bromide and ethylene, using α-radiation from cobalt-60. (2)

Cross-linking of such polymers as polystyrene and polyethylene with either β- or γ-radiation. (3) Vulcanization of rubber with ionizing radiation. (4) Polymerization of methyl methacrylate monomer with cobalt-60 as source of γ-rays. Free radical formation is involved in both cross-linking and polymerization reactions. This technique is also being applied in the textile finishing field for grafting and cross-linking fibers with chemical agents for durable-press fabrics. (5) Processing of various foods (cooking, drying, pasteurizing, etc.) by electromagnetic energy in the microwave region of the spectrum; preservation and sterilization of food products by ionizing radiation (γ- and X-rays). The dosage of radiation is strictly controlled, and FDA approval is required. Irradiation is also effective in inhibiting sprouting and preventing insect infestation of stored grains. (6) Curing or hardening of organic protective coatings (paints, inks) by exposure to infrared, UV, or electron-beam radiation. Required are a monomer or oligomer and a photoinitiator, which induces polymerization by free radical formation.
See "Electrocure."

radiation,   ionizing.   Extremely short-wavelength, highly energetic, penetrating rays of the following types: (1) γ-rays emitted by radioactive elements and isotopes (decay of atomic nuclei); (2) X-rays generated by sudden stoppage of fast-moving electrons; (3) subatomic charged particles (electrons, protons, deuterons) when accelerated in a cyclotron or betatron. The term is restricted to electromagnetic radiation at least as energetic as X-rays, and to charged particles of similar energies. Neutrons also may induce ionization.

Such radiation is strong enough to remove electrons from any atoms in its path, leading to the formation of free radicals. These short-lived but highly reactive particles initiate decomposition of many organic compounds. Thus ionizing radiation can cause mutations in DNA and in cell nuclei; adversely affect protein and amino acid mechanisms; impair or destroy body tissue; and attack bone marrow, the source of red blood cells. Exposure to ionizing radiation for even a short period is highly dangerous, and for an extended period may be lethal. The study of the chemical effects of such radiation is called radiation chemistry or (in the case of body reactions) radiation biochemistry.
See radiation, industrial.

radical.   (1) An ionic group having one or more charges, either positive or negative, e.g., OH^-, NH_4^+, SO_4^{2-}.
(2) See free radical; group (2).

radioactive isotope.   See radioisotope.

radioactive waste.   Disposal of waste containing radioisotopes and of spent nuclear reactor fuel

presents a serious problem for which there is as yet no completely satisfactory solution. Such wastes may remain radioactive for thousands of years and can constitute a long-term hazard that is restraining the development of nuclear-generated electric power. Safe disposal techniques are being intensively studied. Ocean dumping, practiced some years ago, is no longer permissible. Small amounts of low-level wastes containing radioisotopes can be diluted sufficiently with an inert material to reduce its activity to an acceptable point. High-level reactor wastes, for example, at Hanford, are stored in concrete tanks lined with steel and buried under a foot of concrete and 5 or 6 ft of soil. Containers of compressed alumina (corundum) have been recommended, as this material remains impervious to water indefinitely. Storage in the form of calcine (granular solid) and in borosilicate glass is a promising possibility under active investigation. The DOE has recommended disposal in deep geologic formations, although the heat generated by the radioactivity could cause fracturing of the surrounding rock structures; this would admit water that would eventually rise to the surface after being contaminated. A test program involving storage in basalt is being conducted by DOE. Storage in salt formations is under serious consideration because they are self-sealing and free from water.
See waste control.

radioactivity.   Natural or artificial nuclear transformation; discovered by Becquerel in 1895. The energy of the process is emitted in the form of α-, β-, or γ-rays. Thus, radium-226 undergoes radioactive decay by the emission of an α-particle, and the new product is radon-222. The decay series terminates in lead-206. Radioactivity is not affected by the physical state or chemical combination of the element. The radioactivity of a nuclide is characterized by the nature of the radiations, their energy, and the half-life of the process, that is, the time required for the activity to decrease to one-half of the original. Half-lives vary from microseconds to millions of years. Some radioactive elements occur in nature (radium, uranium). Radioactivity can be caused artificially in many stable elements by irradiation with neutrons in a nuclear reactor, or by charged particles from an accelerator.
Amounts of radioactive material are usually expressed in units of activity, the rate of radioactive decay. The accepted unit is the curie (Ci) and its metric multiples and fractions, the mega, kilo, milli-, and microcurie. A curie is 3.73×10^{10} disintegrations per sec. A common unit is millicuries per millimole. Packaging and shipment of radioactive materials, which are highly toxic, must be in accord with official requirements. Consult IATA and DOT shipping regulations for labeling and other instructions.
See rad; rem.

radiocarbon.   See carbon-14; chemical dating.

radiochemistry.   The subdivision of chemistry that deals with the properties and uses of radioactive materials in industry, biology, and medicine, including tracer research and radioactive waste disposal.
See nuclear chemistry.

radiogenic.   Refers to a material produced by radioactive decay. An example is the production of lead from uranium decay.

radioimmunoassay.   A sensitive and quantitative method for detecting trace amounts of a biomolecule, based on its capacity to displace a radioactive form of the molecule from combination with its specific antibody.

radioisotope.   (radionuclide). An isotopic form of an element (either natural or artificial) that exhibits radioactivity. Radioisotopes are used as diagnostic and therapeutic agents in medicine, in biological tracer studies, and for many industrial purposes, from measurement of thickness to initiating polymerization. Artificial radioisotopes are made by neutron bombardment of stable isotopes in a nuclear reactor.
See tracer; isotope.

radionuclide.   (radioactive nuclide; radioisotope). A radioactive unstable isotope of an element that decays or disintegrates spontaneously, emitting radioactive radiation.

radium.   Ra. Radioactive element of group IIA of the periodic table, atomic number 88, aw 226.0254, valence = 2. There are 14 radioactive isotopes but only radium-226 with half-life of 1620 years is usable. Discovered by the Curies in 1898.
Properties: Brilliant-white solid. Mp 700C, bp 1140C, d 5. Luminescent, turns black on exposure to air. Soluble in water with evolution of hydrogen; forms water-soluble compounds. Decays by emission of α-, β-, and γ-radiation. Bone-seeking when taken into the body.
Occurrence: Colorado, Canada, Zaire, France, the former U.S.S.R.
Derivation: Uranium ores (pitchblende and carnotite). The method used for isolating radium is similar to that developed by Mme. Curie and involves coprecipitation with barium and lead, chemical separation with hydrochloric acid, and further purification by repeated fractional crystallization. The metal is separated from its salts by electrolysis and subsequent distillation in hydrogen. Dry salts are stored in sealed glass tubes, opened regularly by experienced workers to relieve pressure. The tubes are kept in lead containers.
Hazard: Highly toxic, emits ionizing radiation. Lead shielding should be used in storage and handling, adequate protective clothing and remote control devices are essential. Destructive to living tissue.

Use: Medical treatment for malignant growths, industrial radiography, source of neutrons and radon.

radium bromide. $RaBr_2$.
Properties: White crystals, becoming yellow or pink, radioactive. D 5.79, mp 728C, sublimes at 900C. Soluble in water and alcohol.
Derivation: Freed from the ores as a bromide mixed with barium bromide.
Method of purification: Fractional crystallization.
Impurities: Barium salts.
Grade: Technical, pure. The purity is determined by the strength of the ionizing power of the salt, i.e., the extent to which it causes air to conduct electricity.
Hazard: As for radium.
Use: Medicine (cancer treatment), physical research.

radium carbonate. $RaCO_3$.
Properties: Amorphous radioactive powder; white when pure, but sometimes yellow, orange, or pink due to impurities. Insoluble in water.
Available forms: Mixture with barium carbonate.
Hazard: As for radium.

radium chloride. $RaCl_2$.
Properties: Yellowish-white crystals becoming yellow or pink on standing. Radioactive. Mp 1000C, d 4.91. Soluble in water and alcohol.
Derivation: Freed from the ores as a chloride mixed with barium chloride.
Method of purification: Fractional crystallization.
Grade: Technical, pure. The purity of radium salts is determined by the strength of their ionizing power, i.e., the extent to which they cause air to conduct electricity.
Hazard: As for radium.
Use: Medicine (cancer treatment), physical research.

radium sulfate. $RaSO_4$.
Properties: White crystals when pure, but sometimes yellow, orange, or pink due to impurities. Radioactive. Insoluble in acids and water.
Hazard: As for radium.

radon.
CAS: 10043-92-2.   Rn.
Gaseous radioactive element. Atomic number 86; noble gas group of periodic table; aw 222; valences = 2, 4, (6); 18 radioactive isotopes, all short-lived. The radon-222 isotope has a half-life of 3.8 days, emits α-radiation.
Properties: Colorless gas. D 9.72 g/L (0C). Soluble in water. Can be condensed to a colorless, transparent liquid (bp −61.8C) and to an opaque, glowing solid. The heaviest gas known.
Derivation: Radioactive decay of radium. Radon is obtained by bubbling air through a radon salt solution and collecting the gas plus air.
Hazard: As for radium.
Use: Medicine (cancer treatment), tracer in leak detection, flow-rate measurement, radiography, chemical research.

raffinate.   The portion of an oil that is not dissolved in solvent refining of lubricating oil.

raffinose.
CAS: 512-69-6. $C_{18}H_{32}O_{16} \cdot 5H_2O$.
A trisaccharide composed of one molecule each of D(+)galactose, D(+)glucose, and D(−)fructose.
Properties: White, crystalline powder; sweet taste. D 1.465, mp (anhydrous) 118–119C, bp decomposes at about 130C, optical rotation +104.5 degrees. Soluble in water; very slightly soluble in alcohol. Split by invertase to melibiose and saccharose. Combustible.
Derivation: Hydrolysis of cottonseed meal, from sugar beet concentrates.
Use: Bacteriology, preparation of other saccharides.

rainfall, induced.   See nucleation.

RAM.   Combination of Rhodamine 6G, Ardox and MBD dyes, which fluoresce when exposed to selected wavelengths of light.
Use: To visualize cyanoacrylate fumed friction ridge detail.

Ramakrishnan, Venkatraman.   (1952–). An Indian-American physicist and molecular biologist who was a co-recipient of the 2009 Nobel Prize with Thomas A. Steitz and Ada E. Yonath. The scientists' research on atomic structure and function of the ribosome provided fundamental understanding of the two ribosomal subunits (small and large). Ramakrishnan made phenomenal discoveries on the small subunit of ribosome's RNA structure and organization, and mapped out how antibiotics bound to the small subunits. He was awarded his Ph.D. by Ohio University. His outstanding career includes awards such as the Jeantet Prize for Medicine (2007), and the Heatley Medal (2008).

Raman spectroscopy.   An analytical technique discovered in 1928 by C. V. Raman, an Indian physicist who received the Nobel Prize in Physics.
See spectroscopy; photometric analysis.

Ramberg–Bäcklund reaction.   Reaction of α-halo sulfones with strong bases to yield alkenes.

ramie.   A natural fiber obtained from the stems of *Boehmeria nivea*, of the hemp family. High wet strength, absorbent but dries quickly, can be spun or woven. Wears well and has great rot and mildew resistance, tensile strength four times that of flax, elasticity 50% greater than flax.
Source: Taiwan, Egypt, Brazil, Florida.
Hazard: Combustible, not self-extinguishing.
Use: High-grade paper (Europe), fabrics (weaving apparel and car seat covers), stern-tube packing in ships, patching water mains (Great Britain).

Ramrod. A selective preemergence herbicide available as a wettable powder (contains 65% 2-chlor-*N*-isopropylacetanilide) and granular form (contains 20% 2-chloro-*N*-isopropylacetanilide).

Ramsay, Sir William. (1852–1916). A British chemist born in Scotland who received the Nobel Prize for Chemistry in 1904. He participated in the discovery of helium, argon, neon, xenon, and krypton. Much of his work concerned investigations of inert gases. He was also known for studies in organic, physical, and inorganic chemistry. Ramsay was educated at the Universities of Glasgow, Heidelberg, and Tubingen, and was a Professor at the Universities of Bristol and London.

Ramsbottom coke test. A laboratory test for carbon residue in petroleum products.

random primed synthesis. If you have a DNA clone and you want to produce radioactive copies of it, one way is to denature it (separate the strands), then hybridize to that template a mixture of all possible 6-mer oligonucleotides. Those oligos will act as primers for the synthesis of labeled strands by DNA polymerase (in the presence of radiolabeled precursors).

Raney nickel.
Properties: Dark-gray powder or crystals, pyrophoric.
Derivation: By leaching the aluminum from an alloy of 50% aluminum-50% nickel with 25% caustic soda solution.
Hazard: Ignites spontaneous in air, store under alcohol or water, dangerous fire risk.
Use: Catalyst for hydrogenation.

ranitidine bismuth citrate.
CAS: 128345-62-0. $C_{13}H_{22}N_4O_3S \cdot C_6H_5BiO_7$.
Hazard: A poison by ingestion.

Rankine. A scale of absolute temperature based on Fahrenheit degrees. Temperatures on the Rankine scale are 9/5 (or 1.8) times those on the Kelvin scale.
See absolute temperature.

Raoult's law. The vapor pressure of a substance in equilibrium with a solution containing the substance is equal to the product of the mole fraction of the substance in the solution and the vapor pressure of the pure substance at the temperature of the solution. The law is not applicable to most solutions, but is often approximately applicable to a mixture of closely similar substances, particularly the substance present in high concentration.

rapeseed. See canola.

rapeseed meal. The ground press-cake left from expression of rapeseed oil. Its major use is as an animal feed ingredient, but the presence of harmful glucosides, which may be goiter-inducing, limits its application for this purpose. Research efforts are being directed to removal of this constituent from meal produced in Canada. The meal also has some use as a fertilizer. If fed to animals, it should be blended with other feeds.
Properties: Pale-yellow liquid. Soluble in chloroform and ether.

rapeseed oil. A vegetable oil derived from rapeseed by expression or solvent extraction; it is now produced chiefly in Canada. It is a viscous, brownish liquid, though when refined it is yellow. D 0.913–0.916, solidifies at 0C, flash p 325F (162C), autoign temp 836F (446C), subject to spontaneous heating. It is high in unsaturated acids, especially oleic, linoleic, and erucic.
Use: Edible oil for salad dressings, margarine, etc.; lubricant additive; substitute for soybean oil; soft soaps; blown oils.

Rapi-Cure. A diluent for radiation curing.

rare-cutter enzyme. See restriction-enzyme cutting site.

rare earth. One of a group of 15 chemically related elements in group IIIB of the periodic table (lanthanide series). Their names and atomic numbers are as follows:

lanthanum	La	57
cerium	Ce	58
praseodymium	Pr	59
neodymium	Nd	60
promethium	Pm	61
samarium	Sm	62
europium	Eu	63
gadolinium	Gd	64
terbium	Tb	65
dysprosium	Dy	66
holmium	Ho	67
erbium	Er	68
thulium	Tm	69
ytterbium	Yb	70
lutetium	Lu	71

The elements 57–62 are known as the cerium subgroup, and 63–71 as the yttrium subgroup. Yttrium, atomic number 39, although not a rare-earth element, is found associated with the rare earths and is separated only with difficulty.
Source: Monazite, bastnasite, and related fluocarbonate minerals as well as minerals of the yttrium group. These ores contain varying percentages of rare-earth oxides, which are often loosely called rare earths. Rare-earth elements also occur as fission products of uranium and plutonium.
Occurrence: (Monazite) India, Brazil, Florida, Carolinas, Australia, South Africa; (bastnasite) California.
See didymium.

rare gas. Any of the six gases composing the extreme right-hand group of the periodic table, namely helium, neon, argon, krypton, xenon, and radon. They are preferably called noble gases or (less accurately) inert gases. The first three have a valence of 0 and are truly inert, but the others can form compounds to a limited extent.

rare metal. A loose term for the less common metallic elements. They include the alkaline-earth metals (barium, calcium, and strontium), beryllium, bismuth, cadmium, cobalt, gallium, germanium, hafnium, indium, lithium, boron, silicon, manganese, molybdenum, rhenium, selenium, tantalum, niobium, tellurium, thallium, thorium, titanium, tungsten, uranium, vanadium, zirconium, and the rare earths.

"Rareox" [Grace]. TM for optical quality cerium oxide for high speed polishing.

Raschig phenol process. Commercial process for the production of phenol by the hydrolysis of chlorobenzene, produced by the chlorination of benzene with hydrochloric acid and air.

Raschig rings. Short sections of metal tubes.
Use: Packing in distillation towers.
See tower, distillation.

Rast method. The melting-point depression method often used for the determination of the molecular weight of organic compounds.

rate-limiting step. The slowest step in a metabolic pathway.

Raticate. 5-(α-hydroxy-α-2-pyridylbenzyl)-7-(α)-2-pyridylbenylidene-5-norbornene-2,3-dicarboximide. A rodenticide specific for rats, supposedly nontoxic to other animals or to humans (U.S. Department of Agriculture).

rauwolfia. The powdered whole root of *Rauwolfia serpentina*, found in India and Indonesia. The plant is of value as a source of alkaloids, especially reserpine.
Properties: Light-tan to light-brown, bitter, fine, amorphous powder; slight aromatic odor. Sparingly soluble in alcohol; very slightly soluble in water.
Grade: NF.
Use: Medicine (antihypertensive agent).

raw material. (1) The basic material from which one or more useful products are derived, e.g., bauxite is the raw material for aluminum, wood for paper, rayon, etc.; petroleum for fuels and chemicals. (2) As commonly used, the term refers to any ingredient or component of a mixture or product before mixing and processing take place, e.g., fillers, colorants, antioxidants.
See storage (1).

"Raykrome" [Rayonier]. (lignin sulfonate). TM for binding and dispersing emulsifiers.
Use: In concrete water reduction, oil-well drilling, dye dispersing, and road stabilization.

γ-ray. See gamma ray.

rayon. Generic name for a semisynthetic fiber composed of regenerated cellulose as well as manufactured fibers composed of regenerated cellulose in which substituents have replaced not more than 15% of the hydrogen of the hydroxyl groups. Rayon was first made by denitration of cellulose nitrate fibers (Chardonnet process), but most rayon is made from wood pulp by the viscose process. "Regular" viscose tenacity = 2 g/denier; "high-tenacity" = 3–6 g/denier (tire cord). Elongation 15–30% (dry) and 20–40% (wet). Swells and weakens when wet. Moisture regain 11–13%, d 1.50.
Modified rayon is made principally of regenerated cellulose and contains nonregenerated cellulose fiber-forming material, for example, a fiber spun from viscose containing casein or other protein (ASTM). This greatly increases both dry and wet strength and also permits mercerization. Rayon is readily dyed by standard methods.
Hazard: Flammable, not self-extinguishing, moderate fire risk.
Use: Nonwoven fabrics, surgical dressings, mechanical rubber goods, coated fabrics, felts and blankets, blends with cotton for home furnishings, etc.
See cellophane; acetate fiber; viscose process.

rayon coning oil. An oil used to lubricate and reduce the static of yarns wound by a coning machine. Usually composed of mineral oils of low viscosity so compounded as to emulsify in water.

"Rayox" [Calgon]. TM for titanium dioxide.

Rb. Symbol for rubidium.

RBR Revitalizer. A liquid naphtha solvent blend.

RCFs. See refractory ceramic fibers.

R&D. Abbreviation for research and development, usually referring to the department or division of a company whose major responsibility is applied research and creative development of new products and processes.

RDA. Abbreviation for recommended dietary allowances of food requirements, including proteins, vitamins and minerals for infants, children, and adults, established by the Food and Nutrition Board of the National Academy of Sciences—National Research Council. They are revised periodically, particularly in reference to certain vitamins (C, B_{12}, and E) and proteins.
See USRDA.

RDGE.   See resorcinol diglycidyl ether.

RDX.   See cyclonite.

Re.   Symbol for rhenium.

ReAct 784.   A two component, no mix, structural adhesive compound.

reaction, chemical.   See chemical reaction.

reaction, fast.   A chemical reaction that has a heat of activation far in excess of critical energy.

reaction, heterogeneous.   A chemical reaction that takes place on the surfaces between various phases in a heterogeneous system.

reaction injection molding.   See injection molding.

reaction, mechanism.   The manner in which a chemical reaction proceeds, expressed in a series of chemical equations.

reaction, nuclear.   See nuclear reaction.

reaction of first order.   A chemical reaction in which the velocity of reaction at a given time depends strictly on the amount per unit volume of the reacting substance at the time, all other parameters staying the same.

reaction of second order.   A chemical reaction in which the rate of reaction is in proportion to the product of the concentrations of two reacting materials.

reaction order.   The total of all the orders with respect to various substances is called the total order or the order of the reaction. Also when a reaction has a rate that is in proportion to some power of the concentration of one of the reacting substances involved.

reaction product imaging.   A beam of hydrogen atoms is crossed with a beam of cold deuterium (D) molecules. The ion images appearing on the detector are two-dimensional projections of the three-dimensional velocity distribution of the D atom products.

reaction step.   The reaction where individual parts of the reaction series are identical in nature with the primary reaction but differ from it only in the order that they occur.

reaction velocity.   The rate at which a chemical reaction takes place as measured by the rate of formation of the product or the rate of disappearance of reactants.

reactive atmosphere.   An atmosphere which causes a change in composition of the material immersed in it.

reading frame.   When mRNA is translated by the cell, the nucleotides are read three at a time. By starting at different positions, the groupings of three that are produced can be entirely different.

reagent.   Any substance used in a reaction for the purpose of detecting, measuring, examining, or analyzing other substances. High purity and high sensitivity are essential requirements of lab reagents. Over 8000 reagent chemicals are commercially available.
See grade.

real gas.   A non-ideal gas that approaches ideal behavior only at high temperature and/or low pressure. In order to bring the behavior of real gases into agreement with the ideal gas law, measurements upon are extrapolated to zero pressure.

Realox. Al_2O_3. Calcined aluminas.
Properties: High-purity, reactive alumina.
Use: In the production of high alumina ceramics and refractories.

rearrangement.   A type of chemical reaction in which the atoms of a single compound recombine, usually under the influence of a catalyst, to form a new compound having the same molecular weight but different properties. Thus ammonium cyanate in solution will rearrange to form urea, in which the four hydrogen atoms are equally distributed between the two nitrogen atoms: $NH_4OCN \rightarrow (NH_2)_2C{=}O$. Many such rearrangements have been named for their discoverers, e.g., Beckmann rearrangement.
See Wohler.

Reaumur.   A temperature scale in which 0 is the freezing point of water and 80 is its boiling point.

rebulose a.   (ribulose; 1,3,4,5-tetrahydroxypentan-2-one). $C_5H_{10}O_5$. A keto-pentose sugar that enters into the carbon dioxide fixation pathway of photosynthesis.

recessive gene.   A gene which will be expressed only if there are 2 identical copies or, for a male, if one copy is present on the X chromosome.

recipe.   A product formula.
Use: Food industry.

reciprocal translocation.   When a pair of chromosomes exchange exactly the same length and area of DNA. Results in a shuffling of genes.

reclaiming.   Recovery and reuse of scrap materials, either in low percentage in new product manufacture or in larger proportions in products in which the highest quality is not essential. Among the materials widely reclaimed in industry are aluminum, steel, paper, rubber, glass, crankcase oil,

greases, etc. Solid materials are comminuted, the contaminants being removed with organic solvents or strong alkali solutions (paint from metals, ink from paper, fabric and metals from tires). In the case of cross-linked elastomers, more intensive solvent and heat treatment is necessary. The resulting product is used as an adulterant in low-quality items. Research on high-temperature conversion of scrap rubber to oil, with recovery of carbon black and metal inserts, indicates that substantial value may be obtained by this method. Devulcanization by means of microwave radiation is also under development. The term *reprocessing* refers specifically to the recovery of nuclear fuels from reactor waste.
See recycling; reprocessing.

recombinant clone. Clone containing recombinant DNA molecules.
See recombinant DNA technology.

recombinant DNA. See genetic engineering; biotechnology.

recombinant DNA molecules. A combination of DNA molecules of different origin that are joined using recombinant DNA technologies.

recombinant DNA technology. Procedure used to join together DNA segments in a cell-free system (an environment outside a cell or organism). Under appropriate conditions, a recombinant DNA molecule can enter a cell and replicate there, either autonomously or after it has become integrated into a cellular chromosome.

recombination. The process by which progeny derive a combination of genes different from that of either parent. In higher organisms, this can occur by crossing over.
See crossing over; mutation.

reconstitution. In food technology, restoration of a dehydrated food product to its original edible condition by adding water to it at the time of use. It is also called *rehydration*.
See dehydration.

rectification. The enrichment or purification of the vapor during the distillation process by contact and interaction with a countercurrent stream of liquid condensed from the vapor.
See reflux.

RectorSeal. A NSF certified pipe repair kit.

recycling. The practice of returning a portion of the reaction products to the start of the system, either for the purpose of more efficient conversion of unreacted components or to reuse auxiliary materials that remain unchanged during processing. In the petroleum refining industry, some of the product stream may be recycled and blended with the fresh input materials to obtain a product of maximum value. In some industries, processing wastes are recycled.
See reclaiming; reprocessing.

red acetate. See aluminum diacetate.

red algae. See algae, red.

red arsenic. See arsenic disulfide.

red brass. Copper–zinc (brass) alloys characterized by their red color and high copper content. The term is used for several different types of brass. One ASTM classification permits 2–8% zinc, tin less than zinc, and lead less than 0.5%. Other alloys referred to as red brass include those with 75–85% copper, up to 20% zinc, and usually very small amounts of lead and tin. In one alloy possessing good machining qualities, the lead content may be as high as 10%, the tin as high as 5%.
Red brasses are widely used for decorative purposes and in plumbing and piping because of their resistance to atmospheric corrosion and dezincification.

Red Dye No. 2. See amaranth; FD&C colors.

red glass. A soda-zinc glass containing a small amount of cadmium and 1% selenium. Red may also be obtained by using cuprous oxide or gold chloride, the latter usually in the form of purple of Cassius.

red iron oxide. See iron oxide, red.

red iron trioxide. See ferric oxide.

red lake C. One of a family of organic acid azo pigments prepared by coupling the diazonium salt of *o*-chloro-*m*-toluidine-*p*-sulfonic acid with β-naphthol. Both sodium and barium salts of the parent dye are used.
Use: Plastics, rubber products, printing inks.

red lead. See lead oxide, red.

red mud. A by-product sludge from aluminum ore processing; it contains 30–60% iron oxide. It may be used for steel making.

red ocher. See ocher.

red oil. A commercial grade of oleic acid comprising 70% oleic and 15% each of linoleic and stearic acids.

redox. Short form of the term *oxidation–reduction*, as in *redox reactions*, *redox conditions*, etc.
See oxidation.

redox pair. An electron donor and its corresponding oxidized form; for example, NADH and NAD+.

redox reaction. An oxidation-reduction reaction.

red oxide. See iron oxide red.

red phosphorus. See phosphorus.

red tide. Yellow-to-reddish discoloration of seawater due to rapid multiplication of various species of plant-like microorganisms, called dinoflagellates, which occurs seasonally in areas of warm water, especially off the coast of Florida and occasionally as far north as New England. Some, though by no means all, of the species are poisonous. Concentration of these organisms (phytoplankton) may be as high as 10^8 units/L. The shellfish are lethal to free-swimming fish. Shellfish are unharmed by them, but are able to store and concentrate the toxin which causes paralytic poisoning when they are eaten by humans. So potent is this poison that death may result from ingestion of milligram amounts.

reduced states. See corresponding states.

reducing agent. (reductant). The electron donor in a redox reaction.

reducing end. The end of a polysaccharide having a terminal sugar with a free anomeric carbon; the terminal residue can act as a reducing sugar during the portion of time it exists in the open-chain form.

reducing equivalent. A general term for an electron or an electron equivalent, usually in the form of a hydrogen atom or a hydride ion.

reducing sugar. A sugar in which the anomeric carbon is not involved in a glycosidic bond and can therefore undergo oxidation during the portion of time it exists in the open-chain form.

reductant. See reducing agent.

reduction. (1) The opposite of oxidation. Reduction may occur in the following ways: (a) acceptance of one or more electrons by an atom or ion, (b) removal of oxygen from a compound, (c) addition of hydrogen to a compound.
(2) Size reduction of materials.
See (1) oxidation; (2) comminution.

"Reductone" *[Olin].* TM for sodium hydrosulfite.
CAS: 7775-14-6.
Grade: Solution.

Use: Continuous vat dyeing and afterscouring; bleaching of clay, ground wood, and thermal mechanical pulp.

Reed reaction. Photochemical sulfonation of paraffins and cycloparaffins by sulfur dioxide and chlorine under irradiation with UV light.

refined bleached shellac. See shellac, bleached, wax-free.

refined mineral oil. See dormant oil.

refinery gas. A mixture of hydrocarbon gases (and often some sulfur compounds) produced in large-scale cracking and refining of petroleum. The usual components are hydrogen, methane, ethane, propane, butanes, pentanes, ethylene, propylene, butenes, pentenes, and small amounts of other components such as butadiene.
Use: Source of raw material for petrochemicals, high-octane gasoline, and organic synthesis of alcohols.

refining. Essentially a separation process whereby undesirable components are removed from various types of mixture to give a concentrated and purified product. Such separation may be effected (1) mechanically, by pressing, centrifuging, filtering, etc.; (2) by electrolysis; (3) by distillation, solvent extraction, or evaporation; and (4) by chemical reaction. One or more of these operations is applied to (1) food products, (2) petroleum, (3) lubricating oils, and (4) metals. As regards petroleum, refining is generally understood to include not only fractional distillation of crude oils to naphthas, low-octane gasoline, kerosene, fuel oil, and asphaltic residues, but also the processes involved in thermal and catalytic cracking (hydroforming, reforming, etc.) for production of high-octane gasoline.

reflux. In distillation processes in which a fractionating column is used, the term *reflux* refers to the liquid that has condensed from the rising vapor and been allowed to flow back down the column toward the still. As it does so, it comes into intimate contact with the rising vapor, resulting in improved separation of the components. The separation resulting from contact of the countercurrent streams of vapor and liquid is called rectification or fractionation.

Reformatskii reaction. Condensation of carbonyl compounds with organozinc derivatives of α-halo esters to yield β-hydroxy esters.

reforming. Decomposition (cracking) of hydrocarbon gases or low-octane petroleum fractions by heat and pressure, either without a catalyst (thermoforming) or with a specific catalyst (molybdenum, platinum). The latter method is the more efficient and is used almost exclusively in the U.S. The chief cracking reactions are (1) dehydrogenation of cyclohexanes to aromatic hydrocarbons; (2) dehydrocyclization of certain paraffins to aromatics; (3) isomerization, i.e., conversion of straight-chain to branched-chain structures, as octane to isooctane. These result in substantial increase in octane number. Steam reforming of natural gas is an important method of producing hydrogen by the reaction $CH_4 + H_2O \rightarrow 3H_2 + CO$; steam reforming of naphtha is used to produce synthetic natural gas. See hydroforming; "Platforming."

Refractaloy. A nickel–cobalt–chromium–molybdenum–iron alloy. Type 26 is a precipitation-hardenable material using titanium as the hardening agent and having high strength, high ductility, and corrosion resistance up to 1450F.
Use: Gas turbine discs, bolting, and blading are typical applications.

refraction. The change in direction (apparent bending) of a light ray passing from one medium to another of different density, as from air to water or glass. The ratio of the sine of the angle of incidence to the sine of the angle of refraction is the index of refraction of the second medium. Index of refraction of a substance may also be expressed as the ratio of the velocity of light in a vacuum to its velocity in the substance. It varies with the wavelength of the incident light, temperature, and pressure. The usual light source is the D line of sodium, the standard temperature being 20C, the expression of refractive index is 20/D.

refractive. Descriptive of a substance having a high refractive index.

refractive index. See refraction.

refractometer. A device for checking the index of refraction.

refractory. (1) An earthy, ceramic material of low thermal conductivity that is capable of withstanding extremely high temperature (1650–2200C) without essential change. There are three broad groups of these: (a) acidic (silica, fireclay), (b) basic (magnesite, dolomite), and (c) amphoteric (alumina, carbon, and silicon carbide). Their primary use is for lining steel furnaces, coke ovens, glass lehrs, and other continuous high-temperature applications. They are normally cast in the form of brick and are sometimes bonded to assure stability. The outstanding property of these materials is their ability to act as insulators. The most important are fireclay (aluminum silicates), silica, high alumina (70–80% Al_2O_3), mullite (clay-sand), magnesite (chiefly MgO), dolomite (CaO–MgO), forsterite (MgO–sand), carbon, chrome ore-magnesite, zirconia, and silicon carbide.
(2) Characterizing the ability to withstand extremely high temperature, e.g., tungsten and tantalum are refractory metals, clay is a refractory earth, ceramics are refractory mixtures.

refractory ceramic fibers. (RCFs). Man-made insulating fibers.
Use: To replace asbestos in coke ovens and industrial furnaces.

"Refrasil" *[Hitco].* TM for a group of materials having outstanding high-temperature resistance. Composed of white, vitreous fibers having up to 99% silicon dioxide content.
Available forms: Bulk fiber, batt, cloth, tape, sleeving, yarn, cordage, and flakes.
Use: Aircraft and missile ablative composites; industrial insulation and filtration applications; protective welding shield for spark and slag containment.

"Refrax" *[Saint-Gobain].* TM for silicon nitride-bonded silicon carbide refractories. Available in brick and precision-formed shapes and parts.
Use: Brazing and furnace fixtures; pumps and pump parts handling corrosive, abrasive slurries; rocket motor components; spray nozzles; burners; pyrometer protection tubes; sinker assemblies in wire aluminizing; bolts and nuts; valve parts; aluminum-melting furnace linings and parts; conveyor parts.

refrigerant. Any substance that by undergoing a change of phase (solid to liquid or liquid to vapor) lowers the temperature of its environment because of its latent heat. Melting ice, with latent heat of 80 calories per gram removes heat and exerts a considerable cooling effect. Most commercial refrigerants are liquids whose latent heat of vaporization results in cooling. Ammonia, sulfur dioxide, and ethyl or methyl chloride were once widely used. The flammability and toxicity of these compounds led to a search for safer refrigerants that resulted in the discovery of halogenated hydrocarbons, especially fluorocarbons, which are nonflammable. Under various trademarks, these are now generally used for domestic refrigeration and air-conditioning. Ice and circulating brine are still used for preservation of fish at sea, and ammonia systems are operated for seafood storage in warehouses.

refrigerant 112a. See 1,1,2-tetrachloro-2,2-difluoroethane.

regenerated cellulose. See cellophane.

regeneration. (1) Restoration of a material to its original condition after it has undergone chemical modification necessary for manufacturing purposes. The most common instance is that of cellulose for rayon production. The wood pulp used must first be converted to a solution by reaction with sodium hydroxide and carbon disulfide; in this form (cellulose xanthate) it can be extruded through spinnerettes. After this operation it is regenerated to cellulose by passing it through acid (viscose process).

Collagen can also be regenerated by acid treatment after it has been purified for use in food products by alkaline solution.

(2) Renewal or reactivation of a catalyst that has accumulated reaction residues such as coke, usually accomplished by passage of steam or reducing gases over the catalyst bed.

(3) Replenishing the sodium ions of a zeolite or similar ion-exchange agent by treatment with sodium chloride solution. Molecular sieves are regenerated by heat removal of the water (200C), followed by treatment with an inert gas.

regulatory enzyme. An enzyme which can be regulated by allosteric mechanisms or by covalent modification.

regulatory gene. A gene that gives rise to a product involved in the regulation of the expression of another gene. An example would be the gene coding for cro, a repressor protein.

regulatory region or sequence. A DNA base sequence that controls gene expression.

regulatory sequence. A DNA sequence involved in regulating the expression of a gene.

regulon. A group of genes that are coordinately regulated.

rehydration. See reconstitution.

Reichert–Meissl number. A measure of volatile, soluble fatty acids derived under arbitrary conditions.

Reich process. A method of purifying carbon dioxide produced in fermentation. The small amounts of organic impurities are oxidized and absorbed, and the gas is then dehydrated with chemicals.

Reimer–Tiemann reaction. Reaction for the formation of phenolic aldehydes by heating a phenol with chloroform in the presence of alkali.

reineckate. Tetrathiocyanato diammino chromium compound.

Reinecke salt. (ammonium tetrathiocyanodiammonochromate; ammonium reineckate).
CAS: 19441-09-9. $NH_4[Cr(NH_3)_2(SCN)_4] \cdot H_2O$.
Properties: Dark-red crystals or crystalline powder. Moderately soluble in cold water; soluble in hot water and alcohol, decomposes in aqueous solution.
Derivation: From fusion of ammonium thiocyanate with ammonium dichromate.
Use: Precipitating agent for organic bases such as choline, amines, for certain amino acids; reagent for mercury.

reinforced plastic. A composite structure comprising of a thermosetting or thermoplastic resin and fibers, filaments, or whiskers of glass, metal, boron, or aluminum silicate. Unless otherwise indicated, this term refers to fiberglass-reinforced plastic (FRP).
Properties: Exceptionally high strength, good electrical resistivity, weather and corrosion resistance, low thermal conductivity, and low flammability.
Derivation: Basic acid glass in the form of fiber (0.0005 inch), strands of 50–200 fibers, filaments, or woven fabric are coated by passing through a bath of molten resin, which acts as a binder. The assembly can be either compression- or injection-molded. Large, high-strength parts are made by filament winding. Resins used are polyester, epoxy, phenolic, polypropylene, polystyrene, nylon, polycarbonate, and polyphenylene oxides.
Use: Automotive body components, ablative coatings on rockets and space vehicles, appliances (air-conditioning and refrigerator cases), electrical equipment, oil-well piping and tubing, large-diameter pipe, industrial piping systems, chemical storage and mixture tanks, unitized cargo containers, marine equipment, pressure vessels, prefabricated building panels and other structural components, blades for wind machines.
See glass fiber; composite; fiber; whiskers.

reinforcing agent. (1) One of numerous fine powders used in rather high percentages to increase the strength, hardness, and abrasion resistance of rubber, plastics, and flooring compositions. The reinforcing effect is in general a function of the particle size of the powder. The finest of all is channel carbon black, whose surface area may be as great as 18 acres per pound. Other widely used reinforcing agents are thermal and furnace blacks, magnesium carbonate, zinc oxide, hard clay (kaolin), and hydrated silicas. Though some reinforcing agents have positive coloring properties, the term should not be used as a general synonym for pigment.
(2) Fibers, fabric, or metal insertions in plastics, rubber, flooring, etc., for the purpose of imparting impact strength and tear resistance.
See pigment; filler; whiskers.

Reinsch test. A test for detecting small amounts of arsenic, silver, bismuth, and mercury.

Reissert indole synthesis. Condensation of an *o*-nitrotoluene with oxalic ester, reduction to the amine, and cyclization to the indole.

Reissert reaction. Formation of 1-acyl-2-cyano-1,2-dihydroquinoline derivatives (Reissert compounds) by reaction of acid chlorides with quinoline and potassium cyanide; hydrolysis of these compounds yields aldehydes and quinaldic acid.

relative error. Ratio of absolute error to exact value.

relaxation time. A measure of the rate at which a disequilibrium distribution decays toward an equilibrium distribution. The electron relaxation time in a metal, for example, describes the time required for a disequilibrium distribution of electron momenta (e.g., in a flowing current) to decay toward equilibrium in the absence of an ongoing driving force and can be interpreted as the mean time between scattering events for a given electron.

relaxin.
CAS: 9002-69-1.
Hazard: A poison. A reproductive hazard.
Use: Hormone.

release. (1) Separation of a cured or baked product from the metal mold or pan in which it is formed. Common release agents for rubber and plastics are waxy or fatty materials such as paraffin and tallow; vegetable oils are used in the baking industry. Such materials are collectively called adherents.
(2) Gradual diffusion of an active ingredient through a permeable or soluble coating. Pelleted products such as fertilizers and medicinals are often covered with a layer of a substance that permits slow and uniform escape of the active principle. Sulfur is used to coat controlled-release fertilizers; gelatin and similar materials serve the same purpose in pharmaceuticals.

release factors. See termination factors.

releasing factors. Hypothalamic hormones that stimulate release of other hormones by the pituitary gland. For example, Luteinizing Hormone Releasing Hormone (LHRH).

rem. The unit of radiation dose equivalent, the dosage in rads multiplied by a factor representing the different biological effects of various types of radiation. The Federal radiation safety standard is 0.5 rem per person per year for non-occupational exposure, and even this is considered too high by some authorities. Occupational exposure is set at 5 rem per year.
See rad; dosimetry; radiation.

Remsen, Ira. (1846–1927). An American chemist born in New York. He began his career in medical practice but abandoned it for chemistry and went to Germany to study. He received his doctorate in chemistry from the University of Gottingen in 1870. Returning to the U.S. he taught physics and chemistry at Williams and was later invited to join the staff of Johns Hopkins University where he became head of the department of chemistry. There he established the first graduate curriculum in chemistry in the U.S. based on the system then in use in Germany. In 1879, saccharin was discovered in his research laboratory. He wrote several widely used textbooks and founded the *American Journal of Chemistry*, which later merged with *JACS*. He was President of Johns Hopkins from 1901 to 1912, and is recognized as one of the great teachers of chemistry.

renaturation. Refolding of a denatured protein so as to restore native structure and protein function.

rendering. The procedure that separates fats from protein connective tissue and other water-insoluble materials by treating small pieces of animal matter with hot water or steam.

renewable resources. See biomass; waste control; gasohol.

"Renite" [Renite]. TM for a series of high-temperature lubricants, swabbing compounds, release agents, and automatic spray equipment for hot-forming glass, hot-working metals, and the lubrication of oven conveyer chains and bearings.
Use: Glassware products, forging, extrusions and die castings; provides lubrication for conveyer chains and bearings in bakery and brick industries.

rennet. See rennin.

rennin. (rennase; chymosin).
CAS: 9001-98-3. A digestive enzyme secreted by the glands of the stomach and causes curdling of milk. It has the power of coagulating 25,000 times its own weight of milk.
Properties: Yellowish-white powder or as yellow grains or scales; characteristic slightly salty taste; peculiar odor. Slightly hygroscopic; partially soluble in water and dilute alcohol.
Derivation: From the glandular layer (inner lining) of the true stomach of the calf. It has been made experimentally by gene-splicing techniques.
Grade: Rennet is the dried commercial extract containing the rennin.
Use: Pharmacy, cheese making, coagulation of casein for plastics, food additive.

repeat sequences. The length of a nucleotide sequence that is repeated in a tandem cluster.

repellent. (1) A substance that causes an insect or animal to turn away from it or reject it as food. Repellents may be in the form of gases (olfactory), liquids, or solids (gustatory). Standard repellents for mosquitos, ticks, etc., are citronella oil, dimethyl phthalate, *n*-butylmesityl oxide oxalate, DEET, and 2-ethyl hexanediol-1,3. Actidione is the most effective rodent repellent, but is too toxic and too costly to use. Thiuram disulfide, amino complexes with trinitrobenzene, and hexachlorophene are successfully used. Copper naphthenate and lime/sulfur mixtures protect vegetation against rabbits and deer. Shark repellents are copper acetate or

formic acid mixed with ground asbestos. Bird repellents are chiefly based on taste, but this sense varies widely with the type of bird so that generalization is impossible. a-Naphthol, naphthalene, sandalwood oil, quinine, and ammonium compounds have been used, with no uniformity of result.
See fumigant.
(2) A substance that, because of its physicochemical nature, will not mix or blend with another substance. All hydrophobic materials have water-repellent properties due largely to differences in surface tension or electric charges, e.g., oils, fats, waxes, and certain types of plastics. Silicone resin coatings can keep water from penetrating masonry by lining the pores, not by filling them; they will not exclude water under pressure.

repetitive DNA.   Sequences of varying lengths that occur in multiple copies in the genome; it represents much of the human genome.

replacement.   See substitution.

replication.   (1) Making a reverse image of a surface by means of an impression on or in a receptive material; usually applied to microscopic techniques for obtaining plastic replicas of observed objects.
(2) In biochemistry, the term refers to reproduction of the DNA molecule, which is composed of two interlocking chains of nucleotides (the double helix structure elucidated by Watson and Crick in 1953). It reproduces itself by forming two identical daughter molecules, each of which receives one of the two chains of the original molecule, the other in each case being synthesized from nucleic acids by enzymes (DNA polymerases). In the oversimplified drawing below the solid lines are the strands of the original molecule, and the broken lines are the synthesized strands:

parent DNA   daughters

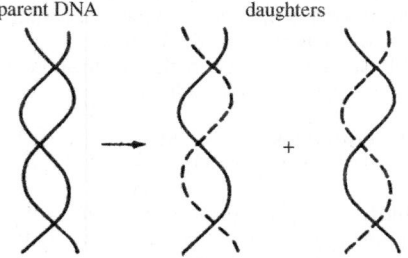

The chemical mechanisms in replication are more complex than previously thought; 12–15 proteins are involved as well as several enzymes.
See deoxyribonucleic acid; genetic code.

replisome.   The multiprotein complex that promotes DNA synthesis at the replication fork.

reporter gene.   See marker.

Reppe process.   Any of several processes involving reaction of acetylene (1) with formaldehyde to produce 2-butyne-1,4-diol which can be converted to butadiene; (2) with formaldehyde under different conditions to produce propargyl alcohol and, from this, allyl alcohol; (3) with hydrogen cyanide to yield acrylonitrile; (4) with alcohols to give vinyl ethers; (5) with amines or phenols to give vinyl derivatives; (6) with carbon monoxide and alcohols to give esters of acrylic acid; (7) by polymerization to produce cyclooctatetraene; and (8) with phenols to make resins. The use of catalysts, pressures up to 30 atm, and special techniques to avoid or contain explosions are important factors in these processes.
See acetylene.

repressible enzyme.   An enzyme whose synthesis is inhibited when its reaction product is readily available to the cell. Occurs only in bacteria.

repression.   A decrease in the expression of a gene in response to a regulatory protein.

repressor.   The protein that binds to the regulatory sequence of a gene, blocking its transcription.

reprocessing.   Treatment of spent nuclear reactor fuel to recover the unconsumed uranium-235 and plutonium by separating them from each other and from the fission products formed in the reactor. The Purex process is the accepted procedure used for this purpose. The spent fuel is dissolved in nitric acid; separation is effected by solvent extraction with tributyl phosphate, ion-exchange reactions, and precipitation. The reclaimed uranium-235 and plutonium are sent to fuel fabrication plants for reuse. The fission product waste is evaporated and stored. Another method, called the Civex process, has been proposed to prevent theft of plutonium; here the mixture of waste products, uranium isotopes, and plutonium is not separated. Since its plutonium content is only 20% it could not be used for weapons; the mixture is suitable for fast breeder reactors. Serious radiation hazards are involved in reprocessing and require use of appropriate shielding and remote-control handling procedures. Storage of the radioactive waste also presents a long-range problem that has not yet been satisfactorily solved. There are no commercial reprocessing plants operational in the U.S., though there are several in Europe.
See radioactive waste; breeder.

reprography.   A coined name for the technique of reproducing drawings, blueprints, and typographic matter by the use of photosensitized papers, or polyester sheeting, which may be coated with diazo dyes. The process has broad potential in the photocopying field and in communications technology. It involves colloid and surface chemistry, ink-paper interactions, and unusual imaging techniques.

Research octane number (RON). See octane number.

resene. The unsaponifiable component of rosin and other natural resins.

reserpine.
CAS: 50-55-5. $C_{33}H_{40}N_2O_9$.
An alkaloid.

Properties: White or pale buff to slightly yellowish; odorless powder; darkens slowly on exposure to light and darkens more rapidly in solution. Mp 264–265C (decomposes). Insoluble in water; slightly soluble in alcohol; soluble in chloroform and benzene.
Derivation: From Rauwolfia serpentina.
Grade: USP.
Hazard: Questionable carcinogen.
Use: Antihypertensive agent, tranquilizer.

Resicure. A series of epoxy curing agents.

residual oil. A liquid or semiliquid product resulting from the distillation of petroleum and containing largely asphaltic hydrocarbons. Also known as asphaltum oil, liquid asphalt, black oil, petroleum tailings, and residuum. Combustible.
Use: Roofing compounds, hot-melt adhesives, friction tape, sealants, heating oil for large buildings, factories, etc.
See fuel oil.
Note: Gasoline of 94 octane can be produced from residual oil in a high-temperature catalytic process, thus increasing the yield of gasoline from a barrel of crude by 33% when full-scale production is achieved.

residue. A single unit within a polymer, for example, an amino acid within a polypeptide chain.

"Resimene" *[INEOS].* TM for melamine and ureaformaldehyde resins. Supplied in organic liquid solutions. The melamine is also available in water-alcohol and soluble, spray-dry powders.
Use: Paint, varnish, lacquer for automobiles, machinery, appliances, construction, electronics, missiles, chemicals, pulp and paper.

resin. A semisolid or solid complex amorphous mix of organic compounds.
Properties: It has no definite melting point and no tendency to crystallize.

Derivation: Resins can be from animal, vegetable, or synthetic origins.

resinamine. $C_{35}H_{42}N_2O_9$. Alkaloid from certain species of *Rauwolfia.*
Properties: White or pale-buff to cream-colored, crystalline powder; odorless; darkens slowly on exposure to light, more rapidly when in solution. Mp 238C (in vacuo). Partially soluble in organic solvents; insoluble in water.
Use: Medicine (antihypertensive).

resinate. A salt of the resin acids found in rosin. They are mixtures rather than pure compounds.
Use: See soap (2).

resiniferatoxin.
CAS: 57444-62-9. $C_{37}H_{40}O_9$.
Hazard: A poison.

resin, ion-exchange. See ion-exchange resin.

resin, liquid. An organic, polymeric liquid that, when converted to its final state for use, becomes solid (ASTM). Example: linseed oil, raw or heat-bodied (partially polymerized).
See drying oil; resinoid.

resin, natural. (1) Vegetable-derived, amorphous mixture of carboxylic acids, essential oils, and terpenes occurring as exudations on the bark of many varieties of trees and shrubs. They are combustible, electrically nonconductive, hard and glassy with conchoidal fracture when cold, and soft and sticky below the glass transition point. Most are soluble in alcohols, ethers and carbon disulfide, and insoluble in water. The best known of these are rosin and balsam, obtained from coniferous trees; these have a high acid content. Of more remote origin are such resins as kauri, congo, dammar, mastic, sandrac, and copal. Their use in varnishes, adhesives, and printing inks is still considerable, though diminishing in favor of synthetic products. (2) Miscellaneous types: shellac, obtained from the secretion of an Indian insect, is still in general use as a transparent coating; amber is a hard, polymerized resin that occurs as a fossil; ester gum is a modified rosin; amorphous sulfur is considered an inorganic natural resin; liquid resins, sometimes called resinoids, are represented by linseed and similar drying oils.
See gum, natural (Note); resin, synthetic (Note).

resinoid. Any thermosetting synthetic resin, either in its initial temporarily fusible state or its final infusible state (ASTM). Heat-bodied linseed oil, partially condensed phenol-formaldehyde and the like are also considered resinoids.

resinol. A coal tar distillation fraction containing phenols. It is the fraction soluble in benzene but insoluble in light petroleum, obtained by solvent extraction of low-temperature tars or similar materials. Resinols are very sensitive to heat and oxidation.

resin, synthetic.   A manufactured high polymer resulting from a chemical reaction between two (or more) substances, usually with heat or a catalyst. This definition includes synthetic rubbers and silicones (elastomers), but excludes modified, water-soluble polymers (often called resins). Distinction should be made between a synthetic resin and a plastic; the former is the polymer itself, whereas the latter is the polymer plus such additives as filters, colorant, plasticizers, etc.
The first truly synthetic resin was developed by Baekeland in 1911 (phenol-formaldehyde). This was soon followed by a petroleum-derived product called coumarone-indene, which did indeed have the properties of a resin. The first synthetic elastomer was polychloroprene (1931) originated by Nieuwland and later called neoprene. Since then many new types of synthetic polymers have been synthesized, perhaps the most sophisticated of which are nylon and its congeners (polyamides, by Carothers), and the inorganic silicone group (Kipping). Other important types are alkyds, acrylics, aminoplasts, polyvinyl halides, polyester, epoxies, and polyolefins.
In addition to their many applications in plastics, textiles, and paints, special types of synthetic resins are useful as ion-exchange media.
See "Cumar" *[Neville]*; plastic; paint; fiber; film; elastomer.
Note: Because the term *resin* is so broadly used as to be almost meaningless, it would be desirable to restrict its application to natural, organo-soluble, hydrocarbon-based products derived from trees and shrubs. But in view of the tendency of inappropriate terminology to "gel" irreversibly, it is a losing battle to attempt to replace "synthetic resin" with the more precise "synthetic polymer."
See gum, natural (Note).

resist.   A material that will prevent the fixation of dye on a fiber, thus making color designs and pattern prints possible. The resist may act mechanically, as a wax, resin, or gel that prevents absorption of the dye or its accompanying mordant. Citric acid, oxalic acid, and various alkalies are among the more common resists of the chemical type.

resistor composition.   A specially treated, semiconducting, metal powder compounded with glass binders and temporary organic carriers. Can be applied to glass or ceramic surfaces by stenciling, spraying, brushing, or dipping; firing range 704–760C. Compositions can be blended with members of the same series to produce intermediate resistance values. Fired resistors have good reproducibility, low temperature and voltage coefficients, and stability to abrasion, moisture, and relatively high (125C) ambient temperature.

Use: To produce fired-on resistor components for electronic circuits.

resite.   See C-stage resin.

resitol.   See B-stage resin.

Resmetal.   A resin–metal composition that when catalyzed converts to metal-like solid. Recommended for mold making, patching, forming, and general repair of metal surfaces and objects.

resol.   See A-stage resin.

resolution.   Degree of molecular detail on a physical map of DNA, ranging from low to high.
See resolving power.

resolving power.   The extent to which a lens can distinguish small particles and minute distances, i.e., fine structure. The human eye can resolve objects of 1/250th inch (100 μm) in any dimension. The compound microscope has a resolving power of 0.5 μm; an electron microscope can resolve fine structure as small as 5 Å, that is, in the molecular range. Two factors determine resolving power, the wavelength of the radiation utilized and the focal depth of the lens. The resolving power of a microscope is much more important than its ability to magnify, for no magnification, however large, can add detail to an image that was not first discerned by the lens system.
See optical microscope; electron microscope.

resonance.   (1) In chemistry, resonance (or mesomerism) is a mathematical concept based on quantum mechanical considerations (i.e., the wave functions of electrons); it is used to describe or express the true chemical structure of certain compounds that cannot be accurately represented by any one valence-bond structure. It was originally applied to aromatic compounds such as benzene, for which there are many possible approximate structures, none of which is completely satisfactory.
See benzene.
The resonance concept indicates that the actual molecular structure lies somewhere between these various approximations, but is not capable of objective representation. This idea can be applied to any molecule, organic or inorganic, in which an electron pair bond is present. The term *resonance hybrid* denotes a molecule that has this property. Such molecules do not vibrate back and forth between two or more structures, nor are they isotopes or mixtures; the resonance phenomenon is rather an idealized expression of an actual molecule that cannot be accurately pictured by any graphic device.
(2) In the terminology of spectroscopy, resonance is the condition in which the energy state of the incident radiation is identical with that of the absorbing

atoms, molecules or other chemical entities. Resonance is applied in various types of instrumental analysis such as nuclear resonance absorption and nuclear magnetic resonance.

See absorption spectroscopy.

Note: The multiple meanings of *nucleus* and *resonance* can be a source of confusion, especially when these terms are closely associated, as in *nuclear magnetic resonance* and *resonance of a molecular nucleus*. In the first of these expressions, *nucleus* is used in the sense of (1) under "nucleus", and *resonance* in sense of (2) under "resonance". In the second expression, *nucleus* is used in the sense of (3) under "nucleus" and *resonance* in the sense of (1) under *resonance*.

resorcinol. (resorcin; *m*-dihydroxybenzene; 3-hydroxyphenol).
CAS: 108-46-3. $C_6H_4(OH)_2$.

Properties: White crystals, becoming pink on exposure to light when not perfectly pure; sweet taste. D 1.2717, mp 110.7, bp 281C, flash p 261F (127C), autoign temp 1126F (607C). Soluble in water, alcohol, ether, glycerol, benzene, and amyl alcohol; slightly soluble in chloroform. Combustible.
Derivation: By fusing benzene-*m*-disulfonic acid with sodium hydroxide.
Grade: USP, powder, resublimed, pure, reagent, technical, crude.
Hazard: Irritant to skin and eyes. Questionable carcinogen.
Use: Resorcinol-formaldehyde resins, dyes, pharmaceuticals, cross-linking agent for neoprene, rubber tackifier, adhesives for wood veneers and rubber-to-textile composites, manufacture of styphnic acid, cosmetics.

resorcinol acetate. (resorcinol monoacetate).
CAS: 102-29-4. $HOC_6H_4OCOCH_3$.
Properties: Viscous, pale-yellow or amber liquid; faint odor; burning taste. Bp 283C (decomposes), boiling range (10 mm Hg) 150–153C, d 1.203–1.207. Saturated solution is acid to litmus; soluble in alcohol and most organic solvents; sparingly soluble in water. Combustible.
Derivation: Action of acetic anhydride on resorcinol.
Grade: CP, NF.
Use: Medicine, cosmetics.

resorcinol bis(diphenyl phosphate).
CAS: 57583-54-7. $C_{30}H_{24}O_8P_2$.
Hazard: Low toxicity by ingestion, inhalation, and skin contact.

resorcinol blue. See lacmoid.

resorcinol dicyanate.
CAS: 1129-88-0. $C_8H_4N_2O_2$.
Hazard: Low toxicity by ingestion and skin contact. A mild skin irritant.

resorcinol diglycidyl ether. (RDGE; 1,3-diglycidyloxybenzene; *m*-bis(2,3-epoxypropoxy-benzene).
CAS: 101-90-6.

$$C_6H_4(OCH_2CHOCH_2)_2.$$

Properties: Straw-yellow liquid. D 1.21 (25C), bp 172C (0.8 mm Hg), refr index 1.541 (25C), viscosity 500 cP (25C), flash p 350F (176C) (COC). Miscible with most organic resins. Combustible.
Use: Epoxy resins.

resorcinol dimethyl ether. (dimethyl resorcinol; 1,3-dimethoxy benzene).
CAS: 151-10-0. $C_6H_4(OCH_3)_2$.
Properties: Pale-straw-tint liquid. Bp 204–212C, d 1.063–1.066 (25/25C), refr index 1.523–1.527 (20C). Combustible.
Use: Organic intermediate, flavoring.

resorcinol-formaldehyde resin. A type of phenol-formaldehyde resin. Permanently fusible; soluble in water, ketones, and alcohols. By dissolving and adjusting the pH to 7, an adhesive base is formed. These adhesives can be used whenever phenolics are used and where fast or room-temperature cure is desired. An important use of these adhesives is in wood gluing, particularly marine plywood.

resorcinol monobenzoate.
CAS: 136-36-7. $C_6H_5COOC_6H_4OH$.
Properties: White, crystalline solid. Mp 132–135C, bp 140C (0.15 mm Hg). Insoluble in water, benzene, di-(2-ethylhexyl)phthalate; soluble in acetone and ethanol. Combustible.
Use: Noncoloring UV inhibitor for various plastics, color stabilizer in cosmetic compositions.

resorcinolphthalein. See fluorescein.

resorcinolphthalein sodium. See uranine.

α-resorcylic acid. (3,5-dihydroxybenzoic acid).
CAS: 99-10-5. $(OH)_2C_6H_3COOH$.
Properties: White crystals. Mp 237C. Soluble in water, ethanol, and ether. Combustible.
Grade: CP.
Use: Intermediate for dyes, pharmaceuticals, light stabilizers, resins.

β-resorcylic acid. (BRA; 2,4-dihydroxybenzene carboxylic acid; 2,4-dihydroxybenzoic acid; 4-hydroxysalicylic acid; 4-carboxyresorcinol).
CAS: 89-86-1. $(OH)_2C_6H_3COOH$.

Properties: White needles. Mp (decomposes) 219–220C, bp decomposes. Almost insoluble in water and benzene; soluble in alcohol, ethyl ether. The sodium, potassium, ammonium, calcium, and barium salts are soluble in water; the silver, lead and copper salts are only slightly soluble. Combustible.
Use: Dyestuff and pharmaceutical intermediate, chemical intermediate in synthesis of fine organic chemicals, reagent for iron.

"Resorsabond" *[Georgia Pacific].* TM for lumber resins.
Available forms: Liquid, powder and slurry.
Use: For lumber laminating and end jointing applications.

respirable dust. (respirable particles). Airborne particulates that are deposited to a greater or lesser in the lungs. The size range of respirable particles varies with the depth and rate of breathing and probably with any of a number of individual and taxonspecific factors.

respiration. (1) In humans and animals, inhalation of oxygen and exhalation of carbon dioxide; the oxygen supports the oxidation (combustion) of organic nutrients in the body, yielding energy, carbon dioxide, and water. (2) In growing plants, respiration occurs in both the presence and the absence of light. Some of the energy produced in respiration is used to form adenosine triphosphate, pyruvic acid, and other metabolic intermediates. Fruits and vegetables continue to respire after harvest, a fact that must be taken into account in transportation and storage.

respiratory chain. The sequence of electroncarrying proteins that transfer electrons from substrates to molecular oxygen in aerobic cells.

response element. By definition, a response element is a portion of a gene which must be present in order for that gene to respond to some hormone or other stimulus. Response elements are binding sites for transcription factors. Certain transcription factors are activated by stimuli such as hormones or heat shock. A gene may respond to the presence of that hormone because the gene has in its promoter region a binding site for hormone-activated transcription factor.

restriction. To "restrict" DNA means to cut it with a restriction enzyme.
See restriction enzyme.

restriction endonucleases. Site-specific endonucleases causing cleavage of both strands of DNA at points within or near the specific site recognized by the enzyme. Many generate sticky ends, making them important tools in genetic engineering.
See restriction enzyme.

restriction enzyme. (restriction endonucleases). A class of enzymes generally isolated from bacteria, which are able to recognize and cut specific sequences (restriction sites) in DNA. Every copy of a plasmid is identical in sequence, so if an enzyme cuts a particular circular plasmid at three sites producing three "restriction fragments", then a million copies of that plasmid will produce those same restriction fragments a million times over. There are more than six hundred known restriction enzymes. Bacteria produce restriction enzymes for protection against invasion by foreign DNA such as phages. The bacteria's own DNA is modified in such a way as to prevent it from being clipped.

restriction-enzyme cutting site. A specific nucleotide sequence of DNA at which a particular restriction enzyme cuts the DNA. Some sites occur frequently in DNA (e.g., every several hundred base pairs); others much less frequently (rare-cutter; e.g., every 10,000 base pairs).

restriction enzyme, endonuclease. A protein that recognizes specific, short nucleotide sequences and cuts DNA at those sites. Bacteria contain over 400 such enzymes that recognize and cut more than 100 different DNA sequences.
See restriction enzyme cutting site.

restriction fragment. (1) The piece of DNA released after restriction digestion of plasmids or genomic DNA. See "Restriction enzyme". One can digest a plasmid and isolate one particular restriction fragment (actually a set of identical fragments). (2) Also describes the fragments detected on a genomic blot which carry the gene of interest.

restriction fragment length polymorphism. (RFLP). Variation between individuals in DNA fragment sizes cut by specific restriction enzymes; polymorphic sequences that result in RFLPs are used as markers on both physical maps and genetic linkage maps. RFLPs usually are caused by mutation at a cutting site.
See marker; polymorphism.

restriction map. A "cartoon" depiction of the locations within a stretch of known DNA where restriction enzymes will cut. The map usually indicates the approximate length of the entire piece (scale on the bottom), as well as the position within the piece at which designated enzymes will cut.

restriction site. See Restriction enzyme.

Reswax. A series of wax-resin blends used as coatings and hot-melt adhesives in paper conversion. The polymers used include butyl rubber, polyisobutylene, chlorinated rubber, polyethylene, and styrene copolymers.

ret. To reduce or digest fibers, especially linen, by enzymatic action.

retarder. See inhibitor.

"Reten" 205 [Hercules]. TM for a strongly cationic, high molecular weight, synthetic, water-soluble polymer. A finely divided white powder, dissolves in either hot or cold water to produce clear, smooth, viscous, nonthixotropic solutions; available in a variety of viscosity grades and cationic functionality. "Reten" 763 [Hercules] is an aqueous solution of a modified polyamide-epichlorohydrin resin.
Use: Flocculant, binder, and viscosifier.

retene. (7-isopropyl-1-methylphenanthrene). $C_{18}H_{18}$.
Properties: Mw 234.36.

retention index. (RI). An indication whereby a compound will appear on a chromatogram with respect to *n*-paraffins.

reticulation. Joining of separate lineages on a phylogenetic tree, generally through hybridization or through lateral gene transfer. Fairly common in certain land plant clades; reticulation is thought to be rare among metazoans.

reticulopodia. Long thread-like pseudopodia that branch apart and rejoin, forming a fine network. They are characteristic of forams.

retinal. Preferred name for retinene.

retinene. (vitamin A aldehyde; retinal). $C_{20}H_{28}O$. A necessary component of rhodopsin, the light-sensitive pigment of the eye. Retinene is the aldehyde form of vitamin A, which is an alcohol.

retinol.
CAS: 68-26-8. $C_{20}H_{29}OH$.
(1) A component of vitamin A.
See carotene.
(2) A resin distillate similar to rosin oil.

retort. A utensil for distilling volatile materials. A globular glass vessel with an extended side tube used for distillation.

retort gas. The gas that results from the heating of coal in retorts, as in the by-product process of coke manufacture.

retorting. (1) A process much used in the early years of chemistry for destructive distillation of heavy organic liquids and for laboratory separations. It involves the use of a cylindrical vessel made of glass (for laboratory work), fireclay, or metal, with a neck bent at a downward angle to facilitate distillation. For gas manufacture the equipment is built on a heavier scale to handle destructive distillation of coal.
(2) Processing shale oil.
(3) Heating canned or pouched foods with steam to stop bacterial growth.
(4) Volatilization of mercury from gold and silver amalgams.

retro Diels–Alder reaction. Thermal dissociation of Diels–Alder adducts, occurring most readily when one or both fragments are particularly stable.

retropinacol rearrangement. Conversion of an alcohol to the rearranged olefin on treatment with acid.

4,14-retro-retinol-14-hydroxy-, (14r)-.
CAS: 139257-77-5. $C_{20}H_{30}O_2$.
Hazard: A reproductive hazard.

***trans*-retrorsine.**
CAS: 15503-87-4. $C_{18}H_{25}NO_6$.
Hazard: A poison.
Source: Natural product.

retrosynthesis. A computer-assisted analysis of an organic molecule that is to be synthesized, i.e., the target molecule, in which the computer works back through the precursors of the target substance to a group of possible starting materials that are readily available from natural sources or as commercial products. Retrosynthetic analysis is thus the opposite of the usual direct approach to laboratory synthesis.
See computational chemistry.

retroviral infection. The presence of retroviral vectors, such as some viruses, which use their recombinant DNA to insert their genetic material into the chromosomes of the host's cells. The virus is then propogated by the host cell.

retrovirus. An RNA virus containing reverse transcriptase.

reumycin. See 1-demethyltoxoflavine.

reverberatory furnace. An ore-roasting kiln having a curved or sloping roof from which the heat is deflected onto the material being treated. The fuel and the charge occupy separate areas in the kiln so that there is no direct contact between them, thus avoiding contamination of the ore with fuel particulates. The heat rising from the ignited fuel impinges on the curved roof and is reflected downward onto the ore. After passing over the ore, the heat escapes through suitably located vents.

Reverdin reaction. Migration of iodine during nitration of iodophenolic ethers.

Reversacol. A line of photochemical dyes.

reversal spectrum. A spectrum that can be observed in intense, white light. The bright lines in the emission spectrum of the gas have traversed a luminous gas reversing the color of the dark spectrum lines.
See emission spectroscopy.

reverse osmosis. See osmosis, desalination.

reverse transcriptase. An RNA-directed DNA polymerase in retroviruses. These viruses are capable of making DNA complementary to an RNA and incorporating it into the genome.

reversible. (1) A chemical reaction that proceeds first to the right and then to the left when the ambient conditions are changed; the product of the first reaction decomposes to the original components as a result of different conditions of temperature or pressure. Examples are $H_2O + CO_2 \leftrightarrow H_2CO_3$, in which the carbonic acid reverts to water and carbon dioxide on heating; $NH_4Cl \leftrightarrow NH_3 + HCl$, in which the ammonium chloride decomposes on heating to ammonia and hydrochloric acid, which recombine on cooling.
(2) A colloidal system such as a gel or suspension that can be changed back to its original liquid form by heating, addition of water, or other method. For example, evaporated egg white can be restored (reconstituted) by addition of water.

reversible colloid. (lyophilic colloid). A colloid that is readily dispersible in a suitable medium and may be redispersed after coagulation.

reversible electrode. An electrode that owes its potential to unit charges used in electroplating that are destroyed during their use.
See chemical reaction; equilibrium constant; irreversible.

reversion. The softening and weakening of a natural rubber vulcanizate when the curing operation has been too long continued.

Reynold's number. The function DUP/μ used in fluid flow calculations to estimate whether flow through a pipe or conduit is streamline or turbulent in nature. D is the inside pipe diameter, U is the average velocity of flow, P is density, and μ is the viscosity of the fluid. Different systems of units give identical values of the Reynold's number, and values much below 2100 correspond to streamline flow, while values above 3000 correspond to turbulent flow.

Rf. Symbol for rutherfordium.

RFLP. See restriction fragment length polymorphism.

RFNA. Abbreviation for red fuming nitric acid. See nitric acid, fuming.

R group. An abbreviation used to denote an organic substituent such as the R groups of amino acids.

Rh. Symbol for rhodium.

rhamnose. $C_6H_{12}O_5$. A deoxyhexose monosaccharide found combined in the form of glycosides in many plants. Two forms exist, α- and β-, of which the α- is the more stable.
Properties: White crystals. Mp 82–92C (α-form converts partially to the β-form on heating). For equilibrium mixture optical rotation is +9.18 degrees. Very soluble in water and in methanol; soluble in absolute alcohol.
Use: Synthetic sweetener research.

rhenium.
CAS: 7440-15-5. Re.
Metallic element, atomic number 75, group VIIB of the periodic table, aw 186.207; valences = −1, 1 through 7; 4, 6, and 7 are most common, the last being the most stable. There are two isotopes.
Properties: Silver-white solid or gray to black powder. D 21.02 (20C), mp 3180C, bp 5630C, tensile strength 80,000 psi, high modulus of elasticity, attacked by strong oxidizing agents (nitric and sulfuric acids). Practically insoluble in hydrochloric acid. Retains its crystalline structure all the way to its mp. Has widest range of valences of any element. Rhenium-molybdenum alloys are superconductive at 10K. Not attacked by seawater.
Source: Principally molybdenite.
Derivation: Solutions from refinery residues (molybdenum ore flue dust, copper ore treatment) are (1) concentrated by a salting-out processes and reduced by hydrogen gas under press to give the metal, or (2) passed through an anionic resin from which pure rhenium can be extracted by a strong mineral acid.
Available forms: Powder that can be consolidated into rods, wires, or strips by powder metallurgy. Annealed metal is very ductile and can be bent, coiled, or rolled. Single crystals 2 inches × 0.05–0.005 inch diameter.
Hazard: Flammable in powder form.
Use: Additive to tungsten- and molybdenum-based alloys, electronic filaments, electrical contact material, high-temperature thermocouples, igniters for flash bulbs, refractory metal components of missiles, catalyst, plating of metals by electrolysis and vapor-phase deposition.

rhenium heptasulfide. Re_2S_7.
Properties: Brown-black solid. D 4.87, decomposes to ReS_2 at 600C. Insoluble in water; dissolves in solutions of alkali sulfides.
Hazard: Ignites on heating in air.
Use: Catalyst.

rhenium heptoxide. Re_2O_7.
Properties: Yellow crystals. D 6.103, mp 297C. Dissolves in water to form perrhenic acid $HReO_4$; very soluble in alcohol.
Derivation: Oxidation of metallic rhenium at 400C.

rhenium pentachloride. $ReCl_5$.
Properties: Dark-green to black solid. D 4.9, decomposes on heating. Decomposes in water; soluble in hydrochloric acid and alkalies.
Derivation: By reacting rhenium heptoxide with carbon tetrachloride at 400C.

rhenium trichloride. $ReCl_3$.
Properties: Dark-red solid, on heating emits green vapor from which the metal may be deposited, nonelectrolyte in solution, soluble in water and glacial acetic acid.
Derivation: Distillation of rhenium pentachloride.

rhenium trioxychloride. ReO_3Cl.
Properties: Colorless liquid. D 3.867 (20/4C), mp 4C, bp 131C. Decomposes in water; soluble in carbon tetrachloride, reacts readily with organic substances.

rheology.   Science of the deformation and flow of materials in terms of stress, strain, and time. Has important bearing on the behavior of viscous liquids in plastic molding.
See liquid, Newtonian; dilatancy; thixotropy; viscosity.

rheometer.   A device that continuously measures the viscosity and elasticity of resin solutions and polymer melts at high shear rates.

rhesus factor.   (Rh factor). A substance present in the red blood cells of the rhesus monkey and of 85% of an average, white, American population. Those whose red cells contain this factor are termed Rh-positive; others, Rh-negative. A negative individual may develop anti-Rh antibodies if Rh-positive red cells enter his blood; such antibodies can then agglutinate Rh-positive cells. Hemolytic reactions may thus occur following transfusion of Rh-positive blood cells into a recipient previously sensitized and having Rh antibodies in the serum. Likewise, an Rh-positive fetus, may give rise to antibodies in the blood of an Rh-negative mother; the antibodies, returning into the fetus may then produce the disease erythroblastosis fetalis. There are many subtypes of the Rh factor; these can be distinguished by serologic tests, and the laws of their inheritance have been determined.

rhizobitoxin.   A broad-spectrum herbicide that attacks young growth and new leaves but has little effect on older growth. Said to have low toxicity to humans (USDA).

rho acid.   See anthraquinone-1,5-disulfonic acid.

rhodamine B.   (CI 45170).
CAS: 81-88-9. $C_{28}H_{31}ClN_2O_3$.
A basic, red fluorescent dye, structurally related to xanthene.
Properties: Green crystals or reddish-violet powder. Very soluble in water and alcohol, forming bluish red, fluorescent solution; slightly soluble in acids or alkalies.
Derivation: By fusion of *m*-diethylaminophenol and phthalic anhydride followed by acidification with hydrochloric acid.
Hazard: Questionable carcinogen.
Use: Red dye for paper, also for wool and silk where brilliant fluorescent effects are desired and lightfastness is of secondary importance; analytical reagent for certain heavy metals, biological stain.

rhodamine toner.   Red to maroon lakes of rhodamine dyes and phosphotungstic or phosphomolybdic acid. They have good lightfastness and are used principally in printing inks.
See phosphomolybdic pigment; phosphotungstic pigment.

rhodanine.   (2-thio-4-keto-thiazolidine).
CAS: 141-84-4.

$$\overline{SCH_2C(O)NHCS}.$$

Properties: Finely crystalline, light-yellow color. D 0.868, bulk d 0.617, mp decomposes (often violently) 166C, pure material 167–168.5C. Soluble in methanol, ethyl ether, and hot water.
Hazard: May explode when heated to 166C. Toxic by ingestion.
Use: Organic synthesis (phenylalanine), laboratory reagent.

rhodinol.
CAS: 106-22-9. A mixture of terpene alcohols consisting principally of *l*-citronellol.
Properties: Colorless liquid; pronounced rose-like odor. D 0.860–0.880 (25C), refr index 1.4630–1.4730 (20C), optical rotation −4 to −9 degrees. Soluble in alcohol and mineral oil; insoluble in water. Combustible.
Derivation: From Reunion geranium oil.
Grade: FCC.
Use: Perfumery, flavoring agent.

rhodinyl acetate.   A mixture of terpene alcohol acetates consisting primarily of *l*-citronellyl acetate.
Properties: Colorless to slightly yellow liquid; rose-like odor. D 0.895–0.908 (25C), refr index 1.4530–1.4580 (20C), optical rotation −2 to 6 degrees. Soluble in alcohol and mineral oil; insoluble in glycerol. Combustible.
Derivation: Action of acetic anhydride on rhodinol in the presence of sodium acetate.
Grade: Technical, FCC.
Use: Perfumery, flavoring agent.

rhodinyl formate. $C_{11}H_{20}O_2$.
Properties: Colorless to slightly yellow liquid; leafy, rose-like odor. Mw 184.28, d 0.901–0.908, refr index 1.453–1.458. Soluble in alcohol, fixed oils; insoluble in glycerin, propylene glycol, water at 200C.
Use: Flavoring agent.

rhodium.
CAS: 7440-16-6.   Rh. Metallic element having atomic number 45, group VIII of the periodic table, aw 102.9055, no isotopes, valence = 3.
Properties: White solid of platinum group. D 12.41 (20C), mp 1966C, bp 4500C. Insoluble in acids and aqua regia; soluble in fused potassium bisulfate. Harder and higher-melting than platinum or palladium, highest in electrical and thermal conductivity of the platinum group. High surface reflectivity. A strong complexing agent.
Occurrence: Ontario, South Africa, Siberia.
Derivation: Occurs with platinum, from which it is recovered during the purification process.
Available forms: Produced as powder that can be fabricated by casting or powder metallurgy techniques. Single crystals are available.
Hazard: Flammable in powder form. Upper respiratory tract irritant. Questionable carcinogen.
Use: Alloy with platinum for high temperature thermocouples, furnace windings, laboratory crucibles, spinnerettes in rayon industry; electrical contacts, jewelry, catalyst, optical instrument mirrors, electrodeposited coatings for metals, vacuum-deposited glass coatings, headlight reflectors.

rhodium carbonyl chloride. $Rh(CO)_2Cl_2$.
Properties: Reddish crystals. Mp 125C. Soluble in alcohol, benzene, acetone, with decomposition; solid material is stable in dry air.
Use: Catalyst for organic reactions.

rhodium chloride.   (rhodium trichloride).
CAS: 10049-07-7. $RhCl_3$.
Properties: Reddish-brown powder. Mp 450–500C, bp 800C (sublimes). Insoluble in water and acids; soluble in solutions of alkalies and cyanides.
Hazard: Toxic by ingestion.
Use: Manufacture of rhodium trifluoride.

rhodium(III), triamminetrinitratato-.   See triamminetrinitratatorhodium(III).

rhodochrosite. $MnCO_3$ with partial replacement by iron, calcium, magnesium, zinc.
Properties: Light-pink, rose-red, brownish-red, or brown mineral; white streak; vitreous to pearly luster; photoluminescent. Found in veins with ores of silver, lead, copper manganese. D 3.3–3.6, Mohs hardness 3–4.
Occurrence: U.S. (Connecticut, New Jersey, Colorado, Montana, Nevada), Europe.
Use: Manganese ore.

rhodonite.   An ore of manganese. See manganous silicate.

rhodopsin.   The red-light-sensitive pigment of the eye (visual purple) consisting of the proteins opsin and retinene (vitamin A aldehyde). It occurs in land and marine vertebrates.

rhodoxanthin. $C_{40}H_{50}O_2$. A natural carotenoid pigment used in the food, drug, and cosmetic industries.
Properties: Soluble in benzene and chloroform; slightly soluble in alcohols.

Rhonite.   Thermosetting modified and unmodified urea-formaldehyde condensates. Supplied as water-clear solutions and aqueous pastes. Reactive with cotton, various grades producing shrink resistance, crease proofing, or modification of hand.
Use: Finishing of natural and synthetic fabrics.

"Rhoplex" [Dow].   TM for aqueous dispersions of acrylic copolymers. White, opaque emulsions; various grades differing in hardness, flexibility, adhesion, and tack of film; some thermosetting. Produce colorless, transparent films with outstanding permanence, durability, adhesion, and pigment-binding capacity.
Use: Emulsion paints, paper coatings and saturation, floor sealers and wax emulsions, textile-backing and finishing, bonding fibers and pigments, clear and pigmented coatings on wood and metals.

Rhothane.   An agricultural insecticide based on 1,1-bis(chlorophenyl)-2,2-dichloroethane and supplied as a wettable powder or emulsion concentrate.

Rhozyme.   Enzyme concentrate with diastatic or proteolytic activity. Buff-colored powders or liquids of fungal or bacterial origin that hydrolyze and soublize proteins and starches.
Use: Desizing textile fabrics; drycleaning; liquefaction of starch paste; fermentation processes; manufacture of corn syrup, fish solubles, septic tank formulations; animal feed; meat tenderizer.

rhubarb yellow.
CAS: 478-43-3. $C_{15}H_8O_6$.
Hazard: Low toxicity by ingestion.

RI.   See retention index.

Rib.   Abbreviation for ribose.

ribavirin.   See 1-β-d-ribofuranosyl-1,2,4-triazole-3-carboxamide.

ribbon mixer.   A mixing or blending machine whose essential components are a steel bowl or trough, jacketed for temperature control, within which rotates an agitating device consisting of two or more metal strips (ribbons) pitched in opposite directions spirally arranged around a central shaft. The curved and reverse-pitched ribbons operate on the principle of the screw; the material is moved forward by one ribbon and backward by another, so that efficient mixing is effected. Continuous operation is possible in some types. Such equipment is used for mixing dry powders, viscous liquids, slurries, etc., as well as for drying, crystallizing, and deaerating. Large sizes have a bowl diameter of 5 ft and a length of 9.5 ft.

riboflavin. (vitamin B_2; 7,8-dimethyl-10-(1'-d-ribityl) isoalloxazine).
CAS: 83-88-5. $C_{17}H_{20}N_4O_6$.
A crystalline pigment, the principal growth-promoting factor of the vitamin B_2 complex. It functions as a flavoprotein in tissue respiration. A syndrome resembling pellagra is thought to be due to riboflavin deficiency.

Properties: Orange-yellow crystals; bitter taste. Mp 282C (decomposes). Slightly soluble in water and alcohols; insoluble in lipid solvents; stable to heat in dry form and in acid solution. Stable to ordinary oxidation, unstable in alkaline solution and quite sensitive to light. In solution, riboflavin has an intense greenish-yellow fluorescence. Amounts are expressed in milligram or micrograms of riboflavin.
Source: (Food) Milk, green leafy vegetables, egg yolk, liver, enriched flour, yeast; (commercial) distiller's residues, fermentation solubles, synthetic production (indirectly from dextrose, lactose, yeast, and whey).
Grade: USP, FCC.
Use: Medicine, nutrition, animal feed supplement, enriched flours, dietary supplement.

riboflavin-5'-phosphate. (FMN; flavin mononucleotide). The phosphate ester of riboflavin in which the phosphate is esterified to the ribityl portion of riboflavin. It functions as a coenzyme for many flavine enzymes. The riboflavin group has the ability to take up hydrogen atoms, thus oxidizing the substrate.
Properties: (Sodium salt) Yellow crystals; quite sensitive to UV light. Much more soluble than riboflavin in water.
Derivation: By treating riboflavin with chlorophosphoric acid.
Use: Dietary supplement, flavor potentiator.

riboflavin 5'-phosphate sodium.
CAS: 130-40-5. $C_{127}H_{20}N_4NaO_9P\cdot2H_2O$.
Properties: Fine orange-yellow crystalline powder; slt odor. Hygroscopic; soluble in water. Decomposed by light when in solution.
Use: Food additive.

9-β-D-ribofuranosyladenine. See adenosine.

1-β-d-ribofuranosyl-1,2,4-triazole-3-carboxamide.
CAS: 36791-04-5. $C_8H_{12}N_4O_5$.
Properties: Mp 174–176C. Soluble in H_2O.
Hazard: Mildly toxic by ingestion. An experimental teratogen.
Use: As an antiviral agent.

ribonome. The complete set of RNA-coding regions of a genome.

ribonuclease. An enzyme that causes splitting of ribonucleic acid. Pancreatic ribonuclease for example, cleaves only phosphodiester bonds that are linked to pyrimidine-3'-phosphates. It is a critical regulator of life processes in the cell. The first enzyme to be synthesized (1969), it is composed of 124 amino acid residues. It is one of the proteins for which the sequence of amino acids has been elucidated (the order or sequence of amino acids is of critical importance in the functioning of enzymes, genes, and nucleotides).
See: Genetic code; RNAse.

ribonucleic acid. (ribose nucleic acid; RNA). Generic term for a group of natural polymers consisting of long chains of alternating phosphate and D-ribose units, with the bases adenine, guanine, cytosine, and uracil bonded to the 1-position of the ribose. Ribonucleic acid is universally present in living cells and has a functional genetic specificity due to the sequence of bases along the polyribonucleotide chain.
Four types are recognized:
(1) Messenger RNA, synthesized in the living cell by the action of an enzyme that carries out the polymerization of ribonucleotides on a DNA template region that carries the information for the primary sequence of amino acids in a structural protein. It is a ribonucleotide copy of the deoxynucleotide sequences in the primary genetic material.
(2) Ribosomal RNA, that exists as a part of a functional unit within living cells called the ribosome, a particle containing protein and ribosomal RNA in roughly 1:2 parts by weight having a particle weight, of about three million.
See ribosome.
Messenger RNA combines with ribosomes to form polysomes containing several ribosome units,

usually 5 (e.g., during hemoglobin synthesis), complexed to the messenger RNA molecule. This aggregate structure is the active template for protein biosynthesis.

(3) Transfer RNA, the smallest and best characterized RNA class. Its molecules contain only 80 nucleotides per chain. Within the class of transfer RNA molecules there are probably at least 20 separate kinds, correspondingly related to each of the 20 amino acids naturally occurring in proteins. Transfer RNA must have at least two kinds of specificity: (a) It must recognize (or be recognized by) the proper amino acid-activating enzyme so that the proper amino acid will be transferred to its free 2′- or 3′-OH group. (b) It must recognize the proper triplet on the messenger RNA-ribosome aggregate. Having these properties, the transfer RNA accepts or forms an intermediate transfer RNA-amino acid that finds its way to the polysome, complexes at a triplet coding for the activated amino acid, and allows transfer of the amino acid into peptide linkage.

(4) Viral RNA, isolated from a number of plant, animal, and bacterial viruses, may be considered as a polycistronic messenger RNA. It has been shown to have molecular weights of one or two million. Generally speaking, there is one molecule of RNA per infective virus particle. The RNA of an RNA virus can be separated from its protein component and is also infective, bringing about the formation of complete virus.

From article by F. J. Bollum in *Encyclopedia of Biochemistry*.

See deoxyribonucleic acid; RNA.

riboprobe.   A strand of RNA synthesized in vitro (usually radiolabeled) and used as a probe for hybridization reactions. An RNA probe can be synthesized at very high specific activity, is single stranded (and therefore will not self-anneal), and can be used for very sensitive detection of DNA or RNA.

ribose.   See ribonucleic acid; deoxyribose.

D-ribose-5-phosphoric acid. $C_5H_9O_4 \cdot H_2PO_4$.
A constituent of nucleotides and nucleic acids.
Properties: The barium salt (5.5 H_2O) is sparingly soluble in cold water and crystallizes in hexagonal plates.
Derivation: From ionosinic acid.
Use: Biochemical research.

ribosomal protein.   Any protein that occurs in the in ribosomes. They may act as catalysts in the reconstitution of the biologically active ribosomal subunits.

ribosome.   A ribonucleoprotein, the smallest organized structure in the cell. Ribosomes occur in all cells including bacteria, fungi, algae, and protozoa. They are the central point of protein synthesis. They contain from 45 to 60% of ribonucleic acid (RNA), the balance being proteins. Ribosome crystals have been produced; in the electron microscope these appear as sheets of black dots, each sheet (one ribosome thick) containing hundreds of ribosomes in recurring groups of four.

See deoxyribonucleic acid; ribonucleic acid.

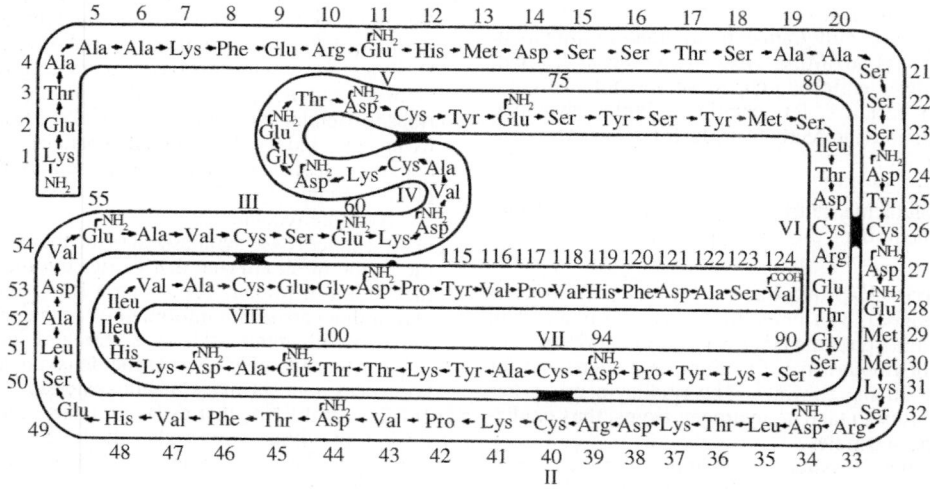

ribonucleotide.   A nucleotide containing D-ribose as its pentose component.
See nucleotide.

ribosome-inactivating protein.   (rip). Any of a class of plant proteins that inhibit normal functioning of ribosomes.

Hazard.   Highly toxic.

D-ribosyl uracil.   See uridine.

ribozyme.   (catalytic RNA; gene shears). A naturally occurring RNA that is enzymatically active and that specifically binds to, cleaves and therefore inactivates mRNA molecules. They offer as a means of inhibiting a gene of interest for target validation studies. They can be engineered to bind to any RNA sequence, with consequent cleavage and inactivation of mRNAs that contain the target sequence.

ribulose biphosphate.   (rubp). A five carbon sugar combined with carbon dioxide to yield a six carbon intermediate in the first stage of the dark reaction of photosynthesis.

rice bran wax.
Properties: Tan to brown hard wax. Mp 75C. Soluble in chloroform, benzene; insoluble in water.
Use: Food additive.

rice paper.   See straw.

Richards, Theodore W.   (1868–1928). An American chemist born in Germantown, PA. He was the first American to receive the Nobel Prize in Chemistry. He studied chemistry at Haverford and Harvard, with a doctorate in chemistry from Harvard where he later became Erving Professor of Chemistry. An outstanding experimental chemist, his major interests were atomic weights, thermochemistry, and thermodynamics. He was also a brilliant teacher. He was president of the ACS in 1914, and the recipient of many honorary awards, including the Davy, Faraday, and Gibbs medals.

Ricilan.   Esters synthesized from castor oil and lanolin components. Amber, viscous liquids.
Use: Hydrophobic emollients, pigment dispersants.

ricin.
CAS: 9009-86-3.
Properties: White powder. The albumin of the castor oil bean is the toxic principle. Extracted from the pressed seeds with 10% solution of sodium chloride followed by precipitation with magnesium sulfate.
Hazard: Very highly toxic by ingestion; small particle in cut, abrasion, eye, or nose may prove fatal.
Use: Reagent for pepsin and trypsin, agriculture.

ricin agglutinin.   Either of two components of ricin mixture that are each comprised of four polypeptide subunits. These components of ricin mixture are inactive orally and affect erythrocytes only when given intravenously.

ricin D.
CAS: 9040-12-4.

Hazard: A poison.
Source: Natural product.

ricinin.
CAS: 524-40-3. $C_8H_8N_2O_2$.
Hazard: A poison by ingestion.
Use: Agricultural chemical.

ricinine.   (1,2-dihydro-4-methoxy-1-methyl-2-oxonicotinonitrile). $C_8H_8O_2N_2$.
Properties: White, crystalline alkaloid. Slightly soluble in water, chloroform, and ether. Mp 201.5C, sublimes 170-180C (20 mm Hg).
Derivation: From castor oil seeds and leaves.
Hazard: Toxic by ingestion.
Use: Biochemical research.

ricinoleic acid.   (*cis*-12-hydroxyoctadec-9-enoic acid; 12-hydroxyoleic acid; castor oil acid).
CAS: 141-22-0. $CH_3(CH_2)_5CH(OH)CH_2CH:CH$ $(CH_2)_7COOH$.
A C_{18} unsaturated fatty acid that comprises 80% of the fatty acid content of castor oil.
Properties: Colorless to yellow, viscous liquid. D 0.940 (27.4/4C), mp 5.5C, bp 226C (10 mm Hg), refr index 1.4697 (20C), dextrorotatory. Soluble in most organic solvents; insoluble in water. Combustible.
Derivation: Saponification of castor oil.
Use: Soaps, Turkey red oil, textile finishing, source of sebacic acid and heptanol, ricinoleate salts, 12-hydroxystearic acid.

ricinoleyl alcohol.   (9-octadecen-1,12-diol). $CH_3(CH_2)_5CH(OH)CH_2CH:CH(CH_2)_7CH_2OH$.
The fatty alcohol derived from ricinoleic acid. It has a long, straight chain with one double bond and one hydroxyl group in a secondary position besides the primary group on the end. Available as a 90% product.
Properties: Colorless, nondrying liquid at room temperature. Iodine value 91.8%, cloud point −12.2C, boiling range 170–328C, viscosity 51 (SSU/21C). Combustible.
Derivation: Reduction of acid made from castor oil.
Impurities: Oleyl and linoleyl alcohols.
Use: Protective coatings, polyesters, plasticizers, organic synthesis, pharmaceuticals, lubricants, surface-active agents.

ricinus oil.   See castor oil.

Riegler's test.   A test for nitrous acid using sodium naphthionate and β-naphthol.

Riehm quinoline synthesis.   Formation of quinoline derivatives by prolonged heating of arylamine hydrochlorides with ketones with or without use of aluminum chloride or phosphorus pentachloride.

Riemschneider thiocarbamate synthesis. The action of concentrated sulfuric acid followed by treatment with ice water serves to transform arylthiocyanates into the corresponding thiocarbamates.

rifametane.
CAS: 94168-98-6. $C_{44}H_{60}N_4O_{12}$.
Hazard: Moderately toxic by ingestion.

rifapentine.
CAS: 61379-65-5. $C_{47}H_{64}N_4O_{12}$.
Hazard: Moderately toxic by ingestion.

Riley oxidations. Oxidations of organic compounds with selenium dioxide, e.g., the oxidation of active methylene groups to carbonyl groups.

"Rilsan" [Arkema]. TM for nylon 11.
See nylon.

RIM. Abbreviation for reaction injection molding.

ring compound. See cyclic compound; alicyclic; aromatic; heterocyclic.

Ringer's solution. Physiologic solution containing 0.650 g sodium chloride, 0.014 g potassium chloride, 0.0129 g calcium chloride, 0.020 g sodium bicarbonate, 0.001 g monosodium phosphate, 0.200 g dextrose in 100 g water. It is isotonic with frog blood serum.

ring whizzer. A fluxional molecule frequently encountered in organometallic chemistry in which rapid rearrangements occur by migrations about unsaturated organic rings.

ripening. (cheese).
See aging; curing.

risedronate sodium.
CAS: 115436-72-1. $C_7H_{10}NO_7P_2$•Na.
Hazard: Human systemic effects.

risk communication. In genetics, a process in which a genetic counselor or other medical professional interprets genetic test results and advises patients of the consequences for them and their offspring.

ristocetin. An antibiotic produced by the fermentation of *Nocardia lurida,* a species of Actinomycetes. The antibiotic has two components, ristocetin A and ristocetin B. The commercial product is a lyophilized preparation representing a mixture of A and B, of which B is no more than 25%.
Properties: White or tan crystals or powder; practically odorless. Freely soluble in water; practically insoluble in organic solvents.
Grade: USP.
Use: Medicine.

"Riston" [Du Pont]. TM for engineering resins and high end specialty polymers.

"Ritalan" [RITA]. TM for ethoxylated, hydrogenated, hydroxylated lanolin.
Use: As an emulsifier, emollient, conditioner, moisturizer, stabilizer in make-up, lipstick, skincare products, bath products, shampoo, soap, shaving cream, ointment, sun preparations, and veterinary products.

ritanserin.
CAS: 87051-43-2. $C_{27}H_{25}F_2N_3OS$.
Hazard: A poison.

Ritter reaction. Formation of amides by addition of olefins or secondary and tertiary alcohols to nitriles in strongly acidic media.

Rittinger's law. The energy required for reduction in particle size of a solid is directly proportional to the increase in surface area.
See Kick's law.

rivastatin.
CAS: 143201-11-0. $C_{26}H_{33}FNO_5$•Na.
Hazard: A poison by ingestion.

Rn. Symbol for radon.

RNA. (ribonucleic acid). A chemical found in the nucleus and cytoplasm of cells; it plays an important role in protein synthesis and other chemical activities of the cell. The structure of RNA is similar to that of DNA. There are several classes of RNA molecules, including messenger RNA, transfer RNA, ribosomal RNA, and other small RNAs, each serving a different purpose.

RNA decoy. An RNA molecule that competitively binds foreign or pathogenic protein molecules and modulates their activity.

RNAi. (RNA interference; RNA silencing). The mechanism by which small double-stranded RNAs can interfere with expression of any mRNA having a similar sequence. Those small RNAs are known as "siRNA", for short interfering RNAs. The mode of action for siRNA appears to be via dissociation of its strands, hybridization to the target RNA, extension of those fragments by an RNA-dependent RNA polymerase, then fragmentation of the target. Importantly, the remnants of the target molecule appears to then act as an siRNA itself; thus the effect of a small amount of starting siRNA is effectively amplified and can have long-lasting effects on the recipient cell.
The RNAi effect has been exploited in numerous research programs to deplete the call of specific messages, thus examining the role of those messages by their absence.

RNA interference. See RNAi.

RNA polymerase. An enzyme essential in imparting the DNA genetic code to ribonucleic acid (RNA).

RNase. Ribonuclease; an enzyme which degrades RNA. It is ubiquitous in living organisms and is exceptionally stable. The prevention of RNase activity is the primary problem in handling RNA.

RNase protection assay. This is a sensitive method to determine (1) the amount of a specific mRNA present in a complex mixture of mRNA and/or (2) the sizes of exons which comprise the mRNA of interest. A radioactive DNA or RNA probe (in excess) is allowed to hybridize with a sample of mRNA (for example, total mRNA isolated from tissue), after which the mixture is digested with single-strand specific nuclease. Only the probe which is hybridized to the specific mRNA will escape the nuclease treatment, and can be detected on a gel. The amount of radioactivity which was protected from nuclease is proportional to the amount of mRNA to which it hybridized. If the probe included both intron and exons, only the exons will be protected from nuclease and their sizes can be ascertained on the gel.

RNA silencing. See RNAi.

RNA splicing. Removal of introns and joining of exons in a primary transcript.

roasting. Heating in the presence of air or oxygen. Most commonly used in converting natural metal sulfide ores to oxides as a first step in recovery of metals such as zinc, lead, copper, etc. Roasting is an oxidation process.
See smelting.

robin.
Derivation: Found in the roots, bark, leaves and seeds of *Robinia pseudoacacia*.
Use: Phytotoxin.

Robinson annellation reaction. Formation of six-membered-ring α,β-unsaturated ketones by condensation of cyclohexanones with methyl vinyl ketone or its equivalents, followed by an intramolecular aldol condensation.

Robinson-Schopf reaction. Synthesis of tropinones from a dialdehyde, methylamine, and acetonedicarboxylic acid.

Robinson, Sir Robert. (1886–1975). An English chemist who won the Nobel Prize for Chemistry in 1947. His work began on plants with biological significance, particularly those containing alkaloids. He synthesized brazilin and haematoxylin. In organic chemistry he discovered an important qualitative electronic theory. He explicated the penicillin structure as well. His education, which began at the University of Manchester, took him throughout Europe.

Rochelle salt. See potassium-sodium tartrate.

rock crystal. See quartz.

rocket fuel. (rocket propellant). A substance or mixture that has the capacity for extremely rapid but controlled combustion that produces large volumes of gas at high pressure and temperature. These fuels may be either liquid, solid, or combinations of both. Liquid monopropellants are hydrogen peroxide and hydrazine, catalyzed by finely divided metals to decompose them into gases. Liquid bipropellants consist of the fuel and an oxidizer; typical fuels of this type are hydrogen, hydrazine, ammonia, and boron hydride, the oxidizers being oxygen, nitric acid, ozone, hydrogen peroxide, and water.
Solid propellants include nitrocellulose plasticized with nitroglycerin or various phthalates and inorganic salts suspended in a plastic or synthetic rubber (e.g., "Thiokol") and containing a finely divided metal. The inorganic oxidizers used are ammonium and potassium nitrates and perchlorates.

rock flour. Mud-sized material ground from coarse grains by the movement of a glacier.

Rockwell hardness. See hardness.

rodenticide. A pesticide used to kill rats and other rodents.
See warfarin.

rod mill. A closed steel cylinder one-third filled with rods of about the same length as the cylinder and 1–2 inches in diameter. As the cylinder rotates, the rods roll over one another, exerting a combination of impact and grinding action on the charge. It gives a product of 50–60 mesh with a minimum of fines. Rod mills are used for pulp grinding in the paper industry and for size reduction of ores, minerals, metal powders, etc.

roentgen. (r). The international unit of quantity or dose for both X-rays and γ-rays. It is defined as the quantity of X- or γ-rays that will produce as a result of ionization one electrostatic unit of electricity of either sign in 1 cc (0.001293 g) of dry air as measured at 0C and standard atmospheric pressure. The use of the roentgen unit has been extended to include particle radiation such as α- and β-particles and protons and neutrons.
See rad; curie.

Rohrbach solution.
Properties: Clear, yellow liquid. Very refractive, d 3.5.
Derivation: An aqueous solution of mercuric barium iodide.
Hazard: Toxic by ingestion and inhalation.
Use: Separating minerals by their specific gravity, microchemical detection of alkaloids.

roll mill. Two chilled steel rolls 48–72 inches wide and 12–24 inches in diameter turning in opposite directions at different speeds to exert shearing action; the separation (or nip) is adjustable by set screws. Because the shearing friction generates considerable heat, the rolls are water-cooled. Such mills are standard equipment in the rubber, plastic, and adhesives industries; several usually rotate on one shaft. They can be used for mixing, but their chief use is for prewarming calender and extruder feed. Mills with three rolls are used for mixing and grinding paints and printing inks. Laboratory sizes of all types are available.

RON. Abbreviation for Research octane number.

ronnell. (O,O-dimethyl-O-(2,4,5 trichlorophenyl)phosphorothioate).
CAS: 299-84-3. $(CH_3O)_2P(S)OC_6H_2Cl_3$.
Properties: Powder or granules. Mp 41C. Insoluble in water; soluble in most organic solvents.
Hazard: Toxic by ingestion and inhalation. Cholinesterase inhibitor, use may be restricted. Questionable carcinogen.
Use: Insecticide.

Röntgen, Wilhelm Conrad. (1845–1923). A German physicist who discovered X-rays in 1895, for which he was awarded the Nobel Prize in 1901. Application of these to a number of important problems in analytical chemistry was developed by Braggs, Moseley, von Laue, and Debye and Scherrer.

room temperature. An ambient temperature from 20 to 25C (68–77F).

ropivacaine.
CAS: 98717-16-9. $C_{17}H_{26}N_2O$.
Hazard: Human systemic effects.

rosanilin dye. (basic fuchsin; 4-[(4-aminophenyl)-4-imino-3-methylcyclohexa-2,5-dien-1-ylidene)methyl]aniline hydrochloride).
CAS: 8053-09-6. $C_{20}H_{20}CLN_3$.
Any of a number of triaminotriphenylmethane dyes.
Properties: The amino groups are unsubstituted and may have methyl groups linked directly to the benzene rings.

rosaniline.
CAS: 632-99-5. $HOC(C_6H_4NH_2)_2C_6H_3(CH_3)NH_2$.
A triphenylmethane dye.

Properties: Reddish-brown crystals. Mp 186C (decomposes). Soluble in acids and alcohol; slightly soluble in water.
Use: Dye (usually as the hydrochloride), fungicide.

roscoelite. $K_2V_4Al_2Si_6O_{20}(OH)_4$.
Properties: A vanadium-bearing species of mica. Formula variable with V_2O_3 up to 28%. Occurs as minute scales with micaceous cleavage, dark green to brown in color, pearly luster, Mohs hardness 2.5, d 3.0.
Occurrence: Colorado, California, Australia.
Use: Source of vanadium.

Rose Bengal. $C_{20}H_4Cl_4I_4O_5$.
Properties: Bluish-pink powder. Soluble in water.
Use: Biological stain; colorant for inks, cellulosics, foods, cosmetics, medicine (diagnostic aid).

Rose, Irwin. (1926–). An American born in Brooklyn, New York who won the Nobel Prize for Chemistry in 2004 for his pioneering work concerning the discovery of ubiquitin-mediated protein degradation. Rose received his B.S. in 1948 from Washington State University and his Ph.D. in biochemistry in 1952 from the University of Chicago. He is currently a distinguished Professor-in-residence at the Department of Physiology and Biophysics of the College of Medicine at the University of California, Irvine.

Rosenheim color test. On addition of nine parts of trichloroacetic acid in water to a solution of ergosterol in chloroform, an immediate red color develops, gradually changing to pure blue. The test is specific to sterols containing a diene system or capable of forming one on acid dehydration.

Rosenmund reaction. Formation of aromatic arsenic acids by heating equimolar amounts of sodium or potassium arsenite with aryl halides in aqueous solution. The reaction is an extension of the Meyer reaction.

Rosenmund reduction. Catalytic reduction of acid chlorides to aldehydes. To prevent further hydrogenation, a poison is added to the catalyst.

Rosenmund—von Braun synthesis. Formation of aromatic nitriles from aryl halides and cuprous cyanide.

rosin.
Properties: Angular, translucent, amber-colored fragments. D 1.08, mp 100–150C, acid number above 150, flash p 370F (187C). Insoluble in water; freely soluble in alcohol, benzene, ether, glacial acetic acid, oils, carbon disulfide, dilute solutions of fixed alkali hydroxides. Hard and friable at room temperature; soft and very sticky when warm. Combustible.

Chief constituents: Resin acids of the abietic and pimaric types, having the general formula $C_{19}H_{29}COOH$ and a phenanthrene nucleus.
See turpentine.

Derivation: From pine trees, chiefly *Pinus palustris* and *Pinus caribaea*. (1) Gum rosin is the residue obtained after the distillation of turpentine oil from the oleoresin tapped from living trees. (2) Wood rosin is obtained by extracting pine stumps with naphtha and distilling off the volatile fraction. (3) Tall-oil rosin is a by-product of the fractionation of tall oil.

Grade: Virgin, yellow dip, hard, NF. Wood rosin grades are B, C, D, E, F, FF, G, H, I, J, K, L, M, N, W-G (window-glass), W-W (water-white). The grading is done by color; B is the darkest and W-W the lightest.

Hazard: Evolves irritating and suffocating fumes on heating.

Use: Hot-melt and pressure-sensitive adhesives, mastics and sealants, varnishes, ester gum, soldering compounds, core oils, insulating compounds, soaps, paper sizing, printing inks, polyesters (formed by reaction of the conjugated acids of rosin with acrylic acid, followed by reaction with a glycol).
See abietic acid.

rosin core solder pyrolysis products.
CAS: 8050-09-7.
Hazard: A poison by ingestion and inhalation route.
Source: Natural product.

rosin oil.
Properties: Water-white to brown liquid, viscous; odorless; strong peculiar taste. D 0.980–1.110, iodine number 112–115. Soluble in ether, chloroform, fatty oils, and carbon disulfide; slightly soluble in alcohol, insoluble in water. Essentially decarboxylated rosin acids.
Derivation: By fractional distillation of rosin, that portion distilling above 360C being rosin oil.
Hazard: Spontaneous heating, fire risk when heated.
Use: Lubricant, adulterant for boiled linseed oil, hot-melt adhesives, printing inks, impregnating paper for wrapping electric cables, core compounds, varnishes.

rosin soap.   See sodium abietate; soap.

rosolic acid.   See aurin.

rotameter.   An instrument for measuring the flow rates of liquids and gases.

Rotax.   A purified grade of 2-mercaptobenzothiazole.
Use: Primary accelerator for rubber.

rotenone.   (tubatoxin).
CAS: 83-79-4. $C_{23}H_{22}O_6$.
A pentacyclic compound.

Properties: White, crystals; odorless. D 1.27 (20C), mp 163C. Strongly levorotatory in solution. Specific rotation for D line 230 degrees in benzene, 62 degrees in ethylene dichloride. Soluble in ether, alcohol, acetone, and other organic solvents; insoluble in water; not compatible with alkalies.
Grade: CP crystals, technical, also as extracts of derris and cube root.
Hazard: Toxic by ingestion, overexposure can be fatal, irritant to skin, eyes and upper respiratory tract. Central nervous system impairment. Questionable carcinogen.
Use: Insecticide, flea powders, fly sprays, mothproofing agents.

Rothemund reactions.   Preparation of *meso*-tetrasubstituted porphyrins by condensation of pyrrole with an aldehyde.

rottenstone.   Soft, friable aluminum silicate.
Use: Polishing agent.

rouge.
(1) CAS: 1309-37-1. A high-grade red pigment used as a polishing agent for glass, jewelry, etc. (2) A cosmetic prepared from dried flowers of the safflower.
Properties: Dark-red powder. Insoluble in water.
Hazard: Questionable carcinogen.
See iron oxide red.

Rowe rearrangement.   Intramolecular rearrangement of pseudophthalazones into phthalazones by heating at 180C in the presence of dilute aqueous acids.

Rowland, Frank Sherwood.   (1927–2012). An American who won the Nobel Prize for Chemistry along with Paul Crutzen and Mario Molina in 1995 for their work in atmospheric chemistry, particularly concerning the formation and decomposition of ozone.
See Crutzen, Paul; Molina, Mario.

Roxel process.   A process for increasing the flame resistance of cotton by chemically modifying the fiber, i.e., by applying a formulation consisting of tetrakis(hydroxymethyl)phosphonium chloride (THPC), triethylolamine, and urea in aqueous solution. These substances cross-link with the hydroxyl groups of the cellulose to form an efficient and durable protective medium. This process is used in the manufacture of blankets, curtains, bed linen, and industrial safety garments.

"Royalac" [Chemtura].   TM for an accelerator for EPDM elastomers.

"Royalene" [Chemtura].   TM for a low-cost, high-performance ethylene-propylene terpolymer.
Properties: Resistance to ozone, heat, weather, sunlight aging, steam, and chemicals. Resilient at low

temperatures, excellent electrical properties and can be easily colored.
Use: Tires, automotive parts, tank cars, hoses, and insulation.

royal jelly. A complex mixture secreted by "worker" honeybees that composes the sole nutrient of the queen bee. It contains 31% protein, 15% carbohydrate, 15% lipid, plus vitamins. The free fatty acid portion of the lipid is a mixture of C_{10} acids.

Royal Society of London. As the U.K. National Academy of Sciences founded in 1660, the Royal Society plays a crucial role as the champion of top quality science and technology. It does this by: funding research, grants, issuing reports, publishing journals, and awarding medals and prizes to scientists throughout the world. By virtue of its independent status and its body of some 1,450 Fellows and Foreign Members covering all scientific disciplines, the Society is uniquely placed to represent the interests of top quality science and technology in its interactions with government, the public and the media. It adopts a high profile on issues which are vital to scientific progress and is taking an increasingly prominent position in furthering the role of science, engineering and technology in society by facilitating constructive dialogue between scientists and non-scientists. Headquarters are at 6–9 Carlton House Terrace, London SW1Y 5AG. Website: http://www.royalsociety.org

RR acid. (2-amino-8-naphthol-3,6-disulfonic acid; 2R acid). $C_{10}H_4NH_2OH(SO_3H)_2$.
Derivation: Fusion of a naphthylamine trisulfonic acid with sodium hydroxide.
Use: Azo dye intermediate.

rRNA. A class of RNA molecules serving as components of ribosomes.

RTV. (room-temperature-vulcanizing). Rubbers that have good physical properties and electrical properties similar to silicone rubber.
Use: Sealing, caulking, encapsulating and flexible mold-making in electronic, aircraft, missile and building industries.
See silicone.

Ru. Symbol for ruthenium.

RU 13. See 9-diethylaminoethyl-2-phenylimidazo(1,2-a)benzimidazole.

rubber. Any of a number of natural or synthetic high polymers having unique properties of deformation (elongation or yield under stress) and elastic recovery after vulcanization with sulfur or other cross-linking agent, which in effect changes the polymer from thermoplastic to thermosetting. The yield or stretch of the vulcanized material ranges

from a few hundred to over 1000%. The deformation after break, called *permanent set*, is usually taken as the index of recovery; it ranges from 5 to 10% for natural rubber to 50% or more for some synthetic elastomers, and varies considerably with the state of vulcanization and the pigment loading. See elastomer.

rubber cement. See adhesive, rubber-based.

rubber, chlorinated. An elastomer (natural rubber or a polyolefin) to which 65% of chlorine has been added to give a solid, film-forming resin. White, amorphous powder available in viscosity grades from 5 to 125 cP, the figures indicating viscosity of a 20% solution in toluene. Decomposes at 125C, soluble in aromatics, insoluble in aliphatics and alcohols. Compatible with almost all natural and synthetic resins. Chief use is in maintenance paints (marine, swimming pool, traffic, masonry, etc.).
See "Hypalon" *[Du Pont Performance]*; rubber hydrochloride.
Hazard: Do not dry-mill chlorinated rubber with zinc oxide; mixture reacts violently at 216C. Do not use in baked enamels.

rubber, cold. See cold rubber.

rubber fiber. Generic name for a manufactured fiber in which the fiber-forming substance comprises natural or synthetic rubber (Federal Trade Commission). Often the rubber is a core around which cotton or other fibers are wrapped to make an elastic yarn used for girdles, swimwear, elastic bands, and tapes.

rubber, hard. A rubber compounded with 30–50% by weight of sulfur and cured until an extremely hard, brittle product is formed. Lime or magnesia is used as activator. The theoretical maximum of sulfur that can combine chemically with rubber hydrocarbon is 32%. Combustible.
Hazard: Flammable in form of dust.
Use: Battery boxes, tank linings, acid- and alkali-resistant equipment, combs. As dust, filler for low-cost rubber products.

rubber hydrochloride. A hydrochloride derivative, as distinct from a chlorine derivative.
Properties: Thermoplastic, white powder or clear film; odorless; tasteless. Chlorine content 29–30.5%. Soluble in aromatic hydrocarbons. Softens at 110–120C. Films are highly resistant to moisture, oils, acids, and alkalies but tend to become brittle on exposure to sunlight. The life of such films is greatly extended by the incorporation of suitable stabilizers and plasticizers. Nonflammable, nontoxic.
Derivation: A solution of natural rubber is treated with anhydrous hydrogen chloride under pressure and at low temperature. After neutralization of

excess hydrochloric acid, the product is precipitated by the addition of ethanol.

Use: Protective coverings for machinery, rain clothing, shower curtains, food packaging.

rubber latex.   See latex.

rubber, liquid.   Any of several proprietary products consisting of high polymers in liquid form for use as coatings, adhesives, etc.

rubber, natural.   (polyisoprene).
CAS: 9006-04-6. $(C_5H_8)_x$.

$$\left(\begin{array}{c} \quad\; H \;\; CH_3 \\ \quad\; | \quad\; | \\ CH_2=C-C=CH_2 \end{array}\right)_x$$

(1) Crude (unvulcanized)

Properties: Chemically unsaturated. D 0.92. Amorphous when unstretched, but has oriented crystalline structure on stretching; not stable to temperature changes (thermoplastic), readily oxidizable by mastication; soluble in acetone, carbon tetrachloride, and most organic solvents; refr index 1.52; dielectric constant 2.5. Processed by calenders and extruders; can be injection molded with low sulfur and high accelerator. Cured by hot-molding or in open steam at temperatures from 120 to 150C after addition of 3% sulfur, 1% organic accelerator, 3% zinc oxide, plus fillers or reinforcing agents. The only factors of significance in vulcanization are the time of exposure to heat and the temperature used.

Derivation: From latex obtained from *Hevea* trees, coagulated with acetic or formic acid. Also made synthetically.

See Coral; "Natsyn" *[Goodyear]*.

Occurrence: Brazil, Malaysia, Indonesia.

Grade: Ribbed, smoked sheets, pale (yellow) crepe; brown crepe.

Use: Cements, adhesives, electrical insulating tapes and cable wrapping.

(2) Cured (vulcanized, i.e., sulfur cross-linkages)

Properties: High tensile strength; relatively low permanent set; insensitive to temperature changes. Attacked by heat, atmospheric oxygen, ozone, hydrocarbons, and unsaturated fats and oils. Insoluble in acetone. Permeable to gases; supports combustion; abrasion resistance poor unless compounded with carbon black; dissipates vibration shock; high electrical resistivity.

Use: Vehicle tires, hose, conveyor-belt covers, footwear, specialized mechanical products, drug sundries, foam rubber, electric insulation, etc.

Note: Gutta percha and balata have similar chemical composition (isomeric) but have very different properties and few commercial uses. Neither can be vulcanized.

See latex; guayule; Appendix II for history of the industry.

rubber sponge.   (foam rubber; cellular rubber). A flexible foam produced by beating air into heat-sensitive latex, with subsequent vulcanization, or by incorporating ammonium carbonate or sodium bicarbonate into a strongly masticated and highly accelerated rubber mixture. As the temperature rises to the curing range, ammonia or carbon dioxide is released, forming uniform pores throughout the mixture just before the onset of vulcanization.

Use: Vibration damping pads and inserts, rug and carpet underlays, mattresses and upholstery, seat cushions.

rubber, synthetic.   Any of a group of manufactured elastomers that approximate one or more of the properties of natural rubber. Some of these are: sodium polysulfide ("Thiokol" *[Toray]*), polychloroprene (neoprene), butadiene-styrene copolymers (SBR), acrylonitrilebutadiene copolymers (nitrile rubber), ethylenepropylene-diene (EPDM) rubbers, synthetic polyisoprene (Coral, "Natsyn" *[Goodyear]*), butyl rubber (copolymer of isobutylene and isoprene), poly-acrylonitrile ("Hycar" *[Lubrizol Advanced]*), silicone (polysiloxane), epichlorohydrin, polyurethane ("Vulkollan" *[Bayer]*).

The properties of these elastomers are widely different. All require vulcanization. In general, sulfur is used only for unsaturated polymers; peroxides, quinones, metallic oxides, or diisocyanates effect vulcanization with saturated types. Many are special-purpose rubbers, some can be used in tires when loaded with carbon black, others have high resistance to attack by heat and hydrocarbon oils and thus are superior to natural rubber for steam hose, gasoline and oil-loading hose. Most are available in latex form.

rubber, thermoplastic.   Any of several block copolymers of propylene/EPDM or styrene/ethylene-butylene. Cross-linking results from crystallization of polypropylene or polystyrene segments. Since this is reversible on heating, the product is thermoplastic. Its chief use is in oil-resistant wire and cable insulation.

rubbing alcohol.   (methyl alcohol; pyroligneous alcohol; propan-2-ol).
CAS: 67-63-0. C_3H_8O.
A 70% aqueous solution of ethanol by volume plus denaturants, or an aqueous solution of isopropyl alcohol).

Properties: Flammable, colorless aqueous solution.
Hazard: Toxic.
Use: In the manufacture of acetone and its derivatives, and as a solvent; topically as an antiseptic.

ruberythric acid. $C_{26}H_{28}O_{14}$. An alizarin glucoside.

rubidium. Rb.
CAS: 7440-17-7. Metallic element of atomic number 37, group IA of the periodic table, aw 85.4678, valence = 1. One stable form, principal natural radioactive isotope is rubidium-87. It is the second most electropositive and the second most alkaline element, has lowest ionization potential. Highly reactive.
Properties: Soft, silvery-white solid. Easily oxidized in air. Decomposes water, d 1.532 (20C as solid), mp 39C, bp 688C, high heat capacity and heat transfer coefficient. Soluble in acids and alcohol.
Derivation: (1) Thermochemical reduction of rubidium chloride with calcium, (2) electrolysis of fused cyanide or chloride.
Source: Lepidolite, carnallite, pollucite, mineral springs, and natural brines.
Grade: Technical, 99+%.
Hazard: Reacts vigorously with air and water, must be stored under kerosene or similar liquid, dangerous fire and explosion risk. Metal causes serious skin burns.
Use: Photocells, catalyst or catalyst promotor.

rubidium alum. See aluminum rubidium sulfate.

rubidium azide.
CAS: 22756-36-1. N_3Rb.
Hazard: A poison by ingestion. May be unstable, especially when heated.

rubidium carbonate. Rb_2CO_3.
Properties: White powder. Mp 837C. Extremely hygroscopic; soluble in water. Dissociates above 900C.
Hazard: Strong irritant to tissue.
Use: Special glass formulations.

rubidium chloride.
CAS: 7791-11-9. RbCl.
Properties: White, crystalline powder; lustrous. D 2.76, mp 715C, bp 1390C. When heated it decrepitates, melts, and volatizes. Soluble in water; almost insoluble in alcohols.
Grade: Technical, CP, single crystals.
Use: Analysis (testing for perchloric acid), source of rubidium metal.

rubidium dichromate.
CAS: 13446-73-6. $Cr_2O_7Rb_2$.
Properties: Crystals. D 3.02–3.13. IDLH Ca [15 mg/m^3 {as Cr(VI)}].
Hazard: A confirmed carcinogen. A poison. A powerful oxidizer.

rubidium fluoride. RbF.
Properties: White crystals. D 3.557, mp 775C, bp 1410C. Soluble in water; insoluble in alcohol.
Available forms: Single large crystals.
Hazard: Strong irritant to tissue.

rubidium hexafluorogermanate. Rb_2GeF_6.
Properties: White, crystalline solid. Mp 696C. Slightly soluble in cold water; very soluble in hot water.

rubidium hydroxide. (rubidium hydrate).
CAS: 1310-82-3. RbOH.
Properties: Grayish-white mass. D 3.2, mp 300C. Extremely hygroscopic. Absorbs carbon dioxide from air. Strong base. Soluble in alcohol, water. Attacks glass at room temperature.
Hazard: Strong irritant to tissue.
Use: Suggested for electrolyte in storage batteries for low-temperature use.

ruby. Synthetic rubies, usually in the form of rods, are made from aluminum oxide containing small amounts of other metals, such as 0.05% chromic oxide, by single-crystal-growing techniques. These are used extensively in masers and lasers.
See corundum.

ruby glass. See glass, ruby.

Rucate. Powder coating resins.

rue oil and herb.
Properties: From steam distillation of fresh blossoming plants *Ruta graveolens* L., *Ruta montana* L., or *Ruta bracteosa* L. (Fam. Rutaceae). Yellow to amber liquid; fatty odor. Soluble in fixed oils, mineral oil; insoluble in glycerin, propylene glycol.
Use: Food additive.

Ruff–Fenton degradation. Shortening of the carbon chain of sugars by the oxidation of aldonic acids (as calcium salts) with hydrogen peroxide and ferric salts.

rug and carpet cleaner. Any formulation used to clean carpeting. They often contain perchloroethylene.
Hazard: Highly toxic; carcinogen.

ruhemann's purple.
Properties: A colored compound.
Derivation: Produced by the reaction ninhydrin with amino acids.

run. (1) The amount of substance processed in a single operation. (2) Complete cycle of operations. (3) Completed procedure.

runner. The secondary feed channel (usually circular) in a multiple-cavity injection mold that connects the sprue with the gate. In certain cases, molding can be performed satisfactorily without runners (runnerless molding).

run-off. see Nuclear run-on.

run-on. see Nuclear run-on.

Rupe reaction. The dehydration of an acetylenic alcohol to a substituted vinyl acetylene followed by hydration of the triple bond to give α,β-unsaturated ketones.

rupture disk. Thin pieces of metal between flanges that break at a certain pressure.

rust. (1) The reddish corrosion product formed by electrochemical interaction between iron and atmospheric oxygen; ferric oxide (Fe_2O_3). The reaction occurs most rapidly in moist air, indicating the catalytic activity of water. (2) A reddish-yellow fungus that attacks plants, especially cereal grains; it is also called smut. It can be controlled by treatment with formaldehyde.

ruthenium.
CAS: 7440-18-8. Ru.
Metallic element of atomic number 44, group VIII of the periodic table, aw 101.07, valences = 3, 4, 5, 6, 8. Seven stable isotopes.
Properties: Silvery-white solid of the platinum metal group. D 12.41 (20C), mp 2310C, bp 3900C, Brinell hardness 220 (cast). Insoluble in acids and in aqua regia. Attacked by alkaline oxidants (concentrated sodium hydroxide solution) and by fused alkalies.
Derivation: Occurs with platinum, from which it is recovered.
Grade: Technical, single crystals.
Use: Hardener for platinum and palladium in jewelry, electrical contact alloys, catalyst, medical instruments, corrosion-resistant alloys, electrodeposited coatings, nitrogen-fixing agent (experimental), solar cells (experimental); the oxide is used to coat titanium anodes in electrolytic production of chloride; the dioxide serves as an oxidizer in photolysis of hydrogen sulfide.

ruthenium chloride. (ruthenic chloride; ruthenium sesquichloride).
CAS: 10049-08-8. $RuCl_3$.
Properties: Black solid; deliquescent. D 3.11. Decomposes above 500C. Insoluble in cold water; decomposes in hot water; slightly soluble in alcohol.
Grade: Technical, CP.
Use: Analysis (testing for sulfur trioxide).

ruthenium red. (ammoniated ruthenium oxychloride). $Ru_2(OH)_2Cl_4 \cdot 7NH_3 \cdot 3H_2O$.
Properties: Brownish-red powder. Soluble in water.
Use: Microscopic stain and reagent for pectin, plant mucin, and gum.

ruthenium tetroxide.
CAS: 20427-56-9. RuO_4.
Properties: Yellow crystals. Mp 25C, bp 100C. Sublimes at room temperature. Strong oxidizing agent. Soluble in carbon tetrachloride.

Derivation: Acidic solutions of ruthenium compounds are heated with strong oxidizing agents.
Hazard: Fire risk in contact with organic materials.
Use: Oxidizing agent (carbon tetrachloride as solvent).

ruthenocene. (dicyclopentadienylruthenium). $(C_5H_5)_2Ru$.
Properties: A stable, light-yellow solid. Mp 199C.
Use: Intermediate for high-temperature compounds and for UV radiation absorbers in paints.

rutherfordium. Rr. A transfermium element. Atomic number 104. Very short half-life.

Rutherford, Sir Ernest. (1871–1937). Born in New Zealand, Rutherford studied under J. J. Thomson at the Cavendish Laboratory in England. His work constituted a notable landmark in the history of atomic research as he developed Becquerel's discovery of radioactivity into an exact and documented proof that the atoms of the heavier elements, which had been thought to be immutable, actually disintegrate (decay) into various forms of radiation. Rutherford was the first to establish the theory of the nuclear atom and to carry out a transmutation reaction (1919) (formation of hydrogen and an oxygen isotope by bombardment of nitrogen with α-particles). Uranium emanations shown to consist of three types of rays, α (helium nuclei) of low penetrating powder, β (electrons), and γ, of exceedingly short wavelength and great energy. Rutherford also discovered the half-life of radioactive elements and applied this to studies of age determination of rocks by measuring the decay period of radium to lead-206.
See Aston; Bohr.

rutile. TiO_2. Natural titanium dioxide. May contain up to 10% iron.
Properties: Red to reddish-brown to black color, adamantine to submetallic luster, streak palebrown. Hardness 6–6.5, d 4.3, mp 1640C, refr index 2.7. Insoluble in acids.
Occurrence: Virginia, West Africa, Australian beach sands.
Use: Source of titanium and titanium compounds; ceramics; steel deoxidizer; welding rod coatings; pigment for paints, enamels, and tile.

Ruzicka large-ring synthesis. Formation of large-ring alicyclic ketones from dicarboxylic acids by thermal decomposition of salts with metals of the second and fourth groups of the periodic table (calcium, thorium, cerium).

Ruzicka, Leopold. (1887–1976). A chemist who won the Nobel Prize in 1939 with Butenandt. His work involved research in organic synthesis including polymethylenes and higher terpenes. He was the first chemist to synthesize musk, androsterone, and testosterone from cholesterol. His medical degree was awarded at the University of Basel, Switzerland, although he was born in Croatia and partially educated in Germany.

ryania.   An insecticidal principle extracted from the wood of a family of tropical American trees (*Ryania speciosa*). In commercial practice the wood is fine-ground and used as a component of certain insecticide formulations.
Hazard: Moderately toxic.

"Rynite" *[Du Pont].*   TM for a glass-reinforced polyester engineering plastic; it is extremely stiff, with glass content of 45%. A modified polyethylene terephthalate featuring high temperature resistance, high tensile and impact strength, and good electrical resistance.

"Ryolex" *[Silbrico].*   TM for industrial perlite insulation.
Use: For coatings, adhesives, and sealants; as a filter aid for purifying chemicals.
See perlite.

S

S. Symbol for (1) sulfur, (2) see Svedberg, (3) see entropy.

s-9 fraction. An enzyme preparation.
Use: In vitro to test a mutagenic chemical to ascertain whether metabolic activation is essential to the expression of mutagenicity.

Sabatier, Paul. (1854–1941). A French chemist who received the Nobel Prize in chemistry in 1912 along with Victor Grignard. His work involved the behavior of oxides as oxidizing catalysts and as agents for dehydrating and dehydrogenating. He received his Ph.D. in Nimes, France, and went on to become lecturer and faculty member in Toulouse, France.

Sabatier–Senderens reduction. Catalytic hydrogenation of organic compounds in the vapor phase by passage over hot, finely divided nickel (the oldest of all hydrogenation methods).

D-saccharic acid. (2,3,4,5-tetrahydroxyhexanedioic acid; tetrahydroxyadipic acid). $COOH(CH_2O)_4COOH$. The 1,6-dicarboxylic acid formed by the oxidation of D-glucose.
Properties: White needles or syrup. Mp 125–126C with decomposition. Very soluble in water, alcohol, or ether; deliquescent. Combustible.
Derivation: Oxidation of cane sugar, glucose, starch by nitric acid.

saccharification. Conversion of wood and other cellulosics (biomass) to glucose by acid hydrolysis or enzymic hydrolysis. An experimental method using anhydrous hydrogen fluoride with vacuum distillation is under development.

saccharin. (o-benzosulfimide; gluside; benzoylsulfonic imide).
CAS: 81-07-2. The anhydride of o-sulfimide benzoic acid. A nonnutritive sweetener.

Properties: White, crystalline powder; exceedingly sweet taste (500 times that of sucrose). Mp 226–230C. Soluble in amyl acetate, ethyl acetate, benzene, and alcohol; slightly soluble in water, chloroform, and ether.
Derivation: A mixture of toluenesulfonic acids is converted into the sodium salt then distilled with phosphorus trichloride and chlorine to obtain the o-toluene sulfonyl chloride which, by means of ammonia, is converted into o-toluenesulfamide. This is oxidized with permanganate, treated with acid, and saccharin crystallized out. In food formulations, saccharin is used in the form of its sodium and calcium salts. Sodium bicarbonate is added to provide water solubility.
Grade: Commercial, CP, USP, FCC.
Hazard: A questionable carcinogen. Products containing it must have a warning label.

saccharin ammonium. See ammonium saccharin.

saccharolactic acid. See mucic acid.

saccharose. See sucrose, carbohydrate.

saccharose unit. See carbohydrate.

Sachsse process. See BASF process.

"SACI" [Daubert]. TM for a rust-, and corrosion-preventive product effective in coatings as thin as 0.5 mil.

S acid. (1-amino-8-naphthol-4-sulfonic acid).
Properties: Gray needles, white when pure. Slightly soluble in water, insoluble in alcohol and ether.
Derivation: Fusion of 1-naphthylamine-4,8-disulfonic acid with sodium hydroxide.
Use: Azo dye intermediate.

2S acid. See Chicago acid.

sacrificial protection. The preferential corrosion of a metal coating for the sake of protecting the substrate metal. For example, when zinc is in contact with a more noble (less reactive) metal in the electromotive series, such as steel, a galvanic cell is created in which electric current will flow in the presence of an electrolyte. Atmospherically contaminated moisture constitutes the electrolyte. Under these conditions, the zinc coating rather than the steel is affected. Thus, galvanic protection with zinc is sometimes called sacrificial protection. See galvanizing.

S-adenosylmethionine. (adoMet; SAM). An enzymatic cofactor involved in methyl group transfers.

SAE. Abbreviation for Society of Automotive Engineers. The initials are applied to its specifications and tests for motor fuels, oils, and steels.

SAE steel. A grade or type of steel indicated by a number system; they are principally plain carbon and of low to medium alloy content, used primarily in machine parts. The first two numerals designate either plain carbon or the alloy grouping and quantity, and the last two the mean carbon content in hundredths of 1%. Thus, 10 indicates carbon steels, 13 manganese steels; 40 and 44, molybdenum steels (the latter of higher alloy content); 50 and 51, chromium steels (the latter of higher alloy content); 41, chromium–molybdenum; 61, chromium–vanadium; 92, silicon and silicon–chromium; 46 and 48, nickel–molybdenum (the latter of higher nickel content); and 81, 94, 86, 87, 88, 47, 43, and 93, nickel–chromium–molybdenum in the order of increasing alloy content. The letter B between the first two and last two numerals indicates the presence of boron in amounts of 0.0005–0.003% as a depth-hardening addition.

SAF black. Abbreviation for super-abrasion furnace black.

safety engineering. Application of engineering principles to chemical plant safety by professionally trained personnel. Following is a checklist of the more important items.
(1) Plant construction: separate buildings or outdoor location of hazardous reaction vessels, storage tanks, etc.; interior fire walls and doors, exterior blow-out walls, sprinkler systems, enclosed stairways, explosion vents and safety valves, scram alarm systems, color-coded pipelines.
(2) Fire and explosion prevention, dust control, proper storage of flammable liquids, grounding of electrical equipment, accessibility of extinguishers and hose lines, leak detection of reaction vessels, adequate ventilation of storage rooms, accumulation of solid wastes, static spark control (metal-free shoes, static bars on friction-generating machinery).
(3) Toxic hazards: workroom concentration of toxic agents must conform to OSHA and ACGIH tolerances.
(4) Protective equipment: goggles and gloves, acid-proof clothing, face masks and respirators, lifelines, eye-wash fountains, flooding showers. For hot labs and plants handling radioactive materials; decontamination equipment, glove boxes, and remote-control devices.
(5) Accident prevention: emergency shutoffs on machines with moving parts; housing on gears, lathes, rotors, etc.; operator-restraining devices; proper handling of chain hoists, carton stackers, pallets; training personnel in safety practice.

safety glass. (shatterproof glass).
A composite or laminate consisting of two sheets of plate glass with an interlayer of polyvinyl butyral.

The plastic is bonded to the glass such that shattering on impact is virtually eliminated. Required for automobile windshields, also used for bulletproof glass, train windows, etc.

Saffil. A group of synthetic inorganic fibers made from alumina and zirconia.

safflower oil. Drying oil from safflower (carthamus) seed, somewhat similar to linseed oil. It is nonyellowing. Contains 78% linoleic acid (unsaturated fatty acid).
Properties: Straw-colored liquid. D 0.923–0.927 (25/25C), refr index 1.4740–1.4745 (25C), acid value 0.6–1.5, iodine no. 140–152, saponification no. 186–193, unsaponifiable 0.3–1.0%. Combustible.
Derivation: Hydraulic or solvent extraction of seeds.
Use: Alkyd resins, paints, varnishes, medicine, dietetic foods, margarine, hydrogenated shortening.

saffron.
Properties: From the dried stigma of *Ceocus saffron* L.
Use: Food additive.

"Saflex" *[Solutia].* TM for polyvinyl butyral adhesive film supplied in clear, translucent, tinted, or graduated color, for the plastic interlayer in safety glass.

safranine. A family of dyes, some of which are useful as biological stains, based on phenazine, having CI Nos. 50200–50375.

safranin *o.* (3,7-dimethyl-10-phenylphenazin-10-ium-2,8-diamine chloride).
CAS: 477-73-6. $C_{20}H_{19}ClN_4$. A red dye that fluoresces when exposed to selected wavelengths of light.
Use: In forensics to visualize details of cyanoacrylate-fumed friction ridges of human digits.

safrole. (4-allyl-1,2-methylenedioxybenzene).
CAS: 94-59-7. $C_3H_5C_6H_3O_2CH_2$.
Properties: Colorless or pale-yellow oil. The odor-giving constituent of sassafras, camphorwood, and other oils. D 1.100–1.107 (15C), mp 11C, bp 233C, optical rotation −0 degrees 30 minutes (15C), refr index 1.5363–1.5385 (20C). Soluble in alcohol; slightly soluble in propylene glycol; insoluble in water and glycerol.
Derivation: From oil of sassafras or camphor oil.
Method of purification: Rectification or freezing.
Grade: Technical.
Hazard: Toxic by ingestion, may not be used in food products (FDA), a possible carcinogen.
Use: Perfumery and soaps, manufacture of heliotropin, medicine (antiseptic).

sage oil. A yellow to green essential oil used in perfumery and flavoring; dextrorotatory. There are two varieties (Clary and Dalmatian) which have different constituents.

"SAG" [Sanchez]. TM for silicone defoamers used in many industrial and chemical processes.

SAIB. Abbreviation for sucrose acetate isobutyrate.

sal ammoniac. See ammonium chloride.

salanine.
Properties: A bitter-tasting alkaloid.
Derivation: Occurs in Irish potatoes.

salicin. (salicyl alcohol glucoside). $HOCH_2C_6H_4OC_6H_{11}O_5$. A glucoside obtained from several species of *Salix* and *Populus*.
Properties: Colorless crystals or white powder. Mp 199–202C, d 1.43. Soluble in water, alcohol, alkalies, glacial acetic acid; insoluble in ether.
Use: Medicine (analgesic), reagent for nitric acid.

salicylal. See salicylaldehyde.

salicyl alcohol. (*o*-hydroxybenzyl alcohol; α-2-dihydroxytoluene; saligenin).
CAS: 90-01-7. $HOC_6H_4CH_2OH$.
Properties: White crystals; pungent taste. Mp 86–87C, d 1.16, sublimes at 100C. Very soluble in alcohol, chloroform, ether; soluble in propylene glycol, benzene, and fixed oils; sparingly soluble in cold water; soluble in hot water. Combustible.
Derivation: Hydrolysis of salicin, heating phenol and methylene chloride with caustic.
Use: Medicine (local anesthetic).

salicyl alcohol glucoside. See salicin.

salicylaldehyde. (salicylal; salicylic aldehyde; *o*-hydroxybenzaldehyde).
CAS: 90-02-8. C_6H_4OHCHO.

Properties: Colorless, oily liquid or dark-red oil; bitter, almondlike odor; burning taste. D 1.165–1.172, fp −7C, bp 196C, flash p 172F (77.7C). Soluble in alcohol, ether, and benzene; slightly soluble in water. Combustible.
Derivation: Interaction of phenol and chloroform in presence of aqueous alkali.
Use: Analytical chemistry, perfumery (violet), synthesis of coumarin, auxiliary fumigant, flavoring.

salicylaldehyde dimethyl acetal carbamate.
CAS: 6884-59-9. $C_{11}H_{15}NO_4$.

Hazard: A poison by ingestion.
Use: Agricultural chemical.

salicylamide. (*o*-hydroxybenzamide).
CAS: 65-45-2. $C_6H_4(OH)CONH_2$.
Properties: White or slightly pink crystals. Mp 139–142C, bp decomposes at 270C. Soluble in hot water, alcohol, ether, chloroform; slightly soluble in cold water, naphtha, and carbon tetrachloride.
Derivation: Treatment of methyl salicylate with dry ammonia gas.
Grade: Technical, NF.
Use: Medicine (analgesic).

salicylanilide.
CAS: 87-17-2. $HOC_6H_4CONHC_6H_5$.
Properties: Odorless, white, or slightly pink crystals. Mp 136–138C, bp decomposes, stable in air. Slightly soluble in water; freely soluble in alcohol, ether, chloroform, and benzene.
Grade: NF.
Hazard: Toxic by ingestion, irritant to skin.
Use: Fungicide, slimicide, antimildew agent, intermediate.

salicylate. (2-hydroxybenzoic acid).
CAS: 69-72-7. $C_7H_6O_3$. Any salt of salicylic acid. They form a class of nonsteroidal anti-inflammatory drugs.
Derivatives: Obtained from the bark of the white willow and wintergreen leaves.
Hazard: Respiratory alkalosis, hyperkalemia, hyperthermia, dehydration, convulsions, shock, respiratory paralysis, respiratory acidosis, lesions and death from respiratory collapse; fetotoxic.

salicylic acid. (*o*-hydroxybenzoic acid).
CAS: 69-72-7. $C_6H_4(OH)(COOH)$.

Properties: White powder; acrid taste. Stable in air but gradually discolored by light, d 1.443 (20/4C), mp 158–161C, bp 211C (20 mm Hg), sublimes at 76C. Soluble in acetone, oil of turpentine, alcohol, ether, benzene; slightly soluble in water. Combustible.
Derivation: Reacting a hot solution of sodium phenolate with carbon dioxide and acidifying the sodium salt thus formed.
Grade: Technical, USP, crude.
Hazard: Dust forms explosive mixture in air. Toxic by ingestion.
Use: Manufacture of aspirin and salicylates, resins, dyestuff intermediate, prevulcanization inhibitor, analytical reagent, fungicide.

salicylic acid, bismuth basic salt.
CAS: 14882-18-9. $C_7H_5BiO_4$.

Hazard: A poison by ingestion. Human systemic effects.
Use: Agricultural chemical.

salicylic acid dipropylene glycol monoester.
See dipropylene glycol monosalicylate.

salicylic aldehyde.   See salicylaldehyde.

salicyloyloxytributylstannane.
CAS: 4342-30-7. $C_{19}H_{32}O_3Sn$.
Hazard: A poison by ingestion. Tributyl tin compounds are extremely toxic to marine life.

saligenin.   (salicyl alcohol). $C_6H_4(OH)CH_2OH$.
Use: Treatment for rheumatism.

saligenin cyclic phenyl phosphate.
CAS: 4081-23-6. $C_{13}H_{11}O_4P$.
Hazard: A poison.
Use: Agricultural chemical.

saline water.   See brine; ocean water.

salinity.   The saltiness of natural water. The salinity of normal seawater is 35 parts salt per 1000 parts water.

salinomycin.
CAS: 53003-10-4. $C_{42}H_{70}O_{11}$.
Properties: Crystals. Mp 112.5–113.5C.
Use: Drug (veterinary); food additive.

saliva.   The secretion of salivary glands.
Properties: Clear, weakly acidic, viscous fluid that contains water, mucin, organic salts and ptyalin.
Use: In the moistening and lubrication of food.

salmine.   A protein specific to the salmon.
Use: Nutritional and biochemical research.

salol.   See phenyl salicylate.

saloquinine.   (salicyl quinine). HOC_6H_4COO $C_{20}H_{23}N_{20}$.
Use: Antipyretic, antiperiodic.

sal soda.   (washing soda; sodium carbonate decahydrate). $Na_2CO_3 \cdot 10H_2O$.
Properties: White crystals. D 1.44, mp 32.5–34.5C (loses water at this temperature). Easily soluble in water; insoluble in alcohol. A pure form of sodium carbonate (soda ash).
Use: Washing textiles, bleaching linen and cotton, general cleanser.

salt.   (1) The compound formed when the hydrogen of an acid is replaced by a metal or its equivalent (e.g., an NH_4^+ radical). Example:
$HCl + NaOH \rightarrow NaCl + H_2O$.

This is typical of the general rule that the reaction of an acid and a base yields a salt and water. Most inorganic salts ionize in water solution.
(2) Common salt, sodium chloride, occurs widely in nature, both as deposits left by ancient seas and in the ocean, where its average concentration is 2.6%.
See sodium chloride; soap.

salt bath.   A molten mixture of sodium, potassium, barium, and calcium chlorides or nitrates to which sodium carbonate and sodium cyanide are sometimes added. Used for hardening and tempering of metals and for annealing both ferrous and nonferrous metals. Temperatures used may be as high as 1315C for hardening high-speed steels. Commercial mixtures are available for a variety of specifications.
See fused salt.

salt bridge.   An ionic bond between charged groups that are part of larger covalent structures; salt bridges occur in many proteins.

salt cake.   Impure sodium sulfate (90–99%).
Properties: For properties and derivation see sodium sulfate.
Grade: Technical, glassmakers (iron-free).
Use: Paper pulp, detergents and soaps, plate and window glass, sodium salts, ceramic glazes, dyes.
See sodium sulfate.

salt, fused.   See fused salt.

salting out.   Reduction in the water solubility of an organic solid or liquid by adding a salt (usually sodium chloride) to an aqueous solution of the substance. Ions of the dissolved salt attract and hold water molecules, thus making them less free to react with the solute. The result of this is to decrease the solubility of the solute molecules with consequent separation or precipitation. Colloidal suspensions of proteins, soaps, and similar substances are precipitated in this way.

salt, molten.   See fused salt.

salt of tartar.   See acid potassium tartrate.

saltpeter.   See niter, potassium nitrate.

salt, rock.   See sodium chloride.

salts of fatty acids.
Properties: Consists of aluminum, calcium, magnesium, potassium, and sodium salts of capric, caprylic, lauric, myristic, oleic, palmitic, and stearic acids manufactured from fats and oils derived from edible sources.
Use: Food additive.

salvage pathway.   Synthesis of a biomolecule from intermediates in the degradative pathway for the biomolecule; a recycling pathway, as distinct from a de novo pathway.

salvarsan. (dihydroxydiaminoarsenobenzene dihydrochloride). $C_{12}H_{14}O_2N_2Cl_2As_2 \cdot 2H_2O$.
Use: To treat syphilis.

salvia oil. The Dalmatian variety of sage oil.

SAM. See S-adenosylmethionine.

samarium. Sm. A rare-earth metal of the lanthanide group (group IIIB of the periodic table), atomic number 62, aw 150.4; valences = 2, 3; seven stable isotopes.
Properties: Hard, brittle metal that quickly develops an oxide film in air. An active reducing agent. Ignites at 150C, liberates hydrogen from water. D 7.53, mp 1072C, bp 1900C, hardness similar to iron, high neutron absorption capacity. Combustible.
Occurrence: Australia, Brazil, southeastern U.S., South Africa; also from bastnasite ore in California.
Derivation: Reduction of the oxide with barium or lanthanum.
Use: Neutron absorber, dopant for laser crystals, metallurgical research, permanent magnets.

samarium chloride. $SmCl_3 \cdot 6H_2O$.
Properties: Faintly yellow, hygroscopic crystals. D 2.383. Loses $5H_2O$ at 110C. Soluble in water.
Derivation: By treating the carbonate or oxide with hydrochloric acid.

samarium oxide. Sm_2O_3.
Properties: Cream-colored powder. D 8.347, mp 2300C. Insoluble in water; soluble in acids. Absorbs moisture and carbon dioxide from the air.
Use: Catalyst in the dehydrogenation of ethanol, infrared-absorbing glass, neutron absorber, preparation of samarium salts.

sampling. The methods and the techniques used in obtaining representative test samples of quantity lots of raw materials, semiprocessed work, and finished product for production and quality control. Rules for sampling procedures for both solid and liquid materials have been established by the National Cottonseed Products Association, Memphis, TN and by the National Institute of Oilseed Products, San Francisco, CA. The techniques of physical sampling are one application of statistical quality control.

SAN. Abbreviation for styrene-acrylonitrile polymer.
See polystyrene.

Sancar, Aziz. (1946–). A Turkish-American biochemist and molecular biologist that received the 2015 Nobel Prize jointly with Thomas Lindahl and Paul Modrich. The scientists mapped how cells repair damaged DNA while safeguarding the genetic information at a genetic level. This work has provided fundamental knowledge of how a living cell functions and is used for the development of new cancer treatments. Sancar's research mapped how cells repair DNA damage from ultraviolet light. His M.D. was awarded by Istanbul University and his Ph.D. was awarded by the University of Texas at Dallas. He is a Distinguished Professor at the University of North Carolina–Chapel Hill School of Medicine.

sand. Sediment particulates ranging in size from 1/16 to 2 mm.
See silica.

sandalwood oil. (santal oil). A pale-yellow, essential oil; strongly levorotatory.
Use: In fragrances, perfumes, and flavoring.

sandalwood oil, West Indian oil. See amyris oil, West Indian type.

sandarac. A natural resin obtained from Morocco. Its commercial form is yellow, brittle, amorphous lumps or powder; soluble in alcohol; insoluble in benzene and water.
Use: Special types of varnishes and lacquers.

sand casting. See foundry sand.

Sandmeyer diphenylurea isatin synthesis. Formation of a cyanoformamidine by treatment of a symmetrical diphenylthiourea with potassium cyanide in alcohol containing lead carbonate, reduction with ammonium sulfide, and ring closure with concentrated sulfuric acid to isatin-2-anil; also formed smoothly by ring closure of the cyanoformamidine with aluminum chloride in benzene or carbon disulfide.

Sandmeyer isonitrosoacetanilide isatin synthesis. Formation of isonitrosoacetodiphenylamidine by condensation of chloral hydrate, hydroxylamine, and aniline; cyclization with concentrated sulfuric acid; and quantitative hydrolysis to isatin on dilution.

Sandmeyer reaction. Replacement of diazonium groups in aromatic compounds by halo or cyano groups in the presence of cuprous salts, copper powder, or cupric salts.

sandstone. A siliciclastic sedimentary rock consisting primarily of sand, usually sand that is predominantly quartz.

sandwich molecule. See metallocene; ferrocene.

Sanger, Frederick. (1918–2013). An English biochemist who won the Nobel Prize for chemistry in 1958. His research was on protein structure. He identified the amino acid sequence of the protein insulin. His Ph.D. was awarded from Cambridge University.

Sanger sequencing. A widely used method of determining the order of bases in DNA. See sequencing; shotgun sequencing.

sanguinarine hydrochloride.
CAS: 5578-73-4. $C_{20}H_{14}NO_4 \cdot Cl$.
Hazard: Moderately toxic by ingestion.

sanitizer. A special class of disinfectant designed for use on food-processing equipment, dairy utensils, dishes, and glassware in restaurants. Among them are the hypochlorites, chloramines, and other organic, chlorine-liberating compounds, and quaternary ammonium compounds, many of which are proprietary.
See antiseptic; disinfectant.

santal oil. See sandalwood oil.

santalol.
CAS: (α) 8006-87-9. $C_{15}H_{24}O$. A sesquiterpene alcohol.
Properties: Colorless liquid; odor of oil of sandalwood. D 0.971–0.973, refr index 1.504–1.508, bp 300C. Soluble in three parts of 70% alcohol; insoluble in water. Combustible.
Derivation: From sandalwood oil.
Use: Perfumery.

santalyl acetate. Acetic acid ester of a mixture of α- and β-santalols.
Properties: Colorless liquid; light odor of sandalwood. D 0.982–0.985, refr index 1.487–1.492.
Derivation: Treatment of sandalwood oil or santalol with acetic anhydride.
Use: Perfumery.

"Santicizer" [Ferro]. TM for a series of plasticizers, including various sulfonamides, phthalates, and glycollates.

"Santicizer 141" [Ferro]. TM for an alkylaryl phosphate ester plasticizer.
Properties: Clear, oily liquid.
Use: Imparts flame resistance, good low temperature properties, light and weathering stability, and stability in vinyls.

"Santicizer 148" [Ferro]. (alkyl diaryl phosphate ester). TM for a flexible, processable, and compatible flame-retardant plasticizer.
Properties: Clear liquid.
Use: For vinyl and nonvinyl resins.

Santochlor. (p-dichlorobenzene).
CAS: 106-46-7. Moth- and mildew-control agents.
Use: For space deodorants.

"Santocure" [Flexsys]. TM for a series of accelerators for natural and synthetic rubbers.

"Santoflex" [Monsanto]. TM for a series of rubber antioxidants and antiozonants.

α-santonin.
CAS: 481-06-1. $C_{15}H_{18}O_3$. A tricyclic structure.
Properties: White powder turning yellow on exposure to light; odorless; tasteless at first, then bitter. D 1.187, mp 170–173C, bp sublimes, specific rotation −170 to −175 degrees (2 g/100 mL alcohol). Soluble in chloroform, alcohol, and most volatile and fatty oils; very slightly soluble in water; solutions are levorotatory.
Derivation: By extraction from Artemisia.
Hazard: Toxic by ingestion, affects color vision.
Grade: Technical.
Use: Medicine (anthelmintic).

"Santoquin" [Novus]. TM for ethoxyquin.
Use: Antioxidant in animal feeds and dehydrated forage crops.

Santovar A. 2,5-di-*tert*-amylhydro-quinone.

"Santowax" R [Solutia]. TM for mixed terphenyls. Yellowish-white, noncrystalline, flaked solid.
Use: Extender for polystyrene.

saponification. The chemical reaction in which an ester is heated with aqueous alkali such as sodium hydroxide to form an alcohol (usually glycerol) and the sodium salt of the acid corresponding to the ester. The process is usually carried out on fats (glyceryl esters of fatty acids). The sodium salt formed is called a soap. A typical saponification reaction is $(C_{17}H_{35}COO)_3C_3H_5 + 3NaOH \rightarrow 3C_{17}H_{35}COONa + C_3H_5(OH)_3$.
See soap.

saponification number. The number of milligrams of potassium hydroxide required to hydrolyze 1 g of a sample of an ester (glyceride, fat) or mixture.

saponin. (1) A general term applied to two groups of plant glycosides that on shaking with water form colloidal solutions giving soapy lathers; they form oil/ester emulsions and are used as protective colloids. They also have the ability to hemolyze red blood corpuscles at very great dilutions. The two groups are triterpenoid and steroid saponins; the latter are used in research on sex hormones.
(2) Specific term refers to saponin derived from Saponaria or Quillaja.
Properties: White, amorphous glucoside; pungent, disagreeable taste and odor. It foams strongly when shaken with water. Soluble in water.
Grade: Crude, purified, highest purity.

Hazard: Highly toxic by injection; destroys red blood cells. Moderately toxic by ingestion.
Use: Foam producer in fire extinguishers, detergent in textile industries, sizing, substitute for soap, emulsification agent for fats and oils.

sapphire. (Al_2O_3). Synthetic sapphire is made by crystal-growing techniques.
Properties: Hard, crystalline solid. D 3.98, Mohs hardness 9.0, mp 2040C, dielectric strength 480 kV/cm, dielectric constant 9.0 (20C), coefficient of friction 0.05 micron, inert to strong acids and alkalies, excellent high-temperature stability. Can be sealed to glass. High transmission in infrared and ultraviolet.
Available forms: Rods, spheres, disks, whiskers, single crystals.
Use: Electron and microwave tubes, optical elements in radiation detectors, substrate for thin-film components and integrated circuits, abrasive, record needles, precision instrument bearings, aluminum composites, micromortars for hand-pulverizing chemicals.
See corundum.

saprozoic. Of or pertaining to organisms that feed on solutions of organic material rather than on solid organic matter.

saquinavir mesylate. See fortovase.

saran. See polyvinylidene chloride; saran fiber.

saran fiber. Generic name for a manufactured fiber in which the fiber-forming substance is any long-chain synthetic polymer composed of at least 80% by weight of vinylidene chloride units ($-CH_2CCl_2-$) (Federal Trade Commission).
Properties: Tenacity 2–2.5 g/denier, elongation 15–30%, softens at 115–137C. Highly resistant to most chemicals and solvents, weather, moths, and mildew. Combustible but self-extinguishing.
Use: Screens, upholstery, curtain and drapery fabrics, rugs and carpets, awnings, filter cloth.
See polyvinylidene chloride.

sarcolysin. See melphalan.

sarcomere. A functional and structural unit of the muscle contractile system.

sarcosine. (methyl glycocoll; methylaminoacetic acid). CH_3NHCH_2COOH.
Properties: Deliquescent crystals; sweet taste. Mp 210–215C (decomposes). Very soluble in water; slightly soluble in alcohol. Combustible.
Derivation: Decomposition of creatine or caffeine.
Grade: Technical.
Use: Synthesis of foaming antienzyme compounds for toothpaste, cosmetics, and pharmaceuticals.

sardine oil. See fish oil.

Sarett oxidation. Oxidation of primary and secondary alcohols to aldehydes and ketones by means of CrO_3^- pyridine complex.

sarin. (methylphosphonofluoride acid, isopropyl ester).
CAS: 107-44-8. $[(CH_3)_2CHO](CH_3)FPO$. A nerve gas.
Hazard: Toxic by inhalation and skin absorption, cholinesterase inhibitor.

Sartoner 213 monomer. 1,4-butanediol diacrylate.
CAS: 1070-70-8.
Available forms: Liquid.
Use: A cross-linking agent with polymerization characteristics. Product is a hard, clear, infusible resin. Uses include inks, adhesives, textile products, polyesters, and photoresists.

SAS. (1) Abbreviation for sodium aluminum sulfate.
See aluminum sodium sulfate.
(2) Abbreviation for sodium alkane sulfonate.

SASOL process. See gasification.

satellite. A chromosomal segment that branches off from the rest of the chromosome but is still connected by a thin filament or stalk.

satellite DNA. Highly repeated, nontranslated segments of DNA whose function is not clear.

satin spar. See calcite.

Satintone. (aluminum silicate). Pigments, coatings, and reinforcements.
Use: In plastics, rubber, adhesives, inks, and pharmaceuticals.

satin white. A high-bulking filler used in paper-coating formulations; a mixture of hydrated lime, potash alum, and aluminum sulfate. Readily hydrated. Particle size range 0.2–2 microns.

saturation. (1) The state in which all available valence bonds of an atom (especially carbon) are attached to other atoms. The straight-chain paraffins are typical saturated compounds. (2) The state of a solution when it holds the maximum equilibrium quantity of dissolved matter at a given temperature.
See unsaturation; solubility (true); supersaturation.

saveall. A device used in the papermaking industry to reclaim wastewater containing suspended solids (fiber, pigment, etc.) for further processing. There are three types: (1) sedimentation, (2) vacuum filter, and (3) flotation. Types 1 and 2 are large tanks made of tile or concrete, type 3 is a rotary vacuum filter drum that separates solids from wastewater.

savory oil.
CAS: 8016-68-0. An essential oil used in flavoring, especially in stuffing for meats and poultry.

saxitoxin.
CAS: 35523-89-8. $C_{10}H_{17}N_7O_4 \cdot 2HCl$. A toxic principle present in certain species of shellfish. It is a paralytic poison that attacks the central nervous system, acting as a muscular nerve block.

Saybolt Universal viscosity. The efflux time in seconds (SUS) of 60 mL of sample flowing through a calibrated Universal orifice in a Saybolt viscometer under specified conditions.
See viscosity.

"Saytex" *[Albermarle Corp.].* TM for flame retardant for engineering plastics used in high temperature applications. Its polymeric structure encourages non-blooming and resists breakdown.
Use: For demanding molding applications such as cool body toasters, PC diskette drives, and electronic connections.

Saytzeff (Zaitsev) rule. The rule predicts that in elimination reactions the olefin predominantly formed will be the one with the largest possible number of substituents on the carbon–carbon double bond.

Sb. Symbol for antimony (from Latin stibium).

SBA. Abbreviation for *sec*-butyl alcohol.

SBG. Abbreviation for standard battery grade, highly purified chemicals manufactured for use in the battery industry.

SBR. Abbreviation for styrene–butadiene rubber.

Sc. Symbol for scandium.

scaffold. In genomic mapping, a series of contigs that are in the right order but not necessarily connected in one continuous stretch of sequence.

scale. (1) A calcareous deposit in water tubes or steam boilers resulting from deposition of mineral compounds present in the water, e.g., calcium carbonate. (2) A type of paraffin or petroleum wax from which all but a few percent of oil has been removed by hydraulic pressing and subsequent processing. (3) A graduated standard of measurement in which the units (degrees) are defined in relation to some property of what is measured, e.g., temperature scale, Brix scale, Baumé scale. (4) The markings indicating such units as on a thermometer or graduate. (5) A weighing device that may be of the beam type in which weights (poises) and lever systems are used, or of the direct-reading spring type in which the gravitational pull of the object being weighed is counterbalanced by a known constant spring force.
See boiler scale; balance (2).

scale-up. A term used in chemical engineering to describe the calculations and planning involved in carrying a chemical processing operation from the pilot plant to large-scale production stage.

scandium. Sc. Metallic element of atomic number 21, group IIIB of the periodic table, aw 44.9559, valence = 3, no stable isotopes.
Properties: Silvery-white solid. Mp 1539C, bp 2727C, d 2.99. Does not tarnish in air, reacts rapidly with acids, strongly electropositive. Not attacked by 1:1 mixture of nitric and 48% hydrofluoric acids. It is chemically similar to the rare earths.
Source: Chief ores are wolframite and thortveitite (Norway, Madagascar).
Derivation: Reduction of scandium fluoride with calcium or with zinc or magnesium alloy. Purification by distillation gives purity of 99+%.
Use: No major industrial use. An artificial radioactive isotope has been used in tracer studies and leak detection, and there is some application in the semiconductor field.

scandium fluoride. ScF_3.
Derivation: Reaction of scandium oxide with ammonium hydrogen fluoride.
Use: Preparation of scandium metal.

scandium oxide. Sc_2O_3.
Properties: White, amorphous powder. D 3.864, specific heat 0.153, a weak base. Soluble in hot acids, less so in cold acids.
Source: Thortveitite.
Use: Preparation of scandium fluoride.

scanning tunneling microscope. Instrument that uses a sharply tipped electrode in close proximity to a surface. As the electrode is moved, the change in electrode signal relates to the shape of the surface being scanned.
Use: Study of atomic-scale structure of surfaces, drawing molecular-sized markings on surfaces and transport of atoms and molecules.
See atomic force microscopy; chemical force microscopy.

scatter diagram. (correlation table). The plotted record of frequency distribution of two different characters showing the tendency for one character to change as the other character or variable changes.

scavenger. (getter).
(1) In chemistry, any substance added to a system or mixture to consume or inactivate traces of impurities. (2) In metallurgy, an active metal added to a molten metal or alloy that combines with oxygen or nitrogen in the melt and causes its removal into the slag. Alloys including such metals as thorium,

zirconium, and cerium as well as carbon and misch metal are used in vacuum tubes to absorb traces of residual gases.

"Scav-Ox" [Arch].   TM for a hydrazine oxygen scavenger.
Use: Prevent corrosion in boilers and oil-wheel casings.

SCC.   Abbreviation for the Society of Cosmetic Chemists.

Schaeffer acid.   (2-naphthol-6-sulfonic acid; β-naphtholsulfonic acid).

Properties: White leaflets. Mp 122C. Soluble in water and alcohol.
Derivation: Sulfonation of β-naphthol with 94% sulfuric acid at 95C, yielding a mixture comprising chiefly Schaeffer acid (56%), R-acid (15%), and G-acid (10%). Separation is effected by dilution with water, boiling to hydrolyze the sulfonic acid group, and addition of metallic salts.
Use: Azo dye intermediate.

Schaeffer's salt.   (sodium salt of 2-naphthol-6-sulfonic acid). $C_{10}H_6OHSO_3Na$.
Use: Intermediate for organic chemicals.

Scheele, C. W.   (1742–1786). One of the outstanding early chemical thinkers and experimenters, Scheele was a Swedish scientist who discovered a number of previously unknown substances, among which were tartaric acid, chlorine, manganese salts, arsine, and copper arsenite (Scheele's green). He also noted the oxidation states of various metals, observed the nature of oxygen two years before Priestley's discovery, and discovered the chemical action of light on silver compounds, thus laying the foundation of photochemistry and photography. Scheelite, or natural calcium tungstate, is named after him.

Scheele's green.   See copper arsenite.

scheelite. $CaWO_4$. A natural calcium tungstate found in igneous rocks, usually with granite. Some molybdenum may replace tungsten.
Use: Ore of tungsten; as a phosphor.

Schiemann reaction.   Formation of diazonium fluoborates by diazotization of aromatic amines in the presence of fluoborates, followed by their thermal decomposition to aryl fluorides.

Schiff base.   A class of compounds derived by chemical reaction (condensation) of aldehydes or ketones with primary amines. The general formula is RR′X=NR″.
Properties: Usually colorless, crystalline solids, although some are dyes. Very weakly basic and hydrolyzed by water and strong acids to carbonyl compounds and amines.
Use: Rubber accelerators; dyes (phenylene blue and naphthol blue); chemical intermediate; liquid crystals in electronic display systems; perfume base.

Schmidlin ketene synthesis.   Formation of ketene by thermal decomposition of acetone over electrically heated wire at 500–750 degrees by a reaction involving radical formation with generation of methane and carbon monoxide.

Schmidt reaction.   Acid catalyzed addition of hydrazoic acid to carboxylic acids, aldehydes, and ketones to give amines, nitriles, and amides, respectively.

Scholler saccharification process.   Industrial saccharification of wood using 0.5% sulfuric acid at 170–180C and 165–180 lb/sq inch pressures. Recovered sugars are fermented to produce about 40 gal alcohol per ton of dry wood.

Scholl reaction.   Coupling of aromatic molecules by treatment with Lewis acid catalysts.

Schorigin (Shorygin) reaction.   Organometallic reactions of the Grignard type, employing sodium in place of magnesium; the reaction of alkyl sodium compounds with carbon dioxide to give monobasic acids is sometimes known as the Wanklyn reaction.

Schotten–Baumann reaction.   Acylation of alcohols with acyl halides in aqueous alkaline solution.

schradan.   (generic name for octamethyl pyrophosphoramide; OMPA).
CAS: 152-16-9. $[(CH_3)_2N]_2P(O)OP(O)[N(CH_3)_2]_2$.
Properties: Viscous liquid. D 1.137, bp 120–125C (0.5 mm Hg), refr index 1.462 (25C). Miscible with water; soluble in most organic solvents; hydrolyzed in the presence of acids, but not by alkalies or water alone.
Hazard: Toxic by ingestion and inhalation, a cholinesterase inhibitor, use may be restricted.
Use: A systemic insecticide that is absorbed by the plant which then becomes toxic to sucking and chewing insects.

Schrock, Richard R.   (1945–). An American Born in Berne, Indiana who won the Nobel Prize for chemistry in 2005 for his pioneering work concerning the development of the metathesis method in organic synthesis. Awarded a B.A. from the University of California, Riverside and a Ph.D. from Harvard University. A member of the American

Academy of Arts and Sciences and the National Academy of Sciences.

Schweitzer's reagent. A solution of copper hydroxide in strong ammonia used in analytical chemistry as a test for wool. It dissolves cotton, silk, and linen.

SCI. Abbreviation for the Society of the Chemical Industry.

scillirubroside.
CAS: 23604-99-1. $C_{30}H_{42}O_{10}$.
Hazard: A poison.
Source: Natural product.

scintillation counter. A device used to detect a pulse of radiation by emitting a flicker of light. α-radiation is counted by inorganic detectors such as sodium iodide, while organic materials such as plastics may be used for β- and γ- particles.
See phosphor.

scirpenetriol triacetate.
CAS: 4297-61-4. $C_{21}H_{28}O_8$.
Hazard: A poison by ingestion.

scission. (1) The rupture of a chemical bond with production of 5-electron-volts of energy. (2) In agricultural technology, the separation of fruit or vegetable products from the tree or vine.

scleroprotein. Any of a large class of proteins that have a supporting or protective function in tendons, bones, cartilages, ligaments, and other hard or tough parts of the animal body. They include the collagens of skin, tendons, and bones, as well as the elastic proteins known as elastins and the keratins. Specific examples are the keratin of hair, hooves, and horns, and fibroin from silk.

scleroscope hardness. See hardness.

scopolamine.
CAS: 51-34-3. $C_{17}H_{21}NO_4$. A drug used to inhibit effects of acetylcholine; viscous liquid, soluble in water and alcohol.

scouring agent. A compound used to remove the natural oils and fats from raw wool, also used to remove lubricants applied to rayon yarns or fabrics during such operations as throwing, winding, weaving, knitting, etc.

SCP. Abbreviation for single-cell protein.
See protein; single-cell.

screen. A woven, fabriclike structure made of intersecting strands of wire or plastic, usually mounted in a steel frame. They are available in a wide range of sizes, weaves, and meshes from as coarse as 25 to as fine as 400. The mesh is the number of apertures per square inch; it is the square of the number of strands of metal or plastic per linear inch. The strands can be made of any suitable metal (copper, nickel), alloy (steel, bronze), or synthetic (nylon, PVC). Some types of screen are mechanically vibrated or gyrated to facilitate solids separation; fine mesh screens require application of force, such as a stream of water, to effect separation. Screens are used for filtration, clarification of suspensions, separation and classification of solids, and removal of contaminants from semisolid materials. Lab sizes are available. A special application is the wire of a fourdrinier papermaking machine; it may be 38–60 inches. wide and 55–85 mesh; it moves continuously over return rolls, the sheet being formed upon it by filtration of wood pulp slurry.
See filter media.

screening. To screen a library (see "library") is to select and isolate individual clones out of the mixture of clones. For example, if you needed a cDNA clone of the pituitary glycoprotein hormone alpha subunit, you would need to make (or buy) a pituitary cDNA library, then screen that library in order to detect and isolate those few bacteria carrying alpha subunit cDNA.
(1) Screening by hybridization involves spreading the mixture of bacteria out on a dozen or so agar plates to grow several ten thousand isolated colonies. Membranes are laid onto each plate, and some of the bacteria from each colony stick, producing replicas of each colony in their original growth position. The membranes are lifted and the adherent bacteria are lysed, then hybridized to a radioactive piece of alpha DNA (the source of which is a story in itself—see "probe"). When X-ray film is laid on the filter, only colonies carrying alpha sequences will "light up." Their position on the membranes show where they grew on the original plates, so you now can go back to the original plate (where the remnants of the colonies are still alive), pick the colony off the plate and grow it up. You now have an unlimited source of alpha cDNA.
(2) Screening by antibody is an option if the bacteria and plasmid are designed to express proteins from the cDNA inserts (see "expression clones"). The principle is similar to hybridization, in that you lift replica filters from bacterial plates, but then you use the antibody (perhaps generated after olde tyme protein purification rituals) to show which colony expresses the desired protein.

screw. (auger; worm).
A simple machine employing the principle of the inclined plane, invented by the Greek scientist Archimedes. It consists of a central shaft around which winds a spiral of ribs (called "flights") that are integral with the shaft. The distance between the flights is the pitch, or angle of inclination of the screw. This distance may be uniform throughout the length of the screw or it may vary from one

point to another, depending on use requirements. The shorter the distance between flights, the more pressure the screw will deliver.

Screws have a number of important applications in industrial operations. (1) Extrusion of rubber, plastics, and food products: the screw rotates in a chamber or barrel, the product being introduced through a port where the flights are farthest apart; it is forced through a die at the opposite end of the barrel, which molds it to the form desired. (2) Mixing of solids: continuous mixing is possible with screws having a wide variety of pitches and contours that impart a back-and-forth motion to the product being mixed without moving it to the discharge end until it is pushed along by added material. Two screws may operate in parallel, their pitches opposing each other in such a way as to effect maximum mixing. Some screws are cored for circulation of cooling water. (3) Conveying of solids: for this purpose screws with uniform and fairly wide pitch are used; as the screw turns the solids are passed along from one flight to the next. These are used for conveying wet and dry solids, wood chips, and similar particulates. (4) An engineering application of the screw is the mechanism known as a worm gear, in which the flights of the screw engage corresponding indentations or notches in a shaft or wheel, causing it to turn.

scrubber. See absorption tower; scrubbing.

scrubbing. Process for removing one or more components from a mixture of gases and vapors by passing it upward and usually countercurrent to and in intimate contact with a stream of descending liquid, the latter being chosen so as to dissolve the desired components and not others. The gas or vapor may be broken into fine bubbles upon entering a tower filled with liquid, but more frequently the tower is filled with coke, broken stone, or other packing, over which the liquid flows while exposing a relatively large surface to the rising gas or vapor. See absorption (1).

scruple. Unit of weight used in pharmacy equivalent to 20 grains or 1/3 dram.

SDA. Abbreviation for (1) specially denatured alcohol, (2) Soap and Detergent Association.

SDDC. Abbreviation for sodium dimethyldithiocarbamate.

SDP. Abbreviation for 4,4'-sulfonyldiphenol.

Se. Symbol for selenium.

Seaboard process. Method of removing hydrogen sulfide from a gas by absorption in sodium carbonate solution. Sodium bicarbonate and sodium hydrosulfide are formed. By blowing air through this solution, hydrogen sulfide is released and carried off, and the sodium carbonate is regenerated.

Seaborg, Glenn Theodore. (1912–1999). An American chemist who won the Nobel Prize for chemistry in 1951 along with McMillan. He did research in nuclear chemistry, physics, and artificial radioactivity. He discovered the elements plutonium, americium, berkelium, californium, einsteinium, fermium, and mendelevium with his colleague. He codiscovered numerous isotopes and radioisotopes. His Ph.D. is from the University of California at Berkeley.

seaborgium. Sg. A transfermium element. Atomic number 106. Very short half-life.

sea coal. Finely ground bituminous coal used in sand mixtures for iron molds to prevent sticking.

Sea Harmony. An organic blend of North Atlantic seaweed and alfalfa.
Use: Food supplement.

sealant. Any organic substance that is soft enough to pour or extrude and is capable of subsequent hardening to form a permanent bond with the substrate. Most sealants are synthetic polymers (silicones, urethanes, acrylics, polychloroprene) that are semisolid before application and later become elastomeric. A few of the best-known sealants are such natural products as linseed oil (putty), asphalt, and various waxes.
See adhesive.

Sealstix. A cement of the deKhotinsky type which adheres to almost any surface; insoluble in all common reagents except alcohol, strong caustics, and chromic acid cleaning solution.

Sealz. A line of thermoplastic rubber compounds that are used as additives for asphalt and tar products to raise softening point, improve low-temperature flexibility, enhance adhesive qualities, and increase elasticity.

Searles Lake brine. See brine; Trona process.

seasoning. Any food additive used in low concentration to contribute to the taste of food. Best known is sodium chloride, but the term also includes various spices, onions, mustards, etc.
See enhancer; potentiator.

SeaWash. A pH neutral floor cleaning solution.
Use: Used in food preparation areas of food manufacturers, hospitals, restaurants, or schools.

SeaWash 70. A parts cleaning solution, certified as clean air solvent.
Use: For degreasing in industrial manufacturing or machine shops.

seawater. See ocean water.

seaweed.   See algae; phycocolloid; carrageenan; kelp.

sebacic acid.   (1,8-octanedicarboxylic acid; sebacylic acid; decanedioic acid).
CAS: 111-20-6. $COOH(CH_2)_8COOH$.
Properties: White leaflets. Mp 133C, d 1.110 (25C), bp 295.0C (100 mm Hg), refr index 1.422 (133.3C). Slightly soluble in water; soluble in alcohol and ether. Combustible.
Derivation: From butadiene via dichlorobutene and its nitrile derivatives, dry distillation of castor oil with alkali.
Grade: CP, purified.
Use: Stabilizer in alkyd resins, maleic and other polyesters, polyurethanes, fibers, paint products, candles and perfumes, low-temperature lubricants and hydraulic fluids, manufacture of nylon 610.

sebaconitrile. $NC(CH_2)_8CN$.
Properties: Straw-colored, oily liquid. Bp 199C (15 mm Hg).
Use: Chemical intermediate for drugs, dyes, and high polymers.

sebacoyl chloride.   (n-octane-1,8-dicarboxylic acid dichloride). $ClOC(CH_2)_8COCl$.
Properties: Liquid. Bp 137–140C (3 mm Hg), 97% pure. Decomposes slowly in cold water; soluble in hydrocarbons and ethers.
Use: Organic intermediate.

sebacoyldioxybis(tributylstannane).   See bis(tributyl(sebacoyldioxy))tin.

sebacylic acid.   See sebacic acid.

sec-.   Abbreviation for secondary.

(1-α,3-β,5z,7e,24r)-9,10-secocholesta-5,7,10(19)-triene-1,3,24-triol.
CAS: 57333-96-7. $C_{27}H_{44}O_3$.
Hazard: A poison by ingestion.

seconal sodium.   (sodium seconal). Proprietary preparation of quinalbarbitone sodium.
Use: Sedative and hypnotic.

secondary.   (1) For chemical meaning refer to "primary." (2) Designates a reversible (rechargeable) electric battery, generally called a storage battery. (3) Designates recovery of metal values from scrap or waste products such as aluminum from cans or steel from car bodies; also recovery of petroleum by chemical flooding and hydraulic fracturing.

secondary air pollutant.   A pollutant formed by chemical reaction in the atmosphere.

secondary alcohol.   An alcohol that contains the CHOH group.

secondary amide.   An amide in which two hydrogen atoms have been replaced by a hydrocarbon group.

secondary arsine.   An arsine in which two hydrogen atoms are replaced by an alkyl radical.

secondary calcium phosphate.   See phosphate, dibasic.

secondary metabolism.   Pathways that lead to specialized products, not found in every living cell. Contrast with intermediary metabolism.

secondary structure.   The localized conformation of a protein.

second law of thermodynamics.   The law stating that the entropy of the universe is increasing.

second messenger.   An effector molecule synthesized within a cell in response to an external signal (first messenger) such as a hormone.

sectrometer.   A potentiometer for electrometric titration using a cathode-ray tube instead of a microammeter.

secular trend.   The persistent tendency of a variable to increase or decrease with passage of time.

sedanolic acid. $C_{12}H_{20}O_3$. A constituent of celery seed oil. Its lactone, also present, which is the source of the characteristic odor, is sedanolide.

sedative.   A natural or synthetic therapeutic agent having the property of inducing relaxation and varying degrees of depression of the central nervous system. The major types of sedatives are (1) chlorine substitution products (chloral hydrate, chlorobutanol); (2) ethanol derivatives (ethyl carbamate, hedonal); (3) specific aldehydes and ketones (paraldehyde, amylene hydrate), (4) barbituric acid derivatives (barbital, barbiturates). Sedatives are almost always prescription drugs.
See tranquilizer.

sedentary.   Living in a fixed location, as with most plants, tunicates, sponges.
See motile.

sedimentation.   The settling out by gravity of solid particles suspended in a liquid, the rate of settling being defined by Stoke's law. This method is used industrially in water purification. It is also an analytical procedure for separation of solids of different particle size, as well as for molecular weight determination. The sedimentation of large molecules in a strong centrifugal field permits

determination of both average molecular weights and the distribution of molecular weights in certain systems. When a solution containing polymer or other large molecules is centrifuged at forces up to 250,000 times gravity, the molecules begin to settle, leaving pure solvent above a boundary that progressively moves toward the bottom of the cell. An optical system is provided for viewing this boundary, and a study as a function of the time of centrifuging yields the rate of sedimentation for the single component, or for each of many components of a polydisperse system. These sedimentation rates may then be related to the corresponding molecular weights of the species present after the diffusion coefficients for each species are determined by independent experiments. The result of this work is the distribution of molecular weights in the sample which is attainable by few other methods.
See precipitate.

sedimentation coefficient.   A physical constant specifying the rate of sedimentation of a particle in a centrifugal field under specified conditions.

seed.   That part of a plant that includes the plant embryo itself, a quantity of stored food (fats, carbohydrates, and proteins in varying proportions), and the enclosing cellulosic coats. The food-storage tissue is called the endosperm. Starch is an important food reserve in the endosperms of cereals and legumes; sugar in sweet corn; protein in soybean and wheat; and fats in such plants as coconut, cacao, castor bean, and most types of palms. Growth substances such as giberellin also occur in seeds; some also contain alkaloids (ricinine, hyoscine, and caffeine). The seeds or bulbs of some plants are highly toxic, e.g., hyacinth, larkspur, castor bean, cashew nut shells. Bitter almond contains as much as 10% of hydrogen cyanide which is removed in processing. Seeds (which in many cases are equivalent to nuts) are an important source of vegetable oils, which are used as food components, in paints and other industrial products, and in medicine. There has been considerable experimentation with sunflower seeds, coconuts, and soybeans as a source of diesel fuel.

seeding.   Trace proportions of a material introduced to initiate a desired reaction where the "seed" acts as a nucleating agent. An example is the addition of a crystal into a supersaturated solution to induce crystallization.
See nucleation.

seeding, cloud.   See nucleation.

Seger cone.   See pyrometric cone.

segregation.   The normal biological process whereby the two pieces of a chromosome pair are separated during meiosis and randomly distributed to the germ cells.

Seidlitz salt.   See epsom salts.

Seismotite.   Trade name for pumice.
Use: Abrasive in scouring agents.

Selectophore.   Products meeting the requirements for the production of reliable ion-selective electrodes and optodes.

"Selectra-Sil" [UCT].   TM for derivatizing reagents of the highest chemical purity.
Use: Reagent.

selenic acid.
CAS: 7783-08-6. H_2SeO_4.
Properties: White, hygroscopic solid. D 3.004 (15/4C), mp 58C (easily undercools), decomposes at 260C. Very soluble in water; decomposes in alcohol; usually available as a liquid.
Hazard: Strong irritant to skin and mucous membranes.

selenide.   (hydrogen selenide).
CAS: 7783-07-5. H_2Se. A metal salt of silane that is an analog of selenium.
Hazard: Toxic.

selenious acid.   See selenous acid.

selenious acid disodium salt pentahydrate.
See sodium selenite pentahydrate.

selenious acid, monosodium salt.
CAS: 7782-82-3. $HO_3Se•Na$.
Hazard: A reproductive hazard.

selenium.
CAS: 7782-49-2.   Se. A nonmetallic element, atomic number 34, group VIA of the periodic table, aw 78.96, valences = 2, 4, 6; 6 stable isotopes.
Properties: Amorphous, red powder becoming black on standing and crystalline on heating; vitreous and colloidal forms may be prepared. Crystalline form has d 4.5, mp 217C, bp 685C; amorphous form softens at 40C and melts at 217C. Crystalline selenium is a p-type semiconductor; electrically it acts as a rectifier and has marked photoconductive and photovoltaic action (converts radiant to electrical energy); the electrical conductivity increases with increasing light irradiation. Soluble in concentrated nitric acid and (in liquid form) in common alkalies; forms binary alloys with silver, copper, zinc, lead, etc. A necessary nutritional factor for animals.
Occurrence: Canada, Japan, Yugoslavia, Mexico; also in certain soils.
Grade: Commercial (powder or lumps), high-purity up to 99.999%.
Hazard: Eye and upper respiratory tract irritant. Questionable carcinogen.

Use: Electronics, xerographic plates, TV cameras, photocells, magnetic computer cores, solar batteries (rectifiers, relays), ceramics (colorant for glass), steel and copper (degasifier and machinability improver), rubber accelerator, catalyst, trace element in animal feeds.

selenium diethyldithiocarbamate.
CAS: 5456-28-0. $Se[SC(S)N(C_2H_5)_2]_4$.
Properties: Orange-yellow powder; characteristic odor. D 1.32 (20/20C), melting range 63–71C. Soluble in carbon disulfide, benzene, chloroform; insoluble in water.
Hazard: Toxic by inhalation, ingestion, and skin absorption.
Use: Vulcanization agent without added sulfur, or as a primary or secondary accelerator with sulfur.

selenium dioxide. (selenous acid anhydride).
CAS: 7446-08-4. SeO_2.
Properties: White or yellowish-white to slightly reddish, lustrous, crystalline powder or needles. D 3.954 (15/15C), mp 340–350C (sublimes). Soluble in alcohol, water.
Hazard: Toxic by inhalation, ingestion, and skin absorption.
Use: Analysis (testing for alkaloids), oxidizing agent, antioxidant in lubricating oils, catalyst.

selenium oxychloride. (seleninyl chloride; selenium chloride oxide). $SeClO_2$.
Properties: Nearly colorless or yellowish, fuming, corrosive liquid, that is miscible with benzene, carbon disulfide, carbon tetrachloride, chloroform, and toluene.
Hazard: Extremely hazardous, toxic irritant, poison, severe vesicant.

selenium oxyfluoride. (seleninyl fluoride; selenyl fluoride). SeF_2O. A liquid that rapidly attacks glass.
Properties: A corrosive fuming liquid.
Hazard: Strong irritant, and vesicant, may cause fatal pulmonary edema.

selenium sulfide. (selenium disulfide).
CAS: 7446-34-6. SeS_2.
Properties: Bright-orange powder. Mp less than 100C. Practically insoluble in water and organic solvents.
Grade: USP.
Hazard: Toxic by ingestion, strong irritant to eyes and skin. Questionable carcinogen.
Use: Medicine (treatment of seborrhea, medicated shampoos).

selenium tetrafluoride.
CAS: 10026-03-6. SeF_4.
Properties: Colorless, fuming liquid. Fp −10C, bp 105C, d 2.75; reacts strongly with phosphorus and with water (hydrolysis). Soluble in alcohol, sulfuric acid, ether, and carbon tetrachloride.

Derivation: Reaction of selenium chloride and silver fluoride.
Hazard: Irritant.
Use: Fluorinating agent.

selenoformaldehyde.
CAS: 6596-50-5. CH_2Se.
Hazard: A poison.
Use: Agricultural chemical.

***dl*-selenomethionine.**
CAS: 2578-28-1. $C_5H_{11}NO_2Se$.
Hazard: A poison.

selenous acid. (selenious acid).
CAS: 7783-00-8. H_2SeO_3.
Properties: Transparent, colorless, deliquescent crystals. D 3.0066, mp 70C (decomposes). Soluble in water and alcohol; insoluble in ammonia.
Derivation: Action of hot nitric acid on selenium.
Hazard: Toxic by inhalation, ingestion, and skin absorption.
Use: Reagent for alkaloids.

selenous acid anhydride. See selenium dioxide.

semantide. A molecule that carries information in the sense that it has the potential to transfer information within a living system.

Semesan. A wettable powder containing 25.3% hydroxymercurichlorophenol.
Hazard: As for mercury compounds.

Semesan Bel. A seed disinfectant containing 12.5% hydroxymercurinitrophenol and 3.8% hydroxymercurichlorophenol.
Hazard: As for mercury compounds.

semicarbazide hydrochloride. (carbamylhydrazine hydrochloride; aminourea hydrochloride).
CAS: 563-41-7. $H_2NCONHNH_2HCl$.
Properties: White crystals. Mp 172–175C (decomposes). Soluble in water; insoluble in absolute alcohol and ether.
Derivation: From hydrazine sulfate, potassium or sodium cyanate, and sodium carbonate, or electrolytically by the reduction of nitrourea.
Grade: CP, technical.
Hazard: Toxic by ingestion. Questionable carcinogen.
Use: Reagent for aldehydes and ketones, isolation of hormones and isolation of certain fractions from essential oils.

semiconductor. An element or compound having an electrical conductivity intermediate between that of conductors and nonconductors (insulators). Most metals have quite high conductivity, while substances like diamond and mica have very low conductivity (high resistance). Between these extremes lie the semiconductors, of which germanium, silicon, silicon carbide, and selenium are examples, with resistivities in the range of 10^{-2} to 10^9 ohms/cm. Slight traces of impurities in the crystalline structure are essential for semiconduction; arsenic is a typical impurity in semiconductor crystals. These impurities function as electron donors or acceptors, and the semiconductor is designated n-type or p-type, depending on the electrical nature of the "holes" or energy deficits in the crystalline lattice.

The functioning of semiconductors involves the science of solid-state physics. Their discovery in the early 1940s made possible the development of transistors, with their manifold applications in electronic devices in which they have largely replaced the vacuum tube.

There are a few organic semiconducting compounds that contain a significant amount of carbon–carbon bonding and are also capable of supporting electronic conduction. Anthracene and Ziegler-catalyzed acetylene polymers (conjugated polyolefins) are examples.

See crystals; impurity; solid; solid-state chemistry.

semimicrochemistry. Any chemical method (usually analytical) in which the weight of the sample used is from 10 to 100 mg.

semipermeable membrane. See membrane, semipermeable.

semisynthetic. A term often used to describe end products that are manufactured from natural materials but do not occur in the free state, for example, paper, glass, soap, cement, rayon, leather, etc.

Semmler–Wolff reaction. Rearrangement of α,β-unsaturated cyclohexenyl ketoximes into aromatic amines under acidic conditions.

Semyonov, Nikolay Nikolayevich. (1896–1986). A Russian chemist and physicist who won the Nobel Prize in 1956. He authored books on the chain reaction and problems of chemical kinetics and reactivity, as well as many articles. His work concerning thermal combustion and explosion is utilized in rockets and jet engines. He received his doctorate at Leningrad State University.

S1 end mapping. A technique to determine where the end of an RNA transcript lies with respect to its template DNA (the gene). Can't be described in a short paragraph. See "RNAse protection assay" for a closely related technique.

Sense strand. The strand of DNA during transcription which is not transcribed into mRNA. This results from its position as collinear, as opposed to complementary, for the RNA sequence. It makes "sense" with the genetic code.

Sentry. (1) Propionic acid. (2) High purity grades of Carbowax polyethane glycol, polyox water-soluble polymers, and polyvinyl chloride. Solvent for flavors and colors.

Use: Fungistats for the control of certain molds and yeast in foods; humectant for baked goods, and plasticizer for cork seals and crowns.

Separan. A series of flocculating agents.

AP30. Synthetic, high molecular weight, anionic polymer.

C-90 and C-120. Synthetic, high molecular weight, cationic polymers.

MGL. Similar to NP10. Used in production of uranium.

NP10. Synthetic, water-soluble, nonionic, high molecular weight polymer of acrylamide. "Separan" NP 10 potable-water-grade flocculant has been accepted, subject to maximum use concentration of 1 ppm, by the U.S. Public Health Service.

NP20 Nonionic polyacrylamide polymer.

PG2 Similar to NP10. Used in paper manufacture.

separation. A collective term including a large number of unit operations that, in one way or another, isolate the various components of a mixture. Chief among these are evaporation, distillation, drying, gas absorption, sedimentation, solvent extraction, press extraction, adsorption, and filtration. Specialized methods include centrifugation, electromagnetic separation (mass spectrograph), gaseous diffusion, and various types of chromatography.

Sephadex [GE Healthcare]. TM for a dry, insoluble powder composed of microscopic beads that are synthetic organic compounds derived from the polysaccharide dextran. The dextran chains are cross-linked to give a three-dimensional network, and the functional ionic groups are attached to the glucose units of the polysaccharide chains by ether linkages. Available in various forms for use in many different phases of chromatography.

sepia. The red-brown color (pigment).

Derivation: Ink of cuttlefish, calcareous portion of the backs of cuttlefish.

septanose ring. A seven-membered ring formed by reaction of a CHO group at one end of the septanose molecule with CH_2OH at the other end.

septicine. A ptomaine produced from hexylamine and amylamiine during putrefaction of red meat.

septiphene. See o-benzyl-p-chlorophenol.

septum. A membranous partition as such that separates two cavities in a plant or animal.

Seqlene. Chelating agents.
Use: In derusting and descaling processes for metal; concrete admixtures to retard set and reduce water required; textile applications; aluminum etching; and caustic bottle washing.

sequence. See base sequence.

sequence assembly. A process whereby the order of multiple sequenced DNA fragments is determined.

sequence tagged site. (STS). Short (200 to 500 base pairs) DNA sequence that has a single occurrence in the human genome and whose location and base sequence are known. Detectable by polymerase chain reaction.

sequencing. Determination of the order of nucleotides (base sequences) in a DNA or RNA molecule, or the order of amino acids in a protein.

sequencing technology. The instrumentation and procedures used to determine the order of nucleotides in DNA.

sequestration. The formation of a coordination complex by certain phosphates with metallic ions in solution so that the usual precipitation reactions of the latter are prevented. Thus, calcium soap precipitates are not produced from hard water treated with certain polyphosphates and metaphosphates (these polyphosphates are often improperly referred to as hexametaphosphates). The term *sequestration* may be used for any instance in which an ion is prevented from exhibiting its usual properties due to close combination with an added material. Two groups of organic sequestering agents (chelates in these examples) of economic importance are the aminopolycarboxylic acids such as ethylenediaminetetraacetic acid, and the hydroxycarboxylic acids such as gluconic, citric, and tartaric acids. In the food industry sequestering agents aid in stabilizing color, flavor, and texture.
See chelate; coordination compound; complex.

Sequestrene. A series of complexing agents and metal complexes consisting of ethylenediaminetetraacetic acid and salts.

Ser. Abbreviation for serine.

serendipity. An unexpected scientific discovery that turns out to be more important than the project being researched. One example is the discovery of the sodium sulfide polymer later known as "Thiokol." White, the researcher, was seeking to develop an improved automotive coolant. Another is the discovery of the reinforcing effect of carbon black on rubber while technologists were using it as a black pigment to counteract the whitening effect of zinc oxide.

serendipity berry. See sweetener, nonnutritive.

SERI. Abbreviation for Solar Energy Research Institute.
See solar energy.

serine. (β-hydroxyalanine; α-α-β-hydroxypropionic acid). $HOCH_2CH(NH_2)COOH$.
A nonessential amino acid occurring naturally in the L($-$) form.
Properties: Colorless crystals. Soluble in water; insoluble in alcohol and ether. Optically active. DL-serine: mp 246C with decomposition; D($+$)-serine: mp 228C with decomposition.
Available forms: Commercially in all three forms.
Derivation: Hydrolysis of protein (especially silk protein); synthesized from glycine.
Use: Biochemical research, dietary supplement, culture media, microbiological tests, feed additive.

dl-**serine.**
CAS: 302-84-1. $C_3H_7NO_3$.
Properties: White crystals from powder, or crystalline powder. Mp 246C (decomp). Soluble in water; insoluble in alc, ether.
Use: Food additive.

Serini reaction. Zinc-promoted rearrangement of 17-hydroxy-20-acetoxysterol derivatives into C-20 ketones. The reaction is applicable to other cyclic, as well as open-chain alcohols.

serolysin. A lytic bactericidal substance in blood serum.

serotonin. (5-hydroxytryptamine; 5-hydroxy-3-(β-aminoethyl)indole).
CAS: 50-67-9. $C_{10}H_{12}N_2O$. A powerful vasoconstrictor occurring in the brain and blood platelets, it is thought to play a part in regulation of blood pressure and also has muscle-contracting properties. It has been isolated from beef serum and may be synthesized from 5-benzyloxyindole. Serotonin is similar to the hallucinogenic drugs mescaline, LSD, psilocin, and psilocybin. Its formation is inhibited by *p*-chlorophenylalanine.

serpentine. A type of asbestos.
See asbestos.

serum. (1) The continuous phase of a biocolloid after the solid or disperse phase has been removed by centrifugation, coagulation, or similar means. In the case of milk, for example, the serum, or whey, is a true solution of sugars, proteins, and mineral compounds in water.
(2) Specifically blood from which corpuscles, platelets, etc., have been removed, especially when prepared with antigenic bacteria for inoculation to effect the cure of a disease.

serum albumin. Blood albumin comprising 60% of the plasma proteins.
Use: As a plasma volume expander.

SES. Abbreviation for sodium-2,4-dichlorophenoxyethyl sulfate.
See sesone.

sesame oil. (benne oil; teel oil).
CAS: 8008-74-0. A bland, yellowish, vegetable oil.
Use: Shortenings, salad oil, margarine, and similar food products.

sesamex. (generic name for 2-(2-ethoxyethoxy) ethyl-3,4-(methylenedioxy)phenyl acetal of acetaldehyde).
$CH_2O_2C_6H_3OCH(CH_3)OC_2H_4OC_2H_4OC_2H_5$.
Properties: Liquid. Bp 137–141C (0.08 mm Hg). Soluble in nonpolar solvents.
Use: Synergist for insecticides.

sesamin. (generic name for 2,6-bis[3,4-(methylenedioxy)phenyl]-3,7-dioxabicyclo[3.3.0] octane). $C_{20}H_{18}O_6$.
Properties: Solid. Mp 122.5C. Combustible.
Use: Synergist for insecticides and fungicides.

sesamolin. (generic name for 6-[3,4-(methylene-dioxy)phenoxy]-2-[3,4-(methylenedioxy)phenyl] 3,7-dioxabicyclo[3.3.0]octane). $C_{22}H_{18}O_7$. Constituent of sesame oil that is a powerful synergist of pyrethrum. A 1:1 mixture of pyrethrum and sesamoline has 31 times the insecticidal activity of pyrethrum alone.

sesin.
CAS: 94-83-7. $C_{15}H_{12}Cl_2O_3$.
Hazard: Moderately toxic by ingestion.
Use: Agricultural chemical.

sesone. (generic name for sodium-2,4-dichlorophenoxy-ethyl sulfate; SES; "Crag").
CAS: 136-78-7. $C_6H_3Cl_2OCH_2CH_2SO_4Na$.
Properties: Crystalline solid. Mp 245C. Soluble in water.
Hazard: Gastrointestinal irritant. Questionable carcinogen.
Use: Preemergence herbicide.

sesqui. A prefix meaning one-and-a-half. Often used for salts in which the proportions of metal oxide to acid anhydride are 2:3, or vice versa, as in sodium sesquicarbonate.

sesquiterpene. A terpene having the formula $C_{15}H_{24}$, one and a half times the standard terpene formula of $C_{10}H_{16}$. An example is cadinene, a constituent of various wood oils (cade oil, cedarleaf oil, etc.).

set point. A value, such as temperature, to be maintained by a controlling instrument for a procedure.

setting point. (1) The temperature at which a liquid sets or congeals. (2) The temperature at which an oil stops flowing when acted on by a small, definite pressure under the proper conditions.

setting up. The cessation of flow in a drying coating such as varnish.

Sevin.
CAS: 63-25-2.
Properties: White crystals. Mp 142C, d 1.232 (20C/20C).
TM for a carbaryl insecticide (1-naphthyl-N-methyl carbamate).
Hazard: Toxic by inhalation.
See carbaryl.

"Sevron" *[Sensient]*. TM for a group of cationic dyes.
Available forms: Liquid and powder.
Use: For acrylic and cationic-dyeable polyester apparel and carpet.

sewage sludge. A mixture of organic materials resulting from purification of municipal waste. There are two types: (1) Imhoff sludge: a low-grade sludge containing 2–3% ammonia and about 1% phosphoric acid.
See Imhoff tank.
(2) Activated sludge: a high-grade sludge containing 5.0–7.5% ammonia and 2.5–4.0% phosphoric acid.
Derivation: (1) By running sewage through settling tanks without access of air. The sludge or solid matter is decomposed by anaerobic bacteria. Lagooning may be used effectively following activated sludge treatment. (2) By running sewage through settling tanks and pumping air (or oxygen) through porous plates at the bottom of the tanks; 20% of the current "make" is also added. The waste acts as nutrient for aerobic bacteria that consume the polluting organic matter. The resulting solids are filtered and dried.
Use: Fertilizer base.

sewer gas.
Properties: Usually flammable, explosive gas, comprised chiefly of methane.
Derivation: Produced by decaying organic matter in sewers.
Hazard: Toxic.

sex chromosome. The X or Y chromosome in human beings that determines the sex of an individual. Females have two X chromosomes in diploid cells; males have an X and a Y chromosome. The sex chromosomes comprise the 23rd chromosome pair in a karyotype.
See autosome.

sex-linked. Traits or diseases associated with the X or Y chromosome; generally seen in males.
See gene; mutation; sex chromosome.

sexual reproduction.   A type of reproduction in which two parents give rise to offspring that have unique combinations of genes inherited through the gametes of the two parents. Sexual reproduction involves meiosis and syngamy.

S-Fatty Acids.   Fatty acids derived from soybean oil. S-210 is a soy fatty acid developed primarily for alkyd resins. S-230 is a soy-type fatty acid having a lower unsaturated acids content and higher titer than S-210.
Properties: S-210 is a light-yellow liquid and S-230 a light-yellow semisolid at ambient temperature.
Use: Chemical intermediate, paints and varnishes, alkyd resins, and soaps.

SFS.   (1) Abbreviation for Saybolt Furol seconds. See Furol viscosity; Saybolt Universal viscosity. (2) Abbreviation for sodium formaldehyde sulfoxylate.

Sg.   Symbol for seaborgium.

SGR.   Abbreviation for steam gas recycle process for shale oil recovery.
See shale oil.

shale oil.
CAS: 68308-34-9.
A mixed-base crude oil extracted from mountains of sedimentary shale in Colorado, Utah, and Wyoming by heating at 425–535C (approximately 800–1000F). Two methods can be used: surface mining and excavation. In the first, the shale is bulldozed from beds, crushed, and fed into retorts of either the vertical or horizontal type. In the second, shafts are driven into the mountain and the shale is heated in situ by direct combustion in an interior chamber excavated for the purpose. The in situ method is the more efficient. The oil-bearing component of the shale is called kerogen. Only 20–30 gallons (less than one barrel) of oil is obtained per ton of shale processed, and less than 33% of the total oil content is recoverable. Major obstacles are the vast earth-moving operations necessary, the need for large volumes of cooling water, and disposal of the spent shale. Though rather extensive pilot plant operations have been carried out in recent years, no large-scale production is likely for the indefinite future for both economic and technical reasons.
See kerogen; oil shale; synfuel.
Hazard: Confirmed carcinogen.

shape-selective catalyst.   See catalyst, shape-selective.

shark liver oil.
Properties: Yellow to red-brown liquid; strong odor. D 0.917–0.928. Soluble in ether, chloroform, benzene, and carbon disulfide.
Derivation: By expression from shark livers.
Method of purification: Chilling and filtration.

Grade: Crude, refined.
Use: Source of vitamin A and squalene, biochemical research.

sharp.   (1) In reference to cheeses, this term denotes length of the curing period, usually at least six months, twice as long as for mild cheeses. (2) Descriptive of a technique for quick-freezing of foods at −20F.

Sharpless, K. Barry.   (1941–). An American who won the Nobel Prize for chemistry in 2001 for his work concerning chirally catalyzed oxidation reactions. Awarded an Undergraduate degree from Dartmouth College and a Ph.D. from Stanford University (E. E. van Tamelen) in 1968. Professorships at Massachusetts Institute of Technology, Stanford and The Scripps Research Institute. Awarded the ACS Arthur C. Cope Award, 1992 and many other awards.

Sharpless reaction.   Metal-catalyzed asymmetric epoxidation of allylic alcohols employing a system consisting of titanium tetraisopropoxide, (+)- or (−)-diethyl tartrate, and *tert*-butyl hydroperoxide. The epoxy alcohols are obtained with a high degree of optical purity (90%), and their absolute stereochemistry is predictable.

Shea butter.   A substance obtained from the nuts of *Bassia*.
Properties: Grey-white solid. D 0.9175.
See "Shebu" *[R. I. T. A.]*.

shear.   The ratio between a stress (force per unit area) applied laterally to a material and the strain resulting from this force. Determination of this ratio is one method of measuring the viscosity of a liquid or semisolid.
See viscosity.

"Shebu" *[R. I. T. A.].*   TM for shea butter.
Derivation: Extracted from fruit of the Karite tree.
Use: To ameliorate dry skin.

Shechtman, Dan.   (1941–). An Israeli chemist who was awarded the 2011 Nobel Prize. Shechtman discovered the quasicrystals, an atomic structure with five-fold symmetry but non-repeating patterns. This discovery was contrary to conventional atomic arrangements. He was criticized by some scientists for his revolutionary discovery until 1987, when scientists from France and Japan made large quasicrystals which could be examined with X-rays. He was educated from the Technion-Israel Institute of Technology where he earned his M.S. and Ph.D. Shechtman had an illustrious career that included awards such as the Wolf Prize in Physics (1999), and the EMET Prize in Chemistry (2002). He is a distinguished professor at the Technion-Israel Institute.

Sheliflex.   A series of petroleum products largely composed of naphthenic and/or paraffinic hydrocarbons.
Properties: Colors ranging from near water-white to black; odor slight to none. Viscosity 3–55 cP at 99C, d 0.85–0.95, very low volatilities.
Hazard: Combustible.
Use: Rubber processing and extending oils, miscellaneous process oil uses.

shell.   (1) In physical chemistry, this term is applied to any of the several sets of, or orbits of, the electrons in an atom as they revolve around its nucleus. They constitute a number of principal quantum paths representing successively higher energy levels. There may be from one to seven shells, depending on the atomic number of the element and corresponding to the seven periods of the periodic table. The shells are usually designated by number, though letter symbols have been used, i.e., K, L, M, N, O, P, Q. The laws of physics limit the number of electrons in the various shells as follows: two in the first (K), eight in the second (L), 18 in the third (M), and 32 in the fourth (N). With the exception of hydrogen and helium, each shell contains two or more orbitals, each of which is capable of holding a maximum of two electrons. See quantum number; orbital theory; Pauli exclusion principle.
(2) The hard integument of mollusks and crustaceans, consisting mostly of calcium carbonate, chitin, etc. See nacre; oyster shells.
(3) The brittle covering of avian eggs, chiefly calcium carbonate, lime, etc. The formation of proper shell structures in certain species of birds is said to be adversely affected by DDT and similar insecticidal contaminants of their food.
(4) The shells of nuts are cellulosic in character. Some contain industrially useful oils.

shellac.   (lac; garnet lac; gum lac; stick lac).
Derivation: A natural resin secreted by the insect *Laccifer lacca (Coccus lacca)* and deposited on the twigs of trees in India. After collection, washing, and purification by melting and filtering, it is formed into thin sheets, which are later fragmented into flakes of orange shellac. This may be dewaxed and bleached to a transparent product. Soluble in alcohol; insoluble in water.
Grade: (1) Orange: TN (impure), fine, superfine, heart, superior; (2) bleached and dewaxed (colorless).
Hazard: (Alcohol solution) Flammable, dangerous fire risk.
Use: Sealer coat under varnish; finish coat for floors, furniture, etc.; dielectric coatings, deKhotinsky cement.

shellac, bleached.
Properties: From the resinous secretion, called lac, of the insect *Llaccifer (Tachardia) lacca Kerr* (fam. *Coccidae*). Off white, amorphous, granular solid.

Soluble in alc; insoluble in water; slightly soluble in acetone, ether.
Use: Food additive.

shellac, bleached, wax-free.
Properties: From the resinous secretion, called lac, of the insect *Llaccifer (Tachardia) lacca Kerr* (fam. *Coccidae*). light yellow, amorphous, granular solid. Soluble in alcohol; insoluble in water; slightly soluble in acetone, ether.
Use: Food additive.

Shell Sol.   A series of aliphatic hydrocarbon solvents composed of a mixture of paraffinic, naphthenic, and aromatic hydrocarbons.

Shellwax.   A series of paraffin waxes derived from distillate lube streams.

sherardizing.   The process by which relatively small articles made of iron or steel are coated with zinc powder. The metal forms an alloy with the steel surface and produces a thin, tightly adherent, corrosion-resistant coating.

Sherbelizer.   An algin derivative-vegetable gum composition.
Use: Stabilizer for sherbets, water ices, frozen fruits, syrups, purees, chocolate ice cream.

Sherbrite.   Brighteners for nickel plating. Active ingredient is 1,2-benzisothiazolin-3-one-1,1-dioxide sodium salt.

sher, extract.   See euphorbia tirucalli l., extract.

shielding.   Protection of personnel from the harmful effects of ionizing radiation and/or neutrons by enclosure of the equipment (reactor, particle accelerator, X-ray generator, etc.) with an absorbing material. The most efficient of these are cadmium, lead, steel, and high-density concrete (3 ft of concrete equals about 1 ft of steel). Among the plastics, polyethylene affords reasonable protection in thick sheets. Water and paraffin wax are good neutron absorbers because of their high hydrogen content. Materials such as graphite and beryllium are able to deflect and retard neutrons and are used for this purpose as moderators in nuclear reactors. Heavy water is also used.

shiga toxin.   See shigella dysenteriae toxin.

shigella dysenteriae toxin.
CAS: 75757-64-1.
Hazard: A poison.
Source: Natural product.

Shimomura, Osamu.   (1928–). A Japanese chemist who was awarded the 2008 Nobel Prize conjointly with Martin Chalfie and Roger Y. Tsien. The scientists discovered and researched the green fluorescent protein (GPF), naturally produced in jellyfish (*Aequorea victoria*), which is used to illuminate certain cells to study their biological processes. Shimomura was primarily interested in understanding the chemistry and biochemistry in production of the GPF bioluminescence. He was awarded his Ph.D. by Nagoya University followed by an outstanding career which includes the Pearse Prize (2004), Emile Chamot Award (2005), Asahi Prize (2006), and the Japan Order of Culture (2008). In 2001, he retired as a professor emeritus from the Marine Biological Laboratory, Massachusetts.

Shine–Dalgarno sequence.   A sequence in an mRNA required for binding prokaryotic ribosomes. The sequence consists of 4–9 purine residues, located 8–13 bp upstream of the initiation codon.

Shirakawa, Hideki.   (1936–). Awarded Nobel Prize in chemistry in 2000 jointly with Alan J. Heeger and Alan G. MacDiarmid for the discovery and development of conductive polymers. He performs his research at University of Tsukuba, Japan.

shiu oil.   (apopino oil).
A Japanese oil with a high linalool content.

shoddy.   Reclaimed scrap wool, rubber, leather, etc., often used in the manufacture of low-quality products.

Shoelkopf acid.   (1-naphthol-4,8-disulfonic acid). $C_{10}H_5OH(SO_3H)_2$.
Properties: Colorless crystals.
Derivation: Decomposition of 1-naphthylamine-4,8-disulfonic acid by diazotization and acidifying with heat.
Use: Azo dye intermediate.

short.   Baked products: crisp, flaky, e.g., shortcake; steel: brittle, friable under certain temperature conditions, e.g., hot-short (above red heat) and cold-short (below red heat). Clay: dry, lacking plasticity.

shortening.   See cooking (5).

shortstopping agent.   A material used in a polymerization reaction to cut off the reaction at a predetermined point, e.g., use of diethylhydroxylamine or sodium dimethyldithiocarbamate to control synthetic rubber polymerization; tetraethylsilanol is used in silicone polymers.

short-term exposure limit.   (STEL). The 15-minute concentration of an air contaminant that should not be exceeded at any time during a workday.
See Threshold Limit Value.

shotgun cloning.   The practice of randomly clipping a larger DNA fragment into various smaller pieces, cloning everything, and then studying the resulting individual clones to determine what occurred.

shotgun method.   Sequencing method that involves randomly sequenced cloned pieces of the genome, with no foreknowledge of where the piece originally came from.
See genomic library.

shotgun sequencing.   See shotgun method.

shot metal.   Lead alloy with less than 3% arsenic.

shuttle vector.   A recombinant DNA vector that can be replicated in two or more different host species, making possible the movement of DNA between organisms.

SI.   Abbreviation for International System of Units (metric system) which is now in use in most countries but not yet officially adopted in U.S.

Si.   Symbol for silicon.

sickle-cell anemia.   A human disease characterized by hemoglobin molecules in which a Glu is replaced by a Val. This creates a "sticky patch" when the hemoglobin is deoxygenated, causing hemoglobin molecules to polymerize and deform the red blood cells.

side chain.   See chain.

siderite.   (chalybite; spathic iron ore). $FeCO_3$ (usually with some calcium, magnesium, or manganese). The term *siderite* is also used for an iron alloy found in meteorites.
Properties: Gray, yellow, brown, green, white, or brownish-red mineral; vitreous inclining to pearly luster; white streak. D 3.83–3.88, Mohs hardness 3.5–4.
Occurrence: U.S. (Vermont, Massachusetts, Connecticut, New York, North Carolina, Pennsylvania, Ohio), Europe.
Use: An ore of iron, when high in manganese used in the manufacture of spiegeleisen.

SiDigester Liquid.   A silicone digester liquid.

sienna.   A yellowish hydrated iron oxide. Raw sienna is a brown-tinted yellow ocher occurring in Alabama, California, Pennsylvania, Cyprus, and Italy. Burnt sienna is an orange-brown pigment made by calcining raw sienna.
See ocher; iron oxide reds.
Use: Colorant in oil paints, stains, pastels, etc.

sieve.   See screen.

Sievert's law.   Refers to the solubility of molecules that dissociate during solution, varies as the square root of the pressure.

siglure. (generic name for *sec*-butyl-6-methyl-3-cyclohexane-1-carboxylate). $CH_3C_6H_8COOCH(CH_3)C_2H_5$.
Properties: Liquid. Bp 113–114C (15 mm Hg). Soluble in most organic solvents; insoluble in water. Combustible.
Use: Insect attractant.

sigma blade. A rotating agitator set horizontally in a kneading bowl or chamber used for mixing doughs and heavy pastes. The blade or arm is shaped somewhat like a Greek capital sigma (Σ) lying on its side; variations of this shape simulate horizontal letters S and Z. Some kneaders have two such blades that overlap as they turn to provide maximum mixing efficiency.
See kneading.

sigma bond. A covalent bond directed along the line joining the centers of two atoms. They are the normal single bonds in organic molecules.
See pi bond.

sigma function. Enthalpy of an air-stream mix, minus the heat of the liquid.

sigma phase. The nonmagnetic, brittle, corundum-hard FeCr constituent in stainless steel.

sigma value. The value of a quantum number, which quantizes the component of angular momentum, of spin about the axis of a diatomic molecule.

signal sequence. An amino-terminal sequence that signals the cellular destination of a newly synthesized protein.

signal transduction. The process by which an extracellular signal is converted to an intracellular response.

sig water. The alkaline solution of soda ash, borax, or ammonia for washing the grain surface of leather before applying color or dye.

silane. (silicon tetrahydride).
CAS: 7803-62-5. SiH_4.
Properties: A gas; repulsive odor. Solidifies at −200C, bp −112C, d 0.68. Decomposes in water; insoluble in alcohol and benzene.
Hazard: Dangerous fire risk, ignites spontaneously in air. Strong irritant to tissue, skin and upper respiratory tract.
Use: Doping agent for solid-state devices, production of amorphous silicon.

silane compounds. Gaseous or liquid compounds of silicon and hydrogen (Si_nH_{2n+2}), analogous to alkanes or saturated hydrocarbons. SiH_3 is called silyl (analogous to methyl), and Si_2H_5 is disilanyl (analogous to ethyl). A cyclic silicon and hydrogen compound having the formula SiH_2 is called a cyclosilane. Organofunctional silanes are noted for their ability to bond organic polymer systems to inorganic substrates.
Hazard: Dangerous fire risk.
See silicone; siloxane.

"Silastic" [Dow Corning]. TM for compositions in physical character comparable to milled and compounded rubber prior to vulcanization but containing organosilicon polymers. Parts fabricated of "Silastic" are serviceable from −73 to +260C, retain good physical and dielectric properties in such service, show excellent resistance to compression set, weathering, and corona. Thermal conductivity is high, water absorption low.
Use: Diaphragms, gaskets and seals, O-rings, hose, coated fabrics, wire and cable, and insulating components for electrical and electronic parts.

"Silcogum" [Grayson]. TM for a rubber additive.
Use: For coloring high consistency silicone rubber.

"Silcolapse 430" [Bluestar]. TM for a 100% non-aqueous antifoam liquid.

"Silcopas" [Grayson]. TM for a coloring liquid.
Use: To color silicone rubber and RTV silicone.

sildenafil citrate. See "Viagra" [Pfizer].

silent mutation. A mutation in a gene that causes no detectable change in the biological characteristics of the gene product.

silica. (silicon dioxide). SiO_2. Occurs widely in nature as sand, quartz, flint, diatomite.
Properties: Colorless crystals or white powder; odorless and tasteless. D 2.2–2.6; thermal conductivity about half that of glass, mp 1710C, bp 2230C, high dielectric constant, high heat and shock resistance. Insoluble in water and acids except hydrogen fluoride; soluble in molten alkali when finely divided and amorphous. Combines chemically with most metallic oxides; melts to a glass with lowest known coefficient of expansion (fused silica). Noncombustible.
Derivation: Can be made from a soluble silicate (water glass) by acidification, washing and ignition. Arc silica is made from sand, vaporized in a 3000C electric arc.
Grade: By purity and mesh size, silica aerogel, hydrated, precipitated.
Hazard: Toxic by inhalation, chronic exposure to dust may cause silicosis. Questionable carcinogen.
Use: (Powder) Manufacture of glass, water glass, ceramics, abrasives, water filtration, microspheres, component of concrete, source of ferrosilicon and elemental silicon, filler in cosmetics, pharmaceuticals, paper, insecticides, hydrated and precipitated grades as rubber reinforcing agent, including silicone rubber, anticaking agent in foods, flattening agents in paints, thermal insulator. (Fused) Ablative

material in rocket engines, spacecraft, etc.; fibers in reinforced plastics; special camera lenses. (Amorphous) Silica gel.
See quartz; silicic acid; silica gel.

silica, amorphous fume.   (amorphous silica fume).
CAS: 69012-64-2.
Hazard: Respirable dust. Questionable carcinogen.

silica, fumed.   A colloidal form of silica made by combustion of silicon tetrachloride in hydrogen–oxygen furnaces. Fine, white powder.
Use: Thickener, thixotropic, and reinforcing agent in inks, resins, rubber, paints, cosmetics, etc. Base material for high-temperature mortars.
See "Aerosil" *[Evonik]*; "Cab-O-Sil" *[Cabot]*.

silica, fused.   (amorphous quartz).
CAS: 60676-86-0. O_2Si.
Properties: Mw 60.09.
Made up of spherical submicroscopic particles under 0.1 micron in size.
Hazard: Questionable carcinogen.
Use: Concrete, grouts, mortars, elastomers, refractory and coating applications.
See silica; quartz, fused.

silica gel.
CAS: 7631-86-9. (silica, amorphous hydrated; silicic acid).
A regenerative adsorbent consisting of amorphous silica. Noncombustible.
Derivation: From sodium silicate and sulfuric acid.
Grade: Commercial grades capable of withstanding temperatures up to 260–315C are supplied in the following mesh sizes: 3–8, 6–16, 14–20, 14–42, 28–200, and through 325.
Use: Dehumidifying and dehydrating agent, air-conditioning, drying of compressed air and other gases and liquids such as refrigerants and oils containing water in suspension, recovery of natural gasoline from natural gas, bleaching of petroleum oils, catalyst and catalyst carrier, chromatography, anticaking agent in cosmetics and pharmaceuticals, in waxes to prevent slipping, in dietary supplements.
See silicic acid; "Britesorb" *[PQ]*.

Silic AR.   Silica-gel-based formulations suitable for various chromatographic applications. The numerical suffixes indicate the approximate pH of a 10% slurry. Letters F, G, or GF indicate that the product contains a fluorescent material, gypsum binder, or both. "TLC" indicates suitability for thin-layer chromatography.

silicate.   Any of the widely occurring compounds containing silicon, oxygen, and one or more metals with or without hydrogen. The silicon and oxygen may combine with organic groups to form silicate esters. Most rocks (except limestone and dolomite) and many mineral compounds are silicates. Typical natural silicates are gemstones (except diamond),

beryl, asbestos, talc, clays, feldspar, mica, etc. Portland cement contains a high percentage of calcium silicates. Best known of the synthetic (soluble) silicates is sodium silicate (water glass).
Hazard: (Natural silicate dusts) Toxic by inhalation.
Use: Fillers in plastics and rubber, paper coatings, antacids, anticaking agents, cements.

silicate garden.   The irregular, colored, tubular growths formed in dilute aqueous silicate solutions by dropping water solutions of heavy metal salts into it.

silicic acid.   (hydrated silica).
CAS: 7699-41-4. $SiO_2 \cdot nH_2O$.
The jellylike precipitate obtained when sodium silicate solution is acidified. The proportion of water varies with the conditions of preparation and decreases gradually during drying and ignition, until relatively pure silica remains. During drying the jelly is converted to a white, amorphous powder or lumps.
Use: Laboratory reagent and reinforcing agent in rubber.
See silica gel.

silicic acid (ortho).
CAS: 10193-36-9. H_4O_4Si.
Hazard: A poison.

silicic acid, sodium salt.
CAS: 1344-09-8.
Hazard: Moderately toxic by ingestion. Low toxicity by skin contact. A severe skin and eye irritant.

silicochloroform.   See trichlorosilane.

silicofluoride.   (hexafluorosilicon(2-)). F_6Si^{-2}. A compound of silicon and fluorine combined with another element.

silicol.   Silicic oxide casein metaphosphate.

silicomanganese.   Alloys consisting principally of manganese, silicon, and carbon.
Use: Low-carbon steel in which silicon is not objectionable. Silicon manganese steels are used for springs and high-strength structural steels.
See ferromanganese.

silicomolybdic acid.   See 12-molybdosilicic acid.

silicon.   (silicon powder, amorphous).
CAS: 7440-21-3.   Si.
Nonmetallic element Atomic number 14, group IVA of the periodic table, aw 28.086, valence = 4, three stable isotopes. It is the second most abundant element (25% of the earth's crust) and is the most important semiconducting element; it can form more compounds than any other element except carbon.

Properties: Dark-colored crystals (the octahedral form in which the atoms have the diamond arrangement). The amorphous form is a dark-brown powder (see silicon, amorphous). D 2.33, mp 1410C, bp 2355C, Mohs hardness 7, dielectric constant 12, coordination number 6. Soluble in a mixture of nitric and hydrofluoric acids and in alkalies; insoluble in water, nitric acid, and hydrochloric acid. Combines with oxygen to form tetrahedral molecules in which one silicon atom is surrounded by four oxygen atoms. In this respect it is similar to carbon. It is also capable of forming –Si=Si– double bonds in organosilicon compounds.

Occurrence: Does not occur free in nature but is a major portion of silica and silicates (rocks, quartz, sand, clays, etc.).

Derivation: Crystalline silicon is made commercially (96–98% pure) in an electric furnace by heating SiO_2 with carbon, followed by zone refining. It can be purified to 99.7% by leaching. The ultra-pure semiconductor grade (99.97%) is obtained by reduction of purified silicon tetrachloride or trichlorosilane with purified hydrogen; the silicon is deposited on hot filaments (800C) of tantalum or tungsten. In a one-step method, sodium fluorosilicate is reacted with sodium, the heat produced being sufficient to form silicon tetrafluoride; this, when reacted with sodium, yields high-purity silicon and sodium fluoride. The process requires no heat except that provided by the original reaction. Single crystals of both n- and p-type are grown by highly specialized techniques.

Grade: Ferrosilicon, regular (97% silicon), semiconductor or hyperpure (99.97% silicon), amorphous.

Hazard: Flammable in powder form.

Use: Semiconductor in solid-state devices (transistors, photovoltaic cells, computer circuitry, rectifiers, etc.); organosilicon compounds; silicon carbide; alloying agent in steels, aluminum, copper, bronze, and iron (ferrosilicon); cermets and special refractories; halogenated silanes; spring steels; deoxidizer in steel manufacture.

See silica; silicate; silicone; ferrosilicon; silicon, amorphous.

silicon, amorphous. A noncrystalline allotrope of silicon that exists in the form of a dark-brown powder. It is made from silane (SiH_4) plus doping agents in a glow-discharge tube at low pressure. A film only a few microns in thickness is deposited on a glass or metal substrate. The amorphous product contains about 20% hydrogen. It has been found superior to crystalline silicon in the manufacture of solar cells.

silicon–bronze. An alloy of copper, tin, and silicon used for telephone and telegraph wires.

silicon carbide.
CAS: 409-21-2.   SiC.
Properties: Bluish-black, iridescent crystals. Mohs hardness 9, d 3.217, sublimes with decomposition at 2700C. Insoluble in water and alcohol; soluble in fused alkalies and molten iron. Excellent thermal conductivity, electrically conductive, resists oxidation at high temperatures. Noncombustible, a nuisance particulate.

Derivation: Heating carbon and silica in an electric furnace at 2000C.

Available forms: Powder, filament, whiskers (3 million psi), single crystals.

Hazard: Upper respiratory tract irritant. Probable carcinogen.

Use: Abrasive for cutting and grinding metals, grinding wheels, refractory in nonferrous metallurgy, ceramic industry and boiler furnaces, composite tubes for steam reforming operations. Fibrous form used in filament-wound structures and heat-resistant, high-strength composites.

See Carborundum; "Carbofrax" *[Saint Gobain]*.

silicon–copper. (copper silicide).
Properties: A hard, tough, bronzelike alloy containing 10–30% silicon.
Derivation: From silica and copper electrolytically.
Use: Manufacture of silicon–bronze.

silicon dioxide. (silica; silicic anhydride; silicon oxide; dioxosilane).
CAS: 7631-86-9. O_2Si.
Properties: A hard crystalline compound.
Derivation: Occurs naturally as sand, quartz, tridymite and crysobalite.
Hazard: Not toxic if ingested, inhaled silica dust can cause silicosis; carcinogen.
Use: In the manufacture of abrasives, glass and ceramics.

silicone. (organosiloxane).
Any of a large group of siloxane polymers based on a structure consisting of alternate silicon and oxygen atoms with various organic radicals attached to the silicon:

$$-OSi\left[\begin{array}{c}CH_3\\ |\\ |\\ CH_3\end{array}\right]-O-Si\left[\begin{array}{c}CH_3\\ |\\ |\\ CH_3\end{array}\right]_n-OSi-\begin{array}{c}CH_3\\ |\\ |\\ CH_3\end{array}$$

Discovered by Kipping in England in 1900.
Properties: Liquids, semisolids, or solids depending on molecular weight and degree of polymerization. Viscosity ranges from less than 1 to more than 1 million centistokes. Polymers may be straight chain, or cross-linked with benzoyl peroxide or other free radical initiator, with or without catalyst. Stable over temperature range from −50 to +250C. Very low surface tension; extreme water repellency; high lubricity; excellent dielectric properties; resistant to oxidation, weathering, and high temperatures; permeable to gases. Soluble in most organic solvents; unhalogenated types are combustible.
Derivation: (1) Silicon is heated in methyl chloride to yield methylchlorosilanes; these are separated and purified by distillation and the desired compound mixed with water. A polymeric silicone results. (2) Reaction of silicon tetrachloride

and a Grignard reagent (RMgCl), with subsequent hydrolysis and polymerization.

Available forms: Fluids, powders, emulsions, solutions, resins, pastes, elastomers.

Use: (Liquid) Adhesives, lubricants, protective coatings, coolants, mold release agents, dielectric fluids, heat transfer, wetting agents and surfactants, foam stabilizer for polyurethanes, diffusion pumps, antifoaming agent for liquids, textile finishes, water repellent, weatherproofing concrete, brake fluids, cosmetic items, polishes, foam shields in solar energy collectors, rust preventives. (Resin) Coatings, molding compounds, laminates (with glass cloth), filament winding sealants, room-temperature curing cements, electrical insulation, impregnating electric coils, bonding agent, modifier for alkyd resins, vibration-damping devices. (Elastomer, or silicone rubber) Encapsulation of electronic parts; electrical insulation; gaskets; surgical membranes and implants; automobile engine components; flexible windows for face masks, air locks, etc.; miscellaneous mechanical products.
See organosilicon.

silicone oil. See silicone.

silicone rubber. See silicone.

silicon–gold alloy. See gold–silicon alloy.

silicon monoxide.
CAS: 10097-28-6. SiO.
Properties: Amorphous, black solid. D 2.15–2.18, sublimes at high temperature, hard and abrasive. Noncombustible.
Grade: Lumps, powders, tablets; optical.
Use: To form thin surface films for protection of aluminum coatings, optical parts, mirrors, dielectrics, or insulators.

silicon nitride. Si_3N_4.
Properties: Gray, amorphous powder (can be prepared as crystals). Sublimes at 1900C, d 3.44, bulk d 70–75 lb/cu ft depending on mesh, Mohs hardness 9+, thermal conductivity 10.83 Btu/in/sq ft/hr/F (400–2400F). Resistant to oxidation, various corrosive media, molten aluminum, zinc, lead, and tin; soluble in hydrogen fluoride.
Derivation: Reaction of powdered silicon and nitrogen in an electric furnace at 1300C.
Use: Refractory coatings, bonding silicon carbide, mortars, abrasives, thermocouple tubes in molten aluminum, crucibles for zone-refining germanium, rocket nozzles, high-strength fibers and whiskers, insulator and passivating agent in transistors and other solid-state devices.

silicon tetrabromide. (tetrabromosilane). $SiBr_4$.
Properties: Fuming, colorless liquid that turns yellow in air; disagreeable odor. Decomposed by water

with evolution of heat, d 2.82 (0C), bp 153C, mp 5C. Noncombustible.
Grade: Purity: 99.999%.
Hazard: Strong irritant to tissue.

silicon tetrachloride. (tetrachlorosilane; silicon chloride).
CAS: 10026-04-7. $SiCl_4$.
Properties: Colorless, exceedingly mobile, fuming liquid; suffocating odor. Corrosive to most metals when water is present; in the absence of water it has practically no action on iron, steel, or the common metals and alloys, and can be stored and handled in metal equipment without danger. D 1.483 (20C), bulk d 12.4 lb/gal, fp −70C, bp 57.6C, refr index 1.412 (20C). Miscible with carbon tetrachloride, tin tetrachloride, titanium tetrachloride, and sulfur mono- and dichlorides; decomposed by water and alcohol with evolution of hydrogen chloride. Noncombustible.
Derivation: Heating silicon dioxide and coke in a stream of chlorine.
Grade: Technical, 99.5%, CP (99.8%), semiconductor.
Hazard: Toxic by ingestion and inhalation, strong irritant to tissue.
Use: Smoke screens; manufacture of ethyl silicate and similar compounds; production of silicones; manufacture of high-purity silica and fused silica glass; source of silicon, silica, and hydrogen chloride; lab reagent.

silicon tetrafluoride. (tetrafluorosilane; silicon fluoride).
CAS: 7783-61-1. SiF_4.
Properties: Colorless gas; suffocating odor similar to hydrogen chloride. Fumes strongly in air, d 3.57 (gas, air = 1) (15C), fp −90C, bp −86C. Absorbed readily in large quantities by water with decomposition; soluble in absolute alcohol. Noncombustible.
Derivation: (1) Action of hydrogen fluoride or concentrated sulfuric acid and a metallic fluoride on silica or silicates; (2) direct synthesis.
Grade: Pure, 99.5% min.
Hazard: Toxic by inhalation, strong irritant to mucous membranes.
Use: Manufacture of fluosilicic acid, intermediate in manufacture of pure silicon, to seal water out of oil wells during drilling.

silicon tetrahydride. See silane.

silicon tetraiodide. SiI_4.
Properties: White crystals. Mp 120C, bp 290C. Noncombustible.
Grade: Up to 99.999%.
Hazard: Toxic by ingestion and inhalation, irritant to tissue.

silicotungstic acid. (12-tungstosilicic acid; silicowolframic acid). $H_4SiW_{12}O_{40}\cdot5H_2O$.

Properties: White, crystalline powder. Very soluble in water and polar organic solvents; relatively insoluble in nonpolar organic solvents. Strong acid. Noncombustible.
Grade: Reagent, technical.
Use: Catalyst for organic synthesis, reagent for alkaloids, additive to plating processes, as precipitant and inorganic ion exchanger, minerals separation, mordant.
See sodium 12-tungstosilicate; tungstosilicates.

silicowolframic acid.   See silicotungstic acid.

silk.   A natural fiber secreted as a continuous filament by the silkworm, *Bombyx mori*; silk consists essentially of the protein fibroin and, in the raw state, is coated with a gum that is usually removed before spinning. D 1.25, elongation at rupture about 20%, tenacity 3–5 g/denier. Combustible but self-extinguishing.

"Sil-Kleer" *[Silbrico].*   TM for perlite filter aid.

sillimanite.   An aluminum silicate, a high-heat-resisting material containing a maximum amount of mullite, developed from the alteration of andalusite during firing. This necessitates firing at above 1550C for the development of a suitable crystalline structure.
Use: Spark plugs, chemical lab ware, pyrometer tubes, special porcelain shapes, furnace patch and refractories.

silo.   See ensilage.

siloxane.   (oxosilane). A straight-chain compound (analogous to paraffin hydrocarbons) consisting of silicon atoms single-bonded to oxygen and so arranged that each silicon atom is linked with four oxygen atoms.

$$
\begin{array}{ccccc}
| & & | & & | \\
O & & O & & O \\
| & & | & & | \\
-Si & -O- & Si & -O- & Si-O- \\
| & & | & & | \\
O & & O & & O \\
| & & | & & |
\end{array}
$$

In some types, hydrogen may replace two or more of the oxygens. Disiloxane and trisiloxane are examples.
See silicone.

Silspan.   Preformed silicone profile seals.

"Siltemp" *[AVS].*   TM for a substantially pure fibrous silica for use in rocket and missile construction and for high-temperature insulation of motor components and similar applications.
Use: As a construction material in laminate form impregnated and bonded with high-temperature resins.

silteplase.
CAS: 131081-40-8.
Hazard: A poison by ingestion.

"Sil-Trode" *[Ampco].*   TM for silicon bronze electrodes and filler rod for use in inert-gas welding.

silumin.   Aluminum–silicon alloys containing 12% silicon.
Properties: D 2.63–2.65.

silver.
CAS: 7440-22-4. Ag.
Metallic element, atomic number 47, group IB of the periodic table, aw 107.868, valence of 1, two stable isotopes.
Properties: Soft, ductile, lustrous, white solid; highest electrical and thermal conductivity of all metals. Excellent light reflector that resists oxidation, but tarnishes in air through reaction with atmospheric sulfur compounds. D 10.53, mp 961C, bp 2212C, thermal conductivity 1.01 cal/cm/sec/C, absorbs oxygen strongly at the melting point. Soluble in nitric acid, hot sulfuric acid, and alkali cyanide solutions; insoluble in water and alkalies. Noncombustible except as powder.
Derivation: By-product of operations on copper, zinc, lead, or gold ores, but some smelters still operate on native silver. The recovery ranges from 166 ounces to a few thousandths of an ounce per ton.
See Parkes process; Pattinson process.
Source: Chief silver ores are native silver, argentite (silver sulfide), and cerargyrite (silver chloride).
Available forms: Pure ("fine"), sterling (7.5% copper), various alloys; plate, ingot, bullion, moss, sheet, wire, tubing, castings, powder, high purity (impurities less than 100 ppm), single crystals, whiskers.
Hazard: Toxic material.
Use: Manufacture of silver nitrate, silver bromide, photographic chemicals; lining vats and other equipment for chemical reaction vessels, water distillation, etc.; mirrors; electric conductors, such as bus bars; silver plating; electronic equipment; sterilant; water purification; surgical cements; hydration and oxidation catalyst; special batteries; solar cells; reflectors for solar towers; low-temperature brazing alloys; table cutlery; jewelry; dental, medical, and scientific equipment; electrical contacts; bearing metal; magnet windings; dental amalgams. Colloidal silver is used as a nucleating agent in photography and in medicine, often combined with protein.
See "Argyrol" *[McKinney].*
Note: A sandwich assembly consisting of a layer of silver between two layers of TiO_2 is used to coat the interior of light bulbs; it is said to reduce power consumption by more than 50% and triple the life of the bulb. Each layer of the coating, applied by the sputtering technique is 180 Å thick.

silver acetate.
CAS: 563-63-3. CH₃COOAg.
Properties: White crystals or powder. D 3.26. Moderately soluble in hot water; soluble in nitric acid.
Hazard: Toxic material.
Use: Lab reagent, oxidizing agent.

silver acetylide.
CAS: 13092-75-6. Ag₂C₂.
Properties: White, unstable powder. A salt of acetylene.
Derivation: Reaction of acetylene with aqueous solution of argentous salts.
Hazard: Severe explosion risk when shocked or heated.
Use: Detonators.

silver bromate. AgBrO₃.
Properties: White powder. Decomposed by heat, d 5.2. Sensitive to light, keep in amber bottle. Soluble in ammonium hydroxide; slightly soluble in hot water.

silver bromide. AgBr.
Properties: Pale-yellow crystals or powder, darkens on exposure to light, finally turning black. D 6.473 (25C), mp 432C, bp decomposes at 700C. Soluble in potassium bromide, potassium cyanide, and sodium thiosulfate solutions; very slightly soluble in ammonia water; insoluble in water.
Derivation: Silver nitrate is dissolved in water and a solution of alkali bromide added slowly. The precipitated silver bromide is washed repeatedly with hot water; the operation must be carried on in a darkroom under a ruby-red light.
Use: Photographic film and plates, photochromic glass, laboratory reagent.

silver carbonate. Ag₂CO₃.
Properties: Yellow to yellowish-gray powder. Contains 78% silver. Light-sensitive. D 6.077; decomposes at 218C. Soluble in ammonium hydroxide, nitric acid; insoluble in alcohol and water.
Use: Lab reagent.

silver chlorate. (argentous chlorate).
CAS: 7783-92-8. AgClO₃.
Properties: White, crystalline solid. Mp 230C, decomposes at 270C, d 4.44. Partially soluble in water, light-sensitive.
Derivation: Reaction of silver nitrate and sodium chlorate.
Hazard: Oxidizing agent, may react violently when shocked or heated, store away from combustible materials. Toxic by ingestion.
Use: Organic synthesis.

silver chloride. AgCl.
Properties: White, granular powder; darkens on exposure to light, finally turning black. Exists in several modifications differing in behavior toward light and in their solubility in various solvents. D 5.56, mp 445C, bp 1550C. Soluble in ammonium

hydroxide, concentrated sulfuric acid and sodium thiosulfate and potassium bromide solutions; very slightly soluble in water. Can be melted, cast, and fabricated like a metal.
Derivation: Silver nitrate solution is heated and hydrochloric acid or salt solution added. The whole is boiled, then filtered, all in the dark or under a ruby-red light.
Method of purification: Resolution in ammonium hydroxide and precipitation by hydrochloric acid.
Grade: Technical; CP; single, pure crystals.
Hazard: As for silver.
Use: Photography, photometry and optics, batteries, photochromic glass, silver plating, production of pure silver, antiseptic. Single crystals are used for infrared absorption cells and lens elements, lab reagent.

silver chromate. Ag₂CrO₄.
Properties: Dark, brownish-red powder. D 5.625. Soluble in acids, ammonium hydroxide, potassium cyanide, solutions of alkali chromates; insoluble in water.
Use: Lab reagent.

silver cyanide.
CAS: 506-64-9. AgCN.
Properties: White powder; odorless, tasteless. Darkens on exposure to light. D 3.95, decomposes at 320C. Soluble in ammonium hydroxide, dilute boiling nitric acid, and potassium cyanide and sodium thiosulfate solutions; insoluble in water.
Derivation: By adding sodium cyanide or potassium cyanide to a solution of silver nitrate.
Hazard: Toxic by ingestion or inhalation.
Use: Silver plating.

silver dichromate. (silver bichromate).
Ag₂Cr₂O₇.
Properties: Dark-red, almost black, crystalline powder. D 4.770. Soluble in ammonium hydroxide and nitric acid; slightly soluble in water.
Hazard: Possible carcinogen.

silver fluoride.
CAS: 7775-41-9. AgF·H₂O.
Properties: Yellow or brownish crystalline mass. Very hygroscopic, becomes dark on exposure to light. D 5.852 (anhydrous), mp 435C, bp 1159C. Soluble in water.
Hazard: Strong irritant.
Use: Medicine (antiseptic), substitution of fluorine for bromine and chlorine in organic compounds.

silver iodide. AgI.
Properties: Pale-yellow powder; odorless, tasteless; darkens on exposure to light. D 5.675, mp 556C. Soluble in hydrogen iodide, potassium iodide, potassium cyanide, ammonium hydroxide, sodium chloride, and sodium thiosulfate solutions. Insoluble in water.
Derivation: Silver nitrate solution is heated, alkali iodide solution added, and the precipitate washed

with boiling water in the dark or under ruby red illumination.

Use: Photography, cloud seeding for artificial rainmaking, lab reagent, antiseptic.

"Silver-Lume" A, B [ATOTECH].   TM for a bright silver electroplating process for use by silversmiths and electronics manufacturers. The materials used are silver cyanide, potassium cyanide, potassium carbonate, and addition agents.

silver mercury iodide.   See mercuric silver iodide.

silver methylarsonate.   (methanearsonic acid, disilver salt). $CH_3AsI_3Ag_2$.

Derivation: Reaction of disodium methylarsonate with silver salts.

Hazard: Very toxic.

Use: Algicides.

silver nitrate.

CAS: 7761-88-8. $AgNO_3$.

Properties: Colorless, transparent, tabular, rhombic crystals; becomes gray or grayish-black on exposure to light in the presence of organic matter; odorless; bitter, caustic metallic taste. Strong oxidizing agent and caustic. D 4.328, mp 212C, bp (decomposes). Soluble in cold water; more soluble in hot water, glycerol, and hot alcohol; slightly soluble in ether.

Derivation: Silver is dissolved in dilute nitric acid and the solution evaporated. The residue is heated to a dull red heat to decompose any copper nitrate, dissolved in water, filtered, and recrystallized.

Grade: Technical, CP, USP.

Hazard: Strong irritant to skin and tissue.

Use: Photographic film, catalyst for ethylene oxide, indelible inks, silver plating, silver salts, silvering mirrors, germicide (as a wall spray), hair dyeing, antiseptic, fused form to cauterize wounds, lab reagent.

silver nitride. Ag_3N.

Properties: Colorless powder.

Derivation: Reaction of silver compounds with ammonia, with or without additives.

Hazard: Severe explosion hazard when shocked. It is unusually sensitive to mechanical action of any kind and can explode even in water suspension.

silver nitrite. $AgNO_2$.

Properties: Small yellow or grayish-yellow needles; becomes gray on exposure to light. Decomposes at 140C, d 4.453 (26C). Contains 70% silver. Decomposed by acids; soluble in hot water; insoluble in alcohol.

Hazard: Highly toxic.

Use: Reagent for alcohols, preparation of aliphatic nitrogen compounds, and standard solutions for water analysis.

silver orthophosphate.   See silver phosphate.

silver oxide.   (argentous oxide).

CAS: 20667-12-3. Ag_2O.

Properties: Dark-brown or black powder; odorless; metallic taste. D 7.14 (16C), mp decomposes when heated above 300C, strong oxidizing agent. Soluble in ammonium hydroxide, potassium cyanide solution, nitric acid, and sodium thiosulfate solution; slightly soluble in water; insoluble in alcohol.

Derivation: Silver nitrate and alkali hydroxide solutions are mixed, the precipitate filtered and washed.

Grade: Technical, up to 99.6% pure, particle size 2–3 microns.

Hazard: Fire and explosion risk in contact with organic materials or ammonia.

Use: Polishing glass, coloring glass yellow, catalyst, purifying drinking water, lab reagent.

silver oxide battery.   See zinc-silver oxide battery.

silver perchlorate.

CAS: 7783-93-9. $AgClO_4$.

Properties: Colorless crystals. Deliquescent, mp 485C (decomposes), d 2.80. Soluble in water; reacts with explosive violence with many organic solvents; explodes on grinding.

Use: Manufacture of explosive compositions.

silver permanganate. $AgMnO_4$.

Properties: Violet, crystalline powder. D 4.27 (25C), mp decomposes; contains 47.5% silver; decomposed by alcohol; light-sensitive, use dark-colored bottles.

Grade: Technical.

Hazard: Dangerous explosion risk, may detonate if shocked or heated.

Use: Gas masks, medicine (antiseptic).

silver peroxide.   (argentic oxide). Ag_2O_2.

Properties: Grayish powder. D 7.44 (25C), decomposes above 100C. Insoluble in water; soluble in sulfuric acid, nitric acid, and ammonium hydroxide.

Hazard: Dangerous fire and explosion risk, strong oxidizing agent. Keep out of contact with organic materials.

Use: Manufacture of silver-zinc batteries.

silver phosphate.   (silver orthophosphate). Ag_3PO_4.

Properties: Yellow powder, turns brown when heated or on exposure to light. D 6.370 (25C), mp 849C. Soluble in acids, potassium cyanide solutions, ammonium hydroxide, ammonium carbonate and acetic acid; very slightly soluble in water.

Derivation: Interaction of silver nitrate and sodium phosphate.

Use: Photographic emulsions, catalyst, pharmaceuticals.

silver picrate.

CAS: 146-84-9. $C_6H_2O(N_2)_3Ag \cdot H_2O$.

Properties: Yellow crystals containing 30% silver, soluble in water. Slightly soluble in alcohol and acetone; insoluble in ether and chloroform.
Hazard: Severe explosion risk. Very toxic.
Use: Medicine (antimicrobial agent).

silver potassium cyanide. (potassium cyanoargentate; potassium argentocyanide).
CAS: 506-61-6. $KAg(CN)_2$.
Properties: White crystals. D 2.36 (25C), sensitive to light. Soluble in water and alcohol; insoluble in acids.
Derivation: By adding silver chloride to a solution of potassium cyanide.
Hazard: Very toxic.
Use: Silver plating, bactericide, antiseptic.

silver sodium chloride. (sodium silver chloride). $AgCl•NaCl$.
Properties: Hard, white crystals. Decomposed by water. Soluble in a concentrated solution of sodium chloride.

silver sodium thiosulfate. (sodium silver thiosulfate). $Ag_2S_2O_3•2Na_2S_2O_3•2H_2O$.
Properties: White to gray, crystalline powder; sweet taste. Soluble in water.

silver, sterling. See sterling silver.

silver sulfate. (silver sulfate, normal). Ag_2SO_4.
Properties: Small, colorless, lustrous crystals or crystalline powder. Contains about 69% silver, turns gray on exposure to light. D 5.45 (29C), bp 1085C (decomposes), mp 652C. Soluble in ammonium hydroxide, nitric acid, sulfuric acid, hot water; insoluble in alcohol.
Grade: Technical, CP.
Hazard: Highly toxic.
Use: Lab reagent.

silver sulfate, normal. See silver sulfate.

silver sulfide. Ag_2S.
Properties: Grayish-black powder. D 7.32 (25C), bp decomposes, mp 825C. Soluble in concentrated sulfuric and nitric acids; insoluble in water.
Derivation: By passing hydrogen sulfide gas into silver nitrate solution, washing and drying.
Use: Inlaying in niello metalwork, ceramics.

silver trinitromethanide.
CAS: 25987-94-4. $CAgN_3O_6$.
Properties: Soft, ductile, malleable, bright silver metal. Tarnishes in air, forming sodium oxides, the carbonates, and the hydroxide. mw 257.90, Mp 97.8C, bp 881C. Soluble in liquid NH_3.
Hazard: An explosive.

Silver Universal Klister. A ski wax.
Use: Wax useful on both sides of freezing as conditions are changing.

silvex. (fenoprop; generic name for 2-(2,4,5-trichlorophenoxy)propionic acid).
CAS: 93-72-1. $Cl_3C_6H_2OCH(CH_3)COOH$.
Properties: Solid. Mp 180.4–181.6C. Slightly soluble in water; freely soluble in acetone, methanol. Combustible.
Hazard: Use has been restricted.
Use: Herbicide and plant growth regulator.

silvichemical. A chemical derived from wood, e.g., lignins, lignosulfonates (from spent sulfite liquor), vanillin, yeast (from fermentation of wood sugars), tall oil, sulfate turpentine, bark extracts, phenolic materials.

silvicide. A nonselective herbicide used to kill or defoliate bushes and small trees, e.g., ammonium sulfamate.

silybum marianum (linn.) gaertn., extract.
CAS: 84604-20-6.
Hazard: Moderately toxic.
Source: Natural product.

silylene. An organosilicon compound containing double-bonded silicon.

simazine. (2-chloro-4,6-bis(ethylamino)-s-triazine).
CAS: 122-34-9. $ClC_3N_3(NHC_2H_5)_2$.
Properties: White solid. Mp 225C. Insoluble in water; slightly soluble in organic solvents. Combustible.
Use: Herbicide.

Simmons–Smith reaction. Stereospecific synthesis of cyclopropanes by treatment of olefins with methylene iodide and zinc–copper couple.

Simonini reaction. The preparation of aliphatic esters by the reaction of two moles of the silver salt of a carboxylic acid and one mole of iodine.

Simonis chromone cyclization. Formation of chromones from phenol and β-keto esters in the presence of phosphorus pentoxide, phosphorus oxychloride, or sulfuric acid. Coumarins may also be formed.

Simons process. An electrochemical fluorination process that makes fluorocarbons by passing an electric current through a mixture of the organic starting compound and liquid anhydrous hydrogen fluoride. The products are hydrogen and the desired fluorocarbon.

simple asphyxiant. A gas or an aerosol that dilutes the air or oxygen that reaches a respiratory surface. It contains high levels of simple asphyxiants and lacks sufficient oxygen to support respiration.

simple diffusion. The unassisted movement of molecules across a membrane to a region of lower concentration.

simple distillation. Distillation in which no appreciable rectification of the vapor occurs, that is, the vapor formed from the liquid in the still is completely condensed in the distillate receiver and does not undergo change in composition due to partial condensation or contact with previously condensed vapor.

simple protein. A protein yielding only amino acids on hydrolysis.

SIMS. Abbreviation for secondary ion mass spectroscopy.

single-cell protein. See protein, single-cell.

single domain antibody. An antibody that has only one protein chain derived from only one of the two domains of the normal antibody. These antibodies can be cloned and expressed into bacteria such that large numbers can be generated and screened in parallel.

single-gene disorder. Hereditary disorder caused by a mutant allele of a single gene. Examples are Duchenne muscular dystrophy, retinoblastoma, and sickle cell disease.
See polygenic disorders.

single nucleotide polymorphism. (SNP). DNA sequence variations that occur when a single nucleotide (A, T, C, or G) in the genome sequence is altered.
See mutation; polymorphism; single-gene disorder.

singlet. Two unpaired electrons of a diradical having antiparallel spins.

singlet oxygen. An electronically stimulated oxygen molecule.

sinigrin. (1-thio-β-d-glucopyranose; 1-[n-(sulfooxy)-3-buteninidate] monopotassium salt; sinigroside; myronate potassium; potassium myronate; allyl glucosinolate; potassium [1-[(2S,3R,4S,5S,6R)-3,4,5-trihydroxy-6-(hydroxymethyl)oxan-2-yl]sulfanylbut-3-enylideneamino]sulfate). $_{10}H_{16}KNO_9S_2$. A crystalline β-glucopyranoside and glycoside of allyl isothiocyanate.
Derivation: Isolated from various species of *Cruciferae* (mustard family, crucifers), most notable from the seeds of *Brassica nigra* and the root of *Alliaria officinalis*.
Hazard: Vesicant; irritant.

sinorphan.
CAS: 112573-73-6. $C_{21}H_{23}NO_4S$.
Hazard: A poison.

sintering. The agglomeration of metal or earthy powders at temperatures below the melting point. Occurs in both powder metallurgy and ceramic firing. While heat and pressure are essential, decrease in surface area is the critical factor. Sintering increases strength, conductivity, and density.
See Rittinger's law.

SIPP. Abbreviation for sodium iron pyrophosphate.

siRNA. (RNAi).
Small Inhibitory RNA.

sisal.
Properties: Hard, strong, light-yellow to reddish fibers obtained from the leaves of Agave sisilana. Strength 4.5 g/denier, fineness ranges from 300–500 denier. Combustible, not self-extinguishing.
Source: Africa, Haiti, Bahama, Indonesia.
Hazard: Dust is flammable, may ignite spontaneously when wet. Toxic by inhalation.
Use: Twine, sacking, upholstery, life preservers, mattress liners, floor covering.

site-directed mutagenesis. A set of methods used to create specific alterations in the sequence of nucleotides in a gene.

site-specific recombination. A type of genetic recombination that occurs only at specific sequences.

Sitol. An oxidizing agent in flake form used in discharge printing.

β-sitosterol. (dihydrostigmasterol). $C_{29}H_{50}O$.

Properties: Waxy-white solid; almost odorless and tasteless. Insoluble in water; soluble in benzene, chloroform, carbon disulfide, and ether. Can be crystallized from ether as anhydrous needles or from aqueous alcohol as leaflets with one molecule of water.
Derivation: Soybeans, tall oil.
Use: Biochemical research, anticholesteremic.
See cholesterol.

sitosterol sulfate.
CAS: 24815-93-8. $C_{29}H_{50}O•HO_4S$.
Hazard: A reproductive hazard.

size oil. See throwing oil.

sizing compound.   (1) A material such as starch, gelatin, casein, gums, oils, waxes, asphalt emulsions, silicones, rosin, and water-soluble polymers applied to yarns, fabrics, paper, leather, and other products to improve or increase their stiffness, strength, smoothness, or weight. (2) A material used to modify the cooked starch solutions applied to warp ends prior to weaving.
See slashing compound.

Skamex.   A fluorocarbon plastic used as a metallurgical additive for the beneficiation of molten metal. Cakes and slugs of it are specifically designed for immersion in hot metal for removal of excessive quantities of dissolved hydrogen. Other forms are added to molds during casting to create a protective atmosphere against the effects of absorbed oxygen. Can be used for ferrous and nonferrous metals.

skatole.   (3-methylindole).
CAS: 83-34-1. C_9H_9N.
Properties: White, crystalline substance; brows upon aging; fecal odor. Mp 93–95C, bp 265C. Soluble in hot water, alcohol, benzene. Gives violet color in potassium ferrocyanide and sulfuric acid.
Derivation: Feces; African civet cat; *Celtis reticulosa*, a Javanese tree.
Use: Perfumery (fixative), artificial civet.

Skou, Jens C.   (1918–). A Danish chemist who won the Nobel Prize in 1997 for his discovery of the first molecular pump, an ion transporting enzyme called Na^+-K^+ ATPase. He is emeritus professor of biophysics at Aarhus University, Denmark. He received his Ph.D. from Aarhus University.
See Boyer, Paul D., Walker, John E.

Skraup synthesis.   Synthesis of quinoline or its derivatives by heating aniline or an aniline derivative, glycerol, and nitrobenzene in the presence of sulfuric acid.

SKY.   See spectral karyotype.

"Skybond" [Monsanto].   TM for polymide resins.
Use: In insulation to provide heat resistance, component strength, fire resistance; in aircraft reinforcement, construction; in foams and adhesives.

"Skydrol" [Solutia].   TM for a series of fire-resistant aircraft hydraulic fluids.
500-A: Used for hydraulic systems in turbo jet and turbo prop aircraft, which must operate at −54C.
7000: Used in aircraft cabin superchargers, expansion turbines for air-conditioning systems and the aircraft hydraulic system itself.

"SkyKleen" [Solutia].   TM for an aviation solvent that is a replacement for chlorinated or ozone depleting solvents such as MEK (methyl ethyl ketone), 1,1,1, trichlorethane, isopropyl alcohol, and acetone. It contains no halogenated materials.

Properties: Biodegradable. Non-combustible.
Use: A sealant and adhesive removal solvent, useful for surface preparation for prepaint and prebonding, general surface cleaning and degreasing.

slack.   (1) Descriptive of a soft paraffin wax resulting from incomplete pressing of the settlings from the petroleum distillate. Though it has some applications in this form, it is actually an intermediate product between the liquid distillate and the scale wax made by expressing more of the oil.
See scale (2).
(2) Specifically, to react calcium oxide (lime) with water to form calcium hydroxide (slaked or hydrated lime), the reaction is $CaO + H_2O \rightarrow Ca(OH)_2$ + heat. The alternate spelling "slake" has the same meaning.

slaframine.   (1-acetoxy-8-aminooctahydroindolizine).
An alkaloid derived from a fungus that infests clover. It is under research development for use as an agent in retarding cystic fibrosis.

slag.   (dross; cinder).
(1) Fused agglomerate (usually high in silicates) that separates in metal smelting and floats on the surface of molten metal. Formed by combination of flux with gangue of ore, ash of fuel, and perhaps furnace lining. Slag is often the medium by means of which impurities may be separated from metal.
(2) The residue or ash from coal gasification processes, it may run as high as 40% depending on the rank of coal used.
Use: Railroad ballast, highway construction, cement and concrete aggregate, raw material for Portland cement, mineral wool, and cinder block.

slake.   See slack.

slaked lime.   See calcium hydroxide; lime, hydrated.

slashing compound.   A textile sizing material applied to cotton or rayon warp ends by a special machine (slasher).

slash pine.   A loblolly pine growing in swampy areas (slashes) in southeastern U.S.
Use: Primarily for manufacture of kraft paper pulp.

slate.   A fine-grained metamorphic rock that cleaves into thin slabs or sheets. Color usually gray to black, sometimes green, yellow, brown, or red. Slates are composed of micas, chlorite, quartz, hematite, clays, and other minerals.
Occurrence: Pennsylvania, Vermont, Maine, Virginia, California, Colorado, Europe.
Use: Roofing, blackboards; (as powder) filler in paint, rubber, abrasive.

slate flour.   Finely divided slate used as a filler and dusting agent in rubber, plastics, etc.

slate, green. See slate flour.

slave. A remote-controlled mechanism or instrument that repeats the action of an identical mechanism that is controlled by an operator in another location; it may be activated by electromagnets or by electronic means. Such devices are used chiefly in handling or processing radioactive materials but also have communication uses, as in the Telautograph.

slimicide. A chemical that is toxic to the types of bacteria and fungi characteristic of aqueous slimes. Examples are chlorine and its compounds, organomercurial compounds, phenols, and related substances.
Use: Largely in paper mills and to some extent in textile and leather industries.
See biocide.

slip clay. A type of clay containing such a high percentage of fluxing impurities and of such a texture that it melts at a relatively low temperature to a greenish or brown glass, thus forming a natural glaze. It must be fine-grained, free from lumps or concretions, show a low air shrinkage, and mature in burning at as low as 704C.

slipicone. Fluid silicone compositions to prevent adhesion of materials to one another.
Use: Food-processing and packaging equipment.

"Slipkote" *[Specialty Lubricants].* TM for an anti-seize lube for engine assembly.

slot blot. Similar to a dot blot, but the analyte is put onto the membrane using a slot-shaped template. The template produces a consistently shaped spot, thus decreasing errors and improving the accuracy of the analysis. See Dot blot.

"Slow-Mag" *[Purdue].* TM for magnesium chloride.
Use: Drug.

sludge. Any thick, viscous mass, usually a sedimented or filtered waste product. Examples are (1) aluminum ore tailings (red mud), (2) asphaltic petroleum residues, and (3) municipal waste (Imhoff and activated sludge). All these may be used as components of useful products, e.g., (1) in steel manufacturing, (2) in roofing and road treatment compositions, and (3) as a base for fertilizers. See sewage sludge.

Sludge Conditioner. A series of polyelectrolytes.
Use: Conditions sludge for dewatering and settling in municipal sewage treatment plants.

slurry. A thin, watery suspension; for example, the feed to a filter press or to a fourdrinier machine; also a stream of pulverized metal ore. A special use of this term refers to a type of explosives

called "slurry blasting agents" based on gelatinized aqueous ammonium nitrate, sensitized with various other explosives.

slushing agent. A nondrying oil, grease, or similar material.
Use: Coat metals to afford temporary protection against corrosion.

slush molding. A method of molding certain toys such as doll parts in which a preheated mold is filled with liquid plastic composition and then heated until the required wall thickness has formed. The remaining liquid plastic is then poured out and the mold heated further at 200–220C until the product has completely set. The mold is then cooled and the product removed.

Sm. Symbol for samarium.

Smalley, Richard E. (1943–2005). An American who won the Nobel Prize for chemistry along with Robert F. Curl, Jr. and Sir Harold W. Kroto in 1996, the 100th anniversary of Alfred Nobel's death. The trio won for the discovery of the C_{60} compound called buckminsterfullerene. He graduated from the University of Michigan and earned a Ph.D. from Princeton University.
See buckminsterfullerene; Curl, Robert F., Jr.; Kroto, Sir Harold W.

small nuclear riboNucleoProtein. See snRNP.

small nuclear RNA (snRNA). Any of several small RNA molecules in the nucleus. Most known snRNAs have a role in the splicing reactions that remove introns from mRNA, tRNA, and rRNA molecules.

smalt.
Properties: Blue powder.
Derivation: A potash-cobalt glass made by fusing pure sand and potash with cobalt oxide, grinding, and powdering.
Use: Paint pigments, ceramic industries (pigment), coloring glass, bluing paper, starch and textiles, coloring rubber.

392 SmartPlate Etch. A prewetting etch.
Use: For use on laser plates.

Smart Plate Fountain Solution. A one step fountain solution.
Use: Formulated for use with laser plates.

394 SmartPlate Prep. A liquid cleaner.
Use: Removes scattered toner particles to allow clean, sharp images.

smectic. A molecular structure (layers or planes) occurring in some liquid crystals; it imparts a soft, soapy property. There are nine types of smectic orientation.

smelting. Heat treatment of an ore to separate the metallic portion with subsequent reduction.
See roasting.

Smiles rearrangement. Intramolecular nucleophilic aromatic substitution in alkaline solution resulting in the migration of an aromatic system from one heteroatom to another.

"Smite" [Zerran]. TM for emulsifiable insecticide containing 12.5% methoxychlor and 12.5% malathion.

Smith, Michael. (1932–2000). A Canadian who won the Nobel Prize for chemistry in 1993 for the development of site-specific mutagenesis.
See Mullis, Kary Banks.

smog. A coined word denoting a persistent combination of smoke and fog occurring under appropriate meteorological conditions in large metropolitan or heavy industrial areas. The discomfort and danger of smog is increased by the action of sunlight on the combustion products in the air, especially sulfur dioxide, nitric oxide, and exhaust gases (photochemical smog). Strongly irritant and even toxic substances may be present, e.g., peroxybenzoyl nitrate. Fatalities have resulted from exposure to particularly severe photochemical smogs.
See air pollution.

smoke. A colloidal or microscopic dispersion of a solid in gas, an aerosol. (1) *Coal smoke*: A suspension of carbon particles in hydrocarbon gases or in air, generated by combustion. The larger particles can be removed by electrostatic precipitation in the stack (Cottrell). Dark color, nauseating odor.
See smog; air pollution; Cottrell. (2) *Wood smoke*: Light-colored particles of cellulose ash, pleasant aromatic odor. Smoke from special kinds of wood (e.g., hickory, maple) is used to cure ham, fish, etc., also to preserve crude rubber. (3) *Chemical smoke*: Generated by chemical means for military purposes (concealment, signaling, etc.). (4) *Metallic smoke (fume)*: An emanation from heated metals or metallic ores, the particles being of specific geometric shapes. Such smoke is particularly damaging to vegetation in the neighborhood of zinc and tin smelters. (5) *Cigarette smoke*: There is conclusive evidence that the tars occurring in cigarette smoke can lead to lung cancer; chief factors are age of individual at initiation of smoking, extent of inhalation, and amount smoked per day. Polonium, a radioactive element, is known to occur in cigarette smoke; more than 100 compounds have been identified, including nicotine, cresol, carbon monoxide, pyridene, and benzopyrene, the latter a carcinogen.
See cigarette tar.

smoke grenade.
Use: Chemical warfare agent that defeats sensors by scattering or absorbing electromagnetic radiation in the visual, infrared, and millimeter wave portions of the electromagnetic spectrum.

Smokehouse Cleaner. A high strength mechanical and recirculation cleaner.
Use: Cleaner.

smokeless powder. Nitrocellulose containing about 13.1% nitrogen, produced by blending material of somewhat lower (12.6%) and slightly higher (13.2%) nitrogen content, converting to a dough with alcohol–ether mixture, extruding, cutting, and drying to a hard, horny product. Small amounts of stabilizers (amines) and plasticizers are usually present, as well as various modifying agents (nitrotoluene, nitroglycerin salts).
Hazard: A low explosive, dangerous fire and explosion risk when exposed to flame or impact.
Use: Sports ammunition, military purposes.

smudge oil. An oil burned in fruit orchards to prevent frost from injuring the trees. No. 3 fuel oil is typical of oils used.

smut. See fungus; rust (2).

Sn. Symbol for tin (from Latin stannum).

snake venom. There are two functional types: (1) those that bring about blood coagulation either by direct action on fibrogen or by converting prothrombin to thrombin; (2) neurotoxins that act on the central nervous system, for example, by inactivation of acetylcholine. Rattlesnake and moccasin venom are examples of (1) and cobra venom of (2). The enzymes of snake venoms are thought to be the actual toxic principles. Solutions of cobra venom have found use in treatment of arthritis and cancer. The chemistry and pharmacological properties of these poisons are not well understood.
Note: A person bitten by a poisonous snake should be carried, *not walked*, to a hospital. *No alcohol* should be administered.

SNDA. Supplemental New Drug Application.

sneezing gas. See diphenylchloroarsine.

snerps. See snRNP.

SNG. Abbreviation for synthetic or substitute natural gas.

snow, artificial. See artificial snow.

"Snowmelt Instant Ice Melter" [Standard]. TM for calcium chloride.
Use: Pellets for snow and ice melter. Flake for concrete accelerator and dust control.

snow point.   Referring to a gas mixture, the temperature at which the vapor pressure of the sublimable component is equal to the actual partial pressure of that component in the gas mixture. This is the gaseous analog to dew point.

SNP.   See single nucleotide polymorphism.

snRNA.   Small nuclear RNA; forms complexes with proteins to form snRNPs; involved in RNA splicing, and polyadenylation reactions.

snRNP.   (small nuclear riboNucleoProtein; snerps). Particles which are complexes between small nuclear RNAs and proteins, and which are involved in RNA splicing and polyadenylation reactions.

SNTA.   Abbreviation for sodium nitrilotriacetate.

S1 nuclease.   An enzyme which digests only single-stranded nucleic acids.

soap.   (1) The water-soluble reaction product of a fatty acid ester and an alkali (usually sodium hydroxide), with glycerol as by-product. For the reaction, see saponification. A soap is actually a specific type of salt, the hydrogen of the fatty acid being replaced by a metal, which in common soaps is usually sodium. Soap lowers the surface tension of water and thus permits emulsification of fat-bearing soil particles. A typical commercial cleansing soap is made by reacting sodium hydroxide with a fatty acid. The lower the hydrogen content of the acid the thinner the soap. The by-product of the reaction is glycerol. Many different carboxyl-containing substances are used, including rosin, tall oil, vegetable and animal oils and fats (stearic, palmitic, and oleic acids). Olive oil is used for Castile soap; transparent soaps are made by decolorized fats. The specific gravity of soaps is slightly more than 1.0; inclusion of air gives a floating product. Water solutions of sodium soaps in bar, chip, or powder form are universally used as mild, emulsifying detergents for washing textiles, skin, paint, etc. Medically, soap is used as an antidote for poisoning by ingestion of mineral acids or heavy metals. Liquid green soap is made with potassium hydroxide and a vegetable oil. (2) Heavy-metal soaps (loosely called metallic soaps) are those formed by metals heavier than sodium (aluminum, calcium, cobalt, lead, and zinc). These soaps are not water soluble; specific types are used in lubricating greases, gel thickeners, and in paints as driers and flatting agents. Napalm is an aluminum soap.
See saponification; detergent.

soap builder.   Any material mixed with soap to improve the cleaning properties, modify the alkali content, or impart water-softening characteristics. The commonest are some of the sodium phosphates and rosin.
See builder detergent.

Soap and Detergent Association.   (SDA). The Soap and Detergent Association is the non-profit trade association representing some 120 North American manufacturers of household, industrial and institutional cleaning products; their ingredients; and finished packaging. SDA member companies produce more than 90% of the cleaning products marketed in the U.S.
Established in 1926, SDA is dedicated to advancing public understanding of the safety and benefits of cleaning products, and protecting the ability of its members to formulate products that best meeting consumer needs. SDA serves both its member and the public by developing and sharing information about industry products with the technical community, policy makers, child care and health professionals, educators, media and consumers. Its headquarters is at 1500 K Street, NW, Suite 300, Washington, DC 20005. Website: http://sdahq.org

soap, heavy metal.   See soap (2).

soap, soft.   See green soap; soap.

soapstone.   See talc.

Society of the Chemical Industry.   (SCIAS, SCI). This society was founded in London in 1881 to advance applied chemistry in all its branches. The American Section was established in 1894. In 1906 on the anniversary of the "birth of the coal tar industry," it founded the famous Perkin Medal, which has since been awarded annually for outstanding achievement in chemistry in the U.S. It has also offered the Chemical Industry Medal since 1932, given for conspicuous service to applied chemistry. Perkin Medalists include L. Baekeland, F. G. Cottrell, Irving Langmuir, Herbert Dow, A. D. Little, R. R. Williams, Glenn T. Seaborg, and Roger Adams. Its address is 14/15 Belgrave Square, London SW1X 8PS, U.K. Website: http://sci.mond.org

Society of Cosmetic Chemists.   (SCC). The society was founded in 1945 to promote high standards of practice in the cosmetic sciences and to serve as a focus for the exchange of ideas and new developments in cosmetic research and technology. The society has a membership of over 4,000 individuals. Its offices are at 120 Wall Street, Suite 2400, New York, NY 10005. Website: http://www.scconline.org

Society of Plastics Engineers.   (SPE). An incorporated technical organization founded in 1942 and devoted primarily to the application of sound engineering principles to the manufacture and use of plastics. This rapidly growing organization has contributed much to the correct evaluation of these versatile materials in many fields. It publishes a monthly journal and sponsors a series of technical books on both theoretical and applied aspects of plastics technology. It can be reached at

13 Church Hill Road, Newtown, CT 06470. Website: http://www.4spe.org

Society of the Plastics Industry.   (SPI). An incorporated technical organization founded in 1937 and serving the needs of the entire plastics industry in the U.S. It establishes standards for the properties and selection of materials and for product design and engineering. Its two major publications are the *Plastics Engineering Handbook* and the *Reinforced Plastics Handbook*. With it are associated the Plastics Pipe Institute and the Reinforced Plastics/Composites Institute. It is located at 1667 K St., NW, Suite 1000, Washington, DC 20006. Website: http://www.plasticsindustry.org

SOCMA.   Abbreviation for Synthetic Organic Chemical Manufacturers Association.

soda.   Any one of the forms of sodium carbonate, also used loosely as equivalent to the word sodium in compounds.

soda alum.   See aluminum sodium sulfate.

soda ash.   (soda, calcined; sodium carbonate, anhydrous).
CAS: 497-19-8. Na_2CO_3. The crude sodium carbonate of commerce.
Properties: Grayish-white powder or lumps containing up to 99% sodium carbonate. Soluble in water; insoluble in alcohol. Noncombustible.
Derivation: For nearly a century most of the soda ash produced in the U.S. has been made synthetically by the ammonia soda (Solvay) process. Largely because of high energy costs and strict pollution controls, most of the synthetic production is being abandoned in favor of the natural product obtained from deposits in Utah, California, Wyoming, etc. One of the largest producing sites of the natural product is in Green River, Wyoming.
Impurities: Sodium chloride, sodium sulfate, calcium carbonate and magnesium carbonate, sodium bicarbonate.
Grade: Dense 58%, light 58%, extra light, natural, refined.
Use: Glass manufacture, chemicals, pulp and paper manufacture, sodium compounds, soaps and detergents, water treatment, aluminum production, textile processing, cleaning preparations, petroleum refining, sealing ponds from leakage (sodium ions bind to clay particles, which swell to seal leaks), catalyst in coal liquefaction.

soda, baking.   See sodium bicarbonate.

soda, calcined.   See soda ash.

soda, caustic.   See sodium hydroxide.

soda crystals.   See sodium carbonate monohydrate.

soda lime.
CAS: 8006-28-8. A mixture of calcium oxide with sodium hydroxide or potassium hydroxide intended for the absorption of carbon dioxide gas and water vapor.
Properties: White or grayish-white granules unless colored by a specified indicator. Must be kept in airtight containers.
Grade: Technical, reagent. Usually percentage moisture and mesh size are stated.
Hazard: Toxic by ingestion and inhalation, strong irritant to tissue.
Use: Drying agent and carbon dioxide absorbent, lab reagent.

soda lime glass.   See glass.

sodalite.   See cage zeolite.

sodamide.   See sodium amide.

soda, modified.   (neutral soda). A combination of soda ash and bicarbonate of soda in definite proportions for purposes where an alkali is needed, ranging in causticity between bicarbonate of soda and soda ash. White, crystalline powders; water-soluble and possessing valuable cleansing and purifying properties. Prepared in various strengths.
Use: Washing powders, laundering, wool scouring powders, bottle cleansers, textile cleaners, mild detergents.

soda monohydrate.   See sodium carbonate, monohydrate.

soda, natural.   See trona.

soda niter.   See sodium nitrate.

"Sodaphos" [ICL].   TM for glassy sodium tetraphosphate.

soda pulp.   See pulp, paper.

soda, washing.   See sal soda.

Soddy, Frederick.   (1877–1956). A English physicist who won the Nobel Prize in chemistry in 1921. His work was concerned with radioactive elements and atomic energy. His concept of isotopes and the displacement law of radioactive change is basic to nuclear physics. His education was at Oxford and Glasgow. He later worked in Canada and Australia.

α-sodio-sodium acetate.   (sodium α-sodio-acetate). $NaCH_2COONa$.
Properties: Free-flowing powder. Stable in dry air, decomposes slowly in moist air, decomposes 280C without melting. Insoluble in ethers and hydrocarbons; reacts mildly with water.

Grade: 80–85% pure. Impurities are sodium acetate, sodium amide, and sodium hydroxide.

Hazard: Toxic by inhalation, irritant to skin and mucous membranes.

Use: Organic intermediate, drying agent for organic solvents.

sodium.   (natrium).
CAS: 7440-23-5.   Na.
Metallic element, atomic number 11, group IA of periodic table, aw 22.98977, valence = 1, no stable isotopes but several radioactive forms, extremely reactive.

Properties: Soft, silver-white solid oxidizing rapidly in air; waxlike at room temperature, brittle at low temperatures. Store in airtight containers or in naphtha or similar liquid that does not contain water or free oxygen. D 0.9674 (25C), mp 97.6C, bp 892C. Decomposes water on contact, with evolution of hydrogen to form sodium hydroxide; insoluble in benzene, kerosene, and naphtha. Has excellent electrical conductivity and high heat-absorbing capacity.

Derivation: Electrolysis of a fused mixture of sodium chloride and calcium chloride.

Method of purification: Distillation.

Grade: Commercial, technical, brick, amalgam, coated powders, dispersions (sodium dispersion), reactor (99.99% pure).

Hazard: Severe fire risk in contact with water in any form, ignites spontaneously in dry air when heated; to extinguish fires use *dry* soda ash, salt, or lime. Forms strong caustic irritant to tissue.

Use: Tetraethyl and tetramethyl lead, titanium reduction, sodium peroxide, sodium hydride, polymerization catalyst for synthetic rubber, lab reagent, coolant in nuclear reactors, electric power cable (encased in polyethylene), nonglare lighting for highways, radioactive forms in tracer studies and medicine, heat transfer agent in solar-powered electric generators.

See sodium dispersions.

sodium abietate.   (rosin soap; sodium resinate).
$C_{19}H_{29}COOONa$.
Properties: White powder. Soluble in water. Combustible.
Derivation: By leaching rosin with sodium hydroxide solution.
Use: Soap making, paper coating.

sodium acetate.
CAS: 127-09-3.   (1) $NaC_2H_3O_2$. (2) $NaC_2H_3O_2 \cdot 3H_2O$.
Properties: Colorless crystals; odorless; efflorescent. (1) D 1.528, mp 324C, (2) d 1.45, mp 58C, autoign temp 1125F (607C). Soluble in water; slightly soluble in alcohol; soluble in ether. Combustible.
Grade: Highest purity, pure fused, CP, NF, technical, FCC.

Use: Dye and color intermediate, pharmaceuticals, cinnamic acid, soaps, photography, purification of glucose, meat preservation, medicine, electroplating, tanning, dehydrating agent, buffer, lab reagent, food additive.

sodium acetone bisulfate.   (acetone-sodium bisulfite). $(CH_3)_2CONaHSO_3$.
Properties: Crystalline material. Soluble in water; decomposed by acids; slightly soluble in alcohol. Combustible.
Derivation: Interaction of sodium bisulfite and acetone.
Use: Chemical (pure acetone), photography, textile (dyeing and printing).

sodium acetylformate.   See sodium pyruvate.

sodium acid carbonate.   See sodium bicarbonate.

sodium acid methanearsonate.   See sodium methanearsonate.

sodium acid phosphate.   See sodium phosphate, monobasic.

sodium acid pyrophosphate.   See sodium pyrophosphate, acid.

sodium acid sulfate.   See sodium bisulfate.

sodium acid sulfite.   See sodium bisulfite.

sodium acid tartrate.   See sodium bitartrate.

sodium alginate.   (sodium polymannuronate).
CAS: 9005-38-3. $(C_6H_7O_6Na)$.
Properties: Colorless or slightly yellow solid occurring in filamentous, granular, and powdered forms. Forms a viscous colloidal solution with water; insoluble in alcohol, ether, and chloroform. Combustible.
Derivation: Extracted from brown seaweeds (alginic acid).
Grade: NF, FCC, technical.
Use: Thickeners, stabilizers, and emulsifiers in foods, especially ice cream; boiler compounds; medicine; experimental ocean-floor covering; textile printing; cement compositions; paper coating; foods; pharmaceutical preparations; water-base paints.

sodium alkane sulfonate.   (SAS). RSO_3Na.
The sodium salt of an alkane sulfonic acid of linear paraffins having chain lengths of 14–18 carbon atoms.
Preparation: By reaction of paraffins with sulfur dioxide and oxygen in the presence of γ-radiation from a cobalt-60 source.
Use: Biodegradable detergent intermediate.

sodium *n*-alkylbenzene sulfonate.
Use: Food additive.

sodium alum.   See aluminum sodium sulfate.

sodium aluminate.
CAS: 11138-49-1. $Na_2Al_2O_3$ or $NaAlO_2$.
Properties: White powder. Mw 81.97, mp 1650C. Soluble in water; insoluble in alcohol; aqueous solution strongly alkaline.
Derivation: By heating bauxite with sodium carbonate and extracting the sodium aluminate with water.
Grade: Technical, reagent, also 27 degrees Bé solution.
Hazard: (Solution) Strong irritant to tissue.
Use: Mordant, zeolites, water purification, sizing paper, manufacture of milk glass, soap and cleaning compounds.

sodium aluminosilicate.   (sodium silicoaluminate).
A series of hydrated sodium aluminum silicates having a $Na_2O:Al_2O_3:SiO_2$ mole ratio of approximately 1:1:13.2.
Properties: Fine, white, amorphous powder or beads; odorless and tasteless. Insoluble in water and in alcohol and other organic solvents; at 80–100C is partially soluble in strong acids and solutions of alkali hydroxides, pH of a 20% slurry is between 6.5 and 10.5. Noncombustible.
Grade: Technical, FCC.
Use: Anticaking agent in food preparations (up to 2%).

sodium aluminum hydride.
CAS: 13770-96-2. $NaAlH_4$.
Properties: White, crystalline material. D 1.24, stable in dry air at room temperature but very sensitive to moisture, begins to melt at 183C with decomposition to evolve hydrogen. Soluble in tetrahydrofuran, dimethyl "Cellosolve" *[Union]*.
Derivation: By reaction of aluminum chloride with sodium hydride.
Hazard: Severe fire and explosion risk in contact with oxidizing agents and water, forms caustic and irritating compounds.
Use: Reducing agent similar to lithium aluminum hydride.

sodium aluminum phosphate.   (sodium aluminum phosphate, acidic).
$NaAl_3H_{14}(PO_4)_8 \cdot 4H_2O$ or $Na_3Al_2H_{15}(PO_4)_n$.
Properties: White powder; odorless. Insoluble in water; soluble in hydrochloric acid.
Grade: Technical, FCC.
Use: Food additive (baked products).

sodium aluminum phosphate, basic.
Properties: White powder; odorless. Insoluble in water; soluble in hydrochloric acid.
Hazard: A nuisance dust.
Use: Food additive.

sodium aluminum silicofluoride.   (sodium aluminum fluosilicate). $Na_5Al(SiF_6)_4$.
Properties: White powder. Somewhat soluble in cold water. Corrosive to galvanized iron.
Hazard: Avoid prolonged skin contact.
Use: Moth-proofing and insecticides, also to obtain acid medium in dyebath.

sodium aluminum sulfate.   See aluminum sodium sulfate.

sodium amalgam. Na_xHg_y.
Properties: A silver-white, porous, crystalline mass containing 2–20% of metallic sodium. Decomposes water.
Derivation: Mercury is heated to about 200C and sodium, in small pieces, is added slowly. Also formed at one stage of the process for making chlorine and sodium hydroxide by the mercury cell process.
Grade: 2, 3, 4, 5, 6, 7, 8, 9, 10, and 20%.
Hazard: Flammable, dangerous fire risk.
Use: Preparation of hydrogen, reduction of metal halogen compounds and organic compounds, reagent in analytical chemistry.

sodium amide.   (sodamide).
CAS: 7782-92-5. $NaNH_2$.
Properties: White, crystalline powder; ammonia odor. Mp 210C, bp 400C. Decomposes in water and hot alcohol.
Derivation: Dry ammonia gas is passed over metallic sodium at 350C.
Hazard: Flammable, dangerous fire risk.
Use: Manufacture of sodium cyanide, organic synthesis, lab reagent, dehydrating agent.

sodium aminophenylarsonate.   See sodium arsanilate.

sodium ammonium hydrogen phosphate.
See sodium ammonium phosphate.

sodium ammonium phosphate.   (microcosmic salt; sodium ammonium hydrogen phosphate; phosphorus salt). $NaNH_4HPO_4 \cdot 4H_2O$.
Properties: Transparent, colorless, monoclinic crystals; odorless; efflorescent. Gives off water and ammonia on heating leaving $NaPO_3$. D 1.57, mp about 79C with decomposition. Soluble in water; insoluble in alcohol.
Derivation: Mixing solutions of sodium phosphate and ammonium chloride.
Grade: Granular, CP, technical.
Use: Analytical reagent.

sodium amytal.   (sodium isoamyl ethyl barbiturate).
Use: Hypnotic.

sodium anilinearsonate.   See sodium arsanilate.

sodium antimonate.   (antimony sodiate).
CAS: 11112-10-0. $NaSbO_3$.
Other forms are sodium metaantimonate, $2NaSbO_3 \cdot 7H_2O$, and sodium pyroantimonate, $Na_2H_2Sb_2O_7 \cdot H_2O$.
Properties: White, granular powder. Slightly soluble in water and alcohol; insoluble in dilute alkalies, mineral acids; soluble in tartaric acid. Noncombustible.
Grade: Technical, glassmakers' grade.
Hazard: Toxic by ingestion and inhalation. TLV: 0.5 mg(Sb)/m³.
Use: Opacifier in enamels for cast iron and glass, ingredient of acid-resisting sheet steel enamels, high-temperature oxidizing agent.

sodium aristolochate I.
CAS: 10190-99-5. $C_{17}H_{10}NO_7 \cdot Na$.
Hazard: Suspected carcinogen.
Source: Natural product.

sodium arsanilate.   (sodium anilinearsonate; sodium aminophenylarsonate).
CAS: 127-85-5. $C_6H_4NH_2(AsO \cdot OH \cdot ONa)$. Often with one or more water.
Properties: White, crystalline powder; odorless; faint salty taste. Soluble in water; slightly soluble in alcohol.
Derivation: By dissolving arsanilic acid in sodium carbonate solution and crystallizing.
Grade: Technical, medicinal.
Hazard: Highly toxic by ingestion and inhalation, may cause blindness.
Use: Medicine (anthelmintic), organic synthesis.

sodium arsenate.
CAS: 7778-43-0. $Na_3AsO_4 \cdot 12H_2O$.
Properties: Clear, colorless crystals; mild alkaline taste. D 1.7593, mp 86C. Soluble in water; slightly soluble in alcohol and glycerol; insoluble in ether.
Derivation: Reaction of arsenic trioxide and sodium nitrate.
Grade: Pure crystals, CP, technical (60% arsenic pentoxide).
Hazard: Toxic by ingestion and inhalation.
Use: Mordant and assist in dyeing and printing, other arsenates, germicide.

sodium arsenite.   (sodium metaarsenite).
CAS: 7784-46-5. $NaAsO_2$.
Properties: Grayish-white powder that absorbs carbon dioxide from the air. D 1.87. Soluble in water; slightly soluble in alcohol.
Derivation: Arsenic trioxide is dissolved in a solution of sodium carbonate or hydroxide and boiled.
Grade: Crude, pure, 75% arsenious oxide (94% solution).
Hazard: Toxic by ingestion and inhalation.

Use: Arsenical soaps for taxidermists, antiseptic, dyeing, insecticides, hide preservation, herbicide.

sodium arsphenamine.   See arsphenamine.

sodium ascorbate.
CAS: 134-03-2. $C_6H_7NaO_6$. The sodium salt of ascorbic acid.
Use: Antioxidant in food products.

sodium azide.
CAS: 26628-22-8. NaN_3.
Properties: Colorless, hexagonal crystals. Decomposes at about 300C, d 1.846 (20C). Soluble in water and in liquid ammonia; slightly soluble in alcohol; hydrolyzes to form hydrazoic acid. Combustible.
Hazard: Highly toxic. Lung damage and cardiac impairment. Questionable carcinogen.
Use: Air-bag inflation, preservative in diagnostic medicinals, intermediate in explosive manufacture.

sodium barbiturate. $C_4H_3N_2O_3Na$.
Properties: White to yellow-tinted powder. Soluble in water and dilute mineral acid, pH of a 1% aqueous solution 10.5.
Hazard: Toxic by ingestion.
Use: Synthetic intermediate, catalyst in ammonium nitrate propellants, wood impregnating solutions. See barbiturate.

sodium benzenesulfonchloramine.   See chloramine-B.

sodium benzoate.
CAS: 532-32-1. C_6H_5COONa.
Properties: White crystals or granular; odorless powder; sweetish, astringent taste. Soluble in water and alcohol. Combustible.
Derivation: Benzoic acid is neutralized with sodium bicarbonate solution, the solution filtered, concentrated, and allowed to crystallize.
Grade: USP, FCC, technical.
Hazard: Use in foods limited to 0.1%.
Use: Food preservative, antiseptic, medicine, tobacco, pharmaceutical preparations, intermediate for manufacture of dyes, rust and mildew inhibitor.

sodium benzosulfimide.   See saccharin.

sodium benzylpenicillin.   See penicillin.

sodium benzyl succinate.
$C_6H_5CH_2O_2C(CH_2)_2COONa$.
Properties: White amorphous or crystalline powder; slight benzyl odor; cool, salty taste. Soluble in water.

sodium beryllium fluoride.   See beryllium sodium fluoride.

sodium bicarbonate.   (baking soda; sodium acid carbonate).
CAS: 144-55-8. $NaHCO_3$.

Properties: White powder or crystalline lumps; cooling, slightly alkaline taste. D 2.159, mp loses carbon dioxide at 270C. Soluble in water; insoluble in alcohol. Stable in dry air, but slowly decomposes in moist air. Noncombustible.

Derivation: Principally by treating a saturated solution of soda ash with carbon dioxide to precipitate the less soluble bicarbonate; also by purifying the crude product from the Solvay process.

Grade: Commercial, pure, highest purity, CP, USP, FCC.

Use: Manufacture of effervescent salts and beverages, artificial mineral water, baking powder; other sodium salts; pharmaceuticals; sponge rubber; gold and platinum plating; treating wool and silk; fire extinguishers; prevention of timber mold; cleaning preparations; lab reagent; antacid; mouthwash.

sodium bichromate.   See sodium dichromate.

sodium bifluoride.   (sodium acid fluoride). CAS: 1333-83-1. $NaHF_2$.

Properties: White, crystalline powder. D 2.08. Soluble in water; decomposes on heating.

Hazard: Highly toxic, strong irritant to tissue.

Use: Tin-plate production, neutralizer in laundry rinsing operations, preservative for zoological and anatomical specimens, etching glass, antiseptic.

sodium biphosphate.   See sodium phosphate, monobasic.

sodium bis(2-methoxyethoxy)-aluminohydride.   An organometallic metal hydride, the alkoxy adduct of sodium aluminum hydride, soluble in benzene and other hydrocarbons; reacts less strongly with water than lithium aluminum hydride.

Use: For reduction of organic compounds.

sodium bismuthate.
CAS: 12232-99-4. $NaBiO_3$.

Properties: Yellow or brown, amorphous powder. Slightly hygroscopic. Insoluble in cold water; decomposes in hot water and acids.

Use: Analysis (testing for manganese in iron, steel, and ores), reagent, pharmaceuticals.

sodium bisulfate.   (sodium acid sulfate; niter cake; sodium hydrogen sulfate).
CAS: 7681-38-1.   (1) $NaHSO_4$ or (2) $NaHSO_4 \cdot H_2O$.

Properties: Colorless crystals or white, fused lumps. Aqueous solution is strongly acid. (1) d 2.435 (13C), mp above 315C, (2) d 2.103 (13C), mp 58.5C. Soluble in water. Noncombustible.

Derivation: A by-product in the manufacture of hydrochloric acid and nitric acid.

Method of purification: Recrystallization.

Grade: Pure crystals, pure fused, pure dry, reagent, crude, CP, technical, FCC.

Available forms: Cakes, powder, prills, pearls.

Hazard: Strong irritant to tissue.

Use: Flux for decomposing minerals; substitute for sulfuric acid in dyeing; disinfectant; manufacture of sodium hydrosulfide, sodium sulfate, and soda alum; liberating CO_2 in carbonic acid baths, in thermophores; carbonizing wool; manufacture of magnesia cements, paper, soap, perfumes, foods, industrial cleaners, metal pickling compounds; lab reagent.

sodium bisulfide.   See sodium hydrosulfide.

sodium bisulfite.   (sodium acid sulfite; sodium hydrogen sulfite).
CAS: 7631-90-5. $NaHSO_3$.

Properties: White crystals or crystalline powder; slight sulfurous odor and taste. Unstable in air, d 1.48, mp decomposes. Soluble in water; insoluble in alcohol. Noncombustible.

Derivation: Sodium carbonate solution is saturated with sulfur dioxide and the solution crystallized.

Grade: Crystals, pure dry, commercial dry, reagent, commercial solution 35 degrees Bé, powder (67% SO_2), CP, USP, FCC.

Hazard: Not permitted in meats and other sources of vitamin B_1, strong irritant to skin and tissue.

Use: Chemicals (sodium salts, cream of tartar), vat dye preparation, textiles (antichlor, mordant, discharge), food preservative. Bleaching groundwood, wool, etc. Reducing agent, fermentation, antiseptic, cask sterilization (brewing), copper and brass plating, color preservative for pale crepe rubber, woodpulp digestion, analytical reagent, dietary supplement.

sodium bitartrate.   (acid sodium tartrate). $NaHC_4H_5O_6 \cdot H_2O$.

Properties: White, crystalline powder. Mp loses water at 100C, bp 219C (decomposes). Soluble in water (aqueous solution in acid); slightly soluble in alcohol. Combustible.

Grade: CP, technical, reagent.

Use: Analysis (testing for potassium), effervescing mixture, nutrient media.

sodium bithionolate.   (USAN; thiobis(4,6-dichloro-*o*-phenylene)oxy disodium; disodium 2,2'-thiobis-(4,6-dichlorophenoxide)). $C_{12}H_4Cl_4Na_2O_2S$.

Properties: Solid. Soluble in water.

Use: Self-sanitizing polishes, waxes, and cleaners; shoe polish and leather conditioners.

sodium borate.   (borax; sodium borate decahydrate; sodium pyroborate; sodium tetraborate).
CAS: 1303-96-4. $Na_2B_4O_7 \cdot 10H_2O$.

See borax, anhydrous; borax pentahydrate.

Properties: White crystals or powder; odorless. D 1.73 (20/4C). Loses water of crystallization when heated, with melting, between 75C and 320C; fuses to a glassy mass at red heat (borax glass); effloresces slightly in warm, dry air. Soluble in water and glycerol, insoluble in alcohol. Noncombustible.

Derivation: Fractional crystallization from Searles Lake brine, solution of kernite ore followed by crystallization. Also from colemanite, natural borax, uxelite, and other borates.

Grade: Crystals, granulated, powdered (refined, USP), CP, technical.

Hazard: Toxic by inhalation.

Use: Heat-resistant glass, porcelain enamel, detergents, herbicides, fertilizers, rust inhibitors, pharmaceuticals, leather, photography, bleaches, paint, boron compounds, flux for smelting, flame-retardant fungicide for wood, soldering flux, cleaning preparations, lab reagent.

sodium borate, anhydrous.   See borax, anhydrous.

sodium boroformate. $NaH_2BO_3•2HCOOH•2H_2O$.

Properties: White, crystalline solid. Mp 110C. Soluble in water.

Use: Buffering agent toward both acid and alkali in the range of pH 8.5, textile treating and tanning baths.

sodium borohydride.

CAS: 16940-66-2. $NaBH_4$.

Properties: White, crystalline powder. D 1.07, hygroscopic, stable in dry air to 300C; decomposes slowly in moist air or in vacuum at 400C. Soluble in water, ammonia, amines, pyridine, and dimethyl ether of diethylene glycol; insoluble in other ethers, hydrocarbons, and alkyl chlorides.

Derivation: By reaction of sodium hydride and trimethyl borate at 250C.

Grade: Technical, powdered, pellets.

Hazard: Reacts with water to evolve hydrogen and sodium hydroxide. Flammable, dangerous fire risk. Store out of contact with moisture.

Note: An explosion can occur by spontaneous ignition of the gases released from a saturated solution of sodium borohydride in dimethylformamide at 17C.

Use: Source of hydrogen and other borohydrides. Reduces aldehydes, ketones and acid chlorides. Bleaching wood pulp, blowing agent for plastics, precipitation of mercury from waste effluent (by reduction), decolorizer for plasticizers, recycling of gold and platinum group metals, inorganic and organometallic reductions, organic synthesis.

sodium bromate.

CAS: 7789-38-0. $NaBrO_3$.

Properties: White crystals or powder; odorless. D 3.339, mp 381C (decomposes). Soluble in water; insoluble in alcohol.

Derivation: By passing bromine into a solution of sodium carbonate, sodium bromide and sodium bromate being formed.

Hazard: Oxidizing material, dangerous fire risk near organic materials. Toxic by ingestion.

Use: Analytical reagent.

sodium bromide.

CAS: 7647-15-6.   (1) NaBr. (2) $NaBr•2H_2O$.

Properties: White, crystalline powder or granules; saline and somewhat bitter taste. Absorbs moisture from the air, becoming very hard. D (1) 3.208, (2) 2.176, mp (1) 757.7C, bp 1390C. Soluble in water; moderately soluble in alcohol.

Derivation: Occurs naturally in some salt deposits. Made synthetically by first causing iron to react with bromine and water. The resulting ferroso-ferric bromide is dissolved in water, sodium carbonate added, the solution filtered and evaporated.

Grade: CP, crystals, powdered, commercial, pure, highest purity, NF.

Hazard: Toxic by inhalation and ingestion.

Use: Photography, medicine (sedative), preparation of bromides.

sodium bromite. $NaBrO_2$.

Use: A desizing compound for textiles.

sodium buthalital.

CAS: 510-90-7. $C_{11}H_{16}N_2O_2S•Na$.

Hazard: A poison.

sodium cacodylate.   (sodium dimethylarsenate).

CAS: 124-65-2. $(CH_3)_2AsOONa•3H_2O$.

Properties: White, amorphous crystals or powder. Deliquescent. Melts about 60C, loses water at 120C. Soluble in water and alcohol.

Derivation: Oxidation and neutralization of cacodyl oxide.

Hazard: Toxic by inhalation and ingestion.

Use: Herbicide.

sodium calcium aluminosilicate, hydrated.

Hazard: A nuisance dust.

Use: Food additive.

sodium carbolate.   See sodium phenate.

sodium carbonate.   (soda).

See soda ash; sal soda; sodium bicarbonate; sodium carbonate monohydrate; sodium sesquicarbonate.

sodium carbonate monohydrate.   (crystal, carbonate; soda monohydrate; soda crystals).

CAS: 497-19-8. $Na_2CO_3•H_2O$.

Properties: White, small crystals or crystalline powder; odorless; alkaline taste. D 1.55, mp 109C (loses water) 851C. Soluble in water and glycerol; insoluble in alcohol. Noncombustible.

Grade: USP, technical, FCC.

Use: Photography, cleaning and boiler compounds, pH control of water, food additive, intermediate in thermochemical reactions, manufacture of glass, cleaning and bleaching textiles, analytical reagent.

sodium carbonate peroxide.   See sodium percarbonate.

sodium carbonate peroxohydrate.   See sodium percarbonate.

sodium carboxymethylcellulose.   See carboxymethylcellulose.

sodium o-carboxymethyltartronate.
CAS: 41999-58-0. $C_5H_3O_7\cdot3Na$.
Hazard: Questionable carcinogen.

sodium caseinate.
CAS: 9005-46-3.
Properties: White, coarse powder; odorless; tasteless. Contains 65% proteins. Soluble in water (usually with turbidity).
Derivation: By dissolving casein in sodium hydroxide and evaporating.
Grade: Edible.
Use: Food additive (binder and extender in sausage, soups, etc.), emulsifier, and stabilizer.

sodium catechol disulfonate.   See disodium-1,2-dihydroxybenzene-3,5-disulfonate.

sodium cellulosate.
Properties: A cellulosic fiber.
Derivation: Reaction of sodium methoxide with cotton or rayon swollen in methanol.
Use: Intermediate in preparing grafted fibers when combined with polyacrylonitrile, there is an average substitution of one polyacrylonitrile chain for every 100 to 300 anhydroglucose units of the cellulose.

sodium chlorate.
CAS: 7775-09-9. $NaClO_3$.
Properties: Colorless crystals; odorless; cooling, saline taste. Must not be triturated with any combustible substance. D 2.490, mp 255C, bp (decomposes). Soluble in water and alcohol.
Derivation: A concentrated acid solution of sodium chloride is heated and electrolyzed, the chlorate crystallizing out.
Grade: Technical, CP, crystals, powder.
Hazard: Dangerous fire risk, strong oxidant, contact with organic materials may cause fire.
Use: Oxidizing agent and bleach (especially to make chlorine dioxide) for paper pulps; ore processing; herbicide and defoliant; substitute for potassium chlorate, being more soluble in water; matches, explosives, flares, and pyrotechnics; recovery of bromine from natural brines; leather tanning and finishing; textile mordant; to make perchlorates.

sodium chloraurate.   See sodium gold chloride.

sodium chloride.   (table salt; sea salt; halite; rock salt). NaCl.
Properties: Colorless, transparent crystals or white, crystalline powder. D 2.165, mp 801C, somewhat hygroscopic. Soluble in water and glycerol; very slightly soluble in alcohol. Essential in diet to maintain chloride balance in body. Noncombustible.
Occurrence: Ocean water (2.6% concentration), deposits in central New York, Southern Michigan, Gulf Coast, Great Salt Lake, Newfoundland.
Derivation: (1) Evaporation and crystallization of natural brines, (2) solar evaporation of seawater, (3) direct mining from underground or surface deposits. Method of purification: Recrystallization.
Impurities: Sulfates, heavy metals, alkaline earths, magnesium salts, ammonium salts.
Grade: Highest purity medicinal, crystals; highest purity, dried; highest purity, fine powder; highest purity, fused; reagent; reagent, fused; sea evaporated; ground; microsized; powdered; table salt; rock salt; CP; USP; FCC; single, pure crystals.
Use: Chemical (sodium hydroxide, soda ash, hydrogen chloride, chlorine, metallic sodium), ceramic glazes, metallurgy, curing of hides, food preservative, mineral waters, soap manufacture (salting out), home water softeners, highway deicing, regeneration of ion-exchange resins, photography, food seasoning, herbicide, fire extinguishing, nuclear reactors, mouthwash, medicine (heat exhaustion), salting out dyestuffs, supercooled solutions. Single crystals are used for spectroscopy, UV and infrared transmissions.
See fused salt.

sodium chlorite.
CAS: 7758-19-2. $NaClO_2$.
Properties: White crystals or crystalline powder. Mp 180–200C (decomposes). Slightly hygroscopic. Soluble in water.
Grade: Technical, reagent.
Hazard: Flammable, strong oxidizing agent, dangerous fire and moderate explosion risk. (Solution) Strong irritant to skin and tissue.
Use: For improving taste and odor of potable water (as an oxidizing agent); bleaching agent for textiles, paper pulp, edible and inedible oils, shellacs, varnishes, waxes and straw products; oxidizing agent; reagent.

sodium chloroacetate.
CAS: 3926-62-3. $ClCH_2COONa$.
Properties: White, nonhygroscopic powder; odorless. Easier to handle than chloroacetic acid. Soluble in water; slightly soluble in methanol; insoluble in acetone benzene, ether, and carbon tetrachloride. Noncombustible.
Use: Manufacture of weed killers, dyes, vitamins, pharmaceuticals; also a defoliant.

sodium chloroaluminate.   See sodium tetrachloroaluminate.

sodium chloroaurate.   See sodium gold chloride.

sodium-*p*-chloro-*m*-cresolate.   A water-soluble preservative for cutting oils.

sodium (r,r)-5-(2-((2-(3-chlorophenyl)-2-hydroxyethyl)amino)propyl)-1,3-benzodioxole-2,2-.
CAS: 138908-40-4. $C_{20}H_{18}ClNO_7 \cdot 2Na$.
Hazard: A poison.

sodium-4-chlorophthalate.   (monosodium-4-chlorophthalate). $ClC_6H_3(COOH)COONa$.
Properties: White to light-gray powder.
Grade: Commercial.
Use: Modifying agent in the manufacture of phthalocyanine pigments.

sodium   chloroplatinate.   (platinic sodium chloride; platinum sodium chloride; sodium platinichloride; sodium hexachloroplatinate).
CAS: 1307-82-0. $Na_2PtCl_6 \cdot 4H_2O$.
Properties: Yellow powder. Soluble in alcohol, water. Noncombustible.
Grade: Technical, CP.
Use: Etching on zinc, indelible ink, microscopy, mirrors, photography, plating, catalyst, determination of potassium.

sodium-*o*-chlorotoluene-*p*-sulfonate.
$NaSO_3C_6H_3(CH_3)Cl$.
Properties: Gray to light-tan powder. Soluble in water and organic solvents.
Derivation: Sulfonation of *o*-chlorotoluene, neutralized to form the sodium salt.
Use: Synthesis of dyes, intermediates and drugs.

sodium chromate.
CAS: 7775-11-3. $NaCrO_4 \cdot 10H_2O$.
Properties: Yellow, translucent, efflorescent crystals. D 1.483, mp 19.92C. Soluble in water; slightly soluble in alcohol.
Derivation: Chrome iron ore is melted in a reverberatory furnace with lime and soda, in presence of air. The melt is dissolved in water, a small amount of sodium carbonate added, the solution decanted, acidified with acetic acid, concentrated, and crystallized.
Available forms: Anhydrous.
Grade: Pure neutral, highest purity, technical, CP, reagent.
Hazard: Toxic material.
Use: Inks, dyeing, paint pigment, leather tanning, other chromates, protection of iron against corrosion, wood preservative.

sodium chromate decahydrate.
CAS: 13517-17-4. $CrO_4 2Na \cdot 10H_2O$.
Properties: Deliquescent yellow monoclinic crystals from H_2O. IDLH Ca [15 mg/m^3 {as Cr(VI)}].
Hazard: Confirmed carcinogen.

sodium chromate tetrahydrate.
$Na_2CrO_4 \cdot 4H_2O$.

Properties: Yellow crystals. Deliquescent. Soluble in water.
Grade: Technical.
Hazard: As for sodium chromate.
Use: Pigment manufacture, corrosion inhibition, leather tanning, other chromium compounds.

sodium citrate.   (trisodium citrate).
CAS: 68-04-2. $C_6H_5O_7Na_3 \cdot 2H_2O$.
Properties: White crystals or granular powder; odorless; pleasant acid taste. Mp loses 2 waters at 150C, bp decomposes at red heat, stable in air. Soluble in water, insoluble in alcohol. Combustible.
Derivation: Sodium sulfate solution is treated with calcium citrate, filtered, concentrated and crystallized.
Grade: Highest purity, medicinal; pure; commercial; CP; USP; FCC.
Use: Soft drinks, photography, frozen desserts, meat products, detergents, special cheeses, electroplating, sequestrant and buffer, nutrient for cultured buttermilk, removal of sulfur dioxide from smelter waste gases, medicine (diuretic, expectorant), anticoagulant for blood withdrawn from body.

sodium clodronate.
CAS: 22560-50-5. $CH_2Cl_2O_6P_2 \cdot 2Na$.
Hazard: Moderately toxic by ingestion.

sodium copper chloride.   See copper sodium chloride.

sodium copper cyanide.   (copper sodium cyanide; sodium cyanocuprate).
CAS: 14264-31-4. $NaCu(CN)_2$.
Properties: White, crystalline, double salt of copper cyanide and sodium cyanide. D 1.013 (20C), decomposes at 100C. Soluble in water.
Hazard: Toxic material.
Use: For preparing and maintaining cyanide copper plating baths based on sodium cyanide.

sodium cyanate.
CAS: 917-61-3.   NaOCN.
Properties: White, crystalline powder. D 1.937 (20C), mp 700C. Soluble in water; insoluble in alcohol and ether.
Use: Organic synthesis, heat treating of steel, intermediate for manufacture of medicinals.

sodium cyanide.
CAS: 143-33-9.   NaCN.
Properties: White, deliquescent, crystalline powder. Mp 563C, bp 149C, the aqueous solution is strongly alkaline and decomposes rapidly on standing. Soluble in water; slightly soluble in alcohol.
Derivation: By absorption of hydrogen cyanide in a solution of sodium hydroxide with subsequent vacuum evaporation.
Grade: 30% solution, 73–75%, 96–98%, reagent, technical, briquettes, granular.
Hazard: Toxic by ingestion and inhalation.

Use: Extraction of gold and silver from ores, electroplating, heat treatment of metals (case-hardening), making hydrogen cyanide, insecticide, cleaning metals, fumigation, manufacture of dyes and pigments, nylon intermediates, chelating compounds, ore flotation.

sodium cyanoaurite.   See sodium gold cyanide.

sodium cyanocuprate.   See sodium copper cyanide.

sodium-2-cyanoethanesulfonate.   (sodium-2-sulfopropionitrile).
$NaO_3S(CH_2)_2CN$.
Properties: Colorless crystals. Mp 240C. Soluble in water and glacial acetic acid; insoluble in benzene, alcohol, and ether.
Use: Introduction of sulfonic acid group into surfactants and other organics.

sodium cyclamate.   (sodium cyclohexylsulfamate).
CAS: 139-05-9. $C_6H_{11}NHSO_3Na$.
Properties: White crystals or powder; practically odorless; sweet taste. Freely soluble in water; practically insoluble in alcohol, benzene, chloroform, and ether; pH (10% solution) 5.5–7.5. Sweetening power about 30 times that of sucrose.
Grade: NF, FCC.
Hazard: Some evidence of causing cancer in laboratory animals. Prohibited by FDA for food use. Questionable carcinogen.
Use: Nonnutritive sweetener.

sodium dehydroacetate.
CAS: 4418-26-2. $C_8H_7NaO_4 \cdot H_2O$.
Properties: Tasteless, white powder. Soluble in water and propylene glycol; insoluble in most organic solvents.
See dehydroacetic acid.
Grade: FCC.
Use: Fungicide, plasticizer, toothpaste, pharmaceutical, preservative in food, mold inhibitor for strawberries and similar fruits.

sodium deoxycholate.   See deoxycholic acid.

sodium dextran sulfate.   (dextran sulfate).
Properties: Solid. Soluble in water.
Derivation: Derivatives of dextran having a molecular weight of 500,000–2,000,000.
Use: Fractionation and separation of biological preparations.

sodium-3,5-diacetamido-2,4,5-triiodobenzoate.   See sodium diatrizoate.

sodium diacetate.
CAS: 126-96-5. $CH_3COONa \cdot x(CH_3COOH)$, anhydrous or $CH_3COONa \cdot x(CH_3COOH) \cdot yH_2O$, technical.

Properties: White crystals; acetic acid odor. Decomposes above 150C. Soluble in water; slightly soluble in alcohol; insoluble in ether. Combustible.
Grade: FCC.
Use: Buffer, mold inhibitor, souring agent, intermediate for acid salts, mordants, varnish hardeners, antitarnishing agents, sequestrant and preservative in foods.

sodium diatrizoate.   (sodium-3,5-diacetamido-2,4,6-triiodobenzoate).
CAS: 737-31-5. $C_6I_3(COONa)(NHCOCH_3)_2$.
Properties: White crystals. Soluble in water; solutions are radiopaque.
Grade: USP (as solution for injection).
Use: Radiopaque medium, medicine.

sodium-1-diazo-2-naphthol-4-sulfonate.
See 1-diazo-2-naphthol-4-sulfonic acid.

sodium dibutyldithiocarbamate.   See Trepidone.

sodium-α,β-dichloroisobutyrate.   A plant growth regulator.

sodium dichloroisocyanurate.   (sodium salt of dichloro-s-triazine-2,4,6-trione).
CAS: 2244-21-5.

$NaNC(O)NClC(O)NClCO$.

Properties: White, slightly hygroscopic, crystalline powder. Loose bulk d about 37 lb/ft^3, granulated 57 lb/ft^3. Active ingredient about 60% available chlorine; decomposes at 230C.
Hazard: Strong oxidizing material, fire risk near organic materials. Toxic by ingestion.
Use: Active ingredient in dry bleaches, dishwashing compounds, scouring powders, detergent-sanitizers, swimming pool disinfectants, water and sewage treatment, replacement for calcium hypochlorite.

sodium-2,4-dichlorophenoxyacetate.
(2,4-D, sodium salt).
CAS: 2702-72-9. $C_6H_5(OCH_2COONa)Cl_2$.
Properties: Crystalline solid. Decomposes at 215C. Slightly soluble in water.
Hazard: Irritant by inhalation.
Use: Herbicide.
See 2,4-D.

sodium-2,4-dichlorophenoxyethyl sulfate.
See sesone.

sodium-2,2-dichloropropionate.   See dalapon.

sodium dichromate.   (sodium bichromate).
CAS: 10588-01-9. $Na_2Cr_2O_7 \cdot 2H_2O$.

Properties: Red or red-orange deliquescent crystals. D 2.52 (13C), mp 357C, decomposes at 400C, loses 2H$_2$O on prolonged heating at 100C. Soluble in water; insoluble in alcohol. Noncombustible.

Derivation: (a) From chromite ore by alkaline roasting and subsequent leaching, (2) action of sulfuric acid on sodium chromate.

Grade: Technical crystals, technical liquor containing 69–70% Na$_2$Cr$_2$O$_7$•2H$_2$O, anhydrous 95C, soluble in water and alcohol.

Use: Colorimetry (copper determination), complexing agent, oxidation inhibitor in ethyl ether.

sodium dihydrogen phosphate. See sodium phosphate, monobasic.

sodium dihydroxyethylglycine.
(*N,N*-bis[2-hydroxyethyl]-glycine).
NaOOCCH$_2$N(CH$_2$CH$_2$OH)$_2$.
Properties: Clear, straw-colored liquid. D 1.204 (25C), fp about −10C.
Use: Complexing agent for the transition metals.

sodium dimethylarsenate. See sodium cacodylate.

sodium dimethyldithiocarbamate. (SDDC).
CAS: 148-18-5. (CH$_3$)$_2$NCS$_2$Na.
Properties: 40% solution is amber to light green. D 1.17–1.20 (25/25C).
Derivation: Reaction of dimethylamine, carbon disulfide, and sodium hydroxide.
Use: Fungicide, corrosive inhibitor, rubber accelerator, intermediate, polymerization shortstop.

sodium 2,2-dimethylpropanoate.
CAS: 1184-88-9. C$_5$H$_9$O$_2$•Na.
Hazard: A reproductive hazard.

sodium dinitro-*o*-cresylate.
CAS: 25641-53-6. CH$_3$C$_6$H$_2$(NO$_2$)$_2$ONa.
Properties: Brilliant-orange-yellow dye that may stain clothing and wood.
Derivation: By treating 4,6-dinitro-*o*-cresol with sodium hydroxide.
Hazard: Toxic by ingestion and inhalation.
Use: Herbicide (control of mustard and other susceptible weeds), fungicide.
See 4,6-dinitro-*o*-cresol.

sodium dioctyl sulfosuccinate. See dioctyl sodium sulfosuccinate.

sodium dioxide. NaO$_2$. Exists as impurity (about 10%) in sodium peroxide, obtained by heating sodium peroxide in oxygen, reacts with water to yield hydrogen peroxide, oxygen, and sodium hydroxide.

sodium dispersion. A stable suspension of microscopic sodium particles in a hydrocarbon or other medium boiling at temperatures above the melting point of sodium (97.5C), e.g., heptane, *n*-octane, toluene, xylene, naphtha, kerosene, mineral oil, *n*-butyl ether. Particles range in size from submicron to 30 μm depending on the method of preparation. Dispersions contain up to 50% (by weight) of sodium metal.

Hazard: Flammable, dangerous fire risk. To extinguish use dry soda ash, sodium chloride, or graphite, followed by carbon dioxide. Do not use carbon tetrachloride.

Use: For chemical reactions where advantages of controlled reaction rate, lower reaction temperature, increased yields, or substitution for more expensive reagent can be achieved, as in, removal of sulfur from hydrocarbons and petroleum, metal powders, sodium hydride, alcohol-free alcoholates, phenylsodium.

sodium dithionate. (sodium hyposulfate).
Na$_2$S$_2$O$_6$•2H$_2$O.
Properties: Large, transparent crystals; bitter taste. D 2.189 (25C), loses 2H$_2$O at 110C, decomposes at 267C. Soluble in water; insoluble in alcohol and concentrated hydrochloric acid. Combustible.
Use: Chemical reagent.

sodium dithionite. (sodium hydrosulfite).
CAS: 7775-14-6. Na$_2$S$_2$O$_4$.
Properties: Light-lemon-colored solid in powder or flake form. Mp 55C (decomposes). Also available in liquid form. Partially soluble in water; insoluble in alcohol; strong reducing agent.
Derivation: (1) Zinc is dissolved in a solution of sodium bisulfite, the zinc-sodium sulfite precipitated by milk of lime, leaving the hydrosulfite in solution; (2) reaction of sodium formate with sodium hydroxide and sulfur dioxide.
Grade: Technical, reagent.
Hazard: Fire risk in contact with moisture. To extinguish fires, flood the reacting mass with water.
Use: Vat dyeing of fibers and textiles; stripping agent for dyes; reagent; bleaching of sugar, soap, oils, and groundwood; oxygen scavenger for synthetic rubbers.

sodium diuranate. (uranium yellow).
Na$_2$U$_2$O$_7$•6H$_2$O.
Properties: Yellow-orange solid. Insoluble in water; soluble in dilute acids.
Derivation: By treating a solution of uranyl salt with sodium hydroxide.
Hazard: Radioactive and probably a poison.
Use: Ceramics to produce colored glazes, manufacture of fluorescent uranium glass.

sodium dodecylbenzenesulfonate.
CAS: 25155-30-0. C$_{12}$H$_{25}$C$_6$H$_4$SO$_3$Na.
Properties: White to light-yellow flakes, granules, or powder; biodegradable. The dodecyl radical may have many isomers and the benzene may be attached to it in many positions. Combustible.

Derivation: Benzene is alkylated with dodecene, to which it attaches itself in any secondary position; the resulting dodecylbenzene is sulfonated with sulfuric acid and neutralized with caustic soda. For ABS (branched-chain alkyl) the dodecene is usually a propylene tetramer, made by catalytic polymerization of propylene. For LAS (straight-chain alkyl), the dodecene may be removed from kerosene or crudes by molecular sieve, may be formed by Ziegler polymerization of ethylene, or by cracking wax paraffins to α-olefins.
See biodegradability; alkyl sulfonate, linear; detergent; alkylate (3).
Use: Anionic detergent.

sodium dodecyldiphenyl oxide disulfonate.
$C_{24}H_{34}O(SO_3)_2Na$.
Properties: Dry form 90% min active solution (45%). Mp 150C, d 1.1. Very soluble in water, strong acids, bases, and electrolytes; stable to oxidation.

sodium dodecyl sulfate.
CAS: 151-21-3. $C_{12}H_{26}NaO_4S$.
An anionic surfactant, usually a mixture of sodium alkyl sulfates that lowers surface tension of aqueous solutions.
Derivation: From petrochemicals.
Uses: Fat emulsifier, wetting agent, detergent in cosmetics, pharmaceuticals and toothpastes, a research tool in protein biochemistry.

sodium edetate.   See tetrasodium EDTA.

sodium enolate monohydrate.   See erythorbic acid sodium salt.

sodium erythorbate.   (sodium isoascorbate).
$NaC_6H_7O_6$. White free-flowing crystals, soluble in water.
Grade: Technical, FCC.
Use: Antioxidant and preservative.

sodium ethoxide.   See sodium ethylate.

sodium ethylate.   (sodium ethoxide; caustic alcohol). C_2H_5ONa.
Properties: White powder, sometimes having brownish tinge; readily hydrolyzes to alcohol and sodium hydroxide.
Derivation: By carefully adding small amounts of sodium to absolute alcohol kept at a temperature of 10C, heating carefully to 37.7C, again carefully adding sodium, cooling to 10C, and adding the same quantity of absolute alcohol as was used originally.
Hazard: As for caustic soda, ethanol. Sodium hydroxide is formed when sodium ethylate is exposed to moisture.
Use: Organic synthesis.
See sodium hydroxide; ethanol.

sodium ethylenebisdithiocarbamate.   See nabam.

sodium 2-ethylhexanoate.
CAS: 19766-89-3. $C_8H_{15}O_2$•Na.
Hazard: A reproductive hazard.

sodium-2-ethylhexylsulfoacetate.
$C_8H_{17}OOCCH_2SO_3Na$.
Properties: Light-cream-colored flakes, water-soluble, good foaming properties and excellent resistance to hard water; solutions practically neutral and stable to mineral acids. Combustible.
Use: Solubilizing agent, particularly for soapless shampoo compositions; electroplating detergent.

sodium ethylmercurithiosalicylate.   See thimerosal.

sodium ethyl oxalacetate.   See ethyl sodium oxalacetate.

sodium ethylxanthate.   (sodium xanthogenate; sodium xanthate). $C_2H_5OC(S)SNa$.
Properties: Yellowish powder. Soluble in water, alcohol.
Use: Ore flotation agent.
See xanthic acids.

sodium ferric pyrophosphate.
$Na_8Fe_4(P_2O_7)_5$•xH_2O.
Properties: White to tan powder; odorless. Insoluble in water; soluble in hydrochloric acid.
Hazard: A nuisance dust.
Use: Food additive.

sodium ferricyanide.   (red prussiate of soda).
$Na_3Fe(CN)_6$•H_2O.
Properties: Ruby-red, deliquescent crystals. Soluble in water; insoluble in alcohol.
Derivation: Chlorine is passed into sodium ferrocyanide solution.
Grade: Technical, CP.
Use: Production of pigments, dyeing, printing.

sodium ferrocyanide.   (yellow prussiate of soda).
CAS: 13601-19-9. $Na_4Fe(CN)_6$•$10H_2O$.
Properties: Yellow, semitransparent crystals. D 1.458. Partially soluble in water; insoluble in organic solvents.
Grade: Technical, FCC.
Use: Manufacture of sodium ferricyanide, blue pigments, blueprint paper, anticaking agent for salt, ore flotation, pickling metals, polymerization catalyst, photographic fixing agent.

sodium fluoborate.   See sodium fluoroborate.

sodium fluophosphate.   See sodium fluorophosphate.

sodium fluorescein. USP name for uranine.

sodium fluoride.
CAS: 7681-49-4. NaF.
Properties: Clear, lustrous crystals or white powder. Insecticide grade frequently dyed blue. D 2.558 (41C), mp 988C, bp 1695C. Soluble in water; very slightly soluble in alcohol.
Derivation: By adding sodium carbonate to hydrogen fluoride.
Grade: Pure; CP; USP; insecticide; technical; single, pure crystals.
Hazard: Toxic by ingestion and inhalation, strong irritant to tissue.
Use: Fluoridation of municipal water (1 ppm), degassing steel, wood preservative, insecticide (not to be used on living plants), fungicide and rodenticide, chemical cleaning, electroplating, glass manufacture, vitreous enamels, preservative for adhesives, toothpastes, disinfectant (fermentation equipment), dental prophylaxis, cryolite manufacture, single crystals used as windows in UV and infrared radiation detecting systems.

sodium fluoroacetate. (also known as 1080).
CAS: 62-74-8. FCH$_2$COONa.
Properties: Fine, white powder; odorless. Nonvolatile, decomposes at 200C. Soluble in water; insoluble in most organic solvents.
Derivation: Ethyl chloroacetate and potassium fluoride form ethyl fluoroacetate, which is then treated with a methanol solution of sodium hydroxide.
Hazard: Toxic by ingestion, inhalation, and skin absorption. For use by trained operators only, use has been restricted. Cardiac impairment, central nervous system impairment, and nausea.
Use: Predator elimination (coyotes), rodenticide.

sodium fluoroborate. NaBF$_4$.
Properties: White powder; bitter acid taste. Slowly decomposed by heat. D 2.47 (20C), mp 384C. Soluble in water; slightly soluble in alcohol.
Derivation: By heating sodium fluoride and hydrofluoboric acid and cooling slowly.
Use: Sand casting of aluminum and magnesium, electrochemical processes, oxidation inhibitor, fluxes for nonferrous metals, fluorinating agent.

sodium fluorophosphate. (sodium fluophosphate). Na$_2$PO$_3$F.
Properties: Colorless crystals. Mp about 625C. Soluble in water.
Grade: 97%.
Hazard: Toxic by ingestion, strong irritant.
Use: Preparation of bactericides and fungicides.

sodium fluorosilicate. (sodium silicofluoride; sodium hexafluorosilicate; sodium fluosilicate). Na$_2$SiF$_6$.

Properties: White, free-flowing powder; odorless; tasteless. D 2.7, mp decomposes at red heat. Partially soluble in cold water; more soluble in hot water; insoluble in alcohol.
Derivation: From fluosilicic acid and sodium carbonate, or sodium chloride.
Grade: Technical, CP.
Hazard: Toxic by ingestion and inhalation, strong irritant to tissue.
Use: Fluoridation, laundry sours, opalescent glass, vitreous enamel frits, metallurgy (aluminum and beryllium), insecticides and rodenticides, chemical intermediate, glue, leather and wood preservative, moth repellent, manufacture of pure silicon.

sodium formaldehyde bisulfite.
HOCH$_2$SO$_3$Na.
Properties: White, water-soluble solid.
Derivation: Action of sodium bisulfite, formaldehyde, and water.
Use: Textile stripping agent.
See hydrosulfite-formaldehyde.

sodium formaldehyde hydrosulfite.
Use: Synthetic rubber polymerization and textile dyeing and stripping agent.
See hydrosulfite-formaldehyde.

sodium formaldehyde sulfoxylate. (sodium sulfoxylate; sodium sulfoxylate formaldehyde; SFS). HCHO•HSO$_2$Na•2H$_2$O.
Properties: White solid. Mp 64C. Soluble in water and alcohol. Usually admixed with a sulfite.
Use: Stripping and discharge agent for textiles, bleaching agent for molasses, soap.
See hydrosulfite-formaldehyde.

sodium formate.
CAS: 141-53-7. HCOONa.
Properties: White, slightly hygroscopic, crystalline powder; slight odor of formic acid. D 1.919, mp 253C. Soluble in water and glycerol; slightly soluble in alcohol; insoluble in ether.
Derivation: Carbon monoxide and sodium hydroxide are heated under pressure, also from pentaerythritol manufacture.
Grade: Technical, CP.
Use: Reducing agent, manufacture of formic acid and oxalic acid, organic chemicals, mordant, manufacture of sodium dithionite, complexing agent, analytical reagent (noble metal precipitant), buffering agent.

sodium glucoheptonate.
HOCH$_2$(CH$_2$O)$_5$COONa.
Properties: Light-tan, crystalline powder. A sequestering agent for polyvalent metals.
Use: Metal cleaning, bottle washing, kier boiling, mercerizing, caustic boiloff, paint stripping, aluminum etching.

sodium gluconate.
CAS: 527-07-1. $NaC_6H_{11}O_7$.
Properties: White to yellowish crystalline powder. Soluble in water; sparingly soluble in alcohol.
Derivation: From glucose by fermentation.
Grade: Purified, technical, FCC.
Use: Foods and pharmaceuticals, sequestering agent, metal cleaners, paint stripper, aluminum deoxidizer, bottle-washing preparations, rust removal, chrome tanning, metal plating, mordant in dyeing.

sodium glucosulfone. (p'-diaminodiphenyl-sulfone-N,N'-di-dextrose sodium sulfonate). $C_{14}H_{34}N_2Na_2O_{18}S_3$.
The USP grade is an aqueous solution (for injection), clear and pale yellow, pH 5.0–6.5.
Use: Medicine.

sodium glutamate. (monosodium glutamate; MSG).
CAS: 142-47-2. $COOH(CH_2)_2CH(NH_2)COONa$.
Sodium salt of glutamic acid, one of the common, naturally occurring amino acids.
Properties: White, crystalline powder. Mp (decomposes). Soluble in water and alcohol. Shows optical activity, most effective between pH 6 and 8.
Derivation: (1) Alkaline hydrolysis of the waste liquor from beet sugar refining, (2) a similar hydrolysis of wheat or corn gluten, (3) organic synthesis based on acrylonitrile.
Grade: Technical, 99%, ND, FCC.
Use: Flavor enhancer for foods in concentration of about 0.3%.
See flavor; glutamic acid.

sodium glycolate. (sodium hydroxyacetate). $NaOOCCH_2OH$.
Properties: White powder.
Use: Buffer in electrodeless plating and textile finishing.

sodium gold chloride. (sodium aurichloride; sodium chloraurate; sodium chloroaurate; gold sodium chloride; gold salts). $NaAuCl_4•2H_2O$.
Properties: Yellow crystals. Soluble in water and alcohol.
Derivation: By neutralizing chloroauric acid with sodium carbonate.
Use: Photography, staining fine glass, decorating porcelain, medicine.

sodium gold cyanide. (sodium cyanoaurite; sodium aurocyanide; gold sodium cyanide). $NaAu(CN)_2$.
Properties: Yellow powder. Soluble in water. Contains 46% gold (min).
Hazard: Toxic.
Use: For gold plating electronic components.

sodium guanylate. (GMP; disodium guanylate).
CAS: 5550-12-9. $Na_2C_{10}H_{12}N_5O_8P•2H_2O$.
A 5'-nucleotide.
Properties: Crystals. Soluble in cold water; very soluble in hot water.
Derivation: From a seaweed or from dried fish.
Use: Flavor potentiator in foods.
See guanylic acid.

sodium heparin. See heparin.

sodium heptametaphosphate. See sodium metaphosphate.

sodium hexachloroosmate. (osmium–sodium chloride; sodium–osmium chloride). Na_2OsCl_6.
Properties: Orange, rhombic prisms. Contains 40.3% osmium, unstable. Soluble in alcohol, water.
Grade: Technical.
Use: Catalyst (oxidation).

sodium hexachloroplatinate. See sodium chloroplatinate.

sodium hexadecanoate. See sodium palmitate.

sodium hexafluorosilicate. See sodium fluorosilicate.

sodium hexametaphosphate. See "Calgon" [Calgon].

sodium hexylene glycol monoborate. $C_6H_{12}O_3BNa$.
Properties: Amorphous, white solid. Bulk d 0.25, mp 426C. Soluble in nonpolar solvents.
Grade: Min 98%.
Use: Corrosion inhibitor in organic systems, additive to lubricating oils, flame-retardant, siloxane resin additive.

sodium hydrate. See sodium hydroxide.

sodium hydride.
CAS: 7646-69-7. NaH.
Properties: Powder; practically odorless. D 0.92, mp 800C (decomposes). Must be kept cool and dry. Particle size range 5–50 µm. Starts to decompose with evolution of hydrogen at about 255C.
Preparation: Reaction of sodium metal with hydrogen. A microcrystalline dispersion of gray powder in oil containing 50 or 25% by weight.
Hazard: Dangerous fire risk, reacts violently with water evolving hydrogen. Irritant.
Use: Condensing or alkylating agent, especially for amines, descaling metals.

sodium hydrogen sulfide. See sodium hydrosulfide; sodium bisulfite.

sodium hydrogen trioxoselenite. See selenious acid, monosodium salt.

sodium hydrosulfide. (sodium sulfhydrate; sodium bisulfide; sodium hydrogen sulfide).
CAS: 16721-80-5. NaSH•2H$_2$O.
Properties: Colorless needles to lemon-colored flakes. 70–72% NaSH, mp 55C, water of crystallization 26–28%. Soluble in water, alcohol, and ether.
Derivation: From calcium sulfide by treating it in the cold with sodium bisulfate.
Grade: Technical, flake, 70–72%, soluble 40–44%.
Hazard: Contact with acids causes evolution of toxic gases.
Use: Paper-pulping dyestuffs processing, rayon and cellophane desulfurizing, unhairing hides, bleaching reagent.

sodium hydrosulfite.   See sodium dithionite.

sodium hydroxide. (caustic soda; sodium hydrate; lye; white caustic).
CAS: 1310-73-2. The most important commercial caustic.
Properties: White, deliquescent solid; chiefly in form of beads or pellets, also 50% and 73% aqueous solutions. Absorbs water and carbon dioxide from the air. D 2.13, mp 318C, bp 1390C. Soluble in water, alcohol, and glycerol.
Derivation: Electrolysis of sodium chloride (brines) (electrolytic cell), reaction of calcium hydroxide and sodium carbonate.
Grade: Commercial, ground, flake, beads, FCC, granulated (60% and 75% Na$_2$O), rayon (low in iron, copper, and manganese), purified by alcohol (sticks, lumps, and drops), reagent, highest purity, CP, USP.
Hazard: Corrosive to tissue in presence of moisture, strong irritant to tissue (eyes, skin, mucous membranes, and upper respiratory tract), poison by ingestion.
Use: Chemical manufacture, rayon and cellophane, neutralizing agent in petroleum refining, pulp and paper, aluminum, detergents, soap, textile processing, vegetable-oil refining, reclaiming rubber, regenerating ion exchange resins, organic fusions, peeling of fruits and vegetables in food industry, lab reagent, etching and electroplating, food additive.

sodium hypochlorite.
CAS: 7681-52-9. NaOCl•5H$_2$O.
Properties: Pale-greenish color; disagreeable, sweetish odor. Mp 18C. Soluble in cold water; decomposed by hot water. Unstable in air unless mixed with sodium hydroxide. Strong oxidizing agent. Usually stored and used in solution.
Derivation: Addition of chlorine to cold dilute solution of sodium hydroxide.
Grade: Technical.
Hazard: Fire risk in contact with organic materials. Toxic by ingestion, strong irritant to tissue.
Use: Bleaching paper pulp, textiles, etc.; intermediate; organic chemicals; water purification; medicine; fungicides; swimming pool disinfectant laundering; reagent; germicide.

sodium hypophosphite.
CAS: 7681-53-0. NaH$_2$PO$_2$•H$_2$O.
Properties: Colorless, pearly, crystalline plates or white, granular powder; saline taste; deliquescent. Soluble in water; partially soluble in alcohol.
Derivation: By neutralizing hypophosphoric acid with sodium carbonate.
Grade: Technical, CP.
Hazard: Explosion risk when mixed with strong oxidizing agents, decomposes to phosphine on heating, store in cool, dry place, away from oxidizing materials.
Use: Pharmaceuticals, reducing agent in electrodeless nickel plating of plastics and metals, lab reagent, substitute for sodium nitrite in smoked meats.

sodium hyposulfate.   See sodium dithionate.

sodium inosinate.
CAS: 4691-65-0. C$_{10}$H$_{11}$Na$_2$N$_4$O$_8$P.
A 5'-nucleotide derived from seaweed or dried fish. Sodium guanylate is a by-product.
Use: Flavor potentiator in foods.
See inosinic acid.

sodium iodate.
CAS: 7681-55-2. NaIO$_3$.
Properties: White crystals. D 4.28. Soluble in water and acetone; insoluble in alcohol.
Derivation: Interaction of sodium chlorate and iodine in presence of nitric acid.
Grade: CP, reagent, technical.
Hazard: Oxidizing agent, fire risk near organic materials.
Use: Antiseptic, disinfectant, feed additive reagent.

sodium iodide.
CAS: 7681-82-5. (1) NaI. (2) NaI•2H$_2$O.
Properties: White, cubical crystals or powder, or colorless crystals; odorless, saline, somewhat bitter taste. D (1) 3.665, (2) 2.448 (21C), mp (1) 653C, bp (1) 1304C. Slowly becomes brown in air, deliquescent. Soluble in water, alcohol, and glycerol.
Grade: Technical, CP, USP, single crystals.
Use: Photography, solvent for iodine, organic chemicals, reagent, feed additive, cloud seeding, scintillation (thallium-activated form), expectorant.

sodium iodide (I-131). (sodium radio-iodide).
A radioactive form of sodium iodide containing iodine-131 which can be used as a tracer.
Grade: USP (as capsules or solution).
See iodine 131.

sodium iodipamide.   C$_{20}$H$_{12}$I$_6$N$_2$NaO$_6$. N,N'-adipolybis(3-amino-2,4,6-triiodobenzoic acid) disodium salt.
Properties: Colorless to pale-yellow, slightly viscous liquid. Water soluble. Radiopaque.
Derivation: By dissolving the free acid in dilute sodium hydroxide and buffering to pH 6.5–7.7.

Available forms: A 20% solution for injection.
Grade: USP.
Use: X-ray contrast medium.

sodium iodomethanesulfonate. See sodium methiodal.

sodium iothalamate. (sodium-5-acetamido-2, 4,6-triiodo-*N*-methylisophthalamate).
CAS: 1225-20-3. $C_6I_3(CONHCH_3)_2COONa$.
Grade: USP (for injection).
Use: Medicine (radiopaque medium).

sodium ipodate. (USAN; sodium-3-(dimethylaminomethyleneamino)-2,4,6-triiodohydrocinnamate).
CAS: 1221-56-3. $NaOOCCH_2CH_2C_6H(I_3)N$: $CHN(CH_3)_2$.
Use: Radiopaque agent.

sodium iron pyrophosphate. (SIPP).
$Na_8Fe_4(P_2O_7)_5 \cdot xH_2O$.
Properties: Tan powder. Insoluble in water; soluble in dilute acid. Min 14.5% iron. Iron is in complex form and will not catalyze oxidation reactions.
Derivation: By reacting tetrasodium pyrophosphate with a soluble iron salt.
Use: For iron enrichment, particularly in flours and cereals.

sodium isoascorbate. See sodium erythorbate.

sodium *d*-isoascorbate. See erythorbic acid sodium salt.

sodium isobutylxanthate. $CH(CH_3)_2CH_2OC$ (S)SNa.
Use: Ore flotation agent.
See xanthic acid.

sodium isopropylxanthate.
CAS: 140-93-2. $(CH_3)_2CHOC(S)SNa$.
Properties: Light-yellow crystals. Soluble in water. Deliquescent, decomposes 150C.
Hazard: Moderate fire risk. Irritant to skin and mucous membranes.
Use: Chemical weed killer, fortifying agent for certain oils, ore flotation.

sodium lactate.
CAS: 72-17-3. $CH_3CH_2OCOONa$.
Properties: Colorless or yellowish, syrupy liquid. Very hygroscopic. Mp 17C, decomposes 140C. Soluble in water. Combustible.
Grade: Technical, USP (solution with pH 6.0–7.3).
Use: Hygroscopic agent, glycerol substitute, plasticizer for casein, corrosion inhibitor in alcohol antifreeze.

sodium-*n*-lauroyl sarcosinate.
$C_{15}H_{28}NO_4Na$.
Use: Dentrifices, hair shampoos, rug shampoos.

sodium lauryl sulfate.
CAS: 151-21-3. $NaC_{12}H_{25}SO_4$.
Properties: Small, white or light-yellow crystals; slight characteristic odor. Soluble in water, forming an opalescent solution.
Grade: USP, technical, FCC.
Use: Wetting agent in textile industry, detergent in toothpaste, food additive (emulsifier and thickener).

sodium–lead alloy. One of several alloys as follows: (1) usually containing 10% sodium and 90% lead, used in the manufacture of lead tetraethyl; (2) containing 2% sodium, used as a deoxidizer and homogenizer in nonferrous metals where lead is a component; (3) used as a stabilizer and deoxidizer for lead in cable sheathing.
Hazard: Moderate fire and explosion risk, reacts with moisture, acids, and oxidizing agents.

sodium lead hyposulfite. See lead sodium thiosulfate.

sodium lead thiosulfate. See lead sodium thiosulfate.

sodium lignosulfonate.
Properties: Tan, free-flowing, spray-dried powder, containing 70–80% total lignin sulfonates, balance wood sugars. Combustible.
Use: Dispersant, emulsion stabilizer, chelating agent. See lignin sulfonate.

sodium liothyronine. (sodium-l-3[4-(4-hydroxy-3-iodophenoxy)-3,5-diiodophenyl]alanine).
$NaOOCCH(NH_2)CH_2C_6H_2I_2OC_6H_3IOH$.
Properties: Light-tan, crystalline powder; odorless. Very slightly soluble in water; slightly soluble in alcohol; insoluble in most other organic solvents.
Grade: USP.
Use: Medicine (a thyroid hormone).

sodium magnesium aluminosilicate. See "Hydrex" *[Energenx]*.

sodium MBT. (NaMBT). $C_7H_4NS_2Na$.
A 50% aqueous solution of sodium mercaptobenzothiazole, light-amber liquid, bulk d 10.5 lb/gal.
Use: Corrosion inhibitor for nonferrous metals, antifreeze, paper-mill systems.

sodium mercaptoacetate. See sodium thioglycolate.

sodium-2-mercaptobenzothiazole. See sodium MBT.

sodium metabisulfite. (sodium pyrosulfite).
CAS: 7681-57-4. $Na_2S_2O_5$.
Chief constituent of commercial dry sodium bisulfite with which most of its properties and uses are practically identical.
Grade: FCC.

Hazard: Toxic by inhalation. Upper respiratory tract irritant. Questionable carcinogen.
Use: In foods, as preservative, lab reagent.

sodium metaborate.
CAS: 7775-19-1. $NaBO_2$.
Properties: White lumps. D 2.464, mp 966C, bp 1434C. Soluble in water. Noncombustible.
Derivation: By fusing sodium carbonate and borax.
Use: Herbicide. Also available commercially as octahydrate and tetrahydrate.

sodium metanilate. $NaSO_3C_6H_4NH_2$.
Derivation: The metasodium sulfonate of aniline sold as a solid or 20% aqueous solution prepared by neutralizing metanilic acid.
Grade: Technical, 99%, also 20% solution.
Use: Manufacture of synthetic dyestuffs and drugs.

sodium metaperiodate. See sodium periodate.

sodium metaphosphate.
CAS: 10361-03-2. $(NaPO_3)_n$. The value of n ranges from 3 to 10 (cyclic molecules) or may be a much larger number (polymers). Cyclic sodium metaphosphate, based on rings of alternating phosphorus and oxygen atoms, range from the trimetaphosphate $(NaPO_3)_3$ to at least the decametaphosphate. So-called sodium hexametaphosphate is probably a polymer where n is between 10 and 20 ("Calgon").
The vitreous sodium phosphates having a Na_2O/P_2O_5 mole ratio near unity are classified as sodium metaphosphates (Graham's salts). The average number of phosphorus atoms per molecule in these glasses ranges from 25 to infinity. The term *sodium metaphosphate* has also been extended to short-chain vitreous compositions, the molecules of which exhibit the polyphosphate formula $Na_{n+2}P_nO_{3n+1}$ with n as low as 4–5. These materials are more correctly called sodium polyphosphates.
Use: Dental polishing agents, detergent builders, water softening, sequestrants, emulsifiers, food additives, textile processing and laundering.

sodium metasilicate, anhydrous.
CAS: 6834-92-0. Na_2SiO_3.
A crystalline silicate.
Properties: Dustless, white granules. Mp 1089C, total Na_2O content 51.5%, total Na_2O in active form 48.6%, bulk d 2.61 or 75 lb/cu ft. Soluble in water; precipitated by acids and by alkaline earths and heavy metal ions, pH of 1% solution 12.6, Noncombustible.
Derivation: Crystallized from a melt of Na_2O and SiO_2 at approximately 1089C.
Use: Laundry, dairy, and metal cleaning; floor cleaning; base for detergent formulations; bleaching aid; deinking paper. Also available as the pentahydrate whose properties are mp 72.2C, total Na_2O content 29.3%, total Na_2O in active form 27.8%, d 1.75 or 55 lb/cu ft.

sodium metavanadate.
CAS: 13718-26-8. $NaVO_3$, often with $4H_2O$.
Properties: Colorless, monoclinic, prismatic crystals or pale-green crystalline powder. Mp 630C. Soluble in water. Noncombustible.
Derivation: Sodium hydrate and vanadium pentoxide in water solution.
Grade: Technical, CP.
Hazard: Toxic by ingestion.
Use: Inks, fur dyeing, photography, inoculation of plant life, mordants and fixers, corrosive inhibitor in gas-scrubbing systems.

sodium methacrylate. $CH_2:C(CH_3)COONa$.
Properties: Water-soluble monomer.
Use: Resins, chemical intermediate.

sodium methanearsonate. (disodium methylarsonate; monosodium methanearsonate; sodium acid methanearsonate; sodium methylarsonate).
CAS: 2163-80-6. $(CH_3AsO(OH)ONa)$.
Properties: White solid. Mp 130–140C. Very soluble in water.
Hazard: Toxic by ingestion and inhalation.
Use: Postemergence herbicide for grassy weeds.

sodium methiodal. (sodium iodomethanesulfonate).
CAS: 126-31-8. ICH_2SO_3Na.
Properties: A white, crystalline powder; odorless; slight salty taste followed by sweetish aftertaste. Decomposes on exposure to light, solutions are neutral to litmus. Soluble in water; very soluble in methanol; slightly soluble in alcohol; practically insoluble in acetone, ether, and benzene.
Derivation: From sodium sulfite and methylene iodide.
Grade: NF.
Use: Radiopaque contrast medium.

sodium methylate. (sodium methoxide).
CAS: 124-41-4. CH_3ONa.
Properties: White, free-flowing powder; sensitive to oxygen; decomposed by water; soluble in methanol and ethanol; decomposes in air above 126C.
Hazard: (Solid) Flammable when exposed to heat or flame. (Solution) Flammable, moderate fire risk.
Use: Condensation reactions; catalyst for treatment of edible fats and oils, especially lard; intermediate for pharmaceuticals; preparation of sodium cellulosate; analytical reagent.

sodium methyl carbonate. $CH_3OCOONa$.
Properties: White powder. Mp 330C (decomposes), d 1.66, purity 90% min.

sodium-*N*-methyldithiocarbamate dihydrate.
CAS: 137-42-8. $CH_3NHC(S)SNa \cdot 2H_2O$.
Properties: White, crystalline solid. Readily soluble in water; moderately soluble in alcohol; stable in concentrated aqueous solution but decomposes in dilute aqueous solution; unstable in moist soil.

Hazard: Irritant to tissue, toxic to plants and vegetation.

Use: Fungicide, nematocide, weed killer, insecticide, soil fumigant.

sodium-*N*-methyl-*N*-oleoyl taurate. (oleyl methyl tauride).
CAS: 137-20-2. $(CH_3)(CH_2)_7CHCH(CH_2)_7CON$ $(CH_3)CH_2CH_2SO_2ONa$.
Properties: Fine, white powder; sweet odor.
Grade: Technical, 32% purity (remainder is mainly sodium sulfate).
Use: Detergent, pesticide aid.

sodium methyl siliconate. Most effective water repellent and cleaner for limestone, concrete, and similar masonry. Reaction product of aqueous sodium hydroxide and a resinous silicone. Total solids about 30%.

sodium-*N*-methyltaurate. See *N*-methyltaurine.

sodium molybdate.
CAS: 7631-95-0. Commercially, the normal molybdate Na_2MoO_4 or its dihydrate (called sodium molybdate crystals). Chemically, a wide variety of complex molybdenum and sodium compounds are known.
Properties: Small, lustrous, crystalline plates. Mp 687C, d 3.28 (18C). Soluble in water. Noncombustible.
Derivation: By the action of sodium hydroxide on molybdenum trioxide. Complex molybdates are prepared by dissolving large amounts of molybdenum trioxide in solutions of normal molybdates.
Grade: Anhydrous, crystals.
Hazard: Irritant.
Use: Reagent in analytical chemistry, paint pigment, production of molybdated toners and lakes, metal finishing, brightening agent for zinc plating, corrosion inhibitor, catalyst in dye and pigment production, additive for fertilizers and feeds, micronutrient.

sodium molybdophosphate.
CAS: 1313-30-0. $Mo_{12}O_{40}P \cdot 3Na$.
Hazard: Moderately toxic by ingestion.

sodium-12-molybdophosphate. (sodium phospho-12-molybdate). $Na_3PMo_{12}O_{40}$.
Properties: Yellow crystals. D 2.83. Soluble in water, although less soluble than the free acid; strong oxidizing action in aqueous solution.
Grade: Technical.
Use: Analysis, neuromicroscopy, catalysts, additives in photographic processes, imparting water resistance to plastics, adhesives and cements, pigments. See heteromolybdate.

sodium-12-molybdosilicate. (sodium silico-12-molybdate). $Na_4SiMo_{12}O_{40} \cdot xH_2O$.

Properties: Yellow crystals. Bulk d 3.44. Soluble in water, acetone, alcohol, ethyl acetate; insoluble in ether, benzene, and cyclohexane.
Grade: Reagent.
Hazard: Water solution is strong oxidizer, store away from organic materials.
Use: Catalysts; reagents; fixing and oxidizing agents in photography; precipitants and ion exchangers in atomic energy; plating processes; imparting water resistance to plastics, adhesives, and cement.

sodium mono- and dimethyl naphthalene sulfonate.
Use: Food additive.

sodium monoxide. (sodium oxide).
CAS: 12401-86-4. Na_2O.
Properties: White powder. D 2.27, sublimes at 1275C. Soluble in molten caustic soda or potash; converted to sodium hydroxide by water.
Hazard: Caustic and strong irritant when wet with water.
Use: Condensing or polymerizing agent in organic reactions, dehydrating agent, strong base.

sodium myristyl sulfate.
CAS: 1191-50-0. $NaC_{14}H_{29}SO$. Anionic detergent.
Use: Foaming, wetting, and emulsifying in cosmetic, household, and industrial uses.

sodium naphthalenesulfonate. $C_{10}H_7SO_3Na$.
Properties: Yellowish, crystalline plates or white odorless scales. Soluble in water; insoluble in alcohol. Combustible.
Grade: Technical.
Hazard: Toxic by ingestion and inhalation.
Use: Organic synthesis, liquefying agent in animal glue preparations, naphthols.

sodium naphthenate. A white paste, the most important of the naphthenic acid salts, commercial samples have consistency of grease, but this will vary with source and manner of processing. Excellent emulsifying and foam-producing properties, low hydrolytic dissociation.
Use: Detergent, emulsifier, disinfectant, manufacture of paint driers.
See naphthenic acid.

sodium naphthionate. (sodium α-naphthylaminesulfonate). $NH_2C_{10}H_6SO_3Na \cdot 4H_2O$.
Properties: White crystals, turns violet on exposure to light. Soluble in water; insoluble in ether. Combustible.
Grade: Technical (paste, crystals).
Use: Riegler's reagent for nitrous acid, manufacture of dyestuffs.

sodium naphthylaminesulfonate. See sodium naphthionate.

sodium niobate.   (sodium columbate).
$Na_2Nb_2O_6 \cdot 7H_2O$.
Important in the purification of niobium materials. The crystalline compound forms when a niobium compound is treated with hot, concentrated sodium hydroxide. It is sparingly soluble in water.

sodium nitrate.   (soda niter).
CAS: 7631-99-4. $NaNO_3$.
Chile saltpeter (caliche) is impure natural sodium nitrate.
Properties: Colorless, transparent crystals; odorless; saline, slightly bitter taste. D 2.267, mp 308C, explodes at 1000F (537C), decomposes at 380C. Soluble in water and glycerol; slightly soluble in alcohol.
Derivation: From nitric acid and sodium carbonate and from Chile saltpeter.
Grade: Granular, sticks, powder; crude, 99.5%, double refined, recrystallized, CP, technical, reagent, diuretic, FCC.
Hazard: Fire risk near organic materials, ignites on friction and explodes when shocked or heated to 1000F (537C). Toxic by ingestion; content in cured meats, fish, and other food products restricted.
Use: Oxidizing agent; solid rocket propellants; fertilizer; flux; glass manufacture; pyrotechnics; reagent; refrigerant; matches; dynamites; black powders; manufacture of sodium salts and nitrates; dyes; pharmaceuticals; an aphrodisiac; color fixative and preservative in cured meats, fish, etc.; enamel for pottery; modifying burning properties of tobacco.

sodium nitrate-nano.
Derivation: A salt recovered from natural deposits in Chile or produced synthetically by reacting nitric acid with sodium carbonate.

sodium nitrilotriacetate.   See nitrilotriacetic acid.

sodium nitrite.
CAS: 7632-00-0. $NaNO_2$.
Properties: Slightly yellowish or white crystals, pellets, sticks, or powder. Oxidizes on exposure to air. D 2.157, mp 271, explodes at 1000F (537C), decomposes at 320C. Soluble in water; slightly soluble in alcohol and ether.
Grade: Reagent, technical, USP, FCC.
Hazard: Dangerous fire and explosion risk when heated to 537C (1000F) or in contact with reducing materials; a strong oxidizing agent. Carcinogen in test animals; its use in curing fish and meat products is restricted to 100 ppm.
Use: Diazotization (by reaction with hydrochloric acid to form nitrous acid); rubber accelerators; color fixative and preservative in cured meats, meat products, fish; pharmaceuticals; photographic and analytical reagent; dye manufacture; antidote for cyanide poisoning.

sodium nitroferricyanide.   (sodium nitroprussiate; sodium nitroprusside). $Na_2Fe(CN)_5NO \cdot 2H_2O$.
Properties: Red, transparent crystals; d 1.72; soluble in water with slow decomposition; slightly soluble in alcohol.
Grade: Reagent, technical.
Use: Analytical reagent.

sodium-*p*-nitrophenolate.   See *p*-nitrophenol, sodium salt.

sodium novobiocin.   See novobiocin.

sodium octyl sulfate.
CAS: 142-31-4. $C_8H_{17}OSO_3Na$. Anionic detergent, available commercially as a 35% solution.
Use: Wetting, dispersing, and emulsifying agent.

sodium oleate.
CAS: 143-19-1. $C_{17}H_{33}COONa$.
Properties: White powder; slight tallowlike odor. Mp 232–235C. Soluble in water with partial decomposition; soluble in alcohol. Combustible.
Derivation: Action of alcoholic sodium hydroxide on oleic acid.
Use: Ore flotation, waterproofing textiles, emulsifier of oil–water systems.

sodium orthophosphate.   See sodium phosphate, monobasic; sodium phosphate, dibasic; sodium phosphate, tribasic.

sodium orthosilicate. $Na_2SiO_3 \cdot 2NaOH$ or other proportions such as $2Na_2O \cdot SiO_2$ (anhydrous) or $2Na_2O \cdot SiO_2 \cdot 5.4H_2O$.
Properties: (Composition $2Na_2O \cdot SiO_2$) Dustless, white, flaked product. Bulk d 75/lb/cu ft. Total Na_2O content 60.8%, percent of total Na_2O in active form 59.0. Soluble in water, pH of a 1% solution 13.0.
Hazard: Strong irritant to skin, eyes, and mucous membranes.
Use: Commercial laundries, metal cleaning, heavy duty cleaning.

sodium orthovanadate.   (sodium vanadate; trisodium orthovanadate; vanadic(II) acid, trisodium salt).
CAS: 13721-39-6. $O_4V \cdot 3Na$.
Properties: Colorless, hexagonal prisms. Mw 183.91, mp 850–866C.
Hazard: Poison by ingestion.

sodium oxalate.
CAS: 62-76-0. $Na_2C_2O_4$.
Properties: White, crystalline powder. D 2.34, mp 250–270C (decomposes). Soluble in water; insoluble in alcohol.
Grade: Reagent, technical, 88%, 99%.
Hazard: Toxic by ingestion.
Use: Reagent, textile finishing, pyrotechnics, leather finishing, blue printing.

sodium oxide. See sodium monoxide.

sodium palconate. The sodium salt of an acid that may be extracted with alkali from redwood dust. The dark reddish-brown material consists mainly of a partially methylated phenolic acid containing aliphatic hydroxyls, phenolic hydroxyls, and carboxyl groups in the ratio of 2:4:3. The viscosity of aqueous solutions rises rapidly with concentration.
Use: To control viscosity and water loss in drilling muds and as a dispersing agent.

sodium palladium chloride. See palladium sodium chloride.

sodium palmitate. (sodium hexadecanoate; sodium pentadecanecarboxylate; sodium salt of hexadecanoic acid).
CAS: 408-35-5. $CH_3(CH_2)_{14}COONa$.
Properties: White to yellow powder.
Hazard: A poison. Combustible.
Use: Polymerization catalyst for synthetic rubbers, laundry and toilet soaps, detergents, cosmetics, pharmaceuticals, printing inks, and as an emulsifier.

sodium pantothenate.
Hazard: A nuisance dust.
Use: Food additive.

sodium paraperiodate. (sodium triparaperiodate). $Na_3H_2IO_6$.
Properties: White, crystalline solid. Very slightly soluble in water; soluble in concentrated sodium hydroxide solutions.
Use: Selectively oxidizes specific carbohydrates and amino acids, wet-strengthens paper, aids combustion of tobacco.

sodium pentaborate decahydrate.
$Na_2B_{10}O_{16} \cdot 10H_2O$.
Properties: White crystals. Free-flowing; stable under ordinary conditions. D 1.72, pH of solution 7.5. Solubility in water 15.4% (20C), increasing with temperature. Noncombustible.
Use: Weed killer, cotton defoliant, fireproofing compositions, glass manufacture, B supplement for tree fruit and truck crops.

sodium pentachlorophenate.
CAS: 131-52-2. C_6Cl_5ONa.
Properties: White or tan powder. Soluble in water, ethanol, and acetone; insoluble in benzene.
Grade: Technical, powder, pellets, or briquettes.
Hazard: Toxic by ingestion, inhalation; skin irritant.
Use: Fungicide; herbicide; slimicide; fermentation disinfectant, especially in finishes and papers.

sodium pentadecanecarboxylate. See sodium palmitate.

sodium pentafluorostannite.
CAS: 22578-17-2. $F_5Sn_2 \cdot Na$.
Properties: Colorless acicular crystals from H_2O. Mw 355.37. Soluble in H_2O; solubility increased in aq HF.
Hazard: A poison by ingestion.
Use: Drug.

sodium pentobarbital. See barbiturate.

sodium "Pentothal" [Hospira]. See "Pentothal" [Hospira]; thiopental sodium.

sodium perborate, anhydrous.
CAS: 7632-04-4. $NaBO_3$.
Properties: White, amorphous powder of unknown constitution containing active oxygen. Evolves oxygen gas when dissolved in water; hygroscopic.
Derivation: By heating sodium perborate tetrahydrate.
Hazard: Strong oxidizing agent, fire risk in contact with organic materials. Toxic by ingestion.
Use: Denture cleaner, oxygen source.

sodium perborate monohydrate. $NaBO_3 \cdot H_2O$, better represented as $Na_2[B_2(O_2)_2(OH)_4]$.
Properties: White, amorphous powder. Rapidly soluble in water giving a solution of H_2O_2 and sodium borate.
Derivation: By partial dehydration of sodium perborate tetrahydrate.
Hazard: Strong oxidizing agent, fire risk in contact with organic materials. Toxic by ingestion.
Use: Denture cleaner, bleaching agent in special detergents.

sodium perborate tetrahydrate. $NaBO_3 \cdot 4H_2O$ better represented as $Na_2[B_2(O_2)_2(OH)_4] \cdot 6H_2O$.
Properties: Colorless crystals. Mp 63C, loses water at 130–150C. Soluble in water 21.5 g/L at 18C, giving a solution of H_2O_2 and sodium borate.
Derivation: By crystallization from solutions of borax and H_2O_2.
Hazard: Oxidizing agent. Toxic by ingestion.
Use: Bleaching agent for domestic detergents and industrial laundries, mild antiseptic, mouthwash (under medical supervision).

sodium percarbonate. (sodium carbonate peroxohydrate, sodium carbonate peroxide).
CAS: 3313-92-6. $Na_2CO_3 \cdot 1.5H_2O_2$.
Properties: Stable, microcrystalline powder. Soluble in water 120 g/kg at 20C giving a solution of H_2O_2 and sodium carbonate.
Derivation: By solution crystallization or by a fluid-bed process, from concentrated solutions of sodium carbonate and hydrogen peroxide, with stabilizers.
Hazard: Oxidizing agent; when intimately mixed with certain organic substances, it may initiate combustion. Toxic by ingestion.

Use: Bleaching agent for domestic and industrial use, denture cleaner, mild antiseptic.

sodium perchlorate.
CAS: 7601-89-0. $NaClO_4$ sometimes with an H_2O.
Properties: White, deliquescent crystals. Mp 482C, bp (decomposes), d 2.02. Soluble in water and alcohol.
Derivation: (1) Sodium chlorate and sodium chloride are mixed and heated until fused. The unchanged chloride is leached out. (2) A cold solution of sodium chlorate is electrolyzed, the solution concentrated and crystallized.
Hazard: Dangerous fire and explosion risk in contact with organic materials and sulfuric acid.
Use: Explosives, jet fuel, analytical reagent.

sodium periodate.   (sodium metaperiodate).
(1) $NaIO_4$. (2) $NaIO_4 \cdot 3H_2O$.
Properties: Colorless crystals. D (1) 3.865 (16C), (2) 3.219 (18C), mp (1) 300C (decomposes), (2) 175C (decomposes). Very soluble in water.
Hazard: Fire risk in contact with organic materials. Toxic by ingestion.
Use: Source of periodic acid, analytical reagent, oxidizing agent.
See sodium paraperiodate.

sodium permanganate.
CAS: 10101-50-5. $NaMnO_4 \cdot 3H_2O$.
Properties: Purple to reddish-black crystals or powder. D 2.47, mp 170C (decomposes). Soluble in water.
Derivation: Sodium manganate is dissolved in water and chlorine or ozone passed in. The solution is concentrated and crystallized.
Grade: Technical, sold commercially in solution.
Hazard: Dangerous fire risk in contact with organic materials, strong oxidizing agent.
Use: Oxidizing agent; disinfectant; bactericide; manufacture of saccharin; antidote for poisoning by morphine, curare, and phosphorus.

sodium peroxide.
CAS: 1313-60-6. Na_2O_2.
Properties: Yellowish-white powder, turning yellow when heated. Absorbs water and carbon dioxide from air. Active oxygen content about 20% by weight. D 2.805, mp 460C (decomposes), bp 657 (decomposes). Soluble in cold water with evolution of heat.
Derivation: Metallic sodium is heated at 300C in aluminum trays in a retort in a current of dry air from which the carbon dioxide has been removed.
Grade: Technical, reagent.
Hazard: Dangerous fire and explosion risk in contact with water, alcohols, acids, powdered metals, and organic materials. Strong oxidizing agent. Keep dry. Irritant.

Use: Oxidizing agent; bleaching of miscellaneous materials including paper and textiles; deodorant; antiseptic; organic chemicals; water purification; pharmaceuticals; oxygen generation for diving bells, submarines, etc.; textile dyeing and printing; ore processing; analytical reagent; calorimetry; germicidal soaps.

sodium persulfate.   (sodium peroxydisulfate).
CAS: 7775-27-1. $Na_2S_2O_8$.
Properties: White, crystalline powder. Soluble in water; decomposed by alcohol; decomposes in moist air.
Hazard: By ingestion, strong irritant to tissue.
Use: Bleaching agent (fats, oils, fabrics, soap), battery depolarizers, emulsion polymerization.

sodium phenate.   (sodium phenolate; sodium carbolate).
CAS: 139-02-6. C_6H_5ONa.
Properties: White, deliquescent crystals. Soluble in water and alcohol; decomposed by carbon dioxide in the air.
Derivation: Phenol is dissolved in caustic soda solution, concentrated, and crystallized.
Hazard: Strong irritant to skin and tissue.
Use: Antiseptic, salicylic acid, organic synthesis.

sodium phenobarbital.   (phenobarbital, solution).
See barbiturate.

sodium phenolate.   Legal label name for sodium phenate.

sodium phenolsulfonate.   (sodium sulfocarbolate). $HOC_6H_4SO_3Na \cdot 2H_2O$.
Properties: Colorless crystals or granules. Slightly efflorescent; chars at high temperature, evolving phenol. Soluble in water, hot alcohol, and glycerol.
Use: Medicine (intestinal antiseptic).

sodium phenylacetate.   (sodium α-toluate). $C_6H_5CH_2 \cdot COONa$.
Properties: Soluble in water; insoluble in alcohol, ether, and ketones; 50% aqueous solution has pH 7.0–8.5 and is pale yellow. Solution tends to crystallize at 15C.
Grade: 50% solution, dry salt.
Use: Precursor in production of penicillin G, intermediate for producing heavy metal salts that act as fungicides.

sodium-*N*-phenylglycinamide-*p*-arsonate.
See tryparsamide.

sodium-*o*-phenylphenate.   (sodium-*o*-phenylphenolate).
CAS: 132-27-4. $C_6H_4(C_6H_5)ONa \cdot 4H_2O$.
Properties: Practically white flakes. Bulk d 38–43 lb/cu ft, pH of saturated solution in water 12.0–13.5. Soluble in water, methanol, acetone.
Hazard: Possible carcinogen.

Use: Industrial preservative (bactericide and fungicide), mold inhibitor for apples and other fruit (postharvest).

sodium phenylphosphinate.
$C_6H_5PH(O)(ONa)$.
Properties: Crystals. Mp 355C (decomposes to give phenylphosphine), stable at room temperature. Soluble in water.
Use: Antioxidant, heat and light stabilizer.

sodium phenyl sulfinate dihydrate.
CAS: 25932-11-0. $C_6H_5O_2S•Na•2H_2O$.
Hazard: A mild eye irritant.

sodium phosphate.   See "Nutrifos" *[ICL]*; sodium metaphosphate; sodium phosphate, dibasic; sodium phosphate, monobasic; sodium phosphate (P-32); sodium phosphate, tribasic; sodium polyphosphate; sodium pyrophosphate; sodium pyrophosphate, acid; sodium tripolyphosphate.

sodium phosphate, dibasic.   (DSP; disodium phosphate; sodium orthophosphate, secondary; disodium orthophosphate; disodium hydrogen phosphate).
CAS: 7558-79-4.   (1) Na_2HPO_4. (2) $Na_2HPO_4•2H_2O$. (3) $Na_2HPO_4•7H_2O$. (4) $Na_2HPO_4•12H_2O$. The dihydrate (2) is also marketed as the duohydrate.
Properties: Colorless, translucent crystals or white powder; saline taste. (1) Hygroscopic; converted to sodium pyrophosphate at 240C; (2) mp loses water at 92.5C, d 2.066 (15C); (3) d 1.679, loses $5H_2O$ at 48C; (4) mp 35C, d 1.5235, readily loses $5H_2O$ on exposure to air at room temperature, loses $12H_2O$ at 100C. Soluble in water; very soluble in alcohol; pH of 1% solution 8.0–8.8. Nonflammable.
Derivation: (1) By treating phosphoric acid with a slight excess of soda ash, boiling the solution to drive off carbon dioxide, and cooling to permit the dodecahydrate to crystallize; (2) by precipitating calcium carbonate from a solution of dicalcium phosphate with soda ash.
Grade: Commercial, NF (1) and (3), FCC (1) or (2).
Use: Chemicals, fertilizers, pharmaceuticals, textiles (weighting silk, dyeing and printing), fireproofing wood and paper; ceramic glazes, tanning, galvanoplastics, soldering enamels, analytical reagent, cheese, detergents, boiler-water treatment, dietary supplement, buffer, sequestrant in foods.

sodium phosphate, monobasic.   (sodium acid phosphate; sodium biphosphate; sodium orthophosphate, primary; MSP, sodium dihydrogen phosphate).
CAS: 7558-80-7.   (1) NaH_2PO_4. (2) $NaH_2PO_4•H_2O$.
Properties: (1) White, crystalline powder. Slightly hygroscopic. Very soluble in water. Has acid reaction; forms sodium acid pyrophosphate at 225–250C and sodium metaphosphate at 350–400C; (2)

large transparent crystals. Mp loses water at 100C, d 2.040. Very soluble in water; insoluble in alcohol; pH of 1% solution 4.4–4.5. Nonflammable.
Derivation: By treating disodium phosphate with proper proportion of phosphoric acid.
Grade: Commercial, food, (2) NF, (1) FCC.
Use: Boiler-water treatment, electroplating, dyeing, acid cleansers, baking powders, cattle feed supplement, buffer, emulsifier, nutrient supplement in food, lab reagent, acidulant.

sodium phosphate (P-32).   (sodium radiophosphate).
A radioactive form of sodium phosphate (which phosphate is not specified) containing phosphorus-32 which can be used as a tracer.
See phosphorus-32.
Grade: USP, as solution.
Use: Biochemical research, medicine (diagnostic aid, antineoplastic).

sodium phosphate, tribasic.   (TSP; trisodium orthophosphate; trisodium phosphate; tertiary sodium phosphate; sodium orthophosphate, tertiary).
CAS: 7601-54-9. $Na_3PO_4•12H_2O$.
Properties: Colorless crystals. D 1.62 (20C), mp 75C (decomposes), loses $12H_2O$ at 100C, pH of 1% solution is 11.8–12.0. Soluble in water. Nonflammable.
Derivation: By mixing soda ash and phosphoric acid in proper proportions to form disodium phosphate and then adding caustic soda.
Grade: Commercial, high purity, CP, FCC (anhydrous), anhydrous salt also available.
Hazard: Toxic by ingestion, irritant to tissue.
Use: Water softeners, boiler-water compounds, detergent, metal cleaner, textiles, manufacture of paper, laundering, tanning, sugar purification, photographic developers, paint removers, industrial cleaners, dietary supplement, buffer, emulsifier, food additive.

sodium phosphate tribasic dodecahydrate.
CAS: 10101-89-0. $O_4P•3Na•12H_2O$.
Hazard: Low toxicity by ingestion.

sodium phosphide.
CAS: 12058-85-4. Na_3P.
Properties: Red solid. Decomposes on heating and in water, forming phosphine.
Hazard: Dangerous fire risk, reacts with water and acids to form phosphine.

sodium phosphite. $Na_2HPO_3•5H_2O$.
Properties: White, crystalline powder. Hygroscopic. Mp 53C, bp 200–250C Soluble in water; insoluble in alcohol. (decomposes).
Use: Antidote in mercuric chloride poisoning.

sodium phosphoaluminate.   White powder composed primarily of sodium aluminate

(hydrated), sodium phosphate (ortho), and small amounts of sodium carbonate and sodium silicate.
Use: Paper industry as a sizing adjunct, as an aid in retention of filler and fiber and in pH control, boiler-feed-water treatment, and as a food additive.

sodium phospho-12-molybdate.   See sodium-12-molybdophosphate.

sodium phospho-12-tungstate.   See sodium-12-tungstophosphate.

sodium phytate.   (USAN; insitol hexaphosphoric ester, sodium salt). $C_6H_9O_{24}P_6Na_9$.
Properties: Hygroscopic powder. Water-soluble.
Use: Chelating agent for trace heavy metals, color improvement, medicine.

sodium picramate.
CAS: 831-52-7. $NaOC_6H_2(NO_2)_2NH_2$.
Derivation: Yellow, water-soluble salt resulting from neutralization of picramic acid with caustic soda.
Hazard: Dangerous fire and explosion hazard when dry. Toxic by ingestion and skin absorption.
Use: Manufacture of dye intermediates, organic synthesis.

sodium platinichloride.   See sodium chloroplatinate.

sodium plumbate. $Na_2PbO_3 \cdot 3H_2O$.
Properties: Fused, light-yellow lumps. Hygroscopic, decomposed by water and acids. Soluble in alkalies.
Hazard: As for lead.

sodium plumbite. Na_2PbO_2.
Derivation: Solution of PbO (litharge) in sodium hydroxide.
Hazard: Highly toxic, corrosive.
See lead.
Use: Doctor solution for improving the odor of gasoline and other petroleum distillates.

sodium polyphosphate. $Na_{n+2}P_nO_{3n+1}$.
The two most important crystalline sodium polyphosphates are the pyrophosphate ($n = 2$) and the tripolyphosphate, also called the triphosphate ($n = 3$). The term *sodium polyphosphate* also includes the system of vitreous sodium phosphates for which the mole ratio of Na_2O/P_2O_5 is between 1 and 2.
Hazard: As for sodium phosphate.
Use: Sequestering and deflocculating agents, primarily in water treatment, food processing, and cleaning compounds; heavy-set detergent builders.
See sodium metaphosphate; sodium pyrophosphate; sodium tripolyphosphate.

sodium polysulfide. Na_2S_x.
Properties: Yellow-brown, granular, free-flowing polymer. Bulk d 56 lb/cu ft. Combustible.

Use: Manufacture of sulfur dyes and colors, insecticides, oil-resistant synthetic rubber ("Thiokol" *[Toray]*), petroleum additives, electroplating.

sodium–potassium alloy.   Legal label name for NaK.

sodium–potassium carbonate.   (potassium–sodium carbonate). $NaKCO_3 \cdot 6H_2O$.
Properties: Colorless crystals. D 1.6344, mp 135C The double salt fuses more readily than the single salts. (decomposes); soluble in water.
Derivation: Mixture of potassium and sodium carbonates.
Use: Analysis (flux).

sodium–potassium phosphate.   (potassium–sodium phosphate). $NaKHPO_4 \cdot 7H_2O$.
Properties: White powder. Stable in air. Soluble in water.

sodium–potassium tartrate.   See potassium–sodium tartrate.

sodium propionate.
CAS: 137-40-6. CH_3CH_2COONa or C_2H_5COO $Na \cdot xH_2O$.
Properties: Transparent crystals or granules; almost odorless. Deliquescent in moist air. Soluble in water and alcohol. Combustible.
Grade: NF, FCC.
Use: Fungicide, mold preventive, food preservative (bread and other bakery products).

sodium prussiate, red.   See sodium ferricyanide.

sodium prussiate, yellow.   See sodium ferrocyanide.

sodium pyroantimonate.   See sodium antimonate.

sodium pyroborate.   See sodium borate.

sodium pyrophosphate.   (tetrasodium pyrophosphate; sodium pyrophosphate, normal; TSPP).
CAS: 7722-88-5.   (1) $Na_4P_2O_7$. (2) $Na_4P_2O_7 \cdot 10H_2O$ (one of the sodium polyphosphates).
Properties: Colorless, transparent crystals or white powder. (1) Mp 880C, d 2.45. Soluble in water; decomposes in alcohol. (2) Mp 94C (loses water), d 1.8. Soluble in water; insoluble in alcohol and ammonia.
Derivation: By fusing sodium phosphate, dibasic.
Grade: Pure crystals, dried, fused, CP, FCC.
Hazard: Toxic by inhalation.
Use: Water softener, soap and synthetic detergent builder, dispersing and emulsifying agent, metal cleaner, boiler-water treatment, viscosity control of drilling muds, de-inking newsprint, synthetic rubber manufacture, textile dyeing, scouring of wool,

buffer, sequestrant, nutrient supplement, food additive.

sodium pyrophosphate, acid. (disodium pyrophosphate; sodium acid pyrophosphate; disodium diphosphate; disodium dihydrogen pyrophosphate; SAPP). $Na_2H_2P_2O_7 \cdot 6H_2O$.
Properties: White, crystalline powder. Mp (decomposes) 220C, d 1.862. Soluble in water.
Derivation: Incomplete decomposition of monobasic sodium phosphate.
Grade: Technical, food, FCC.
Use: Electroplating, metal cleaning and phosphatizing, drilling muds, baking powders and leavening agent, buffer, sequestrant, peptizing agent in cheese and meat products, frozen desserts.

sodium pyrophosphate, normal. See sodium pyrophosphate.

sodium pyrophosphate peroxide.
$Na_4P_2O_7 \cdot 2H_2O_2$.
Properties: White powder. Bulk d 73 lb/cu ft, active oxygen min 9.0% by weight. Water-soluble; mildly alkaline.
Hazard: Fire risk in contact with organic materials, oxidizing agent.
Use: Denture cleansers, dentrifices, household and laundry detergents, antiseptic.

sodium pyroracemate. See sodium pyruvate.

sodium pyrosulfite. See sodium metabisulfite.

sodium pyrovanadate. $Na_4V_2O_7 \cdot 8H_2O$.
Properties: Colorless, six-sided plates. Mp (anhydrous) 654C. Soluble in water; insoluble in alcohol.
Derivation: Sodium hydroxide and vanadium pentoxide in water solution.

sodium pyruvate. (sodium pyroracemate; sodium acetylformate). $NaOOCCOCH_3$. White powder, apparent mp 205C, very soluble in water.
Use: Biochemical research.

sodium resinate. See sodium abietate.

sodium rhodanate. See sodium thiocyanate.

sodium rhodanide. See sodium thiocyanate.

sodium ricinoleate.
CAS: 5323-95-5. $C_{17}H_{32}OHCOONa$.
Properties: White or slightly yellow powder; nearly odorless. Soluble in water or alcohol. Combustible.
Derivation: Sodium salt of the fatty acids from castor oil.
Use: Emulsifying agent in special soaps.

sodium saccharin. (sodium benzosulfimide; gluside, soluble; soluble saccharin).
CAS: 128-44-9. $C_7H_4NNaO_3S \cdot 2H_2O$.

The sodium salt of saccharin.
Properties: White crystals or crystalline powder; odorless or with a faint aromatic odor. In dilute solutions has an intensely sweet taste (500 times sweeter than sugar). Very soluble in water; slightly soluble in alcohol.
Grade: NF, FCC.
Use: Foods (nonnutritive sweetener).
See saccharin.
Hazard: The use of saccharin is being limited due to possible carcinogenicity.

sodium salicylate.
CAS: 54-21-7. HOC_6H_4COONa.
Properties: Lustrous, white, crystalline scales or amorphous powder; saline taste. Soluble in water, alcohol, and glycerol. Combustible.
Grade: Technical, CP, USP.
Use: Production of salicylic acid; preservative for paste, mucilage, glue, and hides; medicine (analgesic).

sodium salt of hexadecanoic acid. See sodium palmitate.

sodium sarcosinate. (sodium sarcosine).
CH_3NHCH_2COONa.
Grade: 33% aqueous solution.
Use: Intermediate, stabilizer for diazonium salts, chelating agent.

sodium secobarbital. See barbiturate.

sodium selenate.
CAS: 13410-01-0. $Na_2SeO_4 \cdot 10H_2O$.
Properties: White crystals. D 1.603–1.620. Soluble in water.
Hazard: Toxic by ingestion.
Use: Reagent, insecticide for nonedible plants.

sodium selenite.
CAS: 10102-18-8. $Na_2SeO_3 \cdot 5H_2O$.
Properties: White crystals. Soluble in water; insoluble in alcohol.
Derivation: By neutralizing selenious acid with sodium carbonate and crystallizing.
Hazard: Toxic by ingestion.
Use: Glass manufacture (color control), reagent in bacteriology, testing germination of seeds, decorating porcelain.

sodium selenite pentahydrate. (selenious acid disodium salt pentahydrate).
CAS: 26970-82-1. $O_3Se \cdot 2Na \cdot 5H_2O$.
Properties: White tetragonal crystals; decomp on heating by H_2O loss. Mw 332.01. Soluble in H_2O; insoluble in EtOH.
Hazard: A poison.

sodium sesquicarbonate.
CAS: 533-96-0. $Na_2CO_3 \cdot NaHCO_3 \cdot 2H_2O$.

Properties: White, needle-shaped crystals. D 2.112, mp decomposes. Soluble in water; less alkaline than sodium carbonate. Noncombustible.

Derivation: Crystallization of a solution containing equimolar quantities of sodium carbonate and sodium bicarbonate, also occurs native (as trona) in desert areas and in Searles Lake brine.

Grade: Technical, FCC.

Hazard: Irritant to tissue.

Use: Detergent and soap builder, mild alkaline agent for general cleaning and water softening, bath crystals, alkaline agent in leather tanning, food additive.

sodium sesquisilicate. $Na_6Si_2O_7$ (anhydrous).

Properties: White, granular powder. Soluble in water; pH of 1% solution 12.7. Noncombustible.

Derivation: Crystals from solutions obtained by heating silica or sodium metasilicate with sodium hydroxide. Intermediate in composition between ortho- and metasilicates, less alkaline than sodium orthosilicate.

Hazard: Toxic by ingestion, protective clothing required.

Use: Heavy-duty cleaning (metals, laundries), textile processing.

sodium silicate. (water glass).

CAS: 6834-92-0. Formulas: Vary from $Na_2O•3.75$ SiO_2 to $2Na_2O•SiO_2$ and with various proportions of water. The simplest form of glass.

Properties: Lumps of greenish glass soluble in steam under pressure, white powders of varying degrees of solubility, or liquids cloudy or clear and varying from highly fluid to extreme viscosity, viscosity range from 0.4 to 600,000 cP, fp slightly below water, miscible with some polyhydric alcohols, partially miscible with primary alcohols and ketones, gels form with acids between pH 3 and pH 9, coagulated by brine, precipitated by alkaline earth and heavy metal ions. Noncombustible.

Derivation: By fusing sand and soda ash.

Grade: (liquid) 40, 47, 52 degrees Bé.

Use: Catalysts and silica gels, soaps and detergents, adhesives (especially sealing and laminating paperboard containers), water treatment, bleaching and sizing of textiles and paper pulp, ore treatment, soil solidification, glass foam, pigments, drilling fluids, binder for abrasive wheels, foundry cores and molds, waterproofing mortars and cements, flame-retardant, chemical equipment lining, enhanced oil recovery.

See sodium metasilicate, anhydrous; sodium metasilicate pentahydrate; sodium sesquisilicate; sodium orthosilicate.

sodium silicoaluminate. See sodium aluminosilicate.

sodium silicofluoride. See sodium fluorosilicate.

sodium silico-12-molybdate. See sodium-12-molybdosilicate.

sodium-12-silicotungstate. See sodium-12-tungstosilicate.

sodium silver chloride. See silver sodium chloride.

sodium silver thiosulfate. See silver sodium thiosulfate.

sodium-α-sodioacetate. See α-sodiosodium acetate.

sodium sorbate.

CAS: 7757-81-5. $CH_3CH:CHCH:CHCOONa$. Combustible.

Use: Food preservative.

sodium stannate. $Na_2SnO_3•3H_2O$ or $Na_2Sn(OH)_6$.

Properties: White to light-tan crystals. Soluble in water; insoluble in alcohol; decomposes in air; aqueous solution slightly alkaline; loses $3H_2O$ at 140C.

Derivation: (1) By fusion of metastannic acid and sodium hydroxide, (2) by boiling tin scrap and sodium plumbate solution.

Hazard: Toxic material.

Use: Mordant in dyeing, ceramics, glass, source of tin for electroplating and immersion plating, textile fireproofing, stabilizer for hydrogen peroxide, blueprint paper, laboratory reagent.

sodium stearate.

CAS: 822-16-2. $NaOOCC_{17}H_{35}$.

Properties: White powder with fatty odor. Soluble in hot water and hot alcohol; slowly soluble in cold water and cold alcohol; insoluble in many organic solvents.

Impurities: Varying quantities of sodium palmitate.

Grade: Technical.

Use: Waterproofing and gelling agent, toothpaste and cosmetics, stabilizer in plastics.

sodium stearoyl lactylate.

CAS: 25383-99-7.

Properties: Cream-colored powder; caramel-like odor. Soluble in hot oil or fat, dispersible in warm water.

Use: Food additive.

See "Emulsilac-S" [ACH].

sodium stearoyl-2-lactylate.

Properties: White powder. Melting range 46–52C.

Derivation: Sodium salt of reaction product of lactic and stearic acids.

Use: Emulsifier; dough conditioner; whipping agent in baked products, desserts, and mixes; complexing agent for starches and proteins.

sodium stearyl fumarate. $C_{22}H_{39}NaO_4$.
Properties: Fine white powder. Soluble in methanol; insoluble in water.
Use: Food additive.

sodium styrenesulfonate. $CH_2:CH_2C_6H_4SO_3$ Na. White, free-flowing powder.
Use: Reactive monomer.

sodium subsulfite. See sodium thiosulfate.

sodium sulfachloropyrazine monohydrate.
Use: Drug (veterinary); food additive.

sodium sulfate, anhydrous.
CAS: 7757-82-6. Na_2SO_4.
Properties: White crystals or powder; odorless; bitter saline taste. D 2.671, mp 888C. Soluble in water and glycerol; insoluble in alcohol. Noncombustible.
Derivation: (1) Purification of natural sodium sulfate from deposits or brines; (2) by-product of hydrochloric acid manufacture from salt and sulfuric acid, $2NaCl + H_2SO_4 \rightarrow 2HCl + Na_2SO_4$; (3) by-product of phenol manufacture (caustic fusion process); (4) Hargreaves process.
Grade: Technical, CP, detergent, rayon, glass maker's.
Use: Manufacture of kraft paper, paperboard, and glass; filler in synthetic detergents; sodium salts; ceramic glazes; processing textile fibers; dyes; tanning; pharmaceuticals; freezing mix; laboratory reagent. food additive.
See salt cake.

sodium sulfate decahydrate. (sodium sulfate, crystals; Glauber's salt). $Na_2SO_4 \cdot 10H_2O$.
Properties: Large, transparent crystals, small needles, or granular powder. D 1.464 (crystals), mp 33C (liquefies). Loses water of hydration at 100C. Energy storage capacity is more than seven times that of water. Soluble in water and glycerol; insoluble in alcohol; solutions neutral to litmus. Nonflammable.
Derivation: Crystallization of sodium sulfate from water solutions (Glauber's salt); also occurs in nature as mirabilite.
Grade: Technical, NF.
Use: Solar heat storage, air-conditioning.

sodium sulfhydrate. See sodium hydrosulfide.

sodium sulfide.
CAS: 1313-82-2. (1) Na_2S. (2) $Na_2S \cdot 9H_2O$.
Properties: Yellow or brick-red lumps or flakes or deliquescent crystals. (1) D 1.856 (14C), mp 1180C; (2) d 1.427 (16C), decomposes at 920C. Soluble in water; slightly soluble in alcohol; insoluble in ether; largely hydrolyzed to sodium acid sulfide and sodium hydroxide.
Derivation: By heating sodium acid sulfate with salt and coal to above 950C, extraction with water, and crystallization.

Grade: Flake, fused, chip sulfide (60% Na_2S), 60% fused and broken, 30% crystals, liquid.
Hazard: Flammable, dangerous fire and explosion risk. Strong irritant to skin and tissue, liberates toxic hydrogen sulfide on contact with acids.
Use: Organic chemicals, dyes (sulfur), intermediates, viscose rayon (sulfur removal), leather (depilatory), paper pulp, hydrometallurgy of gold ores, sulfiding oxidized lead and copper ores preparatory to flotation, sheep dips, photographic reagent, engraving and lithography, analytical reagent.

sodium sulfite.
CAS: 7757-83-7. (1) Na_2SO_3. (2) $Na_2SO_3 \cdot 7H_2O$.
Properties: White crystals or powder; saline, sulfurous taste. D(1) 2.633, (2) d 1.539; mp (1) decomposes, (2) loses 7 H_2O at 150C. Soluble in water; sparingly soluble in alcohol.
Derivation: (1) Sulfur dioxide is reacted with soda ash and water and a solution of the resulting sodium bisulfite is treated with additional soda ash; (2) by-product of the caustic fusion process for phenol.
Grade: Reagent, technical, FCC.
Hazard: Use prohibited in meats and other sources of Vitamin B_1.
Use: Paper industry (semichemical pulp), reducing agent (dyes), water treatment, photographic developer, food preservative and antioxidant, textile bleaching (antichlor).

sodium sulfoacetate derivatives of mono and diglycerides.
Use: Food additive.

sodium sulfobromophthalein.
$C_{20}H_8Br_4O_{10}S_2Na_2$.
Properties: White, crystalline powder; odorless; bitter taste. Hygroscopic; soluble in water; insoluble in alcohol and acetone.
Derivation: From phenol and tetrabromophthalic acid or anhydride.
Grade: USP, technical.
Use: Medicine (diagnostic aid).

sodium sulfocarbolate. See sodium phenolsulfonate.

sodium sulfocyanate. See sodium thiocyanate.

sodium sulfocyanide. See sodium thiocyanate.

sodium sulfonate. Class name for various sulfonates derived from petroleum, e.g., sodium dodecylbenzenesulfonate, sodium xylenesulfonate, sodium toluenesulfonate.
See sodium alkanesulfonate.
Use: Textile processing oils; oils for metal-working (emulsifying and antirust agents); lubricating oils; emulsifiers for insecticides, herbicides, fungicides; preparation of dyes and intermediates; hydrotropic solvent; coatings in food packaging.

sodium-2-sulfopropionitrile. See sodium-2-cyanoethanesulfonate.

sodium sulforicinoleate.
Derivation: Product of successive sulfonation (partial) and saponification of castor oil. Composition indefinite.
Use: Emulsifying and wetting agent.

sodium sulfoxylate. See sodium formaldehyde sulfoxylate.

sodium tartrate. (sal tartar; disodium tartrate).
CAS: 868-18-8. $Na_2C_4H_4O_6 \cdot 2H_2O$.
Properties: White crystals or granules. D 1.794, loses $2H_2O$ at 150C. Soluble in water; insoluble in alcohol.
Derivation: Neutralization of tartaric acid with sodium carbonate, concentration, and crystallization.
Grade: Technical, CP, reagent, FCC.
Use: Reagent, food additive, as sequestrant and stabilizer.

sodium tartrate, acid. See sodium bitartrate.

sodium TCA. See sodium trichloroacetate.

sodium tellurite.
CAS: 10102-20-2. Na_2TeO_3.
Properties: White powder. Soluble in water.
Hazard: Toxic by ingestion.
Use: Bacteriology, medicine.

sodium tetraborate. See sodium borate.

sodium tetraborate pentahydrate.
CAS: 12179-04-3. $B_4O_7 \cdot 2Na \cdot 5H_2O$.
Hazard: A severe eye irritant.

sodium tetrachloroaluminate. (sodium chloroaluminate). $AlCl_4Na$. A water-soluble powder used as catalyst in organic reactions.

sodium-2,3,4,6-tetrachlorophenate.
$C_6HCl_4ONa \cdot H_2O$.
Properties: Buff to light-brown flakes. Bulk d 26–29 lb/cu ft, pH of water-saturated solution 9.0–13.0. Soluble in water; methanol and acetone.
Hazard: Toxic by ingestion.
Use: Industrial preservative (bactericide and fungicide).

sodium tetradecyl sulfate. (sodium-7-ethyl-2-methyl-4-hendecanol sulfate).
CAS: 139-88-8. $C_{14}H_{29}SO_4Na$.
Properties: White, waxy solid; odorless. Soluble in alcohol, ether, and water; 5% solution is clear and colorless; pH (5% solution) 6.5–9.0.
Use: Wetting agent, detergent.

sodium tetrafluoropropionate.
CAS: 22898-01-7. $C_3HF_4O_2 \cdot Na$.
Hazard: Moderately toxic by skin contact. Low toxicity by ingestion.
Use: Agricultural chemical.

sodium tetraphosphate. See sodium polyphosphate.

sodium tetrasulfide. Na_2S_4.
Properties: Yellow, hygroscopic crystals or clear, dark-red liquid. Mp of crystals 275C.
Grade: Aqueous solution containing 40% by weight of compound.
Hazard: Fire risk when exposed to flame. Irritant to skin and tissue.
Use: Reducing organic nitro compounds, manufacture of sulfur dyes, insecticides and fungicides, ore flotation, soaking hides and skins, preparation of metal sulfide finishes.

sodium thimerfonate.
CAS: 5964-24-9. $C_8H_9HgO_3S_2 \cdot Na$.
Hazard: A poison.

sodium thiocyanate. (sodium sulfocyanate; sodium sulfocyanide; sodium rhodanate; sodium rhodanide).
CAS: 540-72-7. NaSCN.
Properties: Colorless, deliquescent crystals or white powder. Mp 287C. Soluble in water and alcohol; hygroscopic and affected by light.
Derivation: By boiling sodium cyanide with sulfur.
Grade: Technical; pure (crystals or dried); CP; reagent; ACS.
Use: Analytical reagent, dyeing and printing textiles, black nickel plating, manufacture of other thiocyanate salts and artificial mustard oil, solvent for polyacrylates, medicine (antihypertensive).

sodium thioglycolate. (sodium mercaptoacetate).
CAS: 367-51-1. $HSCH_2COONa$.
The sodium salt of thioglycolic acid.
Properties: Crystals; characteristic odor; hygroscopic; discolors on exposure to air or iron. Soluble in water; slightly soluble in alcohol. Combustible.
Hazard: Yields toxic hydrogen sulfide on decomposition, may be toxic by skin absorption.
Use: Bacteriology, cold-waving hair, depilatory, analytical reagent.

sodium thiosulfate. (sodium subsulfite; hypo).
CAS: 7772-98-7. $Na_2S_2O_3 \cdot 5H_2O$.
The anhydrous salt is also commercially available.
Properties: White, translucent crystals or powder; cooling taste; bitter aftertaste. Efflorescent above 33C in dry air, d 1.729 (17C), mp 48C, bp decomposes. Soluble in water and oil of turpentine; insoluble in alcohol; deliquescent in moist air.
Derivation: Heating a solution of sodium sulfite with powdered sulfur.

Grade: Technical, crystals, granulated, photographic, CP, pure, USP, FCC.

Hazard: Use in foods restricted to 0.1%.

Use: Photography (fixing agent to dissolve unchanged silver salts from exposed negatives), chrome tanning, removing chlorine in bleaching and papermaking, extraction of silver from its ores, dechlorination of water, mordant, reagent, bleaching, reducing agent in chrome dyeing, sequestrant in salt (up to 0.1%), antidote for cyanide poisoning.

sodium *l*-thyroxine pentahydrate.
CAS: 6106-07-6. $C_{15}H_{10}I_4NO_4 \cdot Na \cdot 5H_2O$.
Hazard: A reproductive hazard.

sodium titanate.   (sodium tritanate). $Na_2Ti_3O_7$.
Properties: White crystals. D 3.35–3.50, mp 1128C. Insoluble in water. Combustible.
Use: Welding.

sodium-α-toluate.   See sodium phenylacetate.

sodium toluenesulfonate.   (*p*-toluenesulfonic acid, sodium salt).
CAS: 657-84-1. $CH_3C_6H_4SO_3Na$.
Properties: Crystals. Soluble in water.
Use: Dye chemistry, hydrotropic solvent.

sodium-*p*-toluenesulfonchloramine.   See chloramine.

sodium trichloroacetate.   (sodium TCA).
CAS: 650-51-1. CCl_3COONa.
Hazard: Toxic by ingestion, irritant to skin and eyes.
Use: Herbicide, pesticide.

sodium-2,4,5-trichlorophenate.
$C_6H_2Cl_3ONa \cdot 1.5H_2O$.
Properties: Buff to light-brown flakes. Bulk d 28–33 lb/cu ft, pH of water-saturated solution 11.0–13.0. Soluble in water, methanol, acetone.
Use: Industrial preservative (bactericide and fungicide).

sodium tridecylbenzenesulfonate.   Not a true compound but a mixture of C_{12} and C_{15} alkyl benzene sulfonates that approximates C_{13}.
See sodium dodecylbenzene sulfonate.

sodium 2,3,5-triiodobenzoate.
CAS: 17274-12-3. $C_7H_2I_3O_2 \cdot Na$.
Hazard: Moderately toxic by ingestion.
Use: Agricultural chemical.

sodium tri-metaphosphate.   See sodium metaphosphate.

sodium triparaperiodate.   See sodium paraperiodate.

sodium triphenyl-*p*-rosaniline sulfonate.
See methyl blue.

sodium triphosphate.   See sodium tripolyphosphate.

sodium tripolyphosphate.   (STFF; sodium triphosphate, tripoly; pentasodium triphosphate).
CAS: 13573-18-7. $Na_5P_3O_{10}$.
Properties: White powder. Two crystalline forms of anhydrous salt (transition p 417C, mp 622C to give melt and sodium pyrophosphate), and a hexahydrate.
Derivation: Controlled calcination of sodium orthophosphate mixture from sodium carbonate and phosphoric acid. May contain up to 10% pyrophosphate and up to 5% tri-metaphosphate.
Grade: Powdered and granular, FCC, food.
Use: Water softening; sequestering, peptizing, or deflocculating agent; food additive and texturizer.

sodium trithiocyanurate.
CAS: 17766-26-6. $C_3N_3S_3 \cdot 3Na$.
Hazard: Moderately toxic by skin contact.

sodium trititanate.   See sodium titanate.

sodium tungstate.   (sodium wolframate).
CAS: 13472-45-2. $Na_2WO_4 \cdot 2H_2O$.
Properties: Colorless crystals. D 3.245, mp loses $2H_2O$ at 100C and then melts at 692C. Soluble in water; insoluble in alcohol and acids. Noncombustible.
Derivation: By dissolving tungsten trioxide or the ground ore in caustic soda solution, concentration, and crystallization.
Grade: Technical, CP, crystals.
Use: Intermediate in preparation of tungsten compounds, (e.g., phosphotungstate), reagent, fireproofing fabrics and cellulose, alkaloid precipitant.

sodium-12-tungstophosphate.   (sodium phosphotungstate; phosphotungstic acid, sodium salt).
CAS: 51312-42-6. $Na_4O_2 \cdot O_5P_2 \cdot O_{36}W_{12} \cdot 18H_2O$.
Properties: Yellowish-white powder. Very soluble in water and alcohols.
Grade: Reagent, technical.
Hazard: An oxidizing agent.
Use: Reagent; manufacture of organic pigments; treatment of furs; antistatic agent for textiles; leather tanning; making plastic films, cements, and adhesives water-resistant.

sodium-12-tungstosilicate.
$Na_4SiW_{12}O_{40} \cdot 5H_2O$.
Properties: White, crystalline powder. Soluble in water and alcohols; although less so than the free acid. Noncombustible.
Grade: Reagent, technical.
Use: Catalyst for organic synthesis, precipitant and inorganic ion-exchanger, additive in plating processes.

sodium undecylenate.
$(CH_2{:}CH(CH_2)_8COONa$.

Properties: White powder. Decomposes above 200C, limite. Solubility in most organic solvents; soluble in water. Combustible.
Use: Bacteriostat and fungistat in cosmetics and pharmaceuticals.

sodium uranate.   See sodium diuranate.

sodium valproate.   A drug that has proved effective in treating epilepsy. Approved (1978) by FDA, successfully used in Europe since 1970.

sodium vanadate.   See sodium orthovanadate; sodium pyrovanadate; sodium metavanadate.

sodium-*p*-vinylbenzene sulfonate.   See sodium styrene sulfonate.

sodium warfarin.   (sodium-(3-α-acetonyl-benzyl)-4-hydroxycoumarin). $C_{19}H_{15}NaO_4$. See warfarin.

sodium xanthate.   See sodium ethylxanthate.

sodium xanthogenate.   See sodium ethylxanthate.

sodium xylenesulfonate.   (dimethylbenzene-sulfonic acid, sodium salt).
CAS: 1300-72-7. $(CH_3)_2C_6H_3SO_3NaH_2O$.
Use: Hydrotropic solvent, used in detergents.

sodium zinc hexa-metaphosphate.   See sodium metaphosphate.

sodium zirconium glycolate.
$NaH_3ZrO(H_2COCOO)_3$.
Properties: Clear, light-straw-colored solution. D 1.28–1.30, containing 35.7–38.6% solids, 12.5–13.5% ZrO_2.
Use: Deodorant, astringent, germicide, sequestrant, fire-retardant.

sodium zirconium lactate.
CAS: 10377-98-7. $NaH_3ZrO(CH_3CHOCOO)_3$.
Properties: Clear, straw-colored solution. D 1.28–1.30, containing 12.5–13.5% ZrO_2, equivalent to 42.5–45.9% sodium zirconium lactate; pH 7.5–8.0.
Use: Deodorant and antiperspirant.

sodium zirconium sulfate.   See zirconium sodium sulfate.

soft.   A nontechnical word used by chemists in several senses, it describes the following: (1) an acid having little or no positive charge and whose valence electrons are easily excited (see acid); (2) water that is relatively free from calcium compounds (see water, hard); (3) wood from coniferous trees (see softwood).

softener.   (1) A substance used when dry powders are added to a polymeric material (e.g., rubber or plastic) to reduce the friction of mechanical mixing and to facilitate subsequent processing. It exerts both lubricating and dispersing action, often by means of emulsification. Examples are vegetable oils, asphaltic materials, and stearic acid, the latter being especially effective with carbon black. It is difficult to distinguish precisely between softeners and plasticizers; in general, softeners do not enter into chemical combination with the polymer, and their softening effect tends to be temporary. (2) A fatliquoring agent used to soften leather. (3) A sulfonated oil, fatty alcohol, or quaternary ammonium compound used in textile finishing to impart superior "hand" to the fabric and facilitate mechanical processing. (4) A substance that reduces the hardness of water by removing or sequestering calcium and magnesium ions; among those used are various sodium phosphates and zeolites.
See water, hard.

"Softigen" *[Cremer]*.   TM for fatty acid esters. See "Witepsol" *[Cremer]*.

"Softisan" *[Cremer]*.   TM for fatty acid esters. See "Witepsol" *[Cremer]*.

soft lithography.   See nanoimprinting.

softwood.   In papermaking terminology, an arbitrary name for the wood from coniferous trees (pine, spruce, fir, hemlock) regardless of the hardness or softness of the wood itself. Softwoods are used for almost all commercial grades of paper.
See pulp, paper.

soil.   (1) A mixture of inorganic matter derived from weathered rocks and organic components resulting from decay of prior vegetation. Eight elements are present in the inorganic component in excess of 1% (oxygen, silicon, aluminum, iron, calcium, potassium, magnesium, and sodium), most in the ionized state. Water and air are also present, either in the voids between the particles or adsorbed on their surfaces. Many other elements occur in lower percentages, including trace elements in concentration of less than 1000 ppm. Some of these, e.g., boron (about 20 ppm), are essential for plant nutrition. Both nitrogen and phosphorus are associated with the organic content. The concentration of these is a fraction of 1% of each, but they play a vital part in plant and animal life. The pH of soils varies widely with location; some soils are as low as pH 4.5 (very acid) and others as high as pH 10 (strongly alkaline). For most crops the pH ranges around the neutral point (6.5–7.5). Texturally, soils are classified on the basis of their content of sand, silt, and clay. Those having 45–50% sand and 20–28% clay are called loams, those with more than 50% sand are called sandy, and those with more than 28% clay are in the clay group. Technologists

consider soil as being made up of layers, known as horizons, each having a characteristic composition and physical properties; the spectrum of these horizons is called the soil profile. Organic matter is usually excluded from the profile.
(2) In textile literature, any foreign matter present in or on fiber or fabric, i.e., dirt, oil, grease, etc. These are usually removable by the action of soap, synthetic detergents, or organic solvents.

Soilbond.   A soil stabilizer.
Use: Dust control for road improvement.

soil conditioner.   (1) A synthetic, long-chain organic molecule having carboxyl groups along the chain whose charges react with the positive charges on the soil particles (aluminum and iron). The conditioners affect the anion exchange capacity of a soil. (2) Loosely, any material added to topsoil to reduce acidity (lime) and promote growth (bone meal).

Soilfume.   A soil fumigant whose active ingredient is ethylene dibromide.
Hazard: As for ethylene dibromide.

soil sterilant.   A chemical agent that is persistent and active in soil for a considerable length of time.
Use: To temporarily or permanently destroy all biota in the soil.

sol.   (1) An abbreviation for soluble. (2) A class of colloid.
See colloidal; colloid chemistry.

solan.   (generic name for 3'-chloro-2-methyl-*p*-valerotoluidide or *N*-(3-chloro-4-methylphenyl)-2-methylpentanamide).
CAS: 2307-68-8. $H_3CC_6H_3(Cl)NHCOCH(CH_3)$ $CH_2CH_2CH_3$.
Properties: Solid. Mp 86C. Insoluble in water; soluble in pine oil, diisobutyl ketone, isophorone, and xylene. Combustible.

solanine.
CAS: 20562-02-1. $C_{45}H_{73}NO_{15}$. An alkaloid found in low percentages in potato and other plants, insoluble in water, soluble in alcohol, decomposes at 285C.

solanoside.
CAS: 4356-33-6. $C_{30}H_{46}O_8$.
Hazard: A poison.
Source: Natural product.

solar cell.   (photovoltaic cell). A battery-like device in which the radiant energy of the sun is converted to electrical energy by means of a semiconductor. The essential component of a solar cell is a thin sheet or wafer of crystalline or amorphous silicon, plus doping agents. The crystalline form is used in 3-inch squares, but the hydrogenated amorphous form is effective in 1-foot squares which are produced as vapor-deposited coatings on glass only a few microns in thickness. A notable advance in the technology of solar cells is the experimental development of a photoelectrochemical cell, that utilizes silicon electrodes by which incident light is converted to energy. The conversion efficiency is sufficient to indicate that this type of cell may have an important future. Its success is due to microthin films of doped oxides of alumina or magnesia deposited on the electrode surfaces, which in turn are coated with platinum a few angstroms thick. The cell can also be operated as a battery for use in remote power sources.

solar collector.   A device for utilizing solar energy either by absorption or reflection of the incident radiation. Several types have been developed for both domestic and industrial application. Some are in the experimental stage. (1) Collection units consisting of a blackened absorber surface enclosed by one or two layers of silvered glass or polished aluminum, which admit the incident radiation but prevent the escape of heat; the energy absorbed is transferred to air or water circulating between the glass plates and the absorber. (2) Energy-focusing systems (solar towers) in which the incident radiation is concentrated on a central receiver from an array of flat or parabolic mirrors. This method is utilized for evaporating low-level radioactive wastewater at Los Alamos, as well as in a power generating plant.
Solar energy has important possibilities for chemical synthesis. A prototype collector for producing ammonia from the reduction of nitrogen and water at low temperature and pressure has been reported. This reaction is

$$N_2 + 3H_2O \xrightarrow[TiO_2]{hv} 2NH_3 + 1.5 O_2$$

solar energy.   The National Renewable Energy Laboratory, 1617 Cole Blvd., Golden, CO 80401 is a center for information on solar energy.
See solar cell; solar collector; solar furnace. For the problem of heat and power storage when sunlight is not available, see storage (4).

solar furnace.   An experimental device for attaining extremely high temperatures for physical and chemical research. Solar radiation is directed by an array of 63 heliostats onto a parabolic reflector that is 2000 m^2 in area and composed of more than 9000 curved mirrors. By this means the radiation is focused on a target area of 0.3 m^2. Three furnaces, each of 1000 kW capacity, have been constructed; one is located in New Mexico, the others are in France and Japan. Calcium carbide is reported to have been made in the French solar furnace from lime and coke at 1980C (3600F), under the sponsorship of the Institute of Gas Technology.

solar pond.   An experimental means of storing solar energy in either fresh- or saltwater. The "pond" is actually a large, suitably lined, open container of varying dimensions. In the freshwater type, the heat absorbed is retained at night by covering the pond with an inflated plastic film, supplemented by a layer of foam applied directly to the water. In the salt-water type, heat loss is prevented by a salt gradient, that minimizes convection currents that would cool the water. The sodium chloride or magnesium chloride used is not uniformly dispersed, but is most highly concentrated in the darker, bottom layers where most of the solar radiation is absorbed. Thus the water is warmest and densest at the bottom, where it remains, and no convection occurs. The salt concentration ranges from zero at the surface to 18% or more at the bottom. An experimental salt-gradient pond at Argonne National Laboratory 100 × 100 × 12 ft containing 2800 tons of water and 700 tons of sodium chloride has attained a bottom temperature approaching 180F.

solder.   A low-melting alloy usually of the lead-tin type used for joining metals at temperatures below 425C. The solder acts as an adhesive and does not form an intermetallic solution with the metals being joined.
See brazing; welding.

solid.   Matter in its most highly concentrated form, that is, the atoms or molecules are much more closely packed than in gases or liquids and thus more resistant to deformation. The normal condition of the solid state is crystalline structure—the orderly arrangement of the constituent atoms of a substance in a framework called a lattice (see crystal). Crystals are of many types and normally have defects and impurities that profoundly affect their applications, as in semiconductors. The geometric structure of solids is determined by X-rays which are reflected at characteristic angles from the crystalline lattices, which act as diffraction gratings.
See crystallography.
Some materials that are physically rigid, such as glass, are regarded as highly viscous liquids because they lack crystalline structure. All solids can be melted (i.e., the attractive forces acting between the crystals are disrupted) by heat and are converted to liquids. For ice, this occurs at 0C; for some metals the melting point may be as high as 3300C. Some solids convert by sublimation directly to a gas.

solid-state chemistry.   Study of the exact arrangement of atoms in solids, especially crystals, with particular emphasis on imperfections and irregularities in the electronic and atomic patterns in a crystal and the effects of these on electrical and chemical properties.
See crystal; semiconductor; impurity.

Solphenyl.   Fast-to-light direct dyes for cellulosic fibers.

solubility.   The ability or tendency of one substance to blend uniformly with another, e.g., solid in liquid, liquid in liquid, gas in liquid, gas in gas. Solids vary from 0–100% in their degree of solubility in liquids, depending on the chemical nature of the substances; to the extent that they are soluble they lose their crystalline form and become molecularly or ionically dispersed in the solvent to form a true solution. Examples are sugar/water, salt/water. Liquids and gases are often said to be miscible with other liquids and gases rather than soluble. Thus nitrogen, oxygen, and carbon dioxide are freely miscible with each other and air is a solution (uniform mixture) of these gases.
The physical chemistry of solutions is an extremely complex mathematical subject in which the principles of electrolytic dissociation, diffusion, and thermodynamics play controlling parts. Raoult's law and Henry's law are also involved.
See miscibility; solution, true.

soluble oil.   An oil (also called emulsifying oil) that, when mixed with water, produces milky emulsions. In some soluble oils the emulsion is so fine that instead of milky solutions in water, amber-colored, transparent solutions are formed. Typical examples are sodium and potassium petroleum sulfonates.
Use: Metal-cutting lubricants, textile lubricants, metal-boring lubricants, emulsifying agents.

soluble starch.   See starch, modified.

"Solulan" [Lubrizol].   TM for an emulsifier and stabilizer emollient, solubilizer, or foam fortifier.
Use: In cold, cleansing or shave creams, lotions, mousses, soaps, and shampoos.

solute.   One or more substances dissolved in another substance, called the solvent; the solute is uniformly dispersed in the solvent in the form of either molecules (sugar) or ions (salt), the resulting mixture is a solution.
See solution, true; solvent.

solution, colloidal.   A liquid colloidal dispersion is often called a solution. Since colloidal particles are larger than molecules it is strictly incorrect to call such dispersions solutions; however, this term is widely used in the literature.
Note: Wolfgang Ostwald stated, "There are no sharp differences between mechanical suspensions, colloidal solutions, and molecular [true] solutions. There is a gradual and continuous transition from the first through the second to the third."
See colloid chemistry.

solution hybridization.   A method closely related to RNase protection (see "RNase protection assay"). Solution hybridization is designed to measure the levels of a specific mRNA species in a complex population of RNA. An excess of radioactive probe is allowed to hybridize to the RNA, then single-strand specific nuclease is used to destroy the remaining unhybridized probe and RNA. The "protected" probe is separated from the degraded fragments, and the amount of radioactivity in it is proportional to the amount of mRNA in the sample which was capable of hybridization. This can be a very sensitive detection method.

solution, true.   A uniformly dispersed mixture at the molecular or ionic level, of one or more substances (the solute) in one or more other substances (the solvent). These two parts of a solution are called phases.
Common types are:
liquid–liquid: alcohol–water
solid–liquid: salt–water
solid–solid: carbon–iron
Solutions that exhibit no change of internal energy on mixing and complete uniformity of cohesive forces are called *ideal*; their behavior is described by Raoult's law. Solutions are involved in most chemical reactions, refining and purification, industrial processing, and biological phenomena.
The proportion of substances in a solution depends on their limits of solution. The solubility of one substance in another is the maximum amount that can be dissolved at a given temperature and pressure. A solution containing such a maximum amount is *saturated*. A state of supersaturation can be created, but such solutions are unstable and may precipitate spontaneously.

solutrope.   A ternary mixture having two liquid phases between which one component is distributed in an apparent ratio varying with concentration from less than 1 to greater than 1. In other words, the solute may be selectively dissolved in one or the other of the phases or solvents, depending on the concentration. This phenomenon has been compared to azeotropic behavior.

solvation.   In the parlance of colloid chemistry, the adsorption of a microlayer or film of water or other solvent on individual dispersed particles of a solution or dispersion. The term *solvated hulls* has been used to describe such particles. It is also applied to the action of plasticizers on resin dispersions in plastisols.
See hydration (2).

Solvay process.   (ammonia soda process). Manufacture of sodium carbonate (soda ash, Na_2CO_3) from salt, ammonia, carbon dioxide, and limestone by an ingenious sequence of reactions involving recovery and reuse of practically all the ammonia and part of the carbon dioxide. Limestone is heated to produce lime and carbon dioxide. The latter is dissolved in water containing the ammonia and salt, with resultant precipitation of sodium bicarbonate. This is separated by filtration, dried, and heated to form normal sodium carbonate. The liquor from the bicarbonate filtration is heated and treated with lime to regenerate the ammonia. Calcium chloride is a major by-product.
Note: Because this process requires much energy and pollutes streams and rivers with chloride effluent, many plants using it have closed, production being obtained from the natural deposits in the western U.S.

"Solvenol" *[Pinova].*   TM for a group of monocyclic terpene hydrocarbons with minor amounts of terpene alcohols and ketones.
Use: General solvent, rubber reclaiming.

solvent.   A substance capable of dissolving another substance (solute) to form a uniformly dispersed mixture (solution) at the molecular or ionic size level. Solvents are either polar (high dielectric constant) or nonpolar (low dielectric constant). Water, the most common of all solvents, is strongly polar (dielectric constant 81), but hydrocarbon solvents are nonpolar. Aromatic hydrocarbons have higher solvent power than aliphatics (alcohols). Other organic solvent groups are esters, ethers, ketones, amines, and nitrated and chlorinated hydrocarbons.
The chief uses of organic solvents are in the coatings field (paints, varnishes and lacquers), industrial cleaners, printing inks, extractive processes, and pharmaceuticals. Since many solvents are flammable and toxic to varying degrees, they contribute to air pollution and fire hazards. For this reason their use in coatings and cleaners has declined in recent years.
See individual compounds.

solvent, aprotic.   A solvent that cannot act as a proton acceptor or donor, i.e., as an acid or base.

solvent drying.   Removal of water from metal surfaces by means of a solvent that displaces it preferentially, as on precision equipment, electronic components, etc. Examples of solvents used are acetone, 1,1,2-trichloro-1,2,2-trifluorethane, 1,1,1-trichloroethane.

solvent dye.   See dye, solvent.

solvent extraction.   A separation operation that may involve three types of mixture: (1) a mixture composed of two or more solids, such as a metallic ore; (2) a mixture composed of a solid and a liquid; (3) a mixture of two or more liquids. One or more components of such mixture are removed (extracted) by exposing the mixture to the action of a solvent in which the component to be removed is soluble. If the mixture consists of two or more solids, extraction is performed by percolation of an appropriate solvent through it. This procedure is also called leaching, especially if the solvent is water; coffee making is an example. Synthetic fuels can be made from coal by extraction with a coal-derived solvent followed by hydrogenation.

In liquid–liquid extraction one or more components are removed from a liquid mixture by intimate contact with a second liquid that is itself nearly insoluble in the first liquid and dissolves the impurities and not the substance that is to be purified. In other cases the second liquid may dissolve, i.e., extract from the first liquid, the component that is to be purified, and leave associated impurities in the first liquid. Liquid–liquid extraction may be carried out by simply mixing the two liquids with agitation and then allowing them to separate by standing. It is often economical to use counter-current extraction, in which the two immiscible liquids are caused to flow past or through one another in opposite directions. Thus fine droplets of heavier liquid can be caused to pass downward through the lighter liquid in a vertical tube or tower.

The solvents used vary with the nature of the products involved. Widely used are water, hexane, acetone, isopropyl alcohol, furfural, xylene, liquid sulfur dioxide, and tributyl phosphate. Solvent extraction is an important method of both producing and purifying such products as lubricating and vegetable oils, pharmaceuticals, and nonferrous metals.

solvent, latent.   (cosolvent).
An organic liquid that will dissolve nitrocellulose in combination with an active solvent. Latent solvents are usually alcohols and are used widely in nitrocellulose lacquers in a ratio of 1 part alcohol to 2 parts active solvent.

solvent naphtha.   See naphtha (2b).

Solvent Red 73.   See 4′,5′-diiodofluorescein.

solvent refining.   See solvent extraction.

Solvent Yellow 3.   See o-aminoazotoluene.

Solvesso 100.
CAS: 63231-51-6. A solvent for active ingredients, typically in emulsifiable concentrates.
Hazard: Moderately toxic. A moderate skin and eye irritant.
Use: In agriculture applications; in coatings; as a solvent in architectural and industrial coatings applications.

solvolysis.   A reaction involving substances in solvent, in which the solvent reacts with the dissolved substance (solute) to form a new substance. Intermediate compounds are usually formed in this process.
See hydrolysis.

soman.   (methylphosphonofluoridic acid-1,2,2-trimethylpropylester).
CAS: 96-64-0. $(CH_3)_3CCH(CH_3)OPF(O)CH_3$.
A nerve gas.
Properties: Colorless liquid. Evolves odorless gas. Bp 167C, fp −70C, d 1.026 (20C).
Hazard: Highly toxic by ingestion, inhalation, and skin absorption; may be fatal on short exposure; cholinesterase inhibitor; military nerve gas; fatal dose (man) 0.01 mg/kg.

somatic cell gene therapy.   Incorporating new genetic material into cells for therapeutic purposes. The new genetic material cannot be passed to offspring.
See gene therapy.

somatic cell genetic mutation.   (acquired mutations). A change in the genetic structure that is neither inherited nor passed to offspring.
See germ line genetic mutation.

somatic cells.   Any cell in the body, except gametes and their precursors.
See gamete.

somatostatin-28 (sheep).
CAS: 73032-94-7. $C_{137}H_{207}N_{41}O_{39}S_3$.
Hazard: A poison.

somatotropic hormone.   (STH; somatotropin).
CAS: 9002-72-6. Hormone secreted by the anterior lobe of the pituitary. It causes an increase in general body growth and also affects carbohydrate and lipid metabolism.

Sommelet–Hauser rearrangement.   Rearrangement of benzyl quaternary ammonium salts to ortho-substituted benzyldialkylamines on treatment with alkali metal amides.

Sommelet reaction.   Preparation of aldehydes from aralkyl or alkyl halides by reaction with hexamethylenetetramine followed by mild hydrolysis of the formed quaternary salt.

Sonn–Muller method.   Preparation of aromatic aldehydes from anilides by conversion of an acid anilide with phosphorus pentachloride to an imido chloride, reduction of the imido chloride with stannous chloride, and hydrolysis of the obtained anil.

sonolysis.   The breaking up (molecular fragmentation) of molecules by ultrasonic radiation. Examples: sonolysis in pure water produces hydrogen atoms, hydroxyl radicals, molecular hydrogen, oxygen, and hydrogen peroxide; acetonitrile in an argon atmosphere produces molecular hydrogen, nitrogen, and methane.

Sonowax. A series of wax emulsions based on microcrystalline wax for use with resins or as a top finish on various textile fabrics.

sorb. See sorption.

sorbent. A substance that has a great capacity for absorbing moisture or other gas or fluid. A compound that absorbs, adsorbs, or entraps something.

sorbic acid. (2,4-hexadienoic acid). CAS: 110-44-1. $CH_3CH:CHCH:CHCOOH$.
Properties: White, crystalline solid. Mp 134.5C, bp 228C (decomposes), 153C (50 mm Hg), flash p (OC) 260F (126C). Slightly soluble in water and many organic solvents. Combustible.
Derivation: Trimerization of acetaldehyde and catalytic air oxidation of the resulting hexadienal. Found in berries of mountain ash.
Grade: FCC, technical.
Use: Fungicide, food preservative (mold inhibitor), alkyd resin coatings, upgrading of drying oils, cold rubber additive, intermediate for plasticizers and lubricants.

sorbide. (dianhydrosorbitol). $C_6H_8O_2(OH)_2$. Generic name for anhydrides (dicyclic ether dihydric alcohols) derivable from sorbitol by removal of two molecules of water. The name is also applied to specific commercial varieties.

sorbitan. (monoanhydrosorbitol; sorbitol anhydride). $C_6H_8O(OH)_4$. Generic name for anhydrides (cyclic ether tetrahydric alcohols) derivable from sorbitol by removal of one molecule of water.

sorbitan fatty acid esters. Mixture of partial esters of sorbitol and its anhydrides with fatty acids.
Properties: Sorbitan monolaurate and sorbitan monooleate are amber liquids; sorbitan monostearate, sorbitan monopalmitate, and sorbitan tristearate are cream-colored, waxy solids with slight odor and bland taste; d of both liquid and solid esters is about 1.0 (25C), mp of solid esters is 54C; they are insoluble in water, somewhat soluble in organic solvents. Combustible.
Derivation: Esterification of sorbitol with fatty acids.
Grade: Technical, FCC (for sorbitan monostearate).
Use: Emulsifiers and stabilizers in foods, cosmetics, drugs, textiles; plastics, agricultural chemicals.
See polysorbate.

sorbitol. (d-sorbite; d-sorbitol; hexahydric alcohol).
CAS: 50-70-4. $C_6H_8(OH)_6$.
Properties: White, odorless, crystalline powder; hygroscopic; faint sweet taste. D 1.47 (-5C), mp (metastable form) 93C, (stable form) 97.5C. Soluble in water, glycerol, and propylene glycol; slightly soluble in methanol, ethanol, acetic acid, phenol,

and acetamide; almost insoluble in most other organic solvents. Approved by FDA for food use.
Derivation: By pressure hydrogenation of dextrose with nickel catalyst. Occurs in small amounts in various fruits and berries.
Grade: Crystals, technical, 70% aqueous solution (USP), resin, powder, FCC (solid and solution).
Use: Ascorbic acid fermentation; in solution form for moisture-conditioning of cosmetic creams and lotions, toothpaste, tobacco, gelatin; bodying agent for paper, textiles, and liquid pharmaceuticals; softener for candy; sugar crystallization inhibitor; surfactants; urethane resins and rigid foams; plasticizer, stabilizer for vinyl resins; food additive (sweetener, humectant, emulsifier, thickener, anticaking agent); dietary supplement.

sorbitol anhydride. See sorbitan.

Sorbo-Cel. A chemical-coated diatomite filter aid used for selectively removing traces of oil from oil-in-water emulsions. Effective for removing other trace components free of emulsified systems.
Use: Conditioning of boiler-feed water.

L(−)-sorbose.
CAS: 87-79-6. $HOCH_2CO(CH_2O)_3CH_2OH$.
Properties: White, crystalline powder; sweet taste. Mp 165C. Soluble in water; almost insoluble in organic solvents. Combustible.
Derivation: From sorbitol by submerged culture aerobic fermentation.
Grade: Technical, reagent.
Use: Manufacture of ascorbic acid (vitamin C), preparation of special diets, and media for the study of metabolism in animals and microorganisms.

Sorel cement. See magnesium oxychloride cement.

Soret effect. The difference in concentration in different parts of a solution when these parts are at different temperatures.

sorghum. A cereal plant cultivated in the temperate zones; its stems are rich in sucrose and can be processed in much the same way as sugarcane and used as a source of sugar and syrups. In the U.S. it is grown in the Midwest and in Texas.

sorption. A surface phenomenon that may be either absorption or adsorption, or a combination of the two. The term is often used when the specific mechanism is not known.

sorrel salt. See potassium binoxalate.

SOS response. In bacteria, a coordinated induction of a variety of genes as a response to high levels of DNA damage.

sour.   (1) Any substance used in textile or laundry operations to neutralize residual alkali or decompose residual hypochlorite bleach. The commonly used sours are sodium bifluoride and sodium fluosilicate. (2) Contaminated with sulfur compounds, e.g., gasoline or natural gas. (3) Taste characteristic of an acidic or fermented substance.
See doctor treatment.

sour dip.   An acid solution of Epsom salts and fermenting corn sugar.

sour gas.   A natural gas containing 30 grains or more of sulfur in the form of hydrogen sulfide and mercaptans. A by-product of gas and oil refining.

sour mash.   An unmalted cereal cooked with the spent liquor of a previously dealcoholized mash resulting from the manufacture of whiskey.

Southern blot.   A technique for analyzing mixtures of DNA whereby the presence and rough size of one particular fragment of DNA can be ascertained. Named for its inventor, Dr. E. M. Southern.
See blotting.

Soxhlet extractor.   A laboratory apparatus consisting of a glass flask and condensing unit used for continuous reflux extraction of alcohol- or ether-soluble components of food products. Named after its inventor, a German chemist.

soybean flour.   A fine-ground powder having a particle size of 100 mesh or less, made by steaming soybeans to inactive enzymes, followed by removal of hulls and mechanical grinding. It contains 40–50% protein, about 20% fat, and 5% moisture. Defatted flours are made by extraction with hexane to remove the oil. The flakes produced are used chiefly for animal feeds but the full-fat flour has become an increasingly important factor in high-protein food products, especially textured proteins and meat analogs.

soybean lectin.
CAS: 90320-57-3.
Hazard: Moderately toxic.
Source: Natural product.

soybean meal.   The crushed residue from the extraction of soybeans. Extraction by the hydraulic or expeller process produces normally a meal with 6% residual oils, while the solvent process yields meal with 1% residual oil. Typical analysis; crude protein 43%, crude fiber 5.5%, nitrogen-free extract 30%, ash 6%, oil content 1–6%. Total digestible nutrients about 75%.
Use: Animal feeds, adhesives, medium for bacitracin production.

soybean oil.   (soya bean oil; Chinese bean oil; soy oil).

Properties: Pale-yellow, fixed drying oil. D 0.924–0.929, mp 22–31C, refr index 1.4760–1.4775, solidifying p −15 to −8C, Hehner value 94–96, saponification value 190–193, iodine value 137–143, flash p 540F (282C), autoign temp 833F (445C), moderate spontaneous heating. Soluble in alcohol, ether, chloroform, and carbon disulfide. Combustible.
Use: Soap manufacture, high-protein foods, paints and varnishes, cattle feeds, margarine and salad dressings, printing inks, source of nylon 9, plasticizer (epoxidized), alkyd resins.
Note: Soybean oil is the most widely used vegetable oil for both edible and industrial use in the U.S.

soybean oil (unhydrogenated).
Properties: From the seed of the legume *Glycine max.* Amber-colored oil.
Use: Food additive.

soy protein, isolated.
Use: Food additive.

space, chemistry in.   Experiments carried out on the space shuttle in the early 1980s indicate that unique types of chemical reactions occur in outer space, and that actual products may result that are not achievable under the terrestrial environment. Several factors are believed to account for this, primarily zero gravity, though absence of oxygen and enhanced magnetic effects may also play a part. Several encouraging results have already been obtained, though until further experiments and operating data have been investigated, the conclusions must be considered tentative. Among projects that have been carried out or are contemplated are the following: (1) Uniform polymer microspheres that are over twice as large as possible on earth have been made due to zero gravity. (2) More effective electrophoresis reactions for making biological materials have been discovered, probably also because of zero gravity. (3) Possibilities exist for (a) making unique alloys in space that are not possible on earth, for example lead–copper, lead–zinc, and aluminum–indium; (b) purer crystals for microelectronics; (c) better glass for fiber optics; (d) new drugs and pharmaceuticals. Future experiments will involve human cells, enzymes, and hormones.

"Spacerite" [Huber]. $Al_2O_3 \cdot 3H_2O$.
TM for a spacer pigment of titanium dioxide in coatings and inks.
Properties: White powder.

space velocity.   The volume of gas or liquid, measured at specified temperature and pressure (usually standard conditions) passing through unit volume in unit time.
Use: Comparing flow processes involving different conditions, rates of flow, and sizes or shapes of containers.

spalling. Chipping an ore for crushing or the cracking, breaking, or splintering of materials due to heat.

spandex. Generic name for a fiber in which the fiber-forming substance is a long-chain synthetic polymer composed of at least 85% of a segmented polyurethane (Federal Trade Commission). Imparts elasticity to garments such as girdles, socks, special hosiery.

Spanish white. (1) Chalk, $CaCO_3$. (2) Bismuth white, $BiO(NO_3)$, basic bismuth white.

spar. (1) A type of crystalline material such as Iceland spar or feldspar, usually containing calcium carbonate or an aluminum silicate; fluorspar is calcium fluoride. Iceland spar has unique optical properties. (2) A weather-resistant varnish originally used for coating wooden spars of sailing ships, which may be the reason for its name.
See spar varnish.

sparger. A perforated pipe through which steam, air, or water is sprayed into a liquid during a fermentation reaction.
Use: Brewing industry to remove traces of wort from the mash.

spar, Greenland. See cryolite.

spar, heavy. See barite.

spar, Iceland. See calcite.

sparking metal. See pyrophoric alloy.

"Sparkolloid" [Scott]. TM for a fining agent for wine making derived from seaweed.
Use: To remove negatively charged particles.

spar, satin. See calcite; gypsum.

spar varnish. A durable, water-resistant varnish for severe service on exterior exposure. It consists of one or more drying oils (linseed, tung, or dehydrated castor oil), one or more resins (rosin, ester gum, phenolic resin, or modified phenolic resin), one or more volatile thinners (turpentine or petroleum spirits), and driers (linoleates, resinates, or naphthenates of lead, manganese, and cobalt). It is classed as a long-oil varnish and generally consists of 45–50 gals of oil for each 100 lb of resin.
See varnish.

SPE. Abbreviation for Society of Plastics Engineers.

spearmint oil. A yellowish essential oil, strongly levorotatory.
Use: Source of carvone and as flavoring for medicines, chewing gum, etc.

special pairs. Antenna molecules which have associated with them proteins required for the initial separation of charge during photosynthesis.

species-specific antigen. An antigen that characterizes a single genus of microorganisms.

specific activity. (1) The activity of a radioelement per unit weight of the element. (2) The activity per unit mass of a pure radionuclide.

specification. A schedule of minimum performance requirements for specialized products such as those established by the various committees of the American Society for Testing and Materials and the Underwriters Laboratories. Such products are subject to inspection and test before acceptance.
See testing.

specific bactericide. A bactericidal immune serum that acts on a single species of bacterium or a group of closely related bacteria.

specific gravity. The ratio of the density of a substance to the density of a reference substance; it is an abstract number that is unrelated to any units. For solids and liquids, specific gravity is numerically equal to density, but for gases it is not, because of the difference between the densities of the reference substances, which are usually water (1 g/cc) for solids and liquids and air (0.00129 g/cc, or 1.29 g/L at 0C and 760 mm Hg) for gases. The specific gravity of a gas is the ratio of its density to that of air; since the *specific gravity* of air = 1.0 (1.29/1.29), this is usually stated to indicate the comparison with the gas under consideration. For example, the density of hydrogen is 0.089 g/L but its specific gravity is 0.069 (i.e., 0.089/1.29). The specific gravity of solids and liquids is the ratio of their density to that of water at 4C, taken as 1.0, as 1 cc of water weighs 1 g. Thus, a solid or liquid with a density of 1.5 g/cc has a specific gravity of 1.5/1 or 1.5.
Since weights of liquids and gases vary with temperature, it is necessary to specify both temperatures involved, except for rough or approximate values. Thus, the specific gravity of alcohol should be given as 0.7893 at 20/4C, the first temperature referring to the alcohol and the latter to the water. At 15.56C the specific gravity of alcohol is 0.816.
See density; API gravity; Baumé.

specific heat. The ratio of the heat capacity of a substance to the heat capacity of water, or the quantity of heat required for a 1 degree temperature change in a unit weight of material. Commonly expressed in Btu/lb/degree F or in cal/g/degree. For gas, the specific heat at constant pressure is greater than that at constant volume by the amount of heat needed for expansion.

specificity. The ability of an enzyme or receptor to discriminate among competing substrates or ligands.

specific rotation. The rotation, in degrees, of the plane of plane-polarized light (D-line of sodium) by an optically active compound at 25C, with a specified concentration and light path.

specific susceptibility. See mass susceptibility.

specific volume. The volume of unit weight of a substance, as cubic feet per pound or gallons per pound, but more frequently milliliters per gram. The reciprocal of density.

specific weight. The weight per unit volume of a substance.

Spectra Flo. Liquid colorants for thermoplastics.
Use: Colorants compatible with polyolefins, styrenics, and engineering resins.

spectral karyotype. (SKY). A graphic of all of the chromosomes of an organism, each labeled with a different color. Useful for identifying chromosomal abnormalities.
See chromosome.

spectrophotometry. See absorption spectroscopy.

spectroquality. A specially prepared chemical of higher purity than those generally available for spectrophotometric use.

spectroscopy. (instrumental analysis).
A branch of analytical chemistry devoted to identification of elements and elucidation of atomic and molecular structure by measurement of the radiant energy absorbed or emitted by a substance in any of the wavelengths of the electromagnetic spectrum in response to excitation by an external energy source. The types of absorption and emission spectroscopy are usually identified by the wavelength involved, namely, γ-ray, X-ray, UV, visible, infrared, microwave, and radiofrequency. The technique of spectroscopic analysis was originated by Fraunhofer who in 1814 discovered certain dark (D) lines in the solar spectrum, which were later identified as characterizing the element sodium. In 1861 Kirchhoff and Bunsen produced emission spectra and showed their relationship to Fraunhofer lines. X-ray spectroscopy was utilized by Moseley (1912) to determine the precise location of elements in the periodic system. Since then, a number of sophisticated and highly specialized techniques have been developed including Raman spectroscopy, nuclear magnetic resonance, nuclear quadrupole resonance, dynamic reflectance spectroscopy, laser, microwave, and γ-ray spectroscopy, and electron paramagnetic resonance.

spectrum. The radiant energy emitted by a substance as a characteristic band of wavelengths by which it can be identified.
See radiation; spectroscopy.

speculum metal. (1) 66% copper, 34% tin with trace of arsenic; (2) 64% copper, 32% tin, 4% nickel.
Properties: D 8.6, mp 750C.
Use: For mirrors for reflecting telescopes.

spelter. Relatively pure zinc as encountered in industrial operations such as galvanizing. Lead and/or iron are common impurities.

spent mixed acid. Mixed acid that has given up part of its nitric acid.
Hazard: Dangerous fire risk. Strong irritant to tissue.

spent oxide. See iron sponge, spent.

Sperry process. An electrolytic process for the manufacture of lead carbonate, basic (white lead) from desilverized lead containing some bismuth. The impure lead forms the anode. A diaphragm separates anode and cathode compartments, and carbon dioxide is passed into the solution. Impurities, including bismuth, remain on the anode as a slime blanket.

sphalerite. (blende, zinc blende). ZnS. Natural zinc sulfide, usually containing some cadmium, iron, and manganese.
Properties: Color yellow, brown, black, or red; luster resinous. Hardness 3.5–4, d 3.9–4.1, good cleavage. Soluble in hydrochloric acid.
Occurrence: Missouri, Kansas, Oklahoma, Colorado, Montana, Wisconsin, Idaho, Australia, Canada, Mexico.
Use: Most important ore of zinc, also a source of cadmium; phosphor; source of sulfur dioxide for production of sulfuric acid.

sphingolipid. An amphipathic lipid with a sphingosine backbone to which are attached a long-chain fatty acid (at the nitrogen) and a polar alcohol.

sphingomyelin. Diaminophosphatides occurring primarily in nervous tissue and containing a fatty acid, phosphoric acid, choline, and sphingosine. They are soluble in hot absolute alcohol and insoluble in ether, acetone, and water.

sphingosine. (1,3-dihydroxy-2-amino-4-octadecene).
$CH_3(CH_2)_{12}CH:CHCH_2OCH(NH_2)CH_2OH$.
Forms part of certain phosphatides, such as cerebrosides and sphingomyelins.
Properties: Waxy crystals. Mp 67C. Soluble in ether.

SPI. Abbreviation for Society of the Plastics Industry.

spider.   (1) A component of the ejector mechanism of a compression molding press. (2) A supporting device used in the die assembly of an extrusion machine.

spin coupling.   (spin–spin interaction). The interaction of fields through electron spins.

spindle oil.   Low-viscosity lubricating oil for textile and other high-speed machinery.

spinel. $MgAl_2O_4$. A natural oxide of magnesium and aluminum with replacement of magnesium by iron, zinc, and manganese and of aluminum by iron and chromium. There are also synthetic spinels, for example, as magnesia-alumina or magnesia-chromia. Their structure is similar to ferrites.
Properties: Color, various shades of red, grading to green, brown, and black; luster vitreous. Hardness 8. There are many varieties.
Occurrence: New York, New Jersey, Massachusetts, Virginia, North Carolina, Ceylon, Thailand, Mozambique.
Use: Crystallography; synthetic spinel is used as a refractory and in electronic applications.

spin moment.   The rotational moment of the momentum of an electron about its axis.

spinnerette.   An extrusion device shaped somewhat like a thimble and containing a number of holes of exceedingly small diameter through which are forced solutions of viscose rayon, nylon, molten glass, and various other materials. It is made of precious metals such as gold or platinum. Spinnerets enable extrusion of filaments of one denier or less, one-denier filament has a diameter of 40 mm. For commercial work 12 to 15 denier fiber is generally used.
See denier.

spinosad.
CAS: 168316-95-8. $C_{41}H_{65}NO_{10} \cdot C_{42}H_{67}NO_{10}$.
Properties: Isolated from *Saccharopolyspora spinosa* (Actinomycetes).
Hazard: A poison by inhalation. Moderately toxic by ingestion.
Use: Agricultural chemical.

spin resonance.   (ESR). The spin state where electrons go from the lower to the upper energy level.

spin–spin interaction.   See spin coupling.

spin state.   Orientations, in a strong magnetic field, of an unpaired electron either parallel or antiparallel to the field direction.

spiramycin.   (foromacidin; kitasamycin; leucomycin; NSC-64393; provamycin; rovamicina; 5337 R.P.; selectomycin; sequamycin; spiramycins;

2-[(4R,5S,6S,7R,9R,10R,11E,13E,16R)-6-[(2S, 3R,4R,5S,6R)-5[(2S,4R,5S,6S)-4,5-dihydroxy-4,6-dimethyloxan-2-yl]oxy-4-(dimethylamino)-3-hydroxy-6-methyloxan-2-yl]oxy-10-[(2R,5S,6S)-5-(dimethylamino)-6-methyloxan-2-yl]oxy-4-hydroxy-5-methoxy-9,16-dimethyl-2-oxo-1-oxoacyclohexadeca-11,13-dien-7-yl]acetaldehyde).
CAS: 8025-81-8. $C_{43}H_{74}N_2O_{14}$.
Properties: An antibiotic substance.
Derivation: Produced by *streptomyces ambofaciens* from soil of northern France.
Hazard: Poison; moderately toxic; teratogen; mutagen; can cause hypermotility, diarrhea, nausea, or vomiting.
Use: Antimicrobial to treat infections caused by bacteria and toxoplasma gondii.

Spiralloy.   Glass- and other filament-wound, resin-bonded internal and external pressure vessels.
Use: Rocket motor cases, underwater and space structures, pressure vessels, radomes, torpedo cases, booms, and tubes.

spirits.   An obsolescent and ambiguous term usually taken to mean the distilled essence of a substance. Mineral or petroleum spirits is a volatile hydrocarbon distillate similar to naphtha; in pharmacy, the term refers to an alcoholic solution of volatile principles, e.g., spirits of ammonia, niter, camphor.
See tincture.

spiro(1,3-benzodioxole-2,1'-cyclohexan)-4-ol, methylcarbamate.
CAS: 22791-18-0. $C_{14}H_{17}NO_4$.
Hazard: A poison by ingestion.
Use: Agricultural chemical.

spirobiindane.
CAS: 1568-80-5. $C_{21}H_{24}O_2$.
Hazard: Low toxicity by ingestion and skin contact. A mild eye irritant.

spirochete.   A type of bacteria characterized by a spiral shape; the infective organism of syphilis.

spirocyclane.   See spiropentane.

spironolactone.   (17-hydroxy-7-α-mercapto-3-oxo-17-α-pregn-4-ene-21-carboxylic   acid-g-lactone-7-acetate).
CAS: 52-01-7. $C_{24}H_{32}O_4S$.
Properties: Light-cream-colored to light-tan, crystalline powder; mild mercaptan-like odor. Mp 198–207C (decomposes). Stable in air. Practically insoluble in water; soluble in ethyl acetate and ethanol; slightly soluble in methanol.
Grade: USP.
Hazard: Questionable carcinogen.
Use: Medicine (diuretic).

spiropentane. (spirocyclane; cyclopropane-spirocyclopropane).

$H_2CH_2CCCH_2CH_2$.

Properties: Colorless liquid. Refr index 1.41220 (20C), d 0.7551 (20/4C), fp −107.05C, bp 39.03C.
Derivation: Heating pentaerythrityl tetrabromide in ethanol with zinc dust.

spiropiperidine alkaloid. Any alkaloid with two piperidine rings.
Deriviation: Occur in the skin secretions of frogs of the genus *Dendrobates*.
Hazard: Toxic.

spiro system. A structural formula consisting of two rings having one carbon atom in common. Most bicyclic compounds such as naphthalene have two carbons in common.
See spiropentane.

splice site. Location in the DNA sequence where RNA removes the noncoding areas to form a continuous gene transcript for translation into a protein.

spodumene. $LiAl(SiO_3)_2$.
Properties: White, pale-green, emerald-green, pink, or purple mineral; white streak; vitreous luster. Contains up to 8% lithium oxide with some replacement by sodium. D 3.13–3.20, Mohs hardness 6.5–7. Insoluble in acids.
Occurrence: U.S. (North Carolina, California, Massachusetts, South Dakota), Brazil, Mozambique.
Use: Source of lithium, in ceramics and glass as a source of lithia and alumina.

sponge. (1) *Metal*: a finely divided and porous form of metal such as iron, platinum, or titanium. May be used in sponge form as a catalyst or pressed into metal ingots. (2) *Plastic*: cellular plastic, foam, rubber sponge. (3) *Natural*: siliceous cells of the Porifera group of sessile sea animals.

sponge iron. See iron sponge.

sponge, iron. (iron, sponge). High-purity, finely divided (dust) of very reactive, unstable metallic iron (Fe0). Made by reducing an iron oxide at such low temperatures that melting does not occur, usually by mixing iron oxide and coke and applying limited increase in temperature.
Hazard: Because of its highly reactive nature, sponge iron is subject to a variety of packaging and shipping restrictions.

spontaneous combustion. See combustion.

sporadic cancer. Cancer that occurs randomly and is not inherited from parents. Caused by DNA changes in one cell that grows and divides, spreading throughout the body.
See hereditary cancer.

spore. A single cell that is dispersed as a reproductive unit, or that encapsulates a cell during unfavorable environmental conditions; in organisms with an alternation of generations.

sporophyte. The diploid stage in the life cycle of an organism undergoing an alternation of generations. The sporophyte is multicellular and develops from a zygote. The mature sporophyte meiotically produces haploid spores that later generate the gametophyte generation.

spot test. Microchemical or semimicrochemical drop test for chemical compounds, ions, or radicals using sensitive chemical reactions.

spray drying. Drying solids by spraying solutions of them into a heated chamber.

sprue. (1) The main feed channel that runs from the outer face of an injection mold to the gate in a single-cavity mold or to the runners of a multiple-cavity mold. The liquid polymer is forced through this orifice from a nozzle till the mold is filled to capacity. Some polymer remains in the sprue after the mold is closed, leaving a projecting piece that must be removed after the product is ejected. The viscosity of the polymer must be low enough to permit it to pass through the sprue readily. (2) A disease attacking the digestive organs as a result of inadequate nutrition, especially in tropical areas.

spun protein. See protein, textured.

sputtered coating. A protective metallic coating applied in a vacuum tube and consisting of metal ions emanating from a cathode and deposited as a film on objects within the tube. The process involves three phases: generation of metal vapor, diffusion of the vapor, and its condensation. Paper, plastics, and similar materials can be coated in this way.

squalane. (perhydrosqualene; 2,6,10,15,19,23-hexamethyltetracosane; spinacane; dodecahydrosqualane).
CAS: 111-01-3. $C_{30}H_{62}$. A saturated hydrocarbon.
Properties: Colorless liquid; odorless; tasteless. D 0.805–0.812 (20C), bp 350C, fp −38C, refr index 1.4520–1.4525 (20C). Miscible with vegetable and mineral oils; organic solvents, and lipophilic substances. Nondrying, nonoxidizing, noncongealing. Combustible.
Derivation: Hydrogenation of squalene, may occur naturally in sebum.
Use: High-grade lubricating oil, vehicle for externally applied pharmaceuticals and cosmetics, perfume fixative, gas chromatographic analysis, transformer oil.

squalene. (spinacene).
CAS: 111-02-4. 2,6,10,15,19,23-hexamethyl-2,6,
10,14,18,22-tetracosahexaene) $C_{30}H_{50}$.
A natural raw material found in human sebum (5%)
and in shark-liver oil. An unsaturated aliphatic
hydrocarbon (carotenoid) with six unconjugated
double bonds.

Properties: Oil with faint odor. D 0.858–0.860
(20C), bp 285C (25 mm Hg), fp −60C, refr index
1.49–1.50, iodine no 360–380, saponification value
0–5. Insoluble in water; slightly soluble in alcohol;
soluble in lipids and organic solvents. Combustible.
Grade: 90% min.
Use: Biochemical and pharmaceutical research, a
precursor of cholesterol in biosynthesis, chemical
intermediate.

squeege oil. A liquid vehicle to mix with colors,
that burns off completely on firing ceramics coated
with it.

squib. A thin tube filled with black powder to
form a slow-burning fuse to fire an explosive.

squirrel-cage disintegrator. See cage mill.

Sr. Symbol for strontium.

SR1. An oil based formula for removing oil based
stains.

SR2. A general purpose spotter for a broad spectrum of stains.

SR3. A concentrated product for removing rust
and iron stains.

SRF black. Abbreviation for semireinforcing
furnace black.
See furnace black.

SS acid. See Chicago acid.

SSR. Simple Sequence Repeat.
See Microsatellite.

STAB2335. A line of organic and inorganic
industrial chemicals, synthetic resins, synthetic rubber, high-compound fertilizers, coating materials,
latexes, pharmaceutical and food additives, explosives, photopolymers and platemaking systems,

separation and ion-exchange membranes, systems,
and equipment.

"Stabaxol P" *[Rhein].*
CAS: 159654-97-4. $(C_{17}H_{22}N_2O_2 \cdot C_{14}H_{16}N_2O_2)_x$.
Hazard: A poison. Moderately toxic by ingestion
and inhalation.
Use: Hydrolysis protection.

"Stabil-9" *[ICL].* TM for sodium aluminum
phosphate.
Grade: Powder.
Use: Primarily in self-rising flours.

Stabileze. A chemical used as thickener and stabilizer for polymers.

Stabilide. Potassium iodide stabilized with calcium stearate.
Use: Animal feeds.

Stabilisal C-1. An organic compound.
Use: An antigellation agent for nitrocellulose-based
bronzing lacquers and for stabilizing dipping lacquers.

stabilizer. Any substance that tends to keep a
compound, mixture, or solution from changing its
form or chemical nature. Stabilizers may retard a
reaction rate, preserve a chemical equilibrium, act
as antioxidants, keep pigments and other components in emulsion form, or prevent the particles in
a colloidal suspension from precipitating.
See inhibitor.

stable transfection. A form of transfection
experiment designed to produce permanent lines
of cultured cells with a new gene inserted into their
genome. Usually this is done by linking the desired
gene with a "selectable" gene, i.e., a gene which
confers resistance to a toxin. Upon putting the toxin
into the culture medium, only those cells which
incorporate the resistance gene will survive, and
essentially all of those will also have incorporated
the experimenter's gene.

Staedel–Rugheimer pyrazine synthesis.
Formation of pyrazines by high-temperature autoclave reaction of α-halogenomethyl ketones with
ammonia.

Staflex Noda. *n*-octyl-*n*-decyl adipate.
Properties: Liquid.
Use: For electrical properties for outdoor wire and
cable applications. In vinyls, combines good heat
and light stability with high compatibility.

stain. (1) An organic protective coating similar to
a paint, but with much lower solids content (pigment
loading).
Use: Exterior and interior coating of wood, furniture,
flooring, etc.

(2) Any compound used to color bacteria for microscopic examination, e.g., osmium tetroxide, phosphotungstic acid, uranyl acetate, and certain chromium compounds.

stainless steel.   See steel, stainless.

standard cell.   See Weston cell.

standard free-energy change.   The free-energy change for a reaction occurring under a set of standard conditions: temperature, 298 K, pressure, 1 atm, and all solutes at 1 M concentration.

standard solution.   A solution of a specifically known concentration used in volumetric analysis.

Stanley, Wendell M.   (1904–1971). An American biochemist who won the Nobel Prize for chemistry in 1946 along with Northrop and Sumner. His work on virus research resulted in isolation of crystals proving the virus to be proteinaceous. In the 1930s, he was concerned with isolating nucleic acid from crystallized virus, and the reproduction of influenza virus. His doctorate was from the University of Illinois. His many accomplishments included membership in the National Advisory Cancer Council of the United States Public Health Service in the 1950s.

stannic.   The designation for quadrivalent tin or compounds containing quadrivalent tin.

stannic acid.   See stannic oxide.

stannic anhydride.   See stannic oxide.

stannic bromide.   (tin bromide; tin tetrabromide).
CAS: 7789-67-5. $SnBr_4$.
Properties: White, crystalline mass. Fumes when exposed to air. D 3.3, bp 203C, mp 31C. Soluble in water, alcohol, and carbon tetrachloride.
Hazard: Irritant to skin and eyes.
Use: Mineral separations.

stannic chloride.   (tin chloride; tin tetrachloride; tin perchloride).
CAS: 7646-78-8. $SnCl_4$. Often sold in the form of the double salt with sodium chloride: $Na_2SnCl_6•H_2O$.
Properties: Colorless, fuming, caustic liquid, that water converts into a crystalline solid, $SnCl_4•5H_2O$. Keep well stoppered. D 2.2788, fp −33C, bp 114C. Soluble in cold water, alcohol, carbon disulfide; decomposed by hot water.
Derivation: Treatment of tin or stannous chloride with chlorine.
Grade: Technical, CP.
Hazard: Evolves heat on contact with moisture. Corrosive liquid.

Use: Electroconductive and electroluminescent coatings, mordant in dyeing textiles, perfume stabilization, manufacture of fuchsin, color lakes, ceramic coatings, bleaching agent for sugar, stabilizer for certain resins, manufacture of blueprint and other sensitized papers, other tin salts, bacteria and fungi control in soaps.

stannic chloride pentahydrate.
CAS: 10026-06-9. $SnCl_4•5H_2O$.
Properties: White solid. Mp 56C. Soluble in water or alcohol.
Hazard: Toxic material.
Use: Substitute for anhydrous stannic chloride where the presence of water is not objectionable.

stannic chromate.   (tin chromate). $Sn(CrO_4)_2$.
Properties: Brownish-yellow, crystalline powder. Partially soluble in water.
Derivation: Action of chromic acid on stannic hydroxide.
Hazard: Toxic material.
Use: Coloring porcelain and china.

stannic oxide.   (stannic anhydride; tin peroxide; tin dioxide; stannic acid). SnO_2 or $SnO_2•xH_2O$.
Properties: White powder; anhydrous or containing variable amounts of water. D 6.6–6.9, mp 1127C, sublimes at 1800–1900C. Soluble in concentrated sulfuric acid, hydrochloric acid; insoluble in water. Noncombustible.
Derivation: (1) Found in nature as in the mineral cassiterite; (2) precipitated from stannic chloride solution by ammonium hydroxide.
Grade: White, pure; white; gray; CP.
Use: Tin salts; catalyst; ceramic glazes and colors; putty; perfume preparations and cosmetics; textiles (mordant, weighting); polishing powder for steel, glass, etc.; manufacture of special glasses.

stannic phosphide.   (tin phosphide).
CAS: 25324-56-5. Sn_2P_2 or SnP.
Properties: Silver-white, hard mass or lumps. D 6.56, mp forms Sn_4P_3 at 415C. Soluble in acids.
Derivation: By heating together tin and phosphorus.
Hazard: Flammable, dangerous fire risk.

stannic sulfide.   (artificial gold; mosaic gold; tin bronze; tin disulfide). SnS_2.
Properties: Yellow to brown powder. D 4.42–4.60, mp decomposes at 600C. Soluble in concentrated hydrochloric acid and alkaline sulfides; insoluble in water.
Derivation: (1) Action of sulfur on a solution of stannic chloride; (2) by heating tin amalgam with sulfur and ammonium chloride, distilling off the mercury sulfide and ammonium chloride.
Grade: Technical, reagent.
Hazard: Irritant to skin and eyes.
Use: Imitation gilding, pigment.

Stannochlor.
CAS: 7772-99-8. Stannous chloride.
Use: For metalizing glass and plastic; catalyst in pharmaceuticals, resins, petrochemicals, solder, tanning agent, and food and beverage preservative.

stannous acetate. $Sn(C_2H_3O_2)_2$.
Properties: Gray to yellow crystals. Mp 182C, d 2.30. Soluble in dilute hydrochloric acid.
Derivation: Refluxing stannous oxide with 50% acetic acid in an atmosphere of nitrogen.
Hazard: Toxic by ingestion.
Use: Reducing agent.

stannous bromide.   (tin bromide; tin dibromide).
$SnBr_2$.
Properties: Yellow powder. D 5.117 (17C), bp 619C, mp 215C. Soluble in hydrochloric acid (dilute); soluble in water, alcohol, ether, and acetone; oxidizes and turns brown in air.
Hazard: Skin irritant.

stannous chloride.   (tin crystals; tin salt; tin dichloride; tin protochloride).
CAS: 7772-99-8.   (1) $SnCl_2$. (2) $SnCl_2 \cdot 2H_2O$.
Properties: White, crystalline mass that absorbs oxygen from the air and is converted into the insoluble oxychloride. (1) D 3.95 (25/4C), mp 246.8C, bp 652C; (2) d 2.71, mp 37.7C, bp (decomposes). Soluble in water, alkalies, tartaric acid, and alcohol.
Derivation: By dissolving tin in hydrochloric acid.
Grade: Technical, CP, reagent, anhydrous, hydrated.
Hazard: Irritant to skin, use in foods restricted to 0.0015%, as tin.
Use: Reducing agent in manufacture of chemicals, intermediates, dyes, polymers, phosphors; manufacture of lakes; textiles (reducing agent in dyeing, discharge in printing); tin galvanizing; reagent in analytical chemistry; silvering mirrors; revivification of yeast sown in must (accelerator); antisludging agent for lubricating oils; food preservative; stabilizer for perfume in soaps; catalyst; soldering flux; sensitizing agent for glass, paper, plastics.

stannous chromate.   (tin chromate). $SnCrO_4$.
Properties: Brown powder. Almost insoluble in water.
Derivation: Interaction of stannous chloride and sodium chromate.
Hazard: Toxic material.
Use: Decorating porcelain.

stannous-2-ethylhexoate.   (stannous octoate; tin octotate). $Sn(C_8H_{15}O_2)_2$.
Properties: Light-yellow liquid. D 1.25, Gardner color 3 (max). Insoluble in water, methanol; soluble in benzene, toluene, petroleum ether; hydrolyzed by acids and bases.
Hazard: Toxic material.
Use: Polymerization catalyst for urethane foams, lubricant, addition agent, stabilizer for transformer oils.

stannous fluoride.   (tin fluoride; tin difluoride).
CAS: 7783-47-3. SnF_2.
Properties: White, lustrous, crystalline powder; bitter, salty taste. Mp 212–214C. Practically insoluble in alcohol, ether, and chloroform; slightly soluble in water.
Grade: NF.
Hazard: Toxic by ingestion, strong irritant to skin and tissue.
Use: Fluoride source in toothpastes.
Note: Stannous hexafluorozirconate is said to be more effective than the fluoride in preventing dental caries.

stannous octoate.   See stannous-2-ethylhexoate.

stannous oleate.   (tin oleate).
CAS: 1912-84-1. $Sn(C_{18}H_{33}O_2)_2$.
Properties: Light-yellow liquid. Insoluble in water and methanol; soluble in benzene, toluene, petroleum ether; hydrolyzed by acids and bases.
Hazard: Absorbed by skin.
Use: Polymerization catalyst, inhibitor.

stannous oxalate.   (tin oxalate).
CAS: 814-94-8. SnC_2O_4.
Properties: Heavy, white, crystalline powder. D 3.56, mp decomposes at 280C. Soluble in acids; insoluble in water and acetone.
Derivation: By the action of oxalic acid on stannous oxide.
Grade: Technical, CP, reagent.
Hazard: Absorbed by skin.
Use: Dyeing and printing textiles, catalyst for esterification reactions.

stannous oxide.   (tin oxide; tin protoxide). SnO.
Properties: Brownish-black powder. Unstable in air. Reacts with acids and strong bases. D 6.3, mp 1080C (600 mm Hg) (decomposes). Insoluble in water. A nuisance particulate.
Derivation: By heating stannous hydroxide in a current of carbon dioxide.
Grade: Technical, CP.
Use: Reducing agent, intermediate in preparation of stannous salts as used in plating and glass industries, pharmaceuticals, soft abrasive (putty powder).

stannous pyrophosphate. $Sn_2P_2O_7$.
Properties: White, free-flowing crystals. D 4.009 (16C). Insoluble in water.
Use: Toothpaste additive.

stannous stearate.
Use: Food additive.

stannous sulfate.   (tin sulfate).
CAS: 7488-55-3. $SnSO_4$.
Properties: Heavy white or yellowish crystals. Mp loses sulfur dioxide at 360C. Soluble in water and sulfuric acid; water solution decomposes rapidly.
Derivation: Action of sulfuric acid on stannous oxide.

Hazard: Toxic material.
Use: Dyeing; tin plating, particularly for plating automobile pistons and steel wire.

stannous sulfide. (tin monosulfide; tin protosulfide; tin sulfide). SnS.
Properties: Dark-gray or black, crystalline powder. D 5.080, bp 1230C, mp 880C. Soluble in concentrated hydrochloric acid (decomposes); insoluble in dilute acids and water.
Hazard: Toxic material.
Use: Making bearing material, catalyst in polymerization of hydrocarbons, analytical reagent.

stannous tartrate. (tin tartrate). $SnC_4H_4O_6$.
Properties: Heavy, white, crystalline powder. Soluble in water; dilute hydrochloric acid.
Derivation: Action of tartaric acid on stannous oxide.
Hazard: Toxic material.
Use: Dyeing and printing fabrics.

stannum. The Latin name for tin, hence the symbol Sn in chemical nomenclature.

staphylococcal enterotoxin b. See enterotoxin b, staphylococcal.

staple. A cotton fiber, usually in reference to length, i.e., short- or long-staple cotton.

starch.
CAS: 9005-25-8. A carbohydrate polymer having the following repeating unit:

It is composed of about 25% amylose (anhydroglucopyranose units joined by glucosidic bonds) and 75% amylopectin, a branched-chain structure.
Properties: White, amorphous, powder or granules; tasteless. Various crystalline forms may be obtained, including microcrystalline. Irreversible gel formation occurs in hot water; swelling of granules can be induced at room temperature with such compounds as formamide, formic acid, and strong bases and metallic salts.
Occurrence: Starch is a reserve polysaccharide in plants (corn, potatoes, tapioca, rice, and wheat are commercial sources).
Grade: Commercial, powdered, pearl, laundry, technical, reagent, edible, USP.
Hazard: Dermatitis. Questionable carcinogen.
Use: Adhesive (gummed paper and tapes, cartons, bags, etc.), machine-coated paper, textile filler and sizing agent, beater additive in papermaking, gelling agent and thickener in food products (gravies, custards, confectionery), oil-well drilling fluids, filler in baking powders (cornstarch), fabric stiffener in laundering, urea-formaldehyde resin adhesives for particle board and fiberboard, explosives (nitrostarch), dextrin (starch gum), chelating and sequestering agent in foods, indicator in analytical chemistry, anticaking agent in sugar, face powders, abherent and mold release agent, polymer base.
See starch-based polymer.

starch-based polymer. (1) A reactive polyol derived from a mixture of a starch with dibasic acids, hydrogen-donating compounds, and catalysts dissolved in water; the slurry is subjected to high temperatures and pressures, yielding a low-viscosity polymer in a 50% solids aqueous solution. A molecular rearrangement takes place, and the polymer formed is completely different from starch in structure and properties. It can be further reacted with acids, bases, and cross-linking agents. Suggested uses are in high-wet-strength papers, as binders in paper coatings, as moisture barriers in packaging, and as water-resistant adhesives. (2) Yeast fermentation of starch to form a biodegradable plastic called "pullulan" (a trigluco polysaccharide) has been reported to be commercially feasible.
See "Pullulan."

starch dialdehyde. Starch in which the original anhydroglucose units have been partially changed to dialdehyde form by oxidation, for example, the product of the oxidation of cornstarch by periodic acid. Available in cationic dispersions up to 15% solids for mixing with paper pulp.
Use: Thickening agent, tanning agent, binder for leaf tobacco, adhesives, wet-strength additive in paper.

starch-iodide paper. Indicator paper made by dipping paper in starch paste containing potassium iodide.
Use: To test for halogens and oxidizers such as hydrogen peroxide.

starch, modified. Any of several water-soluble polymers derived from a starch (corn, potato, tapioca) by acetylation, chlorination, acid hydrolysis, or enzymatic action. These reactions yield starch acetates, esters, and ethers in the form of stable and fluid solutions and films. Modified starches are used as textile-sizing agents and paper coatings. Thin-boiling starches have high gel strength, oxidized starches made with sodium hypochlorite have low gelling tendency. Introduction of carboxyl, sulfonate, or sulfate groups into starch gives sodium or ammonium salts of anionic starches, yielding clear, nongelling dispersions of high viscosity. Cationic starches result from addition of amino groups. The glucose units of starch can be cross-linked with such agents as formaldehyde, soluble metaphosphates, and epichlorohydrin; this increases viscosity and thickening power for adhesives, canned foods, etc.

starch phosphate. An ester made from the reaction of a mixture of orthophosphate salts (sodium dihydrogen phosphate and disodium hydrogen phosphate) with starch.
Properties: Soluble in cold water (unlike regular starch) and has high thickening power. Can be frozen and thawed repeatedly without change in physical properties.
Use: Thickener for frozen foods; taconite ore binder; in adhesives, drugs, cosmetics; substitute for Arabic gum, locust bean gum, and carboxymethyl cellulose.

starch syrup. See glucose.

starch, thin-boiling. See starch, modified.

starch xanthate. A water-insoluble synthetic polysaccharide made by reacting starch with sodium hydroxide and carbon disulfide; biodegradable.
Use: To encapsulate pesticides; the coating, though insoluble, is permeable to water, thus slowly releasing the pesticide. Rubber-reinforcing agent.

starch xanthide. Starch xanthate that has been cross-linked with oxygen.
Use: Strengthening agent in paper and manufacture of powdered rubbers.

Starfol Wax. Synthetic spermaceti wax.
Available forms: Flakes.
Use: As a cosmetic emollient, drawing compounds, finishing aids. lubricants, and leather treatments.

Stark–Einstein law. Every molecule in a chemical reaction induced by exposure to light absorbs one quantum of the radiation causing the reaction.

Stark–Luneland effect. The polarization of light by a beam of moving atoms.

starter distillate.
Properties: Steam distillate of culture of Streptococcus lactis S. cremoris, S. lactis subsp. diacetylactis, Leuconostoc citrovorum, and L. dextranicum.
Use: Food additive.

"Starwax" 100 [Baker Hughes]. TM for a hard, petroleum microcrystalline wax, minimum mp 82.2C.

Sta-Tac. Olefinic hydrocarbon resins used as coating vehicles.

statistical mechanics. Statistical mechanics treats the detailed state of a system (its quantum state or, in classical models, its position in phase space) as unknown and subject to statistical uncertainties. Entropy is a measure of this uncertainty. Statistical mechanics describes the distribution of states in an equilibrium system at a given temperature (describing either the distribution of probabilities of quantum states or the probability density function in phase space), and can be used to derive thermodynamic properties from properties at the molecular level.

Staudinger, Hermann. (1881–1965). A German chemist, winner of a Nobel Prize in 1953 for his pioneer work on the structure of macromolecules and polymerization. A large part of modern high-polymer chemistry is based on his original research.

Staudinger reaction. Synthesis of phosphazo compounds by the reaction of tertiary phosphines with organic azides.

stavudine.
CAS: 3056-17-5. $C_{10}H_{12}N_2O_4$.
Hazard: Moderately toxic by ingestion.

"Staybelite" [Pinova]. TM for hydrogenated rosin, a pale, thermoplastic resin. Acid number 165, USDA color X, softening point 75C, saponification number 167.
Use: Adhesives and protective coatings.

Stayrite. Stabilizers used to provide increased heat and light stability to vinyl resins. Available in range of products, both liquid and solid.

Stead's reagent. A metallographic etching solution consisting of acidified copper chloride. It is useful in detecting phosphorous segregation in steel. Copper tends to deposit on parts of low phosphorous content.

steady state. A condition where properties of any part of a system are constant during a process.

steam. An allotropic form of water formed at 212F (100C) and having a latent heat of condensation of 540 calories per gram. It has a number of industrial uses, one of the most important being the production of hydrogen by the steam-hydrocarbon gas process (reforming), by the steam-water gas process, the steam-iron process, and the steam-methanol process. It is also used in steam cracking of gas oil and naphtha; in food processing; as a cleaning agent; in rubber vulcanization; as a source of heat and power; in distillation of plants for production of essential oils and perfumes; and in secondary oil recovery.
Steam from geothermal sources such as hot springs and fumaroles is utilized as an energy source. Steam for industrial processing is also being generated by solar energy techniques.
See geothermal energy; latent heat.

steam distillation. See hydrodistillation.

steam reforming. See reforming.

steapsin. A lipase in the pancreatic juice. See enzyme.

stearamide. (octadecanamide).
CAS: 124-26-5. $CH_3(CH_2)_{16}CONH_2$.
Properties: Colorless leaflets. Mp 109C, bp 251C (12 mm Hg). Insoluble in water; slightly soluble in alcohol and ether.
Use: Corrosion inhibitor in oil wells.

stearato chromic chloride. A polynuclear complex in the form of a six-membered ring. The two chromium atoms are bridged on one side by a hydroxyl group, and on the other side by the carboxyl oxygens of the stearic acid. The water-soluble complex results from the neutralization of stearic acid with basic chromic chloride. It acts as a water repellent and abherent.

stearic acid. (*n*-octadecanoic acid).
CAS: 57-11-4. $CH_3(CH_2)_{16}COOH$.
The most common fatty acid occurring in natural animal and vegetable fats. Most commercial stearic acid is 45% palmitic acid, 50% stearic acid and 5% oleic acid, but purer grades are increasingly used.
Properties: Colorless, waxlike solid; slight odor and taste suggesting tallow. D 0.8390 (80/4C), mp 69.6C, bp 361.1C, refr index 1.4299 (80C), flash p 385F (196C), autoign temp 743F (395C). Soluble in alcohol, ether, chloroform, carbon disulfide, carbon tetrachloride; insoluble in water. Combustible.
Derivation: (1) From high-grade tallows and yellow grease stearin by washing, hydrolysis with the Twitchell or similar reagent, boiling, distilling, cooling, and pressing; (2) from oleic acid by hydrogenation.
Grade: Saponified, distilled, single-pressed, double-pressed, triple-pressed, USP, FCC, 90% stearic with low oleic, grade free from chick edema factor, 99.8% pure.
Use: Chemicals, especially stearates and stearate driers; lubricants; soaps; pharmaceuticals and cosmetics; accelerator activator; dispersing agent and softener in rubber compounds; shoe and metal polishes; coatings; food packaging; suppositories and ointments.
See Hydrofol.

stearic acid potassium salt. See potassium stearate.

stearin. (tristearin; glyceryl tristearate).
$C_3H_5(C_{18}H_{35}O_2)_3$.
Properties: Colorless crystals or powder; odorless; tasteless. Mp 71.6C, d 0.943 (65C). Insoluble in water; soluble in alcohol, chloroform, carbon disulfide. Combustible.
Derivation: Constituent of most fats.
Grade: Technical, also graded as to source.
Use: Soap, candles, candies, adhesive pastes, metal polishes, waterproofing paper, textile sizes, leather stuffing, manufacture of stearic acid.

stearin pitch. See fatty acid pitch.

stearone. An aliphatic ketone, insoluble in water, stable to high temperatures, acids, and alkalies; compatible with high-melting vegetable waxes, paraffins, and fatty acids; incompatible with resins, polymers and organic solvents at room temperature but compatible with them at high temperature. Combustible.
Use: Antiblocking agent.

N-stearoyl-p-aminophenol.
$HO(C_6H_4)NHCOCH_2(CH_2)_{15}CH_3$.
Properties: White to off-white powder. Mp 131–134C. Insoluble in water; soluble in polar organic solvents (especially when hot) such as alcohol, dioxane, acetone, and dimethylformamide. Combustible.
Use: Antioxidant.

stearoyl chloride. (*n*-octadecanoyl chloride).
$CH_3(CH_2)_{16}COCl$.
Properties: Mp 23C, bp 174–178C (2 mm Hg). Soluble in hydrocarbons and ethers. Combustible.
Use: Preparation of substituted amines and amides, acid anhydrides, esterification of alcohols, synthesis of other organic compounds.

stearoyl propylene glycol hydrogen succinate. See succistearin.

stearyl alcohol. (1-octadecanol; octadecyl alcohol).
CAS: 112-92-5. $CH_3(CH_2)_{16}OH$.
Properties: Unctuous, white flakes or granules; faint odor; bland taste. D 0.8124 (59/4C), bp 210.5C (15 mm Hg), mp 59C. Soluble in alcohol, acetone, and ether; insoluble in water. Combustible.
Derivation: Reduction of stearic acid.
Grade: Commercial, technical, USP.
Use: Perfumery, cosmetics, intermediate, surface active agents, lubricants, resins, antifoam agent.

stearyl citrate.
Use: Food additive.

stearyl-2-lactylic acid.
Use: Food additive.

stearyl mercaptan. (octadecyl mercaptan).
CAS: 2885-00-9. $CH_3(CH_2)_{17}SH$.
Properties: Solid. Mp 25C, bp 205–209C (11 mm Hg), d 0.8420 (25/4C), refr index 1.4591 (34C).
Grade: 95% (min) purity.
Hazard: Strong irritant.
Use: Organic intermediate, synthetic rubber processing.

stearyl methacrylate. Group name for $CH_2:C(CH_3)COO(CH_2)_nCH_3$ ($n = 13$ to 17).
Properties: Bulk d 0.864 (25/15C), boiling range 310–370C.

Use: Monomer for plastics, molding powders, solvent coatings, adhesives, oil additives; emulsions for textile, leather, paper finishing.

stearyl monoglyceridyl citrate.
   Properties: A soft, practically tasteless, off-white to tan, waxy solid having a lardlike consistency; insoluble in water, soluble in chloroform and ethylene glycol.
   Derivation: Reaction of citric acid, monoglycerides of fatty acids (obtained by the glycerolysis of edible fats and oils or derived from fatty acids), and stearyl alcohol.
   Grade: FCC.
   Use: Emulsion stabilizer in foods (not over 0.15%).

steatite.   A mixture of talc, clay, and alkaline-earth oxides.
   Use: Chiefly as a ceramic insulator in electronic devices.

steel.   An alloy of iron and 0.02–1.5% carbon; it is made from molten pig iron by oxidizing out the excess carbon and other impurities. The open-hearth process, which uses air for this purpose, has been supplanted by the basic oxygen process in which pure oxygen is injected into molten iron. A small percentage of steel is also made in the electric furnace, with iron ore as oxidant. Low-carbon (mild) steels contain 0.02–0.3% carbon; medium-carbon grades from 0.3–0.7% carbon; and high-carbon grades 0.7–1.5%.
   There are many special-purpose types of steel in which one or more alloying metals are used, with or without special heat treatment. The most common additives are chromium and nickel, as in the 18–8 stainless steels; these add greatly to corrosion resistance. High-speed and tool steels, designed primarily for efficient cutting, contain such alloying metals as tungsten, molybdenum, manganese, and vanadium as well as chromium. Cobalt and zirconium are used for construction steels.
   Available forms: Rods, bars, sheet, strips, wire, wool.
   Use: Construction, ship hulls, auto bodies, machinery and machine parts, cables, abrasive, chemical equipment, belts for tire reinforcement.

steel, stainless.   Alloy steels containing high percentages of chromium, from less than 10% to more than 25%. There are three groups: (1) Austenitic, which contain both chromium (16% min) and nickel (7% min); a stress-corrosion resistant type contains 2% silicon. (2) Ferritic, which contain chromium only and cannot be hardened by heat treatment. (3) Martensitic, which contain chromium and can be hardened by heat treatment; these last are ferromagnetic. A subgroup of the martensitic steels comprises the precipitation-hardening types.
   See SAE steel.

Steffen process.   A process used in beet sugar manufacture to separate residual sugar from molasses. Based on the formation of insoluble tricalcium saccharate and its subsequent decomposition to sugar in the presence of a weak acid such as carbonic.

Steinbuhl yellow.   See barium chromate.

Steitz, Thomas A.   (1940–). An American biophysicist and biochemist who was a co-recipient of the 2009 Nobel Prize with Venkatraman Ramakrishnan and Ada E. Yonath. The scientists' research on atomic structure and function of the ribosome provided fundamental understanding of the two ribosomal subunits (small and large). Steitz' remarkable discovery was the structure of large subunit and the locations of proteins and RNA inside the subunit. He received his Ph.D. from Harvard University followed by an illustrious career, which includes the Lewis S. Rosenstiel Award (2001), Keio Medical Science Prize (2006), and the Gairdner International Award (2007).

stellarator.   A type of torus developed for fusion research. It differs from tokamaks in respect to the method of confining the plasma.
   See torus; fusion; tokamak.

stem cell.   Undifferentiated, primitive cells in the bone marrow that have the ability to both multiply and to differentiate into specific blood cells.

stem group.   All the taxa in a clade preceding a major cladogenesis event. They are often difficult to recognize because they may not possess synapomorphies found in the crown group.

Stengel process.   A method of making ammonium nitrate fertilizer from anhydrous ammonia and nitric acid. The fertilizer particles can be made in different sizes according to the use.

"Stepantan" [Stepan].   TM for an aqueous solution.
   Use: As a drilling foamer base, fire fighting foam, and other foam-forming applications.

Stephen aldehyde synthesis.   Preparation of aldehydes from nitriles by reduction with stannous chloride in ether saturated with hydrochloric acid. The intermediate aldimine salts have to be hydrolyzed. The best results are obtained in the aromatic series.

stereoblock polymer.   A polymer whose molecule is made of comparatively long sections of identical stereospecific structure, these sections being separated from one another by segments of different structure, as for example, blocks of an isotactic polymer interspersed with blocks of the same polymer with a syndiotactic structure.

See polymer, stereospecific; block polymer.

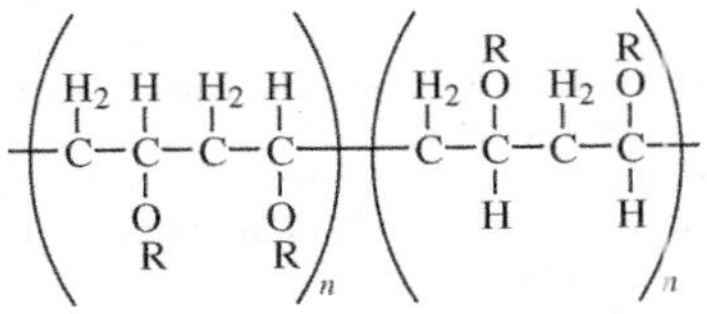

stereoblock polymer

stereochemistry. A subdiscipline of organic chemistry devoted to study of the three-dimensional spatial configurations of molecules. One aspect of the subject deals with stereoisomers—compounds that have identical chemical constitution, but differ as regards the arrangement of the atoms or groups in space. Stereoisomers fall into two broad classes: optical isomers and geometric *(cis-trans)* isomers. Another aspect of stereochemistry concerns control of the molecular configuration of high-polymer substances by use of appropriate (stereospecific) catalysts.
See optical isomer; enantiomer; asymmetry; polymer, stereospecific; Ziegler catalyst; geometric isomer.

stereoisomer. See stereochemistry.

"Stereon" *[Firestone].* TM for styrene–butadiene block copolymers.
Use: In adhesives, plastics, and asphalt.

stereoregular polymer. See polymer, stereospecific.

stereospecific catalyst. See catalyst, stereospecific.

stereospecific polymer. See polymer, stereospecific.

steric. Pertaining to the spatial relationships of atoms in a molecular structure, and in particular, to the space-filling properties of a molecule.

steric hindrance. A characteristic of molecular structure in which the molecules have a spatial arrangement of their atoms such that a given reaction with another molecule is prevented or retarded.

steric strain. The additional energy of molecular bonds due to peculiar molecular geometry.

sterigmatocystin. (3a,12c-dihydro-8-hydroxy-6-methoxy-7H-furo(3′,2′:4,5)furo(2,3-c)xanthen-7-one).
CAS: 10048-13-2. $C_{18}H_{12}O_6$.
A bisfuranoid mycotoxin.
Properties: Pale-yellow needles.

Derivation: Produced by strains of the common molds *Aspergillus versicolor, A. nidulans,* and an unidentified species of *Bipolaris.*
Hazard: Possible carcinogen; tumorigen; poison; mutagen; causes necrosis of the liver and kidney, and an inhibitory effect on orotic acid incorporation into nuclear RNA.

sterilization. (1) Complete destruction of all bacteria and other infectious organisms in an industrial, food, or medical product; it must be followed by aseptic packaging to prevent recontamination, usually by hermetic sealing. The methods used involve either wet or dry heat, use of chemicals such as formaldehyde and ethylene oxide filtration (pharmaceutical products), and irradiation by UV or γ-radiation. Milk is sterilized by heating for 30 seconds at 135C, followed by in-can heating at 115C for several minutes and aseptic packaging.
(2) Rendering a life-form incapable of reproduction by radiation or chemical treatment.
See chemosterilant.

sterling silver. According to U.S. law, this term is restricted to silver alloys containing at least 92.5% pure silver and 7.5% maximum other metal (usually copper).
See silver.

steroid. One of a group of polycyclic compounds closely related biochemically to terpenes. They include cholesterol, numerous hormones, precursors of certain vitamins, bile acids, alcohols (sterols), and certain natural drugs and poisons (such as the digitalis derivatives). Steroids have as a common nucleus a fused, reduced, 17-carbon-atom ring system, cyclopentanoperhydrophenanthrene. Most steroids also have two methyl groups and an aliphatic side chain attached to the nucleus. The length of the side chain varies and generally contains 8, 9, or 10 carbon atoms in the sterols, 5 carbon atoms in the bile acids, 2 in the adrenal cortical steroids, and none in the estrogens and androgens. Steroids are classed as lipids because of their solubility in organic solvents and insolubility in water. Most of the naturally occurring steroids have been synthesized and many new steroids unknown in nature have been synthesized for use in medicine, such as the fluorosteroids (dexamethasone).

steroid alkaloid. An alkaloid that contains a perhydrocyclopentanophenanthrene ring.

sterol. A steroid alcohol. Such alcohols contain the common steroid nucleus, plus an 8- to 10-carbon-atom sidechain and a hydroxyl group. Sterols are widely distributed in plants and animals, both in the free form and esterified to fatty acids. Cholesterol is the most important animal sterol; ergosterol is an important plant sterol (phytosterol).

Stevens rearrangement. Migration of an alkyl group from a quaternary ammonium salt to an adjacent carbanionic center on treatment with strong base. The product is a rearranged tertiary amine, sulfonium, or sulfide.

STH. See somatotropic hormone.

stibic anhydride. See antimony pentoxide.

stibine. See antimony hyride.

stibium. The Latin name for the element antimony; hence the symbol Sb.

stibnite. (gray antimony; antimony glance; antimonite). Sb_2S_3.
Properties: Lead-gray mineral, subject to blackish tarnish, metallic luster. D 4.52–4.62, Mohs hardness 2. Soluble in concentrated boiling hydrochloric acid with evolution of hydrogen sulfide.
Occurrence: Japan, China, Mexico, Bolivia, Peru, South Africa.
Use: The most important ore of antimony.

sticky ends. After digestion of a DNA with certain restriction enzymes, the ends left have one strand overhanging the other to form a short (typically 4 nt) single-stranded segment. This overhang will easily re-attach to other ends like it and become known as "sticky ends."
The overhangs thus produced can still hybridize ("anneal") with each other, even if they came from different parent DNA molecules, and the enzyme ligase will then covalently link the strands. Sticky ends therefore facilitate the ligation of diverse segments of DNA, and allow the formation of novel DNA constructs.

Stieglitz rearrangement. Rearrangement of trityl hydroxylamines to Schiff bases on treatment with phosphorous pentachloride.

stigmasterol. $C_{29}H_{48}O \cdot H_2O$. A plant sterol.

Properties: Insoluble in water; soluble in usual organic solvents. Combustible.
Derivation: From soy or calabar beans.
Use: Preparation of progesterone and other important steroids.

stilbene. (toluelene; *trans* form of α,β-diphenylethylene).

CAS: 103-30-0. $C_6H_5CH:CHC_6H_5$.
Properties: Colorless or slightly yellow crystals. D 0.9707, mp 124–125, bp 306–307C. Soluble in benzene and ether; slightly soluble in alcohol; insoluble in water. Combustible.
Derivation: By passing toluene over hot lead oxide.
Method of purification: Crystallization, zone melting used for very pure crystals.
Grade: Technical, pure.
Use: Manufacture of dyes and optical bleaches, crystals are used as phosphors and scintillators.
Note: The *cis* form of α,β-diphenylethylene (isostilbene) is a yellow oil, bp 145C (13 mm Hg), mp 1C.

stilbene dye. A dye whose molecules contain both the -N=N- and the >C=C< chromophor groups in their structure and whose CI numbers range from 40,000 to 40,999. These are direct cotton dyes.

stilbestrol. See diethylstilbestrol.

still. A device for the conversion of a material from the solid or liquid state to the vapor state, with or without simultaneous chemical decomposition. The vapor is condensed in a separate part of the apparatus becoming liquid or solid again.

stillage. The grain residue from alcohol production used in feeds and feed supplements.

stirrer. See impeller; agitator.

STM. Abbreviation for scanning tunneling microscope.

Stobbe condensation. Condensation of aldehydes or ketones with diethyl succinate in the presence of a strong base to form monoesters of α-alkylidene (or arylidene) succinic acids.

Stock system. See chemical nomenclature.

Stoddard solvent.
CAS: 8052-41-3. A widely used dry cleaning solvent. U.S. Bureau of Standards and ASTM D-484–52 define it as a petroleum distillate clear and free from suspended matter and undissolved water, and free from rancid and objectionable odor. The minimum flash p is 100F (37.7C). Distillation range, more than 50% over at 350F (177C), 90% over at 375F (190C), and the end point below 410F (210C); autoign temp 450F (232C). Combustible.
Hazard: Fire risk. Toxic by ingestion. Eye, skin and kidney damage; nausea; central nervous system impairment.
Use: Dry cleaning, spot and stain removal.

stoichiometry. The branch of chemistry and chemical engineering that deals with the quantities of substances that enter into and are produced by chemical reactions. For example, when methane unites with oxygen in complete combustion, 16 g of methane requires 64 g of oxygen. At the same time 44 g of carbon dioxide and 36 g of water are formed as reaction productions. Every chemical reaction has its characteristic proportions. The methods of obtaining these from chemical formulas, equations, atomic weights, and molecular weights and determination of what and how much is used and produced in chemical processes is the major concern of stoichiometry.

Stokes's law. (1) The rate at which a spherical particle will rise or fall when suspended in a liquid medium varies as the square of its radius; the density of the particle and the density and viscosity of the liquid are essential factors. Stokes's law is used in determining sedimentation of solids, creaming rate of fat particles in milk, etc.
(2) In atomic processes, the wavelength of fluorescent radiation is always longer than that of the exciting radiation.

"Stoko" [Evonik]. TM for a line of skin care and cleansing products.

Stolle synthesis. Formation of indole derivatives by the reaction of arylamines with α-haloacid chlorides or oxalyl chloride, followed by cyclization of the resulting amides with aluminum chloride.

stop bath. An acid solution for stopping photographic development by chemical neutralization of developer.

stop codons. (termination codons; nonsense codons).
UAA, UAG, and UGA. In protein synthesis these codons signal the termination of a polypeptide chain.

storage. Any method of keeping raw materials, chemicals, food products, and energy while awaiting use, transportation, or consumption. The term *storage* is often applied to various types of wastes, but the more accurate word is disposal.
See radioactive waste; chemical waste; waste control.
(1) *Raw materials.* Normally storage is in suitably protected and well-ventilated interior areas at ambient temperature. Outdoor storage is practicable in some cases, e.g., logs for pulpwood, certain bulk solids and liquids received in metal or fiber drums. Storage of flammable liquids in large underground tanks is standard practice (gasoline, fuel oil). Hygroscopic materials (paper, textiles) should be in a humidity-controlled environment. Combustible materials that tend to build up internal heat on long standing at high ambient temperature (cellulosics such as paper, hay, grain; bulk wool; and certain vegetable oils) should be stored in well-ventilated areas.

(2) *Chemicals.* Materials that may react to form hazardous products in case of spillage should be kept well separated. Oxidizing agents (nitrates, peroxides, etc.) should not be stored near reducing or combustible materials. Heat-sensitive materials should be kept away from hot pipes or other heat sources, especially in the case of flammable liquids. Chemicals that will ignite spontaneously in air or react with water vapor require special storage conditions to keep them out of contact with air.
See pyrophoric.
Reactive organic monomers that tend to polymerize at room temperature, e.g., styrene, must contain an inhibitor when stored or shipped.
(3) *Food Products.* Long shelf life at or near room temperature is highly desirable for processed foods. This is achieved partly by the use of antioxidants and other preservatives and partly by processing techniques. Much experimental work has been done in this field. Refrigerated storage at temperatures near 4.5C is used for meats, eggs, and other dairy products. Meats and quick-frozen foods can be stored indefinitely at or below −18C. Controlled-atmosphere storage to retard postharvest ripening is used for unprocessed fruits and vegetables.
See aging (2); atmosphere, controlled.
(4) *Energy.* The conventional method of storing energy is by means of primary and secondary batteries.
See dry cell; storage battery.
The growing need for energy conservation has stimulated research on new and more effective methods, especially in regard to solar and wind energy, the collection of which is intermittent or nonuniform. Two such methods are in limited use. One (for electric power plants) involves compressing air with off-peak electricity and storing it in subterranean cavities from which it can be withdrawn when needed. The other (for domestic use) involves electrically heating refractory bricks at night at off-peak rates; the stored heat is given up during the day with 90% energy recovery. A number of other techniques are in the experimental stage: use of Glauber's salt, which has seven times the heat capacity of water, for storing solar energy; specialized batteries; so-called solar ponds; groundwater heated by solar or industrial process heat and returned to underground storage; and mechanical devices such as flywheel technology.
(5) For information on storage, see chemical data storage.

storage battery. A secondary battery, so called because the conversion of chemical to electrical energy is reversible and the battery is thus rechargeable. An automobile battery usually consists of 12–17 cells with plates (electrodes) made of sponge lead (negative plate or anode) and lead dioxide (positive plate or cathode) that is in the form of a paste. The electrolyte is sulfuric acid. The chemical reaction that yields electric current is $Pb + PbO_2 + 2H_2SO_4 \leftrightarrow 2PbSO_4 + 2H_2O + 2e$. More complicated and expensive types have nickel–iron, nickel–cadmium, silver–zinc, and silver–cadmium as electrode materials. A sodium–liquid sulfur battery for high-temperature operation as well as a chlorine–zinc type using titanium electrodes have also been developed.

As part of the U.S. effort to replace gasoline with another form of energy, DOE is supporting short- and long-term research on batteries for electric vehicles at Argonne National Laboratories. Three types intended to deliver 20–30 kwh are in the short-term program: improved lead–acid, nickel–zinc and nickel–iron. The long-term program includes lithium–metal sulfide, sodium–sulfur (β-battery), zinc–chlorine, and metal–air. Independent research indicates that a zinc–nickel oxide system has encouraging possibilities. A lead–acid battery for storing energy from solar cells has been reported to have a life of 5–7 years.

storax.   USP name for styrax.

"Storite" [Indal].   TM for a post-harvest pesticide.

Stork enamine reaction.   Synthesis of α-alkyl or α-acyl carbonyl compounds from enamines and alkyl or acyl halides.

STP.   Abbreviation for standard temperature and pressure, i.e., 0C and 1 atm pressure.

STP (hallucinogen).
CAS: 15588-95-1. $C_{13}H_{21}NO_2$.
Hazard: A poison by ingestion.

STPP.   Abbreviation for sodium tripolyphosphate.

straight chain.   See aliphatic.

straight-run.   See gasoline.

strain.   The resistance to deformation, that is, the tendency to resume its original shape, of a material subjected to a static or dynamic load. Strain increases as a function of stress; at rupture it represents the maximum strength of the material, usually measured in pounds per square inch or kilograms per square centimeter. The relationship of stress to strain is indicated by a curve obtained by stretching a sample of standard dimensions in a device designed for that purpose.
See stress; modulus of elasticity.

Straits tin.   99.895% pure tin.

strangeness.   See antiparticle.

strategy.   A computer-aided approach to a problem in organic synthesis, e.g., in biomimetic chemistry or recombinant DNA. Strategies may be group oriented, nonoriented, or long range.

straw.   A fibrous, cellulosic component of cereal plants (wheat, rice, etc.). Its fibers are 1–1.5 mm long, similar to those of hardwoods. Straw can be pulped by the alkaline process to yield specialty papers of high quality. Use of straw for papermaking is of limited importance in the U.S. due to the abundance of pulpwood.

streaming electromotive force.   See streaming potential.

streaming potential.   (streaming electromotive force). (1) The current set up between extremes of a diaphragm when a liquid is forced through a porous diaphragm. (2) The electromotive force produced by the flow of a liquid past a solid surface.

Strecker amino acid synthesis.   Synthesis of α-amino acids by simultaneous reaction of aldehydes with ammonia and hydrogen cyanide followed by hydrolysis of the resulting amino nitriles.

Strecker degradation.   Interaction of an α-amino acid with a carbonyl compound in aqueous solution or suspension to give carbon dioxide and an aldehyde or ketone containing one less carbon atom. Inorganic oxidizing agents can also be used to bring about the reaction.

Strecker sulfite alkylation.   Formation of alkyl sulfonates by reaction of alkyl halides with alkali or ammonium sulfites in aqueous solution in the presence of iodide.

streptolin.
Properties: An antibiotic isolated as the hydrochloride. Gummy mass, soluble in water, most stable at pH 3.0–3.5.
Derivation: Produced by *Streptomyces* # 11.
Use: Antibiotic, possible rodenticide.

streptomycin.
CAS: 57-92-1. $C_{21}H_{39}N_7O_{12}$. A specific antibiotic, but the term is also used loosely to designate several chemically related antibiotics produced by actinomycetes. Streptomycin is produced by *Streptomyces griseus* and consists of streptidine attached in glycosidic linkage to the disaccharide, streptobiosamine. It is active against Gram-negative bacteria and the tubercle bacillus. Usually available as trihydrochloride, phosphate, or sesquisulfate.
Properties: A base that readily forms salts with anions. Quite stable but very hygroscopic. One unit equals one microgram of pure crystalline streptomycin base.

Derivation: From *Streptomyces griseus* by aerobic fermentation. The streptomycin is then concentrated by adsorption on activated carbon and purified.

Hazard: Damage to nerves and kidneys may result from ingestion. Use restricted by FDA.

Use: Medicine (antibacterial).

See dihydrostreptomycin.

streptonigrin. (USAN).
CAS: 3930-19-6. $C_{25}H_{22}N_4O_8$. An antibiotic derived from *Streptomyces flocculus*. Dark-brown, rectangular crystals.

stress. The deformation undergone by a material when subjected to a definite load (the force applied per unit area). The load may be static (constant) or dynamic (increasing at a uniform rate). In either case it induces a strain in the material that results in rupture if the deforming force exceeds its strength. See strain; modulus of elasticity.

stress cracking. (tension cracking). Development of transverse cracks in a rubber or plastic product exposed to atmospheric oxygen at low (5–10%) elongation for long periods of time, for example, coiled hose, packaging materials, etc., both in service and during storage. Cracking will occur in the absence of light. It can be minimized in the case of a plastic such as polyethylene by lowering the density and the melt index, and in rubber by use of antioxidants.

stringency. A term used to describe the conditions of hybridization. By varying the conditions (especially salt concentration and temperature), a given probe sequence may be allowed to hybridize only with its exact complement (high stringency), or with any somewhat related sequences (relaxed or low stringency). Increasing the temperature or decreasing the salt concentration will tend to increase the selectivity of a hybridization reaction, and thus will raise the stringency.

stripping. (1) Removal of relatively volatile components from a gasoline or other liquid mixture by distillation, evaporation, or by passage of steam, air, or other gas through the liquid mixture. (2) Rapid removal of color from an improperly dyed fabric or fiber by a chemical reaction. Compounds used for this purpose in vat dyeing or in discharge printing are termed *discharging agents*. Substances commonly used as strippers are sodium hydrosulfite, titanous sulfate, sodium and zinc formaldehyde sulfoxylates.

strobane. (terpene polychlorinates; dichloricide aerosol; dichloride mothproofer; insecticide 3960-x14; (1S,4R)-1,2,2,3,3,4,7-heptachloro-5,5-dimethyl-6-methylidenebicyclo[2.2.1]heptane).
CAS: 8001-35-2. $C_{10}H_9C_7$.

Properties: An amber liquid mixture of polychlorinated terpenes.

Hazard: Carcinogen.

Use: Insecticide.

stroma. The space and aqueous solution enclosed within the inner membrane of a chloroplast.

strontia. See strontium oxide.

strontianite. $SrCO_3$. Natural strontium carbonate.

Properties: White, gray, yellow, green color; vitreous luster. Mohs hardness 3.5–4, d 3.7.

Occurrence: California, New York, Washington, Germany, Mexico.

Use: Source of strontium chemicals.

strontium. Sr. Metallic element of atomic number 38, group IIA of periodic table, aw 87.62, valence = 2, radioactive isotopes strontium-89 and strontium-90. There are four stable isotopes.

Properties: Pale-yellow, soft metal; chemically similar to calcium. D 2.54, mp 752, bp 1390C. Soluble in alcohol and acids, decomposes water on contact.

Occurrence: Ores of strontianite and celestite (Mexico, Spain).

Derivation: (1) Electrolysis of molten strontium chloride in a graphite crucible with cooling of the upper, cathodic space; (2) thermal reduction of the oxide with metallic aluminum (strontium aluminum alloy formed), and distilling the strontium in a vacuum.

Grade: Technical.

Hazard: Spontaneously flammable in powder form, ignites when heated above its mp. Reacts with water to evolve hydrogen. Store under naphtha.

Use: Alloys, "getter" in electron tubes.

strontium-90. Radioactive strontium isotope.

Properties: Half-life is 38 years. Radiation: β.

Derivation: From the fission products of nuclear reactor fuels.

Available forms: A mixture containing strontium-90, yttrium-90, and strontium-89 chlorides in hydrochloric acid solution; also as the carbonate and titanate.

Hazard: Highly toxic radioactive poison; present in fallout from nuclear explosions. Absorbed by growing plants; when ingested attacks bone marrow with possibly fatal results. It may be partially removed from milk by treatment with vermiculite.

Use: Radiation source in industrial thickness gauges, elimination of static charge, treatment of eye diseases, in radio-autography to determine the uniformity of material distribution, in electronics for studying strontium oxide in vacuum tubes, activation of phosphors, source of ionizing radiation in luminous paint, cigarette density control, measuring silk density, atomic batteries, etc.

strontium acetate.
CAS: 543-94-2. $Sr(C_2H_3O_2)_2 \cdot 1/2H_2O$.
Properties: White, crystalline powder. Soluble in water; loses $1/2H_2O$ at 150C.
Derivation: Interaction of strontium hydroxide and acetic acid, followed by crystallization.
Use: Intermediate for strontium compounds, catalyst production.

strontium bromate. $Sr(BrO_3)_2 \cdot H_2O$.
Properties: Colorless or yellowish crystals, lustrous powder, hygroscopic. Soluble in water. D 3.773. Loses water at 120C, decomposes at 240C.
Hazard: Strong oxidant, fire risk in contact with organic materials.

strontium bromide.
CAS: 10476-81-0. $SrBr_2 \cdot 6H_2O$.
Properties: White, hygroscopic crystals or powder. D 2.386 (25/4C), loses $4H_2O$ at 89C, losing remaining water by 180C; mp anhydrous salt 643C. Soluble in water, alcohol, and amyl alcohol; insoluble in ether.
Derivation: Strontium carbonate is treated with bromine or hydrobromic acid.
Grade: Anhydrous powder, crystals, technical, CP.
Hazard: Toxic by ingestion and inhalation.
Use: Medicine (sedative), lab reagent.

strontium carbonate. $SrCO_3$.
Properties: White, impalpable powder. D 3.62, loses carbon dioxide at 1340C. Soluble in acids, carbonated water, and solutions of ammonium salts; slightly soluble in water.
Derivation: Celestite is boiled with a solution of ammonium carbonate or is fused with sodium carbonate.
Grade: Precipitated, technical, natural, reagent.
Use: Catalyst, in radiation-resistant glass for color television tubes, ceramic ferrites, pyrotechnics.

strontium chlorate.
CAS: 7791-10-8.   (1) $Sr(ClO_3)_2$.   (2) $Sr(ClO_3)_2 \cdot 8H_2O$.
Properties: White, crystalline powder. (1) D 3.152, mp decomposes at 120C. Soluble in water; slightly soluble in alcohol.
Derivation: Strontium hydroxide solution is warmed and chlorine passed in, with subsequent crystallization.
Grade: Technical, reagent.
Hazard: Dangerous explosion risk in contact with organic materials; highly sensitive to shock, heat, and friction; strong oxidizing agent.
Use: Manufacture of red-fire and other pyrotechnics, tracer bullets.

strontium chloride.
CAS: 10476-85-4.   (1) $SrCl_2$. (2) $SrCl_2 \cdot 6H_2O$.
Properties: White, crystalline needles; odorless; sharp, bitter taste. (1) D 3.054, mp 872C, bp 1250C;

(2) d 1.964, mp loses $6H_2O$ at 150C. Soluble in water and alcohol.
Derivation: Strontium carbonate is fused with calcium chloride, the melt extracted with water, the solution concentrated and crystallized.
Grade: Reagent, technical, anhydrous.
Use: Strontium salts, pyrotechnics, electron tubes.

strontium chromate.   (chromic acid, strontium salt (1::1)).
CAS: 7789-06-2. $SrCrO_4$.
Properties: Light-yellow pigment. D 3.84. Rust-inhibiting and corrosion-resistant properties. Has good heat and light resistance and low reactivity in highly acid vehicles.
Hazard: Toxic by ingestion. Confirmed carcinogen.
Use: Metal protective coatings to prevent corrosion, colorant in polyvinyl chloride resins; pyrotechnics, sulfate ion control in electroplating baths.

strontium dioxide.   See strontium peroxide.

strontium fluoride.
CAS: 7783-48-4. SrF_2.
Properties: White powder. D 4.2, mp 1463C, bp 2489C. Soluble in hydrochloric acid and hydrogen fluoride; insoluble in water.
Grade: Crystals, 99.2% pure.
Hazard: Toxic material.
Use: Substitute for other fluorides, electronic and optical applications, single crystals for lasers, high-temperature dry-film lubricants.

strontium hydroxide.   (strontium hydrate). (1) $Sr(OH)_2$. (2) $Sr(OH)_2 \cdot 8H_2O$.
Properties: Colorless, deliquescent crystals. (1) D 3.625, mp 375C, decomposes 710C; (2) d 1.90. Soluble in acids and hot water; slightly soluble in cold water; absorbs carbon dioxide from air.
Derivation: Heating the carbonate or sulfide in steam.
Grade: Technical, CP, reagent.
Use: Extraction of sugar from beet sugar molasses, lubricant soaps and greases, stabilizer for plastics, glass, adhesives.

strontium hyposulfite.   See strontium thiosulfate.

strontium iodide.
CAS: 10476-86-5.   (1) $SrI_2 \cdot 6H_2O$. (2) SrI_2.
Properties: (1) White, crystalline plates. Decomposes in moist air; becomes yellow on exposure to air or light. (2) White crystals. Soluble in water and alcohol. (2) D 4.549, mp 515C, bp decomposes. (1) D 2.67 (25C), mp 90C (decomposes).
Derivation: By treating strontium carbonate with hydriodic acid.
Hazard: Toxic by ingestion and inhalation.
Use: Medicine (source of iodine), intermediate.

strontium lactate. $Sr(C_3H_5O_3)_2 \cdot 3H_2O$.
Properties: White crystals or granular powder; odorless. Soluble in water, alcohol, ether.

strontium molybdate. $SrMoO_4$.
Properties: Crystalline powder. D 4, mp 1600C. Insoluble in water.
Grade: 99.39% pure, single crystals.
Use: Anticorrosion pigments, electronic and optical applications, single crystals, solid-state lasers.

strontium monosulfide. See strontium sulfide.

strontium nitrate.
CAS: 10042-76-9. $Sr(NO_3)_2$.
Properties: White powder. D 2.98, mp 570C, bp 645C. Soluble in water; slightly soluble in absolute alcohol.
Derivation: A concentrated solution of strontium chloride is treated with a solution of sodium nitrate.
Grade: Technical, reagent.
Hazard: Strong oxidizing agent, fire risk in contact with organic materials, may explode when shocked or heated.
Use: Pyrotechnics, marine signals, railroad flares, matches.

strontium nitrite. $Sr(NO_2)_2$.
Properties: White or yellowish powder or hygroscopic needles. D 2.8, decomposes at 240C. Soluble in water; insoluble in alcohol.

strontium oxalate. $SrC_2O_4 \cdot H_2O$.
Properties: White powder; odorless. Loses H_2O at 150C. Insoluble in cold water.
Hazard: Toxic by ingestion.
Use: Pyrotechnics, catalyst manufacture, tanning.

strontium oxide. (strontia). SrO.
Properties: Grayish-white, porous lumps. D 4.7, mp 2430C, bp 3000C. Soluble in fused potassium hydroxide, converted to hydroxide by water. Combustible.
Derivation: Decomposition of strontium carbonate or hydroxide.
Grade: Lump, powder, porous carbide-free, reagent.
Use: Manufacture of strontium salts, pyrotechnics, pigments, greases and soaps, a major desiccant.

strontium perchlorate. $Sr(ClO_4)_2$.
Properties: Colorless crystals. Very soluble in water and alcohol.
Hazard: Fire and explosion risk in contact with organic materials.
Use: Oxidizing agent, pyrotechnics.

strontium peroxide. (strontium dioxide).
CAS: 1314-18-7. (1) SrO_2. (2) $SrO_2 \cdot 8H_2O$.
Properties: White powder; odorless; tasteless. (1) D 4.56, mp decomposes at 215C; (2) d 1.951, loses $8H_2O$ at 100C and decomposes when heated to a higher temperature. Soluble in ammonium chloride solution and alcohol; soluble in ammonium chloride solution and alcohol; decomposes in hot water; slightly soluble in cold water.
Derivation: (1) By passing oxygen over heated strontium oxide; (2) reaction of strontium hydroxide and hydrogen peroxide.
Hazard: Dangerous fire and explosion risk in contact with organic materials; sensitive to heat, shock, catalysts, reducing agents; oxidizing agent.
Use: Bleaching, fireworks, antiseptic.

strontium–potassium chlorate. (potassium–strontium chlorate). $Sr(ClO_3)_2 \cdot 2KClO_3$.
Properties: White, crystalline powder. Soluble in water.
Grade: Technical.
Hazard: Fire and explosion risk in contact with organic materials; sensitive to heat, shock, catalysts, reducing agents; oxidizing agent.
Use: Pyrotechnics.

strontium salicylate. $Sr(C_7H_5O_3)_2 \cdot 2H_2O$.
Properties: White crystals or powder; odorless; sweet saline taste. Soluble in water and alcohol; decomposes when heated. Protect from light.
Derivation: Interaction of strontium carbonate and salicylic acid.
Use: Pharmaceutical and fine chemical manufacture.

strontium stearate. $Sr(OOCC_{17}H_{35})_2$.
Properties: White powder. Mp 130–140C. Insoluble in alcohol; soluble (forms gel) in aliphatic and aromatic hydrocarbons.
Use: Grease and wax compounding.

strontium sulfate. $SrSO_4$.
Properties: White precipitate or crystals of the mineral celestite; odorless. D 3.71–3.97, mp 1605C. Slightly soluble in concentrated acids; very slightly soluble in water; insoluble in alcohol and dilute sulfuric acid.
Derivation: (1) Celestite is ground; (2) precipitation of any soluble strontium salt with sodium sulfate.
Grade: Natural, precipitated, air-floated, 90%, 325 mesh, free from sodium salts, CP.
Use: Pyrotechnics, ceramics and glass, paper manufacture.

strontium sulfide. (strontium monosulfide).
CAS: 1314-96-1. SrS.
Properties: Gray powder, has hydrogen sulfide odor when in presence of moist air. D 3.7, mp 2000C. Soluble in acids (decomposes); slightly soluble in water.
Derivation: Reduction of celestite with coke at high temperatures.
Grade: Technical, high purity.
Hazard: Moderate fire and explosion risk. Irritant to skin and tissues.
Use: Depilatory, luminous paints, manufacture of strontium chemicals.

strontium tartrate. $SrC_4H_4O_6 \cdot 4H_2O$.
Properties: White crystals. D 1.966. Slightly soluble in water. Combustible.
Use: Pyrotechnics.

strontium thiosulfate.   (stontium hyposulfite). $SrS_2O_3 \cdot 5H_2O$.
Properties: Fine needles. D 2.17, mp loses $4H_2O$ at 100C. Soluble in water; insoluble in alcohol.

strontium titanate. $SrTiO_3$.
Properties: Powder. D 4.81, mp 2060C. Insoluble in water and most solvents.
Grade: Technical, single crystals.
Use: Electronics, electrical insulation.

strontium zirconate. $SrZrO_3$.
Properties: Powder. Mp 2600C.
Use: Electronics.

structural antagonist.   See antagonist, structural.

structural formula.   See formula, chemical.

structural genomics.   The effort to determine the 3D structures of large numbers of proteins using both experimental techniques and computer simulation.

structure.   (1) In the terminology of the textile industry, any woven fabrics. (2) In chemistry, the arrangement of atoms and groups in a molecule.

strychnidine. $C_{21}H_{24}ON_2$.
Properties: Colorless crystals. Mp 256C.
Use: Reagent for nitrate determination.

strychnine.
CAS: 57-24-9. $C_{21}H_{22}N_2O_2$. An alkaloid.

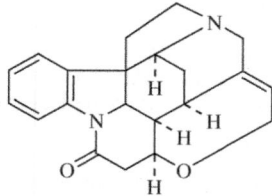

Properties: Hard white crystals or powder; bitter taste. Mp 268–290C, bp 270C (5 mm Hg). Soluble in chloroform; slightly soluble in alcohol and benzene; slightly soluble in water and ether.
Derivation: Extraction of the seeds of *Nux vomica* with acetic acid, filtration, precipitation by alkali and filtration.
Grade: Crystals, powder, technical.
Hazard: Toxic by ingestion and inhalation. Central nervous system impairment.
Use: Pesticide.

STS.   See sequence tagged site.

stuff.   Term used by papermakers to refer to the aqueous pulp suspension fed onto the fourdrinier wire from the headbox.
See furnish.

Stuffer disulfone hydrolysis rule.   (1) All sulfones containing vicinal sulfone group-substituted carbon atoms (γ-disulfones) are hydrolyzed by dilute alkali with formation of β-hydroxysulfone and sulfinate. (2) All sulfones in which adjacent carbon atoms carry a strongly negative group, and one or two sulfone residues can be cleaved into a sulfinic acid and an unsaturated compound.

STXS.
Available forms: Liquid.
Sodium, toluene, xylene, sulfonate.
Use: A hydrotrope or solubilizing agent in detergent formulas.

"Stymer" Vinyl Styrene [Cytec].   TM for resins used as sizes for filament acetates. L.F. A vinyl resin. Styrene copolymer resin.
Properties: Soluble in water, ammonium hydroxide, and sulfur.

S-type synthetic elastomer.   See styrene–butadiene rubber.

styphnic acid.   (2,4,6-trinitroresorcinol).
CAS: 82-71-3. $C_6H(OH)_2(NO_2)_3$.
Properties: Yellow crystals; astringent taste. An initiating explosive. Mp 179–180C. Forms addition compounds with many hydrocarbons; soluble in alcohol and ether; slightly soluble in water.
Derivation: Nitration of resorcinol.
Hazard: Severe explosion risk when heated.
Use: Priming agents in the explosives industry.

styracin.   See cinnamyl cinnamate.

styralyl acetate.   See α-methylbenzyl acetate.

styralyl alcohol.   See α-methylbenzyl alcohol.

styrax.   A type of balsam found in Central America and the Near East.
See balsam.

styrenated oil.   A drying oil whose drying and hardening characteristics have been modified by incorporation of styrene or a similar monomer.

styrene.   See polystyrene; styrene monomer.

styrene–acrylonitrile.   See polystyrene.

styrene–butadiene rubber.   (SBR).

CAS: 9003-55-8. By far the most widely used type of synthetic rubber, its consumption for all applications is four times that of polybutadiene, its nearest competitor, and 1.5 times that of all other elastomers combined. Its manufacture involves copolymerization of three parts butadiene with one part styrene. These materials are suspended in finely divided emulsion form in a large proportion of water, in the presence of a soap or detergent. Also present in small amounts are an initiator or catalyst which is usually a peroxide, and a chain-modifying agent such as dodecyl mercaptan.

Hazard: Questionable carcinogen.

Use: Tires, footwear, mechanical goods, coatings, adhesives, solvent-release sealants, carpet backing. See rubber, synthetic; polymerization; free radical.

styrene glycol.

CAS: 93-56-1. $C_8H_{10}O_2$.

Properties: Acicular crystals. Mp 67C, bp 272C. Soluble in water and organic solvents.

Use: Plasticizers.

styrene monomer.   (vinylbenzene; phenylethylene; cinnamene).

CAS: 100-42-5. $C_6H_5CH:CH_2$.

Properties: Colorless, oily liquid; aromatic odor. Fp −30.63C, bp 145.2C, d 0.9045 (25/25C), bulk d 7.55 lb/gal (20C), flash p 88F (31.1C), autoign temp 914F (490C). Insoluble in water; soluble in alcohol and ether; readily undergoes polymerization when heated or exposed to light or a peroxide catalyst. The polymerization releases heat and may become explosive.

Derivation: From ethylene and benzene in the presence of aluminum chloride to yield ethylbenzene, which is catalytically dehydrogenated at 630C to form styrene.

Grade: Technical 99.2%, polymer 99.6%.

Hazard: Flammable, moderate fire risk, explosive limits in air 1.1–6.1%, must be inhibited during storage. Toxic by ingestion and inhalation. Central nervous system impairment, upper respiratory tract irritant, and peripheral neuropathy. Possible carcinogen.

Use: Polystyrene; SBR, ABS, and SAN resins; protective coatings (styrene–butadiene latex, alkyds); styrenated polyesters; rubber-modified polystyrene; copolymer resins; intermediate.

styrene nitrosite.   A compound resulting from the reaction between styrene and nitrogen dioxide and used as a qualitative or quantitative specific test for monomeric styrene in mixtures with other hydrocarbons.

styrene oxide.

CAS: 96-09-3.

$C_6H_5CHOCH_2$.

Properties: Colorless to pale-straw-colored liquid. Boiling range 194.2–195C (5–95%), fp −36.6C, flash p 180F (82.2C) (COC), refractive index 1.5328 (25C), d 1.0469 (25/4C). Miscible with benzene, acetone, ether, and methanol. Combustible.

Hazard: Toxic by ingestion and inhalation. Possible carcinogen.

Use: Highly reactive organic intermediate.

"Styresol" *[Reichhold].*   TM for a group of styrenated alkyd resins with air-drying and baking properties and high resistance to gasoline, alkalies, acids, and water.

Styrofoam.   Expanded   cellular   polystyrene (available in colors).

Use: Insulating materials; light-weight materials for boats, toys, etc.; separators in packing containers; airport runways; highway construction; battery cases.

"Styron" *[Styron].*   TM for polystyrene resins, general purpose, medium and high impact, heat and impact-heat resistant, and light-stabilized resins ("Styron Verelite" *[Styron]*). Available in wide range of translucent and opaque colors, as well as natural and crystal.

Use: Packaging, toys, appliance parts, bottle closures and containers, hot and cold drinking cups, television cabinet backs, automotive components and machine housings, lighting equipment.

styryl carbinol.   See cinnamic alcohol.

sub-cloning.   If you have a cloned piece of DNA (say, inserted into a plasmid) and you need unlimited copies of only a part of it, you might "subclone" it. This involves starting with several million copies of the original plasmid, cutting with restriction enzymes, and purifying the desired fragment out of the mixture. That fragment can then be inserted into a new plasmid for replication. It has now been subcloned.

suberane.   See cycloheptane.

suberic acid.   (octanedioic acid).

$HOOC(CH_2)_6COOH$.

Properties: Colorless crystals from water. Mp 143C, bp 279C (100 mm Hg). Partially soluble in water and ether; soluble in alcohol. Combustible.

Derivation: Oxidation of oleic acid with nitric acid.

Use: Intermediate for the synthesis of drugs, dyes, and high polymers.

suberone.   See cycloheptanone.

sublimation.   The direct passage of a substance from solid to vapor without appearing in the intermediate (liquid) state. An example is solid carbon dioxide which vaporizes at room temperature; the conversion may also be from vapor to solid under appropriate conditions of temperature.

subnuclear particle.   A particle either found in the nucleus or observed coming from the nucleus as the result of nuclear reaction or rearrangement, i.e., neutrons, mesons, etc.

substance.   Any chemical element or compound. All substances are characterized by a unique and identical constitution and are thus homogeneous. "A material of which every part is like every other part is said to be homogeneous and is called a substance." (Black and Conant, *Practical Chemistry*.) See homogeneous.

substantive dye.   See direct dye.

substituent.   An atom or radical that replaces another in a molecule as the result of a reaction. See substitution.

substituted benzoic acid herbicide.   A family of herbicides whose auxin and herbicidal activity greatly resembles that of the phenoxy auxins. Some of these compounds interfere with geotropic and photorpoic curvatures and also inhibit cell elongation.

substituted urea herbicide.   Any of a number of rather selective pre-emergence herbicides that are absorbed by foliage and roots that penetrate deeply into soil and whose persistence varies considerably.

substitute natural gas.   See synthetic natural gas.

substitution.   The replacement of one element or radical by another as a result of a chemical reaction. Chlorination of benzene to produce chlorobenzene is a typical example; in this case a chlorine atom replaces a hydrogen atom in the benzene molecule.

substitution mutation.   In genetics, a type of mutation due to replacement of one nucleotide in a DNA sequence by another nucleotide; or replacement of one amino acid in a protein by another amino acid.

substrate.   (1) A substance upon which an enzyme or ferment acts. (2) Any solid surface on which a coating or layer of a different material is deposited.

substrate-level phosphorylation.   Phosphorylation of ADP or some other nucleoside 5'-diphosphate coupled to the dehydrogenation of an organic substrate; independent of the electron transport chain.

subtilin.   An antibiotic produced by the metabolic processes of a strain of *Bacillus subtilis*. It is a cyclic polypeptide similar to bacitracin in chemical structure and antibiotic activity, but not as important clinically. Subtilin is active against many Gram-positive bacteria, some Gram-negative cocci, and some species of fungi. It is a surface tension depressant and its antibiotic action is increased by use of wetting agents.
Properties: Soluble in water in pH range 2.0–6.0. Soluble in methanol and ethanol (up to 80%); insoluble in dry ethanol or other common organic solvents. Relatively stable in acid solutions. Inactivated by pepsin and trypsin and destroyed by light.
Use: Seed disinfectant, bacteriostat in foods.

subtilisins.   ("Alk" [*Alko-Abello*]; bacillus subtilis BPN; Maxatase).
CAS: 1395-21-7 and 9014-01-1.
Properties: Light-colored   prills.   Proteolytic enzyme.
Hazard: Skin, upper and lower respiratory tract irritant.
Use: Laundry detergent additive.

succinaldehyde.   (butanedial).
$OHCCH_2CH_2CHO$.
Properties: Liquid. D 1.064 (20/4C), bp 169–170C, refr index 1.4254. Soluble in water, alcohol, and ether. The name succinaldehyde is often incorrectly used in commerce as a synonym for succinic anhydride.

succinic acid.   (butanedioic acid).
CAS: 110-15-6. $CO_2H(CH_2)_2CO_2H$.
Properties: Colorless crystals; odorless; acid taste. D 1.552, mp 185C, bp 235C. Slightly soluble in water; soluble in alcohol and ether. Combustible.
Derivation: Fermentation of ammonium tartrate.
Grade: Technical, CP, FCC.
Use: Organic synthesis, manufacture of lacquers, dyes, esters for perfumes, photography, in foods as a sequestrant, buffer, neutralizing agent.

succinic acid-2,2-dimethylhydrazide.
(diaminozide). $(CH_3)_2NNHCOCH_2CH_2COOH$.
Properties: White crystals. Mp 155C, pH 3.8 (500 ppm). Soluble in water; insoluble in simple hydrocarbons.
Use: Growth retardant used in greenhouses, retards premature fruit drop.

succinic acid peroxide.
CAS: 123-23-9. $(HOOCCH_2CH_2CO)_2O_2$.
Properties: Fine, white powder; odorless. Mp 125C (decomposes). Moderately soluble in water; insoluble in petroleum solvents and benzene.

Hazard: Fire risk in contact with combustible materials, oxidizing agent. Toxic by ingestion and inhalation, irritant to skin.
Use: Polymerization catalyst, deodorants, antiseptics.

succinic anhydride. (2,5-diketotetrahydrofurane; succinyl oxide; butanedoic anhydride).
CAS: 108-30-5.

$H_2CC(O)OC(O)CH_2$.

Properties: Colorless or light-colored needles or flakes. D 1.104 (20/4C), mp 120C, bp 261C, sublimes at 115C (5 mm Hg). Soluble in alcohol and chloroform; insoluble in water. Combustible.
Grade: Distilled.
Hazard: Questionable carcinogen.
Use: Manufacture of chemicals, pharmaceuticals, esters; hardener for resins, starch modifier in foods.

succinimide. (2,5-diketopyrrolidine).
CAS: 123-56-8.

$H_2CC(O)NHC(O)CH_2$ or $C_4H_5O_2N \bullet H_2O$.

Properties: Colorless crystals or thin, light-tan flakes. Nearly odorless; sweet taste. Mp 125–127C, bp 287–288C, d 1.41. Soluble in hot and cold water and in sodium hydroxide solution; slightly soluble in alcohol; insoluble in ether and chloroform.
Derivation: Conversion of succinic acid to succinamide, followed by heating; ammonia splits off to give a diacyl-substituted derivative (succinimide).
Grade: Purified, technical.
Use: Growth stimulants for plants, organic synthesis.

succiniodimide. See *N*-iodosuccinimide.

succinonitrile. See ethylene cyanide.

succinylated monoglycerides.
Properties: Off-white colored waxy solid; bland taste. Mp 60C. Soluble in warm methanol, ether, *n*-propanol.
Use: Food additive.

succinyl oxide. See succinic anhydride.

succistearin.
Use: Food additive.

sucrase. See invertase.

sucroblanc. A mixture used to purify and bleach sugar solution in one operation. Contains high-test calcium hypochlorite, calcium superphosphate, lime, and Filtercel (a grade of Celite).

sucrose. (table sugar; saccharose).
CAS: 57-50-1. $C_{12}H_{22}O_{11}$.

Properties: Hard, white, dry crystals, lumps, or powder; sweet taste; odorless. D 1.5877, decomposes 160–186C, optical rotation +33.6 degrees. Soluble in water; slightly soluble in alcohol; solutions are neutral to litmus. Combustible.
Derivation: By crushing and extraction of sugarcane (Saccharum officinarum) with water or extraction of the sugar beet (*Beta vulgaris*) with water, evaporating, and purifying with lime, carbon, and various liquids. Also obtainable from sorghum by conventional methods. Occurs in low percentages in honey and maple sap.
Grade: Reagent, USP, technical, refined.
Hazard: Dental erosion. Questionable carcinogen.
Use: Sweetener in foods and soft drinks, manufacture of syrups, source of invert sugar, confectionery, preserves and jams, demulcent, pharmaceutical products, caramel, chemical intermediate for detergents, emulsifying agents, and other sucrose derivatives.
See sugar.

sucrose acetate isobutyrate. (SAIB).
$(CH_3COO)_2C_{12}H_{14}O_3[OOCCH(CH_3)_2]_6$.
Properties: Clear sucrose derivative available either as a semisolid (100%) or as a 90% solution in ethanol. D 1.146 (25/25C). Combustible.
Derivation: By controlled esterification of sucrose with acetic and isobutyric anhydrides.
Use: Modifier for lacquers, hot-melt coating formulations, extrudable plastics.

sucrose fatty acid esters.
Use: Food additive.

sucrose monostearate.
Properties: White powder; odorless, tasteless.
Derivation: By reaction of sugar and methyl stearate in a suitable solvent and with potassium carbonate catalyst.
Use: Low-foam nonionic detergent, surfactants.

sucrose octaacetate.
CAS: 126-14-7. $C_{12}H_{14}O_3(OOCCH_3)_8$.
Properties: White, hygroscopic, crystalline material; bitter taste. When once melted does not recrystallize on cooling but becomes a transparent film. Molten film is very adhesive but this property is lost on cooling. Rate of hydrolysis, practically nil. Gives no action with Fehling's solution. D (fused) 1.28 (20/20C), bp 260C (0.1 mm Hg), mp 79–84C, refr index 1.4660 (20C), decomposes at 285C, viscosity

29.5 cP (100C), specific rotation +54.96 degrees (CCl$_4$). Soluble in acetic acid, acetone, benzene; slightly soluble in water. Combustible.

Grade: Technical, reagent, denaturing.

Use: Plasticizer for cellulose esters and plastics, adhesive and coating compositions, insecticide, termite repellent, denaturant in rubbing alcohol formulas, lacquers, flavoring.

sucrose polyester. An experimental, synthetic, noncaloric fat substitute composed of a mixture of 6-, 7-, and 8-fatty acid esters of sucrose. These esters are not decomposed by the digestive enzymes as are the triesters in natural fats; thus they are not metabolized and can be recovered from the excreta. The physical properties of sucrose polyester are similar to those of fats occurring in milk, margarine, and vegetable oils. It has been found effective in notably reducing body weight in preliminary trials when substituted for natural dietary fats in salad dressings, spreads, and milk mixture. Sucrose polyester also has a beneficial effect on serum cholesterol levels. It has met all FDA test requirements satisfactorily.

sudan black.
Use: Dye that stains fats, oils, sebaceous components and contaminants of friction ridge residue.

suffocating gas. Any of a number of chlorinated warfare gases.
Hazard: Acutely irritate the tissues of the trachea, bronchial tubes and lungs, causing pulmonary edema.

sugar. A carbohydrate product of photosynthesis composing one, two, or more saccharose groups. The monosaccharide sugars (often called simple sugars) are composed of chains of 2–7 carbon atoms. One of the carbons carries aldehydic or ketonic oxygen which may be combined in acetal or ketal forms. The remaining carbons usually have hydrogen atoms and hydroxyl groups. Chief among the monosaccharides are glucose (dextrose) and fructose (levulose). These are optical isomers of formula C$_6$H$_{12}$O$_5$, that is, their crystals have the property of rotating to either left or right under polarized light. Hence the alternate names dextrose and levulose.
See D-; optical rotation.
Among the disaccharides are sucrose (cane or beet sugar); lactose, found in milk; maltose, obtained by hydrolysis of starch; and cellobiose from partial hydrolysis of cellulose. High-polymer sugars occur as water-soluble gums such as arabic, tragacanth, etc. Hydrolysis of sucrose yields invert sugar composed of equal parts fructose and glucose. Sugar is an important source of metabolic energy in foods and its formation in plants is an essential factor in the life process.

sugar alcohol.
Derivation: Produced by the reaction of an aldehyde or ketone of a sugar.

sugar–amino acid hybrid. Any of a class of molecules that combines structural attributes of both carbohydrates and amino acids and are biologically active.

sugarcane wax. A hard wax varying from dark green to tan and brown, produced by solvent extraction of cane, mp 76–79C. Combustible.
Grade: Domestic refined, in slabs.
Use: Carbon paper, pigment disperser, castings, lubricant for plastics in food wrappers.

sugar, corn. See dextrose.

sugar, grape. See dextrose.

sugar, invert. See invert sugar, sugar.

sugar of lead. See lead acetate.

sugar of milk. See lactose.

sugar, reducing. Sugars that will reduce Fehling's solution or similar test liquids, with conversion of the blue soluble copper salt to a red, orange or yellow precipitate of cuprous oxide. Glucose and maltose are typical examples of reducing sugars, their molecules containing an aldehyde group that is the basis for this type of reaction.

sugar substitute. See sweetener, nonnutritive.

suicide inhibitor. A relatively inert molecule that is transformed by an enzyme at its active site into a reactive substance that irreversibly inactivates the enzyme.

suicide substrate. A substrate that complexes with an enzyme to yield a reactive metabolite that restricts further activity of the enzyme.

sulfabromomethazine sodium.
CAS: 116-45-0. C$_{12}$H$_{13}$BrN$_4$O$_2$S.
Properties: Crystals. Decomp 250–252C. Soluble in alkaline solutions.
Use: Drug (veterinary); food additive.

sulfachlorpyridazine.
CAS: 80-32-0. C$_{10}$H$_9$ClN$_4$O$_2$S.
Use: Drug (veterinary); food additive.

sulfa drug. Any of a group of 50 or more compounds characterized by the presence of both sulfur and nitrogen, with high specificity for certain bacteria. Because of their toxicity and often serious side effects, their use in treating disease is limited, since the less toxic antibiotics are generally available. Many are used in veterinary medicine.

Among the best known are sulfanilamide, sulfadiazine, sulfapyridine, sulfamerazine, sulfasoxazole, and gantricin.

sulfaethoxypyridazine.
CAS: 963-14-4. $C_{12}H_{14}N_4O_3S$.
Properties: Crystals. Mp 183–184C. Soluble in water at 37°.
Use: Drug (veterinary); food additive.

sulfamic acid.
CAS: 5329-14-6.

$$H-O-\overset{\overset{\displaystyle O}{\|}}{\underset{\underset{\displaystyle O}{\|}}{S}}-NH_2$$

Properties: White, crystalline solid; odorless. Nonvolatile; nonhygroscopic, d 2.1, mp 205C (decomposes). Soluble in water; slightly soluble in organic solvents. Aqueous solutions are highly ionized giving pH values lower than solutions of formic, phosphoric, and oxalic acids. All the common salts (including calcium, barium, and lead) are extremely soluble in water. Combustible.
Grade: Reagent, crystalline, granular.
Hazard: Toxic by ingestion.
Use: Metal and ceramic cleaning, nitrite removal in azo dye operations, gas-liberating compositions, organic synthesis, analytical acidimetric standard, amine sulfamates used as plasticizers and fire-retardants, stabilizing agent for chlorine and hypochlorite in swimming pools, bleaching paper pulp and textiles, catalyst for urea-formaldehyde resins, sulfonating agent, pH control, hard-water-scale removal, electroplating.

sulfanilamide.   See sulfa drug.

sulfanilic acid.   (p-aminobenzenesulfonic acid; p-anilinesulfonic acid).
CAS: 121-57-3. $H_2NC_6H_4SO_3H \cdot H_2O$.
Properties: Grayish-white, flat crystals. Mp chars at 280–300C. Soluble in fuming hydrochloric acid; slightly soluble in water; very slightly soluble in alcohol and ether. Combustible.
Derivation: By heating aniline with weak fuming sulfuric acid and pouring the reaction product into water.
Grade: Technical, pure, reagent.
Use: Dyestuffs, organic synthesis, medicine, reagent.

m-sulfanilic acid.   See metanilic acid.

sulfanitran.
CAS: 122-16-7. $C_{14}H_{13}N_3O_5S$.
Properties: Crystals from alcohol (aq). Mp 239–240C. Soluble in acetone, ethanol, methanol, water, ether.
Use: Drug (veterinary); food additive.

sulfate.   The salt of sulfuric acid as in sodium sulfate.

sulfate mineral.   A mineral in which the basic building block is a sulfur atom linked to four oxygen atoms. Most sulfate minerals are highly soluble in water and form by the evaporation of natural waters.

sulfate process.   The process of producing paper products from woodchips by using a cooking liquor that contains caustic soda, sodium sulfide, soda ash, and sodium sulfate.

sulfate pulp.   See pulp, paper.

sulfate-reducing bacteria.   Fermenting bacteria that obtain energy by converting sulfate compounds into sulfide compounds. These bacteria, which cannot tolerate oxygen, are common in the muds of swamps, ponds, and lagoons. Bacteria of this type seem to have flourished in Archean time, when there was little atmospheric oxygen.

sulfate turpentine.   A by-product turpentine from Kraft paper manufacture.

sulfation.   (1) The formation of insoluble white lead sulfate in the plates of a lead storage battery. (2) The attachment of SO_3H to a carbon or nitrogen atom.

sulfenamide.   A compound having the typical structure $RSNH_2$.

sulfide dye.   (sulfur dye). One of a group of water-insoluble dyes produced by heating various organic compounds with sulfur. The characteristic chromophore groupings are $\equiv$C–S–C$\equiv$ and $\equiv$C–S–S–C$\equiv$; CI numbers range from 53,000–54,999. Like vat dyes, sulfide dyes are reduced to a water-soluble, colorless form before application and are then oxidized to their original colored state. Application is usually to cotton from a sodium sulfide bath. Sulfur black (CI 53,185) is an important example.

1,1′-sulfinylbis(1,2-dichloroethane).
CAS: 23248-53-5. $C_4H_2Cl_4OS$.
Hazard: A poison by ingestion.
Use: Agricultural chemical.

sulfite.   The normal salt of sulfurous acid as in sodium sulfite.
Hazard: Questionable carcinogen.

sulfite acid liquor.   An aqueous solution of calcium bisulfite or calcium and magnesium bisulfite containing a large amount of free sulfur dioxide. It is prepared from sulfur dioxide and limestone, dolomite, or lime by countercurrent extraction.
Use: Manufacture of sulfite pulp for paper.

sulfite process.   The treatment of wood chips with bisulfite under heat and pressure to remove impurities from cellulose to be used in papermaking.

sulfite pulp.   See pulp, paper.

sulfite waste liquor.   A waste liquor produced in the sulfite paper process. Sold in variations from a dilute solution to a solid.
Use: Foam producer, emulsifier, adhesives, tanning agent, binder for briquets, cores, unpaved roads, source of torula yeast.
See lignin sulfonate; waste disposal.

m-sulfobenzoic acid.
$HO_3SC_6H_4COOH \cdot 2H_2O$.
Properties: Grayish-white solid. Stable but hygroscopic in air. Mp 98C, anhydrous form melts at 141C. Soluble in water, alcohol; insoluble in benzene.
Derivation: Direct sulfonation of benzoic acid with sulfur trioxide.
Use: Derivative for surface-active agents.

o-sulfobenzoic acid. $HO_3SC_6H_4COOH$.
Properties: White needles. Mp 68–69C (with $3H_2O$ of crystallization), mp 134C (dry). Soluble in water and alcohol; insoluble in ether.
Derivation: (1) From saccharin and concentrated hydrochloric acid, (2) by the oxidation of thiosalicylic acid with potassium permanganate in alkaline solution.
Use: Manufacture of sulfonaphthalein indicators, dyes.

o-sulfobenzoic anhydride.

$C_6H_4COOSO_2$.
Properties: Solid. Mp 129.5C, bp 184–186C (18 mm Hg). Soluble in hot water, ether, and benzene.
Use: Polymerization inhibitor.

sulfobetaines.   See Crosultaines.

sulfocarbanilide.   See thiocarbanilide.

sulfocarbathion K.
CAS: 114654-31-8.
Hazard: Moderately toxic by ingestion and inhalation.

sulfocarbolic acid.   See phenolsulfonic acid.

sulfodehydroabietic acid monosodium salt.
See ecabet sodium.

sulfoderm.   Silicic acid with 1% colloidal sulfur.

1-(4-sulfo-2,3-dichlorophenyl)-3-methylpyrazolone.

$(Cl_2C_6H_2SO_3H)NNC(CH_3)CH_2CO$.
Properties: White or yellowish powder or crystals. Very soluble in water; soluble in alkalies.
Derivation: By condensation of dichlorophenylhydrazine sulfonic acid with ethylacetoacetate.
Use: Intermediate for dyes.

Sulfogene.   A series of sulfur colors.
Use: Extensively on cotton work clothing and similar fabrics and to a limited extent on rayon and other materials.

sulfolane.   (1,1-dioxide   tetrahydrothiofuran; tetramethylene sulfone).
CAS: 126-33-0.

$CH_2CH_2CH_2CH_2SO_2$.
Properties: Liquid. D 1.2606 at 30/4C, mp 27.4–27.8C, bp 285C, flash p 330F. Miscible with water, acetone, toluene (at 30C). Slightly miscible with octanes, olefins, naphthenes.
Grade: Technical.
Hazard: Combustible. Toxic by ingestion.
Use: Curing agent for epoxy resins, medicine (antibacterial).

n-sulfomethyl-polymyxin   B   sodium   salt.
See sulfomyxin.

sulfomyxin.
Use: Drug (veterinary); food additive.

sulfonamide.   (sulfa drug; sulphoneamide; 4-aminobenzenesulfonamide).

CAS: 63-74-1. $C_6H_8N_2O_2S$. Any of a class of antimetabolices that contain the sulfonamide functional group attached to an aniline.
Hazard: Toxic, cause injury to the urinary tract.
Use: Bacteriostats, herbicide.

sulfonation.   The introduction of the sulfo group into an organic compound whereby it connects to a carbon or nitrogen atom.

sulfonic acid.   An organic compound containing in its molecule one or more sulfo radicals, $-SO_2OH$. The sulfur atom is directly united to a carbon atom.

4,4'-sulfonyldiphenol.   (SDP).
$(C_6H_5OH)_2SO_2$.
Properties: White, crystalline powder. Mp above 247C; 99.5% pure.

Derivation: Oxidation of 4,4'-thiodiphenol.
Use: Intermediate, product modification.

p-1-sulfophenyl-3-methyl-5-pyrazolone.

$(C_6H_4SO_3H)NNC(CH_3)CH_2CO$.

Properties: White or yellowish powder. Very slightly soluble in water; soluble in alkalies.
Derivation: By condensation of phenylhydrazine sulfonic acid with ethylacetoacetate.

4-sulfophthalic anhydride.
$HO_3SC_6H_3(CO)_2O$.
Properties: Reddish-brown syrup. Crystallizes partially on long standing; hygroscopic; fluorescent in solution under UV radiation. D 1.62 (25C). Very soluble in water, alcohol; insoluble in ether, benzene. Combustible.
Use: Esters of 4-sulfophthalic acid used in wetting, cleansing, emulsifying, softening, and equalizing agents with textiles. Derivatives have application as surface-active agents.

5-sulfosalicylic acid.
$C_6H_3OH(SO_2OH)COOH \cdot 2H_2O$.
Properties: Colorless crystals, colored pink by traces of iron. Mp 120C (anhydrous), decomposes at higher temperatures. Very soluble in water.
Derivation: Action of sulfuric acid and salicylic acid.
Use: Reagent for albumin, colorimetric reagent for ferric ion, intermediate for dyes, surfactants, catalysts, grease additives.

sulfotepp.   (ethyl thiopyrophosphate; tetraethyl dithiopyrophosphate).
CAS: 3689-24-5. $(C_2H_5O)_2P(S)OP(S)(OC_2H_5)_2$.
Properties: Yellow liquid. Bp 136–139C (2 mm Hg). Slightly soluble in water; soluble in most organic solvents.
Hazard: Toxic by ingestion, inhalation, and skin absorption; cholinesterase inhibitor; use may be restricted. Questionable carcinogen.
Use: Insecticide.

sulfoxide.   (generic name for 1,2-(methylenedioxy)-4-[2-(octylsulfinyl)propyl]benzene(*n*-octyl sulfoxide isosafrole)).
CAS: 120-62-7. $C_{18}H_{28}O_3S$.
Properties: Brown liquid. Insoluble in water; slightly soluble in oils, miscible with most organic solvents.
Use: Insecticide synergist.

sulfur.
CAS: 7704-34-9. S. Nonmetallic element, atomic number 16, group VIA of periodic table, aw 32.06, valences = 2, 4, 6; four stable isotopes.

Properties: Pure sulfur exists in two stable crystalline forms, α and β, and at least two amorphous (liquid) forms. (1) α-*Sulfur*: rhombic, octahedral, yellow crystals; stable at room temperature. D 2.06, transition to β form 94.5C, mp (rapid heating) 112.8C, refr index 1.957. (2) β-*Sulfur*: monoclinic, prismatic, pale-yellow crystals slowly changing to α form below 94.5C. D 1.96, mp 119C, bp 444.6C, flash p 405F (207C), autoign temp 450F (232C), refr index 2.038. Both forms are insoluble in water, slightly soluble in alcohol and ether, soluble in carbon disulfide, carbon tetrachloride, and benzene. Combustible.
Occurrence: Native in Texas, Louisiana, Sicily, Canada (Alberta), Poland, Saudi Arabia, Mexico, Iraq, offshore deposits in Gulf of Mexico, the former U.S.S.R., Japan.
Derivation: Direct mining by Frasch process, low-grade ores by Chemico process, smelter waste gas, sour natural gas, coal, iron pyrites, gypsum, solvent extraction of volcanic ash, petroleum, coke oven gas, photolysis of hydrogen sulfide.
Grade: Technical (lumps, roll, flour), rubber maker's, NF (sublimed), crude, refined, precipitated (milk of sulfur), high purity (impurities less than 10 ppm), also available in prilled form.
Hazard: Fire and explosion risk in finely divided form.
Use: Sulfuric acid manufacture, pulp and paper manufacture, carbon disulfide, rubber vulcanization, detergents, petroleum refining, dyes and chemicals, drugs and pharmaceuticals, explosives, insecticides, rodent repellents, soil conditioner, fungicide, coating for controlled-release fertilizers, nucleating agent for photographic film, cement sealant, binder and asphalt extender in road paving (up to 40%), base material for low-temperature mortars.

sulfur-35.   Radioactive sulfur of mass number 35.
Properties: Half-life 87.1 days, radiation β.
Derivation: By pile irradiation of elemental sulfur or of various chlorides.
Available forms: Solid elemental sulfur, sulfate in weak hydrochloric acid, barium sulfide in barium hydroxide solution, elemental sulfur in benzene solution, in tagged compounds such as carbon disulfide, chlorosulfonic acid, thiourea, sulfanilamide, thiamine, heparin, insulin, "Sucaryl" *Wampole]*, etc.
Hazard: Radioactive poison.
Use: Research tool in studying mechanism of rubber vulcanization and polymerization of synthetic rubber, role of sulfur in the coking process and in steel, effect of sulfur on engine wear, sulfur removal in the viscose process, behavior of detergents during ashing, sulfur deposition in diesel engines, action of sulfur in silver plating solutions, protein metabolism, surface active agents and surface phenomena, drug actions, etc.

sulfurated lime.   See lime, sulfurated.

sulfurated potash. See potash, sulfurated.

sulfur bromide. (sulfur monobromide). S$_2$Br$_2$.
Properties: Yellow liquid, becomes red when exposed to air. D 2.6 (15C), fp −40C, bp 54C (0.2 mm Hg). Decomposed by water; soluble in carbon disulfide.
Hazard: Avoid inhalation of fumes, strong irritant to tissue.

sulfur chloride. (sulfur subchloride; sulfur monochloride).
CAS: 10025-67-9. S$_2$Cl$_2$.
Properties: Amber to yellowish-red, oily, fuming liquid; penetrating odor. Soluble in alcohol, ether, benzene, carbon disulfide, and amyl acetate; decomposes on contact with water. D 1.690 (15.5C), fp −80C; bp 138C; flash p 266F (130C). Combustible.
Derivation: By passing chlorine over molten sulfur. Method of purification: Distillation.
Hazard: Reacts violently with water in closed vessel. Toxic by inhalation and ingestion, strong irritant to tissue, eyes, skin, and upper respiratory tract.
Use: Chemicals (sulfur solvent, acetic anhydride, thionyl chloride, carbon tetrachloride from carbon disulfide, various chlorohydrins from glycerol, glycol, etc.), analytical reagent, rubber vulcanizing, vulcanized oils, purifying sugar juices, military poison, insecticide, hardening soft woods (by treatment with sulfur chloride dissolved in carbon disulfide), pharmaceuticals, sulfur dyes, extraction of gold from its ores.

sulfur dichloride. SCl$_2$.
Properties: Reddish-brown, fuming liquid; pungent chlorine odor. D 1.638 (15.5C), fp −78C, bp decomposes above 59C on rapid heating, boils near 60C, refr index 1.567 (20C). Decomposes in water and alcohol; soluble in benzene.
Derivation: Chlorine is passed into sulfur monochloride to saturation at 6–10C followed by carbon dioxide to drive off the excess chlorine.
Grade: Technical.
Hazard: Toxic by inhalation and ingestion, strong irritant to tissue.
Use: Chlorine carrier or chlorinating agent, rubber vulcanizing, vulcanized oils, purifying sugar juices, sulfur solvent, chloridizing agent in metallurgy, manufacture of organic chemicals and insecticides.

sulfur dioxide.
CAS: 7446-09-5. SO$_2$.
Properties: Colorless gas or liquid; sharp, pungent odor. D 1.4337, liquid at 0C, fp −76.1C, bp −10C, vap press 3.2 atm (20C), refr index (liquid) 1.410 (24C), an outstanding oxidizing and reducing agent. Soluble in water, alcohol, and ether; forms sulfurous acid (H$_2$SO$_3$). Noncombustible.
Derivation: (1) By roasting pyrites in special furnaces. The gas is readily liquefied by cooling with ice and salt or at a pressure of three atmospheres.

(2) By purifying and compressing sulfur dioxide gas from smelting operations. (3) By burning sulfur.
Grade: Commercial, USP, technical, refrigeration, anhydrous 99.98% min.
Hazard: Toxic by inhalation, strong irritant to eyes and mucous membranes, especially under pressure. Dangerous air contaminant and constituent of smog. Not permitted in meats and other sources of vitamin B$_1$. U.S. atmospheric standard 0.140 ppm. Pulmonary function inhibitor and lower respiratory tract irritant. Questionable carcinogen.
Use: Chemicals (H$_2$SO$_4$, salt cake, sulfites, hydrosulfites of potassium and sodium, thiosulfates, alum from shale, recovery of volatile substances), sulfite paper pulp, ore and metal refining, soybean protein, intermediates, solvent extraction of lubricating oils, bleaching agent for oils and starch, sulfonation of oils, disinfecting and fumigating, food additive (inhibition of browning, of enzyme-catalyzed reactions, bacterial growth), reducing agent, antioxidant.

sulfur dye. See sulfide dye.

sulfuretted hydrogen. See hydrogen sulfide.

sulfur hexafluoride.
CAS: 2551-62-4. SF$_6$.
Properties: Colorless gas; odorless. Fp −64C (sublimes), d (gas) 6.5 g/L, d (liquid) 1.67, specific volume 2.5 cu ft/lb (21.1C). Slightly soluble in water; soluble in alcohol and ether. Noncombustible.
Hazard: Asphyxiant.
Use: Dielectric (gaseous insulator for electrical equipment and radar wave guides).

sulfuric acid. (hydrogen sulfate; battery acid; electrolyte acid).
CAS: 7664-93-9. H$_2$SO$_4$.
By far the most widely used industrial chemical.
Properties: Strongly corrosive, dense, oily liquid; colorless to dark brown depending on purity. Miscible with water. Very reactive, dissolves most metals; concentrated acid oxidizes, dehydrates, or sulfonates most organic compounds, often causes charring; d of pure material 1.84; mp 10.4C; bp varies over range 315–338C due to loss of sulfur trioxide during heating to 300C or higher.
Note: Use great caution in mixing with water due to heat evolution that causes explosive spattering. Always add the acid to water, *never* the reverse.
Derivation: From sulfur, pyrite (FeS$_2$), hydrogen sulfide, or sulfur-containing smelter gases by the contact process (vanadium pentoxide catalyst). The first step is combustion of elemental sulfur or roasting of iron pyrites to yield sulfur dioxide. Then follows the critical reaction, catalytic oxidation of sulfur dioxide to sulfur trioxide

$$\left(SO_2 + \tfrac{1}{2}O_2 \xrightarrow[425°C]{V_2O_5} SO_3 \right)$$

in a four-stage converter at 425–450C with evolution of heat. After cooling to 160C, the sulfur trioxide is absorbed in a circulating stream of 98–99% sulfuric acid, where it unites with the small excess of water in the acid to form more sulfuric acid.

$$(SO_3 + H_2O \xrightarrow{260°C} H_2SO_4)$$

Sulfuric acid can also be made by the "Cat-Ox" process and from gypsum (calcium sulfate).

Grade: Commercial 60 degrees Bé (d 1.71, 77.7% sulfuric acid); 66 degrees Bé (d 1.84, 93.2% sulfuric acid); 98% (d 1.84); 99% (d 1.84); 100% (d 1.84) depending on supplier; reagent ACS, CP.

Hazard: Strong irritant to tissue. Pulmonary function inhibitor. Confirmed carcinogen.

See note above.

Use: Fertilizers, chemicals, dyes and pigments, etchant, alkylation catalyst, electroplating baths, iron and steel, rayon and film, industrial explosives, lab reagent, nonferrous metallurgy.

See sulfuric acid, fuming.

sulfuric acid, fuming. (oleum).

CAS: 8014-95-7. $xH_2SO_4•ySO_3$.

A solution of sulfur trioxide in sulfuric acid.

Properties: Heavy, oily liquid; colorless to dark brown depending on purity. Fumes strongly in moist air. Extremely hygroscopic.

Derivation: Sulfur trioxide produced by the contact process is absorbed in concentrated sulfuric acid.

Grade: Commercial (10–70% free SO_3 depending on supplier), CP.

Hazard: Reacts violently with water. Strong irritant to tissue.

Use: Sulfating and sulfonating agent, dehydrating agent in nitrations, dyes, explosives, lab reagent.

sulfuric anhydride. See sulfur trioxide.

sulfuric chloride. See sulfuryl chloride.

sulfuric oxychloride. See sulfuryl chloride.

sulfur monobromide. See sulfur bromide.

sulfur monochloride. See sulfur chloride.

sulfurous acid.

CAS: 7782-99-2. A solution of sulfur dioxide in water. The formula H_2SO_3 is used, but the acid is known only through its salts.

Properties: Colorless liquid; suffocating sulfur odor. D 1.03, unstable. Soluble in water.

Derivation: Absorption of sulfur dioxide in water.

Grade: Technical, CP.

Hazard: Toxic by ingestion and inhalation, strong irritant to tissue.

Use: Organic synthesis; bleaching straw and textiles, etc.; paper manufacture; wine manufacture; brewing; metallurgy; ore flotation; medicine (antiseptic);

reagent in analytical chemistry; sulfites; as preservative for fruits, nuts, foods, wines; disinfecting ships.

sulfur oxychloride. See thionyl chloride.

sulfur pentachloride. (disulfur decafluoride).

CAS: 5714-22-7. S_2F_{10}.

Properties: Colorless gas; SO_2 odor. Mw 254.11, sp g 2.08 at 0C, mp −92C, bp 29C, vp 561 torr (20C). Insoluble in water.

sulfur subchloride. See sulfur chloride.

sulfur tetrafluoride.

CAS: 7783-60-0. SF_4.

Properties: A gas. Fp −124C, bp −40C. Decomposes in water. Noncombustible.

Grade: Available in cylinders at 90–94% purity.

Hazard: High by inhalation, strong irritant to eyes, mucous membranes, and upper respiratory tract irritant. Lung damage.

Use: Fluorinating agent for making water and oil repellents and lubricity improvers.

sulfur trioxide. (sulfuric anhydride).

CAS: 7446-11-9. SO_3, $(SO_3)_n$.

Properties: Exists in three solid modifications; α mp 62C, β mp 32.5C, γ mp 16.8C. The α form appears to be the stable form but the solid transitions are commonly slow; a given sample may be a mixture of the various forms and its melting point not constant. The solid sublimes easily.

Derivation: Passing a mixture of SO_2 and oxygen over a heated catalyst such as platinum or vanadium pentoxide.

See sulfuric acid.

Hazard: Oxidizing agent, fire risk in contact with organic materials, an explosive increase in vapor pressure occurs when the α form melts. The anhydride combines with water, forming sulfuric acid and evolving heat. Highly toxic, strong irritant to tissue.

Use: Sulfonation of organic compounds, especially nonionic detergents, solar energy collectors. It is usually generated in the plant where it is to be used.

sulfuryl chloride. (chlorosulfuric acid; sulfonyl chloride; sulfuric chloride; sulfuric oxychloride).

CAS: 7791-25-5. SO_2Cl_2.

Properties: Colorless liquid; pungent odor. Rapidly decomposed by alkalies and by hot water; soluble in glacial acetic acid. D 1.667 at 20C, bp 69.2C, fp −54C, vap d 4.6.

Derivation: (1) By heating chlorosulfonic acid in the presence of catalysts. (2) From sulfur dioxide and chlorine in the presence of either activated carbon or camphor.

Grade: Technical.

Hazard: Strong irritant to tissue.

Use: Organic synthesis (chlorinating agent, dehydrating agent, acylating agent) pharmaceuticals, dyestuffs, rubber-base plastics, rayon, solvent, catalyst.

sulfuryl fluoride.
CAS: 2699-79-8. SO_2F_2.
Properties: Colorless gas. Fp −136.7C, bp −55.4C. Slightly soluble in cold water and most organic solvents. Noncombustible.
Hazard: Toxic by inhalation. Central nervous system impairment.
Use: Insecticide, fumigant.

sulphenone.   (*p*-chlorophenyl phenyl sulfone).
CAS: 80-00-2.
Properties: White solid. Mp 98C. Insoluble in water; soluble in hexane, xylene, carbon tetrachloride, and acetone.
Use: A miticide.

Sulphon.   Acid dyestuffs used on wool and silk, fair to good fastness to light, good fastness to washing, etc.; can be used on leather and paper.

sulprofos.   (*O*-ethyl-*O*-[4-(methylthio)phenyl]-*S*-propyl phosphorodithioate; Bolstar).
CAS: 35400-43-2. $C_{12}H_{19}O_2P_3$.
Properties: Tan colored liquid; sulfide odor. Mw 322.43, sp g 1.20 (20C), bp 125C (0.0075 torr). Soluble in organic solvents; insoluble in water.
Hazard: Cholinesterase inhibitor. Questionable carcinogen.
Use: Insecticide.

sultam acid.   (1,8-naphthosultam-2,4-disulfonic acid). $C_{10}H_4(SO_3H)_2NHSO_2$.
Properties: White solid. Soluble in water; slightly soluble in alcohol.
Derivation: Sulfonation of 1-naphthylamine-8-sulfonic acid or 1-naphthylamine-4,8-disulfonic acid.
Use: Intermediate for Chicago acid.

sulvanite.   (ethyl sulfuryl chloride). $ClSO_3C_2H_5$.
Properties: Colorless liquid. Bp 135C.
Use: Poison gas.

sumiseven.
CAS: 83657-22-1. $C_{15}H_{18}ClN_3O$.
Hazard: Moderately toxic by ingestion and skin contact.

Sumner, James B.   (1887–1955). An American biochemist who won the Nobel Prize in chemistry in 1946 along with Northrop and Stanley. His work was mainly concerned with enzymes, and he was the first in the field to isolate and crystallize an enzyme and determined it was a protein. He inspired followers to continue research in virus and enzyme study. His education was at Harvard and he taught for many years at Cornell.

sump.   A chamber or large container into which waste liquids from processing or mechanical operations are allowed to collect and from which they are removed by a pump.

Sun Anti-Chek Wax 55.   General purpose paraffin wax inhibitor for ozone cracking and weathering, nonstaining.
Use: As a process aid and lubricant.

sunflower cake.   The press-cake resulting from hydraulic press expression of sunflower seeds. Typical analysis: proteins 21.0%, fats 8.5%, fiber 48.9%, water 10.2%, ash 11.4%.
Use: Cattle food, fertilizer ingredient.

sunflower meal.   The mealy form assumed by sunflower seeds after crushing and heating. If the oil cake is ground the product again is in this mealy form. It contains 55–60% protein.
Use: Animal feed, fertilizers.

sunflower oil.
Properties: Pale-yellow, semidrying oil; mild taste; pleasant odor. D 0.924–0.926, iodine value 130–135, refr index 1.4611. Soluble in alcohol, ether, chloroform, and carbon disulfide. Combustible.
Chief constituents: Mixed triglycerides of linoleic, oleic, and other fatty acids.
Derivation: By expression from the seeds of *Helianthus annuus*.
Method of purification: Filtration.
Grade: Crude, refined.
Use: Modified alkyd resins, soap, edible oil, margarine, shortening, dietary supplement.

sunflower oil (unhydrogenated).
Properties: From the seed of *Helianthus annuus*. Amber-colored liquid.
Use: Food additive.

"Sunproof" *[Chemtura]*.   TM for a series of protective waxes to prevent static atmospheric cracking in all types of synthetic and natural rubbers.

"Supac" *[Mission]*.   TM for engineered geotextile nonwoven fabric.
Use: Civil engineering applications.

superactinide.   See periodic table.

superalloy.   An iron-base, cobalt-base, or nickel-base alloy that combines high-temperature mechanical properties, oxidation resistance and creep resistance to an unusual degree. Constitutions of these alloys are as follows. Iron-base: 10–45% nickel, 13–19% chromium, 1.3–6% molybdenum, balance iron. Cobalt-base: 0–26% nickel, 0–26% chromium, 0–15% tungsten, balance cobalt. Nickel-base: 55–75% nickel, 10–20% chromium, 0–6% aluminum, 0–5% titanium. All contain less than

0.5% carbon, plus other special ingredients. Super-alloys can be used up to 2500F.
Use: Jet engine parts, turbo-superchargers, extreme high-temperature applications.

Super-Backacite.   Pure phenolic resins used as coating vehicles.

Super-Beckamine.   Melamine-formaldehyde resins used as coating vehicles.

supercalender.   A machine used for finishing paper. It is a vertical stack of from 3 to 12 or more rolls through which the sheet is threaded. The top and bottom rolls are of chilled cast iron and are larger than the intermediate rolls, several of which are of the fiber type. Overall widths may be as much as 15 ft. Operating speeds range from 600 to 2000 ft per minute. The supercalender imparts gloss and smoothness to high-grade papers, both coated and uncoated, as a result of the interaction of the metal and resilient fiber rolls.
See fiber roll; calender; paper.

supercoiled DNA.   (supertwisted DNA). Covalently closed duplex DNA.

superconductivity.   The phenomenon in which certain metals, alloys, and compounds at temperatures usually near absolute zero lose both electrical resistance and magnetic permeability, i.e., have infinite electrical conductivity. Depending upon the substance, the maximum temperature (transition temperature) for the phenomenon as of 1985 was 0.5–28 K. Superconductivity does not occur in alkali metals, noble metals, ferro- and antiferromagnetic metals. It is well known in elements having 3, 5, or 7 valence electrons per atom, and is associated with high room-temperature resistivity. A system for transmitting electric current underground by means of superconducting cables has been developed.
See cryogenics.

supercooled liquid.   A liquid cooled below its normal freezing point without solidification.

supercritical fluid.   A dense gas that is maintained above its critical temperature (the temperature above which it cannot be liquefied by pressure). Such fluids are less viscous and diffuse more readily than liquids, and are thus more efficient than other solvents in liquid chromatography.

"Superfloc A-836" [Kemira].
Properties: Off-white granular powder. D 47–53 lb/ft3· pH of 0.5% solution at 77F 6.5–8.5.
TM for a dry polyacrylamide cationic, flocculated with controlled molecular weight distribution and varying charge densities.

Use: Works effectively to control and substantially prevent silt loss (erosion) in furrow irrigated agriculture.

"Supergro" [Propex].   TM for engineered geotextile fabric.
Use: Erosion control mat.

Superlitefast.   A group of direct dyes.
Available forms: Powder.
Use: For dyeing of cellulosic fibers.

"Superloid" [ISP].   TM for ammonium alginate, a hydrophilic colloid.
Use: Suspending, thickening, emulsifying, and stabilizing agent in creaming and bodying of rubber latex products; protective colloid in resin emulsion paints, adhesives, fire-retarding compositions, ceramics, etc.

"Superlume" [Propper].   TM for a superleveling, bright-nickel electroplating process on steel stampings, brass, copper, zinc die castings, etc. The materials used are nickel sulfate, nickel chloride, boric acid, and addition agents.

supernatant.   A liquid forming a clear layer above a precipitate produced from a solution of the liquid.

superoxide.   A compound characterized by the presence in its structure of the O_2^- ion. The O_2^- ion has an odd number of electrons (17) and as a result all superoxide compounds are paramagnetic. At room temperature they have a yellowish color. At low temperature many of them undergo reversible phase transitions accompanied by a color change to white. The stable superoxides are

sodium superoxide	NaO_2
potassium superoxide	KO_2
rubidium superoxide	RbO_2
cesium superoxide	CsO_2
calcium superoxide	$Ca(O_2)_2$
strontium superoxide	$Sr(O_2)_2$
barium superoxide	$Ba(O_2)_2$
tetramethylammonium superoxide	$(CH_3)_4NO_2$

In these compounds each oxygen atom has an oxidation number of −1/2 instead of −2, as a normal oxide.

superoxide dismutase.   Any of a type of antioxidant metalloenzymes that occur in aerobic and facultatitive bacteria and in eukarotes. They catalyze a reaction in which two molecules of the highly toxic, highly reactive, superoxide anion is converted into one molecule each of hydrogen peroxide and molecular oxygen.

superpalite. (diphosgene; green cross gas; trichloromethylchloroformate). ClCOOCCl₃.
Properties: D 1.6525, bp 127.5–128C.

Superpax. 92–94.5% zirconium silicate with bulk d 68 lb/cu ft, average particle size 5 microns max.
Use: Ceramic glazes and as a filler for resins and rubbers.

superphosphate. (acid phosphate). The most important phosphorus fertilizer, made by the action of sulfuric acid on insoluble phosphate rock (essentially calcium phosphate, tribasic) to form a mixture of gypsum and calcium phosphate, monobasic. A typical composition is CaH₄(PO₄)₂•H₂O 30%, CaHPO₄ 10%, CaSO₄ 45%, iron oxide, alumina, silica 10%, water 5%. Typical analysis: Moisture 10–15%, available phosphoric acid (as P₂O₅) 18–21%, insoluble phosphoric acid 0.3–2%, total phosphoric acid (as P₂O₅) 19–23%.
Grade: Based on available P₂O₅.
Use: Fertilizer.
See triple superphosphate; nitrophosphate.

superphosphoric acid. See polyphosphoric acid.

supersaturation. The condition in which a solvent contains more dissolved matter (solute) than is present in a saturated solution of the same components at equivalent temperature. Such solutions may occur or can be made when a saturated solution cools gradually so that nucleating crystals do not form. They are extremely unstable and will precipitate upon addition of even one or two crystals of the solute or upon shaking or other slight agitation. Supersaturated solutions occur in the confectionery industry, e.g., in fudges, maple sugar, etc.

"Super Shock" *[Avon].* TM for a shock system for swimming pools.
Use: Fast dissolving weekly pool treatment.

Super-Sta-Bac. Hydrocarbon resins that are polymers of mixed olefins.

suppressor gene. A gene that can suppress the action of another gene.

suppressor mutation. A mutation that totally or partially restores a function lost in a previous mutation. Must be located at a site different from the site of the first mutation.

"Supralan" *[Dystar Colours].* TM for metallized acid colors of good fastness and level dyeing properties.

Supramine XA. A leather chemical, solubilized sulfur phenol condensate, 75% active.

"Supranol" *[Dystar Textilefarben].* TM for dyestuffs used on wool and silk; good fastness to light, washing, and seawater; can also be used on leather.

surface. In physical chemistry the area of contact between two different phases or states of matter, e.g., finely divided solid particles and air or other gas (solid–gas); liquids and air (liquid–gas); insoluble particles and liquid (solid–liquid). Surfaces are the sites of the physicochemical activity between the phases that is responsible for such phenomena as adsorption, reactivity, and catalysis. The depth of a surface is of molecular order of magnitude. The term *interface* is approximately synonymous with *surface*, but it also includes dispersions involving only one phase of matter, i.e., solid–solid or liquid–liquid.
See interface; surface area; surface chemistry.

surface-active agent. (surfactant). Any compound that reduces surface tension when dissolved in water or water solutions, or that reduces interfacial tension between two liquids, or between a liquid and a solid. There are three categories of surface-active agents: detergents, wetting agents, and emulsifiers; all use the same basic chemical mechanism and differ chiefly in the nature of the surfaces involved.
See interface; surface chemistry.

surface area. The total area of exposed surface of a finely divided solid (powder, fiber, etc.) including irregularities of all types. Since activity is greatest at the surface, that is, the boundary between the particle and its environment, the larger the surface area of a given substance, the more reactive it is. Thus reduction to small particles is a means of increasing the efficiency of both chemical and physical reactions; for example, the coloring effect of pigments is increased by maximum size reduction. Carbon black is notable among solids for its huge surface area (as much as 18 acres/lb for some types); the activity of its surface accounts for its outstanding ability to increase the strength and abrasion resistance of rubber. The capacity of activated carbon to adsorb molecules of gases is due to this factor. Surface area is measured most accurately by nitrogen adsorption techniques.

surface chemistry. The observation and measurement of forces acting at the surfaces of gases, liquids and solids or at the interfaces between them. This includes the surface tension of liquids (vapor pressure, solubility); emulsions (liquid–liquid interfaces); finely divided solid particles (adsorption, catalysis); permeable membranes and microporous materials; and biochemical phenomena such as osmosis, cell function, and metabolic mechanisms in plants and animals. Surface chemistry has many industrial applications, a few of which are air pollution, soaps and synthetic detergents, reinforcement

of rubber and plastics, behavior of catalysts, color and optical properties of paints, aerosol sprays of all types, monolayers and thin films, both metallic and organic. Outstanding names in the development of this science are Graham, Freundlich, and W. Ostwald in the 19th Century and Harkins, Langmuir, LaMer, and McBain in the 20th.
See colloid chemistry.

surface tension. In any liquid, the attractive force exerted by the molecules below the surface upon those at the surface–air interface, resulting from the high molecular concentration of a liquid compared to the low molecular concentration of a gas. An inward pull, or internal pressure, is thus created which tends to restrain the liquid from flowing. Its strength varies with the chemical nature of the liquid. Polar liquids have high surface tension (water = 73 dynes/cm at 20C); nonpolar liquids have much lower values (benzene = 29 dynes/cm, ethanol = 22.3 dynes/cm), thus they flow more readily than water. Mercury, with the highest surface tension of any liquid (480 dynes/cm) does not flow, but disintegrates into droplets.
See interface; surface-active agent.

surfactant. See surface-active agent.

"Sur-Gard" [Nalco]. TM for chemicals used for treatment of boiler water to inhibit scale and corrosion and to remove oxygen from the boiler water.

"Surlyn" [Du Pont]. TM for a group of ionomer resins.
Properties: ("Surlyn" A) Thermoplastic produced as a granular material; flexible, transparent, grease resistant; very light weight but tough. Izod impact strength 5.7–14.6 ft-lb/in (higher than any other polyolefin), tensile strength 3,500–5,500 psi, elongation 300–400%, softening point 71. Insoluble in any commercial solvent. Subject to slow swelling by hydrocarbons, to slow attack by acids.
Use: Coatings, packaging films, products made by injection or blow molding, or by thermoforming.

SUS. Abbreviation for Saybolt Universal Seconds.
See Saybolt Universal viscosity.

suspension. A system in which very small particles (solid, semisolid, or liquid) are more or less uniformly dispersed in a liquid or gaseous medium. If the particles are small enough to pass through filter membranes, the system is a colloidal suspension (or solution). Examples of solid-in-liquid suspensions are comminuted wood pulp in water, which becomes paper on filtration; the fat particles in milk; and the red corpuscles in blood. A liquid-in-gas suspension is represented by fog or by an aerosol spray. If the particles are larger than colloidal dimensions

they will tend to precipitate if heavier than the suspending medium, or to agglomerate and rise to the surface if lighter. This can be prevented by incorporation of protective colloids. Polymerization is often carried out in suspension, the product being in the form of spheres or beads.
See solution; colloidal; dispersion; emulsion; colloid chemistry.

"Sustane" [Eastman]. TM for a low viscosity blend weighting agent for liquids.

Sustane. Synthetic, food-grade antioxidant product line including BHA, BHT, TBHQ, propyl gallate, and liquid blends.
Use: To preserve vegetable oils, animal fats, spices, baked goods, nuts, pet foods, dressing oils, confections, cereals, sausage, cosmetics, and dehydrated potatoes.

Suzuki, Akira. (1930–). A Japanese chemist who was awarded the 2010 Nobel Prize jointly with Richard F. Heck and Ei-ichi Negishi. The scientists discovered the use of palladium as a catalyst to produce organic molecules. Heck began the process in the late 1960s, and Negishi and Suzuki refined palladium catalysis for widespread use. Suzuki was awarded his Ph.D. (1959) by Hokkaido University.

Svedberg. (S). A unit of measure of the rate at which a particle sediments in a centrifugal field.

Svedberg, Theodor. (1884–1971). A Swedish chemist who won the Nobel Prize in 1926. Author of *Die Methoden zur Herstellung Kolloider Losungen anorganischer Stoffe.* His work included research in colloidal chemistry, molecular size determination, and methods of electrophoresis, as well as the development of the ultracentrifuge for separation of colloidal particles in solution. His education was in Sweden with later work done at the University of Wisconsin before returning to Uppsalla.

Swarts reaction. Fluorination of organic polyhalides with antimony trifluoride (or zinc and mercury fluorides) in the presence of a trace of a pentavalent antimony salt.

Sweet Design. A sweetener system.
Use: In food and beverage manufacturing to produce better tasting frozen desserts and sweetened products with little or no sugar.

sweeten. (1) To add sugar or a synthetic product to foods or beverages to provide a sweet taste (flavor). (2) To deodorize and purify petroleum products by removing sulfur compounds (doctor treatment). (3) In industrial slang, to increase the quality of a low-cost product by adding more expensive ingredients.

sweetener, nonnutritive. A food additive, either natural or synthetic, usually having much greater sweetness intensity than sugar (sucrose), but without its caloric value. In some cases it acts as an enhancer or potentiator of sweetness. Chief among them is saccharin, a benzoic acid derivative. The cyclamate group was removed from the market by FDA in 1970 because of animal carcinogenicity, though the evidence in respect to human toxicity is controversial.

Increasing research has developed several new noncaloric sweeteners: the dihydrochalcone group of disaccharide, glycyrrhizin (licorice extract), especially its ammoniated derivative; dulcin (4-ethoxyphenylurea), the glycoprotein of "miracle fruit;" and a polypeptide from a tropical fruit called the "serendipity berry," said to be 3000 times sweeter than sucrose. "Aspartame" ($C_{14}H_{18}N_2O_5$) has been approved by FDA.

sweet oil. See olive oil.

sweet water. (1) The glycerin and water mixture obtained when fats are split (or hydrolyzed) with water to give fatty acid and glycerin. (2) The washings from char used in sugar refining. (3) In engineering terminology, plain water cooled to just below the freezing point and used to preserve milk and other food products.

swep. (generic name for methyl-3,4-dichlorocarbanilate; methyl-*n*-3,4-dichlorophenylcarbamate). $CH_3OOCNHC_6H_3Cl_2$.
Properties: Crystals. Mp 113C. Insoluble in water and kerosene; soluble in acetone and dimethylformamide.
Hazard: Toxic by ingestion.
Use: Herbicide.

swertianolin.
CAS: 23445-00-3. $C_{20}H_{20}O_{11}$.
Hazard: A poison.
Source: Natural product.

swiss-prot.
A curated, heavily annotated, largely nonredundant protein sequence database. Annotations include the description of protein function, domain structures, post-translational modifications and variants.

Sybicort. A drug.
Use: For treatment of asthma and pulmonary diseases.

"Sylgard" [Dow Corning]. TM for a series of silicone resin encapsulants used in electronic assemblies.

"Syloid" [Grace]. TM for a series of micron-sized silicas.
Use: Flatting agents, lacquer, baking finishes, inks, paper reproduction, adhesives, pharmaceuticals, insulation.

"Syltherm" [Dow Corning]. TM for dimethylsiloxane polymer.
Use: (561) Dielectric coolant for transformers; (800) high temperature, liquid-phase, heat-transfer liquid. Also used in industrial applications, glass and lens coating, cosmetics, penetrating oils, and personal-care products.

sylvan. (α-methylfuran). C_5H_6O. Constituent of wood tar.

sylvic acid. See abietic acid.

sylvite. (sylvine). KCl. A natural potassium chloride, contains 43% potassium chloride, 57% sodium chloride, sometimes with up to 0.26% bromide.
Properties: Colorless or white, bluish or reddish in color, white streak, vitreous luster, resembles rock salt in appearance. D 1.97–1.99, Mohs hardness 2.
Occurrence: West Texas, New Mexico, Europe.
Use: Major source of potassium compounds in the U.S., fertilizers.

symbiosis. Two (or more) organisms that are mutually interdependent and live in physical association.

"Symmetrel" [Endo]. (amantadine hydrochloride). TM for a synthetic antiviral drug specific for A_2-type virus infection (one type of Asian flu). It is taken orally and acts by preventing the virus from entering cells. It is inactive against B-type viruses. Reported to be effective in treatment of Parkinson's disease.

symmetrical compound. A compound derived from benzene in which the hydrogen atom on three alternate carbon atoms has been replaced with another element or group, for example, at the 1, 3, and 5 positions.

symmetry. Arrangement of the constituents of a molecule in a definite and continuously repeated space pattern or coordinate system. It is described in terms of three parameters called elements of symmetry (1) The center of symmetry, around which the constituent atoms are located in an ordered arrangement; there is only one such center in a molecule, which may or may not be an atom. (2) Planes of symmetry, which represent division of a molecule into mirror-image segments. (3) Axes of symmetry, represented by lines passing through the center of symmetry; if the molecule is rotated it will have the same position in space more than once in a complete 360 degree turn, for example, the benzene molecule with 6 axes of symmetry requires a 60-degree rotation to return to its identical position.
See stereochemistry; asymmetry.

symport. Cotransport of solutes across a membrane in the same direction.

synapomorphy. A character which is derived, and because it is shared by the taxa under consideration, is used to infer common ancestry.

synaptase. See emulsion.

"Syncal" [PMC]. TM for synthetic sweetener, saccharin, including sodium and calcium salts and insoluble form, as well as several different particle size distributions.

syndet. Abbreviation form for synthetic detergent.
See detergent.

syndiotactic. A type of polymer molecule in which groups of atoms that are not part of the backbone structure are located in some symmetrical and recurring fashion above and below the atoms in the backbone chain, when the latter are arranged so as to be in a single plane.
See polymer, stereospecific.

Syndiotactic polymer

syndrome. The group or recognizable pattern of symptoms or abnormalities that indicate a particular trait or disease.

syneresis. The contraction of a gel on standing, with exudation of liquid. The separation of serum from a blood clot, or of the whey in milk souring or cheese making are examples.

synergism. A chemical phenomenon (the opposite of antagonism) in which the effect of two active components of a mixture is more than additive, that is, it is greater than the equivalent volume or concentration of either component alone. A substance that induces this result when added to another substance is called a synergist.

synfuel. Any gaseous or liquid fuel made by the synthesis gas reaction; synthetic natural gas (SNG) and methanol are examples, the latter being the only synfuel commercially available in the U.S. (Othmer, 1983).
See semisynthetic.

syngamy. The process of union of two gametes; sometimes called fertilization. It encompasses both plasmogamy and karyogamy.

syngas. Shortened form of synthesis gas.

syngeneic. Genetically identical members of the same species.

Synge, Richard L. M. (1914–1994). An Irish mathematician and physicist who won the Nobel Prize for chemistry in 1952 along with Archer. His research was on the application of methods of physical chemistry to isolate and analyze proteins, with special attention to antibiotic peptides and higher plants. He received his doctorate from Cambridge.

Synoca. A group of cleaning compounds including wetting agents, detergents, or alkaline hexametaphosphate.
Use: Continuous conditioning and batch washing of paper machine felts. Paper-mill system boil-outs. Deinking of secondary fiber.

Synotol. A series of alkanolamides of fatty acids.
Use: Foam builders, viscosity improvers, wetting agents, detergents, and opacifiers.

synperonic-n.
Use: In the preparation of the detergent solution in physical developer.

"Synpro" [Ferro]. TM for aluminum, calcium, lithium, magnesium, and zinc stearates.

syntan. A synthetic organic tanning material made from phenolsulfonic acids and formaldehyde.
Use: Chrome- and vegetable-tanned leathers.
See tanning.

synteny. Genes occurring in the same order on chromosomes of different species.
See linkage; conserved sequence.

synthane process. See Fischer–Tropsch process; gasification.

synthases. Enzymes that catalyze reactions in which no NTP is required as an energy source.

Synthe-Copal. Pale-colored rosin ester gums used as coating vehicles, and in chewing gums and rubber adhesives.

"Synthemul" [Reichhold]. TM for a series of acrylate-acrylonitrile, acrylic,and styrene acrylic emulsions.

synthesis. Creation of a substance that either duplicates a natural product or is a unique material not found in nature, by means of one or more chemical reactions, or (for elements) by a nuclear change. High temperature and pressures as well as a catalyst are usually required. Though syntheses are more readily achieved with organic compounds because of the great versatility of the carbon atom, extremely important syntheses of inorganic compounds were made in the early years of chemistry. A list of noteworthy syntheses with approximate dates would include the following:

Inorganic			Organic	
sulfuric acid			urea (Wohler)	1828
(chamber)	1746		mauveine (Perkin)	1857
(contact)	1890		celluloid (Hyatt)	1869
soda ash			ethylbenzene (Friedel Crafts)	1877
(LeBlanc)	1800		rayon (Chardonnet)	1884
(Solvay)	1861		phenolic resins (Baekeland)	1910
ammonia (Haber–Bosch)	1912		neoarsphenamine (Ehrlich)	1910
plutonium	1940		aldehydes, alcohols (Oxo synthesis)	1920
			methanol	1927
			neoprene (Nieuwland)	1930
			nylon (Carothers)	1935
			SBR rubber	1940
			polyisoprene	1950

The tremendous proliferation of synthetic materials in recent years, especially in the high-polymer field, was made possible by the increasingly sophisticated use of catalysts, particularly the Ziegler and Natta stereospecific type.

Synthesis of elements has also occurred; since 1940 all the transuranic elements from 93 to 106, as well as a vast array of radioisotopes of natural elements, have been created by nuclear bombardment of various types.

See resin, synthetic; semisynthetic.

synthesis gas. Any of several gaseous mixtures used for synthesizing a wide range of compounds, both organic and inorganic, especially ammonia. Such mixtures result from reacting carbon-rich substances with steam (steam reforming) or steam and oxygen (partial oxidation); they contain chiefly carbon monoxide and hydrogen, plus low percentages of carbon dioxide and usually less than 2.0% nitrogen. The organic source materials may be biomass, natural gas, methane, naphtha, heavy petroleum oils, or coke (coal). The reactions are nickel-catalyzed steam-cracking (reforming) of methane or natural gas ($CH_4 + H_2O \rightarrow CO + 3H_2$); partial oxidation of methane, naphtha, or heavy oils; and the water-gas reaction with coke ($C + H_2O \rightarrow CO + H_2$). With transition-metal catalysts synthesis gas yields alcohols, aldehydes, acrylic acid, etc., and has many useful reactions with acetylene; it is the basis of the Oxo and Fischer–Tropsch reactions. Synthesis gas from that the carbon monoxide has been removed and which has been adjusted to a ratio of 3 parts hydrogen and 1 part nitrogen is used for ammonia synthesis (nitrogen fixation). Processing with a catalyst at high temperatures and pressures yields ammonia.

See ammonia, anhydrous; Oxo process; Haber, Fritz; water gas; gasification; Fischer–Tropsch process.

synthetases. Enzymes that catalyze reactions using ATP or another NTP as an energy source.

synthetic alkaloid. Any synthetic compound that is structurally similar to a plant alkaloid.

synthetic dye.
Derivation: From coal-tar, petroleum-based intermediates, and synthesized from benzene and its derivatives.
Use: Brightly colored permanent dyes.

synthetic natural gas. (SNG; syngas). Any manufactured fuel gas of approximately the same composition and Btu value as that obtained naturally from oil fields (85% methane and 15% ethane). There are two major methods of synthesis involving catalysts, high temperature, and high pressure: (1) direct hydrogenation of coal; and (2) methanation of synthesis gas, obtained by hydrogenolysis of coal or steam-reforming of naphtha or similar petroleum distillate. Other methods involve pyrolysis of solid wastes and extraction from oil shale and manures.
See gasification; hydrogenolysis.

Synthetic Organic Chemical Manufacturers Association. (SOCMA). Manufacturers of synthetic organic chemicals that are products manufactured from coal, natural gas, crude petroleum and certain natural substances such as vegetable oils, fats protein, carbohydrates, rosin, grain, and their derivatives. The address is 1850 M St N.W., Suite 700, Washington, DC 20036. Website: http://www.socma.com

synthetic paraffin and succinic derivatives.
Use: Food additive.

synthetic resin. See resin, synthetic.

synthine process. See Fischer–Tropsch process.

synthol. Liquid fuel containing fatty acids, water-soluble alcohols, aldehydes, ketones, and esters. Obtained from reduction of carbon monoxide in water-gas at high temperature and pressure with iron as catalyst.

Synthorich NC. A softener.
Use: To give value added finish to fabrics.

Synthorm RLS. A mould releasing agent.
Use: For rubber moulds.

Synthosoft SWNI. A nonionic softener with extra whiteners.

synton. Any of several isomers of prostaglandin from which prostaglandin analogs and intermediates may be derived.

syntopherol acetate. See *dl*-α-tocopheryl acetate.

sypro ruby. A non-destructive fluorescent protein stein that has a sensitivity of 1–2 ng.

syrosingopine.
CAS: 84-36-6. $C_{35}H_{42}N_2O_{11}$. Methyl reserpate ester of syringic acid ethyl carbonate. An analog of reserpine.

Properties: White or slightly yellowish powder. Practically insoluble in water; slightly soluble in methanol; soluble in chloroform and acetic acid.
Grade: NF.
Use: Medicine.

syrup. Commercial name for an aqueous solution of cane or beet sugar (sucrose) sold in tank car lots to manufacturers of candy, soft drinks, soda-fountain goods, etc. USP grade is an aqueous solution of cane sugar (85 g/100 mL). A viscous liquid with d 1.313.

systematics. Field of biology that deals with the diversity of life. Systematics is usually divided into phylogenetics and taxonomy.

systemic herbicide. One that is absorbed and transported internally, killing the plant by acting on tissues that may be remote from the point of contact.

systemic pesticide. A chemical absorbed by an organism that is translocated within the organism that makes the organism toxic to pests.

Systox. Demeton.

"Syton" *[Du Pont Air].* TM for a series of colloidal silicas dispersed in water.
Use: Antisoil and antislip agents.

T

T. Symbol for (1) tritium; (2) tera-; (3) thymine.

2,4,5-T. Abbreviation for 2,4,5-trichlorophenoxyacetic acid.

2,4,6-T. Abbreviation for 2,4,6-trichlorophenol.

Ta. Symbol for tantalum.

tabun. (dimethylphosphoramidocyanidic acid, ethyl ester).
CAS: 77-81-6. $(CH_3)_2NP(O)(C_2H_5O)(CN)$.
A nerve gas.
Properties: Liquid. Fp $-50C$, bp 240C, flash p 172F (77.7C), d 1.4250 (20/4C). Readily soluble in organic solvents; miscible with water but readily hydrolyzed; destroyed by bleaching powder, generating cyanogen chloride. Combustible.
Hazard: Very toxic by inhalation, cholinesterase inhibitor, a military nerve gas, fatal dose (man) 0.01 mg/kg.

"TAC" [Mallinckrodt]. TM for tested additive chemical items, satisfactory for food additives and medical uses.

tachysterol. $C_{28}H_{44}O$.
Properties: Oil; levorotatory. Insoluble in water; soluble in most organic solvents. Protect from air.
Use: Medicine, as the dihydrotachysterol.

tackifiers. Refers to compounds used for making an adhesive stickier.

tackiness. (tack). Property of being sticky or adhesive.

taconite. A low-grade iron ore consisting essentially of a mixture of hematite and silica. It contains 25% iron. Found in the Lake Superior district and western states.

tacrolimus hydrate.
CAS: 109581-93-3. $C_{44}H_{69}NO_{12} \cdot H_2O$.
Hazard: A poison by ingestion.

tacticity. The regularity or symmetry in the molecular arrangement or structure of a polymer molecule. Contrasts with random positioning of substituent groups along the polymer backbone, or random position with respect to one another of successive atoms in the backbone chain of a polymer molecule.
See polymer, stereospecific; isotactic.

Tafel rearrangement. Rearrangement of the carbon skeleton of substituted acetoacetic esters to hydrocarbons with the same number of carbon atoms by electrolytic reduction at a lead cathode in alcoholic sulfuric acid.

Tag Closed Cup. See TCC.

tagetes meal and extract.
Properties: Extracted from the dried ground flower petals of the Aztec marigold *Tagetes erecta L.* Mp: 53.5–55.0C. A permissible food additive used to increase the yellow color of the skin and eggs of poultry.
Use: Food additive.

tagged atom. A radioactive isotope used in tracing the behavior of a substance in both biochemical and engineering research, e.g., C-14 or I-131.
See tracer; label (2).

Tagliabue Closed Cup. A standard method of determining flash points.

Tagliabue Open Cup. A standard method of determining flash points.

Tag Open Cup. See Tagliabue Open Cup.

tailings. (1) In flour milling, the product left after grinding and bolting middlings. (2) Impurities remaining after the extraction of useful minerals from an ore. (3) In general, any residue from a mechanical refining or separation process.

tailored molecule. A molecule that has been modified chemically to give it certain properties.

tails. Refers to high-boiling impurities that are less volatile than the solvent being distilled.

talc. (talcum; soapstone; steatite).
CAS: 14807-96-6. $Mg_3Si_4O_{10}(OH)_2$ or $3MgO \cdot 4SiO_2 \cdot H_2O$.
A natural hydrous magnesium silicate. Compact, massive varieties may be called steatite in distinction from the foliated varieties, which are called talc. Soapstone is an impure variety of steatite.

Properties: White, apple-green, gray powder; pearly or greasy luster, greasy feel. Mohs hardness 1–1.5 (may be harder when impure), high resistance to acids, alkalies, and heat; d 2.7–2.8.

Grade: Crude, washed, air-floated, USP, fibrous (99.5%, 99.95%).

Hazard: Toxic by inhalation. Effects pulmonary function and causes pulmonary fibrosis. Questionable carcinogen.

Use: Ceramics; cosmetics and pharmaceuticals; filler in rubber, paints, soap, putty, plaster, oilcloth; abherent; dusting agent; lubricant; paper; slate pencils and crayons; electrical insulation.

See magnesium silicate.

tall oil. (tallol; liquid rosin).
CAS: 8002-26-4.

A mixture of rosin acids, fatty acids, and other materials obtained by acid treatment of the alkaline liquors from the digesting (pulping) of pine wood; flash p 360F (182C). Combustible.

Derivation: The spent black liquor from the pulping process is concentrated until the sodium salts (soaps) of the various acids separate out and are skimmed off. These are acidified by sulfuric acid. Composition and properties vary widely, but average 35–40% rosin acids, 50–60% fatty acids.

Grade: Crude, refined.

Use: Paint vehicles, source of rosin, alkyd resins, soaps, cutting oils and emulsifiers, driers, flotation agents, oil-well drilling muds, core oils, lubricants and greases, asphalt derivatives, rubber reclaiming, synthesis of cortisone and sex hormones, chemical intermediates.

tall oil imidazoline.
CAS: 61791-36-4.
Hazard: A severe skin and eye irritant.

tallow. An animal fat containing C_{16} to C_{18}.

Properties: The solidifying points of the different tallows are as follows: from 20–45C for horse fat, 27–38C for beef tallow, 54–56C for stearin and oleo, 32–41C for mutton tallow; d 0.86; refr index 46–49 (40C) (Zeiss); iodine value 193–202; flash p 509F (265C). Combustible.

Derivation: Extracted from the solid fat or "suet" of cattle, sheep, or horses by dry or wet rendering.

Chief constituents: Stearin, palmitin, and olein.

Grade: Edible; inedible; beef tallow; mutton tallow; horse fats; acidless; edible, extra.

Use: Soap stock, leather dressing, candles, greases, manufacture of stearic and oleic acids, animal feeds, abherent in tire molds.

tallow oil. See Peacock.

taloximine hydrochloride.
CAS: 20230-77-7. $C_{12}H_{16}N_4O_2 \cdot ClH$.
Hazard: Moderately toxic by ingestion.

"Tamol" [Dow]. TM for anionic, polymer-type dispersing agents. Supplied as light-colored powders or aqueous solutions. Effective dispersant for aqueous suspensions of insoluble dyestuffs, polymers, clays, tanning agents, and pigments.

Use: Manufacture of dyestuff pastes, textile printing and dyeing, pigment dispersion in textile backings, latex paints and paper coatings, retanning and bleaching of leather, dye resist in leather dyeing, dispersion of pitch in paper manufacture, prefloc prevention in the manufacture of synthetic rubber.

"Tamol 850" [Dow]. TM for carboxylic homopolymer.

Use: For water treatment, as a scale inhibitor, and sludge dispersant.

tamuslosin hydrochloride. See r-(−)-5-(2-((2-(2-ethoxyphenoxy)ethyl)amino)propyl)-2.

Tanacol CG. (isopropylbiphenyl).
CAS: 25640-78-2.
A solvent.
Use: In carbonless-copy-paper systems, replaces PCB in capacitors, heat transfer fluid.

Tanaka, Koichi. (1959–). Born in Japan, Tanaka won the Nobel Prize for chemistry in 2002 for his pioneering work concerning the development of methods for identification and structure analyses of biological macromolecules. Tanaka received a degree in Electrical Engineering from Tohoku University and undertakes his research at the Shimadzu Corporation.

Tanak MRX. A melamine–formaldehyde resin tanning agent used to make pure-white leather and for bleaching and filling chrome leather.

Tanamer. A sodium polyacrylate adhesive for use during the drying of leather.
See acrylate.

Tandem 552. Emulsifiers.
Use: As pan release agents for frozen desserts, ice cream, baked goods; antibrowning agent for fresh-cut fruits and vegetable.

tandem repeat sequences. Multiple copies of the same base sequence on a chromosome; used as markers in physical mapping.
See physical map.

tangerine oil. See citrus peel oil.

tankage. (animal tankage; tankage rough ammoniate).

The product obtained in abattoir by-product plants from meat scraps and bones, that are boiled under pressure and allowed to settle. The grease is removed from the top and the liquor drawn off. The scrap is then pressed, dried, and sold for fertilizer.

Grade: Based on percentage of ammonia and bone phosphate. A medium grade has 10% ammonia and 20% bone phosphate. Concentrated tankage has had the boiled-down tank liquor and press water added to it before drying and runs 15–17% ammonia.
Hazard: Flammable, may ignite spontaneously.

tankage, garbage.   (tankage fertilizer).
Garbage treated with steam under pressure, the water and some of the grease removed by pressing, and further grease removed by solvent extraction. Contains 3–4% ammonia, 2–5% phosphoric acid, and 0.50–1.00% potash.
Hazard: Flammable, may ignite spontaneously.
Use: Fertilizer.

tannic acid.   (gallotannic acid; described as a penta-(*m*-digalloyl)-glucose).
CAS: 1401-55-4. $C_{76}H_{52}O_{46}$.
Natural substance widely found in nutgalls, tree barks, and other plant parts. Tannins are known to be gallic acid derivatives. A solution of tannic acid will precipitate albumin. Tannins are classified according to their behavior on dry distillation into two groups, (1) condensed tannins, which yield catechol, and (2) hydrolyzable tannins, which yield pyrogallol; (2) comprises two groups on the basis of its products of hydrolysis, glucose, and (a) ellagic acid or (b) gallic acid.
Properties: Lustrous, faintly yellowish, amorphous powder, glistening scales, or spongy mass; darkens on exposure to air; odorless; strong, astringent taste. Mp decomposes at 210C. Soluble in water, alcohol, and acetone; almost insoluble in benzene, chloroform, and ether. Flash p 390F (198C), autoign temp 980F (526C). Combustible.
Derivation: Extraction of powdered nutgalls with water and alcohol.
Grade: Technical, CP, NF, fluffy, FCC.
Hazard: Toxic by ingestion and inhalation. Questionable carcinogen.
Use: Chemicals (tannates, gallic acid, pyrogallic acid, hydrosols of the noble metals); alcohol denaturant; tanning; textiles (mordant and fixative); electroplating; galvanoplastics (gelatin precipitant); clarification agent in wine manufacture, brewing and foods, writing inks; pharmaceuticals; deodorization of crude oil; photography; paper (sizing, mordant for colored papers); treatment of minor burns.

tannin.   Any of a broad group of plant-derived phenolic compounds characterized by their ability to precipitate proteins. Some are more toxic than others, depending on their source. Those derived from nutgalls are believed to be carcinogens, while those found in tea and coffee may be virtually nontoxic.
See tannic acid.

tanning.   The preservation of hides or skins by use of a chemical that (1) makes them immune to bacterial attack; (2) raises the shrinkage temperature; and (3) prevents the collagen fibers from sticking together on drying, so that the material remains porous, soft, and flexible. Vegetable tanning is used mostly for sole and heavy-duty leathers. The chief vegetable tannins are water extracts of special types of wood or bark, especially quebracho and wattle. The main active constituent is tannic acid. The tannins penetrate the skin or hides after long periods of soaking, during which the molecular aggregates of the tannin form cross-links between the polypeptide chains of the skin proteins; hydrogen bonding is an important factor.
In mineral or chrome tanning, the sulfates of chromium, aluminum, and zirconium are used (the last two for white leather); here the reaction is of a coordination nature between the carboxyl groups of the skin collagen and the metal atom. Syntans are also used; these are sulfonated phenol or naphthols condensed with formaldehyde. Condensation products other than phenol having strong hydrogen-bonding power have been developed. Tannage by any method is a time-consuming and exacting process, requiring careful control of pH, temperature, humidity, and concentration factors.
For further information refer to Tanners' Council, University of Cincinnati, Cincinnati, Ohio.
See leather.

tantalic acid anhydride.   See tantalum oxide.

tantalic chloride.   See tantalum chloride.

tantalite.   (Fe, Hg)(TaNb)$_2$O$_6$. The most important ore of tantalum, found in Canada, Africa, Brazil, Malaysia. When niobium content exceeds that of tantalum, the ore is called columbite.

tantalum.
CAS: 7440-25-7.
Ta. Element of atomic number 73 in group VB of the periodic table, aw 180.9479, valences of 2, 3, 5; no stable isotopes.
Properties: (1) Black powder. (2) Steel-blue-colored metal when unpolished, nearly a platinum-white color when polished. D (1) 14.491, (2) 16.6 (worked metal), mp 2996C, bp 5425C, tensile strength of drawn wire may be as high as 130,000 psi, refr index 2.05, expansion coefficient 8×10^{-6} over range 20–1500C. Electrical resistance 13.6 $\mu\Omega$·cm (0C), 32.0 (500C). Soluble in fused alkalies; insoluble in acids except hydrofluoric and fuming sulfuric acids.
Occurrence: Canada, Thailand, Malaysia, Brazil.
Derivation: From tantalum potassium fluoride by heating in an electric furnace, by sodium reduction, or by fused salt electrolysis. The powdered metal is converted to a massive metal by sintering in a vacuum. Foot-long crystals can be grown by arc fusion. Corrosion resistance: 99.5% pure tantalum is resistant to all concentrations of hot and cold sulfuric acid (except concentrated boiling), hydrochloric acid, nitric and acetic acids, hot and cold dilute

sodium hydroxide, all dilutions of hot and cold ammonium hydroxide, mine and seawaters, moist sulfurous atmospheres, aqueous solutions of chlorine.
Grade: Powder 99.5% pure, sheet, rods, wire, ultrapure, single crystals.
Hazard: Dust or powder may be flammable. Toxic by inhalation.
Use: Capacitors, chemical equipment, dental and surgical instruments, rectifiers, vacuum tubes, furnace components, high-speed tools, catalyst, getter alloys, sutures and body implants, electronic circuitry, thin-film components.

tantalum alcoholate.
(pentaethoxytantalum). $(C_2H_5O)_5Ta$.
Use: Catalyst, intermediate for pure tantalates, preparing thin dielectric films.

tantalum carbide.   TaC.
Properties: Hard, heavy, refractory, crystalline solid. Mp 3875C, bp 5500C, d 14.5, hardness 1800 kg/sq mm, resistivity 30 $\mu\Omega\cdot$cm (room temperature). Extremely resistant to chemical action except at elevated temperature.
Derivation: Tantalum oxide and carbon heated at high temperatures.
Use: Cutting tools and dies, cemented carbide tools.

tantalum chloride.   (tantalic chloride; tantalum pentachloride).
CAS: 7721-01-9. $TaCl_5$.
Properties: Pale-yellow, crystalline powder; highly reactive. Decomposed by moist air. D 3.7, bp 242C, mp 221C. Soluble in alcohol and potassium hydroxide solution. Keep well stoppered.
Grade: Technical.
Use: Chlorination of organic substances, intermediate, production of pure metal.

tantalum disulfide. TaS_2.
Properties: Black powder or crystals. Mp above 3000C. Insoluble in water.
Available forms: 40 micron size.
Use: Solid lubricant.

tantalum fluoride.   (tantalum pentafluoride).
CAS: 7783-71-3. TaF_5.
Properties: Deliquescent crystals. Mp 97C, d 4.74. Soluble in water and nitric acid.
Use: Catalyst in organic reactions.

tantalum nitride.   TaN.
Properties: Hexagonal, brown, bronze, or black crystals. D 16.3, mp 3310–3410C. Insoluble in water; slightly soluble in aqua regia, nitric acid, hydrogen fluoride.
Grade: Technical, powder.

tantalum oxide.   (tantalic acid anhydride; tantalum pentoxide).
CAS: 1314-61-0. Ta_2O_5.

Properties: Rhombic, crystalline prisms. D 7.6, mp 1800C. Insoluble in water and acids except hydrogen fluoride.
Derivation: From tantalite, by removal of other metals.
Grade: Technical, optical (99.995% pure), single crystals.
Hazard: TLV: 5 mg/m^3.
Use: Production of tantalum; "rare-element" optical glass; intermediate in preparation of tantalum carbide; piezoelectric, maser, and laser applications; dielectric layers in electronic circuits.

tantalum pentachloride.   See tantalum chloride.

tantalum pentafluoride.   See tantalum fluoride.

tantalum pentoxide.   See tantalum oxide.

tantalum potassium fluoride.   (potassium tantalum fluoride; potassium fluotantalate).
CAS: 16924-00-8. K_2TaF_7.
Properties: White, silky needles. Slightly soluble in cold water; quite soluble in hot water.
Hazard: Toxic by inhalation.
Use: Intermediate in preparation of pure tantalum.

tantiron.   A ferrous alloy containing 84.8% iron, 13.5% silicon, 1% carbon, 0.4% manganese, 0.18% phosphorus, 0.05% sulfur. It is resistant to acids.
Use: Chemical equipment.

tapioca.   A product rich in starch, obtained from the tuberous roots of the cassava plant. It is used both directly as a food and as a thickening agent in emulsions for food products and industrial applications (lithographic inks). In many uses it is a satisfactory substitute for gum arabic.

TAPPI.   Abbreviation for Technical Association of the Pulp and Paper Industry.

taq polymerase.   A DNA polymerase isolated from the bacterium *Thermophilis aquaticus* which is very stable to high temperatures. It is used in PCR procedures and high temperature sequencing.

tar.   See coal tar; cigarette tar; pine tar; tar, wood.

tar acid.   Any mixture of phenols present in tars or tar distillates and extractable by caustic soda solutions. Usually refers to tar acids from coal tar and includes phenol, cresols, and xylenols. When applied to the products from other tars it should be qualified by the appropriate prefix, e.g., wood tar acid, lignite tar acid.
Properties: Soluble in alcohol and coal tar hydrocarbons. Combustible.
Grade: 15–18%, 25–28%, and 50–53% phenol.

Hazard: Toxic by ingestion, inhalation, and skin absorption; strong irritant to tissue.
Use: Wood preservative, insecticide for cattle and sheep dipping (dip oils or sheep dip), manufacture of disinfectants.

tar base. A basic nitrogen compound derived from coal tar, such as pyridine, picoline, lutidine, and quinoline.

tar camphor. See naphthalene.

tar camphor, chlorinated. See chloronaphthalene.

Tarcel. DLC form of Tarene.
See Tarene.

tar, dehydrated. (tar, refined).
Properties: Dark-brown, thick, viscid liquid. Combustible.
Derivation: Tar from which the water has been driven off.
Grade: Technical.
Hazard: Strong irritant, absorbed by skin.
Use: Waterproofing compounds, roads, medicine.

tare. (1) The weight of a container, wrapper, or liner that is deducted in determining the net weight of a material. (2) A weight used in analytical work to offset the weight of a container.

Tarene. Mixed naval stores (pine tar type).
Properties: Dark-brown, viscous liquid. D 0.99–1.02, ash 0.5%. Benzene soluble. Combustible.
Use: Plasticizer and softener.

targeted mutagenesis. Deliberate change in the genetic structure directed at a specific site on the chromosome. Used in research to determine the targeted region's function.
See mutation; polymorphism.

Tarmac. Proprietary preparation of blast furnace slag, refined tar, and other materials.
Use: Road dressing.

tarnish. A reaction product that occurs readily at room temperature between metallic silver and sulfur in any form. The well-known black film that appears on silverware results from reaction between atmospheric sulfur dioxide and metallic silver, forming silver sulfide. It is easily removable with a cleaning compound and is not a true form of corrosion. Plating with a mixture of silver and indium will increase tarnish resistance. Gold will also tarnish in the presence of a high concentration of sulfur in the environment.

tar oil. See creosote; coal tar.

tar oil, wood. (pine tar oil).

Properties: Almost colorless liquid when freshly distilled, turns dark reddish-brown; strong odor and taste. D 0.862–0.872. Soluble in ether, chloroform, alcohol, and carbon disulfide; flash p 144F (62.2) (CC). Combustible.
Derivation: Obtained by the destructive distillation of the wood of *Pinus palustris*.
Method of purification: Rectification.
Grade: Technical, rectified, refined.
Use: Paints, waterproofing paper, rubber reclaiming, varnishes, stains, ore flotation, cattle dips, insecticides.

tar, refined. See tar, dehydrated.

tar sands. See oil sands.

tar soap. Soap containing juniper tar oil.

tartar, cream of. See potassium bitartrate.

tartar emetic. See antimony potassium tartrate.

tartaric acid. (dihydroxysuccinic acid).
CAS: 87-69-4. $HOOC(CH_2O)_2COOH$.
Properties: Colorless, transparent crystals or white, crystalline powder; odorless; acidic taste; stable in air. Soluble in water, alcohol, and ether. D 1.76, mp 170C. It has unusual optical properties. The molecule has two asymmetric carbon atoms that result in four isomeric forms, three of which are the natural D-form and the corresponding L-form, which are optically active, and the inactive *meso* form:

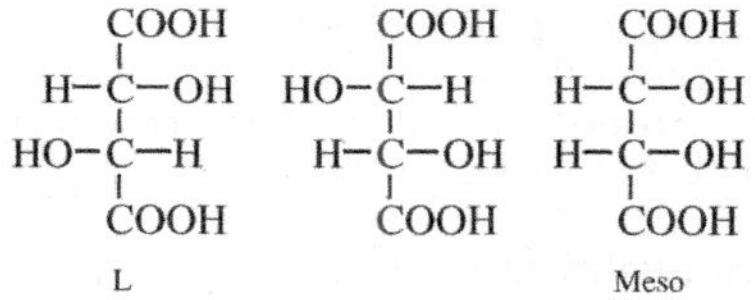

The occurrence of some DL-tartaric acid along with natural D-tartaric acid in the wine industry is explained on the basis of partial racemization. The L-form is the commercial product.
Derivation: Occurs naturally in wine lees; made synthetically from maleic anhydride and hydrogen peroxide and by an enzymatic reaction with a succinic acid derivative.
Grade: Technical, CP, crystals, powder, granular, NF, FCC.
Use: Chemicals (cream of tartar, tartar emetic, acetaldehyde), sequestrant, tanning, effervescent beverages, baking powder, fruit esters, ceramics, galvanoplastics, photography (printing and developing, light-sensitive iron salts), textile industry, silvering mirrors, coloring metals, acidulant in foods.

tartrazine. (FD&C Yellow No. 5; (3-carboxy-5-hydroxy)-1-*p*-sulfophenyl-4-*p*-sulfophenylazopyrazole trisodium salt; CI 19140).

CAS: 1934-21-0. $C_{16}H_9N_4O_9S_2$.
Properties: Bright orange-yellow powder. Soluble in water.
Hazard: An allergen.
Use: Dye, especially for foods, drugs, and cosmetics.

tar, wood. (tar, hardwood).
Properties: Black, syrup-like, viscous fluid.
Derivation: A by-product of the destructive distillation of wood.
Grade: Technical.
Hazard: Flammable, moderate fire risk.
Use: Hardwood pitch, wood creosote, heavy high-boiling wood oils, wood-preserving oils, paint thinners.

taste. See flavor.

TATA box. A sequence found in the promoter (part of the 5′ flanking region) of many genes. Deletion of this site (the binding site of transcription factor TFIID) causes a marked reduction in transcription, and gives rise to heterogeneous transcription initiation sites.

Taube, Henry. (1915–2005). A Canadian-born chemist who won the Nobel Prize for chemistry in 1983 for his pioneering work in inorganic chemistry and the study of electron-transfer reactions, particularly of metal complexes. Known as an outstanding teacher, he was admired and respected by students and colleagues for his work at Stanford University.

taurin. See gracilin.

taurine. (2-aminoethanesulfonic acid).
CAS: 107-35-7. $NH_2CH_2CH_2SO_3H$.
A crystallizable amino acid found in combination with bile acids; its combination with cholic acid is called taurocholic acid.
Properties: Solid. Decomposes at 300C. Soluble in water; insoluble in alcohol.
Derivation: Isolated from ox bile, organic synthesis.
Hazard: Toxic by ingestion.
Use: Biochemical research, pharmaceuticals, wetting agents.

taurocholic acid. (cholaic acid; cholyltaurine).
CAS: 81-24-3. $C_{26}H_{45}NO_7S$.
Occurs as sodium salt in bile. It is formed by the combination of the sulfur-containing amino acid taurine and cholic acid as the sodium salt. It aids in digestion and absorption of fats.
Properties: Crystals. Stable in air. Mp 125C (decomposes.) Freely soluble in water; soluble in alcohol, almost insoluble in ether and ethyl acetate.
Derivation: Isolation from bile.
Use: Biochemical research, emulsifying agent in foods (not over 0.1%).

tautomerism. A type of isomerism in which migration of a hydrogen atom results in two or more structures, called tautomers. The two tautomers are in equilibrium. For example, acetoacetic ester has the properties of both an unsaturated alcohol and a ketone. The tautomers are called enol and keto. See enol; isomer (1).

"Taxol." *[Bristol-Myers].* TM for 5-β,20-epoxy-1,2-α,4,7-β,10-β,13-α-hexahydroxy-tax-11-en-9-one 4,10-diacetate 2-benzoate 13-ester with (2r,3s)-*n*-benzoyl-3-phenylisoserine.
See paclitaxel.

taxon. Any named group of organisms, not necessarily a clade. A taxon may be designated by a Latin name or by a letter, number, or any other symbol.

taxonomy. The science of naming and classifying organisms.

taxotere.
CAS: 114977-28-5. $C_{43}H_{53}NO_{14}$.
Hazard: A poison. Human systemic effects.

Tb. Symbol for terbium.

TBC. (tri-*n*-butyl citrate).
CAS: 77-94-1.
Aqueous-based pharmaceutical coatings, direct food ingredient.

TBDMS chloride. See *tert*-butylchlorodimethylsilane.

TBH. Abbreviation for technical benzene hexachloride.
Use: An insecticide.
See 1,2,3,4,5,6-hexachlorocyclohexane.

TBP. Abbreviation for tributyl phosphate.

TBSM. See *p-tert*-butoxycarbonyloxystyrene monomer.

TBT. Abbreviation for tetrabutyl titanate.

TBTO. Abbreviation for bis(tributyltinoxide).

TC. Abbreviation for trichloroacetic acid or its sodium salt.

Tc. Symbol for technetium.

TCA cycle. (tricarboxylic acid cycle; Krebs cycle; citric acid cycle). A series of enzymatic reactions occurring in living cells of aerobic organisms, the net result of which is the conversion of pyruvic acid, formed by anaerobic metabolism of carbohydrates, into carbon dioxide and water. The metabolic intermediates are degraded by a combination of decarboxylation and dehydrogenation. It is the major terminal pathway of oxidation in animal, bacterial, and plant cells. Recent research indicates that the TCA cycle may have predated life on earth and may have provided the pathway for formation of amino acids.

TCB.   Abbreviation for tetracarboxybutane.

TCBO.   See trichlorobutylene oxide.

TCC.   Abbreviation for Tagliabue closed cup, a standard method of determining flash points.

TCDD.   See dioxin.

TCEO.   See epoxy-1,1,2-trichloroethane.

TCHH.   See cyhexatin.

TCHTH.   See cyhexatin.

TCP.   Abbreviation for (1) 2-hydroxy-3,5,6-trichloropyridine, (2) tricresyl phosphate.

TDE.   (generic name for 1,1-dichloro-2,2-bis(*p*-chlorophenyl)ethane;   tetrachlorodiphenylethane; DDD).
CAS: 72-54-8. $(ClC_6H_4)_2CHCHCl_2$.
Properties: Colorless crystals. Mp 109–110C. Soluble in organic solvents; insoluble in water; not compatible with alkalies. Combustible.
Derivation: Chlorination of ethanol and condensation with chlorobenzene.
Grade: Technical.
Hazard: Toxic by ingestion, inhalation, and skin absorption; use restricted in some states. Questionable carcinogen.
Use: Dusts, emulsions, and wettable powders for contact control of leaf rollers and other insects.

TDI.   Abbreviation for toluene diisocyanate.

TDP.   Abbreviation for 4,4'-thiodiphenol.

TDQP.   Abbreviation for trimethyldihydroquinoline polymer.

Te.   Symbol for tellurium.

TEA.   (1) Abbreviation for triethanolamine. (2) Abbreviation for triethylaluminum.

TEAC.   Abbreviation for tetraethylammonium chloride.

tea catechin.   See epigallocatechin 3-gallate.

tear gas.   (2-chloro-1-phenylethanone).
C_8H_7ClO. An irritant gas that can be readily stored and placed in artillery shells, bombs, or grenades that detonate and release as an aerosol.
Hazard: Causes profuse tearing and temporary blindness.
Use: In crowd control and warfare.

tear resistance.   The amount of force necessary to cause additional damage, in a cut made in a sample, by force on the sample, where the force is primarily normal to the plane of the cut.

tebufenozide.
CAS: 112410-23-8. $C_{22}H_{28}N_2O_2$.
Hazard: Low toxicity by ingestion, inhalation, and skin contact.
Use: Agricultural chemical.

TEC.   (triethyl citrate).
CAS: 77-93-0.
Aqueous-based pharmaceutical coatings.

techneplex.
CAS: 65454-61-7. $C_{14}H_{17}N_3O_{10}Tc•Na$.
Hazard: Moderately toxic.

technetium.   Tc. Element with atomic number 43, group VIIB of the periodic table, aw 98.9062, valences of 4, 5, 6, 7; three radioactive isotopes of half-life more than 105 years, also several of relatively short half-life, some of which are β emitters. Technetium was first obtained by the deuteron bombardment of molybdenum, but since has been found in the fission products of uranium and plutonium. The chemistry of technetium has been studied by tracer techniques and is similar to that of rhenium and manganese. The free metal is obtained from reactor fission products by solvent extraction followed by crystallization as ammonium pertechnetate, which is reduced with hydrogen. The metal is silver-gray in appearance, mp 2200C (4000F), d 11.5, slightly magnetic. Compounds of the types TcO_2, Tc_2O_7, NH_4TcO_4, etc. have been prepared. The pertechnetate ion has strong anticorrosive properties. Technetium and its alloys are superconductors and can be used to create high-strength magnetic fields at low temperature. Tc-99 (metastable) is the most widely used isotope in nuclear medicine.
Use: Metallurgical tracer, cryochemistry, corrosion resistance, nuclear medicine.

Technical Association of the Pulp and Paper Industry.   (TAPPI). A professional group of scientists devoted to the interests of pulp and paper chemistry and technology. Founded in 1915, it has seven sections, each concerned with a specific phase of the industry. It also has ten local sections that hold monthly meetings. The association publishes its own journal, as well as industry data sheets, bibliographies, technical monographs on subjects relating to the paper industry. It establishes standards of quality and testing procedures. The address is 15 Technology Parkway South, Suite 115, Peachtree Corners, GA 30092. Website: http://www.tappi.org

technology transfer.   The process of transferring scientific findings from research laboratories to the commercial sector.

"Tecto" *[American].*   TM a flowable SC fungicide.

"Tedlar" *[Du Pont].*   TM for polyvinylfluoride film.

TEDP.   Abbreviation for tetraethyl dithiopyrophosphate.
See sulfotepp.

"Tedur" *[Albis].*   TM for polysulfide polymers.
Available forms: Glass, mineral, and mineral/glass grades.
Use: Injection molding for high-temperature and performance electronic and automotive parts.

tefestrol.
CAS: 129407-28-9.
Hazard: A reproductive hazard.

Teflon.   Tetrafluoroethylene (TFE) fluorocarbon polymers available as molding and extrusion powders, aqueous dispersion, film, finishes, and multifilament yarn or fiber. The name also applies to fluorinated ethylene–propylene (FEP) resins available in the same forms. The no-stick cookware finishes may be of either type. Fibers are monofilaments made from copolymer of TFE and FEP.
Use: Packing, bearings, filters, electrical insulation, high-temperature industrial plastics, cooking utensils, plumbing sealants, coating glass fiber for architectural structure composites, bonding industrial diamonds to metal in the manufacture of grinding wheels.
See fluorocarbon polymer.

Teflon EFTE.   A co-polymer of ethylene and tetrafluroethylene, also known as Tefzel.
Use: Non-stick surfaces.

Teflon FEP.   Non-stick melt and flow system.
Use: During backing to provide non-porous films.

Teflon PTFE.   A non-stick double coating (primer, top coat).
Use: Coating for the highest operating temperatures.

TEG.   Abbreviation for tetraethylene glycol and triethylene glycol.

Tego.   Thin tissue impregnated with heat-convertible phenol–formaldehyde resin, supplied in rolls. Produces waterproof bond with plywood veneers.
Use: Hot-press bonding of furniture veneers, premium wall paneling.

"Tekflame" *[Exxon].*   TM for a firefighting fuel.

TEL.   Abbreviation for tetraethyl lead.

telluric acid.   (hydrogen tellurate).
CAS: 7803-68-1. $H_2TeO_4 \cdot 2H_2O$ or H_6TeO_6.

Properties: White, heavy crystals. D 3.07, mp 136C. Soluble in hot water and alkalies; slightly soluble in cold water.
Derivation: Action of sulfuric acid on barium tellurate.
Hazard: As for tellurium.
Use: Chemical reagent.

telluric bromide.   See tellurium tetrabromide.

tellurium.
CAS: 13494-80-9.   Te. A nonmetallic element with many properties similar to selenium and sulfur. Atomic number 52; group VIA of the period table; aw 127.60; valences of 2, 4, 6; eight stable isotopes.
Properties: Silvery-white, lustrous solid with metal characteristics. D 6.24 g/cc (30C), Mohs hardness 2.3, mp 450C, bp 990C. Soluble in sulfuric acid, nitric acid, potassium hydroxide, and potassium cyanide solutions; insoluble in water. Imparts garlic-like odor to breath, can be depilatory. It is a p-type semiconductor and its conductivity is sensitive to light exposure.
Source: From anode slime produced in electrolytic refining of copper and lead.
Derivation: Reduction of telluric oxide with sulfur dioxide; by dissolving the oxide in a caustic soda solution and plating out the metal.
Grade: Powder, sticks, slabs, and tablets, 99.5% pure, crystals up to 99.999% pure.
Hazard: (Metal and compounds as tellurium) Toxic by inhalation. Halitosis.
Use: Alloys (tellurium lead, stainless steel, iron castings), secondary rubber vulcanizing agent, manufacture of iron and stainless steel castings, coloring agent in glass and ceramics, thermoelectric devices, catalysts, with lithium in storage batteries for spacecraft.

tellurium bromide.   See tellurium dibromide and tellurium tetrabromide.

tellurium chloride.   See tellurium dichloride.

tellurium dibromide.   (tellurium bromide; tellurous bromide). $TeBr_2$.
Properties: Blackish-green, crystalline mass or gray to black needles; very hygroscopic. Mp 210C, bp 339C. Decomposed by water; soluble in ether; violet vapor.
Hazard: As for tellurium.

tellurium dichloride.   (tellurium chloride; tellurous chloride). $TeCl_2$.
Properties: Amorphous, black mass, greenish-yellow when powdered. Decomposed by water. D 6.9, bp 327C, mp 209C.
Hazard: As for tellurium.

tellurium dioxide.   (tellurous acid anhydride).
CAS: 7446-07-3. TeO_2.

Properties: Heavy, white, crystalline powder; odorless. D 5.89, mp 733C, bp 1245C. Soluble in concentrated acids, alkalies; slightly soluble in dilute acids, water.
Hazard: As for tellurium.

tellurium disulfide.   (tellurium sulfide). TeS_2.
Properties: Red powder, turns in time to a dark-brown amorphous powder, fuses to gray lustrous mass. Soluble in alkali sulfides; insoluble in acids, water.
Hazard: As for tellurium.

tellurium lead.   See lead; tellurium.

tellurium sulfide.   See tellurium disulfide.

tellurium tetrabromide.   (telluric bromide; tellurium bromide).
CAS: 10031-27-3. $TeBr_4$.
Properties: Yellow crystals. D 4.3, mp 363C, bp 420C (decomposes into bromine and dibromide). Soluble in a little water (decomposes in excess water).
Hazard: As for tellurium.

tellurous acid. H_2TeO_3.
Properties: White, crystalline powder. D 3.053, mp 40C (decomposes). Soluble in dilute acids, alkalies; slightly soluble in water, alcohol.
Hazard: As for tellurium.

tellurous acid anhydride.   See tellurium dioxide.

tellurous bromide.   See tellurium dibromide.

tellurous chloride.   See tellurium dichloride.

telodrin.   (1,3,4,5,6,7,8,8-octachloro-3a,4,7,7a-tetrahydro-4,7-methanonaphthalan).
CAS: 297-78-9.
Properties: Vap press 3 mm Hg (20C). Soluble in acetone, benzene, carbon tetrachloride, fuel oil, toluene, xylene.

telomer.   Polymeric products formed from a telomerization reaction.

telomerase.   The enzyme that directs the replication of telomeres.

telomere.   The end of a chromosome. This specialized structure is involved in the replication and stability of linear DNA molecules.
See DNA replication.

telomerization reactions.   In telomerization reactions, a polymerizable unsaturated compound (the taxogen) is reacted under polymerization conditions in the presence of radical-forming catalysts or promoters with a so-called telogen. During the reaction, the telogen is split into radicals that attach to the ends of the polymerizing taxogen and in some instances add on to the double bond of the taxogen and thereby form chains whose terminal groups are formed of the radicals formed from the telogen. Organic compounds containing an olefinic double bond, such as ethylene, propylene, hexene, octene, or styrene, are normally employed as taxogens. Many different types of compounds can be employed as telogens, for example, halogenated hydrocarbons, such as chloroform or carbon tetrachloride, halogen derivatives of cyanogen, such as cyanogen chloride, aldehydes, alcohols, and the like. Radical-forming catalysts, such as organic peroxides, hydrogen peroxide, aliphatic azo compounds of the type of azoisobutyric acid nitrile, and redox systems are employed for telomerization reactions. Telomerization reactions are as a rule carried out at an elevated temperature up to 250C. When volatile reactants are used, the reaction is carried out under elevated pressures, i.e., between 20 and 1000 atm.

"Telone" *[Dow Agrosciences].*   TM for fumigants containing 1,3-dichloropropene and related C_3 hydrocarbons.

"Tel-Tale" *[Grace].*   TM for silica gel that is impregnated with cobalt chloride and turns from blue to pink as the relative humidity increases.

TEM.   Abbreviation for triethylene melamine.

temephos.   (O,O,O',O'-tetramethyl-O,O'-thio-di-p-phenylene phosphorothionate).
CAS: 3383-96-8.
Properties: Mp 30–30.5C.
Hazard: Cholinesterase inhibitor. Questionable carcinogen.
Use: Larvicide for mosquito and blackfly.

Temfibre.   Wood-based chemicals.

tempalgin.
CAS: 39296-38-3. $C_{13}H_{17}N_3O_4S•C_9H_{17}NO•C_7H_8O_3S•Na$.
Hazard: Moderately toxic by ingestion.

temper.   To increase the hardness and strength of a metal by quenching or heat treatment.

temperate phage.   A phage whose DNA may be incorporated into the host-cell genome without being expressed. Contrast with a virulent phage, which destroys the host cell.

temperature.   The thermal state of a body considered with reference to its ability to communicate heat to other bodies (J. C. Maxwell). There

is a distinction between temperature and heat, as is evidenced by Helmholtz's definition of heat as "energy that is transferred from one body to another by a thermal process," whereby a thermal process is meant radiation, conduction, and/or convection. Temperature is measured by such instruments as thermometers, pyrometers, thermocouples, etc., and by scales such as centigrade (Celsius), Fahrenheit, Rankine, Reaumur, and absolute (Kelvin).
See absolute temperature; thermodynamics.

template.   A macromolecular pattern for the synthesis of another molecule. For example, DNA is a template for RNA synthesis.

Temprite CPVC.   Chlorinated polyvinyl chloride resins formulated for process piping, valves, tanks, ductwork, and protective covers.
Use: For electrical conductors, residential hot and cold water-supply piping.

tenacity.   Strength per unit weight of a fiber or filament, expressed as g/denier. It is the rupture load divided by the linear density of the fiber.
See tensile strength; denier.

tenderization.   (1) Treatment of meats with UV radiation or with certain enzyme preparations to accelerate softening of the collagen fibers and thus reduce the time necessary to "hang" the meat. (2) The degradation and mechanical weakening undergone by textile fibers under excessive wet abrasion during laundering.

"Tenox" *[Eastman].*   TM for food-grade antioxidants containing one or more of the following ingredients: butylated hydroxyanisole, butylated hydroxytoluene, and/or propyl gallate with or without citric acid. Some formulas are supplied in solvents such as propylene glycol.

"Tenox GT-1" *[Eastman].*   TM for natural mixed tocopherols.
Use: Antioxidant in foods, cosmetics, and ointments.

tensile strength.   The rupture strength (stress–strain product at break) per unit area of a material subjected to a specified dynamic load; it is usually expressed in pounds per square inch (psi). This definition applies to elastomeric materials and to certain metals.
See tenacity.

tension cracking.   See stress cracking.

tenter.   A machine used for holding a processed fabric taut as it is fed into a wind-up or to a cutter. It consists of a frame along the inner sides of which travel continuous chains to which gripping devices are attached at intervals of a few inches; these may be either hooks or clamps of various kinds. As the fabric moves into the machine, the edges are engaged by the grippers and are automatically released at the end of the frame.

tepa.   (generic name for tris(1-aziridinyl)phosphine oxide).
See triethylenephosphoramide.

Tepidone. $(C_4H_9)_2NC(S)SNa$. A water solution of sodium dibutyldithiocarbamate.
Properties: Amber liquid. D 1.08.
Use: To accelerate vulcanization of natural and synthetic rubber and latex compounds.

TEPP.   (ethyl pyrophosphate; tetraethyl pyrophosphate).
CAS: 107-49-3. $(C_2H_5)_4P_2O_7$.
Properties: Water-white to amber liquid depending on purity; hygroscopic. D 1.20, refr index 1.420, bp (pure compound) 135–138C (1 mm Hg). Miscible with water and all organic solvents except aliphatic hydrocarbons; hydrolyzed in water with formation of mono-, di-, and triethyl orthophosphates; water solutions attack metals; commercial material contains 40% TEPP.
Derivation: From phosphorus oxychloride and ethanol or phosphorus oxychloride and triethyl phosphate.
Grade: 40%.
Hazard: Toxic by skin contact, inhalation, or ingestion; rapidly absorbed through skin; repeated exposure may, without symptoms, be increasingly hazardous; cholinesterase inhibitor, use may be restricted.
Use: Insecticide for aphids and mites, rodenticide.

tera-.   Prefix meaning 10^{12} units (symbol T), 1 Tg = 1 teragram = 10^{12} grams.

teratogen.   An agent that causes growth abnormalities in embryos, genetic modifications in cells, etc.; ionizing radiation can have this effect.

teratogenic.   Substances such as chemicals or radiation that cause abnormal development of an embryo.
See mutatgen.

terbacil.
CAS: 5902-51-2. $C_9H_{13}ClN_2O_2$.
Properties: Colorless crystals. Mp 175C. Soluble in dimethylacetamide and cyclohexanone; partially soluble in xylene and butyl acetate.
Use: Herbicide.

Terbec.   A wet strength improver for soils; based on 4-*tert*-butylcatechol.

terbia.   See terbium oxide.

terbium.   Tb. Atomic number 65, group IIIB of the periodic table, a rare-earth element of the

yttrium subgroup (lanthanide series), aw 158.9254, valences of 3, 4; no stable isotopes.
Properties: Metallic luster. D 8.332, mp 1356C, bp 2800C. Reacts slowly with water; soluble in dilute acids. Salts are colorless, highly reactive, handled in inert atmosphere or vacuum.
Source: See rare earths.
Derivation: Reduction of fluoride with calcium.
Grade: Regular, 99.9+% purity (ingots, lumps), single crystals.
Use: Phosphor activator, dope for solid-state devices.

terbium chloride, hexahydrate.
$TbCl_3 \cdot 6H_2O$.
Properties: Transparent, colorless, prismatic crystals. D 4.35, mp (anhydrous) 588C, very hygroscopic. Readily soluble in water or alcohol.
Derivation: By treatment of carbonate or oxide with hydrochloric acid in an atmosphere of dry hydrogen chloride.

terbium earths.   Elements with atomic numbers 63 through 66 that are: 63 europium (eu), 64 gadolinium (gd), 65 terbium (tb), 66 dysprosium (dy).

terbium fluoride. $TbF_3 \cdot 2H_2O$.
Properties: Solid. Mp 1172C, bp 2280C. Insoluble in water.
Hazard: Strong irritant.
Use: Source of terbium.

terbium nitrate. $Tb(NO_3)_3 \cdot 6H_2O$.
Properties: Colorless, monoclinic needles or white powder. Mp 89.3C. Soluble in water.
Derivation: By treatment of oxide, carbonate, or hydroxide with nitric acid.
Hazard: Strong oxidant, fire risk in contact with organic materials.

terbium oxide.   (terbia). Tb_2O_3.
Properties: Dark-brown powder. Soluble in dilute acids; slightly hygroscopic, absorbs carbon dioxide from air.
Derivation: By ignition of hydroxides or salts of oxy-acids.
Grade: 98–99%.
See rare earth.

terbium sulfate. $Tb_2(SO_4)_3 \cdot 8H_2O$.
Properties: Colorless crystals. Lose $8H_2O$ at 360C. Soluble in water.

terbufos.   (*O,O*-diethyl-*S*-(*tert*-butyl)methyl phosphorodithioate).
CAS: 13071-79-9. $C_9H_{21}O_2PS_3$. An ester of phosphoric acid.
Properties: Yellowish liquid. D 1.10, bp 70C (0.01 mm Hg), fp −29C, flash p 88C (TOC). Soluble in alcohol, acetone.
Hazard: Moderate fire risk. Toxic by ingestion. Cholinesterase inhibitor. Questionable carcinogen.
Use: Soil insecticide.

terbutol.
CAS: 1918-11-2. $C_{17}H_{27}NO_2$.
Hazard: Low toxicity by ingestion and skin contact.
Use: Agricultural chemical.

terebene.   A mixture of terpenes, chiefly dipentene and terpinene.
Properties: Colorless liquid. D 0.862–0.866, inactive optical rotation, Bp 160–172C. Soluble in alcohol; insoluble in water.
Derivation: From oil of turpentine.
Hazard: Flammable, moderate fire risk. Toxic by ingestion and inhalation.
Use: To impart water and oil resistance to cellulosics.

terephthalic acid.   (*p*-phthalic acid; TPA; benzene-*p*-dicarboxylic acid).
CAS: 100-21-0. $C_6H_4(COOH)_2$.
Properties: White crystals or powder. D 1.51, sublimes above 300C. Insoluble in water, chloroform, ether, acetic acid; slightly soluble in alcohol; soluble in alkalies. Combustible.
Derivation: (1) Oxidation of *p*-xylene or of mixed xylenes and other alkyl aromatics (phthalic anhydride). (2) Reacting benzene and potassium carbonate over a cadmium catalyst.
Grade: Commercial, fiber.
Use: Production of linear, crystalline polyester resins, fibers, and films by combination with glycols; reagent for alkali in wool; additive to poultry feeds.

2-terephthaloylbenzoic acid. $HOOCC_6H_4$ COC_6H_4COOH.
Properties: White to gray powder. Mp 233–237C.
Use: Chemical intermediate.

terephthaloyl   chloride.   (1,4-benzenedicarbonyl chloride). $C_6H_4(COCl)_2$.

Properties: Colorless needles. Mp 82–84C, bp 259C, flash p 356F (180C). Decomposes in water and alcohol; soluble in ether. Combustible.
Hazard: Skin irritant.
Use: Dye manufacture; synthetic fibers, resins, films; UV absorption; pharmaceuticals; rubber chemicals; cross-linking agent for polyurethanes and polysulfides.

"Tergitol" *[Union].*   TM for a series of nonionic and anionic surfactants.

Use: Detergents, wetting agents, emulsifiers in water systems, leveling and spreading agents.

terminal transferase. An enzyme that catalyzes the addition of nucleotide residues of a single kind to the 3′ end of DNA chains, e.g., polyA synthetase.

termination codons. XH.
See stop codons.

termination factors. (release factors). Protein factors of the cytoplasm required in releasing a completed polypeptide chain from a ribosome.

termination sequence. A DNA sequence that appears at the end of a transcriptional unit and signals the end of transcription.

ternary. Descriptive of a solution or alloy having three components, or of a chemical compound having three constituent elements or groups.

ternary acid. An acid that contains three different elements. An example would be acetic acid, CH_3COOH.

ternary diagram. A constitution diagram for a ternary alloy system.

terneplate. (roofing tin).
A lead-tin alloy used for coating iron or steel; its composition is 75% lead and 25% tin. It has a dull finish.
Use: Roofing, deep stamping, gasoline tanks, etc.

1,4(8)-terpadiene. See terpinolene.

terpene. $C_{10}H_{16}$. An unsaturated hydrocarbon occurring in most essential oils and oleoresins of plants. The terpenes are based on the isoprene unit

C_5H_8, and may be either acyclic or cyclic with one or more benzenoid groups. They are classified as monocyclic (dipentene), dicyclic (pinene), or acyclic (myrcene), according to the molecular structure. Many terpenes exhibit optical activity. Terpene derivatives (camphor, menthol, terpineol, borneol, geraniol, etc.) are called terpenoids; many are alcohols.
See polyterpene resin.

terpene alcohol. A generic name for an alcohol related to or derived from a terpene hydrocarbon, such as terpineol or borneol.

terpene hydrochloride. See 2-bornyl chloride.

terpeneless oil. An essential oil from which the terpene components have been removed by extraction and fractionation, either alone or in combination. The optical activity of the oil is thus reduced. The terpeneless grades are much more highly concentrated than the original oil (15–30 times). Removal of terpenes is necessary to inhibit spoilage, particularly of oils derived from citrus sources. On atmospheric oxidation the specific terpenes form compounds that impair the value of the oil; for example, *d*-limonene oxidizes to carvone and γ-terpinene to *p*-cymene. Terpeneless grades of citrus oils are commercially available.

terpene resin, natural.
Properties: Extracted from wood. Mp: 155C
Use: Food additive.

terpene resin, synthetic.
Properties: Polymers of α-pinene, β-pinene, and/or dipentene (FCC III). Mp: 112–118C. Soluble in benzene.
Use: Food additive.

terpenoid. See terpene.

1,3-terphenyl. See *m*-terphenyl.

***m*-terphenyl.** (*m*-diphenylbenzene; isodiphenylbenzene; 1,3-terphenyl).
CAS: 92-06-8. $C_{18}H_{14}$.
Properties: Needles. Mw 230.32, mp 86–87C, bp 379C.
Hazard: Moderately toxic by ingestion. Combustible. Eye and upper respiratory tract irritant.

***o*-terphenyl.** (1,2-diphenylbenzene).
CAS: 84-15-1. $C_{18}H_{14}$.
Properties: Prisms from MeOH. Mp 58–59C, bp 337C, mw 230.32, flash p >230F.
Hazard: Combustible. Eye and upper respiratory tract irritant.

***p*-terphenyl.** (1,4-diphenylbenzene).
CAS: 92-94-4. $(C_6H_5)_2C_6H_4$.
Properties: Liquid. D 1.234 (0C), mp 213C, bp 405C, flash p 405F (207C). Combustible.
Derivation: From *p*-dibromobenzene or bromobenzene and sodium.
Method of purification: Zone-melting.
Grade: Technical, scintillation.
Hazard: Toxic by ingestion and inhalation. Eye and upper respiratory tract irritant.
Use: Polymerized with styrene to make a plastic phosphor. Single crystals used as scintillation counters.

terpilenol. See terpineol.

terpinene. $C_{10}H_{16}$. A mixture of three isomeric cyclic terpenes, α-, β-, and γ-terpinene. α-terpinene has a bp of 180–182C and d 0.8484 (14C). It is found in cardamom, marjoram, and coriander oils.

β-terpinene has a bp of 183C and d 0.853 (15C). It is found in coriander, lemon, cumin, and ajawan oils.
Use: Synthetic flavors.

terpineol.   (α-terpineol; β-terpineol; γ-terpineol; terpilenol).
CAS: (α) 8006-39-1. $C_{10}H_{17}OH$.
Properties: Colorless liquid or low-melting transparent crystals; lilac odor. D 0.930–0.936 (25C), solidification p 2C, optical rotation between −0 degrees 10 min and +0 degrees 10 min, boiling range 214–224C, 90% within 5C, refr index 1.4825–1.4850 (20C). Soluble in two volumes of 70% alcohol; slightly soluble in water, glycerol. Combustible.
Available forms: Mixture of the three isomers.
Derivation: By heating terpin hydrate with phosphoric acid and distilling or with dilute sulfuric acid, using an azeotropic separation; fractional distillation of pine oil. Occurs naturally in several essential oils.
Grade: Technical, perfumery, extra, prime, FCC.
Use: Solvent for hydrocarbon materials, mutual solvent for resins and cellulose esters and ethers, perfumes, soaps, disinfectant, antioxidant, flavoring agent.

Terpineol 318.   A highly refined mixture of tertiary terpene alcohols, predominantly α-terpineol; 97% total terpene alcohols; colorless liquid; fp below 10C; d 0.937 (15.6/15.6C); ASTM distillation range 5–95%, (216–220C).

terpin hydrate.   (dipentene glycol).
$CH_3(OH)C_6H_9C(CH_3)_2OH \cdot H_2O$.
Properties: Colorless, lustrous, rhombic, crystalline prisms or white powder; slight characteristic odor; slightly bitter taste. Efflorescent, mp 115–117C, anhydrous mp 105C, bp 258C. Soluble in alcohol and ether; slightly soluble in water. Combustible.
Derivation: From turpentine oil or *d*-limonene.
Grade: Technical, NF.
Use: Pharmaceuticals, source of terpineol, medicine (expectorant).

terpinolene.   (1,4(8)-terpadiene).
CAS: 586-62-9. $C_{10}H_{16}$.
Properties: Water-white to pale-amber liquid. D 0.864 (15.5/15.5C), bp 183–185C, flash p 99F (37.2C) (CC), bulk d 7.2 lb/gal (15.5C). Insoluble in water; soluble in alcohol, ether, glycol.
Derivation: Fractionation of wood turpentine.
Hazard: Flammable, moderate fire risk.
Use: Solvent for resins, essential oils; manufacture of synthetic resins, synthetic flavors.

terpinyl acetate.
CAS: 80-26-2. $C_{10}H_{17}OOCCH_3$.
Properties: Colorless liquid; odor suggestive of bergamot and lavender. D 0.958–0.968 (15C), refr index 1.4640–1.4660 (20C), optical rotation varies around 0 degrees, fp −50C, bp 220C. Soluble in five

or more volumes of 70% alcohol; slightly soluble in water and glycerol. Combustible.
Derivation: By heating terpineol with acetic acid or anhydride in the presence of sulfuric acid and subsequent distillation.
Grade: Technical, prime, extra, FCC.
Use: Perfumes, flavoring agent.

terpolymer.   A polymer made from three monomers, e.g., ABS polymers.

"Terramycin" *[Pfizer].*   Proprietary preparation containing oxytetracycline and oxytetracycline hydrochloride.
Use: Antibiotic.

Terra XPS.   Fiber optic coatings.

terreic acid.   (5,6-epoxy-3-hydroxy-p-toluquinone; 2-hydroxy-3-methyl-1,4-benzoquinone 5,6-epoxide; 3-hydroxy-4-methyl-7-oxabicyclo(4,1,0)hept-3-ene-2,5-dione; (1R-*cis*)-3-hydroxy-4-methyl-7-oxabicyclo(4.1.0)hept-3-ene-2,5-dione; (1S,6R)-4-hydroxy-3-methyl-7-oxabicyclo[4.1.0]hept-3-ene-2,5-dione).
CAS: 121-40-5. $C_7H_6O_4$. An epoxide.
Derivation: Produced by *aspergillus terreus*.
Hazard: Mycotoxin; poison; mutagen.

Tersan 75.   A turf fungicide containing 75% thiram. Tersan OM contains 45% thiram and 10% hydroxymercurichlorophenol.
Hazard: Toxic by ingestion and inhalation.

tert-.   Abbreviation for tertiary.

dl-**tertatalol.**
CAS: 34784-64-0. $C_{16}H_{25}NO_2S$.
Hazard: Human systemic effects.

tertiary.   (1) For chemical meaning see primary. (2) In petroleum extraction technology, the term refers to recovery of petroleum by pumping detergents, high polymers, silicates, etc. into the rock structure. These techniques are generally known as enhanced oil recovery.
See chemical flooding; hydraulic fracturing.

tertiary alcohol.   An alcohol in which the carbon atom to which the hydroxyl group is attached is also attached to three other carbon atoms.

tertiary amine.   An amine in which three carbon atoms are attached to the amino nitrogen.

tertiary carbon.   Carbon atom that is joined to three other carbon atoms.

tertiary structure.   The three-dimensional conformation of a protein in its native folded state.

testing, chemical.   Identification of a substance by means of reagents, chromatography,

spectroscopy, melting and boiling point determination, etc.
See analytical chemistry.

testing, physical. Application of any procedure whose object is to determine the physical properties of a material. There are four major categories of tests: (1) Those that are direct measurements of a property, e.g., tensile strength. (2) Those that subject the material to actual service conditions; these often require a long period of time, e.g., shelf life of foods and corrosion of metals. (3) Accelerated tests, which require specially designed equipment that simulates service conditions on an exaggerated scale; in these, only a few hours are necessary to duplicate years of service life, e.g., oxygen bomb aging of elastomers. (4) Nondestructive testing by X-ray or radiography. Elaborate standard testing procedures are established by the American Society for Testing and Materials. The more common types of tests are as follows:

abrasion (elastomers, textiles)
adhesion (glues, resins)
aging (elastomers, plastics, leather, food products)
color stability (pigments, organic dyes) (exposure)
corrosion (metals, alloys) (exposure)
dielectric (electrical tapes, plastics, glass)
flammability (textiles, fibers, paper, plastics)
flash point of combustible liquids (Tag closed cup TCC, Cleveland open cup COC, open cup OC)
hardness (metals, elastomers, plastics) (Brinell, Rockwell, Shore penetration)
high temperature (elastomers, adhesives)
impact strength (composites, glass cement)
sun-cracking (paints, varnishes, elastomers) (exposure)
tear (paper, rubber, textiles)
tensile strength (fibers, elastomers, paper, textiles, metals)
viscosity (lubricants) (Saybolt, Engler)
See exposure testing; nondestructive testing; aging.

testing, physiological. Determination of the toxicity of a substance or product by administering it to laboratory animals in controlled dosages, by mouth, skin application, or injection. Materials commonly subjected to such evaluation are pharmaceuticals, pesticides, and foods. Extensive testing programs are required before such products are approved for human use.
See LD_{50}.

testosterone.
CAS: 58-22-0. $C_{19}H_{28}O_2$.

An androgenic steroid; the male sex hormone produced by the testis. It has six times the androgenic activity of its metabolic product, androsterone.
Properties: White or slightly cream-white crystals or crystalline powder; odorless; stable in air. Mp 153–157C. Dextrorotatory in dioxane solution; very soluble in chloroform; soluble in alcohol, dioxane, and vegetable oils; slightly soluble in ether; insoluble in water.
Derivation: Isolation from extract of testis, synthesis from cholesterol or from the plant steroid diosgenin.
Grade: NF.
Hazard: A confirmed carcinogen.
Use: Medicine, biochemical research.
See methyltestosterone.

testosterone-17-β-estradiol mixt.
CAS: 8055-33-2. $C_{19}H_{28}O_2 \cdot C_{18}H_{24}O_2$.
Hazard: Low toxicity by ingestion. A reproductive hazard.
Use: Hormone.

TETD. Abbreviation for tetraethylthiuram disulfide.

tetraamminecopper sulfate. $N_4H_{12}CuSO_4$.
Derivation: Dissolving copper sulfate in ammonia water, with precipitation by alcohol.
Use: Mordant in textile printing, fungicide.

1,5,8,12-tetraazadodecane.
CAS: 10563-26-5. $C_8H_{22}N_4$.
Hazard: A poison by skin contact. Moderately toxic by ingestion.

tetrabromobisphenol A. See 4,4′-(1-methylethylidene)bis(2,6-dibromophenol).

tetrabromo-*m*-cresolphthalein sulfone.
CAS: 76-60-8. $C_{21}H_{14}Br_4O_5S$.
Hazard: Moderately toxic.

tetrabromo-*o*-cresol.
CAS: 576-55-6. $C_7H_4Br_4O$.
Properties: Fine, white crystals. Mp 205C (decomposes). Insoluble in water; soluble in alcohol and ether.
Derivation: Bromination of *o*-cresol.
Hazard: Irritant to skin and mucous membranes.
Use: Fungicide.

tetrabromo-*m*-cresolphthalein sulfone. See bromocresol green.

1,1,1,2-tetrabromoethane.
CAS: 630-16-0. $C_2H_2Br_4$.
Hazard: Low toxicity by inhalation.

***sym*-tetrabromoethane.** See acetylene tetrabromide.

(1r,3s)3[(1′rs)(1′,2′,2′,2′-tetrabromoethyl)]-2,2-dimethylcyclopropanecarboxylic acid (s)-α-cyano-3-phenoxybenzyl ester.
Use: Food additive; insecticide.

tetrabromoethylene. C_2Br_4.
Properties: Colorless crystals. Mp 55–56C, bp 227C.
Derivation: Bromination of dibromoacetylene.
Use: Organic synthesis.

tetrabromofluorescein.   See eosin.

tetrabromomethane.   See carbon tetrabromide.

tetrabromophthalic anhydride. $C_6Br_4C_2O_3$.
Properties: Pale-yellow, crystalline solid. Mp 280C.
Use: Flame-retardant for plastics, paper, and textiles.

3,4,5,6-tetrabromophthalic anhydride.
CAS: 632-79-1. $C_8Br_4O_3$.
Hazard: Low toxicity by ingestion, inhalation, and skin contact. A mild eye irritant.

tetrabromosilane.   See silicon tetrabromide.

tetra-*n*-butylammonium chloride.
  $(C_4H_9)_4NCl$
Properties: Deliquescent, light-tan powder. Mp 50C.
Use: Substitute in solution for glass-calomel electrode systems, coagulant for silver iodide solutions, stereospecific catalyst.

tetrabutyl dichlorostannoxane.
CAS: 10428-19-0. $C_{16}H_{36}Cl_2OSn_2$.
Properties: White crystals from Me_2CO. Mp: 112.5C. Soluble in organic solvents.
Hazard: A poison.

tetra(2-butylisopropyl) orthosilicate.
CAS: 63449-47-8. $C_{28}H_{60}O_4Si$.
Hazard: Low toxicity by ingestion and skin contact. A moderate skin and eye irritant.

tetrabutylphosphonium phenylphosphinate.
See tetra-*n*-butylphosphonium phenylphosphinate.

tetra-*n*-butylphosphonium phenylphosphinate.
CAS: 60767-85-3. $C_{16}H_{36}P•C_6H_6O_2P$.
Hazard: A poison by ingestion. A severe skin irritant.

tetrabutylthiuram disulfide.
  $[(C_4H_9)_2NCH]_2S_2$.
Properties: Amber-colored liquid; slightly sweet odor. D 1.03–1.06 (20/20C), solidifies at −30C. Soluble in carbon disulfide, benzene, chloroform, and gasoline; insoluble in water and 10% caustic. Combustible.
Use: Vulcanizing and accelerating agent.

tetrabutylthiuram monosulfide.
  $[(C_4H_9)_2NCS]_2S$.
Properties: Brown, free-flowing liquid. D 0.99. Soluble in acetone, benzene, gasoline, and ethylene dichloride; insoluble in water. Combustible.
Use: Rubber accelerator.

tetrabutyltin.
CAS: 1461-25-2. $(C_4H_9)_4Sn$.
Properties: Colorless or slightly yellow, oily liquid. D 1.0572 (20/4C), fp −97C, bp 145C (10 mm Hg), decomposes at 265C. Insoluble in water; soluble in most common organic solvents. Combustible.
Derivation: Reaction of tin tetrachloride with butyl magnesium chloride.
Hazard: Irritant.
Use: Stabilizing and rust-inhibiting agent for silicones, lubricant and fuel additive, polymerization catalyst, hydrogen chloride scavenger.

tetrabutyl titanate.   (TBT; butyl titanate; titanium butylate).
CAS: 5593-70-4. $Ti(OC_4H_9)_4$.
Properties: Colorless to light-yellow liquid. Bp 310–314C, forms a glass −55C, d 0.996, refr index 1.486, flash p 170F (76.6C). Decomposes in water; soluble in most organic solvents except ketones. Combustible.
Derivation: Reaction of titanium tetrachloride with butanol.
Use: Ester exchange reactions; heat-resistant paints (up to 500C); improving adhesion of paints, rubber, and plastics to metal surfaces; cross-linking agent; condensation catalyst.

tetrabutyl urea. $(C_4H_9)_2NCON(C_4H_9)_2$.
Properties: Liquid. D 0.880, refr index 1.4535, vap press below 0.01 mm Hg, bp 305C, fp below −60C, flash p 200F (93.3C). Insoluble in water. Combustible.
Use: Plasticizer.

tetrabutyl zirconate. $(C_4H_9O)_4Zr$.
Properties: White solid from reaction of zirconium tetrachloride with butanol.
Use: Condensation catalyst and cross-linking agent.

Tetracaine.   (2-dimethylaminoethyl-*p*-butylaminobenzoate).
CAS: 94-24-6. $CH_3(CH_2)_3NHC_6H_4COOCH_2CH_2N(CH_3)_2$.
Properties: White or light-yellow, waxy solid. Mp 41–46C. Very slightly soluble in water; soluble in alcohol, ether, benzene, chloroform.
Grade: USP.
Use: Medicine (anesthetic).

tetracalcium aluminoferrate.   An ingredient of Portland cement.

tetracarbonylhydrocobalt).   See cobalt hydrocarbonyl.

1,2,3,4-tetracarboxybutane. (TCB; 1,2,3,4-butanetetracarboxylic acid). HOOCCH$_2$CH(COOH)CH(COOH)CH$_2$COOH.
Use: Alkyd resins, epoxy curing agent, sequestrant.

tetracene. See naphthacene.

1,2,3,4-tetrachlorobenzene.
CAS: 634-66-2. C$_6$H$_2$Cl$_4$.
Properties: White crystals. Mp 46.6, bp 254C. Insoluble in water. Combustible.
Use: Component of dielectric fluids, synthesis.

1,2,4,5-tetrachlorobenzene.
CAS: 95-94-3. C$_6$H$_2$Cl$_4$.
Properties: White flakes. Mp 137.5–140C, flash p 311F (155C), distillation range 240–246C. Combustible.
Use: Intermediate for herbicides and defoliants, insecticide, impregnant for moisture resistance, electrical insulation.

tetrachloro-p-benzoquinone. See chloranil.

tetrachlorobisphenol.
(C$_6$H$_2$Cl$_2$OH)$_2$C(CH$_3$)$_2$. A monomer for flame-retardant epoxy, polyester, and polycarbonate resins.

tetrachlorocyclopentane.
CAS: 59808-78-5. C$_5$H$_6$Cl$_4$.
Hazard: Moderately toxic by ingestion, skin contact, and inhalation. A moderate eye irritant.

2,3,7,8-tetrachlorodibenzo-p-dioxin. See dioxin.

1,1,2-tetrachloro-2,2-difluoroethane. (halocarbon 112a; refrigerant 112a).
CAS: 76-11-9. CCl$_2$CF$_2$Cl.
Properties: Colorless, liquid or solid; slight ether odor. Mw 202.83, mp 40.56C, bp 91.67C, vap press 40 torr (20C). Insoluble in water; soluble in ether, chloroform, and alcohol. Noncombustible.
Hazard: Central nervous system impairment, liver and kidney damage.
Use: Refrigerant.

sym-tetrachlorodifluoroethane. (1.2-difluoro-1,1,2,2-tetrachloroethane; freon 112).
CAS: 76-12-0. CCl$_2$FCCl$_2$F.
Properties: White solid or colorless liquid; slightly camphor-like odor when concentrated. Bp 92.8C, mp 26C, critical temp 278C, d 1.6447 (25C), refr index 1.413 (25C), bulk d 13.8 lb/gal. Insoluble in water; soluble in alcohol. Nonflammable.
Grade: Purified, solvent.
Hazard: Toxic by inhalation.
Use: Degreasing solvent.

tetrachlorodinitroethane.
O$_2$NCCl$_2$CCl$_2$NO$_2$.

Properties: White crystals. Decomposes at 130C to nitrogen peroxide.
Hazard: Toxic by ingestion and inhalation, strong irritant.

tetrachlorodiphenylethane. See TDE.

2′,3,4,6′-tetrachlorodiphenyl ether. See 1,3-dichloro-2-(3,4-dichlorophenoxy)benzene.

tetrachlorodiphenyl sulfone. See tetradifon.

sym-tetrachloroethane. (acetylene tetrachloride).
CAS: 79-34-5. CHCl$_2$CHCl$_2$.
Properties: Heavy, colorless, corrosive liquid; chloroform-like odor. D 1.593 (25/25C), bp 146.5C, fp −43C, bulk d 13.25 lb/gal (25C), refr index 1.4918 (25C), flash p none. Soluble in alcohol and ether; slightly soluble in water. Nonflammable.
Derivation: Reaction of acetylene and chlorine and subsequent distillation.
Grade: Technical.
Hazard: Toxic by ingestion, inhalation, skin absorption. Questionable carcinogen.
Use: Solvent, cleansing and degreasing metals, paint removers, varnishes, lacquers, photographic film, resins and waxes, extraction of oils and fats, alcohol denaturant, organic synthesis, insecticides, weed killer, fumigant, intermediate in manufacture of other chlorinated hydrocarbons.

tetrachloroethylene. See perchloroethylene.

cis-n-[1,1,2,2-tetrachloroethyl)thio]-4-cyclohexene-1,2-dicarboximide. ("Difolatan"; "captafol").
CAS: 2425-06-1. C$_{10}$H$_9$Cl$_4$NO$_2$S.
Properties: White solid. Mp 160C. Slightly soluble in most organic solvents; insoluble in water.
Hazard: Absorbed by skin. Probable carcinogen.
Use: Fungicide.

n-((tetrachloro-2-fluoroethyl)thio) methanesulfoanilide. See n-(2-fluoro-1,1,2,2-tetrachloroethylthio).

tetrachloroisophthalonitrile. (1,3-dicyano-2,4,5,6-tetrachlorobenzene). C$_8$Cl$_4$N$_2$.
Properties: Colorless crystals. Mp 245C, bp 350C, d 1.70, insoluble in water, almost insoluble in organic solvents.
Use: Bactericide, nematocide.

tetrachloromethane. See carbon tetrachloride.

tetrachloronaphthalene. (Halowax).
CAS: 1335-88-2. C$_{10}$H$_4$Cl$_4$.
Properties: Colorless to pale-yellow solid; aromatic odor. Mw 265.96, sp g 1.59–1.65, mp 115C, bp 311.5–360C, flash p 410F. Insoluble in water.

Hazard: Combustible. Toxic by inhalation and skin contact. Liver damage.
Use: Electrical insulating materials; in resins and polymers for coating textiles, wood, and paper; additive in cutting oils.
See chlorinated naphthalene.

2,3,4,6-tetrachlorophenol.
CAS: 58-90-2. C_6HCl_4OH.
Properties: Brown flakes or sublimed mass; strong odor. Mp 69–70C, bp 164C (23 mm Hg), d 1.839 (25/4C). Soluble in acetone, benzene, ether, and alcohol. Nonflammable.
Hazard: Toxic by ingestion and inhalation, strong irritant.
Use: Fungicide.

2,4,5,6-tetrachlorophenol.
CAS: 58-90-2. C_6HCl_4OH.
Properties: Brown solid; phenol odor. D 1.65 (60/4C), mp above 50C. Soluble in sodium hydroxide solutions and most organic solvents; insoluble in water.
Hazard: Toxic by ingestion and inhalation, strong irritant.
Use: Fungicide, wood preservative.

tetrachlorophthalic acid. $C_6Cl_4(CO_2H)_2$.
Properties: Colorless, crystalline plates. Soluble in hot water; sparingly soluble in cold water.
Derivation: By passing a stream of chlorine through a mixture of phthalic anhydride and antimony pentachloride.
Use: Dyes, intermediates.

tetrachlorophthalic anhydride.
CAS: 117-08-8. $C6Cl_4(CO)_2O$.
Properties: White, odorless, free-flowing, non-hygroscopic powder. Mp 254–255C, bp 371C. Slightly soluble in water.
Use: Intermediate in dyes, pharmaceuticals, plasticizers, and other organic materials; flame-retardant in epoxy resins.

tetrachloroquinone.   See chloranil.

tetrachlorosalicylanilide. $C_{13}H_7Cl_4NO_2$.
Properties: Crystalline solid. Mp 160C. Insoluble in water; soluble in common organic solvents and alkaline solutions.
Use: Bacteriostat; preservative (textile finishes, cutting oils, plastics); use in foods, drugs, and cosmetics may be restricted.

tetrachlorosilane.   See silicon tetrachloride.

3,3,4,4-tetrachlorotetrahydrothiophene-1,1-dioxide.
CAS: 3737-41-5. $C_4H_4Cl_4O_2S$.
Hazard: A poison by ingestion and skin contact. Moderately toxic by inhalation. A severe skin irritant.

tetrachlorothiophene.
CAS: 6012-97-1.

CClCClCClCClS.
Properties: Liquid. Mp 29–30, bp 104C (10 mm Hg). Soluble in benzene, hexane, alcohols, chlorocarbons.
Use: Agricultural chemicals, lubricants.

tetrachlorvinphos.
CAS: 22248-79-9. $C_{10}H_9Cl_4O_4P$.
Properties: Powder. Mp 97C. Partially soluble in chloroform; slightly soluble in water.
Hazard: Cholinesterase inhibitor. Questionable carcinogen.
Use: Insecticide.

tetracine.   See tetrazene.

tetracosamethylhendecasiloxane.
$C_{24}H_{72}O_{10}Si_{11}$.
Properties: Colorless liquid. Bp 200C (5 mm Hg), d 0.924, refr index 1.399, flash p 188C. Soluble in benzene and low molecular weight hydrocarbons; viscosity unaffected by temperature. Combustible.
Use: Silicone oils, foam inhibitor in lubricating oils.

tetracosane. $C_{24}H_{50}$ or $CH_3(CH_2)_{22}CH_3$.
Properties: Crystals. D 0.779 (51/4C), bp 324.1C, mp 51.5C. Soluble in alcohol; insoluble in water. Combustible.
Use: Organic synthesis.

n-tetracosanoic acid.   See lignoceric acid.

tetracyanoethylene.
CAS: 670-54-2. $(CN)_2C:C(CN)_2$. The first member of a class of compounds called cyanocarbons.
Properties: Colorless crystals. Sublimes above 120C, mp 198–200C, bp 223C, high thermal stability. Burns in oxygen with a hotter flame than acetylene.
Hazard: Hydrolyzes in moist air to hydrogen cyanide.
Use: Organic synthesis, dyes, to make colored solutions with aromatics.

tetradecamethylhexasiloxane. $C_{14}H_{42}O_5Si_6$.
Properties: Colorless liquid. Bp 140C (20 mm Hg), fp below −100C, d 0.89, refr index 1.4, flash p 118C. Viscosity unaffected by temperature. Soluble in benzene and low molecular weight hydrocarbons; slightly soluble in alcohol. Combustible.
Use: Silicone oil base, foam inhibitor in lubricating oils.

n-tetradecane.
CAS: 629-59-4. $C_{14}H_{30}$ or $CH_3(CH_2)_{12}CH_3$.
Properties: Colorless liquid. D 0.7653 (20/4C), mp 5.5C, bp 253.5C, refr index 1.4302 (20C), flash p

212F (100C), autoign temp 396F (202C). Soluble in alcohol; insoluble in water. Combustible.
Grade: 95%, 99%.
Hazard: A fire hazard. Lower flammable limit in air 0.5%.
Use: Organic synthesis, solvent standardized hydrocarbon, distillation chaser.

tetradecanoic acid.　　See myristic acid.

tetradecanol.　　See myristyl alcohol; 7-ethyl-2-methyl-4-undecanol.

9-tetradecenal, (z)-.
CAS: 53939-27-8. $C_{14}H_{26}O$.
Hazard: Low toxicity by ingestion.
Source: Natural product.

11-tetradecenal, (e)-.
CAS: 35746-21-5. $C_{14}H_{26}O$.
Hazard: Low toxicity by ingestion, inhalation, and skin contact.
Source: Natural product.

1-tetradecene.　　(α-tetradecylene).
$CH_2:CH(CH_2)_{11}CH_3$.
Properties: Colorless liquid. D 0.775 (20/4C), fp −12C, bp 256C, flash p 230F (110C). Insoluble in water; very slightly soluble in alcohol and ether. Combustible.
Use: Solvent in perfumes, flavors, medicines, dyes, oils, resins.

cis-tetradec-9-enoic acid.　　See myristoleic acid.

(z)-9-tetradecenol acetate.
CAS: 16725-53-4. $C_{16}H_{30}O_2$.
Hazard: Low toxicity by ingestion.
Source: Natural product.

tetradecylamine.　　$C_{14}H_{29}NH_2$.
Properties: White solid; odor of ammonia. Mp 37C, bp 291.2C. Insoluble in water; soluble in alcohol and ether. Combustible.
Grade: 90% purity.
Use: Intermediate for manufacture of cationic surface-active agents, germicides.

tetradecylbenzyldimethylammonium chloride monohydrate.　　$[C_{14}H_{29}N(CH_3)_2CH_2C_6H_5]Cl•H_2O$. A quaternary ammonium salt.
Properties: (50% solution in aqueous isopropanol) Viscous liquid that may gelatinize on standing. D 0.978 (16C). Miscible with water, alcohol, glycerol and acetone; pH of a 10% solution in distilled water 7–8.
Grade: 50% solution in aqueous isopropanol.
Use: Production of bacteriostatic and fungistatic paper.

tetradecyl chloride.　　(myristyl chloride).
$CH_3(CH_2)_{13}Cl$.
Properties: Water-white distilled liquid; mild odor. D 0.8590, fp −0.2C, bp 154–155C (15 mm Hg), 15.2% chloride, subject to mild hydrolysis on standing.
Grade: 97% min.

α-tetradecylene.　　See 1-tetradecene.

tetradecyl thiol.　　(myristyl mercaptan).
$CH_3(CH_2)_{13}SH$.
Properties: Liquid; strong odor. Mp 6.5C, bp 176–180C (22 mm Hg), d 0.8398 (25/4C), refr index 1.4612 (20C). Combustible.
Grade: 95% (min) purity.
Hazard: Toxic by inhalation, strong irritant.
Use: Organic intermediate, synthetic rubber processing.

tetradifon.　　(chlorophenyl-2,4,5-trichlorophenyl sulfone; 2,4,4′,5-tetrachlorodiphenylsulfone).
CAS: 116-29-0. $Cl_3C_6H_2SO_2C_6H_4Cl$.
Properties: White, crystalline powder. Mp 147C. Soluble in chloroform and aromatic hydrocarbons; insoluble in water.
Use: Insecticide, acaricide, ovicide.

tetra(diphenylphosphito)pentaerythritol.
See pentaerythritol tetrakis(diphenyl phosphite).

tetradotoxin.　　A highly toxic venom found in puffer-like fish.

tetraethanolammonium hydroxide.
CAS: 77-98-5. $(HOCH_2CH_2)_4NOH$.
Properties: White, crystalline solid. Mp 123C, vap press below 0.01 mm Hg (20C). Completely soluble in water. A strong base, approaching sodium hydroxide in alkalinity. Aqueous solutions are stable at room temperature but decomposes on heating to weakly basic polyethanolamines.
Grade: Commercial grade is a 40% water solution.
Hazard: Strong irritant to skin and tissue.
Use: Alkaline catalyst, solvent for certain types of dyes, metal-plating solutions.

2,4,6,8-tetraethenyl-2,4,6,8-tetramethylcyclotetrasiloxane.
CAS: 2554-06-5. $C_{12}H_{24}O_4Si_4$.
Hazard: Low toxicity by ingestion and inhalation.

tetraethylammonium perfluoro-1-octanesulfonate.
CAS: 56773-42-3. $C_8H_{20}N•C_8F_{17}O_3S$.
Hazard: A poison by ingestion.

1,1,3,3-tetraethoxypropane.
CAS: 122-31-6. $[(C_2H_5O)_2CH]_2CH_2$.

Properties: Liquid. Bp 105C, refr index 1.410 (20C), d 0.920 (20/20C), flash p 190F (87.7C). Slightly soluble in water; soluble in ether and alcohol. Combustible.
Use: Organic synthesis.

tetraethylammonium chloride.   (TEAC; TEA chloride). $(C_2H_5)_4NCl$.
Properties: (Anhydrous) Colorless, crystals; odorless. D 1.080. Hygroscopic; freely soluble in water, alcohol, chloroform, acetone; slightly soluble in benzene and ether. (Tetrahydrate) $(C_2H_5)_4NCl \cdot 4H_2O$. Crystals. Mp 37.5C, d 1.084.
Use: Medicine (nerve-blocking agent).

tetraethylammonium   hexafluorophosphate. $(C_2H_5)_4NPF_6$.
Properties: Solid. Mp 255C. Nonhygroscopic but soluble in water; stable to heat, can be stored in solution without decomposition of the PF_6 ion.
Hazard: Irritant to skin.
Use: Maintenance of fluoride atmospheres; preparation of bactericides and fungicides.

tetra-(2-ethylbutyl) silicate. $[(C_2H_5)C_4H_8O]_4Si$.
Properties: Colorless liquid. D 0.8920–0.9018 (20/20C), fp −100C, bp 238C (50 mm Hg). Insoluble in water; slightly soluble in methanol; miscible with most organic solvents.
Use: Heat-transfer medium, hydraulic fluid, wide-temperature-range lubricant.

tetraethyl dithiopyrophosphate.   (TEDP).
See sulfotepp.

tetraethylene glycol.   (TEG). $HO(C_2H_4O)_3C_2H_4OH$.
Properties: Colorless liquid; hygroscopic. D 1.1248 (20/20C), fp −4C, bp 327.3C, vap press above 0.001 mm Hg (20C), refr index 1.4577 (20C), flash p 345F (174C), bulk d 9.4 lb/gal (20C). Soluble in water; insoluble in benzene, toluene, or gasoline. Combustible.
Use: Solvent for nitrocellulose, plasticizer, lacquers, coating compositions.

tetraethylene glycol dibutyl ether.   See dibutoxytetraglycol.

tetraethylene glycol dimethacrylate.
CAS: 109-17-1.
Properties: Water-white to pale-straw liquid. Bp 200C (1 mm Hg), d 1.075 (20/20C), refr index 1.4620 (20C), viscosity 12 cP. Insoluble in water; soluble in styrene, many esters, and aromatics; limited solubility in aliphatic hydrocarbons. Combustible.
Hazard: Irritant to skin and eyes.
Use: Plasticizer.

tetraethylene glycol dimethyl ether.   See dimethoxytetraglycol.

tetraethylene glycol distearate. $(C_{17}H_{35}COOCH_2CH_2OCH_2CH_2)_2O$.
Properties: Liquid. Mp 32–33C. Insoluble in water. Combustible.
Use: Plasticizer.

tetraethylene glycol monostearate. $C_{17}H_{35}COO(CH_2CH_2O)_4H$.
Properties: Liquid. D 0.971, mp 30–31C. Insoluble in water. Combustible.
Use: Plasticizer.

tetraethylenepentamine.
CAS: 112-57-2 $NH_2(CH_2CH_2NH)_3CH_2CH_2NH_2$.
Properties: Viscous, hygroscopic liquid. D 0.9980 (20/20C), fp −30C, bp 333C, vap press above 0.01 mm Hg (20C), bulk d 8.3 lb/gal (20C), flash p 325F (162.7C). Soluble in most organic solvents and water. Combustible.
Hazard: Strong irritant to eyes and skin.
Use: Solvent for sulfur, acid gases, and various resins and dyes; saponifying agent for acidic materials; manufacture of synthetic rubber; dispersant in motor oils; intermediate for oil additives.

tetra-(2-ethylhexyl) silicate. $[C_4H_9CH(C_2H_5)CH_2O]_4Si$.
Properties: Colorless liquid. D 0.8838, bp 350–370C, fp −90C, flash p 390F (198C). Solubility in water below 0.01, 7.4 lb/gal. Combustible.
Use: Synthetic lubricants and functional fluids.

tetra-(2-ethylhexyl)   titanate.   (tetrakis(2-ethylhexyl)titanate). $[C_4H_9CH(C_2H_5)CH_2O]_4Ti$.
Properties: Light-yellow, viscous liquid from the transesterification of isopropyl titanate with 2-ethylhexanol. Combustible.
Use: Cross-linking agent, condensation catalyst, adhesion promoter, water repellents.

tetraethyl lead.   (TEL).
CAS: 78-00-2. $Pb(C_2H_5)_4$.
Properties: Colorless, oily liquid; pleasant odor. D 1.65, bp 198–202C, 75–85C (13–14 mm Hg), fp −136C, decomposes slowly at room temperature, rapidly at 125–150C. Soluble in all organic solvents; insoluble in water and dilute acids or alkalies. Combustible.
Derivation: (1) Alkylation of lead–sodium alloy with ethyl chloride; (2) electrolysis of an ethyl Grignard reagent with an anode of lead pellets.
Grade: One grade only, about 98% pure.
Hazard: Toxic by ingestion, inhalation, and skin absorption. Central nervous system impairment. Questionable carcinogen.
Use: Antiknock agent. Leaded gasoline contains 1.10 g lead per gallon. TEL has been largely replaced by MBTE.

tetraethyl orthosilicate.   See ethyl silicate.

tetraethyl pyrophosphate.   Legal label name for tepp.

tetraethylthiuram disulfide.   (disulfiram; TTD; TETD; bis[diethylthiocarbamyl] disulfide).
CAS: 97-77-8. $[(C_2H_5)_2NCS]_2S_2$.
Properties: Light-gray powder; slight odor. D 1.27, melting range 65–70C. Soluble in carbon disulfide, benzene, and chloroform; insoluble in water.
Hazard: Toxic symptoms when ingested with alcohol; animal teratogen. Vasodilation and nausea. Questionable carcinogen.
Use: Fungicide, ultraaccelerator for rubber.

tetraethylthiuram sulfide.   [bis-(diethyl thiocarbamyl)sulfide]. $[(C_2H_5)_2NCS]_2S$.
Properties: Dark-brown powder; slight odor. D 1.12 (20/20C), boiling range 225–240C (3 mm Hg).
Hazard: Toxic by ingestion and inhalation.
Use: Pharmaceutical ointments, fungicide, insecticide.

tetraethyltin.
CAS: 597-64-8. $Sn(C_2H_5)_4$.
Properties: Colorless liquid. D 1.187 (23C), bp 181C, fp−112C. Insoluble in water; soluble in alcohol and ether.
Hazard: Toxic material.

tetrafluorodichloroethane.   See dichlorotetrafluoroethane.

tetrafluoroethylene.   (perfluoroethylene; TFE).
CAS: 116-14-3. $F_2C:CF_2$.
Properties: Colorless gas. Fp −142C, bp −78.4C. Insoluble in water; much heavier than air.
Derivation: By passing chlorodifluoromethane through a hot tube.
Hazard: Flammable, dangerous fire risk. Kidney and liver damage; kidney and liver cancer. Possible carcinogen.
Use: Monomer for polytetrafluoroethylene polymers.

tetrafluoroethylene epoxide.   (TFEO).

Derivation: Oxidation of tetrafluoroethylene at 120C with UV light; reaction proceeds by free-radical mechanism.
Use: Monomer for products ranging from dimers to polymers of dp 35.
See FreonE; "Krytox" [Du Pont].

tetrafluorohydrazine.
CAS: 13847-65-9. F_2NNF_2.

Properties: Colorless, mobile liquid or colorless gas. Bp (calc) −73C, heat of vaporization 3170 cal/mole, critical temp 36C.
Hazard: Explodes on contact with reducing agents and at high pressures. Irritant.
Use: Organic synthesis; oxidizer in fuels for rockets, missiles, etc.

tetrafluoromethane.   (fluorocarbon 14; carbon tetrafluoride).
CAS: 75-73-0. CF_4.
Properties: Colorless gas. D (liquid) 1.96 (−184C), sp vol 4.4 cu ft/lb (70F), fp −184C, bp −128C. Slightly soluble in water. Nonflammable.
Grade: 95% min purity.
Hazard: Toxic by inhalation.
Use: Refrigerant, gaseous insulator.

3-(4,6-(tetrafluoromethoxy)pyrimidin-2-yl)-1-((2-methoxycarbonyl)phenylsulfonyl) urea.
CAS: 86209-51-0. $C_{15}H_{12}F_4N_4O_7S$.
Hazard: Moderately toxic by skin contact. Low toxicity by ingestion.

tetrafluorosilane.   See silicon tetrafluoride.

tetraformaltrisazine.   See octahydro-(1,2,4,5)tetrazino(1,2-a)(1,2,4,5)tetrazine.

tetragastrin.   See gastrin tetrapeptide amide.

tetraglycol dichloride. $(ClCH_2CH_2OCH_2 CH_2)_2O$.
Properties: Colorless liquid. D 1.186, bp 114 (2 mm Hg). Slightly soluble in water. Combustible.
Use: High-boiling solvent and extractant for oils, fats, waxes, and greases; chemical intermediate.

tetraglyme.   See dimethoxytetraglycol.

1,2,3,6-tetrahydrobenzaldehyde.   Legal label name for 3-cyclohexene-1-carboxaldehyde.

1,2,3,4-tetrahydrobenzene.   See cyclohexene.

tetrahydrobiopterin.   The fully reduced coenzyme form of biopterin.

tetrahydrocannibol. $C_{21}H_{30}O_2$. The active principle of marijuana, a hallucinatory drug. It has been synthesized and is available in laboratory quantities, subject to legal restrictions. Animal tests have indicated that it can retard cancer growth and also may promote the acceptance of organ transplants in the human body.

1,4,5,6-tetrahydro-3-(10,11-dihydrodibenzo (a,d)cyclohepten-5-yl)-1-methyl-as-triazine hydrobromide.
CAS: 22187-44-6. $C_{19}H_{21}N_3 \cdot BrH$.
Hazard: A poison.

α-(tetrahydro-4,6-dioxo-2-thioxo-5(2h) pyrimidinylidene)-*o*-toluic acid.
CAS: 73909-20-3. $C_{12}H_8N_2O_4S$.
Hazard: A poison.

3,4,5,6-tetrahydro-2-(α-ethoxybenzyl)-5-ethyl-5-methylpyrimidine.
CAS: 33236-07-6. $C_{16}H_{24}N_2O$.
Hazard: A poison by ingestion.

tetrahydrofolate.   The reduced, active coenzyme form of the vitamin folate which functions as an antioxidant.

tetrahydrofuran.   (THF).
CAS: 109-99-9.

$\overline{CH_2CH_2CH_2CH_2O}$.

Properties: Water-white liquid; ethereal odor. D 0.888 (20C), refr index 1.4070 (20C), fp −65C, bp 66C, flash p 5F (−15C) (OC), autoign temp 610F (321C). Soluble in water and organic solvents.
Derivation: (1) Catalytic hydrogenation of furan with nickel catalyst. (2) Acid-catalyzed dehydration of 1,4-butanediol.
Grade: Technical, spectrophotometric.
Hazard: Flammable, dangerous fire risk. Flammable limits in air 2–11.8%. Toxic by ingestion and inhalation. Upper respiratory tract irritant, central nervous system impairment, and kidney damage. Possible carcinogen.
Use: Solvent for natural and synthetic resins (particularly vinyls), in top-coating solutions, polymer coating, cellophane, protective coatings, adhesives, magnetic tapes, printing inks, etc. Grignard reactions, lithium aluminum hydride reductions, and polymerizations; chemical intermediate and monomer.

tetrahydro-3-furancarboxaldehyde.   See 3-formyltetrahydrofuran.

2,5-tetrahydrofurandimethanol.   (2,5-bis[hydroxymethyl]tetrahydrofuran). $C_6H_{12}O_3$.
Properties: Colorless liquid. D 1.154, mp below −50C, bp 265C, refr index 1.47. Soluble in water, alcohol, acetone, benzene; hygroscopic.
Hazard: Strong irritant to tissue.
Use: Organic synthesis, humectant, solvent.

(tetrahydro-2-furanyl)methyl 2-propenoate.
See tetrahydrofurfuryl acrylate.

tetrahydrofurfuryl acetate. $C_4H_7OCH_2$ $OOCCH_3$.
Properties: Colorless liquid. D 1.061 (20/0C), bp 194–195C (753 mm Hg). Soluble in water, alcohol, ether, and chloroform. Combustible.
Derivation: By treatment of tetrahydrofurfuryl alcohol with acetic anhydride.
Use: Flavoring.

tetrahydrofurfuryl acrylate.
CAS: 2399-48-6. $C_8H_{12}O_3$.
Hazard: A moderate skin irritant.

tetrahydrofurfuryl alcohol.   (tetrahydrofuryl carbinol).
CAS: 97-99-4. $-C_4H_7OCH_2OH$.

Properties: Colorless liquid; mild odor; hygroscopic. D 1.0543 (20/20C), bp 178C, refr index 1.4520 (20C), flash p 167F (75C) (OC), viscosity 5.49 cP (25C), autoign temp 540F (282C). Miscible with water. Combustible.
Derivation: Catalytic hydrogenation of furfural.
Grade: Commercial, industrial (80%).
Use: Solvent for vinyl resins; dyes for leather; chlorinated rubber; cellulose esters; solvent softener for nylon; vegetable oils; coupling agent; organic synthesis.

tetrahydrofurfurylamine.
CAS: 4795-29-3. $C_4H_7OCH_2NH_2$.
Properties: Colorless to light-yellow liquid. Refr index 1.4520–1.4535 (25C), distilling range 150–156C, d 0.977 (20C/20C). Combustible.
Use: Chemical intermediate, fine-grain photographic development, vulcanization accelerator.

tetrahydrofurfuryl benzoate. $C_4H_7OCH_2$ $OOCC_6H_5$.
Properties: Colorless liquid. D 1.137 (20/0C), bp 300–302C, 138–140C (2 mm Hg). Insoluble in water; soluble in alcohol, ether, and chloroform. Combustible.
Derivation: Tetrahydrofurfuryl alcohol and benzoic acid by esterification.
Use: Chemical intermediate.

tetrahydrofurfuryl laurate. $C_4H_7OCH_2$ $OOCC_{11}H_{23}$.
Properties: Colorless liquid. D 0.930 (25C). Insoluble in water. Combustible.
Use: Plasticizer.

tetrahydrofurfuryl levulinate. CH_3CO $(CH_2)_2COOCH_2C_4H_7O$.
Properties: Colorless liquid. Mp 59–62C. Soluble in water. Combustible.
Use: Plasticizer.

tetrahydrofurfuryl oleate. $C_{17}H_{33}COOCH_2$ C_4H_7O.
Properties: Colorless liquid. D 0.923 (25C), bp 240C (5 mm Hg), fp −30C, flash p 329F (165C). Insoluble in water. Combustible.
Use: Plasticizer.

tetrahydrofurfuryl phthalate.
$C_6H_4(COOCH_2C_4H_7O)_2$.
Properties: Colorless liquid. D 1.194 (25C), mp below 15C. Insoluble in water. Combustible.
Use: Plasticizer.

tetrahydrofuryl carbinol.   See tetrahydrofuryl alcohol.

3-(2-tetrahydrofuryl)-5-fluorouracil.   See 5-fluoro-3-(tetrahydro-2-furyl)uracil.

tetrahydrogeraniol.   See 3,7-dimethyl-1-octanol.

4-(1,2,3,4-tetrahydro-4-(4-hydroxy-2-oxo-2h-1-benzopyran-3-yl)-2-naphthalenyl) benzonitrile.
CAS: 90035-01-1. $C_{26}H_{19}NO_3$.
Hazard: A poison by ingestion.
Use: Agricultural chemical.

4′-(1,2,3,4-tetrahydro-4-(4-hydroxy-2-oxo-2h-1-benzopyran-3-yl)-2-naphthalenyl)(1,1′-biphenyl)-4-carbonitrile, cis-.
CAS: 90034-98-3. $C_{32}H_{23}NO_3$.
Hazard: A poison by ingestion.
Use: Agricultural chemical.

4′-(1,2,3,4-tetrahydro-4-(4-hydroxy-2-oxo-2h-1-benzopyran-3-yl)-2-naphthalenyl)(1,1′-biphenyl)-4-carbonitrile, trans-.
CAS: 90034-97-2. $C_{32}H_{23}NO_3$.
Hazard: A poison by ingestion.
Use: Agricultural chemical.

4-((4-(1,2,3,4-tetrahydro-4-(4-hydroxy-2-oxo-2h-1-benzopyran-3-yl)-2-naphthalenyl)phenoxy)benzonitrile.
CAS: 90035-05-5. $C_{32}H_{23}NO_4$.
Hazard: Moderately toxic by ingestion.
Use: Agricultural chemical.

4,5,6,7-tetrahydroisoxazolo(4,5-c)pyridin-3-ol.
CAS: 64603-91-4. $C_6H_8N_2O_2$.
Hazard: A poison.

tetrahydrolinalool.   (3,7-dimethyl-3-octanol).
CAS: 78-69-3. $C_{10}H_{21}OH$.
Properties: Colorless liquid; floral odor. D 0.832–0.837, optically inactive. Combustible.
Use: Perfumery, flavoring.

1-((1,2,3,4-tetrahydro-1-(4-(4-(2-methoxyphenyl)-1-piperazinyl)butyl)-7-me   thyl-2,4-dioxo-3-phenylpyrido(2,3-d)pyrimidin-5-yl)carbonyl)pyrrolidine.
CAS: 272774-82-0. $C_{34}H_{40}N_6O_4$.
Hazard: A poison.

1-((1,2,3,4-tetrahydro-7-methyl-2,4-dioxo-3-phenyl-1-(4-(4-phenyl-1-piperazinyl)   butyl)pyrido(2,3-d)pyrimidin-5-yl)carbonyl)yrrolidine.
CAS: 272774-79-5. $C_{33}H_{38}N_6O_3$.
Hazard: A poison.

1-((1,2,3,4-tetrahydro-7-methyl-1-(4-(4-methyl-1-piperazinyl)butyl)-2,4-di oxo-3-phenylpyrido(2,3-d)pyrimidin-5-yl)carbonyl)pyrrolidine.
CAS: 272774-83-1. $C_{28}H_{36}N_6O_3$.
Hazard: A poison.

1,4,5,6-tetrahydro-3-(α-methylphenethyl)-as-triazine hydrochloride.
CAS: 21038-21-1. $C_{12}H_{17}N_3 \cdot ClH$.
Hazard: A poison.

1,2,3,6-tetrahydro-1-methyl-4-phenylpyridine.
CAS: 28289-54-5. $C_{12}H_{15}N$.
Hazard: A poison. A reproductive hazard.

1,2,3,4-tetrahydro-6-methylquinoline.
$C_{10}H_{13}N$.
Properties: Yellowish crystals; strong, civet-like odor. Mp 33C. Soluble in 2 parts of 80% alcohol. Combustible.
Use: Perfumery.

tetrahydronaphthalene.
CAS: 119-64-2. $C_{10}H_{12}$.

Properties: Colorless liquid; pungent odor. D 0.981 (13C), bp 206C, refr index 1.540–1.547, flash p 160F (71.1C), fp −25C, moisture content none, residue on evaporation none, acidity neutral, bulk d 8 lb/gal, autoign temp 723F (384C). Miscible with most solvents; compatible with natural and synthetic vehicles; insoluble in water. Combustible.
Derivation: Hydrogenation of naphthalene in the presence of a catalyst at 150C.
Grade: Technical.
Hazard: Irritant to eyes and skin; narcotic in high concentration.
Use: Chemical intermediate; solvent for greases, fats, oils, waxes; substitute for turpentine.

tetrahydrophthalic anhydride.
CAS: 85-43-8. $C_6H_8(CO)_2O$.
Properties: White, crystalline powder. Solidification p 99–101C, d 1.20 (105C), flash p 315F (157C) (OC). Slightly soluble in petroleum ether and ethyl ether; soluble in benzene. Combustible.
Derivation: Diels–Alder reaction of butadiene and maleic anhydride.

Use: Chemical intermediate for light-colored alkyds, polyesters, plasticizers, and adhesives; intermediate for pesticides; hardener for resins.

(+)-(4ar,10br)-3,4,4a,10b-tetrahydro-4-propyl-2h,5h-(1)benzopyrano(4,3-b)-1,4-oxazin-9-ol.
CAS: 123671-92-1. $C_{14}H_{19}NO_3$.
Hazard: A poison.

tetrahydropyran-2-methanol.
CAS: 100-72-1.

$OCH_2CH_2CH_2CH_2CHCH_2OH$.

Properties: Liquid. D 1.0272 (20C), bp 187.2C, fp sets to glass below −70C, flash p 200F (93.3C) (OC). Miscible with water. Combustible.
Use: Chemical intermediate.

1,2,3,3-tetrahydro-3h-pyrano-(3,2-f) quinoline-8(7h)-one.
CAS: 128202-32-4. $C_{12}H_{11}NO_2$.
Hazard: Moderately toxic by ingestion.

1,2,5,6-tetrahydropyridine. C_5H_9N.
Properties: Colorless liquid. D 0.912–0.914 (20/4C), fp −44C, bp 115.5–120.0C. Combustible.
Grade: 96% min.
Use: Organic intermediate.

tetrahydrothiophene.   (thiophane).
CAS: 110-01-0.

$CH_2CH_2CH_2CH_2S$.

Properties: Water-white liquid. D 1.00 (15.6/15.6C), boiling range 115–124.4C. Combustible.
Use: Solvent, intermediate, fuel gas odorant.

tetrahydrothiophene-1,1-dioxide.   See sulfolane.

7,8,9,10-tetrahydro-6,8,11-trihydroxy-8-(hydroxyacetyl)-1-methoxy-10-((2,3,6-trideoxy-3-(2,3-dihydro-1h-pyrrol-1-yl)-αlyxo-hexopyranosyl)oxy)-,   (8s,10s)-5,12-naphthacenedione.
CAS: 175795-76-3. $C_{31}H_{33}NO_{11}$.
Hazard: A poison.

tetrahydroxybutane.   See erythritol.

tetrahydroxydiphenyl.   See diresorcinol.

tetrahydroxyethylethylenediamine.
[*N,N,N',N'*-tetrakis-(2-hydroxyethyl)ethylenediamine]. $(HOCH_2CH_2)_2NCH_2CH_2N(CH_2CH_2OH)_2$.
Properties: Clear, viscous liquid. Good heat stability. Combustible.

Use: Organic intermediate, cross-linking of rigid polyurethane foams, chelating agent, humectant, gas absorbent, resin formation, detergent processing.

3',4',5,7-tetrahydroxyflavone.
CAS: 491-70-3. $C_{15}H_{10}O_6$.
Hazard: A poison.

2,3,4,5-tetrahydroxyhexanedioic acid.   See saccharic acid.

2,3,9,10-tetrahydroxyprotoberberine.
CAS: 162854-37-7. $C_{17}H_{14}NO_4$.
Hazard: A poison by ingestion.

tetrahydroxysilane.   See silicic acid (ortho).

tetraiodoethylene.   (iodoethylene). $I_2C:CI_2$.
Properties: Light-yellow, crystals; odorless; turns brown on exposure to light. Mp 187C, d 2.98. Insoluble in water; soluble in most organic solvents.
Derivation: Iodine on diiodoacetylene obtained from calcium carbide and iodine.
Use: Surgical dusting powder, antiseptic ointment, fungicide.

tetraiodofluorescein.   See iodeosin.

tetraisopropylstannane.   See tetraisopropyltin.

tetraisopropylthiuram disulfide. $[(CH_3CH_3CH)_2NCS]S_2$.
Properties: Tan powder; amine odor. D 1.12 (20/20C), melting range 95–99C. Soluble in benzene, chloroform, gasoline; insoluble in water, 10% caustic, carbon disulfide.
Use: Rubber accelerator.

tetraisopropyltin.
CAS: 2949-42-0. $C_{12}H_{28}Sn$.
Hazard: A poison.

tetraisopropyl titanate.   (TPT; titanium isopropylate; isopropyl titanate). $Ti[OCH(CH_3)_2]_4$.
Properties: Light-yellow liquid that fumes in moist air. Bp 102–104C (10 mm Hg), mp 14.8C, d 0.954, refr index 1.46, apparent viscosity 2.11 cP (25C). Decomposes rapidly in water; soluble in most organic solvents.
Derivation: Reaction of titanium tetrachloride with isopropanol.
Use: Ester exchange reactions; adhesion of paints, rubber, and plastics to metals; condensation catalyst.

tetraisopropyl zirconate. $Zr[OCH(CH_3)_2]_4$.
Properties: White solid. Decomposes before melting.

Derivation: By reaction of zirconium tetrachloride with isopropanol.
Use: Condensation catalyst, cross-linking agent.

tetrakis(2-chloroethyl) ethylene diphosphate.
CAS: 33125-86-9. $C_{10}H_{20}Cl_4O_8P_2$.
Hazard: A reproductive hazard.

tetrakis(dimethylamino)silane.   See octamethylsilanetetramine.

tetrakis(1,1-dimethylpentyl) silicate.   See tetra(2-butylisopropyl) orthosilicate.

tetrakis(hydroxymethyl)phosphonium chloride.   (THPC).
CAS: 124-64-1. $(HOCH_2)_4PCl$.
A crystalline compound made by the reaction of phosphine, formaldehyde, and hydrochloric acid.
Hazard: Liver damage. Questionable carcinogen.
Use: Flame-retarding agent for cotton fabrics. May be used in combination with triethylolamine and urea (Roxel process) or with triethanolamine and tris(1-aziridinyl) phosphine oxide.

***N,N,N′,N′*-tetrakis(2-hydroxypropyl) ethylenediamine.**   (ethylenedinitrilotetra-2-propanol). $[-CH_2N(CH_2CH_2OCH_3)_2]_2$.
Properties: Viscous, water-white liquid. Miscible with water; soluble in ethanol, toluene, ethylene glycol, and perchloroethylene. Combustible.
Use: Cross-linking agent and catalyst in urethane foams, epoxy resin curing, metal complexes, intermediate.

tetrakis(*p*-phenoxyphenyl)tin.
CAS: 6452-62-6. $C_{48}H_{36}O_4Sn$.
Hazard: A poison.

"Tetralin" *[Sigma-Aldrich].*   TM for tetrahydronaphthalene.

tetralite.   See tetryl.

"Tetralol" *[Millenium].*   TM for an additive.
Use: In floral fragrances and soap perfumes to import fresh floral and citrus aroma.

1-tetralone.

CHCHCHCHCCCH$_2$CH$_2$CH$_2$CO.

A ketone of tetrahydronaphthalene.
Properties: Liquid. D 1.090–1.095 (20/20C), bp 120–125C (10 mm Hg), vap press 0.02 mm Hg (20C), mp 5.3–6.0C, flash p 265F (129.5C). Insoluble in water. Combustible.
Grade: Solvent and intermediate.

tetram.   (*O,O*-diethyl-*S*-(β-diethylamino)-ethyl phosphorothioate hydrogen oxalate).
CAS: 78-53-5. $(C_2H_5O)_2POSCH_3CH_2N(C_2H_5)_2$.
Hazard: Cholinesterase inhibitor. Use may be restricted.
Use: Insecticide.

tetramer.   An oligomer whose molecule is composed of four molecules of the same chemical composition.
See polymer.

1,1,3,3-tetramethoxypropane.
$[(CH_3O)_2CH]_2CH_2$.
Properties: Liquid. Bp 183C, refr index 1.408 (20C), d 0.995 (20/20C). Soluble in water, hexane, ether, and alcohol. Combustible.
Use: Organic synthesis.

tetramethoxy silane.   See methyl silicate.

3,3′,4,4′-tetramethoxystilbene.
CAS: 18513-98-9. $C_{18}H_{20}O_4$.
Hazard: Moderately toxic by ingestion.

tetramethylammonium chloride.
$(CH_3)_4NCl$. A quaternary ammonium compound.
Properties: White, crystalline solid. D 1.1690 (20/4C), mp (decomposes). Soluble in water and alcohol; insoluble in ether, also available as a 50% solution.
Use: Chemical intermediate, catalyst, inhibitor.

tetramethylammonium chlorodibromide.
$(CH_3)_4NClBr_2$.
Properties: Powder. Mp 118–126C. Soluble in water and other polar solvents.
Hazard: Evolves bromine on contact with water.
Use: Dry brominating agent, ingredient in formulation of sanitizers.

tetramethylammonium hydroxide.
CAS: 75-59-2. $(CH_3)_4NOH$.
Properties: A strong base available in 10% solution.
Hazard: Strong irritant to skin and tissue.

1,2,3,5-tetramethylbenzene.   See isodurene.

***sym*-tetramethylbenzene.**   See durene.

***N,N,N′,N′*-tetramethyl-1,3-butanediamine.**
$CH_3CHN(CH_3)_2CH_2CH_2N(CH_3)_2$.
Properties: Colorless, stable liquid. Fp −100C, bp 165.0C, d 0.8020 (20/20C), vap press 1.64 mm Hg (20C), miscible with water, viscosity 1.0 cP (20C), flash p 114F (45.5C) (TOC). Combustible.
Use: Catalyst for polyurethane foams and epoxy resins; high-energy fuels.

***n,n′,o,o′*-tetramethylcurinium diiodide.**
CAS: 16240-52-1. $C_{40}H_{48}N_2O_6 \cdot 2I$.
Hazard: A poison.

2,2,4,4-tetramethyl-1,3-cyclobutanediol.
$(CH_3)_4C_4H_2(OH)_2$.
Properties: White solid. Mp 124–135C, bp 220–225C, isomer composition 50% *cis*, 50% *trans*, flash p 125F (51.6C).
Hazard: Moderate fire risk. Irritant.
Use: Chemical intermediate, lubricants.

tetramethyldiamidophosphoric fluoride.
See dimefox.

tetramethyldiaminobenzhydrol. (tetramethyldiaminodiphenylcarbinol; Michler's hydrol; hydrol). $(CH_3)_2NC_6H_4CH(OH)C_6H_4N(CH_3)_2$.
Properties: Colorless prisms. Forms a colorless solution in ether or benzene and a blue solution in alcohol or acetic acid. Mp 96C. Soluble in alcohol, ether, benzene, and acetic acid Combustible.
Derivation: Reaction of tetramethyldiaminodiphenylmethane, hydrochloric acid, and glacial acetic acid; oxidized with lead peroxide.
Grade: Technical.
Use: Dye intermediate, organic synthesis.

tetramethyldiaminobenzophenone. (Michler's ketone; 4,4'-bis[dimethylamino]benzophenone).
CAS: 90-94-8. $CO[C_6H_4N(CH_3)_2]_2$.
Properties: Crystalline leaflets. Mp 172C, bp decomposes at 360C. Soluble in alcohol, ether, and water. Combustible.
Derivation: From dimethylaniline by reaction with phosgene.
Hazard: Possible carcinogen.
Use: Synthesis of dyestuffs, especially auramine derivatives.

4,4'-tetramethyldiaminodiphenylmethane.
(tetra base).
CAS: 101-61-1. $H_2C[C_6H_4N(CH_3)_2]_2$.
Properties: Yellowish leaflets or glistening plates. Mp 90–91C, sublimes with decomposition, bp 390C. Insoluble in water; soluble in benzene, ether, carbon disulfide, and acids.
Derivation: By heating dimethylaniline with hydrochloric acid and formaldehyde.
Use: Dye intermediate.

tetramethyldiaminodiphenylsulfone. (4,4'-bis(dimethylamino)diphenylsulfone). $[(CH_3)_2NC_6H_4]_2SO_2$.
Properties: Solid. Mp 259–260C. Combustible.
Grade: Technical, reagent.
Use: Intermediate in making dyestuffs and medicinal chemicals; analytical reagent for lead.

1,1,3,3-tetramethyl-1,3-divinyldisilazane.
CAS: 7691-02-3. $C_8H_{19}NSi_2$.
Hazard: A poison by ingestion.

***sym*-tetramethyldivinyldisiloxane.**
CAS: 2627-95-4. $C_8H_{18}OSi_2$.
Hazard: Low toxicity by ingestion and inhalation.

tetramethylene. See cyclobutane.

tetramethylenediamine.
CAS: 110-60-1. $H_2N(CH_2)_4NH_2$.
Properties: Colorless crystals; strong odor. Mp 27C, bp 158–159C. Soluble in water with strongly basic reaction. Combustible.
Use: Chemical intermediate, complexing agent, catalyst in resin technology, synthesis of quaternary ammonium compounds.

tetramethylene dichloride. See 1,4-dichlorobutane.

3,3'-(tetramethylenedioxy)bis(propylamine).
See 1,4-bis(3-aminopropoxy)butane.

tetramethylene glycol. See 1,4-butylene glycol.

tetramethylene sulfone. See sulfolane.

1,1,4,4-tetramethyl-6-ethyl-7-acetyl-1,2,3,4-tetrahydronaphthalene. $C_{18}H_{26}O$. A polycyclic musk.
Properties: Colorless crystals. Mp 45C, bp 130C (2 mm Hg). Insoluble in water; soluble in alcohol.
Use: Perfumes, cosmetics, soaps.

tetramethylethylenediamine. (TMEDA; N,N,N',N'-tetramethylethylenediamine).
CAS: 110-18-9. $(CH_3)_2NCH_2CH_2N(CH_3)_2$.
Properties: Colorless liquid; slight ammoniacal odor. Bp 121–122C, d 0.7765 (20/4C), refr index 1.4170 (25C), fp −55.1C. Soluble in water and most organic solvents. Combustible.
Grade: Anhydrous (100%), aqueous (65%).
Use: Preparation of epoxy curing agents, polyurethane formation, corrosion inhibitor, textile finishing agents, intermediate for quaternary ammonium compounds.

tetramethylguanidine.
$(CH_3)_2NC(NH)N(CH_3)_2$.
Properties: Liquid; slight ammoniacal odor. Bp 159–160C. Soluble in both water and organic solvents. A strong base. Combustible.

2,6,10,14-tetramethylhexadecane. See phytane.

tetramethyl lead. (TML).
CAS: 75-74-1. $(CH_3)_4Pb$.
Properties: Colorless liquid. D 1.995, fp −27.5C, bp 110C (10 mm Hg), flash p 100F (37.7C). Insoluble in water; slightly soluble in benzene, petroleum ether, alcohol.
Derivation: As for tetraethyl lead. Methyl reagents used instead of ethyl.
Hazard: Flammable, moderate fire risk. Toxic by ingestion, inhalation, and skin absorption. Lower

explosion level 1.8%. Central nervous system impairment. Questionable carcinogen.

tetramethylmethane.   See neopentane.

3,3′-tetramethylnonyl   thiodipropionate.
See ditridecyl thiodipropionate.

tetramethyl phosphonosuccinate.   See phosphonosuccinic acid tetramethyl ester.

tetramethylsilane.
CAS: 75-76-3. $(CH_3)_4Si$.
Properties: Colorless, volatile liquid. Bp 26.5C, d 0.646 (20/4C), flash p 0F (−17.7C). Insoluble in water and cold concentrated sulfuric acid; soluble in most organic solvents.
Derivation: By Grignard reaction of silicon tetrachloride and methylmagnesium chloride.
Grade: Technical, purified.
Hazard: Flammable, high fire risk.
Use: Aviation fuel, internal standard for NMR analytical instruments.

tetramethyl succinonitrile.   (TMSM).
CAS: 3333-52-6. $C_8H_{12}N_2$.
Properties: Colorless, solid; odorless. Mw 136.19, sp g 1.070, mp 170.5C, bp (sublimes). Insoluble in water.
Hazard: Toxic by inhalation and skin contact. Headache, nausea, and central nervous system convulsions.
Use: Blowing agent for vinyl foam production.

tetramethylthiourea.
CAS: 2782-91-4. $C_5H_{12}N_2S$.
Properties: Crystals from H_2O. Mp: 79–80C, bp: 245C.
Hazard: Moderately toxic by ingestion.
Use: As a drug.

1,1,3,3-tetramethylthiourea.   See tetramethylthiourea.

tetramethylthiuram disulfide.   See thiram.

tetramethylthiuram   monosulfide.   (bisdimethylthiocarbamyl sulfide).
CAS: 97-74-5. $[(CH_3)_2NCH]_2S$.
Properties: Yellow powder. D 1.40, mp 104–107C. Soluble in acetone, benzene, and ethylene dichloride; insoluble in water and gasoline. Combustible.
Use: Ultraaccelerator for rubber, fungicide, insecticide.

tetramethylurea.
CAS: 632-22-4. $C_5H_{12}N_2O$.
Properties: Liquid. Bp 176C, flash p 75C (167F), d 1.45. Soluble in water and organic solvents.
Use: Solvent, analytical reagent.

tetranitroaniline.   (TNA).
CAS: 3698-54-2. $C_6H(NO_2)_4NH_2$. A nitration product of aniline.
Properties: Mp 170C; explodes at 237C.
Hazard: Dangerous fire and explosion risk.
Use: Manufacture of detonators and primers.

tetranitromethane.
CAS: 509-14-8. $C(NO_2)_4$.

$$O_2N-\overset{\overset{\displaystyle NO_2}{|}}{\underset{\underset{\displaystyle NO_2}{|}}{C}}-NO_2$$

Properties: Colorless liquid; pungent odor. Bp 125.7C, mp 12.5C, d 1.650 (13C). Miscible with alcohol and ether; insoluble in water; decomposed by alcoholic solution of potassium hydroxide. Powerful oxidizing agent.
Derivation: By action of fuming nitric acid on benzene, acetic anhydride, or acetylene.
Hazard: Dangerous fire and explosion risk. Toxic by ingestion, inhalation, skin absorption. Eye and upper respiratory tract irritant. Upper respiratory tract cancer. Possible carcinogen.
Use: Rocket fuel, as an oxidant or monopropellant; qualitative test for unsaturated compounds; diesel fuel booster; organic reagent.

tetra(octylene glycol) titanate.   See octylene glycol titanate.

tetraphene.   See benz(a)anthracene.

1,1,4,4-tetraphenylbutadiene.
(TPB). $(C_6H_5)_2C:CHCH:C(C_6H_5)_2$.
Properties: White crystals. Two forms, mp 194–196C and mp 202–204C. Insoluble in water; soluble in most organic solvents. Combustible.
Grade: Purified.
Use: Primary fluor or wavelength shifter in soluble scintillators.

tetraphenylresorcinol   diphosphate.   See resorcinol bis(diphenyl phosphate).

tetraphenylsilane. $(C_6H_5)_4Si$.
Properties: White solid. Mp 237C, bp 428C. Very stable and inert. Combustible.
Derivation: By Grignard reaction of silicon tetrachloride and phenylmagnesium chloride.
Grade: Technical.
Use: Heat-transfer medium, polymers.

tetraphenyltin. $(C_6H_5)_4Sn$.
Properties: White powder. D 1.490, mp 225–228C, bp above 420C. Insoluble in water; soluble in hot benzene, toluene, xylene.
Derivation: Reaction of tin tetrachloride with phenylmagnesium bromide.
Hazard: Skin irritant.

Use: Stabilizer in chlorinated transformer oils, mothproofing agent, scavenger in dielectric fluids, intermediate.

tetraphosphoric acid. See polyphosphoric acid.

tetraphosphorus heptasulfide. See phosphorus heptasulfide.

tetraphosphorus hexasulfide. See phosphorus trisulfide.

tetraphosphorus trisulfide. See phosphorus sesquisulfide.

tetrapotassium ethylenediaminetetraacetate. See ethylenediaminetetraacetic acid (note).

tetrapotassium pyrophosphate. (TKPP). See potassium pyrophosphate.

tetrapropenylsuccinic anhydride. See dodecenylsuccinic anhydride. Many isomers are possible.

tetra-*n*-propyl dithionopyrophosphate.
CAS: 3244-90-4. $(C_3H_7O)_2P(S)OP(S)(OC_3H_7)_2$.
Properties: Amber liquid. Bp 148C (2 mm Hg). Miscible with most organic solvents; insoluble in water.
Hazard: Highly toxic.
Use: Insecticide, acaricide.

tetrapropylene. (dodecene; propylene tetramer).
CAS: 6842-15-5. $C_{12}H_{24}$. A mixture of C_{12} monoolefins.
Properties: Liquid. D 0.770 (20/20C), boiling range 183–218C, bulk d 6.44 lb/gal (15.5). Combustible.
Derivation: Olefin fraction obtained from catalytic polymerization of propylene.
Use: Detergents (dodecylbenzene), lubricant additives, plasticizers.

tetrapropylenepentamine.
CAS: 13274-42-5. $C_{12}H_{31}N_5$.
Hazard: A poison by ingestion and skin contact. A moderate skin and severe eye irritant.

tetrapropylthiuram disulfide.
$[(C_3H_7)_2NCS]_2S_2$.
Properties: Light-cream color; musty odor. D 1.13 (20/20C), melting range 49–51.5C. Soluble in carbon disulfide, benzene, chloroform, and gasoline; insoluble in water and 10% caustic soda. Combustible.

tetrasilane. Si_4H_{10}.
Properties: Colorless liquid. Fp −93.5C, bp 109C, d 0.825 (0C).
Hazard: Severe fire and explosion risk, can ignite or explode in air.

tetrasodium diphosphate. See sodium pyrophosphate; sodium polyphosphate.

tetrasodium EDTA. (ethylenediaminetetraacetic acid; tetrasodium salt; EDTA Na$_4$; sodium edetate).
CAS: 64-02-8. $C_{10}H_{12}N_2Na_4O_8$ anhydrous or 2H$_2$O.
Properties: White powder. Freely soluble in water.
Use: General-purpose chelating agent.

See ethylenediaminetetraacetic acid (note).

tetrasodium hydrogen 2-phosphonatobutane-1,2,4-tricarboxylate.
CAS: 66669-53-2. $C_7H_7O_9P•4Na$.
Hazard: Moderately toxic by ingestion and skin contact.

tetrasodium monopotassium tripolyphosphate. $Na_4KP_3O_{10}$.
Properties: White, crystalline solid. Mp 580–600C, d 2.55. Solubility in water 30 g/100 mL (26C).
Use: Sequestrant.

tetrasodium pyrophosphate. See sodium pyrophosphate; sodium polyphosphate.

tetrastearyl titanate. Organic intermediate, adhesion promoter, pigment dispersant.

tetrathiin. See oxidimethiin.

tetrazene. (4-amidino-1-[nitrosamino-amidino]-1-tetrazene).
CAS: 109-27-3. $H_2NC(:NH)NHNHN:NC(:NH)NHNHNO$.
Properties: Colorless or pale-yellow, fluffy solid. Apparent d 0.45 but yields a pellet of d 1.05 under pressure of 3000 psi. Practically insoluble in water, alcohol, ether, benzene, and carbon tetrachloride; slightly hygroscopic.
Derivation: Interaction of an aminoguanidine salt with sodium nitrite in the absence of free mineral acid.
Use: Initiating explosive.

1h-tetrazol-5-amine. See aminotetrazole.

tetrazolium chloride. (tetrazolium salt; TTC; 2,4,5-triphenyltetrazolium chloride). $CN_4Cl(C_6H_5)_3$.
Properties: White to pale-yellow crystalline powder that darkens on exposure to light. Mp (with decomposition) 245C. Readily soluble in water.

Use: In germination and viability tests. Viable parts of seed are stained red by deposition of red insoluble triphenyl formazan.

tetrazolium nitro blue. See *p*-nitro blue tetrazolium chloride.

tet resistance. See: Antibiotic resistance.

tetrol. See furan.

"Tetrone" A *[Du Pont Performance].* TM for dipentamethylenethiuram tetrasulfide rubber accelerator.

"Tetronic" *[BASF].* TM for a nonionic tetrafunctional series of polyether block polymers ranging in physical form from liquids through pastes to flakable solids. They are polyoxyalkylene derivatives of ethylenediamine. Physical state varies with molecular weight and oxyethylene content, 100% active.
Use: Low-foaming detergent formulations; defoaming agents; flexible and rigid polyurethane foams; emulsifying and demulsifying agents; textile processing.

tetryl. (tetralite and nitramine are common commercial names for trinitrophenylmethylnitramine).
CAS: 479-45-8. $(NO_2)_3C_6H_2N(NO_2)CH_3$.
Properties: Yellow crystals. Mp 130–132C, d 1.57 (19C), explodes at 187C. Insoluble in water; soluble in alcohol, ether, benzene, glacial acetic acid.
Hazard: Dangerous fire and explosion risk. Skin irritant, absorbed by skin. Upper respiratory tract irritant.
Use: Detonating agent for less-sensitive high explosives, indicator (colorless at pH 10.8, dull red at pH 13.0).

textile oil. Any of various specially compounded oils used to condition raw fibers, yarns, or fabric for manufacturing, bleaching, dyeing, and finishing operations.

"Textone" *[Ahlstrom].* TM for sodium chlorite.
Available forms: Liquid solutions and powder.
Use: Used for water disinfection, as a bleaching agent for textiles, and for algae control.

textryl. Generic name for nonwoven structures that may be manufactured by wet-processing from staple fibers and fibrid binder.

texture. The physical structure of a solid or semisolid material that results from the shape, arrangement, and proportions of its components. The term is used in the textile industry to characterize fabrics of various types and in the food industry to describe quality characteristics of bakery products, margarines, meats, spun proteins, etc. It is also

regarded by geologists as a property of rocks and soils.
See protein, textured.

textured protein. See protein, textured.

TFE. Abbreviation for tetrafluoroethylene.
See polytetrafluorethylene.

TFEO. Abbreviation for tetrafluoroethylene epoxide.

TGA. Abbreviation for (1) triglycollamic acid; (2) thermogravimetric analysis.
See thermogravimetric analysis; nitrilotriacetic acid.

TGDR. See thioguanine deoxyriboside.

Th. Symbol for thorium.

thallium.
CAS: 7440-28-0. Tl. Metallic element of atomic number 81, group IIIA of the periodic table, aw 204.37, valences = 1, 3; two stable isotopes.
Properties: Bluish-white, lead-like metal. D 11.85, mp 302C, bp 1457C, oxidizes in air at room temperature. Soluble in nitric and sulfuric acids; insoluble in water but readily forms soluble compounds when exposed to air or water.
Derivation: Flue dusts from lead and zinc smelting. The thallium compounds recovered are treated to obtain the metal by electrolysis, precipitation, or reduction.
Grade: Technical, high purity.
Hazard: Forms toxic compounds on contact with moisture; keep from skin contact. Gastrointestinal damage and peripheral neuropathy.
Use: Thallium salts, mercury alloys, low-melting glasses, rodenticides, photoelectric applications, electrodes in dissolved oxygen analyzers.

thallium acetate. (thallous acetate).
CAS: 563-68-8. TlOCOCH₃.
Properties: White, deliquescent crystals. Mp 131, d 3.68. Soluble in water and alcohol.
Derivation: Interaction of acetic acid and thallium carbonate.
Hazard: As for thallium.
Use: High specific gravity solutions used to separate ore constituents by flotation.

thallium amalgam.
Properties: Reported to have fp –60C (30C below that of mercury).
Hazard: As for thallium.
Use: Substitute for mercury in electrical switches, thermometers, etc. for extremely low-temperature service.

thallium bromide. (thallous bromide).
CAS: 7789-40-4. TlBr.

Properties: Yellowish-white, crystalline powder. D 7.557, bp 815C, mp 460C. Soluble in alcohol. Slightly soluble in water. Insoluble in acetone.
Hazard: As for thallium.
Use: Mixed crystals with thallium iodide for infrared radiation transmitters used in military detection devices.

thallium carbonate. (thallous carbonate).
CAS: 6533-73-9. Tl_2CO_3.
Properties: Heavy, shiny, colorless or white crystals. D 7, mp 272C. Highly refractive, melts to dark-gray mass, slightly alkaline taste. Soluble in water; insoluble in alcohol.
Hazard: As for thallium.
Use: Analysis (testing for carbon disulfide), artificial diamonds.

thallium chloride. (thallous chloride).
CAS: 7791-12-0. TlCl.
Properties: White, crystalline powder; becomes violet on exposure to light. D 7.004 (30/4C), mp 430C, bp 720C. Slightly soluble in water; insoluble in alcohol, ammonium hydroxide.
Hazard: As for thallium.
Use: Catalyst (chlorination), suntan lamp monitors.

thallium hydroxide. (thallous hydroxide).
$TlOH•H_2O$.
Properties: Yellow needles. Bp (dehydrated) 139C, (decomposes). Soluble in alcohol, water.
Hazard: As for thallium.
Use: Analysis, indicator.

thallium iodide. (thallous iodide).
CAS: 7790-30-9. TlI.
Properties: Yellow powder. D 7.09, bp 824C, mp 440C, becomes red at 170C. Insoluble in alcohol; slightly soluble in water; soluble in aqua regia.
Hazard: As for thallium.
Use: Mixed crystals with thallium bromide for infrared radiation transmitters.

thallium monoxide. (thallium oxide; thallous oxide).
CAS: 1314-32-5. Tl_2O.
Properties: Black powder. Mp 300C, bp 1080C, d 9.52(16C). Soluble in alcohol, water (decomposes). Oxidizes when exposed to air, keep well stoppered.
Hazard: As for thallium.
Use: Analysis (testing for ozone), artificial gem, optical glass of high refractive index.

thallium nitrate. (thallous nitrate).
CAS: 10102-45-1. $TlNO_3$.
Properties: Colorless crystals. D 5.5, mp 206C (solidifies to a glass-like solid), decomposes at 450C. Soluble in hot water; insoluble in alcohol.
Grade: Technical.
Hazard: A poison. Strong oxidizing agent, fire and explosion risk. TLV: 0.1 mg(Tl)/m^3. Toxic by skin absorption.
Use: Analysis, pyrotechnics (green fire).

thallium(III) nitrate.
CAS: 13746-98-0. N_3O_9Tl.
Hazard: A poison by skin contact.

thallium oxide. (thallous oxide).
CAS: 1314-12-1. Tl_2O.
Properties: Finely divided black solid. Mp about 300C. Soluble in water and alcohol. Oxidizes when exposed to air.
Hazard: Toxic by ingestion.
Use: Optical glass, synthetic gemstones.

thallium sesquichloride. (thallo-thallic chloride). $TlCl_3•3TlCl$ or Tl_2Cl_3.
Properties: Yellow, crystalline powder. D 5.9, mp 400–500C. Slightly soluble in water.
Hazard: As for thallium.

thallium sulfate. (thallous sulfate).
CAS: 7446-18-6. Tl_2SO_4.
Properties: Colorless crystals. D 6.77, mp 632C. Soluble in water.
Grade: Technical, 99%.
Hazard: As for thallium.
Use: Analysis (testing for iodine in the presence of chlorine), ozonometry, rodenticides, pesticide.

thallium sulfide. (thallous sulfide). Tl_2S.
Properties: Blue-black, lustrous, microscopic crystals or amorphous powder. D 8.46, mp 448C. Soluble in mineral acids; insoluble in water, alcohol, or ether.
Hazard: As for thallium.
Use: Infrared-sensitive photocells.

thallous compound. See corresponding thallium compound.

thalrugosamine. See homoaromoline.

THAM. See tris(hydroxymethyl)aminomethane.

thebaine. (p-morphine).
CAS: 115-37-7. $C_{19}H_{21}NO_3$.
Properties: White, crystalline alkaloid. Mp 193C, d 1.30. Slightly soluble in water; soluble in alcohol and ether. It is extracted from poppies of a different species from morphine-producing opium poppies.
Hazard: Toxic by ingestion, may induce addiction.

theca. General term for any stiff outer covering of a unicellular protist, and usually made up of interlocking plates. Dinoflagellates and diatoms are examples of protists with thecae.

theine. See caffeine.

Thenard's blue. See cobalt blue.

The National Dendrimer and Nanotechnology Center. (NDNC). Serves as a catalyst for new dendrimer-based research yielding a growing body of valuable intellectual property. Working coopera-tively on a broad range of research projects. Located at 625 Denison Drive, Mount Pleasant, MI 48858. Website: http://www.dendrimercenter.org/research .html

thenyl alcohol.　　(2-thienylmethanol; 2-hydro-xymethylthiophene; 2-thiophenecarbinol). $C_4H_4SCH_2OH$.
Properties: Colorless liquid. Bp 207C. Insoluble in water; soluble in alcohol and ether.
Derivation: A heterocyclic alcohol made by reaction of 2-thienylmagnesium iodide and formaldehyde.
Use: No commercially developed applications.

thenyldiamine.　　(coined name for 2-[2-dime-thylaminoethyl]-3-thenylaminopyridine).
CAS: 91-79-2.　　$(C_4H_3SCH_2)N(C_5H_4N)CH_2CH_2$ $N(CH_3)_2$.
Properties: Liquid. Bp 169–172C (1.0 mm Hg).
Derivation: Condensing N,N-dimethylaminoethyl-α-aminopyridine with 3-thenyl bromide.
Use: Medicine (as base for various salts, especially the hydrochloride).

theobroma oil.　　(cacao butter; cocoa butter).
Properties: Yellowish-white solid; chocolate-like taste and odor. D 0.858–0.864 (100/25C), mp 30–35C, refr index 1.4537–1.4585 (40C), saponifica-tion number 188–195, iodine number 35–43. Insol-uble in water; slightly soluble in alcohol; soluble in boiling dehydrated alcohol; freely soluble in ether and chloroform. Combustible.
Derivation: From the cacao bean, by expression, decoction, or extraction by solvent.
Chief constituents: Glycerides of stearic, palmitic, and lauric acids.
Grade: Crude, refined, USP.
Use: Confectionery, suppositories and pharmaceuti-cals, soaps, cosmetics.

theobromine.　　(3,7-dimethylxanthine).
CAS: 83-67-0.　$C_7H_8N_4O_2$. The alkaloid found in cocoa and chocolate products. A purine base closely related to caffeine. Also occurs in tea and cola nuts.

Hazard: Toxic by ingestion. Questionable carcino-gen.

theophylline.　　(1,3-dimethylxanthine).
CAS: 58-55-9.　$C_7H_8N_4O_2$.

Properties: White, crystalline alkaloid; odorless; bitter taste. Mp 270–274C. Slightly soluble in water and alcohol; more soluble in hot water; soluble in alkaline solutions.
Derivation: (1) By extraction from tea leaves, (2) synthetically from ethyl cyanoacetate.
Grade: Technical, NF.
Hazard: Questionable carcinogen.
Use: Medicine (diuretic, muscle relaxant).

theoretical plate.　　Any contacting device in a fractionating column, such as packing, grids, or screens, that effects the same degree of separation of vapor from liquid as one simple distillation. A col-umn that gives the same separation as 10 successive simple distillations is considered to have 10 theo-retical plates. The effectiveness of a fractionating column is measured in terms of theoretical plates. As many as 100 theoretical plates are used in lab-oratory and industrial operations. The total column height divided by the number of theoretical plates is known as HETP (height equivalent to a theoretical plate). This concept is also used in chromatographic techniques.

"Therabloat" *[Zoetis].*　　(poloxalene).
CAS: 9003-11-6.　$HO(CH_2CH_2O)_n[CH(CH_3)$ $CH_2O]_n(CH_2CH_2O)_nH$.
Properties: Liquid nonionic surfactant polymer. TM for a veterinary drug used to treat bloat.
Use: Drug (veterinary); food additive.

therm.　　A unit of heat equal to 100,000 Btu. It has also been used to mean one Btu or one small calorie, but these uses have been abandoned in the U.S.

thermal black.　　Carbon black made from natural gas by the thermatomic process or by pyrolysis of bituminous coal. It is no longer widely used.

thermal cracking.　　See cracking.

thermal decomposition.　　See decomposition (6); pyrolysis.

thermal expansion coefficient.　　The change in volume per unit volume per degree change in tem-perature (cubical coefficient). For isotropic solids the expansion is equal in all directions, and the cubi-cal coefficient is about three times the linear coef-ficient of expansion. These coefficients vary with temperature, but for gases at constant pressure the coefficient of volume expansion is nearly constant and equals 0.00367 for each degree Celsius at any temperature.

thermal neutron.   A slow neutron.
See neutron.

thermal pollution.   Heat introduced into rivers or estuaries by power plants or other industrial cooling waters or chemical wastes, which has adverse effects on estuarine ecology.
See water pollution.

thermatomic process.   Methane or natural gas is cracked over hot bricks at 870C to form amorphous carbon (carbon black) and hydrogen.
See thermal black; carbon black.

"Therm-Chek" *[Ferro].*   TM for heat stabilizers.

"Therminol" FR *[Solutia].*   TM for a family of heat-transfer fluids thermally stable to 400C.

thermite.   A mixture of ferric oxide and powdered aluminum, usually enclosed in a metal cylinder and used as an incendiary bomb, invented by the German chemist Hans Goldschmidt around 1900. On ignition by a ribbon of magnesium, the reaction produces a temperature of 2200C, that is sufficient to soften steel. This is typical of some oxide/metal reactions which provide their own oxygen supply and thus are very difficult to stop.
Hazard: Dangerous fire risk.

thermochemistry.   That branch of chemistry comprising the measurement and interpretation of heat changes that accompany changes of state and chemical reactions. It is closely related to chemical thermodynamics. The heat of formation of a compound is the heat absorbed when it is formed from its elements in their standard states. An exothermic reaction evolves heat, an endothermic reaction requires heat for initiation. Application of thermochemical principles to generation of hydrogen by water splitting is being extensively researched. The advantage of this method over electrolysis is its greater net efficiency. Many techniques have been explored, but only a few have practical potential. One of the more promising is the S-I cycle under study at Lawrence Livermore Laboratory. The heat for this would be obtained from either solar receivers or nuclear reactors. The reactions involved are
$$2H_2O + SO_2 + I_2 \rightarrow H_2SO_4 + 2HI$$
$$H_2SO_4 \rightarrow H_2O + SO_2 + \tfrac{1}{2}O_2$$
$$2HI \rightarrow H_2 + I_2$$
See hydrogen (Note 2).

thermocouple.   (thermoelectric thermometer). An instrument composed of two wires made of dissimilar metals or semiconducting materials that are joined at one end (the measuring junction), the other end being the reference junction, which is maintained at a known temperature (usually 0C).

The difference in temperature between the measuring junction and the reference junction generates an electromotive force that is proportional to the temperature difference. Thermocouples are applicable over a range of −200C to 1800C. The most suitable conducting materials are iron–constantan, platinum–platinum–rhodium, copper–constantan, and Chromel–Alumel; graphite–silicon carbide is used in the metallurgical field. Thermocouples are essential for determinations of extreme temperatures that are beyond the range of liquid-in-glass thermometers. Their industrial applications include molten metals, fuel beds, ceramic kilns, furnaces, etc.; in laboratories they are used for both high-temperature and cryogenic research. They are also applicable to intermediate temperatures in cases where conventional thermometers are impractical.
See thermoelectricity.

thermodynamic potential.   See Nernst potential.

thermodynamics.   A rigorously mathematical analysis of energy relationships (heat, work, temperature, and equilibrium), the principles of which were first elaborated by J. Willard Gibbs in the mid-19th century. It describes systems whose states are determined by thermal parameters, such as temperature, in addition to mechanical and electromagnetic parameters. A *system* is a geometric section of the universe, whose boundaries may be fixed or varied, and that may contain matter or energy or both. The *state* of a system is a reproducible condition, defined by assigning fixed numerical values to the measurable attributes of the system. These attributes may be wholly reproduced as soon as a fraction of them has been reproduced. In this case the fractional number of attributes determines the state, and is referred to as the *number of variables of state* or *the number of degrees of freedom* of the system.
The concept of *temperature* can be evolved as soon as a means is available for determining when a body is "hotter" or "colder." Such means might involve the measurement of a physical parameter such as the volume of a given mass of the body. When a "hotter" body, A, is placed in contact with a "colder" body, B, it is observed that A becomes "colder" and B "hotter." When no further changes occur, and the joint system involving the two bodies has come to equilibrium, the two bodies are said to have the same temperature. Thus temperature can only be measured at equilibrium. Therefore thermodynamics is a science of equilibrium, and a thermodynamic state is necessarily an equilibrium state. Thermodynamics is a macroscopic discipline, dealing only with the properties of matter and energy in bulk, and does not recognize atomic and molecular structure. Although severely limited in this respect, it has the advantage of being completely insensitive to any change in our ideas concerning molecular

phenomena, so that its laws have broad and permanent generality. Its chief service is to provide mathematical relations among the measurable parameters of a system in equilibrium so that, for example, a variable-like pressure may be computed when the temperature is known, and vice versa.

thermodynamics, chemical. See chemical thermodynamics.

thermoelectricity. Electricity produced directly by applying a temperature difference to various parts of electrically conducting or semiconducting materials. Usually two dissimilar materials are used, and the points of contact are kept at different temperatures (Peltier effect). Many temperature-measuring devices (thermocouples, thermopiles) work on this principle, since the voltage is proportional to the temperature difference. Metallic conductors are usually used for these "thermometers, " which produce a rather small current. A newer use for the effect is as a source of electrical energy, i.e., a means of direct conversion of heat into electricity (or vice versa) without the use of a generator (or motor). The materials used for these thermoelectric couples are semiconductors (e.g., tellurium; zinc antimonide; lead, bismuth, and germanium tellurides; samarium sulfide) or thermoelectric alloys, all of which produce relatively large currents. Several of these "cells" are then hooked in series much like the cells of a battery.

"Thermoflex" A [Du Pont]. TM for a rubber antioxidant containing 25% di-*p*-methoxydiphenylamine $(CH_3OC_6H_4)_2NH$; 25% diphenyl-*p*-phenylenediamine $C_6H_4(NHC_6H_5)_2$; and 50% phenyl-β-naphthylamine $C_{10}H_7NHC_6H_5$.
Properties: Dark-gray pellets. D 1.21, fp above 67C.
Use: Tire carcasses, transmission belts, etc., to promote resistance to flexing at operating temperatures. See antioxidant.

thermofor. A heat-transfer medium. See coolant.

thermoforming. (1) See reforming. (2) Forming or shaping a thermoplastic sheet by heating the sheet above its melting point, fitting it along the contours of a mold with pressure supplied by vacuum or other force, and removing it from the mold after cooling below its softening point. The method is applied to polystyrenes, acrylics, vinyls, polyolefins, cellulosics, etc.

thermofor process. A moving-bed catalytic cracking process in which petroleum vapor is passed up through a reactor countercurrent to a flow of small beads or catalyst. The deactivated catalyst then passes through a regenerator and is recirculated.

thermogravimetric analysis. (TGA). The weight of a substance heated or cooled at a controlled rate, in thermogravimetry, which is recorded as a function of time or temperature. Frequently the rate of weight change also is measured electronically by taking the first derivative of the weight change with time.

"Thermoguard" [Thermoguard]. TM for antimony-based materials for incorporation in PVC and other chlorine-containing plastics for flame-proofing properties.

thermometer. An instrument for measuring temperature. The liquid-in-glass thermometer consists of a graduated glass tube and a bulb containing a suitable liquid whose expansion and contraction indicate the temperature. Its range is from −130 to 600C. For scientific purposes the most widely used liquid is mercury down to its freezing point at −40C; below this, alcohol gives readings to −100C and pentane to −130C. Colored alcohol is generally used in household thermometers. Mercury thermometers ranging up to 600C are available; the mercury is prevented from vaporizing by a pressurized inert gas inserted above the mercury column. Metal protection tubes for stem and bulb are necessary. The softening point of the glass is of primary importance; borosilicate glasses are satisfactory up to 500C, but Jena glass is required for higher temperatures. Minimum and maximum thermometers are so made as to retain their lowest and highest readings indefinitely; the latter are used for oil-well and other geothermal measurements.
   There are several other types of thermometers: (1) Gas in which either the pressure at constant volume or the volume at constant pressure measure the temperature; these are used for extremely accurate thermodynamic determinations. The gases used are helium, nitrogen, and hydrogen. (2) Bimetallic, in which the sensing element consists of two strips of metals having different expansion coefficients; its range is from −185 to 425C. (3) Thermoelectric (thermocouple), in which measurement is made by the electromotive force generated by two dissimilar metals; its range is from −200 to 1800C. (4) Resistance, in which temperature is measured by change in the electrical resistance of a metal, usually platinum; its range is from −163 to 660C. (5) An optical fiber thermometer developed by NBS Center for Chemical Engineering has a range of up to 2000C. It is made from a single crystalline sapphire and is much more accurate than the existing standard. Based on fundamental radiation principles, it measures thermodynamic temperatures directly.
   See thermocouple; bimetal.

thermonuclear reaction. See fusion.

thermoplastic. A high polymer that softens when exposed to heat and returns to its original condition when cooled to room temperature. Natural substances that exhibit this behavior are crude rubber and a number of waxes; however, the term is usually applied to synthetics such as polyvinyl chloride, nylons, fluorocarbons, linear polyethylene, polyurethane prepolymer, polystyrene, polypropylene, and cellulosic and acrylic resins. See thermoset.

thermoset. A high polymer that solidifies or "sets" irreversibly when heated. This property is usually associated with a cross-linking reaction of the molecular constituents induced by heat or radiation, as with proteins, and in the baking of doughs. In many cases, it is necessary to add "curing" agents such as organic peroxides or (in the case of rubber) sulfur. For example, linear polyethylene can be cross-linked to a thermosetting material by either radiation or chemical reaction. Phenolics, alkyds, amino resins, polyesters, epoxides, and silicones are usually considered to be thermosetting, but the term also applies to materials in which additive-induced cross-linking is possible, e.g., natural rubber.

THF. Abbreviation for tetrahydrofuran.

thia-. Prefix indicating the presence of sulfur in a heterocyclic ring.

thiabendazole. (4-[2-benzimidazolyl]thiazole). CAS: 148-79-8. $C_{10}H_7N_3S$.
Properties: White to tan crystals. Mp 304C. Slightly soluble in water, alcohols, and chlorinated hydrocarbons; soluble in dimethylformamide.
Use: Fungicide effective on citrus fruits, anthelmintic.

thiacloprid. See (3-((6-chloro-3-pyridinyl) methyl)-2.

thiaminase. A plant enzyme that catalyzes the hydrolysis of thiamine into pyrimidine and thiazole derivatives.
Derivation: Occurs in raw fish, certain bacteria, and an occasional plant.
Hazard: Toxic.

thiamine. (3-(4-amino-2-methylpyrimidyl-5-methyl)-4-methyl-5, β-hydroxy-ethylthiazolium chloride; vitamin B_1). $C_{12}H_{17}ClN_4OS$. The antineuritic vitamin, essential for growth and the prevention of beriberi. It functions in intermediate carbohydrate metabolism in coenzyme form in the decarboxylation of α-keto acids. Deficiency symptoms: emotional hypersensitivity, loss of appetite, susceptibility to fatigue, muscular weakness, and polyneuritis.

Source: Enriched and whole-grain cereals, milk, legumes, meats, yeast. Most of the thiamine commercially available is synthetic.
Use: Medicine, nutrition, enriched flours. Isolated usually as the chloride (see formula above). Available as thiamine hydrochloride and thiamine mononitrate.

thiamine pyrophosphate. The active coenzyme form of vitamin B1 which functions in aldehyde transfer reactions.

1,4-thiazane.

CH$_2$SCH$_2$CHNHCH$_2$.
Properties: Colorless liquid; pyridine-like odor. Bp 169C (758 mm Hg). Fumes in air. Absorbs carbon dioxide from the air. Soluble in alcohol, benzene, ether, water. Combustible.
Derivation: Interaction of alcoholic ammonia and dichlorodiethyl sulfide.
Grade: Technical.
Use: Organic synthesis.

thiazin dye. Any of a group of important biological stains that are similar to azin dyes except that one of the connecting nitrogen atoms is replaced by sulfur.
Use: Hematology.

thiazole.
CAS: 288-47-1.

SCH:NCH:CH.
Properties: Colorless or pale-yellow liquid; odor resembles that of pyridine. D 1.18, bp 116.8C. Soluble in alcohol and ether; slightly soluble in water.
Use: Organic synthesis of fungicides, dyes, and rubber accelerators.

thiazole dye. A dye whose molecular structure contains the thiazole ring. The chromophore groups are =C=N–, –S–C=, but the conjugated double bonds are also of importance. The members of the class are mainly used as direct or developed dyes for cotton, though some find use as union dyes. One example is primuline, CI 49000.
See thiazole.

3-(n-2-thiazolylformimidoyl)indole.
CAS: 22394-38-3. $C_{12}H_9N_3S$.
Hazard: A poison.

thiazophyr.
CAS: 117718-60-2. $C_{16}H_{17}F_5N_2O_2S$.
Hazard: Questionable carcinogen.
Use: Agricultural chemical.

2-thiazylamine.   See 2-aminothiazole.

thickened oil.   See blown oil.

thickener.   A circular or cylindrical tank equipped with revolving rakes or plows so designed as to increase the concentration of solids in a suspension. The slurry enters through a feed trough or well, which distributes it uniformly around the circumference of the tank; the rakes move the thickened sediment toward the center where it is discharged through a collecting cone. Thickeners are applicable to the metallurgical and food industries, ore flotation, and municipal water treatment.
See clarification.

thickening agent.   Any of a variety of hydrophilic substances used to increase the viscosity of liquid mixtures and solutions and to aid in maintaining stability by their emulsifying properties. Four classifications are recognized: (1) starches, gums, casein, gelatin, and phycocolloids; (2) semisynthetic cellulose derivatives (carboxymethyl-cellulose, etc.); (3) polyvinyl alcohol and carboxy-vinylates (synthetic); and (4) bentonite, silicates, and colloidal silica. The first group is widely used in the food industries, especially in ice creams, confectionery, gravies, etc.; other major consumers are the paper, adhesives, textiles, and detergent fields.
See emulsion; colloid, protective; gel.

Thiele reaction.   Formation of triacetoxy aromatic compounds by the reaction of quinones with acetic anhydride catalyzed by sulfuric acid or boron trifluoride.

(thienyl)hydrazine.
CAS: 64059-33-2. $C_4H_6N_2S$.
Hazard: A poison.

thimerosal.   (sodium ethylmercurithiosalicylate; ethylmercurithiosalicylic acid, sodium salt).
CAS: 54-64-8. $NaOOCC_6H_4SHgC_2H_5$.
Properties: Light-cream-colored, crystalline powder; slight characteristic odor. Affected by light. Soluble in water and alcohol; almost insoluble in ether and benzene; pH (1% solution) 6.7.
Derivation: Reaction between ethylmercuric chloride and thiosalicylic acid in alcoholic sodium hydroxide.
Grade: NF.
Hazard: Toxic by ingestion, inhalation.
Use: Medicine, bacteriostat, fungistat.

thin.   A nontechnical word used by scientists with a variety of meanings. (1) In electronic metallurgy, a thin film is a vapor-deposited coating having a thickness of only a single atom; such *monatomic* films, e.g., thorium on tungsten, are used in electronic devices such as cathodes. (2) A coating or film of a fatty acid on water that is one molecule thick (about 200 Å) is called a *monomolecular* film. (3) In thin-layer chromatography the term applies to a specially prepared mixture of adsorbents spread on a glass slide to a thickness of 1/100 inch. (4) The word is also used in the sense of a liquid of low viscosity, as in paint thinner and thin-boiling starch.

thin-boiling starch.   See starch, modified; thin.

thin film.   In electronic engineering a film having a "thickness" of a single atom and consisting of a metal deposited on a metallic substrate either externally by vapor deposition or internally by diffusion. The base metal is usually tungsten (for a cathode), the film being any of a number of other metals (thorium, cesium, zirconium, barium, or cerium). "The greatest benefit is obtained when the films are of a monatomic nature; in the case of thorium on tungsten, the optimum coverage is 0.67 monatomic layer, that is, the thorium atoms do not completely cover the tungsten surface. Such films are very tightly bound to the base metal by atomic forces" (W. H. Kohl).
See thin; film.

thin-layer chromatography.   (TLC). A micro type of chromatography. The thin layer (0.01 inch) is the adsorbent, usually a special silica gel spread on glass or incorporated in a plastic film. Single drops of the solutions to be investigated are placed along one edge of the glass plate, and this edge then dipped into a solvent. The solvent carries the constituents of the original test drops up the thin layer in a selective separation, so that a comparison with known standards and various identifying tests may be made on the spots formed.
See thin.

thinner.   A hydrocarbon (naphtha) or oleoresinous solvent (turpentine) used to reduce the viscosity of paints to appropriate working consistency, usually just before application. In this sense a thinner is a liquid diluent, except that it has active solvent power on the dissolved resin.

thio-.   A prefix used in chemical nomenclature to indicate the presence of sulfur in a compound, usually as a substitute for oxygen.
See thiol.

thioacetamide.
CAS: 62-55-5. CH_3CSNH_2.
Properties: Colorless leaflets. Mp 115C. Stable in solution. Soluble in water, alcohol, ether, benzene. Combustible.

Hazard: Toxic by ingestion and inhalation, a possible carcinogen.

Use: To replace gaseous hydrogen sulfide in qualitative analysis.

thioacetic acid.   (thiacetic acid; ethanethiolic acid).
CAS: 507-09-5. CH_3COSH.
Properties: Clear, yellow liquid; strong unpleasant odor. D 1.05 (25C), fp −17C, bp 81.8C (630 mm Hg). Soluble in water, alcohol, and ether. Combustible.
Derivation: By heating glacial acetic acid and phosphorus pentasulfide, with subsequent distillation.
Hazard: Toxic by ingestion and inhalation.
Use: Chemical reagent, lachrymator.

thioallyl ether.   See allyl sulfide.

thioanisole. $C_6H_5CH_3$.
Properties: Colorless liquid; strong unpleasant odor. D 1.053 (25C), fp −15.5C, bp 188C, refr index 1.5842 (25C). Insoluble in water; soluble in most organic solvents. Combustible.
Use: Intermediate, solvent for polymeric systems.

thiobenzoic acid. C_6H_5COSH.
Properties: Yellow oil or crystals. D 1.1825–1.1835 (20/4C), mp 24C, bp 77.5C (5 mm Hg), 122C (30 mm Hg), refr index 1.602–1.604 (20C). Insoluble in water; miscible with organic solvents. Combustible.
Grade: 95% min.
Use: Organic intermediate.

2-thio-2h-1,3-benzoxazine-2,4(3h)-dione.
CAS: 10021-35-9. $C_8H_5NO_2S$.
Hazard: A poison by ingestion.
Use: Agricultural chemical.

4,4′-thiobis(6-*tert*-butyl-*m*-cresol).
CAS: 96-69-5.
Properties: Light-gray to tan powder. Mp 150C min, d 1.10 (25C).
Hazard: Toxic by inhalation. Upper respiratory tract irritant. Questionable carcinogen.
Use: Protection of light-colored rubber from oxidation and of nonstaining neoprene compounds against deterioration.

2,2′-thiobis(chlorophenol).
CAS: 97-18-7. $[ClC_6H_3(OH)]_2S$.
Properties: White, crystalline solid; odorless. Mp 175.8–186.8C. Insoluble in water.
Hazard: Toxic by ingestion.
Use: Bacteriostat for cosmetics, fungicide.

2,2′-thiobis(4,6-di-*sec*-amylphenol).   (2,2′-thiobis[4,6-bis-(1-methylbutyl)phenol]).
$[(CH_3[CH_2]_2CH[CH_3])_2OHC_6:H_2]_2S$.
Properties: A dark, viscous liquid. Softening p 0C, d 0.99 (50C).
Use: Rubber antioxidant.

thiocarbamate herbicide.   Any of a small family of herbicides that are toxic to germinating seeds that interfere with a variety of processes such as photosynthesis, respiration, oxidative phosphorylation, protein synthesis, and nucleic acid metabolism.
Hazard: Volatile and nonpersistent.

thiocarbamide.   See thiourea.

thiocarbanil.   See phenyl mustard oil.

thiocarbanilide.   (*N,N′*-diphenylthiourea; sulfocarbanilide).
CAS: 102-08-9. $CS(NHC_6H_5)_2$.
Properties: Gray powder. Mp 148C, d 1.32. Soluble in alcohol and ether; insoluble in water. Combustible.
Derivation: Interaction of aniline and carbon disulfide and alcohol in the presence of sulfur.
Use: Intermediates, dyes (sulfur colors, indigo, methyl indigo), vulcanization accelerator, synthetic organic pharmaceuticals, flotation agent, acid inhibitor.

thiocarbonyl chloride.   See thiophosgene.

4-thiochromanyl-*o,o*-dimethyl dithiophosphate.
CAS: 41219-25-4. $C_{11}H_{15}O_2PS_3$.
Hazard: A poison by ingestion.
Use: Agricultural chemical.

thioctic acid.   See *dl*-α-lipoic acid.

thiocyanate.   (rhodanate; sulfocyanate).
CAS: 302-04-5.   CNS−.
Any compound that contains the radical −SCN and are derivatives of thiocyanic acid, HSCN, in which the hydrogens have been replaced by hydrocarbon moieties.
Hazard: Rapid-acting poisons, thyrotoxic.
Use: Fumigants.

thiocyanic acid, phenyl ester.
CAS: 5285-87-0. C_7H_5NS.
Hazard: A poison by ingestion.

thiocyanic acid, trimethylstannyl ester.   See trimethyltin thiocyanate.

thiodiethylene glycol.   See thiodigylcol.

thiodiglycol.   (thiodiethylene glycol; β-bis-hydroxyethyl sulfide; dihydroxyethyl sulfide).
$(CH_2CH_2OH)_2S$.
Properties: Syrupy, colorless liquid; characteristic odor. D 1.1852 (20C), bp 283C, fp −10C, viscosity 0.652 cP (20C), flash p 320F (160C), bulk d 9.85 lb/gal, refr index 1.5217 (20C). Soluble in acetone, alcohol, chloroform, water; slightly soluble in benzene, carbon tetrachloride, and ether. Combustible.

Derivation: Hydrolysis of dichloroethyl sulfide, interaction of ethylene chlorohydrin and sodium sulfide.

Hazard: Do not use with hydrochloric acid.

Use: Intermediate for elastomers and antioxidants; solvent for dyes in textile printing.

thiodiglycolic acid.
CAS: 123-93-3. $HOOCCH_2SCH_2COOH$. A dicarboxylic acid.

Properties: Colorless crystals. Mp 128C. Soluble in water and alcohol. Combustible.

Use: Analytical reagent.

4,4′-thiodiphenol. (TDP). $(C_6H_5OH)_2S$.
Properties: White, crystalline powder. Mp above 151C, 99.5% pure.

Use: Intermediate, flame-retardant, antioxidant, engineering plastics.

thiodiphenylamine. See phenothiazine.

thiodipropionic acid.
CAS: 111-17-1. $HOOCCH_2CH_2 \cdot S \cdot CH_2CH_2COOH$. A dicarboxylic acid.

Properties: Leaflets. Mp 135. Soluble in water and alcohol.

Hazard: Use in foods restricted to 0.02% of fat and oil content, including essential oils.

Use: Antioxidant in food packaging, soaps, plasticizers, lubricants, fats, and oils.

β,β-thiodipropionitrile.
CAS: 111-97-7. $S(CH_2CH_2CN)_2$.
Properties: White crystals or light-yellow liquid. D 1.1095 (30C), mp 28.65C. Slightly soluble in water and alcohol; soluble in acetone, chloroform, and benzene.

Use: Preservative, selective solvent, chromatography.

thio-1,3-dithio[4,5-b]quinoxaline. See thioquinox.

thioester. An ester of a carboxylic acid with a thiol instead of an alcohol.

thioethanolamine. See 2-aminoethanethiol.

thioflavine T. (CI 49005). $CH_3C_6H_3N(HCl)$ $SCC_6H_4N(CH_3)_2)$.
Properties: A yellow basic dye of the thiazole class, fluoresces yellow to yellowish-green when excited by UV.

Derivation: By heating p-toluidine with sulfur in the presence of lead oxide.

Use: Textile dyeing, fluorescent sign paints, in combination with green or blue pigments to produce brilliant greens, phosphotungstic pigments.

thiofuran. See thiophene.

1-thioglycerol.
CAS: 96-27-5. $CH_2(OH)CH(OH)CH_2SH$.
Properties: Water-white liquid. Bp 118C (5 mm Hg), d 1.295 (14.4C). Soluble in water, alcohol, and ether. Combustible.

Use: Reducing agent for cystine molecule in human hair and wool, for stabilization of acrylonitrile polymers, medicine.

thioglycolic acid. (mercaptoacetic acid).
CAS: 68-11-1. $HSCH_2COOH$.
Properties: Colorless liquid; strong unpleasant odor. D 1.325, fp −16.5C, bp 123C (29 mm Hg). Miscible with water, alcohol, or ether. Combustible.

Derivation: Heating chloracetic acid with potassium hydrogen sulfide.

Hazard: Toxic by ingestion and inhalation, strong irritant to tissue, eyes, and skin.

Use: Reagent for iron, manufacture of thioglycolates, permanent-wave solutions and depilatories, vinyl stabilizer, manufacture of pharmaceuticals.

thioguanine deoxyriboside.
CAS: 789-61-7. $C_{10}H_{13}N_5O_3S$.
Hazard: Questionable carcinogen.

2-thiohydantoin. (glycolylthiourea).
$NHC(S)NHC(O)CH_2$.
Properties: Crystals or tan powder. Mp 230C. Slightly soluble in water; insoluble in alcohols and ethers.

Grade: 99% min.

Use: Intermediate for pharmaceuticals, rubber accelerators, copper-plating brighteners, and dyestuffs.

2-thio-4-keto-thiazolidine. See rhodanine.

-thiol. (mercaptan). Suffix indicating that a substance belongs to the group of organic compounds resembling alcohols but having the oxygen of the hydroxyl group replaced by sulfur, as in ethanethiol (C_2H_5SH). Many thiols are characterized by strong and repulsive odors.

Hazard: Aliphatic thiols are flammable. Toxic by inhalation.

Use: Warning agents in fuel gas lines, chemical intermediates.

Note: Adoption of the name *thiol* to replace *mercaptan* has been officially approved as more consistent with the molecular constitution of these compounds. The older term, which literally means "mercury seizing," is inappropriate.

thiolactic acid. (2-mercaptopropionic acid).
CAS: 79-42-5. $CH_3CH(SH)COOH$.
Properties: Oily liquid; unpleasant odor. Becomes crystalline at 10C, d 1.22, bp 116C (16 mm Hg), refr index 1.482. Soluble in water, alcohol, and acetone. Readily forms salts with numerous metals that have quite different properties.

Derivation: Reaction of sodium sulfide, sulfur, and bromopropionic acid.

Use: Depilatory, hair-waving preparations.

thiol protease inhibitor.
CAS: 66701-25-5. $C_{15}H_{27}N_5O_5$.
Hazard: A reproductive hazard.

thiomalic acid. (mercaptosuccinic acid).
CAS: 70-49-5. $HOOCCH(SH)CH_2COOH$.
Properties: White crystals or powder; sulfuric odor. Mp 149–150C. Soluble in water, alcohol, acetone, and ether; slightly soluble in benzene. Combustible.
Use: Biochemical research, intermediate, rust inhibitor, antidarkening agent for crepe rubber, tackifier for synthetic rubber.

thiomerin.
CAS: 20223-84-1. $C_{16}H_{25}HgNO_6S \cdot 2H$.
Hazard: A poison.

thionazin. (generic name for O,O-diethyl-O-2-pyrazinyl phosphorothioate).
CAS: 297-97-2.

NCHCHNCHCOPS$(C_2H_5O)_2$.
Properties: Amber liquid. Mp -1.7C, bp 80C (0.001 mm Hg). Slightly soluble in water; miscible with most organic solvents.
Hazard: Toxic by ingestion, inhalation, and skin absorption; cholinesterase inhibitor.
Use: Insecticide, fungicide, nematocide.

"Thionex" [Makhteshim]. TM for tetramethylthiuram monosulfide.

thionyl chloride. (sulfurous oxychloride; sulfur oxychloride).
CAS: 7719-09-7. $SOCl_2$.
Properties: Pale-yellow to red liquid; suffocating odor. D 1.638, fp -105C, bp 79C, decomposes at 140C, decomposes (fumes) in water. Soluble in benzene, carbon tetrachloride.
Grade: 93%, 97.5%.
Hazard: Strong irritant to skin, tissue, and upper respiratory tract.
Use: Pesticides, engineering plastics, chlorinating agent, catalyst.

thiopental sodium. ("Pentothal" [Hospira]; "Sodium Pentothal" [Hospira]; sodium-5-ethyl-5 (1-methylbutyl)-2-thiobarbiturate). ($C_{11}H_{17}N_2$ O_2SNa). A rapidly acting barbiturate administered intravenously for general anesthesia and hypnosis. Commonly known as "truth serum."
Hazard: May cause respiratory failure; use only with physician in attendance.

thiophane. See tetrahydrothiophene.

thiophene. (thiofuran).
CAS: 110-02-1.

Properties: Colorless liquid. Refr index 1.5285 (20C), d 1.0644 (20/4C), fp -38.5C, bp 84C, flash p 30F (-1.1C). Soluble in alcohol and ether; insoluble in water.
Derivation: From coal tar (benzene fraction) and petroleum, synthetically from heating sodium succinate with phosphorus trisulfide.
Hazard: Flammable, dangerous fire risk.
Use: Organic synthesis (condenses with phenol and formaldehyde, copolymerizes with maleic anhydride), solvent, dye, and pharmaceutical manufacture.

α-thiophenealdehyde. (2-thiophenecarboxaldehyde). C_4H_3SCHO.
Properties: Oily liquid; almond-like odor. Bp 198C, 90C (20 mm Hg), d 1.210–1.220. Very soluble in alcohol, benzene, ether; slightly soluble in water. Combustible.
Grade: 95%.
Use: Thiophene derivatives, introducing thenyl group into organic compounds.

thiophenol. (phenyl mercaptan).
CAS: 108-98-5. C_6H_5SH.
Properties: Water-white liquid; repulsive odor. Bp 168.3C, bp 71C (15 mm Hg), refr index 1.5891, d 1.075 (25/25C). Insoluble in water; soluble in alcohol and ether. Combustible.
Derivation: Reduction of benzenesulfonyl chloride with zinc dust in sulfuric acid.
Grade: 99%.
Hazard: Skin irritant.
Use: Pharmaceutical synthesis.

thiophosgene. (thiocarbonyl chloride).
CAS: 463-71-8. $CSCl_2$.
Properties: Reddish liquid. D 1.5085 (15C), bp 73.5C. Decomposes in water and alcohol; soluble in ether.
Hazard: Toxic by ingestion and inhalation.
Use: Organic synthesis.

thiophosphoryl chloride.
CAS: 3982-91-0. $PSCl_3$.
Properties: Colorless liquid; penetrating odor. D 1.635, bp 126C, fp -35C, decomposed by water, flash p none. Soluble in carbon disulfide, carbon tetrachloride. Nonflammable.
Hazard: Strong irritant to skin and tissue.

thiophosphoryl chlorodifluoride. See thiophosphoryl difluoride monochloride.

thiophosphoryl difluoride monochloride.
CAS: 2524-02-9. ClF_2PS.
Hazard: Low toxicity by inhalation.

thioquinox.   (generic name for 2,3-quinoxa-
linedithiol cyclic trithiocarbonate). $C_6H_4N_2$-
C_2S_2CS (tricyclic).
Properties: Yellow solid. Mp 180C. Insoluble in
water; slightly soluble in acetone, kerosene, and
alcohol.
Hazard: Toxic by ingestion.
Use: Fungicide, acaricide.

thioridazine.   (USAN for 2-methylthio-10-[2-
(*N*-methyl-2-piperidyl)-ethyl]phenothiazine).
CAS: 50-52-2. $C_{21}H_{26}N_2S_2$.
Properties: Colorless crystals. Mp 158C. Soluble in
water and alcohol.
Grade: ND (as the hydrochloride).
Use: Medicine (tranquilizer).

thiosalicylic acid.   (2-mercaptobenzoic acid).
CAS: 147-93-3. $HOOCC_6H_4SH$.
Properties: Yellow solid. Mp 164–165C, sublimes.
Slightly soluble in hot water; soluble in alcohol,
ether, and acetic acid. Combustible.
Grade: Reagent, technical 80%, pharmaceutical.
Use: Dyes, reagent for iron determination, interme-
diate.

thiosemicarbazide.   (aminothiourea).
CAS: 79-19-6. $NH_2CSNHNH_2$.
Properties: White, crystalline powder; no odor. Mp
180–184C. Soluble in water and alcohol.
Derivation: From potassium thiocyanate and
hydrazine salts.
Grade: Technical and pure.
Use: Reagent for ketones and certain metals, pho-
tography, rodenticide.

thiosinamine.   See allyl thiourea.

thiosorbitol. $CH_2OH(CH_2O)_4CH_2SH$.
Properties: Colorless, nonhygroscopic crystals. Mp
92C, strong reducing agent. Soluble in water, ethylene glycol, and formamide; insoluble in benzene,
carbon tetrachloride, and carbon disulfide.
Use: Corrosion protection in pickling and plating
baths.

Thiostat-B.   40% aqueous solution of sodium
dimethyldithiocarbamate.
Use: Bactericide in oil-well water flooding, paper
slimicide, specialty bactericide.

Thiostop.   Aqueous solutions of the sodium and
potassium salts of dimethyl dithiocarbamate.
Properties: Clear, yellow to amber liquid. D 1.18–
1.23, good storage stability, potassium salt will
crystallize out at −6.6C and the sodium salt at 0C.
Avoid long storage in partially filled containers.
Use: Short stop in SBR polymerization.

thiotepa.
CAS: 52-24-4.
USP name for triethylenethiophosphoramide.
Hazard: Confirmed carcinogen.

thiourea.   (thiocarbamide).
CAS: 62-56-6. $(NH_2)_2CS$.

$$S=C\overset{\displaystyle NH_2}{\underset{\displaystyle NH_2}{\diagdown}}$$

Properties: White, lustrous crystals; bitter taste. D
1.406, mp 180–182C, bp sublimes in vacuo at
150–160C. Soluble in cold water, ammonium thiocyanate solution, and alcohol, nearly insoluble in
ether.
Derivation: (1) By heating dry ammonium thiocyanate, extraction with a concentrated solution
of ammonium thiocyanate, with subsequent crystallization; (2) action of hydrogen sulfide on
cyanamide.
Grade: Technical, reagent.
Hazard: A questionable carcinogen. May not be
used in food products (FDA); skin irritant (allergenic).
Use: Photography and photocopying papers, organic
synthesis (intermediate, dyes, drugs, hair preparations), rubber accelerator, analytical reagent, amino
resins, mold inhibitor.

Thioxin.   A series of rubber-to-metal bonding
adhesives. Available in water- and solvent-based
formulations for bonding a wide variety of elastomers and rubbers to metal and engineered plastics.
Use: Antivibration components, transmission seals,
hoses, bushings, automotive weatherstrip, solid
tires, storage tanks, and bridge bearing pads.

thioxylenol.   (mixed isomers). $C_6H_3(CH_3)_2SH$.
Properties: Colorless liquid. D 1.03 (15.6/15.6C),
bp 122–133C (50 mm Hg). Oxidizes on exposure
to air, supplied under nitrogen atmosphere. Combustible.
Use: Chemical intermediate, peptizer for natural and
SBR rubbers.

thiram.   (tetramethylthiuram disulfide; bis-(di-
methylthiocarbamyl)disulfide; thiuram; TMTD).
CAS: 137-26-8. $[(CH_3)_2NCH]_2S_2$.
Properties: White, crystalline powder; characteristic
odor. D 1.29 (20C), melting range 155–156C. Soluble in alcohol, benzene, chloroform, carbon disulfide; insoluble in water, dilute alkali, gasoline.
Grade: 75% wettable powder, 95% technical powder, ND.
Hazard: Toxic by ingestion and inhalation, irritant
to skin and eyes. Body weight and hematologic
effects. Questionable carcinogen.
Use: Vulcanizing agent for rubber, especially for
steam hose and other heat-resistant uses; fungicide;
insecticide; seed disinfectant; lubricating oil additive; bacteriostat; animal repellent.

Thiramad.   A 75% thiram formulation. A turf
fungicide.
Hazard: See thiram.

thiuram. A compound containing the group R_2NCS. Most are disulfides. The most common monosulfide is tetramethylthiuram monosulfide. See thiram.

thixotropy. The ability of certain colloidal gels to liquefy when agitated (as by shaking or ultrasonic vibration) and to return to the gel form when at rest. This is observed in some clays, paints, and printing inks that flow freely on application of slight pressure, as by brushing or rolling. Suspensions of bentonite clay in water display this property, which is desirable in oil-well drilling fluids. See rheology; dilatancy.

"Thompson's Concrete Care Cleaner and Degreaser" [Thompson]. TM for a liquid concrete cleaner.
Use: To remove oil, dirt, algae, and mildew.

"Thompson's Water Seal Maximum Deck Stripper" [Thompson]. TM for a liquid deck stripper.
Use: To remove oil and latex-based solid and semi-transparent stains and waterproofers from decks and other surfaces.

"Thompson's Water Seal No Drip Gel Stain" [Thompson]. TM for a gel wood stain.
Use: A gel stain for outdoor furniture and shutters without the mess of ordinary stain.

Thomson, J. J., Sr. (1856–1940). A native of England, Thomson entered Cambridge University in 1876 and remained there permanently as a professor of physics, especially in the field of electrical phenomena. His observations and calculations of cathode ray experiments led to proof of the existence of the electron as the lightest particle of matter (1896). This proof was announced at the Royal Institution in the following year. This was the keystone of the theory of atomic structure and one of the most notable discoveries in the history of science. See Rutherford, Sir Ernest.

thomsonite. $(NaCa_2Al_5(SiO_4)_5 \cdot 6H_2O)$. A mineral, one of the zeolites.

thonzylamine hydrochloride. (2-[(2-di-methylaminoethyl)(p-methoxybenzyl)amino]-pyrimidinehydrochloride).
CAS: 63-56-9. $C_{16}H_{22}N_4O \cdot HCl$.
Properties: White, crystalline powder; faint odor. Mp 173–176. Very soluble in water; freely soluble in alcohol and chloroform; practically insoluble in ether and benzene; pH 5.0–6.0 (2% solution).
Grade: NF.
Use: Medicine (antihistamine).

thoria. See thorium dioxide.

thorin. $HOC_{10}H_4(SO_3H)_2NNC_6H_4AsO_3H_2$.
A reagent for the colorimetric determination of microgram quantities of thorium.
Hazard: Toxic by ingestion.

thorite. $ThSiO_4$. A natural thorium silicate, usually impure, found in pegmatites.
Properties: Black to orange color, vitreous to resinous luster. Mohs hardness 4.5–5, d 4.4–5.2. Radioactive.
Occurrence: Norway, Ceylon.
Use: Source of thorium.

thorium.
CAS: 7440-29-1. Th. Metallic element of atomic number 90, a member of the actinide series (group IIIB of periodic table), aw 232.0381, valence of 4; radioactive, no stable isotopes.
Properties: Soft metal with bright silvery luster when freshly cut, similar to lead in hardness when pure. Can be cold-rolled, extruded, drawn, and welded. D about 11.7, mp 1700C, bp 4500C. Soluble in acids; insoluble in alkalies and water. Some alloys may ignite spontaneously, the metal in massive form is not flammable.
Source: Monazite, thorite. It is about as abundant as lead.
Derivation: (1) Reduction of thorium dioxide with calcium; (2) fused salt electrolysis of the double fluoride $ThF_4 \cdot KF$. The product of both processes is thorium powder, fabricated into the metal by powder metallurgy techniques. Hot surface decomposition of the iodide produces crystal bar thorium.
Available forms: Powder, unsintered bars, sintered bars, sheets.
Hazard: Flammable and explosive in powder form. Dusts of thorium have very low ignition points and may ignite at room temperature. Radioactive decay isotopes are dangerous when ingested.
Use: Nuclear fuel (thorium-232 is converted to uranium-233 on neutron bombardment after several decay steps), sun lamps, photoelectric cells, target in X-ray tubes, alloys.

thorium acetylacetonate. $Th[OC(CH_3):CHCOCH_3]_4$.
Properties: Crystalline powder. Slightly soluble in water; resistant to hydrolysis. A chelating, nonionizing compound.

thorium anhydride. See thorium dioxide.

thorium carbide. ThC_2.
Properties: Yellow solid. D 8.96 (18C), mp 2630–2680C, bp 5000C. Decomposes in water.
Use: Nuclear fuel.

thorium chloride. (thorium tetrachloride).
CAS: 10026-08-1. $ThCl_4$.
Properties: Colorless or white, lustrous needles (light-yellow color caused by iron trace); hygroscopic; partially volatile; crystallizes with variable

water of crystallization. D 4.59, bp 928C (decomposes), mp 820C. Soluble in alcohol, water.
Grade: Technical; as 50% ThO_2.
Use: Incandescent lighting.

thorium decay series. The series of radioactive elements produced as successive intermediate products when thorium undergoes spontaneous natural radioactive disintegration into lead. Many of these are severe radioactive poisons when ingested or inhaled as thorium dust particles.

thorium dioxide. (thorium anhydride; thorium oxide; thoria). ThO_2.
Properties: Heavy white powder. D 9.7, mp 3300C (highest of all oxides), bp 4400C, Mohs hardness 6.5, very refractory. Soluble in sulfuric acid; insoluble in water.
Derivation: Reduction of thorium nitrate.
Grade: Technical and purities to 99.8% ThO_2; granular particles; crystals.
Use: Ceramics (high temperature), gas mantles, nuclear fuel, flame spraying, crucibles, medicine, nonsilica optical glass, catalyst, thoriated tungsten filaments.

thorium disulfide. ThS_2.
Properties: Dark-brown crystals. D 7.30 (25/4C), mp 1875–1975C (in vacuo). Insoluble in water.
Use: Solid lubricant.

thorium fluoride. ThF_4.
Properties: White powder. Dehydrated between 200 and 300C. D (anhydrous) 6.32 (24C), mp 1111C, above 500C reacts with atmospheric moisture to form thorium oxyfluoride, $ThOF_2$, and finally the oxide, ThO_2. Forms a series of compounds with other metallic fluorides such as NaF and KF.
Grade: 79–80% ThO_2.
Hazard: Toxic material.
Use: Production of thorium metal and magnesium–thorium alloys, high temperature ceramics. $ThOF_2$ is used as a protective coating on reflective surfaces.

thorium nitrate.
CAS: 13823-29-5. $Th(NO_3)_4 \cdot 4H_2O$.
Properties: White, crystalline mass. Mp 500C (decomposes). Soluble in water and alcohol.
Grade: Technical, CP.
Hazard: Dangerous fire and explosion risk in contact with organic materials, strong oxidizing agent. Radioactive.
Use: Reagent for determination of fluorine, thoriated tungsten filaments.

thorium oxalate. $Th(C_2O_4)_2 \cdot 2H_2O$.
Properties: White powder. D (anhydrous) 4.637 (16C). Insoluble in water and most acids. Soluble in solutions of alkali and of ammonium oxalates. Above 300–400C decomposes to thorium oxide, ThO_2.
Grade: Purities to 99.9%; as 59% ThO_2.
Use: Ceramics.

thorium oxide. See thorium dioxide.

thorium sulfate. (thorium sulfate, normal $Th(SO_4)_2 \cdot 8H_2O$.
Properties: White, crystalline powder. D 2.8, mp loses $4H_2O$ at 42C, remainder at 400C. Slightly soluble in water; soluble in ice water.
Grade: As 43% ThO_2.

thorium tetrachloride. See thorium chloride.

Thorpe reaction. Base-catalyzed condensation of two molecules of nitrile to yield imines that tautomerize to enamines.

thortveitite. An ore containing 37–42% scandium oxide. A basic material in the production of scandium.

THPC. See tetrakis(hydroxymethyl)phosphonium chloride.

Thr. Abbreviation of threonine.

3′ end. The end of a nucleic acid that lacks a nucleotide bound at the 3′ position of the terminal residue.

three-phase equilibrium. The state of equilibrium of vapor, liquid, and solid phases of a pure substance at a definite pressure and temperature.

threonine. (α-amino-β-hydroxybutyric acid).
CAS: 72-19-5. $CH_3CH(OH)CH(NH_2)COOH$. An essential amino acid.
Properties: Colorless crystals. (*dl*-threonine) Mp 228–229C with decomposition; (*l*(−)-threonine) (naturally occurring) mp 255–257C with decomposition; (*dl*-allo-threonine), mp 250–252C. Soluble in water. Optically active.
Derivation: Hydrolysis of protein (casein), organic synthesis.
Use: Nutrition and biochemical research, dietary supplement.

threshold energy. The energy limit for an incident particle or photon below which a particular endothermic reaction will not occur.

Threshold Limit Value. (TLV). A set of standards established by the American Conference of Governmental Industrial Hygienists for concentrations of airborne substances in workroom air. They are time-weighted averages based on conditions that it is believed that workers may be repeatedly exposed to day after day without adverse effects. The TLV values are revised annually and provide the basis for the safety regulations of OSHA. They are intended to serve as guides in control of health hazards rather than definitive marks between safe and dangerous concentration. In this book, these are indicated by TLV.

See air pollution; American Conference of Governmental Industrial Hygienists.

thrombin. A proteolytic enzyme that catalyzes the conversion of fibrinogen to fibrin and thus is essential in the clotting mechanism of blood. It is present in the blood in the form of prothrombin under normal conditions; when bleeding begins, the prothrombin is converted to thrombin, which in turn activates the formation of fibrin.

thrombocyte. See platelet.

thromboxane. See prostaglandin.

throwing oil. An oil applied to prepare raw silk and filament rayon for "throwing," the operation by which the filaments are twisted into threads. Applied by a bath, the oils condition the filaments and yarns for subsequent weaving or knitting. Usually compounded to be self-emulsifying and may contain a sizing agent such as dextrin, gelatin, etc. See slashing compound.

throwing power. A term denoting the effectiveness of an electrolytic cell for depositing metal uniformly over a surface being electroplated, particularly in irregular and recessed areas. The throwing power is the weight of deposition per unit distance between the electrodes.

thuja oil. (arbor vitae oil).
CAS: 8007-20-3.
Properties: Pale-yellow essential oil; camphor-like odor. D 0.910–0.920, refr index 1.459 (20C), optical rotation −10 to −13 degrees in 100-mm tube. Soluble in alcohol, ether, chloroform, carbon disulfide, fixed oils, and mineral oil. Combustible. Chief known constituents: Dextro-pinene, levofenchone, thujone, should contain more than 60% ketones calculated as thujone.
Derivation: Distilled from the leaves of the white cedar, *Thuja occidentalis.*
Grade: Technical, FCC (as cedar leaf oil).
Use: Perfumery, flavoring.

α-thujaplicin.
CAS: 1946-74-3. $C_{10}H_{12}O_2$.
Properties: A minor component of *Thujopsis dolabrata* Sieb. et Zucc. var. *hondai* Makino.
Hazard: A poison.

thujone.
CAS: 546-80-5. $C_{10}H_{16}O$.
A terpene-type ketone contained in thuja oil and the oils of sage, tansy, and wormwood.
Properties: Colorless liquid. D 0.915–0.919 (20/20C), bp 203C. Insoluble in water; soluble in alcohol. Combustible.
Hazard: Toxic by ingestion.
Use: Solvent.

thulia. See thulium oxide.

thulium. Tm. Atomic number 69, group IIIB of the periodic table, a rare-earth element of the lanthanide groups, aw 168.9342, valence of 3; no stable isotopes.
Properties: Metallic luster. D 9.318, mp 1550C, bp 1727C. Reacts slowly with water; soluble in dilute acids. Salts colored green.
Derivation: Isolated by reduction of the fluoride with calcium.
Grade: Regular high purity (ingots, lumps).
Hazard: Fire risk in form of dust.
Use: Ferrites, X-ray source.
See rare earth.

thulium-170. Radioactive thulium of mass number 170.
Use: X-ray source in portable units.

thulium chloride. $TmCl_3 \cdot 7H_2O$.
Properties: Green, deliquescent crystals. Mp 824C, bp 1440C. Very soluble in water and alcohol.

thulium nitrate.
CAS: 14985-19-4. $N_3O_9 \cdot Tm$.
Hazard: A poison.

thulium oxalate. $Tm_2(C_2O_4)_3 \cdot 6H_2O$.
Properties: Greenish-white precipitate, loses H_2O at 50C. Soluble in aqueous alkali oxalates.
Derivation: Precipitation of a solution containing a thulium salt and a mineral acid by addition of oxalic acid.
Use: Analytical separation of thulium (and other rare-earth metals) from the common metals.

thulium oxide. (thulia). Tm_2O_3.
Properties: Dense white powder with greenish tinge, slightly hygroscopic. Absorbs water and carbon dioxide from the air. D 8.6. Exhibits a reddish incandescence on heating, changing to yellow and then white on prolonged heating. Slowly soluble in strong acids.
Derivation: By ignition of thulium oxalate, salt of other oxyacids, or hydroxide.
Grade: 99–99.9%.
Use: Source of thulium metal.

thylakoid. Closed cisterna, or disc, formed by the pigment-bearing internal membranes of chloroplasts.

Thylate. A wettable off-white powder containing 65% thiram.

thymic acid. See thymol.

thymidine. (thymine-2-deoxyriboside).
CAS: 50-89-5. $C_{10}H_{14}N_2O_5$. The nucleoside (deoxyriboside) of thymine. Occurs in DNA.

Properties: Crystalline needles. Mp 185C. Dextro-rotatory in solution; soluble in water, methanol, hot ethanol, hot acetone, and hot ethyl acetate; sparingly soluble in hot chloroform; soluble in pyridine and glacial acetic acid.
Use: Biochemical research.
Also available as trityl thymidine and as tritiated thymidine in a radioactive form.

thymidylic acid.   The nucleotide of thymine, i.e., the phosphate ester of thymidine.

thymine.   (5-methyluracil; 5-methyl-2,4-dioxy-pyrimidine; T).
CAS: 65-71-4.

CH$_3$CC(O)NHC(O)NHCH.

One of the pyrimidine bases of living matter.
Properties: White, crystalline powder. Decomposes at 335–337C. Slightly soluble in hot water; insoluble in cold water, alcohol; sparingly soluble in ether; readily soluble in alkalies.
Derivation: Hydrolysis of deoxyribonucleic acid, from methylcyanoacetylurea by catalytic reduction.
Use: Biochemical research.
See base pair; nucleotide.

thymine-2-deoxyriboside.   See thymidine.

thymine dimer.   See pyrimidine dimer.

thymol.   (isopropyl-*m*-cresol; thyme camphor; thymic acid).
CAS: 89-83-8.   (CH$_3$)$_2$CHC$_6$H$_3$(CH$_3$)OH.

Properties: White crystals; aromatic odor and taste. D 0.979, mp 48–51, bp 233C. Soluble in alcohol, carbon disulfide, chloroform, glacial acetic acid, ether, and fixed or volatile oils; slightly soluble in water and glycerol. Combustible.
Derivation: From thyme oil or other oils, synthetically from *m*-cresol and isopropyl chloride by the Friedel-Crafts method at −10C.
Grade: Technical, NF, reagent.
Use: Perfumery, mold and mildew preventive, microscopy, preservative, antioxidant, flavoring, laboratory reagent, synthetic menthol.

thymol blue.   (thymolsulfonphthalein).
CAS: 76-61-9.

C$_6$H$_4$SO$_2$OC[C$_6$H$_2$(CH$_3$)(OH)CH(CH$_3$)$_2$]$_2$.

Properties: Brown-green powder or crystals. Mp 223C (decomposes). Insoluble in water; soluble in alcohol or dilute alkali.
Use: Acid-base indicator in pH range 1.5 (pink) to 2.8–8 (yellow) to 9.6 (blue).
See indicator.

thymol iodide.   [C$_6$H$_2$(CH$_3$)(OI)(C$_3$H$_7$)]$_2$. Principally dithymol diiodide.
Properties: Red-brown powder or crystals; slight aromatic odor; affected by light. Soluble in ether, chloroform, and fixed or volatile oils; slightly soluble in alcohol; insoluble in water. Combustible.
Derivation: Interaction of thymol and potassium in alkaline solution.
Grade: Technical.
Use: Feed additive, antifungal agent.

thymolphthalein.

C$_6$H$_4$COOC[C$_6$H$_2$(CH$_3$)(OH)CH(CH$_3$)$_2$]$_2$.

Properties: White powder. Mp 245C. Insoluble in water; soluble in alcohol and acetone, and dilute alkali and acids.
Use: Acid-base indicator in pH range 9.3 (colorless) to 10.5 (blue).
See indicator.

thymolsulfonphthalein.   See thymol blue.

thymotoxic serum.   A serum that is selectively toxic to thymus tissue.

***p*-thymoquinone.**   (2-isopropyl-5-methylben-zoquinone).
CAS: 490-91-5.   C$_6$H$_2$O$_2$(CH$_3$)CH(CH$_3$)$_2$.
Properties: Bright-yellow crystals; penetrating odor. Mp 45.5C, bp 232C. Slightly soluble in water; soluble in alcohol and ether. Combustible.
Derivation: From diazonium salt of aminothymol and nitrous acid.
Use: Fungicide.

thyrocalcitonin.   A thyroid hormone having significant effect on the calcium content of bone and blood. It can retard bone damage and may be involved in the incidence of dental caries and the form of dental calculus known as tartar.

thyroid.   See thyroxine.

thyrolytic serum.   (thyrotoxic serum). A serum that is selectively toxic to the thyroid gland.

thyronine.   (desiodothyroxine).
HOC$_6$H$_4$OC$_6$H$_4$CH$_2$CH(NH$_2$)COOH.   The *p*-hydroxyphenyl ether of tyrosine. Thyronine and its iodinated derivatives are used in biochemical research on the thyroid gland and its activity.

thyrotropic hormone. (TSH; thyrotropin). A hormone secreted by the anterior lobe of the pituitary gland. It increases the rate of removal of iodine from the blood by the thyroid gland, synthesis of the thyroid hormone, and its release into the bloodstream. The thyrotropic hormone is a protein with a low molecular weight (about 10,000) that contains some carbohydrate.

thyroxine. (3,5,3′,5′-tetraiodothyronine). CAS: 51-48-9. $HOC_6H_2I_2OC_6H_2I_2CH_2CH(NH_2)COOH$.

The hormone produced by the thyroid gland. It is an amino acid and a derivative of tyrosine. It increases the metabolic rate and oxygen consumption of animal tissues.

Properties: Optically active, the *l*-isomer is the natural and physiologically active form.

(*dl*-thyroxine) Needles. Decomposes at 231–233C. Insoluble in water, alcohol, and the common organic solvents; soluble in alcohol in the presence of mineral acids or alkalies. (*l*-thyroxine) Crystals. Decomposes at 235–236C. (*d*-thyroxine) Crystals. Decomposes at 237C.

Derivation: Obtained from the thyroid glands of animals; also made synthetically.

Use: Medicine (hormone), biochemical research. See triiodothyronine.

t lymphocyte. (t-cell; t-cell lymphocyte; thymus-dependent lymphocyte; thymus-derived lymphocyte).

Any of a type of lymphocyte that develops bone marrow stem cells that migrate to and develop in the thymus. The mature cells are transported to other lymphoid tissues where they interact with B-cells and accessory cells to stimulate antibody production. T-cells also interact directly with antigen-bearing cells, thereby providing cell-mediated immunity. They bear specific receptors on the outer surface of the cell membrane that bind substances or even microorganisms that are identified as foreign and initiates immune response.

Ti. Symbol of titanium.

TIBAL. Abbreviation for triisobutylaluminum.

"TIC Pretested Guarcel 302" *[TIC].* TM for a powder developed with bland guar gum.
Use: For dairy-based dry mix beverages.

"TIC Pretested Nutriloid Arabic" *[TIC].* TM for a sprayable dry powder food additive.

Use: As a mouthfeel aid providing a creamy texture without viscosity.

Tiemann rearrangement. Rearrangement of amide oximes (available from nitriles and hydroxylamine) to monosubstituted ureas by treatment with benzenesulfonyl chloride and water.

Tiffeneau-Demjanov ring expansion. Rearrangement of β-amino alcohols on diazotization with nitrous acid to give ring-expanded carbonyl compounds.

tiglic acid. (methylcrotonic acid; crotonolic acid; *trans*-2-methyl-2-butenoic acid). CAS: 80-59-1. $CH_3CH:C(CH_3)COOH$. The *trans* isomer of angelic acid.
Properties: Thick, syrupy liquid or colorless crystals; spicy odor. D 0.9641, mp 65C, bp 198.5C. Soluble in alcohol, ether, hot water. Combustible.
Derivation: From croton oil. Also occurs in English chamomile oil.
Use: Perfumes, flavors, emulsion breakers.

tiglium oil. See croton oil.

time-weighted average. (TWA). The average concentration of an air contaminant for a conventional 8-hour workday and a 40-hour work. See Threshold Limit Value.

timolol. $C_{13}H_{24}N_4O_3S$.

Properties: Colorless crystals. Mp 72C.
Use: Adrenergic blocker used in treatment of hypertension, glaucoma, and for reducing risk of second heart attack. FDA approved.

tin. (stannum). CAS: 7440-31-5. Sn. Metallic element of atomic number 50, group IVA of the periodic system, aw 118.69, valences of 2, 4; 10 isotopes.
Properties: Silver-white, ductile solid (β form). D 7.29 (20C), mp 232C, bp 2260C. Changes to brittle, gray (α) tin at temperature of 18C, but the transition is normally very slow. Soluble in acids and hot potassium hydroxide solution; insoluble in water. Elemental tin has low toxicity, but most of its compounds are toxic.
Derivation: By roasting the ore (cassiterite) to oxidize sulfates and to remove arsine, then reducing with coal in a reverberatory furnace or by smelting in an electric furnace; secondary recovery from tin plate.
Occurrence: Malaysia (called Straits tin), Indonesia, Thailand, Bolivia.

Grade: By percentage purity; spectrographic grade is 99.9999%; block tin is a common designation for pure tin.

Available forms: Sheet, wire, tape, pipe, bar, ingot, powder, single crystals.

Hazard: All organic tin compounds are toxic. Eye and upper respiratory tract irritant, headache, nausea, central nervous system and immune effects. Questionable carcinogen.

Use: Tin plate, terneplate, Babbitt metal, pewter, bronze, corrosion-resistant coatings, collapsible tubes, anodes for electrotin plating, hot-dipped coatings, cladding, solders, low-melting alloys for fire control, organ pipes, dental amalgams, diecasting. White, type, and casting metal. Manufacture of chemicals, tinned wire (all copper wire that is to be rubber covered). Block tin is used to coat copper cooking utensils and lead sheet, or to line lead pipe for distilled water, beer, carbonated beverages, and some chemicals.

Note: In speaking of fabricated articles, "tin" is often used when tinplate (thin sheets of steel coated with tin) is meant, e.g., "a tin can." To distinguish, articles (such as condenser coils) made of solid tin are said to be made of "block tin."

tin anhydride. See stannic oxide.

tin ash. See stannic oxide.

tin bis(diethyldithiocarbamate).
CAS: 16248-90-1. $C_{10}H_{20}N_2S_4$•Sn.
Hazard: A poison.
Use: Agricultural chemical.

tin bisulfide. See stannic sulfide.

tin bronze. See stannic sulfide.

tincal. See borax.

tincture. An alcoholic or aqueous alcoholic solution of an animal or vegetable drug or a chemical substance. The tincture of potent drugs is essentially a 10% solution. Tinctures are more dilute than fluid extracts and less volatile than spirits.

tin dibutylditrifluoroacetate. See bis(trifluoroacetoxy)dibutyltin.

tin, dichlorodidodecyl–. See didodecyltin dichloride.

tin, dioctyldiphenyl–. See dioctyldiphenyltin.

"Tinopal" *[BASF].* TM for a group of fluorescent whitening agents, which absorb UV light in the near visible range and reemit the energy as visible light.
Use: Heavy-duty detergents to whiten fabrics, paper, plastics, fibers, coatings, waxes, etc.

tin oxide. See stannous oxide.

tin perchloride. See stannic chloride.

tin plating. The process of covering steel, iron, or other metal with a layer of tin by dipping it in the molten metal, by electroplating, or by immersion in solutions that deposit tin by chemical action of their components. Ingredients for the chemical process are cream of tartar and stannous chloride. The metal being plated also takes part in the process. The objective is to utilize the superior corrosion resistance of tin and in some cases to improve appearance.
See terneplate.

tin protochloride. See stannous chloride.

tin protosulfide. See stannous sulfide.

tin protoxide. See stannous oxide.

tin resinate. See soap (2).

tin salt. See stannous chloride.

tin spirits. Solutions of tin salts used in dyeing.

tin stearate. See stannous stearate.

tin tetrabromide. See stannic bromide.

tin tetrachloride. Legal label name (Rail) for stannic chloride.

tintometer. A device for comparison of colors of solutions.

"Tinuvin" *[BASF].* TM for a group of UV absorbers, substituted hydroxyphenyl benzotriazoles, e.g., 2-(2'-hydroxy-5'-methylphenyl)benzotriazole.

tiocarbazil.
CAS: 36756-79-3. $C_{16}H_{25}NOS$.
Hazard: Moderately toxic by skin contact. Low toxicity by ingestion.
Use: Agricultural chemical.

Tipersul. Fibrous potassium titanate, crystalline fibers 1 micron in diameter melting at 1371C, useful to 1204C.
Available forms: Lumps, blocks, and loose fibers.
Use: High-temperature thermal, acoustical, and electrical insulation; filter media.

Tiron. Disodium-1,2-dihydroxybenzene-3,5-disulfonate.
Use: Colorimetric determination of ferric iron, titanium, or molybdenum.

Tiselius, Arne W. K.   (1902–1971). A Swedish biochemist who won the Nobel Prize for chemistry in 1948. Renowned for research in separation methods of biochemical matter, in particular electrophoresis and chromatography. Work also involved virus isolation and synthesis of blood plasma. He earned degrees from the University of Upsala and Princeton University, as well as a multitude of honorary degrees.

Tishchenko reaction.   Formation of esters from aldehydes by an oxidation–reduction process in the presence of aluminum or sodium alkoxides.

tissue culture.   Method by which cells derived from multicellular organisms are grown in liquid media.

tissue-specific expression.   Gene function which is restricted to a particular tissue or cell type. For example, the glycoprotein hormone alpha subunit is produced only in certain cell types of the anterior pituitary and placenta, not in lungs or skin; thus, expression of the glycoprotein hormone alpha-chain gene is said to be tissue-specific. Tissue specific expression is usually the result of an enhancer which is activated only in the proper cell type.

titanellow.   See titanium trioxide (not to be confused with titan yellow, an organic dye containing no titanium).

titania.   See titanium dioxide.

titanic acid.   (titanic hydroxide; metatitanic acid). H_2TiO_3 or $Ti(OH)_4$. Water content variable.
Properties: White powder. Insoluble in mineral acids and alkalies except when freshly precipitated; insoluble in water.
Derivation: From hydrochloric acid solution of titanates by treating with ammonia and then drying over concentrated sulfuric acid or by boiling titanium sulfate solution.
Grade: Technical.
Use: Mordant.

titanium.
CAS: 7440-32-6.   Ti. Metallic element of atomic number 22, group IVB of the periodic table, aw 47.90, valences of 2, 3, 4; five isotopes.
Properties: Silvery solid or dark-gray, amorphous powder. D 4.6 (20C), mp 1675C, bp 3260C, sp heat 0.13 Btu/lb/F, thermal conductivity 105 Btu/ft^2/F/hour, as strong as steel but 45% lighter, Vickers hardness 80–100, excellent resistance to atmospheric and seawater corrosion and to corrosion by chlorine, chlorinated solvents, and sulfur compounds; reactive when hot or molten. Insoluble in water, inert to nitric acid but attacked by concentrated sulfuric acid and hydrochloric acid. Unaffected by strong alkalies.

Source: Ilmenite, rutile, titanite, titanium slag from certain iron ores.
Derivation: (1) Reduction of titanium tetrachloride with magnesium (Kroll process) or sodium (Hunter process) in an inert atmosphere of helium or argon. The titanium sponge is consolidated by melting. (2) Electrolysis of titanium tetrachloride in a bath of fused salts (alkali or alkaline-earth chlorides).
Grade: Technical (powder), commercially pure (sheets, bars, tubes, rods, wire, and sponge), single crystals.
Hazard: Flammable, dangerous fire and explosion risk. (Metal) Ignites in air at 1200C, will burn in atmosphere of nitrogen. Do not use water or carbon dioxide to extinguish.
Use: Alloys (especially ferrotitanium); as structural material in aircraft, jet engines, missiles, marine equipment, textile machinery, chemical equipment (especially as anode in chloride production), desalination equipment, surgical instruments, orthopedic appliances, food-handling equipment; X-ray tube targets; abrasives; cermets; metal-ceramic brazing, especially in nickel-cadmium batteries for space vehicles; electrodeposited and dipped coatings on metals and ceramics; electrodes in chlorine battery.

titanium acetylacetonate.   See titanylacetylacetonate.

titanium ammonium oxalate.   See ammonium titanium oxalate.

titanium boride.   (titanium diboride). TiB_2.
Properties: Extremely hard solid with oxidation resistance up to 1400C. Mp 2980C, d 4.50, Mohs hardness 9+, low electrical resistivity.
Use: Metallurgical additive, high-temperature electrical conductor, refractory, cermet component, coatings resistant to attack by molten metals, aluminum manufacture, super alloys.

titanium butylate.   See tetrabutyl titanate.

titanium carbide.   TiC.
Properties: Extremely hard, crystalline solid with gray metallic color. Mp 3140C, bp 4300C, d 4.93, resistivity 60 $\mu\Omega\cdot$cm (room temperature). Insoluble in water; soluble in nitric acid and aqua regia.
Use: Additive (with tungsten carbide) in making cutting tools and other parts subjected to thermal shock, arc-melting electrodes, cermets, coating dies for metal extrusion (0.2 mil film by vapor deposition).

titanium chelate. $(HOYO)_2Ti(OR)_2$ or $(H_2NYO)_2Ti(OR)_2$. A series of titanium compounds where Y and R are hydrocarbon groups, e.g., octylene glycol titanate, triethanolamine titanate.
Use: Surface-active agents, corrosive inhibitors, cross-linking agents.

titanium diboride.   See titanium boride.

titanium dichloride.
CAS: 10049-06-6. $TiCl_2$.
Properties: Black powder. D 3.13. Decomposed by water; decomposes at 475C in vacuo; hygroscopic. Soluble in alcohol; insoluble in chloroform, ether, carbon disulfide.
Derivation: Reduction of titanium tetrachloride with sodium amalgam, dissolving titanium metal in hydrochloric acid.
Hazard: Flammable, dangerous fire risk, ignites in air, store under water or inert gas.

titanium dioxide.   (titanic anhydride; titanic acid anhydride; titanic oxide; titanium white; titania).
CAS: 13463-67-7. TiO_2.
Properties: White powder in two crystalline forms, anatase and rutile. It has the greatest hiding power of all white pigments. Noncombustible.
Derivation: From ilmenite or rutile. (1) Ilmenite is treated with sulfuric acid and the titanium sulfate further processed. The product is primarily the anatase form. (2) Rutile is chlorinated and the titanium tetrachloride converted to the rutile form by vapor-phase oxidation. Papermakers are using this form to an increasing extent in preference to the anatase form.
Grade: Technical, of many variations, pure, USP, single crystals, whiskers.
Hazard: Lower respiratory tract irritant. Possible carcinogen.
Use: White pigment (in paints, paper, rubber, plastics, etc.), opacifying agent, cosmetics, radioactive decontamination of skin, floor coverings, glassware and ceramics, enamel frits, delustering synthetic fibers, printing inks, welding rods. Single crystals are high-temperature transducers.

titanium disilicide. $TiSi_2$.
Use: In special alloy applications, as a flame or blast impingement-resistant coating material.

titanium disulfide. TiS_2.
Properties: Yellow solid. D 3.22 (20C). Decomposed by steam.
Use: Solid lubricant.

titanium ester.   A series of titanium compounds whose general formula is $Ti(OR)_4$ where R is a hydrocarbon group, e.g., tetraisopropyl titanate.
Use: Adhesion promoters, ester-exchange catalysts, cross-linking agents, heat-resistant paints.
See tetrabutyl titanate; tetra(2-ethylhexyl) titanate.

titanium ferrocene.   See titanocene dichloride.

titanium hydride. TiH_2.
Properties: Black, metallic powder that dissociates above 288C. The evolution of hydrogen is gradual and practically complete at 650C. D 3.8, attacked by strong oxidizing agents.
Derivation: Direct combination of titanium with hydrogen, reduction of titanium oxide with calcium hydride in the presence of hydrogen above 600C.
Hazard: Flammable, dangerous fire risk, dust may explode in presence of oxidizing agents.
Use: Powder metallurgy, production of pure hydrogen (can contain 1800 cc (STP) H/cc of hydride), production of foamed metals, solder for metal-glass composites, electronic getter, reducing atmosphere for furnaces, hydrogenation agent, refractories.

titanium isopropylate.   See tetraisopropyl titanate.

titanium monoxide.   TiO. A weakly basic oxide of titanium, prepared by reduction of the dioxide at 1500C; it has no important industrial uses.

titanium nitride.   TiN.
Properties: Golden-brown, hard, brittle plates. Mp 2927C, d 5.24, sp heat 8.86 cal/mole at 25C, electrical resistivity 21.7 $\mu\Omega\cdot$cm.
Use: High-temperature bodies, cermets, alloys, rectifiers, semiconductor devices.

titanium ore.   See rutile; ilmenite.

titanium oxalate.   (titanous oxalate). $Ti_2(C_2O_4)_3\cdot10H_2O$.
Properties: Yellow prisms. Soluble in water; insoluble in alcohol and ether.
Derivation: Action of oxalic acid on titanous chloride.

titanium peroxide.   See titanium trioxide.

titanium potassium fluoride.   (potassium titanium fluoride). TiK_2F_6.
Properties: White leaflets. Soluble in hot water.
Hazard: Toxic material.
Use: Titanic acid, titanium metal.

titanium potassium oxalate.   (potassium titanium oxalate). $TiO(C_2O_4K)_2\cdot2H_2O$.
Properties: Colorless, lustrous crystals. Soluble in water.
Derivation: By treating titanium hydroxide with potassium oxalate and oxalic acid.
Grade: Technical, pure, 22% TiO_2 (min).
Use: Mordant in cotton and leather dyeing, sensitization of aluminum for photography.

titanium sesquisulfate.   See titanous sulfate.

titanium sponge.   The metal in crude form as reduced from the tetrachloride; ingots are made from it by consumable-electrode refining (arc-melting).
See titanium.

titanium sulfate. (titanic sulfate; basic titanium sulfate; titanyl sulfate).
CAS: 13825-74-6. Ti(SO$_4$)$_2$•9H$_2$O, also TiOSO$_4$•H$_2$SO$_4$•8H$_2$O. A commercial material, possibly a mixture.
Properties: White, cake-like solid. Highly acidic, similar to 50% sulfuric acid, typical composition 20% TiO$_2$, 50% sulfuric acid, 30% water, hygroscopic. D 1.47. Soluble in water; solutions hydrolyze readily unless protected from heat and dilution.
Derivation: Action of sulfuric acid on ilmenite ore.
Hazard: Irritant to skin and tissue.
Use: Treatment of chrome yellow and other colors, production of titanous sulfate used as reducing agent or stripper for dyes, laundry chemical.

titanium tetrachloride. (titanic chloride).
CAS: 7550-45-0. TiCl$_4$.
Properties: Colorless liquid. Fumes strongly when exposed to moist air, forming a dense and persistent white cloud. (Pure) d 1.7609 (0C), bp 136.4C, fp −30C. Soluble in dilute hydrochloric acid; soluble in water with evolution of heat; concentrated aqueous solutions are stable and corrosive; dilute solutions precipitate insoluble basic chlorides.
Derivation: By heating titanium dioxide or the ores and carbon to redness in a current of chlorine.
Grade: Technical, CP.
Hazard: Toxic by inhalation, strong irritant to skin and tissue.
Use: Pure titanium and titanium salts, iridescent effects in glass, smoke screens, titanium pigments, polymerization catalyst.

titanium trichloride. (titanous chloride).
CAS: 7705-07-9. TiCl$_3$.
Properties: Dark-violet, anhydrous, deliquescent crystals. D 2.6, decomposes above 440C, decomposes in air and water with heat evolution. Soluble in alcohol, acetonitrile, certain amines; slightly soluble in chloroform; insoluble in ether and hydrocarbons.
Hazard: Fire risk in the presence of oxidizing materials. Irritant to skin and tissue; open container only in oxygen-free or inert atmosphere.
Use: Reducing agent, organic synthesis, cocatalyst for polyolefin polymerization, organometallic synthesis involving titanium, laundry stripping agent.

titanium trioxide. (titanium peroxide; titanello; pertitanic acid). TiO$_3$.
Properties: Yellow powder. Soluble in acids.
Use: Dental porcelain and cements, yellow tile.

titanocene dichloride. (dicyclopentadienyltitanium dichloride; titanium ferrocene). (C$_5$H$_5$)$_2$TiCl$_2$.
Properties: Red crystals. Mp 287–289C. Moderately soluble in toluene, chloroform; sparingly soluble in ether and water. Stable in dry air, slowly hydrolyzed in moist air.

Hazard: Toxic by inhalation, irritant to skin and mucous membranes.
Use: Ziegler polymerization catalyst (with aluminum alkyls).
See metallocene.

titanous chloride. See titanium trichloride.

titanous oxalate. See titanium oxalate.

titanous sulfate. (titanium sesquisulfate). Ti$_2$(SO$_4$)$_3$.
Properties: Green, crystalline powder. Insoluble in water, alcohol, concentrated sulfuric acid; soluble in dilute hydrochloric acid or sulfuric acid, giving violet solutions.
Grade: Commercial grade made and supplied as a dark-purple solution containing 15% Ti$_2$(SO$_4$)$_3$.
Use: Textile industry as reducing agent for stripping or discharging colors.

Titanox. A series of white pigments comprising titanium dioxide in both anatase and rutile crystalline forms and also extended with calcium sulfate (titanium dioxide-calcium pigment). Noncombustible.
Use: Paints, paper, rubber, plastics, leather and leather finishes, inks, coated textiles, textile delustering, ceramics, roofing granules, welding rod coatings, and floor coverings.

titanyl acetylacetonate. (titanium acetylacetonate). TiO[OC(CH$_3$):CHCOCH$_3$]$_2$.
Properties: Crystalline powder. Slightly soluble in water; resistant to hydrolysis. A chelating, nonionizing compound.
Derivation: Reaction of titanium oxychloride with acetylacetone and sodium carbonate.
Use: Cross-linking agent for cellulosic lacquers.

titanyl sulfate. See titanium sulfate.

titer. In solutions (1) the concentration of a dissolved substance as determined by titration; (2) the minimum amount or volume needed to bring about a given result in titration; or (3) the solidification point of fatty acids that have been liberated from the fat by hydrolysis.

titration. Any of a number of methods for determining volumetrically the concentration of a desired substance in solution by adding a standard solution of known volume and strength until the reaction is completed, usually as indicated by a change in color due to an indicator. Organic solvents are used in titrating water-insoluble substances (nonaqueous titration). Several types of electrical measurement techniques are used, among which are amperometric, conductometric, and coulometric titration methods, and spectrophotometry is another possible procedure. Titration has been successfully automated, the extent depending on the type of measurement used.

titration curve.   A plot of the pH versus the equivalents of acid or base, added during titration of a base or an acid.

TKP.   Abbreviation for tripotassium phosphate. See potassium phosphate, tribasic.

TKPP.   Abbreviation   for   tetrapotassium pyrophosphate. See potassium pyrophosphate.

Tl.   Symbol for thallium.

TLV.   See Threshold Limit Value.

Tm.   Symbol for thulium.

Tm.   The melting point for a double-stranded nucleic acid. Technically, this is defined as the temperature at which 50% of the strands are in double-stranded form and 50% are single-stranded, that is, midway in the melting curve. A primer has a specific Tm because it is assumed that it will find an opposite strand of appropriate character.

TMA.   Abbreviation for trimethylamine and trimellitic anhydride.

TMEDA.   Abbreviation for tetramethylethylenediamine.

TML.   Abbreviation for tetramethyllead.

TMSM.   See tetramethyl succinonitrile.

TMTD.   Abbreviation for tetramethylthiuram disulfide. See thiram.

TMTU.   See tetramethylthiourea.

TNA.   Abbreviation for tetranitroaniline.

TNB.   Abbreviation for trinitrobenzene.

TNBS.   See 2,4,6-trinitrobenzenesulfonic acid.

TNT.   Abbreviation for trinitrotoluene.

tobacco.   Cured leaves of plants of the family *Solanaceae*, genus *Nicotiiana*; the species *N. tabacum* is the most important domestic source. Curing consists of drying and long aging. The leaves are dehydrated by hanging in warm air to terminal moisture content of 20–30%, during which time starches reduce to sugars, the chlorophyll is discharged, and the color darkens. The product is then aged from 1 to 5 years to remove unpleasant odor; cigar tobacco is "fermented" with water, with resulting hydrolysis, deamination, and decarboxylation. Aging fermentation and oxidizers improve taste, aroma, and smoking qualities. Humectants and flavoring agents are added to some cigarette blends. Scores of chemical compounds have been identified in unburned tobacco. Basically it is cellulose. The cured product contains acids (citric, oxalic, formic), alkaloids (nicotine, anabasine, myosmine), and carbohydrates (lignin, pentosans, starch, sucrose), as well as tannin, ammonia, glutamine, and micro amounts of zinc, iodine, copper, manganese, and polonium-210.
See cigarette tar; smoke (4).
A nicotine-free tobacco substitute, TM "Cytrel" *[Cleanese]*, has been developed.
See "Cytrel" *[Celanese]*.
Hazard: A confirmed carcinogen. Many other deleterious effects from chronic inhalation of cigarette smoke.

tobacco mosaic virus.   The first virus to be obtained in crystalline form (from diseased tobacco plants). The protein portion of this virus (95% of each particle) contains about 2300 peptide chains, each having 150 amino acids and ending with threonine. Molecular weight of each chain is about 17,000, total molecular weight 40 million. The complete sequence of the amino acids has been determined.

tobermorite gel. $3CaO \cdot 2SiO_2 \cdot 3H_2O$. A calcium silicate hydrate; a main cementing ingredient of concrete because of its great surface area.

Tobias acid.   (2-naphthylamine-1-sulfonic acid; 2-amino-1-naphthalene-sulfonic acid).

Properties: White needles. Soluble in hot water.
Derivation: Sulfonation of β-naphthol with chlorosulfonic acid in nitrobenzene at 0C followed by heating the resulting 1-sulfonic acid with ammonium hydrogen sulfite and ammonia at 145–150C. Purification is by precipitation from dilute solution of the sodium salt.
Use: Azo dye intermediate, optical brighteners.

TOC.   Abbreviation for Tagliabue open cup, a standard method for determining flash points.

tocamphyl.   See hepasynthyl.

tocopherol.   (vitamin E). Any of a group of related substances (α-, β-, γ-, and Δ-tocopherol) that constitute vitamin E. The α-form (which occurs naturally as the *d*-isomer) is the most potent. Occurs naturally in plants, especially wheat germ. All are derivatives of dihydrobenzo-γ-pyran and differ from each other only in the number and position of methyl groups. Vitamin E is required by certain rodents for normal reproduction. Muscular and central nervous system depletion along with generalized edema are deficiency symptoms in all animals. It is not required as a dietary supplement for humans.

Properties: Viscous oils. Soluble in fats; insoluble in water. Stable to heat in the absence of oxygen, to strong acids, and to visible light; unstable to UV light, alkalies, and oxidation.

Use: Medicine, nutrition, antioxidants for fats, animal feed additive.

dl-α-tocopherol. (dl-2,5,7,8-tetramethyl-2-(4′,8′,12′-trimethyltridecyl)-6-chromanol).
CAS: 59-02-9. $C_{29}H_{49}OOH$. The most important of the vitamin E group.

Properties: Clear yellow, viscous oil. D 0.947–0.958 (25/25C), refr index 1.5030–1.5070 (20C).
Derivation: Synthetic.
Grade: NF, FCC.
Use: Biological antioxidant, meat curing (nitrosamine blocker), nutrient, medicine. Also available as the acetate.
See tocopherol.

tocophersolan. (USAN for *d*-α-tocopheryl polyethylene glycol 1000 succinate). $C_{29}H_{49}O \cdot OOC(CH_2)_2COO(CH_2CH_2O)_nH$. $n = 22$. A water-miscible vitamin E.
Use: Dietary food supplement, medicine.

d-α-tocopheryl acetate.
CAS: 58-95-7. $C_{31}H_{52}O_3$.
Properties: From vacuum steam distillation and acetylation of edible vegetable oil products. (FCC III) Colorless to yellow oil; odorless. Mp: 26.5–27.5C, bp: 205C @ 0.02 mm (bath). Insoluble in water, soluble in alc; misc with acetone, chloroform, ether, vegetable oil.
Use: Food additive; drug.

dl-α-tocopheryl acetate.
CAS: 7695-91-2. $C_{31}H_{52}O_3$.
Properties: From vacuum steam distillation and acetylation of edible vegetable oil products. (FCC III) Colorless to yellow oil; odorless. Insoluble in water; soluble in alc; misc with acetone, chloroform, ether, vegetable oil.
Hazard: A reproductive hazard.
Use: Food additive.

dl-α-tocopheryl acid succinate.
CAS: 4345-03-3. $C_{33}H_{54}O_5$.
Properties: From vacuum steam distillation and succinylation of edible vegetable oil products. (FCC III) Colorless to white crystalline powder or needles from pet ether; odorless and tasteless. Mp: 76–77C. Insoluble in water; very soluble in chloroform; soluble in acetone, alc, ether, vegetable oil.
Use: Food additive.

Todd, Sir Alexander R. (1907–1997). A British chemist who won the Nobel Prize for chemistry in 1957. His diverse research and accomplishments involved phosphorylation and mechanisms of biological reactions concerning phosphates. Many of his studies concerned the structure of nucleic acids, nucleotides, nucleotidic coenzymes, as well as vitamins B_1, B_{12}, and E. Work in biological organic chemistry indicated that hemp plant could be used for production of narcotics. Todd had degrees awarded from Oxford, Frankfurt, and Glasgow, among others.

Tofranil. A preparation of imipramine hydrochloride.
Use: Antidepressant.

toilet bowl cleaner. Any of a number of caustic formulations of varying strength designed to use on toilet bowls and other surfaces that are hard to clean.
Hazard: All are toxic and should be considered hazardous.

toiletries. Liquid cosmetic preparations usually consisting of an essential oil–alcohol mixture used chiefly for application to hair and scalp, as aftershave lotions, etc.

Tok. 2,4-dichlorophenyl-4-nitrophenyl ether product.
Use: Herbicide.

tokamak. A transliterated Russian word for an assembly for producing nuclear fusion. It consists essentially of a doughnut-shaped evacuated chamber called a torus, of 18-inch cross section, through which the plasma moves. Surrounding the torus is an electromagnetic field powerful enough to confine the energized plasma sufficiently to achieve the required density of 10^{14} particles/cc/sec and a temperature well above 44 million centigrade. Up to 74 million centigrade has been obtained experimentally, but 100 million centigrade will be necessary

for power production. The vacuum chamber and the magnetic field simulate conditions on the sun, i.e., absence of air and immense gravitational forces. Several experimental tokamaks are in operation in the U.S., the largest being at Princeton, NJ. It is utilizing hydrogen and deuterium as a test plasma and thus can attain temperatures of only 100,000C. After further experimentation and modification, it should be capable of reaching the 100 million centigrade necessary for power production when tritium is used.
See fusion.

tokuthion oxon. See ethaphos.

tolan. (diphenylacetylene). $C_6H_5C:CC_6H_5$.
Properties: Monoclinic crystals. Mp 59–61C, bp 300C, 170C (19 mm Hg), d 0.966 (100/4C). Insoluble in water; soluble in ether or hot alcohol.
Grade: Technical, purified.
Use: Organic synthesis, purified grade, primary fluor or wavelength shifter in soluble scintillators.

o-tolidine. (3,3'-dimethylbenzidine; diaminoditolyl).
CAS: 119-93-7. $[C_6H_3(CH_3)NH_2]_2$.
Properties: Glistening plates, white to reddish. Mp 129–131C. Soluble in alcohol and ether; sparingly soluble in water. Affected by light. Combustible.
Derivation: Reduction of o-nitrotoluene with zinc dust and caustic soda and conversion of the hydrazotoluene by boiling with hydrochloric acid.
Grade: Technical, dry, or paste.
Hazard: Eye, bladder and kidney irritant, bladder cancer, and methemoglobinemia. Possible carcinogen.
Use: Dyes, sensitive reagent for gold (1:10 million detectable) and for free chlorine in water, curing agent for urethane resins (also available as the dihydrochloride).

Tollens' reagent. Solution of ammoniacal silver nitrate containing free sodium bicarbonate.
Hazard: May explode on standing from the formation of silver nitride.
Use: To test for aldehydes.

tolnaftate. (USAN for o-2-naphthyl-meta-N-dimethylthiocarbanilate).
CAS: 2398-96-1. $C_{10}H_7OC(S)N(CH_3)C_6H_4CH_3$.
Use: Antifungal treatment (medicine).

tolualdehdye. See tolyl aldehydes.

toluazotoluidine. See o-aminoazotoluene.

tolu balsam. See balsam.

toluene. (methylbenzene; phenylmethane).

CAS: 108-88-3. $C_6H_5CH_3$.

Properties: Colorless liquid; benzene-like odor. D 0.866 (20/4C), fp −94.5C, bp 110.7C, refr index 1.497 (20C), aniline equivalent 15, flash p 40F (4.4C) (CC), autoign temp 997F (536C). Soluble in alcohol, benzene, and ether; insoluble in water.
Derivation: (1) By catalytic reforming of petroleum. (2) By fractional distillation of coal tar light oil.
Method of purification: Rectification.
Grade: (Usually defined in terms of boiling ranges) Pure, commercial, straw-colored, nitration, scintillation, industrial.
Hazard: Flammable, dangerous fire risk. Explosive limits in air 1.27–7%. Toxic by ingestion, inhalation, and skin absorption. Visual impairment, female reproductive effects, and pregnancy loss. Questionable carcinogen.
Use: Aviation gasoline and high-octane blending stock; benzene, phenol, and caprolactam; solvent for paints and coatings, gums, resins, most oils, rubber, vinyl organosols; diluent and thinner in nitrocellulose lacquers; adhesive solvent in plastic toys and model airplanes; chemicals (benzoic acid, benzyl and benzoyl derivatives, saccharin, medicines, dyes, perfumes); source of toluenediisocyanates (polyurethane resins); explosives (TNT); toluene sulfonates (detergents); scintillation counters.

toluene-2,4-diamine. (m-tolylenediamine; MTD; m-toluylenediamine; diaminotoluene).
CAS: 95-80-7. $CH_3C_6H_3(NH_2)_2$.
Properties: Colorless crystals. Mp 99C, bp 280C. Soluble in water, alcohol, and ether.
Derivation: Reduction of m-dinitrotoluene with iron and hydrochloric acid.
Hazard: A possible carcinogen.
Use: Chain extender and cross-linker, intermediate in organic synthesis of dyes, polymers, especially polyurethanes.

toluene-2,4-diisocyanate. (2,4-tolylene diisocyanate; m-tolylene diisocyanate; TDI).
CAS: 584-84-9. $CH_3C_6H_3(NCO)_2$.

Properties: Water-white to pale-yellow liquid; sharp pungent odor. Bp 251C, 120C (10 mm Hg), flash p 270F (132C), mp 19.4–21.5C (pure isomer), d 1.22

(25/15.5C), vap press 0.01 mm Hg (20C). Reacts with water, producing carbon dioxide; reacts with compounds containing active hydrogen (may be violent). Soluble in ether, acetone, and other organic solvents. Combustible.
Derivation: Reaction of 2,4-diaminotoluene with phosgene.
Method of purification: Distillation to remove hydrogen chloride.
Grade: 100% 2,4-isomer; 80% and 65% 2,4-isomer both mixed with 2,6-isomer.
Hazard: Toxic by ingestion and inhalation; strong irritant to skin and tissue, especially to eyes. Respiratory sensitization.
Use: Polyurethane foams, elastomers and coatings, cross-linking agent for nylon 6.

toluene-2,6-diisocyanate.   See   toluene-2,4-diisocyanate.

p-toluenesulfamine.   See   p-toluenesulfonamide.

p-toluenesulfanilide. $CH_3C_6H_4SO_2C_6H_4NH_2$.
Properties: White to pink crystalline solid. Mp 103C. Soluble in most lacquer solvents. Combustible.
Derivation: p-Toluene sulfonchloride treated with aniline in the presence of lime or carefully regulated amounts of alkalies.
Grade: Technical.
Use: Softener for acetylcellulose in proportions up to 50%, dyestuff intermediate.

o-toluenesulfonamide.
CAS: 88-19-7. $CH_3C_6H_4SO_2NH_2$.
Properties: Colorless crystals. Mp 156.3C. Soluble in alcohol; slightly soluble in water and ether. Combustible.
Use: Plasticizer.

p-toluenesulfonamide.   (p-toluenesulfamine; PTSA).
CAS: 70-55-3. $CH_3C_6H_4SO_2NH_2$.
Properties: White leaflets. Mp 137C. Soluble in alcohol; very slightly soluble in water. Combustible.
Derivation: By amination of p-toluene sulfonchloride.
Use: Organic synthesis, plasticizers and resins, fungicide and mildewicide in paints and coatings.

p-toluenesulfondichloroamide.   See   dichloramine-T.

o-toluenesulfonic acid.   (o-toluenesulfonate).
CAS: 104-15-4. $C_6H_4(SO_3H)(CH_3)$.
Properties: Colorless crystals. Mp 67.5C, bp 129C. Soluble in alcohol, water, and ether. Combustible.
Derivation: By sulfonating toluene with concentrated sulfuric acid at 100C.
Grade: Anhydrous, monohydrate, 40% aqueous solution.

Hazard: Toxic by ingestion and inhalation, strong irritant to tissue.
Use: Dyes, organic synthesis, acid catalyst.

p-toluenesulfonic acid.   (p-toluenesulfonate).
CAS: 104-15-4. $C_6H_4(SO_3H)(CH_3)$.
Properties: Colorless leaflets. Mp 107C, bp 140C (20 mm Hg). Soluble in alcohol, ether, and water. Combustible.
Derivation: By action of chlorosulfonic acid on toluene at a low temperature.
Grade: Anhydrous, monohydrate; 40% aqueous solution.
Hazard: Skin irritant.
Use: Dyes, organic synthesis, organic catalyst.

o-toluenesulfonyl chloride.   (o-toluenesulfochloride; o-toluenesulfonchloride).
CAS: 98-59-9. $H_3CC_6H_4SO_2Cl$.
Properties: Oily liquid. D 1.3383 (20/4C), mp 10.2C, bp 154C (36 mm Hg). Insoluble in water; soluble in hot alcohol and in ether and benzene. Combustible.
Derivation: Action of chlorosulfonic acid on toluene.
Use: Intermediate in the synthesis of saccharin and dyestuffs.

p-toluenesulfonyl chloride.   (tosyl chloride; p-toluenesulfochloride; p-toluenesulfonchloride).
CAS: 98-59-9. $H_3CC_6H_4SO_2Cl$.
Properties: Solid. Mp 71C, bp 145–146C (15 mm Hg). Insoluble in water; soluble in alcohol, ether, and benzene. Combustible.
Use: Organic synthesis.

4-toluenesulfonyl isocyanate.   See 4-methylbenzenesulfonyl isocyanate.

p-toluenesulfonyl semicarbazide.
$C_8H_{11}N_3O_3S$.
Properties: Fine, white powder. D 1.428, decomposes at 440F (226C dry), 415–430F (212–221C compounded).
Use: Blowing agent for polyolefins, impact polystyrene, polypropylene, ABS, etc.

toluenethiol.   (thiocresol; tolyl mercaptan).
CAS: 100-53-8. $CH_3C_6H_4SH$.
Properties: Cream to white, moist crystals; musty odor. Bp 195C. Insoluble in water; soluble in alcohol or ether. There are three isomers with different boiling points.
Hazard: Skin irritant.
Use: Intermediate, bacteriostat.

α-toluenethiol.   See benzyl thiol.

toluene trichloride.   See benzotrichloride.

toluene trifluoride. See benzotrifluoride.

toluhydroquinone. $CH_3C_6H_3(OH)_2$.
Properties: Pink to white crystals. Mp 125–127C. Slightly soluble in water; soluble in alcohol and acetone.
Use: Antioxidant, polymerization inhibitor.

α-toluic acid. See phenylacetic acid.

m-toluic acid. (m-toluylic acid; 3-methylbenzoic acid). $C_6H_4CH_3COOH$.
Properties: White to yellowish crystals. D 1.0543, mp 109C, bp 263C, ionization constant 5.3×10^{-5}. Slightly soluble in water; soluble in alcohol and ether. Combustible.
Derivation: Oxidation of m-xylene with nitric acid.
Use: Organic synthesis, to form N,N-diethyl-m-toluamide, a broad-spectrum insect repellent.

o-toluic acid. (o-toluylic acid; 2-methylbenzoic acid). $C_6H_4CH_3COOH$.
Properties: White crystals. D 1.0621, mp 103.5–104C, bp 256C, refr index 1.512 (114.6C), ionization constant 1.2×10^{-5}. Slightly soluble in water; soluble in alcohol and chloroform. Combustible.
Derivation: Oxidation of o-xylene with dilute nitric acid.
Use: Bacteriostat.

p-toluic acid. (p-toluylic acid; 4-methylbenzoic acid). $C_6H_4CH_3COOH$.

Properties: Transparent crystals. Mp 180C, bp 275C, ionization constant 4.3×10^{-5}. Slightly soluble in water; soluble in alcohol and ether. Combustible.
Derivation: By treating cymene or turpentine with nitric acid.
Use: Agricultural chemicals, animal feed supplement.

α-toluic aldehyde. See phenylacetaldehyde.

m-toluidine. (m-aminotoluene).
CAS: 108-44-1. $CH_3C_6H_4NH_2$.
Properties: Colorless liquid. D 0.980, fp −31.5C, bp 203.3C, flash p 187F (86.1C). Slightly soluble in water; soluble in alcohol or ether. Combustible.
Derivation: Reduction of m-nitrotoluene.
Hazard: Toxic by inhalation and ingestion, absorbed by skin. Eye, bladder, and kidney irritant and methemoglobinemia. Questionable carcinogen.
Use: Dyes, manufacture of organic chemicals.

o-toluidine. (o-aminotoluene).
CAS: 95-53-4. $CH_3C_6H_4NH_2$.

Properties: Light-yellow liquid, becomes reddish-brown on exposure to air and light. Volatile with steam, d 1.008 (20/20C), fp −16C, bp 200C, flash p 185F (85C). Soluble in alcohol and ether; very slightly soluble in water. Combustible.
Derivation: By the reduction of o-nitrotoluene or obtained mixed with p-toluidine by the reduction of crude nitrotoluene.
Grade: Technical.
Hazard: Toxic by inhalation and ingestion, absorbed by skin. Confirmed carcinogen.
Use: Textile printing dyes, vulcanization accelerator, organic synthesis.

p-toluidine. (p-aminotoluene).
CAS: 106-49-0. $CH_3C_6H_4NH_2$.
Properties: White lustrous plates or leaflets. D 1.046 (20/4C), mp 45C, bp 200.3C, flash p 189F (87.2C). Soluble in alcohol and ether; very slightly soluble in water. Combustible.
Derivation: By the reduction of p-nitrotoluene with iron and hydrochloric acid.
Grade: Technical, flake, or cast.
Hazard: Toxic by inhalation and ingestion, absorbed by skin. Methemoglobinemia. Possible carcinogen.
Use: Dyes, organic synthesis, reagent for lignin, nitrite, chloroglucinol.

toluidine maroon. $CH_3C_6H_3NO_2N_2C_{10}H_5$ $OHCONHC_6H_4NO_2$. An organic azo pigment obtained by the azo coupling of m-nitro-p-toluidine with the m-nitroanilide of β-hydroxynaphthoic acid.
Properties: Good lightfastness and weather resistance, excellent acid and alkali resistance, poor resistance to bleeding.
Use: Air-dried and baked enamels, truck body finishes.

toluidine red. Any of a class of pigments based on couplings of β-naphthol and m-nitro-p-toluidine. See toluidine maroon.

tolu oil. (tolu balsam oil).
Properties: Yellow liquid; hyacinth-like odor. D 0.945–1.09. Soluble in alcohol, ether, chloroform, and carbon disulfide. Chief known constituents: A terpene, $C_{10}H_{16}$, and esters of cinnamic and benzoic acid.
Derivation: From tolu balsam by distillation.
Grade: Technical.
Use: Medicine (expectorant), cough syrups. See balsam.

toluol. Obsolete name for toluene.

toluqinone. (2-methylquinone; p-toluqinone).
$CH_3C_6H_3O_2$.
Properties: Yellow leaflets or needles. Mp 65–67C. Soluble in hot water; very soluble in alcohol, ether, acetone, ethyl acetate, and benzene.

Tolu-Sol. A series of solvents, predominantly C$_7$ hydrocarbons, low in naphthenic hydrocarbons, and containing 3–50% aromatics, the balance being essentially paraffinic.
Use: Lacquer diluents and gravure ink solvents.
Hazard: May be flammable.

toluyl aldehyde. See tolylaldehyde.

toluylene. See stilbene.

m-**toluylenediamine.** See toluene-2,4-diamine.

toluylene red. See neutral red.

m-, *o*-, **and** *p*-**toluylic acid.** See corresponding toluic acid.

tolyl acetate. See cresyl acetate.

p-**tolylaldehyde.** (*p*-toluylaldehyde; *p*-tolualdehyde; *p*-methylbenzaldehyde).
CAS: 1334-78-7.   CH$_3$C$_6$H$_4$CHO.
Properties: Colorless liquid. Refr index 1.54693 (16.6C), d 1.020, bp 204C. Slightly soluble in water; soluble in alcohol and ether. There are also *o*- and *m*- isomers. Combustible.
Grade: Technical, pure.
Use: Perfumes, pharmaceutical and dyestuff intermediate, flavoring agent.

α-**tolylaldehyde dimethylacetal.** See phenylacetaldehyde dimethylacetal.

4-*o*-**tolylazo-*o*-diacetotoluide.** See diacetylaminoazotoluene.

o-**tolyl biguanide.** NH$_2$(CNHNH)$_2$C$_6$H$_4$CH$_3$.
Properties: White to off-white powder. Mp 138C (min). Combustible.
Use: Antioxidant for soaps produced from animal or vegetable oil.

p-**tolyldiethanolamine.**
(HOC$_2$H$_4$)$_2$NC$_6$H$_4$CH$_3$.
Properties: Crystals. Mp 62C, bp 297.1C, vap press <0.1 (20C), d 1.0723 (20/20C), solubility in water 1.67% by weight (20C), viscosity 155 cP (20C), flash p 385F (196C). Very soluble in acetone, ethanol, ethyl acetate, benzene. Combustible.
Use: Emulsifier, dyestuff intermediate.

2-*p*-**tolyl-5,6-dihydroimidazo(2,1-a)isoquinoline.**
CAS: 61001-19-2.   C$_{18}$H$_{16}$N$_2$.
Hazard: A reproductive hazard.

m-**tolylenediamine.** See toluene-2,4-diamine.

m-**tolylenediaminesulfonic acid.** [4,6-diamino-*m*-toluenesulfonic acid (SO$_3$H = 1)].
CH$_3$C$_6$H$_2$(NH$_2$)$_2$SO$_3$H.

Properties: White, crystalline solid. Soluble in alkalies.
Derivation: By addition of *m*-toluylenediamine sulfate to oleum and heating.
Use: Dyes.

m-**tolylenediisocyanate.** See toluene-2,4-diisocyanate.

2-*p*-**tolylimidazo(2,1-a)isoquinoline.**
CAS: 61001-04-5.   C$_{18}$H$_{14}$N$_2$.
Hazard: A reproductive hazard.

p-**tolyl isobutyrate.** See *p*-cresyl isobutyrate.

2-**tolyl isocyanate.** See 1-isocyanato-2-methylbenzene.

4-**tolyl isocyanate.** See 1-isocyanato-4-methylbenzene.

p-**tolyl-1-naphthylamine-8-sulfonic acid.**
(tolylperi acid). C$_{17}$H$_{15}$NO$_3$S.
Properties: Greenish-gray needles. Soluble in alcohol; rather insoluble in water. Combustible.
Derivation: Arylation of 1-naphthylamine-8-sufonic acid with *p*-toluidine.
Grade: Technical, mostly as sodium salt.
Use: Azo colors.

4-(*p*-**tolyl**)-**6h-1,3,5-oxathiazine.** See 4-(4-methylphenyl)-6h-1,3,5-oxathiazine.

p-**tolyl phenylacetate.** See *p*-cresyl phenylacetate.

3-(*p*-**tolylsulfonyl)acrylonitrile.**
CAS: 1424-48-2.   C$_{10}$H$_9$NO$_2$S.
Hazard: Moderately toxic by ingestion. A moderate eye irritant.

tolyl-*p*-toluenesulfonate. See cresyl-*p*-toluenesulfonate.

tomatidine. C$_{27}$H$_{45}$NO$_2$. A steroid secondary amine, the nitrogenous aglycone of tomatine. Isolated from the roots of the Rutgers tomato plant as the hydrochloride, C$_{27}$H$_{45}$NO$_2$•HCl.
Properties: Crystals. Decomposes at 275–280C.

tomatine.
CAS: 17406-45-0.   C$_{50}$H$_{83}$NO$_{21}$.
A glycosidal alkaloid prepared from the dried leaves and stems of the tomato plant. White crystals used as plant fungicide and as a specific precipitating agent for cholesterol and other sterols. The crude extract is referred to as tomatin.

tomatine hydrochloride.
CAS: 17605-83-3.   C$_{50}$H$_{83}$NO$_{21}$•ClH.
Hazard: A poison. An eye irritant.

Tone. Caprolactone-based monomers and polymers.
Use: For high-performance urethane elastomers and adhesives.

toner. An organic pigment that does not contain inorganic pigment or inorganic carrying base (ASTM). Important toners are Pigment Green 7, Pigment Blue 15, Pigment Yellow 12, and Pigment Blue 19.
See lake.

tonka. (tonka bean; coumarouna bean; dipteryx).
Properties: Black-brownish seeds with wrinkled surface and brittle shining or fatty skins, aromatic bitterish taste, balsamic vanilla-like odor, efflorescences of coumarin often appear on the surface. Combustible.
Use: Production of natural coumarin, flavoring extracts, toilet powders.

tonka bean camphor. See coumarin.

"Tonox" *[Chemtura].* TM for a blend of aromatic primary amines, the main component of which is *p,p*-diaminodiphenylmethane.
Use: Epoxy and urethane curative.

Topfer's reagent. Dimethylaminoazobenzene 0.5 g in 100 cc 95% ethanol.
Use: To test acidity of stomach contents.

topical. A medical term meaning "applied to the surface of the skin."

TOPO. Abbreviation for trioctylphosphinic oxide.

topochemical reaction. Any chemical reaction that is not expressible in stoichiometric relationships. Such reactions are characteristic of cellulose; they can take place only at certain sites on the molecule where reactive groups are available, i.e., in the amorphous areas or on the surfaces of the crystalline areas.

topoisomerases. Enzymes that introduce positive or negative supercoils in closed, circular duplex DNA.

"Tordon" *[Dow Agrosciences].* TM for picloram.

torr. A pressure unit used chiefly in vacuum technology; it is the pressure required to support 1-mm Hg at 0C.

torsion balance. A balance having equal arms and using horizontal steel bands for pivots and bearings.

torula yeast. A yeast that utilizes fermentable sugar in industrial wastes, such as fruit cannery refuse and sulfite liquor from pulp mills. The dried yeast is high in protein and vitamin content, enabling it to be used for enriching animal feeds. The enzymes present are destroyed during drying. It is now being made by a new process utilizing petroleum-derived ethanol.

torus. A doughnut-shaped vacuum chamber that is an essential part of a nuclear fusion reactor.
See tokamak; JET.

tosyl. (Ts).
CAS: 302-17-0. $CH_3C_6H_4SO_2{}^-$.
The *p*-toluenesulfonyl radical. Esters of *p*-toluenesulfonic acid are known as tosylates.

total ash. The residue of the mineral matter obtained by controlled incineration, as of coal.

Toth process. A process for production of aluminum metal that utilizes kaolin and other high-alumina clays. The clay is chlorinated after calcination, and the aluminum chloride resulting is reacted with metallic manganese to yield aluminum and manganese chloride. The reaction occurs at the comparatively low temperature of 260C. The manganese chloride is recovered as manganese metal and chlorine by oxidation and subsequent reduction, the manganese being recycled. This is a much cheaper and more efficient method than the Hall process, because it requires less energy input and does not utilize imported bauxite.

tourmaline. $(Na,Ca)(Al,Fe)B_3Al_3(AlSi_2O_9)$ $(O,OH,F)_4$. A complex borosilicate of aluminum.
Use: Pressure gauges, optical equipment, oscillator plates, source of boric acid.

tower, distillation. A metal cylinder from 6 inches to 20 ft in diameter located between the boiler and the condenser in distillation units. The vapors rise through the tower, some of the liquid condensate flowing back down through the tower (reflux). Horizontal plates at intervals of 2 ft are used to achieve contact between countercurrent liquid and the vapor stream. Vapor passes through the liquid on the plate through several apertures, each covered by a cap, an arrangement called a bubble cap plate. Various types of packings are used in smaller columns, e.g., metal chains, tubes, glass beads, Raschig rings.
See HETP (2).
Possibly supplanting distillation towers in the future is a circular metal drum that rotates at 1800 rpm and contains conventional packing materials. The centrifugal force separates the liquid and vapor components as they pass through the drum. This novel distillation technique has been developed by ICI in England and is called Higee, i.e., high gravity.

toxaphene. (generic name for technical chlorinated camphene).
CAS: 8001-35-2. $C_{10}H_{10}Cl_8$.
Formula is approximate, contains 67–69% chlorine.
Properties: Amber, waxy solid; mild odor of chlorine and camphor. Melting range 65–90C, d 1.66 (27C). Soluble in common organic solvents.
Hazard: Toxic by ingestion, inhalation, skin absorption; most uses prohibited. Central nervous system convulsions and liver damage. Possible carcinogen.

toxenzyme. Any poisonous enzyme.

toxicity. The ability of a substance to cause damage to living tissue, impairment of the central nervous system, severe illness, or, in extreme cases, death when ingested, inhaled, or absorbed by the skin. The amounts required to produce these results vary widely with the nature of the substance and the time of exposure to it. "Acute" toxicity refers to exposure of short duration, i.e., a single brief exposure; "chronic" toxicity refers to exposure of long duration, i.e., repeated or prolonged exposures.
The toxicity hazard of a material may depend on its physical state and on its solubility in water and acids. Some metals that are harmless in solid or bulk form are quite toxic as fume, powder, or dust. Many substances that are intensely poisonous are actually beneficial when administered in micro amounts, as in prescription drugs, e.g., strychnine.
Toxicity is objectively evaluated on the basis of test dosages made on experimental animals under controlled conditions. Most important of these are the LD_{50} (lethal dose, 50%) and the LC_{50} (lethal concentration, 50%) tests, which include exposure of the animal to oral ingestion and inhalation of the material under test. A substance having an LD_{50} of less than 400 mg/kg of body weight is considered very toxic.

toxicogenomics. The study of how genomes respond to environmental stressors or toxicants. Combines expression profiling of genome-wide mRNA with protein expression patterns using bioinformatics to understand the role of gene-environment interactions in disease and dysfunction.

toxicology. The branch of medical science devoted to the study of poisons, including their mode of action, effects, detection, and countermeasures. The subject may be subdivided into (1) clinical, (2) environmental, (3) forensic, and (4) occupational (industrial). The Institute of Chemical Toxicology was formed in 1975, its members comprising a number of the larger chemical companies.
See toxicity; poison (1).

toxic substances. The following list includes a number of chemical individuals and groups that are generally regarded as having toxic properties by ingestion, inhalation, or absorption via the skin.

There is considerable variation in the degree of toxicity among these, and the listing is by no means exhaustive.
See toxicity.

Individuals	Groups
aniline	aldehydes
asbestos (carcinogen)	alkaloids
benzidine (carcinogen)	allyl compounds
benzpyrene (carcinogen)	arsenic and compounds
carbon monoxide	barium and soluble
chlorine	compounds
coal tar pitch	barbiturates
(carcinogen)	
cresol	beryllium and soluble
hydrogen peroxide	compounds
hydrogen sulfide	chlorinated
	hydrocarbons
methanol	chromium (hexavalent)
nickel carbonyl	carcinogenic
	compounds
osmium tetroxide	corrosive materials
oxalic acid	cyanides
ozone	cadmium and
	compounds
phenol	fluorine and compounds
pyrene	lead compounds
pyridine	mercury and
	compounds
phosphine	organic phosphate esters
stibine	radioactive substances
sulfur dioxide	selenium and
	compounds
vinyl chloride monomer	thallium and
(carcinogen)	compounds tin
	(organic compounds)

The Toxic Substances Control Act (TSCA) passed by the U.S. Congress in 1976 provides the legal basis for regulations concerning all aspects of the manufacture of such products. Establishment and enforcement of such regulations are carried out by the Environmental Protection Agency (EPA). The American Conference of Governmental Industrial Hygienists issues a periodically revised list of Threshold Limit Values for substances in workroom air, upon which the industrial safety standards of the Occupational Safety and Health Administration (OSHA) are based. The Food and Drug Administration (FDA) is responsible for the enforcement of the Food, Drug, and Cosmetic Act. Decisions made by these agencies are arrived at only after extensive testing by both manufacturers and independent groups. Effective control of toxic materials has assumed increasing importance in recent years and may be expected to become still more rigorous.
See poison (1); NIOSH.

A Toxicology Data Bank (TDB) has been established by the National Library of Medicine of NIH. The data are set up in a computer-based online file available for public use. Extensive physical data are included. Additional information can be found in *Dangerous Properties of Industrial Materials, 11th Edition*, Richard J. Lewis, Sr., Editor, published by John Wiley & Sons, Inc. This book contains 25,000 entries, each of which gives physical, chemical, and toxicological data about potentially hazardous materials.

Toxic Substances Control Act.　(TSCA). See toxic substances; Environmental Protection Agency.

toxins.　Proteins produced by some organisms and toxic to some other species.

TPA.　Abbreviation for terephthalic acid.

TPB.　Abbreviation for tetraphenylbutadiene.

TPG.　Abbreviation for triphenylguanidine.

TPN.　Abbreviation for triphosphopyridine nucleotide. See nicontinamide adenine dinucleotide phosphate.

TPO rubber.　Abbreviation for thermoplastic polyolefin rubber. See elastomer.

Tpp.　(1) Abbreviation for triphenyl phosphate. (2) Abbreviation for thiamine pyrophosphate. See cocarboxylase.

TPT.　(1) Abbreviation for triphenyltetrazolium chloride. See tetrazolium chloride. (2) Abbreviation for tetraisopropyl titanate.

trace element.　(micronutrient). An element essential to plant and animal nutrition in trace concentration, i.e., minute fractions of 1% (1000 ppm or less). Plants require iron, copper, boron, zinc, manganese, potassium, molybdenum, sodium, and chlorine. Animals require iron, copper, iron, manganese, cobalt, selenium, and potassium. Such elements are also called micronutrients. Do not confuse with *tracer*.

tracer.　A chemical entity (almost invariably radioactive and usually an isotope) added to the reacting elements or compounds in a chemical process, which can be traced through the process by appropriate detection methods, e.g., Geiger counter. Compounds containing tracers are often said to be "tagged" or "labeled." Carbon-14 is a commonly used tracer, and radioactive forms of iodine and sodium are also used. Many complex biochemical reactions have been examined in this way (e.g.,

photosynthesis). Nonradioactive deuterium (hydrogen isotope) is sometimes used, the detection being by molecular weight determination. Radioactive enzymes are also available for tracer studies, e.g., ribonuclease, pepsin, trypsin, and others. See labeling (2).

Tra-Cide.　Testing service. **Use:** To determine biocide toxicity and microbiological activity in industrial process waters.

trademark.　(TM). A word, symbol, or insignia designating one or more proprietary products or the manufacturer of such products, that has been officially registered with the government trademark agency. The accepted designation is a superior capital *R* enclosed in a circle; however, quotation marks may also be used, as in this dictionary. The term *trade name*, though widely used, is not applicable to such products; according to the U.S. Trade Mark Association, a trade name is the name under which a company does business, e.g., the Blank Chemical Company. Use of a trademark without proper indication of its proprietary nature places the name in jeopardy; a number of trademarks have been invalidated as a result of this practice.

trade name.　See trademark.

trade sales.　In the paint industry this term is applied to paints intended for sales to the general public, as in hardware stores and similar outlets.

tragacanth gum.
CAS: 9000-65-1.
Properties: Dull white, translucent plates or yellowish powder. Soluble in alkaline solutions, aqueous hydrogen peroxide solution; strongly hydrophilic; insoluble in alcohol. Combustible.
Chief constituents: Polysaccharides of galactose, fructose, xylose, and arabinose with glucuronic acid.
Occurrence: Southwestern Europe, Greece, Turkey, Iran.
Grade: USP, FCC, No. 1, 2, 3.
Use: Pharmacy (emulsions), adhesives, leather dressing, textile printing and sizing, thickener and emulsifier, dyes, food products (ice cream, desserts), toothpastes, coating soap chips and powders, hairwave preparations, confectionery, printing inks, tablet binder.

tranquilizer.　See psychotropic drug.

trans-.　See *cis-*.

transalkylation.　A type of disproportionation reaction by which toluene is hydrogenated to benzene and mixed xylene isomers free from ethylbenzene, avoiding the formation of methane resulting from the conventional hydrodealkylation process. Transalkylation of toluene to benzene involves the use of a catalyst; the yield is claimed to be 97%, based on toluene feed.

transaminase. (aminopherase). An enzyme that facilitates the reversible transfer of an amino group from an α-amino acid to an α-keto acid.
See aminotransferases.

transamination. The use of nitrogen of one amino acid for synthesizing another amino acid in vivo.

transcription. The enzymatic process whereby the genetic information contained in one strand of DNA is synthesized into a complementary sequence of bases in an mRNA chain.

transcriptional control. The regulation of the rate of a protein's synthesis by regulation of the formation of its mRNA.

transcription factor. A protein which is involved in the transcription of genes. These usually bind to DNA as part of their function (but not necessarily). A transcription factor may be general (i.e., acting on many or all genes in all tissues) or tissue-specific (i.e., present only in a particular cell type and activating the genes restricted to that cell type). Its activity may be constitutive or may depend on the presence of some stimulus; for example, the glucocorticoid receptor is a transcription factor which is active only when glucocorticoids are present.

transcription unit. The sequence from the transcriptional start site to the site of termination, inclusive of introns.

transcriptome. The full complement of activated genes, mRNAs, or transcripts in a particular tissue at a particular time.

transduction. (1) The conversion of energy from one form to another (ATP synthesis by a chemiosmotic mechanism). (2) The conversion of information from one form to another (epinephrine outside a liver cell causing cAMP inside the cell). (3) The transfer of genetic information from one cell to another by means of a viral vector.

trans effect. Bond holding a group trans to the more electronegative or other labilizing group is weakened.

transesterification. See ester interchange.

transfection. The introduction of foreign DNA into a host cell.
See cloning vector; gene therapy.

transferase. An enzyme whose activity causes a transfer of a radical from one molecule to another. Examples are transaminases, transacetylases, and transmethylases, which effect the transfer of amino, acetyl, and methyl groups respectively.
See enzyme.

transference number. That portion of the total current carried by any species of ion in an electrolyte in the fluid state. The symbol t^+ is usually used for a positive ion and t^- for a negative ion.

transfermium element. An element with an atomic number greater than 100, the atomic number of fermium.
See periodic table; atomic number.

transfer mold. A chamber in which a thermosetting plastic is softened by heat and pressure and from which it is forced by high pressure through a suitable orifice into a closed mold for final curing.

transferrin. A 90 kDa protein responsible for the transport of iron and uptake by the liver.

transfer RNA. See ribonucleic acid.

transformation. A process by which the genetic material carried by an individual cell is altered by incorporation of exogenous DNA into its genome.

transformer oil. A liquid having the property of insulating the coils of transformers, both electrically and thermally. There are two broad classes, (1) natural and (2) synthetic. The natural type includes refined mineral oils (petroleum fractions), which have low viscosities and high chemical and oxidative stability. The synthetic types are (1) chlorinated aromatics (chlorinated biphenyls and trichlorobenzene) known collectively as askarels; (2) silicone oils; and (3) ester liquids such as dibutyl sebacate. All these types are nonflammable but combustible. Flash points are 250–300F (121–148C).
See dielectric.
Note: Use of PCB (chlorinated biphenyls) has been discontinued because of their ecologically damaging effects.

transgenic. An organism that has genes from another organism incorporated within its genome as a result of recombinant DNA procedures.

transgenic mouse. A mouse which carries experimentally introduced DNA. The procedure by which one makes a transgenic mouse involves the injection of DNA into a fertilized embryo at the pro-nuclear stage. The DNA is generally cloned and may be experimentally altered. It will become incorporated into the genome of the embryo. That embryo is implanted into a foster mother, who gives birth to an animal carrying the new gene. Various experiments are then carried out to test the functionality of the inserted DNA.

transient transfection. When DNA is transfected into cultured cells, it is able to stay in those cells for about 2–3 days, but then it will be lost unless steps are taken to ensure that it is retained—see stable transfection. During those 2–3 days, the DNA is functional and any functional genes it contains will be expressed. Investigators take advantage of this transient expression period to test gene function.

Transist AR. A grade of reagent-quality chemicals, specially controlled for use in the manufacture of semiconductors and other electronic devices and precision instruments.

transition element. (transition metal). Any of a number of elements in which the filling of the outermost shell to eight electrons within a period is interrupted to bring the penultimate shell from 8 to 18 or 32 electrons. Only these elements can use penultimate shell orbitals as well as outermost shell orbitals in bonding. All other elements, called "major group" elements, can use only outermost shell orbitals in bonding. Transition elements include elements 21 through 29 (scandium through copper), 39 through 47 (yttrium through silver), 57 through 79 (lanthanum through gold), and all known elements from 89 (actinium) on. All are metals. Many are noted for exhibiting a variety of oxidation states and forming numerous complex ions, as well as possessing extremely valuable properties in the metallic state.
See orbital theory.

transition point. The temperature at which two crystalline forms of a polymorphic substance are in equilibrium.

transition state. An activated form of a molecule in which the molecule has undergone a partial chemical reaction and has bond characteristics of both reactant and product.

transition state theory. (TSS). Gives approximate descriptions of chemical reaction rates based on the properties of two potential wells and a transition state between them.

transition-type system. See peritectic system.

translation. The process of decoding a strand of mRNA, thereby producing a protein based on the code. This process requires ribosomes (which are composed of rRNA along with various proteins) to perform the synthesis, and tRNA to bring in the amino acids.

translational control. The regulation of the rate of a protein's synthesis by regulation of the rate of translocation of the ribosome.

translational repressor. A repressor that blocks translation of an mRNA.

translocated herbicide. A herbicide that is distributed throughout the plant following absorption through the leaves or roots.

transmix. See 3-(2,2-dichloroethenyl)-2,2-dimethylcyclopropanecarboxylic.

transmutation. The natural or artificial transformation of atoms of one element into atoms of a different element as the result of a nuclear reaction. The reaction may be one in which two nuclei interact, as in the formation of oxygen from nitrogen and helium nuclei (β-particles), or one in which a nucleus reacts with an elementary particle such as a neutron or proton. Thus a sodium atom and a proton form a magnesium atom. Radioactive decay, e.g., of uranium, can be regarded as a type of transmutation. The first transmutation was performed by the English physicist Rutherford in 1919.

transpiration. Passage of water from the roots of a plant to the atmosphere via the vascular system and the stomata of the leaves.

transposable element. A class of DNA sequences that can move from one chromosomal site to another.
See transposon.

transposition. The movement of a gene or set of genes (transposon) from one site in the genome to another.

transportation label. See label (1).

transporters. Proteins that span a membrane and transport specific molecules across the membrane.

Transport-Plus. Boiler water treatment program.

transposon. (transposable element). A segment of DNA that can move from one position in the genome to another.

transuranic element. An element of higher atomic number than uranium, not found naturally, and produced by nuclear bombardment. The highest discovered are 105 and 106. They are all radioactive.
See actinide; periodic table.

"Trasar" [Nalco]. TM for technology that uses information from a chemical trace to enhance program value such as automation and diagnostics.

Traube purine synthesis. Preparation of an appropriate 4,5-diaminopyrimidine by introduction of the amino group into the 5 position of 4-amino-6-hydroxy- or 4,6-diaminopyrimidines by nitrosation and ammonium sulfide reduction, followed by ring closure with formic acid or chlorocarbonic ester.

Traube's rule. The adsorption from aqueous solution on carbon increases with increase in molecular weight of a homologous series of fatty acids.

tremolite.
CAS: 1332-21-4. $Ca_2Mg_5Si_8(OH)_2$.
A variety of asbestos. Some tremolite is sold as "fibrous talc."
Properties: White to light-green color, vitreous to silky luster. Mohs hardness 5–6, d 3.0–3.3, resistant to acids. Noncombustible.
Occurrence: New York, California, Maryland, South Africa.
Hazard: Inhalation of dust or fine particles is dangerous. Carcinogenic.
Use: As asbestos, particularly in acid-resisting applications, ceramics, paint.

tremorine dichlorohydrate.
CAS: 300-68-5. $C_{12}H_{20}N_2 \cdot 2ClH$.
Hazard: A poison.

treemoss concrete.
CAS: 68648-41-9.
Hazard: Low toxicity by ingestion and skin contact. A mild skin irritant.
Source: Natural product.

trenbolone. $C_{18}H_{22}O_2$.
Properties: Crystals. Mp: 186C.
Use: Drug (veterinary); food additive.

Trendcheck. Real-time statistical process control system.

tretamine. (generic name for 2,4,6-tris(1-aziridinyl)-s-triazine).
See triethylenemelamine.

triacetate fiber. See acetate fiber.

triacetin. (glyceryl triacetate).
CAS: 102-76-1. $C_3H_5(OCOCH_3)_3$.
Properties: Colorless liquid; slight fatty odor; bitter taste. D 1.160 (20C), bp 258–260C, sets to a glass at −37C, refr index 1.4312 (20C), flash p 300F (149C), bulk d 9.7 lb/gal. Slightly soluble in water; very soluble in alcohol, ether, and other organic solvents. Combustible.
Derivation: Action of acetic acid on glycerol.
Method of purification: Vacuum distillation.
Grade: Technical, CP, ND, FCC.
Use: Plasticizer, fixative in perfumery, manufacture of cosmetics, specialty solvent, to remove carbon dioxide from natural gas, medicine (topical antifungal).

triacontanoic acid. See melissic acid.

1-triacontanol. $CH_3(CH_2)_{28}CH_2OH$. A 30-carbon, straight-chain fatty alcohol.
Properties: Colorless needles from ether. Mp 85–88C, d at mp 0.777. Soluble in most organic solvents; insoluble in water. Combustible.
Occurrence: Beeswax, carnauba wax, leaf wax.
Use: Biochemical research, growth promoter, fertilizer supplement.

triacyiglycerol. An ester of glycerol with three molecules of fatty acid; also called a triglyceride or neutral fat.

Triadine. Industrial bactericide/fungicide.
Use: As a preservative in inhibiting growth of bacteria and fungi in aqueous-based metal working fluids, aqueous analytical and diagnostic reagents.

trialkyl boranes. R_3B. (where R = alkyl radical).
See tributylborane; triethylborane.

trialkylsilanol. An alcohol derivative of silane.
Hazard: May be flammable.
Use: Short-stopping agent for silicone polymers.

triallylamine.
CAS: 102-70-5. $(H_2C{:}CHCH_2)_3N$.
Properties: Liquid. D 0.800 (20/4C), fp −70C, bp 150–151C, refr index 1.4501 (20C), flash p 103F (39.4C) (TOC).
Hazard: Fire risk. Irritant.
Use: Intermediate.

triallyl cyanurate.
CAS: 101-37-1. $(CH_2{:}CHCH_2OC)_3N_3$.
Properties: Colorless liquid or solid. Mp 27.32C, flash p above 176F (80C) (TOC), d 1.1133 (30C), refr index 1.5049 (25C). Miscible with acetone, benzene, chloroform, dioxane, ethyl acetate, ethanol, and xylene. Combustible.
Hazard: Toxic by ingestion and inhalation.
Use: Polymers as monomer and modifier, organic intermediate.

triallyl cyanurate coagent. See Perkalink.

triallyl phosphate. $(CH_2{:}CHCH_2O)_3PO$.
Properties: Water-white liquid. Fp −50C, bp 80C (0.5 mm Hg), refr index 1.448 (25C), d 1.064 (25/15C). Combustible.
Use: Intermediate.

triallyl phosphite. See phosphorous acid, triallyl ester.

triamcinolene. (α-fluoro-16(α)-hydroxyprednisolone).
CAS: 124-94-7. $C_{21}H_{27}FO_6$.

Properties: White, crystalline powder. Mp 264–268C. Insoluble in water; slightly soluble in usual organic solvents; soluble in dimethylformamide.
Grade: ND.
Use: Medicine. Also available as the acetonide.

1,3,5-triaminobenzene.
CAS: 108-72-5. $C_6H_3(NH_2)_3$.
Properties: Solid. Mp (anhydrous) 129C, (hydrate) 84–86C (1.5 moles water). Soluble in water, acetone, and alcohol; insoluble in ether, cold benzene, carbon tetrachloride, and petroleum ether. Combustible.
Use: Ion-exchange resin intermediate, wetting and frothing agents, photographic developers, organic reactions.

2,4,6-triaminotoluene trihydrochloride.
$C_6H_2(NH_2HCl)_3CH_3 \cdot H_2O$.
Properties: Light-tan to cream crystals. Mp 119C (free base). Very soluble in water; soluble in alcohol and acetone; insoluble in benzene.
Use: Nongelatin photographic emulsion with ethylenediamine for fixation, ion-exchange resins, wetting and frothing agents, photographic developers, intermediate for varnishes and rubber chemicals.

2,4,6-triamino-*sym*-triazine. See melamine.

s-triaminotrinitrobenzene. See 2,4,6-trinitro-1,3,5-benzenetriamine.

triamminetrinitratatorhodium (III).
CAS: 41762-18-9. $H_9N_6O_9Rh$.
Hazard: A poison.

triamylamine. $(C_5H_{11})_3N$.
Properties: Colorless to yellow liquid. D 0.79–0.80 (20C), triamylamine content at least 98.0%, 95% boils between 225 and 260C, flash p 215F (101.6C), refr index 1.4374 (18C). Insoluble in water; soluble in gasoline. Combustible.
Derivation: Reaction of amyl chloride and ammonia.
Hazard: Irritant.
Use: Corrosion inhibitor, insecticidal preparations.

triamylbenzene. $(C_5H_{11})_3C_6H_3$.
Properties: Colorless liquid; odor faintly aromatic. D 0.87 (20C), boiling range 300–320C, flash p 270F (132.2C). Combustible.
Use: Chemical intermediate.

triamyl borate. $(C_5H_{11})_3BO_3$.
Properties: Colorless liquid; odor faintly alcoholic. D 0.845 (20C), boiling range 220–280C, flash p 180F (82.2C). Soluble in alcohol and ether. Combustible.
Derivation: direct heating of boric acid and amyl alcohol.
Use: Varnishes.

tri-*p-tert*-amylphenyl phosphate.
$(C_5H_{11}C_6H_4)_3PO$.
Properties: White solid; odorless. Liquid. Boiling range 305–345C (5 mm Hg), mp 62–63C. Insoluble in water. Combustible.
Use: Plasticizer.

triarylmethane dye. Any of a group of dyes whose molecular structure involves a central carbon atom joined to three aromatic nuclei. CI numbers range from 42000 to 44999. The color is due in part to the aromatic rings and to the chromophore groups =C=NH and =C=N–. The members of this class function as basic dyes for cotton, using tannin as a mordant, or if they contain sulfonic acid groups, as acid dyes for wool and silk. Examples are malachite green and methyl violet.
See triphenylmethane dye.

s-triazine derivatives. See ammelide; ammeline; melamine.

s-triazine-3,5(2H,4H)dione riboside. See 6-azauridine.

triazine herbicide.
Hazard: Can cause abdominal pain, impaired adrenal function, anemia, dermatitis, diarrhea, eye irritation, mucous membrane irritation, nausea, disturbed thiamine and riboflavin function, or vomiting.
Use: To control annual grasses and broadleaf weeds.

s-triazine-2,4,6-triol. See cyanuric acid.

triazole. $C_2H_3N_3$. A five-membered ring compound containing three nitrogens in the ring.
Use: Suggested as photoconductors in copying systems.

(1,2,4)triazolo(1,5-a)pyrimidine.
CAS: 275-02-5. $C_5H_4N_4$.
Hazard: A poison.

(1,2,4)triazolo(4,3-a)pyrimidine.
CAS: 274-98-6. $C_5H_4N_4$.
Hazard: A poison by ingestion.

(1,2,4)triazolo(4,3-a)quinoline.
CAS: 235-06-3. $C_{10}H_7N_3$.
Hazard: A poison by ingestion.

(1h-1,2,4-triazolyl-1-yl)tricyclohexylstannane.
CAS: 41083-11-8. $C_{20}H_{35}N_3Sn$.
Hazard: A poison by ingestion. Moderately toxic by skin contact.

triazone resin. One of a class of amino resins produced from urea, formaldehyde, and a primary amine.
Use: Textile and fabric treatment.
See dimethylolethyltriazone.

tribasic. See monobasic.

tribenzoside. See ethyl-3,5,6-tri-*o*-benzyl-*d*-glucofuranoside.

tribenzylchlorostannane. See chlorotribenzylstannane.

tribenzyltin chloride. See chlorotribenzylstannane.

tribenzyltin formate. See formyloxytribenzylstannane.

tribrominated polystyrene. See ethenylbenzene tribromo deriv. homopolymer.

tribromoacetaldehyde. (bromal). CBr_3CHO.
Properties: Oily, yellowish liquid. D 2.66, bp 174C. Soluble in water, alcohol, or ether. Combustible.
Derivation: (1) By adding bromine to a solution of paraldehyde in ethylacetate. (2) By adding bromine to absolute alcohol, fractionating, treating the fraction boiling at 165C with water, and distilling.
Hazard: As for bromine.
Use: Organic synthesis.

tribromoacetic acid. CBr_3COOH.
Properties: Colorless crystals. Mp 135C, bp 245–250C. Soluble in water, alcohol, or ether. Combustible.
Derivation: By oxidizing tribromoacetaldehyde with nitric acid.
Hazard: As for bromine.
Use: Organic synthesis.

tribromo-*tert*-butyl alcohol. (acetone–bromoform). $CBr_3C(CH_3)_2OH$.
Properties: Fine, white, prismatic crystals; camphor odor and taste. Mp 176C. Slightly soluble in water; soluble in alcohol and ether. Combustible.
Derivation: Reaction of acetone and bromoform with solid potassium hydroxide.
Hazard: As for bromine.
Use: Vinyl chloride polymerization.

tribromoethanol. (1,1,1-tribromoethyl alcohol).
CAS: 75-80-9. CBr_3CH_2OH.
Properties: White crystals or powder; slight aromatic odor and taste. Mp 79–82C, bp 94C (11 mm Hg), unstable in air and light. Slightly soluble in water; soluble in alcohol, ether, benzene, and amylene hydrate; aqueous and alcoholic solutions decompose on exposure to light. Combustible.
Grade: NF.
Derivation: By reduction of tribromoacetaldehyde with aluminum isopropylate.
Use: Medicine (basal anesthetic).

tribromomethane. See bromoform.

1,1,1-tribromo-2-methyl-2-propanol.
$CBr_3C(CH_3)_2OH$
Properties: Fine, white crystals. Mp 176–177C. Soluble in water, methanol, ether. Combustible.
Use: Organic synthesis.

tribromonitromethane. See bromopicrin.

tribromophenol. See bromol.

1,2,3-tribromopropane. (allyl tribromide).
CAS: 96-11-7. $BrCH_2CHBrCH_2Br$.
Properties: Colorless liquid. D 2.43, mp 16C, bp 220C, refr index 1.584. Soluble in alcohol and ether; insoluble in water.
Derivation: Gamma-ray initiated reaction of bromotrichloromethane with allyl bromide.
Use: Nematocide.

3,4′,5-tribromosalicylanilide. (tribromasalan).
CAS: 87-10-5. $Br_2C_6H_2(OH)C(O)NHC_6H_4Br$.
An active antiseptic.
Use: Soaps.
Hazard: A suspected carcinogen. Use in cosmetics prohibited (FDA).

tributoxyethyl phosphate.
$[CH_3(CH_2)_3O(CH_2)_2O]_3PO$.
Properties: Slightly yellow, oily liquid. D 1.020 (20C), fp −70C (viscous liquid), boiling range 215–228C (4 mm Hg), flash p 435F (223C), refr index 1.434 (25C). Insoluble or limited solubility in glycerol, glycols, and certain amines; soluble in most organic liquids. Combustible.
Use: Primary plasticizer for most resins and elastomers, floor finishes and waxes, flame-retarding agent.

tri-*n*-butyl aconitate. $C_3H_3(COOC_4H_9)_3$.
Properties: Colorless, odorless liquid. D 1.018 (20C), refr index 1.4500–1.4530 (25C), bp 190C (3 mm Hg). Insoluble in water; soluble in organic solvents. Combustible.
Use: Plasticizer–stabilizer for vinylidene chloride polymers, synthetic rubbers and cellulosic lacquers, insecticides.

tri-*n*-butylaluminum. $(CH_3CH_2CH_2CH_2)_3Al$.
Properties: Colorless, pyrophoric liquid. Bulk d 0.823 g/mL (20C), fp −26.7C.
Derivation: Exchange reaction of butene-1 and isobutyl aluminum.
Hazard: Highly flammable, dangerous fire risk, ignites spontaneously.
Use: Production of organo-tin compounds.

tri-*n*-butylamine.
CAS: 102-82-9. $(C_4H_9)_3N$.
Properties: Pale-yellow liquid; amine odor. Bp 214C, fp −70C, d 0.8 (20/20C), bulk d 6.5 lb/gal, refr index 1.4297 (20C), flash p 185F (85C) (OC).

Slightly soluble in water; soluble in most organic solvents. Combustible.
Derivation: By reaction of butanol or butyl chloride with ammonia.
Grade: Technical.
Hazard: Skin irritant, CNS stimulant.
Use: Solvent, inhibitor in hydraulic fluids, intermediate.

tri-*n*-butylborane.　(tri-*n*-butylborine).
$(CH_3CH_2CH_2CH_2)_3B$.
Properties: Colorless pyrophoric fluid. Fp $-34C$, bp 170C (222 mm Hg), d 0.747 (25C), vap press 0.1 mm Hg (20C), refr index 1.4285 (20C), flash p $-32F$ ($-35.5C$). Insoluble in water; soluble in most organic solvents.
Hazard: Flammable, dangerous fire risk, store and use in inert atmosphere, ignites spontaneously in air.
Use: Petrochemical industry, organic reactions, catalyst.

tributyl borate.　(butyl borate). $(C_4H_9)_3BO_3$.
Properties: Water-white liquid. D 0.8550–0.8570, bp 232.4C, distillation range, 85% distills between 135C and 140C (40 mm Hg), refr index 1.4071 (25C), fp below $-70C$, viscosity 1.601 cP (25C), flash p 200F (93.3C) (OC). Hydrolyzes rapidly; miscible with common organic liquids. Combustible.
Derivation: From butanol and boric acid.
Use: Welding fluxes, intermediate in preparation of borohydrides, flame retardant for textiles (with boric acid).

tri-*n*-butylborine.　See tri-*n*-butylborane.

tributylchlorostannane.　See chlorotributylstannane.

tri-*n*-butylchlorostannate.　See tributyltin chloride.

tributyl citrate.　(butyl citrate). C_3H_5O
$(COOC_4H_9)_3$.
Properties: Colorless or pale-yellow, stable, odorless, nonvolatile liquid. Fp $-20C$, bp 233.5C (22.5 mm Hg), flash p 315F (157C) (COC), refr index 1.4453 (20C), d 1.042 (25/25C), bulk d 8.7 lb/gal at 20C, pour p viscosity 31.9 cP (25C). Practically insoluble in water. Combustible.
Grade: Technical.
Use: Plasticizer, antifoam agent, solvent for nitrocellulose.

tri-*n*-butyl citrate.　See tributyl citrate; "Citroflex 4" *[Vertellus]*; TBC.

n,n,n-tributyl-2,4-dichlorobenzenemethanaminium chloride.
CAS: 3278-43-1.　$C_{19}H_{32}Cl_2N•Cl$.
Hazard: A severe eye irritant.

tributyl(2,4-dichlorobenzyl)phosphonium chloride.
CAS: 115-78-6.　$Cl_2C_6H_3CH_2P(C_4H_9)_3Cl$.
Properties: White, crystalline solid; mild aromatic odor. Technical grade melts at 114–120C. Soluble in water, acetone, ethanol, isopropanol and hot benzene; insoluble in hexane and ether.
Use: Growth retardant for ornamental plants.

tributyl(glycoloyloxy)tin.　See (glycoloyloxy) tributylstannane.

tributyliosocyanatostannane.
CAS: 681-99-2.　$C_{13}H_{27}NOSn$.
Hazard: A poison. Flammable liquid.

tributyl(methacryloxy)stannane.
CAS: 2155-70-6.　$C_{16}H_{32}O_2Sn$.
Hazard: A poison by ingestion and intravenous routes.

tributyl(oleoyloxy)stannane.
CAS: 3090-35-5.　$C_{30}H_{60}O_2Sn$.
Hazard: A poison by ingestion.

2,4,6-tri-*tert*-butylphenol.
$[(CH_3)_3C]_3C_6H_2OH$.
Properties: Solid. Mp 131C. Insoluble in water; soluble in most organic solvents. Combustible.
Use: Permissible antioxidant for aviation gasolines (when mixed with other butylphenols) (ASTM).

tributyl-*o*-phenylphenoxytin.　See (2-biphenyloxy)tributyltin.

tri-*p*-tert-butylphenyl phosphate.　$[(CH_3)_3$
$CC_6H_4O]_3PO$.
Properties: Solid. Bp 320C (5 mm Hg), mp 102–105C, flash p 545F (285C). Insoluble in water. Combustible.
Use: Plasticizer.

tributyl phosphate.　(TBP).
CAS: 126-73-8.　$(C_4H_9)_3PO_4$.
Properties: Stable, colorless liquid; odorless. Refr index 1.4226 (25C), bp 292C, latent heat of vaporization 55.1 cal/g (289C), fp below $-80C$, flash p 295F (146C) (COC), Saybolt viscosity 38.6 sec at 29.4C, bulk d 8.19 lb/gal, d 0.978 (20/20C). Miscible with most solvents and diluents; soluble in water. Combustible.
Grade: Technical.
Hazard: Toxic by ingestion and inhalation, irritant to skin, bladder, eye and upper respiratory tract. Possible carcinogen.
Use: Heat-exchange medium, solvent extraction of metal ions from solution of reactor products, solvent for nitrocellulose, cellulose acetate, plasticizer, pigment grinding assistant, antifoam agent, dielectric.

tributylphosphine. (phosphine, tributyl-; tributylfosfin; *tri-n*-butylphosphine).
CAS: 998-40-3. $C_{12}H_{27}P$.
An agent that can reduce disulfide bonds.
Hazard: Moderately toxic.

tri-*n*-butyl phosphine. $(CH_3CH_2CH_2CH_2)_3P$
Properties: Colorless liquid; garlic odor. D 0.8100 (min at 25/4C), fp −60 to −65C, bp 240C, flash p 104F (40C), autoign temp 392F (200C), refr index 1.463 (20C). Almost insoluble in water; miscible with ether, methanol, ethanol, benzene. An organic base and strong reducing agent. Combustible.
Use: Polymerization cross-linking catalyst, organic intermediate, fuels.

tributyl phosphite.
CAS: 102-85-2. $(C_4H_9O)_3P$.
Properties: Water-white liquid. Bp 120C (8 mm Hg), d 0.911 (25C), refr index 1.4301 (25C), flash p 250F (121C). Decomposes in water; soluble in common organic solvents. Combustible.
Use: Additive for greases and extreme-pressure lubricants, stabilizer for fuel oils and polyamides, gasoline additive.

O,O,O-tributyl phosphorothioate. (tributyl thiophosphate). $(C_4H_9O)_3PS$.
Properties: Colorless liquid; characteristic odor. Bp 142–145C (4.5 mm Hg), d 0.987, flash p 295F (146C) (COC). Insoluble in water; soluble in most organic solvents. Combustible.
Hazard: Highly toxic, cholinesterase inhibitor.
Use: Plasticizer, lubricant additive, antifoam agent, hydraulic fluid, intermediate.

S,S,S-tributyl phosphorotrithioate. (DEF).
CAS: 78-48-8. $(C_4H_9S)_3PO$.
Properties: Liquid. Bp 150C (0.3 mm Hg). Insoluble in water; soluble in aliphatic, aromatic, and chlorinated hydrocarbons.
Hazard: Cholinesterase inhibitor.
Use: Cotton defoliant.

tributylphosphorotrithioite.
CAS: 150-50-5. $(C_4H_9S)_3P$.
Properties: Nearly colorless liquid. Bp 115–134C (0.08 mm Hg), d 1.02 (20C), refr index 1.542 (25C). Insoluble in water; soluble in a variety of organic solvents.
Derivation: Reaction of butyl mercaptan with phosphorus trichloride.
Hazard: Cholinesterase inhibitor.
Use: Cotton defoliant.

n-(tributylplumbyl)benzimidazole.
CAS: 23188-89-8. $C_{19}H_{32}N_2Pb$.
Hazard: A poison.

tributyl(8-quinolinolato)tin.
CAS: 5488-45-9. $C_{21}H_{33}NOSn$.
Hazard: A poison.

tributylstannanecarbonitrile.
CAS: 2179-92-2. $C_{13}H_{27}NSn$.
Hazard: A poison.

tri-*n*-butylstannane hydride.
CAS: 688-73-3. $C_{12}H_{28}Sn$.
Properties: A liquid. D: 1.103 @ 20C, bp: 112.5–113.5C @ 8 mm.
Hazard: Moderately toxic by inhalation.

tributylstannic hydride. See tri-*n*-butylstannane hydride.

tributyl thiophosphate. See tributyl phosphorothioate.

tributyltin-*p*-acetamidobenzoate.
CAS: 2857-03-6. $C_{21}H_{35}NO_3Sn$.
Hazard: A poison.

tributyltin acetate.
CAS: 56-36-0. $(C_4H_9)_3SnOOCCH_3$.
Properties: White, crystalline solid.
Derivation: Reaction of sodium acetate with tributyltin chloride.
Hazard: Toxic material.
Use: Fungicide and bactericide.

tributyltin benzoate. See benzoyloxytributylstannane.

tri-*n*-butyltin bromide.
CAS: 1461-23-0. $C_{12}H_{27}Sn \cdot Br$.
Properties: Liquid. D: 1.3365, bp: 163C/12 mm, refr index: 1.5000.
Hazard: A poison by ingestion. Moderately toxic by skin contact and inhalation.

tributyltin chloride. (tri-*n*-butylchlorostannate). $(C_4H_9)_3SnCl$.
Properties: Colorless liquid. D 1.20 (20/4C); refr index 1.4903 (25C); bp 145–147C (5 mm Hg). Soluble in the common organic solvents; including alcohol, heptane, benzene, and toluene; insoluble in cold water, but hydrolyzes in hot water.
Derivation: Reaction of tetrabutyltin with dibutyltin chloride.
Hazard: Toxic material.
Use: Rodenticide, intermediate, rodent-repellent cable coatings.

tributyltin chloroacetate.
CAS: 5847-52-9. $C_{14}H_{29}ClO_2Sn$.
Hazard: A poison.

tributyltin-γ-chlorobutyrate.
CAS: 33550-22-0. $C_{16}H_{33}ClO_2Sn$.
Hazard: A poison.

tributyltin cyanate. See cyanatotributylstannane.

tri-*n*-butyltin cyanide. See tributylstannanecarbonitrile.

tributyltin cyclohexanecarboxylate.
CAS: 2669-35-4. $C_{19}H_{38}O_2Sn$.
Hazard: A poison.

tributyltin-s,s'-dibutyldithiocarbamate.
See 2,2-dibutyl-1,3,2-oxathiastannolane.

tributyltin dimethyldithiocarbamate.
CAS: 20369-63-5. $C_{15}H_{33}NS_2Sn$.
Hazard: A poison.

tributyltin-2-ethylhexanoate.
CAS: 5035-67-6. $C_{20}H_{42}O_2Sn$.
Hazard: Moderately toxic by ingestion.

tributyltin hydroxide.
CAS: 1067-97-6. $C_{12}H_{28}OSn$.
Properties: Waxy solid or oil. Mp: 15–16C.
Hazard: Moderately toxic.
Use: Agricultural chemical.

tri-*n*-butyltin iodide. See iodotributylstannane.

tributyltin iodoacetate.
CAS: 73927-91-0. $C_{14}H_{29}IO_2Sn$.
Hazard: A poison.

tributyltin-*o*-iodobenzoate.
CAS: 73927-93-2. $C_{19}H_{31}IO_2Sn$.
Hazard: A poison.

tributyltin-*p*-iodobenzoate.
CAS: 73940-88-2. $C_{19}H_{31}IO_2Sn$.
Hazard: A poison.

tributyltin-β-iodopropionate.
CAS: 73927-95-4. $C_{15}H_{31}IO_2Sn$.
Hazard: A poison.

tributyltin isopropylsuccinate.
CAS: 53404-82-3. $C_{19}H_{38}O_4Sn$.
Hazard: A poison.

tributyltin isothiocyanate. See tributyliso-
cyanatostannane.

tributyltin laurate.
CAS: 3090-36-6. $C_{24}H_{50}O_2Sn$.
Hazard: A poison by ingestion.

tributyl tin linoleate. See (linoleoyloxy)
tributylstannane.

tributyltin methacrylate. See tributyl
(methacryloxy)stannane.

tri-*n*-butyltin methanesulfonate.
CAS: 13302-06-2. $C_{13}H_{30}O_3SSn$.
Hazard: A poison.

tributyltin neodecanoate.
CAS: 28801-69-6. $C_{22}H_{46}O_2Sn$.
Hazard: Moderately toxic by ingestion.

tributyltin nonanoate.
CAS: 4027-14-9. $C_{21}H_{44}O_2Sn$.
Hazard: A poison.

tri-*n*-butyltin oleate. See tributyl(oleoyloxy)
stannane.

tributyltin-*o*-phenylphenoxide. See (2-bip-
henyloxy)tributyltin.

tributyltin salicylate. See salicyloyloxytributylstannane.

tributyltin sulfide. See 1,1,1,3,3,3-hexabuty-
ldistannthiane.

tri-*n*-butyltin undecylate. See tributyl
(undecanoyloxy)stannane.

**tributyltin-α-(2,4,5-trichlorophenoxy)
propionate.**
CAS: 73940-89-3. $C_{21}H_{33}Cl_3O_3Sn$.
Hazard: A poison.

tri-*n*-butyl tricarballylate. $(C_4H_9OCOCH_2)_2$
$CHCOOC_4H_9$.
Properties: Liquid. D 1.004 (24C), refr index 1.4388
(26.5C). Insoluble in water. Combustible.
Use: Plasticizer.

tributyl(2,4,5-trichlorophenoxy)tin.
CAS: 73927-98-7. $C_{18}H_{29}Cl_3OSn$.
Hazard: A poison.

tributyl(undecanoyloxy)stannane.
CAS: 69226-47-7. $C_{23}H_{48}O_2Sn$.
Hazard: A poison by ingestion.

tributyrin. See glyceryl tributyrate.

tricalcium aluminate. See calcium aluminate.

tricalcium citrate. See calcium citrate.

tricalcium orthoarsenate. See calcium arsenate.

tricalcium orthophosphate. See calcium
phosphate, tribasic.

tricalcium phosphate. See calcium phosphate,
tribasic.

tricalcium silicate.
Use: Anticaking agent in foods (up to 2%).
See cement, Portland.

tricamba. (3,5,6-trichloro-*o*-anisic acid).
CAS: 2307-49-5. $C_6HCl_3(COOH)(OCH_3)$.
Properties: Crystals. Mp 137–139C. Very slightly soluble in water; moderately soluble in xylene; freely soluble in alcohol.
Hazard: Toxic by ingestion.
Use: Herbicide.

tricarbimide. See cyanuric acid.

tricarboxylic acid cycle. See TCA cycle.

trichlohexyltin hydroxide. See cyhexatin.

trichlorfon. (*O,O*-dimethyl[2,2,2-tri-chloro-1-hydroxyethyl]phosphonate).
CAS: 52-68-6. $(CH_3O)_2P(O)CH(OH)CCl_3$.
Properties: White, crystalline solid. Mp 83–84C, bp 100C (1 mm Hg), d 1.73 (20/4C). Soluble in water, benzene, chloroform, ether; insoluble in oils.
Hazard: Cholinesterase inhibitor, absorbed by skin, use may be restricted.
Use: Systemic insecticide, medicine (anthelmintic).

trichloroacetaldehyde. See chloral.

trichloroacetaldehyde, hydrated. See chloral hydrate.

trichloroacetic acid. (TCA).
CAS: 76-03-9. CCl_3COOH.
Properties: Deliquescent, colorless crystals; sharp, pungent odor. D 1.6298, mp 57.5C, bp 197.5C, flash p none. Soluble in water, alcohol, and ether. Nonflammable.
Derivation: (1) By treating chloral hydrate with fuming nitric acid. (2) From glacial acetic acid by the action of chlorine in the presence of sunlight, UV radiation, or catalysts.
Grade: Technical, CP, USP.
Hazard: Toxic by ingestion and inhalation, strong irritant to skin, tissue, eyes, and upper respiratory tract. Possible carcinogen.
Use: Organic synthesis, reagent for detection of albumin, medicine, pharmacy, herbicides.

trichloroacetic acid tripropylstannyl ester.
See tripropyltin trichloroacetate.

trichloroacetonitrile.
CAS: 545-06-2. CCl_3CN.
Properties: Colorless liquid. Bp 85C, d 1.44, refr index 1.440.
Derivation: Reaction of methylnitrile, hydrochloric acid, and chlorine.
Hazard: Strong irritant to tissue. Questionable carcinogen.
Use: Insecticide.

trichloroacetyl chloride. CCl_3COCl.
Properties: Liquid. D 1.654 (0/4C), bp 118C. Decomposes in water; soluble in alcohol.
Hazard: Toxic by ingestion and inhalation, strong irritant to skin and tissue.

***S*-2,3,3-trichloroallyl-*N,N*-diisopropylthiol-carbamate.**
CAS: 2303-17-5. $[(CH_3)_2CH]_2NC(O)SCH_2$ $CCl:CCl_2$.
Properties: Oily liquid. Bp 148–149C (9 mm Hg), mp 29–30C. Practically insoluble in water; soluble in alcohol, acetone, ether, and heptane. Combustible.
Use: Herbicide.

trichloro-*o*-anisic acid. See tricamba.

2,4,6-trichloroanisole. $C_7H_5Cl_3O$.
Properties: Acicular crystals. Mp 60C, bp 240C (238 mm Hg). Insoluble in water; soluble in benzene, methanol, and dioxane; gradually sublimes at room temperature.
Use: Dyeing auxiliary for polyester fabrics.

1,2,3-trichlorobenzene.
CAS: 87-61-6. $C_6H_3Cl_3$.
Properties: White crystals. D (solid) 1.69, refr index 1.5776 (19C), bp 221C, mp 52.6C, flash p 235F (112.7C) (CC). Insoluble in water; slightly soluble in alcohol; soluble in ether. Combustible.
Hazard: Toxic by ingestion and inhalation.
Use: Organic intermediate.

1,2,4-trichlorobenzene.
CAS: 120-82-1. $C_6H_3Cl_3$.
Properties: Colorless, stable liquid; odor similar to that of *o*-dichlorobenzene. D 1.4634 (25C), bp 213C, mp 17C, flash p 210F (98.9C). Miscible with most organic solvents and oils; insoluble in water. Combustible.
Derivation: Chlorination of monochlorobenzene.
Grade: Technical, 99%, mixture of 1,2,4- and 1,2,3-isomers distilling at 213–219C.
Hazard: Toxic by ingestion and inhalation. Eye and upper respiratory tract irritant.
Use: Solvent in chemical manufacturing, dyes and intermediates, dielectric fluid, synthetic transformer oils, lubricants, heat-transfer medium, insecticides.

2,3,6-trichlorobenzoic acid. (benzabar; benzac; fen-all; HC 1281; NCI-C60242; T-2; 2,3,6-TBA; 2,3,6-TCB; 2,3,6-TCBA; tribac; 2,3,6-trichlorobenzoesaeure; trichlorobenzoic acid; tryben; trysben 200; zobar).
CAS: 50-31-7. $C_7H_3Cl_3O_2$.
A substituted benzoic acid that is a potent synthetic auxin.
Properties: A solid.
Hazard: Poison; moderately toxic.
Use: An all-purpose, non-selective, pre-emergent and early post-emergent herbicide.

2,3,6-trichlorobenzyloxypropanol.
$(Cl)_3C_6H_2CH_2OCH_2CH(CH_3)OH$.
Properties: Liquid. Bp 121–124C (0.1 mm Hg). Almost insoluble in water; soluble in most organic solvents. Combustible.
Hazard: Toxic by ingestion and inhalation.
Use: Herbicide.

((2,3,6-trichlorobenzyl)oxy)-2-propanol.
CAS: 34314-31-3. $C_{10}H_{11}C_{13}O_2$.
Hazard: Moderately toxic by ingestion.
Use: Agricultural chemical.

1,1,1-trichloro-2,2-bis(chlorophenyl)ethane.
See DDT.

1,1,1-trichloro-2,2-bis(_p_-methoxyphenyl)ethane.   See methoxychlor.

B-trichloroborazole.

BClNHBClNHBClNH.
Properties: White, crystalline solid. Mp 84.5–85.5C, bp 96.5–98C (37 mm Hg). Soluble in many organic solvents; highly reactive.
Use: Intermediate, gelling agent, catalyst complexing agent.

trichlorobromomethane.   See bromotrichloromethane.

trichlorobutylene oxide.   (TCBO).

$$Cl-\underset{\underset{Cl}{|}}{\overset{\overset{Cl}{|}}{C}}-CH_2-CH-CH$$

A reactive liquid epoxide used as an organic solvent and surfactant intermediate; its polymers can be used for polyester, polyurethane, and polyacrylic resins, polyether polyols, flame-retardants, etc.

3,4,4′-trichlorocarbanilide.
$C_6H_3Cl_2NHCONHC_6H_4Cl$.
Properties: Heat-resistant white powder. Mp 250C.
Use: Bacteriostat in soaps and detergents, plastics.

1,2,4-trichloro-5-(2,6-dichlorophenoxy) benzene.
CAS: 130892-67-0. $C_{12}H_5Cl_5O$.
Hazard: A reproductive hazard.

trichloroethane.   See "Extrema" _[JJISCO]_.

1,1,1-trichloroethane.   (methyl chloroform).
CAS: 71-55-6. CH_3CCl_3.
Properties: Colorless liquid. D 1.325, bp 75C, fp −38C, flash p none. Insoluble in water; soluble in alcohol and ether. Nonflammable.

Hazard: Irritant to eyes and tissue. Central nervous system impairment and liver damage. Questionable carcinogen.
Use: Solvent for cleaning precision instruments, metal degreasing, pesticide, textile processing.
See Aerothenet TT; "Tri-Ethane" _[Axiall]_.

1,1,2-trichloroethane.   (vinyl trichloride; β-trichloroethane).
CAS: 79-00-5. $CHCl_2CH_2Cl$.
Properties: Clear, colorless liquid; characteristic sweet odor. Bp 113.7C, d 1.4432 (20C/4C), refr index 1.4458, vap press 16.7 mm Hg (20C), bulk d 12.0 lb/gal (20C), fp −36.4C, flash p none. Miscible with alcohols, ethers, esters, and ketones; insoluble in water. Nonflammable.
Grade: Technical.
Hazard: Irritant, absorbed by skin. Central nervous system impairment and liver damage. Questionable carcinogen.
Use: Solvent for fats, oils, waxes, resins, other products; organic synthesis.

trichloroethanol.
CAS: 115-20-8. CCl_3CH_2OH.
Properties: Viscous liquid; ether-like odor; hygroscopic. Bp 150C, fp 13C, d 1.541 (25/4C). Slightly soluble in water; miscible with alcohol, ether, and carbon tetrachloride. Combustible.
Use: Intermediate, anesthetic.

trichloroethylene.   (tri).
CAS: 79-01-6. $CHCl:CCl_2$.
Properties: Stable, low-boiling, colorless, photoreactive liquid; chloroform-like odor. Will not attack the common metals even in the presence of moisture. Bp 86.7C, fp −73C, d 1.456–1.462 (25/25C), refr index 1.4735 (27C). Miscible with common organic solvents; slightly soluble in water. Nonflammable.
Derivation: From tetrachloroethane by treatment with lime or alkali in the presence of water, or by thermal decomposition, followed by steam distillation.
Grade: USP, technical, high purity, electronic, metal degreasing, extraction.
Hazard: Toxic by inhalation. Use as solvent not permitted in some states. FDA has prohibited its use in foods, drugs, and cosmetics. Central nervous system impairment, cognitive decrements and renal toxicity. Confirmed carcinogen.
Use: Metal degreasing; extraction solvent for oils, fats, waxes; solvent dyeing; dry-cleaning; refrigerant and heat-exchange liquid; fumigant; cleaning and drying electronic parts; diluent in paints and adhesives; textile processing; chemical intermediate; aerospace operations (flushing liquid oxygen).

trichloroethylene epoxide.   See epoxy-1,1,2-trichloroethane.

trichloroethyltin.   See ethyltin trichloride.

trichlorofluoroethylene.
CAS: 359-29-5. C_2Cl_3F.
Hazard: Moderately toxic by ingestion and inhalation.

trichlorofluoromethane. (fluorotrichloromethane; fluorocarbon-11).
CAS: 75-69-4. CCl_3F.
Properties: Colorless, volatile liquid; nearly odorless. Bp 23.7C, fp –111C, d 1.494 (17.2C), critical press 43.2 atm. Noncombustible.
Derivation: From carbon tetrachloride and hafnium, in the presence of fluorinating agents such as antimony tri- and pentafluoride.
Grade: Technical, 99.9% min.
Hazard: Cardiac sensitization. Questionable carcinogen.
Use: Solvent, fire extinguishers, chemical intermediate, blowing agent.

3,5,6-trichloro-2-hydroxypyridine. See 2-hydroxy-3,5,6-trichloropyridine.

trichloroisocyanuric acid. (1,3,5-trichloro-*s*-triazine-2,4,6-trione).
CAS: 87-90-1.

OCNClCONClCONCl.
Properties: White, slightly hygroscopic, crystalline powder or granules. Loose bulk d 31 lb/cu ft, granular 60 lb/cu ft; available chlorine 85%; decomposes at 225C.
Hazard: Fire risk in contact with organic materials, strong oxidizing agent. Toxic by ingestion.
Use: Active ingredient in household dry bleaches, dishwashing compounds, scouring powders, detergent sanitizers, commercial laundry bleaches, swimming-pool disinfectant, bactericide, algicide, bleach, and deodorant.

trichloroisocyanuric acid-potassium dichloroisocyanurate (1:4). See monotrichlorotetra (monopotassium dichloro)-penta-s.

trichloromelamine. (*N*,*N'*,*N'*-trichloro-2,4,6-triamine-1,3,5-triazine).

NC(NHCl)NC(NHCl)NC(NHCl).
Properties: Fine, white powder. Autoign temp 320F (160C). Slightly soluble in water and glacial acetic acid; insoluble in carbon tetrachloride and benzene; pH of saturated aqueous solution 4.
Derivation: By chlorination of melamine.
Grade: 89% available chlorine.
Hazard: Dangerous fire risk, can ignite spontaneously in contact with reactive organic materials.
Use: Chlorine bleach and bactericide.

trichloromethane. See chloroform.

α-(trichloromethyl)benzyl acetate. See trichloromethylphenylcarbinol acetate.

trichloromethyl chloroformate. (diphosgene). $ClCOOCCl_3$.
Properties: Colorless liquid; odor similar to phosgene (newly mown hay). D 1.65 (15C), bp 127–128C, fp –57C, vap d 6.9 (air = 1), refr index 1.45664 (22C). Decomposed by heat, porous substances, activated carbons (with evolution of phosgene), also by alkalies, hot water. Soluble in alcohol, benzene, and ether. Noncombustible.
Derivation: (1) By chlorinating methyl formate, (2) by chlorinating methyl chloroformate. In both methods the mixture of chloro-derivatives is then separated by fractionation.
Grade: Technical.
Hazard: Toxic by inhalation and ingestion, strong irritant to tissue.
Use: Organic synthesis, military poison gas.

trichloromethyl ether. $CHCl_2OCH_2Cl$.
Properties: Liquid; pungent odor. D 1.5066 (10C), bp 130–132C. Soluble in alcohol, benzene, and ether; insoluble in water.
Hazard: Strong irritant to eyes and skin, evolves lachrymatory fumes.

N-(trichloromethylmercapto)-tetrahydrophthalimide. See captan.

trichloromethylphenylcarbinyl acetate. (α-[trichloromethyl]benzyl acetate).
CAS: 90-17-5. $C_6H_5CH(CCl_3)OOCCH_3$.
Properties: White, crystalline solid; intense rose odor. Mp 86–88C. Soluble in 18 parts of 95% alcohol.
Use: Perfumes, fixative for essential oils and perfumes.

trichloromethylphosphonic acid. CCl_3PO (OH)$_2$. Strong dibasic acid.
Properties: Soluble in water and alcohol; insoluble in benzene and hexane.
Use: Catalyst and condensation agent.

1,1,1-trichloro-2-methyl-2-propanol. See chlorobutanol.

trichloromethylsulfenyl chloride. (perchloromethyl mercaptan).
CAS: 594-42-3. $ClSCCl_3$.
Properties: Yellow, oily liquid; disagreeable odor. Mildly decomposed by moist air, subject to the action of oxidizing agents, reducing agents, chlorine, etc. D 1.722 (0C), bp 148–149C (decomposes), vap d 6.414, volatility 18,000 mg/cu m (20C). Insoluble in water. Nonflammable but supports combustion.
Derivation: Chlorination of carbon disulfide, thiophosgene, or methyl thiocyanate.
Grade: Technical.

Hazard: Toxic by ingestion and inhalation, strong irritant to eyes, skin, and upper respiratory tract irritant.

Use: Organic synthesis, dye intermediate, fumigant.

3-trichloromethylthiobenzothiazolone.
CAS: 3567-79-1. $C_8H_4Cl_3NOS_2$.
Hazard: Moderately toxic by ingestion.
Use: Agricultural chemical.

trichloronaphthalene.   (halowax).
CAS: 1321-65-9. $C_{10}H_5Cl_3$.
Properties: A white solid. Mw 231.50.
Hazard: A poison. Liver damage and chloracne.
Use: Wire coating, electrical insulations.

trichloronitromethane.   See chloropicrin.

trichloronitrosomethane. CCl_3NO.
Properties: Dark-blue liquid; unpleasant odor. D 1.5 (20C); bp 5C (70 mm Hg). Slowly decomposes, but is more stable in solution. Soluble in alcohol, benzene, ether; insoluble in water.
Derivation: Interaction of sulfuric acid, sodium trichloromethylsulfinate, and sodium nitrate.
Grade: Technical.
Hazard: Strong irritant to eyes and tissue.
Use: Organic synthesis, military poison gas (lachrymator).

trichlorononylsilane.   See nonyl trichlorosilane.

trichlorooctadecylsilane.   See octadecyltrichlorosilane.

trichlorooctylsilane.   See octyl trichlorosilane.

2,4,5-trichlorophenol.
CAS: 95-95-4. $C_6H_2Cl_3OH$.
Properties: Gray flakes in sublimed mass; strong phenolic odor. D 1.678 (25/4C), bp 252C, mp 68–70C, no flash p. Soluble in alcohol, ether, and acetone. Nonflammable.
Hazard: May cause skin irritation.
Use: Fungicide, bactericide.

2,4,6-trichlorophenol.   (2,4,6-T).
CAS: 88-06-2. $C_6H_2Cl_3OH$.
Properties: Yellow flakes; strong phenolic odor. D 1.675 (25/4C), fp 61C, bp 248–249C. Soluble in acetone, alcohol, and ether. Nonflammable.
Hazard: May cause skin irritation.
Use: Fungicide, herbicide, defoliant.

3,4,5-trichlorophenol.
CAS: 609-19-8. $C_6H_3Cl_3O$.
Hazard: A poison.

2,4,5-trichlorophenoxyacetic acid.   (2,4,5-T).
CAS: 93-76-5. $C_6H_2Cl_3OCH_2CO_2H$.

Properties: Light-tan solid. Mp 151–153C. Soluble in alcohol; insoluble in water; available as sodium and amine salts.
Hazard: Use has been restricted. Peripheral nervous system impairment. Questionable carcinogen.
Use: Plant hormone, herbicide, defoliant.
See dioxin.

2-(2,4,5-trichlorophenoxy)ethyl-2,2-dichloropropionate.   See erbon.

2-(2,4,5-trichlorophenoxy)propionic acid.
See silvex.

2,4,6-trichlorophenyl acetate.
CAS: 23399-90-8. $C_6H_2Cl_3OOCCH_3$.
Use: Fungicide, especially on cotton seed.

1-((2,3,6-trichlorophenyl)methoxy)-2-propanol.
CAS: 1861-44-5. $C_{10}H_{11}Cl_3O_2$.
Hazard: A mild eye irritant.

1,2,3-trichloropropane.
CAS: 96-18-4. $CH_2ClCHClCH_2Cl$.
Properties: Colorless liquid. D 1.3888 (20/4C), fp −15C, bp 156.17C, refr index 1.4822 (20C), flash p 180F (82.2C) (COC). Slightly soluble in water; dissolves oils, fats, waxes, chlorinated rubber, and numerous resins. autoign temp 580F (304C). Combustible.
Derivation: Chlorination of propylene.
Hazard: Toxic by inhalation and skin absorption; strong irritant to eyes and upper respiratory tract. Liver and kidney damage. Probable carcinogen.
Use: Paint and varnish remover, solvent, degreasing agent.

tri(chloropropyl) phosphate.
CAS: 26248-87-3. $C_9H_{18}Cl_3O_4P$.
Hazard: Moderately toxic by ingestion. A reproductive hazard.
See "Antiblaze" [Hickory].

trichlorosilane.   (1) (silicochloroform).
CAS: 10025-78-2. $SiHCl_3$.
Properties: Colorless volatile liquid. D 1.336, fp −127C, bp 32C, refr index 1.3990, flash p 7F (−13.9C). Soluble in benzene, ether, heptane, perchloroethylene; decomposed by water. Purity of 99.9999% is commercially attainable.
Hazard: Flammable, dangerous fire risk.
Use: Intermediate, purification of silicon.
(2) Generic name for compounds of the formula $RSiCl_3$ of which methyl trichlorosilane, CH_3SiCl_3 is most important.

N,N′,N″-trichloro-2,4,6-triamine-1,3,5-triazine.   See trichloromelamine.

2,4,6-trichloro-1,3,5-triazine.   See cyanuric chloride.

1,3,5-trichloro-*s*-triazine-2,4,6-trione.   See trichloroisocyanuric acid.

trichloromethyltin.   See methyltrichlorostannane.

trichlorotrifluoroacetone.   (1,1,3-trichloro-1,3,3-trifluoroacetone). $CCl_2FCOCClF_2$.
Properties: Colorless liquid. Bp 84.5C, fp −78C. Miscible with water and most organic solvents; stable to acid but not alkalies. Nonflammable.
Hazard: Strong irritant to eyes.
Use: Solvent in acid media, complexing agent.

1,1,2-trichloro-1,2,2-trifluoroethane.   (trifluorotrichloroethane).
CAS: 76-13-1. CCl_2FCClF_2.
Properties: Colorless, volatile liquid; nearly odorless. Bp 47.6C, fp −35C, critical press 33.7 atm, d 1.42 (25C). Noncombustible.
Derivation: From perchloroethylene and hafnium.
Grade: Technical, spectrophotometric.
Hazard: Central nervous system impairment. Questionable carcinogen.
Use: Dry-cleaning solvent, fire extinguishers, to make chlorotrifluoroethylene, blowing agent, polymer intermediate, solvent drying, drying electronic parts and precision equipment.

trichocyst.   Organelle in ciliates and dinoflagellates which releases long filamentous proteins when the cell is disturbed. Used as a defense against would-be predators.

tricholine citrate.   (tris[2-hydroxyethyl]trimethylammonium citrate). $[(CH_3)_3NCH_2CH_2OH]_3 \cdot C_6H_5O_7$.
Use: Medicine, nutrition.

trichotecene.   See mycotoxin.

tricobalt tetraoxide.   See cobalto-cobaltic oxide.

tricosane. $CH_3(CH_2)_{21}CH_3$.
Properties: Glittering leaflets. D 0.779 (48C), bp 234C (15 mm Hg), mp 48C. Soluble in alcohol; insoluble in water. Combustible.
Grade: Technical.
Use: Organic synthesis.

***n*-tricosanoic acid.** $CH_3(CH_2)_{21}COOH$. A saturated fatty acid not normally found in natural fats or oils.
Properties: Synthetic compound is a white crystalline solid. Mp 79.1C.
Use: Purified product is used in medical research and as reference standard for gas chromatography.

tri-*m,p*-cresyl borate. $(CH_3C_6H_4)_3BO_3$.
Properties: Light-amber liquid. D 1.065 (25C), bp 385–395C, refr index 1.5480 (24C), flash p 240F (115.5C) (COC). Miscible with acetone, benzene, chloroform; hydrolyzes on contact with water. Combustible.
Use: Plasticizer, organic synthesis.

tricresyl phosphate.   (tritolyl phosphate; TCP). CAS: 78-30-8. $(CH_3C_6H_4O)_3PO$. A mixture of isomers.
Properties: Practically colorless liquid; odorless. Stable, nonvolatile. Bp 420C, refr index 1.556 (25C), d 1.162 (25/25C), bulk d 9.7 lb/gal, crystallizing p −35C, flash p 437F (225C), autoign temp 770F (410C). Miscible with all the common solvents and thinners, also with vegetable oils; insoluble in water. Combustible.
Derivation: From cresol and phosphorus oxychloride.
Hazard: Toxic by ingestion and skin absorption. The *o*- isomer is highly toxic. TLV: 0.1 mg/m^3 (skin); not classifiable as a Human Carcinogen.
Use: Plasticizer for polyvinyl chloride, polystyrene, nitrocellulose; fire-retardant for plastics; air-filter medium; solvent mixtures; waterproofing; additive to extreme pressure lubricants; hydraulic fluid; heat-exchange medium.

tri-*o*-cresylphosphate.   See tricresyl phosphate.

tricresyl phosphite. $(CH_3C_6H_4O)_3P$.
Properties: Colorless liquid; slight phenolic odor. Bp 191C (0.11 mm Hg), d 1.115 (20/4C), flash p 440F (226.6C) (OC). Insoluble in water; miscible with acetone, alcohol, benzene, ether, and kerosene. Combustible.
Grade: Technical.
Use: Stabilizer and plasticizer for plastics and resins.

tricyanic acid.   See cyanuric acid.

tricyclic.   An organic compound composed of only three-ring structures, which may be identical or different, e.g., anthracene.

***sym*-tricyclodecane.**   See adamantane.

tricyclodecanedimethanol.
CAS: 26160-83-8. $C_{12}H_{20}O_2$.
Hazard: Moderately toxic by ingestion.

tricyclohexyl borate.   See boric acid ester.

1-(tricyclohexylstannyl)-1h-1,2,4-triazole.
See (1h-1,2,4-triazolyl-1-yl)tricyclohexylstannane.

***n*-tridecane.** $CH_3(CH_2)_{11}CH_3$.
Properties: Colorless liquid. D 0.755 (20/4C), bp 225.5C, fp −5.45C, refr index 1.4250 (20C), flash p 175F (79.4C). Soluble in alcohol; insoluble in water. Combustible.
Grade: 95%, 99%, research.
Use: Organic synthesis, distillation chaser.

n-**tridecanoic acid.** (tridecylic acid; tridecoic acid). $CH_3(CH_2)_{11}COOH$. A saturated fatty acid usually prepared synthetically.
Properties: Colorless crystals. Mp 44.5C, d 0.8458 (80/4C), bp 312.4C, 192.2C (16 mm Hg), refr index 1.4328 (50C). Slightly soluble in water; soluble in alcohol and ether. Combustible.
Grade: 99% pure.
Use: Organic synthesis, medical research.

tridecanol. See tridecyl alcohol.

2-tridecenal.
CAS: 7774-82-5. $C_{13}H_{24}O$.
Properties: White to yellow liquid; oily, citrus odor. D: 0.842–0.862, refr index: 1.457. Soluble in alc, fixed oils; insoluble in water.
Hazard: Low toxicity by ingestion and skin contact.
Use: Food additive.

tridecoic acid. See *n*-tridecanoic acid.

tridecyl alcohol. (tridecanol). $C_{12}H_{25}CH_2OH$. A commercial mixture of isomers.
Properties: Low-melting white solid; pleasant odor. Bp 274C, mp 31C, d 0.845 (20/20C), bulk d 7.0 lb/gal, flash p 180F (82.2C) (TOC). Combustible.
Derivation: Oxo process from C_{15} hydrocarbons.
Grade: Technical.
Use: Esters for synthetic lubricants, detergents, antifoam agent, other tridecyl compounds, perfumery.

tridecylbenzene. (1-phenyltridecane). $C_6H_5(CH_2)_{12}CH_3$.
Properties: Colorless liquid. D 0.85–0.86 (60/60F), refr index 1.4815–1.4830. Combustible.
Use: Detergent intermediate.

tridecylic acid. See *n*-tridecanoic acid.

tri(decyl)orthoformate. $CH(OC_{10}H_{21})_3$.
Properties: Liquid. Bp 194C, fp −15 to −20C, refr index 1.448. Insoluble in water; soluble in benzene, naphtha, ether, and alcohol.
Use: To remove small quantities of water from ethers or other solvents in which acid catalysts can be employed.

tri(decyl) phosphite. $(C_{10}H_{21}O)_3P$.
Properties: Water-white liquid; decyl alcohol odor. D 0.892 (25/15.5C), mp below 0C, refr index 1.4565 (25C), flash p 455F (235C). Combustible.
Use: Chemical intermediate, stabilizer for polyvinyl and polyolefin resins.

2,4,6-tri(dimethylaminomethyl)phenol.
$[(CH_3)_2NCH_2]_3C_6H_2OH$.
Properties: Liquid. Refr index 1.5181. Combustible.
Use: Antioxidants, acid neutralizers, stabilizers, and catalysts for epoxy and polyurethane resins.

tri(dimethylphenyl)phosphite. (trixylenyl phosphate). $[(CH_3)_2C_6H_3O]_3PO$.
Properties: Liquid. D 1.155, refr index 1.5535, bp 243–265C (10 mm Hg), flash p 450F (232C). Solubility in water 0.002% (85C) by weight. Combustible.
Use: Plasticizer.

tridodecyl amine. See trilauryl amine.

tridodecyl borate. See boric acid ester.

tridymite. (christensenite; crystalline silica).
CAS: 15468-32-3. SiO_2.
A vitreous, colorless, or white native form of pure silica, found variously but not so commonly as quartz. Quartz will change into tridymite with a 16.2% increase in volume at 870C. Unlike quartz, it is soluble in boiling sodium carbonate solution; d 2.28–2.3; Mohs hardness 7.

trietazine. (generic name for 2-chloro-4-diethylamino-6-ethylamino-*s*-triazine).
CAS: 1912-26-1. $ClC_3N_3[N(C_2H_5)_2]NHC_2H_5$.
Properties: Solid. Practically insoluble in water; partially soluble in benzene and chloroform.
Use: Herbicide, plant growth regulator.

"Tri-Ethane" *[Axiall]*.
CAS: 71-55-6.
TM for 1,1,1-trichloroethane solvent.
Use: Cold cleaning, vapor degreasing, resins application, dry-film photoresist processing, adhesive solvent, in aerosols as solvent and vapor-pressure depressant.

triethanolamine. (TEA; tri[2-hydroxyethyl] amine).
CAS: 102-71-6. $(HOCH_2CH_2)_3N$.
Properties: Colorless, viscous, hygroscopic liquid; slight ammoniacal odor. Mp 21.2C, bp 335C (decomposes), vap press below 0.01 mm Hg (20C), d 1.126, flash p 375F (190.5C) (OC), bulk d 9.4 lb/gal. Miscible with water, alcohol; soluble in chloroform; slightly soluble in benzene and ether; slightly less alkaline than ammonia, commercial product contains up to 25% diethanolamine and up to 5% monoethanolamine. Combustible.
Derivation: Reaction of ethylene oxide and ammonia.
Grade: Technical, regular, 98%, USP.
Hazard: Eye and skin irritant. Questionable carcinogen.
Use: Fatty acid soaps used in dry-cleaning, cosmetics, household detergents, and emulsions. Wool scouring, textile antifume agent and water-repellent, dispersion agent, corrosion inhibitor, softening agent, emulsifier, humectant and plasticizer, chelating agent, rubber accelerator, pharmaceutical alkalizing agent.

triethanolamine dodecylbenzene sulfonate.
CAS: 27323-41-7. $C_{18}H_{20}O_3S \cdot C_6H_{15}NO_3$.
Hazard: Low toxicity by ingestion and skin contact.
Use: Food additive.

triethanolamine lauryl sulfate.
CAS: 139-96-8. $(HOC_2H_4)_3NOS(O)_2OC_{12}H_{25}$.
A liquid or paste.
Use: Detergent; wetting, foaming, and dispersing agent for industrial, cosmetic and pharmaceutical applications, especially shampoos.

triethanolamine methanearsonate.
CAS: 5902-97-6. $C_6H_{15}NO_3 \cdot xCH_5AsO_3$.
Hazard: Low toxicity by ingestion.
Use: Agricultural chemical.

triethanolamine oleate.   See trihydroxyethylamine oleate.

triethanolamine stearate.   See trihydroxyethylamine stearate.

triethanolamine titanate.   See titanium chelate.

1,1,3-triethoxybutane.
CAS: 5870-82-6. $C_{10}H_{22}O_3$.
Hazard: A poison by ingestion and skin contact. Low toxicity by inhalation. A mild skin irritant.

1,1,3-triethoxyhexane.
$CH(OC_2H_5)_2CH_2CH(OC_2H_5)C_3H_7$.
Properties: Colorless liquid. D 0.8746 (20/20C), bp 133C (50 mm Hg), fp −100C, bulk d 7.3 lb/gal, flash p 210F (98.9C). Insoluble in water. Combustible.
Use: Synthesis of aldehydes, acids, esters, chlorides, amines, etc.

triethoxymethane.   See triethyl-*o*-formate.

1,1,3-triethoxy-3-methoxypropane.   (triethylmethyl malonaldehyde diacetal). (CH_3O) $(C_2H_5O)CHCH_2CH(OC_2H_5)_2$.
Properties: Colorless liquid. D 0.9300 (25/4C), bp 86C (6 mm Hg). Combustible.
Grade: 99%.
Use: Intermediate, cross-linking, and insolubilizing agent.

triethylaconitate.
$C_2H_5OOCCHC(COOC_2H_5)CH_2COOC_2H_5$.
Properties: Liquid. D 1.096 (25C), refr index 1.4517 (26C), bp 154–156 (5 mm Hg). Combustible.
Use: Plasticizer.

triethylaluminum.   (ATE; TEA; aluminum triethyl).
CAS: 97-93-8. $(C_2H_5)_3Al$.

Properties: Colorless liquid. D 0.837, fp −52.5C, bp 194C, sp heat 0.527 (33C), flash p −63F (−53C). Miscible with saturated hydrocarbons.
Derivation: By introduction of ethylene and hydrogen into an autoclave containing aluminum. The reaction proceeds at moderate temperature and varying pressures.
Grade: 88–94%.
Hazard: Flammable, dangerous fire risk, ignites spontaneously in air. Reacts violently with water, acids, alcohols, halogens, and amines. Destructive to tissue.
Use: Catalyst intermediate for polymerization of olefins, especially ethylene; pyrophoric fuels; production of α-olefins and long-chain alcohols; gas plating of aluminum.

triethylamine.
CAS: 121-44-8. $(C_2H_5)_3N$.
Properties: Colorless liquid; strong ammoniacal odor. Bp 89.7C, fp −115.3C, d 0.7293 (20/20C), bulk d 6.1 lb/gal, flash p 10F (−6.67C) (OC). Soluble in water and alcohol.
Derivation: From ethyl chloride and ammonia with heat and pressure.
Hazard: Flammable, dangerous fire risk, explosive limits in air 1.2–8.0%. Toxic by ingestion and inhalation, strong irritant to tissue and upper respiratory tract. Visual impairment. Questionable carcinogen.
Use: Catalytic solvent in chemical synthesis; accelerator activators for rubber; wetting, penetrating, and waterproofing agents of quaternary ammonium types; curing and hardening of polymers (e.g., core-binding resins); corrosion inhibitor; propellant.

triethyl(3-aminopropyl)silane.   See (3-aminopropyl)triethylsilane.

triethylborane.   (triethylborine; boron triethyl). $(C_2H_5)_3B$.
Properties: Colorless liquid. D (25C), flash p −32F (35.5C), fp −93C, bp 95C, refr index 1.3971, heat of combustion 20,000 Btu/lb. Miscible with most organic solvents; immiscible with water.
Derivation: Reaction of triethylaluminum and boron halide or diborane and ethylene.
Hazard: Flammable, dangerous fire risk, ignites spontaneously in air. Reacts violently with water and oxidizing materials. Toxic by inhalation, strong irritant.
Use: Igniter or fuel for jet and rocket engines, fuel additive, olefin polymerization catalyst, intermediate.

triethyl borate.   (ethyl borate). $(C_2H_5)_3BO_3$.
Properties: Colorless liquid; mild odor. Hydrolyzes rapidly depositing boric acid in finely divided crystalline form. Bp 120C, d 0.863–0.864 (20/20C), flash p 51.8F (11C) (CC), bulk d 7.20 lb/gal (20C), refr index 1.37311 (20C).
Hazard: Flammable, dangerous fire risk.
Use: Antiseptics, disinfectants, antiknock agent.

triethylborine.　　See triethylborane.

triethyl citrate.　　(ethyl citrate).
CAS: 77-93-0.　$C_3H_5(COOC_2H_5)_3$.
Properties: Colorless, mobile liquid; bitter taste. Bp 294C, bp 126–127C (1 mm Hg), d 1.136 (25C), pour p −46C, flash p 303F (150.5C) (COC). Solubility in water 6.5 g/100 cc; solubility in oil 0.8 g/100 cc. Combustible.
Derivation: Esterification of citric acid.
Grade: Technical, refined, FCC.
Use: Solvent and plasticizer for nitrocellulose and natural resins, softener, paint removers, agglutinant, perfume base, food additive (not over 0.25%).
See "TEC" *[Reilly]*.

triethylene diamine.　　(1,4-diazobicyclo[2,2,2] octane).
CAS: 280-57-9.　$N(CH_2CH_2)_3N$.
Properties: Colorless, hygroscopic crystals. Mp 158C, bp 174C.
Hazard: Skin irritant.
Use: Catalyst for polyurethane foams, oxidation and polymerization catalyst, chemical intermediate (metal complexes, quaternary ammonium compounds, etc.), bromine and iodine addition compounds.

triethylene glycol.　　(TEG).
CAS: 112-27-6.　$HO(C_2H_4O)_3H$.
Properties: Colorless, hygroscopic liquid; practically odorless. D 1.1254 (20/20C), bp 287.4C, vap press below 0.01 mm Hg (20C), flash p 350F (176.6C) (CC), bulk d 9.4 lb/gal (20C), fp −7.2C, viscosity 0.478 cP (20C), autoign temp 700F (371C). Soluble in water; immiscible with benzene, toluene, and gasoline. Combustible.
Derivation: From ethylene and oxygen as a by-product of ethylene glycol manufacture.
Grade: Technical, CP.
Use: Solvent and plasticizer in vinyl, polyester, and polyurethane resins; dehydration of natural gas; humectant in printing inks; extraction solvent.

triethylene glycol diacetate.
CAS: 111-21-7.　$CH_3COOCH_2CH_2OCH_2$ $CH_2OCH_2CH_2OOCCH_3$.
Properties: Colorless liquid. D 1.112 (25C), refr index 1.437 (25C), bp 300C, fp below −60C. Combustible.
Use: Plasticizer.

triethylene glycol dibenzoate.
$C_6H_5CO(OCH_2CH_2)_3OOCC_6H_5$.
Properties: Crystals. Bp 210–223C, mp 46C, flash p 457F (236C) (TOC), d 1.168. Combustible.
Use: Plasticizer for vinyl resins, adhesives.

triethylene glycol dicaprylate.　　(triethylene glycol dioctoate).　$C_7H_{15}COO(CH_2CH_2O)_3$ OCC_7H_{15}.
Properties: Clear liquid. D 0.973 (20C), acidity 0.3% max (caprylic), moisture 0.05% max, fp −3C, bp 243C (5 mm Hg), soluble in most organic solvents. Combustible.
Use: Low-temperature plasticizer for elastomers.

triethylene glycol dichloride.　　See triglycol dichloride.

triethylene glycol didecanoate.
$C_9H_{19}COO(C_2H_4O)_3OCC_9H_{19}$.
Properties: Colorless liquid. Bp 237 (2.0 mm Hg), d 0.9584 (20/20C), viscosity 28.6 cP (20C). Combustible.
Use: Plasticizer.

triethylene glycol di(2-ethylbutyrate).
$C_5H_{11}OCOCH_2(CH_2OCH_2)_2CH_2OCOC_5H_{11}$.
Properties: Light-colored liquid. D 0.9946 (20/20C), bulk d 8.3 lb/gal (20C), bp 196C (5 mm Hg), vap press 5.8 mm Hg (200C), viscosity 10.3 cP (20C), flash p 385F (196C). Solubility in water 0.02% by weight (20C). Combustible.
Use: Plasticizer.

triethylene glycol di(2-ethylhexoate).
$C_7H_{15}OCOCH_2(CH_2OCH_2)_2CH_2OCOC_7H_{15}$.
Properties: Light-colored liquid. D 0.9679 (20/20C), bulk d 8.1 lb/gal (20C), bp 219C (5 mm Hg), vap press 1.8 mm Hg (200C), viscosity 15.8 cP (20C), flash p 405F (207C). Insoluble in water. Combustible.
Use: Plasticizer.

triethylene glycol dihydroabietate.
$C_{19}H_{31}COO(C_2H_4O)_3OCC_{19}H_{31}$.
Properties: Liquid. D 1.080–1.090 (25C), refr index 1.5180 (20C), vap press 2.5 (225C), flash p 438F (226C). Insoluble in water. Combustible.
Use: Plasticizer.

triethylene glycol dimethyl ether.　　(triglyme).
$CH_3(OCH_2CH_2)_3OCH_3$.
Properties: Water-white liquid; mild ether odor. D 0.9862 (20/20C), refr index 1.4233 (20C), flash p 232F (111C), bp 216.0C (760 mm Hg), 153.6C (100 mm Hg), fp −46C. Completely soluble in water and hydrocarbons at 20C. May contain peroxides. Combustible.
Use: Solvent for gases, coupling immiscible liquids.

triethylene glycol dioctoate.　　See triethylene glycol dicaprylate.

triethylene glycol dipelargonate.
$C_8H_{17}COO(C_2H_4O)_3OCC_8H_{17}$.
Properties: Clear liquid. D 0.964 (20/20C), bp 251C (5 mm Hg), fp +1 to −4C, refr index 1.4470 (23C),

flash p 410F (210C). Almost insoluble in water; soluble in most organic solvents. Combustible.
Use: Plasticizer.

triethylene glycol dipropionate.
$C_2H_5CO(OCH_2CH_2)_2OOCC_2H_5$.
Properties: Colorless liquid. D 1.066 (25C), refr index 1.436 (25C), bp 138–142C (2 mm Hg), fp below −60C. Solubility in water 6.70% by weight. Combustible.
Use: Plasticizer.

triethylene glycol monobutyl ether.   See butoxytriglycol.

triethylene glycol monohexyl ether.   See 2-(2-(hexyloxy)ethoxy)ethanol.

triethylenemelamine.   (tretamine; TEM; 2,4,6-tris(1-aziridinyl)-s-triazine).
CAS: 51-18-3.

NC[N(CH$_2$)$_2$]NC[N(CH$_2$)$_2$]NC[N(CH$_2$)$_2$].

Properties: White, crystalline, powder; odorless. Mp 160C (polymerizes); polymerizes readily with heat or moisture. Soluble in alcohol, water, methanol, chloroform, and acetone.
Grade: NF.
Hazard: Highly toxic.
Use: Medicine (antineoplastic), insecticide, chemosterilant.

triethylenephosphoramide.   (tepa; tris-(1-aziridinyl)-phosphine oxide; APO).
CAS: 545-55-1.

(NCH$_2$CH$_2$)$_3$PO.

Properties: Colorless crystals. Mp 41C. Soluble in water, alcohol, and ether. Combustible.
Derivation: From ethyleneimine.
Hazard: Highly toxic, strong irritant to skin and tissue.
Use: Medicine (antineoplastic), insect sterilant. Also used with tetrakis(hydroxymethyl)phosphonium chloride (THPC) to form a condensation polymer suitable for flame-proofing cotton.
See tris[1-(2-methyl)aziridinyl]phosphine oxide.

triethylenetetramine.
CAS: 112-24-3. $NH_2(C_2H_4NH)_2C_2H_4NH_2$.
Properties: Moderately viscous, yellowish liquid, less volatile than diethylenetriamine but resembles it in many other properties. Bp 277.5C, d 0.9818 (20/20C), mp 12C, flash p 275F (135C) (CC), bulk d 8.2 lb/gal (20C), autoign temp 640F (337.7C). Soluble in water. Combustible.
Grade: Technical, anhydrous.
Hazard: Strong irritant to tissue, causes skin burns and eye damage.

Use: Detergents and softening agents; synthesis of dyestuffs, pharmaceuticals, and rubber accelerators.

tri(2-ethylhexyl)phosphate.
CAS: 78-42-2. $[C_4H_9CH(C_2H_5)CH_2]_3PO_4$.
Properties: Light-colored liquid. D 0.9260 (20/20C), bulk d 7.70 lb/gal (20C), bp 220C (5 mm Hg), vap press 1.9 mm Hg (200C), viscosity 14.1 cP (20C), pour p −74C, flash p 405F (207C). Insoluble in water. Combustible.
Use: Plasticizer.

tri(2-ethylhexyl) phosphite. $(C_8H_{17}O)_3P$.
Properties: Straw-colored liquid. D 0.897 (25/15C), mp sets to a glass at low temperature, refr index 1.451 (25C), flash p 340F (171C) (COC). Combustible.
Use: Plasticizer, intermediate.

tri(2-ethylhexyl) trimellitate.
$C_6H_3(COOC_8H_{17})_3$.
Properties: Clear liquid; mild odor. D 0.992 (20/20C), distillation range 278–284C (3 mm Hg) (5–95%), fp a gel at −35C, refr index 1.4846 (23C), bulk d 8.26 lb/gal (20C). Combustible.
Use: Plasticizer.

triethylhydroxytin sulfate.   See bis(triethyltin) sulfate.

triethylmethane.   See 3-ethylpentane.

triethylmethyl malonaldehyde diacetal.
See 1,1,3-triethoxy-3-methoxypropane.

triethylorthoformate.   (triethoxymethane).
CAS: 122-51-0. $CH(OC_2H_5)_3$.
Properties: Colorless liquid; pungent odor. Bp 145.9C, refr index 1.39218 (18.8C), d 0.895 (20/20C), flash p 86F (30C) (CC). Soluble in alcohol, ether; decomposes in water.
Derivation: Reaction of sodium ethylate with chloroform or reaction of hydrochloric acid with hydrocyanic acid in ethanol solution.
Hazard: Flammable, moderate fire risk. Toxic.
Use: Organic synthesis, pharmaceuticals.

triethyl phosphate.   (TEP).
CAS: 78-40-0. $(C_2H_5)_3PO_4$.
Properties: Colorless, high-boiling liquid; mild odor. Fp −56.4C, bp 216C, flash p 240F (115.5C), refr index 1.4055 (20C), bulk d 8.90 lb/gal (20C). Very stable at ordinary temperatures, compatible with many gums and resins, soluble in most organic solvents, miscible with water. When mixed with water is quite stable at room temperature, but at elevated temperatures it hydrolyzes slowly. Combustible.
Grade: Technical, 97%.
Hazard: May cause nerve damage but to a lesser extent than other cholinesterase-inhibiting compounds.

Use: Solvent; plasticizer for resins, plastics, gums; manufacture of pesticides; catalyst; lacquer remover.

triethyl phosphite.
CAS: 122-52-1. $(C_2H_5)_3PO_3$.
Properties: Colorless liquid. D 0.9687 (20C), bp 156.6C, refr index 1.413 (25C), flash p 130F (54.4C). Insoluble in water; soluble in alcohol and ether. Combustible.
Hazard: Moderate fire risk.
Use: Synthesis, plasticizers, stabilizers, lubricant and grease additives.

***O,O,O*-triethyl phosphorothioate.**   (triethyl thiophosphate).
CAS: 126-68-1. $(C_2H_5O)_3PS$.
Properties: Colorless liquid; characteristic odor. Bp 93.5–94C (10 mm Hg), d 1.074, flash p 225F (107.2C) (COC). Combustible.
Hazard: Toxic by ingestion, cholinesterase inhibitor.
Use: Plasticizer, lubricant additive, antifoam agent, hydraulic fluid, intermediate.

triethylstannium bromide.   See bromotriethylstannane.

triethyltin acetate.   See acetoxytriethylstannane.

triethyltin bromide.   See bromotriethylstannane.

triethyltin bromide-2-pipecoline.
CAS: 73926-90-6. $C_6H_{15}BrSn•C_6H_{13}N$.
Hazard: A poison.

triethyltin chloride.
CAS: 994-31-0. $C_6H_{15}ClSn$.
Properties: Colorless liquid. D: 1.440 @ 20C/4C, mp: 15.5C, bp: 210C. Insoluble in water; soluble in organic solvents.
Hazard: A poison.

triethyltin phenoxide.
CAS: 1529-30-2. $C_{12}H_{20}OSn$.
Hazard: A poison.

triethyltin sulphate.   See bis(triethyltin) sulfate.

triethyl tricarballylate. $(C_2H_5OCOCH_2)_2$ $CHCOOC_2H_5$.
Properties: Colorless liquid. D 1.087 (20C), refr index 1.4234 (26C), bp 158–160C (5 mm Hg). Solubility in water 0.62% (20C) by weight. Combustible.
Use: Plasticizer.

triethyl(trifluoroacetoxy) stannane.   See trifluoroacetic acid triethylstannyl ester.

trifluoroacetic acid.
CAS: 76-05-1. CF_3COOH.
Properties: Colorless, fuming liquid; hygroscopic; pungent odor. Bp 72.4C, d 1.535, fp −15.25C, index of refr 1.2850 (20C). Very soluble in water. Nonflammable.
Hazard: Irritant to skin.
Use: Strong nonoxidizing acid, laboratory reagent, solvent, catalyst.

trifluoroacetic acid triethylstannyl ester.
CAS: 429-30-1. $C_8H_{15}F_3O_2Sn$.
Hazard: A poison.

trifluoroamine oxide.
CAS: 13847-65-9. (F_3NO).
A perfluorated amine obtained by fluorination of nitrosyl fluoride with UV light or high temperature or pressure. An alternate process is by burning nitric oxide and fluorine, rapidly quenching the gaseous mixture as it leaves the flame zone.

trifluorobromomethane.   (bromotrifluoromethane; Halon 1301).
CAS: 75-63-8. $CBrF_3$.
Properties: Nonflammable colorless gas. Mw 148.92, mp −167.7C, bp −58C, vap d 5 (air = 1) at −58C. Very soluble in chloroform.
Hazard: Central nervous system and cardiac impairment.
Use: Fire-extinguisher agent.

trifluorochloroethylene.   Legal label name (Rail) for chlorotrifluoroethylene.

trifluorochloromethane.   See chlorotrifluoromethane.

1,1,1-trifluoro-2,6-dinitro-*N,N*-dipropyl-*p*-toluidine.   See trifluralin.

trifluoroethene.
CAS: 359-11-5. C_2HF_3.
Hazard: Low toxicity by inhalation.

trifluoromethane.   See fluoroform.

4-trifluoromethoxy-*n*-chlorocarboxyphenylurethan.   See methyl (chlorocarbonyl)(4-(trifluoromethoxy)phenyl)carbamate.

trifluoromethylbenzene.   See benzotrifluoride.

2-trifluoromethyl benzimidazole.
CAS: 312-73-2. $C_8H_5F_3N_2$.
Hazard: A poison by ingestion.
Use: Agricultural chemical.

3-trifluoromethyl-4-nitrophenol.   (1,1,1-trifluoro-4-nitro-*m*-cresol). $CF_3C_6H_3(NO_2)OH$.
Properties: Crystals. Mp 74–76C.

Use: To exterminate lampreys, especially in the Great Lakes. It is placed in tributary streams, where it kills the lamprey larvae.

4-(3-(trifluoromethyl)phenyl)-1-piperazinee-thanol mono(2-(acetyloxy)benzoate) (salt).
CAS: 54851-13-7. $C_{13}H_{17}F_3N_2O \cdot C_9H_8O_4$.
Hazard: Moderately toxic by ingestion.

trifluoronitrosomethane. CF_3NO.
Properties: Midnight blue, fairly stable gas; disagreeable odor. Bp −84C, fp −150C. Nonflammable.
Derivation: (1) Interaction of fluorine and silver cyanide in the presence of silver nitrate; (2) from nitric oxide and iodotrifluoromethane or bromotrifluoromethane in the presence of UV light.
Hazard: Strong irritant to mucous membranes and tissue.
Use: Monomer for nitroso rubber.

trifluorostyrene. $C_8H_5F_3$. A monomer designed for the production of polytrifluorostyrene and for copolymerization with vinyl monomers.
Properties: Liquid. Bp 68C, fp −23C, refr index 1.474, d 1.22, dipole moment 1.98 (D). The polymer is soluble in toluene, chloroform, and methyl ethyl ketone and has dielectric constant of 2.56. Nonflammable.
Use: Membranes for fuel tanks and water purification.

trifluorotrichloroethane.   See trichlorotrifluoroethane.

4,4′-((2,2,2-trifluoro-1-(trifluoromethyl) ethylidene)bis(4,1-phenyleneoxy))bisbenzenamine.
CAS: 69563-88-8. $C_{27}H_{20}F_6N_2O_2$.
Hazard: Moderately toxic by ingestion and skin contact. An eye irritant.

trifluorovinylchloride.   See chlorotrifluoroethylene.

trifluralin.   (generic name for 1,1,1-trifluoro-2,6-dinitro-N,N-dipropyl-p-toluidine).
CAS: 1582-09-8. $F_3C(NO_2)_2C_6H_2N(C_3H_7)_2$.
Properties: Yellowish-orange solid. Mp 48.5–49C, bp 139–140C (4.2 mm Hg). Insoluble in water; soluble in xylene, acetone, and ethanol.
Hazard: Toxic by ingestion. Questionable carcinogen.
Use: Herbicide, especially for cotton plant.

triforine.
CAS: 26644-46-2. $C_{10}H_{14}Cl_6N_4O_2$.
Properties: White crystals. Mp: 155C. Soluble in water, CMF, DMSO.
Hazard: Low toxicity by ingestion, inhalation, and skin contact. Human systemic effects.
Use: food additive; fungicide; agricultural chemical.

triformol.   See sym-trioxane.

triglyceride.   Any naturally occurring ester of a normal acid (fatty acid) and glycerol. The chief constituents of fats and oils, they have the general formula: $CH_2(OOCR_1)CH(OOCR_2)CH_2(OOCR_3)$, where R_1, R_2, and R_3 are usually of different chain length. Refining processes often yield commercial products in which the R chain lengths are the same.
Derivation: Extraction from animal, vegetable, and marine matter.
Use: Fatty acids and derivatives, manufacture of edible oils and fats, manufacture of monoglycerides.

triglycerol monolinolenate.   See polyglycerol ester.

triglycerol trilinoleate.   See polyglycerol ester.

triglycine.   See nitrilotriacetic acid.

triglycol dichloride.   (triethylene glycol dichloride). $Cl(C_2H_4O)_2C_2H_4Cl$.
Properties: Colorless liquid. D 1.1974 (20/20C), bp 241.3C, flash p 250F (121C), bulk d 10.0 lb/gal (20C), fp −31.5C. Insoluble in water. Combustible.
Grade: Technical.
Use: Solvent for hydrocarbons, oils, etc.; extractant; intermediate for resins and insecticides; organic synthesis.

triglycollamic acid.   See nitrilotriacetic acid.

triglyme.   See triethylene glycol dimethyl ether.

trigonelline.   (coffearine; caffearine; gynesine; N-methylnicotinic acid betaine). $C_5H_4NCOOCH_3 \cdot H_2O$. A base formed in the seeds of many plants.
Properties: Colorless prisms. Mp 218C (decomposes). Very soluble in water; soluble in alcohol; nearly insoluble in ether, benzene, and chloroform.
Derivation: Plant seeds, coffee beans, synthetically by heating nicotinic acid with methyl iodide and treatment with silver oxide.
Use: Biochemical research.

"Trigonox 22-8880" [Akzo].   TM for mixture of 80% 1,1-bis(tert-butylperoxy)cyclohexane in butyl benzoyl phthalate.
Available forms: Liquid.
Use: Initiator for SMC and BMC formulations.

tri-n-hexylaluminum. $(C_6H_{13})_3Al$.
Properties: Colorless, pyrophoric liquid. Bp 105C (0.001 mm Hg).
Derivation: Exchange reaction between hexene and isobutyl aluminum.
Hazard: Ignites in air at room temperature.
Use: Polyolefin catalyst.

trihexylene glycol biborate. $(C_6H_{12}O_2)_3B_2$. A cyclic borate.
Properties: Colorless liquid. D 0.982 (21C), boiling range 314–326C, refr index 1.4375 (25C), flash p 345F (173.9C). Soluble in most organic solvents; hydrolyzes slowly in water. Combustible.
Derivation: Reaction of hexylene glycol with boric oxide.
Use: Gasoline additive, chemical intermediate.

trihexyl phosphite. $(C_6H_{13}O)_3P$.
Properties: Mobile, colorless liquid; characteristic odor. D 0.897 (20/4C), bp 135–141C (0.2 mm Hg), flash p 320F (160C) (COC). Miscible with most common organic solvents; insoluble in water; hydrolyzes very slowly, high degree of thermal stability; exposure to air should be minimum. Combustible.
Use: Intermediate for insecticides, component of vinyl stabilizers, lubricant additive, specialty solvent.

trihexyltin acetate. See acetoxytrihexylstannane.

tri-*n*-hexyl trimellitate. See "Morflex 560" *[vertellus]*.

trihydric. Any alcohol in which three hydroxyl groups are present.
See polyol; glycerol.

trihydrochloride.
CAS: 146714-97-8. $C_{25}H_{34}N_4O_2$.
Hazard: A poison.

1,2,3-trihydroxyanthraquinone. See anthragallol.

1,2,4-trihydroxyanthraquinone. See purpurin.

1,2,7-trihydroxyanthraquinone. See anthrapurpurin.

1,2,3-trihydroxybenzene. See pyrogallol.

1,3,5-trihydroxybenzene. See phloroglucinol.

3,4,5-trihydroxybenzoic acid. See gallic acid.

2,4,5-trihydroxybutyrophenone.
$C_6H_2(OH)_3COC_3H_7$.
Properties: Yellow-tan crystals. Mp 149–153C, bulk d 6.0 lb/gal (20C). Very slightly soluble in water; soluble in alcohol and propylene glycol.
Use: Antioxidant for polyolefins and paraffin waxes, food additive.

4,5,7-trihydroxycoumarin.
CAS: 17575-26-7. $C_9H_6O_5$.
Hazard: Moderately toxic by ingestion.

tri(2-hydroxyethyl)amine. See triethanolamine.

trihydroxyethylamine oleate. (triethanolamine oleate). $(HOCH_2CH_2)_3 \cdot HOOCC_{17}H_{33}$. Surfactant made by reaction of triethanolamine with oleic acid. Combustible.
Use: Emulsifying agent.

trihydroxyethylamine stearate. (triethanolamine stearate). $(HOCH_2CH_2)_3N \cdot HOOCC_{17}H_{35}$.
Properties: Cream-colored, wax-like solid; faint fatty odor. D 0.968, pH 8.8–9.2 (25C) (5% aqueous dispersion), mp 42–44C. Soluble in methanol, ethanol, mineral oil, vegetable oil; dispersible in hot water. Combustible.
Use: Emulsifying agent for cosmetic and pharmaceutical industries.

1,3,8-trihydroxy-6-methylanthraquinone. See emodin.

2,4,6-trihydroxytoluene. See methylphloroglucinol.

2,3,5-triiodobenzoic acid. $C_6H_2I_3COOH$. A plant growth regulator used as a growth retardant.

2,3,5-triiodobenzoic acid sodium salt. See sodium 2,3,5-triiodobenzoate.

triiodoisopropylgermane.
CAS: 21342-26-7. $C_3H_7GeI_3$.
Hazard: A poison.

triiodomethane. See iodoform.

triiodopropylgermane.
CAS: 13904-39-7. $C_3H_7GeI_3$.
Hazard: A poison by ingestion.

triiodothyronine. (liothyronine; 3,5,3'-triiodothyronine). $HOC_6H_3IOC_6H_2I_2CH(NH_2)COOH$. Either a derivative or precursor of thyroxine. Triiodothyronine increases the metabolic rate and oxygen consumption of animal tissues.
Use: Biochemical research, medicine (metabolic insufficiency).

triisobutylaluminum. (TIBAL).
$[(CH_3)_2CHCH_2]_3Al$.
Properties: Colorless liquid. D 0.7876 (20C), fp −5.6C, bp 114C (30 mm Hg), flash p 32F (0C), autoign temp 39F (4C).
Derivation: Reaction of isobutylene and hydrogen with aluminum under moderate temperature and varying pressures.
Hazard: Highly toxic, destroys tissue. Flammable; dangerous fire risk; ignites spontaneously in air; reacts violently with water, acids, alcohols, amines, and halogens.

Use: Polyolefin catalyst, manufacture of primary alcohols and olefins, pyrophoric fuel.

triisobutylene. $(C_4H_8)_3$. A mixture of isomers readily prepared by polymerizing isobutylene. A typical mixture is 2,2,4,6,6-pentamethylheptane-3 and 2-neopentyl-4,4-dimethylpentene-1. May be depolymerized to simpler isobutylene derivatives.
Properties: Liquid. D 0.764 (60F), boiling range 175.5–178.9C. Combustible.
Use: Synthesis of resins, rubbers, and intermediate organic compounds; lubricating-oil additive; raw material for alkylation in producing high-octane motor fuels.

triisobutyltin chloride.   See chloro(triisobutyl) stannane.

triisodecyl phosphite.   See isodecyl phosphite.

triisooctyl phosphite. $(C_8H_{17}O)_3P$.
Properties: Colorless liquid; characteristic odor. D 0.891 (20/4C), bp 161–164C (0.3 mm Hg), flash p 385F (196C) (COC). Miscible with most common organic solvents; insoluble in water, hydrolyzes very slowly; exposure to air should be minimum; high thermal stability. Combustible.
Use: Intermediate for insecticides, component of vinyl stabilizers, lubricant additive, specialty solvent.

O,O,O-triisooctyl phosphorothioate. (triisooctyl thiophosphate). $(C_8H_{17}O)_3PS$.
Properties: Colorless liquid; characteristic odor. Bp 160–170C (0.2 mm Hg), d 0.933, flash p 410F (210C) (COC). Insoluble in water; soluble in most organic solvents. Combustible.
Hazard: Highly toxic, cholinesterase inhibitor.
Use: Plasticizer, lubricant additive, hydraulic fluid, intermediate.

triisooctyl trimellitate. $C_6H_3(COOC_8H_{17})_3$.
Properties: Clear liquid; mild odor. D 0.992 (20/20C), distillation range 272–286C (5–95%), fp a gel at −45C, refr index 1.4852 (23C). Practically insoluble in water. Combustible.
Use: Plasticizer.

triisopropanolamine. $N(C_3H_6OH)_3$.
Properties: Crystalline, white solid. Mild base (a mixture of isopropanolamines that has density of 1.004–1.010 and is liquid at room temperature is also marketed). D 0.9996 (50/20C), mp 45C, bp 305C, vap press below 0.01 mm Hg (20C), viscosity 1.38 cP (60C), flash p 320F (160C) (OC). Soluble in water. Combustible.
Grade: Technical.
Hazard: Irritant to skin and eyes.
Use: Emulsifying agents.

triisopropyl borate. $[(CH_3)_2CH]_3BO_3$.

Properties: Colorless liquid. Bp 138–140C, fp −59C, d 0.8138, flash p 82F (27.7C) (TCC).
Derivation: Reaction of isopropyl alcohol with boric oxide.
Hazard: Flammable, moderate fire risk.

triisopropyl phosphite. $[(CH_3)_2CH]_3PO_3$.
Properties: Colorless liquid; characteristic odor. D 0.914 (20/4C), bp 94–96C (50 mm Hg), flash p 165F (73.9C) (COC). Miscible with most common organic solvents; insoluble in water; hydrolyzes slowly in water; exposure to air should be minimum; high thermal stability. Combustible.
Use: Intermediate for insecticides, component of vinyl stabilizers, lubricant additive, specialty solvent.

triisopropyltin acetate.
CAS: 19464-55-2. $C_{11}H_{24}O_2Sn$.
Hazard: A poison by ingestion.

triisopropyltin chloride.
CAS: 14101-95-2. $C_9H_{21}ClSn$.
Hazard: A poison by ingestion.

triisopropyltin undecylenate.
CAS: 73928-00-4. $C_{20}H_{42}O_2Sn$.
Hazard: A poison.

triketohydrindene hydrate.   See ninhydrin.

trilaurin.   The glyceride of lauric acid (glyceryl trilaurate).

trilaurylamine. (tridodecyl amine). $(C_{12}H_{25})_3N$.
Properties: Colorless liquid. D 0.82, mp 14C. Soluble in organic solvents; insoluble in water. Combustible.
Use: Chemical intermediate, metal complexes.

trilauryl phosphite. $(C_{12}H_{25}O)_3P$.
Properties: Water-white liquid. D 0.866 (25/15C), refr index 1.456 (25C), mp 10C. Combustible.
Use: Stabilizer in polymers, chemical intermediate.

trilauryl trithiophosphite. $(C_{12}H_{25}S)_3P$.
Properties: Pale-yellow liquid. D 0.915 (25/15C), mp 20C, refr index 1.502 (25C), flash p 430F (221C) (COC). Combustible.
Use: Stabilizer, lubricant, chemical intermediate.

"Trilene" [Lion].   TM for liquid EPDM.

trimagnesium phosphate.   See magnesium phosphate, tribasic.

trimec.
CAS: 8077-38-1. $C_{10}H_{11}ClO_3 \cdot C_8H_6Cl_2O_3 \cdot C_8H_6Cl_2O_3 \cdot C_2H_7N$.
Hazard: Moderately toxic by ingestion.
Use: Agricultural chemical.

trimedlure. (generic name for *tert*-butyl 4(or 5)-chloro-2-methylcyclohexanecarboxylate). H$_3$C(Cl)C$_6$H$_9$COOC(CH$_3$)$_3$.
Properties: Liquid. Bp 90–92C (0.6 mm Hg). Soluble in most organic solvents; insoluble in water.
Use: Insect attractant.

trimellitic acid. (1,2,4-benzenetricarboxylic acid).
CAS: 528-44-9. C$_9$H$_6$O$_6$.
Properties: Colorless crystals. Mp 220C. Partially soluble in DMF and alcohol; insoluble in benzene and carbon disulfide; slightly soluble in water.
Derivation: Oxidation of pseudocumene.
Use: Organic synthesis (plasticizers, polymers, and similar products).

trimellitic anhydride. (TMA; 1,2,3-benzenetricarboxylic acid-1,2-anhydride).
CAS: 552-30-7.

HOCOC$_6$H$_3$COOCO.
Properties: Solid. Mp 164–166C. Combustible.
Derivation: From pseudocumene.
Hazard: Toxic by inhalation. Respiratory sensitization.
Use: Plasticizer for polyvinylchloride, alkyd coating resins, high-temperature plastics, wire insulation, gaskets, automotive upholstery.

"Trimene Base" [Chemtura]. TM for a reaction product of ethyl chloride, formaldehyde, and ammonia.
Properties: Dark-brown, viscous liquid. D 1.10. Soluble in water and acetone; insoluble in gasoline and benzene.
Use: Rubber accelerator.

trimer. An oligomer whose molecule is comprised of three molecules of the same chemical composition. Examples are trioxane and tripropylene. See polymer; dimer.

trimercuric orthophosphate. See mercuric phosphate.

trimercurous orthophosphate. See mercurous phosphate.

trimesoyl trichloride. (benzene-1,3,5-tricarboxylic acid chloride). C$_6$H$_3$(COCl)$_3$.
Use: Specialty organic.

trimethadione. (3,5,5-trimethyl-2,4-oxazolidinedione).
CAS: 127-48-0.

C$_6$H$_9$NO or OC(O)N(CH$_3$)C(O)C(CH$_3$)$_2$.
Properties: White, granular, crystalline substance; camphor-like odor. Mp 45–47C. Soluble in water;

freely soluble in alcohol, chloroform, and ether; pH 6.0 (5% solution).
Grade: USP.
Hazard: May have adverse side effects; toxic in overdose.
Use: Medicine (anticonvulsant).

trimethoxyborine. See trimethyl borate.

trimethoxyboroxine. (methyl metaborate). (CH$_3$O)$_3$B$_3$O$_3$. A cyclic compound.
Properties: Colorless liquid. Mp 10–11C, bp (dissociates), d 1.216 (25C), refr index 1.3986. Nonflammable.
Derivation: Reaction of methyl borate with boric acid.
Grade: 99%.
Use: Metal-fire extinguishing fluid.

trimethoxymethane. See methyl orthoformate.

3,4,5-trimethoxyphenethylamine. See mescaline.

2,4,5-trimethoxy-1-propenylbenzene. (asarone). (CH$_3$O)$_3$C$_6$H$_2$CH:CHCH$_3$.
Properties: Crystals. Mp 67C. Insoluble in water; soluble in alcohol.
Derivation: Either extracted from calamus oil or synthesized (Wittig reaction).
Use: Grain fumigant, insect chemisterilant.

3,4,5-trimethoxy-*n*-(4-propylcyclohexyl) benzamide.
CAS: 315706-65-1. C$_{19}$H$_{29}$NO$_4$.
Hazard: A poison by ingestion.

3-(trimethoxysilyl)propyl chloride.
CAS: 2530-87-2. C$_6$H$_{15}$ClO$_3$Si.
Hazard: A poison by ingestion and skin contact.

***n*-(3-(trimethoxysilyl)propyl)-1,3-propanediamine.**
CAS: 25147-91-5. C$_9$H$_{24}$N$_2$O$_3$Si.
Hazard: A poison by ingestion and skin contact. A mild skin and severe eye irritant.

3-(trimethoxysilyl)-*n*-(3-(trimethoxysilyl) propyl)-1-propanamine.
CAS: 82985-35-1. C$_{12}$H$_{31}$NO$_6$S$_{12}$.
Hazard: A poison by ingestion and skin contact. A moderate skin and severe eye irritant.

trimethylacethydrazide ammonium chloride. See Girard's T Reagent.

trimethylacetic acid. (pivalic acid; neopentanoic acid).
CAS: 75-98-9. (CH$_3$)$_3$CCOOH.
Properties: Colorless crystals. D 0.905 (50C), refr index 1.3931 (36.5C), mp 35.5C, bp 163.8C. Soluble in water, alcohol, and ether. Combustible.

Use: Intermediate, as a replacement for some natural materials.

trimethylacetic acid sodium salt. See sodium 2,2-dimethylpropanoate.

trimethyladipic acid. $C_9H_{16}O_4$.
Properties: Powder. Combustible.
Use: Esterification agent for plasticizers, lubricants, alkyd resins, polyurethane, polyester, special polyamides; intermediate for the production of glycols.

trimethylaluminum. (aluminum trimethyl; ATM).
CAS: 75-24-1. $(CH_3)_3Al$.
Properties: Colorless, pyrophoric liquid. Bp 126C, mp 15.4C, d 0.752.
Derivation: By sodium reduction of dimethylaluminum chloride.
Hazard: Highly flammable; dangerous fire risk; flames instantly on contact with air; reacts violently with water, acids, halogens, alcohols, and amines.
Use: Catalyst for olefin polymerization, pyrophoric fuel, manufacture of straight-chain primary alcohols and olefins, to produce luminous trails in upper atmosphere to track rockets.

trimethylamine. (TMA).
CAS: 75-50-3. $(CH_3)_3N$.
Properties: Colorless gas at room temperature; fishy ammoniacal odor; readily liquefied. Anhydrous form shipped as liquefied compressed gas. D 0.6621 (−5C), bp −4C, fp −117.1C, autoign temp 374F (190C), flash p 10F (−12.2C) (CC), flash p (25% solution) 38F (3.3C) (TOC). Soluble in water, alcohol, and ether.
Derivation: Interaction of methanol and ammonia over a catalyst at high temperature. The mono-, di-, and trimethylamines are produced, and yields are regulated by conditions.
Method of purification: Azeotropic or extractive distillation.
Grade: Anhydrous 99% min, aqueous solution 25, 30, 40%.
Hazard: Flammable, dangerous fire risk, explosive limits in air 2–11.6%. Toxic by inhalation, vapor highly irritating. Eye, skin and upper respiratory tract irritant.
Use: Organic synthesis, especially of choline salts, warning agent for natural gas, manufacture of disinfectants, flotation agent, insect attractant, quaternary ammonium compounds, plastics.

trimethylamine sulfur trioxide.
$(CH_3)_3N\cdot SO_3$.
Properties: White powder. Mp 232–238C (decomposes). Soluble in hot water, ethanol; soluble with difficulty in cold water and acetone; not dissociated in benzene and chloroform solutions; distinctly different from the isomeric adduct of trimethylamine oxide and sulfur dioxide.

Use: Separation of isomers, soil sterilant, catalyst for thermosetting resins.

2,4,5-trimethylaniline. See pseudocumidine.

n,n,4-trimethylaniline.
CAS: 99-97-8. $C_9H_{13}N$.
Hazard: A poison.

trimethyl benzene.
CAS: 25551-13-7. $(CH_3)_3C_6H_3$.
Properties: Liquid. Mw 120.9. Insoluble in water.
Hazard: Central nervous system impairment, asthma and hematologic effects.
Use: Raw material for chemical synthesis.

1,2,3-trimethylbenzene. See hemimellitene.

1,2,4-trimethylbenzene. See pseudocumene.

1,3,5-trimethylbenzene. See mesitylene.

(trimethylbenzyl)dodecyldimethyl ammonium chloride. $[(CH_3)_3C_6H_2CH_2N(CH_3)_2 C_{12}H_{25}]Cl$. A quaternary ammonium salt.
Properties: White to slightly yellow, crystalline powder; mild odor and taste. Mp 162–163C, bulk d 4.62 lb/gal. Soluble in water, alcohol, glycerol, acetone; pH of 10% solution in distilled water 4.4.
Use: Germicide.

trimethyl borate. (methyl borate; trimethoxyborine).
CAS: 121-43-7. $(CH_3O)_3B$.
Properties: Water-white liquid. Bp 67–68C, d 0.915, fp −29C. Miscible with ether, methanol, hexane, tetrahydrofuran; decomposes in presence of water; flash p 80F (26.6C).
Derivation: Reaction of boric acid and methanol.
Hazard: Flammable, fire risk, reacts with water and oxidizing agents.
Use: Solvent, dehydrating agent, fungicide for citrus fruit, neutron scintillation counters, brazing flux, boron compounds, catalyst.

2,2,3-trimethylbutane. (isopropyltrimethylmethane; triptane). $CH_3C(CH_3)_2C(CH_3)CH_3$.
Properties: Colorless liquid. D 0.691, bp 81.0C, fp −24.96C, refr index 1.3895 (20C). Soluble in alcohol; insoluble in water.
Hazard: Flammable, moderate fire risk.
Use: Organic synthesis, aviation fuel.

β,γ,γ-trimethylcaproaldehyde thiosemicarbazone.
CAS: 63884-77-5.
Hazard: A poison by ingestion.
Use: Agricultural chemical.

trimethyl carbinol.   See *tert*-butyl alcohol.

1,1,3-trimethyl-5-carboxy-3-(*p*-carboxy-phenyl)indane.   An aromatic di-acid used as an intermediate in the manufacture of polyester fibers, polyamides, and alkyd resins; hot-melt adhesives; engineering thermoplastics.

trimethylchlorosilane.
CAS: 75-77-4. $(CH_3)_3SiCl$.
Properties: Colorless liquid. Bp 57C, d 0.854 (25/25C), refr index 1.3893 (25C), flash p −18F (−27.7C). Readily hydrolyzed with liberation of hydrogen chloride; soluble in benzene, ether and perchloroethylene.
Derivation: By Grignard reaction of silicon tetrachloride and methylmagnesium chloride.
Hazard: Flammable, dangerous fire risk, reacts violently with water. Strong irritant to tissue.
Use: Intermediate for silicone fluids, as a chain-terminating agent, imparting water repellency.

trimethylchlorostannane.   See chlorotrimethylstannane.

trimethylchlorotin.   See chlorotrimethylstannane.

3,3,5-trimethyl-5-cyanocyclohexanone.
See isophoronenitrile.

trimethylcyclododecatriene.   (TMCDT).   A cyclic hydrocarbon.
Use: Intermediate in making derivatives useful in the perfume and pharmaceutical industries, as well as catalysts.

3,3,5-trimethylcyclohexanol-1.
CAS: 116-02-9. $C_6H_8(CH_3)_3OH$.
Properties: Colorless liquid. D 0.878 (40/20C), mp 35.7C, bp 198C, flash p 165F (73.9C) (OC). Soluble in most organic solvents, hydrocarbons, oils; insoluble in water. Combustible.
Hazard: Toxic by inhalation, strong irritant.
Use: Menthol and camphor substitute, antifoaming agent, hydraulic fluids and textile soaps, odor masking, esterification agent, pharmaceuticals, wax additive, printing inks.

3,5,5-trimethyl-2-cyclohexen-1-one.   See isophorone.

3,3,5-trimethylcyclohexyl   salicylate.   See homomenthyl salicylate.

1,7,7-trimethyl-*o*-(3-(diethylamino)-2-hydroxypropyl)oxime   (1r,4r)-bicyclo(2.2.1)heptan-2-one.
CAS: 314238-30-7. $C_{17}H_{32}N_2O_2$.
Hazard: A poison by ingestion.

1,7,7-trimethyl-*o*-(3-(2,6-dimethyl-4-morpholinyl)-2-hydroxypropyl)oxime (1r,4r)-bicyclo(2.2.1)heptan-2-one.
CAS: 316172-59-5. $C_{19}H_{34}N_2O_3$.
Hazard: A poison by ingestion.

1,7,7-trimethyl-*o*-(3-(3,5-dimethyl-1-piperidinyl)-2-hydroxypropyl)oxime (1r,4r)-bicyclo(2.2.1)heptan-2-one.
CAS: 316172-58-4. $C_{20}H_{36}N_2O_2$.
Hazard: A poison.

trimethyl dihydroquinoline polymer.
(TDQP). $(C_{12}H_{15}N)_n$ (probably three or more quinoline groups).
Properties: Amber pellets. D 1.08, softening p 75C. Insoluble in water; miscible with ethanol, acetone, benzene, monochlorobenzene, isopropyl acetate, and gasoline.
Use: Antioxidant, stabilizer, or polymerization inhibitor.

3,7,11-trimethyl-1,6,10-dodecatrien-3-ol.
See nerolidol.

3,7,11-trimethyl-2-6-10-dodecatrien-1-ol.
See farnesol.

trimethylene.   See cyclopropane.

trimethylene bromide.   (1,3-dibromopropane).
CAS: 109-64-8. $CH_2BrCH_2CH_2Br$.
Properties: Colorless liquid; sweet odor. D 1.979 (20/4C), bp 166C, fp −34.4C. Insoluble in water; soluble in organic solvents. Combustible.
Grade: Technical, CP.
Use: Intermediate for dyestuff and pharmaceutical industries, cyclopropane manufacture.

trimethylene chlorobromide.   See 1-bromo-3-chloropropane.

trimethylene   chlorohydrin.   (3-chloro-1-propanol). $ClCH_2CH_2CH_2OH$.
Properties: Colorless to pale-yellow liquid; characteristic odor. D 1.130–1.150 (25/25C), refr index 1.445–1.447 (25C). Soluble in water, alcohols, and ethers; insoluble in hydrocarbons. Combustible.
Use: Intermediate.

trimethylenedicyanide.   See glutaronitrile.

trimethylene   glycol.   (1,3-propylene glycol; 1,3-propanediol).
CAS: 504-63-2. $CH_2OHCH_2CH_2OH$.
Properties: Colorless liquid; odorless. D 1.0537 (25C), bp 210–211C, autoign temp 752F (400C). Soluble in water, alcohol, and ether. Combustible.
Derivation: From acrolein.
Grade: Technical 95%, pure 99%.
Use: Intermediate, especially for polyesters.

trimethylene oxide.   See oxetane.

***sym*-trimethylene trinitramine.**   See cyclonite.

trimethylethylene.   See 3-methyl-2-butene.

trimethylglycine.   See betaine.

trimethylheptanoic acid.   See isodecanoic acid.

1,7,7-trimethyl-*o*-(3-(hexahydro-1h-azepin-1-yl)-2-hydroxypropyl)oxime (1r,4r)-bicyclo (2.2.1)heptan-2-one.
CAS: 314238-35-2. $C_{19}H_{34}N_2O_2$.
Hazard: A poison by ingestion.

2,2,5-trimethylhexane.
CAS: 3522-94-9. $(CH_3)_3CCH_2CH_2CH(CH_3)_2$.
Properties: Colorless liquid. Fp $-105.84C$, bp 124.06C, d 0.711 (15.5/15.5C), refr index 1.399 (20C), flash p 55F (12.7C).
Grade: 95%, 99%; research.
Hazard: Flammable, dangerous fire risk; the 2,3,3-, 2,3,4-, and 3,3,4-isomers are less flammable.
Use: Synthesis, motor-fuel additive.

3,5,5-trimethylhexan-1-ol. $C_9H_{20}O$.
Properties: Colorless liquid; mild odor. Bp 194C, d 0.8236 (25/4C), bulk d 6.86 lb/gal (25C), refr index 1.4300 (25C), flash p 200F (93.3C) (OC). Insoluble in water. Combustible.
Derivation: High-pressure synthesis.
Use: Synthetic lubricants, additives to lubricating oils, wetting agent, softener in manufacture of various plastics, disinfectants and germicides.

1,7,7-trimethyl-*o*-(2-hydroxy-3-((1-methyl-ethyl)amino)propyl)oxime (1r,4r)-bicyclo (2.2.1)heptan-2-one.
CAS: 314238-31-8. $C_{16}H_{30}N_2O_2$.
Hazard: A poison by ingestion.

1,7,7-trimethyl-*o*-(2-hydroxy-3-(4-morpho-linyl)propyl)oxime (1r,4r)-bicyclo(2.2.1) heptan-2-one.
CAS: 314238-34-1. $C_{17}H_{30}N_2O_3$.
Hazard: A poison by ingestion.

1,7,7-trimethyl-*o*-(2-hydroxy-3-(1-piperi-dinyl)propyl)oxime (1r,4r)-bicyclo(2.2.1) heptan-2-one.
CAS: 314238-33-0. $C_{18}H_{32}N_2O_2$.
Hazard: A poison by ingestion.

1,7,7-trimethyl-*o*-(2-hydroxy-3-(1-pyrroli-dinyl)propyl)oxime (1r,4r)-bicyclo(2.2.1) heptan-2-one.
CAS: 314238-32-9. $C_{17}H_{30}N_2O_2$.
Hazard: A poison by ingestion.

trimethylmethane.   See isobutane.

trimethyl nitrilotripropionate. $N(CH_2CH_2 COOCH_3)_3$.
Properties: Weakly basic tertiary amine and organic ester.
Use: Plasticizer for PVC, intermediate for pharmaceutical and agricultural chemicals, high-boiling solvents, low-temperature lubricants.

trimethylnonanol.   See 2,6,8-trimethylnonyl-4 alcohol.

2,6,8-trimethyl-4-nonanone.   (isobutyl heptyl ketone).
CAS: 123-17-1. $C_{12}H_{26}O$.
Properties: Water-white liquid; pleasant odor. D 0.8165 (20/20C), bulk d 6.8 lb/gal, fp $-75C$, bp 211–219C, viscosity 1.91 cP (20C), flash p 195F (90.5C) (COC). High solvent power for vinyl resins, the cellulose esters and ethers and many difficultly soluble substances; insoluble in water. Combustible.
Use: Solvent, dispersant, intermediate, lubricating-oil dewaxing.

2,6,8-trimethylnonyl-4-alcohol.   (trimethylnonanol).
$(CH_3)_2CHCH_2CH_2OCH_2CH(CH_3)CH_2CH(CH_3)_2$.
Properties: Colorless liquid; characteristic odor. D 0.8913 (20/20C), bulk d 6.9 lb/gal (20C), bp 225.2C, fp $-60C$ (sets to a glass), viscosity 21.4 cP, flash p 200F (93.3C) (COC). Insoluble in water. Combustible.
Derivation: Oxo process.
Use: Surface-active and flotation agents, lubricant additives, rubber chemicals.

trimethylolethane.   (pentaglycerine; methyltrimethylolmethane). $CH_3C(CH_2OH)_3$.
Properties: Colorless, hygroscopic crystals. Soluble in water and alcohol. Combustible.
Use: Conditioning agent, manufacture of varnishes, alkyd and polyester resins, synthetic drying oils.

trimethylolmelamine.
CAS: 1017-56-7. $C_3N_3(NHCH_2OH)_3$.
The first stage in making melamine resins.

trimethylolpropane.   (hexaglycerol). $C_2H_5C (CH_2OH)_3$.
Properties: Colorless, hygroscopic crystals. Soluble in water and alcohol. Combustible.
Use: Conditioning agent, manufacture of varnishes, alkyd resins, synthetic drying oils, urethane foams and coatings, silicone lubricant oils, lactone plasticizers, textile finishes, surfactants, epoxidation products.

trimethylolpropane monooleate. $C_2H_5C (CH_2OH)_2CH_2OOCC_{17}H_{33}$ (theoretically). The commercial product is a mixture of mono-, di-, and triesters, free polyol, and free oleic acid. Combustible.

Properties: Oily liquid. D 0.954 (25C), fp <−20C. Insoluble in water; soluble in most organic solvents.

Use: Water-in-oil emulsifier, corrosion inhibitor, low-temperature plasticizer, deicing agent for gasoline.

trimethylolpropane triacetoacetate. See 2-ethyl-2-(hydroxymethyl)-1,3-propanediol triacetoacetate.

trimethylolpropane tris(mercaptopropionate). $C_2H_5C(CH_2OOCCH_2CH_2SH)_3$.
Properties: Liquid. D 1.21 (25C), refr index 1.5151 (25C), insoluble in water and hexane, soluble in acetone, benzene, and alcohol. Combustible.

Use: Curing or cross-linking agents for polymers, especially epoxy resins, intermediate for stabilizers and antioxidants.

trimethylolpropane trithioglycolate.
$CH_3CH_2C(CH_2OOCCH_2SH)_3$.
Properties: Liquid. D 1.28, refr index 1.5292 (25C). Combustible.

Use: Curing or cross-linking agent for polymeric systems, especially epoxy resins; intermediate for stabilizers and antioxidants.

2,4,5-trimethyl δ-3-oxazoline. $C_6H_{11}NO$.
Properties: Yellow-orange liquid; powerful, musty, nut-like odor. D: 0.911–0.932, refr index: 1.414–1.435. Soluble in alc, propylene glycol, water; insoluble in fixed oils.

Use: Food additive.

3,5,5-trimethyl-2,4-oxazolidinedione.   See trimethadione.

2,4,8-trimethyl-5-oxo-6-oxa-3,9-dithia-2,4,7-triazadec-7-enoic acid ethyl ester.
CAS: 64028-99-5. $C_9H_{17}N_3O_4S_2$.
Hazard: A poison by ingestion.
Use: Agricultural chemical.

2,2,4-trimethylpentane.
CAS: 540-84-1. $(CH_3)_2CHCH(CH_3CH)CH_3CH_3$.
Properties: Liquid. Fp −109.43C, bp 113C, d 0.723 (60/60F), refr index 1.4042 (20C), flash p 41F (5C).
Grade: 95%, 99%; research.
Hazard: Flammable, dangerous fire risk.
Use: Intermediate, azeotropic distillation entrainer.

2,2,4-trimethyl-1,3-pentanediol.
$(CH_3)_2CHCH(OH)C(CH_3)_2CH_2OH$.
Properties: (96% pure): White solid. D 0.928 (55/15C), mp 46–55C, bp 215–235C, flash p 235F (112.7C). Slightly soluble in water; soluble in alcohol, acetone, ether, and benzene. Combustible.

Use: Polyester resins, plasticizers, lubricants, surface coatings and printing inks, insect repellent.

2,2,4-trimethyl-1,3-pentanediol monoisobutyrate.
$(CH_3)_2CHCH(OH)C(CH_3)_2CH_2OOCCH(CH_3)_2$.
Properties: Liquid. D 0.945–0.955 (20/20C), bp 180–182C (125 mm Hg), pour p −57C, refr index 1.4423 (20C), flash p 245F (118.3C) (COC). Insoluble in water; soluble in benzene, alcohol, acetone, and carbon tetrachloride. Combustible.

Use: Intermediate in the manufacture of plasticizers, surfactants, pesticides, and resins.

2,4,4-trimethylpentene-1.   (α-diisobutylene).
$H_2C:C(CH_3)CH_2C(CH_3)_3$.
Properties: Colorless liquid. Bp 101.44C, fp −93.5C, refr index 1.4086 (20C), d 0.7150 (20C), flash p 35F (1.67C).
Derivation: Polymerization of isobutene.
Grade: 95%, 99%; research.
Hazard: Flammable, dangerous fire risk.
Use: Organic synthesis; motor-fuel synthesis, particularly isooctane; peroxide reactions.

2,4,4-trimethylpentene-2.   (β-diisobutylene).
$H_3CC(CH_3):CHC(CH_3)_3$.
Properties: Colorless liquid. Bp 104.55C, d 0.724 (60/60F), fp −106.4C, refr index 1.416 (20C), flash p 35F (1.67C) (TOC).
Grade: 95%.
Hazard: Flammable, dangerous fire risk. Irritant and narcotic in high concentration.
Use: Organic synthesis.

tri-2-methylpentylaluminum.
$[(CH_3)_2CH(CH_2)_3]_3Al$.
Properties: Colorless liquid.
Derivation: Reaction of 2-methylpentene and isobutylaluminum.
Hazard: Flammable.
Use: Polyolefin catalyst.

n,n,α-trimethylphenethylamine.
CAS: 4075-96-1. $C_{11}H_{17}N$.
Hazard: Moderately toxic by ingestion.

2,4,6-trimethyl phenol.   See mesitol.

trimethyl phosphate.
CAS: 512-56-1. $(CH_3O)_3PO$.
Properties: Colorless liquid, 22.1% phosphorus. D 1.210 (68F), flash p above 300F (148C), bp 193C, refr index 1.397 (20C), pour p −46C. Soluble in both gasoline and water. Combustible.
Hazard: Toxic by ingestion and inhalation, strong irritant to skin and eyes.
Use: For controlling spark plug fouling, surface ignition and rumble in gasoline engines.

trimethyl phosphite.
CAS: 121-45-9. $(CH_3O)_3P$.
Properties: Colorless liquid. Bp 108–108.5C, pour p below −60C, d 1.046 (20/4C), flash p 100F (37.7C)

(COC). Insoluble in water; soluble in hexane, benzene, acetone, alcohol, ether, carbon tetrachloride, and kerosene.
Hazard: Flammable, moderate fire risk. Eye irritant and cholinesterase inhibitor.
Use: Chemical intermediate, especially for insecticides.

trimethyl phosphorotrithioate.
CAS: 816-80-8. $C_3H_9PS_3$.
Properties: Water-white liquid. Bp 78C (12 mm Hg).
Hazard: A poison by ingestion. Moderately toxic by skin contact. A severe eye irritant.
Use: Extraction of mineral salts from alkyl acid phosphate solvent solutions, plasticizer.

1,2,4-trimethylpiperazine. $(CH_3)_3C_4H_7N_2$.
Properties: Colorless liquid. D 0.851 (25/25C), fp 44C, bp 151C, refr index 1.4480 (20C), flash p 257F (125C) (OC). Miscible with water. Combustible.
Use: Polymerization catalyst.

trimethylpropylmethane.   See 2,2-dimethylpentane.

2,4,6-trimethylpyridine.   See 2,4,6-collidine.

trimethylstannane sulphate.   See trimethyltin sulphate.

3,3,6-trimethyl-2,5-thiomorpholinedione-2-(o-((methylamino)carbonyl)oxime).
CAS: 66637-26-1. $C_9H_{15}N_3O_3S$.
Hazard: A poison by ingestion.
Use: Agricultural chemical.

trimethyltin acetate.
CAS: 1118-14-5. $C_5H_{12}O_2Sn$.
Properties: White crystals. Mp: 37.5C. Spar soluble in $CHCl_3$ and CCl_4.
Hazard: A poison by ingestion.

trimethyltin cyanate.
CAS: 73940-86-0. C_4H_9NOSn.
Hazard: A poison.

trimethyltin hydroxide.
CAS: 56-24-6. $C_3H_{10}OSn$.
Properties: Colorless, white crystals. Mp: 118C (decomp). Soluble in water and many organic solvents.
Hazard: A poison.
Use: Drug.

trimethyltin iodide.   See iodotrimethyltin.

trimethyltin isothiocyanate.   See isothiocyanatotrimethyltin.

trimethyltin sulphate.
CAS: 63869-87-4. $C_3H_{10}O_4SSn$.
Hazard: A poison by ingestion.

trimethyltin thiocyanate.
CAS: 4638-25-9. C_4H_9NSSn.
Hazard: A poison.

2,4,6-trimethyl-1,3,5-trioxane.   See paraldehyde.

1,3,5-trimethyl-2,4,6-tris(3,5-di-tert-butyl-4-hydroxybenzyl)benzene.
Properties: Free-flowing, white, crystalline powder; no odor. Mp 244C. Partially soluble in benzene and methylene chloride; insoluble in water. Permissible in contact with food products. Combustible.
Use: Antioxidant for polypropylene, high-density polyethylene, spandex fibers, polyamides, and specialty rubbers.

2,6,10-trimethyl-9-undecen-1-al.
$(CH_3)_2C{:}CHCH_2H_4CH(CH_3)C_3H_6CH(CH_3)CHO$
Properties: Clear, yellow liquid; strong, pungent, ozone-like odor. D 0.850–0.860 (25/25C), refr index 1.4530–1.4630 (20C). Combustible.
Use: Perfume.

trimethylvinylammonium hydroxide.   See neurine.

trimethylxanthine.   See caffeine.

(+)-trimipramine.
CAS: 3564-66-7. $C_{20}H_{26}N_2$.
Hazard: A poison by ingestion.

"Trimmit" [Syngenta].
CAS: 76738-62-0. $C_{15}H_{20}ClN_3O$. TM for a plant growth regulator for turfgrass.
Hazard: Moderately toxic by ingestion. A reproductive hazard.
Use: Agricultural chemical.

trimyristin.   The glyceride of myristic acid, glyceryl trimyristate.

trineophyltin acetate.
CAS: 1636-70-0. $C_{32}H_{42}O_2Sn$.
Hazard: Moderately toxic by ingestion.

trinickelous orthophosphate.   See nickel phosphate.

trinitite.   Green glazed glass material produced by the 100 million degree fahrenheit heat from an atomic explosion. It was named for Trinity, New Mexico, a sand-covered site 80 miles from the Almogordo atomic bomb test site.

trinitroaniline.   (picramide).
CAS: 489-98-5. $C_6H_2NH_2(NO_2)_3$.
Properties: Orange-red crystals. Mp 188C, bp (explodes), d 1.762.

Derivation: Nitrating aniline in glacial acetic acid solution or by the use of mixed nitric-sulfuric acid in limited amounts.
Hazard: Dangerous, explodes by heat or shock.
Use: Explosive compositions.

trinitroanisole. (methyl picrate; 2,4,6-trinitrophenyl methyl ether). $CH_3OC_6H_2(NO_2)_3$.
Properties: Crystals, mp 68.4C, d 1.408 (20/4C).
Hazard: High. Dangerous, explodes by heat or shock.
Derivation: Interaction of methyl iodide and silver picrate, nitration of anisic acid.
Use: Explosive compositions.

1,3,5-trinitrobenzene. (TNB).
CAS: 99-35-4. $C_6H_3(NO_2)_3$.
Properties: Yellow crystals. D 1.688 (20/4C), mp 122C. Soluble in alcohol and ether; insoluble in water.
Derivation: From trinitrotoluene by removal of the methyl group.
Hazard: Dangerous, explodes by heat or shock.
Use: Explosive compositions.

2,4,6-trinitrobenzoic acid. (trinitrobenzoic acid).
CAS: 129-66-8. $C_6H_2(NO_2)_3COOH$.
Properties: Orthorhombic crystals. Mp 228.7C. Sublimes with decomposition, forming carbon dioxide and trinitrobenzene; slightly soluble in water and benzene; soluble in alcohol, ether, and acetone.
Derivation: Oxidation of 2,4,6-trinitrotoluene with chromic acid.
Hazard: Dangerous, explodes by heat or shock.
Use: Explosive compositions.

2,4,6-trinitrobenzenesulfonic acid.
CAS: 2508-19-2. $C_6H_3N_3O_9S$.
Hazard: A poison.

2,4,6-trinitro-1,3,5-benzenetriamine.
CAS: 3058-38-6. $C_6H_6N_6O_6$.
Hazard: Low toxicity by ingestion and inhalation. A mild eye irritant.

4,4,4-trinitrobutyric acid.
CAS: 5029-46-9. $C_4H_5N_3O_8$.
Hazard: A poison by inhalation. Moderately toxic by ingestion.

2,4,6-trinitro-*m*-cresol. (cresolite; cresylite). $(NO_2)_3C_6H(CH_3)OH$.
Properties: Yellow crystals. Mp 106C. Readily soluble in alcohol, ether, and acetone.
Derivation: Prepared from *m*-cresol by a process similar to that in picric acid is prepared from phenol.
Hazard: Explodes at 300F (148.8C), severe explosion risk when shocked or heated.
Use: Bursting charges and other high explosive uses.

2,4,7-trinitrofluoren-9-one.
CAS: 129-79-3. $C_{13}H_5N_3O_7$.
Properties: Pale-yellow needles from AcOH or C_6H_6. Mp: 176C.
Hazard: Suspected carcinogen. Mildly toxic by ingestion. A skin and eye irritant.

trinitroglycerin. See nitroglycerin.

trinitromethane.
CAS: 517-25-9. $CH(NO_2)_3$.
Properties: White crystals. Mp 15C, d 1.469 (25C), decomposes above 25C, heat of combustion 746 cal/g. Soluble in water.
Derivation: Reaction of acetylene with nitric acid.
Hazard: Explodes on heating, concentrations above 50% in air may explode.
Use: Manufacture of propellants and explosives.

1,3,5-trinitronaphthalene. (naphite). $C_{10}H_5(NO_2)_3$. Commercial preparation is a mixture of isomers that melts at 110C.
Hazard: Explosion risk when shocked or heated.
Use: Explosive, stabilizer for nitrocellulose.

trinitrophenol. See picric acid.

2,4,6-trinitrophenyl methyl ether. See trinitroanisole.

trinitrophenylmethylnitramine. See tetryl.

2,4,6-trinitroresorcinol. Legal label name for styphnic acid.

2,4,6-trinitrotoluene. (TNT; methyltrinitrobenzene).
CAS: 118-96-7. $CH_3C_6H_2(NO_2)_3$.

Properties: Yellow, monoclinic needles. D 1.654, mp 80.9C. Soluble in alcohol and ether; insoluble in water.
Derivation: Nitration of toluene with mixed acid. Small amounts of the 2,3,4- and 2,4,5-isomers are produced that may be removed by washing with aqueous sodium sulfite solution.
Grade: Technical.
Hazard: Flammable, dangerous fire risk, moderate explosion risk, will detonate only if vigorously shocked or heated to 450F (232C). Toxic by ingestion, inhalation, and skin absorption. Methemoglobinemia, liver damage, and cataracts. Questionable carcinogen.
Use: Explosive, intermediate in dyestuffs and photographic chemicals.

trinitrotrimethylenetriamine.   See cyclonite.

trioctadecyl phosphite. $(C_{18}H_{37}O)_3P.$
Properties: White, waxy solid. Mp 45–47C, d 0.940 (25/25C).
Use: Stabilizer in polymers; intermediate.

tri-*n*-octylaluminum. $(C_8H_{17})_3Al.$
Properties: Colorless, pyrophoric liquid.
Derivation: Reaction between octene and isobutylaluminum.
Hazard: Flammable, ignites in air.
Use: Polyolefin catalyst.

trioctyl(butylthio)stannane.   See (butylthio)trioctylstannane.

trioctyl(ethylthio)stannane.   See (ethylthio)trioctylstannane.

trioctyl phosphate.   (octyl phosphate). $(C_8H_{17})_3$ $PO_4.$
Properties: Liquid, d 0.924 (26C), bp 220–230C (8 mm Hg). Soluble in alcohol, acetone, and ether. Combustible.
Hazard: Toxic by ingestion and inhalation.
Use: Solvent, antifoaming agent, plasticizer.

tri-*n*-octylphosphine oxide.
CAS: 78-50-2. $C_{24}H_{51}OP.$
Hazard: A severe eye and moderate skin irritant.

trioctylphosphinic oxide.   (TOPO). $(C_8H_{17})_3$ PO.
Properties: Solid. Mp 55C, min purity 95%.
Use: Reagent for extraction of metals from aqueous and nonaqueous solutions, including fissionable actinide elements.

trioctyl phosphite.   See tris-2-ethyl-hexyl phosphite.

tri-*n*-octyltin chloride.
CAS: 2587-76-0. $C_{24}H_{51}ClSn.$
Hazard: Moderately toxic by ingestion.

triolein.   See olein.

triose.   A simple sugar with a backbone containing three carbon atoms.

***sym*-trioxane.**   (triformol; trioxin; metaformaldehyde).

┌─────────┐
$(CH_2O)_3$ or $CH_2OCH_2OCH_2O.$

A trimer of formaldehyde, not to be confused with paraformaldehyde, which consists of eight or more formaldehyde units.
Properties: White crystals; formaldehyde odor. Mp 62C, sublimes at 115C, flash p 113F (45C) (OC),

autoign temp 777F (413C). Soluble in water, alcohol, and ether.
Derivation: By distillation of formaldehyde with an acid catalyst and extraction with solvent.
Hazard: Moderate fire risk, explosive limits in air 3.6–29%.
Use: Organic synthesis; disinfectant; nonluminous, odorless fuel.
See formaldehyde.

2,6,8-trioxypurine.   See uric acid.

tripalmitin.   (palmitin; glyceryl tripalmitate). $C_3H_5(OOCC_{15}H_{31})_3.$
Properties: White, crystalline powder. Mp 65.5C, d 0.866 (80/4C). Soluble in ether and chloroform; insoluble in water. Combustible.
Derivation: From glycerol and palmitic acid.
Grade: Technical.
Use: Soap, leather dressing.

trip balance.   Stabilized pan balance in which plates or pans are positioned above an equal arm beam.

tripelennamine citrate.
CAS: 91-81-6. $C_{16}H_{21}N_3 \cdot C_6H_8O_7.$ 2-[benzyl-(2-dimethylaminoethyl)amino]pyridine   dihydrogen citrate.
Properties: White, bitter, crystalline powder. Solutions are acid to litmus. Mp 107C. Soluble in water and alcohol; very slightly soluble in ether; practically insoluble in chloroform and benzene; 1% solution in water has a pH of 4.3.
Grade: USP.
Hazard: Toxic by ingestion.
Use: Medicine (antihistamine, sunburn treatment).

tripentaerythritol. $C_{15}H_{32}O_{10}.$
Properties: White to ivory powder; has eight primary hydroxyl groups, all esterifiable. Melting range 225–240C. Combustible.
Use: Hard resins, varnishes, and fast-drying tall-oil vehicles.

tri-*n*-pentyltin bromide.   See bromotripentylstannane.

triphenol phosphorus.   See 1,1,3-tris(hydroxyphenyl)-propane.

triphenylamine.
CAS: 603-34-9. $C_{18}H_{15}N.$
Properties: Monoclinic crystals from EtOAc. D 0.774 @ 0°/0°C, mp 127°C, bp 195–205°C @ 10–22 mm, mw 245.34.

triphenylantimony.   (triphenylstibine).
CAS: 603-36-1. $Sb(C_6H_5)_3.$
Properties: White, crystalline solid. D 1.434 (25C), mp 46–53C, bp below 360C. Insoluble in water;

slightly soluble in alcohol; soluble in most organic solvents. Combustible.

Derivation: Reaction of antimony trichloride with phenyl magnesium bromide or phenyl sodium.

Use: Stibonium salts, cocatalyst in converting trienes to aromatics and hydroaromatics, react with nitric-sulfuric acid to give trinitro derivatives, polymerization inhibitor, lubricating-oil additive.

triphenylborane.
CAS: 960-71-4. $C_{18}H_{15}B$.
Hazard: A severe eye irritant.

triphenylboron. $B(C_6H_5)_3$. A type of Lewis acid used as catalyst and intermediate.

triphenyl formazan. $CN_4H(C_6H_5)_3$. Red, insoluble derivative of tetrazolium chloride, formed when the latter comes into contact with viable portions of a seed.
Use: Germination and viability tests.

triphenylguanidine. (TPG).
$C_6H_5NC(C_6H_5NH)_2$.
Properties: White, crystalline powder. D 1.10, mp 144C. Soluble in alcohol. Combustible.
Derivation: Desulfurization of thiocarbanilide in the presence of aniline.
Use: Accelerator for vulcanization of rubber.

triphenylmethane dyes. Any of a group of dyes whose molecular structure is basically derived from $(C_6H_5)_3CH$, usually by substitution of NH_2, OH, HSO_3, or other groups or atoms for some of the hydrogen of the C_6H_5 groups. Many coal tar and synthetic dyes are of this class, including rosaniline, fuchsin, malachite green, and crystalline violet.
See triarylmethane dye.

triphenylmethane triisocyanate. Available as a brown 20% solution in methylene chloride.
Use: Bonding uncured rubber to metal or other surfaces.

triphenylmethylhexafluorophosphate.
Properties: Orange-colored, free-flowing, crystalline powder.
Use: Catalyst in the manufacture of polyoxymethylenes (polymers of formaldehyde and trioxane).
See acetal resin.

triphenyl phosphate. (TPP).
CAS: 115-86-6. $PO(OC_6H_5)_3$.
Properties: Colorless, crystalline powder; odorless. Mp 50C, bp 245C (11 mm Hg mercury), d 1.268 (60C), bulk d 10.5 lb/gal, refr index 1.550 (60C), flash p 428F (220C) (CC). Soluble in most lacquers, solvents, thinners, oils; insoluble in water. Combustible.
Derivation: Interaction of phenol and phosphorus oxychloride.
Grade: Technical.

Hazard: Toxic by inhalation. Cholinesterase inhibitor. Questionable carcinogen.
Use: Fire-retarding agent, plasticizer for cellulose acetate and nitrocellulose.

triphenylphosphine. See triphenylphosphorus.

triphenylphosphine monosulfide.
CAS: 3878-45-3. $C_{18}H_{15}PS$.
Hazard: A poison.

triphenyl phosphite.
CAS: 101-02-0. $(C_6H_5O)_3P$.
Properties: Water-white to pale-yellow solid or oily liquid; pleasant odor. D 1.184 (25/25C), mp 22–25C, bp 155–160C (0.1-mm Hg), refr index 1.589 (25C), flash p 425F (218.3C) (COC). Combustible.
Use: Chemical intermediate, stabilizer systems for resins, metal scavenger, diluent for epoxy resins.

triphenylphosphorus. (triphenylphosphine).
CAS: 603-35-0. $(C_6H_5)_3P$.
Properties: White, crystalline solid. Mp 79–82C, bp above −360C, d 1.132 (25C), flash p 356F (180C) (OC). Insoluble in water; slightly soluble in alcohol; soluble in benzene, acetone, carbon tetrachloride. Combustible.
Derivation: By a modified Grignard synthesis.
Use: Synthesis of organic compounds, phosphonium salts, other phosphorus compounds, polymerization initiator.

triphenyl-2-propenyl-stannane. See allyltriphenyltin.

triphenylstannane sulfate (2::1). See bis(triphenyltin)sulfate.

triphenylstannyl benzoate.
CAS: 910-06-5. $C_{25}H_{20}O_2Sn$.
Hazard: A poison.

triphenylstibine. See triphenylantimony.

triphenyltetrazolium chloride. See tetrazolium chloride.

triphenylthiocyanatostannane.
CAS: 7224-23-9. $C_{19}H_{15}NSSn$.
Hazard: A poison.

triphenyltin *p*-acetamidobenzoate.
CAS: 2847-65-6. $C_{27}H_{23}NO_3Sn$.
Hazard: A poison.

triphenyltin acetate.
CAS: 900-95-8. $(C_6H_5)_3SnOOCCH_3$. An agricultural biocide, white crystalline solid, made by reaction of sodium acetate with triphenyltin chloride.
Hazard: Irritant to skin.

triphenyltin benzoate.   See triphenylstannyl benzoate.

triphenyltin chloride.
CAS: 639-58-7. $(C_6H_5)_3SnCl$.
Properties: White, crystalline solid. Mp 106C, bp 240C (13.5 mm Hg). Insoluble in water; soluble in organic solvents.
Derivation: Reaction of tin tetrachloride with phenylmagnesium bromide.
Hazard: An irritant to skin.
Use: Biocidal intermediate.

triphenyltin cyanoacetate.
CAS: 73927-89-6. $C_{21}H_{17}NO_2Sn$.
Hazard: A poison.

triphenyltin hydroxide.
CAS: 76-87-9. $(C_6H_5)_3SnOH$.
Properties: White solid. Mp 118–120C. Insoluble in water; soluble in ether, benzene, and alcohol.
Hazard: Irritant to skin.
Use: Insect chemisterilant, fungicide.

triphenyltin iodide.   See iodotriphenylstannane.

triphenyltin levulinate.
CAS: 23292-85-5. $C_{23}H_{22}O_3Sn$.
Hazard: A poison.

triphenyltin methanesulfonate.
CAS: 13302-08-4. $C_{19}H_{18}O_3SSn$.
Hazard: A poison.

triphenyltin propiolate.
CAS: 67410-20-2. $C_{21}H_{16}O_2Sn$.
Hazard: A poison.

triphenyltin thiocyanate.   See triphenylthiocyanatostannane.

triphenyl-1h-1,2,4-triazol-1-yl tin.
CAS: 974-29-8. $C_{20}H_{17}N_3Sn$.
Hazard: A poison.

triphosgene.   See hexachloromethylcarbonate.

triphosphoric acid.   See polyphosphoric acid.

triple bond.   A highly unsaturated linkage between the two carbon atoms of acetylenic compounds (alkynes), typified by acetylene (HC≡CH). See double bond; chemical bonding.

triple point.   The temperature and pressure at which the solid, liquid, and vapor of a substance are in equilibrium with one another. Also applied to similar equilibrium between any three phases, i.e., two solids and a liquid, etc. The triple point of water is +0.072C at 4.6 mm Hg; it is of special importance because it is the fixed point for the absolute scale of temperature.
See Plait point.

triple superphosphate.   A dry, granular, free-flowing product, gray in color. Produced by addition of phosphoric acid to phosphate rock, thus avoiding formation of insoluble gypsum, as in superphosphate, and achieving three times the amount of available phosphate (as P_2O_5). Typical analysis: moisture 2%, available P_2O_5 50%, water solution P_2O_5 45%, free phosphoric acid 1%, also minor ingredients.
Use: Fertilizer.
See superphosphate; nitrophosphate.

triplet.   An electronic state of a molecule in which two spins are aligned. This term is derived from spectroscopy: a system of two aligned spins has three possible orientations with respect to a magnetic field; each has a different energy, resulting in sets of three field-dependent spectral lines.
See doublet; singlet.

tripoli.   (amorphous silica; cristobalite; rottenstone; silicon dioxide).
CAS: 1317-95-9.
Properties: Soft, porous granules resulting from natural decomposition of siliceous rock.
Grade: Various grades according to fineness for polishing; rose, cream, white.
Hazard:
Use: Abrasive, polishing powder, filtering material, absorbent for insecticidal chemicals, paints (inert filler, wood filler), rubber filler, base for scouring soaps and powders, oil-well drilling muds.

tripolyphosphate.   See sodium tripolyphosphate.

tripotassium dicitratobismuthate.   See bismuth subcitrate.

tripotassium orthophosphate.   See potassium phosphate, tribasic.

tripotassium phosphate.   See potassium phosphate, tribasic.

tripropionin.   See glyceryl tripropionate.

tri-*n*-propylaluminum. $(C_3H_7)_3Al$.
Properties: Colorless, pyrophoric liquid. D 0.820; fp −84C.
Derivation: Reaction of propylene and isobutylaluminum.
Hazard: Flammable, dangerous fire risk, ignites spontaneously in air.
Use: Polyolefin catalyst.

tripropylamine.
CAS: 102-69-2. $(CH_3CH_2CH_2)_3N$.

Properties: Water-white liquid; amine odor. Fp −94C, boiling range 150–156C, d 0.754 (20/20C), refr index 1.417 (20C), flash p 105F (40.5) (OC). Combustible.
Hazard: Moderate fire risk. Toxic by inhalation and ingestion.

tripropyl(butylthio)stannane.
CAS: 67445-50-5. $C_{13}H_{30}SSn$.
Hazard: A poison.
Use: Drug.

tripropylene.   (propylene trimer). $(C_3H_6)_3$.
Properties: Colorless liquid. D 0.738 (20/20C), boiling range 133.3–141.7C, bulk d 6.17 lb/gal (60F), flash p 75F (23.9C) (TOC).
Derivation: Catalytic polymerization of propylene.
Hazard: Flammable, dangerous fire risk.
Use: Oxo feed stock, lubricant additive, plasticizers, nonyl phenol.

tripropylene glycol.
CAS: 24800-44-0. $HO(C_3H_6O)_2C_3H_6OH$.
Properties: Colorless liquid. Supercools instead of freezing, bp 268C, d 1.019 (25/25C), bulk d 8.51 lb/gal, refr index 1.442 (25C), flash p 285F (140.5C). Soluble in water, methanol, ether. Combustible.
Use: Intermediate in resins, plasticizers, pharmaceuticals, insecticides, dyestuffs, mold lubricants.

tripropylene glycol monomethyl ether.
CAS: 20324-33-8. $HO(C_3H_6O)_2C_3H_6OCH_3$.
Properties: Colorless liquid. D 0.961 (25C), bp 242C, 116C (10 mm Hg), viscosity 5.5 cP (25C), refr index 1.427 (25C), flash p 250F (121C). Miscible with water, VM&P naphtha, acetone, ethanol, benzene, carbon tetrachloride, ether, methanol, and monochlorobenzene. Combustible.
Use: Ingredient in hydraulic fluids.

tripropyltin acetate.
CAS: 3267-78-5. $C_{11}H_{24}O_2Sn$.
Hazard: A poison by ingestion.

tri-*n*-propyltin bromide.   See bromotripropylstannane.

tripropyltin chloride.   See chlorotripropylstannane.

tripropyltin iodide.
CAS: 7342-45-2. $C_9H_{21}ISn$.
Properties: Colorless liquid. D: 1.692 @ 16C, mp: −53C, bp: 262C. Soluble in organic solvents.
Hazard: A poison.

tripropyltin iodoacetate.
CAS: 73927-92-1. $C_{11}H_{23}IO_2Sn$.
Hazard: A poison.

tripropyltin-*o*-iodobenzoate.   See (*o*-iodobenzoyloxy)tripropylstannane.

tripropyltin isothiocyanate.
CAS: 31709-32-7. $C_{10}H_{21}NSSn$.
Hazard: A poison.

tripropyltin trichloroacetate.
CAS: 73927-99-8. $C_{11}H_{21}Cl_3O_2Sn$.
Hazard: A poison.

triptane.   See 2,2,3-trimethylbutane.

triptolide 12,13-chlorhydrin.
CAS: 132368-08-2. $C_{20}H_{25}ClO_6$.
Hazard: A reproductive hazard.

tris-.   A prefix indicating that a certain chemical grouping occurs three times in a molecule, e.g., tris(hydroxymethyl)aminomethane $(CH_2OH)_3CNH_2$.
See bis-.

tris amine buffer.   See tris(hydroxymethyl)aminomethane.

tris(aminoethyl)amine.   See 2,2′,2″-nitrilotris(ethylamine).

tris(1-aziridinyl)phosphine   oxide.   Legal label name for triethylenephosphoramide.

2,4,6-tris(1-aziridinyl)-*s*-triazine.   See triethylenemelamine.

trisazo dye.   One of the four kinds of azo dyes, characterized by the presence of three azo couplings (−N=N−) in each molecule. (CI 30000–34999).

tris(2-chloroethyl) phosphate.
CAS: 115-96-8. $(ClC_2H_4O)_3PO$.
Properties: Clear, transparent liquid. D 1.425 (20/20C), bp 214C (25 mm Hg), refr index 1.4721 (20C), flash p 421F (216C) (COC). Combustible.
Use: Flame-retardant plasticizer.

tris(2-chloroisopropyl) thionophosphate.
$[CH_3(CH_2Cl)CHO]_3PS$.
Properties: Liquid. Phosphorus content 9.0%, d 1.282 at (20C), flash p above 347F (175C) (OC), pour p below −50C. Readily soluble in gasoline; insoluble in water. Combustible.
Use: To extend spark plug life, to control deposit-induced knocking in gasoline engines.

tris(2-chloroethyl) phosphite.
CAS: 140-08-9. $(ClC_2H_4O)_3P$.
Properties: Colorless liquid; characteristic odor. D 1.353 (20/4C), bp 119C (0.15 mm Hg), flash p 280F (137.7C) (OC). Miscible with most common organic solvents; insoluble in water and hydrolyzes in water. Undergoes intramolecular isomerization

at higher temperatures, exposure to air should be minimum. Combustible.
Use: Intermediate, component of vinyl stabilizers, grease additives, flameproofing compositions.

tris(*p*-chlorophenyl)tin fluoride.
CAS: 427-45-2. $C_{18}H_{12}Cl_3FSn$.
Hazard: A poison.

tris(2,3-dibromopropyl) phosphate.
$(CH_2BrCHBrCH_2O)_3PO$.
Properties: Viscous, pale-yellow liquid. Bulk d 18.5 lb/gal, refr index 1.5772 (20C). Combustible.
Hazard: A carcinogen; use restricted.
Use: Flame-retardant for plastics and synthetic fibers.

tris(2,3-dichloropropyl) phosphate.
$(CH_2ClCHClCH_2O)_3PO$. Combustible.
Use: Flame-retardant in plastics and as a secondary plasticizer.
Hazard: A carcinogen.

tris(diethyldithiocarbamato)iron.   See iron(III) diethyldithiocarbamate.

tris(diethylene glycol monoethyl ether) citrate. $C_{19}H_{42}O_3$.
Properties: Solid. D 1.28 (25C), mp 16–19C. Soluble in water. Combustible.
Use: Plasticizer.

tris((2-ethylhexanoyl)oxy)methylsilane.
CAS: 70682-61-0. $C_{25}H_{48}O_6Si$.
Hazard: Moderately toxic by ingestion.

tris(2-ethylhexyl)   phosphate. $[C_4H_9CH(C_2H_5)CH_2O]_3PO$.
Properties: Colorless liquid. D 0.9260 (20/20C), fp −90C (sets to a glass), pour p −74C, mid-boiling p 216C (4 mm Hg), refr index 1.441 (25C), flash p 420F (215.5C). Insoluble in water; soluble in mineral oil and gasoline. Combustible.
Use: Low-temperature plasticizer for PVC resins, imparting flame and fungus resistance.

tris-2-ethylhexyl phosphite.   (trioctyl phosphite). $[C_4H_9CH(C_2H_5)CH_2O]_3P$.
Properties: Colorless liquid. D 0.902, bp 163–164C (0.3 mm Hg). Insoluble in water; soluble in alcohol and ether, flash p 340F (171C). Combustible.
Use: Synthesis, plasticizers, stabilizers, lubricant and grease additives, flameproofing compositions.

tris(2-hydroxyethyl)isocyanurate.   (THEIC; tris(2-hydroxyethyl)-*s*-triazine-2,4,6-trione).
$C_3N_3O_3(CH_2CH_2OH)_3$.
Properties: White solid. Mp 135C, bp (dissociates) 180C (3 mm Hg). Very soluble in water; somewhat soluble in alcohol and acetone; insoluble in chloroform and benzene. Combustible.

Use: Additive to plastics, especially to impart thermal stability.

tris(hydroxymethyl)acetic acid. $HOCH_2$ $C(CH_2OH)_2COOH$. A photographic chemical made by bacterial oxidation of pentaerythritol.

tris(hydroxymethyl)aminomethane.
(tri[hydroxymethyl]-aminomethane;   THAM; 2-amino-2-hydroxymethyl-1,3-propanediol;   tris amine buffer). $(CH_2OH)_3CNH_2$.
Properties: White, crystalline solid. Mp 171–172C, bp 219–220C (10 mm Hg), pH (0.1M aqueous solution) 10.6, corrosive to copper, brass, aluminum. Solubility in water 80 g/100 cc (20C). Combustible.
Hazard: Irritant to eyes and skin.
Use: Emulsifying agent (in soap form) for oils, fats, and waxes; absorbent for acidic gases; chemical synthesis; buffer; medicine.

tris(hydroxymethyl)nitromethane.   (2-hydroxymethyl-2-nitro-1,3-propanediol).
CAS: 126-11-4. $(CH_2OH)_3CNO_2$.
Properties: White crystals or amorphous solid. Mp 175C (decomposes). Soluble in water and alcohol.
Hazard: Moderate fire risk. Irritant to skin and eyes.
Use: Bactericide and slimicide for aqueous systems, cutting oil emulsions, industrial water systems, drilling muds.

tris(hydroxymethyl)phosphine.
CAS: 2767-80-8. $C_3H_9O_3P$.
Hazard: A poison by ingestion. A severe skin and eye irritant.

tris(4-hydroxyphenyl)ethane.   See 4,4′,4″-ethylidynetrisphenol.

1,1,3-tris(hydroxyphenyl)propane.   (triphenol P). $(C_6H_4OH)_2CHCH_2CH_2C_6H_4OH$.
Properties: White solid. Mp 84C, d 1.226 (20/20C), fp 90–110C, sets to glass below this temperature. Combustible.
Use: Antioxidant, intermediate for polyester and alkyd resins.

trisilane. Si_3H_8.
Properties: Colorless liquid. Bp 53C, mp −117C, d 0.74.
Hazard: Explodes on contact with air, reacts violently with carbon tetrachloride and chloroform.

trisiloxane.   See siloxane.

1,1,1-tris(4-isocyanatophenoxymethyl) propane.
CAS: 121514-80-5. $C_{27}H_{23}N_3O_6$.
Hazard: Moderately toxic by ingestion. A mild eye irritant.

tris[1-(2-methyl)aziridinyl]phosphine oxide.
$(C_3H_6N)_3PO$.
See metepa.

tris(nictinato)aluminum. See aluminum nicotinate.

Tris Nitro. 2-hydroxymethyl-2-nitro-1,3-propanediol.
Available forms: Solid and liquid.
Use: A water-soluble antimicrobial agent for use in cutting fluids and other aqueous alkaline systems; formaldehyde donor with controlled rate of release and minimum odor.

trisodium (carboxymethoxy)propanedioate.
See sodium o-carboxymethyltartronate.

trisodium citrate. See sodium citrate.

trisodium dipotassium tripolyphosphate.
$Na_3K_2P_3O_{10}$.
Properties: White, crystalline solid. Mp 620–640C, d 2.48. Solubility in water 80 g/100 cc (26C).
Use: Sequestrant.

trisodium EDTA. (ethylenediaminetetraacetic acid trisodium salt).
CAS: 150-38-9. $C_{10}H_{13}N_2Na_3O_8 \cdot H_2O$.
Properties: White powder. Freely soluble in water.
Use: Chelating agent.

trisodium hydroxyethylethylenediaminetriacetate. $(NaOOCCH_2)_2NC_2H_4N(CH_2COONa)C_2H_4OH$.
Properties: Liquid. D 1.285, fp below −5C.
Use: Chelating agent.

trisodium nitrilotriacetate.
CAS: 5064-31-3. $N(CH_2COONa)_3$.
A sodium salt of nitrilotriacetic acid.
Use: Sequestrant.

trisodium orthophosphate. See sodium phosphate, tribasic.

trisodium orthovanadate. See sodium orthovanadate.

trisodium phosphate. See sodium phosphate, tribasic.

trisodium phosphate, chlorinated.
$4(Na_3PO_4 \cdot 12H_2O) \cdot NaOCl$. Active ingredients 3.25% min sodium hypochlorite and 91.75% min trisodium phosphate dodecahydrate. Inert ingredients below 5% sodium chloride.
Properties: White, crystalline solid. Water soluble; stable under normal storage conditions; in solution has the properties of both trisodium phosphate and sodium hypochlorite.

Derivation: By reacting sodium phosphate, caustic soda, and sodium hypochlorite.
Hazard: Irritant to skin and eyes.
Use: Cleaner and bactericide in dairies, food plants, dishwashing compounds, and scouring powders.

trisodium phosphate monohydrate. See sodium phosphate, tribasic, monohydrate.

trisodium trithiocyanuric acid. See sodium trithiocyanurate.

trisomy. Possessing three copies of a particular chromosome instead of the normal two copies.
See cell; gene; gene expression; chromosome.

tristearin. See stearin.

tris(tetrahydrofurfuryl) phosphate.
$(C_4H_7OCH_2O)_3PO$.
Properties: Yellow liquid. Fp −75C, refr index 1.4759 (20C). Soluble in water and aromatic solvents.
Use: Plasticizer, solvent.

tris(tridecyl)amine.
CAS: 5910-77-0. $C_{39}H_{81}N$.
Hazard: A poison by ingestion and skin contact.

Tritac. A group of herbicides whose active ingredient is 2,3,6-trichlorobenzyloxypropanol.

triticonazole.
CAS: 131983-72-7. $C_{17}H_{20}ClN_3O$.
Hazard: A reproductive hazard.
Use: Agricultural chemical.

tritium. T. Radioactive isotope of hydrogen, mass number 3, isotopic weight 3.017 (two neutrons and one proton in the nucleus).
Properties: Half-life 12.5 years, radiation β, radiotoxic material.
Derivation: Bombardment of lithium with low-energy neutrons.
Available forms: Gas packaged in ampules and in tagged compounds such as water, streptomycin, cortisone, epinephrine, octadecane, stearic acid, etc.
Use: Bombarding particle in cyclotrons, activator in self-luminous phosphors, in cold cathode tubes, tracer in biochemical research and various special problems in chemical analysis, luminous instrument dials, thermonuclear power research.
See fusion.

tritolyl phosphate. See tricresyl phosphate.

"Triton" *[Dow].* TM for specialty surfactants that are alcohol and ethoxylates, alkoxylates, sulfates, sulfonates, sulfosuccinates, or phosphate esters.

Use: Industrial and institutional cleaners, various detergents, emulsions polymerization processes, paper, and textile processing.

"Triton RW 20" *[Dow].*
CAS: 119823-35-7.
TM for a surfactant.
Hazard: Moderately toxic by ingestion and skin contact. A severe skin irritant.

"Triton RW 50" *[Dow].*
CAS: 119823-36-8.
TM for a surfactant.
Hazard: Moderately toxic by ingestion and skin contact. A moderate skin and severe eye irritant.

"Triton RW 75" *[Dow].*
CAS: 86924-47-2.
TM for a surfactant.
Hazard: Low toxicity by ingestion and skin contact. A severe eye irritant.

"Triton RW 100" *[Dow].*
CAS: 83383-01-1.
TM for a surfactant.
Hazard: Moderately toxic by ingestion. Low toxicity by skin contact. A severe eye irritant.

"Triton RW 125" *[Dow].*
CAS: 86924-48-3.
TM for a surfactant.
Hazard: Moderately toxic by ingestion. Low toxicity by skin contact. A severe eye irritant.

"Triton RW 150" *[Dow].*
CAS: 86924-49-4.
TM for a surfactant.
Hazard: Moderately toxic by ingestion. Low toxicity by skin contact. A severe eye irritant.

triton. The nucleus of the T atom (one proton, two neutrons).

tritopine. See laudanidine.

triturate. To reduce to a powder by rubbing or grinding, to pulverize.

triuranium octoxide. (uranous-uranic oxide; uranyl uranate). U_3O_8.
Properties: Olive-green to black solid, crystals, or granules. D 8.39, decomposes when heated to 1300C to uranium dioxide. Insoluble in water; soluble in nitric acid and sulfuric acid.
Source: The naturally occurring uranium oxide found in pitchblende.
Derivation: (1) As one of the forms of uranium produced from the ores, often by a solvent extraction process. The solvent used is dodecylphosphoric acid. (2) A common form of triuranium octoxide is yellow cake, the powder obtained by evaporating an ammonia solution of the oxide.

Hazard: Radioactive poison. Use appropriate protection in handling.
Use: Nuclear technology, preparation of other uranium compounds.

trivial name. The name applied by early chemists to a number of simple organic compounds, usually based on their sources or properties, e.g., acetone and acetic acid, from Latin *acetum* (vinegar), urea from urine, glucose, and glycerol from Greek *glyc-* (sweet). Such names remained in common use regardless of the systematic nomenclature later developed.

trivinyltin chloride. See chloro(trivinyl) stannane.

trixylenyl phosphate. See tri(dimethylphenyl) phosphate.

troclosene potassium. See potassium dichloroisocyanurate.

trona. (urao).
CAS: 497-19-8. $Na_2CO_3 \cdot NaHCO_3 \cdot 2H_2O$.
A natural sodium sesquicarbonate and the most important of the natural sodas.
Properties: White, gray, or yellow with vitreous, glistening luster. Noncombustible.
See trona process.
Occurrence: Hungary, Egypt, Africa, Venezuela, and U.S. (Wyoming, California, especially Searles Lake, Owens Lake).
Use: Source of sodium compounds, especially the sodium carbonates.

trona process. The method used for separation and purification of soda ash, anhydrous sodium sulfate, boric acid, borax, potassium sulfate, bromine, and potassium chloride from Searles Lake (California) brine.

tropacocaine hydrochloride. $C_{15}H_{19}NO_2 \cdot$ HCl. An alkaloidal salt.
Properties: White crystals. Mp 271C. Soluble in water, alcohol, and ether.
Derivation: From a variety of *Erythroxylon coca.*
Hazard: Toxic by ingestion and inhalation.
Use: Medicine (spinal anesthetic).

tropaeolin D. See methyl orange.

tropaeolin OO. (Orange IV). $NaSO_3C_6H_4$ $NNC_6H_4NHC_6H_5$. (*p*-diphenylaminoazobenzene-sodium sulfonate). A biological stain and acid–base indicator, red at pH 1.4, yellow at pH 2.6. CI 13080.

tropisetron.
CAS: 89565-68-4. $C_{17}H_{20}N_2O_2$.
Hazard: Human systemic effects.
Use: A substance that is used to prevent nausea and vomiting caused by chemotherapy.

Trouton's rule. The molal heat of vaporization of normal liquids, at the boiling point and under atmospheric pressure, divided by the absolute boiling temperature is a constant, about 22.

"Troysan 395" *[Troy].* TM for 1,3-bis (hydroxymethyl)-5,5-dimethyl hydantoin.
Available forms: Liquid.
Use: Bactericide for water-based systems, paints and coatings, adhesives, joint cements, caulks, and printing inks.

Trp. Abbreviation for tryptophan.

true solution. See solution (true).

truth serum. See scopolamine.

truttine. A protein obtained from fish of the trout family.

"Trycite" *[Dow].* TM for an oriented polystyrene film.
Use: Packaging and in envelope windows.

tryparsamide. (sodium-*N*-phenylglycine-amide-*p*-arsonate).
CAS: 618-25-7. NaOAs(OOH)C$_6$H$_4$NHCH$_2$CONH$_2$•1/2H$_2$O.
Properties: White, crystalline powder; odorless. Contains 24.6% arsenic, may affect eyes. Soluble in water; almost insoluble in alcohol; insoluble in ether, chloroform, benzene.
Grade: Medicinal, USP.
Hazard: Highly toxic by ingestion and inhalation.
Use: Medicine (treatment of syphilis).

trypsin.
CAS: 9002-07-7.
The proteolytic enzyme of pancreatic juice, yellow to grayish powder, soluble in water, insoluble in alcohol or glycerol. It acts on albuminoid material, producing amino acids. The maximum result is obtained in a neutral or slightly alkaline medium. Trypsins or similar materials are found not only in the pancreas but also in the spleen, leucocytes, and urine, as well as in beer yeast, molds, and bacteria.
Grade: NF (crystallized).
Hazard: Irritant to skin.
Use: Dehairing of hides.

trypsinogen. An inactive precursor of trypsin.

tryptophan. (indole-α-aminopropionic acid; 1-α-amino-3-indolepropionic acid).
CAS: 54-12-6.

One of the essential amino acids occurring naturally in the L(−)-form.
Properties: (DL-) White crystals. Slightly soluble in water; stable in alkaline solution; decomposed by strong acids.
(D(+)-) Characteristic sweet taste. Mp 275–290C (decomposes). Soluble in water, hot alcohol, alkali hydroxides; insoluble in chloroform.
(L(−)-) Flat taste (other properties identical with D(+)-tryptophan).
Derivation: (1) Synthetic tryptophan can be made by the conversion of indole to gramine followed by methylation, interaction with acetylaminomalonic ester and hydrolysis, (2) hydrolysis of proteins.
Grade: Reagent, technical, FCC.
Use: Nutrition and research, medicine, dietary supplement, cereal enrichment.
Available commercially in all three forms, as well as acetyl-DL-tryptophan.

Trysben 200. A weed killer based on aqueous solution of the dimethylamine salt of trichlorobenzoic acid, containing two pounds of acid equivalent per gallon.
Use: Control of broadleaf weeds.

Ts. Abbreviation for tosyl.

TSA. Abbreviation for toluenesulfonic acid.

Tscherniac–Einhorn reaction. Introduction of the amidomethyl group into aromatic rings or activated methylene groups in the presence of sulfuric acid.

TSH. See thyrotropic hormone.

Tsien, Roger Y. (1952–). An American chemist who was awarded the 2008 Nobel Prize conjointly with Osamu Shimomura and Martin Chalfie. The scientists discovered and researched the green fluorescent protein (GPF), naturally produced in jellyfish (*Aequorea victoria*), which is used to illuminate certain cells to study their biological processes. Tsien mapped out the structure of GPF and how it can be used to understand the activities of calcium ions in living cells. He was awarded his Ph.D. from the University of Cambridge. Apart from the Nobel Prize, Tsien has won numerous awards for his remarkable work including the Gairdner Foundation International Award (1995), Wolf Prize in Medicine (2004), and Foreign Member of Royal Society Award (2006).

TSP. Abbreviation for trisodium phosphate.
See sodium phosphate, tribasic.

TSPA. Abbreviation for triethylenethiophosphoramide.

TSPP. Abbreviation for tetrasodium pyrophosphate.
See sodium pyrophosphate.

TSS.   See transition state theory.

TTC.   Abbreviation for tetrazolium chloride.

TTD.   Abbreviation for tetraethylthiuram disulfide.

tubatoxin.   See rotenone.

tube mill.   A fine-grinding machine composed of a rotating steel cylinder that may be from 15 to 50 ft long and 6 to 8 ft in diameter, within which are steel balls from 1 to 5 inches in diameter. Depending on its construction it may be either batch or continuous. Some types have several compartments, each containing balls of different sizes. Its function is finish-grinding of particulates; it is usually fed with 20-mesh material, which it reduces to 325-mesh. See ball mill.

tuberose oil.
Properties: Colorless to very-light-colored oil. D 1.007–1.035 (15C).
Method of purification: Taken from *Polianthes tuberosa*, by enfleurage.
Use: Perfume, flavoring.

***d*-tubocurarine chloride.** $C_{38}H_{44}Cl_2N_2O_6 \cdot 5H_2O$.
Properties: White to light-tan crystalline alkaloid; odorless. Mp 270C with decomposition. Soluble in water and alcohol; insoluble in acetone, chloroform, and ether; aqueous solution is strongly dextrorotatory (specific rotation for 1% solution of anhydrous −208 to +218 degrees).
Grade: USP.
Hazard: Highly toxic.
Use: Medicine (muscle relaxant).

"Tuex" [Chemtura].   TM for tetramethylthiuram disulfide.
Available forms: Pellets as "Tuex Naugets."
Use: Sulfur donor accelerator for EV and semi-EV systems.
See thiram.

"Tuffak" [Bayer Materialscience].   TM for colored or colorless polycarbonate resin.
Available forms: Sheet material of various thickness; film.
Use: Skylights, display fixtures, signs, lighting fixtures, safety guards, car mats, and chair mats.

"Tuflin" [Union].   TM for linear low-density polyethylene film resins.
Use: For trash bags, packaging and stretch wrap, and shipping sacks.

tumor suppressor.   A gene that inhibits progression toward neoplastic transformation. The best known examples of tumor suppressors are the proteins p53 and Rb.

tung oil.   (China-wood oil).
Properties: Yellow drying oil. D 0.9360–0.9432, saponification value 193, iodine value 150–165, refr index 1.5030, flash p 552F (288.9C), autoign temp 855F (457C). Soluble in chloroform, ether, carbon disulfide, and oils. Combustible.
Derivation: From the seeds of *Aleurites cordata*, a tree indigenous to China. It is now produced in China, Argentina, and Paraguay; U.S. production has been virtually discontinued.
Chief constituents: Eleostearic acid (80%).
Hazard: Toxic by ingestion. Contact causes dermatitis. Ingestion causes nausea, vomiting, cramps, diarrhea and tenesmus, thirst, dizziness, lethargy, and disorientation. Large doses can cause fever, tachycardia, and respiratory effects. Combustible when exposed to heat or flame. Can react with oxidizing materials.
Use: Food additive; wood finish.

tungstated pigment.   See phosphotungstic pigment.

tungstate white.   See barium tungstate.

tungsten.   (wolfram).
CAS: 7440-33-7.   W.
Metallic element, atomic number 74, group VIB of periodic system, aw 183.85, valences of 2, 4, 5, 6; five isotopes.
Properties: Hard, brittle, gray solid; not found native; the ores are scheelite and wolframite. D 19.3 (20C), mp 3410C (highest of all the metals), bp 5927C, high electrical conductivity, soluble in a mixture of nitric acid and hydrogen fluoride, corroded by seawater, oxidizes in air at 400C, the rate increases rapidly with temperature.
Derivation: (1) Aluminothermic reduction of tungstic oxide, (2) hydrogen reduction of tungstic acid or its anhydride.
The metal can be plated onto objects by vapor deposition from tungsten hexafluoride or hexacarbonyl and can bond metal parts together. Tungsten powder (produced by carbon reduction) is converted into solid metal by powder metallurgical techniques. Large single crystals are grown by an arc-fusion process. Granules obtained by reduction of the hexafluoride.
Occurrence: Canada, Bolivia, Peru, Thailand, China, the former U.S.S.R., U.S. (Arizona, California, Colorado, Nebraska, Nevada, New Mexico, and Texas).
Grade: Technical, powder, single crystals, ultrapure granules of 50–600 microns.
Hazard: Finely divided form is highly flammable and may ignite spontaneously. Central nervous system impairment, pulmonary fibrosis, and lower respiratory tract irritant.
Use: High-speed tool steel, ferrous and nonferrous alloys, ferrotungsten, filaments for electric light bulbs, contact points, X-ray and electron tubes, welding electrodes, heating elements in furnaces

and vacuum-metallizing equipment, rocket nozzles and other aerospace applications, shell steel, chemical apparatuses, high-speed rotors as in gyroscopes, solar energy devices (as vapor-deposited film that retains heat at 500C).

tungsten boride. WB_2.
Properties: Silvery solid. D 10.77, mp 2900C. Insoluble in water; soluble in aqua regia and concentrated acids; decomposed by chlorine at 100C.
Derivation: Heating tungsten and boron in electric furnace.
Use: Refractory.

tungsten carbide.
CAS: 12070-12-1. WC.
A ditungsten carbide, W_2C, with similar physical properties is also known. WC is said to be the strongest of all structural materials.
Properties: Gray powder. D 15.6, mp 2780C, bp 6000C, Mohs hardness of 9+ in solid form. Insoluble in water but readily attacked by a nitric acid–hydrofluoric acid mixture. Stable to 400C with chlorine, burns in fluorine at room temperature, oxidizes on heating in air.
Derivation: Heating tungsten and lampblack at 1500–1600C.
Hazard: Toxic by inhalation.
Use: Cemented carbide, dies and cutting tools, wear-resistant parts, cermets, electrical resistors. An abrasive in liquids.

tungsten carbide, cemented.
A mixture consisting of tungsten carbide 85–95% and cobalt 5–15%.
Properties: Hardness about that of corundum; not affected by high temperatures. D 12–16.
Derivation: Ball milling of powdered tungsten carbide with metallic cobalt, followed by sintering.
Hazard: As for tungsten carbide.
Use: Machine tools and abrasives for machining and grinding metals, rocks, molded products, porcelain, and glass; in gauges, blast nozzles, knives, drill bits.

tungsten carbonyl.
See tungsten hexacarbonyl.

tungsten diselenide. WSe_2.
Lamellar-structured, dry, solid lubricant with exceptionally high temperature and high vacuum stability. Retains its lubricity to temperatures as low as −265C.
Grade: 1–2 micron and 40 micron.
Hazard: TLV: 5 mg(W)/m^3.

tungsten disulfide. WS_2.
Properties: Grayish-black solid. Mp above 1480C, can lubricate at temperatures above 2400F (1316C). Attacked by fluorine and hot sulfuric and hydrogen fluoride.
Grade: 0.40, 0.70, and 1–2 micron grades.
Use: Solid lubricant for many applications, including use as an aerosol.

tungsten hexacarbonyl. (tungsten carbonyl). $W(CO)_6$.
Properties: White, volatile, highly refractive, crystalline solid. Decomposes without melting at 150C. One of the more stable carbonyls. D 2.65, vap press 0.1 mm Hg (20C). Insoluble in water; soluble in organic solvents.
Derivation: Reaction of tungsten with carbon monoxide at high pressures, reduction of tungsten hexachloride with iron alloy powders in carbon monoxide atmosphere.
Use: Tungsten coatings on base metals by deposition and decomposition of the carbonyl.

tungsten hexachloride. WCl_6.
Properties: Dark-blue or violet hexagonal crystals, volatile. Mp 275C, bp 347C, d 3.52, vap press 43 mm Hg (215C). Electrical conductivity (fused state) poor. Soluble in organic solvents including ligroin and ethanol; decomposed by moist air and water; reduced by hydrogen to the metal.
Derivation: Heating tungsten with dry chlorine at red heat.
Use: Tungsten coatings on base metals, vapor deposition for bonding metals, single crystal tungsten wire, additive to tin oxide to produce electrically conducting coating for glass, catalyst for olefin polymers.

tungsten hexafluoride.
CAS: 7783-82-6. WF_6.
Properties: Colorless gas or light-yellow liquid. D (liquid) 3.44, mp 2.5C, bp 19.5C. Decomposes in water.
Derivation: Direct fluorination of powdered tungsten, purified by distillation under pressure.
Hazard: Toxic by ingestion and inhalation, strong irritant to tissue.
Use: Vapor-phase deposition of tungsten, fluorinating agent.

tungsten lake.
See phosphotungstic pigment.

tungsten oxychloride. $WOCl_4$.
Properties: Dark-red, acicular crystals. Decomposed by water and moist air. Keep in sealed glass container. Bp 227.5C, mp 211C, d 11.92. Soluble in carbon disulfide.
Derivation: By the action of chlorine on tungsten or tungstic oxide at elevated temperatures.
Method of purification: Vacuum distillation.
Grade: Technical.
Hazard: Irritant.
Use: Incandescent lamps.

tungsten silicide.
A ceramic, probably WSi_2.
Properties: Blue-gray, very hard solid. D 9.4, mp above 900C. Insoluble in water; attacked by fused alkalies and mixture of nitric and hydrofluoric acids.

Grade: Cylindrical shapes, lumps, standard sieve sizes.
Hazard: Dust flammable.
Use: Oxidation-resistant coatings, electrical resistance and refractory applications.

tungsten steel. In many of its alloying effects, tungsten is similar to molybdenum. Tungsten increases the density of alloys to which it is added.
Use: To obtain steels with great wear resistance and special resistance to tempering, as in high-speed steels, hot-work steels, finishing steels that maintain keen cutting edge and great wear resistance, creep-resisting steels, and oxidation-resistant, high-temperature, high-strength alloys.

tungsten trioxide. See tungstic oxide.

tungstic acid. (wolframic acid; orthotungstic acid).
CAS: 11105-11-6. H_2WO_4.
Properties: Yellow powder. D 5.5. Insoluble in water; soluble in hydrogen fluoride and alkalies. A white form of tungstic acid exists, having the formula $H_2WO_4 \cdot H_2O$. This is formed by acidifying tungsten solutions in the cold.
Derivation: Decomposition of sodium tungstate with hot sulfuric acid.
Grade: Technical, CP, reagent.
Hazard: Toxic material.
Use: Textiles (mordant, color resist), plastics, tungsten metal, wire, etc.

tungstic acid anhydride. See tungstic oxide.

tungstic anhydride. See tungstic oxide.

tungstic oxide. (tungstic acid anhydride; tungstic anhydride; tungsten trioxide; wolframic acid, anhydrous). WO_3.
Properties: Canary-yellow, heavy powder, dark-orange when heated and regains original color on cooling. Mp 1473C, d 7.16. Insoluble in water, soluble in caustic alkalies, soluble with difficulty in acids. Noncombustible.
Derivation: Scheelite ore is treated with hydrochloric acid and the resulting product dissolved out with ammonia. The complex ammonium tungstate can then be ignited to tungstic oxide.
Hazard: Toxic material.
Use: To form metal by reduction, alloys, preparation of tungstates for X-ray screens, fireproofing fabrics, yellow pigment in ceramics.

tungstophosphate. Structure and properties are similar to molybdophosphate.

12-tungstophosphoric acid. See phosphotungstic acid.

tungstosilicate. One of a group of complex inorganic compounds of high molecular weight, containing a central silicon atom surrounded by tungsten oxide octahedra. They have high molecular weight, high degree of hydration, strong oxidizing action in aqueous solutions; decompose in strongly basic aqueous solutions to give simple tungstate solutions; form highly colored anions or reaction products.
See silicotungstic acid; sodium-12-tungstosilicate.

12-tungstosilicic acid. See silicotungstic acid.

turinabol. See 4-chlorotestosterone 17-acetate.

turkey brown. (turkey umber). Natural earth used as permanent pigment. Contains iron and manganese oxide with some clay.

Turkey red. See iron oxide red.

Turkey red oil. (castor oil, sulfonated; castor oil, soluble).
CAS: 72-48-0.
It is also known as alizarin assistant and alizarin oil because of its use in dyeing with alizarin.
Properties: Viscous liquid. D 0.95, iodine No. 82.1, acid No. 174.3, saponification No. 189.3, autoign temp 833F (445C). Soluble in water. Combustible.
Derivation: By sulfonating castor oil with sulfuric acid and washing.
Grade: Sulfonated castor oil graded as to moisture and color.
Use: Textiles, leather, manufacture of soaps, alizarin dye assistant, paper coatings.

Turnbull's Blue. An inorganic blue pigment made by the reaction of a ferrous salt and potassium ferricyanide $[Fe_3Fe(CN)_6]_2$. One of its important uses is in making blueprints, in which sensitized paper containing ferric ammonium citrate and potassium ferricyanide is exposed to light, the ferric ions being thus reduced to ferrous ions.
See iron blue.

turnover number. The number of times an enzyme molecule transforms a substrate molecule per sec, under conditions giving maximal activity at substrate concentrations that are saturating.

turpentine (gum). The oleoresin or pitch obtained from living pine trees.
Properties: Sticky, viscous, balsamic liquid comprising a mixture of rosin and turpentine oil; strong, piney odor. Soluble in alcohol, ether, chloroform, and glacial acetic acid. Combustible.
Use: Source of turpentine oil and gum rosin.
See rosin.

turpentine (oil).
CAS: 8006-64-2. $C_{10}H_{16}$.

A volatile essential oil whose chief constituents are pinene and diterpene.

Derivation: Steam-distillation of the turpentine gum exuded from living pine trees (gum turpentine), naphtha-extraction of pine stumps (wood turpentine), destructive distillation of pine wood.

See rosin; rosin oil.

Properties: Colorless liquid; penetrating odor. Immiscible with water; lighter than water. Considerable variation appears in constants reported; the following are based on tests made by the Forest Products Laboratory: d 0.860–0.875 (15C), refr index 1.463–1.483 (20C), flash p 90–115F (32–46C) (CC), no acidity, bulk d 7.18 lb/gal, autoign temp 488F (253C).

Hazard: Moderate fire risk. Toxic by ingestion. Lung irritant. Questionable carcinogen.

Use: Solvent; thinner for paints, varnishes, and lacquers; rubber solvent and reclaiming agent; insecticide; synthesis of camphor and menthol; wax-based polishes; medicine (liniments); perfumery.

tutocaine hydrochloride.   (γ-dimethylamino-α,β-dimethylpropyl-p-aminobenzoate hydrochloride).

CAS: 532-62-7. $C_{14}H_{22}O_2N_2 \cdot HCl$.

Use: Local anesthetic.

tuyere.   A duct or pipe through which a stream of hot air is introduced into a blast furnace or cupola to support combustion.

Tw.   Abbreviation for Twaddell, used in reporting specific gravities for densities greater than water, as degrees Tw. A twaddell reading, multiplied by five and added to 1000, gives specific gravity with reference to water as 1000.

TWA.   See time-weighted average.

twinning axis.   Axis around which one part of a twin crystal may be viewed as having been rotated 180 degrees relative to the other part.

Twitchell process.   Commercial process for splitting fats to glycerol and fatty acids by heating the sulfuric acid-washed fat 20–48 hours in an open tank with steam in a mixture of 25–50% water, 0.5% sulfuric acid, and 0.75–1.25% Twitchell reagent. The original Twitchell catalyst was prepared by sulfonation of a mixture of fatty acid and benzene, but toway sulfonated petroleum products are used.

Twitchell reagent.   Catalyst for the Twitchell process (acid hydrolysis of fats). It is a sulfonated addition product of naphthalene and oleic acid, a naphthalenestearosulfonic acid.

"Tygon" [Saint-Gobail].   TM for a series of vinyl compounds used as linings, coatings, adhesives, tubing, and extruded shapes applied to chemical process equipment as corrosion protection.

"Tylac" [Mallard].   TM for a series of synthetic latexes and elastomers.

(BL) High-strength, film-forming, nitrile rubber latexes and rubbers and carboxylated polymers characterized by exceptional oil, solvent, and abrasion resistance.

(SBL) Butadiene/styrene and carboxylated butadiene/styrene and high styrene latexes of various monomer ratios. Modified versions are available.

Use: Paper, textiles, adhesives, Portland cement, molded products, and polyvinyl chloride, and phenolic resin blends.

"Tylenol" [Johnson & Johnson].   TM for p-acetylaminophenol (acetaminophen).

tylosin and sulfamethazine.

Use: Drug (veterinary); food additive.

tyloxapol.   (USAN).

CAS: 25301-02-4.

An oxyethylated-*tert*-octylphenolpolymethylene polymer and low toxicity, nonionic surfactant.

tyloxin.   Antibiotic substance isolated from a strain of *Streptomyces fradiae*.

Properties: Crystals. Mp 128–130C. Solutions are stable between pH 4 to 9.

Use: Veterinary medicine.

Tyndall effect.   A colloidal phenomenon in which particles too small to be resolved in an optical microscope suspended in a gas or liquid reveal their presence by scattering a beam of light as it passes through the suspension, the extent of reflection being dependent on the position of the irregularly shaped particles relative to the incident light. The effect causes the appearance of a visible cone of light through the suspension. This principle is utilized in the ultramicroscope.

"Tynex" [Du Pont].   TM for nylon filament. Available tapered, with an essentially uniform taper from butt to tip, and also level, that is, in a wide range of constant diameters. The tapered form is used primarily in paint brushes, the level form in other brushes.

type metal.   Alloy of 75–95% lead, 2.5–18% antimony, with a little tin and sometimes copper, which expands slightly upon solidification and produces sharp castings.

Tyr.   Abbreviation for tyrosine.

Tyrer sulfonation process.   Sulfonation of benzene in the vapor phase with sulfuric acid at 170–180C with passage of benzene vapor and azeotropic removal of the water of reaction.

Tyrez.   A series of impact modifiers for plastics.

tyrian brilliant blue i 3g.   See 6,15-dihydrohydroxy-5,9,14,18-anthrazinetetrone.

"Tyril" *[Styron].*   TM for a group of styrene–acrylonitrile copolymers.

"Tyrilfoam" 80 *[Styron].*   TM for expanded styreneacrylonitrile for flotation uses under conditions of gasoline spillage, petroleum scum from outboard motors, and stagnant water.

tyrocidine.   Antibiotic produced by the metabolic processes of the bacteria *Bacillus brevis.* It is a cyclic polypeptide that is active against most Gram-positive pathogenic bacteria. It is one of the two antibiotic components of tyrothricin but has been isolated and used alone.
   Properties: (Probably the hydrochloride) Fine, crystalline needles that decompose at 240C; soluble in 95% alcohol, acetic acid, and pyridine; slightly soluble in water, acetone, and absolute alcohol; insoluble in ether, chloroform, and hydrocarbons; depresses surface tension; forms fairly stable colloidal emulsion in distilled water.
   Use: Medicine (usually as component of tyrothricin), possible fungistat and bacteriostat.

tyrosinase.   An enzyme containing copper that occurs in plant and animal tissue and is responsible for turning peeled potatoes black when exposed to air.
   Use: Medicine (antihypertensive).

tyrosine.   (β-*p*-hydroxyphenylalanine; α-amino-β-*p*-hydroxyphenylpropionic acid).
CAS: 60-18-4. $C_6H_4OHCH_2CHNH_2COOH$.
A nonessential amino acid.
   Properties: White crystals. *dl*-Tyrosine mp 316C, d(+)−tyrosine mp 310–314C, l(−)-tyrosine mp 295C (with decomposition), d 1.456 (20/4C). Soluble in water; slightly soluble in alcohol; insoluble in ether. Optically active. Readily oxidized by animal organisms.
   Derivation: Hydrolysis of protein (casein), organic synthesis.
   Grade: FCC.
   Use: Growth factor in nutrition, biochemical research, dietary supplement.
   Available commercially as *dl*-tyrosine.

tyrothricin.   An antibiotic produced by growth of *Bacillus brevis.* It consists of a mixture of antibiotics, principally gramicidin and tyrocidine. Gramicidin is the more active component. Use is generally limited to local external applications. It is active against some Gram-positive bacteria, including species of pneumococci, streptococci, and staphylococci.

"Tyzor" *[Du Pont].*   TM for a group of simple and chelated esters of ortho-titanic acid, such as tetrabutyltitanate, of varying reactivity.
   Use: Chemical intermediates, primers for adhesion promotion in extrusion coating, dispersants, scratch-resistant finishes on glass, masonry water repellents, cross-linking agents, esterification and olefin polymerization catalyst.

U

U. Symbol for uranium.

"Ucar" [Union]. TM for various synthetic organic chemicals including solvents, coatings, resins, and deicing fluids. Also applied to synthetic latexes and water-soluble polymers.

"Ucarcide" [Union]. TM for glutaraldehyde-based antimicrobials.
Use: Preservative and sanitizing applications; oil fields.
See glutaraldehyde.

"Ucare" [Union]. TM for catatonic hydroxyethel cellulose, conditioning polymer.
Use: Shampoos and body washes.

"Ucarsan" [Union]. TM for a blend of glutaraldehyde and a surfactant.
Use: As a sanitizer for agricultural equipment, such as poultry hatcheries and farm-animal housing facilities; industrial equipment.
See glutaraldehyde.

"Ucon" [Union]. (1) TM for a series of non-flammable fluorocarbon solvents and solvent blends with high chemical stability and extremely low residue levels.
Use: Cleaning electronic and mechanical instruments and controls, degreasing motors, cleaning liquid oxygen equipment, and motion-picture and television film and magnetic tape.
(2) TM for polyalkylene glycols and diesters. Available as water-soluble or insoluble products.
Grade: LB-Series, 50-HB Series, 75-H Series, DLB Series, and Hydrolube Series.
Use: High-temperature lubricants, low-temperature fluids, compressor lubricants, hydraulic brake fluids, quenchant fluid, heat-transfer fluids, textile lubricants, rock drill lubricants, leather and paper-treating compounds, rubber lubricants, plasticizers and solvents, chemical intermediate, stationary phase in gas chromatography.
See fluorocarbon.

"Uconex" [Union]. TM for a biocide that is an aqueous solution of glutaraldehyde.
Use: In controlling bacteria and fungi in aqueous metalworking fluids.
See glutaraldehyde.

Ucure. Polyvinyl acetate reactive modifiers.

Use: Promoting reinforcement and improving dispersion, dimensional control, and surface quality in molded reinforced plastic automotive exterior panels, bathroom fixtures, and business machine housings.

UDP. Abbreviation for uridine diphosphate. See uridine phosphate.

UDPG. Abbreviation for uridine diphosphate glucose.

"Uformite" [Allnex]. TM for synthetic resins based on urea-formaldehyde, melamine-formaldehyde, and triazine condensates. Supplied as colorless or light-colored aqueous solutions or solutions in volatile solvents. Solvent type produces hard, alkali-resistant, colorless coatings on curing with adhesion to a variety of surfaces.
Use: With alkyd resins in coatings, industrial finishes on appliances, automobiles, etc.; adhesives for paperboard boxes, paper coatings, wet-strength paper, textile pigment binding.

"Ulano Fotocoat" [Ulano]. TM for a solvent resistant emulsion.
Use: For use with weak light sources.

ulexite. (cotton balls). $NaCaB_5O_9 \cdot 8H_2O$. A natural hydrated borate of sodium and calcium.
Properties: White color; silky luster. Mohs hardness 1–2.5, d 1.96. Usually found as rounded, loose-textured masses of fine crystals.
Occurrence: Chile, Argentina, California, Nevada.
Use: Source of borax.

ullage. (1) The loss by evaporation or leaking. (2) The amount a vessel lacks of being full.
See head space.

Ullman reaction. Synthesis of biaryls by copper-induced coupling of aryl halides. Similar coupling of aryl halides with aroxides yields diaryl ethers. (A modification of the Fittig synthesis in which copper powder is used instead of sodium.)

ulmin. One of a class of amorphous substances resulting from the decomposition of the cellulose and lignite tissues of plants. Ulmins represent one of the initial changes by which vegetable matter is converted into coal.

ulmin brown. See Van Dyke brown.

Hawley's Condensed Chemical Dictionary, Sixteenth Edition. Michael D. Larrañaga, Richard J. Lewis, Sr., and Robert A. Lewis.
© 2016 John Wiley & Sons, Inc. Published 2016 by John Wiley & Sons, Inc.

Ultimate.   A lime calcium and rust remover.

"Ultimer" *[Nalco].*   TM for oil-free, water-soluble polymers.
Use: Water and waste-treatment processes.

"Ultimet" *[Haynes].*   TM for a high-strength, cobalt-based alloy designed with excellent resistance to pitting and general corrosion, especially in oxidizing acids, coupled with exceptional wear resistance (cavitation erosion, galling, and abrasion).
Available forms: Sheet, plate, castings, bars, rods, welding electrodes and wire.
Use: Fabrication into all types of process equipment.

"ULTI-PRO 100" *[Betaseed].*   TM for a rubber processing aid.
Use: A reactive viscosity modifier for difficult-to-process rubber compounds.

"Ultisil" *[Ametek].*   TM for specially processed silica textiles having improved temperature properties of higher strength retention and flexibility after long exposures to temperatures as high as 2000F (1050C).
Use: Furnace curtains, stress-relief blankets, expansion joints, insulations involving encapsulation of refractory wools, aerospace composites, seals, and gaskets.

ultraaccelerator.   An unusually powerful accelerator of rubber vulcanization, typified by thiuram sulfides and dithiocarbamates.

ultracentrifuge.   A high-speed rotational separating device, usually of laboratory size, capable of developing a force of 250,000 times gravity. Its major uses are in research on molecular weight distribution, macromolecular structure and properties (proteins, nucleic acids, viruses) and separation of solutes from solutions.
See centrifugation.

"Ultra and Super Shield" *[Ultrashield].*   TM for gelcoat liquid.
Use: For boats, recreational vehicles, trucks, sanitary ware, pools.

Ultra-Clor.   A turf fungicide whose active ingredients are mercuric dimethyldithiocarbamate, potassium chromate, and cadmium succinate.
Hazard: Highly toxic by ingestion.

"Ultraflex" *[AVS].*   TM for microcrystalline wax.
Available forms: Prills, slab, or molten.
Use: Hot-melt adhesives for coatings and packaging, chewing-gum base, cosmetics, pharmaceuticals, lubricants, ceramics, processing aids, laminants, textiles, paints, rubber and elastomers, and waterproofing.

ultramarine blue.   (CI 77007).
Properties: Inorganic pigment, blue powder, good alkali and heat resistance, low hiding power, poor acid resistance, poor outdoor durability. Noncombustible.
Derivation: Heating a mixture of sulfur, clay, alkali, and a reducing agent to high temperatures.
Use: Colorant for machinery and toy enamels, white baking enamels, printing inks, rubber products, soaps and laundry blues, cosmetics, textile printing.
Note: Used in very low percentages to intensify whiteness of white enamels, rubber compounds, laundered clothing, etc., by offsetting yellowish undertones, giving a "blue" rather than a "yellow" white.

ultramicroscope.   A development of the compound optical microscope invented in 1903 by Zsigmondy and Siedentopf. Its essential feature is a strong light beam from an arc lamp, focused by passing through two lenses, which illuminates the specimen at right angles to the axis of observation. The presence of suspended colloidal particles as small as 5 microns is detectable because of the light-scattering effect of the particles as they move about in the suspension (Tyndall effect). Since the light reflected or scattered by the particles is the only light that enters the microscope, the particles appear as points of light against a dark background (dark-field illumination). There is no resolution of individual particle shape or size; the instrument shows only that particles are present. The ultramicroscope has been of great value in the study of colloidal suspensions, such as rubber latex, and of various biological phenomena; its usefulness has diminished since the advent of the electron microscope.
See Tyndall effect.

Ultraray.   Laser pure gases.
Use: Excimen, chemical, and gas lasers are available as ultraray surgical lasers.

"Ultra-San" *[Ecolab].*   TM for antibacterial in mats used by workers in the food processing or preparation, food service or related industry.
Use: To prevent microbial, unsanitary and odorous build up.

ultrasonics.   The science of effects of sound vibrations beyond the limit of audible frequencies. Used for dust, smoke, and mist precipitation; preparation of colloidal dispersions; cleaning of metal parts, precision machinery, fabrics, etc. Friction welding, formation of catalysts, degassing and solidification of molten metals, extracting flavor oils in brewing, electroplating, drilling hard materials, fluxless soldering, nondestructive testing. Also used for investigation of physical properties, determination of molecular weights of liquid polymers, degree of association of water, and for inducing chemical reaction. A developing application is the use of ultrasonic vibration in diagnostic medicine.

ultrastructure. (fine structure). The detailed structure of a specimen, such as a cell, tissue, or organ, that can be observed only by electron microscopy.

"Ultratex" *[AMF].* TM for a group of silicone elastomeric finishes for natural and synthetic textile fabrics.

"Ultrathene" *[Equistar].* TM for a series of ethylenevinyl acetate copolymer resins for adhesives, conversion coatings, and thermoplastic modifiers. Wide range of melt indexes. Improves specific adhesion of hot-melt, solvent-based, and pressure-sensitive adhesives.

ultraviolet. (UV). Radiation in the region of the electromagnetic spectrum including wavelengths from 100 to 3900 Å. UVA covers the region from 315 to 400 Å. UVB covers the region from 280 to 315 Å.
See radiation.
Hazard: Dangerous to eyes, overexposure causes severe skin burns (sunburn).
Use: Air sterilization in hospitals, microscopy.

ultraviolet absorber. A substance that absorbs radiant energy in the wavelength of UV. The radiant energy absorbed is converted to heat (thermal energy). UV absorbers are added to unsaturated substances (plastics, rubbers, etc.) to decrease light sensitivity and consequent discoloring and degradation. Among compounds used are benzophenones, benzotriazoles, substituted acrylonitriles, and phenol–nickel complexes.
See absorption (2).

"Ultrawet" *[Vitax].* TM for a series of biodegradable linear alkylate sulfonate (LAS) anionic detergents or surface-active agents (linear dodecylbenzene type).
Available forms: Sodium or triethanolamine salts in liquid, slurry, flake, or bead form.

"Ultrion" *[Nalco].* TM for coagulent program.
Use: In water and wastewater treatment.

umbellic acid. See anisic acid.

umbelliferone. CAS: 93-35-6. $C_9H_6O_3$.

Properties: Colorless needles. Mp 225C. Soluble in alcohol, acetic acid, weak alkalies, and boiling water.
Derivation: Distillation of plant resins (Umbelliferae).
Use: Cosmetics, sunscreen preparations.

umber. A naturally occurring brown earth containing ferric oxide together with silica, alumina, manganese oxides, and lime. Raw umber is umber that is ground and then levitated. Burnt umber is umber calcined at low heat.
Use: Paint pigment, lithographic inks, wallpaper (pigment), artists' color.

UMP. Abbreviation for uridine monophosphate. See uridine phosphates; uridylic acid.

uncertainty principle. The conclusion of Heisenberg based on quantum mechanical theory that the precise position of a specific electron in an atomic orbit cannot be determined and that consequently the ultimate nature of matter is not susceptible to objective measurement. The result of this concept was development of the orbital theory in which electron behavior is dealt with on a statistical basis. Its validity was confirmed by 1930 by the work of other mathematical physicists such a DeBroglie, Fermi, and Schrödinger.
See orbital theory; Heisenberg, Werner, P.

uncoupling agent. A substance that uncouples phosphorylation of ADP from electron transfer by disrupting the pH gradient; for example, 2,4-dinitrophenol.

γ-undecalactone. (peach aldehyde; γ-undecyl lactone; 4-hydroxy-undecanoic acid, γ-lactone). CAS: 710-04-3.

$$CH_3(CH_2)_6CHCH_2CH_2COO.$$

Properties: Colorless to light-yellow liquid; peach-like odor. D 0.941–0.944, refr index 1.450–1.454. Soluble in 4–5 volumes of 60% alcohol; soluble in benzyl alcohol, benzyl benzoate, and most fixed oils. Combustible.
Derivation: By heating undecylenic acid in the presence of sulfuric acid.
Grade: Chlorine-free, FCC.
Use: Perfumery, flavoring agent.

undecanal. (n-undecylic aldehyde; hendecanal). CAS: 112-44-7. $CH_3(CH_2)_9CHO$.
Properties: Colorless liquid; sweet odor. D 0.825–0.832 (25C), refr index 1.4310–1.4350 (20C), flash p 235F (112.7C). Soluble in oils and alcohol; insoluble in glycerol and water. Combustible.
Derivation: By oxidation of 1-undecanol or reduction of undecanoic acid.

Grade: FCC.
Hazard: Toxic by ingestion and inhalation, irritant to tissue.
Use: Perfumery, flavors.

_n_-undecane.   (hendecane).
CAS: 1120-21-4. $CH_3(CH_2)_9CH_3$.
Properties: Colorless liquid. D 0.7402 (20/4C), fp −25.75C, bp 195.6C, refr index 1.41725 (20C), flash p 149F (65C). Combustible.
Grade: 95%, 99%, research.
Use: Petroleum research, organic synthesis, distillation chaser.

undecanoic acid.   (*n*-undecylic acid; hendecanoic acid). $CH_3(CH_2)_9COOH$. Small amounts occur in castor oil. It is best derived from undecylenic acid by hydrogenation.
Properties: Colorless crystals. D 0.8505 (80/4C), mp 28.5C, bp 284.0C, 222.2C (128 mm Hg), refr index 1.4319 (40C). Insoluble in water; soluble in alcohol and ether.
Grade: Technical, 99%.
Use: Organic synthesis.

1-undecanol.   (*n*-undecyl   alcohol;   decyl carbinol; 1-hendecanol).
CAS: 112-42-5. $CH_3(CH_2)_9CH_2OH$.
Properties: Colorless liquid; citrus odor. D 0.829–0.834, refr index 1.435–1.443, mp 19C, flash p 200F (93.3C). Soluble in 60% alcohol. Combustible.
Use: Perfumery, flavoring.

2-undecanol.   (2-hendecanol).
$CH_3(CH_2)_8CH_2OCH_3$.
Properties: Colorless liquid. D 0.8363 (20C), mp 12C, bp 228–229C, flash p 235F (112.7C). Insoluble in water; soluble in alcohol and ether. Combustible.
Use: Antifoaming agent, intermediate, perfume fixatives, plasticizer.

2-undecanone.   See methyl nonyl ketone.

undecenal.   (undecylenic aldehyde; hendecen-1-al). Listed by different authorities as 10-undecenal, $CH_2{:}CH(CH_2)CHO$, and 9-undecenal. $CH_3CH{:}CH(CH_2)_7CHO$.
Properties: Colorless liquid; strong odor suggesting rose. D 0.840–0.850 (25/25C), refr index 1.4410–1.4470 (20C). Soluble in 80% alcohol. Combustible.
Use: Perfumery, flavoring.

10-undecenoic acid.   See undecylenic acid.

2-undecenol. $C_{11}H_{22}O$.
Properties: White to slightly yellow liquid; oily, sweet, floral odor. D 0.847, refr index 1.450 @ 22°F. Insoluble in water.
Use: Food additive.

10-undecen-1-ol.   See undecylenic alcohol.

_n_-undecyl alcohol.   See 1-undecanol.

undecylenic acid.   (10-undecenoic acid).
CAS: 112-38-9. $CH_2{:}CH(CH_2)_8COOH$.
Properties: Light-colored liquid; fruity-rosy odor. Congealing p 21C, d 0.910–0.913 (25/25C), refr index 1.4475–1.4485 (25C), flash p 295F (146C). Almost insoluble in water; miscible with alcohol, chloroform, ether, benzene, and fixed and volatile oils. Combustible.
Derivation: Destructive distillation of castor oil.
Grade: Technical, NF.
Use: Perfumery, flavoring, plastics, modifying agent (plasticizer, lubricant additive, etc.), medicine (antifungal agent).

undecylenic alcohol.   (*n*-undecylenic alcohol; 10-undecen-1-ol; alcohol C-11).
CAS: 112-43-6. $CH_2{:}CH(CH_2)_8CH_2OH$.
Properties: Colorless liquid; citrus odor. D 0.842–0.847, refr index 1.449–1.454, fp −3.0C. Soluble in 70% alcohol. Combustible.
Use: Perfumes.

undecylenic aldehyde.   See undecenal.

undecylenyl acetate.   (10-hendecenyl acetate).
CAS: 112-19-6. $CH_3COO(CH_2)_9CH{:}CH_2$.
Properties: Colorless liquid; floral-fruity odor. D 0.876–0.883, refr index 1.438–1.442. Soluble in 80% alcohol. Combustible.
Use: Perfumery, flavoring.

_n_-undecylic acid.   See undecanoic acid.

γ-undecyl lactone.   See γ-undecalactone.

undulipodium.   See flagellum.

UNH.   Abbreviation for uranyl nitrate hydrated. See uranyl nitrate.

unhairing.   (dehairing). Removal of hair from hides and skins as practiced on a commercial scale in the leather industry. Several methods are used, involving application of hydrated lime, dimethylamine, trypsin, and other enzymes.

"Unicarb" *[Union].*   TM for spray-coatings system employing supercritical fluid technology that significantly reduces VOC and HAP emissions.

"Unicel" *[Beckman].*   TM for blowing agents for natural and synthetic rubber sponge.

"Unichrome" *[ATOTECH].*   (copper pyrophosphate). TM for electroplating.
Use: Printed circuit boards.

Unilin.   An alcohol.

unimolecular. Occurring to or within a single molecule like intramolecular, but can refer to fragmentation reactions.

"Unipol" [Jason]. TM for low pressure, gas-phase process technology.
Use: Production of polyethylene and polypropylene.

uniport. A transport system that carries only one solute. Contrast with cotransport.

"UniSeal" [Uniseal]. TM for a high-end hardening and dustproofing compound that is chemically reactive.
Use: To permanently bond concrete.

United States Department of Agriculture. (USDA). The federal regulatory authority for meats and meat products. It attempts to support production of agriculture, ensure a safe, affordable, nutritious, and accessible food supply, care for agricultural, forest and range lands, support sound development of rural communities, provide economic opportunities for farm and rural residents, expand global markets for agricultural and forest products and services, and work to reduce hunger in America and throughout the world. Its central office is at 1400 Independence Ave., S.W., Washington, DC 20250. Website: http://www.usda.gov

United States Pharmacopeia. (USP). An independent, science-based public health organization and the official public standard-setting authority for all prescription and over-the-counter medicines, dietary supplements, and other health-care products manufactured and sold in the U.S. USP sets standards for the quality of these products and works with healthcare providers to help them reach the standards. USP's standards are also recognized in more than 140 countries. These standards have been helping to ensure good pharmaceutical care for people throughout the world for more than 194 years.
The USP is a self-sustaining nonprofit organization and is funded through revenues from the sale of products and services that help to ensure good pharmaceutical care. USP's contributions to public health are enriched by the participation and over-sight of volunteers representing pharmacy, medicine, and other healthcare professions as well as academia, government, the pharmaceutical industry, health plans, and consumer organizations. Its headquarters is located at 12601 Twinbrook Parkway, Rockville, MD 20852. Website: http://www.usp.org

"Unithox" [Baker Hughes]. TM for ethoxylated grade alcohol.

unit operation. A particular kind of physical change used in the industrial production of various chemicals and related materials. Filtration, evaporation, distillation, fluid flow, and heat transfer are examples.
See chemical engineering.

unit process. A process characterized by a particular kind of chemical reaction; oxidation, hydrolysis, esterification, and nitration are examples.
See kinetics, chemical; chemical engineering.

"Unival" [Union]. TM for high-density polyethylene resins.
Use: Milk, juice, and water bottles.

univalent antibody. (incomplete antibody; inhibiting antibody). A form of antibody that may coat an antigen but does not have a second receptor that could attach to another molecule of antigen.

universal antidote. A hypothetical antidote that would be therapeutically effective against all poisons. No such antidote is known to exist.

uns-. (unsym). Abbreviation for unsymmetrical. A prefix denoting the structure of organic compounds in which substituents are disposed unsymmetrically with respect to the carbon skeleton or to a functional group, such as a double bond. For example, unsdichloroethane is CH_3CHCl_2.

unsaponifiable matter. The weight of matter extracted by ether from the aqueous alkaline solution after complete saponification of a fat or oil.

unsaturated fatty acid. A fatty acid containing one or more double bonds.

unsaturation. Of a chemical compound, the state in which not all the available valence bonds along the alkyl chain are satisfied; in such compounds the extra bonds usually form double or triple bonds (chiefly with carbon). Thus, unsaturated compounds are more reactive than saturated compounds because other elements readily add to the unsaturated linkage. An unsaturated compound (ethylene, C_2H_4; butadiene, C_4H_6; benzene, C_6H_6) has fewer hydrogen atoms or equivalent groups than the corresponding saturated compound (ethane, C_2H_6; butane, C_4H_{10}; cyclohexane, C_6H_{12}).
In structural formulas unsaturation may be represented by parallel lines joining the carbon atoms (ethylene, $H_2C{=}CH_2$; butadiene $H_2C{=}CHCH{=}CH_2$) or by colons or triple dots, $H_2C{:}CH_2H_2C{:}CH_2$ (ethylene) and $HC\ldots CH$ (acetylene).
See carboxyl group.

unstable compound system. See peritectic system.

upholstery cleaner. (upholstery shampoo). A product designed for use in cleaning upholstery.

Properties: Common components are per-chloroethylene, naphthalene, ethanol, ammonia, and detergents.
Hazard: Toxic.

UPS. Abbreviation for ultraviolet photoelectron spectroscopy.

upstream activator sequence. A binding site for transcription factors, generally part of a promoter region. A UAS may be found upstream of the TATA sequence (if there is one), and its function is (the same as an enhancer) to increase transcription. Unlike an enhancer, it cannot be positioned just anywhere or in any orientation.

upstream/downstream. In an RNA, anything toward the 5′ end of a reference point is "upstream" of that point. This orientation reflects the direction of both the synthesis of mRNA and its translation—from the 5′ end to the 3′ end. In DNA, the situation is a bit more complicated. In the vicinity of a gene (or in a cDNA), the DNA has two strands, but one strand is virtually a duplicate of the RNA, so its 5′ and 3′ ends determine upstream and downstream, respectively.
Note: In genomic DNA, two adjacent genes may be on different strands and thus oriented in opposite directions. Upstream or downstream is only used on conjunction with a given gene.

"Urac" [Allnex]. TM for products based on urea-formaldehyde condensates used mainly as adhesives for the production of moisture-proof bonds in plywood manufacture, plywood assembly, and furniture manufacture.

uracil. (2,4-dioxypyrimidine).
CAS: 66-22-8. $HNC(O)NHC(O)CHCH$. A pyrimidine that is a constituent of ribonucleic acids and the coenzyme, uridine diphosphate glucose.
Properties: Crystalline needles. Mp 335C (decomposes). Soluble in hot water, ammonium hydroxide, and other alkalies; insoluble in alcohol and ether.
Derivation: Hydrolysis of nucleic acids, precipitation from urea and ethyl formylacetate. Radioactive forms available.
Use: Biochemical research.

uracil-6-carboxylic acid. See orotic acid.

Uracil Mustard.
CAS: 66-75-1. 5-[bis(2-chloroethyl)amino]uracil. An antineoplastic.
Hazard: Possible carcinogen.

uracil, D-ribosyl. See uridine.

Uramite. A fertilizer derived from ureaformaldehyde containing 38% nitrogen.

Uramon. Solutions of urea in aqueous ammonia.
Use: Manufacture of mixed fertilizers.

urania. See uranium dioxide.

urania-thoria. Crystals of the mixed oxides of uranium and thorium are available. The crystals are denser and cheaper than the pellet form.
Use: Nuclear fuel.

uranic chloride. See uranium tetrachloride.

uranic oxide. See uranium dioxide.

uranine. (uranine yellow; sodium fluorescein; resorcinolphthalein sodium; CI 45350).
CAS: 518-47-8. $Na_2C_{20}H_{10}O_5$.
Properties: Orange-red powder; odorless. Hygroscopic, soluble in water; sparingly soluble in alcohol.
Derivation: By treatment of fluorescein with sodium carbonate solution and crystallizing.
Method of purification: Recrystallization.
Grade: Technical, USP (as sodium fluorescein).
Use: Dyeing silk and wool yellow; tracing subterranean waters; marking water for air-sea rescues; clinical test solution.

uraninite. UP_2. A natural phosphide of uranium usually partly oxidized to UO_3 with variable amounts of lead, radium, thorium, rare-earth metals, helium, argon, nitrogen. Pitchblende is an important variety.
Occurrence: Colorado, Utah, South Africa, Canada, Europe, the former U.S.S.R., Australia, Zaire.
Use: Source of uranium and radium.

uranium.
CAS: 7440-61-1. U. Metallic element number 92; a member of the actinide series; aw 238.029; valences of 3, 4, 6; three natural radioactive isotopes: uranium-234 (0.006%), uranium-235 (0.7%), and uranium-238 (99%).
Properties: Dense, silvery solid. D 19.0, mp 1132C, bp 3818C, heat of fusion 4.7 kcal/mol, heat capacity 6.6 cal/mol/C. Strongly electropositive, ductile and malleable, poor conductor of electricity. Forms solid solutions (for nuclear reactors) with molybdenum, niobium, titanium, and zirconium. The metal reacts with nearly all nonmetals. It is attacked by water, acids, and peroxides, but is inert toward alkalies. Green tetravalent uranium and yellow uranyl ion (UO_2^{++}) are the only species that are stable in solution.
Occurrence: Pitchblende (essentially UO_2), a variety of uraninite, coffinite ($USiO_4$), and carnotite (Colorado, New Mexico, France, Zaire, Canada, South Africa, Australia, the former U.S.S.R.). U.S. resources of uranium oxide in the 1970s were estimated at 150,000–175,000 tons; there is about the same amount in Canada.
Derivation: Finely ground ore is leached under oxidizing conditions to give uranyl nitrate solution. The uranyl nitrate, purified by solvent extraction (ether, alkyl phosphate esters), is then reduced with

hydrogen to uranium dioxide. This is treated with hydrogen fluoride to obtain uranium tetrafluoride, followed by either electrolysis in fused salts or by reduction with calcium or magnesium. Uranium can also be recovered from phosphate sand.

Available forms: Solid pure metal, alloys, powder (99.7%).

Hazard: (Powder) Dangerous fire risk, ignites spontaneously in air. Highly toxic, radioactive material, source of ionizing radiation. Kidney damage. Confirmed carcinogen.

Use: Source of fissionable isotope uranium-235, source of plutonium by neutron capture, electric power generation.

See enrichment (2); uranium compounds.

uranium-233.   (^{233}U). A fissionable isotope of uranium produced artificially by bombarding thorium-232 with neutrons. Used as an atomic fuel in molten salt reactor and is a possible fuel in breeder reactors. Half-life 1.62×10^5 years.

uranium-234.   (^{234}U). A natural isotope of uranium with half-life of 2.48×10^5 years; it is separated by extraction with trioctylphosphine oxide.

Use: Nuclear research, with potential use in fission detectors for counting fast neutrons.

uranium-235.   (^{235}U). The readily fissionable isotope of uranium used to enrich natural uranium in nuclear fuels. It is present in uranium only to the extent of 0.7% and can be separated from it by any of several methods: the gaseous diffusion process using uranium hexafluoride, the gas centrifuge process, and the electromagnetic separation method. Its half-life is 7.13×10^8 years. It was the energy source used in the original atomic bomb. Its critical mass is about 33 lb.

See fission; uranium.

uranium-238.   (^{238}U). The abundant isotope of uranium of which it comprises 99%. It is not fissionable but will form plutonium-239 as a result of bombardment by neutrons in a reactor. Its half-life is 4.51×10^9 years. It will be used in breeder reactors, together with plutonium, where its energy potential will be exploited by transmuting it to fissionable plutonium.

See breeder.

uranium carbide.   See uranium dicarbide.

uranium compounds.   Before the advent of nuclear energy, uranium had very limited uses. It had been suggested for filaments of lamps. A small tube of uranium dioxide, UO_2, connected in series with the tungsten filaments of large incandescent lamps used for photography and motion pictures, tends to eliminate the sudden surge of current through the bulbs when the light is turned on, thereby extending their lives. Compounds of uranium have been used in photography for toning,

and in the leather and wood industries uranium compounds have been used for stains and dyes. Uranium salts are mordants of silk or wool. In making special steels, a little ferrouranium has been utilized, but its value is questionable in this connection. Such alloys have not proved commercially attractive. In the production of ceramics, sodium and ammonium diuranates have been used to produce colored glazes.

Uranium carbide has been suggested as a good catalyst for the production of synthetic ammonia. Uranium salts in small quantities are claimed to stimulate plant growth, but large quantities are clearly poisonous to plants.

By far the most important use of uranium lies in its application for nuclear (or atomic) energy.

uranium decay series.   (uranium-radium series). The series of elements produced as successive intermediate products when the element uranium undergoes spontaneous natural radioactive disintegration into lead. Radium and radon are members of this series.

uranium, depleted.   Uranium from which most of the uranium-235 isotope has been removed.

Properties: Uranium metal wherein uranium-235 isotopic content has been reduced below 0.7% as found in normal uranium. Has high structural strength coupled with high density of 18.9 g/cc.

Derivation: Processed from uranium hexafluoride as tailing from gaseous diffusion plants.

Use: Balance weights for aircraft, high-speed rotors in gyro-compasses, γ-radiation shielding, radio-isotope transportation casks and fuel element transfer casks, in general as structural material applicable to radiation shielding.

See uranium-238.

uranium dicarbide.   (uranium carbide). UC_2.

Properties: Gray crystals. D 11.28 (18C), mp 2350C, bp 4370C. Decomposes in water; slightly soluble in alcohol.

Hazard: Highly toxic, radiation risk.

Use: Crystals, pellets, or microspheres for nuclear reactor fuel.

uranium dioxide.   (uranium oxide; uranic oxide; urania; yellowcake).

CAS: 1344-57-6. UO_2.

Properties: Black crystals. D 10.9, mp 3000C. Insoluble in water; soluble in nitric acid and concentrated sulfuric acid.

Derivation: (pure oxide) Powdered uranium ore is digested with hot nitric-sulfuric acid mixture and filtered to remove the insoluble portion. Sulfate is precipitated from the solution with barium carbonate, and uranyl nitrate is extracted with ether. After re-extraction into water, it is heated to drive off nitric acid, leaving uranium trioxide. The latter is reduced with hydrogen to the dioxide. Can be prepared from uranium hexafluoride by treating

with ammonia and subsequent heating of the ammonium diuranate. It is also recovered from phosphoric acid.
Hazard: High radiation risk. Ignites spontaneously in finely divided form.
Use: A crystalline (or pellet) form is used to pack nuclear fuel rods.

uranium, enriched. Natural uranium to which a few percent of the fissionable 235-isotope has been added.
See enrichment.

uranium hexafluoride.
CAS: 7783-81-5. UF_6.
Properties: Colorless, volatile crystals. Sublimes, triple point 64.0C (1134 mm Hg), mp 64.5C (2 atm), d 5.06 (25C). Soluble in liquid bromine, chlorine, carbon tetrachloride, *sym*-tetrachloroethane, and fluorocarbons. Reacts vigorously with water, alcohol, ether, and most metals. Vapor behaves as nearly perfect gas.
Derivation: (1) Triuranium octoxide (U_3O_8) and nitric acid react to form a solution of uranyl nitrate; this is decomposed to UO_3 and reduced to the dioxide with hydrogen. The dioxide as a fluidized bed is reacted with hydrogen fluoride. The resulting tetrafluoride is fluorinated to the hexafluoride. (2) Triuranium octoxide is converted directly to the hexafluoride with hydrogen fluoride and fluorine, then purified by fractional distillation.
Hazard: Highly corrosive; radiation risk.
Use: Gaseous diffusion process for separating isotopes of uranium.

uranium hydride. UH_3.
Properties: Brown-gray to black powder. D 10.92, conductor of electricity.
Derivation: Action of hydrogen on hot uranium.
Hazard: Highly toxic. Ignites spontaneously in air.
Use: Preparation of finely divided uranium metal by decomposition, separation of hydrogen isotopes, reducing agent, laboratory source of pure hydrogen.

uranium monocarbide. UC.
Properties: Lumps or powder that can be formed into desired shapes by powder metallurgy or arc-melt casting. Mp 2375C, d 13.63, thermal conductivity 0.08 cal/sec/cm^2/C/cm. Must be stored in inert atmosphere.
Hazard: Radioactive poison.

uranium nitrate.
CAS: 15905-86-9. $HNO_3 \cdot xU$.
Hazard: A poison.

uranium oxide. See uranium dioxide.

uranium tetrafluoride. (green salt).
CAS: 10049-14-6. UF_4.

Properties: Green, nonvolatile, crystalline powder. D 6.70, mp 1036C. Insoluble in water.
Derivation: Treatment of uranium dioxide with hydrogen fluoride.
Hazard: Highly corrosive, radioactive poison.
Use: Intermediate in preparation of uranium metal. See uranium hexafluoride.

uranocene. (bis-cyclooctatetraenyluranium). A synthetic metallocene "sandwich" molecule whose unique feature is that the pi-molecular orbitals of the large organic rings share electrons with f-atomic orbitals of the uranium.
See metallocene.

uranyl acetate.
CAS: 541-09-3. $UO_2(C_2H_3O_2)_2$.
Properties: (Dihydrate) Yellow crystals. D 2.89. Soluble in water plus acetic acid; slightly soluble in alcohol.
Use: Bacterial oxidation activator, copying inks, reagent.

uranyl nitrate. (uranium oxynitrate; UNH; yellow salt).
CAS: 10102-06-4. $UO_2(NO_3)_2 \cdot 6H_2O$.
Properties: Yellow, rhombic crystals. D 2.807, mp 60.2C, bp 118C. Soluble in water, alcohol, and ether.
Derivation: Action of nitric acid on uranium octoxide (U_3O_8).
Hazard: Strong oxidizer. Corrosive and irritating.
Use: Source of uranium dioxide, extraction of uranium into nonaqueous solvents.

urban waste. Solid waste materials including garbage, cellulosics, glass, metals, etc., but not sewage (municipal waste). Garbage is being used both directly as fuel and fermented to yield proteins; the cellulosic portion can be hydrolyzed to glucose, which is in turn converted to methane by anaerobic fermentation. A continuous process for this operation utilizing newspapers or sawdust has been reported.
See waste control; biomass.

urea. (carbamide).
CAS: 57-13-6. $CO(NH_2)_2$.

$$O = C \underset{NH_2}{\overset{NH_2}{\big<}}$$

Occurs in urine and other body fluids. The first organic compound to be synthesized (Wohler, 1824).
Properties: White crystals or powder; almost odorless; saline taste. D 1.335, mp 132.7C, decomposes before boiling. Soluble in water, alcohol, and benzene; slightly soluble in ether; almost insoluble in chloroform. Noncombustible.

Derivation: Liquid ammonia and liquid carbon dioxide at 1750–3000 psi and 160–200C react to form ammonium carbamate, $NH_4CO_2NH_2$, which decomposes at lower pressure (about 80 psi) to urea and water. Several variations of the process include once-through, partial recycle, and total recycle.
Method of purification: Crystallization.
Grade: Technical, CP, USP, fertilizer (45–46% nitrogen), feed grade (about 42% nitrogen).
Use: Fertilizer, animal feed, plastics, chemical intermediate, stabilizer in explosives, medicine (diuretic), adhesives, separation of hydrocarbons (as urea adducts), pharmaceuticals, cosmetics, dentifrices, sulfamic acid production, flameproofing agents, viscosity modifier for starch or casein-based paper coatings, preparation of biuret.
See urea-formaldehyde resin.

urea adduct. See inclusion complex.

urea ammonia liquor. A solution of crude urea in aqueous ammonia containing ammonium carbamate.
Use: Reaction with superphosphate in preparation of fertilizers, furnishing combined nitrogen.

urea-ammonium orthophosphate. A fertilizer developed especially for food-deficient regions, particularly rice-dependent areas. Several grades contain all three primary plant nutrients (nitrogen, phosphorus, and potassium). Contains up to 60% nitrogen, phosphoric anhydride, and potassium oxide.

urea-ammonium polyphosphate. A fertilizer similar to urea-ammonium orthophosphate except that about half the phosphorus is in polyphosphate form, which gives improved sequestering action and solubility. It is excellent for use as a liquid fertilizer.

urea cycle. A metabolic pathway in vertebrates for the synthesis of urea from amino groups and carbon dioxide; occurs in mitochondrial and cytoplasmic compartments of the liver.

ureaform. A urea-formaldehyde reaction product that contains more than one molecule of urea per molecule of formaldehyde. It can be used as a fertilizer because of its high nitrogen content, its insolubility in water, and its gradual decomposition in the soil during the growing season to yield soluble nitrogen.

urea-formaldehyde resin. An important class of amino resin. Urea and formaldehyde are united in a two-stage process in the presence of pyridine, ammonia, or certain alcohols with heat and control of pH to form intermediates (methylolurea, dimethylolurea) that are mixed with fillers to produce molding powders. These are converted to thermosetting resins by further controlled heating and pressure in the presence of catalysts. These were

the first plastics that could be made in white, pastel, and colored products.
Use: Scale housings, dinnerware, interior plywood, foundry core binder, flexible foams, insulation.
See amino resin; melamine resin.

urea hydrogen peroxide. See urea peroxide.

urea nitrate.
CAS: 124-47-0. $CO(NH_2)_2 \cdot HNO_3$.
Properties: Colorless crystals. Decomposes at 152C. Slightly soluble in water; soluble in alcohol.
Derivation: By adding an excess of nitric acid to a strong aqueous solution of urea.
Hazard: Dangerous fire and explosion risk.
Use: Explosives, manufacture of urethane.

urea peroxide. (urea hydrogen peroxide; percarbamide; carbamide peroxide).
CAS: 124-43-6. $CO(NH_2)_2 \cdot H_2O_2$.
Properties: White crystals or crystalline powder. Mp (decomposes) 75–78C, decomposed by moisture at temperatures around 40C. Soluble in water, alcohol, and ethylene glycol; solvents such as ether and acetone extract hydrogen peroxide and may form explosive solutions. Active oxygen (min) 16%.
Grade: Technical, sometimes compounded with waxes in pellet form.
Hazard: Dangerous fire risk in contact with organic materials; strong oxidizing agent. Irritant.
Use: Source of water-free hydrogen peroxide, bleaching disinfectant, cosmetics, pharmaceuticals, blueprint developer, modification of starches.

urea phosphoric acid. See carbamide phosphoric acid.

urease. Enzyme present in low percentages in jack bean and soybean; water soluble, its action is inhibited by heavy-metal ions. Its principal use is in the determination of urea in urine, blood, and other body fluids; it splits urea into ammonia and carbon dioxide or ammonium carbonate.

Urech cyanohydrin method. Cyanohydrin formation by addition of alkali cyanide to the carbonyl group in the presence of acetic acid (Urech); or by reaction of the carbonyl compound with anhydrous hydrogen cyanide in the presence of a basic catalyst (Ultee).

Urech hydantoin synthesis. Formation of hydantoins from α-amino acids by treatment with potassium cyanate in aqueous solution and heating of the salt of the intermediate hydantoic acid with 25% hydrochloric acid.

ureotelic. Organisms which excrete excess nitrogen in the form of urea.

urethane. (ethyl carbamate; ethyl urethane).
CAS: 51-79-6. $CO(NH_2)OC_2H_5$. Its structure is typical of the repeating unit in polyurethane resins.

Properties: Colorless crystals or white powder; odorless, saltpeter-like taste. D 0.9862, mp 49C, bp 180C. Solutions neutral to litmus. Soluble in water, alcohol, ether, glycerol, and chloroform; slightly soluble in olive oil. Combustible.

Derivation: (1) By heating ethanol and urea nitrate to 120–130C; (2) by action of ammonia on ethyl carbonate or ethyl chloroformate.

Grade: Technical, NF.

Hazard: Toxic by ingestion. Probable carcinogen.

Use: Intermediate for pharmaceuticals, pesticides, and fungicides; biochemical research; medicine (antineoplastic).

See polyurethane.

urethane alkyd. (uralkyd). A urethane resin modified with a drying oil (linseed or safflower) for use in high-quality paints.

"UrethHall" *[Hallstar].* TM for polyester polyols for urethane systems that are fluid and pourable at room temperature and are based on chemistry completely different from the traditional adipate polyesters. Available in 25 different variations of glycol base, hydroxyl number, or molecular weight.

Use: All forms of polyurethanes made from polyesters, especially those where improved hydrolytic stability and flexibility without external plasticizers are valued.

Urey, Harold Clayton. (1893–1981). An American chemist who received the Nobel Prize in chemistry in 1934 for his discovery of the heavy isotopes of hydrogen and oxygen. His discovery became an important factor in the development of nuclear fission and fusion and made possible the production of the first transuranic element Pu. He was one of the leaders of the Manhattan Project, which constructed the first nuclear reactor at the University of Chicago and eventually produced the first atomic bomb. After obtaining his doctorate at the University of California in 1923, he taught at several leading universities, including Columbia, where he discovered deuterium (D) oxide (heavy water), used as a moderator in early types of nuclear reactors. Later he devoted much study to the origin of the universe and the origin of life on earth. He was the author of many scientific treatises and made notable contributions to the cosmological theories.

uric acid. (lithic acid; uric oxide; 2,6,8-trioxypurine).
CAS: 69-93-2.

OCNHC(O)NHCCNHC(O)NH (keto form).

May also be written in enolic form. The end product of purine metabolism in human and other primates, birds, and some dogs and reptiles.

Properties: Odorless, white crystals; tasteless. D 1.855–1.893, mp (decomposes). Soluble in hot concentrated sulfuric acid; very slightly soluble in water; insoluble in alcohol and ether; soluble in glycerol, solutions of alkali hydroxides, sodium acetate, and sodium phosphate.

Derivation: From guano.

Grade: Technical, reagent.

Hazard: Evolves a highly toxic hydrogen cyanide when heated.

Use: Organic synthesis.

uricotelic. Organisms which excrete excess nitrogen in the form of uric acid (urate at physiological conditions).

uridine. (D-ribosyl uracil).
CAS: 58-96-8. $C_9H_{12}N_2O_6$. The nucleoside of uracil. It is a constituent of ribonucleic acid and some coenzymes (such as uridine diphosphate glucose).

Properties: White powder; odorless; slightly acrid and faintly sweet taste. Mp 165C. Soluble in water; slightly soluble in dilute alcohol; insoluble in strong alcohol.

Derivation: From nucleic acid hydrolyzates from yeast. Radioactive forms available.

Use: Biochemical research.

uridine diphosphate glucose. (UDPG). A coenzyme that acts in the transfer of glucose from the coenzyme to another chemical compound during the reaction for which the coenzyme is a catalyst.

Use: Biochemical research.

uridine monophosphate. See uridylic acid.

uridine phosphate. A nucleotide used by the body in growth processes; important in biochemical and physiological research. Those isolated and commercially available (as sodium salts) are the monophosphate (UMP), the diphosphate (UDP), and the triphosphate (UTP).

See uridine diphosphate glucose (UDPG).

uridylic acid. (uridine phosphoric acid; UMP; uridine monophosphate).
CAS: 58-97-9. $(C_9H_{13}N_2O_9P)$. The monophosphoric ester of uracil, i.e., the nucleotide containing uracil-d-ribose and phosphoric acid. The phosphate may be esterified to either the 2, 3, or 5 carbon of ribose, yielding uridine-2'-phosphate, uridine-3'-phosphate, and uridine-5'-phosphate, respectively.

Properties: (Uridine-3'-phosphate) Crystallizes in prisms from methanol. Mp 202C (decomposes). Freely soluble in water and alcohol; dextrorotatory in solution.

Derivation: (of commercial product) From yeast ribonucleic acid. Also made synthetically; radioactive forms available.

Use: Biochemical research.

uroformine.   (urotropin, formin, crystogen, cystamine).
See hexamethyltetramine.

uronic acid.   Any of a class of compounds similar to sugars but differing from them in that the terminal carbon has been oxidized from an alcohol to a carboxyl group. The most common are galacturonic acid and glucuronic acid.

ursin.   See arbutin.

urushiol.   Mixture of catechol derivatives.
  Properties: Pale-yellow liquid. D 0.968, bp 200C. Soluble in alcohol, ether, and benzene.
  Derivation: Poison ivy (*Rhus toxicodendron*).
  Hazard: The toxic principle of poison ivy. Causes severe allergenic dermatitis.

urylon.   A polyurea synthetic fiber made by condensation of nonamethylenediamine and urea.
  Properties: D 1.07, softening p 205C, mp 235C, weakens on heating to 150C or long exposure to light.
  Use: Fiber blends, fishing gear, and the like.

USAN.   Abbreviation for United States Adopted Name, a nonproprietary name approved by the American Pharmaceutical Association, American Medical Association, and the U.S. Pharmacopeia. Such names applied to pharmaceutical products do not imply endorsement; their use in advertising and labeling is required by law.

USDA.   Abbreviation for United States Department of Agriculture.

usnic acid.   (usninic acid).
  CAS: 125-46-2. $C_{18}H_{16}O_7$. A tricyclic compound. A constituent of many lichens. Known in *d*, *l*, and *dl* forms.
  Properties: Crystalline, yellow solid. Melting range 192–203C. Insoluble in water; slightly soluble in alcohol and ether.

Derivation: From *Usnea barbata*, a lichen growing on trees.
Use: Medicine (antibiotic).

USP.   Abbreviation for United States Pharmacopeia, the official publication for drug product standards.

USRDA.   Abbreviation for United States Recommended Dietary Allowances for food and nutrition, established by the Food and Drug Administration (FDA) to serve as a basis for regulations on nutritional labeling of food products. They are based on the Recommended Dietary Allowances previously established by the Food and Nutrition Board, and the National Research Council of the National Academy of Sciences, and they are revised periodically.
See RDA.

UTP.   See uridine phosphate.

UV.   See ultraviolet.

UVA.   See ultraviolet.

UVB.   See ultraviolet.

"UV-Chek" *[Ferro].*   TM for light stabilizers.

"Uvinul" *[BASF].*   TM for 2-ethylhexyl-2-cyano-3′,3′-diphenyl acrylate.
  Use: In sunscreen products.

"Uvitex" *[Huntsman].*   TM for a group of fluorescent whitening agents used on natural and synthetic textile fabrics and yarn.

17-α-uzarigenin.
  CAS: 663-97-8. $C_{23}H_{34}O_4$.
  Hazard: A poison.
  Source: Natural product.

V

V. Symbol for vanadium.

vacancy. (1) A defect in a crystal due to the absence of an atom in the lattice. See hole. (2) The absence of one or more electrons in the outer shell of an atom.

vacuole. Membrane-bound fluid-filled space within a cell. In most plant cells, there is a single large vacuole filling most of the cell's volume. Some bacterial cells contain gas vacuoles.

vacuum deposition. The process of coating a base material by evaporating a metal under high vacuum and condensing it on the surface of the material to be coated, which is usually another metal or a plastic. Aluminum is most commonly used for this purpose. The coatings obtained range in thickness from 0.01 to as many as 3 mils. A vacuum of about one-millionth atmosphere is necessary. The process is used for jewelry, electronic components, decorative plastics, etc. Thermally evaporated metals and dielectric coatings can be effectively applied to glass by this method. It is also called vacuum coating and vacuum metallizing.

vacuum distillation. Distillation at a pressure below atmospheric but not so low that it would be classed as molecular distillation. Since lowering the pressure also lowers the boiling point, vacuum distillation is useful for distilling high-boiling and heat-sensitive materials such as heavy distillates in petroleum, fatty acids, vitamins, etc.

vacuum forming. See thermoforming (2).

val. Abbreviation for value.

valacidin.
CAS: 3930-19-6. $C_{26}H_{24}N_4O_8$.
Properties: Brown to red solid. Soluble in polar solvents and alkaline solutions; insoluble in most nonpolar solvents and acid solutions.
Use: Antibiotic (used as a preservative for biological specimens); embalming fluids.

Valclene. Clear fluorocarbon formulations with slight ethereal odor. Nonflammable.
Use: Dry-cleaning fluids.

valence. A whole number that represents or denotes the combining power of one element with another. By balancing these integral valence numbers in a given compound, the relative proportions of the elements present can be accounted for. If hydrogen and chlorine both have a valence of 1, oxygen a valence of 2, and nitrogen 3, the valence-balancing principle gives the formulas HCl, H_2, NH_3, Cl_2O, NCl_3, and N_2O_3, which indicate the relative numbers of atoms of these elements in compounds that they form with each other. In inorganic compounds it is necessary to assign either a positive or negative value to each valence number, so that valence balancing will give a zero sum by algebraic addition. Negative numbers are called polar valence numbers (-1 and -2). The valence of chlorine may be -1, $+1$, $+3$, $+5$, or $+7$, depending on the type of compound in which it occurs. In organic chemistry, only nonpolar valence numbers are used.
See chemical bonding; oxidation; oxidation number; coordination number.

valence electrons. The electrons in the outermost shell of an atom determining chemical properties.

valentinite. (antimony trioxide [ortho-rhombic]; white antimony).
CAS: 1309-64-4. Sb_2O_3.
Properties: White or gray mineral, sometimes pale red, white streak and adamantine or silky luster. D 5.57–5.76, Mohs hardness 2–3.
Occurrence: Algeria, the former Yugoslavia, Italy, Germany.
Use: Ore of antimony.

valepotriate.
CAS: 18296-44-1. $C_{22}H_{30}O_8$.
Hazard: A poison.

valeral. See valeraldehyde.

***n*-valeraldehyde.** (valeric aldehyde; valeral; amyl aldehyde; pentanal).
CAS: 110-62-3. $CH_3(CH_2)_3CHO$.
Properties: Colorless liquid. D 0.8095 (20/4C), fp -91C, bp 102–103C, refr index 1.3944 (20C), flash p 54F (12.2C) (CC). Slightly soluble in water; soluble in alcohol and ether.
Derivation: Oxidation of amyl alcohol, also by the Oxo process.
Hazard: Flammable, dangerous fire risk. Eye, skin, and upper respiratory tract irritant.
Use: Flavoring, rubber accelerators.
See isovaleraldehyde.

Hawley's Condensed Chemical Dictionary, Sixteenth Edition. Michael D. Larrañaga, Richard J. Lewis, Sr., and Robert A. Lewis.
© 2016 John Wiley & Sons, Inc. Published 2016 by John Wiley & Sons, Inc.

valeranone.
  CAS: 55528-90-0. $C_{15}H_{26}O$.
  Hazard: A poison.

valerenal.
  CAS: 4176-16-3. $C_{15}H_{22}O$.
  Hazard: A poison.

valerenic acid.
  CAS: 3569-10-6. $C_{15}H_{22}O_2$.
  Hazard: A poison.

valerianic acid.   See valeric acid.

valerian oil.
  Properties: Yellowish or brownish liquid; penetrating odor. D 0.903–0.960, refr index 1.486 (20C). Soluble in alcohol, ether, chloroform, acetone, benzene, and carbon disulfide. Combustible.
  Chief constituents: Pinene, camphene, borneol, and esters of borneol and valeric acid.
  Derivation: Distilled from roots and rhizome of *Valeriana officinalis*.
  Use: Tobacco perfume, industrial odorant, flavors.

valeric acid.   (valerianic acid; *n*-pentanoic acid).
  CAS: 109-52-4. $CH_3(CH_2)_3COOH$.
  Properties: Colorless liquid; penetrating odor and taste. D 0.9394 (20/4C), bp 185.4C, refr index 1.4081 (20C), vap press 0.08 mm Hg (20C), fp −34C, flash p 205F (96C) (OC). Soluble in water, alcohol and ether. Undergoes reactions typical of normal monobasic organic acids. Combustible.
  Derivation: With other C_5 acids by distillation from valerian, by oxidation of *n*-amyl alcohol, numerous essential oils.
  Grade: Technical, reagent.
  Hazard: Toxic by ingestion, strong irritant to skin and tissue.
  Use: Intermediate for flavors and perfumes, ester-type lubricants, plasticizers, pharmaceuticals, vinyl stabilizers.
  See isovaleric acid.

valeric aldehyde.   See *n*-valeraldehyde.

γ-valerolactone.
  CAS: 108-29-2.

$$H_2C-CH_2-CH_2-CH-CH_3$$
$$\;\;\;|\qquad\qquad\qquad\quad|$$
$$\;\;\;O\text{————————}C=O$$

  Properties: Colorless liquid. D 1.0518 (25/25C), bp 205–206.5C, crystallizing point −37C, flash p 205F (96C) (COC), refr index 1.4301 (25C), surface tension 30 dynes/cm (25C), viscosity 2.18 cP (25C), pH (anhydrous): 7. pH (10% solution in distilled water): 4.2. Miscible with water and most organic solvents, resins, waxes, etc.; slightly miscible with zein, beeswax, petrolatum; immiscible with

anhydrous glycerin, glue, casein, arabic gum, and soybean protein. Combustible.
  Use: Dye baths (coupling agent), brake fluids, cutting oils, and as solvent for adhesives, insecticides, and lacquers.

valine.   (α-aminoisovaleric acid).
  CAS: 72-18-4. $(CH_3)_2CHCH(NH_2)COOH$. An essential amino acid.
  Properties: White, crystalline solid. Soluble in water; very slightly soluble in alcohol; insoluble in ether; shows the following optical isomers:
  DL-valine: Mp 298C with decomposition.
  D-valine (natural isomer): Mp 315C with decomposition.
  L-valine: Mp 293C with decomposition.
  Derivation: Hydrolysis of proteins, synthesized by the reaction of ammonia with α-chloroisovaleric acid. Available commercially as D-, L-, or DL-valine.
  Use: Dietary supplement, culture media, biochemical and nutritional investigations.

"Valium" [Roche].   TM for diazepam.

valone.   (2-isovaleryl-1,3-indanedione).
  $C_{14}H_{14}O_3$.
  Properties: Yellow, crystalline solid. Mp 68C. Insoluble in water; soluble in common organic solvents. A blood anticoagulant.
  Hazard: Toxic by ingestion.
  Use: Pesticide, rodenticide.

vanadic acid.   (1) meta-HVO_3, (2) ortho-H_3VO_4, (3) pyro-$H_4V_2O_7$. These acids apparently do not exist in the pure state but are represented in various alkali and other metal vanadates. Ordinarily, vanadic acid implies vanadium pentoxide (vanadic acid anhydride).

vanadic acid anhydride.   See vanadium pentoxide.

vanadic(II) acid, trisodium salt.   See sodium orthovanadate.

vanadic sulfate.   See vanadyl sulfate.

vanadic sulfide.   See vanadium sulfide.

vanadinite. $Pb_5Cl(VO_3)$. A natural chlorovanadate of lead.
  Properties: Ruby-red, orange-red, brown, yellow solid; resinous to adamantine luster. Mohs hardness 3, d 6.7–7.1. Soluble in strong nitric acid.
  Occurrence: New Mexico, Arizona, Africa, Scotland, the former U.S.S.R.
  Use: Ore of vanadium and lead.

vanadium.
  CAS: 7440-62-2. V. Metallic element having atomic, number 23, group VB of the periodic table, aw

50.9414, valences of 2, 3, 4, 5; two natural isotopes.

Properties: Silvery-white ductile solid. D 6.11, mp 1900C, bp 3000C. Insoluble in water; resistant to corrosion, but soluble in nitric, hydrofluoric, and concentrated sulfuric acids; attacked by alkali, forming water-soluble vanadates. Acts as either a metal or a nonmetal and forms a variety of complex compounds.

Source: Not found native; principal ores are patronite, roscoelite, carnotite and vanadinite. Also from phosphate rock (Idaho, Montana, Arkansas).

Occurrence: Colorado, Utah, New Mexico, Arizona, Mexico, and Peru.

Derivation: (1) Calcium reduction of vanadium pentoxide yields 99.8+% pure ductile vanadium; (2) aluminum, cerium, etc. reduction produces a less pure product; (3) solvent extraction of petroleum ash or ferrophosphorus slag from phosphorus production (4) electrolytic refining using a molten salt electrolyte containing vanadium chloride.

Grade: 99.99% pure (electrolytic process), single crystals.

Use: Target material for X rays, manufacture of alloy steels, vanadium compounds, especially catalyst for sulfuric acid and synthetic rubber.

See ferrovanadium.

vanadium acetylacetonate.

Properties: Blue to blue-green crystals. Decomposes before melting.

Derivation: Reaction of vanadyl sulfate with acetylacetone and sodium carbonate.

Use: Catalyst.

vanadium carbide.   VC.

Properties: Crystal. Mohs hardness 2800 kg/sq mm, d 5.77, mp 2800C, bp 3900C, resistivity 150 $\mu\Omega\cdot$cm (room temperature).

Use: Alloys for cutting tools, steel additive.

vanadium dichloride.   (vanadous chloride).

CAS: 10213-09-9. VCl_2.

Properties: Apple-green hexagonal plates. D 3.23 (18C). Soluble in alcohol and ether; decomposes in hot water.

Derivation: From vanadium trichloride by heating in atmosphere of nitrogen.

Method of purification: Sublimation in nitrogen.

Grade: CP.

Hazard: Strong irritant to tissue.

Use: Strong reducing agent, purification of hydrogen chloride from arsenic.

vanadium disulfide. V_2S_2.

Properties: Solid. D 4.20, mp (decomposes). Soluble in hot sulfuric or nitric acid; insoluble in alkalies.

Use: Solid lubricant, electrode in lithium-based batteries (experimental).

vanadium ethylate. $(C_2H_5O)_4V$.

Properties: Dark-reddish-brown solid.

Derivation: Reaction of vanadium chloride with sodium ethylate.

Use: Polymerization catalyst.

vanadium hexacarbonyl. $V(CO)_6$.

Properties: Blue-green powder. Sublimes easily at 50C (15 mm Hg), paramagnetic, decomposes without melting at 60–70C.

Hazard: Strong irritant to tissue; store under inert gas.

Use: Chemical intermediate, production of plating compounds and fuel additives.

vanadium hexacarbonyl, sodium salt.

(bisdiglymesodium hexacarbonyl vanadate). $Na(C_6H_{14}O_3)_2V(CO)_6$. A convenient though incomplete name for the complex compound.

Properties: Yellow solid. Mp 173–176C (with decomposition). Soluble in water, alcohol, and ether; slightly soluble in hydrocarbons; relatively inert to air but is stored and shipped under nitrogen.

Hazard: Strong irritant to tissue.

Use: Source of vanadium hexacarbonyl by treatment with phosphoric acid under special conditions.

vanadium nitride.   VN.

Properties: Black solid. D 6.13, mp 2320C. Insoluble in water; slightly soluble in aqua regia.

Use: Refractory.

vanadium oxydichloride.   See vanadyl chloride.

vanadium oxytrichloride.

CAS: 7727-18-6. $VOCl_3$.

Properties: Lemon-yellow liquid. D 1.811 (32C), fp −78.9C, bp 125–127C, nonionizing solvent. Dissolves most nonmetals; dissolves and/or reacts with many organic compounds; hydrolyzes in moisture.

Hazard: Strong irritant to tissue.

Use: Catalyst in olefin polymerization (ethylene-propylene rubber) organovanadium synthesis.

vanadium pentoxide.   (vanadic acid anhydride).

CAS: 1314-62-1. V_2O_5.

Properties: Yellow to red crystalline powder. D 3.357 (18C), mp 690C, bp (decomposes) 1750C. Soluble in acids and alkalies; slightly soluble in water.

Derivation: (1) Alkali or acid extraction from vanadium minerals. (2) By igniting ammonium metavanadate. (3) From concentrated ferrophosphorus slag by roasting with sodium chloride, leaching with water and purification by solvent extraction followed by precipitation and heating.

Method of purification: Alkali solution, precipitation as ammonium metavanadate and ignition to V_2O_5.

Grade: Commercial air-dried, commercial fused, CP air-dried, CP fused.

Hazard: Toxic by inhalation. TLV: 0.05 mg $(V_2O_5)/m^3$; not classifiable as a human carcinogen.

Use: Catalyst for oxidation of sulfur dioxide in sulfuric acid manufacture, ferrovanadium, catalyst for many organic reactions, ceramic coloring material, vanadium salts, inhibiting UV transmission in glass, photographic developer, dyeing textiles. See contact process.

vanadium sesquioxide. See vanadium trioxide.

vanadium sulfate. See vanadyl sulfate.

vanadium sulfide. (vanadium pentasulfide; vanadic sulfide). V_2S_5.
Properties: Black-green powder. D 3.0, decomposes on heating. Soluble in acids, alkali-metal sulfides, and alkalies; insoluble in water.
Derivation: Action of hydrogen sulfide on vanadium chloride solution.
Hazard: Toxic by inhalation (especially to animals).
Use: Vanadium compounds.

vanadium tetrachloride.
CAS: 7632-51-1. VCl_4.
Properties: Red liquid. D 1.816 (20C), fp −28C, bp 154C. Decomposes slowly to vanadium trichloride and chlorine below 63C. Soluble in absolute alcohol and ether. Nonflammable.
Derivation: Chlorination of ferrovanadium.
Method of purification: Distillation and fractionation.
Hazard: Toxic by ingestion, inhalation, and skin absorption. Open containers only in dry, oxygen-free atmosphere or inert gas; wear goggles and protective clothing; corrosive to tissue.
Use: Preparation of vanadium trichloride, vanadium dichloride, and organovanadium compounds.

vanadium tetraoxide. V_2O_4.
Properties: Blue-black powder. D 4.339, mp 1967C. Insoluble in water; soluble in alkalies and acids.
Derivation: (1) From vanadium pentoxide by oxalic acid reduction. (2) From vanadium pentoxide by carbon reduction.
Hazard: Toxic and irritating.
Use: Catalyst at high temperature.

vanadium trichloride.
CAS: 7718-98-1. VCl_3.
Properties: Pink, deliquescent crystals. D 3.0 (18C), decomposes on heating. Soluble in absolute alcohol and ether; decomposes in water.
Derivation: From vanadium tetrachloride boiling under reflux condenser.
Hazard: Irritant.
Use: Preparation of vanadium dichloride and organovanadium compounds.

vanadium trioxide. (vanadium sesquioxide).
CAS: 1314-34-7. V_2O_3.
Properties: Black crystals. D 4.87 (18C), mp 1970C. Soluble in alkalies and hydrogen fluoride; slightly soluble in water.

Derivation: From vanadium pentoxide by either hydrogen or carbon reduction.
Hazard: Irritant.
Use: Catalyst for conversion of ethylene to ethanol.

vanadous chloride. See vanadium dichloride.

vanadyl chloride. (vanadium oxydichloride; vanadyl dichloride; divanadyl tetrachloride).
CAS: 10213-09-9. $V_2O_2Cl_4•5H_2O$.
Properties: Green, deliquescent crystal. Usual technical product is a dark-green, syrupy mass, 76–82% pure, or a solution. Slowly decomposed by water; soluble in water, alcohol, and acetic acid; may react violently with potassium.
Grade: Technical.
Hazard: Irritant.
Use: Mordanting textiles.

vanadyl sulfate. (vanadic sulfate; vanadium sulfate).
CAS: 27774-13-6. $VOSO_4•2H_2O$.
Properties: Blue crystals. Soluble in water.
Derivation: Reduction of cold solution of concentrated sulfuric acid and vanadium pentoxide by sulfur dioxide gas.
Hazard: Irritant.
Use: Mordant, catalyst, aniline black preparation, reducing agent, colorant in glasses and ceramics.

Vanaquat. A cleaner and disinfectant for food processing equipment.

Vanason. A tank sanitizer for dairy farm equipment.

vancomycin hydrochloride.
Properties: Tan to brown powder; odorless; bitter taste. Soluble in water; moderately soluble in dilute methanol; insoluble in higher alcohols, acetone, ether.
Derivation: Produced by *Streptomyces orientalis* from Indonesian and Indian soil.
Grade: USP.
Use: Medicine (antibiotic).

"Vancote 2K" [Vanderbilt]. TM for a waterborne epoxy fast-dry shop primer.

van der Waals forces. (London dispersion force). Weak attractive forces acting between molecules. They are somewhat weaker than hydrogen bonds and far weaker than interatomic valences. They are involved in the van der Waals equation of state for gases, which compensates for the actual volume of the molecules and the forces acting between them. "Information regarding the numerical values of van der Waals forces is mostly semiempirical, derived with the aid of theory from an analysis of chemical or physical data. Attempts to calculate the forces from first principles have had a measure of success only for the simplest systems,

such as H–H, He–He, and a few others. When judging the difficulties of such calculations, one must bear in mind that the energies sought are of the same order of magnitude as those in the best atomic energy calculations." (Henry Margenau).
See hydrogen bond; chemical bonding.

Van Dyke brown. (Cassel brown; Cologne brown; Cologne earth; ulmin brown). A naturally occurring pigment.
Derivation: Indefinite mixture of iron oxide and organic matter. Obtained from bog-earth, peat deposits, or ochers that contain bituminous matter.
Use: Pigment for artists' colors and stains.

Van Dyke red. A brownish-red pigment consisting of copper ferrocyanide; sometimes used to refer to red varieties of ferric oxide.
Use: Pigment.
See iron oxide red.

vanillin. (3-methoxy-4-hydroxybenzaldehyde; vanillic aldehyde).
CAS: 121-33-5. $(CH_3O)(OH)C_6H_3CHO$. The methyl ether of protocatechuic aldehyde.
Properties: White, crystalline needles; sweetish smell. D 1.056, mp 81–83C, bp 285C. Soluble in 125 parts water, in 20 parts glycerol, and in 2 parts 95% alcohol; soluble in chloroform and ether. Combustible.
Derivation: (1) By extraction from the vanilla bean, (2) from lignin contained in sulfite waste pulp liquor.
Method of purification: Crystallization.
Grade: Technical, USP, FCC.
Use: Perfumes, flavoring, pharmaceuticals, laboratory reagent, source of L-dopa.

van Leeuwenhoek, Anton. (1632–1723). A native of Holland, van Leeuwenhoek was a professional lens grinder. He developed the compound optical microscope, which had been invented in 1590, to a point where he was able to obtain magnifications up to 275 times. He was the first to observe bacteria, spermatozoa, and other unicellular animals, which he described to the Royal Society of London.
See optical microscope.

"Vansil" *[Vanderbilt].* TM for a calcium metalsilicate filler mineral.

Van Slyke determination. Treatment of primary aliphatic amines and α-amino acids with nitrous acid and volumetric determination of evolved nitrogen.

van't Hoff, Jacobus H. (1852–1911). A Dutch chemist who received the first Nobel Prize for chemistry in 1901. A father of physical chemistry, he did research on decomposition and formation of double salts. He related optically active carbon compounds to three-dimensional and asymmetrical

molecular structure. This led to the development of stereochemistry. He was educated at the Universities of Paris and Utrecht, where he received a doctorate in 1874. He was a professor at Amsterdam, Leipzig, and Berlin.

"Vapona" *[Bayer Healthcare].* TM for an insecticide that contains more than 93% 2,2-dichlorovinyl dimethyl phosphate and less than 7% active, related compounds.
See DDVP.
Hazard: As for dichlorovos.

vapor. An air dispersion of molecules of a substance that is liquid or solid in its normal state, i.e., at standard temperature and pressure. Examples are water vapor and benzene vapor. Vapors of organic liquids are also loosely called fumes.
See evaporation; gas.

vapor density. Weight of a vapor per unit volume at any given temperature and pressure.

vapor-phase chromatography. See gas chromatography.

vapor pressure. (vap press; vap p; v.p.). The pressure (often expressed in millimeters of mercury, mm Hg) characteristic at any given temperature of a vapor in equilibrium with its liquid or solid form.

vapor tension. See vapor pressure.

varnish. (1) An organic protective coating similar to paint except that it does not contain a colorant. It may be composed of a vegetable oil (linseed, tung, etc.) and solvent or of a synthetic or natural resin and solvent. In the first case the formation of the film is due to polymerization of the oil and in the second, due to evaporation of the solvent. "Long-oil" varnishes such as spar varnish have a high proportion of drying oil; "short-oil" types have a lower proportion, i.e., furniture varnishes. Spirit varnishes contain solvents such as methanol, toluene, ketones, etc., and often also thinners such as naphtha or other light hydrocarbon. Flammable. (2) A hard, tightly adherent deposit on the metal surfaces of automobile engines resulting from resinous oxidation products of gasoline and lubricating oils.

varnish remover. See paint remover.

"Varonic" *[Evonik].* TM for ethoxylated primary amines. 2- to 30-mole adducts of primary fatty amines based on coco, soya, tallow, and oleic acid.
Properties: Oil and water soluble.
Use: In textile industry and other emulsifier applications.

Varrentrapp reaction. Cleavage of oleic acid into palmitic and acetic acids by heating with molten potassium hydroxide. The procedure has been extended to olefinic acids in general.

vasopressin. (β-hypophamine; antidiuretic hormone). One of the hormones secreted by the posterior lobe of the pituitary gland. It causes increased blood pressure and an increase in water retention by the kidney. Vasopressin is an octapeptide consisting of eight amino acids.
Derivation: Synthetic, or from the posterior lobe of the pituitary of food animals.
Grade: USP as an aqueous solution for injection.
Use: Medicine (antidiuretic).

vasopressin, 8-*l*-arginine-. See arginine vasopressin.

vat dye. A class of water-insoluble dyes that can be easily reduced, i.e., vatted to a water-soluble and usually colorless leuco form in which they can readily impregnate fibers. Subsequent oxidation then produces the insoluble colored form that is remarkably fast to washing, light, and chemicals. Examples are indigo (CI 73000) and Indanthrene Blue BFP (CI 69825). The reducing agents are usually an alkaline solution of sodium hydrosulfite ($Na_2S_2O_4$). Oxidation is by air, perborate, dichromate, etc.
Use: For cotton, wool, and cellulose acetate.

vat printing assistant. A mixture of gums, and reducing and wetting agents used to carry the dye in printing fabrics with vat dyes. They assist in securing penetration of the fabric and in converting the dyes from a semileuco to a leuco state.

Vatsol. A series of wetting agents made in several different grades and types: OS, sodium isopropyl naphthalene sulfonic acid; OT, sodium dioctyl sulfosuccinate.

vattin. The process of solubilizing vat dyes in an aqueous solution of caustic soda and sodium hydrosulfite.

Vauquelin's salt. $(Pd(NH_3)_4)Cl_2 \cdot PdCl_2$. A compound obtained by treating palladium chloride with ammonia.

VC. Abbreviation for vinyl chloride or vinylidene chloride.

vector. (1) In biochemistry, an animal (insect, rodent, etc.) that carries or transports infectious microorganisms. Typical vectors are rats and mosquitoes. (2) A DNA molecule known to replicate autonomously in a host cell, to which a segment of DNA may be spliced to allow its replication. The current types are plasmids, cosmids, yeast artificial chromosomes or a temperate-phage DNA.

vegetable black. In general, any form of more or less pure carbon produced by incomplete combustion or destructive distillation of vegetable matter, wood, vines, wine lees.

vegetable dye. A colorant derived from a vegetable source, i.e., logwood, indigo, madder, etc.

vegetable gum. See dextrin.

vegetable oil. An oil extracted from the seeds, fruit, or nuts of plants and generally considered to be a mixture of glycerides (e.g., cottonseed, linseed, corn, coconut, babassu, olive, tung, peanut, perilla, oiticica). Many types are edible. Being plant-derived products, vegetable oils are a form of biomass. Some are reported to be convertible to liquid fuels by passing them over zeolite catalysts.
Use: Paints (as drying oils), shortenings, salad dressings, margarine, soaps, rubber softeners, dietary supplements, pesticide carriers.

vegetable tanning. The tanning of leather by plant extracts.
See tannic acid; tanning; wattle bark; quebracho.

vehicle. A term used in paint technology to indicate the liquid portion of a paint, composed of drying oil or resin, solvent, and thinner, in which the solid components are dissolved or dispersed.
See paint.

Venetian red. A high-grade ferric oxide pigment of a pure red hue. It is obtained either native as a variety of hematite red or more often artificially by calcining copperas (ferrous sulfate) in the presence of lime. The composition ranges from 15 to 40% ferric oxide and from 60 to 80% calcium sulfate. The 40% ferric oxide is the "pure" grade and has a d 3.45.
Grade: 20–40% ferric oxide.
See iron oxide red.

venturi. A type of flowmeter used for liquids or fine particulates. It is a tube like device having broad, flaring ends and a narrow central portion, or throat; this so constricts the passage of the fluid that its rate of flow increases while the pressure decreases. The difference in pressure thus created is a measure of the flow. Venturis are used in scrubbers, liquid and solid conveying systems, pipelines, and aircraft instrument control, as well as in numerous chemical process techniques such as hydrogenation, chlorination, and oxidation. The hydrogenation technology involves hydrogen entrainment by rapid flow of a liquid catalyst through the venturi nozzle.

veratraldehyde. (vertraldehyde; 3,4-dimethoxy benzaldehyde).
CAS: 120-14-9. $(CH_3O)_2C_6H_3CHO$.
Properties: Mp 42–45C, bp 281C, flash p 235F (112C), mw 166.18.

veratric acid. (3,4-dimethoxybenzoic acid).
CAS: 93-07-2. $(CH_3O)_2C_6H_3COOH$.
Properties: Mw 182.18, mp 179–82C.

veratrole. (1,2-dimethoxybenzene; pyrocatechol dimethyl ether).
CAS: 91-16-7. $C_6H_4(OCH_3)_2$.
Properties: Colorless crystals or liquid. Mp 21–22C, bp 206–207C, d 1.084 (25/25C). Soluble in alcohol and ether; slightly soluble in water.
Derivation: Treatment of catechol in methanol with dimethyl sulfate and caustic.
Use: Medicine (antiseptic).

veratrum alkaloid. One of a group of alkaloids used in medicine to relieve hypertension. They include *Veratrum viride* (American hellebore).
Hazard: May have severe side effects.
See alkaloid.

verdigris. See copper acetate, basic.

vermiculite. Hydrated magnesium–iron–aluminum silicate capable of expanding 6–20 times when heated to 1093C. The platelets exhibit an active curling movement when heated, hence the name.
Occurrence: Montana, North Carolina, South Carolina, Wyoming, Colorado, South Africa.
Properties: Platelet-type crystalline structure. High porosity, high void volume to surface area ratio, low density, large range of particle size. Insoluble in water and organic solvents; soluble in hot concentrated sulfuric acid. Water vapor adsorption capacity of expanded vermiculite less than 1%, liquid adsorption dependent on conditions and particle size, ranges 200–500%. Noncombustible.
Grade: Unexpanded (ore concentrate), expanded (also called exfoliated), flake, activated.
Use: Lightweight concrete aggregate, insulation, sound conditioning, fireproofing, plaster, soil conditioner, additive for fertilizers, seed bed for plants, refractory, lubricant, oil-well drilling mud. Filler in rubber, paint, plastics. Wallpaper printing, removal of strontium-90 from milk, absorption of oil spills on seawater, animal feed additive, packing, carrier for insecticides, catalyst and catalyst support, litter for hatcheries, adsorbent.
See verxite.

vermifuge. An agent used in veterinary medicine to eliminate intestinal worms; an anthelmintic.

vermilion, natural. See cinnabar.

vernolepin. A sesquiterpene dilactone extracted from leaves of an African plant *Veronia hymenolepis.*
Use: Biochemical research (inhibits tumor growth in rats and reversibly retards plant growth).

"Versaflex" *[VersaFlex].* TM for polymer emulsion.
Use: In latex paint formulations.

Versa Guma. A xantham gum specialty ingredient.

Use: In food formulation in a broad range of pH and temperatures.

"Versamag" *[Martin].* See "Novogel" *[Organovo].*

"Versene" *[Dow].*
CAS: 62-33-9. TM for calcium disodium EDTA chelating agent.
Use: Pharmaceuticals, foods, and biological applications to control oxidation, and chelate metal ions.

"Versene Fe-3 Specific" *[Dow].* TM for an iron specialty chelating agent; active ingredient is sodium dihydroxyethylglycine; available as straw-colored liquid or white powder.

"Versenex" **80** *[Dow].* TM for pentasodium salt of diethylenetriaminepentaacetic acid.

"Versenol" **120** *[Dow].* TM for the trisodium salt of *N*-hydroxyethylethylenediaminetriacetic acid ($C_{10}H_{15}O_7N_2Na_3$).
Use: Organic chelating agent.

verxite. (exfoliated hydrobiotite).
Properties: Thermally expanded (exfoliated) magnesium-iron-aluminum silicate having a minimum of 98% purity and a bulk d of 5–7 lb/cu ft. Expansion occurs by heating at 793C. Spongelike structure that absorbs liquids and permits re-expansion after compression to 70–80% of the original heat-expanded volume.
Use: Poultry feed in quantities less than 5% as a nonnutritive bulking agent, pelleting or anti-caking agent and nutrient carrier in dog and ruminant feeds. For dog feeds the maximum permitted is 1.5%. Labeling must state content when in excess of 1%.
See vermiculite.

"Vespel" *[Du Pont].* TM for fabricated parts, based on polyimide resin and diamond abrasive wheels formulated with a high-temperature polyimide binder.

vetiver oil. An essential oil with violetlike odor, strongly dextrorotatory.
Use: Perfumery and fragrances.

"Vexar" *[Conwed].* TM for both low and high-density polyethylene and polypropylene plastic netting. Available in a wide variety of forms and colors for use in packaging.

V-G-B. A reaction product of acetaldehyde and aniline.
Properties: Brown resinous powder. D 1.152, mp 60–80C. Soluble in acetone, benzene, and ethylene dichloride; insoluble in water and gasoline.
Use: Rubber antioxidant.

"Viagra" [Pfizer].   (sildenafil citrate).
CAS: 171599-83-0. $C_{22}H_{30}N_6O_4S \cdot C_6H_8O_7$.
Use: A treatment for male erectile dysfunction.

"Vibrathane" [Chemtura].   TM for a liquid castable, urethane prepolymer.

"Vibrin" [Aoe].   TM for resin compositions of polyesters and cross-linking monomers that when catalyzed, will polymerize to infusible solid resins without evolving water or other by-products.
Use: Molding, laminating, impregnating, casting, automotive and aircraft structural parts, wall panels, table tops, coating for paper, boat hulls, chemically inert tanks, large-diameter pipe.

vic-.   Prefix meaning vicinal.

vicariance.   Speciation which occurs as a result of the separation and subsequent isolation of portions of an original population.

vicinal.   (vic-). Neighboring or adjoining positions on a carbon ring or chain; the term is used in naming derivatives with substituting groups in such locations in a structural formula or molecule. For example, vicinal locations in the molecule shown are occupied by the hydrogen atoms and the hydroxyl groups:

$$
\begin{array}{ccc}
\text{H} & & \text{H} \\
| & & | \\
-\text{C} & - & \text{C}- \\
| & & | \\
\text{HO} & & \text{OH}
\end{array}
$$

Victoria blue.   (CI 44045). $C_{33}H_{31}N_3 \cdot HCl$.
Properties: Bronze-colored, crystalline powder. Soluble in hot water, alcohol, or ether.
Derivation: Michler's ketone is condensed with phenyl-α-naphthylamine.
Use: Dyeing silk, wool, and cotton; biological stain; dye intermediate for complex acid pigment toners.

Victoria green.   See malachite green.

Victory.   A plastic grade of petroleum microcrystalline wax.

vicuna.   A soft, wool-like fiber obtained from a South American animal similar to the llama.
Use: Specialty high-grade coats, sweaters, etc. Combustible.

vidarabine.   Generic name for ara-A.

Vigofac 6.   An unidentified growth factor for addition to animal feeds. Derived from dried streptomyces fermentation solubles.

"Vikane" [Dow Agrosciences].   TM for sulfuryl fluoride.

Vilsmeier-Haack reaction.   Formulation of activated aromatic or heterocyclic compounds with disubstituted formamides and phosphorus oxychloride.

vinal fiber.   Generic name for a manufactured fiber in which the fiber-forming substance is any long-chain synthetic polymer composed of at least 50% by weight of vinyl alcohol units, $-CH_2CH_2O-$, and in which the total of the vinyl alcohol units and any one or more of the various acetal units is at least 85% by weight of the fiber (Federal Trade Commission). It has good chemical resistance, low affinity for water, good resistance to mildew and fungi. Combustible.
Use: Fishing nets, stockings, gloves, hats, rainwear, swimsuits.

vinegar.
Properties: Brownish or colorless liquid. Dilute aqueous solution containing 4–8% acetic acid, depending on source. Legal minimum is 4%. Also contains low percentages of alcohols and mineral salts. Nonflammable.
Derivation: (1) Bacterial fermentation of apple cider, wine, or other fruit juice. (2) Fermentation of malt or barley. The fermenting agent is usually a mold, e.g., *Mycoderma aceti*, generally known as "mother." Either type can be distilled to remove color and other impurities and is then called white vinegar.
Use: Mayonnaise, salad dressings, pickled foods, preservative, medicine (antifungal agent), latex coagulant.

Vinoflex.   Vinyl chloride homo- and copolymers.

"Vinsol" [Pinova].   TM for a series of low-cost, dark, brittle, thermoplastic resins; ruby-red by transmitted light, dark-brown by reflected light. Available in solid form, flakes, fine powder, and aqueous dispersion.
Use: Adhesives, asphalt emulsions, electrical insulation, inks, plastics.

vinyl.   See vinyl compound.

vinyl acetate.
CAS: 108-05-4. $CH_3COOCH{:}CH_2$. Raw material for polyvinyl resins.
Properties: Colorless liquid stabilized with either hydroquinone or diphenylamine inhibitors. The hydroquinone stabilized material can be polymerized without redistillation. The DPA-stabilized material must be distilled before polymerization. D 0.9345 (20/20C), fp −100.2C, bp 73C, refr index 1.3941, bulk d 7.79 lb/gal, flash p 30F (−1.1C) (TOC), autoign temp 800F (426.6C). Soluble in most organic solvents including chlorinated solvents; insoluble in water.
Derivation: (1) Vapor-phase reaction of ethylene, acetic acid, and oxygen, with a palladium catalyst.

(2) Vapor-phase reaction of acetylene, acetic acid, and oxygen, with zinc acetate catalyst. (3) From synthesis gas.

Grade: Technical.

Hazard: Flammable, dangerous fire risk. Flammable limits in air 2.6–13.4%. Toxic by inhalation and ingestion. Eye, skin, and upper respiratory tract irritant and central nervous system impairment.

Use: Polyvinyl acetate, polyvinyl alcohol, polyvinyl butyral, and polyvinyl chloride-acetate resins, used particularly in latex paints, paper coating, adhesives, textile finishing, safety glass interlayers. A vinyl acetate-ethylene copolymer is available for specialty products.

See polyvinyl acetal; polyvinyl acetate; polyvinyl alcohol; polyvinyl chloride-acetate.

vinylacetonitrile.   See allyl cyanide.

vinylacetylene. C_4H_4 or $H_2C{:}CHC{:}CH$. The dimer of acetylene, formed by passing it into a solution of cuprous and ammonium chloride in hydrochloric acid.

Properties: Colorless gas or liquid. D 0.6867 (20C), bp 5C. Combustible.

Use: Intermediate in manufacture of neoprene and for various organic syntheses.

See divinyl acetylene.

vinyl acrylate.   See ethenyl 2-propenoate.

vinyl alcohol.   (ethenol). $CH_2{:}CH_2O$. Unstable liquid; isolated only in the form of its esters or the polymer, polyvinyl alcohol.

vinylation.   The formation of a vinyl derivative by reaction with acetylene. Thus vinylation of alcohols yields vinyl ethers such as vinyl ethyl ether, $C_2H_5OC_2H_3$.

vinyl bromide.

CAS: 593-60-2. CH_2CHBr.

Properties: Gas. Fp −138C, bp 15.6C, d 1.51.

Hazard: Liver cancer. A probable carcinogen.

Use: Flame-retarding agent for acrylic fibers.

vinyl-*n*-butyl ether.   (*n*-butyl vinyl ether; BVE).

CAS: 111-34-2. $CH_2{:}CHOC_4H_9$.

Properties: Liquid. D 0.7803 (20C), bp 94.1C, fp −113C, refr index 1.3997, flash p 15F (−9.4C) (OC), bulk d 7.45 lb/gal (20C). Slightly soluble in water; soluble in alcohol and ether.

Derivation: Reaction of acetylene with *n*-butyl alcohol.

Grade: Technical (98%).

Hazard: Flammable, dangerous fire risk.

Use: Synthesis, copolymerization.

vinyl butyrate.

CAS: 123-20-6. $CH_2{:}CHOOCC_3H_7$.

Properties: Liquid. D 0.9022 (20/20C), bp 116.7C, fp −86.8C, flash p 68F (20C) (OC). Very slightly soluble in water.

Hazard: Flammable, dangerous fire risk.

Use: Polymers, emulsion paints.

n-vinylcarbazole.

CAS: 1484-13-5. $C_2H_8NHC{:}CH_2$.

Properties: Liquid. Combustible.

Derivation: From acetylene and carbazole.

Use: Polymerizes to form heat-resistant and insulating resins somewhat similar to mica in dielectric properties.

See polyvinyl carbazole.

vinyl cetyl ether.   See cetyl vinyl ether.

vinyl chloride.   (VC; chloroethene; chloroethylene).

CAS: 75-01-4. $CH_2{:}CHCl$. The most important vinyl monomer.

Properties: Compressed gas, easily liquefied, ethereal odor, usually handled as liquid. Phenol is added as a polymerization inhibitor. D 0.9121 (liquid at 20/20C), bp −13.9C, fp −159.7C, vap press 2300 mm Hg (20C), flash p −108F (−77C), autoign temp 882F (472C). Slightly soluble in water; soluble in alcohol and ether.

Derivation: (1) Dehydrochlorination of ethylene dichloride; (2) reaction of acetylene and hydrogen chloride, either as liquids or gases.

Grade: Technical, pure 99.9%.

Hazard: Explosive limits in air 4–22% by volume. An extremely toxic and hazardous material by all avenues of exposure. A carcinogen. Use in aerosol sprays is prohibited. Lung cancer and liver damage. Confirmed carcinogen.

Use: Polyvinyl chloride and copolymers, organic synthesis, adhesives for plastics.

vinyl-2-chloroethyl ether.

CAS: 110-75-8. $CH_2{:}CHOCH_2CH_2Cl$.

Properties: Liquid. D 1.0498 (20C), bp 109.1C, fp −69.7C, flash p 80F (26.6C) (OC). Very slightly soluble in water.

Hazard: Flammable, dangerous fire risk.

vinyl compound.   A compound having the vinyl grouping ($CH_2{=}CH{-}$), specifically vinyl chloride, vinyl acetate, and similar esters but also referring more generally to compounds such as styrene $C_6H_5CH{=}CH_2$, methyl methacrylate $CH_2{=}C(CH_3)COOCH_3$, and acrylonitrile $CH_2{=}CHCN$. The vinyl compounds are highly reactive, polymerize easily, and the basis of a number of important plastics.

vinyl cyanide.   See acrylonitrile.

vinylcyclohexene.   (1-vinylcyclohexene-3; 4-vinylcyclohexene-1; cyclohexenylethylene).

CAS: 100-40-3. CH_2:$CHCHCH_2CH$:$CHCH_2CH_2$. A butadiene dimer.

Properties: Liquid. D 0.8303 (20/4C), fp −108.9C, bp 128C, refr index 1.464 (20C), flash p 70F (21.2C) (TOC), autoign temp 517F (269C). Temperatures above 80F (26.6C) and prolonged exposure to oxygen-containing gases should be avoided because these conditions lead to discoloration and gum formation.

Grade: Technical 95%, pure 99%, research.

Hazard: Flammable, dangerous fire risk. Narcotic in high concentration. Female and male reproductive damage. Possible carcinogen.

Use: Polymers, organic synthesis.

vinylcyclohexene dioxide.   (vinylcyclohexane dioxide).
CAS: 106-87-6.

$CH_2CHOC_6H_9O$.

Properties: Colorless liquid. D 1.098 (20/20C), bp 227C, refr index 1.4782 (20C), viscosity 7.77 cP (20C), flash p 230F (110C). Combustible.

Hazard: Toxic by ingestion and skin absorption, strong irritant to skin and tissue. Female and male reproductive damage. Possible carcinogen.

Use: Polymers, organic synthesis.

vinylcyclohexene monoxide.   (vinylcyclohexane monoxide). CH_2:CHC_6H_9O.

Properties: Liquid. D 0.9598 (20/20C), bp 169C, fp −100C, flash p 136F (57.8C), viscosity 1.69 cP (20C). Very slightly soluble in water. Combustible.

Hazard: Moderate fire risk. Irritant.

Use: Polymers, organic synthesis.

vinyl ether.   (divinyl ether; divinyl oxide).
CAS: 109-93-3. CH_2:$CHOCH$:CH_2.

Properties: Colorless liquid; characteristic odor. D 0.769, bp 39C, refr index 1.3989 (20C), flash p −22F (−30C), autoign temp 680F (360C). Slightly soluble in water; miscible with alcohol, acetone, chloroform, and ether. Must be protected from light.

Derivation: Treatment of dichloroethyl ether with alkali.

Grade: NF (contains 96–97% vinyl ether, remainder dehydrated alcohol).

Hazard: Flammable, severe fire and explosion risk; explosive limits in air 1.7–27%. Toxic by inhalation, overexposure may be fatal.

Use: Copolymer with 3–5% polyvinyl chloride for plastic products such as clear plastic bottles; medicine (anesthetic, for brief operations only).

vinyl-β-ethoxyethyl sulfide. CH_2: $CHSCH_2CH_2OC_2H_5$.

Properties: Colorless liquid; pungent, camphorlike odor. D 0.9532 (15C), bp 65C (8 mm Hg).

Use: Organic synthesis.

vinylethylene.   See butadiene.

vinyl ethyl ether.   (ethyl vinyl ether; EVE).
CAS: 109-92-2. CH_2:$CHOC_2H_5$.

Properties: Colorless liquid. Extremely reactive, can be polymerized in liquid or vapor phase. D 0.754 (20/20C), bulk d 6.28 lb/gal (20C), fp −115.0C, viscosity 0.22 cP (20C), refr index 1.3739, flash p −50F (−46C), autoign temp 395F (201.6C). Slightly soluble in water (0.9% by weight). Commercial material contains inhibitor to prevent premature polymerization. Often stored underground to minimize vapor losses.

Derivation: Reaction of acetylene with ethanol.

Grade: Technical.

Hazard: Highly flammable, severe fire and explosion risk, explosive limits in air 1.7–28%.

Use: Copolymerization, intermediate.

vinyl ethyl ether polymer.   See 5-(3 or 6-oxo-1-cyclohexen-1-yl)-5-ethylbarbituric acid.

vinyl-2-ethylhexoate.
CAS: 94-04-2. CH_2=$CHOOCCH(C_2H_5)C_4H_9$.

Properties: Liquid. D 0.8751 (20/20C), bp 185.2C, fp −90C, flash p 165F (73.9C) (OC). Insoluble in water. Combustible.

Use: Polymers, emulsion paints.

vinyl-2-ethylhexyl ether.
CAS: 103-44-6. CH_2=$CHOCH_2CH(C_2H_5)C_4H_9$.

Properties: Liquid. D 0.8102 (20/20C), bp 177.7C, fp −100C, flash p 135F (57.2C) (OC), autoign temp 395F (201.6C). Insoluble in water. Combustible.

Hazard: Moderate fire and explosion risk.

Use: Intermediate for pharmaceuticals, insecticides, adhesives, viscosity index improver.

2-vinyl-5-ethylpyridine.
$(CH_2$=$CH)C_5H_3N(C_2H_5)$.

Properties: Liquid. D 0.9449 (20/20C), bp 138C (100 mm Hg), vap press 0.1 mm Hg (20C), fp −50.9C, flash p 200F (93.3C) (COC). Insoluble in water. Combustible.

Use: Copolymer, synthesis.

vinyl fluoride.   (fluoroethylene).
CAS: 75-02-5. CH_2=CHF.

Properties: Colorless gas. Bp −72C. Insoluble in water; soluble in alcohol and ether.

Hazard: Flammable, dangerous fire and explosion risk. Toxic by inhalation. Liver damage and liver cancer. Probable carcinogen.

Use: Monomer.
See polyvinyl fluoride.

vinylheptacyclotetrasiloxane.
CAS: 3763-39-1. $C_9H_{24}O_4Si_4$.

Hazard: Low toxicity by ingestion. A severe skin irritant.

vinylidene chloride.   (VC).
CAS: 75-35-4. CH_2=CCl_2.

Properties: Colorless liquid. Fp −122.53C, bp 37C, flash p 14F (−10C) (OC), autoign temp 856F (457C), readily polymerizes. Insoluble in water. Commercial product contains small proportion of inhibitor.

Hazard: Flammable, dangerous fire risk, explosive limits in air 5.6–11.4%. Toxic by inhalation. Liver and kidney damage. Questionable carcinogen.

Use: Copolymerized with vinyl chloride or acrylonitrile to form various kinds of saran. Other copolymers are also made. Adhesives; component of synthetic fibers.

See saran.

vinylidene chloride-butyl acrylate copolymer.

CAS: 9011-09-0. $(C_7H_{12}O_2 \cdot C_2H_2Cl_2)_x$.

Hazard: Low toxicity by ingestion.

vinylidene fluoride.   (1,1-difluoroethylene).

CAS: 75-38-7. $H_2C=CF_2$. A monomer.

Properties: Colorless gas; faint ethereal odor. Bp −83C (1 atm), fp −144C (1 atm), d (liquid) 0.617 (24C). Slightly soluble in water; soluble in alcohol and ether.

Derivation: Interaction of hydrogen with dichlorodifluoroethane.

Grade: 99% min purity.

Hazard: Flammable, dangerous fire risk, explosive limits in air 5.5–21%. Toxic by inhalation. Liver damage. Questionable carcinogen.

Use: Polymers and copolymers, chemical intermediate.

See polyvinylidene fluoride.

vinylidene resin.   (polyvinylidene resin). A polymer in which the unit structure is $(-H_2CCX_2-)$, in which X is usually chlorine, fluorine, or cyanide radical. Examples are saran and "Vitron" A *[Chemtura]*.

vinyl isobutyl ether.   (isobutyl vinyl ether; IVE).

CAS: 109-53-5. $CH_2=CHOCH_2CH(CH_3)_2$.

Properties: Colorless liquid. D 0.7706 (20/20C), bp 83.3C, vap press 68 mm Hg (20C), fp −132C, refr index 1.3938, flash p 15F (−9.4C) (OC). Very slightly soluble in water; soluble in alcohol and ether; easily polymerized.

Derivation: Catalytic union of acetylene and isobutyl alcohol.

Method of purification: Washing with water, drying in the presence of alkali, and distillation from metallic sodium.

Grade: Technical.

Hazard: Flammable, dangerous fire risk.

Use: Polymer and copolymers used in surgical adhesives, coatings, and lacquers; modifier for alkyd and polystyrene resins; plasticizer for nitrocellulose and other plastics; chemical intermediate.

vinylmagnesium chloride. $CH_2=CHMgCl$.

Usually supplied dissolved in tetrahydrofuran.

Use: Grignard reagent.

vinyl methyl ether.   (methyl vinyl ether; methoxyethylene; MVE).

CAS: 107-25-5. $CH_2=CHOCH_3$.

Properties: Colorless compressed gas, or colorless liquid. D 0.7500 (20/20C), bp 6.0C, vap press 1052 mm Hg (20C), flash p −60F (−51C), fp −121.6C. Slightly soluble in water; soluble in alcohol and ether; easily polymerized. Commercial material contains a small proportion of polymerization inhibitor.

Derivation: Catalytic reaction of acetylene and methanol.

Grade: Technical (95% min), pure.

Hazard: Highly flammable, severe fire and explosion risk, explosive limits in air 2.6–39%.

Use: Copolymers used in coatings and lacquers; modifier for alkyl, polystyrene, and ionomer resins; plasticizer for nitrocellulose and adhesives.

See polyvinyl methyl ether.

vinyl methyl ketone.   (3-buten-2-one; methyl vinyl ketone).

CAS: 78-94-4. $CH_3COCH=CH_2$.

Properties: Colorless liquid. D 0.8636 (20/4C), bp 80C, flash p 20F (−6.6C) (CC). Soluble in water and alcohols.

Hazard: Flammable, dangerous fire risk. Skin, eye and upper respiratory tract irritant. Central nervous system impairment.

Use: Monomer for vinyl resins, component of ionomer resins, intermediate in steroid and vitamin A synthesis, alkylating agent.

2-(vinyloxy)ethyl nitrate.   See ethylene glycol mononitrate vinyl ether.

vinyl plastics.   See polyvinyl resins.

vinyl propionate. $CH_2=CHOOCC_2H_5$.

Properties: Liquid. D 0.9173 (20/20C), bp 95.0C, fp −81.1C, flash p 34F (1.1C) (OC). Almost insoluble in water.

Hazard: Flammable, dangerous fire risk.

Use: Polymers, emulsion paints.

vinylpyridine. $C_5H_4NCH=CH_2$.

Properties: Colorless liquid. Boils with resinification at 159C, d 0.9746 (20C), refr index 1.5509 (20C). Dissolves in water to extent of 2.5%; water dissolves in it to 15%; soluble in dilute acids, hydrocarbons, alcohols, ketones, esters. Commercial material contains inhibitor. Combustible.

Hazard: Irritant to skin and eyes.

Use: Elastomers and pharmaceuticals. Latex used as tire cord binder (41% solids). The latex is copolymerized with butadiene-styrene.

n-vinyl-2-pyrrolidone.

CAS: 88-12-0.

Properties: Colorless liquid. Bp 148C (100 mm Hg), mp 13.5C, flash p 209F (98.3C) (COC), d 1.04. Combustible.
Derivation: From acetylene and formaldehyde by high-pressure synthesis.
Hazard: Irritant and narcotic. Liver damage. Questionable carcinogen.
Use: Polyvinylpyrrolidone, organic synthesis.

1-vinyl-2-pyrrolidone crosslinked insoluble polymer. See polyvinylpolypyrrolidone.

vinyl stabilizer. A substance added to vinyl chloride resins during compounding, to retard the rate of deterioration caused by formation of hydrogen chloride. Many combine readily with hydrogen chloride but do not otherwise interfere with the properties and uses of the final product. Amines, basic oxides, and metallic soaps are commonly used.

vinyl stearate. $CH_3(CH_2)_{16}COOCH=CH_2$.
Properties: White, waxy solid. Mp 28–30C, bp 175C (3 mm Hg), d 0.9037 (20/20C), refr index 1.4355–1.4362 (55C), iodine no. 80–82. Insoluble in water and alcohol; moderately soluble in ketones and vegetable oils; soluble in most hydrocarbon and chlorinated solvents. Combustible.
Use: Plasticizer (copolymerizer), lubricant.

vinylstyrene. See divinyl benzene.

vinyl toluene. (methyl styrene).
CAS: 25013-15-4. $CH_2=CHC_6H_4CH_3$.
Properties: Colorless liquid. Fp −76.8C, bp 170–171C, d 0.890 (25/25C), bulk d 7.41 lb/gal, refr index 1.534 (34C), flash p 130F (54.4C) (CC), autoign temp 921F (494C). Very slightly soluble in water; soluble in methanol, ether. Combustible.
Hazard: Moderate fire risk. Eye and upper respiratory tract irritant. Questionable carcinogen.
Use: Solvent, intermediate.

vinyl trichloride. See 1,1,2-trichloroethane.

vinyltrichlorosilane.
CAS: 75-94-5. $CH_2CHSiCl_3$.
Properties: Colorless or pale-yellow liquid. Bp 90.6C, d 1.265 (25/25C), refr index 1.432 (20C), flash p 16F (−8.89C). Readily hydrolyzed with liberation of hydrogen chloride; polymerizes easily; soluble in most organic solvents; reacts with alcohol.
Derivation: Reaction of acetylene and trichlorosilane (peroxide catalyst), reaction of trichlorosilane with vinyl chloride.
Grade: Technical.
Hazard: Highly toxic by ingestion and inhalation, strong irritant to tissue. Flammable, dangerous fire risk.
Use: Intermediate for silicones, coupling agent in adhesives and bonds.

vinyon. Generic name for a manufactured fiber in which the fiber-forming substance is any long-chain synthetic polymer composed of at least 85% by weight of vinyl chloride units, $-CH_2CHCl-$ (Federal Trade Commission). It has good resistance to chemicals, bacteria, moths; unaffected by water and sunlight; low softening point. Tenacity 3.1 g/denier, difficult to ignite, self-extinguishing.
See polyvinyl chloride.

vioform. (iodochloroxyquinoline). Odorless, nonirritant, sterilizable substitute for iodoform.

violanthrone. See dibenzanthrone.

Violet #1. An FD&C color used for meat-grading, cosmetics, beverages, etc. It is a triphenylmethane dye banned by the FDA in 1973 because of carcinogenic risk.

violet gentian. See methyl violet.

viomycin. An antibiotic produced by *Streptomyces puniceus*. Unique among antibiotics in that it is more active against acid-fast organisms than against other groups. Mycobacteria are most sensitive to viomycin, and the antibiotic is active against strains of *Mycobacterium tuberculosis*, which are resistant to other antibiotics. Available commercially as sulfate.

viosterol. Irradiated ergosterol.

"Vipex 19" [LFL]. TM for a specialty formulated high viscosity oil.
Use: In asphalt pavement emulsified rejuvenation.

viral vector. A viral DNA altered so that it can act as a vector for recombinant DNA.

viridogrisein. See etamycin.

virion. A single virus particle.

Virtanen, Arrturi I. (1895–1973). A Finnish biochemist who won the Nobel Prize in 1945. His work was primarily concerned with research in nutrition and agriculture. He made important discoveries regarding prevention of fodder spoilage and bacterial fermentation as well as nitrogen metabolism in plants. His Ph.D. was awarded at the University of Helsinki and followed by an illustrious career that included awards throughout Scandinavia.

virus. An infectious agent composed almost entirely of protein and nucleic acids (nucleoprotein). Viruses can reproduce only within living cells and are so small that they can be resolved only with an electron microscope. Since they pass through filters that retain bacteria, they are often called filterable viruses. Tobacco mosaic was the first virus to be crystallized and isolated (Dr. W. M. Stanley,

1935); it contains some 2000 protein molecules in a sequence of 158 amino acids (mw 40,000,000). Bushy stunt virus found in tomato plants has a molecular weight of 7,600,000. First synthesis of a virus was reported in 1967.

Viruses differ from organisms in that they are only half alive; they lack metabolism, are unable to utilize oxygen, to synthesize macromolecules, to grow, or to die. They are parasites, relying on a living host cell. They account for many diseases, including mumps, measles, scarlet fever, smallpox, influenza, and possibly the common cold. Their shapes are similar to those of bacteria (rods, spheres, filaments). They have the ability to mutate; they are also antigenic and thus initiate formation of antibodies. Some act as bacteriophages. A direct relation between virus and cancer has been shown, the DNA of the virus becoming irreversibly bound to the DNA of the affected cells.
See bacteria; deoxyribonucleic acid.

viscometer. (viscosimeter). A device for measuring the viscosity of a liquid. The types most widely used are the Engler, Saybolt, and Redwood, which indicate viscosity by the rate of flow of the test liquid through an orifice of standard diameter or the flow rate of a metal ball through a column of the liquid; other types utilize the speed of a rotating spindle or vane immersed in the test liquid. The liquids commonly measured are lubricating oils and the like; heavier (non-Newtonian) liquids such as paints and paper coatings require more complex devices, e.g., Brookfield and Krebs-Stormer.
See viscosity.

viscose process. The best-known process for making regenerated cellulose (rayon) by converting cellulose to the soluble xanthate, which can be spun into fibers and then reconverted to cellulose by treatment with acid. Wood pulp is steeped with 17–20% caustic soda; the resulting alkali cellulose is pressed to remove excess liquor and the soluble β- and γ-cellulose, and then shredded and aged. It is then treated with carbon disulfide and sodium hydroxide to form an orange, viscous solution of cellulose xanthate. After filtration and deaeration, this solution (viscose) is forced through minute spinnerette openings (or long slit dies in the case of cellophane) into a bath containing sulfuric acid and various salts such as sodium and zinc sulfate. The salts cause the viscose to gel immediately, forming a fiber or film of sufficient strength to permit it to be drawn through the bath under tension. At the same time the sulfuric acid decomposes the xanthate, converting the fibers to cellulose, in which form they are washed and dried.
See rayon; cellophane.

viscosification. Increasing the viscosity of a liquid.

viscosimeter. See viscometer.

viscosity. The internal resistance to flow exhibited by a fluid; the ratio of shearing stress to rate of shear. A liquid has a viscosity of one poise if a force of one dyne per square centimeter causes two parallel liquid surfaces one square centimeter in area and one centimeter apart to move past one another at a velocity of one centimeter per second. One poise equals 100 centipoises. Viscosity in centipoises divided by the liquid density at the same temperature gives kinematic viscosity in centistokes (cs). One hundred centistokes equal one stoke. To determine kinematic viscosity, the time is measured for an exact quantity of liquid to flow by gravity through a standard capillary.

Water is the primary viscosity standard with an accepted viscosity at 20C of 0.010019 poise. Hydrocarbon liquids such as hexane are less viscous. Molasses may have a viscosity of several hundred centistokes, while for a very heavy lubrication oil the viscosity may be 100 centistokes. There are many empirical methods for measuring viscosity.
See Saybolt Universal Viscosity; viscometer.

viscosity index improver. A lubricating-oil additive that has the effect of increasing the viscosity of the oil in such a way that it is greater at high temperature than at low temperature. Agents used for this purpose are polymers of alkyl esters of methacrylic acid, polyisobutylenes, etc.

viscosity, kinematic. See viscosity.

"Vistac A" [Chevron Intellectual]. TM for a series of synthetic hydrocarbon polymers.
Use: Rubber-base adhesives and cements; latex and asphalt emulsions.

visual purple. See rhodopsin.

vital dye. One that is absorbed by living organisms and colors certain structures without apparent serious injury to the structure or organism.

vitamin. Any of a number of complex organic compounds, present in natural products or made synthetically, which are essential in small proportions in the diet of animals and humans. Some are fat soluble (A, D, K); others are water soluble (B complex, C). Their precursors are called provitamins. A normal diet usually contains sufficient vitamins for health, although older people, the ill, young, or infirm may require different standards. Their usual use in medicine is restricted to correction of specific metabolic deficiencies. Some authorities believe that habitual intake of standard vitamin preparations readily available on the market is of little, if any, nutritional benefit. The following list of cross-references will serve to locate technical information about the various vitamin entries in this book:
Vitamin A: See carotene; cryptoxanthin; retinol; 3-dehydroretinol; provitamin.

Vitamin B: See vitamin B complex; thiamine; riboflavin; niacin; panthothenic acid; biotin; cyanocobalamin; pyridoxine; folic acid; inositol.

Vitamin C: See ascorbic acid.

Vitamin D: See ergosterol; ergocalciferol; cholecalciferol.

Vitamin E: See tocopherol.

Vitamin K: See phytonadione; menadione; phthiocol.

vitamin A_1, anhydrous. See all-*trans*-anhydroretinol.

vitamin B complex. A group of closely interrelated vitamins found in rice bran, yeast, wheat germ, etc., originally thought to be one substance. Studies carried out by R. J. Williams and associates later revealed the astonishing complexity of this group. He states as follows: "The physiological activity originally observed was due to the *additive* effect of a considerable number of substances, *each one* of which is of itself essential. If and when the designations B_1, B_2, B_3, etc., are used, they have an entirely different meaning from the parallel use of D_1, D_2, D_3 or K_1 and K_2 because in the case of the D and K vitamins one form can replace another. In the case of the B vitamins each form is a distinctly different substance with different functions, and each member of the family is separately indispensable. No one B vitamin can replace any other."

vitamin B_6 tripalmitate. See pyridoxine tripalmitate.

vitamin E acetate. See d-α-tocopheryl acetate.

vitexin. See 8-β-d-glucopyranosyl-apigenin.

"Viton" [Du Pont]. TM for a series of fluoroelastomers based on the copolymer of vinylidene fluoride and hexafluoropropylene with the repeating structure possibly $-CF_2-CH_2-CF_2-CF(CF_3)-$.
Properties: White transparent solid. D 1.72–1.86. Resistant to corrosive liquids and chemicals up to 315C. Useful continuous service at 204–232C. Resistant to ozone, weather, flame, oils, fuels, lubricants, many solvents; radiation resistance good. Nonflammable.
Use: Gaskets, seals, diaphragms, tubing, aerospace and automotive components, high-vacuum equipment, low-temperature and radiation equipment.

vitreous. Descriptive of a material having the appearance and properties of a glass, i.e., a hard, amorphous, brittle structure, as in porcelain enamel. See vitrification; glass.

vitreous enamel. See porcelain enamel.

vitrification. The process of converting a siliceous material into an amorphous, glassy form by melting and cooling. As applied to radioactive waste disposal, it refers to incorporation of the waste in glassy materials for permanent storage.

vitriol. An obsolete term once used to refer to a number of sulfates (lead, copper, zinc) because of their glassy appearances. Sulfuric acid was called oil of vitriol. Derived from *vitrum* (glass).

Vmax. The maximum velocity of an enzymatic reaction when the binding site is saturated with substrate.

VM&P naphtha. See naphtha (la).

VOC. See volatile organic compound.

voids. Empty spaces of molecular dimensions occurring between closely packed solid particles, as in powder metallurgy. Their presence permits barriers made by powder metallurgy techniques to act as diffusion membranes for separation of uranium isotopes in the gaseous diffusion process. See diffusion, gaseous.

Voight amination. Amination of benzoins with amines in the presence of phosphorus pentoxide or hydrochloric acid.

"Volan" [Zaclon]. (methacrylatochromic chloride). TM for bonding agent.
Use: Applied to glass fibers used in reinforced plastic laminates to improve adhesion between glass and resin.

volatile organic compound. (VOC). Any hydrocarbon, except methane and ethane, with vapor pressure equal to or greater than 0.1 mm Hg.

volatility. The tendency of a solid or liquid material to pass into the vapor state at a given temperature. Specifically the vapor pressure of a component divided by its mole fraction in the liquid or solid.

Volhard-Erdmann cyclization. Synthesis of alkyl and aryl thiophenes by cyclization of disodium succinate or other 1,4-difunctional compounds (g-oxo acids; 1,4-diketones; chloroacetyl-substituted esters) with phosphorus heptasulfide.

Volhard's solution. A solution of potassium thiocyanate.
Use: Analytical chemistry.

voltaic cell. Two conductive metals of different potentials, in contact with an electrolyte, which generate an electric current. The original voltaic cell was composed of silver and zinc, with brine-moistened paper as electrolyte. Semisolid pastes are now used; electrodes may be lead, nickel, zinc, or cadmium.
See solar cell.

volumetric analysis. See titration.

von Baeyer, Adolf. (1835–1917). A German chemist who received the Nobel Prize for chemistry in 1905. His work concerned organic dyes and hydroaromatic compounds. He was educated in Berlin under the direction of Bunsen and Kekule. He was a professor in Strasbourg and Munich. Von Baeyer made many discoveries, including barbituric acid and the molecular structure of indigo.

Von Braun reaction. The reaction between phosphorous pentahalide and an N-substituted benzamide to give an alkylene dihalide and benzonitrile.

von Richter reaction. Carboxylation of substituted aromatic nitro compounds with ethanolic potassium cyanide at 120–270C, the carboxyl group entering a position ortho to that previously occupied by the eliminated nitro group.

von Richter synthesis. Formation of cinnoline derivatives by diazotization of *o*-aminoarylpropiolic acids or *o*-aminoarylacetylenes followed by hydration and cyclization. The method is applicable to preparation of cinnolines substituted in the benzenoid ring.

"Voranate" *[Dow].* TM for a series of urethane intermediates that are the reaction products of polyols and isocyanates. They are adducts or quasiprepolymers to be used in combination with "Voranol" products to obtain rigid urethane foams.

Vorane. A group of polyurethane chemicals, raw materials for polyurethane elastomers, coatings, and foams.

"Voranol" *[Dow].* TM for a series of polyols used as intermediates in urethane elastomers, coatings, and flexible and rigid foams.

VPC. Abbreviation for vapor-phase chromatography.
See gas chromatography.

VPM. A solution containing 32.7% sodium methyldithiocarbamate used as a soil fumigant.

vulcanization. A physicochemical change resulting from cross-linking of the unsaturated hydrocarbon chain of polyisoprene (rubber) with sulfur, usually with application of heat. The precise mechanism that produces the network structure of the cross-linked molecules is still not completely known. Sulfur is also used with unsaturated types of synthetic rubbers; some types require use of peroxides, metallic oxides, chlorinated quinones, or nitrobenzenes. Natural rubber can be vulcanized with selenium, organic peroxides, and quinone derivatives but these have limited industrial use; high-energy radiation curing is an important innovation.

Vulcanization can be effected with sulfur alone in high percentage, but the time required is too long to be economical and the properties obtained are poor. Inorganic accelerators and metallic oxides (usually zinc) are essential for satisfactory cure. Organic accelerators, introduced in the early 1920s, notably shortened vulcanization time.
Three factors affect the properties of a vulcanizate: (1) the percentage of sulfur and accelerator used, (2) the temperature, and (3) the time of cure. Sulfur is usually from 1 to 3%; with strong acceleration the time can be as short as 3 minutes at high temperature (150C). Vulcanization can also occur at room temperature with specific formulations (self-curing cements).
Vulcanization was discovered in 1846 by Charles Goodyear in the U.S. and simultaneously by Thomas Hancock in England. Its overall effect is to convert rubber hydrocarbon from a soft, tacky, thermoplastic to a strong, temperature-stable thermoset having unique elastic modulus and yield properties.
See rubber; rubber, synthetic.

Vulcosal. The industrial grade of salicylic acid.
Use: Stabilizer and retarder of vulcanization.

Vulklor. Tetrachloro-*p*-benzoquinone.
Use: Vulcanizing agent.

Vultex. A vulcanized rubber or synthetic rubber latex.
Use: Health care workers' gloves, drug sundries, dipped products, adhesives.

"Vybar" *[Baker Hughes].* TM for a grade of petroleum microcrystalline wax.
Use: Hardening and upgrading paraffins, coatings, polishes, adhesives, cosmetics, color dispersants, paper coatings, lubricants, floor wax, ski wax inks, ozone barriers for elastomers, and anticaking treatments.

"Vycel P" *[Crowley].* TM for a polyester modifier.
Use: Improves impact resistance.

"Vycor" Brand Glass No. 7900 *[Corning].* TM for a glass made by a process in which an article fabricated by conventional methods is chemically leached to remove substantially all of the ingredients except silica. When fired at high temperatures a transparent glass of high softening point and extremely low expansion coefficient is produced.
Properties: Softening point 1500C, temp limit in service 900C, linear coefficient of expansion per degree C = 0.0000008, d 2.18, refr index 1.458 (a similar glass, No. 7910, will transmit over 60% radiation at 254 millimicrons in a 2-mm section).
Use: Laboratory and industrial glassware, including beakers, crucibles, flasks, dishes, tubes, cylinders, containers, flat glass rods.

W.   Symbol for tungsten.
See tungsten, wolfram.

Wacker reaction.   The oxidation of ethylene to acetaldehyde in the presence of palladium chloride and cupric chloride.

Wagner-Jauregg reaction.   Addition of maleic anhydride to diarylethylenes with formation of bis adducts that can be converted to aromatic ring systems.

Wagner–Meerwin rearrangement.   Carbon-to-carbon migration of alkyl, aryl, or hydride ions. The original example is the acid-catalyzed rearrangement of camphene hydrochloride to isobornyl chloride.

Waksman, Selman A.   (1888–1973). American microbiologist (Nobel Prize winner in 1952) and professor at Rutgers University. He was the first to use the term *antibiotic* to designate the mold-produced antibacterial substances discovered by Fleming in 1928. He became the outstanding authority in this field.

Walden inversion.   Inversion of configuration of a chiral center in bimolecular nucleophilic substitution reactions.

Walker, John E.   (1941–). An English chemist who won the Nobel Prize in 1997 for his work on elucidating the enzymatic mechanism by which ATP synthase (ATPase) catalyzes the synthesis of adenosine triphosphate (ATP), the energy source of living cells. He is a senior scientist at the Medical Research Council Laboratory of Molecular Biology, Cambridge, England. He received his Ph.D. from Oxford University, Great Britain.
See Boyer, Paul D., Skou, Jens C.

Wallach, Otto.   (1847–1931). A German chemist who received the Nobel Prize for chemistry in 1910 for his work in alicyclic compounds. His mentors were Hofmann and Wahler, and he worked at the University of Bonn under Kekule. He studied pharmacy and did work on terpenes, camphors, and essential oils. This was followed by research in aromatic oils, perfumes, and spices. He studied thuja oil and fenchone. His research on terpenes revealed their significance in sex hormones and vitamins. Ethereal oils and industrial uses were made possible by his work.

Wallach rearrangement.   Acid-catalyzed rearrangement of azoxybenzenes to *p*-hydroxyazobenzenes.

Wallpol.   A series of vinyl acetate/acrylate and vinyl acetate emulsions.

Warfarin.   (3-(α-acetonylbenzyl)-4-hydroxycoumarin).
CAS: 81-81-2. $C_6H_4C_3O(OH)(O)CH(CH_2COCH_3)C_6H_5$.
Properties: Colorless crystals; odorless; tasteless. Mw 308.35, mp 161C. Soluble in acetone, dioxane; slightly soluble in methanol, ethanol; very soluble in alkaline aqueous solution; insoluble in water and benzene.
Hazard: A poison. A coagulant.
Use: Anticoagulant rodenticide.

warning odor.   A distinctive odor imparted to fuel gases for safety purposes, because they have little or no odor of their own. Ethanethiol and other malodorous mercaptan derivatives are added to natural gas, SNG, LPG, etc., for this purpose.

warp.   The lengthwise threads or strands of a textile fabric.

Warshel, Arieh.   (1940–). An American-Israeli chemist who was awarded the 2013 Nobel Prize jointly with Martin Karplus and Michael Levitt. The scientist developed accurate computer models to understand and predict chemical processes. As the computer simulations could produce the exact output, this created an efficient method to study and test effect of drugs on living molecules, leading to drug design for cancer therapy. Warshel was interested in enzymes reaction, and he and Levitt developed the first generalized computer model of enzymatic reactions, and this model proved significant as it could be used to model other molecules. He was awarded his M.S. and Ph.D. by the Weizmann Institute of Science, Israel. He was an associate professor at the Wiseman Institute of Science and the University of Southern California, and later became a professor (1984) and a distinguished professor of chemistry (2011) at the University of Southern California, Los Angles.

Hawley's Condensed Chemical Dictionary, Sixteenth Edition. Michael D. Larrañaga, Richard J. Lewis, Sr., and Robert A. Lewis.
© 2016 John Wiley & Sons, Inc. Published 2016 by John Wiley & Sons, Inc.

wash-and-wear fabric.   See aminoplast resin.

washing soda.   (soda crystals). $Na_2CO_3 \cdot 10H_2O$. See soda ash; sodium carbonate monohydrate.

waste, chemical.   See chemical waste.

waste control.   Waste materials can be classified by type, as gaseous, liquid, solid, and radioactive; and by source, as chemical, municipal, agricultural, urban, and nuclear. Many methods of treating wastes, either by converting them to useful by-products or by disposing of them, are in operation or under experimentation. Dumping in streams and rivers has long been illegal and ocean dumping has been prohibited since 1973. Methods of disposal or treatment include incineration (garbage, paper, plastics), precipitation (smoke, solid-in-liquid suspensions), adsorption (gases and vapors), chemical treatment (neutralization, ion-exchange, chlorination), reclamation of sulfite liquors (paper industry), reclaiming (paper, metals, rubber), compaction (urban wastes), bacterial digestion (sewage), comminution, and melting (glass, metals). Some cases may involve a combination of these methods.

The following procedures should also be mentioned: (1) Flash pyrolysis of certain solid wastes yields synthetic fuel oil and other useful products. (2) Urban, animal, and agricultural wastes can be fermented by anaerobic bacteria to yield proteins, fuel oil, etc.; the possibilities of converting cannery and other food-processing wastes into protein-rich foods and feeds are being investigated. (3) Catalytic oxidation of exhaust emission gases; devices have been installed in cars since 1973 for air-pollution control. (4) Recovery of methane from manures. (5) Incorporation of special additives to certain plastics (polystyrene bottles) to render them biodegradable. (6) High-pressure hydrogenation of garbage to yield a low-sulfur combustible oil. (7) Deactivating radioactive wastes by adsorption or ion exchange, as well as by solidification and hydraulic fracturing; high-activity wastes are buried in steel-lined concrete tanks. (8) Catalytic oxidation of waste chlorinated hydrocarbons, with partial recovery of chlorine. (9) Incineration of semisolid and liquid wastes at sea in ships designed for that purpose.

See radioactive waste; chemical waste; sewage sludge; urban waste.

waste, hazardous.   See chemical waste; radioactive waste.

waste, radioactive.   See radioactive waste.

waste wool, wet.   See wool waste.

Watchung.   Precipitated diazo red pigments.

water.   (ice, steam). H_2O or HOH.
  Properties: Colorless liquid; odorless; tasteless. Allotropic forms are ice (solid) and steam (vapor).

Water is a polar liquid with high dielectric constant (81 at 17C), which largely accounts for its solvent power. It is a weak electrolyte, ionizing as H_3O^+ and OH^-. At atmospheric pressure it has d 1.00 (4C), fp 0C (32F), and expands about 10% on freezing. Viscosity 0.010019 poise (20C), sp heat 1 cal/g, vap press 760 mm Hg (100C), triple p 273.16 K (4.6 mm), surface tension 73 dynes/cm (20C), latent heat of fusion (ice) 80 cal/g, latent heat of condensation (steam) 540 cal/g, bulk d 8.337 lb/gal, 62.3/lb/cu ft, refr index 1.333. Water may be superheated by enclosing in an autoclave and increasing pressure; it may be supercooled by adding sodium chloride or other ionizing compound. It has definite catalytic activity, especially of metal oxidation. Physiologically water is classed as a nutrient substance.
Derivation: (1) Oxidation of hydrogen; (2) end product of combustion; (3) end product of acid-base reaction; (4) end product of condensation reaction.
Purification: (1) Distillation; (2) ion exchange reaction (zeolite); (3) chlorination; (4) filtration.
Use: Suspending agent (papermaking, coal slurries), solvent (extraction, scrubbing), diluent, beer and carbonated beverages, hydration of lime, paper coatings, textile processing, moderator in nuclear reactors, debarking logs, industrial coolant, filtration, washing and scouring, sulfur mining, hydrolysis, Portland cement, hydraulic systems, power source, steam generation, food industry, source of hydrogen by electrolysis and thermochemical decomposition.
See hydrogen; ice; steam; heavy water; ocean water; water, hard.

Water Based Ink Wash.   A liquid cleaner.
  Use: To remove all solvent based inks from screen after printing, leaving film stencil intact.

water, bound.   See bound water.

water gas.   (blue gas). A mixture of gases made from coke, air, and steam. The steam is decomposed by passing it over a bed of incandescent coke or by high-temperature reaction with natural gas or similar hydrocarbons. Approximate composition: carbon monoxide 40%, hydrogen 50%, carbon dioxide 3%, and nitrogen 3%.
  Hazard: Flammable, dangerous fire and explosion risk. Explosive limits 7–72% in air. Toxic by inhalation.
  Use: Organic synthesis, fuel gas, ammonia synthesis. See synthesis gas.

water glass.   See sodium silicate.

water, hard.   Water containing low percentages of calcium and magnesium carbonates, bicarbonates, sulfates, or chlorides as a result of long contact with rocky substrates and soils. Degree of hardness is expressed either as grains per gallon or parts per million (ppm) of calcium carbonate (1 grain of $CaCO_3$ per gal is equivalent to 17.1 ppm). Up

to 5 grains is considered soft; more than 30 grains is very hard. Hardness may be temporary (carbonates and bicarbonates) or permanent (sulfates, chlorides). Treatment with zeolites is necessary to soften permanently hard water. Temporary hardness can be reduced by boiling. These impurities are responsible for boiler scale and corrosion of metals on long contact. Hard waters require use of synthetic detergents for satisfactory sudsing.
See zeolite.

watermark.   See dandy roll.

water of crystallization.   Water chemically combined in many crystallized substances; it can be removed at or near 100C, usually with loss of crystalline properties.

water pollution.   Contamination of fresh- or saltwater with materials that are toxic, noxious, or otherwise harmful to fish and other animals and to humans, including thermal pollution. Disposal of untreated chemical and municipal wastes in streams and rivers has been illegal since the early 1900s; in 1973 the EPA prohibited dumping of all types of wastes into the ocean. Unintentional pollution results from runoff containing toxic insecticidal residues. Oil spills at sea are a continual problem and probably will remain so.
See waste treatment; Environmental Protection Agency; environmental chemistry.

waterproofing agent.   (1) Any film-forming substance that coats a substrate with a water-repellent layer, such as paint, a rubber or plastic film, a wax, or an asphaltic compound. These are used on a wide variety of surfaces, including cement, masonry, metals, textiles, etc.
(2) Any metal salt or other chemical that impregnates textile fibers in such a way as to give an air-permeable, water-resistant product. There are three types of these: renewable, semidurable, and durable. Renewable repellents are water dispersions containing aluminum acetate or formate, emulsifying agents and protective colloids in the continuous phase, and a blend of waxes in the disperse phase. Semidurable repellents involve precipitation of insoluble metal salts on the fibers; water-soluble soaps and waxes are usually added to the mixture, which is especially effective on synthetic fibers. Durable repellents coat each fiber with a protective film without bonding them together or sealing the apertures.

water purification.   (water conditioning). Any process whereby water is treated in such a way as to remove or reduce undesirable impurities. The following methods are used: (1) sedimentation, in which coarse suspended matter is allowed to settle by gravity in special tanks or reservoirs; (2) coagulation of aggregates (called "floc") by means of

aluminum sulfate, ferric sulfate, or sodium aluminate (the aggregates are formed from colloidally dispersed impurities activated by the coagulant); (3) filtration through a bed of fine sand, either by gravity flow or by pressure, to remove suspended particles; (4) chlorination, which is effective in sterilizing potable water, swimming pools, etc.; (5) adsorption on activated carbon for removal of organic contaminants causing unpleasant taste and odor; (6) hardness removal by ion-exchange or zeolite process. The USDA has reported that mercury can be removed from water by treatment with low concentrations of black liquor from kraft papermaking.

water-soluble gum.   See gum, natural.

water-soluble oil.   Ammonia, potash, or sodium soaps of oleic, rosin, or naphthenic acids dissolved in mineral oils.
Use: Boring, lathe-cutting, milling, polishing lubricants, dressing textile fibers, dust laying.
See soluble oil.

water-soluble resin.   See polymer, water-soluble.

wattle bark.   (Australian bark; mimosa bark).
Derivation: From the Australian and South African wattles and South African acacias. Bark contains 25–35% tannin.
Grade: Based on tannin content.
Use: Source of wattle bark extract, used in vegetable tanning of leather, especially retannage of upper leathers and production of heavy leathers.

wave function.   In quantum mechanics, a complex function extending over the configuration space of a system; its complex conjugate yields the probability density function, and other mathematical operations yield other physical quantities.

wax.   A low-melting organic mixture or compound of high molecular weight, solid at room temperature and generally similar in composition to fats and oils except that it contains no glycerides. Some are hydrocarbons, others are esters of fatty acids and alcohols. They are classed among the lipids. Waxes are thermoplastic, but since they are not high polymers, are not considered in the family of plastics. Common properties are water repellency, smooth texture, low toxicity, freedom from objectionable odor and color. They are combustible and have good dielectric properties; soluble in most organic solvents; insoluble in water. The major types are as follows:
I. Natural
(1)   Animal (beeswax, lanolin, shellac wax, Chinese insect wax)
(2)   Vegetable (carnauba, candelilla, bayberry, sugar cane)

(3) Mineral
   (a) Fossil or earth waxes (ozocerite, ceresin, montan)
   (b) Petroleum waxes (paraffin, microcrystalline) (slack or scale wax)

II. Synthetic
(1) Ethylenic polymers and polyol ether-esters ("Carboxwax," sorbitol)
(2) Chlorinated naphthalenes ("Halowax")
(3) Hydrocarbon type, i.e., Fischer–Tropsch synthesis

Use: Polishes, candles, crayons, sealants, suncracking protection of rubber and plastic products, cosmetics, paper coating, packaging food products, electrical insulation, waterproofing and cleaning compounds, carbon paper, precision investment casting.

wax, chloronaphthalene.

Properties: Translucent, black, light and varied colors. D 1.40–1.7 (300F), mp 87.7–129.4C, bp 287–371C. Soluble in many organic solvent liquids and oils (when heated together).
Derivation: By chlorinating naphthalene.
Hazard: Toxic by ingestion and skin contact.
Use: Condenser impregnation, moisture-, flame-, acid-, insect-proofing of wood, fabric, and other fibrous bodies; moisture- and flameproofing covered wire and cable; solvent (for rubber, aniline and other dyes, mineral and vegetable oils, varnish gums and resins, and other waxes when mixed in the molten state).

wax, microcrystalline.
A wax, usually branched-chain paraffins, characterized by a crystalline structure much smaller than that of normal wax and also by much higher viscosity. Obtained by dewaxing tank bottoms, refinery residues, and other petroleum waste products; they have an average molecular weight of 500–800 (twice that of paraffin). Viscosity 45–120 sec (SUS at 98.9C), penetration value 3–33. Petroleum-derived products are used for adhesives, paper coating, cosmetic creams, floor wax, electrical insulation, heat-sealing, glass fabric impregnation, leather treatment, emulsions, etc. Some natural products, notably chlorophyll, are classed as microcrystalline waxes.

wax, polymethylene.
White, odorless solid with congealing point of 96.1C. Offered in flaked form. Approved by FDA.

wax tailings.
Brown, sticky, semiasphalt product obtained in the destructive distillation of petroleum tar just before formation of coke.
Use: Wood preservative, roofing paper.

weaponized anthrax toxin.
An modified filtrate from *Bacillus antrhacis*.
Use: A biological weapon.

Weatherometer.
See aging (c).

web.
A roll of paper as it comes from the fourdrinier machine and used to feed a rotary printing press.

weedkiller.
See herbicide.

Weerman degradation.
Formation of an aldose with one less carbon atom from an aldonic acid by a Hoffmann-type rearrangement of the corresponding amide. This is a general reaction of α-hydroxy carboxylic acids.

weight.
See mass.

weighting agent.
(1) In soft drink technology, an oil or oil-soluble compound of high specific gravity, such as a brominated olive oil, which is added to citrus flavoring oils to raise the specific gravity of the mixture to about 1.00, so that stable emulsions with water can be made for flavoring. (2) In the textile industry a compound used both to deluster and lower the cost of a fabric, at the same time improving its "hand" or feeling. Zinc acetylacetonate, clays, chalk, etc. are used.

welding.
Joining or bonding of metals or thermoplastics by application of temperatures high enough to melt the materials so that they fuse to a permanent union on cooling. In general, the temperatures used for thermoplastics are considerably lower than required for metals. The following methods are used for metals: (1) An oxyacetylene flame is applied with a torch to the butted ends or edges of the pieces to be joined. (2) A method called brazing is similar to (1), except that a nonferrous filler alloy is inserted between the pieces. A number of alloys are used, e.g., Ag/Cu/Zn; the filler cannot be remelted. It forms an intermetallic compound at the interfaces. (3) In resistance welding, the heat is provided by the resistance to an electric current as it passes through the material. No filler metal is used. (4) In ultrasonic welding, the heat source is the friction resulting from ultrasonic vibrations. It is a type of friction welding. (5) Electron-beam welding is a comparatively recent technique in which energy is supplied by a stream of electrons focused by a magnetic field under high vacuum. It is used for complicated weldments of tool steels.

The following methods are used for welding such thermoplastics as polyvinyl chloride, HDPE, polypropylene, and polycarbonates: (1) Hot gas technique, in which an electrically or gas-heated "gun" melts a rod of the same material as the parts to be joined. (2) Friction welding, in which heat is generated by rapid rubbing together of the two surfaces, one of which is held stationary while the other is rubbed against it at a speed great enough to cause softening. (3) Ultrasonic welding, which is also used for metals. See (4) above.
See solder.

"Wellbrom" *[Albemarle].* TM for sodium bromide solution, completion fluid.
Use: Completion, work-over and packer fluid in oilfield applications.

Werner, A. (1866–1919). A native of Switzerland, Werner was awarded the Nobel Prize for his development of the concept of the coordination theory of valence, which he advanced in 1893. His ideas revolutionized the approach to the structure of inorganic compounds and in recent years have permeated this entire area of chemistry. The term *Werner complex* has largely been replaced by "coordination compound."

Wessely–Moser rearrangement. Rearrangement of flavones and flavanones possessing a 5-hydroxyl group, through fission of the heterocyclic ring and reclosure of the intermediate diaroyl-methanes in the alternate direction.

Western blot. A technique for analyzing mixtures of proteins to show the presence, size, and abundance of one particular type of protein.
See DNA; Northern blot; protein; RNA; Southern blot.

Weston cell. An electrical cell used as a standard that consists of an amalgamated cadmium anode covered with crystals of cadmium sulfate dipping into a saturated solution of the salt, and a mercury cathode covered with solid mercury sulfate.

Westphalen–Lettré rearrangement. Dehydration of 5-hydroxycholesterol derivatives accompanied by C-10 to C-5 methyl migration in compounds with a β-substituent in C-6.

wet deposition. See acid precipitation.

wetting agent. A surface-active agent that, when added to water, causes it to penetrate more easily into, or to spread over the surface of, another material by reducing the surface tension of the water. Soaps, alcohols, and fatty acids are examples.
See detergent.

WFNA. Abbreviation for white fuming nitric acid.
See nitric acid, fuming.

Wharton reaction. Reduction of α,β-epoxy ketones by hydrazine to allylic alcohols.

wheat germ oil. Light-yellow, fat-soluble oil extracted from wheat germ.
Use: A food supplement.
See tocopherol.

whey. The serum remaining after removal of the solids (fat and casein) from milk. Dried whey contains about 13% protein, 71% lactose, 2.3% lactic acid, 4.5% water, and 8% ash, including a low concentration of phosphoric anhydride. Besides its value as an inexpensive source of protein for animal feeds, whey is used as a source of lactose and lactic acid, as well as for the synthesis of riboflavin, acetone, butanol, and fuel-grade ethanol by fermentation processes. Some types of cheese are made from whey, and it is also a possible culture medium. Dried whey may be used to replace up to 75% of the polyol component of rigid polyurethane foams.
See lactose.

whiskers. Single, axially oriented, crystalline filaments of metals (iron, cobalt, aluminum, tungsten, rhenium, nickel, etc.), refractory materials (sapphire, aluminum oxide, silicon carbide), carbon, boron, etc. They have tensile strengths of 3–6 million psi and very high elastic moduli. Their upper temperature limit in oxidizing atmospheres may be as high as 1700C and in inert atmospheres up to 2000C. Length may be up to two inches, with diameter up to 10 μm. Their chief use is in the manufacture of composite structures with plastics, glass, or graphite that have many applications in the aircraft and space vehicle field, where their high heat capacity and tremendous strength are invaluable, especially as ablative agents.

whiskey.
Properties: Light-yellow to amber liquid. D 0.923–0.935 (15.56C), 47–53% alcohol by volume, flash p 26.6C (CC).
Derivation: Distillation of fermented malted grains (corn, rye, or barley). After distillation, whiskey is aged in wooden containers for several years. The following changes occur during aging: extraction of wood components (acids and esters), oxidation of the components of the liquid, and reaction between organic compounds in the liquid, forming new flavors.
Hazard: Flammable, moderate fire risk. A non-cumulative poison; usually harmless in moderate amounts but may be toxic when habitually taken in large amounts.
Use: Beverage, medicine (stimulant, antiseptic, vasodilator).
See ethanol; proof.

white acid. A mixture of ammonium bifluoride and hydrogen fluoride used for etching glass.
Hazard: Strong irritant to skin and tissue.

white arsenic. See arsenic trioxide.

white copperas. See zinc sulfate.

white dye. An optical bleach or, in general, any substance, such as bluing, that may be added to a white article to increase its apparent whiteness.

white gasoline.   See gasoline.

white gold.   Alloy of 90% gold and 10% palladium, or 59% nickel and 41% gold.

white lead.   Name primarily applied to lead carbonate, basic, but also used for lead sulfate, basic (white lead sulfate), and lead silicate, basic (white lead silicate).
Hazard: Toxic by ingestion and skin absorption. Content in paints limited to 0.05% (FDA).
Use: Paint pigment.

white liquor.   See liquor (c).

white metal.   (1) Any of a group of alloys having relatively low melting points. They usually contain tin, lead, or antimony as the chief component. Type metal, Babbitt, pewter, and Britannia metal are of this group. (2) Copper matte containing about 75% copper, as obtained in copper smelting operations. See copper.

whitener.   (1) Any of several oil-in-water emulsions in powder form that are dried and concentrated; when added to an aqueous medium they form stable emulsions, giving a white color and a cream-like body to coffee. A typical formulation contains vegetable fat, protein, sugar, corn syrup solids, plus emulsifier and stabilizers. (2) A white pigment or colorant used in the paper and textile industries.

white oil.   Any of several derivatives of paraffinic hydrocarbons having moderate viscosity, low volatility, and high flash point. Available in USP grade. Combustible.
Use: Textile lubricant and finishing agent, plastics modifier.

white phosphorus.   See phosphorus.

white precipitate.   See mercury, ammoniated.

white shellac.   See shellac, bleached.

white vitriol.   See zinc sulfate.

whitewash.   A suspension of hydrated lime or calcium carbonate in water used for temporary antiseptic coatings of interiors of chicken houses and the like.
See calcimine.

whiting.   Finely ground, naturally occurring calcium carbonate, $CaCO_3$, 98% pure, contaminated with silica, iron, aluminum, or magnesium.
Properties: White or off-white powder. D 2.7. Insoluble in water; soluble in acids. It has no tinctorial power and hence is not a pigment.
Derivation: Traditionally from chalk, obtained from England, France, Belgium. A pure limestone or calcite is the principal commercial source. Crude chalk

or limestone is ground dry or wet, air- or water-floated, and sieved. Grades are based on particle size, softness, and light reflectance. Dry ground, air-floated limestone whiting can be as fine as 99% through 300-mesh.
Grade: Various. Paris white is the finest; coarser grades are extra gilders whiting, gilders whiting, and commercial, the last being quite coarse and of poor color. A putty grade is also sold.
Use: Filler in rubber, plastics, and paper coatings; putty (with linseed oil); whitewashes; sealants.
See calcium carbonate; chalk.

Whiting reaction.   Alkynediols are reduced by lithium aluminum hydride in ether or tertiary amines to dienes.

Wichterle reaction.   Modification of the Robinson annulation reaction in which 1,3-dichloro-*cis*-2-butene is used instead of methyl vinyl ketone.

Widman–Stoermer synthesis.   Synthesis of cinnolines by cyclization of diazotized *o*-aminoarylethylenes at room temperature.

Wieland, Heinrich O.   (1877–1957). A German chemist who won the Nobel Prize for chemistry in 1927. His research included work on bile acids, organic radicals, nitrogen compounds, toxic substances, and chemical oxidation, as well as the discovery of the structure of cholesterol. He received his Ph.D. from the University of Munich.

Wij's solution. ICl_3 and I_2.
Source: Iodine chloride (9.4 g) and iodine (7.2 g) are dissolved separately in glacial acetic acid and the solutions added together.
Use: Determination of iodine values of fats and oils.

wild type.   The form of an organism that occurs most frequently in nature.

Wilkinson, Geoffrey.   (1921–1996). A British organic chemist who won the Nobel Prize for chemistry in 1973 with Fischer. Their work involved research on organometallic compounds. He was a professor at the University of California and Harvard before returning to University of London as professor of inorganic chemistry.

willemite. Zn_2SiO_4. Natural zinc orthosilicate. Troostite is a manganese-bearing variety.
Properties: Color: yellow, green, red, brown, white; luster vitreous to resinous; sometimes fluoresces in UV light. Mohs hardness 5.5, d 3.3.
Occurrence: New Jersey, New Mexico, Africa, Greenland.
Use: Ore of zinc, a phosphor.

Willgerodt reaction.   Discovered in 1887, this reaction involves heating a ketone, e.g., $ArCOCH_3$, with an aqueous solution of yellow ammonium

sulfide (sulfur dissolved in ammonium sulfide). It results in the formation of an amide derivative of an arylacetic acid and in some reduction of the ketone: $ArCOCH_3 + (NH)_4S_x \rightarrow ArCH_2CONH_2 + ArCH_2CH_3$.
The dark reaction mixture usually is refluxed with alkali to hydrolyze the amide, the arylacetic acid being recovered from the alkaline solution (L. & M. Fieser).

Williamson synthesis.   An organic method for preparing ethers by the interaction of an alkyl halide with a sodium alcoholate (or phenolate).

Willstätter, Richard Martin.   (1872–1942). A German chemist who won the Nobel Prize for chemistry in 1915 for his work with plant pigmentation. His education was at the University of Munich where he studied and taught before going to Zurich, Switzerland. He researched chlorophylls and pigments of plants and the relationship of cornflower blue to rose red. Work included the study of alkaloids including cocaine, tropine, and atropine. His work perfected the process of chromatographic partition.

"Wilube 139" *[Wilube].*   TM for a semisynthetic machine and grinding coolant with lubricating and pressure additives.
Use: As a machine and grinding coolant.

"Wilube 655" *[Wilube].*   TM for a synthetic fluid grinding lubrication.
Use: Provides hard water stability and low foam characteristics.

"Wilube Blue" *[Wilube].*   TM for a premium grade oil containing extreme pressure additives.
Use: To inhibit growth of bacteria and fungus in hard water.

Wilzbach procedure.   Exposure of organic compounds to tritium gas yields tritiated products of high activity without extensive radiation damage. Concentrations of tritium ranging from 1 to 90 millicuries per gram have been obtained with quite varied compounds.

Windaus, Adolf.   (1876–1959). A German chemist who won the Nobel Prize for chemistry in 1928. His work involved the study of steroids and the effect of ultraviolet light activity, ergosterol, and vitamin D_2. He also researched digitalis and histamine. Although he studied medicine, he received his doctorate in chemistry at the University of Freiburg.

wine.   The fermented juice of grapes or other fruits or plants. Contains 7–20% alcohol (by volume). The higher percentages are obtained by addition of pure alcohol (fortifying). Coloring matter, sugars, and small amounts of acetic acid, salts, higher fatty acids, etc. give wines their distinctive appearance and flavor.
Hazard: Flammable, moderate fire risk.

wine ether.   See ethyl pelargonate.

wine gallon.   Same as U.S. gallon.

wine lees.   A deposit or sediment formed in the bottom of wine casks during fermentation. Wine lees vary greatly in quality, but usually contain 20–35% potassium acid tartrate and up to 20% calcium tartrate. They also contain yeast cells, proteins, and other solid matter that was suspended in the grape juice.
Use: Source of tartaric acid and tartarates.
See Pasteur, Louis; tartaric acid.

"Wingstay 29" *[Elikem].*   TM for *p*-oriented styrenated diphenyl amine.
Available forms: Liquid or powder.
Use: Stabilizer for SBR and NBR, back-loaded compounds using diene polymer systems.

wintergreen oil.   See methyl salicylate.

winterize.   A process of refrigerating edible and lubricating oils to crystallize the saturated glycerides, which are then removed by filter pressing.

wire cloth.   See screen; filter media.

Wiswesser   Line-Formula   notation.   See WLN.

"Witepsol" *[Cremer].*   TM for fatty acid esters.
Derivation: Products from coconut oil.
Grade: Liquid, hard fats, flake, and powder form.
Use: As dispersants, cream bases, emollients, refatting agents, emulsifiers, suppository bases, and consistency regulators.

witherite. $BaCO_3$. A natural barium carbonate usually found in veins with lead ores.
Properties: White, yellowish, or grayish; vitreous, inclining to resinous, luster. D 4.27–4.35, Mohs hardness 3–3.75.
Occurrence: U.S. (Kentucky, Lake Superior region), England (most important source).
Hazard: Toxic by ingestion, skin irritant.
Use: Chemicals (barium dioxide, barium hydroxide, blanc fixe), plate glass and porcelain, brick making, rodenticide.

Wittig, Georg.   (1897–1987). A University of Heidelberg professor who won the Nobel Prize for chemistry in 1979 along with Brown of Purdue. Wittig's research showed that phosphorous ylides react with ketones and aldehydes to form alkenes. This reaction is used a great deal in the synthesis of pharmaceuticals and other complex organic substances.

Wittig reaction. Preparation of olefins from alkylidene phosphoranes (ylids) and carbonyl compounds.

Wittig rearrangement. Rearrangement of ethers with alkyl lithiums to yield alcohols via a 1,2 shift.

Wizard. An aqueous solution for degreasing commercial kitchens.

WLN. (Wiswesser Line-Formula notation). A system for converting chemical structural diagrams to a linear string of symbols of letters and numbers.

wobble. The relatively loose base pairing between the base at the $3'$ end of a codon and the complementary base at the $5'$ end of the anticodon.

Wohl degradation. Method for the conversion of an aldose into an aldose with one less carbon atom by the reversal of the cyanohydrin synthesis. In the Wohl method, the nitrile group is eliminated by treatment with ammoniacal silver oxide.

Wohler, Friedrich. (1800–1882). A native of Germany, Wohler, working with Berzelius, placed the qualitative analysis of minerals on a firm foundation. In the early 19th century, chemists still thought it impossible to synthesize organic compounds. In 1828, Wohler publicized his laboratory synthesis of urea, which he had obtained four years previously by heating the inorganic substance ammonium cyanate; the synthesis is a result of intramolecular rearrangement:

$$(NH_4)OCN \longrightarrow NH_2-\overset{\overset{\text{O}}{\|}}{C}-NH_2$$

This was the beginning of the science of synthetic organic chemistry.

Wohler synthesis. Classical synthesis of urea by heating an aqueous solution of ammonium cyanate extended to preparation of urea derivatives.

Wohlwill process. The official process of the U.S. mints for refining gold. It consists in subjecting gold anodes to electrolysis in a hot solution of hydrochloric acid containing gold chloride, the solution being continuously agitated with compressed air.

Wohl–Ziegler reaction. Allylic bromination of olefins with N-bromosuccinimide. Peroxides or ultraviolet light is used as initiators.

Wolffenstein–Böters reaction. Simultaneous oxidation and nitration of aromatic compounds to nitrophenols with nitric acid or the higher oxides of nitrogen in the presence of a mercury salt as catalyst. Hydroxynitration of benzene yields picric acid.

Wolff–Kishner reaction. This reduction reaction was discovered independently in Germany (Wolff, 1912) and in Russia (Kishner, 1911). A ketone (or aldehyde) is converted into the hydrazone, and this derivative is heated in a sealed tube or an autoclave with sodium ethoxide in absolute ethanol.

$$\underset{/}{\overset{\backslash}{}}C=O \xrightarrow{H_2NNH_2} \underset{/}{\overset{\backslash}{}}C=NNH_2 \xrightarrow{NaOC_2H_5,\ 200°} \underset{}{\overset{\backslash}{}}CH_2 + N_2$$

After preliminary technical improvements, Huang Minlon (1946) introduced a modified procedure by which the reduction is conducted on a large scale at atmospheric pressure with efficiency and economy (L. & M. Fieser).

Wolff rearrangement. Rearrangement of diazoketone to ketenes by action of heat, light or some metallic catalysts. The rearrangement is the key step in the Arndt–Eistert synthesis.

wolfram. Wolfram is the official international alternate name for tungsten. Tungsten is preferred in the U.S.
See tungsten.

wolframic acid, anhydrous. See tungstic oxide.

wolframite. $(Fe,Mn)WO_4$. A natural tungstate of iron and manganese. Ferberite is the iron-rich member of the series, and huebnerite is the manganese-rich member.
Properties: Black to brown color, submetallic to resinous luster, black to brown streak. Mohs hardness 5–5.5, d 7.0–7.5.
Occurrence: Colorado, South Dakota, Nevada, Australia, Bolivia, Europe.
Use: Chief ore of tungsten.

wolfram white. See barium tungstate.

wollastonite.
CAS: 13983-17-0. $CaSiO_3$.
A natural calcium silicate found in metamorphic rocks.
Properties: White to brown, red, gray, yellow solid; vitreous to pearly luster. Mohs hardness 4.5–5, d 2.8–2.9.
Occurrence: New York, California.
Grade: Fine, medium paint grades.
Hazard: Questionable carcinogen.
Use: Ceramics; paint extender; welding rod coatings; rubber filler; silica gels; paper coating; filler in plastics, cements, and wallboard; mineral wool; soil conditioner.

Women Plus.   A rich source of soybean protein essence with a wealth of vitamins.

wood.   A mixture composed of 67–80% holocellulose and 17–30% lignin, together with low percentages of resins, sugars, a variable amount of water, and potassium compounds. Its fuel value varies widely around 3000–6000 Btu/lb according to variety, moisture, etc. Combustible.
Use: Pulp and paper, construction, packaging and cooperage, furniture, destructive distillation products (charcoal), extraction products (turpentine, rosin, tall oil, pine oil, etc.), methanol, plywood, fuel, rayon and cellophane, flock.
Note: Catalytic conversion of wood chips to fuel oil and methanol is under development.
See biomass.

wood alcohol.   See methanol.

wood ash.   The inorganic residue of wood combustion.
Use: Fertilizer for its potash content, which averages about 4% K_2O.

wood flour.   (wood flock). Pulverized dried wood from either soft- or hardwood wastes. Graded according to color and fineness.
Grade: Domestic standard, domestic fine, imported 40–60 mesh, 70–80 mesh, etc.
Hazard: Flammable, dangerous fire risk, especially suspended in air.
Use: Extender and filler in dynamite, plastics, rubber, paperboard; fur cleaning; polishing agents; Sorel cement.

wood, indurated.   A wood hardened by impregnation with a phenol-formaldehyde product.
Use: Storage batteries.

wood, petrified.   Wood in which the original chemical components have been replaced by silica. The change occurs in such a way that the original form and structure of the wood are preserved. The most famous instance is the Petrified Forest of Arizona.

wood pulp.   See pulp, paper.

wood rosin.   See rosin.

Wood's metal.   A four-component fusible alloy used largely in sprinkler systems, it melts at 70C, the composition being bismuth 50%, cadmium 10%, tin 13.3%, lead 26.7%.
See alloy, fusible; eutectic.

wood sugar.   See D(+)-xylose.

wood tar.   See pine tar; creosote, coal tar; woodtar.

wood turpentine.   See turpentine (oil).

Woodward–Hoffmann rules.   Rules that predict the stereochemical course of concerted reactions in terms of the symmetry of the interacting molecular orbitals.

Woodward *cis*-hydroxylation.   The hydroxylation of an olefin with iodine and silver acetate in acetic acid to give *cis*-glycols. The method involves the trans addition of iodine and silver acetate to give a *trans*-iodo-acetoxy derivative. The latter is hydrolyzed to a *cis*-mono-glycol acetate with acetic acid and water. Alkaline hydrolysis results in the final *cis*-glycol.

Woodward, Robert B.   (1917–1979). An American chemist born in Quincy, MA, and widely regarded as one of the world's leading synthetic organic chemists. After receiving his doctorate from M.I.T. (the youngest student in the history of the institute to do so), he joined the Harvard faculty in 1937 as instructor and attained full professorship in 1950. He was recipient of the Nobel Prize in 1965 for his brilliant work in synthesizing complex organic compounds, among them quinine, cholesterol, chlorophyll, reserpine, and cobalamin (vitamin B_{12}). When he died in 1979, his synthesis of the antibiotic erythromycin was virtually complete; it was finished by his associates two years later. He was director of the Woodward Research Institute in Basel, Switzerland, and a member of the governing board of the Weizmann Institute in Israel.

wool.   Staple fibers, usually 2–8 in long, obtained from the fleece of sheep (and also alpaca, vicuna, and certain goats). Physically, wool differs from hair in fineness and by the presence of prominent cortical scales and a natural crimp. The latter properties are responsible for the felting properties of wool and the ability of the fibers to cling together when spun into yarns. Chemically, wool consists essentially of protein chains (keratin) bound together by disulfide cross linkages.
Properties: Tenacity ranges from 1 to 2 g/denier; elongation 25–50%, d 1.32, moisture regain 16% (21.2C, 65% relative humidity), decomposes at 126C, scorches at 204C, resistant to most acids except hot sulfuric, destroyed by alkalies and chlorine bleach, resistant to mildew but attacked by insects, amphoteric to dyes. Combustible.
Source: Australia, Argentina, U.S., New Zealand, Uruguay, the former U.S.S.R., England.
Use: Outerwear, blankets, carpets, upholstery, felt, clothing source of lanolin.

wool fat.   See lanolin.

wool waste.   Wool from scrap materials cut up for remaking into cloth. Used as a fertilizer. Wool waste usually contains from 4 to 7% ammonia. Pure

wool shoddy may contain as much as 15% nitrogen and is particularly valued by hop growers.
Hazard: Wet wool waste is flammable and a dangerous fire risk.

work function. The energy needed to transport electrons, ions, and molecules from the inside of one medium into an adjoining medium.

working draft DNA sequence. See Draft DNA Sequence.

work, internal. The work done by a system against internal forces or between its parts or upon the system by these forces.

work, theoretical. Ideal reversible work calculated from properties.

wormseed oil. See chenopodium oil.

wormwood. See absinthium.

wort. A clear infusion of grain extract (usually malt) used as the basis of fermentation in the brewing industry.
See brewing.

wove. (paper).
See dandy roll.

writing ink. A solution of colorant in water, usually containing also low concentrations of tannic or gallic acid. Washable inks contain glycerol. Various water-soluble dyes are used for colored inks; dispersions of carbon black stabilized with a protective colloid are used for drawing inks. Fountain pen inks retain the fluidity of water; for ballpoint pens the mixture is of a paste-like consistency.

wrought iron. A ferrous material, aggregated from a solidifying mass of pasty particles of highly refined metallic iron into which 1–4% of slag is uniformly dispersed without subsequent melting. Wrought iron is distinguished by its low carbon and manganese contents. Carbon seldom exceeds 0.035%, and manganese content is held at 0.06% maximum. Phosphorus usually ranges from 0.10% to 0.15%; it adds about 1000 psi for each 0.01% above 0.10%. Sulfur content is normally low, ranging from 0.006% to less than 0.015%. Silicon content ranges from 0.075% to 0.15%; silicon content of base metals is 0.015% or less. Residuals such as chromium, nickel, cobalt, copper, and molybdenum are generally low, totaling less than 0.05%. Wrought iron is readily fabricated by standard methods and is quite corrosion resistant.

WS-Extra 400. A cooling agent blend for oral products.

wulfenite. $PbMoO_4$, sometimes with calcium, chromium, vanadium.
Properties: Yellow, orange, or bright-orange-red mineral of resinous luster. D 6.7–7.0, Mohs hardness 2.75–3.
Occurrence: Found in veins with ores of lead in U.S. (Massachusetts, New York, Pennsylvania, Nevada, Utah, New Mexico, Arizona), Europe, Australia.
Use: Ore of molybdenum.

Wulff process. Production of acetylene by treating a hydrocarbon gas with superheated steam in a regenerative type of refractory furnace at 1148–1371C. Contact times are very short. Acetylene and ethylene are produced in 1:1 ratio.
See acetylene.

Wurtz reaction. A method of synthesizing hydrocarbons discovered by Wurtz in 1855. It consists in treatment of an alkyl halide with metallic sodium, which has a strong affinity for bound halogen and acts on methyl iodide in such a way as to strip iodine from the molecule and produce sodium iodide. The reaction involves two molecules of methyl iodide and two atoms of sodium.
The reaction probably proceeds through the formation of methylsodium, which interacts with methyl iodide:

$$CH_3I \xrightarrow{Na} CH_3Na \xrightarrow{CH_3I} CH_3CH_3$$

The Wurtz reaction can be applied generally to synthesis of hydrocarbons by the joining together of hydrocarbon residues of two molecules of an alkyl halide (usually the bromide or iodide). With halides of high molecular weight the yields are often good and the reaction has been serviceable in the synthesis of higher hydrocarbons starting with alcohols found in nature.

Wuthrich, Kurt. (1938–). Born in Aarberg, Switzerland, Wuthrich won the Nobel Prize for chemistry in 2003 for his pioneering work concerning for the development of nuclear magnetic resonance spectroscopy for determining the three-dimensional structure of biological macromolecules in solution. He earned an undergraduate degree from the University of Bernborn in Aarberg and a Ph.D. in chemistry from the University of Basel, Switzerland. Currently a Professor of Biophysics in Eidgenrich, Switzerland and Cecil H. and Ida M. Green Visiting Professor of Structural Biology, The Scripps Research Institute, La Jolla, CA.

X

xanthan. A synthetic, water-soluble bipolymer made by fermentation of carbohydrates; it is a thickening and suspending agent that is heat-stable, with good tolerance for strongly acidic and basic solutions. Viscosity remains stable over a wide temperature range.
Use: In drilling fluids, ore flotation, and the food and pharmaceutical fields, especially in prepared mixed and high-protein fast foods. An experimental use is to flood oil fields as an aid to enhanced recovery.

xanthate. A salt (usually potassium or sodium) of a xanthic acid. Available as yellow pelletized solids having a pungent odor; soluble in water. Examples are potassium amyl xanthate, potassium ethyl xanthate, sodium isobutyl xanthate, sodium isopropyl xanthate.
Use: Collector agents in the flotation of sulfide minerals, metallic elements such as copper, silver, gold, and some oxidized minerals of lead and copper.
See cellulose xanthate; viscose process.

xanthene. (dibenzopyran, tricyclic). $CH_2(C_6H_4)_2O$. The central structure of the fluoroscein, eosin, and rhodamine dyes.
Properties: Yellowish, crystalline leaflets. Mp 100.5C, bp 315C. Soluble in ether; slightly soluble in alcohol; very slightly soluble in water.
Derivation: By the condensation of phenol and o-cresol by means of aluminum chloride.
Use: Organic synthesis, fungicide.

xanthene dye. A group of dyes whose molecular structure is related to that of xanthene. The aromatic (C_6H_4) groups constitute the chromophore. CI ranges from 45000 to 45999. The dyes are closely related structurally to diaryl methane dyes. Eosin (CI 45380) is an example.

xanthenol. See xanthydrol.

xanthic acid. (xanthogenic acid). A substituted dithiocarbonic acid of the type ROC(S)SH, in which R is ordinarily an alkyl radical. Xanthic acid salts are called xanthates. Unless otherwise designated, xanthic acid is understood to be the ethyl derivative $(C_2H_5OC(S)SH)$, also called ethylxanthic acid.
Properties: Liquid. Mp −53C, decomposing at room temperature.

xanthine. (dioxopurine).
CAS: 69-89-6. $C_5H_4N_4O_2$.

A purine base occurring in the blood and urine and in some plants. Theophylline and theobromine, the alkaloids of tea and cocoa, respectively, are both dimethylxanthines; caffeine in coffee is a trimethylxanthine.
Properties: Yellowish-white powder. Sublimes with partial decomposition; soluble in potassium hydroxide; insoluble in water and acid.
Derivation: By the action of nitrous acid on guanine.
Grade: Technical, CP, monohydrate, sodium salt, radioactive forms available.
Hazard: Toxic by ingestion.
Use: Organic synthesis, medicine.

xanthine oxidase. An enzyme found in animal tissues that acts upon hypoxanthine, xanthine, aldehydes, reduced coenzyme I, etc., producing, respectively, xanthine, uric acid, acids, oxidized coenzyme I, etc.
Use: Biochemical research.

xanthogenic acid. See xanthic acid.

xanthone. (benzophenone oxide; dibenzopyrone; xanthene ketone).
CAS: 90-47-1. $CO(C_6H_4)_2O$.
Tricyclic; occurs in some plant pigments.
Properties: White needles or crystalline powder. Mp 173–174C, bp 350C, sublimes. Insoluble in water; soluble in alcohol, chloroform, and benzene, especially when hot.
Use: Larvicide; intermediate for dyes, perfumes, and pharmaceuticals.

xanthophyll. $C_{40}H_{56}O_2$.
Properties: Yellow pigment. An oxygenated carotenoid occurring in green vegetation and in some animal products, notably egg yolk. Mp 190–193C. Insoluble in water; slightly soluble in alcohol and ether.
See lutein; carotenoid.

xanthopterin. (2-amino-4,6-dihydroxypteridine). $C_6H_5N_5O_2 \cdot H_2O$. Pigment found in the wings of butterflies; can be converted by yeast into folic acid.
Properties: Orange-yellow crystals. Sinters at 360C, decomposes above 410C. Practically insoluble in water; freely soluble in dilute ammonium or sodium hydroxide giving yellow solutions and in $2N$ hydrochloric acid giving colorless solutions.
Use: Biochemical research.

Hawley's Condensed Chemical Dictionary, Sixteenth Edition. Michael D. Larrañaga, Richard J. Lewis, Sr., and Robert A. Lewis.
© 2016 John Wiley & Sons, Inc. Published 2016 by John Wiley & Sons, Inc.

xanthydrol.   (xanthenol).
  CAS: 90-46-0.   HCOH(C_6H_4)$_2$O.
  A derivative of xanthene.
  Properties: White powder. Mp 123C. Insoluble in water; soluble in alcohol.
  Derivation: Reduction of xanthone with alcohol and sodium.
  Grade: CP (analytical).
  Use: Determination of urea and DDT.

xatral.
  CAS: 81403-68-1. $C_{19}H_{27}N_5O_4 \cdot ClH$.
  Hazard: A poison.

X chromosome.   One of the two sex chromosomes, X and Y.
  See Y chromosome; sex chromosome.

Xe.   Symbol for xenon.

xenobiotic.   Any substance that enters and interacts with an organism and is neither endogenously produced as part of a normal metabolic pathway of the organism.

xenograft.   Tissue or organs from an individual of one species transplanted into or grafted onto an organism of another species, genus, or family. A common example is the use of pig heart-valves in humans.

3-(2-xenolyl)-1,2-epoxypropane.
  Use: Food additive; plasticizer.

xenon.   Xe. Element of atomic number 54, noble gas group of the periodic table, aw 131.30, valences = 2, 4, 6, 8; nine stable isotopes.
  Properties: Colorless, odorless gas or liquid. Gas (at STP) has d 5.8971 g/L (air = 1.29 g/L), dielectric constant 1.0012 (25C) (1 atm); liquid has bp −108.12C (1 mm Hg), d (at bp) of 1.987 g/cc; liquefaction temp −106.9C. Chemically unreactive but not completely inert. Noncombustible.
  Derivation: Fractional distillation of liquid air.
  Use: Luminescent tubes, flash lamps in photography, fluorimetry, lasers, tracer studies, anesthesia.
  See noble gas; xenon compounds.

xenon compounds.   Xenon tetrafluoride, XeF_4, is easily prepared by mixing fluorine and xenon in gaseous form, heating in a nickel vessel to 400C, and cooling. The product forms large, colorless crystals. The difluoride and hexafluoride, XeF_2 and XeF_6, also colorless crystals, can be obtained somewhat similarly. The hexafluoride melts to a yellow liquid at 50C and boils at 75C. Many xenon fluorine complexes with other compounds are also known. Xenon oxytetrafluoride, $XeOF_4$, a volatile liquid at room temperature, is obtained from the reaction of xenon hexafluoride and silica. Gram amounts have been isolated and studied.

All these fluorides must be protected from moisture to avoid formation of xenon trioxide, XeO_3, a colorless, nonvolatile solid that is dangerously explosive when dry. Its solution, the so-called xenic acid, is a stable weak acid but a strong oxidizing agent, which will even liberate chlorine from hydrochloric acid.
In an alkaline solution, xenon trioxide reacts to give free xenon and perxenate, such as $Na_4XeO_6 \cdot 8H_2O$. Perxenates are probably the most powerful oxidizing agents known, just as the xenon fluorides are extremely effective fluorinating agents.
Complex compounds containing nitrogen bonded to xenon have also been prepared.
  Hazard: Toxic by inhalation; oxidizing agents, strong irritant.

xenyl.   The biphenyl group $C_6H_5C_6H_4-$.

p-xenylamine.   See p-aminodiphenyl.

xerography.   A "dry" method of photography or photocopying. A metal plate is covered with a layer of photoconductive powder, such as selenium; the surface of this plate is given an electric charge by passing it under a series of charged wires. An image of the material to be photographed is projected onto the charged plate through a camera lens. The electric charges disappear in the areas exposed to light, but elsewhere the surface retains its charge. A powder consisting of a coarse carrier and a fine developing resin is then spread over the plate. Adhesion between powder and plate occurs only at the charged areas. Elsewhere developing resin and carrier are not retained on the plate, which thus has become a negative of the original image. A positive is obtained by placing a piece of paper against the plate and applying an electric charge as in the first stage of the process. This causes adhesion of developing resin and its carrier to the paper. This positive print is fixed by heating in a press for a few seconds to melt the developing resin and fuse it to the paper. Colored prints are possible by use of suitable developing resins. Various materials other than paper can thus be printed.

xibornol.   (6-isobornyl-3,4-xylenol). $C_{18}H_{26}O$.
  Properties: Crystalline solid or viscous, yellowish liquid. D 1.02, refr index 1.538.
  Use: Antibacterial agent, rubber antioxidant.

X-linked.   Shortened form for cross-linked, e.g., X-linked polyethylene.

XLPE.   Trade abbreviation for cross-linked polyethylene.

XPS.   Abbreviation for X-ray photoelectron spectroscopy.

X-radiation.   (Roentgen rays; X-rays). Electromagnetic radiation of extremely short wavelength (0.06–120 Å), emitted as the result of electron transitions in the inner orbits of heavy atoms bombarded by cathode rays in a vacuum tube. Those of the shortest wavelength have the highest intensity and are called "hard" X-rays. X-radiation was discovered by Roentgen in 1898. Its properties are (1) penetration of solids of moderate density, such as human tissue but retardation by bone, barium sulfate, lead, and other dense materials. (2) Action on photographic plates and fluorescent screens. (3) Ionization of the gases through which they pass. (4) Ability to damage or destroy diseased tissue; there is also a cumulative deleterious effect on healthy tissue.

Hazard: Overexposure can permanently damage cells and tissue structures; effect is cumulative.

Use: Spectrometry; structure determination of molecules, cancer therapy, diagnostic medicine, nondestructive testing of metals, identification of original paintings, preservation of foods.

See radiation, ionizing; diffraction; roentgen.

"Xydar" *[Solvay Advanced].*   TM for liquid crystal polymer glass/mineral combination product.

Use: Microwave cabinets; cookware; automotive and electronic components.

xylene.   (dimethylbenzene).
CAS: 1330-20-7. $C_6H_4(CH_3)_2$.
A commercial mixture of the three isomers, *o*-, *m*-, and *p*-xylene. The last two predominate.

Properties: Clear liquid. D about 0.86. See under Grade for boiling range, flash p 81–115F (27.2–46.1C) (TOC). Soluble in alcohol and ether; insoluble in water.

Derivation: (1) Fractional distillation from petroleum (90%), coal tar, or coal gas, (2) by catalytic reforming from petroleum, followed by separation of *p*-xylene by continuous crystallization, (3) from toluene by transalkylation.

Grade: Nitration (bp range 137.2–140.5C), 4 degrees (bp range 138–134C), 5 degrees (bp range 137–142C, high in *m*-isomer), 10 degrees (bp range 135–145C), industrial (bp 90% 40C, complete 160C). Also other grades depending on use. In some cases one or another of the industrial isomers is partially removed for use in chemical production.

Hazard: Flammable, moderate fire risk. Toxic by ingestion and inhalation. Eye and upper respiratory tract irritant. Central nervous system impairment. Questionable carcinogen.

Use: Aviation gasoline; protective coatings; solvent for alkyd resins, lacquers, enamels, rubber cements; synthesis of organic chemicals.

m-xylene.   (1,3-dimethylbenzene).
CAS: 108-38-3. $1,3\text{-}C_6H_4(CH_3)_2$.

Properties: Clear, colorless liquid. D 0.8684 (15C), fp −47.4C, bp 138.8C, refr index 1.4973 (20C), flash p 85F (29.4C), autoign temp 982F (527.7C). Soluble in alcohol and ether; insoluble in water.

Derivation: Selective crystallization or solvent extraction of *m*, *p*-mixture.

Grade: 95% (technical), 99%, 99.9% (research).

Hazard: Flammable, moderate fire risk. Questionable carcinogen.

Use: Solvent; intermediate for dyes and organic synthesis, especially isophthalic acid; insecticides; aviation fuel.

m-xylene-α,α′-diamine.   (MXDA; *m*-phenylenebis(methylamine)).
CAS: 1477-55-0. $C_8H_{12}N_2$.

Properties: A liquid. Mw 136.22, bp 245–248C.

Hazard: Moderately toxic by skin contact and ingestion. Mildly toxic by inhalation. A severe skin, eye, and gastrointestinal irritant.

Use: Manufacture of polyamide fibers and resins, as a curing agent.

o-xylene.   (1,2-dimethylbenzene).
CAS: 95-47-6. $1,2\text{-}C_6H_4(CH_3)_2$.

Properties: Clear, colorless liquid. D 0.880 (20/4C), fp −25C, bp 144C, refr index 1.505 (20C), flash p 115F (46.1C) (TOC), autoign temp 867F (463.8C). Soluble in alcohol and ether; insoluble in water. Combustible.

Grade: 99%, free of hydrogen sulfide and sulfur dioxide, technical 95%, research 99.9%.

Hazard: Moderate fire risk. Questionable carcinogen.

Use: Manufacture of phthalic anhydride, vitamin and pharmaceutical syntheses, dyes, insecticides, motor fuels

p-xylene.   (1,4-dimethylbenzene).
CAS: 106-42-3. $1,4\text{-}C_6H_4(CH_3)_2$.

Properties: Colorless liquid. D 0.8611 (20C), mp 13.2C, bp 138.5C, refr index 1.5004 (21C), flash p 81F (27.2C) (TOC). Crystallizes at low temperature. Soluble in alcohol and ether; insoluble in water.

Derivation: Selective crystallization or solvent extraction of *m*-, *p*-mixture; separation from mixed-xylene feedstocks by adsorption.

Hazard: Flammable, dangerous fire risk. Questionable carcinogen.
Use: Synthesis of terephthalic acid for polyester resins and fibers ("Dacron" *[Invista]*, Mylar, Terylene), vitamin and pharmaceutical syntheses, insecticides.

p-xylene-α,α′-diol. $C_6H_4(CH_2OH)_2$.
Properties: White, crystalline solid. Mp 118C, bp 138–144C (0.8–1.0 mm Hg). Slightly soluble in water (25C).
Grade: 98%.
Use: Polyester resins.

xylenol.   (dimethylphenol; hydroxydimethylbenzene; dimethylhydroxybenzene).
CAS: 1300-71-6. $(CH_3)_2C_6H_3OH$.
There are five isomers (2,4-; 2,5-; 2,6-; 3,4-; 3,5-). This entry describes the commercially offered mixture.

Properties: White, crystalline solid. D 1.02–1.03 (15C), mp 20–76C, bp 203–225C. Only slightly soluble in water; soluble in most organic solvents and caustic soda solution. Combustible.
Derivation: Cresylic acid or tar acid fraction of coal tar.
Hazard: Toxic by ingestion and skin absorption.
Use: Disinfectants, solvents, pharmaceuticals, insecticides and fungicides, plasticizers, rubber chemicals, additives to lubricants and gasolines, manufacture of polyphenylene oxide (2,6-isomer only), wetting agents, dyestuffs.

xylidine.   (aminodimethylbenzene; aminoxylene).
CAS: 1300-73-8. $(CH_3)_2C_6H_3NH_2$.
A varying mixture of isomers (2,3-; 2,4-; 2,5-; 2,6-).

Properties: Liquid. D 0.97–0.99, bp 213–226C, flash p 206F (96.6C) (CC). Slightly soluble in water; soluble in alcohol and ether. Combustible.
Derivation: Nitration of xylene and subsequent reduction.
Hazard: Toxic by ingestion, inhalation, and skin absorption. Liver damage. Methemoglobinemia. Possible carcinogen.
Use: Dye intermediate, organic syntheses, pharmaceuticals.

3,4-xylidino-2-oxazoline.
CAS: 23420-61-3. $C_{11}H_{14}N_2O$.
Hazard: A poison.

xylocaine.   Preparation of lignocaine hydrochloride.
Use: Local anesthetic.

xylol.   Commercial-grade xylene, a mixture of isomers and other benzene derivatives.

xylopic acid.
Properties: *Xylopia aethiopica.* Family Annonacea.
CAS: 6619-97-2. $C_{22}H_{32}O_4$.
Hazard: A poison by ingestion.

D(+)-xylose.   (wood sugar).
CAS: 58-86-6. $C_5H_{10}O_5$.
(not to be confused with phenylosazone of the same name).
Properties: White, crystalline, dextrorotatory powder; sweet taste. D 1.525 (20C), mp 144C (also given as 153C). Soluble in water and alcohol. Combustible.
Derivation: Hydrolysis with hot dilute acids of wood, straw, corncobs, etc.; wood pulp wastes.
Grade: Reagent, technical.
Use: Dyeing, tanning, diabetic food, source of ethanol.

xylyl bromide.   (α-bromoxylene).
CAS: 35884-77-6. $CH_3C_6H_4CH_2Br$.
Mixed *o*-, *m*-, and *p*-isomers.
Properties: Colorless liquid; pleasant aromatic odor. Decomposed slowly by water, d 1.4, bp 210–220C. Combustible.
Derivation: Bromination of xylene.
Grade: Technical.
Hazard: Toxic by inhalation and ingestion; strong irritant to eyes, skin, and tissue.
Use: Organic synthesis, tear gas.

m-xylyl chloride. $CH_3C_6H_4CH_2Cl$.
Properties: Colorless liquid. Bp 196, d 1.064. Combustible.
Hazard: Toxic by ingestion and inhalation, strong irritant to eyes and skin.
Use: Intermediate.

p-xylylene. $CH_2{:}C_6H_4{:}CH_2$. A monomer.
See parylene.

n-2,3-xylylmaleimide.
CAS: 31581-09-6. $C_{12}H_{11}NO_2$.
Hazard: Low toxicity by ingestion and inhalation. A severe eye irritant.

((2,6-xylyl)sulfinyl)methylcarbamic acid-2,3-dihydro-2,2-dimethyl-7-benzofuranyl ester.
CAS: 77248-44-3. $C_{20}H_{23}NO_5S$.
Hazard: A poison by ingestion.
Use: Agricultural chemical.

Y

Y.   Symbol for yttrium.

YAC.   Abbreviation for yeast artificial chromosome.

Yankee machine.   A papermaking machine similar to a fourdrinier but with much shorter wire. It is designed for light-weight papers such as toilet and facial tissues and operates at comparatively high speeds. It has a specialized cylindrical drying roll that effects drying of the web much more efficiently than felts.
See fourdrinier.

"Yarmor" [Pinova].   TM for series of pine oils for widely varied uses. Total terpene alcohol contents from 55 to 91%.
Use: Disinfectants, textile specialties, wetting and flotation agents, special solvents, household and industrial cleaners, odorants.

Yb.   Symbol for ytterbium.

Y chromosome.   One of the two sex chromosomes, X and Y.
See X chromosome; sex chromosome.

yeast.   (barm). Unicellular organisms known as Saccharomycetaceae. The following description applies to the cultured commercial product and not to various wild varieties.
Properties: Yellowish white, viscid liquid or soft mass, flakes, or granules, consisting of cells and spores of *Saccharomyces cerevisiae*.
Derivation: A ferment obtained in brewing. Yeasts induce fermentation by enzymes (zymases) that convert glucose and other carbohydrates into carbon dioxide and water in the presence of oxygen or into alcohol and carbon dioxide (or lactic acid) in the absence of oxygen.
Grade: Technical, brewers', cooking, compressed (contains about 74% moisture), dried, NF (contains no starch or filler, not more than 7% moisture nor more than 8% ash). Also graded according to vitamin B_1 content.
Use: Fermentation of sugars, molasses, and cereals for alcohol; brewing; baking; food supplement; protein biosynthesis from many carbonaceous and nitrogenous materials, including petroleum; source of vitamins, enzymes, nucleic acids, etc.; biochemical research.
See bacteria; fermentation.

yeast adenylic acid.   See adenylic acid.

yeast artificial chromosome.   (YAC). Constructed from yeast DNA, it is a vector used to clone large DNA fragments; a method for cloning very large fragments of DNA. Genomic DNA in fragments of 200–500 kb are linked to sequences which allow them to propagate in yeast as a mini-chromosome (including telomeres, a centromere, and an ARS—an autonomous replication sequence). This technique is used to clone large genes and intergenic regions, and for chromosome walking.
See cloning vector; cosmid.

Yellotone.
CAS: 7704-34-9.   A sulfur product.
Use: As a rubber vulcanizing agent, insecticide, and fungicide.

yellow AB.   (1-(phenylazo)-2-naphthylamine; CI 11380).
CAS: 85-84-7. $C_6H_5N_2C_{10}H_6NH_2$.
Properties: Orange or red platelets. Mp 102–104C. Insoluble in water; soluble in alcohol and oils.
Hazard: Consult regulations before using in food products. Questionable carcinogen.
Use: Biological stain.

yellow brass.   A brass containing 34–37% zinc; it has excellent fabrication properties and corrosion resistance.
Use: Structural and decorative purposes.
See brass.

yellowcake.   See uranium dioxide.

yellow enzyme.   See flavin enzyme.

yellow glass.   A soda-lime glass.

yellow lake.   Any of several pigments made by precipitating soluble yellow dyes on an aluminum hydrate base. They are transparent in oil and lacquer vehicles and are used for metal decorating.

yellow OB.   (1-o-tolueneazonaphthylamine-2; CI 11390).
CAS: 131-79-3. $CH_3C_6H_4N_2C_{10}H_6NH_2$.
Properties: Orange or yellow powder. Mp 122–125C. Insoluble in water; soluble in alcohol and oils.

Hawley's Condensed Chemical Dictionary, Sixteenth Edition. Michael D. Larrañaga, Richard J. Lewis, Sr., and Robert A. Lewis.
© 2016 John Wiley & Sons, Inc. Published 2016 by John Wiley & Sons, Inc.

Hazard: Consult regulations before using in food products. Questionable carcinogen.
Use: Biological stain.

yellow phosphorus. See phosphorus.

yellow precipitate. See mercuric oxide, yellow.

yellow prussiate of potash. See potassium ferrocyanide.

yellow prussiate of soda. See sodium ferrocyanide.

yellow rain. See mycotoxin.

yellow salt. See uranyl nitrate.

-yl. The suffix of univalent radicals derived from aromatic hydrocarbons or heterocyclic compounds by the removal of one hydrogen atom from the ring.

ylang ylang oil. An essential, yellowish, volatile oil distilled from the flowers of *Cananga odorata*. Strongly levorotatory.
Use: Perfumery.
See cananga oil.

ylid. A substance in which a carbanion is attached to a heteratom with a high degree of positive charge, i.e., $>C^--X^+$. It is similar to a zwitterion and related to the Wittig reaction.

-ylidene. (-ylidyne).
A suffix for bivalent or trivalent radicals derived from saturated hydrocarbons by the removal of two or three hydrogen atoms from the same carbon atom.

-ylidyne. See -ylidene.

yohimbine. (aprodine; corynine; quebrachine).
CAS: 146-48-5. $C_{21}H_{26}O_3N_2$.
Properties: Glistening, needle-like alkaloid, mp 234C, soluble in alcohol and ether, very slightly soluble in water.
Derivation: By extraction from the bark of *Corynanthe yohimbe*, found in the Cameroons.
Hazard: Said to be an aphrodisiac. Toxic by ingestion.

β-yohimbine hydrochloride.
CAS: 75444-63-2. $C_{21}H_{26}N_2O_3 \cdot ClH$.
Hazard: A poison by ingestion.

Yonath, Ada E (1939–). An Israeli protein crystallographer who was a co-recipient of the 2009 Nobel Prize with Venkatraman Ramakrishnan and Thomas A. Steitz. The scientists' research on atomic structure and function of the ribosome provided fundamental understanding of the two ribosomal subunits (small and large). Her research concentrated in studying ribosomes using X-ray crystallography, and she identified the atomic structure of the subunits and the atomic structure of antibiotics. She received her Ph.D. from the Weizmann Institute of Science followed by an outstanding career which includes awards such as Louisa Gross Horwitz Prize for Biology or Biochemistry (2005), The Paul Ehrlich and Ludwig Darmstaedter Prize (2007), and the Albert Einstein World Award of Science (2008).

Young's modulus. See modulus of elasticity.

ytterbium.
CAS: 7440-64-4. Yb.
A metallic element. A rare-earth metal of yttrium subgroup, atomic number 70, aw 173.04, valence of 2, 3; exists in α and β forms, the latter being semiconductive at pressures above 16,000 atm. There are seven natural isotopes.
Properties: Metallic luster, malleable. Mp 824C, bp 1427C, d 7.01. Reacts slowly with water; soluble in dilute acids and liquid ammonia.
Source: See rare-earth metals.
Derivation: Reduction of the oxide with lanthanum or misch metal.
Grade: Regular high purity (ingots and lumps).
Use: Lasers, dopant for garnets, portable X-ray source, chemical research.

ytterbium chloride. $YbCl_3 \cdot 6H_2O$.
Properties: Green crystals. D 2.575, loses $6H_2O$ at 180C, mp 865C. Very soluble in water; hygroscopic.

ytterbium fluoride. YbF_3.
Properties: Solid. Mp 1157C, bp 2200C. Insoluble in water; hygroscopic.

ytterbium oxide. (ytterbia). Yb_2O_3.
Properties: Colorless mass when free of thulia, tinted brown or yellow when containing thulia. D 9.2, mp 2346C. The weakest base of the yttrium group, except scandia and lutetia. Slightly hygroscopic, absorbs water and carbon dioxide from the air. Soluble in hot, dilute acids; less so in cold acids.
Use: Special alloys, dielectric ceramics, carbon rods for industrial lighting, catalyst, special glasses.

ytterbium sulfate. $Yb_2(SO_4)_3 \cdot 8H_2O$.
Properties: Crystalline solid. D 3.286. Soluble in water.

yttria. See yttrium oxide.

yttrium.
CAS: 7440-65-5. Y. Metallic element of atomic number 39, group IIIB of the periodic table, aw 88.9059, valence of 3; no stable isotopes.

Properties: Dark-gray metal. D 4.47, mp 1500C, bp 2927C. Soluble in dilute acids and potassium hydroxide solution; decomposes water. Known only in the tripositive state. Low neutron capture cross section.
Source: See rare-earth metals.
Derivation: Reduction of the fluoride with calcium.
Grade: Regular high purity (ingots, lumps, turnings), metallurgical, low-oxygen, crystal sponge, powder.
Hazard: Flammable in finely divided form. Pulmonary fibrosis.
Use: Nuclear technology, iron and other alloys, deoxidizer for vanadium and other nonferrous metals, microwave ferrites, coating on high-temperature alloys, special semiconductors.

yttrium acetate. $Y(C_2H_3O_2)_3 \cdot 9H_2O$.
Properties: Colorless crystals. Soluble in water.
Derivation: Action of acetic acid on yttrium oxide.
Use: Analytical chemistry.

yttrium antimonide. YSb. A high-purity binary semiconductor.

yttrium arsenide. YAs. A high-purity binary semiconductor.
Hazard: Highly toxic.

yttrium bromide. $YBr_3 \cdot 9H_2O$.
Properties: Colorless crystals. Hygroscopic, mp (anhydrous) 904C. Soluble in water; slightly soluble in alcohol; insoluble in ether.

yttrium chloride.
CAS: 10361-92-9. $YCl_3 \cdot 6H_2O$.

Properties: Reddish-white, transparent, deliquescent prisms. D 2.18, decomposes at 100C. Soluble in water and alcohol; insoluble in ether.
Derivation: By the action of hydrochloric acid on yttrium oxide.
Grade: Purities to 99%.
Use: Analytical chemistry.

yttrium oxide. (yttria).
CAS: 1314-36-9. Y_2O_3.
Properties: Yellowish-white powder. D 4.84, mp 2410C. Soluble in dilute acids; insoluble in water.
Derivation: By the ignition of yttrium nitrate.
Grade: Purities to 99.8%, electronic grade 99.999%.
Use: Phosphors for color TV tubes (alloy with europium), yttrium-iron garnets for microwave filters, stabilizer for high-temperature service materials (zirconia and silicon nitride refractories).

yttrium phosphide. YP.
Use: High-purity binary semiconductor.

yttrium sulfate. $Y_2(SO_4)_3 \cdot 8H_2O$.
Properties: Small, reddish-white, monosymmetric crystals. D 2.558, loses $8H_2O$ at 120C, decomposes 700C. Soluble in concentrated sulfuric acid; sparingly soluble in water; insoluble in alkalies.
Derivation: Action of sulfuric acid on monazite sand.
Method of purification: Fractional crystallization.
Use: Reagent.

yttrium vanadate. YVO_4.
Properties: White, crystalline solid.
Use: With europium vanadate as red phosphor in color television tubes.

Z

Zaclon. A series of fluxes based on active zinc ammonium chloride combined with additives; available in various forms.

Zalba. A rubber antioxidant containing a hindered phenol.
Properties: Yellow or cream-colored powder. D 1.30.
Use: Nondiscoloring antioxidant for natural and synthetic rubbers and latex; stabilizer in SBR manufacture.

Zanzibar gum. A hard, usually fossil type of copal.
Source: Found on the island of Zanzibar and the adjacent African mainland.
Properties: D 1.062–1.068, mp 240–250C. Insoluble in most solvents. Combustible.
Use: Varnishes.

zaprinast.
CAS: 37762-06-4. $C_{13}H_{13}N_5O_2$.
A specific cGMP-phosphodiesterase inhibitor.
Properties: White solid. Mp 235–237C.
Hazard: A poison.

"Zar Exterior Polyurethane" [United]. TM for liquid additive containing ultra violet radiation absorbers.
Use: To resist UV rays.

"Zar Tung Oil" [United]. TM for oil coating for wood.
Use: For durable bartops to resist marks caused by hot and cold liquids.

"Zar Wood Patch" [United]. TM for a latex compound.
Use: To fill nail holes, cracks, or gouges. It accepts stains better than conventional wood.

ZDP. Abbreviation for zinc dithiophosphate.

Zectran. Insecticides containing 4-dimethylamino-3,5-xylyl-*n*-methylcarbamate.

zedoarondiol.
CAS: 98644-24-7. $C_{15}H_{24}O_3$.
Hazard: A poison by ingestion.

zein. The protein of corn, a prolamin.
Properties: White to slightly yellow powder; odorless; tasteless. D 1.226. Free of cystine, lysine, and tryptophan. A resinous material dispersible in water with neutral sulfonated castor oil; soluble in dilute alcohol; insoluble in water, dilute acids, anhydrous alcohols, turpentine, esters, oils, fats. Nontoxic protein of the prolamine class, derived from corn; contains 17 amino acids. Combustible.
Derivation: By-product of corn processing, by extraction of gluten meal with 85% isopropanol, extraction of the zein from the extract with hexane, precipitation by water, and spray drying.
Use: Paper coating, grease-resistant coating, label varnishes, laminated board, solid-color prints, printing inks, food coatings, microencapsulation, fibers.

Zeisel determination. Cleavage of methoxy and ethoxy groups with boiling hydriodic acid, distillation of the alkyl iodide into alcoholic silver nitrate, and gravimetric determination as silver iodide. Alternatively, the alkyl iodide is oxidized to form iodate and the iodine liberated by addition of potassium iodide is titrated.

Zeise's salt. $(Pt(C_2H_4)Cl_3)K$. Formed by adding potassium chloride to a solution of platinous chloride saturated with ethylene.

"Zelan" [Du Pont]. TM for a line of durable, water-repellent textile finishes.

"Zelcon" SL [Stepan]. TM for a quaternized, long-chain, complex amine condensation product used as a fabric softener and conditioner for compounding into home laundry softeners.

"Zelec" [Stepan]. TM for a series of antistatic agents, both durable and nondurable.

"Zenite" [Ticona]. TM for a group of rubber accelerators based on zinc salt of 2-mercaptobenzothiazole, with or without various modifying agents.

zeolite.
CAS: 1318-02-1.
A natural hydrated silicate of aluminum and either sodium or calcium or both, of the type $Na_2O \cdot Al_2O_3 \cdot xSiO_2 \cdot xH_2O$. Both natural and artificial zeolites are used extensively for water

softening, as detergent builders, and cracking catalysts. For the former purpose the sodium or potassium compounds are required, since their usefulness depends on the cationic exchange of the sodium of the zeolite for the calcium or magnesium of hard water. When the zeolite has become saturated with calcium or magnesium ion it is flooded with strong salt solution, a reverse exchange of cations takes place and the material is regenerated. The natural zeolites are analcite, chabazite, heulandite, natrolite, stilbite, and thomsonite.

Synthetic zeolites are made either by a gel process (sodium silicate and alumina) or by a clay process (kaolin), which forms a matrix to which the zeolite is added. These processes are quite complex, involving substitution of various rare-earth oxides. The effectiveness of zeolites depends on their pore size, which may be as small as 4–5 Å. Other applications of zeolites are as adsorbents, desiccants, and in solar collectors, where they function as both heating and cooling agent.
Hazard: Questionable carcinogen.
See molecular sieve; aluminosilicate; ion-exchange resin; cage zeolite.

Zepar BP. A reducing agent based on sodium hydrosulfite.
Use: Bleaching agent for groundwood pulp, unbleached pulp, and old papers.

Zepel. A fluorocarbon textile finish used as a durable oil and water repellent.

Zerewitinoff determination. The reaction of methylmagnesium iodide with an active hydrogen-containing compound provides methane, which can be measured quantitatively by the increase in volume of the system at constant pressure.

Zerewitinoff reagent. Solution of methylmagnesium iodide in purified n-butyl ether. A clear, light-colored liquid that reacts rapidly with moisture and oxygen.
Use: Analytical reagent for active hydrogen atoms in organic compounds; also to determine water, alcohols, and amines in inert solvents.

Zerlate. A wettable fungicide powder containing 75% ziram.
Hazard: Irritant to eyes and mucous membranes.

"Zerlon" [Anchor]. TM for a methyl methacrylate-styrene copolymer.
Use: Plastic molding material.

zero gravity. See space, chemistry in.

zero group. See group (1).

zeroth law (of thermodynamics). Two bodies that have been shown to be individually in equilibrium with a third body will be in equilibrium when placed in contact with each other, that is, they will have the same temperature (H. Reiss).

"Zeset" [Du Pont]. TM for a fiber-reactant resin for crease resistance and dimensional stabilization of textiles.

zeta potential. (electrokinetic potential). The potential across the interface of all solids and liquids. Specifically, the potential across the diffuse layer of ions surrounding a charged colloidal particle, which is largely responsible for colloidal stability. Discharge of the zeta potential, accompanied by precipitation of the colloid, occurs by addition of polyvalent ions of sign opposite to that of the colloidal particles. Zeta potentials can be calculated from electrophoretic mobilities, that is, the rates at which colloidal particles travel between charged electrodes placed in the solution.
See electric double layer.

Zewail, Ahmed H. (1946–). Awarded Nobel Prize in Chemistry in 1999 for showing that it is possible with rapid laser technique to see how atoms in a molecule move during a chemical reaction using femtosecond spectroscopy. He performs his research at the California Institute of Technology in Pasadena.

Ziegler catalyst. A type of stereospecific catalyst, usually a chemical complex derived from a transition metal halide and a metal hydride or a metal alkyl. The transition metal may be any of those in groups IV to VIII of the periodic table; the hydride or alkyl metals are those of groups I, II, and III. Typically, titanium chloride is added to aluminum alkyl in a hydrocarbon solvent to form a dispersion or precipitate of the catalyst complex. These catalysts usually operate at atmospheric pressure and are used to convert ethylene to linear polyethylene and also in stereospecific polymerization of propylene to crystalline polypropylene (Ziegler process).
See Natta catalyst.

Ziegler, Karl. (1898–1973). A German chemist who won the Nobel Prize for chemistry in 1963 with Natta. A great deal of work was concerned with the chemistry of carbon compounds and development of plastics. A recipient of the Swinburne medal from the Plastics Institute of London in 1964. After studying at Marburg he was a professor at Heidelberg.

Ziegler method. Cyclization of dinitriles at high dilution in dialkyl ether in the presence of ether-soluble metal alkylanilide and hydrolysis of the resultant imino-nitrile with formation of macrocyclic ketones in good yields.

Ziegler–Natta polymerization. Polymerization of vinyl monomers under mild conditions using aluminum alkyls and $TiCl_4$ (or other transition element halide) catalyst to give a stereoregulated, or tactic, polymer. These polymers, in which the stereochemistry of the chain is not random, have very useful physical properties.

"Zimate" *[Vanderbilt].*   TM for zinc dimethyldithiocarbamate. See ziram.

Zimmermann reaction.   The reaction that occurs between methylene ketones and aromatic polynitro compounds in the presence of alkali. When applied to 17-oxosteroids, the colored compounds formed can be used for the quantitative determination of 17-oxosteroids.

zinc.
CAS: 7440-66-6.   Zn.
Metallic element of atomic number 30, group IIB of the periodic table, aw 65.38, valence of 2; five stable isotopes.
Properties: Shining white metal with bluish-gray luster (called spelter). Not found native. D 7.14, mp 419C, bp 907C, malleable at 100–150C, strongly electropositive, zinc foil will ignite in the presence of moisture. Soluble in acids and alkalies; insoluble in water.
Source: British Columbia, Mexico, U.S. (Colorado), Australia, Belgium.
Derivation: Extracted from ores by two distinct methods, both starting with zinc oxide formed by roasting the ores: (1) the pyrometallurgical or distillation process wherein the zinc oxide is reduced with carbon in retorts from which the resultant zinc is distilled and condensed; and (2) the hydrometallurgical or electrolytic process wherein the zinc oxide is leached from the roasted or calcined material with sulfuric acid to form zinc sulfate solution, which is electrolyzed in cells to deposit zinc on cathodes.
Grade: Special high-grade (99.990%), high-grade (99.95%), intermediate (99.5%), brass special (99%), prime western (98%).
Available forms: Slab, rolled (strip, sheet, rod, tubing), wire, mossy zinc, zinc dust powder (99% pure), single crystals, zinc anodes.
Hazard: (dust) Flammable, dangerous fire and explosion risk.
See zinc dust.
Use: Alloys (brass, bronze, and die-casting alloys), galvanizing iron and other metals, electroplating, metal spraying, automotive parts, electrical fuses, storage and dry-cell batteries, fungicides, nutrition (essential growth element), roofing, gutters, engravers' plates, cable wrappings, organ pipes.
See calamine; franklinite; hydrozincite; sphalerite; willemite.

zinc-65.   Radioactive isotope of mass number 65.
Properties: Half-life 250 days, radiation β and γ.
Derivation: Pile irradiation of zinc metal and, in the cyclotron, by bombarding copper-65 with deuterons.
Available forms: Zinc metal and zinc chloride in hydrochloric acid solution.
Hazard: A radioactive poison.
Use: Tracer nuclide in study of wear in alloys, the nature of phosphor activators, galvanizing, function of traces of zinc in body metabolism, the functions of oil additives in lubricating oils, etc.

zinc abietate.   See zinc resinate.

zinc acetate.
CAS: 557-34-6. $Zn(C_2H_3O_2)_2 \cdot 2H_2O$.
Properties: White, monoclinic, crystalline plates; pearly luster; faint acetous odor; astringent taste. D 1.735, loses $2H_2O$ at 100C, mp 200C (decomposes). Soluble in water and alcohol.
Derivation: Action of acetic acid on zinc oxide.
Use: Medicine (astringent), preserving wood, textile dyeing (mordant and resist), zinc chromate, laboratory reagent, cross-linking agent for polymers, ingredient of dietary supplements (up to 1 mg daily), feed additive, ceramic glazes.

zinc acetylacetonate.
$Zn[OC(CH_3):CHCO(CH_3)]_2$.
Properties: Crystalline solid. Mp 138C, bp (sublimes). Very soluble in benzene, acetone; decomposes in water.
Use: Catalyst in synthesis of long-chain alcohols and aldehydes, textile weighting agent.

"Zincalume" *[Bluescope].*   TM for a bright zinc electroplating process for consumer's goods and military materials. The bath contains zinc cyanide, sodium cyanide, sodium hydroxide, and addition agents.

zinc ammonium chloride. $ZnCl_2 \cdot 2NH_4Cl$. A complex salt; double salts with 3–6 molecules of ammonium chloride have also been prepared.
Properties: White powder or crystals. D 1.8. Soluble in water.
Grade: Technical (foaming and nonfoaming).
Use: Welding, soldering flux, dry batteries, galvanizing.

zinc ammonium nitrite.
CAS: 63885-01-8. $ZnNH_4(NO_2)_3$.
Properties: White powder. Strong oxidizing agent.
Hazard: Dangerous fire risk in contact with organic materials.

zinc antimonide. $ZnSb_2$. Silvery-white crystals, d 6.33, mp 570C, decomposes in water.
Use: Thermoelectric devices.
Hazard: May be irritating to skin.

zinc arsenite.   (zinc metaarsenite; ZMA).
CAS: 10326-24-6. $Zn(AsO_2)_2$.
Properties: Colorless powder. Soluble in acids; insoluble in water. Federal specification TT-W-581 describes the composition of the solution used for wood preservation.
Hazard: Toxic by ingestion and inhalation.
Use: Timber preservative, insecticide.

zinc arsenite.   (zinc metaarsenite; ZMA).
CAS: 10326-24-6. $Zn(AsO_2)_2$.
Properties: Colorless powder. Soluble in acids; insoluble in water. Federal specification TT-W-581 describes the composition of the solution used for wood preservation.
Hazard: Toxic by ingestion and inhalation. TLV: 10 mg/m^3 of total dust when toxic impurities are not present, e.g., quartz <1%.
Use: Timber preservative, insecticide.

zinc bacitracin.
Properties: Creamy-white powder. Almost insoluble in water. Good thermal stability, usually has 50–60 units/mg of bacitracin activity.
Derivation: Action of zinc salts on bacitracin broth.
Grade: USP.
Use: Antibacterial agent in ointments, suppositories, etc.

zinc benzenesulfinate.
CAS: 24308-84-7. $C_{12}H_{12}O_4S_2 \cdot Zn$.
Hazard: Moderately toxic by ingestion. A mild eye irritant.

zinc borate.   Typical composition: zinc oxide 45%, B_2O_3 34%, may have 20% water of hydration.
Properties: White, amorphous powder. Zinc borate of composition $3ZnO \cdot 2B_2O_3$ has d 3.64, mp 980C. Soluble in dilute acids; slightly soluble in water. Nonflammable.
Derivation: Interaction of the oxides at 500–1000C or of zinc oxide slurries with solutions of boric acid or borax.
Use: Medicine, fireproofing textiles, fungistat and mildew inhibitor, flux in ceramics.
See Firebrake ZB.

zinc bromate. $Zn(BrO_3)_2 \cdot 6H_2O$.
Properties: White solid. D 2.566, mp 100C. Deliquescent, loses $6H_2O$ at 200C; very soluble in water.
Hazard: Dangerous fire risk in contact with organic materials, strong oxidizing agent.

zinc bromide. $ZnBr_2$.
Properties: White, hygroscopic, crystalline powder. D 4.219, mp 394C, bp 650C. Soluble in water, alcohol, and ether.
Derivation: Interaction of solutions of barium bromide and zinc sulfate, with subsequent crystallization.

Use: Photographic emulsions, manufacture of rayon. A solution of 80% zinc bromide is used as a radiation viewing shield.

zinc butylxanthate. $Zn(C_4H_9OCS_2)_2$.
Properties: White powder. D 1.45. Decomposes when heated; moderately soluble in benzene and ethylene dichloride; slightly soluble in acetone, insoluble in water and gasoline.
Use: Ultraaccelerator used in self-curing rubber cements.
See xanthate.

zinc cadmium sulfide.   A fluorescent pigment, a phosphor.
Hazard: As for cadmium.

zinc caprylate.   (zinc octanoate).
CAS: 557-09-5. $Zn(C_8H_{15}O_2)_2$.
Properties: Lustrous scales. Mp 136C, decomposes in moist atmosphere giving off caprylic acid. Slightly soluble in boiling water; fairly soluble in boiling alcohol.
Derivation: By precipitation from a solution of ammonium caprylate with zinc sulfate.
Use: Fungicide.

zinc carbolate.   See zinc phenate.

zinc carbonate.
CAS: 3486-35-9. $ZnCO_3$.
Properties: White, crystalline powder. D 4.42–4.45, evolves carbon dioxide at 300C. Soluble in acids, alkalies, and ammonium salt solutions; insoluble in water.
Derivation: (1) Grinding the mineral smithsonite, (2) action of sodium bicarbonate on a solution of a zinc salt.
Use: Ceramics, fireproofing filler for rubber and plastic compositions exposed to flame temperature, cosmetics and lotions, pharmaceuticals (ointments, dusting powders), zinc salts, medicine (topical antiseptic).

zinc chlorate.
CAS: 10361-95-2. $Zn(ClO_3)_2 \cdot 4H_2O$.
Properties: Colorless to yellowish crystals. Deliquescent, d 2.15, decomposes at 60C. Soluble in water, alcohol, glycerol, and ether.
Hazard: Dangerous fire risk in contact with organic materials, strong oxidizing agent.

zinc chloride.
CAS: 7646-85-7. $ZnCl_2$.
Properties: White, granular, deliquescent crystals or crystalline powder. D 2.91, mp 290C, bp 732C, a 10% solution is acid to litmus. Soluble in water, alcohol, glycerol, and ether.
Derivation: Action of hydrochloric acid on zinc or zinc oxide.
Method of purification: Recrystallization.

Grade: CP, technical; fused, crystals, granulated; 62.5% solution, 50% solution, USP.

Hazard: (Solid) skin irritant; (solution) severe irritant to skin and tissue. Lower and upper respiratory irritant.

Use: Catalyst, dehydrating and condensing agent in organic synthesis, fireproofing and preserving food, soldering fluxes, burnishing and polishing compounds for steel, electroplating, antiseptic and deodorant preparations (up to 2% solution), textiles (mordant; carbonizing agent; mercerizing, sizing, and weighting compositions; resist for sulfur colors, albumin colors, and para red), adhesives, dental cements, glass etching, petroleum refining, parchment, dentifrices, embalming and taxidermists' fluids, medicine (astringent), antistatic, denaturant for alcohol.

zinc chloride, chromated.   A mixture of zinc chloride and sodium dichromate used as a wood preservative. Federal specification TT-W-551 requires that it contain no less than 77.5% zinc chloride and 17.5% sodium dichromate dihydrate.

zinc chloroiodide.   Mixture of zinc chloride and iodide.

Properties: White powder. Soluble in water.

Use: Disinfectant, pharmaceutical preparations.

zinc chromate. $ZnCrO_4 \cdot 7H_2O$.

Properties: Solid yellow pigment. Mw 307.6.

Hazard: Nasal cancer, Confirmed carcinogen.

Use: Pigments.

zinc cyanide.

CAS: 557-21-1. $Zn(CN)_2$.

Properties: White powder. D 1.852, mp 800C (decomposes). Soluble in dilute mineral acids with production of hydrogen cyanide; soluble in alkalies; insoluble in water and alcohol.

Derivation: By precipitation of a solution of zinc sulfate or chloride with potassium cyanide.

Grade: Technical.

Hazard: Toxic by ingestion and inhalation.

Use: Metal plating, chemical reagent, insecticide.

zinc dialkyldithiophosphate.   A lubricating-oil additive for corrosion resistance, wear resistance, antioxidant.

zinc dibenzyldithiocarbamate.

$Zn[SCSN(C_7H_7)_2]_2$.

Properties: White powder. D 1.41, melting range 165–175C. Moderately soluble in benzene and ethylene dichloride; insoluble in acetone, gasoline, and water.

Use: Accelerator for latex dispersions and cements.

zinc dibutyldithiocarbamate.

CAS: 136-23-2. $Zn[SC(S)N(C_4H_9)_2]_2$.

Properties: White powder; pleasant odor. D 1.24 (20/20C), melting range 104–108C. Soluble in carbon disulfide, benzene, and chloroform; insoluble in water.

Use: Accelerator for latex dispersions and cements, etc; ultra-accelerator for lubricating oil additive.

zinc _o,o_-dibutyl dithiophosphate.

CAS: 6990-43-8. $C_{16}H_{36}O_4P_2S_4Zn$.

Hazard: Moderately toxic by ingestion. A mild skin and severe eye irritant.

zinc dichromate.

CAS: 7789-12-0. $ZnCr_2O_7 \cdot 3H_2O$.

Properties: Orange-yellow powder. Soluble in acids and hot water; insoluble in alcohol and ether.

Derivation: Action of chromic acid on zinc hydroxide.

Hazard: Toxic by ingestion and inhalation.

Use: Pigment.

zinc diethyl.   See diethylzinc.

zinc diethyldithiocarbamate.

CAS: 14324-55-1. $Zn[SC(S)N(C_2H_5)_2]_2$.

Properties: White powder. D 1.47 (20/20C), melting range 172–176C. Soluble in carbon disulfide, benzene, and chloroform; insoluble in water.

Hazard: Strong irritant to eyes and mucous membranes.

Use: Rubber vulcanization accelerator, especially for latex foam; heat stabilizer for polyethylene.

zinc dimethyldithiocarbamate.   See ziram.

zinc dimethyldithiocarbamate cyclohexylamine complex.   (zinc dithioamine complex).

Properties: White powder or slurry of low solubility.

Hazard: Toxic by ingestion.

Use: Fungicide, rodent poison, deer and rabbit repellent.

zinc dioxide.   See zinc peroxide.

zinc dithionite.   See zinc hydrosulfite.

zinc dust.   A gray powder.

Grade: Commercial, pigment.

Hazard: Dangerous fire risk; may form explosive mixture with air; in bulk when damp, may heat and ignite spontaneously on exposure to air.

Use: Zinc salts and other zinc compounds, reducing agent, precipitating agent, purifier, catalyst, rust-resistant paints, bleaches, pyrotechnics, soot removal, pipe-thread compounds, sherardizing, decorative effect in resins, autobody coatings.

zinc ethyl.   See diethylzinc.

zinc ethylenebisdithiocarbamate.   See zineb.

zinc-2-ethylhexoate.   See zinc octoate.

zinc ethylsulfate. $Zn(C_2H_5SO_4)_2 \cdot 2H_2O$.
Properties: Colorless, hygroscopic, crystalline leaflets. Soluble in water and alcohol.
Derivation: Interaction of zinc hydroxide and diethyl sulfate.
Use: Organic synthesis.

zinc-finger protein.   A secondary feature of some proteins containing a zinc atom; a DNA-binding protein.

zinc fluoride.
CAS: 7783-49-5. ZnF_2.
Properties: White powder. D 4.84 (15C), mp 872C, bp about 1500C. Soluble in hot acids; slightly soluble in water; insoluble in alcohol.
Derivation: (1) Action of hydrogen fluoride on zinc hydroxide, (2) addition of sodium fluoride to a solution of zinc acetate.
Grade: Technical, 95% pure.
Hazard: Toxic material.
Use: Phosphors, ceramic glazes, wood preserving, electroplating, organic fluorination.

zinc fluoroborate. $Zn(BF_4)_2$.
Properties: Colorless liquid. Handled as 40 or 48% solution.
Use: Plating and bonderizing, resin curing.

zinc fluorosilicate.   (zinc silicofluoride).
CAS: 16871-71-9. $ZnSiF_6 \cdot 6H_2O$.
Properties: White crystals. D 2.104. Decomposes on heating; soluble in water.
Derivation: Reaction of zinc oxide and fluosilicic acid.
Use: Concrete hardener, laundry sour, preservative, moth-proofing agents.

zinc formaldehyde sulfoxylate.
$Zn(HSO_2 \cdot CH_2O)_2$ (normal); $Zn(OH)(HSO_2 \cdot CH_2O)$ (basic).
Properties: Rhombic prisms. Very soluble in water (normal) (basic is insoluble in water); insoluble in alcohol; decomposes in acid.
Derivation: Reaction of formaldehyde and zinc sulfoxylate.
Grade: Basic, normal.
Hazard: Toxic by ingestion.
Use: Stripping and discharging agent for textiles.
See hydrosulfite-formaldehyde.

zinc formate. $Zn(CH_2O)_2 \cdot 2H_2O$.
Properties: White crystals. D 2.207 (20C). Loses $2H_2O$ at 140C. Soluble in water; insoluble in alcohol.
Derivation: Action of formic acid on zinc hydroxide.
Hazard: Toxic by ingestion.
Use: Catalyst for production of methanol, waterproofing agent, textiles, antiseptic.

zinc gluconate.
CAS: 468-02-4. $C_{12}H_{22}O_4Zn$.

Properties: White granular or crystalline powder. Sol in water; very slightly soluble in alcohol.
Use: Dietary supplement and food additive, vitamin tablets.

zinc green.   One of a group of brilliant green pigments consisting of a mixture of prussian blue and zinc yellow. They are permanent to light but not to alkali or water.
Use: Flat wall paints and interior work.

zinc hydrosulfite.   (zinc dithionite).
CAS: 7779-86-4. (ZnS_2O_4).
Properties: White, amorphous solid. Soluble in water.
Grade: Technical.
Use: Brightening groundwood, kraft, and other paper pulps; to treat beet and cane sugar juices; depressant in mining flotations; bleaching textiles, vegetable oils, straw, hemp, vegetable tannins, animal glue, etc.

zinc hydroxide. $Zn(OH)_2$.
Properties: Colorless crystals. D 3.053. Decomposes at 125C. Almost insoluble in water; forms both zinc salts and zincates.
Derivation: Addition of a strong alkali to a solution of a zinc salt.
Use: Intermediate, absorbent in surgical dressings, rubber compounding.

zinc hypophosphite. $Zn(H_2PO_2)_2 \cdot H_2O$.
Properties: White, hygroscopic crystals. Soluble in water and alkalies.
Derivation: Action of hypophosphorous acid on zinc hydroxide.

zinc iodide. ZnI_2.
Properties: Hygroscopic, white, crystalline powder; sharp saline taste. Turns brown on exposure to light or air. D 4.67, mp 446C, bp 625C (decomposes). Soluble in water, alcohol, and alkalies.
Derivation: Interaction of barium iodide and zinc sulfate, with subsequent crystallization.
Use: Medicine (topical antiseptic), analytical reagent.

Zincke disulfide cleavage.   Formation of sulfenyl halides by three essentially similar methods involving the action of chlorine or bromine on aryl disulfides, thiophenols, or arylbenzyl sulfides.

Zincke nitration.   Replacement of o- or p-bromine or iodine atoms (but not fluorine or chlorine atoms) in phenols by a nitro group on treatment with nitrous acid or a nitrite in acetic acid.

Zincke–Suhl reaction.   Phenol-dienone rearrangement of p-cresols by addition of carbon tetrachloride in the presence of aluminum chloride, with formation of 4-methyl-4-trichloromethylcyclohexa-2,5-dienone.

zinc lactate. $Zn(C_3H_5O_3)_2 \cdot 3H_2O$.
Properties: White crystals. Soluble in water. Combustible.
Derivation: Action of lactic acid on zinc hydroxide.

zinc laurate. $Zn(C_{12}H_{23}O_2)_2$.
Properties: White powder. Mp 128C. Almost insoluble in water and alcohol. Combustible.
Derivation: Precipitation of a soluble coconut oil soap with a solution of a zinc salt.
Use: Paints, varnishes, rubber compounding (softener and activator).

zinc linoleate. $Zn(C_{17}H_{31}COO)_2$.
Properties: Brown solid containing 8.5–9.5% zinc. Combustible.
Derivation: Precipitation from solutions of sodium linoleate and soluble zinc salt, or by fusion of the fatty acid and zinc oxide.
Use: Paint drier, especially with cobalt and manganese soaps.

zinc malate. $Zn(OOCCH_2CH_2OCOO) \cdot 3H_2O$.
Properties: White, crystalline powder. Soluble in water. Combustible.
Derivation: Action of malic acid on zinc hydroxide.

zinc-2-mercaptobenzothiazole.
$Zn(C_7H_4NS_2)_2$.
Use: Rubber accelerator, fungicide.

zinc methionine sulfate.
CAS: 56329-42-1.
Properties: White powder. Freely sol in water.
Use: Food additive.

zinc molybdate. $ZnMoO_4 \cdot 2H_2O$.
Properties: Solid. D 3.3, mp 1650C. Insoluble in water.
Use: Anticorrosion agent, starting material for growing single crystals.

zinc naphthenate. $Zn(C_6H_5COO)_2$.
Properties: Amber, viscous, basic liquid or basic solid. The liquid contains 8–10% zinc, the solid contains 16% zinc. Very soluble in acetone. Combustible.
Derivation: Fusion of zinc oxide or hydroxide and naphthenic acid or precipitation from mixture of soluble zinc salts and sodium naphthenate.
Use: Drier and wetting agent in paints, varnishes, resins; insecticide; fungicide; mildew preventive; wood preservative; waterproofing textiles; insulating materials.

zinc nitrate.
CAS: 7779-88-6. $Zn(NO_3)_2 \cdot 6H_2O$.
Properties: Colorless lumps or crystals. D 2.065 (13C), mp 36.4C, loses water of crystallization between 105 and 131C. Soluble in water and alcohol.

Derivation: Action of nitric acid on zinc or zinc oxide.
Hazard: Dangerous fire and explosion risk, strong oxidizing agent.
Use: Acidic catalyst, latex coagulant, reagent, intermediate, mordant.

zinc octanoate. See zinc caprylate.

zinc octoate. (zinc-2-ethylhexoate).
$Zn(OOCCH(C_2H_5)C_4H_9)_2$.
Properties: Light-straw-colored, viscous liquid. D 1.16. Insoluble in water; soluble in hydrocarbon solvents. Combustible.
Use: Catalyst.

zinc oleate.
CAS: 557-07-3. $Zn(C_{17}H_{33}COO)_2$.
Properties: Dry, white to tan, greasy, granular powder containing 8.5–10.5% zinc. Mp 70C. Soluble in alcohol, ether, carbon disulfide, ligroin; insoluble in water. Combustible.
Derivation: Interaction of solutions of zinc acetate and sodium oleate or by fusion of zinc oxide and oleic acid.
Use: Paints, resins, and varnishes (drier).

"Zinc Omadine" [Arch].
CAS: 13463-41-7.
TM for zinc pyrithione.
Properties: Powder.
Use: Additive in antidandruff preparations, preservation of cosmetics, metalworking fluids, inhibition of bacterial and fungal growth on fabric.

"Zincon" [Ultimark]. TM for 2-carboxy-2'-hydroxy-5'-sulfoformazylbenzene.
Use: In colorimetric determination of zinc and copper.

zinc orthoarsenate. See zinc arsenate.

zinc orthophosphate. See zinc phosphate.

zinc orthosilicate. See zinc silicate.

zinc oxalate. $ZnC_2O_4 \cdot 2H_2O$.
Properties: White powder. D 2.562 (24C), mp 100C. Soluble in acids and alkalies; slightly soluble in water. (decomposes), combustible.
Derivation: Interaction of zinc sulfate and sodium oxalate.
Use: Zinc oxide, organic synthesis.

zinc oxide. (Chinese white; zinc white).
CAS: 1314-13-2. ZnO.
Properties: Coarse white or grayish powder; odorless; bitter taste. Absorbs carbon dioxide from the air. Has greatest UV absorption of all commercial pigments. D 5.47, mp 1975C. Soluble in acids and alkalies; insoluble in water and alcohol. Noncombustible.

Derivation: (1) Oxidation of vaporized pure zinc (French process), (2) roasting of zinc oxide ore (franklinite) with coal and subsequent oxidation with air, (3) similar treatment starting with other ores, (4) oxidation of vapor-fractionated die castings.

Grade: American process, lead-free; French process; lead-free; green seal; red seal; white seal (according to fineness); leaded (white lead sulfate); USP; single crystals.

Hazard: Zinc oxide fume is harmful by inhalation. Zinc oxide powder reacts violently with chlorinated rubber at 215C. Metal fume fever.

Use: Accelerator activator, pigment and reinforcing agent in rubber, ointments, pigment and mold-growth inhibitor in paints, UV absorber in plastics, ceramics, floor tile, glass, zinc salts, feed additive, dietary supplement, seed treatment, cosmetics, photoconductor in office copying machines and in color photography, piezoelectric devices, artists' colorant.

zinc oxychloride. A saturated solution of zinc chloride and zinc oxide.
Use: Dentistry.

zinc palmitate. $Zn(C_{16}H_{31}O_2)_2$.
Properties: White, amorphous powder. D 1.121, mp 100C. Insoluble in water and alcohol; slightly soluble in benzene and toluene. Combustible.
Use: Flatting agent in lacquer, pigment suspending agent for paints, rubber compounding, lubricant in plastics.

zinc perborate. $Zn(BO_3)_2$ with water of hydration.
Properties: Amorphous white powder. Insoluble in water but slowly decomposed by it, liberating hydrogen peroxide.
Derivation: Interaction of sodium peroxide, boric acid, and zinc salt, or of boric acid and zinc peroxide.
Hazard: Fire risk when wet, in contact with organic materials.
Use: Medicine, oxidizing agent.

zinc permanganate.
CAS: 23414-72-4. $Zn(MnO_4)_2 \cdot 6H_2O$.
Properties: Violet-brown or black, hygroscopic crystals. D 2.47, loses $5H_2O$ at 100C. Decomposes on exposure to light and air; soluble in water and acids; decomposes in alcohol.
Grade: Technical (95% pure).
Hazard: Dangerous fire risk in contact with organic materials, strong oxidizing agent.
Use: Oxidizing agent, medicine (antiseptic).

zinc peroxide. (zinc dioxide).
CAS: 1314-22-3. ZnO_2.
Properties: White powder containing 45–60% ZnO_2, balance ZnO. D 1.571, decomposes rapidly

above 150C. Decomposes in acids, alcohol, acetone; insoluble in water but decomposed by it.
Derivation: Action of barium peroxide on zinc sulfate solution, followed by filtration.
Grade: USP (mixture of peroxide, carbonate, and hydroxide), technical 50–60%.
Hazard: Severe explosion risk when heated; explosive range 190–212C. Fire risk in contact with organic materials; strong oxidizing agent.
Use: Curative for rubber and elastomers, pharmaceuticals, high-temperature oxidation.

zinc-*o*-phenanthroline complex.
CAS: 16561-55-0. $C_{12}H_8N_2 \cdot Zn$.
Hazard: A reproductive hazard.

zinc phenate. (zinc carbolate; zinc phenolate). $Zn(C_6H_5O)_2$. (May be only a mixture of zinc oxide and phenol).
Properties: White powder. Soluble in alcohol; slightly soluble in water. Combustible.
Derivation: By heating zinc hydroxide with phenol and extracting with alcohol.
Hazard: Toxic by ingestion.
Use: Insecticide.

zinc-1,4-phenolsulfonate. (zinc sulfophenate; zinc sulfocarbolate).
CAS: 127-82-2. $Zn(SO_3C_6H_4OH)_2 \cdot 8H_2O$.
Properties: Colorless, transparent crystals or white granular powder; odorless; astringent metallic taste. Effloresces in air, turns pink on exposure to air and light. Loses water of crystallization at 120C; soluble in water and alcohol.
Derivation: By heating zinc hydroxide with *p*-phenolsulfonic acid.
Grade: Technical.
Hazard: Toxic by ingestion.
Use: Insecticide, medicine (antiseptic).

zinc phosphate. (zinc orthophosphate; zinc phosphate, tribasic). $Zn_3(PO_4)_2$.
Properties: White powder. D 3.998 (15C), mp 900C. Soluble in acids and ammonium hydroxide; insoluble in water.
Derivation: Interaction of zinc sulfate and trisodium phosphate.
Grade: Technical, 98% pure.
Use: Dental cements; phosphors; conversion coating of steel, aluminum, and other metal surfaces.

zinc phosphide.
CAS: 1314-84-7. Zn_3P_2.
Properties: Dark gray, gritty powder. D 4.55 (15C), mp above 420C, stable if dry. Insoluble in alcohol; soluble in acids; decomposes in water.
Derivation: By passing phosphine into a solution of zinc sulfate.
Grade: Technical, 80–85% pure.

Hazard: Reacts violently with oxidizing agents, produces toxic and flammable phosphine by reaction with acids. Toxic by ingestion. A deadly poison.
Use: Rodenticide.

zinc potassium chromate.   See zinc yellow.

zinc potassium iodide.   See potassium zinc iodide.

zinc propionate. $Zn(OOCC_2H_5)_2$.
Properties: Platelets, tablets, or needle-like crystals. Fairly soluble in water; slightly soluble in alcohol; decomposes in moist atmosphere liberating propionic acid. Combustible.
Derivation: By dissolving zinc oxide in dilute propionic acid and concentrating the solution.
Use: Fungicide on adhesive tape.

zinc-1,2-propylene bisdithiocarbamate.
See propineb.

zinc pyrithione.   See "Zinc Omadine" *[Arch]*.

zinc pyrophosphate. $Zn_2P_2O_7$.
Properties: White powder. D 3.756 (23C). Soluble in acids and alkalies; insoluble in water.
Use: Pigment.

zinc resinate.
Properties: Powder, clear amber lumps, or yellowish liquid. May be acidic, basic, or neutral; soluble in some organic solvents (ether, amyl alcohol). Combustible.
Chief constituent: zinc abietate.
Derivation: By fusion of zinc oxide and rosin or by precipitation from solutions of zinc salts and sodium resinate.
Use: Wetting, dispersing, and hardening agent; drier in paints, varnishes, and resins.

zinc rhodanide.   See zinc thiocyanate.

zinc ricinoleate.
$Zn[(CH_3CH_2)_5CH_2OCH_2CH:CH(CH_2)_7CO_2]_2$.
Properties: Fine white powder; faint fatty acid odor. Mp 92–95C, d 1.10 (25/25C). Combustible.
Use: Fungicide, emulsifier, greases, lubricants, waterproofing, lubricating-oil additive, stabilizer in vinyl compounds.

zinc salicylate. $Zn[C_6H_4(OH)COO]_2 \cdot 3H_2O$.
Properties: White crystalline needles or powder. Soluble in water and alcohol. Combustible.
Derivation: By heating zinc hydroxide and salicylic acid.
Use: Medicine (antiseptic).

zinc selenide.   ZnSe.
Properties: Yellowish to reddish crystals. D 5.42 (15/4C), mp above 1100C. Insoluble in water.
Hazard: Fire risk in contact with water or acids.

Use: Windows in infrared optical equipment, phosphor.

zinc silicate.   (zinc orthosilicate). Zn_2SiO_4.
Properties: White crystals. D 4.103, mp 1509C. Insoluble in water.
Use: Phosphors, spray ingredients, to remove traces of copper from gasoline.
See willemite.

zinc silicofluoride.   See zinc fluorosilicate.

zinc-silver oxide battery.   Primary or secondary battery used where space and weight are critical, i.e., in missiles. The battery has large energy output for its weight, but the components are expensive and the cycle life is short. To avoid deterioration, potassium hydroxide electrolyte is added just before use.
See battery.

zinc stearate.
CAS: 557-05-1. $Zn(C_{18}H_{35}O_2)_2$.
Percentage of zinc may vary according to intended use, some products being more basic than others.
Properties: (Pure substance) White, hydrophobic powder free from grittiness; faint odor. D 1.095, mp 130C. Soluble in acids and common solvents when hot; insoluble in water, alcohol, and ether. Combustible.
Derivation: Action of sodium stearate on solution of zinc sulfate.
Grade: USP, technical, available free from chick edema factor.
Hazard: Questionable carcinogen.
Use: Cosmetics, lacquers, ointments, dusting powder, lubricant, mold-release agent, filler, antifoamer, heat and light stabilizer, medicine (dermatitis), tablet manufacture, dietary supplement.

zinc sulfate.   (white vitriol; white copperas; zinc vitriol).
CAS: 7733-02-0. $ZnSO_4 \cdot 7H_2O$.
Properties: Colorless crystals, small needles, or granular, crystalline powder; without odor; astringent, metallic taste. Efflorescent in air. Solutions acid to litmus. D 1.957 (25/4C), mp 100C, loses $7H_2O$ at 280C. Soluble in water and glycerol; insoluble in alcohol.
Derivation: (1) Roasting zinc blend and lixiviating with subsequent purification, (2) action of sulfuric acid on zinc or zinc oxide.
Grade: Technical, USP, reagent.
Use: Rayon manufacture, dietary supplement, animal feeds, mordant, wood preservative, analytical reagent.

zinc sulfate monohydrate. $ZnSO_4 \cdot H_2O$.
Properties: White, free-flowing powder. Soluble in water; insoluble in alcohol.
Use: Rayon manufacture, agricultural sprays, chemical intermediate, dyestuffs, electroplating.

zinc sulfide.   ZnS. Exists in two crystalline forms, α (wurtzite) and β (sphalerite).
Properties: Yellowish-white powder. Stable if kept dry. α: d 3.98. β: d 4.102, changes to α form at 1020C. Sublimes at 1180C. Soluble in acids; insoluble in water.
Derivation: By passing hydrogen sulfide gas into a solution of a zinc salt.
Grade: Technical, CP, fluorescent or luminous, single crystals.
Use: Pigment, white and opaque glass, base for color lakes, rubber, plastics, dyeing (hydrosulfite process), ingredient of lithopone, phosphor in X-ray and television screens, luminous paints, fungicide.

zinc sulfite. $ZnSO_3 \cdot 2H_2O$.
Properties: White, crystalline powder. Absorbs oxygen from the air to form sulfate. Loses $2H_2O$ at 100C; decomposes at 200C. Soluble in sulfurous acid; insoluble in cold water and alcohol; decomposes in hot water.
Derivation: Action of sulfurous acid on zinc hydroxide.
Use: Preservative for anatomical specimens.

zinc sulfocarbolate.   See zinc phenolsulfonate.

zinc sulfocyanate.   See zinc thiocyanate.

zinc sulfophenate.   See zinc phenolsulfonate.

zinc sulfoxylate. $ZnSO_2$.
Properties: White, crystalline powder. Decomposed by heat; salt of unstable sulfoxylic acid (H_2SO_2). A strong reducing agent.
Derivation: Action of zinc and sulfuryl chloride in ethereal solution or of sulfur dioxide on granulated zinc in absolute alcohol.
Use: Stripping agent in dyeing.

zinc telluride.   ZnTe.
Properties: Reddish crystals. D 6.34 (15C), mp 1238C. Decomposes in water. Single crystals available for phosphors.
Derivation: Reaction of zinc oxide and tellurium powder in alkaline solution.
Use: Semiconductor research, photoconductor.

zinc thiocyanate.   (zinc rhodanide; zinc sulfocyanate). $Zn(CNS)_2$.
Properties: White, hygroscopic powder or crystals. Soluble in water, alcohol, and ammonium hydroxide.
Derivation: Interaction of zinc hydroxide and ammonium thiocyanate.
Grade: Technical, solution, reagent, ACS.
Use: Analytical chemistry, swelling agent for cellulose esters, dyeing assistant.

zinc thiophenate.   A coined name for a class of peptizing agents for natural and synthetic rubbers; a typical example is the *tert*-butylphenyl sulfide $[(CH_3)_3CC_6H_4S]_2Zn$.

zinc undecylenate. $[CH_2{:}CH(CH_2)_8COO]_2Zn$.
Properties: White, amorphous powder. Mp 115–116C. Nearly insoluble in water and alcohol. Combustible.
Grade: NF.
Use: Medicine (fungistat), cosmetics, chemical intermediate.

zinc white.   See zinc oxide.

zinc yellow.   (citron yellow; buttercup yellow; zinc potassium chromate; zinc chrome).
CAS: 11103-86-9. $4ZnO \cdot 4CrO_3 \cdot K_2O \cdot 3H_2O$.
Properties: Greenish-yellow pigment of comparatively low tinting strength. Partially water soluble.
Derivation: Reaction of a solution of potassium dichromate with zinc oxide and sulfuric acid.
Hazard: Toxic by ingestion.
Use: Rust-inhibiting paints, artists' color.

zinc zirconium silicate. $ZnO \cdot ZrO_2 \cdot SiO_2$.
Properties: White powder. D 4.8, bulk d 115 lb/cu ft, mp 2100C. Soluble in hydrogen fluoride; insoluble in water and alkalies; slightly soluble in mineral acids and hot concentrated sulfuric acid. Noncombustible.
Use: Opacifier for ceramic glazes.

zineb.   (zinc ethylenebis[dithiocarbamate]).
CAS: 12122-67-7. $Zn(CS_2NHCH_2)_2$.
Properties: Light-tan solid. Insoluble in water; soluble in pyridine; decomposes on heating.
Derivation: Reaction of sodium ethylenebisdithiocarbamate with zinc sulfate or other zinc salts. In practical application as a fungicide these reactants are mixed in the presence of lime; the zineb is not formed until after reaction of the carbon dioxide of the air with the film of the other chemicals on the leaf or fruit.
Grade: Commercial dusts and wettable powders usually contain 65% active material.
Hazard: Toxic by inhalation and ingestion; irritant to eyes and mucous membranes. Questionable carcinogen.
Use: Insecticide and fungicide.

zingerone.   (4-[4-hydroxy-3-methoxyphenyl]-2-butanone).
CAS: 122-48-5. $HOC_6H_3(OCH_3)CH_2CH_2COCH_3$.
Properties: Crystals. Mp 40–41C. Soluble in ether; sparingly soluble in water and petroleum ether.
Use: Flavoring.

"Zin-O-Lyte" *[Du Pont].*   TM for a series of electroplating products for use in zinc cyanide plating baths.

Zinophos. A soil insecticide and nematocide whose active ingredient is thionazin.

"Zinpol" *[Lubrizol].* TM for shellac.
Available forms: Flake, powder, and solution.
Use: Ink, coatings, confectionary, pharmaceutical, and industrial processing.

ziram. (zinc dimethyldithiocarbamate).
CAS: 137-30-4. $Zn(SCSNCH_3CH_3)_2$.
Properties: White and odorless when pure. D 1.71, mp 246C. Almost insoluble in water; soluble in acetone, carbon disulfide, chloroform, dilute alkalies, and concentrated hydrochloric acid.
Derivation: Reaction of sodium dimethyldithiocarbamate with a soluble zinc salt in aqueous solution.
Grade: 76% wettable powder, 90% technical powder.
Hazard: Strong irritant to eyes and mucous membranes. Questionable carcinogen.
Use: Fungicide, rubber accelerator.

Zircaloy. Alloys of zirconium with low percentages of antimony, iron, chromium, and nickel.
Use: Cladding for nuclear fuel elements and other reactor applications.

Zircat.
CAS: 7440-67-7.
A zirconium catalyst.
Use: Waterborne coatings.

Zirco. An oil-soluble polymeric zirconyl complex in odorless mineral spirits. Not a paint drier in itself; has synergistic action on metallic driers.

Zircofrax. Superrefractory products from zirconium oxide and zirconium silicate.
Properties: High heat resistance, great strength, high thermal conductivity, high resistance to attack by acids and acid slags, porosity about 25%, low permeability.
Use: Bricks and special shapes for ceramic kiln furniture and in chemical and metallurgical furnaces.

zircon.
CAS: 14940-68-2. $ZrSiO_4$ or $ZrO_2 \cdot SiO_2$.
A natural zirconium silicate, represents 60–70% of all zirconium used.
Properties: Brown, gray, red, or colorless; adamantine luster. Mohs hardness 7.5, d 4.68. Insoluble in acids.
Occurrence: Georgia, Florida, Australia, Brazil.
Use: Source of zirconium oxide, metallic zirconium, and hafnium; abrasive; refractories; enamels; refractory porcelain; catalyst; silicone rubbers; foundry cores.

zirconia. See zirconium oxide.

zirconic anhydride. See zirconium oxide.

zirconium.
CAS: 7440-67-7. Zr.
Metallic element of atomic number 40, group IVB of the periodic table, aw 91.22, valences of 2, 3 (halogens only) 4; five stable isotopes.
Properties: Hard, lustrous, grayish, crystalline scales or gray amorphous powder. D 6.4, mp 1850C, bp 4377C. Soluble in hot, very concentrated acids; insoluble in water and cold acids. Corrosion resistant, low neutron absorption.
Source: Zircon, baddeleyite (zirconia).
Derivation: The ore is converted to a cyanonitride and is chlorinated to obtain zirconium tetrachloride. This is reduced with magnesium (Kroll process) in inert atmosphere. The metal can be prepared in a highly pure and ductile form by vapor-phase decomposition of the tetraiodide. Hafnium must be removed for uses in nuclear reactors.
See hafnium.
Grade: Plate, strip, bars, wire, sponge and briquettes, powder, foil, technical, pure (hafnium free), single crystals.
Hazard: Flammable and explosive as dust or powder and in form of borings, shavings, etc. Not permitted in cosmetics (FDA). Powder should be kept wet in storage and protective clothing should be worn. Questionable carcinogen.
Use: Coating nuclear fuel rods, corrosion-resistant alloys, photo flashbulbs (foil), pyrotechnics, metal-to-glass seals, special welding fluxes, getter in vacuum tubes, explosive primers, acid manufacturing plants, deoxidizer and scavenger in steel manufacturing, laboratory crucibles, spinnerets.

zirconium-95. Radioactive zirconium of mass number 95.
Properties: Half-life 63 days, radiation β and γ.
Derivation: Obtained in a mixture with niobium from the fission products of nuclear reactor fuels.
Available forms: Zirconium oxalate complex in oxalic acid solution.
Hazard: Radioactive poison.
Use: To trace the flow of petroleum products in pipelines, to measure rate of catalyst circulation in petroleum cracking plants, to study the cracking and polymerization of hydrocarbons with various catalysts, etc.

zirconium acetate. $H_2ZrO_2(C_2H_3O_2)_2$.
Properties: (1) Available as aqueous solution, 22% ZrO_2. Clear to pale-amber liquid. D 1.46, pH 3.8–4.2 (20C), fp −7C, stable at room temperature. (2) Available as 13% ZrO_2 (aqueous solution). D 1.20 (approx.), pH 3.3–4.0 (20C), stable at room temperature but temperature of hydrolysis decreases with pH; undergoes exchange with anion-exchange resins but not with cation exchangers.
Use: Preparation of water repellents, other chemicals.

zirconium acetylacetonate. See zirconium tetraacetylacetonate.

zirconium ammonium fluoride.   (ammonium zirconifluoride). $Zr(NH_4)_2F_6$.
Properties: White crystals. Soluble in water.
Hazard: Irritant.

zirconium anhydride.   See zirconium oxide.

zirconium boride.   (zirconium diboride). ZrB_2.
Properties: Gray metallic crystals or powders. D 6.085, mp 3000C, Mohs hardness 8, electrical resistivity 9.2 micro-ohm-cm (20C), excellent thermal shock resistance, poor oxidation resistance above 1100C.
Hazard: Toxic.
Use: Refractory for aircraft and rocket applications, thermocouple protection tubes, high temperature electrical conductor, cutting tool component, coating tantalum, cathode in high-temperature electrochemical systems.

zirconium carbide. ZrC.
Properties: Gray, crystalline solid. D 6.78, Mohs hardness 8+, mp 3400C, bp 5100C. Insoluble in water and hydrochloric acid; soluble in oxidizing acids and attacked by oxidizers.
Derivation: By heating zirconium oxide and coke in an electric furnace.
Grade: Technical.
Hazard: Powder or dust will ignite spontaneously.
Use: Incandescent filaments, abrasive, cermet component, high-temperature electrical conductor, refractory, metal cladding, cutting tool component.

zirconium carbonate, basic.   (zirconyl carbonate; zirconium carbonate). $ZrOCO_3$ or $ZrOCO_3 \cdot xH_2O$.
Properties: White, amorphous powder. Soluble in acids; insoluble in water.
Derivation: By adding sodium carbonate to a solution of zirconium salt.
Use: Preparation of zirconium oxide.

zirconium chloride.   See zirconium tetrachloride.

zirconium chloride, basic.   See zirconium oxychloride.

zirconium diboride.   See zirconium boride.

zirconium dioxide.   See zirconium oxide.

zirconium disilicide.   (zirconium silicide). $ZrSi_2$.
Properties: Gray solid. D 4.88 (22C). Soluble in hydrogen fluoride; insoluble in water and aqua regia.
Use: Coatings resistant to flame or blast impingement, special alloys.

zirconium disulfide. ZrS_2.
Properties: Gray, crystalline solid. D 3.87, mp 1550C. Insoluble in water.
Use: A solid lubricant.

zirconium fluoride.   See zirconium tetrafluoride.

zirconium glycolate. $H_2ZrO(C_2H_2O_3)_3$.
Properties: Solid. Decomposes without melting on heating to 220C. Insoluble in water and organic solvents; soluble in alkali and sulfuric acid solutions. One or more of the acidic hydrogens may be replaced by alkali metals or ammonium to give water-soluble salts.
Use: Cosmetic (deodorant), medicine, sequestrant.

zirconium hydride.
CAS: 7704-99-6. ZrH_2.
Contains 1.7–2.1% combined hydrogen which can be driven off in a vacuum above 600C.
Properties: Gray-black metallic powder. Stable toward air and water. D 5.6, autoign temp 518F (270C).
Derivation: Reduction of zirconia with calcium hydride or magnesium in the presence of hydrogen, direct combination of hydrogen and zirconium.
Grade: Commercial (contains hafnium), reactor (hafnium free), electronic.
Hazard: Flammable, dangerous fire risk, especially in the presence of oxidizers.
Use: Vacuum-tube getter, powder metallurgy, source of hydrogen, metal-foaming agent, nuclear moderator, reducing agent, hydrogenation catalyst.

zirconium hydroxide. $Zr(OH)_4$.
Properties: White, bulky, amorphous powder. D 3.25, decomposes to ZrO_2 at 550C. Soluble in dilute mineral acids; insoluble in water and alkalies.
Derivation: Action of sodium hydroxide (solution) on a zirconium salt solution.
Use: Source of zirconium oxide and zirconium sulfate, glass colorants.

zirconium hydroxychloride solution.   See zirconyl hydroxychloride solution.

zirconium lactate.
CAS: 63919-14-2. $H_4ZrO(CH_3CHOCO_2)_3$.
Properties: White, slightly moist pulp. Decomposes without melting. Very slightly soluble in water and the common organic solvents; soluble in aqueous alkalies with formation of salts, decomposes to hydrous zirconia above pH 10.5, efficient odor absorber. Combustible.
Grade: Zirconia 25% (min).
Use: Body deodorants, source of zirconia.

zirconium naphthenate.
Properties: Amber-colored, heavy, transparent liquid. D 1.05. Viscosity equivalent to that of heavy lubricating oil. Very stable, unlike other metallic

naphthenates, possesses no drying properties. Soluble in all common solvents. Combustible.

Derivation: By heating a mixture of naphthenic acid and zirconium sulfate.

Use: Ceramics (enamels, glazes); lubricants; paints and varnishes (antichalking agent, minimizer of moisture and solar radiation effects).

zirconium nitrate.
CAS: 13746-89-9. $Zr(NO_3)_2 \cdot 5H_2O$.
Properties: White, hygroscopic crystals. Decomposes at 100C. Soluble in water and alcohol.
Derivation: Action of nitric acid on zirconium oxide.
Hazard: Dangerous fire and explosion risk in contact with organics, strong oxidizing agent.
Use: Preservative.

zirconium nitride. ZrN. A brassy-colored powder produced by heating the metal in nitrogen.
Properties: Mohs hardness 8+, d 7.09, mp 2930C. Slightly soluble in dilute hydrochloric acid and sulfuric acid; soluble in concentrated acids.
Use: Special crucibles, cermets, refractories.

zirconium orthophosphate. See zirconium phosphate.

zirconium oxide. (zirconia; zirconium dioxide; zirconic anhydride; zirconium anhydride). ZrO_2. Occurs in nature as baddeleyite.
Properties: Heavy, white, amorphous powder. D 5.73, mp 2700C, Mohs hardness 6.5, refr index 2.2. Insoluble in water and most acids or alkalies at room temperature; soluble in nitric acid and hot concentrated hydrochloric, hydrofluoric, and sulfuric acids. Most heat resistant of commercial refractories; dielectric.
Derivation: By heating zirconium hydroxide or zirconium carbonate.
Grade: Reagent, technical, crystals, fused, whiskers, CP (99% zirconia), hydrous. The fused grade is reported to be harder than diamond (11 on Mohs scale).
Use: (Unstabilized) Production of piezoelectric crystals, high-frequency induction coils, colored ceramic glazes, special glasses, source of zirconium metal, heat-resistant fibers, (hydrous) odor absorbent, to cure dermatitis caused by poison ivy. (Stabilized with CaO refractory furnace linings, crucibles, solid electrolyte for batteries operating at high temperature.

zirconium oxychloride. (zirconium chloride, basic; zirconyl chloride). $ZrOCl_2 \cdot 8H_2O$.
Properties: White, silky crystals. Loses $6H_2O$ at 150C, $8H_2O$ at 210C, d 44 lb/cu ft. Soluble in water, methanol, and ethanol; insoluble in other organic solvents; aqueous solutions are acidic.
Derivation: Action of hydrochloric acid on zirconium oxide.
Grade: Technical, 36% ZrO_2, HP.

Use: Textile, cosmetic and grease additive, antiperspirant, water repellents, chemical reagent, zirconium salts, in lakes and toners of acidic and basic dyes, oil-field acidizing aid.

zirconium phosphate. (zirconium phosphate, basic; zirconium orthophosphate). $ZrO(H_2PO_4)_2 \cdot 3H_2O$.
Properties: White, dense, amorphous powder. Decomposes on heating. Soluble in acids; insoluble in water and organic solvents; extensively hydrolyzed in basic solution.
Derivation: Action of phosphoric acid on zirconium hydroxide.
Use: Chemical reagent, cation scavenger, coagulant, carrier for radioactive phosphorus.

zirconium potassium chloride. (potassium zirconium chloride). $ZrCl_4 \cdot KCl$. A source of zirconium for magnesium alloys, to remove iron in an insoluble form.

zirconium potassium fluoride. (potassium fluozirconate; potassium zirconifluoride).
CAS: 16923-95-8. ZrK_2F_6.
Properties: White crystals. Soluble in water (hot).
Hazard: Irritant.
Use: Grain refiner in magnesium and aluminum, welding fluxes, catalyst, optical glass.

zirconium potassium sulfate. (potassium zirconium sulfate). $2K_2SO_4 \cdot Zr(SO_4)_2 \cdot 3H_2O$.
Properties: White, crystalline powder. Slightly soluble in water.

zirconium pyrophosphate. ZrP_2O_7.
Properties: White solid. Stable to 1550C. Insoluble in water and dilute acids other than hydrogen fluoride, coefficient of thermal expansion 5×10^{-6} at 1000C.
Use: Refractory, olefin polymerization catalyst, phosphor.

zirconium silicate. See zircon.

zirconium silicide. See zirconium disilicide.

"Zirconium Spinel". Trade designation for a synthetic complex containing 39–41% zirconium oxide, 20-22% silicon dioxide, 18.5–20.5% aluminum oxide, and 17–21% zinc oxide. Mp 1704C.
Use: Glaze opacifier in the ceramic industry.

zirconium sulfate.
CAS: 14644-61-2. $Zr(SO_4)_2 \cdot 4H_2O$.
Properties: White, crystalline powder. Bulk d 70 lb/cu ft, decomposes to monohydrate at 100C. Soluble in water; slightly soluble in alcohol; insoluble in hydrocarbons. Aqueous solutions are strongly acidic, will precipitate potassium ions and amino acids from solution, and are decomposed by bases and heat.

Derivation: Action of sulfuric acid on zirconium hydroxide.
Use: Chemical reagent, lubricants, catalyst support, protein precipitation, tanning of white leather.

zirconium sulfate, basic. (zirconyl sulfate). $Zr_5O_8(SO_4)_2 \bullet xH_2O$. Similar in properties to the oxychloride and is prepared in a similar fashion, the end result being in cake form.
Use: Textile treatment and white leather tanning and retannage.

zirconium tetraacetylacetonate. (zirconium acetylacetonate). $Zn[OC(CH_3):CHCO(CH_3)]_4$.
Properties: A colorless, crystalline tetrachelate. D 1.415, mp 194–195C (decomposition begins at 125C). Soluble in pyridine, acetone, benzene, and other organic solvents having some polarity; slightly soluble in water.
Derivation: Reaction among zirconyl chloride, acetylacetone, and sodium carbonate.
Use: Cross-linking agent for polyol, polyester, and polyalkyloxy resins; lubricant and grease additive; reagent; catalyst.

zirconium tetrachloride. (zirconium chloride).
CAS: 10026-11-6. $ZrCl_4$.
Properties: White, lustrous crystals. D 2.8, sublimes above 300C. Soluble in alcohol; decomposes in water.
Derivation: Action of hydrochloric acid on zirconium hydroxide.
Grade: Technical.
Hazard: Irritant.
Use: Source of the pure metal (formed as intermediate in process), analytical chemistry, water repellents for textiles, tanning agent, zirconium compounds, special catalysts of the Friedel–Crafts and Ziegler types.

zirconium tetrafluoride. (zirconium fluoride).
CAS: 7783-64-4. ZrF_4.
Properties: White powder. D 4.43, mp 600C (sublimes), slightly soluble in water and hydrogen fluoride.
Hazard: Strong irritant. TLV: 2.5 mg(F)/m^3.
Use: Component of molten salts for nuclear reactors.

zirconocene dichloride. (dicyclopentadienyl-zirconium dichloride). $(C_5H_5)_2ZrCl_2$.
Properties: White crystals. Mp 244C. Soluble in polar organic solvents. Stable in dry air, very slowly hydrolyzes in moist air.
Hazard: Toxic by inhalation and skin contact, irritant to eyes and mucous membranes.
Use: Rubber accelerator, component of a catalyst system for polymerization of vinyl monomers, curing agent for water-repellent silicone materials, agent for plating with zirconium.
See metallocene.

zircon porcelain. See porcelain, zircon.

zircon sand. A sand containing considerable zirconium, titanium, and related metals.
Use: Source of these elements and also in foundries for casting of alloys.

zirconyl acetate.
CAS: 20645-04-9. $Zr(OH)_2(C_2H_3O_2)_2$.
Properties: (22% ZrO_2 solution) Tacky, resinous, amorphous mass. D 1.46, fp −7C. Becomes solid on heating.
Derivation: Addition of acetic acid to water suspension of carbonated zirconia.
Use: Waterproofing textiles, precipitation of proteins, starch, etc.; for textile and paper coatings.

zirconyl carbonate. See zirconium carbonate, basic.

zirconyl chloride. See zirconium oxychloride.

zirconyl hydroxychloride.
CAS: 10119-31-0. $ZrOOHCl \bullet xH_2O$.
Properties: Colorless or slightly amber liquid (aqueous solution). D 1.26. Forms a soluble glass on evaporation, pH of solution 0.8, reacts with alkalies to form hydrous zirconia; contains 20% zirconia.
Use: Pharmaceuticals, deodorants, precipitation of acid dyes, water repellents for textiles.

zirconyl nitrate (basic). (zirconyl hydroxynitrate). $ZrO(OH)NO_3$.
Properties: Aqueous solution. D 1.35 (25C).
Hazard: Fire risk in contact with organic materials.
Use: Gelatins and improving lamination bonds of polyvinyl alcohol.

zirconyl sulfate. See zirconium sulfate, basic.

Zircotan. Zirconium tanning agents that produce through-white leather.
Use: Tannage of white kid suede, glove leathers; retannage of chrome leather.

ZMA. Abbreviation for zinc metaarsenite. See zinc arsenite.

Zn. Symbol for zinc.

zoalene. (dinitolmide; 3,5-dinitro-o-toluamide).
CAS: 148-01-6. $(O_2N)_2C_6H_2(CH_3)CONH_2$.
Properties: Yellowish solid. Mp 177C. Very slightly soluble in water; soluble in acetone, acetonitrile, dioxane, and dimethylformamide.
Hazard: Liver damage. Questionable carcinogen.
Use: Coccidiostat, food additive.

"Zoamix" *[Zoetis].* TM for poultry coccidiostats containing 3,5-dinitro-o-toluamide.

zoapatanolide a.
CAS: 84886-38-4. $C_{20}H_{26}O_6$.
Hazard: A reproductive hazard.

Zobar. A weed killer based on an aqueous solution of the dimethylamine salts of polychlorobenzoic acids, containing four pounds of acid equivalent per gallon.

"Zoldine" [Dow]. TM for bicyclic oxazolidines.
Grade: Industrial.
Available forms: Liquid.
Use: Catalysts, resin reactions, formaldehyde substitute, cross-linking agents, and tanning agents.

zone refining. A purification process that involves repeated melting and crystallization. The sample to be purified is placed in a relatively long, narrow tube and then passed slowly through a furnace having short, alternate hot and cold zones. Melting occurs opposite the hot zones and crystallization opposite the cold zones. As the rod moves through the furnace the zones move along the rod. Impurities remain in the molten zones and so are carried to one end of the rod. The process has been most used for relatively high-cost materials used in small quantities at very high purities, as for solid-state electronic purposes.

Zonester. Rosin esters.
Available forms: Solid or beaded grades.
Use: As a gum base, tackifier, resin for adhesives, printing inks and coatings.

Zonolite. Verxite (expanded hydrobiotite).

"Zonyl" [Du Pont]. TM for a fluorosurfactant wetting agent that is superior to hydrocarbon surfactants because of greater surface tension reduction.

Zoxamide technical.
CAS: 156052-68-5. $C_{14}H_{16}Cl_3NO_2$.
A fungicide.

Hazard: Low toxicity by ingestion, inhalation, and skin contact.

Zr. Symbol for zirconium.

Zsigmondy, Richard. (1865–1929). A native of Austria, Zsigmondy received the Nobel Prize in Chemistry in 1925 for his work in the field of colloid chemistry, which was initiated by his interest in ruby glass (a colloidal gold suspension). His most important contribution to chemistry was his invention of the ultramicroscope (with Siedentopf) in 1903.
See ultramicroscope; Tyndall effect.

zwitterion. See isoelectric point.

zygote. The product of gamete fusion. In organisms with a haploid life cycle, the zygote immediately undergoes meiosis, but in organisms with a multicellular diploid stage, the zygote is merely the first stage in the diploid portion of the life cycle.

zym-. (zymo-). Relating to a ferment or fermentation.

zymase. The enzyme present in yeast that converts sugars to alcohol and carbon dioxide.

zymo-. See zym-.

zymogen. An inactive precursor of an enzyme; for example, trypsinogen—the precursor of trypsin.

zymohexase. See aldolase.

zymose. Enzyme found in yeast. It hydrolyzes glucose to alcohol (fermentation).

zyprexa.
CAS: 132539-06-1. $C_{17}H_{20}N_4S$.
Hazard: Human systemic effects.

"Zytel" [DuPont]. TM for engineering resins and high-end specialty polymers.

APPENDIX I

Origin of Some Chemical Terms

The etymology of chemical terms is not only of historical interest but is often of value in illuminating their meanings. Those listed here were selected chiefly on the basis of these two criteria.

The names of many elements (e.g., tin, sulfur, gold, zinc, lead) are merely transliterations from older languages; others are in honor of famous scientists (curium, fermium, einsteinium) or of countries (germanium, francium, polonium); still others are in honor of their place of discovery (berkelium, yttrium, strontium). The predominance of Latin (L.) and Greek (Gk.) origins and the many references to mythology reflect the importance of the classical tradition in the 19th century. The dates given in parentheses indicate the approximate year in which the term was introduced. Many common chemical words were in literary use long before the development of modern chemistry in the late 1700s. For dates before 1750, the editor is indebted to the *Oxford Universal Dictionary*; however, some of these (e.g., petroleum, 1526) appear open to question.

abrasive (1656) L. *ab-* (away) + *ras-* (scrape). From the same root as **erase**.

abscission L. *ab-* (from) + *sciss-* (cut). To detach, as fruit from a tree. From the same root as **abscissa**.

absorption See **sorb**.

accelerator (1900) L. *ad-* (*ac-*) (to) + *celer-* (speed). An agent that adds speed to a reaction.

acetic (1808) L. *acetum* (vinegar, a dilute solution of acetic acid). **Acetone** is from the same root.

acid (1626) L. *acidum*, from *acer* (sour, sharp tasting), referring to one of the characteristic properties of acids. **Acrid** is from the same root.

adhesive (1670) L. *ad-* (to) + *haes-* (stick, cling), referring to the clinging together of two dissimilar materials. See **coherent**.

adsorption See **sorb**.

agglomeration L. *ad-* (*ag-*) + *glomer-* (ball, sphere), i.e., a material made up of small, spherical aggregates.

aggregation (1547) L. *ad-* (to, together) + *greg-* (group, herd), i.e., of particulates. From the same root are **gregarious** (group-seeking) and **egregious** (out of the herd), i.e., noteworthy, outstanding in a deprecatory sense, as in an egregious error.

albumin L. *alba* (white). In the 1600s this term (spelled albumen) referred to egg white, as well as to a substance found in plants and seeds. Modern use as a class of proteins began in 1869.

alchemy (1514) Arabic *al* (the) + Gk. *kemia* (transmutation of lead to gold, which was the primary concern of the alchemists). See **chemistry**.

alcohol (1543) Arabic *al* (the) + *koh'l*, a cosmetic paste (antimony sulfide) used to paint the eyelids; later the term referred to any fine metal powder. By some inexplicable semantics, probably originating with the alchemists, it came to mean distilled spirits

of wine (1753). The present sense originated about 1850.

aldehyde (1850) Coined from *al* (cohol) and *dehyd* (rogenation), describing the chemical derivation.

aliphatic Gk. *aliphatos* (oil, fat) from *aleiphein* (to rub or smear with oil). Originated with early chemists who were familiar with animal and vegetable fats whose properties were determined by high molecular weight hydrocarbon groups.

alkali (1813) Arabic *al* (the) + *qalay* (pan-roast, calcine). As early as 1612 it referred to any soda like substance obtained by calcining plants and reducing them to ashes.

alkaloid (1831) See **alkali**. So-called because of their superficial resemblance to alkalis, i.e., bitter taste.

alkyd (1847) Coined from the words *al* (cohol) and *acid* (cid = kyd), describing their chemical derivation. First prepared by Berzelius.

aluminum L. *alumen* (alum). So named by Sir Humphry Davy in 1812.

amalgam (1471) L. *amalgma*, from Gk. *malessein* (a soft mixture or poultice); hence, a mix softened with mercury.

ammonia (1799) From the name of the Egyptian deity Ammon, to whom a temple was built in the North African desert. It refers to the gas evolved by camel excreta in the vicinity of the temple. The ammonium group, NH_4^+ was named by Berzelius in 1808.

amorphous (1801) Gk. *a-* (not, without) + *morph-* (shape), referring to a material that lacks crystal structure, usually a liquid.

amphoteric (1849) Gk. *ampho-* (both), i.e., both acidic and alkaline.

analysis (1581) Gk. *ana-* (up) + *lyein* (loosen, set free), i.e., to separate. It was applied to chemistry in 1655. **Dialysis** is from the same root.

Hawley's Condensed Chemical Dictionary, Sixteenth Edition. Michael D. Larrañaga, Richard J. Lewis, Sr., and Robert A. Lewis.
© 2016 John Wiley & Sons, Inc. Published 2016 by John Wiley & Sons, Inc.

anesthesia (1846) Gk. *an-* (not) + *aisthesis* (feeling, sense). This term was coined by Dr. Oliver Wendell Holmes soon after the first successful use of ether.

anhydride (1863) Gk. *an-* (not, without) + *hydr-* (water), in the sense of a compound that forms an acid on combining with water, e.g., SO_3 is the anhydride of H_2SO_4.

anhydrous (1819) Same roots as for anhydride but used in the sense of absence of water of crystallization in a salt or similar compound.

aniline (1850) Arabic *anil* (indigo), from which aniline was originally made.

anthracene (1863) *anthrac-* (coal), from which it is obtained by distillation.

anthracite See anthracene.

antibiotic Term coined by Selman A. Waksman for mold-derived antibacterial agents.

argon (1894) Gk. *argos* (inactive), referring to the chemical nature of this element.

aromatic (1869) L. *aroma* (spice, fragrance). Applied to benzenoid compounds derived from balsams because of their characteristic odor.

asbestos Gk. *a-* (not) + *sbestos* (quench, extinguish). The meaning as derived is the exact opposite of the unique property of this material.

asphalt (1650?) Gk. *asphaltos*, used in ancient times as an illuminant.

atom Gk. *a* (not) + *tom* (divide, cut). The original use goes back to the Greek philosopher Democritus (465 BC) who first conceived of matter as composed of small indivisible particles. In about 1650, Boyle came to a similar conclusion, but it was not until 1807 that Dalton advanced his atomic theory, and the word entered the scientific vocabulary. However, the etymological meaning was proved to be inapplicable by Rutherford's discovery of atomic disintegration in 1910.

auxiliary L. *auxilia* (help, aid).

azo Gk. *a* (not) + *zo-* (alive), referring to the presence of nitrogen, which Lavoisier had named azote in 1791 because of its nonreactive nature. Related words are zoology, zoo.

barbiturate L. *barba* (beard), presumably in reference to the original plant source of these substances, namely, a lichen characterized by a beard-like structure (*Usnea barbata*).

barium, barytes (1808, 1789) Gk. *barus* (heavy, referring to the characteristic property of these substances). The same occurs in barometer.

base (1810) Gk. *basis* from root *ba-* (go). The reason for the selection of this term in its present chemical sense is not clear.

bauxite Named for a town in southern France, Les Baux, where it was discovered.

bentonite Named for Fort Benton, Wyoming, where it was discovered.

benzene See benzoin.

benzoin (1560) Probably from French *benjoin*, said to be derived from Arabic *ban jawi*, an aromatic gum of Java (frankincense?). The term was later corrupted to *benjamin*. *Benz-* is the root responsible for the word benzol (Liebig, 1838), later modified to benzene, and for benzoic, an acid present in gum benzoin.

bromine (1827) Gk. *bromos* (stench), after one of its characteristic properties.

butyric (1826) L. *butyrum* (butter). The term butyl, from the same source, was introduced in 1868.

cadmium (1917) L. *cadmia* (calamine = zinc ore); this in turn was derived from the legendary hero Cadmus, founder of the Greek city of Thebes, though the connection is obscure.

calcium (1808) L. *calx* (lime). Related is the term *calcine*, used by the alchemists to mean reduction of a metal to a powder (calx) by heat; also to desiccate by fire (1640).

calender (1688) L. *cylindrus* (cylinder), descriptive of the major feature of this machine.

carbon (1789) L. *carbo* (coal); so named by Lavoisier, as the chief constituent of coal.

casein (1841) L. *caseus* (cheese), of which it is a major component.

catalysis Gk. *kata-* (completely) + *lyein* (loosen, set free). Adopted by Berzelius (1836) to describe this type of reaction. The term dates back to 1660 in the nontechnical sense of dissolution or destruction.

cement (1490) L. *caed-* (cut), in reference to the comminuted or crushed materials that are the basis of this product. Cementation (finely divided metal powders) has the same origin.

ceramic (1850) Gk. *keramos* (pottery, potters' clay) Contrary to general belief, this term is *not* derived from Latin *cera* (wax).

ceresin See ozocerite.

cerium (1804) Named by Berzelius from the then recently discovered asteroid Ceres, which in turn was named after the Greek goddess of the harvest. Cereal has the same derivation.

cesium (1860) L. *caesius* (sky-blue). So named by Kirchhoff and Bunsen because of the blue lines in its spectrum.

chelate (1930) Gk. *chela* (claw of a crab or lobster). So named by G. T. Morgan and H. D. K. Drew because of the tenacity with which the coordinating group holds the metal ion.

chemistry Gk. *kemia* (transmutation). *Oxford Universal Dictionary* states that this relates to "the Egyptian art" associated with the "black earth" (*khem*) of the Nile delta, but the exact etymology is obscure. Early writers referred to chemistry as an art rather than a science. Use in its present sense probably began with Robert Boyle about 1650. See alchemy.

chiral (1950?) Gk. *chiro* (hand), referring to the right- and left-handed structure of optical isomers. See enantiomorph.

chlorine (1810) Gk. *chloros* (green). So named from its greenish-yellow color by Sir Humphry Davy.

chlorophyll (1819) Gk. *chloros* (green) + *phyl-* (leaf).

cholesterol (1827) Gk. *chol-* (bile) + *stereo-* (hard, solid).

chromatography (1906) Gk. *chroma* (color) + *graph-* (write, draw). So named by Tswett, its inventor, who used it to analyze plant pigments, which produced bands of characteristic color.

chromium (1797) Gk. *chroma* (color), in reference to the distinctive colors of its compounds. Many other chemical terms have the same derivation, e.g., **chromosome**, **chromatin**, **chromatography**; also **chromatic**.

chromophore Gk. *chroma* (color) + *phor-* (carry, bear), i.e., a chemical group that imparts color to a compound.

chromosome (1890) Gk. *chroma* (color) + *soma* (body), the reference to color being due to the extreme ease with which this nucleoprotein accepts biological stains. The sense of the word *body* here is probably the same as that in *antibody*.

clathrate Gk. *klathra* (a grating of metal bars), suggestive of the cagelike structure of these compounds (also called inclusion compounds). The term was known in the 19th century in a botanical sense, but its chemical use did not appear till about 1940.

coagulate (1610) L. *co-* (together) + *ag-* (drive), i.e., a natural forcing together of dissolved substances to form a clot.

cobalt (1735) German *kobold* (goblin). The extreme difficulty of separating Co from Ni ore during extraction operations led miners in Saxony to endow it with an evil spirit that interfered with obtaining the nickel, which they considered the valuable material. See **nickel**.

coherent L. *co-* (together) + *haer-* (stick, cling), referring to sticking together of identical or closely similar materials, e.g., coherent light. See **adhesive**.

colloid (1861) Gk. *kolla* (glue). Introduced by Thomas Graham to describe suspensions whose disperse phase did not pass readily through a parchment membrane. **Collagen** and **collodion** have the same derivation.

component L. *con-* (together) + *pon-* (place), i.e., a material that is blended with other materials in a mixture. Compound has the same derivation but a different chemical meaning.

composite L. *con-* (together) + *pos-* (place), i.e., a combination or placing together of two or more materials, usually in a laminar structure. See **component**.

compound See **component**.

conjugated L. *con-* (together) + *jugum* (yoke for oxen). An organic compound "yoked together" by alternate double bonds.

copper L. *cuprum* or *Cyprium*, because originally found in Cyprus. Also German *Kupfer*.

corrode (1610) L. *con-* (completely) + *rod-* (gnaw). From the same root as erode and rodent.

crucible (1460) L. *cruc-* (cross), referring to the practice of the alchemists of placing a cross on their experimental equipment to ward off evil spirits. **Crucil** has the same derivation, e.g., a crossroad.

desiccate L. *de-* (out, away) + *siccus* (dry). From the same root as **siccative**, a drying agent.

detergent (1620) L. *de-* (away) + *terg-* (wipe), no doubt because of the cleaning action of these compounds.

dextrose See **levulose**.

dialysis See **analysis**.

diffusion L. *dis-* (*dif-*) (apart) + *fusus* (melt), i.e., gradual dispersal of one substance in another.

disintegrate L. *dis-* (apart) + *in-* (not) + *teg-* (touch), i.e., division of an untouched whole (integer) into many parts. An integer is an "untouched" whole number.

dissociation L. *dis-* (apart) + *socius* (friend, ally), i.e., a separation of substances previously united. Association is from the same root but with opposite meaning.

distillation L. *de-* (down) + *still-* (to fall as drops). The exact date is unknown, but the term was familiar to the alchemists in the Middle Ages. It was also used by poets in the 16th century, e.g., "leperous distilment" (Shakespeare).

dysprosium (1886) Gk. *dysprositos* (hard to get at), because of the difficulty of its isolation.

electrode Gk. *elektron* (amber) + *-ode* (road, pathway), i.e., a highway for electricity.

electron (1890) Gk. *elektron* (amber), referring to the ability of amber to accumulate a static charge by being rubbed. Similarly, electric and related terms.

enantiomorph (about 1900) Gk. *enantio-* (opposite, contrary) + *morph-* (form, shape), descriptive of the mirror-image structure of optical isomers. Also called **enantiomer**.

enzyme (1881) Gk. *zyme* (ferment, leaven), in reference to the catalytic activity of these compounds.

equilibrium (1608) L. *aequus* (the same) + *libra* (weight, balance); hence, any stabilized condition regardless of whether weight is involved.

ether (1587) Gk. *aether* (the sky or upper atmosphere), from a root meaning to glow or burn. The term was long used to mean simply "the air"; it was also applied to a supposed elastic medium that filled all space and by which light was transmitted. This was disproved by Michelson and Morley in 1887. The chemical sense originated about 1838, doubtless because of the highly volatile (airlike) nature of ether.

ethyl See **ether**.

eutectic (1884) Gk. *eu-* (well) + *tek-* (melt), i.e., well or ideally melting.

evolve, evolution L. *e-* (out) + *volv-* (roll), e.g., a gas given off by a chemical reaction. In general, the "rolling out" of any development over a long time span.

extract L. *ex-* (out) + *tract-* (draw, drag). From the same root as **contract** (draw together) and **retract** (draw back).

extrude L. *ex-* (out) + *trud-* (thrust), descriptive of the action of an extrusion machine.

ferment (1605) L. *fermentum* (yeast), from the root *ferv-* (boil), in reference to the apparent similarity to boiling. The chemical meaning did not appear until Pasteur's studies, about 1850. From the same root as **effervescent**.

filter (1791) L. *filtrum* (felt, which was originally used for liquid–solid separations).

flavone L. *flavus* (yellow), for the predominant color produced in flowers by this class of plant pigments.

flocculation (1877) L. *floccus* (a tuft of wool), referring to the appearance of such aggregates.

fluorine (1813) L. *flu-* (flow). So named by Ampere, after Scheele's earlier naming of hydrofluoric acid.

flux (1800) L. *flu-* (flow), originally applied to excretory processes of the body; later it referred to flow-inducing materials in ceramics and metallurgy, as well as to neutron emanations.

fossile L. *foss-* (dig), e.g., fossil fuels.

fractionate L. *fract-* (break), i.e., separation of a liquid mixture into its components by distillation. From the same root as **refraction** ("breaking" of a light ray).

fumigant, fume L. *fumus* (smoke).

fuse, fusion L. *fusus* (melt). The original meaning (1681) was to melt by application of heat, e.g., fused salts, electric fuse. This was later (1817) extended to include blending or joining, as in welding. Since 1940 the term has been applied to the union of hydrogen nuclei with release of energy. In this sense the connection with melting has disappeared.

gas (1600) Adapted by van Helmont from Gk. *chaos*, the letter *g* replacing the *ch*. The Gk. term meant "a yawning gulf or abyss; the formless void of primordial matter" (*Oxford Universal Dictionary*). Modern usage dates from the late 1700s.

gasoline (1871) Standard dictionaries give the etymology as gas + ol + ine, but no clue as to how the term originated. It was probably coined from "gas oil", "a heavy petroleum fraction from which the first" gasoline was made.

glycerol (1838) Gk. *glyc-* (sweet). From the same root are glycol, glycogen (sweet-maker), glycine, etc.

graphite (1789) Gk. *graph-* (write, draw); so named by Werner in reference to its marking ability, as in pencils. **Graph** is from the same root.

halogen Gk. *halo-* (salt) + L. *gen-* (create), from the tendency of these elements to form salts with metals.

helium (1868) Gk. *helios* (the sun), where it was originally discovered by spectrographic analysis.

heterogeneous (1630) Gk. *hetero-* (different) + L. *gen-* (kind). Descriptive of mixtures and solutions, which are made up of two or more different components. Its opposite is homogeneous (*homo-* = the same).

homologous (1850) Gk. *homo-* (the same) + *log-* (proportion), i.e., compounds having the same proportion of certain constituents in an ascending series.

hormone (1906) Gk. *ormon* (to excite, stimulate), in reference to the physiological activity of these compounds.

hydrogen (1783) Gk. *hydro-* (water) + L. *gen-* (create), i.e., water producer. Named by Lavoisier.

hydrolysis (1880) Gk. *hydro-* (water) + *lysis-* (loosen, dissolve), in reference to the cleavage of the water molecule in this reaction.

hydrophilic Gk. *hydro-* (water) + *phil-* (love). The term describes the chief property of vegetable gums, seaweeds, gelatin, etc., which absorb many times their weight of water. Hydrophobic (*phob* = hate) has the opposite meaning.

hygroscopic (1775) Gk. *hygro-* (moisture) + *skopos* (observe, measure). Hygroscopic materials "observe" the presence of atmospheric moisture by absorbing it.

-iferous L. *fer-* (carry, bear). A common suffix as in **carboniferous, coniferous**, etc.

inhibitor L. *in-* (in) + *hab-* (*hib-*) (have, hold), i.e., an agent that holds in or restrains a reaction.

insulin (1925) L. *insula* (island). So named by Banting, its discoverer, because it is formed in the isles of Langerhans in the pancreas. From the same word as **insulation**.

iodine (1814) Gk. *iodos* (violet-colored), so named from the color of its vapor by Gay-Lussac.

ion (1834) L. *i-* (go, move), in reference to the extreme mobility of these particles.

iridium (1804) Named by Tennant from Iris, goddess of the rainbow, because of "the striking color it gives while dissolving in marine acid."

isomer (1838) Gk. *iso-* (the same) + *mer-* (part, share), i.e., compounds that (usually) share their constituent atoms equally, though their arrangements are different.

isotope (1913) Gk. *iso-* (the same) + *topos* (place), i.e., in the periodic table.

kerosene (1854) L. *cer-* (wax), referring to the Pennsylvania paraffinic crude from which it was derived.

ketone (1851) A German modification of **acetone**.

krypton (1898) Gk. *kryptos* (hidden). Named by Sir William Ramsay in reference to the difficulty of its separation.

laser Coined from the first letters of "light amplification by stimulated emission."

latent L. *lat-* (lie hidden), e.g., a latent solvent.

levulose L. *laev-* (the left) + *-ose* (sugar suffix), descriptive of optical isomers that rotate light to the left. The opposite meaning is dextrose from L. *dextro-* (the right).

ligand L. *lig-* (bind), referring to the capacity of certain molecules to attach themselves to metal atoms in coordination compounds. Discovered by Werner in 1893.

lignin, lignite (1820) L. *lignus* (wood). Lignin is a component of wood, and lignite is formed from it.

linoleic L. *linum* (flax) + *oleum* (oil). The principal fatty acid in linseed (flaxseed) oil.

litharge (1800) Gk. *lithos* (stone) + *argyros* (silver); so named because it was once thought to be a product of the separation of silver from lead.

lithium (1818) Gk. *lithos* (stone), probably because of the rocklike nature of petalite, the mineral complex in which it was discovered.

-lysis Gk. *lysis* (decompose, loosen, break down). A suffix in many chemical terms, e.g., **analysis, hydrolysis, catalysis**, etc.

magnesia (1610) Gk. for Magnesian Stone, which was thought to have magnetic properties and which the alchemists believed to be a component of the Philosopher's Stone. The element magnesium was so named by Sir Humphry Davy in 1808. See **manganese**.

manganese (1775) L. *magnes* (magnet) = German *Mangan*, because of the magnetic nature of its ore, pyrolusite. A corrupted form of magnesium.

mercaptan (1835) L. *mer* (curium) (mercury) + *captans* (seizing), in reference to the ability of such compounds to form relatively insoluble salts of mercury and other heavy metals. Has been replaced by *thiol*.

mercury L. *Mercurius*, in Roman mythology the messenger of the gods, noted for his quickness and agility.

microscope (1656) Gk. *micro-* (small) + *skopos* (observe).

molecule (1794) L. *moles* (mass) + diminutive suffix *-cule*, i.e., a little mass.

molybdenum (1816) Gk. *molybdos* (a soft, leadlike mineral). Scheele had used the term *molybdic acid* in 1778.

mordant L. *morsus* (bite), i.e., an agent that bites (binds) a dye to a fiber.

nano Gk. *nanos* – dwarf, *Lithos* – rock, and *grapho* – to write.

naphtha (1572) Transliteration of Gk. *naphtha*; a component of asphalt. Both naphtha and asphalt were used as illuminants in ancient times.

nascent (1790?) L. *nasc-* (arise, be born), referring to the active state of an element as it is liberated from a compound, e.g., nascent oxygen. Now obsolete.

neon (1898) Named by Sir William Ramsay from Gk. *neos* (new).

neutron (1932) L. *neuter* (neither). So named by its discoverer Chadwick because it has neither a positive nor a negative charge. Neutral has the same derivation (neither acidic nor basic).

nickel (1751) Swedish *koppernickel*, from German *Kupfer* (copper) + *nickel* (demon, devil). So called because the ore niccolite (NiAs), being reddish in color, resembled copper, and was thus supposed the trick of an evil spirit. See **cobalt**.

niobium (1644) Named by H. Rose after the lachrymose goddess Niobe, though his reason for doing so is not clear.

nitrogen (1790) L. *nitrum* (nitre) + *gen-* (create). The term was used by Chaptal to indicate that the element is a constituent of nitre. It had previously been recognized as an element by Lavoisier.

nucleus (1762) L. *nuc-* (nut, kernel).

olefin (1860) L. *ole-* (oil). Ethylene was originally called olefin gas because on combination with chlorine it formed an oily liquid, ethylene dichloride, also called Dutch oil. See **aliphatic**.

oleum L. *ole-* (oil). Fuming H_2SO_4, probably so called because of its oily appearance.

oligomer Gk. *oligo-* (a few) + *mer-* (part). Originally used in botany; adopted by chemists to describe polymers composed of fewer than five monomer units.

organic (1807) First used by Berzelius to designate substances derived from living organisms. The Greek word *organon* meant implement, instrument, machine, part of the body. The semantics of its transfer from inanimate matter to living entities is not clear.

oxygen (1790) Gk. *oxy-* (sharp, acidic) + L. *gen-* (create). Thus the literal but erroneous meaning is "acid maker" because of the belief of Lavoisier, who named it, that oxygen is the characteristic element of acids. In 1774, it had been discovered by Priestely, who called it dephlogisticated air.

ozocerite (1834) Gk. *ozo-* (bad odor) + *keros* (wax), i.e., a malodorous earth wax. From the latter root is **ceresin**.

ozone (1840) Gk. *ozo-* (bad odor), a characteristic property of this gas.

palladium (1803) Named by its discoverer, Wollaston, after the then recently discovered asteroid Pallas, which in turn was so named because of its pale-white color associated with the goddess Pallas.

paper (1670) Gk. and L. *papyrus*, a plant used by the ancients for writing. Thin strips were cut from the stem, soaked in water, and pressed together after drying.

paraffin (1830) L. *parum* (little) + L. *affinis* (relationship). The literal meaning is "having little affinity," in reference to the low reactivity of saturated hydrocarbons. Introduced by the German chemist Reichenbach.

periodic (1870) Gk. *peri-* (around) + *odos* (way, road), i.e., a circuit; any system having repeated beginnings and endings, e.g., the periodic table.

petroleum (1526) L. *petr-* (rock) + *ole-* (oil), from its early discovery in rock formations.

pH (1909) Abbreviation of *pouvoir hydrogene* (hydrogen power), the term used by Sorenson, who invented this scale. The word power is used in its mathematical sense.

phen- The root that characterizes derivatives of benzene. Gk. *phaino* (shine, show, appear). First used by Laurent (1841) in *hydrate de phenyle* and *acid phenique*, which were names for a coal tar product (later called phenol) from which illuminating gas could be made. Hence, this root was used to name all compounds derived from benzene, e.g., phenanthrene *phen* + *anthrac-* (coal); **phenol**; **phenyl**; **phenacetin**, etc. **Phenomenon** (i.e. "what appears") is from the same root.

phosphorus (1669) Gk. *Phos-* (light) + *phor-* (bear, carry). So named by H. Brand because of its property of glowing in the dark.

phthalic (1857) Coined from na*phthal*ene, from which phthalic anhydride is made.

pigment L. *pigmentum* (paint) from the root *ping-* (paint a picture), implying use of a colorant.

platinum (1803) Originally called *platina* from the Spanish *plata* (silver), because of its color.

plutonium (1941) Named after the former most remote planet, Pluto, because plutonium was the last element in the periodic system at the time of its discovery. See **uranium**.

polymer (1866) Gk. *poly-* (many) + *mer-* (part); similarly for monomer, dimer, etc.

polyunsaturated See **saturated**.

precipitation L. *prae-* (first) + *caput* (head). Literally, headfirst, in the sense of falling directly downward.

protein (1868) Gk. *proteios* (first, primary), in reference to its fundamental part in nutrition. **Proton** is from the same root.

racemic (1890) L. *racemus* (grapes). Refers back to Pasteur's discovery of optical isomers in the tartaric acid of wine lees.

radical (1560) L. *radic-* (root). First used in the chemical sense in 1816, but the connection is obscure.

reaction, reagent (1797) L. *re-* (back) + *act-, ag-* (do, drive). The same root is responsible for such common words as **action**, **activity**, **agent**, etc. See **coagulate**.

retort (1605) L. *re-* (back) + *tort-* (twist), referring to the shape of a type of simple distillation unit characterized by a glass neck twisted downward.

rhodium (1803) Gk. *rhodon* (rose), so named by Wollaston from the rosy color of its chloro salt. Rhodopsin, the red pigment of the eye, is from the same root.

rubber (1788) So named by Joseph Priestley, when he found that the crude product obtained by coagulation of latex was able to erase pencil marks by rubbing.

salt L. *sal*; present chemical meaning (acid–base reaction), 1790. Its original use cannot be dated accurately; it goes back at least to the alchemists and Chaucer ("salte teeres"), about 1400. Historically its economic value was unique. Roman soldiers accepted salt as payment for their services. Hence, the word *salary* and the expression "earn your salt."

saponification (1821) L. *sapo-* (soap). The term *soap* has the same derivation, with vowels transposed.

saturated L. *satis* (enough), i.e., an element with which enough constituents are united to fill (satisfy) its valence bonds. Thus *polyunsaturated* literally means "many times not enough" constituents (i.e., two or more double bonds). Similarly, any material that has absorbed all the liquid it can retain.

scan L. *scand-* (climb over), i.e., as a light beam moves over an irregular surface.

scintillation (1809) L. *scintilla* (spark).

scleroscope Gk. *sclero-* (hard) + *skopos* (measure, observe), i.e., a type of metal-hardness tester.

selenium (1818) Named by Berzelius from Gk. *selenon* (the moon), because its chemical similarity to tellurium (from L. *tellus*, the earth) suggested an analogical relationship between the two elements similar to that of the moon to the earth.

silicon (1823) L. *silic-* (flint).

silvicide L. *silva* (trees, forest) + *caed-* (kill), i.e., a special type of herbicide.

solder L. *solid-* (to make firm), descriptive of the metal-joining function of these alloys.

solvent L. *solv-* (free, loosen), i.e., a liquid that "frees" solids from their confined state. Related words from the same root are **solve**, **solution**, **dissolve**, **resolve**.

sorb, sorption L. *sorb-, sorp-* (suck in, swallow). **Absorption** describes the swallowing action of a porous material; **adsorption** means attraction to the surface only (at, *ad* = to, toward). **Sorb** is used when the distinction is not clear.

sparger (1839) L. *sparg-* (sprinkle). **Sparse** (thinly scattered) is from the same root.

ster-, stereo- Gk. *stereo* (solid), i.e., three-dimensional. This root occurs in many chemical terms, e.g., **sterol** (solid alcohol), **stearic**, **steroid**, **steric**, **stereospecific**, **cholesterol** (hard bile alcohol).

synthesis (1611) Gk. *syn-* (together) + *tithenai* (place). The chemical meaning did not appear until 1733.

tacticity Gk. *tax-* (arrange), referring to the arrangement of atoms in the molecular structure of high polymers. This and related terms (**syndiotactic**, **atactic**, etc.) were introduced in the 1930s when stereospecific catalysts were developed by Ziegler and Natta.

tall Derived from the Swedish word for pine, to distinguish tall oil from the U.S. meaning of **pine oil**.

tall oil See **tall**.

tantalus (1802) For the mythological punishment of Tantalus for revealing the secrets of the gods. He was condemned to stand in water, which always receded as he tried to drink. Isolation of the element was apparently attended by equally frustrating difficulties.

tellurium See **selenium**.

tension, tensile L. *ten-* (hold, as against strain). From the same root as *retension* (holding back) and *tenter* (a machine for holding fabric taut).

tetrahedral Gk. *tetra* (four) + *hedron* (side), i.e., a pyramidal structure characteristic of certain molecules.

thallium (1861) L. *thallus* (budding leaf). Named by Sir William Crookes because of a pronounced green line in its spectrum.

thiol (1899) Gk. *thio-* (sulfur) + *-ol*, a sulfur alcohol in which SH has replaced the OH group. From the same root as **thiamine**, **dithiocarbamate**, etc.

tincture (1648) L. *tinct-* (color, dye) The term is obsolete in this sense; the connection with alcoholic solutions is not clear, though possibly some were colored. From the same root are **tinctorial** and **tint**.

titanium (1800) So named by Klaproth after the Titans, legendary giants supposed to have originally inhabited the earth. His reason for naming this substance thus is not clear.

transfer, transference L. *trans-* (across) + *fer-* (carry, bear). There is no connection between this root and the L. *ferrum* (iron) from which **ferric** is derived.

transition L. *trans-* (across) + *i-* (go, move), e.g., transition elements that "move across" the periodic table, making a bridge from electropositive to electronegative.

transmutation L. *trans-* (across) + *mut-* (change), i.e., changing one element into another. In the Middle Ages, alchemists tried to change lead into gold with little success.

tungsten (1781) Swedish *tung* (heavy) + *sten* (stone); named in honor of Swedish chemist Scheele.

turpentine Gk. *terebenthos* (turpentine tree). **Terpene** is undoubtedly related.

uranium (1789) Named by Klaproth after the then recently discovered planet Uranus, believed to be the most distant member of the solar system. This planet in turn was named from the Greek word for sky (heavens). See **plutonium**.

valence (1869) L. *val-* (be strong), referring to the binding strength of interatomic attraction. From the same root as **valid**.

vanadium (1830) For the Scandinavian goddess Vanadis, because of its multicolored compounds. So named by its discoverer, Sefstrom.

vermiculite (1824) L. *vermis* (worm), so called because of the worm-like writhing motion of the material as it expands on exposure to high temperature.

vitamin (1921) L. *vita* (life) + amin(e). Originally named vitamine by Funk (1911); the terminal *-e* was later dropped when the nonnitrogenous carotene (vitamin A) was discovered.

vitriol L. *vitrum* (glass). An old name for concentrated H_2SO_4, because of its glassy appearance. **Vitrification** (conversion to a glassy state) is from the same root.

vulcanization (1846) Named by Charles Goodyear after Vulcan, the Roman god of fire and forging, to describe Goodyear process for heat curing of rubber with sulfur.

xenon (1898) Gk. *xenos* (strange); so named by Sir William Ramsay.

X-ray (1896) Named by Roentgen, the discoverer of this form of radiation, because its nature was at that time unknown.

xylene Gk. *hyle*, a suffix meaning stuff or material, as in **ethyl** = ether material (Liebig).

zein L. *zea* (grain). The protein occurring in corn.

zymase See **enzyme**.

Highlights in the History of Chemistry

A. Chronology of Notable Achievements

Democritus (460–370 B.C.). First to conceive matter in the form of particles, which he called atoms.

Alchemists (about 1000–1650). Attempted to (1) change lead and other base metals to gold; (2) discover a universal solvent; and (3) discover a life-prolonging elixir. Used plant products and arsenic compounds to treat diseases.

Torricelli, Evangelista (1608–1647). Developed the first true barometer.

Pascal, Blaise (1623–1662). Repeated Torricelli's experiments and confirmed that the height of mercury barometer decreased with increasing altitude, creating the foundation for meteorology.

Boyle, Sir Robert (1627–1691). Formulated fundamental gas laws. First, to conceive the possibility of small particles combining to form molecules; distinguished between compounds and mixtures; studied air and water pressures, desalination, crystals, and electrical phenomena.

Brand, Henning (1630–1692). An alchemist that accidentally discovered phosphorous while attempting to condense metal fumes into gold. He was the first to isolate an element. He kept the discovery a secret for some time, later selling the method.

von Guericke, Otto (1645). Constructed the first vacuum pump.

Leibniz, Gottfried (1646–1716). Proposed a precursor to the modern concept of energy, which he called the "living force." Developed differential and integral calculus.

Fahrenheit, Daniel (1686–1736). Created the temperature scale that bears his name and identified the freezing and boiling points of water in 1724.

Celsius, Anders (1701–1714). Created a temperature scale that set the freezing point of water at 0 degrees and the boiling point at 100 to simplify meteorological measurements.

Bradley, James (1728). Used aberration of starlight to determine the speed of light to within 5% accuracy.

Cavendish, Henry (1731–1810). Accurately calculated the ratio of hydrogen to oxygen in water (2.02:1) and calculated the mean density of Earth to within one standard deviation of today's accepted value.

Priestley, Joseph (1733–1804). Discovered oxygen, carbon monoxide, and nitrous oxide. Proposed the electrical inverse-square law (1767).

Scheele, Carl Wilhelm (1742–1786). Discovered chlorine, tartaric acid, sensitivity of silver compounds to light (photochemistry), and oxidation of metals.

Leblanc, Nicolas (1742–1806). Invented a process for making soda ash from sodium sulfate, limestone and coal.

Lavoisier, Antoine-Laurent de (1743–1794). Discovered nitrogen, studied acids, and described the composition of many organic compounds. Generally regarded as the Father of Chemistry.

Volta, Alessandro Giuseppe Antonio Anastasio (1745–1827). Invented the electric battery, a series of "piles" or stacks of alternating layers of silver and zinc, or copper and zinc, separated by paper soaked in brine (electrolyte). See activity (1).

Berthollet, Claude Louis (1748–1822). Corrected Lavoiser's theory of acids; discovered bleaching power of chlorine; studied combining weights of atoms (stoichiometry).

Jenner, Edward (1749–1823). Discoverer of vaccination for prevention of smallpox (1776).

Franklin, Benjamin (1752). Demonstrated that lightning is electricity.

Dalton, John (1766–1844). The first great chemical theorist; proposed atomic theory (1807); stated the law of partial pressure of gases. His ideas led to the

Hawley's Condensed Chemical Dictionary, Sixteenth Edition. Michael D. Larrañaga, Richard J. Lewis, Sr., and Robert A. Lewis.
© 2016 John Wiley & Sons, Inc. Published 2016 by John Wiley & Sons, Inc.

laws of multiple proportions, constant composition, and conservation of mass.

Avogadro di Quaregna e di Cerreto, Lorenzo Romano Amedeo Carlo (1776–1856). Proposed principle that equal volumes of gases contain the same number of molecules. The number (6.02 × 10^{23} for 22.41 liters of any gas) is a fundamental constant that applies to all chemical units.

Davy, Sir Humphry (1778–1829). Laid foundation of electrochemistry; studied electrolysis of salts in water and other electrochemical phenomena; isolated Na and K.

Gay-Lussac, Joseph Louis (1778–1850). Discovered boron and iodine; studied acids and bases and discovered indicators (litmus); improved production method for H_2SO_4; did basic research on behavior of gases versus temperature and on the rations of gas volumes in chemical reactions.

Berzelius, Jöns Jacob (1779–1848). Classified minerals chemically; discovered and isolated many elements (Se, Th, Si, Ti, Zr); coined the terms isomer and catalyst; noted existence of radicals; anticipated discovery of colloids.

Faraday, Michael (1791–1867). Extended Davy's work in electrochemistry; he developed theories of electrical and mechanical energy, electrolysis, corrosion, batteries, and electrometallurgy.

Coulomb, Charles (1795). Introduced the inverse-square law of electrostatics.

Count Rumford (1798). Thought that heat was a form of energy.

Goodyear, Charles (1800–1860). Discovered vulcanization of rubber (1844) by sulfur, inorganic accelerator, and heat. Hancock in England made a parallel discovery.

Wöhler, Friedrich (1800–1882). First to synthesize an organic compound—a rearrangement reaction (urea, 1828). This discovery was the beginning of synthetic organic chemistry.

Young, Thomas (1801). Demonstrated the wave nature of light and the principle of interference.

Liebig, Justus Freiherr Von (1803–1873). Fundamental investigator of plant life (photosynthesis) and soil chemistry; first to propose use of fertilizers. Discovered chloroform and cyanogen compounds.

Graham, Thomas (1805–1869). Studied diffusion of solutions through membranes; established principles of colloid chemistry.

Oersted, Hans (1820). Observed that a current in a wire can deflect a compass needle; provided first concrete evidence of the connection between electricity and magnetism.

Pasteur, Louis (1822–1895). (1) First to recognize infective bacteria as disease-causing agents; (2) developed concept of immunochemistry; (3) initiated heat sterilization of wine and milk (pasteurization); (4) observed optical isomers (enantiomers) in tartaric acid.

Sturgeon, William (1823). Invented the electromagnet.

Carnot, Sadi (1824). Analyzed heat engines.

Ohm, Simon (1826). Stated law of electrical resistance.

Brown, Robert (1827). Discovered Brownian motion.

Lister, Joseph (1827–1912). Initiated use of antiseptics in surgery, e.g., phenols, carbolic acid, and cresols.

Kekulé, Friedrich August (1829–1896). Laid foundations of aromatic chemistry; conceived of four-valent carbon and structure of benzene ring; predicted isomeric substitutions (ortho-, meta-, para-). Considered the Father of Aromatic Chemistry.

Nobel, Alfred (1833–1896). Invented dynamite, smokeless powder, and blasting gelatin. Established international awards for achievements in chemistry, physics, and medicine (Nobel Prize).

Mendeleev, Dmitri Ivanovich (1834–1907). Discovered periodicity of the elements and compiled the first periodic table with elements arranged into seven groups (1869).

Hyatt, John Wesley (1837–1920). Initiated plastics industry (1869) by invention of celluloid (nitrocellulose modified with camphor).

Beilstein, Friedrich Konrad (1838–1906). Compiled *Handbuch der Organischen Chemie*, a multivolume compendium of properties and reactions of organic chemicals.

Lord Kelvin (1838). Described the absolute zero point of temperature.

Perkin, Sir William Henry (1838–1907). Synthesized first organic dye (mauveine, 1856) and first synthetic perfume (coumarin). His work on dyes was continued and expanded by Hofmann in Germany.

Gibbs, Josiah Willard (1839–1903). Stated three principle laws of thermodynamics; expounded on the nature of entropy and phase rule and the relation between chemical, electric, and thermal energy.

Chardonnet, Hilaire Bernigaud, count de (1839–1924). First to produce a synthetic fiber (nitrocellulose) with properties similar to rayon.

Joule, James (1843–1849). Experimentally demonstrated that heat is a form of energy.

Boltzmann, Ludwig Eduard (1844–1906). Developed kinetic theory of gases; their viscosity and diffusion properties are summarized in Boltzmann's law.

Röntgen, Wilhelm Conrad (1845–1923). Discovered X-radiation (1895); awarded the Nobel Prize in 1901.

Le Chatelier, Henry-Louis (1850–1936). Fundamental research on equilibrium reactions (Le Chatelier's Law), combustion of gases, and metallurgy of iron and steel.

Moissan, Ferdinand Frederick Henri (1852–1907). Developed electric furnace for making carbides and preparing pure metals; isolated fluorine (1886); was awarded the Nobel Prize in 1906.

Becquerel, Antoine-Henri (1852–1908). Discovered radioactivity, deflection of electrons by magnetic fields, and gamma radiation; awarded the Nobel Prize in 1903 with the Curies. While studying effects of X-rays on photographic film in 1896, he discovered that some chemicals spontaneously decompose and emit very penetrating rays.

Fischer, Hermann Emil Louis (1852–1919). Conducted basic research on sugars, purines, uric acid, enzymes, nitric acid, and ammonia; pioneered work in stereochemistry; awarded the Nobel Prize in 1902.

Thomson, Sir Joseph John (1856–1940). Conducted research on cathode rays; resulting in proof of existence of electrons (1886); was awarded the Nobel Prize in 1906. Used a cathode ray tube to experimentally determine the charge to mass ratio of an electron. Found that "canal rays" were associated with the proton H+.

Arrhenius, Svante August (1859–1927). Conducted fundamental research on rates of reaction versus temperature, expressed by the Arrhenius equation, and on electrolytic dissociation; was awarded the Nobel Prize in 1903.

Maxwell, James Clerk (1859). Described the mathematical distribution of the velocities of molecules of a gas.

Plucker, Julius (1859). Built one of the first gas discharge tubes (cathode ray tubes).

Hall, Charles Martin (1863–1914). Invented a method of aluminum manufacture by electrochemical reduction of alumina. Parallel discovery by Héroult in France.

Baekeland, Leo Hendrik (1863–1944). Invented phenolformaldehyde plastic (1907), the first completely synthetic resin (Bakelite).

Nernst, Walther Hermann (1864–1941). Awarded the Nobel Prize in 1920 for his work in thermochemistry; did basic research in electrochemistry and thermodynamics.

Werner, Alfred (1866–1919). Introduced concept of coordination theory of valence (complex chemistry); was awarded the Nobel Prize in 1913.

Curie, Marie (1867–1934). Discovered and isolated radium; research on radioactivity of uranium; was awarded the Nobel Prize in 1903 (with Becquerel) in physics; in 1911 in chemistry.

Ipatieff, Vladimir Nikolayevich (1867–1952). Basic research and development of catalytic alkylation and isomerization of hydrocarbons (with Herman Pines).

Haber, Fritz (1868–1934). Synthesized ammonia from nitrogen and hydrogen, the first industrial fixation of atmospheric nitrogen (the process was further developed by Bosch); was awarded the Nobel Prize in 1918.

Rutherford, Sir Ernest (1871–1937). First to prove radioactive decay of heavy elements and to carry out a transmutation reaction (1919). Discovered half-life of radioactive elements and he was awarded the Nobel Prize in 1908.

Maxwell, James Clerk (1873). Proposed that electric and magnetic fields filled space.

Lord Kelvin (1874). Stated the second law of thermodynamics.

Stoney, George Johnstone (1874). Proposed that electricity consisted of discrete negative particles that he named "electrons."

Lewis, Gilbert Newton (1875–1946). Proposed electron-pair theory of acids and bases; authority on thermodynamics.

Aston, Francis William (1877–1945). Pioneer work on isotopes and their existence and separation by mass spectrography; was awarded the Nobel Prize in 1922.

Crookes, Sir William (1879). Discovered that cathode rays travel in straight lines, impart a negative charge, are deflected by electric and magnetic fields (indicating negative charge), cause glass to fluoresce, and cause pinwheels in their path to spin (indicating mass).

Fischer, Hans (1881–1945). Basic research on porphyrins, chlorophyll, carotene; synthesized hemin; was awarded the Nobel Prize in 1922.

Fleming, Sir Alexander (1881–1955). Discovered penicillin (1928); initiated antibiotics; was awarded the Nobel Prize in 1945. Selman A. Waksman developed the science in the U.S.

Langmuir, Irving (1881–1957). Fundamental research on surface chemistry, monomolecular films, emulsion chemistry, electric discharges in gases, cloud seeding, etc. Langmuir was awarded the Nobel Prize in 1932.

Staudinger, Hermann (1881–1965). Fundamental research on high-polymer structure, catalytic synthesis, polymerization mechanisms, resulting eventually in development of stereospecific catalysts by Ziegler and Natta (stereospecific catalysts by Ziegler and Natta (stereoregular polymers); was awarded the Nobel Prize in 1963.

Goldstein, Eugen (1886). Used cathode ray tube to study "canal rays," which possessed electrical and magnetic properties opposite those an electron.

Hertz, Heinrich (1887). Discovered the photoelectric effect.

Moseley, Henry Gwyn Jeffreys (1887–1915). Discovered the relation between frequency of X-rays emitted by an element and its atomic number, thus indicating the element's true position in the periodic table.

Midgley, Thomas Jr (1889–1944). Discovered tetraethyl-lead and antiknock treatment for gasoline (1921) and fluorocarbon refrigerants; early research on synthetic rubber.

Adams, Roger (1889–1971). Noted educator and contributor to industrial research in catalysis and structural analysis; was awarded the Priestley Medal.

Banting, Sir Frederick Grant (1891–1941). Isolated the insulin molecule; was awarded the Nobel Prize in 1923.

Chadwick, Sir James (1891–1974). Discovered the neutron (1932) and was awarded the Nobel Prize in 1935.

Urey, Harold Clayton (1893–1981). Discovered a heavy isotope of hydrogen (deuterium) and was awarded the Nobel Prize in 1934. A leader of the Manhattan Project. Made original contributions to theories of the origin of the universe and of life processes.

Roentgen, Wilhelm (1895). Discovered that certain chemicals near a cathode ray tube glowed. Found highly penetrating rays that were not deflected by a magnetic field, which he named "X-rays."

Carothers, Wallace Hume (1896–1937). Polymerization research resulting in synthesis of neoprene (polychloroprene) and of nylon (polyamide).

Kistiakowsky, George Bogdanovich (1900–1982). Developed the detonating device used in the first atomic bomb.

Plank, Max (1900). Stated radiation law and Planck's constant.

Soddy, Frederick (1900). Observed spontaneous disintegration of radioactive elements into "isotopes" or new elements, described "half-life," and made calculations of the energy of decay.

Fermi, Enrico (1901–1954). First to achieve a controlled nuclear fission reaction (1939); basic research on subatomic particles. Formulated a theory on Beta decay (1934). Enrico was awarded the Nobel Prize in 1938.

Lawrence, Ernest Orlando (1901–1958). Invented the cyclotron in which the first synthetic elements were created and was awarded the Nobel Prize in 1939.

Heisenberg, Werner Karl (1901–1976). Researched in quantum mechanics resulting in development of the orbital theory of chemical bonding; stated uncertainty principle and was awarded the Nobel Prize in 1932.

Nagaoka, Hantaro (1903). Postulated a "Saturnian" atom model with flat rings of electrons revolving about a positively charged particle.

Abegg, Richard (1904). Discovered that inert gases have a stable electron configuration (8 electrons in the outermost shell), which results in their chemical inactivity.

Geiger, Hans (1906). Developed an electrical device which made an audible "click" when hit with alpha particles.

Libby, Willard Frank (1908–1980). Developed a radiocarbon dating technique based on carbon-14 and was awarded the Nobel Prize in 1960.

Rutherford, Ernest and Royds, Thomas (1909). Demonstrated that alpha particles are doubly ionized helium atoms.

Bohr, Niels (1913). Devised quantum model of the atom in which atoms had orbital shells of electrons.

Milliken, Robert (1913). Experimentally determined the charge and mass of an electron using an oil drop.

Crick, Francis Harry Compton (1916–2004). Elucidated the structure of the DNA molecule (1953) with Watson, J. D., resulting in the development of gene-splicing (recombinant DNA) techniques.

Woodward, Robert Burns (1917–1979). Was awarded the Nobel Prize in 1965 for his brilliant syntheses of such compounds as cholesterol, quinine, chlorophyll, and cobalamin. Woodward and Roald Hoffmann put forth the Woodward–Hoffmann rules to explain stereochemistry in chemical reactions.

Brønsted, Johannes N., Lowry, Thomas N., and Lewis, Gilbert Newton (1923). Produced refined definitions of acids and bases.

Chemical & Engineering News (1923). Begun as the News Edition of Industrial & Engineering Chemistry.

de Broglie, Louis (1923). Described the particle/wave duality of electrons.

IG Farben Chemical Conglomerate (1925). Six German chemical firms merge to form the world's largest chemical company. The allies broke up IG Farben after World War II, creating Bayer, Hoechst, and BASF.

Pauli, Wolfgang E., Heisenberg, Warner, and Schrodinger, Erwin (1925–1927). Work from these scientists ushers in the field of quantum mechanics.

Tacke, Ida, Noddack, Walter, and Berg, Otto (1925). Preliminary discovered element 43 (now named Technetium), which was earlier predicted to exist, by Russian chemist Dmitri Mendeleev, creator of the periodic table. In 1937, Carlo Perrier and Emilio Segre artificially produced Technetium, and the pair has historically been attributed to discovering Technetium. However, a 1999 study by Dave Curtis and colleagues at Los Alamos National Laboratory reproduced Taccke, Noddack, and Berg's original research experiment and obtained similar results, confirming that the 1925 research likely discovered Technetium.

Imperial Chemical Industries (1926). Four U.K. chemical companies merge to produce the British Empire's largest chemical company. Imperial Chemical Industries patented the process for producing polyethylene (1935), one of today's most important and commonly used plastics.

Dirac, Paul (1930). Proposed anti-particles and discovered the anti-electron (positron) in 1932. (Segre and Chamberlain detected the anti-proton in 1955).

IG Farben–Bayer AG (1930s). Sulfanomides (sulfa drugs) were the first antibiotics systematically used and commercialized.

Anderson, Carl (1932). Discovered the positron.

King, Charles G. and Szent-Györgyi, Albert (1932). Each independently became the first to isolate Vitamin C.

Pauli, Wolfgang (1933). Proposed the existence of neutrinos as a means of accounting for what was considered a violation of the law of conservation of energy in some nuclear reactions.

Berchet, Gerard and Carothers, Wallace H (1935). While working at DuPont, Carothers and Berchet synthesized nylon.

Houdry, Eugene J (1935). Develops industrial-scale petroleum catalytic cracking, leading to the development of the modern crude oil refinery.

Food, Drug & Cosmetic Act (1938). Mass poisonings from an improperly prepared sulfa drug (Elixir Sulfanilamide) killed over 100 people in 15 states in the U.S., prompting legislation for additional safety regulation of foods, drugs, and cosmetics.

Hahn, Otto, Meitner, Lise, and Strassmann, Fritz (1938). Recognized that heavy elements capture electrons to form fissionable products in a process that ejects neutrons, enabling a chain reaction. They were the first to recognize that the uranium atom split when bombarded by neutrons. News of the splitting of the atom was brought by Niels Bohr to scientists in the U.S., resulting in the Manhattan Project. Hahn, Meitner, and Strassmann were not engaged in nuclear weapons research during World War II. Hahn was awarded the Nobel Prize in 1944.

Plunkett, Roy J (1938). Plunkett, of DuPont, accidentally discovered Teflon after having trouble with a gas cylinder, which emitted a white powder that Plunkett characterized and later became known as Teflon, which had wide industrial and consumer use.

Muller, Paul Hermann (1939). Muller discovered that dichlorodiphenyltrichloroethane (DDT) is an effective insecticide.

Pauling, Linus C (1939). Pauling published an instant classic text titled "The Nature of the Chemical

Bond." Proposed the α–helix structure for proteins in 1951.

Seaborg, Glenn (1941–1951). Synthesized several transuranium elements and suggested a revision to the layout of the periodic table.

Bloch, Felix and Purcell, Edward M (1945). Each independently discovered nuclear magnetic resonance (NMR).

Cremer, Erika and Prior, Fritz (1947). Produced the first gas chromatograph.

Dow Chemicals (1949). Dow introduced Saran Wrap, a thin and clingy clear plastic film for preserving foods.

National Science Foundation (1949). The United States Congress established the National Science Foundation.

Kendrew, John C. and Perutz, Max F (1958). Kendrew and Perutz determined myoglobin's structure with X-ray crystallography, producing the first high-resolution protein structure.

Feynmnan, Richard (1959). Proposed a method to manipulate individual atoms and molecules. His work is considered the beginning of the field of nanotechnology.

Djerassi, Carl and Zaffaroni, Alejandro (1960). The birth control is approved for use in the U.S. Djerassi and Zaffaroni developed the birth control pill in the 1950s from yam extracts in Mexico.

Maiman, Theodore H (1960). Demonstrated the first operational laser.

Jevons, M. Patricia (1961). Identified strains of *Staphylococcus aureus* resistant to the antibiotic methicillin.

Thalidomide (1961). Countries around the world begun to remove Thalidomide from the market because it causes birth defects.

Kwolek, Stephanie L (1965). Kwolek, while working at DuPont, invented the thin, light, and strong polymer Kevlar, which has seen widespread commercial and consumer use.

Anderson, Weston A. and Ernst, Richard R (1966). Developed Fourier transform nuclear magnetic resonance (NMR), paving the way for the development of NMR spectroscopy.

Cotzias, George C (1967). Identified relief of Parkinson's disease symptoms using the psychoactive drug levodopa.

Wall, Monroe E. and Wani, Mansukh C (1967). Isolated the cancer drug paclitaxel, which later became Taxol, from the bark of a Pacific yew tree.

Berg, Paul (1970). Conducts the first successful recombinant DNA experiment by assembling DNA molecules from different organisms.

Clean Air Act (1970). The United States Congress passed the Clean Air act, paving the way for additional environmental regulatory laws in the 1970s. The Clean Air Act Amendments of 1990 amended the Act, made air pollution standards significantly tougher and created the United States Chemical Safety and Hazard Investigation Board (CSB). The CSB is an independent safety board, which investigates accidents resulting from the production, processing, handling or storage of chemical substances causing death, serious injury, or substantial property damage (including damage to natural resources).

Environmental Protection Agency (1970). President Richard Nixon created the United States Environmental Protection Agency in 1970.

Taniguchi, Norio (1974). Coined the term nanotechnology to describe semi-conductor processes such as thin film deposition and ion beam milling.

Ames, Bruce N. and Blum, Arlene (1977–1978). Showed two flame-retardants in children's pajamas to have mutagenic properties.

Love Canal (1978). Leaking chemical waste from a former industrial dump forces the evacuation of the Love Canal, N.Y. neighborhood.

Hull, Charles W (1983). Invented the 3D printer and started the field of commercial rapid prototyping using 3D printing systems. 3D printers have gained widespread use and have printed biological materials, ceramics, metals, plastics, composites, and other materials to make a variety of products.

Kroto, Harry, Smalley, Richard, and Curl, Robert (1985). Discovered Carbon-60, a form of carbon nanoscale material that is shaped like a soccer ball. The scientists named the discovery buckminsterfullerene, which was later shortened in popular culture to buckyballs. The three were awarded the Nobel Prize in 1996.

Mullis, Kary B (1985). Succeeded with the first polymerase chain reaction.

Drexler, K. Eric (1986). Building on Richard Feynman's work, Drexler proposed the idea of a nanoscale assembler which would build a copy of itself and other items. Drexler is considered the father of molecular nanotechnology, also called molecular manufacturing.

Eigler, Don (1989). The first to manipulate atoms using a scanning tunneling microscope.

Human Genome Project (1989). The U.S. launched the Human Genome Project, an international scientific research effort to determine the sequence of chemical base pairs which make up human DNA, and to identify and map all of the genes of the human genome from both a physical and functional standpoint. In 2001, the Human Genome Project and Celera Genomics independently report the complete sequencing of the human genome. The project was successful and culminated in 2003.

Iijima, Sumio (1991). Discovered multi-walled carbon nanotube, and is widely considered the inventor of carbon nanotubes.

Dolly, a sheep (1996). Dolly is the first successfully cloned animal.

Leaded Gasoline (1996). The United States Environmental Protection Agency completes the phase-out of leaded gasoline.

Bisphenol A (1997). Low doses of bisphenol A are determined to be toxic to laboratory animals, prompting controversy over the use of bisphenol A in plastics, especially children's products.

Anastas, Paul and Warner, John C (1998). Published the 12 Principles of Green Chemistry.

Campbell, Sir Simon (1998). Campbell invented sildenafil, marketed as Viagra, which was approved by the United States Food and Drug Administration to treat impotence.

Fire, Andrew Z. and Craig, C. Mello (1998). Demonstrated that small RNA molecules inhibit expression in the worm *C. elegans*.

Ribosomal Structure (2000). Three research groups identified the atomic resolution structures of the ribosome independently.

Cipla Pharmaceuticals (2001). Cipla supplies AIDS drugs to Africa for a small fraction of market prices.

Gleevec (2001). The United States Food and Drug Administration approves Gleevec (imatinib) for treating leukemia caused by a specific gene defect.

Humira (2002). The United States Food and Drug Administration approved Humira (adalimumab), the first fully human monoclonal antibody drug to treat rheumatoid arthritis.

Embryonic Stem Cell Therapy (2009). The United States Food and Drug Administration approved the first human clinical trial of embryonic stem-cell therapy.

IBM (2009). IBM visualized all of the atom position and bonds of a single molecule for the first with improvements in atomic force microscopy.

Kobilka, Brian K. and Sunahara, Roger (2011). Determined the structure of a G protein-coupled receptor with its G protein partner.

Higgs boson (2012). An international research team discovered a particle, the Higgs boson, a particle believed to exist and give other particles their mass.

McAlpine, Michael (2013). McAlpine's research team printed a bionic ear into a 3D shape with embedded electronics, which is believe to be the first printed device to integrate biological tissue with electronic circuitry.

Cairns, Elton (2014). Cairns and colleagues at Lawrence Berkeley National Laboratory developed a lithium–sulfur battery that can hold more than double the energy of ordinary lithium-ion batteries.

Wang, Lai-Sheng (2014). Observed the first experimental evidence of borospherene, a box-like boron fullerene (B-40).

B. American Chemical Society (ACS)
www.acs.org

(The following is based in part on an article prepared by Alden H. Emery, then secretary of the ACS, for the first edition of *Encyclopedia of Chemistry*.)

At a meeting held a Priestley's old home in Northumberland, Pennsylvania, in 1874 to celebrate the centennial of the discovery of oxygen, plans were laid to organize an American Chemical Society. The first official meeting was held two years later in New York City. It was attended by 35 chemists, out of a membership list of 133. Contrasted with these figures are the current attendance of about 15,000 at national meetings, from a membership of more than 161,000.

The society was originally incorporated in New York State; a national character was granted by Congress as of January 1, 1938. The objectives of the Society are:

"to encourage in the broadest and most liberal manner the advancement of chemistry in all its branches; the promotion of research in chemical science and industry; the improvement of the qualifications and usefulness of chemists . . . ; the increase and diffusion of chemical knowledge; and to promote scientific interests and inquiry; thereby fostering public welfare and education, aiding the development of our country's industries, and adding to the material prosperity and happiness of our people."

These objectives are pursued by several means: (1) Meetings at the national, regional, and local levels, at which papers devoted to advanced chemical theory and practice are presented; several hundred of these are delivered at a single national meeting. (2) Publication of over 60 journals in major areas of chemistry, including *Chemical Abstracts*, *Journal of the American Chemical Society*, *Industrial and Engineering Chemistry*, *Chemical Reviews*, and *Chemical and Engineering News*. The society also publishes reports of chemical symposia, Advances in Chemistry, and numerous chemical treatises, including the Monograph Series. (3) Strong and positive interest in chemical education, evidenced by a series of intensive short courses, internet broadcasts on special topics, recorded symposia availability, and the development of educational resources for K-12 education. (4) Maintenance of the general well-being of its membership by conservative management and the highest ethical and professional standards.

The Priestley Medal, considered the highest award in chemistry, was established by the ACS in 1923. The Society also administers or sponsors many other awards.

C. Chemical Abstracts Service (CAS)
http://www.cas.org

CAS originated in the publication *Chemical Abstracts*[TM] (*CA*), a journal of the American Chemical Society first published in 1907. The purpose of *CA* was to help scientists benefit from the published work of their colleagues around the world by monitoring, abstracting, and indexing the world's chemistry-related literature. Over the years, CAS has evolved into the world's authority for chemical information, providing the global scientific community with access to the most current chemical and related scientific information available immediately through databases such as CAS REGISTRY[SM] and CAplus[SM]. CA originally came into being primarily because U.S. chemists felt that European abstracting journals were neglecting U.S. chemical research. From the beginning, the American Chemical Society charged CA with abstracting the complete world's literature of chemistry. The exponential growth of chemical research and publication in the ensuing years led to a parallel growth in CA and the CAS. The first year's issues of CA contained just under 12,000 abstracts. It now abstracts, indexes, or cites more than 1.5 million scientific papers and patents annually. The 75 millionth substance was added to the CAS REGISTRY in 2013.

The early issues were edited by William A. Noyes, Sr., chairman of the ACS committee on papers and publications. Noyes was assisted by two part-time editors, a secretary, and 129 unpaid volunteer abstractors. Currently, staff members review and intellectually analyze published research from more than 10,000 scientific journals and patent documents from 63 patent authorities.

Document analysts abstract papers and patents published in more than 50 languages, and staff thoroughly index each paper or patent using both key words and phrases from titles and abstracts of the document as well as highly controlled and structured index entries to subjects covered and chemical substances mentioned. CA has grown in influence and become international in terms of both the information it covers and the audience it serves, as nearly two-thirds of its circulation is abroad.

Noyes was succeeded as editor of CA in 1908 by Austin M. Patterson. Patterson retired in 1914 but made a major contribution after his retirement. When it was decided to publish a 10-year collective index to CA in 1916, it became evident that some systematic means of naming and indexing chemical substances was necessary. Patterson and Carleton E. Curran devised a naming scheme that had a profound effect on chemical nomenclature in general.

In 1915, E. J. Crane took over as editor and remained at the helm for 43 years, becoming the first director of Chemical Abstracts Service when the CA editorial organization was renamed and made a division of ACS in 1956. Crane made CA the model and pacesetter for all scientific abstracting and indexing services. He was succeeded as director of CAS in 1958 by Dale B. Baker, who led it through a difficult transition to financial self-sufficiency, and a major growth in staff, facilities, and technology, by moving CAS into the computer age. An extensive, in-depth systems approach to the storage and handling of chemical information has enabled CAS to assemble a vast computerized body of knowledge, typified by its Chemical Compound Registry.

Today, CAS offers a CAS REGISTRY database containing more than 75 million patent and journal article references in all areas of organic, macromolecular, applied, physical, inorganic, and analytical chemistry and biochemistry. CAS also offers several products and services including SciFinder, a leading research tool for chemists and researchers that offers instant access to the most comprehensive collection of chemical substance and reaction information including the full CAS Registry; STN, an online database service that provides global access to published research, journal literature, patents, structures, sequences, properties, and other data; Science IP, a CAS search service staffed by experienced researchers who provide high-quality information retrieval in many areas of science as well as comprehensive patent searches; and Common Chemistry, a free web-based resource offered to the public containing approximately 7900 chemicals of widespread and general interest as well as all 118 elements form the periodic table.

D. The Chemical Heritage Foundation (CHF) https://www.chemheritage.org

The Center for the History of Chemistry was launched in 1982 by the University of Pennsylvania and the

American Chemical Society. By 1987 the center was incorporated as a nonprofit organization called the National Foundation for the History of Chemistry and was renamed the Chemical Heritage Foundation in 1992. The purpose of CHF is to collect, preserve, and exhibit historical artifacts; engage communities of scientists and engineers; and tell the stories of the people behind breakthroughs and innovations. It is located in the heart of Independence National Historical Park in Philadelphia's Old City.

Manufacturers of Trademarked Products (Alphabetical List)

20/20 GeneSystems Inc.
9430 Key West Avenue
Rockville, Maryland 20850
www.2020gene.com

3M Company
3M Corporate Headquarters
3M Center
St. Paul, Minnesota 55144
www.solutions.3m.com

Abbott Laboratories
100 Abbott Park Road
Abbott Park, Illinois 60064
www.abbott.com

AbbVie Biotherapeutics Inc.
A Subsidiary of AbbVie Inc.
1500 Seaport Boulevard
Redwood City, California 94063
www.abbvie.com

Accurate Chemical & Scientific
  Corporation
300 Shames Drive
Westbury, New York 11590
www.accuratechemical.com

Active Organics
1097 Yates Street
Lewisville, Texas 75057
www.activeorganics.com

Adfast Corporation
2750 Paulus
Saint-Laurent, QC H4S 1G1
Canada
www.adfastcorp.com

Agilent Technologies, Inc.
5301 Stevens Creek Boulevard
Santa Clara, California 95051
www.home.agilent.com

Agrisel USA, Inc.
P.O. Box 3528
Suwanee, Georgia 30024
www.agrisel.com

Ahlstrom Corporation
Alvar Allon katu 3 C
P.O. Box 329
00101 Helsinki
Finland
www.ahlstrom.com

AI Chem & CY US AcquiCo., Inc.
1 Heilman Avenue
Willow Island, West Virginia
  26134

Ajinomoto Company
15-1, Kyobashi 1-Chrome
Chuo-ku, Tokyo 104-8315
Japan
www.ajinomoto.com

Akros Chemicals Limited
500 Jersey Avenue
New Brunswick, New Jersey
  08901
www.ackros.com

Akzo Nobel Corporation
Strawinskylaan 2555
1077 ZZ Amsterdam
Netherlands
www.akzonobel.com

Albaugh Inc.
4900 Packers Avenue
St. Joseph, Missouri 64504
www.albaughinc.com

Albemarle Corporation
451 Florida Street
Baton Rouge, Louisiana 70801
www.albemarle.com

Albis Plastic GmbH
Muhlenhagen 35
20539 Hamburg
Germany
www.albis.com

Alconox, Inc.
30 Gleen Street
Suite 309
White Plains, New York 10603
www.alconox.com

Alk-Abello A/S Corporation
Boge Alle 6
2970 Horsholm
Denmark
www.alk-abello.com

Allnex USA Inc.
Corporate Services Company
2711 Centerville Road, Suite 400
Wilmington, County of New Castle, Delaware 19808
www.allnex.com

Amedra Pharmaceuticals LLC
2 Walnut Grove Drive
Suite 190
Horsham, Pennsylvania 19044
www.amedrapharma.com

Amerchol Corporation
A Subsidiary of The Dow Chemical Company
13875 Highway 43 South
Greensburg, Louisiana 70441
www.dow.com

American Metal Chemical Corporation
835 West Smith Road
Medina, Ohio 44265
www.amcor-usa.com

Ametek Corporation
1100 Cassatt Road
Berwyn, Pennsylvania 19312
www.amtek.com

AMF Support Surfaces Corpora-
  tion
A Subsidiary of Tridien Medical
  Inc.
1691 North Delilah Street
Corona, California 92879

AMVAC Chemical Corporation
4695 MacArthur Court
Newport Beach, California
  92660
www.amvac-chemical.com

Hawley's Condensed Chemical Dictionary, Sixteenth Edition. Michael D. Larrañaga, Richard J. Lewis, Sr., and Robert A. Lewis.
© 2016 John Wiley & Sons, Inc. Published 2016 by John Wiley & Sons, Inc.

Anchor Wire Corporation
425 Church Street
Goodlettsville, Tennessee
    37072
www.hillmangroup.com.

AOC, LLC
955 Highway 57 East
Collierville, Tennessee 38017
www.aoc-resins.com

Aquasol Corporation
80 Thompson Street
North Tonawanda, New York
    14120
www.aquasolwelding.com

Arch Chemicals Inc.
A Subsidiary of Lonza Group, Ltd.
Muenchensteinerstrasse 38
CH-4002 Basel
Switzerland
www.lonza.com

Arch Wood Protection, Inc.
5660 New Northside Drive
Atlanta, Georgia 30328
www.wolmanizedwood.com

Arkema France Corporation
420 Rue d'Estienne d'Orves
92705 Colombes Cedex
France
www.arkema.com

Arkema, Inc.
900 First Avenue
King of Prussia, Pennsylvania
    19406
www.arkema-americas.com

Arthur H. Thomas Company
1654 High Hill Road
Swedesboro, New Jersey 08085
www.thomassci.com

Ashland Licensing and Intellectual
    Property LLC
50 East RiverCenter Boulevard
Covington, Kentucky 41012
www.ashland.com

Atotech Deutschland GmbH
Erasmusstrasse 20
10553 Berlin
Germany
www.atotech.com

Avon Products Inc.
777 Third Avenue
New York, New York 10017
www.avoncompany.com

AVS Industries
21 Bellecor Drive
Unit C
New Castle, Delaware 19720
www.avsind.com

Avantor Materials
222 Red School Lane
Phillipsburg, New Jersey 08865
https://www.avantormaterials
    .com/

Axalta Coating Systems IP
    Company, LLC
A Subsidiary of E. I. du Pont
    de Nemours and Company
1007 Market Street
Wilmington, Delaware 19898
www.dupont.com

Axel Plastics Research
    Laboratories Inc.
5820 Broadway
Flushing, New York 11377
www.axelplastics.com

Axiall Ohio Inc.
A Subsidiary of Axiall Corporation
1000 Abernathy Road
Suite 1200
Atlanta, Georgia 30328
www.axiall.com

Babco Europe Limited
    Corporation
Eglinton House 8 Englinton Road
Bray Co. Wicklow
Ireland
www.babcoeurope.com

Baker Hughes Incorporated
2929 Allen Parkway
Suite 2100
Houston, Texas 77019
www.bakerhughes.com

BASF
100 Park Avenue
Florham Park, New Jersey
    07932
www.basf.com

Bayer AG
51368 Leverkusen
Germany
www.bayer.com

Bayer Healthcare LLC
555 White Plains Road
Tarrytown, New York 10591
www.healthcare.bayer.com

Bayer Materialscience NAFTA
100 Bayer Road
Pittsburgh, Pennsylvania 15205
www.bayermaterialsciencenafta
    .com

Beckman Coulter, Inc.
250 South Kraemer Boulevard
Brea, California 92821
www.beckmancoulter.com

Betaseed, Inc.
1788 Marshall Road
Shakopee, Minnesota 55379
www.betaseed.com

Bimeda Inc.
One Tower Lane
Suite 2250
Oakbrook Terrace, Illinois 60181
www.bimedaus.com

Blanver Farmoquimica Ltda
Rua Doutor Jose Alexandre Cros-
    gnac
715, Itapevi-SP
06680-035, Brazil
www.blanver.com

Bluescope Steel Limited Company
1540 Genessee Street
Kansas City, Missouri 64102
www.bluescopesteel.com

Bluestar Silicones USA Corp.
Two Towers Center Boulevard
Suite 1601
East Brunswick, New Jersey 08816
www.bluestarsilicones.com

Boston Scientific Scimed Inc.
A Subsidiary of Boston Scientific
    Corporation
One Scimed Place
Maple Grove, Minnesota 55311
www.bostonscientific.com

Brandt Consolidated Inc.
2935 South Koke Mill Road
Springfield, Illinois 62711
www.brandt-inc.com

Bristol-Myers Squibb Company
345 Park Avenue
New York, New York 10154
www.bms.com

Buckman Laboratories Interna-
    tional, Inc.
1256 North McLean Boulevard
Memphis, Tennessee 38108
www.buckman.com

Cabot Corporation
Two Seaport Lane
Suite 1300
Boston, Massachusetts 02210
www.cabot-corp.com

Calgon Carbon Corporation
400 Carbon Drive
Pittsburgh, Pennsylvania 15205
www.calgoncarbon.com

Calumet Penreco, LLC
A Subsidiary of Calumet Specialty
  Products Partners LP
2780 Waterfront Parkway East
  Drive
Suite 200
Indianapolis, Indiana 46214
www.calumetspecialty.com

Camtex Fabrics, Ltd.
Blackwood Road, Lillyhall
Lillyhall Industrial Estate
Workington, Cumbria CA14 4JJ
United Kingdom
www.cambrelle.com

Cargill Food Ingredients US,
  LLC
Cargill, Incorporated
P.O. Box 9300
Minneapolis, Minnesota 55440
www.cargill.com

Cargill Texturizing Solutions
  Deutschland GmbH &
  Company KG
Ausschlager Elbdeich 62
20539 Hamburg
Germany
www.cargilltexturizing.com

Carus Corporation
315 5th Street
Peru, Illinois 61354
www.caruscorporation.com

Catalent Pharma Solutions, Inc.
14 Schoolhouse Road
Somerset, New Jersey 08873
www.catalent.com

Celanese
AM Unisyspark 1
65843 Sulzbach
Germany
www.celanese.com

CelChem, LLC
19123 Hickory Bay Court
Baton Rouge, Louisiana 70817
http://www.celchemllc.com/

Chemtura Corporation
1818 Market Street
Suite 3700
Philadelphia, Pennsylvania
  19103
www.chemtura.com

Chevron Intellectual Property LLC
6001 Bollinger Canyon Road
San Ramon, California 94583

Chevron Phillips Chemical Company LP
10001 Six Pines Drive
The Woodlands, Texas 77380
www.cpchem.com

Ciba Specialty Chemical Corporation
A Subsidiary of BASF
100 Park Avenue
Florham Park, New Jersey 07932
www.basf.com

Clariant International Ltd.
Rothausstrasse 61
4132 Muttenz
Switzerland
www.clariant.com

Cleanese Corporation
222 West Las Colinas Boulevard
Suite 900N
Irving, Texas 75039
www.celanese.com

Cognis IP Management GmbH
A Subsidiary of BASF
100 Park Avenue
Florham Park, New Jersey 07932
www.basf.com

Consolidated Plastics Company
4700 Prosper Road
Stow, Ohio 44224
www.consolidatedplastics.com

Conwed Plastics LLC
2810 Weeks Avenue South East
Minneapolis, Minnesota 55414
www.conwedplastics.com

Corning Incorporated
One Riverfront Plaza
Corning, New York 14831
www.corning.com

Cranston Print Works Company
1381 Cranston Street
Cranston, Rhode Island 02920
www.cpw.com

Cray Valley HSC
Oaklands Corporate Center
468 Thomas Jones Way
Suite 100
Exton, Pennsylvania 19341
www.crayvalley.com

C. R. Bard, Inc.
730 Central Avenue
Murray Hill, New Jersey 07974
www.crbard.com

Cremer Oleo GmbH & Company
Glockengiesserwall 3
20095 Hamburg
Germany
www.cremeroleo.de

Cristy Corporation
260 Authority Drive
Fitchburg, Massachusetts 01420
www.drygas.com

Croda International PLC
Cowick Hall
Snaith, Goole
East Yorkshire DN14 9AA
United Kingdom
www.croda.com

Crompton Corporations
A Subsidiary of Chemtura Corporation
1818 Market Street
Suite 3700
Philadelphia, Pennsylvania 19103
www.chemtura.com

Crowley Chemical Company
One Grand Central Place
305 Madison Avenue
Suite 1035
New York, New York 10165
www.crowleychemical.com

Cultor Corporation
Kyllikinportti 2
00241 Helsinki
Finland

Cytec Surface Specialties Ger-
  many GmbH & Co. KG
A Subsidiary of Cytec Industries,
  Inc.
Helbingstrasse 46
22047 Hamburg
Germany
www.cytec.com

Cytec Technology Corporation
5 Garret Mountain Plaza
Woodland Park, New Jersey 07424
www.cytec.com

Dallas Group of America, Inc.
1425 Production Drive
Jeffersonville, Indiana 47130
www.dallasgrp.com

Danisco US Inc.
A Subsidiary of E. I. du Pont de
  Nemours and Company
925 Page Mill Road
Palo Alto, California 94304
www.biosciences.dupont.com

Daubert Chemical Company, Inc.
4700 South Central Avenue
Chicago, Illinois 60638
www.daubertchemical.com

Dover Chemical Corporation
3676 Davis Road
Dover, Ohio 44622
www.doverchem.com

Dow Agrosciences LLC
9330 Zionsville Road
Indianapolis, Indiana 46268
www.dowagro.com

Dow Corning Corporation
Corporate Center
P.O. Box 994
Midland, Michigan 48686
www.dowcorning.com

DSM IP Assets
P.O. Box 6500
6401 JH Heerlen
Netherlands
www.dsm.com

Dupont Air Products Nanomateri-
  als, LLC
A Subsidiary of Air Products &
  Chemicals, Inc.
2507 West Erie Drive
Tempe, Arizona 85282
www.airproducts.com

Dupont Performance Elastomers
  LLC
A Subsidiary of E. I. du Pont de
  Nemours and Company
4417 Lancaster Pike
Suite 300
Wilmington, Deleware 19805
www.dupont.com/products-and-
  services/plastics-polymers-
  resins/elastomers.html

Duracool Refrigerants, Inc.
2695 Slough Street
Mississauga, ON L4T 1G2
Canada
www.duracool.com

Durferrit GmbH
Industriestrasse 3
68169 Mannheim
German
www.durferrit.com

Dyno Nobel Inc.
2795 East Cottonwood Parkway
#500
Salt Lake City, Utah 84121
www.dynonobel.com

Dystar Colours Distribution
  GmbH
AM Prime Park 10-12
65479 Raunheim
Germany
www.dystar.com

Dystar Textilfarben GmbH & Co.
A Subsidiary of Kiri Industries
  Limited and Zhejiang Long-
  sheng Group Co., Ltd.
Industriepark Hochst
Building B 598
65926 Frankfurt am Main
Germany

Eastman Chemical Company
200 South Wilcox Drive
Kingsport, Tennessee 37660
www.eastman.com

Eastman Kodak Company
343 State Street
Rochester, New York 14650
www.kodak.com

Ecolab USA Inc.
370 Wabasha Street North
#100
St. Paul, Minnesota 55102
www.ecolab.com

E. I. du Pont de Nemours and Company
1007 Market Street
Wilmington, Delaware 19898
www.dupont.com

Elcat, Inc.
163 Washington Valley Road
Warren, New Jersey 07059
http://www.el-cat.com/

Electro Abrasives, LLC
701 Willet Road
Buffalo, New York 14218
www.electroabrasives.com

Eliokem, Inc.
1452 East Archwood Avenue
Suite 240
Akron, Ohio 44306
https://www.omnova.com

Eli Lilly and Company
Lilly Corporate Center
Indianapolis, Indiana 46285
www.lilly.com

Emerald Bioagriculture
  Corporation
2123 University Park Drive
Suite 105
Okemos, Michigan 48864
www.emeraldbio.com

EMS-CHEMIE Holding AG
  Corporation
Fuederholzstrasse 34
8704 Herrliberg
Switzerland
www.ems-group.com

Endo Pharmaceuticals Inc.
Glandore Business Centeres
No. 33, Fitzwilliam Square
Dublin 2
Ireland
www.endo.com

Energenx LLC
10183 North Aero Drive
Suite 2
Hayden, Idaho 83835
www.energenx.com

Enpath Medical Inc.
A Subsidiary of Greatbatch
  Medical
2300 Berkshire Lane North
Minneapolis, Minnesota 55441
www.greatbatchmedical.com

EP Minerals, LLC
9785 Gateway Drive
Reno, Nevada 89521
www.epminerals.com

Equistar Chemicals, LP
A Subsidiary of Lyondell-
  Basell Industries
1501 McKinzie Road
Corpus Christi, Texas 78410
www.lyondellbasell.com

EuroChem Agro GmbH
Rechskanzler-Muller-Strasse 23
68165 Mannheim
Germany
www.intl.eurochemagro.com

Evonik Degussa GmbH
A Subsidiary of Evonik
  Industries AG
Rodenbacher Chaussee, 4
63457 Hanau-Wolfgang
Germany
www.evonik.com

Evonik Goldschmidt Rewo
GmbH
Max-Wolf-Strasse 7
36396 Steinau an der Strasse
Germany
www.evonik.com

ExcelAG Corporation
7300 North Kendall Drive
Miami, Florida 33156
www.excelag.com

Exxon Mobil Corporation
5959 Las Colinas Boulevard
Irving, Texas 75039
www.corporate.exxonmobil.
com

Ferro Corporation
6060 Parkland Boulevard
Suite 250
Mayfield Heights, Ohio 44124
www.ferro.com

F. Hoffmann-La Roche AG
Konzern Hauptsitz
Grenzacherstrasse 124
CH-4070 Basel
Schweiz
www.roche.com

Fiberstar, Inc.
713 Saint Croix Street
River Falls, Wisconsin 54022
www.fiberstar.net

Firestone Natural Rubber Company
Bridgestone Americas, Inc.
535 Marriott Drive
P.O. Box 140990
Nashville, Tennessee 37214
www.firestonetire.com

Firestone Polymers, LLC
381 West Wilberth Road
Akron, Ohio 44301
www.firesyn.com

Flexsys America LP
A Subsidiary of Eastman Chemical
Co.
260 Springside Drive
P.O. Box 5444
Akron, Ohio 44333
www.eastman.com

Flow International Corporation
23500 64th Avenue South
Kent, Washington 98032
www.flowwaterjet.com

FMC Corporation
1735 Market Street
Philadelphia, Pennsylvania 19103
www.fmc.com

Gabriel Performance Products,
LLC
725 State Road
Ashtabula, Ohio 44004
www.gabepro.com

Galata Chemicals, LLC
464 Heritage Road
Southbury, Connecticut 06488
www.galatachemicals.com

Gayson Silicone Dispersions, Inc.
A Subsidiary of ColorMatrix Corporation
30 Second Street SW
Barberton, Ohio 44203
www.gsdi.com

GC Corporation
3-2-14 Hongo
Bunkyo-ku, Tokyo 113-0033
Japan
www.gcdental.co.jp

G. D. Searle LLC
A Subsidiary of Pfizer
5200 Old Orchard Road
Skokie, Illinois 60077
www.pfizer.com

Geberit International AG
Schachenstrassee 77
CH-8645 Rapperswil-Jona
Switzerland
www.geberit.com

GE Healthcare Bio-sciences AB
Bjorkgatan 30
751 84 Uppsala
Sweden
www.gehealthcare.com

GelTech Solutions, Inc.
1460 Park Lane South
Suite 1
Jupiter, Florida 33458
www.geltechsolutions.com

Georgia-Pacific
133 Peachtree Street North East
Atlanta, Georgia 30303
www.gp.com

Geo Specialty Chemicals, Inc.
340 Mathers Road
Ambler, Pennsylvania 19002
www.geosc.com

Gibbs Sports Amphibians, Inc.
50 Corporate Drive
Auburn Hills, Michigan 48326
www.gibbssports.com

Gilead Sciences, Inc.
333 Lakeside Drive
Foster City, California 94404
www.gilead.com

Gilman, Michael (Private Owner)
13700 Marina Pointe Drive #907
Marina Del Rey, California 90292

Glanbia Nutritionals, Inc.
1603 Orrington
Suite 1000
Evanston, Illinois 60201
www.glanbianuritionals.com

Glaxo Group Limited Corporation
A Subsidiary of GlaxoSmithKline
PLC
980 Great West Road
Brentford TW8 9GS
United Kingdom
www.gsk.com

Glyptal, Inc.
305 Eastern Avenue
Chelsea, Massachusetts 02150
www.glyptal.com

Goodyear Tire & Rubber Company
200 Innovation Way
Akron, Ohio 44316
www.goodyear.com

Gowan Company, LLC
13200 Metcalf Avenue
Overland Park, Kansas 66213
www.gowanco.com

Grayson O. Company
6509 Grayson Lane
Kannapolis, North Carolina 28081

Great Lakes Solutions
1801 U.S. Highway 52 West
West Lafayette, Indiana 47906
www.greatlakes.com

Gun EI Chemical Industry Co.
700 Shukuroui-machi
Takasaki-shi, Gunma 370-0032
Japan
www.gunei-chemical.co.jp.eng

HA International LLC
630 Oakmont Lane
Westmont, Illinois 60559
www.ha-international.com

Haiyan Hatehui Machinery Hardware Company, Ltd.
Fenghuang Industrial Zone
Yuantong Town, Haiyan County
Zhejlang Province, China
www.hth-hardware.com

Halliburton Energy Services, Inc.
3000 North Sam Houston Parkway E
Houston, Texas 77032
www.halliburton.com

Hallstar Innovations Corporation
120 South Riverside Plaza
Chicago, Illinois 60606
www.hallstar.com

Hangsterfer's Laboratories Inc.
175 Ogden Road
West Deptford, New Jersey 08051
www.hangsterfers.com

Haynes International Inc.
1020 West Park Avenue
Kokomo, Indiana 46901
www.haynesintl.com

H. B. Fuller IP Licensing GmbH
Subsidiary of H. B. Fuller Deutschland GmbH
1200 Willow Lake Boulevard
P.O. Box 64683
St. Paul, Michigan 55164
www.hbfuller.com

Henkel AG & Company
One Henkel Way
Rocky Hill, Connecticut 06067
www.henkel.com

Hercules Incorporated
A Subsidiary of Ashland Inc.
50 East River Center Boulevard
Covington, Kentucky 41011
www.ashland.com

Heritage Pharmaceuticals Inc.
105 Fieldcrest Avenue
Edison, New Jersey 08837
www.heritagepharma.com

Hickory Springs Manufacturing Company
235 2nd Avenue North West
Hickory, North Carolina 28601
www.hsmsolutions.com

Hitco Carbon Composites, Inc.
1600 West 135th Street
Gardena, California 90249
www.hitco.com

Homeland Tobacco Company
888 East Las Olas Boulevard
#504
Fort Lauderdale, Florida 33301
http://www.homelandtobacco.com/

Honeywell International Inc.
101 Columbia Road
Morristown, New Jersey 07960
www.honeywell.com

Hospira, Inc.
275 North Field Drive
Lake Forest, Illinois 60045
www.hospira.com

Howmet Corporation
475 Steamboat Road
Greenwich, Connecticut 06836

HSAIO-Products
Huber Specialty Hydrates, LLC
A Subsidiary of the J. M. Humber Corporation
499 Thornall Street
Suite 8
Edison, New Jersey 08837
www.huber.com

Huber Specialty Hydrates, LLC
A Subsidiary of J. M. Huber Corporation
4750 Alcoa Road
Bauxite, Arkansas 72011
www.huber.com

Huntington Alloys Corporation
3200 Riverside Drive
Huntington, West Virginia 25705
www.specialmetals.com

Huntsman Corporation
100003 Woodloch Forest Drive
The Woodlands, Texas 77380
www.huntsman.com

Huntsman International, LLC
A Subsidiary of Huntsman Corporation
500 Huntsman Way
Salt Lake City, Utah 84108
www.huntsman.com

Ichthyol-Gesellschaft
Sportallee 85
22335 Hamburg
Germany
www.ichthyol.de

ICL Performance Products LP
622 Emerson Road
Suite 500
St. Louis, Missouri 63141
www.icl-pp.com

Imerys Minerals California, Inc.
2500 Miguelito Canyon Road
Lompoc, California 93436
www.worldminerals.com

Immuno-Mycologics, Inc.
2700 Technology Plaza
Norman, Oklahoma 73071
www.immy.com

Indal Enterprises, LLC
23480 Park Sorrento
Calabasas, California 91302

Indspec Chemical Corporation
133 Main Street
Petrolia, Pennsylvania 16050
www.indspec-chem.com

INEOS Melamines GmbH
Alt-Fechenheim 34
60386 Frankfurt am Main
Germany
www.ineos.com

INEOS USA LLC
2600 South Shore Boulevard
#500
League City, Texas 77573
www.ineos.com

Innovia Films Limited
290 Interstate North Circle SE
Suite 100
Atlanta, Georgia 30339
www.innoviafilms.com

International Nutrition, Inc.
7706 I Plaza
Omaha, Nebraska 68127
www.ini-agworld.com

International Paper Company
6400 Poplar Avenue
Memphis, Tennessee 38119
www.internationalpaper.com

Interplastic Corporation
1225 Willow Lake Boulevard
Saint Paul, Minnesota 55110
www.interplastic.com

Invista
Invista Building
4123 East 37th Street North
Wichita, Kansas 67220
www.invista.com

ISP Investments, Inc.
A Subsidiary of Ashland, Inc.
300 Delaware Avenue
Wilmington, Delaware 19801
www.ashland.com

ITC Gums, Inc.
A8 Workshop
Clifford Industrial Park
No. 288 Juhuashi Street
Huashan Town, Huadu
    District
Guangzhou 510800
China
www.ticgums.com

Jason GmbH
A Subsidiary of Jason Incorporated
Eisenhammerstr. 9
Sulzbach-Rosenberg
92224 Amberg-Sulzbach
Germany
www.jasoninc.com

JJISCO Inc.
1815 Executive Drive
Oconomowoc, Wisconsin
    53066
http://www.jjisco.com/

J. M. Huber Corporation
499 Thornall Street
Suite 8
Edison, New Jersey 08837
www.huber.com

Johns Manville Corporation
P.O. Box 5108
Denver, Colorado 80217
www.jm.com

Johnson & Johnson Corporation
One Johnson & Johnson Plaza
New Brunswick, New Jersey
    08933
www.jnj.com

Johnson Matthey
5th Floor
25 Farringdon Street
London EC4A 4AB
United Kingdom
www.matthey.com

J. P. Morgan Chase Bank
270 Park Avenue
New York, New York 10017
www.jpmorganchase.com

J. R. Simplot Company
P.O. Box 27
Boise, Idaho 83707
www.simplot.com

Kasten Masonry Sales, Inc.
713 Kasten Drive
Jackson, Missouri 63755
www.kastenmasonry.net

Kemira OYJ Company
Prokkalankatu 3
P.O. Box 330
00101 Helsinki
Finland
www.kemira.com

Kenrich Petrochemicals, Inc.
140 East 22nd Street
Bayonne, New Jersey 07002
www.4kenrich.com

Kentucky Equine Research, Inc.
3910 Delaney Ferry Road
Versailles, Kentucky 40383
www.ker.com

King Industries, Inc.
1 Science Road
Norwalk, Connecticut 06855
www.kingindustries.com

King Pharmaceuticals Research
    and Development, Inc.
A Subsidiary of Pfizer
235 East 42nd Street
New York, New York 10017
www.pfizer.com

Kraton Polymers U.S. LLC
15710 John F. Kennedy Boulevard
Suite 300
Houston, Texas 77032
www.kraton.com

Krokeide, Gunnar (Private Owner)
Bahnhofstrase 4 D-22885
Barsbutel Fed Rep Germany
Kydex LLC
6685 Low Street
Bloomsburg, Pennsylvania
    17815
www.kydex.com

Lanxess Deutschland GmbH
Kennedyplatz 1
50569 Cologne
Germany
www.lanxess.com

Lafayette Pharmaceuticals, Inc.
A Subsidiary of Covidien, Ltd.
526 North Earl Avenue
Lafayette, Indiana 47904
www.covidien.com

Lawson Products
8770 W. Bryn Mawr Avenue
Suite 900
Chicago, Illinois 60631
www.lawsonproducts.com

LFL Investments, LLC
5130 Isleworth Country Club
    Drive
Windermere, Florida 34786

Lion Copolymers Geismar, LLC
P.O. Box 397
36191 Highway 30
Geismar, Louisiana 70734
www.lioncopolymer.com

Lipo Chemicals, Inc.
207 19th Avenue
Paterson, New Jersey 07504
www.lipochemicals.com

Loders Croklaan BV Corporation
Hogeweg 1
1521 AZ Wormerveer
Netherlands
www.corklaan.com

Lonza Inc.
90 Boroline Road
Allendale, New Jersey 07401
www.lonza.com

Loveland Products, LLC
3005 Rocky Mountain Avenue
Loveland, Colorado 80538
www.lovelandproducts.com

Lucite International
6350 North Twin City Highway
Nederland, Texas 77627
www.leuciteinternational.com

LyondellBassel
1221 McKenney Street
Suite 700
Houston, Texas 77010
www.lyondellbasell.com

MAG Instrument Inc.
2001 South Hellman Avenue
P.O. Box 50600
Ontario, California 91761
www.maglite.com

Makhteshim Chemical Works, Ltd.
A Subsidiary of ADAMA
Derech Hebron, Industrial Zone
Beer Sheva, 84100, Israel
www.adama.com

Mallard Creek Polymers, Inc.
8901 Research Drive
Charlotte, North Carolina
    28262
www.mcpolymers.com

Marigold Works, LLC
1000 Urban Center Drive
#350
Birmingham, Alabama 35242
http://www.marigoldworks.com/

Martin Marietta Magnesia Specialties, LLC
8140 Corporate Drive
Suite 220
Baltimore, Maryland 21236
www.magnesiaspecialties.com

McKinney, Christine Ann (Private
Owner)
1715 Galatea Terrace
Corona del Mar, California
92625
MeadWestvaco Corporation
501 South 5th Street
Richmond, Virginia 23219
www.mwv.com

Merck KGAA
Frankfurter Str. 250
64293 Darmstadt
Germany
www.emdgroup.com

MetalTek International
905 East Saint Paul Avenue
Waukesha, Wisconsin 53188
www.metaltek.com

Metglas Inc.
440 Allied Drive
Conway, South Carolina
29526
www.metglas.com

Milacron, LLC
3010 Disney Street
Cincinnati, Ohio 45209
www.milacron.com

Millennium Specialty Chemicals
Inc.
A Subsidiary of LyondellBasell
Industries
601 Crestwood Street
Jacksonville, Florida 32208
www.lyondellbasell.com

Milwaukee Metropolitan Sewer-
age District
260 West Seeboth Street
Milwaukee, Wisconsin 53204
www.mmsd.com

Mission Pharmacal Company
10999 West IH 10
San Antonio, Texas 78230
www.missionpharmacal.com

Mitsubishi Gas Chemical Company Inc.
Mitsubishi Building
5-2, Marunouchi 2-chome
Chiyoda-ku, Tokyo 100-8324
Japan
www.mgc.co.jp

Momentive Specialty Chemicals
180 East Broad Street
Columbus, Ohio 43215
www.momentive.com

Mona Industries Inc.
Cedarhurst, New York 11516
www.monaindustries.com

Monsanto Company
800 North Lindbergh Boulevard
St. Louis, Missouri 63167
www.monsanto.com

Monument Chemical Kentucky
LLC
2450 Olin Road
Brandenburg, Kentucky 40108
www.monumentchemical.com

Nalco Chemical Company
A Subsidiary of Ecolab
Company
1601 W. Diehl Road
Naperville, Illinois 60563
www.nalco.com

Nano-Tex LLC
38500 Woodward Avenue
Suite 201
Bloomfield Hills, Michigan
48304
www.nano-tex.com

NEASC Corporation
New England Association of
Schools and Colleges, Inc.
3 Burlington Woods Drive
Suite 100
Burlington, Massachusetts
01803
www.neasc.org

Neoperl International AG
Pfeffingerstrasse 21
4153 Reinach
Switzerland
www.neoperl.ch/

Neville Chemical Company
2800 Neville Road
Pittsburgh, Pennsylvania
15225
www.nevchem.com

Nicon Construction Company
5425 North 59th Street
Tampa, Florida 33610
http://niconllc.com/

Novus International Inc.
20 Research Park Drive
Saint Charles, Missouri 63304
www.novusint.com

Nu Mark LLC
Attention: Quality Department
P.O. Box 18583
Pittsburgh, Pennsylvania 15236
www.nu-mark.com

Nutrasweet Property Holdings Inc.
222 West Merchandise Mart Plaza
Chicago, Illinois 60654
www.nutrasweet.com

O2 Partners, LLC
9 Research Drive
Suite 7
Amherst, Massachusetts 01002
www.ortholite.us

Occidental Chemical Corporation
5005 LBJ Freeway
Dallas, Texas 75244
www.oxy.com

OHP, Inc.
1722 Sumneytown Pike
Mainland, Pennsylvania 19451
www.ohp.com

Oil-Dri Corporation
410 North Michigan Avenue
Suite 352
Chicago, Illinois 60611
www.oildri.com

Olin Corporation
190 Carondelet Plaza
Suite 1530
Clayton, Missouri 63105
www.olin.com

Onyx Specialty Papers Inc.
40 Willow Street
South Lee, Massachusetts 01260
www.onyxpapers.com

Organovo, Inc.
6275 Nancy Ridge Drive
San Diego, California 92121
www.organovo.com

Orica Explosives Technology Pty.
Ltd.
L 9 1 Nicholson Street
Melbourne East, Victoria 3002
Australia
www.orica.com

Owensboro Specialty Polymers Inc.
5529 US60 East
Owensboro, Kentucky 42303
www.onssp.com

Paratec Elastomers LLC
14550 Torrey Chase Boulevard
Suite 199
Houston, Texas 77014
www.paratecelastomers.com

Parker Intangibles LLC
6035 Parkland Boulevard
Cleveland, Ohio 44124
www.parker.com

Par Pharmaceutical
300 Tice Boulevard
Woodcliff Lake, New Jersey
   07677
www.parpharm.com

Par Sterile Products, LLC
870 Parkdale Road
Rochester, Michigan 48307
www.jhppharma.com

Particle Dynamics International, LLC
2629 South Hanley Road
St. Louis, Missouri 63144
www.particledynamics.com

PBI-Gordon Corporation
1217 West 12th Street
Kansas City, Missouri 64101
www.pbigordon.com

PBM Pharmaceuticals, Inc.
200 Garrett Street
Charlottesville, Virginia 22902
www.pbmpharma.net

Pfizer
235 East 42nd Street
New York, New York 10017
www.pfizer.com

Pines International, Inc.
P.O. Box 927
Lawrence, Kansas 66044
www.wheatgrass.com

Pinova Inc.
2801 Cook Street
Brunswick, Georgia 31520
www.pinovasolutions.com

PMC Biogenix Inc.
1231 Pope Street
Memphis, Tennessee 38108
www.pmcbiogenix.com

PMC Group
1288 Route 73
Suite 401
Mount Laurel, New Jersey
   08054
www.pmc-group.com

Porocel International LLC
1 Landy Lane
Reading, Ohio 45215
www.porocel.com

PPG Industries
One PPG Place
Pittsburgh, Pennsylvania
   15272
www.ppg.com

PQ Corporation
1700 Kansas Avenue
Kansas City, Kansas 66105
www.pqcorp.com

Precision Combustion, Inc.
410 Sackett Point Road
North Haven, Connecticut 06473
www.precision-combustion.com

Prestige Brands
660 White Plains Road
Suite 250
Tarrytown, New York 10591
www.prestigebrands.com

ProDox, LLC
2400 North Central Avenue
Phoenix, Arizona 85004
www.prodox.com

Produits Chimiques Magnus Ltee
   Corporation
1271 Ampere Suite 101
Boucherville, QC J4B 5Z5
Canada
www.magchem.com

Propex Operating Company, LLC
6025 Lee Highway
Chattanooga, Tennessee 37421
www.propexglobal.com

Propper Manufacturing Company
3604 Skillman Avenue
Long Island City, New York
   11101
www.proppermfg.com

Purdue Pharma L.P.
One Streamford Forum
201 Tresser Boulevard
Stamford, Connecticut 06901
www.purduepharma.com

QOL Labs, LLC
Quality of Life
2975 Westchester Avenue
Purchase, New York 10577
www.qualityoflife.net

Raymond G. and Ruth Perelman
   Education Foundation Trust
   Organization
Bala Cynwyd, Pennsylvania 19004
www.guidestar.org/organizations/
   23-2819735/raymond-
   ruth-perelman-education-
   foundations.aspx

Rectorseal Corporation
2601 Spenwick Drive
Houston, Texas 77055
www.rectorseal.com

Reheis, Inc.
235 Snyder Avenue
Berkeley Heights, New Jersey
   07922
www.reheis.lookchem.com

Reichhold, Inc.
1035 Swabia Court
Durham, North Carolina 27703
www.reichhold.com

Renite Company
P.O. Box 30830
Columbus, Ohio 43230
www.renite.com

Rhein Chemie Rheinau GmbH
145 Parker Court
Chardon, Ohio 44024
www.rheinchemie.com

Rhodia Operations
25 Rue de Clichy
75009 Paris
France
www.solvay.com

Rhodia UK Ltd.
Oak House
Reeds Crescent
Watford WD24 4QP
United Kingdom
www.solvay.com

RITA Corporation
850 South Illinois Route 31
Crystal Lake, Illinois 60014
www.ritacorp.com

Roche Products Inc.
F. Hoffmann-La Roche AG
Konzern Hauptsitz

Grenzacherstrasse 124
CH-4070 Basel
Schweiz
www.roche.com

Rockland Timber Co.
305 South 4th Street
Springfield, Oregon 97477

Rohm & Haas is Dow Chemical
100 South Independence Mall
    West
Philadelphia, Pennsylvania
http://www.dow.com/

R. R. Street & Co.
215 Shuman Boulevard
Suite 403
Naperville, Illinois 60563
www.4streets.com

Sabinsa Corporation
20 Lake Drive
East Windsor, New Jersey 08520
www.sabinsa.com

Saint-Gobain Advanced Ceramics
Saint-Gobain Ceramic Materials
168 Creekside Drive
Amherst, New York 14228
www.bn.saint-gobain.com

Sanchez Agular, Guillermo (Private Owner)
#723, 4701 Monterey Oaks Boulevard
Austin, Texas 78749
Sanofi
54 Rue La Boetie
75008 Paris
France
www.sanofi.us

Sasol Wax Americas, Inc.
21325 Cabot Boulevard
Hayward, California 94545
www.sasolwax.com

S. C. Johnson & Sons, Inc.
1525 Howe Street
Racine, Wisconsin 53403
www.scjohnson.com

Scott Laboratories, Inc.
2220 Pine View Way
Petaluma, California 94954
www.scottlab.com

Sealed Air Corporation
200 Riverfront Boulevard
Elmwood Park, New Jersey 07407
www.sealedair.com

Sensient Colors LLC
2515 North Jefferson

St. Louis, Missouri 63106
www.sensientfoodcolors.com

SePRO Corporation
11550 North Meridian Street
Carmel, Indiana 46032
www.sepro.com

Shell Trademark Management BV
A Subsidiary of Royal Dutch Shell
    PLC
Shell UK Limited
Shell Centre
London SE1 7NA
United Kingdom
www.royaldutchshellplc.com

Sigma-Aldrich Biotechnology LP
Sigma-Aldrich
3050 Spruce Street
St. Louis, Missouri 63103
www.sigmaaldrich.com

Silbrico Corporation
6300 River Road
Hodgkins, Illinois 60525
www.silbrico.com

Solutia Inc.
575 Maryville Center Drive
St. Louis, Missouri 63141
www.solutia.com

Solvay Advanced Polymers LLC
A Subsidiary of Solvay SA
4500 McGinnis Ferry Road
Alpharetta, Georgia 30005
www.solvayplastics.com

Solvin S. A. Corporation
Rue de Ransbeek 310
Brussels, 1120
Belgium
www.solvinpvc.com

Sonneborn Inc.
100 Sonneborn Lane
Perolia, Pennsylvania 16050
www.sonneborn.com

Specialty Lubricants Corporation
8300 Corporate Park Drive
Macedonia, Ohio 44056
www.speclubes.com

SQM North America Corporation
2727 Paces Ferry Road South East
Atlanta, Georgia 30339
www.sqm.com

Stainless Foundry & Engineering
    Inc.
5110 North 35th Street
Milwaukee, Wisconsin 53209
www.stainlessfoundry.com

Standard Process Inc.
1200 West Royal Lee Drive
Palmyra, Wisconsin 53516
www.standardprocess.com

Stepan Company
22 Frontage Road
Northfield, Illinois 60093
www.stepan.com

Stollberg Group
4111 Witmer Road
Niagara Falls, New York
    14305
www.sollberg.com

Storopack, Inc.
4758 Devitt Drive
Cincinnati, Ohio 45246
www.storopack.us

Stromdahl, Andrew W. (Private
    Owner)
8390 Driftwood Place
Minocqua, Wisconsin 54548

Styron Europe GmbH LLC
Zugerstrasse 231
8810 Horgen
Switzerland
http://www.trinseo.com/

Sumitomo Bakelite North American, Inc.
46820 Magellan Drive
Suite C
Novi, Michigan 48377
www.sbna-inc.com

Symrise AG
300 North Street
Teterboro, New Jersey 07608
www.symrise.com

Syngenta Participations AG
Schwarzwaldallee 215
4058 Basel
Switzerland
www.syngenta-rfx.com

SynPep Corporation
6905 Sierra Court
Dublin, California 94568
Sysmex Corporation
1-5-1 Wakinohama-Kaigandori
Chuo-ku
Kobe 651-0073
www.sysmex.com

Technopharma Ltd.
2-6-9 Hiranomachi
Chuo-ku, Osaka 541-0-046
Japan

TechPac LLC
A Subsidiary of Central Garden &
  Pet Company
2365 Harrodsburg Road
Suite B230
Lexington, Kentucky 40504
www.gardentech.com

Tessenderlo Kerley Inc.
2255 North 44th Street
#300
Phoenix, Arizona 85008
www.tkinet.com

TEVA Pharmaceuticals
1090 Horsham Road
North Wales, Pennsylvania 19454
www.tevausa.com

The Dow Chemical Company
2030 Dow Center
Midland, Michigan 48674
www.dow.com

The Euclid Chemical Company
  Corporation
19215 Redwood Road
Cleveland, Ohio 44110
www.euclidchemical.com

The Gleason Works Corporation
1000 University Avenue
Rochester, New York 14607
www.gleason.com

The Gorilla Glue Company
4550 Red Bank Road
Cincinnati, Ohio 45227
www.gorillaglue.com

The International Group, Inc.
50 Salome Drive
Toronto, ON M1S 2A8
Canada
www.igiwax.com

The Lubrizol Corporation
29400 Lakeland Boulevard
Wickliffe, Ohio 44092
www.lubrizol.com

The Proctor & Gamble Company
1 Proctor & Gamble Plaza
Cincinnati, Ohio 45202
www.pg.com

Thermoguard Deutschland
Johannes Schmitt Kuhlwetter-
  strasse 34
40239 Dusseldorf
Germany
www.thermoguard.com

The Thompson's Company
A Subsidiary of The Sherwin-
  Williams Company
825 Crossover Lane
Memphis, Tennessee 38117
www.thompsonwaterseal.com

The Tintometer Limited
Lovibond House
Solar Way
Solstice Park
Amesbury SP4 7SZ
United Kingdom
www.lovibondcolour.com

TIC Gums, Inc.
1-552 Philadelphia Road
White Marsh, Maryland 21162
www.ticgums.com

Ticona LLC
8040 Dixie Highway
Florence, Kentucky 41042
www.ticona.com

TIFA International LLC
109 Stryker Lane
Hilssborough Township, New Jersey 08844
www.tifausa.com

Timminco Corporation
3595 Moline Street
Aurora, Colorado 80010

Toray Fine Chemicals Co., Ltd.
2-3-1 Kanda-Sudacho
Chiyoda-ku, Tokyo 101-0041
Japan
www.torayfinechemicals.com

T-Rex
8100 Washington Avenue
Suite 200
Houston, Texas 77007
www.trexec.com

Tricor Refining, LLC
1134 Manor Street
Bakersfield, California 93308
www.tricorrefining.com

Troy Chemical Corporation
One Avenue L
Newark, New Jersey 07105
www.troycorp.com

Tyco Fire Products LP
A Subsidiary of Tyco International
  Ltd.
451 North Cannon Avenue
Lansdale, Pennsylvania 19446
www.tycofsbp.com

UCT, Inc.
2731 Bartram Road
Bristol, Pennsylvania 19056
www.unitedchem.com

Ulano Corporation
110 Third Avenue
Brooklyn, New York 11217
www.ulano.com

Ultimark Products, LLC
Funston Media Management &
  Ultimark Products
One Belmont Avenue
Suite 602
Bala Cynwyd, Pennsylvania
  19004
www.ultimarkproducts.com

Ultrashield
3702 North West 16th Street
Lauderhill, Florida 33311
www.ultrashield.com

Unimin Corporation
258 Elm Street
New Canaan, Connecticut 06840
www.unimin.com

Union Carbide Corporation
A Subsidiary of The Dow Chemical Company
1254 Enclave Parkway
Houston, Texas 77077
www.unioncarbide.com

Uniseal, Inc.
1014 Uhlhorn Street
Evansville, Indiana 47710
www.uniseal.com

United Gilsonite Laboratories
P.O. Box 70
Scraton, Pennsylvania 18501
www.ugl.com

Upsher-Smith Laboratories
6701 Evenstad Drive
Maple Grove, Minnesota 55369
www.upsher-smith.com

U. S. Borax Inc.
8051 East Maplewood Avenue
Building 4
Greenwood   Village,   Colorado
  80111
www.borax.com

Vanderbilt Minerals, LLC
30 Winfield Street
Norwalk, Connecticut 06855
www.vanderbiltminerals.com

Vantage Specialty Chemicals
Vantage Oleochemicals
4650 South Racine Avenue
Chicago, Illinois 60609
www.vantagespecialties.com

VersaFlex Inc.
686 South Adams Street
Kansas City, Kansas 66105
www.versaflex.com

Vertellus Specialties Inc.
201 North Illinois Street
Indianapolis, Indiana 46204
www.vertellus.com

Veyance Technologies, Inc.
703 South Cleveland Massillon
    Road
Fairlawn, Ohio 44333
www.goodyearep.com

Vintage Specialties, Inc.
Vantage Oleochemicals
4650 South Racine Avenue
Chicago, Illinois 60609
www.vantagespecialties.com

Vitax Ltd.
Owen Street
Coalville, Leicester LE67 3DE
United Kingdom
www.vitak.co.ul

Virbac
13 e rue LID-BP 27
06511 Carros Cedex
France
www.virbac.com

W.A. Hammond Drierite Com-
    pany, Ltd.
138 Daytona Avenue
Xenia, Ohio 45385
www.drierite.com

Wampole, Inc.
A Subsidiary of Pharmex Indus-
    tries Inc.
1255 West Pender Street
Vancouver, BC V6E 2V1
Canada
http://www.wampole.ca/

Washington Pacific Bio Control
    Corporation
14615 North East 13th Court
#A
Vancouver, Washington 98685
www.pacificbiocontrol.com

White Square Chemical, Inc.
171 Hood Avenue
Unit 2
Tavernier, Florida 33070
http://www.wsqchem.com/

Wilube Chemical Inc.
21845 Mason Street
Edwardsburg, Michigan 49112
www.wilube.com

Windmoller & Holscher Corpora-
    tion
23 New England Way
Lincoln, Rhode Island 02865
www.whcorp.com

W.R. Grace & Company
7500 Grace Drive
Columbia, Maryland 21044
www.grace.com

Wyeth LLC
A Subsidiary of Pfizer Inc.
5 Giralda Farms
Madison, New Jersey 07940
www.pfizer.com

Yorkshire Group PLC
27 Kirkstall Road
Leeds LS3 1LL
United Kingdom

Zaclon LLC
2981 Independence Road
Cleveland, Ohio 44115
www.zalcon.com

Zerran International Corporation
12880 Pierce Street
Pacoima, California 91331
www.zerran.com

Zoetis Products, LLC
100 Campus Drive
Florham Park, New Jersey 07932
www.zoetis.com

APPENDIX IV

Chemical Abstract (CAS) Number Index

Hawley's Condensed Chemical Dictionary, Sixteenth Edition. Michael D. Larrañaga, Richard J. Lewis, Sr., and Robert A. Lewis.
© 2016 John Wiley & Sons, Inc. Published 2016 by John Wiley & Sons, Inc.

54-31-9	4-chloro-*N*-furfuryl-5-sulfamoylanthranilic acid	57-24-9	strychnine
54-31-9	furosemide	57-27-2	morphine
54-35-3	procaine penicillin G	57-39-6	metepa
54-64-8	thimerosal	57-44-3	barbital
54-47-4	pyridoxal phosphate	57-47-6	physostigmine
54-95-5	pentylenetetrazole	57-50-1	sucrose
55-18-5	*n*-nitrosodiethylamine	57-52-3	bis(triethyltin) sulfate
55-21-0	benzoylglucuronic acid	57-53-4	meprobamate
55-38-9	fenthion	57-55-6	1,2-propylene glycol
55-63-0	nitroglycerin	57-57-8	β-propiolactone
55-86-7	mechlorethamine hydrochloride	57-63-6	ethinylestradiol
55-91-4	diisopropyl fluorophosphate	57-74-9	chlordane
55-98-1	busulfan	57-83-0	progesterone
56-12-2	*a*-aminobutyric acid	57-88-5	cholesterol
56-12-2	aminobutyric acid	57-92-1	streptomycin
56-18-8	3,3′-iminobispropylamine	57-97-6	9,10-dimethyl-1,2-benzanthracene
56-23-5	carbon tetrachloride	58-00-4	apomorphine
56-24-6	trimethyltin hydroxide	58-05-9	*n*-[*p*-[[(2-amino-5-formyl-5,6,7,8-tetrahydro-4-hydroxy-6-pteridinyl)methyl]amino]benzoyl]glutamic acid
56-25-7	cantharidine		
56-29-1	hexobarbital	58-08-2	caffeine
56-35-9	bis(tri-*n*-butyltin) oxide	58-18-4	17-methyltestosterone
56-36-0	tributyltin acetate	58-22-0	testosterone
56-40-6	glycine	58-25-3	chlordiazepoxide
56-49-5	methylcholanthrene	58-27-5	menadione
56-53-1	diethylstilbestrol	58-39-9	perphenazine
56-54-2	quinidine	58-55-9	theophylline
56-55-3	benz(*a*)anthracene	58-56-0	pyridoxal hydrochloride
56-65-5	adenosine phosphate	58-61-7	adenosine
56-65-5	adenosine triphosphate	58-63-9	inosine
56-72-4	coumaphos	58-64-0	adenosine diphosphate
56-73-5	glucose-6-phosphate	58-74-2	papaverine
56-75-7	chloramphenicol	58-82-2	bradykinin
56-81-5	glycerol	58-86-6	D(+)-xylose
56-82-6	glyceraldehyde	58-89-9	lindane
56-84-8	*l*-aspartic acid	58-90-2	2,3,4,6-tetrachlorophenol
56-85-9	glutamine	58-90-2	2,4,5,6-tetrachlorophenol
56-86-0	glutamate	58-95-7	*d*-α-tocopheryl acetate
56-86-0	glutamic acid	58-96-8	uridine
56-87-1	lysine	58-97-9	5′-uridylic acid
56-89-3	cystine	59-02-9	*dl*-α-tocopherol
57-06-7	allyl isothiocyanate	59-05-2	methotrexate
57-06-7	isothiocyanic acid allyl ether	59-06-3	ethopabate
57-09-0	cetyltrimethylammonium bromide	59-23-4	galactose
57-10-3	palmitic acid	59-50-7	4-chloro-3-methylphenol
57-11-4	hydrofol	59-51-8	methionine
57-11-4	stearic acid	59-52-9	2,3-dimercaptopropanol
57-12-5	cyanide	59-67-6	niacin
57-13-6	isourea	59-69-1	phenoxybenzamine
57-13-6	urea	59-92-7	dopa
57-14-7	1,1-dimethylhydrazine	60-00-4	edetic acid
57-15-8	chlorobutanol	60-00-4	ethylenediaminetetraacetic acid

70-49-5	thiomalic acid	75-09-2	methylene chloride
70-51-9	deferoxamine	75-10-5	difluoromethane
70-55-3	*p*-toluenesulfonamide	75-11-6	methylene iodide
71-00-1	histidine	75-12-7	formamide
71-23-8	propyl alcohol	75-15-0	carbon disulfide
71-30-7	cytosine	75-18-3	dimethyl sulfide
71-36-3	*n*-butyl alcohol	75-19-4	cyclopropane
71-41-0	*n*-amyl alcohol, primary	75-20-7	calcium carbide
71-43-2	benzene	75-21-8	ethylene oxide
71-48-7	cobaltous acetate	75-21-8	oxirane
71-55-6	aerothenet TT	75-24-1	trimethylaluminum
71-55-6	1,1,1-trichloroethane	75-25-2	bromoform
71-55-6	"Tri-Ethane"	75-26-3	isopropyl bromide
71-58-9	mecoprop	75-28-5	isobutane
71-63-6	digitalin	75-29-6	2-chloropropene
71-63-6	digitoxin	75-29-6	isopropyl chloride
72-17-3	sodium lactate	75-30-9	isopropyl iodide
72-18-4	D-valine	75-31-0	isopropylamine
72-19-5	threonine	75-33-2	isopropyl mercaptan
72-20-8	endrin	75-34-3	ethylidene chloride
72-43-5	methoxychlor	75-35-4	vinylidene chloride
72-48-0	turkey red oil	75-36-5	acetyl chloride
72-54-8	TDE	75-37-6	1,1-difluoroethane
72-55-9	DDE	75-38-7	vinylidene fluoride
72-89-9	acetylcoenzyme A	75-43-4	dichloromonofluoromethane
73-24-5	adenine	75-44-5	phosgene
73-32-5	isoleucine	75-45-6	chlorodifluoromethane
73-40-5	guanine	75-46-7	fluoroform
74-31-7	*N,N′*-diphenyl-*p*-phenylenediamine	75-47-8	iodoform
74-82-8	methane	75-50-3	trimethylamine
74-83-9	methyl bromide	75-52-5	nitromethane
74-84-0	ethane	75-54-7	methyldichlorosilane
74-85-1	ethylene	75-55-8	propyleneimine
74-86-2	acetylene	75-56-9	propylene oxide
74-87-3	methyl chloride	75-59-2	tetramethylammonium hydroxide
74-88-4	methyl iodide	75-60-5	cacodylic acid
74-89-5	methylamine	75-61-6	dibromodifluoromethane
74-90-8	hydrocyanic acid	75-62-7	bromotrichloromethane
74-93-1	methanethiol	75-63-8	trifluorobromomethane
74-95-3	methylene bromide	75-64-9	*tert*-butylamine
74-96-4	ethyl bromide	75-65-0	*tert*-butanol
74-97-5	bromochloromethane	75-65-0	*tert*-butyl alcohol
74-98-6	propane	75-68-3	1-chloro-1,1-difluoroethane
74-99-7	methyl acetylene	75-69-4	cfc-11
75-00-3	blastoderm	75-69-4	trichlorofluoromethane
75-00-3	ethyl chloride	75-71-8	dichlorodifluoromethane
75-01-4	vinyl chloride	75-72-9	chlorotrifluoromethane
75-02-5	vinyl fluoride	75-73-0	tetrafluoromethane
75-03-6	ethyl iodide	75-74-1	tetramethyl lead
75-04-7	ethylamine	75-75-2	methanesulfonic acid
75-05-8	acetonitrile	75-76-3	tetramethylsilane
75-07-0	acetaldehyde	75-77-4	trimethylchlorosilane
75-08-1	ethanethiol	75-78-5	dimethyldichlorosilane

79-34-5	sym-tetrachloroethane	83-59-0	propyl isomer
79-37-8	oxalyl chloride	83-63-6	diacetylaminoazotoluene
79-38-9	chlorotrifluoroethylene	83-67-0	theobromine
79-39-0	methacrylamide	83-79-4	rotenone
79-41-4	methyacrylic acid	83-86-3	phytic acid
79-42-5	thiolactic acid	83-88-5	lactoflavin
79-43-6	dichloroacetic acid	83-88-5	riboflavin
79-44-7	dimethylcarbamoyl chloride	83-89-6	intercalating agent
79-46-9	NiPar S-20	84-11-7	phenanthrenequinone
79-46-9	2-nitropropane	84-15-1	o-terphenyl
79-57-2	oxytetracycline	84-16-2	hexestrol
79-68-5	γ-irone	84-36-6	syrosingopine
79-69-6	α-irone	84-61-7	Morflex 150
79-70-9	β-irone	84-65-1	anthraquinone
79-92-5	camphene	84-66-2	diethyl phthalate
79-94-7	4,4′-(1-methylethylidene)bis(2,6- dibromophenol)	84-69-5	diisobutyl phthalate
		84-74-2	dibutyl phthalate
80-00-2	sulphenone	84-77-5	didecyl phthalate
80-05-7	bisphenol A	85-01-8	phenanthrene
80-08-0	dapsone	85-34-7	fenac
80-10-4	diphenyldichlorosilane	85-35-2	5′-guanylic acid
80-13-7	halazone	85-41-6	phthalimide
80-15-9	cumene hydroperoxide	85-43-8	memtetrahydrophthalic anhydride
80-26-2	terpinyl acetate	85-43-8	tetrahydrophthalic anhydride
80-32-0	sulfachlorpyridazine	85-44-9	phthalic anhydride
80-33-1	ovex	85-68-7	butylbenzyl phthalate
80-38-6	fenson	85-70-1	Morflex 190
80-40-0	ethyl-p-toluenesulfonate	85-79-0	dibucaine
80-46-6	p-tert-pentylphenol	85-84-7	yellow AB
80-47-7	p-menthane-8-hydroperoxide	85-91-6	dimethyl anthranilate
80-48-8	methyl-p-toluenesulfonate	86-50-0	azinphos methyl
80-51-3	p,p′-oxybis(benzenesulfonyl- hydrazide)	86-51-1	2,3-dimethoxybenzaldehyde
		86-57-7	α-nitronaphthalene
80-52-4	menthanediamine	86-73-7	fluorene
80-59-1	tiglic acid	86-74-8	carbazole
80-62-6	methyl methacrylate	86-75-9	8-hydroxyquinoline benzoate
81-07-2	saccharin	86-85-1	methylmercury quinolinolate
81-13-0	pantothenol	86-87-3	α-naphthaleneacetic acid
81-24-3	taurocholic acid	86-88-4	ANTU
81-25-4	cholate	86-88-4	α-naphthylthiourea
81-25-4	cholic acid	87-00-3	homotropine
81-77-6	indanthrone	87-10-5	3,4′,5-tribromosalicylanilide
81-81-2	"Warfarin"	87-17-2	samicylanilide
81-88-9	rhodamine B	87-18-3	p-tert-butylphenyl salicylate
82-66-6	diphenadione	87-19-4	isobutyl salicylate
82-68-8	pentachloronitrobenzene	87-20-7	isoamyl salicylate
82-71-3	styphnic acid	87-25-2	ethyl anthranilate
83-12-5	phenindione	87-29-6	cinnamyl anthranilate
83-26-1	2-pivaloyl-1,3-indanedione	87-31-0	diazodinitrophenol
83-32-9	acenaphthene	87-47-8	pyrolan
83-34-1	skatole	87-51-4	3-indoleacetic acid
83-43-2	methylprednisolone	87-52-5	gramine
83-44-3	deoxycholic acid	87-61-6	1,2,3-trichlorobenzene

94-17-7	Cadox PS	96-45-7	ethylene thiourea
94-17-7	*p*-chlorobenzoyl peroxide	96-47-9	2-methyltetrahydrofuran
94-24-6	tetracaine	96-48-0	butyrolactone
94-36-0	benzoyl peroxide	96-49-1	ethylene carbonate
94-46-2	isoamyl benzoate	96-64-0	soman
94-51-9	dipropylene glycol dibenzoate	96-69-5	4,4′-thiobis(6-*tert*-butyl-*m*-cresol)
94-52-0	6-nitrobenzimidazole	96-76-4	2,4-di-*tert*-butylphenol
94-59-7	safrole	96-83-3	iopanoic acid
94-63-3	2-pyridine aldoxime methiodide	96-91-3	picramic acid
94-70-2	*o*-phenetidine	97-00-7	1-chloro-2,4-dinitrobenzene
94-74-6	MCPA	97-02-9	2,4-dinitroaniline
94-75-7	2,4-D	97-17-6	dichlofenthion
94-78-0	phenazopyridine	97-18-7	bithionol
94-81-5	MCPB	97-18-7	2,2′-thiobis(chlorophenol)
94-83-7	sesin	97-23-4	dichlorophene
94-96-2	2-ethylhexanediol-1,3	97-39-2	di-*o*-tolylguanidine
95-13-6	indene	97-53-0	eugenol
95-33-0	*n*-cyclohexyl-2-	97-54-1	isoeugenol
	benzothiazolesulfenamide	97-56-3	*o*-aminoazotoluene
95-39-6	5-norbornene-2-mthyl acrylate	97-61-0	2-methylpentanoic acid
95-47-6	*o*-xylene	97-62-1	ethyl isobutyrate
95-48-7	*o*-cresol	97-63-2	ethyl methacrylate
95-49-8	*o*-chlorotoluene	97-64-3	ethyl lactate
95-50-1	*o*-dichlorobenzene	97-72-3	isobutyric anhydride
95-51-2	*o*-chloroaniline	97-74-5	tetramethylthiuram monosulfide
95-53-4	*o*-toluidine	97-77-8	tetraethylthiuram disulfide
95-54-5	*o*-phenylenediamine	97-85-8	isobutyl isobutyrate
95-55-6	*o*-aminophenol	97-86-9	isobutyl methacrylate
95-57-8	*o*-chlorophenol	97-88-1	di-butyl methylacrylate
95-63-6	psedocumene	97-93-8	triethylaluminum
95-69-2	2-amino-4-chlorotoluene	97-95-0	2-ethylbutyl alcohol
95-69-2	4-amino-2-chlorotoluene	97-96-1	2-ethylbutyraldehyde
95-76-1	3,4-dichloroaniline	97-99-4	tetrahydrofurfuryl alcohol
95-80-7	toluene-2,4-diamine	98-00-0	furfuryl alcohol
95-94-3	1,2,4,5-tetrachlorobenzene	98-01-1	furfural
95-95-4	2,4,5-trichlorophenol	98-05-5	phenylarsonic acid
96-08-2	dipentene dioxide	98-06-6	*tert*-butylbenzene
96-09-3	styrene oxide	98-07-7	benzotrichloride
96-10-6	diethylaluminum chloride	98-12-4	cyclohexyl richlorosilane
96-11-7	1,2,3-tribromopropane	98-13-5	phenyltrichlorosilane
96-13-9	dibromopropanol	98-15-7	*m*-chlorobenzotrifluoride
96-14-0	3-methylpentane	98-16-8	*m*-aminobenzotrifluoride
96-18-4	1,2,3-trichloropropane	98-46-4	*m*-nitrobenzotrifluoride
96-20-8	2-amino-1-butanol	98-47-5	*m*-nitrobenzenesulfonic acid
96-22-0	diethyl ketone	98-51-1	*p-tert*-butyltoluene
96-23-1	α-dichlorohydrin	98-54-4	*p-tert*-butylphenol
96-24-2	chlorohydrin	98-56-6	*p*-chlorobenzotrifluoride
96-27-5	1-thioglycerol	98-59-9	*o*-toluenesulfonyl chloride
96-31-1	*N,N′*-dimethylurea	98-59-9	*p*-toluenesulfonyl chloride
96-32-2	methyl bromoacetate	98-82-8	cumene
96-33-3	methyl acrylate	98-83-9	α-methylstyrene
96-34-4	methyl chloroacetate	98-84-0	α-methylbenzylamine
96-37-7	methylcyclopentane	98-86-2	acetophenone

103-48-0	2-phenylethyl isobutyrate		105-87-3	geranyl acetate
103-50-4	benyl ether		106-11-6	diglycol monostearate
103-50-4	dibenzyl ether		106-14-9	12-hydroxystearic acid
103-64-0	β-bromostyrene		106-22-9	citronellol
103-65-1	*n*-propylbenzene		106-22-9	rhodinol
103-69-5	*n*-ethylaniline		106-24-1	geraniol
103-71-9	phenyl isocyanate		106-25-2	nerol
103-73-1	phenetole		106-27-4	isoamyl butyrate
103-76-4	*N*-hydroxyethyl piperazine		106-31-0	butyric anhydride
103-81-1	phenylacetamide		106-32-1	ethyl caprylate
103-82-2	phenylacetic acid		106-34-3	quinhydrone
103-85-5	1-phenyl-2-thiourea		106-35-4	butyl ethyl ketone
103-90-2	acetaminophen		106-35-4	ethyl butyl ketone
103-90-2	*p*-acetylaminophenol		106-36-5	*n*-propyl propionate
103-95-7	cyclamen aldehyde		106-40-1	4-bromoaniline
104-03-0	*p*-nitrophenylacetic acid		106-41-2	*p*-bromophenol
104-15-4	*o*-toluenesulfonic acid		106-42-3	*p*-xylene
104-15-4	*p*-toluenesulfonic acid		106-43-4	*p*-chlorotoluene
104-51-8	*n*-butylbenzene		106-44-5	*p*-cresol
104-53-0	phenylpropyl aldehyde		106-46-7	*p*-dichlorobenzene
104-54-1	cinnamic alcohol		106-46-7	santochlor
104-55-2	cinnamic aldehyde		106-47-8	*p*-chloroaniline
104-75-6	2-ethylhexylamine		106-48-9	*p*-chlorphenol
104-76-7	2-ethylhexyl alcohol		106-49-0	*p*-toluidine
104-83-6	chlorobenzyl chloride		106-50-3	*p*-phenylenediamine
104-90-5	2-methyl-5-ethylpyridine		106-51-4	quinone
104-91-6	*p*-nitrosophenol		106-54-7	*p*-chlorothiophenol
104-93-8	*p*-methylanisole		106-63-8	isobutyl acrylate
104-94-9	*p*-anisidine		106-69-4	1,2,6-hexanetriol
105-13-5	anise alcohol		106-70-7	methyl caproate
105-21-5	γ-heptalactone		106-87-6	vinylcyclohexene dioxide
105-30-6	amyl methyl alcohol		106-88-7	1,2-butylene oxide
105-36-2	ethyl bromoacetate		106-89-8	epichlorohydrin
105-37-3	ethyl propionate		106-90-1	glycidyl acrylate
105-39-5	ethyl chloroacetate		106-92-3	allyl glycidal ether
105-45-3	acetoacetate		106-93-4	ethylene dibromide
105-45-3	methyl acetoacetate		106-94-5	*n*-propyl bromide
105-46-4	*sec*-butyl acetate		106-95-6	allyl bromide
105-52-2	di-*n*-hexyl maleate		106-96-7	propargyl bromide
105-53-3	ethyl malonate		106-97-8	butane
105-54-4	ethyl butyrate		106-99-0	1,3-butadiene
105-55-5	1,3-diethylthiourea		106-99-0	butadiene
105-56-6	ethyl cyanoacetate		107-00-6	ethylacetylene
105-57-1	acetal		107-02-8	acrolein
105-58-8	diethyl carbonate		107-03-9	1-propanethiol
105-59-9	methyldiethanolamine		107-05-1	allyl chloride
105-60-2	caprolactam		107-06-2	ethylene dichloride
105-65-7	isopropyl xanthogen disulfide		107-07-3	ethylene chlorohydrin
105-66-8	propyl butyrate		107-10-8	*n*-propylamine
105-74-8	lauroyl peroxide		107-12-0	ethyl cyanide
105-76-0	dibutyl maleate		107-13-1	acrylonitrile
105-85-1	citronellyl formate		107-14-2	chloroacetonitrile
105-86-2	geranyl formate		107-15-3	ethylenediamine

109-87-5	methylal	111-14-8	*n*-heptanoic acid
109-89-7	diethylamine	111-15-9	ethylene glycol monoethyl ether acetate
109-90-0	ethyl isocyanate		
109-92-2	ethylene bis(dithiocarbamate)	111-16-0	pimelic acid
109-92-2	vinyl ethyl ether	111-17-1	thiodipropionic acid
109-93-3	vinyl ether	111-20-6	sebacic acid
109-94-4	ethyl formate	111-26-2	*n*-hexylamine
109-97-7	pyrrole	111-27-3	hexyl alcohol
109-99-9	tetrahydrofuran	111-29-5	1,5-pentanediol
110-00-9	furan	111-30-8	glutaraldehyde
110-01-0	tetrahydrothiophene	111-34-2	vinyl-*n*-butyl ether
110-02-1	thiophene	111-36-4	*n*-butylisocyanate
110-05-4	di-*tert*-butyl peroxide	111-40-0	diethylenetriamine
110-12-3	methyl isoamyl ketone	111-41-1	hydroxyethylethylenediamine
110-13-4	acetonylacetone	111-42-2	diethanolamine
110-15-6	succinic acid	111-44-4	*sym*-dichloroethyl ether
110-16-7	maleic acid	111-46-6	diethylene glycol
110-17-8	fumaric acid	111-49-9	hexamethyleneimine
110-18-9	tetramethylethylenediamine	111-55-7	ethylene glycol diacetate
110-19-0	isobutyl acetate	111-60-4	ethylene glycol monostearate
110-22-5	acetyl peroxide	111-65-9	*n*-octane
110-27-0	isopropylmyristate	111-66-0	1-octene
110-30-5	acrawax	111-69-3	adiponitrile
110-38-3	ethyl caprate	111-70-6	heptyl alcohol
110-43-0	methyl-*n*-amylketone	111-71-7	heptanal
110-44-1	sorbic acid	111-76-2	ethylene glycol monobutyl ether
110-45-2	isoamyl formate	111-77-3	diethylene glycol monomethyl ether
110-49-6	ethylene glycol monomethyl ether acetate	111-83-1	*n*-octyl bromide
		111-84-2	nonane
110-54-3	*n*-hexane	111-87-5	*n*-octyl alcohol, primary
110-60-1	tetramethylenediamine	111-88-6	*n*-octyl mercaptan
110-61-2	ethylene cyanide	111-90-0	diethylene glycol monoethyl ether
110-62-3	*n*-valeraldehyde	111-92-2	di-*n*-butylamine
110-63-4	1,4-butylene glycol	111-96-6	diethylene glycol dimethyl ether
110-65-6	butynediol	111-97-7	β,β-thiodipropionitrile
110-66-7	pentanethiol	112-00-5	lauryltrimethylammonium chloride
110-68-9	*N*-methylbutylamine	112-02-7	hexadecyltrimethylammonium chloride
110-74-7	propyl formate		
110-75-8	2-chloroethyl vinyl ether	112-05-0	pelargonic acid
110-75-8	vinyl-2-chloroethyl ether	112-06-1	1-heptyl acetate
110-80-5	ethylene glycol monoethyl ether	112-12-9	methyl nonyl ketone
110-82-7	cyclohexane	112-15-2	diethylene glycol monoethyl ether acetate
110-83-8	cyclohexene		
110-85-0	piperazine	112-18-5	lauryldimethylamine
110-86-1	pyridine	112-19-6	undecylenyl acetate
110-89-4	piperidine	112-21-7	triethylene glycol diacetate
110-91-8	morpholine	112-24-3	triethylenetetramine
110-94-1	glutaric acid	112-27-6	triethylene glycol
110-96-3	diisobutylamine	112-32-3	octyl formate
111-01-3	squalane	112-34-5	diethylene glycol monobutyl ether
111-02-4	squalene	112-36-7	diethylene glycol diethyl ether
111-12-6	methyl heptyne carbonate	112-38-9	undecylenic acid
111-13-7	methyl hexyl ketone	112-40-3	*n*-dodecane

120-83-2	2,4-dichlorophenol	123-25-1	diethyl succinate
120-92-3	cyclopentanone	123-30-8	*p*-aminophenol
121-14-2	2,4-dinitrotoluene	123-31-9	hydroquinone
121-32-4	ethyl vanillin	123-33-1	maleic hydrazide
121-33-5	vanillin	123-38-6	propionaldehyde
121-43-7	trimethyl borate	123-42-2	diacetone alcohol
121-44-8	triethylamine	123-51-3	isoamyl alcohol, primary
121-45-9	trimethyl phosphite	123-54-6	acetylacetone
121-47-1	metanilic acid	123-56-8	succinimide
121-57-3	sulfanilic acid	123-62-6	propionic anhydride
121-69-7	*N*,*N*-dimethylaniline	123-63-7	paraldehyde
121-73-3	*m*-chloronitrobenzene	123-66-0	2-ethylbutyl acetate
121-75-5	malathion	123-66-0	ethyl caproate
121-79-9	propyl gallate	123-72-8	butyraldehyde
121-82-4	cyclonite	123-75-1	pyrrolidine
121-87-9	2-chloro-4-nitroaniline	123-76-2	levulinic acid
121-90-4	*m*-nitrobenzoyl chloride	123-86-4	*n*-butyl acetate
121-91-5	isophthalic acid	123-91-1	1,4-dioxane
121-92-6	*m*-nitrobenzoic acid	123-92-2	isoamyl acetate
122-03-2	cuminic aldehyde	123-93-3	thiodiglycolic acid
122-04-3	*p*-nitrobenzoyl chloride	124-02-7	diallylamine
122-14-5	fenitrothion	124-07-2	octanoic acid
122-16-7	sulfanitran	124-07-2	caprylic acid
122-31-6	1,1,3,3-tetraethoxypropane	124-09-4	hexamethylenediamine
122-34-9	simazine	124-13-0	1-octanal
122-37-2	*p*-hydroxydiphenylamine	124-17-4	diethylene glycol monobutyl ether acetate
122-39-4	diphenylamine		
122-42-9	IPC	124-18-5	*n*-decane
122-48-5	zingerone	124-19-6	nonanal
122-51-0	triethylorthoformate	124-26-5	stearamide
122-52-1	triethyl phosphite	124-38-9	carbon dioxide
122-59-8	pheoxyacetic acid	124-40-3	dimethylamine
122-60-1	phenoxypropylene oxide	124-41-4	sodium methylate
122-60-1	phenyl glycidyl ether	124-43-6	urea peroxide
122-62-3	di(2-ethylhexyl) sebacate	124-47-0	urea nitrate
122-64-5	phenylmercuric lactate	124-58-3	methylarsenic acid
122-66-7	hydrazobenzene	124-64-1	tetrakis(hydroxymethyl)phosphonium chloride
122-67-8	isobutyl cinnamate		
122-69-0	cinnamyl cinnamate	124-65-2	sodium cacodylate
122-72-5	phenylpropyl acetate	124-76-5	isoborneol
122-73-6	isoamyl benzyl ether	124-87-8	picrotoxin
122-78-1	phenylacetaldehyde	124-94-7	triamcinolene
122-79-2	phenyl acetate	125-46-2	usnic acid
122-88-3	chlorophenoxyacetic acid	125-64-4	methylprylon
122-97-4	phenylpropyl alcohol	126-07-8	griseofulvin
122-99-6	ethylene glycol monophenyl ether	126-11-4	tris(hydroxymethyl)nitromethane
123-04-6	2-ethylhexyl chloride	126-14-7	sucrose octaacetate
123-08-0	*p*-hydroxybenzaldehyde	126-30-7	neopentyl glycol
123-11-5	*p*-anisaldehyde	126-31-8	sodium methiodal
123-17-1	2,6,8-trimethyl-4-nonanone	126-33-0	sulfolane
123-19-3	dipropyl ketone	126-68-1	*O*,*O*,*O*-triethyl phosphorothioate
123-20-6	vinyl butyrate	126-73-8	tributyl phosphate
123-23-9	succinic acid peroxide	126-96-5	sodium diacetate

140-89-6	potassium xanthate	144-02-5	barbital sodium
140-93-2	sodium isopropylxanthate	144-21-8	disodium methylarsonate
140-95-4	oxymethurea	144-21-8	DSMA
141-01-5	ferrous fumarate	144-49-0	fluoroacetic acid
141-02-6	di(2-ethylhexyl) fumarate	144-55-8	sodium bicarbonate
141-04-8	diisobutyl adipate	144-62-7	oxalic acid
141-05-9	diethyl maleate	145-73-3	endothall
141-10-6	pseudoionone	146-14-5	flavin adenine dinucleotide
141-22-0	ricinoleic acid	146-48-5	yohimbine
141-23-1	methyl-12-hydroxystearate	146-84-9	silver picrate
141-32-2	*n*-butyl acrylate	147-93-3	thioaslicylic acid
141-43-5	ethanolamine	147-94-4	cytosine arabinoside
141-46-8	glycolic aldehyde	148-01-6	zoalene
141-53-7	sodium formate	148-18-5	dithiocarb sodium
141-57-1	*n*-propyltrichlorosilane	148-18-5	sodium dimethyldithiocarbamate
141-66-2	bidrin	148-24-3	8-hydroxyquinoline
141-78-6	ethyl acetate	148-79-8	thiabendazole
141-79-7	mesityl oxide	148-82-3	mephalan
141-82-2	malonic acid	149-30-4	2-mercaptobenzothiazole
141-84-4	rhodanine	149-32-6	erythritol
141-91-3	2,6-dimethylmorpholine	149-91-7	gallic acid
142-16-5	di(2-ethylhexyl) maleate	150-13-0	*p*-aminobenzoic acid
142-22-3	diethylene glycol bis(allyl carbonate)	150-38-9	edetate trisodium
142-22-3	nouryset 200	150-38-9	trisodium EDTA
142-29-0	cyclopentene	150-39-0	hydroxyethylethylenediaminetriacetic acid
142-31-4	sodium octyl sulfate		
142-47-2	sodium glutamate	150-50-5	tributylphosphorotrithioite
142-59-6	nabam	150-68-5	monuron
142-62-1	caproic acid	150-69-6	dulcin
142-64-3	piperazine dihydrochloride	150-76-5	hydroquinone monomethyl ether
142-71-2	copper acetate	150-84-5	citronellyl acetate
142-72-3	magnesium acetate	150-86-7	phytol
142-73-4	iminodiacetic acid disodium salt hydrate	151-10-0	resorcinol dimethyl ether
		151-21-3	sodium dodecyl sulfate
142-82-5	*n*-heptane	151-21-3	sodium lauryl sulfate
142-84-7	di-*n*-propylamine	151-38-2	2-methoxyethylmercury acetate
142-90-5	lauryl methacrylate	151-50-8	potassium cyanide
142-92-7	hexyl acetate	151-56-4	ethyleneimine
142-96-1	butyl ether	151-67-7	halothane
143-07-7	lauric acid	152-16-9	schradan
143-08-8	*n*-nonyl alcohol	152-72-7	acenocoumarin
143-13-5	*n*-nonyl acetate	154-02-9	5-hydroxy-1h-indole-3-ethanol
143-18-0	potassium oleate	154-21-2	lincomycin
143-19-1	sodium oleate	154-41-6	propadrine hydrochloride
143-22-6	butoxytriglycol	154-93-8	carmustine
143-24-8	dimethoxytetraglycol	156-08-1	benzphetamine
143-28-2	oleyl alcohol	156-10-5	*p*-nitrosodiphenylamine
143-33-9	"Cyanobrik"	156-43-4	*p*-phenetidine
143-33-9	sodium cyanide	156-57-0	cysteamine hydrochloride
143-50-0	chlorodecone	156-57-0	β-mercaptoethylamine hydrochloride
143-50-0	kepone	156-62-7	calcium cyanamide
143-62-4	digitoxigenin	170-30-2	methylchloromethyl ether
143-74-8	phenolsulfonephthalein	193-39-5	indeno(1,2,3-*cd*)pyrene

422-56-0	3,3-dichloro-1,1,1,2,2-pentafluoropropane	485-35-8	cytisine
		485-47-2	ninhydrin
427-45-2	tris(*p*-chlorophenyl)tin fluoride	485-50-7	*l*-capnoidine
429-30-1	trifluoroacetic acid triethylstannyl ester	485-51-8	(+)-corlumine
		486-25-9	9-fluorenone
431-03-8	diacetyl	486-47-5	ethaverine
431-63-0	1,1,1,2,3,3-hexafluoropropane	487-79-6	kainic acid
434-03-7	ethisterone	488-10-8	jasmone
434-13-9	lithocholic acid	489-98-5	trinitroaniline
434-16-2	7-dehydrocholesterol	490-79-9	gentisic acid
437-38-7	fentanyl	490-91-5	*p*-thymoquinone
443-79-8	*dl*-isoleucine	491-26-9	nicotine-*n'*-oxide
446-86-6	azathioprine	491-35-0	lepidine
457-60-3	neosalvarsan	491-70-3	3',4',5,7-tetrahydroxyflavone
460-19-5	cyanogen	492-17-1	2,4'-diphenyldiamine
461-72-3	hydantoin	492-80-8	auramine
462-06-6	fluorobenzene	493-52-7	methyl red
463-04-7	amyl nitrite	494-38-2	acridine orange
463-40-1	linolenic acid	494-47-3	hydrofuramide
463-49-0	allene	495-48-7	azoxybenzene
463-51-4	ketene	495-69-2	hippuric acid
463-58-1	carbonyl sulfide	496-11-7	indan
463-71-8	thiophosgene	497-19-8	soda ash
463-82-1	neopentane	497-19-8	sodium carbonate monohydrate
463-88-7	neurine	497-19-8	trona
464-41-5	2-bornyl chloride	497-23-4	butenolide
465-12-3	3-episarmentogenin	498-40-8	cysteic acid
465-42-9	capsanthin	499-75-2	carvacrol
465-73-6	isodrin	500-38-9	nordihydroguaiaretic acid
468-02-4	zinc gluconate	501-30-4	kojic acid
469-62-5	propoxyphene	502-37-4	hypoglycine B
470-82-6	eucalyptol	502-42-1	cycloheptanone
470-90-6	chlorfenvinphos	502-55-6	dixanthogen
470-90-6	chlorofenvinophos	502-85-2	β-hydroxybutyric acid
471-46-5	oxamide	503-30-0	oxetane
475-25-2	hematein	503-74-2	isopentanoic acid
477-30-5	demecolcine	503-74-2	isovaleric acid
477-47-4	picropodophyllotoxin	504-15-4	orcin
477-73-6	safranin o	504-20-1	phorone
478-43-3	rhubarb yellow	504-24-5	4-aminopyridine
479-18-5	dyphylline	504-29-0	2-aminopyridine
479-20-9	atranorin	504-63-2	trimethylene glycol
479-45-8	tetryl	505-29-3	1,4-dithiane
479-61-8	chlorophyll A	505-60-2	dichlorodiethyl sulfide
479-97-3	nornicotine	506-12-7	*n*-heptadecanoic acid
480-16-0	morin	506-32-1	arachidonic acid
480-68-2	5-nitrobarbituric acid	506-61-6	silver potassium cyanide
480-78-4	platiphillin	506-64-9	silver cyanide
481-06-1	α-santonin	506-68-3	cyanogen bromide
481-39-0	5-hydroxy-1,4-naphthoquinone	506-77-4	cyanogen chloride
482-27-9	isopimpinellin	506-87-6	ammonium carbonate
482-89-3	indigo	506-96-7	acetyl bromide
483-18-1	emetrine	507-02-8	acetyl iodide

553-54-8	lithium benzoate	585-99-9	allolactose
553-79-7	5-nitro-2-*n*-propoxyaniline	586-62-9	terpinolene
554-12-1	methyl propionate	587-65-5	*n*-phenylchloroacetamide
554-13-2	lithium carbonate	589-18-4	α-methylbenzyl alcohol
554-84-7	*m*-nitrophenol	589-38-8	3-hexanone
555-21-5	*p*-nitrobenzyl cyanide	589-41-3	2-hydroxyethyl carbamate
555-43-1	"Neustrene 064"	589-82-2	3-heptanol
556-22-9	glyodin	590-00-1	potassium sorbate
556-52-5	glycidol	590-01-2	*n*-butyl propionate
556-61-6	methyl isothiocyanate	590-28-3	potassium cyanate
556-67-2	octamethylcyclotetrasiloxane	590-54-5	acetyl phosphate
556-88-7	nitroquanidine	590-86-3	isopentaldehyde
556-89-8	nitrourea	590-86-3	isovaleraldehyde
557-04-0	magnesium stearate	590-88-5	1,3-diaminobutane
557-05-1	zinc stearate	591-27-5	*m*-aniophenol
557-07-3	zinc oleate	591-60-6	butyl acetoacetate
557-09-5	zinc caprylate	591-78-6	methyl butyl ketone
557-19-7	nickel cyanide	591-87-7	allyl acetate
557-20-0	diethylzinc	591-89-9	mercuric potassium cyanide
557-21-1	zinc cyanide	592-01-8	calcium cyanide
557-34-6	zinc acetate	592-04-1	mercuric cyanide
558-23-5	carbon tetrabromide	592-05-2	lead cyanide
561-27-3	diacetylmorphone	592-31-4	*n*-*n*-butylurea
563-12-2	ethion	592-41-6	1-hexene
563-25-7	dibutyldifluorostannane	592-84-7	butyl formate
563-41-7	semicarbazide hydrochloride	592-85-8	mercuric thiocyanate
563-45-1	3-methyl-1-butene	592-87-0	lead thiocyanate
563-46-2	2-methyl-1-butene	593-29-3	potassium stearate
563-47-3	chloromethylpropane	593-53-3	fluoromethane
563-47-3	β-methylallyl chloride	593-60-2	vinyl bromide
563-63-3	silver acetate	593-84-0	guanidine thiocyanate
563-68-8	thallium acetate	593-89-5	methyldichloroarsine
563-71-3	ferrous carbonate	594-42-3	trichloromethylsulfenyl chloride
563-80-4	methyl isopropyl ketone	594-72-9	1,1-dichloro-1-nitroethane
564-00-1	erythritol anhydride	595-33-5	medroxyprogesterone acetate
569-61-9	basic fuchsin	597-64-8	tetraethyltin
569-61-9	pararosaniline	597-71-7	pentaerythritol tetraacetate
569-64-2	malachite green	598-14-1	ethyldichloroarsine
573-58-0	Congo red	598-31-2	bromoacetone
576-55-6	tetrabromo-*o*-cresol	598-58-3	methyl nitrate
577-11-7	dioctylsodium sulfosuccinate	598-78-7	2-chloropropionic acid
578-54-1	*o*-ethylaniline	600-14-6	acetyl propionyl
578-94-9	diphenylamine chloroarsine	600-25-9	korax
580-48-3	chlorazine	603-34-9	triphenylamine
582-22-9	2-phenylpropylamine	603-35-0	triphenylphosphours
583-15-3	mercuric benzoate	603-36-1	tripheylantimony
583-52-8	potassium oxalate	606-68-8	NADH
583-60-8	*o*-methylcyclohexanone	608-10-6	6-(2-carboxyethylthio)purine
584-02-1	*sec*-isoamyl alcohol	608-73-1	1,2,3,4,5,6-hexachlorocyclohexane
584-02-1	3-pentanol	608-73-1	tetra-hexachlorocyclohexane
584-08-7	potassium carbonate	609-19-8	3,4,5-trichlorophenol
584-42-9	metachrome yellow	609-72-3	*n*,*n*-dimethyl-*o*-toluidine
584-84-9	toluene-2,4-diisocyanate	611-13-2	methyl-2-furoate

1317-35-7	manganese tetroxide	1345-05-7	lithopone
1317-35-7	manganous oxide	1345-25-1	ferrous oxide
1317-36-8	litharge	1395-21-7	bacillus subtilis BPS
1317-37-9	ferrous sulfide	1395-21-7	subtilisins
1317-39-1	copper oxide black	1401-55-4	tannic acid
1317-43-7	brucite	1402-68-2	aflatoxins
1317-60-8	hematite	1403-29-8	carzinophillin A
1317-60-8	hematite, red	1403-66-3	gentamicin
1317-65-3	agstone	1404-04-2	neomycin
1317-65-3	calcium carbonate	1405-87-4	bacitracin
1317-65-3	limestone	1406-05-9	penicillin
1317-95-9	tripoli	1406-11-7	polymyxin
1318-02-1	zeolite	1407-03-0	ammoniated glycyrrhizin
1319-46-6	basic lead carbonate	1414-45-5	nisin preparation
1319-69-3	potassium glycerophosphate	1424-48-2	3-(p-tolysulfonyl)acrylonitrile
1319-77-3	cresol	1459-93-4	Morflex 1129
1321-64-8	pentachloronaphthalene	1461-22-9	chlorotributylstannane
1321-65-9	trichloronaphthalene	1461-23-0	tri-n-butyltin bromide
1321-74-0	divinylbenzene	1461-25-2	tetrabutyltin
1322-90-3	dypnone	1477-55-0	m-xylene-α,α'-diamine
1324-28-3	6,15-dihydrohydroxy-5,9,14,18-	1478-61-1	hexafluoroacetone bisphenol A
	anthrazinetetrone	1484-13-5	n-vinylcarbazole
1327-53-3	arsenic trioxide	1496-02-2	p-fluorobenzoic acid
1330-20-7	xylene		2-phenylhydrazide
1330-43-4	anhydrous borax	1529-30-2	triethyltin phenoxide
1332-21-4	asbestos	1563-66-2	"Furadan"
1332-21-4	tremolite	1568-80-5	spirobiindane
1332-37-2	iron oxide red	1575-72-0	2-allylpentanoic acid
1332-58-7	kaolin	1582-09-8	trifluralin
1333-74-0	hydrogen	1589-47-5	2-methoxy-1-propanol
1333-83-1	sodium bifluoride	1590-87-0	disilane
1333-86-4	carbon black	1592-23-0	calcium stearate
1334-78-7	p-tolyaldehyde	1596-09-4	2-cyclohexyl-4-methylphenol
1335-31-5	mercuric oxycyanide	1600-27-7	mercuric acetate
1335-32-6	lead subacetate	1609-06-9	n'-cyano-n,n-dimethylguanidine
1335-87-1	hexachloronaphthalene	1610-18-0	2,4-bis(isopropylamino)-6-
1335-88-2	tetrachloronaphthalene		methoxy-s-triazine
1336-21-6	ammonia water	1622-61-3	clonazepam
1336-36-3	polychlorinated biphenyl	1624-01-7	octamethylsilanetetramine
1338-02-9	copper naphthenate	1624-02-8	bis(triphenylsilyl)chromate
1338-23-4	methyl ethyl ketone peroxide	1629-74-4	4-methyl-α-(1-methylethyl)
1341-49-7	ammonium bifluoride		benzeneacetic acid, (5-(2-
1344-09-8	silicic acid, sodium salt		furanylmethyl)-2-thienyl)methyl
1344-28-1	aluminum oxide		ester
1344-28-1	alunite	1634-04-4	methyl-tert-butyl ether
1344-28-1	dycron	1636-70-0	trineophyltin acetate
1344-40-7	lead phosphite, dibasic	1642-54-2	diethylcarbamazine citrate
1344-48-5	mercuric sulfide, black	1643-20-5	lauryldimethylamine oxide
1344-48-5	mercuric sulfide, red	1656-48-0	β,β'-oxydipropionitrile
1344-57-6	uranium dioxide	1663-39-4	tert-butyl acrylate
1344-67-8	copper chloride	1689-84-5	bromoxynil
1344-95-2	calcium metasilicate	1694-09-3	benzyl violet
1345-04-6	antimony trisulfide	1694-20-8	trans-4-nitrostilbene

2425-06-1	cis-n-[1,1,2,2-tetrachloroethyl)thio]-4-cyclohexene-1,2-dicarboximide
2426-08-6	n-butyl glycidyl ether
2429-82-5	C.I. direct brown 2
2439-01-2	oxythioquinox
2439-10-3	dodline
2440-45-1	ethylmercuric phosphate
2444-90-8	bisphenol A disodium salt
2451-62-9	glycidyl isocyanurate
2465-27-2	auramine hydrochloride
2508-19-2	2,4,6-trinitrobenzenesulfonic acid
2511-00-4	ethyl 2-cyclohexylpropionate
2514-52-5	2,3,4,4,5,5-hexachloro-2-cyclopenten-1-one
2524-02-9	thiophosphoryl difluoride monochloride
2524-03-0	O,O-dimethyl phosphorochloridothioate
2524-04-1	O,O-diethyl phosphorochloridothioate
2528-36-1	dibutyl phenyl phosphate
2530-87-2	3-(trimethoxysilyl)propyl chloride
2540-82-1	formothion
2551-62-4	sulfur hexafluoride
2554-06-5	2,4,6,8-tetraethenyl-2,4,6,8-tetramethylcyclotetrasiloxane
2564-35-4	deoxyguanosine-5'-triphosphate
2578-28-1	dl-selenomethionine
2587-76-0	tri-n-octyltin chloride
2595-54-2	mecarbam
2616-10-1	bisaniline-p
2623-22-5	(3-α)-12,13-epoxytrichothec-9-ene-3,4,15-triol 15-acetate
2627-95-4	sym-tetramethyldivinyldisiloxane
2631-37-0	promecarb
2636-26-2	cyanophos
2638-94-0	4,4'-azobis(4-cyanovaleric acid)
2641-56-7	diethylbis(octanoyloxy)stannane
2650-18-2	peacock blue
2654-47-9	ethylmercury toluenesulfonate
2669-35-4	tributyltin cyclohexanecarboxylate
2675-35-6	bis(4-hydroxyphenyl)methanone (2,4-dintrophenyl)hydrazone
2696-92-6	nitrosyl chloride
2698-41-1	o-chlorobenzylidene malononitrile
2699-79-8	sulfuryl fluoride
2702-72-9	sodium-2,4-dichlorophenoxyacetate
2738-06-9	n-ethyl-1,2-dimethylpropylamine
2764-72-9	diquat
2767-41-1	dichlorodihexylstannane
2767-47-7	dimethyltin dibromide
2767-54-6	bromotriethylstannane
2767-55-7	diethyldiiodostannane
2767-61-5	bromotripropylstannane
2767-80-8	tris(hydroxymethyl)phosphine
2778-04-3	endothion
2781-10-4	dibutyltin di-2-thylhexoate
2782-91-4	tetramethylthiourea
2787-93-1	1,1,1,3,3,3-hexaoctyldistannoxane
2788-26-3	phosphonosuccinic acid tetramethyl ester
2788-93-1	n-butyl acid phosphate
2835-39-4	alyl isovalerate
2837-89-0	chlorotetrafluoroethane
2847-65-6	triphenyltin p-acetamidobenzoate
2857-03-6	tributyltin-p-acetamidobenzoate
2858-66-4	pelletierine
2865-19-2	dibutyldiiodostannane
2867-47-2	dimethylaminoethyl methacrylate
2885-00-9	stearyl mercaptan
2897-46-3	acetoxytrihexylstannane
2905-62-6	3,5-dichlorobenzoyl chloride
2917-26-2	cetyl mercaptan
2921-88-2	dursban
2949-22-6	ethyl isocyaatoacetate
2949-42-0	tetraisopropyltin
2952-70-7	5-α-cyprinol
2955-88-6	1-pyrrolidineethanol
2971-38-2	krotiline
2971-90-6	clopidol
2978-58-7	2-methyl-3-butyn-2-amine
2984-64-7	o-ethyl s-4-chlorophenyl ethylphosphonodithioate
3011-89-0	aklomide
3021-41-8	bis(triphenytin)sulfate
3056-17-5	stavudine
3058-38-6	2,4,6-trinitro-1,3,5-benzenetriamine
3060-89-7	metobromuron
3068-88-0	β-butyrolacetone
3073-87-8	1,4-bis[2-(4-methyl-5-phenyloxazolyl)]benzene
3090-35-5	tributyl(oleoyloxy)stannane
3090-36-6	tributyltin laurate
3091-18-7	bromotripentylstannane
3091-25-6	octyltrichlorostannane
3131-23-5	4-hydroxyestrone
3151-41-5	chlorotribenzylstannane
3164-29-2	ammonium tartrate
3165-93-3	4-chloro-o-toluidine hydrochloride
3173-53-3	cyclohexyl isocyanate
3173-72-6	1,5-naphthalene diisocyanate
3237-50-1	5,5-dihydroxy barbituric acid
3240-78-6	paraquat dibromide
3244-90-4	tetra-n-propyl dithionopyrophosphate
3248-28-0	propionyl peroxide
3251-23-8	copper nitrate

4083-64-1	4-methylbenzenesulfonyl isocyanate	4795-29-3	tetrahydrofurfurylamine
4097-89-6	2,2',2''-nitrilotris(ethylamine)	4808-30-4	1,1,1,3,3,3-hexabutyldistannthiane
4098-71-9	isophorone diisocyanate	4848-63-9	bis(phenylthio)dimethyltin
4114-31-2	ethyl carbazate	4914-36-7	n-acetylloline
4147-05-1	1-hydroxyestradiol	4949-20-6	2,4-pentadien-1-ol
4147-10-8	18-norestrone methyl ether	4968-09-6	progesterone 16,17-acetonide
4170-30-3	crotonaldehyde	5001-51-4	calcium lactobionate
4176-16-3	valerenal	5008-52-6	luciculine
4208-80-4	C.I. basic yellow 11	5016-18-2	1-demethyltoxoflavine
4253-22-9	dibutyltin sulfide	5029-46-9	4,4,4-trinitrobutyric acid
4261-68-1	2-chloroethyldiisopropylamine hydrochloride	5035-67-6	tributyltin-2-ethylhexanoate
		5064-31-3	trisodium nitrilotriacetate
4281-67-8	cyclohexyl phenyl phosphate	5124-30-1	methylene bis(4-cyclohexylisocyanate)
4297-61-4	scirpenetriol triacetate		
4337-86-4	oxochlorpromazine	5188-42-1	bis(guanidinium) chromate
4342-30-7	salicyloyloxytributylstannane	5189-40-2	4-(cyclohexylidene(4-hydroxy-phenyl)methyl)phenol
4342-36-3	benoyloxtributylstanane		
4345-03-3	dl-α-tocopheryl acid succinate	5216-25-1	p-chlorobenzotrichloride
4351-70-6	(1-(((2-chloroethoxy)(2-chloroethyl)phosphinyl)oxy)ethyl)-phosphonic acid, 1-(bis(2-chloroethoxy)phosphinyl)ethyl 2-chloroethyl ester	5219-17-0	4-ethyl-7-hydroxy-3-(p-methoxyphenyl) coumarin
		5230-87-5	(−)-anisatin
		5283-66-9	octyl trichlorosilane
4354-73-8	cyclohexylidenemalononitrile	5285-87-0	thiocyanic acid, phenyl ester
4356-33-6	solanoside	5323-95-5	sodium ricinoleate
4372-46-7	pyridoxine tripalmitate	5329-14-6	sulfamic acid
4409-11-4	4-(4-chlorobenzyl)pyridine	5332-73-0	3-methoxypropylamine
4418-26-2	sodium dehydroacetate	5339-43-5	1,3-dimethyl-2-tetradecyl-2-thiopseudourea hydriodide
4418-61-5	aminotetrazole		
4420-65-9	mono-o-acetylsolanoside		
4455-26-9	n-methyl-n-octyl-1-octanamine	5339-67-3	n-ethylbenzenesulfonamide
4471-27-6	ethylene azelate	5392-40-5	citral
4474-24-2	C.I. acid blue 80	5412-01-1	isopropyl carbitol
4484-72-4	dodecyltrichlorosilane	5441-63-4	diglycolic acid, diallyl ester
4485-12-5	lithium stearate	5443-89-0	6-(pentylthio)purine
4486-44-6	isopropyl phosphorothioate	5456-28-0	selenium diethyldithiocarbamate
4525-33-1	dimethyl dicarbonate	5459-93-8	N-ethylcyclohexylamine
4536-23-6	2-hexanecarboxylic acid	5470-02-0	1-propylpiperidine
4564-87-8	carbomycin	5470-11-1	hydroxylamine hydrochloride
4569-88-4	C.I. basic blue 16	5486-03-3	buquinolate
4570-11-0	β-dolabrin	5488-45-9	tributyl(8-quinolinolato)tin
4602-84-0	farnesol	5538-94-3	n,n-dimthyl-b-octyl-1-octanaminium chloride
4620-70-6	t-butylethanolamine		
4638-25-9	trimethyltin thiocyanate	5550-12-9	sodium guanylate
4678-44-8	grayanotoxin II	5555-13-5	α-(4-biphenylyloxy)propionic acid
4678-45-9	grayanotoxin III	5578-73-4	sanguinarine hydrochloride
4685-14-7	paraquat	5593-70-4	tetrabutyl titanate
4685-14-7	paraquat dihydride	5606-24-6	n-butylmelamine
4685-18-1	furyltriazine	5672-84-4	4-bromophenyl trifluoroacetate
4691-65-0	sodium inosinate	5683-33-0	2-dimethylaminopyridine
4713-59-1	dibromodiphenylstannane	5687-22-9	9-arsafluoreninic acid
4731-77-5	bis(octanoyloxy)di-n-butyl stannane	5707-69-7	drazoxolon
4759-48-2	accutane	5714-22-7	sulfur pentachloride

7342-47-4	iodotributylstannane	7440-48-4	cobalt
7392-96-3	dibutyl(diformyloxy)stannane	7440-50-8	copper
7400-08-0	*p*-coumaric acid	7440-54-2	gadolinium
7428-48-0	lead stearate	7440-55-3	gallium
7429-90-5	aluminum	7440-57-5	gold
7439-88-5	iridium	7440-58-6	hafnium
7439-89-6	iron	7440-59-7	helium
7439-90-9	krypton	7440-61-1	uranium
7439-91-0	lanthanum	7440-62-2	vanadium
7439-92-1	lead	7440-64-4	ytterbium
7439-93-2	lithium	7440-65-5	yttrium
7439-95-4	magnesium	7440-66-6	zinc
7439-96-5	manganese	7440-67-7	zircat
7439-97-6	mercury	7440-70-2	calcium
7439-98-7	molybdenum	7440-74-6	indium
7440-00-8	neodymium	7440-76-7	zirconium
7440-01-9	neon	7446-07-3	tellurium dioxide
7440-02-0	nickel	7446-08-4	selenium dioxide
7440-03-1	niobium	7446-09-5	sulfur dioxide
7440-04-2	osmium	7446-11-9	sulfur trioxide
7440-05-3	palladium	7446-14-2	lead sulfate
7440-06-4	platinum	7446-18-6	thallium sulfate
7440-06-4	platinum foil	7446-27-7	lead phosphate
7440-09-7	potassium	7446-32-4	antimony sulfate
7440-15-5	rhenium	7446-34-6	selenium sulfide
7440-16-6	rhodium	7447-39-4	cupric chloride
7440-17-7	rubidium	7447-40-7	potassium chloride
7440-18-8	ruthenium	7447-41-8	lithium chloride
7440-21-3	silicon	7473-98-5	α-hydroxy-α-methylpropiophenone
7440-22-4	silver	7487-94-7	mercuric chloride
7440-23-5	sodium	7488-55-3	stannous sulfate
7440-25-7	tantalum	7506-77-6	diethylphospinic acid anhydride with diethyl phosphorothionate
7440-28-0	thallium		
7440-29-1	thorium	7526-26-3	methanephosphonic acid, diphenyl ester
7440-31-5	tin		
7440-32-6	titanium	7532-85-6	divinyltin dichloride
7440-33-7	tungsten	7550-35-8	lithium bromide
7440-36-0	antimony	7550-45-0	titanium tetrachloride
7440-37-1	argon	7553-56-2	iodine
7440-38-2	arsenic	7558-79-4	sodium phosphate, dibasic
7440-39-3	barium	7558-80-7	sodium phosphate, monobasic
7440-41-7	beryllium	7568-93-6	phenylethanolamine
7440-42-8	boron	7572-29-4	dichloroacetylene
7440-42-8	boron compound	7580-67-8	lithium hydride
7440-43-9	cadmium	7585-39-9	β-cavitron
7440-44-0	buckytube	7587-88-9	magnesium sulfate
7440-44-0	carbon	7601-54-9	"Emulsiphos"
7440-44-0	carbon-12	7601-54-9	"Nutrifos"
7440-44-0	graphite	7601-54-9	"Nutriphos STP"
7440-45-1	cerium	7601-54-9	sodium phosphate, tribasic
7440-46-2	cesium	7601-89-0	sodium perchlorate
7440-47-3	chrome	7601-90-3	perchloric acid
7440-47-3	chromium	7616-94-6	perchloryl fluoride

7775-11-3	sodium chromate		7783-60-0	mercurous sulfate
7775-14-6	"Reductone"		7783-60-0	sulfur tetrafluoride
7775-14-6	sodium dithionite		7783-61-1	silicon tetrafluoride
7775-19-1	sodium metaborate		7783-64-4	zirconium tetrafluoride
7775-27-1	sodium persulfate		7783-66-6	iodine pentafluoride
7775-41-9	silver fluoride		7783-70-2	antimony pentafluoride
7778-39-4	arsenic acid		7783-71-3	tantalum fluoride
7778-43-0	sodium arsenate		7783-81-5	uranium hexafluoride
7778-44-1	calcium arsenate		7783-82-6	tungsten hexafluoride
7778-50-9	potassium dichromate		7783-92-8	silver chlorate
7778-53-2	potassium phosphate, tribasic		7783-93-9	silver perchlorate
7778-54-3	calcium hypochlorite		7784-24-9	aluminum potassium sulfate
7778-54-3	"Pittclor"		7784-33-0	arsenic tribromide
7778-74-7	potassium perchlorate		7784-34-1	arsenic trichloride
7778-80-5	potassium sulfate		7784-35-2	arsenic trifluoride
7779-86-4	zinc hydrosulfite		7784-37-4	mercuric arsenate
7779-88-6	zinc nitrate		7784-40-9	lead arsenate
7782-41-4	fluorine		7784-41-0	potassium arsenate
7782-44-7	molecular oxygen		7784-42-1	arsine
7782-44-7	oxygen		7784-44-3	ammonium arsenate
7782-49-2	selenium		7784-46-5	sodium arsenite
7782-50-5	chlorine		7785-87-7	manganous sulfate
7782-65-2	germanium tetrahydride		7786-30-3	magnesium chloride
7782-68-5	iodic acid		7786-34-7	mevinphos
7782-75-4	magnesium phosphate, dibasic		7786-67-6	isopulegol
7782-77-6	nitrous acid		7786-81-4	nickel sulfate
7782-78-7	mitrosylsulfuric acid		7787-32-8	barium fluoride
7782-79-8	hydrazoic acid		7787-47-5	beryllium chloride
7782-82-3	selenious acid, monosodium salt		7787-68-0	bismuth trisulfate
7782-87-8	potassium hypophosphite		7787-71-5	bromine trifluoride
7782-89-0	lithium amide		7788-97-8	chromic fluoride
7782-92-5	sodium amide		7788-98-9	ammonium chromate
7782-99-2	sulfurous acid		7789-00-6	potassium chromate
7783-00-8	selenous acid		7789-04-0	chromic phosphate
7783-06-4	hydrogen sulfide		7789-06-2	strontium chromate
7783-07-5	hydrogen selenide		7789-12-0	zinc dichromate
7783-07-5	selenide		7789-18-6	cesium nitrate
7783-08-6	selenic acid		7789-21-1	fluorosulfonic acid
7783-18-8	ammonium thiosulfate		7789-23-3	potassium fluoride
7783-20-2	ammonium sulfate		7789-24-4	lithium fluoride
7783-30-4	mercurous iodide		7789-25-5	nitrosyl fluoride
7783-33-7	Nessler's reagent		7789-26-6	fluorine nitrate
7783-35-9	mercuric sulfate		7789-29-9	potassium bifluoride
7783-40-6	magnesium fluoride		7789-30-2	bromine petafluoride
7783-41-7	oxygen fluoride		7789-33-5	iodine monobromide
7783-46-2	lead fluoride		7789-38-0	sodium bromate
7783-47-3	stannous fluoride		7789-40-4	thallium bromide
7783-48-4	strontium fluoride		7789-41-5	calcium bromide
7783-49-5	zinc fluoride		7789-42-6	cadmium bromide
7783-50-8	ferric fluoride		7789-43-7	cobaltous bromide
7783-54-2	nitrogen fluoride		7789-47-1	mercuric bromide
7783-54-2	nitrogen trifluoride		7789-59-5	phosphorus oxybromide
7783-56-4	antimony trifluoride		7789-60-8	phosphorus tribromide

9001-16-5	cytochrome oxidase	9007-13-0	calcium resinate
9001-54-1	hyaluronidase	9007-34-5	collagen
9001-62-1	lipase	9007-43-6	cytochrome c
9001-66-5	monoamine oxidase	9007-83-4	immune endoglobulin
9001-73-4	papain	9009-54-5	polyurethane
9001-75-6	pepsin	9009-86-3	ricin
9001-84-7	phospholipase A2	9010-85-8	isobutylene-isoprene copolymer
9001-86-9	phospholipase C	9010-98-4	neoprene
9001-87-0	phospholipase D	9011-05-6	cascamite
9001-98-3	rennin	9011-09-0	vinylidene chloride-butyl acrylate copolymer
9002-07-7	trypsin		
9002-60-2	ACTH	9012-25-3	methyltransferase
9002-68-0	follicle-stimulating hormone	9012-36-6	agarose
9002-69-1	relaxin	9012-37-7	n-acylamino acid
9002-72-6	somatotropic hormone	9012-56-0	amidase
9002-84-0	polytetrafluoroethylene	9013-38-1	dopamine hydroxylase
9002-86-2	geon	9014-89-5	polyethylene glycol lauryl thioether
9002-86-2	polyvinyl chloride	9015-68-3	L-asparaginase
9002-88-4	polythylene	9016-18-6	carboxylic esterase
9002-89-5	polyvinyl alcohol	9016-87-9	polymethylene polyphenylisocyanate
9002-93-1	p-$tert$-octylphenoxy polyethoxyethanol	9027-96-7	citrate synthase
9003-07-0	polypropylene	9028-86-8	acetaldehyde dehydrogenase
9003-11-6	purina bloat block	9028-86-8	aldehyde dehydrogenase
9003-11-6	"Therabloat"	9031-37-2	ceruloplasmin
9003-20-7	polyvinyl acetate	9032-73-9	arylesterase
9003-36-5	formaldehyde, polymer with (chloromethyl)oxirane and phenol	9032-73-9	neurotoxic esterase
		9032-88-6	fumarase
9003-39-8	polyvinylpyrrolidone	9040-12-4	ricin D
9003-53-6	polystyrene	9050-36-6	maltodetrin
9003-55-8	styrene-butadiene rubber	9054-89-1	dismutase
9004-32-4	akucell	9056-38-6	nitrostarch
9004-32-4	carboxymethylcellulose	9067-32-7	actimoist
9004-32-4	carbodymethyl cellulose sodium	9067-32-7	hyalure
9004-34-6	cellulose	10008-90-9	chloro(trivinyl)stannane
9004-62-0	natrosol	10021-35-9	2-thio-2h-1,3-benzoxazine-2,4(3h)-dione
9004-64-2	hydroxypropyl cellulose		
9004-67-5	benecel	10022-31-8	barium nitrate
9004-67-5	methylcellulose	10022-50-1	nitryl fluoride
9004-70-0	nitrocellulose	10024-97-2	anesthetic gas
9004-83-5	polyoxyethylated (C9-C10) alhyl thioether	10024-97-2	nitrous oxide
		10025-65-7	platinum chloride
9004-99-3	polyoxyethylene (40) monostearate	10025-67-9	sulfur chloride
9004-99-3	polyoxyethylene (8) stearate	10025-73-7	chromic chloride
9005-08-7	polyglycol distearate	10025-76-0	europium chloride
9005-25-8	starch	10025-78-2	trichlorosilane
9005-38-3	sodium alginate	10025-82-8	indium chloride
9005-42-4	heparin	10025-82-8	indium trichloride
9005-46-3	casein	10025-85-1	nitrogen trichloride
9005-46-3	sodium caseinate	10025-87-3	phosphorus oxychloride
9005-79-2	glycogen	10025-91-9	antimony trichloride
9006-04-6	rubber, natural	10026-03-6	selenium tetrafluoride
9006-26-2	ethylene-maleic anhydride copolymer	10026-04-7	silicon tetrachloride
		10026-06-9	stannic chloride pentahydrate

10476-86-5	strontium iodide	12165-69-4	phosphorus trisulfide
10526-15-5	bis(2-ethylhexyl)thiodipropionate	12174-11-7	attapulgite
10563-26-5	1,5,8,12-tetraazadodecane	12179-04-3	sodium tetraborate pentahydrate
10584-98-2	dibutyldi(2-ethylhexyloxycarbonylmethyl-thio)stannane	12232-99-4	sodium bismuthate
		12244-57-4	gold sodium thiomalate
		12401-86-4	sodium monoxide
10588-01-9	sodium dichromate	12427-38-2	maneb
10599-70-9	3-acetyl-2,5-dimethylfuran	12550-82-2	bis(l-cysteinato)mercury
10599-90-3	chloramine	12604-58-9	ferrovanadium
10691-18-2	dioctyltin maleate	12645-45-3	iridium trichloride
11024-24-1	digitonin	12656-85-8	molybdate orange
11028-71-0	concanavalin A	13007-92-6	chromium hexacarbonyl
11032-79-4	α-bungarotoxin	13047-38-3	bis(tribenylstannyl)sulfide
11041-12-6	cholestyramine resin	13092-75-6	silver acetylide
11052-72-5	oligomycin C	13121-70-5	cyhexatin
11056-06-7	bleomycin	13138-45-9	nickel nitrate
11056-06-7	bleomycin A2	13171-21-6	phosphamidon
11075-17-5	carboxypeptidase	13189-98-5	s-(3-hydroxypropyl)-l-cysteine
11080-14-1	neurotoxin A (naja naja reduced)	13252-14-7	2-(1,1,2,3,3,3-hexafluoro-2-(heptafluoro-propoxy)propoxy)-2,3,3,3-tetrafluoro-propanoic acid
11096-82-5	aroclor 1260		
11097-21-9	chlorodiphenyl		
11097-69-1	aroclor 1254		
11100-45-1	enterotoxin b, staphylococcal	13253-44-6	methionine hydantoin
11103-86-9	zinc yellow	13274-42-5	tetrapropylenepentamine
11105-11-6	tungstic acid	13275-42-8	2-(2-bromophenyl)-1h-benzimidazole
11112-10-0	sodium antimonate		
11138-49-1	dynaflock	13302-06-2	tri-n-butyltin methanesulfonate
11138-49-1	sodium aluminate	13302-08-4	triphenyltin methanesulfonate
12001-26-2	mica	13327-32-7	beryllium hydroxide
12001-89-7	dicumene chromium	13400-13-0	cesium fluoride
12007-97-5	molybdenum boride	13410-01-0	sodium selenate
12018-95-0	copper indium selenide	13413-18-8	hexaoctyldistannthiane
12031-80-0	lithium peroxide	13422-55-4	methylcobalamin
12037-82-0	phosphorus heptasulfide	13424-46-9	lead azide
12054-48-7	nickelic hydroxide	13444-71-8	periodic acid
12057-24-8	lithium oxide	13444-85-4	nitrogen triiodide
12058-85-4	sodium phosphide	13444-90-1	nitryl chloride
12067-99-1	phosphotungstic acid	13446-18-9	nitric acid, magnesium salt, hexahydrate
12068-85-8	pyrite		
12069-69-1	copper carbonate	13446-73-6	rubidium dichromate
12070-12-1	tungsten carbide	13453-07-1	gold trichloride
12071-83-9	propineb	13455-36-2	cobaltous phosphate
12079-65-1	manganese cyclopentadienyl tricarbonyl	13463-39-3	nickel carbonyl
		13463-40-6	iron pentacarbonyl
12108-13-3	methylcyclopentadienyl manganese tricarbonyl	13463-41-7	"Zinc Omadine"
		13463-67-7	titanium dioxide
12122-67-7	metiram	13464-82-9	indium sulfate
12122-67-7	zineb	13466-78-9	δ-3-carene
12124-99-1	ammonium sulfide	13472-08-7	2,2'-azobisisovaleronitrile
12125-01-8	ammonium fluoride	13472-45-2	sodium tungstate
12125-02-9	ammonium chloride	13477-00-4	barium chlorate
12125-03-0	potassium stannate	13483-18-6	bis-1,2-(chlromethoxy)ethane
12125-56-3	nickelous hydroxide	13494-80-9	tellurium

15422-00-1	3-(diethylamino)propyl isopropyl(phenyl)glycolate hydrochloride
15468-32-3	tridymite
15503-87-4	*trans*-retrorsine
15520-10-2	dytek
15535-69-0	dibutyltin maleate
15535-79-2	dioctylthioacetoxystannane
15541-45-4	bromate
15546-11-9	bis(methoxymaleoyloxy) dioctylstannane
15546-12-0	bis((2-ethyl)hexyloxy)maleoyloxy) di(*n*-butyl)stannane
15546-16-4	bis(butoxymaleoyloxy) dibutylstannane
15571-58-1	di-*n*-octyltin bis(2-ethylhexyl) mercapto-acetate
15588-95-1	STP (hallucinogen)
15597-43-0	isothiocyanatotrimethyltin
15652-38-7	decafentin
15687-27-1	ibuprofen
15699-18-0	nickel ammonium sulfate
15710-39-1	pentacarbonyl(piperidine)chromium
15825-70-4	mannitol hexanitrate
15829-53-5	mercurous oxide
15879-93-3	α-chloralose
15905-86-9	uranium nitrate
15972-60-8	alachlor
15978-91-3	*cis*-diamminedibromoplatinum(II)
16034-77-8	iocetamic acid
16079-88-2	1-bromo-3-chloro-5,5-dimethylhydantoin
16128-42-0	(1-imidazolyl)tributylplumbane
16219-75-3	5-ethylidene-2-norbornene
16219-99-1	*n*-nitroso-*n*-methyl-4-aminopyridine
16240-52-1	*n,n′,o,o′*-tetramethylcurinium diiodide
16248-90-1	tin bis(diethyldithiocarbamate)
16377-01-8	indospicine monohydrochloride monohydrate
16409-45-3	menthyl acetate
16409-46-4	methyl 3-menthylbutyrate
16423-68-0	erythrosine
16470-24-9	C.I. fluorescent brightener 220
16478-59-4	hemicholinium
16481-54-2	podophyllotoxin 4-o-glucoside
16484-86-9	1,2-diethoxy ethylene
16503-95-0	7,8-dimethoxyisoquinoline
16518-17-5	potassium nicotinate
16561-55-0	zinc-*o*-phenanthroline complex
16721-80-5	sodium hydrosulfide
16725-53-4	(*z*)-9-tetradecenol acetate
16726-46-8	chloroxyphenamine

16731-55-8	potassium metabisulfite
16752-77-5	methomyl
16758-26-2	diphenylaminochloroarsine
16816-67-4	pantethine
16842-03-8	cobalt hydrocarbonyl
16853-85-3	lithium aluminum hydride
16871-71-9	zinc fluorosilicate
16871-90-2	potassium fluorosilicate
16872-09-6	1,2-dicarbandodecaborane(12)
16872-11-0	fluoroboric acid
16881-77-9	methyldimethoxysilane
16899-81-3	3′,4′-dihydroxy-2-(isopropylamino)acetophenone hydrochloride
16919-19-0	ammonium fluorosilicate
16919-19-0	ammonium hexafluorosilicate
16919-58-7	ammonium chloroplatinite
16921-30-5	potassium chloroplatinate
16923-95-8	zirconium potassium fluoride
16924-00-8	tantalum potassium fluoride
16940-66-2	sodium borohydride
16940-81-1	hexafluorophosphoric acid
16949-15-8	lithium borohydride
16961-83-4	fluorosilicic acid
16971-82-7	dicatechol borate 1,3-di(*o*-tolyl)guanidine salt
16984-48-8	fluoride
17003-79-1	formaldehyde bis(2-fluoro-2,2-dinitroethyl)acetal
17008-69-4	pseudobufarenogin
17014-71-0	potassium peroxide
17058-53-6	2,5-bis(phenylthio(benzoquinone))
17091-40-6	2-isobutylaminoethanol
17099-70-6	aluminum hexafluorosilicate
17126-65-7	ethyl 4-isothiocyanatobutanoate
17132-74-0	homoaromoline
17146-95-1	pentazocine lactate
17226-43-6	glycerin 1-isopropyl ether
17243-39-9	benzoctamine
17247-12-3	sodium 2,3,5-triiodobenzoate
17256-39-2	1-chloro-*n,n*-dimethyl-2-propanamine hydrochloride
17297-82-4	oxypendyl hydrochloride
17360-35-9	oxotremorine sesquifumarate
17381-88-3	diallyldibromo stannane
17406-45-0	tomatine
17431-55-9	7-chloroincomycin hydrochloride
17465-86-0	δ-cavitron
17476-04-9	lithium aluminum-tri-*tert*-butoxyhydride
17489-40-6	malloside
17523-77-2	niobium potassium oxyfluoride

21709-44-4	*o*-(1,3-dithiolan-2-yl)phenyl dimethylcarbamate	22936-20-5	*o,o*-diethyl *o*-(4-(1-(((((butylamino) caronyl)oxy)imino)ethyl)phenyl) phosphorothioi acid ester
21725-46-2	cyanazine		
21727-09-3	(pentachlorophenyl)glycolonitrile	22936-34-1	*o*-(4-(1-(((ethoxycarbonyl)oxy) imino)ethyl)phenyl) *o,o*-diethyl phosphorothioate
21800-49-7	haratavic acid		
21908-53-2	mercuric oxide, red		
21908-53-2	mercuric oxide, yellow	22936-44-3	*o*-(4-(1-((acetyloxy)imino)ethyl)-3-methylphenyl) *o,o*-diethyl-phosphorothioate
21961-30-8	3-amino-5-chlorobenzoic acid		
22041-28-7	1-carboxymethyl-1-methylpyrrolidinium iodide methyl ester	22941-83-9	*o,o*-diethyl *o*-(4-(1-(((((methylamino)carbonyl)oxy) imino)ethyl)phenyl) phosphorothioic acid ester
22041-39-0	*n,n*-dimethyl-3-(pyrrolidin-1-yl)propionamide		
22047-25-2	2-acetyl pyrazine	22942-02-5	4-(*o*-(*o,o*-diethylphosphorothioyl)) benzal-doximino-*n*-butylcarbamate
22128-43-4	4-methyl-4-hydroxy-1-octyne		
22143-50-6	*n*-methylloline		
22187-44-6	1,4,5,6-tetrahydro-3-(10,11-dihydro-dibenzo(*a,d*) cyclohepten-5-yl)-1-methyl-*as*-triazine hydrobromide	22942-43-4	*o,o*-diethyl *o*-(4-(1-(((((dimethylamino)carbonyl)oxy) imino)ethyl)phenyl) phosphorothioic acid ester
22208-25-9	2-ethyl-2-(hydroxymethyl)-1,3-propanediol triacetoacetate	22942-83-2	1-keto-2,3-dihydrocyclopentindole oxime
22224-92-6	fenamiphos	22967-92-6	methylmercury
22248-79-9	tetrachlorvinphos	23031-36-9	*d,d*-t80-parallethrin
22287-69-0	*p*-(phenylazo)phenyl isocyanide	23110-15-8	fumagillin
22374-89-6	1-phenyl-3-aminobutane	23125-28-2	diethyl 2,6-dimethyl-4(2-pyridyl)-1,4-dihydro-3,5-pyridine-dicarboxylate
22393-63-1	*cis*-broparestrol		
22394-38-3	3-(*n*-2-thiazolylformimidoyl)indole		
22398-80-7	indium phosphide	23155-02-4	fosfomycin
22560-50-5	sodium clodronate	23181-80-8	7-aminoheptanenitrile
22578-17-2	sodium pentafluorostannate	23184-66-9	butachlor
22708-05-0	barium reineckate	23188-89-8	*n*-(tributylplumbyl)benzimidazole
22729-75-5	*n*-(2-fluoro-1,1,2,2-tetrachloroethylthio)-methanesulfonanilide	23192-42-9	(*z*)-7-hexadecen-1-ol acetate
		23319-66-6	phenylmercuritriethanolammonium lactate
22756-36-1	rubidium azide	23248-53-5	1,1'-sulfinylbis(1,2-dichloroethane)
22791-18-0	spiro(1,3-benzodioxole-2,1'-cyclohexan)-4-ol, methylcarbamate	23292-85-5	triphenyltin levulinate
		23383-11-1	ferrous citrate
		23399-90-8	2,4,6-trichlorophenyl acetate
22791-23-7	1,3-benzodioxol-4-yl methylcarbamate	23414-72-4	zinc permanganate
		23420-61-3	3,4-xylidino-2-oxazoline
22791-33-9	1,3-benzodioxol-4-ol, 2,2-dimethyl-, acetylmethylcarbamate	23422-53-9	formetanate
		23445-00-3	swertianolin
22898-01-7	sodium tetrafluoropropionate	23522-05-6	gracilin
22935-72-4	4-(*o*-(*o,o*-diethylphosphorothioyl)) benzaldoximino-*n*-morpholinylcarbamate	23535-89-9	dibutyl(tetrachlorophthalato) stannane
		23572-32-9	9-diethylaminoethyl-2-phenylimidazo(1,2-*a*)benzimidazole dihydrochloride
22936-03-4	2-methoxy-4-(*o*-(*o,o*-diethylphosphoro-thioyl))benzaldoximino-*n*-methylcarbamate		
		23590-99-0	inosine dialdehyde
22936-17-0	*o*,-diethyl *o*-(4-(1-((((((1-methylethyl) amino)carbonyl)oxy)imino)ethyl) phenyl)phosphorothioic acid ester	23597-98-0	5-(*p*-chlorophenyl)-2,3,5,6-tetrahydro-imidazo(1,2-*c*)quinazolin

26134-62-3	lithium nitride	28140-60-5	*n*-(3-phenylimino-1-propenyl)anilinehydrochloride
26155-31-7	morantel tartrate		
26160-83-8	tricyclodecanedimethanol	28249-77-6	benthiocarb
26183-52-8	polyoxyethylated (6) isodecyl alcohol	28289-54-5	1,2,3,6-tetrahydro-1-methyl-4-phenylpyridine
26225-79-6	ethofumesate	28553-12-0	phthalic acid, diisononyl ester
26248-87-3	tri(chloroporpyl)phosphate	28660-63-1	di-*n*-butyl(dibutyryloxy)stannane
26249-12-7	*o*-dibromobenzene	28660-67-5	bis(tetradecanoyloxy)dibutylstannane
26264-05-1	benzenesulfonic acid, dodecyl-, compd. with isopropylamine	28801-69-6	tributyltin neodecanoate
26401-97-8	*d*(*n*-octyl)tin-*S,S'*-bis-(isooctylmercaptoacetate)	28846-35-7	9-(2,2-dimethylhydrazino)acridine
		28846-37-9	4-(9-acridinyl)-2-methyl-3-thiosemicarbazide
26419-72-7	2-methyl-*m*-dithiane-2-carboxaldehyde *o*-(methylcarbamoyl)oxime	28846-38-0	9-(2,2-dimethylhydrazino)acridine mono(methyl sulfate)
26543-09-9	monocrotalic acid	28846-39-1	9-(2,2-dibutylhydrazino)acridine monohydrochloride
26544-23-0	isodecyl diphenyl phosphite		
26571-79-9	chlorophenyltrichlorosilane	28846-40-4	9-(piperidinoamino)acridine
26628-22-8	sodium azide	28846-41-5	9-(morpholinoamino)acridine
26636-01-1	bis(isooctylcarbonylmethylthio)dimethylstannane	28846-42-6	9-(morpholinoamino)acridine mono(methylsulfate)
26640-60-8	*n*-isopropyl-α-methylphenethylamine hydrochloride	28846-43-7	9-(3'-methyl-4'-ethylidene-thiosemicarb-azido)acridine
26644-46-2	triforine	28846-44-8	2-(1-(9-acridnyl)hydrazino)ethanol monohydrochloride
26691-08-7	9-methoxyellipticine lactate	28894-74-8	deacetyllyoniatoxin
26761-40-0	diisodecyl phthalate	29023-82-3	2-chloro-9-(2,2-dimethylhydrazino)acridine
26896-20-8	neodecanoic acid		
26913-06-4	polyethylene imine	29023-83-4	9-(2,2-bis(2-chloroethyl)hydrazino)acridine monohydrochloride
26952-21-6	isooctyl alcohol		
26970-82-1	sodium selenite pentahydrate		
27013-91-8	α-hederine	29023-84-5	4-(9-acridinyl)-3-thiosemicarbazide
27074-70-0	α-(diamylaminomethyl)-1,2,3,4-tetrahydro-9-phenanthrenemethanol	29023-85-6	4-(9-acridinyl)-2-methyl-3-thiosemicarbazone acetone
27107-89-7	octyltris(2-ethylhexyloxycarbonyl-methylthio)stannane	29171-20-8	linalool, dehydro-
		29575-02-8	bis(butoxymaleoyloxy)dioctylstannane
27193-86-8	dodecylphenol		
27205-24-9	3-(ethylthio)butyraldehyde	29714-87-2	ocimene
27215-10-7	diisooctyl acid phosphate	30099-72-0	bis(tributyl(sebacoyldioxy))tin
27252-69-3	polyethylene glycol 400, dichloride	30460-34-5	grayanotoxin III 6,14-diacetate
27323-41-7	triethanolamine dodecylbenzene sulfonate	30460-36-7	grayanotoxin VI
		30525-89-4	paraformaldehyde
27371-95-5	2,2-dibutyl-1,3,2-oxathiastannolane	30560-19-1	acephate
27478-34-8	dinitronaphthalene	30586-10-8	dichloropentane
27554-26-3	diisooctyl phthalate	30618-84-9	glyceryl monothioglycolate
27774-13-6	vanadyl sulfate	30638-08-5	cobalt phthalocyaninesulfonate
27896-84-0	5-nitrobenzimidazole nitrate	30846-35-6	formaldehyde, polymer with 4-nonylphenol and oxirane
27953-64-6	17-α,21-dihydroxy-14-α-pregn-4-ene-3,20-dione 21-iodoacetate	30947-30-9	((3,5-bis(1,1-dimethylethyl)-4-hydroxy-phenyl)methyl) phosphonic acid, monoethyl ester, nickel(2+) salt (2:1)
27955-94-8	4,4',4''-ethylidynetrisphenol		
27970-32-7	methyl orange		
28069-72-9	*trans,cis*-2,6-nonadienol	31078-10-1	isovaltrate

37139-99-4	olealkonium chloride
37231-28-0	melittin
37321-09-8	apramycin
37326-33-0	hyaluronoglucosaminidase
37430-50-5	2,3-dihydro-2,2-dimethyl-7-benzofuranyl-methyl ((trichloromethyl)thio)carbamate
37677-14-8	isohexenyl cyclohexenyl carboxaldehyde
37762-06-4	zaprinast
37902-85-5	methyl 3-((ethyl(propylamino) phosphino-thioyl)oxy)-2-butenoate
37924-13-3	perfluidone
37971-35-0	2-phosphonobutanetricarbonic acid pentamethyl ester
38083-17-9	1-(p-chlorophenoxy)-3,3-dimethyl-1-(1-imidazolyl)-2-butanone
38090-84-5	o,o-diethyl s-((5-methoxy-1,3,4-thiadiazol-2-yl)methyl) phosphorothioate
38090-92-5	o,o-dimethyl s-((5-(methylthio)-1,3,4-thiadiazol-2-yl) methyl)phosphorodithioate
38230-32-9	dehydrocurdione
38302-26-0	leucothol B
38527-91-2	ethaphos
38700-88-8	m-chlorophenyl phenyl sulfide
38802-82-3	diisopropyltin dichloride
38951-85-8	1-chloro-2,2,5,5-tetramethyl-4-imidazolidinone
38998-91-3	bis(1,3-dithiocyanato-1,1,3,3,-(tetrabutyldistannoxane)
39236-46-9	"Germall 115"
39296-38-3	tempalgin
39324-65-7	herbifert
39464-66-9	polyoxyethylene lauryl ether phosphate
39562-22-6	2-methoxyethyl 2-(m-nitrobenzylidene) acetoacetate
39637-16-6	(2,4-dichlorophenoxy) tributylstannane
39807-15-3	5-tert-butyl-3-(2,4-dichloro-5-propargyloxyphenyl)1,3,4-oxadiazol-2(3h)-one
40199-26-6	(4-chlorophenyl)methyl dodecyl 3-pyridinylcarbonimidodithioate
40596-69-8	methoprene
40626-35-5	heterophos
40630-63-5	1-octanesulfonyl fluoride
41083-11-8	(1h-1,2,4-triazolyl-1-yl) tricyclohexylstannane

41219-25-4	4-thiochromanyl-o,o-dimethyldithiophosphate
41219-30-1	6-chloro-4-thiochromanyl o,o-dimethyl dithiophosphate
41219-31-2	6-chloro-4-thiochromanyl-o,o-diethyl dithiophosphate
41643-22-5	(4-chlorophenyl)methyl tetradecyl 3-pyridinylcarbonimidodithioate
41643-23-6	(4-chlorophenyl)methyl hexadecyl 3-pyridinylcarbonimidodithioate
41643-24-7	(4-chlorophenyl)methyl octadecyl 3-pyridinylcarbonimidodithioate
41663-84-7	4-nitro-n-methylphthalimide
41762-18-9	triamminetrinitratatorhodium(III)
41920-59-6	p-(benzyloxy)phenyl bis(1-aziridinyl)phosphinate
41941-50-8	epichlorohydrin-bis(3-aminopropyl)methylamine copolymer
41999-58-0	sodium o-carboxymethyltartronate
42296-74-2	1,4-hexadiene
42615-29-2	benzenesulfonic acid
42723-79-5	3-pyridinylcarbonimidothioic acid, s-((4-(1,1-dimethylethyl)phenyl) methyl)o-(3-methylcyclohexyl) ester
42754-15-4	3-pyridinylcarbonimidothioic acid, o-cyclopentyl s-((4-(1,1-dimethylethyl)phenyl)methyl) ester
42754-16-5	3-pyridinylcarbonimidothioic acid, o-cyclohexyl s-((4-(1,1-dimethylethyl)phenyl)methyl) ester
42754-17-6	3-pyridinylcarbonimidothioic acid, s-((4-(1,1-dimethylethyl)phenyl) methyl)o-(2-methylcyclohexyl) ester
42754-18-7	3-pyridinylcarbonimidothioic acid, s-((4-(1,1-dimethylethyl)phenyl) methyl)o-(4-methylcyclohexyl) ester
42754-19-8	3-pyridinylcarbonimidothioic acid, o-(2,5-dimethylcyclohexyl) s-((4-(1,1-dimethylethyl)phenyl) methyl)ester
42754-20-1	C3-pyridinylarbonimidothioic acid, o-(3,4-dimethylcyclohexyl) s-((4-(1,1dimethylethyl)phenyl)methyl) ester
42754-21-2	3-pyridinylcarbonimidothioic acid, o-(3,5-dimethylcyclohexyl) s-((4-(1,1-dimethylethyl)phenyl) methyl)ester

51308-71-5 s-((4-(1,1-dimethylethyl)phenyl) methyl)o-octyl-3-pyridinylcarbonimidothioate

51308-72-6 methyl (4-(1-methylethyl)phenyl) methyl 3-pyridinylcarbonimido-dithioate

51308-74-8 heptyl (4-(1-methylethyl)phenyl) methyl 3-pyridinylcarbonimido-dithioate

51308-75-9 bis((4-(1-methylethyl)phenyl) methyl) 3-pyridinylcarbonimido-dithioate

51308-76-0 bis((4-(1,1-dimethylethyl)phenyl) methyl) 3-pyridinylcarbonimido-dithioate

51308-77-1 o-((4-chlorophenyl)methyl) s-(2-methylpropyl)-3-pyridinylcarbonimidothioate

51308-78-2 s-((4-chlorophenyl)methyl) o-(2-methylpropyl)-3-pyridinylcarbonimidothioate

51308-79-3 s-((4-bromophenyl)methyl) o-butyl 3-pyridinylcarbonimidothioate

51308-80-6 (4-bromophenyl)methyl butyl 3-pyridinylcarbonimidodithioate

51312-42-6 sodium-12-tungstophosphate

51317-24-9 lead mononitroresorcinate

51338-27-3 illoxan

51379-04-5 (4-(1,1-dimethylethyl)phenyl) methyl pentyl-3-pyridinylcarbonimidodithioate

51379-05-6 3-pyridinylcarbonimidodithioic acid, (4-(1,1-dimethylethyl)phenyl)methyl 1,1-dimethylpropyl ester

51628-36-5 5'-benyl-3'-furylmethyl α-ethyl-phenylacetate

51628-37-6 3'-phenoxybenzyl α-ethylphenylacetate

51628-56-9 5'-benyl-3'-furylmethyl α-isopropyl-4-methoxyphenyl acetate

51628-95-6 3'-phenoxybenzyl α-isopropyl-4-methoxyphenylacetate

51628-96-7 5'-phenoxyfurfuryl α-isopropyl-4-methoxyphenylacetate

51629-13-1 3'-phenoxybenzyl α-ethyl-4-methoxyphenylacetate

51629-15-3 3'-phenoxybenzyl α-isopropenyl-4-methoxyphenylacetate

51629-37-9 3'-phenoxybenzyl α-isopropyl-4-methylphenylacetate

51629-48-2 3'-phenoxybenzyl α-isopropyl-4-chlorophenylacetate

51629-54-0 3'-phenoxybenzyl α-isopropenyl-4-methylphenylacetate

51629-58-4 4-chloro-α-(1-methylethyl) benzeneacetic acid, (2,6-dimethyl-4-(2-propynyl)phenyl) methyl ester

51629-79-9 4-methyl-α-(1-methylethyl) benzeneacetic acid, (4,5,6,7-tetrahydrobenzo(b)thien-2-yl) methyl ester

51630-04-7 (3-phenoxyphenyl)methyl α-ethyl-4-methylbenzeneacetate

51630-12-7 (5-(phenylmethyl)-3-furanyl)methyl 4-chloro-α-ethylbenzeneacetate

51630-33-2 (3-phenoxyphenyl)methyl 4-chloro-α-(1-methylethyl)benzeneacetate

51630-58-1 fenvalerate

51715-75-4 5'-propargyl-α'-ethynyl-3'-furylmethyl α-ethylphenyl acetate

51961-45-6 n-(3-aminopropyl)-1,3-propanediamine polymer with (chloromethyl)oxirane

52032-20-9 n-perfluorooctanesulfonyl-n-methyl-carbamoyl ((nonadecaethoxy) butoxy)butyl ether

52112-09-1 bis(trifluoroacetoxy)dibutyltin

52207-48-4 disodium S,S'-(2-dimethylamino-1,3-propanediyl)bis(thiosulfate)

52236-29-0 molinate sulfoxide

52315-07-8 cypermethrin

52334-99-3 n-methyl 3-mercaptopropionamide

52557-31-0 grayanol A

52611-78-6 grayanol B

52645-53-1 permethrin

52659-57-1 aplysiatoxin

52740-16-6 calcium arsonate (1:1)

52918-63-5 decamethrin

53003-10-4 salinomycin

53061-10-2 1,1'-(oxybis(methylenesulfonyl)) bis(2-chloroethane)

53083-27-5 1-oxo-2-(2,4,6-trimethylphenyl)-1h-inden-3-yl dodecanoate

53120-27-7 (z,z)-3,13-octadecadien-1-ol acetate

53369-07-6 glufosinate

53403-68-2 potassium dinitrotetrachloroplatinate

53404-82-3 tributyltin isopropylsuccinate

53413-47-1 dicyclohexylphenyltin hydroxide

53469-21-9 aroclor 1242

53744-50-64 acetoxyphenyl methyl carbinol-4-acetoxyphenyl methyl carbinol

53775-55-6 potassium 5-(2,4-dichlorophenoxy)-2-nitrobenzoate

60735-64-0 4'-(n,n-diethylalanyl) methanesulfonanilide hydrochloride

60767-85-3 tetra-n-butylphosphonium phenylphosphinate

60837-57-2 anoxomer

61001-04-5 2-p-tolyimidazo(2,1-a)isoquinoline

61001-06-7 2-(p-bromophenyl)imidazo(2,1-a)isoquinoline

61001-09-0 2-(4-fluorophenyl)imidazo(2,1-a)isoquinoline

61001-11-4 2-(3-chlorophenyl)imidazo(2,1-a)isoquinoline

61001-12-5 2-(3,4-dichlorophenyl)imidazo(2,1-a)isoquinoline

61001-13-6 2-(4-nitrophenyl)imidazo(2,1-a)isoquinoline

61001-14-7 2-(m-ethoxyphenyl)imidazo(2,1-a)isoquinoline

61001-15-8 2-(3-propoxyphenyl)imidazo(2,1-a)isoquinoline

61001-16-9 2-(3-aalyloxyphenol)imidazo(2,1-a)isoquinoline

61001-19-2 2-p-tolyl-5,6-dihydroimidazo(2,1-a)isoquinoline

61001-21-6 imidazo(1,2-a)isoquinoline, 5,6-dihydro-2-(m-methoxyphenyl)-

61262-53-1 1,1'-(1,2-ethanediylbis(oxy))bis(2,3,4,5,6-pentabromobenzene)

61288-13-9 bromkal 80

61319-99-1 2-(1,1-dimethylethyl)pyrimidine

61379-65-5 rifapentine

61532-76-1 2-(nitromethylene)-1-(phenylmethyl)imidazolidine

61718-82-9 fluvoxamine maleate

61788-32-7 hydrogenated terphenyls

61789-40-0 coco amido betaine

61790-14-5 lead naphthenate

61790-53-2 diatomaceous earth

61791-36-4 tall oil imidazoline

61825-94-3 oxalatoplatinum

61827-42-7 polyoxyethylated (4) isodecyl alcohol

61947-30-6 diisobutyloxostannane

62078-98-2 n-(2-chloroethyl)-n-ethyl-2-bromobenzylamine

62232-46-6 2-(4-methylaminobutoxy) diphenylmethanehydrochloride

62382-21-2 2-methyl-n-(n'-methyl-n'-(4-morpholino-sulfenyl) carbamoyloxy)thio-acetimidate

62382-23-4 2-methyl-2-(methylthio)propanol-o-((n-methyl-n-morpholinosulfenyl) carbamoyl)oxime

62524-93-0 7-(3,5-bis((tetrahydro-2h-pyran-2-yl)oxy)-2-(4-phenoxy-3-((tetrahydro-2h-pyran-2-yl)oxy)-1-butenyl)cyclopentyl)-2-(phenylseleno)-5-heptenoic acid, methyl ester

62601-71-2 protopolygonatoside g

62602-94-2 optimax

63091-06-5 katapol vp-532

63123-39-7 2-(acetylthioglycolic amide) benzothiazole

63166-73-4 phylanthoside

63231-51-6 solvesso 100

63333-35-7 bromethalin

63428-83-1 nylon

63449-47-8 tetra(2-butylisopropyl) orthosilicate

63680-10-4 2-aminoethylisothiouronium diacetate

63697-52-9 1-(cyclohexylcarbonyl)-1,2,3,6-tetra-hydropyridine

63716-40-5 n-butoxypropanol

63782-90-1 flamprop-m-isopropyl

63868-81-5 1,1'-(p-phenylenebis(carbonylmethyl))di-3-picolinium dibromide

63869-03-4 ethylmercuric dicyandiamide

63869-06-7 methylmercury toluenesulfonate

63869-12-5 copper methane arsonate

63869-87-4 trimethyltin sulfate

63884-77-5 β,γ,γ-trimethylcaproaldehyde thiosemicarbazone

63885-01-8 zinc ammonium nitrite

63901-83-7 5-fluoro-3-(tetrahydro-2-furyl)uracil

63906-07-0 (4,4'-biphenylylenebis(2-oxoethylene))bis(3-iodopyridinium) dibromide

63917-25-9 bis(2-methylallyl) diglycolate

63918-97-8 lead trinitroresorcinate

63919-09-5 dimethyl diisopropyl pyrophosphate

63919-14-2 zirconium lactate

63919-17-5 diammine(benzylmalonato) platinum(II)

63938-62-5 3-(2-(4-chlorobenzylamino) ethyl)indole monohydrochloride

63942-42-7 n,n'-bis(2-methylsulfonyl-2-methylpropionaldehyde-o-(n-methylcarbamoyl)oxime)sulfide

63942-43-8 n,n'-bis(2-cyano-2-methylpropionaldeheyo(n-methylcarbamoyl)oxime)sulfide

63942-44-9 n,n'-bis(1-methylthio-1-(n,n-dimethyl-carbonyl)formaldehyde-o-(n-methyl-carbamoyl)oxime) sulfide

67287-95-0	6-chloro-2,3,4,5-tetrahydro-3-methyl-1-(3-methylphenyl)-1h-3-benzazepine-7,8-diol, hydrobromide
67293-51-0	α-heptyl-3,4,5-trimethoxyphenethylamine
67306-00-7	fenpropidine
67410-20-2	triphenyltin propiolate
67445-50-5	tripropyl(butylthio)stanane
67465-66-1	10-(2-(diethylamino)propyl)-10h-pyrido(3,2-b)(1,4)benzothiazine
67469-78-7	1-(2-(bis(4-fluorophenyl)methoxy)ethyl)-4-(3-phenylpropyl)piperazine dihydrochloride
67481-15-6	metofos
67704-68-1	metazine
68000-78-2	isepamicin disulfate
68002-18-6	isobutylated urea formaldehyde
68070-99-5	polyoxyethylene(4)docyl alcohol phosphate potassium salt
68083-14-7	polydimethyldiphenylsiloxane
68131-39-5	alcohols, C12-C15, ethoxylated
68154-34-7	coconut amine oil condensate
68308-34-9	shale oil
68333-98-2	coconut oil, esters with polyethylene glycol nonylphenyl ether
68334-67-8	1-isopropoxypentachlorobutadiene
68424-04-4	polydextrose
68439-49-6	alcohols, C16-C18, ethoxylated
68476-34-6	diesel oil
68476-85-7	liquefied petroleum gas
68515-41-3	1,2-benzenedicarboxylic acid, di-c7-c9-branched alkyl esters
68515-48-0	diisononyl phthalate
68515-49-1	di-(C9-C11 alkyl) phthalate
68551-12-1	alcohols, C12-C6, ethoxylated
68586-17-4	ethylene glycol mono-dicyclopentenyl ether
68603-25-8	alcohols, C8-C10, ethoxylated propoxylated
68603-42-9	coconut oil acid diethanolamine
68648-41-9	treemoss concrete
68648-87-3	linear alkylbenzene A-315
68789-89-9	n,n'-bis-(2-(o-(n-methylcarbamoyl)oximino)-1,4-dithiane)disulfide
68789-93-5	n,n'-bis-(1-(2-cyanoethylthioacetaldehyde-o-(n-methylcarbamoyl)oxime)disulfide
68789-90-2	n,n'-bis(1-methylthioacetaldehyd-o-(n-methylcarbamoyl)oxim)-disulfid

68815-56-5	α-(3-carboxy-1-oxosulfopropyl)-ω-hydroxy-poly(oxy-1,2-ethanediyl), C10-C16 alkyl ethers, disodium salts
68848-64-6	lithium silicon
68890-66-4	piroctone olamine
68891-01-0	melamine, formaldehyde, toluenesulfonamide polymer, butylated
68891-47-4	2-propene-nitrile, polymer with 1,3-butadiene, 3-carboxy-1-cyano-1-methylpropyl-terminated,2-hydroxy-3-((1-oxo-2-propenyl)oxy)propyl ester
68906-88-7	indenolol hydrochloride
68909-26-2	α-hydro-ω-hydroxypoly(oxy(methyl-1,2-ethanediyl)) ether with bis((2-(hydroxyethyl)amino)methyl)phenol (3:1)
68917-09-9	ocotea oil
68951-67-7	alcohols, C14-C15, ethoxylated
68955-53-3	C12-C14-tert-alkyl amines
69012-64-2	silica, amorphous fume
69226-43-3	di-n-octyltin bis(laurylthioglycolate)
69226-44-4	di-n-octyltin ethyleneglycol dithioglycolate
69226-47-7	tributyl(undecanoyloxy)stannane
69239-37-8	di-n-butyltin di(monononyl)maleate
69327-76-0	buprofezine
69381-94-8	fenprostalene
69408-81-7	mafidimide
69521-64-8	(2-hydroxyethyl)-2-propenenitrile
69563-88-8	4,4'-((2,2,2-trifluoro-1-(trifluoro-methyl)ethylidene)bis(4,1-phenyleneoxy))bisbenzenamine
69766-47-8	azabicyclooctanol methyl bromide diphenylacetate
69806-50-4	fluazifop-butyl
69840-61-5	1-((4-chlorophenyl)methyl)-2-(nitromethylene)imidazolidine
70303-46-7	(ethylthio)trioctylstannane
70303-47-8	(butylthio)trioctylstannane
70441-84-8	2-formyl-2'-methyl-1,1'-(oxydimethylene)dipyridinium, dichloride oxime
70601-54-6	n-((phenylmethoxy)carbonyl)-o-(phenylmethyl)-l-seryl-l-tryptophyl-l-seryl-l-tyrosyl-2,3,4,5,6-pentamethyl-d-phenylalanyl-l-leucyl-l-arginyl-n-ethyl-l-prolinamide, monoacetate (salt), hydrate

74381-53-6   leutenizing hormone-releasing factor (pig), 6-d-leucine-9-(*n*-ethyl-*l*-prolinamide)-10-deglycinamide-, monoacetate (salt)

74578-69-1   ceftriaxone sodium hydrate

74749-74-9   ginsenoside

74764-40-2   3-isobutoxy-2-pyrrolidino-*n*-phenyl-*n*-benylpropylamine hydrochloride hydrate

74806-04-5   (+)car-3-ene

74861-59-9   ammonium boranecarboxylate

74938-11-7   7-hydroxy-2-(dipropylamino)tetraline

74940-61-7   *o*,*o*-dimethyl(2,2,2-trichloro-1-(chloroacetoxy)ethyl)phosphonate

75184-71-3   5-(propylsulfonyl)-2-benzimidazolecarbmic acid methyl ester

75318-64-8   2-(4-biphenylyl)-5,6-dihydro-*s*-traizolo(5,1-*a*)isoquinoline

75318-65-9   2-(4-biphenylyl)-5h-*s*-triazolo(5,1-*a*)isoindole

75444-63-2   β-yohimbine hydrochloride

75464-11-8   10-butyryldithranol

75464-12-9   1,8-dihydroxy-10-(1-oxopentyl)-9(10h)-anthracenone

75625-24-0   1,2,3,4,6,7-hexabromonaphthalene

75754-64-1   shigella dysenteriae toxin

76306-40-6   3,4,3′,4′-dimethylenedioxystilbene

76306-39-3   3,4-dimethoxy-3′,4′-methylenedioxystilbene

76547-98-3   lisinopril

76706-59-7   6-aminonicotinohydroxamic acid

76706-97-3   *s*-(4-chloro-2-butynyl) *o*,*o*-diethyl phosphorothioate

76706-98-4   *s*-(4-chloro-2-butynyl) *o*-ethyl phenylphosphonothioate

76706-99-5   *s*-(4-chloro-2-butynyl) diphenylphosphinothioate

76738-62-0   "Trimmit"

76849-19-9   (n^{10})-propargyl-5,8-dideazafolic acid

76858-53-2   2-((*o*-(*n*-methyl-*n*-(*tert*-butylthiosulfenyl)carbamoyl)oximin))-1,3-dithiolane

77227-69-1   5-(2-chloro-6-fluoro-4-(trifluoromethyl)phenoxy)-*n*-(ethylsulfonyl)-2-nitrobenzamide

77248-43-2   ((hexyloxy)sulfinyl)methylcarbamic acid-2,3-dihydro-2,2-dimethyl-7-benzofuranyl ester

77248-44-3   ((2,6-xylyl)sulfinyl)methylcarbamic acid-2,3-dihydro-2,2-dimethyl-7-benzofuranyl ester

77248-45-4   2-methyl-2-(methylthio)propionaldehyde-ω-((methyl)(decoxysulfinyl)carbamoyl)oxime

77267-47-1   2,3-dihydro-2,2-dimethyl-7-benzofuranyl methyl (phenoxysulfinyl)carbamate

77267-48-2   3-isopropylphenyl (methyl)(*n*-hexoxy-sulfinyl)carbamate

77267-49-3   2-isopropoxyphenyl (methyl)(*n*-hexoxy-sulfinyl)carbamate

77267-52-8   1-naphthylenyl ((hexyloxy)sulfinyl)methyl-carbamate

77267-59-5   2,3-dihydro-2,2-dimethylbenzofuranyl-7-(methyl)(*t*-butoxyfulfinyl)carbamate

77267-60-8   2,3-dihydro-2,2-dimethyl-7-benzofuranyl (butoxysulfinyl)methylcarbamate

77276-08-5   2-isopropoxyphenyl (methyl)(*t*-butoxysulfinyl)carbamate

77405-29-9   ethyldiphenyltin acetate

78194-09-9   2-chloro-*n*-(2-methoxy-3,6-dimethyl-phenyl)-*n*-((1-methylethoxy)methyl)acetamide

78308-32-4   aromex

78649-41-9   iomeprol

78950-78-4   8-hydroxy-2-(di-*n*-propylamino)tetraline

79006-76-1   5-oxo-2,4,8-trimethyl-6-oxa-3,9-dithia-2,4,7-triazadec-7-enoic acid, (1-methylethylidene)di-4,1-phenylene ester

79547-78-7   levocabastine hydrochloride

79622-59-6   3-chloro-*n*-(5-chloro-2,6-dinitro-4-trifluoro-methylphenyl)-5-trifluoromethyl-2-pyridinamine

79967-32-1   4-((*p*-(bis(2-chloroethyl)amino)phenyl)imino)methyl)-5-hydroxy-6-methyl-3-pyidinemethanol

80756-85-0   (2-mercaptobenzothiazolyl)-2-(2-aminothiazol-4-yl)-2-methoxyiminoacetate (*syn*)

81103-11-9   6-*o*-methylerythromycin

81131-70-6   lipostat

81147-92-4   methyl 4-(2-hydroxy-3-((1-methylethyl)amino)propoxy)benzenepropanoate

81226-60-0   4(4-chlorophenyl)-1-(1h-indol-3-ylmethyl)piperidin-4-ol

85850-93-7   3-((10-ethyl-11-(*p*-hydroxyphenyl)
              dibenz(*b*,*f*)oxypin-3-yl)oxy)-1,2-
              propanediol hydrate(4:1)

86209-51-0   3-(4,6-(tetrafluoromethoxy)
              pyrimidin-2-yl)-1-((2-
              methoxycarbonyl)phenyl-
              sulfonyl)urea

86329-79-5   cefodizime disodium
86408-72-2   ecabet sodium
86438-78-0   lauramidopropyl betaine
86924-47-2   "Triton RW 75"
86924-48-3   "Triton RW 125"
86924-49-4   "Triton RW 150"
89565-68-4   tropisetron
87051-43-2   ritanserin
87056-78-8   *n*,*n*-diethyl-*n*-(1,2,3,4,4*a*,5,10,10*a*-
              octahydro-6-hydroxy-1-
              propylbenzo(*g*)qui nolin-3-yl)-
              sulfamide, (3-α, 4*a*-α,
              10*a*-β)-(+−)

87130-20-9   ispropyl 3,4-diethoxycarbanilate
87177-09-1   1,3-diisocyanatomethylbenzene
              polymer with Niax E 488

87188-51-0   *p*-*tert*-butoxycarbonyloxystyrene
              monomer

87233-62-3   emedastine difumarate
87767-48-4   (((((1,4-dithian-2-ylideneamino)
              oxy)carbonyl)methylamino)thio)
              methylcarbamic acid,
              (1-methylethylidene)di-4,
              1-phenylene ester

88069-49-2   pilsicainide hydrochloride
              hemihydrate

88150-42-9   amlodipine
88255-01-0   netobimin
88671-89-0   myclobutanil
89365-50-4   (+−)-4-hydroxy-α-1-(((6-(4-
              phenylbutoxy)hexyl)amino)
              methyl)-1,3-benzene-dimethanol

89398-07-2   7-benzyl-3-thia-7-
              azabicyclo(3.3.1)nonane
              perchlorate

89985-01-3   *n*-(2,4-dichlorophenyl)-*n*-(4,5-
              dihydro-2-thiazolyl)-3-
              pyridinemethanamine

90034-97-2   4'-(1,2,3,4-tetrahydro-4-(4-hydroxy-
              2-oxo-2h-1-benzopyran-3-yl)-2-
              naphthalenyl)(1,1'-biphenyl)-4-
              carbonitrile, trans-

90034-98-3   4'-(1,2,3,4-tetrahydro-4-(4-hydroxy-
              2-oxo-2h-1-benzopyran-3-yl)-2-
              naphthalenyl)(1,1;-biphenyl)-4-
              carbonitrile, cis-

90034-99-4   4-hydroxy-3-(1,2,3,4-tetrahydro-3-
              (4-(trifluoromethyl)phenyl)-1-
              naphthalenyl)2h-benzopyran-2-
              one

90035-01-1   4-(1,2,3,4-tetrahydro-4-(4-hydroxy-
              2-oxo-2h-1-benzopyran-3-yl)-2-
              naphthalenyl)benzonitrile

90035-05-5   4-((4-(1,2,3,4-tetrahydro-4-(4-
              hydroxy-2-oxo-2h-1-benzopyran-
              3-yl)-2-naphthalenyl)phenoxy)
              benzonitrile

90035-06-6   3-(3-(4-((4-bromophenyl)
              methoxy)phenyl)-1,2,3,4-
              tetrahydro-1-naphthalenyl)-4-
              hydroxy2h-1-benzopyran-2-one

90035-11-3   3-(3-(4-(2-(4-bromophenyl)ethyl)
              phenyl)1,2,3,4-tetrahydro-1-
              naphthalenyl)-4-hydroxy2h-1-
              benzopyran-2-one

90035-12-4   3(3-(4-(2-(4-chlorophenyl)ethyl)
              phenyl)-1,2,3,4-tetrahydro-1-
              naphthalenyl)-4-hydroxy2h-1-
              benzopyran-2-one

90035-14-6   4-hydroxy-3-(1,2,3,4-tetrahydro-3-
              (4-(4-(trifluoromethyl)phenoxy)
              phenyl)-1-naphthalenyl)2h-1-
              benzopyran-2-one

90045-36-6   *Ginkgo biloba* L., root extract
90293-48-4   *n*,5-dimethyl-4-((dimethylamino)
              carbonyl)-*n*-((4-(1,1-
              dimethylethyl)phenyl)thio)-2,7-
              dioxa-3,6-diazaocta-3,5-
              dienamide

90293-50-8   9-(1-(methoxyimino)ethyl)-6-oxo-
              n,*n*,2,2,5-pentamethyl-7-oxa-3,4-
              dithia-5,8-diazadec-8-en-10-
              amide

90293-52-0   *n*-((8-((dimethylamino)carbonyl)-
              2,4,9-trimethyl-1,5-dioxo-6,11-
              dioxa-3-thia-2,4,7,10-
              tetraazadodeca-7,9-dien-1-
              yl)oxy)-ethanimidothioic acid,
              methyl ester

90293-53-1   8-((dimethylamino)carbonyl)-5-
              oxo-2,4,9-trimethyl-6,11-dioxa-3-
              thia-2,4,7,10-tetraazadodeca-7,9-
              dienoic acid, 1-naphthalenyl
              ester

90293-54-2   8-((dimethylamino)carbonyl)-5-
              oxo-2,4,9-trimethyl-6,11-dioxa-3-
              thia-2,4,7,10-tetraazadodeca-7,9-
              dienoic acid, 2,3-dihydro-2,2-
              dimethyl-7-bezofuranyl ester

106602-80-6 Otto Fuel II

106897-63-6 2,7-bis(phenylmethyl)benzo (*lmn*)(3,8)phenanthroline-1,3,6,8(2h,7h)-tetrone

107097-80-3 loxiglumide

107133-36-8 peridopril *tert*-butylamine

107359-39-7 4-chloro-2-(4-chlorophenyl)-5-((4-chlorophenyl)methoxy)-3(2h)-pyridazinone

107359-42-2 5-((4-bromophenyl)methoxy)-4-chloro-2-(4-chlorophenyl)3-(2h)-pyridazinone

107359-69-3 2-(4-bromophenyl)-4-chloro-5-((4-chlorophenyl)methoxy)-3(2h)-pyridazinone

107359-74-0 4-chloro-2-(4-chloro-2-fluoropjenyl)-5-((4-chlorophenyl)methoxy)-3(2h)-pyridazinone

107359-76-2 5-((4-bromophenyl)methoxy)-4-chloro-2-(4-chloro-2-fluorophenyl)-3(2h)-pyridazinone

107746-52-1 *n*-methoxy-3-(3,5-di-*tert*-butyl-4-hydroxybenzylidene)-2-pyrrolidone

107868-30-4 6-methylenadrosta-1,4-diene-3,17-dione

108910-63-0 AI3-36161

108944-67-8 neocurdione

109293-97-2 2-(1-(((((3,5-difluorophenyl)amino)carbonyl)hydrazono)ethyl)-3-pyridinecarboxylic acid, 98.1%

109581-93-3 tacrolimus hydrate

110690-43-2 3-(3-(6-benzoyloxy-3-cyano-2-pyridyloxy-carbonyl)benzoyl)-1-ethoxymethyl-5-fluorouracil

111109-77-4 dipropylene glycol dimethyl ether

111841-85-1 isopropyl 6-benzyloxy-4-methoxymethyl-β-carboline-3-carboxylate

111974-72-2 quetiapine fumarate

111988-43-3 (1-((6-chloro-3-pyridinyl)methyl)-4,5-dihydro-1h-imidazol-2-yl)cyanamide

111988-49-9 (3-((6-chloro-3-pyridinyl)methyl)-2-thiazolidinylidene)cyanamide

112410-23-8 tebufenozide

112573-73-6 sinorphan

112885-41-3 mosapride

113136-77-9 cyclanilide

113852-37-2 (*s*)-1-(3-hydroxy-2-phosphonylmethoxy-propyl)cytosine

114088-58-3 9-((2-phosphonylmethoxy)-ethyl)guanine

114654-31-8 sulfocarbathion K

114949-22-3 activin

114977-28-5 taxotere

115086-54-9 1-((6-chloro-3-pridinyl)methyl)-*n*-nitro-1h-imidazol-2-amine

115436-72-1 risedronate sodium

115659-47-7 perfluoro(2-butoxypropyl vinyl ether)

115722-23-1 Naja mossambica mossambica α-neurotoxin I

115970-17-7 4,5-dihydro-((6-chloro-3-pyridinyl)methyl)-1h-imidazol-2-amine

116397-83-2 murotox

116425-35-5 aerugidiol

116580-64-4 neem extract

116929-00-1 methyl-3-((5-(2-chloro-4-(trifluoro-methyl)phenoxy)-2-nitrophenyl)amino)butyrate

117568-24-8 10-acetyldithranol

117718-60-2 thiazophyr

117900-35-3 2,2′-diiododiaceamide

117906-15-7 1-((6-chloro-3-pyridinyl)methyl)-3-methyl-*n*-niro-2-imidazolidinimine

118359-59-4 7-*n*-((2-((2-(glutamylamino)ethyl)dithio)ethyl))mitomycin c

119515-38-7 1-(1-methylpropoxycarbonyl)-2-(2-hydroxyethyl)piperidine

119823-35-7 "Triton RW 20"

119823-36-8 "Triton RW 50"

120162-55-2 azimsulfuron

120373-24-5 isopropyl unoprostone

120720-15-2 glycoprotease

120868-66-8 1–(6-chloro-3-prindinyl)methyl)-2-imidazolidinone

121514-80-5 1,1,1-tris(4-isocyanatophenoxy-methyl)propane

121776-33-8 furilazole

122130-63-6 *s*-nitrosocaptopril

122322-18-3 4-chloro-2-(3,4-dichlorophenyl)-5-((6-iodo-3-pyridinyl)methoxy)-3(2h)-pyridazinone

122322-19-4 4-chloro-2-(4-chlorophenyl)-5-((6-iodo-3-pyridinyl)methoxy)-3(2h)-pyridazinone

122322-20-7 5-((6-bromo-3-pyridinyl)methoxy)-4-chloro-2-(4-chlorophenyl)-3(2h)-pyridazinone

122322-21-8 5-((6-bromo-3-pyridinyl)methoxy)-4-chloro-2-(3,4-dichlorophenyl)-3(2h)-pyridazinone

139257-77-5	4,14-retro-retinol-14-hydroxy-, (14r)-	156482-87-0	n-(2-(5-ethyl-2-methoxyphenyl) ethyl)propionamide
139906-72-2	(r)-2-ethylhexanoic acid sodium salt	156768-16-0	peroxisomicin A2
140456-78-6	nitronitramide ammonium salt	157810-81-6	indinavir sulfate
140695-21-2	osutidine	158440-71-2	6-(hydroxymethyl)acylfulvene
141206-73-7	1,3-bis(trifluoromethyl)-5-isocyanobenzene	158681-13-1	carboxamide hydrochloride
		159081-23-9	2-(1h-imidazol-4-ylmethyl)-8h-indeno(1,2-d)thiazole monofumarate
141807-57-0	3-β-(4-chlorophenyl)tropane-2-β-carboxylic acid phenyl eser hydrochloride		
		159654-97-4	"Stabaxol P"
143201-11-0	rivastatin	161050-58-4	methoxyfenozide
143390-89-0	methyl (E)-2-methoxyimino-(2-(o-tolyloxymethyl)phenyl)acetate	161326-34-7	fenamidone
		162280-52-6	8-chloro-3,4-dihydrospiro (naphthalene-2(1h),4'(5'h)-oxazol)-2'-amine
143563-20-6	n'-dimethylaminoacetylparticin A dimethylaminoethylamide diaspartat		
		162750-10-9	n-((3s)-2,3-dihydro-6-(2,6-difluorophenyl)methoxy)-3-benzofuranyl)-n-hydrourea, xy-
143621-35-6	3-amino-pyridine-2-carboxyaldehyde		
146714-97-8	trihydrochloride	162854-37-7	2,3,9,10-tetrahydroxyprotoberberine
147526-32-7	NK 104 (acid)	162885-01-0	(+−)-epibatidine dihydrochloride
147536-97-8	bosentan	164150-85-0	carboxylic acid hydrochloride
149845-06-7	fortovase	168316-95-8	spinosad
149950-60-7	emivirine	171599-83-0	"Viagra"
150155-61-6	d-phenylalanyl-l-phenylalanyl-l-tyrosyl-d-tryptophyl-l-lysyl-l-valyl-l-phenylalanyl-3-(2-naphthalenyl)-d-alaninamide	173381-90-3	3,5-dimethyl-n-(2-methylphenyl)-1h-pyrazole-1-acetamide
		173584-44-6	indoxacarb
		173903-15-6	methyl (chlorocarbonyl)(4-(triluoromethoxy)phenyl) carbamate
150403-89-7	l-N⁶-(1-iminoethyl)lysine hydrochloride		
150785-53-8	8,9-amhydro-4''-deoxy-3'-n-desmethyl-3'-n-ethylerythromycin b-6,9-hemiacetal	174689-39-5	3-(2-ethylphenoxy)-1-((1s)-1,2,3,4-tetrahydronaphth-1-ylamino)-(2s)-2-propanol oxalate
151776-26-0	s-desmethylzopiclone	175442-95-2	2-(r,s)-(di-n-propylamino)-6-(4-methoxy-phenylsulfonylmethyl)-1,2,3,4-tetra-hydronaphthalene
153049-45-7	dithiadenoxide		
153436-22-7	4,6-dichloro-3-((1c)-3-oxo-3-(phenylamino)-1-propenyl)-1h-indole-2-carboxylic acid		
		175795-76-3	7,8,9,10-tetrahydro-6,8,11-trihydroxy-8-(hydroxyacetyl)-1-methoxy-10-((2,3,6-trideoxy-3-(2,3-dihydro-1h-pyrrol-1-yl)-alphalyxo-hexopyranosyl)oxy)-, (8s,10s)-5,12-naphthacenedione
153857-27-3	(+−)-dibenzo(a,1)pyrene-11,12-dihydrodiol		
155536-45-1	dihydroetorphine hydrochloride		
155802-65-6	4,5-dihydro-1-((6-chloro-3-pyridinyl)methyl)-2-(nitroamino)-1h-imidazole-4,5-diol		
		179710-86-4	3-formyltetrahydrofuran
		181274-17-9	flucarbazone-sodium
156052-68-5	"Zoxamide technical"	182620-63-9	2,6-dimethyl-3,5-pyridine dicarboylic acid, dimethylester
156482-69-8	n-(2-(2-methoxy-1-naphthyl) ethyl)cyclo-propylcarboxamide		
		182912-58-9	clenoliximab
156482-71-2	n-(2-(2-methoxynaphthyl)ethyl) butyramide	188364-50-3	inol k65
		189624-85-9	(−)-nigaldipine hydrochloride
156482-72-3	n-(2-(2-methoxy-1-naphthyl) ethyl)pentanamide	190650-04-5	n-(2-(5-methoxy-4-indolyl) ethyl)acetamide
156482-84-7	n-(2-(2-methoxy-5-methylphenyl) ethyl)propionamide	193551-21-2	trans-1-(4-hydroxycyclohexyl)-4-(4-fluorophenyl)-5-(2-methoxpyrimidin-4-yl)imidazole
156842-85-5	n-(2-(2-methoxy-5-methylphenyl) ethyl)trifluoroacetamide		
		194205-01-1	5,6-dehydro-n-acetylloline

301644-23-5   3-((4-chlorophenyl)amino)-
n-(2-ethoxyethyl)-4,5-dihydro-
2h-benz(g)indazole-2-acetamide

301644-24-6   3-((4-chlorophenyl)amino)-4,5-
dihydro-n-(phenylmethyl)-2h-
benz(g)indazole-2-acetamide

301644-25-7   4-((3-((4-chlorophenyl)amino)-4,5-
dihydro-2h-benz(g)indazol-2-
yl)acetyl)morpholine

301644-26-8   3-((4-bromophenyl)amino)-n-(2-
ethoxyethyl)-4,5-dihydro-2h-
benz(g)indazole-2-acetamide

301644-27-9   4-((3-((4-bromophenyl)amino)-4,5-
dihydro-2h-benz(gndazol-2-
yl)acetyl)morpholine

302542-40-1   n-(2-methylphenyl)-1h-pyrazole-1-
acetamide

302542-42-3   4-iodo-3,5-dimethyl-n-(2-
methylphenyl)-1h-pyrazole-1-
acetamide

302542-44-5   3,5-dimethyl-n-(2-methylphenyl)-4-
nitro-1h-pyrazole-1-acetamide

302542-49-0   n-(3-methylphenyl)-1h-pyrazole-1-
acetamide

302542-50-3   3,5-dimethyl-n-(3-methylphenyl)-
1h-pyrazole-1-acetamide

302542-51-4   4-iodo-3,5-dimethyl-n-(3-
methylphenyl)-1h-pyrazole-1-
acetamide

302542-57-0   n-(4-methylphenyl)-1h-pyrazole-1-
acetamide

302542-60-5   3,5-dimethyl-n-(4-methylphenyl)-
1h-pyrazole-1-acetamide

302542-63-8   4-iodo-3,5-dimethyl-n-(4-
methylphenyl)-1h-pyrazole-1-
acetamide

302561-65-5   1-(2,4-dimethylphenyl)-3-(4-(2-
methoxyphenyl)-1-piperazinyl)-
1-propanone

302959-28-0   3-(6-fluoro-2-naphthalenyl)-1,2-
dimethyl(2r,3s)-rel-3-pyrrolidinol
hydrochloride

302959-30-4   3(6-methoxy-2-naphthalenyl)-1,2-
dimethyl(2r,3s)-rel-3-pyrrolidinol
hydrochloride

302959-32-6   1,2-dimethyl-3-(2-naphthalenyl)
(2r,3s)-rel-3-pyrrolidinol
drochloride

314238-30-7   1,7,7-trimethyl-o-(3-(diethylamino)-
2-hydroxypropyl)oxime (1r,4r)-
bicyclo(2.2.1)heptan-2-one

314238-31-8   1,7,7-trimethyl-o-(2-hydroxy-3-((1-
methylethyl)amino)propyl)oxime
(1r,4r)-bicyclo(2.2.1)heptan-2-
one

314238-32-9   1,7,7-trimethyl-o-(2-hydroxy-3-(1-
pyrrolidinyl)propyl)oxime
(1r,4r)-bicyclo(2.2.1)heptan-2-one

314238-33-0   1,7,7-trimethyl-o-(2-hydroxy-3-(1-
piperidinyl)propyl)oxime
(1r,4r)-bicyclo(2.2.1)heptan-2-one

314238-34-1   1,7,7-trimethyl-o-(2-hydroxy-3-(4-
morpholinyl)propyl)oxime
(1r,4r)-bicyclo(2.2.1)heptan-2-one

314238-35-2   1,7,7-trimethyl-o-(3-(hexahydro-1h-
azepin-1-yl)-2-hydroxypropyl)
oxime (1r,4r)-bicyclo(2.2.1)
heptan-2-one

315706-65-1   3,4,5-trimethoxy-n-(4-
propylcyclohexyl)benzamide

315706-66-2   4-chloro-n-(4-propylcyclohexyl)
benzamide

315706-67-3   4-fluoro-n-(4-propylcyclohexyl)
benzamide

315706-68-4   4-bromo-n-(4-propylcyclohexyl)
benzamide

315706-69-5   2,4-dichloro-n-(4-propylcyclohexyl)
benzamide

315706-70-8   n-(4-propylcyclohexyl)benzamide

315706-72-0   4-nitro-n-(4-propylcyclohexyl)
benzamide

315706-73-1   n-(4-propylcyclohexyl)-3-(3,4,5-
trimethoxyphenyl)-2-propenamide

315706-74-2   3-(4-chlorophenyl)-n-(4-
propylcyclohexyl)-2-propenamide

315706-75-3   3-(4-fluorophenyl)-n-(4-
propylcyclohexyl)-2-propenamide

315706-76-4   3-(4-bromophenyl)-n-(4-
propylcyclohexyl)-2-propenamide

315706-77-5   3-(2,4-dichlorophenyl)-n-(4-
propylcyclohexyl)-2-propenamide

315706-78-6   3-phenyl-n-(4-propylcyclohexyl)-2-
propenamide

315706-79-7   3-(4-methylphenyl)-n-(4-
propylcyclohexyl)-2-propenamide

315706-80-0   3-(4-nitrophenyl)-n-(4-
propylcyclohexyl)-2-propenamide

315706-81-1   n-(4-propylcyclohexyl)-4-
morpholine-carboxamide

326800-75-3   6-(2,5-dihydro-1,2-dimethyl-1h-
pyrrol-3-yl)-2-naphthalenol
hydrochloride

APPENDIX V
Tables

Hawley's Condensed Chemical Dictionary, Sixteenth Edition. Michael D. Larrañaga, Richard J. Lewis, Sr., and Robert A. Lewis.
© 2016 John Wiley & Sons, Inc. Published 2016 by John Wiley & Sons, Inc.

Los Alamos National Laboratory Chemistry Division
Periodic Table of the Elements

Legend:
- New notation
- Previous IUPAC from
- CAS version

Group 1 / IA

#	Sym	Config	Name	Mass
1	H	$1s^1$	hydrogen	1.008
3	Li	[He]$2s^1$	lithium	6.94
11	Na	[Ne]$3s^1$	sodium	22.99
19	K	[Ar]$4s^1$	potassium	39.10
37	Rb	[Kr]$5s^1$	rubidium	85.47
55	Cs	[Xe]$6s^1$	cesium	132.9
87	Fr	[Rn]$7s^1$	francium	(223)

Group 2 / IIA

#	Sym	Config	Name	Mass
4	Be	[He]$2s^2$	beryllium	9.012
12	Mg	[Ne]$3s^2$	magnesium	24.31
20	Ca	[Ar]$4s^2$	calcium	40.08
38	Sr	[Kr]$5s^2$	strontium	87.62
56	Ba	[Xe]$6s^2$	barium	137.3
88	Ra	[Rn]$7s^2$	radium	(226)

Transition metals (Groups 3–12)

#	Sym	Config	Name	Mass
21	Sc	[Ar]$4s^23d^1$	scandium	44.96
22	Ti	[Ar]$4s^23d^2$	titanium	47.88
23	V	[Ar]$4s^23d^3$	vanadium	50.94
24	Cr	[Ar]$4s^13d^5$	chromium	52.00
25	Mn	[Ar]$4s^23d^5$	manganese	54.94
26	Fe	[Ar]$4s^23d^6$	iron	55.85
27	Co	[Ar]$4s^23d^7$	cobalt	58.93
28	Ni	[Ar]$4s^23d^8$	nickel	58.69
29	Cu	[Ar]$4s^13d^{10}$	copper	63.55
30	Zn	[Ar]$4s^23d^{10}$	zinc	65.39
39	Y	[Kr]$5s^24d^1$	yttrium	88.91
40	Zr	[Kr]$5s^24d^2$	zirconium	91.22
41	Nb	[Kr]$5s^14d^4$	niobium	92.91
42	Mo	[Kr]$5s^14d^5$	molybdenum	95.96
43	Tc	[Kr]$5s^24d^5$	technetium	(98)
44	Ru	[Kr]$5s^14d^7$	ruthenium	101.1
45	Rh	[Kr]$5s^14d^8$	rhodium	102.9
46	Pd	[Kr]$4d^{10}$	palladium	106.4
47	Ag	[Kr]$5s^14d^{10}$	silver	107.9
48	Cd	[Kr]$5s^24d^{10}$	cadmium	112.4
72	Hf	[Xe]$6s^24f^{14}5d^2$	hafnium	178.5
73	Ta	[Xe]$6s^24f^{14}5d^3$	tantalum	180.9
74	W	[Xe]$6s^24f^{14}5d^4$	tungsten	183.9
75	Re	[Xe]$6s^24f^{14}5d^5$	rhenium	186.2
76	Os	[Xe]$6s^24f^{14}5d^6$	osmium	190.2
77	Ir	[Xe]$6s^24f^{14}5d^7$	iridium	192.2
78	Pt	[Xe]$6s^14f^{14}5d^9$	platinum	195.1
79	Au	[Xe]$6s^14f^{14}5d^{10}$	gold	197.0
80	Hg	[Xe]$6s^24f^{14}5d^{10}$	mercury	200.5
104	Rf	[Rn]$7s^25f^{14}6d^2$	rutherfordium	(265)
105	Db	[Rn]$7s^25f^{14}6d^3$	dubnium	(268)
106	Sg	[Rn]$7s^25f^{14}6d^4$	seaborgium	(271)
107	Bh	[Rn]$7s^25f^{14}6d^5$	bohrium	(270)
108	Hs	[Rn]$7s^25f^{14}6d^6$	hassium	(277)
109	Mt	[Rn]$7s^25f^{14}6d^7$	meitnerium	(276)
110	Ds	[Rn]$7s^15f^{14}6d^9$	darmstadtium	(281)
111	Rg	[Rn]$7s^15f^{14}6d^{10}$	roentgenium	(280)
112	Cn		copernicium	(285)

Groups 13–18

#	Sym	Config	Name	Mass
5	B	[He]$2s^22p^1$	boron	10.81
6	C	[He]$2s^22p^2$	carbon	12.01
7	N	[He]$2s^22p^3$	nitrogen	14.01
8	O	[He]$2s^22p^4$	oxygen	16.00
9	F	[He]$2s^22p^5$	fluorine	19.00
2	He	$1s^2$	helium	4.003
10	Ne	[He]$2s^22p^6$	neon	20.18
13	Al	[Ne]$3s^23p^1$	aluminum	26.98
14	Si	[Ne]$3s^23p^2$	silicon	28.09
15	P	[Ne]$3s^23p^3$	phosphorus	30.97
16	S	[Ne]$3s^23p^4$	sulfur	32.06
17	Cl	[Ne]$3s^23p^5$	chlorine	35.45
18	Ar	[Ne]$3s^23p^6$	argon	39.95
31	Ga	[Ar]$4s^23d^{10}4p^1$	gallium	69.72
32	Ge	[Ar]$4s^23d^{10}4p^2$	germanium	72.64
33	As	[Ar]$4s^23d^{10}4p^3$	arsenic	74.92
34	Se	[Ar]$4s^23d^{10}4p^4$	selenium	78.96
35	Br	[Ar]$4s^23d^{10}4p^5$	bromine	79.90
36	Kr	[Ar]$4s^23d^{10}4p^6$	krypton	83.79
49	In	[Kr]$5s^24d^{10}5p^1$	indium	114.8
50	Sn	[Kr]$5s^24d^{10}5p^2$	tin	118.7
51	Sb	[Kr]$5s^24d^{10}5p^3$	antimony	121.8
52	Te	[Kr]$5s^24d^{10}5p^4$	tellurium	127.6
53	I	[Kr]$5s^24d^{10}5p^5$	iodine	126.9
54	Xe	[Kr]$5s^24d^{10}5p^6$	xenon	131.3
81	Tl	[Xe]$6s^24f^{14}5d^{10}6p^1$	thallium	204.4
82	Pb	[Xe]$6s^24f^{14}5d^{10}6p^2$	lead	207.2
83	Bi	[Xe]$6s^24f^{14}5d^{10}6p^3$	bismuth	209.0
84	Po	[Xe]$6s^24f^{14}5d^{10}6p^4$	polonium	(209)
85	At	[Xe]$6s^24f^{14}5d^{10}6p^5$	astatine	(210)
86	Rn	[Xe]$6s^24f^{14}5d^{10}6p^6$	radon	(222)
113	Uut		ununtrium	(284)
114	Fl		flerovium	(289)
115	Uup		ununpentium	(288)
116	Lv		livermorium	(293)
117	Uus		ununseptium	(294)
118	Uuo		ununoctium	(294)

Lanthanide Series*

#	Sym	Config	Name	Mass
57	La	[Xe]$6s^25d^1$	lanthanum	138.9
58	Ce	[Xe]$6s^24f^15d^1$	cerium	140.1
59	Pr	[Xe]$6s^24f^3$	praseodymium	140.9
60	Nd	[Xe]$6s^24f^4$	neodymium	144.2
61	Pm	[Xe]$6s^24f^5$	promethium	(145)
62	Sm	[Xe]$6s^24f^6$	samarium	150.4
63	Eu	[Xe]$6s^24f^7$	europium	152.0
64	Gd	[Xe]$6s^24f^75d^1$	gadolinium	157.2
65	Tb	[Xe]$6s^24f^9$	terbium	158.9
66	Dy	[Xe]$6s^24f^{10}$	dysprosium	162.5
67	Ho	[Xe]$6s^24f^{11}$	holmium	164.9
68	Er	[Xe]$6s^24f^{12}$	erbium	167.3
69	Tm	[Xe]$6s^24f^{13}$	thulium	168.9
70	Yb	[Xe]$6s^24f^{14}$	ytterbium	173.0
71	Lu	[Xe]$6s^24f^{14}5d^1$	lutetium	175.0

Actinide Series**

#	Sym	Config	Name	Mass
89	Ac	[Rn]$7s^26d^1$	actinium	(227)
90	Th	[Rn]$7s^26d^2$	thorium	232
91	Pa	[Rn]$7s^25f^26d^1$	protactinium	231
92	U	[Rn]$7s^25f^36d^1$	uranium	238
93	Np	[Rn]$7s^25f^46d^1$	neptunium	(237)
94	Pu	[Rn]$7s^25f^6$	plutonium	(244)
95	Am	[Rn]$7s^25f^7$	americium	(243)
96	Cm	[Rn]$7s^25f^76d^1$	curium	(247)
97	Bk	[Rn]$7s^25f^9$	berkelium	(247)
98	Cf	[Rn]$7s^25f^{10}$	californium	(251)
99	Es	[Rn]$7s^25f^{11}$	einsteinium	(252)
100	Fm	[Rn]$7s^25f^{12}$	fermium	(257)
101	Md	[Rn]$7s^25f^{13}$	mendelevium	(258)
102	No	[Rn]$7s^25f^{14}$	nobelium	(259)
103	Lr	[Rn]$7s^25f^{14}$	lawrencium	(262)

Los Alamos NATIONAL LABORATORY — CHEMISTRY

Elements Table

Element	Symbol	Atomic number	Element	Symbol	Atomic number
Actinium	Ac	89	Mendelevium	Md	101
Aluminum	Al	13	Mercury	Hg	80
Americium	Am	95	Molybdenum	Mo	42
Antimony	Sb	51	Neodymium	Nd	60
Argon	Ar	18	Neon	Ne	10
Arsenic	As	33	Neptunium	Np	93
Astatine	At	85	Nickel	Ni	28
Barium	Ba	56	Niobium (Columbium)	Nb	41
Berkelium	Bk	97	Nitrogen	N	7
Beryllium	Be	4	Nobelium	No	102
Bismuth	Bi	83	Osmium	Os	76
Bohrium	Bh	107	Oxygen	O	8
Boron	B	5	Palladium	Pd	46
Bromine	Br	35	Phosphorus	P	15
Cadmium	Cd	48	Platinum	Pt	78
Calcium	Ca	20	Plutonium	Pu	94
Californium	Cf	98	Polonium	Po	84
Carbon	C	6	Potassium	K	19
Cerium	Ce	58	Praseodymium	Pr	59
Cesium	Cs	55	Promethium	Pm	61
Chlorine	Cl	17	Protactinium	Pa	91
Chromium	Cr	24	Radium	Ra	88
Cobalt	Co	27	Radon	Rn	86
Copernicium	Cn	112	Rhenium	Re	75
Copper	Cu	29	Rhodium	Rh	45
Curium	Cm	96	Roentgenium	Rg	111
Darmstadtium	Ds	110	Rubidium	Rb	37
Dubnium	Db	105	Ruthenium	Ru	44
Dysprosium	Dy	66	Rutherfordium	Rf	104
Einsteinium	Es	99	Samarium	Sm	62
Erbium	Er	68	Scandium	Sc	21
Europium	Eu	63	Seaborgium	Sg	106
Fermium	Fm	100	Selenium	Se	34
Flerovium	Fl	114	Silicon	Si	14
Fluorine	F	9	Silver	Ag	47
Francium	Fr	87	Sodium	Na	11
Gadolinium	Gd	64	Strontium	Sr	38
Gallium	Ga	31	Sulfur	S	16
Germanium	Ge	32	Tantalum	Ta	73
Gold	Au	79	Technetium	Tc	43
Hafnium	Hf	72	Tellurium	Te	52
Hassium	Hs	108	Terbium	Tb	65
Helium	He	2	Thallium	Tl	81
Holmium	Ho	67	Thorium	Th	90
Hydrogen	H	1	Thulium	Tm	69
Indium	In	49	Tin	SN	50
Iodine	I	53	Titanium	Ti	22
Iridium	Ir	77	Tungsten (Wolfram)	W	74
Iron	Fe	26	Ununoctium*	Uuo	118
Krypton	Kr	36	Ununpentium*	Uup	115
Lanthanum	La	57	Ununseptium*	Uus	117
Lawrencium	Lr	103	Ununtrium*	Uut	113
Lead	Pb	82	Uranium	U	92
Lithium	Li	3	Vanadium	V	23
Livermorium	Lv	116	Xenon	Xe	54
Lutetium	Lu	71	Ytterbium	Yb	70
Magnesium	Mg	12	Yttrium	Y	39
Manganese	Mn	25	Zinc	Zn	30
Meitnerium	Mt	109	Zirconium	Zr	40

*Officially unnamed and temporary symbol for an element with a specific atomic number.

Temperature Conversion Table

		°C = (°F − 32) × 5/9 (or 0.55)		°F = °C × 9/5 (or 1.8) + 32	
°F	°C	°F	°C	°F	°C
−40	−40	180	82.2	400	204.4
−35	−37.2	185	85.0	405	207.2
−30	−34.4	190	87.8	410	210.0
−25	−31.6	195	90.5	415	212.8
−20	−28.9	200	93.3	420	215.5
−15	−26.1	205	96.1	425	218.3
−10	−23.3	210	98.9	430	221.1
−5	−20.5	212	100.0	435	223.9
0	−17.78	215	101.6	440	226.6
+5	−15.0	220	104.4	445	229.4
10	−12.2	225	107.2	450	232.2
15	−9.4	230	110.0	455	235.0
20	−6.6	235	112.8	460	237.8
25	−3.9	240	115.5	465	240.5
30	−1.1	245	118.3	470	243.3
32	0	250	121.1	475	246.1
35	+1.6	255	123.9	480	248.9
40	4.4	260	126.6	485	251.6
45	7.2	265	129.4	490	254.4
50	10.0	270	132.2	495	257.2
55	12.8	275	135.0	500	260.0
60	15.5	280	137.8	550	287.8
65	18.3	285	140.5	600	315.5
70	21.1	290	143.3	650	343.3
75	23.9	295	146.1	700	371.1
80	26.6	300	148.9	750	398.9
85	29.4	305	151.6	800	426.6
90	32.2	310	154.4	850	454.4
95	35.0	315	157.2	900	482.2
100	37.8	320	160.0	950	510.0
105	40.5	325	162.8	1000	537.8
110	43.3	330	165.5	1100	593.3
115	46.1	335	168.3	1200	648.9
120	48.9	340	171.1	1300	704.4
125	51.6	345	173.9	1400	760.0
130	54.4	350	176.6	1500	815.5
135	57.2	355	179.4	1600	871.1
140	60.0	360	182.2	1700	926.6
145	62.8	365	185.0	1800	982.2
150	65.5	370	187.8	1900	1037.8
155	68.3	375	190.5	2000	1093.3
160	71.1	380	193.3		
165	73.9	385	196.1		
170	76.6	390	198.9		
175	79.4	395	201.6		

This table is public domain information provided by NIST (http://www.nist.gov/pml/wmd/metric/common-conversion.cfm).